CHILTON'S

TRANSMISSION SERVICE MANUAL

DOMESTIC CARS & TRUCKS

CHILTON BOOK COMPANY
ONE OF THE ***ABC PUBLISHING COMPANIES,***
A PART OF ***CAPITAL CITIES/ABC, INC.***
Manufactured in USA • Chilton Way, Radnor, Pa. 19089
ISBN 0-8019-7959-5 Library of Congress Card Catalog Number 88-43180
1234567890 8765432109

ACKNOWLEDGEMENTS

Chilton Book Company expresses appreciation to the following associations and firms for their cooperation and technical assistance:

ATSG—Automatic Transmission Service Group, Miami, Florida
Special Thanks to: Mr. Robert D. Cherrnay—President and Technical Director
Mr. Dale England—Technical Consultant
Ms. Dora Miller—Technical Assistant

ATRA—Automatic Transmission Rebuilders Association, Ventura, California

ASA—Automotive Service Association, Bedford, Texas
Special Thanks to: Mr. Allen Ritchey—President
Mr. John F. Mullins, Jr., Manager, Special Projects, Retired Chairman of the Board
Mr. Edward Anderson—Manager, Transmission Division

Brandywine Transmission Service, Inc., Wilmington, Delaware
Special Thanks to: Mr. Mark Mullins

Borg-Warner Automotive Inc., Muncie, Indiana

Chrysler Corportion, Detroit, Michigan

Component Remanufacturing Specialists, Inc., Ramsey, New Jersey
Special Thanks to: Mr. Michael LaPore

Ford Motor Company, Dearborn, Michigan

General Motors Corporation, Flint, Michigan

Hydra-Matic, Division of General Motors, Ypsilanti, Michigan

American Isuzu Motors, Inc., Whittier, California

Lee's Auto Service, Trainer, Chester, Pennsylvania

Mazda Motors of America, Inc., Montvale, New Jersey

MGW Manufacturing Corp., Bohemia, New York
Special Thanks to: The Technical Advisement Committee
Mr. Ralph Raymond and Staff

Pat's Transmission Service, Broomall, Pennsylvania

Ralph's Garage, Linwood, Pennsylvania

RAM Automotive, King of Prussia, Pennsylvania
Special Thanks to: Mr. Bob Alekna
Mr. Tony Kidd
Mr. Chet Freas

Rockland Standard Gear, Inc., Nanuet, New York
Special Thanks to: Mr. Michael Weinberg

Seuro Transmissions, Incorporated, Pittsburgh, Pennsylvania

Transmission By Lucille, Pittsburgh, Pennsylvania

Wayne Ford, Wayne, Pennsylvania
Special Thanks to: Mr. John Lanzar
Mr. Larry Devereaux
Mr. Steve Pancoe

ZF of North America, Inc., Chicago, Illinois

AUTOMATIC TRANSMISSION/TRANSAXLE SERVICE DATA LOCATOR

			Professional Transmission Manual 1984–89 Domestic #7959		Professional Automatic Transmission Manual 1980–84 #7390		Professional Automatic Transmission Manual 1974–80 #6927	
Manufacturer	Transaxle	Transmission	Service Info	Modifications Info	Service Info	Modifications Info	Service Info	Modifications Info
Chrysler Corp.	AR4		Yes	Yes	–	–	–	–
AMC/Jeep-Eagle	A404, A413, A415, A470		–	Yes	Yes	Yes	–	–
	A604		Yes	Yes	–	–	–	–
	KM171, KM172		–	Yes	Yes	Yes	–	–
	MJ3		Yes	Yes	–	–	–	–
	ZF–4HP–18		Yes	Yes	–	–	–	–
		AW4	Yes	Yes	–	–	–	–
		A500	Yes	Yes	–	–	–	–
		A727, A904, A998, A999	–	Yes	–	Yes	Yes	Yes
		KM148, AW372	Yes	Yes	–	–	–	–
		Loadflite	–	–	–	–	Yes	Yes
		Torque Command	–	–	–	–	Yes	Yes
Ford Motor Co.	ATX		–	Yes	Yes	Yes	–	–
	AXOD		Yes	Yes	–	–	–	–
	F3A		Yes	Yes	–	–	–	–
	4EAT		Yes	Yes	–	–	–	–
		AOD	Yes	Yes	–	Yes	–	–
		A4LD	Yes	Yes	–	–	–	–
		C3	–	Yes	–	Yes	Yes	Yes
		C4	–			Yes	Yes	Yes
		C5	–	Yes	Yes	Yes	–	–
		C6	–	Yes	–	Yes	Yes	Yes
		E4OD	Yes	Yes	–	–	–	–
		FIOD	–	–	–	Yes	Yes	Yes
		ZF–4HP–22	–	Yes	Yes	Yes	–	–
		FMX	–	–	–	–	Yes	Yes
General Motors	AW131L		Yes	Yes	–	–	–	–
	A240E		Yes	Yes	–	–	–	–
	KF100		Yes	Yes	–	–	–	–
	Sprint/Metro		Yes	Yes	–	–	–	–
	THM 125		–	–	–	–	Yes	Yes
	THM 125C		–	Yes	Yes	Yes	–	–
	THM 325		–	–	–	–	Yes	Yes
	THM 325–4L		–	Yes	Yes	Yes	–	–
	THM 425		–	–	–	–	Yes	Yes
	THM 440–T4, F7		Yes	Yes	–	Yes	–	–
		Powerglide	–	–	–	–	Yes	Yes
		THM 180	–	–	–	–	Yes	Yes
		THM 180C	–	Yes	Yes	Yes	–	–
		THM 200	–	Yes	–	Yes	Yes	Yes
		THM 200C	–	Yes	–	Yes	Yes	Yes
		THM 200–4R	–	Yes	Yes	Yes	–	–
		THM 250	–	Yes	–	Yes	Yes	Yes
		THM 250C	–	Yes	–	Yes	Yes	Yes
		THM 350	–	Yes	–	Yes	Yes	Yes
		THM 350C	–	Yes	–	Yes	Yes	Yes
		THM 375B	–	–	–	Yes	Yes	Yes
		THM 400, 475	–	Yes	–	Yes	Yes	Yes
		THM 700–R4	–	Yes	Yes	Yes	–	–

– No coverage in this publication

SAFETY NOTICE

Proper service and repair procedures are vital to the safe, reliable operation of all motor vehicles, as well as the personal safety of those performing repairs. This manual outlines procedures for servicing and repairing vehicles using safe, effective methods. The procedures contain many NOTES, CAUTIONS and WARNINGS which should be followed along with standard safety procedures to eliminate the possibilty of personal injury or improper service which could damage the vehicle or compromise its safety.

It is important to note that the repair procedures and techniques, tools and parts for servicng motor vehicles, as well as the skill and experience of the individual performing the work vary widely. It is not possible to anticipate all of the conceivable ways or conditions under which vehicles may be serviced, or to provide cautions as to all of the possible hazards that may result. Standard and accepted safety precautions and equipment should be used when handling toxic or flammable fluids, and safety goggles or other protection should be used during cutting, grinding, chiseling, prying, or any other process that can cause material removal or projectiles.

Some procedures require the use of tools specially designed for a specific purpose. Before substituting another tool or procedure, you must be completely satisfied that neither your personal safety, nor the performance of the vehicle will be endangered

PART NUMBERS

Part numbers listed in this reference are not recomendations by Chilton for any product by brand name. They are references that can be used with interchange manuals and aftermarket supplier catalogs to locate each brand supplier's discrete part number.

Although information in this manual is based on industry sources and is complete as possible at the time of publication, the possibilty exists that some car manufacturers made later changes which could not be included here. While striving for total accuracy, Chilton Book Company cannot assume responsibity for any errors, changes or omissions that may occur in the compilation of this data.

Section 1

Automatic Transmissions and Transaxles

General Information

GENERAL INFORMATION

DIAGNOSING MALFUNCTIONS

GENERAL INFORMATION

Introduction

With this edition of Chilton's Professional Transmission Manual- domestic vehicles, we continue to assist the professional transmission repair trade to perform quality repairs and adjustments for that "like new" dependability of the transmission/transaxle assemblies.

This concise, but comprehensive service manual places emphasis on diagnosing, troubleshooting, adjustments, testing, disassembly and assembly of the automatic transmission/transaxle.

Metric Fasteners and Inch System Fasterners

Metric bolt sizes and thread pitches are more commonly used for all fasteners on the automatic transmissions/transaxles now being manufactured. The metric bolt sizes and thread pitches are very close to the dimensions of the similar inch system fasteners and for this reason, replacement fasteners must have the same measurement and strength as those removed.

Do not attempt to interchange metric fasteners for inch system fasteners. Mismatched and incorrect fasteners can result in

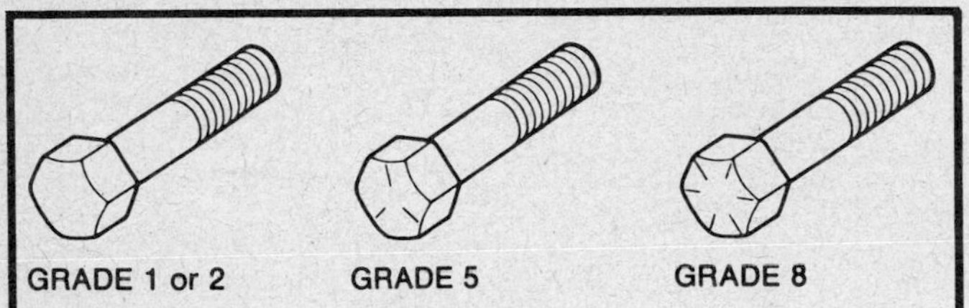
GRADE 1 or 2
GRADE 5
GRADE 8

English (Inch) Bolts—Identification Marks Correspond To Bolt Strength—Increasing Number Of Slashes Represent Increasing Strength.

Typical SAE bolt head identification marks

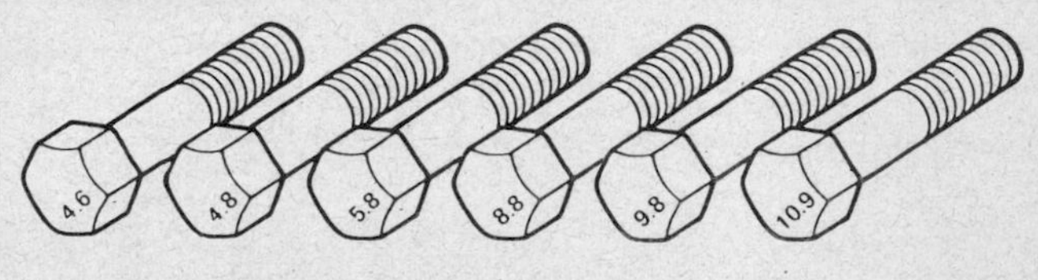
4.6
4.8
5.8
8.8
9.8
10.9

Metric Bolts—Identification Class Numbers Correspond To Bolt Strength—Increasing Numbers Represent Increasing Strength. Common Metric Fastener Bolt Strength Property Are 9.8 And 10.9 With The Class Identification Embossed On The Bolt Head.

Typical metric bolt head identification marks

damage to the transmission/transaxle unit through malfunction, breakage or possible personal injury. Care should be exercised to reuse the fasteners in their same locations as removed when every possible. If any doubt exists in the reuse of fasteners, install new ones.

To avoid stripped threads and to prevent metal warpage, the use of the torque wrench becomes more important, as the gear box assembly and internal components are being manufactured from light weight material. The torque conversion charts should be understood by the repairman, to properly service the requirements of the torquing procedures. When in doubt, refer to the specifications for the transmission/transaxle being serviced or overhauled.

Critical Measurements

With the increase use of transaxles and the close tolerances needed throughout the drive train, more emphasis is placed upon making the critical bearing and gear measurements correctly and being assured that correct preload and turning torque exists before the unit is reinstalled in the vehicle. Should a comeback occur because of the lack of proper clearances or torque, a costly rebuild can result. Rather than rebuilding a unit by "feel", the repairman must rely upon precise measuring

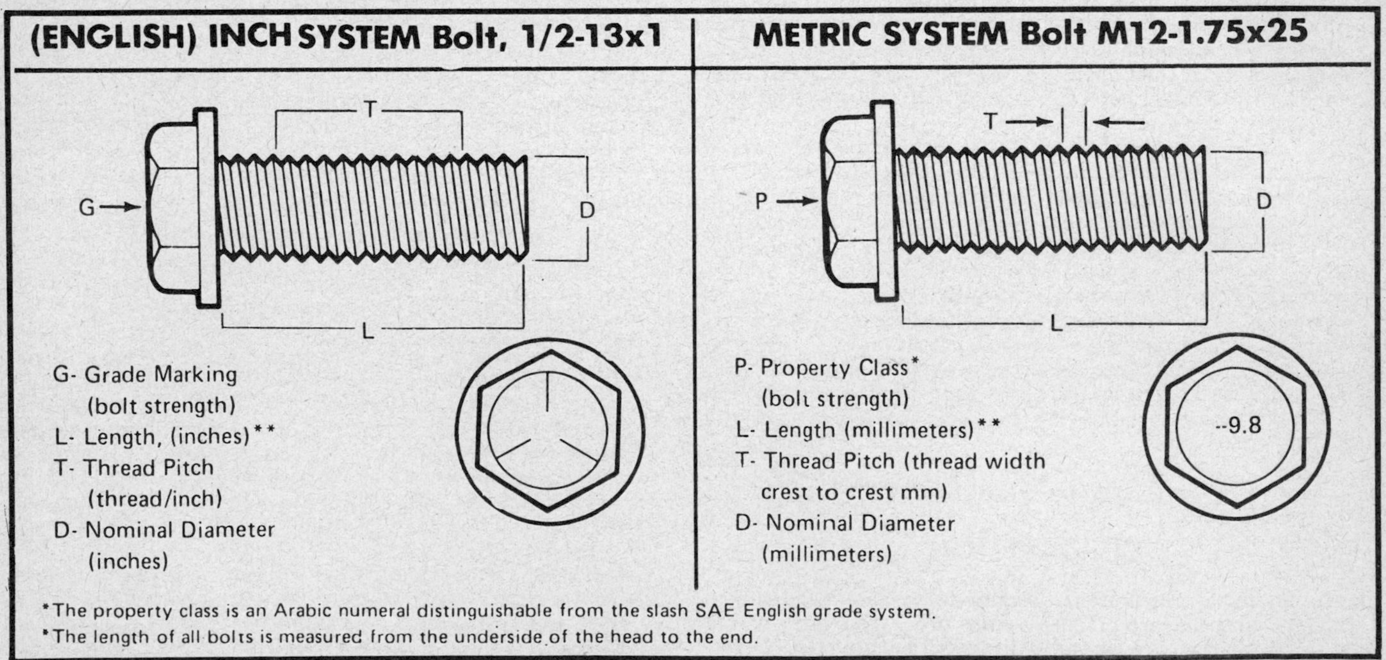
(ENGLISH) INCH SYSTEM Bolt, 1/2-13x1
T
G
D
L
G- Grade Marking (bolt strength)
L- Length, (inches)**
T- Thread Pitch (thread/inch)
D- Nominal Diameter (inches)
METRIC SYSTEM Bolt M12-1.75x25
T
P
D
L
P- Property Class* (bolt strength)
L- Length (millimeters)**
T- Thread Pitch (thread width crest to crest mm)
D- Nominal Diameter (millimeters)
--9.8
*The property class is an Arabic numeral distinguishable from the slash SAE English grade system.
*The length of all bolts is measured from the underside of the head to the end.

SAE and metric system bolt and thread nomenclature

(ENGLISH) INCH SYSTEM		METRIC SYSTEM	
Grade	Identification	Class	Identification
Hex Nut Grade 5	 3 Dots	Hex Nut Property Class 9	 9 Arabic 9
Hex Nut Grade 8	 6 Dots	Hex Nut Property Class 10	 10 Arabic 10
Increasing dots represent increasing strength.		May also have blue finish or paint daub on hex flat. Increasing numbers represent increasing strength.	

SAE and metric system strength identification marks

tools, such as the dial indicator gauge, micrometers, torque wrenches and feeler gauges to insure that correct specifications are adhered to. At the end of each transmission/transaxle section specification data is provided so that the repairman can measure important clearances that will effect the outcome of the transmission/transaxle rebuild.

Electronically Controlled Units

Today transmissions/transaxles are being developed and manufactured with electronically controlled components. The demand for lighter, smaller and more fuel efficient vehicles has resulted in the use of electronics to control both the engine spark and fuel delivery. Certain transmission/transaxle assemblies are a part of the electronic controls, by sending signals of vehicle speed and throttle opening to an on-board computer, which in turn computes these signals, along with others from the engine assembly, to determine if spark occurrence should be changed or the delivery of fuel should be increased or decreased. The computer signals are then sent to their respective controls and/or sensors as required.

Automatic transmissions/transaxles with microcomputers to determine gear selections are now in use. Sensors are used for engine and road speeds, engine load, gear selector lever position, kickdown switch and a status of the driving program to send signals to the microcomputer to determine the optimum gear selection, according to a preset program. The shifting is accomplished by solenoid valves in the hydraulic system. The electronics also control the modulated hydraulic system during shifting, along with regulating engine torque to provide smooth shifts between gear ratio changes. This type of system can be designed for different driving programs, such as giving the operator the choice of operating the vehicle for either economy or performance.

Lockup Torque Converter Units

DESCRIPTION

Most all vehicle transmissions/transaxles are equipped with a lockup torque converter. The lockup torque converter clutch should apply when the engine has reached near normal operating temperature in order to handle the slight extra load and when the vehicle speed is high enough to allow the operation of the clutch to be smooth and the vehicle to be free of engine pulses.

When the converter clutch is coupled to the engine, the engine pulses can be felt through the vehicle in the same manner as if equipped with a clutch and standard transmission. Engine condition, engine load and engine speed determines the severity of the pulsations.

The converter clutch should release when torque multiplication is needed in the converter, when coming to a stop, or when the mechanical connection would affect exhaust emissions during a coasting condition.

The electrical control components consists of the brake release switch, the low vacuum switch and the governor switch. Some transmission/transaxles have a thermal vacuum switch, a relay valve and a delay valve. Diesel engines use a high vacuum switch in addition to certain above listed components. These various components control the flow of current to the apply valve solenoid. By controlling the current flow, these components activate or deactivate the solenoid, which in turn engages or disengages the transmission/transaxle converter clutch, depending upon the driving conditions. The components have 2 basic circuits, electrical and vacuum.

ELECTRICAL CURRENT FLOW

All of the components in the electrical circuit must be closed or grounded before the solenoid can open the hydraulic circuit to engage the converter clutch. The circuit begins at the fuse panel and flows to the brake switch as long as the brake pedal is not depressed. The current will flow to the low vacuum switch on gasoline engines and to the high vacuum switch on diesel engines. These switches open or close the circuit path to the solenoid, dependent upon the engine or pump vacuum. If the low vacuum switch is closed (high switch on diesel engines), the current continues to flow to the transmission/transaxle case connector and then into the solenoid and to the governor pressure switch. When the vehicle speed is approximately 35–50 mph, the governor switch grounds to activate the solenoid. The solenoid, in turn, opens a hydraulic circuit to the converter clutch assembly, engaging the unit.

It should be noted that external vacuum controls include the thermal vacuum valve, the relay valve, the delay valve, the low vacuum switch and a high vacuum switch (used on diesel engines). Keep in mind that all of the electrical or vacuum components may not be used on all engines, at the same time.

VACUUM FLOW

The vacuum relay valve works with the thermal vacuum valve to keep the engine vacuum from reaching the low vacuum valve switch at low engine temperatures. This action prevents the clutch from engaging while the engine is still warming up. The delay valve slows down the response of the low vacuum switch to changes in engine vacuum. This action prevents the low vacuum switch from causing the converter clutch to engage and disengage too rapidly. The low vacuum switch deactivates the converter clutch when engine vacuum drops to a specific low level during moderate acceleration just before a part-throttle transmission downshift. The low vacuum switch also deactivates the clutch while the vehicle is coasting because it receives no vacuum from its ported vacuum source.

The high vacuum switch, on diesel engines, deactivates the converter clutch while the vehicle is coasting. The low vacuum switch used on diesel engines only deactivates the converter clutch during moderate acceleration, just prior to a part-throttle downshift. Because the diesel engine's vacuum source is a vacuum pump, rather than from a carburetor port, diesel engines require both the high and the low vacuum switch to achieve the same results as the low vacuum switch used on gasoline engines.

COMPUTER CONTROLLED CONVERTER CLUTCH

With the use of microcomputers governoring the engine fuel and spark delivery, the converter clutch electronic control has been changed to provide the grounding circuit for the solenoid valve through the microcomputer, rather than the governor pressure switch. Sensors are used in place of the formerly used switches. These sensors send signals back to the microcomputers to indicate if the engine is in its proper mode to accept the mechanical lockup of the converter clutch.

Normally a coolant sensor, a throttle position sensor, an engine vacuum sensor and a vehicle speed sensor are used to signal the microcomputer when the converter clutch can be applied. Should a sensor indicate the need for the converter clutch to be deactivated, the grounding circuit to the transmission/transaxle solenoid valve would be interrupted and the converter clutch would be released.

HYDRAULIC CONVERTER CLUTCH OPERATION

Numerous automatic transmissions/transaxles rely upon hydraulic pressures to sense and determine when to apply the converter clutch function. This type of automatic transmission/transaxle unit is considered to be a self-contained unit with only the shift linkage, throttle cable or modulator valve being external. Specific valves, located within the valve body or oil pump housing, are put into operation when a sequence of events occur within the unit. For example, to engage the converter clutch, most all automatic transmissions require the gear ratio to be in the top gear before the converter clutch control valves can be placed in operation. The governor and throttle pressures must maintain specific fluid pressures at various points within the hydraulic circuits to aid in the engagement or disengagement of the converter clutch. In addition, check valves must properly seal and move the exhaust pressurized fluid at the correct time to avoid "shudders" or "chuckles" during the initial application and engagement of the converter clutch.

CENTRIFUGAL CONVERTER CLUTCH

Transmissions/transaxles also use a torque converter that mechanically locks up centrifugally without the use of electronics or hydraulic pressure. At specific input shaft speeds, brake-like shoes move outward from the rim of the turbine assembly, to

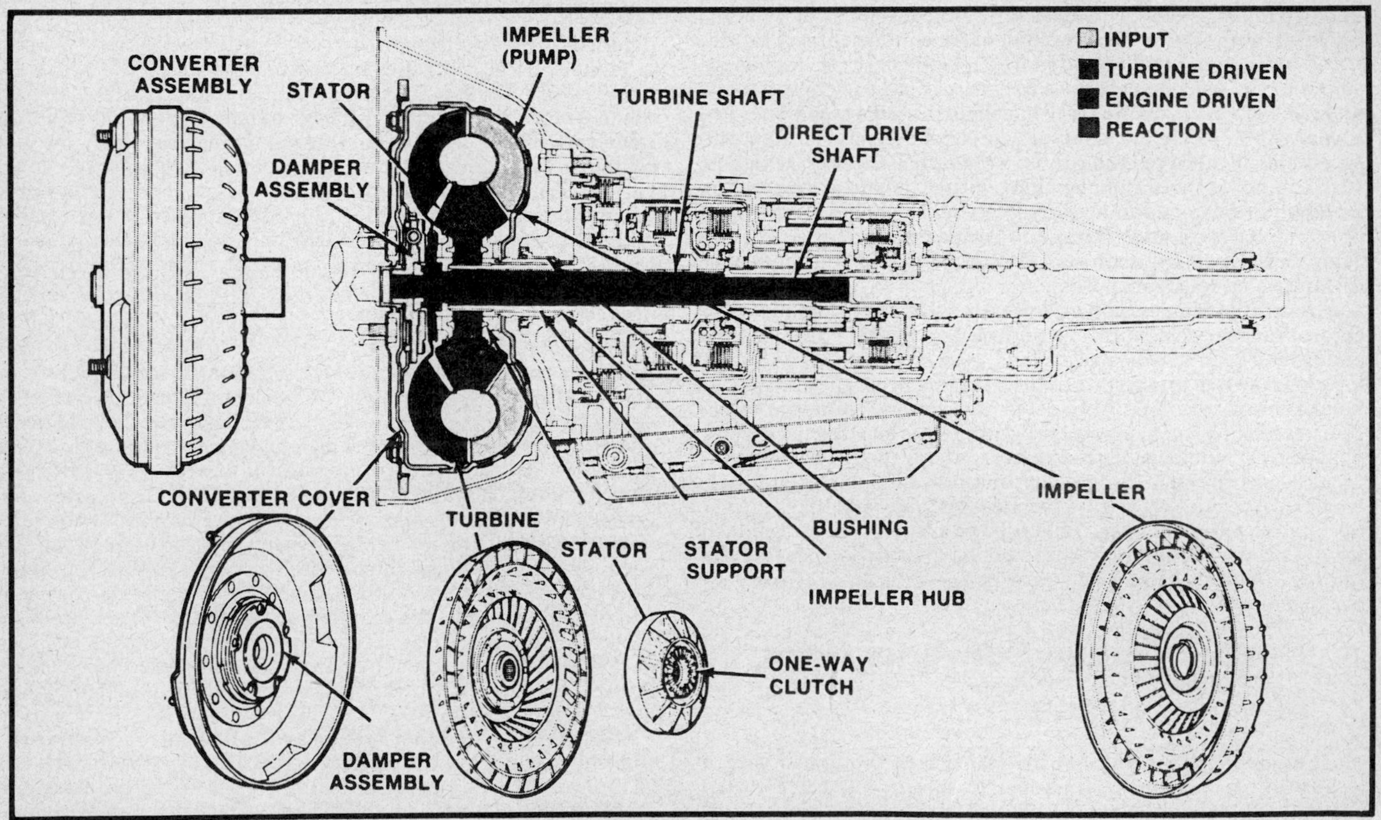
INPUT
TURBINE DRIVEN
ENGINE DRIVEN
REACTION
CONVERTER ASSEMBLY
STATOR
IMPELLER (PUMP)
TURBINE SHAFT
DIRECT DRIVE SHAFT
DAMPER ASSEMBLY
CONVERTER COVER
TURBINE
STATOR
STATOR SUPPORT
BUSHING
IMPELLER HUB
IMPELLER
ONE-WAY CLUTCH
DAMPER ASSEMBLY

Torque converter and direct driveshaft—AOD transmission

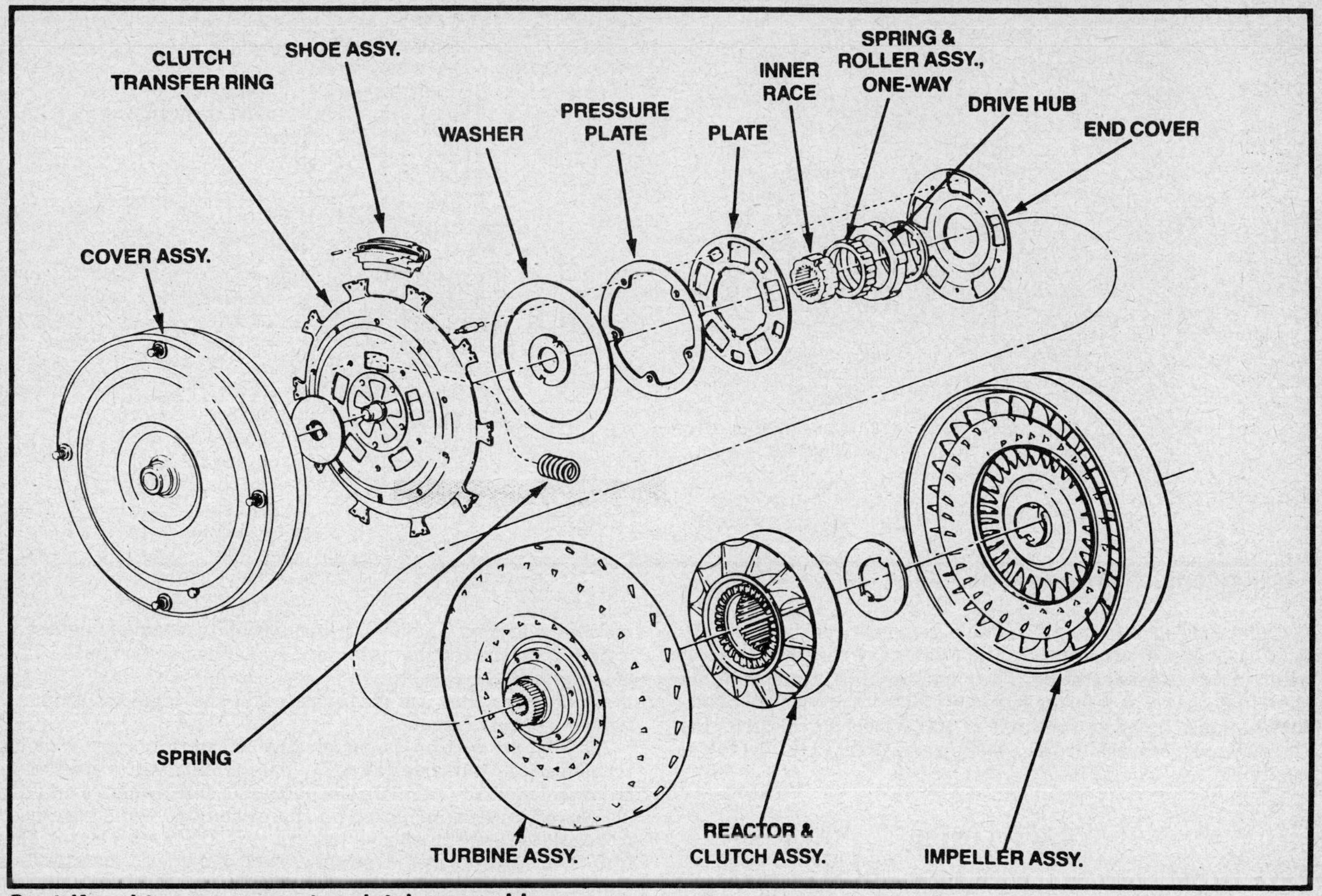
CLUTCH
TRANSFER RING
SHOE ASSY.
INNER
RACE
SPRING &
ROLLER ASSY.,
ONE-WAY
DRIVE HUB
END COVER
WASHER
PRESSURE
PLATE
PLATE
COVER ASSY.
SPRING
TURBINE ASSY.
REACTOR &
CLUTCH ASSY.
IMPELLER ASSY.

Centrifugal torque converter clutch assembly

engage the converter housing, locking the converter unit mechanically together for a 1:1 ratio. Slight slippage can occur at the low end of the rpm scale, but the greater the rpm, the tighter the lockup. Again, it must be mentioned, that when the converter has locked up, the vehicle may respond in the same manner as driving with a clutch and standard transmission. This is considered normal and does not indicate converter clutch or transmission/transaxle problems. Keep in mind if the engine is in need of a tune-ups or repairs, the lockup "shudder" or "chuckle" feeling may be greater.

MECHANICAL CONVERTER LOCKUP

Another type of converter lockup is the Ford Motor Co. AOD automatic overdrive transmission, which uses a direct drive input shaft splined to the damper assembly of the torque converter cover to the direct clutch,thereby bypassing the torque converter reduction components. A second shaft encloses the direct drive input shaft and is coupled between the converter turbine and the reverse clutch or forward clutch, depending upon their applied phase. With this type of unit, when in third gear, the input shaft torque is split, 30% hydraulic and 70% mechanical. When in the overdrive or fourth gear, the input torque is completely mechanical and the transmission is locked mechanically to the engine.

CONFIRMING CONVERTER LOCKUP

To confirm that the lockup function of the torque converter has occurred, check the engine rpm with a tachometer while the vehicle is being driven. If the torque converter is locked up, the engine rpm will decrease approximately 200–400 rpm, at the time of lockup.

Overdrive Units

With need for greater fuel economy, the automatic transmission/transaxles were among the many vehicle components that have been modified to aid in this quest. Internal changes have been made and in some cases, additions of a fourth gear to provide the overdirect or overdrive gear ratio. The reasoning for adding the overdrive capability is that an overdrive ratio enables the output speed of the transmission/transaxle to be greater than the input speed, allowing the vehicle to maintain a given speed with less engine speed. This results in better fuel economy and a slower running engine.

The automatic overdrive unit usually consists of an overdrive planetary gear set, a roller one-way clutch assembly and 2 friction clutch assemblies, one as an internal clutch pack and the second for a brake clutch pack. The overdrive carrier is splined to the turbine shaft, which in turn, is splined into the converter turbine.

Another type of overdrive assembly is a separation of the overdrive components by having them at various points along the gear transassembly and also utilizing them for other gear ranges. Instead of having a brake clutch pack, an overdrive band is used to lock the planetary sun gear. In this type of transmission, the converter cover drives the direct driveshaft clockwise at engine speed, which in turn drives the direct clutch. The direct clutch then drives the planetary carrier assembly at engine

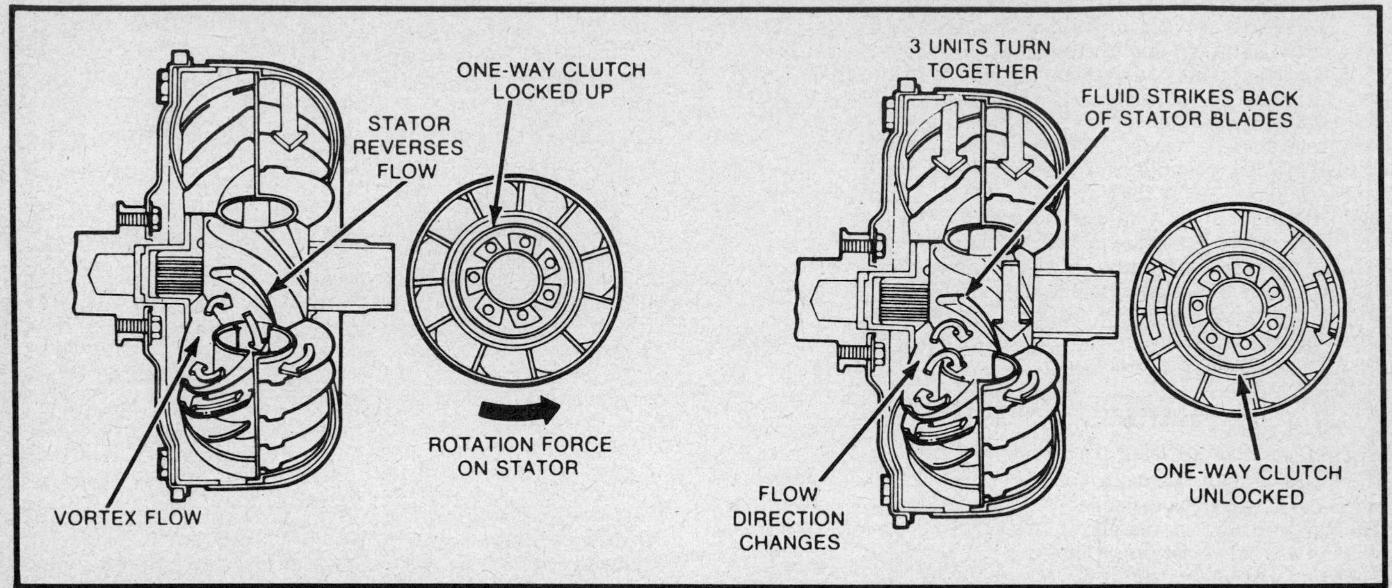
ONE-WAY CLUTCH LOCKED UP
STATOR REVERSES FLOW
ROTATION FORCE ON STATOR
VORTEX FLOW
3 UNITS TURN TOGETHER
FLUID STRIKES BACK OF STATOR BLADES
FLOW DIRECTION CHANGES
ONE-WAY CLUTCH UNLOCKED

Lockup torque converter operation

speed in a clockwise direction. The pinion gears of the planetary gear assembly "walk around" the stationary reverse sun gear, again in a clockwise rotation. The ring gear and output shaft are therefore driven at a faster speed by the rotation of the planetary pinions. Because the input is 100% mechanical drive, the converter can be classified as a lockup converter in the overdrive position.

CVT Units

The continuously variable transmission/transaxle (CVT) is a new type of automatic transmission/transaxle. The CVT offers a vehicle drive ratio that is equal to that of a 5 speed (or more) manual gear box. The CVT transmission/transaxle is lighter in weight than the regular automatic transmission/transaxle. This is because 2 variable sheave pulleys eliminate the need for the mass of gears that are incorporated in the regular automatic transmission/transaxle.

Also in production is an electronic continuously variable transmission/transaxle (ECVT). This transmission uses electronic solenoids to regulate the hydraulic shift controls and has electronic controls mounted on the valve body, which are used to regulate how the sheave pulleys vary their diameters. This unit also incorporates a powder electromagnetic clutch. This clutch consists of a chamber filled with very fine stainless steel powder. The clutch spins free until the coils are energized and in turn magnetize the powder and lockup the clutch.

DIAGNOSING AUTOMATIC TRANSMISSION/TRANSAXLE MALFUNCTIONS

Introduction

Diagnosing automatic transmission/transaxle problems is simplified following a definite procedure and understanding the basic operation of the individual transmission/transaxle that is being inspected or serviced. Do no attempt to short-cut the procedure or take for granted that another technician has performed the adjustments or the critical checks. It may be an easy task to locate a defective or burned-out unit, but the technician must be skilled in locating the primary reason for the unit failure and must repair the malfunction to avoid having the same failure occur again.

Each automatic transmission/transaxle manufacturer has developed a diagnostic procedure for their individual transmissions/transaxles. Although the operation of the units are basically the same, many differences will appear in the construction, method of unit application and the hydraulic control system.

The same model transmission/transaxle can be installed in different makes of vehicles and are designed to operate under different load stresses, engine applications and road conditions. Each make of vehicle will have specific adjustments or use certain outside manual controls to operate the individual unit, but may not interchange with another transmission/transaxle vehicle application from the same manufacturer.

The identification of the transmission/transaxle is most important so that the proper preliminary inspections and adjustments may be done and if in need of a major overhaul, the correct parts may be obtained and installed to avoid costly delays.

Systematic Diagnosis

Transmission/transaxle manufacturers have compiled diagnostic aids to use when diagnosing malfunctions through oil pressure tests or road test procedures. Diagnostic symptom charts, operational shift speed charts, oil pressure specifications, clutch and band application charts and oil flow schematics are some of the aids available.

Numerous manufacturers and re-manufacturers require a diagnosis check sheet be filled out by the diagnostician, pertaining to the operation, fluid level, oil pressure (idling and at various speeds), verification of adjustments and possible causes and the need correction of the malfunctions. In certain cases, authorization must be obtained before repairs can be done, with the diagnostic check sheet accompanying the request for payment or warranty claim, along with the return of defective parts.

It is a good policy to use the diagnostic check sheet for the evaluation of all transmission/transaxle diagnosis and include the complete check sheet in the owners service file, should future reference be needed.

Many times, a rebuilt unit is exchanged for the defective unit, saving down time for the owner and vehicle. However, if the diagnostic check sheet would accompany the removed unit to the rebuilder, more attention could be directed to verifying and repairing the malfunctioning components to avoid costly comebacks of the rebuilt unit, at a later date. Most large volume rebuilders employ the use of dynamometers, as do the new unit manufacturers, to verify proper build-up of the unit and its correct operation before it is put in service.

General Diagnosis

Should the diagnostician not use a pre-printed check sheet for the diagnosing of the malfunctioning unit, a sequence for diagnosis of the gear box is needed to proceed in an orderly manner. During the road test, use all the selector ranges while noting any differences in operation or changes in oil pressure, so that the defective unit or hydraulic circuit can be isolated and the malfunction corrected. A suggested sequence is as follows:

1. Inspect and correct the fluid level.
2. Inspect and adjust the throttle or kickdown linkage.
3. Inspect and adjust the manual linkage.
4. Be sure to properly install a pressure gauge to the transmission/transaxle as instructed in the individual repair section.
5. Road test the vehicle (with owner if possible).

Road Test Diagnosis

Prior to driving the vehicle on a road test, have the vehicle operator explain the malfunction of the transmission/transaxle as fully and as accurate as possible. Because the operator may not have the same technical knowledge as the diagnostician, ask questions concerning the malfunction in a manner that the operator can understand. It may be necessary to have the operator drive the vehicle, with the technician, on a road test and to identify the problem. The diagnostician can observe the manner in which the transmission/transaxle is being operated and can point out constructive driving habits to the operator to improve operation reliability.

Many times, an actual transmission/transaxle malfunction can occur without the operator's knowledge, due to slight slippages occurring and increasing in duration while the vehicle is being driven. Had the operator realized that a malfunction existed, minor adjustments possibly could have been done to avoid costly repairs.

Be aware of the engine's performance. For example, if a vacuum modulator valve is used to control the throttle pressure, an engine performing poorly cannot send the proper vacuum signals to the transmission/transaxle for proper application of the throttle pressure and control pressure, in the operation of the bands and clutches. Slippages and changes in shift points can occur.

During the road test, the converter operation must be considered. Related converter malfunctions affecting the road test could be that the stator assembly free wheels or the stator assembly remains locked up.

When the stator roller clutch freewheels in both directions, the vehicle will have poor acceleration from a standstill. At speeds above approximately 45 mph, the vehicle will act normally. A check to make on the engine is to accelerate to a high rpm in neutral. If the engine responds properly, this is an indication that the engine is operating satisfactorily and the problem may be with the stator.

When the stator remains locked up, the engine rpm and the vehicle speed will be restricted at higher speeds, although the vehicle will accelerate from a standstill normally. Engine overheating may be noticed and visual inspection of the converter may reveal a blue color, resulting from converter overheating.

Clutch and Band Application Diagnosis

During the road test, operate the transmission/transaxle in each gear position and observe the shifts for signs of any slippage, variation, sponginess or harshness. Note the speeds at which the upshifts and downshifts occur. If slippage and engine flare-up occurs in any gear, clutch band or overrunning clutch problems are indicated and depending upon the degree of wear, a major overhaul may be indicated.

The clutch and band application chart in each transmission/transaxle section provides a basis for road test analysis to determine the internal units applied or released in a specific gear ratio.

NOTE: Some transmissions/transaxles use brake and clutches in place of bands and are usually indicated at B1 and B2 on the unit application chart. These components are diagnosed in the same manner as one would diagnose a band equipped gearbox.

TRANSMISSION/TRANSAXLE NOISE DIAGNOSIS

In diagnosisng transmission/transaxle noises, the diagnostician must be alert to any abnormal noises from the transmission/transaxle area or any excessive movement of the engine or the transmission/transaxle assembly during torque application or transmission/transaxle shifting.

CAUTION

Before attempting to diagnose automatic transmission/transaxle noises, be sure the noises do not originate from the engine components, such as the water pump, alternator, air conditioner compressor, power steering or the air injection pump. Isolate these components by removing the proper drive belt and operate the engine. Do not operate the engine longer than 2 minutes at a time to avoid overheating.

1. Whining or siren type noises can be considered normal if occurring during a stall speed test, due to the fluid flow through the converter.
2. A continual whining noise with the vehicle stationary and if the noise increases and decreases with the engine speed, the following defects could be present:
   a. Oil level low
   b. Air leakage into pump (defective gasket, O-ring or porosity of a part)
   c. Pump gears damaged or worn
   d. Pump gears assembled backward
   e. Pump crescent interference
3. A buzzing noise is normally the result of a pressure regulator valve vibrating or a sealing ring broken or worn out and will usually come and go, depending upon engine and the transmission/transaxle speed.
4. A constant rattling noise that usually occurs at low engine speed can be the result of the vanes stripped from the impeller or turbine face or internal interference of the converter parts.
5. An intermittent rattleing noise reflects a broken flywheel or flex plate and usually occurs at low engine speed with the transmission/transaxle in gear. Placing the transmission/transaxle in **N** or **P** will change the rattling noise or stop it for a short time.
6. Gear noise (1 gear range) will normally indicate a defective planetary gear unit. Upon shifting into another gear range, the noise will cease. If the noise carries over to the next gear range, but at a different pitch, defective thrust bearings or bushings are indicated.

7. Engine vibration or excessive movement can be caused by transmission/transaxle filler or cooler lines vibrating due to broken or disconnected brackets. If excessive engine or transmission/transaxle movement is noted, look for broken engine or transmission/transaxle mounts.

CAUTION

When necessary to support an engine equipped with metal safety tabs on the mounts, be sure the metal tabs are not in contact with the mount bracket after the engine or transmission/transaxle assembly is again supported by the mounts. A severe vibration can result.

8. Squeal at low vehicle speeds can result from a speedometer driven gear seal, a front pump seal or rear extension seal being dry.

9. The above list of noises can be used as a guide. Noises other than the ones listed can occur around or within the transmission assembly. A logical and common sense approach will normally result in the source of the noise being detected.

Fluid Diagnosis

FLUID INSPECTION AND LEVEL

Most automatic transmissions/transaxles are designed to operate with the fluid level between the **ADD** or **ONE PINT** and **FULL** marks on the dipstick indicator, with the fluid at normal operating temperature. The normal operating temperature is attained by operating the engine assembly for at least 8–15 miles of driving or its equivalent. The fluid temperature should be in the range of 150–200°F when normal operating temperature is attained.

NOTE: If the vehicle has been operated for long periods at high speed or in extended city traffic during hot weather, an accurate fluid level check cannot be made until the fluid cools, normally 30 minutes after the vehicle has been parked, due to fluid heat in excess of 200°F.

The transmission fluid can be checked during 2 ranges of temperatures.

1. Transmission/transaxle at normal operating temperature.
2. Transmission/transaxle at room temperature.

During the checking procedure and adding of fluid to the transmission/transaxle, it is most important not to overfill the reservoir in order to avoid foaming and loss of fluid through the breather, which can cause slippage and transmission/transaxle failure.

Transmission/Transaxle at Room temperature

65–95°F – DIPSTICK COOL TO TOUCH

CAUTION

The automatic transmissions/transaxles are sometimes overfilled because the fluid level is checked when the transmission/transaxle has not been operated and the fluid is cold and contracted. As the transmission/transxle is warmed to normal operating temperature, the fluid level can change as much as ¾ in.

1. With the vehicle on a level surface, engine idling, wheels blocked or parking brake applied, move the selector lever through all the ranges to fill the passages with fluid.
2. Place the selector lever in the **P** position and remove the dipstick from the transmission/transaxle. Wipe clean the re-insert it back into the dipstick tube.
3. Remove the dipstick and observe the fluid level mark on the dipstick stem. The fluid should be directly below the **FULL** indicator.

NOTE: Most dipsticks will have either one mark or two marks, such as dimples or holes in the stem of the dipstick, to indicate the cold level, while others may be marked HOT or COLD levels.

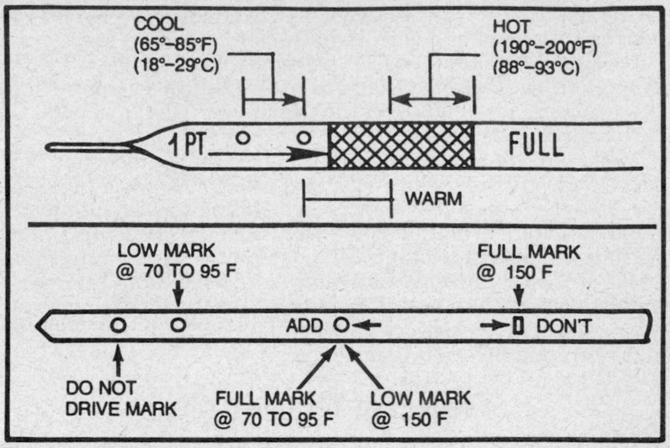
COOL
(65°–85°F)
(18°–29°C)
HOT
(190°–200°F)
(88°–93°C)
1 PT
FULL
WARM
LOW MARK
@ 70 TO 95 F
FULL MARK
@ 150 F
ADD
DON'T
DO NOT
DRIVE MARK
FULL MARK
@ 70 TO 95 F
LOW MARK
@ 150 F

Typical fluid level indicators

4. Add enough fluid, as necessary, to the transmission/transaxle, but do not overfill.

CAUTION

This operation is most critical, due to the expansion of the fluid under heat.

Transmission/Transaxle at Normal Operating Temperature

150–200°F – DIPSTICK HOT TO THE TOUCH

1. With the vehicle on a level surface, engine idling, wheels blocked or parking brake applied, move the gear selector lever through all the ranges to fill the passages with fluid.
2. Place the selector lever in the **P** position and remove the dipstick from the transmission/transaxle. Wipe clean and reinsert and dipstick to its full length into the dipstick tube.
3. Remove the dipstick and observe the fluid level mark on the dipstick stem. The fluid level should be between the **ADD** and the **FULL** marks. If necessary, add fluid through the filler tube to bring the fluid level to its proper height.
4. Reinstall the dipstick and be sure it is sealed to the dipstick filler tube to avoid the entrance of dirt or water.

FLUID TYPE SPECIFICATIONS

The automatic transmission fluid is used for numerous functions such as a power-transmitting fluid in the torque converter, a hydraulic fluid in the hydraulic control system, a lubricating agent for the gears, bearings and bushings, a frictoin-controlling fluid for the bands and clutches and a heat transfer medium to carry the heat to an air or cooling fan arrangement.

Because of the varied automatic transmission/transaxle designs, different frictional characteristics of the fluids are required so that one fluid cannot assure freedom from chatter or squawking from the bands and clutches. Operating temperatures have increased sharply in many new transmissions/transaxles and the transmission drain intervals have been extended or eliminated completely. It is therefore most important to install the proper automatic transmission fluid into the automatic transmission/tranaxle design for its use.

FLUID CONDITION

During the checking of the fluid level, the fluid condition should be inspected for color and odor. The normal color of the fluid is deep red or orange-red and should not be a burned brown or black color. If the fluid color should turn to a green/brown shade at an early stage of transmission/transaxle operation and have an offensive odor, but not a burned odor, the fluid condition is considered normal and not a positive sign of required maintenance or transmission/transaxle failure.

With the use of absorbent white paper, wipe the dipstick and examine the stain for black, brown or metallic specks, indicating clutch, band or bushing failure, and for gum or varnish on the dipstick or bubbles in the fluid, indicating either water or antifreeze in the fluid.

Should there be evidence of water, antifreeze or specks of residue in the fluid, the oil pan should be removed and the sediment inspected. If the fluid is contaminated or excessive solids are found in the removed oil pan, the transmission/transaxle should be disassembled, completely cleaned and overhauled. In addition to the cleaning of the transmission/transaxle, the converter and transmission/transaxle cooling system should be cleaned and tested.

SYSTEM FLUSHING

Much reference has been made to the importance of flushing the transmission/transaxle fluid coolers and lines during an overhaul. With the increased use of converter clutch units and the necessary changes to the internal fluid routings, the passage of contaminated fluid, sludge or metal particles to the fluid cooler is more predominate. In most cases, the fluid returning from the fluid cooler is directed to the lubrication system and should the system be deprived of lubricating fluid due to blockage premature unit failure will occur.

Procedure

1. Disconnect both fluid lines from the transmission/transaxle assemblies, leaving the lines attached to the cooler.
2. Add a length of hose to the return line and place in a container. Flush both lines and the cooler at the same time.

NOTE: When flushing the cooling components, use a commercial flushing fluid or its equivalent. Reverse flush the lines and cooler with the flushing fluid and pulsating air pressure. Continue the flushing process until clean flushing fluid appears. Remove the flushing fluid by the addition of transmission fluid through the lines and cooler.

Special Tools

There are an unlimited amount of special tools and accessories available to the transmission rebuilder to lessen the time and effort required in performing the diagnosing and overhaul of the automatic transmission/transaxles. Specific tools are necessary during the disassembly and assembly of each unit and its subassemblies. Certain tools can be fabricated, but it becomes the responsibility of the repair shop operator to obtain commercially manufactured tools to insure quality rebuilding and to avoid costly "come backs".

The commercial labor saving tools range from puller sets, bushing and seal installer sets, compression tools and presses (both mechanically and hydraulically operated), holding fixtures, oil pump aligning tools, degreaser tanks, steam cleaners, converter flushing machines, transmission/transaxle jacks and lifts, to name a few. For specific information concerning the various tools, a parts and tool supplier should be consulted.

The use of the basic measuring tools has become more critical in the rebuilding process. The increased use of front drive transaxles, in which both the automatic transmission/transaxle and the final drive gears are located, has required the rebuilder to adhere to specifications and tolerances more closely than ever before.

Bearings must be torqued or adjusted to specific preloads in order to meet the rotating torque drag specifications. The end play and backlash of the varied shafts and gears must be measured to avoid excessive tightness or looseness. Critical tensioning bolts must be torqued to specification.

Dial indicators must be protected and used as a delicate measuring instrument. A mutilated or un-calibrated dial indicator invites premature unit failure and destruction. Torque wrenches are available in many forms, some cheaply made and others, accurate and durable under constant use. To obtain accurate readings and properly applied torque, recalibration should be applied to the torque wrenches periodically, regardless of the type used. Micrometers are used as precise measuring tools and should be properly stored when not in use. Instructions on the recalibration of the micrometers and a test bar usually accompany the tool when it is purchased.

Other measuring tools are available to the rebuilder and each in their own way, must be protected when not in use to avoid causing mis-measuring in the fitting of a component to the unit.

Pan Gasket Identification

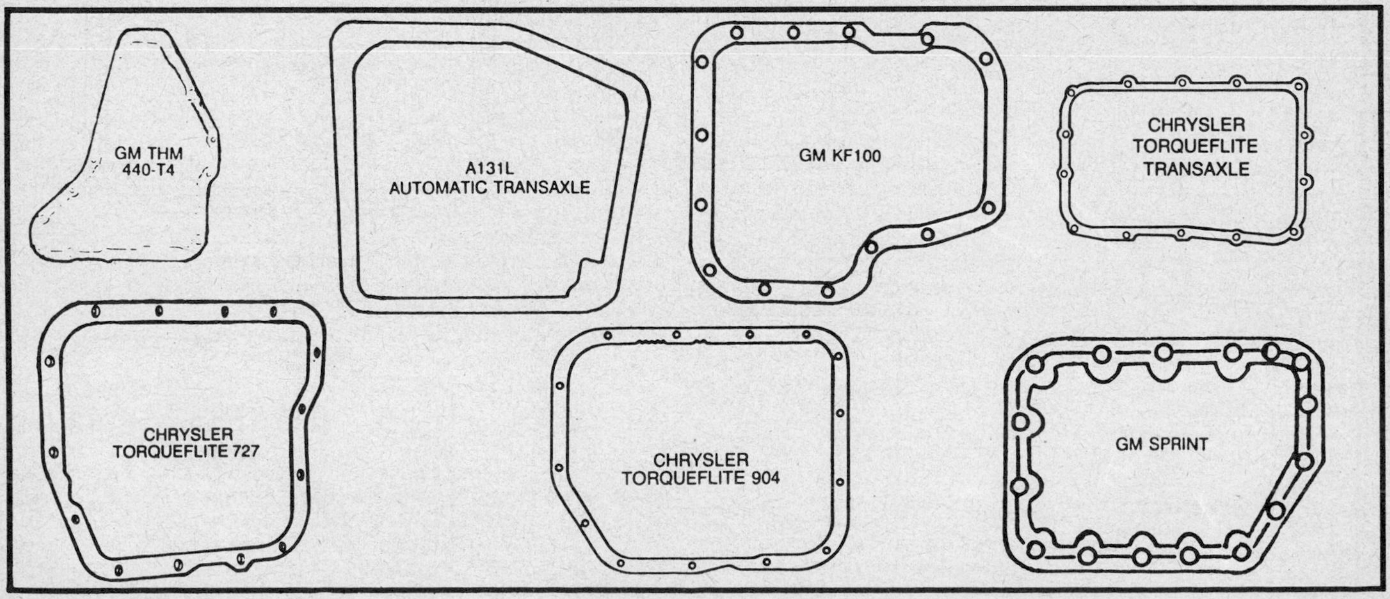
GM THM 440-T4
A131L AUTOMATIC TRANSAXLE
GM KF100
CHRYSLER TORQUEFLITE TRANSAXLE
CHRYSLER TORQUEFLITE 727
CHRYSLER TORQUEFLITE 904
GM SPRINT

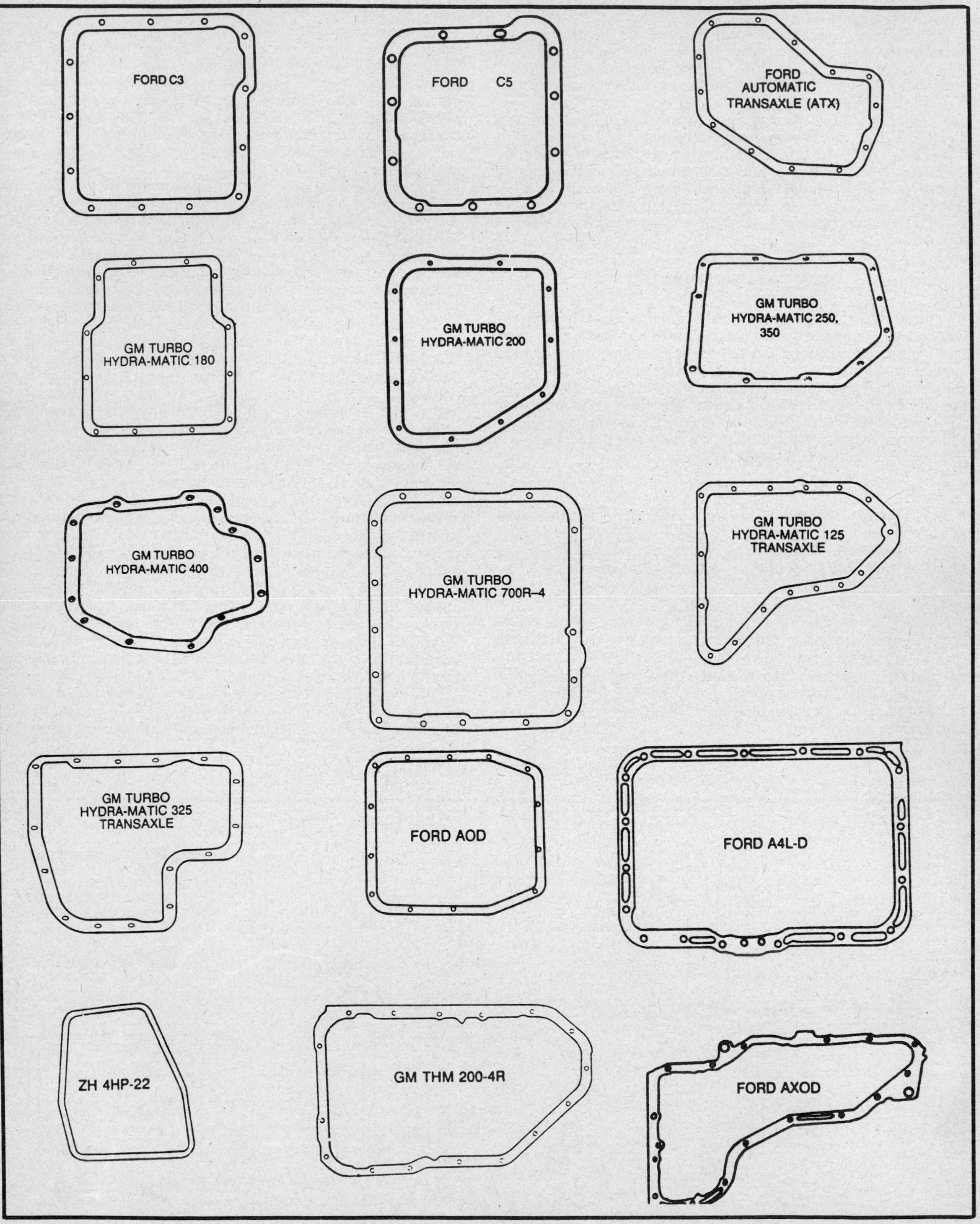
FORD C3
FORD C5
FORD AUTOMATIC TRANSAXLE (ATX)
GM TURBO HYDRA-MATIC 180
GM TURBO HYDRA-MATIC 200
GM TURBO HYDRA-MATIC 250, 350
GM TURBO HYDRA-MATIC 400
GM TURBO HYDRA-MATIC 700R–4
GM TURBO HYDRA-MATIC 125 TRANSAXLE
GM TURBO HYDRA-MATIC 325 TRANSAXLE
FORD AOD
FORD A4L-D
ZH 4HP-22
GM THM 200-4R
FORD AXOD

Section 2

Automatic Transmission/ Transaxle Applications

AMC-JEEP-EAGLE TRANSMISSION APPLICATIONS

Transmission	Year	Engine	Vehicle	Transmission Code
727	1984	2.5L	CJ	89-33-000-914, 89-33-000-915, 89-33-000-918, 89-53-001-836
			Wagoneer	89-33-000-914, 89-33-000-915, 89-33-000-918, 89-53-001-836
		2.8L	Wagoneer	89-33-000-914, 89-33-000-915, 89-33-000-918, 89-53-001-836
		4.2L	CJ	89-33-000-914, 89-33-000-915, 89-33-000-918, 89-53-001-836
			J10	89-33-000-914, 89-33-000-915, 89-33-000-918, 89-53-001-836
		5.9L	Grand Wagoneer	89-33-000-914, 89-33-000-915, 89-33-000-918, 89-53-001-836
			J10	89-33-000-914, 89-33-000-915, 89-33-000-918, 89-53-001-836
			J20	89-33-000-914, 89-33-000-915, 89-33-000-918, 89-53-001-836
727	1985	2.1L	Cherokee	89-33-000-914, 89-33-000-915, 89-33-000-918, 89-53-001-836
		2.5L	CJ	89-33-000-914, 89-33-000-915, 89-33-000-918, 89-53-001-836
		4.2L	CJ	89-33-000-914, 89-33-000-915, 89-33-000-918, 89-53-001-836
			Grand Wagoneer	89-33-000-914, 89-33-000-915, 89-33-000-918, 89-53-001-836
			J10	89-33-000-914, 89-33-000-915, 89-33-000-918, 89-53-001-836
		5.9L	Grand Wagoneer	89-33-000-914, 89-33-000-915, 89-33-000-918, 89-53-001-836
			J10	89-33-000-914, 89-33-000-915, 89-33-000-918, 89-53-001-836
			J20	89-33-000-914, 89-33-000-915, 89-33-000-918, 89-53-001-836
727	1986	2.5L	CJ	89-33-000-914, 89-33-000-915, 89-33-000-918, 89-53-001-836
		4.2L	CJ	89-33-000-914, 89-33-000-915, 89-33-000-918, 89-53-001-836
			Grand Wagoneer	89-33-000-914, 89-33-000-915, 89-33-000-918, 89-53-001-836
			J10	89-33-000-914, 89-33-000-915, 89-33-000-918, 89-53-001-836
		5.9L	Grand Wagoneer	89-33-000-914, 89-33-000-915, 89-33-000-918, 89-53-001-836
			J10	89-33-000-914, 89-33-000-915, 89-33-000-918, 89-53-001-836
			J20	89-33-000-914, 89-33-000-915, 89-33-000-918, 89-53-001-836
727	1987	4.2L	J10	53005-019
		5.9L	Grand Wagoneer	53005-019
			J10	53005-019
			J20	53005-019
727	1988	4.2L	J10	53005-019
		5.9L	J10	53005-019
			J20	53005-019

AMC-JEEP-EAGLE TRANSMISSION APPLICATIONS

Transmission	Year	Engine	Vehicle	Transmission Code
727	1989	5.9L	Grand Wagoneer	53005-019
998	1984	2.5L	Eagle	33000-916
		4.2L	Eagle	33000-916
998	1985	4.2L	Eagle	33002-366
998	1986	4.2L	Eagle	33002-366
998	1987	4.2L	Eagle	33004-126
998	1988	4.2L	Eagle	33004-126 Type I, 83505-566 Type II
999	1984	2.5L	CJ	89-33-000-913, 89-33-000-917
		4.2L	CJ	89-33-000-913, 89-33-000-917
			Grand Wagoneer	89-33-000-913, 89-33-000-917
			J10	89-33-000-913, 89-33-000-917
	1985	2.5L	CJ	89-33-000-913, 89-33-000-917
		4.2L	CJ	89-33-000-913, 89-33-000-917
			Grand Wagoneer	89-33-000-913, 89-33-000-917
			J10	89-33-000-913, 89-33-000-917
	1986	2.5L	CJ	89-33-000-913, 89-33-000-917
		4.2L	CJ	89-33-000-913, 89-33-000-917
			Grand Wagoneer	89-33-000-913, 89-33-000-917
			J10	89-33-000-913, 89-33-000-917
	1987	2.5L	Wrangler	53003-074, 53005-018, 83505-55
		4.2L	Wrangler	53003-074, 53005-018, 83505-55
	1988	2.5L	Wrangler	53003-074, 53005-018, 83505-55
		4.2L	Wrangler	53003-074, 53005-018, 83505-55
	1989	4.2L	Wrangler	53003-074, 53005-018, 83505-55
AR4	1987	2.2L	Medallion	AR4-xxx ①
		2.5L	Premier	AR4-xxx ①
		3.0L	Premier	AR4-xxx ①
AR4	1988	2.2L	Medallion	AR4-xxx ①
		2.5L	Premier	AR4-xxx ①
		3.0L	Premier	AR4-xxx ①
AR4	1989	2.2L	Medallion	AR4-xxx ①
		2.5L	Premier	AR4-xxx ①
		3.0L	Premier	AR4-xxx ①
AW4	1984	2.5L	Cherokee	530001-141, 530001-334, 53002-430
			Wagoneer	530001-141, 530001-334, 53002-430
		2.8L	Cherokee	530001-141, 530001-334, 53002-430
			Wagoneer	530001-141, 530001-334, 53002-430
AW4	1985	2.1L	Cherokee	530001-334
			Wagoneer	530001-334
		2.5L	Cherokee	530001-141, 530001-334, 53002-430
			Wagoneer	530001-141, 530001-334, 53002-430
		2.8L	Cherokee	530001-141, 530001-334, 53002-430
			Wagoneer	530001-141, 530001-334, 53002-430

① Serial numbers in sequence

AMC-JEEP-EAGLE TRANSMISSION APPLICATIONS

Transmission	Year	Engine	Vehicle	Transmission Code
AW4	1986	2.1L	Cherokee	53001-672, 53003-810, 811, JR775-140, 1
			Commanche	53001-672, 53003-810, 811, JR775-140, 141
			Wagoneer	53001-672, 53003-810, 811, JR775-140, 141
		2.5L	Cherokee	53001-672, 53003-810, 811, JR775-140, 141
			Commanche	53001-672, 53003-810, 811, JR775-140, 141
			Wagoneer	53001-672, 53003-810, 811, JR775-140, 141
		2.8L	Cherokee	53002-840, 839, JR775-142, 143
			Commanche	53002-840, 839, JR775-142, 143
			Wagoneer	53002-840, 839, JR775-142, 143
AW4	1987	2.1L	Cherokee	53001-672, 53003-810, 811, JR775-140, 141
			Commanche	53001-672, 53003-810, 811, JR775-140, 141
			Wagoneer	53001-672, 53003-810, 811, JR775-140, 141
		2.5L	Cherokee	53001-672, 53003-810, 811, JR775-140, 141
			Commanche	53001-672, 53003-810, 811, JR775-140, 141
			Wagoneer	53001-672, 53003-810, 811, JR775-140, 141
		2.8L	Cherokee	53002-840, 839, JR775-142, 143
			Commanche	53002-840, 839, JR775-142, 143
		2.8L	Wagoneer	53002-840, 839, JR775-142, 143
		4.0L	Cherokee	53002-840, 839, JR775-142, 143
			Commanche	53002-840, 839, JR775-142, 143
			Wagoneer	53002-840, 839, JR775-142, 143
AW4	1988	2.5L	Cherokee	53001-672, 53003-810, 811, JR775-140, 141
			Commanche	53001-672, 53003-810, 811, JR775-140, 141
			Wagoneer	53001-672, 53003-810, 811, JR775-140, 141
		2.8L	Cherokee	53002-840, 839, JR775-142, 143
			Commanche	53002-840, 839, JR775-142, 143
			Wagoneer	53002-840, 839, JR775-142, 143
		4.0L	Cherokee	53002-840, 839, JR775-142, 143
			Commanche	53002-840, 839, JR775-142, 143
			Wagoneer	53002-840, 839, JR775-142, 143
AW4	1989	2.5L	Cherokee	53001-672, 53003-810, 811, JR775-140, 141
			Commanche	53001-672, 53003-810, 811, JR775-140, 141
			Wagoneer	53001-672, 53003-810, 811, JR775-140, 141
		4.0L	Cherokee	53002-840, 839, JR775-142, 143
			Commanche	53002-840, 839, JR775-142, 143
			Wagoneer	53002-840, 839, JR775-142, 143
MJ3	1987	2.2L	Medallion	MJ3-xxx ①
MJ3	1988	2.2L	Medallion	MJ3-xxx ①
MJ3	1989	2.2L	Medallion	MJ3-xxx ①
ZF-4HP-18	1988	2.5L	Premier	ZF-xxx ①
		3.0L	Premier	ZF-xxx ①
ZF-4HP-18	1989	2.5L	Premier	ZF-xxx ①
		3.0L	Premier	ZF-xxx ①

① Serial numbers in sequence

CHRYSLER-DODGE-PLYMOUTH TRANSMISSION APPLICATIONS

Transmission	Year	Engine	Vehicle	Body Code	Transmission Code
727	1984	3.7L	Pick-Up and Ram	D, W	4058-384, 4295-941, 4329-468, 4329-482, 4329-458, 4329-488
			Van	B	4295-941, 4329-438, 4329-821
		5.2L	Pick-Up and Ram	D, W	4058-384, 4295-941, 4329-468, 4329-482, 4329-458, 4329-488
			Van	B	4295-941, 4329-438, 4329-821
		5.9L	Pick-Up and Ram	D, W	4058-384, 4295-941, 4329-468, 4329-482, 4329-458, 4329-488
			Van	B	4295-941, 4329-438, 4329-821
		7.2L	Pick-Up and Ram	D, W	4058-384, 4295-941, 4329-468, 4329-482, 4329-458, 4329-488
727	1985	3.7L	Pick-Up and Ram	D, W	4058-384, 4295-941, 4329-468, 4329-482, 4329-458, 4329-488
			Van	B	4295-941, 4329-438, 4329-821
		5.2L	Pick-Up and Ram	D, W	4058-384, 4295-941, 4329-468, 4329-482, 4329-458, 4329-488
			Van	B	4295-941, 4329-438, 4329-821
		5.9L	Pick-Up and Ram	D, W	4058-384, 4295-941, 4329-468, 4329-482, 4329-458, 4329-488
			Van	B	4295-941, 4329-438, 4329-821
		7.2L	Pick-Up and Ram	D, W	4058-384, 4295-941, 4329-468, 4329-482, 4329-458, 4329-488
727	1986	3.7L	Pick-Up and Ram	D, W	4377-823, 4331-552, 4377-824, 4431-552, 4348-785, 4348-718
			Van	B	4377-823, 4331-552, 4377-824, 4431-552
		5.2L	Diplomat	M	4412-560
			Fifth Avenue	M	4412-560
			Gran Fury	M	4412-560
			Pick-Up and Ram	D, W	4377-823, 4331-552, 4377-824, 4431-552, 4348-785, 4348-718, 4348-783, 4431-562
			Van	B	4377-823, 4331-552, 4377-824, 4431-552
		5.9L	Pick-Up and Ram	D, W	4377-823, 4331-552, 4377-824, 4431-552, 4348-785, 4348-718, 4348-786
			Van	B	4377-823, 4331-552, 4377-824, 4431-552
727	1987	3.7L	Pick-Up and Ram	D, W	4377-823, 4331-552, 4377-824, 4431-552, 4348-783, 4431-562, 4348-786
			Van	B	4377-823, 4331-552, 4377-824, 4431-552
		5.2L	Diplomat	M	4431-560, 4431-574
			Fifth Avenue	M	4431-560, 4431-574
			Gran Fury	M	4431-560, 4431-574
			Pick-Up and Ram	D, W	4377-823, 4331-552, 4377-824, 4431-552, 4348-783, 4431-562, 4348-786
			Van	B	4377-823, 4331-552, 4377-824, 4431-552
		5.9L	Pick-Up and Ram	D, W	4377-823, 4331-552, 4377-824, 4431-552, 4348-783, 4431-562, 4348-786, 4412-004, 4412-517
			Van	B	4377-823, 4331-552, 4377-824, 4431-552

CHRYSLER-DODGE-PLYMOUTH TRANSMISSION APPLICATIONS

Transmission	Year	Engine	Vehicle	Body Code	Transmission Code
727	1988	3.7L	Pick-Up and Ram	D, W	4431-563, 4431-552, 4471-406
			Van	B	4431-563, 4431-552
		3.9L	Pick-Up and Ram	D, W	4431-563, 4431-552, 4471-406
		5.2L	Diplomat	M	4431-574, 4505-209
			Fifth Avenue	M	4505-209, 4431-574
			Gran Fury	M	4431-574, 4505-209
			Pick-Up and Ram	D, W	4431-563, 4471-406
			Van	B	4431-563, 4431-552
		5.9L	Pick-Up and Ram	D, W	4431-552, 4471-406
			Van	B	4431-563, 4431-552
727	1989	3.7L	Van	B	4431-563, 4431-552
		5.2L	Diplomat	M	4505-209
			Fifth Avenue	M	4505-209
			Pick-Up and Ram	D, W	4431-563, 4471-406
			Van	B	4431-563, 4431-552
		5.9L	Pick-Up and Ram	D, W	4431-552, 4471-406
			Van	B	4431-563, 4431-552
904	1984	3.7L	Pick-Up and Ram	D, W	4058-383
			Van	B	4058-383
		5.2L	Diplomat	M	4058-398, 4329-436, 4295-887, 4329-631
			Fifth Avenue	M	4058-398, 4329-436, 4295-887, 4329-631
			Gran Fury	M	4058-398, 4329-436, 4295-887, 4329-631
904	1985	3.7L	Pick-Up and Ram	D, W	4058-383
			Van	B	4058-383
		5.2L	Diplomat	M	4058-398, 4329-436, 4295-887
			Fifth Avenue	M	4058-398, 4329-436, 4295-887
			Gran Fury	M	4058-398, 4329-436, 4295-887
			Newport	M	4058-398, 4329-436, 4295-887
904	1986	3.7L	Pick-Up and Ram	D, W	4329-633, 4431-566, 4348-782, 4431-555
			Van	B	4329-633, 4431-566, 4348-782, 4431-555
		5.2L	Diplomat	M	4412-001, 4412-002, 4348-703
			Fifth Avenue	M	4412-001, 4412-002, 4348-703
			Gran Fury	M	4412-001, 4412-002, 4348-703
904	1987	3.7L	Pick-Up and Ram	D, W	4329-633, 4431-566, 4348-782, 4431-555
			Van	B	4329-633, 4431-566, 4348-782, 4431-555
904	1988	3.7L	Van	B	4329-633, 4431-566, 4348-782, 4431-555
998	1986	2.2L	Dakota	N	4329-608, 4431-412
		2.6L	Dakota	N	4329-608, 4431-412
998	1987	2.2L	Dakota	N	4446-367
		2.6L	Dakota	N	4446-367
		3.9L	Dakota	N	4446-367

CHRYSLER-DODGE-PLYMOUTH TRANSMISSION APPLICATIONS

Transmission	Year	Engine	Vehicle	Body Code	Transmission Code
998	1988	2.2L	Dakota	N	4446-367, 4431-424, 4429-628
		2.6L	Dakota	N	4446-367, 4431-424, 4429-628
		3.9L	Dakota	N	4446-367, 4431-424, 4429-628
			Pick-Up and Ram	D, W	4471-406, 4431-425
			Van	B	4471-406, 4431-425
998	1989	2.2L	Dakota	N	4446-367, 4431-424, 4429-628
		2.6L	Dakota	N	4446-367, 4431-424, 4429-628
		3.9L	Dakota	N	4446-367, 4431-424, 4429-628
			Pick-Up and Ram	D, W	4471-406, 4431-425
			Van	B	4471-406, 4431-425
999	1984	5.2L	Pick-Up and Ram	D, W	4058-398
			Van	B	4058-398
999	1985	5.2L	Van	B	4058-398
999	1986	5.2L	Pick-Up and Ram	D, W	4348-715, 4431-554, 4329-632, 4431-554
			Van	B	4348-715, 4431-554
999	1987	5.2L	Diplomat	M	4431-501, 4431-572, 4431-503, 4431-573
			Fifth Avenue	M	4431-501, 4431-572, 4431-503, 4431-573
			Gran Fury	M	4431-501, 4431-572, 4431-503, 4431-573
			Pick-Up and Ram	D, W	4348-715, 4431-554, 4329-632, 4431-554
			Van	B	4348-715, 4431-554
999	1988	5.2L	Diplomat	M	4471-529, 4471-533, 4431-572, 4446-365, 4505-207, 4471-537, 4471-531, 4471-535, 4505-171
			Fifth Avenue	M	4471-529, 4471-533, 4431-572, 4446-365, 4505-207, 4471-537, 4471-531, 4471-535, 4505-171
			Gran Fury	M	4471-529, 4471-533, 4431-572, 4446-365, 4505-207, 4471-537, 4471-531, 4471-535, 4505-171
			Pick-Up and Ram	D, W	4431-556, 4431-428, 4431-554, 4431-427, 4431-441
			Van	B	4431-556, 4431-428
999	1989	5.2L	Diplomat	M	4471-533, 4505-207, 4505-171, 4446-368
			Fifth Avenue	M	4471-533, 4505-207, 4505-171, 4446-368
			Gran Fury	M	4471-533, 4505-207, 4505-171, 4446-368
			Pick-Up and Ram	D, W	4431-556, 4431-428, 4431-554, 4431-427, 4431-441
			Van	B	4431-556-4431-428
A413, A470, A670	1984	2.2L	600	K	4295-512, 4295-513
			Aries	K	4295-512, 4295-513
			Caravan	S	4295-763
			Charger	L	4295-512, 4295-513
			Charger Shelby	L	4295-512, 4295-513
			Daytona	G	4295-512, 4295-513
			E-Class	E	4295-512, 4295-513
			Horizon	L	4295-512, 4295-513
			Laser	G	4295-512, 4295-513
			LeBaron	K	4295-512, 4295-513
			Omni	L	4295-512, 4295-513
			Reliant	K	4295-512, 4295-513

CHRYSLER-DODGE-PLYMOUTH TRANSMISSION APPLICATIONS

Transmission	Year	Engine	Vehicle	Body Code	Transmission Code
A413, A470, A670	1984	2.2L	Town and Country	K	4295-512, 4295-513
			Turismo	L	4295-512, 4295-513
			Voyager	S	4295-763
		2.2L ①	600	E	4329-827
			Daytona	G	4329-827
			E-Class	E	4329-827
			Laser	G	4329-827
			Town and Country	K	4329-827
		2.6L	600	E	4295-515, 4329-544, 4295-517, 4329-556, 4329-819
			Aries	K	4295-515, 4329-544, 4295-517, 4329-556, 4329-819
			E-Class	E	4295-515, 4329-544, 4295-517, 4329-556, 4329-819
			LeBaron	K	4295-515, 4329-544, 4295-517, 4329-556, 4329-819
			New Yorker	E	4295-515, 4329-544, 4295-517, 4329-556, 4329-819
			Reliant	K	4295-515, 4329-544, 4295-517, 4329-556, 4329-819
			Town and Country	K	4295-515, 4329-544, 4295-517, 4329-556, 4329-819
			Voyager	S	4295-515, 4329-544, 4295-517, 4329-556, 4329-819
A413, A470, A670	1985	2.2L	600	E, K	4207-905, 4329-506
			Aries	K	4207-905, 4329-506
			Caravan	S	4329-546
			Caravelle	E	4207-905, 4329-506
			Charger	L	4207-905, 4329-506
			Charger Shelby	L	4207-905, 4329-506
			Daytona	G	4207-905, 4329-506
			Horizon	L	4207-905, 4329-506
			Lancer	H	4207-905, 4329-506
			Laser	G	4207-905, 4329-506
			LeBaron	K, H	4207-905, 4329-506
			Omni	L	4207-905, 4329-506
			Reliant	K	4207-905, 4329-506
			Turismo	L	4207-905, 4329-506
			Voyager	S	4329-546
		2.2L ①	600	E, K	4329-538
			Caravelle	E	4329-538
			Charger Shelby	L	4329-538
			Daytona	G	4329-538
			Laser	G	4329-538
			LeBaron	K, H	4329-538
			New Yorker	E	4329-538
			Omni	L	4329-538
			Town and Country	K	4329-538
			Lancer	H	4329-538
		2.6L	600	E, K	4329-547, 4329-564
			Aries	K	4329-547, 4329-564
			Caravan	S	4329-565
			New Yorker	E	4329-547, 4329-564

CHRYSLER·DODGE·PLYMOUTH TRANSMISSION APPLICATIONS

Transmission	Year	Engine	Vehicle	Body Code	Transmission Code
A413, A470, A670	1985	2.6L	Reliant	K	4329-547, 4329-564
			Town and Country	K	4329-547, 4329-564
			Voyager	S	4329-565
A413, A470, A670	1986	2.2L	600	E, K	4377-906, 4377-954, 4446-779, 4377-902, 4377-951, 4446-780, 4377-907, 4377-955, 4446-782
			Aries	K	4377-906, 4377-954, 4446-779, 4377-902, 4377-951, 4446-780, 4377-907, 4377-955, 4446-782
			Caravan	S	4377-903, 4377-952
			Caravelle	E	4377-906, 4377-954, 4446-779, 4377-902, 4377-951, 4446-780, 4377-907, 4377-955, 4446-782
			Charger	L	4377-906, 4377-954, 4446-779, 4377-902, 4377-951, 4446-780, 4377-907, 4377-955, 4446-782
			Daytona	G	4377-906, 4377-954, 4446-779, 4377-902, 4377-951, 4446-780, 4377-907, 4377-955, 4446-782
			Horizon	L	4377-906, 4377-954, 4446-779, 4377-902, 4377-951, 4446-780, 4377-907, 4377-955, 4446-782
			Lancer	H	4377-906, 4377-954, 4446-779, 4377-902, 4377-951, 4446-780, 4377-907, 4377-955, 4446-782
			Laser	G	4377-906, 4377-954, 4446-779, 4377-902, 4377-951, 4446-780, 4377-907, 4377-955, 4446-782
			LeBaron	K, H	4377-906, 4377-954, 4446-779, 4377-902, 4377-951, 4446-780, 4377-907, 4377-955, 4446-782
			Omni	L	4377-906, 4377-954, 4446-779, 4377-902, 4377-951, 4446-780, 4377-907, 4377-955, 4446-782
			Reliant	K	4377-906, 4377-954, 4446-779, 4377-902, 4377-951, 4446-780, 4377-907, 4377-955, 4446-782
			Turismo	L	4377-906, 4377-954, 4446-779, 4377-902, 4377-951, 4446-780, 4377-907, 4377-955, 4446-782
			Voyager	S	4377-903, 4377-952
		2.2L ①	600	E, K	4377-907, 4377-955, 4446-782
			Caravelle	E	4377-907, 4377-955, 4446-782
			Charger Shelby	L	4377-907, 4377-955, 4446-782
			Daytona Shelby	G	4377-907, 4377-955, 4446-782
			Lancer	H	4377-907, 4377-955, 4446-782
			New Yorker	E	4377-907, 4377-955, 4446-782
			Omni	L	4377-907, 4377-955, 4446-782
		2.5L	600	E, K	4377-903, 4377-952, 4377-907, 4377-955, 4377-951, 4446-780
			Aries	K	4377-903, 4377-952, 4377-907, 4377-955, 4377-951, 4446-780
			Caravelle	E	4377-903, 4377-952, 4377-907, 4377-955, 4377-951, 4446-780
			Daytona	G	4377-903, 4377-952, 4377-907, 4377-955, 4377-951, 4446-780
			Lancer	H	4377-903, 4377-952, 4377-907, 4377-955, 4377-951, 4446-780
			Laser	G	4377-903, 4377-952, 4377-907, 4377-955, 4377-951, 4446-780

① Turbocharged engine

CHRYSLER-DODGE-PLYMOUTH TRANSMISSION APPLICATIONS

Transmission	Year	Engine	Vehicle	Body Code	Transmission Code
A413, A470, A670	1986	2.5L	LeBaron	K, H	4377-903, 4377-952, 4377-907, 4377-955, 4377-951, 4446-780
			New Yorker	E	4377-903, 4377-952, 4377-907, 4377-955, 4377-951, 4446-780
			Reliant	K	4377-903, 4377-952, 4377-907, 4377-955, 4377-951, 4446-780
		2.6L	Caravan	S	4377-911, 4377-958, 4446-784
			Voyager	S	4377-911, 4377-958, 4446-784
A413, A470, A670	1987	2.2L	600	E, K	4377-954, 4431-483, 4471-483, 4431-484, 4431-485, 4431-485
			Aries	K	4377-954, 4431-483, 4471-483, 4431-484, 4431-485, 4431-485
			Caravan	S	4431-486, 4471-486
			Caravelle	E	4377-954, 4431-483, 4471-483, 4431-484, 4431-485, 4431-485
			Charger	L	4377-954, 4431-483, 4471-483, 4431-484, 4431-485, 4431-485
			Horizon	L	4377-954, 4431-483, 4471-483, 4431-484, 4431-485, 4431-485
			Lancer	H	4377-954, 4431-483, 4471-483, 4431-484, 4431-485, 4431-485
			LeBaron	K, H	4377-954, 4431-483, 4471-483, 4431-484, 4431-485, 4431-485
			Omni	L	4377-954, 4431-483, 4471-483, 4431-484, 4431-485, 4431-485
			Reliant	K	4377-954, 4431-483, 4471-483, 4431-484, 4431-485, 4431-485
			Shadow	P	4377-954, 4431-483, 4471-483, 4431-484, 4431-485, 4431-485
			Sundance	P	4377-954, 4431-483, 4471-483, 4431-484, 4431-485, 4431-485
			Town and Country	K	4377-954, 4431-483, 4471-483, 4431-484, 4431-485, 4431-485
			Turismo	L	4377-954, 4431-483, 4471-483, 4431-484, 4431-485, 4431-485
			Voyager	S	4431-486, 4471-486
		2.2L ①	600	E, K	4377-955, 4431-485, 4471-485
			Caravelle	E	4377-955, 4431-485, 4471-485
			Charger Shelby	L	4377-955, 4431-485, 4471-485
			Lancer	H	4377-955, 4431-485, 4471-485
			LeBaron	J,K,H	4377-955, 4431-485, 4471-485
			New Yorker	E	4377-955, 4431-485, 4471-485
			Omni	L	4377-955, 4431-485, 4471-485
			Shadow	P	4377-955, 4431-485, 4471-485
			Shadow	P	4377-955, 4431-485, 4471-485
			Sundance	P	4377-955, 4431-485, 4471-485
			Town and Country	K	4377-955, 4431-485, 4471-485
		2.5L	600	E, K	4431-491, 4431-492

① Turbocharged engine

CHRYSLER·DODGE·PLYMOUTH TRANSMISSION APPLICATIONS

Transmission	Year	Engine	Vehicle	Body Code	Transmission Code
A413, A470, A670	1987	2.5L	Aries	K	4431-491, 4431-492
			Caravan	S	4431-492, 4471-486
			Caravelle	E	4431-491, 4431-492
			Daytona	G	4431-491, 4431-492
			Lancer	H	4431-491, 4431-492
			LeBaron	K, H	4431-491, 4431-492
			New Yorker	E	4431-491, 4431-492
			Reliant	K	4431-491, 4431-492
			Voyager	S	4431-492, 4471-486
		3.0L	Caravan	S	4431-493, 4471-493
			Voyager	S	4431-493, 4471-493
A413, A470, A670	1988	2.2L	600	E	4471-568, 4531-085, 4471-491
			Aries	K	4471-568, 4531-085, 4471-491
			Caravelle	E	4471-568, 4531-085, 4471-491
			Horizon	L	4471-568, 4531-085, 4471-491
			Lancer	H	4471-568, 4531-085, 4471-491
			LeBaron	K, H	4471-568, 4531-085, 4471-491
			Omni	L	4471-568, 4531-085, 4471-491
			Reliant	K	4471-568, 4531-085, 4471-491
			Shadow	P	4471-568, 4531-085, 4471-491
			Sundance	P	4471-568, 4531-085, 4471-491
		2.2L ①	600	E, K	4471-494, 4531-099
			Caravelle	E	4471-494, 4531-099
			Daytona Shelby	G	4471-494, 4531-099
			Lancer	H	4471-494, 4531-099
			LeBaron	J,K,H	4471-494, 4531-099
			New Yorker	E	4471-494, 4531-099
			Omni	L	4471-494, 4531-099
			Shadow	P	4471-494, 4531-099
			Sundance	P	4471-494, 4531-099
			Town and Country	K	4471-494, 4531-099
		2.5L	600	E, K	4531-115, 4471,491
			Aries	K	4531-115, 4471,491
			Caravan	S	4471-486, 4531-087
			Caravelle	E	4531-115, 4471-491
			Daytona	G	4531-115, 4471-491
			Lancer	H	4531-115, 4471-491
			LeBaron	K, H	4531-115, 4471-491
			New Yorker	E	4531-115, 4471-491
			Reliant	K	4531-115, 4471-491
			Sundance	P	4531-115, 4471-491
		2.5L	Voyager	S	4471-486, 4531-087
			Caravan	S	4471-493, 4471-527, 4531-097
			Dynasty	C	4471-569, 4531-027

① Turbocharged engine

CHRYSLER-DODGE-PLYMOUTH TRANSMISSION APPLICATIONS

Transmission	Year	Engine	Vehicle	Body Code	Transmission Code
A413, A470, A670	1988	3.0L	New Yorker	E	4471-569, 4531-027
			Voyager	S	4471-493, 4471-527, 4531-097
A604	1989	2.5L	Caravan	S	4471-895
			Voyager	S	4471-895
		3.0L	Caravan	S	4471-895
			Voyager	S	4471-895
AW372	1987	2.0L	D-50 Pick Up		MD72-4505
		2.6L	D-50 Pick Up 2WD		MD72-4506
AW372	1988	2.0L	D-50 Pick Up		MD72-4505
		2.6L	D-50 Pick Up 2WD		MD72-4506
AW372	1989	2.0L	D-50 Pick Up		MD72-4505
		2.6L	D-50 Pick Up 2WD		MD72-4506
KM148	1987	2.6L	D-50 Pick Up 4WD		MD72-4507
			Raider		MD72-4508
KM148	1988	2.6L	D-50 Pick Up 4WD		MD72-4507
			Raider		MD72-4508
KM148	1989	2.6L	D-50 Pick Up 4WD		MD72-4507
			Raider		MD73-1530
KM171	1985	1.5L	Colt		MD99-6026
KM171	1986	1.5L	Colt		MD99-6026
KM171	1987	1.5L	Colt		MD99-6091
KM171	1988	1.5L	Colt		MD99-6091
KM171	1989	1.5L	Colt		MD99 6183, MD99 6184
KM172	1984	1.6L	Colt		MD70-7058
		2.0L	Colt Vista		MD99-6047, MD70-4959
KM172	1985	1.6L	Colt		MD99-6049
		2.0L	Colt Vista		MD99-6047
KM172	1986	1.6L	Colt		MD99-6049
		2.0L	Colt Vista		MD99-6047
KM172	1987	1.6L	Colt		MD99-6092
		2.0L	Colt Vista		MD99-6047
KM172	1988	1.6L	Colt		MD99-6092
		2.0L	Colt Vista		MD99-6047
KM172	1989	1.6L	Colt		MD99-6161
		2.0L	Colt Vista		MD99-6187, MD99-6186
KM175	1989	2.0L	Summit		MD99-6192

FORD-LINCOLN-MERCURY TRANSMISSION APPLICATIONS

Transmission	Year	Engine	Vehicle	Transmission Code
4EAT	1989	2.2L	Probe	NONE
A4DL	1985	2.3L	Aerostar	85GT-ACA
			Ranger	85GT-ABA
		2.8L	Aerostar	85GT-AMA, BCA
			Bronco II	85GT-AGA, BAA
			Ranger 2WD	85GT-AEA, ALA
			Ranger 4WD	85GT-AGA, BAA
A4DL	1986	2.3L	Aerostar	85GT-ACA
			Ranger	86GT-AAA
		2.8L	Aerostar	85GT-AMA, BCA
		2.9L	Bronco II 2WD	86GT-ABA, ACA
			Bronco II 4WD	86GT-DAA, EAA
			Ranger 2WD	86GT-BAA, CAA
			Ranger 4WD	86GT-DAA, EAA
		3.0L	Aerostar	86GT-KAA, LAA
A4DL	1987	2.3L	Cougar	87GT-AAA, BAA
			Mustang	87GT-ABA
			Ranger	87GT-CAA, CAB
			Thunderbird	87GT-AAA, BAA
		2.9L	Bronco II	87GT-KAA, KAG, LAA, HAE
			Ranger	87GT-DAA, DAC, FAA, FAD, HAA, HAE, KAA, KAG
		3.0L	Aerostar	87GT-MAA, NAA
A4DL	1988	2.3L	Mustang	88GT-HAA, NAA
			Ranger	88GT-ABB
			Thunderbird	88GT-GAA, KAA
		2.9L	Bronco II	88GT-DAA, EAA
			Ranger	88GT-BAA, CAA
		3.0L	Aerostar	88GT-LAA, LAB, MAA, MAB
A4DL	1989	2.3L	Mustang	88GT-HAA, NAA
			Ranger	89GT-AAB, BBB
			Thunderbird	88GT-GAA, KAA
		2.9L	Bronco II	89GT-BAA, BBA, CAA, DAA
			Ranger 2WD	89GT-AAA, BAA
			Ranger 4WD	89GT-CAA, DAA
		3.0L	Aerostar	89GT-EAA, GAA, TAA, TBA
AOD	1984	3.8L	Capri	PKA-BZ1, 2, 3, 4, CD1, 2, 3, 4
			Cougar	PKA-BT6, 7, 8, 9, 10, CB6, 7, 8, 9, 10
			LTD	PKA-BT6, 7, 8, 9, 10, CB6, 7, 8, 9, 10
			Marquis	PKA-BT6, 7, 8, 9, 10, CB6, 7, 8, 9, 10
			Mustang	PKA-BZ1, 2, 3, 4, CD1, 2, 3, 4
			Thunderbird	PKA-BT6, 7, 8, 9, 10, CB6, 7, 8, 9, 10
		4.9L	F-100/250	PKB-E6, E7, E8, E9, E10, F4, F5, F6, F7, F8, F9
		5.0L	Capri	PKA-BW1, 2, 3, 4, BZ1, 2, 3, 4
			Continental	PKA-BD18, 19, 20, 21, 22
			Cougar	PKA-K6, 7, 8, 9, 10

FORD-LINCOLN-MERCURY TRANSMISSION APPLICATIONS

Transmission	Year	Engine	Vehicle	Transmission Code
AOD	1984	5.0L	Crown Victoria	PKA-AG23, 24, 25, 26, 27, AU23, 24, 25, 26, 27, AY18, 19, 20, 21, 22, BB18, 19, 20, 21, 22
			F-150/250	PKB-A26, A27, A28, A29, A30, A31
			Gran Marquis	PKA-AG23, 24, 25, 26, 27, AU23, 24, 25, 26, 27, AY18, 19, 20, 21, 22, BB18, 19, 20, 21, 22
			LTD	PKA-CE1, 2, 3, CF1, 2, 3
			Mark VII	PKA-BV1, 2, 3, 4
			Marquis	PKA-CE1, 2, 3, CF1, 2, 3
			Mustang	PKA-BW1, 2, 3, 4
			Thunderbird	PKA-K6, 7, 8, 9, 10
			Town Car	PKA-M31, 32, 33, 34, 35, BC12, 13, 14, 15, 16
		5.8L	Crown Victoria	PKA-C31, 32, 33, 34, 35, AS23, 24, 25, 26, 27
			Gran Marquis	PKA-C31, 32, 33, 34, 35, AS23, 24, 25, 26, 27
AOD	1985	3.8L	Capri	PKA-CD7
			Cougar	PKA-CB13
			LTD	PKA-CB13
			Marquis	PKA-CB13
			Mustang	PKA-CD7
			Thunderbird	PKA-CB13
		4.9L	E-150/250 Van	PKB-F12
			F-150	PKB-E12, F12
		5.0L	Bronco	PKB-K, M
			Capri	PKA-BW7
			Continental	PKA-BD25
			Cougar	PKA-K13, CJ
			Crown Victoria	PKA-AG30, AY25
			E-150/250 Van	PKB-A33, G
			F-150/250	PKB-A33, J, K, L
			F-150/250 4WD	PKB-M
			Gran Marquis	PKA-AG30, AY25
			LTD	PKA-BW17, CE6, CF6
			Mark VII	PKA-BD25
			Mustang	PKA-BW7
			Thunderbird	PKA-K13, CJ
			Town Car	PKA-M38, BC19
		5.8L	Crown Victoria	PKA-C38
			Gran Marquis	PKA-C38
AOD	1986	2.3L	Capri	PKA-CD9
		3.8L	Cougar	PKA-CB15
			LTD	PKA-CB15
			Marquis	PKA-CB15
			Mustang	PKA-CB15
		4.9L	E-150/250 Van	PKB-F17
			F-150	PKB-E17, E18, F17, F18
		5.0L	Bronco	PKB-K5, K6, K7, M5, M6, M7
			Capri	PKA-CY4

FORD-LINCOLN-MERCURY TRANSMISSION APPLICATIONS

Transmission	Year	Engine	Vehicle	Transmission Code
AOD	1986	5.0L	Continental	PKA-CS4, CT4
			Cougar	PKA-CZ4
			Crown Victoria	PKA-CL4, CM4, DB, CN4
			E-150/250	PKB-J5, J6, J7, N1, N2, N3
			F-150/250	PKB-J5, J6, J7, J8, L5, L6, L7, L8
			F-150/250 4WD	PKB-K5, K6, K7, M5, M6, M7
			Gran Marquis	PKA-CL4, CM4, DB
			Mark VII	PKA-CU4, CV4, CW4
			Mustang	PKA-CY4
			Thunderbird	PKA-CZ4
			Town Car	PKA-CP5, DC2
		5.8L	Crown Victoria	PKA-C38, AS30
			Gran Marquis	PKA-C38
AOD	1987	3.8L	Cougar	PKA-CB15, DK
			Thunderbird	PKA-CB15, DK
		4.9L	E-150 Van	PKB-T2, T3, U2, U3
		5.0L	Bronco	PKB-P2, R2
			Continental	PKA-CS5, CT5
			Cougar	PKA-CZ5, DE1
			Crown Victoria	PKA-CL4, 5, CM4, 5, CN4, 5, DB, 1
			E-150/250 Van	PKB-J10, N5
			F-150/250	PKB-J10, L9, P2, R2, T2, T3
			Gran Marquis	PKA-CL4, 5, CM4, 5, DB, 1
			Mark VII	PKA-CU5, CV4, CW5, DG, DJ
			Mustang	PKA-DL, 1, DD3
			Thunderbird	PKA-CZ5, DE1
			Town Car	PKA-CP5, DC2
		5.8L	Crown Victoria	PKA-C40, AS32
			Gran Marquis	PKA-C40
AOD	1988	3.8L	Cougar	PKA-DK1, 2
			Thunderbird	PKA-DK1, 2
		4.9L	E-150 Van	PKB-AA, AB
			F-150	PKB-AA
		5.0L	Bronco	PKB-Y, Z
			Cougar	PKA-DR2, DS2
			Crown Victoria	PKA-CM8, CN7, CN8, CN9, DB3, DB4, DU, DU1, EM1
			E-150 Van	PKB-N
			F-150/250	PKB-J12, L11
			F-150/250 4WD	PKB-Y, Z
			Gran Marquis	PKA-CM8, DB3, DB4
			Mark VII	PKA-DG4, DG5, DG7, DJ4, DJ6
			Mustang	PKA-DD4, DD5, DL2, DL3
			Thunderbird	PKA-DK1, 2, DR2, DS2
			Town Car	PKA-CP7, CP8, CP9, DC4, DC5
		5.8L	Gran Marquis	PKA-C42, C43

FORD-LINCOLN-MERCURY TRANSMISSION APPLICATIONS

Transmission	Year	Engine	Vehicle	Transmission Code
AOD	1989	3.8L	Thunderbird	PKA-DK1, 2
		4.9L	E-150 Van	PKB-AC, AD
		5.0L	Bronco	PKB-AG, AJ
			Cougar	PKA-DR2, DS2
			Crown Victoria	PKA-CM8, CN7, CN8, CN9, DB3, DB4, DU, DU1, EM1
			E-150 Van	PKB-AF
			F-150/250	PKB-AC, AE, AH
			F-150/250 4WD	PKB-AG, AJ
			Gran Marquis	PKA-CM8, DB3, DB4
			Mark VII	PKA-DG4, DG5, DG7, DJ4, DJ6
			Mustang	PKA-DD4, DD5, DL2, DL3
			Thunderbird	PKA-DK1, 2, DR2, DS2
			Town Car	PKA-CP7, CP8, CP9, DC4, DC5
		5.8L	Gran Marquis	PKA-C42, C43
ATX	1984	1.6L	Escort/EXP	PMA-V3, U1, 2, Z3, PMB-C2, D, R
			Lynx/LN-7	PMA-V3, U1, 2, Z3, PMB-C2, D, R
		2.3L	Tempo	PMA-N1, 2, 3, AA, AA1, 2, 3, AE
			Topaz	PMA-N, N1, 2, 3, AA, AA2, 3, AE
ATX	1985	1.6L	Escort/EXP	PMA-V5, V6, 7, Z6, Z7, Z8, U4, U5, 6
			Lynx/LN-7	PMA-U4, 5, 6, V5, 6, 7, Z6, 7, 8
		1.9L	Lynx	PMA-U4, 5, 6, V5, 6, 7, Z6, 7, 8
			Topaz	PMA-AD
		2.3L	Tempo	PMA-N5, 6, 7, AA5, AA6
			Topaz	PMA-N5, 6, 7, AA5, AA6, AS
ATX	1986	1.9L	Escort/EXP	PMA-AD, AD1, AD2, AM, AP
			Lynx	PMA-AD, AD1, AD2, AM, AP
		2.3L	Tempo	PMA-N8, N8A, N9, 10, AA8, AA8A, AA9, 10
			Topaz	PMA-N8, N8A, N9, 10, AA8, AA8A, AA9, 10
		2.5L	Sable	PMA-AK, 1, 2, 3
			Taurus	PMA-AK, 1, 2, 3
ATX	1987	1.9L	Excort/EXP	PMA-AU, BE, PMP-W, X
			Lynx	PMA-AU, BE, PMB-W, X
		2.3L	Tempo AWD	PMA-AW
			Topaz	PMA-AV, BD
			Topaz AWD	PMA-AW
		2.5L	Sable	PMA-AK, 1, 2, 3
			Taurus	PMA-AK, 1, 2, 3
ATX	1988	1.9L	Escort/EXP	PMA-K3, 4, P, P1
			Escort/EXP	PMA-BR1, BS1, PMB-Y, PMB-Z
		2.3L	Tempo	PMA-BJ, BM, CB
			Tempo AWD	PMA-BX
			Topaz	PMA-BJ, BM, CB
			Topaz AWD	PMA-BX
		2.5L	Sable	PMT-BT

FORD-LINCOLN-MERCURY TRANSMISSION APPLICATIONS

Transmission	Year	Engine	Vehicle	Transmission Code
ATX	1989	1.9L	Escort/EXP	PMA-BR1, BS1, PMB-Y, PMB-Z
		2.3L	Tempo	PMA-BJ, BM, CB
			Tempo AWD	PMA-BX
			Topaz	PMA-BJ, BM, CB
			Topaz AWD	PMA-BX
AXOD	1986	3.0L	Sable	PNA-C
			Taurus	PNA-C
AXOD	1987	3.0L	Sable	PNA-C
			Taurus	PNA-C
AXOD	1988	3.8L	Continental	PNA-V, W
		3.0L	Sable	PNA-W, Y
			Taurus	PNA-W, Y
AXOD	1989	3.8L	Continental	PNA-V, W
		3.0L	Sable	PNA-W, Y
			Taurus	PNA-W, Y
C3	1984	2.3L	Capri	84DT-AAA, ACA, AJA
			Cougar	84DT-AEA, AFA
			LTD	84DT-ADA, AKA
			Marquis	84DT-ADA, AKA
			Mustang	84DT-AAA, ACA, AJA
			Ranger	83DT-ALB, AMB
			Thunderbird	84DT-AEA, AFA
		2.8L	Ranger 2WD	83DT-DLC, DMC
C3	1985	2.3L	Capri	85DT-AAA, EAA, JAA
			Cougar	85DT-KAA, LAA
			LTD	85DT-BAA, DAA, FAA
			Marquis	85DT-BAA, DAA, FAA
			Mustang	85DT-AAA, EAA, JAA
			Thunderbird	85DT-KAA, LAA
C3	1986	2.3L	Capri	86DT-AAA, 85DT-ACA
			Cougar	86DT-AEA, AFA
			LTD	86DT-ABA, ADA
			Marquis	86DT-ABA, ADA
			Mustang	86DT-AAA, 85DT-ACA
			Thunderbird	86DT-AEA, AFA
C5	1984	2.8L	Bronco II	PEJ-AJ
			Ranger 4WD	PEJ-AJ
		3.8L	Capri	PEP-AF
			Cougar	PEP-AD, AE
			LTD	PEP-Z, AC, AE, AK
			Marquis	PEP-Z, AC, AE, AK
			Mustang	PEP-AF
			Thunderbird	PEP-AE, K6, 7, 8, 9, 10
		4.9L	F-150	PEA-CU
		5.0L	F-150	PEA-CW

FORD-LINCOLN-MERCURY TRANSMISSION APPLICATIONS

Transmission	Year	Engine	Vehicle	Transmission Code
C5	1985	3.8L	Capri	PEP-AF1
			Cougar	PEP-AD1, AE1, AN, AP
			LTD	PEP-Z1, AC1, AE1, AL, A
			Marquis	PEP-Z1, AC1, AE1, AL, AM
			Mustang	PEP-AF1
			Thunderbird	PEP-AD1, AE1, AN, AP
		4.9L	F-150	PEA-CU
		5.0L	F-150	PEA-CW
C5	1986	2.3L	Capri	PEP-AF1, AF2
		3.8L	Cougar	PEP-AD1, AP
			LTD	PEP-Z1, AC1, AE1
			Marquis	PEP-Z1, AC1, AE1
			Mustang	PEP-AD1, AP
			Thunderbird	PEP-AD1, AP
		4.9L	F-150	PEA-CU
C6	1984	4.9L	Bronco	PGD-EG2, EG3
			E-100/350	PGD-A32, AW26, EK10, DB17
			F-100/250 4WD	PGD-EA12, EA13, EG2, EG3, EG4
			F-100/350 2WD	PGD-EK10
		5.0L	Bronco	PGD-EA12, EA13
			E-150/250	PGD-BF27, DC18, EP1, EU
			F-150/250	PGD-BF27, EP1
		5.8L	Bronco	PGD-EY1, EY2
			E-150/350 Van	PGD-DK13, DP14, DW11, EC11, EV, EW, EZ, FA
			F-150/350 4WD	PGD-DL13, DL14, DL15, EY, EY1, EY2
		6.9L	F-250/350	PJE-A, B
			F-250/350 4WD	PJE-C, C1
		7.5L	F-250/350	PJD-BB1, BC1
			F-250/350 4WD	PJD-BA3
C6	1985	4.9L	Bronco	PGD-EG4
			E-100/350	PGD-FB, AW27, FC10, F12
			F-100/250 4WD	PGD-EG4
			F-150	PEA-FE, FF
		5.8L	Bronco	PGD-DL15
			E-150/350 Van	PGD-EV1, DW12, FD
			F-150/350	PGD-DW12, EV1, FD
			F-150/350 4WD	PGD-DL15, EY2
		6.9L	E-250/350 Van	PJE-B1
			F-250/350	PJE-A, B1, C1
		7.5L	E-250/350 Van	PJD-BB1, BC2
			F-250/350	PJD-BA3, BA4, BB1, BB2, BC2, BC3
C6	1986	4.9L	Bronco	PGD-EG4
			E-150/350 Van	PGD-AW27, EK10, EK11, FB1, FC1, FC2, FF1
			F-150 4WD	PGD-EG4, EG5
			F-150/250 2WD	PGD-AW27, AW28, EK10, EK11, FE, FE1, FF, FF1

FORD-LINCOLN-MERCURY TRANSMISSION APPLICATIONS

Transmission	Year	Engine	Vehicle	Transmission Code
C6	1986	5.0L	F-150	PGD-FG, FG1
		5.8L	Bronco	PGD-EY2
			E-150/350 Van	PGD-EV1, EV2, FD1, FD2
			F-150/350	PGD-FD, FD1, EV1, EV2
			F-150/350 4WD	PGD-EY2, EY3
		6.9L	E-250/350 Van	PJE-B2, B3, B4
			F-250/350	PJE-A, A1, A2, B2, B3, B4
			F-250/350 4WD	PJE-C2, C3, C4
		7.5L	E-250/350 Van	PJD-BB2, BB3, BC3, BC4
			F-250/350	PJD-BB2, BB3, BC3, BC4
			F-250/350 4WD	PJD-BA4, BA5
C6	1987	4.9L	Bronco	PGD-FR, FU
			E-150/350 Van	PGD-FN, FP, FW, FZ
			F-150/350	PGD-FM, FN, FP, FR, FZ, GD, GD1, GD2
		5.0L	F-150/250 2WD	PGD-FM, GA, GA1
		5.8L	Bronco	PGD-FK, FT
			E-150/350 Van	PGD-FV, FY, FY1
			F-150/350 2WD	PGD-FK, FV, FY, FY1
		6.9L	E-250/350 Van	PJE-G, G1
			F-250/350	PJE-F, G, G1
		7.5L	E-250/350 Van	PJD-BF, BG, BH
			F-250/350	PJD-BE, BF, BF1, BG, BH
C6	1988	4.9L	Bronco	PGD-GN1
			E-150/350 Van	PGD-FN2, FP2
			F-150 4WD	PGD-GM1
			F-150/350 2WD	PGD-FN2, FP2, GD2, GW
		5.0L	Bronco	PGD-GU
			F-150 4WD	PGD-GT
			F-150/250 2WD	PGD-FM1, GA1
		5.8L	Bronco	PGD-GH, GJ
			E-150/350 Van	PGD-GE, GF
			F-150/350	PGD-GE, GF, GY
			F-150/350 4WD	PGD-GG, GR
		7.3L	E-250/350 Van	PJE-G1
			F-250/350	PJE-G1, H, J
		7.5L	E-250/350 Van	PJD-BF1
			F-250/350	PJD-BF1, BR, BT
C6	1989	4.9L	Bronco	PGD-JD, HP
			E-150/350 Van	PGD-HA, HB, HZ
			F-150 4WD	PGD-HN, JC
			F-150/350	PGD-HA, HD, HF, HZ
		5.0L	Bronco	PGD-HV
			F-150 2WD	PGD-GZ, HD
			F-150 4WD	PGD-HU

FORD-LINCOLN-MERCURY TRANSMISSION APPLICATIONS

Transmission	Year	Engine	Vehicle	Transmission Code
C6	1989	5.8L	Bronco	PGD-HL, HM
			E-150/350 Van	PGD-HG, HH, JA, JE
			F-250/350	PGD-HG, HH, HL, HM, HS, JA, JE
			F-250/350 4WD	PGD-HJ
		7.3L	E-250/350 Van	PJE-K, L
			F-250/350	PJE-K, L
			F-250/350 4WD	PJE-M
		7.5L	E-250/350 Van	PJD-BV, CB
			F-250/350 2WD	PJD-BV, CB
			F-250/350 4WD	PJD-BZ
E40D	1989	5.8L	E-250 Van	PRA-E, E1, S
			F-250/350	PRA-E, E1, S
			F-250/350 4WD	PRA-H, H1
		7.3L	E-250 Van	PRA-L, L1, U
			F-250/350	PRA-L, L1, U
			F-250/350 4WD	PRA-G, G1
			F-450 Super Duty	PRA-C, C1
		7.5L	E-250 Van	PRA-J
			F-250/350	PRA-J, T
			F-250/350 4WD	PRA-F
F3A	1987	1.6L	Tracer	NONE
	1988	1.6L	Tracer	NONE
F3A	1989	1.3L	Festiva	NONE
		1.6L	Tracer	NONE
ZF 4 HP-22	1984	2.4L Diesel	Continental	E4LP-CA
			Mark VII	E4LP-CA
ZF 4 HP-22	1985	2.4L Diesel	Continental	E4LP-CA
			Mark VII	E4LP-CA

GENERAL MOTORS TRANSMISSION APPLICATIONS

Transmission	Year	Engine	Engine Code	Vehicle	Body Code	Transmission Code
125C	1984	1.8L	J	Cavalier	J	PE, P4, PR
			J	Firenza	J	PE, P4, PR
			J	Skyhawk	J	PE, P4, PR
			J	Sunbird, J2000	J	PE, P4, PR
		2.0L	B, P	Cavalier	J	HC, HY, CA
			B, P	Cimarron	J	CA, C4, HY, C3, CB, CI, CM
			B, P	Firenza	J	HC, HY, CA
			B, P	Skyhawk	J	HC, HY, CA
			B, P	Sunbird, J2000	J	HC, HY, CA
		2.5L	R	6000	A	PD, PW
			R	Celebrity	A	PD, PW
			R	Century	A	PD, PW
			R	Ciera	A	PD, PW
			R	Citation	X	5PD, 5PW
			R	Fiero	P	6PF
			R	Omega	X	5PD, 5PW
			R	Phoenix	X	5PD, 5PW
			R	Skylark	X	5PD, 5PW
		2.8L	Z	6000	A	CL, CC, HS
			X	6000	A	CL, CC, HS
			Z	Celebrity	A	CL, CC, HS
			X	Celebrity	A	CL, CC, HS
			X	Century	A	CL, CC, HS
			X	Ciera	A	CL, CC, HS
			X, Z	Citation	X	5CE, CC, CT
			Z	Omega	X	5CE, CC, CT
			Z	Phoenix	X	5CE, CC, CT
			X, Z	Skylark	X	5CE, CC, CT
		3.0L	E	Century	A	4BF, 5BF, BBZ
			E	Ciera	A	4BF, 5BF, BBZ
		4.3L	T	6000	A	OP
			T	Celebrity	A	OP
			T	Century	A	OP
			T	Ciera	A	OP
125C	1985	1.8L	J	Cavalier	J	PE, P4, PR, 5P5, 5P2, 5PJ
			J	Cavalier	J	PE, P4, PR, 5P5, 5P2, 5PJ
			J	Firenza	J	PE, P4, PR, 5P5, 5P2, 5PJ
			J	Skyhawk	J	PE, P4, PR, 5P5, 5P2, 5PJ
			J	Skyhawk	J	PE, P4, PR, 5P5, 5P2, 5PJ
			J	Sunbird, J2000	J	PE, P4, PR, 5P5, 5P2, 5PJ
			J	Sunbird, J2000	J	PE, P4, PR, 5P5, 5P2, 5PJ
		2.0L	P	Cavalier	J	6CA, 6CC, CA, MI, CM
			P	Cimarron	J	CA, C4, HY, C3, CB, CI, CM, 5CA, 5CC
			P	Firenza	J	5CA, 5CC, CA, MI, CM
			P	Skyhawk	J	6CA, 6CC, CA, MI, CM

GENERAL MOTORS TRANSMISSION APPLICATIONS

Transmission	Year	Engine	Engine Code	Vehicle	Body Code	Transmission Code
125C	1985		P	Sunbird, J2000	J	6CA, 6CC, CA, MI, CM
		2.5L	R	6000	A	5PD, 5PW
			U	Calais	N	6PN
			R	Celebrity	A	5PD, 5PW
			R	Century	A	5PD, 5PW
			R	Ciera	A	5PD, 5PW
			R	Citation	X	5PD, 5PW
			R	Fiero	P	6PF
			U	Grand Am	N	6PN
			R	Skylark	X	5PD, 5PW
			U	Somerset, Skylark	N	6PN, PN
		2.8L	X	6000	A	HS, 5CT, 5CL
			W	6000	A	HS, 5CT, 5CL
			X	Celebrity	A	HS, 5CT, 5CL
			W	Celebrity	A	HS, 5CT, 5CL
			X	Century	A	HS, 5CT, 5CL
			X	Ciera	A	HS, 5CT, 5CL
			W	Cimarron	J	CJ
			W	Citation	X	5CE, CC, CT
			9	Fiero	P	CD, 6CP
			W	Firenza	J	CJ
			W	Skylark	X	5CE, CC, CT
		3.0L	L	Calais	N	BD, BP
			E	Century	A	4BF, BBZ
			E	Ciera	A	4BF, BBZ
			L	Grand Am	N	BD, BP
			L	Somerset, Skylark	N	BD, BP, 6BA
			L	Somerset, Skylark	N	BD, BP
		4.3L	T	6000	A	50P
			T	Celebrity	A	50P
			T	Century	A	50P
			7	Ciera	A	50P
125C	1986	1.8L	J	Cavalier	J	PE, P4, PR, 5P5, 5P2, 5PJ, 6PA, 6PJ
			J	Cavalier	J	PE, P4, PR, 5P5, 5P2, 5PJ, 6PA, 6PJ
			J	Firenza	J	PE, P4, PR, 5P5, 5P2, 5PJ, 6PA, 6PJ
			J	Firenza	J	PE, P4, PR, 5P5, 5P2, 5PJ, 6PA, 6PJ
			J	Skyhawk	J	PE, P4, PR, 5P5, 5P2, 5PJ, 6PA, 6PJ
			J	Skyhawk	J	PE, P4, PR, 5P5, 5P2, 5PJ, 6PA, 6PJ
			J	Sunbird, J2000	J	PE, P4, PR, 5P5, 5P2, 5PJ, 6PA, 6PJ
		2.0L	P	Cavalier	J	6CA, 6CC, CB, CI, CM
			P	Cimarron	J	6CA, 6CC, CB, CI, CM, CA, C4, NY, CB
			P	Firenza	J	6CA, 6CC, CB, CI, CM
			P	Skyhawk	J	6CA, 6CC, CB, CI, CM
			P	Sunbird, J2000	J	6CA, 6CC, CB, CI, CM

GENERAL MOTORS TRANSMISSION APPLICATIONS

Transmission	Year	Engine	Engine Code	Vehicle	Body Code	Transmission Code
125C	1986	2.5L	R	6000	A	6PD, 6PW
			U	Calais	N	6PN
			R	Celebrity	A	6PD, 6PW
			R	Century	A	6PD, 6PW
			R	Ciera	A	6PD, 6PW
			R	Fiero	P	6PP
			U	Grand Am	N	6PN
			U	Somerset, Skylark	N	6PN
		2.8L	W	6000	A	HS, 6CT, 6CL, 6CU, CX
			X	6000	A	HS, 6CT, 6CL, 6CU, CX
			X	Celebrity	A	HS, 6CT, 6CL, 6CU, CX
			W	Celebrity	A	HS, 6CT, 6CL, 6CU, CX
			X	Century	A	HS, 6CT, 6CL, 6CU, CX
			W	Ciera	A	HS, 6CT, 6CL, 6CU, CX
			X	Ciera	A	HS, 6CT, 6CL, 6CU, CX
			W	Cimarron	J	CJ, 6CJ
			9	Fiero	P	CD, 6CP
			W	Firenza	J	CJ, 6CJ
		3.0L	L	Calais	N	BD, 6BD, 6BP
			L	Grand Am	N	BD, 6BD, 6BP
			L	Somerset, Skylark	N	BD, 6BD, 6BP
			L	Somerset, Skylark	N	BD, BP, 6BA
125C	1987	2.0L	M	Cavalier	J	7PKC, 7PFC, 7PHC, 7PPC
			K	Cavalier	J	7PKC, 7PFC, 7PHC, 7PPC
			1	Corsica, Beretta	L	7CRC, 7C8C
			K	Firenza	J	7PKC, 7PFC, 7PHC, 7PPC, 7CBC
			1	Firenza	J	7PKC, 7PFC, 7PHC, 7PPC, 7CBC
			M	Grand Am	N	7PMC
			1	Skyhawk	J	7PKC, 7PFC, 7PHC, 7PPC
			M	Skyhawk	J	7PKC, 7PFC, 7PHC, 7PPC
			K	Skyhawk	J	7PKC, 7PFC, 7PHC, 7PPC
			M	Sunbird, J2000	J	7PKC, 7PFC, 7PHC, 7PPC
			K	Sunbird, J2000	J	7PKC, 7PFC, 7PHC, 7PPC
		2.3L	D	Calais	N	8KDC
			D	Grand Am	N	8KDC
			D	Somerset, Skylark	N	8KDC
		2.5L	R	6000	A	7PDC
			U	Calais	N	7PNC
			R	Celebrity	A	7PDC
			R	Century	A	7PDC
			R	Ciera	A	7PDC
			R	Fiero	P	7PSC
			U	Grand Am	N	7PNC
			U	Somerset, Skylark	N	7PNC

GENERAL MOTORS TRANSMISSION APPLICATIONS

Transmission	Year	Engine	Engine Code	Vehicle	Body Code	Transmission Code
125C	1987	2.8L	W	6000	A	7CXC, 7CTC
			W	Cavalier	J	7CJC
			W	Celebrity	A	7CXC, 7CTC
			W	Century	A	7CXC, 7CTC
			W	Ciera	A	7CXC, 7CTC
			W	Cimarron	J	7CJC
			W	Corsica, Beretta	L	7CVC
			9	Fiero	P	7CPC
			W	Firenza	J	7CJC
			W	Sunbird	J	7CJC
		3.0L	L	Calais	N	7BPC, 7BDC, 7BHC, 7BJC
			L	Grand Am	N	7BPC, 7BDC, 7BHC, 7BJC
			L	Somerset, Skylark	N	7BPC, 7BDC, 7BHC, 7BJC
			L	Somerset, Skylark	N	7BPC, 7BDC, 7BJC
125C	1988	1.6L	6	Lemans	T	7, 8, 9PTC
		2.0L	K	Cavalier	J	8TRC, 8PPC
			1	Corsica, Beretta	L	8CRC
			1	Firenza	J	8TRC, 8PPC, 8CBC, 8PKC
			K	Firenza	J	8TRC, 8PPC, 8CBC, 8PKC
			M	Grand Am	N	8PMC
			K	Lemans	T	8PRC
			1	Skyhawk	J	8TRC, 8PPC
			K	Skyhawk	J	8TRC, 8PPC
			M	Sunbird, J2000	J	8TRC, 8PPC
			K	Sunbird, J2000	J	8TRC, 8PPC
		2.3L	D	Calais	N	8KDC
			D	Grand Am	N	8KDC
			D	Somerset, Skylark	N	8KDC
			D	Somerset, Skylark	N	8KDC
		2.5L	R	6000	A	8PDC
			U	Calais	N	8PNC
			R	Celebrity	A	8PDC
			R	Ciera	A	8PDC
			R	Fiero	P	7PSC, 8PSC
			U	Grand Am	N	8PNC
			U	Somerset, Skylark	N	8PNC
		2.8L	W	6000	A	8CTC, 8LSC
			W	Cavalier	J	8TNC, 8CJC
			W	Celebrity	A	8CTC, 8LSC
			W	Century	A	8CTC, 8LSC
			W	Ciera	A	8CTC, 8LSC
			W	Cimarron	J	8CTC, 8LSC
			W	Corsica, Beretta	L	8CVC
			9	Fiero	P	8CPC

GENERAL MOTORS TRANSMISSION APPLICATIONS

Transmission	Year	Engine	Engine Code	Vehicle	Body Code	Transmission Code
125C	1988	3.0L	L	Calais	N	8BHC, 8BJC
			L	Somerset, Skylark	N	8BHC, 8BJC
125C	1989	1.6L	6	Lemans	T	7, 8, 9PTC
		2.0L	1	Cavalier	J	9TRC, 9PPC, 9CBC
			K	Cavalier	J	9TRC, 9PPC
			1	Corsica, Beretta	L	9CRC
			1	Firenza	J	9TRC, 9PPC, 9CBC, 9PKC
			M	Grand Am	N	9PMC
			K	Lemans	T	8PRC
			K	Skyhawk	J	9TRC, 9PPC
			1	Skyhawk	J	9TRC, 9PPC, 9CBC
			K	Sunbird, J2000	J	PTRC, 9PPC, PCBC
			M	Sunbird, J2000	J	9TRC, 9PPC, 9CBC
		2.3L	D	Grand Am	N	9KDC, 9KCC
			D	Somerset, Skylark	N	9KDC
		2.5L	R	6000	A	8PDC
			U	Calais	N	9PNC
			R	Celebrity	A	8PDC
			R	Century	A	8PDC
			R	Ciera	A	8PDC
			U	Grand Am	N	9PNC
			U	Somerset, Skylark	N	9PNC
		2.8L	W	6000	A	8CTC, 8LSC
			W	Cavalier	J	9TNC, 9CJC
			W	Celebrity	A	8CTC, 8LSC
			W	Century	A	8CTC, 8LSC
			W	Ciera	A	8CTC, 8LSC
			W	Corsica, Beretta	L	9CVC, 9CRC
		3.1L	T	6000	A	9CTC
			T	Celebrity	A	9CTC
		3.3L	N	Calais	N	9BUC
			N	Somerset, Skylark	N	9BUC
180C	1984	1.6L	C	Chevette	T	PR, TP, TN, VQ, JY
			C	T1000	T	PR, TP, TN, VQ, JY
180C	1985	1.6L	C	Chevette	T	PR
			C	T1000	T	PR
180C	1986	1.6L	C	Chevette	T	PR, UG, UR
			C	T1000	T	PR, UG, UR
180C	1987	1.6L	C	Chevette	T	UG, UR
			C	T1000	T	UG, UR
180C	1989	1.6L		Geo Tracer	J1	None
200-4R	1984	3.8L	A	Cutlass	G	4CH
			A	Grand Prix	G	4CH
			A	Monte Carlo	G	4CH
			9	Regal	G	4CH

GENERAL MOTORS TRANSMISSION APPLICATIONS

Transmission	Year	Engine	Engine Code	Vehicle	Body Code	Transmission Code
200-4R	1984	4.1L		98	D	4AP, 4AA, 4BT
				Electra	D	4AP, 4AA, 4BT
			8	Fleetwood DeVille	D	4AP, 4AA, 4BT
			4	LeSabre, Estate Wagon	B	4BY, 4BT
			4	Regal	G	4GH
		4.3L	V	Regal	G	4OF
			V	Regal	G	4OF
		5.0L	Y	88, Custom Cruiser	B	4KC, KJ
			Y	98	D	4OJ, 4OG
			Y	Cutlass	G	4OZ, 4OG
			H	Cutlass	G	4HG, 4CR
			Y	Electra	D	4OJ, 4OG
			H	Grand Prix	G	4HG, 4CR
			Y	LeSabre, Estate Wagon	B	4OG
			G	Monte Carlo	G	5CQ
			H	Monte Carlo	G	4HG, 4CR
			H	Regal	G	4HG, 4CR
		5.7L	N	88, Custom Cruiser	B	4OM
			N	98	D	4OM
			N	Caprice	B	4OM
			N	Cutlass	G	4OM
			N	Electra	D	4OM
			N	Fleetwood DeVille	D	4OM
			N	Grand Prix	G	4OM
			N	LeSabre, Estate Wagon	B	4OM
			N	Monte Carlo	G	4OM
			N	Parisienne	B	4OM
200-4R	1985	3.8L	A	Cutlass	G	5CH, 5OK
			9	Regal	G	5BQ
		4.1L	8	Fleetwood Brgm	D	5AA, 5AO, 6AB, AP
			4	LeSabre, Estate Wagon	B	4BY, 4BT
		5.0L	Y	88, Custom Cruiser	B	5OG, 5OM
			Y	88, Custom Cruiser	B	5OG, 5OJ
			H	Caprice	B	KC, KJ
			Y	Cutlass	G	6KA, 6KB
			H	Cutlass	G	5HG, 5CR, 6CR
			H	Grand Prix	G	5HG, 5CR, 6CR
			Y	LeSabre, Estate Wagon	B	4OG
			H	Monte Carlo	G	5HG, 5CR, 6CR
			H	Parisienne	B	KC, KJ
		5.7L	N	88, Custom Cruiser	B	5OM
			N	Caprice	B	5OM
			N	Cutlass	G	4OM
			N	Fleetwood Brgm	D	4OM
			N	LeSabre, Estate Wagon	B	4OM

GENERAL MOTORS TRANSMISSION APPLICATIONS

Transmission	Year	Engine	Engine Code	Vehicle	Body Code	Transmission Code
200-4R	1985	5.7L	N	Parisienne	B	50M
200-4R	1986	3.8L	7	Regal	G	6HH
		4.3L	Z	Grand Prix	G	6CH
			Z	Monte Carlo	G	6CH
		5.0L	Y	Caprice	B	6KC, 6KJ
			H	Caprice	B	6CR, 6HC, 6CA
			9	Cutlass	G	6KZ
			Y	Estate Wagon	B	6KC, 6KB
			Y	Fleetwood Brgm	D	5AJ
			H	Parisienne	B	6CR, 6HC, 6CA
			Y	Parisienne	B	6KC, 6KJ
			H	Regal	G	6HL
			Y	Regal	G	6KB, 6KC
200-4R	1987	3.8L	7	Regal	G	7BHB
		4.3L	Z	Grand Prix	G	7CHF, 7CYF
			Z	Monte Carlo	G	7CHF, 7CYF
		5.0L	Y	Caprice	B	7HFF, 7KCF, 7KJF
			Y	Custom Cruiser	B	6KC, 6KY
			H	Cutlass	G	7CRF, 7CCF, 7HTF
			Y	Cutlass	G	7KJF, 8KJF
			Y	Estate Wagon	B	7HFF, 7KCF, 7KJF
			Y	Fleetwood Brgm	D	7KJF, 7KCF
			H	Grand Prix	G	7CRF, 7CCF, 7HTF
			G	Monte Carlo	G	7CZF
			H	Regal	G	7HTF
			Y	Regal	G	7KJF, 7KTF, 8KJF
			Y	Safari	B	7HFF, 7KCF, 7KJF
200-4R	1988	5.0L	H	Caprice	B	8CTF
			Y	Custom Cruiser	B	8KJF, 8KTF
			Y	Cutlass	G	8KJF
			Y	Estate Wagon	B	8KJF, 8KTF
			Y	Fleetwood Brgm	D	8KJF, 8KTF
			H	Safari	B	8CTF
			Y	Safari	B	8KJF, 8KTF
200-4R	1989	3.8L	7	Firebird	F	9TAF
		5.0L	E	Custom Cruiser	B	9CTF, 9CUF
			Y	Custom Cruiser	B	9KJF, 9KTF
			Y	Estate Wagon	B	9KJF, 9KTF
			Y	Fleetwood Brgm	D	9KJF, 9KTF
			Y	Safari	B	9KJF, 9KTF
			E	Safari	B	9CTF, 9CUF
200C	1984	1.6L	C	Chevette	T	JY
			C	T1000	T	JY
		3.8L	A	88, Custom Cruiser	B	4BH
			A	Caprice	B	4BH

GENERAL MOTORS TRANSMISSION APPLICATIONS

Transmission	Year	Engine	Engine Code	Vehicle	Body Code	Transmission Code
200C	1984	1.6L	A	LeSabre, Estate Wagon	B	4BH
			A	Parisienne	B	4BH
			A	Regal	G	4XE, 4WK
		4.3L		Cutlass	G	5C6, 5C5, 5CH, 5CY
		5.0L	Y	88, Custom Cruiser	B	40I
			H	Caprice	B	5HL
			9, Y	Cutlass	G	40I
			Y	LeSabre, Estate Wagon	B	40I
			H	Parisienne	B	5HL
			Y	Parisienne	B	40U
			H	Regal	G	5CO, 5CV, 9CU
		5.7L	N	88, Custom Cruiser	B	40U
			N	Caprice	B	40U
			N	Cutlass	G	40U
			N	Grand Prix	G	40U
			N	LeSabre, Estate Wagon	B	40U
			N	Monte Carlo	G	40U
			N	Parisienne	B	40U
200C	1985	3.8L	A	88, Custom Cruiser	B	5HH, 5BH
			A	Caprice	B	5HH, 5BH
			A	Cutlass	G	6BH
			A	Grand Prix	G	5HH
			A	LeSabre, Estate Wagon	B	4BH
			A	Monte Carlo	G	5HH
			A	Parisienne	B	5HH, 5BH
			A	Regal	G	6BH
		4.3L	Z	Caprice	B	5CS
			V	Cutlass	G	5C6, 5C5, 5CH, 5CY
			V	Monte Carlo	G	5C6, 5C5, 5CH, 5CY
			Z	Parisienne	B	5CS
		5.0L	H	Caprice	B	5HL
			Y	Custom Cruiser	B	50I
			9, Y	Cutlass	G	50I
			H	Cutlass	G	5CO, 5CV, 9CU
			H	Grand Prix	G	5CO, 5CV, 9CU
			Y	LeSabre, Estate Wagon	B	40I
			H	Monte Carlo	G	5CO, 5CV, 9CU
			H	Parisienne	B	5HL
			H	Regal	G	5HL
		5.7L	N	88, Custom Cruiser	B	50U
			N	Caprice	B	50U
			N	Cutlass	G	50U
			N	LeSabre, Estate Wagon	B	40U
			N	Parisienne	B	50U

GENERAL MOTORS TRANSMISSION APPLICATIONS

Transmission	Year	Engine	Engine Code	Vehicle	Body Code	Transmission Code
200C	1986	3.8L	A	Cutlass	G	6HH
			A	Grand Prix	G	6HH
			A	Monte Carlo	G	6HH
			A	Regal	G	6KJ
		4.3L	Z	Caprice	B	6CA
			Z	Parisienne	B	6CA
		5.0L	H	Caprice	B	6HL
			Y	Custom Cruiser	B	50I
			9, Y	Cutlass	G	6KA, 6KB
			H	Cutlass	G	6HL, 6CU
			Y	Estate Wagon	B	6KA, 6KB
			H	Grand Prix	G	6HL, 6CU, 6BR
			H	Monte Carlo	G	6HL, 6CU
			H	Parisienne	B	6HL
			H	Regal	G	6BR
			Y	Regal	G	6HC, 6KA
200C	1987	3.8L	A	Cutlass	G	7HHB
			A	Grand Prix	G	7HHB
			A	Monte Carlo	G	7HHB
			A	Regal	G	7HHB, 7BHB
		4.3L	Z	Grand Prix	G	7CAB
			Z	Monte Carlo	G	7CAB
		5.0L	H	Cutlass	G	7HLB, 7CUB
			H	Grand Prix	G	7HLB, 7CUB
			H	Monte Carlo	G	7HLB, 7CUB
			H	Regal	G	7HLB
			Y	Regal	G	7KBB
			Y	Cutlass	G	8KBB
250C	1984	3.8L	A	88, Custom Cruiser	B	4XP
			A	Caprice	B	4XD
			A	Grand Prix	G	4WK, 4XE
			A	LeSabre, Estate Wagon	B	4XD
			A	Monte Carlo	G	4WK, 4XE
			A	Parisienne	B	4XD
250C	1985	3.8L	A	LeSabre, Estate Wagon	B	4XD
3254L	1984	2.8L	L	Riviera	E	5BJ
		4.1L	8	Eldorado, Seville	E, K	4AJ, 5AB, 4AA, 5AJ
			4	Riviera	E	4BE
			4	Toronado	E	4BE
		5.0L	Y	Riviera	E	50E, 50Q
			Y	Toronado	E	50E, 50Q
		5.7L	N	Riviera	E	50K
			N	Toronado	E	50K

GENERAL MOTORS TRANSMISSION APPLICATIONS

Transmission	Year	Engine	Engine Code	Vehicle	Body Code	Transmission Code
3254L	1985	5.0L	N	Riviera	E	50K
			Y	Riviera	E	50E, 50Q
			Y	Toronado	E	50E, 50Q
		5.7L	N	Toronado	E	50K
3254L	1986	4.1L	8	Eldorado, Seville	E, K	4AJ, 5AB, 4AA, 5AJ
350C	1984	5.0L	H, F	Light Truck	C, K	XX, 5XX
			H	Light Truck	C, K	4MF, 4MD, 4MK, 4TE, 4ME, 4TK
			H	Light Truck	C, K	5MF, 5MD, 4MK, 4MD, 5TE, 5ME, TK, 5TK
			F, H	Van	G	4XX, 5XX, XX
350C	1985	5.0L	H	Light Truck	C, K	5MF, 5MD, 5MK, 5MDM, 5TE, 5ME, 6TKM, 5TK
			F, H	Van	G	6XX, 5XX
350C	1986	5.0L	F, H	Van	G	6XX
400	1984	4.8L	T	Light Truck	C, K	FC
		5.7L	M	Light Truck	C, K	5TH, 5FZ, TH, FZ, FM, 4FA, 5FA
		6.0L		Fleetwood Limo	D	40M
		6.2L	J	Light Truck	C, K	5TZ, 5FX, FD, 5FD, FX
			J	Van	G	FD, 4FD, FX
		7.4L	W	Light Truck	C, K	FJ, 5FJ, FB, 5FB, FK, 5FK, FN, 4FN
400	1985	4.8L	T	Light Truck	C, K	5FC
		5.7L	M	Light Truck	C, K	5VH, 6MDM, 5VH, 6TKM, 5TK, 5TH, 6FMA, 5FM
		6.2L	J	Light Truck	C, K	5T, 5FX, 6FDA, 5FD
			J	Van	G	6FDA, 5FD, 5FX
		7.4L	W	Light Truck	C, K	6FJA, 5FJ, 6FBA, 5FB, 6FKA, 5FK, 6FNA, 5FN
400	1986	4.3L	N	Light Truck	C, K	6LSA
			Z	Van	G	6LSA
		4.8L	T	Light Truck	C, K	6FQA, 6FCA, 5FC
		5.0L	9	Fleetwood Brgm	D	6AH
		5.7L	M	Light Truck	C, K	6FAA, 6FMA, 6FZA, 6FWA
			M	Van	G	6FMA, 6FAA
		6.2L	J	Light Truck	C, K	6FDA, 6FXA, 6FFA
			J	Van	G	6FDA
		7.4L	W	Light Truck	C, K	6FPA, 6FJA, 6FKA, 6FNA, 6FBA
400	1987	4.3L	Z	Light Truck	R, V	7LAA, 7LSA
			Z	Van	G	7LSA
		5.0L	9	Fleetwood Brgm	D	7AH
			H	Light Truck	R, V	7LCA, 7LJA, 7LDA
		5.7L	K	Light Truck	R, V	7FTA, 7TKA
			M	Light Truck	R, V	7FAA
			M	Van	G	7FAA
			K	Van	G	7FTA, 7LTA, 6TKA
		6.2L	J	Light Truck	R, V	7FZA, 7LZA, 8LKA, 7FXA
			J	Van	G	7FZA, 7LZA, 7FXA
		7.4L	W	Light Truck	R, V	7FBA
			N	Light Truck	R, V	7TCA, 7FUA
			N	Van	G	7FYA

GENERAL MOTORS TRANSMISSION APPLICATIONS

Transmission	Year	Engine	Engine Code	Vehicle	Body Code	Transmission Code
400	1988	4.3L	Z	Light Truck	C, K	7LFA, 8LFA, 7LRA, 8LRA, 7FHA, 7FCA, 8FCA
		4.8L	T	Light Truck	R, V	8FQA, 8TLA
		5.0L	9	Fleetwood Brgm	D	8AHA
			H	Light Truck	C, K	7LHA, 8LBA, 7LBA, 8LHA, 7LDA, 8LDA
			H	Van	G	8LJA
		5.7L	K	Light Truck	R, V	7FWA, 8LWA, 7LWA, 7TDA, 8LTA, 8TKA, 8TDA, 8FWA
			M	Light Truck	R, V	7TAA, 8FAA, 8TAA
			K	Light Truck	C, K	7FMA, 8FMA, 7LMA, 8LMA, 87MA, 7TMA, 7FWA, 8FWA, 8LWA, 7LWA, 7TDA, 8TDA
			K	Van	G	8FTA, 8LTA, 8KTA
			M	Van	G	8FAA
		6.2L	J	Light Truck	C, K	7FOA, 8FDA, 7LVA, 8LVA, 7FFA, 8FFA, 8KLA, 7LKA
			J	Light Truck	R, V	7TDA, 7FFA, 7FRA, 7LKA, 8LZA, 8FZA, 8FXA, 8FFA, 8FRA, 8LKA
			J	Van	G	8FZA, 8LZA, 8FXA
		7.4L	N	Light Truck	C, K	7FJA, 8FJA, 8FPA, 7FPA, 8TNA, 7TNA, 8FNA, 7FNA, 8TFA, 7TFA
			N, W	Light Truck	R, V	7TBA, 8FBA, 7FNA, 8FBA, 8TPA, 8FUA, 8TBA, 9FNA, 9TFA
			N	Van	G	9LSA
400	1989	4.3L	Z	Light Truck	C, K	8LFA, 8LRA, 8FCA, 9LFA, 9MXM, 9MFM, 9LRA, 9FCA
		4.8L	T	Light Truck	R, V	9FQA
		5.0L	9	Fleetwood Brgm	D	9AHA
			H	Light Truck	C, K	8LBA, 8LHA, 8LDA, 9LBA, 9LHA, 9LDA
			H	Van	G	9LJA
		5.7L	K	Light Truck	C, K	8FMA, 8LMA, 87MA, 8FWA, 8LWA, 8TDA, 9TMA, 9FMA, 9LMA, 9TDA, 9FWA, 9LWA
			K	Light Truck	R, V	9TUA, 9TYA, 9TKA, 9FAA, 9FWA, 9TDA, 9TAA, 9LWA
			K	Van	G	9LTA, 9TKA, 9TFA
		6.2L	J	Light Truck	C, K	8FDA, 8LVA, 8FFA, 8KLA, 9FDA, 9LVA, 9FFA, 9LKA
			J	Light Truck	R, V	9FZA, 9LZA, 9FXA, 9FFA, 9LKA, 9FRA
			J	Van	G	9FZA, 9LZA, 9FXA
		7.4L	N	Light Truck	C, K	8FJA, 8FPA, 8TNA, 8FNA, 8TFA, 9FJA, 9FPA, PTNA, 9FNA, 9TFA
			N, W	Light Truck	R, V	9FUA, 9FYA, 9TCA, 9TPA, 9TBA
			N	Van	G	9LTA, 9FUA
440T4	1984	2.8L	Z	6000	A	CN, CW
			X	6000	A	CN, CW
			X	Celebrity	A	CN, CW
			Z	Celebrity	A	CN, CW
			X	Century	A	CN, CW
			X	Ciera	A	CN, CW
		3.0L	L	Ciera	A	BS, BU, BV, BN, BV
		3.8L	3	Century	A	BR, BC, BA
		5.7L	N	Eldorado, Seville	E, K	6AA, 6AT, 6ATH, 6ADH, 7AHH, 7ADH

GENERAL MOTORS TRANSMISSION APPLICATIONS

Transmission	Year	Engine	Engine Code	Vehicle	Body Code	Transmission Code
440T4	1985	2.8L	X	6000	A	CN, CWCM, CP, 5HT, HJ
			W	6000	A	CN, CWCM, CP, 5HT, HJ
			W	Celebrity	A	CN, CWCM, CP, 5HT, HJ
			X	Celebrity	A	CN, CWCM, CP, 5HT, HJ
			X	Century	A	CN, CWCM, CP, 5HT, HJ
			X	Ciera	A	CN, CWCM, CP, 5HT, HJ
		3.0L	E	98	C	BS, BU, BV, 5BV, BN, BY, 5BY
			L	Ciera	A	BS, BU, BV, 5BV, BN
			E	Electra	C	BS, BU, BV, 5BV, BN, BY, 5BY
			E	Fleetwood, DeVille FWD	C	BS, BU, BV, 5BV, BN, BY, 5BY
			E	Fleetwood, DeVille FWD	C	BS, BU, BV, 5BV, BN, BY, 5BY
		3.8L	—	98	C	5BA, 5BC, 5BR, 5BW, 5BX
			3	Century	A	BR, BC, BA
			—	Electra	C	5BA, 5BC, 5BR, 5BW, 5BX
			—	Fleetwood, DeVille FWD	C	5BA, 5BC, 5BR, 5BW, 5BX
			—	Fleetwood, DeVille FWD	C	5BA, 5BC, 5BR, 5BW, 5BX
		4.1L	8	Eldorado, Seville	E, K	6AA, 6AT, 6ATH, 6ADH, 7AHH, 7ADH
		4.3L	T	98	C	0Y, 0B
			T	Electra	C	0Y, 0B
			T	Fleetwood, DeVille FWD	C	0Y, 0B
			T	Fleetwood, DeVille FWD	C	0Y, 0B
		5.7L	N	Eldorado, Seville	E, K	6AA, 6AT, 6ATH, 6ADH, 7AHH, 7ADH
440T4	1986	2.8L	W	6000	A	6CF, 6CM, CN, CF, HJ
			X	6000	A	6CF, 6CM, CN, CF, HJ
			X	Celebrity	A	6CF, 6CM, CN, CF, HJ
			W	Celebrity	A	6CF, 6CM, CN, CF, HJ
			X	Century	A	6CF, 6CM, CN, CF, HJ
			X	Ciera	A	6CF, 6CM, CN, CF, HJ
			W	Ciera	A	6CF, 6CM, CN, CF, HJ
		3.0L	L	Electra	C	BS, BU, BV, 5BV, BN, BY, 5BY
			L	Fleetwood, DeVille FWD	C	BS, BU, BV, 5BV, BN, BY, 5BY
		3.8L	B, 3	6000	A	6BS, 6BAH, 6BC
			3	88 FWD	H	6BHH, 6BB, 6BD, 6BA
			B	88 FWD	H	6BHH, 6BB, 6BD, 6BA
			—	98	C	6BA, BM, BS, 6BAH, 6BL, 6FTH, 6BT, 6BB
			B, 3	Celebrity	A	6BS, 6BAH, 6BC
			3	Century	A	6BS, 6GAH, 6BC
			3	Ciera	A	6BS, 6BAH, 6BC
			B	Ciera	A	6BS, BGAH, 6BC
			—	Electra	C	6BA, BM, BS, 6BAH, 6BL, 6FTH, 6BT, 6BB
			—	Fleetwood, DeVille FWD	C	6BA, BM, BS, 6BAH, 6BL, 6FTH, 6BT, 6BB
			C	LeSabre	H	6BHH, 6BB, 6BD, 6BAHCH
			3	LeSabre	H	6BHH, 6BB, 6BD, 6BA
			B	Riviera	E	6FZH, 6BZ, 6FYH, 6BY
			B	Toronado	E	6FZH, 6BZ, 6FYH, 6BY

GENERAL MOTORS TRANSMISSION APPLICATIONS

Transmission	Year	Engine	Engine Code	Vehicle	Body Code	Transmission Code
440T4	1987	2.8L	W	6000	A	7CAH, 7CBH, 7CAH
			W	Celebrity	A	7CAH, 7CBH, 7CAH
			W	Century	A	7CAH, 7CBH, 7CAH
			W	Ciera	A	7CAH, 7CBH, 7CAH
		3.8L	3	6000	A	7FCH, YFLH, 7FSH
			3	88 FWD	H	7FBH, 7FKH, 7FJH, 7HAH, 7HCH
			—	98	C	7FBH, 7FKH, 7FNH, 7TFH, 7FSH
			3	Bonneville	H	7FBH, 7FKH, 7FHJ, 7HAH, 7HCH
			3	Celebrity	A	7FCH, YFLH, 7FSH
			3	Century	A	7FCH, 7FLH, 7FSH, 7FSH
			3	Ciera	A	7FCH, FFLH, 7FSH, 7FSH
			—	Electra	C	7FBH, 7FKH, 7FNH, 7FTH, 7FSH
			—	Fleetwood, DeVille FWD	C	7FBH, 7FKH, 7FNH, 7FTH, 7FSH
			3	LeSabre	H	7FBH, 7FKH, 7FJH, 7HAH, 7HCH
			C	LeSabre	H	7FBH, 7FKH, 7FHJ, 7HCH
			3	Riviera	E	7FRH, 7FZH
			3	Toronado	E	7FRH, 7FZH
		4.1L	7	Allante	V	7APZ
			8	Eldorado, Seville	E, K	6AA, 6AT, 6ATH, 6ADH, 7AHH, 7ADH
440T4	1988	2.8L	W	Ciera	A	8CFH, 8CMN, 8CWH
			W	Cutlass FWD	W	8CRH
			W	Grand Prix	W	8CRH, 8CTH
			W	Regal	W	8CRH
		3.8L	C	88 FWD	H	8FBH, 8FJH, 8FSH, 8BJH, 8BKH
			3	88 FWD	H	8FBH, 8FJH, 8FSH, 8BJH, 8BKH
			C	98	C	8FBH, 8FJH, 8FSH, 8BJH, 8BKH
			3	Bonneville	H	8FBH, 8FJH, 8FSH, 8BJH, 8BKH
			C	Bonneville	H	8FBH, 8FJH, 8FSH, 8BJH, 8BKH
			3	Century	A	8FCH
			3	Ciera	A	8FCH
			C	Electra	C	8FBH, 8FJH, 8FJH, 8FTH, 8FSH
			3	LeSabre	H	8FBH, 8FJH, 8FSH, 8FJH, 8FKH
			C	LeSabre	H	8FBH, 8FJH, 8FSH, 8FJH, 8FKH
			C	Riviera	E	8BRH, 8BYH
			C	Toronado	E	8BRH, 8BYH
		4.1L	7	Allante	V	8APZ
		4.5L	5	Eldorado, Seville	E, K	8AJH, 8ATH
			5	Fleetwood, DeVille FWD	C	8FBH, 8FKH, 8FJH, 8FTH, 8FSH
440T4	1989	2.8L	W	6000	A	8CHF, 8CMN, 8CWH
			W	Celebrity	A	8CHF, 8CMN, 8CWH
			W	Century	A	8CHF, 8CMN, 8CWH
			W	Ciera	A	8CHF, 8CMN, 8CWH
			W	Cutlass FWD	W	9CDH
			W	Grand Prix	W	9CDH, 9CYH
			W	Regal	W	9CDH

GENERAL MOTORS TRANSMISSION APPLICATIONS

Transmission	Year	Engine	Engine Code	Vehicle	Body Code	Transmission Code
440T4	1989	3.1L	T	Cutlass FWD	W	0CHH
			T	Grand Prix	W	0CJH, 0LAH
			T	Regal	W	0CHH
		3.8L	C	88 FWD	H	9FJH, 9BHH, 9PAH, 9BWH
			C	98	C	9BHH, 9BWH
			C	Bonneville	H	9BJH, 9BHH, 9PAH, 9BWH
			C	Electra	C	9BHH, PBWH
			C	LeSabre	H	9BJH, 9BHH, 9PAH, PBWH
			C	Riviera	E	9BPH, 9BWH
			C	Toronado	E	9BPH, PBWH
		4.5L	—	Allante	V	9AZH
			5	Eldorado, Seville	E, K	9AJH, 9ATH, 9ABH
			5	Fleetwood, DeVille FWD	C	9BHH, 9BWH
700R4	1984	2.0L	Y	S-10, S-15	S, T	T2
		2.5L	2	Camaro	F	PQ, 6PL
			2	Firebird	F	PQ, 6PL
		2.8L	1	Camaro	F	Y7, 4Y7
			1	Firebird	F	Y7, 4Y7
			B	S-10, S-15	S, T	5T7, 4MP, 4ML, 5ML, 5MP, 4MS
		4.3L	N	Light Truck	C, K	5VR, VR, 5MX, VF, VZ, TW, 5TW
		5.0L	G, H	Camaro	F	4YP, Y8, YF, 6YF, 4YG
			H	Caprice	B	4YK, Y6, YL
			G, H	Firebird	F	4YP, Y8, YF, 6YF, 4YG
			F	Light Truck	C, K	5TE, TE
			H	Parisienne	B	4YK, Y6, YL
			F, H	Van	G	5MF, 5ME, MF, ME
		5.7L	8	Corvette	Y	Y9
			M	Van	G	FA, 5FA
			L	Van	G	5MK, 5ME, MD, TE
		6.2L	C	Light Truck	C, K	MH, GH, 5VJ, 5MH, 5MG, 5VJ, 5VE, 5TL, 5T8, 5TM, 5TZ, VJ, VE, TL
			C	Light Truck	C, K	5MH, 5MG, 5VJ, MH, 5MH, 5MG, 5VJ, 5VE, 5TL, 5T8, 5TM, TR, 4TZ
			J	Van	G	5TL, 5VJ, 5VE, TL, VJ, MG, VE
700R4	1985	2.5L	2	Camaro	F	5PQ
			2	Firebird	F	5PQ
			E	S-10, S-15	S, T	6PRM, 5PR, 6MTM, 5MT
			S	Camaro	F	YX, 6YX
			S	Firebird	F	YX, 6YX
			B	S-10, S-15	S, T	5T7, 5ML, 5MP, 5MS
		4.3L	N	Light Truck	C, K	6MMM, 5VR, 6MXM, 5MX, 5VF, 5VZ, 5TW, 6TWM
		4.3L	Z	Caprice	B	5YT
			Z	Parisienne	B	5YT
		5.0L	H	Camaro	F	Y8, YF, 6YF, 5YP
			F	Camaro	F	5YZ, YS, YW, YZ, 6YW

GENERAL MOTORS TRANSMISSION APPLICATIONS

Transmission	Year	Engine	Engine Code	Vehicle	Body Code	Transmission Code
700R4	1985	2.5L	H	Caprice	B	YK, 6YK, 5Y6, 5YK
			F	Firebird	F	5YZ, YS, YW, YZ, 6YW
			H	Firebird	F	Y8, YF, 6YF, 5YP
			F	Light Truck	C, K	5TE, 6TJM
			H	Parisienne	B	YK, 6YK, 5Y6, 5YK
			F, H	Van	G	5MF, 5ME
		5.7L		Caprice	B	5TS
			8	Corvette	Y	Y9
			L	Van	G	5MK, 5ME, 6MDM, 6TJM
			M	Van	G	6FAA, 5FA
		6.2L	C	Light Truck	C, K	5MH, 5MG, 5VJ, 6MHM, 5MH, 5MG, 5VJ, 5VE, 5TL, 5T8, 5TM, 6TRM, 5TZ
			J	Van	G	5TL, 5VJ, 5VE, 6MHM, 5MG, 5VE
700R4	1986	2.5L	2	Camaro	F	YX, 6YX
			2	Firebird	F	YX, 6YX
			E	S-10, S-15	S, T	6PRM, 6MTM
		2.5L	S	Camaro	F	YX, 6YX
			S	Firebird	F	YX, 6YX
			R	S-10, S-15	S, T	6TB, 6TA
		4.3L	Z	Caprice	B	5YT, 6YT, YT
			N	Light Truck	C, K	6MMM, 6MAM, 6MXM, 6MRM, 6TWM
			Z	Parisienne	B	5YT, 6YT, YT
			N	Van	G	6MAM, 6MMM, 6MXM, 6MFM, 6MHM
		5.0L	H	Camaro	F	YPM, 6YP
			H	Caprice	B	YL, Y6, 6YL
				Firebird	F	YPM, 6YP
			F	Light Truck	C, K	6TJM, 6MPM
			H	Light Truck	C, K	6MDM, 6TKM
			H	Parisienne	B	YL, Y6, 6YL
			F, H	Van	G	6TJM, 6MPM, 6MDM, 6MKM
		5.7L	8	Corvette	Y	GYA
			L	Light Truck	C, K	6MDM, 6MKM, 6TJM, 6MPM, 6MWM, 5MW
			L	Van	G	6MDM, 6MKM, 6TJM, 6MPM
		6.2L	C	Light Truck	C, K	6MHM, 6TNM, 6TRM
			C	Van	G	6MHM, 6TNM
700R4	1987	2.5L	E	S-10, S-15	S, T	8PRM, 7PRM, 7MTM
		2.8L	S	Camaro	F	7YXM
			S	Firebird	F	7YXM
			R	S-10, S-15	S, T	7TBM, 7TAM
		4.3L	Z	Light Truck	R, V	7MMM, 8MAM, 7MAM
			Z	Van	G	8MAM, 7MAM
		5.0L	H	Camaro	F	7YFM, 7YPM
			F	Camaro	F	7YSM, Y7WM, 7YZM
			F	Firebird	F	7YSM, Y7WM, 7YZM
			H	Firebird	F	7YFM, 7YPM

GENERAL MOTORS TRANSMISSION APPLICATIONS

Transmission	Year	Engine	Engine Code	Vehicle	Body Code	Transmission Code
700R4	1987	5.0L	H	Light Truck	R, V	7MUM, 8TJM, 7TJM
			H	Van	G	8TUM, 7TUM, 8TJM, 7TJM
		5.7L	8	Camaro	F	7YMM
			8	Corvette	Y	7YAM, 7YCM, 7YDM
			8	Firebird	F	7YMM
			K	Light Truck	R, V	8TUM, 7TUM, 8MWM, 7MWM, 8TXM, 7TXM
		6.2L	C	Light Truck	R, V	7MHM, 7TNM
			C	Van	G	7TNM
700R4	1988	2.5L	E	S-10, S-15	S, T	8PRM, 8MTM
		2.8L	S	Camaro	F	8YXM
			S	Firebird	F	8YXM
			R	S-10, S-15	S, T	8TBM, 8TAM
		4.3L	Z	Caprice	B	8YTM
			Z	Light Truck	C, K	8MXM, 7MXM, 7MFM, 8MFM, 8MRM, 7MRM
			Z	S-10, S-15	S, T	8THM, 9TLM, 8TLM
			Z	Safari	B	8YTM
			Z	Van	G	8LSA
		5.0L	E, F	Camaro	F	8YPM, 8YWM, 8YZM
			H	Caprice	B	8YKM
			E, F	Firebird	F	8YPM, 8YWM, 8YZM
			H	Light Truck	R, V	8TUM
			H	Light Truck	C, K	7MRM, 8MRM, 7MDM, 8MDM, 8MLM, 7MLM, 8TRM
			H	Safari	B	8YKM
			H	Van	G	8MAM, 8TJM
		5.7L	8	Camaro	F	8YMM
			8	Caprice	B	9YNM
			8	Corvette	Y	8YDM
			8	Firebird	F	8YMM
			K	Light Truck	C, K	8MPM, 8MZM, 7MZM, 8TXM, 7TXM, 7FWA
			K	Light Truck	R, V	8TXM, 7TXM, 8MWM
			8	Safari	B	8YNM
		6.2L	C	Light Truck	R, V	8MHN, 8TNM, 9PCM, 8PCM
			C	Light Truck	C, K	7PAM, 8MPM, 7MPM, 7PBM, 8PBM, 8PAM, 7PAM, 7PCM, 8PCM
			C	Van	G	8TNM
700R4	1989	2.5L	E	S-10, S-15	S, T	9PRM
		2.8L	S	Firebird	F	9YPM, 9YWM, 9YZM
		4.3L	Z	Caprice	B	9YTM
			Z	Light Truck	C, K	8MXM, 8MFM, 8MRM, 9MPM
			Z	S-10, S-15	S, T	9THM, 9TLM
			Z	Van	G	9LSA
		5.0L	Y	Caprice	B	9YKM
			H	Caprice	B	9YKM

GENERAL MOTORS TRANSMISSION APPLICATIONS

Transmission	Year	Engine	Engine Code	Vehicle	Body Code	Transmission Code
700R4	1989	5.0L	E, F	Firebird	F	9YPM, 9YZM
			H	Light Truck	C, K	8MRM, 8MDM, 8MLM, 8TRM, 9MLM, 9TRM, 9MKM, 9MDM, 9MLM
			H	Van	G	9TJM
		5.7L	8	Camaro	F	9FKM, 9FXM
			8	Corvette	Y	9YDM
			8	Firebird	F	9FKM, 9FXM
			K	Light Truck	C, K	8MPM, 8MZM, 8TXM, PMZM, 9TXM
			K	Light Truck	R, V	9TUM, 9MWM, 9TXM
		6.2L	C	Light Truck	C, K	8MPM, 8PBM, 8PAM, 8PCM, 9PCM, 9PBM, 9PAM
			C	Light Truck	R, V	9PCM, 9TNM, 9MHM
			C	Van	G	9TNM
A131L	1985	1.6L	4	Nova	S	None
A131L	1986	1.6L	4	Nova	S	None
A131L	1987	1.6L	4	Nova	S	None
A131L	1988	1.6L	4	Nova	S	None
A131L	1989	1.6L	6	Geo Prizm	S	None
A240E	1988	1.6L	5	Nova	S	MS7
KF100	1985	1.5L	K	Spectrum	R	04-260
KF100	1986	1.5L	7	Spectrum	R	04-260
KF100	1987	1.5L	7	Spectrum	R	04-260
KF100	1988	1.5L	7	Spectrum	R	04-260
KF100	1989	1.5L	7	Geo Spectrum	R	04-260
Sprint	1985	1.0L	M	Sprint	M	04-012
Sprint	1986	1.0L	5	Sprint	M	04-012
Sprint	1987	1.0L	5	Sprint	M	04-012
Sprint	1988	1.0L	5	Sprint	M	04-012
Sprint	1989	1.0L	5, 6	Geo Metro	0	04-034

Section 3

MJ3 Transaxle
AMC/Jeep-Eagle

APPLICATION

1987–89 Medallion

GENERAL DESCRIPTION

The MJ3 automatic transaxle enables 3 foward speeds to be engaged one after the other with continous action. There are 3 main components, the torque converter, the differential and the rear case. The transaxle is electronically controlled, with the aid of an electronic microprocessor (computer) that interprets information from the road speed sensor, the engine load potentiometer and the malfunction switch. It then converts the information received from these sensors into electrical instruction (inputs) to the solenoid valves in the valve body assembly to change gears. The solenoid operated ball valves open or close hydraulic channels to change gears. These solenoid valves are controlled by the computer. The governor is small low wattage alternator (appoximately 1 watt). It supplies to the computer an alternating current (AC) which varies according to the vehicle speed and engine load.

The torque converter provides a smooth coupling for transmitting the engine torque to the rear case components with automatic clutch action. Increased torque is provided for moving from a standard position.

The differential transmits power from the rear case components to the wheels. It consists of a step-down gear cluster that lowers the drive centerline and a ring gear and pinion that drive the differential housing.

The rear case components provide 3 reduction ratios for forward movement and 1 for reverse movement. The rear case components include an epicyclic gear train with 3 different control elements; mechanical, hydraulic and electric. The epicyclic gear train is an assembly of helical gears that enable the different ratios to be obtained (3 forward, 1 reverse) depending on the hydraulic pressure to the receivers.

Transaxle and Converter Identification

TRANSAXLE

The transaxle has a stamped identification plate that is located on the top center rear of the transaxle. The identification number is stamped on the plate with the following data, the first part of the number is the transaxle type, the second part of the number is the transaxle model number and the last identification number is the transaxle fabrication number.

CONVERTER

The torque converter is a completely sealed unit and is not equipped with a drain plug and therefore cannot be flushed if contaminated, only replaced. The torque converter may vary depending on the engine and transaxle combination being used. The converter may have an identifying decal attached to the front cover. The decal is circular in shape and states converter type and stall ratio.

Electronic Controls

The transaxle is electronically controlled, with the aid of an electronic microprocessor (computer) that interprets information from the road speed sensor, the engine load potentiometer and the malfunction switch. It then converts the information received from these sensors into electrical instruction (inputs) to the solenoid valves in the valve body assembly to change gears. The solenoid operated ball valves open or close hydraulic channels to change gears. These solenoid valves are controlled by the computer.

The potentiometer provides variable voltage based on the throttle position. The road speed sensor is a winding located opposite the park ring that senses vehicle speed.

Metric Fasteners

The metric fastener dimensions are very close to the dimensions of the familiar inch system fasteners and for this reason, replacement fasteners must have the same measurement and strength as those removed.

Do not attempt to interchange metric fasteners for inch system fasteners. Mismatched or incorrect fasteners can result in damage to the transaxle unit through malfunctions, breakage or possible personal injury.

Care should be taken to reuse the fasteners in the same locations as removed.

Capacities

Vehicles equipped with extra transaxle coolers will all vary slightly in their capacity. Therefore, check fluid level carefully. The transaxle capacity is 6.4 quarts dry fill capacity and 2.6 quarts refill capacity.

Checking Fluid Level

The dipstick is located on the right side of the vehicle, on top of the transaxle.

1. With the vehicle on a level surface, engine idling, wheels blocked and the parking brake applied, move the selector lever through all the gear positions and return to the **P** position. Allow the engine to idle.
2. Clean the dipstick area of dirt and remove the dipstick from the filler tube. Wipe the dipstick indicator clean and reinsert the dipstick back into the filler tube and seat firmly.
3. Remove the dipstick from the filler tube again and check the fluid level as indicated on the dipstick indicator. The fluid level should be at the between the **ADD**and the **FULL** mark when the vehicle is at 160°–170°F. With the temperature at 100°F the fluid level should be at the **ADD** mark.
4. If necessary, add fluid through the filler tube to bring the fluid level to its proper height.
5. When the fluid level is correct, fully seat the dipstick in the filler tube to avoid entrance of dirt or other foreign matter.

NOTE: It takes only a ½ pint of transmission fluid to raise the level from the ADD mark to the FULL mark. Do not overfill.

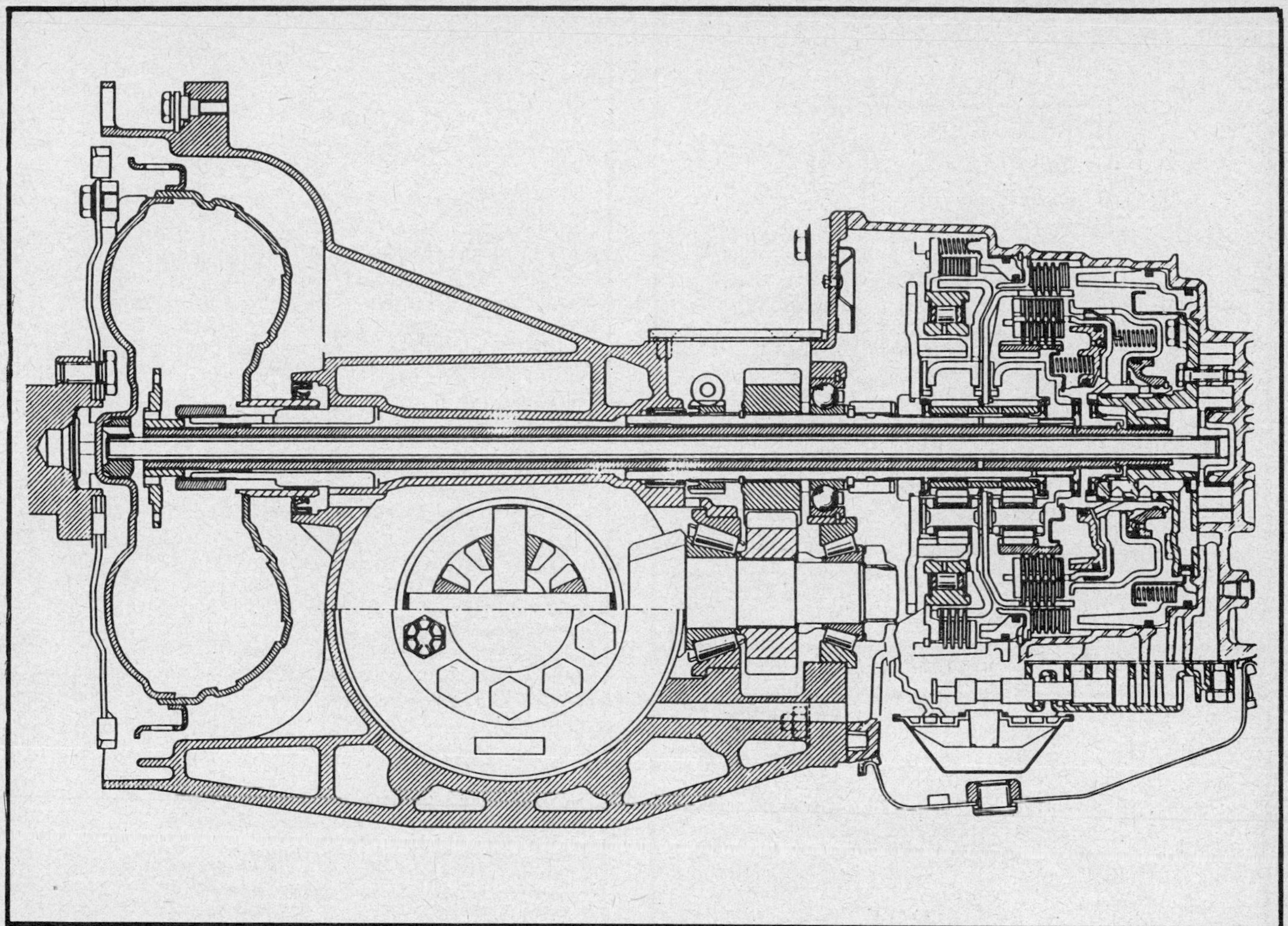

Cross section of the MJ3

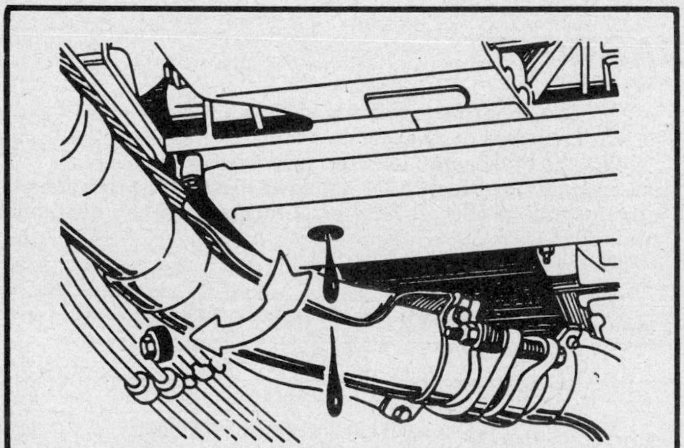

Fluid drain plug location

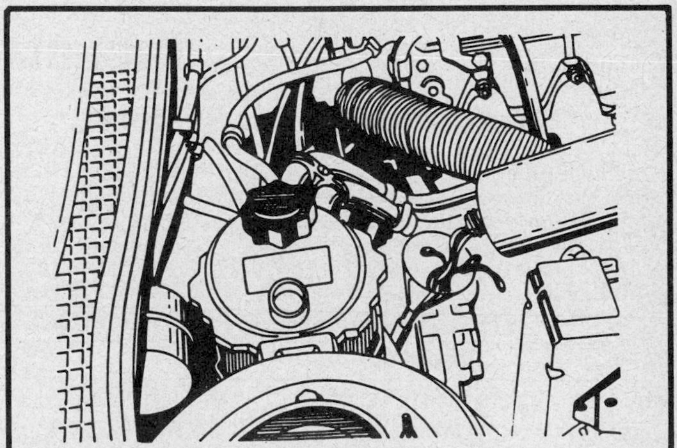

Fluid fill location

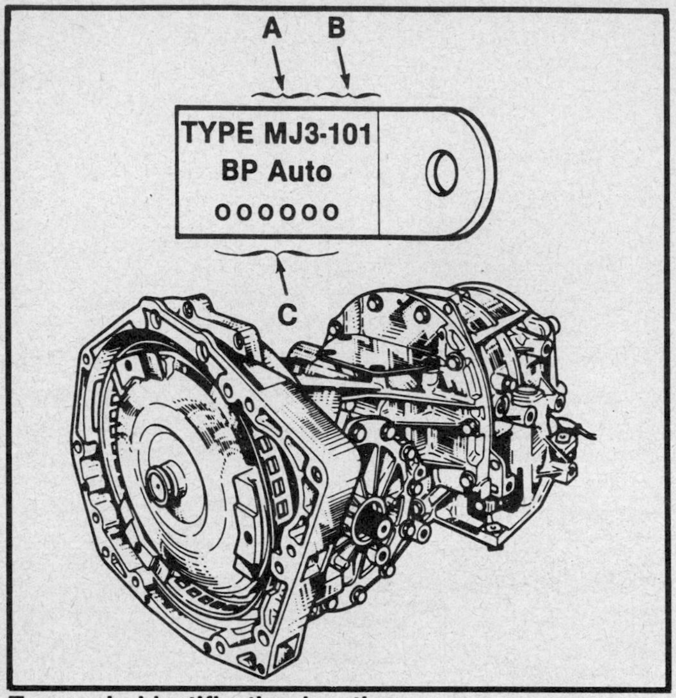
A
B
TYPE MJ3-101
BP Auto
000000
C
Transaxle identification location

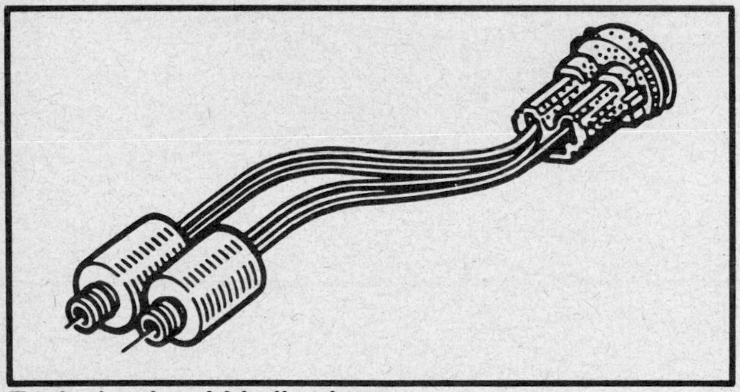
Typical solenoid ball valves

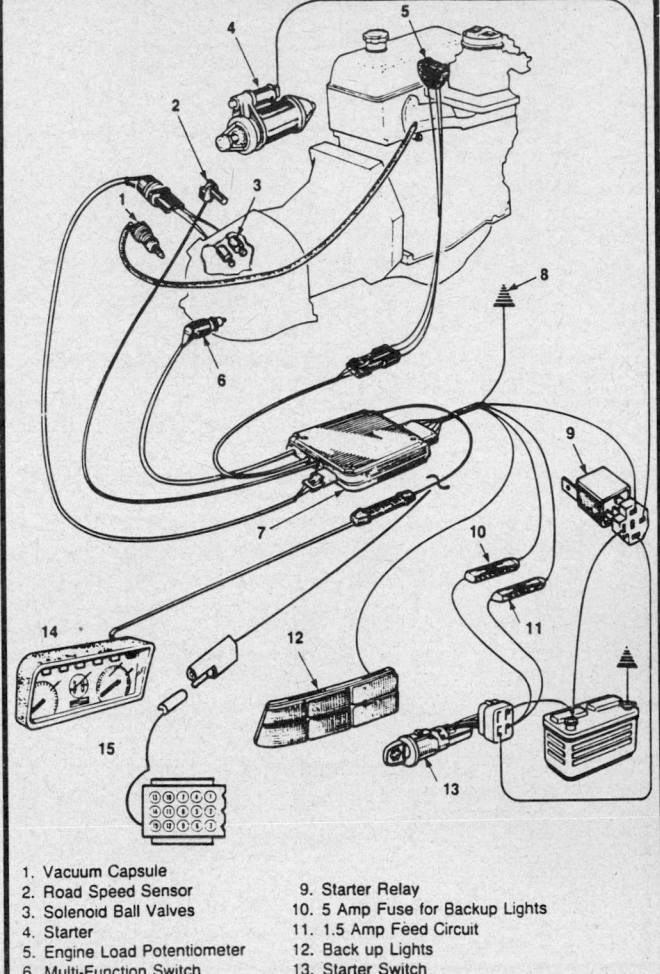

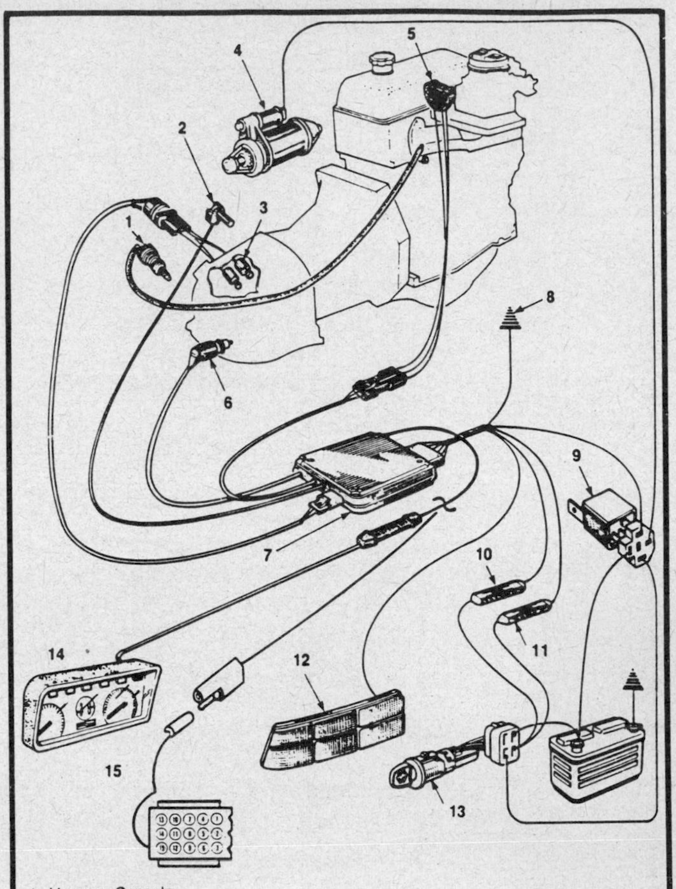
1. Vacuum Capsule
2. Road Speed Sensor
3. Solenoid Ball Valves
4. Starter
5. Engine Load Potentiometer
6. Multi-Function Switch
7. Computer
8. Automatic Transaxle Ground
9. Starter Relay
10. 5 Amp Fuse for Backup Lights
11. 1.5 Amp Feed Circuit
12. Back up Lights
13. Starter Switch
14. Instrument Panel Indicator (Early Production)
15. Diagnostic Connector (Late Production)
Electronic transaxle wire harness and electrical components

TROUBLE DIAGNOSIS

Hydraulic Control System

The hydraulic control system has 4 important functions to perform in order to make the transaxle fully automatic. These functions are grouped into basic functional systems.

a. The pressure supply system.
b. The pressure regulating system.
c. The valve body system.
d. The brakes and clutches system.

An explanation of each system follows to assist the technician in understanding the oil flow circuitry within the automatic transaxles.

CHILTON'S THREE "C'S" AUTOMATIC TRANSMISSION DIAGNOSIS CHART

Condition	Cause	Correction
Engine stalls and has uneven idle	a) Engine idle b) Ignition system c) Accelerator control d) Vacuum modulator and/or hoses	a) Correct engine Idle b) Correct ignition system malfunction c) Repair accelerator control d) Repair or renew vacuum modulator and/or hoses
Creeps in "N" position	a) Gear selector lever b) E-1/E-2 clutches	a) Adjust gear selector lever b) Overhaul as required

CHILTON'S THREE "C'S" AUTOMATIC TRANSMISSION DIAGNOSIS CHART

Condition	Cause	Correction
Excessive creep in "D"	a) Engine Idle b) Accelerator control c) Converter	a) Correct engine idle b) Correct accelerator control c) Renew converter assembly
Slippage when starting in "D" or "R"	a) Fluid level b) Fluid pressure c) Valve body d) Converter	a) Correct fluid level b) Adjust fluid pressure c) Clean, repair or renew valve body assembly d) Renew converter assembly
Slippage when starting off in "D" only	a) Fluid level b) E-1/E-2 clutches c) Overrunning clutch	a) Correct fluid level b) Correct or renew clutches c) Renew overrunning clutch
Slippage during shift	a) Fluid pressure b) Modulator c) Valve body d) Oil pump screen e) E-1/E-2 or overrunning clutches	a) Correct fluid pressure b) Adjust or renew modulator assembly c) Clean, repair or renew valve body d) Clean or renew oil pump screen e) Overhaul and renew as required
No 1st gear hold	a) Selector lever b) Harness, plugs, grounds c) Computer control unit d) Multifunction switch e) Valve body	a) Adjust selector lever b) Clean, repair, or renew components as required c) Test and/or renew computer control unit d) Clean, repair or renew components as required e) Clean, repair or renew valve body assembly
No 2nd gear hold	a) Selector lever b) Harness, plugs, grounds c) Computer control unit d) Multifunction switch e) Valve body	a) Adjust selector lever b) Clean, repair, or renew components as required c) Test and/or renew computer control unit d) Clean, repair or renew components as required e) Clean, repair or renew valve body assembly
Remains in 1st gear when in "D" position	a) Harness, plugs, grounds b) Computer control unit c) Solenoid valves d) Road speed indicator e) Valve body	a) Clean, repair or renew components as required b) Test and/or renew computer control unit c) Clean, repair or renew solenoid valves d) Test, repair or renew indicator assembly e) Clean, repair or renew valve body assembly
Remains in 3rd gear	a) Fuses b) Harness, plugs, grounds c) Computer control assembly d) Oil pump e) Valve body	a) Test circuits and renew fuses b) Clean, repair, or renew components as required c) Test and/or renew computer control unit d) Repair or renew oil pump e) Clean, repair or renew valve body assembly
Some gear ratios unobtainable and selector lever out of position	a) Selector lever b) Selector control c) Manual valve mechanical control	a) Adjust selector lever b) Adjust or repair selector control c) Repair or renew components as required
Park position not operating	a) Selector lever b) Broken or damaged components	a) Adjust selector lever b) Repair or renew components as required
No 1st in "D" position, operates—2nd to 3rd to 2nd	a) El-1 solenoid ball valve stays open	a) Test and/or renew El-1 solenoid ball valve
No 2nd in "D" position, operates—1st to 3rd to 1st	a) El-1 solenoid ball valve stays closed b) Solenoid ball valves reversed	a) Test and/or renew El-1 solenoid ball valve b) Reverse solenoid ball valves

CHILTON'S THREE "C'S" AUTOMATIC TRANSMISSION DIAGNOSIS CHART

Condition	Cause	Correction
Operates in 3rd only	a) El-2 solenoid ball valve stays open	a) Test and/or renew El-2 solenoid ball valve
No 3rd, operates 1st to 2nd to 1st	a) El-2 solenoid ball valve stays closed	a) Test and/or renew El-2 solenoid ball valve
Surge when starting off	a) Idle speed b) Accelerator controls c) Fluid level	a) Correct idle speed b) Correct the accelerator controls c) Correct fluid level
Surge during shifting	a) Modulator valve and/or hoses b) Valve body assembly	a) Adjust or renew modulator assembly and/or hoses b) Clean, repair or renew valve body assembly
Incorrect shifting speeds	a) Accelerator controls b) Load potentiometer setting c) Harness, plugs or grounds d) Kickdown switch e) Computer control units f) Road speed	a) Correct accelerator controls b) Adjust setting of load potentiometer c) Clean, repair or renew components as required d) Adjust or renew kickdown switch e) Test and/or renew computer control f) Bring vehicle to correct road speed
No drive	a) Selector lever ad b) Fluid level c) Valve body d) Oil pump e) Oil pump screen f) Oil pump shaft broken g) Turbine shaft broken h) Final drive i) Converter drive plate broken j) Converter k) E-1/E-2 clutches	a) Adjust selector lever b) Correct fluid level c) Clean, repair or renew valve body assembly d) Repair or renew oil pump e) Clean or renew oil pump screen f) Renew oil pump shaft g) Replace turbine shaft h) Correct final drive malfunction i) Replace converter drive plate j) Renew converter k) Correct or renew clutches
No drive in 1st gear hold or in "D" position	a) Valve body b) E-1/E-2 clutches c) Overrunning clutch	a) Clean, repair or renew valve body assembly b) Correct or renew clutches c) Renew overrunning clutch
No drive in "R" or 3rd gear	a) Valve body b) E-1/E-2 clutches	a) Clean, repair or renew valve body assembly b) Correct or renew clutches
No reverse or engine braking in 1st gear hold	a) Multifunction switch b) Valve body c) F-1 brake	a) Test, repair or renew multifunction switch b) Clean, repair or renew valve body assembly c) Correct or renew brake
No 1st gear in "D" position	a) Harness, plugs or grounds b) Solenoid valves c) Overrunning clutch	a) Clean, repair or renew components as required b) Clean, repair or renew solenoid valves c) Renew overrunning clutch
No 2nd gear in "D" position	a) Harness, plugs or grounds b) Valve body assembly c) F-2 brake	a) Clean, repair or renew components as required b) Clean, repair or renew valve body assembly c) Correct or renew brake
No 3rd gear in "D" position	a) Harness, plugs or grounds b) Computer control unit c) Solenoid valves d) Multifunction switch e) Valve body	a) Clean, repair or renew components as required b) Test and/or renew computer control c) Clean, repair or renew solenoid valves d) Test, repair or renew multifunction switch e) Clean, repair or renew valve body assembly

CLUTCH AND BAND APPLICATION CHART
MJ3 Automatic Transaxle

Lever Position	Free-Wheel	Clutches E1	Clutches E2	Brakes F1	Brakes F2	Solenoid Valves EL1	Solenoid Valves EL2
P	—	—	—	—	—	—	Applied
R	—	—	Applied	Applied	—	—	Applied
N	—	—	—	—	—	—	Applied
A1	Applied	Applied	—	—	—	—	Applied
A2	—	Applied	—	—	Applied	Applied	Applied
A3	—	Applied	Applied	—	—	—	—
2nd Hold	—	Applied	—	—	Applied	Applied	Applied
1st Hold	—	Applied	—	Applied	—	—	Applied

E1—Clutch 1
E2—Clutch 2
F1—Brake 1
F2—Brake 2
EL1—Solenoid valve 1
EL2—Solenoid valve 2

THE PRESSURE SUPPLY SYSTEM

The pressure supply system consists of the involute oil pump which is located in the back of the rear case. The pump is driven directly by the engine through the torque converter drive plate and supplies pressurized fluid for the converter, gear lubrication and the brakes and clutches.

THE PRESSURE REGULATING SYSTEM

The pressure regulating system consists of the vacuum capsule and pressure regulator which provides pressure that, depending on the engine load, determines the fluid pressure to the receivers and as a result controls the gear change quality.

THE VALVE BODY SYSTEM

The valve body ensures that the fluid pressure regulation is taking into account the engine load. It controls the pressurized fluid supply or release to or from the clutches and brakes. Ratio changes are determined by the operation of the 2 solenoid ball valves (EL1 and EL2). They receive the electrical instructions from the computer.

THE BRAKES AND CLUTCHES

Clutches E1 and E2 and brakes F1 and F2 are the multi-disc oil bath type. They are hydraulic receivers that, depending on their feed, lock or release units in the epicy clic gear train to engage the various forward gear ratios.

Diagnostic Tests

FLUID PRESSURE CHECK AND ADJUSTMENT

This test must be performed when the transaxle is hot and during a road test. The fluid pressure varies with the fluid temperature. Low fluid pressure will cause excessive slip during gear shifting, overheating of the clutches and brakes and eventual damage to the transaxle.

High fluid pressure will cause harsh gear shifting that is harmful to the transaxle. The fluid pressure testing gauge B.Vi. 466.04 or equivalent, must be handled carefully and calibrated regularly.

1. Verify the correct accelerator pedal cable adjustment. Adjust the cable as necessary.
2. Remove the oil pressure test port.
3. Connect pressure test gauge B.Vi.466.07 or equivalent to the pressure test port. Position the gauge inside the vehicle so it can be easily read while the vehicle in in motion.
4. Drive the vehicle a few miles to warm the transaxle fluid to normal operating temperature. Then, check fluid pressure as follows:
   a. Shift the transaxle into 2nd gear.
   b. Simultaneously press the accelerator pedal to the floor and apply the brakes to hold the vehicle speed at 50 mph.
   c. Note the oil pressure. The oil pressure reading should be 66.7 psi (4.6 bars). Release the accelerator pedal and brakes and shift the transaxle into the **D** position. If the pressure is above or below specifications, go on to the next step.
   d. Remove the vacuum capsule retaining bolt and bracket.
   e. Rotate the vacuum capsule clockwise to increase the pressure and counterclockwise to decrease the pressure.

NOTE: The pressure is increased or decreased by 1.2 psi (0.08 bars) for every 2 notches of capsule rotation.

5. Install the bracket and retaining bolt. Place a white paint mark on the vacuum capsule to indicate the pressure has been asjusted correctly.

VACUUM CAPSULE CHECK

1. Apply an extended vacuum of 16 in. Hg. to the vacuum capsule.
2. If the test gauge needle remains steady, check the fluid pressure. If the needle falls, either the capsule or its pipe must be replaced.
3. Make sure that the union on the inlet manifold is sound. Check that the pipe is tight at both capsule and union ends.
4. An air leak into the capsule or into its pipe will cause whistling, unsteady idling and rough gear shifting on a light load.

NOTE: The vacuum capsule should be replaced it it is the cause of a fluid leak. The capsule cannot be repaired.

HYDRAULIC PRESSURE TEST

Before starting the pressure test, make sure the fluid level is correct and linkage is adjusted properly. A good pressure test is an important part of diagnosing transaxle problems. Oil pressure checks are made once it is obivious that the vacuum capsule and vacuum circuit are in good condition. Regulated oil prerssure varies with the ATF fluid temperature. It is therefore normal to find, in the case of a fluid temperature of less than 176°F pressure values higher than the specifications.

1. Check fluid level and linkage adjustments. Remember that fluid must be at operating temperature (176–200°F) during tests.
2. Connect a suitable fluid pressure gauge to the transaxle. Apply the parking brake and block the drive wheels. Connect a suitable tachometer to the engine.

High Pressure Check in Park

1. Move the selector lever to the **P** position.
2. The fluid pressure must be 68 psi (5 bars) minimum at 800 rpm; it should then rise rapidly as the engine rpm is increased until it reaches a maximum pressure of 183–203 psi (13–14 bars).

Light Throttle Pressure Check

1. Disconnect the vacuum capsule.
2. Move the selector lever to the **N** position.
3. The light throttle pressure should be 36–39 psi (2.45–2.7 bars).
4. If necessary, adjust the pressure by turning the vacuum capsule; 1 notch is equal to 1.5 psi (0.1 bar).
5. The pressure will increase as the vacuum capsule is screwed in. The initial adjustment is as follows:
   a. Place the selector lever to the **N** position.
   b. Run the engine at 3800 rpm.
   c. The pressure obtained must be closed to 68 psi (5 bars).
   d. The final test and adjustment must then be made by road testing at 176°F.

ROAD TEST

The road test will be done at full throttle pressure.

1. Be sure to reconnect the vacuum capsule.
2. Drive a few miles to bring the ATF fluid up to operating temperature.
3. With the selector lever in the **D** position, fully depress the accelerator pedal. Just before the transaxle shifts from 1st to 2nd, read the maximum pressure, which should be 68 psi (5 bars).
4. If the pressure is abnormal, check the vacuum capsule and vacuum circuit. Replace the capsule if necessary and adjust the full throttle pressure.
5. If the pressure is still not correct, the pressure regulator or the transaxle is at fault.
6. Too low a pressure produces a considerable slippage during gear shifts, overheating of the clutches and brakes and consequently damage these parts.
7. Too high a pressure causes gear shifts that are too harsh and jerky, both for driver comfort and for the long life of the transaxle.
8. Frequently have the fluid pressure gauge calibration checked, especially in the 68 psi (5 bars) zone and if the gauge has been mistreated.

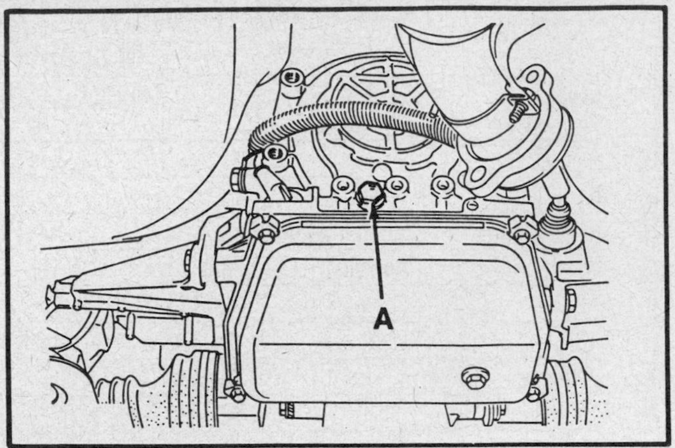
A
Location of the oil pressure test port plug (A)

Converter Clutch Operation and Diagnosis

The torque converter provides a smooth coupling for transmitting the engine torque to the rear case components with automatic clutch action. Increased torque is provided for moving from a standard position. The torque converter used with this transaxle is not a lockup torque converter.

The torque converter is a completely sealed unit, therefore if contaminated it must be replaced. The torque converter may vary depending on the the engine and transaxle combination being used. The converter may have an identifying decal attached to the front cover. The decal is circular in shape and states converter type and stall ratio.

TROUBLESHOOTING THE TRANSAXLE COMPUTER SYSTEM

There are 2 testers available for testing the transaxle electrical components. There is the diagnostic tester MS1700 and the diagnostic tester B.Vi.958. Both testers should be used according to the manufacturers instructions supplied with the testers.

The B.Vi.958 tester has a self checking feature. Check the tester by connecting terminal 14 to the battery and switching the tester to the test position. The test lights and the (+) red zone should light up. If they do not light, the tester is defective. It is also possible to test this system with a suitable volt/ohmmeter. Use the charts provided and a volt/ohmmeter to test the system.

Checking the 6-Way Connector

1. Unplug the connector from the computer and make the following checks:
   a. With the ignition switch in the **OFF** position, check the **B** terminal of the connector. The reading should be 4 ohms ± 3 ohms. This is used to diagnosis the backup lamps.
   b. With the ignition switch in the **ON** position, check the **A** terminal of the connector. The reading should be 12 volts ± 2 volts. This is used to diagnosis the backup lamps.
   c. With the ignition switch in the **OFF** position, check the **E** terminal of the connector. The reading should be 0 ohms. This is used to diagnosis the computer ground.
   d. With the ignition switch in the **ON** position, check the **F** terminal of the connector. The reading should be 12 volts ± 2 volts. This is used to diagnosis the current feed to the computer.
   d. With the ignition switch in the **ON** position, check the **C**

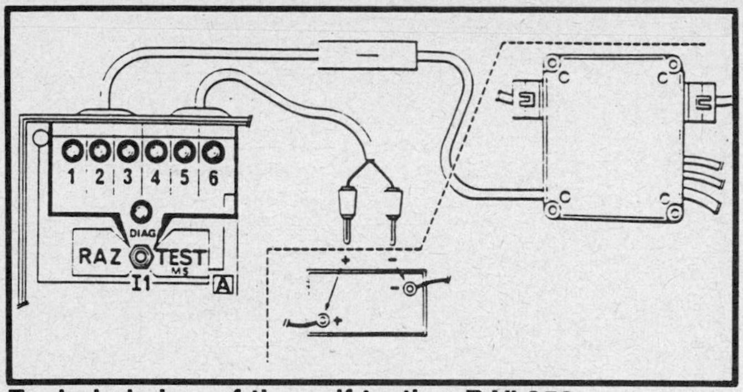
1 2 3 4 5 6
DIAG
RAZ
TEST
MS
I1
A

Exploded view of the self testing B.VI.958 diagnostic tester

VEHICLE STOPPED (ENGINE RUNNING)					
Checks	Check light(s)	Good	Bad	Faulty components	Oper-ation
Solenoid Valves	1	○	⊛	Solenoid valves Harness	VII-IX-X
Road Speed Sensor	2	○	⊛	Faulty road speed sensor	
Potentio-meter	3	○	⊛	Load potentiometer harness	XI - XIV

ENGINE NOT RUNNING - IGNITION SWITCH ON				
Position of the control lever	Check light 2 is on - (do not take it into consideration)			
	Check light(s)	Good	Faulty components	Oper-ation
2nd hold	4	[4]⊛ [5]○	If bad, multifunction switch and harness	V - XIII
1st hold	4 and 5	[4]⊛ [5]⊛	If bad, check multifunction switch and harness	V - XIII
P R N D	4 and 5	[4]○ [5]○	If bad, check multifunction switch and harness	III - IV XII- XIII
P - N	6	⊛	If bad, check selector lever adjustment and multifunction switch operation and harness	II

⊛ Light 'ON' ○ Light 'OFF'

Diagnostic charts to be used with diagnostic tester B.VI.958

terminal of the connector. The reading should be 12 volts ± 2 volts. This is used to diagnosis the starter circuit.

Checking The 3–Way Connector

1. Unplug the 3–way connector from the computer and make the following checks:

a. With the ignition switch in the **ON** position, check the **B** terminal of the connector. The reading should be 4.3 volts ± 0.5 volts.

b. If this test proves a problem exists, check connector 7. Replace the computer if connector 7 checks out good.

Checking the Solenoid Valves and Harness

1. Unplug the 3–way connector from the computer and make the following checks:

a. Check the continuity between terminals **A** and **C**, the reading should be 30 ohms ± 10 ohms. If there is a reading of 0, replace the wiring or solenoid valves.

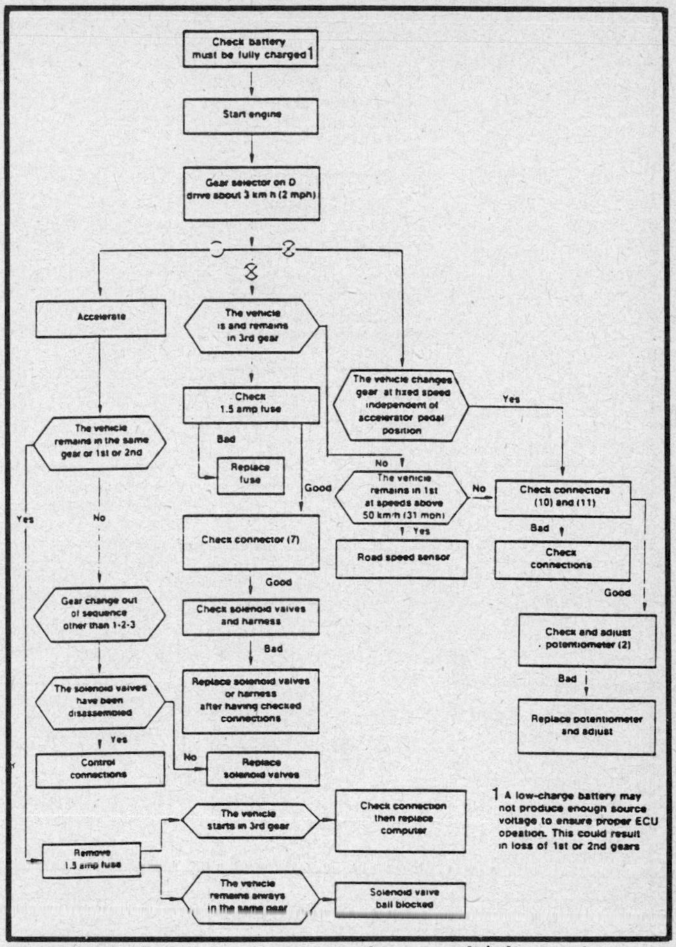
Check battery must be fully charged 1
Start engine
Gear selector on D drive about 3 km h (2 mph)
Accelerate
The vehicle is and remains in 3rd gear
Check 1.5 amp fuse
Bad
Replace fuse
Good
The vehicle changes gear at fixed speed independent of accelerator pedal position
Yes
No
The vehicle remains in the same gear or 1st or 2nd
The vehicle remains in 1st at speeds above 50 km/h (31 mph)
Check connectors (10) and (11)
Check connector (7)
Road speed sensor
Check connections
Check solenoid valves and harness
Gear change out of sequence other than 1-2-3
Check and adjust potentiometer (2)
Replace solenoid valves or harness after having checked connections
The solenoid valves have been disassembled
Replace potentiometer and adjust
Control connections
Replace solenoid valves
The vehicle starts in 3rd gear
Check connection then replace computer
1 A low-charge battery may not produce enough source voltage to ensure proper ECU opeation. This could result in loss of 1st or 2nd gears
Remove 1.5 amp fuse
The vehicle remains always in the same gear
Solenoid valve ball blocked

Electrical check sequence using a volt/ohmmeter

b. Check the continuity between terminals **B** and **C**, the reading should be 30 ohms ± 10 ohms. If the reading is higher than specified, there is a poor connection.

c. Check the continuity between terminal **C** and ground, the reading should be infinity. If the reading is not infinity, there is a short circuit between the solenoid valve windings and ground. Replace the wiring or solenoid valves.

Checking Engine Load Potentiometer

1. Unplug the connector from the potentiometer:

a. Check the continuity between terminals **C** and **B**, the reading should be 4 kilo-ohms ± 1.

b. Check the continuity between terminals **A** and **B**, the reading should be 2.5 kilo-ohms ± 1. Be sure to open the throttle slowly, the ohmmeter should never show infinite resistance.

c. If the readings are different, the potentiometer is faulty or incorrectly adjusted.

Partial Check of the Multi-Function Switch

1. Unplug the 6–way connector from the computer and make the following checks at the computer socket:

a. Check the continuity between terminals **A** and **B**, the reading should be 0 ohms with the vehicle in **R**.

b. Check the continuity between terminals **E** and **C**, the reading should be 0 ohms with the vehicle in **P** or **N**.

c. If there is no continuity at either of these check points, then the multi-function switch must be replaced.

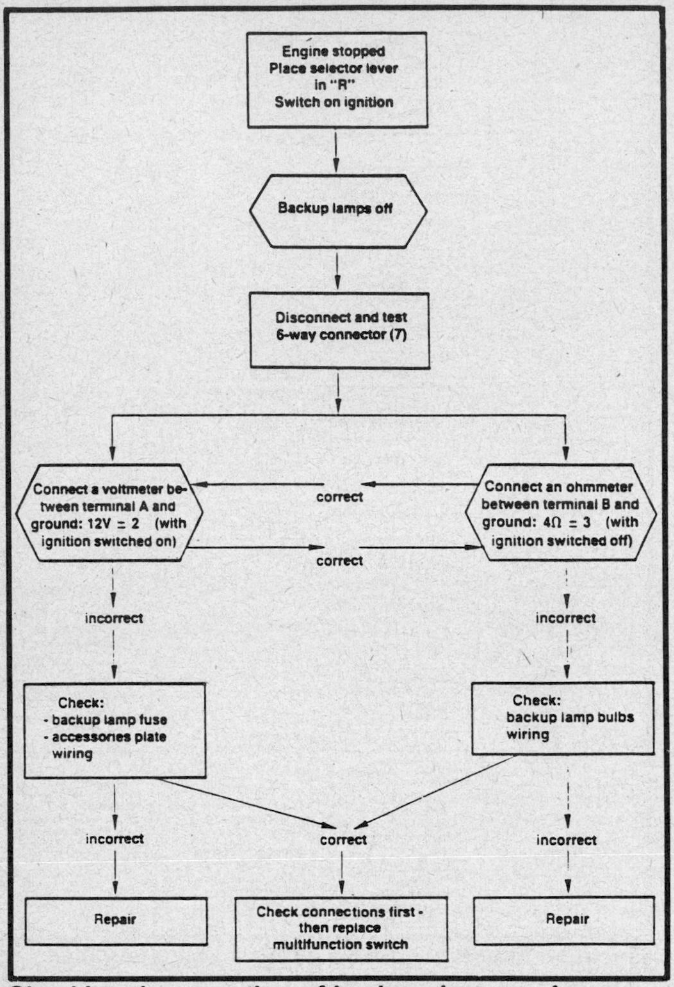
Engine stopped
Place selector lever in "R"
Switch on ignition
Backup lamps off
Disconnect and test 6-way connector (7)
Connect a voltmeter between terminal A and ground: 12V = 2 (with ignition switched on)
correct
correct
Connect an ohmmeter between terminal B and ground: 4Ω = 3 (with ignition switched off)
incorrect
incorrect
Check:
- backup lamp fuse
- accessories plate wiring
Check:
backup lamp bulbs
wiring
incorrect
correct
incorrect
Repair
Check connections first - then replace multifunction switch
Repair

Checking the operation of back-up lamps using a volt/ohmmeter

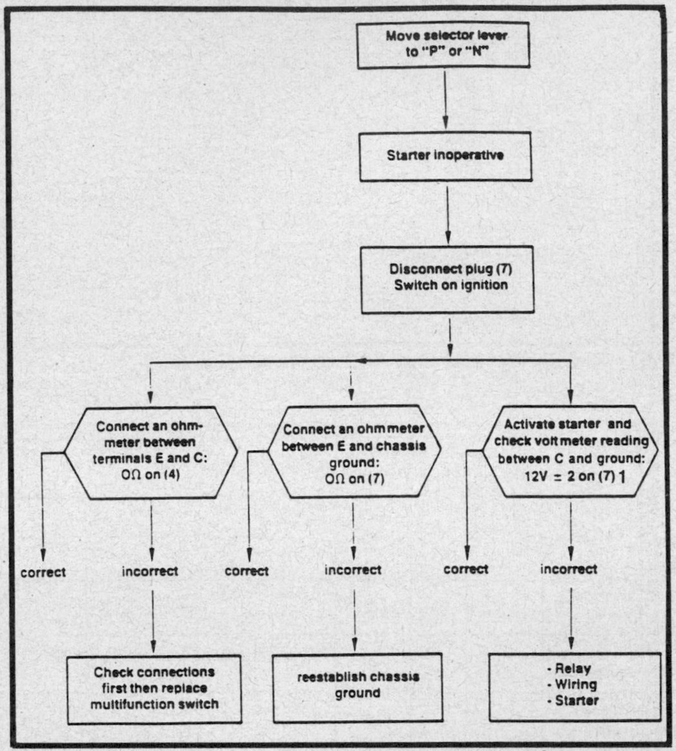
Move selector lever to "P" or "N"
Starter inoperative
Disconnect plug (7)
Switch on ignition
Connect an ohmmeter between terminals E and C: 0Ω on (4)
Connect an ohmmeter between E and chassis ground: 0Ω on (7)
Activate starter and check voltmeter reading between C and ground: 12V = 2 on (7) 1
correct
incorrect
correct
incorrect
correct
incorrect
Check connections first then replace multifunction switch
reestablish chassis ground
- Relay
- Wiring
- Starter

Checking the operation of starter circuit using a volt/ohmmeter

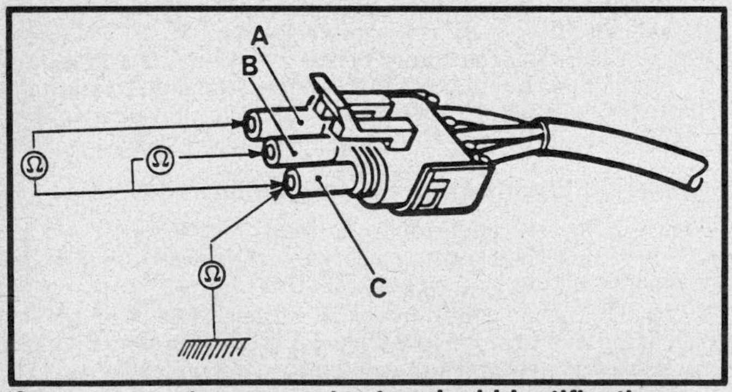
A
B
C
Ω
Ω
Ω

3-way computer connector terminal identifications

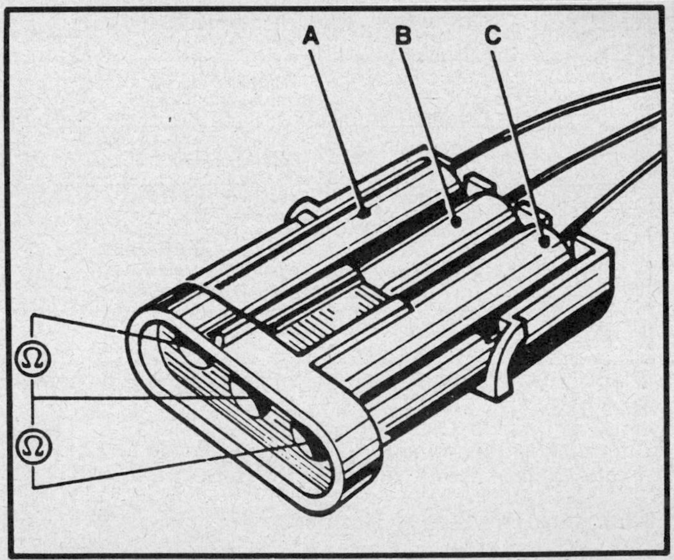
A
B
C
Ω
Ω

Pontentiometer connector terminal identifications

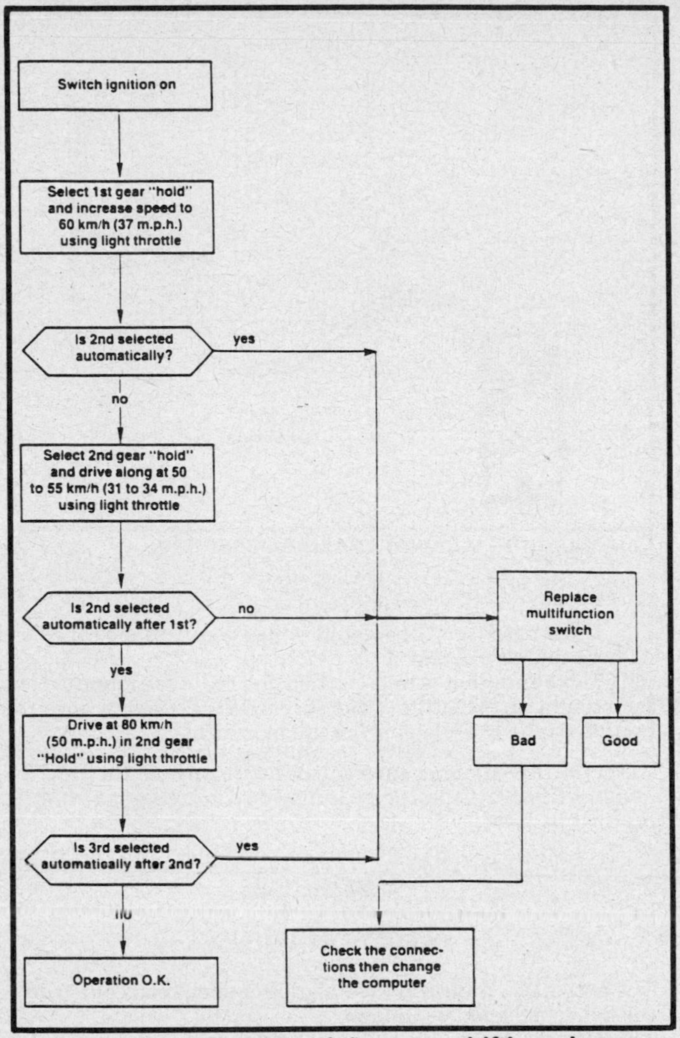
Switch ignition on
Select 1st gear "hold" and increase speed to 60 km/h (37 m.p.h.) using light throttle
Is 2nd selected automatically?
yes
no
Select 2nd gear "hold" and drive along at 50 to 55 km/h (31 to 34 m.p.h.) using light throttle
Is 2nd selected automatically after 1st?
no
yes
Replace multifunction switch
Bad
Good
Drive at 80 km/h (50 m.p.h.) in 2nd gear "Hold" using light throttle
Is 3rd selected automatically after 2nd?
yes
no
Operation O.K.
Check the connections then change the computer

Checking the operation of the gear shifting phases using a volt/ohmmeter

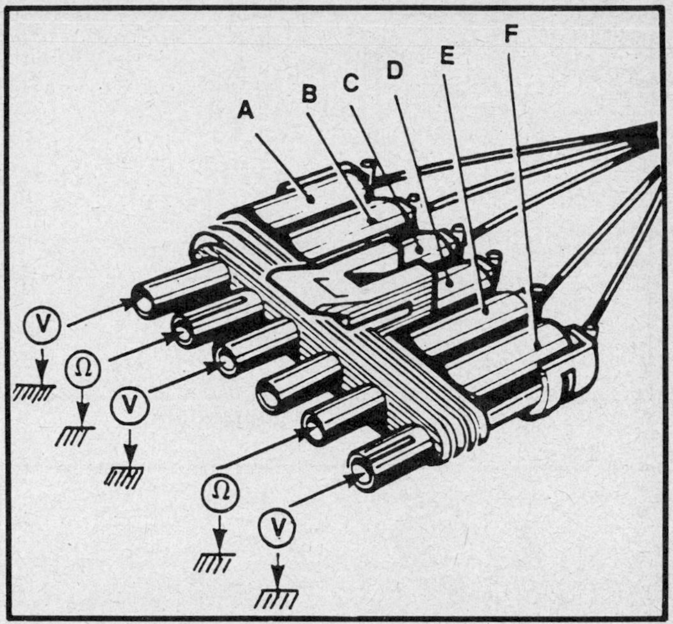
A
B
C
D
E
F
V
Ω
V
Ω
V

6-way computer connector terminal identifications

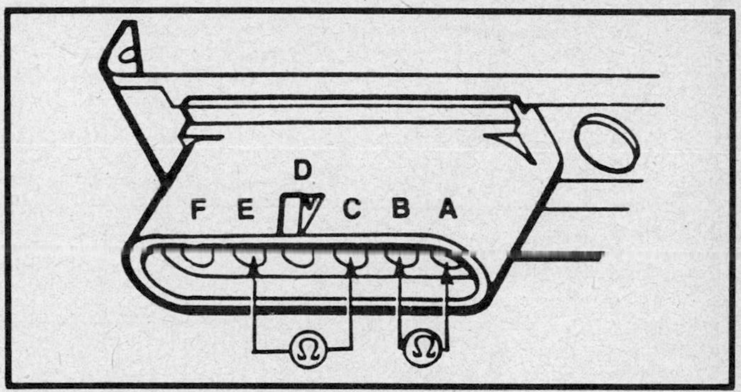
D
F E C B A
Ω
Ω

Computer socket terminal identifications

ON CAR SERVICES

Adjustments

KICKDOWN SWITCH

The kickdown switch adjustment is made with the accelerator cable.

1. Be sure that the accelerator cable has sufficient play to allow a $\frac{1}{16}$ in. (3–4mm) movement in the stop sleeve when the accelerator pedal is completely depressed.
2. Make sure that the cover is in position to prevent tarnishing of the contacts.

NOTE: The accelerator pedal travel, kickdown switch adjustments and governor control cable adjustments are all closely related, it is therefore wise to check and adjust them at the same time.

3. To check the kickdown switch, connect a test lamp between the kickdown switch and the positive terminal of the battery. When the accelerator pedal is depressed the test lamp should light showing that the kickdown switch is making contact.

GOVERNOR CABLE

Before making the governor cable adjustment, be certain that the accelerator cable has been adjusted properly.

1. Depress the accelerator pedal to the wide open throttle position (WOT).
2. Adjust the accelerator cable to obtain 0.080 in. (2mm) compression of the spring in the cable stop. Make certain that the kickdown switch is operating properly.
3. Adjust the governor cable adjusters on both the governor and the throttle sides to mid way position.
4. Adjust the cable stop to obtain a clearance of 0.008–0.028

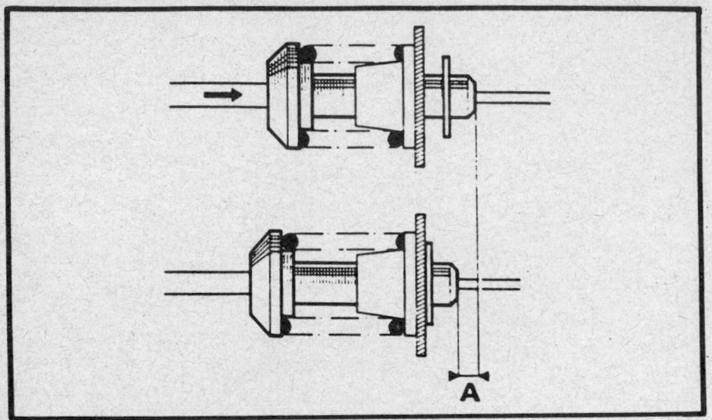
A
Making the accelerator cable adjustment

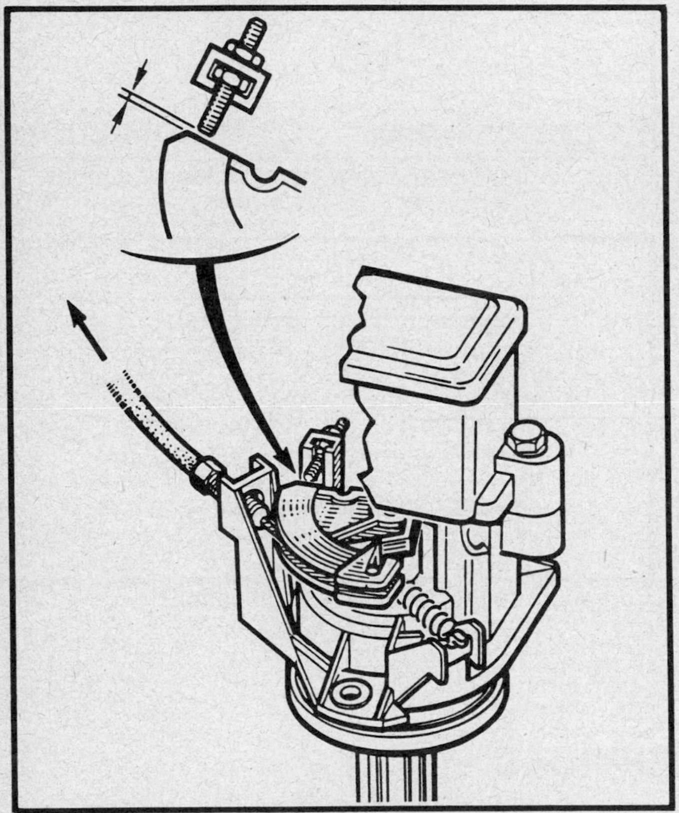
Making the governor cable adjustment

in. (0.2–0.7mm) between the screw and the lever with the throttle fully open. The screw has been preset at the factory and must not be adjusted under any circumstances.

5. Verify that the length of the governor cable is approximately 0.788 in. (20mm) between the wide open throttle position and the closed position.

CHECKING THE VACUUM CAPSULE

The vacuum capsule is checked with the engine stopped.

1. Connect a suitable vacuum gauge to the vacuum hose. Apply a vacuum of approximately 15.7 in. Hg to the capsule.
2. If the needle on the vacuum gauge does not move, check the full and light throttle pressure.

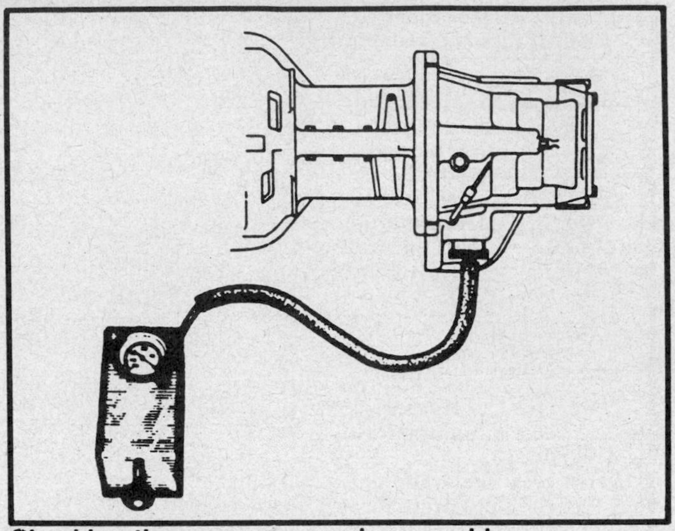
Checking the vacuum capsule assembly

3. If the needle on the vacuum gauge does fall, the capsule or its hose must be changed.
4. Make sure that the connection on the intake manifold is in good condition. Check to make sure that the vacuum hose connection to the capsule and the union is tight.

NOTE: An air leak into the capsule or into its pipe will cause whistling, unsteady idling and rough gear shifting on a light load.

Services

FLUID CHANGES

The automatic transaxle fluid and filter should be changed and the bands adjusted as follows:

a. Normal usage—every 30,000 miles.
b. Severe usage—every 15,000 miles.

Severe usage would be any prolonged operation with heavy loading especially in hot weather.

When refilling the transaxle, make sure only AMC/JEEP/RENAULT© ATF. The oil filter change should be made at the time of the oil change.

Drain and Refill

The fluid must be drained when hot, immediately after the engine has been turned off. This will remove all impurities suspended in the fluid.

1. Remove the transaxle dipstick. Raise and safely support vehicle. Place a container with a large opening under the transaxle oil pan.
2. Unscrew the drain plug from the transaxle oil pan.
3. Let the fluid drain for as long as possible.
4. Reinstall the drain plug into the oil pan. Torque the drain plug to 11 ft. lbs.
5. Refill the transaxle with 2⅔ quarts of the proper transaxle fluid through the dipstick tube.
6. Lower the vehicle and insert the dipstick.
7. Start the engine and allow it to idle for at least 2 minutes. Apply the parking brake and block the drive wheels. Move the selector lever momentarily to each position, ending in the **P** position.
8. Recheck the ATF fluid level, if necessary, add sufficient fluid to bring level to the the proper level.

9. Do not overfill. Make sure the dipstick is properly seated to seal against dirt and water entering system.

NOTE: If there is evidence of contamination or if trouble shooting indicates a problem in the converter, it must be replaced. Whenever the transmission fluid is replaced, automatic transmission fluid additive, part number 8983-100-034 or equivalent should be added. This will minimize the fluid foaming condition which occurs during normal vehicle operation. Do not add more than one bottle of automatic tranmission fluid additive per transmission oil change.

OIL PAN

Removal and Installation

1. Remove the transaxle dipstick. Raise and safely support vehicle. Place a container with a large opening under the transaxle oil pan.
2. Unscrew the drain plug from the transaxle oil pan.
3. Let the fluid drain for as long as possible.
4. Reinstall the drain plug into the oil pan. Torque the drain plug to 11 ft. lbs.
5. Loosen oil pan bolts and gently pull one corner down so any fluid left in the pan will drain. If transaxle is hot, be careful of spilling oil.
6. Remove oil pan bolts and pan.
7. Carefully inspect filter and pan bottom for a heavy accumulation of friction material or metal particles. A little accumulation can be considered normal, but a heavy concentration indicates damaged or worn parts.
8. Filter replacement is recommended when a the pan is removed. To remove the filter, remove the filter retaining bolts and and slip the filter out from under the third bolt holding it in place, do not remove the third bolt after removing the filter.
9. Be sure not damage the suction line seal. When installing the new filter be certain to install in the same position as it was removed from and torque the retaining bolts to 80 inch lbs. (9 Nm).
10. Check pan carefully for distortion, straightening the edges with a straight block of wood and a rubber mallet if necessary. Clean the pan and make sure that the magnets are positioned correctly in the pan.
11. Install a new gasket on the pan and install the pan. Torque bolts to 54 inch lbs. (6 Nm).
12. Refill the transaxle with 2⅔ quarts of the proper transaxle fluid through the dipstick tube.
13. Lower the vehicle and insert the dipstick.
14. Start the engine and allow it to idle for at least 2 minutes. Apply the parking brake and block the drive wheels. Move the selector lever momentarily to each position, ending in the **P** position.
15. Recheck the ATF fluid level, if necessary, add sufficient fluid to bring level to the the proper level.
16. Do not overfill. Make sure the dipstick is properly seated to seal against dirt and water entering system.

VALVE BODY ASSEMBLY

Removal and Installation

1. Remove the transaxle dipstick. Raise and safely support vehicle. Place a container with a large opening under the transaxle oil pan.
2. Unscrew the drain plug from the transaxle oil pan.
3. Let the fluid drain for as long as possible.
4. Reinstall the drain plug into the oil pan. Torque the drain plug to 11 ft. lbs.
5. Loosen oil pan bolts and gently pull one corner down so any fluid left in the pan will drain. If transaxle is hot, be careful of spilling oil.

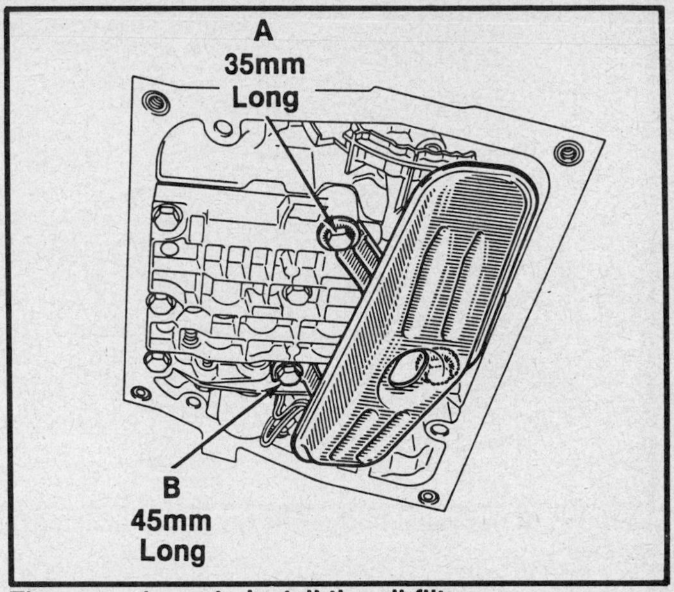
A
35mm
Long
B
45mm
Long
The correct way to install the oil filter

6. Remove oil pan bolts and pan.
7. Disconnect the solenoid valve plug from the sealed plug connector. Disconnect the solenoid ball valve wires after removing the clips. Mark the solenoid ball valves, indicating the color of the wire to which each was attached. Remove the 2 bolts holding the solenoid ball valve supporting plate and remove the valves.
8. Place large opening drain pan under transaxle, then remove the 6 bolts holding the valve body assembly to the transaxle. Hold the valve body in position while the bolts are being removed. Remove the valve body assembly from the transaxle.

CAUTION

Do not clamp any portion of the valve body assembly in a vise. Even the slightest distortion will result in stuck valves and leakage or fluid cross leakage. Handle the valve body with care at all times.

9. Installation is the reverse order of the removal procedure and torque the valve body bolts to 54 inch lbs. (6 Nm).
10. Install a new gasket on the pan and install the pan. Torque bolts to 54 inch lbs. (6 Nm).
11. Refill the transaxle with 2⅔ quarts of the proper transaxle fluid through the dipstick tube.
12. Lower the vehicle and insert the dipstick.
13. Start the engine and allow it to idle for at least 2 minutes. Apply the parking brake and block the drive wheels. Move the selector lever momentarily to each position, ending in the **P** position.
14. Recheck the ATF fluid level, if necessary, add sufficient fluid to bring level to the the proper level.
15. Do not overfill. Make sure the dipstick is properly seated to seal against dirt and water entering system.

SHIFT MECHANISM

Removal and Installation

1. Raise and support ther vehicle safely. Remove gear shift handle by pulling straight up on it. Remove the bezel by pressing on the left hand side and lift to disengage the prongs.
2. Remove the 4 nuts on the exhaust pipe flange. Remove the 2 bolts connecting the front and rear exhaust pipe sections.
3. Remove the exhaust pipe hanger nut. Remove the front exhaust pipe section from the vehicle.

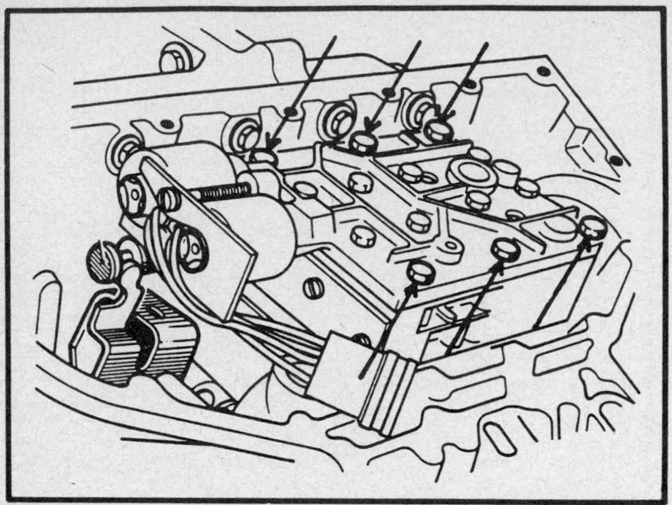

Location of the valve body assembly and retaining screws

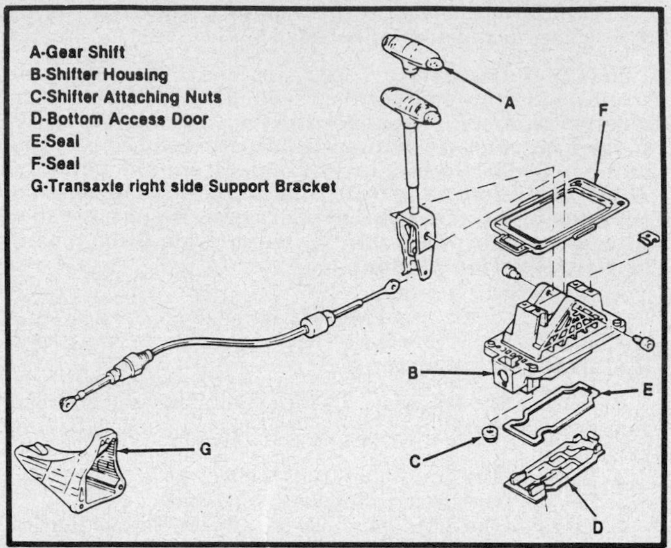
A-Gear Shift
B-Shifter Housing
C-Shifter Attaching Nuts
D-Bottom Access Door
E-Seal
F-Seal
G-Transaxle right side Support Bracket
A
F
B
E
G
C
D

Exploded view of the shift lever assembly

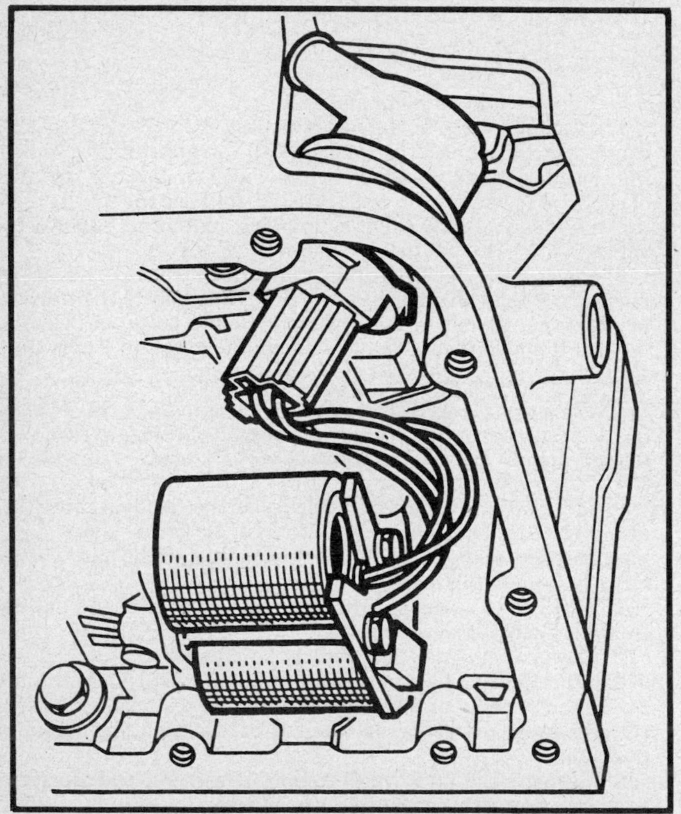

Location of the solenoid ball valves and support plate

4. Support the engine and transaxle assembly with a suitable lifting device. Snap the external control cable end off the shift lever ball stud. Lower the engine-transaxle cradle approximately ⅛ in. by loosening the rear bolts.
5. Support the transaxle assembly with a suitable transaxle jack. Remove the automatic transaxle right side support bracket.
6. Remove the shifter attaching nuts from under the vehicle.
7. Install the shift mechanism with the seal in the proper position.
8. Position the shift mechanism into the cutout in the floor pan (board).
9. Install the retainer nuts and tighten securely.
10. Tighten the nuts on the spring loaded clamping bolts, at the catalytic converter, until the coils are completely collapsed, then back off each nut 1½ turns.
11. Tighten all fasteners securely. Snap the trim panel back into the shifter console.
12. Install the shift handle by lightly rapping it with a soft faced mallet.

SHIFT CABLE

Removal and Installation

1. Raise and support the vehicle safely. The shift mechanism must be removed to perform this procedure.
2. Unsnap the cable from the transaxle right side support bracket.
3. Using a suitable tool, open the bottom access door of the shift housing.
4. Unsnap the cable from the shifter lower end. Remove the cable lock pin.
5. Disengage the cable from the shifter housing by depressing the 2 locking tangs in the housing.
6. Snap the new cable into the shifter housing and make sure the locking tangs are in the lock position.
7. Snap the cable end onto the lower end of the shifter. Snap the access door shut with the seal in the correct position.
8. Snap the other end of the cable into the transaxle right side support bracket.
9. Adjust the shift cable as follows:
   a. Install the transaxle right side support bracket and torque the bolts to 30 ft. lbs. (40 Nm).
   b. Tighten the 2 rear engine to transaxle cradle bolts to 62 ft. lbs. (85 Nm).
   c. Unlock the cable adjuster by popping up the lock tab.
   d. Set the shift selector lever to **D** position.
   e. Shift the transaxle outside lever all the way foward and back 2 detent positions (**D**).
   f. Snap the cable end onto the transaxle outer lever ball stud. In this position and with the shift lever hanging under the vehicle, snap the cable adjustment lock into the lock position.

REMOVAL AND INSTALLATION

TRANSAXLE REMOVAL

The transaxle can be removed separately from the engine from underneath the vehicle, using the special transaxle jack Desvil 701 ST or equivalent with 4 studs. Install a suitable engine support tool across the front of the engine compartment. Secure the chain around the front of the exhaust port and take up the slack in the chain. This will keep the engine from tilting forward when the transaxle is removed.

1. Disconnect the negative battery cable. Raise and support the vehicle safely. Drain the transaxle fluid from the transaxle.
2. Disconnect the oxygen sensor electrical connector. Remove the hose clamp and tube from the lower end of the heat tube.
3. Remove the heat tube bracket bolt located at the rear of the engine. Remove the remaining heat tube bracket bolts and nuts, which are located under the intake manifold. Remove the heat tube.
4. Remove the bolts attaching the top dead center sensor to the converter housing and remove the sensor.
5. Remove the steering bracket. The bracket is attached to the steering rack and the steering tie rod with nuts and bolts.
6. On vehilces equipped with air conditioning, discharge the refrigerant from the A/C system. Disconnect the A/C lines at the expansion valve and retainer.
7. Remove the bolts attaching the crossmember to the side still and body.
8. Remove the front wheels. Disconnect the passenger side tie rod ball stud from the knuckle with tool T.Av. 476 or equivalent.

NOTE: Run the tie rod ball joint nut to the end of the ball stud before installing the removal tool. This protects the stud threads when loosening the stud.

9. Remove the passenger side steering tie rod.
10. Loosen the coolant expansion tank retaining strap. Pull the tank out of the strap and move the tank aside. Move the tank far enough away to permit the A/C lines foward and away from the crossmember.
11. Remove the nuts attaching the exhaust pipe to the exhaust manifold.
12. Remove the 2 bolts connecting the front and rear exhaust pipe sections. Remove the hanger nut. Remove the front exhaust pipe section from the vehicle.
13. Turn the crossmember and remove it through the wheel well opening on the passenger side of the vehicle.
14. Remove the steering knuckle upper mounting bolt. Then loosen (but do not remove) the lower bolt.

NOTE: The bolts are splined just below the head so it is necessary to remove the nut and tap the bolt out with a brass or lead mallet.

15. Remove the driveshaft roll pins with a suitable pin drift. Swing each rotor and steering knuckle outward. Then slide the driveshafts off of the transaxle output shafts.
16. Remove the mounting bracket by removing the bracket bolts.
17. Remove the wiring from the starter solenoid. Remove the rear mounting bolts. Remove the starter mounting bolts and remove the starter and locating bushing.
18. Install the drive plate tool Mot. 582 or equivalent, to hold the driveplate stationary. Remove the torque converter bolts. Remove the computer module from the bracket. Do not remove the computer module from the transaxle.
19. Disconnect the coolant lines from the heat exchanger and remove the line clamp on the side rail.
20. Disconnect the speedometer cable. Support the transaxle with a suitable transaxle jack.

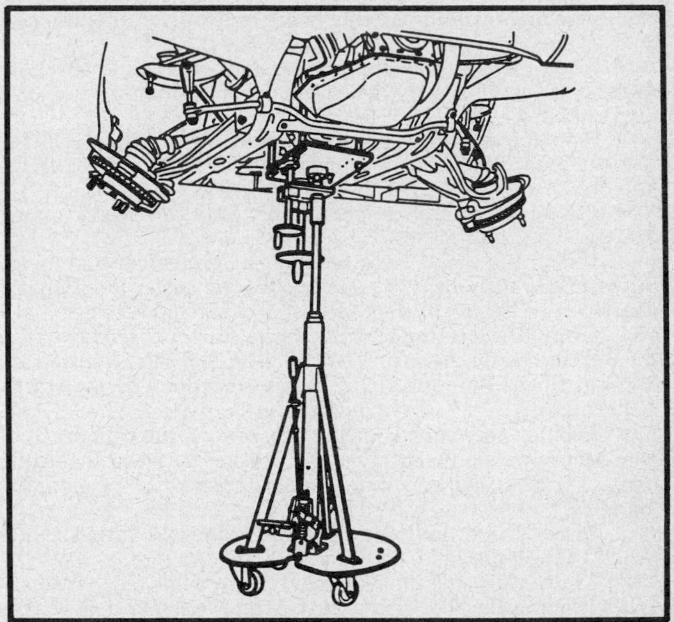
Installing the transaxle jack for removal purposes

21. Lower the the engine transaxle cradle by loosening the 2 rear bolts. Lower the cradle by 0.6 in. (15mm).
22. Remove the shift cable from the shift lever ball stud. Leave the shift cable in the right side transaxle support bracket but tie the bracket out of the way with mechanics wire.
23. Remove the ground strap from the transaxle case. Remove all transaxle to engine retaining bolts.
24. Remove the nuts that attach the transaxle support cushions to the vehicle. Remove the bolts tahat attach the 2 transaxle mounting brackets to the transaxle. Then remove the brackets.
25. Lower the transaxle carefully and guide the computer module and the cooler lines out.

NOTE: As soon as the transaxle is removed, use a suitable converter retaining plate to prevent the torque converter from separating.

TRANSAXLE INSTALLATION

If the transaxle was removed because of malfunction caused by sludge, accumulated friction material or metal particles, the oil cooler and lines must be flushed thoroughly. The transaxle and converter must be installed as an assembly; otherwise, the converter drive plate, pump bushing and oil seal will be damaged.

1. Apply a small amount of grease to the torque converter pilot area in the back of the crankshaft. Make sure that the alignment dowels are in position in the rear block face.
2. Remove the converter holding tool. Using the transaxle jack, raise the transaxle up into position while guiding the cooler lines and the computer module into position.
3. The drive plate and torque converter must be properly mated. The painted marks on the driveplate align with the marks on the torque converter.
4. Position the transaxle onto the engine dowels. Install the engine to transaxle bolts and torque them to 37 ft. lbs. (50 Nm).
5. Check the torque converter to see that it is free. Install the

torque converter bolts and torque them to 22 ft. lbs. (30 Nm). Install the drive plate tool Mot. 582 or equivalent, to hold the driveplate stationary, while tightening the bolts.

6. Install the grounding strap to the transaxle case. Raise the transaxle and install the 2 mounting brackets on the transaxle and torque the bolts to 30 ft. lbs. (40 Nm).

7. Align the transaxle support cushion studs and lower the transaxle. Install and tighten the support cushion nuts to 30 ft. lbs. (40 Nm).

8. Snap the shift cable end onto the transaxle outside shift lever ball stud. Torque the 2 rear engine cradle bolts to 62 ft. lbs. (85 Nm).

9. Install the computer module into its bracket and connect the electrical connector.

10. Install the cooler lines to the heat exchanger and torque them to 175 inch lbs. (20 Nm). Install the cooler line clamp of the side rail. Install the speedometer cable and retainer.

11. Install the locating bushing in the starter assembly flange. Install the starter mounting bolts on the converter housing and tighten them. Reconnect the wire connections to the starter solenoid.

12. Position the rear mounting brackets on the cylinder block and secure the bolts. Position the bracket between the engine block and the intake manifold and secure with the retaining bolts.

13. Install the driveshafts onto the transaxle output shafts. Align the roll pin holes in the shafts.

14. Tilt the steering knuckle and rotor assemblies inward and align them in the shock bracket. Install the upper bolt. Torque the upper and lower bolt to 148 ft. lbs. (200 Nm).

15. Line up the driveshaft using tool B.Vi. 31-01 or equivalent and install the roll pins with a suitable drift pin. Place a small amount of silicone sealant at each end of the roll pins.

16. Install the crossmember through the wheel well opening on the passenger side of the vehicle. Position the crossmember on the side sills and body and install the crossmember attaching bolts and nuts. Tighten the crossmember to body bolts first and tighten the remaining bolts.

17. On the vehicles with A/C, connect the A/C lines to the connector block on the dash panel. Secure the lines to the side sill with the retainer.

18. Connect the steering tie rods to the knuckles. Torque the tie rod nuts to 30 ft. lbs. (40 Nm).

19. Connect the steering tie rods to the steering gear bracket. Tighten the attaching bolts and nuts to 25 ft. lbs. (35 Nm). Connect the steering gear bracket to the steering gear rack. Torque the bracket bolts to 30 ft. lbs. (40 Nm). Install the locknuts and washers on the bolts and torque the nuts to 25 ft. lbs. (35 Nm).

20. Install the front wheels and torque the bolts to 66 ft. lbs. (90 Nm).

21. Install the top dead center sensor. Install the warm up tube, brackets and hoses. Connect the exhaust head pipe to the exhaust manifold.

22. Install a replacement seal on the exhaust head pipe and connect the pipe to the converter. Tighten the bolts that attach the head pipe and converter until the spring coils are touching then back off each nut 1½ turns.

23. Connect the oxygen sensor wire. Install the vacuum capsule tube. Remove the engine support tool.

24. Adjust the shift cable as necessary. Connect the negative battery cable.

25. Fill the transaxle with 3 quarts of the proper automatic transmission fluid before starting the engine. On vehicles equipped with A/C recharge the system with fresh freon. Road test the vehicle and make all and any necessary adjustments.

BENCH OVERHAUL

Before Disassembly

Cleanliness during disassembly and assembly is necessary to avoid further transaxle trouble after overhaul. Before removing any of the transaxle subassemblies, plug all the openings and clean the outside of the of the transaxle thoroughly. Steam cleaning or car wash type high pressure equipment is preferable. During disassembly, clean all parts in suitable solvent and dry each part. Do not use cloth or paper towels to dry parts. Use compressed air only. Before disassembling the transaxle, remove the torque converter, dipstick tube, wiring connections on the transaxle, the governor computer and the malfunction switch.

Converter Inspection

The torque converter is removed by simply sliding the unit out of the transaxle off the input and reaction shaft. If the converter is to be reused, set aside so it will not be damaged. Since the unit is welded and does not have a drain plugs converters subject to burnt fluid or other contamination must be replaced.

Transaxle Disassembly

OIL PAN AND FILTER

Removal

1. Make sure transaxle is held securely either in a stand or fixture.

2. Remove the pan to case bolts and gently tap pan loose. Do not insert a tool between pan and case as a prying tool or case damage may result.

3. Check pan carefully for distortion, straightening the edges of the pan with a straight block of wood and a rubber mallet if necessary. A power driven wire wheel is useful in removing glued on gaskets.

4. Carefully inspect filter and pan bottom for a heavy accumulation of friction material or metal particles. A little accumulation can be considered normal, but a heavy concentration indicates damaged or worn parts.

5. To remove the filter, remove the filter retaining bolts and remove the filter, be sure not damage the suction line seal. When installing the new filter be certain to install in the same position as it was removed from and torque the retaining bolts to 6 ft. lbs.

6. When the pan is eventually reinstalled torque the bolts evenly to 54 inch lbs. (6 Nm).

VALVE BODY

Removal

1. Place large opening drain pan under transaxle, then remove the 6 bolts holding the valve body assembly to the transaxle.

2. Hold the valve body in position while the bolts are being removed. Remove the valve body assembly from the transaxle.

REAR CASE

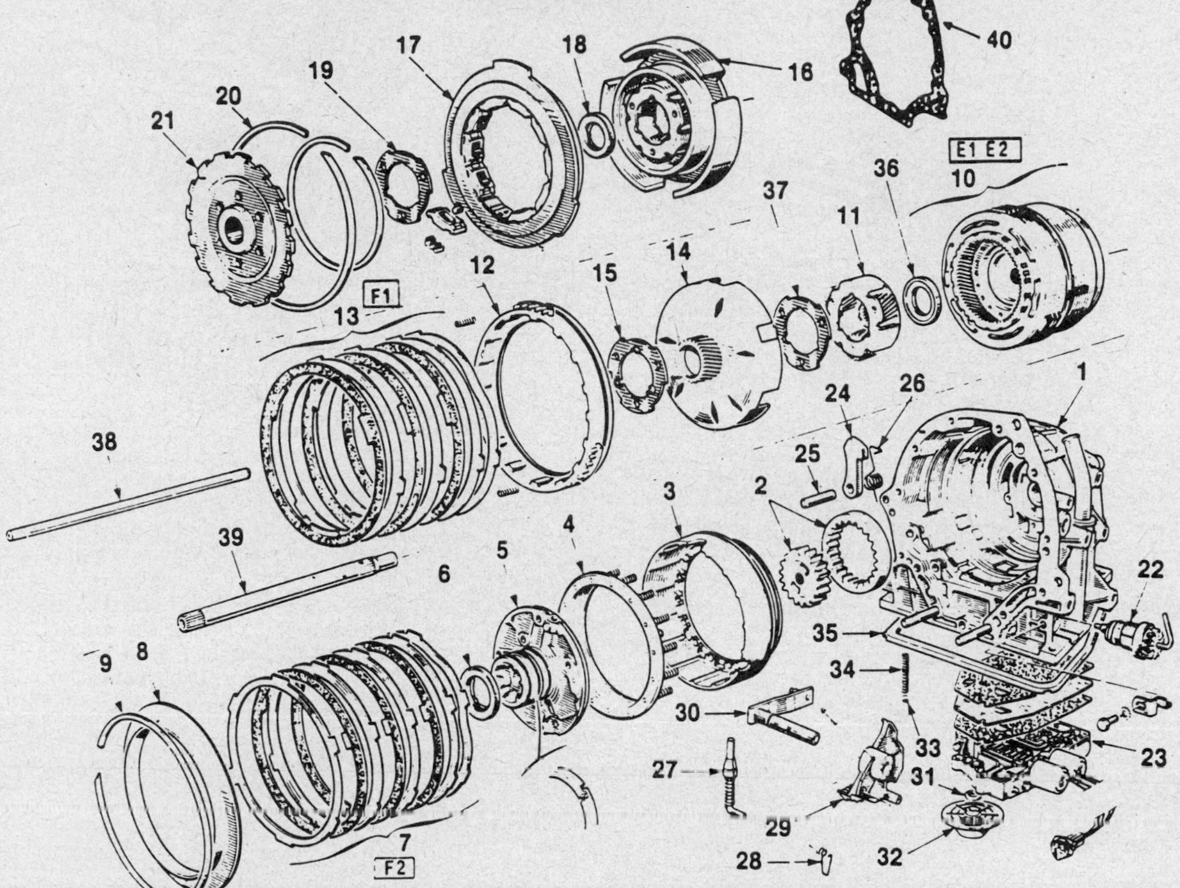
40
16
18
17
19
20
21
E1 E2
10
36
37
11
14
12
15
F1
13
1
24
26
38
25
3
2
39
4
5
6
22
35
34
9
8
30
33
23
27
31
29
28
32
7
F2

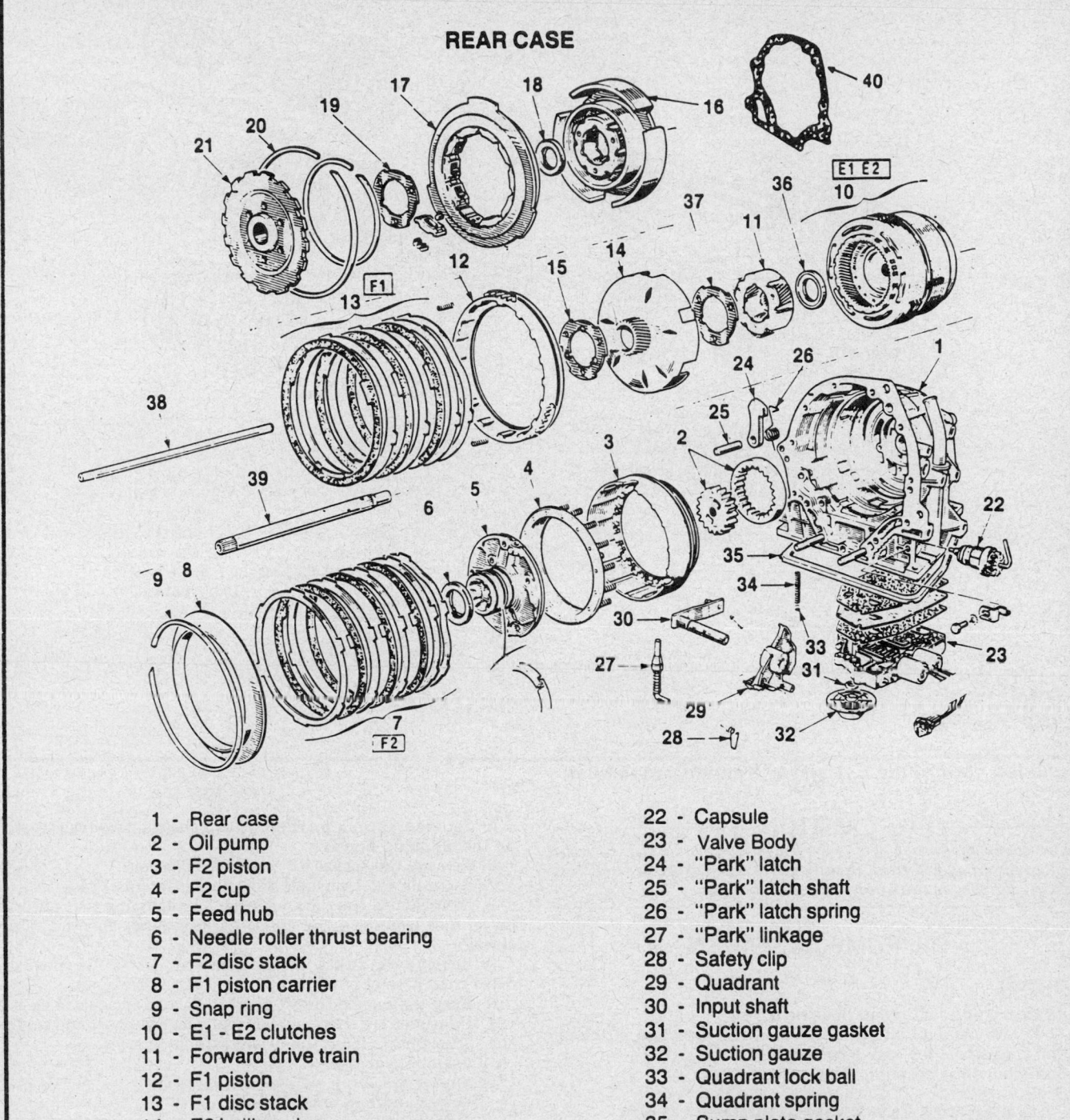

1 - Rear case	22 - Capsule
2 - Oil pump	23 - Valve Body
3 - F2 piston	24 - "Park" latch
4 - F2 cup	25 - "Park" latch shaft
5 - Feed hub	26 - "Park" latch spring
6 - Needle roller thrust bearing	27 - "Park" linkage
7 - F2 disc stack	28 - Safety clip
8 - F1 piston carrier	29 - Quadrant
9 - Snap ring	30 - Input shaft
10 - E1 - E2 clutches	31 - Suction gauze gasket
11 - Forward drive train	32 - Suction gauze
12 - F1 piston	33 - Quadrant lock ball
13 - F1 disc stack	34 - Quadrant spring
14 - E2 bellhousing	35 - Sump plate gasket
15 - Friction washer (1.5 mm thick)	36 - Needle roller bearing
16 - Reverse drive train	37 - Friction washer (1.5 mm thick)
17 - Freewheel	38 - Pump shaft
18 - Needle roller thrust bearing	39 - Turbine shaft
19 - Friction washer (thickness to be determined)	40 - Rear Case-to-Intermediary Case gasket
20 - Snap ring	
21 - "Park" wheel	

Exploded view of the rear case and its components

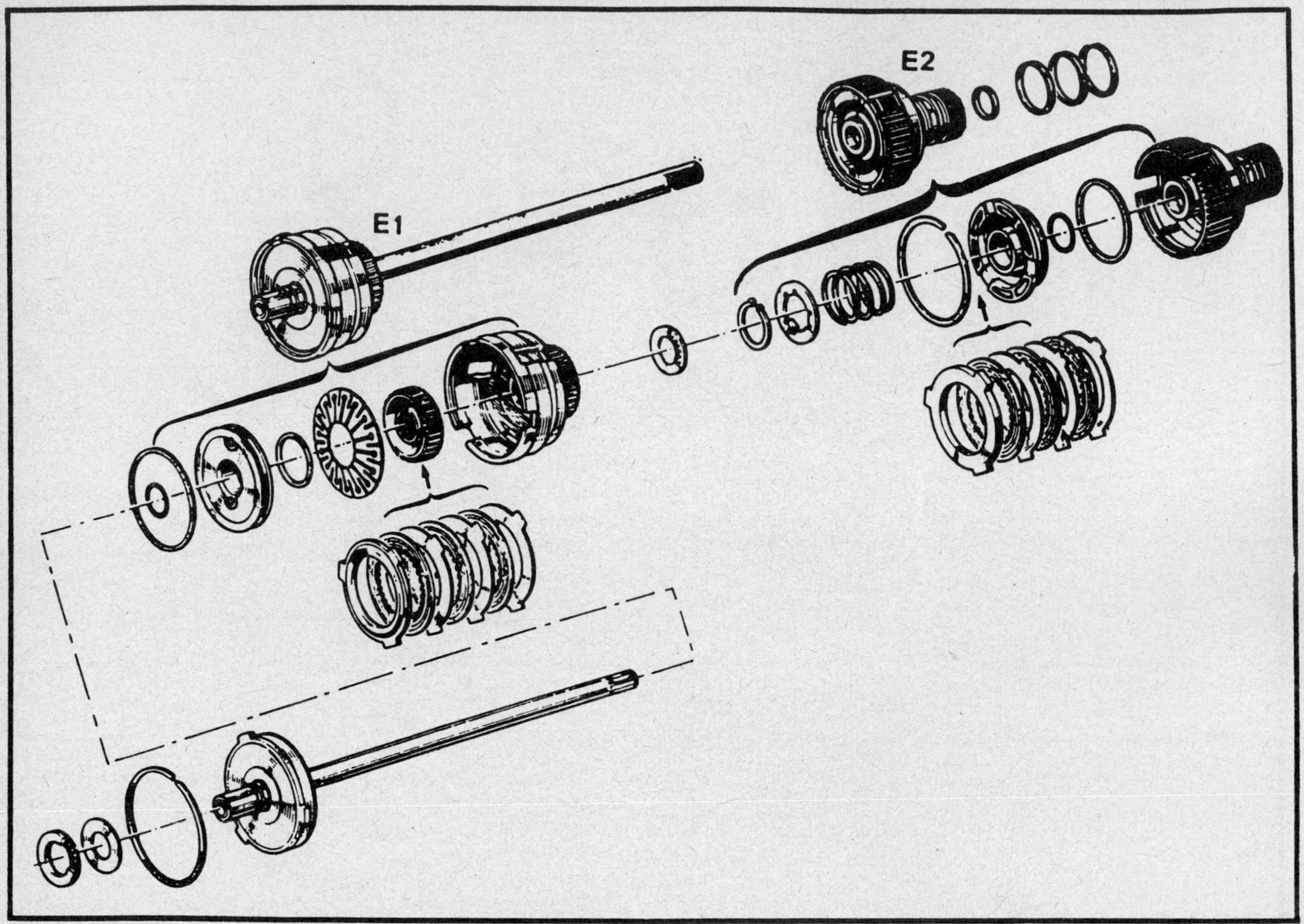
E2
E1

Exploded view of the E–1 and E–2 clutch assemblies

CAUTION

Do not clamp any portion of the valve body assembly in a vise. Even the slightest distortion will result in stuck valves and leakage or fluid cross leakage. Handle the valve body with care at all times.

OIL PUMP ASSEMBLY

Removal

1. Remove the oil pump housing cover.
2. Remove the oil pump shaft. Mark the direction of the oil pump ring wheel for easy reassembly.
3. Remove the oil pump gear.

TRANSAXLE COMPONENTS AND DIFFERENTIAL

Removal

1. Remove the 4 inner differential assembly bolts in the transaxle case.
2. Remove the 2 roll pins with a suitable drift pin.
3. Remove the parking latch and remove the connecting arm without separating the ball joints.
4. Remove the shaft and save the tooth quadrant. Remove the gear control linkage. The socket containing the locking ball must not be removed unless it is to be changed.
5. Set the transaxle on the end with the transaxle case resting on the oil pump housing.
6. Remove the transaxle case assembly bolts.
7. Separate the transaxle from the differential case.
8. Disassemble the parking latch by, removing the centering dowel (use a puller), the shaft, parking latch and the return spring.
9. Remove the brake mechanism fixing bolts. Remove the drive train assembly while holding the turbine shaft.
10. Keep the needle thrust bearing located inside the case.
11. Position the drive train vertically on a support and remove the various parts from the drive train as follows.
   Planetary gear train
   Sun gear assembly
   The F–1 and F–2 assembly
   The E–2 clutch
   The E–1 clutch and the turbine shaft
12. Take note of the location of the needle thrust bearings, so as to keep them in their proper location during reassembly.

Unit Disassembly and Assembly

PLANETARY GEAR TRAIN

Disassembly

1. Remove the sprag clutch.
2. Remove the adjusting shim.

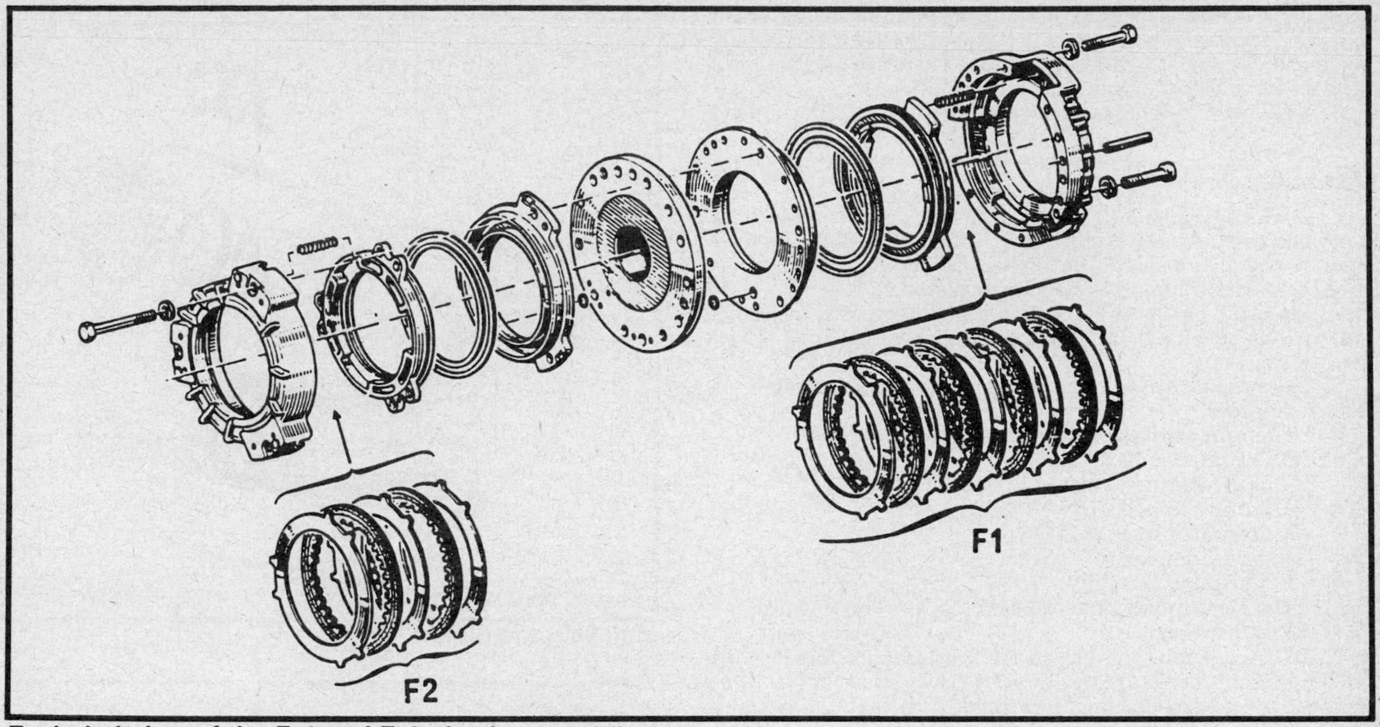
F1
F2

Exploded view of the F–1 and F–2 clutch assemblies

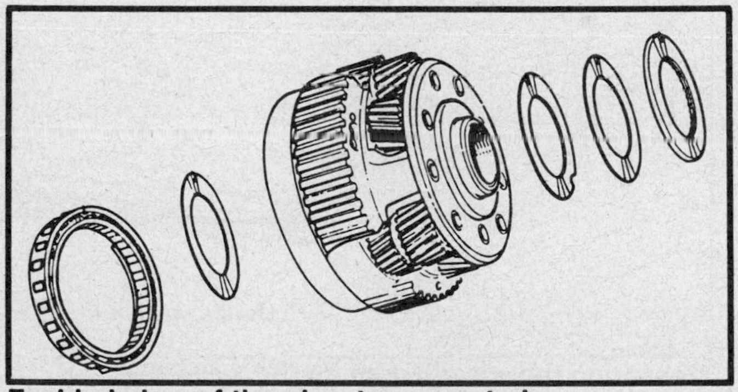

Expided view of the planetary gear train

3. Remove the needle thrust bearing plate, needle thrust bearing and the last needle thrust bearing plate.

NOTE: The thrust bearing inside the planetary gear train cannot be disassembled and the assembly must be replaced as a unit.

Inspection

1. Inspect all oil passages in the shaft and make sure they are open and clean.
2. Inspect the bearing surfaces nicks, burrs, scores or other damage. Light scratches, small nicks or burrs can be removed with crocus cloth or a fine stone.
3. Inspect all thrust plates for wear and scores, replace if damaged or worn below specifications.
4. Inspect the thrust faces of the planetary gear carriers for wear, scores or other damage, replace as required. Inspect the planetary gear carrier for cracks and pinions for broken or worn gear teeth and for broken pinion shaft welds.

Assembly

1. Install the first needle thrust bearing plate, needle thrust bearing and the last needle thrust bearing plate.
2. Install the sprag clutch on the planetary gear train. The shoulder should face towards the inside (bottom) of the carrier.
3. Check to see that the thrust bearing plate is correctly positioned.
4. Install the sun gear so that it centers and holds the inner thrust bearing.

E–1 CLUTCH

Disassembly

1. Remove the compresion ring from the needle thrust bearing before removing the needle thrust bearing.
2. The turbine shaft and the hub are one piece. Push down on the E–1 pistion housing to remove the retaining ring.
3. Remove the housing. Push the piston out by applying compressed air to the piston housing input hole.
4. Remove the diaphragm spring, thrust plate, 3 lined discs, 3 intermediate flat discs, hub and bell housing.

Inspection

1. If plates show any sign of deterioration or wear, they must be replaced. The plates and disc should be flat, they must not be warped or coned shaped.
2. Inspect the facing material on all driving discs. Replace the discs that are charred, glazed or heavily pitted. Discs should also be replaced if they show any evidence of material flaking off or if the facing material can be scrapped off easily. Inspect the driving disc splines for wear or other damage.
3. Inspect the steel plate and pressure plate surface for burning, scoring or damaged driving lugs. Replace if necessary.
4. Check to see that the discs slide easily on the hub splines and that the plates slide easily in the bell housing.
5. Check to see that the wave discs are correct with a feeler

gauge. Lay the wave disc flat on a surface plate without exerting pressure on it and slip the feeler gauge between the surface plate and the disc at 3 different points. The wave height should be between 0.010–0.018 in. (0.25–0.45mm).

6. If the wave height is out of specifications, replace the wave discs.

Assembly

1. Place the thrust ring onto the piston. Lubricate the 2 piston seals and place the rectangular seal on the piston and to O-ring on the piston sleeve. Be sure to lightly lubricate the bore and the part of the hub in which the piston slides.

2. Place the piston (2 balls) into its housing with the flange upward, by pushing it in with the thumbs while tilting it back and forth.

3. Place the following into the connecting bell housing in the following order:

The hub, recessed part facing upward
A flat intermediate disc
A lubricated lined disc
A flat intermediate disc
A lubricated lined disc
A flat intermediate disc
A lubricated lined disc
The thrust plate, smooth side towards the line disc
The diaphragm.

4. Join the piston and bell housing assemblies together. Install the circlip, be sure to position the circlip ends between the 2 lugs on the piston housing and make sure that the circlip is properly seated in the groove all the way around.

5. Install the bearing plate for the needle thrust bearing. The seal ring after checking the ring gap play.

6. Check the operation of the E–1 clutch components by applying compressed air to the piston through the hole provided.

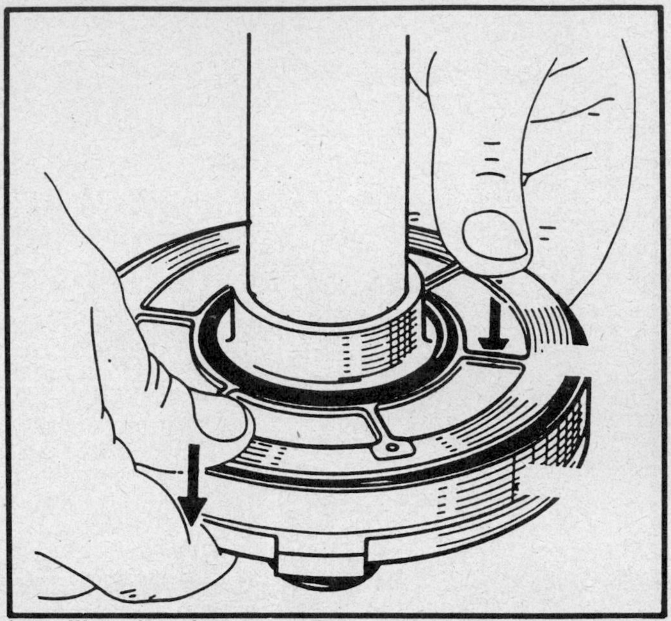

Installing the piston

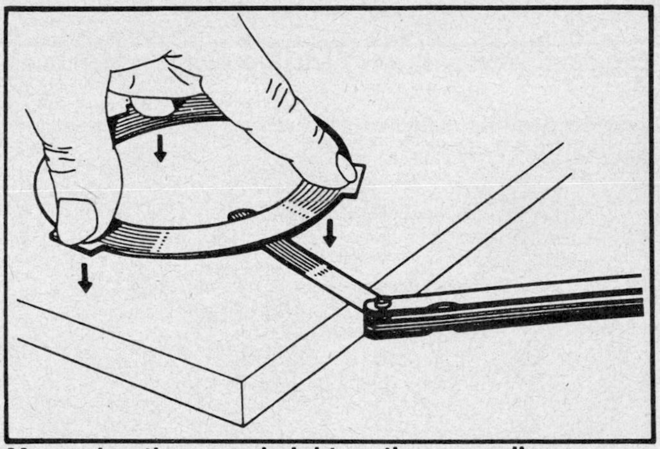

Measuring the wave height on the wave disc

E–2 CLUTCH

Disassembly

1. Using a suitable tool and a arbor press, slightly compress the clutch return spring.

2. Remove the circlip, spring retainer, the spring and the 3 seal rings.

3. Remove the thrust plate circlip, the plate, 3 lined discs, 2 wave discs, the flat disc.

4. Push the piston out by applying compressed air to the piston housing input hole.

Inspection

1. If plates show any sign of deterioration or wear, they must be replaced. The plates and disc should be flat, they must not be warped or coned shaped.

2. Inspect the facing material on all driving discs. Replace the discs that are charred, glazed or heavily pitted. Discs should also be replaced if they show any evidence of material flaking off or if the facing material can be scrapped off easily. Inspect the driving disc splines for wear or other damage.

3. Inspect the steel plate and pressure plate surface for burning, scoring or damaged driving lugs. Replace if necessary.

4. Check to see that the discs slide easily on the hub splines and that the plates slide easily in the bell housing.

5. Check to see that the wave discs are correct with a feeler gauge. Lay the wave disc flat on a surface plate without exerting pressure on it and slip the feeler gauge between the surface plate and the disc at 3 different points. The wave height should be between 0.010–0.018 in. (0.25–0.45mm).

6. If the wave height is out of specifications, replace the wave discs.

Assembly

1. Install the 3 seal rings on the E–2 bell housing after checking the ring gap play, by placing the rings in the sprag clutch hub and the condition of the 3 grooves in the bell housing.

2. Lubricate the 2 piston seals and install the rectangular seal on the piston and the O-ring on the piston hub in the E–2 bell housing. Make sure that the seals fit tightly in their bores.

3. Install the piston in its place in the bell housing.

4. Using a suitable tool and a arbor press, slightly compress the clutch return spring and install the circlip, being careful to guide the spring retainer.

5. The slots in the piston and the bell housing must be lined up. Install one after the other on the piston after lubricating:

The thrust plate
A lubricated lined disc
A wave disc
A lubricated lined disc
A wave disc
A lubricated lined disc
The thrust plate with the punch-marked surface towards the outside.

6. Install the circlip.

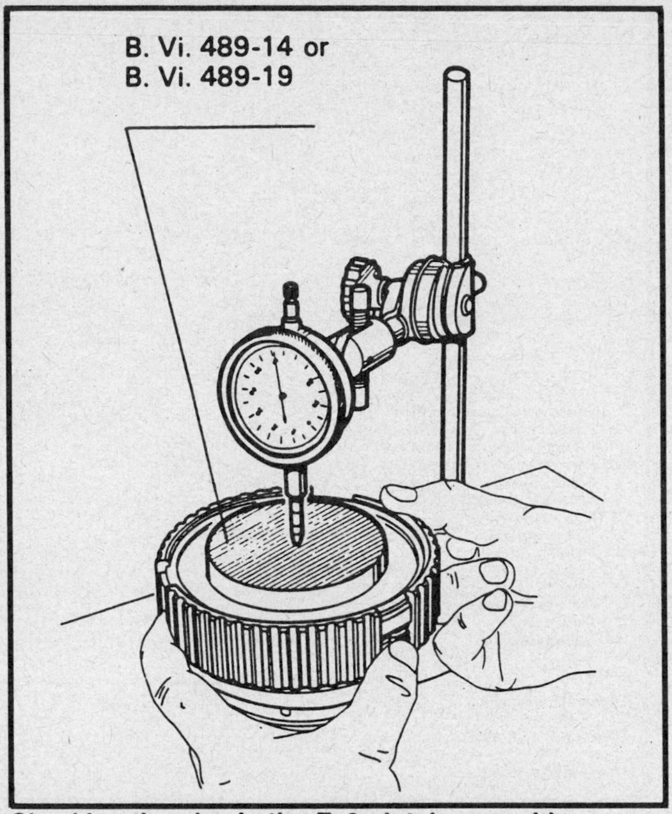
B. Vi. 489-14 or
B. Vi. 489-19

Checking the play in the E-2 clutch assembly

7. Check the operation of the E-2 clutch components by applying compressed air to the piston through the hole provided.

8. Check the endplay of the clutch as follows:

a. Place the E-2 clutch and a dial indicator gauge with bracket on a flat surface.

b. Place the B.Vi. 489-14 or B.Vi. 489-19 reference tool on the pile of the clutch discs.

c. Set the dial indicator to 0, raise the disc pile (circlip pushed up against the top groove but without compressing the waved discs) and read the off the play on the dial.

d. The play should be between 0.043-0.083 in. (1.1-2.1mm)

e. If the play is more than specified, insert a thrust plate (spacer) 0.098 in. (2.5mm) thick.

F-1 AND F-2 BRAKES

Disassembly

1. On the F-1 brake remove the following:

a. Unscrew the 3 bell housing bolts.

b. Remove the bell housing, the 6 springs and the steel discs and lined discs.

c. Save the O-ring located between the sprag clutch bearing and the piston housing.

2. On the F-2 brake remove the following

a. Unscrew the 3 bell housing bolts.

b. Remove the bell housing, the 6 springs and the steel discs and lined discs.

c. Save the O-ring located between the sprag clutch bearing and the piston housing.

d. Push the piston out by applying compressed air to the piston housing input hole.

e. Remove the piston seals.

Inspection

1. If plates show any sign of deterioration or wear, they must be replaced. The plates and disc should be flat, they must not be warped or coned shaped.

2. Inspect the facing material on all driving discs. Replace the discs that are charred, glazed or heavily pitted. Discs should also be replaced if they show any evidence of material flaking off or if the facing material can be scrapped off easily. Inspect the driving disc splines for wear or other damage.

3. Inspect the steel plate and pressure plate surface for burning, scoring or damaged driving lugs. Replace if necessary.

4. Check to see that the discs slide easily on the hub splines and that the plates slide easily in the bell housing.

5. Check to see that the wave discs are correct with a feeler gauge. Lay the wave disc flat on a surface plate without exerting pressure on it and slip the feeler gauge between the surface plate and the disc at 3 different points. The wave height should be between 0.010-0.018 in. (0.25-0.45mm).

6. If the wave height is out of specifications, replace the wave discs.

Checking and Adjusting the F-2 Brake Operating Play

This operation consists of measuring the play for the disc pile in the housing. The play is limited by, on one end the piston at the end of its travel and on the other end the bell housing.

1. Insert the piston with a seal in the F-2 piston housing.

2. Stack up a flat disc 0.059 in. (1.5mm) thick, a lined disc, the wave disc 0.079 in, (2mm) thick, which is marked with 2 notches, a lined disc and a flat disc 0.059 in. (1.5mm) thick.

3. Install the bell housing and attach the assembly to sprag the clutch hub.

4. With the dial indicator gauge point resting on a spline of the first lines disc, set the gauge to 0.

5. Raise the disc assembly so that the it makes contact with the bell housing (against the top).

6. Take measurements at several points and average the reading.

7. The play should be between 0.028-0.067 in. (0.70-1.70mm).

8. If the play is more than specified, check all parts which could affect the value, such as the piston, wave disc, line disc and bell housing. Repair or replace as necessary.

Assembly

1. Lubricate the 4 seals for the F-1 and F-2 pistons and install them in their respective housings. Check beforehand that the seals fit tightly in the brake bell housing bores.

2. Place the O-ring between the sprage clutch hub and the F-1 housing. Insert the F-1 piston in its housing, being careful not to damage the seals.

3. Stack up the flat discs and the lined discs. Place the 6 springs in their housings and cap the assembly with the F-1 bell housing.

4. Screw the 3 F-1 assembly bolts into the sprag clutch hub. Turn the assembly over and rest it on the F-1 bell housing.

5. Place the O-ring between the sprag clutch hub and the F-2 housing. Insert the F-2 piston into its housing, being careful not to damage the seals.

6. Stack up the flat discs, a lined disc and the wave disc marked with 2 notches, a lined disc and a flat disc. Place the 6 springs in their housings and cap the assembly with the F-2 bell housing.

7. Screw the 3 F-2 assembly bolts into the F-1 sprag clutch hub.

8. Check the operation of the F-1 and F-2 components by applying compressed air to the piston through the holes provided.

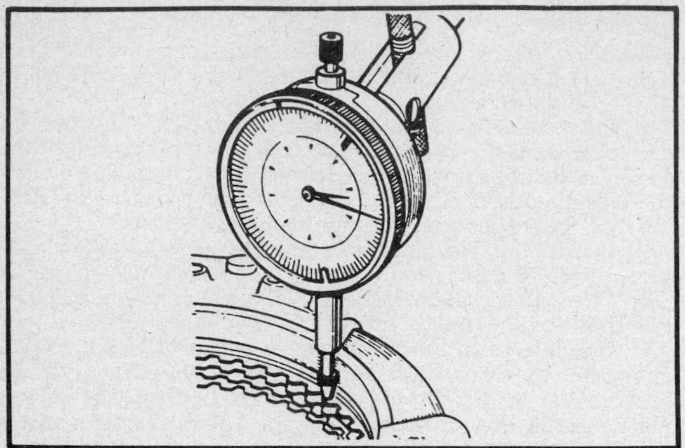
Checking the play in the F-2 brake assembly

VALVE BODY

NOTE: The valve body and regulator assembly should not be disassembled. Only the solenoid ball valves should be changed. Whenever the transaxle is disassembled due to damaged clutches or brakes or because of poor quality gear shifting, the valve body and its regulator must be changed as an assembly.

Disassembly

Do not clamp any part of the valve body or plate in a vise, for this will cause sticking valves or excessive leakage or both. When removing and installing valves or plugs, slide them in or out very carefully. Do not use force to remove or install the valves.

When disassembling the valve body identify all valve springs with a tag for assembly reference later. Also, the oil filter screws may be longer than transfer plate screws, so do not mix them up.

1. Remove the 2 closure plate bolts, remove the manual valve and turn the unit over.
2. Remove the pressure limiting valve ball, pressure limiting valve medium spring, pressure limiting valve and pressure limiting valve seal.
3. Remove the sequence valve small spring and sequence valves.
4. Remove the pilot valve, plungers and the second pilot valve.
5. Remove the pressure regulating valve and valve spring.

Inspection

1. Thoroughly wash and blow dry all parts. Be sure all passages are clean and free from dirt or other obstructions.
2. Check the manual and throttle levers and shaft for looseness or being bent or worn excessively. If bent, replace the assembly. Check all parts for burrs or nicks. Slight imperfections can be removed with crocus cloth.
3. Using a straightedge, inspect all mating surfaces for warpage or distortion. Slight distortion can be corrected by abrading the mating surfaces on a sheet of crocus cloth on a flat piece of glass, using very light pressure. Be sure all metering holes are open in both the valve body and separator plate.
4. Use a penlight to inspect valve body bores for scratches, burrs, pits or scores. Inspect valve springs for distortion. Check valves and plugs for burrs, nicks and scores. Remove slight irregularities with crocus cloth, but do not round off the sharp edges. The sharpness of these edges is vitally important because it prevents foreign matter from lodging between the valve and the bore.

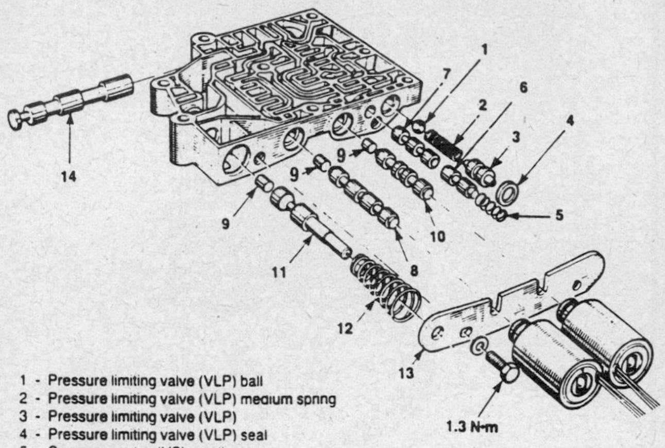
1.3 N·m

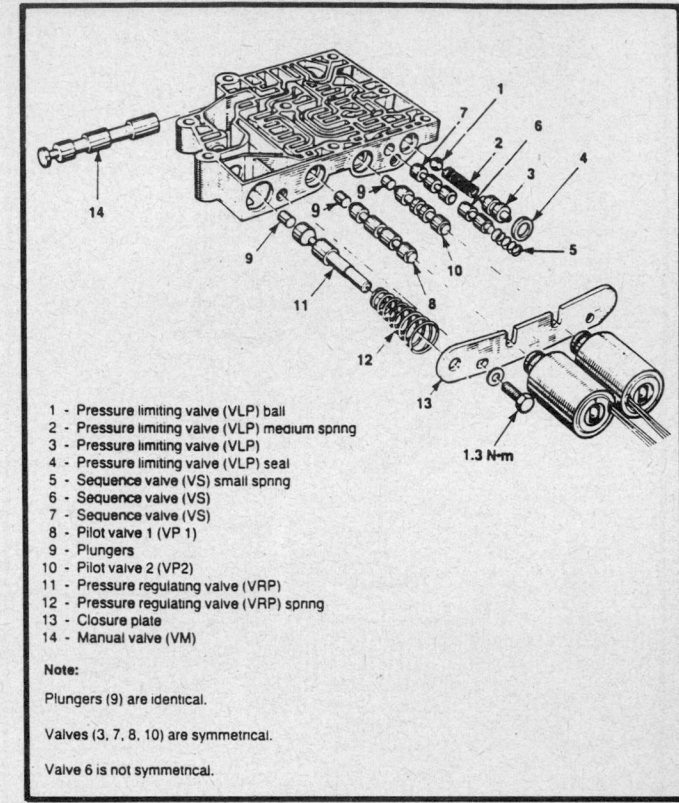

1 - Pressure limiting valve (VLP) ball
2 - Pressure limiting valve (VLP) medium spring
3 - Pressure limiting valve (VLP)
4 - Pressure limiting valve (VLP) seal
5 - Sequence valve (VS) small spring
6 - Sequence valve (VS)
7 - Sequence valve (VS)
8 - Pilot valve 1 (VP 1)
9 - Plungers
10 - Pilot valve 2 (VP2)
11 - Pressure regulating valve (VRP)
12 - Pressure regulating valve (VRP) spring
13 - Closure plate
14 - Manual valve (VM)

Note:

Plungers (9) are identical.

Valves (3, 7, 8, 10) are symmetrical.

Valve 6 is not symmetrical.

Exploded view of the disassembled valve body assembly

5. When valves, plugs and bores are clean and dry, they should fall freely in the bores.

Asssembly

All screws used in the valve body are tightened to the same torque, 54 inch lbs. (6 Nm).

1. Install the pressure regulating valve and valve spring.
2. Install the pilot valve, plungers and the second pilot valve.
3. Install the sequence valve small spring and sequence valves.
4. Install the pressure limiting valve ball, pressure limiting valve medium spring, pressure limiting valve and pressure limiting valve seal.
5. Install the 2 closure plate bolts, remove the manual valve and turn the unit over.

OIL PUMP

Disassembly

1. Remove the oil pump housing cover.
2. Remove the oil pump shaft. Mark the direction of the oil pump ring wheel for easy reassembly.
3. Remove the oil pump gear.

Inspection

1. Inspect the pump rotors for scoring or pitting and clean.
2. Inspect the machined surfaces on the pump body and reaction shaft support for nicks and burrs. Inspect the pump body and reaction shaft support bushings for wear or scores.
3. Inspect the pump gears for scoring and pitting. With gears cleaned and installed in the pump body, Check the oil pump shaft play.

Removing the oil pump shaft

4. The oil pump shaft play should be 0.014–0.031 in. (0.35–0.80 mm).

Assembly

1. Lubricate all components.
2. Install the ring wheel and be sure to use tha marks made during disassembly for wheel direction, or position the chamfer to face downward into the housing.
3. Install the gear and the pump driveshaft.
4. The pump oil seal can be replaced without removing the pump and reaction shaft support assembly from the transaxle case.
6. Using a suitable seal removal tool, remove the seal.
7. To install a new seal, lightly lubricate the new seal place the seal in the opening of the pump housing and push it into place with a suitable seal driver tool as it is tap gently.
8. Drive the new seal into the housing until the tool bottoms out.

DIFFERENTIAL COMPONENTS

Disassembly

1. Remove the endplay adjustment shim, spacer, closure plate bolts and the closure plate.
2. Remove the snapring, tapered washer, speedometer drive spindle, speedometer drive gear, speedometer drive seal and the snapring. Open the snapring and pull on the output shaft at the same time.
3. Remove the ball bearings. Push the speedometer worm gear toward the converter and pull the output shaft.
4. Remove the speedometer worm gear, step-down driving gear, thick washer and output shaft.
5. Using tool B.Vi. 953 or equivalent, block the step-down driven gear and unlock the lock plates, remove the screws and remove the final drive pinion nut.
6. Use tool B.Vi. 905 or equivalent to remove the speedometer drive oil seal it it was not removed previously.
7. Push out the final drive pinion using a mallet. Remove the step-down driven gear and spacer.
8. Remove the pinion bearing outer races. The needle roller bearing on the output shaft cannot be removed from the differential case.
9. The taper roller bearings are removed with a bearing splitter extractor. The band ring collar will have to be destroyed in order to work inside the differential housing.
10. When removing the large bearing from the differential housing, insert a steel bar into the case and place it on the flat of the bearing. Use a piece of tubing and an arbor press to remove the bearing.
11. When removing the small bearing from the differential housing, discard the snapring holding the bearing and using a 1.97 in. (50mm) diameter tube and an arbor press, press the bearing into the case.

Inspection

Inspect and clean all parts. Repair or replace the components as necessary. Be sure to dip each diffential component in automatic transaxle fluid prior to assembly.

Assembly

1. When installing the small bearing into the differential housing, position the new bearing over the opening and install using an arbor press and a piece of 2.560 in. (65mm) tube. Install a new retaining snapring.
2. When installing the large bearing into the differential housing, install it using an arbor press and a slightly shouldered steel bar 5.118 in. (130mm) long, or a 4.291–5.040 in. (125–128mm) diameter tube.
3. To install the converter and differential bearing, be sure to clean the seat area for the bearing throughly. Use emery cloth and compressed air to remove all burrs and dust.
4. Using an arbor press, install the bearing flush with the inside face of the case. Stake the bearing in place using a narrow cold chisel. The depth should be 0.051 in. (1.3mm).

NOTE: This bearing must be replaced whenever the complete transaxle or final drive is overhauled.

5. Install the taper roller bearings, using a suitable bearing installer.
6. Install the pinion bearing outer races. Install final drive pinion using a mallet and install the step-down driven gear and spacer.
7. Install a new speedometer drive oil seal.
8. Install the step-down driven gear and lock the lock plates. Install the screws and the final drive pinion nut. The final drive pinion bolts should be torqued to 118 ft. lbs.
9. Install the speedometer worm gear, step-down driving gear, thick washer and output shaft.
10. Install the ball bearings. Install the snapring, tapered washer, speedometer drive spindle, speedometer drive gear, speedometer drive seal and the snapring.
11. Install the endplay adjustment shim, spacer, closure plate bolts and the closure plate. Be sure to place a suitable sealant on the closure plate and torque the bolts to 18 ft. lbs. (25 Nm).
12. The collar must be crimped in position using an arbor press and tool B.Vi. 883 or equivalent. The tapered roller bearing must not be in position at this time. Install the ring gear using new replacement bolts. Torque the bolts to 18 ft. lbs. (25 Nm).

NOTE: The differential preload must be checked and adjusted if necessary whenever the final drive components are disassembled. The differential preload adjustment is performed without the lip seals or the pinion in place. When the differential preload shim increases in thickness, the preload increases and vice versa.

13. The differential components must be rotate under a load of 3–7 ft. lbs. (15–30 Nm) pull when new bearings are installed. These components must rotate freely and without play when bearings are reused.

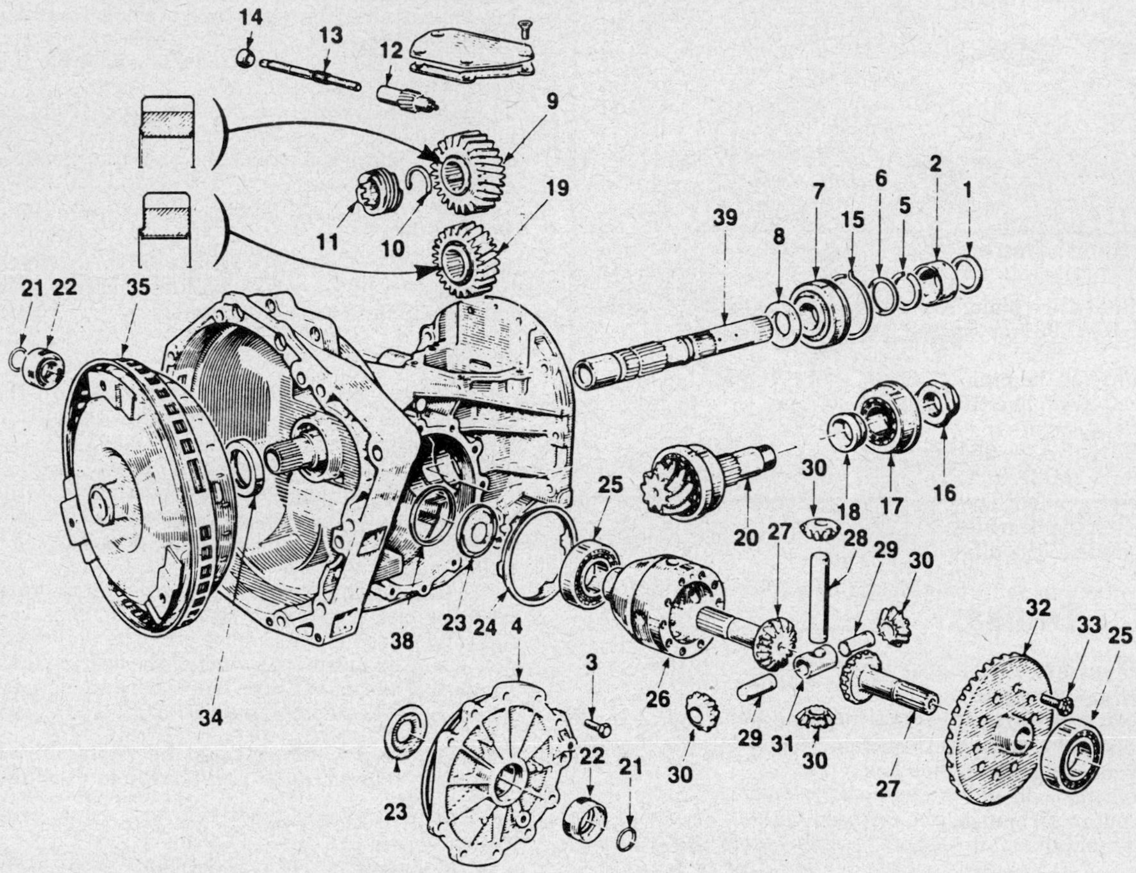

1 - End play adjusting shim
2 - Spacer
3 - Closure plate bolt
4 - Closure plate
5 - Snap ring
6 - Tapered washer
7 - Ball bearing
8 - Thick washer
9 - Step-down driving gear
10 - Snap ring
11 - Speedo. worm gear
12 - Speedo. drive gear
13 - Speedo. drive spindle
14 - Speedo. drive spindle seal
15 - Snap ring
16 - Final drive pinion nut
17 - Tapered roller bearing
18 - Spacer
19 - Step-down driven gear
20 - Final drive pinion
21 - "O" ring
22 - Lip-type oil seal
23 - Deflector
24 - Collar
25 - Tapered roller bearing
26 - Differential housing
27 - Side gears
28 - Long shaft
29 - Short shaft
30 - Spider gears
31 - Core
32 - Ring gear
33 - Ring gear bolts
34 - Converter oil seal
35 - Converter
38 - Differential preload adjusting shim
39 - Output shaft
40 - Spacers

Exploded view of the differential case and its components

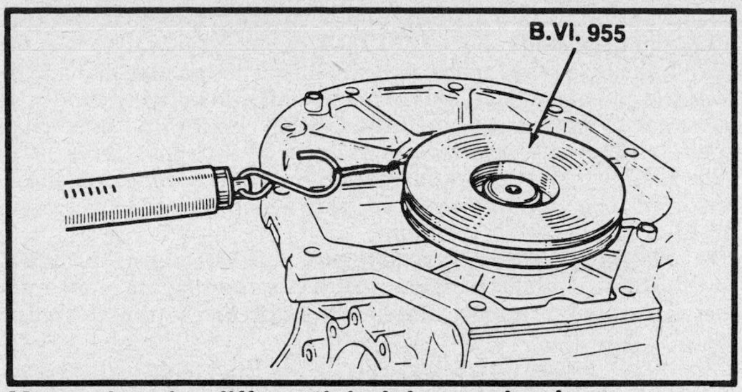
B.Vi. 955
Measuring the differential pinion preload

14. The final drive pinion outer track rings should be refitted using tool B.Vi. 961 or equivalent, if new bearings are to be installed.
15. The differential pinion preload is checked without the differential installed. The thickness of the spacer determines the pinion preload.
16. Install tool B.Vi. 955 on the diifferntial pinion nut. The pinion should rotate under a load of 5–8 lbs. pull when new nearings are installed. The pinion should rotate freely and without play when the bearings are being reused. When this adjustment is complete, the differential assembly may be installed.

Transaxle Assembly

Do not use force to assemble mating parts. If parts do not assemble freely, invesitgate the cause and correct the trouble before proceeding. Always use new gaskets during assembly operations. Use only the specified automatic transaxle fluid to lubricate the transaxle parts during assembly.

With the transaxle case thoroughly cleaned and the various sub assemblies overhauled and assembled, the assembly procedure is as follows:

1. Place the E–1 assembly and the turbine shaft on a tube approximatly 4 in. (100mm) in diameter. Place the needle thrust bearing into the E–1 clutch assembly with the pins facing down.
2. Roughly center the E–2 discs and slide this assembly onto the splines in the E–1 connecting bell housing. Turn gently, without forcing, to avoid damaging the discs.
3. When all the discs are properly in place, there is a play of approximately ⅛ in (4mm) between E–1 and E–2.
4. Lubricate the E–2 rings and their bearing surfaces on the sprag clutch hub.
5. Roughly center the F–2 discs and slide this assembly slowly onto the E–2 clutch assembly. Place the thrust bearing between the E–2 clutch and the sun gear assembly. Be sure the pins are facing down.
6. Center the F–1 discs and slowly lower the sun gear and planetary carrier assembly onto the F–1 brake. Make sure that all the discs are correctly positioned.
7. Install the transaxle case thrust bearing and 2 0.276 in. (7mm) diameter studs to allow the mechanism to be lowered more easily. Lubricate the seal ring housing and the location for the sprag clutch hub. Be sure that there is no burr that might interfere with the assembly.
8. Position the assembly holding it by the turbine shaft. Slowly lower the turbine shaft, making sure that the sprag clutch hub is lined up with the E–1 and E–2 clutch assemblies. Check to make sure that the sprag clutch hub fits well up against the case. The play between the E–1 and E–2 clutch should be 0.118–0.197 in. (3–5mm).
9. Remove the 2 studs and install the retaining bolts, torque them to 15 ft. lbs.

NOTE: The F–1 operating clearance should be between 0.043–0.122 in. (1.1–3.1mm).

10. Install the parking latch return spring on its shaft, the parking latch, the shaft (threaded hole upward), the centering dowel and the circlip.
11. Install the main shaft needle bearing. Install the needle thrust bearing plate onto the planetary gear carrier.
12. Install the needle thrust bearing on the output shaft in the differential case.
13. Install the endplay adjusting shim and the needle thrust bearing plate.
14. Before assembling the 2 housings, be sure that the oil seal is correctly positioned on the transaxle case.
15. Install 0.276 in. (7mm) diameter studs on the transaxle case. The 2 centering dowels the smear the sealing surface with a suitable sealant. Lubricate the turbine shaft and slowly lower the differential case onto the transaxle case.
16. Connect the 2 cases with a few bolts and torque them to 22 ft. lbs. Install the lower cover plate with a new gasket and torque the retaining bolts to 18 ft. lbs.
17. Place a dial indicator gauge bracket on the cover plate and set the pointer of the gauge on the E–1 shaft.
18. Pull on the turbine shaft, set the gauge to 0 and push the turbine shaft back again. Read off the endplay on the dial indicator gauge. The reading should be between 0.016–0.032 in. (0.4–0.8mm).
19. Once the endplay is correct, finish assembling the 2 housings, taking care to arrange the wiring clips correctly. Install the sealed junction plug equipped with a new O-ring that has been lubricated.
20. Install the assembled control shaft making sure that is has its O-ring, but do not push it all the way in because the assembled parking rod with its end in the housing and the control selector must be installed on it.
21. Install the roll pin.
22. Before installing the valve body perform the air pressure test as follows:
    a. Applying air to passage **A** will apply the F–1 brake.
    b. Applying air to passage **B** will apply the F–2 brake.
    c. Applying air to passage **C** will apply the E–2 brake.
    d. Applying air to passage **D** will apply the E–1 brake.

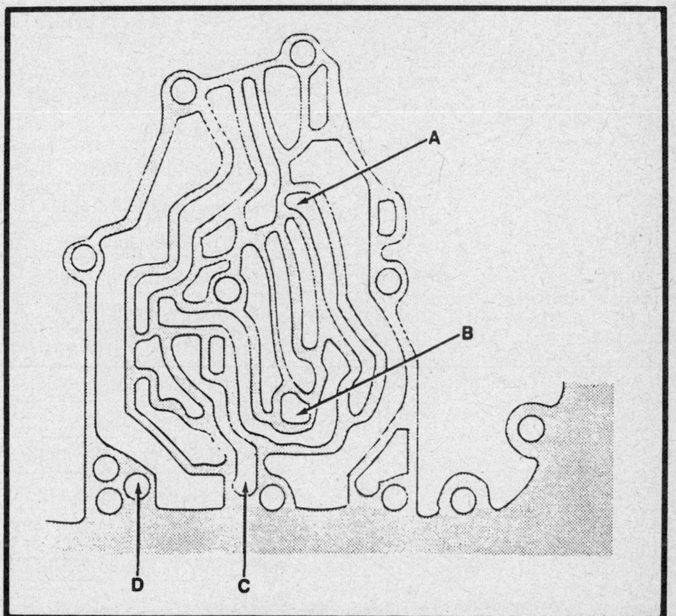
A
B
D
C
Performing the air pressure test

NOTE: Air pressure testing is used as a method of confirming proper clutch and brake operation after a repair. The tests involves substituting air pressure for fluid pressure. By applying air pressure to the appropriate case passages, movement of the piston can be felt and a soft thud may be heard as the clutch or break is applied. Check for excessive air leakage. Use dry filtered compressed air. Pressures of 30–100 psi are required to perform the tests.

23. Install the valve body assembly, as follows:

a. The sealing surfaces on the housing and the valve body must be cleaned and free of any burrs.

b. Be sure that the 2 centering dowels are positioned on the valve body.

c. Be sure that the 2 toothed quadrants are properly meshed in the **P** position.

24. Install the valve body assembly onto the case housing, engaging the manual valve on the ball joint. Torque the 6 retaining bolts in progressive steps to 54 inch lbs. (6 Nm).

25. Connect the sealed junction plug and check that the marks on the valve body, solenoid ball valves and the plugs all match. Position the magnet on the solenoid ball valve retaining clamp.

26. Check that the oil pump housing is clean, lubricate and install the ring wheel (be sure to use the marks made at disassembly for the proper wheel direction or position the chamfer to face downward into the housing). Install the oil pump gear and the oil pump driveshaft.

27. Lubricate the O-ring on the oil filter suction pipe and slip it over the end of the pipe. Push the pipe into its housing carefully so as not to damage or pinch the O-ring. Secure the filter with the retaining bolts and torque them to 54 inch lbs. (6 Nm).

28. Install the oil pan and gasket. Rotate the oil pump shaft and check the oil pump shaft endplay which should be between 0.014–0.031 in. (0.35–0.80mm).

29. Place a 0.276 in. (7mm) diameter stud into one of the holes in the stator support location. Lubricate the turbine shaft and install the stator support. Remove the stud and torque the bolts to 13 ft. lbs.

30. Install a new converter seal, lightly lubricate the new seal place the seal in the opening of the pump housing and push it into place with a suitable seal driver tool as it is tap gently.

31. Drive the new seal into the housing until the tool bottoms out.

32. Install the converter after lubricating the white metal sleeve, the turbine shaft and the pump shaft splines.

33. Install the governor computer with its seal, the vacuum capsule, the wiring and the dipstick.

34. Install the transaxle assembly into the vehicle. Refill to the proper level with automatic transaxle fluid. Road test the vehicle, make all adjustments as necessary.

SPECIFICATIONS

TORQUE SPECIFICATIONS

Component	ft. lbs.	Nm
Converter to drive plate bolt	22	30
Cooler line fitting	175 ①	20
Drive plate to crankshaft bolt	51	70
Lower shock bracket nut	148	200
Oil pan bracket bolt	54 ①	6
Steering arm bracket bolt	30	40
Steering arm bracket nut	25	35
Steering link ball joint nut	30	40
Tie rod bracket nuts	25	35
Transmission cushion stud nut	30	40
Transmission to engine bolt	37	50
Vacuum capsule holddown bolt	132 ①	15
Transaxle mount bracket to transaxle case bolt	30	40
Valve body to transmission case bolt	80 ①	9
Wheel nut	66	90

① Inch lbs.

SPECIAL TOOLS

Tool Ref.	Description
B.Vi. 31-01	Set of 3 Roll Pin Drifts
B.VI. 946	Planetary Snap Ring Installer
B.Vi. 952	Feed Hub Alignment Dowels and Front Piston Removing Tool
Mot. 50	Torque Wrench or Equivalent Beam Type Torque Wrench
Mot. 53	Drain Plug Wrench (98 mm square drive)
B. Vi. 465	Converter Oil Seal Replacement Tool and Converter Holding Lug
B. Vi. 466-04	Oil Pressure Gauge (or J-24027 with adapter 8981 320759)
B. Vi. 466-06	Oil Pressure Gauge for B. Vi. 466-04
B. Vi. 715	Tool From B. Vi. 710 Kit or use B. Vi. LM
B. Vi. 883	Differential Outside Band Installer
B. Vi. 905	Speedometer Shaft Seal Replacement Tool
B. Vi. 945	Planetary Oil Seal Installing Mandrel
B. Vi. 947	Intermediary Case Bearing Installer
B. V. 951	Differential Oil Seal Installer
B. Vi. 953	Step Down Driven Gear Holding Tool
B. Vi. 955	Differential Pinion Bearing Preload Measuring Tool
B. Vi. 958	Diagnostic Tester (or MS 1700 With Adapter)
B. Vi. 959	Output Shaft Circlip Installing Tool
B. Vi. 961	Differential Pinion Bearing Race Installing Tool
B. Vi. 962	Converter Oil Seal Installing Tool
B.Vi. 31-01	Roll Pin Punch Set
B.Vi. 465	Torque Converter Tool Set
B.Vi. 466-07	Pressure Gauge Set
B.Vi. 905-02	Speedometer Seal Tool Set
Mot. 582	Drive Plate Locking Tool
MS 1700	Electronic Diagnostic Tester
MS 1900	Engine Support Tool
T. Av. 476	Ball Joint Extractor

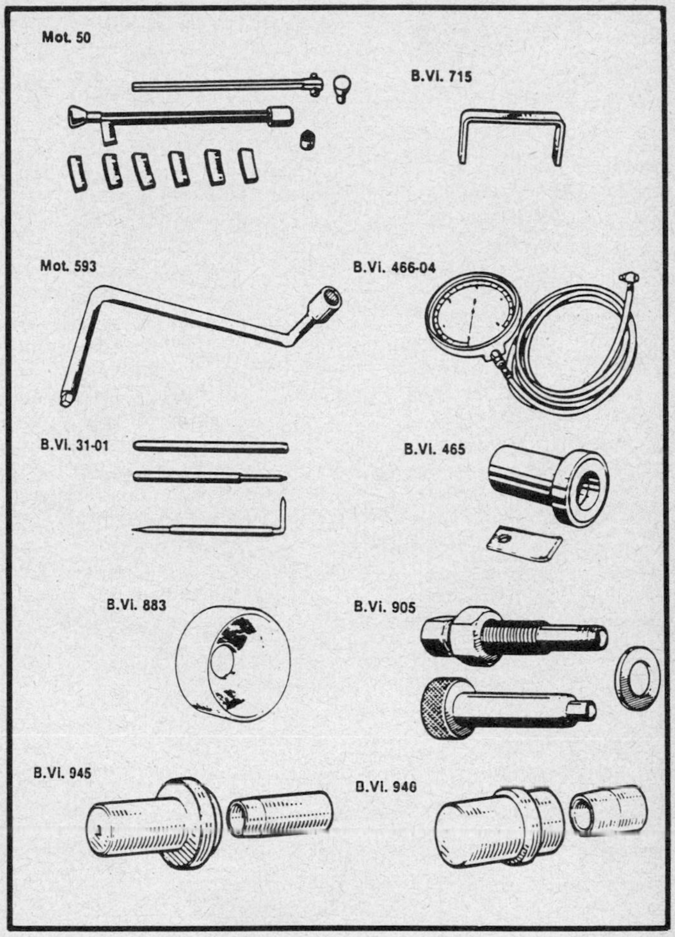
Mot. 50
B.VI. 715
Mot. 593
B.Vi. 466-04
B.VI. 31-01
B.VI. 465
B.VI. 883
B.Vi. 905
B.VI. 945
B.VI. 946

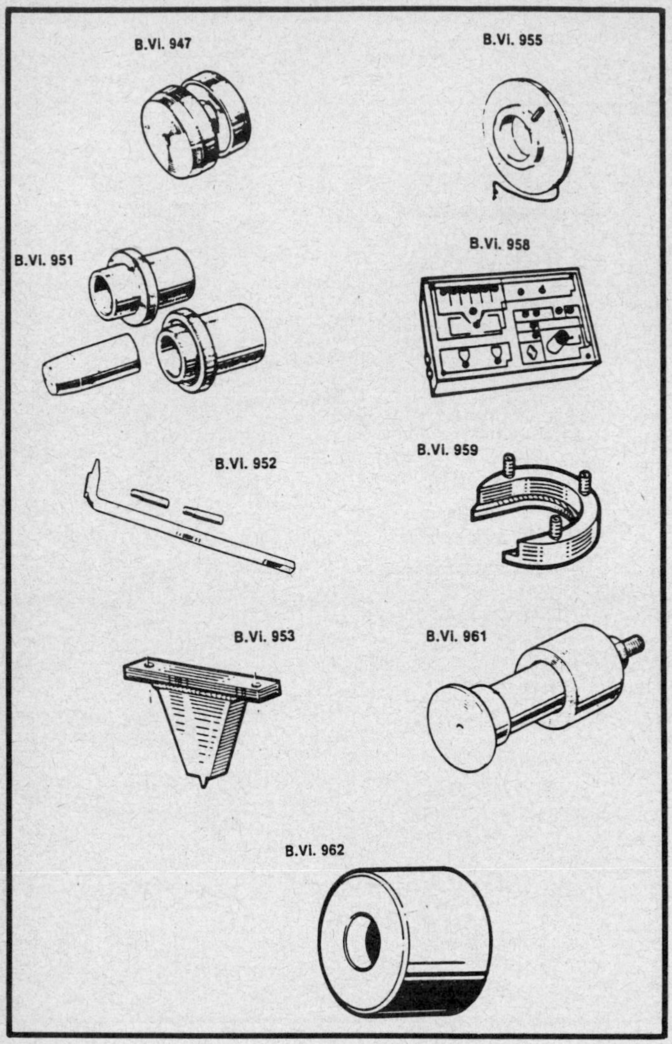
B.Vi. 947
B.Vi. 955
B.Vi. 951
B.Vi. 958
B.Vi. 952
B.Vi. 959
B.Vi. 953
B.Vi. 961
B.Vi. 962

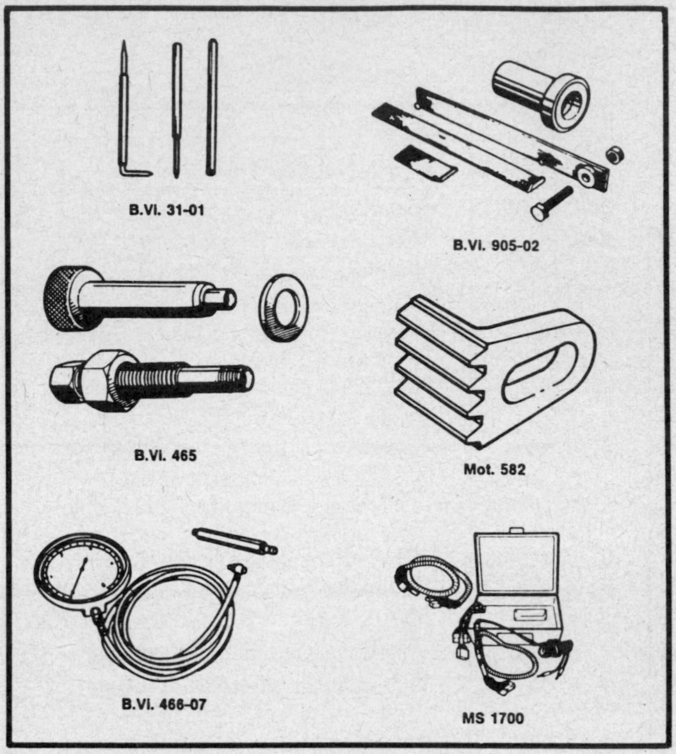
B.Vi. 31-01
B.Vi. 905-02
B.Vi. 465
Mot. 582
B.Vi. 466-07
MS 1700

Section 3

ZF-4HP-18 Transaxle
AMC/Jeep-Eagle

APPLICATION

1988–89 Premier

GENERAL DESCRIPTION

The ZF 4HP–18 is a 4 speed, automatic transaxle. Third gear ratio is 1:1. Fourth gear is an overdrive range providing an 0.74:1 gear ratio. Shifting is controlled by a governor valve, a line pressure valve, a throttle pressure regulator valve and a modulator valve. Valve operation is dependant on shift lever position, vehicle speed and throttle position.

Transaxle and Converter Identification

TRANSAXLE

The ZF 4HP–18 transaxle can be identified by a tag on the left side of the transaxle case just above the oil pan. The information on the plate consists of the builders sequence number, manufacturers part number and the transaxle type.

CONVERTER

The torque converter diameter is 10.2 in. Identification numbers or symbols, are either stamped into the cover or ink stamped on the cover surface. The torque converter used in this transaxle is not serviceable. If the converter is damaged in any way, it must be replaced.

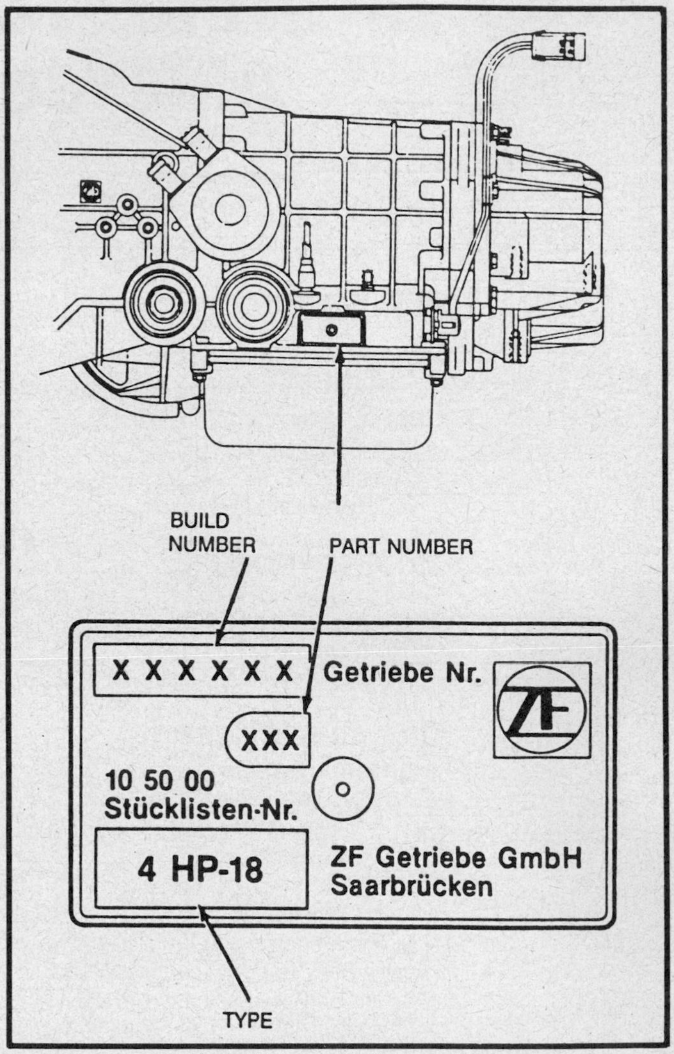
BUILD NUMBER
PART NUMBER
X X X X X X
Getriebe Nr.
XXX
10 50 00
Stücklisten-Nr.
4 HP-18
ZF Getriebe GmbH
Saarbrücken
TYPE

Transmission identification label and location

Metric Fasteners

The ZF 4HP–18 transaxle is designed and assembled using metric fasteners. Metric measurements are used to determine clearances within the unit during servicing. Metric fasteners are required to service the unit and torque specifications must be strictly adhered to.

The metric thread is extremely close to the dimensions of the standard inch system threads and for this reason, extreme care must be taken to prevent the interchanging of inch system bolts or screws with metric bolts or screws. Mismatched or incorrect fasteners can result in damage to the transaxle unit. The fasteners should be used in the same location as they were removed from or replaced with fasteners of the same size and grade.

Capacities

NOTE: The transmission and differential sections are not integral in the ZF 4HP–18. They are separate and require different lubricants.

The ZF 4HP–18 model transaxle requires Mercon® automatic transmission fluid. This is the fluid to be used in this transaxle. Do not substitute any other type of fluid. The transaxle uses 7.4 qts. of fluid for a dry fill and 2.8 qts. for a fluid change.

The differential requires a synthetic-type SAE grade 75W–140 gear lubricant. It is the only lubricant recommended. The synthetic lubricant is designed to last the life of the differential under normal conditions. Periodic lubricant changes are not required. The capacity of the differential is 0.66 qts.

Checking Transaxle Fluid Levels

1. Check the level when the fluid is cold (at ambient temperature). If the fluid temperature is more than 125°F, allow it to cool before checking the level.
2. Position the vehicle on a level surface. Start and run the engine at idle.
3. Shift the transaxle through every gear range, return it to the **P** position and check the fluid level.
4. Correct level is to the cold fill mark on the dipstick.

TRANSAXLE MODIFICATIONS

No transaxle modifications have been published by the manufacturer at the time of this publication.

TROUBLE DIAGNOSIS

The ZF model 4HP–18 automatic transaxle assembly are not to be overhauled at this time. If diagnosis determines damage or malfunctions are present in the converter or transaxle, the entire assembly must be removed from the vehicle and replaced through the dealership parts exchange program.

CLUTCH AND BAND APPLICATION
ZF 4HP–18 Automatic Transaxle

	Gear Drive				
Component	First	Second	Third	Fourth	Reverse
1–3 clutch	Applied	Applied	Applied	—	—
Reverse clutch	—	—	—	—	Applied
Forward brake	—	Applied	—	—	—
2–4 band	—	Applied	—	Applied	—
First/reverse brake	Applied	—	—	—	Applied
3–4 clutch	—	—	Applied	Applied	—
Roller clutch	Applied	—	—	—	—
Sprag clutch	—	Applied	—	—	—

CHILTON'S THREE C's TRANSMISSION DIAGNOSIS
ZF 4HP–18 Automatic Transaxle

Condition	Cause	Correction
Will not engage or hold in park	a) Shift cable Incorrectly adjusted b) Excess clearance on detent plate c) Detent segment out of position d) Park pawl damaged	a) Adjust cable b) Adjust or replace plate c) Correct position or replace segment and rod d) Replace pawl and pin
Engine will not start	a) Neutral start switch inoperative b) Excess clearance on selector shaft	a) Replace switch b) Adjust or replace shaft
No reverse	a) Shift cable incorrectly adjusted b) Oil screen plugged c) Reverse clutch damaged d) First reverse brake damaged) Also engine will not decelerate in position one, 1st gear e) Governor sticking f) Lock-up valve 1 and reverse gear sticking	a) Adjust cable b) Replace oil screen c) Replace transmission d) Replace transmission e) Replace governor f) Replace valve body
Slips/vibrates when accelerating from stop	a) 1–2–3 clutch damaged b) First-reverse brake damaged c) Turbine shaft O-ring or pump starter malfunction d) Oil leaking into reverse clutch or seat piston ring has scored center plate	a) Replace transmission b) Replace transmission c) Replace transmission
d) Replace transmission		
Harsh engagement P-to-R, or N-to-R may be accompanied by improper 2–1 downshift	a) Accumulator inoperative	a) Replace valve body
Back-up lights not functioning (electrical feed/ground ok)	a) Neutral start switch malfunction	a) Replace switch
Engine will not start	a) Neutral start switch defective	a) Replace switch

CHILTON'S THREE C's TRANSMISSION DIAGNOSIS
ZF 4HP–18 Automatic Transaxle

Condition	Cause	Correction
Vehicle creeps in park or neutral	a) Shift cable incorrectly adjusted	a) Adjust cable
No power, poor acceleration in D range	a) Converter valve operation	a) Replace transmission
	b) Oil screen plugged	b) Replace oil screen
	c) 1–2–3 clutch defective	c) Replace transmission
	d) Roller clutch slips	d) Replace transmssion
	e) Shift cable incorrectly adjusted	e) Correct adjustment
	f) Throttle or shift valve sticking	f) Replace valve body
Harsh engagement during N-to-D shift	a) Accumulator sticking or spring broken	a) Replace valve body
	b) 1–2–3 clutch damaged	b) Replace transmission
No shift in cold or warm condition		
No 1–2 or 2–1 shift	a) Governor sticking	a) Replace governor
	b) 1–2 shift valve sticks	b) Replace valve body
No 1–2 shift	c) Forward brake or 2–4 band malfunction	c) Replace transmission
No 2–3 or 3–2 shift	d) Governor sticking	d) Replace governor
	e) 2–3 shift valve sticks	e) Replace valve body
No 2–3 shift	f) 3–4 clutch damaged	f) Replace transmission
	g) Oil supply for 3–4 clutch leaking	g) Replace transmission
No 3–4 or 4–3 shift	h) Governor sticking	h) Replace governor
	i) 3–4 shift valve sticks	i) Replace valve body
No 3–4 shift	j) Forward brake inoperative (1–2 up-shift not ok)	j) Replace transmission
	k) 2–4 band loose	k) Replace transmission
	l) 2–3–4 up-shift valve sticks	l) Replace valve body
	m) Position 3 valve sticks	m) Replace valve body
Vehicle takes off in 2nd gear (will not downshift to 1st)	a) Governor piston sticks	a) Replace governor
	b) 1–2 shift valve sticks	b) Replace valve body
	c) 2–4 band binds	c) Correct adjustment
	d) 2–4 band will not release	d) Replace transmission
Vehicle takes off in 3rd gear (will not downshift to 1st or 2nd)	e) Center ring of governor flange defective	e) Replace transmission
	f) Governor piston sticking	f) Replace governor
	g) 1–2 and 2–3 shift valve sticking	g) Replace valve body
	h) Closing cap in center plate leaking (reverse clutch always filled with oil	h) Replace transmission
Transmission shifts 1–3 (no 2nd gear)	i) 2–3 shift valve sticks	i) Replace valve body
	j) 2–3–4 shift valve sticks	j) Replace valve body
	k) 1–2–3 shift valve sticks	k) Replace valve body
Transmission shifts 1–4 (no 1–2; or 2–3 shift)	l) Engine will not accelerate	l) Replace valve body
Will not downshift to 1st at idle or no kickdown shift) Will not return to 1st gear at idle speed when stopped	a) Governor sticking	a) Replace governor
	b) Leakage in governor assembly	b) Replace transmission
	c) Shift valve binding	c) Replace valve body
No full throttle kickdown	d) Throttle valve cable incorrectly adjusted	d) Adjust cable
No kickdown shifts	e) Throttle valve cable incorrectly adjusted	e) Adjust cable
	f) Governor sticking	f) Replace governor
Harsh engagement at idle speeds	a) Accumulator malfunction	a) Replace valve body
	b) Modulator pressure too high	b) Replace valve body
	c) Clutch pack damage	c) Replace transmission
Full throttle kickdown shifts too long	d) Accumulator malfunction	d) Replace valve body
	e) Clutch pack damage	e) Replace transmission

CHILTON'S THREE C's TRANSMISSION DIAGNOSIS
ZF 4HP–18 Automatic Transaxle

Condition	Cause	Correction
Full throttle and kickdown shifts too harsh	f) Modulator pressure not OK g) Accumulator malfunction	f) Replace valve body g) Replace valve body
Engine overspeed at 3–4 shift	h) Orifice control valve sticking i) 3–4 traction valve binding j) 2–4 band slips	h) Replace valve body i) Replace valve body j) Replace transmission
Engine overspeed at 3–4 downshift	k) Time control valve and 4–3 downshift valves not coordinated l) 1–2–3 clutch damaged m) Damper function of 1–2–3 clutch and 4–3 traction valve not functioning properly	k) Replace valve body l) Replace transmission m) Replace valve body
Manual 2nd gear downshift incorrect or downshift early or late	a) Lockup valve 2 binding b) Replace governor	a) Replace valve body b) Governor piston binding
No overrun braking in D1	a) 2–4 band inoperative b) 2–4 band damaged	a) Check/replace band piston and cover O-rings if required b) Replace transmission
Manual 2–1 downshift incorrect	a) Lockup valve 1 and reverse gear binding b) Governor piston binding	a) Replace valve body b) Replace governor
No overrun braking in 1st gear	a) 1st/reverse brake damaged	a) Replace transmission
Throttle valve cable sticks	a) Cable not attached to cam b) Internal friction in cable c) Throttle pressure piston sticks	a) Connect cable to cam b) Replace cable c) Replace valve body
After long drive, noise develops and vehicle will not move in drive or reverse	a) Valve body oil screen plugged	a) Replace oil screen
Transmission noisy) Will not move in drive or reverse	a) Converter driveplate damaged b) Oil pump gears worn or damaged	a) Replace driveplate b) Replace transmission
Oil leaking from converter housing seam	a) Torque converter leaking at welded b) Pump seal leaking	a) Replace converter b) Replace pump seal
Leakage between transmission and oil pan	a) Oil pan bolts loose or pan warped b) Oil pan gasket damaged	a) Tighten bolts or replace pan b) Replace gasket
Leakage between transmission housing and differential cover	a) Cover bolts loose	a) Tighten bolts
Leakage at transmission cooler	a) Cooler attaching bolt loose b) Gasket damaged c) Cooler cracked or split	a) Tighten bolt b) Replace gasket c) Replace cooler
Leakage at 2–4 band piston cover	a) Cover O-rings worn or damaged	a) Replace O-rings
Leakage from 2–4 band retaining shaft	a) Retaining shaft O-ring damaged	a) Remove valve body and replace shaft O-ring
Leakage at output shaft	a) Bolts loose b) Seal rings damaged	a) Tighten bolts b) Replace seal rings
Oil leakage at throttle cable connection in case	a) Cable connector O-ring damaged	a) Replace O-ring; if necessary, replace cable
Leakage at differential	a) Output shaft seals or cover seal damaged	a) Replace seals
Leakage at speedometer sensor	a) Sensor or O-ring damaged	a) Replace O-ring or sensor
Leakage at breather vents in transmission or differential	a) Transmission or differential overfilled b) Incorrect fluid or lubricant	a) Correct oil level b) Replace transmission
Leakage at selector shaft	a) Seal ring damaged	a) Replace seal ring
Noise in all positions	a) Fluid level low b) Valve body leaking internally c) Oil screen plugged	a) Correct level b) Replace valve body c) Replace oil screen

CHILTON'S THREE C's TRANSMISSION DIAGNOSIS CHART
ZF 4HP-18 Automatic Transaxle

Condition	Cause	Correction
Noise at certain speeds	a) Bearing adjustment of differential b) Bearing adjustment of differential incorrectly set	a) Replace transmission pinion gear incorrectly set b) Replace transmission

ON CAR SERVICES

Adjustments

THROTTLE VALVE CABLE

1. Loosen the cable locknuts and lift the threaded cable shank out of the engine bracket.
2. Place the throttle lever in idle position.
3. Pull cable wire forward and place a 1.55 in. (39.5mm) long gauge block on the wire between the cable connector and cable end.

NOTE: Vernier calipers can be substituted for the gauge block.

4. Pull cable shank rearward to the detent position (but not to the wide open throttle position).
5. The detent position will provide a definite feel, similar to a stop, when it is reached.
6. Hold the shank at the detent position and insert it in the engine bracket. Tighten the locknuts to lock it in place.
7. Remove the gauge block and verify adjustments. Detent position should be reached when travel of the cable wire is 1.55 in. (39.5mm).

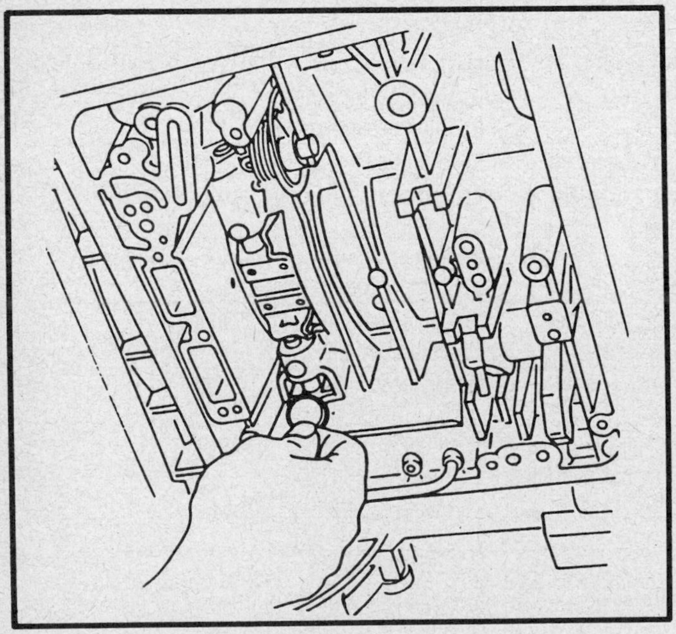
Installing band adjusting shims

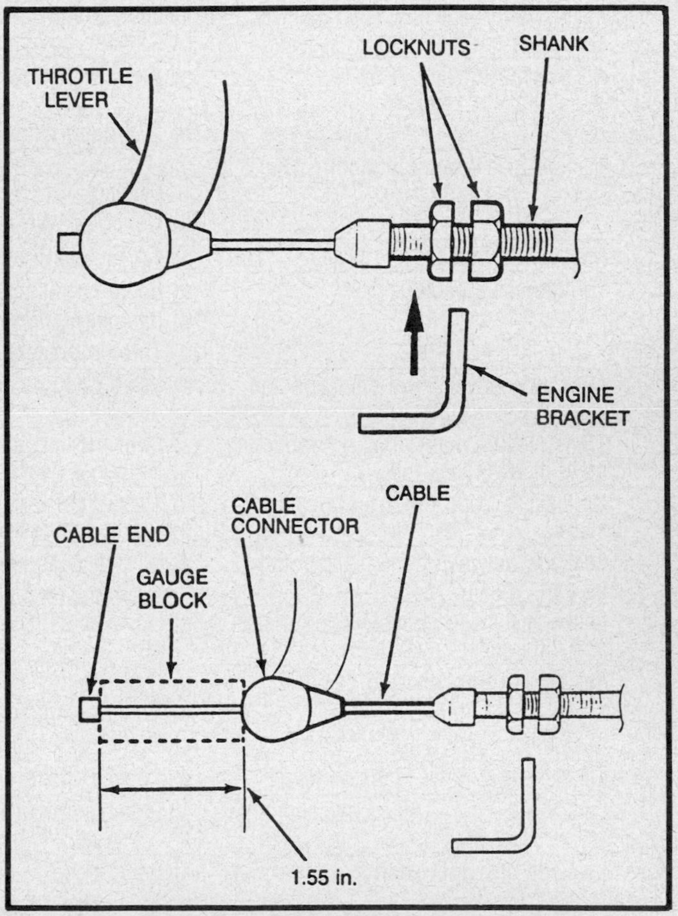
THROTTLE LEVER
LOCKNUTS
SHANK
ENGINE BRACKET
CABLE END
CABLE CONNECTOR
CABLE
GAUGE BLOCK
1.55 in.
Throttle valve cable adjustment

SHIFT CABLE

1. Shift the transmission into **P**.
2. Raise and safely support the vehicle.
3. Unlock the shift cable by releasing the cable adjuster clamp. Move the clamp outward to release it.
4. Move the transmission lever rearward into the park detent. Be sure the lever is centered in the detent.

NOTE: The park detent is the last rearward position.

5. Verify positive engagement of the park lock by attempting

to rotate the driveshafts. The shafts cannot be turned if the park lock is properly engaged.

6. Lock the shift cable by pressing the adjuster clamp back into position, it should lock into position. Lower the vehicle.
7. Turn the ignition key to the **LOCK** position and verify that the shift lever remains locked in **P**. It should not be possible to move the lever.
8. Turn the ignition key to the **ON** position.
9. Verify that the engine only starts when the shift lever is in the **P** or **N** positions.

NOTE: If the engine starts in any position other than park or neutral, the shift cable adjustment is incorrect or a problem exists with the multifunction switch.

10. Shift the transmission into **PARK** and verify that the key can be returned to the **LOCK** position.

2ND AND 4TH GEAR BAND

1. Raise and safely support the vehicle.
2. Remove the under-body splash shield. Loosen the nut attaching the the fill tube to the oil pan and drain the fluid.
3. Remove nuts attaching the oil pan clamps and remove the oil pan.
4. Remove the valve body bolts and remove the valve body.
5. Remove the adjusting shim from behind the nut on the band pin.
6. Using a feeler gauge, check the clearance between the band pin and the case. Correct clearance is should be between 0.049–0.059 in. (1.25–1.50mm).
7. If the clearance is not in the correct range, adjust it be adding either a thicker or thinner shim.
8. Reinstall the valve body and the oil pan.
9. Fill the transaxle with the correct grade and quantity of fluid.

Services

FLUID CHANGES

NOTE: The transmisssion and differential sections are not integral in the ZF 4HP-18. They are separate and require different lubricants.

TRANSAXLE

The manufacturer recommends that the transaxle fluid and filter be changed at 30,000 miles.

1. Raise and safely support the vehicle.
2. Remove the underbody splash shield and loosen the nut that attaches the filler tube to the oil pan. Drain the fluid.
3. When all of the fluid is drained, tighten the nut on the filler tube to 74 ft. lbs. Install the splash shield and lower the vehicle.
4. Remove the transmission dipstick and add 2.36 qts. of Mercon® transmission fluid.
5. Check and adjust the fluid as necessary.

DIFFERENTIAL

NOTE: The differential requires a synthetic-type SAE grade 75W-140 gear lubricant. It is the only lubricant recommended. The synthetic lubricant is designed to last the life of the differential under normal conditions. Periodic lubricant changes are not required. The fluid level should be checked at the regular service interval. The capacity of the differential is 0.66 qts.

1. Raise and safely support the vehicle. Remove the underbody splash shield.
2. Remove the differential drain plug and drain the fluid.
3. Install a replacement washer on the drain plug and reinstall the plug, tightening to 18 ft. lbs.

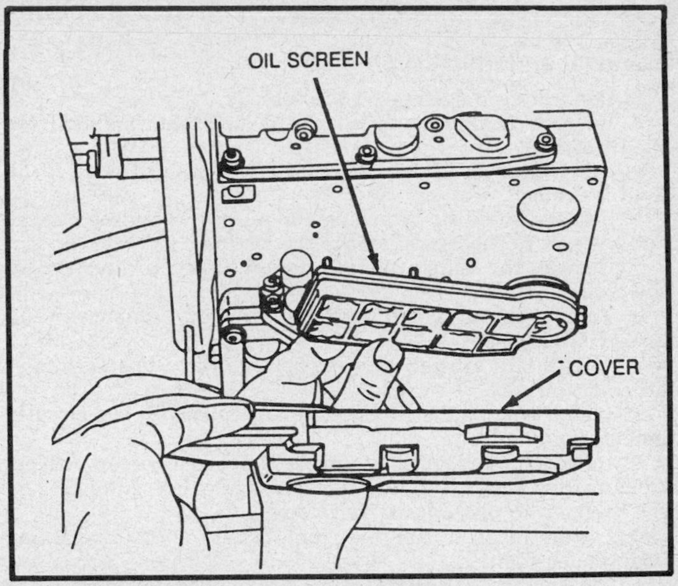
OIL SCREEN
COVER

Removing the oil screen and cover

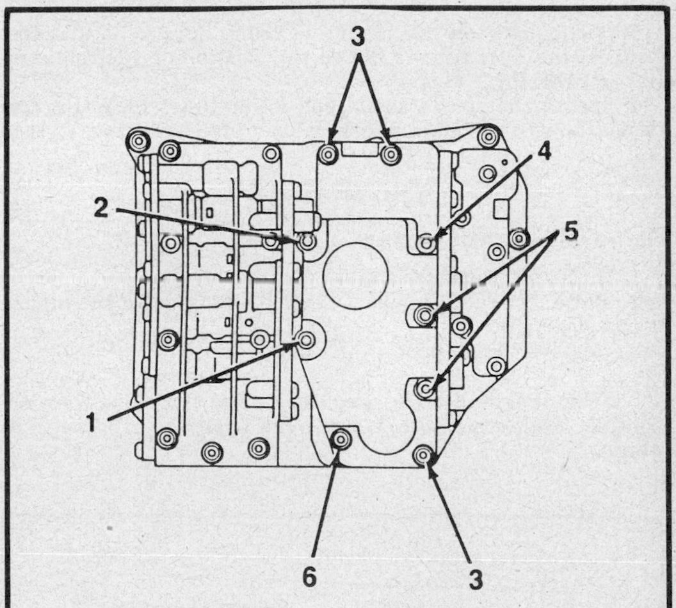
3
4
2
5
1
6
3

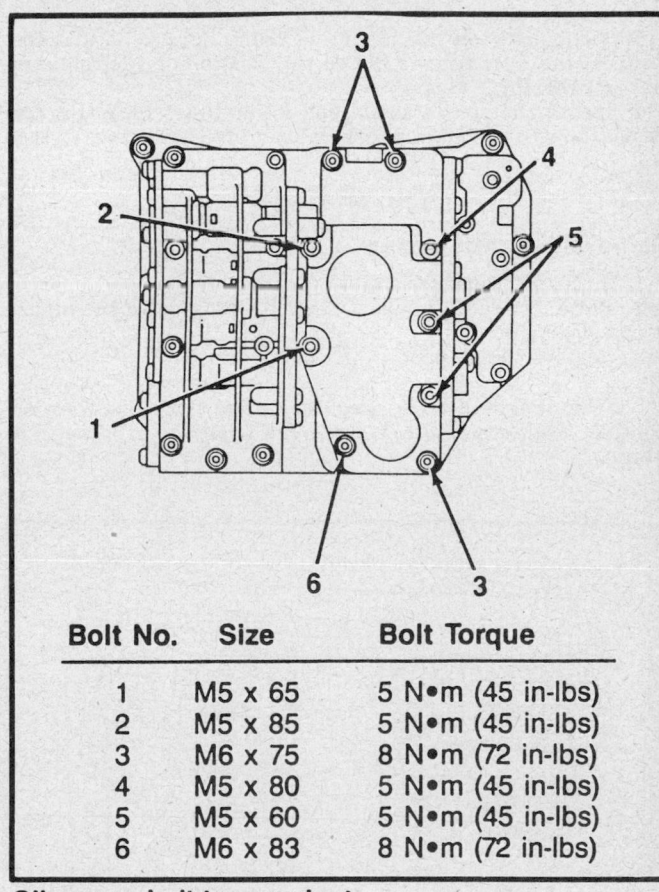

Bolt No.	Size	Bolt Torque
1	M5 x 65	5 N•m (45 in-lbs)
2	M5 x 85	5 N•m (45 in-lbs)
3	M6 x 75	8 N•m (72 in-lbs)
4	M5 x 80	5 N•m (45 in-lbs)
5	M5 x 60	5 N•m (45 in-lbs)
6	M6 x 83	8 N•m (72 in-lbs)

Oil screen bolt torque chart

4. Remove the differential fill plug and fill the differential with 75W-140 synthetic-type hypoid gear lubricant.
5. Continue adding lubricant until it starts to flow out of the fill hole. Install a new washer on the fill plug and install it. Tighten the fill plug to 37 ft. lbs.

OIL PAN

Removal and Installation

1. Raise and safely support the vehicle.
2. Remove the underbody splash shield. Drain the fluid from the transaxle.
3. Disconnect the filler tube from the transaxle after all of the fluid is drained.
4. Remove the nuts attaching the oil pan retaining clamps and the pan to the case. Remove the oil pan.
5. Remove the bolts retaining the oil screen cover. Remove the cover.
6. Remove the oil screen from the valve body. Remove the oil screen cover gasket.
7. Clean the oil pan and screen, replace the screen if necessary.
8. Install the magnet in the oil pan, it goes in the circular indentation.
9. Install a new O-ring on the screen. Coat the screen with petroleum jelly and install it in the valve body. Press the tabs on the screen into place in the valve body.
10. Install the oil screen cover and install the cover retaining bolts finger tight.
11. Tighten the oil screen cover bolts to specification.
12. Position the oil pan on the case and install the oil pan mounting clamps.
13. Tighten the oil pan mounting clamp nuts to 54 inch lbs. Connect the filler tube to the oil pan, tightening the retaining nut to 74 ft. lbs.
14. Install the splash shield and lower the vehicle. Fill the transaxle with the appropriate fluid and to the correct level.

VALVE BODY

Removal and Installation

1. Shift the transmission into manual 1st gear.
2. Raise and safely support the vehicle. Remove the underbody splash shield.
3. Remove the oil pan.

NOTE: It is not necessary to remove the oil screen assembly. The valve body is removed with the oil screen in place.

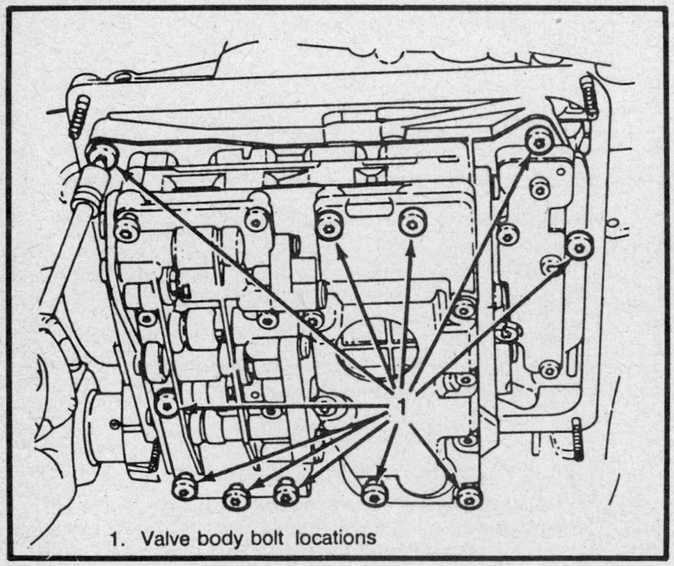
1
1. Valve body bolt locations

Valve body bolt locations

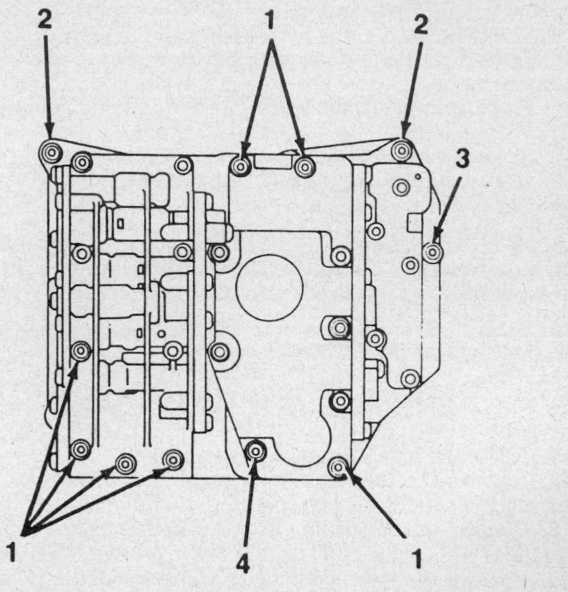
2
1
2
3
1
4
1

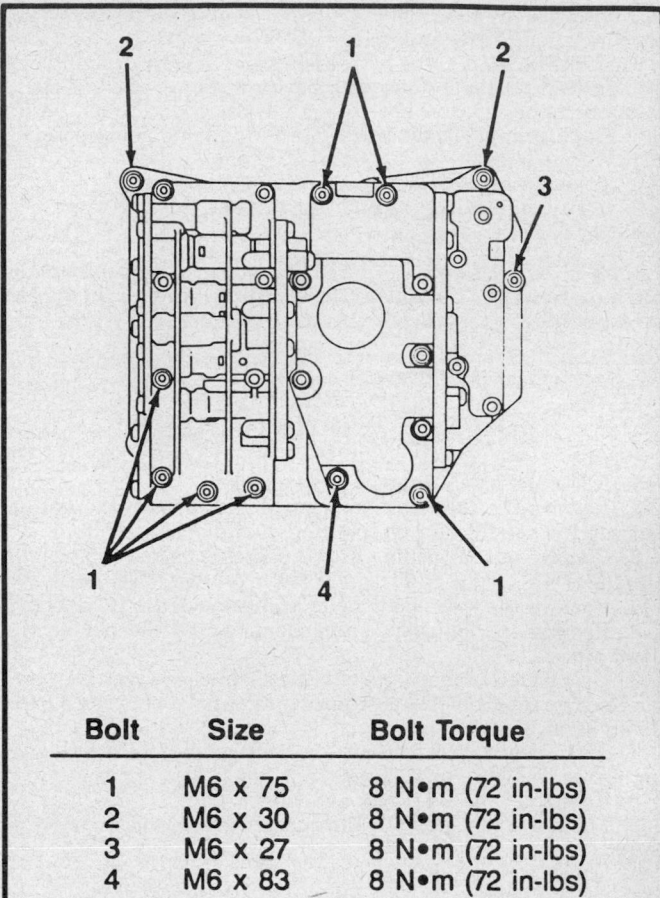

Bolt	Size	Bolt Torque
1	M6 x 75	8 N•m (72 in-lbs)
2	M6 x 30	8 N•m (72 in-lbs)
3	M6 x 27	8 N•m (72 in-lbs)
4	M6 x 83	8 N•m (72 in-lbs)

Valve body bolt torque chart

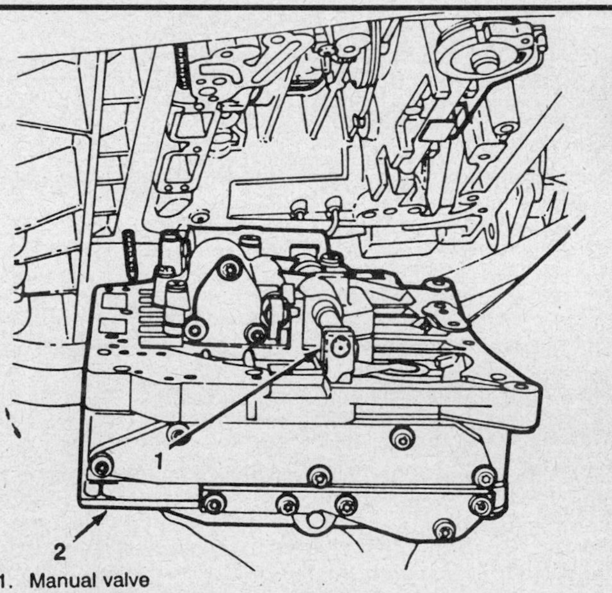
1
2
1. Manual valve
2. Valve body

Aligning the valve body for installation

4. Remove the valve body retaining bolts and remove the valve body.
5. Move the selector lever into the 1st gear detent. This is the last detent in the counterclockwise direction.

6. Pull the throttle cable to wide open throttle position to avoid jamming the throttle cam and piston during the valve body installation.
7. Push the manual valve all the way in to the 1st gear position.
8. Align and install the valve body on the transmission case.
9. Install and tighten the valve body bolts to 72 inch lbs. Be sure to install the correct length bolt in each position.
10. Install the oil pan. Install the underbody splash shield.
11. Fill the transmission with the correct level and grade of fluid. Adjust the throttle valve cable.

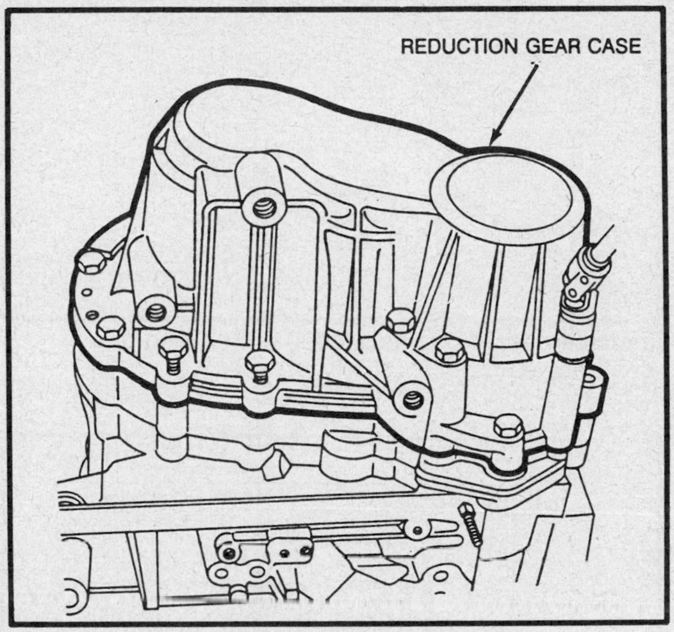
REDUCTION GEAR CASE

Removing the reduction gear case

REDUCTION GEAR CASE AND GOVERNOR

Removal and Installation

NOTE: The reduction gear case must be removed for access to the governor and governor support.

1. Raise and safely support the vehicle. Remove the underbody spalsh shield.
2. Disconnect the exhaust pipes. Remove the differential drain plug and drain the lubricant.
3. Remove the bolts attaching the reduction gear case to the transaxle case and remove the assembly.
4. Remove the housing flange bolts and remove the flange from the reduction gear housing.
5. Remove the spring washers from in front of the governor cover on the transaxle. Note the position of the washers for ease of reassembly.
6. Remove the governor cover and remove the governor assembly.
7. Remove the bolts attaching the valves to the governor and remove the valves.
8. Remove the 1 metal and 2 rubber seal rings from the governor body.
9. Remove the governor support retaining bolt from inside the transaxle case. Remove the governor support using an puller tool. Note the position of the 2 oil seals at the forward end of the governor support for assembly reference.
10. Remove the seals and seal rings from the support.
11. Install replacement oil seals on the governor support.

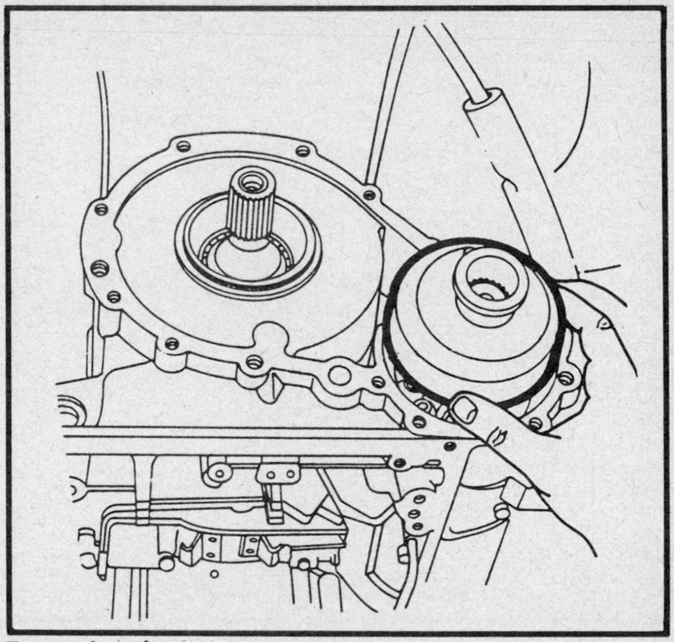

Removing the governor cover

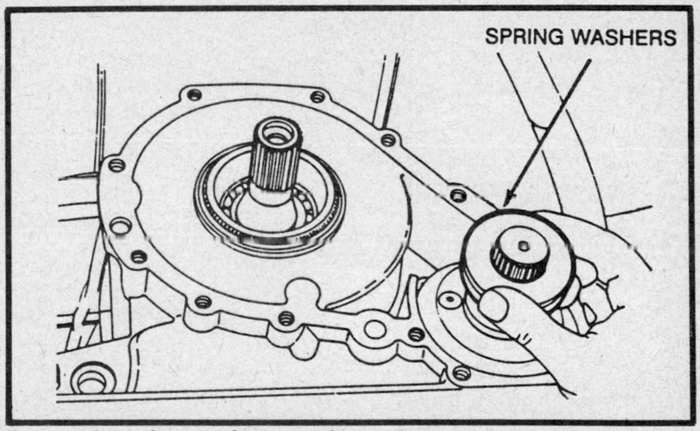
SPRING WASHERS

Removing the spring washers from the governor

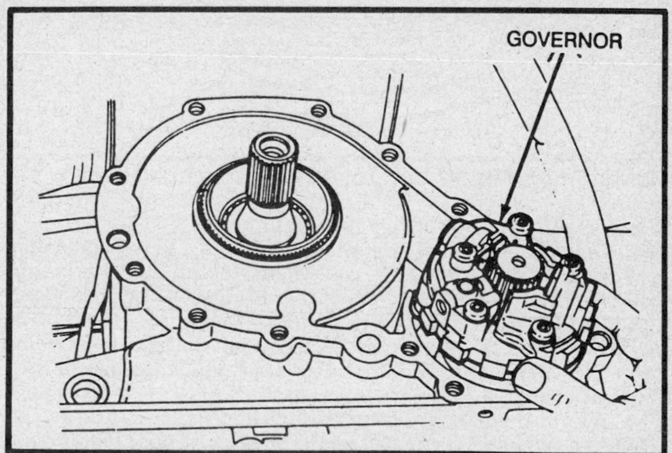
GOVERNOR

Removing the governor assembly

NOTE: Seal position is important. Install the small diameter seal with the seal lip toward the rear of the support. Install the large diameter seal with the seal lip facing out.

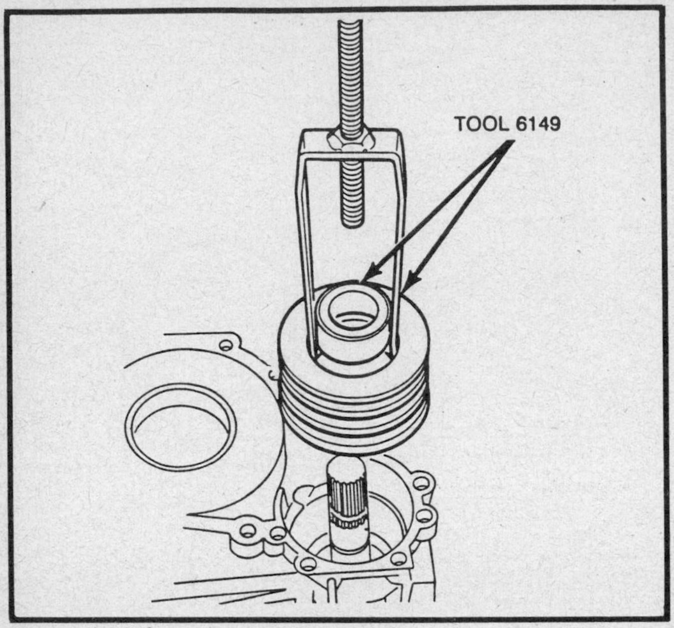
TOOL 6149

Removing the governor support

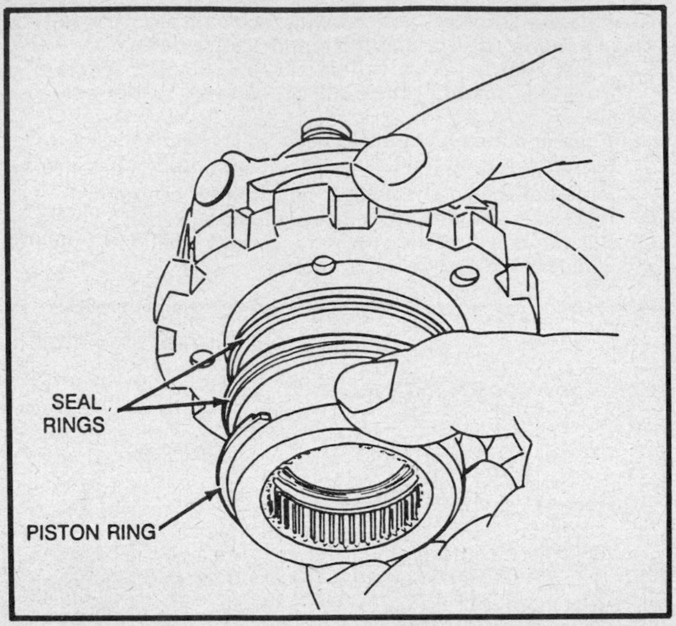
SEAL RINGS
PISTON RING

Installing seal rings and metal piston ring on governor support

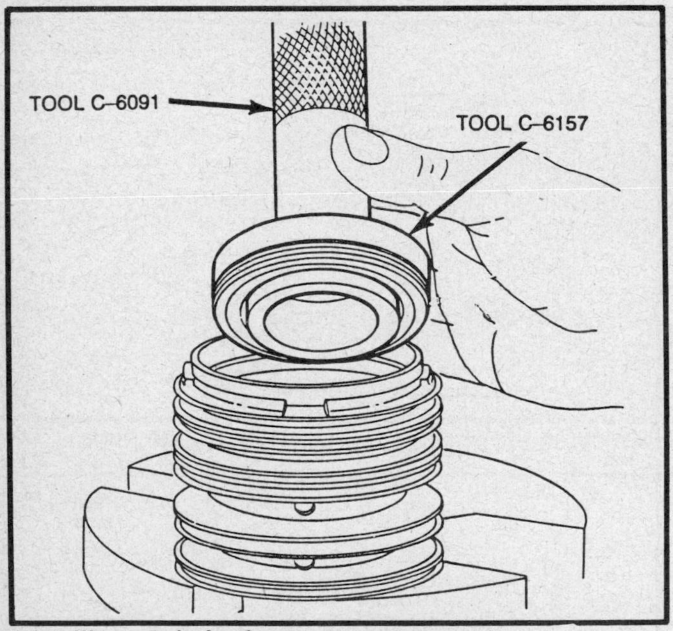
TOOL C-6091
TOOL C-6157

Installing seals in the governor support

12. Install the small diameter seal with tool 6157 or equivalent and driver handle C–6091 or equivalent. Install the large diameter seal with tool 6158 and driver handle C–6091 or equivalent.
13. Install replacement O-rings on the support.
14. Lubricate the seals and O-rings with petroleum jelly and install the support in the case. Use a soft mallet to tap the support into the case.
15. Install the support retainer bolt and tighten to 17 ft. lbs.
16. Install the governor valves on the governor. Tighten the valve attaching bolts to 8 ft. lbs.
17. Install replacement rubber seal rings and a replacement metal piston ring on the governor body.
18. Install the governor assembly in the case. Install the governor cover and assemble and install the 4 spring washers in sets.
19. Coat a replacement flange-to-case gasket with petroleum

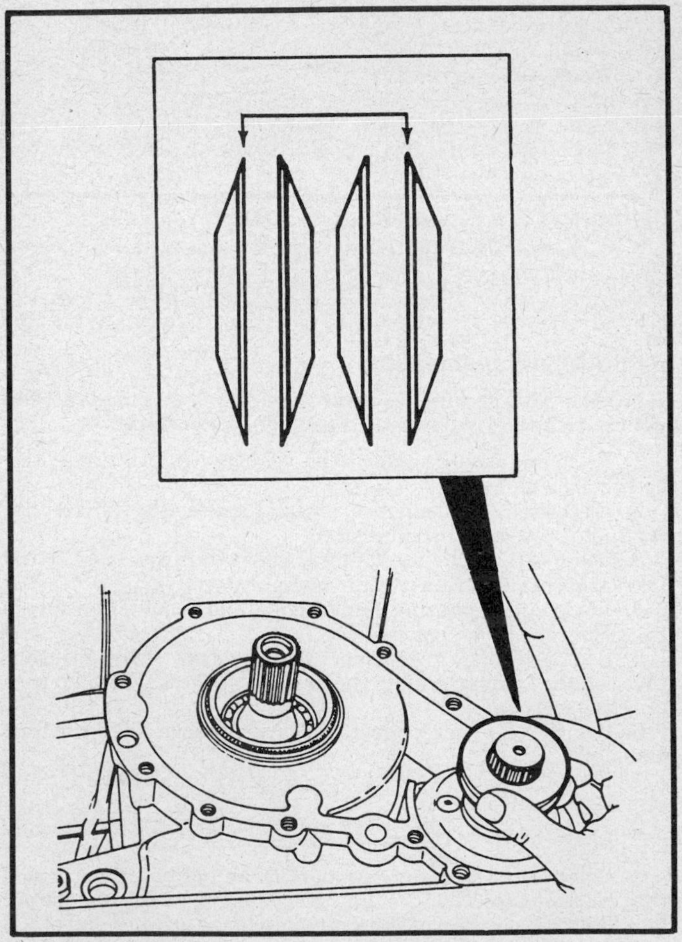

Proper installation of spring washers on the governor

jelly and position the gasket on the case. Install the flange on the case tightening the attaching bolts to 6 ft. lbs.

20. Lubricate a replacement case gasket with petroleum jelly and position it on the transmission.

21. Install the reduction gear case on the transmission. Tighten the case attaching bolts to 17 ft. lbs.

22. Install the drain plug. Remove the differential fill plug and fill the differential with the correct grade and quantity of lubricant.

23. Install the exhaust pipes and install the underbody splash shield. Lower the vehicle.

DIFFERENTIAL OUTPUT SHAFT SEAL AND BEARING

Removal and Installation

1. Raise and safely support the vehicle.
2. Remove the underbody splash shield and drain the differential lubricant.
3. Remove the roll pins attaching the driveshafts to the transaxle output shafts using a pin punch or equivalent.
4. Remove the output shaft dust cover using a small pry bar.
5. Loosen the bolt and pull the the short shaft and bearing out of the cover.
6. Pry the shaft outer seal out of the differential cover. Remove the differential fill plug from the cover.
7. Remove the differential cover bolts. Raise the transaxle as far as possible using a suitable lifting device.
8. Loosen the engine cradle nuts until there is about ½–⅞ in. space between the cradle and side sill. Do not remove the nuts completely. Lowering the cradle will allow easy cover removal.
9. Disconnect the oil filler tube and remove the cover from the case.
10. Remove the differential ring gear and case.
11. Remove the dust cover from the long output shaft. Remove the long output shaft seal using an appropriate tool.
12. Remove the long shaft retaining snapring and remove the long shaft from the case.
13. Remove the output shaft inner seals, use care not to damage the seal bores.
14. Place the long shaft in a press. Remove the snapring that retains the bearing on the shaft and press the shaft out of the bearing.
15. Press the replacement bearing onto the long shaft and install the snapring. Pack both sides of the bearing with grease. Install a replacement O-ring in the long shaft groove.

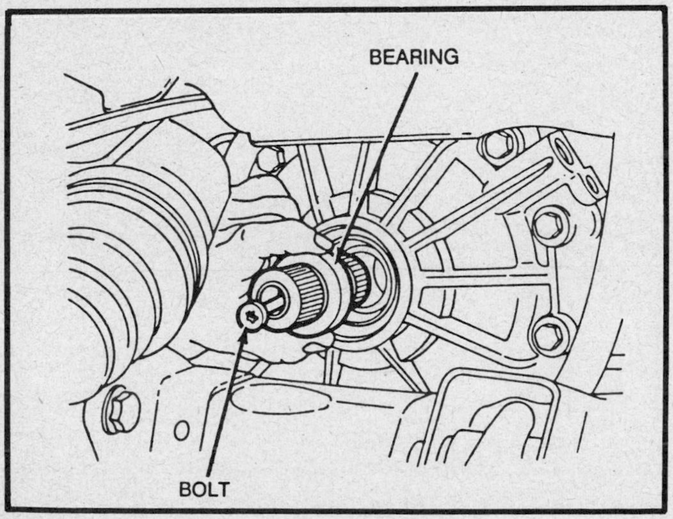
BEARING
BOLT

Removing the short shaft assembly

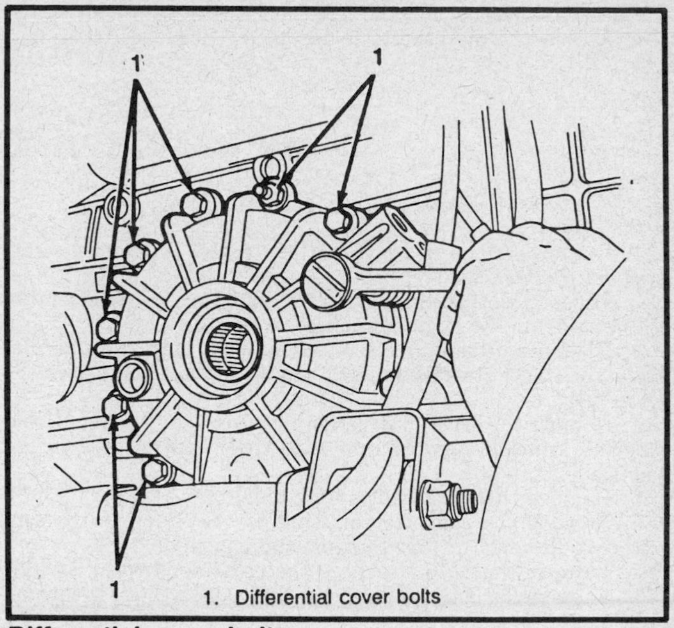
1
1
1
1. Differential cover bolts

Differential cover bolts

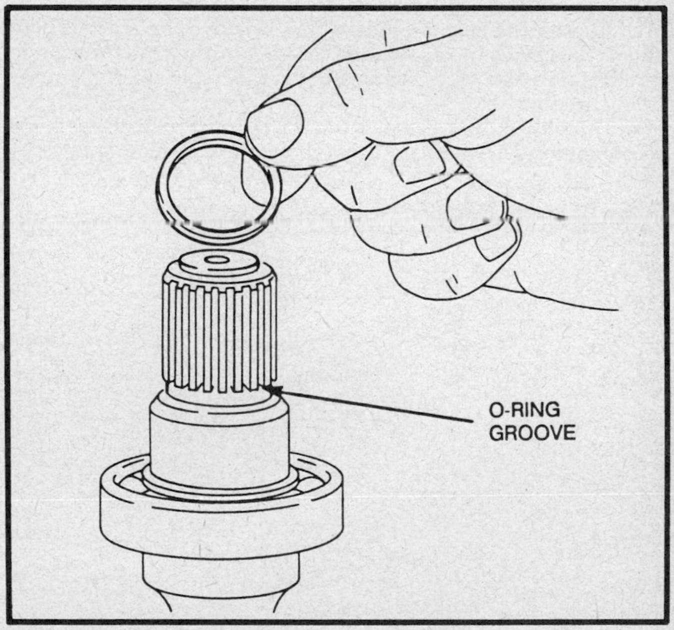
O-RING
GROOVE

Installing the O-ring on the long shaft

16. Install a replacement O-ring in the short shaft groove. Install the inner shaft seals using a seal driver.
17. Install the long output shaft in the case and secure the shaft with the snapring.
18. Install the long shaft outer seal and dust cover.
19. Install the differential ring gear assembly. Install a replacement seal on the differential cover and install the differential cover on the case.
20. Install the differential cover bolts and tighten to 17 ft. lbs.
21. Tighten the engine cradle nuts and remove the jack used to support the transaxle.
22. Install the short output shaft in the case. Tighten the shaft bolt to 18 ft. lbs.
23. Install the short output outer seal and dust cover. Install

the differential plug. Fill the differential with lubricant.

24. Attach the driveshafts to the differential output shafts. Use a pin punch to install the roll pins.

25. Install the underbody splash shield and lower the vehicle.

REMOVAL AND INSTALLATION

TRANSAXLE REMOVAL

1. Disconnect the negative battery cable. Raise and safely support the vehicle.
2. Loosen the throttle valve cable adjusting nut and remove the cable from the engine bracket.
3. Disengage the shift cable and support it to the side. Remove the upper steering knuckle mounting bolt and loosen the lower bolt.
4. Remove the drive shaft retaining pin. Swing each rotor and steering knuckle outward and slide the driveshafts from the transaxle.
5. Remove the underbody splash shield.
6. Remove the converter housing covers. Remove the converter-to-driveplate bolts. Support the transaxle.
7. Remove the nuts attaching the crossmember to the side sills. Remove the large bolt and nut that attach the rear cushion to the support bracket.
8. Remove the support bracket and rear cushion.
9. Disconnect the header pipes from the exhaust manifolds and the catalytic converter.
10. Loosen the engine cradle bolts. Remove the starter, plate and dowel. Disconnect the shift cable from the transmission lever. Remove the cable bracket bolts and separate the bracket from the case.
11. Disconnect and remove the TDC sensor, disconnect the speedometer sensor. Disconnect the transaxle cooling lines.
12. Remove the transaxle-to-engine bolts, pull the transaxle back and away from the engine.

TRANSAXLE INSTALLATION

1. Position the transaxle to the engine. Install the transaxle-to-engine bolts and tighten to 31 ft. lbs.
2. Connect all electrical leads, install the TDC sensor. Connect the speedometer. Install the transaxle cooler lines.
3. Attach the shift bracket to the case and tighten the bolts to 125 inch lbs. Install the shift cable into the bracket.
4. Install the starter. Connect the exhaust head pipes to the manifolds and he converter.
5. Install the rear support and cushion, install the mounting bolts and tighten to 49 ft. lbs.
6. Tighten the engine cradle bolts to 92 ft. lbs. Connect the driveshafts.
7. Install the converter-to-driveplate bolts and tighten to 24 ft. lbs. Install the converter housing covers.
8. Tilt the steering knuckles in and install the top bolts, tighten all to 148 ft. lbs.
9. Install the front wheels. Install the under body splash shield. Attach the throttle valve cable.
10. Connect the negative battery cable. Check the fluid level and check the transaxle operation.

Section 3

AR-4 Transaxle
AMC/Jeep-Eagle

APPLICATION

1987–89 Premier

GENERAL DESCRIPTION

The AR-4 automatic transaxle is 4 speed, electronically controlled fully automatic transaxle with 3 clutches and 2 brakes. The 4th gear is an overdrive range providing a 0.068:1 ratio. The transaxle and differential sections are not intergral and require different lubricants. Shifting is controlled electronically by solenoids in the valve body, speed and throttle sensors and by the transmission computer unit (TCU).

The 3 element torque converter couples the engine to the planetary gears and overdrive unit through oil and hydraulically provides additional torque multiplication when required. The converter torque multiplication feature is operational in **R**, **D1** and **D2** ranges only.

Transaxle and Converter Identification

TRANSAXLE

The AR-4 automatic transaxle can be identified by an identification tag next to the transaxle oil cooler on the passenger side of the vehicle. The information on the tag provides the transaxle type, suffix, fabrication number and plant of manufacture. The identification number "T" is used in the VIN to designate the AR-4 transaxle.

CONVERTER

The torque converter is a welded unit and cannot be disassembled for service. The torque converter diameter is 9.8 in. (250mm). Any internal malfunctions require the replacement of the converter assembly. No specific identification is available for matching the converter to the transaxle for the average repair shop.

Metric Fasteners

Metric bolt sizes and thread pitches are used for all fasteners on the AR-4 automatic transaxle. The use of metric tools is mandatory in the service of this transaxle.

Do not attempt to interchange metric fasteners for inch system fasteners. Mismatched or incorrect fasteners can result in damage to the transaxle unit through malfunctions, breakage or possible personal injury. Care should be taken to reuse the fasteners in the same location as removed, whenever possible. Due to the large number of alloy parts used, torque specifications should be strictly observed. Before installing capscrews into aluminum parts, always dip screws into oil to prevent the screws from galling the aluminum threads and to prevent seizing.

Capacities

If the pan was removed, the approximate fluid needed to fill the transaxle is 2.8 qts. (2.6L). A completely overhauled transaxle will require 5.6 qts. (5.3L) of transaxle fluid. The fluid capacities are approximate and the correct fluid level should be determined by the dipstick indicator. Only Mercon® automatic transmission fluid should be used when adding fluid or servicing the AR-4 automatic transaxle.

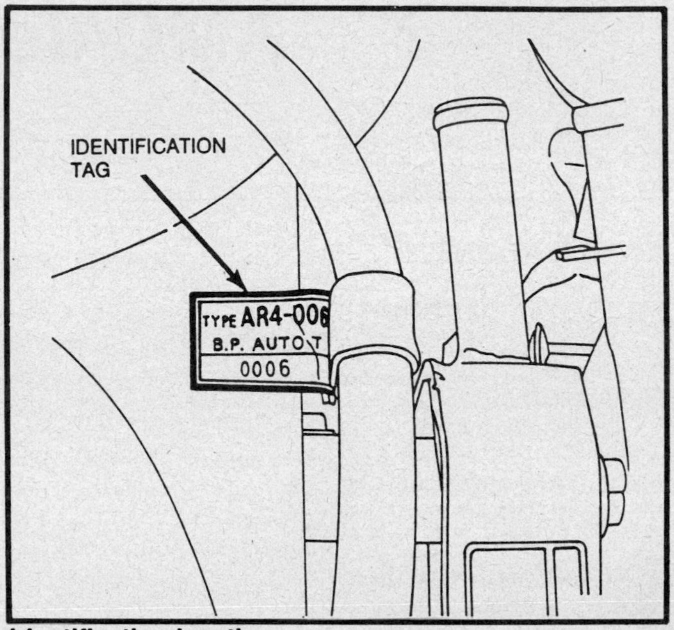
IDENTIFICATION TAG
TYPE AR4-006
B.P. AUTO T
0006

Identification location

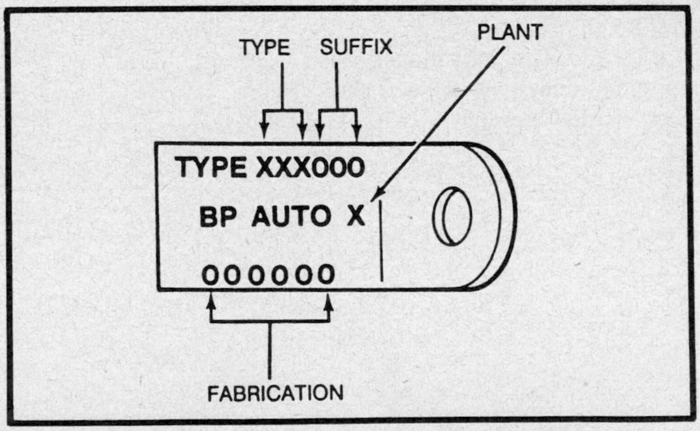
TYPE
SUFFIX
PLANT
TYPE XXX000
BP AUTO X
000000
FABRICATION

Identification tag

NOTE: The differential is a separate unit. The 4 cylinder model's differential require .89 qts (.85L) of SAE 75W-140 gear lubricant and the 6 cylinder model's differential require .73 qts (.70L) of SAE 75W-140 gear lubricant.

Checking Fluid Level

TRANSAXLE

The AR-4 transaxle is designed to operate at the hot **FULL** mark on the dipstick at normal operating temperatures, which

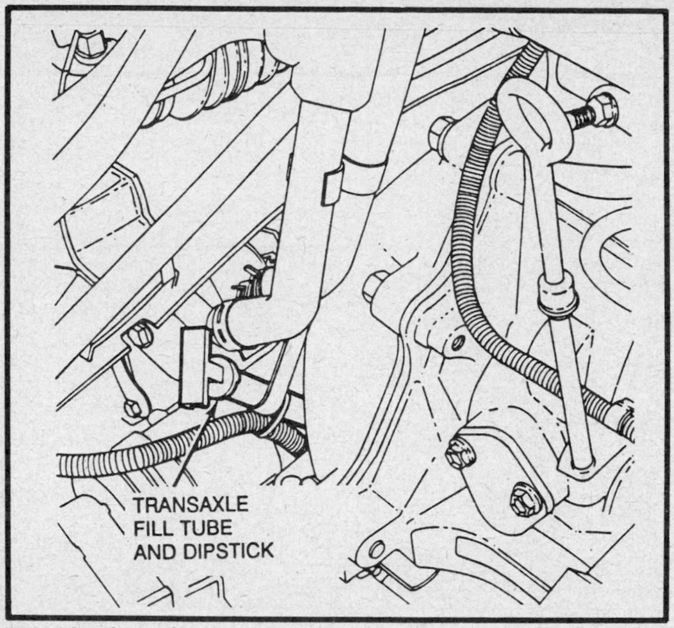
TRANSAXLE
FILL TUBE
AND DIPSTICK

Fill tube location

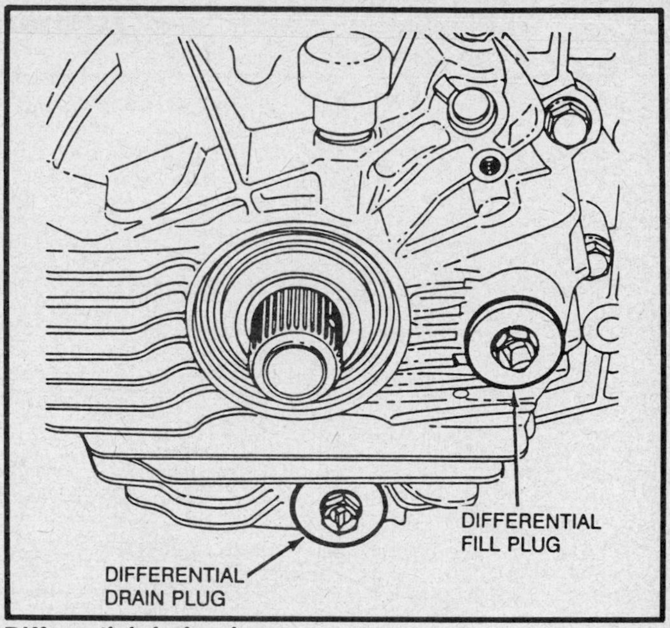
DIFFERENTIAL
FILL PLUG
DIFFERENTIAL
DRAIN PLUG

Differential drain plug

range from 190–200°F. Automatic transaxles are frequently overfilled because the fluid level is checked when cool and the dipstick level reads low. However, as the fluid warms up, the level of the fluid will rise, as much as ¾in. Note that if the transmission fluid is too hot, as it might be when operating under city traffic conditions, trailer towing or extended high speed driving, an accurate fluid level cannot be determined until the fluid has cooled somewhat, perhaps 30 minutes after shutdown. It requires 0.28 qts. (0.3L) of fluid to increase level from the **ADD** to the **FULL** mark, on the dipstick.

To determine proper fluid level under normal operating temperatures, proceed as follows:

1. Make sure vehicle is parked level.
2. Apply parking brake; move selector to **P**.
3. Start but do not race engine. Allow to idle.
4. Move selector through each range, then back to **P**, then check level. The fluid should read **FULL**.

Do not overfill the transaxle. Overfilling can cause foaming and loss of fluid from the vent. Overheating can also be a result of overfilling since heat will not transfer as readily. Notice the condition of the fluid and whether there seems to be a burnt smell or metal particles on the end of the dipstick. A milky appearance is a sign of water contamination, possibly from a damaged cooling system. All this can be a help in determining transaxle problems and their source.

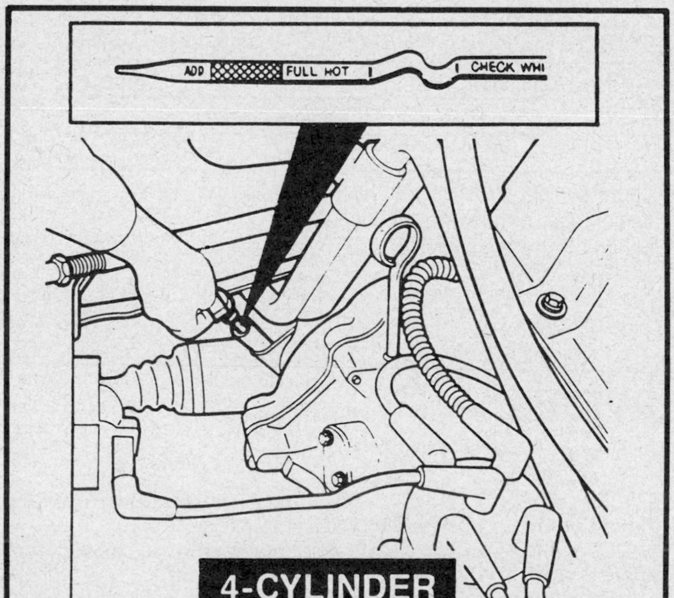
ADD
FULL HOT
CHECK WHI
4-CYLINDER

Transaxle dipstick location—4 cylinder engine

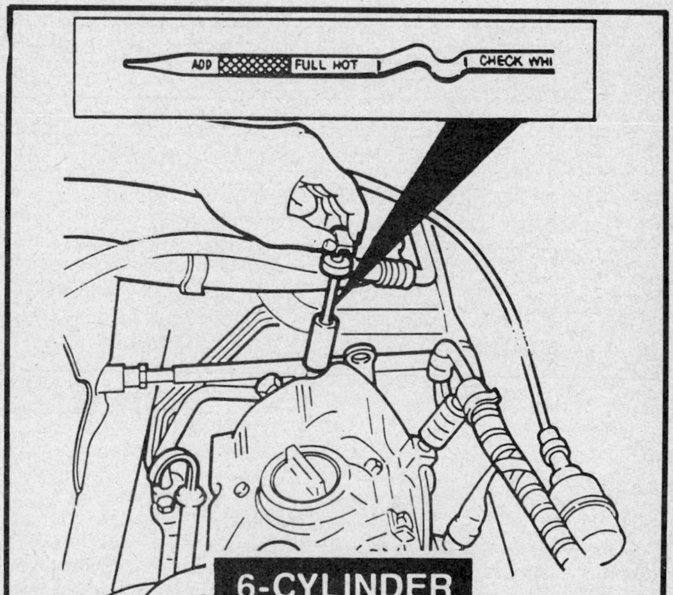
ADD
FULL HOT
CHECK WHI
6-CYLINDER

Transaxle dipstick location—6 cylinder engine

TROUBLE DIAGNOSIS

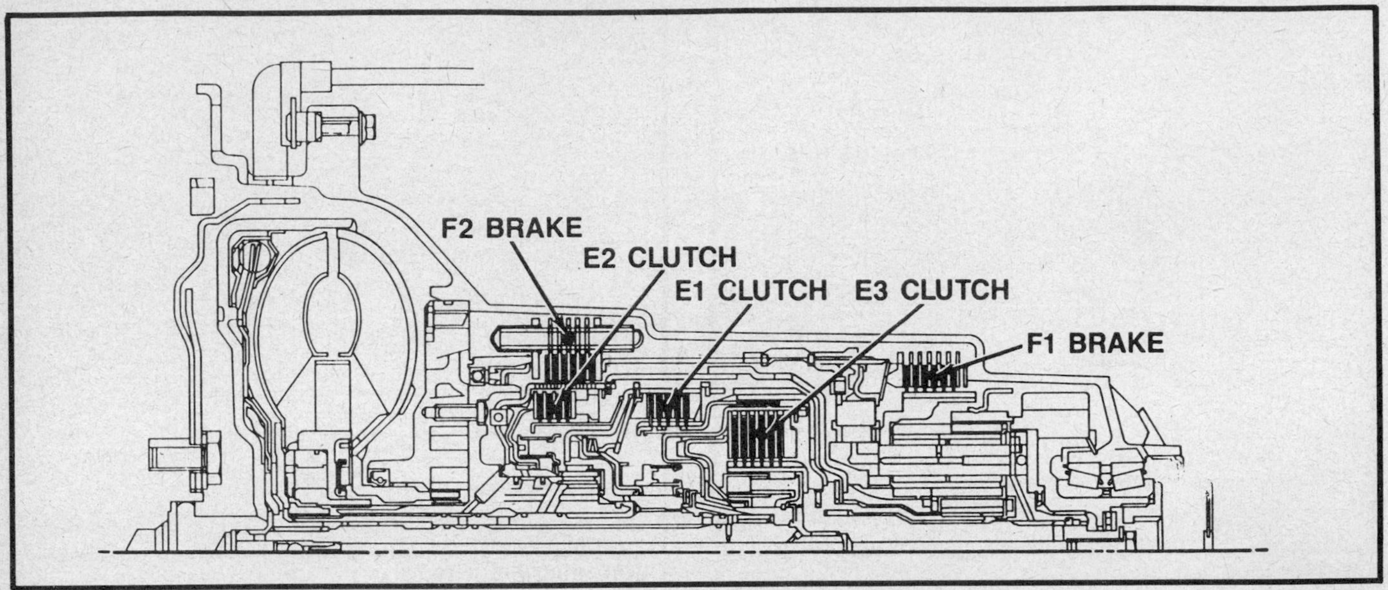
F2 BRAKE
E2 CLUTCH
E1 CLUTCH
E3 CLUTCH
F1 BRAKE

Clutch and brake locations

CLUTCH AND BAND APPLICATION

Shift Lever Position	Line Pressure Level	Clutch No. 1	Clutch No. 2	Clutch No. 3	Brake No. 1	Brake No. 2	Roller Clutch	Solenoid Valve No. 1	Solenoid Valve No. 2	Solenoid Valve No. 3
Park	Low	–	– –	–	–	–	–	–	–	–
Reverse	High	–	Applied	–	Applied	–	–	–	–	–
Neutral	Low	–	–	–	–	–	–	–	–	–
Drive–1	1st gear	Applied	–	–	–	–	Applied	–	Applied	–
Drive–2	2nd gear	Applied	–	–	–	Applied	–	–	–	Applied
Drive–3	3rd gear	Applied	Applied	Applied	–	–	–	Applied	–	–
Drive–4	4th gear	–	–	Applied	–	Applied	–	Applied	Applied	Applied
1st gear hold	1st gear	Applied	–	–	Applied	–	Applied	–	Applied	Applied
2nd gear hold	2nd gear	Applied	–	–	–	Applied	–	–	–	Applied
3rd gear hold	3rd gear	Applied	Applied	Applied	–	–	–	Applied	–	–
3rd gear default①	High	Applied	Applied	–	–	–	–	–	–	–

① If an electrical malfunction causes the valve body solenoids to become inoperative, the transmission will still operate in the 3rd gear range. The vehicle can be driven at reduced speeds to a repair facility

ON CAR SERVICES

Adjustments

SHIFT CABLE

1. Shift the transaxle into **P**.
2. Raise and support vehicle safely.
3. Unlock shift cable by releasing the cable adjusting clamp. Move the clamp outward to release it.
4. Move the transaxle shift lever rearward into the **P** detent. Make certain the lever is centered in the detent.
5. Verify the positive engagement of the park lock by attempting to rotate the driveshafts. The shafts cannot be turned if the park lock is properly engaged.

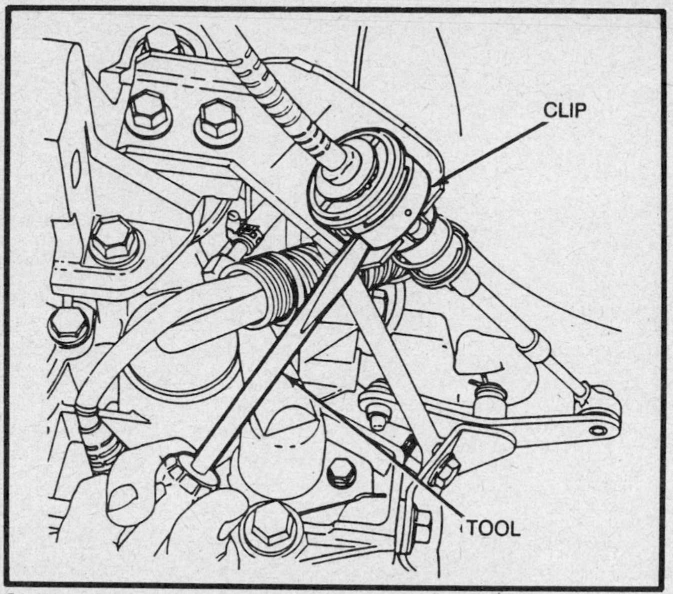
CLIP
TOOL
Shift cable—releasing the adjusting clamp

6. Lock the shift cable by pressing the adjuster clamp back into position. Be sure the clamp snaps into place.
7. Lower the vehicle.
8. Turn the ignition key to **LOCK** and verify that the shift lever remains locked in **P**. It should not nbe possible to move the lever out of **P**.
9. Turn the ignition key to **ON**.
10. Make certain the engine starts only when the shift lever is in **P** or **N** positions.
11. If the engine starts in any other position the cable adjustment is incorrect.
12. Shift the transaxle back into **P** and make certain the key can be returned to **LOCK** and then removed.

Services

FLUID CHANGES

The main considerations in establishing fluid change intervals are the type of driving that is done and the heat levels that are generated by such driving. Normally, the fluid and strainer would be changed at 30,000 miles. However, if the vehicle is driven under severe conditions, it is recommended that the fluid be changed and the filter screen serviced at 15,000 mile intervals. Be sure not to overfill the unit.

NOTE: The differential and transaxle sections are not integral, the differential uses SAE 75W-140 gear lubricant.

OIL PAN

Removal and Installation

1. Raise and safely support vehicle.
2. Remove the underbody splash shield bolts and remove the shield.
3. Place drain pan under transaxle oil pan and remove the drain plug.
4. Remove the oil pan bolts and remove pan.
5. Clean pan in solvent and dry with compressed air.
6. Remove the filter screen bolts and remove the filter screen from the valve body. Discard old screen and gasket.

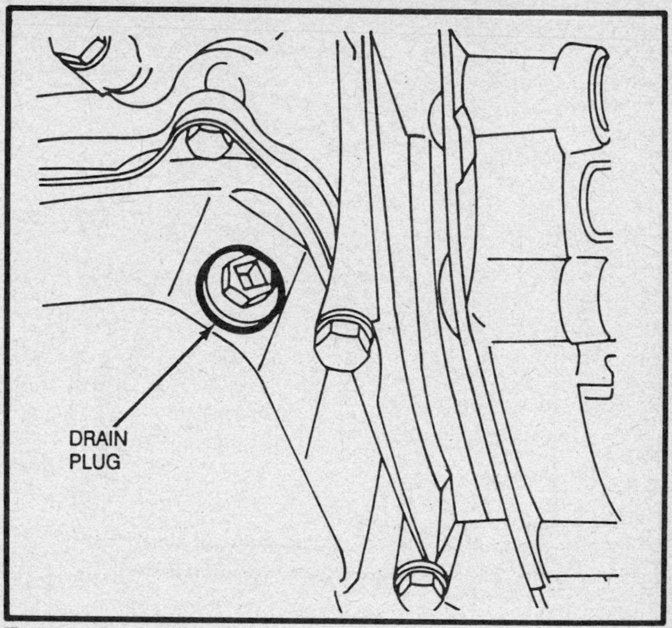
DRAIN
PLUG
Transaxle drain plug

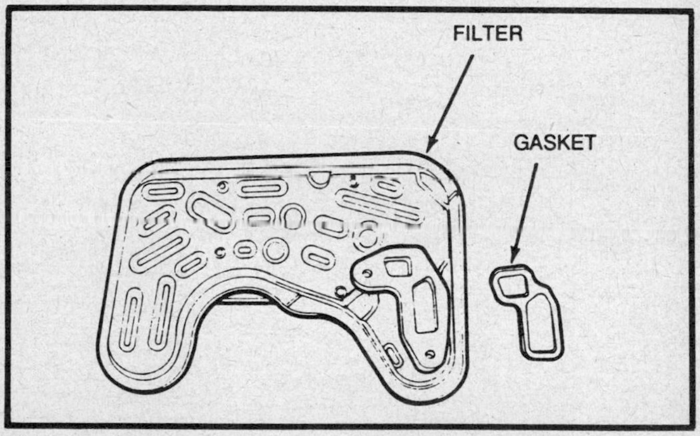
FILTER
GASKET
Filter and gasket

7. Install new screen and gasket, if necessary hold gasket in place on screen with petroleum jelly.
8. Torque screen bolts to 46 inch lbs. (5 Nm).
9. Install new gasket on pan dry, do not use any type of sealer, make certain to replace all gasket mounting spacers.
10. Install the pan and torque bolts to 90 inch lbs. (10 Nm).
11. Install drain plug with new seal ring and torque to 177 inch lbs. (20 Nm).
12. Replace the underbody splash shield.
13. Lower the vehicle.
14. Remove dipstick and fill transaxle through the fill tube with Mercon® automatic transmission fluid to the proper level.
15. With selector in **P** start engine and idle. Apply parking brake. Do not race engine.
16. Move selector through all ranges and return to **P**. Check fluid level. Add transmission fluid as necessary to bring to proper level.

VALVE BODY

Removal and Installation

1. Raise and safely support vehicle.

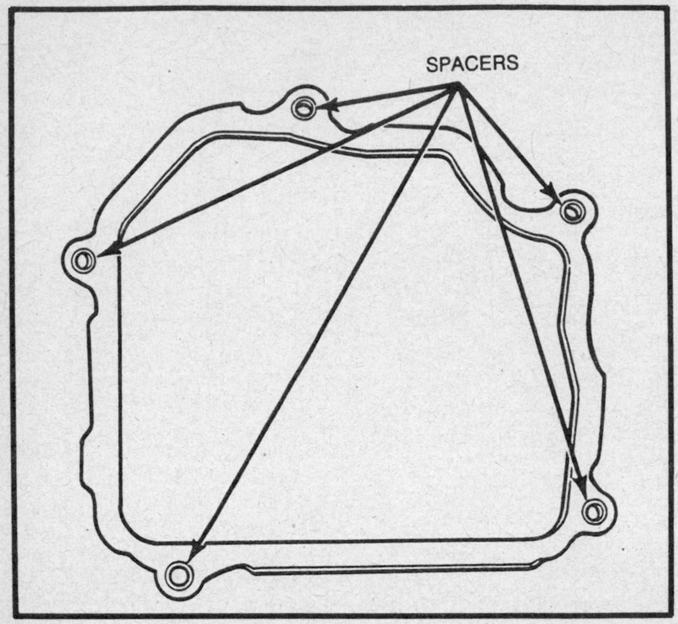
SPACERS

Transaxle oil pan spacers

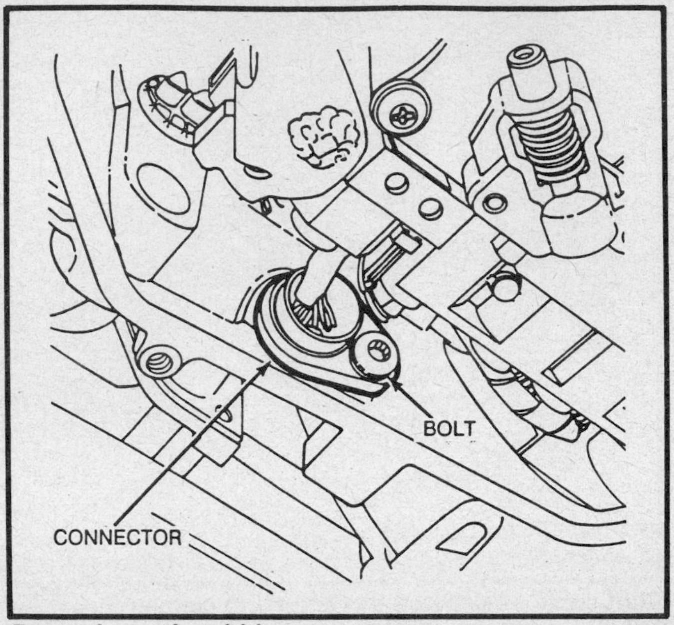
BOLT
CONNECTOR

Removing solenoid harness connector

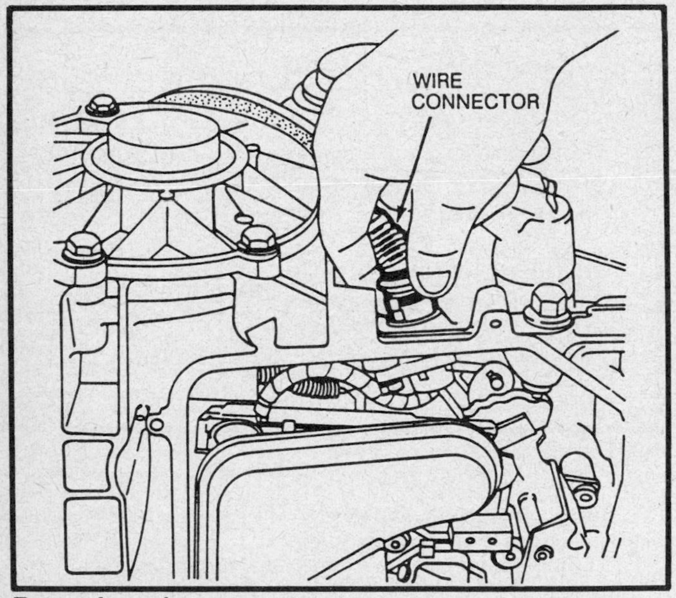
WIRE
CONNECTOR

Removing wire connector

2. Remove the underbody splash shield bolts and remove the shield.
3. Place drain pan under transaxle oil pan and remove the drain plug.
4. Remove the oil pan bolts and remove pan.
5. Remove the filter from the valve body.
6. Squeeze the lock ring on the solenoid wire harness connector and remove the wire connector. Do not use pliers to squeeze the lock ring.
7. Remove the bolt attaching the valve body solenoid harness connector to the case.
8. Disconnect the shift rod from the shift lever.
9. Rotate the shift arm outward fully.
10. Disengage the caliper from the manual valve and remove the caliper and valve.
11. Remove the valve body attaching bolts.

NOTE: Do not remove the 2 smaller bolts in the valve body.

12. Remove the valve body solenoid connector from the case.
13. Lower and remove the valve body.
14. Check the position of the 7 valve body baffles. If any are loose, install them in their correct locations using petroleum jelly to hold them in place.

NOTE: If any baffles fall out, make certain they are installed with the tabs facing the valve body.

15. Install a replacement O-ring on the solenoid connector and lubricate with transmission fluid.
16. Raise the valve body into position and push the solenoid connector into the transaxle case. Then align the valve body on the case and install the valve body bolts finger tight.
17. Tighten the valve body bolts to 46 inch lbs. (5 Nm) in the proper sequence.
18. Install the caliper on the manual valve, insert the metal end of the caliper first.
19. Swing the shift arm over into the channel in the caliper.
20. Install the bolt that attaches the solenoid harness connector to the case and tighten to 46 inch lbs. (5 Nm).
21. Connect the external wire harness to the solenoid connector. Squeeze the lock ring on the harness connector to install it. Listen for the connector to click into place.
22. Install new screen and gasket, if necessary hold gasket in place on screen with petroleum jelly.
23. Torque screen bolts to 46 inch lbs. (5 Nm).
24. Install new gasket on pan dry, do not use any type of sealer, make certain to replace all gasket mounting spacers.
25. Install the pan and torque bolts to 90 inch lbs. (10 Nm).
26. Install drain plug with new seal ring and torque to 177 inch lbs. (20 Nm).
27. Connect the shift cable rod to the transaxle shift lever.
28. Replace the underbody splash shield.
29. Lower the vehicle.
30. Remove dipstick and fill transaxle through the fill tube with Mercon® automatic transmission fluid to the proper level.
31. With selector in **P** start engine and idle. Apply parking brake. Do not race engine.
32. Move selector through all ranges and return to **P**. Check fluid level. Add transmission fluid as necessary to bring to proper level.

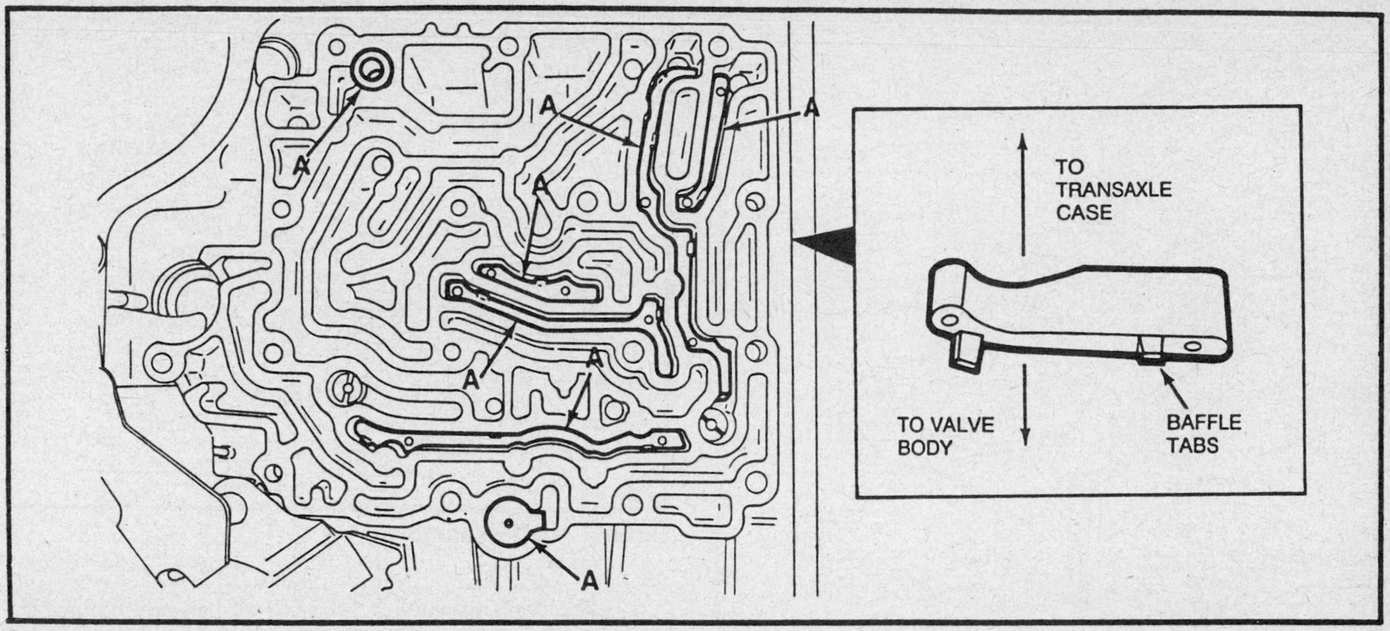
A
A
A
A
A
A
A
TO TRANSAXLE CASE
TO VALVE BODY
BAFFLE TABS

Baffle positions

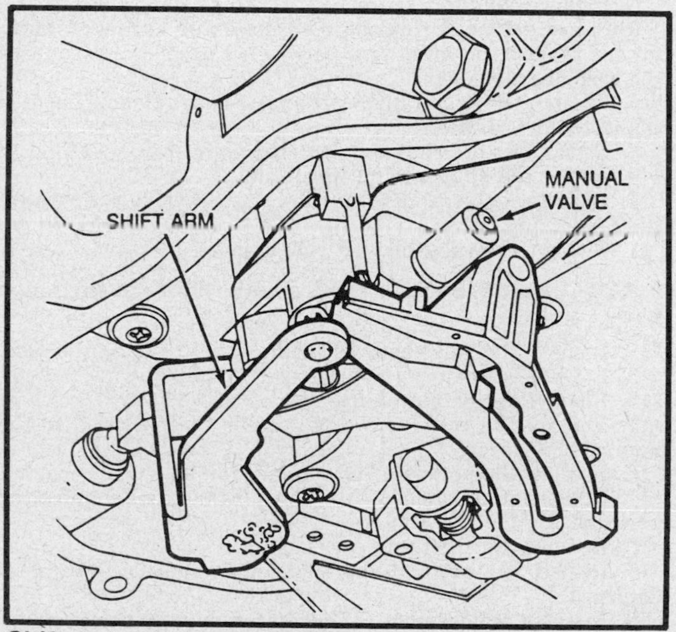
MANUAL VALVE
SHIFT ARM

Shift rod and manual valve

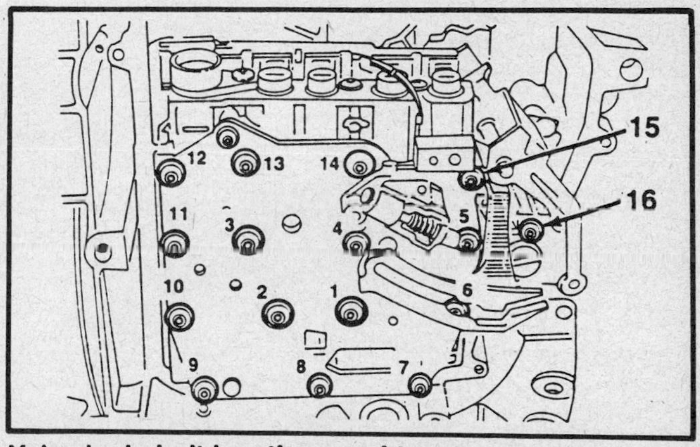
15
16
12
13
14
11
3
4
5
10
2
1
6
9
8
7

Valve body bolt locations and torque sequence

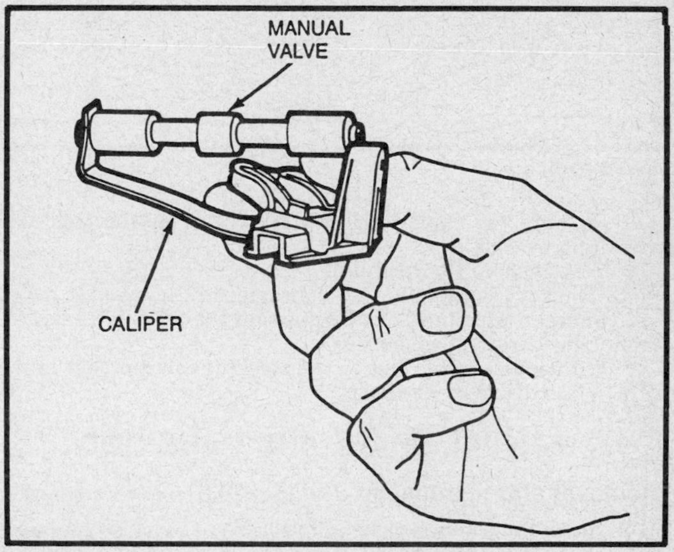
MANUAL VALVE
CALIPER

Installing the caliper onto the manual valve

SHIFT CABLE

Removal and Installation

1. Disconnect the shift cable from the steering column shift arm, under the dash.
2. Squeeze lock tabs to release cable from the bracket.
3. Raise and support the vehicle safely.
4. Disconnect the cable from shift lever, under vehicle.
5. Squeeze lock tabs to release cable from the bracket.
6. Pull cable grommet out of the dash and remove the cable.
7. From under the vehicle, insert the cable into the drivers compartment through the cable grommet hole in the dash panel.

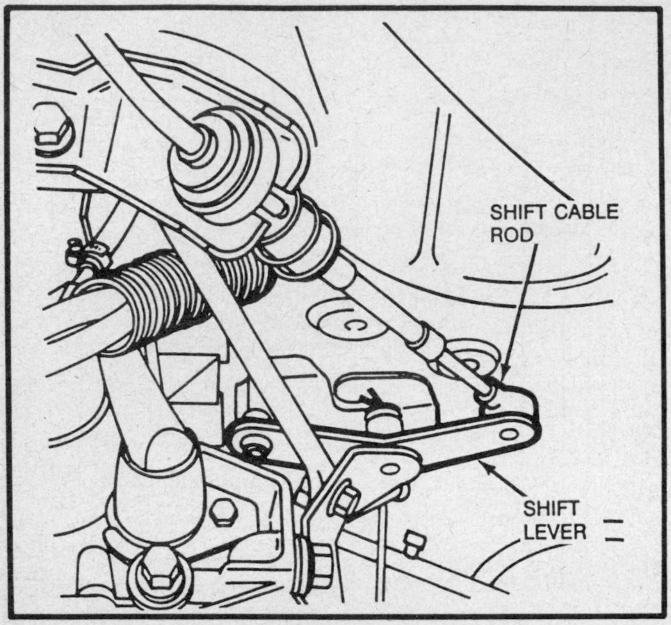
SHIFT CABLE ROD
SHIFT LEVER

Connecting shift cable rod

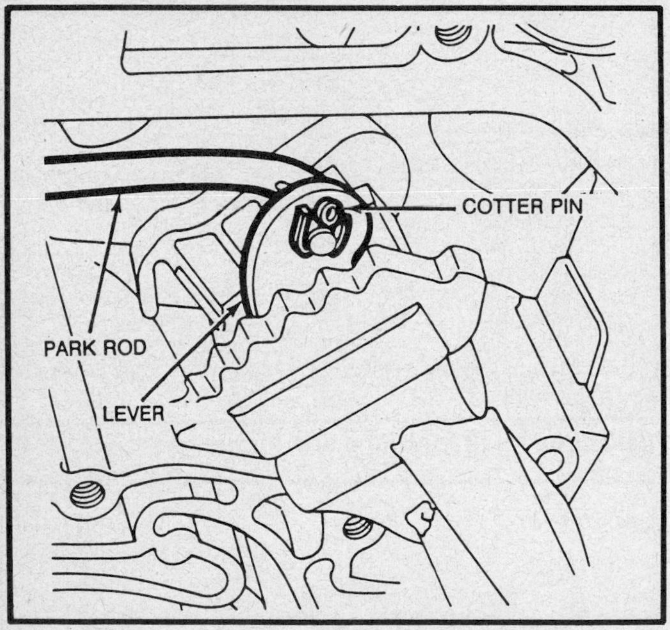
COTTER PIN
PARK ROD
LEVER

Park rod

8. Squeeze the cable lock and seat cable in the transaxle mounting bracket.
9. Seat cable in floorpan hole.
10. Connect the cable to the transaxle shift lever.
11. Squeeze the cable lock tabs and seat the cable in the instrument panel bracket, under dash.
12. Connect the cable end to the steering column shift arm.
13. Adjust cable as necessary.

MANUAL SHAFT AND PARK ROD

Removal and Installation

1. Raise and safely support vehicle.
2. Remove the underbody splash shield bolts and remove the shield.

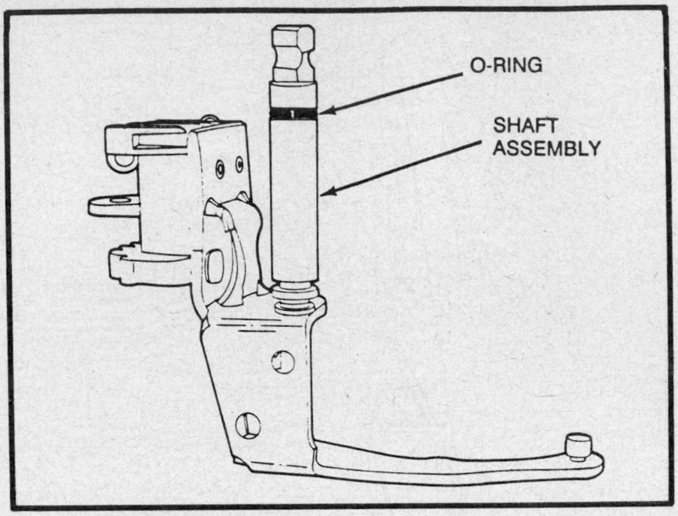
O-RING
SHAFT ASSEMBLY

Manual shaft assembly

3. Place drain pan under transaxle oil pan and remove the drain plug.
4. Remove the oil pan bolts and remove pan.
5. Remove the filter screen.
6. Squeeze the lock ring on the solenoid wire harness connector and remove the wire connector. Do not use pliers to squeeze the lock ring.
7. Remove the bolt attaching the valve body solenoid harness connector to the case.
8. Disconnect the shift rod from the shift lever.
9. Rotate the shift arm outward fully.
10. Disengage the caliper from the manual valve and remove the caliper and valve.
11. Remove the valve body attaching bolts.

NOTE: Do not remove the 2 smaller bolts in the valve body.

12. Remove the valve body solenoid connector from the case.
13. Lower and remove the valve body.
14. Remove the detent plunger.
15. Remove the cotter pin that secures the park rod to the manual shift lever.
16. Remove the manual shaft assembly bolt.
17. Pull the manual shaft assembly down and disconnect the park rod.
18. Remove the shaft and park rod.
19. Inspect the manual shaft assembly, replace if worn of damaged.
20. Install a replacement O-ring on the manual shaft and lubricate with automatic transmission fluid.
21. Replace park rod if damaged or distorted and replace shaft bushing if damaged, cracked or worn.
22. Install the park rod in correct position.
23. Install the manual shaft assembly into the case and engage the park rod in the manual shaft lever.

NOTE: Take care not to damage the multi-function switch plunger buttons when installing the manual shaft assembly.

24. Install the retainer plate and position the plate in the manual shaft groove.
25. Install and tighten the retainer bolt to 80 inch lbs. (9 Nm).
26. Secure the park rod to the manual shaft lever with a replacement cotter pin.
27. Install a new dust seal on the manual shaft.
28. Install a new O-ring, lubricated with transmission fluid, on

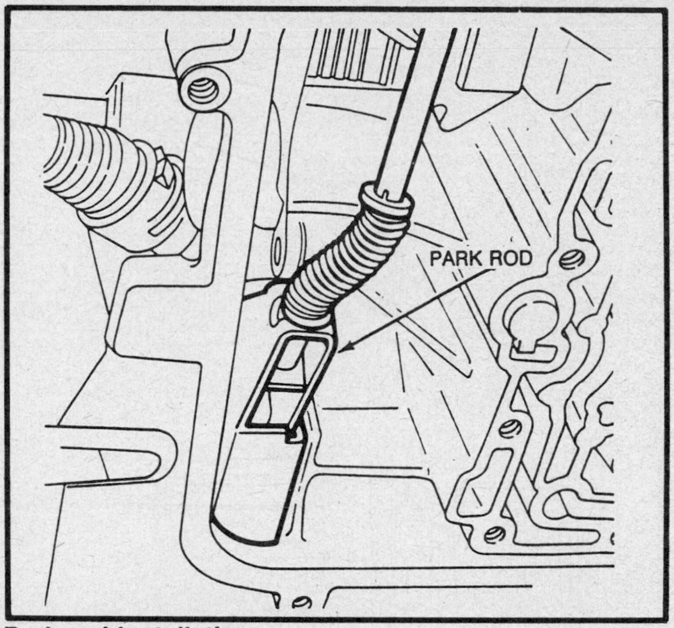
PARK ROD
Park rod installation

the detent plunger and install plunger. Tighten plunger to 150 inch lbs. (17 Nm).

29. Install the valve body and filter screen.
30. Check the position of the 7 valve body baffles. If any are loose, install them in their correct locations using petroleum to hold them in place.

NOTE: If any baffles fall out, make certain they are installed with the tabs facing the valve body.

31. Install a replacement O-ring on the solenoid connector and lubricate with transmission fluid.
32. Raise the valve body into position and push the solenoid connector into the transaxle case. Then align the valve body on the case and install the valve body bolts finger tight.
33. Tighten the valve body bolts to 46 inch lbs. (5 Nm) in the proper sequence.
34. Install the caliper on the manual valve, insert the metal end of the caliper first.
35. Swing the shift arm over into the channel in the caliper.
36. Install the bolt that attaches the solenoid harness connector to the case and tighten to 46 inch lbs. (5 Nm).
37. Connect the external wire harness to the solenoid connector. Squeeze the lock ring on the harness connector to install it. Listen for the connector to click into place.
38. Install new screen and gasket, if necessary hold gasket in place on screen with petroleum jelly.
39. Torque screen bolts to 46 inch lbs. (5 Nm).
40. Install new gasket on pan dry, do not use any type of sealer, make certain to replace all gasket mounting spacers.
41. Install the pan and torque bolts to 90 inch lbs. (10 Nm).
42. Install drain plug with new seal ring and torque to 177 inch lbs. (20 Nm).
43. Connect the shift cable rod to the transaxle shift lever.
44. Lower vehicle and replace the underbody splash shield.
45. Remove dipstick and fill transaxle through the fill tube with Mercon® automatic transmission fluid to the proper level.
46. With selector in **P** start engine and idle. Apply parking brake. Do not race engine.
47. Move selector through all ranges and return to **P**. Check fluid level. Add transmission fluid as necessary to bring to proper level.

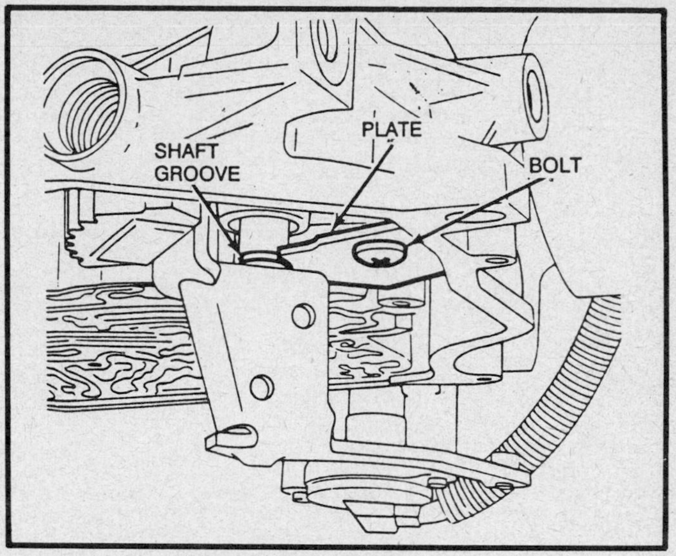
PLATE
SHAFT
GROOVE
BOLT
Positioning manual shaft groove

MULTI-FUNCTION SWITCH

Removal and Installation

1. Disconnect the negative battery cable.
2. Raise the vehicle and support it safely.
3. Remove the switch attaching bolt and pull the multi-function switch out of the case.
4. Lower vehicle and remove windshield washer bottle.
5. Disconnect the multi-function switch harness from the transmission control unit.
6. Remove the old harness.
7. Install the new harness, making certain wiring is clear from any hot or moving parts.
8. Connect harness to transmission control unit and replace windshield washer bottle.
9. Install new O-ring on switch, lubricate with transmission fluid and install switch. Make certain the switch ground wire is on the attaching bolt.
10. Connect the negative battery cable.

SPEED SENSOR

Removal and Installation

TRANSMISSION CONTROL UNIT SENSOR

1. Disconnect the negative battery cable.
2. Disconnect the sensor electrical connector.
3. Remove the sensor bracket bolt and remove the sensor.
4. Install a new lubricated O-ring on the sensor and install the sensor.
5. Tighten the sensor mounting bracket bolt and connect the sensor electrical connector.
6. Connect the negative battery cable.

ROAD SENSOR

NOTE: The road speed sensor is the electronic pickup unit for the vehicle speedometer. It is mounted on the differential case just above the driveshaft.

1. Disconnect the negative battery cable.
2. Raise and support vehicle safely.
3. Remove the sensor bolt and pull sensor from the case.
4. Disconnect the sensor electrical connector.
5. Install a new lubricated O-ring on the sensor and reconnect the electrical connector.

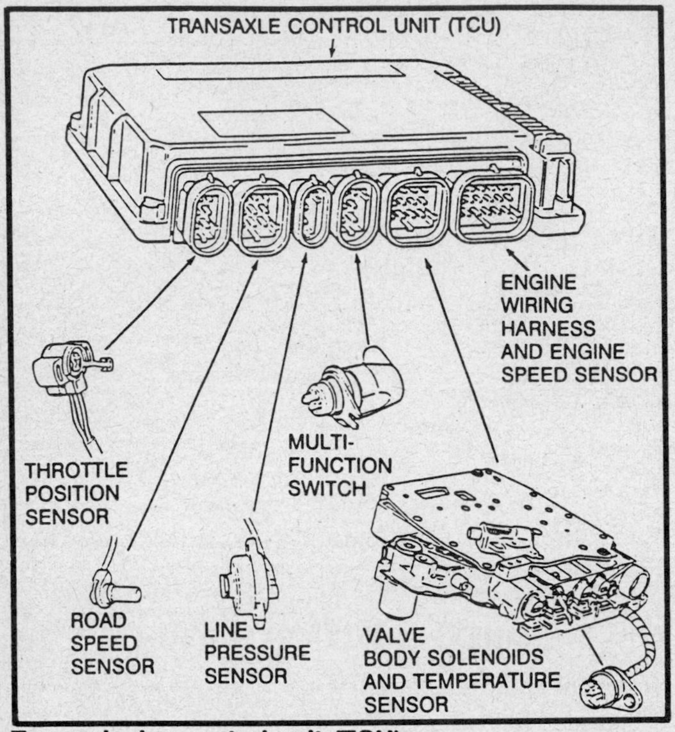
TRANSAXLE CONTROL UNIT (TCU)
ENGINE WIRING HARNESS AND ENGINE SPEED SENSOR
THROTTLE POSITION SENSOR
MULTI-FUNCTION SWITCH
ROAD SPEED SENSOR
LINE PRESSURE SENSOR
VALVE BODY SOLENOIDS AND TEMPERATURE SENSOR

Transmission control unit (TCU)

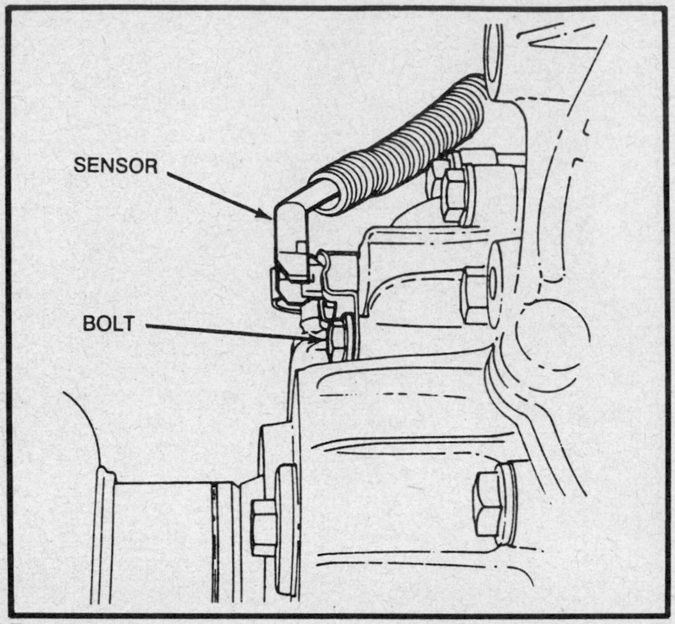
SENSOR
BOLT

Road speed sensor

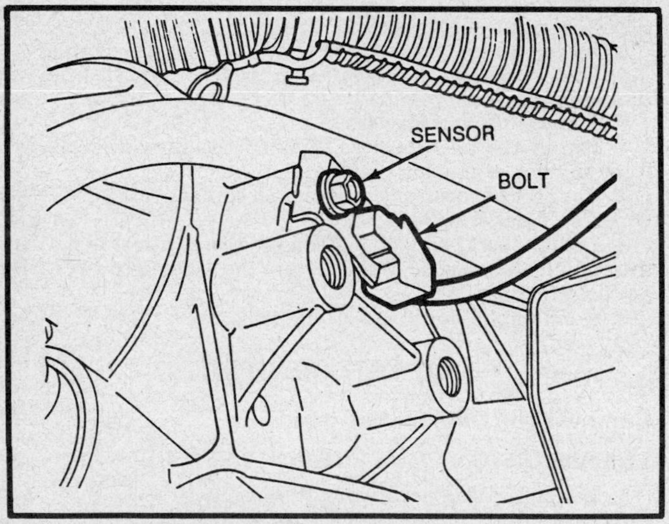
SENSOR
BOLT

TCU speed sensor

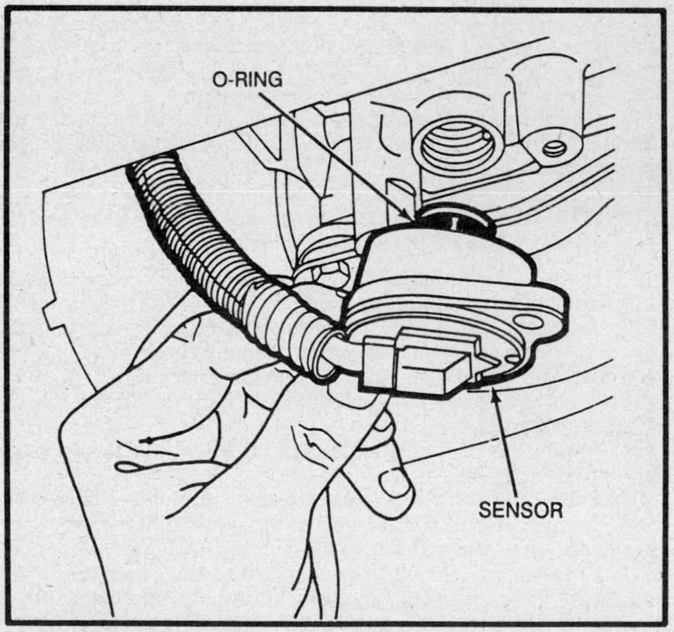
O-RING
SENSOR

Line pressure sensor

6. Replace sensor and tighten the sensor mounting bolt.
7. Connect the negative battery cable.

LINE PRESSURE SENSOR

Removal and Installation

1. Disconnect the negative battery cable.
2. Raise and support vehicle safely.
3. Remove the underbody splash shield.
4. Remove the sensor attaching screws and pull sensor from the case.
5. Lower vehicle and remove the windshield washer bottle.
6. Disconnect the pressure line harness from the transmission control unit.
7. Remove sensor and harness from underneath vehicle.
8. Install new sensor and harness, making certain wiring is clear from any hot or moving parts.
9. Connect harness to transmission control unit and install windshield washer bottle.
10. Raise the vehicleand support safely, using a lubricated O-ring, install new sensor and tighten retaining screws.
11. Install underbody cover and lower vehicle.
12. Connect the negative battery cable.

TRANSMISSION CONTROL UNIT

Removal and Installation

1. Disconnect the negative battery cable.

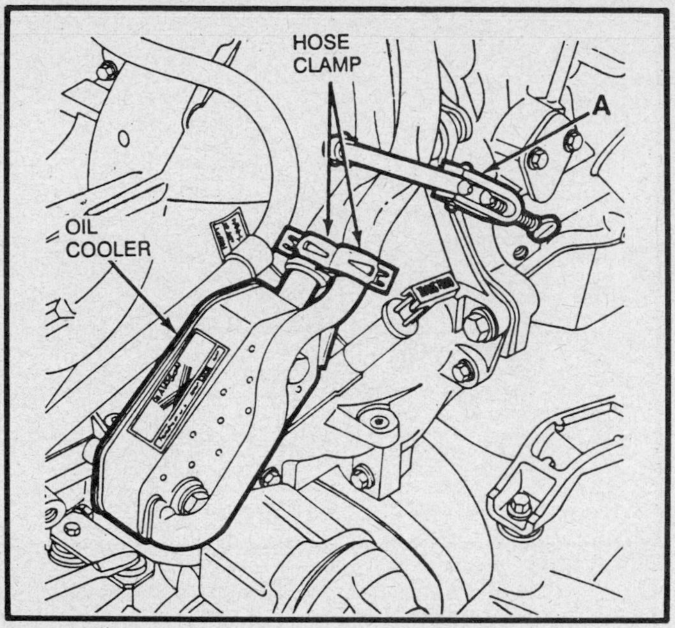
HOSE
CLAMP
A
OIL
COOLER
Oil cooler

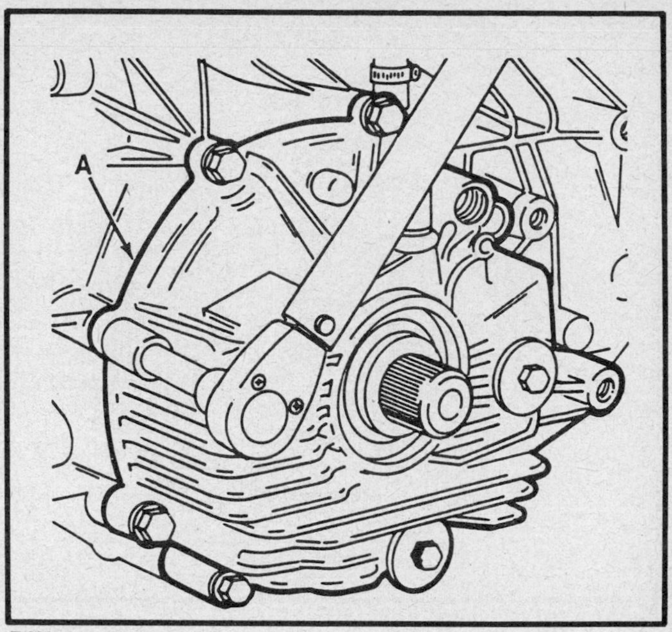
A
Differential cover

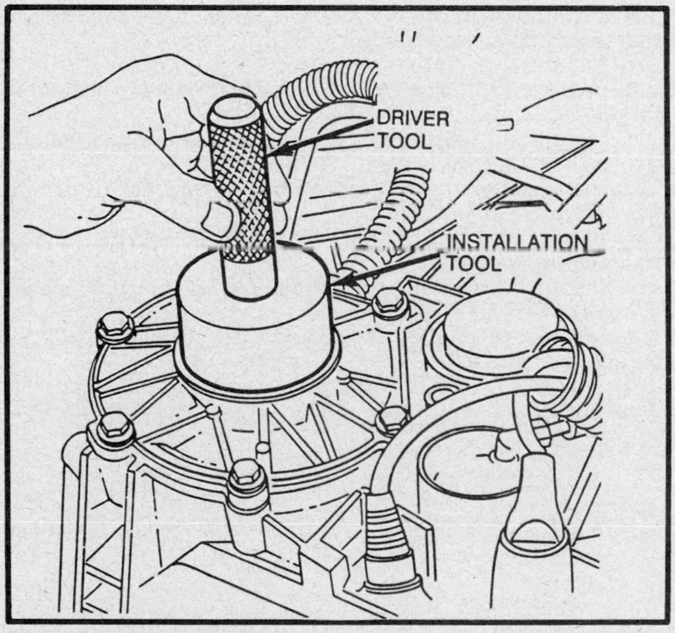
DRIVER
TOOL
INSTALLATION
TOOL
Rear plug replacement

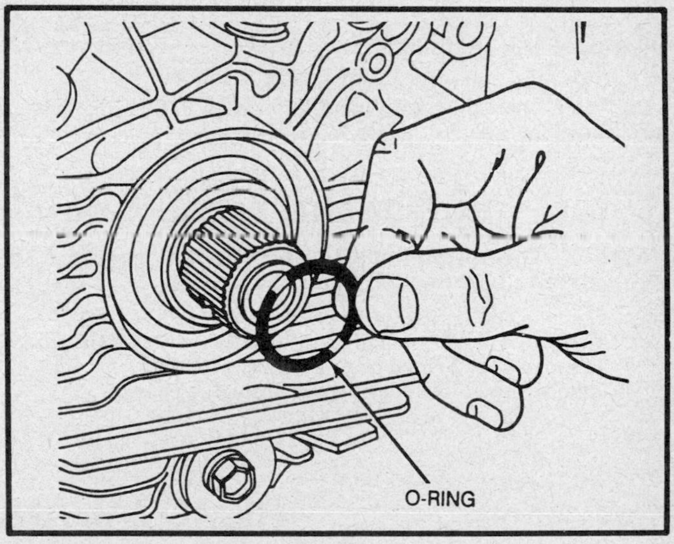
O-RING
Removing output shaft O-ring

2. Remove windshield washer bottle.
3. Unclip the strap that secures the transmission control unit to the inner fender panel.
4. Mark or tag the sensor harnesses for installation reference.
5. Disconnect the sensor harnesses from the transmission control unit.
6. Connect harnesses to new transmission control unit.
7. Replace the transmission control unit and windshield washer bottle.

OIL COOLER

Removal and Installation

1. Clamp off oil cooler hoses.
2. Remove the oil cooler bolts and remove cooler.
3. Install new O-rings on replacement cooler and new seals on cooler bolts.
4. Install cooler and torque bolts to 24 ft. lbs. (32 Nm).
5. Install oil cooler hoses and remove clamps.
6. Place hose clamps ends at a 6 o-clock position to avoid contacting the crossmember.
7. Check and add coolant as needed.

BREATHER VENT

Removal and Installation

1. Grip each vent base with pliers, twist and pull upward to remove it.
2. Unsnap cap off of new vent and start vent into case.
3. Tap vent into case with a hammer and small socket.
4. Snap cap back onto vent.

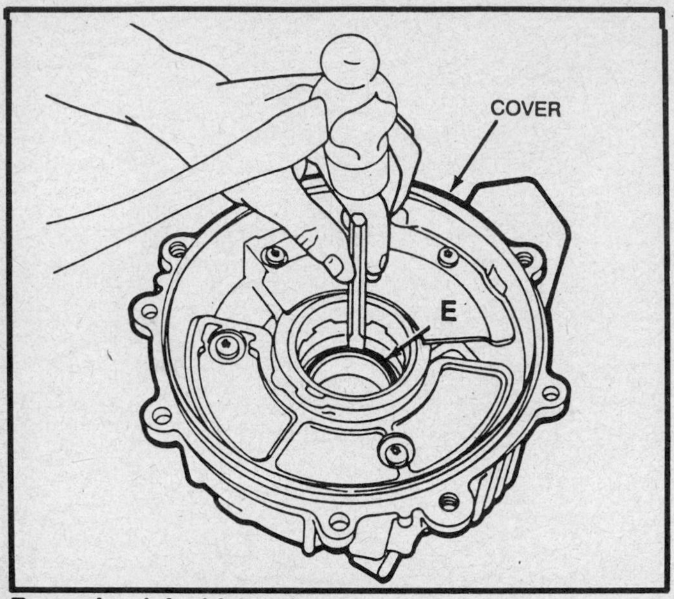
COVER
E

Removing left-side seal

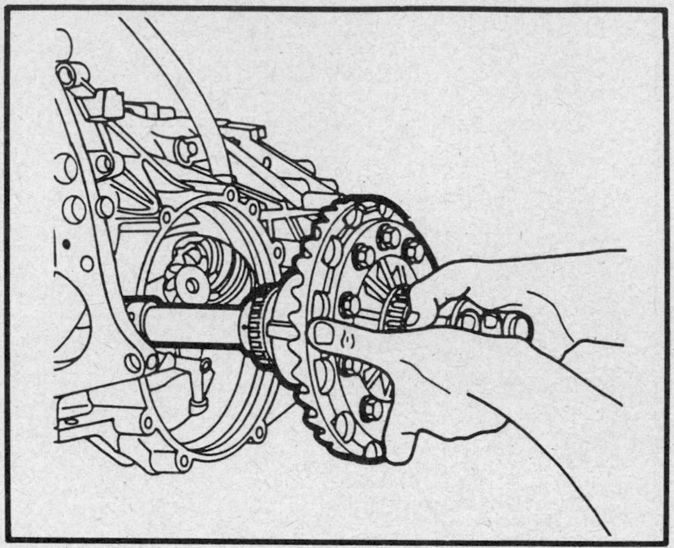

Differential

REAR PLUG

Removal and Installation

1. Tap plug out with a drift punch.
2. Install new plug with installation tool 6184 or equivalent and driver handle.

OUTPUT SHAFT SEAL

NOTE: The output shaft seal removal requires the transaxle to be removed from the vehicle.

1. Remove the transaxle from the vehicle.
2. Remove the bolts holding differential cover to the case.
3. Remove the O-rings from the output shafts.
4. Remove the differential cover from the case.
5. Remove the left-side shaft seal by tapping it out of the cover, using a hammer and drift.
6. Install new lubricated seal using a hammer and installation tool 6186 or equivalent.
7. Remove differential from case.
8. Remove right-side seal using slide hammer and tool J–29369–2 or an equivalent puller.
9. Install new lubricated seal using a hammer and installation tool 6185 or equivalent.
10. Install differential into case, taking care not to damage new seal.
11. Install new O-ring, lubricated with petroleum jelly, on right-side output shaft.
12. Install differential cover and tighten bolts to 150 inch lbs. (17 Nm).
13. Install new O-ring, lubricated with petroleum jelly, on leftside output shaft.
14. Install transaxle into vehicle, making all necessary adjustments.

REMOVAL AND INSTALLATION

TRANSAXLE REMOVAL

NOTE: Transaxle can be removed without removing the engine from the vehicle.

1. Disconnect the negative battery cable and all transaxle electrical connections.
2. Disconnect all the electrical connectors at the transmission control unit.

NOTE: Do not remove the sensors, these components will remain in place for transaxle removal.

3. Disconnect and plug transaxle cooler lines.
4. Remove the timing sensor.
5. Raise and safely support the vehicle.
6. Remove the upper strut bolt and only loosen the lower bolt. Tilt the steering knuckle outward.

NOTE: The strut bolts are splined just under the bolt head. Do not turn the bolt. Hold the bolt with a wrench and loosen the nuts as required.

7. Remove the underbody splash shield.
8. Remove drain plug and drain the transaxle. Replace plug when drained.
9. Remove the driveshaft retaining pin. Swing each rotor and steering knuckle outward and slide the driveshafts from the transaxle.
10. Remove the starter and heat shield.
11. Remove the converter housing covers. Remove the torque converter-to-driveplate bolts. Support the transaxle.
12. Remove the exhaust bracket.
13. Using a transmission jack, support the transaxle and remove the crossmember.
14. Disconnect the header pipes from the exhaust manifolds and the catalytic converter.
15. Disconnect the shift cable from the lever.
16. Remove the brace rod and the manual shift lever.

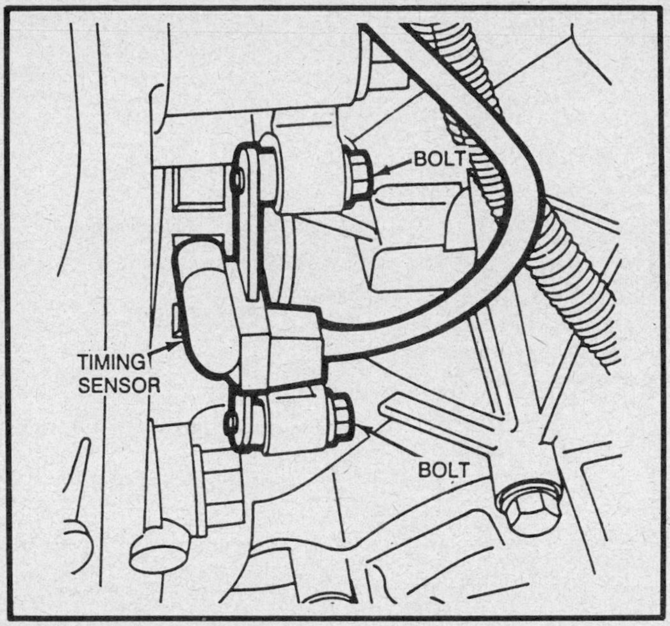
BOLT
TIMING SENSOR
BOLT
Engine timing sensor

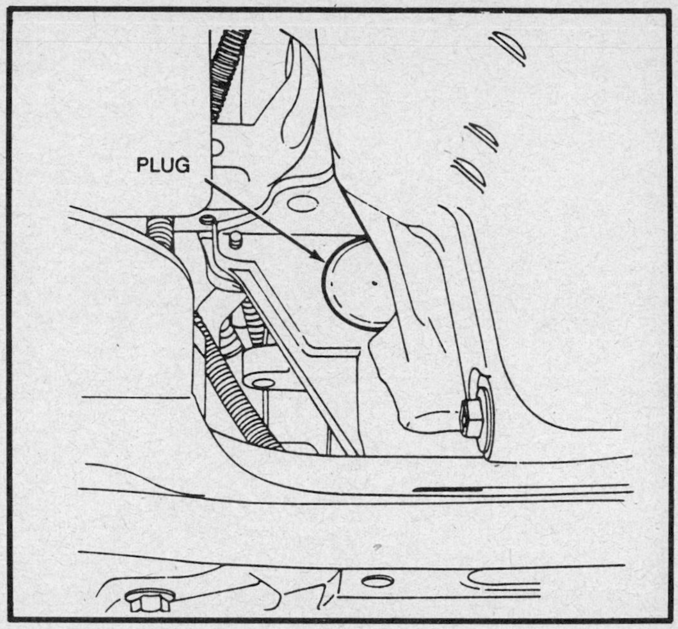
PLUG
Converter housing access plug

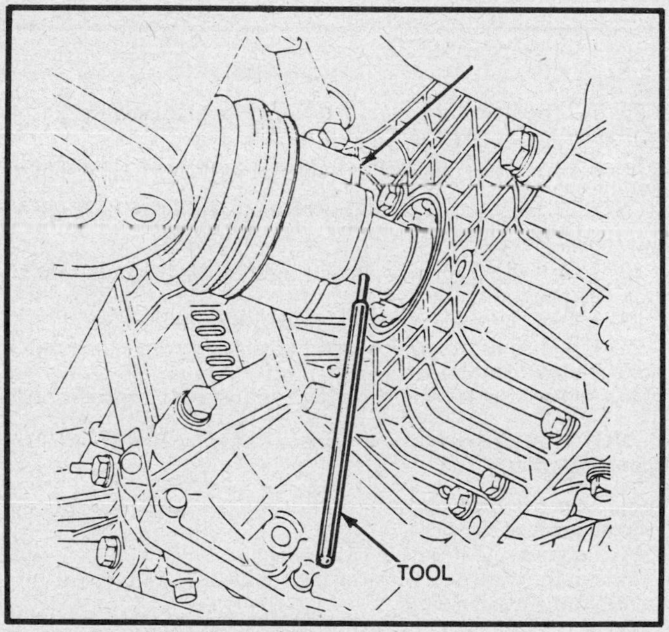
TOOL
Removing driveshaft roll pin

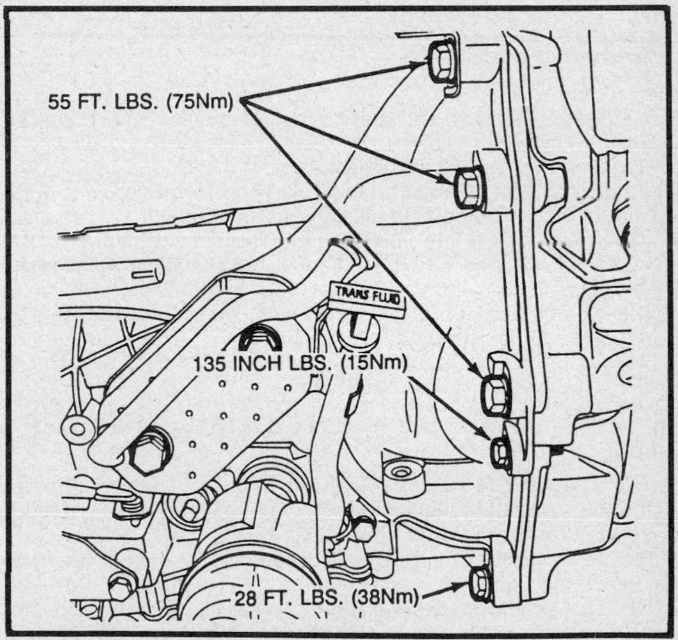
55 FT. LBS. (75Nm)
TRANS FLUID
135 INCH LBS. (15Nm)
28 FT. LBS. (38Nm)
Transaxle-to-engine bolt torque

17. Remove the brace rod bracket.
18. Remove the transaxle-to-engine bolts and pull the transaxle rearward and lower from vehicle.

TRANSAXLE INSTALLATION

1. Position the transaxle to the engine. Install the transaxle-to-engine bolts and tighten top bolts to 55 ft. lbs. (75 Nm).
2. Tighten lower bolt to 28 ft. lbs. (38 Nm) and tighten small bolt to 135 inch lbs. (15 Nm).

NOTE: Make certain the dowel pins are seated in the converter housing before tightening any bolts. Also be sure the converter is aligned in the driveplate trigger wheel.

3. Check that the torque converter rotates freely.
4. Install rear mount bracket and torque to 29 ft. lbs. (40 Nm).
5. Install the shift lever bracket and torque bolts to 32 ft. lbs. (43 Nm).
6. Install the manual shift lever and torque the bolt to 110 inch lbs. (12.5 Nm).
7. Install the shift cable bracket to case. Do not tighten at this time.
8. Install the brace rod.
9. Tighten the shift cable bracket bolts to 32 ft. lbs. (43 Nm) and tighten the the brace rod bolts to 185 inch lbs. (21 Nm).
10. Snap the shift cable onto the shift lever.
11. Install the crossmember and tighten the rear mount bolt

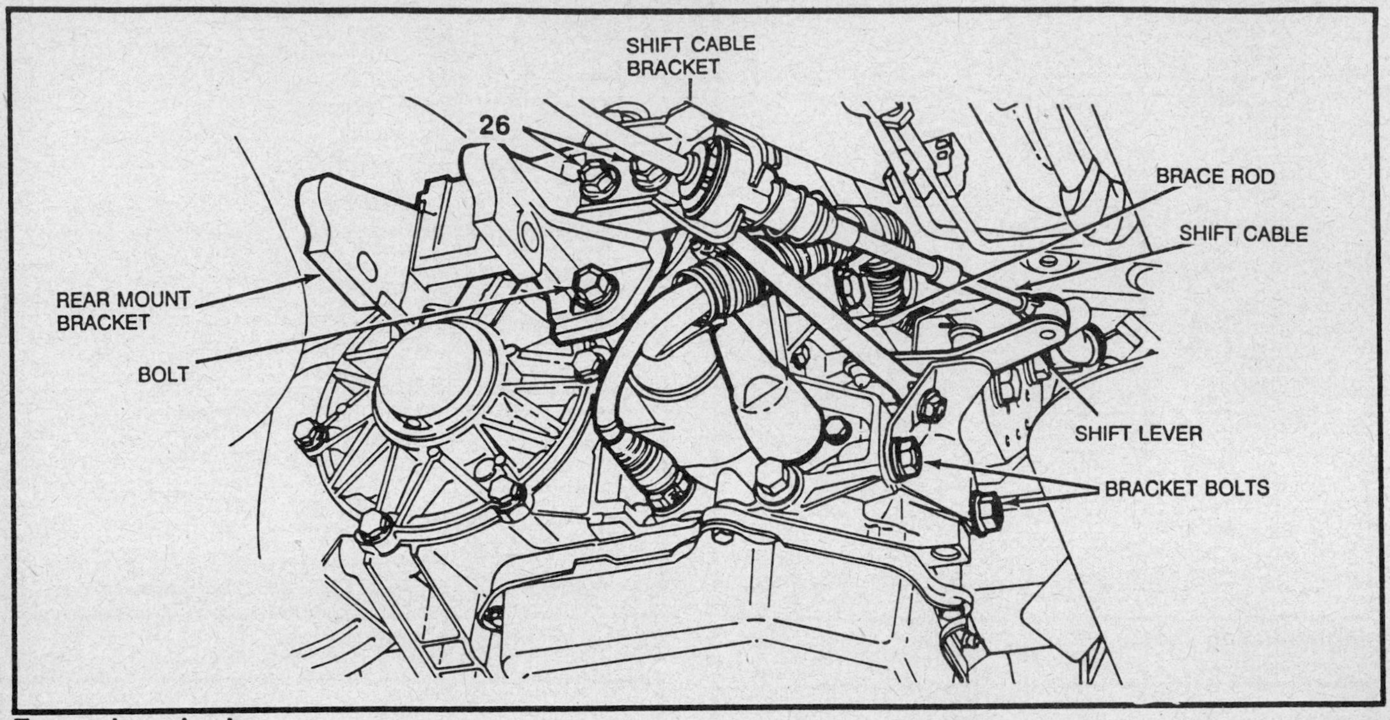
SHIFT CABLE
BRACKET
26
BRACE ROD
SHIFT CABLE
REAR MOUNT
BRACKET
BOLT
SHIFT LEVER
BRACKET BOLTS

Transaxle underview

to 49 ft. lbs. (67 Nm) and the crossmember bolts to 31 ft. lbs. (43 Nm).

12. Remove the transaxle jack.
13. Install the exhaust pipe bracket and torque bolts to 31 ft. lbs. (43 Nm). Clamp exhaust pipe to bracket.
14. Apply Loctite® to converter-to-driveplate bolts and tighten bolts to 25 ft. lbs. (33 Nm). Make certain to replace the access hole plug.
15. Before installing the heat shield, install and tighten the housing bolt to 135 inch lbs. (15 Nm).
16. Install the starter, starter heat shield and connect the exhaust head pipes to the manifolds.

NOTE: Connect the ground cable to the center starter bolt.

17. Tighten the starter mounting bolts to 31 ft. lbs. (43 Nm) and the wire nuts to 80 inch lbs. (9 Nm). Tighten the heat shield attaching nuts to 96 inch lbs. (11 Nm).
18. Make certain the transaxle output shafts O-rings are in the correct position.
19. Install the driveshafts, aligning the roll pin hole in the output shaft with the hole in the driveshaft and install roll pins.
20. Tilt the steering knuckles inward into place and tap the upper steering knuckle bolt and nuts into place.
21. Tighten steering knuckle bolts to 148 ft. lbs. (200 Nm).

NOTE: Do not turn the bolt head. Hold the bolt with a wrench and tighten bolt to specifications.

22. Install the front wheels and tighten the lug nuts to 62 ft. lbs. (85 Nm).
23. Install the coolant hoses and tighten the clamps.
24. Fill the differential section with the proper amount of 75W-140 gear lubricant.
25. Replace the transaxle wiring and install the splash shield.

NOTE: Make certain wiring is not near hot or rotating components.

26. Lower the vehicle and connect the transaxle transmission control unit wire connectors.
27. Connect the negative battery cable.
28. Fill the transaxle with the proper amount and type of automatic transmission fluid.
29. Check and adjust the shift cable.

BENCH OVERHAUL

At the time of this publication transaxle overhaul information was not available from the manufacturer.

SPECIFICATIONS

E1 Clutch	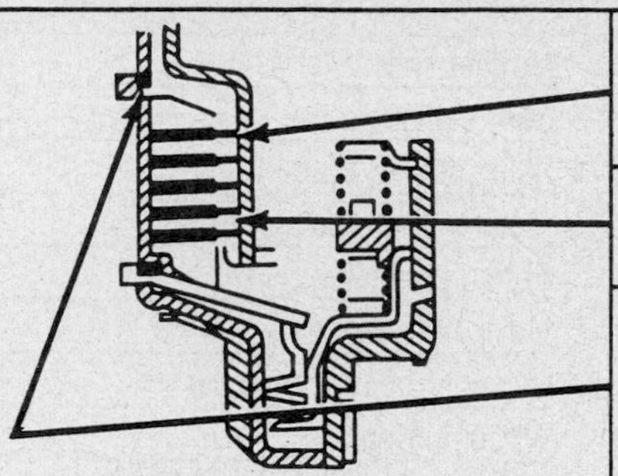	Lined Discs	4
		Intermediate flat discs	3
		End Play 1.4 - 1.6 mm (0.055 - 0.063 in) with spacer washer	
E2 Clutch	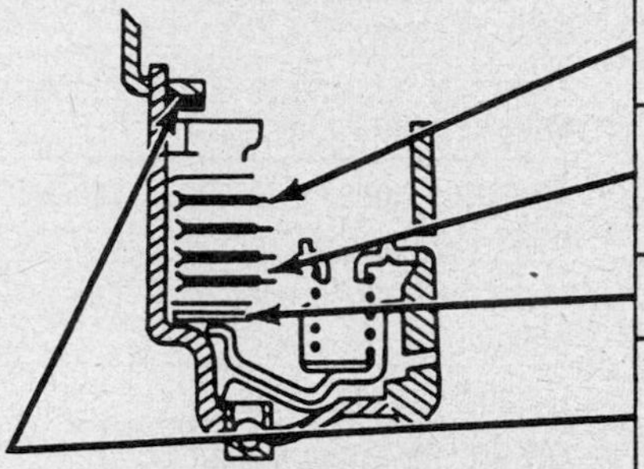	Lined discs	4
		Intermediate flat discs	5
		Spring disc	1
		End Play 1.0 - 1.4 mm (0.039 - 0.055 in) with spacer washer	
E3 Clutch	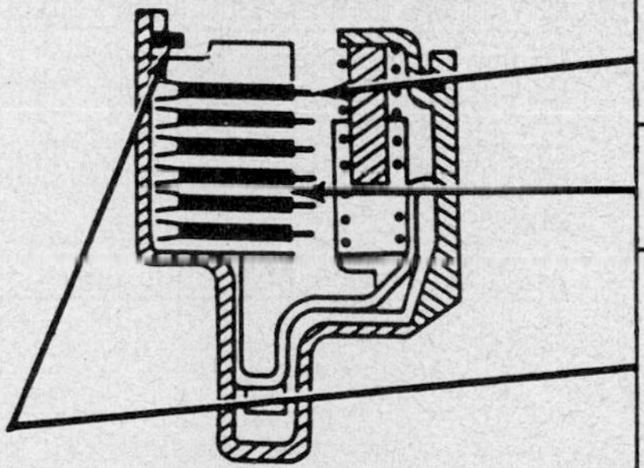	Lined discs	6
		Intermediate flat discs	6
		End Play 1.6 - 2.05 mm (0.063 - 0.081 in) with spacer washer	
F1 Brake	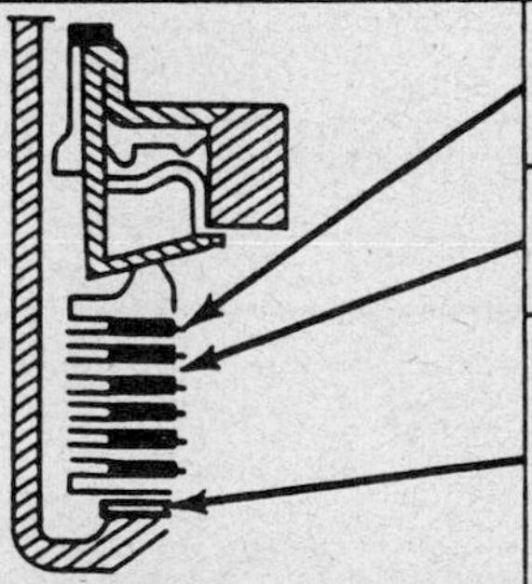	Lined discs	6
		Intermediate flat discs	6
		End Play 1.4 - 1.8 mm (0.055 - 0.070 in) with steel washer on housing side	
F2 Brake	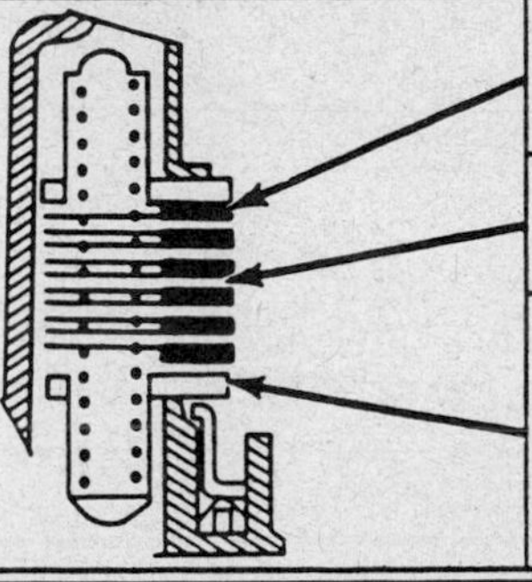	Lined discs	5
		Intermediate flat discs	4
		End Play 1.4 - 1.85 mm (0.055 - 0.073 in) with spacer plate on piston side	

TORQUE SPECIFICATIONS

Component	ft. lbs.	Nm
Differential housing drain full plugs	170–184①	19–21
Rear support bracket to transaxle case bolts	28–30	38–42
Starter shield nuts	90–102①	10–12
Rear support bracket to rear cushion bolt/nut	46–52	64–70
Exhaust pipe flange to manifold nuts	21–25	29–33
Starter wire harness nuts	75–85①	8.5–9.5
Exhaust pipe flange to catalytic cconverter nut and bolt	28–32	39–43
Starter attaching bolts	29–33	40–44
Oil cooler attaching bolts	22–26	30–34
Transmission oil pan drain plug	170–184①	19–21
Lower shock bracket nut	140–156	190–210
Rear cushion to rear crossmember nuts	22–26	30–34
Transaxle housing to engine block bolcks	52–58	71–79
Fill tube bracket to transaxle case bolt	142–158①	16–18
Starter to transaxle case bolts	29–33	41–44
Wiring harness clamp bolt	—	—
Oil screen retaining bolts	43–49①	4.5–5.5
TCU speed sensor barcket bolt	84–96①	9–11
Rear coverplate bolts	142–158①	16–18
Differential housing cover bolts	142–158①	16–18

Component	ft. lbs.	Nm
Shift bracket to transaxle case bolts	29–33	41–44
Manual shift lever bolt	96–124①	11–14
Shift cable bracket bolts	29–33	41–44
Shift bracket brace bolts	168–202①	19–22
Manual shaft lock plate bolt	84–96①	9–11
Detent plunger	170–184①	19–21
Valve body bolts	43–49①	4.5–5.5
Solenoid connector bolts	84–96①	9–11
Wheel lug nuts	59–65	81–89
Transaxle housing to engine block bolts (12mm × 1.25)	52–58	71–79
Transaxle housing to engine block bolts (10mm × 1.75)	26–30	35–41
Transaxle housing to engine block bolts (8m × 1.50)	152–168①	17–19
Crossmember to engine cradle bolts/nuts	29–33	41–44
Exhaust bracket bolts	29–33	41–44
Drive plate to crankshaft bolts	37–43 ±60°	51–57
Drive plate to torque converter bolts	23–27	31–35
Engine timing sensor bolts	68–76①	7.5–8.5
Transmission oil pan bolts	84–96①	9–11
Differential housing cover baffle plate bolt	84–96①	9–11
Road speed sensor bracket bolt	84–96①	9–11

① inch lbs.

SPECIAL TOOLS

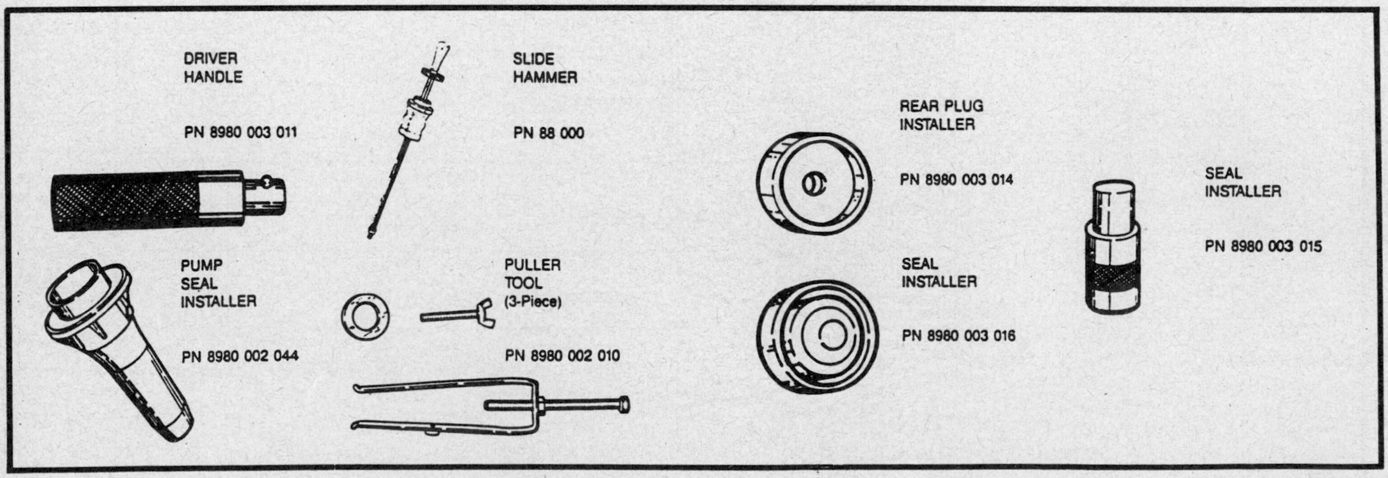
DRIVER HANDLE
PN 8980 003 011
SLIDE HAMMER
PN 88 000
REAR PLUG INSTALLER
PN 8980 003 014
SEAL INSTALLER
PN 8980 003 015
PUMP SEAL INSTALLER
PN 8980 002 044
PULLER TOOL (3-Piece)
PN 8980 002 010
SEAL INSTALLER
PN 8980 003 016

Section 3

A413, A415, A470 & A670 Transaxles

Chrysler Corp.

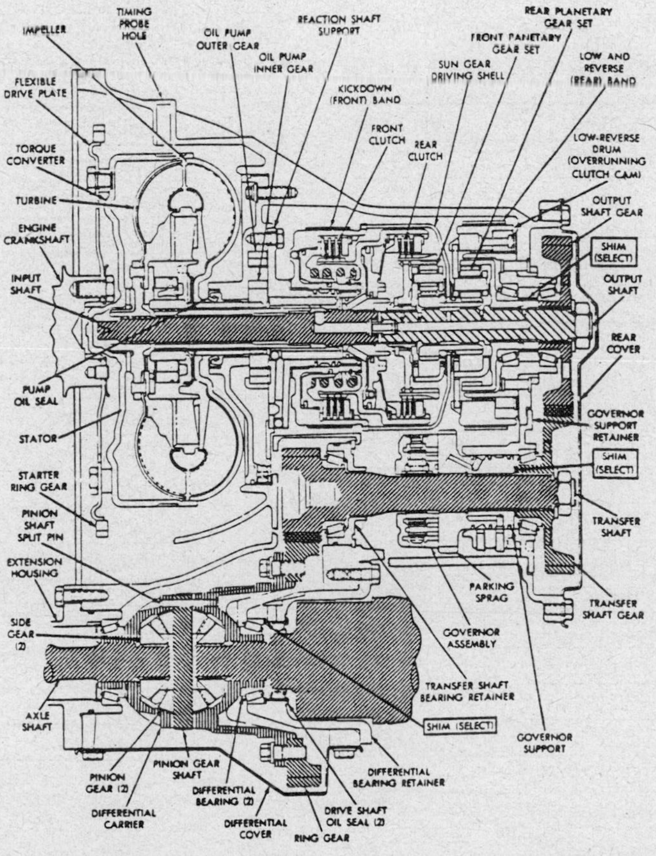

IMPELLER
TIMING PROBE HOLE
OIL PUMP OUTER GEAR
OIL PUMP INNER GEAR
REACTION SHAFT SUPPORT
KICKDOWN (FRONT) BAND
FRONT CLUTCH
REAR CLUTCH
SUN GEAR DRIVING SHELL
FRONT PANETARY GEAR SET
REAR PLANETARY GEAR SET
LOW AND REVERSE (REAR) BAND
FLEXIBLE DRIVE PLATE
TORQUE CONVERTER
TURBINE
ENGINE CRANKSHAFT
INPUT SHAFT
LOW-REVERSE DRUM (OVERRUNNING CLUTCH CAM)
OUTPUT SHAFT GEAR
SHIM (SELECT)
OUTPUT SHAFT
REAR COVER
PUMP OIL SEAL
STATOR
GOVERNOR SUPPORT RETAINER
SHIM (SELECT)
STARTER RING GEAR
PINION SHAFT SPLIT PIN
TRANSFER SHAFT
EXTENSION HOUSING
PARKING SPRAG
TRANSFER SHAFT GEAR
SIDE GEAR (2)
GOVERNOR ASSEMBLY
TRANSFER SHAFT BEARING RETAINER
AXLE SHAFT
SHIM (SELECT)
GOVERNOR SUPPORT
PINION GEAR SHAFT
PINION GEAR (2)
DIFFERENTIAL BEARING (2)
DIFFERENTIAL BEARING RETAINER
DIFFERENTIAL CARRIER
DIFFERENTIAL COVER
DRIVE SHAFT OIL SEAL (2)
RING GEAR

Cross section of Chrysler A–413 automatic transaxle

REFERENCES

APPLICATION

1980–84 Chilton's Professional Automatic Transmission Manual #7390
1984–89 Chilton's Professional Transmission Manual—Domestic Vehicles #7959

GENERAL DESCRIPTION

1980–84 Chilton's Professional Automatic Transmission Manual #7390

MODIFICATIONS

1980–84 Chilton's Professional Automatic Transmission Manual #7390
1984–89 Chilton's Professional Transmission Manual—Domestic Vehicles #7959

TROUBLE DIAGNOSIS

1980–84 Chilton's Professional Automatic Transmission Manual #7390

ON CAR SERVICES

1980–84 Chilton's Professional Automatic Transmission Manual #7390

REMOVAL AND INSTALLATION

1980–84 Chilton's Professional Automatic Transmission Manual #7390
1984–89 Chilton's Professional Transmission Manual—Domestic Vehicles #7959

BENCH OVERHAUL

1980–84 Chilton's Professional Automatic Transmission Manual #7390

SPECIFICATIONS

1980–84 Chilton's Professional Automatic Transmission Manual #7390
1984–89 Chilton's Professional Transmission Manual—Domestic Vehicles #7959

SPECIAL TOOLS

1980–84 Chilton's Professional Automatic Transmission Manual #7390

APPLICATION

A-413, A-415 AND A-470 AUTOMATIC TRANSAXLE

Year	Vehicle	Engine	Vin Code
1984	Aries	2.2L/2.6L	D
	Caravan	2.2L	K
	Carvelle	2.2L/2.6L	E
	Charger	1.6L/2.2L	Z
	Daytona	2.2L/2.6L	V
	E-Class	2.2L/2.6L	T
	Horizon	1.6L/2.2L	M
	Laser	2.2L/2.6L	C
	LeBaron	2.2L/2.6L	C
	Mini Ram Van	2.2L/2.6L	K
	New Yorker	2.2L/2.6L	T
	Omni	1.6L/2.2L	Z
	Rampage	1.6L/2.2L	Z
	Reliant	2.2L/2.6L	P
	Turismo	1.6L/2.2L	M
	Voyager	2.2L	H
	600	2.2L/2.6L	E
1985	Aries	2.2L/2.6L	D
	Caravan	2.2L	K
	Carvelle	2.2L/2.6L	J
	Carvelle	2.2L/2.6L	L
1985	Charger	1.6L/2.2L	Z
	Daytona	2.2L/2.6L	V
	Horizon	1.6L/2.2L	M
	Lancer	2.2L/2.6L	D
	Laser	2.2L/2.6L	C
	LeBaron	2.2L/2.6L	C
	Mini Ram Van	2.2L/2.6L	K
	New Yorker	2.2L/2.6L	T
	Omni	1.6L/2.2L	Z
	Reliant	2.2L/2.6L	P
	Turismo	1.6L/2.2L	M
	Voyager	2.2L	H
	600	2.2L/2.6L	E
	600	2.2L/2.6L	V
1986	Aries	2.2L/2.6L	D
	Caravan	2.2L	K
	Carvelle	2.2L/2.6L	J
	Charger	1.6L/2.2L	Z
	Daytona	2.2L/2.6L	V
	Horizon	1.6L/2.2L	M
	Lancer	2.2L/2.6L	D

A-413, A-415 AND A-470 AUTOMATIC TRANSAXLE

Year	Vehicle	Engine	Vin Code
1986	Laser	2.2L/2.6L	C
	LeBaron	2.2L/2.6L	C
	Mini Ram Van	2.2L/2.6L	K
	New Yorker	2.2L/2.6L	T
	Omni	1.6L/2.2L	Z
	Reliant	2.2L/2.6L	P
	Turismo	1.6L/2.2L	M
	Voyager	2.2L	H
	600	2.2L/2.6L	E
	600	2.2L/2.6L	V
1987	Aries	2.2L/2.5L	D
	Caravan	2.2L	K
	Carvelle	2.2L/2.5L	J
	Charger	2.2L	Z
	Daytona	2.2L/2.5L	V
	Horizon	2.2L	M
	Lancer	2.2L/2.5L	D
	LeBaron	2.2L/2.5L	C
	Mini Ram Van	2.2L/2.6L	K
	New Yorker	2.2L/2.5L	T
	Omni	2.2L	Z
	Reliant	2.2L/2.5L	P
	Shadow	2.2L	D
	Sundance	2.2L	P
	Turismo	2.2L	M
	Voyager	2.2L	H
	600	2.2L/2.5L	E
1988	Aries	2.2L/2.5L	D
	Caravan	2.5L/3.0L	K
	Carvelle	2.2L/2.5L	J

Year	Vehicle	Engine	Vin Code
1988	Daytona	2.2L/2.5L	V
	Dynasty	2.5L/3.0L	D
	Horizon	2.2L	M
	Lancer	2.2L/2.5L	D
	LeBaron	2.2L/2.5L	C
	Mini Ram Van	2.5L/3.0L	K
	New Yorker	2.5L/3.0L	C
	Omni	2.2L	Z
	Reliant	2.2L/2.5L	P
	Shadow	2.2L	D
	Sundance	2.2L	P
	Voyager	2.5L/3.0L	H
	600	2.2L/2.5L	E
1989	Acclaim	2.5L/3.0L	P
	Aries	2.2L/2.5L	D
	Caravan	2.5L/3.0L	K
	Daytona	2.2L/2.5L	V
	Dynasty	2.5L/3.0L	D
	Horizon	2.2L	M
	Lancer	2.2L/2.5L	D
	LeBaron	2.2L/2.5L	C
	Mini Ram Van	2.5L/3.0L	K
	New Yorker	2.5L/3.0L	C
	Omni	2.2L	Z
	Reliant	2.2L/2.5L	P
	Shadow	2.2L/2.5L	D
	Spirit	2.5L/3.0L	D
	Sundance	2.2L/2.5L	P
	Voyager	2.5L/3.0L	H

TRANSAXLE MODIFICATIONS

1987–88 Harsh or Abrupt Engagement In Drive

When the vehicle selector lever is placed into the **D** detent from any other position, the engagement is noticeably harsh. This condition can be aggravated by idle speeds higher than specified on the emissions label for warm curb idle.

Diagnosis

1. Park the vehicle on a level surface. Set the parking brake and block the drive wheels.
2. Connect a suitable tachometer to the engine. Start the vehicle and let it run until it reaches normal operating temperature.
3. Verify the correct engine idle speed per vehicle emissions label.
4. Check the transaxle oil level. Check for proper engine mount isolator adjustment.
5. Shift the vehicle into **D** at curb idle speed several times. Allow 30 seconds of idle in **N** between engagements.
6. If unusually harsh or abrupt shift engagements are evident, use the following repair procedure.

Procedure

After using the diagnosis procedure, the condition is caused by the rear clutch discs (green or olive, drab in color) which are not up to specifications. These clutch discs must be replaced to correct the condition. The parts required for this repair are new rear clutch discs (brown friction material only) 3 or 4 depending on the transaxle application.

1. Using this section, remove and disassemble the transaxle to access the rear clutch assembly.
2. Remove the rear clutch pack and replace with new clutch pack (part number 5224084 brown friction material only). Reset the rear clutch clearance to 0.026–0.043 in. (0.67–1.10mm).
3. Using this section, reassemble the transaxle assembly.
4. Install the transaxle into the vehicle. Lower the vehicle and check the engine and transaxle alignment.
5. Refill the transaxle to the correct level with the proper automatic transmission fluid.
6. Check the vehicle for smooth engagements into drive. Road test the vehicle to verify proper transaxle operation.

1987–88 Except the Horizon/Turismo And Omni/Charger Equipped With 2.5L and 3.0L Engine

LOCKUP TORQUE CONVERTER SURGE/ BUCK AT 33–35 MPH

A surge may be noticed at steady cruising speeds above 35 mph due to the lockup torque converter. In addition, bucking may occur under light deceleration conditions, especially when cresting a hill.

Diagnosis

This condition may be diagnosed by disconnecting the transaxle lockup torque converter solenoid and driving the vehicle at a steady cruising speed above 35 mph. If the problem goes away, use the following repair procedure.

Procedure

The following procedure involves disconnecting the lockup torque converter solenoid. It should be noted that after this repair has been made, a loss in fuel economy of 1–2 miles per gallon may be noticed, depending on the driving conditions. The following parts (Driveability Kit part number 4419447) will be required to insure a proper repair.

a. Wiring jumper assembly, part number 4400768
b. Lockup solenoid cap, part number 4419446
c. Tie wrap, part number 6015756
d. Authorized modification label, part number 4275086.

1. Disconnect the wiring connector from the lockup torque converter solenoid. This is located directly behind the transaxle dipstick.
2. Install the wiring jumper assembly to the lockup solenoid wiring connector that was previously removed.
3. On all models equipped with the 2.5L engine, reroute the new wiring assembly (jumper plugged into the original lockup wiring) under the battery so that it loops back along the original lockup wiring. Tie wrap the loose harness to the original lockup wiring in 2 places.
4. On all LeBaron GTS and Vans equipped with the 3.0L engine, reroute the new wiring assembly (jumper plugged into the original lockup wiring) so that it loops back along the original lockup wiring. Tie wrap the loose harness to the original lockup wiring in 2 places.
5. Install supplied lockup solenoid cap, onto the lockup solenoid.
6. Type the information on the authorized modification label and attach it next to the vehicle emissions control information label.

REMOVAL AND INSTALLATION

TRANSAXLE REMOVAL

While the removal of the transaxle does not require the removal of the engine, it should be noted that care must be used to prevent damage to the converter drive plate. The drive plate will not support any weight, so the transaxle and converter must be removed as an assembly. Do not let any of the weight of the transaxle or converter rest.

1. Disconnect the negative battery cable.
2. Disconnect the throttle and shift linkage from the transaxle levers.
3. If equipped with a lockup torque converter, disconnect the lockup solenoid connector, located near the dipstick tube.
4. With the vehicle on the floor and foot brakes applied, loosen the hub and wheel nuts. Remove the upper and lower cooler lines at the transaxle.
5. Position an engine support fixture to the engine and tighten to equalize the engine weight from the mounts.

NOTE: This step can be done later in the procedure, at the discretion of the technician.

6. Remove the upper bolts of the bell housing.
7. Raise and support the vehicle safely. Remove the wheel and tire assemblies and the hub nut with cotter pin, lock and spring washer from the driveshaft.
8. From under the left front fender, remove the inner fender splash panel.
9. Remove the front halfshaft from the vehicle as follows:
   a. For removal of the right halfshaft, the speedometer pinion must be removed, before the shaft removal.
   b. Remove the clamp bolt securing the ball joint stud into the steering knuckle.
   c. Separate the ball joint stud from the steering knuckle by prying against the knuckle leg and control arm. Do not damage the ball joint or CV-joint boots.
   d. Separate the outer CV-joint splined shaft from the hub by holding the CV housing while moving the knuckle (hub) assembly away. Do not pry on or other wise damage wear sleeve on the outer CV-joint.
   e. Support assembly at the CV-joint housing. Remove by pulling outward on the inner joint housing.

The driveshaft, when installed, acts as a bolt and secures the hub/bearing assembly. If the vehicle is to be supported or moved on its wheels, install a bolt through the hub to insure that the hub bearing assembly cannot loosen.

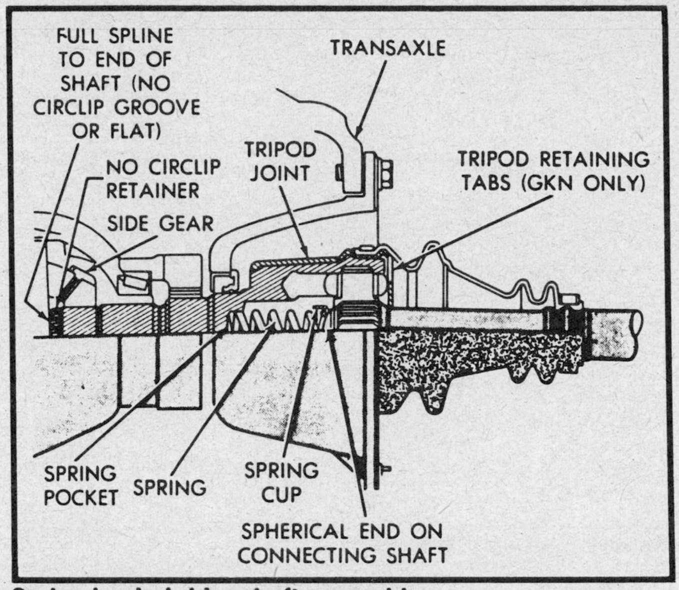
FULL SPLINE
TO END OF
SHAFT (NO
CIRCLIP GROOVE
OR FLAT)
TRANSAXLE
NO CIRCLIP
RETAINER
TRIPOD
JOINT
TRIPOD RETAINING
TABS (GKN ONLY)
SIDE GEAR
SPRING
POCKET
SPRING
SPRING
CUP
SPHERICAL END ON
CONNECTING SHAFT
Spring loaded driveshaft assembly

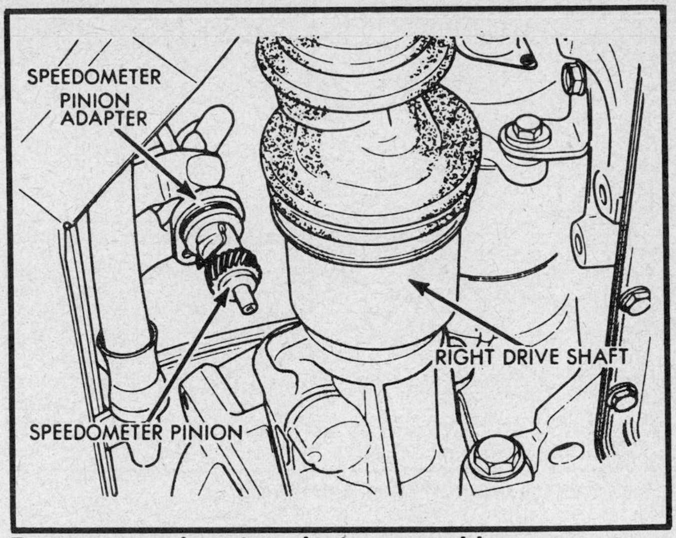
SPEEDOMETER
PINION
ADAPTER
RIGHT DRIVE SHAFT
SPEEDOMETER PINION
Remove speedometer adapter assembly

CAUTION

Mishandling of the driveshaft assemblies, such as allowing the assemblies to dangle unsupported, pulling or pushing the ends can result in pinched rubber boots or damage to the CV joints. Boot sealing is vital to retaining special joint lubricants and to prevent contaminents from entering the joint areas.

10. Remove the converter dust cover, mark the torque converter and drive plate relationship and remove the torque converter mounting bolts.
11. To rotate the engine, remove the access plug located in the right fender splash panel, and use a suitable socket/rachet assembly to turn the engine crankshaft.
12. Remove the connector from the neutral starter switch. Certain vehicles may require the removal of the lower cooler pipe at this time, if not accomplished earlier.
13. Remove the engine mount bracket from the front cross member.
14. Remove the front mount insulator through bolt and front engine mount bolts. Remove the upper bell housing bolts.

NOTE: It will be necessary to adjust the engine support to obtain the zero clearance needed to relieve the pressure from the mounts and bolts.

18. Place the removal jack under the transaxle, remove the left engine mount and the long bolt through the mount.
19. Remove the starter assembly and remove the lower bell housing bolts. Pry against the engine and lower transaxle, being careful of the torque converter.

TRANSAXLE INSTALLATION

Installation of the transaxle is a reversal of the removal procedure but remember to fill the differential with the proper automatic transmission fluid, if equipped with separate sump, before lowering the vehicle. Install the halfshafts as follows:
1. Hold the inner joint assembly at the housing while aligning and guiding the inner joint spline into the transaxle or intermediate shaft assembly.

NOTE: On turbocharged models, be sure that the rubber washer seal is in place on the right inner CV-joint. On all vehicles be sure to clean and lubricate all seals and the wear sleeve.

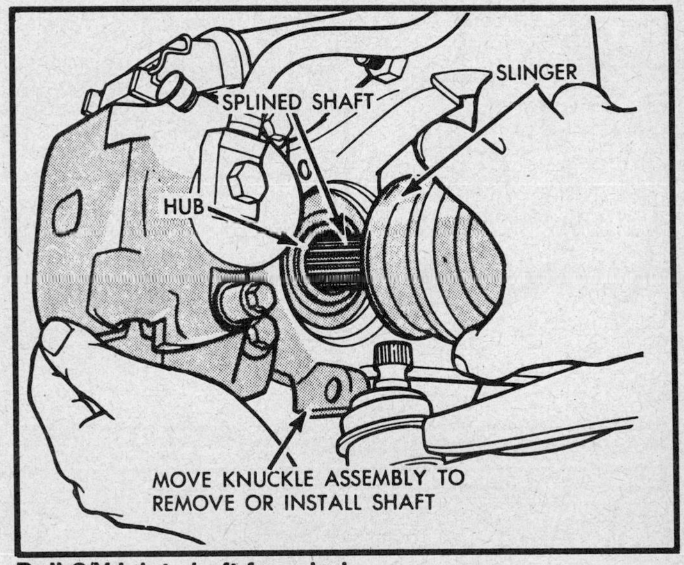
SLINGER
SPLINED SHAFT
HUB
MOVE KNUCKLE ASSEMBLY TO
REMOVE OR INSTALL SHAFT
Pull C/V joint shaft from hub

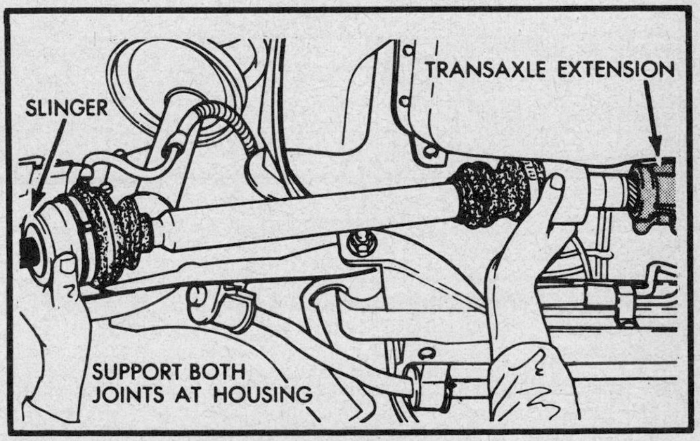
TRANSAXLE EXTENSION
SLINGER
SUPPORT BOTH
JOINTS AT HOUSING
Removing driveshaft assembly

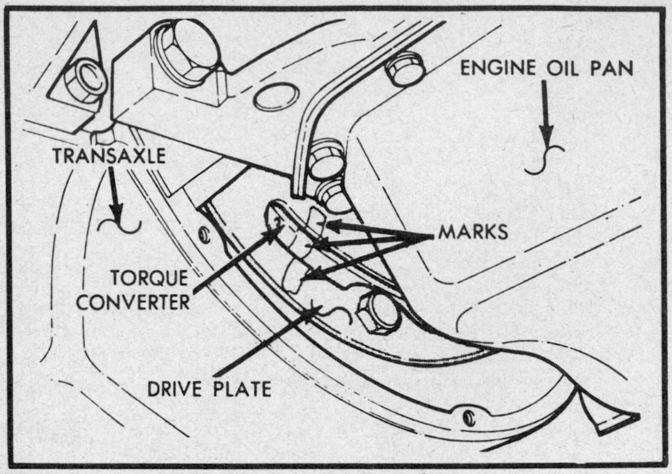
ENGINE OIL PAN
TRANSAXLE
MARKS
TORQUE CONVERTER
DRIVE PLATE

Mark torque converter and drive plate

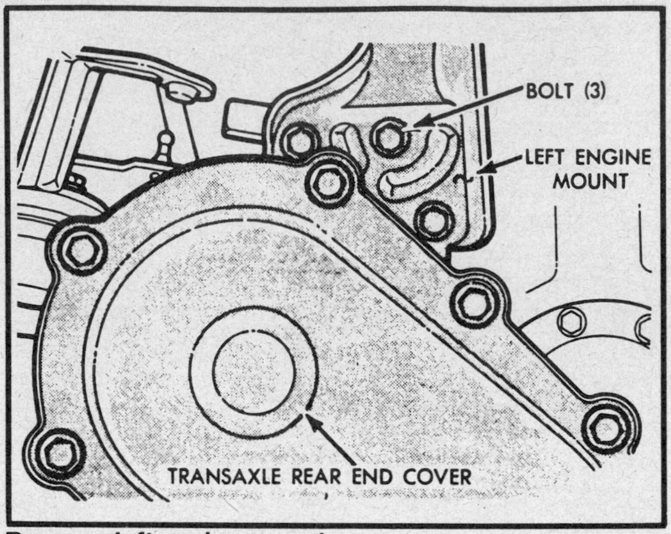
BOLT (3)
LEFT ENGINE MOUNT
TRANSAXLE REAR END COVER

Remove left engine mount

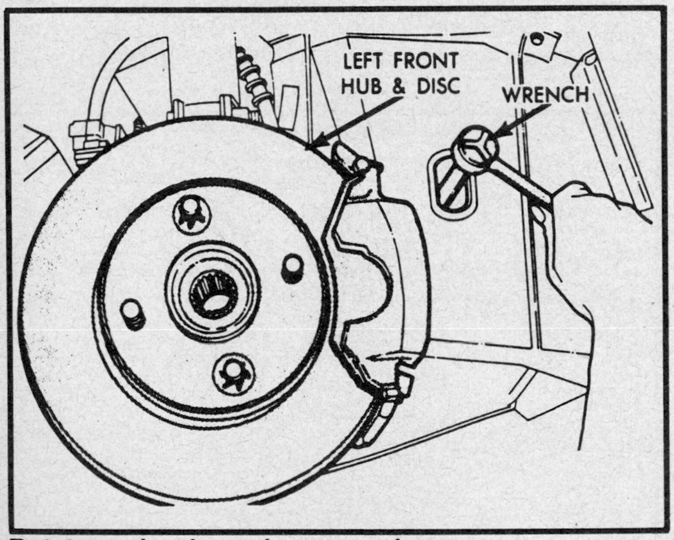
LEFT FRONT HUB & DISC
WRENCH

Rotate engine through access plug

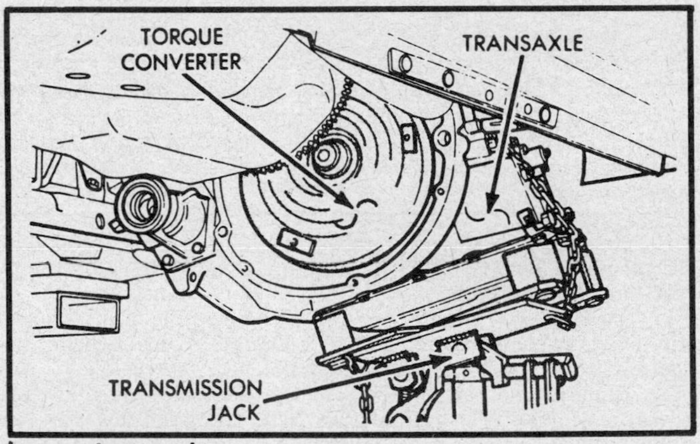
TORQUE CONVERTER
TRANSAXLE
TRANSMISSION JACK

Lower transaxle

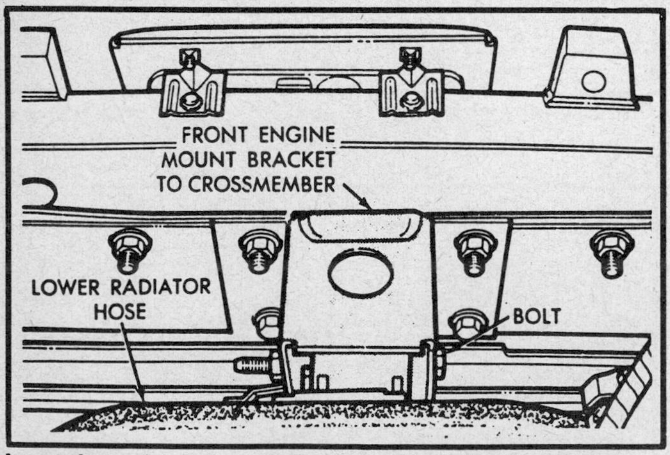
FRONT ENGINE MOUNT BRACKET TO CROSSMEMBER
LOWER RADIATOR HOSE
BOLT

Location of front engine mount bracket to cross member

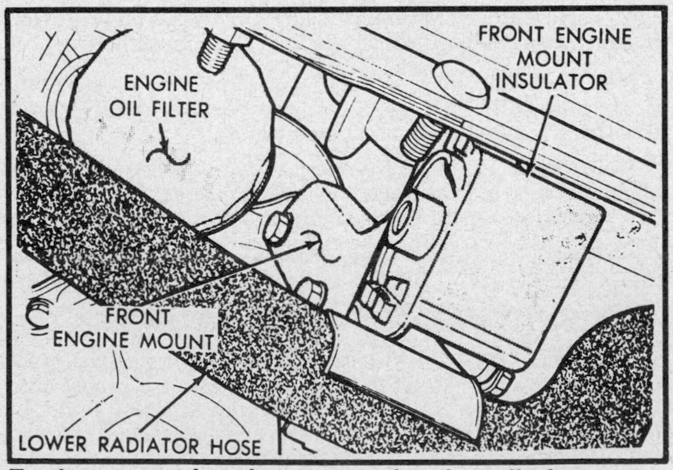
FRONT ENGINE MOUNT INSULATOR
ENGINE OIL FILTER
FRONT ENGINE MOUNT
LOWER RADIATOR HOSE

Engine mount bracket removal or installation

2. Apply a bead ¼ in. (6 mm) wide of multi purpose lubriant to the full circumference of the wear sleeve to seal the contact area. Fill the lip to the housing cavity on the seal, complete circumference and wet the seal ip with lubricant.

3. Push the knuckle hub assembly out and install the splined outer CV-joint into the hub.

d. Reinstall the knuckle assembly onto the ball joint stud. The steering knuckle clamp bolt is a prevailing torque type,

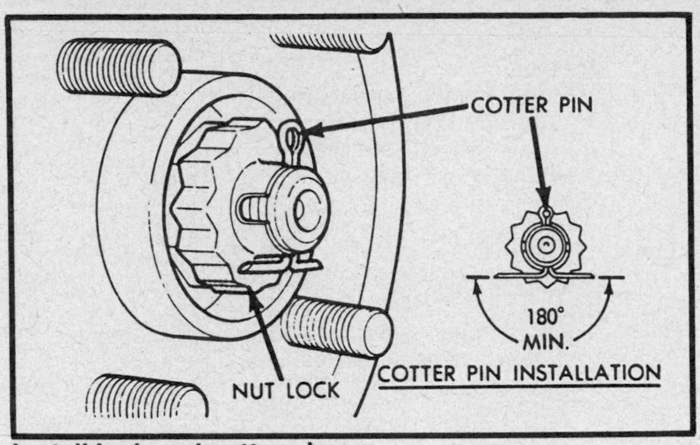
COTTER PIN
180° MIN.
NUT LOCK
COTTER PIN INSTALLATION

Install lock and cotter pin

original or equivalent bolt must be reinstalled during reassembly.

4. Install and torque the steering knuckle clamp bolt to 70 ft. lbs (95 Nm).

5. Install the speedometer pinion. Be sure the differential is filled with the proper lubricant.

6. Install the hub nut assembly.

7. If after installing the halfshaft assembly the inboard boot appears collasped or deformed, vent the inner boot by inserting, a round tipped small diameter rod between the boot and the shaft. If necessary, massage the boot to remove all wrinkles being careful not to allow dirt to enter or grease to the leave the boot cavity. If the boot is clamped to the shaft with a rubber garter clamp, it need not be removed to perform this venting operation.

8. If the boot is clamped to the shaft using a metal clamp, the clamp must be removed and discarded before the rod can be inserted. After venting, install a new service clamp, (part number 5212720) or equivalent.

9. Install the hub nut washer and nut and torque the hub nut to 180 ft.lbs. (245 Nm). Install the lock, spring washer and new cotter pin. Wrap the cotter pin prongs tightly around the nut lock.

The differential cover is not installed with a gasket, but with RTV sealant in a ribbon about ⅛ in. wide. The screws are torqued to 165 inch lbs. (19 Nm). Adjust the gearshift and throttle linkage. Refill transaxle with the proper automatic transmission fluid. Road test the vehicle.

SPECIFICATIONS

Item	Qty.	Thread Size	Torque Newton-meters	Torque Inch-Pounds	Torque Foot-Pounds
Automatic Transaxle:					
Bolt—Bell Housing Cover	3	9.8-M6-1-10	12	105	—
Bolt—Flex Plate to Crank	8	M10×1.5×18	95	—	70
Bolt—Flex Plate to Torque Converter	4	10.9-M10×1.5×13.2	74	—	55
Screw Assy. Transaxle to Cyl. Block	3	9.8A-M12-1.75-65	95	—	70
Screw Assy. Lower Bell Housing Cover	3	9.8-M6-1-10	12	105	—
Screw Assy. Manual Control Lever	1	9.8A-M6-1-35	12	105	—
Screw Assy. Speedometer to Extension	1	9.8A-M6-1-14	7	60	—
Connector, Cooler Hose to Radiator	2	1/8-27 NPTF	12	110	—
Bolt—Starter to Transaxle Bell Housing	3	M10-1.5-30	54	—	40
Bolt—Throttle Cable to Transaxle Case	1	M6-1.0-14	12	105	—
Bolt—Throttle Lever to Transaxle Shaft	1	M6-1-25	12	105	—
Bolt—Manual Cable to Transaxle Case	1	M8-1.75-30	28	250	—
Bolt—Front Motor Mount	2	M10	54	—	40
Bolt—Left Motor Mount	3	M10-1.5-25	54	—	40
Dress Up:					
Connector Assembly, Cooler Line	2	M12-1.75-122	28	250	—
Plug, Pressure Check	7	1/16-27NPTF	5	45	—
Switch, Neutral Safety	1	3/4-16UNF	34	—	25
Differential Area:					
Ring Gear Screw	12	12.9-M10-1.5-25	95	—	70
Bolt, Extension to Case	4	9.8-M8-1.25-28	28	250	—
Bolt, Differential Bearing Retainer to Case	6	9.8-M8-1.25-28	28	250	—
Screw Assy. Differential Cover to Case	10	9.8-M8-1.25-16	19	165	—
Transfer & Output Shaft Areas:					
Nut, Output Shaft	1	M20-1.5	271	—	200
Nut, Transfer Shaft	1	M20-1.5	271	—	200
Bolt, Gov to Support	2	9.8-M5-0.8-20	7	60	—
Bolt, Gov to Support	1	9.8-M5-0.8-30	7	60	—
Screw Assy., Governor Counterweight	1	M8-1.25-35	28	250	—
Screw Assy., Rear Cover to Case	10	9.8-M8-1.25-16	19	165	—
Plug, Reverse Band Shaft	1	1/4-18-NPTF	7	60	—
Pump & Kickdown Band Areas:					
Bolt, Reaction Shaft Assembly	6	9.8-M8-1.25-19	28	250	—
Bolt Assy., Pump to Case	7	10.9-M8-1.25-25	31	275	—
Nut, Kickdown Band Adjustment Lock	1	M12-1.75	47	—	35
Valve Body & Sprag Areas:					
Bolt, Sprag Retainer to Transfer Case	2	9.8-M8-1.25-23	28	250	—
Screw Assy., Valve Body	16	9.8A-M5-0.8-11	5	40	—
Screw Assy., Transfer Plate	16	9.8A-M5-0.8-25	5	40	—
Screw Assy., Filter	2	9.8A-M5-0.8-30	5	40	—
Screw, Transfer Plate to Case	7	9.8-M6-1-30	12	105	—
Screw Assy., Oil Pan to Case	14	9.8-M8-1.25-16	19	165	—
Nut, Reverse Band Adjusting Lock	1	M8-1.25	14	120	—

		Metric Measure	U.S. Measure
Type		Automatic Three Speed With Torque Converter and Integral Differential	
Torque Converter Diameter		241 millimeters	9.48 inches
Oil Capacity—Transaxle and Torque Converter:			
except fleet		8.4 Liters	8.9 qts.
fleet only		8.7 Liters	9.2 qts.
Use MOPAR ATF Automatic Transmission Fluid Type 7176 (or DEXRON II)			
Cooling Method		Water-Heat Exchanger and/or oil-to-air heat exchanger	
Lubrication		Pump (Internal-External Gear Type)	
Gear Ratios:			
Transmission Portion: First		2.69	
Second		1.55	
Third		1.00	
Reverse		2.10	
Pump Clearances:		**(Millimeter)**	**(Inch)**
Outer Gear to Pocket		.045-.141	.0018-.0056
Outer Gear Side Clearance		.020-.046	.0008-.0018
Inner Gear Side Clearance		.020-.046	.0008-.0018
End Play:		**(Millimeter)**	**(Inch)**
Input Shaft		.19-1.50	.008-.060
Front Clutch Retainer		.76-2.69	.030-.106
Front Carrier		.89-1.45	.007-.057
Front Annulus Gear		.09-0.50	.0035-.020
Planet Pinion		.15-0.59	.006-.023
Reverse Drum		.76-3.36	.030-.132
Clutch Clearance and Selective Snap Rings:		**(Millimeter)**	**(Inch)**
Front Clutch (Non-Adjustable) Measured from Reaction Plate to "Farthest" Wave	3 Disc	2.22-3.37	.087-.133
	4 Disc	2.29-3.71	.090-.146
Rear Clutch (3 and 4 Disc) Adjustable	3 Disc	.67-1.10	.026-.043
	4 Disc	.67-1.10	.026-.043
Selective Snap Rings (5)		1.22-1.27	.048-.050
		1.52-1.57	.060-.062
		1.73-1.78	.068-.070
		1.88-1.93	.074-.076
		2.21-2.26	.087-.089
Band Adjustment:			
Kickdown, Backed off from 8 N·m (72 in. lbs.)		2-1/2 Turns	
Low-Reverse		3-1/2 Turns backed off from 5 N·m (41 in. lbs.)	
Thrust Washers:		**(Millimeter)**	**(Inch)**
Reaction Shaft Support (Phenolic)	No. 1	1.55-1.60	.061-.063
Rear Clutch Retainer (Phenolic)	No. 2	1.55-1.60	.061-.063
Output Shaft, Steel Backed Bronze	(Select) No. 3	1.98-2.03	.077-.080
		2.15-2.22	.085-.087
		2.34-2.41	.092-.095
Front Annulus, Steel Backed Bronze	No. 4	2.95-3.05	.116-.120
Front Carrier, Steel Backed Bronze	Nos. 5, 6	1.22-1.28	.048-.050
Sun Gear (Front)	No. 7	.85-0.91	.033-.036
Sun Gear (Rear)	No. 8	.85-0.91	.033-.036
Rear Carrier, Steel Backed Bronze	Nos. 9, 10	1.22-1.28	.048-.050
Rev. Drum, Phenolic	No. 11	1.55-1.60	.061-.063
Tapered Roller Bearing Settings:		**(Millimeter)**	**(Inch)**
Output Shaft		.0-.07 Preload	.0-.0028 Preload
Transfer Shaft		.05-.25 End Play	.002-.010 End Play
Differential		.15-.29 Preload	.006-.012 Preload

Section 3

A604 Transaxle
Chrysler Corp.

APPLICATION

1989

Acclaim, Spirit LE, Dynasty, Dynasty LE
New Yorker, New Yorker Landau,
Voyager LE, Grand Voyager SE,
Grand Voyager LE, Caravan LE,
Grand Caravan SE, Grand Caravan LE

GENERAL DESCRIPTION

Transaxle and Converter Indentification

TRANSAXLE

The A-604 Ultradrive electronic 4-speed FWD transaxle (transaxle assembly No. 4471895) makes use of fully-adaptive controls. Adaptive controls are those which perform their functions based on real-time feedback sensor information, just as is done by electronic antilock brake controls. Although the transaxle is conventional in that it uses hydraulically-applied clutches to shift a planetary geartrain, its use of electronics to control virtually all functions is unique. The overall top gear ratio in overdrive is 2.36 and is equipped in several Chrysler models with the 3.0L V6 engine.

Operation

The transaxle provides forward ratios of 2.84, 1.57, 1.00, and 0.69 with torque converter lockup available in 2nd, direct, or overdrive gear; the Reverse ratio is 2.21. The shift lever is conventional with 6 positions: **P, R, N, OD, D,** and **L**. When in **OD** is selected, the transaxle shifts normally through all 4 speeds with lockup in overdrive; this position is recommended for most driving. The **D** position is tailored for use in hilly or mountainous driving. When **D** is selected, the transaxle uses only 1st, 2nd, and direct gears with 2-direct shift delayed to 40 mph or greater. When operating in **D** or **L** positions torque converter lockup occurs in direct gear for improved transaxle cooling when towing trailers and steep grades. If high engine coolant temperature occurs, the torque converter will also lock up in 2nd gear. The **L** position provides maximum engine braking for descending steep grades. Unlike most current transaxles, upshifts are provided to 2nd or direct at peak engine speeds if the accelerator is depressed. This provides engine over-speed protection and maximum performance.

CONVERTER

The converter is a welded unit and cannot be disassembled. The torque converter is a fluid drive coupling between the engine and transaxle. It is designed to slip at low speeds, such as when the engine is idling. As engine speed increases, the torque converter will engage the engine to the transmission. There is also a hydraulically controlled mechanical clutch inside the torque converter. This clutch is controlled by the electronic control module (ECM).

Electronic Controls

SOLENOIDS

Since the solenoid valves perform virtually all control functions, these valves must be extremely durable and tolerant of normal dirt particles. For that reason hardened-steel poppet and ball valves are used. These are free from any close operating clearances, and the solenoids operate the valves directly without any intermediate element. Direct operation means that these units must have very high output so that they can close against the sizeable flow areas and high line pressures. Fast response is also required to meet the control requirements.

Two of the solenoids are normally-venting and 2 are normally-applying; this was done to provide a default mode of operation. With no electrical power, the transaxle provides 2nd gear in **OD, D,** or **L** shift lever positions, neutral in **N**, reverse in **R**, and park in **P**. The choice of 2nd gear was made to provide adequate breakaway performance while still accommodating highway speeds.

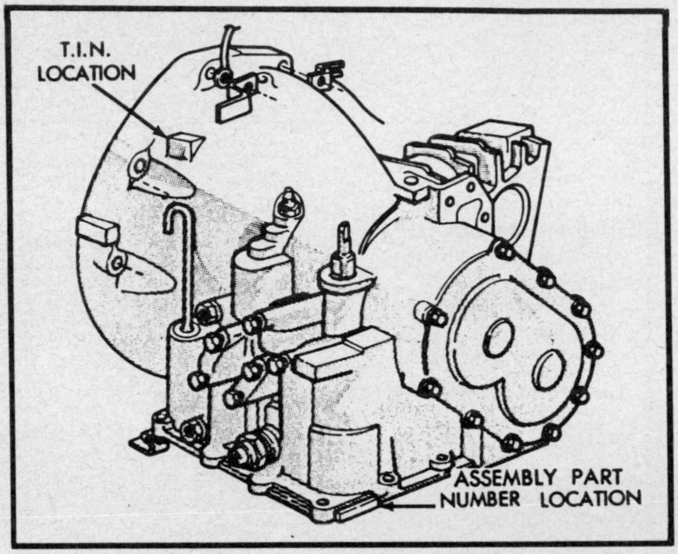
T.I.N.
LOCATION
ASSEMBLY PART
NUMBER LOCATION
Transaxle Indentification Number (TIN) location

SENSORS

Other electrical components include: 3 pressure switches to identify solenoid application, 2 speed sensors to read input (torque converter turbine) and output (parking sprag) speeds, and position switches to indicate the manual shift lever position. The pressure switches are incorporated in an assembly with the solenoids. Engine speed, throttle position, temperature, etc., are also observed. Some of these signals are read directly from the engine control sensors; others are read from a C^2D multiplex circuit with the engine controller.

ELECTRONICS

The control electronic unit is located underhood in a potted, diecast aluminum housing with a sealed, 60-way connector.

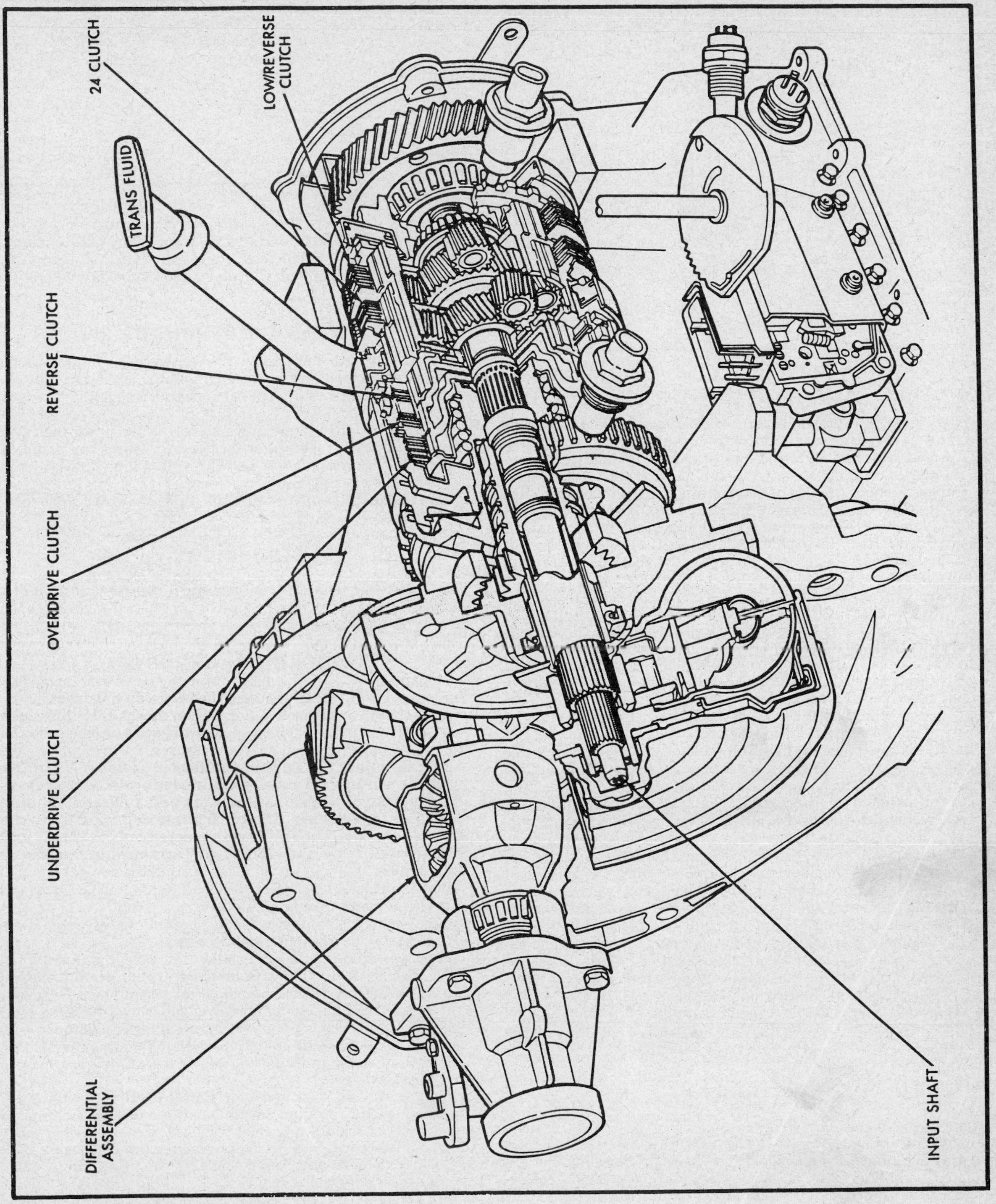
24 CLUTCH
LOW/REVERSE CLUTCH
TRANS FLUID
REVERSE CLUTCH
OVERDRIVE CLUTCH
UNDERDRIVE CLUTCH
DIFFERENTIAL ASSEMBLY
INPUT SHAFT

A604 transaxle cut-away view

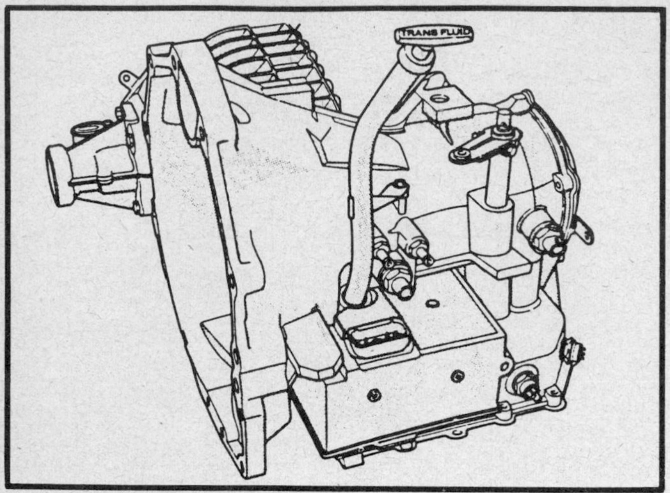
TRANS FLUID
Oil level indicator location

ADAPTIVE CONTROLS

These controls function by reading the input and output speeds over 140 times a second and responding to each new reading. This provides the precise and sophisticated friction element control needed to make smooth clutch-to-clutch shifts for all gear changes without the use of overrunning clutches or other shift quality aids. As with most automatic transaxles, all shifts involve releasing 1 element and applying a different element. In simplified terms, the upshift logic allows the releasing element to slip backwards slightly to ensure that it does not have excess capacity; the apply element is filled until it begins to make the speed change to the higher gear; its apply pressure is then controlled to maintain the desired rate of speed change until the shift is complete. The key to providing excellent shift quality is precision; for example, as mentioned, the release element for upshifts is allowed to slip backwards slightly; the amount of that slip is typically less than a total of 20 degrees. To achieve that precision, the controller learns the characteristics of the particular transaxle that it is controlling; it learns the release rate of the releasing element, the apply time of the applying element, the rate a which the apply element builds pressure sufficient to begin to make the speed change, and so on. This method achieves more precision than would be possible with exacting tolerances and it can adapt to any changes that occur with age or environment, for example, altitude, temperature, engine output, etc.

For kickdown shifts, the control logic allows the releasing element to slip and then controls the rate at which the input (and engine) accelerate; when the lower gear speed is achieved, the releasing element reapplies to maintain that speed until the apply element is filled. This provides quick response since the engine begins to accelerate immediately and a smooth torque exchange since the release element can control the rate of torque increase. This control can make any powertrain feel more responsive without increasing harshness.

Since adaptive controls respond to input speed changes, they inherently compensate for changes in engine or friction element torque and provide good, consistent shift quality for the life of the transaxle.

DIAGNOSTICS

These controls also provide comprehensive, on-board transaxle diagnostics, and, thanks to the learning of individual characteristics, the information available can be truly revealing. For example, apply element buildup rate indicates solenoid performance. Also included are self-diagnostic functions which allow the technician to test the integrity of the electronic controls

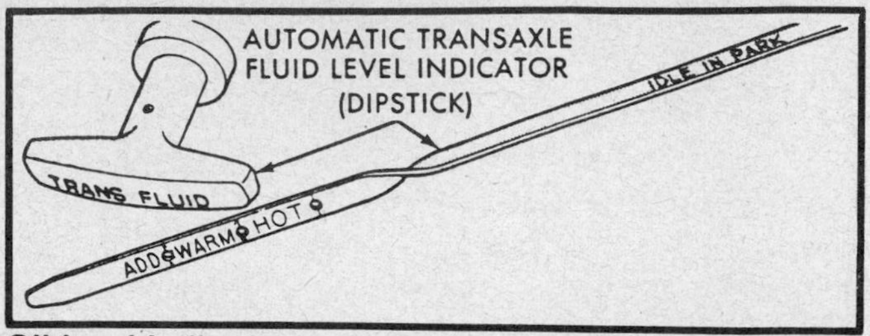
AUTOMATIC TRANSAXLE
FLUID LEVEL INDICATOR
(DIPSTICK)
IDLE IN PARK
TRANS FLUID
ADD WARM HOT
Oil level indicator

without requiring a road test. Moreover, the controller continuously monitors its critical functions, records any malfunctions, and the number of engine starts since the last malfunction so that the technician can use the information in the event of a customer complaint.

Metric Fasteners

The metric fastener dimensions are very close to the dimensions of the familiar inch system fasteners. For this reason, replacement fasteners must have the same measurement and strength as those removed.

Do not attempt to interchange metric fasteners for inch system fasteners. Mismatched or incorrect fasteners can result in damage to the transmission unit through malfunctions, breakage or possible personal injury.

Care should be taken to reuse the fasteners in the same locations as removed.

Capacities

The capacity of the A-604 transaxle and converter is 18.25 pints/ 9 quarts (8.6 liters). The oil type used in the transaxle is Mopar ATF Type 7176 or Dexron®II.

Checking Fluid Level

The transaxle fluid level should be inspected on the indicator every time other underhood service are preformed. To check the fluid level, start the engine and allow it to idle, make sure the transaxle is in **P** or **N**.

NOTE: Allow the engine to idle for at least 1 minute with the vehicle on level ground. This will assure complete oil level stabilization between differential and transaxle. A properly filled transaxle will read near the ADD mark when the fluid temperature is 70°F (21°C) and in the HOT region at 180°F (82°C) (average operating temperature).

Low fluid level can cause a variety of conditions because it allows the pump to take in air along with the fluid. As in any hydraulic system, air bubbles make the fluid spongy, therefore, pressures will be low and build up slowly.

Improper filling can also raise the fluid level too high. When the transaxle has too much fluid, the gears churn up foam and cause the same conditions which occur with a low fluid level.

In either case, the air bubbles can cause over-heating, fluid oxidation, and varnishing, which can interfere with normal valve, clutch, and accumulator operation. Foaming can also result in fluid escaping from the transaxle vent where it may be mistaken for a leak.

Along with fluid level, it is important to check the condition of the fluid. When the fluid smells burned, and is contaminated with metal or friction material particles, a complete transaxle overhaul is needed. Be sure to examine the fluid on the dipstick closely. If there is any doubt about its condition, drain out a sample for a double check.

After the fluid has been checked, seat the dipstick fully to seal out water and dirt.

TRANSAXLE MODIFICATIONS

Momentary Deceleration, Default to Second Gear (Limp-In) Mode or Excessive Clutch Slippage During 3–4 Upshift

On some 1989 Dynasty, New Yorker, Landau, Caravan and Voyager vehicles equipped with A604 transaxle, a default to 2nd gear (limp-in) mode or excessive clutch slippage during a 3–4 upshift are complaints that may be caused by reaction shaft support seal ring hang-up. Dirt, debris and imperfections in the area of the reaction shaft support seal rings can cause this condition. Hang-up of this seal ring may cause underdrive and/or overdrive clutch failures.

An improperly functioning seal ring can direct hydraulic pressure between the overdrive and underdrive clutches during a 3–4 upshift. Under some conditions, this may result in momentary vehicle deceleration and/or clutch drag when the clutch normally should be venting hydraulic fluid.

REPAIR PROCEDURES

CAUTION: Do not perform this procedure unless the vehicle has 3-4 shift problems as described.

Fault codes other than 46 and 39 must be diagnosed using regular diagnostic tests, starting with Test #1.

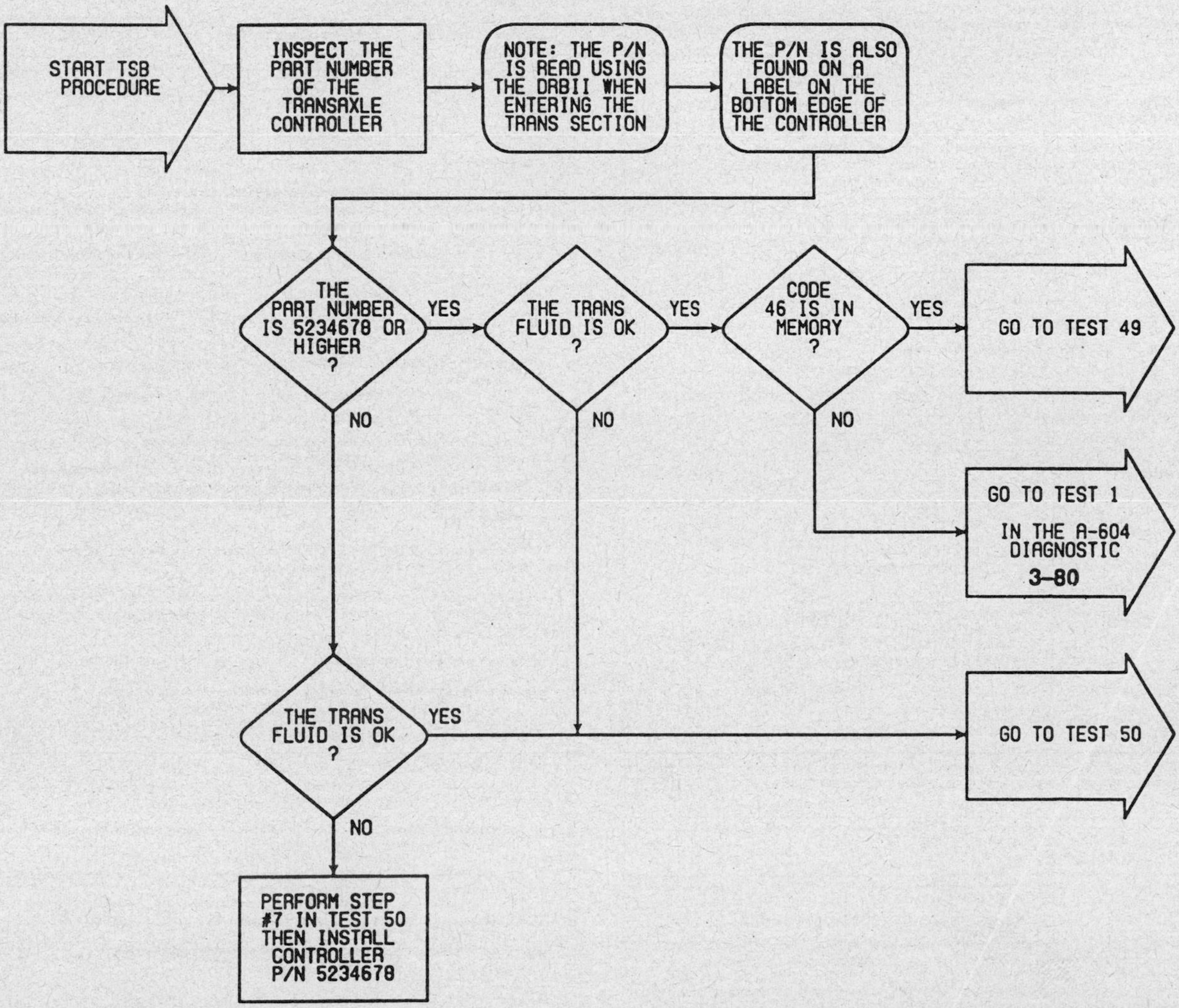
START TSB PROCEDURE
INSPECT THE PART NUMBER OF THE TRANSAXLE CONTROLLER
NOTE: THE P/N IS READ USING THE DRBII WHEN ENTERING THE TRANS SECTION
THE P/N IS ALSO FOUND ON A LABEL ON THE BOTTOM EDGE OF THE CONTROLLER
THE PART NUMBER IS 5234678 OR HIGHER ?
YES
THE TRANS FLUID IS OK ?
YES
CODE 46 IS IN MEMORY ?
YES
GO TO TEST 49
NO
NO
NO
GO TO TEST 1 IN THE A-604 DIAGNOSTIC 3–80
THE TRANS FLUID IS OK ?
YES
GO TO TEST 50
NO
PERFORM STEP #7 IN TEST 50 THEN INSTALL CONTROLLER P/N 5234678

TEST 49 — CODE 46 TEST PROCEDURE

Code 46 can only be generated by an upgraded transaxle controller P/N 5234678 or later. If you find code 46 in memory, the vehicle already has the upgrade - DO NOT REPLACE THE CONTROLLER.

Code 46 is primarily set in memory after the controller has made 3 unsuccessful attempts at a proper 3-4 upshift. Controller logic then prevents the transaxle from any further upshift attempts and causes it to remain in third gear until downshifts into first or second gears have been obtained.

Code 46 will **not** cause a limp-in (default to second) condition.

Faulty connections at the Turbine or Output speed sensors can result in aborted 3-4 upshifts and set code 46 in memory. Carefully inspect the Turbine and Output speed sensor connectors for anything that could cause an intermittent connection to the sensor (i.e. spread, bent or misaligned terminals). If the connectors are OK, proceed with the road test below.

Road Test

Road test to confirm that code 46 can be repeated by using the following guidelines:

A. Use the DRBII to clear any fault codes before road testing.

B. Leaving the DRBII connected, road test the vehicle using the guidelines found in step #3 of Test 50, then continue to road test with an emphasis on the 3-4 shift.

C. If code 46 occurs early during the road test, erase the code and continue the road test. (A false code 46 can sometimes set during the shift learning process)

D. If transaxle performance is judged to be acceptable, return the vehicle to the owner.

E. If code 46 continues to repeat, go to Test 50 Step 7.

TEST 50 — TESTING THE A-604 CLUTCH VOLUMES

Transaxles that have 3-4 shift problems could have normal or slightly discolored fluid but no serious clutch distress. The DRBII diagnostic tool can be used to help determine the amount of clutch wear in all A-604 clutches with the exception of reverse.

NOTE: A transmission which has experienced a **sudden** clutch failure may be incapable of learning the correct clutch volume index values. A road test of a transmission with this condition may exhibit an upshift runaway condition, but have acceptable clutch volumes. In this case, a clutch is obviously failing.

1. Finding A-604 clutch volumes

Connect the DRBII to the vehicle's bus diagnostic connector as shown in Transaxle Test #1. (Entering A-604 Diagnostics and Bus Tests)

After selecting the Transmission Section, drop down to the "SELECT TEST" level using the "YES" key. The first select test is the "DISPLAY FAULT CODES" display. Using the "F1" key, move to the left past the "CLEAR FAULT CODES" display until you see . . .

```
SELECT TEST
CONTINUE PRESS F1/F2
A604 TRANSAXLE
PRESS F1, F2 FOR MORE
```

. . . displayed on the DRBII. (**NOTE:** This display is not shown on the DRBII functional flow diagram found at the beginning of this book.)

Use the "YES" key to drop down to the display shown below.

```
CONDITIONS SHOWN
OCCUR IN REAL TIME
AND MAY BE NORMAL
```

Press "YES" again . . .

```
SELECT TEST
ENGINEERING DATA 1
A604 TRANSAXLE
PRESS F1, F2 FOR MORE
```

Press the "F2" key . . .

```
SELECT TEST
ENGINEERING DATA 2
A604 TRANSAXLE
PRESS F1, F2 FOR MORE
```

Press the "YES" key again.

2. Understanding clutch volume values

The display on the DRBII shows the four A-604 clutches that can be examined for clutch wear. Next to each clutch is a number that represents the volume required to fill and pressurize that clutch circuit. The initial values that will be displayed following a battery disconnect are shown on the next page.

TEST 50 — TESTING THE A-604 CLUTCH VOLUMES

```
LR - 64    OD - 89
2-4 - 48   UD - 45
```

NOTE: These are start up values and are **not** to be used in the actual diagnosis of the transaxle.

These numbers will change as the A-604 controller "learns" or updates clutch fill volumes due to the normal usage and wear that occurs during the life of the transaxle.

New clutch discs will have maximum friction material present and tend to take less fluid to fill and apply - hence the value for these clutch circuits is smaller. As the clutch ages, the clutch clearance increases, increasing the amount of oil required to fill and apply the clutch - hence the value for these clutch circuits will be larger.

NOTE: To obtain useful clutch information on a vehicle which may have recently had its battery disconnected, the transaxle controller must be allowed to "re-learn" each clutch circuit.

3. "Teaching" clutch volume values to the A-604 controller

A. The transaxle fluid level must be properly set.
B. The transaxle must be at normal operating temperatures.
C. At least three constant throttle upshifts at approximately half throttle from a standing start through the 2-3 upshift must be made. (These learn the overdrive and 2-4 clutch)
D. At least three heavy throttle downshifts to 1st at 15 mph must be made. (These learn the low/reverse clutch)
E. At least three part throttle 4-3 downshifts between 40 and 50 mph must be made. (These learn the underdrive clutch.

4. Acceptable transaxle clutch volume value guidelines

The range of clutch volume values which have been found to be within **normal** wear limits are shown below:

LR - 35 TO 85	OD - 75 TO 135
2-4 - 20 TO 77	UD - 24 TO 70

5. Other factors affecting clutch volume values

1. Incorrect fluid level
2. High transaxle temperature level
3. Restricted clutch feed circuit
4. Restricted solenoid feed circuit
5. Valve body leakage
6. Leaky lip seals
7. Leaky seal rings
8. Case porosity
9. Circuit leak in pump housing
10. Circuit leak in reaction shaft support
11. Damaged accumulator seal ring
12. Clogged oil filter
13. Aerated fluid
14. Faulty oil pump
15. Insufficient clutch pack clearance
16. Sticky regulator valve
17. Failed clutch return spring

NOTE: The above factors will normally be accompanied by transaxle symptoms not always related to this procedure.

TEST 50 — TESTING THE A-604 CLUTCH VOLUMES

6. Interpreting clutch volume values

If all clutch volume values are still within the acceptable guidelines found in step 4, the vehicle may be serviced by only replacing the A-604 controller with P/N 5234678. Road test the vehicle using the instructions found in step #8.

If any clutch volume value exceeds the **upper** range limit, a worn clutch pack is likely. Perform step #7.

7. Repair guidelines

Begin the repair by upgrading the transaxle controller with P/N 5234678 (if not already done). (**NOTE:** 60-way connector torque is 35-45 in./lbs.)

Using the cautions listed below, disassemble, clean and rebuild the transaxle as outlined in the 1989 front wheel drive service manual.

CAUTION:

A. The cooler & lines should be REVERSE flushed using mineral spirits and short blasts of compressed air, followed by rinsing with 1 qt. of ATF as described in the service manual.

B. Inspect the input clutch hub and reaction shaft support assemblies. It is critical that these parts be free of debris and burrs. The seal rings must be removed, inspected and the rings and mating parts must be washed.

C. Replace distressed clutch components as required and replace the valve body filter. Make sure parts are installed properly as outlined in the service manual.

D. During reassembly, input shaft end play **must** fall between .005" and .025". To achieve this, a selection from a set of #4 thrust plates may be necessary. (**NOTE:** These end play specifications are different from the service manual.)

E. Replace gaskets and seals as required.

F. The torque converter must be changed if major clutch failure has occurred.

G. The solenoid assembly can be re-used unless the transaxle has experienced a geartrain failure. In this case a complete replacement of the solenoid assembly is required.

8. Proper road test procedure

A. Use the DRBII to clear any fault codes before road testing.

B. Leaving the DRBII connected, road test the vehicle using the guidelines found in step #3, then continue to road test with an emphasis on the 3-4 shift.

C. If code 46 occurs early during the road test, erase the code and continue the road test. (A false code 46 can sometimes set during the shift learning process)

D. If transaxle performance is judged to be acceptable, return the vehicle to the owner.

E. If Code 46 continues to repeat, contact Automatic Transmission Hotline.

TROUBLE DIAGNOSIS

CLUTCH APPLICATION CHART

Shift Lever Position	Start Safety	Park Sprag	Clutches				
			Underdrive	Overdrive	Reverse	2/4	Low/ Reverse
P-Park	Applied	Applied	—	—	—	—	—
R-Reverse	—	—	—	—	Applied	—	Applied
N-Neutral	Applied	—	—	—	—	—	Applied
OD-Overdrive							
First	—	—	Applied	—	—	—	Applied
Second	—	—	Applied	—	—	Applied	—
Direct	—	—	Applied	Applied	—	—	—
Overdrive	—	—	—	Applied	—	Applied	—
D-Drive ①							
First	—	—	Applied	—	—	—	Applied
Second	—	—	Applied	—	—	Applied	—
Direct	—	—	Applied	Applied	—	—	—
L-Low ①							
First	—	—	Applied	—	—	—	Applied
Second	—	—	Applied	—	—	Applied	—
Direct	—	—	Applied	Applied	—	—	—

① Vehicle upshift and downshift speeds are increased when in these selector positions.

CHILTON THREE "C" TRANSAXLE DIAGNOSIS

Condition	Cause	Correction
Harsh engagement from N to D	a) Poor engine performance b) Underdrive clutch worn or faulty c) Low/reverse clutch worn or faulty d) Accumulator seal rings worn or damaged e) Valve body malfunction or leakage f) Hydraulic pressures too high g) Engine idle speed too high	a) Check engine tuneup b) Overhaul c) Overhaul d) Replace seal rings e) Clean or overhaul f) Adjust to specifications g) Adjust idle speed
Harsh engagement from N to R	a) Poor engine performance b) Reverse clutch worn or faulty c) Low/reverse clutch worn or faulty d) Accumulator seal rings worn or damaged e) Valve body malfunction or leakage	a) Check engine tuneup b) Overhaul c) Overhaul d) Replace seal rings e) Clean or overhaul

CHILTON THREE "C" TRANSAXLE DIAGNOSIS

Condition	Cause	Correction
Harsh engagement from Neutral to D	f) Hydraulic pressures too high	f) Adjust to specifications
	g) Engine idle speed too high	g) Adjust idle speed
Delayed engagement from N to D	a) Damaged clutch seal	a) Replace seal
	b) Underdrive clutch worn or faulty	b) Overhaul
	c) Incorrect gearshift control linkage adjustment	c) Adjust gearshift control linkage
	d) Accumulator seal rings worn or damaged	d) Replace seal rings
	e) Valve body malfunction or leakage	e) Clean or overhaul
	f) Reaction shaft support seal rings worn or broken	f) Replace seal ring
	g) Input shaft seal rings worn or damaged	g) Replace seal ring
	h) Hydraulic pressure too low	h) Adjust to specification
	i) Oil pump faulty	i) Overhaul pump
	j) Oil filter clogged	j) Replace filter
	k) Fluid level low	k) Add as required
	l) Fluid aerated	l) Check for overfill
	m) Engine idle speed too low	m) Adjust idle speed
Delayed engagement from N to R	a) Damaged clutch seal	a) Replace seal
	b) Reverse clutch worn or faulty	b) Overhaul
	c) Incorrect gearshift control linkage adjustment	c) Adjust gearshift control linkage
	d) Accumulator seal rings worn or damaged	d) Replace seal rings
	e) Valve body malfunction or leakage	e) Clean or overhaul
	f) Reaction shaft support seal rings worn or broken	f) Replace seal ring
	g) Input shaft seal rings worn or damaged	g) Replace seal ring
	h) Hydraulic pressure too low	h) Adjust to specification
	i) Oil pump faulty	i) Overhaul pump
	j) Oil filter clogged	j) Replace filter
	k) Fluid level low	k) Add as required
	l) Fluid aerated	l) Check for overfill
	m) Engine idle speed too low	m) Adjust idle speed
Poor shift quality	a) Reaction shaft support seal rings worn or broken	a) Replace seal ring
	b) Hydraulic pressure too low	b) Adjust to specification
	c) Oil pump faulty	c) Overhaul pump
	d) Oll filter clogged	c) Replace filter
	e) Fluid level low	e) Add as required
	f) Fluid aerated	f) Check for overfill
Shifts erratic	a) Poor engine performance	a) Check engine tuneup
	b) Clutches worn or faulty	b) Overhaul
	c) Incorrect gearshift control linkage adjustment	c) Adjust gearshift control linkage
	d) Valve body malfunction or leakage	d) Clean or overhaul
	e) Reaction shaft support seal rings worn or broken	e) Replace seal ring
	f) Hydraulic pressure too low	f) Adjust to specification
	g) Oil pump faulty	g) Overhaul pump
	h) Oil filter clogged	h) Replace filter
	i) Fluid level low	i) Add as required
	j) Fluid aerated	j) Check for overfill
Drives in Neutral	a) Underdrive clutch worn or faculty	a) Overhaul
	b) Overdrive clutch worn or faulty	b) Overhaul
	c) Reverse clutch worn or faulty	c) Overhaul
	d) Clutches dragging	d) Check clearance
	e) Clutch plate clearance insufficient	e) Check clearance

CHILTON THREE "C" TRANSAXLE DIAGNOSIS

Condition	Cause	Correction
Drives in Neutral	f) Incorrect gearshift control linkage adjustment g) Valve body malfunction or leakage	f) Adjust gearshift control linkage g) Clean or overhaul
Drags or locks	a) Clutches worn or faulty b) Gear teeth chipped or damaged c) Planetary gearsets broken or seized d) Bearings worn or damaged	a) Overhaul b) Replace gear teeth c) Replace gearsets d) Replace bearings
Grating, scraping, growling noise	a) Gear teeth chipped or damaged b) Planetary gearsets broken or seized c) Bearings worn or damaged d) Driveshaft(s) bushing(s) worn or damaged	a) Replace gear teeth b) Replace gearsets c) Replace bearings d) Replace bushing(s)
Buzzing noise	a) Valve body malfunction or leakage b) Fluid level low c) Fluid aerated	a) Clean or overhaul b) Add as required c) Check for overfill
Buzzing noise during shifts only	a) Normal solenoid operation b) Solenoid sound cover loose	a) No correction b) Tighten cover
Hard to fill, oil blows out filler tube	a) High fluid level b) Oil filter clogged c) Fluid aerated	a) Remove as required b) Change filter c) Check for overfill
Transaxle overheats	a) Clutch plate clearance insufficient b) Incorrect gearshift control linkage adjustment c) Cooling system faulty d) Hydraulic pressure too low e) Oil pump faulty f) Fluid level low g) Fluid aerated h) Fluid level high i) Engine idle speed too high	a) Check clearance b) Adjust gearshift control linkage c) Service system d) Adjust to specification e) Overhaul pump f) Add as required g) Check for overfill h) Remove as required i) Adjust idle speed
Harsh upshift	a) Poor engine performance b) Overdrive clutch worn or faulty c) 2/4 clutch worn or faulty d) Hydraulic pressure too low e) Hydraulic pressure too high	a) Check engine tuneup b) Overhaul c) Overhaul d) Adjust to specification e) Adjust to specification
No upshift into overdrive	a) Overdrive clutch worn or faulty b) Engine coolant temperature too low	a) Overhaul b) Service cooling system
No lockup	a) Engine coolant temperature too low b) Valve body malfunction or leakage c) Input shaft seal rings worn or damaged d) Hydraulic pressure too low e) Oil pump faulty f) Fluid level low g) Fluid aerated	a) Service cooling system b) Clean or overhaul c) Replace seal ring d) Adjust to specification e) Overhaul pump f) Add as required g) Check for overfill
Harsh downshifts	a) Poor engine performance b) Underdrive clutch worn or faulty c) 2/4 clutch worn or faulty d) Low/reverse clutch worn or faulty e) Damaged clutch seal f) Accumulator seal ringe worn or damaged g) Valve body malfunction or leakage h) Reaction shaft support seal rings worn or broken i) Hydraulic pressure too high	a) Check engine tuneup b) Overhaul d) Overhaul d) Overhaul d) Replace seal f) Replace seal rings g) Clean or overhaul h) Replace seal ring i) Adjust to specification

CHILTON THREE "C" TRANSAXLE DIAGNOSIS

Condition	Cause	Correction
Harsh downshifts	j) Fluid level low k) Fluid aerated l) Engine idle speed too high	d) Add as required k) Check for overfill l) Adjust idle speed
High shift efforts	a) Shift linkage damaged b) Valve body malfunction or leakage	a) Check/repair linkage b) Clean or overhaul
Harsh lockup shift	a) Lockup piston sticking	a) Clean or overhaul

PRESSURE CHECK SPECIFICATIONS CHART

		Pressure Taps					
Shift Lever Position	Actual Gear	Under-Drive Clutch	Over-Drive Clutch	Reverse Clutch	Lockup Off	2/4 Clutch	Low/ Reverse Clutch
Park ① 0 mph	Park	0–2	0–5	0–2	60–110	0–2	115–145
Reverse ① 0 mph	Reverse	0–2	0–7	165–235	50–100	0–2	165–235
Neutral ① 0 mph	Neutral	0–2	0–5	0–2	60–110	0–2	115–145
L ② 20 mph	First	110–145	0–5	0–2	60–110	0–2	115–145
D ② 30 mph	First	110–145	0–5	0–2	60–110	115–145	0–2
D ② 45 mph	Direct	75–95	75–95	0–2	60–90	0–2	0–2
OD ② 30 mph	Overdrive	0–2	75–95	0–2	60–90	75–95	0–2
OD ② 50 mph	OD Lockup	0–2	75–95	0–2	0–5	75–95	0–2

① Engine speed at 1500 rpm
② Both front wheels must be turning at the same speed

Hydraulic Control System

NOTE: Please refer to Section 9 for all oil flow circuits.

CLUTCH AND GEAR

The A-604 transaxle consists of 3 multiple-disc input clutches, 2 multiple-disc grounded clutches, 4 hydraulic accumulators, and 2 planetary gearsets to provide 4 speeds forward and a reverse ratio. Since this transaxle is expected to operate properly with today's high-speed engines, its clutch-apply pistons were designed with centrifugally-balanced oil cavities so that quick response and good control can be achieved at any speed. A unique push/pull piston is incorporated for 2 of the 3 input clutches with out any added pressure seals; the 3rd clutch requires 1 additional seal.

HYDRAULICS

The hydraulics of the new transaxle provide the manual shift lever select function, main line pressure regulation, and torque converter and cooler flow control. Oil flow to the friction elements is controlled directly by 4 solenoid valves. The hydraulics also include a unique logic-controlled "solenoid switch valve" which locks out the 1st gear reaction element with the application of 2nd, direct, or overdrive gear elements, and redirects the 1st gear solenoid output so that it can control torque converter lockup operation. To regain access to 1st gear, a special sequence of solenoid commands must be used to uplock and move the solenoid switch valve. This precludes any application of the 1st gear reaction element with other elements applied unless specifically commanded by a properly functioning controller; it also allows 1 solenoid to control 2 friction elements.

Small, high-rate accumulators are provided in each controlled friction element circuit. These serve to absorb the pressure responses, and allow the controls to read and respond to changes that are occurring.

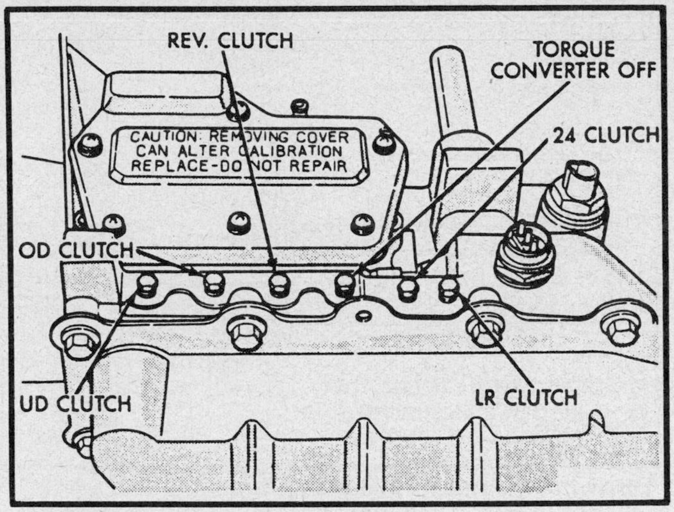
REV. CLUTCH
TORQUE CONVERTER OFF
CAUTION: REMOVING COVER CAN ALTER CALIBRATION REPLACE-DO NOT REPAIR
24 CLUTCH
OD CLUTCH
UD CLUTCH
LR CLUTCH

Transaxle pressure taps location

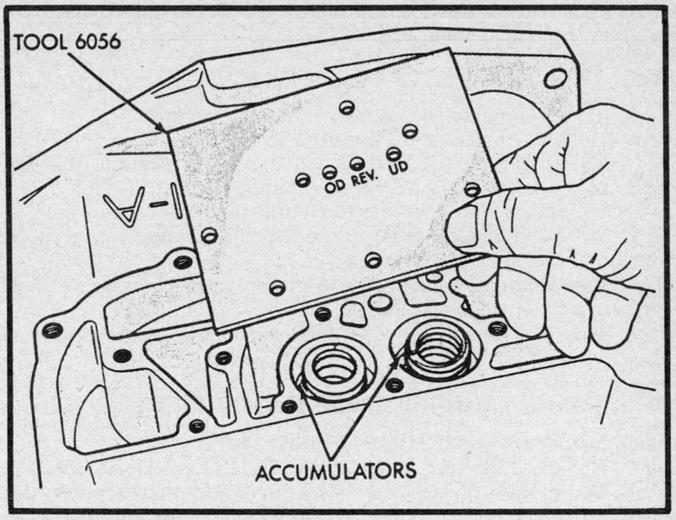
TOOL 6056
OD REV. UD
ACCUMULATORS

Air pressure test plate tool

Diagnosis Tests

GENERAL DIAGNOSIS

NOTE: Before attempting any repair on the A-604 Electronic Automatic Transaxle, always check for fault codes with the DRB II.

Automatic transaxle malfunctions may be caused by 4 general conditions: poor engine malfunctions, mechanical malfunctions, and electronic malfunctions. Diagnosis of these problems should always begin by checking the easily accessible variables: fluid level and condition, gear shift cable adjustment. Then perform a road test to determine if the problem has been corrected or that more diagnosis is necessary. If the problem exists after the preliminary tests and corrections are completed, hydraulic pressure checks should be preformed.

CONTROL PRESSURE TEST

Pressure testing is a very important step in the diagnostic procedure. These tests usually reveal the cause of most transaxle problems.

Before performing pressure tests, be certain that the fluid level and the condition, and shift cable adjustments have been checked and approved. Fluid must be at operating temperature (150–200°F).

Install an engine tachometer, raise the vehicle on a hoist which allows the front wheels to turn, and position the tachometer so it can be read.

Attach 150 psi gauges to the ports as required for test being conducted. A 300 psi gauge (C–3293) is required for reverse pressure test.

Test One (Selector In L–1st Gear)

1. Attach pressure gauge to the low/reverse clutch tap.
2. Move the selector lever to the **L** position.
3. Allow the vehicle wheels to turn and increase throttle opening to achieve an indicated vehicle speed to 20 mph.
4. Low/reverse clutch pressure should read 115–145 psi.
5. This test checks the pump output, pressure regulation and condition of the low/reverse clutch hydraulic circuit and shift schedule.

Test Two (Selector In D–2nd Gear)

1. Attach a gauge to the underdrive clutch tap.

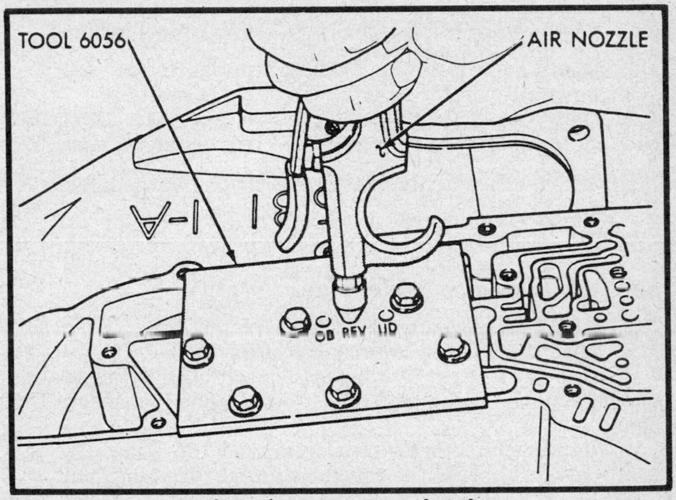
TOOL 6056
AIR NOZZLE
OD REV UD

Air pressure testing the reverse clutch

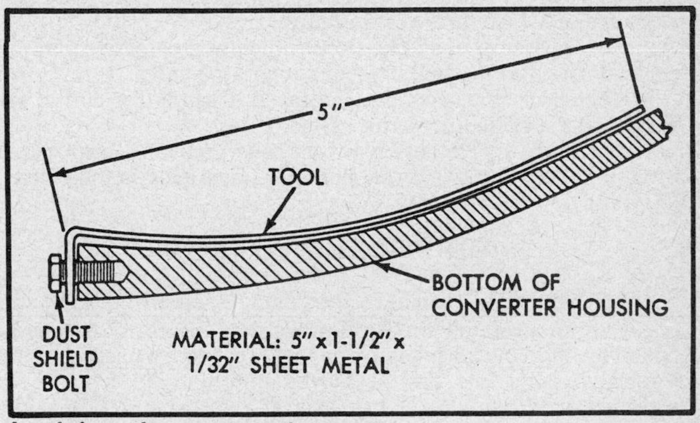
5"
TOOL
BOTTOM OF CONVERTER HOUSING
DUST SHIELD BOLT
MATERIAL: 5"x1-1/2"x 1/32" SHEET METAL

Leak locating test probe tool

2. Move the selector lever to the **D** position.
3. Allow the vehicle wheels to turn and increase throttle opening to achieve an indicated vehicle speed of 30 mph.
4. Underdrive clutch pressure should read 110–145 psi.

5. This test checks the underdrive clutch hydraulic circuit as well as the shift schedule.

Test Three (Overdrive Clutch Check)

1. Attach a gauge to the overdrive clutch tap.
2. Move the selector lever to the circle **D** position.
3. Allow the vehicle wheels to turn and increase throttle opening to achieve an indicated vehicle speed of 20 mph.
4. Overdrive clutch pressure should read 74–95 psi.
5. Move the selector lever to the **D** position and increase indicated vehicle speed to 30 mph.
6. The vehicle should be in 2nd gear and overdrive clutch pressure should be less than 5 psi.
7. This test checks the overdrive clutch hydraulic circuit as well as the shift schedule.

Test Four (Selector In Circle D–Overdrive Gear)

1. Attach a gauge to the 2–4 clutch tap.
2. Move the selector lever to the circle **D** position.
3. Allow the vehicle wheels to turn and increase throttle opening to achieve an indicated vehicle speed of 30 mph.
4. The 2–4 clutch pressure should read 74–95 psi.
5. This test checks the 2–4 clutch hydraulic circuit.

Test Five (Selector In Circle D–Overdrive Lockup)

1. Attach a gauge to the lockup off pressure tap.
2. Move the selector lever to the circle **D** position.
3. Allow the vehicle wheels to turn and increase throttle opening to achieve an indicated vehicle speed of 50 mph.

NOTE: Both wheels must turn at the same speed.

4. The lockup off pressure should be less than 5 psi.
5. This test checks the lockup clutch hydraulic circuit.

Test Six (Selector In Reverse)

1. Attach the gauge to the reverse clutch tap.
2. Move the selector lever to the **R** position.
3. Read the reverse clutch pressure with the output stationary (foot on the brake) and the throttle opened to achieve 1500 rpm.
4. Reverse clutch pressure should read 165–235 psi.
5. This test checks the reverse clutch hydraulic circuit.

Test Result Indications

1. If proper line pressure is found in any 1 test, the pump and pressure regulator are working properly.
2. Low pressure in all positions indicates a defective pump, a clogged filter, or a stuck pressure regulator valve.
3. Clutch circuit leaks are indicated if pressures do not fall within the specified pressure range.
4. If the overdrive clutch pressure is greater than 5 psi in Step 6 of Test 3, a worn reaction shaft seal ring is indicated.

AIR PRESSURE TEST

Inoperative clutches can be located using a series of tests by substituting air pressure for fluid pressure. The clutches may be tested by applying air pressure to their respective passages after the valve body has been removed and tool 6056 has been installed.

NOTE: The compressed air supply must be free of all dirt and moisture. Use a pressure of 30 psi.

Overdrive Clutch Check

Apply air pressure to the overdrive clutch apply passage and watch for the push/pull piston to move forward. The piston should return to its starting position when the air pressure is removed.

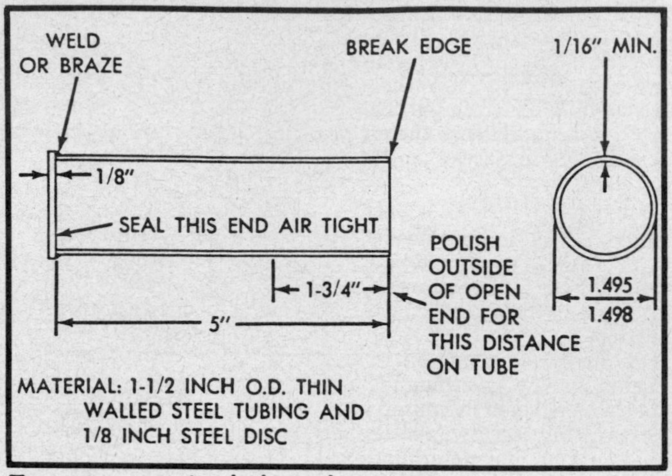
WELD
OR BRAZE
BREAK EDGE
1/16" MIN.
1/8"
SEAL THIS END AIR TIGHT
POLISH
OUTSIDE
OF OPEN
END FOR
THIS DISTANCE
ON TUBE
1-3/4"
5"
1.495
1.498
MATERIAL: 1-1/2 INCH O.D. THIN
WALLED STEEL TUBING AND
1/8 INCH STEEL DISC

Torque converter hub seal cup tool

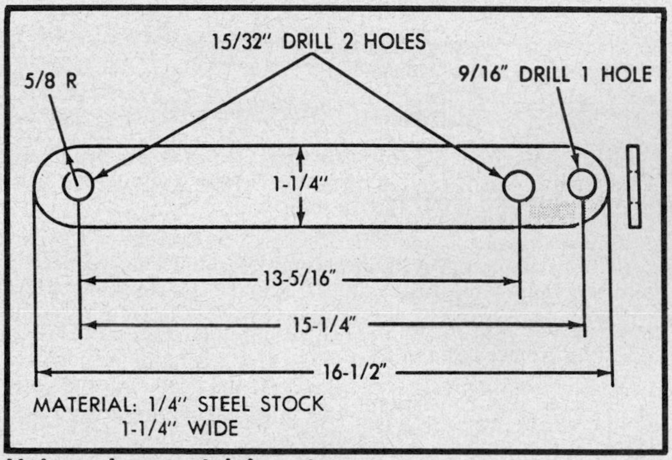
15/32" DRILL 2 HOLES
5/8 R
9/16" DRILL 1 HOLE
1-1/4"
13-5/16"
15-1/4"
16-1/2"
MATERIAL: 1/4" STEEL STOCK
1-1/4" WIDE

Hub seal cup retaining strap

Reverse Clutch Check

Apply air pressure to the reverse clutch apply passage and watch for the push/pull piston to move rearward. The piston should return to its starting position when the air pressure is removed.

2–4 Clutch Check

Apply air pressure to the feed hole located on the 2–4 clutch retainer. Look in the area where the 2–4 piston contacts the first separator plate and watch carefully for the 2–4 piston to move rearward. The piston should return to its original position after the air pressure is removed.

Low/Reverse Clutch Check

Apply air pressure to the low/reverse clutch feed hole (rear of the case, between 2 bolts holes). Then, look in the area where the low/reverse piston contacts the first separator plate and watch carefully for the piston to move forward. The piston should return to its original position after the air pressure is removed.

Underdrive Clutch Check

Because this clutch piston can not be seen, its operation is checked by function. Air pressure is applied to the low/reverse and the 2–4 clutches. This locks the output shaft. Use a piece of rubber hose wrapped around the input shaft and a pair of clamp-on pliers to turn the input shaft. Next apply air pressure to the underdrive clutch. The input shaft should not rotate with hand torque. Release the air pressure and confirm that the input shaft will rotate.

Transaxle Test

Fabricate equipment needed for the test.

The transaxle should be prepared for pressure test as follows after removal of the torque converter:

1. Install a dipstick bore plug and plug the oil cooler line fitting.
2. With rotary motion, install the converter hub seal cup over input shaft, and through the converter hub seal until the cup bottoms against the pump gear lugs. Secure with cup retainer strap using starter upper hole and opposite bracket hole.
3. Attach and clamp hose from the nozzle of tool C–4080 to the upper cooler line fitting position in the case.

CAUTION

Do not, under any circumstances, pressurize a transaxle to more than 10 psi.

4. Pressurize the transaxle using tool C–4080 until the pressure gauge reads 8 psi. Position the transaxle to that the pump housing and the case front may be covered with soapy solution of water. Leaks are sometimes caused by porosity in the case or the pump housing.

If a leak source is located, that part and all associated seals, O-rings, and gaskets should be replaced with new parts.

STALL SPEED TEST

CAUTION

Do not let anyone stand in front of the vehicle during this test.

The stall test consists of determining the engine speed obtained at full throttle in **D** position only, with the front wheels blocked. This test checks the torque converted stator clutch operation, and the holding ability of the transaxle clutch. The transaxle oil level should be checked and the engine brought to normal operating temperature before stall operation.

NOTE: Both the parking and service brakes must be fully applied and front wheels blocked while making this test.

Do not hold the throttle open any longer than is necessary to obtain a maximum engine speed reading, and never longer than 5 seconds at a time. If more than 1 stall check is required, operate the engine at approximately 1000 rpm in **N** for 20 seconds to cool the transaxle fluid between runs. If the engine speeds exceeds the maximum limits, release the accelerator immediately since transaxle clutch slippage is indicated.

Stall Speed Above Specification

If the stall speeds exceeds the maximum specified in the chart by more than 200 rpm, the transaxle clutch slippage is indicated. Follow the transaxle oil pressure and air pressure checks to determine the cause of the slippage.

Stall Speed Below Specification

Low stall speeds with a properly tuned engine indicate torque converter stator clutch problems. A road test will be necessary to identify the exact problem.

The stall speeds are 250–350 rpm below the minimum specification, and the vehicle operates properly at highway speeds, the stator overrunning clutch is slipping.

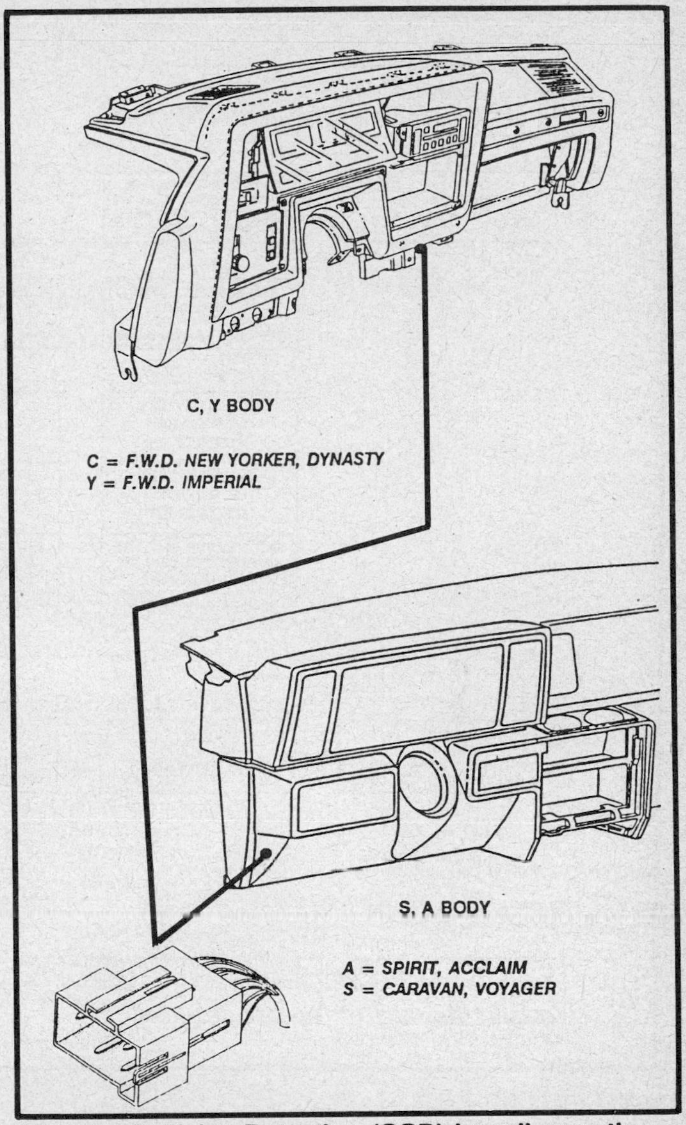
C, Y BODY
C = F.W.D. NEW YORKER, DYNASTY
Y = F.W.D. IMPERIAL
S, A BODY
A = SPIRIT, ACCLAIM
S = CARAVAN, VOYAGER

Chrysler Collision Detection (CCD) bus diagnostic connector location

If stall speed and acceleration are normal, but abnormally high throttle opening is required to maintain highway speeds, the stator clutch has seized.

Both of these stator defects require replacement of the torque converter.

Noise Diagnosis

A whining or siren-like noise due to fluid flow is normal during the stall operation with some torque converters; however, loud metallic noises from loose parts or interference within the assembly indicate a defective torque converter. To confirm that the noise originates with the torque converter, operate the vehicle at light throttle in **D** and **N** on a hoist and listen under the transaxle bell housing.

ROAD TEST

Prior to performing a road test, be certain that the fluid level and condition, and control cable adjustment have been checked and approved.

During the road test, the transaxle should be operated in each

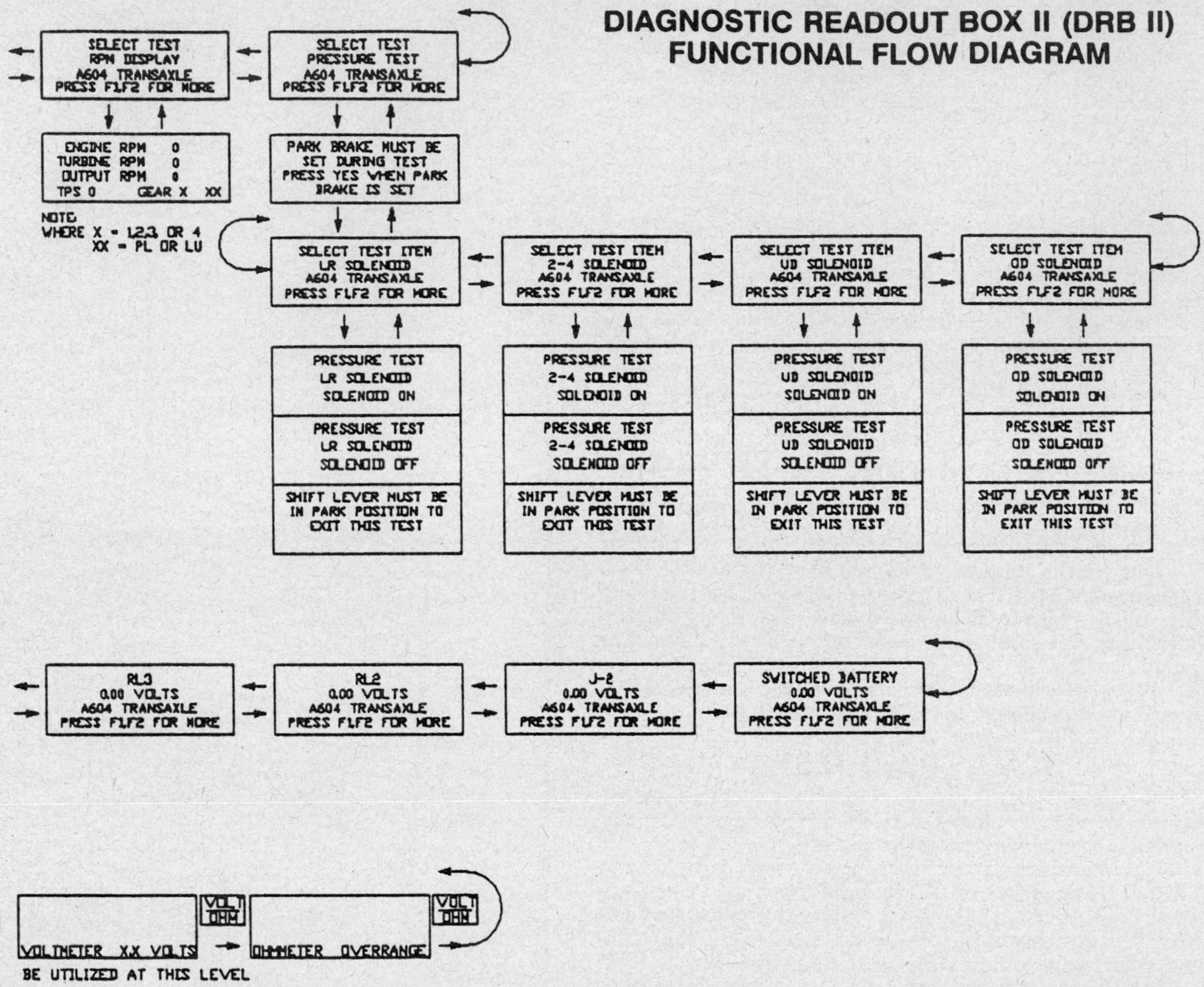
SELECT TEST
RPM DISPLAY
A604 TRANSAXLE
PRESS F1,F2 FOR MORE
SELECT TEST
PRESSURE TEST
A604 TRANSAXLE
PRESS F1,F2 FOR MORE
ENGINE RPM 0
TURBINE RPM 0
OUTPUT RPM 0
TPS 0 GEAR X XX
NOTE:
WHERE X = 1,2,3, OR 4
XX = PL OR LU
PARK BRAKE MUST BE
SET DURING TEST
PRESS YES WHEN PARK
BRAKE IS SET
SELECT TEST ITEM
LR SOLENOID
A604 TRANSAXLE
PRESS F1,F2 FOR MORE
SELECT TEST ITEM
2-4 SOLENOID
A604 TRANSAXLE
PRESS F1,F2 FOR MORE
SELECT TEST ITEM
UD SOLENOID
A604 TRANSAXLE
PRESS F1,F2 FOR MORE
SELECT TEST ITEM
OD SOLENOID
A604 TRANSAXLE
PRESS F1,F2 FOR MORE
PRESSURE TEST
LR SOLENOID
SOLENOID ON
PRESSURE TEST
LR SOLENOID
SOLENOID OFF
SHIFT LEVER MUST BE
IN PARK POSITION TO
EXIT THIS TEST
PRESSURE TEST
2-4 SOLENOID
SOLENOID ON
PRESSURE TEST
2-4 SOLENOID
SOLENOID OFF
SHIFT LEVER MUST BE
IN PARK POSITION TO
EXIT THIS TEST
PRESSURE TEST
UD SOLENOID
SOLENOID ON
PRESSURE TEST
UD SOLENOID
SOLENOID OFF
SHIFT LEVER MUST BE
IN PARK POSITION TO
EXIT THIS TEST
PRESSURE TEST
OD SOLENOID
SOLENOID ON
PRESSURE TEST
OD SOLENOID
SOLENOID OFF
SHIFT LEVER MUST BE
IN PARK POSITION TO
EXIT THIS TEST
RL3
0.00 VOLTS
A604 TRANSAXLE
PRESS F1,F2 FOR MORE
RL2
0.00 VOLTS
A604 TRANSAXLE
PRESS F1,F2 FOR MORE
J-2
0.00 VOLTS
A604 TRANSAXLE
PRESS F1,F2 FOR MORE
SWITCHED BATTERY
0.00 VOLTS
A604 TRANSAXLE
PRESS F1,F2 FOR MORE
VOLT
OHM
VOLTMETER X.X VOLTS
VOLT
OHM
OHMMETER OVERRANGE
BE UTILIZED AT THIS LEVEL

DIAGNOSTIC READOUT BOX II (DRB II) FUNCTIONAL FLOW DIAGRAM

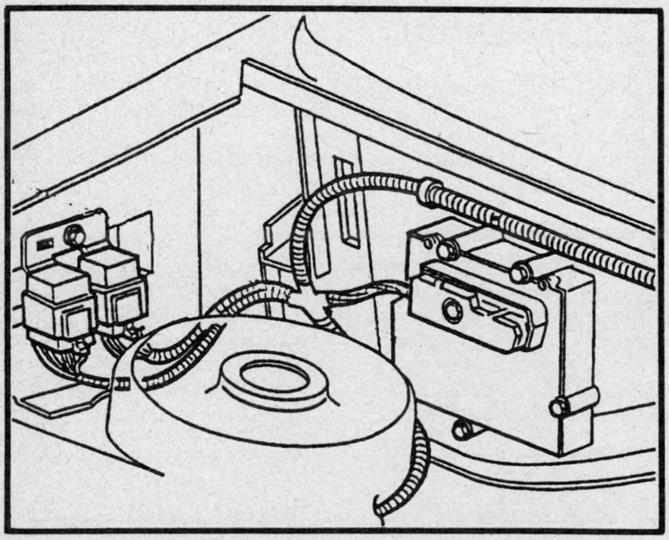

Transaxle controller, EATX relay and reverse lamp relay location—minivan

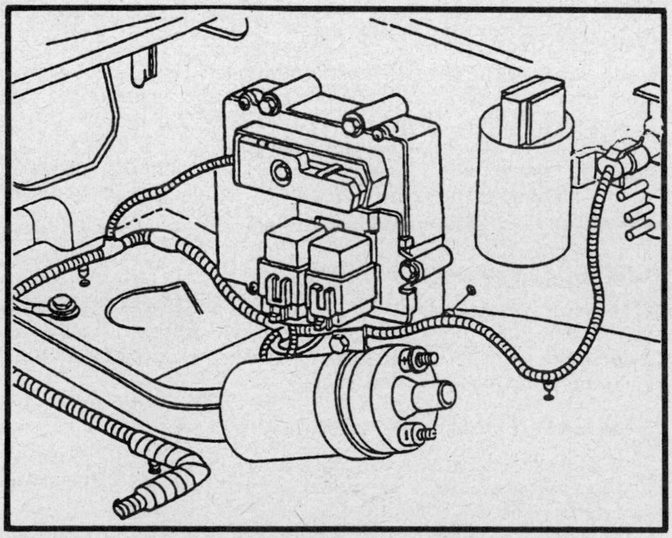

Transaxle controller, EATX relay and reverse lamp relay location—passenger car

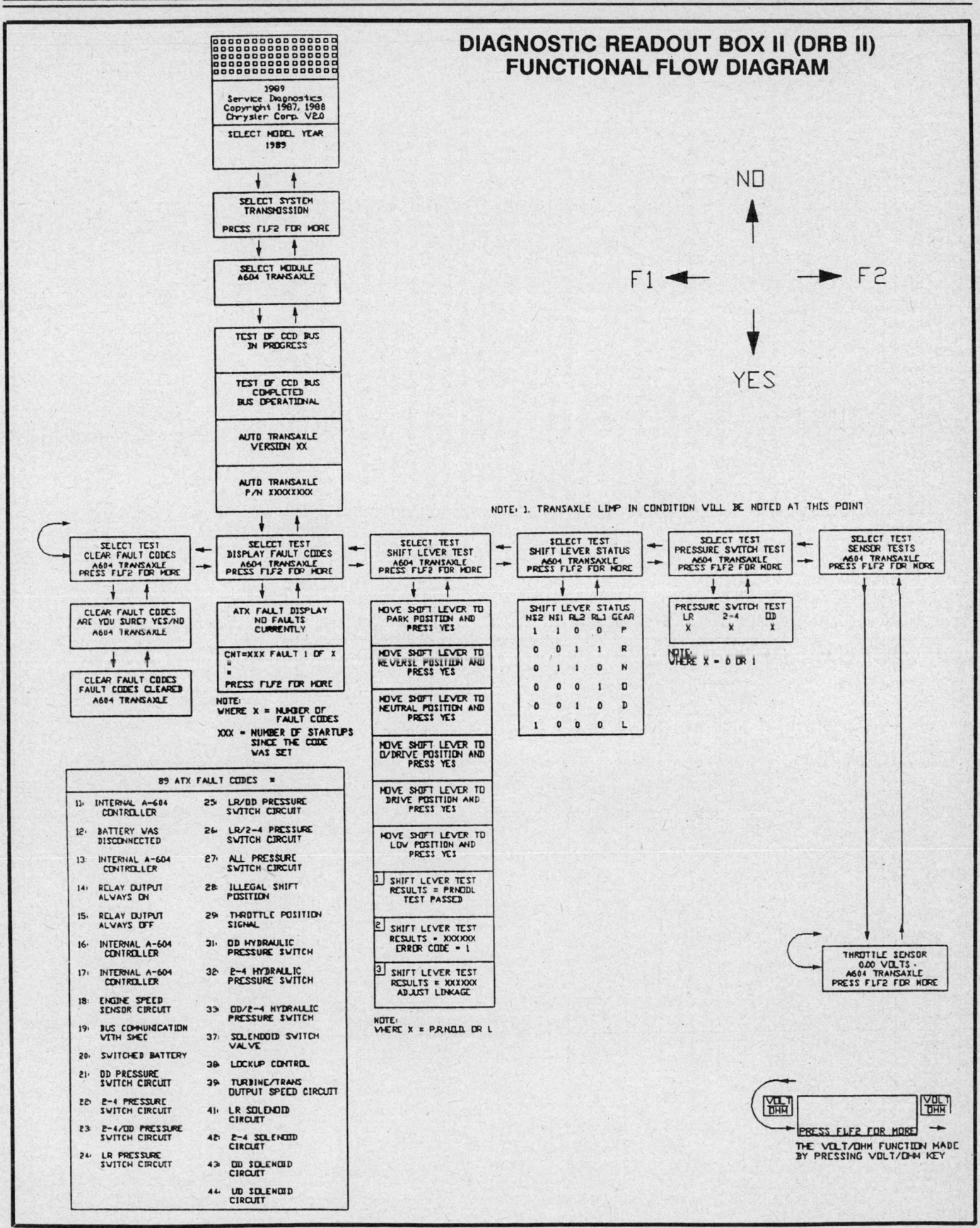

DIAGNOSTIC READOUT BOX II (DRB II) FUNCTIONAL FLOW DIAGRAM

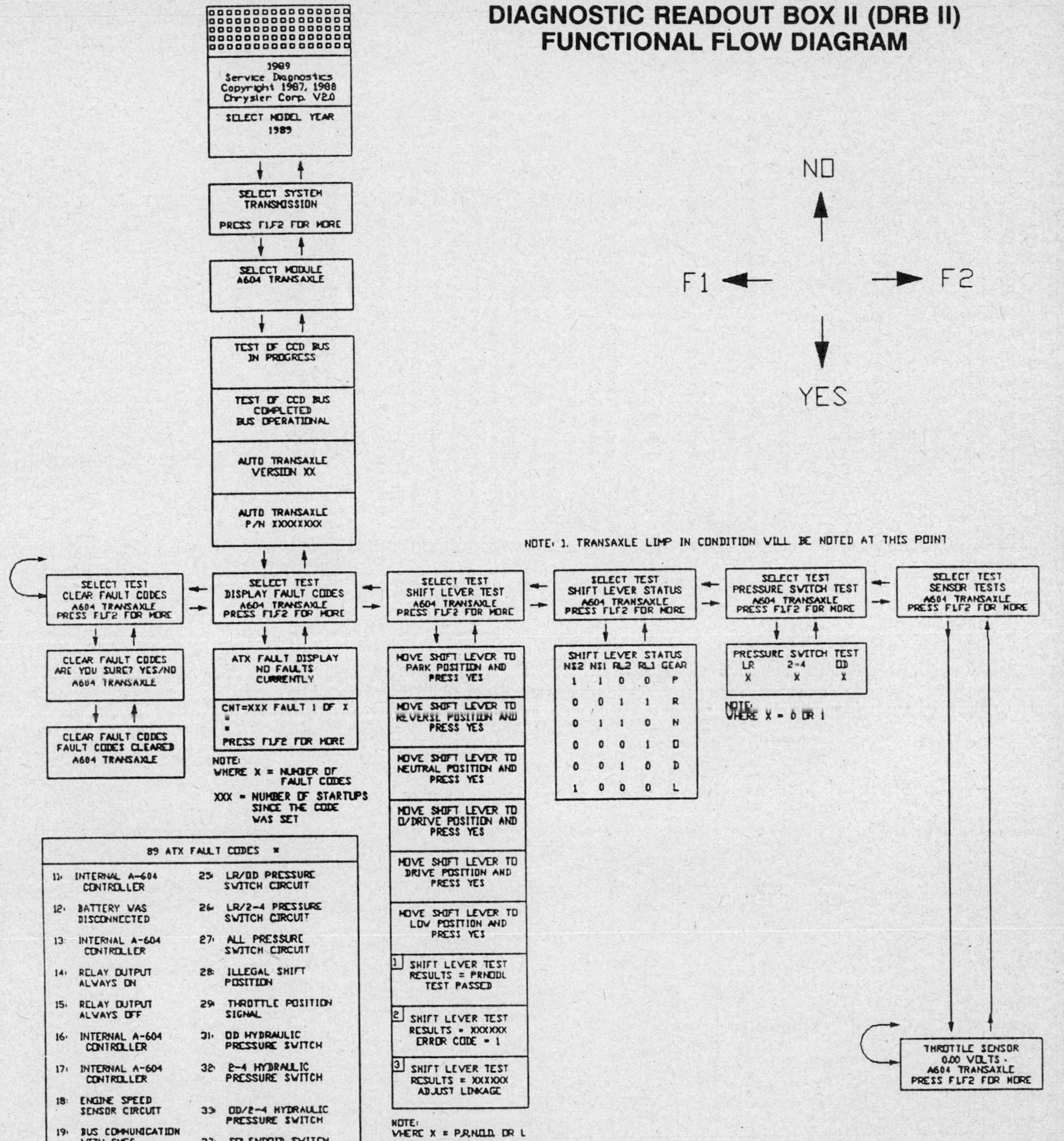
1989 Service Diagnostics Copyright 1987, 1988 Chrysler Corp. V2.0
SELECT MODEL YEAR 1989
SELECT SYSTEM TRANSMISSION PRESS F1,F2 FOR MORE
SELECT MODULE A604 TRANSAXLE
TEST OF CCD BUS IN PROGRESS
TEST OF CCD BUS COMPLETED BUS OPERATIONAL
AUTO TRANSAXLE VERSION XX
AUTO TRANSAXLE P/N XXXXXXXX
NO
F1
F2
YES
NOTE: 1. TRANSAXLE LIMP IN CONDITION WILL BE NOTED AT THIS POINT
SELECT TEST CLEAR FAULT CODES A604 TRANSAXLE PRESS F1,F2 FOR MORE
SELECT TEST DISPLAY FAULT CODES A604 TRANSAXLE PRESS F1,F2 FOR MORE
SELECT TEST SHIFT LEVER TEST A604 TRANSAXLE PRESS F1,F2 FOR MORE
SELECT TEST SHIFT LEVER STATUS A604 TRANSAXLE PRESS F1,F2 FOR MORE
SELECT TEST PRESSURE SWITCH TEST A604 TRANSAXLE PRESS F1,F2 FOR MORE
SELECT TEST SENSOR TESTS A604 TRANSAXLE PRESS F1,F2 FOR MORE
CLEAR FAULT CODES ARE YOU SURE? YES/NO A604 TRANSAXLE
CLEAR FAULT CODES FAULT CODES CLEARED A604 TRANSAXLE
ATX FAULT DISPLAY NO FAULTS CURRENTLY
CNT=XXX FAULT 1 OF X
PRESS F1,F2 FOR MORE
NOTE: WHERE X = NUMBER OF FAULT CODES
XXX = NUMBER OF STARTUPS SINCE THE CODE WAS SET
MOVE SHIFT LEVER TO PARK POSITION AND PRESS YES
MOVE SHIFT LEVER TO REVERSE POSITION AND PRESS YES
MOVE SHIFT LEVER TO NEUTRAL POSITION AND PRESS YES
MOVE SHIFT LEVER TO O/DRIVE POSITION AND PRESS YES
MOVE SHIFT LEVER TO DRIVE POSITION AND PRESS YES
MOVE SHIFT LEVER TO LOW POSITION AND PRESS YES
1 SHIFT LEVER TEST RESULTS = PRNODL TEST PASSED
2 SHIFT LEVER TEST RESULTS = XXXXXX ERROR CODE = 1
3 SHIFT LEVER TEST RESULTS = XXXXXX ADJUST LINKAGE
NOTE: WHERE X = P,R,N,O,D, OR L
SHIFT LEVER STATUS
NS2 NS1 RL2 RL1 GEAR
1 1 0 0 P
0 0 1 1 R
0 1 1 0 N
0 0 0 1 O
0 0 1 0 D
1 0 0 0 L
PRESSURE SWITCH TEST
LR 2-4 OD
X X X
NOTE: WHERE X = 0 OR 1
THROTTLE SENSOR 0.00 VOLTS A604 TRANSAXLE PRESS F1,F2 FOR MORE
VOLT OHM
PRESS F1,F2 FOR MORE
VOLT OHM
THE VOLT/OHM FUNCTION MADE BY PRESSING VOLT/OHM KEY

89 ATX FAULT CODES

11: INTERNAL A-604 CONTROLLER	25: LR/OD PRESSURE SWITCH CIRCUIT
12: BATTERY WAS DISCONNECTED	26: LR/2-4 PRESSURE SWITCH CIRCUIT
13: INTERNAL A-604 CONTROLLER	27: ALL PRESSURE SWITCH CIRCUIT
14: RELAY OUTPUT ALWAYS ON	28: ILLEGAL SHIFT POSITION
15: RELAY OUTPUT ALWAYS OFF	29: THROTTLE POSITION SIGNAL
16: INTERNAL A-604 CONTROLLER	31: OD HYDRAULIC PRESSURE SWITCH
17: INTERNAL A-604 CONTROLLER	32: 2-4 HYDRAULIC PRESSURE SWITCH
18: ENGINE SPEED SENSOR CIRCUIT	33: OD/2-4 HYDRAULIC PRESSURE SWITCH
19: BUS COMMUNICATION WITH SMEC	37: SOLENOID SWITCH VALVE
20: SWITCHED BATTERY	38: LOCKUP CONTROL
21: OD PRESSURE SWITCH CIRCUIT	39: TURBINE/TRANS OUTPUT SPEED CIRCUIT
22: 2-4 PRESSURE SWITCH CIRCUIT	41: LR SOLENOID CIRCUIT
23: 2-4/OD PRESSURE SWITCH CIRCUIT	42: 2-4 SOLENOID CIRCUIT
24: LR PRESSURE SWITCH CIRCUIT	43: OD SOLENOID CIRCUIT
	44: UD SOLENOID CIRCUIT

TEST 1–ENTERING DIAGNOSTICS AND BUS TESTS

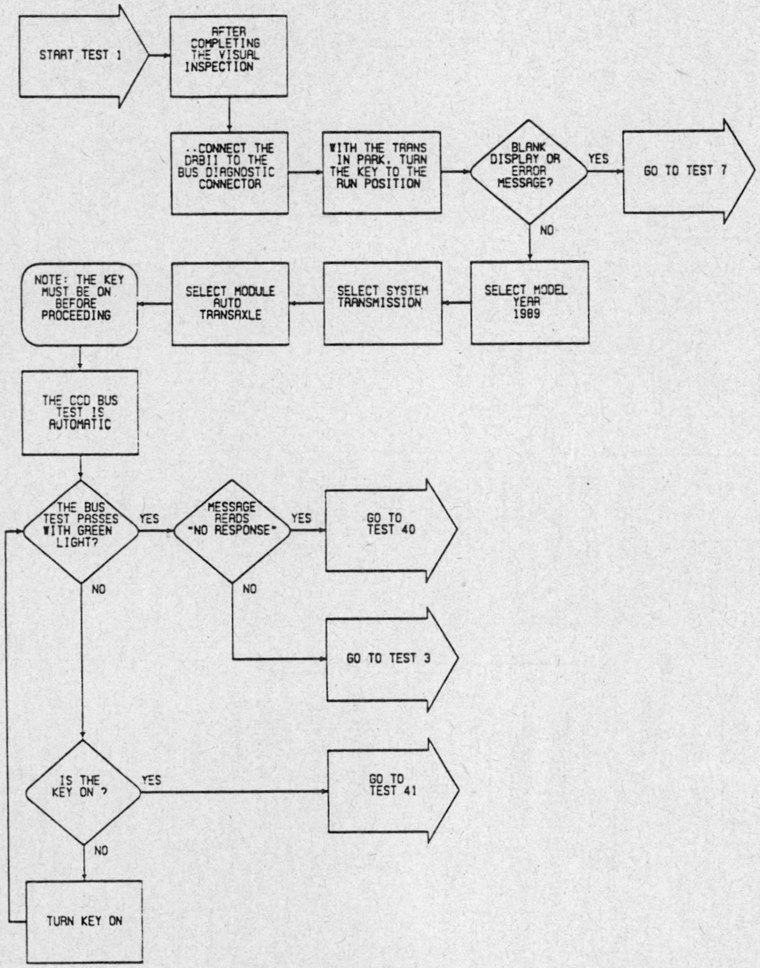
START TEST 1
AFTER COMPLETING THE VISUAL INSPECTION
..CONNECT THE DRBII TO THE BUS DIAGNOSTIC CONNECTOR
WITH THE TRANS IN PARK, TURN THE KEY TO THE RUN POSITION
BLANK DISPLAY OR ERROR MESSAGE?
YES
GO TO TEST 7
NO
SELECT MODEL YEAR 1989
SELECT SYSTEM TRANSMISSION
SELECT MODULE AUTO TRANSAXLE
NOTE: THE KEY MUST BE ON BEFORE PROCEEDING
THE CCD BUS TEST IS AUTOMATIC
THE BUS TEST PASSES WITH GREEN LIGHT?
YES
MESSAGE READS "NO RESPONSE"
YES
GO TO TEST 40
NO
GO TO TEST 3
NO
IS THE KEY ON ?
YES
GO TO TEST 41
NO
TURN KEY ON

TEST 2–CONDITIONS REQUIRED TO SET FAULT CODES PERFORM TEST 3 BEFORE PROCEEDING

A. Write down, then erase any fault code(s) in memory.

B. Find the code that you erased in the list below. Following the instructions next to that code, attempt to make the code reappear on the DRBII display.

CODE	REPEAT REQUIREMENTS
11	Turn ignition from off to on, then start engine. Move slowly through each shift lever position.
12	Status Code. Requires no action.
13	Same as Code 14.
14	Turn ignition from off to on. If no Limp-In occurs, start engine.
15	Same as Code 14.
16	Turn ignition from off to on.
17	Turn ignition from off to on.
18	Start engine and allow to run in park for a minimum of 3 seconds.
19	Start engine and wait for 15 seconds.
20	Turn ignition from off to on. If no code appears, start engine.
21 to 27	Run vehicle in N,R,1,2,3,4 for a minimum of 30 seconds in each gear.
28	Start engine and move shift lever slowly from Park through Low.
29	Turn ignition on, move throttle from closed to W.O.T.
31	Transaxle at normal operating temperature, drive in 1st and 2nd gears for a minute each.
32	Transaxle at normal operating temperature, drive in 1st and 3rd gears for a minute each.
33	Transaxle at normal operating temperature, drive in 1st gear for at least 1 minute.
37	Drive in 2nd, apply brake until 2-1 downshift occurs. Do this at least 3 times.
38	Transaxle at normal operating temperature, drive in 4th gear at light throttle for 17 to 20 seconds. Verify lockup by monitoring the bottom of the DRBII RPM display.
41 to 44	Start and run the engine for a minimum of 30 seconds.

C. If the fault code comes back, it is considered a "Hard Fault". GO TO TEST 4.

D. If the fault code does not come back, it is an intermittent code. GO TO TEST 9.

TEST 3—READING FAULT CODES PERFORM TEST 1 BEFORE PROCEEDING

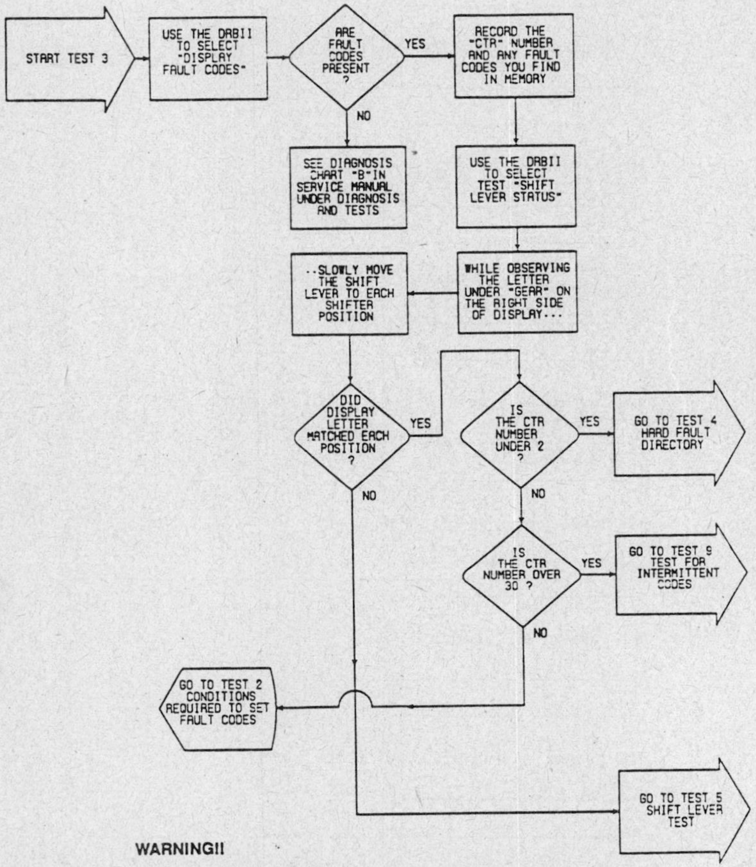
START TEST 3
USE THE DRBII TO SELECT "DISPLAY FAULT CODES"
ARE FAULT CODES PRESENT ?
YES
NO
RECORD THE "CTR" NUMBER AND ANY FAULT CODES YOU FIND IN MEMORY
SEE DIAGNOSIS CHART "B" IN SERVICE MANUAL UNDER DIAGNOSIS AND TESTS
USE THE DRBII TO SELECT TEST "SHIFT LEVER STATUS"
WHILE OBSERVING THE LETTER UNDER "GEAR" ON THE RIGHT SIDE OF DISPLAY...
..SLOWLY MOVE THE SHIFT LEVER TO EACH SHIFTER POSITION
DID DISPLAY LETTER MATCHED EACH POSITION ?
YES
NO
IS THE CTR NUMBER UNDER 2 ?
YES
NO
GO TO TEST 4 HARD FAULT DIRECTORY
IS THE CTR NUMBER OVER 30 ?
YES
NO
GO TO TEST 9 TEST FOR INTERMITTENT CODES
GO TO TEST 2 CONDITIONS REQUIRED TO SET FAULT CODES
GO TO TEST 5 SHIFT LEVER TEST

WARNING!!

DO NOT USE FAULT CODES TO DIAGNOSE TRANSMISSION PROBLEMS UNTIL AFTER YOU HAVE PERFORMED AND PASSED THE "SHIFT. LEVER STATUS" TEST.

TEST 4—"HARD FAULT" CODE DIRECTORY

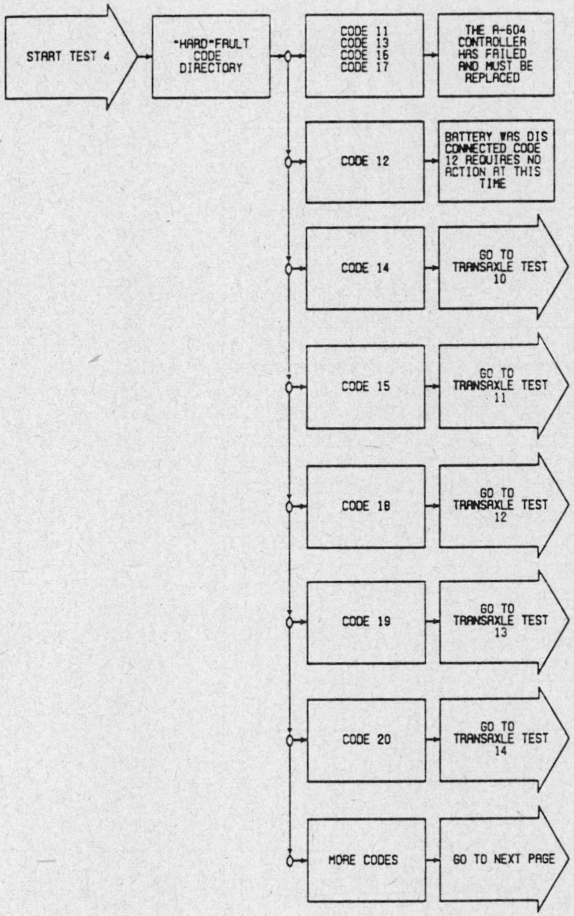
START TEST 4
"HARD" FAULT CODE DIRECTORY
CODE 11 CODE 13 CODE 16 CODE 17
THE A-604 CONTROLLER HAS FAILED AND MUST BE REPLACED
CODE 12
BATTERY WAS DISCONNECTED CODE 12 REQUIRES NO ACTION AT THIS TIME
CODE 14
GO TO TRANSAXLE TEST 10
CODE 15
GO TO TRANSAXLE TEST 11
CODE 18
GO TO TRANSAXLE TEST 12
CODE 19
GO TO TRANSAXLE TEST 13
CODE 20
GO TO TRANSAXLE TEST 14
MORE CODES
GO TO NEXT PAGE

TEST 4—"HARD FAULT" CODE DIRECTORY

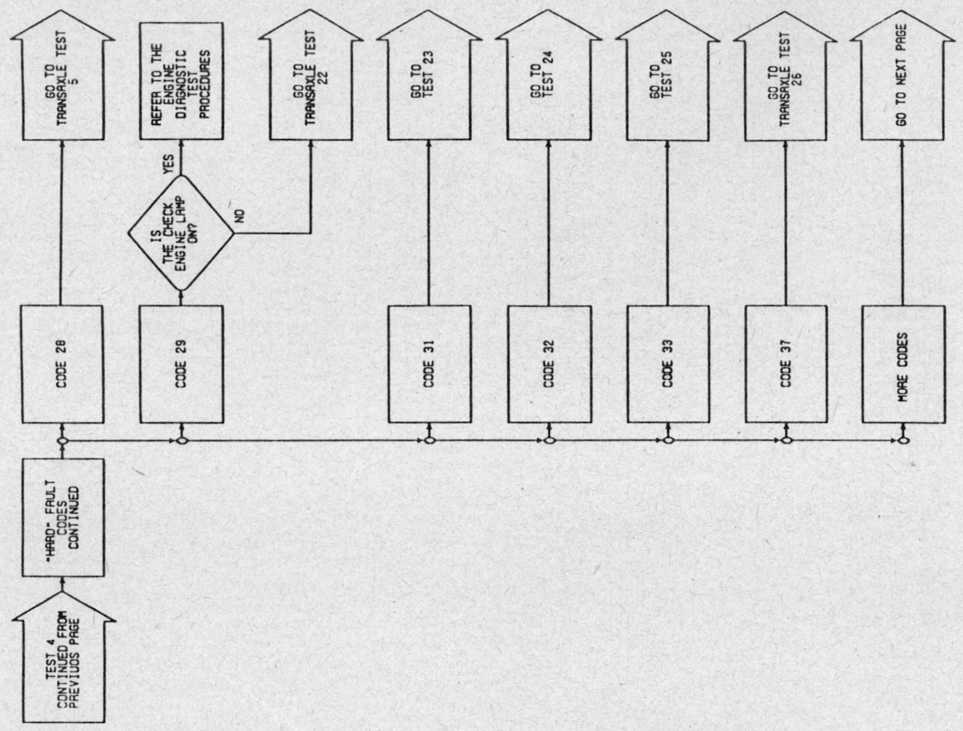
TEST 4 CONTINUED FROM PREVIOUS PAGE
"HARD" FAULT CODES CONTINUED
CODE 28
GO TO TRANSAXLE TEST 5
CODE 29
IS THE CHECK ENGINE LAMP ON?
YES
REFER TO THE ENGINE DIAGNOSTIC TEST PROCEDURES
NO
GO TO TRANSAXLE TEST 22
CODE 31
GO TO TEST 23
CODE 32
GO TO TEST 24
CODE 33
GO TO TEST 25
CODE 37
GO TO TRANSAXLE TEST 26
MORE CODES
GO TO NEXT PAGE

TEST 4—"HARD FAULT" CODE DIRECTORY

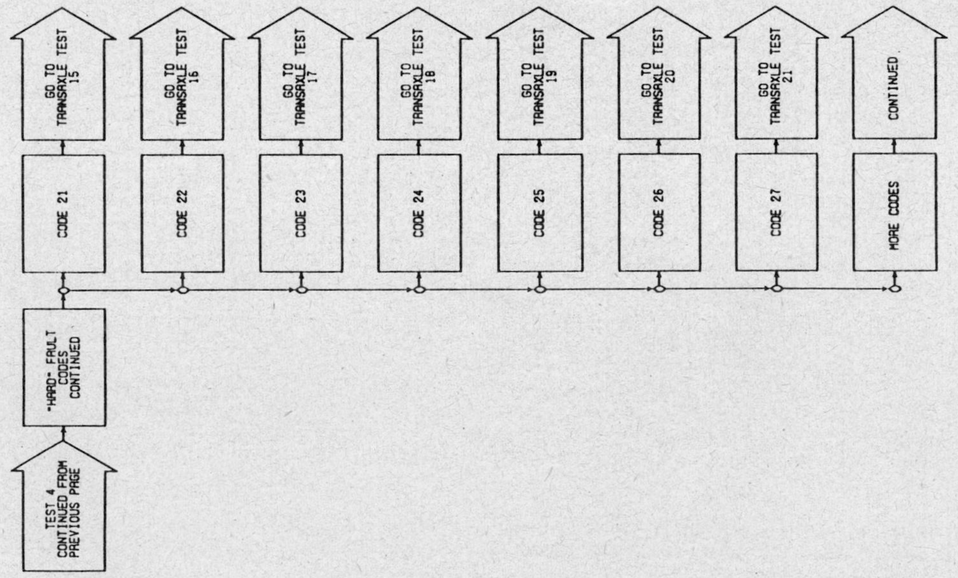
TEST 4 CONTINUED FROM PREVIOUS PAGE
"HARD" FAULT CODES CONTINUED
CODE 21
GO TO TRANSAXLE TEST 15
CODE 22
GO TO TRANSAXLE TEST 16
CODE 23
GO TO TRANSAXLE TEST 17
CODE 24
GO TO TRANSAXLE TEST 18
CODE 25
GO TO TRANSAXLE TEST 19
CODE 26
GO TO TRANSAXLE TEST 20
CODE 27
GO TO TRANSAXLE TEST 21
MORE CODES
CONTINUED

TEST 4—"HARD FAULT" CODE DIRECTORY

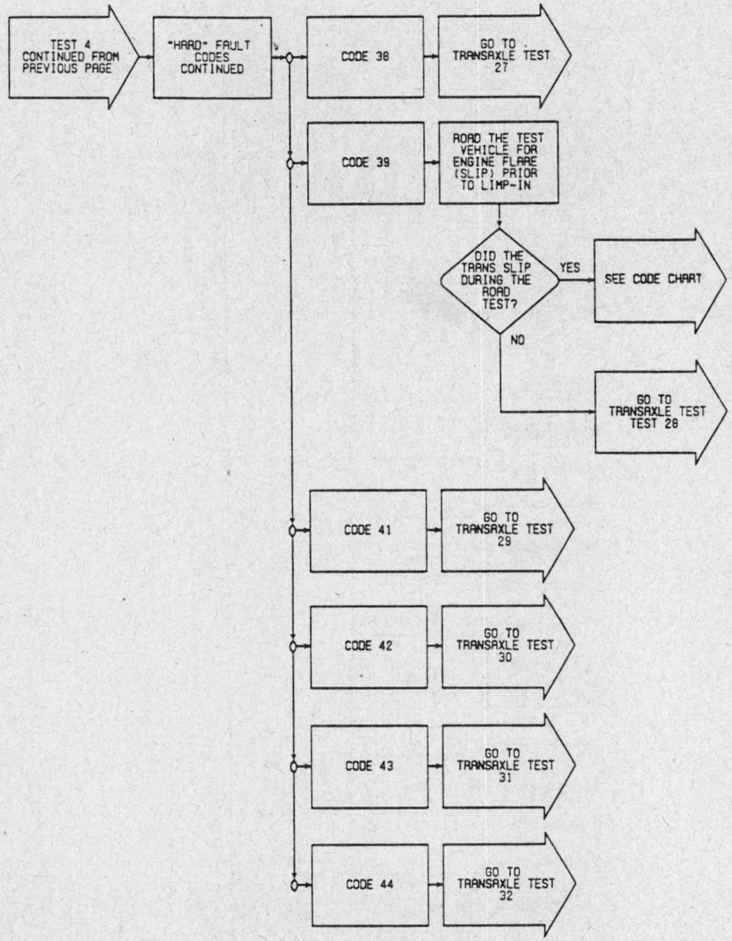
TEST 4 CONTINUED FROM PREVIOUS PAGE
"HARD" FAULT CODES CONTINUED
CODE 38
GO TO TRANSAXLE TEST 27
CODE 39
ROAD THE TEST VEHICLE FOR ENGINE FLARE (SLIP) PRIOR TO LIMP-IN
DID THE TRANS SLIP DURING THE ROAD TEST?
YES
SEE CODE CHART
NO
GO TO TRANSAXLE TEST TEST 28
CODE 41
GO TO TRANSAXLE TEST 29
CODE 42
GO TO TRANSAXLE TEST 30
CODE 43
GO TO TRANSAXLE TEST 31
CODE 44
GO TO TRANSAXLE TEST 32

TEST 5—SHIFT LEVER TEST
PERFORM TEST 2 BEFORE PROCEEDING

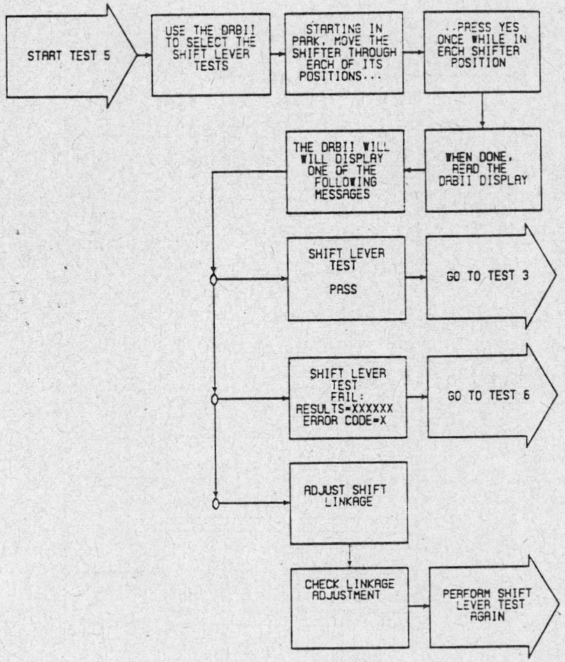
START TEST 5
USE THE DRBII TO SELECT THE SHIFT LEVER TESTS
STARTING IN PARK, MOVE THE SHIFTER THROUGH EACH OF ITS POSITIONS...
..PRESS YES ONCE WHILE IN EACH SHIFTER POSITION
WHEN DONE, READ THE DRBII DISPLAY
THE DRBII WILL WILL DISPLAY ONE OF THE FOLLOWING MESSAGES
SHIFT LEVER TEST PASS
GO TO TEST 3
SHIFT LEVER TEST FAIL: RESULTS=XXXXXX ERROR CODE=X
GO TO TEST 6
ADJUST SHIFT LINKAGE
CHECK LINKAGE ADJUSTMENT
PERFORM SHIFT LEVER TEST AGAIN

TEST 6—FAILED SHIFT LEVER TESTS PERFORM TEST 5 BEFORE PROCEEDING

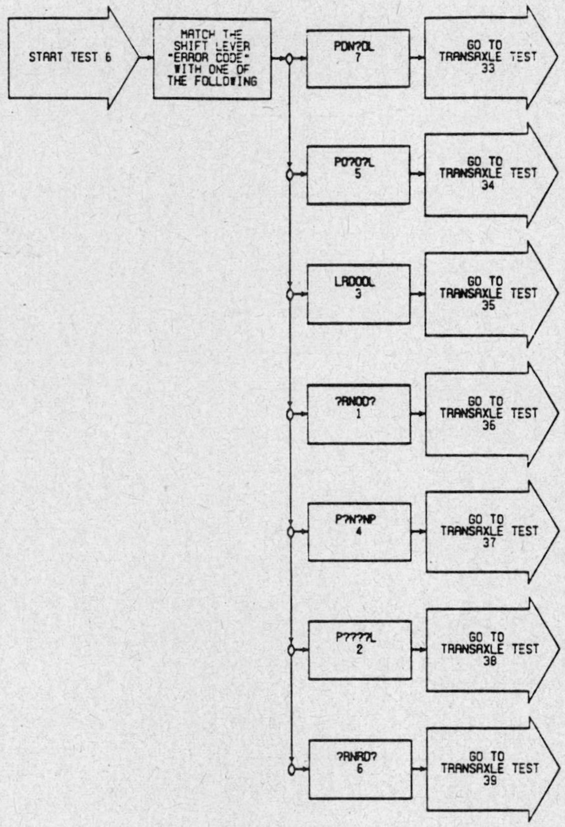
START TEST 6
MATCH THE SHIFT LEVER "ERROR CODE" WITH ONE OF THE FOLLOWING
PDN?DL 7
GO TO TRANSAXLE TEST 33
PD?D?L 5
GO TO TRANSAXLE TEST 34
LRDODL 3
GO TO TRANSAXLE TEST 35
?RNOD? 1
GO TO TRANSAXLE TEST 36
P?N?NP 4
GO TO TRANSAXLE TEST 37
P????L 2
GO TO TRANSAXLE TEST 38
?RNRD? 6
GO TO TRANSAXLE TEST 39

TEST 7—DRBII ERROR MESSAGES PERFORM TEST 1 BEFORE PROCEEDING

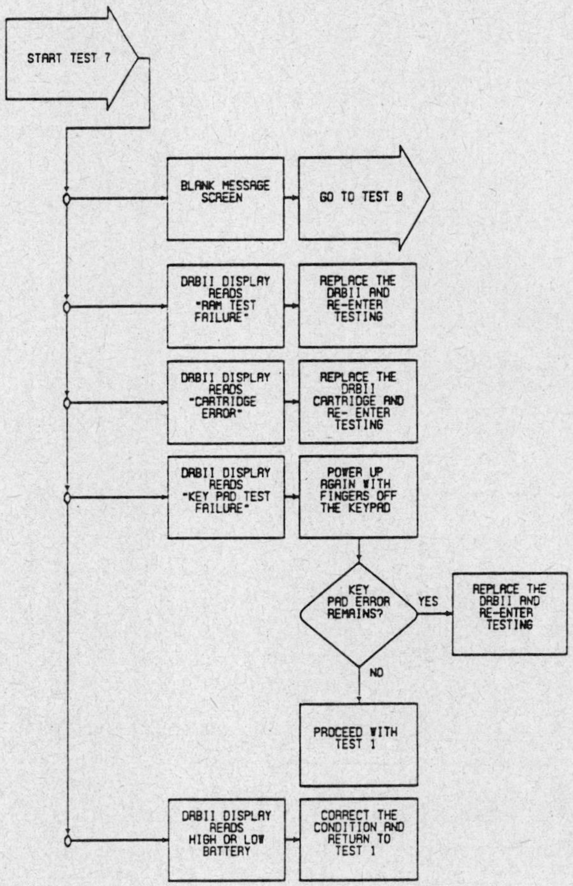
START TEST 7
BLANK MESSAGE SCREEN
GO TO TEST 8
DRBII DISPLAY READS "RAM TEST FAILURE"
REPLACE THE DRBII AND RE-ENTER TESTING
DRBII DISPLAY READS "CARTRIDGE ERROR"
REPLACE THE DRBII CARTRIDGE AND RE- ENTER TESTING
DRBII DISPLAY READS "KEY PAD TEST FAILURE"
POWER UP AGAIN WITH FINGERS OFF THE KEYPAD
KEY PAD ERROR REMAINS?
YES
REPLACE THE DRBII AND RE-ENTER TESTING
NO
PROCEED WITH TEST 1
DRBII DISPLAY READS HIGH OR LOW BATTERY
CORRECT THE CONDITION AND RETURN TO TEST 1

TEST 8—TEST FOR CAUSE OF BLANK MESSAGE SCREEN PERFORM TEST 1 BEFORE PROCEEDING

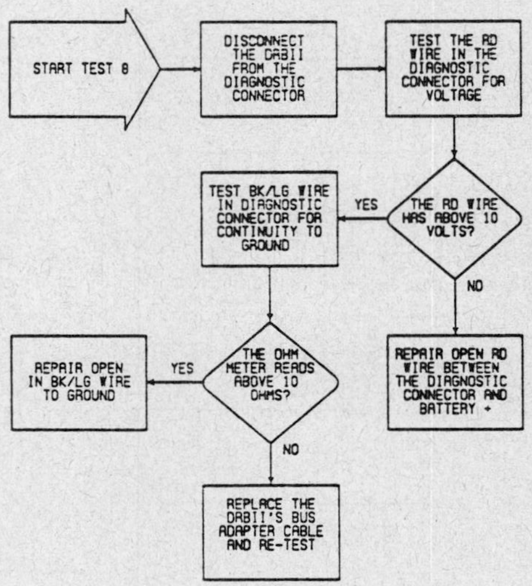
START TEST 8
DISCONNECT THE DRBII FROM THE DIAGNOSTIC CONNECTOR
TEST THE RD WIRE IN THE DIAGNOSTIC CONNECTOR FOR VOLTAGE
THE RD WIRE HAS ABOVE 10 VOLTS?
YES
NO
TEST BK/LG WIRE IN DIAGNOSTIC CONNECTOR FOR CONTINUITY TO GROUND
REPAIR OPEN RD WIRE BETWEEN THE DIAGNOSTIC CONNECTOR AND BATTERY +
THE OHM METER READS ABOVE 10 OHMS?
YES
NO
REPAIR OPEN IN BK/LG WIRE TO GROUND
REPLACE THE DRBII'S BUS ADAPTER CABLE AND RE-TEST

TEST 9—TEST FOR INTERMITTENT FAULT CODES PERFORM TEST 2 BEFORE PROCEEDING

1. Check the following for push-outs or flaired connectors:
   A. Engine ground connection
   B. 60-way ground pins 53, 54, 57 & 58
   C. 60-way battery feed (J-11) pin 56
   D. Battery feed to EATX relay from I.O.D. connector
   E. 60-way J-2 feed pin 11

   If the above connections are OK . . .

2. Select the intermittent code from the list below. Carefully inspect the pins and connector cavities that are listed next to the code for push-outs or terminal damage.

INTERMITTENT CODE	60-WAY PIN #	8-WAY PIN #	OTHER
11,13,16,17			Replace the A-604 controller
12			Requires no action
14	16 and 17		EATX connector/relay
15	15 and 16		EATX connector/relay
18	45		
19	4 and 43		Engine controller pin 46 and 26 (Pin 46 and 6 on Minivan)
20	16 and 17		EATX connector/relay
21	9	3 and 4	
22	47	1, 3 and 4	
23	47	1 and 4	
24	50	2 and 4	
25	50	2, 3 and 4	
26	50	2 and 4	
27	47	1 and 4	
28	1, 2, 3, 41, 42		NS connector/switch, PRNODL connector/switch
29	12 and 51		Throttle position sensor connector
31			
32			
33			
37			
38			
39	13, 14, 52		Turbine or output speed sensor connector
41	16, 20, 57, 58	4 pin 7	
42	16, 19, 57, 58	4 pin 8	
43	16,60	4 pin 6	
44	16, 59	4 pin 5	

3. If steps 1 and 2 fail to turn up a defective condition, use the DRBII to erase all fault codes and reset the start counter to zero. Road test the vehicle for proper transaxle function. If the transaxle function is acceptable, take no action at this time.

TEST 10
TEST FOR CODE 14—RELAY OUTPUT ALWAYS ON
PERFORM TEST 3 BEFORE PROCEEDING

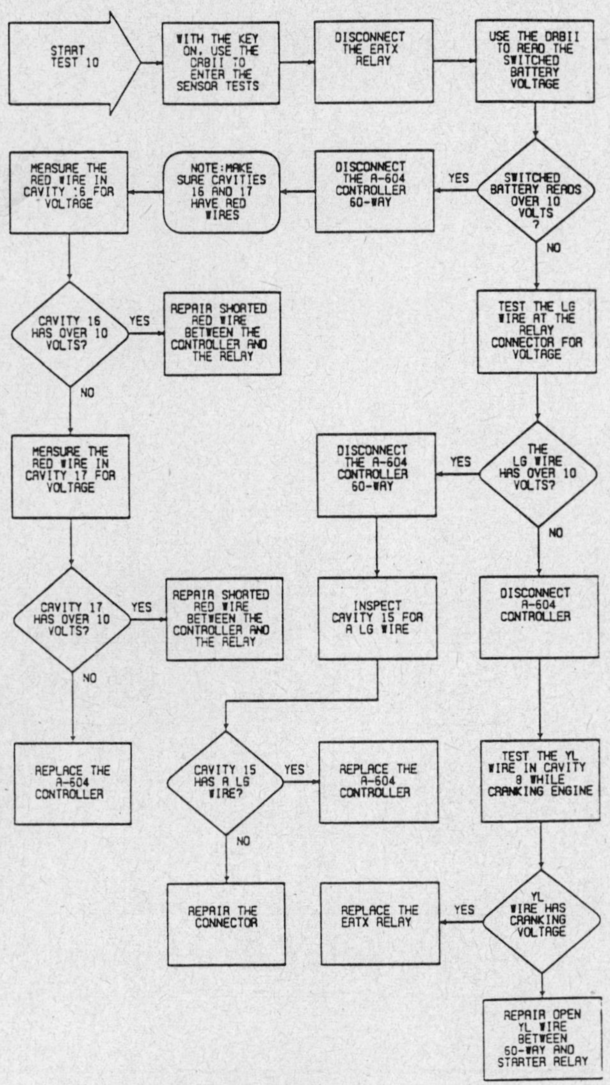
START TEST 10
WITH THE KEY ON, USE THE DRBII TO ENTER THE SENSOR TESTS
DISCONNECT THE EATX RELAY
USE THE DRBII TO READ THE SWITCHED BATTERY VOLTAGE
SWITCHED BATTERY READS OVER 10 VOLTS ?
YES
NO
DISCONNECT THE A-604 CONTROLLER 60-WAY
NOTE: MAKE SURE CAVITIES 16 AND 17 HAVE RED WIRES
MEASURE THE RED WIRE IN CAVITY 16 FOR VOLTAGE
CAVITY 16 HAS OVER 10 VOLTS?
YES
REPAIR SHORTED RED WIRE BETWEEN THE CONTROLLER AND THE RELAY
NO
MEASURE THE RED WIRE IN CAVITY 17 FOR VOLTAGE
CAVITY 17 HAS OVER 10 VOLTS?
YES
REPAIR SHORTED RED WIRE BETWEEN THE CONTROLLER AND THE RELAY
NO
REPLACE THE A-604 CONTROLLER
TEST THE LG WIRE AT THE RELAY CONNECTOR FOR VOLTAGE
THE LG WIRE HAS OVER 10 VOLTS?
YES
DISCONNECT THE A-604 CONTROLLER 60-WAY
INSPECT CAVITY 15 FOR A LG WIRE
CAVITY 15 HAS A LG WIRE?
YES
REPLACE THE A-604 CONTROLLER
NO
REPAIR THE CONNECTOR
NO
DISCONNECT A-604 CONTROLLER
TEST THE YL WIRE IN CAVITY 8 WHILE CRANKING ENGINE
YL WIRE HAS CRANKING VOLTAGE
YES
REPLACE THE EATX RELAY
REPAIR OPEN YL WIRE BETWEEN 60-WAY AND STARTER RELAY

TEST 11
TEST FOR CODE 15—RELAY OUTPUT ALWAYS OFF
PERFORM TEST 3 BEFORE PROCEEDING

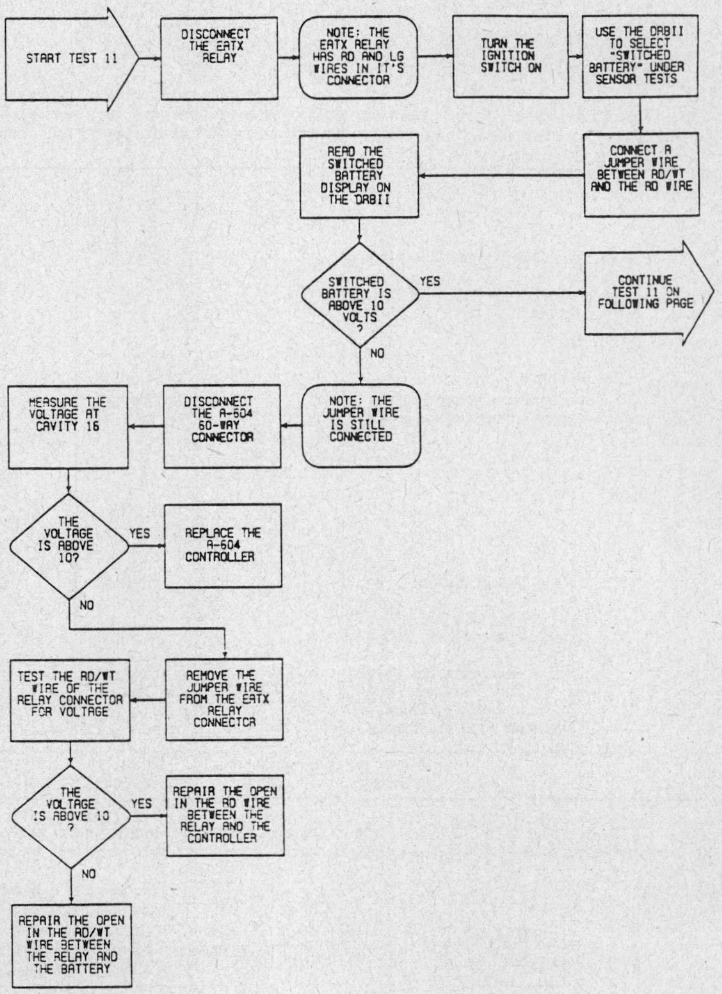
START TEST 11
DISCONNECT THE EATX RELAY
NOTE: THE EATX RELAY HAS RD AND LG WIRES IN IT'S CONNECTOR
TURN THE IGNITION SWITCH ON
USE THE DRBII TO SELECT "SWITCHED BATTERY" UNDER SENSOR TESTS
CONNECT A JUMPER WIRE BETWEEN RD/WT AND THE RD WIRE
READ THE SWITCHED BATTERY DISPLAY ON THE DRBII
SWITCHED BATTERY IS ABOVE 10 VOLTS ?
YES
CONTINUE TEST 11 ON FOLLOWING PAGE
NO
NOTE: THE JUMPER WIRE IS STILL CONNECTED
DISCONNECT THE A-604 60-WAY CONNECTOR
MEASURE THE VOLTAGE AT CAVITY 16
THE VOLTAGE IS ABOVE 10?
YES
REPLACE THE A-604 CONTROLLER
NO
REMOVE THE JUMPER WIRE FROM THE EATX RELAY CONNECTOR
TEST THE RD/WT WIRE OF THE RELAY CONNECTOR FOR VOLTAGE
THE VOLTAGE IS ABOVE 10 ?
YES
REPAIR THE OPEN IN THE RD WIRE BETWEEN THE RELAY AND THE CONTROLLER
NO
REPAIR THE OPEN IN THE RD/WT WIRE BETWEEN THE RELAY AND THE BATTERY

TEST 11
TEST FOR CODE 15—RELAY OUTPUT ALWAYS OFF
PERFORM TEST 3 BEFORE PROCEEDING

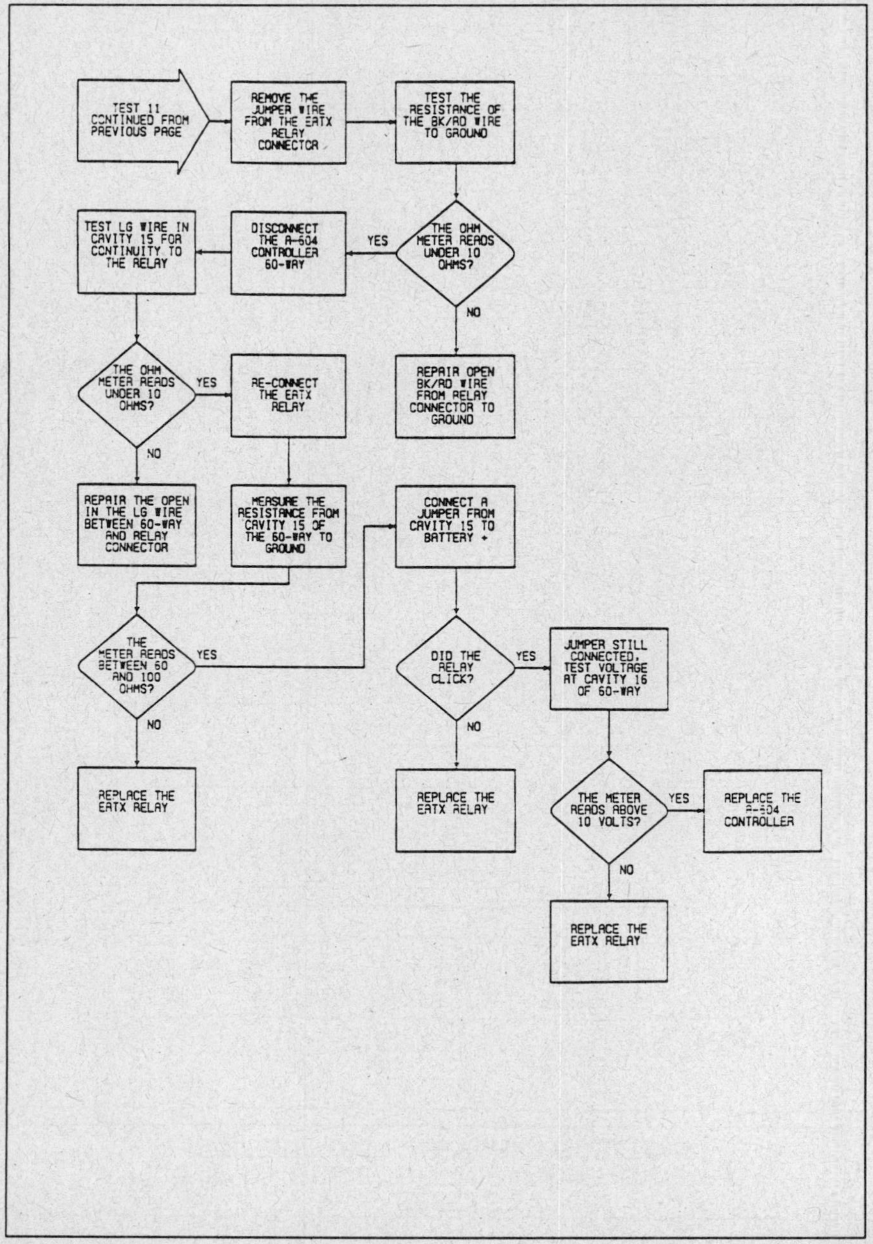
TEST 11 CONTINUED FROM PREVIOUS PAGE
REMOVE THE JUMPER WIRE FROM THE EATX RELAY CONNECTOR
TEST THE RESISTANCE OF THE BK/RD WIRE TO GROUND
THE OHM METER READS UNDER 10 OHMS?
YES
NO
DISCONNECT THE A-604 CONTROLLER 60-WAY
REPAIR OPEN BK/RD WIRE FROM RELAY CONNECTOR TO GROUND
TEST LG WIRE IN CAVITY 15 FOR CONTINUITY TO THE RELAY
THE OHM METER READS UNDER 10 OHMS?
YES
NO
RE-CONNECT THE EATX RELAY
REPAIR THE OPEN IN THE LG WIRE BETWEEN 60-WAY AND RELAY CONNECTOR
MEASURE THE RESISTANCE FROM CAVITY 15 OF THE 60-WAY TO GROUND
THE METER READS BETWEEN 60 AND 100 OHMS?
YES
NO
REPLACE THE EATX RELAY
CONNECT A JUMPER FROM CAVITY 15 TO BATTERY +
DID THE RELAY CLICK?
YES
NO
REPLACE THE EATX RELAY
JUMPER STILL CONNECTED, TEST VOLTAGE AT CAVITY 16 OF 60-WAY
THE METER READS ABOVE 10 VOLTS?
YES
NO
REPLACE THE A-604 CONTROLLER
REPLACE THE EATX RELAY

TEST 12
TEST FOR CODE 18—ENGINE SPEED SENSOR CIRCUIT
PERFORM TEST 3 BEFORE PROCEEDING

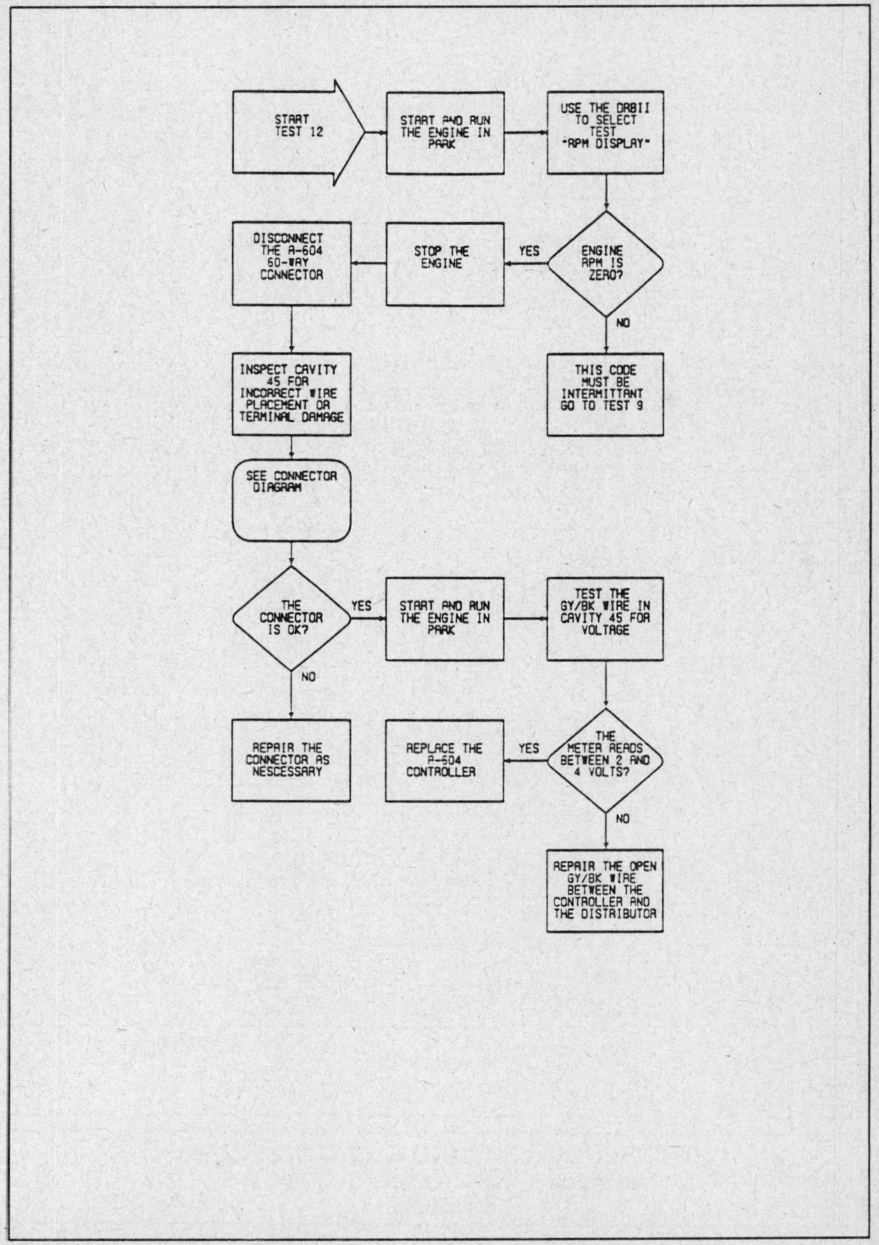
START TEST 12
START AND RUN THE ENGINE IN PARK
USE THE DRBII TO SELECT TEST "RPM DISPLAY"
ENGINE RPM IS ZERO?
YES
NO
THIS CODE MUST BE INTERMITTANT GO TO TEST 9
STOP THE ENGINE
DISCONNECT THE A-604 60-WAY CONNECTOR
INSPECT CAVITY 45 FOR INCORRECT WIRE PLACEMENT OR TERMINAL DAMAGE
SEE CONNECTOR DIAGRAM
THE CONNECTOR IS OK?
YES
NO
REPAIR THE CONNECTOR AS NESCESSARY
START AND RUN THE ENGINE IN PARK
TEST THE GY/BK WIRE IN CAVITY 45 FOR VOLTAGE
THE METER READS BETWEEN 2 AND 4 VOLTS?
YES
NO
REPLACE THE A-604 CONTROLLER
REPAIR THE OPEN GY/BK WIRE BETWEEN THE CONTROLLER AND THE DISTRIBUTOR

TEST 13
CODE 19—BUS COMMUNICATION WITH SMEC
PERFORM TEST 3 BEFORE PROCEEDING

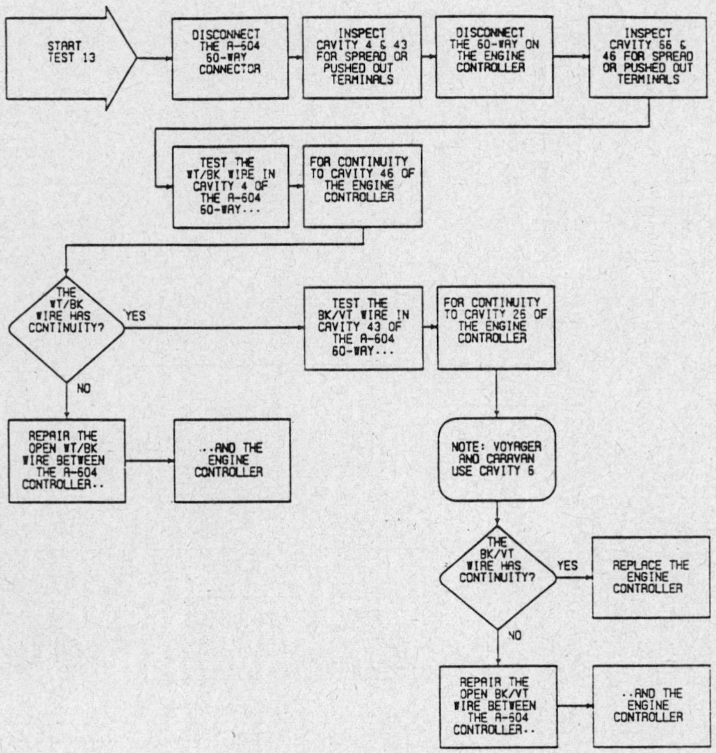
START TEST 13
DISCONNECT THE A-604 60-WAY CONNECTOR
INSPECT CAVITY 4 & 43 FOR SPREAD OR PUSHED OUT TERMINALS
DISCONNECT THE 60-WAY ON THE ENGINE CONTROLLER
INSPECT CAVITY 56 & 46 FOR SPREAD OR PUSHED OUT TERMINALS
TEST THE WT/BK WIRE IN CAVITY 4 OF THE A-604 60-WAY...
FOR CONTINUITY TO CAVITY 46 OF THE ENGINE CONTROLLER
THE WT/BK WIRE HAS CONTINUITY?
YES
NO
REPAIR THE OPEN WT/BK WIRE BETWEEN THE A-604 CONTROLLER..
..AND THE ENGINE CONTROLLER
TEST THE BK/VT WIRE IN CAVITY 43 OF THE A-604 60-WAY...
FOR CONTINUITY TO CAVITY 25 OF THE ENGINE CONTROLLER
NOTE: VOYAGER AND CARAVAN USE CAVITY 6
THE BK/VT WIRE HAS CONTINUITY?
YES
REPLACE THE ENGINE CONTROLLER
NO
REPAIR THE OPEN BK/VT WIRE BETWEEN THE A-604 CONTROLLER..
..AND THE ENGINE CONTROLLER

TEST 14
CODE 20—SWITCHED BATTERY
PERFORM TEST 3 BEFORE PROCEEDING

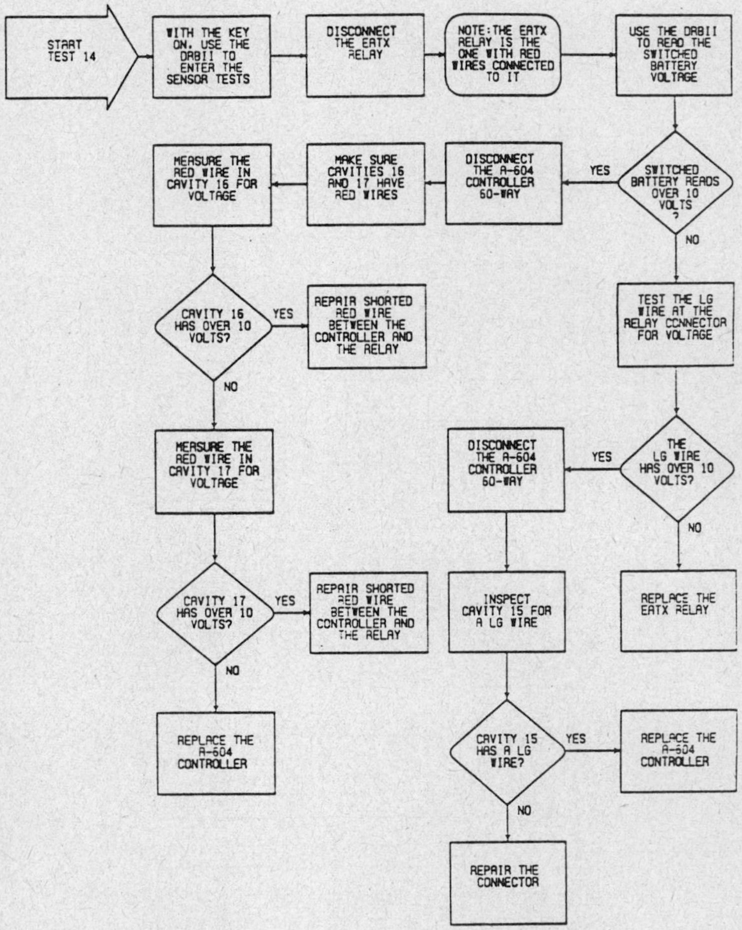
START TEST 14
WITH THE KEY ON, USE THE DRBII TO ENTER THE SENSOR TESTS
DISCONNECT THE EATX RELAY
NOTE: THE EATX RELAY IS THE ONE WITH RED WIRES CONNECTED TO IT
USE THE DRBII TO READ THE SWITCHED BATTERY VOLTAGE
SWITCHED BATTERY READS OVER 10 VOLTS ?
YES
NO
DISCONNECT THE A-604 CONTROLLER 60-WAY
MAKE SURE CAVITIES 16 AND 17 HAVE RED WIRES
MEASURE THE RED WIRE IN CAVITY 16 FOR VOLTAGE
CAVITY 16 HAS OVER 10 VOLTS?
YES
REPAIR SHORTED RED WIRE BETWEEN THE CONTROLLER AND THE RELAY
NO
MEASURE THE RED WIRE IN CAVITY 17 FOR VOLTAGE
CAVITY 17 HAS OVER 10 VOLTS?
YES
REPAIR SHORTED RED WIRE BETWEEN THE CONTROLLER AND THE RELAY
NO
REPLACE THE A-604 CONTROLLER
TEST THE LG WIRE AT THE RELAY CONNECTOR FOR VOLTAGE
THE LG WIRE HAS OVER 10 VOLTS?
YES
NO
REPLACE THE EATX RELAY
DISCONNECT THE A-604 CONTROLLER 60-WAY
INSPECT CAVITY 15 FOR A LG WIRE
CAVITY 15 HAS A LG WIRE?
YES
REPLACE THE A-604 CONTROLLER
NO
REPAIR THE CONNECTOR

TEST 15
CODE 21—OD PRESSURE SWITCH CIRCUIT
PERFORM TEST 3 BEFORE PROCEEDING

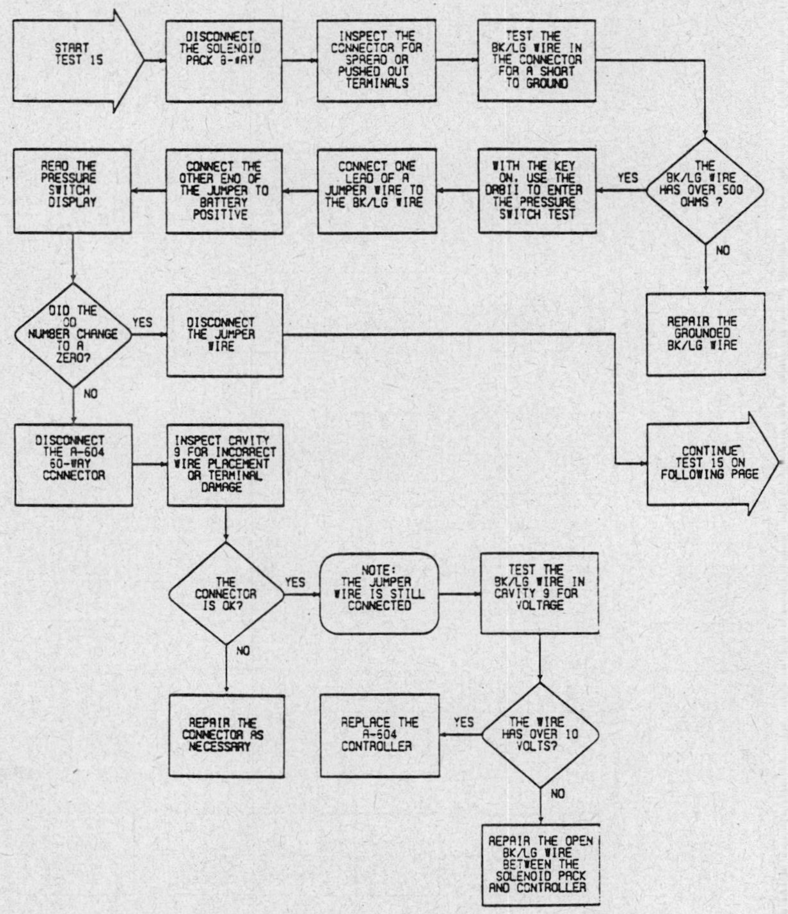
START TEST 15
DISCONNECT THE SOLENOID PACK 8-WAY
INSPECT THE CONNECTOR FOR SPREAD OR PUSHED OUT TERMINALS
TEST THE BK/LG WIRE IN THE CONNECTOR FOR A SHORT TO GROUND
THE BK/LG WIRE HAS OVER 500 OHMS ?
YES
NO
REPAIR THE GROUNDED BK/LG WIRE
WITH THE KEY ON, USE THE DRBII TO ENTER THE PRESSURE SWITCH TEST
CONNECT ONE LEAD OF A JUMPER WIRE TO THE BK/LG WIRE
CONNECT THE OTHER END OF THE JUMPER TO BATTERY POSITIVE
READ THE PRESSURE SWITCH DISPLAY
DID THE OD NUMBER CHANGE TO A ZERO?
YES
DISCONNECT THE JUMPER WIRE
CONTINUE TEST 15 ON FOLLOWING PAGE
NO
DISCONNECT THE A-604 60-WAY CONNECTOR
INSPECT CAVITY 9 FOR INCORRECT WIRE PLACEMENT OR TERMINAL DAMAGE
THE CONNECTOR IS OK?
YES
NOTE: THE JUMPER WIRE IS STILL CONNECTED
TEST THE BK/LG WIRE IN CAVITY 9 FOR VOLTAGE
THE WIRE HAS OVER 10 VOLTS?
YES
REPLACE THE A-604 CONTROLLER
NO
REPAIR THE OPEN BK/LG WIRE BETWEEN THE SOLENOID PACK AND CONTROLLER
NO
REPAIR THE CONNECTOR AS NECESSARY

TEST 15
CODE 21—OD PRESSURE SWITCH CIRCUIT
PERFORM TEST 3 BEFORE PROCEEDING

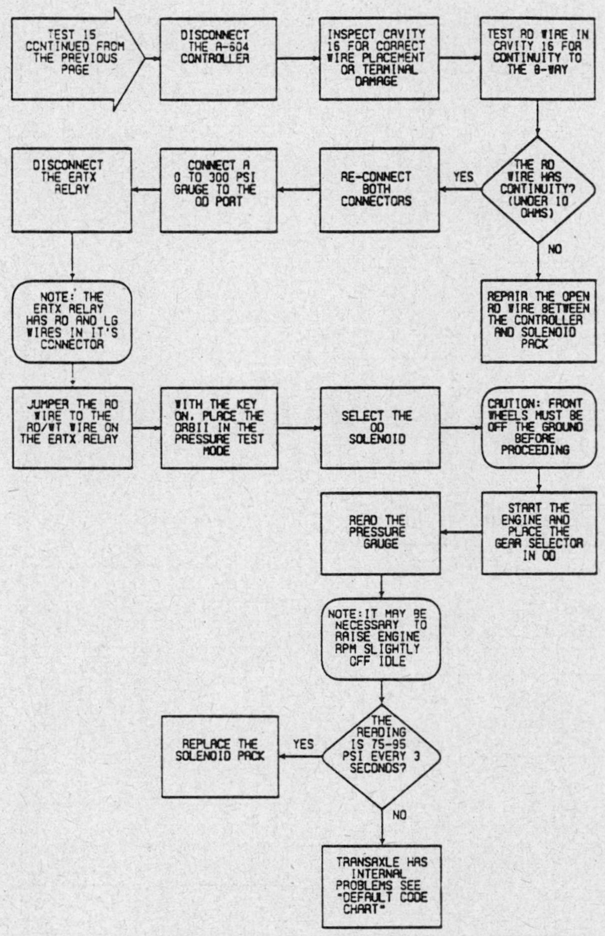
TEST 15 CONTINUED FROM THE PREVIOUS PAGE
DISCONNECT THE A-604 CONTROLLER
INSPECT CAVITY 16 FOR CORRECT WIRE PLACEMENT OR TERMINAL DAMAGE
TEST RD WIRE IN CAVITY 16 FOR CONTINUITY TO THE 8-WAY
THE RD WIRE HAS CONTINUITY? (UNDER 10 OHMS)
NO
REPAIR THE OPEN RD WIRE BETWEEN THE CONTROLLER AND SOLENOID PACK
YES
RE-CONNECT BOTH CONNECTORS
CONNECT A 0 TO 300 PSI GAUGE TO THE OD PORT
DISCONNECT THE EATX RELAY
NOTE: THE EATX RELAY HAS RD AND LG WIRES IN IT'S CONNECTOR
JUMPER THE RD WIRE TO THE RD/WT WIRE ON THE EATX RELAY
WITH THE KEY ON, PLACE THE DRBII IN THE PRESSURE TEST MODE
SELECT THE OD SOLENOID
CAUTION: FRONT WHEELS MUST BE OFF THE GROUND BEFORE PROCEEDING
START THE ENGINE AND PLACE THE GEAR SELECTOR IN OD
READ THE PRESSURE GAUGE
NOTE: IT MAY BE NECESSARY TO RAISE ENGINE RPM SLIGHTLY OFF IDLE
THE READING IS 75-95 PSI EVERY 3 SECONDS?
YES
REPLACE THE SOLENOID PACK
NO
TRANSAXLE HAS INTERNAL PROBLEMS SEE "DEFAULT CODE CHART"

TEST 16
CODE 22—2–4 PRESSURE SWITCH CIRCUIT
PERFORM TEST 3 BEFORE PROCEEDING

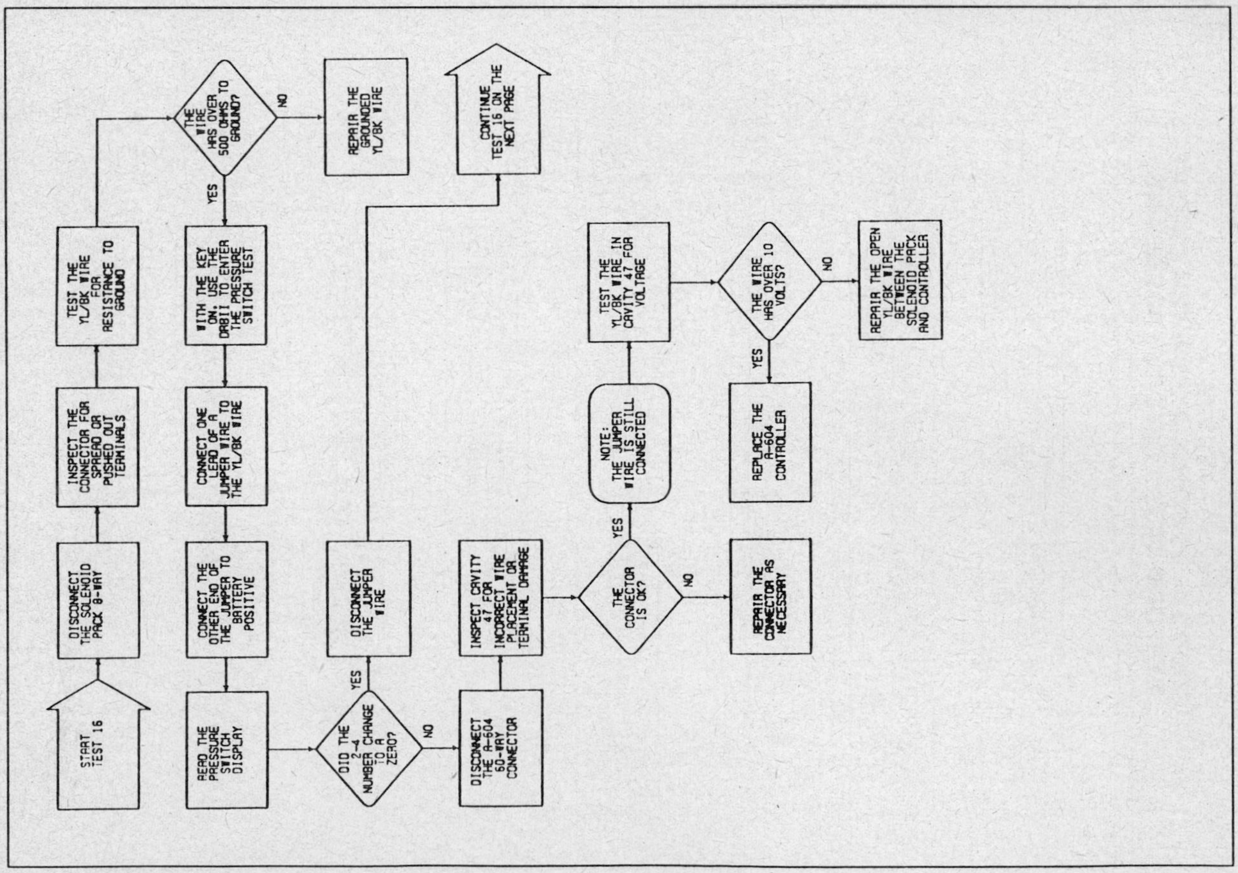
START TEST 16
DISCONNECT THE SOLENOID PACK 8-WAY
INSPECT THE CONNECTOR FOR SPREAD OR PUSHED OUT TERMINALS
TEST THE YL/BK WIRE FOR RESISTANCE TO GROUND
THE WIRE HAS OVER 500 OHMS TO GROUND?
NO
REPAIR THE GROUNDED YL/BK WIRE
YES
WITH THE KEY ON, USE THE DRBII TO ENTER THE PRESSURE SWITCH TEST
CONNECT ONE LEAD OF A JUMPER WIRE TO THE YL/BK WIRE
CONNECT THE OTHER END OF THE JUMPER TO BATTERY POSITIVE
READ THE PRESSURE SWITCH DISPLAY
DID THE 2-4 NUMBER CHANGE TO A ZERO?
YES
DISCONNECT THE JUMPER WIRE
CONTINUE TEST 16 ON THE NEXT PAGE
NO
DISCONNECT THE A-604 60-WAY CONNECTOR
INSPECT CAVITY 47 FOR INCORRECT WIRE PLACEMENT OR TERMINAL DAMAGE
THE CONNECTOR IS OK?
NO
REPAIR THE CONNECTOR AS NECESSARY
YES
NOTE: THE JUMPER WIRE IS STILL CONNECTED
TEST THE YL/BK WIRE IN CAVITY 47 FOR VOLTAGE
THE WIRE HAS OVER 10 VOLTS?
YES
REPLACE THE A-604 CONTROLLER
NO
REPAIR THE OPEN YL/BK WIRE BETWEEN THE SOLENOID PACK AND CONTROLLER

TEST 16
CODE 22—2–4 PRESSURE SWITCH CIRCUIT
PERFORM TEST 3 BEFORE PROCEEDING

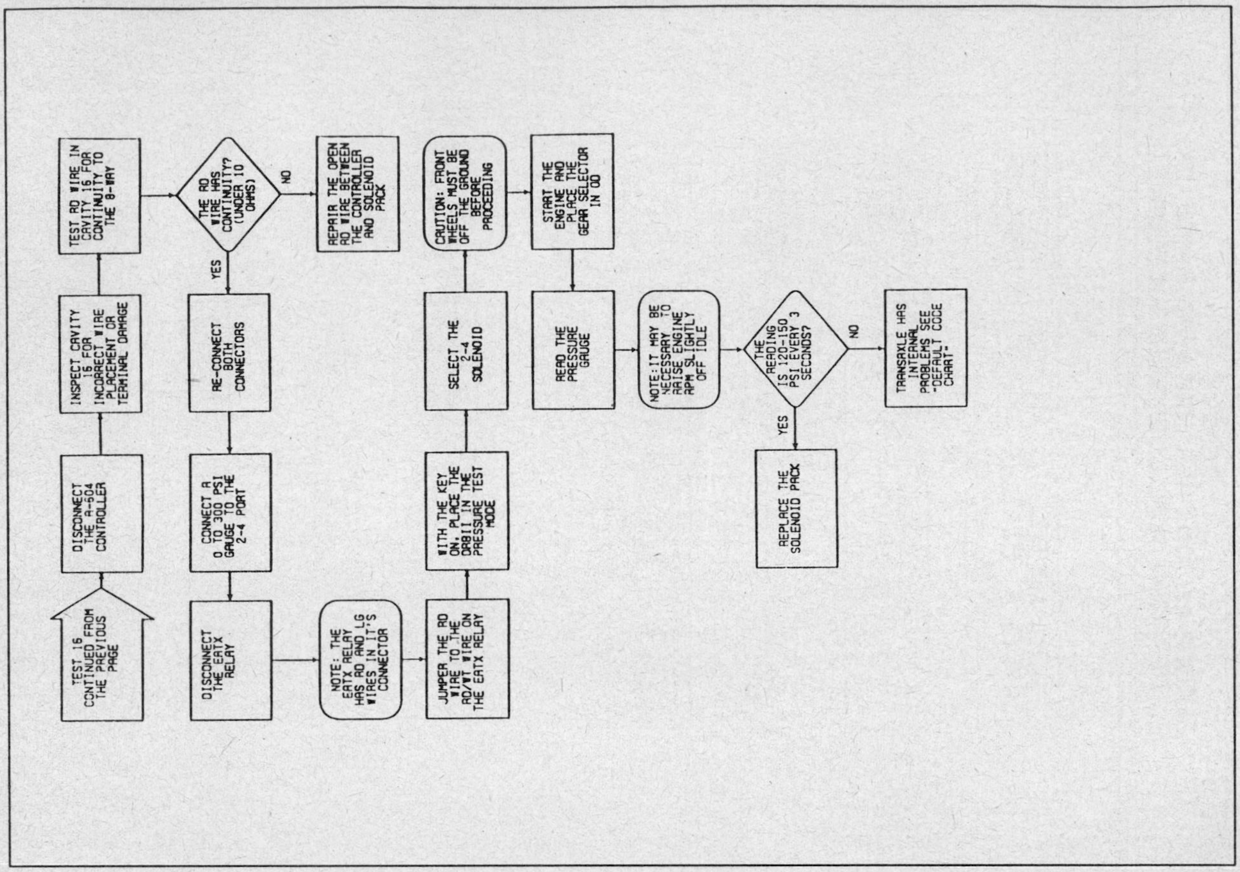
TEST 16 CONTINUED FROM THE PREVIOUS PAGE
DISCONNECT THE A-604 CONTROLLER
INSPECT CAVITY 16 FOR INCORRECT WIRE PLACEMENT OR TERMINAL DAMAGE
TEST RD WIRE IN CAVITY 16 FOR CONTINUITY TO THE 8-WAY
THE RD WIRE HAS CONTINUITY? (UNDER 10 OHMS)
NO
REPAIR THE OPEN RD WIRE BETWEEN THE CONTROLLER AND SOLENOID PACK
YES
RE-CONNECT BOTH CONNECTORS
CONNECT A 0 TO 300 PSI GAUGE TO THE 2-4 PORT
DISCONNECT THE EATX RELAY
NOTE: THE EATX RELAY HAS RD AND LG WIRES IN IT'S CONNECTOR
JUMPER THE RD WIRE TO THE RD/WT WIRE ON THE EATX RELAY
WITH THE KEY ON, PLACE THE DRBII IN THE PRESSURE TEST MODE
SELECT THE 2-4 SOLENOID
CAUTION: FRONT WHEELS MUST BE OFF THE GROUND BEFORE PROCEEDING
START THE ENGINE AND PLACE THE GEAR SELECTOR IN OD
READ THE PRESSURE GAUGE
NOTE: IT MAY BE NECESSARY TO RAISE ENGINE RPM SLIGHTLY OFF IDLE
THE READING IS 120-150 PSI EVERY 3 SECONDS?
YES
REPLACE THE SOLENOID PACK
NO
TRANSAXLE HAS INTERNAL PROBLEMS SEE "DEFAULT CODE CHART"

TEST 17
CODE 23—2-4/OD PRESSURE SWITCH CIRCUIT
PERFORM TEST 3 BEFORE PROCEEDING

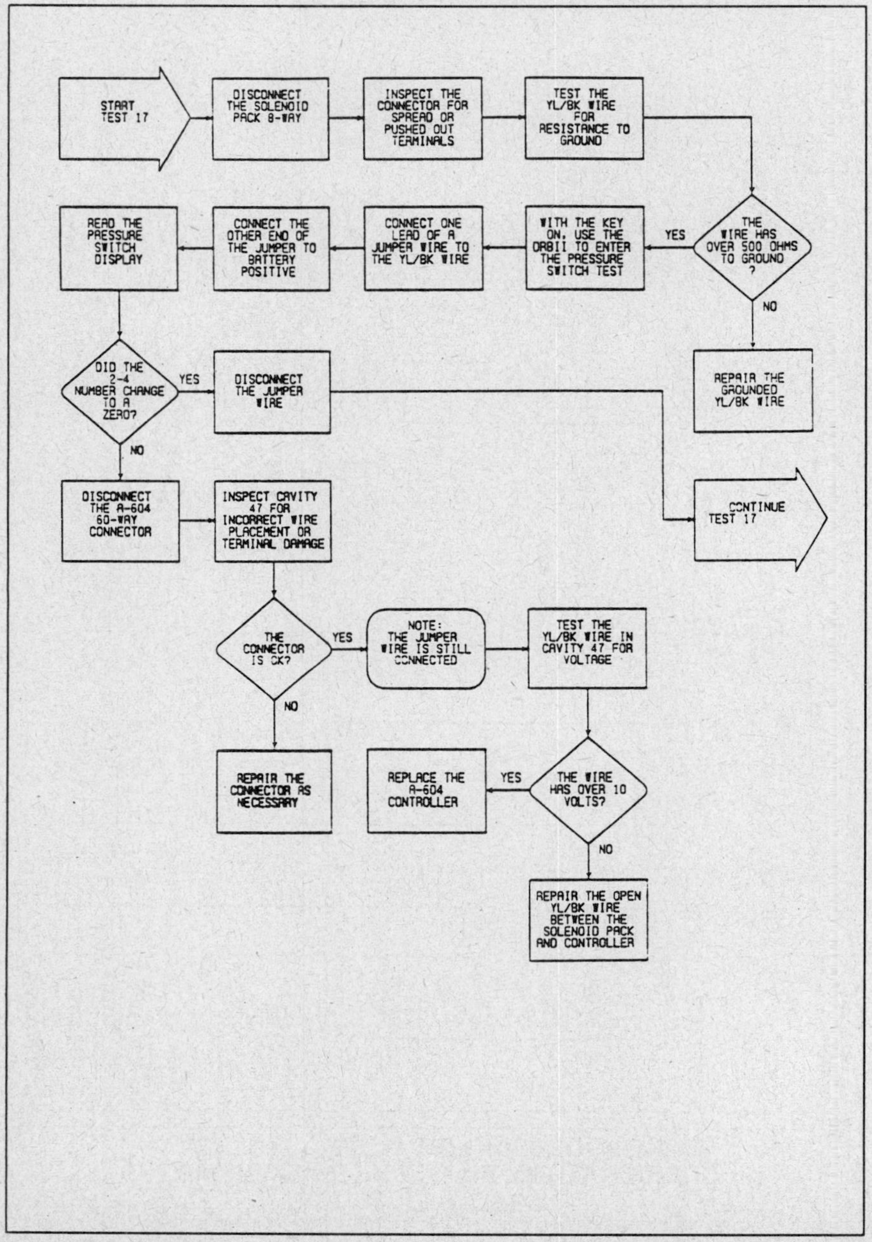
START TEST 17
DISCONNECT THE SOLENOID PACK 8-WAY
INSPECT THE CONNECTOR FOR SPREAD OR PUSHED OUT TERMINALS
TEST THE YL/BK WIRE FOR RESISTANCE TO GROUND
THE WIRE HAS OVER 500 OHMS TO GROUND ?
YES
NO
REPAIR THE GROUNDED YL/BK WIRE
WITH THE KEY ON, USE THE DRBII TO ENTER THE PRESSURE SWITCH TEST
CONNECT ONE LEAD OF A JUMPER WIRE TO THE YL/BK WIRE
CONNECT THE OTHER END OF THE JUMPER TO BATTERY POSITIVE
READ THE PRESSURE SWITCH DISPLAY
DID THE 2-4 NUMBER CHANGE TO A ZERO?
YES
DISCONNECT THE JUMPER WIRE
CONTINUE TEST 17
NO
DISCONNECT THE A-604 60-WAY CONNECTOR
INSPECT CAVITY 47 FOR INCORRECT WIRE PLACEMENT OR TERMINAL DAMAGE
THE CONNECTOR IS OK?
YES
NOTE: THE JUMPER WIRE IS STILL CONNECTED
TEST THE YL/BK WIRE IN CAVITY 47 FOR VOLTAGE
NO
REPAIR THE CONNECTOR AS NECESSARY
THE WIRE HAS OVER 10 VOLTS?
YES
REPLACE THE A-604 CONTROLLER
NO
REPAIR THE OPEN YL/BK WIRE BETWEEN THE SOLENOID PACK AND CONTROLLER

TEST 17
CODE 23—2-4/OD PRESSURE SWITCH CIRCUIT
PERFORM TEST 3 BEFORE PROCEEDING

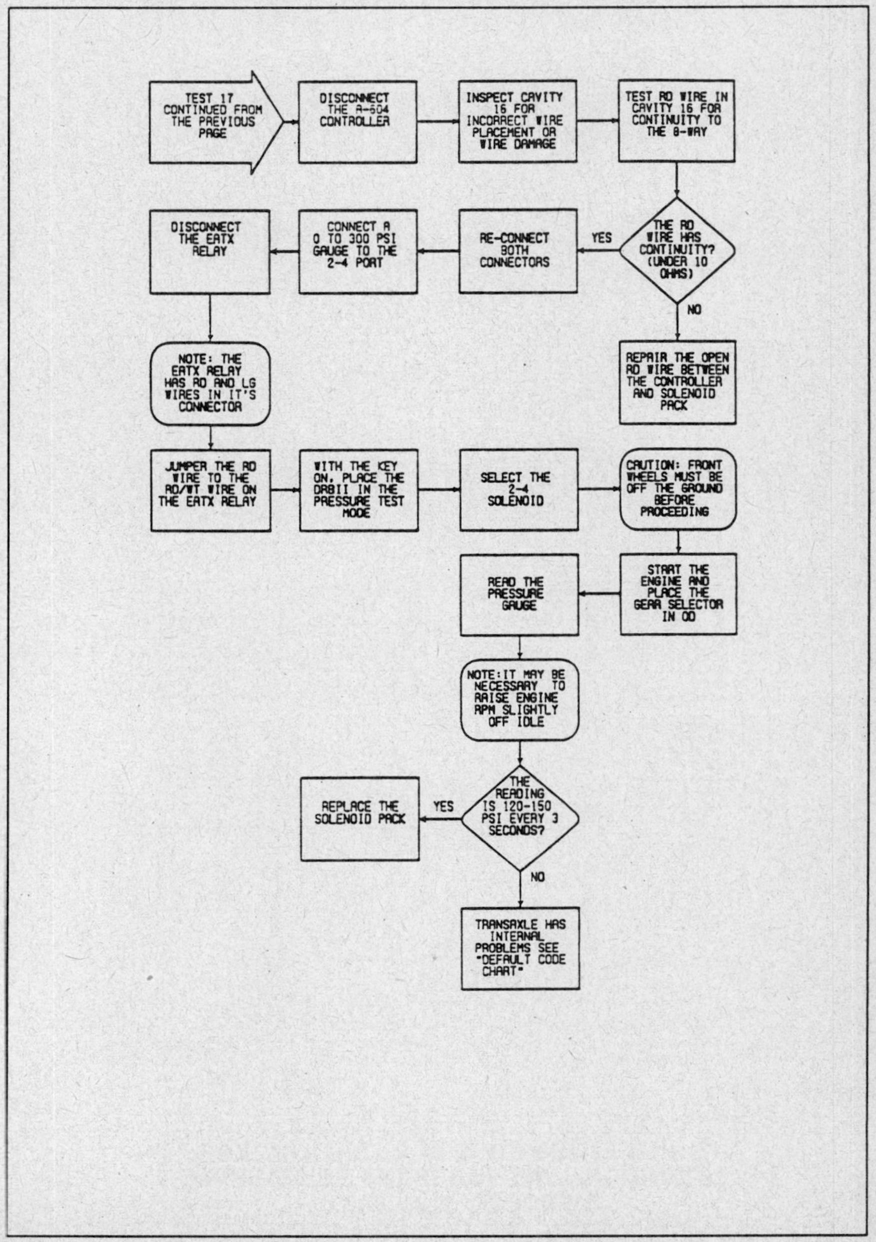
TEST 17 CONTINUED FROM THE PREVIOUS PAGE
DISCONNECT THE A-604 CONTROLLER
INSPECT CAVITY 16 FOR INCORRECT WIRE PLACEMENT OR WIRE DAMAGE
TEST RD WIRE IN CAVITY 16 FOR CONTINUITY TO THE 8-WAY
THE RD WIRE HAS CONTINUITY? (UNDER 10 OHMS)
YES
NO
REPAIR THE OPEN RD WIRE BETWEEN THE CONTROLLER AND SOLENOID PACK
RE-CONNECT BOTH CONNECTORS
CONNECT A 0 TO 300 PSI GAUGE TO THE 2-4 PORT
DISCONNECT THE EATX RELAY
NOTE: THE EATX RELAY HAS RD AND LG WIRES IN IT'S CONNECTOR
JUMPER THE RD WIRE TO THE RD/WT WIRE ON THE EATX RELAY
WITH THE KEY ON, PLACE THE DRBII IN THE PRESSURE TEST MODE
SELECT THE 2-4 SOLENOID
CAUTION: FRONT WHEELS MUST BE OFF THE GROUND BEFORE PROCEEDING
START THE ENGINE AND PLACE THE GEAR SELECTOR IN OD
READ THE PRESSURE GAUGE
NOTE: IT MAY BE NECESSARY TO RAISE ENGINE RPM SLIGHTLY OFF IDLE
THE READING IS 120-150 PSI EVERY 3 SECONDS?
YES
REPLACE THE SOLENOID PACK
NO
TRANSAXLE HAS INTERNAL PROBLEMS SEE "DEFAULT CODE CHART"

TEST 18 CODE 24
LOW/REVERSE PRESSURE SWITCH CIRCUIT
PERFORM TEST 3 BEFORE PROCEEDING

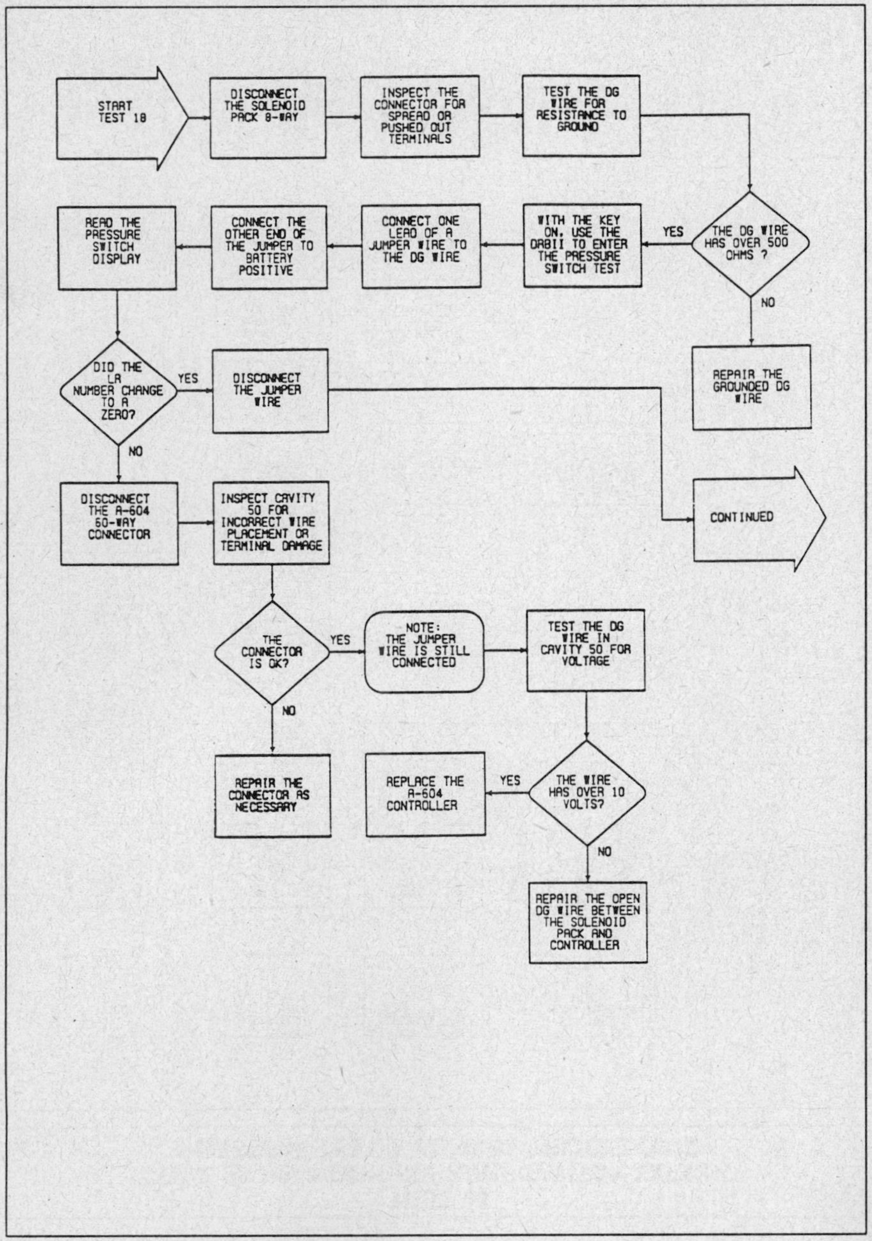
START TEST 18
DISCONNECT THE SOLENOID PACK 8-WAY
INSPECT THE CONNECTOR FOR SPREAD OR PUSHED OUT TERMINALS
TEST THE DG WIRE FOR RESISTANCE TO GROUND
THE DG WIRE HAS OVER 500 OHMS ?
YES
NO
REPAIR THE GROUNDED DG WIRE
WITH THE KEY ON, USE THE DRBII TO ENTER THE PRESSURE SWITCH TEST
CONNECT ONE LEAD OF A JUMPER WIRE TO THE DG WIRE
CONNECT THE OTHER END OF THE JUMPER TO BATTERY POSITIVE
READ THE PRESSURE SWITCH DISPLAY
DID THE LA NUMBER CHANGE TO A ZERO?
YES
DISCONNECT THE JUMPER WIRE
CONTINUED
NO
DISCONNECT THE A-604 60-WAY CONNECTOR
INSPECT CAVITY 50 FOR INCORRECT WIRE PLACEMENT OR TERMINAL DAMAGE
THE CONNECTOR IS OK?
YES
NOTE: THE JUMPER WIRE IS STILL CONNECTED
TEST THE DG WIRE IN CAVITY 50 FOR VOLTAGE
NO
REPAIR THE CONNECTOR AS NECESSARY
THE WIRE HAS OVER 10 VOLTS?
YES
REPLACE THE A-604 CONTROLLER
NO
REPAIR THE OPEN DG WIRE BETWEEN THE SOLENOID PACK AND CONTROLLER

TEST 18 CODE 24
LOW/REVERSE PRESSURE SWITCH CIRCUIT
PERFORM TEST 3 BEFORE PROCEEDING

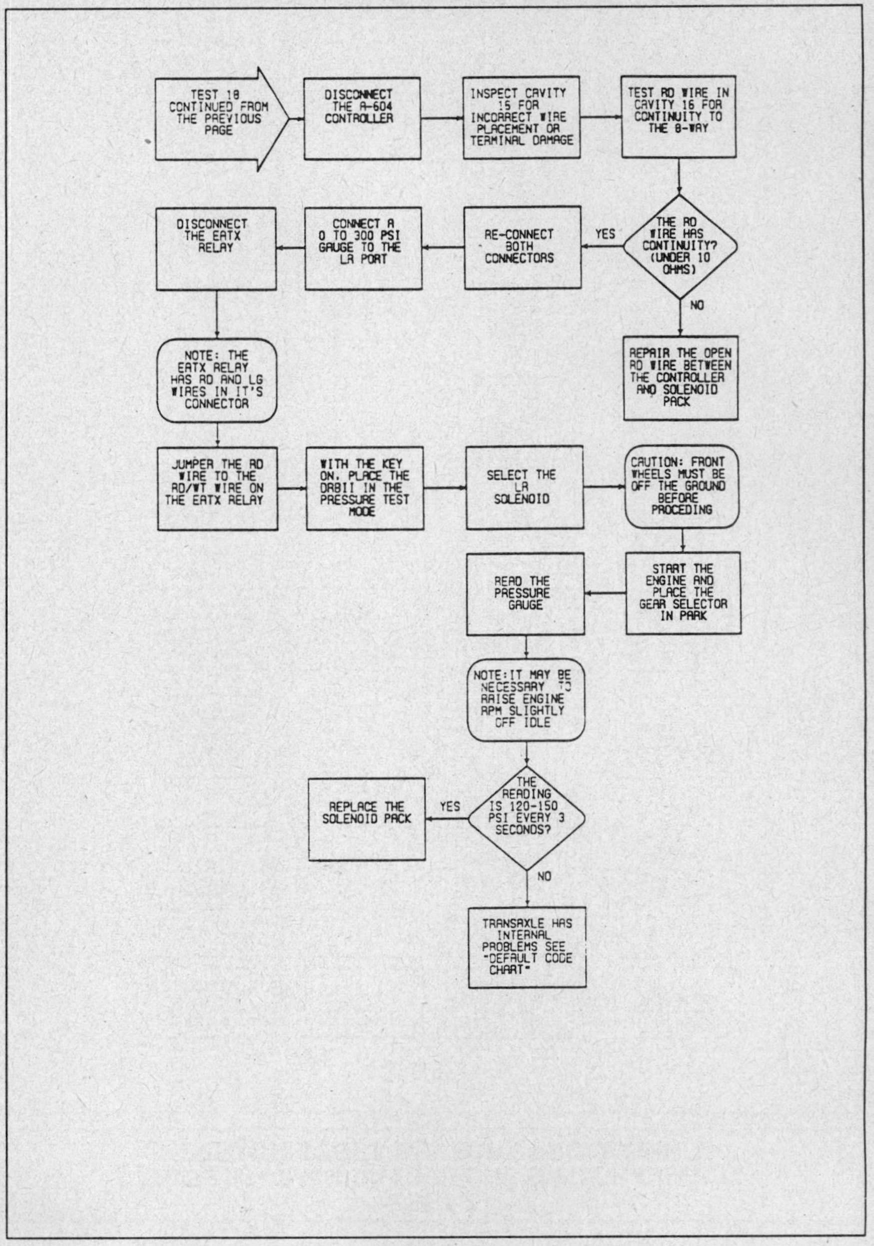
TEST 18 CONTINUED FROM THE PREVIOUS PAGE
DISCONNECT THE A-604 CONTROLLER
INSPECT CAVITY 15 FOR INCORRECT WIRE PLACEMENT OR TERMINAL DAMAGE
TEST RD WIRE IN CAVITY 16 FOR CONTINUITY TO THE 8-WAY
THE RD WIRE HAS CONTINUITY? (UNDER 10 OHMS)
NO
REPAIR THE OPEN RD WIRE BETWEEN THE CONTROLLER AND SOLENOID PACK
YES
RE-CONNECT BOTH CONNECTORS
CONNECT A 0 TO 300 PSI GAUGE TO THE LR PORT
DISCONNECT THE EATX RELAY
NOTE: THE EATX RELAY HAS RD AND LG WIRES IN IT'S CONNECTOR
JUMPER THE RD WIRE TO THE RD/WT WIRE ON THE EATX RELAY
WITH THE KEY ON, PLACE THE DRBII IN THE PRESSURE TEST MODE
SELECT THE LR SOLENOID
CAUTION: FRONT WHEELS MUST BE OFF THE GROUND BEFORE PROCEDING
START THE ENGINE AND PLACE THE GEAR SELECTOR IN PARK
READ THE PRESSURE GAUGE
NOTE: IT MAY BE NECESSARY TO RAISE ENGINE RPM SLIGHTLY OFF IDLE
THE READING IS 120-150 PSI EVERY 3 SECONDS?
YES
REPLACE THE SOLENOID PACK
NO
TRANSAXLE HAS INTERNAL PROBLEMS SEE "DEFAULT CODE CHART"

TEST 19
CODE 25—LR/OD PRESSURE SWITCH CIRCUIT
PERFORM TEST 3 BEFORE PROCEEDING

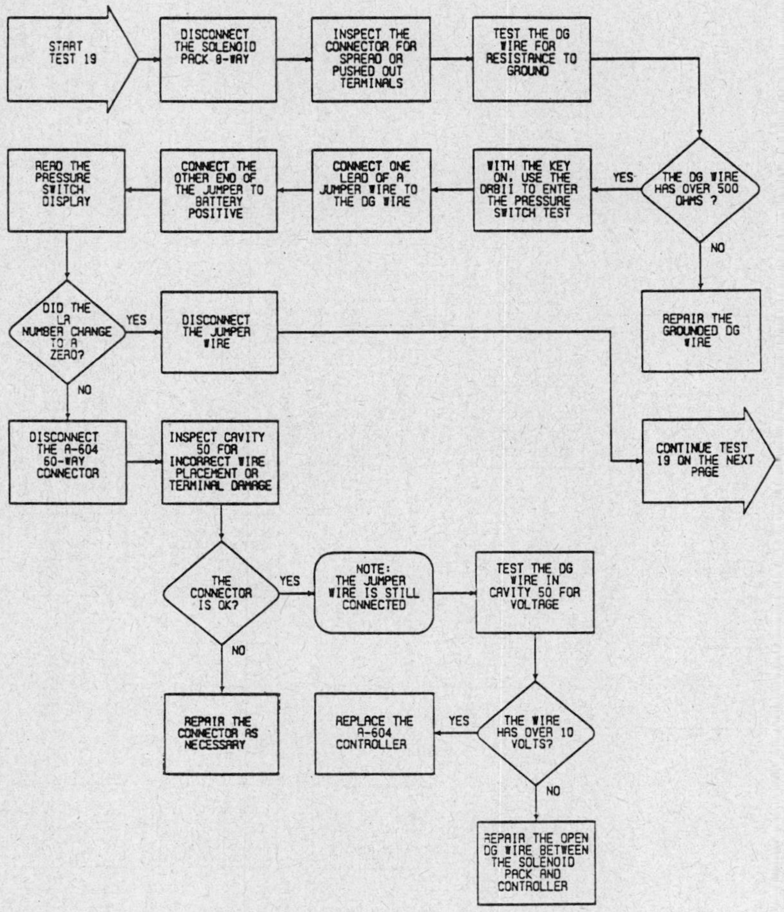
START TEST 19
DISCONNECT THE SOLENOID PACK 8-WAY
INSPECT THE CONNECTOR FOR SPREAD OR PUSHED OUT TERMINALS
TEST THE DG WIRE FOR RESISTANCE TO GROUND
THE DG WIRE HAS OVER 500 OHMS ?
YES
NO
REPAIR THE GROUNDED DG WIRE
WITH THE KEY ON, USE THE DRBII TO ENTER THE PRESSURE SWITCH TEST
CONNECT ONE LEAD OF A JUMPER WIRE TO THE DG WIRE
CONNECT THE OTHER END OF THE JUMPER TO BATTERY POSITIVE
READ THE PRESSURE SWITCH DISPLAY
DID THE LR NUMBER CHANGE TO A ZERO?
YES
DISCONNECT THE JUMPER WIRE
CONTINUE TEST 19 ON THE NEXT PAGE
NO
DISCONNECT THE A-604 60-WAY CONNECTOR
INSPECT CAVITY 50 FOR INCORRECT WIRE PLACEMENT OR TERMINAL DAMAGE
THE CONNECTOR IS OK?
YES
NOTE: THE JUMPER WIRE IS STILL CONNECTED
TEST THE DG WIRE IN CAVITY 50 FOR VOLTAGE
THE WIRE HAS OVER 10 VOLTS?
YES
REPLACE THE A-604 CONTROLLER
NO
REPAIR THE OPEN DG WIRE BETWEEN THE SOLENOID PACK AND CONTROLLER
NO
REPAIR THE CONNECTOR AS NECESSARY

TEST 19
CODE 25—LR/OD PRESSURE SWITCH CIRCUIT
PERFORM TEST 3 BEFORE PROCEEDING

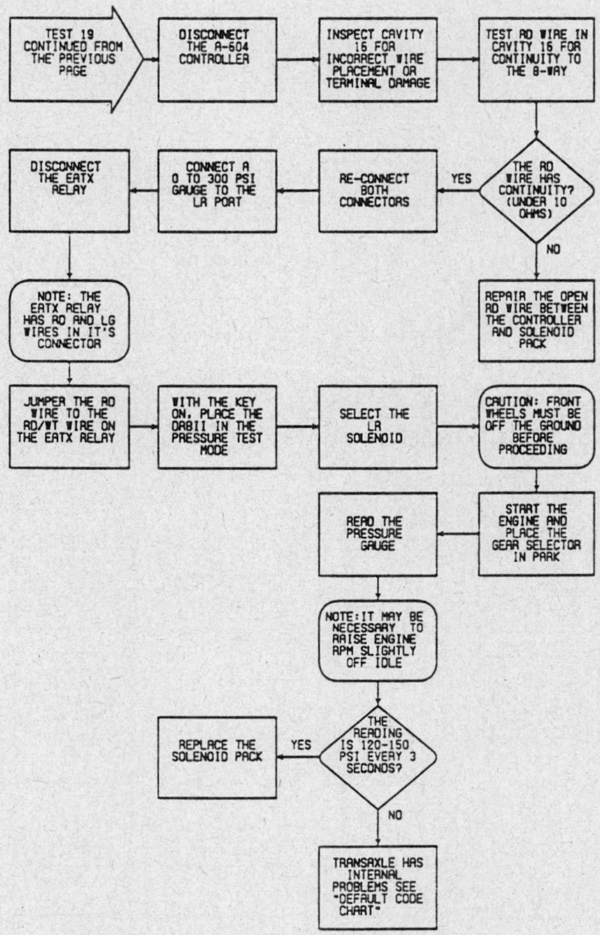
TEST 19 CONTINUED FROM THE PREVIOUS PAGE
DISCONNECT THE A-604 CONTROLLER
INSPECT CAVITY 16 FOR INCORRECT WIRE PLACEMENT OR TERMINAL DAMAGE
TEST RD WIRE IN CAVITY 16 FOR CONTINUITY TO THE 8-WAY
THE RD WIRE HAS CONTINUITY? (UNDER 10 OHMS)
NO
REPAIR THE OPEN RD WIRE BETWEEN THE CONTROLLER AND SOLENOID PACK
YES
RE-CONNECT BOTH CONNECTORS
CONNECT A 0 TO 300 PSI GAUGE TO THE LR PORT
DISCONNECT THE EATX RELAY
NOTE: THE EATX RELAY HAS RD AND LG WIRES IN IT'S CONNECTOR
JUMPER THE RD WIRE TO THE RD/WT WIRE ON THE EATX RELAY
WITH THE KEY ON, PLACE THE DRBII IN THE PRESSURE TEST MODE
SELECT THE LR SOLENOID
CAUTION: FRONT WHEELS MUST BE OFF THE GROUND BEFORE PROCEEDING
START THE ENGINE AND PLACE THE GEAR SELECTOR IN PARK
READ THE PRESSURE GAUGE
NOTE: IT MAY BE NECESSARY TO RAISE ENGINE RPM SLIGHTLY OFF IDLE
THE READING IS 120-150 PSI EVERY 3 SECONDS?
YES
REPLACE THE SOLENOID PACK
NO
TRANSAXLE HAS INTERNAL PROBLEMS SEE "DEFAULT CODE CHART"

TEST 20
CODE 26—LR/2-4 PRESSURE SWITCH CIRCUIT
PERFORM TEST 3 BEFORE PROCEEDING

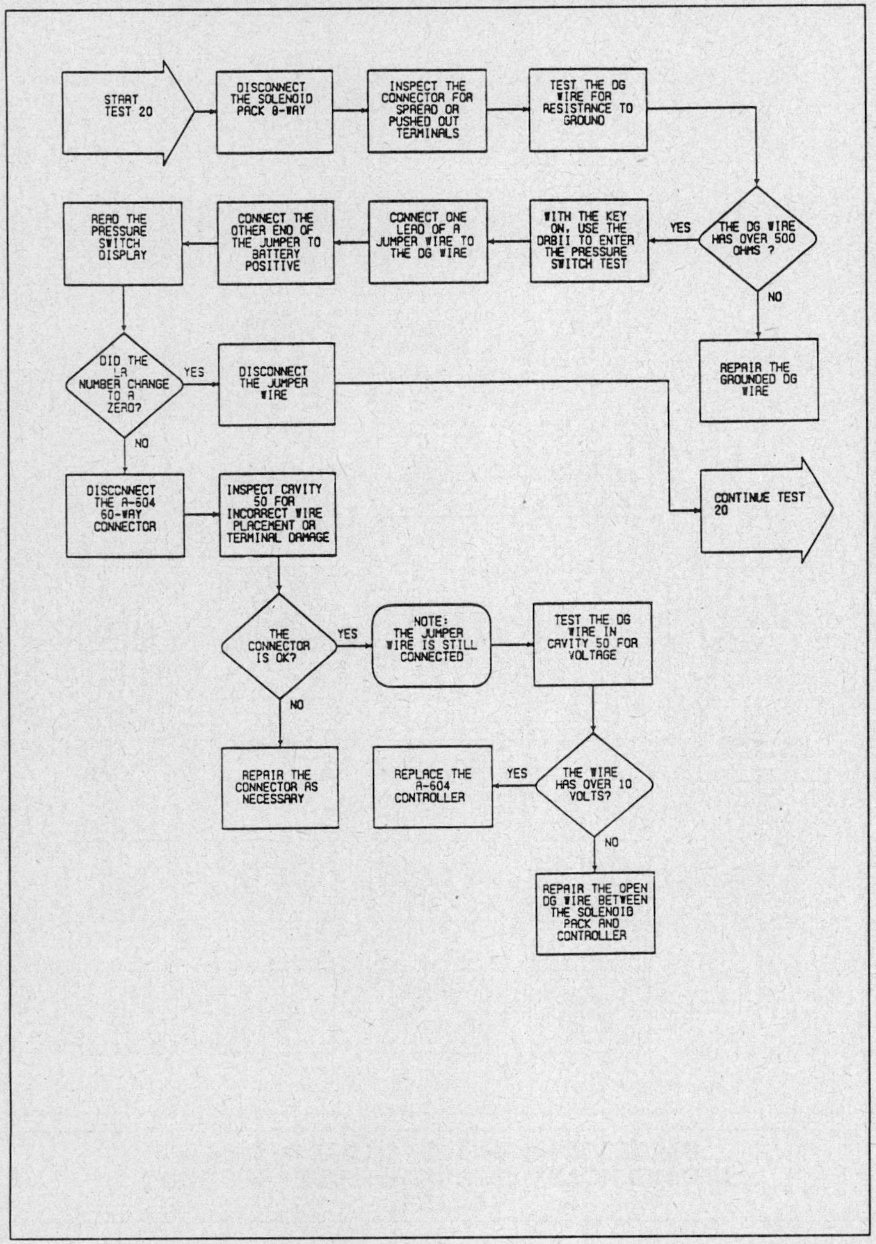
START TEST 20
DISCONNECT THE SOLENOID PACK 8-WAY
INSPECT THE CONNECTOR FOR SPREAD OR PUSHED OUT TERMINALS
TEST THE DG WIRE FOR RESISTANCE TO GROUND
THE DG WIRE HAS OVER 500 OHMS ?
YES
NO
REPAIR THE GROUNDED DG WIRE
WITH THE KEY ON, USE THE DRBII TO ENTER THE PRESSURE SWITCH TEST
CONNECT ONE LEAD OF A JUMPER WIRE TO THE DG WIRE
CONNECT THE OTHER END OF THE JUMPER TO BATTERY POSITIVE
READ THE PRESSURE SWITCH DISPLAY
DID THE LR NUMBER CHANGE TO A ZERO?
YES
DISCONNECT THE JUMPER WIRE
CONTINUE TEST 20
NO
DISCONNECT THE A-604 60-WAY CONNECTOR
INSPECT CAVITY 50 FOR INCORRECT WIRE PLACEMENT OR TERMINAL DAMAGE
THE CONNECTOR IS OK?
YES
NOTE: THE JUMPER WIRE IS STILL CONNECTED
TEST THE DG WIRE IN CAVITY 50 FOR VOLTAGE
NO
REPAIR THE CONNECTOR AS NECESSARY
THE WIRE HAS OVER 10 VOLTS?
YES
REPLACE THE A-604 CONTROLLER
NO
REPAIR THE OPEN DG WIRE BETWEEN THE SOLENOID PACK AND CONTROLLER

TEST 20
CODE 26—LR/2-4 PRESSURE SWITCH CIRCUIT
PERFORM TEST 3 BEFORE PROCEEDING

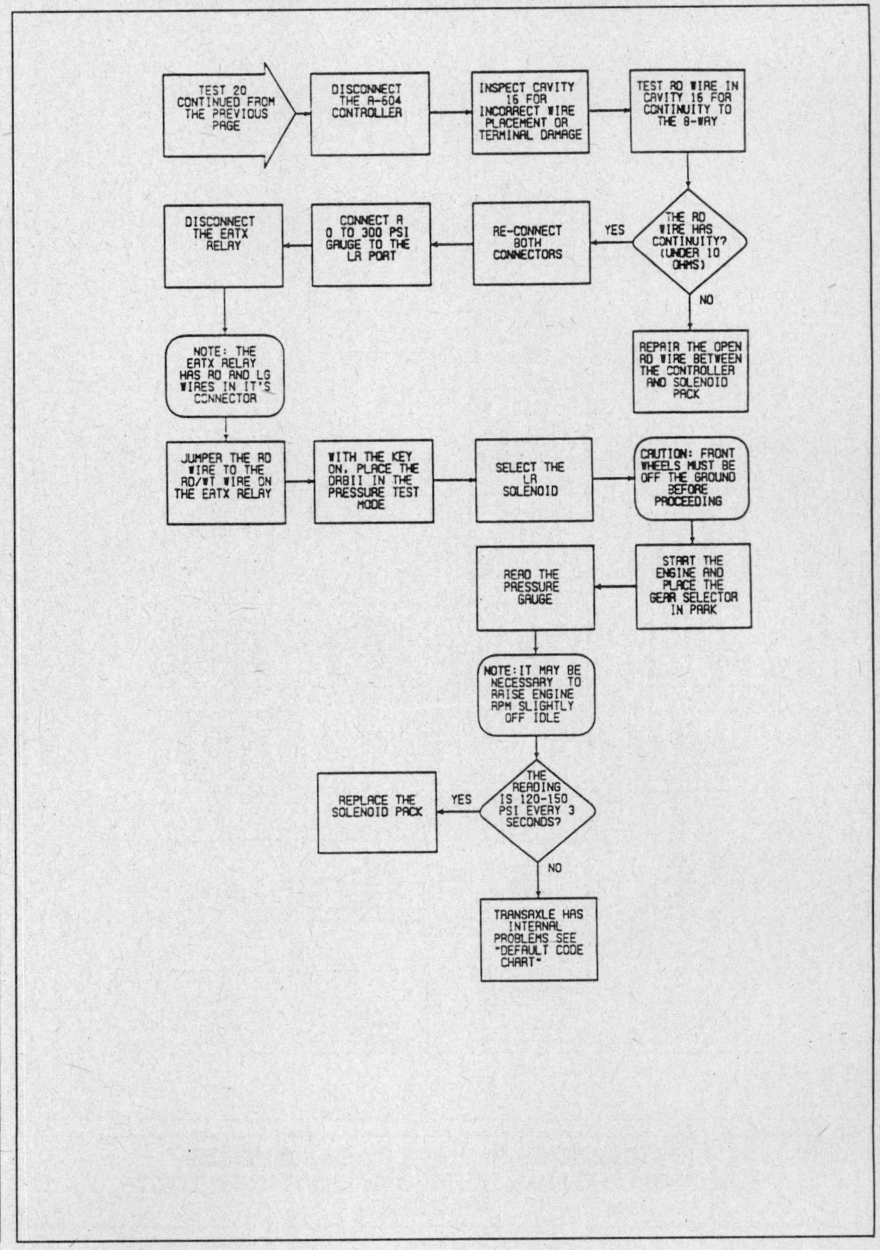
TEST 20 CONTINUED FROM THE PREVIOUS PAGE
DISCONNECT THE A-604 CONTROLLER
INSPECT CAVITY 16 FOR INCORRECT WIRE PLACEMENT OR TERMINAL DAMAGE
TEST RD WIRE IN CAVITY 16 FOR CONTINUITY TO THE 8-WAY
THE RD WIRE HAS CONTINUITY? (UNDER 10 OHMS)
YES
RE-CONNECT BOTH CONNECTORS
CONNECT A 0 TO 300 PSI GAUGE TO THE LR PORT
DISCONNECT THE EATX RELAY
NO
REPAIR THE OPEN RD WIRE BETWEEN THE CONTROLLER AND SOLENOID PACK
NOTE: THE EATX RELAY HAS RD AND LG WIRES IN IT'S CONNECTOR
JUMPER THE RD WIRE TO THE RD/WT WIRE ON THE EATX RELAY
WITH THE KEY ON, PLACE THE DRBII IN THE PRESSURE TEST MODE
SELECT THE LR SOLENOID
CAUTION: FRONT WHEELS MUST BE OFF THE GROUND BEFORE PROCEEDING
START THE ENGINE AND PLACE THE GEAR SELECTOR IN PARK
READ THE PRESSURE GAUGE
NOTE: IT MAY BE NECESSARY TO RAISE ENGINE RPM SLIGHTLY OFF IDLE
THE READING IS 120-150 PSI EVERY 3 SECONDS?
YES
REPLACE THE SOLENOID PACK
NO
TRANSAXLE HAS INTERNAL PROBLEMS SEE "DEFAULT CODE CHART"

TEST 21
CODE 27—ALL PRESSURE SWITCH CIRCUITS
PERFORM TEST 3 BEFORE PROCEEDING

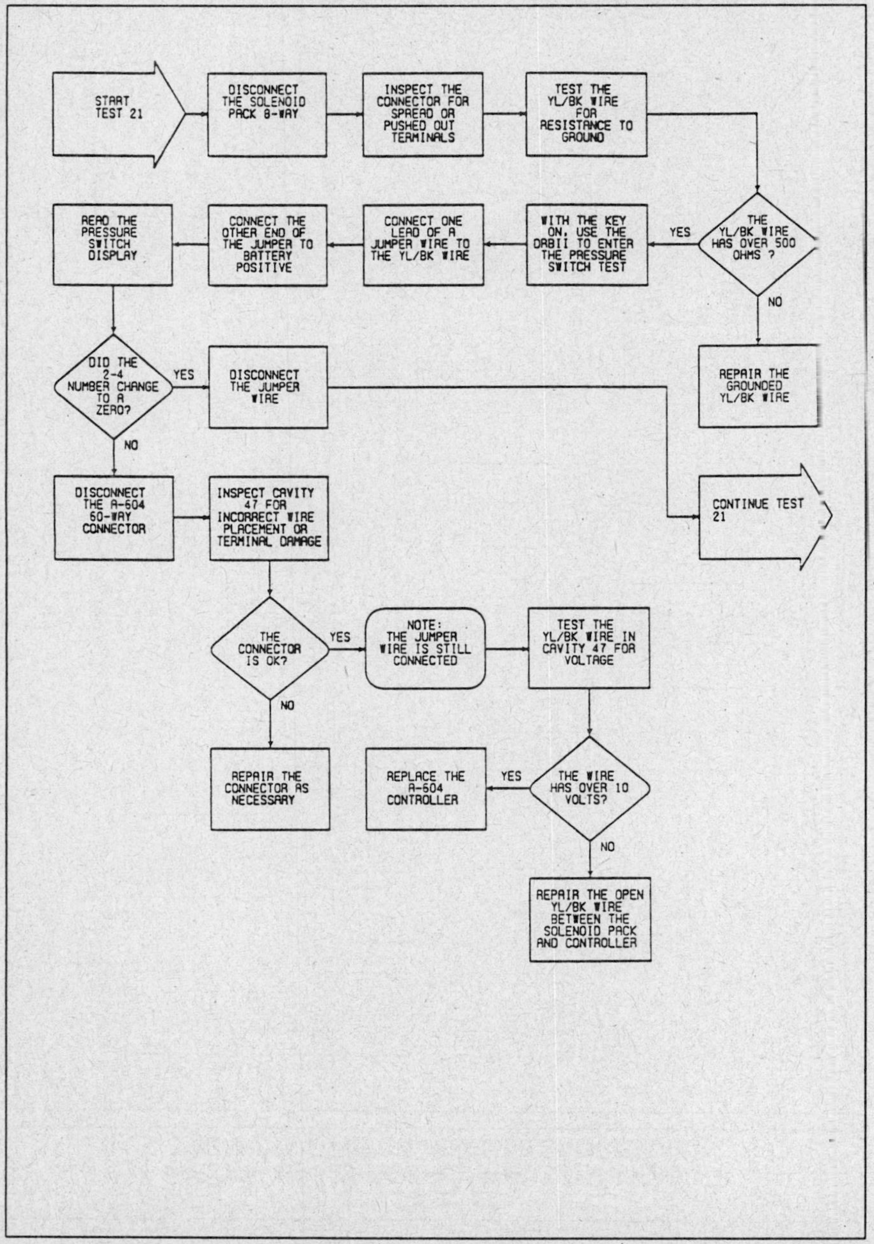
START TEST 21
DISCONNECT THE SOLENOID PACK 8-WAY
INSPECT THE CONNECTOR FOR SPREAD OR PUSHED OUT TERMINALS
TEST THE YL/BK WIRE FOR RESISTANCE TO GROUND
THE YL/BK WIRE HAS OVER 500 OHMS ?
YES
NO
REPAIR THE GROUNDED YL/BK WIRE
WITH THE KEY ON, USE THE DRBII TO ENTER THE PRESSURE SWITCH TEST
CONNECT ONE LEAD OF A JUMPER WIRE TO THE YL/BK WIRE
CONNECT THE OTHER END OF THE JUMPER TO BATTERY POSITIVE
READ THE PRESSURE SWITCH DISPLAY
DID THE 2-4 NUMBER CHANGE TO A ZERO?
YES
DISCONNECT THE JUMPER WIRE
CONTINUE TEST 21
NO
DISCONNECT THE A-604 60-WAY CONNECTOR
INSPECT CAVITY 47 FOR INCORRECT WIRE PLACEMENT OR TERMINAL DAMAGE
THE CONNECTOR IS OK?
YES
NOTE: THE JUMPER WIRE IS STILL CONNECTED
TEST THE YL/BK WIRE IN CAVITY 47 FOR VOLTAGE
THE WIRE HAS OVER 10 VOLTS?
YES
REPLACE THE A-604 CONTROLLER
NO
REPAIR THE OPEN YL/BK WIRE BETWEEN THE SOLENOID PACK AND CONTROLLER
NO
REPAIR THE CONNECTOR AS NECESSARY

TEST 21
CODE 27—ALL PRESSURE SWITCH CIRCUITS
PERFORM TEST 3 BEFORE PROCEEDING

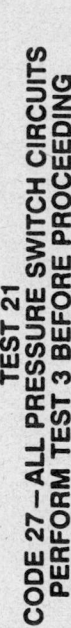

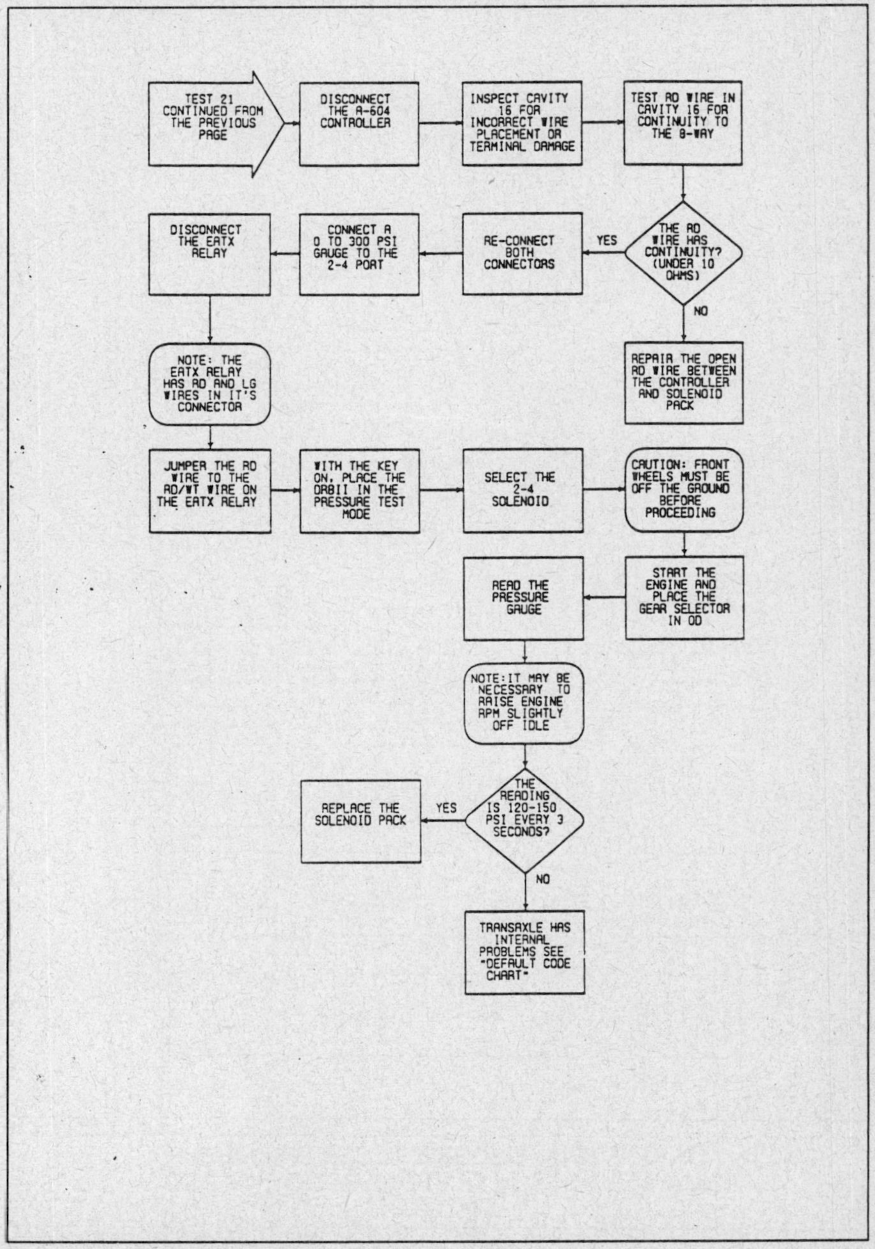
TEST 21 CONTINUED FROM THE PREVIOUS PAGE
DISCONNECT THE A-604 CONTROLLER
INSPECT CAVITY 16 FOR INCORRECT WIRE PLACEMENT OR TERMINAL DAMAGE
TEST RD WIRE IN CAVITY 16 FOR CONTINUITY TO THE 8-WAY
THE RD WIRE HAS CONTINUITY? (UNDER 10 OHMS)
NO
REPAIR THE OPEN RD WIRE BETWEEN THE CONTROLLER AND SOLENOID PACK
YES
RE-CONNECT BOTH CONNECTORS
CONNECT A 0 TO 300 PSI GAUGE TO THE 2-4 PORT
DISCONNECT THE EATX RELAY
NOTE: THE EATX RELAY HAS RD AND LG WIRES IN IT'S CONNECTOR
JUMPER THE RD WIRE TO THE RD/WT WIRE ON THE EATX RELAY
WITH THE KEY ON, PLACE THE DRBII IN THE PRESSURE TEST MODE
SELECT THE 2-4 SOLENOID
CAUTION: FRONT WHEELS MUST BE OFF THE GROUND BEFORE PROCEEDING
START THE ENGINE AND PLACE THE GEAR SELECTOR IN OD
READ THE PRESSURE GAUGE
NOTE: IT MAY BE NECESSARY TO RAISE ENGINE RPM SLIGHTLY OFF IDLE
THE READING IS 120-150 PSI EVERY 3 SECONDS?
YES
REPLACE THE SOLENOID PACK
NO
TRANSAXLE HAS INTERNAL PROBLEMS SEE "DEFAULT CODE CHART"

TEST 22
CODE 29—THROTTLE POSITION SIGNAL
PERFORM TEST 3 BEFORE PROCEEDING

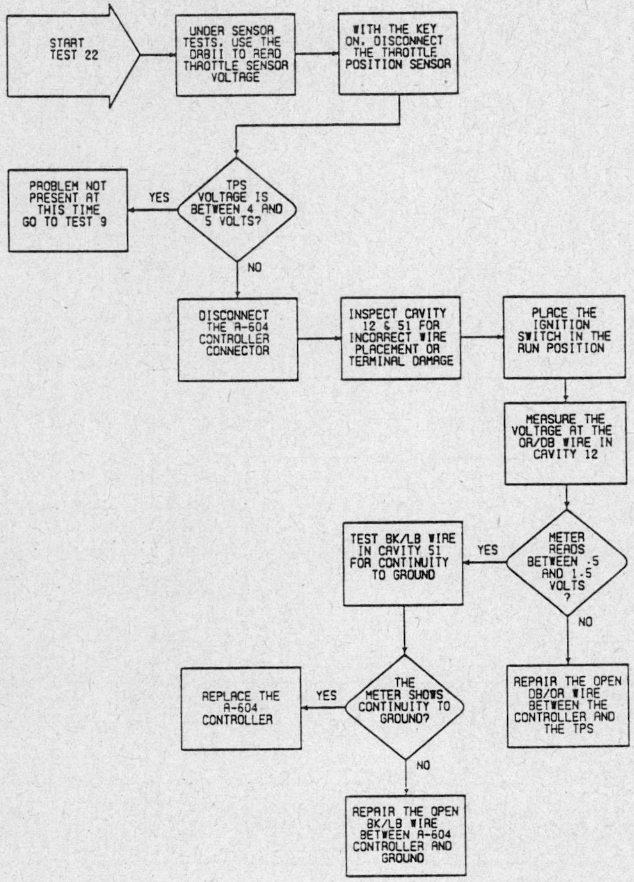
START TEST 22
UNDER SENSOR TESTS, USE THE DRBII TO READ THROTTLE SENSOR VOLTAGE
WITH THE KEY ON, DISCONNECT THE THROTTLE POSITION SENSOR
TPS VOLTAGE IS BETWEEN 4 AND 5 VOLTS?
YES
PROBLEM NOT PRESENT AT THIS TIME GO TO TEST 9
NO
DISCONNECT THE A-604 CONTROLLER CONNECTOR
INSPECT CAVITY 12 & 51 FOR INCORRECT WIRE PLACEMENT OR TERMINAL DAMAGE
PLACE THE IGNITION SWITCH IN THE RUN POSITION
MEASURE THE VOLTAGE AT THE OR/DB WIRE IN CAVITY 12
METER READS BETWEEN .5 AND 1.5 VOLTS ?
YES
TEST BK/LB WIRE IN CAVITY 51 FOR CONTINUITY TO GROUND
NO
REPAIR THE OPEN DB/OR WIRE BETWEEN THE CONTROLLER AND THE TPS
THE METER SHOWS CONTINUITY TO GROUND?
YES
REPLACE THE A-604 CONTROLLER
NO
REPAIR THE OPEN BK/LB WIRE BETWEEN A-604 CONTROLLER AND GROUND

TEST 23
CODE 31—OD HYDRAULIC PRESSURE SWITCH
PERFORM TEST 3 BEFORE PROCEEDING

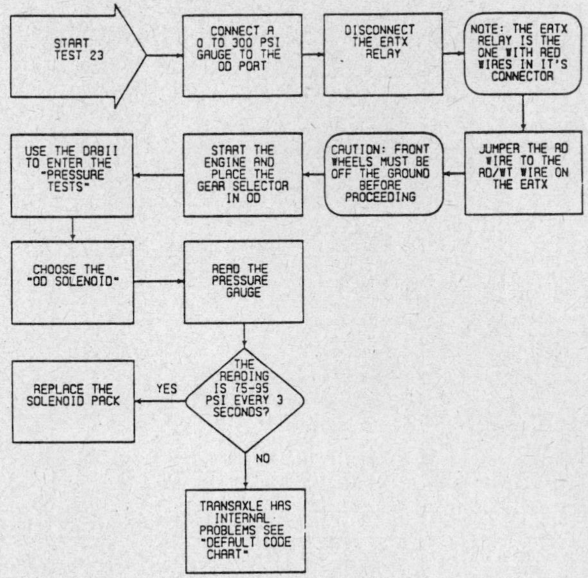
START TEST 23
CONNECT A 0 TO 300 PSI GAUGE TO THE OD PORT
DISCONNECT THE EATX RELAY
NOTE: THE EATX RELAY IS THE ONE WITH RED WIRES IN IT'S CONNECTOR
JUMPER THE RD WIRE TO THE RD/WT WIRE ON THE EATX
CAUTION: FRONT WHEELS MUST BE OFF THE GROUND BEFORE PROCEEDING
START THE ENGINE AND PLACE THE GEAR SELECTOR IN OD
USE THE DRBII TO ENTER THE "PRESSURE TESTS"
CHOOSE THE "OD SOLENOID"
READ THE PRESSURE GAUGE
THE READING IS 75-95 PSI EVERY 3 SECONDS?
YES
REPLACE THE SOLENOID PACK
NO
TRANSAXLE HAS INTERNAL PROBLEMS SEE "DEFAULT CODE CHART"

TEST 24
CODE 32—2-4 HYDRAULIC PRESSURE SWITCH
PERFORM TEST 3 BEFORE PROCEEDING

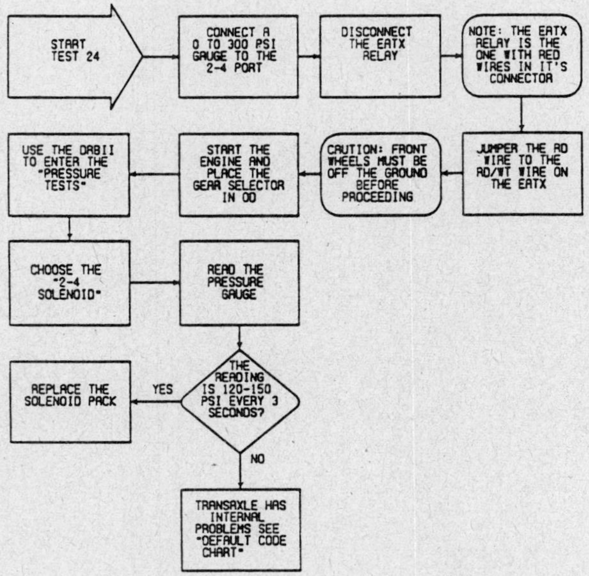
START TEST 24
CONNECT A 0 TO 300 PSI GAUGE TO THE 2-4 PORT
DISCONNECT THE EATX RELAY
NOTE: THE EATX RELAY IS THE ONE WITH RED WIRES IN IT'S CONNECTOR
JUMPER THE RD WIRE TO THE RD/WT WIRE ON THE EATX
CAUTION: FRONT WHEELS MUST BE OFF THE GROUND BEFORE PROCEEDING
START THE ENGINE AND PLACE THE GEAR SELECTOR IN OD
USE THE DRBII TO ENTER THE "PRESSURE TESTS"
CHOOSE THE "2-4 SOLENOID"
READ THE PRESSURE GAUGE
THE READING IS 120-150 PSI EVERY 3 SECONDS?
YES
REPLACE THE SOLENOID PACK
NO
TRANSAXLE HAS INTERNAL PROBLEMS SEE "DEFAULT CODE CHART"

TEST 25
CODE 33—OD/2-4 HYDRAULIC PRESSURE SWITCH
PERFORM TEST 3 BEFORE PROCEEDING

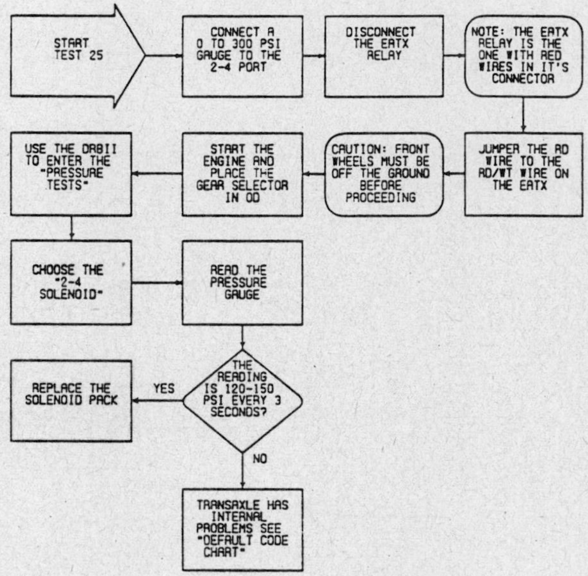
START TEST 25
CONNECT A 0 TO 300 PSI GAUGE TO THE 2-4 PORT
DISCONNECT THE EATX RELAY
NOTE: THE EATX RELAY IS THE ONE WITH RED WIRES IN IT'S CONNECTOR
JUMPER THE RD WIRE TO THE RD/WT WIRE ON THE EATX
CAUTION: FRONT WHEELS MUST BE OFF THE GROUND BEFORE PROCEEDING
START THE ENGINE AND PLACE THE GEAR SELECTOR IN OD
USE THE DRBII TO ENTER THE "PRESSURE TESTS"
CHOOSE THE "2-4 SOLENOID"
READ THE PRESSURE GAUGE
THE READING IS 120-150 PSI EVERY 3 SECONDS?
YES
REPLACE THE SOLENOID PACK
NO
TRANSAXLE HAS INTERNAL PROBLEMS SEE "DEFAULT CODE CHART"

TEST 26
CODE 37—SOLENOID SWITCH VALVE
PERFORM TEST 3 BEFORE PROCEEDING

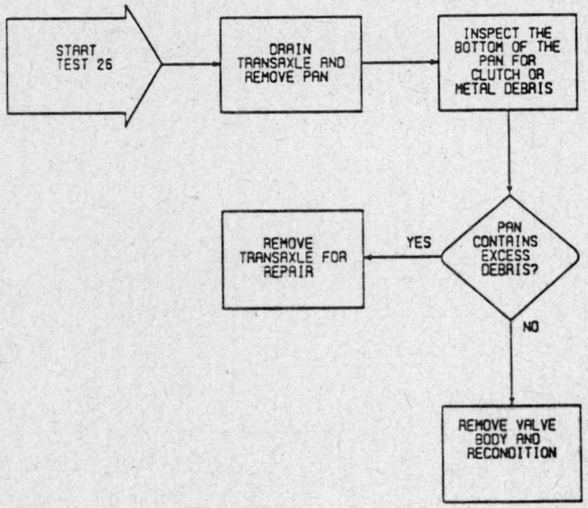
START TEST 26
DRAIN TRANSAXLE AND REMOVE PAN
INSPECT THE BOTTOM OF THE PAN FOR CLUTCH OR METAL DEBRIS
PAN CONTAINS EXCESS DEBRIS?
YES
REMOVE TRANSAXLE FOR REPAIR
NO
REMOVE VALVE BODY AND RECONDITION

TEST 27
CODE 38—LOCK-UP CONTROL
PERFORM TEST 3 BEFORE PROCEEDING

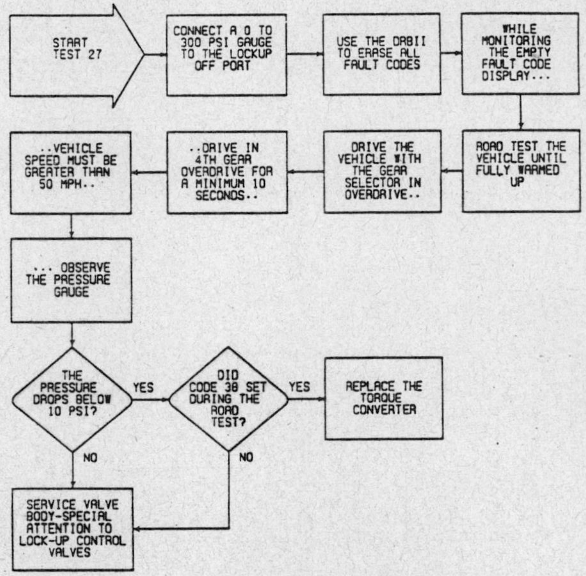
START TEST 27
CONNECT A 0 TO 300 PSI GAUGE TO THE LOCKUP OFF PORT
USE THE DRBII TO ERASE ALL FAULT CODES
WHILE MONITORING THE EMPTY FAULT CODE DISPLAY...
ROAD TEST THE VEHICLE UNTIL FULLY WARMED UP
DRIVE THE VEHICLE WITH THE GEAR SELECTOR IN OVERDRIVE..
..DRIVE IN 4TH GEAR OVERDRIVE FOR A MINIMUM 10 SECONDS..
..VEHICLE SPEED MUST BE GREATER THAN 50 MPH..
... OBSERVE THE PRESSURE GAUGE
THE PRESSURE DROPS BELOW 10 PSI?
YES
DID CODE 38 SET DURING THE ROAD TEST?
YES
REPLACE THE TORQUE CONVERTER
NO
NO
SERVICE VALVE BODY-SPECIAL ATTENTION TO LOCK-UP CONTROL VALVES

TEST 28
CODE 39—TURBINE/TRANS OUTPUT SPEED CIRCUIT
PERFORM TEST 3 BEFORE PROCEEDING

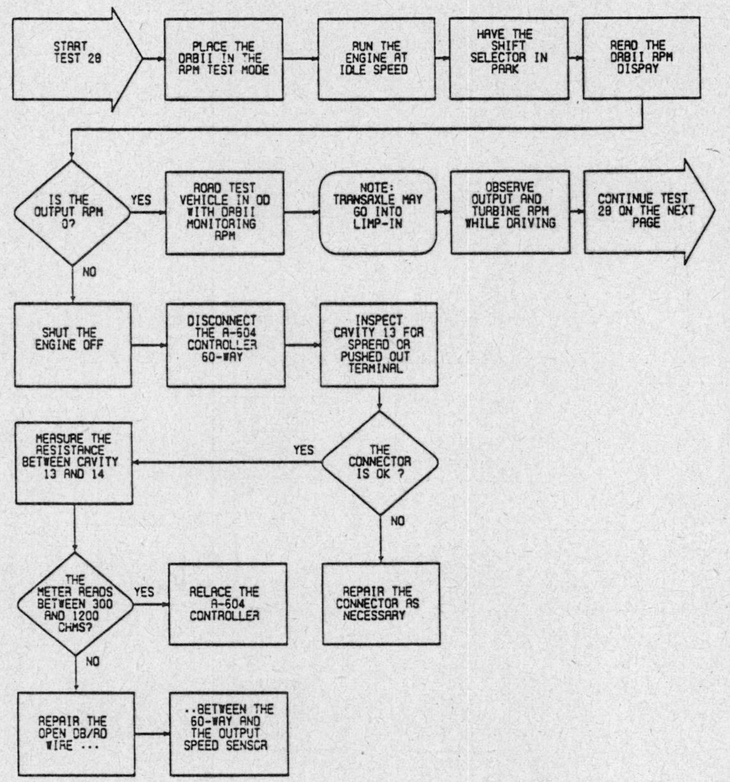
START TEST 28
PLACE THE DRBII IN THE RPM TEST MODE
RUN THE ENGINE AT IDLE SPEED
HAVE THE SHIFT SELECTOR IN PARK
READ THE DRBII RPM DISPAY
IS THE OUTPUT RPM 0?
YES
ROAD TEST VEHICLE IN OD WITH DRBII MONITORING RPM
NOTE: TRANSAXLE MAY GO INTO LIMP-IN
OBSERVE OUTPUT AND TURBINE RPM WHILE DRIVING
CONTINUE TEST 28 ON THE NEXT PAGE
NO
SHUT THE ENGINE OFF
DISCONNECT THE A-604 CONTROLLER 60-WAY
INSPECT CAVITY 13 FOR SPREAD OR PUSHED OUT TERMINAL
THE CONNECTOR IS OK ?
YES
MEASURE THE RESISTANCE BETWEEN CAVITY 13 AND 14
NO
REPAIR THE CONNECTOR AS NECESSARY
THE METER READS BETWEEN 300 AND 1200 OHMS?
YES
RELACE THE A-604 CONTROLLER
NO
REPAIR THE OPEN DB/RD WIRE ...
..BETWEEN THE 60-WAY AND THE OUTPUT SPEED SENSOR

TEST 28
CODE 39—TURBINE/TRANS OUTPUT SPEED CIRCUIT
PERFORM TEST 3 BEFORE PROCEEDING

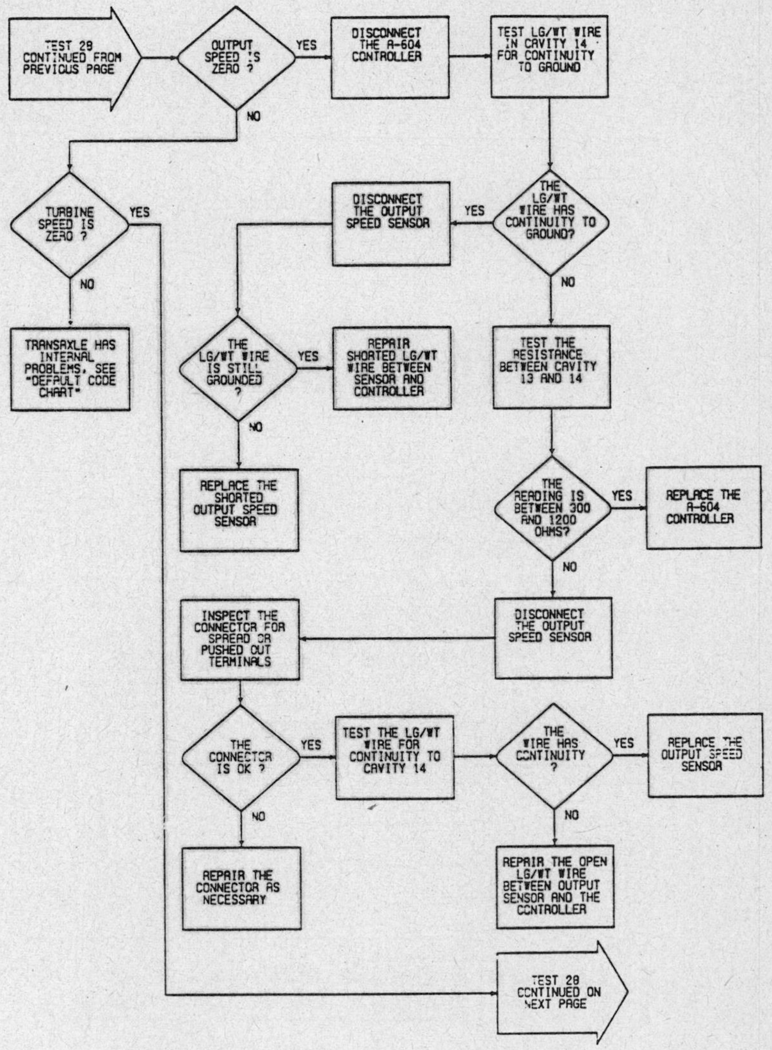
TEST 28 CONTINUED FROM PREVIOUS PAGE
OUTPUT SPEED IS ZERO ?
YES
DISCONNECT THE A-604 CONTROLLER
TEST LG/WT WIRE IN CAVITY 14 FOR CONTINUITY TO GROUND
NO
TURBINE SPEED IS ZERO ?
YES
NO
TRANSAXLE HAS INTERNAL PROBLEMS, SEE "DEFAULT CODE CHART"
THE LG/WT WIRE HAS CONTINUITY TO GROUND?
YES
DISCONNECT THE OUTPUT SPEED SENSOR
NO
TEST THE RESISTANCE BETWEEN CAVITY 13 AND 14
THE LG/WT WIRE IS STILL GROUNDED ?
YES
REPAIR SHORTED LG/WT WIRE BETWEEN SENSOR AND CONTROLLER
NO
REPLACE THE SHORTED OUTPUT SPEED SENSOR
THE READING IS BETWEEN 300 AND 1200 OHMS?
YES
REPLACE THE A-604 CONTROLLER
NO
DISCONNECT THE OUTPUT SPEED SENSOR
INSPECT THE CONNECTOR FOR SPREAD OR PUSHED OUT TERMINALS
THE CONNECTOR IS OK ?
YES
TEST THE LG/WT WIRE FOR CONTINUITY TO CAVITY 14
THE WIRE HAS CONTINUITY ?
YES
REPLACE THE OUTPUT SPEED SENSOR
NO
REPAIR THE CONNECTOR AS NECESSARY
NO
REPAIR THE OPEN LG/WT WIRE BETWEEN OUTPUT SENSOR AND THE CONTROLLER
TEST 28 CONTINUED ON NEXT PAGE

TEST 28
CODE 39—TURBINE/TRANS OUTPUT SPEED CIRCUIT
PERFORM TEST 3 BEFORE PROCEEDING

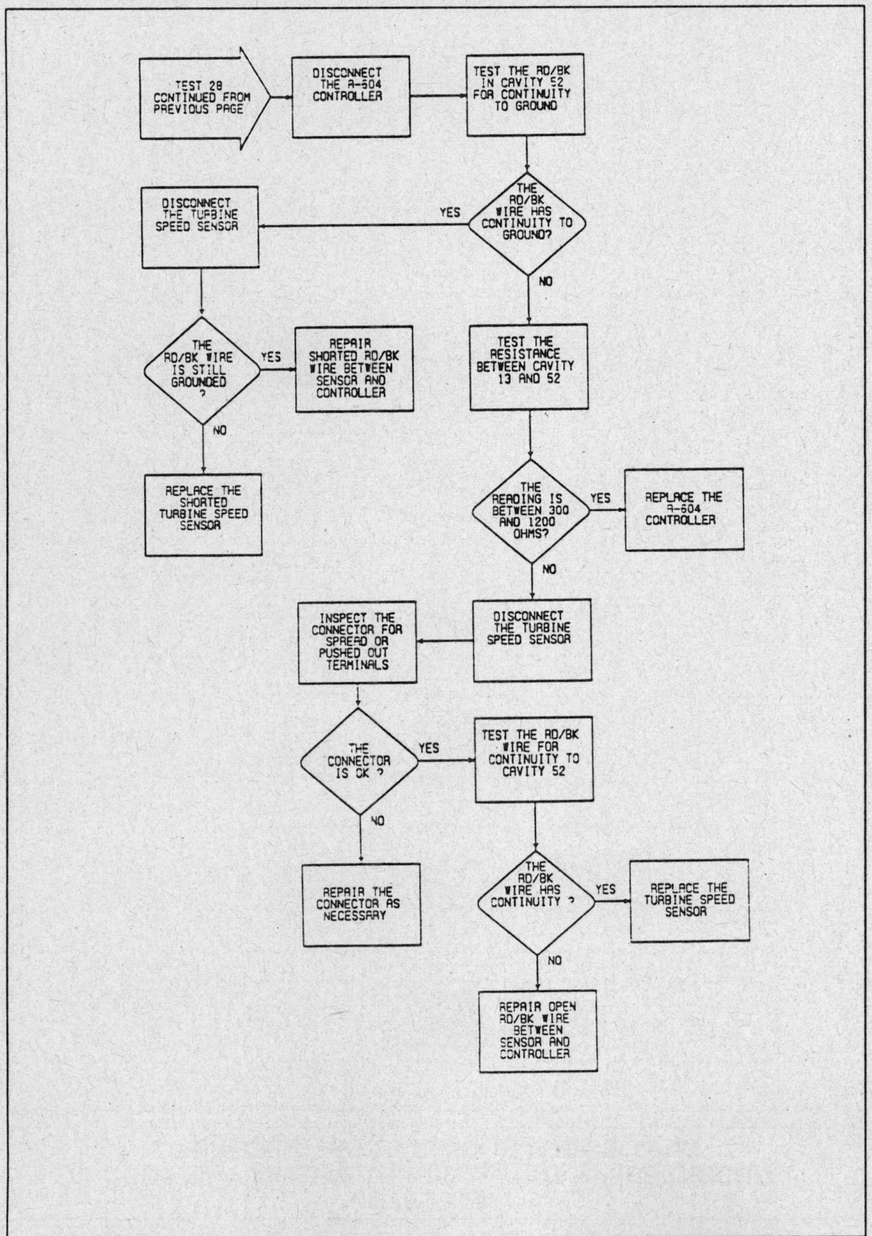
TEST 28 CONTINUED FROM PREVIOUS PAGE
DISCONNECT THE A-604 CONTROLLER
TEST THE RD/BK IN CAVITY 52 FOR CONTINUITY TO GROUND
THE RD/BK WIRE HAS CONTINUITY TO GROUND?
YES
NO
DISCONNECT THE TURBINE SPEED SENSOR
THE RD/BK WIRE IS STILL GROUNDED ?
YES
NO
REPAIR SHORTED RD/BK WIRE BETWEEN SENSOR AND CONTROLLER
REPLACE THE SHORTED TURBINE SPEED SENSOR
TEST THE RESISTANCE BETWEEN CAVITY 13 AND 52
THE READING IS BETWEEN 300 AND 1200 OHMS?
YES
NO
REPLACE THE A-604 CONTROLLER
DISCONNECT THE TURBINE SPEED SENSOR
INSPECT THE CONNECTOR FOR SPREAD OR PUSHED OUT TERMINALS
THE CONNECTOR IS OK ?
YES
NO
REPAIR THE CONNECTOR AS NECESSARY
TEST THE RD/BK WIRE FOR CONTINUITY TO CAVITY 52
THE RD/BK WIRE HAS CONTINUITY ?
YES
NO
REPLACE THE TURBINE SPEED SENSOR
REPAIR OPEN RD/BK WIRE BETWEEN SENSOR AND CONTROLLER

TEST 29
CODE 41—LR SOLENOID CIRCUIT
PERFORM TEST 3 BEFORE PROCEEDING

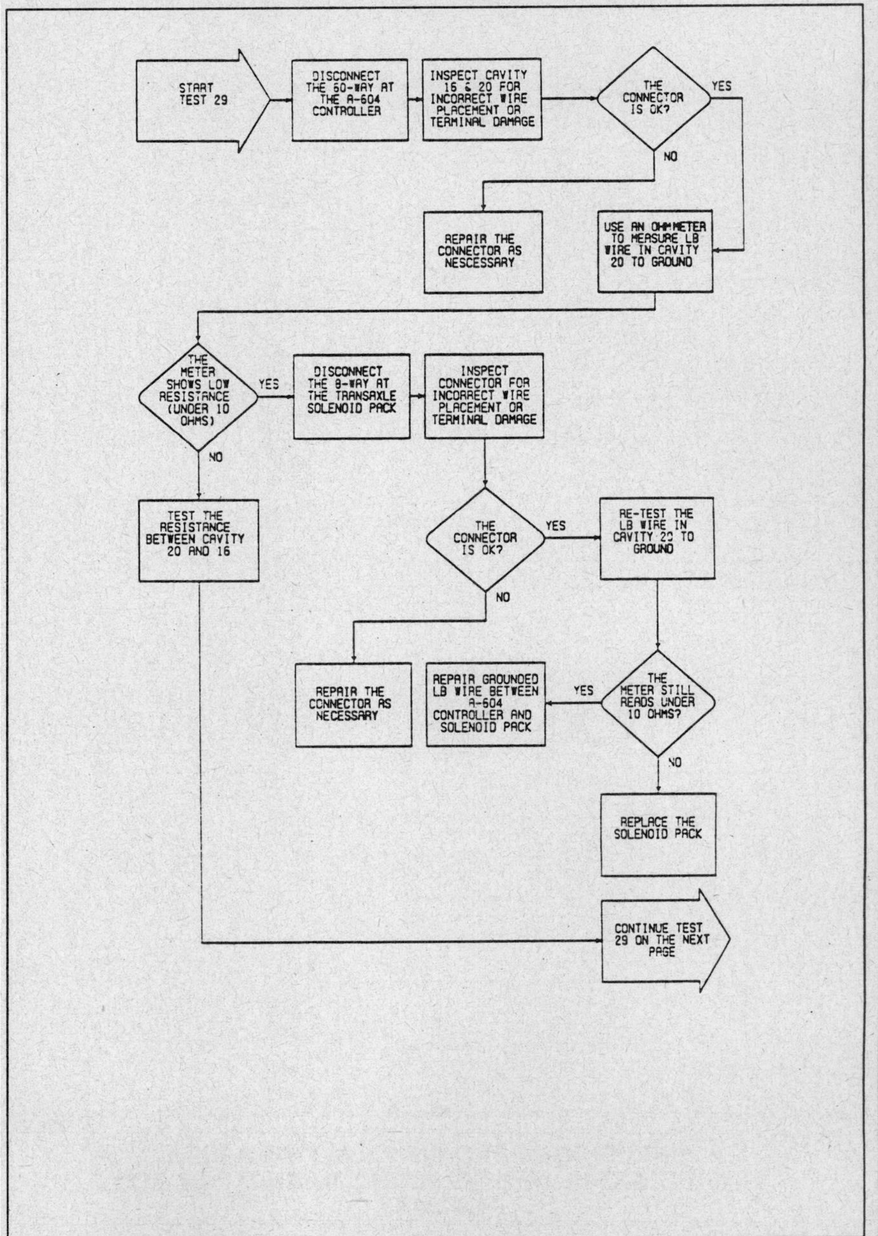
START TEST 29
DISCONNECT THE 60-WAY AT THE A-604 CONTROLLER
INSPECT CAVITY 16 & 20 FOR INCORRECT WIRE PLACEMENT OR TERMINAL DAMAGE
THE CONNECTOR IS OK?
YES
NO
REPAIR THE CONNECTOR AS NESCESSARY
USE AN OHMMETER TO MEASURE LB WIRE IN CAVITY 20 TO GROUND
THE METER SHOWS LOW RESISTANCE (UNDER 10 OHMS)
YES
NO
DISCONNECT THE 8-WAY AT THE TRANSAXLE SOLENOID PACK
INSPECT CONNECTOR FOR INCORRECT WIRE PLACEMENT OR TERMINAL DAMAGE
TEST THE RESISTANCE BETWEEN CAVITY 20 AND 16
THE CONNECTOR IS OK?
YES
NO
RE-TEST THE LB WIRE IN CAVITY 20 TO GROUND
REPAIR THE CONNECTOR AS NECESSARY
THE METER STILL READS UNDER 10 OHMS?
YES
NO
REPAIR GROUNDED LB WIRE BETWEEN A-604 CONTROLLER AND SOLENOID PACK
REPLACE THE SOLENOID PACK
CONTINUE TEST 29 ON THE NEXT PAGE

TEST 29
CODE 41—LR SOLENOID CIRCUIT
PERFORM TEST 3 BEFORE PROCEEDING

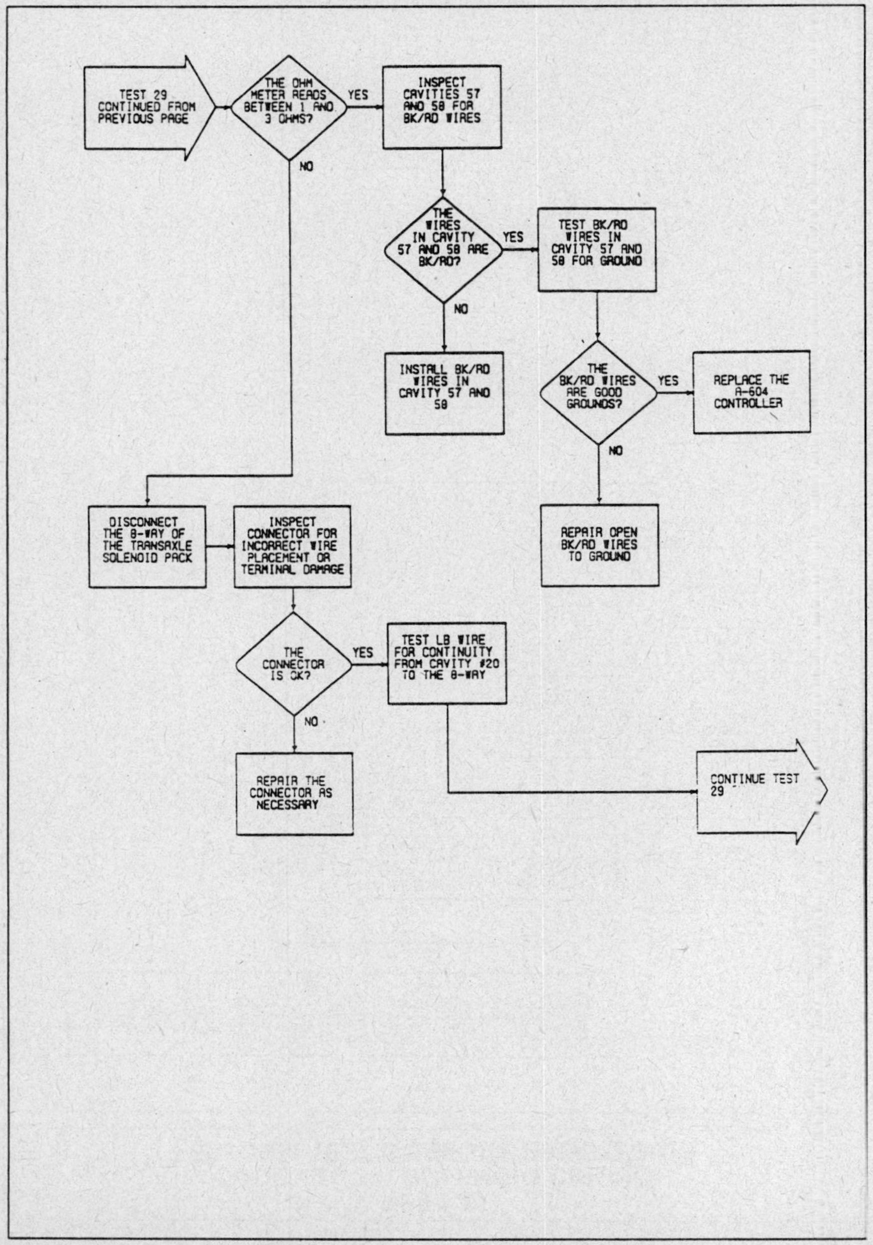
TEST 29 CONTINUED FROM PREVIOUS PAGE
THE OHM METER READS BETWEEN 1 AND 3 OHMS?
YES
NO
INSPECT CAVITIES 57 AND 58 FOR BK/RD WIRES
THE WIRES IN CAVITY 57 AND 58 ARE BK/RD?
YES
NO
TEST BK/RD WIRES IN CAVITY 57 AND 58 FOR GROUND
INSTALL BK/RD WIRES IN CAVITY 57 AND 58
THE BK/RD WIRES ARE GOOD GROUNDS?
YES
NO
REPLACE THE A-604 CONTROLLER
REPAIR OPEN BK/RD WIRES TO GROUND
DISCONNECT THE 8-WAY OF THE TRANSAXLE SOLENOID PACK
INSPECT CONNECTOR FOR INCORRECT WIRE PLACEMENT OR TERMINAL DAMAGE
THE CONNECTOR IS OK?
YES
NO
TEST LB WIRE FOR CONTINUITY FROM CAVITY #20 TO THE 8-WAY
REPAIR THE CONNECTOR AS NECESSARY
CONTINUE TEST 29

TEST 29
CODE 41—LR SOLENOID CIRCUIT
PERFORM TEST 3 BEFORE PROCEEDING

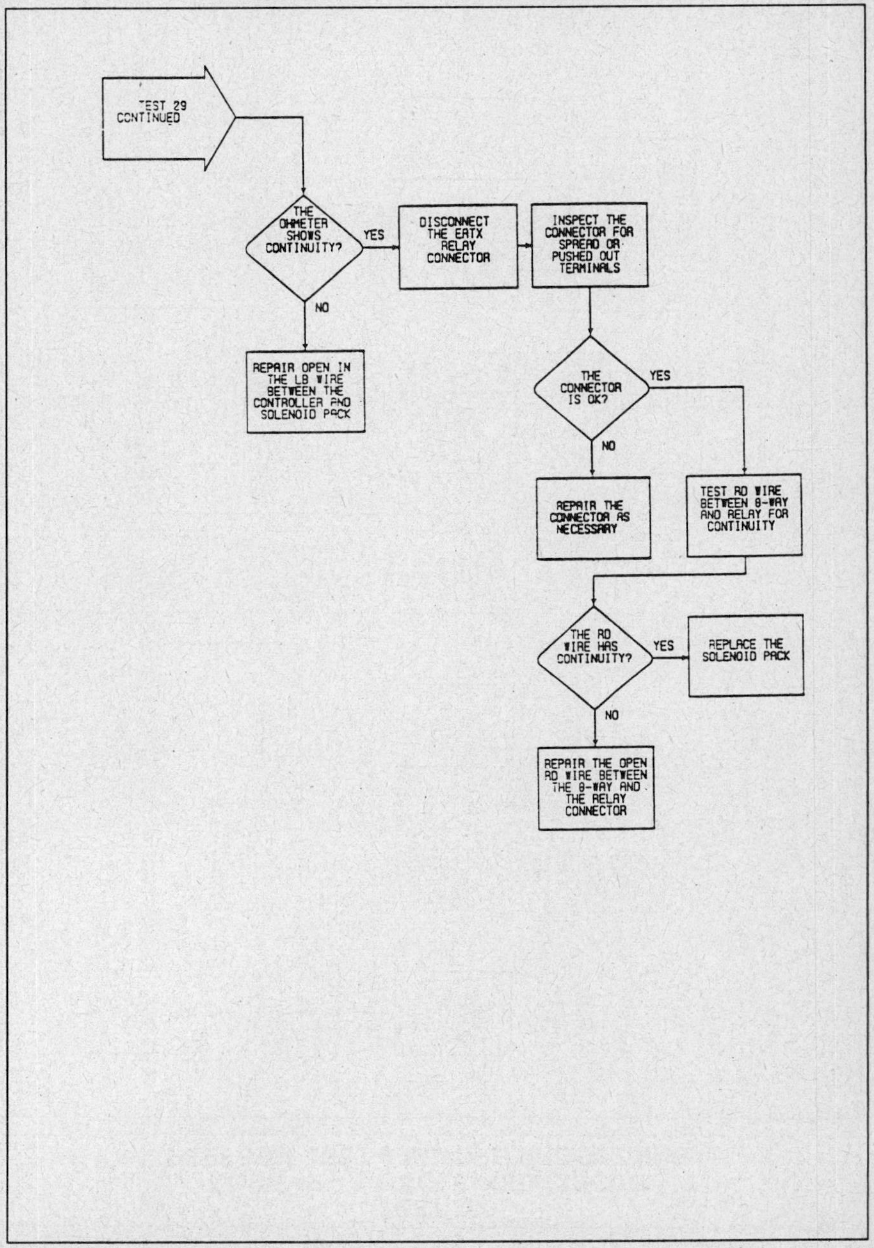
TEST 29 CONTINUED
THE OHMETER SHOWS CONTINUITY?
YES
NO
DISCONNECT THE EATX RELAY CONNECTOR
INSPECT THE CONNECTOR FOR SPREAD OR PUSHED OUT TERMINALS
REPAIR OPEN IN THE LB WIRE BETWEEN THE CONTROLLER AND SOLENOID PACK
THE CONNECTOR IS OK?
YES
NO
REPAIR THE CONNECTOR AS NECESSARY
TEST RD WIRE BETWEEN 8-WAY AND RELAY FOR CONTINUITY
THE RD WIRE HAS CONTINUITY?
YES
NO
REPLACE THE SOLENOID PACK
REPAIR THE OPEN RD WIRE BETWEEN THE 8-WAY AND THE RELAY CONNECTOR

TEST 30
CODE 42—2-4 SOLENOID CIRCUIT
PERFORM TEST 3 BEFORE PROCEEDING

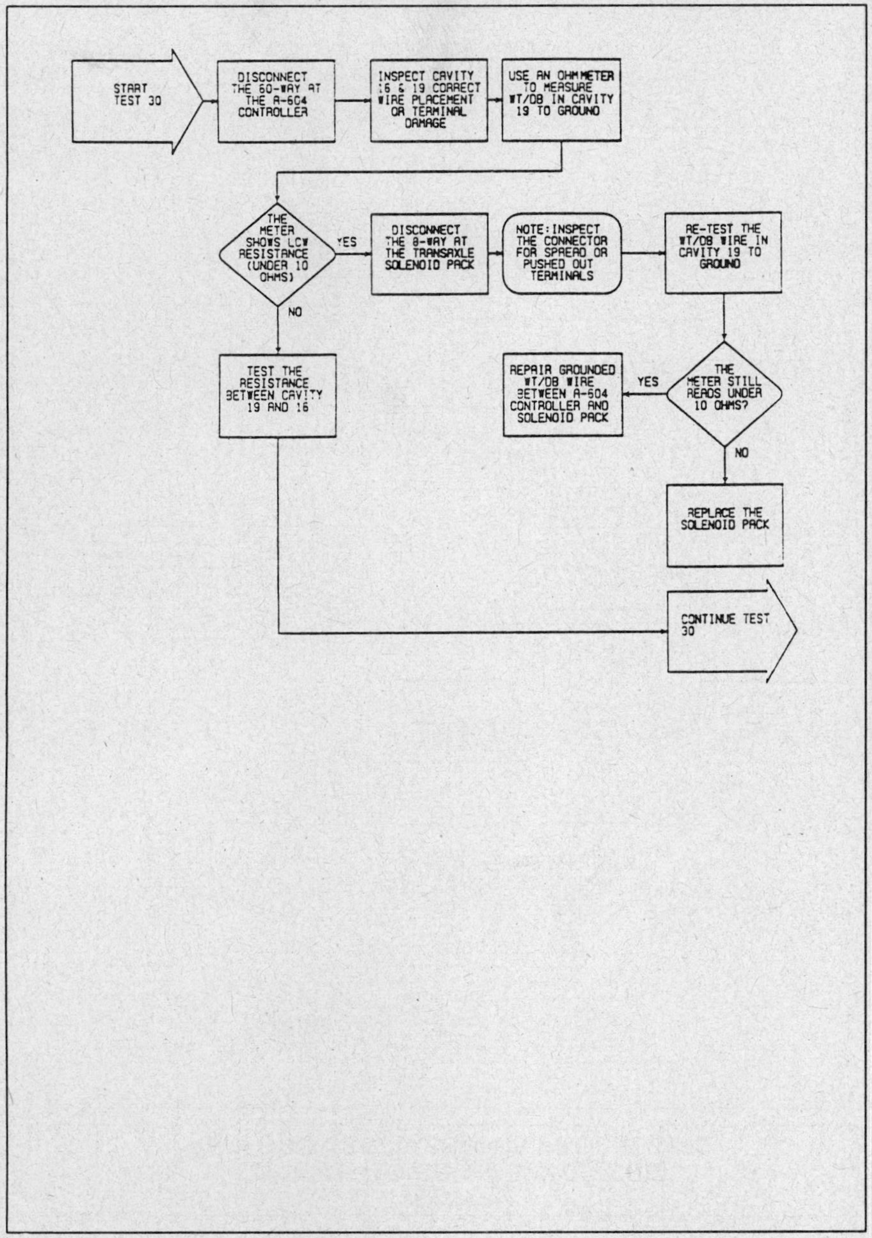
START TEST 30
DISCONNECT THE 60-WAY AT THE A-604 CONTROLLER
INSPECT CAVITY 16 & 19 CORRECT WIRE PLACEMENT OR TERMINAL DAMAGE
USE AN OHMMETER TO MEASURE WT/DB IN CAVITY 19 TO GROUND
THE METER SHOWS LOW RESISTANCE (UNDER 10 OHMS)
YES
NO
DISCONNECT THE 8-WAY AT THE TRANSAXLE SOLENOID PACK
NOTE: INSPECT THE CONNECTOR FOR SPREAD OR PUSHED OUT TERMINALS
RE-TEST THE WT/DB WIRE IN CAVITY 19 TO GROUND
TEST THE RESISTANCE BETWEEN CAVITY 19 AND 16
REPAIR GROUNDED WT/DB WIRE BETWEEN A-604 CONTROLLER AND SOLENOID PACK
YES
THE METER STILL READS UNDER 10 OHMS?
NO
REPLACE THE SOLENOID PACK
CONTINUE TEST 30

TEST 30
CODE 42—2-4 SOLENOID CIRCUIT
PERFORM TEST 3 BEFORE PROCEEDING

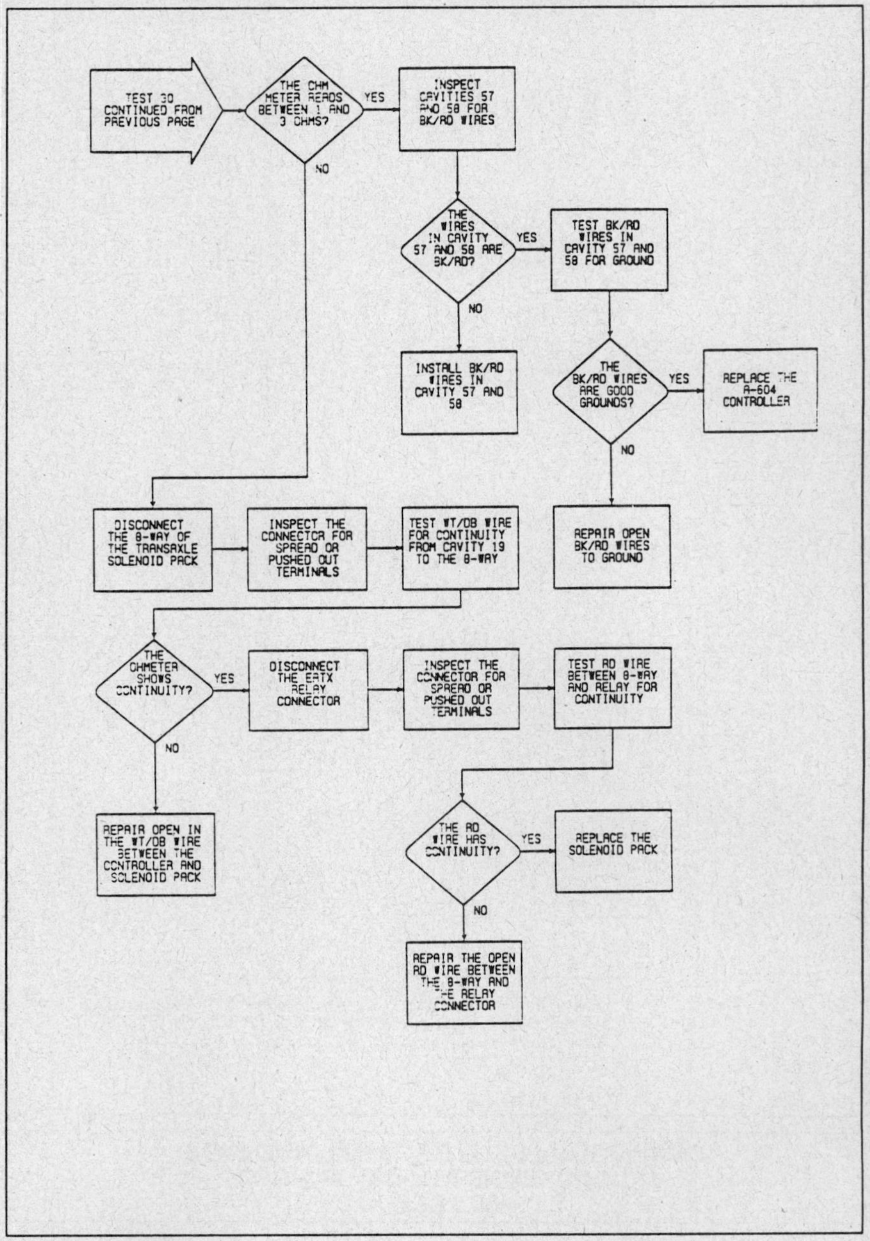
TEST 30 CONTINUED FROM PREVIOUS PAGE
THE OHM METER READS BETWEEN 1 AND 3 OHMS?
YES
NO
INSPECT CAVITIES 57 AND 58 FOR BK/RD WIRES
THE WIRES IN CAVITY 57 AND 58 ARE BK/RD?
YES
NO
TEST BK/RD WIRES IN CAVITY 57 AND 58 FOR GROUND
INSTALL BK/RD WIRES IN CAVITY 57 AND 58
THE BK/RD WIRES ARE GOOD GROUNDS?
YES
NO
REPLACE THE A-604 CONTROLLER
REPAIR OPEN BK/RD WIRES TO GROUND
DISCONNECT THE 8-WAY OF THE TRANSAXLE SOLENOID PACK
INSPECT THE CONNECTOR FOR SPREAD OR PUSHED OUT TERMINALS
TEST WT/DB WIRE FOR CONTINUITY FROM CAVITY 19 TO THE 8-WAY
THE OHMMETER SHOWS CONTINUITY?
YES
NO
DISCONNECT THE EATX RELAY CONNECTOR
INSPECT THE CONNECTOR FOR SPREAD OR PUSHED OUT TERMINALS
TEST RD WIRE BETWEEN 8-WAY AND RELAY FOR CONTINUITY
REPAIR OPEN IN THE WT/DB WIRE BETWEEN THE CONTROLLER AND SOLENOID PACK
THE RD WIRE HAS CONTINUITY?
YES
NO
REPLACE THE SOLENOID PACK
REPAIR THE OPEN RD WIRE BETWEEN THE 8-WAY AND THE RELAY CONNECTOR

TEST 31
CODE 43—OD SOLENOID CIRCUIT
PERFORM TEST 3 BEFORE PROCEEDING

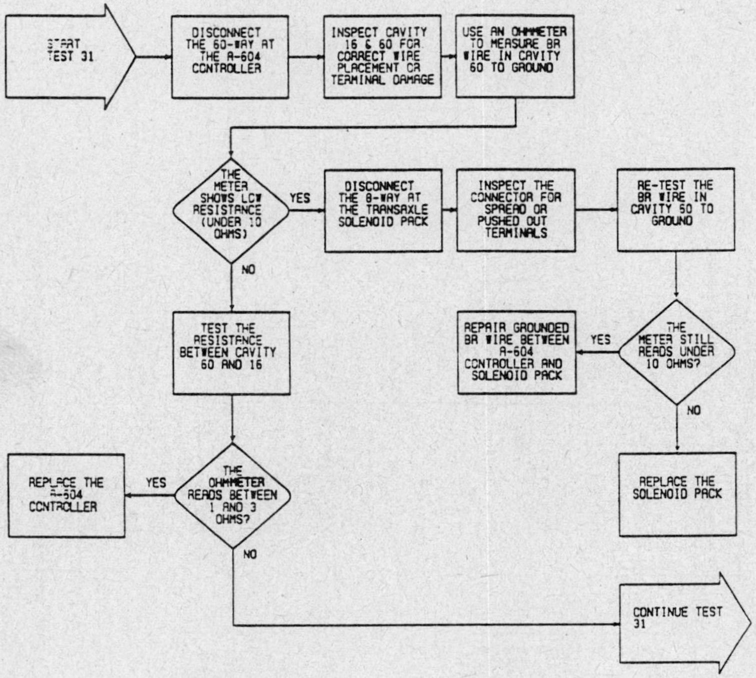
START TEST 31
DISCONNECT THE 60-WAY AT THE A-604 CONTROLLER
INSPECT CAVITY 16 & 60 FOR CORRECT WIRE PLACEMENT OR TERMINAL DAMAGE
USE AN OHMMETER TO MEASURE BR WIRE IN CAVITY 60 TO GROUND
THE METER SHOWS LOW RESISTANCE (UNDER 10 OHMS)
YES
DISCONNECT THE 8-WAY AT THE TRANSAXLE SOLENOID PACK
INSPECT THE CONNECTOR FOR SPREAD OR PUSHED OUT TERMINALS
RE-TEST THE BR WIRE IN CAVITY 50 TO GROUND
THE METER STILL READS UNDER 10 OHMS?
YES
REPAIR GROUNDED BR WIRE BETWEEN A-604 CONTROLLER AND SOLENOID PACK
NO
REPLACE THE SOLENOID PACK
NO
TEST THE RESISTANCE BETWEEN CAVITY 60 AND 16
THE OHMMETER READS BETWEEN 1 AND 3 OHMS?
YES
REPLACE THE A-604 CONTROLLER
NO
CONTINUE TEST 31

TEST 31
CODE 43—OD SOLENOID CIRCUIT
PERFORM TEST 3 BEFORE PROCEEDING

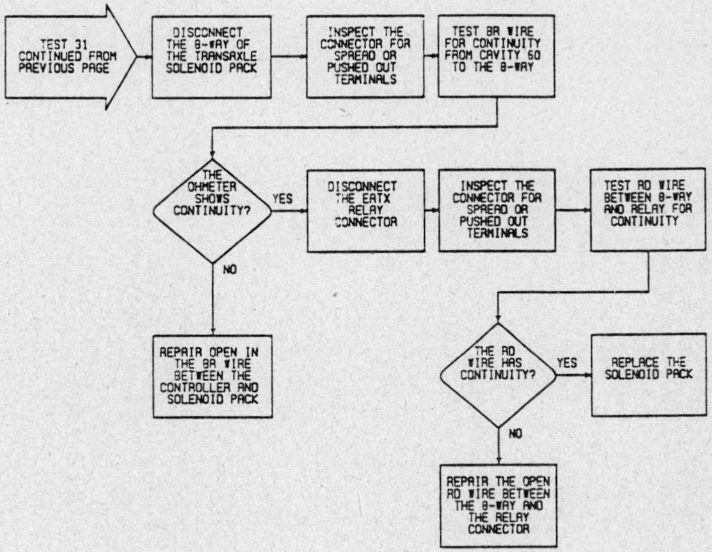
TEST 31 CONTINUED FROM PREVIOUS PAGE
DISCONNECT THE 8-WAY OF THE TRANSAXLE SOLENOID PACK
INSPECT THE CONNECTOR FOR SPREAD OR PUSHED OUT TERMINALS
TEST BR WIRE FOR CONTINUITY FROM CAVITY 50 TO THE 8-WAY
THE OHMMETER SHOWS CONTINUITY?
YES
DISCONNECT THE EATX RELAY CONNECTOR
INSPECT THE CONNECTOR FOR SPREAD OR PUSHED OUT TERMINALS
TEST RD WIRE BETWEEN 8-WAY AND RELAY FOR CONTINUITY
THE RD WIRE HAS CONTINUITY?
YES
REPLACE THE SOLENOID PACK
NO
REPAIR THE OPEN RD WIRE BETWEEN THE 8-WAY AND THE RELAY CONNECTOR
NO
REPAIR OPEN IN THE BR WIRE BETWEEN THE CONTROLLER AND SOLENOID PACK

TEST 32
CODE 44—UD SOLENOID CIRCUIT
PERFORM TEST 3 BEFORE PROCEEDING

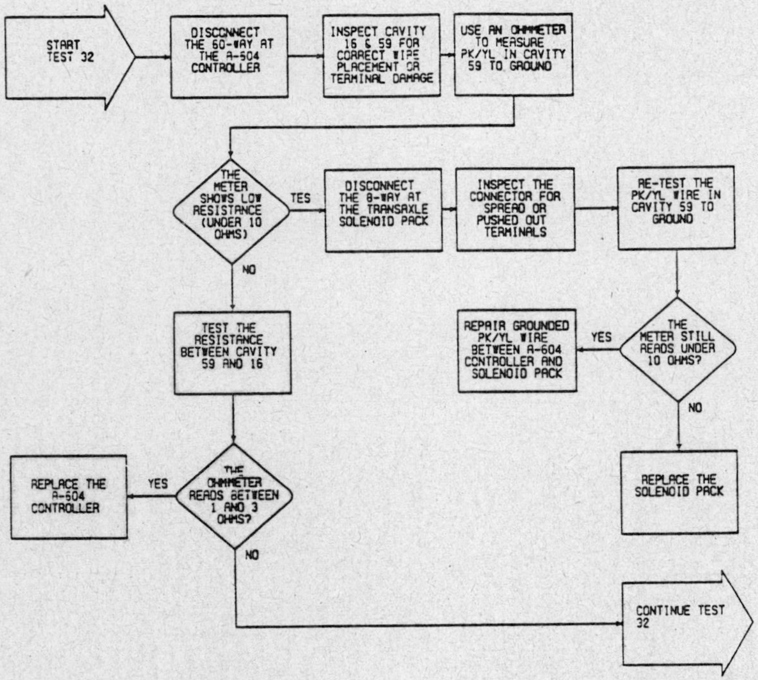
START TEST 32
DISCONNECT THE 60-WAY AT THE A-604 CONTROLLER
INSPECT CAVITY 16 & 59 FOR CORRECT WIRE PLACEMENT OR TERMINAL DAMAGE
USE AN OHMMETER TO MEASURE PK/YL IN CAVITY 59 TO GROUND
THE METER SHOWS LOW RESISTANCE (UNDER 10 OHMS)
TES
DISCONNECT THE 8-WAY AT THE TRANSAXLE SOLENOID PACK
INSPECT THE CONNECTOR FOR SPREAD OR PUSHED OUT TERMINALS
RE-TEST THE PK/YL WIRE IN CAVITY 59 TO GROUND
THE METER STILL READS UNDER 10 OHMS?
YES
REPAIR GROUNDED PK/YL WIRE BETWEEN A-604 CONTROLLER AND SOLENOID PACK
NO
REPLACE THE SOLENOID PACK
NO
TEST THE RESISTANCE BETWEEN CAVITY 59 AND 16
THE OHMMETER READS BETWEEN 1 AND 3 OHMS?
YES
REPLACE THE A-604 CONTROLLER
NO
CONTINUE TEST 32

TEST 32
CODE 44—UD SOLENOID CIRCUIT
PERFORM TEST 3 BEFORE PROCEEDING

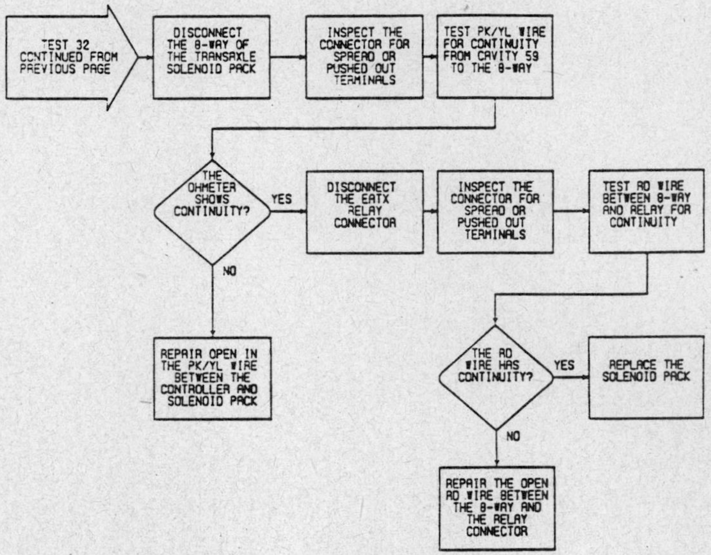
TEST 32 CONTINUED FROM PREVIOUS PAGE
DISCONNECT THE 8-WAY OF THE TRANSAXLE SOLENOID PACK
INSPECT THE CONNECTOR FOR SPREAD OR PUSHED OUT TERMINALS
TEST PK/YL WIRE FOR CONTINUITY FROM CAVITY 59 TO THE 8-WAY
THE OHMMETER SHOWS CONTINUITY?
YES
DISCONNECT THE EATX RELAY CONNECTOR
INSPECT THE CONNECTOR FOR SPREAD OR PUSHED OUT TERMINALS
TEST RD WIRE BETWEEN 8-WAY AND RELAY FOR CONTINUITY
THE RD WIRE HAS CONTINUITY?
YES
REPLACE THE SOLENOID PACK
NO
REPAIR THE OPEN RD WIRE BETWEEN THE 8-WAY AND THE RELAY CONNECTOR
NO
REPAIR OPEN IN THE PK/YL WIRE BETWEEN THE CONTROLLER AND SOLENOID PACK

TEST 33
TEST FOR FAILED SHIFT LEVER TEST
PERFORM TEST 5 BEFORE PROCEEDING

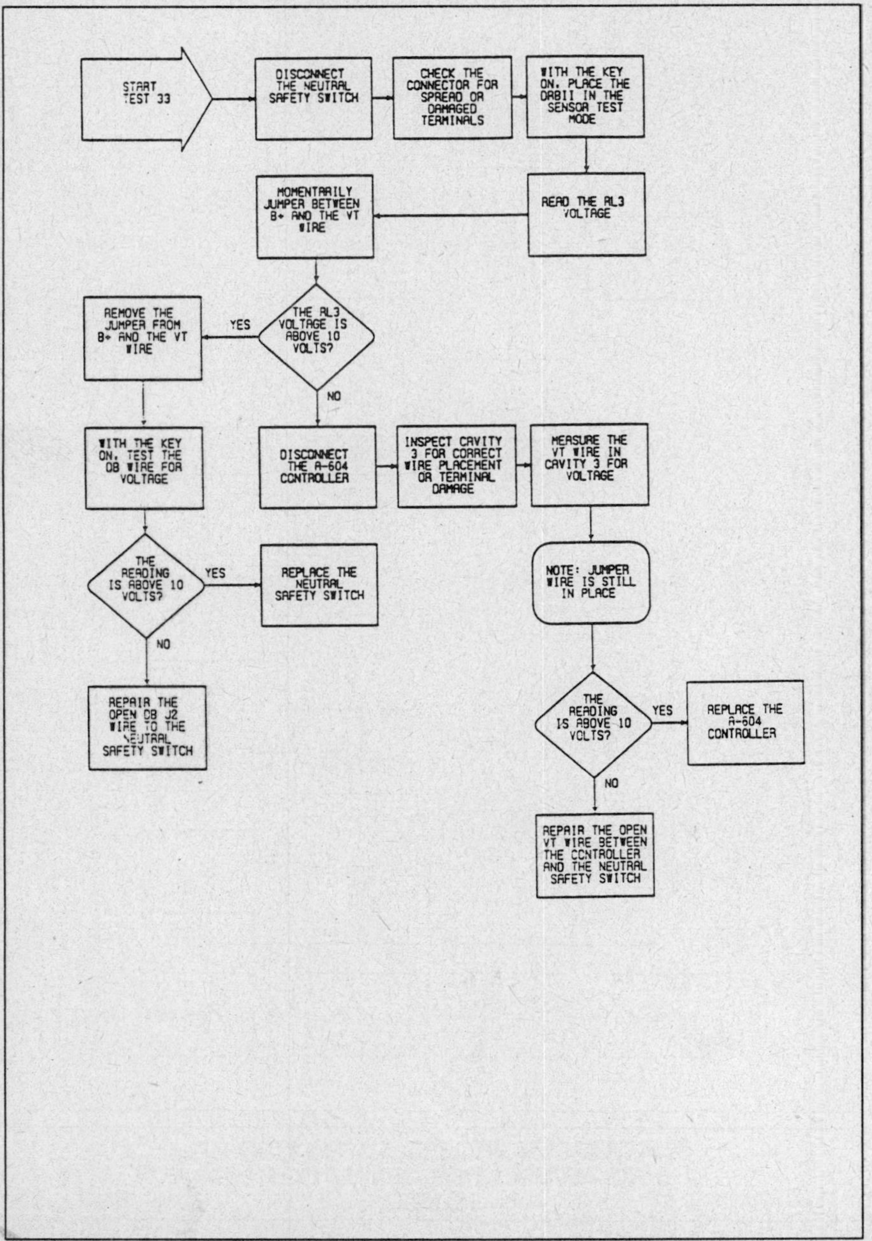
START TEST 33
DISCONNECT THE NEUTRAL SAFETY SWITCH
CHECK THE CONNECTOR FOR SPREAD OR DAMAGED TERMINALS
WITH THE KEY ON, PLACE THE DRBII IN THE SENSOR TEST MODE
READ THE RL3 VOLTAGE
MOMENTARILY JUMPER BETWEEN B+ AND THE VT WIRE
THE RL3 VOLTAGE IS ABOVE 10 VOLTS?
YES
REMOVE THE JUMPER FROM B+ AND THE VT WIRE
WITH THE KEY ON, TEST THE DB WIRE FOR VOLTAGE
THE READING IS ABOVE 10 VOLTS?
YES
REPLACE THE NEUTRAL SAFETY SWITCH
NO
REPAIR THE OPEN DB J2 WIRE TO THE NEUTRAL SAFETY SWITCH
NO
DISCONNECT THE A-604 CONTROLLER
INSPECT CAVITY 3 FOR CORRECT WIRE PLACEMENT OR TERMINAL DAMAGE
MEASURE THE VT WIRE IN CAVITY 3 FOR VOLTAGE
NOTE: JUMPER WIRE IS STILL IN PLACE
THE READING IS ABOVE 10 VOLTS?
YES
REPLACE THE A-604 CONTROLLER
NO
REPAIR THE OPEN VT WIRE BETWEEN THE CONTROLLER AND THE NEUTRAL SAFETY SWITCH

TEST 34
TEST FOR FAILED SHIFT LEVER TEST
PERFORM TEST 5 BEFORE PROCEEDING

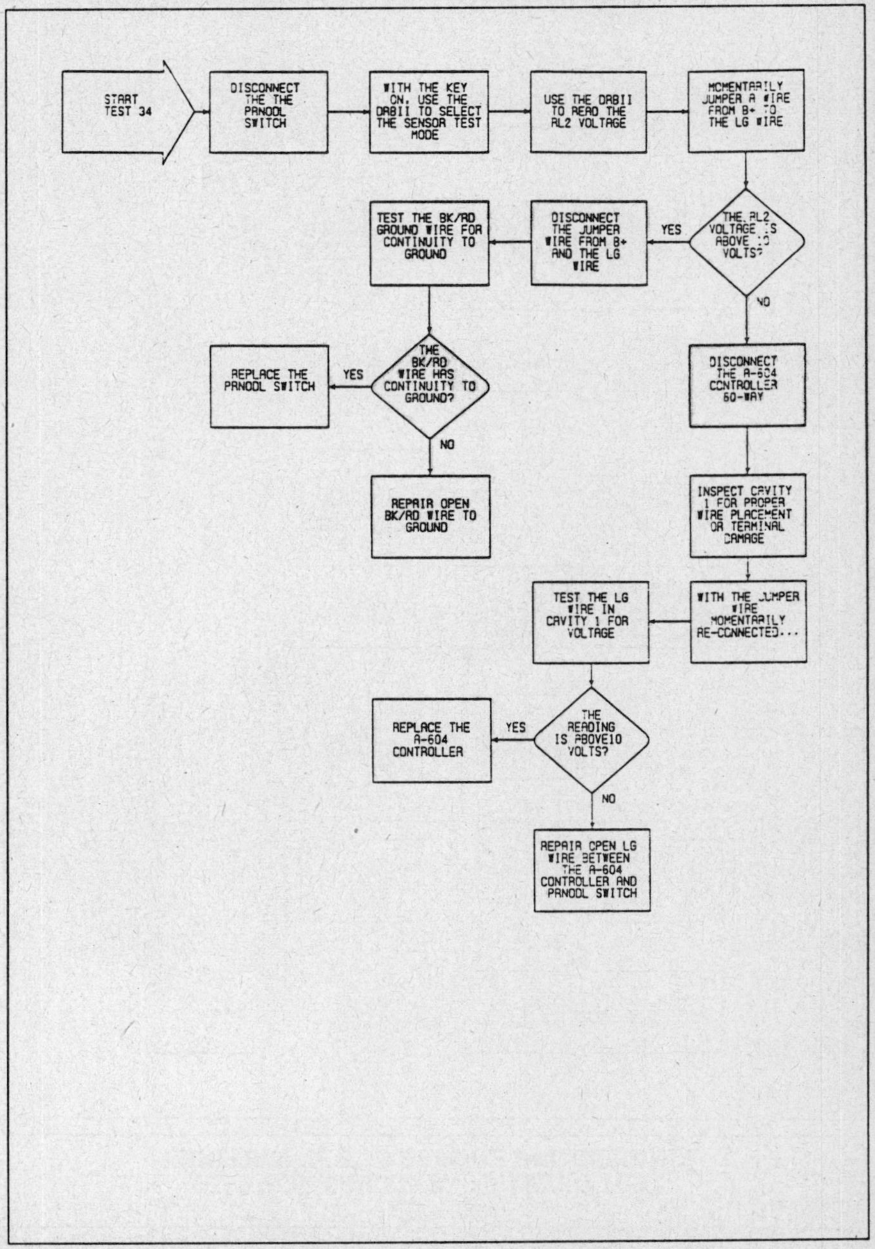
START TEST 34
DISCONNECT THE THE PRNODL SWITCH
WITH THE KEY ON, USE THE DRBII TO SELECT THE SENSOR TEST MODE
USE THE DRBII TO READ THE RL2 VOLTAGE
MOMENTARILY JUMPER A WIRE FROM B+ TO THE LG WIRE
THE RL2 VOLTAGE IS ABOVE 10 VOLTS?
YES
DISCONNECT THE JUMPER WIRE FROM B+ AND THE LG WIRE
TEST THE BK/RD GROUND WIRE FOR CONTINUITY TO GROUND
THE BK/RD WIRE HAS CONTINUITY TO GROUND?
YES
REPLACE THE PRNODL SWITCH
NO
REPAIR OPEN BK/RD WIRE TO GROUND
NO
DISCONNECT THE A-604 CONTROLLER 60-WAY
INSPECT CAVITY 1 FOR PROPER WIRE PLACEMENT OR TERMINAL DAMAGE
WITH THE JUMPER WIRE MOMENTARILY RE-CONNECTED...
TEST THE LG WIRE IN CAVITY 1 FOR VOLTAGE
THE READING IS ABOVE10 VOLTS?
YES
REPLACE THE A-604 CONTROLLER
NO
REPAIR OPEN LG WIRE BETWEEN THE A-604 CONTROLLER AND PRNODL SWITCH

TEST 35
TEST FOR FAILED SHIFT LEVER TEST
PERFORM TEST 5 BEFORE PROCEEDING

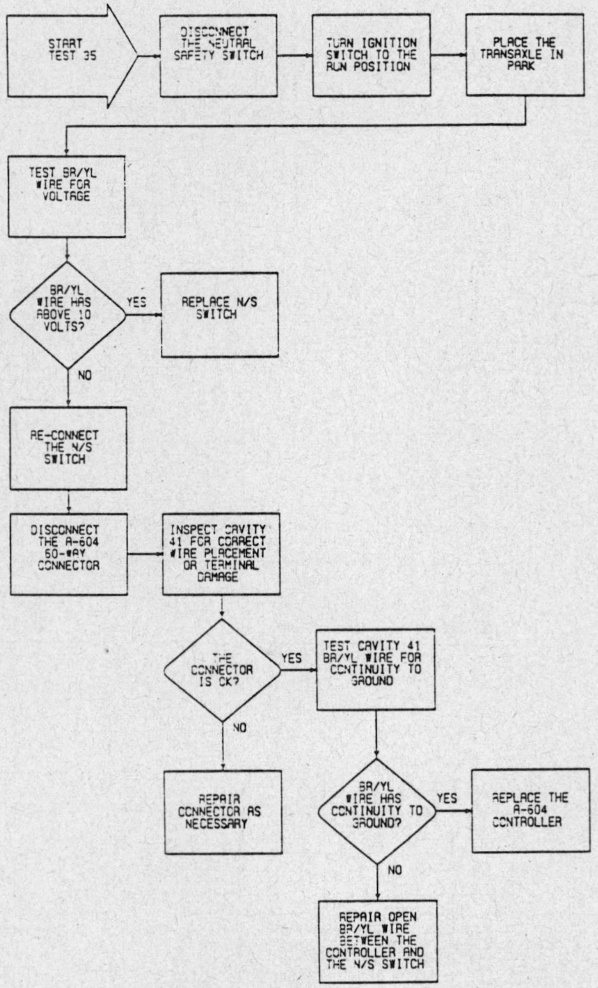
START TEST 35
DISCONNECT THE NEUTRAL SAFETY SWITCH
TURN IGNITION SWITCH TO THE RUN POSITION
PLACE THE TRANSAXLE IN PARK
TEST BR/YL WIRE FOR VOLTAGE
BR/YL WIRE HAS ABOVE 10 VOLTS?
YES
REPLACE N/S SWITCH
NO
RE-CONNECT THE N/S SWITCH
DISCONNECT THE A-604 60-WAY CONNECTOR
INSPECT CAVITY 41 FOR CORRECT WIRE PLACEMENT OR TERMINAL DAMAGE
THE CONNECTOR IS OK?
YES
TEST CAVITY 41 BR/YL WIRE FOR CONTINUITY TO GROUND
NO
REPAIR CONNECTOR AS NECESSARY
BR/YL WIRE HAS CONTINUITY TO GROUND?
YES
REPLACE THE A-604 CONTROLLER
NO
REPAIR OPEN BR/YL WIRE BETWEEN THE CONTROLLER AND THE N/S SWITCH

TEST 36
TEST FOR FAILED SHIFT LEVER TEST
PERFORM TEST 5 BEFORE PROCEEDING

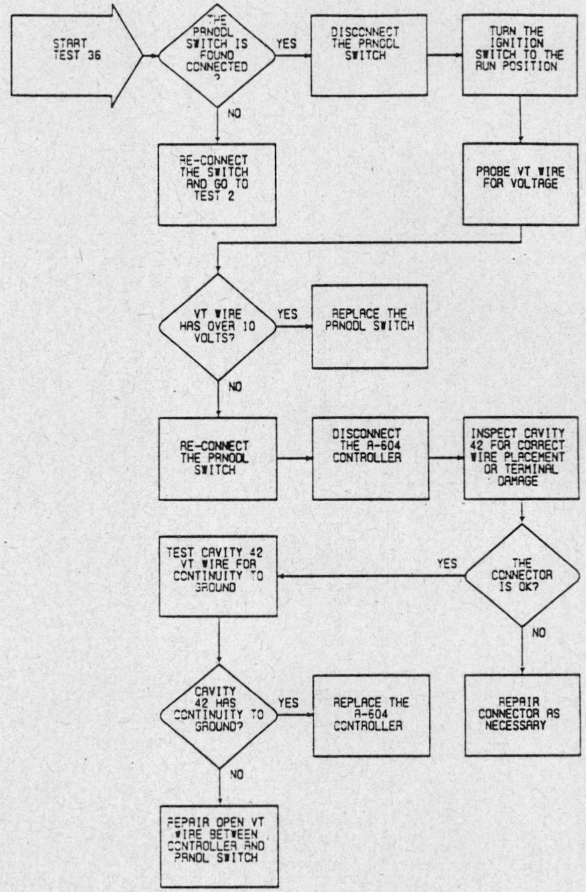
START TEST 36
THE PANDOL SWITCH IS FOUND CONNECTED ?
YES
DISCONNECT THE PANDOL SWITCH
TURN THE IGNITION SWITCH TO THE RUN POSITION
NO
RE-CONNECT THE SWITCH AND GO TO TEST 2
PROBE VT WIRE FOR VOLTAGE
VT WIRE HAS OVER 10 VOLTS?
YES
REPLACE THE PANDOL SWITCH
NO
RE-CONNECT THE PANDOL SWITCH
DISCONNECT THE A-604 CONTROLLER
INSPECT CAVITY 42 FOR CORRECT WIRE PLACEMENT OR TERMINAL DAMAGE
THE CONNECTOR IS OK?
YES
TEST CAVITY 42 VT WIRE FOR CONTINUITY TO GROUND
NO
REPAIR CONNECTOR AS NECESSARY
CAVITY 42 HAS CONTINUITY TO GROUND?
YES
REPLACE THE A-604 CONTROLLER
NO
REPAIR OPEN VT WIRE BETWEEN CONTROLLER AND PANDL SWITCH

TEST 37
TEST FOR FAILED SHIFT LEVER TEST
PERFORM TEST 5 BEFORE PROCEEDING

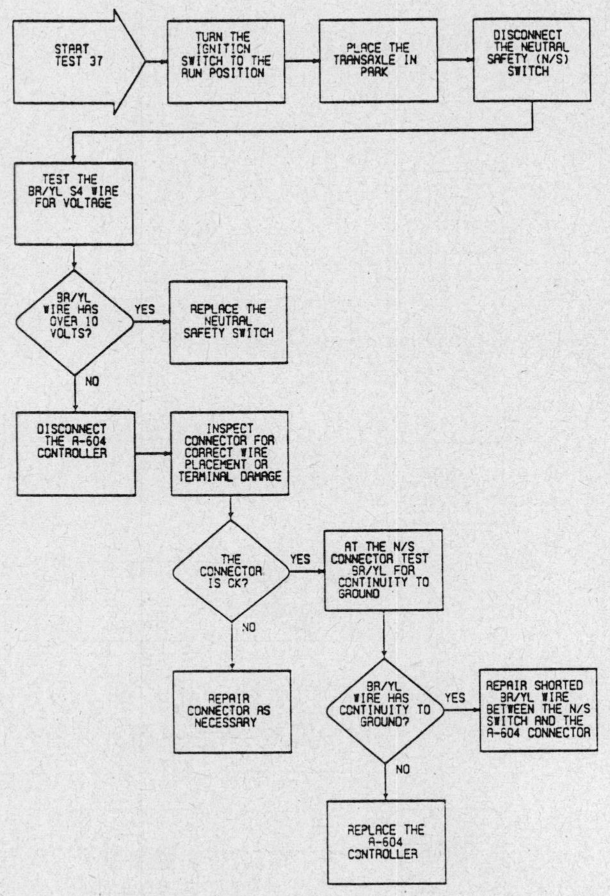
START TEST 37
TURN THE IGNITION SWITCH TO THE RUN POSITION
PLACE THE TRANSAXLE IN PARK
DISCONNECT THE NEUTRAL SAFETY (N/S) SWITCH
TEST THE BR/YL S4 WIRE FOR VOLTAGE
BR/YL WIRE HAS OVER 10 VOLTS?
YES
REPLACE THE NEUTRAL SAFETY SWITCH
NO
DISCONNECT THE A-604 CONTROLLER
INSPECT CONNECTOR FOR CORRECT WIRE PLACEMENT OR TERMINAL DAMAGE
THE CONNECTOR IS OK?
YES
AT THE N/S CONNECTOR TEST BR/YL FOR CONTINUITY TO GROUND
NO
REPAIR CONNECTOR AS NECESSARY
BR/YL WIRE HAS CONTINUITY TO GROUND?
YES
REPAIR SHORTED BR/YL WIRE BETWEEN THE N/S SWITCH AND THE A-604 CONNECTOR
NO
REPLACE THE A-604 CONTROLLER

TEST 38
TEST FOR FAILED SHIFT LEVER TEST
PERFORM TEST 5 BEFORE PROCEEDING

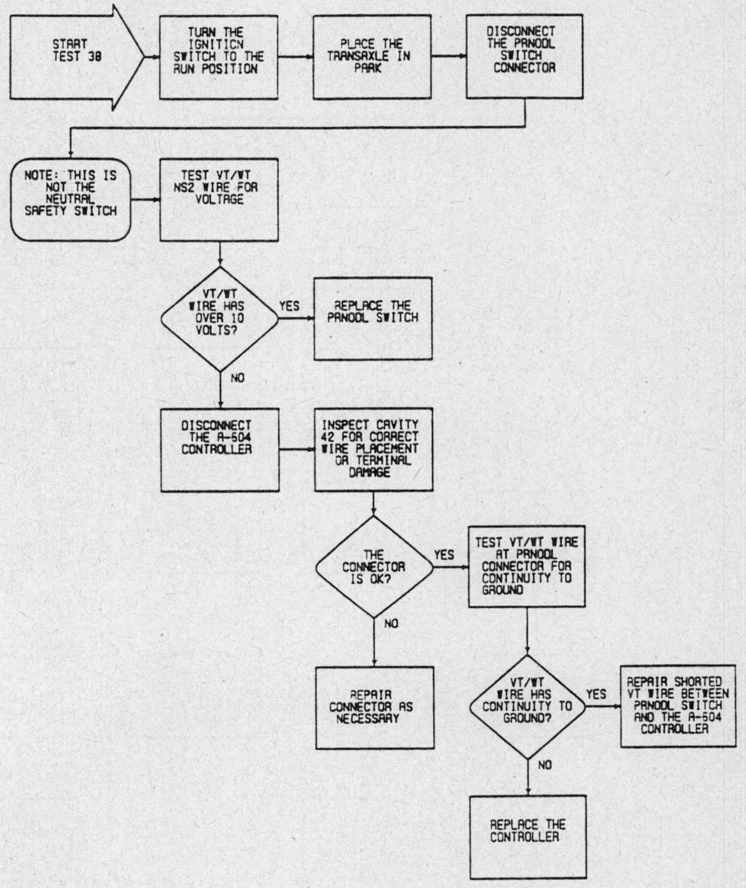
START TEST 38
TURN THE IGNITION SWITCH TO THE RUN POSITION
PLACE THE TRANSAXLE IN PARK
DISCONNECT THE PRNODL SWITCH CONNECTOR
NOTE: THIS IS NOT THE NEUTRAL SAFETY SWITCH
TEST VT/WT NS2 WIRE FOR VOLTAGE
VT/WT WIRE HAS OVER 10 VOLTS?
YES
REPLACE THE PRNODL SWITCH
NO
DISCONNECT THE A-604 CONTROLLER
INSPECT CAVITY 42 FOR CORRECT WIRE PLACEMENT OR TERMINAL DAMAGE
THE CONNECTOR IS OK?
YES
TEST VT/WT WIRE AT PRNODL CONNECTOR FOR CONTINUITY TO GROUND
NO
REPAIR CONNECTOR AS NECESSARY
VT/WT WIRE HAS CONTINUITY TO GROUND?
YES
REPAIR SHORTED VT WIRE BETWEEN PRNODL SWITCH AND THE A-604 CONTROLLER
NO
REPLACE THE CONTROLLER

TEST 39
TEST FOR FAILED SHIFT LEVER TEST
PERFORM TEST 5 BEFORE PROCEEDING

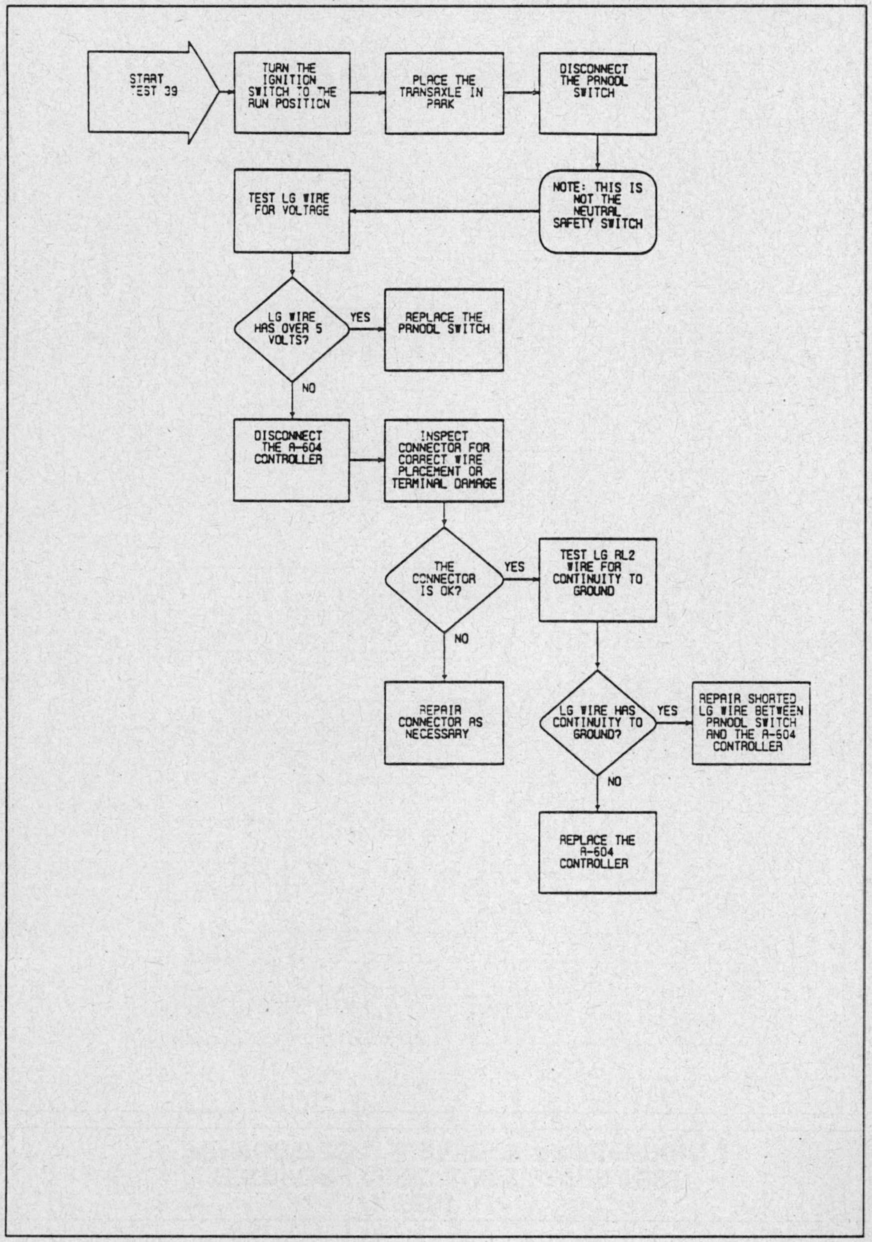
START TEST 39
TURN THE IGNITION SWITCH TO THE RUN POSITION
PLACE THE TRANSAXLE IN PARK
DISCONNECT THE PRNODL SWITCH
NOTE: THIS IS NOT THE NEUTRAL SAFETY SWITCH
TEST LG WIRE FOR VOLTAGE
LG WIRE HAS OVER 5 VOLTS?
YES
REPLACE THE PRNODL SWITCH
NO
DISCONNECT THE A-604 CONTROLLER
INSPECT CONNECTOR FOR CORRECT WIRE PLACEMENT OR TERMINAL DAMAGE
THE CONNECTOR IS OK?
YES
TEST LG RL2 WIRE FOR CONTINUITY TO GROUND
NO
REPAIR CONNECTOR AS NECESSARY
LG WIRE HAS CONTINUITY TO GROUND?
YES
REPAIR SHORTED LG WIRE BETWEEN PRNODL SWITCH AND THE A-604 CONTROLLER
NO
REPLACE THE A-604 CONTROLLER

TEST 40
TEST FOR "NO RESPONSE" BUS MESSAGE
PERFORM TEST 1 BEFORE PROCEEDING

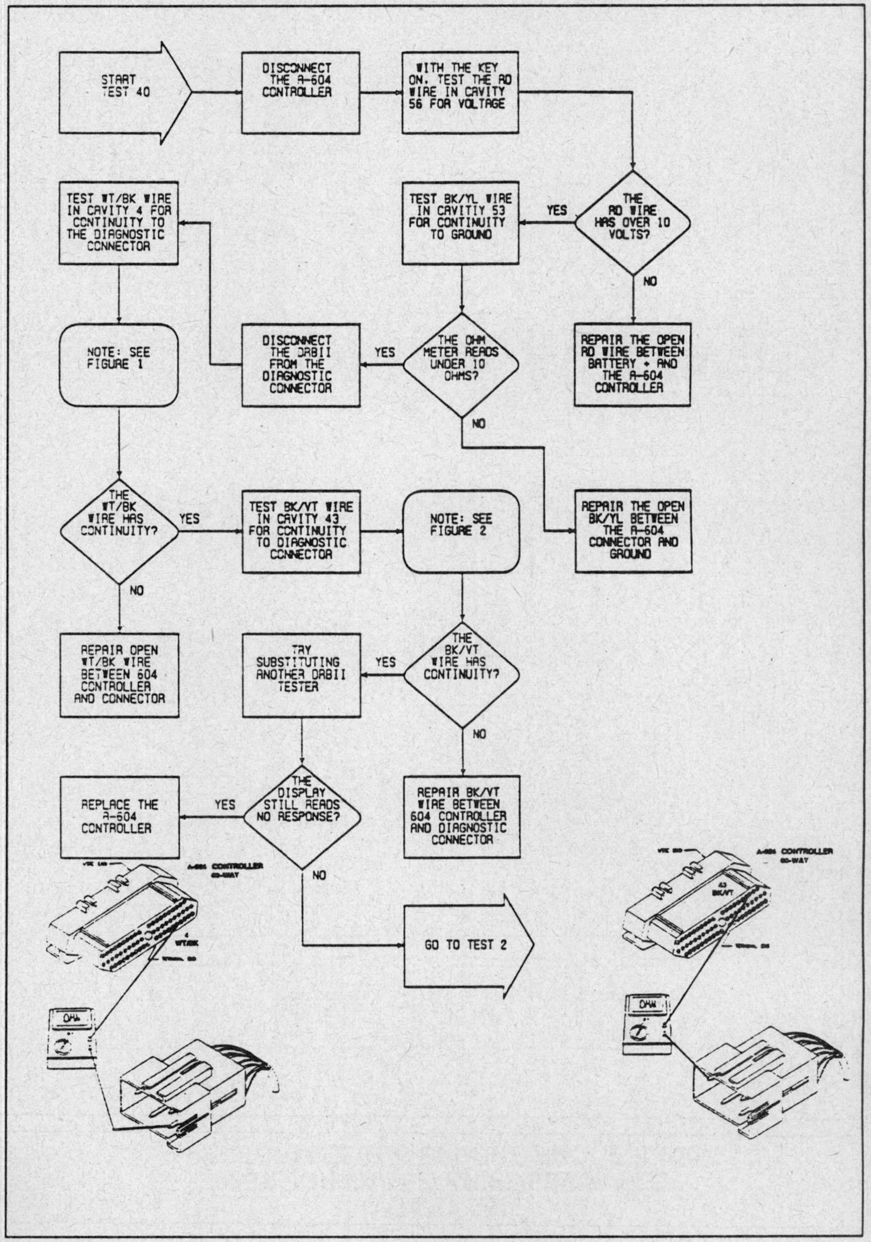
START TEST 40
DISCONNECT THE A-604 CONTROLLER
WITH THE KEY ON, TEST THE RD WIRE IN CAVITY 56 FOR VOLTAGE
THE RD WIRE HAS OVER 10 VOLTS?
YES
NO
REPAIR THE OPEN RD WIRE BETWEEN BATTERY + AND THE A-604 CONTROLLER
TEST BK/YL WIRE IN CAVITY 53 FOR CONTINUITY TO GROUND
THE OHM METER READS UNDER 10 OHMS?
YES
NO
REPAIR THE OPEN BK/YL BETWEEN THE A-604 CONNECTOR AND GROUND
DISCONNECT THE DRBII FROM THE DIAGNOSTIC CONNECTOR
TEST WT/BK WIRE IN CAVITY 4 FOR CONTINUITY TO THE DIAGNOSTIC CONNECTOR
NOTE: SEE FIGURE 1
THE WT/BK WIRE HAS CONTINUITY?
YES
NO
REPAIR OPEN WT/BK WIRE BETWEEN 604 CONTROLLER AND CONNECTOR
TEST BK/YT WIRE IN CAVITY 43 FOR CONTINUITY TO DIAGNOSTIC CONNECTOR
NOTE: SEE FIGURE 2
THE BK/YT WIRE HAS CONTINUITY?
YES
NO
REPAIR BK/YT WIRE BETWEEN 604 CONTROLLER AND DIAGNOSTIC CONNECTOR
TRY SUBSTITUTING ANOTHER DRBII TESTER
THE DISPLAY STILL READS NO RESPONSE?
YES
REPLACE THE A-604 CONTROLLER
NO
GO TO TEST 2
A-604 CONTROLLER 60-WAY
OHM

TEST 41
DETERMINING BODY CONFIGURATION
PERFORM TEST 1 BEFORE PROCEEDING

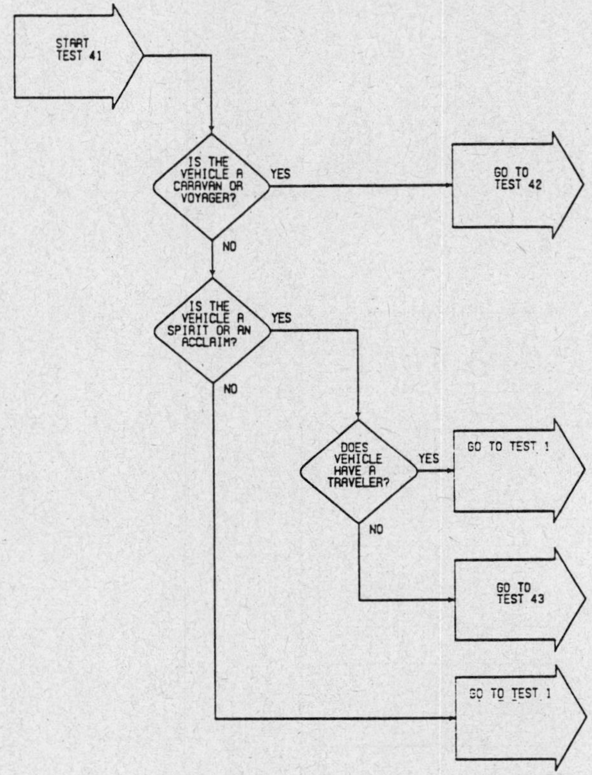
START TEST 41
IS THE VEHICLE A CARAVAN OR VOYAGER?
YES
GO TO TEST 42
NO
IS THE VEHICLE A SPIRIT OR AN ACCLAIM?
YES
DOES VEHICLE HAVE A TRAVELER?
YES
GO TO TEST 1
NO
GO TO TEST 43
NO
GO TO TEST 1

TEST 42
S-BODY FAILED BUS MESSAGE INDEX
PERFORM TEST 41 BEFORE PROCEEDING

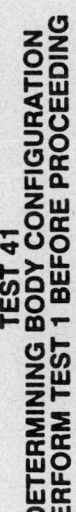

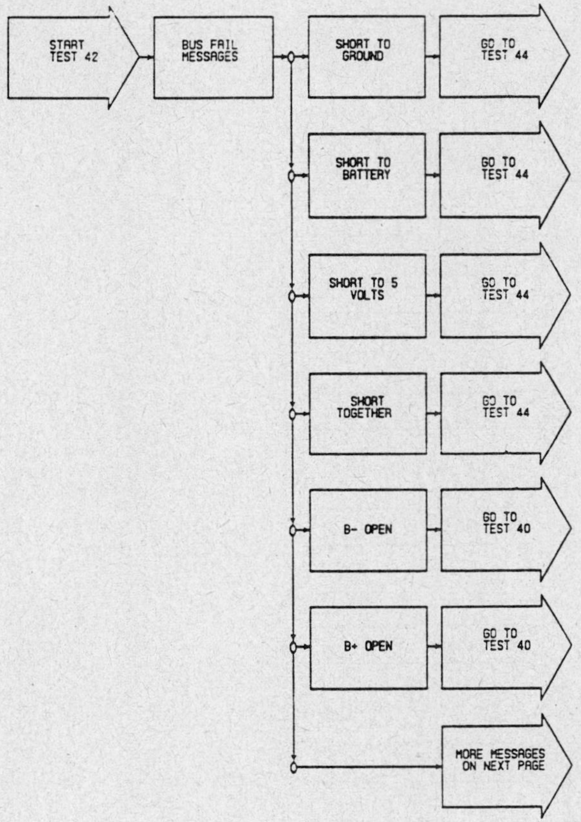
START TEST 42
BUS FAIL MESSAGES
SHORT TO GROUND
GO TO TEST 44
SHORT TO BATTERY
GO TO TEST 44
SHORT TO 5 VOLTS
GO TO TEST 44
SHORT TOGETHER
GO TO TEST 44
B- OPEN
GO TO TEST 40
B+ OPEN
GO TO TEST 40
MORE MESSAGES ON NEXT PAGE

TEST 42
S-BODY FAILED BUS MESSAGE INDEX
PERFORM TEST 41 BEFORE PROCEEDING

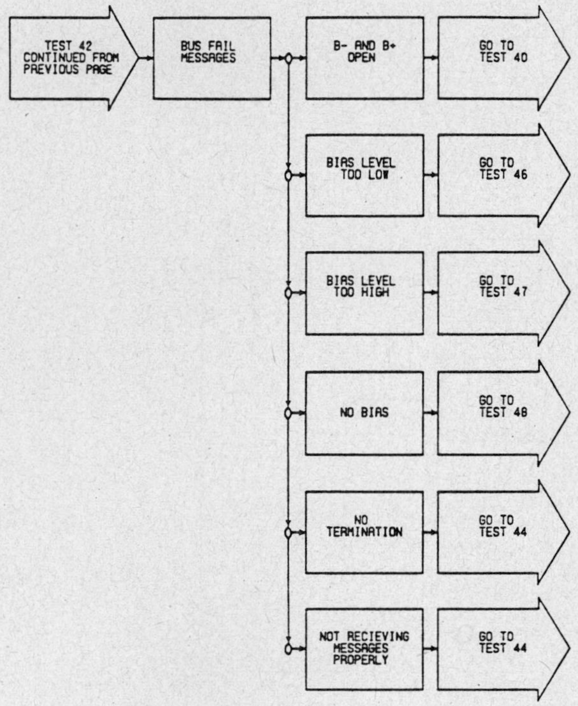
TEST 42 CONTINUED FROM PREVIOUS PAGE
BUS FAIL MESSAGES
B- AND B+ OPEN
GO TO TEST 40
BIAS LEVEL TOO LOW
GO TO TEST 46
BIAS LEVEL TOO HIGH
GO TO TEST 47
NO BIAS
GO TO TEST 48
NO TERMINATION
GO TO TEST 44
NOT RECIEVING MESSAGES PROPERLY
GO TO TEST 44

TEST 43
A-BODY FAILED BUS MESSAGE INDEX
PERFORM TEST 41 BEFORE PROCEEDING

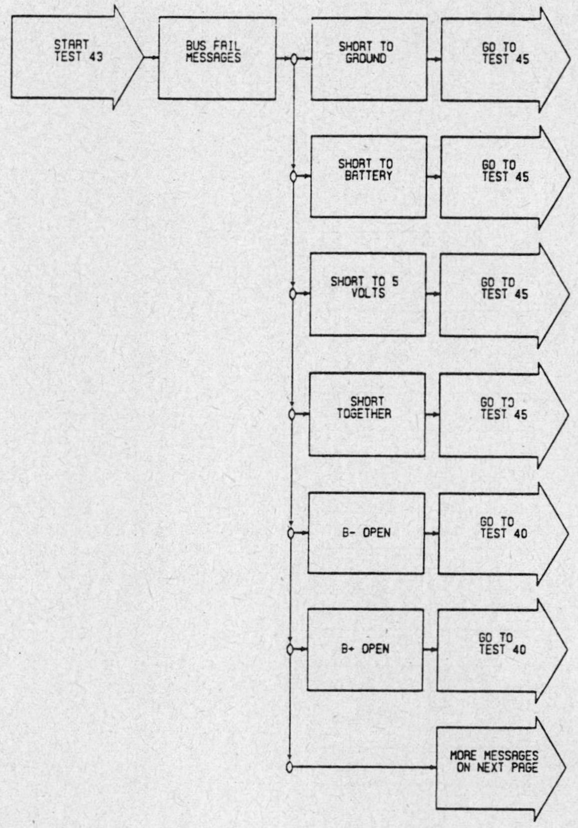
START TEST 43
BUS FAIL MESSAGES
SHORT TO GROUND
GO TO TEST 45
SHORT TO BATTERY
GO TO TEST 45
SHORT TO 5 VOLTS
GO TO TEST 45
SHORT TOGETHER
GO TO TEST 45
B- OPEN
GO TO TEST 40
B+ OPEN
GO TO TEST 40
MORE MESSAGES ON NEXT PAGE

TEST 43
A-BODY FAILED BUS MESSAGE INDEX
PERFORM TEST 41 BEFORE PROCEEDING

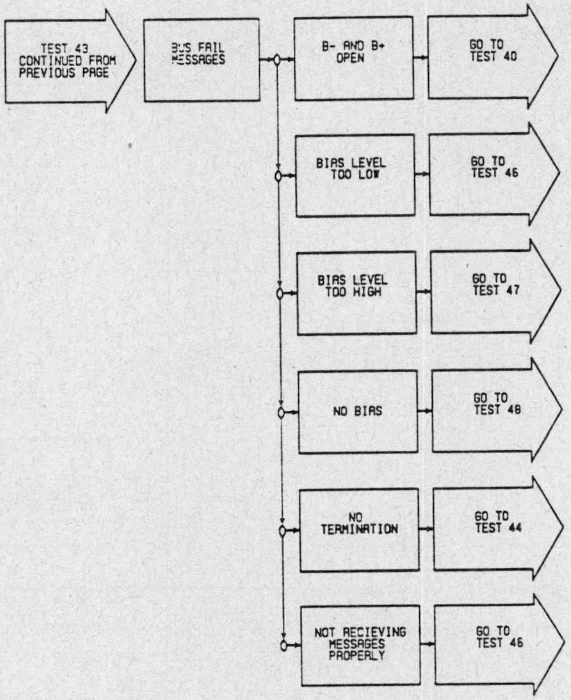
TEST 43 CONTINUED FROM PREVIOUS PAGE
BUS FAIL MESSAGES
B- AND B+ OPEN
GO TO TEST 40
BIAS LEVEL TOO LOW
GO TO TEST 46
BIAS LEVEL TOO HIGH
GO TO TEST 47
NO BIAS
GO TO TEST 48
NO TERMINATION
GO TO TEST 44
NOT RECIEVING MESSAGES PROPERLY
GO TO TEST 46

TEST 44
S-BODY TESTS FOR SHORTED MESSAGES

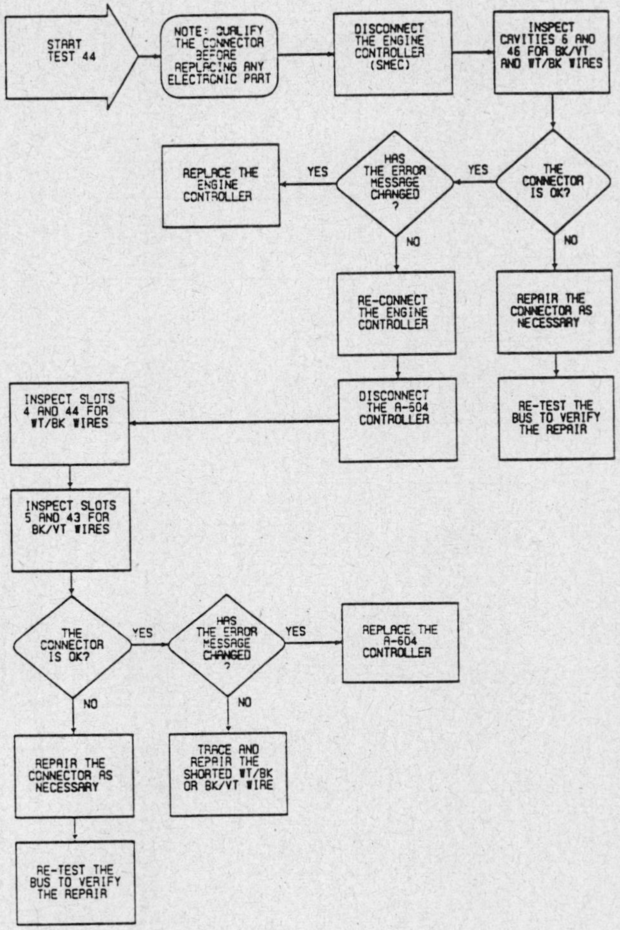
START TEST 44
NOTE: QUALIFY THE CONNECTOR BEFORE REPLACING ANY ELECTRONIC PART
DISCONNECT THE ENGINE CONTROLLER (SMEC)
INSPECT CAVITIES 6 AND 46 FOR BK/VT AND WT/BK WIRES
THE CONNECTOR IS OK?
YES
NO
HAS THE ERROR MESSAGE CHANGED ?
YES
NO
REPLACE THE ENGINE CONTROLLER
REPAIR THE CONNECTOR AS NECESSARY
RE-TEST THE BUS TO VERIFY THE REPAIR
RE-CONNECT THE ENGINE CONTROLLER
DISCONNECT THE A-604 CONTROLLER
INSPECT SLOTS 4 AND 44 FOR WT/BK WIRES
INSPECT SLOTS 5 AND 43 FOR BK/VT WIRES
THE CONNECTOR IS OK?
YES
NO
HAS THE ERROR MESSAGE CHANGED ?
YES
NO
REPLACE THE A-604 CONTROLLER
REPAIR THE CONNECTOR AS NECESSARY
RE-TEST THE BUS TO VERIFY THE REPAIR
TRACE AND REPAIR THE SHORTED WT/BK OR BK/VT WIRE

TEST 45
A-BODY TESTS FOR SHORTED MESSAGES

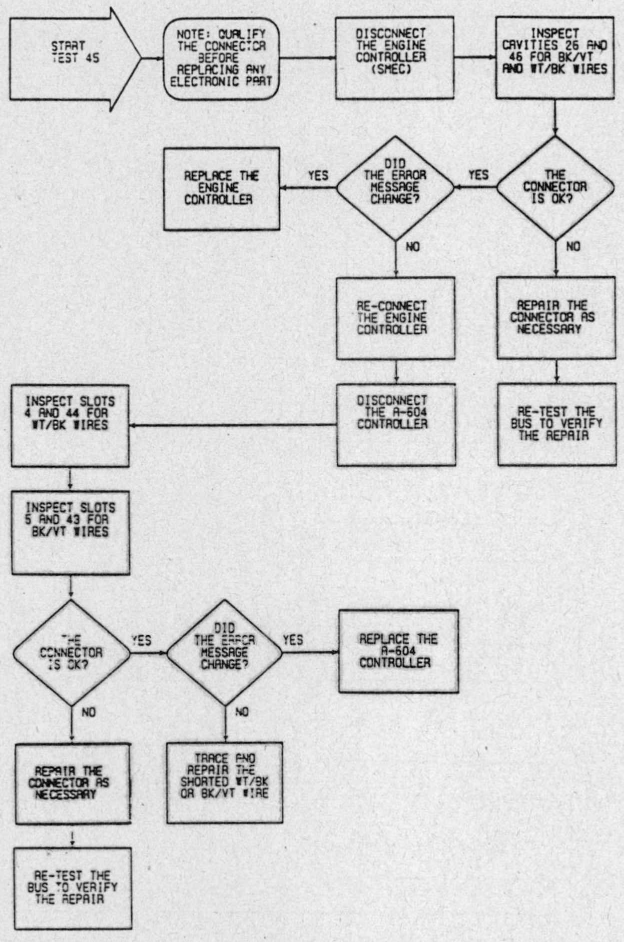
START TEST 45
NOTE: QUALIFY THE CONNECTOR BEFORE REPLACING ANY ELECTRONIC PART
DISCONNECT THE ENGINE CONTROLLER (SMEC)
INSPECT CAVITIES 26 AND 46 FOR BK/YT AND WT/BK WIRES
THE CONNECTOR IS OK?
YES
NO
DID THE ERROR MESSAGE CHANGE?
YES
NO
REPLACE THE ENGINE CONTROLLER
REPAIR THE CONNECTOR AS NECESSARY
RE-TEST THE BUS TO VERIFY THE REPAIR
RE-CONNECT THE ENGINE CONTROLLER
DISCONNECT THE A-604 CONTROLLER
INSPECT SLOTS 4 AND 44 FOR WT/BK WIRES
INSPECT SLOTS 5 AND 43 FOR BK/VT WIRES
THE CONNECTOR IS OK?
YES
NO
DID THE ERROR MESSAGE CHANGE?
YES
NO
REPLACE THE A-604 CONTROLLER
REPAIR THE CONNECTOR AS NECESSARY
RE-TEST THE BUS TO VERIFY THE REPAIR
TRACE AND REPAIR THE SHORTED WT/BK OR BK/VT WIRE

TEST 46
TESTING FOR "BUS BIAS LEVEL TOO LOW" MESSAGE

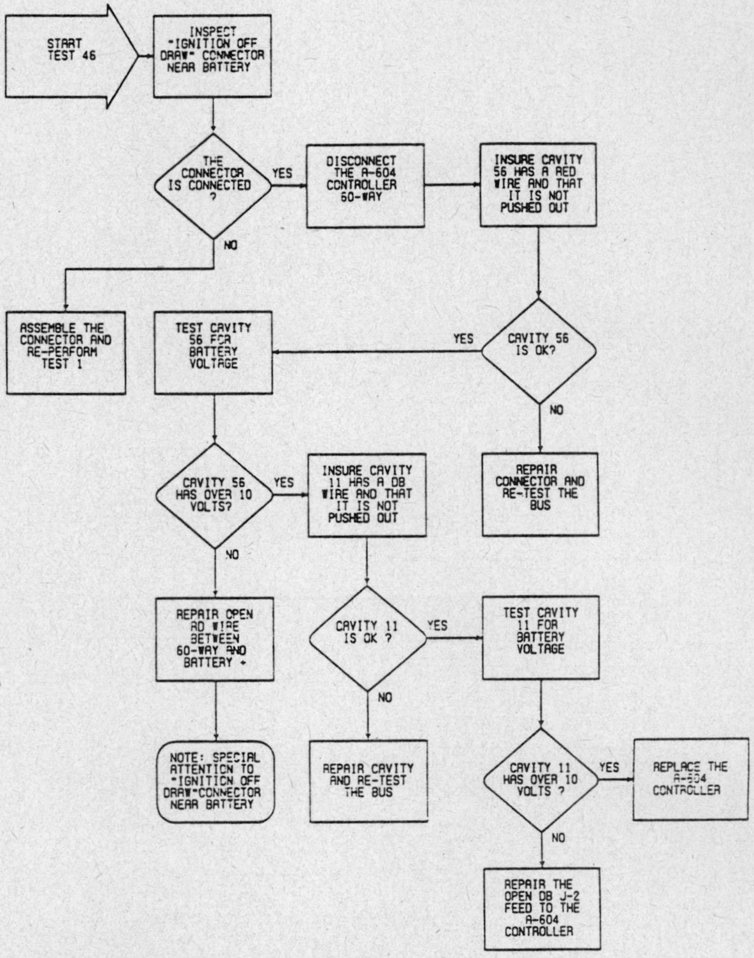
START TEST 46
INSPECT "IGNITION OFF DRAW" CONNECTOR NEAR BATTERY
THE CONNECTOR IS CONNECTED ?
YES
NO
DISCONNECT THE A-604 CONTROLLER 60-WAY
INSURE CAVITY 56 HAS A RED WIRE AND THAT IT IS NOT PUSHED OUT
ASSEMBLE THE CONNECTOR AND RE-PERFORM TEST 1
CAVITY 56 IS OK?
YES
NO
REPAIR CONNECTOR AND RE-TEST THE BUS
TEST CAVITY 56 FOR BATTERY VOLTAGE
CAVITY 56 HAS OVER 10 VOLTS?
YES
NO
INSURE CAVITY 11 HAS A DB WIRE AND THAT IT IS NOT PUSHED OUT
REPAIR OPEN RD WIRE BETWEEN 60-WAY AND BATTERY +
NOTE: SPECIAL ATTENTION TO "IGNITION OFF DRAW" CONNECTOR NEAR BATTERY
CAVITY 11 IS OK ?
YES
NO
REPAIR CAVITY AND RE-TEST THE BUS
TEST CAVITY 11 FOR BATTERY VOLTAGE
CAVITY 11 HAS OVER 10 VOLTS ?
YES
NO
REPLACE THE A-604 CONTROLLER
REPAIR THE OPEN DB J-2 FEED TO THE A-604 CONTROLLER

TEST 47
TESTING FOR "BUS BIAS LEVEL TOO HIGH" MESSAGE

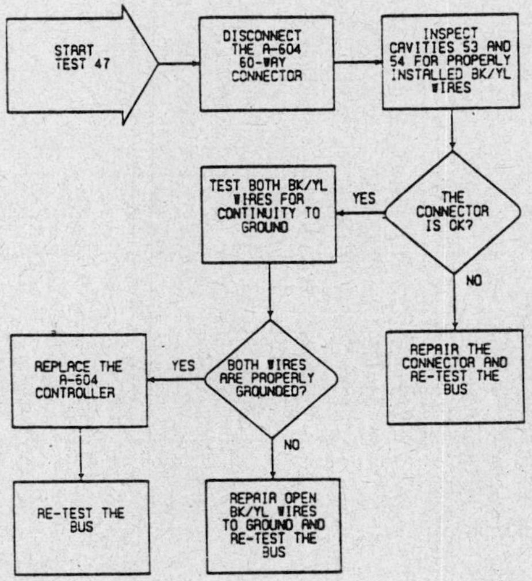
START TEST 47
DISCONNECT THE A-604 60-WAY CONNECTOR
INSPECT CAVITIES 53 AND 54 FOR PROPERLY INSTALLED BK/YL WIRES
THE CONNECTOR IS OK?
YES
NO
TEST BOTH BK/YL WIRES FOR CONTINUITY TO GROUND
REPAIR THE CONNECTOR AND RE-TEST THE BUS
BOTH WIRES ARE PROPERLY GROUNDED?
YES
NO
REPLACE THE A-604 CONTROLLER
REPAIR OPEN BK/YL WIRES TO GROUND AND RE-TEST THE BUS
RE-TEST THE BUS

TEST 48
TESTING FOR A "NO BIAS" MESSAGE

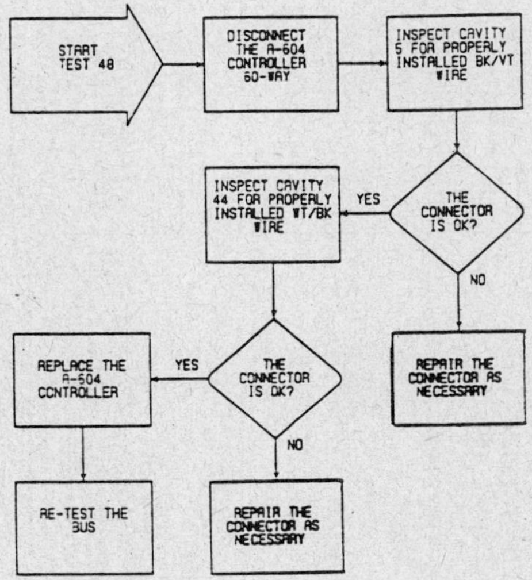
START TEST 48
DISCONNECT THE A-604 CONTROLLER 60-WAY
INSPECT CAVITY 5 FOR PROPERLY INSTALLED BK/VT WIRE
THE CONNECTOR IS OK?
YES
NO
INSPECT CAVITY 44 FOR PROPERLY INSTALLED WT/BK WIRE
REPAIR THE CONNECTOR AS NECESSARY
THE CONNECTOR IS OK?
YES
NO
REPLACE THE A-604 CONTROLLER
REPAIR THE CONNECTOR AS NECESSARY
RE-TEST THE BUS

VEHICLE WILL NOT MOVE DIAGNOSIS

CHECK THE TRANSAXLE FLUID LEVEL BEFORE STARTING THE ENGINE. IF NO FLUID IS VISIBLE ON THE DIPSTICK, ADD FLUID TO THE "ADD" MARK BEFORE STARTING THE ENGINE. THEN START THE ENGINE WITH THE TRANSAXLE IN NEUTRAL AND LISTEN FOR NOISE.

ABNORMAL NOISE, STOP ENGINE IMMEDIATELY, REMOVE THE TRANSAXLE AND CONVERTER AS AN ASSEMBLY. DISASSEMBLE, CLEAN AND INSPECT ALL PARTS. CLEAN VALVE BODY; INSTALL ALL NEW SEALS, RINGS AND GASKETS; REPLACE WORN OR DEFECTIVE PARTS.

NO ABNORMAL NOISE, MOVE THE SELECTOR TO A FORWARD DRIVE RANGE AND OBSERVE THE FRONT WHEELS FOR TURNING

DRIVE SHAFTS TURN BUT FRONT WHEELS DO NOT TURN, INSPECT FOR BROKEN DRIVE SHAFT PARTS.

DRIVE SHAFTS DO NOT TURN

REMOVE ALL THREE OIL PANS. INSPECT FOR DEBRIS AND IF AXLE SHAFTS ARE PROPERLY INSTALLED.

NO DEBRIS. REMOVE VALVE BODY. DISASSEMBLE, CLEAN AND INSPECT ALL PARTS. REASSEMBLE, INSTALL AND CHECK PRESSURES AND OPERATION.

DEBRIS IS PRESENT. REMOVE TRANSAXLE AND CONVERTER AS AN ASSEMBLY; DISASSEMBLE, CLEAN AND INSPECT ALL PARTS; CLEAN THE VALVE BODY. INSTALL ALL NEW SEALS, RINGS, AND GASKETS; REPLACE WORN OR DEFECTIVE PARTS.

REPLACE TORQUE CONVERTER FLUSH COOLER AND LINES

FLUID LEAKS DIAGNOSIS

VISUALLY INSPECT FOR SOURCE OF LEAK. IF THE SOURCE OF LEAK CANNOT BE READILY DETERMINED, CLEAN THE EXTERIOR OF THE TRANSAXLE. CHECK TRANSAXLE FLUID LEVEL. CORRECT IF NECESSARY.

THE FOLLOWING LEAKS MAY BE CORRECTED WITHOUT REMOVING THE TRANSAXLE:

MANUAL LEVER SHAFT OIL SEAL
PRESSURE GAUGE PLUGS
NEUTRAL START SWITCH
OIL PAN RTV
OIL COOLER FITTINGS
EXTENSION HOUSING TO CASE BOLTS
SPEEDOMETER ADAPTER "O" RING
FRONT BAND ADJUSTING SCREW
EXTENSION HOUSING AXLE SEAL
DIFFERENTIAL BEARING RETAINER AXLE SEAL
REAR END COVER RTV
DIFFERENTIAL COVER RTV
EXTENSION HOUSING "O" RING
DIFFERENTIAL BEARING RETAINER RTV

THE FOLLOWING LEAKS REQUIRE REMOVAL OF THE TRANSAXLE AND TORQUE CONVERTER FOR CORRECTION.

TRANSAXLE FLUID LEAKING FROM THE LOWER EDGE OF THE CONVERTER HOUSING; CAUSED BY FRONT PUMP SEAL, PUMP TO CASE SEAL, OR TORQUE CONVERTER WELD.

CRACKED OR POROUS TRANSAXLE CASE.

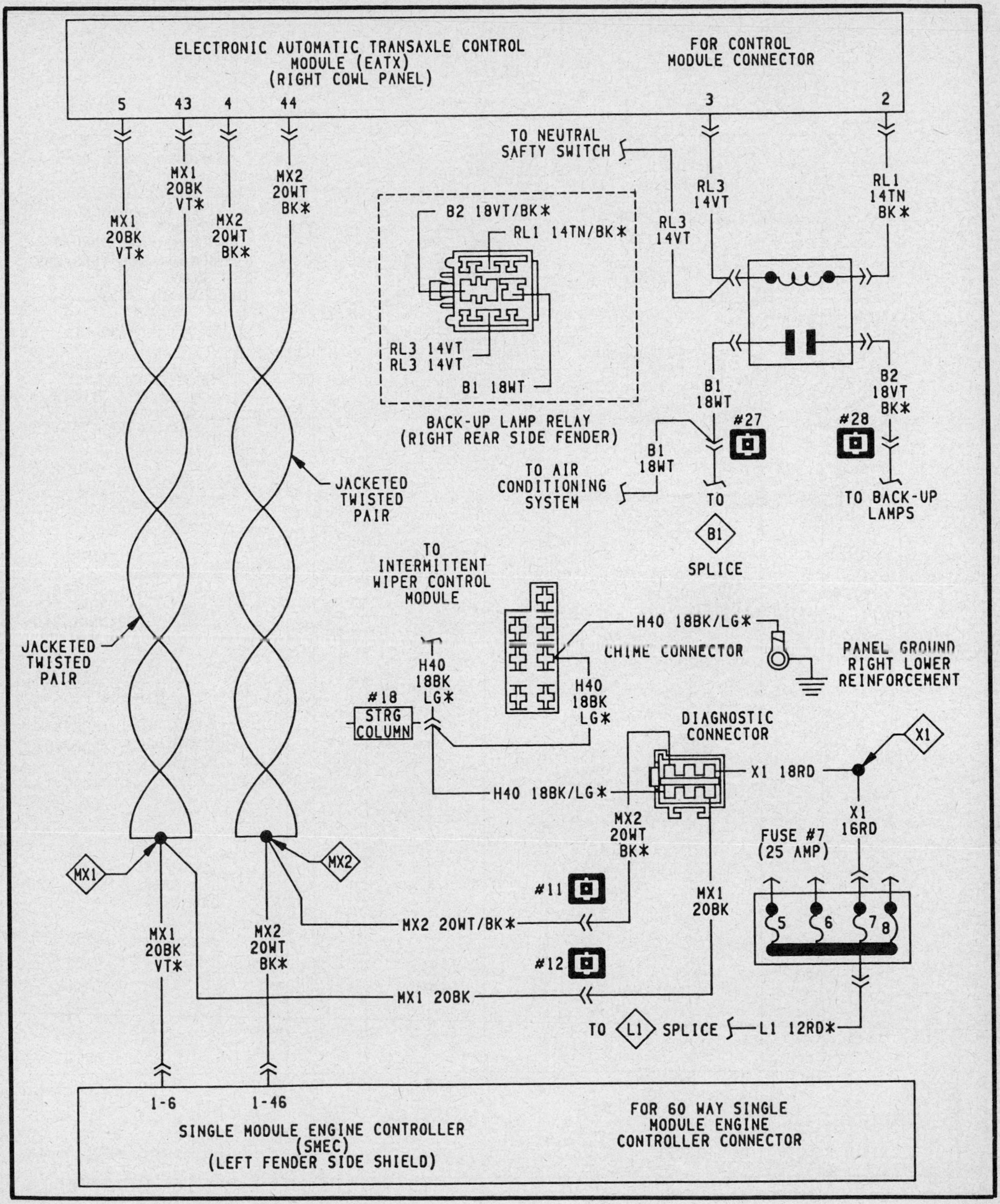
ELECTRONIC AUTOMATIC TRANSAXLE CONTROL MODULE (EATX) (RIGHT COWL PANEL)
FOR CONTROL MODULE CONNECTOR
5
43
4
44
3
2
TO NEUTRAL SAFTY SWITCH
MX1 20BK VT*
MX2 20WT BK*
RL3 14VT
RL1 14TN BK*
B2 18VT/BK*
RL1 14TN/BK*
RL3 14VT
RL3 14VT
B1 18WT
BACK-UP LAMP RELAY (RIGHT REAR SIDE FENDER)
B1 18WT
B2 18VT BK*
#27
#28
JACKETED TWISTED PAIR
TO AIR CONDITIONING SYSTEM
TO
B1
SPLICE
TO BACK-UP LAMPS
TO INTERMITTENT WIPER CONTROL MODULE
H40 18BK/LG*
CHIME CONNECTOR
PANEL GROUND RIGHT LOWER REINFORCEMENT
H40 18BK LG*
#18
STRG COLUMN
DIAGNOSTIC CONNECTOR
X1
X1 18RD
H40 18BK/LG*
X1 16RD
FUSE #7 (25 AMP)
5
6
7
8
MX1
MX2
MX2 20WT BK*
#11
MX2 20WT/BK*
MX1 20BK
#12
MX1 20BK
TO
L1
SPLICE
L1 12RD*
1-6
1-46
SINGLE MODULE ENGINE CONTROLLER (SMEC) (LEFT FENDER SIDE SHIELD)
FOR 60 WAY SINGLE MODULE ENGINE CONTROLLER CONNECTOR

Electronic automatic transaxle wiring diagram

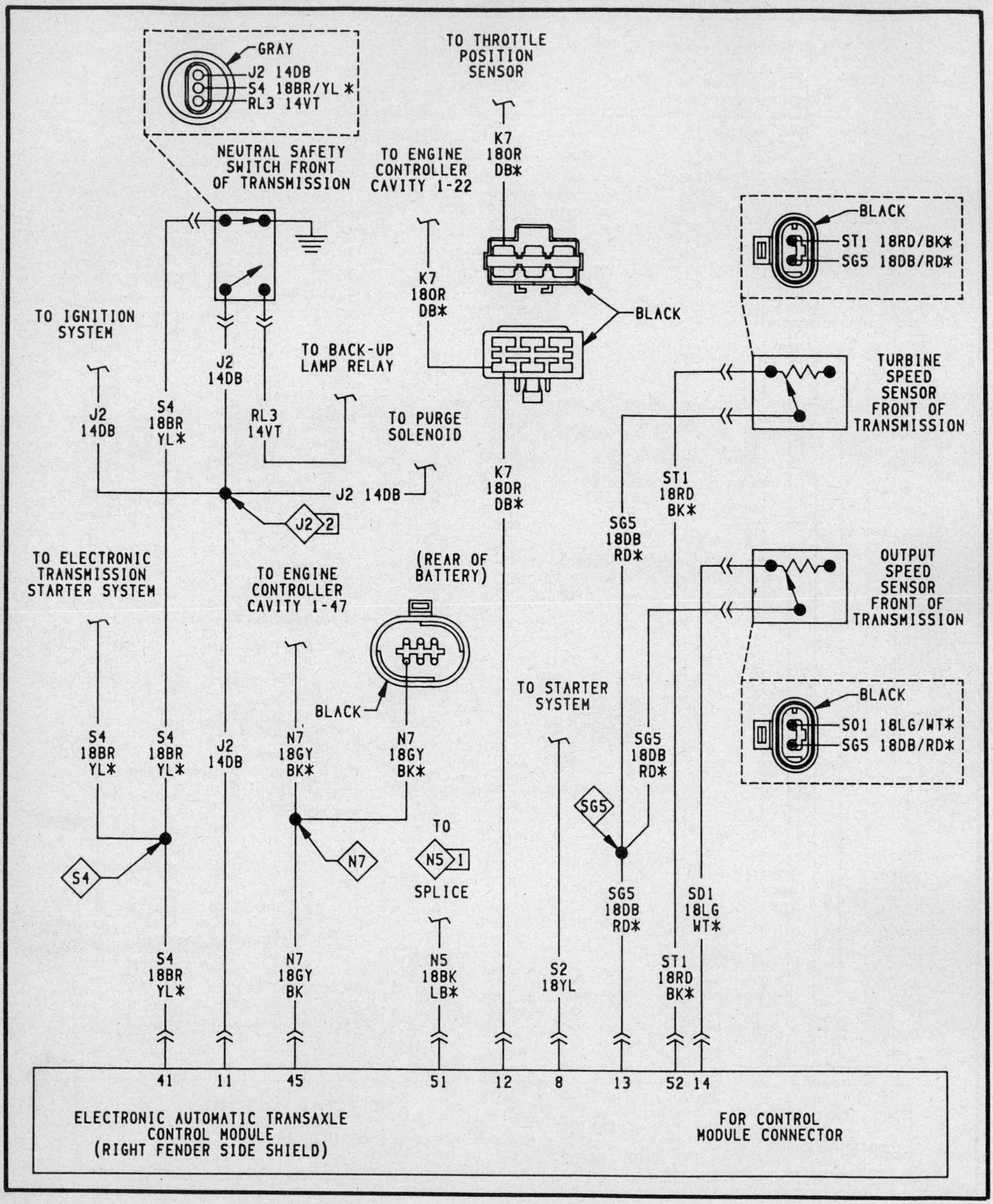
GRAY
J2 14DB
S4 18BR/YL *
RL3 14VT
NEUTRAL SAFETY SWITCH FRONT OF TRANSMISSION
TO THROTTLE POSITION SENSOR
K7 18OR DB*
TO ENGINE CONTROLLER CAVITY 1-22
K7 18OR DB*
BLACK
ST1 18RD/BK*
SG5 18DB/RD*
BLACK
TO IGNITION SYSTEM
J2 14DB
S4 18BR YL*
J2 14DB
RL3 14VT
TO BACK-UP LAMP RELAY
TO PURGE SOLENOID
J2 14DB
J2 2
K7 18DR DB*
TURBINE SPEED SENSOR FRONT OF TRANSMISSION
ST1 18RD BK*
SG5 18DB RD*
TO ELECTRONIC TRANSMISSION STARTER SYSTEM
TO ENGINE CONTROLLER CAVITY 1-47
(REAR OF BATTERY)
OUTPUT SPEED SENSOR FRONT OF TRANSMISSION
BLACK
SO1 18LG/WT*
SG5 18DB/RD*
BLACK
TO STARTER SYSTEM
S4 18BR YL*
S4 18BR YL*
J2 14DB
N7 18GY BK*
N7 18GY BK*
SG5 18DB RD*
SG5
S4
N7
TO
N5 1
SPLICE
SG5 18DB RD*
SD1 18LG WT*
S4 18BR YL*
N7 18GY BK
N5 18BK LB*
S2 18YL
ST1 18RD BK*
41
11
45
51
12
8
13
52
14
ELECTRONIC AUTOMATIC TRANSAXLE CONTROL MODULE (RIGHT FENDER SIDE SHIELD)
FOR CONTROL MODULE CONNECTOR

Electronic automatic transaxle wiring diagram

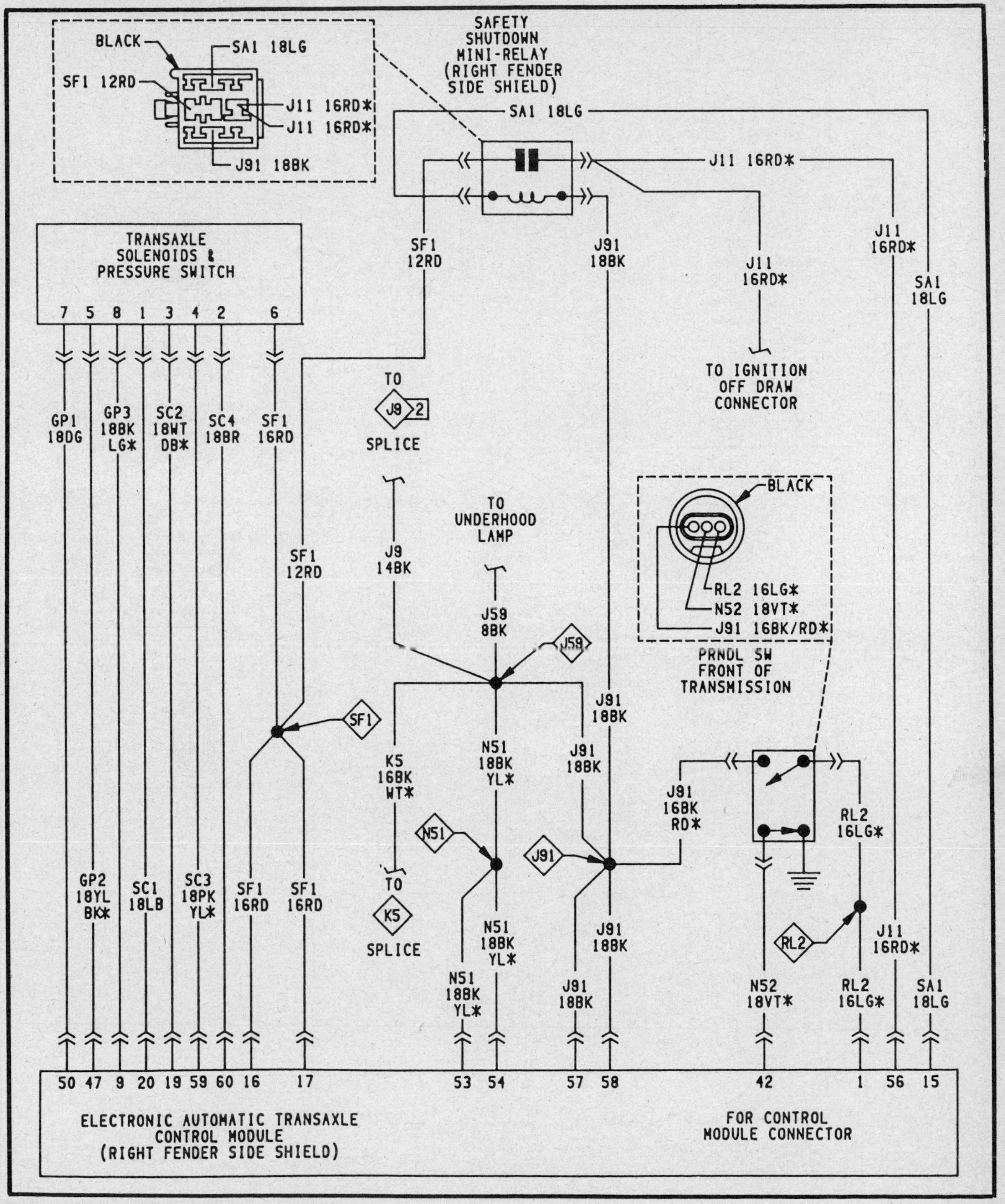
BLACK
SA1 18LG
SF1 12RD
J11 16RD*
J11 16RD*
J91 18BK
SAFETY SHUTDOWN MINI-RELAY (RIGHT FENDER SIDE SHIELD)
SA1 18LG
J11 16RD*
TRANSAXLE SOLENOIDS & PRESSURE SWITCH
7 5 8 1 3 4 2 6
SF1 12RD
J91 18BK
J11 16RD*
J11 16RD*
SA1 18LG
TO IGNITION OFF DRAW CONNECTOR
TO
J9 2
SPLICE
GP1 18DG
GP3 18BK LG*
SC2 18WT DB*
SC4 18BR
SF1 16RD
TO UNDERHOOD LAMP
BLACK
SF1 12RD
J9 14BK
J59 8BK
J59
RL2 16LG*
N52 18VT*
J91 16BK/RD*
PRNDL SW FRONT OF TRANSMISSION
SF1
J91 18BK
K5 16BK WT*
N51 18BK YL*
J91 18BK
J91 16BK RD*
RL2 16LG*
N51
J91
TO
K5
SPLICE
GP2 18YL BK*
SC1 18LB
SC3 18PK YL*
SF1 16RD
SF1 16RD
N51 18BK YL*
N51 18BK YL*
J91 18BK
J91 18BK
RL2
J11 16RD*
N52 18VT*
RL2 16LG*
SA1 18LG
50 47 9 20 19 59 60 16 17 53 54 57 58 42 1 56 15
ELECTRONIC AUTOMATIC TRANSAXLE CONTROL MODULE (RIGHT FENDER SIDE SHIELD)
FOR CONTROL MODULE CONNECTOR

Electronic automatic transaxle wiring diagram

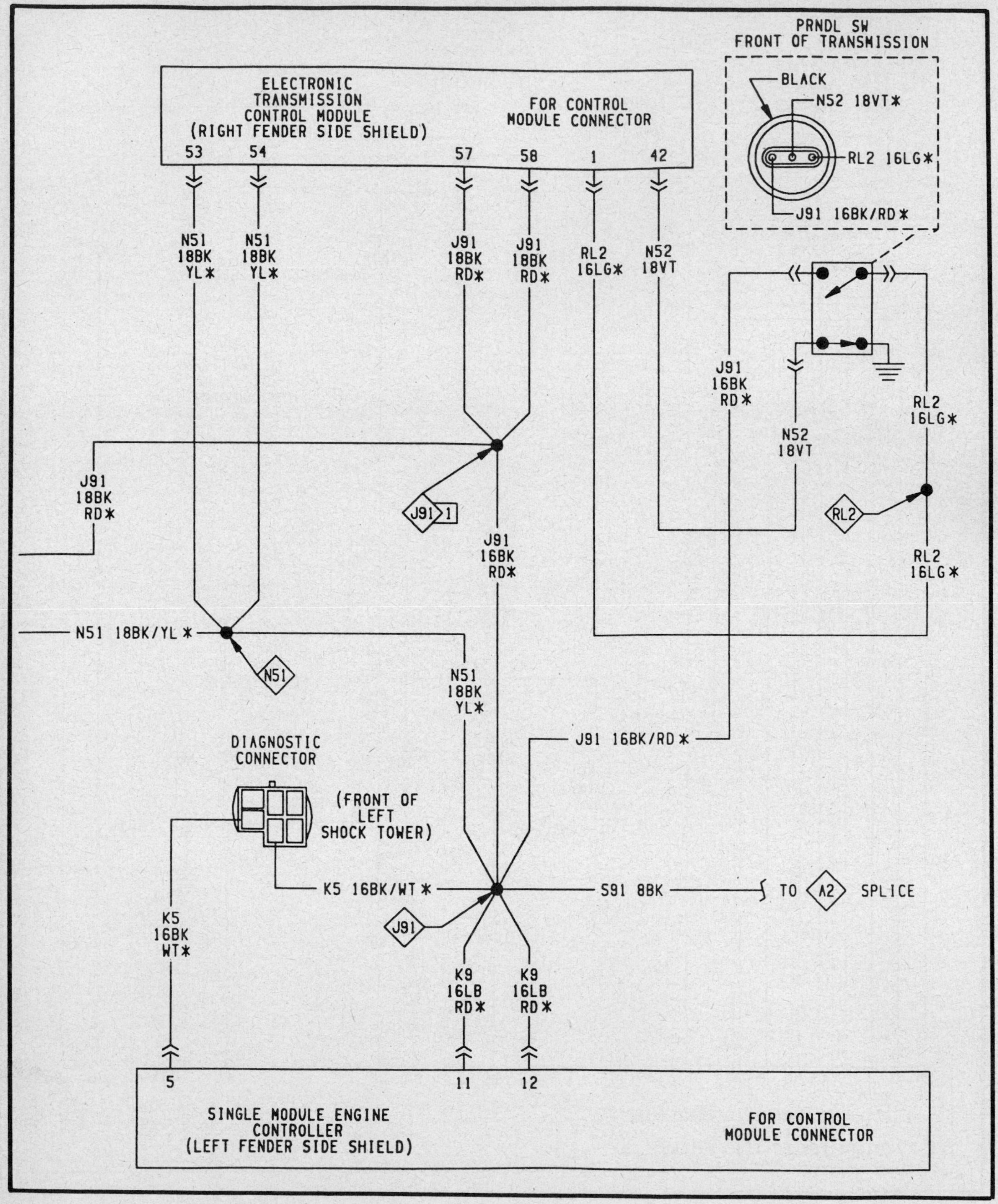
PRNDL SW
FRONT OF TRANSMISSION
BLACK
N52 18VT*
RL2 16LG*
J91 16BK/RD*
ELECTRONIC
TRANSMISSION
CONTROL MODULE
(RIGHT FENDER SIDE SHIELD)
FOR CONTROL
MODULE CONNECTOR
53
54
57
58
1
42
N51
18BK
YL*
N51
18BK
YL*
J91
18BK
RD*
J91
18BK
RD*
RL2
16LG*
N52
18VT
J91
16BK
RD*
RL2
16LG*
N52
18VT
J91
18BK
RD*
J91
1
J91
16BK
RD*
RL2
RL2
16LG*
N51 18BK/YL*
N51
N51
18BK
YL*
J91 16BK/RD*
DIAGNOSTIC
CONNECTOR
(FRONT OF
LEFT
SHOCK TOWER)
K5 16BK/WT*
S91 8BK
TO
A2
SPLICE
K5
16BK
WT*
J91
K9
16LB
RD*
K9
16LB
RD*
5
11
12
SINGLE MODULE ENGINE
CONTROLLER
(LEFT FENDER SIDE SHIELD)
FOR CONTROL
MODULE CONNECTOR

Electronic automatic transaxle wiring diagram

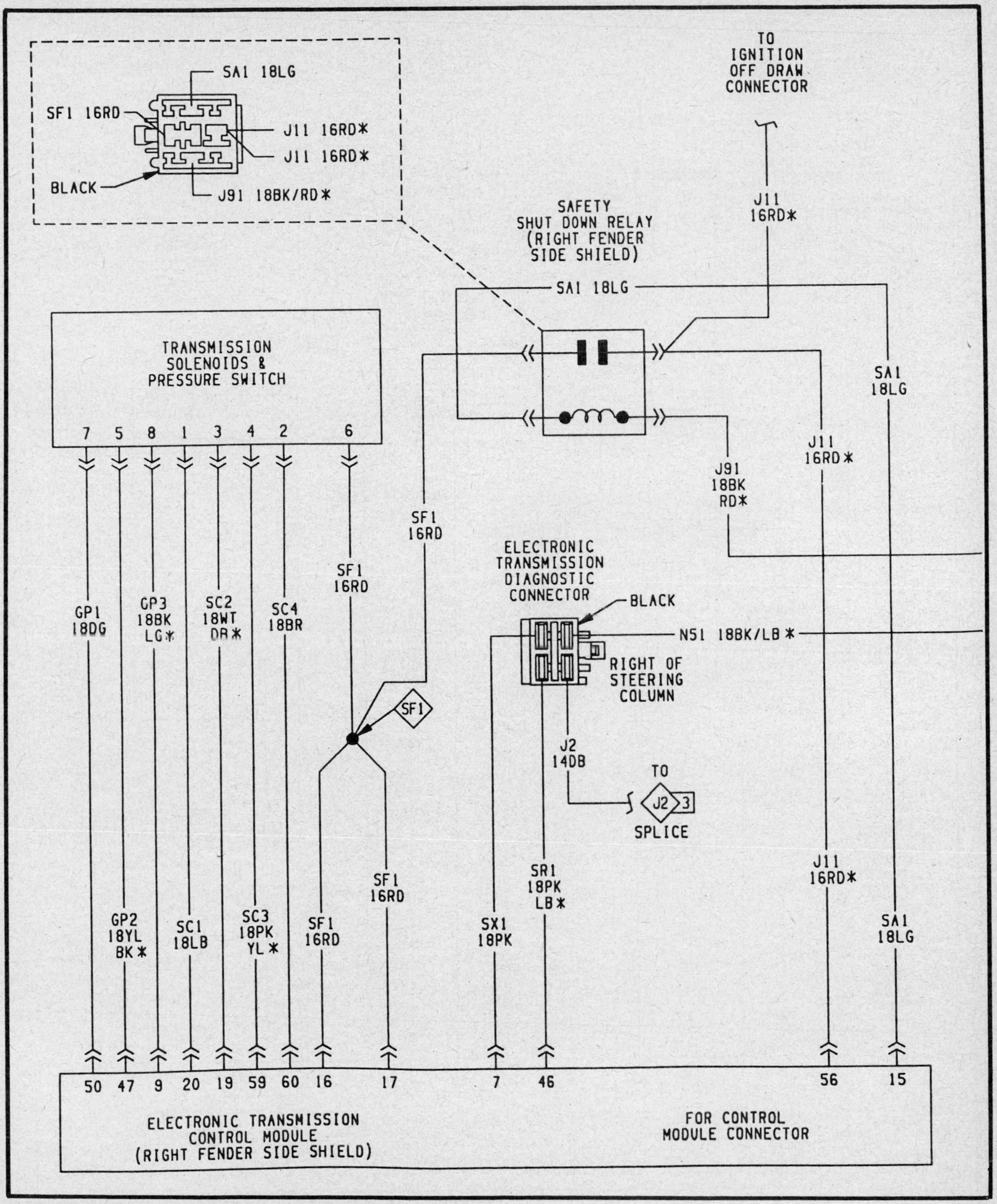
SA1 18LG
SF1 16RD
J11 16RD*
J11 16RD*
BLACK
J91 18BK/RD*
TO IGNITION OFF DRAW CONNECTOR
J11 16RD*
SAFETY SHUT DOWN RELAY (RIGHT FENDER SIDE SHIELD)
SA1 18LG
TRANSMISSION SOLENOIDS & PRESSURE SWITCH
7 5 8 1 3 4 2 6
SA1 18LG
J11 16RD*
J91 18BK RD*
SF1 16RD
SF1 16RD
ELECTRONIC TRANSMISSION DIAGNOSTIC CONNECTOR
BLACK
N51 18BK/LB*
RIGHT OF STEERING COLUMN
GP1 18DG
GP3 18BK LG*
SC2 18WT DB*
SC4 18BR
SF1
J2 14DB
TO
J2 3
SPLICE
GP2 18YL BK*
SC1 18LB
SC3 18PK YL*
SF1 16RD
SF1 16RD
SX1 18PK
SR1 18PK LB*
J11 16RD*
SA1 18LG
50 47 9 20 19 59 60 16 17 7 46 56 15
ELECTRONIC TRANSMISSION CONTROL MODULE (RIGHT FENDER SIDE SHIELD)
FOR CONTROL MODULE CONNECTOR

Electronic automatic transaxle wiring diagram

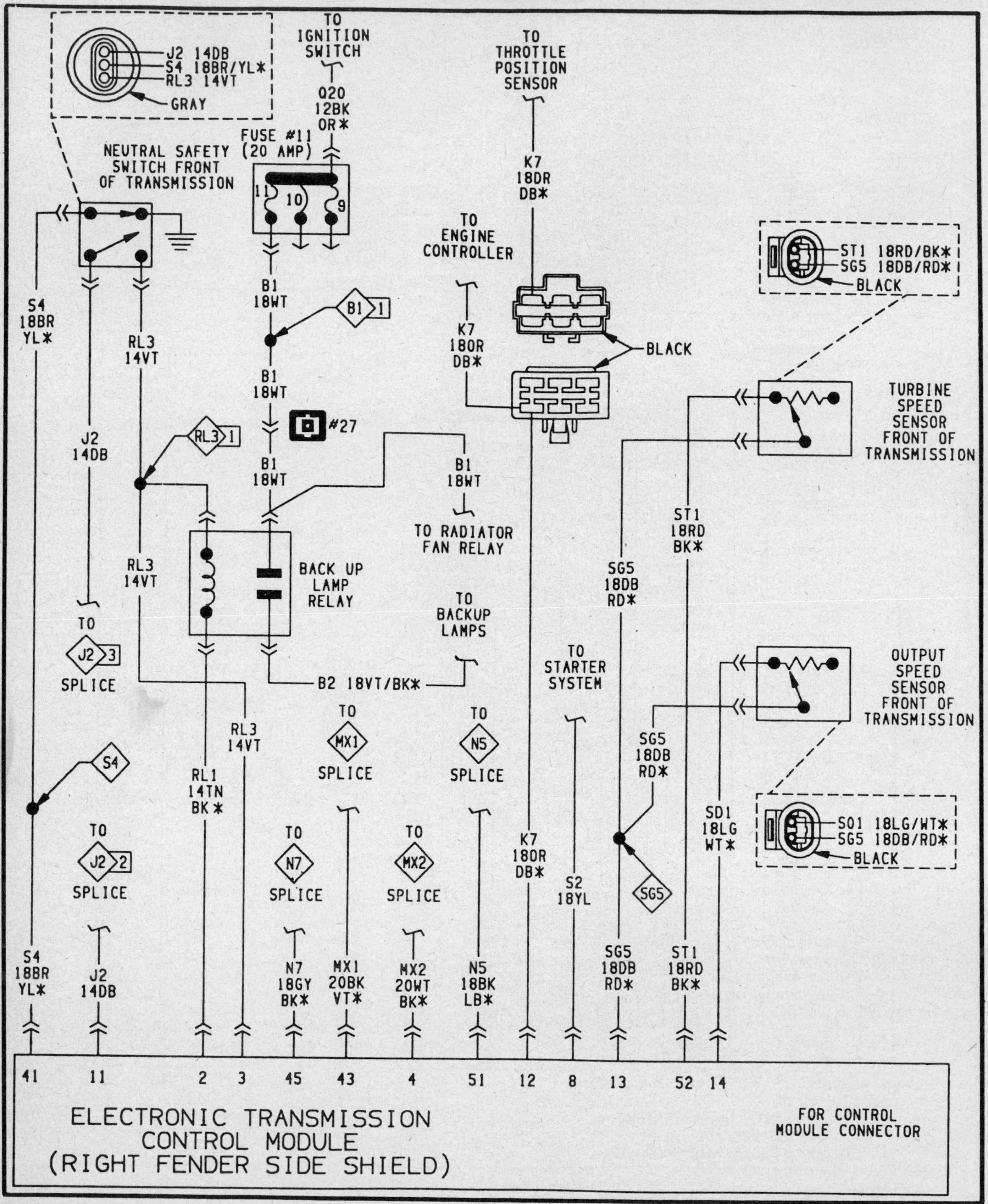
J2 14DB
S4 18BR/YL*
RL3 14VT
GRAY
TO IGNITION SWITCH
Q20 12BK OR*
FUSE #11 (20 AMP)
NEUTRAL SAFETY SWITCH FRONT OF TRANSMISSION
11
10
9
TO THROTTLE POSITION SENSOR
K7 18DR DB*
TO ENGINE CONTROLLER
K7 18OR DB*
BLACK
ST1 18RD/BK*
SG5 18DB/RD*
BLACK
TURBINE SPEED SENSOR FRONT OF TRANSMISSION
S4 18BR YL*
RL3 14VT
B1 18WT
B1 1
B1 18WT
#27
J2 14DB
RL3 1
B1 18WT
B1 18WT
TO RADIATOR FAN RELAY
ST1 18RD BK*
SG5 18DB RD*
RL3 14VT
BACK UP LAMP RELAY
TO BACKUP LAMPS
TO
J2 3
SPLICE
B2 18VT/BK*
TO STARTER SYSTEM
OUTPUT SPEED SENSOR FRONT OF TRANSMISSION
RL3 14VT
TO
MX1
SPLICE
TO
N5
SPLICE
SG5 18DB RD*
S4
RL1 14TN BK*
SD1 18LG WT*
S01 18LG/WT*
SG5 18DB/RD*
BLACK
TO
J2 2
SPLICE
TO
N7
SPLICE
TO
MX2
SPLICE
K7 18OR DB*
S2 18YL
SG5
S4 18BR YL*
J2 14DB
N7 18GY BK*
MX1 20BK VT*
MX2 20WT BK*
N5 18BK LB*
SG5 18DB RD*
ST1 18RD BK*
41
11
2
3
45
43
4
51
12
8
13
52
14
ELECTRONIC TRANSMISSION CONTROL MODULE (RIGHT FENDER SIDE SHIELD)
FOR CONTROL MODULE CONNECTOR

Electronic automatic transaxle wiring diagram

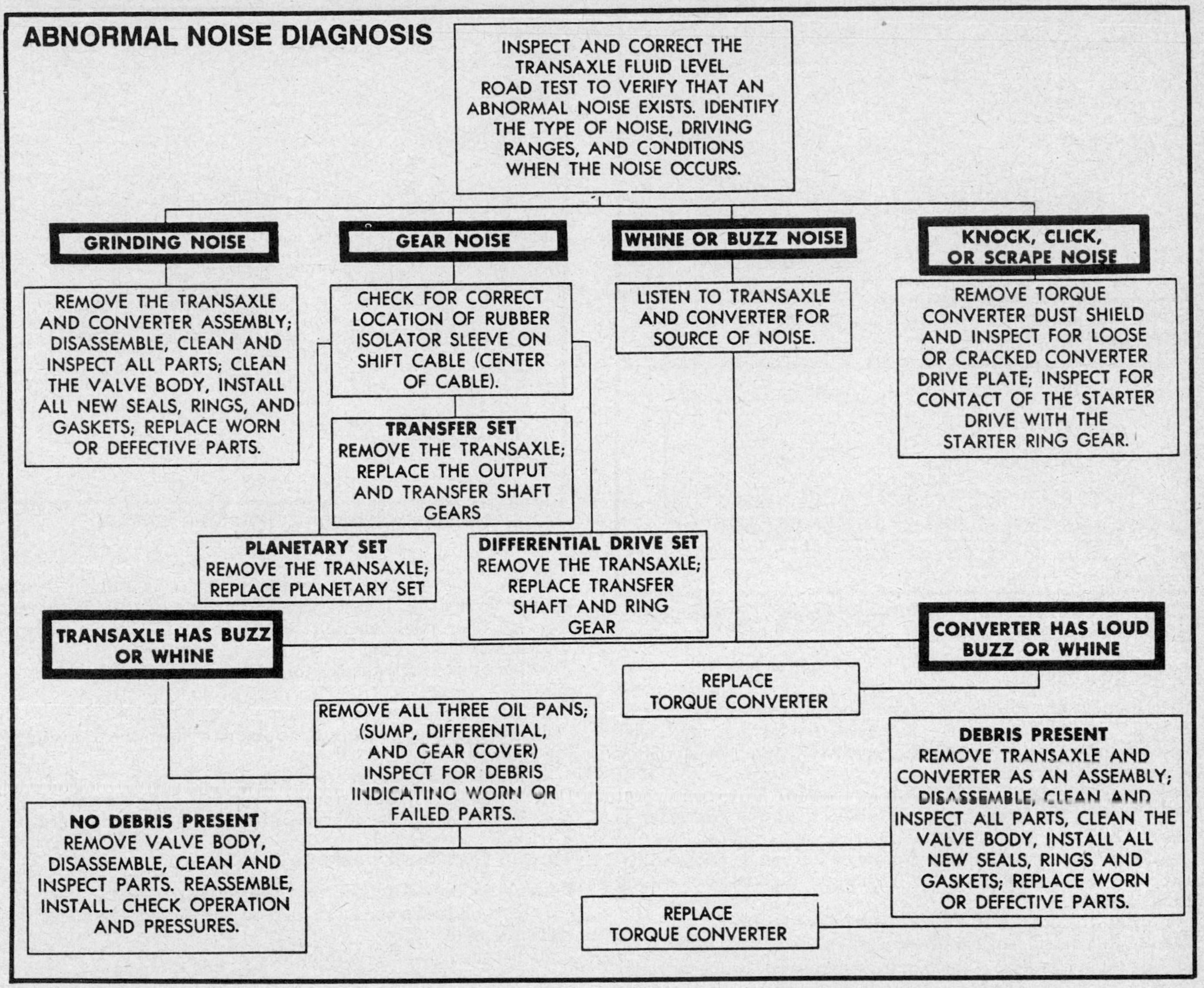
ABNORMAL NOISE DIAGNOSIS
INSPECT AND CORRECT THE TRANSAXLE FLUID LEVEL. ROAD TEST TO VERIFY THAT AN ABNORMAL NOISE EXISTS. IDENTIFY THE TYPE OF NOISE, DRIVING RANGES, AND CONDITIONS WHEN THE NOISE OCCURS.
GRINDING NOISE
GEAR NOISE
WHINE OR BUZZ NOISE
KNOCK, CLICK, OR SCRAPE NOISE
REMOVE THE TRANSAXLE AND CONVERTER ASSEMBLY; DISASSEMBLE, CLEAN AND INSPECT ALL PARTS; CLEAN THE VALVE BODY, INSTALL ALL NEW SEALS, RINGS, AND GASKETS; REPLACE WORN OR DEFECTIVE PARTS.
CHECK FOR CORRECT LOCATION OF RUBBER ISOLATOR SLEEVE ON SHIFT CABLE (CENTER OF CABLE).
LISTEN TO TRANSAXLE AND CONVERTER FOR SOURCE OF NOISE.
REMOVE TORQUE CONVERTER DUST SHIELD AND INSPECT FOR LOOSE OR CRACKED CONVERTER DRIVE PLATE; INSPECT FOR CONTACT OF THE STARTER DRIVE WITH THE STARTER RING GEAR.
TRANSFER SET REMOVE THE TRANSAXLE; REPLACE THE OUTPUT AND TRANSFER SHAFT GEARS
PLANETARY SET REMOVE THE TRANSAXLE; REPLACE PLANETARY SET
DIFFERENTIAL DRIVE SET REMOVE THE TRANSAXLE; REPLACE TRANSFER SHAFT AND RING GEAR
TRANSAXLE HAS BUZZ OR WHINE
CONVERTER HAS LOUD BUZZ OR WHINE
REPLACE TORQUE CONVERTER
REMOVE ALL THREE OIL PANS; (SUMP, DIFFERENTIAL, AND GEAR COVER) INSPECT FOR DEBRIS INDICATING WORN OR FAILED PARTS.
DEBRIS PRESENT REMOVE TRANSAXLE AND CONVERTER AS AN ASSEMBLY; DISASSEMBLE, CLEAN AND INSPECT ALL PARTS, CLEAN THE VALVE BODY, INSTALL ALL NEW SEALS, RINGS AND GASKETS; REPLACE WORN OR DEFECTIVE PARTS.
NO DEBRIS PRESENT REMOVE VALVE BODY, DISASSEMBLE, CLEAN AND INSPECT PARTS. REASSEMBLE, INSTALL. CHECK OPERATION AND PRESSURES.
REPLACE TORQUE CONVERTER

position to check for slipping and any variation in shifting.

In most cases, the clutch that is slipping can be determined by noting the transaxle operation in all selector positions and by comparing which internal units are applied in those positions.

The process of eliminating can be used to detect any unit which slip and to confirm proper operation of good units. However, although road test analysis can usually diagnose slipping units, the actual cause of the malfunction usually cannot be decided. Practically any condition can be caused by leaking hydraulic circuits or sticking valves.

ON-BOARD DIAGNOSTICS

The transaxle controller monitors critical input and output circuits relating to the control of the transaxle. Some of these circuits are tested continuously, and others are checked only during normal driving conditions.

If the controller senses a problem in the system, a fault code will be stored in the controller's memory. Each monitored circuit has its own designated fault code. Any stored fault code will remain in memory until erased or until displaced by more recent codes.

Converter Clutch Operation and Diagnosis

TORQUE CONVERTER CLUTCH

Fluid Leakage—Transaxle Torque Converter Housing Area

Since fluid leakage at or around the torque converter area may originate from an engine oil leak, the area should be examined closely. Factory fill fluid is dyed red and, therefore, can be distinguished from engine oil.

Prior to removing the transaxle, check the following:

1. When leakage is determined to originate from the transaxle, check fluid level prior to removal of the transaxle and torque converter.
2. High oil level can result in oil leakage out the vent in the manual shaft. If the fluid level is high, adjust to proper level.
3. After performing this operation, inspect for leakage. If a leak persists, determine if it is the torque converter or the transaxle that is leaking.

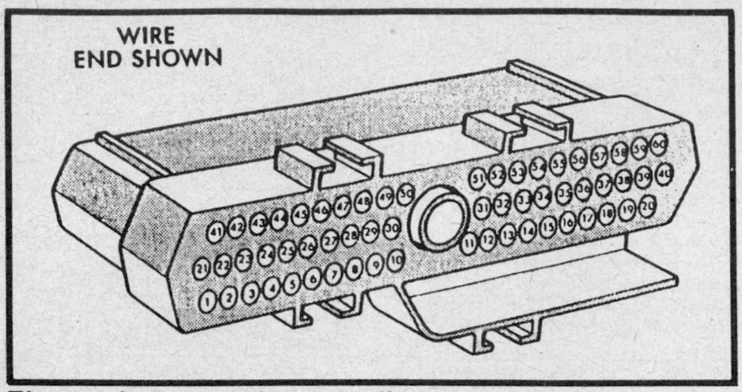
WIRE
END SHOWN

Electronic automatic controller connector terminal identification

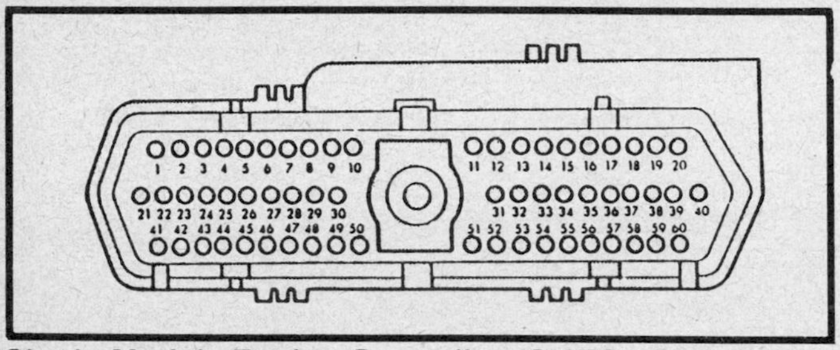

Single Module Engine Controller (SMEC) connector terminal identification

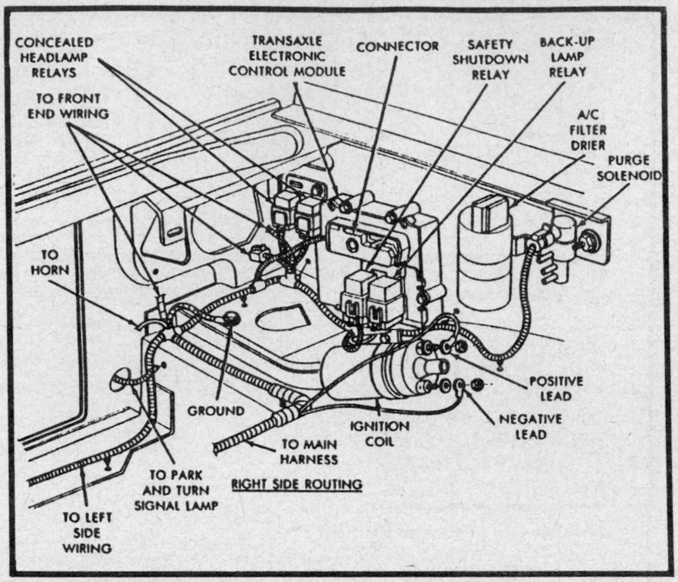
CONCEALED HEADLAMP RELAYS
TRANSAXLE ELECTRONIC CONTROL MODULE
CONNECTOR
SAFETY SHUTDOWN RELAY
BACK-UP LAMP RELAY
TO FRONT END WIRING
A/C FILTER DRIER
PURGE SOLENOID
TO HORN
POSITIVE LEAD
GROUND
NEGATIVE LEAD
IGNITION COIL
TO MAIN HARNESS
TO PARK AND TURN SIGNAL LAMP
RIGHT SIDE ROUTING
TO LEFT SIDE WIRING

Transaxle electronic control module location

Leakage Test Probe

1. Remove the torque converter housing dust shield.
2. Clean the inside of the torque converter housing (lower area) as dry as possible. A solvent spray followed by compressed air drying is preferable.
3. Fabricate and fasten the test probe securely to convenient dust shield bolt hole. Make certain the torque converter is cleared by the test probe. Tool must be clean and dry.
4. Run the engine at approximately 2500 rpm with the transaxle in **N**, for about 2 minutes. Transaxle must be at operating temperature.
5. Stop the engine and carefully remove the tool.
6. If the upper surface of the test probe is dry, there is no torque converter leak. A path of fluid across the probe indicates a torque converter leak. Oil leaking under the probe is coming from the transaxle torque converter area.
7. Remove transaxle and torque converter assembly from the vehicle for further investigation. The fluid should be drained from the transaxle. Reinstall the oil pan (with RTV sealant) and torque attaching bolts.

Possible sources of transaxle torque converter area fluid leakage are:

1. Torque converter hub seal.
2. Fluid leakage at the outside diameter from the pump housing O-ring.
3. Fluid leakage at the front pump to case bolts. Check condition of washers on bolts and use new bolts, if necessary.
4. Fluid leakage due to case or front pump housing porosity.

Torque Converter Leakage

Possible sources of torque converter leakage are:

1. Torque converter weld leaks at the outside diameter (peripheral) weld.
2. Torque converter hub weld.

NOTE: Hub weld is inside and not visible. Do not attempt to repair. Replace the torque converter.

ON CAR SERVICES

Adjustments

MANUAL LINKAGE

Gearshift Linkage

Normal operation of the PRNDL and neutral safety switch provides a quick check to confirm proper manual linkage adjustment.

Move the selector lever slowly upward until it clicks into the **P** notch in the selector gate. If the starter will operate the **P** position is correct.

After checking **P** position move the selector slowly toward **N** position until lever drops at the end of the **N** stop in the selector gate. If the starter will also operate at this point the gearshift linkage is properly adjusted. If required, adjustment gearshift linkage as follows:

1. Set the parking brake and place the gearshift lever in **P** position.
2. Loosen the clamp bolt on the gearshift cable bracket.
3. On column shift, insure that the preload adjustment spring engages the fork on the transaxle bracket.
4. Pull the shift lever by hand all the way to the front detent position (**P**) and tighten the lock screw to 100 inch lbs. (11 Nm). Gearshift linkage should now be properly adjusted.
5. To check the adjustment, the detent position for neutral and drive should be within limits of hand lever gate stops, and the key start must occur only when the shift lever is in the park or neutral positions.

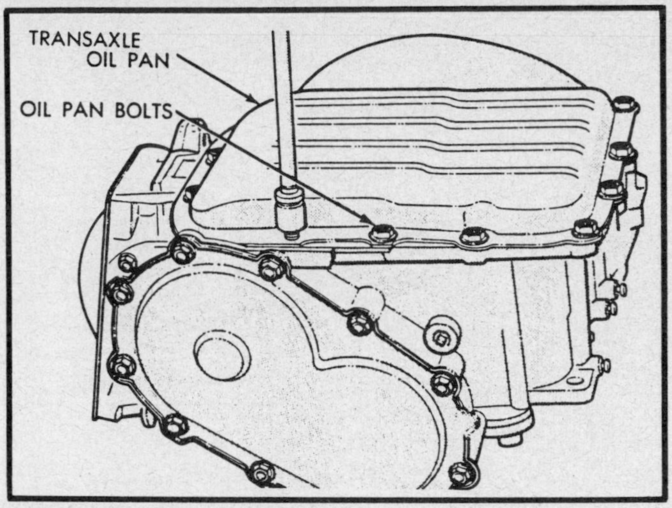
TRANSAXLE
OIL PAN
OIL PAN BOLTS

Transaxle oil pan servicing

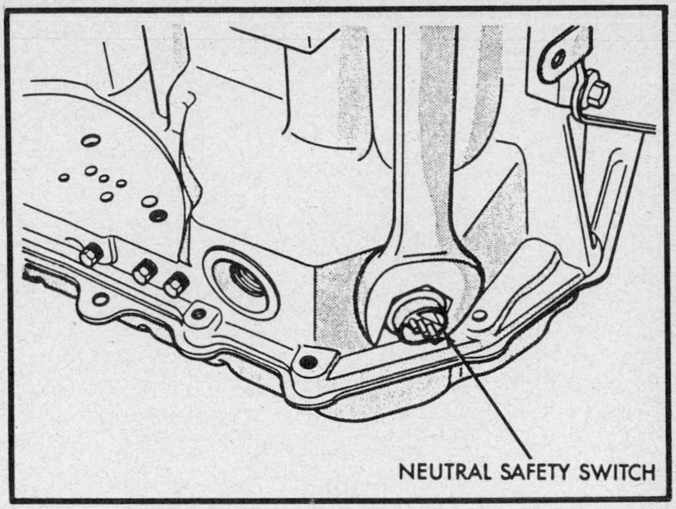
NEUTRAL SAFETY SWITCH

Neutral safety switch servicing

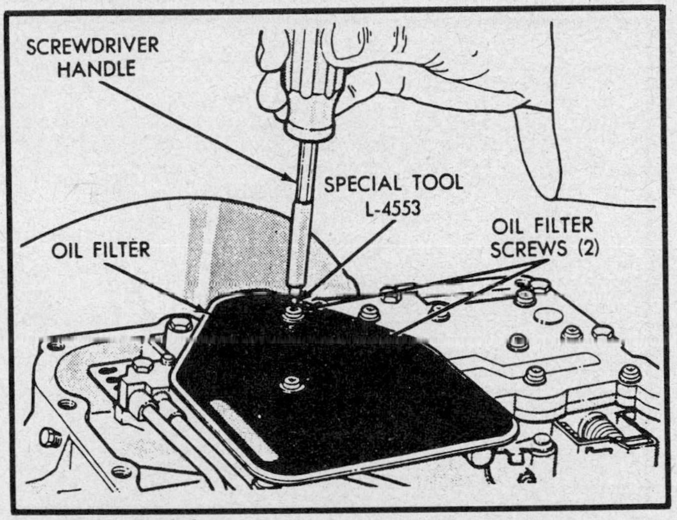
SCREWDRIVER
HANDLE
SPECIAL TOOL
L-4553
OIL FILTER
OIL FILTER
SCREWS (2)

Oil filter servicing

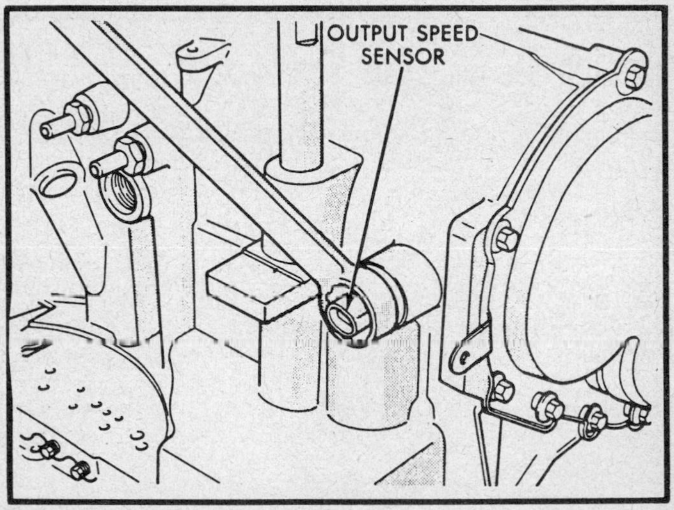
OUTPUT SPEED
SENSOR

Output speed sensor servicing

SHIFT QUALITY QUICK-LEARN PROCEDURE

NOTE: This procedure will quickly optimize the shift quality after the battery has been disconnected.

The transaxle operating temperature must be warm before learning is allowed. To warm up the transaxle fluid, idle the engine for approximately 10 minutes before proceeding.

Upshift Learn Procedure

1. Maintain constant throttle opening during shifts.

NOTE: Do not move the accelerator pedal during the upshifts.

2. Accelerate the vehicle with the throttle opening angle in the range of 10–50 degrees.
3. Make approximately 15–20 upshifts—1st to 2nd, 2nd to 3rd, and 3rd to 4th upshifts.

NOTE: Accelerating from stop to approximately 45 mph each time at moderate throttle angle (20–25 degrees) is sufficient.

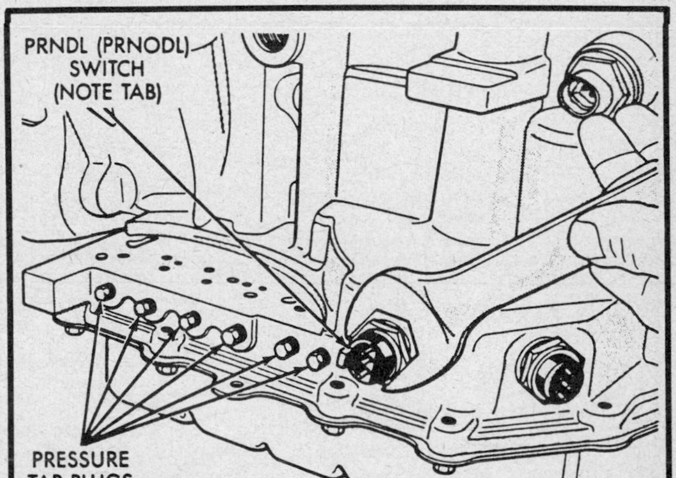
PRNDL (PRNODL)
SWITCH
(NOTE TAB)
PRESSURE
TAP PLUGS

PRNDL switch servicing

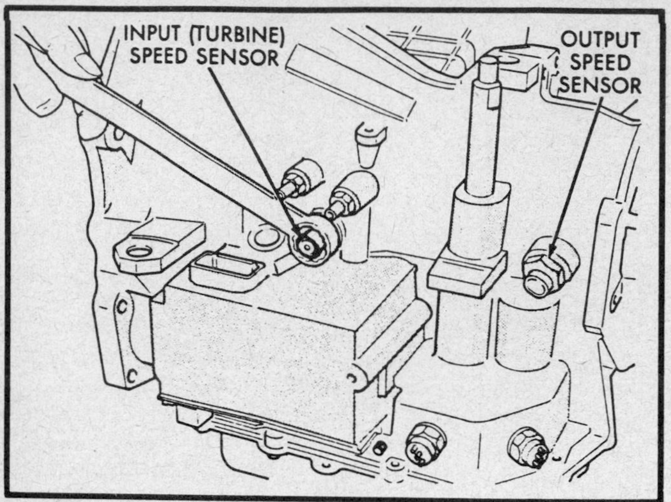
INPUT (TURBINE)
SPEED SENSOR
OUTPUT
SPEED
SENSOR

Input speed sensor servicing

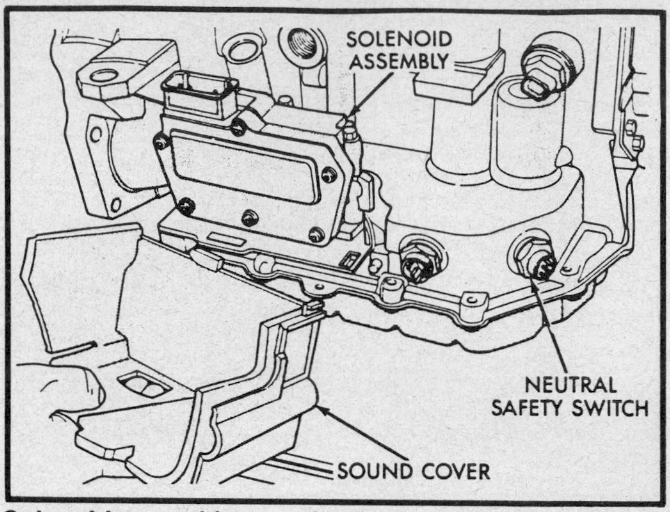
SOLENOID
ASSEMBLY
NEUTRAL
SAFETY SWITCH
SOUND COVER

Solenoid assembly sound cover servicing

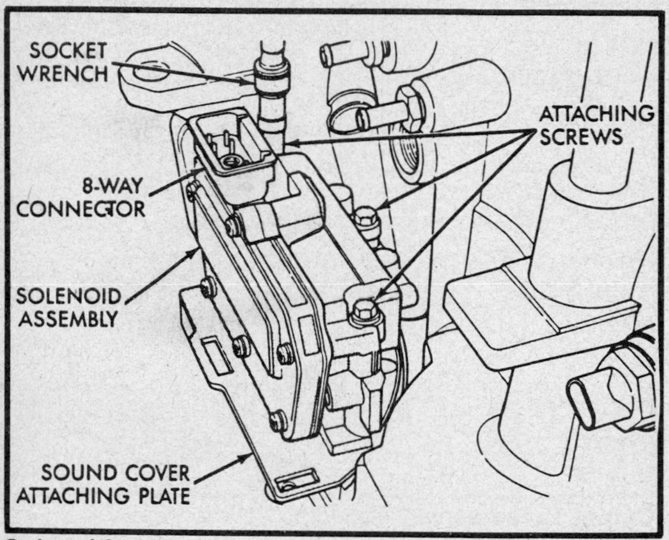
SOCKET
WRENCH
ATTACHING
SCREWS
8-WAY
CONNECTOR
SOLENOID
ASSEMBLY
SOUND COVER
ATTACHING PLATE

Solenoid assembly attaching bolts location

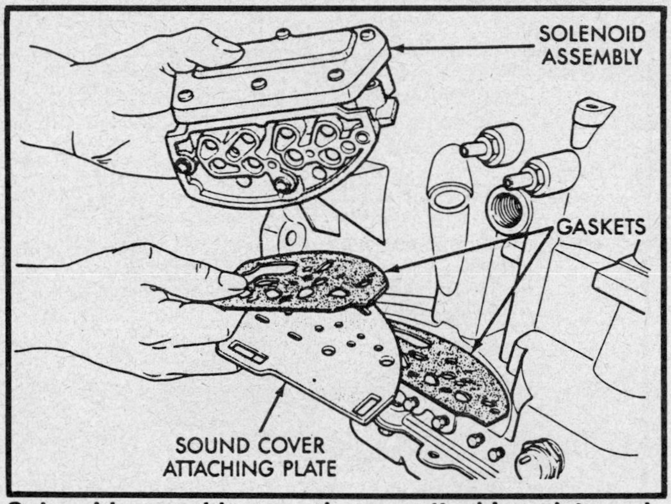
SOLENOID
ASSEMBLY
GASKETS
SOUND COVER
ATTACHING PLATE

Solenoid assembly, sound cover attaching plate and gaskets

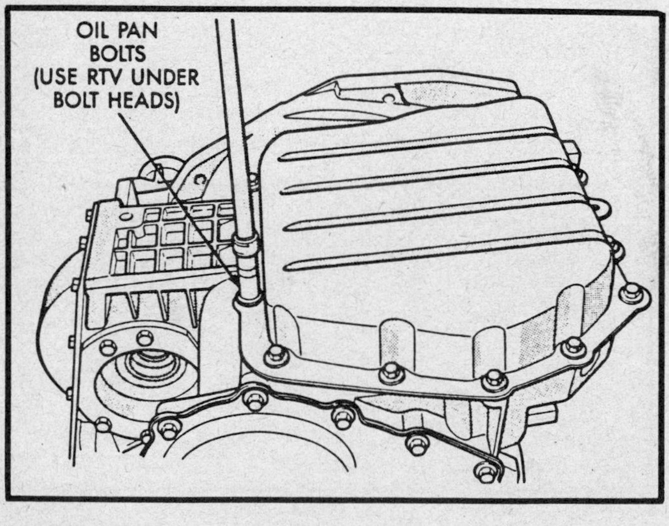
OIL PAN
BOLTS
(USE RTV UNDER
BOLT HEADS)

Transaxle oil pan servicing

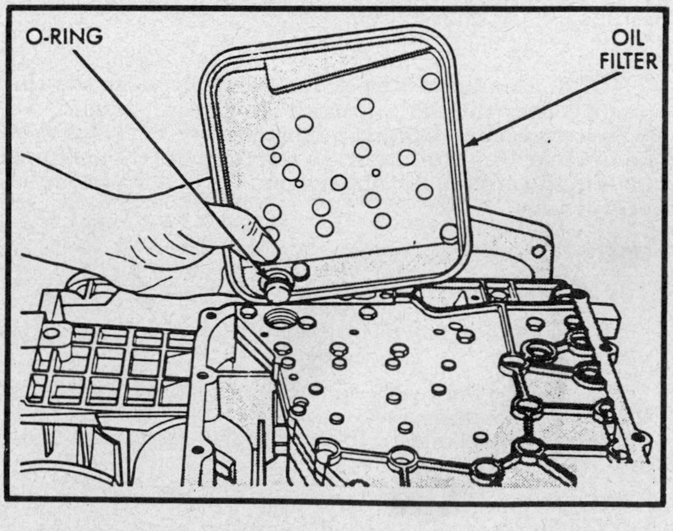
O-RING
OIL
FILTER

Oil filter servicing

Kickdown Learn Procedure

1. With the vehicle speed below 25 mph, make 5–8 WOT kickdowns to 1st gear from either 2nd or 3rd gear. Allow for 5 seconds or more of operation in 2nd or 3rd prior to the kickdown.
2. With the vehicle speed greater than 25 mph, make 5–8 part throttle to WOT kickdowns to either 3rd or 2nd gear from 4th gear (for example, 4–3 or 4–2 kickdowns). Allow for 5 seconds or more of operation in 4th, preferably at road-load throttle, prior to performing the kickdown.

Services

FLUID CHANGES

Fluid and filter changes are not required for average passenger vehicle usage.

Only severe usage, such as more than 50% operation in heavy city traffic during hot weather above 90°F (32°C) or police, taxi, commercial type operation and trailer towing, requires that the fluid and filter be changed, the magnet (on the inside of the oil pan) should be cleaned with a clean, dry cloth every 15,000 miles (24,000 km).

If the transaxle is disassembled for any reason, the fluid and filter should be changed.

1. Raise the vehicle and support safely. Place a drain container with a large opening under the transaxle oil pan.
2. Loosen the pan bolts and tap the pan at 1 corner to break it loose allowing fluid to drain, then remove the oil pan.
3. Install a new filter and O-ring on the bottom of the valve body.
4. Clean the oil pan and magnet. Reinstall the pan using new RTV sealant. Tighten the oil pan bolts to 165 inch lbs. (19 Nm).
5. Install 4 quarts of Mopar® ATF type 7176, or Dexron®II, through the filler tube.
6. Start the engine and allow to idle for at least 1 minute. Then, with the parking and service brakes applied, move the selector lever momentarily to each position, ending in the **P** or **N** position.
7. Add sufficient fluid to bring level to ⅛ in. below the **ADD** mark.
8. Recheck the fluid level after the transaxle is at normal operating temperature. The level should be in the **HOT** region.

NOTE: To prevent dirt from entering the transaxle, make certain that the dipstick is seated into the dipstick fill tube.

OIL PAN

Removal and Installation

1. Raise the vehicle and support safely. Place a drain container with a large opening under the transaxle oil pan.
2. Loosen the pan bolts and tap the pan at 1 corner to break it loose allowing fluid to drain, then remove the oil pan.
3. To install, reverse the removal procedure.
4. Reinstall the pan using new RTV sealant. Tighten the oil pan bolts to 165 inch lbs. (19 Nm).
5. Install 4 quarts of Mopar® ATF type 7176, or Dexron®II, through the filler tube.
6. Start the engine and allow to idle for at least 1 minute. Then, with the parking and service brakes applied, move the selector lever momentarily to each position, ending in the **P** or **N** position.
7. Add sufficient fluid to bring level to ⅛ in. below the **ADD** mark.
8. Recheck the fluid level after the transaxle is at normal operating temperature. The level should be in the **HOT** region.

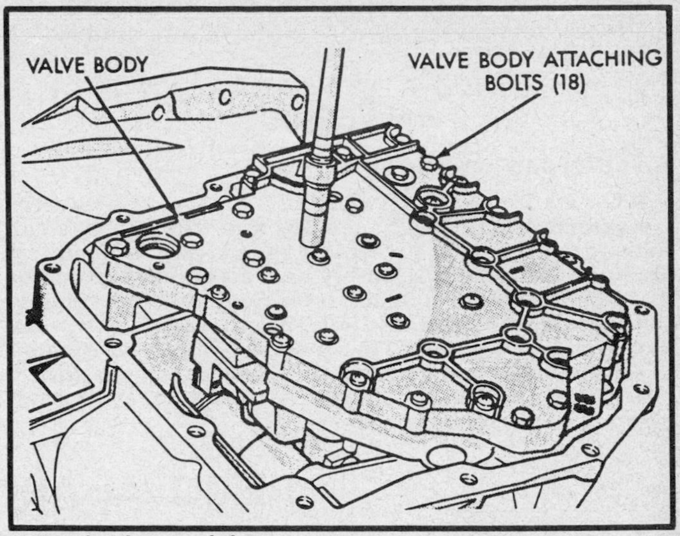
VALVE BODY
VALVE BODY ATTACHING BOLTS (18)

Valve body servicing

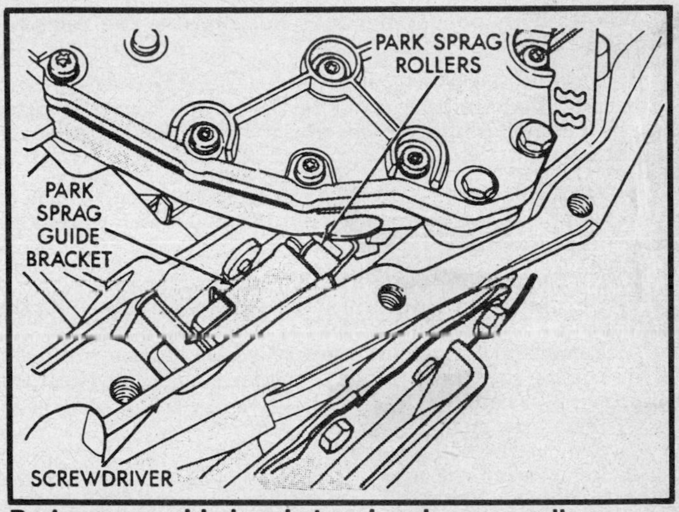
PARK SPRAG ROLLERS
PARK SPRAG GUIDE BRACKET
SCREWDRIVER

Park sprag guide bracket and park sprag rollers location

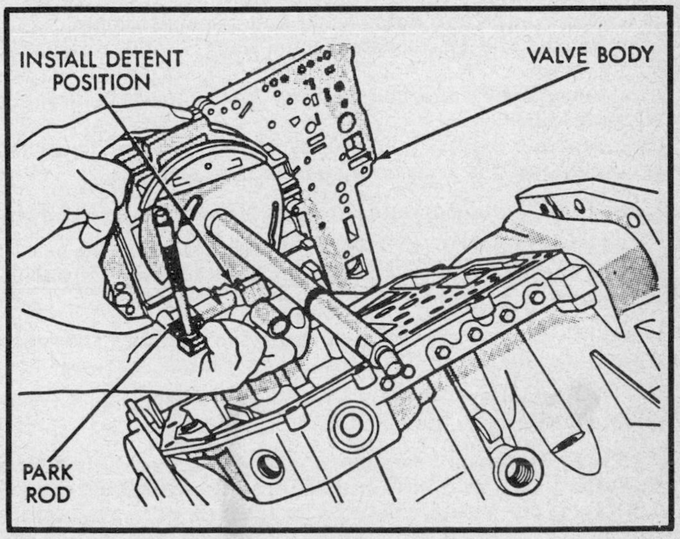
INSTALL DETENT POSITION
VALVE BODY
PARK ROD

Valve body assembly servicing

NOTE: To prevent dirt from entering the transaxle, make certain that the dipstick is seated into the dipstick fill tube.

VALVE BODY

Removal and Installation

NOTE: Prior to removing any transaxle subassemblies, plug all openings and thoroughly clean exterior of the unit, preferably with steam. Cleanliness through entire disassembly and assembly cannot be overemphasized. When disassembling, each part should be washed in a suitable solvent, then dried by compressed air. Do not wipe parts with shop towels. All mating surfaces in the transaxles are accurately machined; therefore, careful handling of all parts must be exercised to avoid nicks and burrs.

1. Raise the vehicle and support safely.
2. Remove the oil pan bolts and remove the oil pan.
3. Remove the oil filter.
4. Remove the valve body attaching bolts.
5. Push the park rod rollers from the guide bracket.
6. Remove the valve body.
7. To install, reverse the removal procedure.

SOLENOID ASSEMBLY

Removal and Installation

1. Raise the vehicle and support safely.
2. Remove the input speed sensor.
3. Remove the sound cover.
4. Remove the solenoid assembly attaching bolts.
5. Remove the solenoid assembly and gaskets.
6. To install, reverse the removal procedure.

REMOVAL AND INSTALLATION

TRANSAXLE REMOVAL

NOTE: This procedure does not require the removal of the engine.

1. The torque converter and transaxle must be removed as an assembly. Damage may result to the torque converter drive plate, pump bushing, or oil seal if the units are removed separately.

NOTE: The drive plate will not support a load. None of the weight of the transaxle should be allowed to rest on the plate during removal.

2. Disconnect the negative battery cable.
3. Disconnect the throttle linkage and shift linkage from the transaxle.
4. Remove the upper and lower oil cooler hoses.
5. Unplug the lockup torque converter plug, located near the dipstick, if so equipped.
6. Install the engine support fixture.
7. Remove the bell housing upper bolts.
8. Remove the cotter pin, lock and spring washer from the hub assembly.
9. Loosen the hub nut and wheel nuts while the vehicle is on the floor and the brakes are applied.
10. Raise and support the vehicle safely.
11. Remove the hub nut, washer and wheel assembly.
12. Remove the speedometer pinion (for the right driveshaft) clamp.
13. Remove the clamp bolt securing the ball joint stud into the steering knuckle.
14. Separate the ball joint stud from the steering knuckle by prying against the knuckle leg and control arm.

NOTE: Do not damage the ball joint or CV joint boots.

15. Separate the outer CV joint splined shaft from the hub by holding the CV housing while moving the knuckle (hub) assembly away.

NOTE: Do not pry on or otherwise damage the wear sleeve on the outer CV joint.

16. Support the assembly at the CV joint housings. Remove by pulling outward on the inner joint housing.

NOTE: Do not pull on the shaft. The driveshaft, when installed, acts as a bolt and secures the hub/bearing assembly. If the vehicle is to be supported or moved on its wheels, install a bolt through the hub to insure that the hub bearing assembly cannot loosen.

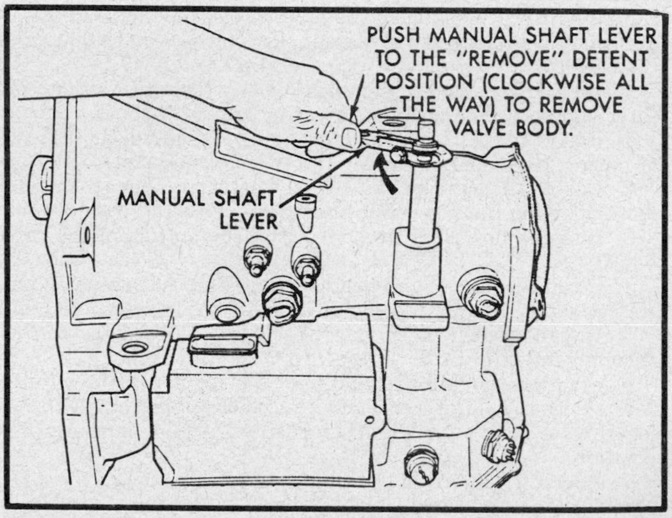
PUSH MANUAL SHAFT LEVER TO THE "REMOVE" DETENT POSITION (CLOCKWISE ALL THE WAY) TO REMOVE VALVE BODY.
MANUAL SHAFT LEVER

Positioning manual shaft lever for transaxle removal

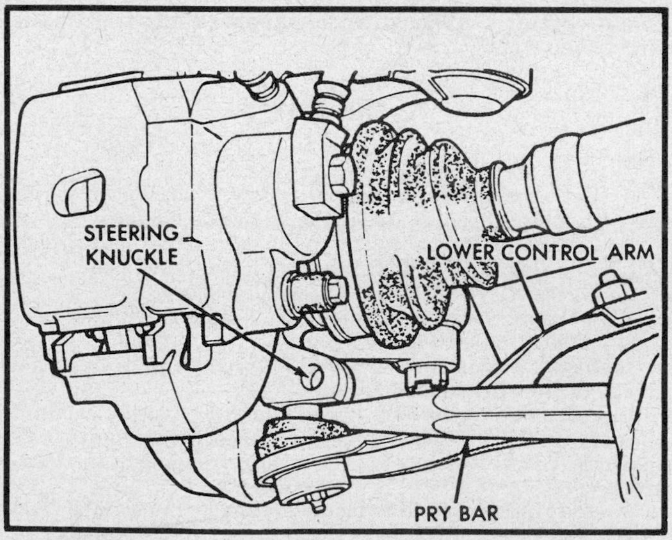
STEERING KNUCKLE
LOWER CONTROL ARM
PRY BAR

Separating lower control arm from steering knuckle

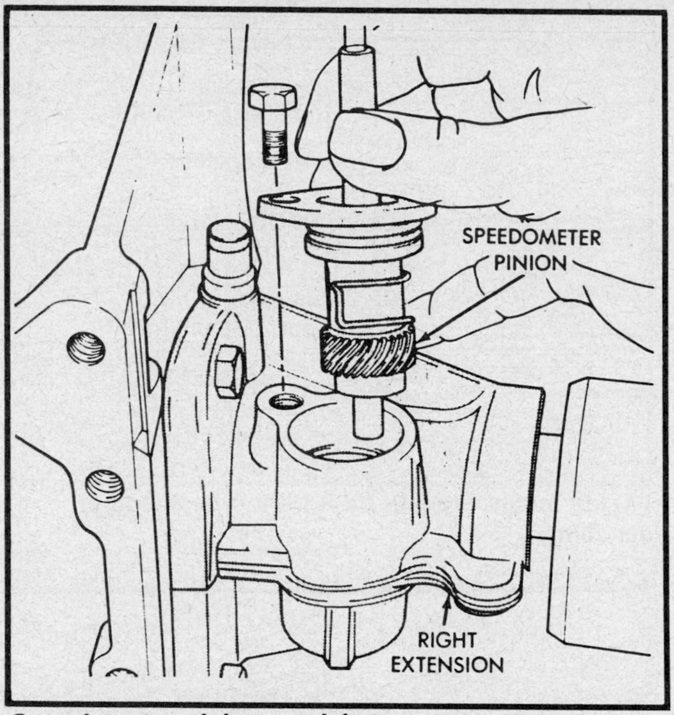
SPEEDOMETER PINION
RIGHT EXTENSION

Speedometer pinion servicing

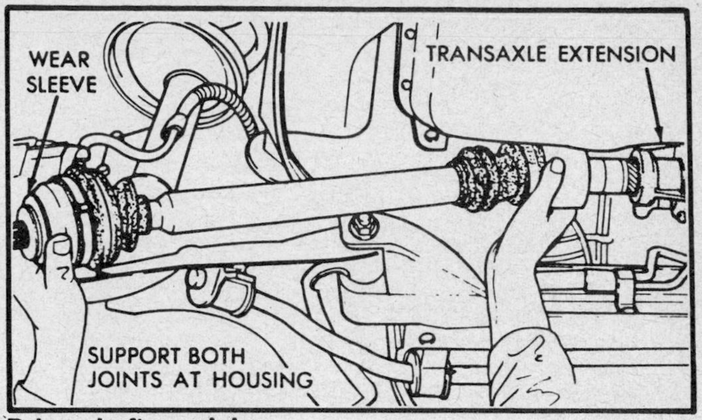
WEAR SLEEVE
TRANSAXLE EXTENSION
SUPPORT BOTH JOINTS AT HOUSING

Drive shaft servicing

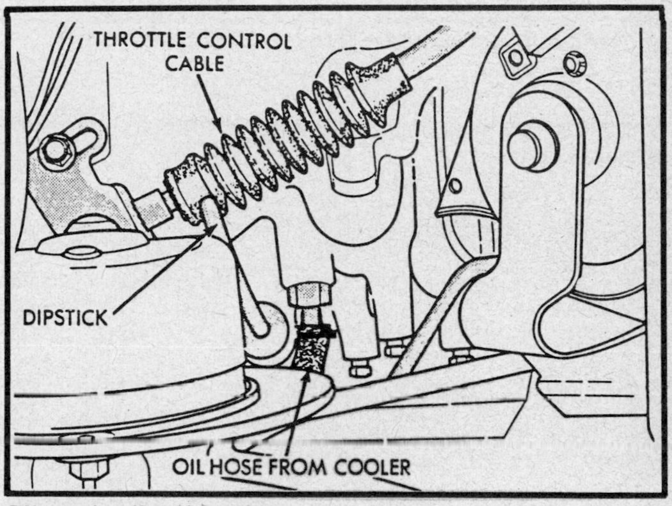
THROTTLE CONTROL CABLE
DIPSTICK
OIL HOSE FROM COOLER

Oil cooler line location

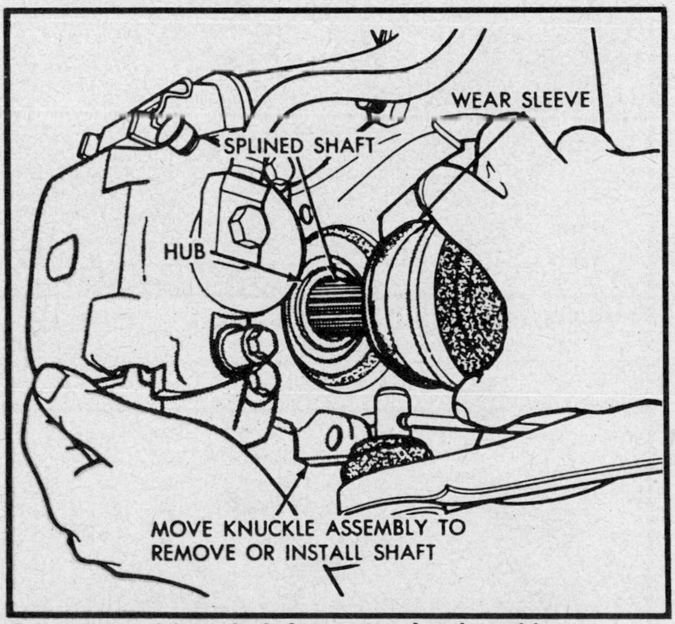
WEAR SLEEVE
SPLINED SHAFT
HUB
MOVE KNUCKLE ASSEMBLY TO REMOVE OR INSTALL SHAFT

Separating drive shaft from steering knuckle

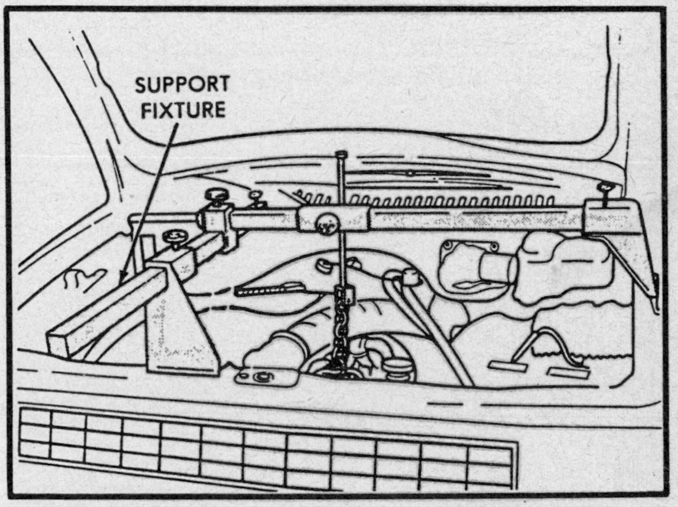
SUPPORT FIXTURE

Engine support fixture—Caravan and Voyager

17. Remove both driveshafts.
18. Remove the torque converter dust cover. Mark the torque converter and drive plate with chalk, for reassembly. Remove the torque converter mounting bolts.
19. Remove the access plug in the right splash shield to rotate the engine crankshaft.
20. Remove the wire to the neutral/park safety switch.
21. Remove the engine mount bracket from the front crossmember.
22. Remove the front mount insulator through bolt and the bell housing bolts.
23. Position the transmission jack.
24. Remove the left engine mount.
25. Remove the starter. Remove the lower bell housing bolts.
26. Pry the engine for clearance and lower the transaxle.

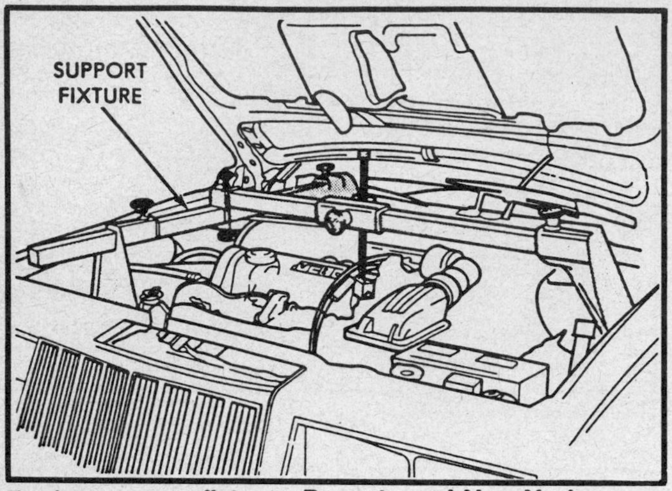
SUPPORT
FIXTURE

Engine support fixture—Dynasty and New Yorker

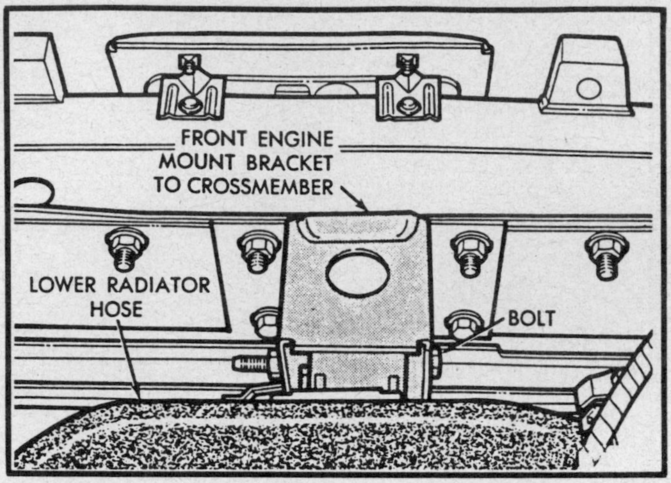
FRONT ENGINE
MOUNT BRACKET
TO CROSSMEMBER
LOWER RADIATOR
HOSE
BOLT

Engine mount bracket from front crossmember servicing

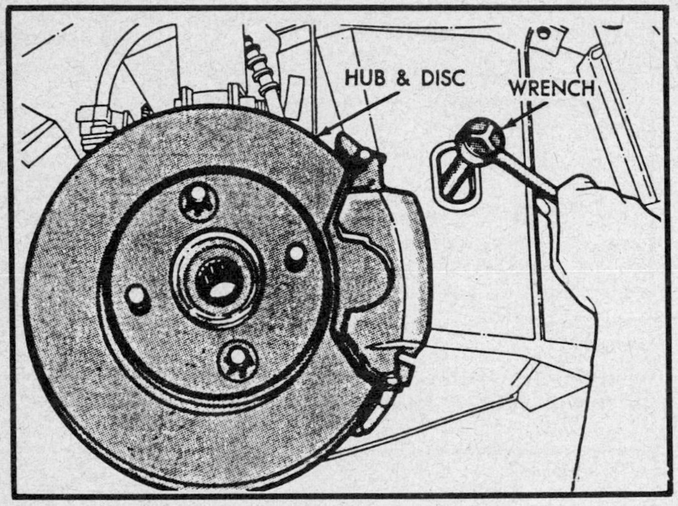
HUB & DISC
WRENCH

Right splash shield access plug location

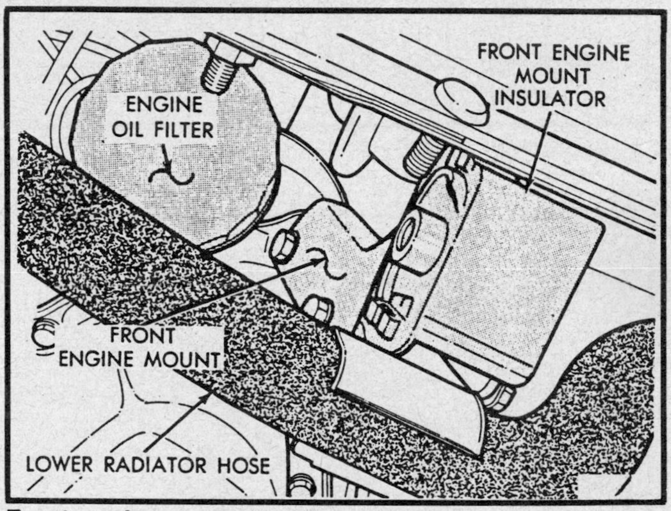
FRONT ENGINE
MOUNT
INSULATOR
ENGINE
OIL FILTER
FRONT
ENGINE MOUNT
LOWER RADIATOR HOSE

Front engine mount insulator location

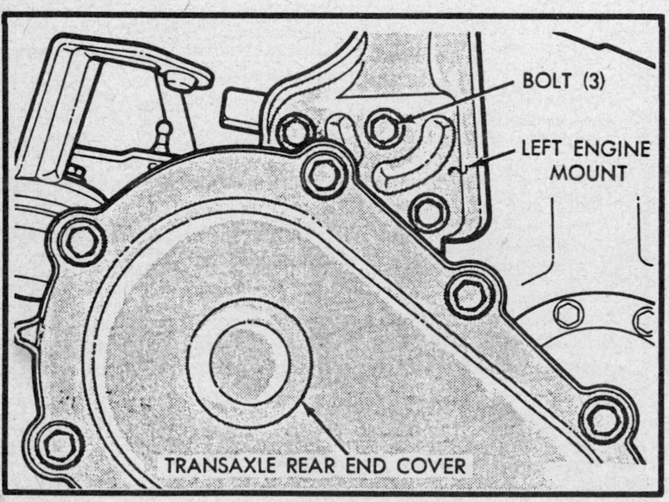
BOLT (3)
LEFT ENGINE
MOUNT
TRANSAXLE REAR END COVER

Left engine mount bolts location

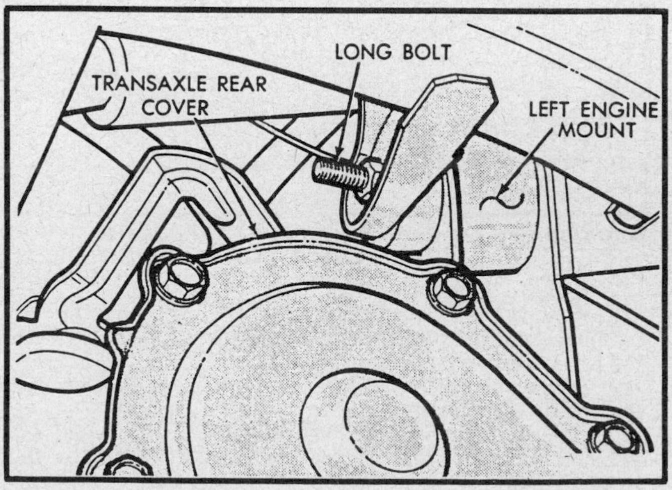
LONG BOLT
TRANSAXLE REAR
COVER
LEFT ENGINE
MOUNT

Left engine mount-from-engine bolt location

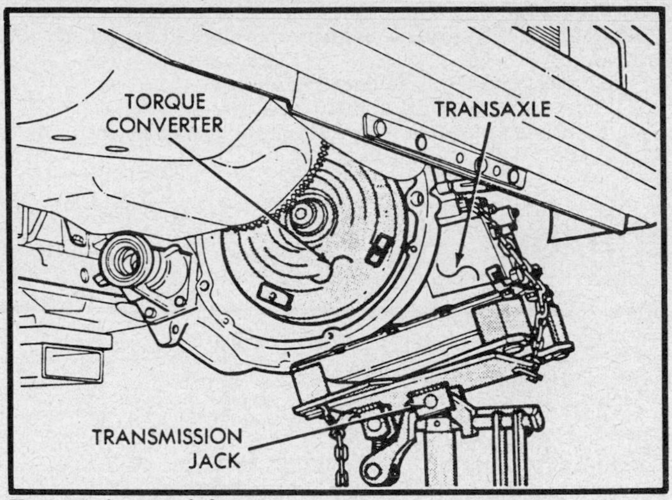
TORQUE CONVERTER
TRANSAXLE
TRANSMISSION JACK
Transaxle servicing

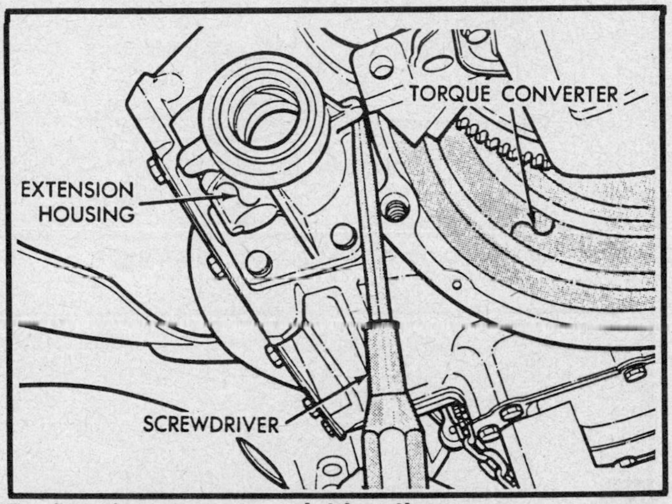
TORQUE CONVERTER
EXTENSION HOUSING
SCREWDRIVER
Engine clearance pry point location

TRANSAXLE INSTALLATION

1. Raise the transaxle into position.
2. Install the lower bell housing bolts. Install the starter.
3. Install the left engine mount to the engine.
4. Install the left engine mount.
5. Remove the transmission jack.
6. Install the front mount insulator through bolt and bell housing bolts.
7. Install the engine mount bracket to the front crossmember.
8. Install the wire to the neutral/park safety switch.
9. Install the access plug in the right splash shield.
10. Turn the torque converter to matchmark the driveplate made on disassembly, and install torque converter mounting bolts.
11. Install the torque converter dust cover.
12. To install the driveshafts, hold the inner joint assembly at the housing while aligning and guiding the inner joint spline into the transaxle or intermediate shaft assembly.

NOTE: When installing an A.C.I. type shaft, be sure that the tripod is engaged in the housing and the boot is not twisted.

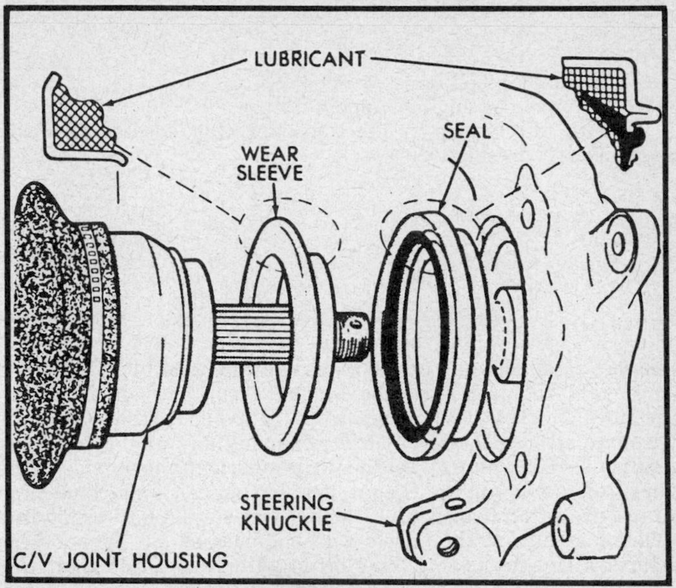
LUBRICANT
SEAL
WEAR SLEEVE
STEERING KNUCKLE
C/V JOINT HOUSING
Wear sleeve and seal servicing

13. Install the inner shaft into the transaxle.

NOTE: During any service procedures where the knuckle and driveshaft are separated, thoroughly clean the seal and wear sleeve with a suitable solvent (solvent must not touch the boots) and relubricate both components prior to reinstalling driveshaft. Lubricate wear sleeve and seal with Mopar Multi-Purpose Lubricant, part number 4318063 or equivalent.

14. Apply lubricate on the full circumference of the wear sleeve and bead of lubricate that is ¼ in. (6mm) wide to seal the contact area. Fill the lip to housing cavity on the seal, complete circumference, and wet the seal lip with lubricate.
15. Push the knuckle (hub) assembly out and install the splined outer CV joint shaft in the hub.
16. Reinstall the knuckle assembly on the ball joint stud.

NOTE: The original or equivalent steering knuckle clamp bolt must be reinstalled during reassembly.

17. Install and tighten the clamp bolt to 70 ft. lbs. (95 Nm).
18. Install speedometer pinion.
19. Fill the differential with proper lubricant.
20. Install the hub nut assembly.

NOTE: If the inboard boot appears collapsed or deformed after installing the driveshaft assembly, vent the inner boot by inserting a round tipped, small diameter rod between the boot and shaft. If necessary, massage the boot to remove all puckers being careful not to allow dirt to enter or grease to leave the boot cavity. If the boot is clamped to the shaft with a rubber "garter" clamp, it need not be removed to perform this venting operation. If the boot is clamped to the shaft using a metal clamp, the clamp must be removed and discarded before the rod can be inserted. After venting, install a new service clamp, part number 5212720, or equivalent and special tool C–4653.

21. Install the washer and hub nut after cleaning foreign matter from the threads.
22. With the brakes applied, tighten the hub nut to 180 ft. lbs. (245 Nm).
23. Install the lock, washer and new cotter pin. Wrap the cotter pin prongs tightly around the nut lock.

24. Install the wheel assembly and tighten wheel nuts to 95 ft. lbs. (129 Nm).
25. Lower the vehicle.
26. Install the bell housing upper bolts.
27. Remove the engine support fixture.
28. Plug the lockup torque converter plug, located near the dipstick, if so equipped.
29. Install the upper and lower oil cooler hoses.
30. Connect the throttle linkage and shift linkage from the transaxle.
31. Connect the negative battery cable.
32. Adjust the gearshift and throttle cables.
33. Refill the transaxle with automatic transmission fluid.

BENCH OVERHAUL

Before Disassembly

Before removing any of the transaxle subassemblies for bench overhaul, the unit should be cleaned. Cleanliness during disassembly and assembly is necessary to avoid further transaxle trouble after assembly. Before removing any of the transaxle subassemblies, plug all the openings and clean the outside of the transaxle thoroughly. Steam cleaning or car wash type high pressure equipment is preferable. Parts should be washed in a cleaning solvent, then dried with compressed air. Do not wipe parts with shop towels. The lint and fibers will find their way into the valve body and other parts and cause problems later. The case assembly was accurately machined and care must be used to avoid damage. Pay attention to the torque values to avoid case distortion.

Transaxle Disassembly

CONVERTER

Removal

The torque converter is removed by sliding the unit out of the transaxle input and reaction shaft. If the converter is to be reused, set it aside so it will not be damaged. Since the units are welded and have no drain plugs, converters subject to burnt fluid or other contamination should be replaced.

OIL PAN

Removal

Loosen the pan bolts and tap the pan at a corner to break it loose allowing fluid to drain, then remove the oil pan.

VALVE BODY

Removal

1. Remove the oil filter.
2. Remove the valve body attaching bolts.
3. Push the park rod rollers from the guide bracket.
4. Remove the valve body.

ACCUMULATORS

Removal

1. Remove the underdrive clutch accumulator and return spring.
2. Remove the overdrive clutch accumulator and return spring.
3. Remove the low/reverse accumulator snapring.
4. Remove the low/reverse accumulator plug with adjustable pliers.

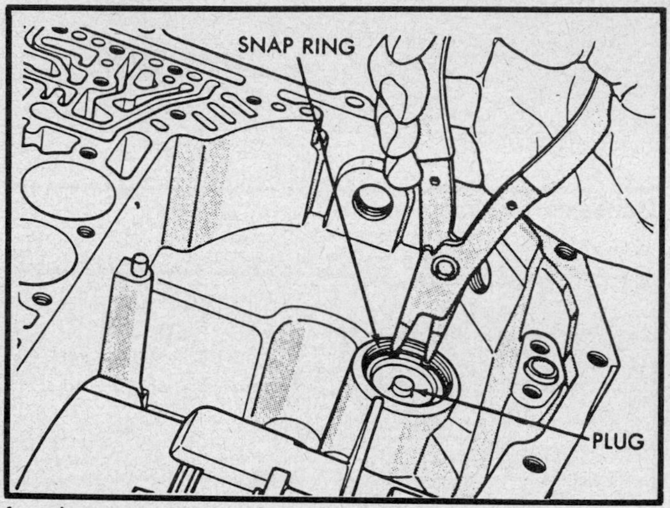
SNAP RING
PLUG
Low/reverse accumulator snapring servicing

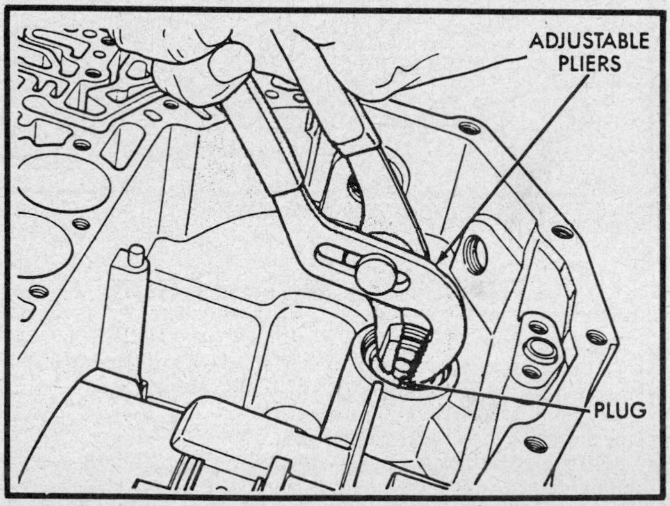
ADJUSTABLE PLIERS
PLUG
Low/reverse accumulator plug removal

5. Remove the low/reverse accumulator piston with a suitable dowel tool and petroleum jelly.
6. Remove the low/reverse accumulator piston return springs.

GOVERNOR

Removal

This transaxle utilizes electronic sensors and solenoids in place of the governor. Therefore, there is nc governor assembly.

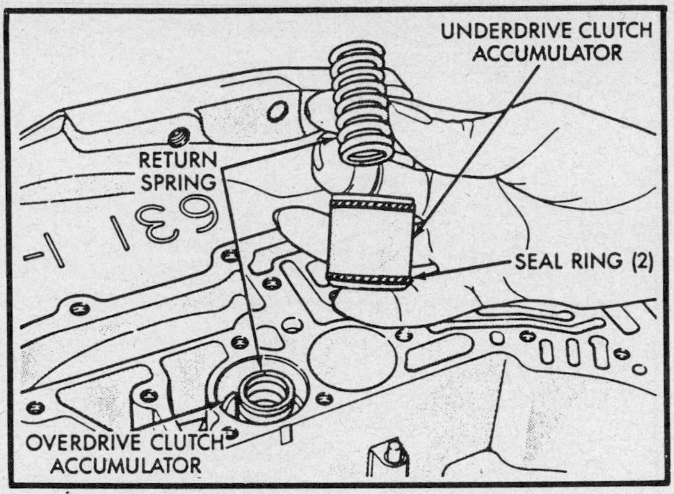
UNDERDRIVE CLUTCH
ACCUMULATOR
RETURN
SPRING
SEAL RING (2)
OVERDRIVE CLUTCH
ACCUMULATOR
Underdrive clutch accumulator servicing

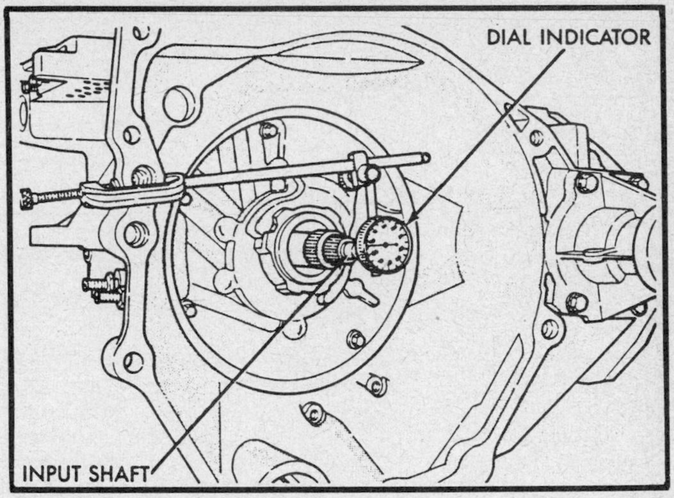
DIAL INDICATOR
INPUT SHAFT
Checking input shaft endplay

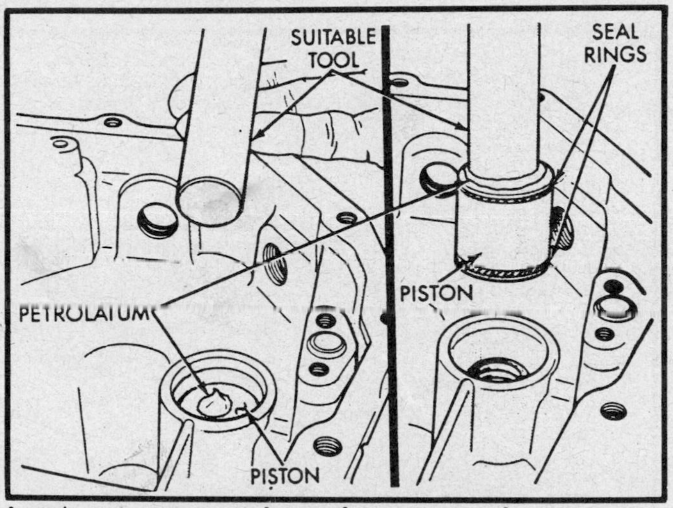
SUITABLE
TOOL
SEAL
RINGS
PETROLATUM
PISTON
PISTON
Low/reverse accumulator piston removal

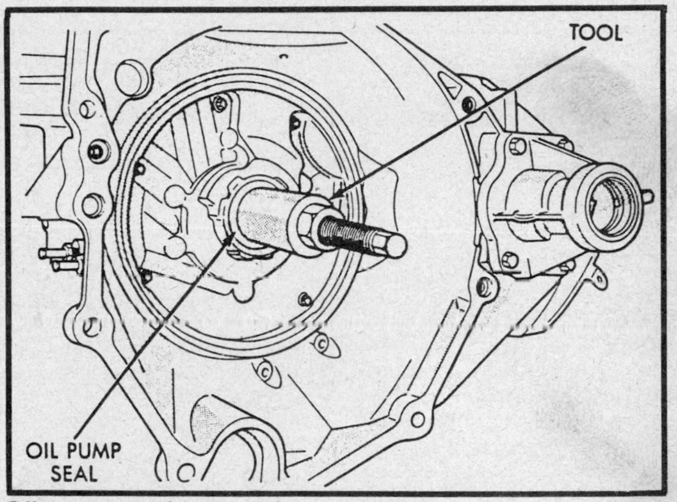
TOOL
OIL PUMP
SEAL
Oil pump seal removal

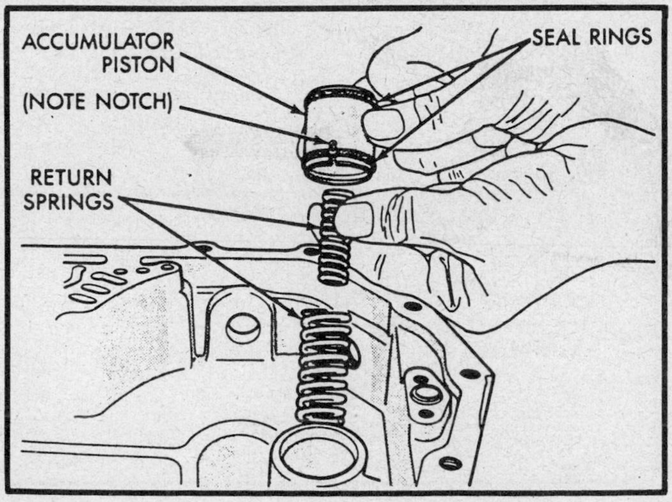
ACCUMULATOR
PISTON
(NOTE NOTCH)
SEAL RINGS
RETURN
SPRINGS
Low/reverse accumulator and return springs

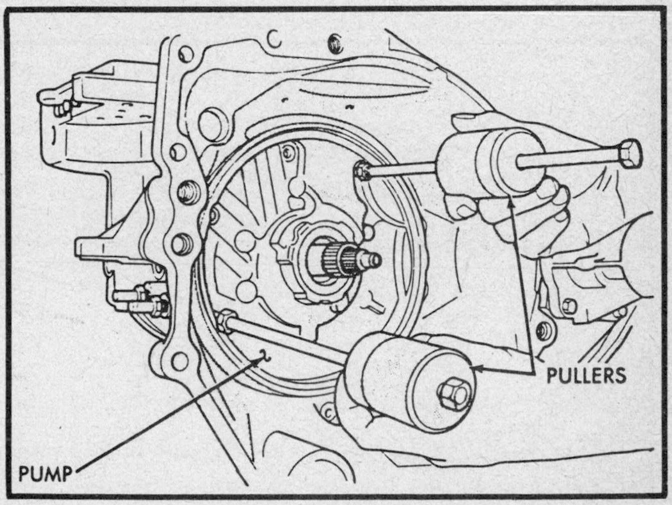
PULLERS
PUMP
Oil pump puller tools

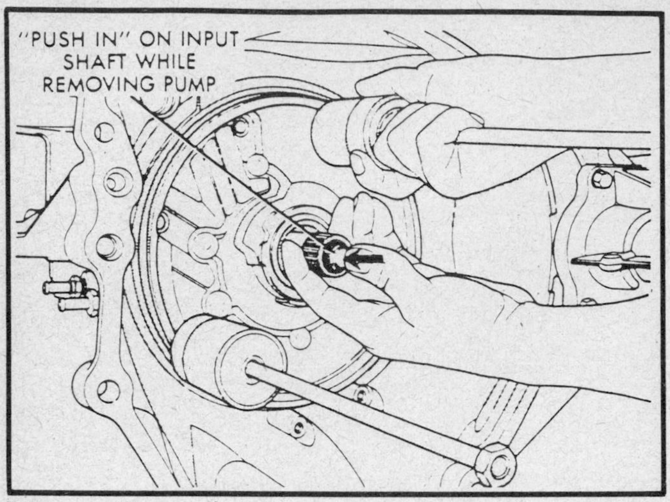
"PUSH IN" ON INPUT SHAFT WHILE REMOVING PUMP
Oil pump removal

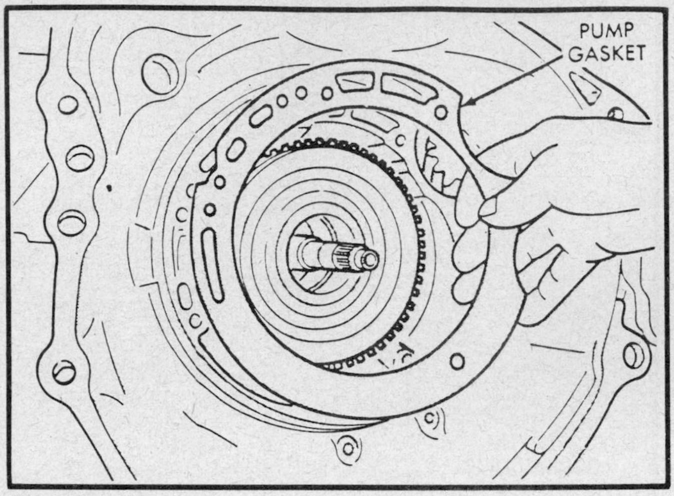
PUMP GASKET
Oil pump gasket

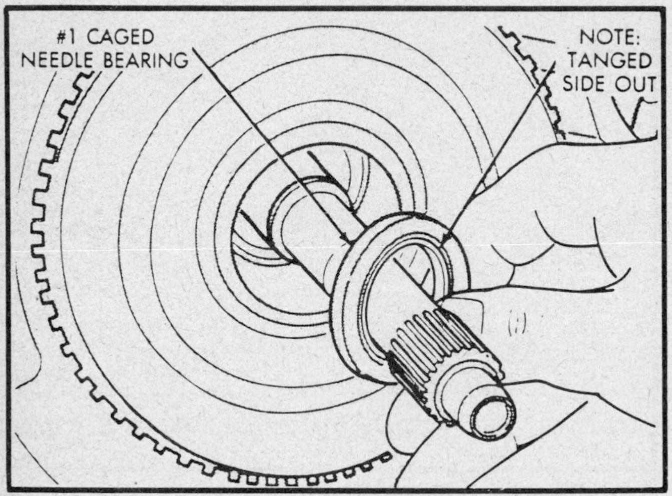
#1 CAGED NEEDLE BEARING
NOTE: TANGED SIDE OUT
Caged needle bearing #1 assembly

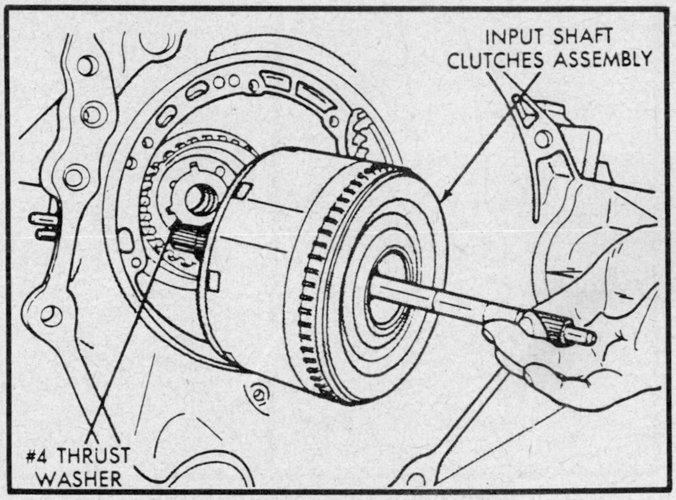
INPUT SHAFT CLUTCHES ASSEMBLY
#4 THRUST WASHER
Input shaft clutches assembly

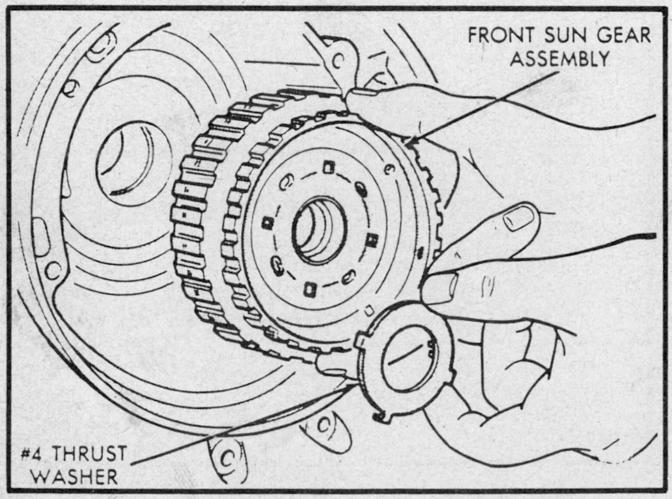
FRONT SUN GEAR ASSEMBLY
#4 THRUST WASHER
Front sun gear assembly and #4 thrust washer

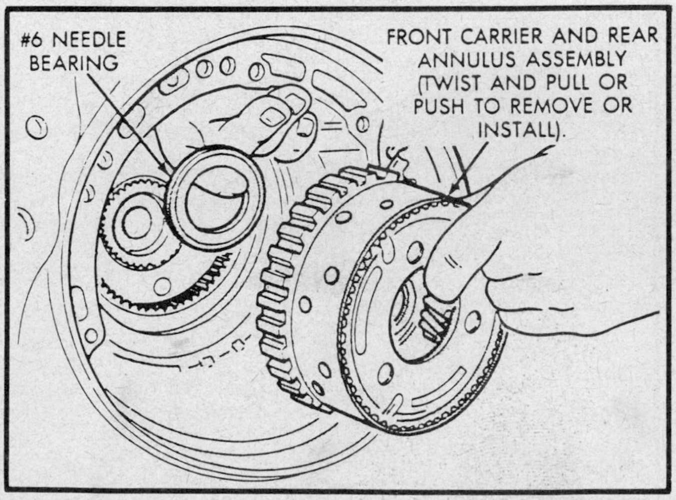
#6 NEEDLE BEARING
FRONT CARRIER AND REAR ANNULUS ASSEMBLY (TWIST AND PULL OR PUSH TO REMOVE OR INSTALL).
Front carrier/rear annulus assembly and #6 needle bearing

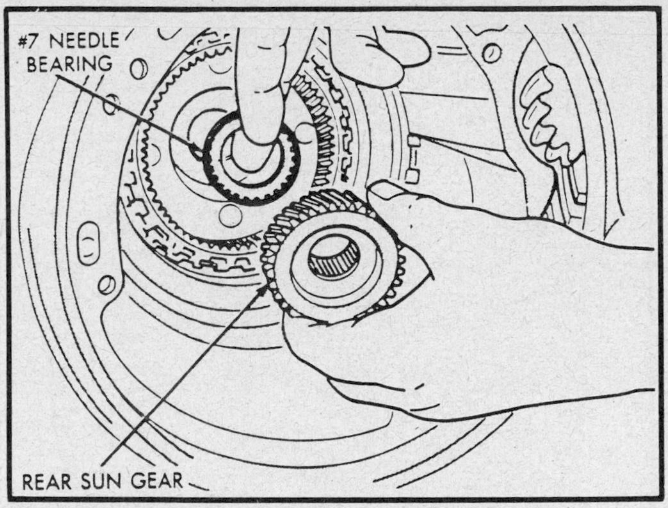
#7 NEEDLE
BEARING
REAR SUN GEAR

Rear sun gear and #7 needle bearing

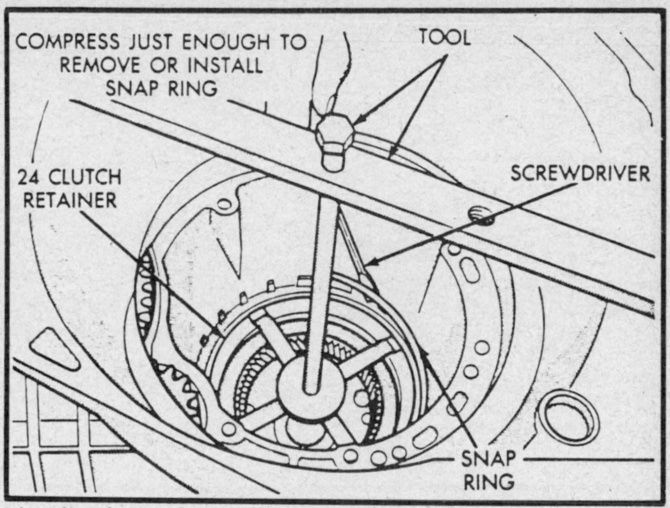
COMPRESS JUST ENOUGH TO
REMOVE OR INSTALL
SNAP RING
TOOL
24 CLUTCH
RETAINER
SCREWDRIVER
SNAP
RING

2/4 clutch retainer snapring compression tool

INPUT SHAFT ENDPLAY

Removal

Measuring input shaft endplay before disassembly will usually indicate when a No. 4 thrust plate change is required (except when major parts are replaced). The thrust washer is located behind the input shaft.

1. Attach a dial indicator to the transaxle bell housing with its plunger seated against the end of the input shaft.
2. Move the shaft in and out to obtain endplay reading. Endplay specifications are 0.012–0.030 in. (0.31–0.76mm).
3. Record the indicator reading for reference when reassembling the transaxle.

OIL PUMP

Removal

1. Remove the input speed sensor.
2. Remove the oil pump seal using tool C–3981.
3. Remove the oil pump attaching bolts.
4. Install pullers tool C–3752 to pump.
5. Push in on the input shaft while removing the pump.
6. Remove the oil pump and oil pump gasket.

INPUT SHAFT CLUTCHES ASSEMBLY

Removal

1. Remove the No. 1 caged needle bearing.
2. Remove the input shaft clutches assembly.

FRONT SUN GEAR ASSEMBLY

Removal

1. Remove the No. 4 thrust washer.
2. Remove the front sun gear assembly.

FRONT CARRIER AND REAR ANNULUS ASSEMBLY

Removal

1. Remove the front carrier and rear annulus assembly. Twist and pull to remove the assembly.
2. Remove the No. 6 needle bearing.

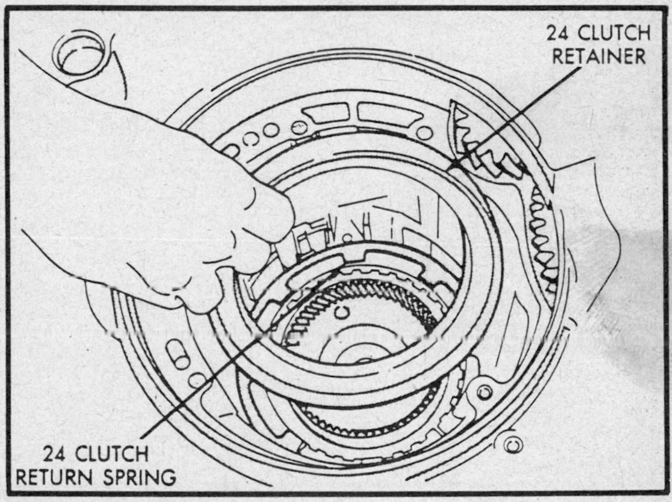
24 CLUTCH
RETAINER
24 CLUTCH
RETURN SPRING

2/4 clutch retainer

REAR SUN GEAR

Removal

1. Remove the rear sun gear.
2. Remove the No. 7 needle bearing.

2–4 CLUTCH PACK

Removal

1. Install tool 5058 and compress the 2–4 clutch retainer spring just enough to remove the 2–4 clutch retainer snapring. Remove the 2–4 clutch retainer snapring.
2. Remove the 2–4 clutch retainer.
3. Remove the 2–4 clutch return spring.
4. Remove the 2–4 clutch pack. Tag and identify clutch packs to assure original replacement.

LOW/REVERSE CLUTCH PACK

Removal

1. Remove the tapered snapring. Remove in sequence.
2. Remove the low/reverse reaction plate.

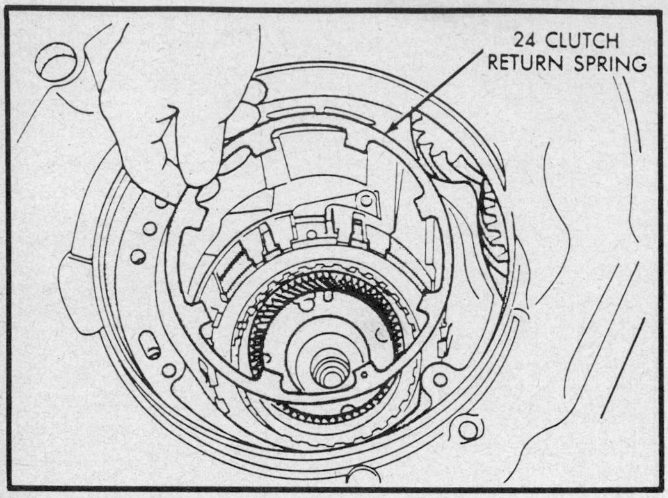
24 CLUTCH
RETURN SPRING

2/4 clutch return spring

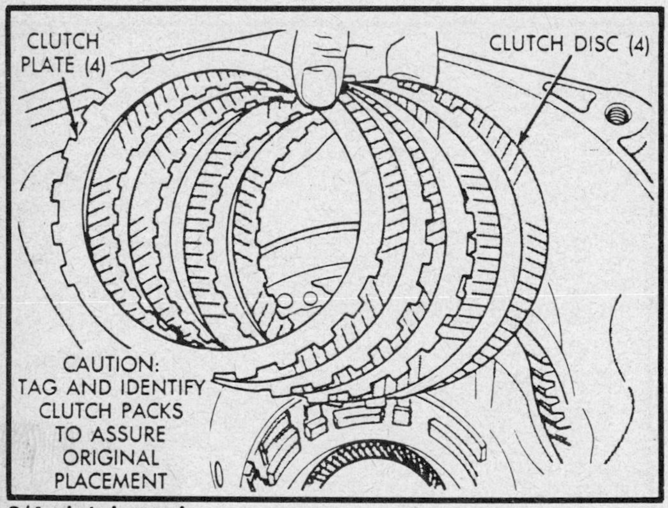
CLUTCH
PLATE (4)
CLUTCH DISC (4)
CAUTION:
TAG AND IDENTIFY
CLUTCH PACKS
TO ASSURE
ORIGINAL
PLACEMENT

2/4 clutch pack

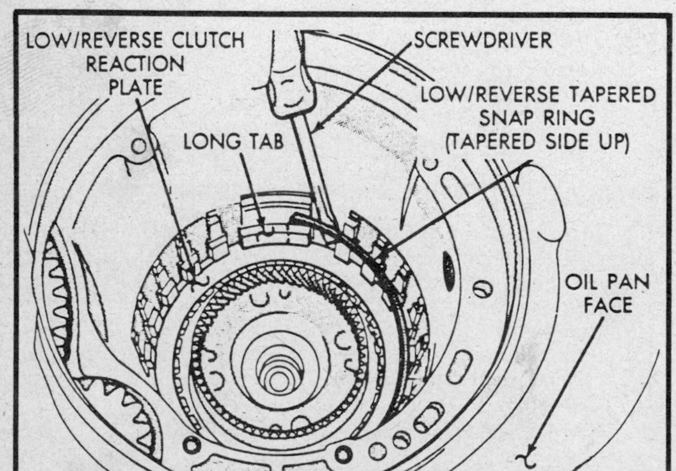
LOW/REVERSE CLUTCH
REACTION
PLATE
SCREWDRIVER
LONG TAB
LOW/REVERSE TAPERED
SNAP RING
(TAPERED SIDE UP)
OIL PAN
FACE

Low/reverse tapered snapring

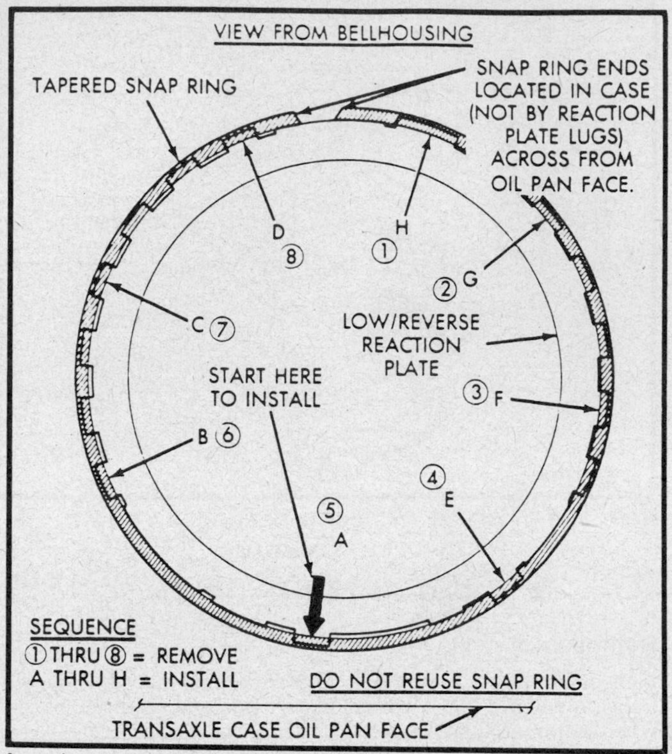
VIEW FROM BELLHOUSING
TAPERED SNAP RING
SNAP RING ENDS
LOCATED IN CASE
(NOT BY REACTION
PLATE LUGS)
ACROSS FROM
OIL PAN FACE.
D ⑧
H ①
② G
C ⑦
LOW/REVERSE
REACTION
PLATE
START HERE
TO INSTALL
③ F
B ⑥
④ E
⑤ A
SEQUENCE
① THRU ⑧ = REMOVE
A THRU H = INSTALL
DO NOT REUSE SNAP RING
TRANSAXLE CASE OIL PAN FACE

Low/reverse tapered snapring removal/installation sequence instructions

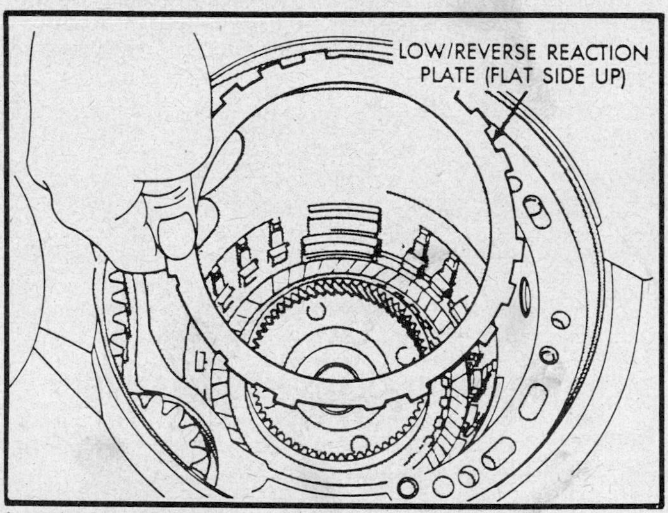
LOW/REVERSE REACTION
PLATE (FLAT SIDE UP)

Low/reverse reaction plate

3. Remove the 1 disc from the low/reverse clutch.
4. Remove the low/reverse reaction plate snapring. Do not scratch the clutch plate.
5. Remove the low/reverse clutch pack. Tag and identify clutch packs to assure original replacement.

TRANSFER SHAFT GEAR

Removal

1. Remove the rear cover bolts.
2. Remove the rear cover.
3. Remove the transfer shaft gear nut and lockwasher using tool 6259.

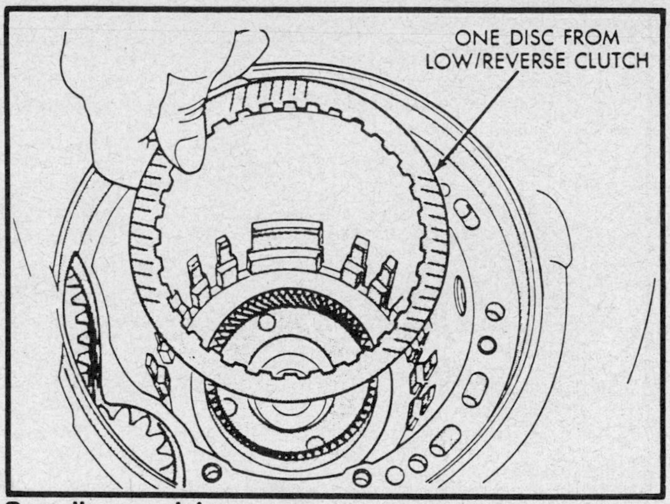
ONE DISC FROM
LOW/REVERSE CLUTCH
One disc servicing

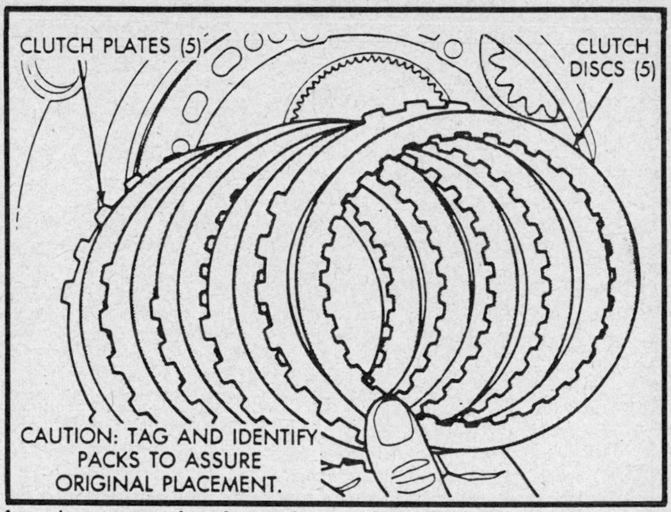
CLUTCH PLATES (5)
CLUTCH
DISCS (5)
CAUTION: TAG AND IDENTIFY
PACKS TO ASSURE
ORIGINAL PLACEMENT.
Low/reverse clutch pack

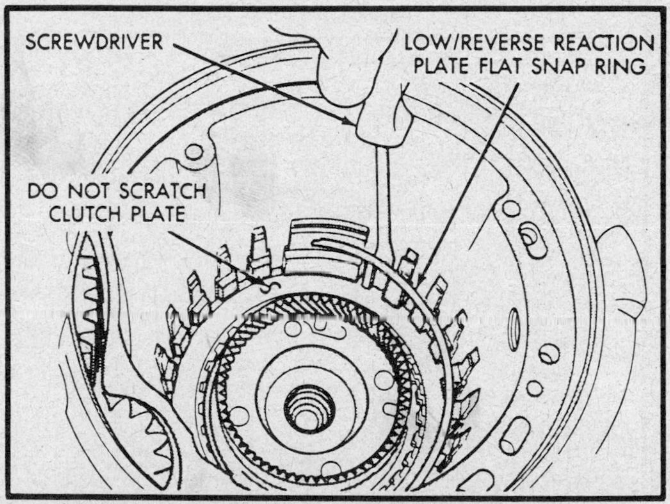
SCREWDRIVER
LOW/REVERSE REACTION
PLATE FLAT SNAP RING
DO NOT SCRATCH
CLUTCH PLATE
Low/reverse reaction plate snapring

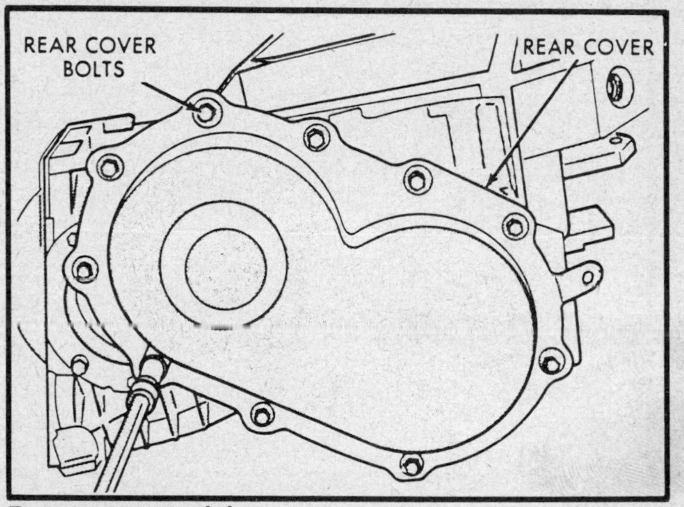
REAR COVER
BOLTS
REAR COVER
Rear cover servicing

4. Remove the transfer shaft gear and shim using puller tool L–4407 and bolts tool L–4407–6.

NOTE: If necessary, the transfer shaft bearing cone can be removed from the gear by using tool 5048 with jaws tool 5048–4 and button tool 6055.

TRANSFER SHAFT

Removal

1. Remove the transfer shaft bearing cup retainer.

NOTE: If necessary, the transfer shaft bearing cup can be removed from the transfer shaft bearing cup retainer by using tool 6062.

2. Remove the transfer shaft bearing snapring using snapring pliers tool 6051.
3. Remove the transfer shaft using tool 5049–A.

NOTE: If necessary, the transfer shaft bearing cone can be removed from the transfer shaft with the use of tool P–334 and arbor press.

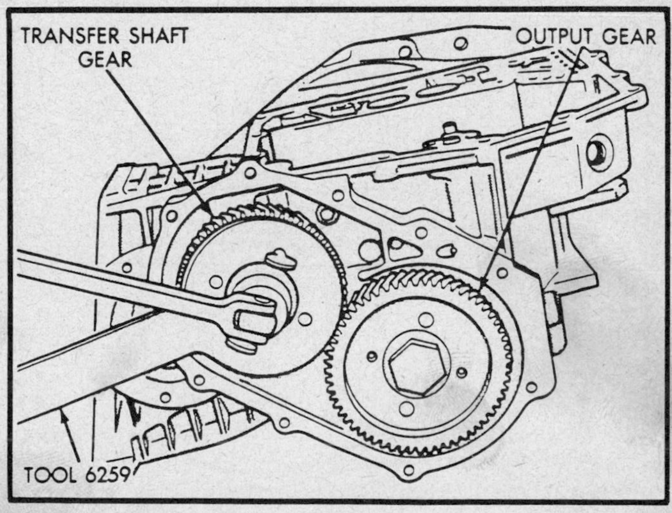
TRANSFER SHAFT
GEAR
OUTPUT GEAR
TOOL 6259
Transfer shaft gear nut removal

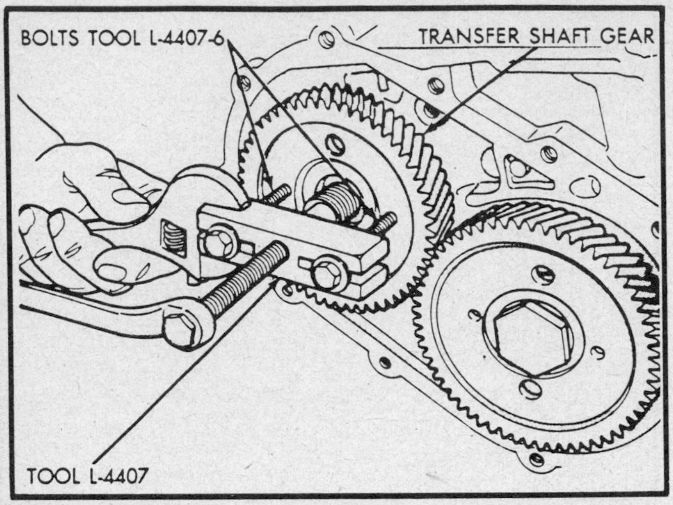
BOLTS TOOL L-4407-6
TRANSFER SHAFT GEAR
TOOL L-4407
Transfer shaft gear removal tool

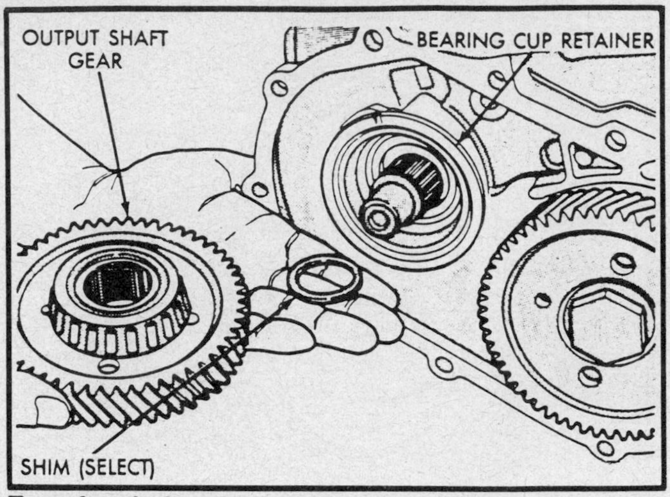
OUTPUT SHAFT GEAR
BEARING CUP RETAINER
SHIM (SELECT)
Transfer shaft gear and shim

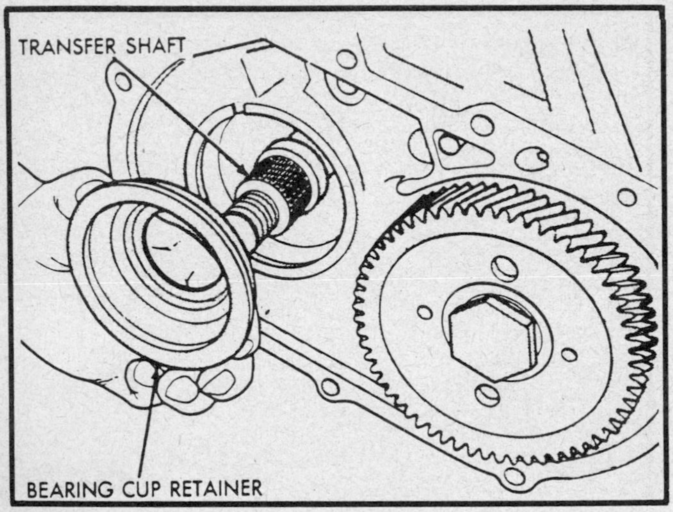
TRANSFER SHAFT
BEARING CUP RETAINER
Transfer shaft bearing cup retainer

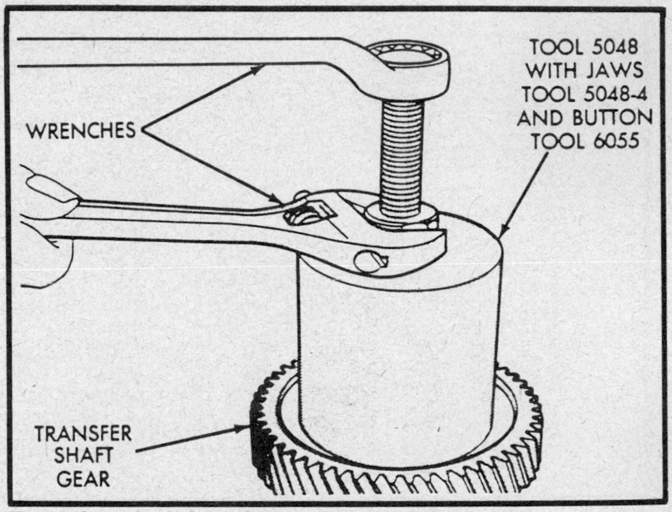
WRENCHES
TOOL 5048 WITH JAWS TOOL 5048-4 AND BUTTON TOOL 6055
TRANSFER SHAFT GEAR
Transfer shaft bearing gear cone removal

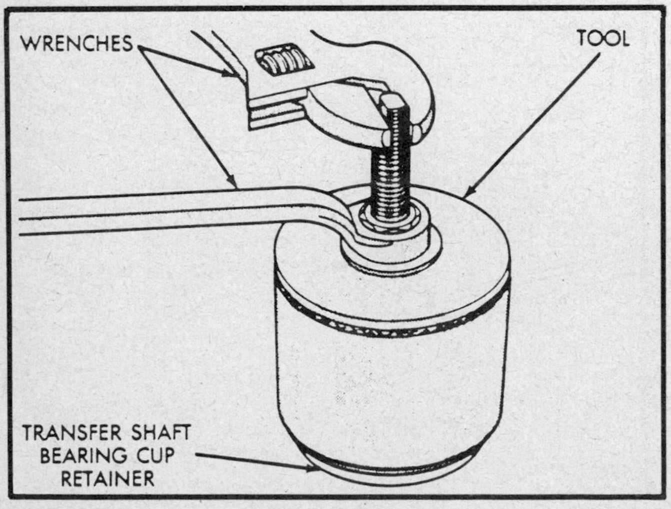
WRENCHES
TOOL
TRANSFER SHAFT BEARING CUP RETAINER
Transfer shaft bearing cup removal

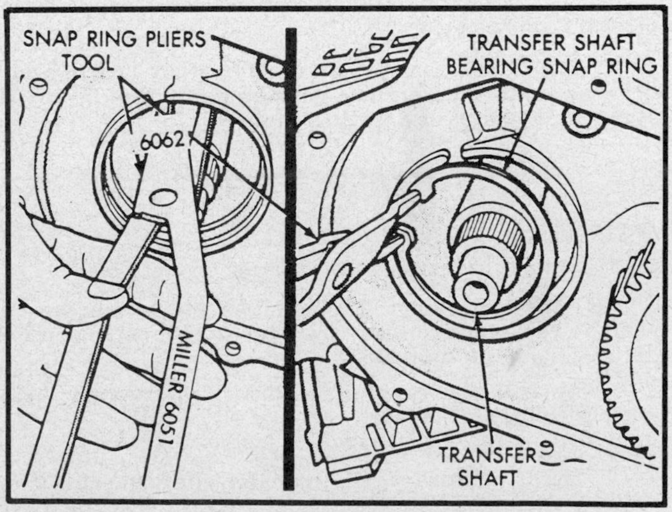
SNAP RING PLIERS
TOOL
6062
MILLER 6051
TRANSFER SHAFT BEARING SNAP RING
TRANSFER SHAFT
Transfer shaft bearing snapring servicing

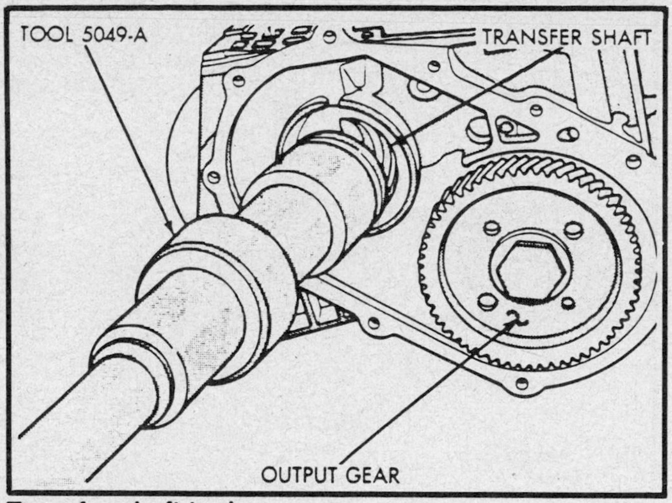
TOOL 5049-A
TRANSFER SHAFT
OUTPUT GEAR

Transfer shaft tool

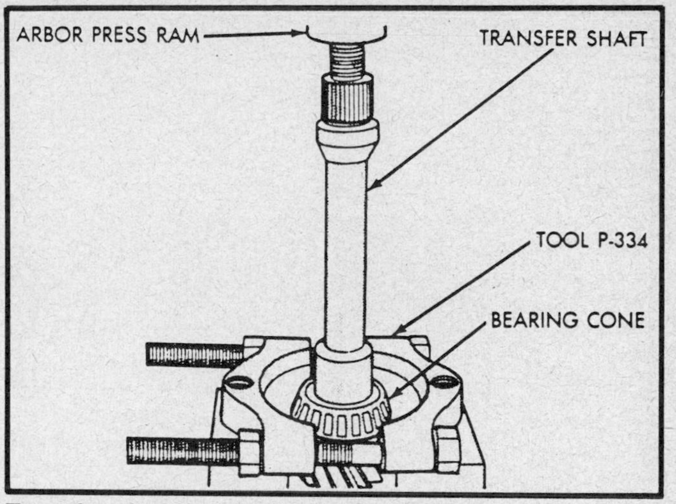
ARBOR PRESS RAM
TRANSFER SHAFT
TOOL P-334
BEARING CONE

Transfer shaft bearing cone removal

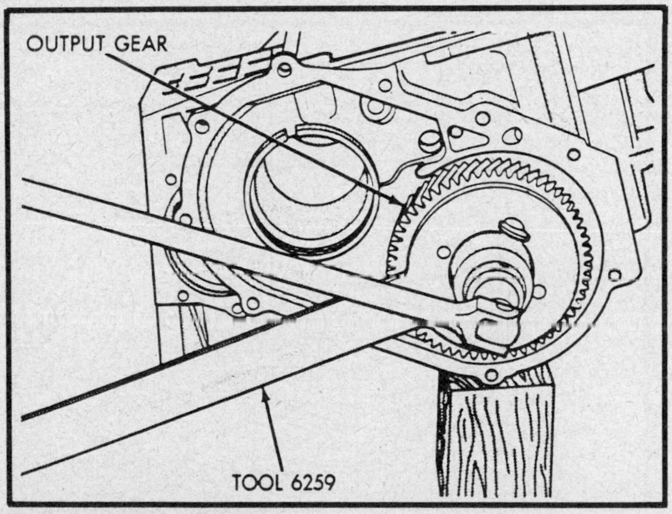
OUTPUT GEAR
TOOL 6259

Output gear bolt removal

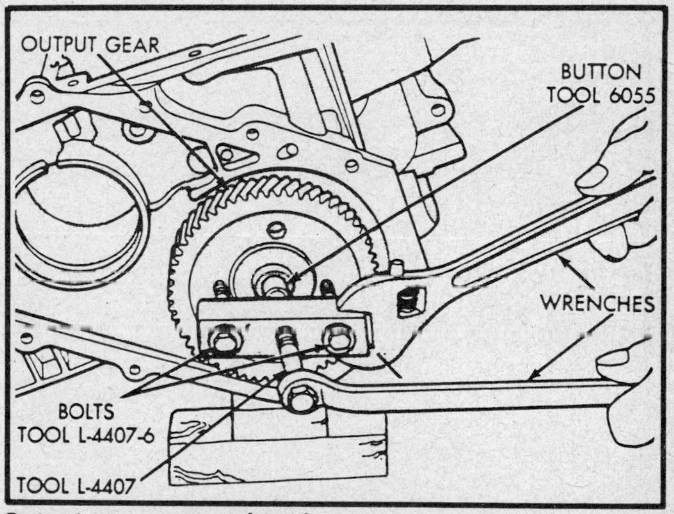
OUTPUT GEAR
BUTTON
TOOL 6055
WRENCHES
BOLTS
TOOL L-4407-6
TOOL L-4407

Output gear removal tool

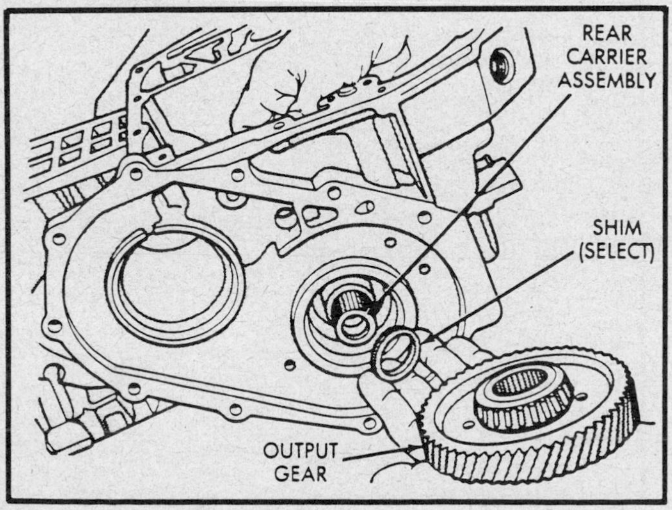
REAR
CARRIER
ASSEMBLY
SHIM
(SELECT)
OUTPUT
GEAR

Output gear and shim

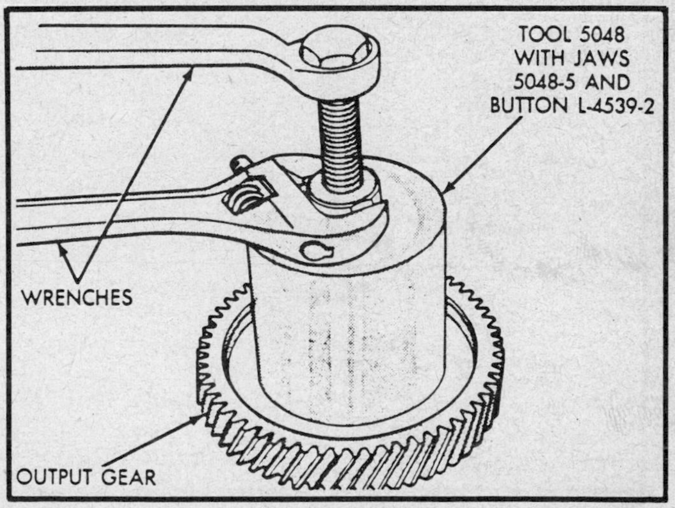
TOOL 5048
WITH JAWS
5048-5 AND
BUTTON L-4539-2
WRENCHES
OUTPUT GEAR

Output gear bearing cone removal

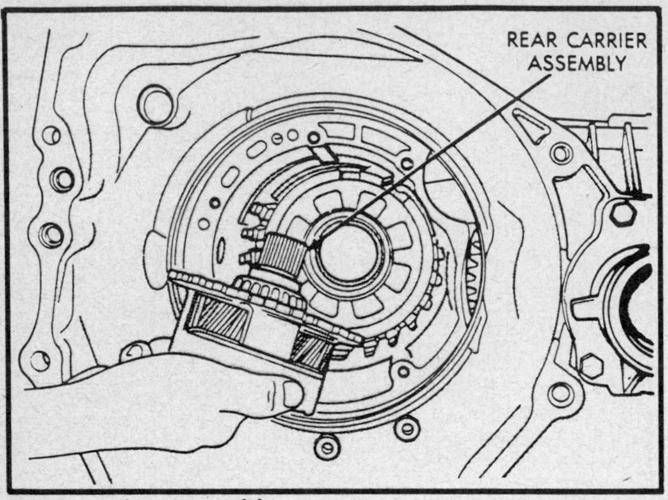
REAR CARRIER ASSEMBLY

Rear carrier assembly

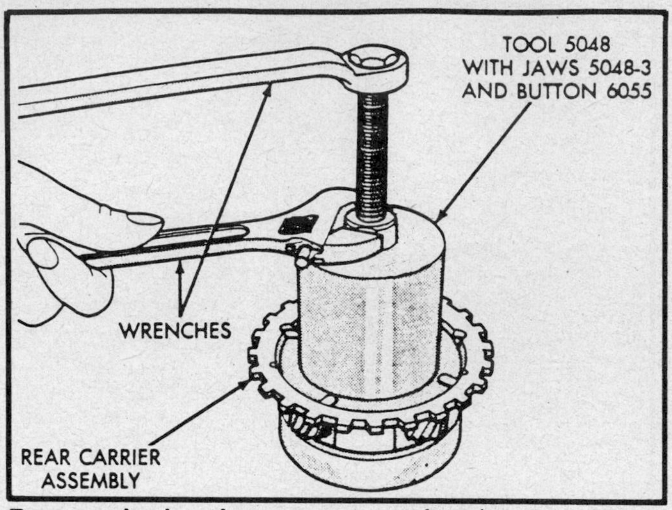
TOOL 5048 WITH JAWS 5048-3 AND BUTTON 6055
WRENCHES
REAR CARRIER ASSEMBLY

Rear carrier bearing cone removal tool

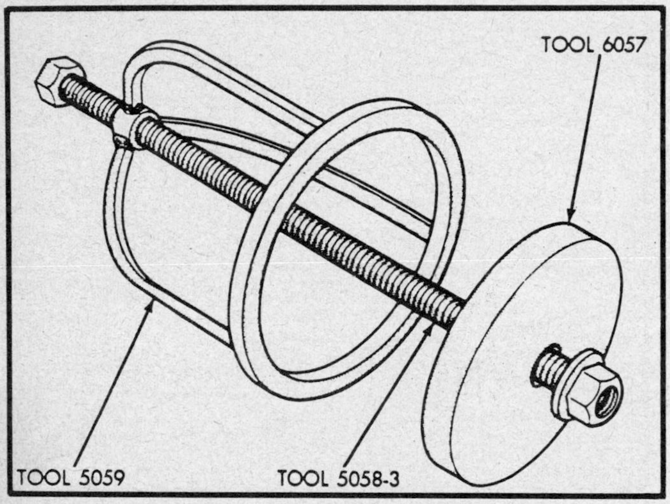
TOOL 6057
TOOL 5059
TOOL 5058-3

Low/reverse spring compressor tool

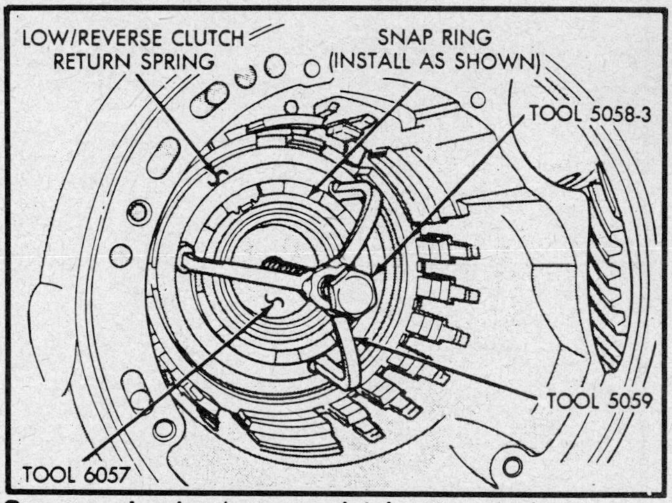
LOW/REVERSE CLUTCH RETURN SPRING
SNAP RING (INSTALL AS SHOWN)
TOOL 5058-3
TOOL 5059
TOOL 6057

Compressing low/reverse clutch return spring

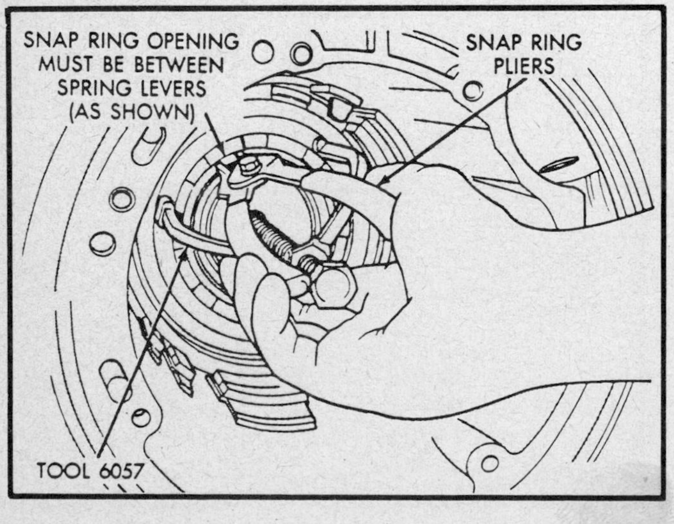
SNAP RING OPENING MUST BE BETWEEN SPRING LEVERS (AS SHOWN)
SNAP RING PLIERS
TOOL 6057

Low/reverse clutch return spring snapring servicing

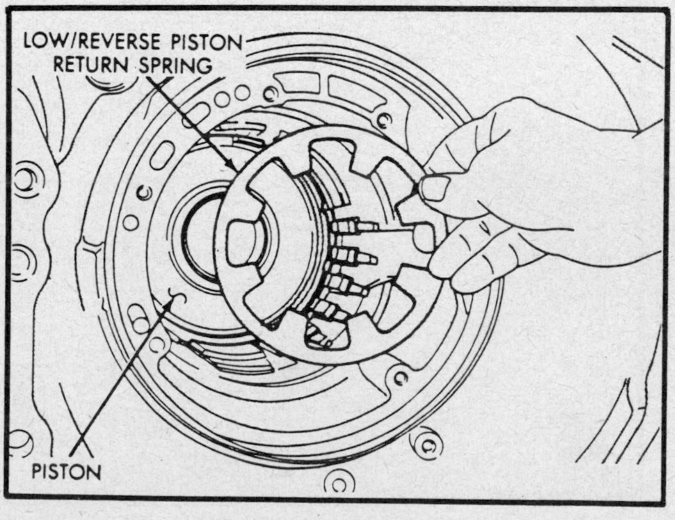
LOW/REVERSE PISTON RETURN SPRING
PISTON

Low/reverse clutch return spring

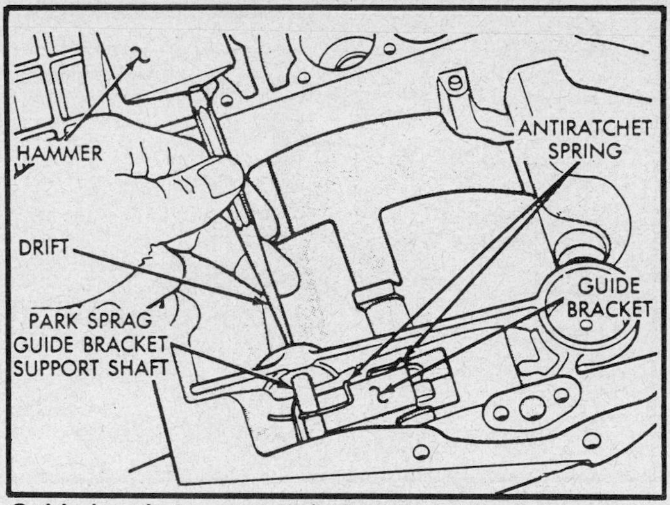
HAMMER
DRIFT
PARK SPRAG
GUIDE BRACKET
SUPPORT SHAFT
ANTIRATCHET
SPRING
GUIDE
BRACKET
Guide bracket support shaft removal

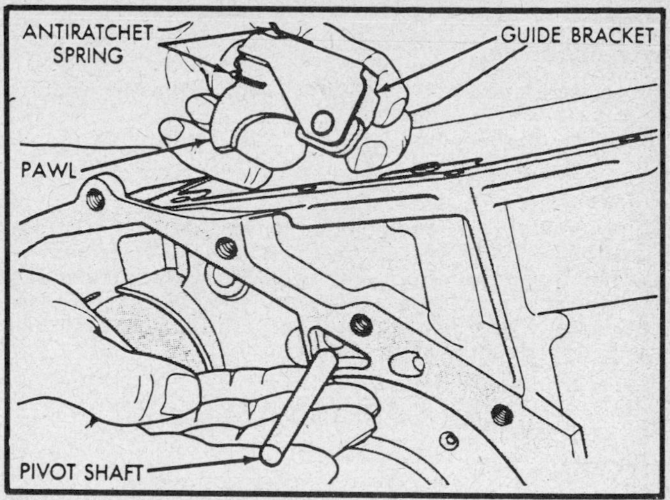
ANTIRATCHET
SPRING
GUIDE BRACKET
PAWL
PIVOT SHAFT
Guide bracket pivot shaft

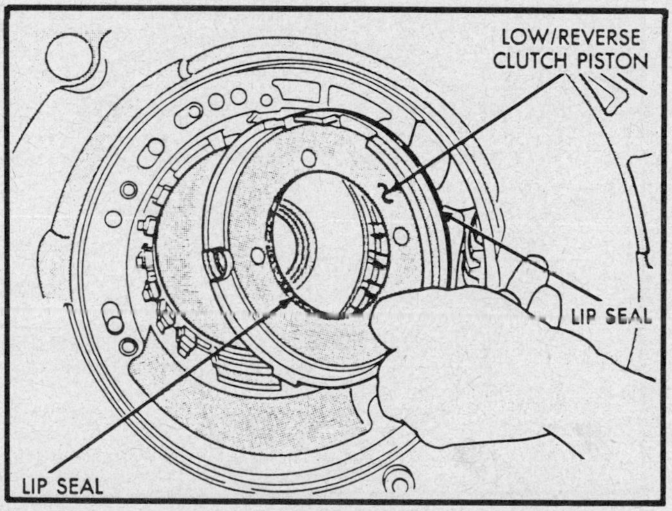
LOW/REVERSE
CLUTCH PISTON
LIP SEAL
LIP SEAL
Low/reverse clutch piston

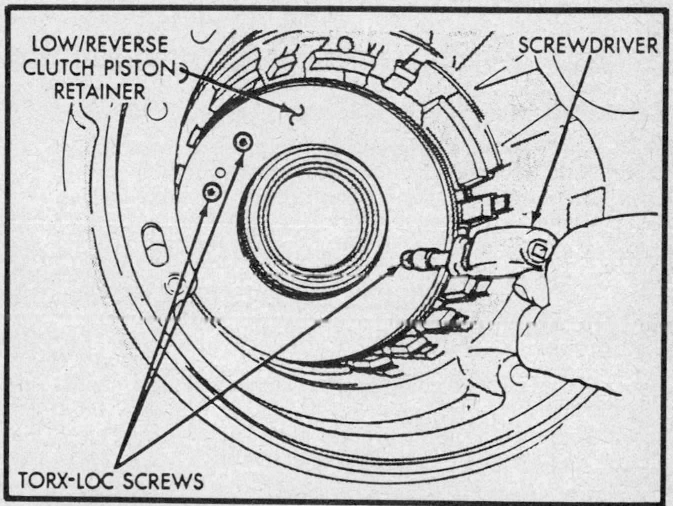
LOW/REVERSE
CLUTCH PISTON
RETAINER
SCREWDRIVER
TORX-LOC SCREWS
Low/reverse clutch piston retainer attaching screws

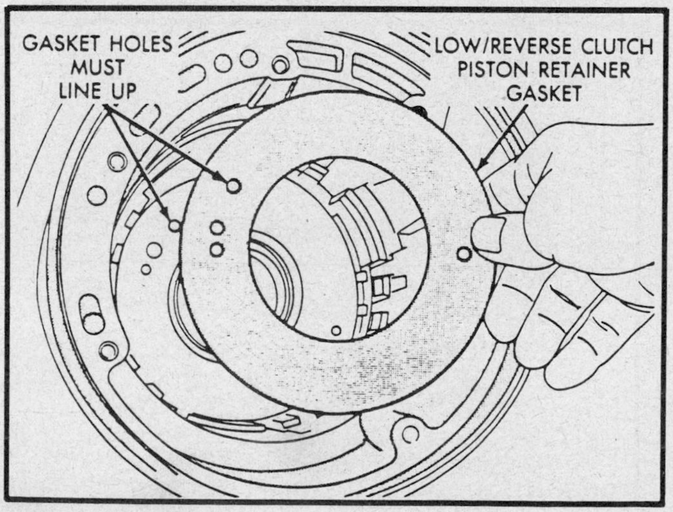
GASKET HOLES
MUST
LINE UP
LOW/REVERSE CLUTCH
PISTON RETAINER
GASKET
Low/reverse clutch piston retainer gasket

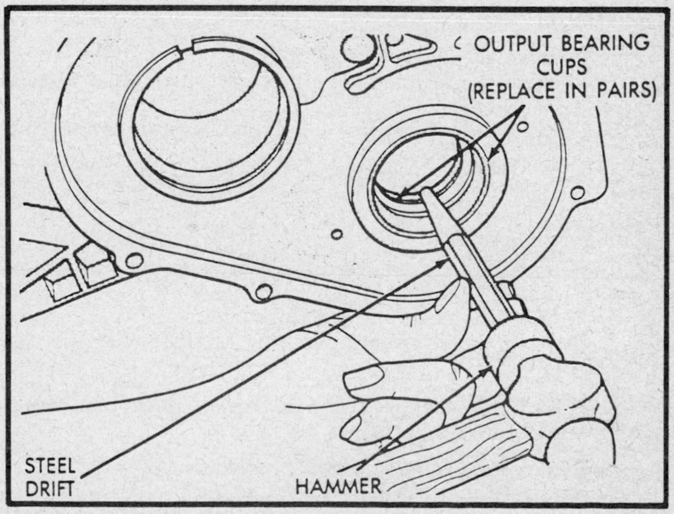
OUTPUT BEARING
CUPS
(REPLACE IN PAIRS)
STEEL
DRIFT
HAMMER
Output gear bearing cups removal

OUTPUT SHAFT GEAR

Removal

1. Remove the output gear bolt and washer by using tool 6259.
2. Remove the output gear and shim with the use of puller tool L–4407, bolts tool L–4407–6 and button tool 6055.

NOTE: If necessary, the output gear bearing cone can be removed from the gear by using tool 5048 with jaws tool 5048–5 and button tool L–4539–2. If necessary, remove the output shaft gear bearing cups with a drift after the rear carrier assembly, and the low/reverse piston and return spring are removed. Make sure to drift bearing cups all the way around.

REAR CARRIER ASSEMBLY

Removal

From the bell housing side, remove the rear carrier assembly.

NOTE: If necessary, the rear carrier bearing cone can be removed from the rear carrier using tool 5048 with jaws tool 5048–3 and button tool 6055.

LOW/REVERSE PISTON RETURN SPRING

Removal

1. Install low/reverse spring compressor tools 5059, 5058–3 and 6057 to compress low/reverse piston return spring.
2. Remove low/reverse piston return spring snapring.
3. Remove low/reverse spring compressor tool.
4. Remove low/reverse piston return spring.

GUIDE BRACKET

Removal

1. Drive out the guide bracket support shaft and plug with a drift.
2. Remove the guide bracket pivot shaft with a pair of pliers.
3. Remove the guide bracket assembly.

LOW/REVERSE CLUTCH PISTON AND RETAINER

Removal

1. Remove the low/reverse clutch piston.
2. Remove the low/reverse clutch piston retainer attaching screws.
3. Remove the low/reverse clutch piston retainer and gasket.

DIFFERENTIAL ASSEMBLY

Removal

1. Remove the oil seal from the extension housing.
2. Remove the differential cover bolts.
3. Remove the differential retainer bolts.
4. Remove the differential bearing retainer with the use of tool L–4435.

NOTE: If necessary, remove the oil seal from the differential bearing retainer.

5. Remove the extension housing bolts.
6. Remove the extension housing using tool L–4435.
7. Remove the differential assembly from the transaxle assembly.

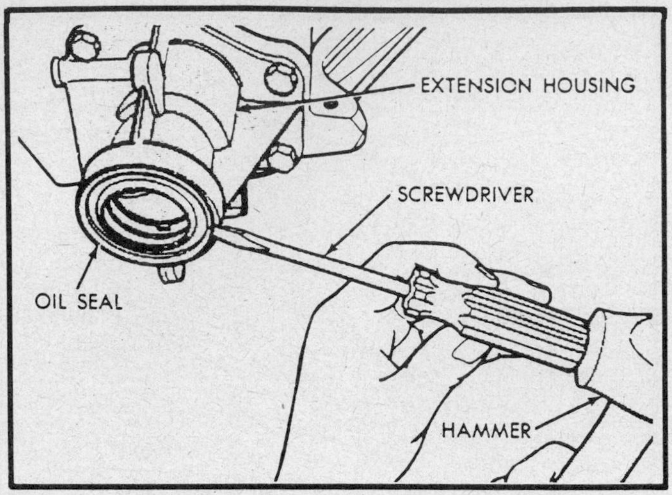
EXTENSION HOUSING
SCREWDRIVER
OIL SEAL
HAMMER

Extension housing oil seal removal

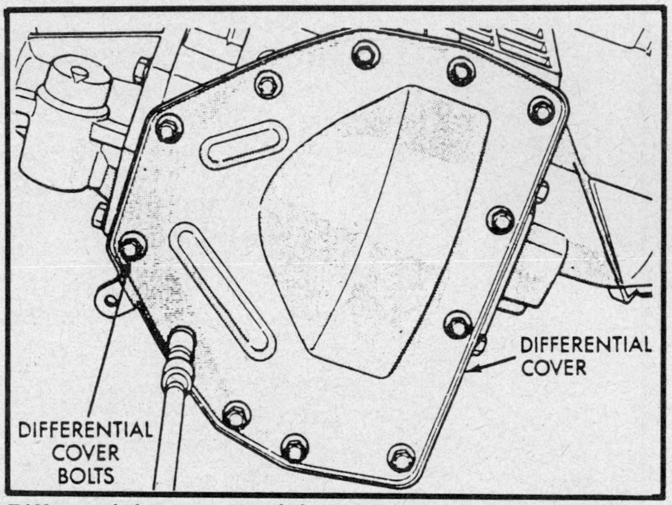
DIFFERENTIAL COVER
DIFFERENTIAL COVER BOLTS

Differential cover servicing

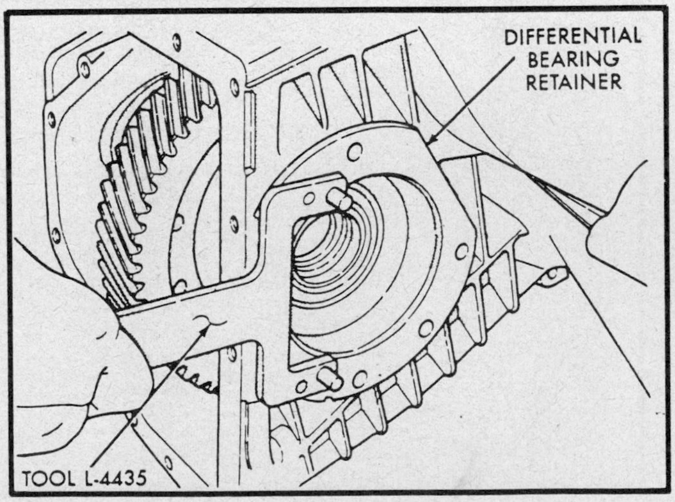
DIFFERENTIAL BEARING RETAINER
TOOL L-4435

Differential bearing retainer servicing

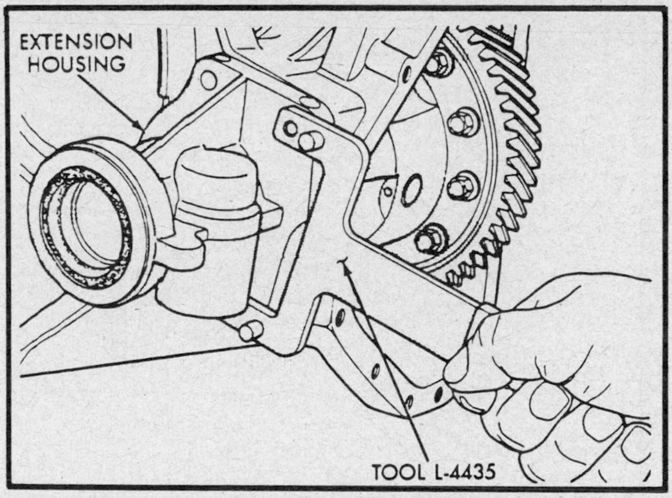
EXTENSION
HOUSING
TOOL L-4435

Differential extension housing servicing

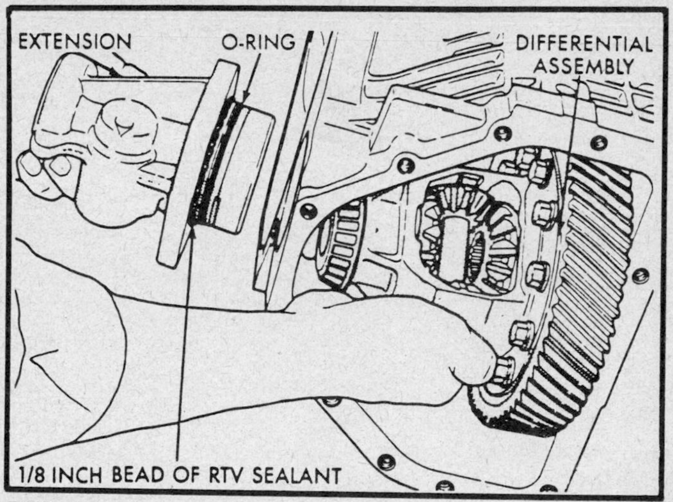
EXTENSION
O-RING
DIFFERENTIAL
ASSEMBLY
1/8 INCH BEAD OF RTV SEALANT

Differential assembly servicing

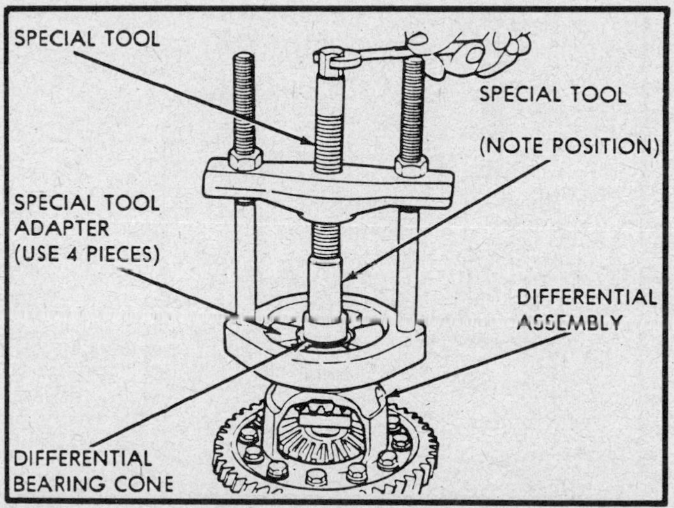
SPECIAL TOOL
SPECIAL TOOL
(NOTE POSITION)
SPECIAL TOOL
ADAPTER
(USE 4 PIECES)
DIFFERENTIAL
ASSEMBLY
DIFFERENTIAL
BEARING CONE

Differential assembly bearing cone (side gear side) removal

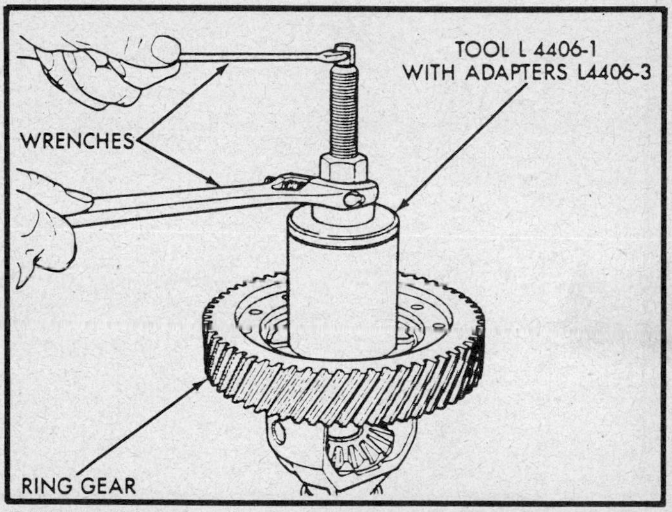
TOOL L 4406-1
WITH ADAPTERS L4406-3
WRENCHES
RING GEAR

Differential assembly bearing cone (ring gear side) removal

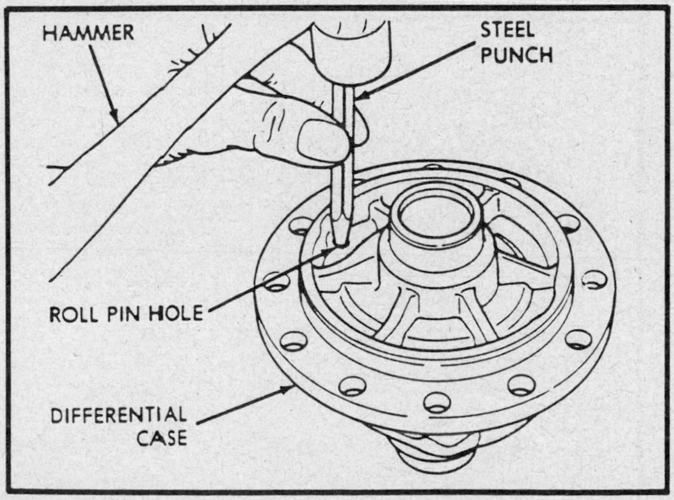
HAMMER
STEEL
PUNCH
ROLL PIN HOLE
DIFFERENTIAL
CASE

Differential pinion shaft roll pin removal

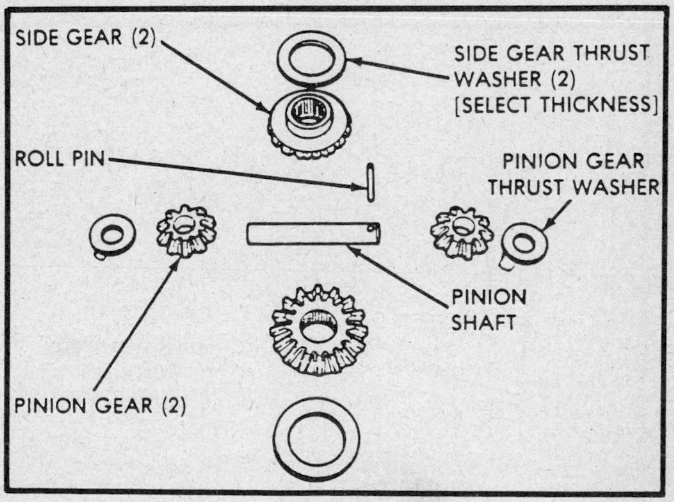
SIDE GEAR (2)
SIDE GEAR THRUST
WASHER (2)
[SELECT THICKNESS]
ROLL PIN
PINION GEAR
THRUST WASHER
PINION
SHAFT
PINION GEAR (2)

Differential gears

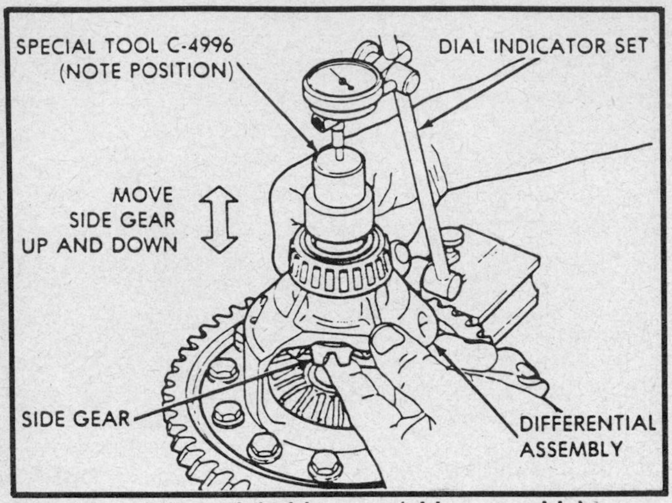
SPECIAL TOOL C-4996
(NOTE POSITION)
DIAL INDICATOR SET
MOVE
SIDE GEAR
UP AND DOWN
SIDE GEAR
DIFFERENTIAL
ASSEMBLY

Checking differential side gear (side gear side) endplay

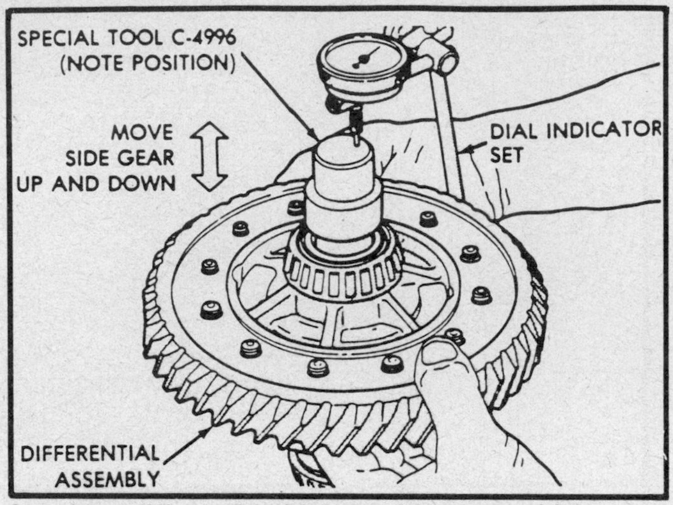
SPECIAL TOOL C-4996
(NOTE POSITION)
MOVE
SIDE GEAR
UP AND DOWN
DIAL INDICATOR
SET
DIFFERENTIAL
ASSEMBLY

Checking differential side gear (ring gear side) endplay

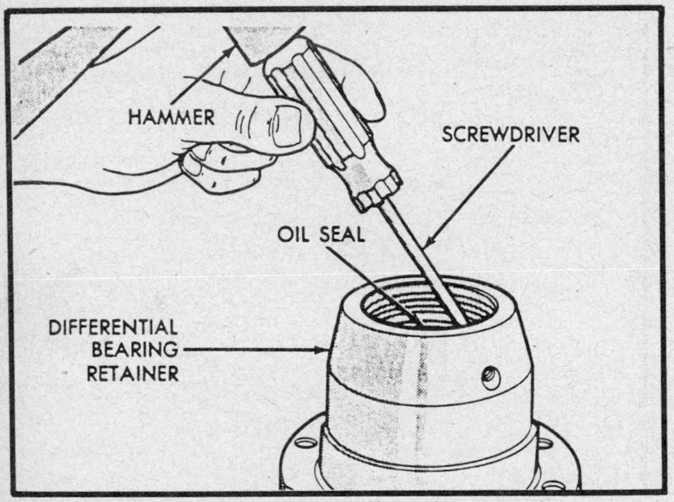
HAMMER
SCREWDRIVER
OIL SEAL
DIFFERENTIAL
BEARING
RETAINER

Differential bearing retainer oil seal removal

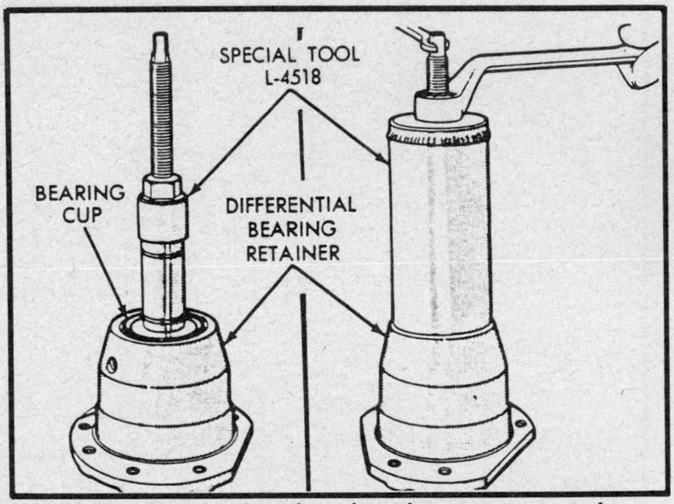
SPECIAL TOOL
L-4518
BEARING
CUP
DIFFERENTIAL
BEARING
RETAINER

Differential bearing retainer bearing cup removal

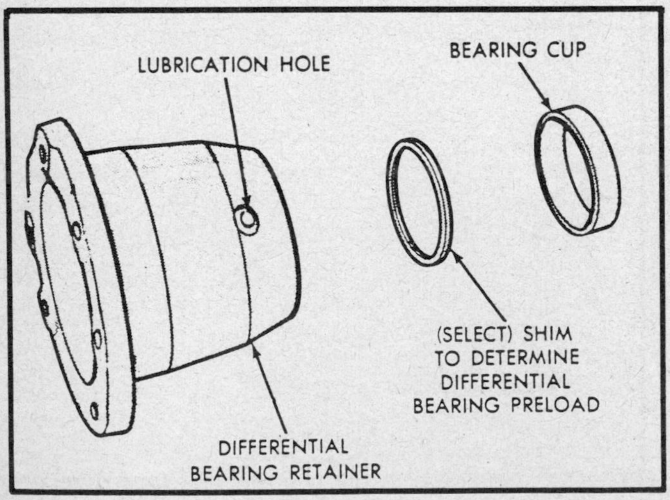
LUBRICATION HOLE
BEARING CUP
(SELECT) SHIM
TO DETERMINE
DIFFERENTIAL
BEARING PRELOAD
DIFFERENTIAL
BEARING RETAINER

Differential bearing retainer bearing cup and shim

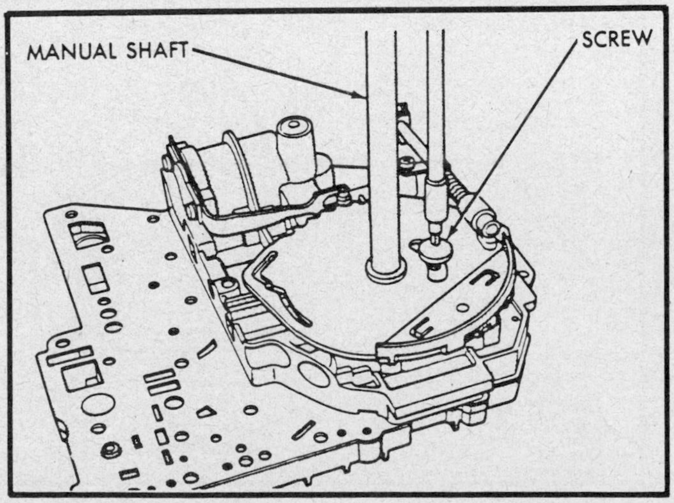
MANUAL SHAFT
SCREW

Manual shaft servicing

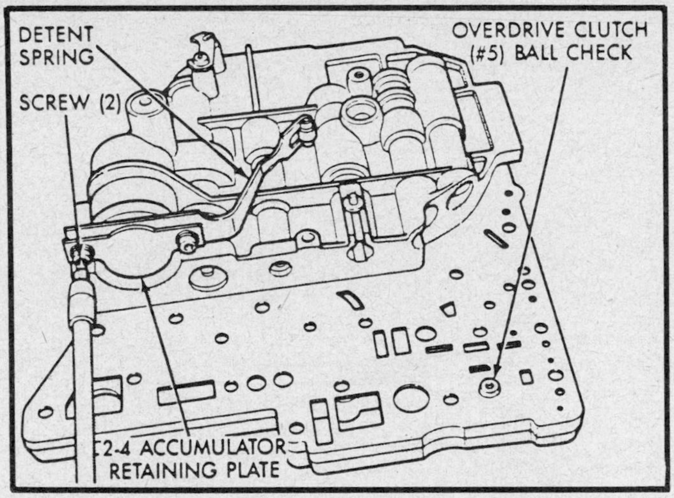
DETENT SPRING
OVERDRIVE CLUTCH (#5) BALL CHECK
SCREW (2)
2-4 ACCUMULATOR RETAINING PLATE
2/4 accumulator retainer plate screws location

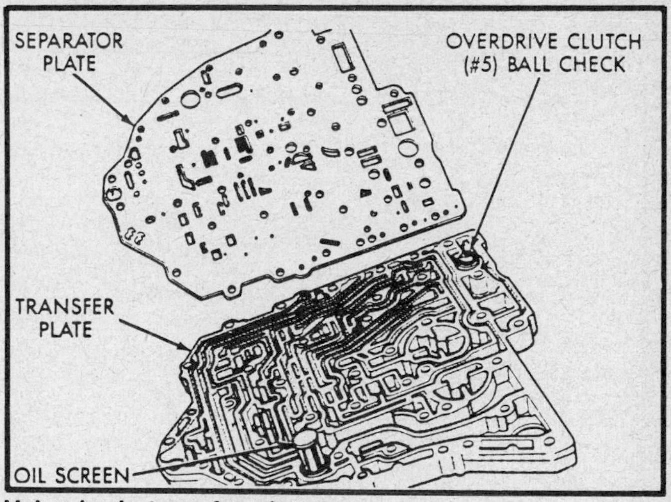
SEPARATOR PLATE
OVERDRIVE CLUTCH (#5) BALL CHECK
TRANSFER PLATE
OIL SCREEN
Valve body transfer plate

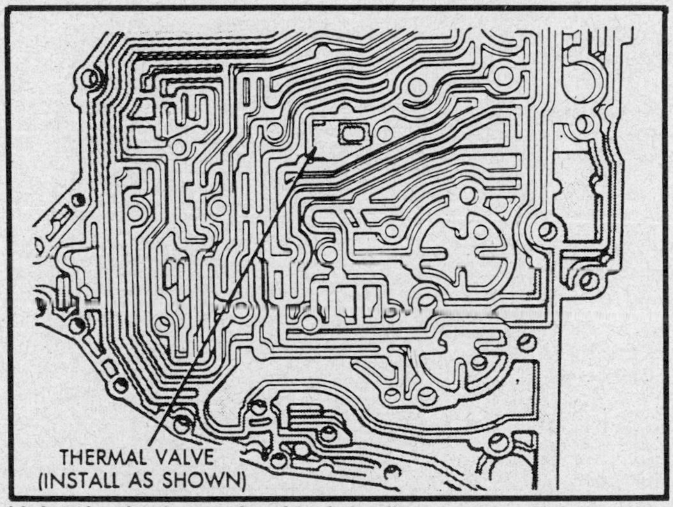
THERMAL VALVE (INSTALL AS SHOWN)
Valve body thermal valve location

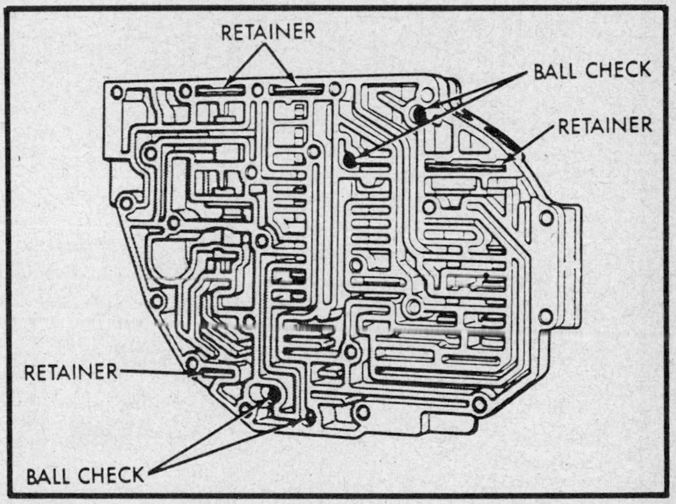
RETAINER
BALL CHECK
RETAINER
RETAINER
BALL CHECK
Valve body ball check and retainer locations

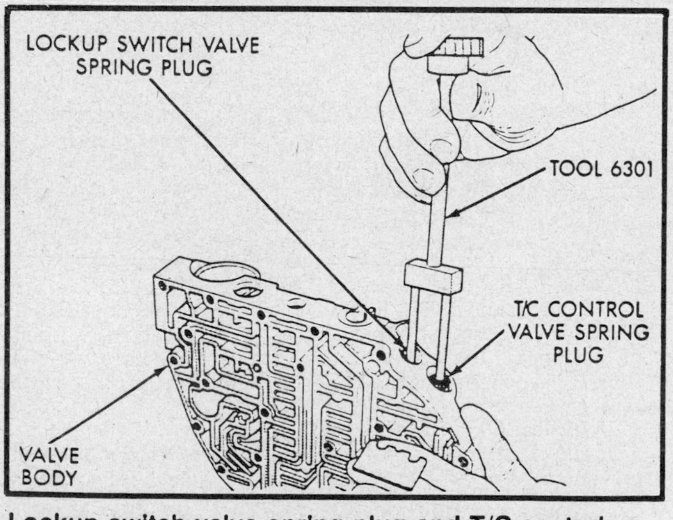
LOCKUP SWITCH VALVE SPRING PLUG
TOOL 6301
T/C CONTROL VALVE SPRING PLUG
VALVE BODY
Lockup switch valve spring plug and T/C control valve spring plug servicing

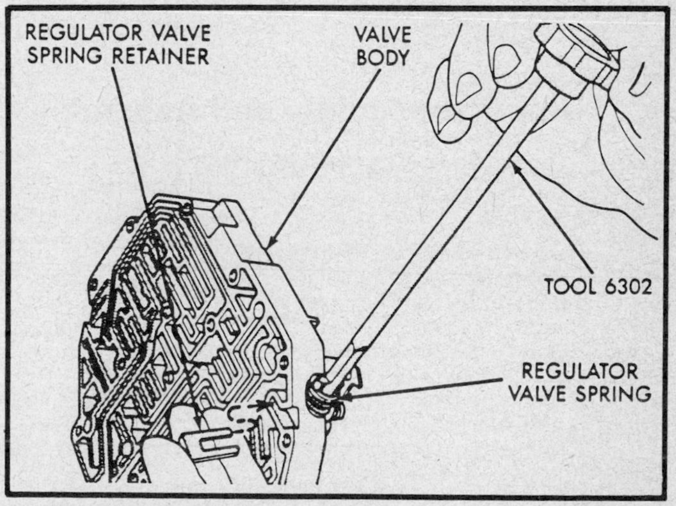
REGULATOR VALVE SPRING RETAINER
VALVE BODY
TOOL 6302
REGULATOR VALVE SPRING
Regulator valve spring servicing

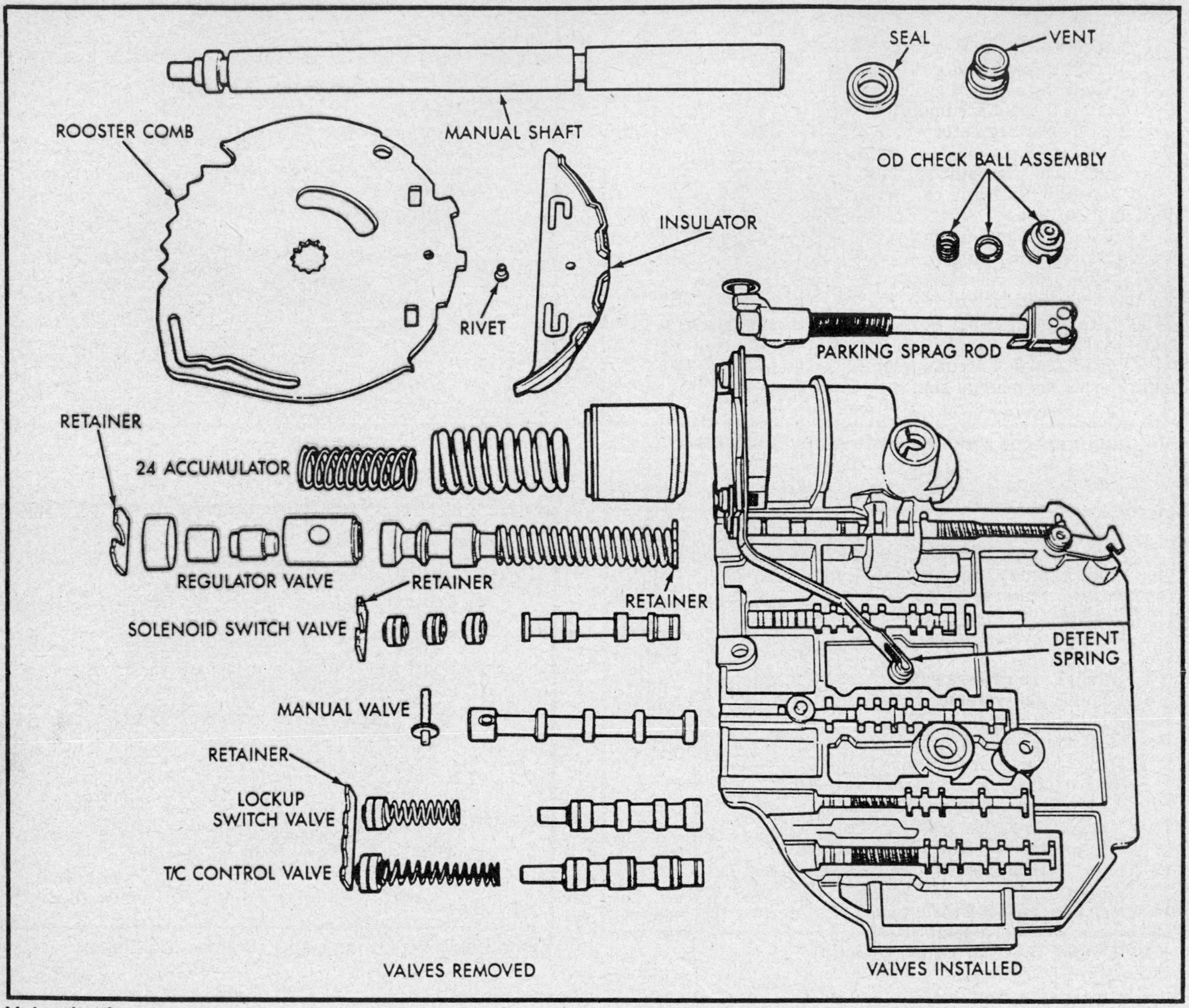
SEAL
VENT
ROOSTER COMB
MANUAL SHAFT
OD CHECK BALL ASSEMBLY
INSULATOR
RIVET
PARKING SPRAG ROD
RETAINER
24 ACCUMULATOR
REGULATOR VALVE
RETAINER
RETAINER
SOLENOID SWITCH VALVE
DETENT SPRING
MANUAL VALVE
RETAINER
LOCKUP SWITCH VALVE
T/C CONTROL VALVE
VALVES REMOVED
VALVES INSTALLED

Valve body components

Unit Disassembly and Assembly

VALVE BODY

Disassembly

1. Remove the manual shaft screw.
2. Remove the 2–4 accumulator retaining plate screws.
3. Separate the transfer plate and the separator plate.
4. Remove the thermal valve from the transfer plate.
5. Remove the dual retainer plate from the torque converter control valve and lockup switch valve.
6. Remove the torque converter control valve and lockup switch valve spring plugs using tool 6301.
7. Remove the torque converter control valve and lockup switch valve.
8. Remove the regulator valve retainer plate.
9. Remove the regulator valve and valve spring using tool 6302.
10. Remove the 2–4 accumulator and accumulator springs.
11. Remove the solenoid switch valve retainer plate.
12. Remove the solenoid switch valve.
13. Remove the manual valve.

Inspection

1. Inspect all valves and plug bores for scores. Check all fluid passages for obstructions. Inspect the check valves for freedom of movement. Inspect all mating surfaces for burrs or distortion. If needed, polish the valves and plugs with crocus cloth to remove minor burrs or scores.
2. Inspect all springs for distortion.
3. Check all valves and plugs in their bores for freedom of movement. Valves and plugs, when dry, must fall from their own weight in their respective bores.
4. Roll the manual valve on a flat surface to check for a bent condition.

Assembly

1. Install the manual valve.
2. Install the solenoid switch valve.
3. Install the solenoid switch valve retainer plate.
4. Install the 2–4 accumulator and accumulator springs.
5. Install the regulator valve and valve spring using tool 6302.
6. Install the regulator valve retainer plate.
7. Install the torque converter control valve and lockup switch valve.
8. Install the torque converter control valve and lockup switch valve spring plugs using tool 6301.
9. Install the dual retainer plate from the torque converter control valve and lockup switch valve.
10. Install the thermal valve from the transfer plate.
11. Separate the transfer plate and the separator plate.
12. Install the 2–4 accumulator retaining plate screws.
13. Install the manual shaft screw.

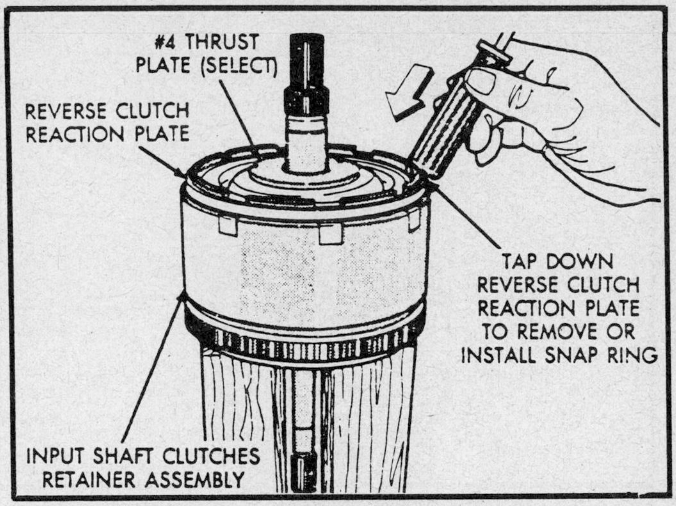
#4 THRUST
PLATE (SELECT)
REVERSE CLUTCH
REACTION PLATE
TAP DOWN
REVERSE CLUTCH
REACTION PLATE
TO REMOVE OR
INSTALL SNAP RING
INPUT SHAFT CLUTCHES
RETAINER ASSEMBLY

Reverse clutch reaction plate tap down

INPUT SHAFT CLUTCHES RETAINER ASSEMBLY

Disassembly

1. With the input shaft clutches retainer assembly in a suitable holding fixture, tap down the reverse clutch reaction plate to remove the reverse clutch snapring.
2. Remove the reverse clutch snapring.
3. Pry the reverse clutch reaction plate up with a small pry tool.
4. Remove the reverse clutch reaction plate.
5. Remove the reverse clutch pack. Tag and identify the clutch packs to assure original replacement.
6. Remove the overdrive/reverse pressure plate snapring.
7. Remove the overdrive/reverse pressure plate.
8. Remove the overdrive/reverse clutch waved snapring.
9. Remove the overdrive shaft assembly and overdrive clutch pack with No. 3 thrust plate.
10. Remove the overdrive clutch pack from the overdrive shaft assembly. Tag and identify the clutch packs to assure original replacement.
11. Remove the No. 3 thrust plate and No. 4 thrust plate from the overdrive shaft assembly.
12. Remove the No. 3 thrust washer and underdrive shaft assembly.
13. Remove the No. 2 needle bearing.
14. Remove the overdrive/underdrive clutches reaction plate tapered snapring with a small pry tool. Do not scratch the reaction plate.
15. Remove the overdrive/underdrive clutch reaction plate.
16. Remove the underdrive clutch disc.
17. Remove the underdrive clutch reaction plate flat snapring.
18. Remove the underdrive clutch pack. Tag and identify clutch packs to assure original replacement.
19. Install tool 5059 with arbor press and remove underdrive retainer snapring.
20. Remove the underdrive return spring retainer and the piston return spring.
21. Remove the underdrive clutch piston.
22. Remove the input hub tapered snapring.
23. With a plastic hammer, tap on the input hub and remove the input shaft and hub assembly.
24. Remove the input clutches retainer from the overdrive/reverse piston.
25. With an arbor press ram, compress the return spring on the overdrive/reverse piston just enough to remove the snapring.
26. Remove the snapring and return spring from the overdrive/reverse piston.

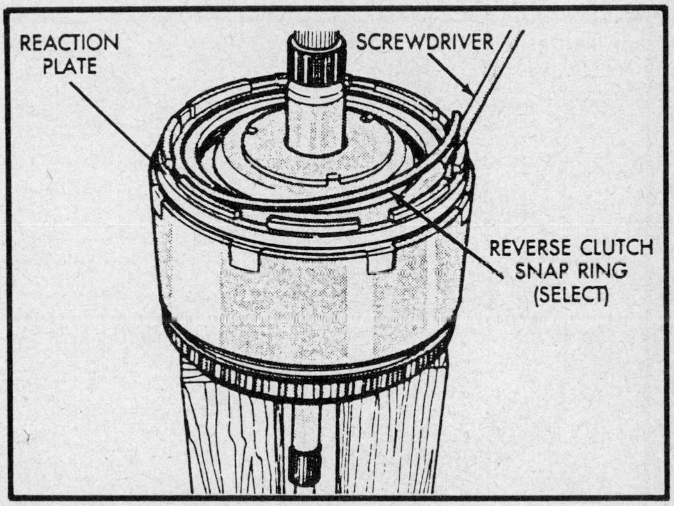
REACTION
PLATE
SCREWDRIVER
REVERSE CLUTCH
SNAP RING
(SELECT)

Reverse clutch snapring servicing

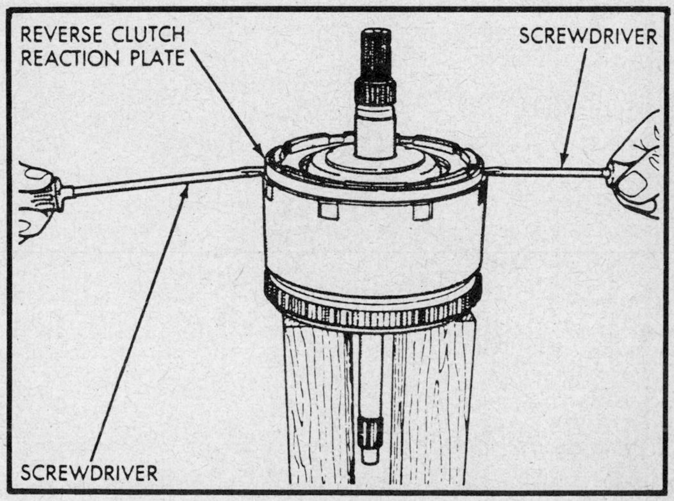
REVERSE CLUTCH
REACTION PLATE
SCREWDRIVER
SCREWDRIVER

Reverse clutch reaction plate servicing

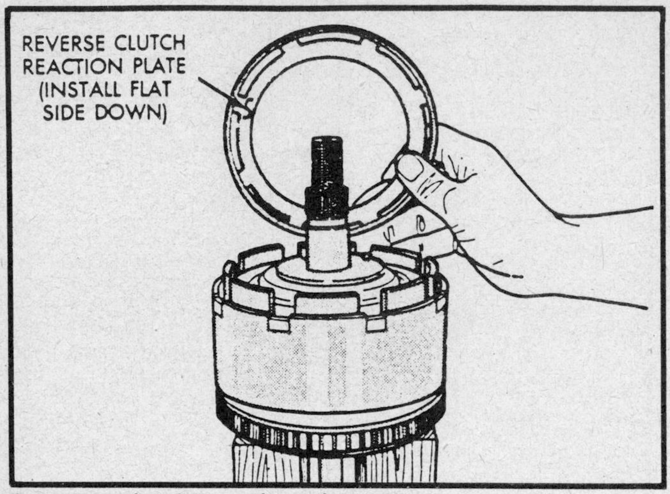
REVERSE CLUTCH
REACTION PLATE
(INSTALL FLAT
SIDE DOWN)

Reverse clutch reaction plate

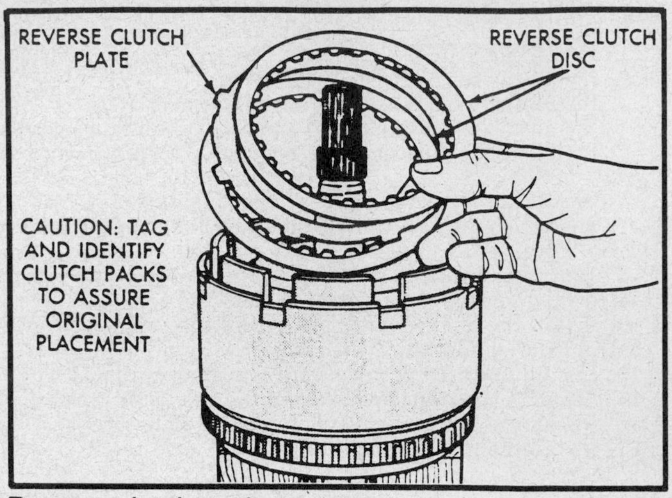
REVERSE CLUTCH
PLATE
REVERSE CLUTCH
DISC
CAUTION: TAG
AND IDENTIFY
CLUTCH PACKS
TO ASSURE
ORIGINAL
PLACEMENT

Reverse clutch pack servicing

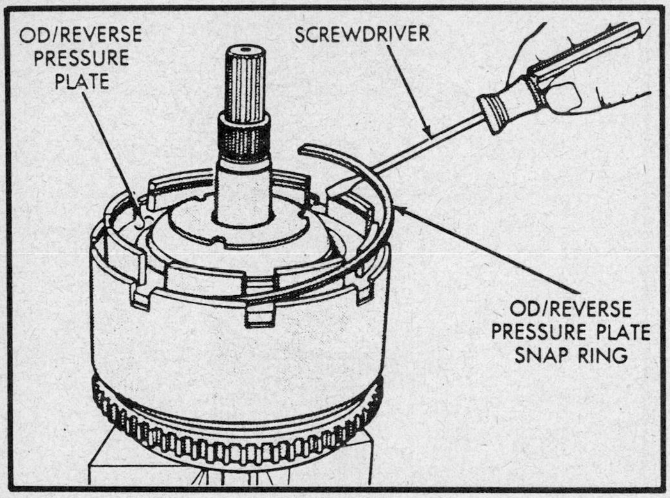
OD/REVERSE
PRESSURE
PLATE
SCREWDRIVER
OD/REVERSE
PRESSURE PLATE
SNAP RING

Overdrive/reverse pressure plate snapring servicing

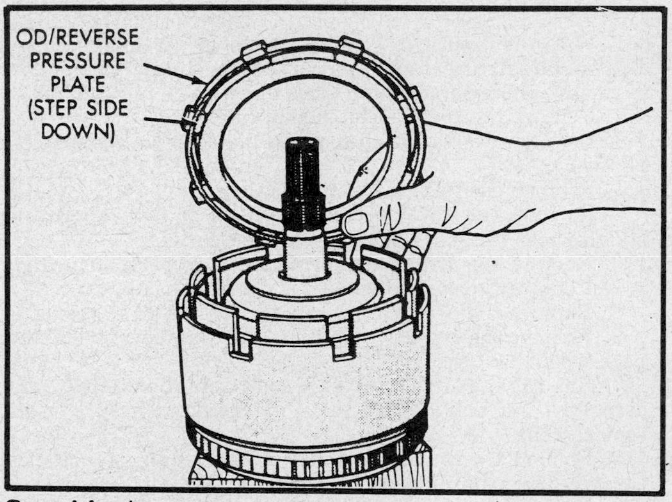
OD/REVERSE
PRESSURE
PLATE
(STEP SIDE
DOWN)

Overdrive/reverse pressure plate

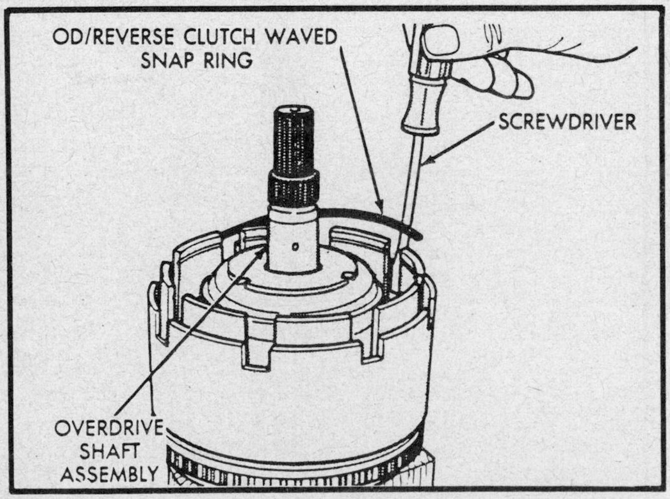
OD/REVERSE CLUTCH WAVED
SNAP RING
SCREWDRIVER
OVERDRIVE
SHAFT
ASSEMBLY

Overdrive/reverse clutch waved snapring servicing

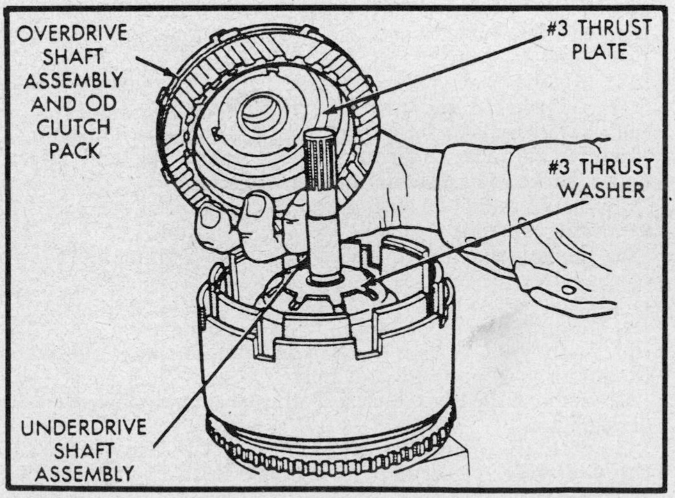
OVERDRIVE
SHAFT
ASSEMBLY
AND OD
CLUTCH
PACK
#3 THRUST
PLATE
#3 THRUST
WASHER
UNDERDRIVE
SHAFT
ASSEMBLY

Overdrive shaft assembly and overdrive clutch pack servicing

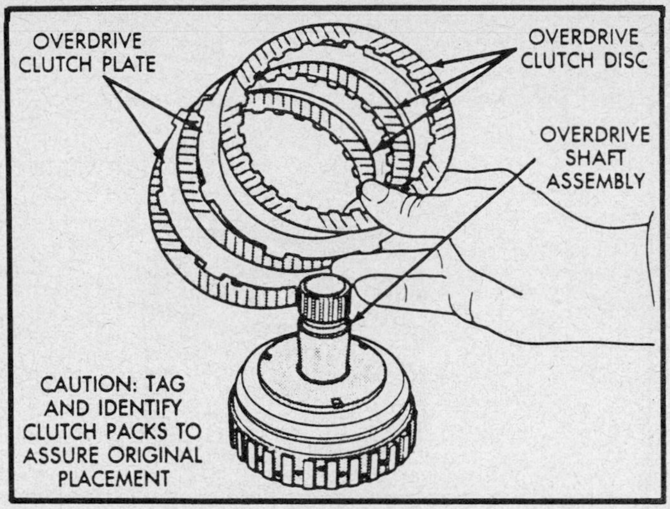
OVERDRIVE CLUTCH PLATE
OVERDRIVE CLUTCH DISC
OVERDRIVE SHAFT ASSEMBLY
CAUTION: TAG AND IDENTIFY CLUTCH PACKS TO ASSURE ORIGINAL PLACEMENT

Overdrive clutch pack servicing

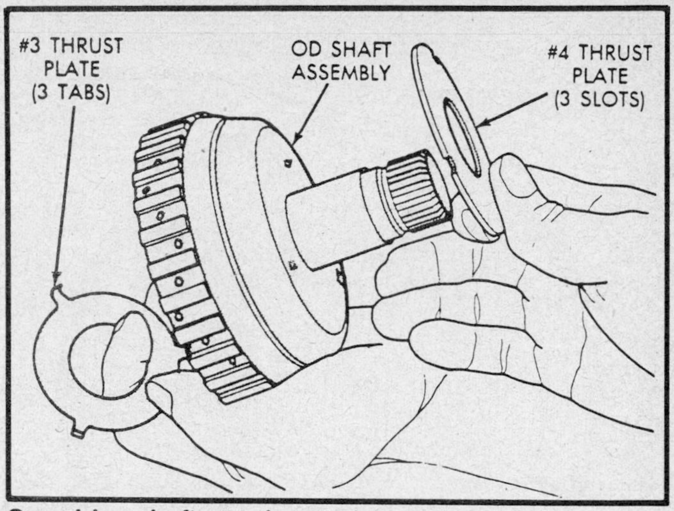
#3 THRUST PLATE (3 TABS)
OD SHAFT ASSEMBLY
#4 THRUST PLATE (3 SLOTS)

Overdrive shaft, #4 thrust plate and #3 thrust plate

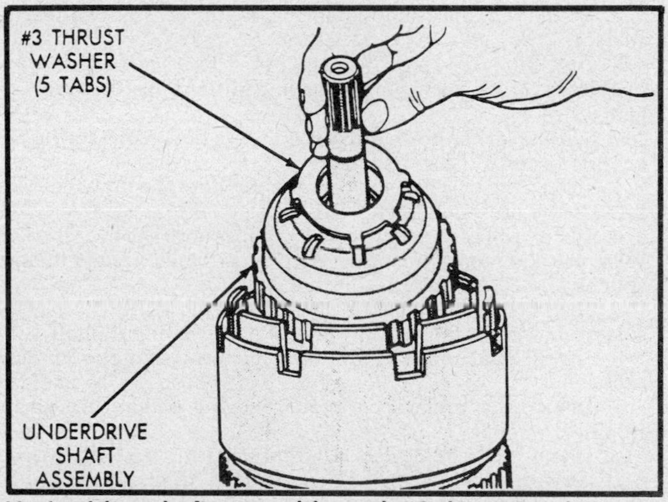
#3 THRUST WASHER (5 TABS)
UNDERDRIVE SHAFT ASSEMBLY

Underdrive shaft assembly and #3 thrust washer

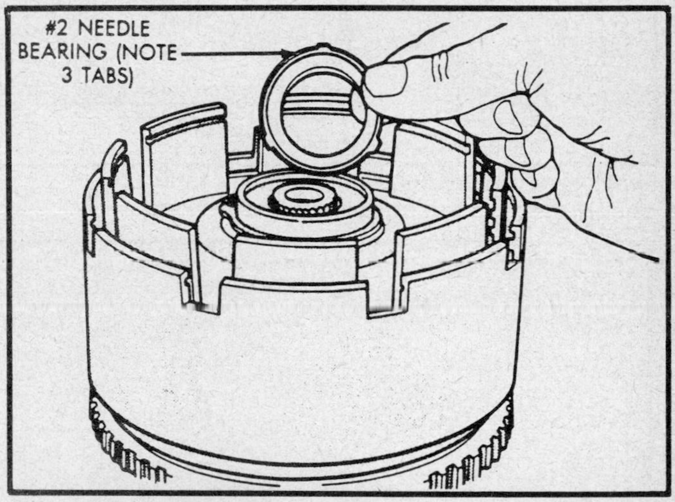
#2 NEEDLE BEARING (NOTE 3 TABS)

#2 needle bearing assembly

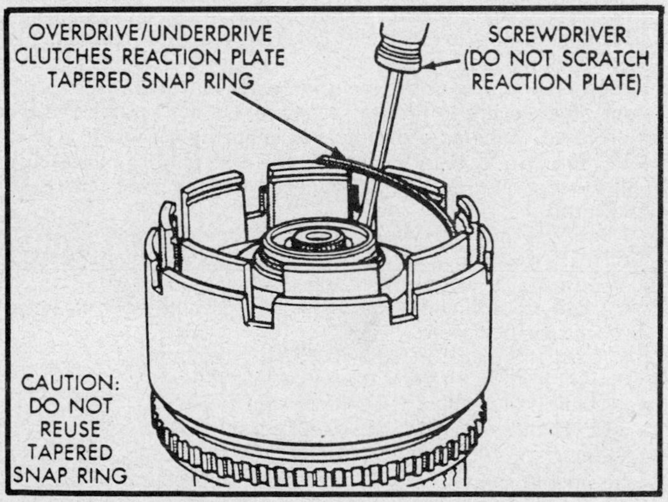
OVERDRIVE/UNDERDRIVE CLUTCHES REACTION PLATE TAPERED SNAP RING
SCREWDRIVER (DO NOT SCRATCH REACTION PLATE)
CAUTION: DO NOT REUSE TAPERED SNAP RING

Overdrive/underdrive reaction plate tapered snapring servicing

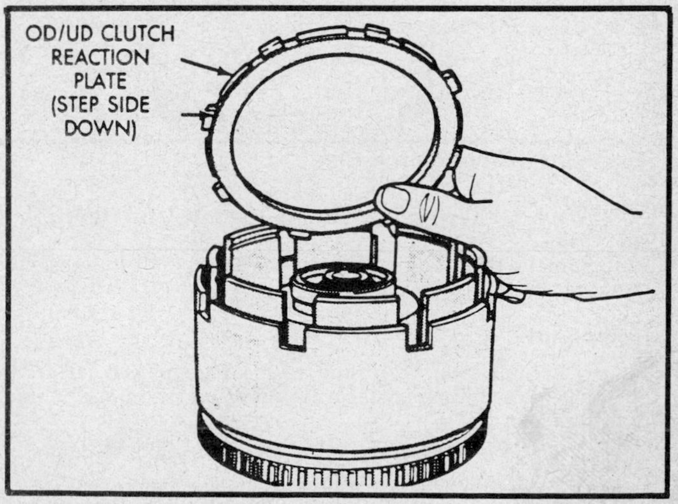
OD/UD CLUTCH REACTION PLATE (STEP SIDE DOWN)

Overdrive/underdrive reaction plate

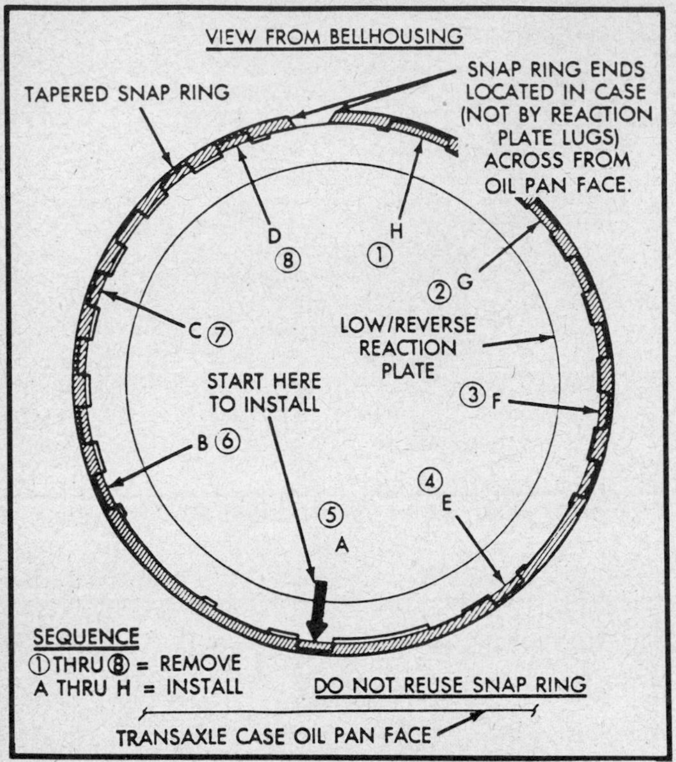
VIEW FROM BELLHOUSING
TAPERED SNAP RING
SNAP RING ENDS LOCATED IN CASE (NOT BY REACTION PLATE LUGS) ACROSS FROM OIL PAN FACE.
D ⑧
H ①
② G
C ⑦
LOW/REVERSE REACTION PLATE
START HERE TO INSTALL
③ F
B ⑥
④ E
⑤ A
SEQUENCE
① THRU ⑧ = REMOVE
A THRU H = INSTALL
DO NOT REUSE SNAP RING
TRANSAXLE CASE OIL PAN FACE

Overdrive/underdrive reaction plate tapered snapring servicing sequence instructions

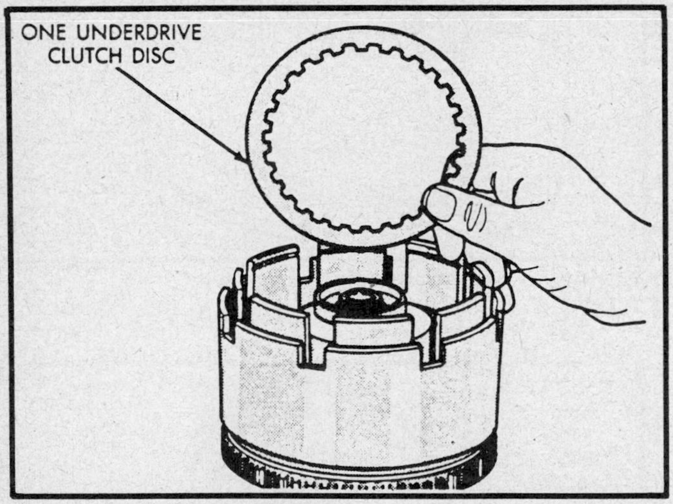
ONE UNDERDRIVE CLUTCH DISC

One underdrive clutch disc

27. With a sharp pointed tool, remove the input shaft snapring.
28. Remove the input shaft from the input shaft hub assembly with an arbor press ram.

Inspection

Inspect splines, gear teeth and all mating surfaces for scores, pitting, chips and abnormal wear. Replace the components as required.

Assembly

NOTE: To ease assembly of components, use petrolatum on all seals.

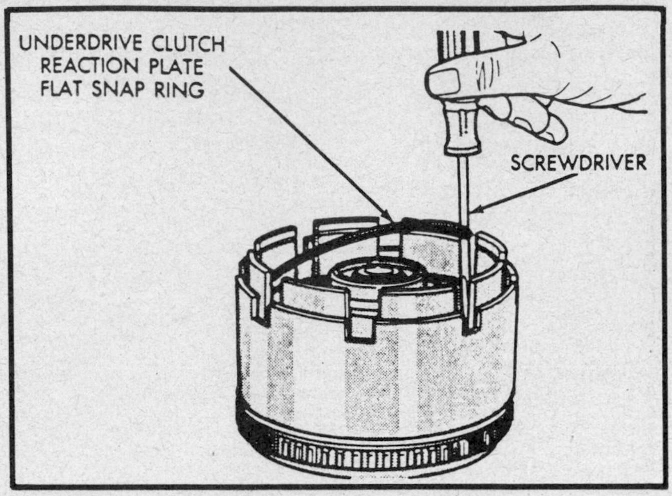
UNDERDRIVE CLUTCH REACTION PLATE FLAT SNAP RING
SCREWDRIVER

Underdrive clutch flat snapring servicing

1. Press the input shaft into the input shaft hub assembly with a arbor press ram.
2. Install the input shaft snapring. Do not scratch the bearing surface.
3. Install new O-rings into the overdrive/reverse piston and install the return spring and snapring into the overdrive/reverse piston.
4. With an arbor press ram to compress the return spring just enough to install the snapring.
5. Install the overdrive/reverse piston onto the input clutches retainer by pressing down with both hands.
6. Install the input shaft hub assembly onto the overdrive/reverse piston by pressing down with both hands. Rotate to align splines.
7. Install the input hub tapered snapring (with the tapered side up and with tabs in the cavity) onto the input shaft.
8. Install the underdrive clutch piston. Note the antispin tabs.
9. Install the underdrive return spring and return spring retainer.
10. Install tools 5067 and 5059 into the input shaft clutches retainer assembly and press down with a arbor press ram. Install the underdrive return spring retainer snapring.
11. Install the underdrive clutch pack in proper order.
12. Install the underdrive clutch reaction plate flat snapring.
13. Install the underdrive clutch disc.
14. Install the overdrive/underdrive clutches reaction plate with the step side down.
15. Install a new overdrive/underdrive clutches reaction plate tapered snapring in proper sequence. Do not scratch the reaction plate. Make sure the tapered snapring is seated.
16. Set up a dial indicator on the assembly to check clutch clearance. Compress the clutch pack with fingers to zero the dial indicator.
17. Use a hook to raise the 1 clutch disc and check the underdrive clutch pack clearance. Clearance must be 0.036–0.058 in. (0.91–1.47mm).
18. Use the following reaction plate part number to achieve proper specifications:
    Part number 4377185 – 0.257 in. (6.52mm)
    Part number 4377186 – 0.276 in. (7.01mm)
    Part number 4377187 – 0.295 in. (7.50mm)
    Part number 4377188 – 0.315 in. (7.99mm)
19. Remove the dial indicator set-up.
20. Install the overdrive clutch pack in proper order.
21. Install the overdrive reaction plate waved snapring.
22. Install the overdrive/reverse pressure plate with the step side down.

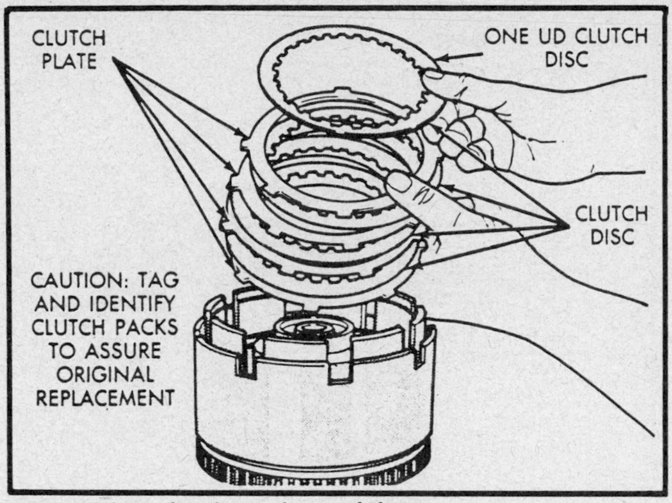
CLUTCH PLATE
ONE UD CLUTCH DISC
CLUTCH DISC
CAUTION: TAG AND IDENTIFY CLUTCH PACKS TO ASSURE ORIGINAL REPLACEMENT

Underdrive clutch pack servicing

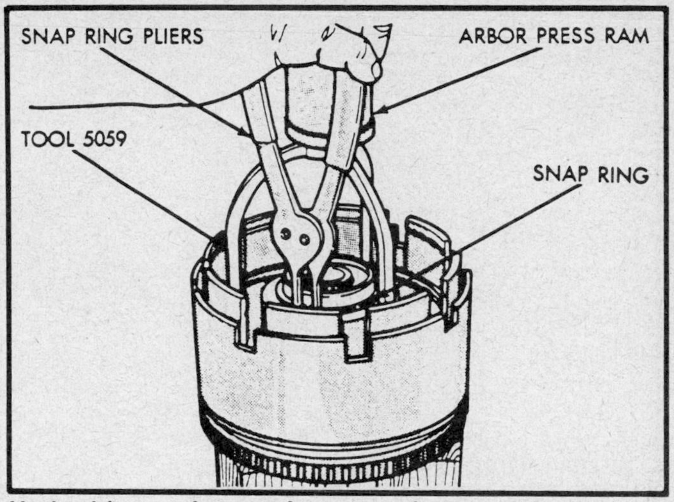
SNAP RING PLIERS
ARBOR PRESS RAM
TOOL 5059
SNAP RING

Underdrive spring retainer snapring servicing

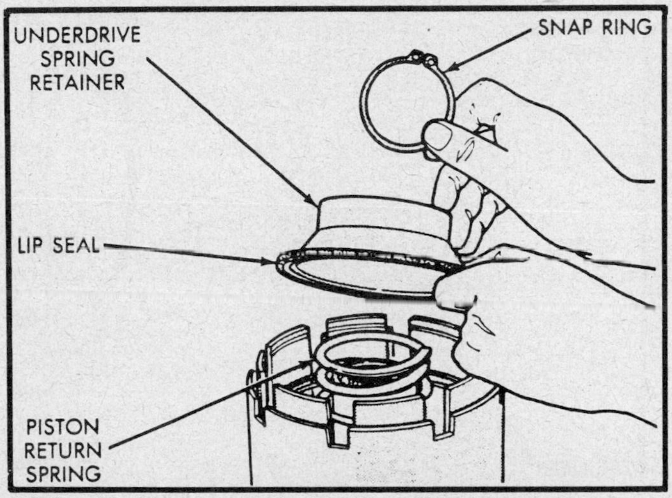
UNDERDRIVE SPRING RETAINER
SNAP RING
LIP SEAL
PISTON RETURN SPRING

Underdrive return spring and retainer servicing

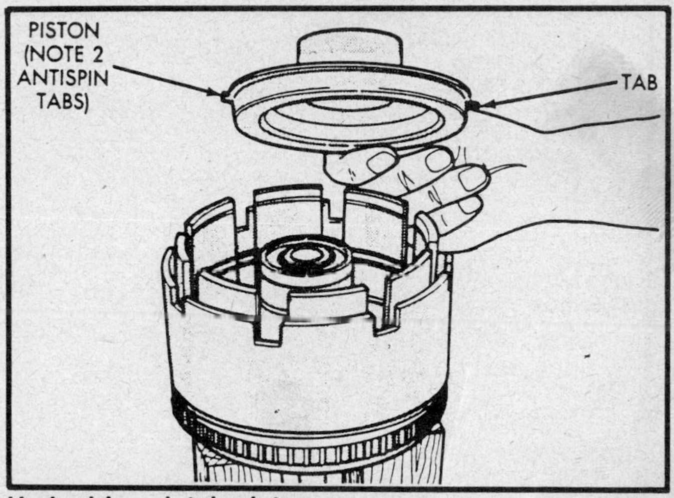
PISTON (NOTE 2 ANTISPIN TABS)
TAB

Underdrive clutch piston

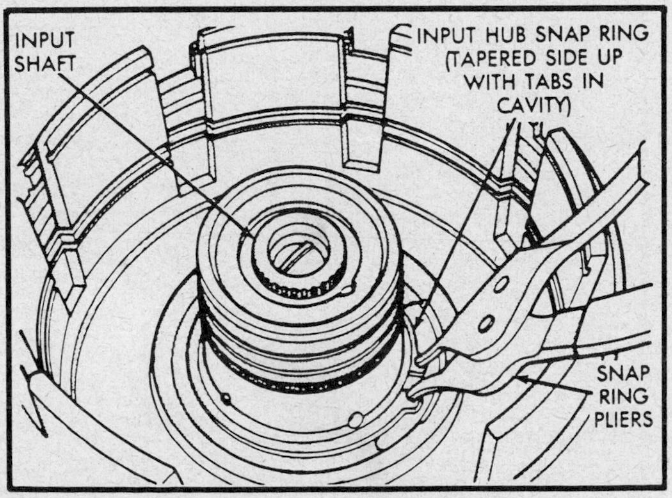
INPUT SHAFT
INPUT HUB SNAP RING (TAPERED SIDE UP WITH TABS IN CAVITY)
SNAP RING PLIERS

Input hub tapered snapring servicing

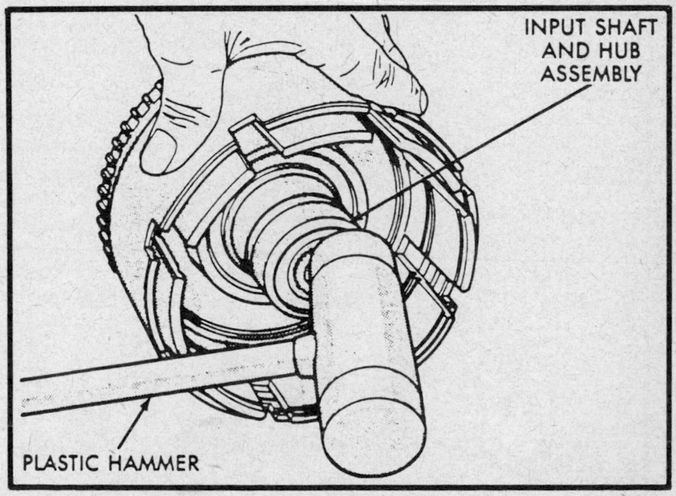
INPUT SHAFT AND HUB ASSEMBLY
PLASTIC HAMMER

Input shaft/hub assembly servicing

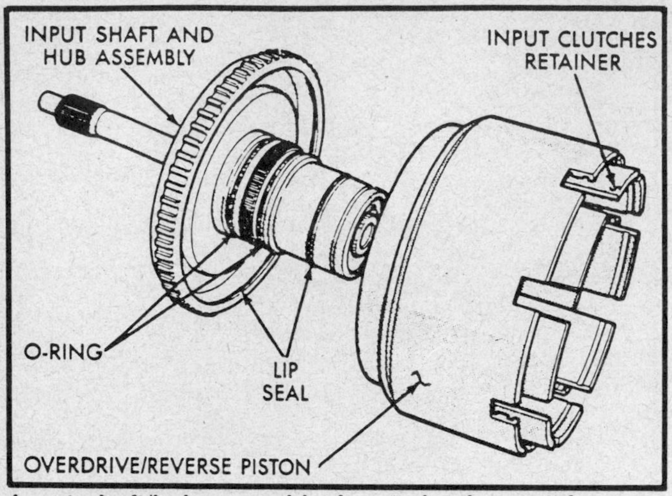
INPUT SHAFT AND
HUB ASSEMBLY
INPUT CLUTCHES
RETAINER
O-RING
LIP
SEAL
OVERDRIVE/REVERSE PISTON

Input shaft/hub assembly, input clutches retainer and overdrive/reverse piston

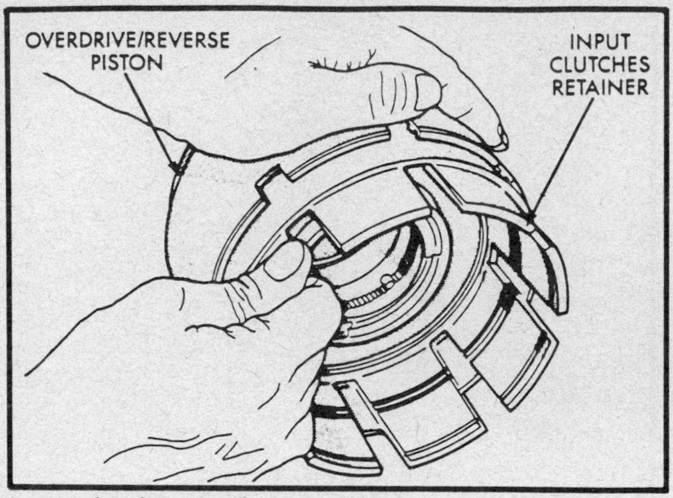
OVERDRIVE/REVERSE
PISTON
INPUT
CLUTCHES
RETAINER

Input clutches retainer servicing

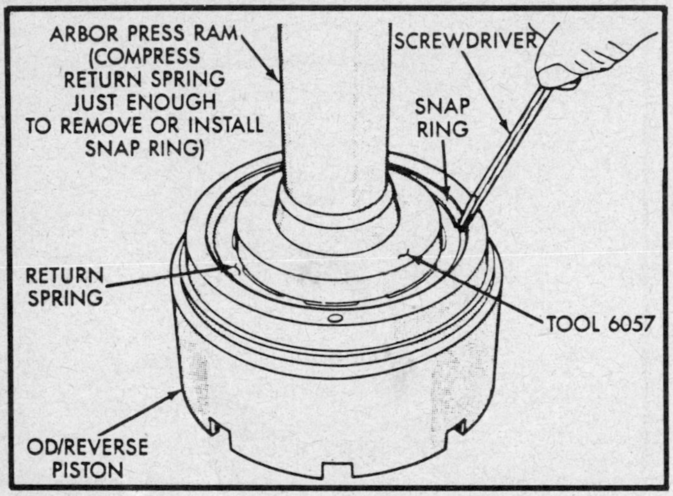
ARBOR PRESS RAM
(COMPRESS
RETURN SPRING
JUST ENOUGH
TO REMOVE OR INSTALL
SNAP RING)
SCREWDRIVER
SNAP
RING
RETURN
SPRING
TOOL 6057
OD/REVERSE
PISTON

Overdrive/reverse piston snapring servicing

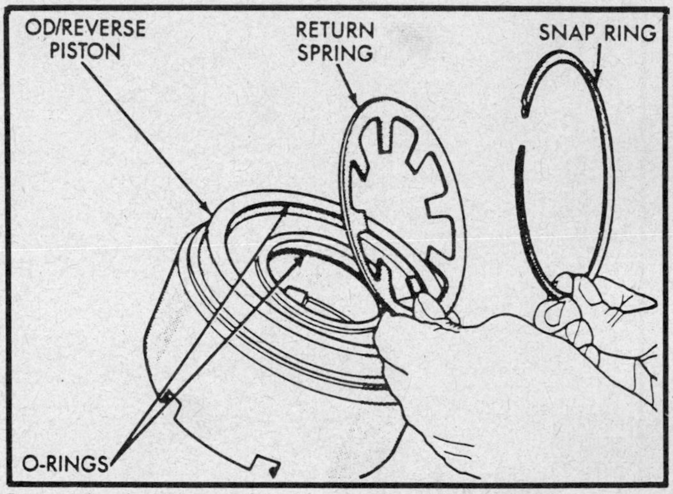
OD/REVERSE
PISTON
RETURN
SPRING
SNAP RING
O-RINGS

Overdrive/reverse piston return spring servicing

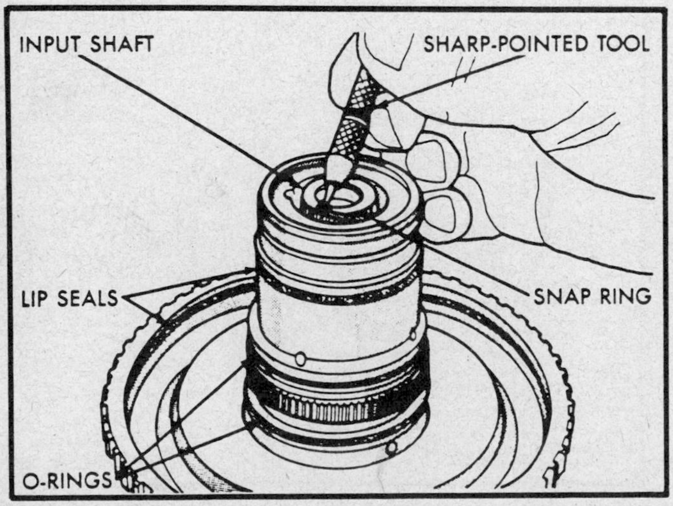
INPUT SHAFT
SHARP-POINTED TOOL
LIP SEALS
SNAP RING
O-RINGS

Input shaft snapring servicing

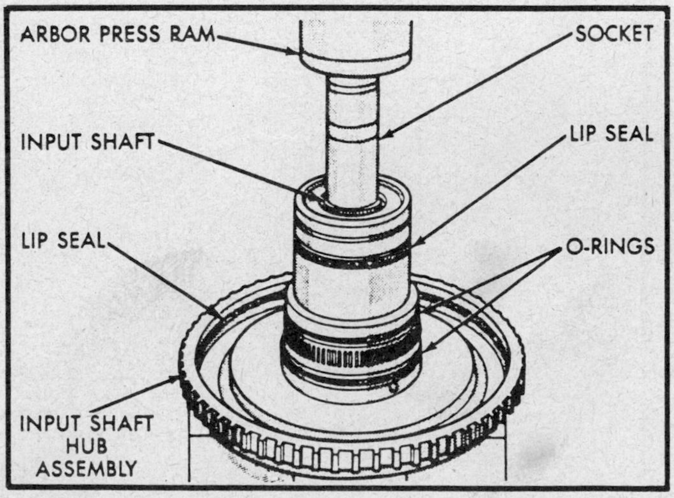
ARBOR PRESS RAM
SOCKET
INPUT SHAFT
LIP SEAL
LIP SEAL
O-RINGS
INPUT SHAFT
HUB
ASSEMBLY

Input shaft removal

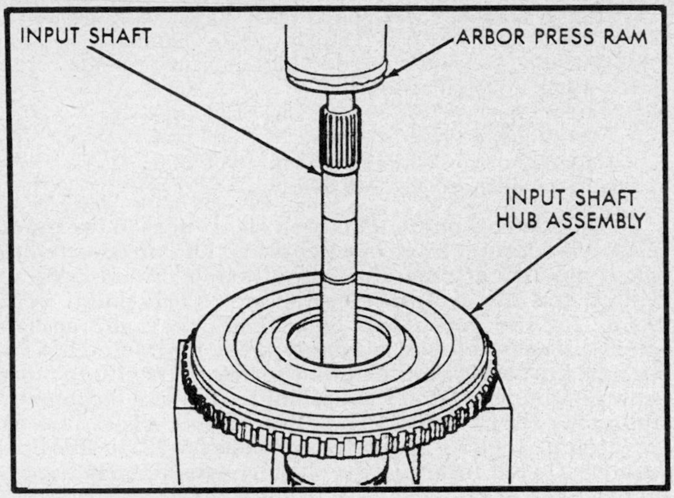
INPUT SHAFT
ARBOR PRESS RAM
INPUT SHAFT
HUB ASSEMBLY

Input shaft installation

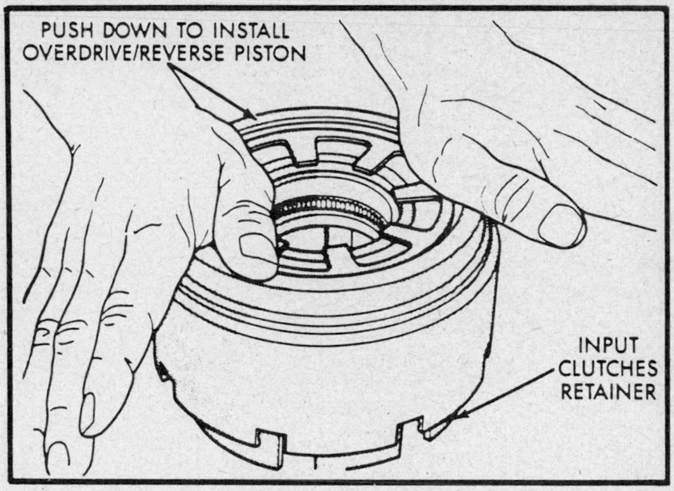
PUSH DOWN TO INSTALL
OVERDRIVE/REVERSE PISTON
INPUT
CLUTCHES
RETAINER

Overdrive/reverse piston installation

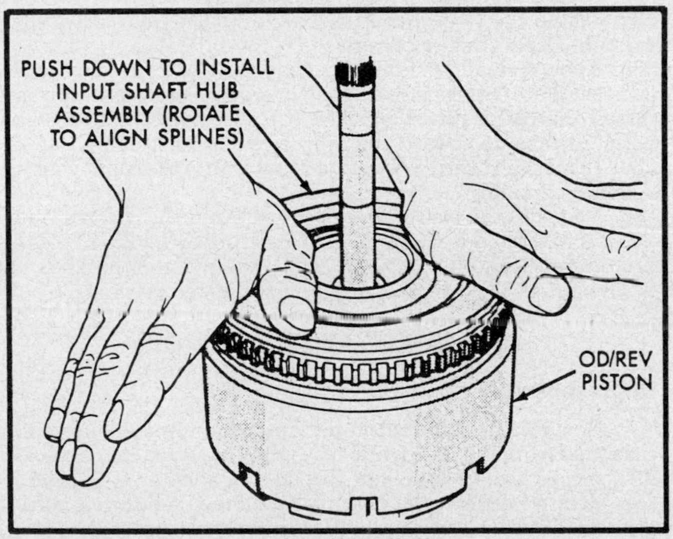
PUSH DOWN TO INSTALL
INPUT SHAFT HUB
ASSEMBLY (ROTATE
TO ALIGN SPLINES)
OD/REV
PISTON

Input shaft hub assembly installation

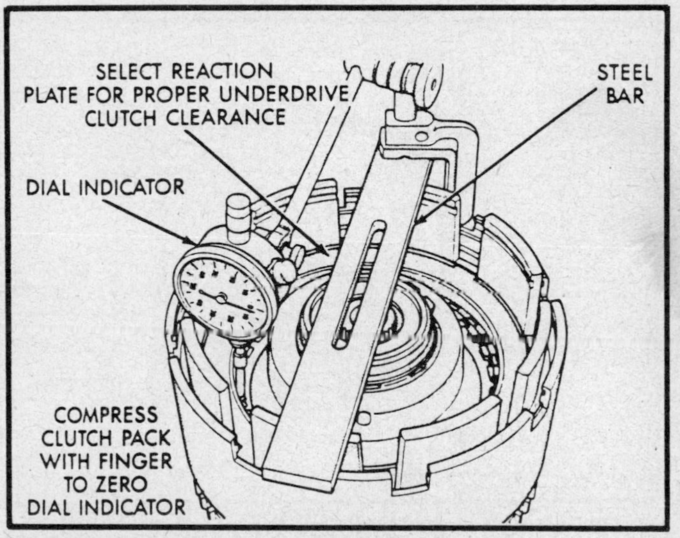
SELECT REACTION
PLATE FOR PROPER UNDERDRIVE
CLUTCH CLEARANCE
STEEL
BAR
DIAL INDICATOR
COMPRESS
CLUTCH PACK
WITH FINGER
TO ZERO
DIAL INDICATOR

Checking underdrive clutch clearance set-up

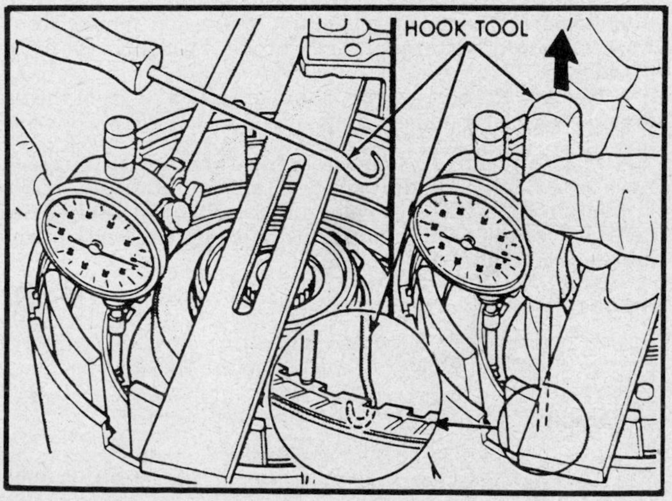
HOOK TOOL

Checking underdrive clutch pack clearance

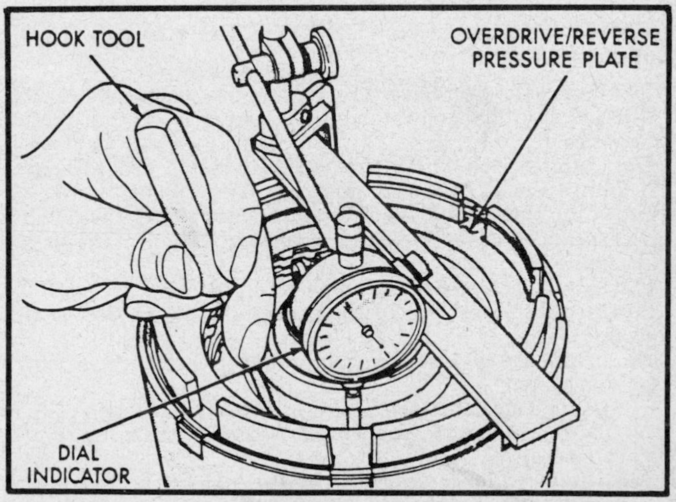
HOOK TOOL
OVERDRIVE/REVERSE
PRESSURE PLATE
DIAL
INDICATOR

Checking overdrive clutch pack clearance

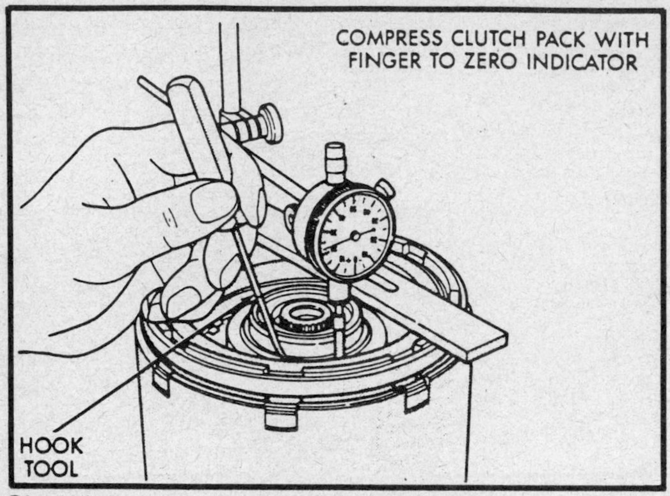
COMPRESS CLUTCH PACK WITH
FINGER TO ZERO INDICATOR
HOOK
TOOL
Checking reverse clutch pack clearance

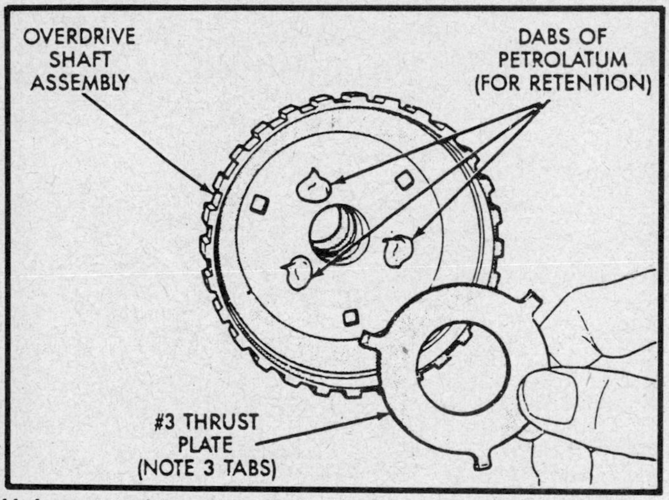
OVERDRIVE
SHAFT
ASSEMBLY
DABS OF
PETROLATUM
(FOR RETENTION)
#3 THRUST
PLATE
(NOTE 3 TABS)
Using petrolatum to retain #3 thrust plate

23. Install tool 5059 on the overdrive/reverse pressure plate and press down with an arbor press ram.

NOTE: Press down JUST ENOUGH to expose the snapring groove.

24. Install the overdrive/reverse pressure plate flat snapring.
25. Set up a dial indicator on the assembly to check clutch clearance.
26. Using a hook, check the overdrive clutch pack clearance. Clearance should be 0.042–0.096 in.
27. If the clutch pack is not within specifications, the clutch is not assembled properly. The overdrive clutch clearance is not adjustable.
28. Install the reverse clutch pack in proper order.
29. Install the reverse clutch reaction plate with the flat side down.
30. Install the reverse clutch snapring onto the reverse clutch reaction plate.
31. Using a small pry tool, lift up the reverse reaction plate to seat the reverse clutch snapring and to determine the reverse clutch clearance.
32. Set up a dial indicator on the assembly to check clutch clearance. Compress the clutch pack with fingers to zero the dial indicator.
33. Using a hook, check the reverse clutch pack clearance. Clearance should be 0.030–0.039 in. (0.76–0.99mm).
34. Use the following reverse clutch snapring part number to achieve proper specifications:
    Part number 4377195 – 0.061 in. (1.56mm)
    Part number 4412871 – 0.071 in. (1.80mm)
    Part number 4412872 – 0.081 in. (2.05mm)
    Part number 4412873 – 0.090 in. (2.30mm)

NOTE: At this point, all clutch clearances in the input clutches retainer have been checked and/or adjusted to meet specifications. The installation of the reverse clutch and the overdrive clutch, up to this point, were solely for the purpose of clearance check and adjustment. To complete the assembly of the input clutches retainer, the reverse clutch and the overdrive clutch must now be removed from the retainer. This can be done by following Steps 1-10 in the Disassembly Procedure or reversing Steps 20–31 (omitting Steps 25–27) in this procedure. Do not intermix the clutch parts. Keep in the exact same order.

35. Install the No. 2 needle bearing with the small tabs up.
36. Install the underdrive shaft assembly.
37. Install the No. 3 thrust washer onto the underdrive shaft assembly with the tabs down.
38. Apply a small amount of petroleum jelly (to retain the No. 3 thrust plate) to the inside of the overdrive shaft assembly and install the No. 3 thrust plate to the overdrive shaft assembly with the tabs up toward the overdrive shaft assembly.
39. Install the overdrive shaft assembly into the input clutches retainer assembly.
40. Reinstall the overdrive clutch pack and the reverse clutch pack as described in Steps 20–31 (omitting Steps 25–27) of this procedure. Rechecking clearances of these clutch packs is not necessary, as they were checked and adjusted previously.

DIFFERENTIAL ASSEMBLY

Disassembly

1. Remove the differential bearing cone (on the differential case side) from the differential assembly by using special tool C–293, special tool C–4996 and special tool adapter C–293–45.
2. Remove differential bearing cone (on the differential ring gear side) from the differential assembly by using tool L–4406–1 with adapters L–4406–3.
3. Remove the ring gear bolts.
4. Remove the pinion shaft roll pin from the pinion shaft.
5. Remove the pinion shaft from the differential case.
6. Remove the pinion gears, side gears and tabbed thrust washers by rotating the pinion gears to the opening in the differential case.
7. Remove the bearing cup, shim and oil baffle from the differential bearing retainer using special tool L–4518.

NOTE: The differential bearing retainer shim thickness need only be determined and/or changed if the transaxle case, differential carrier, differential bearing retainer, extension housing or differential bearing cups/cones are replaced.

Inspection

Inspect splines, gear teeth and all mating surfaces for scores, pitting, chips and abnormal wear. Replace the components as required.

Assembly

1. Install a 0.020 in. (0.50mm) gauging shim and bearing cup into the differential bearing retainer using special tool L–4520, handle C–4171 and press. The oil baffle is not required when making shim selection.

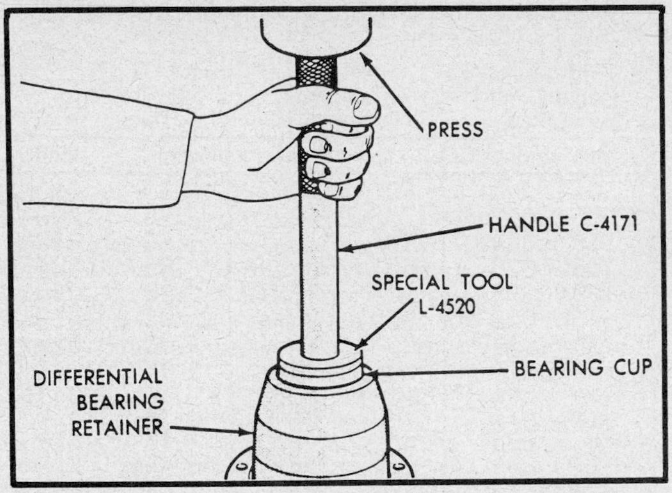
PRESS
HANDLE C-4171
SPECIAL TOOL
L-4520
BEARING CUP
DIFFERENTIAL
BEARING
RETAINER
Differential bearing retainer bearing cup installation

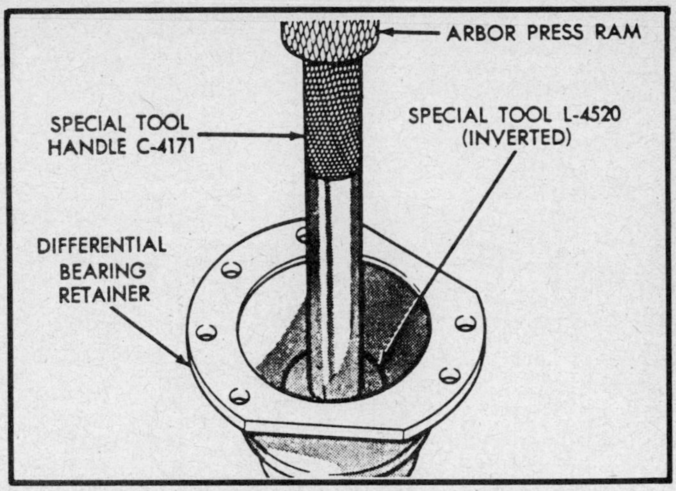
ARBOR PRESS RAM
SPECIAL TOOL
HANDLE C-4171
SPECIAL TOOL L-4520
(INVERTED)
DIFFERENTIAL
BEARING
RETAINER
Differential bearing retainer oil seal installation

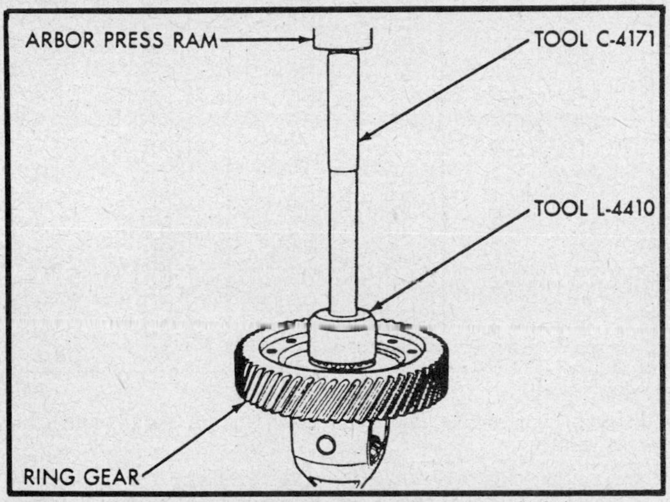
ARBOR PRESS RAM
TOOL C-4171
TOOL L-4410
RING GEAR
Differential bearing retainer bearing cone (ring gear side) installation

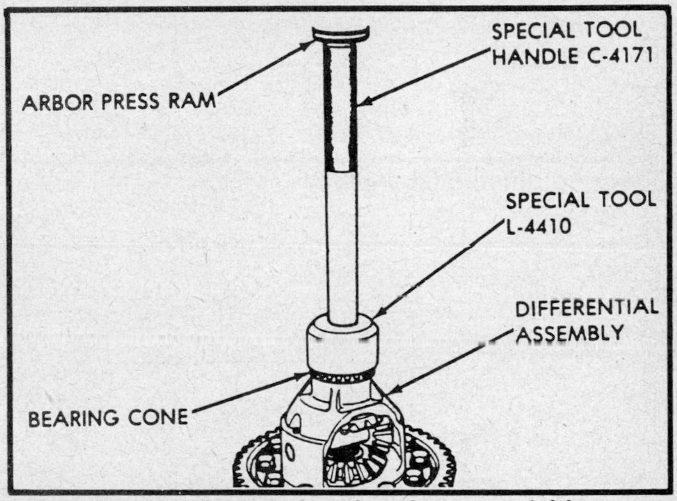
SPECIAL TOOL
HANDLE C-4171
ARBOR PRESS RAM
SPECIAL TOOL
L-4410
DIFFERENTIAL
ASSEMBLY
BEARING CONE
Differential bearing retainer bearing cone (side gear side) installation

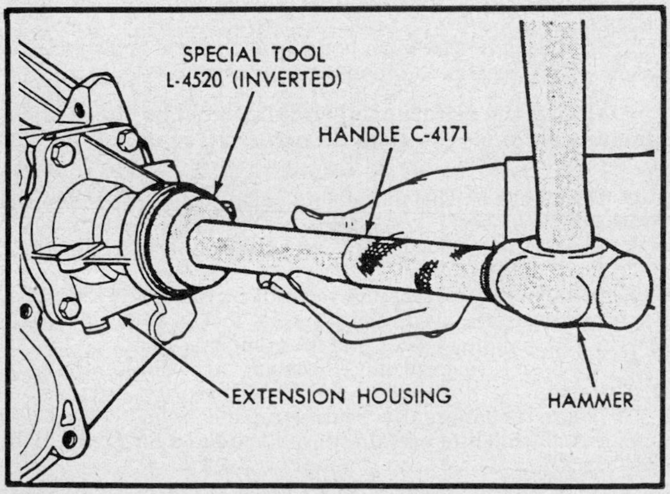
SPECIAL TOOL
L-4520 (INVERTED)
HANDLE C-4171
EXTENSION HOUSING
HAMMER
Differential extension seal installation

NOTE: The differential bearing retainer shim thickness need only be determined and/or changed if the transaxle case, differential carrier, differential bearing retainer, extension housing or differential bearing cups/cones are replaced.

2. Install pinion gear, side gears and tabbed thrust washers by rotating the pinion gears to the opening in the differential case.
3. Install the pinion shaft into the differential case. Do not install the pinion shaft roll pin at this time. If necessary, tape can be place around the differential case to hold the pinion shaft in place.
4. Install new ring gear bolts and torque to 70 ft. lbs. (95 Nm).

NOTE: Always use new ring gear bolts and torque properly.

5. Install a dial indicator and special tool C–4996 to check side gear endplay to the differential assembly (differential case side). Check the side gear endplay by moving the side gear up and down. Side gear endplay must be within 0.001–0.013 in. Thrust washers are available in 0.032, 0.037, 0.042 and 0.047 in.

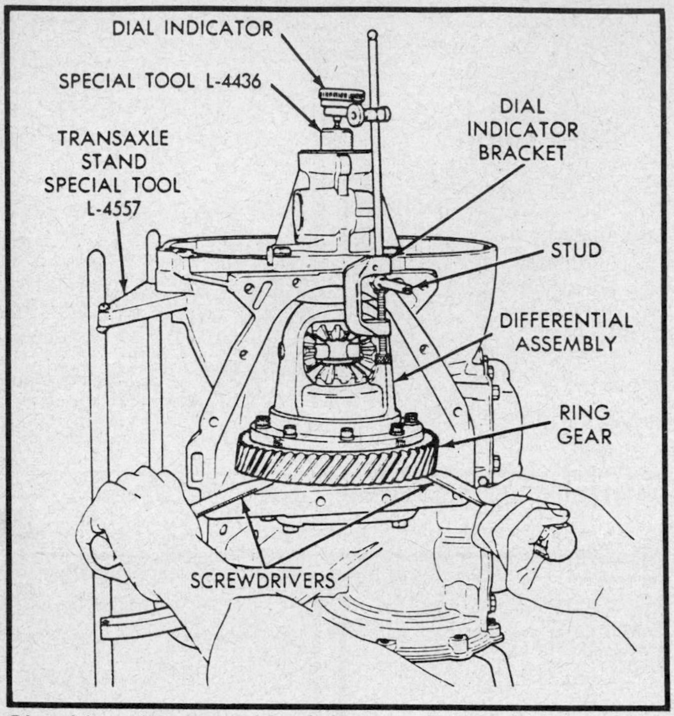
DIAL INDICATOR
SPECIAL TOOL L-4436
TRANSAXLE STAND SPECIAL TOOL L-4557
DIAL INDICATOR BRACKET
STUD
DIFFERENTIAL ASSEMBLY
RING GEAR
SCREWDRIVERS

Checking differential endplay

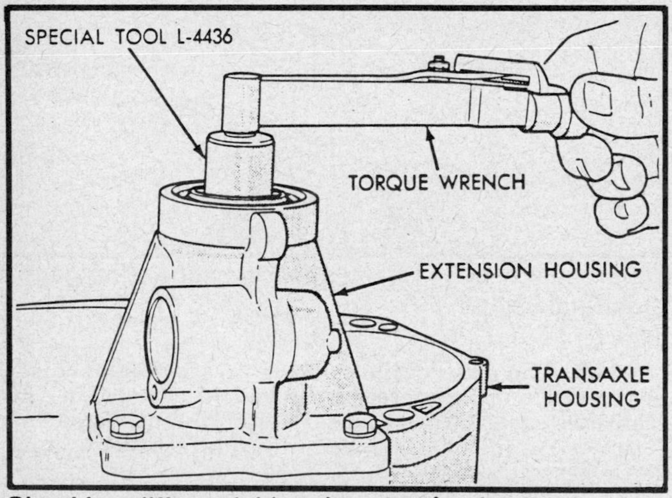
SPECIAL TOOL L-4436
TORQUE WRENCH
EXTENSION HOUSING
TRANSAXLE HOUSING

Checking differential bearings turning torque

6. Install a dial indicator and special tool C–4996 to check side gear endplay to the differential assembly (ring gear side). Check the side gear endplay by moving the side gear up and down. Side gear endplay must be within 0.001–0.013 in.. Thrust washers are available in 0.032, 0.037, 0.042 and 0.047 in..
7. When side gear endplay is within specifications, install the pinion shaft roll pin.

CAUTION

Make sure the pinion shaft roll pin is installed or the pinion shaft could drift and cause damage.

8. Install differential bearing cone onto differential assembly (ring gear side) using tool L–4410, tool C–4171 and arbor press ram.
9. Install differential bearing cone onto differential assembly (differential case side) using tool L–4410, tool C–4171 and arbor press ram.

DIFFERENTIAL BEARING SHIM SELECTION

End Play (with .50mm gauging shim Installed) mm	Inch	Required Shim Combination mm	Total Thickness mm	Inch
.0	.0	.50	.50	.020
.05	.002	.75	.75	.030
.10	.004	.80	.80	.032
.15	.006	.85	.85	.034
.20	.008	.90	.90	.035
.25	.010	.95	.95	.037
.30	.012	1.00	1.00	.039
.35	.014	1.05	1.05	.041
.40	.016	.50 + .60	1.10	.043
.45	.018	.50 + .65	1.15	.045
.50	.020	.50 + .70	1.20	.047
.55	.022	.50 + .75	1.25	.049
.60	.024	.50 + .80	1.30	.051
.65	.026	.50 + .85	1.35	.053
.70	.027	.50 + .90	1.40	.055
.75	.029	.50 + .95	1.45	.057
.80	.031	.50 + 1.00	1.50	.059
.85	.033	.50 + 1.05	1.55	.061
.90	.035	1.00 + .60	1.60	.063
.95	.037	1.00 + .65	1.65	.065
1.00	.039	1.00 + .70	1.70	.067
1.05	.041	1.00 + .75	1.75	.069
1.10	.043	1.00 + .80	1.80	.071
1.15	.045	1.00 + .85	1.85	.073
1.20	.047	1.00 + .90	1.90	.075
1.25	.049	1.00 + .95	1.95	.077
1.30	.051	1.00 + 1.00	2.00	.079
1.35	.053	1.00 + 1.05	2.05	.081
1.40	.055	1.05 + 1.05	2.10	.083

Transaxle Assembly

DIFFERENTIAL ASSEMBLY

Installation

1. Install the differential assembly into the transaxle assembly.
2. Install a new extension housing O-ring and apply an ⅛ in. bead of RTV sealant to the extension housing.

NOTE: If the differential endplay must be check and a gauging shim is installed, do not use O-ring and RTV at this time.

3. Install the extension housing using tool L–4435 onto the transaxle assembly.
4. Install the extension housing bolts and torque bolts to 21 ft. lbs. (28 Nm).
5. Install a new oil seal into the differential bearing retainer, if the seal was removed, using special tool L–4520 (inverted), special tool handle C–4171 and an arbor press ram.
6. Apply an ⅛ in. bead of RTV sealant to the differential bearing retainer.
7. Install the differential bearing retainer using tool L–4435.
8. Install the differential retainer bolts and torque to 21 ft. lbs. (28 Nm).
9. Apply an ⅛ in. bead of RTV sealant to the differential cover and install the differential cover.

10. Install the differential cover bolts and torque to 165 inch lbs. (19 Nm).
11. Install a new oil seal into the extension housing using special tool L–4520 (inverted) and tool handle C–4171.

Checking Differential Endplay

1. Install the transaxle assembly vertically on the support stand and install tool L–4436 into the extension.
2. Rotate the differential at least 1 full revolution to ensure the tapered roller bearings are fully seated.
3. Install a dial indicator to the case and zero the dial indicator. Place the indicator tip on the end of tool L–4436.
4. Place a small pry tool to each side of the ring gear and lift. Check the dial indicator for the amount of endplay.

NOTE: Do not damage the transaxle case and/or the differential cover sealing surface.

5. Record the endplay and refer to the Differential Bearing Shim Chart for the correct shim combination to obtain the proper bearing setting.
6. Remove the differential bearing retainer. Remove the bearing cup and the 0.020 in. (0.50mm) gauging shim.
7. Install the proper shim combination under the bearing cup. Make sure the oil baffle is installed properly in the bearing retainer, below the bearing shim and cup.
8. Install a new extension housing O-ring and apply an ⅛ in. bead of RTV sealant to the extension housing.
9. Install the extension housing and torque the extension housing bolts to 21 ft. lbs. (28 Nm).

Checking Differential Bearings Turning Torque

1. Install the transaxle assembly vertically on the support stand and install tool L–4436 into the extension.
2. Rotate the differential at least 1 full revolution to ensure the tapered roller bearings are fully seated.
3. Using an inch lbs. torque wrench installed on tool L–4436, check the turning torque of the differential. The turning torque should be between 5–18 inch lbs..
4. If the turning torque is too high, install a 0.002 in. (0.05mm) thinner shim. If the turning torque is too low, install a 0.002 in. (0.05mm) thicker shim. Repeat this procedure until 5–18 inch lbs. turning torque is obtained.

LOW/REVERSE CLUTCH PISTON AND RETAINER

Installation

1. Install the low/reverse clutch piston retainer gasket and low/reverse clutch piston retainer.

NOTE: Low/reverse clutch piston retainer gasket holes must line up.

2. Install the low/reverse clutch piston retainer attaching screws. Torque the screws to 40 inch lbs. (5 Nm).
3. Install the low/reverse clutch piston.

GUIDE BRACKET

Installation

1. Install the guide bracket assembly.
2. Install the guide bracket pivot shaft and plug.
3. Install the guide bracket support shaft and plug.

NOTE: Be sure the guide bracket and split sleeve touch the rear of the transaxle case.

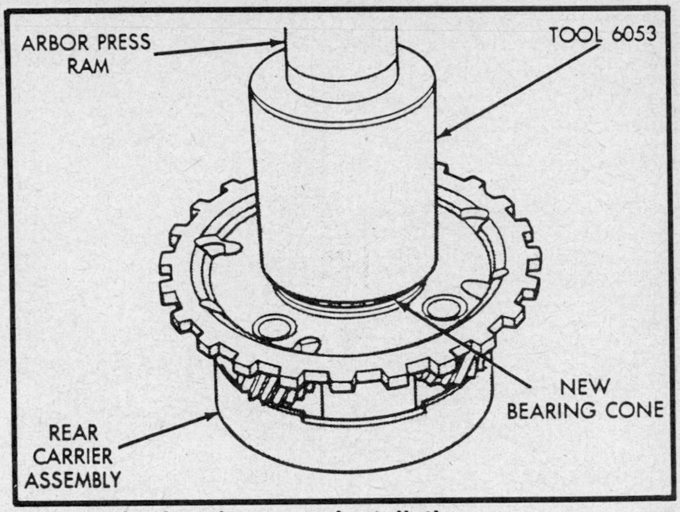
ARBOR PRESS RAM
TOOL 6053
NEW BEARING CONE
REAR CARRIER ASSEMBLY

Rear carrier bearing cone installation

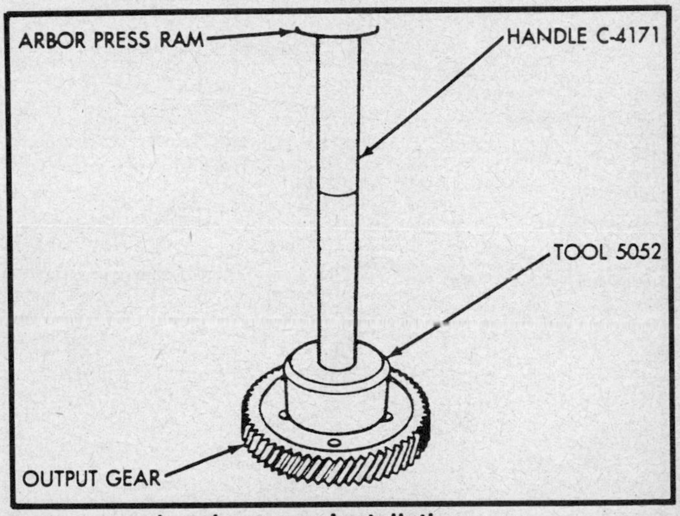
ARBOR PRESS RAM
HANDLE C-4171
TOOL 5052
OUTPUT GEAR

Output gear bearing cone installation

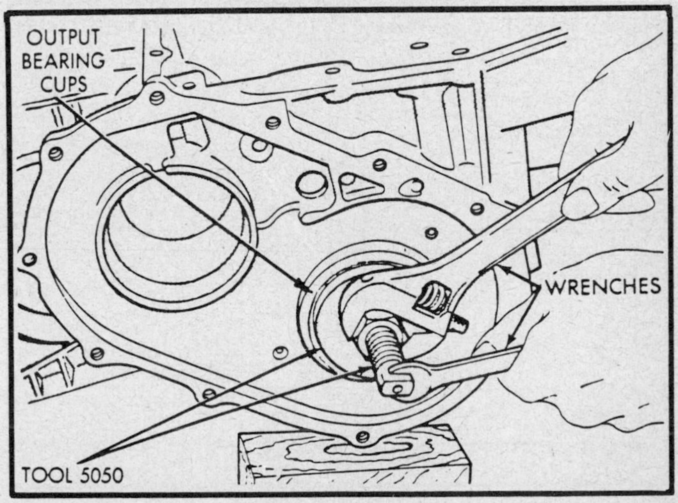
OUTPUT BEARING CUPS
WRENCHES
TOOL 5050

Output gear bearing cups installation

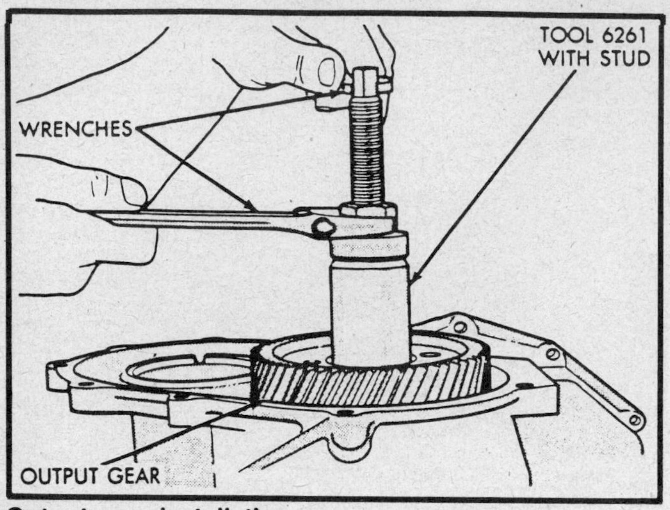
TOOL 6261
WITH STUD
WRENCHES
OUTPUT GEAR

Output gear installation

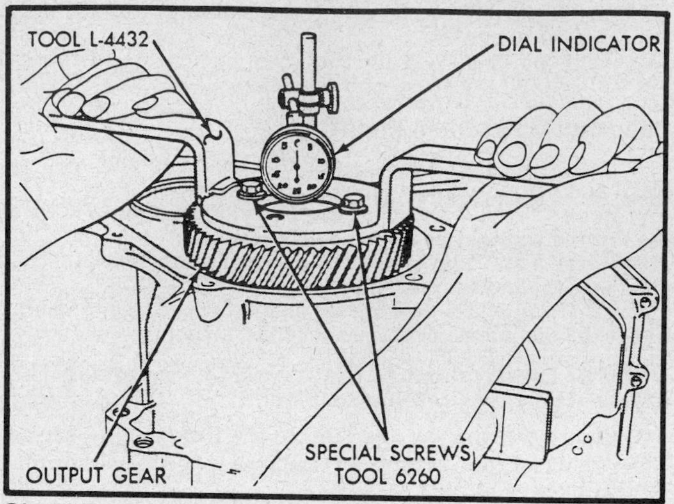
TOOL L-4432
DIAL INDICATOR
OUTPUT GEAR
SPECIAL SCREWS
TOOL 6260

Checking output gear bearings endplay

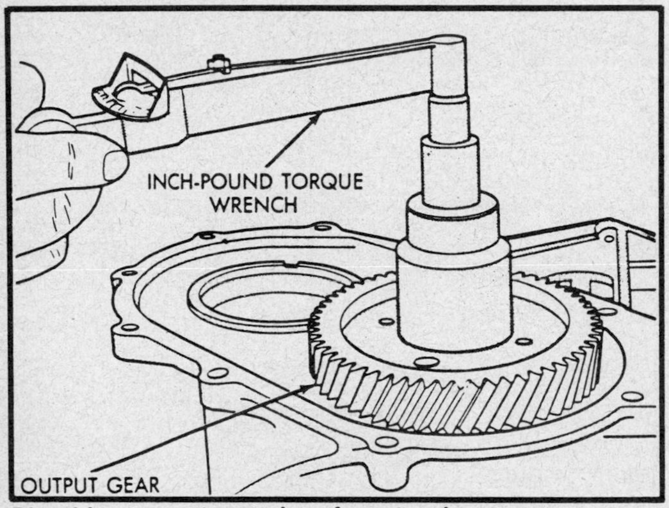
INCH-POUND TORQUE
WRENCH
OUTPUT GEAR

Checking output gear bearings turning torque

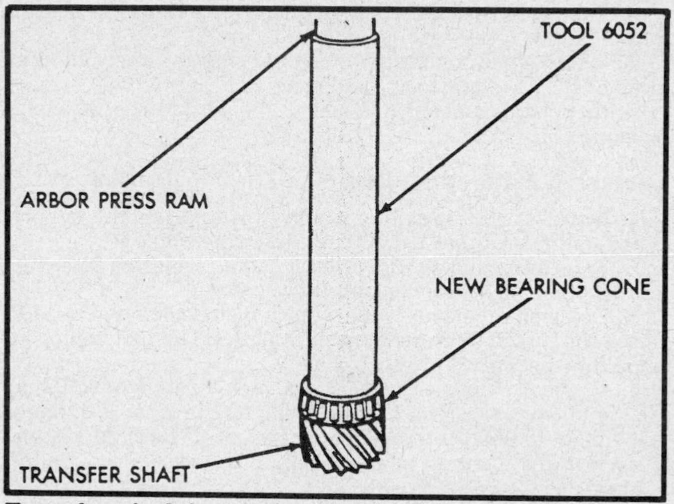
TOOL 6052
ARBOR PRESS RAM
NEW BEARING CONE
TRANSFER SHAFT

Transfer shaft bearing cone (small gear side) installation

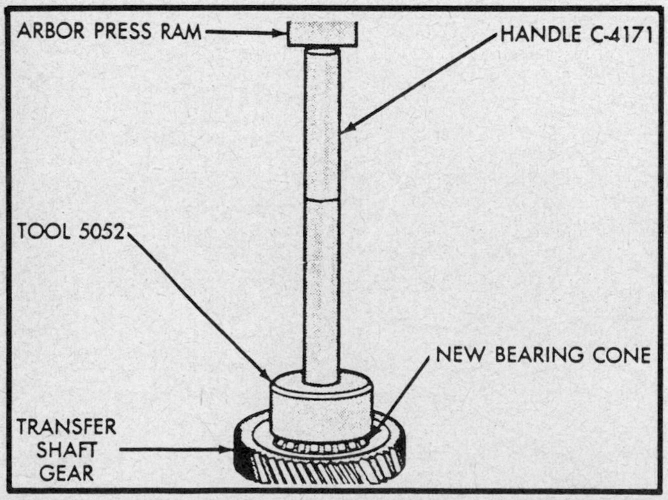
ARBOR PRESS RAM
HANDLE C-4171
TOOL 5052
NEW BEARING CONE
TRANSFER
SHAFT
GEAR

Transfer shaft bearing cone (large gear side) installation

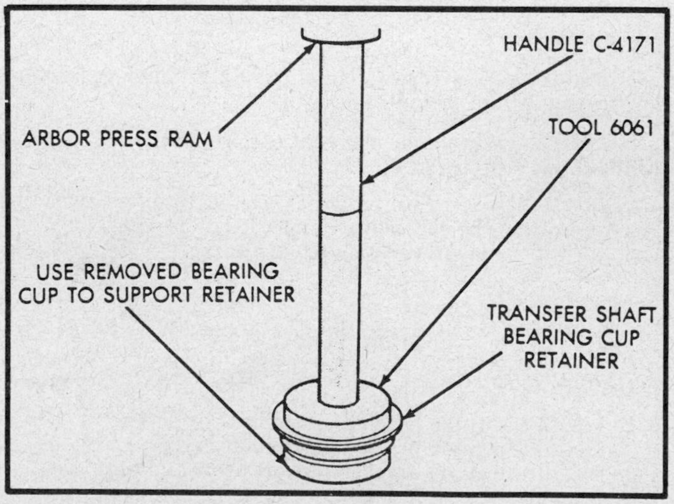
HANDLE C-4171
ARBOR PRESS RAM
TOOL 6061
USE REMOVED BEARING
CUP TO SUPPORT RETAINER
TRANSFER SHAFT
BEARING CUP
RETAINER

Transfer shaft bearing cup installation

LOW/REVERSE PISTON RETURN SPRING

Installation

1. Install the low/reverse piston return spring.
2. Install low/reverse spring compressor tool 5059, 5058–3 and 6057. Compress the low/reverse piston spring.
3. Install the low/reverse piston return spring snapring.
4. Remove the low/reverse spring tools.

REAR CARRIER ASSEMBLY

Installation

1. If removed, install the rear carrier bearing cone to the rear carrier using tool 6053 and an arbor press ram.
2. Install the rear carrier assembly into the transaxle assembly (bell housing side).

OUTPUT SHAFT GEAR

Installation

1. If removed, install the output gear bearing cone using tool 5052, tool handle C–4171 and an arbor press ram.
2. If removed, install the output gear bearing cups using tool 5050.
3. Install the output gear shim and output gear.
4. Install the output gear bolt and washer using tool 6261 with stud. Use tool 6259 and torque output gear bolt to 200 ft. lbs. (271 Nm).

Checking Bearing Endplay

1. Install tool L–4432, special screws tool 6260 and a dial indicator to the output gear.
2. Move gear up and down to measure bearing endplay. Output gear endplay must have a preload of 0.0008–0.002 in. (0.02–0.05mm).
3. If the output gear endplay is not within specifications, remove the output gear following the Output Gear Removal procedure and install the proper output gear shim. Install by following the Output Gear Installation procedure and recheck the endplay.

Checking Bearing Turning Torque

1. Install proper socket onto the output gear attaching bolt. Install an inch lbs. torque wrench onto the socket.
2. Check the turning torque of the output gear. The torque should be 3–8 inch lbs..
3. If the output gear turning torque is not within specifications, remove the output gear following the Output Gear Removal procedure and install the proper output gear shim. Install by following the Output Gear Installation procedure and recheck the turning torque.

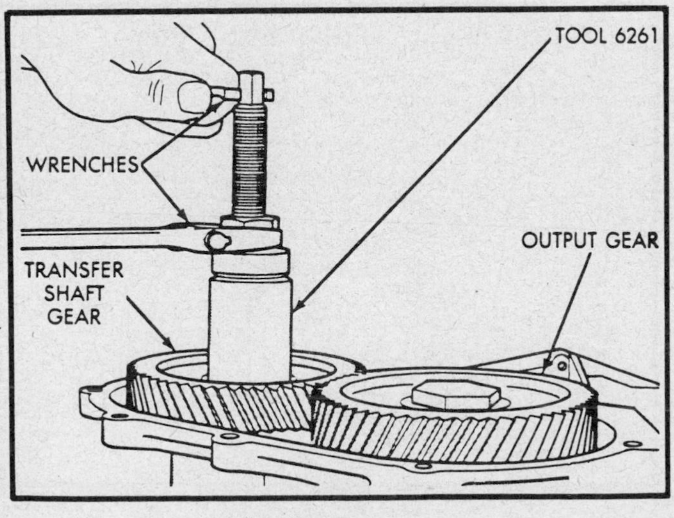
TOOL 6261
WRENCHES
OUTPUT GEAR
TRANSFER SHAFT GEAR

Transfer shaft gear installation

TRANSFER SHAFT

Installation

1. If removed, install the transfer shaft bearing cone using tool 6052 and an arbor press ram.
2. Install the transfer shaft using tool 5049–A.
3. Install the transfer shaft bearing snapring using snapring pliers tool 6051.
4. If removed, install the transfer shaft bearing cone onto the transfer shaft gear using tool 5052, tool handle C–4171 and an arbor press ram.
5. Install the transfer shaft bearing cup retainer.

TRANSFER SHAFT GEAR

Installation

1. If removed, install the transfer shaft bearing cone using tool 5052, tool handle C–4171 and an arbor press ram.
2. Install the transfer shaft gear and shim using tool 6261.
3. Install transfer shaft gear lockwasher and nut. Using tool 6259 to hold the transfer shaft gear, torque the transfer shaft gear nut to 200 ft. lbs. (271 Nm).
4. Check the transfer shaft endplay. The transfer shaft must have endplay of 0.002–0.004 in. (0.05–0.10mm). Install proper shim to meet specifications.
5. Apply an ⅛ in. bead of RTV sealant on the rear cover.
6. Install the rear cover.
7. Install the rear cover bolts and torque to 14 ft. lbs. (19 Nm).

LOW/REVERSE CLUTCH PACK

Installation

1. Install the low/reverse clutch pack in the proper order.
2. Install the low/reverse reaction plate snapring. Do not scratch the clutch plate.
3. Install the 1 disc onto the low/reverse clutch.
4. Install the low/reverse reaction plate (flat side up).
5. Install a new tapered snapring in sequence according to the illustration.
6. To check the low/reverse clutch clearance, install dial indicator and dial indicator tip tool 6268. Using a hook tool to raise the 1 clutch disc, check the clearance. Low/reverse clutch pack clearance is 0.042–0.065 in. (1.04–1.65mm).
7. If the low/reverse clutch pack is not within specifications, select the proper low/reverse reaction plate to achieve specifications.

Part number 4377150 – 0.273 in. (6.92mm)
Part number 4377149 – 0.262 in. (6.66mm)
Part number 4377148 – 0.252 in. (6.40mm)
Part number 4412268 – 0.242 in. (6.14mm)
Part number 4412267 – 0.232 in. (5.88mm)
Part number 4412266 – 0.221 in. (5.62mm)
Part number 4412265 – 0.211 in. (5.36mm)

2–4 CLUTCH PACK

Installation

1. Install the 2–4 clutch pack in proper order.

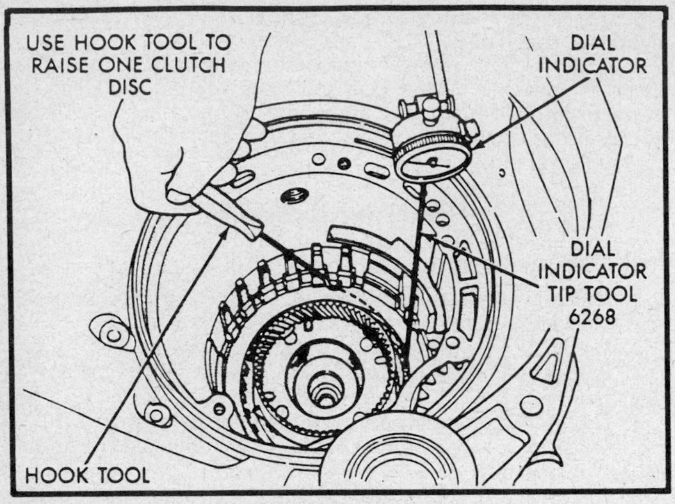
USE HOOK TOOL TO RAISE ONE CLUTCH DISC
DIAL INDICATOR
DIAL INDICATOR TIP TOOL 6268
HOOK TOOL

Checking low/reverse clutch clearance

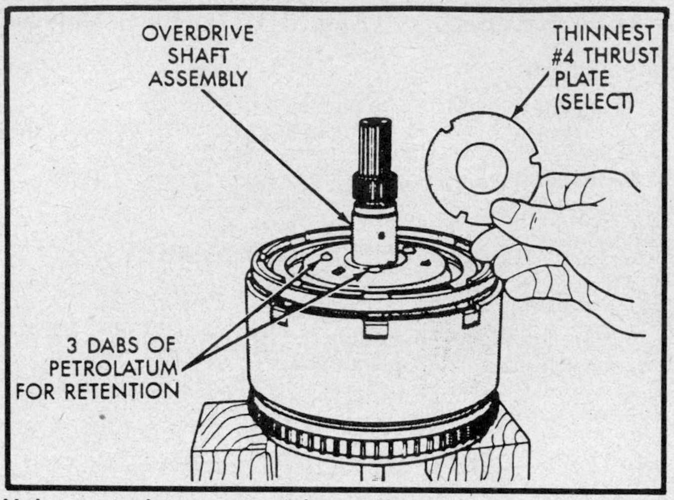
OVERDRIVE SHAFT ASSEMBLY
THINNEST #4 THRUST PLATE (SELECT)
3 DABS OF PETROLATUM FOR RETENTION

Using petrolatum to retain #4 thrust plate

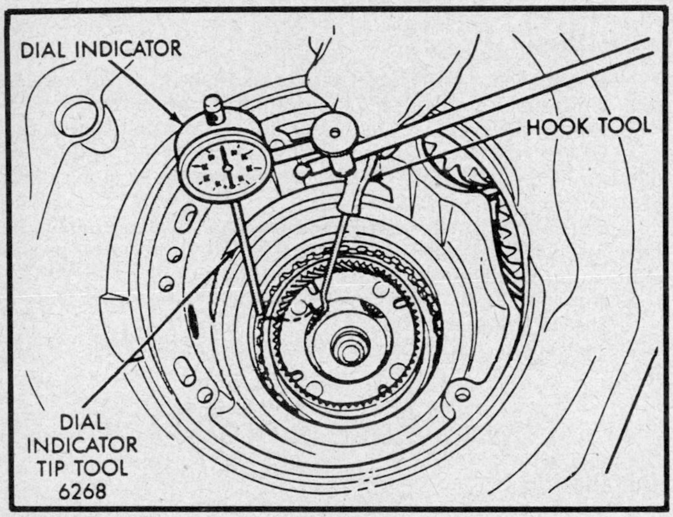
DIAL INDICATOR
HOOK TOOL
DIAL INDICATOR TIP TOOL 6268

Checking 2/4 clutch clearance

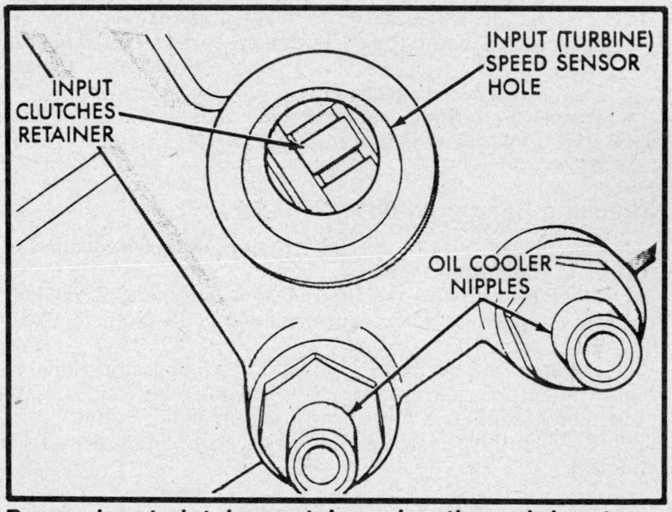
INPUT CLUTCHES RETAINER
INPUT (TURBINE) SPEED SENSOR HOLE
OIL COOLER NIPPLES

Proper input clutches retainer view through input speed sensor hole

2. Install the 2–4 clutch return spring onto the 2–4 clutch retainer. Note the position of the return spring so it is indexed properly onto the retainer.
3. Install the 2–4 clutch return spring and retainer into the transaxle assembly.
4. Install tool 5058 and compress the 2–4 clutch return spring just enough to install the 2–4 clutch retainer snapring.
5. Install the 2–4 clutch retainer snapring.
6. Install a dial indicator and dial indicator tip tool 6268 into the transaxle assembly. Press down on the clutch pack with fingers and zero the dial indicator.
7. With a hook tool, check the 2–4 clutch pack clearance. Clearance should be 0.030–0.104 in. (0.76–2.64mm). If the clutch clearance is not within specification, the 2–4 clutch pack is not assembled properly. There is no adjustment for the 2–4 clutch clearance.

REAR SUN GEAR

Installation

1. Install the No. 7 needle bearing assembly.
2. Install the rear sun gear.

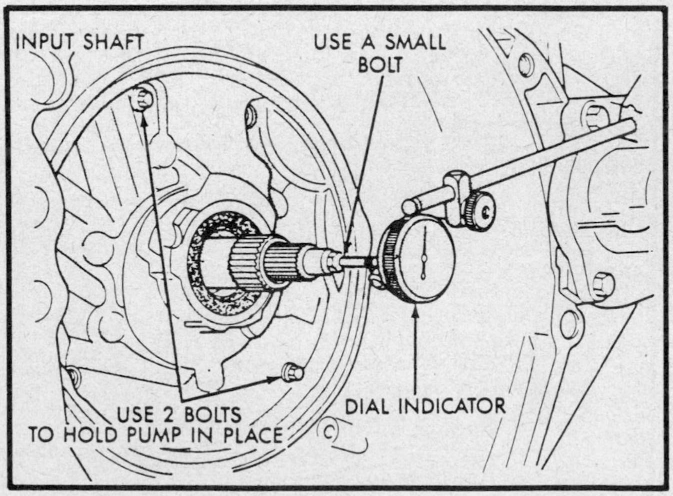
INPUT SHAFT
USE A SMALL BOLT
USE 2 BOLTS TO HOLD PUMP IN PLACE
DIAL INDICATOR

Checking input shaft endplay

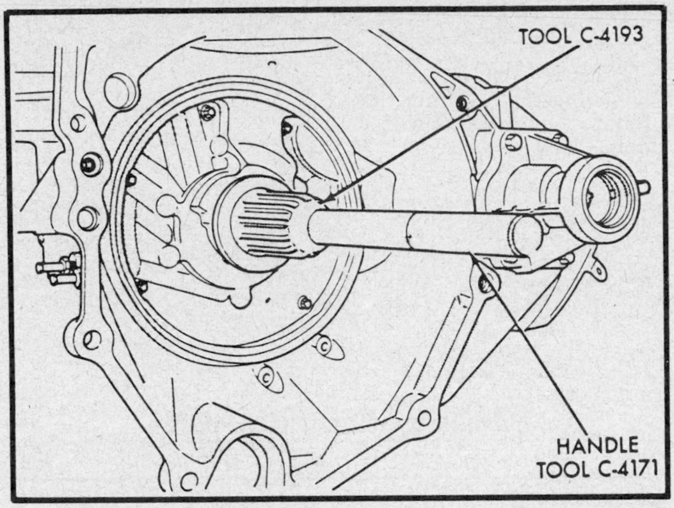
TOOL C-4193
HANDLE
TOOL C-4171
Oil pump seal installation

FRONT CARRIER AND REAR ANNULUS ASSEMBLY

Installation

1. Install the No. 6 needle bearing assembly.
2. Install the front carrier and rear annulus assembly. Twist and push to install the assembly.

FRONT SUN GEAR ASSEMBLY

Installation

1. Install the front sun gear assembly.
2. Install the No. 4 thrust washer onto the front sun gear assembly.

NOTE: If the input shaft endplay was checked before disassembly and the No. 4 thrust washer shim size is known, install the proper shim and continue assembling the transaxle unit. If not, follow the procedure to determine the No. 4 thrust plate (washer) thickness.

DETERMINING NO. 4 THRUST PLATE (WASHER) THICKNESS

1. Select the thinnest No. 4 thrust plate.
2. Apply a small amount of petroleum jelly to the input clutches retainer to hold the thrust plate in place.
3. Install the No. 4 thrust plate to the input clutches retainer.
4. Install the input clutches retainer into the transaxle unit. Make sure the input clutches retainer is completely seated.
5. To ease installation and removal, remove the oil pump O-ring.
6. Install the oil pump and install 2 bolts to hold the oil pump in place.
7. Install a dial indicator and check the input shaft endplay. Input shaft endplay must be 0.012–0.030 in. (0.31–0.76mm).
8. Once the input shaft endplay has been recorded, remove the oil pump bolts and remove the oil pump.
9. Install a oil pump O-ring onto the oil pump.
10. Remove the input shaft clutches retainer and No. 4 thrust plate.
11. Install the proper size No. 4 thrust plate and continue assembling the transaxle unit.

Part number 4431662—0.037–0.039 in. (0.93–1.00mm)
Part number 4431663—0.045–0.048 in. (1.15–1.22mm)
Part number 4431664—0.054–0.057 in. (1.37–1.44mm)
Part number 4431665—0.063–0.066 in. (1.59–1.66mm)
Part number 4431666—0.071–0.074 in. (1.81–1.88mm)
Part number 4431667—0.080–0.083 in. (2.03–2.10mm)
Part number 4431668—0.089–0.091 in. (2.25–2.32mm)
Part number 4431669—0.097–0.100 in. (2.47–2.54mm)
Part number 4446670—0.106–0.109 in. (2.69–2.76mm)
Part number 4446671—0.114–0.117 in. (2.91–2.98mm)
Part number 4446672—0.123–0.126 in. (3.13–3.20mm)
Part number 4446601—0.132–0.135 in. (3.35–3.42mm)

INPUT SHAFT CLUTCHES ASSEMBLY

Installation

1. Install the input shaft clutches assembly. Make sure the input shaft clutches assembly is completely seated.
2. Install the No. 1 caged needle bearing. Install with the tanged side out.

OIL PUMP

Installation

1. Install the oil pump gasket. Make sure the hole line up.
2. Install a new O-ring onto the oil pump and install the pump.
3. Install the oil pump attaching bolts. Torque the bolts to 23 ft. lbs. (32 Nm).
4. Install a new oil pump seal using tool C–4193 and handle tool C–4171.

GOVERNOR

Installation

This transaxle utilizes electronic sensors and solenoids in place of the governor. Therefore, there is no governor assembly.

ACCUMULATORS

Installation

1. Install the low/reverse accumulator piston return springs.
2. Install the low/reverse accumulator piston. Note the piston notch so the piston is installed correctly.
3. Install the low/reverse accumulator plug.
4. Install the low/reverse accumulator plug snapring.
5. With the seal ring installed on the overdrive clutch accumulator, install the accumulator and return spring.
6. With the seal ring installed on the underdrive clutch accumulator, install the accumulator and return spring.

VALVE BODY

Installation

1. Install the valve body by guiding the park rod rollers into the guide bracket while shifting the manual lever assembly.
2. Install the valve body attaching bolts and torque to 40 inch lbs. (5 Nm).
3. Install a new oil filter and O-ring.

OIL PAN

Installation

1. Apply an ⅛ in. bead of RTV sealant on the oil pan.
2. Install the oil pan.
3. Apply RTV sealant under the oil pan bolt heads and install the oil pan bolts. Torque the bolts to 14 ft. lbs. (19 Nm).

CONVERTER

Installation

The torque converter is installed by sliding the unit into the transaxle input and reaction shaft. Rotate the torque converter to seat the converter onto the oil pump.

SPECIFICATIONS

AUTOMATIC TRANSAXLE SPECIFICATIONS

Item	in.	mm
Pump Clearance		
Outer gear to pocket	0.0018–0.0056	0.045–0.141
Outer gear side clearance	0.0008–0.0018	0.020–0.046
Inner gear side clearance	0.0008–0.0018	0.020–0.046
Input shaft endplay	0.012–0.030	0.31–0.76
Differential side gear clearance	0.001–0.013	0.025–0.330
Clutch pack clearance		
Underdrive clutch	0.036–0.058	0.91–1.47
Overdrive clutch	0.042–0.096	1.07–2.44
Reverse clutch	0.030–0.039	0.76–0.99
Low/reverse clutch	0.042–0.065	1.07–1.65
2/4 clutch	0.030–0.104	0.76–2.64
Tapered roller bearing settings		
Output gear (preload)	0.008–0.002	0.02–0.05
Transfer shaft (endplay)	0.002–0.004	0.05–0.10
Differential (preload)	0.006–0.012	0.15–0.29

TORQUE SPECIFICATIONS

Item	Ft. lbs.	Nm
Cooler line fittings	110 ①	12
Differential cover	165 ①	19
Differential ring gear	70	95
Differential bearing retainer	21	28
Rear end cover	14	19
Extension housing	21	28
Input speed sensor	20	27
Low/Reverse clutch retainer	40 ①	5
Neutral safety switch	25	34
Oil pan to case	14	19
Output gear bolt (1.5 in. hex)	200	271
Output speed sensor	20	27
Pressure taps	45 ①	5
PRNDL switch	25	34
Pump to case	23	32
Reaction shaft to pump	23	32
Solenoid assy. to case	105 ①	12
Transfer plate to case	105 ①	12
Transfer gear nut (1.25 in. hex)	200	271
Valve body and transfer plate	40 ①	5
Vent assembly	110 ①	12
8-way solenoid connector	38 ①	4
60-way EATX connector	38 ①	4

① inch lbs.

Section 3

KM171L and KM172 Transaxles
Chrysler Corp.

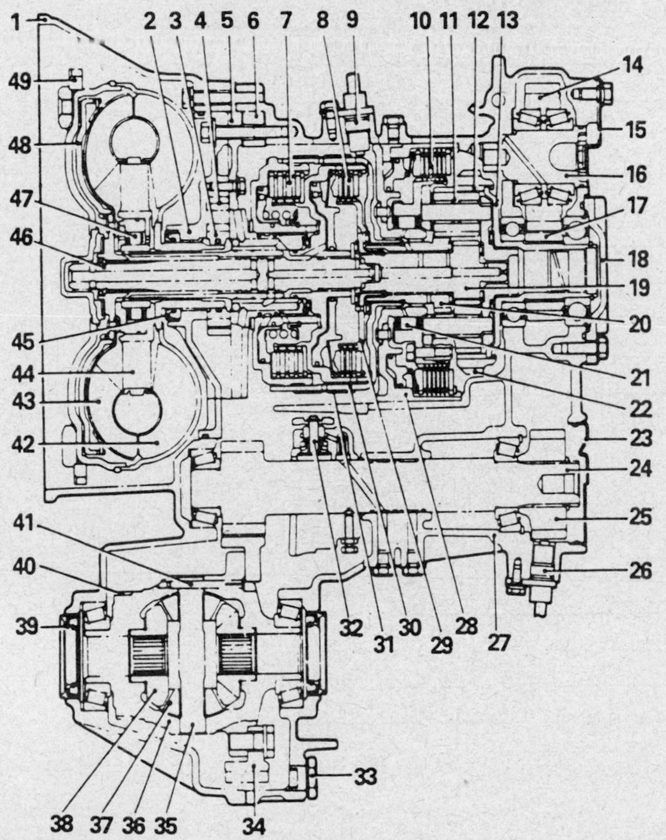

1
2 3 4 5 6 7 8 9
10 11 12 13
14
15
16
17
18
19
20
21
22
23
24
25
26
27
28
29
30
31
32
33
34
35
36
37
38
39
40
41
42
43
44
45
46
47
48
49

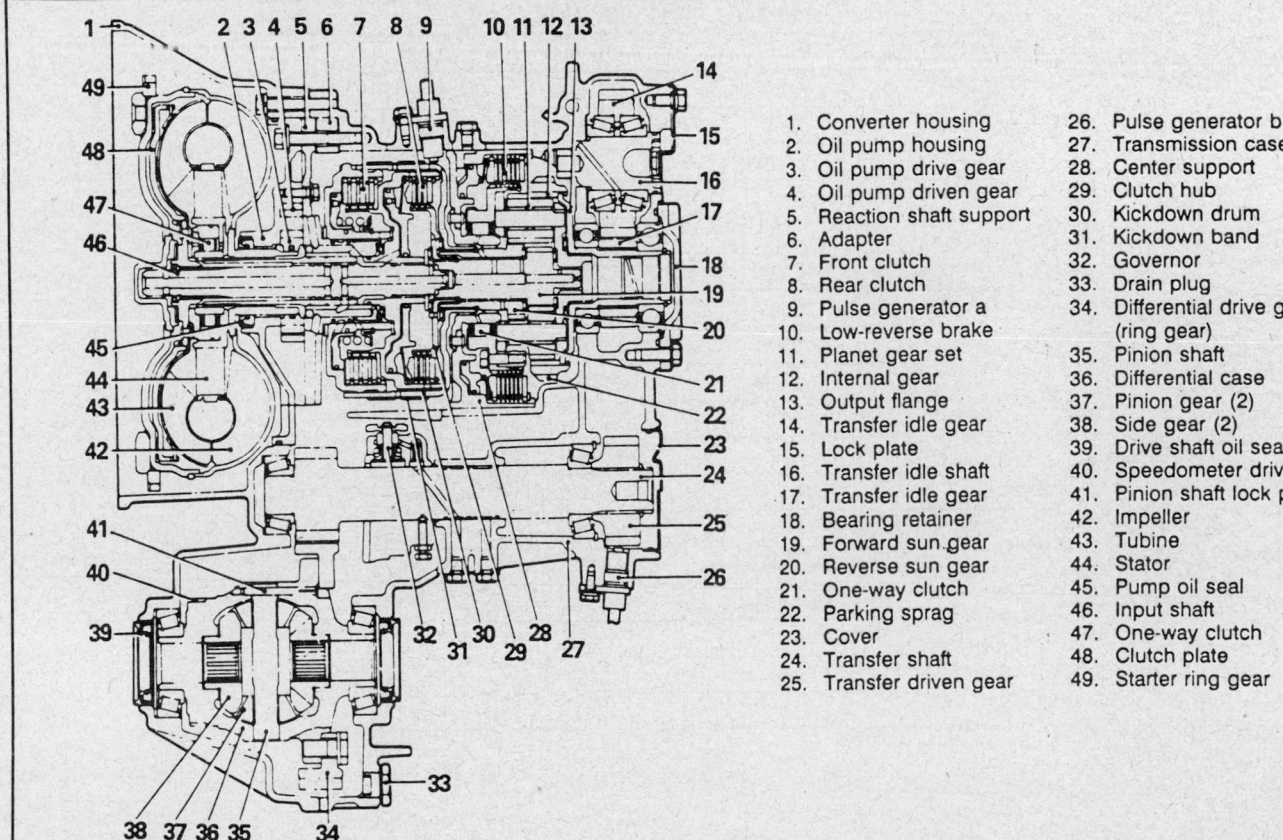

1. Converter housing
2. Oil pump housing
3. Oil pump drive gear
4. Oil pump driven gear
5. Reaction shaft support
6. Adapter
7. Front clutch
8. Rear clutch
9. Pulse generator a
10. Low-reverse brake
11. Planet gear set
12. Internal gear
13. Output flange
14. Transfer idle gear
15. Lock plate
16. Transfer idle shaft
17. Transfer idle gear
18. Bearing retainer
19. Forward sun gear
20. Reverse sun gear
21. One-way clutch
22. Parking sprag
23. Cover
24. Transfer shaft
25. Transfer driven gear
26. Pulse generator b
27. Transmission case
28. Center support
29. Clutch hub
30. Kickdown drum
31. Kickdown band
32. Governor
33. Drain plug
34. Differential drive gear (ring gear)
35. Pinion shaft
36. Differential case
37. Pinion gear (2)
38. Side gear (2)
39. Drive shaft oil seal (2)
40. Speedometer drive gear
41. Pinion shaft lock pin
42. Impeller
43. Tubine
44. Stator
45. Pump oil seal
46. Input shaft
47. One-way clutch
48. Clutch plate
49. Starter ring gear

Cross section of automatic transaxle

REFERENCES

APPLICATION

1980–84 Chilton's Professional Automatic Transmission Manual #7390
1984–89 Chilton's Professional Transmission Manual—Domestic Vehicles #7959

GENERAL DESCRIPTION

1980–84 Chilton's Professional Automatic Transmission Manual #7390

MODIFICATIONS

1980–84 Chilton's Professional Automatic Transmission Manual #7390
1984–89 Chilton's Professional Transmission Manual—Domestic Vehicles #7959

TROUBLE DIAGNOSIS

1980–84 Chilton's Professional Automatic Transmission Manual #7390

ON CAR SERVICES

1980–84 Chilton's Professional Automatic Transmission Manual #7390

REMOVAL AND INSTALLATION

1980–84 Chilton's Professional Automatic Transmission Manual #7390
1984–89 Chilton's Professional Transmission Manual—Domestic Vehicles #7959

BENCH OVERHAUL

1980–84 Chilton's Professional Automatic Transmission Manual #7390

SPECIFICATIONS

1980–84 Chilton's Professional Automatic Transmission Manual #7390
1984–89 Chilton's Professional Transmission Manual—Domestic Vehicles #7959

SPECIAL TOOLS

1980–84 Chilton's Professional Automatic Transmission Manual #7390

APPLICATION

KM 171 AND KM 172

Year	Vehicle	Engine	Transaxle
1984	Colt	1.5L	KM 171
	Colt	1.6L	KM 172
	Colt Vista	2.0L	KM 172
1985	Colt	1.5L	KM 171
	Colt	1.6L	KM 172
	Colt Vista	2.0L	KM 172
1986	Colt	1.5L	KM 171
	Colt	1.6L	KM 172
	Colt Vista	2.0L	KM 172
1987	Colt	1.5L	KM 171
	Colt	1.6L	KM 172
	Colt Vista	2.0L	KM 172
1988	Colt	1.5L	KM 171
	Colt	1.6L	KM 172
	Colt Vista	2.0L	KM 172
1989	Colt	1.5L	KM 171
	Colt	1.6L	KM 172
	Colt Vista	2.0L	KM 172

TRANSAXLE MODIFICATIONS

New Planetary Carrier Bolt

There was a change made in the 1985 Colt Vista automatic transaxle planet carrier bolts and their torque specifications begining in February of 1985 this change started with transaxle serial number AL1771. The new bolt, part number MD720857, can be identified by the under cut made on the back side of the bolt head. If necessary, the new bolt is fully interchangeable with the previously used bolt.

With this change in the bolt, the torque specification has also changed. The new torque specification for either the new or the old bolt has been increased to 10 ft. lbs. (13.5 Nm). This change should be made to the previous bolt specification used on the Colt Vista.

Gasket Kit Installation

A new heavy duty torque converter housing to transaxle gasket kit is available. Included in the kit is a new front pump O-ring and various thickness differential and transfer shaft spacers. When installing the new gasket, it will be necessary to check the differential and transfer shaft spacer thickness and replace, if necessary.

The KM-171 transaxles built prior to December 1986 (up to transaxle CK0520), use kit, part number MD726244. KM-172 transaxles built prior to May 1985 (up to transaxle AQ4209), use kit, part number MD726246. The procedure involves replacing the gasket and proper spacer installation.

Engine Surge at Steady Speeds When at Normal Operating Temperature

The 1985–87 Colt equipped with the KM-171 or the KM-172 transaxle may encounter an engine surge at full operating temperature with the vehicle at a steady cruise/light load condition.

If a surge from 100–200 rpm can be felt when the engine speed is between 1800–2600 rpm, disconnect the automatic transaxle control unit. If the surge is gone, a transaxle modification kit part number MD728159 must be installed.

The modification kit includes the transaxle parts.

Damper clutch control valve—part number MD727246
Damper clutch control sleeve—part number MD723456
Lower valve body—part number MD727244
Lower valve spring—part number MD725203
Solenoid valve— part number MD727245
Oil pan gasket— part number MD707183
O-ring— part number MD707603
O-ring— part number MD707752

This procedure involves replacing the transmission lower valve body damper clutch control valve, sleeve, spring and the lockup solenoid valve. This will correct a lockup transaxle induced engine surge.

Relocation Of Transaxle Breather

On 1986–87 Colt vehicles built after May 1986 and equipped with the KM-171 and KM-172 transaxles, have the transaxle breather relocated from the front of the bell housing through the front pump assembly, to the top of the transaxle case adjacent to the pulse generator.

Manual Control Lever Change

On 1987 Colt and Colt Vista vehicles equipped with the KM-171 and KM-172 transaxle have had the groove in the manual control lever changed. This change is to improve the shifting lever feel. The change requires a new O-ring. The new shaft and O-ring have been installed on these transaxles since August of 1986. If it should become necessary to change the manual control lever or the O-ring on a transaxle prior to this date, both components should be replaced as a set with these new parts.

The new manual control shaft part number is MD726962 and the new O-ring part number is MD726966.

Transaxle Dipstick Improvement

A new dipstick and filler tube has been released, for use on the 1988 Colt Wagon equipped with the KM-171 transaxle. This new dipstick and filler tube is much longer and are positioned for easier access when checking or adding transaxle fluid. This change was effective in February 1988 for transaxle number EF9730 and later.

The new dipstick and filler tube is completely interchangeable with the earlier units. However the dipstick and filler tube must be replaced as a set along with a new O-ring. The new dipstick part number is MD732523, the new filler tube part number is MD731610 and the new O-ring part number is MD966612.

REMOVAL AND INSTALLATION

TRANSAXLE REMOVAL

1. Disconnect the control cable from the transaxle. Remove the center floor console and remove the floor shift assembly.
2. Disconnect the throttle control cable.
3. Disconnect the negative battery cable. Remove the battery and the battery tray.
4. Remove the air cleaner case.
5. Remove the reservoir tank and the windshield washer tank.
6. Disconnect the inhibitor switch connector, pulse generator connectors, lockup solenoid valve connector, oil cooler hoses, and speedometer cable from transaxle.

NOTE: Be sure to plug the oil cooler lines, so that no foreign matter enters the lines. It will also keep the oil from draining out of the oil cooler.

7. Remove the starter motor
8. Raise and support the vehicle safely. Remove the front wheels and engine splash shield. Drain the transmission fluid.
9. Remove the strut bars and the stabilizer bar from the lower arms.
10. Remove the right and left driveshafts from the transaxle as follows:
    a. Remove the castle nut cotter pin, wheel bearing nut and washer.
    b. Remove the lower control arm self locking nut and dis-

connect the lower control arm from the steering knuckle using a suitable ball joint spreader tool.

c. Remove the tie rod cotter pin and separate the tie rod from the steering knuckle using a suitable ball joint spreader.

d. Remove the stabilzer bars locknut and disconnect the stabilizer bar from the steering knuckle.

e. When pulling out the driveshafts from the transaxle case, do so by using 2 suitable pry tools and pry the driveshafts out of the transaxle case. Never attempt to pull out the driveshaft by using the dust cover part of the driveshaft as a fulcrum for leverage. Do not insert the pry bar or the oil seal will be damaged.

g. Force the driveshaft from the hub using tool CT-1003 or equivalent.

11. Remove the bell housing cover and then remove the 3 bolts connecting the torque converter to the drive plate.

12. After removing the bolts, turn and force the torque converter toward the transaxle to prevent the converter from remaining on the engine side.

13. Remove the 5 upper bolts connecting the transaxle with the engine.

14. Using a transmission jack or equivalent, support the lower part of the transaxle and remove the remaining bolts that connect the engine to the transaxle.

NOTE: Support a wide area of the transaxle so that the oil pan is not distorted when supported.

15. Remove the transaxle mount insulator bolts and the mount bracket.

16. Remove the blank cap from inside the right fender shield and remove installation bolts.

17. Slide the transaxle assembly to the right and lower it to remove it. Do not allow the converter to fall from the unit.

TRANSAXLE INSTALLATION

The transaxle and converter should be installed as a unit. Connecting the torque converter to engine first could cause damage to the oil seal on the transaxle and the drive plate.

1. Place the transaxle on the removing jack or equivalent and raise it into position in the vehicle.
2. Replace the transaxle mounting bolts and torque them to the proper specification.
3. Install the 3 bolts that connect the torque converter to the drive plate.
4. Reinstall the bell housing cover and torque the bolts.
5. Install starter motor and torque the bolts.
6. Install the speedometer cable at the transaxle, and reinstall the oil cooler lines.
7. Reinstall the driveshafts as follows:
   a. Install the driveshaft into the hub.
   b. Install the stabilizer bar to the steering knuckle and tighten the locknut.
   c. Install the tie rod to the steering knuckle and install the new cotter pin.
   d. Install the lower control arm to the steering knuckle and torque the self locking nut to 43–52 ft. lbs.
   e. Install the wheel bearing, washer and nut torque the nut to 144–188 ft. lbs. Install the castle nut cotter pin.
   f. Reinstall the front wheels.
8. Fill the transaxle with Dexron® II ATF fluid.
9. Reinstall all the disconnected electrical connectors.
10. Reinstall the throttle control cable to the carburetor. Adjust the cable.
11. Install the control cable to the transaxle, and adjust the slack out of the cable.
12. Install the battery, the battery tray and the air cleaner case.
13. Reinstall the reservoir tank and the windshield washer tank.
14. Confirm ignition switch starter action in the **P** and **N** positions. Make sure that the starter doesn't engage in any other positions.
15. Check to see that the selector lever operates smoothly and is properly shifting into every selector position.
16. Refill the assembly using the correct transmission fluid.

SPECIFICATIONS

TORQUE SPECIFICATIONS
Colt and Colt Vista

Component	ft. lbs.	Nm
Air cleaner (1.6L Engine)		
Nut	5.8–6.5	8–9
Bolt	5.8–7.2	8–10
Lever to bracket assembly	10–14	14–20
Transaxle drain plug	22–25	30–35
Starter motor to transaxle	20–25	27–34
Transaxle mount bracket to body	22–29	30–40
Transaxle mount bracket to transaxle bracket	65–80②	90–110②
Transaxle bracket to transaxle	65–80②	90–110②
Bell housing cover to transaxle	7–9	10–12
Tie rod end to knuckle—Colt	11–25	15–34
Lower arm ball joint to knuckle—Colt	43–52	60–72

TORQUE SPECIFICATIONS
Colt and Colt Vista

Component	ft. lbs.	Nm
Transaxle mount bolt	31–40	43–55
Transaxle mount bolt	22–25	30–35
Drive plate	94–101	130–140
Drive plate-to-converter tightening bolt	34–38	46–53
Transaxle mounting bolt (10mm diameter bolt)	32–39	43–55
Transaxle mounting bolt (8mm diameter bolt)	22–25①	30–35①
Bell housing cover bolt	7–9	10–12
Drain plug	22–25	30–35
Pressure check plug	6–7	8–10
Pulse generator mounting bolt	7.5–8.5	10–12
Bearing retainer bolt	11–15	15–22

TORQUE SPECIFICATIONS
Colt and Colt Vista

Component	ft. lbs.	Nm
Lock plate bolt	15–19	20–27
Oil cooler connector	11–15	15–22
Converter housing bolt	14–16	19–23
Oil pan bolt	7.5–8.5	10–12
Kickdown servo piston plate screw	4–6	6–8
Center support bolt	15–19	20–27
One-way clutch outer race bolt	6–7	8–10
Differential drive gear bolt	94–101	130–140
Governor bolt	6–7	8–10
Governor bolt lock nut	3–4	4–6
Manual control lever nut	13–15	17–21
Manual control shaft set screw	6–7	8–10
Neutral safety switch	7.5–8.5	10–12
Sprag rod support bolt	15–19	20–27
Pump housing-to-reaction shaft support bolt	7.5–8.5	10–12
Oil pump assembly mounting bolt	11–15	15–22
Valve body bolt	3–4	4–6
Throttle cam bolt	6–7	8–10
Valve body assembly mounting bolt	7.5–8.5	10–12
Oil filter bolt	4–5	5–7
Speedometer sleeve locking plate bolt	2.5–3.5	3–5

SERVICE SPECIFICATIONS
Colt Vista

Component	in.	mm
Differential case preload	0–.006	0–0.15
Differential side gear and pinion backlash	.001–.006	0.025–0.150
Input shaft endplay	.012–.039	0.3–1.0
Transfer shaft preload	.004–.006	0.1–0.15
Oil pump gear side clearance	.001–.002	0.03–0.05
Front clutch snapring clearance	.028–.035	0.7–0.9
Rear clutch snapring clearance	.016–.024	0.4–0.6
Low reverse brake snapring clearance	.031–.043	0.78–1.09
Output flange bearing endplay	0–.002	0–0.06
Transfer idler gear bearing preload	0.6 ①	0.8 ②

① ft. lbs.
② Nm

VALVE BODY SPRING IDENTIFICATION
Colt Vista

Component	Free height in. (mm)	Outside diameter in. (mm)	Number of Loops	Diameter of Wire in. (mm)
Throttle valve spring	1.262 (32.05)	.374 (9.5)	11.8	.039 (1.0)
Kickdown valve spring	1.029 (24.14)	.232 (5.9)	19	.050 (0.5)
Range control valve spring	0.921 (23.4)	.331 (8.4)	11	.039 (1.0)
Torque converter control valve spring	1.039 (26.4)	.346 (8.8)	12	.043 (1.1)
Regulator valve spring	2.024 (51.4)	.606 (15.4)	12	.055 (1.4)
1–2 shift valve spring	1.232 (31.3)	.299 (76)	9.9	.055 (1.4)
2–3 control valve spring	1.897 (48.19)	.260 (6.6)	22.31	.035 (0.9)
2–3 shift valve spring	0.933 (23.71)	.283 (7.2)	14	.035 (0.9)
Line relief spring	0.681 (17.3)	.276 (7.0)	10	.039 (1.0)
Low relief spring	0.491 (12.46)	.260 (6.6)	8.5	.024 (0.6)
Reducing valve spring	1.589 (40.35)	.374 (9.5)	—	.031 (0.8)
Damper clutch control valve spring	0.618 (15.7)	.244 (6.2)	10.5	.028 (0.7)
Kickdown servo control valve spring	1.011 (25.69)	.291 (7.4)	14	.024 (0.6)
N–D accumulator valve spring	2.100 (53.34)	.307 (7.8)	27.8	.031 (0.8)
N–D accumulator plug spring	1.511 (38.38)	.535 (13.6)	10.4	.055 (1.4)

SNAPRING SELECTION
Colt Vista

Component	Thickness in. (Nm)	Identification Symbol	Part Number
Snapring (for output flange preload adjustment)	.0717 (1.82)	None	MD722538
	.0740 (1.88)	Blue	MD721014
	.0764 (1.94)	Brown	MD721015
	.0787 (2.00)	None	MD721016
	.0811 (2.06)	Blue	MD721017
	.0835 (2.12)	Brown	MD722539
Spacer (for transfer shaft preload adjustment)	.0323 (0.82)	82	MD712638
	.0335 (0.85)	85	MD712639
	.0346 (0.88)	88	MD712640
	.0358 (0.91)	91	MD712641
	.0370 (0.94)	94	MD712642
	.0382 (0.97)	97	MD712643
	.0394 (1.00)	00	MD712644
	.0406 (1.03)	03	MD712645
	.0417 (1.06)	06	MD712646
	.0429 (1.09)	09	MD712647
	.0441 (1.12)	12	MD712648
	.0453 (1.15)	15	MD712649
	.0465 (1.18)	18	MD712650
	.0476 (1.21)	21	MD712651
	.0488 (1.24)	24	MD712652
	.0500 (1.27)	27	MD712653
	.0512 (1.30)	30	MD712654
	.0524 (1.33)	33	MD712655
	.0535 (1.36)	36	MD712656
	.0547 (1.39)	39	MD712657
	.0559 (1.42)	42	MD712658
	.0571 (1.45)	45	MD712659
	.0583 (1.48)	48	MD712660
	.0594 (1.51)	51	MD712661
	.0606 (1.54)	54	MD712662
	.0618 (1.57)	57	MD712663
	.0630 (1.60)	60	MD712664
	.0642 (1.63)	63	MD712665
	.0654 (1.66)	66	MD712666
	.0665 (1.69)	69	MD712667
	.0677 (1.72)	72	MD722807
	.0713 (1.81)	81	MD722808
Spacer (for differential case preload adjustment)	.043 (1.10)	J	MD710454
	.044 (1.13)	D	MD700270
	.046 (1.16)	K	MD710455
	.047 (1.19)	L	MD710456
	.048 (1.22)	G	MD700271
	.049 (1.25)	M	MD710457

SNAPRING SELECTION
Colt Vista

Component	Thickness In (Nm)	Identification Symbol	Part Number
Spacer (for differential case preload adjustment)	.050 (1.28)	N	MD710458
	.052 (1.31)	E	MD706574
	.053 (1.34)	O	MD710459
	.054 (1.37)	P	MD710460
	.055 (1.40)	None	MD706573
	.056 (1.43)	Q	MD710461
	.057 (1.46)	R	MD710462
	.059 (1.49)	C	MD706572
	.060 (1.52)	S	MD710463
	.061 (1.55)	T	MD710464
	.062 (1.58)	B	MD706571
	.063 (1.61)	U	MD710465
	.065 (1.64)	V	MD710466
	.066 (1.67)	A	MD706570
	.067 (1.70)	W	MD710467
	.068 (1.73)	X	MD710468
	.069 (1.76)	F	MD706575
	.070 (1.79)	Y	MD710469
	.072 (1.82)	Z	MD710470
	.073 (1.85)	H	MD700272
	.074 (1.88)	AA	MD710471
	.075 (1.01)	BB	MD715955
	.076 (1.94)	CC	MD715956
	.078 (1.97)	DD	MD715957
	.079 (2.00)	EE	MD715958
	.080 (2.03)	FF	MD715959
	.081 (2.06)	GG	MD715960
	.082 (2.09)	HH	MD715961
	.083 (2.12)	II	MD715962
	.085 (2.15)	JJ	MD715963
	.086 (2.18)	KK	MD715964
	.087 (2.21)	LL	MD715965
	.088 (2.24)	MM	MD715966
	.089 (2.27)	NN	MD715967
	.091 (2.30)	OO	MD715968
	.092 (2.33)	PP	MD715969
	.093 (2.36)	QQ	MD715970
	.094 (2.39)	RR	MD715971
	.095 (2.42)	SS	MD722734
	.096 (2.45)	TT	MD722735
	.098 (2.48)	UU	MD722736
Snapring (front, rear clutch snapring clearance adjustment)	.0630 (1.6)	None	MD955630
	.0709 (1.8)	Blue	MD955631
	.0787 (2.0)	Brown	MD955632

SNAPRING SELECTION
Colt Vista

Component	Thickness In (Nm)	Identification Symbol	Part Number
Snapring (front, rear clutch snapring clearance adjustment)	.0866 (2.2)	None	MD955633
	.0945 (2.4)	Blue	MD955634
	.1024 (2.6)	Brown	MD955635
	.1102 (2.8)	None	MD955636
	.1181 (3.0)	Blue	MD955637
Spacer (for adjustment of differential side gear and pinion backlash)	.0295–.0323 (0.75–0.82)	—	MA180862
	.0327–.0362 (0.83–0.92)	—	MA180861
	.0366–.0394 (0.93–1.00)	—	MA180860
	.0398–.0425 (1.01–1.08)	—	MA180875
	.0429–.0457 (1.09–1.16)	—	MA180876

Section 3

ATX Transaxle
Ford Motor Co.

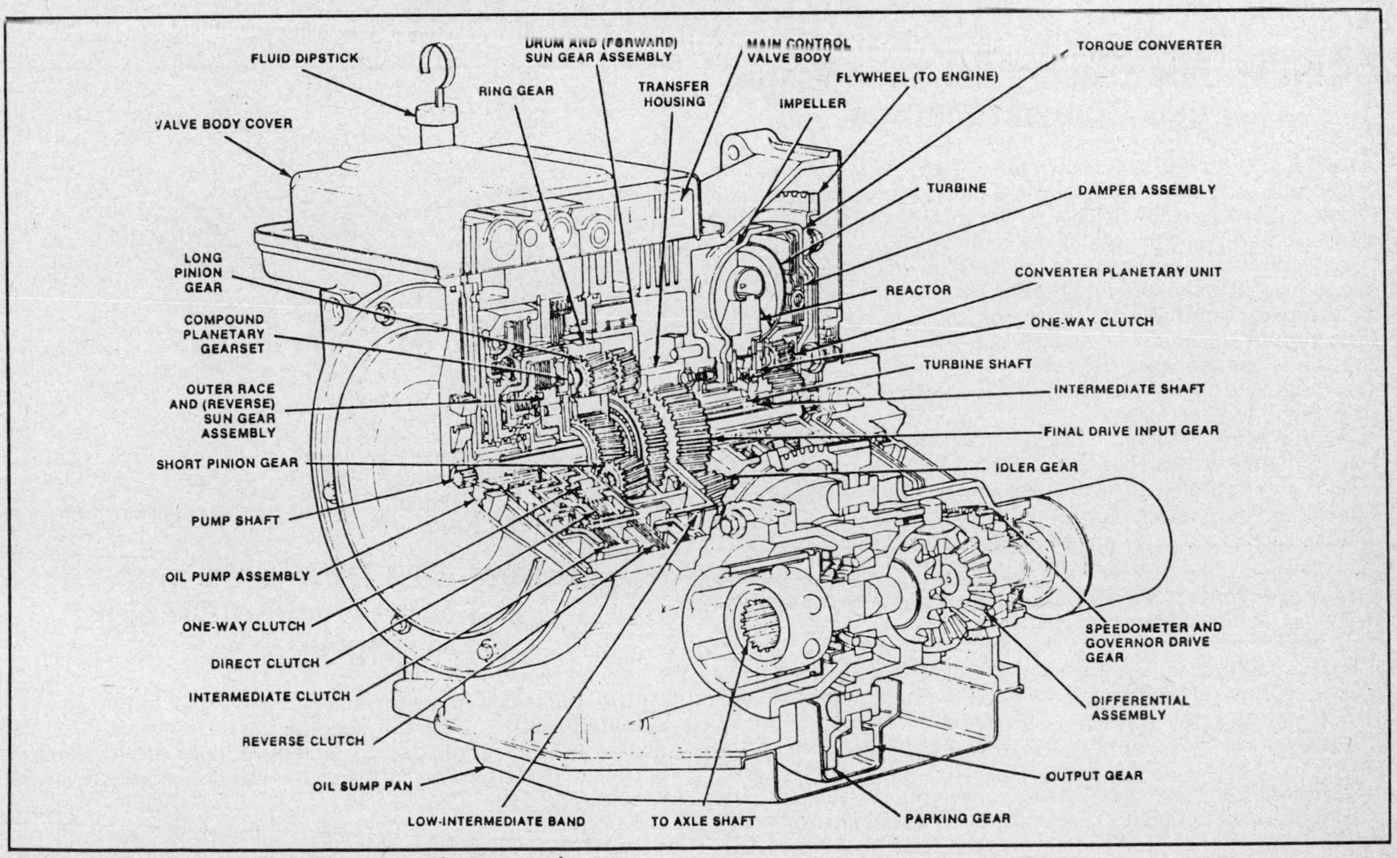

FLUID DIPSTICK
DRUM AND (FORWARD) SUN GEAR ASSEMBLY
MAIN CONTROL VALVE BODY
TORQUE CONVERTER
RING GEAR
TRANSFER HOUSING
FLYWHEEL (TO ENGINE)
IMPELLER
VALVE BODY COVER
TURBINE
DAMPER ASSEMBLY
LONG PINION GEAR
CONVERTER PLANETARY UNIT
REACTOR
COMPOUND PLANETARY GEARSET
ONE-WAY CLUTCH
TURBINE SHAFT
OUTER RACE AND (REVERSE) SUN GEAR ASSEMBLY
INTERMEDIATE SHAFT
FINAL DRIVE INPUT GEAR
SHORT PINION GEAR
IDLER GEAR
PUMP SHAFT
OIL PUMP ASSEMBLY
ONE-WAY CLUTCH
SPEEDOMETER AND GOVERNOR DRIVE GEAR
DIRECT CLUTCH
INTERMEDIATE CLUTCH
DIFFERENTIAL ASSEMBLY
REVERSE CLUTCH
OIL SUMP PAN
OUTPUT GEAR
LOW-INTERMEDIATE BAND
TO AXLE SHAFT
PARKING GEAR

Cross section of Ford ATX automatic transaxle

REFERENCES

APPLICATION

1980–84 Chilton's Professional Automatic Transmission Manual #7390
1984–89 Chilton's Professional Transmission Manual—Domestic Vehicles #7959

GENERAL DESCRIPTION

1980–84 Chilton's Professional Automatic Transmission Manual #7390

MODIFICATIONS

1980–84 Chilton's Professional Automatic Transmission Manual #7390
1984–89 Chilton's Professional Transmission Manual—Domestic Vehicles #7959

TROUBLE DIAGNOSIS

1980–84 Chilton's Professional Automatic Transmission Manual #7390

ON CAR SERVICES

1980–84 Chilton's Professional Automatic Transmission Manual #7390

REMOVAL AND INSTALLATION

1980–84 Chilton's Professional Automatic Transmission Manual #7390
1984–89 Chilton's Professional Transmission Manual—Domestic Vehicles #7959

BENCH OVERHAUL

1980–84 Chilton's Professional Automatic Transmission Manual #7390

SPECIFICATIONS

1980–84 Chilton's Professional Automatic Transmission Manual #7390
1984–89 Chilton's Professional Transmission Manual—Domestic Vehicles #7959

SPECIAL TOOLS

1980–84 Chilton's Professional Automatic Transmission Manual #7390

APPLICATION

1984–89 Tempo/Topaz and Escort
1984–87 Lynx
1986–89 Taurus and Sable

TRANSAXLE MODIFICATIONS

Cooler Line Disconnect Tool Usage Push Connect Fittings

To service the transaxle cooler lines, tool T82L–9500–AH or equivalent, is required. The purpose of the tool is to spread the "duck bill" retainer to disengage the tube bead. The following steps are necessary for use of the tool:

To facilitate use of the tool, clean the road dirt from the fitting before inserting the tool into the fitting. Also, it is important to avoid any contamination of the fitting and transaxle, dirt in the fitting could cause an O-ring leak.

1. Slide the tool over the tube.
2. Align the opening of the tool with 1 of the 2 tabs on the fitting "duck bill" retainer.
3. Firmly insert tool into fitting until it seats against the tube bead (a definite click should be hard).
4. With a thumb held against the tool, firmly pull back on the tube until it disengages from the fitting.

CAUTION

Do not attempt to separate the cooler line from the fitting by prying with another tool. This will break the plastic insert in fitting and bend the cooler lines at the junction to the fitting.

Before assembly of the lines in the fitting, visually inspect the plastic retainer in the fitting for a broken tab. If a tab is broken, the fitting must be replaced. Also visually inspect the cooler lines to make sure they are not bent at the junction of the fitting.

Tube assembly is accomplished by inserting the tube into the fitting until the retainer engages the tube head (a definite click should be heard). Pull back on the tube to ensure full engagement.

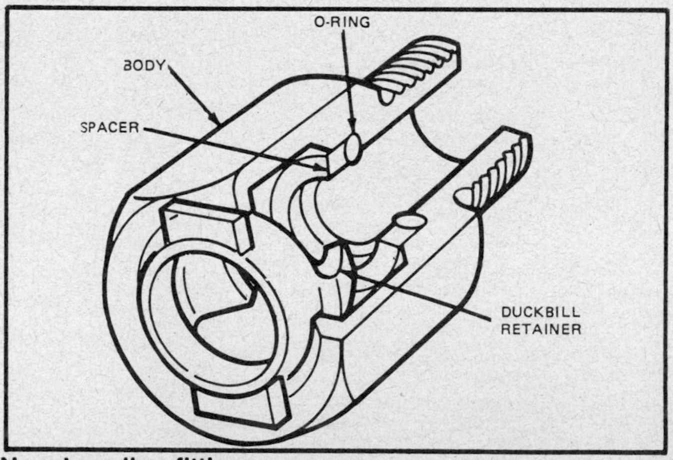
O-RING
BODY
SPACER
DUCKBILL RETAINER
New type line fitting

Sensitive Downshift and 3–2/2–3 Shift Cycling On Light Throttle

On the 1984–85 Tempo/Topaz, concerns of 3–2 torque demand sensitivity and 3–2/2–3 shift hunting can be resolved by discarding the 3–2 control valve spring and installing a check ball at the bottom of the bore. This repair is for vehicles with the following calibration codes:

4–26E–RO
4–26D–R18
4–26S–R13

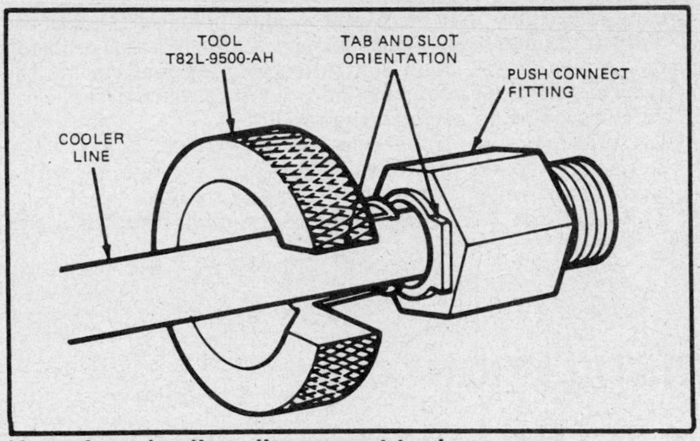
TOOL
T82L-9500-AH
TAB AND SLOT
ORIENTATION
PUSH CONNECT
FITTING
COOLER
LINE

Use of cooler line disconnect tool

To perform the repair the main control assembly must be removed and disassembled. Revised assembly of the 3–2 control valve is as follows:

1. Remove main control assembly and disassemble.
2. Locate and remove the 3–2 control valve components.
3. Discard the control valve spring (yellow or purple in color).
4. Install check ball (EOAZ–7E195–B) at bootom of bore.
5. Replace 3–2 control valve with main control, do not install spring, and install retainer.
6. Assemble and install the main control assembly.

3–2/2–3 Shift Cycling on Light Throttle

On the 1984–85 Tempo/Topaz, concerns of light throttle 2–3/3–2 shift hunting at approximately 24mph can be resolved by installing a new design governor spring. This repair is for vehicles with the following calibration codes:

4–26D–R18
4–26S–R13
4–26G–R11

To perform the repair, the governor assembly must be removed from the transaxle.

1. Remove the governor assembly.
2. Remove and discard the governor spring (pink in color) and replace with service spring (E43Z–7E467–A—brown in color).

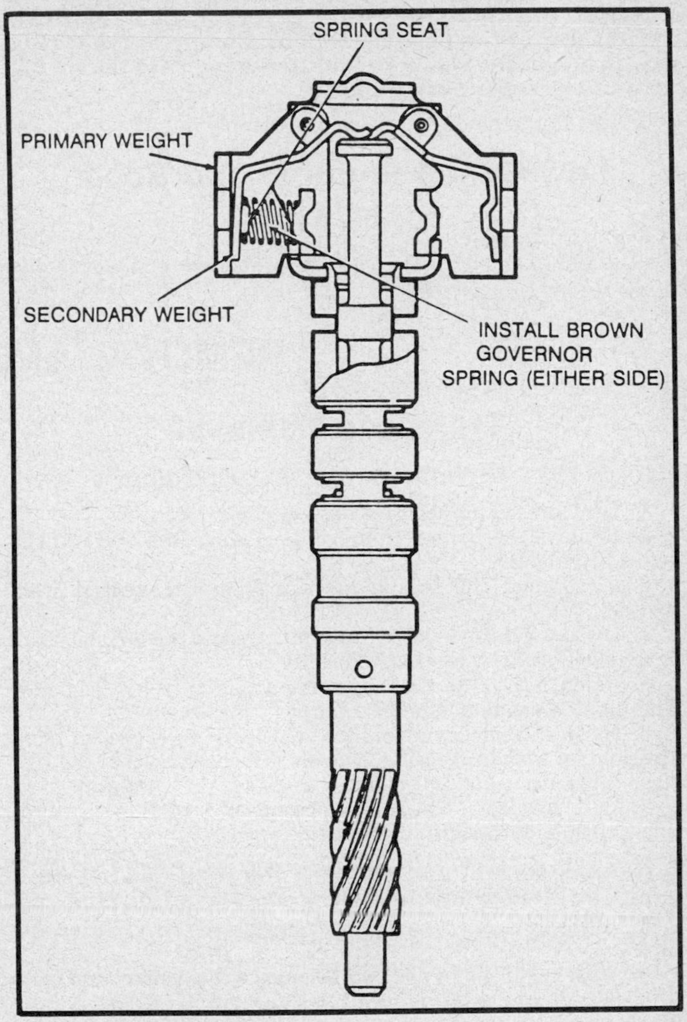
SPRING SEAT
PRIMARY WEIGHT
SECONDARY WEIGHT
INSTALL BROWN
GOVERNOR
SPRING (EITHER SIDE)

Governor modification

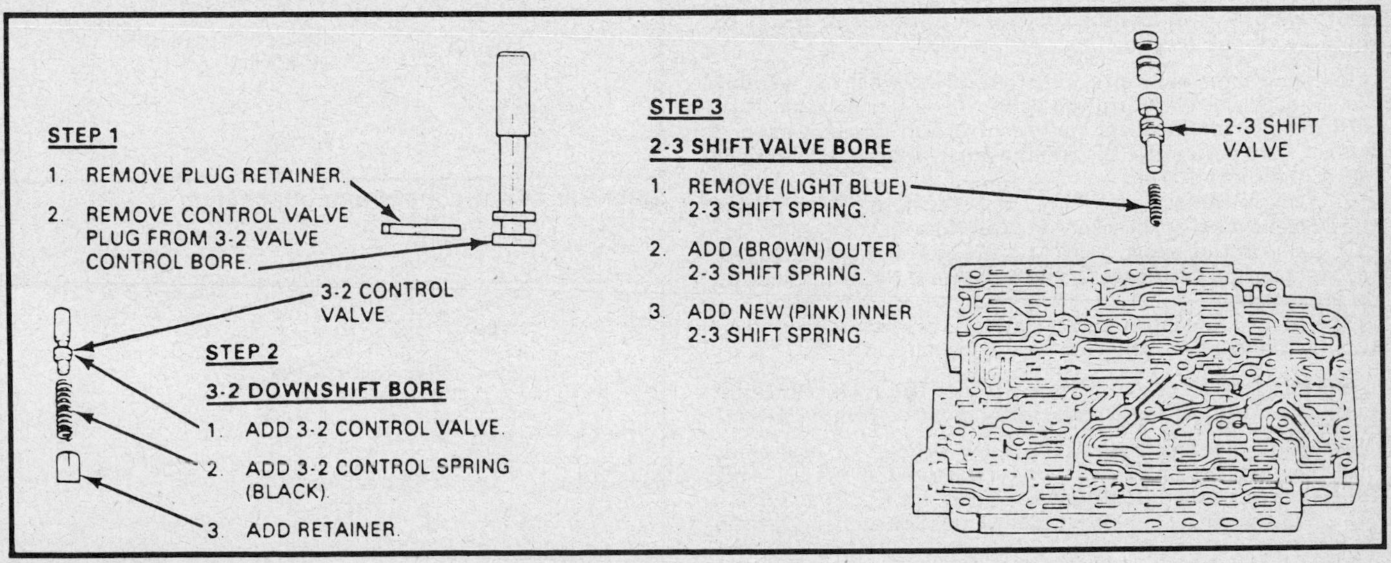
STEP 1
1. REMOVE PLUG RETAINER.
2. REMOVE CONTROL VALVE PLUG FROM 3-2 VALVE CONTROL BORE.
3-2 CONTROL VALVE
STEP 2
3-2 DOWNSHIFT BORE
1. ADD 3-2 CONTROL VALVE.
2. ADD 3-2 CONTROL SPRING (BLACK).
3. ADD RETAINER.
STEP 3
2-3 SHIFT VALVE BORE
1. REMOVE (LIGHT BLUE) 2-3 SHIFT SPRING.
2. ADD (BROWN) OUTER 2-3 SHIFT SPRING.
3. ADD NEW (PINK) INNER 2-3 SHIFT SPRING.
2-3 SHIFT VALVE

Modifying the valve body

NOTE: Position 1 end of the governor spring onto the spring seat of the primary weight. Compress the spring and position the other end of the spring onto the spring seat of the secondary weight.

3. Install the modified governor assembly into the transaxle.

Low Speed Shudder and Boom in 3rd Gear

On the 1986 Tempo/Topaz, a low speed shudder and boom while driving in 3rd gear may be caused by shift point variance in the 2–3 upshift and 3–2 downshift pattern. To correct this problem, install a new design main control service kit that revises the part throttle 2–3 upshift and 3–2 downshift pattern. Use the following procedure to perform the repair:

1. Obtain main control service kit (E6FZ–7A230–A).
2. Replace the 3–2 control valve bore plug and retainer with a 3–2 downshift valve, spring and flat plate retainer.
3. Replace the 2–3 spring with the assembly from the repair kit.

REMOVAL AND INSTALLTION

Transaxle Removal

Except 1984–85 Vehicles with 2.3L HSC Engine

1. Disconnect the negative battery cable.
2. Remove the bolts retaining the managed air valve to the valve body cover.
3. Disconnect the wiring harness from the neutral start switch.
4. Disconnect the throttle valve linkage and the manual valve lever cable at their respective levers.
5. Remove the transaxle to engine upper bolts, located below and on either side of the distributor.
6. Raise and safely support the vehicle. Remove the nut from the control arm to steering knuckle attaching bolt, at the ball joint, both right and left sides.
7. With the use of a punch and hammer, drive the bolt out of the knuckle, both right and left sides.

NOTE: The bolt and nut from the right and left sides must be discarded and new ones used during the assembly.

8. With the use of a pry bar, disengage the control arm from the steering knuckle, on both the right and left sides.

NOTE: The plastic shield installed behind the rotor contains a molded pocket into which the lower control arm ball joint fits. When disengaging the control arm from the knuckle, clearance for the ball joint can be provided by bending the shield back towards the rotor. Failure to provide clearance for the ball joint can result in damage to the shield.

9. Remove the bolts attaching the stabilizer bar to the frame rail on both sides. Discard the bolts.
10. Remove the stabilizer bar to control arm attaching nut and washer from both sides. Discard the nuts. Pull the stabilizer bar out of the control arms.
11. Remove the bolt attaching the brake hose routing clip to the suspension strut bracket, on both sides.
12. Remove the steering gear tie rod to steering knuckle attaching nut and disengage the tie rod from the steering knuckle on both sides.
13. Pry the halfshaft from the right side of the transaxle. Position the halfshaft on the transaxle housing.

NOTE: Due to the configuration of the ATX transaxle case, the right halfshaft assembly must be removed first.

14. Insert the differential tool, T81P–4026–A or equivalent into the right side halfshaft bore and drive the left halfshaft from the transaxle differential side gear.
15. Pull the left halfshaft from the transaxle and support the end of the shaft by wiring the shaft to the underbody of the vehicle.

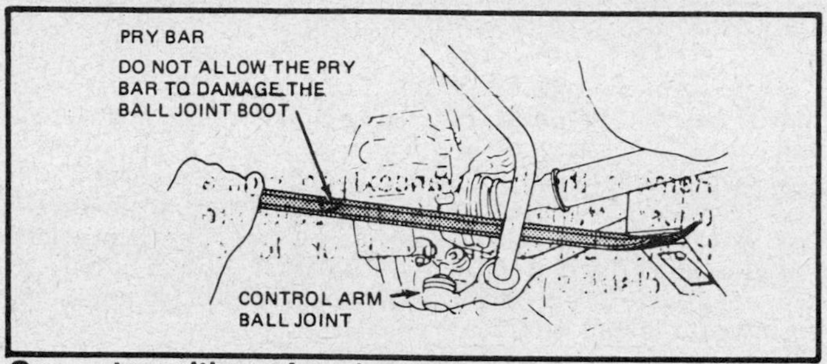
PRY BAR
DO NOT ALLOW THE PRY BAR TO DAMAGE THE BALL JOINT BOOT
CONTROL ARM BALL JOINT
Correct position of prybar

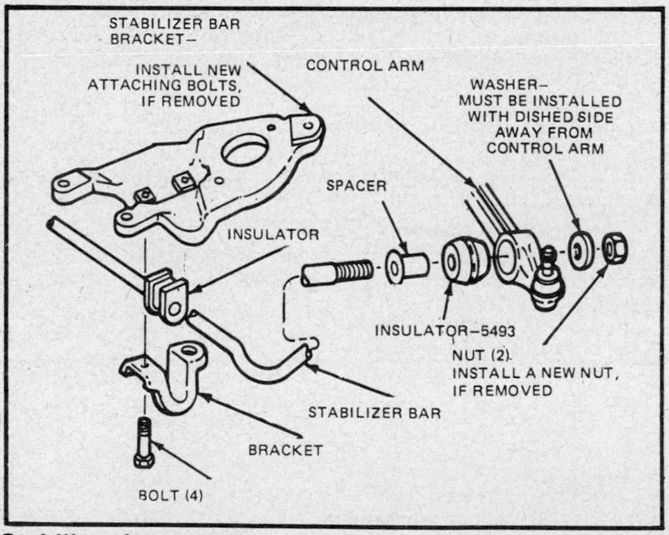
STABILIZER BAR BRACKET—
INSTALL NEW ATTACHING BOLTS, IF REMOVED
CONTROL ARM
WASHER— MUST BE INSTALLED WITH DISHED SIDE AWAY FROM CONTROL ARM
SPACER
INSULATOR
INSULATOR–5493
NUT (2). INSTALL A NEW NUT, IF REMOVED
STABILIZER BAR
BRACKET
BOLT (4)
Stabilizer bar to control arm attachment

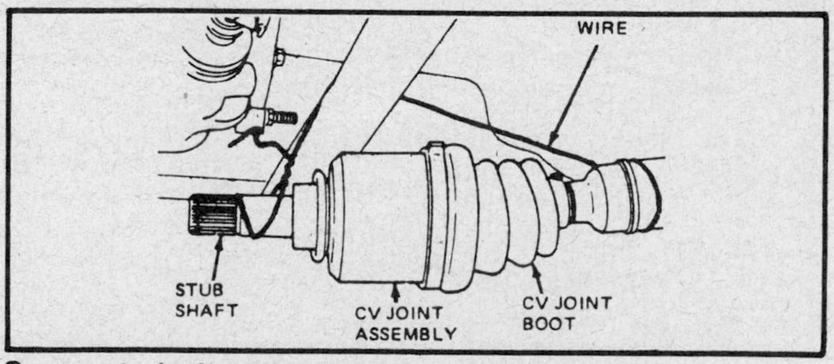
WIRE
STUB SHAFT
CV JOINT ASSEMBLY
CV JOINT BOOT
Support shaft and CV joint with wire

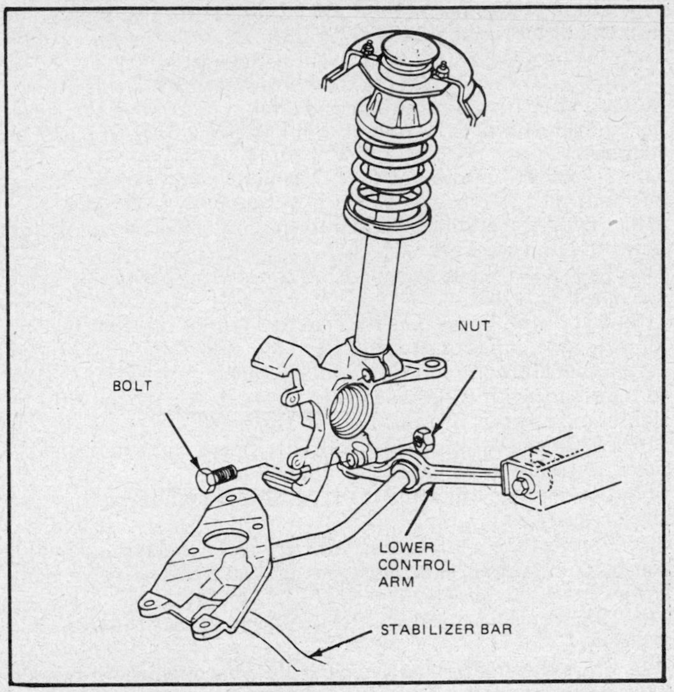
NUT
BOLT
LOWER
CONTROL
ARM
STABILIZER BAR

Steering knuckle attachment

NOTE: Never allow the halfshaft to hang unsupported as damage to the outboard CV joint may result.

16. Install seal plugs into the bores of the left and right halfshafts.
17. Remove the starter support brackets and disconnect the starter cable. Remove the starter attaching bolts and remove the starter.
18. Remove the transaxle support bracket and the outer dust cover from the torque converter housing.
19. Remove the torque converter to flywheel retaining nuts. Matchmark torque converter to flywheel. Turn the crankshaft pulley bolt to bring the retaining nuts into an accessible position.
20. Remove the nuts attaching the left front insulator mount to the body bracket.
21. Remove the bracket to body attaching bolts and remove the bracket.
22. Remove the left rear insulator mount bracket attaching nut.
23. Disconnect the transaxle cooler lines and remove the bolts attaching the manual lever bracket to the transaxle case.
24. Position a transmission jack or other lifting device under the transaxle and remove the remaining transaxle to engine attaching bolts.
25. Separate the transaxle from the engine enough for the torque converter studs to clear the flywheel and lower the transaxle approximately 3 in.
26. Disconnect the speedometer cable and continue lowering the transaxle from the vehicle.

NOTE: When moving the transaxle away from the engine, if the insulator mount contacts the body before the converter studs clear the flywheel, remove the insulator mount.

1984–85 Vehicles with 2.5L HSC Engine

NOTE: In 1984–85 vehicles equipped with the 2.5L HSC engine, the engine and transaxle must be removed as an assembly.

1. Mark the position of the hood and remove the hood from the vehicle.
2. Disconnect the negative battery cable and remove the air cleaner.
3. Position a drain pan under the lower radiator hose and remove the lower hose. Allow the coolant to drain into the pan.

CAUTION

Do not drain the cooling system at this point if the coolant is at normal operation temperature. Personal injury can result, due to excessive heat of the coolant.

4. Remove the upper radiator hose from the engine.
5. Disconnect the oil cooler lines at the rubber hoses below the radiator.
6. Remove the coil assembly from the cylinder head.
7. Diconnect the coolant fan electrical connector, remove the radiator shroud and cooling fan as an assembly. Remove the radiator.
8. If equipped with air conditioning, discharge the system and remove the pressure and suction lines from the air conditioning compressor.

CAUTION

Refrigerant R-12 is contained in the air conditioning system under high pressure. Extreme care must be used when discharging the system, personal injury can result.

9. Identify and disconnect all electrical and vacuum lines as necessary.
10. Disconnect the accelerator linkage, the fuel supply and return hoses on the engine and the thermactor pump discharge hose at the pump. Disconnect T.V. linkage at transaxle.
11. If equipped with power steering, disconnect the pressure and return lines at the power steering pump. Remove the power steering lines bracket at the cylinder head.
12. Install an engine holding or support tool device to the engine lifting eye. Raise and safely support the vehicle.
13. Remove the starter cable from the starter.
14. Remove the hose from the catalytic converter.
15. Remove the bolt attaching the exhaust pipe bracket to the oil pan.
16. Remove the exhaust pipes to exhaust manifold retaining nuts. Pull the exhaust system from the rubber insulating grommets.
17. Remove the speedometer cable from the transaxle.
18. Position a coolant drain pan under the heater hoses and remove the heater hose from the water pump inlet tube. Remove the remaining heater hoses from the steel tube on the intake manifold.
19. Remove the water pump inlet tube clamp attaching bolt at the engine block and remove the 2 clamp attaching bolts at the underside of the oil pan. Remove the inlet tube.
20. Remove the bolts retaining the control arms to the body. Remove the stabilizer bar brackets retaining bolts and remove the brackets.
21. Remove the bolt retaining the brake hose routing clip to the suspension strut.
22. From the right and left sides, remove the nut from the ball joint to steering knuckle attaching bolt. Drive the bolt out of the steering knuckle with a punch and hammer. Discard the bolt and nut.
23. Separate the ball joint from the steering knuckle by using a pry bar. Position the end of the pry bar outside of the bushing pocket, to avoid damage to the bushing or ball joint boot.

NOTE: The lower control arm ball joint fits into a pocket formed in the plastic disc brake shield. this shield must be bend back, away from the ball joint while prying the ball joint out of the steering knuckle.

24. Due to the configuration of the ATX transaxle housing, the

right side halfshaft must be removed first. Position the pry bar between the case and the shaft and pry outward.

NOTE: Use extreme care to avoid damaging the differential oil seal or the CV-joint boot.

25. Support the end of the shaft by suspending it from a convenient underbody component with a length of wire.

NOTE: Do not allow the halfshaft to hang unsupported; damage to the outboard CV-joint may occur.

26. Install driver tool T81P–4026–A or equivalent, in the right halfshaft bore of the transaxle and tap the left halfshaft from its circlip retaining groove in the differential side gear splines. Support the left halfshaft in the same manner as the right halfshaft. Install plugs in the left and right halfshaft bores.
27. Disconnect the manual shift cable clip from the lever on the transaxle. Remove the manual shift linkage bracket bolts from the transaxle and remove the bracket.
28. Remove the left hand rear insulator mount bracket from the body bracket by removing the 2 retaining nuts.
29. Remove the left hand front insulator to transaxle mounting bolts.
30. Lower the vehicle and attach the lifting equipment to the existing lifting eyes on the engine. Remove the engine holding or support tool.

NOTE: Do not allow the front wheels to touch the floor.

31. Remove the right hand insulator intermediate bracket to engine bracket bolts, intermediate bracket to insulator attaching nuts and the nut on the bottom of the double ended stud which attaches the intermediate bracket to the engine bracket. Remove the bracket.
32. Carefully lower the engine/transaxle assembly to the floor. Raise the vehicle from over the assembly. Separate the engine from the transaxle and do the necessary repair work to the transaxle assembly.

TRANSAXLE INSTALLATION

Except 1984–85 Vehicles with 2.3L HSC ENGINE

1. With the transaxle on a lifting device or transmission jack, position the assembly under the vehicle and slowly raise the unit into position to mate with the engine.
2. With the unit almost in position, attach the speedometer cable to the transaxle assembly. Rotate the converter or the flywheel until the matchmarks made during the removal, are in alignment. Raise the transaxle and position to the engine. Install the 4 lower transaxle to engine retaining bolts. Torque to 40–50 ft. lbs. (54–68 Nm).
3. Install the oil cooler lines and bolt the manual lever bracket to the transaxle case.
4. Install the left rear insulator mount bracket attaching nut.
5. Install the left front insulator mount bracket attaching bolts and install the nuts attaching the left front insulator mount to body bracket.
6. Install the torque converter to flywheel retaining nuts and torque to 17–29 ft. lbs. (23–39 Nm).
7. Install the transaxle support bracket and the dust cover to the torque converter housing.
8. Install the starter and retaining bolts. Attach the starter cable and install the starter support brackets.
9. Remove the left halfshaft bore seal plug, install a new circlip on the CV-joint stub shaft and insert the stub shaft into the differential. Carefully align the splines of the stub axle with those of the differential gear. Push the CV joint until the circlip is felt to seat in the differential gear.

NOTE: Use care during the stub axle installation, not to damage the differential oil seat.

10. Using the same method, install the right side stub shaft into the differential side gear splines. Be sure the new circlip seats in the groove of the differential side gear.
11. Install the steering gear tie rod to steering knuckle attaching nut, after installing the tie rod stud in its bore on the steering knuckle, on both the right and left sides. Lock in place, as required.
12. Install the brake hose routing clip retaining bolts on the left and right side suspension.
13. Install the stabilizer bar into the control arms and install new nuts and washers.
14. Install new bolts and attach the stabilizer bar to the frame rails on both sides.
15. Install the lower control arm ball joints into the steering knuckle assemblies on the right and left sides. Using new bolts, install them into the steering knuckle, locking the ball joint stud to the steering knuckle. Torque to 37–44 ft. lbs. (50–60 Nm) by tightening the nut. Do not tighten the bolt.
16. Lower the vehicle and install the upper engine to transaxle bolts, torquing to 40–50 ft. lbs. (54–68 Nm).
17. Connect the linkages, the throttle and manual controls, to their respective levers.
18. Connect the neutral start switch wire connector and install the bolts retaining the managed air valve to the valve body cover.
19. Connect the negative battery cable and verify the transaxle fluid level.
20. Start the engine and cycle the fluid through the transaxle by moving the manual valve control lever.
21. Recheck the assembly, the fluid level and road test as required.

1984–85 Vehicles with 2.3L HSC Engine

1. Raise and safely support the vehicle.
2. Position the assembled engine/transaxle assembly directly under the engine compartment.
3. Slowly and carefully, lower the vehicle over the engine/transaxle assembly.

NOTE: Do not allow the front wheels to touch the floor.

4. With lifting equipment in place and attached to the lifting eyes on the engine, raise the engine/transaxle assembly up through the engine compartment and position it to be bolted.
5. Install the right hand insulator intermediate attaching nuts and intermediate bracket to the engine bracket bolts. Install the nut on the bottom of the double ended stud that attaches intermediate bracket to the engine bracket. Tighten to 75–100 ft. lbs. (100–135 Nm).
6. Install an engine support fixture to an engine lifting eye to support the engine/transaxle assembly. Remove the lifting equipment.
7. Raise the vehicle and position a lifting device under the engine. Raise the engine and transaxle assembly into its operating position.
8. Install the insulator to bracket nut and tighten to 75–100 ft. lbs. (100–135 Nm).
9. Tighten the left hand rear insulator bracket to body bracket nuts to 75–100 ft. lbs. (100–135 Nm).
10. Install the starter cable to the starter.
11. Install the lower radiator hose and install the remaining bracket and bolts. Tighten to specifications.
12. Install the manual shift linkage bracket bolts to the transaxle. Install the cable clip to the lever on the transaxle.
13. Connect the lower radiator hose to the radiator. Install the thermactor pump discharge hose at the pump.
14. Install the speedometer cable to the transaxle.
15. Position the exhaust system up and into the insulating grommets, located at the rear of the vehicle.

16. Install the exhaust pipe to the exhaust manifold bolts and tighten to specifications.
17. Connect the gulp valve hose to the catalytic converter.
18. Position the stabilizer bar and the control arm assemblies in position and install the attaching bolts. Tighten all fasteners to specifications.
19. Install new circlips in the sub axle inboard spline grooves on both the left and right halfshafts. Carefully align the splines of the stub axle with the splines of the differential side gears and with some force, push the halfshafts into the differential unit until the circlips can be felt to seat in their grooves in the differential side gears.
20. Connect the control arm ball joint stud into its bore in the steering knuckle and install new bolts and nuts.
21. Tighten the new bolt and nut to 37–44 ft. lbs. (50–60 Nm).
22. Position the brake hose routing clip on the suspension components and install their remaining bolts.
23. Lower the vehicle and remove the engine support tool.
24. Connect the vacuum and electrical lines that were disconnected during the removal procedure.
25. Install the disconnected air conditioning components.
26. Connect the fuel supply and return lines to the engine and connect the accelerator cable.
27. Install the power steering pressure and return lines. Install the brackets.
28. Connect the T.V. linkage the transaxle.
29. Install the radiator shroud and the cooling fan assembly. Tighten the bolts to specifications.
30. Install the coil and connect the coolant fan electrical connector.
31. Install the upper radiator hose to the engine and connect the transaxle cooler lines to the rubber hoses under the radiator. Fill the radiator and engine with coolant.
32. Install the negative battery cable and the air cleaner assembly.
33. Install the hood in its original position.
34. Check all fluid levels and correct as required.
35. Start the engine and check for leakage.
36. Charge the air conditioning system and road test the vehicle as necessary.

SPECIFICATIONS

CLUTCH PACK PLATE USAGE AND CLEARANCE

Steel	Friction	Clearance	Selective Snap Ring Thickness
		REVERSE CLUTCH	
3*	3*	0.76-1.40mm (0.030-0.055 in.)	1.89-1.99mm (0.074-0.078 in.)
			2.33-2.43mm (0.092-0.096 in.)
			2.77-2.87mm (0.109-0.113 in.)
			3.21-3.31mm (0.126-0.130 in.)
*With Cushion Spring			
		DIRECT CLUTCH	
4	4	1.01-1.43mm (0.040-0.056 in.)	1.26-1.36mm (0.050-0.054 in.)
			1.58-1.68mm (0.062-0.066 in.)
			1.90-2.00mm (0.075-0.079 in.)
		INTERMEDIATE CLUTCH	
3	3	0.75-11.12mm (0.030-0.044 in.)	1.24-1.34mm (0.049-0.053 in.)
			1.51-1.61mm (0.060-0.064 in.)
			1.78-1.88mm (0.071-0.075 in.)

SERVO PISTON TRAVEL

Acceptable Travel ①	Available Rod Lengths ②	Identification
5.15-7.04mm (0.203-0.277 inch)	160.22-160.52mm (6.313-6.324 inch)	0 Groove
	159.61-159.90mm (6.289-6.300 inch)	1 Groove
	159.00-159.30mm (6.265-6.276 inch)	2 Groove
	158.39-158.69mm (6.240-6.252 inch)	3 Groove
	157.78-158.08mm (6.216-6.189 inch)	4 Groove
	157.17-157.47mm (6.197-6.209 inch)	5 Groove

① Rod Stroke—not piston stroke
② Measured from far end of snap ring groove to end of rod

TRANSAXLE ENDPLAY

Measured Depth	Thrust Washer Required	Identification Code
2.00-1.77mm (0.079-0.070 inch)	1.40-1.45mm (0.055-0.057 inch)	AA
2.20-2.00mm (0.087-0.079 inch)	1.60-1.65mm (0.063-0.065 inch)	BA
2.41-2.20mm (0.095-0.087 inch)	1.80-1.85mm (0.071-0.073 inch)	CA
1.77-1.46mm (0.070-0.057 inch)	1.15-1.20mm (0.045-0.047 inch)	EA

LINE PRESSURE

Transaxle Model	Range	Pressure (At Idle)		Pressure (WOT Stall)	
		kPa	PSI	kPa	PSI
PMA-N	D-2-1-P-N	338-420	49-61	655-765	95-111
	R	455-689	66-100	1413-1772	205-257
PMA-U, V	D-2-1-P-N	—	54-66	—	106-122
PMB-C, D	R	—	77-111	—	232-284

NOTE: Governor Pressure is at zero (vehicle stationary). Transaxle is at operating temperature.

TORQUE CHART

Description	ft. lbs.	N•m
Reactor Support to Case	6-8	8-11
Separator Plate to Valve Body	6-8	8-11
Filler Tube Bracket to Case	7-9	9-12
Filter to Case	7-9	9-12
Valve Body Cover to Case	7-9	9-12
Pump Support to Pump Body	6-8	8-11
Natural Safety Switch to Case	7-9	9-12
Pump Assembly to Case	7-9	9-12
Valve Body to Case	72-96 (in. lbs.)	8-11
Oil Pan to Case	15-19	20-26
Lower Ball Joint to Steering Knuckle	37-44	50-60
Transfer Housing to Case	18-23	24-32
Differential Retainer to Case (with Sealant)	15-19	20-26
Pressure Test Port Plugs to Case	4-8	5-11
Cooler Tube Fitting to Case	18-23	24-31
Outer Throttle Lever to Shaft Nut	7.5-9.5	10-13
Inner Manual Lever to Shaft Nut	32-48	43-65
Idler Shaft Attaching	80-100	108-136
Converter Drain Plug	8-12	10-16
Valve Body Retaining Bolts	6-8	8-11
T.V. Adjuster Locknut	24-36 (in. lbs.)	2.7-4.1

PISTON ROD SIZES
With Paint Identification

I.D. Color①	Rod Length②	
	MM	Inch
Yellow	160.52-160.22	6.319/6.307
Green	159.91-159.61	6.295/6.283
Red	159.30-159.00	6.271/6.259
Black	158.69-158.39	6.247/6.235
Orange	158.08-157.78	6.223/6.211
Blue	157.47-157.17	6.199/6.187

① Daub of paint on tip or rod.
② From far end of snap ring groove to end ot rod.

For This Dial Indicator Reading		Install A New Piston Rod That Is . . .
MM	Inch	
2.45-2.98	.096-.117	5th Size Shorter
3.05-3.58	.120-.141	4th Size Shorter
3.65-4.18	.144-.165	3rd Size Shorter
4.25-4.78	.167-.188	2nd Size Shorter
5.12-5.14	.202-.202	1 Size Shorter
5.15-6.28	.203-.247	NO CHANGE
6.29-6.31	.248-.248	1 Size Longer
6.65-7.18	.262-.283	2nd Size Longer
7.25-7.78	.285-.306	3rd Size Longer
7.85-8.38	.309-.330	4th Size Longer
8.45-8.98	.333-.353	5th Size Longer

NOTE: For readings not on Table, select nearest one and repeat measurement with the new rod.

PISTON ROD SIZES
With Groove Identification

I.D	Rod Length(1)	
	mm	Inch
0 Groove	160.22-160.52	6.313-6.324
1 Grooves	159.61-159.90	6.289-6.300
2 Grooves	159.00-159.30	6.265-6.276
3 Grooves	158.39-158.69	6.240-6.252
4 Grooves	157.78-158.08	6.216-6.189
5 Grooves	157.17-157.47	6.197-6.209

(1) From far end of snap ring groove to end of rod.

If the dial indicator reads:

Less than 5.15mm (used) Less than 5.15mm (new) (.203 in.)	More than 2.04mm (used) More than 6.28mm (new) (.247 in.)	5.15-2.04mm (used) 5.15-6.28mm (new) (.203-.247 in.)
The piston rod is too long. A shorter rod (more grooves) will have to be installed.	The piston rod is too short. A longer rod (less grooves) will have to be installed.	The piston rod is the correct length and no change is required.

REVERSE CLUTCH
Selective Snap Rings

Thickness	Part No.
1.89-1.94 mm (.074-.076 inch)	N800654
2.33-2.43 mm (.092-.096 inch)	N800655
2.77-2.87 mm (.109-.113 inch)	N800656
3.21-3.31 mm (.126-.130 inch)	N800657
For This Clearance Reading	**Install a New Ring That Is . . .**
0.12-0.46 mm (.005-.018 inch)	One Size THINNER
1.02-1.32 mm (.040-.052 inch)	One Size THICKER
1.45-1.75 mm (.057-.069 inch)	2nd Size THICKER
1.88-2.26 mm (.074-.089 inch)	3rd Size THICKER

DIFFERENTIAL BEARING END PLAY SHIMS

Part No.	Inch	MM
E1FZ-4067-A	0.012	0.30
B	0.014	0.35
C	0.016	0.40
D	0.018	0.45
E	0.020	0.50
F	0.022	0.55
G	0.024	0.60
H	0.026	0.65
J	0.028	0.70
K	0.030	0.75
L	0.032	0.80
M	0.033	0.85
N	0.035	0.90
P	0.037	0.95
R	0.039	1.00
S	0.041	1.05
T	0.043	1.10
U	0.045	1.15
V	0.047	1.20
W	0.049	1.25
X	0.051	1.30

END PLAY ADJUSTMENT MEASUREMENTS

For This Reading	Use This Washer Part ID
.779-.796 inch (19.78-20.22 mm)	AA
.789-.804 inch (20.04-20.42 mm)	BA
.797-.812 inch (20.24-20.62 mm)	CA
.807-.825 inch (20.50-20.95 mm)	CA

WASHER THICKNESS

Inch	MM	ID
.055-.057	1.40-1.45	AA
.063-.065	1.60-1.65	BA
.071-.073	1.80-1.85	CA
.081-.083	2.05-2.10	DA

STALL SPEED SPECIFICATION
ATX Transaxle

Vehicle Application	Engine Disp.	Converter Size	Converter ID	Stall Speed Min.	Stall Speed Max.
Escort/Lynx/EXP	1.6L (EFI)	9¼″	E04	2627	3095
Tempo/Topaz	2.3L	9¼″	E05	2272	2664
Escort/Lynx/EXP	1.6L (HO)	9¼″	E04	2655	3147

SHIFT POINTS

Drive Range		PMA-N 2.3L HSC Engine km/h	PMA-N 2.3L HSC Engine mph
Idle:	1-2	17-29	11-18
	2-3	25-50	16-31
	3-2	26-40	16-25
	2-1	15-23	9-14
Part Throttle:	1-2	19-44	12-27
	2-3	42-70	26-43
	3-2	36-45	22-41
WOT:	1-2	38-66	23-41
	2-3	84-112	52-69
	3-2	76-102	49-64
	2-1	24-51	15-32
Manual Low:	2-1	32-61	20-38

Axle Ratio: 3.3:1
Tire Size: P165/70R13, P165/80R13

Section 3

AXOD Transaxle
Ford Motor Co.

APPLICATION

1988–89 Continental
1986–89 Taurus and Sable

GENERAL DESCRIPTION

The AXOD transaxle is a 4-speed fully automatic overdrive transmission. The unit consists of 2 planetary gearsets, a combination planetary/differential gearset, 4 multiple plate clutches, 2 band assemblies and 2 one-way clutches. The lockup converter transmits engine power to the geartrain by means of the drive chain.

The converter clutch is controlled by the EEC-IV computer system. The fully automatic shift control responds to the road speed and engine torque demand. Operation of the EEC-IV system's electronic controls, along with the valve body hydraulic controls, operate a piston plate clutch in the torque converter to eliminate converter slippage and improve fuel economy.

Transaxle Identification

Code letters and numbers are stamped on the identification tag located on top of the converter housing. The tag denotes the serial number and the date of manufacture. The numbers are important when ordering service replacement parts.

Electronic Controls

The transaxle, being fully controlled by the EEC-IV computer system, is equipped with pressure switches and a solenoid; it responds to road speed and engine torque demand.

The internal equipment, attached to the main control assembly, consists of 3 pressure switches, an oil temperature switch (3.8L only) and a bypass clutch solenoid. A bulkhead connector/wiring assembly, attached to the middle, rear side of the chain cover assembly, provides an electrical path to the EEC-IV system.

Metric Fasteners

The transaxle is of a metric design; all bolt sizes and thread pitches are metric. Metric fastener dimensions are very close to the customary inch system fastener dimensions; replacement of the fasteners must be of the same measurement and strength as those removed.

Do not attempt to interchange metric fasteners with the customary inch system fasteners. Mismatched or incorrect fasteners can result in damage to the transaxle. Care should be taken to reuse the same fasteners in the location from which they were removed.

UD
ASSY E6DP-BA
BD-6D015
BUILD DATE
SN-000001
SERIAL NO.
AAAB1000001

View of the transaxle identification tag

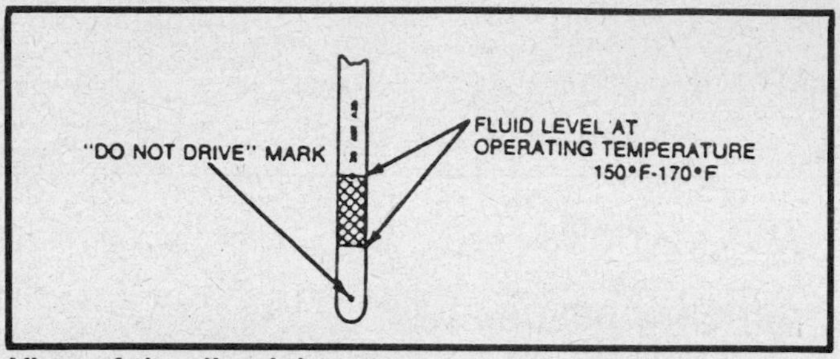
"DO NOT DRIVE" MARK
FLUID LEVEL AT
OPERATING TEMPERATURE
150°F-170°F

View of the dipstick

NOTE: Be sure to check the case holes for the quality of their threads. It is rather difficult to rethread bolt holes after the transaxle is installed in the vehicle.

Capacities

The fluid quantities are approximate and the correct fluid level should be determined by the dipstick indicator for the correct level. Dry fill is 13.1 qts. (12.46L)

Checking Fluid Level

NOTE: When checking the fluid level, the vehicle must be at normal operating temperatures and positioned on a flat surface. The vehicle should not be driven if the fluid level is below the dipstick's DO NOT DRIVE hole.

1. If the transaxle's oil is below 150–170°F (21–35°C), drive the vehicle until normal operating temperatures are reached.

NOTE: If the outside temperature is above 50°F (10°C), drive the vehicle 15–20 miles (24–32 km) of city driving. If the vehicle has been operated under extreme conditions in hot weather, turn the engine OFF and allow the fluid to cool for at least 30 minutes.

2. With the vehicle on a level surface, place the transaxle in **P**, idle the engine and apply the footbrake. Move the transaxle selector lever through each range; allow time in each range for the transaxle to engage.
3. Position the selector lever in **P**, fully apply the parking brake, block the drive wheels and allow the vehicle to idle.
4. Before removing the dipstick, wipe the dirt from the dipstick cap.
5. Remove the dipstick and wipe it clean. Insert it into the dipstick tube and be sure it fully seats.
6. Remove the dipstick again and observe the fluid level; it must be between the indicators.
7. If necessary to add fluid, add it through the oil filler (dipstick) tube. Add enough fluid to raise the level. Do not overfill; overfilling will cause foaming, fluid loss and transaxle malfunction.

TRANSAXLE MODIFICATIONS

Temperature Switch—3.8L Engine

A temperature switch sensor is installed on the main control assembly, near the pressure switches when this transaxle is used with the 3.8L engine.

TROUBLE DIAGNOSIS

CLUTCH AND BAND APPLICATION

Gear	Lo-Int Band	Overdrive Band	Forward Clutch	Intermediate Clutch	Direct Clutch	Reverse Clutch	Low One-Way Clutch	Direct One-Way Clutch
1st Gear Manual Low	Applied	—	Applied	—	Applied	—	Applied	Applied
1st Gear (Drive)	Applied	—	Applied	—	—	—	Applied	—
2nd Gear (Drive)	Applied	—	Applied	Applied	—	—	Holding	—
3rd Gear (Drive)	—	—	Applied	Applied	Applied	—	—	—
4th Gear (Overdrive)	—	Applied	—	Applied	Applied	—	—	Holding
Reverse (R)	—	—	Applied	—	—	Applied	Holding	—
Neutral (N)	—	—	—	—	—	—	—	—
Park (P)	—	—	—	—	—	—	—	—

CHILTON'S THREE C'S TRANSMISSION DIAGNOSIS
Ford AXOD

Condition	Cause	Correction
Oil leak	a) Damaged gasket or pan rail	a) Replace gasket, repair pan rail
	b) Distorted pan	b) Replace pan
	c) TV cable, fill tube or electrical bulkhead connector. Loose fit/damaged case. External seal damage/missing	c) Reseal TV cable, fit tube or connector. Repair or replace damaged case. Replace damaged or missing seal
	d) Manual shaft, seal damaged	d) Replace seal
	e) Governor cover and servo covers. O-ring seal damaged	e) Replace O-ring seal
	f) Cooler fittings or pressure taps Low torque, damaged threads	f) Repair damaged threads
	g) Converter or converter seal, damaged, garter spring missing. Converter hub scored or weld seam leaking	g) Inspect CV joint journal for damage. Replace seal or spring. Weld seam
	h) Halfshaft seal damaged or garter spring missing	h) Replace seal
	i) Speedometer cable or speed sensor O-ring seal damaged	i) Replace O-ring seal
Oil venting or foaming	a) Transaxle overfilled	a) Drain and fill transaxle to proper level
	b) Transmission fluid contaminated with antifreeze or engine overheating	b) Determine source of leak and repair leak
	c) Bi-metallic element stuck open	c) Replace element
	d) Oil filter plugged, damaged or missing O-rings	d) Replace filter O-rings and filter

CHILTON'S THREE C'S TRANSMISSION DIAGNOSIS
Ford AXOD

Condition	Cause	Correction
High or low oil pressure (verify with gauge)	a) Oil level too low or too high	a) Drain or fill transaxle as necessary
	b) Improper T.V. cable/linkage actuation (travel)	b) Inspect for broken or disconnected component
	c) Pressure regulator valve or spring damaged	c) Replace valve or spring
	d) Pressure relief valve damaged. Missing ball or spring	d) Replace or repair pressure relief valve
	e) Oil pump ring stuck, seals damaged, vanes damaged	e) Determine source of damage. Repair seals
	f) Oil pump driveshaft broken or damaged	f) Replace oil pump
No 1–2 shift (first gear only)	a) Governor assembly weights binding	a) Perform governor test. Replace springs
	b) Governor springs damaged, misaligned or missing	b) Replace springs
	c) Governor gears damaged	c) Replace gears
	d) Governor shaft seal damaged or missing	d) Replace seal
	e) Governor value (ball) stuck or missing	e) Replace valve ball
	f) Governor tube leaking/damaged	f) Replace governor tube
	g) Intermediate clutch plates damaged/missing	g) Replace clutch plates
	h) Intermediate clutch piston or seals damaged	h) Replace piston seals
	i) Intermediate clutch ball check stuck/damaged or missing	i) Replace damaged ball
	j) Intermediate clutch cylinder damaged	j) Replace clutch cylinder
	k) Direct/intermediate clutch hub seals damaged, missing or holes blocked	k) Replace seals or unblock holes
	l) Driven sprocket support seals damaged, missing or holes blocked	l) Replace seals or unblock holes
	m) 1–2 shift valve stuck, nicked or damaged	m) Replace valve
	n) 1–2 throttle delay valve stuck, nicked or damaged	n) Replace valve
	o) 1–2 Accumulator capacity modulator valve stuck, nicked or damaged	o) Clean or replace valve
	p) No. 9 check ball missing or damaged	p) Replace check ball
	q) Control assembly bolts too loose or too tight	q) Tighten bolts to specification
	r) Front carrier damaged	r) Inspect welds and repair
	s) Intermediate clutch tap plug loose/missing. (Located on oil pump body)	s) Tighten or replace plug
	t) T.V. cable damaged/disconnected	t) Replace cable
1–2 shift feels harsh or soft	a) High or low oil pressure	a) Perform control pressure test
	b) 1–2 Accumulator regulator valve stuck, nicked, spring missing or damaged	b) Replace valve or spring
	c) Improper T.V. cable/linkage actuation	c) Check for broken or disconnected parts
	d) 1–2 Accumulator capacity modulator valve stuck, nicked, spring missing or damaged	d) Replace valve or spring
	e) 1–2 Accumulator assembly piston stuck, seal damaged, springs damaged or missing	e) Replace piston assembly or springs
1–2 shift speed high or low	a) Governor weights binding, spring damaged, misaligned or missing	a) Perform governor test
	b) Governor gear damaged	b) Replace gear
	c) Governor shaft seal, damaged or missing	c) Replace seal
	d) Governor lube tube leaking/damaged	d) Replace tube

CHILTON'S THREE C'S TRANSMISSION DIAGNOSIS
Ford AXOD

Condition	Cause	Correction
1–2 shift speed high or low	e) Governor valve balls damaged, stuck or missing	e) Replace balls
	f) Improper T.V. cable/linkage actuation	f) Check for broken or disconnected parts
	g) T.V. control valve, T.V. plunger, T.V. line modulator valve, 1–2 throttle delay valve	g) Check and/or replace T.V. valve parts
	h) Control assembly valve(s) stuck, nicked or damaged	h) Replace valve
	i) Control assembly spring(s)—missing or damaged	i) Replace spring(s)
	j) Control assembly valve balls damaged, stuck or missing	j) Replace balls
No 2–3 shift (1–2 shift OK)	a) Low/intermediate servo apply rod (too long)	a) Install correct apply rod, if required
	b) Low/intermediate servo bore or piston damaged	b) Replace piston
	c) Low/intermediate servo piston seals damaged/missing	c) Replace seals
	d) Low/intermediate servo missing/broken return spring or retaining clip	d) Replace spring or clip
	e) Direct clutch assembly plates damaged/missing	e) Replace clutch plates
	f) Direct clutch assembly piston cylinder, or seals damaged	f) Replace piston or seals
	g) Direct clutch assembly ball check assembly stuck or missing	g) Replace check ball assembly
	h) Direct/intermediate clutch hub seals damaged or missing or holes blocked	h) Replace seals or unblock holes
	i) Driven sprocket support seals damaged or missing or holes blocked	i) Replace seals or unblock holes
	j) Direct one-way clutch assembly cage/rollers/springs damaged	j) Disassemble and inspect. Replace parts
	k) Direct one-way clutch assembly rollers missing or misassembled on inner race	k) Replace rollers
	l) Control assembly bolts too loose or too tight	l) Tighten to specification
	m) 2–3 shift valve. Valve stuck, nicked or damaged	m) Replace valve
	n) No. 4 check ball missing/damaged	n) Replace check ball
	o) Bypass solenoid not energized during wide open throttle upshift	o) Refer to Electrical System Diagnosis
	p) Case servo release passage blocked	p) Determine source of blockage
	q) Servo release tube leaking or improperly installed	q) Seal or seat tube
	r) Direct clutch pressure tap plug loose/missing on oil pump body	r) Tighten or replace
	s) TTS temperature switch on 3.8L engine	s) Refer to Electrical diagram
2–3 shift feels harsh or soft	a) Low or high oil pressure	a) Perform control pressure test
	b) Wrong low/intermediate servo apply rod length	b) Install correct apply rod, if required
	c) Low/intermediate servo piston, seal, springs or rod damaged	c) Replace piston, seal, spring or rod
	d) Backout valve stuck, nicked or spring damaged	d) Determine source of contamination or damage and replace valve or spring
2–3 shift speed high or low (1–2 shift OK)	a) Governor weights binding, springs damaged, shaft seal or valve damaged	a) Perform governor pressure test
	b) Governor tube leaking/damaged	b) Repair tube leak or replace tube

CHILTON'S THREE C'S TRANSMISSION DIAGNOSIS
Ford AXOD

Condition	Cause	Correction
2–3 shift speed high or low (1–2 shift OK)	c) Governor valve balls damaged d) Improper T.V. cable/linkage actuation e) T.V. control valve, T.V. plunger, T.V. line modulator valve, 2–3 throttle modulator valve stuck, nicked or damaged	c) Replace valve balls d) Service T.V. cable e) Determine source of contamination or damage and repair or replace damaged part
No 3–4 shift (1–2 and 2–3 OK)	a) Overdrive band assembly not holding b) Overdrive servo apply rod (too long) c) Overdrive servo bore, piston, piston seals damaged d) Overdrive servo assembly return spring or retaining clip missing or broken e) Forward clutch assembly return springs/ piston damaged f) Control assembly bolts too loose or too tight g) 3–4 shift valve stuck, nicked or spring damaged h) 3–4 modulator valve stuck, nicked or spring missing i) 4–3 scheduling valve stuck, nicked or spring missing	a) Perform air pressure test and replace defective part b) Install correct apply rod, if required c) Determine source of contamination or damage. Repair damaged part d) Replace spring or clip e) Determine source of damage. Replace springs or piston f) Tighten bolts to specification g) Determine source of contamination or damage. Repair or replace valve or spring h) Determine source of contamination or damage. Repair or replace valve or spring i) Determine source of contamination. Repair or replace valve or spring
3–4 shift feels harsh or soft	a) Oil pressure too high or too low b) 3–4 Accumulator assembly piston stuck, piston seal missing or damaged c) 3–4 Accumulator assembly springs missing or damaged d) No. 14 check ball missing/damaged	a) Perform control pressure test b) Determine source of damage or contamination. Replace piston or seal c) Replace springs d) Replace ball
3–4 shift speed high or low (1–2 and 2–3 OK)	a) Governor weights binding, spring damaged or misaligned b) Governor gear, shaft seal or valve damaged c) Governor tube leaking d) T.V. control valve, T.V. plunger, T.V. line modulator valve, 3–4 modulator valve stuck, nicked or spring(s) missing or damaged	a) Perform governor test. Replace spring or weights b) Replace gear, seal or valve c) Seal or seat tube d) Determine source of contamination or damage. Replace valve(s) or spring(s)
No converter clutch apply	a) No lock-up signal b) By-pass solenoid damaged or inoperative c) Bulkhead connector damaged d) Pinched wires e) 4–3 pressure switch, 3–2 pressure switch inoperative f) Turbine shaft, seals damaged or missing g) Bypass clutch control valve stuck h) Bypass plunger stuck i) Missing or damaged pump shaft seals or cup plug j) Valve body pilot sleeve damaged/ misaligned	a) Repair or replace wiring or component. Refer to Electrical System Diagnosis b) Repair or replace wiring or component. Refer to Electrical System Diagnosis c) Repair or replace wiring or component. Refer to Electrical System Diagnosis d) Repair or replace wiring or component. Refer to Electrical System Diagnosis e) Repair or replace wiring or component. Refer to Electrical System Diagnosis f) Replace seals g) Determine source of contamination. Repair or replace valve h) Repair or replace plunger i) Determine source of contamination or damage. Replace seals or plug j) Determine source of damage. Replace pilot sleeve

CHILTON'S THREE C'S TRANSMISSION DIAGNOSIS
Ford AXOD

Condition	Cause	Correction
Converter clutch does not release	a) No unlock signal	a) Repair or replace wiring or component. Refer to Electrical System Diagnosis
	b) Bypass solenoid damaged or inoperative	b) Repair or replace wiring or component. Refer to Electrical System Diagnosis
	c) Bulkhead connector wires damaged	c) Repair or replace wiring or component. Refer to Electrical System Diagnosis
	d) Bypass clutch control valve or plunger valve stuck, nicked or damaged	d) Determine source of contamination. Repair or replace valve
	e) Solenoid filter plug (in main control)	e) Tighten or replace solenoid filter plug (in main control)
4–3 downshifts harsh	a) Incorrect overdrive servo apply rod length	a) Install correct apply rod, if required
	b) Damaged overdrive servo piston, springs or seal	b) Determine source of contamination or damage. Replace piston, seal or springs
	c) No converter clutch release	c) Refer to Electrical System Diagnosis
3–2 downshift harsh	a) Damaged or missing low/intermediate servo assembly springs	a) Determine source of contamination or damage. Replace springs
	b) Incorrect low/intermediate servo apply rod length	b) Install correct apply rod, if required
	c) 3–2 Control valve stuck, nicked or damaged	c) Determine source of contamination. Repair or replace valve
	d) No. 5 check ball missing	d) Replace check ball
	e) Intermediate clutch return spring retaining ring out of position	e) Reposition spring
3–1, 2–1 downshift harsh	a) Damaged low/intermediate servo piston, springs, or seal	a) Determine source of damage. Replace piston, seal or springs
	b) Incorrect low/intermediate servo apply rod length	b) Install correct apply rod, if required
	c) No. 9 check ball missing (3–1 only)	c) Replace check ball
No drive in drive range and no reverse in reverse range	a) Oil level low	a) Add oil
	b) Oil pressure too low	b) Perform control pressure test. Repair system
	c) Manual linkage misadjusted, disconnected, damaged, broken, bent	c) Adjust, replace or repair linkage
	d) Oil pump assembly worn or damaged	d) Determine source of damage. Replace oil pump
	e) Drive chain assembly damaged or broken	e) Determine source of damage. Replace chain
	f) Drive sprocket shaft to converter turbine spline damaged	f) Determine source of damage. Replace drive sprocket shaft
	g) Driven sprocket shaft to direct/intermediate clutch hub damaged	g) Determine source of damage. Replace drive sprocket shaft
	h) Oil filter damaged/missing O-rings or plugged	h) Clean oil filter or replace O-rings
	i) Forward clutch assembly's clutch plates burned or missing	i) Replace clutch plates
	j) Forward clutch assembly's damaged piston seals or pistons damaged	j) Replace seals or pistons
	k) Forward clutch assembly's forward clutch ball check assembly missing or damaged	k) Replace clutch ball
	l) Forward clutch assembly's driven sprocket support seals or direct intermediate clutch hub seals damaged/missing or holes blocked	l) Clean blocked holes or replace seals
	m) Gearset's front sun, front/rear carriers, ring gear and/or final drive assembly	m) Replace damaged component

CHILTON'S THREE C'S TRANSMISSION DIAGNOSIS
Ford AXOD

Condition	Cause	Correction
No drive in drive range and no reverse in reverse range	n) Low one-way clutch has two way rotation o) Damaged output shaft splines/misassembled with axles p) Halfshaft splines damaged or disengaged from transaxle	n) Replace clutch o) Determine source of damage. Replace shaft or align with axles p) Service halfshaft and CV-joints
No drive. Reverse OK	a) Low/intermediate band assembly burned or broken ends b) Low/intermediate servo assembly apply rod (too short) c) Piston/seal/rod damaged d) Low/intermediate servo oil tubes or case bores damaged (leaking oil) e) 2–3 Servo regulator valve stuck	a) Determine source of damage. Replace bond assembly b) Install correct apply rod, if required c) Determine source of contamination. Replace damaged component d) Unblock tubes or case bores. Repair oil tube leaks e) Replace valve
No reverse. Drive OK	a) Reverse clutch plates burned or missing b) Reverse apply tube leaking or improperly installed	a) Determine source of damage. Replace clutch plates b) Seat or seal tube
No park range	a) Chipped or broken parking pawl or park gear b) Broken park pawl return spring c) Bent or broken actuating rod d) Manual linkage misadjusted	a) Replace pawl or gear b) Replace spring c) Replace rod d) Adjust linkage
Harsh neutral to reverse or harsh neutral to drive	a) Damaged or missing low/intermediate servo assembly springs b) Incorrect servo apply rod length c) 3–2 Control valve stuck, nicked or damaged d) No. 5 ball check. Ball missing e) Neutral-drive accumulator assembly piston stuck or seal and/or springs damaged or missing f) No. 1 check ball missing/damaged (harsh reverse) g) Main control separator plate thermal elements do not close when warm	a) Determine source of contamination or damage. Replace springs b) Install correct apply rod, if required c) Determine source of contamination. Replace valve d) Replace ball check e) Determine source of contamination. Replace piston, seal or springs f) Replace ball g) Replace thermal elements
Transaxle overheats	a) Excessive tow loads b) Improper fluid level c) Incorrect engine idle or performance d) Improper clutch or band application or oil pressure control system e) Restriction in cooler or lines f) Seized converter one-way clutch g) Dirty or sticking valve body	a) Check Owner's Manual for tow restriction b) Perform fluid level check c) Perform engine tune-up d) Perform control pressure test e) Service restriction f) Replace converter g) Clean, service or replace valve body
Transaxle fluid leaks	a) Improper fluid level b) Leakage at gaskets, seals and/or etc.	a) Perform fluid level check b) Remove all traces of lubrication on exposed surfaces of transaxle. Check the vent for free-breathing. Operate transaxle at normal temperatures and inspect for leakage

Electrical System Diagnosis

These test should only be conducted if a problem has been detected with the transaxle. If any of the following service codes appear during the Self Test, perform the AXOD Drive Cycle Test.

CODE 39: Transaxle converter bypass clutch not working properly.

CODE 59: Transaxle Hydraulic Switch (THS) 4–3 pressure switch circuit has failed open.

CODE 62: Transaxle Hydraulic Switch (THS) 4–3 and/or 3–2 pressure switch(es) has failed closed. If code appears in the "Key On, Engine Off" test, the 3–2 circuit has failed. If the code appears in the "Engine Running" test, the 4–3 circuit has failed. If the code appears in both tests, check both circuits.

CODE 69: Transaxle Hydraulic Switch (THS) 3–2 circuit has failed open.

CODE 89: Transaxle converter bypass clutch solenoid has failed open or closed.

The following codes are not transaxle related but can affect operation of the converter clutch bypass. These components should be serviced before servicing the transaxle codes.

CODE 21: Engine Coolant Temperature (ECT) sensor out of range.

CODE 22: Manifold Absolute Pressure (MAP) sensor out of range.

CODE 23: Throttle Position Sensor (TPS) out of range.

CODE 24: Air Charge Temperature (ACT) sensor out of range.

CODE 29: Vehicle Speed Sensor (VSS) nonfunctioning.

CODE 74: Brake ON/OFF (BOO) switch always open or brake not applied during Engine Running On-Demand Self Test.

CODE 75: Brake ON/OFF (BOO) switch always closed.

The following service code is transaxle related and may cause faulty engine idle speed control if not working properly.

CODE 57: Neutral Pressure Switch (NPS) failed in **N** (open). The NPS is a normally open switch that closes with hydraulic pressure; its failure will not allow the transaxle to engage in **D** or **R**. Before testing the electrical components, check for proper hydraulic operation.

NOTE: After performing the Self Test, perform the AXOD Drive Cycle Test to check for continuous codes; this test must be performed on flat terrain or a slight upgrade.

AXOD DRIVE CYCLE TEST

1. Record and zero the Self Test codes.
2. Operate the engine until normal operating temperatures are reached.
3. Place the transaxle in the **D** range, slowly accelerate to 40 mph (64 km/h) until the transaxle shifts into 3rd gear and hold the speed/throttle steady for 15 seconds or 30 seconds (above 4,000 ft.).
4. Shift the transaxle into **OD**, accelerate from 40–50 mph (64–80 km/h) until the transaxle shifts into 4th gear and hold the speed/throttle steady for 15 seconds.
5. With the transaxle in 4th gear, the speed steady and the throttle opening, lightly apply/release the brakes (to operate the brakelamps) and maintain the steady speed for the next 15 seconds.
6. Apply the brakes, come to a stop and remain stopped for the next 20 seconds with the transaxle still in **OD**.
7. Perform the Self Test and record any continuous codes.

NOTE: If any other continuous codes appear, service them first, for they could affect the transaxle's electrical operation.

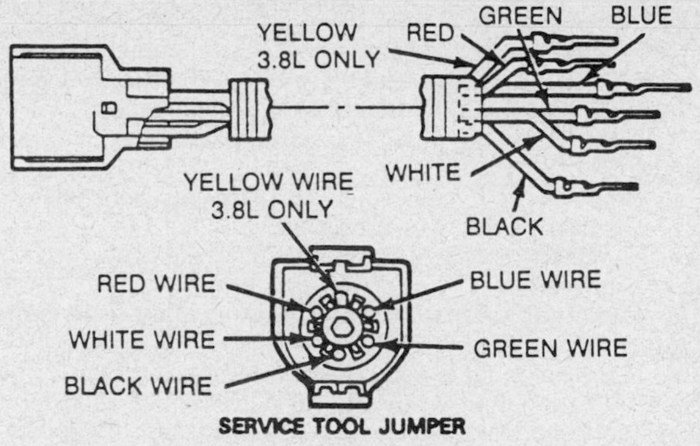
GREEN
BLUE
YELLOW
3.8L ONLY
RED
YELLOW WIRE
3.8L ONLY
WHITE
BLACK
RED WIRE
BLUE WIRE
WHITE WIRE
GREEN WIRE
BLACK WIRE
SERVICE TOOL JUMPER
CAUTION: DO NOT CONNECT A POWER SUPPLY WITH REVERSED POLARITY TO THESE WIRES OR SOLENOID DIODE WILL BE DAMAGED.

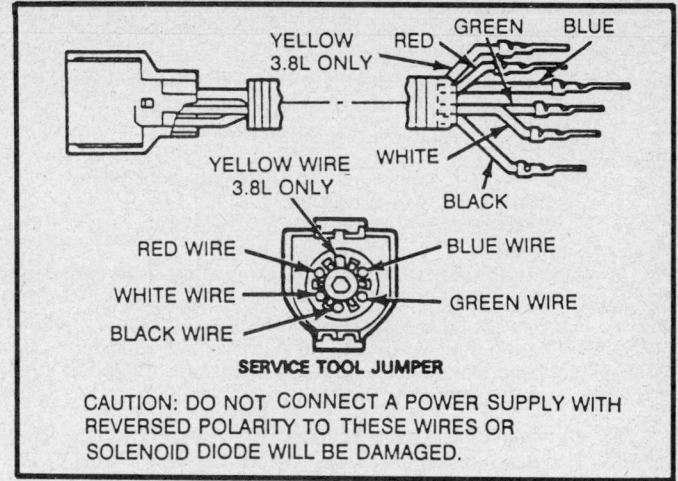

View of the service jumper harness

DIAGNOSIS CHART INDEX

Service Code	Pinpoint Test
39	A
59	B
62	C
69	D
89	E
57	F

CODE 39

The following procedure is used to determine why the converter bypass clutch is not applying properly.

1. Make sure the vehicle harness connector is fully seated and the terminals engaged with the transaxle's bulkhead connector. If not OK, repair the connector(s) or terminal(s) and repeat the Self Test. If OK, check the solenoid resistance.
2. To check the solenoid resistance, perform the following procedures:
   a. Using a service jumper harness, connect it to the transaxle's bulkhead connector.
   b. Using an ohmmeter, connect it to the service jumper harness leads; the positive (+) lead to the red wire and the negative (−) lead to the black wire. Check the resistance, it should be 20–40 ohms. Disconnect the ohmmeter.
   c. If the reading is not OK, repeat the Self Test and service the codes as required. If the reading is OK, check the bypass clutch application.
3. To check the bypass clutch application, perform the following procedures:
   a. Using the service jumper harness, connect the Red wire to the positive (+) battery terminal.
   b. With the engine running and transaxle in 3rd gear, connect the Black wire (jumper harness) to ground; this will energize the bypass clutch solenoid.

NOTE: When connecting a power supply to the wiring, do not reverse the polarity for it will damage the solenoid diode.

   c. If the clutch applies (engine speed drops slightly), there is no electrical component failure, the bypass clutch solenoid is operating properly.
   d. Any error code produced may be caused by a slipping bypass clutch. Inspect the torque converter or refer to the transaxle diagnosis "No Converter Clutch Apply".

e. If the clutch does not apply (engine speed does not drop slightly), check the bypass solenoid valve (next step).

4. Inspect the main control bypass clutch control valve for a sticking condition. If it is sticking, service the spool valve. If the valve is OK, check the bypass solenoid (next step).

5. To check the bypass solenoid, perform the following procedures:

a. Remove the bypass solenoid from the transaxle.

b. Check the O-ring's condition.

c. Shake the solenoid to check for a free armature and internal valve.

d. Check for solenoid contamination; make sure the small hole in the valve is open.

e. If the solenoid is defective, replace it and repeat the Self Test. If the solenoid is OK and service is still required, inspect the torque converter, refer to the transaxle diagnosis "No Converter Clutch Apply" and repeat the Self Test.

NOTE: The Transaxle Hydraulic Switch (THS) 4-3 and 3-2 pressure switches are normally open and close with hydraulic pressure. With the shift selector in D and the transaxle fails to engage, a service code 59 will appear. When the transaxle fails to shift to 3rd gear, a service code 69 will appear. Before performing electrical part inspection, be sure the hydraulic system is operating correctly.

CODE 59

The following procedure is used to check the 4–3 pressure switch's failure to open.

1. Make sure the vehicle harness connector is fully seated and the terminals engaged with the transaxle's bulkhead connector. If not OK, repair the connector(s) or terminal(s) and repeat the Self Test. If OK, check the switch for continuity.

2. To check the switch for continuity, perform the following procedures:

a. Using the service jumper harness connector, attach it to the transaxle's bulkhead connector.

b. Using a ohmmeter, attach 1 lead to the Blue wire and the other to ground.

c. Start the engine and place the shift selector in **N**; the ohmmeter should show no continuity.

d. Move the shift selector to **D**; the switch should close and the ohmmeter should read less than 10 ohms.

e. When the transaxle shifts through the 1st, 2nd and 3rd gears, the switch should stay closed; when shifting to 4th gear, the switch should open.

f. If the circuit is OK, repeat the Self Test and service any codes. If the circuit is not OK, check the internal connections.

3. To check internal connections, perform the following procedures:

a. Remove the side cover from the transaxle and check that the Blue wire is firmly attached to the 4–3 pressure switch.

b. Using an ohmmeter, remove the connector from the pressure switch and check the resistance of the wire; it should be less than 2.0 ohms.

c. If the connector/wire is not OK, replace the bulkhead connector/wiring assembly and repeat the Self Test. If the connector/wire is OK, check the 4–3 pressure switch.

4. To check the 4–3 pressure switch, perform the following procedures:

a. Remove the 4–3 pressure switch and install it into a ⅛–27 pipe fitting. Connect the pipe fitting to low pressure air (do not use water) supply line (for testing purposes).

b. Apply 50 psi (345 kPa) to the switch and check for ruptured diaphram.

c. Submerge the switch in transaxle fluid and check for bubbles at the small vent hole near the switch terminal.

d. If bubbles appear, replace the 4–3 pressure switch and repeat the Self Test. If no bubbles appear, check the switch's resistance.

5. To check the switch's resistance, perform the following procedures:

a. Apply 50 psi (345 kPa) of air pressure to the switch.

b. Using an ohmmeter, measure the resistance between the switch's terminal and the case; it should be less than 8.0 ohms.

c. If the resistance is greater than 8.0 ohms, replace the switch, repeat the switch continuity check and repeat the Self Test; codes 39, 59, 62, 69 and 89 should no longer appear. If the resistance is less than 8.0 ohms, inspect the hydraulic circuit supplying pressure to the switch for excessive leakage and/or main control assembly operation.

CODE 62

The following procedure is used to check the 4–3 and/or 3–2 pressure switches for a failed closed condition.

NOTE: Code 62 will appear under the following Self Test situations: "Engine Running" when the 4–3 circuit has failed closed to ground, "Key On, Engine Off" when the 3–2 circuit has failed closed to ground or under both testing situations.

1. To check the wiring, perform the following procedures:

a. Remove the vehicle harness connector from the bulkhead connector.

b. Using the service jumper harness connector, attach it to the bulkhead connector.

c. Using a ohmmeter (engine Off), attach 1 lead to the White wire and the other to ground; the resistance should be infinite.

d. Using a ohmmeter (engine **OFF** and/or **RUNNING** in **N**), attach 1 lead to the Blue wire and the other to ground; the resistance should be infinite.

e. If the reading(s) are infinite, repeat the Self Test and service any code. If the reading(s) are are not infinite, check the internal wiring.

2. To check the internal wiring, perform the following procedures:

a. Remove the side cover and check for pinched, cut or grounded wiring.

b. If the wiring is pinched, cut or grounded, replace the bulkhead wiring assembly and repeat the Self Test.

c. If the wiring for the 3–2 switch is OK, inspect the 3–2 switch continuity. If the wiring for the 4–3 switch is OK, inspect the 4–3 switch's continuity.

3. To check the 3–2 and 4–3 switches continuity, perform the following procedures:

a. From each switch, remove the wiring connector.

b. Using an ohmmeter, connect 1 lead to the pressure switch terminal and the other to valve body.

c. If there is continuity, replace the defective switch and repeat the Self Test. If there is no continuity, check the internal wiring.

4. To check the internal wiring of both switches, perform the following procedures:

a. Remove the wiring connectors from both switches.

NOTE: When performing this test, make sure the wiring terminals are not contacting any metal surfaces.

b. Using an ohmmeter, connect 1 lead to the White wire and the other to ground; the resistance should be infinite.

c. Using an ohmmeter, connect 1 lead to the Blue wire and the other to ground; the resistance should be infinite.

d. If the resistance is infinite, repeat the Self Test and service any codes. If there is continuity, replace the bulkhead connector/wiring assembly and repeat the Self Test.

NOTE: The Transaxle Hydraulic Switch (THS) 4-3 and 3-2 pressure switches are normally open and close with hydraulic pressure. With the shift selector in D and the transaxle fails to engage, a service code 59 will ap-

pear. When the transaxle fails to shift to 3rd gear, a service code 69 will appear. Before performing electrical part inspection, be sure the hydraulic system is operating correctly.

CODE 69

The following procedure is used to check the 3–2 pressure switch's failure to open.

1. Make sure the vehicle harness connector is fully seated and the terminals engaged with the bulkhead connector. If not OK, repair the connector(s) or terminal(s) and repeat the Self Test. If OK, check the switch for continuity.
2. To check the switch for continuity, perform the following procedures:
   a. Using the service jumper harness connector, attach it to the bulkhead connector.
   b. Using a ohmmeter, attach 1 lead to the White wire and the other to ground.
   c. Start the engine and place the shift selector in **D**, **1** or **2**; the ohmmeter should show no continuity.
   d. When the transaxle shifts to **3rd** or **4th** gears, the switch should close; resistance should be less than 10 ohms.
   e. If the circuit is OK, repeat the Self Test and service any codes. If the circuit is not OK, check the internal connections.
3. To check internal connections, perform the following procedures:
   a. Remove the side cover from the transaxle and check that the White wire is firmly attached to the 3–2 pressure switch.
   b. Using an ohmmeter, remove the connector from the pressure switch and check the resistance of the wire; it should be less than 2.0 ohms.
   c. If the connector/wire is not OK, replace the bulkhead connector/wiring assembly and repeat the Self Test. If the connector/wire is OK, check the 3–2 pressure switch.
4. To check the 3–2 pressure switch, perform the following procedures:
   a. Remove the 3–2 pressure switch and install it into a ⅛—27 pipe fitting. Connect the pipe fitting to low pressure air (do not use water) supply line (for testing purposes).
   b. Apply 50 psi (345 kPa) to the switch and check for ruptured diaphram.
   c. Submerge the switch in transaxle fluid and check for bubbles at the small vent hole near the switch terminal.
   d. If bubbles appear, replace the 3–2 pressure switch and repeat the Self Test. If no bubbles appear, check the switch's resistance.
5. To check the switch's resistance, perform the following procedures:
   a. Apply 50 psi (345 kPa) of air pressure to the switch.
   b. Using an ohmmeter, measure the resistance between the switch's terminal and the case; it should be less than 8.0 ohms.
   c. If the resistance is greater than 8.0 ohms, replace the switch, repeat the switch continuity check and repeat the Self Test; codes 39, 59, 62, 69 and 89 should no longer appear. If the resistance is less than 8.0 ohms, inspect the hydraulic circuit supplying pressure to the switch for excessive leakage and/or main control assembly operation.

CODE 89

The following procedure is used to check the Bypass Clutch Solenoid circuit's failure.

NOTE: Code 39 may also be present; if so, refer to the diagnosis procedures.

1. Make sure the vehicle harness connector is fully seated and the terminals engaged with the bulkhead connector. If not OK, repair the connector(s) or terminal(s) and repeat the Self Test. If OK, check the solenoid's resistance.
2. To check the solenoids resistance, perform the following procedures:
   a. Using the service jumper harness attach it to the transaxle's bulkhead connector.
   b. Using an ohmmeter, connect the positive (+) lead to the Red wire and the negative (−) lead to the Black wire. Check the resistance, it should be 20–40 ohms.

NOTE: When connecting a power supply to the wiring, do not reverse the polarity for it will damage the solenoid diode.

   c. If the resistance is OK, repeat the Self Test and service any code. If the resistance is not OK, check the internal connection.
3. To check the internal connection, remove the side cover and make sure the internal connector is fully engaged with the solenoid. If the connection is not OK, fully engage the connector, check the continuity, recheck the resistance and repeat the Self Test. If the connection is OK, check the solenoid continuity.
4. To check the solenoid's continuity, perform the following procedures:
   a. Disconnect the wires from the solenoid connector.
   b. Using an ohmmeter, connect the positive (+) lead to the solenoid's positive (+) terminal and the negative (−) lead to the solenoid's negative (−) terminal; the resistance should be 20–40 ohms.
   c. If the resistance is infinite (open circuit), replace the solenoid and repeat the Self Test.
   d. If the solenoid is OK, replace the bulkhead connector/wiring assembly, reconnect the internal connectors, recheck the resistance, replace the side cover and repeat the Self Test "Key On, Engine Off On Demand"; code 89 should no longer appear.

CODE 57

The following procedure is used to check the Neutral Park Switch (NPS) for failure in **N**.

NOTE: The NPS is normally open and closes with hydraulic pressure. If the transaxle fails to engage in D or R, a service code 57 will appear. Before testing the electrical components, be sure to inspect the hydraulic system functions.

1. Make sure the vehicle harness connector is fully seated and the terminals engaged with the bulkhead connector. If not OK, repair the connector(s) or terminal(s) and repeat the Self Test. If OK, check the switch's continuity.
2. To check the switch for continuity, perform the following procedures:
   a. Using the service jumper harness connector, attach it to the bulkhead connector.
   b. Using a ohmmeter, attach 1 lead to the Green wire and the other to ground.
   c. Start the engine and place the shift selector in **N** or **P**; the resistance should be infinite.
   d. Shift the transaxle into to **R** and **D**; the switch should close and the resistance should be less than 10 ohms in both ranges.
   e. If the circuit is OK, repeat the Self Test and service any codes. If the circuit is not OK, check the internal connections.
3. To check internal connections, perform the following procedures:
   a. Remove the side cover from the transaxle and check that the Green wire is firmly attached to the neutral pressure switch.
   b. Using an ohmmeter, remove the connector from the pressure switch and check the resistance of the wire; it should be less than 2.0 ohms.
   c. If the connector/wire is not OK, replace the bulkhead connector/wiring assembly and repeat the Self Test. If the connector/wire is OK, check the neutral pressure switch.

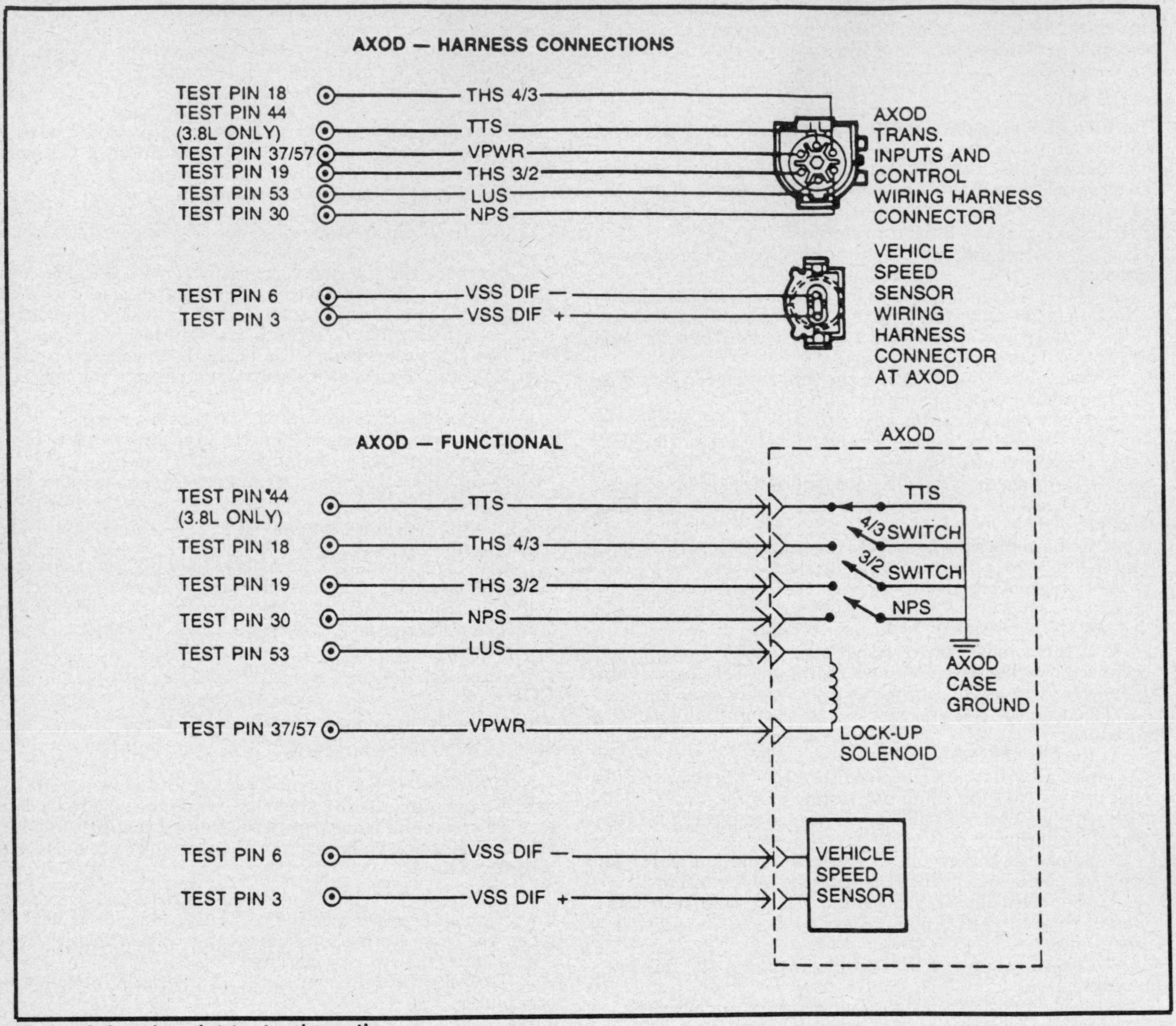
AXOD — HARNESS CONNECTIONS
TEST PIN 18
TEST PIN 44
(3.8L ONLY)
TEST PIN 37/57
TEST PIN 19
TEST PIN 53
TEST PIN 30
THS 4/3
TTS
VPWR
THS 3/2
LUS
NPS
AXOD
TRANS.
INPUTS AND
CONTROL
WIRING HARNESS
CONNECTOR
TEST PIN 6
TEST PIN 3
VSS DIF −
VSS DIF +
VEHICLE
SPEED
SENSOR
WIRING
HARNESS
CONNECTOR
AT AXOD
AXOD — FUNCTIONAL
AXOD
TEST PIN 44
(3.8L ONLY)
TEST PIN 18
TEST PIN 19
TEST PIN 30
TEST PIN 53
TEST PIN 37/57
TTS
THS 4/3
THS 3/2
NPS
LUS
VPWR
4/3 SWITCH
3/2 SWITCH
AXOD
CASE
GROUND
LOCK-UP
SOLENOID
TEST PIN 6
TEST PIN 3
VSS DIF −
VSS DIF +
VEHICLE
SPEED
SENSOR

View of the pinpoint test schematic

4. To check the neutral pressure switch, perform the following procedures:
   a. Remove the neutral pressure switch and install it into a ⅛–27 pipe fitting. Connect the pipe fitting to low pressure air (do not use water) supply line (for testing purposes).
   b. Apply 50 psi (345 kPa) to the switch and check for ruptured diaphram.
   c. Submerge the switch in transaxle fluid and check for bubbles at the small vent hole near the switch terminal.
   d. If bubbles appear, replace the neutral pressure switch and repeat the Self Test. If no bubbles appear, check the switch's resistance.
5. To check the switch's resistance, perform the following procedures:
   a. Apply 50 psi (345 kPa) of air pressure to the switch.
   b. Using an ohmmeter, measure the resistance between the switch's terminal and the case; it should be less than 8.0 ohms.
   c. If the resistance is greater than 8.0 ohms, replace the switch, repeat the switch continuity check and repeat the Self Test; codes 39, 59, 62, 69 and 89 should no longer appear. If the resistance is less than 8.0 ohms, inspect the hydraulic circuit supplying pressure to the switch for excessive leakage and/or main control assembly operation.

Pinpoint Test

NOTE: Perform this test only when the "Key On/Engine Off Self Test" service codes 62, 67 and 89 are present; when "Engine Running Self Test" service code 62 is displayed; or when "Continuous Self Test" service codes 29, 39, 57, 59 and/or 69 are displayed. Make sure that all components are connected before performing the test.

AXOD DRIVE CYCLE TEST

1. Record and zero the Self Test codes.
2. Operate the engine until normal operating temperatures are reached.
3. Place the transaxle in the **D** range, slowly accelerate to 40

mph (64 km/h) until the transaxle shifts into 3rd gear and hold the speed/throttle steady for 15 seconds (under 4000 ft.) or 30 seconds (above 4000 ft.).

4. Shift the transaxle into **OD**, accelerate from 40 mph (64 km/h) to 50 mph (80 km/h) until the transaxle shifts into 4th gear and hold the speed/throttle steady for 15 seconds.

5. With the transaxle in 4th gear, the speed steady and the throttle opening, lightly apply/release the brakes (to operate the brake lights) and maintain the steady speed for the next 15 seconds.

6. Apply the brakes, come to a stop and remain stopped for the next 20 seconds with the transaxle still in **OD**.

7. Turn the engine **OFF**. Perform the "Run Key On/Engine Off Self Test" and record any continuous codes.

NOTE: If any other continuous codes appear, service them first, for they could affect the electrical operation.

CODE 29

This procedure is performed in an attempt to generate Code 29.

1. Perform the AXOD Drive Cycle Test. If Code 29 appears, check the Vehicle Speed Sensor (VSS) harness for continuity (next step). If codes other than 29 appear, service them first. If no codes appear, the test is complete.

2. To check the continuity of the Vehicle Speed Sensor (VSS) harness, perform the following procedures:
   a. Turn the key **OFF** and wait 10 seconds.
   b. Disconnect the VSS harness connector.
   c. From the ECA processor, disconnect the 60 pin connector. Inspect the pins for damage, corrosion, loose wires and/or etc.; repair as necessary.
   d. While leaving the processor disconnected, install the breakout box.
   e. Using a DVOM, place it on the 200 ohm scale. Measure the resistance between the VSS harness connector (Pin 3) and the breakout box. Measure the resistance between the VSS harness connector (Pin 6) and the breakout box.
   f. If the resistance is over 5 ohms, remove the breakout box, reconnect the components, service the open circuits and repeat the AXOD Drive Cycle Test for Code 29.
   g. If the resistance is under 5 ohms, inspect the VSS harness for shorts to power or ground (next step).

3. To check the VSS harness for shorts, perform the following procedures:
   a. Turn the key **OFF**. Disconnect the processor and the VSS.
   b. Using a DVOM, place it on the 200k scale. Measure the resistance between the VSS harness connector (Pin 3) and the breakout box (Pins 37, 40 and 6). Measure the resistance between the VSS harness connector (Pin 6) and the breakout box.
   c. If the resistance is under 10k ohms, remove the breakout box, reconnect the components, service the short circuits and repeat the AXOD Drive Cycle Test for Code 29.
   d. If the resistance is over 10k ohms, remove the breakout box, install a new VSS, reconnect the components and repeat the AXOD Drive Cycle Test for Code 29.
   e. If Code 29 appears, replace the processor and repeat the AXOD Drive Cycle Test. If Code 29 did not appear, replace the VSS and repeat the AXOD Drive Cycle Test for Code 29.

CODE 69

This portion of the procedure is used in an attempt to generate Code 69.

1. Perform the AXOD Drive Cycle Test. If Code 69 appears, check the 3–2 circuit for continuity (next step). If codes other than 69 appear, service them first. If no codes appear, the test is complete.

2. To check the continuity of the 3–2 circuit, perform the following procedures:
   a. Turn the key **OFF** and wait 10 seconds.
   b. Disconnect the AXOD harness connector.
   c. From the ECA processor, disconnect the 60 pin connector. Inspect the pins for damage, corrosion, loose wires and/or etc.; repair as necessary.
   d. While leaving the processor disconnected, install the breakout box.
   e. Using a DVOM, place it on the 200 ohm scale. Measure the resistance between the AXOD connector (Pin 19) and the breakout box.
   f. If the resistance is over 5 ohms, remove the breakout box, reconnect the components, service the 3–2 open circuit and repeat the AXOD Drive Cycle Test for Code 69.
   g. If the resistance is under 5 ohms, inspect the 3–2 circuit a for power short (next step).

3. To check the 3–2 circuit for a short, perform the following procedures:
   a. Turn the key **OFF**. Disconnect the processor and the AXOD harness connector.
   b. Using a DVOM, place it on the 200k scale. Measure the resistance between the 3–2 circuit connector (Pin 19) and the breakout box (Pin 37).
   c. If the resistance is under 10k ohms, remove the breakout box, reconnect the components, service the short in the 3–2 circuit and repeat the AXOD Drive Cycle Test for Code 69.
   d. If the resistance is over 10k ohms, check the processor (next step).

4. To inspect the processor, perform the following procedures:
   a. Turn the key **OFF**. Install the breakout box. Reconnect the electrical harness connector to the processor and AXOD.
   b. At the breakout box, install a jumper wire between Pin 19 and Pin 40.
   c. Perform the "Run Key On/Engine Off Test".
   d. If Code 62 or 69 does not appear, remove the breakout box and the jumper wire. Replace the processor and repeat the AXOD Drive Cycle Test for Code 69.
   e. If Code 62 or 69 appears, remove the breakout box and the jumper wire.

This portion of the test is used to check the AXOD harness and the 3–2 circuit for shorts.

1. To verify the working order of the AXOD harness, perform the following procedures:
   a. Turn the key **OFF**. Disconnect the AXOD electrical harness connector.
   b. Perform the "Run Key On/Engine Off Self Test".
   c. If Code 69 does not appear, reconnect the AXOD harness connector; test is complete.
   d. If Code 69 does appear, check the 3–2 circuit for a short to ground (next step).

2. To check the 3–2 circuit for a short to ground, perform the following procedures:
   a. Turn the key **OFF**.
   b. Disconnect the AXOD harness connector.
   c. From the ECA processor, disconnect the 60 pin connector. Inspect the pins for damage, corrosion, loose wires and/or etc.; repair as necessary.
   d. While leaving the processor disconnected, install the breakout box.
   e. Using a DVOM, place it on the 200k ohm scale. At the breakout box, measure the resistance of Pin 19 between Pin 40 and 60.
   f. If both resistances are under 10k ohms, remove the breakout box, reconnect the AXOD harness and the processor, service the short to ground and repeat the AXOD Drive Cycle Test for Code 69.
   g. If both resistances are over 10k ohms, remove the break-

out box, reconnect AXOD harness, replace the processor and repeat the AXOD Drive Cycle Test for Code 69.

CODE 59

Code Generation

This portion of the procedure is used in an attempt to generate Code 59.

1. Perform the AXOD Drive Cycle Test. If Code 59 appears, check the 4–3 circuit for continuity (next step). If codes other than 59 appear, service them first. If no codes appear, the test is complete.
2. To check the continuity of the 4–3 circuit, perform the following procedures:
   a. Turn the key **OFF** and wait 10 seconds.
   b. Disconnect the AXOD harness connector.
   c. From the ECA processor, disconnect the 60 pin connector. Inspect the pins for damage, corrosion, loose wires and/or etc.; repair as necessary.
   d. While leaving the processor disconnected, install the breakout box.
   e. Using a DVOM, place it on the 200 ohm scale. Measure the resistance between the AXOD connector (Pin 18) and the breakout box.
   f. If the resistance is over 5 ohms, remove the breakout box, reconnect the components, service the 4–3 open circuit and repeat the AXOD Drive Cycle Test for Code 59.
   g. If the resistance is under 5 ohms, inspect the 4–3 circuit a for power short (next step).
3. To check the 4–3 circuit for a short, perform the following procedures:
   a. Turn the key **OFF**. Install the breakout box.
   b. Disconnect the processor and the AXOD harness connector.
   c. Using a DVOM, place it on the 200k scale. Measure the resistance between the 4–3 circuit connector (Pin 18) and the breakout box (Pin 37).
   d. If the resistance is under 10k ohms, remove the breakout box, reconnect the components, service the short in the 4/3 circuit and repeat the AXOD Drive Cycle Test for Code 59.
   e. If the resistance is over 10k ohms, check the processor (next step).
4. To inspect the processor, perform the following procedures:
   a. Turn the key **OFF**. Install the breakout box. Reconnect the electrical harness connector to the processor and AXOD.
   b. At the breakout box, install a jumper wire between Pin 18 and Pin 40.
   c. Perform the "Run Key On/Engine Off Self Test".
   d. If Code 62 or 59 does not appear, remove the breakout box and the jumper wire. Replace the processor and repeat the AXOD Drive Cycle Test for Code 59.
   e. If Code 62 or 59 appears, remove the breakout box and the jumper wire.

Short Test

This portion of the test is used to check the AXOD harness and the 4–3 circuit for shorts.

1. To verify the working order of the AXOD harness, perform the following procedures:
   a. Turn the key **OFF**. Disconnect the AXOD electrical harness connector.
   b. Perform the "Run Key On/Engine Off Self Test".
   c. If Code 59 does not appear, reconnect the AXOD harness connector; test is complete.
   d. If Code 59 does appear, check the 4–3 circuit for a short to ground (next step).
2. To check the 4–3 circuit for a short to ground, perform the following procedures:
   a. Turn the key **OFF**.
   b. Disconnect the AXOD harness connector.
   c. From the ECA processor, disconnect the 60 pin connector. Inspect the pins for damage, corrosion, loose wires and/or etc.; repair as necessary.
   d. While leaving the processor disconnected, install the breakout box.
   e. Using a DVOM, place it on the 200k ohm scale. At the breakout box, measure the resistance of Pin 18 between Pin 40 and 60.
   f. If both resistances are under 10k ohms, remove the breakout box, reconnect the AXOD harness and the processor, service the short to ground and repeat the AXOD Drive Cycle Test for Code 59.
   g. If both resistances are over 10k ohms, remove the breakout box, reconnect AXOD harness, replace the processor and repeat the AXOD Drive Cycle Test for Code 59.

CODE 39

This procedure is used in an attempt to generate Code 39.

NOTE: Should Code 59 be present, go directly to Code 59 and perform the series of checks.

Perform the AXOD Drive Cycle Test. If Code 39 appears, go to Electrical System Diagnosis and perform checks on the components. If codes other than 39 appear, service them first. If no codes appear, the test is complete.

CODE 57

This procedure is used in an attempt to generate Code 57.

1. Perform the AXOD Drive Cycle Test. If Code 57 appears, check the NPS harness circuit for continuity (next step). If codes other than 57 appear, service them first. If no codes appear, the test is complete.
2. To check the continuity of the NPS harness circuit, perform the following procedures:
   a. Turn the key **OFF** and wait 10 seconds.
   b. Disconnect the AXOD harness connector.
   c. From the ECA processor, disconnect the 60 pin connector. Inspect the pins for damage, corrosion, loose wires and/or etc.; repair as necessary.
   d. While leaving the processor disconnected, install the breakout box.
   e. Using a DVOM, place it on the 200 ohm scale. Measure the resistance between the AXOD connector (Pin 30) and the breakout box.
   f. If the resistance is over 5 ohms, remove the breakout box, reconnect the components, service the NPS open circuit and repeat the AXOD Drive Cycle Test for Code 57.
   g. If the resistance is under 5 ohms, remove the breakout box and reconnect the components.

CODE 89

This procedure is used in an attempt to generate Code 89.

1. To check the continuity of the VPWR circuit, perform the following procedures:
   a. Turn the key **OFF** and wait 10 seconds.
   b. Disconnect the AXOD harness connector.
   c. From the ECA processor, disconnect the 60 pin connector. Inspect the pins for damage, corrosion, loose wires and/or etc.; repair as necessary.
   d. While leaving the processor disconnected, install the breakout box.
   e. Using a DVOM, place it on the 200 ohm scale. Measure the resistance between the AXOD connector (Pin 37) and the breakout box.
   f. If the resistance is over 5 ohms, remove the breakout box, reconnect the components, service the LUS open circuit and repeat the this test for Code 89.
   g. If the resistance is under 5 ohms, check the LUS circuit continuity (next step).

2. To check the continuity of the LUS circuit, perform the following procedures:
a. Turn the key **OFF**. Install the breakout box.
b. Disconnect the processor and the AXOD harness connector.
c. Using a DVOM, place it on the 200 ohm scale. Measure the resistance between the AXOD harness connector (Pin 53) and the breakout box.
d. If the resistance is over 5 ohms, remove the breakout box, reconnect the components, service the open in the LUS circuit and repeat this test.
e. If the resistance is under 5 ohms, check the LUS circuit for a short to power or ground (next step).
3. To check the LUS circuit for a short to power or ground, perform the following procedures:
a. Turn the key **OFF**. Install the breakout box.
b. Disconnect the processor and the AXOD harness connector.
c. Using a DVOM, place it on the 200k scale. At the breakout box, measure the resistance Pin 53 between Pins 37 and 40.
d. If both resistances are under 10k ohms, remove the breakout box, reconnect the components, service the short in the LUS circuit. Repeat the AXOD Drive Cycle Test for Code 89; if it is still present, replace the processor and repeat the AXOD Drive Cycle Test for Code 89.
e. If both resistances are over 10k ohms, check the total circuit resistance (next step).
4. To check the total circuit resistance, perform the following procedures:
a. Turn the key **OFF**. Install the breakout box.
b. Disconnect the processor and the AXOD harness connector.
c. Using a DVOM, place it on the 200 ohm scale. At the breakout box, measure the resistance Pin 53 and Pin 57.
d. If the resistance is 20–40 ohms, remove the breakout box, replace processor and repeat the AXOD Drive Cycle Test for Code 89.
e. If the resistance is not 20–40 ohms, remove the breakout box and reconnect the processor.

CODE 62

This code is used to check the AXOD harness, the 3–2 and 4–3 circuits for shorts.
1. To verify the working order of the AXOD harness, perform the following procedures:
a. Turn the key **OFF**. Disconnect the AXOD electrical harness connector.
b. Perform the "Run Key On/Engine Off Self Test".
c. If Code 62 does not appear, reconnect the AXOD harness connector; test is complete.
d. If Code 62 does appear, check the 3–2 and 4–3 circuits for a short to ground (next step).
2. To check the 3–2 and the 4–3 circuits for a short to ground, perform the following procedures:
a. Turn the key **OFF**.
b. Disconnect the AXOD harness connector.
c. From the ECA processor, disconnect the 60 pin connector. Inspect the pins for damage, corrosion, loose wires and/or etc.; repair as necessary.
d. While leaving the processor disconnected, install the breakout box.
e. Using a DVOM, place it on the 200k ohm scale. At the breakout box, measure the resistance of Pin 18 between Pin 40 and 60.
f. At the breakout box, measure the resistance of Pin 19 between Pin 40 and 60.
g. If all resistances are under 10k ohms, remove the breakout box, reconnect all components, service the short(s) to ground and repeat the AXOD Drive Cycle Test for Code 62.
h. If all resistances are over 10k ohms, remove the breakout box, reconnect all components, replace the processor and repeat the AXOD Drive Cycle Test for Code 62.

TEMPERATURE TIMED SWITCH (TTS) – 3.8L ONLY

1. To check the TTS harness circuit continuity, perform the following procedures:
a. Turn the key **OFF** and wait 10 seconds.
b. Disconnect the AXOD harness connector.
c. From the ECA processor, disconnect the 60 pin connector. Inspect the pins for damage, corrosion, loose wires and/or etc.; repair as necessary.
d. While leaving the processor disconnected, install the breakout box.
e. Using a DVOM, place it on the 200 ohm scale. Measure the resistance between the AXOD connector (Pin 44) and the breakout box.
f. If the resistance is over 5 ohms, remove the breakout box, reconnect the components, service the TTS open circuit and drive the vehicle to verify that the drive complaint is eliminated.
g. If the resistance is under 5 ohms, check the TTS circuit for a short to the power or ground (next step).
2. To check the TTS harness circuit for a short to power or ground, perform the following procedures:
a. Turn the key **OFF**.
b. Disconnect the AXOD harness connector.
c. Disconnect the processor disconnected and install the breakout box.
d. Using a DVOM, place it on the 200k ohm scale. At the breakout box, measure the resistance between Pin 44 and Pin 37.
e. At the breakout box, measure the resistance between Pin 40 and Pin 44.
f. If all resistances are under 10k ohms, remove the breakout box, reconnect all components, service the TTS circuit's short(s) and drive the vehicle to verify that the drive complaint is eliminated.
g. If all resistances are over 10k ohms, check the processor's operation (next step).
3. To check the processor operation, perform the following procedures:
a. Turn the key **OFF** and install the breakout box.
b. Reconnect the processor and the AXOD harness.
c. Using a jumper wire at the breakout box, connect Pin 44 to Pin 40.
d. Drive the vehicle to verify the drive complaint.
e. If the drive complaint was eliminated, remove the breakout box and the jumper wire.
f. If the drive complaint was not eliminated, remove the breakout box and the jumper wire. Replace the processor.

CODE 67

1. To check the NPS input to processor voltage, perform the following procedures:
a. Turn the key **ON** and the engine **OFF**.
b. From the ECA processor, disconnect the 60 pin connector. Inspect the pins for damage, corrosion, loose wires and/or etc.; repair as necessary.
c. Reconnect the processor and install the breakout box.
d. Using a DVOM, place it on the 20 volt scale. At the breakout box, measure the voltage between Pin 30 and Pin 46.
e. If the voltage over 4 volts, check the A/C input of the neutral drive switch. Using a DVOM on the 20 volt scale, measure the breakout box Pin 10 to ground voltage; the voltage should be greater than 1.0 volt, if not replace the processor.
f. If the voltage is under 4 volts, check the NPS harness circuit for a short to ground (next step).
2. To check the NPS harness circuit for a short to ground,

perform the following procedures:

a. Turn the key **OFF**.

b. Disconnect the AXOD harness connector.

c. Disconnect the processor disconnected and install the breakout box.

d. Using a DVOM, place it on the 200k ohm scale. At the breakout box, measure the resistance of Pin 30 between Pin 40 and Pin 60.

e. If both resistances are under 10k ohms, remove the breakout box, reconnect all components, service the NPS circuit's short and repeat the AXOD Drive Cycle Test.

g. If both resistances are over 10k ohms, check the processor's operation (next step).

3. To check the processor's operation, perform the following procedures:

a. Turn the key **OFF**.

b. Install the breakout box and reconnect the processor.

c. Disconnect the AXOD harness connector.

d. Perform the "Run Key On/Engine Off Self Test".

e. If Code 67 is not present, remove the breakout box, reconnect the components.

f. If Code 67 is present, remove the breakout box, reconnect the components, replace the processor and repeat the AXOD Drive Cycle test for code 67.

Hydraulic Control System

The hydraulic shifting operation is monitored by the EEC-IV system's computer which operates a bypass clutch solenoid within the transaxle to eliminate converter slippage. Signals from the NPS Neutral, 3–2 and 4–3 pressure switches inform the computer of transaxle gear shifts so the computer may control engine operations.

The hydraulic system consists of a main control assembly (valve body), oil pump, overdrive servo, low/intermediate servo, governor and 2 reservoir areas.

Main Control Assembly (Valve Body)

The main control assembly controls the transaxle operation by directing pressurized fluid to the torque converter, band servos, clutches and governor.

Oil Pump

The oil pump, located in the control valve/pump assembly, is a variable capacity vane/rotor pump which provides pressurized fluid (proportional to demand) to operate, lubricate and cool the transaxle.

Low/Intermediate Servo

The low/intermediate servo applies the low/intermediate band in the manual low, 1st and 2nd gears.

Overdrive Servo

The overdrive servo applies the overdrive band in the 4th gear.

Governor

The governor, driven by a differential assembly gear, provides a road speed signal to the hydraulic control for shift control.

Reservoirs

Upper and lower reservoirs, dependent upon fluid temperature, are used to control the oil level. As fluid temperature rises in the lower sump, a thermostatic element closes, retaining fluid in the upper reservoir.

Diagnosis Tests

CONTROL PRESSURE TEST

1. Firmly set the parking brake and block the drive wheels.
2. Remove the pressure line tap plug from the transaxle and install the pressure gauge.
3. Start the engine, move the shift lever through the various selector positions and check the fluid pressures; refer to the control pressure chart.
4. If the fluid pressures are not within specifications, proceed to the air pressure test and/or service the main control system.

AIR PRESSURE TEST

Because of inoperative bands or clutches, a no drive condition can exist, even if the fluid pressures are correct. By substituting air pressure for fluid pressure, an erratic shift condition location can be determined.

NOTE: An inoperative forward clutch, low/intermediate one-way or low intermediate band may cause a NO DRIVE condition when the shift lever is positioned in

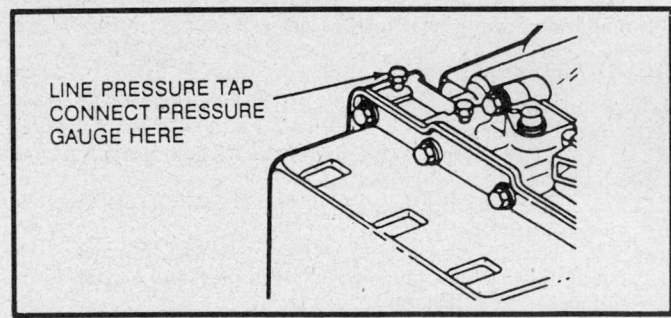

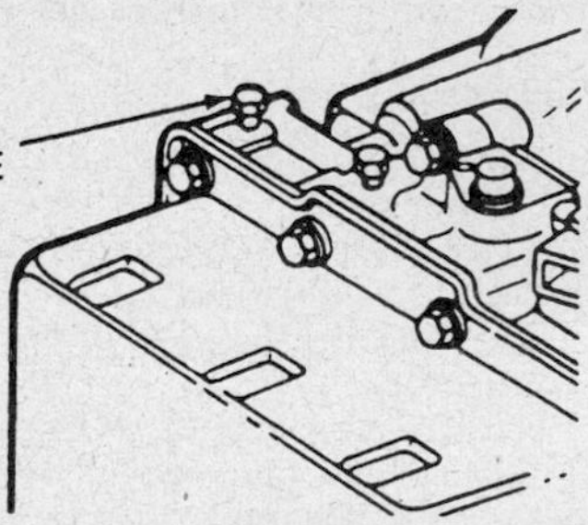
LINE PRESSURE TAP
CONNECT PRESSURE
GAUGE HERE

Location of the transaxle line pressure tap

CONTROL PRESSURE TEST

Engine	Range	Idle psi	Idle kPa	Stall (WOT) psi	Stall (WOT) kPa
2.5L and 3.0L	P, N	81–95	558–655	—	—
	R	93–152	641–1048	242–279	1669–1924
	OD, D	81–95	558–655	158–183	1089–1262
	L	112–169	772–1165	158–183	1089–1262
3.8L	P, N	80–91	551–627	—	—
	R	93–152	641–1048	248–289	1709–1992
	OD, D	80–91	551–627	182–213	1254–1468
	L	112–169	772–1165	158–183	1089–1261

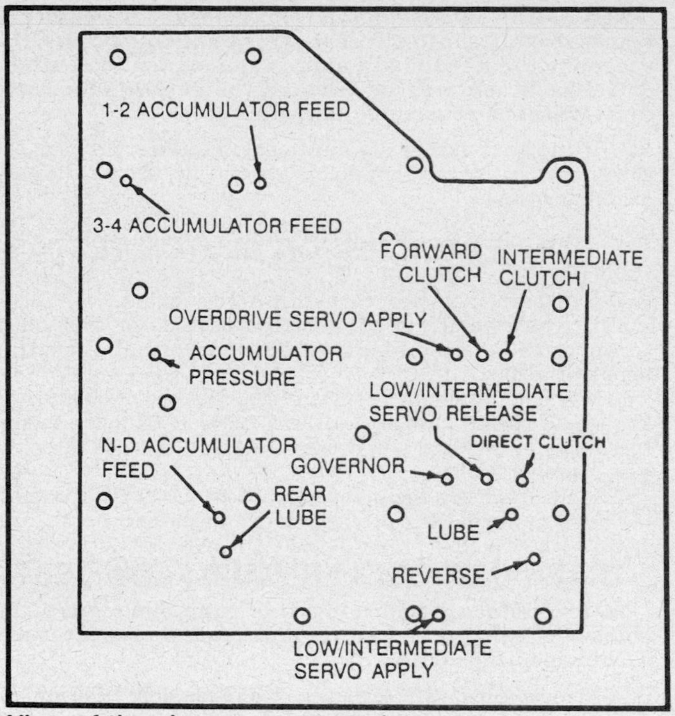
1-2 ACCUMULATOR FEED
3-4 ACCUMULATOR FEED
FORWARD CLUTCH
INTERMEDIATE CLUTCH
OVERDRIVE SERVO APPLY
ACCUMULATOR PRESSURE
LOW/INTERMEDIATE SERVO RELEASE
DIRECT CLUTCH
N-D ACCUMULATOR FEED
GOVERNOR
REAR LUBE
LUBE
REVERSE
LOW/INTERMEDIATE SERVO APPLY

View of the air pressure test plate

OD, D or 1st gears. An inoperative direct clutch or direct one-way clutch may cause a NO COAST condition in 1st gear. A malfunctioning reverse clutch, forward clutch or low/intermediate one-way clutch may cause a NO DRIVE condition in reverse gear.

1. Raise and support the vehicle on jackstands.
2. Drain the fluid from the transaxle and remove the oil pan.
3. Remove the main control cover, the oil pump and the main control assembly.
4. Using the air pressure test plate and the chain cover gasket, install it to the main control assembly.
5. Using air pressure, introduce it to the various test plate passages as follows:

FORWARD CLUTCH

When applying air pressure to the forward clutch test port, a dull thud should be heard or movement from the piston be felt. If a hissing sound is noticed, the clutch seal(s) is leaking.

GOVERNOR

When applying air pressure to the governor test port, listen for a whistling or a sharp clicking noise; the noise indicates proper operation.

OVERDRIVE SERVO

When applying air pressure to the overdrive servo test port, the band should tighten around the overdrive drum. Due to the servo release spring's cushioning effect, band application may not be heard or felt. The servo should not leak (while holding pressure) and a dull thud (piston returning to original position) should be heard when the pressure is removed.

DIRECT CLUTCH

When applying air pressure to the direct clutch test port, a dull thud should be heard or movement of the piston should be felt. If hissing is noticed, the clutch seal(s) is leaking.

INTERMEDIATE CLUTCH

When applying air pressure to the intermediate clutch test port, a dull thud should be heard or movement of the piston should be felt. If hissing is noticed, the clutch seal(s) is leaking.

LOW/INTERMEDIATE SERVO

When applying air pressure to the low/intermediate servo apply test port, the band should tighten around the rear planetary gearset's sun gear. Due to the servo release spring's cushioning effect, band application may not be heard or felt. The servo should not leak (while holding pressure) and a dull thud (piston returning to original position) should be heard when the pressure is removed.

While applying air pressure to the low/intermediate servo apply test port, introduce air pressure to the low/intermediate release test port; the band should loosen, a dull thud should be heard and the piston move to the release position. Remove air pressure to the apply test port, the release test port should hold air pressure without leakage. The servo requires service if leakage occurs or the piston fails to move.

LUBE AND REAR LUBE

When applying air pressure to the lube and rear lube test ports, air should move freely through the ports. If a blockage occurs, remove the test plate and check for obstructions or damage.

1-2, 3-4 AND N-D ACCUMULATORS

When applying air pressure to each accumulator feed test port, the accumulator should apply, holding air pressure. Due to the accumulator release spring's cushioning effect, accumulator application may not be heard or felt. The accumulator should not leak (while holding pressure) and a dull thud (accumulator returning to original position) should be heard when the pressure is removed.

STALL SPEED TEST

The stall test is used to test the converter's one-way clutch, forward clutch, low one-way clutch, reverse clutch, reverse clutch, low/intermediate band and engine performance.

NOTE: Be sure the engine and transaxle are at normal operating temperatures before performing this test.

1. Using a tachometer, connect it to the engine.
2. Block the drive wheels and firmly apply the parking brake. While performing the test, firmly apply the service brakes.

STALL SPEED HIGH (SLIP)

Range	Possible Source
OD, D, 1	Forward clutch Low/intermediate one-way clutch Low/intermediate band or servo
R	Forward clutch Low/intermediate one-way clutch Reverse clutch

CAUTION

Do not maintain wide open throttle in any gear range for more than a few seconds.

3. Start the engine and place the shift selector in a gear range (one at a time). Press the accelerator to the floor (wide open throttle) and record the engine speed reached in each range; the speeds should be within 1950–2275 rpm.

NOTE: After testing a gear range, place the shift selector in N and run the engine for 15 seconds to allow the converter to cool before testing the next range.

CAUTION

If the engine speed exceeds 2275 rpm, release the accelerator immediately; clutch or band slippage is indicated.

If the stall speeds are too low, check the engine tune-up. If the stall speeds are too high refer to the Stall Speed Diagnosis chart for possible source of the problem. If the engine is OK, remove the torque converter and check the one-way clutch for slippage.

SHIFT POINT TEST

Road Test

1. Drive the vehicle until the engine and transaxle are at normal operating temperatures.
2. Shift the transaxle to **OD** range. Apply minimum throttle pressure and note the upshift speeds and speed which the converter clutch applies.
3. Stop the vehicle and move the shift selector into **D**. Apply minimum throttle pressure and note the upshift speeds and speed which the converter clutch applies; the transaxle should make all upshifts (except 3–4 and the converter clutch apply should occur above 27 mph (46 km/h).
4. Fully depress the accelerator to wide open throttle. Depending upon vehicle speed, the transaxle should shift from 3rd-to-2nd or 3rd-to-1st and the converter clutch should release.

NOTE: If the shift lever is placed in the OD range and the accelerator is in the wide open throttle position, a 4th-to-3rd downshift can be obtained regardless of the vehicle speed.

5. When the vehicle speed is above 30 mph (48 km/h), move the shift selector from **D** to **L** and release the accelerator; the transaxle should immediately downshift to **2nd** gear. When the vehicle speed drops below 20 mph, the transaxle should downshift to **1st** gear.
6. If the transaxle does not perform according to the proceeding tests, refer to governor pressure and shift control valve diagnosis.

In-Shop Test

This test is designed to check the governor circuits, the shift delay pressures and the throttle boost.

1. Raise and support the front of the vehicle by placing supports under the suspension so the wheel are off the floor.

CAUTION

Do not exceed the speedometer reading of 60 mph (97 km/h) for the tire speed is actually twice the speedometer reading. Do not allow the suspension to hang free; damage to the velocity seals and joints may occur and heavy vibrations will be emitted.

2. Start the engine, place the shift lever in **OD** and apply minimum throttle pressure.
3. Note the shift speeds and the speed when the converter locks up; 1–2, 2–3, converter lockup and 3–4.

NOTE: The converter will remain locked up when the transaxle shifts into 4th gear. At the shifting points, the speedometer needle will surge (momentarily), a slight driveline bump may be felt and the engine speed will drop (without releasing the pedal).

4. If the shift speeds are not to specifications, perform the governor check. If the shift points are too low, inspect the shift modulator valves.

CONVERTER PISTON CLUTCH TEST

1. Using a tachometer, connect it to the engine.
2. Drive the vehicle in **OD** at highway speeds for 15 minutes to warm the engine and transaxle to normal operating temperatures.
3. While maintaining a constant vehicle speed of 50 mph (80 km/h) and tap the brake pedal with the left foot; the engine speed should increase and decrease about 5 seconds after the pedal is released.
4. If the transaxle does not perform according to the proceeding tests, perform the converter clutch diagnosis.

TRANSAXLE COOLER FLUID FLOW TEST

NOTE: Before performing this test, make sure the transaxle linkage, fluid level and control pressures are within specifications.

1. Remove the dipstick from the transaxle filler tube and insert a funnel.
2. Raise and support the vehicle in a level position.
3. Remove the fluid cooler tube from the lower (return) transaxle fitting.
4. Using a hose, connect it to the fluid cooler tube and insert the other end into the funnel.
5. Firmly set the parking brake. Position the transaxle in **N**, start the engine and adjust the idle to 1000 rpm. Observe the fluid flowing into the funnel; it should be liberal and solid.
6. If the flow is not liberal, stop the engine and install the lower tube to the transaxle. Remove the upper transaxle tube, connect it to the hose and repeat the flow test.
7. If the flow is still not liberal, inspect the fluid pump for low capacity, the main circuit system for leakage and the converter drain valve or regulator valve for a sticking condition.

Converter Clutch Operation and Diagnosis

TORQUE CONVERTER CLUTCH

The torque converter clutch is a one-way clutch located within the torque converter's stator. It operates in conjunction with the piston plate clutch/damper assembly, when the EEC-IV system energizes the bypass clutch solenoid, and is designed to hold the stator stationary in the **LOCK-UP** mode. The result is improved fuel economy by eliminating converter clutch slippage.

ON CAR SERVICES

Adjustments

THROTTLE VALVE CABLE

Normally, the throttle valve (T.V.) cable does not require adjustment. Only, if the main control assembly, the T.V. cable, the T.V. cable engine mounting bracket, the throttle control lever link/lever assembly, the throttle body and/or transaxle assembly have been replaced, should adjustment be necessary.

1. Connect the T.V. cable's eye to the transaxle throttle control lever link and the cable boot to the chain cover.
2. With the T.V. cable attached to the engine bracket, be sure the threaded shank is fully retracted. Using the index fingers, pull the spring rest upward and wiggle the threaded shank top, while pressing the shank through the spring with the thumbs.
3. Connect the T.V. cable end to the throttle body.

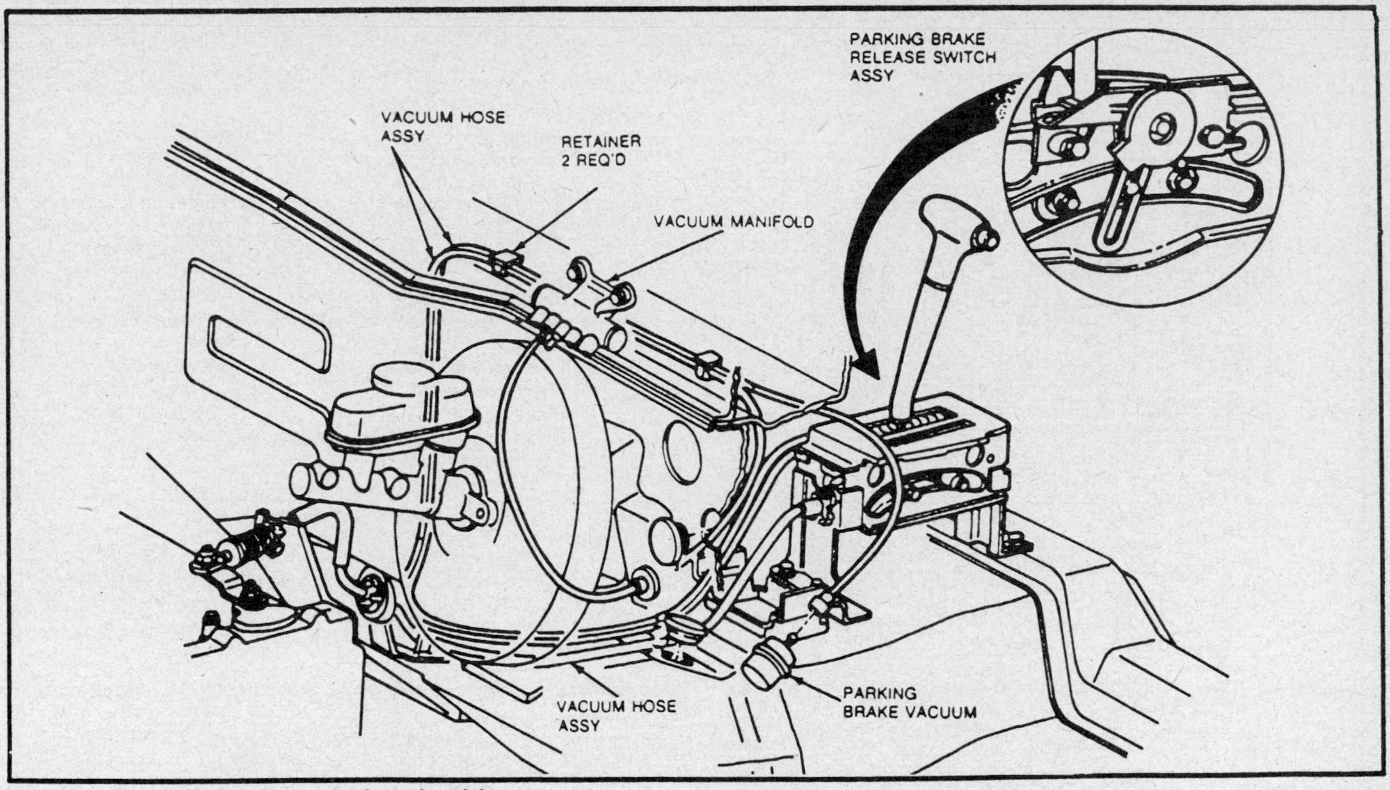
PARKING BRAKE
RELEASE SWITCH
ASSY
VACUUM HOSE
ASSY
RETAINER
2 REQ'D
VACUUM MANIFOLD
VACUUM HOSE
ASSY
PARKING
BRAKE VACUUM
View of the shift lever control and cable

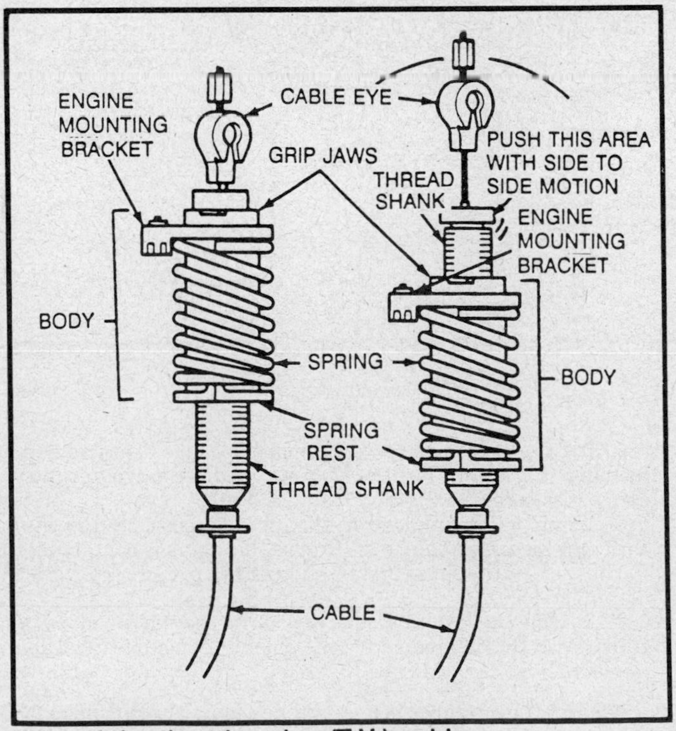
ENGINE
MOUNTING
BRACKET
CABLE EYE
GRIP JAWS
THREAD
SHANK
PUSH THIS AREA
WITH SIDE TO
SIDE MOTION
ENGINE
MOUNTING
BRACKET
BODY
SPRING
BODY
SPRING
REST
THREAD SHANK
CABLE
View of the throttle valve (T.V.) cable

4. Rotate the throttle lever to the wide open throttle (WOT) position and release it.

NOTE: The threaded shank must move or ratchet from the grip jaws. If movement is not noticed, inspect the system for broken or disconnected parts.

MANUAL LINKAGE

TAURUS AND SABLE

The following procedure is used for both floor and column mounted shifters.

1. Move the shift selector into the **OD** position against the rear stop.

NOTE: While the linkage is being adjusted, the shift lever must be held in the rearward position.

2. Loosen the manual lever-to-control cable retaining nut and move the lever to the **OD** position (2nd detent from the most rearward position).
3. Tighten the manual lever-to-control cable nut to 10–15 ft. lbs. (13.5–20 Nm).
4. Check the operation in each shift lever position; ensure that the park/neutral switch is functioning properly.

CONTINENTAL

This vehicle is equipped with a column shift only.

1. From the transaxle lever's pivot ball, remove the cable plastic terminal.
2. At the cable trunnion, mounted on the retaining bracket, loosen the adjusting bolt and free the cable in the trunnion.
3. From the passenger's compartment, move the shift selector into the **OD** position. Using an 8 lb. weight, suspend it from the shift lever.
4. At the transaxle, rotate the shift lever clockwise to the **L** position and counterclockwise to the **OD** position.
5. Install the shift cable plastic terminal onto the transaxle lever's pivot ball, from the flat side of the terminal.
6. Torque the trunnion's cable adjustment screw to 11–14 ft. lbs. (14–20 Nm).
7. Check the operation in each shift lever position; ensure that the park/neutral switch is functioning properly.

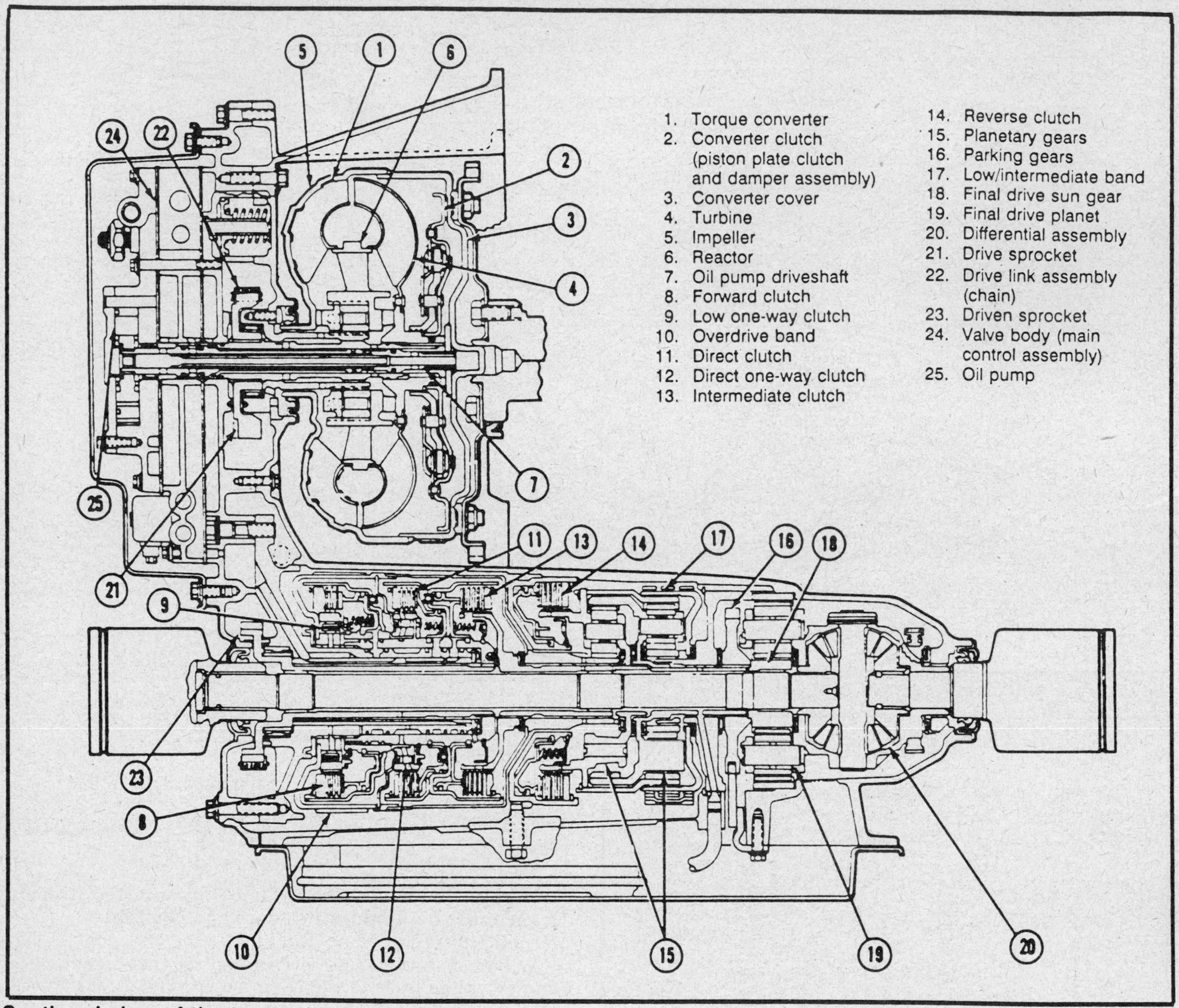
1. Torque converter
2. Converter clutch (piston plate clutch and damper assembly)
3. Converter cover
4. Turbine
5. Impeller
6. Reactor
7. Oil pump driveshaft
8. Forward clutch
9. Low one-way clutch
10. Overdrive band
11. Direct clutch
12. Direct one-way clutch
13. Intermediate clutch
14. Reverse clutch
15. Planetary gears
16. Parking gears
17. Low/intermediate band
18. Final drive sun gear
19. Final drive planet
20. Differential assembly
21. Drive sprocket
22. Drive link assembly (chain)
23. Driven sprocket
24. Valve body (main control assembly)
25. Oil pump

Sectional view of the transaxle components

LOW/INTERMEDIATE AND OVERDRIVE BANDS

The bands do not require adjustment but when the transaxle has been overhauled, the servo travel check must be performed and the servo piston rod(s) possibly be replaced.

1. If the servo covers are installed, remove the servo cover, the piston and rod.
2. Using the spring from the overdrive servo rod tool, low/intermediate servo rod tool or equivalent, install it in the case bore.
3. Install the servo piston and rod in the case bore.

NOTE: On the low/intermediate servo, the piston must be installed without the seal.

4. Using the overdrive servo rod tool, low/intermediate servo rod tool or equivalent, install it in the case bore. Using the servo cover bolts, torque the tool(s)-to-case bore to 7–9 ft. lbs. (9–12 Nm).
5. For the overdrive servo, torque the gauge disc screw to 10 inch lbs. (1.13 Nm). For the low/intermediate servo, torque the gauge disc screw to 30 inch lbs. (3.4 Nm).
6. Using a dial indicator, mount and position the stylus through the disc gauge hole. Be sure the stylus is contacting the flat surface of the servo piston and not the piston step. Zero the dial indicator.
7. Loosen the gauge disc screw until the piston movement stops. Read the dial indicator; the amount of piston travel determines the rod length to be installed.

NOTE: The overdrive servo reading should be 0.070–0.149 in. (1.8–3.8mm); the low/intermediate servo reading should be 0.216–0.255 in. (5.5–6.5mm). If a new low/intermediate band has been installed, the reading should be 0.196–0.236 in. (5–6mm).

8. Using the measurement acquired, select and install a new piston rod. Recheck the piston movement to verify the amount of piston travel.

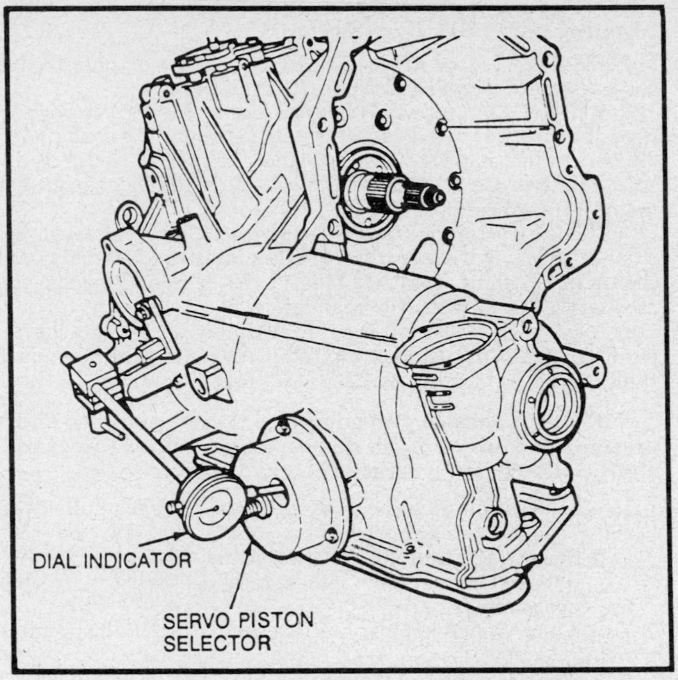
DIAL INDICATOR
SERVO PISTON
SELECTOR

Using a dial indicator to check the servo piston movement

9. On the low/intermediate servo piston, install the seal(s).
10. Install the servo pistons and springs; be sure they are fully seated.
11. Using new gaskets, install the servo-to-case bore covers and torque the bolts to 7–9 ft. lbs. (9–12 Nm).

NOTE: When installing the low/intermediate servo cover, be sure the tab aligns with the case port. Tighten the bolts 2–3 turns at a time to prevent cocking the case cover.

Services

FLUID CHANGES

Under normal vehicle usage, the transaxle necessitates partial drain and refill procedures. Only, when operated under severe/continuous conditions or part replacement, should the transaxle be removed from service, totally drained, cleaned and refilled; at this time, the converter, cooler and cooler lines should be throughly flushed.

1. Raise and support the vehicle safely.
2. Place a drain pan under the transaxle.
3. Loosen the oil pump/valve body cover bolts and the lower pan bolts and drain the fluid into the drain pan; if necessary, use a rubber mallet to bump the cover or pan loose from the transaxle.
4. When the fluid has drained from the transaxle (except from the lower pan), remove the remaining pan bolts (working from the right side), allow the pan to drop and drain slowly. Remove the oil pump/valve body cover.
5. Clean the pan and cover, discard the gasket.
6. Using new gaskets and sealant (if necessary), install the pan and cover-to-case bolts and torque to 10–12 ft. lbs. (14–16 Nm).
7. Using the correct lubricant, refill the transaxle to the correct dipstick Level; do not overfill or foaming will occur.
8. Operate the transaxle to distribute fluid to the upper reservoir and recheck/refill to dipstick levels.

OIL PAN

Removal and Installation

1. Raise and support the vehicle safely.
2. Place a drain pan under the transaxle.
3. Loosen the lower pan bolts and drain the fluid into the drain pan; if necessary, use a rubber mallet to bump the pan loose from the transaxle.
4. When the fluid has drained from the transaxle, remove the remaining pan bolts (working from the right side), allow the pan to drop and drain slowly.
5. Clean the pan and discard the gasket.
6. Using new gaskets and sealant (if necessary), install the pan-to-case bolts and torque to 10–12 ft. lbs. (14–16 Nm).
7. Using the correct lubricant, refill the transaxle to the correct dipstick level; do not overfill, for foaming will occur.
8. Operate the transaxle to distribute fluid to the upper reservoir and recheck/refill to dipstick levels.

SIDE COVER

Removal and Installation

The side cover is located on the left-side of the transaxle and is a secondary reservoir.

1. Disconnect the battery cables, the negative cable first.
2. Remove the battery, the battery tray and the air cleaner.
3. Secure the necessary hoses, vacuum lines and wiring away from the side case.
4. If necessary, raise and support the vehicle safely.
5. Place an oil catch pan under the side cover.
6. Loosen the side cover bolts and drain the fluid. After the fluid has drained, remove the cover and gasket, if necessary, use a rubber mallet to bump the cover loose from the transaxle.
7. Remove the side cover-to-case bolts and the cover.
8. Clean the cover and discard the gasket.
9. Using new gaskets and sealant (if necessary), install the cover-to-case bolts and torque to 10–12 ft. lbs. (14–16 Nm).
10. Using the correct lubricant, refill the transaxle to the correct dipstick; do not overfill, for foaming will occur.

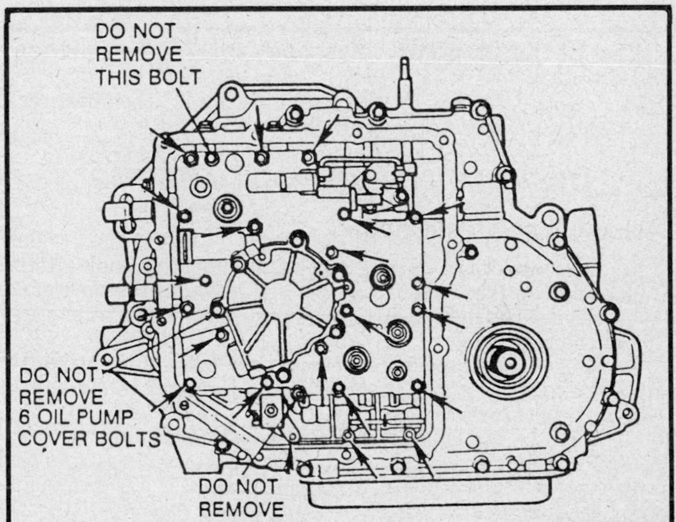
DO NOT
REMOVE
THIS BOLT
DO NOT
REMOVE
6 OIL PUMP
COVER BOLTS
DO NOT
REMOVE
THIS BOLT

View of the oil pump/valve body assembly-to-transaxle bolts. Remove only the bolts indicated

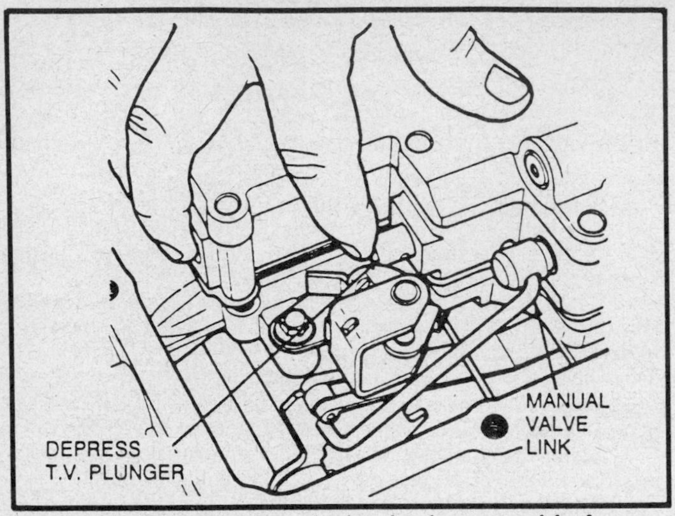
MANUAL
VALVE
LINK
DEPRESS
T.V. PLUNGER

Separating the oil pump/valve body assembly from the manual valve link

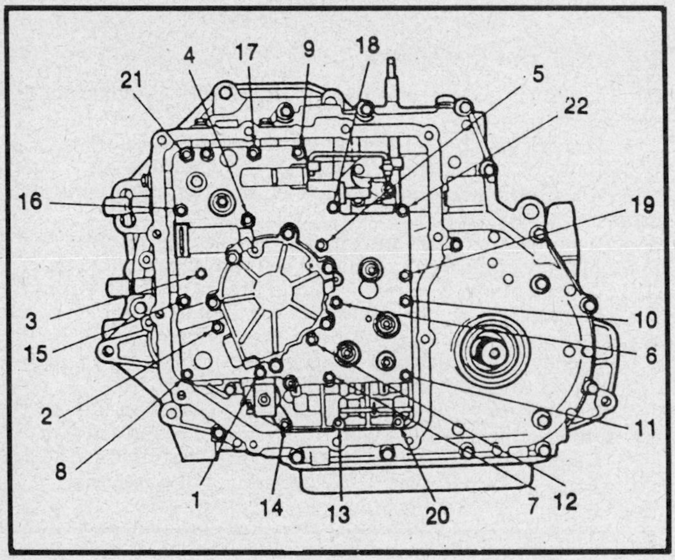
4
17
9
18
21
5
22
16
19
3
10
15
6
2
11
8
1
14
13
20
7
12

View of the oil pump/valve body torquing sequence

11. Operate the transaxle to distribute fluid to the upper reservoir and recheck/refill to dipstick levels.

OIL PUMP/VALVE BODY ASSEMBLY

Removal and Installation

1. Disconnect the battery cables, the negative cable first.
2. Remove the battery, the battery tray and the air cleaner.
3. Secure the necessary hoses, vacuum lines and wiring away from the side case.
4. Raise and support the vehicle on a hoist or jackstands.
5. Using the engine support bar tool, install it and support the engine/transaxle assembly.
6. Remove the left-side engine mounts and supports.
7. Place an oil catch pan under the side cover. Loosen the side cover bolts and drain the fluid. After the fluid has drained, remove the cover and gasket.
8. Disconnect the electrical connectors from the pressure switches and the solenoid.
9. Remove the oil pump/valve body assembly-to-transaxle bolts.

NOTE: Do not remove the oil pump cover bolts or the 2 oil pump-to-valve body bolts.

10. Push the T.V. plunger inward and pull the oil pump/valve body assembly outward to clear the bracket.
11. Pull the oil pump/valve body assembly forward enough to clear the throttle valve bracket, rotate it toward the dash panel, disconnect the manual valve link and remove the assembly.
12. Before installing the assembly, be sure to clean the gasket mounting surfaces.
13. Using a new oil pump/valve body-to-chain cover gasket, position it on the mating surface. While sliding the assembly onto the oil pump shaft, rotate it toward the dash and engage the manual valve link with the manual valve.
14. Slightly jiggle or rotate the assembly to engage the oil pump splines with the oil pump rotor; it should slide flush onto the chain cover without force.

NOTE: If necessary to complete the oil pump-to-shaft engagement, use a ⅞ in. deep well socket on the crankshaft pulley bolt to rotate the crankshaft.

15. If the oil pump/valve body assembly will not fully seat, perform the following procedures:
    a. Remove the manual valve from the valve body.
    b. Rotate the assembly (360 degrees, if necessary) to allow full engagement.
    c. When the assembly has engaged, install the manual valve.
16. If the assembly-to-chain cover bolt holes do not align, use the valve body guide pin tool or equivalent align the them; do not use the retaining bolts to perform alignment.
17. Install the oil pump/valve body assembly-to-chain cover bolts and torque them to 7–9 ft. lbs. (9–12 Nm), using the specified torquing sequence.
18. Connect the electrical connectors to the pressure switches and the solenoid.
19. Using a new side cover gasket and sealant (if necessary), install the side cover and torque the bolts to 10–12 ft. lbs. (14–16 Nm).
20. Install the left-side engine mounts and supports, remove the engine/transaxle assembly support bar and lower the vehicle to the ground.
21. Reposition the hoses, vacuum lines and wiring. Install the air cleaner, the battery tray and the battery. Reconnect the battery cables.
22. Using the correct fluid, fill the transaxle to the specified level.
23. Start the engine, move the shift selector through all ranges and check for oil leaks around the side cover.

LOW/INTERMEDIATE AND OVERDRIVE SERVO ASSEMBLIES

The servo assemblies, located on the rear side of the transaxle assembly, may be removed easily, without removing oil pans.

Removal and Installation

1. Remove the servo assembly-to-transaxle bolts and the servo assembly.
2. Check the servo body's for cracks, the piston bore for scores and the servo spring(s) for distortion.
3. Check the fluid passages for obstructions.
4. Check the band ends for cracks and the band lining for excessive wear and/or bond to the metal band. Check the band and struts for distortion.
5. Install new seals and lubricate them with petroleum jelly.
6. and the assembly into the case. Torque the servo cover screws to 7–9 ft. lbs. (9–12 Nm).

GOVERNOR

Removal and Installation

The governor is located on the top right-side of the lower case.

1. Remove the governor cover-to-transaxle bolts, the cover and seal (discard it).
2. Remove the governor, speedometer drive gear assembly and bearing (located on top of the speedometer gear) from the case.
3. Check the governor shaft seal for cracks, scoring and/or cuts.
4. Check the balance weight retaining pin for wear and the spring for distortion, damage or misalignment.
5. Check the pressure balls for scoring and free movement.
6. Check the governor drive, the driven gear and the speedometer drive gear for broken, chipped or worn teeth; replace, if necessary.
7. Using a new seal, install it onto the governor cover.
8. Position the governor assembly in the case bore, align the driven gear with the speedometer gear teeth and seat the assembly in the bore.
9. Torque the governor cover-to-case bolts to 7–9 ft. lbs. (9–12 Nm).

REMOVAL AND INSTALLATION

TRANSAXLE REMOVAL

1. Raise and support the vehicle safely. Raise the hood.
2. Using fender covers, place them on the fenders. Disconnect the negative battery cable.
3. Remove the air cleaner, the hoses and tubes. Remove the shift cable/bracket assembly-to-transaxle bolts.

NOTE: It may be necessary to place a small pry bar in the bracket slot to keep it from moving.

4. Disconnect the electrical connectors from the neutral safety switch and bulkhead connector.
5. To disconnect the throttle valve cable, perform the following procedures:
   a. Pull the cable upward and unsnap it from the throttle body lever.
   b. Remove the throttle valve cable-to-transaxle bolt, carefully lift the cable and slide it from the T.V. link.

NOTE: Be careful not to pull the cable too hard, for the internal T.V. bracket may bend.

6. From the left engine support strut, remove the nut and bolt. From the top to the transaxle, remove the torque converter housing-to-engine bolts.
7. Using an engine lifting bracket tool or equivalent and a bolt, attach it to the left-rear cylinder head; the engine lifting eye should still be connected to the front right cylinder head.
8. Using the engine support bar tool or equivalent, position it over the rocker arm covers and attach the bar chains to the lifting brackets.

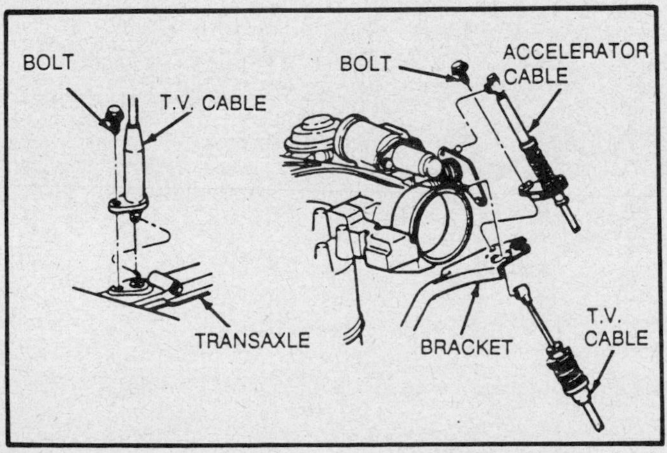
BOLT
BOLT
ACCELERATOR CABLE
T.V. CABLE
TRANSAXLE
BRACKET
T.V. CABLE

Disconnecting the throttle valve cable from the transaxle and throttle lever—3.0L engine

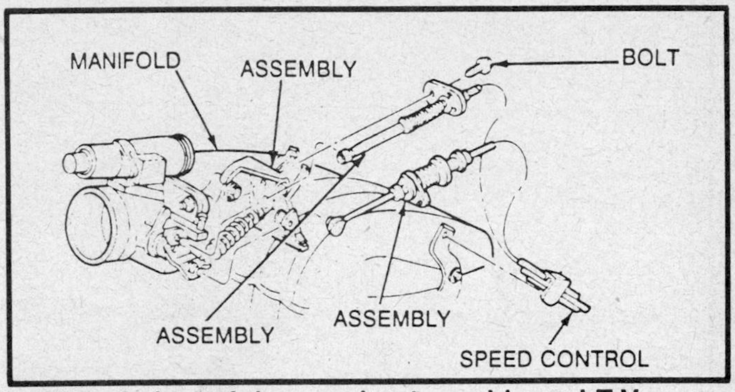
MANIFOLD
ASSEMBLY
BOLT
ASSEMBLY
ASSEMBLY
SPEED CONTROL

Exploded view of the accelerator cable and T.V. cable—3.8L engine

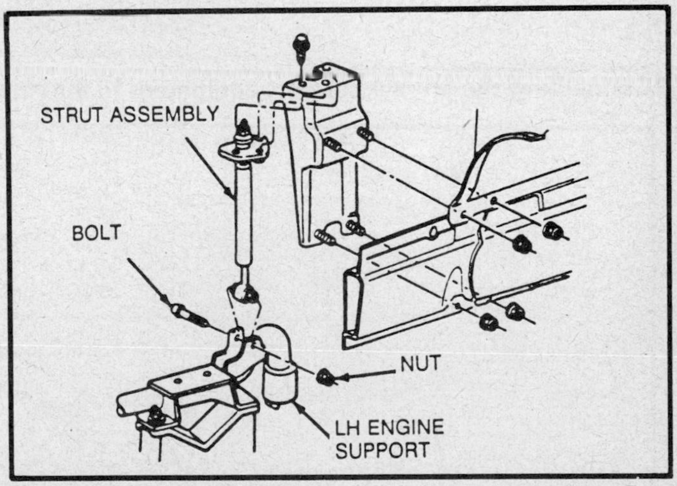
STRUT ASSEMBLY
BOLT
NUT
LH ENGINE SUPPORT

Removing the nut/bolt from the left engine support strut—3.0L engine only

NOTE: When using the 2 support points, the engine assembly will hang slightly lower at the rear (transaxle attached) or slightly lower at the front (transaxle removed). To eliminate the forward tilt, attach the left support bar chain to the No. 4 exhaust runner stud; to eliminate the rearward tilt, attach the right support bar chain between No. 2 and No. 3 exhaust manifold runners. At the front attaching point, the chain hook must face forward. Do not run the chain across the throttle cable or T.V. mechanism, for damage may occur to them.

9. Raise and support the vehicle so the wheels are off the ground. Remove both front wheels.
10. Disconnect the tie rod end from each steering knuckle. Re-

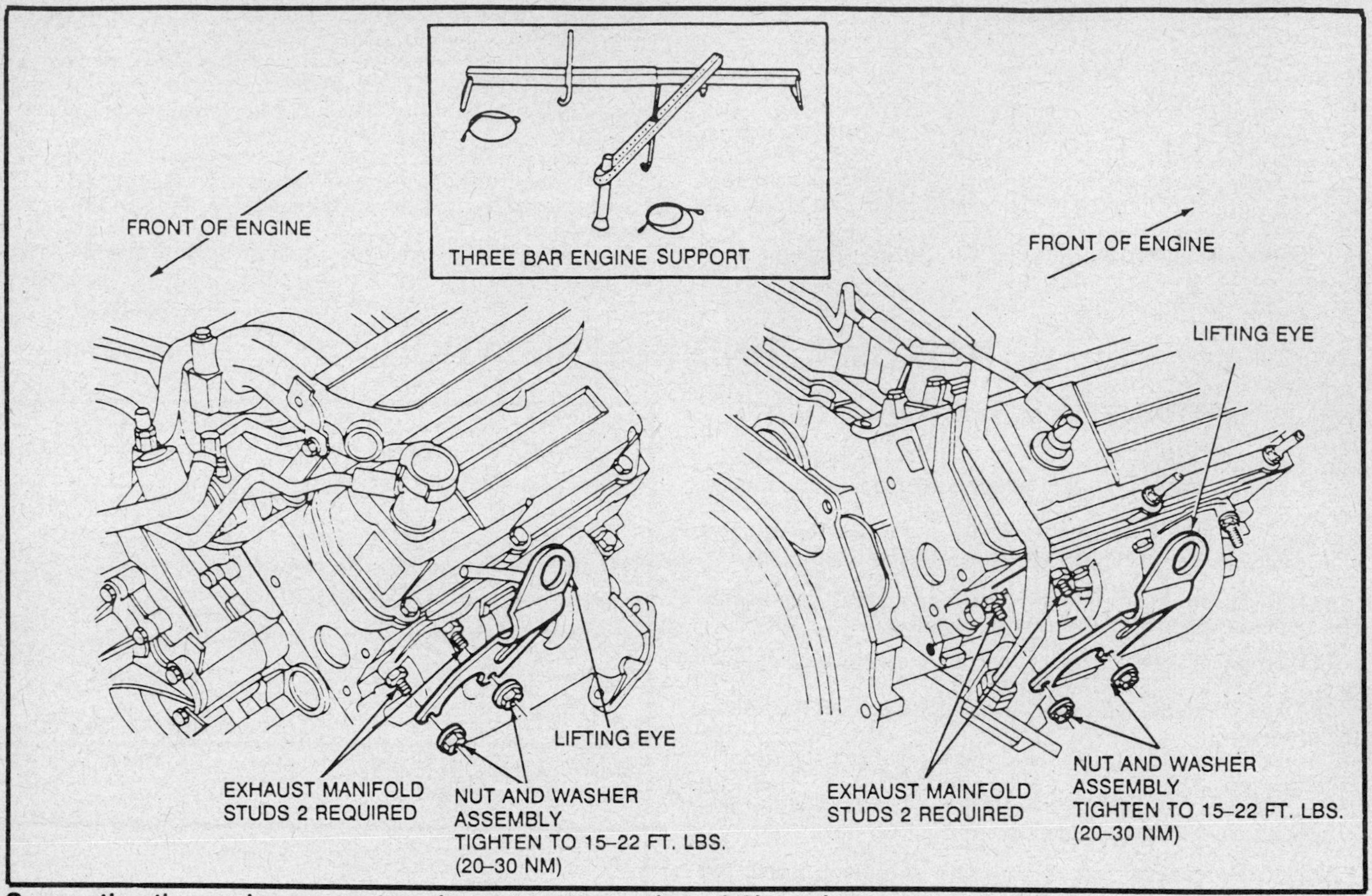
THREE BAR ENGINE SUPPORT
FRONT OF ENGINE
FRONT OF ENGINE
LIFTING EYE
LIFTING EYE
EXHAUST MANIFOLD
STUDS 2 REQUIRED
NUT AND WASHER
ASSEMBLY
TIGHTEN TO 15–22 FT. LBS.
(20–30 NM)
EXHAUST MAINFOLD
STUDS 2 REQUIRED
NUT AND WASHER
ASSEMBLY
TIGHTEN TO 15–22 FT. LBS.
(20–30 NM)

Connecting the engine support equipment to the engine—3.8L engine

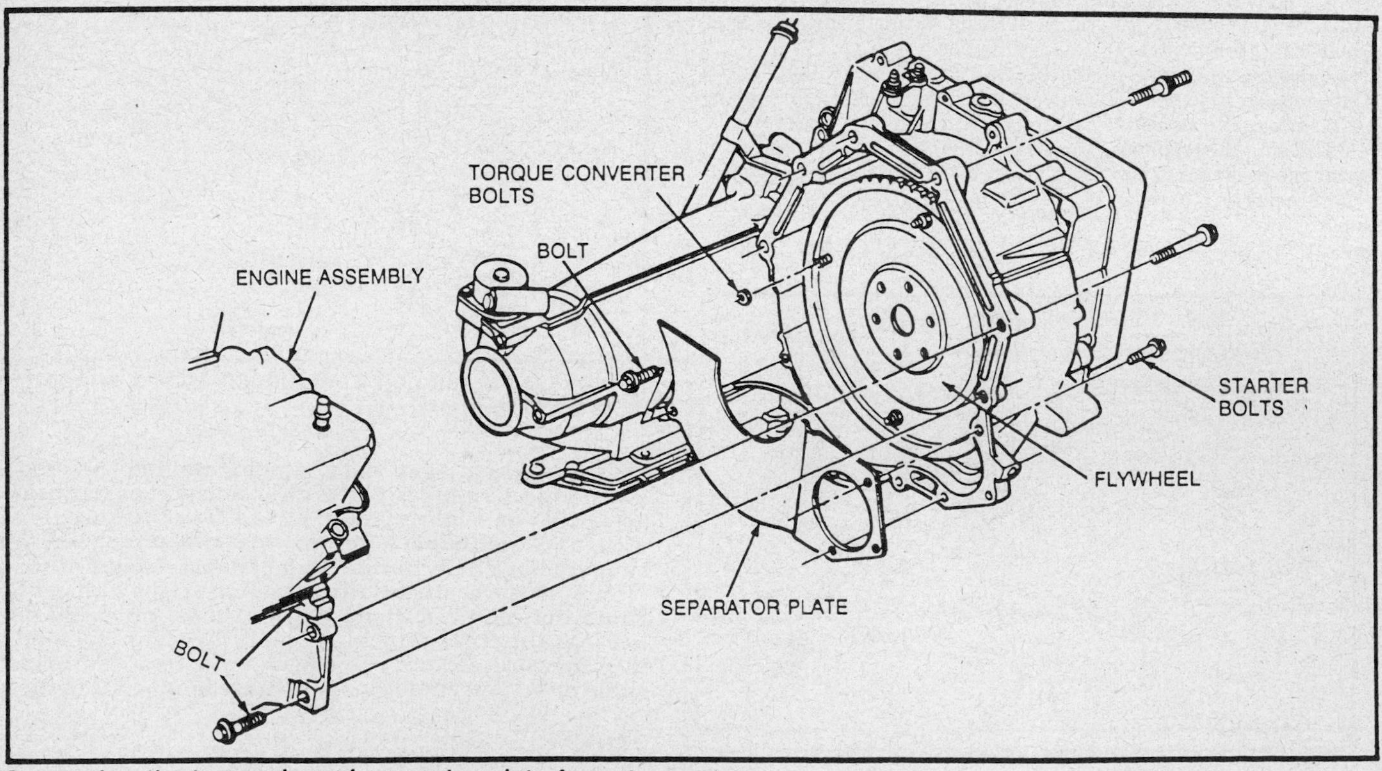
TORQUE CONVERTER
BOLTS
BOLT
ENGINE ASSEMBLY
STARTER
BOLTS
FLYWHEEL
SEPARATOR PLATE
BOLT

Separating the transaxle and separator plate from the engine

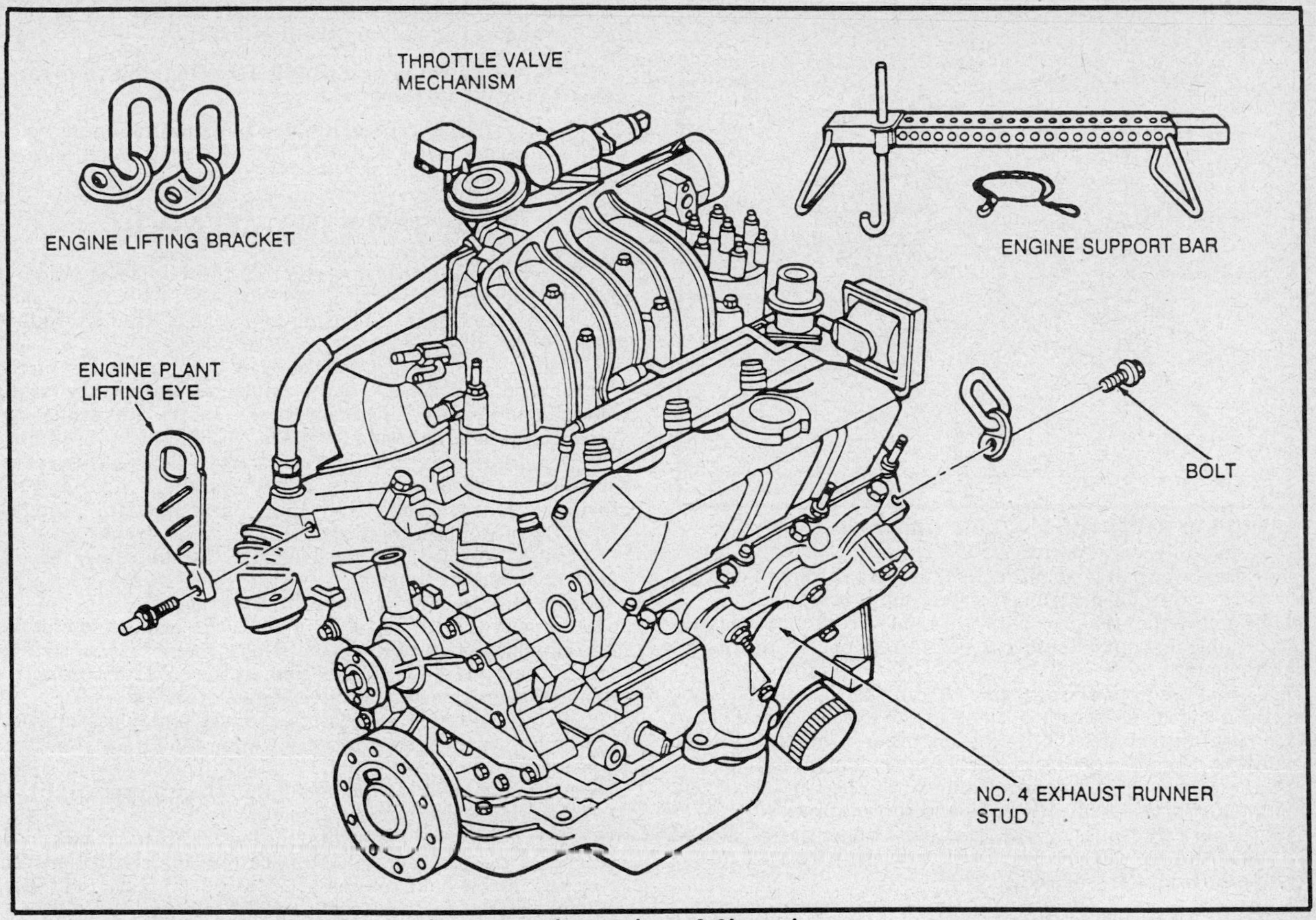
THROTTLE VALVE
MECHANISM
ENGINE LIFTING BRACKET
ENGINE SUPPORT BAR
ENGINE PLANT
LIFTING EYE
BOLT
NO. 4 EXHAUST RUNNER
STUD

Connecting the engine support equipment to the engine—3.0L engine

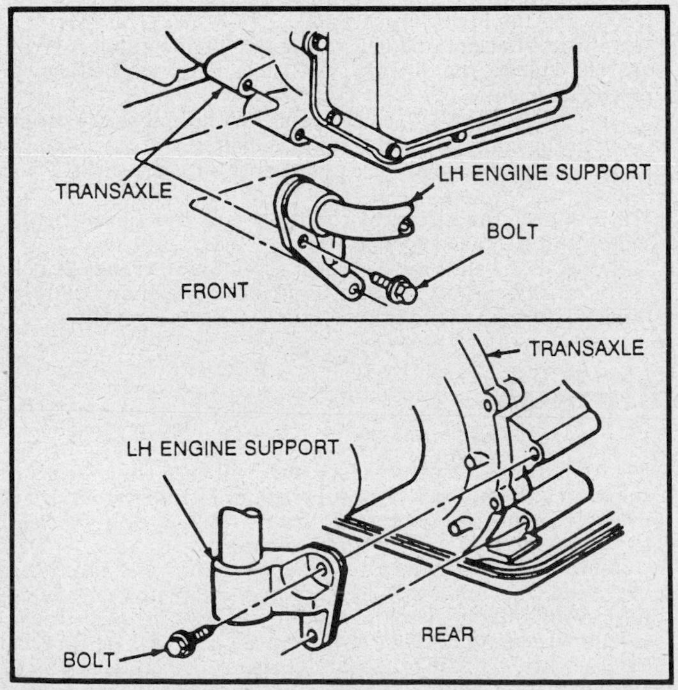
TRANSAXLE
LH ENGINE SUPPORT
BOLT
FRONT
TRANSAXLE
LH ENGINE SUPPORT
BOLT
REAR

Exploded view of the engine supports

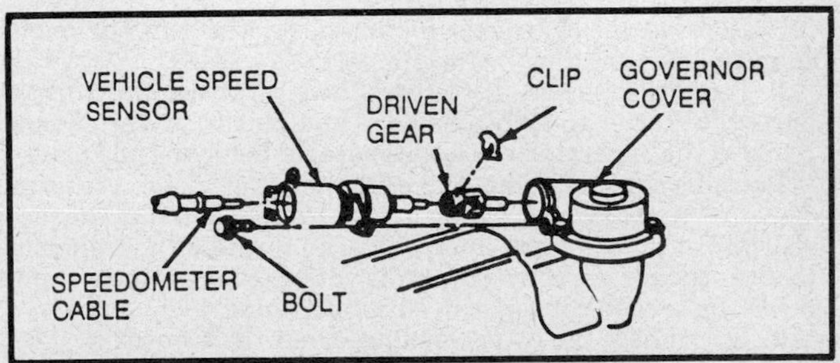
VEHICLE SPEED
SENSOR
CLIP
GOVERNOR
COVER
DRIVEN
GEAR
SPEEDOMETER
CABLE
BOLT

Exploded view of the vehicle speed sensor assembly

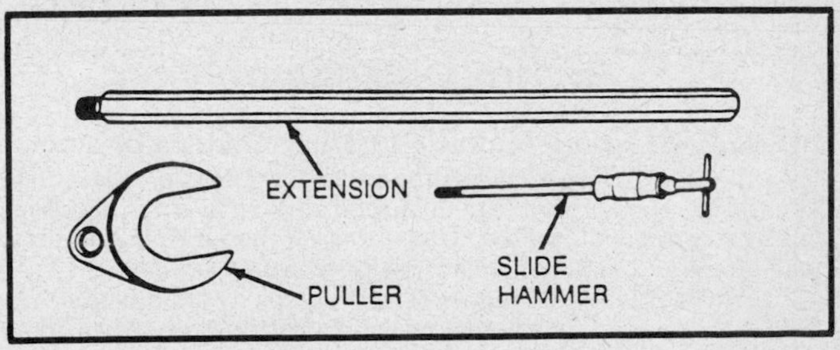
EXTENSION
PULLER
SLIDE
HAMMER

View of the halfshaft removal tools

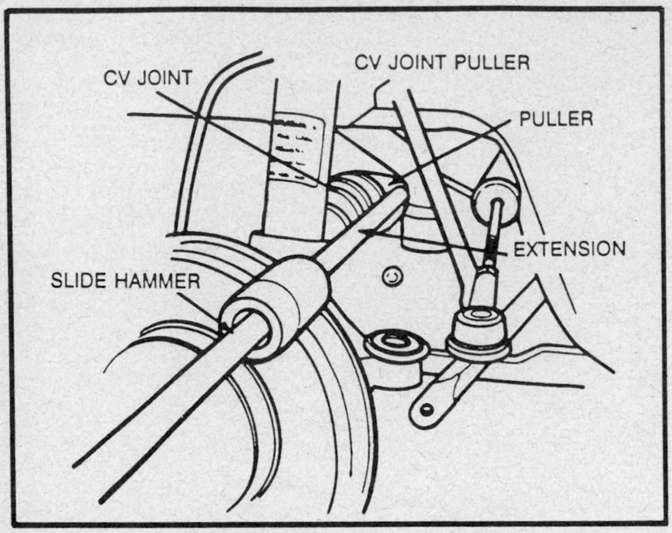
CV JOINT
CV JOINT PULLER
PULLER
EXTENSION
SLIDE HAMMER

Pulling the halfshafts from the transaxle

move the lower ball joint nut/bolts and the ball joints; separate the lower control arms from the steering knuckles.

11. Remove the stabilizer bar nuts and the rack/pinion assembly-to-subframe nuts. Remove the front/rear engine mounts-to-subframe nuts.

12. From the front of the engine, disconnect the electrical connector from the O_2 sensor, located in the exhaust manifold.

13. Disconnect the "Y" exhaust pipe from the exhaust mainfolds and the rear section from the exhaust pipe flange.

14. Remove the subframe-to-frame bolts, the left-side engine mount support-to-subframe bolts and the subframe.

15. Using a transmission jack, position it under the oil pan and support the weight. Remove the vehicle speed sensor (if equipped) from the transaxle.

NOTE: Vehicles equipped with electronic instrument clusters do not use a speedometer cable.

16. Remove the transaxle mount bolts, the left-side engine support bolts and the engine support.

17. From the starter, disconnect the electrical connectors, remove the starter-to-transaxle bolts and move the starter aside. Remove the separator plate-to-transaxle bolts and the plate.

18. Using chalk or paint, make alignment marks on the torque converter and flywheel. Using a ½ in. ratchet and the ⅞ in. deep well socket on the crankshaft pulley bolt, rotate the crankshaft to align the torque converter bolts with the starter hole and remove the torque converter-to-flywheel bolts.

19. Disconnect the oil cooler lines from the transaxle.

20. Assemble the CV-joint puller tool or equivalent, a screw extension tool or equivalent, and a slide hammer puller tool or equivalent, position the puller assembly behind the CV-joint and pull the halfshafts from the transaxle.

NOTE: When removing the halfshafts from the transaxle, do not pry against the case.

21. Remove the torque converter housing-to-engine bolts. Carefully separate the transaxle from the engine and lower it from the vehicle.

TRANSAXLE INSTALLATION

1. Carefully raise the transaxle into the vehicle and align it with the engine.
2. Install the transaxle housing-to-engine bolts and torque to 41–50 ft. lbs. (55–68 Nm).
3. Using a ½ in. ratchet and the ⅞ in. deep well socket on the crankshaft pulley bolt, rotate the crankshaft to align the torque converter-to-flywheel alignment marks. Insert the torque converter-to-flywheel bolts and torque to 23–39 ft. lbs. (31–53 Nm).
4. Install the separator plate and starter. Torque the separator plate-to-transaxle bolts to 7–9 ft. lbs. (9–12 Nm) and the starter-to-transaxle bolts to 30–40 ft. lbs. (41–54 Nm). Connect the electrical connectors to the starter.
5. Install the left-side engine support, the subframe, the left-side engine mount support.
6. Connect the oil cooler lines to the transaxle.
7. Connect the "Y" exhaust pipe to the exhaust manifolds and the rear section to the exhaust pipe flange.
8. At the front of the engine, connect the electrical connector to the O_2 sensor, located in the exhaust manifold.
9. Install the stabilizer bar nuts, the rack/pinion assembly-to-subframe nuts and the front/rear engine mounts-to-subframe nuts.
10. Align and push the halfshafts into the transaxle until the retaining ring snaps into position.
11. Install the lower ball joints, the lower control arms and tie rod end to the steering knuckles. Torque the control arm-to-steering knuckle bolts to 36–44 ft. lbs. (50–60 Nm) and the tie rod end-to-steering knuckle nut to 23–35 ft. lbs. (31–47 Nm).
12. Install the wheels and lower the vehicle to the ground.
13. Remove the engine lifting bar tool or equivalent, the bar chains and the lifting brackets.
14. Install the nut and bolt to the left engine support strut.
15. To connect the throttle valve cable, perform the following procedures:
    a. Connect the T.V. cable to the T.V. link, seat the connector to the transaxle and install the bolt.
    b. Pull the cable upward and snap it onto the throttle body lever.
16. Connect the electrical connectors to the neutral safety switch and bulkhead connectors.
17. Install the shift cable/bracket assembly-to-transaxle bolts.
18. Connect the negative battery cable. Check and refill the fluid. Drive the vehicle to check the transaxle operation.

BENCH OVERHAUL

Before Disassembly

When servicing the transaxle it is important to be aware of cleanliness. Before disassembling the transaxle, the outside should be throughly cleaned, preferably with a high-pressure spray cleaning equipment. Dirt entering the unit may negate all the effort and time spent on the overhaul.

During inspection and reassembly, all parts should be cleaned with solvent and dried with compressed air; do not use wiping rags or cloths for lint may find its way into the valve body passages. Lubricate the seals with Dexron® II and use petroleum jelly to hold the thrust washers; this will ease the assembly of the seals and not leave harmful residues in the system. Do not use solvent on neoprene seals, friction plates or composition thrust washers, if they are to be reused.

Before installing bolts into aluminum parts, dip the threads into clean transmission fluid. Anti-seize compound may be used to prevent galling the aluminum or seizing. Be sure to use a torque wrench to prevent stripping the threads. Be especially careful when installing the seals (O-rings), the smallest nick can cause a leak. Aluminum parts are very susceptible to damage; great care should be used when handling them. Reusing

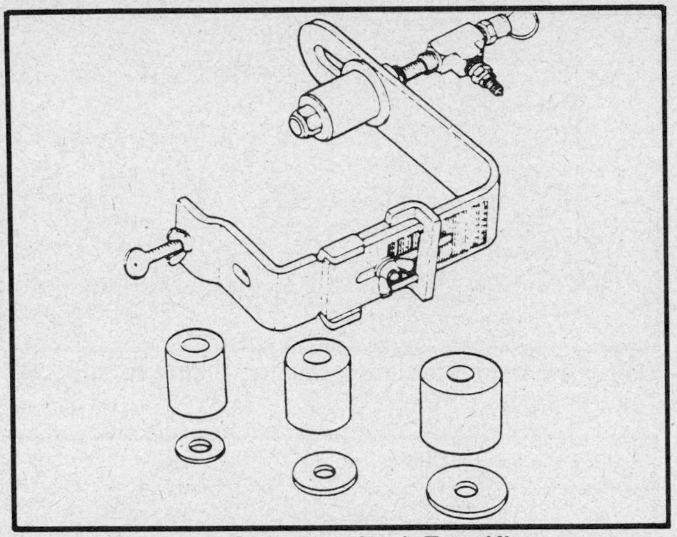

View of a Torque Converter Leak Test Kit

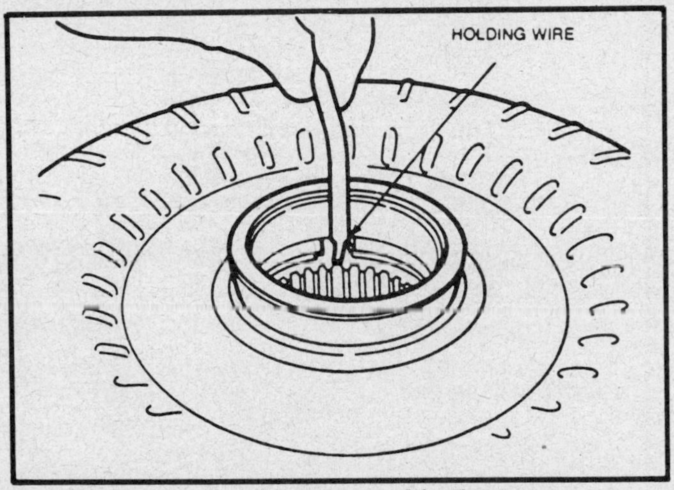
HOLDING WIRE

Positioning the holding wire tool in the torque converter's thrust washer slot

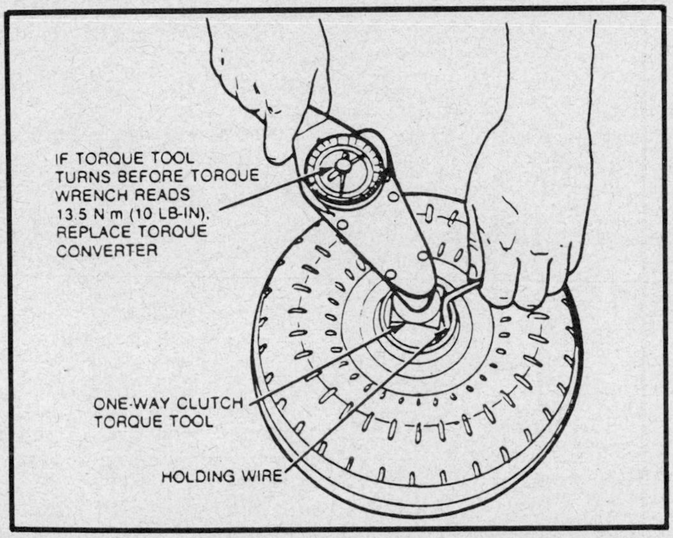
IF TORQUE TOOL TURNS BEFORE TORQUE WRENCH READS 13.5 N m (10 LB-IN), REPLACE TORQUE CONVERTER
ONE-WAY CLUTCH TORQUE TOOL
HOLDING WIRE

Using a torque wrench and a holding wire to check the torque converter's one-way clutch operation

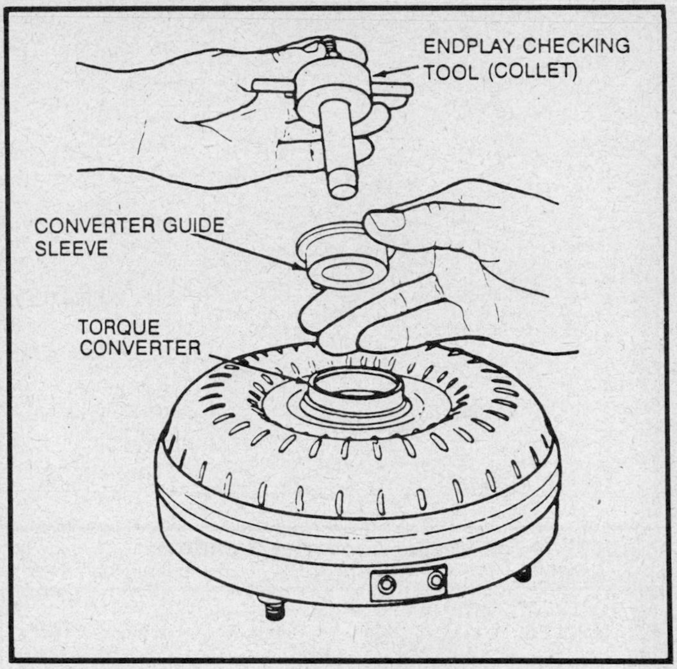
ENDPLAY CHECKING TOOL (COLLET)
CONVERTER GUIDE SLEEVE
TORQUE CONVERTER

Installing the Endplay Checking and Converter Guide Sleeve tool into the torque converter

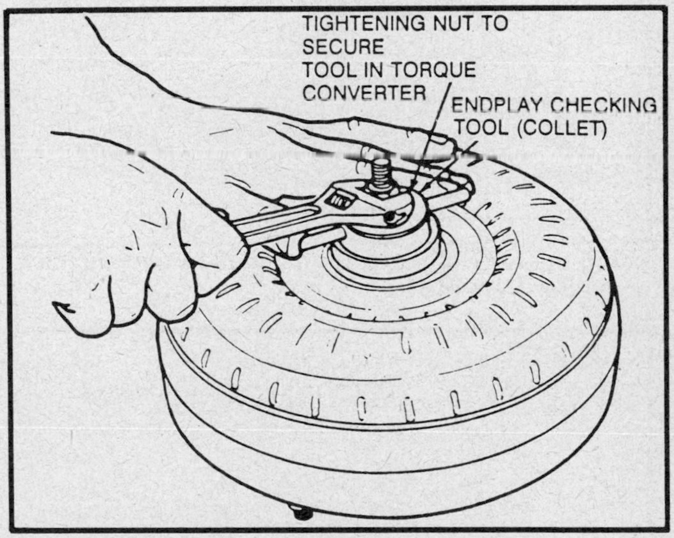
TIGHTENING NUT TO SECURE TOOL IN TORQUE CONVERTER
ENDPLAY CHECKING TOOL (COLLET)

Torquing the Endplay Checking tool nut

snaprings is not recommended but should they be: compress the internal ones and compress the external ones.

Converter Inspection

The torque converter used a sealed, welded design that cannot be disassembled for service or repair; there are a few checks that can be made.

The torque converter contains approximately 2 quarts of transmission fluid. Since there is no drain plug on the unit, the fluid can only be drained through the hub. To drain the converter, invert it over a catch pan and drain the fluid. Fluid that is drained can help diagnosis the converter's condition.

1. If the fluid is discolored but does not contain metal bits or particles, the converter is usable and need not be replaced.

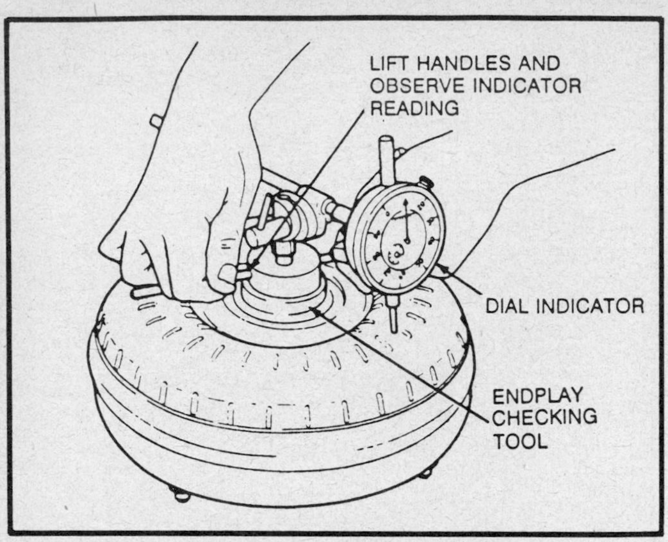
LIFT HANDLES AND
OBSERVE INDICATOR
READING
DIAL INDICATOR
ENDPLAY
CHECKING
TOOL

Inspecting the torque converter's endplay

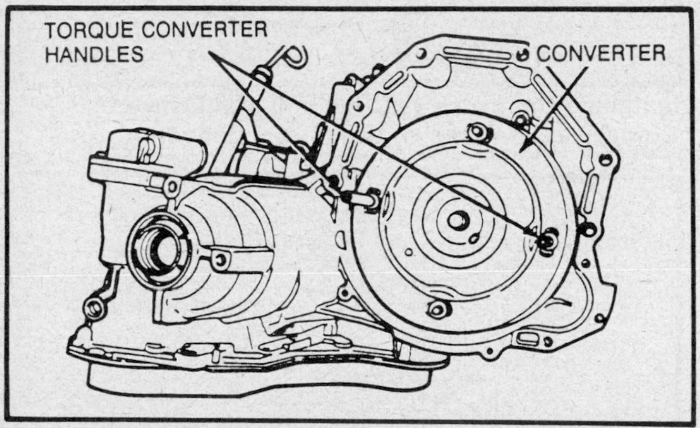
TORQUE CONVERTER
HANDLES
CONVERTER

Using the torque converter handles to remove the converter from the transaxle

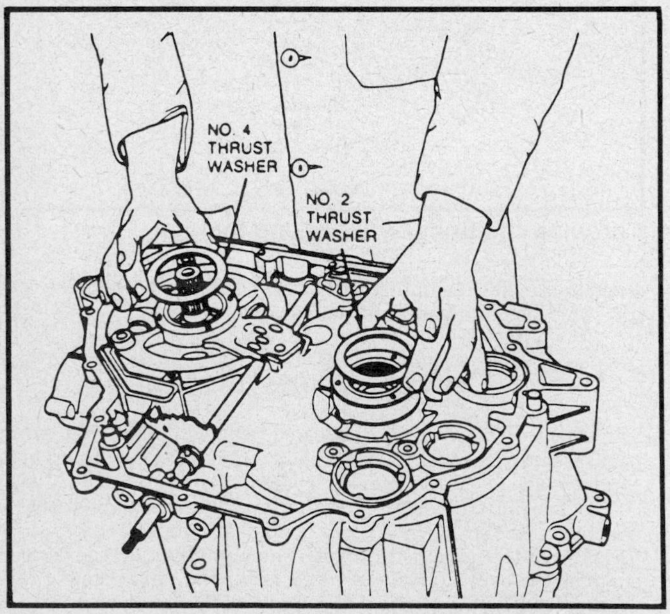
NO. 4
THRUST
WASHER
NO. 2
THRUST
WASHER

Removing the No. 4 and No. 2 thrust washers from the chain case

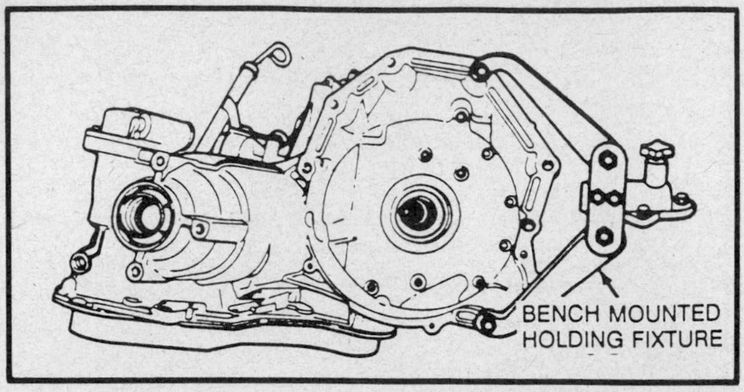
BENCH MOUNTED
HOLDING FIXTURE

Using the bench mounted holding fixture to support the transaxle

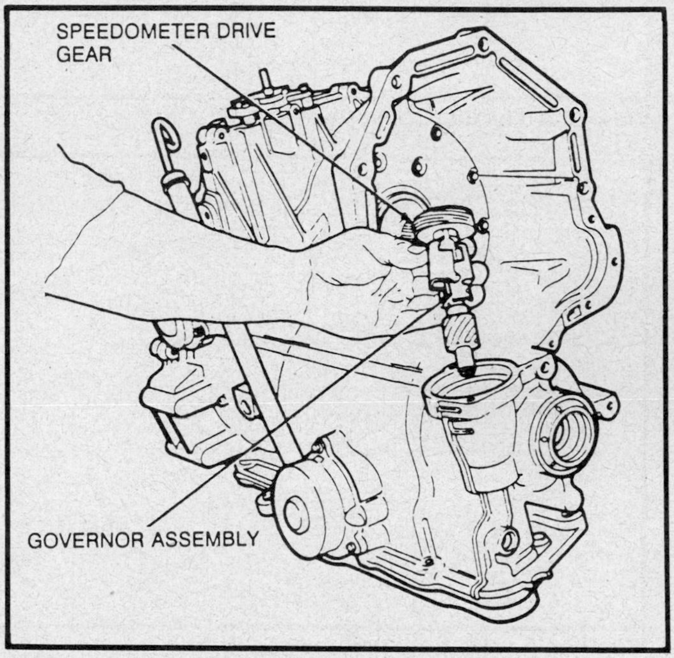
SPEEDOMETER DRIVE
GEAR
GOVERNOR ASSEMBLY

Removing the governor from the transaxle

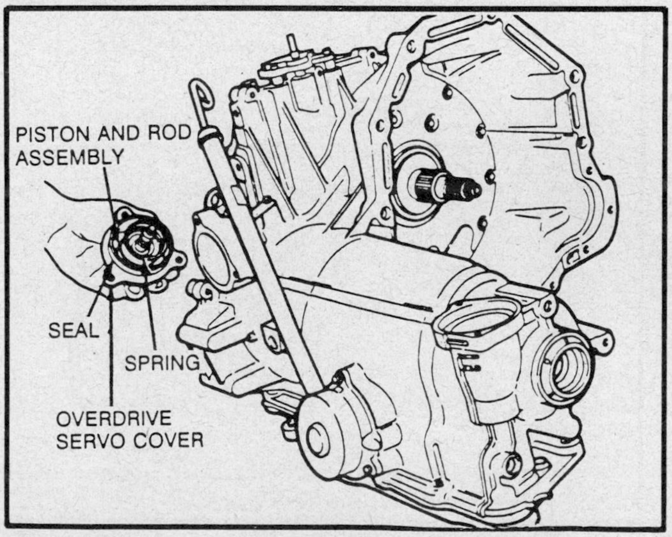
PISTON AND ROD
ASSEMBLY
SEAL
SPRING
OVERDRIVE
SERVO COVER

Removing the overdrive servo cover from the transaxle

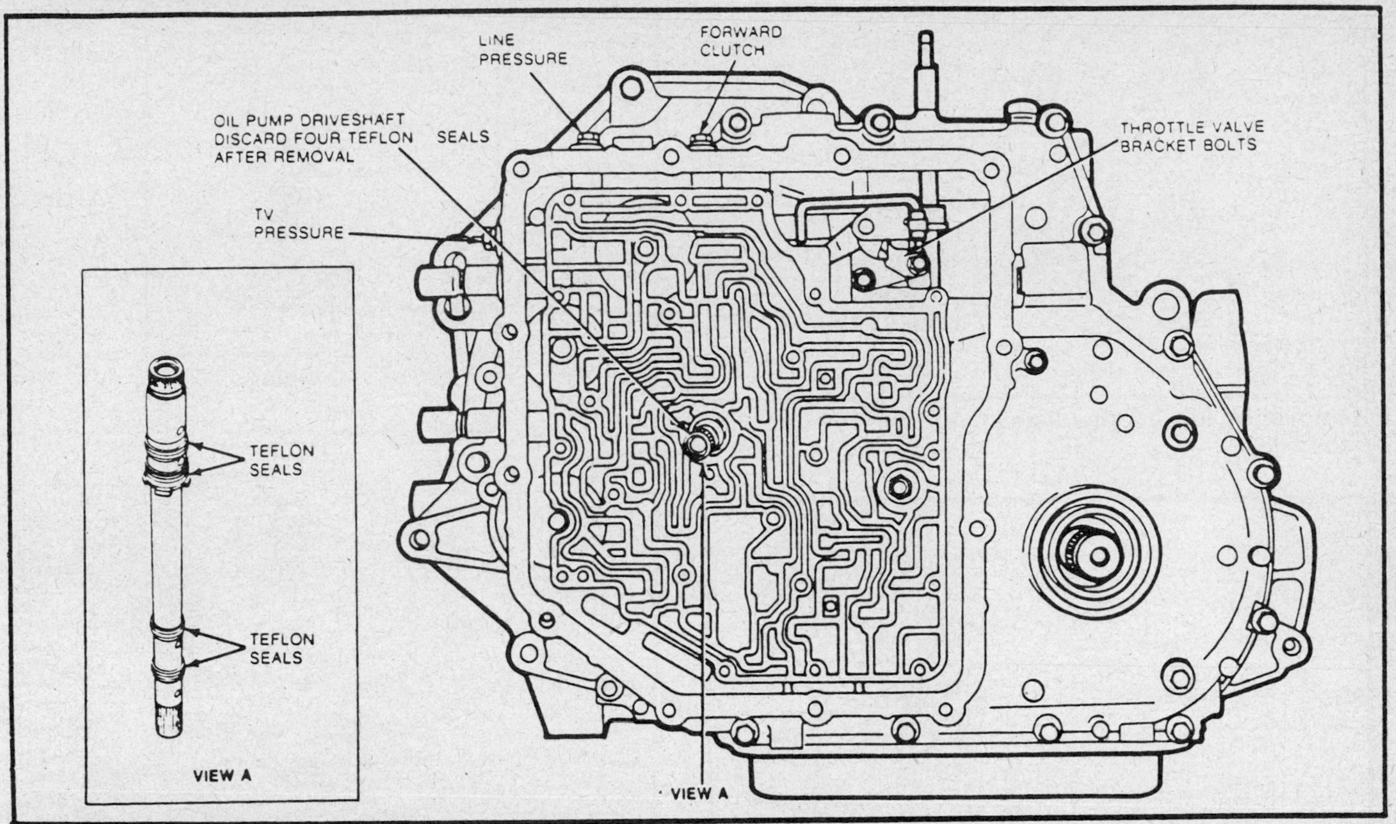
LINE PRESSURE
FORWARD CLUTCH
OIL PUMP DRIVESHAFT DISCARD FOUR TEFLON SEALS AFTER REMOVAL
THROTTLE VALVE BRACKET BOLTS
TV PRESSURE
TEFLON SEALS
TEFLON SEALS
VIEW A
VIEW A

View of the oil pump driveshaft and chain case cover

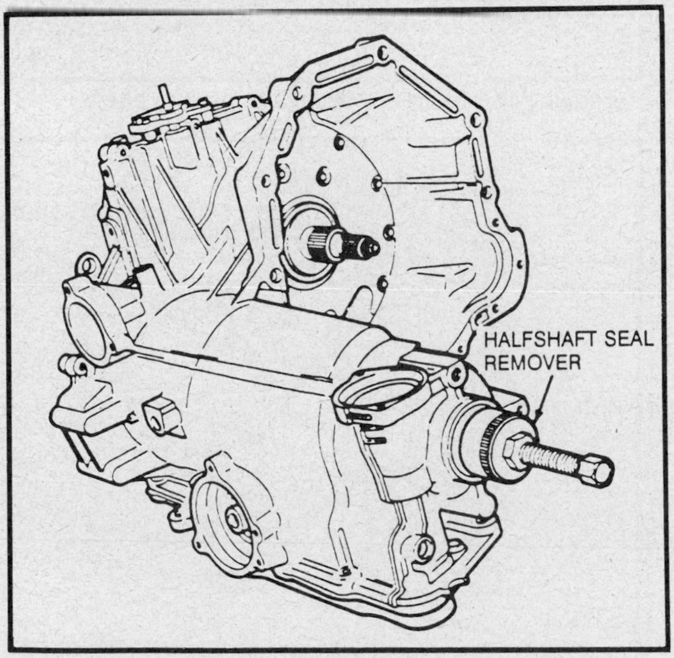
HALFSHAFT SEAL REMOVER

View of the halfshaft seal removal tool installed on the transaxle

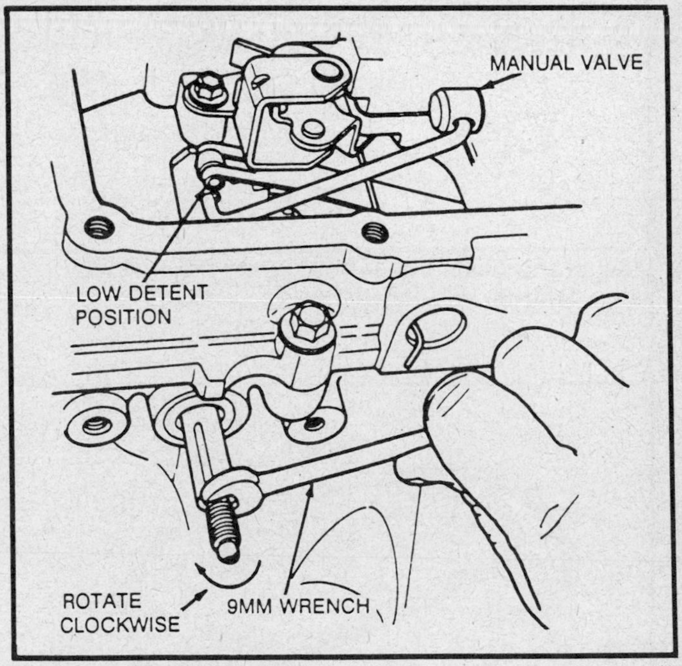
MANUAL VALVE
LOW DETENT POSITION
ROTATE CLOCKWISE
9MM WRENCH

Rotating the manual shift shaft

NOTE: Remember the fluid color is not longer a good indicator of the fluid condition. In the past, the dark color would indicate overheated transaxle fluid; with the newer fluids, this is not a positive sign of transaxle failure.

2. Metal particles in the fluid, having an aluminum paint appearance, indicating converter damage and replacement.
3. If fluid contamination is due to burned clutch plates, overheated oil or antifreeze, the converter should be cleaned or replaced.

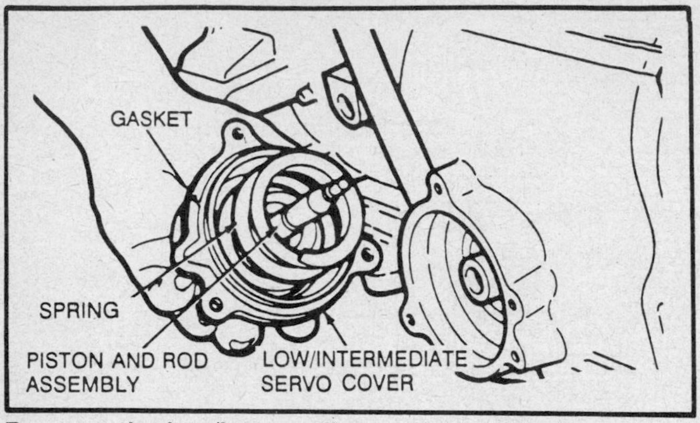
GASKET
SPRING
PISTON AND ROD ASSEMBLY
LOW/INTERMEDIATE SERVO COVER

Remove the low/intermediate cover from the transaxle

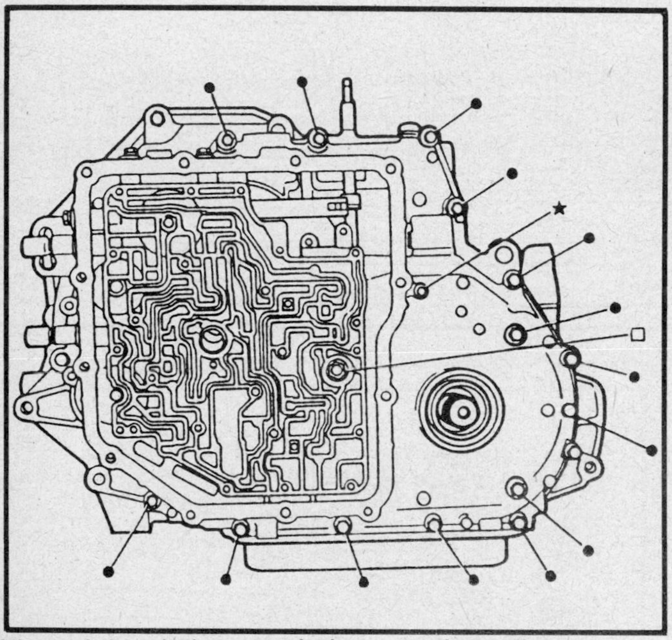

View of the chain cover bolts; note the various sizes, lengths and locations

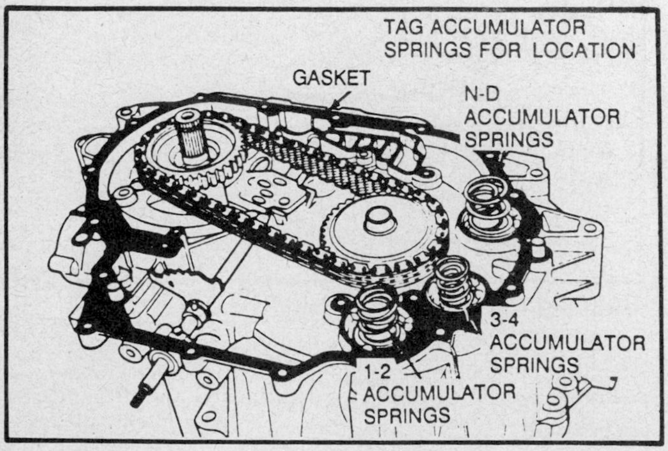
TAG ACCUMULATOR SPRINGS FOR LOCATION
GASKET
N-D ACCUMULATOR SPRINGS
3-4 ACCUMULATOR SPRINGS
1-2 ACCUMULATOR SPRINGS

View of the accumulator springs

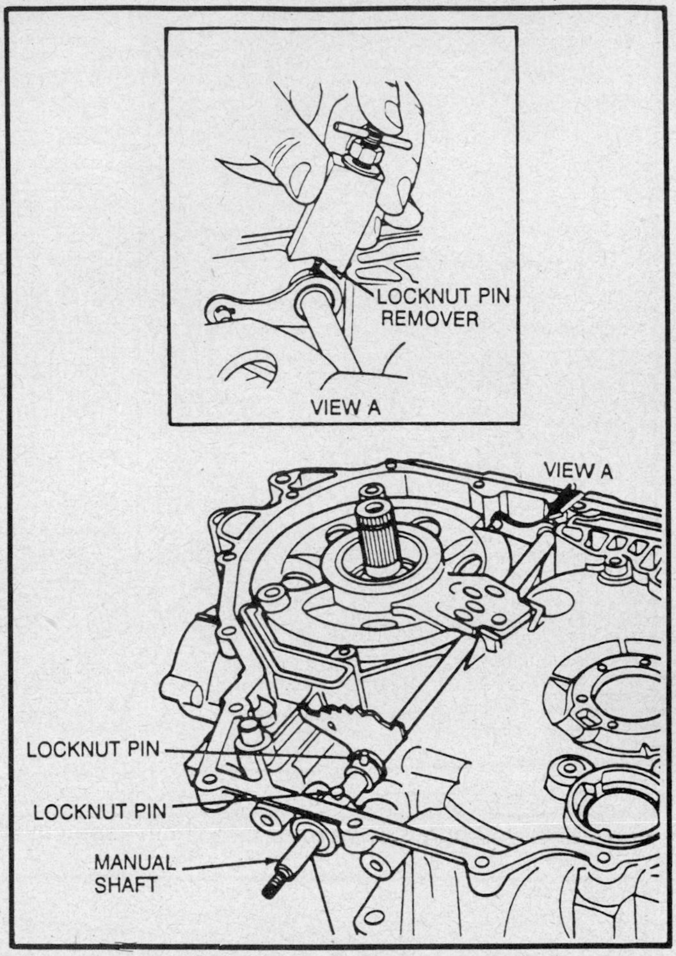
LOCKNUT PIN REMOVER
VIEW A
VIEW A
LOCKNUT PIN
LOCKNUT PIN
MANUAL SHAFT

Removing the lockpins from the manual shaft

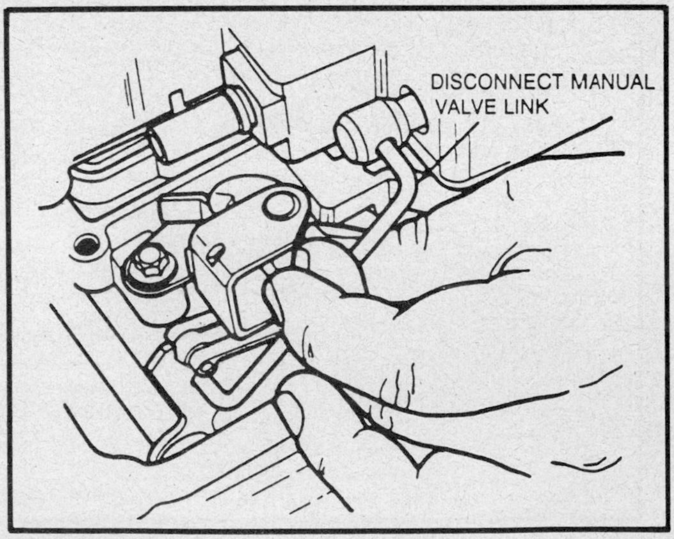
DISCONNECT MANUAL VALVE LINK

Disconnecting the manual valve link

The converter should be checked carefully for damage, especially around the seal area; remove any sharp edges or burrs from the seal surface. Do not expose the seal to any type of solvent. If the converter is to be washed with solvent, the seal must be removed.

Leakage Check

If the torque converter is suspected of leaking at the welded seams, remove it and perform the leakage check. Using a torque converter leak test Kit or equivalent, pressurize the converter and inspect the seams.

Reactor One-Way Clutch Check

To perform this test, the torque converter must be removed from the transaxle.

1. Using a holding wire, position it in the torque converter's thrust washer slot.
2. Using a torque wrench and a one-way clutch torque tool, install the assembly into the torque converter.
3. While supporting the holding wire stationary, rotate the torque wrench counterclockwise to check the torque. The torquing tool should not turn under 10 ft. lbs. (13.55 Nm); if it turns, replace the torque converter.

Endplay Check

1. Using the endplay checking tool or equivalent, and the converter guide sleeve tool or equivalent, install them into the torque converter hub.
2. Tighten the endplay checking tool nut.
3. Using a dial indicator, mount it onto the endplay checking tool. Using the indicator's stylus, contact the converter shell and zero the indicator.
4. Lift the endplay checking tool handles and note the indicator reading; if the reading changes more than 0.05 in. (1.27mm), replace the converter.

Transaxle Disassembly

TRANSAXLE UNIT

1. Using 2 torque converter handle tools or equivalent, install them onto the torque converter and pull the converter from the transaxle.
2. Using the bench mounted holding fixture tool or equivalent, mount the transaxle to it. If necessary to drain the fluid, rotate the transaxle to the vertical position (right halfshaft side down), drain the fluid into a catch pan and return it to the horizontal position.
3. Remove the governor cover-to-transaxle bolts, the cover and seal (discard it). Remove the governor, speedometer drive gear assembly and bearing (located on top of the speedometer gear) from the case.
4. Remove the overdrive servo cover-to-transaxle bolts, the cover, piston assembly and spring; discard the O-ring seal.
5. Remove the low/intermediate servo cover-to-transaxle bolts, the cover, piston assembly and spring; discard the gasket.
6. To remove the right-side output shaft seal, perform the following procedures:

NOTE: The halfshaft shaft seal is a 2-piece construction; inner rubber seal and outer metal protector.

a. Using a shaft protector tool or equivalent, install it into the output shaft opening.
b. Using the output shaft seal remover tool or equivalent, screw it into the metal seal protector. Tighten the tool's screw until the metal seal protector is removed and remove the protector from the tool.
c. Reinsert the tool into the rubber seal, tighten the tool's screw and pull the seal from the transaxle.

7. From the top of the transaxle, remove the neutral safety switch-to-transaxle bolts and the switch.
8. Remove the dipstick tube-to-transaxle bolt and pull out the tube.
9. From inside the torque converter housing, remove the torque converter-to-chain cover bolts.
10. Using the seal remover tool or equivalent and a slide hammer puller, pull the oil seal from the torque converter housing shaft.
11. Remove the side cover (upper reservoir) bolts, the bolts and the gasket (discard it).
12. Disconnect the electrical connectors from the pressure switches and the solenoid.

NOTE: When disconnecting the electrical connectors, grasp the connector with one hand and push against it with the finger from the other hand.

13. At the bulkhead connector, compress the tabs (on both sides) and remove the connector from the chain cover.

NOTE: When removing the bulkhead connector, do not pull on the wiring or the connector.

14. Place a 9mm wrench on the manual shaft flats and rotate it clockwise to position the linkage in the **L** detent (valve all the way in).
15. Remove the oil pump/valve body assembly-to-chain case bolts.

NOTE: When removing the oil pump/valve body assembly, do not remove the 2 oil pump-to-valve body retaining bolts or the oil cover bolts.

16. Push the T.V. plunger inward and pull the oil pump/valve body assembly outward to clear the bracket. Rotate the assembly clockwise and remove the link from the manual valve. Disconnect the manual valve link from the detent lever and remove the oil pump/valve body assembly.
17. Remove the throttle valve bracket-to-chain cover bolts and the bracket. After pulling the oil pump driveshaft from the chain case, remove the Teflon® seals (discard them) from the shaft.
18. Rotate the transaxle to the vertical position and remove the left output shaft circlip.
19. To remove the left output shaft seal, perform the following procedures:

NOTE: The halfshaft shaft seal is a 2-piece construction; inner rubber seal and outer metal protector.

a. Using a shaft protector tool or equivalent, install it into the output shaft opening.
b. Using the output shaft seal remover tool or equivalent, screw it into the metal seal protector. Tighten the tool's screw until the metal seal protector is removed and remove the protector from the tool.
c. Reinsert the tool into the rubber seal, tighten the tool's screw and pull the seal from the transaxle.

20. Remove the chain cover-to-transaxle bolts, the chain cover and the gasket (discard it).

NOTE: When removing the chain cover, be sure to note the location and length of the bolts. Be sure to tag the actuator springs so they may be installed in their correct locations during assembly.

21. From the chain cover, remove the No. 1 and No. 3 thrust washers.
22. Using both hands, grasp and lift both chain sprockets and chain assembly from the chain case. Remove the No. 2 thrust washer from the drive sprocket support and the No. 4 thrust washer from the driven sprocket support.

NOTE: The No. 4 thrust washer may stay on the driven sprocket.

23. Determine if the drive sprocket support bearing (No. 2

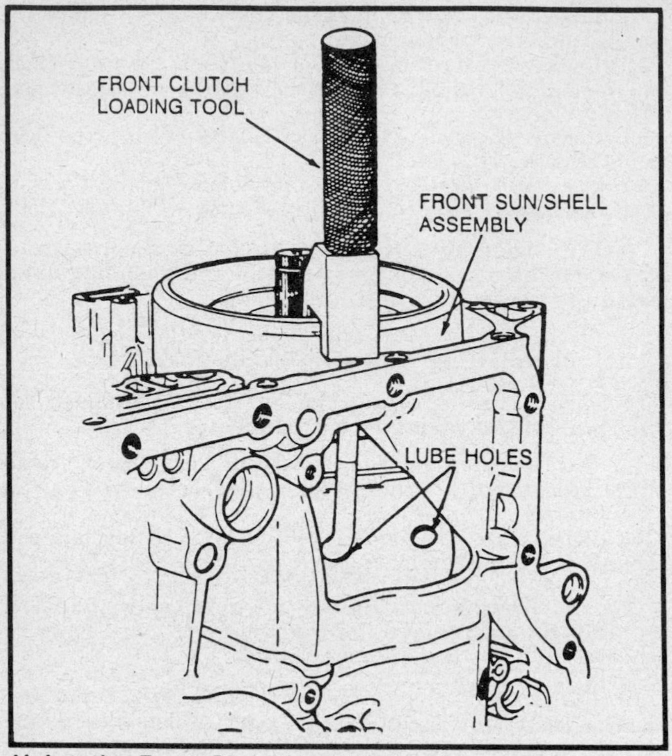
FRONT CLUTCH
LOADING TOOL
FRONT SUN/SHELL
ASSEMBLY
LUBE HOLES

Using the Front Clutch Loading tool to remove the sun/shell assembly

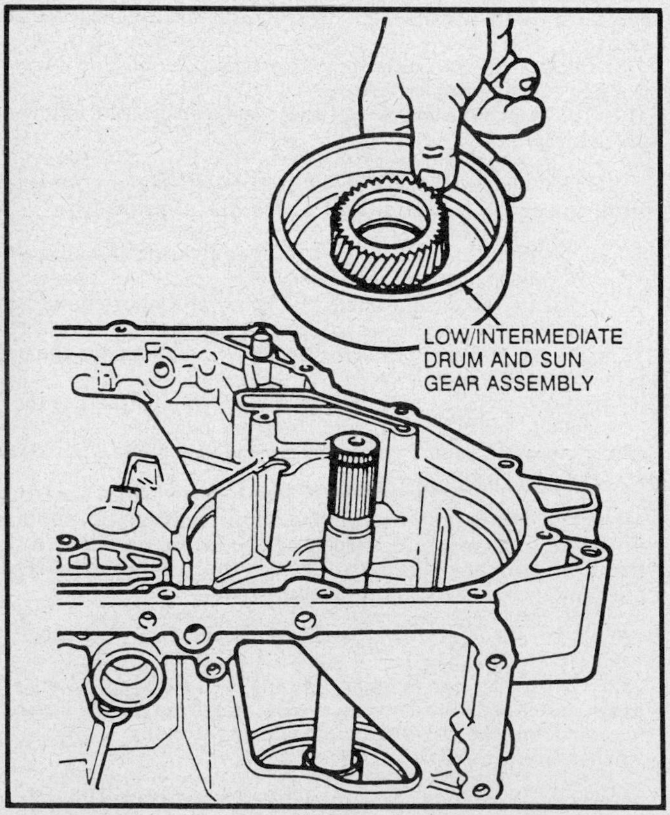
LOW/INTERMEDIATE
DRUM AND SUN
GEAR ASSEMBLY

Removing the low/intermediate drum/sun gear assembly

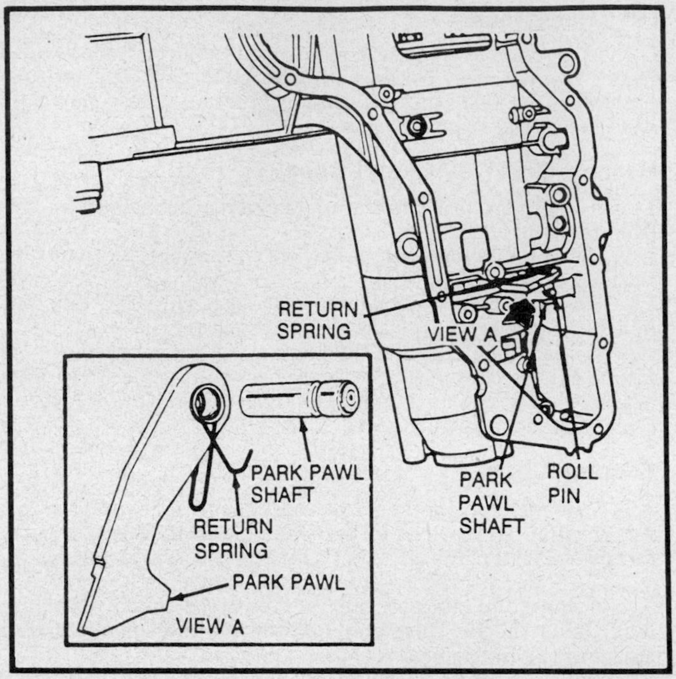
RETURN
SPRING
VIEW A
PARK PAWL
SHAFT
RETURN
SPRING
PARK PAWL
VIEW A
PARK
PAWL
SHAFT
ROLL
PIN

Exploded view of the park pawl assembly

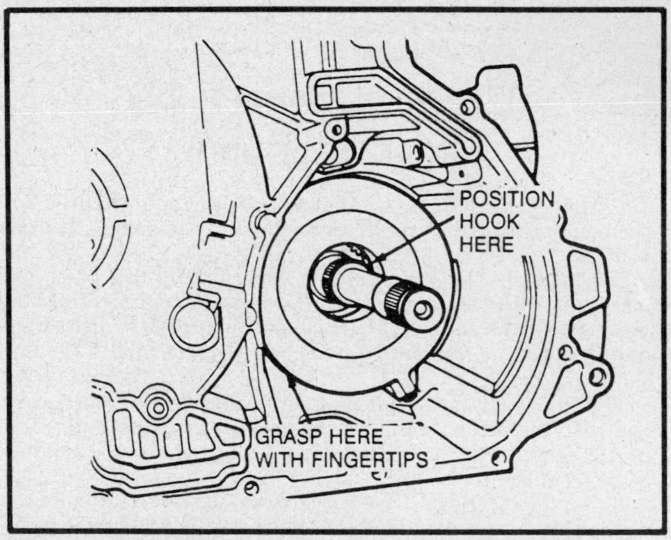
POSITION
HOOK
HERE
GRASP HERE
WITH FINGERTIPS

View of the reverse clutch cylinder's inner diameter

thrust washer) needs replacement; if the bearing is OK, remove the driven sprocket support-to-case housing Torx® bolts, from the torque converter housing side.

24. Using the locknut pin remover tool or equivalent, from the chain case side, remove and discard the manual shaft's lockpin and 2 roll pins; be careful not to damage the machined surfaces. Slide the manual shaft linkage from the case and pry the seal from the case.

25. Using a flat block or a straight edge, determine if the driven sprocket support's machined bolt hole surfaces are above or below the case's machined surface; this is for reassembly purposes.

26. From the chain case, remove the driven sprocket support assembly, the Teflon® seals (from the support assembly shaft) and the thrust washer (it may stay with the driven sprocket support). If the No. 8 selective thrust washer and No. 9 needle bearing were not removed with the driven sprocket support assem-

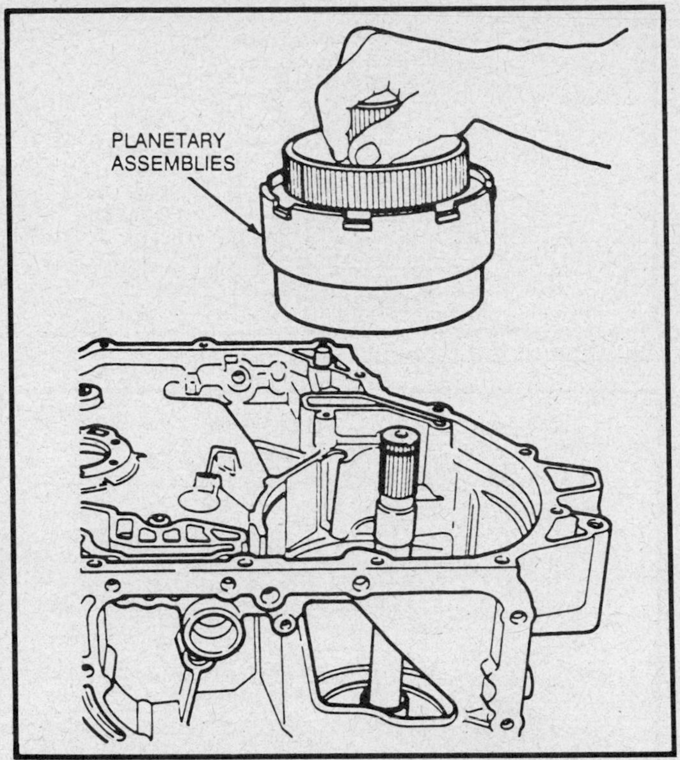
PLANETARY ASSEMBLIES

Removing the front planetary assembly from the case

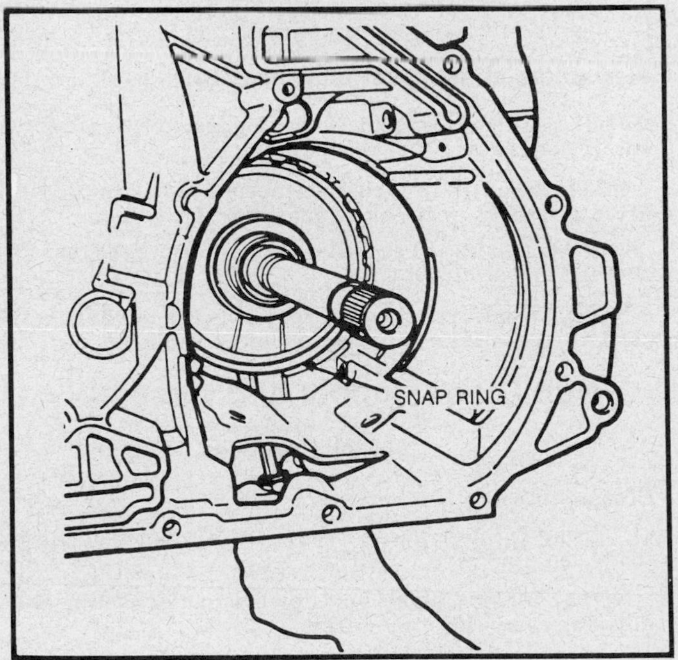
SNAP RING

Removing the snapring from the final drive gear assembly

bly, use a wire hook to remove them from the bottom of the chain case cylinder.

NOTE: If the driven sprocket support assembly is binding in the chain case housing, it may be necessary to back out the reverse clutch anchor bolt.

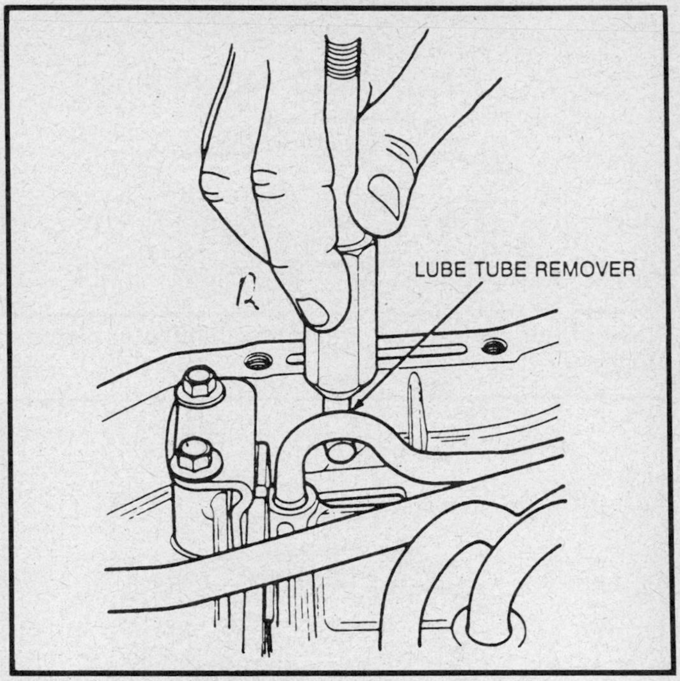
LUBE TUBE REMOVER

Removing the lube tubes from the transaxle

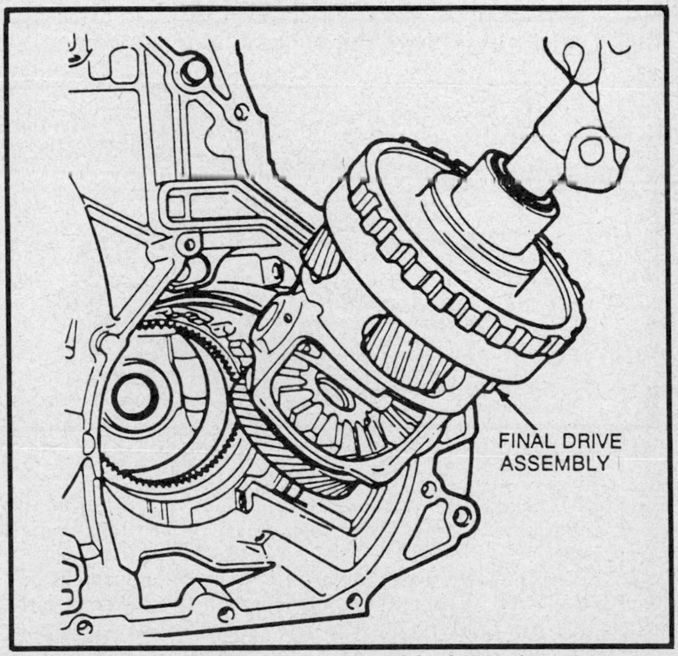
FINAL DRIVE ASSEMBLY

Removing the final drive assembly from the side case

27. At the overdrive band, remove the plastic retainer and the overdrive band.
28. Install the hooked end of the front clutch loading tool or equivalent, into 1 of the sun/shell assembly's 6 holes, position the notched block over the assembly's edge, tighten the handle (do not overtighten) and lift the assembly from the case.
29. Remove the oil pan-to-transaxle cover bolts, the oil pan and gasket (discard it). Remove the reverse apply tube/oil filter bracket bolt, the bracket and the oil filter screen (discard both O-rings).
30. From the lower case, remove the lube tube bracket bolts

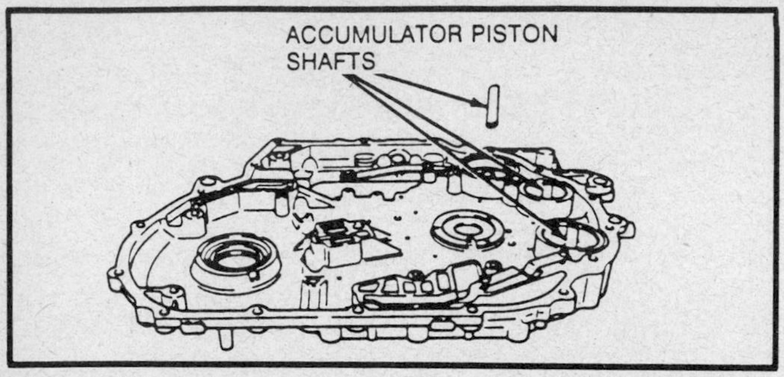
ACCUMULATOR PISTON
SHAFTS

View of the chain cover and the accumulator piston shafts

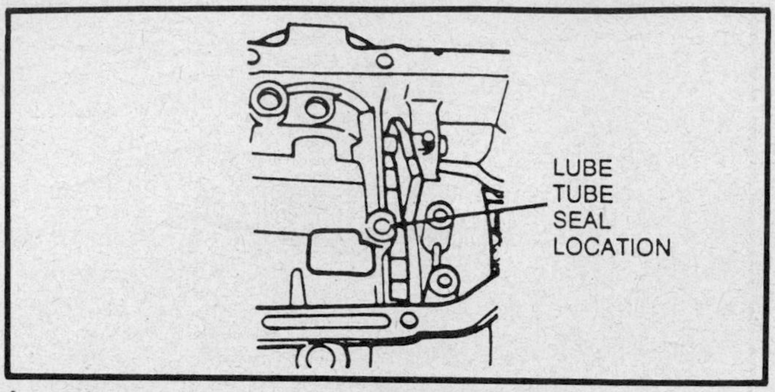
LUBE
TUBE
SEAL
LOCATION

Location of the lube tube seal

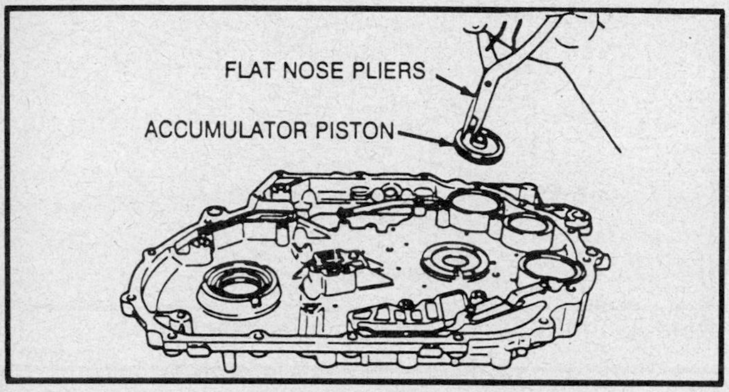
FLAT NOSE PLIERS
ACCUMULATOR PISTON

Using pliers to remove the accumulator pistons

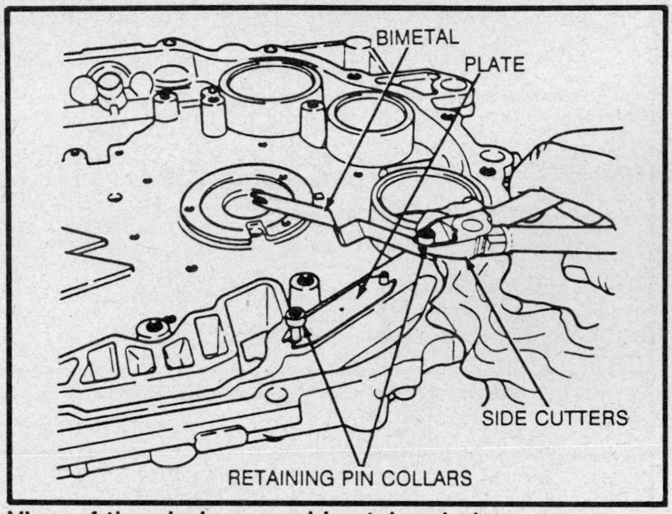
BIMETAL
PLATE
SIDE CUTTERS
RETAINING PIN COLLARS

View of the chain cover bimetal and plate

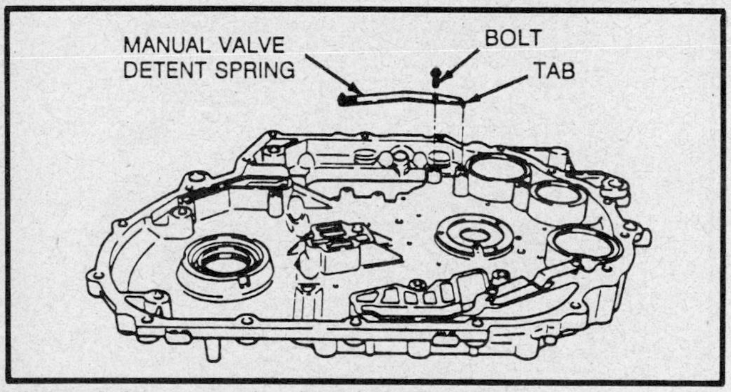
MANUAL VALVE
DETENT SPRING
BOLT
TAB

View of the manual valve detent spring

and the brackets. Using the lube tube remover tool or equivalent, and a slide hammer puller tool, pull the lube tubes from the transaxle.

31. From the lower case, remove the park rod abutment bolts and lift the rod to clear the abutment and the lower case. Remove the park pawl roll pin, the park pawl shaft (use a magnet), the park pawl and the spring.

32. At the reverse clutch band, loosen the reverse clutch anchor pin nut and remove the Allen bolt. Rotate the transaxle to the horizontal position.

33. From the side case, locate the inner diameter of the reverse clutch cylinder, using the hooked portion of the front clutch loading tool or equivalent, grasp the cylinder's outer diameter (with fingertips) and slide the clutch assembly from the case.

34. Rotate the transaxle to the vertical position. From the side case, grasp the front planetary shaft and lift the front/rear planetary assembly from the case. Remove the low/intermediate drum/sun gear assembly and the low/intermediate band.

35. Insert a small pry bar through the lower case to remove the snapring from the final drive gear assembly. Grasp the output shaft and lift the final drive assembly from the side case. From the side case, remove the final drive ring gear, the No. 18 thrust washer and the No. 19 needle bearing.

NOTE: The No. 18 thrust washer may stick to the final drive assembly next to the governor drive gear.

36. Using a 3/8 in. rod at the lower case, remove the rear lube tube seal and discard it.

NOTE: Whenever the differential is removed from the case, the rear lube tube seal must be replaced.

Unit Disassembly and Assembly

CHAIN COVER

Disassembly

1. Using flat-nose pliers, remove the 3 accumulator piston shafts and pistons.

NOTE: Do not place any objects in the piston shaft bore for removal procedures.

2. Using a pair of side cutters, carefully remove the bimetal pin collars, the bimetal strip and the plate.
3. Pull the pins from the cover.
4. Remove the manual valve detent spring bolt and spring.
5. If necessary, use the stator/driven sprocket bearing remover tool or equivalent, and slide hammer puller to remove the drive sprocket support needle bearing.

Inspection

1. Using solvent, clean the chain cover.

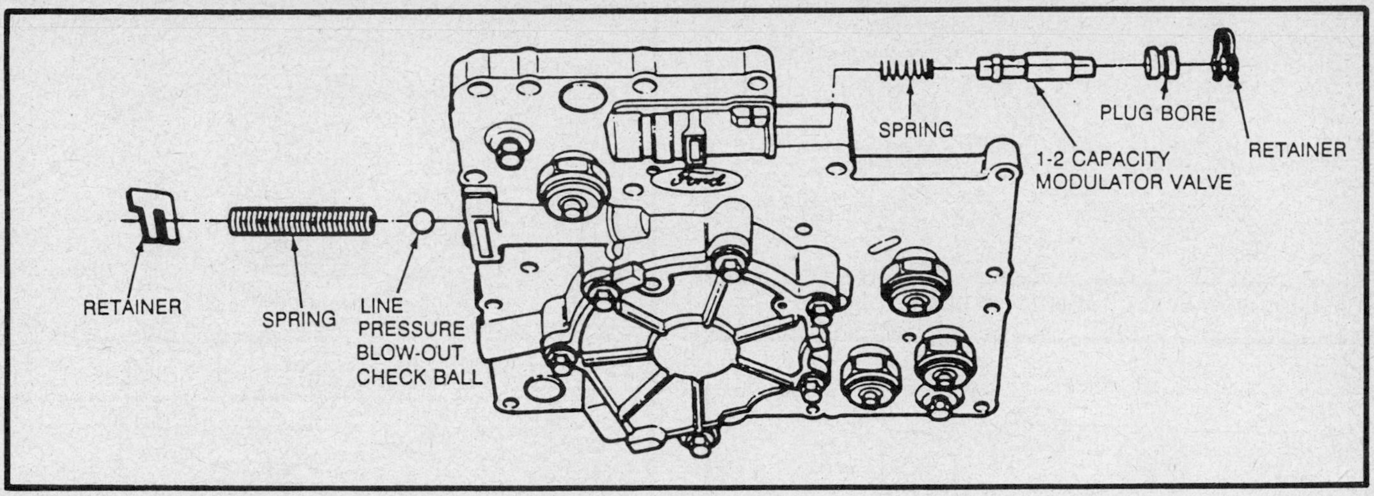
SPRING
PLUG BORE
RETAINER
1-2 CAPACITY
MODULATOR VALVE
RETAINER
SPRING
LINE
PRESSURE
BLOW-OUT
CHECK BALL

Exploded view of the oil pump's pressure and modulator valves

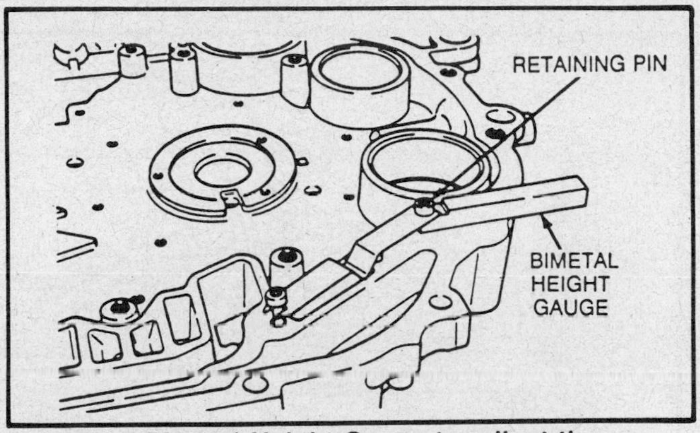
RETAINING PIN
BIMETAL
HEIGHT
GAUGE

Using the Bimetal Height Gauge to adjust the retaining pin height

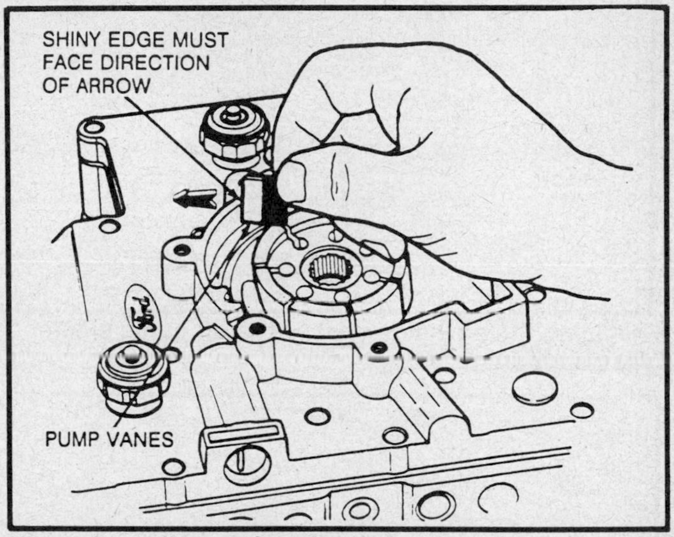
SHINY EDGE MUST
FACE DIRECTION
OF ARROW
PUMP VANES

View of an oil pump's rotor vane

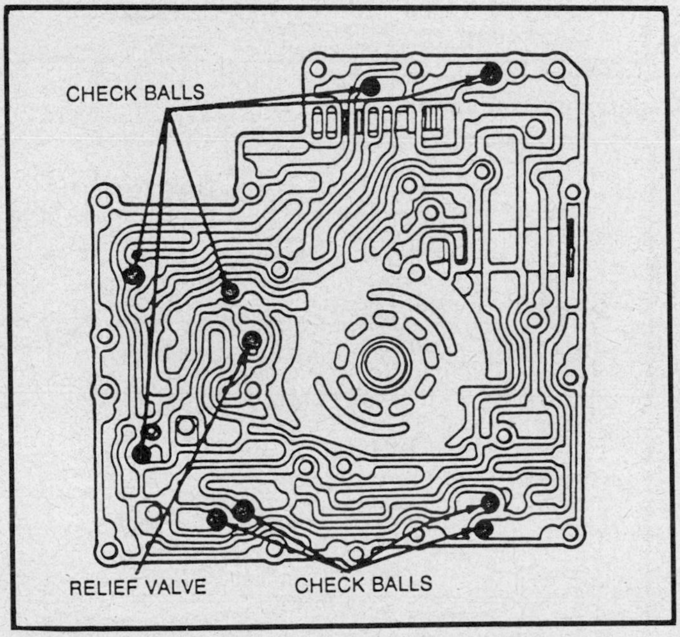
CHECK BALLS
RELIEF VALVE
CHECK BALLS

Location of the oil pump's check balls and relief valve

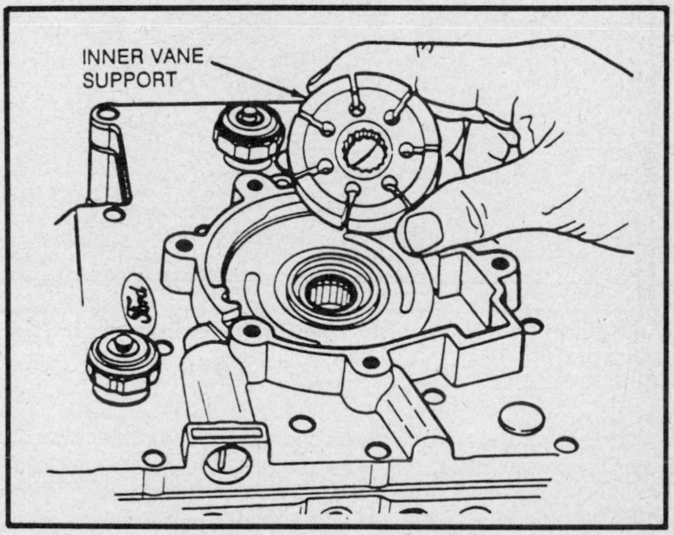
INNER VANE
SUPPORT

View of the oil pump's inner vane support

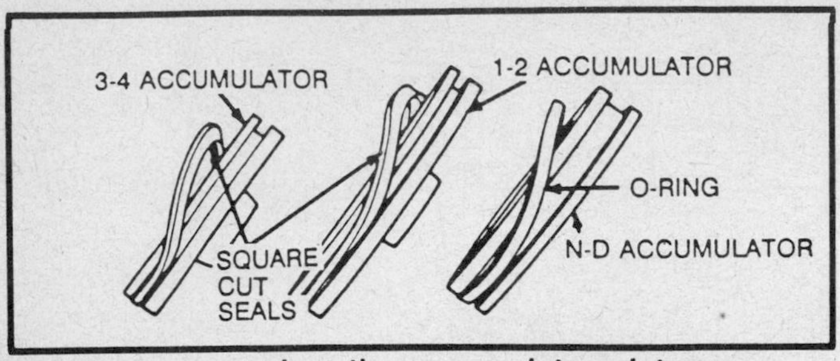
3-4 ACCUMULATOR
1-2 ACCUMULATOR
O-RING
N-D ACCUMULATOR
SQUARE
CUT
SEALS

Installing new seal on the accumulator pistons

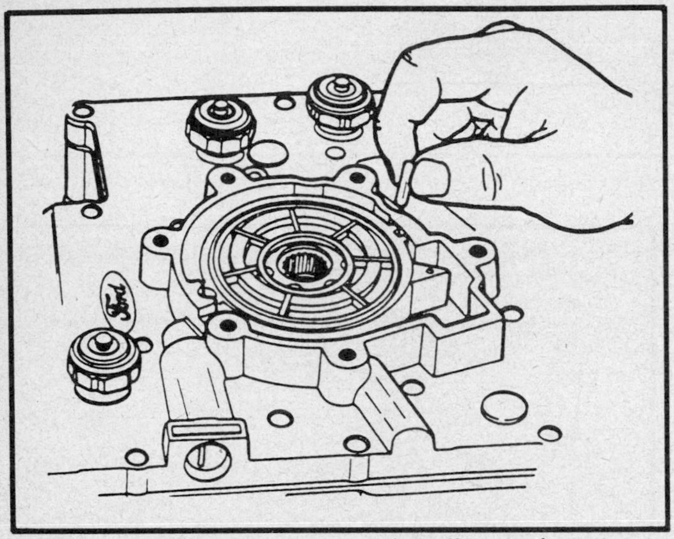

Installing the new side seal to the oil pump's outer vane support

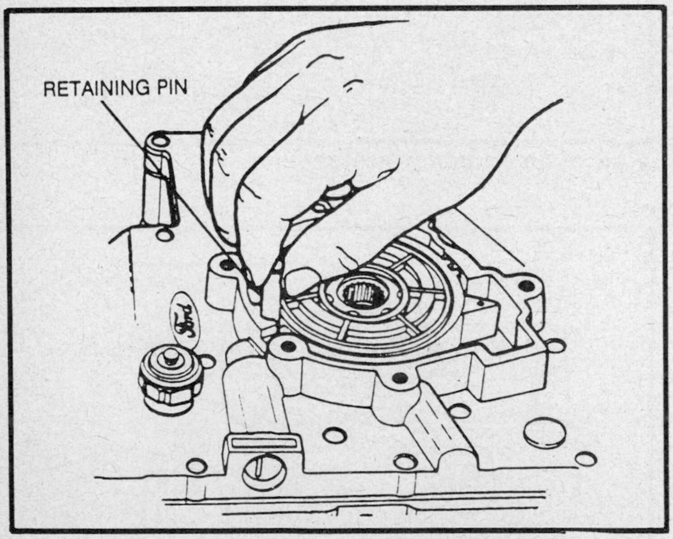
RETAINING PIN

Installing the outer vane support retaining pin to the oil pump

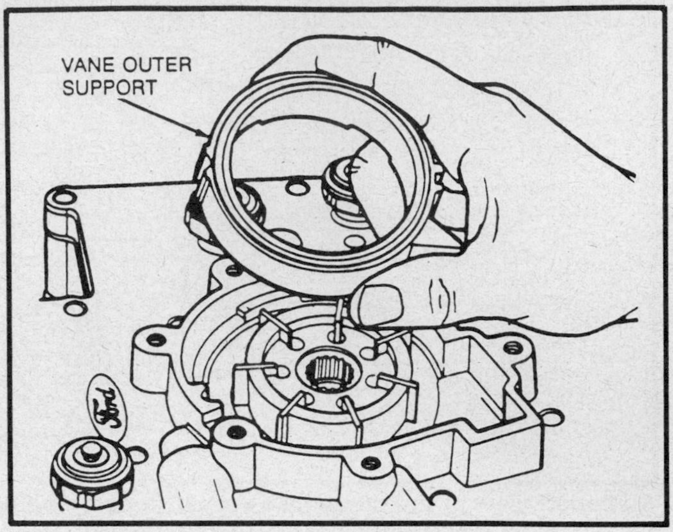
VANE OUTER
SUPPORT

View of the oil pump's outer vane support

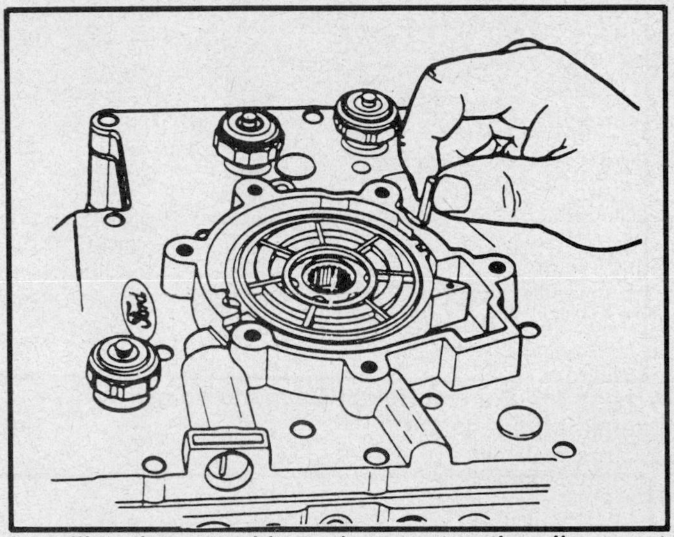

Installing the new side seal support to the oil pump's outer vane support

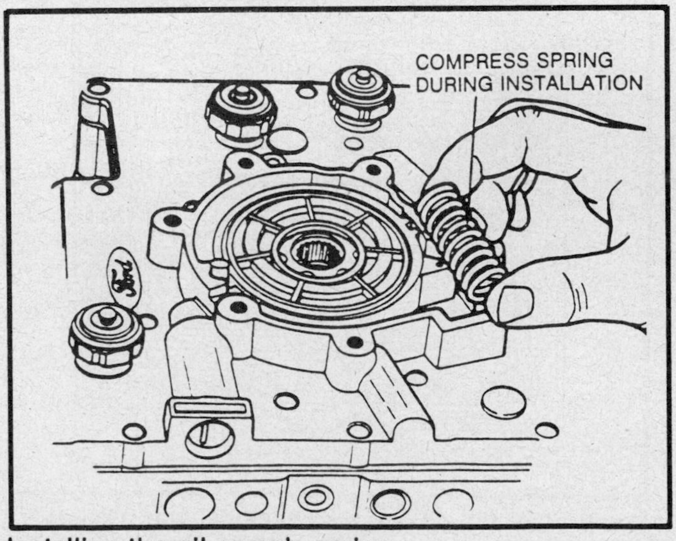
COMPRESS SPRING
DURING INSTALLATION

Installing the oil pump's spring

2. Inspect the case for cracks and/or stripped threads. Inspect the gasket mounting surfaces for burrs. Check the vent for obstructions. Check the fluid passages for obstructions and leakage.

3. Check the parking linkage parts for wear and/or damage. Inspect the case bushings for scores.

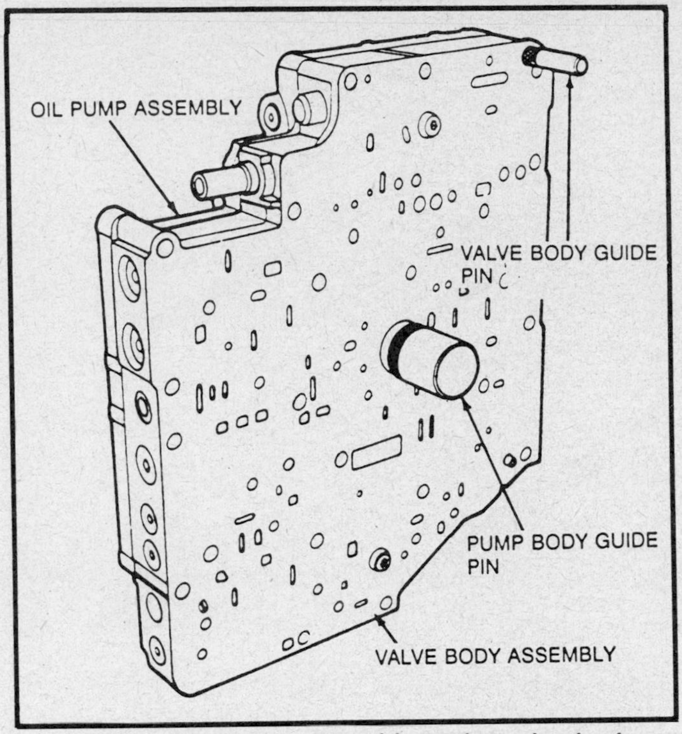
OIL PUMP ASSEMBLY
VALVE BODY GUIDE PIN
PUMP BODY GUIDE PIN
VALVE BODY ASSEMBLY
Installing the oil pump assembly to the valve body assembly

Assembly

1. If the drive sprocket support needle bearing was removed, use the stator/driven sprocket bearing replacer tool or equivalent, and an arbor press to the bearing into the chain cover until it seats.
2. Position the manual valve detent spring's tab in the chain cover's locator hole and install the bolts. Torque the bolt to 7–9 ft. lbs. (9–12 Nm).
3. If the bimetal retaining pins were removed, start them into the cover.
4. To install the bimetal, perform the following procedures:
   a. Position the bimetal over the retaining pins and start the retaining collars.
   b. Position the bimetal height gauge (slotted end) or equivalent, under the bimetal/retaining collar (hole side).
   c. Tap the retaining collar onto the pin until the assembly seats against the tool. Repeat the height procedure for the other bimetal retaining collar.
   d. Disconnect the bimetal's slotted end and adjust the center pin's height with the slotted end of the tool.
   e. Install the bimetal plate on the center and rear pins. Reinstall the bimetal.
5. Using new seals and O-ring, install them on the 3 accumulator pistons; use the square cut seals for the 3-4 accumulator and 1-2 accumulator pistons, while the O-ring is for the N-D accumulator.
6. Install the pistons (into the proper cylinder) and the piston shafts.

OIL PUMP

Disassembly

1. Remove the oil pump assembly-to-valve body bolts and the oil pump assembly from the valve body.
2. Remove the gasket and discard it. Place the oil pump assembly with the separator plate facing upwards.

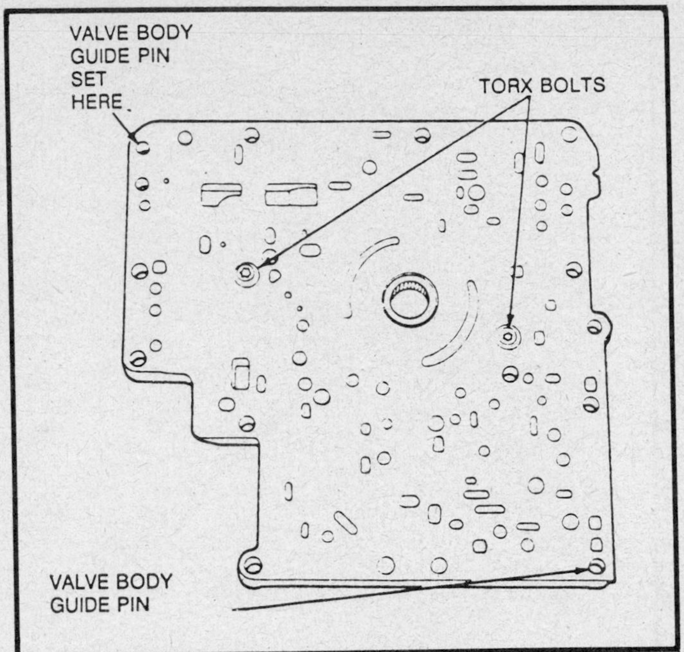
VALVE BODY GUIDE PIN SET HERE
TORX BOLTS
VALVE BODY GUIDE PIN
Installing the separator plate onto the oil pump housing

3. Remove the separator plate-to-oil pump housing Torx® bolts and the separator plate. Remove the check ball and the relief valve; be sure to note the location of each for installation purposes.
4. Remove the oil pump cover-to-oil pump housing bolts and the cover. Using a pry bar, pry the bore spring from the oil pump housing; be careful not to damage the gasket mounting surface.

CAUTION

Be careful when removing the spring for it is under pressure and can cause personal injury.

5. From the oil pump bore's outside vane support, remove the retaining pin, the metal O-ring, the soft O-ring (discard it), the side seal (discard it), the seal support, the top vane positioning ring and the outer vane support.
6. From the oil pump rotor, remove the 7 vanes, the inner vane support and the bottom vane positioning ring.

Inspection

The only oil pump parts that are servicable are the seals. If any part is worn or damaged, replace the entire pump assembly.

Assembly

1. Using clean transmission fluid, lubricate the pump's parts.
2. Into the oil pump's bore, install the bottom vane positioning ring, the inner vane support (face the small inner diameter counterbore upward) and the vanes into the inner vane support (the shiny surface must face outwards).
3. Install the outer vane support over the inner vane support assembly, the top vane positioning ring, a new side seal support, new side seal and the outer vane support retaining pin.
4. Compress the oil pump's spring and install it between the case and outer vane tab.
5. Install the new rubber O-ring and the the metal O-ring retainer into the outer vane support groove.
6. Install the oil pump cover onto the oil pump housing and torque the bolts to 7–9 ft. lbs. (9–12 Nm). If the line pressure blow-out ball/spring and the 1–2 modulator valve/spring/plug

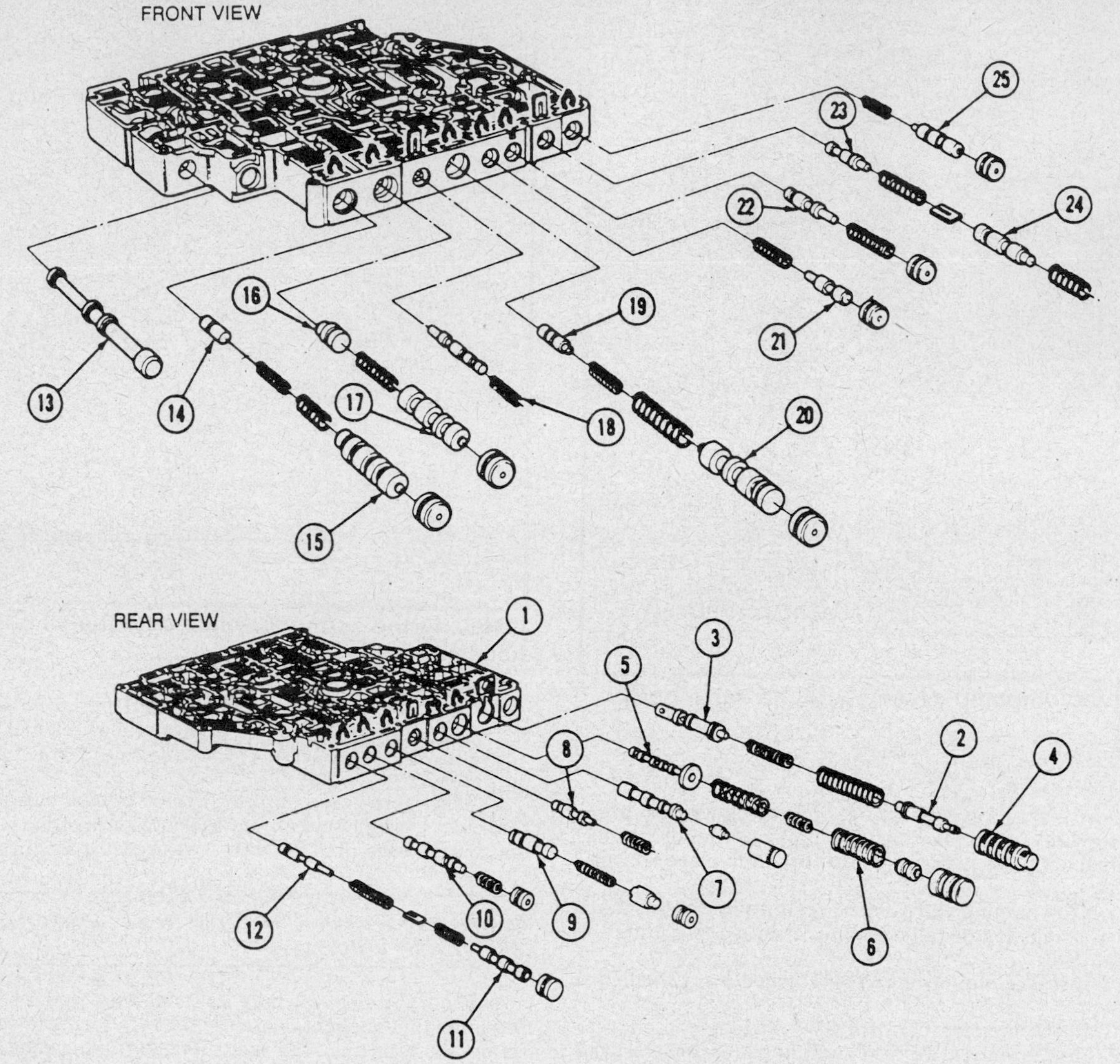
FRONT VIEW
REAR VIEW

1. Valve body
2. Throttle valve (T.V.)
3. T.V. plunger
4. T.V. valve sleeve
5. Main regulator valve
6. Main regulator boost valve
7. Converter clutch control valve
8. Converter regulator valve
9. Accumulator regulator valve
10. Backout valve
11. T.V./line modulator valve
12. 4-3 scheduling valve
13. Manual valve
14. 2-3 T.V. modulator valve
15. 2-3 shift valve
16. 1-2 throttle delay valve
17. 1-2 shift valve
18. 2-1 scheduling valve
19. 3-4 T.V. modulator valve
20. 3-4 shift valve
21. 2-4 inhibit valve
22. 3-2 control valve
23. N-D Engagement valve
24. T.V. limit valve
25. 2-3 servo regulator valve

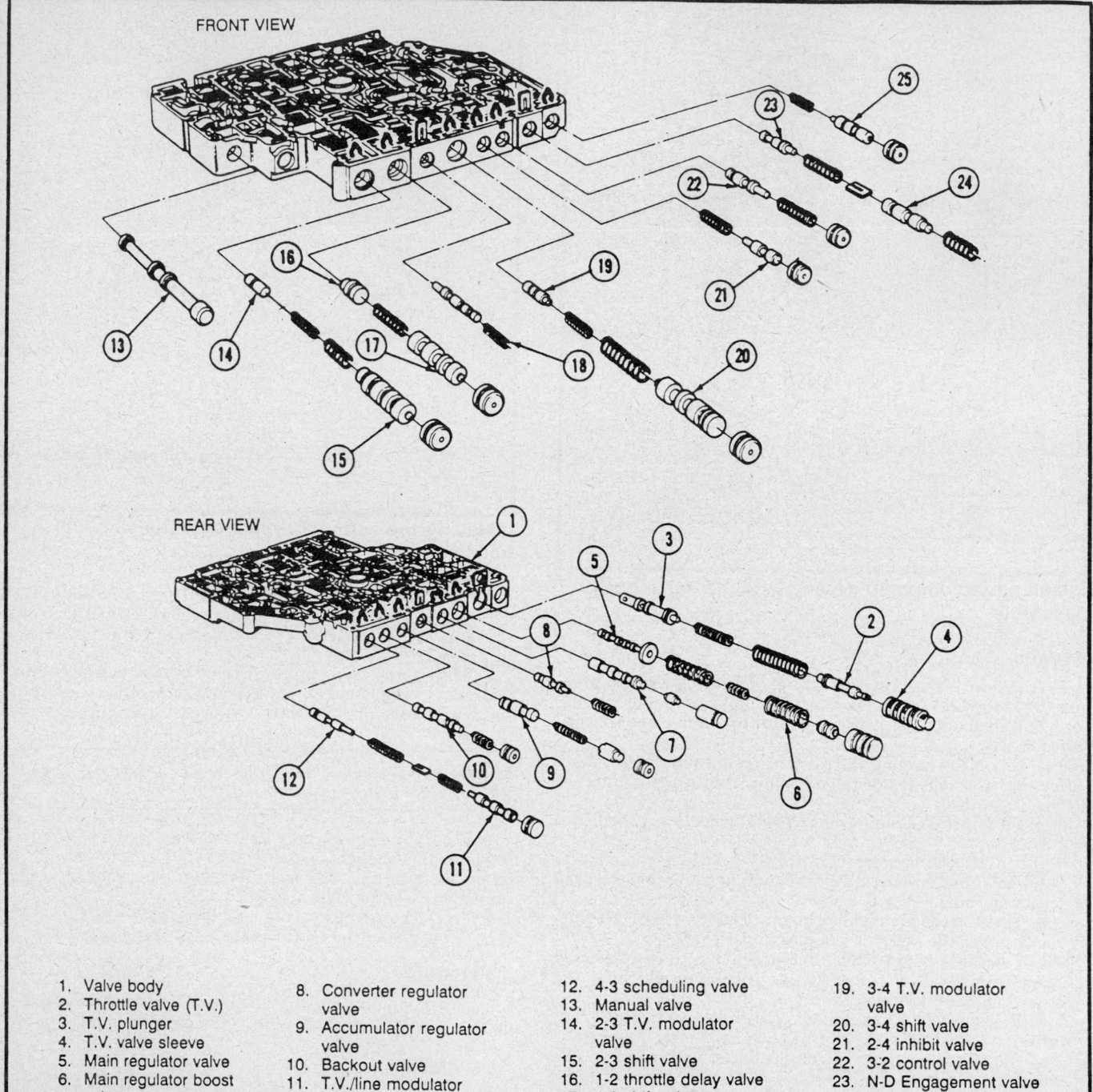

Exploded view of the valve body components

have been removed from the housing, install them and the retaining clips.

7. Turn the oil pump housing over and install the check balls/relief valve into their proper locations.

8. Using a new gasket, position the separator plate onto the oil pump housing. Using the insert valve body guide pin set tool or equivalent, and the valve body guide pin tool or equivalent, insert them through the separator plate into the oil pump housing and torque the Torx® to 7–9 ft. lbs. (9–12 Nm); remove the pins.

9. Using a new gasket, position the oil pump assembly onto the valve body.

10. Using the pump body guide pin tool or equivalent, and the valve body guide pin tool or equivalent, install them through the valve body and into the oil pump assembly; this will align the oil pump assembly-to-valve body.

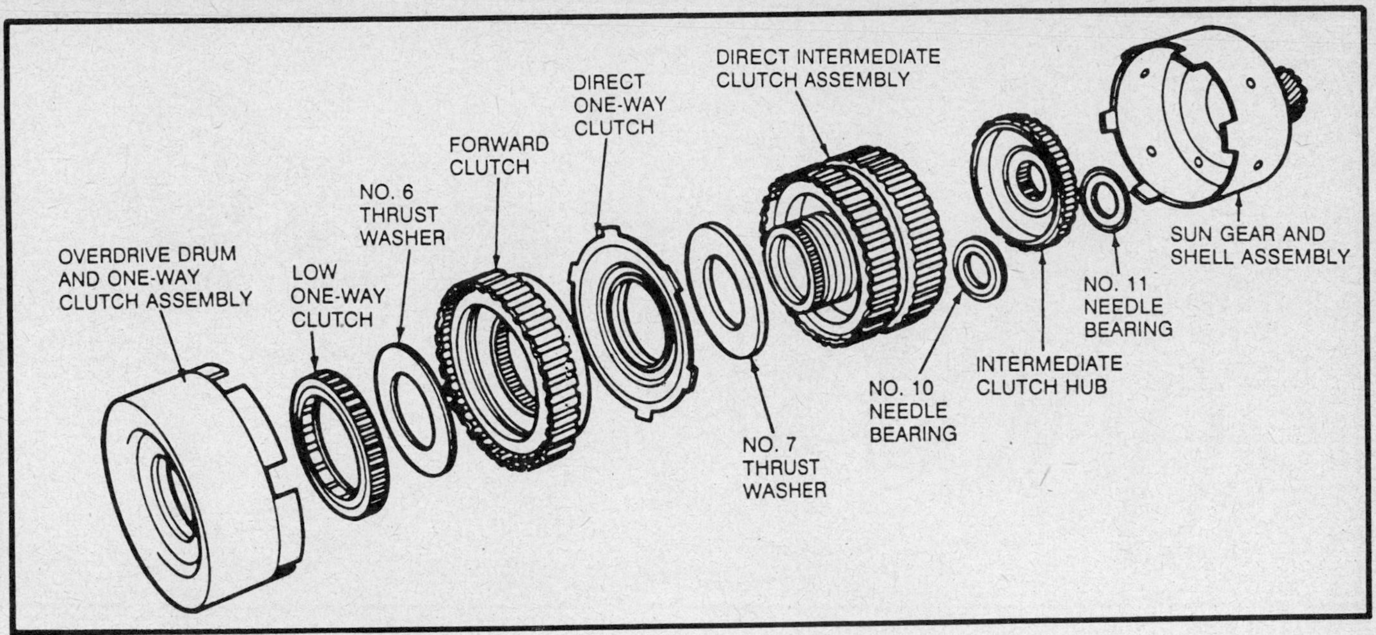
DIRECT INTERMEDIATE
CLUTCH ASSEMBLY
DIRECT
ONE-WAY
CLUTCH
FORWARD
CLUTCH
NO. 6
THRUST
WASHER
OVERDRIVE DRUM
AND ONE-WAY
CLUTCH ASSEMBLY
LOW
ONE-WAY
CLUTCH
SUN GEAR AND
SHELL ASSEMBLY
NO. 11
NEEDLE
BEARING
INTERMEDIATE
CLUTCH HUB
NO. 10
NEEDLE
BEARING
NO. 7
THRUST
WASHER

Exploded view of the shell assembly

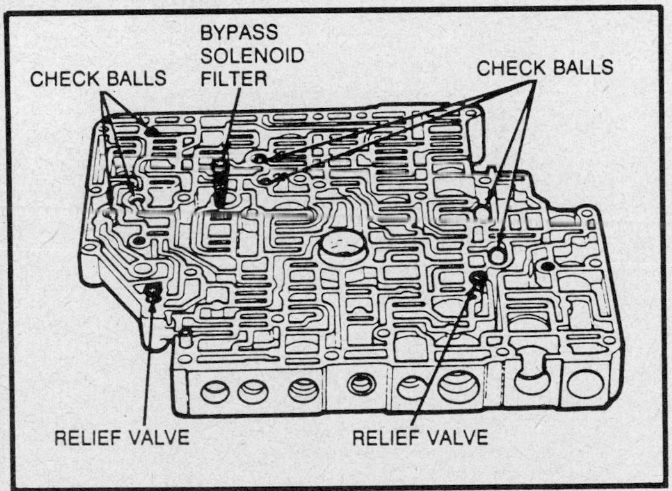
BYPASS
SOLENOID
FILTER
CHECK BALLS
CHECK BALLS
RELIEF VALVE
RELIEF VALVE

Location of the valve body's check balls, relief valves and solenoid bypass filter

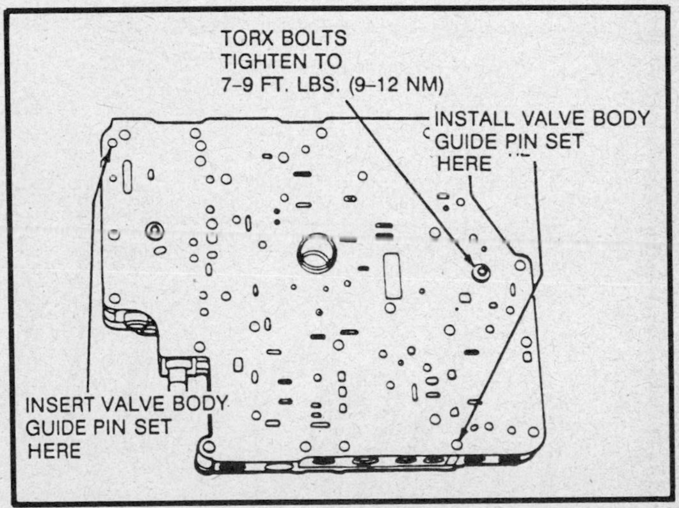
TORX BOLTS
TIGHTEN TO
7–9 FT. LBS. (9–12 NM)
INSTALL VALVE BODY
GUIDE PIN SET
HERE
INSERT VALVE BODY
GUIDE PIN SET
HERE

Installing the separator plate onto the valve body housing

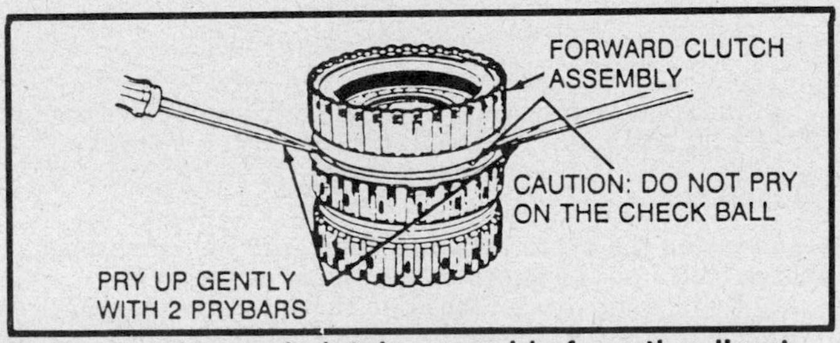
FORWARD CLUTCH
ASSEMBLY
CAUTION: DO NOT PRY
ON THE CHECK BALL
PRY UP GENTLY
WITH 2 PRYBARS

Prying the forward clutch assembly from the direct clutch hub

11. Install the oil pump assembly-to-valve body bolts and torque them to 7–9 ft. lbs. (9–12 Nm). Remove the alignment tools.

VALVE BODY

Disassembly

1. Remove the valve body-to-oil pump assembly bolts and the valve body from the oil pump assembly.
2. Remove the gasket and discard it. Place the valve assembly with the separator plate facing upwards.
3. Remove the separator plate-to-valve body Torx® bolts and the separator plate. Remove the check balls, the relief valves and bypass solenoid filter (clean it); be sure to note the location of each for installation purposes.
4. From the valve body, remove the retaining clips, the bore plugs, the valves and springs; be sure to note their locations for installation purposes.

NOTE: Since most valve are made of aluminum, they cannot be removed with a magnet. It will be necessary to remove them by tapping the valve body with the palm of

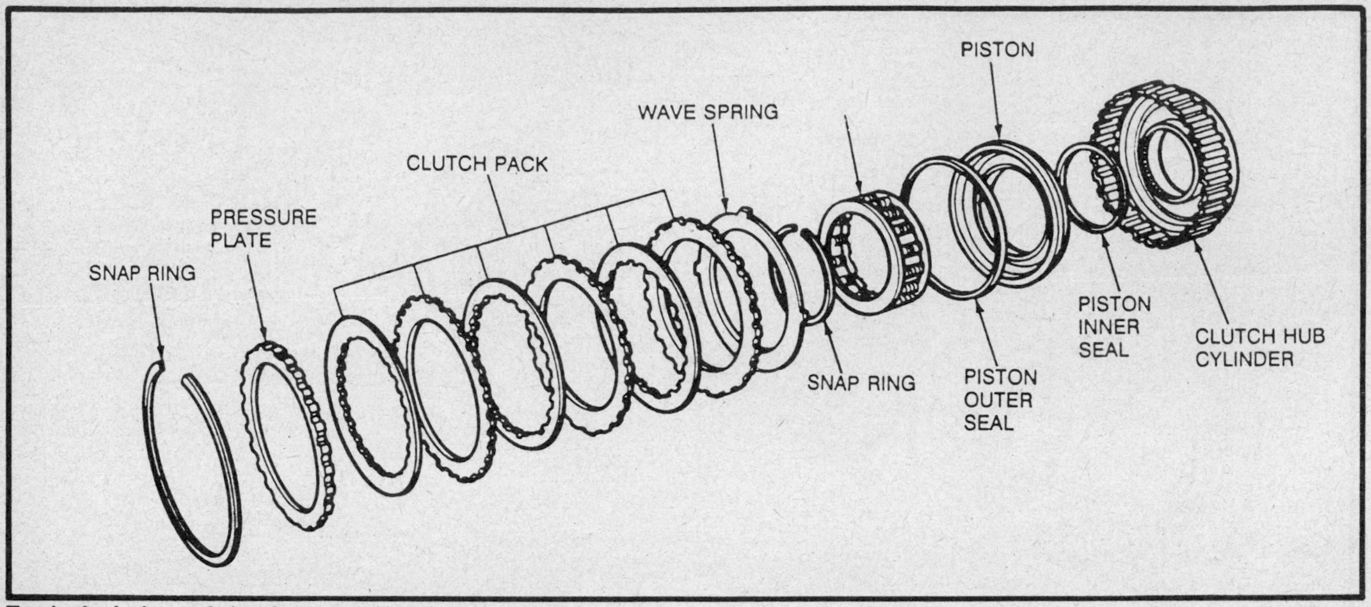
PISTON
WAVE SPRING
CLUTCH PACK
PRESSURE PLATE
SNAP RING
PISTON INNER SEAL
CLUTCH HUB CYLINDER
SNAP RING
PISTON OUTER SEAL

Exploded view of the forward clutch assembly

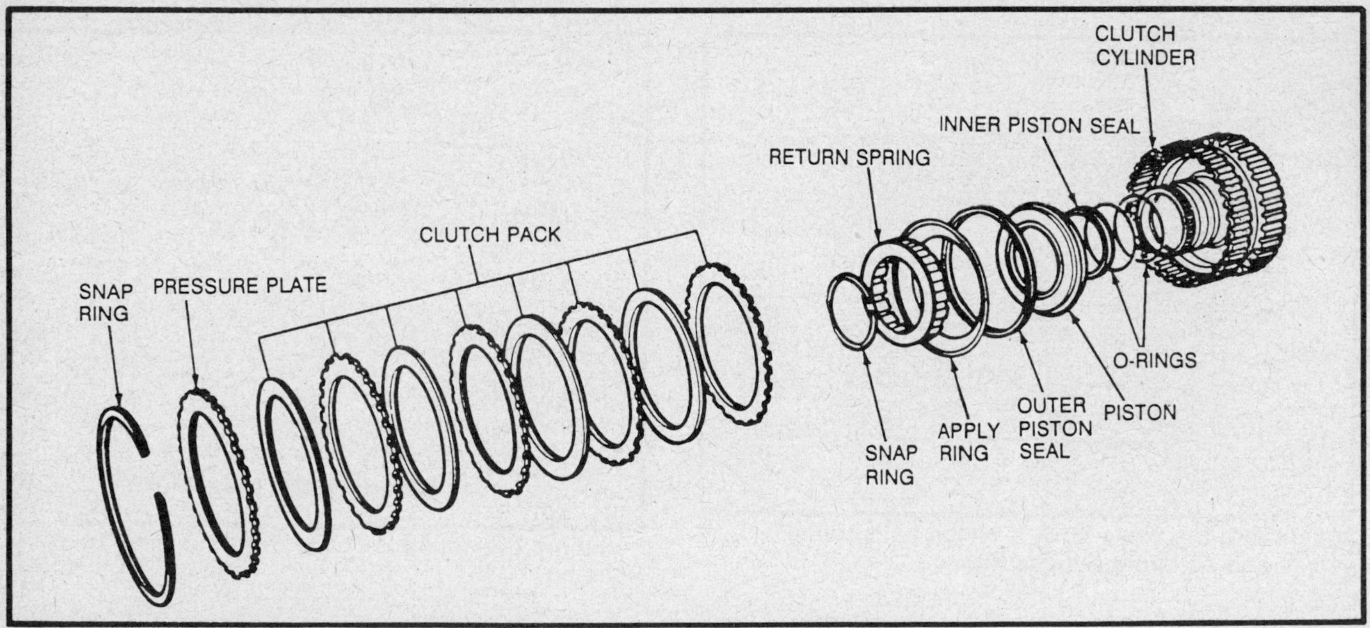
CLUTCH CYLINDER
INNER PISTON SEAL
RETURN SPRING
CLUTCH PACK
SNAP RING
PRESSURE PLATE
O-RINGS
OUTER PISTON SEAL
PISTON
APPLY RING
SNAP RING

Exploded view of the direct clutch assembly

the hand or a tooth pick. If using a tooth pick, be careful not to damage the valves or the valve bore. Do not turn the throttle valve adjusting screw.

Inspection

1. Using clean solvent, wash the valves, springs and valve body; do not clean the check balls with solvent. Using moisture-free compressed air, blow dry the parts.
2. Check the valve/plug bores for scores, all passage ways for obstructions and the mounting surfaces for scores and/or burrs. If necessary, polish the valves and plugs with crocus cloth; be careful not to round their sharp edges.
3. Check the spring for distortion and the valves/plugs for free-bore movement in their bores; the valves/plugs must be free to move through their own weight.
4. Place the manual valve on a flat surface and roll it to check for a bent condition; replace it, if necessary.

Assembly

1. Install the bypass solenoid filter, the valves, springs and check balls into their correct valve body positions.
2. Using a new gasket, position the separator plate onto the valve body housing. Using the valve body guide pin set tools or equivalent, insert them through the separator plate into the valve body housing and torque the Torx® to 7–9 ft. lbs. (9–12 Nm); remove the pins.
3. Using a new gasket, position the valve body assembly onto the oil pump housing assembly.
4. Using the pump body guide pin tool or equivalent, and the valve body guide pin tool or equivalent, install them through the

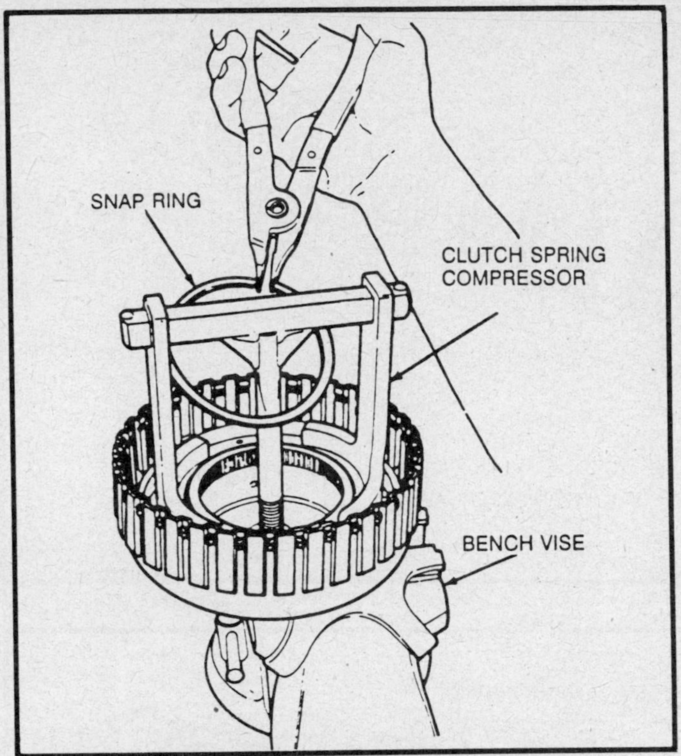
SNAP RING
CLUTCH SPRING
COMPRESSOR
BENCH VISE

Compressing the piston assembly into the clutch hub cylinder to remove the snapring

valve body and into the oil pump assembly; this will align the oil pump assembly-to-valve body.

5. Install the oil pump assembly-to-valve body bolts and torque them to 7–9 ft. lbs. (9–12 Nm). Remove the alignment tools.

CLUTCH/SHELL ASSEMBLY

The shell assembly consists of the overdrive drum, the one-way clutch, the forward clutch, the direct clutch, the intermediate clutch and the sun gear.

Disassembly

SHELL ASSEMBLY

1. Position the shell assembly on the work bench with the sun gear facing upwards. Remove the sun gear/shell assembly from the overdrive drum.
2. Remove the No. 11 needle bearing, the intermediate clutch hub and the No. 10 needle bearing.
3. Turn the assembly over, onto the intermediate cylinder hub.
4. Remove the overdrive drum/one-way clutch assembly and the No. 6 thrust washer.
5. Using 2 pry bars, one on each side of the forward clutch assembly, pry the forward clutch upward.

NOTE: Since the forward clutch assembly is retained to the hub by O-ring seals, be sure to pry evenly. Do not position the pry bars near the check ball.

6. Remove the direct one-way clutch and the No. 7 thrust washer.

FORWARD CLUTCH ASSEMBLY

1. From the forward clutch assembly, remove the snapring, the pressure plate, the clutch pack and the wave spring.
2. Using the clutch spring compressor tool or equivalent, compress the clutch hub cylinder's piston assembly and remove the snapring.
3. Remove the compressor tool and the piston assembly.
4. From the piston, remove the outer seal. From the clutch hub cylinder, remove the inner seal.

DIRECT CLUTCH ASSEMBLY

The direct clutch assembly occupies ½ of the direct/intermediate clutch cylinder assembly.

1. From the direct clutch cylinder, remove the snapring, pressure plate and clutch pack.
2. Using the clutch spring compressor tool or equivalent, compress the clutch hub cylinder's piston assembly and remove the snapring.
3. Remove the compressor tool and the 2-piece piston assembly.
4. Separate the apply ring from the piston. From the piston, remove the outer seal. From the clutch hub cylinder, remove the inner seal.

INTERMEDIATE CLUTCH ASSEMBLY

The intermediate clutch assembly occupies ½ of the direct/intermediate clutch cylinder assembly.

1. From the intermediate clutch cylinder, remove the snapring, pressure plate and clutch pack.
2. Using the clutch spring compressor tool or equivalent, compress the clutch cylinder's piston assembly and remove the snapring.
3. Remove the compressor tool and the piston assembly.
4. From the piston, remove the outer seal. From the clutch cylinder, remove the inner seal.

REVERSE CLUTCH ASSEMBLY

1. From the reverse clutch cylinder, remove the snapring, pressure plate, clutch pack and wave spring.
2. Using the clutch spring compressor tool or equivalent, compress the clutch cylinder's piston assembly and remove the snapring.
3. Remove the compressor tool and the piston assembly.
4. From the piston, remove the outer seal. From the clutch cylinder, remove the inner seal.

Inspection

CLUTCHES

The following inspection is used to check the forward, direct, intermediate and reverse clutches.

1. Check the clutch cylinder thrust surfaces, clutch plate serrations and piston bore for burrs and/or scores. If the clutch cylinder is badly damaged or scored, replace it; otherwise, use crocus cloth to remove minor scores and/or burrs.
2. Inspect the clutch cylinder's fluid passages for obstructions; if necessary, clean them. Inspect the check balls for proper seating and freedom of movement. Check the clutch pistons for scoring; replace them, if necessary.
3. Inspect the clutch release spring for cracks and/or distortion; replace the spring, if necessary.
4. Check the steel clutch plates, composition clutch plates and pressure plate for scored or worn bearing surfaces; replace the plates, if necessary.
5. Inspect the clutch plates for fit on the clutch hub serrations and flatness; replace the plates, if they do not slide freely on the serrations.
6. Inspect the clutch hub thrust surfaces for scoring and the clutch hub splines for wear.

ONE-WAY CLUTCHES

1. Check the outer/inner races surface areas for damage and/or scoring.
2. Check the rollers, sprags and springs for wear and/or damage.

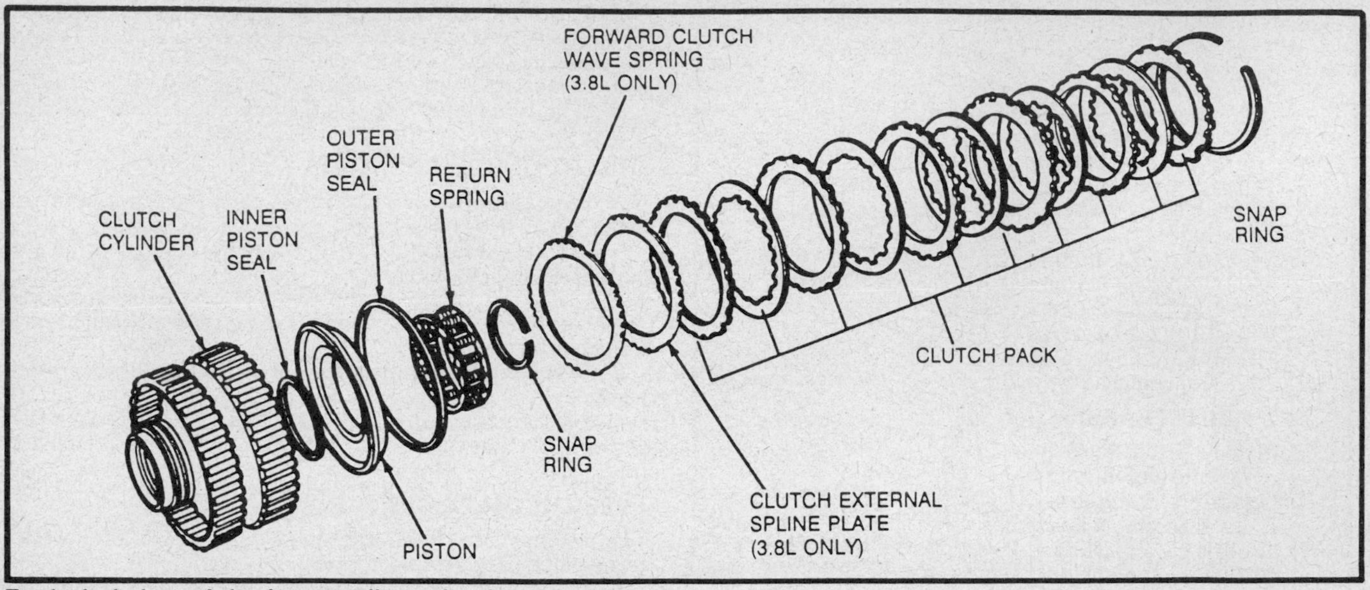
FORWARD CLUTCH
WAVE SPRING
(3.8L ONLY)
OUTER
PISTON
SEAL
RETURN
SPRING
CLUTCH
CYLINDER
INNER
PISTON
SEAL
SNAP
RING
CLUTCH PACK
SNAP
RING
CLUTCH EXTERNAL
SPLINE PLATE
(3.8L ONLY)
PISTON

Exploded view of the intermediate clutch assembly

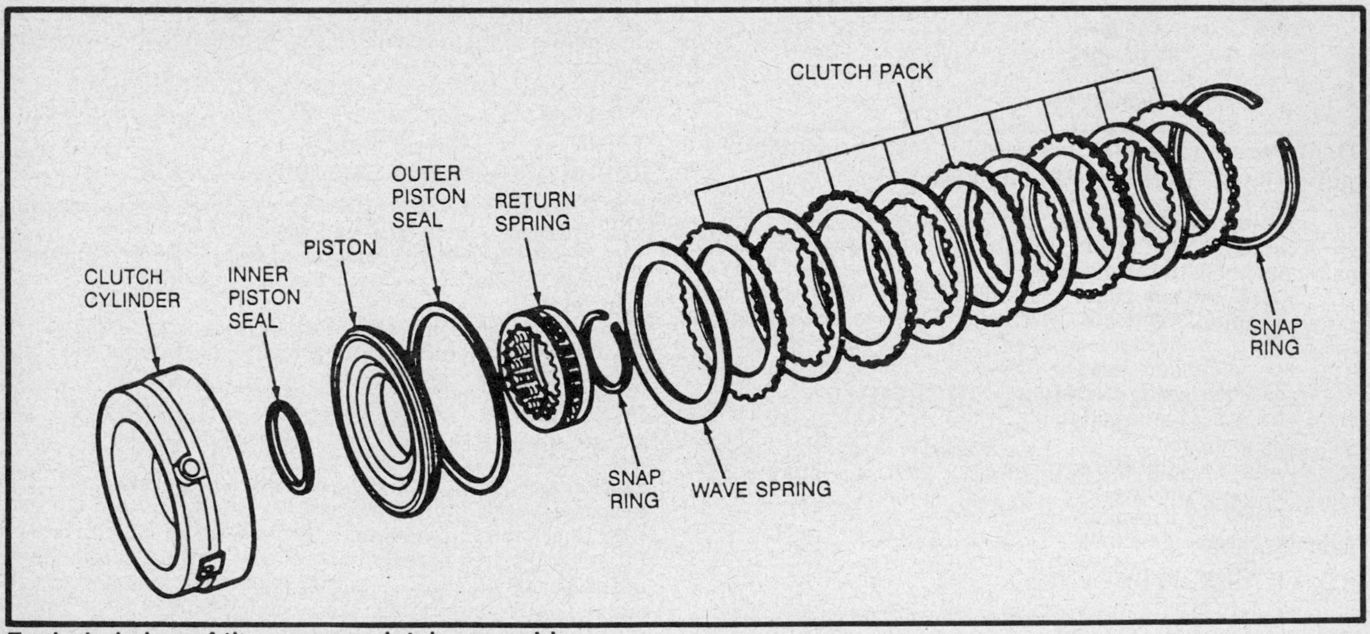
CLUTCH PACK
OUTER
PISTON
SEAL
RETURN
SPRING
PISTON
CLUTCH
CYLINDER
INNER
PISTON
SEAL
SNAP
RING
SNAP
RING
WAVE SPRING

Exploded view of the reverse clutch assembly

3. Check the spring/case for damaged and/or bent spring retainers.

THRUST WASHERS

1. Using clean solvent, throughly clean the thrust bearings. Using compressed air, blow dry the bearings.
2. Check the bearings for pitting and/or roughness; replace them, if necessary.
3. Before installation, lubricate the cleaned bearings with clean transmission fluid.

Assembly

FORWARD CLUTCH ASSEMBLY

1. Using new piston seals, install them with their lips facing the bottom of the cylinder.
2. Using the forward clutch seal lip protector tool or equivalent, place the piston assembly inside the tool and the tool/piston assembly into the clutch hub cylinder. Remove the installation tool.
3. Using the clutch spring compressor tool or equivalent, compress the piston assembly into the clutch hub cylinder and install the snapring. Remove the installation tool.
4. Install the wave spring, the clutch pack, the pressure plate and snapring into the clutch hub cylinder.
5. To check the clutch pack clearance, perform the following procedures:
   a. Using a dial indicator, connect it to a rigid mount.
   b. Position the forward clutch assembly on a flat surface with the clutch package facing upwards.
   c. Using at least 10 lbs. pressure, force the clutch package downward and release it.
   d. Position the dial indicator's stylus on the forward clutch's pressure plate and zero the indicator.

e. While securing the assembly on the flat surface, lift the pressure plate to the bottom of the snapring and note the indicator's reading.

f. Move the indicator to the opposite side (180 degrees) and repeat this procedure.

g. Using the 2 readings, determine the reading average; the clearance should be 0.055–0.075 in. (1.40–1.89mm).

h. If the reading is not within specifications, the snapring must be changed.

Snapring sizes:

0.049–0.053 in. (1.24–1.34mm)
0.063–0.067 in. (1.60–1.70mm)
0.077–0.081 in. (1.95–2.05mm)
0.091–0.094 in. (2.30–2.40mm)
0.104–0.108 in. (2.65–2.75mm)

i. If the snapring was replaced, recheck the clearance.

DIRECT CLUTCH ASSEMBLY

1. Using new piston seals, install them with their lips facing the bottom of the clutch cylinder.
2. Using the direct clutch seal lip protector tool or equivalent, place the piston assembly inside the tool and the tool/piston assembly into the clutch cylinder. Remove the installation tool.
3. Install the piston apply ring on top of the piston. Be sure the piston's check ball has free movement.
4. Install the return spring into the clutch cylinder by aligning the notch with the piston's check ball.
5. Using the clutch spring compressor tool or equivalent, compress the piston assembly into the clutch cylinder and install the snapring. Remove the installation tool.
6. Install the clutch pack, the pressure plate and snapring into the clutch cylinder.
7. To check the clutch pack clearance, perform the following procedures:

a. Using a dial indicator, connect it to a rigid mount.

b. Position the direct/intermediate clutch assembly on a flat surface with the direct clutch package facing upwards.

c. Using at least 10 lbs. pressure, force the clutch package downward and release it.

d. Position the dial indicator's stylus on the direct clutch's pressure plate and zero the indicator.

e. While securing the assembly on the flat surface, lift the pressure plate to the bottom of the snapring and note the indicator's reading.

f. Move the indicator to the opposite side (180 degrees) and repeat this procedure.

g. Using the 2 readings, determine the reading average; the clearance should be 0.031–0.051 in. (0.78–1.29mm).

h. If the reading is not within specifications, the snapring must be changed.

Snapring sizes:

0.047–0.051 in. (1.20–1.30mm)
0.065–0.069 in. (1.67–1.76mm)
0.084–0.088 in. (2.14–2.24mm)
0.102–0.106 in. (2.61–2.71mm)
0.119–0.123 in. (3.04–3.14mm)

i. If the snapring was replaced, recheck the clearance.

INTERMEDIATE CLUTCH ASSEMBLY

1. Using new piston seals, install them with their lips facing the bottom of the cylinder. Inspect the clutch cylinder for free check ball movement.
2. Using the forward clutch seal lip protector tool or equivalent, place the piston assembly inside the tool and the tool/piston assembly into the clutch cylinder. Remove the installation tool.
3. Using the clutch spring compressor tool or equivalent, compress the piston assembly into the clutch cylinder and install the snapring. Remove the installation tool.
4. Install the clutch pack, the pressure plate and snapring into the clutch cylinder.
5. To check the clutch pack clearance, perform the following procedures:

a. Using a dial indicator, connect it to a rigid mount.

b. Position the direct/intermediate clutch assembly on a flat surface with the intermediate clutch package facing upwards.

c. Using at least 10 lbs. pressure, force the clutch package downward and release it.

d. Position the dial indicator's stylus on the intermediate clutch's pressure plate and zero the indicator.

e. While securing the assembly on the flat surface, lift the pressure plate to the bottom of the snapring and note the indicator's reading.

f. Move the indicator to the opposite side (180 degrees) and repeat this procedure.

g. Using the 2 readings, determine the reading average; the clearance should be 0.040–0.061 in. (1.04–1.55mm).

h. If the reading is not within specifications, the snapring must be changed.

Snapring sizes:

0.049–0.053 in. (1.24–1.34mm)
0.065–0.069 in. (1.66–1.76mm)
0.081–0.085 in. (2.08–2.18mm)
0.098–1.020 in. (2.50–2.60mm)
0.114–0.118 in. (2.92–3.02mm)

i. If the snapring was replaced, recheck the clearance.

REVERSE CLUTCH ASSEMBLY

1. Using new piston seals, install them with their lips facing the bottom of the cylinder.
2. Using the reverse clutch seal lip protector tool or equivalent, place the piston assembly inside the tool and the tool/piston assembly into the clutch cylinder. Remove the installation tool. Install the return spring into the cylinder.
3. Using the clutch spring compressor tool or equivalent, compress the piston assembly into the clutch cylinder and install the snapring. Remove the installation tool.
4. Install the wave spring, the clutch pack, the pressure plate and snapring into the clutch cylinder.
5. To check the clutch pack clearance, perform the following procedures:

a. Using a dial indicator, connect it to a rigid mount.

b. Position the reverse clutch assembly on a flat surface with the clutch package facing upwards.

c. Using at least 10 lbs. pressure, force the clutch package downward and release it.

d. Position the dial indicator's stylus on the reverse clutch's pressure plate and zero the indicator.

e. While securing the assembly on the flat surface, lift the pressure plate to the bottom of the snapring and note the indicator's reading.

f. Move the indicator to the opposite side (180 degrees) and repeat this procedure.

g. Using the 2 readings, determine the reading average; the clearance should be 0.038–0.064 in. (0.97–1.63mm).

h. If the reading is not within specifications, the snapring must be changed.

Snapring sizes:

0.059–0.064 in. (1.52–1.62mm)
0.078–0.081 in. (1.98–2.08mm)
0.096–0.100 in. (2.45–2.55mm)
0.115–0.118 in. (2.92–3.02mm)

i. If the snapring was replaced, recheck the clearance.

SHELL ASSEMBLY

1. Position the intermediate clutch cylinder on the work bench with shaft end facing upwards.
2. Position the No. 7 thrust washer so its tabs (facing downward) align with the direct clutch slots.
3. Install the direct one-way clutch into the intermediate

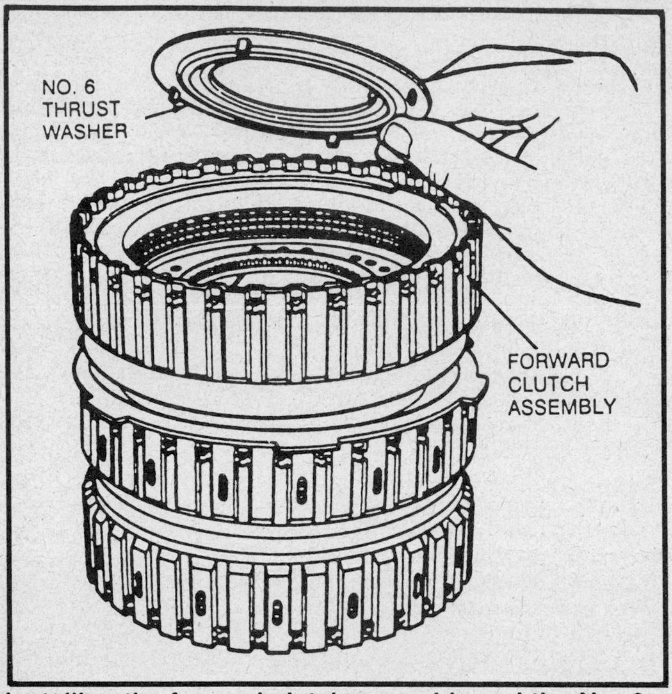
NO. 6
THRUST
WASHER
FORWARD
CLUTCH
ASSEMBLY

Installing the forward clutch assembly and the No. 6 thrust washer onto the intermediate clutch cylinder

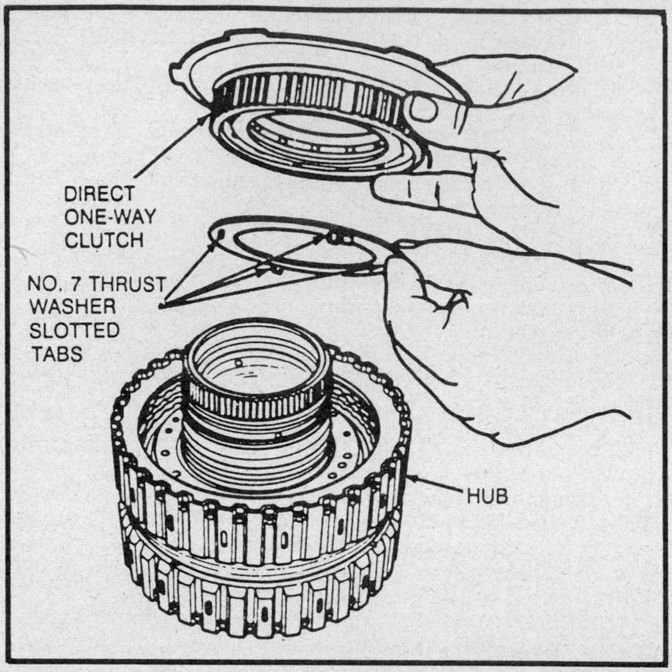
DIRECT
ONE-WAY
CLUTCH
NO. 7 THRUST
WASHER
SLOTTED
TABS
HUB

Installing the No. 7 thrust washer and the direct 1-way clutch onto the intermediate clutch cylinder

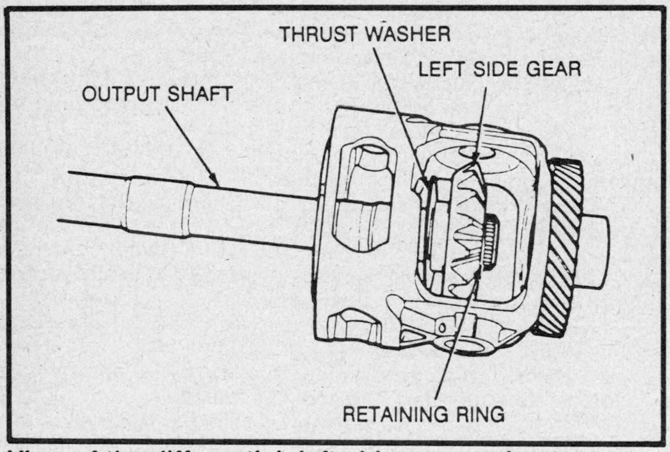
THRUST WASHER
LEFT SIDE GEAR
OUTPUT SHAFT
RETAINING RING

View of the differential, left side gear and output shaft assembly

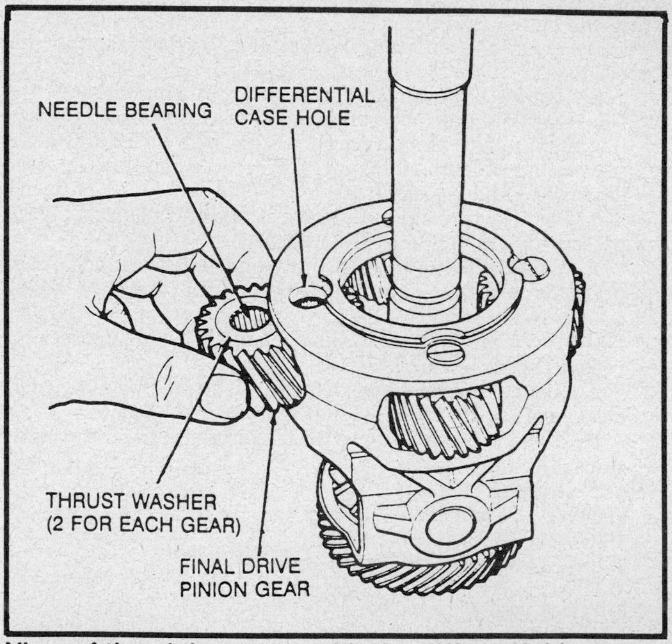
NEEDLE BEARING
DIFFERENTIAL
CASE HOLE
THRUST WASHER
(2 FOR EACH GEAR)
FINAL DRIVE
PINION GEAR

View of the pinion gears with the differential housing

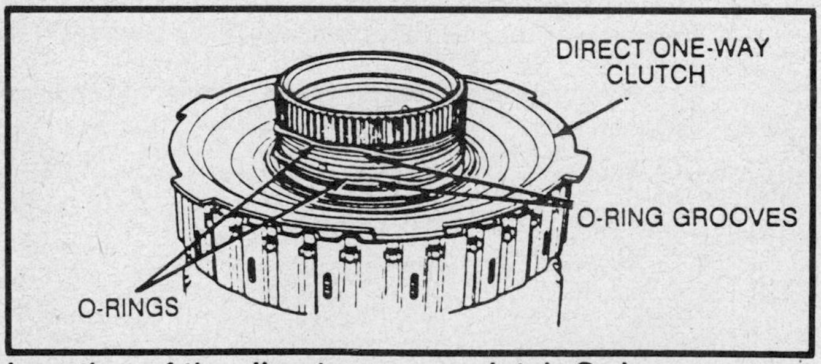
DIRECT ONE-WAY
CLUTCH
O-RING GROOVES
O-RINGS

Location of the direct one-way clutch O-rings

clutch cylinder by aligning its serrations with the clutch pack splines.

4. Install new O-ring seals onto the direct one-way clutch.
5. When installing the forward clutch assembly onto the direct one-way clutch, be careful not to damage the O-ring seals. Install the No. 6 thrust washer (tabs facing downward) onto the forward clutch.
6. Install the overdrive drum/one-way clutch assembly over the forward clutch.

NOTE: When installing the overdrive drum/one-way clutch assembly, be sure the one-way clutch outer race groove is visible; if not, rotate the clutch counterclockwise to expose it.

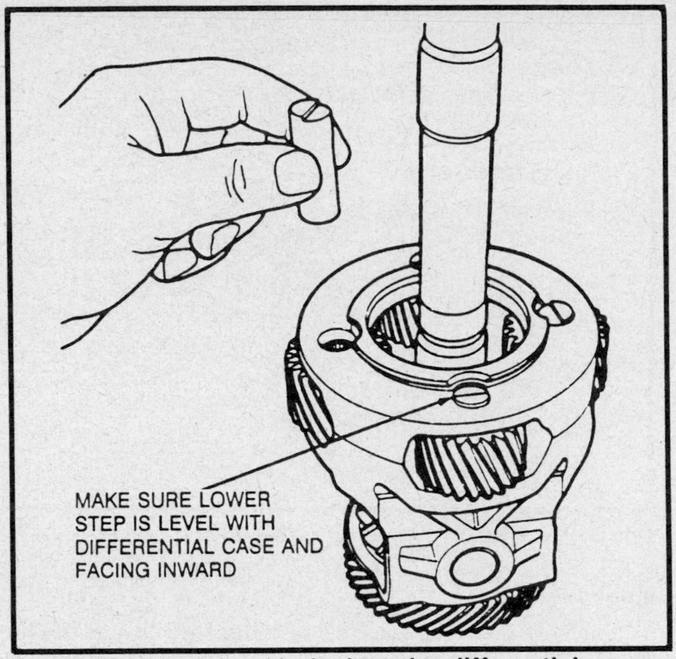
MAKE SURE LOWER
STEP IS LEVEL WITH
DIFFERENTIAL CASE AND
FACING INWARD

Installing the pinion shafts into the differential case

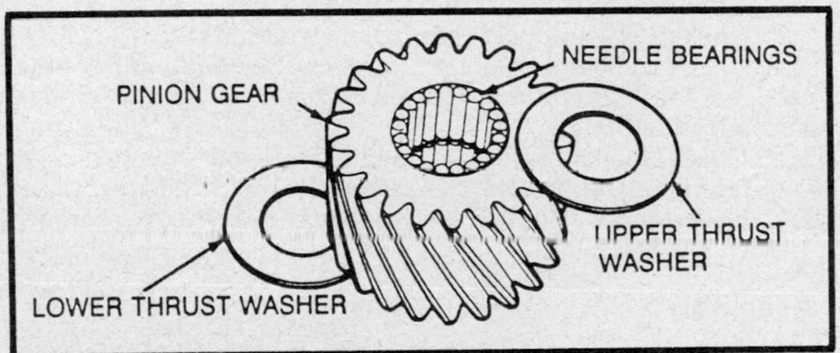
NEEDLE BEARINGS
PINION GEAR
UPPER THRUST
WASHER
LOWER THRUST WASHER

View of the pinion gear assemblies

7. Turn the shell assembly over (rest it on the overdrive drum). Grease and install the No. 10 needle bearing into the intermediate clutch hub. Install the intermediate clutch hub (with No. 10 needle bearing) into the intermediate clutch cylinder; be sure the clutch plates align with the hub's serrations.
8. Install the No. 11 needle bearing (outer lip facing downward) onto the intermediate clutch hub.
9. Install the sun gear/shell assembly onto the overdrive drum assembly.

DIFFERENTIAL AND GEARSET (FINAL DRIVE)

Disassembly

1. Position the differential assembly in the vertical position (shaft facing upward).
2. Remove the snapring from the planetary pinion shaft.
3. Using a magnet, lift the planetary pinion shafts from the differential case housing. Remove the pinion gears and thrust washers from the differential case.
4. Inspect the pinion gear's needle bearings and shafts for wear and/or damage; replace them, if necessary.
5. From the top of the differential planetary assembly, remove the No. 17 needle bearing.
6. Move the differential assembly to the horizontal position. Using a drift punch, remove the roll pin from the differential

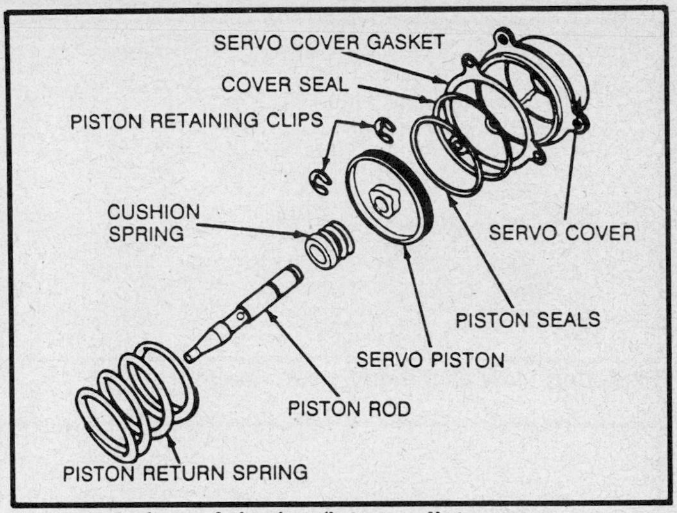
SERVO COVER GASKET
COVER SEAL
PISTON RETAINING CLIPS
CUSHION
SPRING
SERVO COVER
PISTON SEALS
SERVO PISTON
PISTON ROD
PISTON RETURN SPRING

Exploded view of the low/intermediate servo

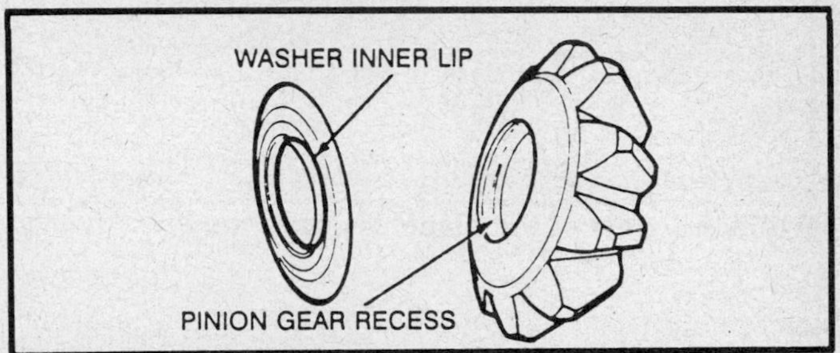
WASHER INNER LIP
PINION GEAR RECESS

View of the pinion gear and thrust washer

pinion shaft and the pinion shaft from the differential assembly.
7. Rotate the output shaft and remove the pinion gears and thrust washers.
8. Remove the right-hand side gear and thrust washer.
9. Move the output shaft toward the center of the differential housing and the left-hand side gear upward on the shaft (to gain access to the retaining ring). Remove the retaining ring and the output shaft from the differential housing. Remove the pinion gear and the thrust washer.

Inspection

OUTPUT SHAFT

1. Check the output shaft bearing surfaces for scoring and/or wear; if necessary, replace the shaft.
2. Check the output shaft splines for wear; if necessary, replace the shaft.
3. Inspect the shaft bushings for scoring and/or wear; if necessary, replace them.

PINION GEARS AND SHAFTS

1. Check the pinion gear and shafts for looseness and/or complete disengagement; be sure to check the shaft welds.
2. Check the pinion gears for freedom of rotation, damage and/or excessively worn teeth.

THRUST WASHERS

1. Using clean solvent, throughly clean the thrust bearings. Using compressed air, blow dry the bearings.
2. Check the bearings for pitting and/or roughness; replace them, if necessary.
3. Before installation, lubricate the cleaned bearings with clean transmission fluid.

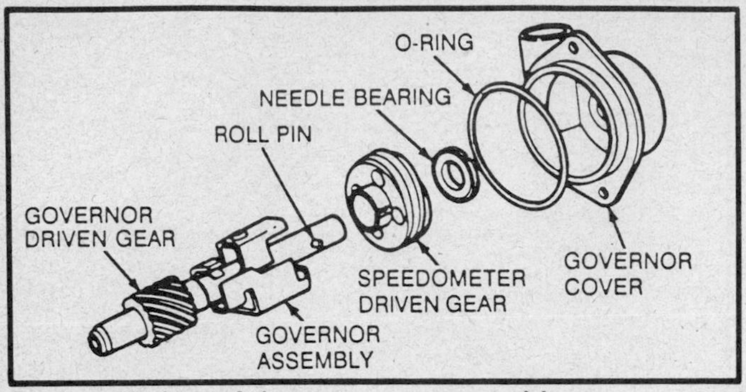
O-RING
NEEDLE BEARING
ROLL PIN
GOVERNOR DRIVEN GEAR
SPEEDOMETER DRIVEN GEAR
GOVERNOR COVER
GOVERNOR ASSEMBLY

Exploded view of the governor assembly

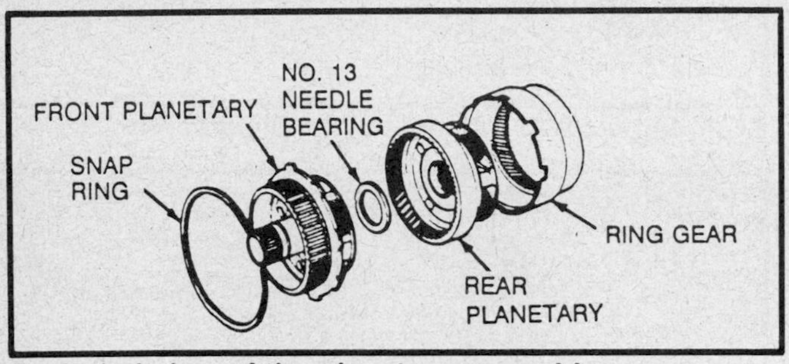
NO. 13 NEEDLE BEARING
FRONT PLANETARY
SNAP RING
RING GEAR
REAR PLANETARY

Exploded view of the planetary assembly

Assembly

1. Insert the output shaft into the differential case. Install the thrust washer, the left-hand side gear and the retaining ring onto the output shaft.
2. Install the thrust washer and the right-hand side gear into the differential case.
3. Seat the thrust washers (lips facing pinion gear) onto the pinion gears. Engage the pinion gear teeth with the side gears and rotate the output shaft to install the pinion gear assemblies into the differential housing.
4. After walking (turning the output shaft) the pinion gears into position, install the pinion shaft (tap it into position) and align the retaining pin hole with the differential housing. Using a drift punch, tap the roll pin into the differential housing/pinion shaft.
5. Install the No. 17 needle bearing (tabs facing upward) over the output shaft and seat it onto the differential housing.
6. Using a small amount of grease, install the needle bearings and the thrust washers onto the pinion gears.
7. Position the pinion gear assemblies into the differential housing and align the case holes. Install the pinion shafts through the differential case/gears until the lower shaft step is level with the differential case.
8. Install the retaining ring into the differential case grooves to hold the pinion shafts in place.

LOW/INTERMEDIATE SERVO

Disassembly

1. Remove the servo-to-transaxle cover bolts, the cover and the return spring.
2. From the cover, remove the piston and the rod.
3. From the piston assembly, remove the retaining clips, the rod and the cushion spring.
4. From the piston, remove the seal. From the cover, remove the seal and the gasket.

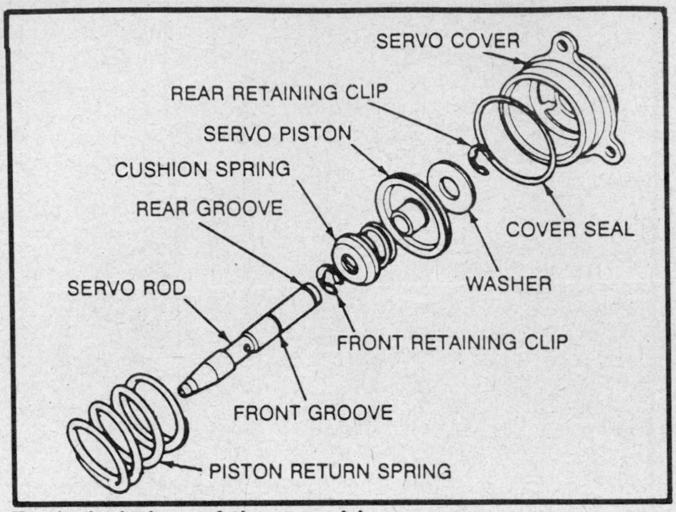
SERVO COVER
REAR RETAINING CLIP
SERVO PISTON
CUSHION SPRING
REAR GROOVE
COVER SEAL
SERVO ROD
WASHER
FRONT RETAINING CLIP
FRONT GROOVE
PISTON RETURN SPRING

Exploded view of the overdrive servo

Inspection

1. Remove the servo assembly-to-transaxle bolts and the servo assembly.
2. Check the servo body's for cracks, the piston bore for scores and the servo spring(s) for distortion.
3. Check the fluid passages for obstructions.
4. Check the band ends for cracks and the band lining for excessive wear and/or bond to the metal band. Check the band and struts for distortion.
5. Check and/or replace the damaged seals.
6. To install, lubricate the piston seal with petroleum jelly and the assembly into the case. Torque the servo cover screws to 7–9 ft. lbs. (9–12 Nm).

Assembly

1. Onto the piston rod, install the front retaining clip, the cushion spring and the piston.
2. Compress the assembly and install the rear retaining clip.
3. Using a new piston seal, install it onto the piston. Using a new cover seal and gasket, install them onto the servo cover.
4. Using petroleum jelly, lubricate the piston seals.
5. Install the piston assembly and return spring into the servo cover.
6. Install the servo cover assembly into the transaxle case; make sure the return spring is positioned correctly.
7. Install the servo cover-to-transaxle bolts and torque them to 7–9 ft. lbs. (9–12 Nm).

OVERDRIVE SERVO

Disassembly

1. Remove the overdrive servo cover-to-transaxle bolts and the cover.
2. From the cover, remove the servo piston.
3. From the piston rod, remove the rear retaining clip, the washer, the piston, the seal (from the piston), the cushion spring and the front retaining clip.
4. Remove the seal from the cover.

Inspection

1. Remove the servo assembly-to-transaxle bolts and the servo assembly.
2. Check the servo body's for cracks, the piston bore for scores and the servo spring(s) for distortion.

3. Check the fluid passages for obstructions.
4. Check the band ends for cracks and the band lining for excessive wear and/or bond to the metal band. Check the band and struts for distortion.
5. Check and/or replace the damaged seals.
6. To install, lubricate the piston seal with petroleum jelly and the assembly into the case. Torque the servo cover screws to 7–9 ft. lbs. (9–12 Nm).

Assembly

1. Onto the piston rod, install the front retaining clip, the cushion spring, the piston and the washer.
2. Compress the assembly and install the rear retaining clip.
3. Using a new piston seal, install it onto the piston. Using a new cover seal, it onto the servo cover.
4. Using petroleum jelly, lubricate the piston seal.
5. Install the piston assembly and return spring into the transaxle; make sure the return spring is positioned correctly.
6. Install the servo cover-to-transaxle bolts and torque them to 7–9 ft. lbs. (9–12 Nm).

GOVERNOR

Disassembly

1. Remove the governor-to-transaxle bolts and the cover.
2. Remove the cover seal and discard it.
3. Remove the governor assembly from the transaxle and disassemble the speedometer drive gear bearing, the speedometer drive gear and the governor assembly.

Inspection

1. Remove the governor-to-transaxle bolts and the governor assembly from the transaxle assembly.
2. Check the governor shaft seal for cracks, scoring and/or cuts.
3. Check the balance weight retaining pin for wear and the spring for distortion, damage or misalignment.
4. Check the pressure balls for scoring and free movement.
5. Check the governor drive, the driven gear and the speedometer drive gear for broken, chipped or worn teeth; replace, if necessary.
6. To install, place the assembly into the case bore, align the driven gear with the speedometer gear teeth and seat the assembly in the bore. Torque the governor cover-to-case bolts to 7–9 ft. lbs. (9–12 Nm).

Assembly

1. Onto the governor shaft, push the speedometer drive gear and align its slots with the roll pin.
2. Onto the speedometer drive gear, install the speedometer drive gear bearing with the outer race (black side) facing upward.
3. Using a new seal, install it onto the cover.
4. Install the assembly into the transaxle and torque the cover-to-transaxle bolts to 7–9 ft. lbs. (9–12 Nm).

PLANETARY ASSEMBLY

Disassembly

1. Remove the snapring, the front planetary and the No. 13 needle bearing from the planetary assembly.
2. From the shell/ring gear assembly, remove the rear planetary.

Inspection

The individual planetary carrier parts are not servicable, except for the differential.

1. Check the planetary assembly's pins/shafts for looseness and/or complete disengagement; be sure to check the shaft welds. If replacement is necessary, install a new planetary assembly.
2. Check the pinion gears for freedom of rotation, damage and/or excessively worn teeth.

Assembly

1. Install the rear planetary assembly into the shell/ring gear assembly.
2. Install the No. 13 needle bearing, the front planetary and the snapring into the planetary assembly.

DRIVEN SPROCKET SUPPORT

Disassembly

Using the stator/driven sprocket bearing remover tool or equivalent, and a slide hammer, pull the needle bearing from the driven sprocket support.

Assembly

Using the Stator/Driven Sprocket Bearing Replacer tool or equivalent, press the needle bearing into the driven sprocket support.

Transaxle Assembly

UNIT ASSEMBLY

1. Place the transaxle case if the horizontal position.
2. If the drive sprocket support was removed from the torque converter housing, install it and torque the bolts to 7–9 ft. lbs. (9–12 Nm).

NOTE: Since the drive sprocket support bolts holes are offset, it can only be aligned in one direction.

3. Using the converter oil seal replacer tool or equivalent, install a new oil seal into the front of the torque converter housing; be sure the seal is equipped with a garter spring.
4. Using the output shaft seal replacer tool or equivalent, install a new oil seal into the right-side halfshaft opening; be sure the seal is equipped with a garter spring.
5. Using the AXOD endplay tool or equivalent, the step plate adapter tool or equivalent, and 2 bolts, mount the tool assembly of the right-side halfshaft opening; this assembly will be used later to perform selective thrust washer checks.
6. Turn the case so the left-side is facing upward. Install the No. 19 needle bearing (flat side facing upward and outer lip facing down) on the case boss.
7. Install the final drive ring gear (external splines facing upward) into the case; it may be necessary to use a hammer handle (tap gently) to seat the gear into the case splines.
8. Onto the final drive assembly, assemble the governor drive gear, the final drive sun gear, the paring gear, the No. 16 needle bearing, the rear planetary support, the No. 15 needle bearing and the No. 18 thrust washer.
9. Install the final drive assembly into the case and secure with the snapring; align the snapring end with the low/intermediate band anchor pin.
10. To perform the end clearance check on the No. 18 thrust washer, perform the following procedures:
   a. Using a small pry bar, place it under the differential assembly and pry upward.
   b. Using a dial indicator, attach it to the transaxle case, rest the stylus on the final drive's output shaft.
   c. Locate the AXOD tool assembly (right-side output opening) and back off the adjusting bolt until it no longer touches the shaft.
   d. Zero the dial indicator.

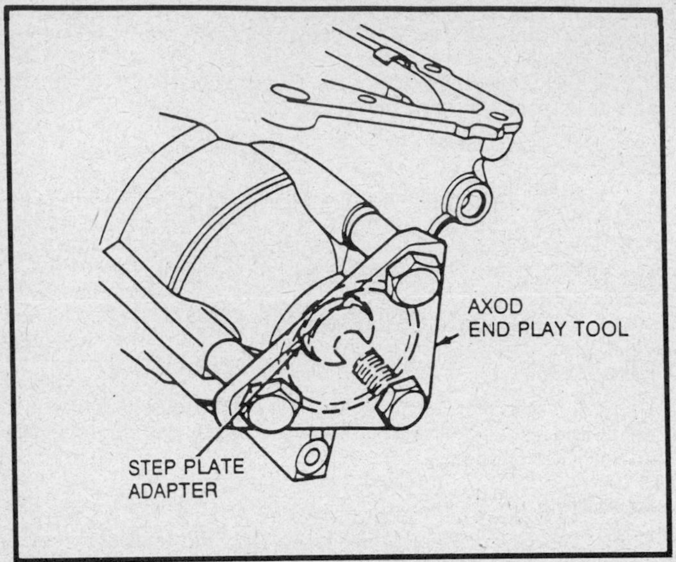
AXOD
END PLAY TOOL
STEP PLATE
ADAPTER

View of the adjustment tools installed in the right-side halfshaft opening

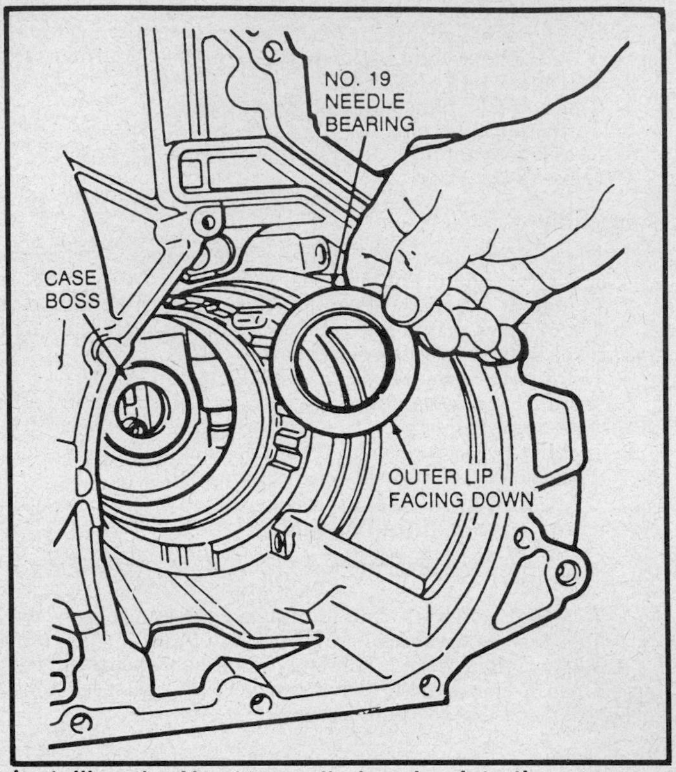
NO. 19
NEEDLE
BEARING
CASE
BOSS
OUTER LIP
FACING DOWN

Installing the No. 19 needle bearing into the case

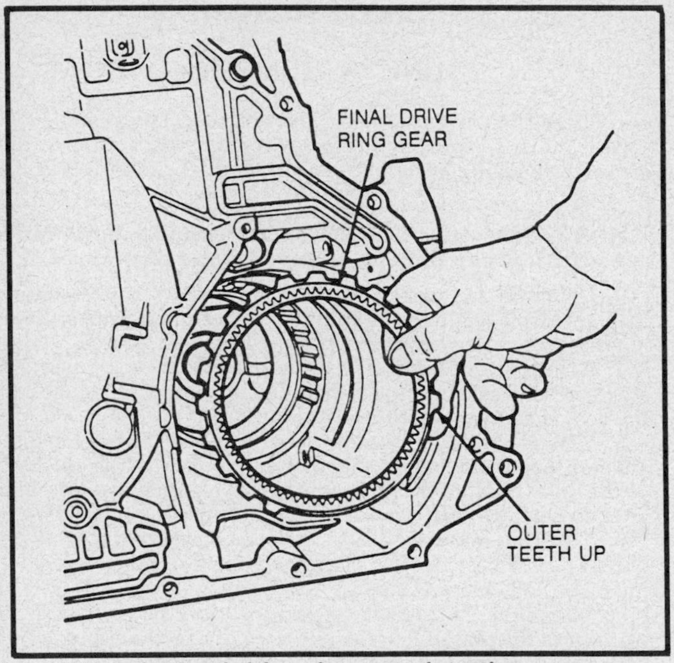
FINAL DRIVE
RING GEAR
OUTER
TEETH UP

Installing the final drive ring gear into the case

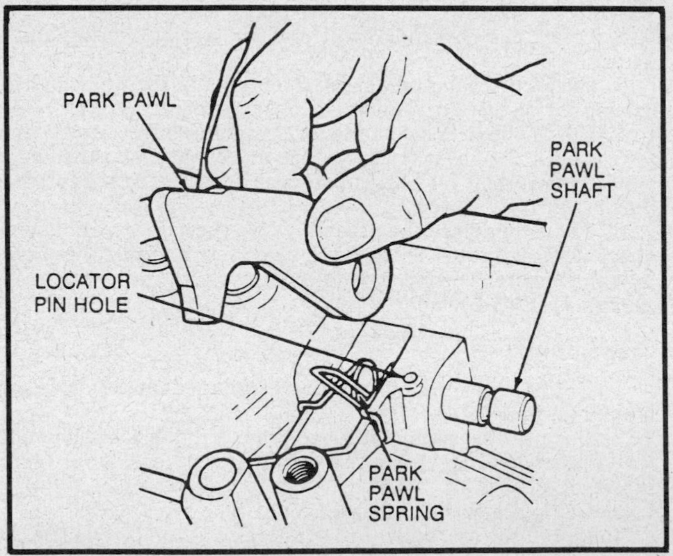
PARK PAWL
PARK
PAWL
SHAFT
LOCATOR
PIN HOLE
PARK
PAWL
SPRING

Installing the park pawl assembly

e. Torque the tool's bolt to 35–44 inch lbs. (4–5 Nm) and observe the dial indicator; the clearance should be 0.004–0.025 in. (0.1–0.65mm). If the measurement is not within specifications, the thrust washer must be replaced.

Thrust washer sizes:
Orange – 0.045–0.049 in. (1.15–1.25mm)
Purple – 0.055–0.059 in. (1.50–1.65mm)
Yellow – 0.065–0.069 in. (1.65–1.75mm)

f. After installing another thrust washer, recheck the end clearance.

11. At the parking gear, install the park pawl, the return spring, the park pawl shaft and the locator pin; be sure the park pawl engages the park gear and returns freely.

12. Install the park rod actuating lever/park rod into the case and the park rod abutment (start the bolts). Push the park pawl inward and locate the rod between the pawl and abutment.

13. Using a 3/8 in. drift (rod), push the lube tube seal (rubber end first) into the case (oil pan side) until it is flush.

14. At the side cover end of the case, install the low/intermediate band and align the anchor pin pocket with the anchor pin. Install the low/intermediate drum and sun gear assembly over the output shaft.

15. Assemble the planetary assembly components: the ring gear/shell assembly, the rear planetary, the No. 13 needle bearing, the front planetary and snapring; carefully slide the planetary assembly over the output shaft.

16. Install the reverse clutch into the case and engage the

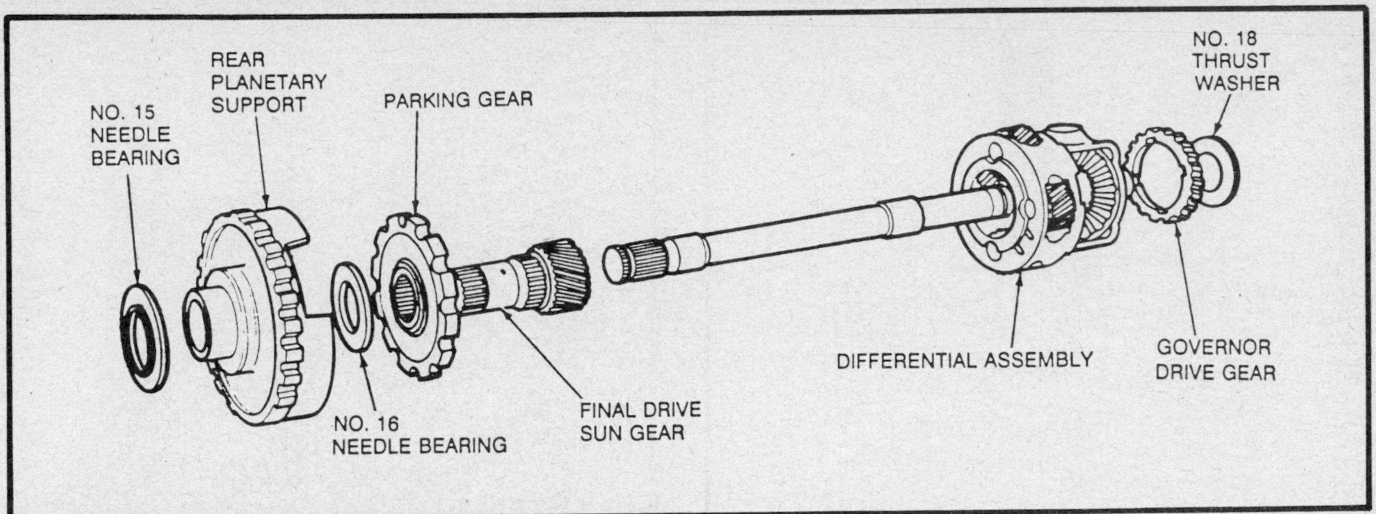
NO. 15
NEEDLE
BEARING
REAR
PLANETARY
SUPPORT
PARKING GEAR
NO. 18
THRUST
WASHER
DIFFERENTIAL ASSEMBLY
GOVERNOR
DRIVE GEAR
NO. 16
NEEDLE BEARING
FINAL DRIVE
SUN GEAR
Exploded view of the final drive/planetary assembly

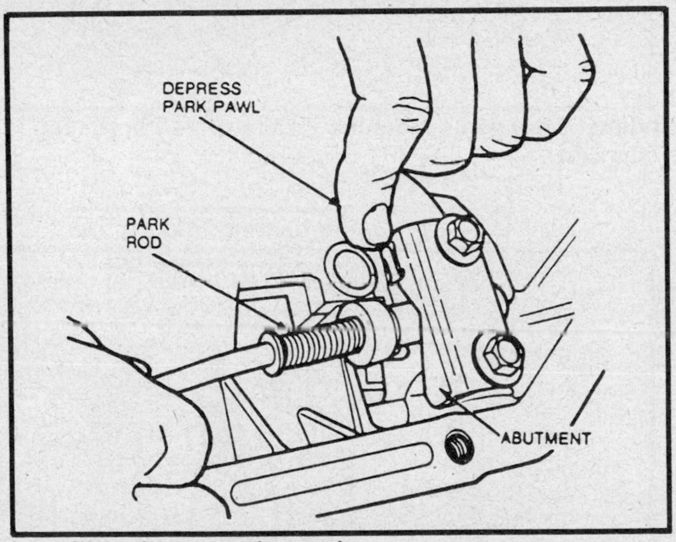
DEPRESS
PARK PAWL
PARK
ROD
ABUTMENT
Installing the actuating rod

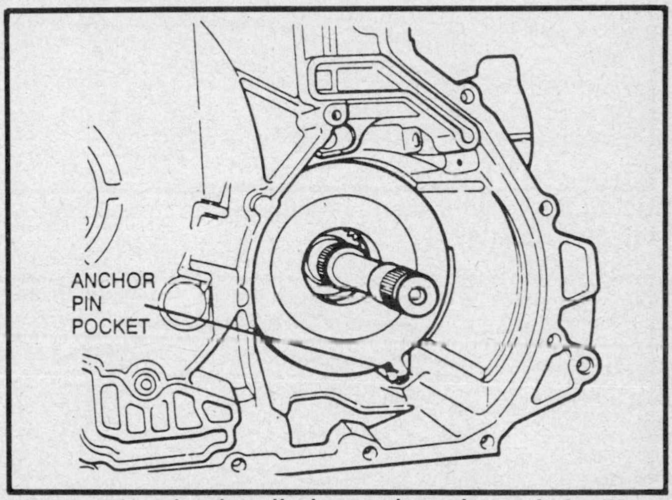
ANCHOR
PIN
POCKET
Aligning the clutch cylinder anchor pin pocket with the anchor pin case hole

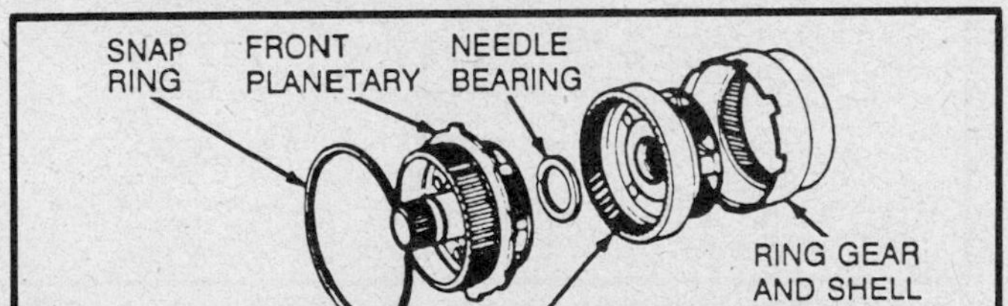
SNAP
RING
FRONT
PLANETARY
NEEDLE
BEARING
RING GEAR
AND SHELL
ASSEMBLY
REAR
PLANETARY
View of the planetary assembly components

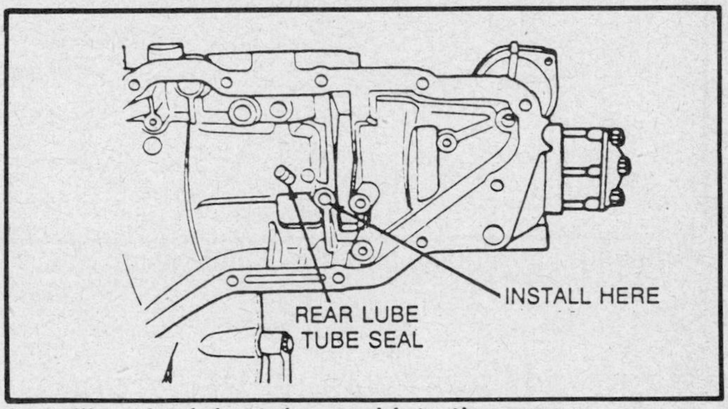
INSTALL HERE
REAR LUBE
TUBE SEAL
Installing the lube tube seal into the case

clutch plate by aligning the clutch cylinder anchor pin pocket with the anchor pin case hole.

17. While installing the intermediate clutch hub (when engaged with the planetary shaft splines), rotate the hub to seat the reverse clutch.

18. From the oil pan side, start the reverse anchor pin bolt but do not tighten.

19. After assembling the forward, direct and intermediate clutch assembly, attach the front clutch loading tool or equivalent, to the assembly (using a lube hole). Lower the assembly into the case and align the shell/sun gear splines with the forward planetary; be sure the assembly is fully assembled before removing the tool.

20. Install the overdrive band and the plastic retainer (cross hairs facing upward).

21. To check the drive sprocket end clearance for the No. 5 and

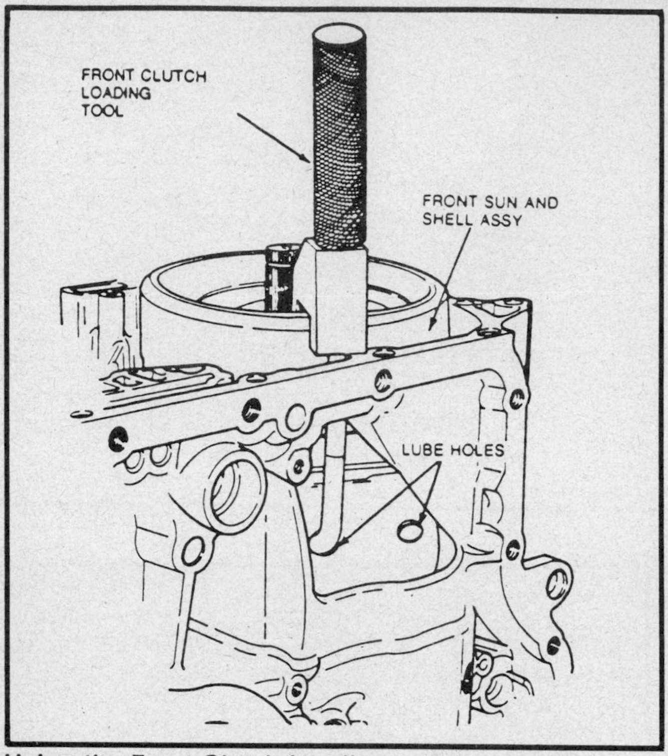
FRONT CLUTCH LOADING TOOL
FRONT SUN AND SHELL ASSY
LUBE HOLES
Using the Front Clutch Loading tool to install the clutch assembly

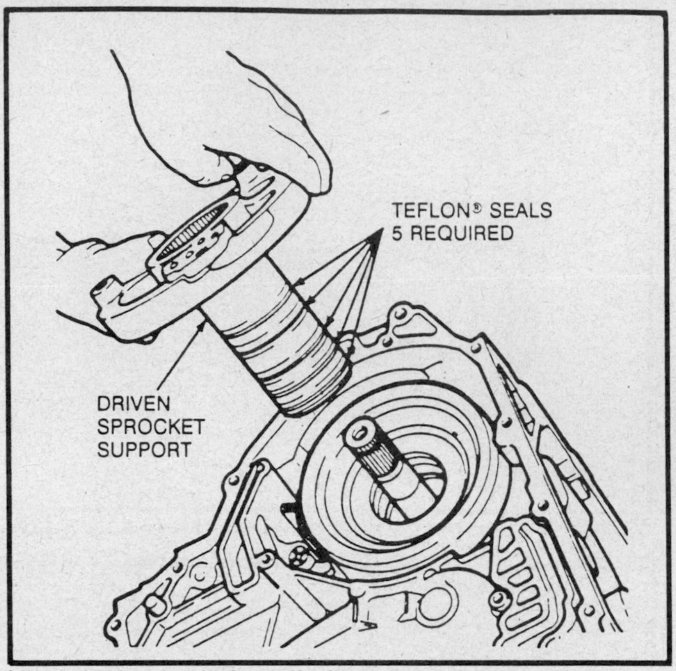
TEFLON® SEALS 5 REQUIRED
DRIVEN SPROCKET SUPPORT
View of the seals installed on the driven sprocket support

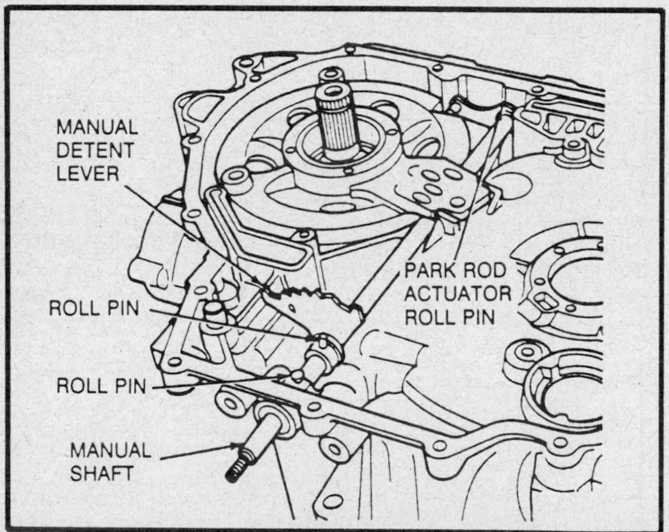
MANUAL DETENT LEVER
PARK ROD ACTUATOR ROLL PIN
ROLL PIN
ROLL PIN
MANUAL SHAFT
Installing the roll pins into the manual shaft

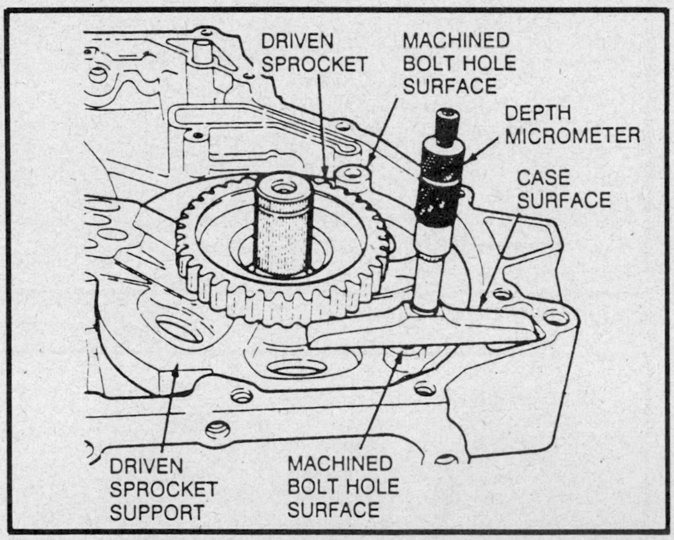
DRIVEN SPROCKET
MACHINED BOLT HOLE SURFACE
DEPTH MICROMETER
CASE SURFACE
DRIVEN SPROCKET SUPPORT
MACHINED BOLT HOLE SURFACE
Using a depth micrometer to check difference of the driven sprocket support-to-case machined surfaces

NO. 8 THRUST WASHER SELECTION

Thrust Washer Thickness		
Inches	mm	Color
0.060–0.056	1.53–1.43	Natural
0.070–0.066	1.78–1.68	Dark green
0.079–0.075	2.02–1.92	Light blue
0.089–0.085	2.27–2.17	Red

NO. 5 THRUST WASHER SELECTION

Thrust Washer Thickness		
Inches	mm	Color
0.090–0.086	2.28–2.18	Green
0.099–0.095	2.53–2.43	Black
0.109–0.105	2.77–2.67	Natural
0.118–0.115	3.02–2.92	Red

NO. 4
THRUST
WASHER
NO. 2
THRUST
WASHER

Installing the No. 2 and No. 4 thrust washers onto the case supports

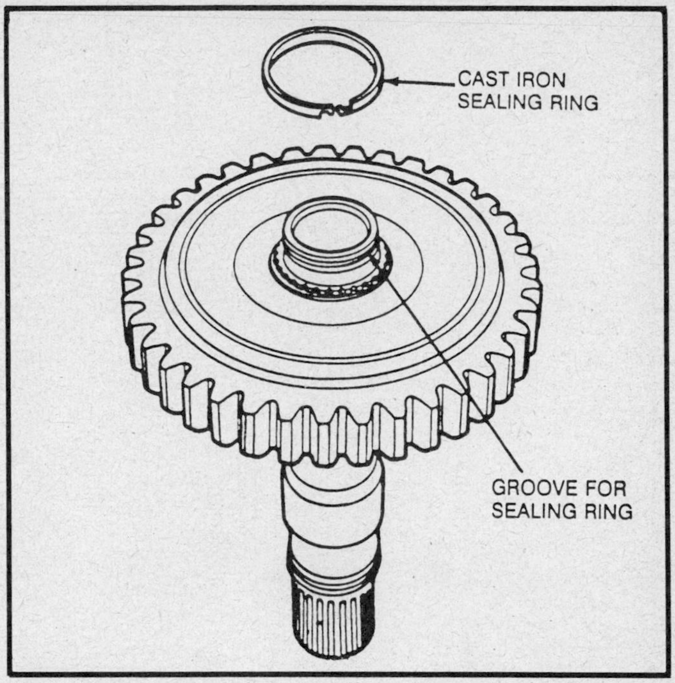
CAST IRON
SEALING RING
GROOVE FOR
SEALING RING

Installing the case iron sealing ring onto the input shaft

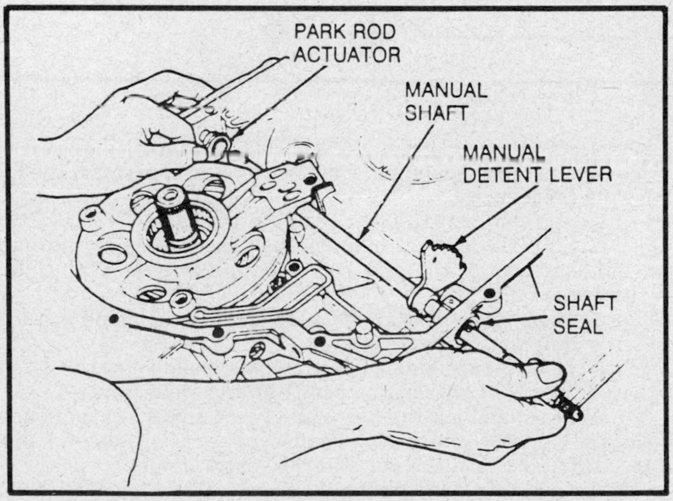
PARK ROD
ACTUATOR
MANUAL
SHAFT
MANUAL
DETENT LEVER
SHAFT
SEAL

Installing the manual shaft into the transaxle case

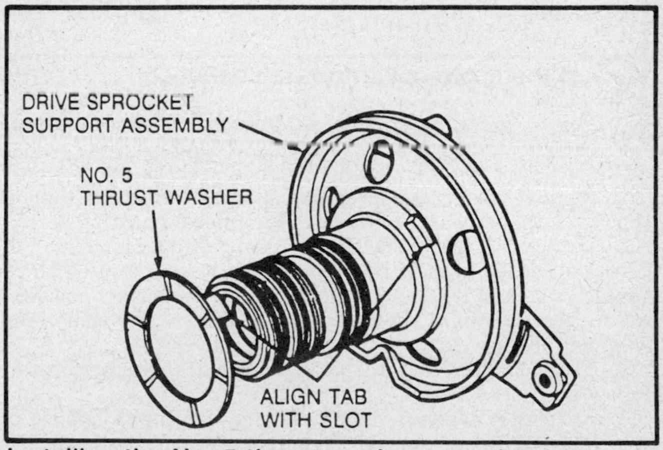
DRIVE SPROCKET
SUPPORT ASSEMBLY
NO. 5
THRUST WASHER
ALIGN TAB
WITH SLOT

Installing the No. 5 thrust washer onto the drive sprocket support assembly

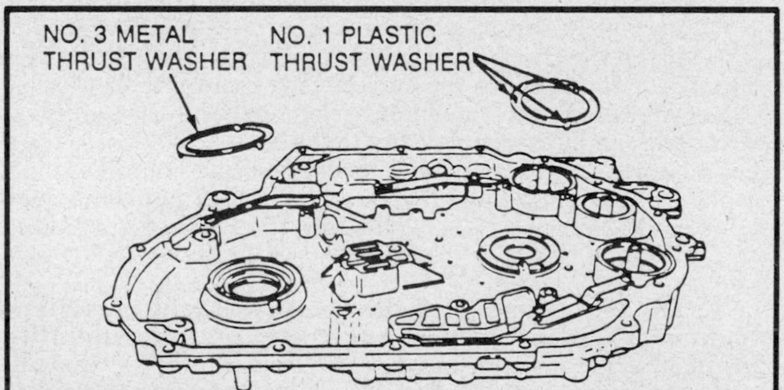
NO. 3 METAL
THRUST WASHER
NO. 1 PLASTIC
THRUST WASHER

Installing the No. 1 and No. 3 thrust washers onto the chain cover

No. 8 thrust washers, perform the following procedures:

a. At the right-side output shaft, tighten the endplay check tool's screw.

b. From the driven sprocket support assembly, remove the 5 Teflon® seals (if not already removed).

c. Over the output shaft, install the No. 9 needle bearing (outer lip facing upward) and the No. 8 thrust washer.

d. If the No. 5 thrust washer is attached to the sprocket support, remove it.

e. Install the driven sprocket support and the driven sprocket to the case.

f. Determine if the machined bolt hole surfaces of the driven sprocket support are above or below the case. Using a depth micrometer, position it on the machined bolt hole surface (support hole surfaces above case) or on the case (support hole surfaces below case) and measure the averaged difference of the 2 surfaces (at both machined hole surfaces); if the average exceeds (support hole surfaces above case) 0.008 in.

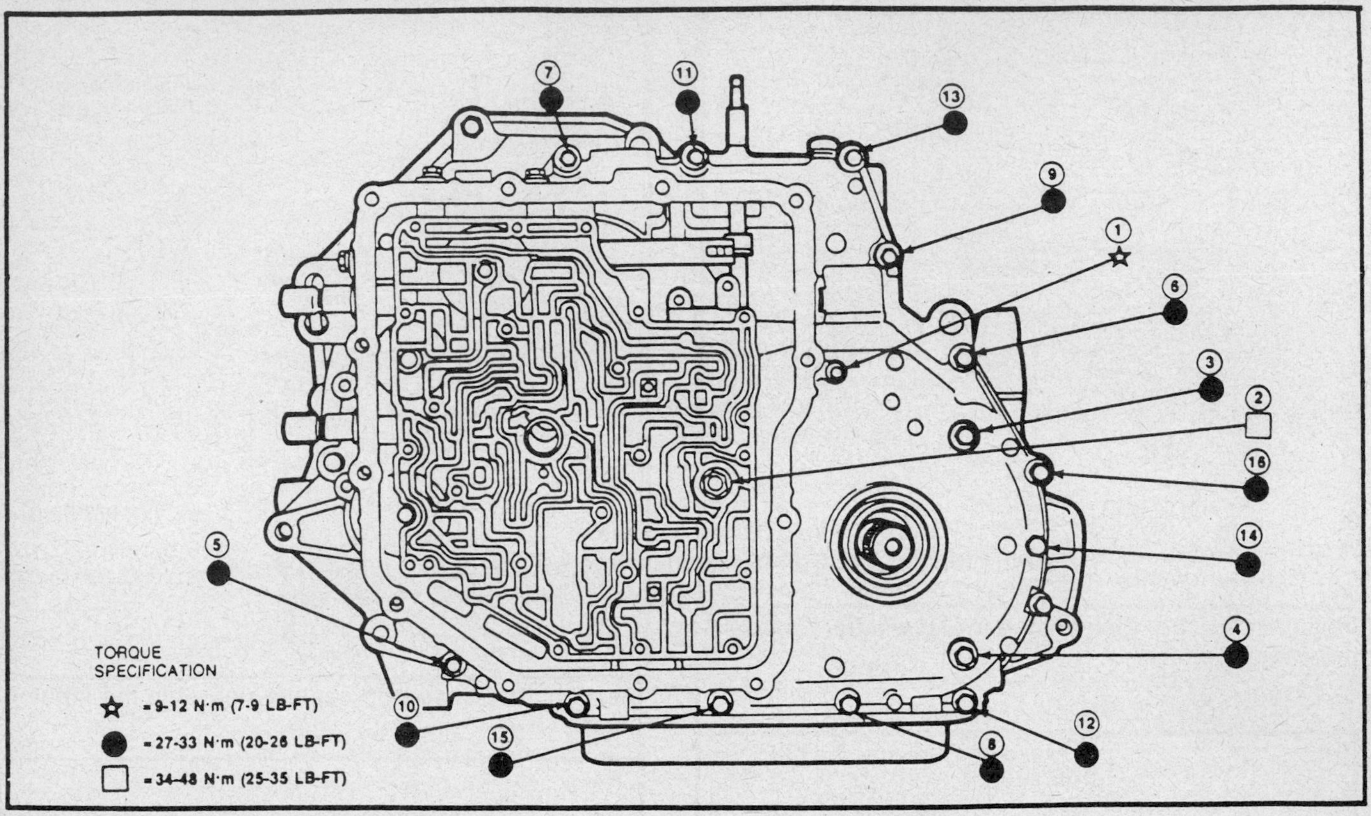
TORQUE
SPECIFICATION
= 9-12 N·m (7-9 LB-FT)
= 27-33 N·m (20-26 LB-FT)
= 34-48 N·m (25-35 LB-FT)

View of the chain cover torque sequence

(0.21mm) or (support hole surfaces below case) 0.018 in. (0.46mm), replace the No. 8 thrust washer.

g. After replacing the No. 8 washer, recheck the clearance.

22. Remove the driven sprocket, the driven sprocket support, the No. 8 thrust washer and the No. 9 needle bearing.

23. Using grease, coat the No. 5 thrust washer, install it on the driven sprocket support by aligning the washer's tab with the support's slot. While leaving the No. 8 thrust washer and No. 9 needle bearing out of the assembly, install the driven sprocket support into the case.

24. To check the No. 5 thrust washer thickness, perform the following procedures:

a. Using a depth micrometer, measure the difference between the driven sprocket support surface (at both machined bolt holes) and the case surface; average the readings.

NOTE: The following procedure should only be used if the driven sprocket support surface is below the case surface.

b. Using the averaged reading of the No. 5 thrust washer, subtract the averaged reading of the No. 8 thrust washer; the difference should be 0–0.033 in. (0–85mm). If the difference exceeds specifications, select the correct thrust washer from the No. 5 thrust washer chart and install it.

NOTE: The following procedure should only be used if the driven sprocket support surface is above the case surface.

c. Using the averaged reading of the No. 5 thrust washer, add the averaged reading of the No. 8 thrust washer; the difference should be 0–0.033 in. (0–85mm). If the difference exceeds specifications, select the correct thrust washer from the No. 5 thrust washer chart and install it.

25. Pull the driven sprocket support from the case. Install the No. 9 needle bearing, the correct No. 8 thrust washer, the Teflon® seals and the correct No. 5 thrust washer (using grease).

26. To install the manual shaft into the case, perform the following procedures:

a. Tap the manual shaft seal into the case.

b. Slide the manual shaft through the shaft seal, the manual detent lever, the park rod actuating lever and tap it into the case hole.

c. Align the shaft's groove with the case hole and install a new lock pin.

d. Using new manual shaft roll pins, install 1 at the detent lever and the other at the park rod actuating lever.

27. When installing the No. 2 and No. 4 thrust washers onto the drive and driven sprocket supports, grease them and align the their tabs with the holes in the sprocket supports.

28. Lubricate and install the case iron sealing ring onto the input shaft. Assemble the chain onto the drive and driven sprockets and install the assembly into the sprocket supports; when installing, rotate the sprockets to be sure they are fully seated.

29. Onto the chain cover, align (tabs with slots) and install the No. 1 and No. 3 thrust washers.

30. Install the chain cover gasket onto the case. Install the accumulator springs into the correct case positions.

31. Carefully lower the chain cover onto the case, align the cover pins, apply gentle pressure to the cover to compress the accumulator springs and install the cover-to-case bolts. Torque (in sequence) the 8mm bolts to 7–9 ft. lbs. (9–12 Nm), the 10mm bolts to 20–26 ft. lbs. (27–33 Nm) and the 13mm bolts to 25–35 ft. lbs. (34–48 Nm).

NOTE: When installing the chain cover, be careful not to damage the input shaft's cast iron sealing ring. If the input shaft does not have some endplay after installation, remove the cover and inspect the cast iron seal for damage.

32. Torque the park rod abutment bolts to 20–22 ft. lbs. (27–

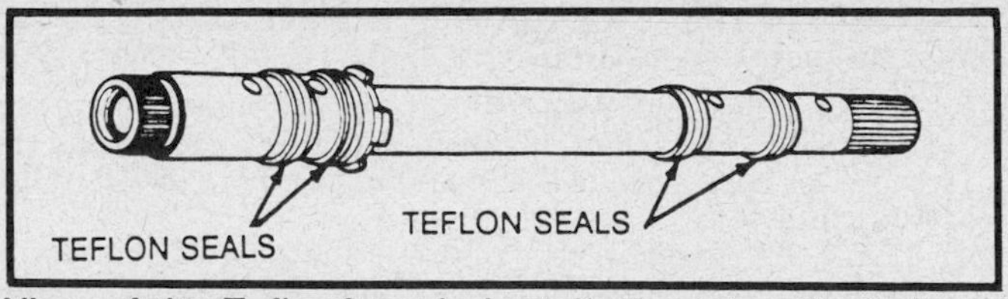
TEFLON SEALS
TEFLON SEALS
View of the Teflon® seals installed on the oil shaft

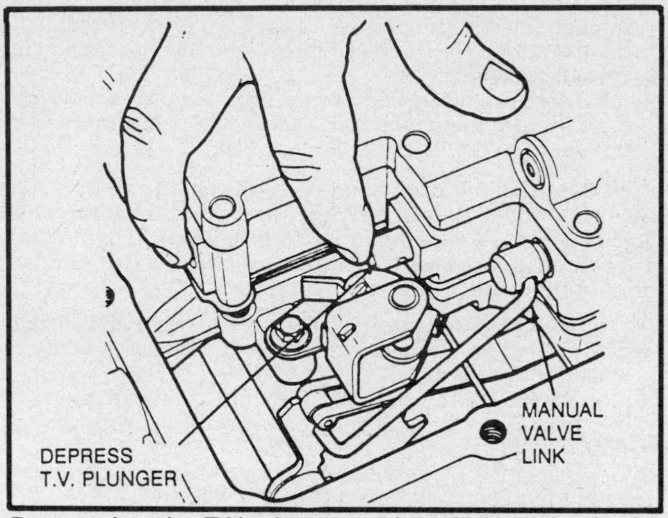
DEPRESS
T.V. PLUNGER
MANUAL
VALVE
LINK
Depressing the T.V. plunger to install the valve body

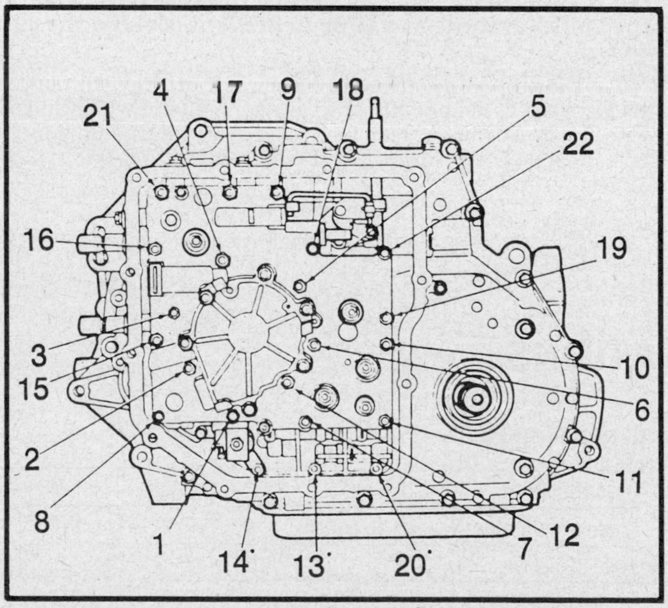
4 17 9 18
21
5
22
16
19
3
15
10
6
2
11
8
1
14
13
20
7
12
View of the valve body bolt torquing sequence

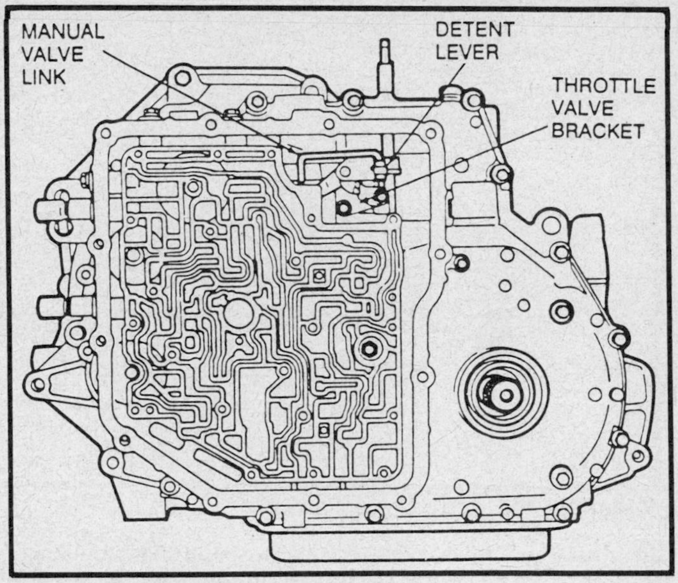
MANUAL
VALVE
LINK
DETENT
LEVER
THROTTLE
VALVE
BRACKET
View of the throttle valve assembly

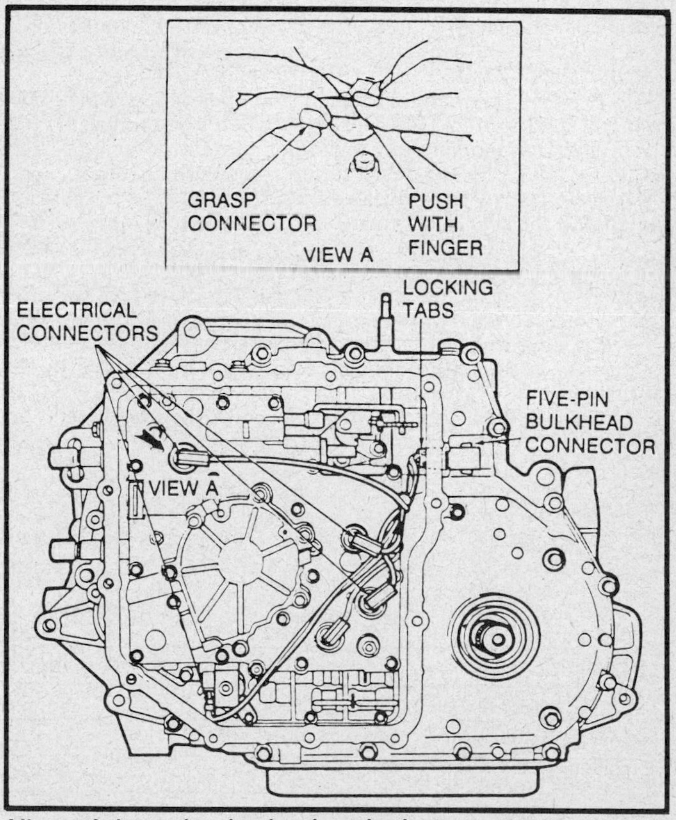
GRASP
CONNECTOR
PUSH
WITH
FINGER
VIEW A
ELECTRICAL
CONNECTORS
LOCKING
TABS
FIVE-PIN
BULKHEAD
CONNECTOR
VIEW A
View of the valve body electrical connectors

30 Nm), the reverse drum 6mm Allen® anchor bolt to 7.5–9 ft. lbs. (10–12 Nm) and the 19mm locknut to 25–35 ft. lbs. (34–47 Nm).

33. At the oil pan side, lightly tap the lube tubes into position; using threadlock compound, coat the lube tube-to-case surfaces. Install the tube retaining brackets.

34. Install 2 new O-rings onto the oil filter and push the filter into the case. Install the reverse apply tube/oil filter bracket.

35. To install the oil pan, use a new gasket and the pan. Torque the pan-to-case bolts to 10–12 ft. lbs. (14–16 Nm).

36. Onto the oil pump shaft, install new Teflon® seals and install the shaft.

37. Through the case hole near the side cover, install the T.V. bracket with the T.V. link and torque the bolts to 7–9 ft. lbs. (9–12 Nm). Connect the manual valve link to the detent lever.

38. While installing the oil pump/valve body over the oil pump shaft, attach the manual valve link to the manual valve. Depress the T.V. plunger (to clear the T.V. bracket) and seat the oil pump/valve body.

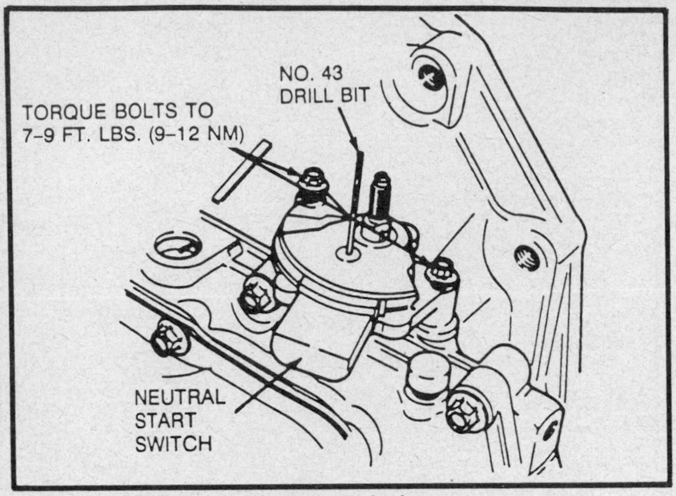
NO. 43
DRILL BIT
TORQUE BOLTS TO
7–9 FT. LBS. (9–12 NM)
NEUTRAL
START
SWITCH

Installing the neutral start switch

39. Install the valve body-to-chain case bolts and torque (in sequence) to 7–9 ft. lbs. (9–12 Nm); be sure the short bolts are installed in their correct location.
40. To the chain cover, install the bulkhead connector and secure with the locking tabs. Attach the electrical connectors to the switches and solenoid; the connectors are secure when a slight "click"is felt.
41. Position the manual shaft in the **N** position and install the neutral start switch. Using a No. 43 drill bit or equivalent, position it in the neutral start switch alignment hole. Torque the bolts to 7–9 ft. lbs. (9–12 Nm) and remove the drill bit.
42. Using a new self-adhesive gasket, install it on the side cover. Install the side cover and torque the bolts to 7–9 ft. lbs. (9–12 Nm).
43. Move the transaxle to the horizontal position.
44. To inspect, adjust and install the overdrive and low/intermediate servo travel, perform the following procedures:
  a. Using the spring from the overdrive servo rod tool, low/intermediate servo rod tool or equivalent, install it in the case bore.
  b. Install the servo piston and rod in the case bore.

NOTE: On the low/intermediate servo, the piston must be installed without the seal.

  c. Using the overdrive servo rod tool, low/intermediate servo rod tool or equivalent, install it in the case bore. Using the servo cover bolts, torque the tool(s)-to-case bore to 7–9 ft. lbs. (9–12 Nm).
  d. For the overdrive servo, torque the gauge disc screw to 10 inch lbs. (1.13 Nm). For the low/intermediate servo, torque the gauge disc screw to 30 inch lbs. (3.4 Nm).
  e. Using a dial indicator, mount and position the stylus through the disc gauge hole. Be sure the stylus is contacting the flat surface of the servo piston and not the piston step. Zero the dial indicator.
  f. Loosen the gauge disc screw until the piston movement stops. Read the dial indicator; the amount of piston travel determines the rod length to be installed.

NOTE: The overdrive servo reading should be 0.070–0.149 in. (1.8–3.8mm); the low/intermediate servo reading should be 0.216–0.255 in. (5.5–6.5mm). If a new low/intermediate band has been installed, the reading should be 0.196–0.236 in. (5–6mm).

  g. Using the measurement acquired, select and install a new piston rod. Recheck the piston movement to verify the amount of piston travel.
  h. On the low/intermediate servo piston, install the seal(s).
  i. Install the servo pistons and springs; be sure they are fully seated.
  j. Using new gaskets, install the servo-to-case bore covers and torque the bolts to 7–9 ft. lbs. (9–12 Nm).

NOTE: When installing the low/intermediate servo cover, be sure the tab aligns with the case port. Tighten the bolts 2–3 turns at a time to prevent cocking the case cover.

45. To install the governor, use a new seal on the governor cover, place the assembly into the case bore, align the driven gear with the speedometer gear teeth and seat the assembly in the bore. Torque the governor cover-to-case bolts to 7–9 ft. lbs. (9–12 Nm).

SPECIFICATIONS

TORQUE SPECIFICATIONS

Description	ft. lbs.	Nm
Separator plate to main control	7–9	9–12
Separator plate to pump body	7–9	9–12
Detent spring to chain cover	7–9	9–12
Dust cover to case	7–9	9–12
T.V. control lever to chain cover	7–9	9–12
Solenoid to main control	7–9	9–12
Low-intermediate servo cover to case	7–9	9–12
Overdrive servo cover to case	7–9	9–12
Pump cover to pump body	7–9	9–12
Filler tube to case	7–9	9–12
Governor cover to case	7–9	9–12

Description	ft. lbs.	Nm
Case to stator support	7–9	9–12
Case to chain cover (10mm)	7–9	9–12
Oil pump assembly to main control	7–9	9–12
Neutral start switch to case	7–9	9–12
Valve body/solenoid to chain cover	7–9	9–12
Bracket tubes to case	7–9	9–12
T.V. cable to case	6–9	8–12
Chain cover to case (10mm)	7–9	9–12
Pump body to chain cover	7–9	9–12
Oil pan to case (lower reservoir)	10–12	14–16

TORQUE SPECIFICATIONS

Description	ft. lbs.	Nm
Main control cover to chain cover (upper reservoir)	10–12	14–16
Manual lever to manual shaft	12–16	16–22
Park abutment to case	20–22	27–30
Chain cover to case (13mm)	20–22	27–30
Case to chain cover (13mm)	24.3–26.6	33–36
Chain cover to front support (13mm)	20–22	27–30
Chain cover to front support (7mm)	25–35	34–48
Differential brace to case	25–35	34–48
Engine to case/case to engine	41–50	55–68
Case to reverse clutch screw	7-9	10–12
Case to reverse clutch nut	25–35	34–47
Pressure tap plug for chain cover and pump body	6-9	8–12

Description	ft. lbs.	Nm
Pressure switch to pump body	6–9	8–12
Transaxle to engine	41–50	55–68
Control arm to knuckle	36–44	50–60
Stabilizer U-clamp to bracket	60–70	81–95
Stabilizer to control arm	98–125	133–169
Brake hose routing clip	8	11
Tie rod to knuckle	23–35	31–47
Manual cable bracket	10–20	14–27
Starter	30–40	41–54
Dust cover	7–9	9–12
Torque converter to flywheel	23–39	31–53
Insulator to bracket	55–70	75–90
Insulator bracket to frame	40–50	55–70
Insulator mount to transmission	25–33	34–45

SPECIAL TOOLS

Tool Number	Description
D79P-100-A	Slide hammer—universal
T59L-100-B	Impact slide hammer
T58L-101-A	Impact slide hammer
T57L-500-B	Bench Mount Holding Fixture
D80L-515-S	Puller screw
D80L-522-A	Gear and pully support bar
D80L-625-A	Shaft protector
D80L-630-3	Step plate adapter
T00L-1175-AC	Seal remover
T86P-1177-B	Output shaft seal replacer
D81P-3504-N	Locknut pin remover
T86P-3514-A1	C.V. joint puller
T86P-3514-A2	Screw extension
T00L-4201-C	Dial indicator
D79P-6000-A	Engine support bar
D81L-6001-D	Engine lifting bracket
T74P-6700-A	Output shaft seal remover
T77L-7902-A	Holding wire
T80L-7902-A	End play checking tool
T80L-7902-C	End play checking tool
T81P-7902-B	One-way clutch torque tool
T81P-7902-C	Torque converter handles
T86P-7902-A	Converter guide sleeve tool
T86P-70001-A	Lube tube remover tool

Tool Number	Description
T86P-70023-B	Overdrive servo rod tool
T86P-70023-A	Low/Intermediate servo rod tool
T86P-70043-A	Stator and driven sprocket bearing remover
T86P-70043-B	Stator and driven sprocket bearing replacer
T86P-70100-A	Valve body guide pin set
T86P-70100-B	Guide pin
T86P-70100-C	Valve body guide pin
T86P-70234-A	Direct clutch lip seal protector
T86P-70234-A	Output shaft seal replacer
T86P-70370-A	Pump body guide pin
T86P-70373-A	Direct/intermediate clutch bushing replacer
T86P-70389-A	Front clutch loading tool
T86P-70401-A	Converter oil seal replacer
T86P-70403-A	Reverse clutch outer lip seal protector
T86P-70422-A	Bimetal height gauge
T86P-70423-A	Direct clutch bearing replacer
T86P-70548-A	Forward clutch seal lip protector
T86P-77265-AH	Cooler line disconnect tool
T65L-77515-A	Clutch spring compressor
T81P-78103-A	Slide hammer adapter
	ROTUNDA EQUIPMENT
021-00047	Torque converter leak test kit
014-00737	Automatic transmission tester kit
014-00028	Torque converter and oil cooler cleaner

THRUST WASHER AND NEEDLE BEARING LOCATION

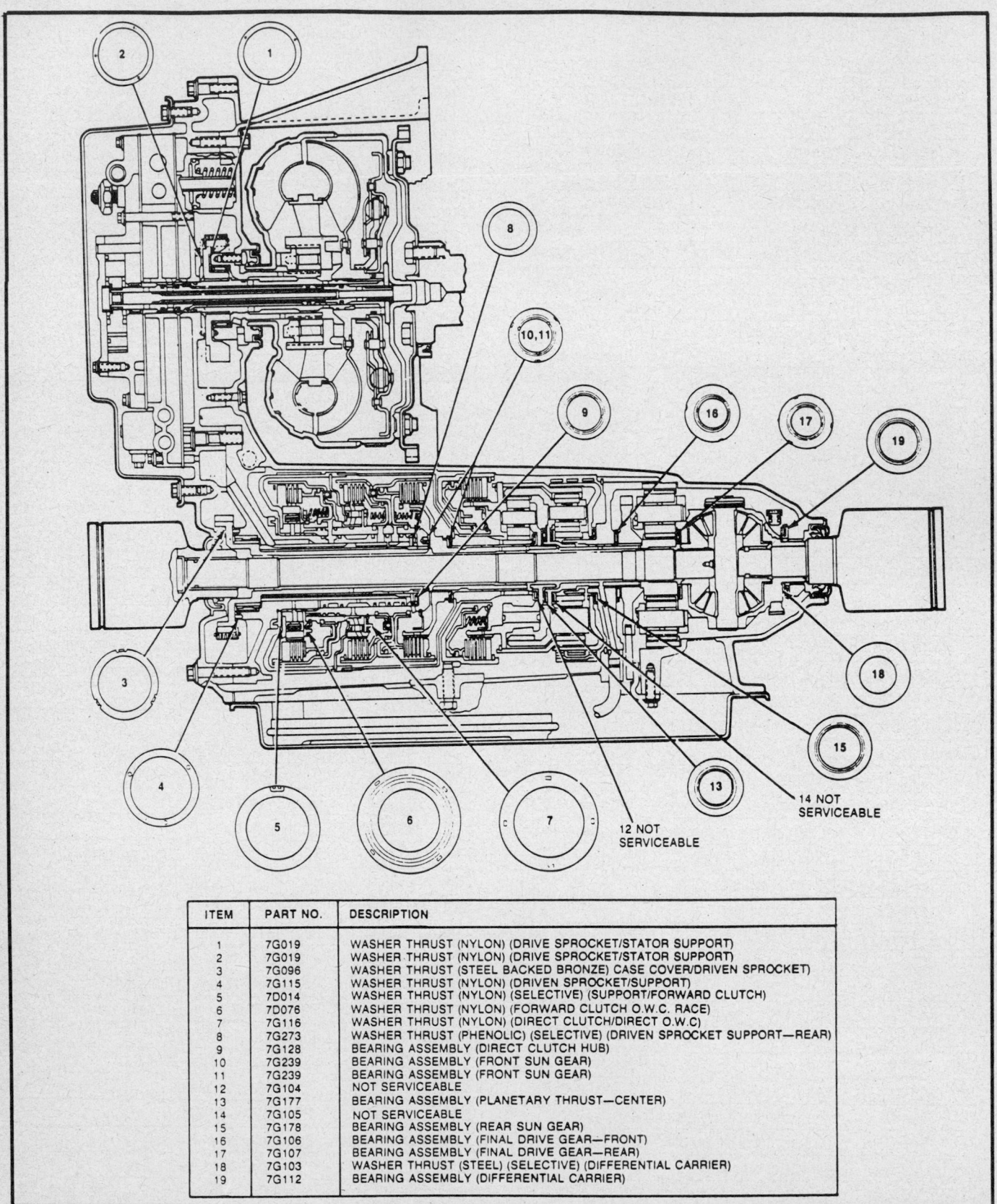

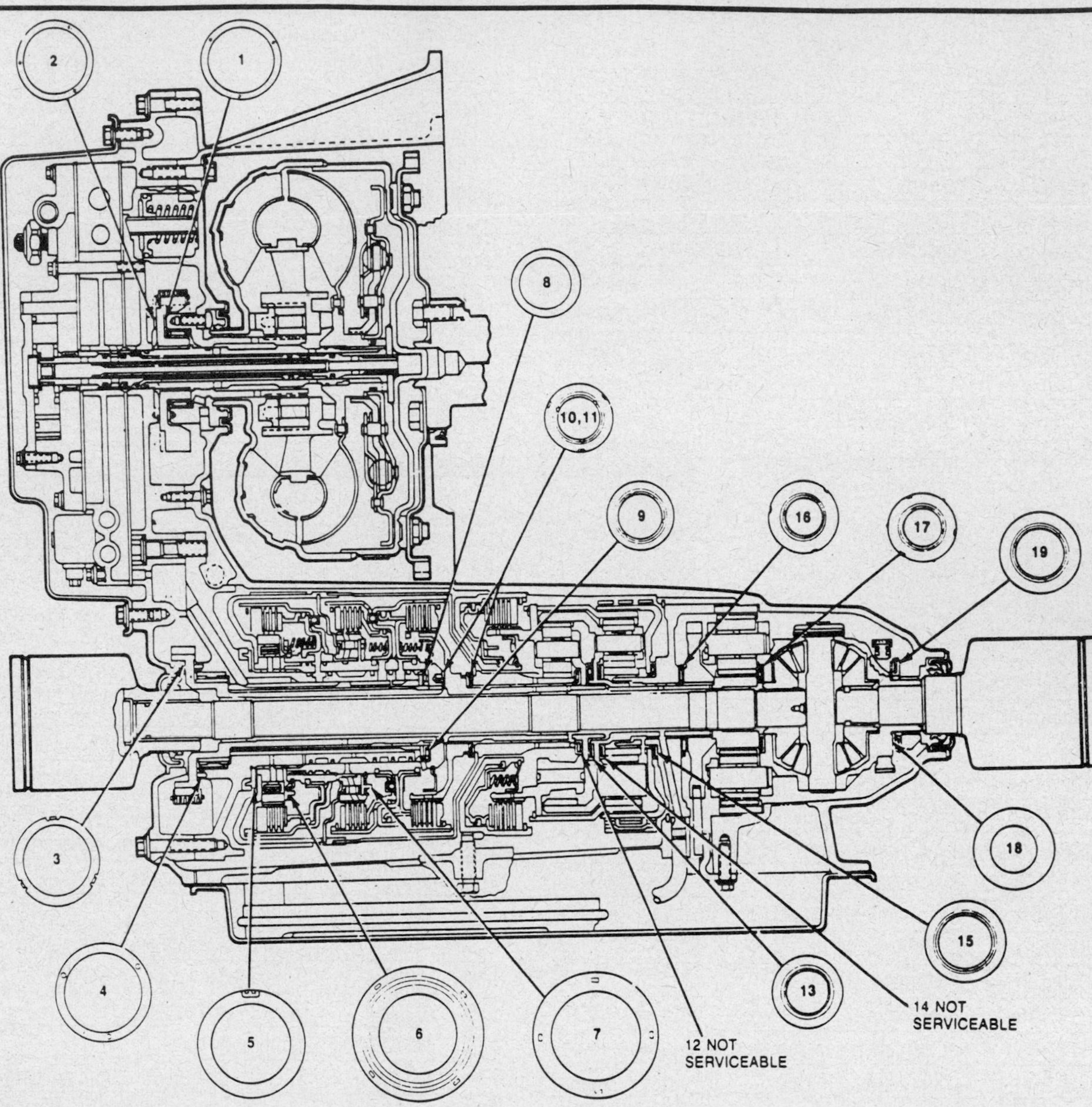
1
2
8
10,11
9
16
17
19
18
15
14 NOT SERVICEABLE
13
12 NOT SERVICEABLE
7
6
5
4
3

ITEM	PART NO.	DESCRIPTION
1	7G019	WASHER THRUST (NYLON) (DRIVE SPROCKET/STATOR SUPPORT)
2	7G019	WASHER THRUST (NYLON) (DRIVE SPROCKET/STATOR SUPPORT)
3	7G096	WASHER THRUST (STEEL BACKED BRONZE) CASE COVER/DRIVEN SPROCKET)
4	7G115	WASHER THRUST (NYLON) (DRIVEN SPROCKET/SUPPORT)
5	7D014	WASHER THRUST (NYLON) (SELECTIVE) (SUPPORT/FORWARD CLUTCH)
6	7D076	WASHER THRUST (NYLON) (FORWARD CLUTCH O.W.C. RACE)
7	7G116	WASHER THRUST (NYLON) (DIRECT CLUTCH/DIRECT O.W.C)
8	7G273	WASHER THRUST (PHENOLIC) (SELECTIVE) (DRIVEN SPROCKET SUPPORT—REAR)
9	7G128	BEARING ASSEMBLY (DIRECT CLUTCH HUB)
10	7G239	BEARING ASSEMBLY (FRONT SUN GEAR)
11	7G239	BEARING ASSEMBLY (FRONT SUN GEAR)
12	7G104	NOT SERVICEABLE
13	7G177	BEARING ASSEMBLY (PLANETARY THRUST—CENTER)
14	7G105	NOT SERVICEABLE
15	7G178	BEARING ASSEMBLY (REAR SUN GEAR)
16	7G106	BEARING ASSEMBLY (FINAL DRIVE GEAR—FRONT)
17	7G107	BEARING ASSEMBLY (FINAL DRIVE GEAR—REAR)
18	7G103	WASHER THRUST (STEEL) (SELECTIVE) (DIFFERENTIAL CARRIER)
19	7G112	BEARING ASSEMBLY (DIFFERENTIAL CARRIER)

Section 3

4EAT Transaxle
Ford Motor Co.

APPLICATION

1989 Ford Probe

GENERAL DESCRIPTION

The 4EAT electronically-controlled automatic transaxle is a 4 speed overdrive transaxle with a lockup torque converter. The 4EAT differs from most Ford transaxles because it is controlled by both mechanical and electronic systems. Several sensors and switches allow the 4EAT system to constantly monitor driving conditions. Signals from these sensors are sent to the 4EAT control unit, which uses the input to control shift pattern, gear position and lockup timing. The 4EAT control unit has built-in self-diagnosis, fail-safe and warning code display functions for the main input sensors and solenoid valves.

The 4 solenoid valves are located on the valve body. These valves actuate shifting and lockup by switching the oil flow through passages within the valve body. The valve body utilizes hydraulic pressure to control the application of the friction elements. The friction elements transmit power from the engine to the planetary gear unit.

A manual switch is located on the selector lever. Below it, on the selector console, is the the shift mode switch. The manual and shift mode switches provide a number of shifting and engine braking options.

The unique mechanical features of the 4EAT include a single compact combination-type planetary gear (4 speed capability) instead of the usual 2 planetary gears and a new variable capacity oil pump.

Converter Identification

CONVERTER

The torque converter is identified by either a reference or part number stamped on the converter body and is matched to a specific engine. The torque converter is a welded unit and is not repairable. If internal problems exists, the torque converter must be replaced.

Electronic Controls

The 4EAT electronic control system consists of several sensors and switches which send information to the 4EAT control unit. The control unit utilizes the input to operate the 4EAT transaxle. The governor, used in conventional transmissions to perform shifiting and lockup, is replaced in the 4EAT system by solenoid valves. These valves maintain or drain hydraulic pressure by actuating the shift and lockup control valves. The solenoid valves are controlled by the 4EAT control unit.

In the POWER mode, the transaxle shifts and the torque converter lockups occur at higher vehicle speeds to permit faster acceleration and improved performance feel. In the NORMAL mode, the shift points are selected to produce smoother engine operations and optimum fuel efficiency.

In the MANUAL SHIFT mode, the driver can use the selector lever to manually select and hold each of the lower 3 gears. A small push button on the gear selector is used to activate this mode.

Metric Fasteners

All metric fasteners are used on the 4EAT transaxle. Metric fastener dimensions are very close to the dimensions of the familiar inch system fasteners. For this reason, replacement fasteners must have the same measurement and strength as those removed.

Do not attempt to interchange metric fasteners for inch system fasteners. Mismatched or incorrect fasteners can result in damage to the transmission unit through malfunctions, breakage or possible personal injury.

NOTE: Care should be taken to reuse the fasteners in the same locations as removed.

Capacities

The 4EAT automatic transaxle use Motorcraft Mercon automatic transmission fluid. The capacity of the 4EAT automatic transaxle is 7.2 U.S. quarts or 6.8 liters.

Checking Fluid Level

1. Apply the parking brake and position wheel chocks to prevent the vehicle from rolling.

NOTE: Place the vehicle on a flat level surface. Use the low temperature scale when the fluid temperature is 148°F (68°C) or lower. Use the high temperature scale when the fluid temperature is 149°F (65°C) or higher.

4EAT SYSTEM ELECTRONIC COMPONENTS

Components	4EAT Control Unit Input/Output
4EAT Control Unit	———
Vehicle Speed Sensor	Input
Pulse Generator	Input
Throttle Position Sensor	Input
Idle Switch	Input
Coolant Temperature Switch	Input
Fluid Temperature Switch (ATF)	Input
Brake Light Switch	Input
Neutral Safety Switch	Input
Mode Switch	Input
Manual Switch	Input
Solenoid Valve 1–2 Shift	Output
Solenoid Valve 2–3 Shift	Output
Solenoid Valve 3–4 Shift	Output
Solenoid Valve Lockup	Output

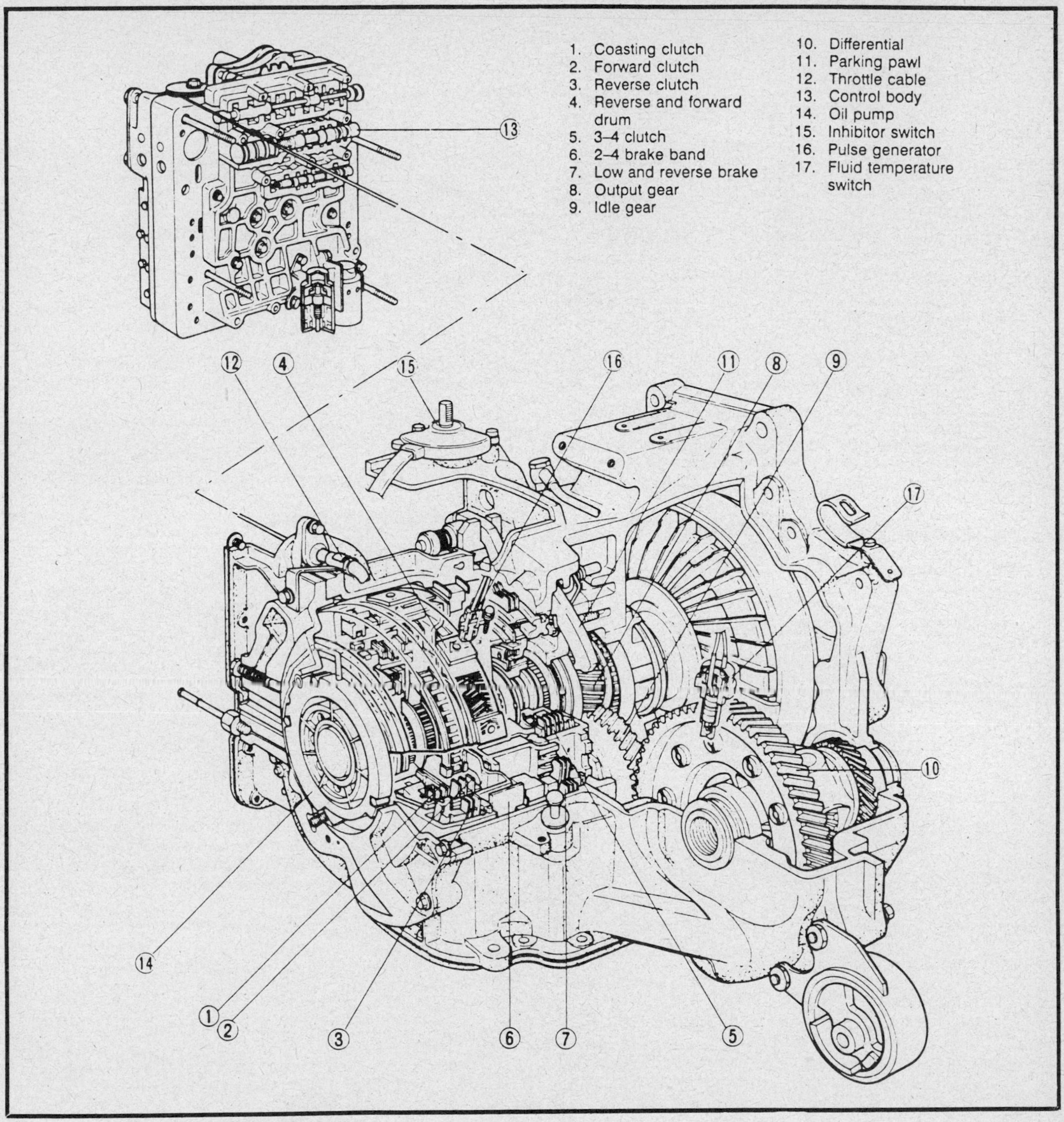
1. Coasting clutch
2. Forward clutch
3. Reverse clutch
4. Reverse and forward drum
5. 3–4 clutch
6. 2–4 brake band
7. Low and reverse brake
8. Output gear
9. Idle gear
10. Differential
11. Parking pawl
12. Throttle cable
13. Control body
14. Oil pump
15. Inhibitor switch
16. Pulse generator
17. Fluid temperature switch

Structural view of the 4EAT transaxle

2. Start the engine to allow the transmission fluid to warm up to specification.

3. With the engine idling, shift the gear selector lever from **P** to **L** and back again.

4. Let the engine idle.

5. Make sure the transmission fluid level is between the **F** and **L** marks. Add the correct fluid as required.

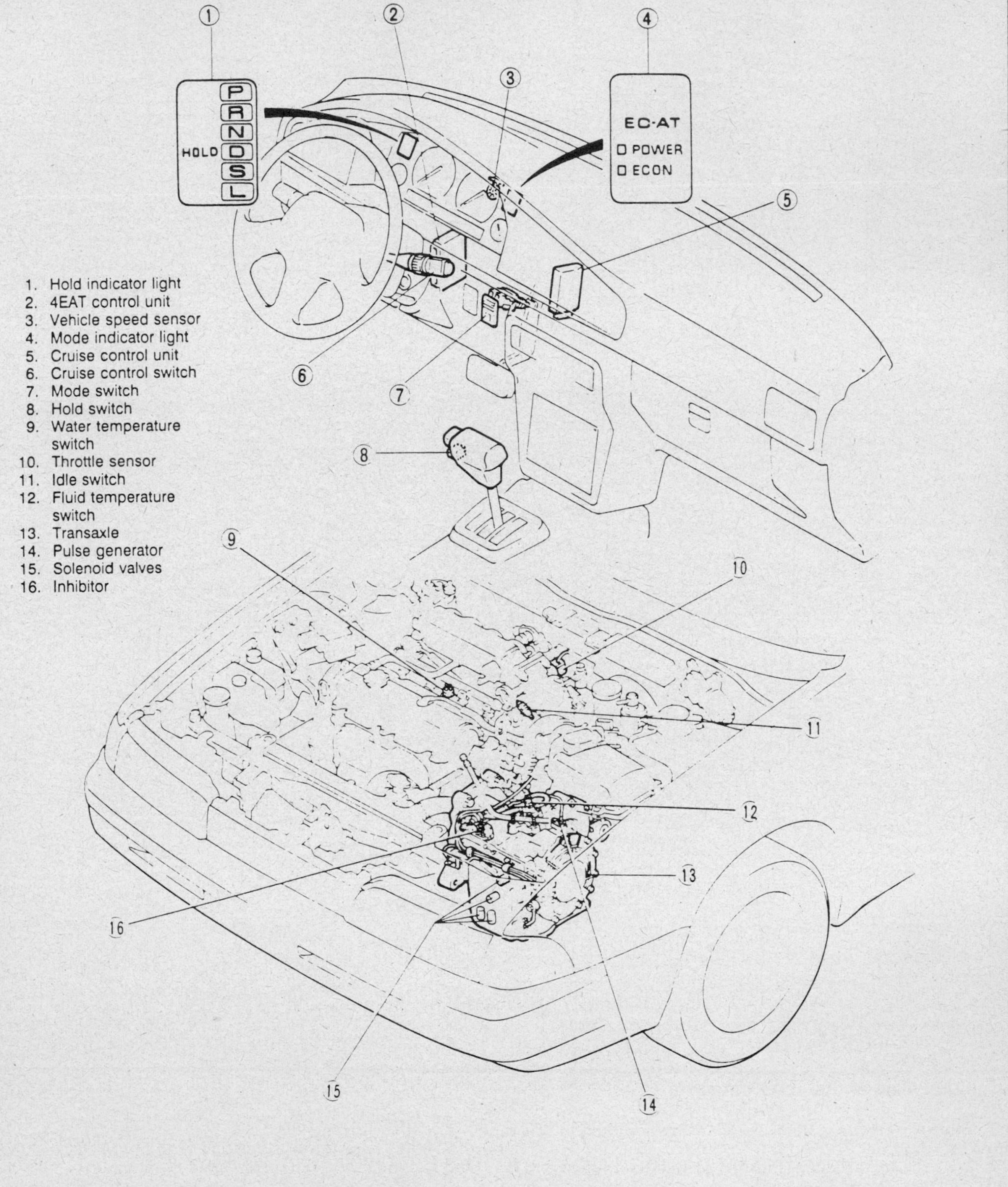
HOLD
P
R
N
D
S
L
EC-AT
POWER
ECON

1. Hold indicator light
2. 4EAT control unit
3. Vehicle speed sensor
4. Mode indicator light
5. Cruise control unit
6. Cruise control switch
7. Mode switch
8. Hold switch
9. Water temperature switch
10. Throttle sensor
11. Idle switch
12. Fluid temperature switch
13. Transaxle
14. Pulse generator
15. Solenoid valves
16. Inhibitor

Electrical component location

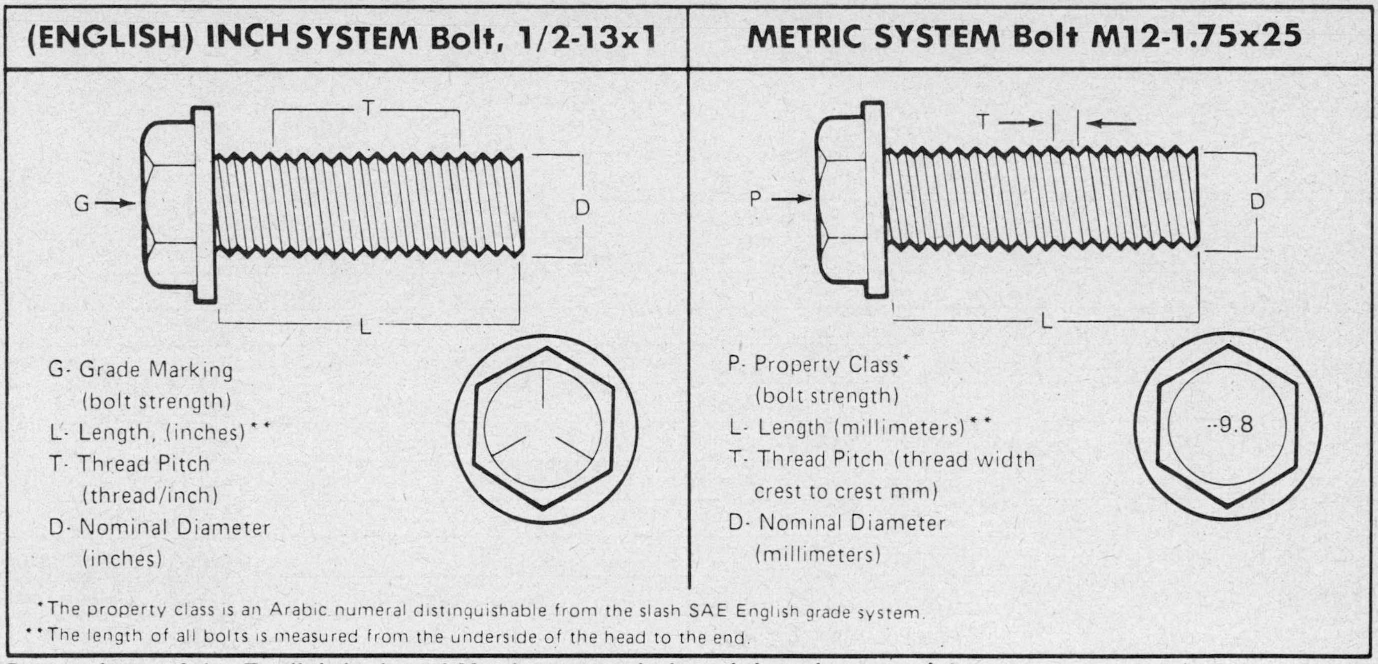
(ENGLISH) INCH SYSTEM Bolt, 1/2-13x1
METRIC SYSTEM Bolt M12-1.75x25
T
G
D
L
G- Grade Marking (bolt strength)
L- Length, (inches)**
T- Thread Pitch (thread/inch)
D- Nominal Diameter (inches)
P
P- Property Class* (bolt strength)
L- Length (millimeters)**
T- Thread Pitch (thread width crest to crest mm)
D- Nominal Diameter (millimeters)
9.8
*The property class is an Arabic numeral distinguishable from the slash SAE English grade system.
**The length of all bolts is measured from the underside of the head to the end.

Comparison of the English Inch and Metric system bolt and thread nomenclature

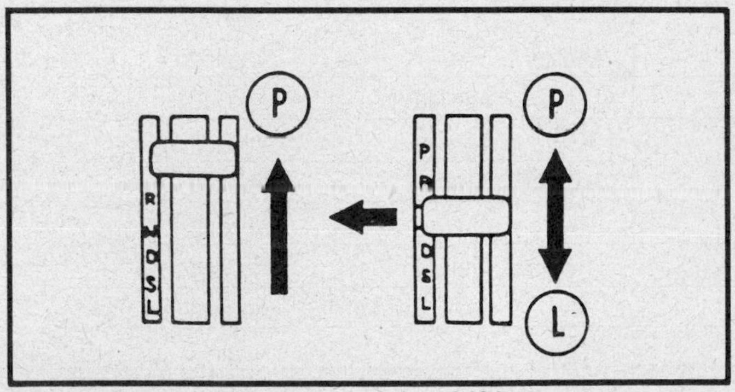
P
R
N
D
S
L
P
L

Checking transaxle fluid level—shifting gear selector

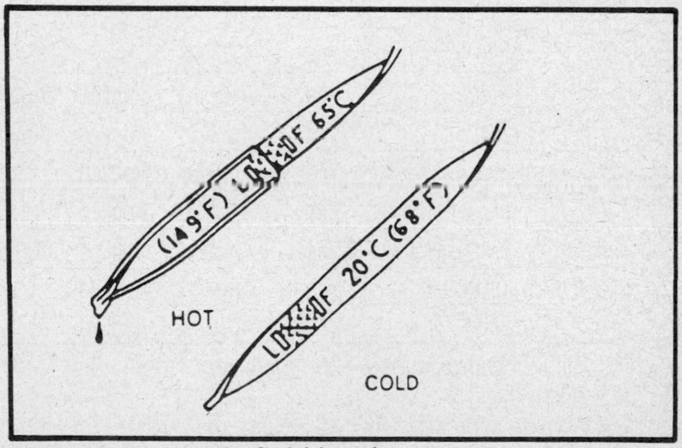
(149°F) LO 65°C
20°C (68°F)
HOT
COLD

Checking transaxle fluid level

TRANSAXLE MODIFICATIONS

There have been no modifications to the 4EAT transaxle at the time of this printing.

TROUBLE DIAGNOSIS

A logical and orderly diagnosis outline and charts are provide to assist the repairman in diagnosing the problems, causes and the extent of repairs needed to bring the automatic transaxle back to its acceptable level of operation.

Preliminary checks and adjustments should be made to all electrical related components, idle speed, selector lever, kickdown cable and throttle cable. Transaxle oil level should be checked, both visually and by smell, to determine whether the fluid level is correct and to observe any foreign material in the fluid, if present. Smelling the fluid will indicate if any of the bands or clutches have been burned through excessive slippage or overheating of the transaxle.

It is most important to locate the defect, its cause and to properly repair them to avoid having the same problem re-occur.

In order to more fully understand the 4EAT automatic transaxle and to diagnose possible defects more easily, the clutch and band application chart and a general description of the hydraulic and electrical control systems are given.

CLUTCH AND BAND APPLICATION
4EAT Automatic Transaxle

Range	Gear	Engine Braking Effect	Forward Clutch	Coasting Clutch	3–4 Clutch	Reverse Clutch	2–4 Brake		Low & Reverse Brake	One-way Clutch One	One-way Clutch Two
							Applied	Released			
P		—	—	—	—	—	—	—	—	—	—
R	Reverse	Yes	—	—	—	Applied	—	—	Applied	—	—
N		—	—	—	—	—	—	—	—	—	—
D	1st	No	Applied	—	—	—	—	—	—	Applied	Applied
	2nd	No	Applied	—	—	—	Applied	—	—	Applied	—
	3rd①	Yes	Applied	Applied	Applied	—	—	Applied	—	Applied	—
	3rd②	Yes	Applied	Applied	Applied	—	③	Applied	—	Applied	—
	OD	Yes	Applied	—	Applied	—	Applied	—	—	—	—
S	1st	No	Applied	—	—	—	—	—	—	Applied	Applied
	2nd	No	Applied	—	—	—	Applied	—	—	Applied	—
	3rd①	Yes	Applied	Applied	Applied	—	—	Applied	—	Applied	—
	3rd②	Yes	Applied	Applied	Applied	—	①	Applied	—	Applied	—
L	1st	No	Applied	—	—	—	—	—	Applied	Applied	Applied
	2nd	Yes	Applied	Applied	—	—	Applied	—	—	Applied	—
Hold D	2nd	No	Applied	—	—	—	Applied	—	—	Applied	—
	3rd①	Yes	Applied	Applied	Applied	—	—	Applied	—	Applied	—
	3rd②	Yes	Applied	Applied	Applied	—	①	Applied	—	Applied	—
S	2nd	Yes	Applied	Applied	—	—	Applied	—	—	Applied	—
	3rd①	Yes	Applied	Applied	Applied	—	—	Applied	—	Applied	—
	3rd②	Yes	Applied	Applied	Applied	—	③	Applied	—	Applied	—
L	1st	Yes	Applied	Applied	—	—	—	—	Applied	Applied	Applied
	2nd	Yes	Applied	Applied	—	—	Applied	—	—	Applied	—

① Below approximately 25 mph or 40km/h
② Above approximately 25 mph or 40km/h
③ Fluid pressure to servo but band not applied due to pressure difference in servo

CHILTON'S THREE C's TRANSAXLE DIAGNOSIS
4EAT Automatic Transaxle

Condition	Cause	Correction
Engine will not crank in any shift lever position	a) Neutral start switch stuck or failed b) Neutral start switch damaged or disconoected c) 4EAT control module	a) Go to Quick Test b) Go to Quick Test c) Go to Quick Test
Engine does not crank in P	a) Selector lever and linkage out of adjustment b) Neutral start switch not correctly aligned to transmission	a) Confirm selector or linkage adjustment and operation b) Adjust neutral start switch

CHILTON'S THREE C's TRANSAXLE DIAGNOSIS
4EAT Automatic Transaxle

Condition	Cause	Correction
Engine starts in shift lever positions other than P or N	a) Shift linkage damaged, out of adjustment b) Neutral start switch short circuit c) 4EAT control module	a) Confirm selector linkage adjustment and operation b) Go to Quick Test c) Go to Quick Test
Vehicle moves in P or parking gear not disengaged when P is disengaged	a) Selector lever and linkage out of adjustment b) Parking pawl	a) Confirm selector linkage adjustment and operation b) Inspect parking pawl
Vehicle moves in N	a) Selector lever and linkage out of adjustment b) Control valve damaged	a) Confirm selector linkage adjustment and operation b) Inspect control valve. Service or replace as required
Vehicle does not move in Overdrive, D, L or R	a) Control valves b) Improper fluid level c) Oil pump dirty, broken or bad seals d) Torque converter damaged	a) Go to Quick Test b) Check and fill c) Inspect oil pump d) Inspect torque converter
Vehicle does not move in any forward shift position. Reverse OK	a) Control valves b) Forward clutch worn or damaged c) One-way clutch No. 1 worn or damaged d) Oil flow to forward clutch blocked	a) Go to Quick Test b) Inspect clutches c) Go to Operational Test d) Go to Operational Test
Vehicle does not move in reverse. Forward OK	a) Reverse clutch worn or damaged b) Low and reverse clutch slipping	a) Go to Operational Test b) Inspect clutch and clutch adjustment
Noise severe under acceleration or deceleration. Ok in P, N or steady speed	a) Speedometer cable b) Torqur converter failure c) Gear or clutch failure d) Selector cable grounding out e) Engine mounts grounding out	a) Service or replace b) Examine/service c) Examine or service d) Install and route cable as necessary e) Neutralize engine mounts
Noise in P or N. Does not stop in D at stall	a) Loose flywheel to converter bolts b) Oil pump worn c) Torque converter failure	a) Torque to specification b) Examine/servcie engine c) Examine/servcie converter or Go to Operational Test (Stall Test)
Noise in all gears. Changes power to coast	a) Final drive gearset worn b) CV joints	a) Examine/service final drive gearset b) Servcie as required
Noise in all gears—does not change power to coast	a) Defective speedometer gears gear b) Bearings worn or damaged c) Planetary gearset noisy	a) Examine/replace speed drive or driven b) Examine/replace c) Service planetary gearset
Harsh shifts (any gear)	a) Kickdown cable out of adjustment b) Valve body c) Sticking accumulators d) CV joints e) Engine mounts loose f) Throttle valve sticking g) Band adjustment h) Band servo	a) Check kickdown cable adjustment b) Inspect valve body. Go to Quick Test c) Inspect accumulators d) Service as required e) Service as required f) Inspect throttle valve g) Check band adjustment h) Inspect band servo
Soft shifts (any gears)	a) Kickdown cable b) Band adjustment c) Band servo d) Pressure regulator damaged e) ATF level f) Valve body g) Sticking accumulators h) Throttle valve sticking	a) Check kickdown cable b) Check band adjustment c) Inspect band servo d) Inspect pressure regulator e) Check and fill f) Inspect valve body. Go to Quick Test g) Inspect accumulators h) Inspect throttle valve

CHILTON'S THREE C's TRANSAXLE DIAGNOSIS
4EAT Automatic Transaxle

Condition	Cause	Correction
Erratic shifting, incorrect shift points, incorrect shift sequence	a) Kickdown cable b) Control valves c) Band adjustment d) Clutches slipping e) Fluid level and quality	a) Check kickdown cable adjustment b) Go to Quick Test c) Check band adjustment d) Inspect clutches e) Check and fill
Improper lockup	a) Control valves b) Torque converter	a) Go to Quick Test b) Inspect torque converter
Skipping gears (shift 1st to 3rd or 2nd to OD, for example)	a) Control valves b) Valve body c) 2-4 band	a) Go to Quick Test b) Inspect valve body c) Check band adjustment
Transaxle overheating	a) Impoper fluid level b) Poor engine performance c) Worn clutch, incorrect band application or poor oil pressure control d) Restriction in cooler lines e) Clogged cooler	a) Check fluid level b) Adjust according to specifications c) Go to Operational Test (Stall Test) d) Check cooler lines for kinks and damage. Clean, service or replace cooler lines e) Inspect cooler for plugging. Service as required
Drags in R like parking brake is applied	a) 2-4 band adjustment	a) Inspect band adjustment
Drags in forward gears	a) Band adjustment	a) Inspect band adjustment
Engine runaway on upshift	a) Fluid level low b) Valve body c) Oil pump d) Damaged bypass valve e) Clutches slipping	a) Check fluid level b) Inspect valve body, solenoid valve c) Inspect oil pump d) Inspect bypass valve e) Inspect clutches
Engine runaway on downshift	a) Coasting bypass valve sticking b) Clutches slipping c) Fluid level d) Oil pump	a) Go to Operational Test (Stall Test) b) Inspect clutches c) Check fluid level d) Inspect oil pump
Manual light flashing	a) Control module b) Sensors c) Circuit	a) Go to Quick Test b) Go to Quick Test c) Go to Quick Test
Mode will not switch from manual to automatic or from automatic to manual	a) Control module b) Sensors c) Circuit	a) Go to Quick Test b) Go to Quick Test c) Go to Quick Test
Excessive creep	a) Torque converter b) Kickdown cable c) Ignition timing and idle speed	a) Inspect torque converter b) Inspect kickdown cable adjustment c) Correct or adjust
No creep	a) AFT level and condition b) Kickdown cable c) Selector level d) Valve body e) Control valves f) Forward clutch g) Reverse clutch h) Oil pump	a) Check level and condition b) Inspect kickdown cable adjustment c) Confirm selector linkage adjustment and operation d) Inspect valve body e) Inspect control valves f) Inspect clutches g) Inspect clutches h) Inspect oil pump

Hydraulic Control System

NOTE: Please refer to Section 9 for all oil flow circuits.

TORQUE CONVERTER

The torque converter consists of an impeller, stator, turbine, converter clutch and converter cover. The converter clutch couples the turbine hub and the spline. During lockup, the converter clutch slides on the turbine hub and is pressed against the converter cover. Torsional damper springs are provided in the converter clutch to absorb engine torque and pulsations during lockup.

OIL PUMP

The pump consists of a cam ring, rotor vanes, guide ring, pivot roller, seal pins and oil pump flange, incorporated within the pump housing. The oil pump flange and shaft are coupled to the rotor. The other end of the shaft is coupled to the torque converter so that the rotation is the same as the engine speed. The cam ring, which rotates eccentrically with the pivot roller as a fulcrum, regulates the discharge quantity. A valve and spring are provided at the discharge port to regulate the hyddraulic pressure in the variable chamber.

FLUID PASSAGES

The fluid passages are located in the oil pump, converter housing and transaxle case.

VALVE BODY

The valve body is mounted on the side of the transaxle which faces the radiator. It can be removed without removing the transaxle. The valve body consists of 4 subsections: the front, premain, main and rear control bodies.

THROTTLE VALVE

The throttle valve produces throttle pressure according to the depression of the accelerator pedal.

THROTTLE-MODULATOR VALVE

The throttle-modulator valve produces throttle-modulated pressure from throttle pressure.

PRESSURE-REGULATOR VALVE

The pressure-regulator valve adjusts the line pressure produced by the oil pump to match the current driving condition.

MANUAL VALVE

The manual valve is moved by the range selector and switches the passage for line pressure produced by the oil pump. The manual valve determines where line pressure is distributed for each selector position.

SHIFT VALVES

The 1–2 shift valve is activated by the 1–2 solenoid valve. It controls the automatic shifting between first and second gear. The 2–3 shift valve is activated by the 2–3 solenoid valve. It controls the automatic shifting between second and third gear. The 3–4 shift valve is activated by the 3–4 solenoid valve. It controls the automatic shifting between third and fourth gear.

LOW REDUCING VALVE

The low reducing valve reduces the low and reverse clutch engagement pressure, which in turn reduces low-first gear shift shock.

ACCUMULATORS

The N – OD accumulator reduces shift shock when shifting to overdrive range. The N – R accumulator reduces shift shock when shifting to reverse range. The 1–2 accumulator reduces shift shock when shifting from first to second gear. Unlike the other valves which are incorporated within the valve body, the 2–3 accumulator is located beside the bearing housing. This accumulator functions to reduce the shift shock when shifting from second to third gear.

TIMING VALVES

The 2–3 timing valve controls the 3–4 clutch engagement timing during a second to third gear shift. This timing is determined by the amount of throttle opening. The 3–2 timing valve controls the engagement timing of the 2–4 band. This timing is determined by the amount of the throttle opening during an overdrive-third to drive-second gear shift.

BYPASS VALVES

The bypass valve controls the 3–4 clutch engagement timing and shift shock during a second to third gear shift. The coasting bypass valve controls the engagement of the coasting clutch.

3–2 CAPACITY VALVE

The 3–2 capacity valve adjusts the rate at which the release pressure of the 2–4 band drains during a downshift to second gear in drive range.

LOCKUP CONTROL VALVE

The lockup control valve is activated by the lockup solenoid valve. This valve controls torque converter lockup operation.

Electronic Control System

PULSE GENERATOR

The pulse generator is a magnetic pickup sensor located on the transaxle housing. It detects the reverse-forward drum speed.

VEHICLE SPEED SENSOR

The speed sensor acts as a substitute for the pulse generator, if the pulse generator malfunctions. There are 2 types of vehicle speed sensors used on the 4EAT transaxle. One type is present on vehicles equipped with an analog cluster (sensor located within speedometer assembly) the other type is present on vehicles equipped with a digital cluster (sensor located on the transaxle housing).

THROTTLE POSITION SENSOR

The throttle position sensor is a variable resistor attached to the throttle body. The sensor detects the opening of the throttle and relays these signals to the 4EAT control unit.

IDLE SWITCH

The idle switch is attached to the throttle body, when the throttle plate is fully closed the switch sends a signal to the 4EAT control unit. The idle switch is preset at the factory and should not be adjusted.

NEUTRAL SAFETY SWITCH

The neutral safety switch is located on the transaxle case. It sends signals to the 4EAT control unit indicating the position of the manual valve (P, R, N, OD, D or L). This switch has a 0.079 in. hole which permits alignment of the switch during installation.

BRAKE LIGHT SWITCH

The brake light switch sends a signal to the 4EAT control unit when the brake pedal is depressed.

ENGINE COOLANT TEMPERATURE SWITCH

The coolant temperature switch is located on the lower portion of the intake manifold. It sends a signal to the 4EAT control unit when the engine coolant temperature is below 149°F (65°C).

FLUID TEMPERATURE SWITCH

The fluid temperature switch is located on the switch box which connects the oil cooler fluid pipe to the transaxle. It sends a signal to the 4EAT control unit when the transaxle fluid temperature is above 302°F (150°C).

SOLENOID VALVES

There are 3 solenoid valves for shifting and 1 for lockup. Located on the valve body, these valves are switched on and off by electrical signals from the control unit.

CONTROL UNIT

The control unit has internal self-diagnosis, fail-safe and warning code display functions for the the main input sensors and solenoid valves.

Diagnosis Tests

These diagnostic procedures are to be used on 4EAT equipped vehicles only. To help locate problems with the transaxle, the following sequence should be followed.

PRELIMINARY ROAD TEST

This road test, if possible, should be performed to verify the complaint. No special test equipped is used at this time.

PRELIMINARY INSPECTIONS

Visually inspect all 4EAT related components, fuses, sensors, switches etc. also inspect CV joints, engine mounts, oil cooler, halfshafts and any other external component. Inspect fluid level for burnt, discoloration or contamination of fluid. Check and adjust idle speed, if required. Check selector lever for smooth operation of the button and clicks in each position. Check transaxle for any fluid leaks from seals, lines and gaskets. Check for smooth operation of kickdown cable from idle to wide open throttle, service or replace the kickdown cable, as required. Check for proper operation of throttle cable. Adjust or replace the throttle cable as necessary. Check and inflate all tires to the proper level. After preliminary inspections review the diagnosis charts which will provide basic direction or test procedures.

QUICK TEST

This step will find fault codes that may indicate 4EAT input or output device failure. Follow Pinpoint Test step direction given in the Quick Test before continuing with the procedure. When directed to perform Quick Test, Operational Tests and Road Test for the same symptom, always perform Quick Test first.

OPERATIONAL TESTS

This step determines the causes of most basic problems that may exist. When directed to perform Operational Tests and Road Test for the same symptom always perform Operational Tests first. This action will prevent causing possible damage to the transaxle during driving.

ROAD TEST

This step isolates problems that are evident during driving. The road test is an evaluation of the 4EAT system while driving with the 4EAT tester in service. Repair or inspection of the transaxle during this test may involve major disassembly, therefore secondary road test should always be done last. The powertrain may also show problems during the test that can cause transaxle problems. If no problems are found during the road test it is likely that the problem is intermittent. Since the problem may not reoccur, the symptom should be evaluated again.

NOTE: After any repair is made, re-test the transaxle to verify if symptom is still present. If the symptom reoccurs, further testing must be performed to isolate problem. Any time fluid is drained from the transaxle, be certain the proper type and amount of fluid is replaced.

4EAT QUICK TEST

Description

The Quick test is the procedure to activate the 4EAT electronic control module self-test. The self-test is divided into 3 specialized tests: Key On- Engine Off Test, Continuous Test, Switch Monitor Test.

The processor stores the self-test program in its permanent memory. When activated, it checks the 4EAT control system by testing its memory integrity and processing capability and verifies that various sensors and actuators are connected and operating properly.

Any time a repair is made, clear memory by disconnecting the small (16-pin) connector on the control unit. Remove the 4EAT tester. Turn ignition switch **OFF**. Then repeat the test to ensure that the repair was effective.

NOTE: The Quick Test procedure should be used only when the preliminary inspection steps result in a PASS condition. If all phases of the Quick Test result in a PASS CONDITION, it is likely that the problem will be found elsewhere. Proceed to the Operational Test for further evaluation of the transaxle.

Test Steps

VEHICLE PREPARATION

NOTE: It may be necessary to disconnect or disassemble harness connector assemblies to do some of the inspections. Pin locations should be noted before disassembly.

1. Place shift lever firmly into the **P** postion and block drive wheels.
2. Start engine. Observe manual shift light.
3. Proceed to equipment hookup step.

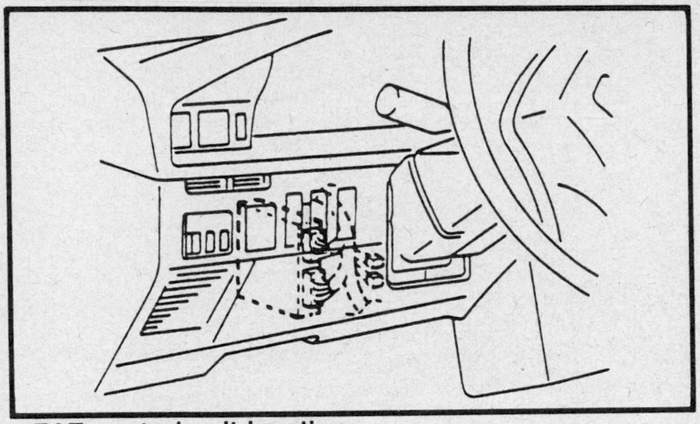

4EAT control unit location

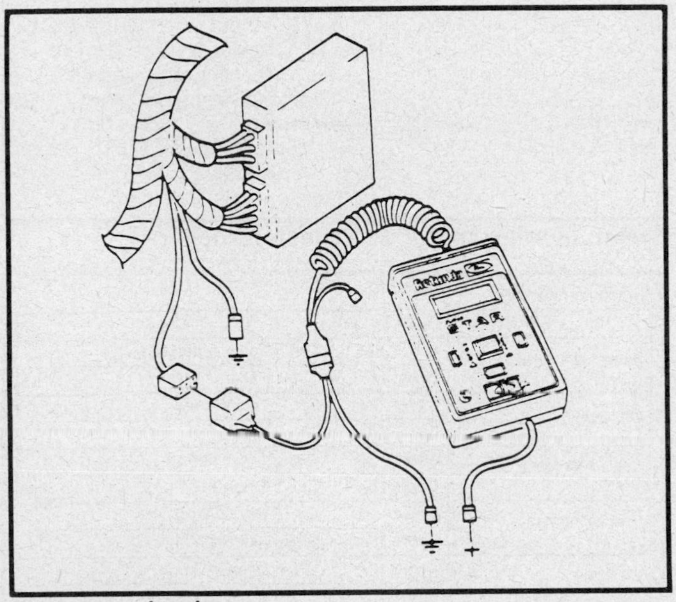

Star tester hookup

EQUIPMENT HOOKUP

1. Turn ignition key **OFF**.
2. Locate the service connector.
3. Using the Star tester
   a. Ground the Star tester.
   b. Connect the Star tester to the 6-pin 4EAT Star tester output (STO) connector.
   c. Connect the Star tester to the 1-pin 4EAT Star tester input (STI) connector.
4. Using the VOM
   a. Set the VOM on a DC voltage range to read from 0–20V.
   b. Ground the VOM lead.
   c. Connect the VOM lead to the red wire (6-pin 4EAT STO connector).
   d. Ground the single-pin 4EAT STI connector with a jumper wire.
5. Go to Key On- Engine Off Test.

KEY ON ENGINE OFF TEST

Is a system to display service codes which are present or past failures. The Star tester will display service codes and the VOM needle will pulse across the dial face representing service codes. If no codes are indicated on the Star tester or VOM and the manual shift flashed during the vehicle preperartion step, then proceed to Pinpoint Test PPQ. If the manual shift light did not flash, proceed to the Switch Monitor Test.

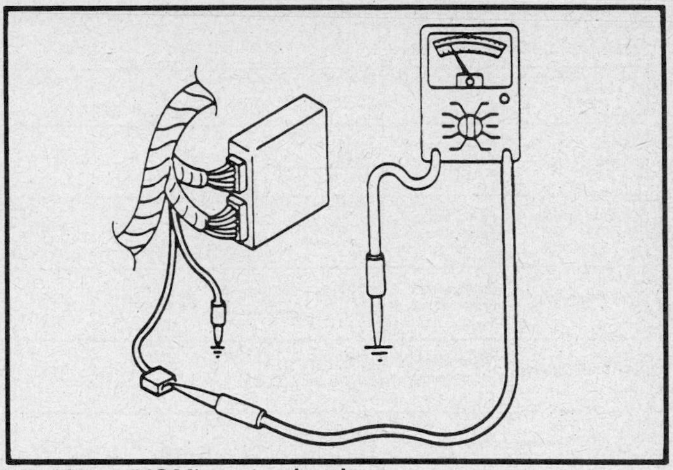

Volt/Ohm (VOM) meter hookup

1. To perform Key On- Engine Off Test verify that the vehicle has been properly prepared per vehicle preperation and equipment hookup steps.
2. Place ignition key in the **ON** position.
3. Record codes indicated by Star tester or VOM.
4. Refer to Key On- Engine Off Test passenger vehicle service code chart troubleshooting guide.

NOTE: If the manual shift light flashed during vehicle preparation step, use Key On- Engine Off passenger vehicle service code chart troubleshooting guide. If the manual shift light did not flash during vehicle preperation step, proceed to Continuous Test

CONTINUOUS TEST

Continuous memory codes are issued as a result of information stored during Continuous Test, while the vehicle was in normal operation. These codes are displayed during testing and should be used for diagnosis only when a continuous code results from previous test steps.

NOTE: Verify that the manual shift light did not flash during the vehicle preparation step before continuing with this test. It is necessary to clear the codes in memory before continuing this test. Only service those codes which are recreated by Continuous Test.

1. To clear continuous memory codes turn ignition switch **OFF**.
2. Disconnect the small (16-pin) connector on the 4EAT control unit.
3. Remove the Star tester or VOM.
4. Connect Star tester or VOM.
5. Start the engine and do not shut the engine down.
6. The system is now in the Engine Running Continuous Monitor mode.

NOTE: The Continuous Monitor mode (wiggle test) allow the technician to attempt to recreate an intermittent fault. It is necessary to drive the vehicle each time the suspect sensor and or harness is tapped, moved or wiggled. If a fault is detected, a service code will be stored in memory.

7. Refer to the Continuous Test passenger vehicle service code chart troubleshooting guide.

SWITCH MONITOR TEST

Is a check of the 4EAT control module inputs using the 4EAT tester. If a switch fails, proceed to Pinpoint test PPM. If a PASS code is received, an indication that the 4EAT control systems are OK, proceed to the Operational Test.

Key On Engine Off Test
Passenger Car Service Code Chart

Defective System	Possible Cause	Diagnosis Indication Mode		Pinpoint Test
		VOM	Code No.	
Vehicle Speed Sensor	Failed sensor Faulty wiring	ON OFF	06	PPA
Throttle Position Sensor	Failed sensor Faulty wiring		12	PPB
Pulse Generator	Failed sensor Faulty wiring		55	PPC
1–2 Shift Solenoid	Failed solenoid Faulty wiring		60	PPD
2–3 Shift Solenoid	Failed solenoid	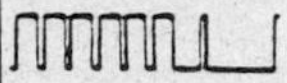	61	PPE
3–4 Shift Solenoid	Failed solenoid Faulty wiring	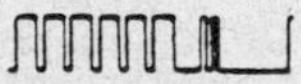	62	PPF
Lockup Solenoid	Failed solenoid Faulty wiring	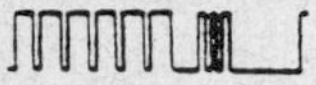	63	PPG

Diagnostic test

Continuous Test
Passenger Car Service Code Chart

Defective System	Possible Cause	Diagnosis Indication Mode		Pinpoint Test
		VOM	Code No.	
Vehicle Speed Sensor	Failed sensor Faulty wiring	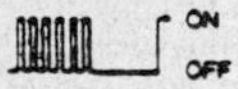ON OFF	06	PPA
Throttle Position Sensor	Failed sensor Faulty wiring		12	PPB
Pulse Generator	Failed sensor Faulty wiring		55	PPC
1–2 Shift Solenoid	Failed solenoid Faulty wiring		60	PPD
2–3 Shift Solenoid	Failed solenoid	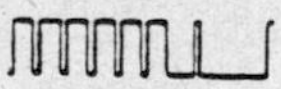	61	PPE
3–4 Shift Solenoid	Failed solenoid Faulty wiring	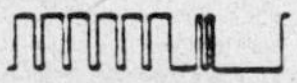	62	PPF
Lockup Solenoid	Failed solenoid Faulty wiring	1.2 SEC. 1.8 SEC.	63	PPG

Diagnostic test

1. To perform test disconnect vehicle harness from 4EAT control module.
2. Connect the 4EAT tester adaptor between the harness and the module.
3. Turn the 4EAT tester on.
4. Turn ignition switch **ON**.
5. Test all switches under conditions specified in the Switch Monitor Test chart.

Manual Shift Light Operation

SYSTEM OK

The manual shift light will illuminate when using manual shift mode.

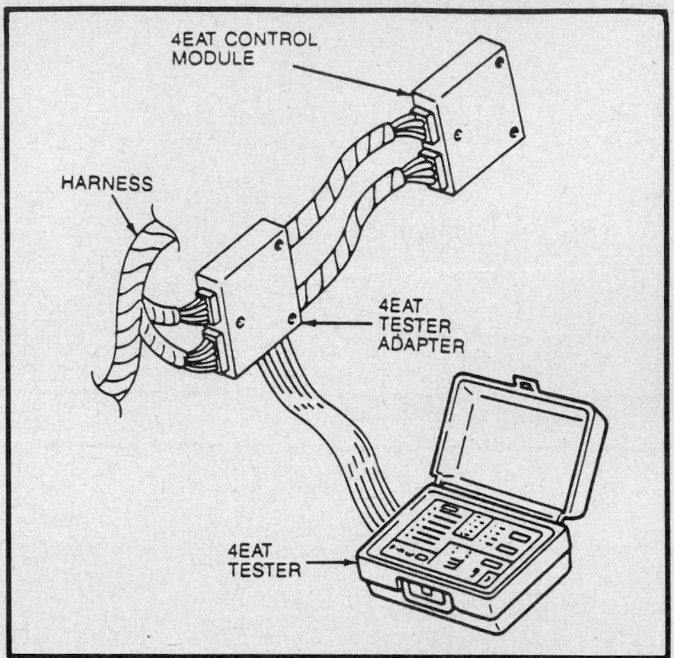
4EAT CONTROL MODULE
HARNESS
4EAT TESTER ADAPTER
4EAT TESTER

Switch monitor test— 4EAT tester hookup

Switch Monitor Test

Switch	VOM or LED		Condition
Brake Light Switch	Above 10v	ON	Pedal depressed
	Below 1.5v	OFF	Pedal released
Idle Switch	Above 10v	ON	Other speeds
	Below 1.5v	OFF	At idle
Coolant Temperature Switch	Above 10v	ON	Above 72°C (162°F)
	Below 1.5v	OFF	Below 65°C (149°F)
Check Connect	Below 1.5v	ON	Key OFF
	Above 10v	OFF	Key ON
L	Above 10v	ON	L range
	Below 1.5v	OFF	Other ranges
D	Above 10v	ON	D range
	Below 1.5v	OFF	Other ranges
Ⓓ	Above 10v	ON	Ⓓ range
	Below 1.5v	OFF	Other ranges
N or P	Below 1.5v	ON	N or P range
	Above 10v	OFF	Other ranges
Mode Switch	Above 10v	ON	Normal mode
	Below 1.5v	OFF	Power mode
Mode Indicator	Above 4.5v	OFF	Manual mode
	Below 1.5v	ON	Other mode
Manual Switch	Above 10v	ON	Switch depressed
	Below 1.5v	OFF	Switch released
Manual Indicator	Below 1.5v	ON	Manual mode
	Above 10v	OFF	Other mode
No Load Signal	Above 10v	ON	Drum speed below 80 rpm
	Below 1.5v	OFF	Drum speed above 640 rpm and N or P range
Throttle Position Sensor	Above 4.3v	ON	Throttle fully open
	Below 0.5v	OFF	Throttle closed
	Changes 0.5v		Every 1/8 position change
ATF Temperature Switch	Above 10v	OFF	ATF Temperature below 143°C (289°F)
	Below 0.5v	ON	ATF Temperature above 150°C (302°F)

Diagnostic test

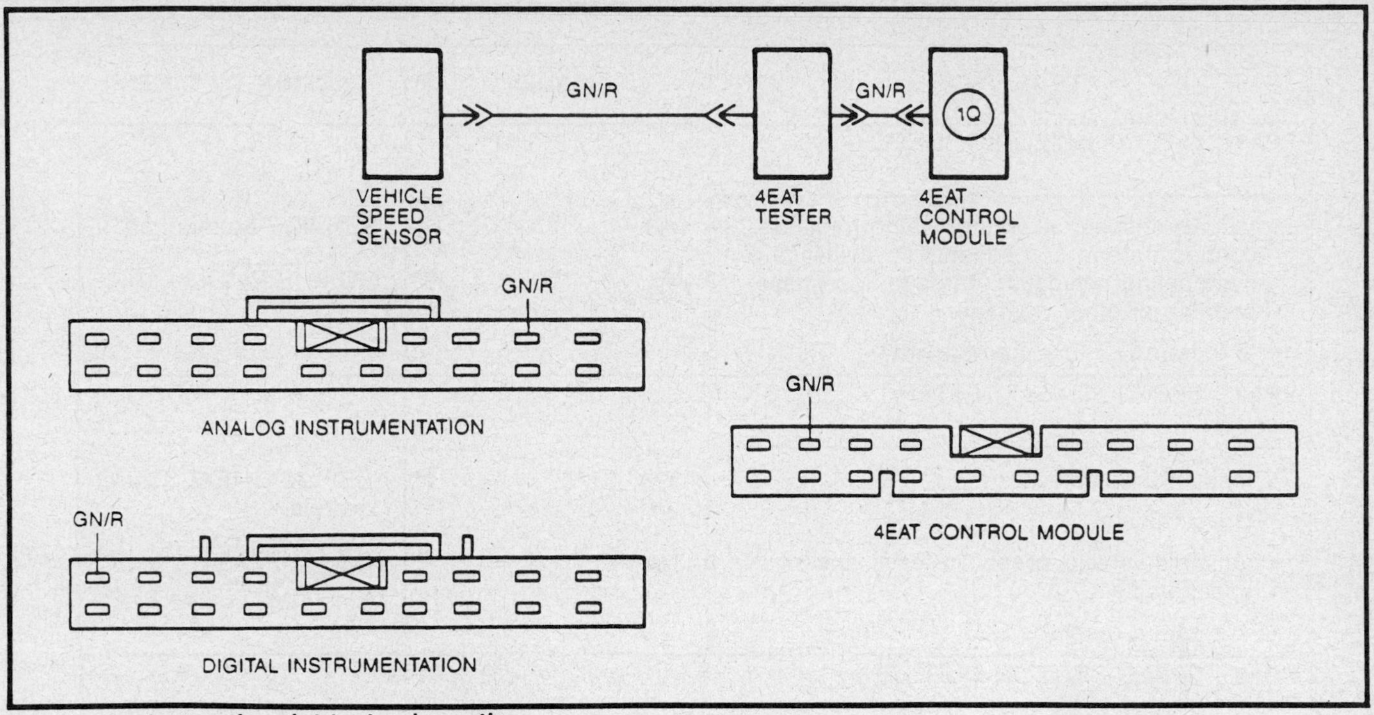
GN/R
GN/R
1Q
VEHICLE SPEED SENSOR
4EAT TESTER
4EAT CONTROL MODULE
GN/R
ANALOG INSTRUMENTATION
GN/R
4EAT CONTROL MODULE
GN/R
DIGITAL INSTRUMENTATION

Service code 6 – pinpoint test schematic

BK Black	N Natural
BL Blue	O Orange
BR Brown	PK Pink
DB Dark Blue	P Purple
DG Dark Green	R Red
GY Gray	T Tan
GN Green	W White
LB Light Blue	Y Yellow
LG Light Green	

Standard Ford color abbreviations

SYSTEM NOT OK

If the manual shift light flashes during driving, run Key On- Engine Off Test to completion. If the manual shift light never comes on go, to Pinpoint Test PPQ.

Pinpoint Test

Each Pinpoint Test assumes that a fault has been detected in the system with direction to enter a specific repair routine. Doing any Pinpoint Test without direction may produce incorrect results and replacement of non-defective components.

In using the Pinpoint Tests, follow each step in order, starting from the first step in the appropriate test. Follow each step until the fault is found.

After completing any repairs to the 4EAT system, verify all components are properly reconnected and repeat the Quick test.

SERVICE CODE 6/PPA TEST

Enter this Pinpoint Test only when this service code is received in the Quick Test steps. This Pinpoint Test is intended to diagnose only the following: speed sensor, wiring harness, 4EAT control module.

SERVICE CODE 12/PPB TEST

Enter this Pinpoint Test only when this service code is received in the Quick Test steps. This Pinpoint Test is intended to diagnose only the following: TP sensor, sensor harness circuit, 4EAT control module.

SERVICE CODE 55/PPC TEST

Enter this Pinpoint Test only when this service code is received in the Quick Test steps. This Pinpoint Test is intended to diagnose only the following: pulse generator, wiring harness, 4EAT control module.

SERVICE CODE 60/PPD TEST

Enter this Pinpoint Test only when this service code is received in the Quick Test steps. This Pinpoint Test is intended to diagnose only the following: 1–2 Shift Solenoid, harness circuit (BL)

SERVICE CODE 61/PPE TEST

Enter this Pinpoint Test when this service code is received in the Quick Test steps. This Pinpoint Test is intended to diagnose only the following: 2–3 Shift Solenoid, harness circuit (BL/BK)

SERVICE CODE 62/PPF TEST

Enter this Pinpoint Test when this service code is received in the Quick Test steps. This Pinpoint Test is intended to diagnose only the following: 3–4 Shift Solenoid, harness circuit (BL/O)

SERVICE CODE 63/PPG TEST

Enter this Pinpoint Test only when this service code is received in the Quick Test steps. This Pinpoint Test is intended to diagnose only the following: lockup solenoid, harness circuit (BL/W)

NO SERVICE CODE/PPM TEST

Enter this Pinpoint Test only when directed here by Quick Test steps. This Pinpoint Test is intended to diagnose only malfunctioning switches.

NO SERVICE CODE/PPQ TEST

Enter this Pinpoint Test only when directed here by Quick Test steps. This Pinpoint Test is intended to diagnose only the following: wiring harness problems such as STI and STO connectors, VPWR, KAPWR, ground and 4EAT control module.

TEST STEP		RESULT ►	ACTION TO TAKE
PPA1	SYSTEM INTEGRITY CHECK		
• Visually inspect all wiring, wiring harness, connectors and components for evidence of overheating, insulation damage, looseness, shorting or other damage. • Is there any cause for concern?		Yes ► No ►	SERVICE as required. GO to PPA2
PPA2	SPEED AT 4EAT TESTER		
• Install 4EAT tester. • Drive vehicle. • Observe vehicle speed on tester and on speedometer. • Are speed readings OK?		Yes ► No ►	REPLACE 4EAT control module. GO to PPA3
PPA3	SPEED SENSOR OUTPUT		
• Key on. • VOM on 20 volt scale. • Measure voltage between speed sensor "GN/R" wire and ground. **Analog Instrumentation:** Slowly turn speedometer cable one turn. Does voltage reading 4.5 volts show four times? 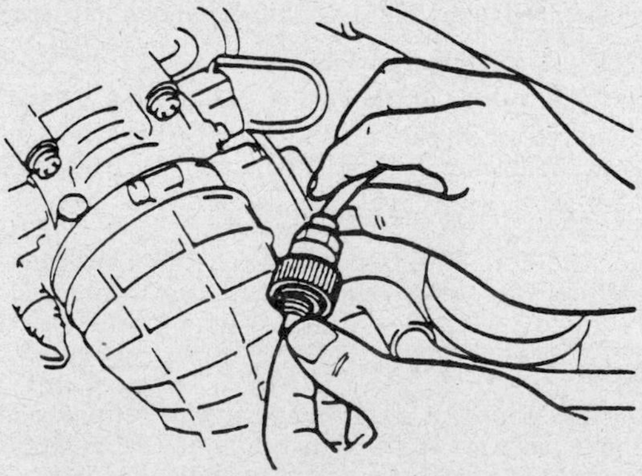 **Digital Instrumentation:** Driving .... Above 4.5v Vehicle stopped .... 4.5v or below 1.5v • Are voltage readings OK?		Yes ► No ►	REPAIR "GN/R" wire from 4EAT module to speed sensor. REPLACE speed sensor. NOTE: To prevent replacement of a good speed sensor be aware that: – The rotor may be damaged – Installation of rotor or sensor may be incorrect.

Service code 6 test information

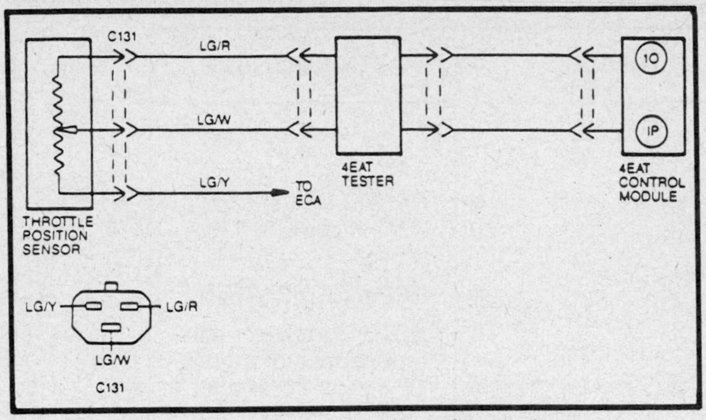
C131
LG/R
LG/W
LG/Y
TO ECA
4EAT TESTER
10
IP
4EAT CONTROL MODULE
THROTTLE POSITION SENSOR
LG/Y
LG/R
LG/W
C131

Service code 12 —pinpoint test schematic

OPERATIONAL TESTS

Description

These test are used to determine the cause of (and provide the corrective actions for) malfunctions most likely to occur. These include the torque converter, the powertrain, the friction elements (clutches and bands), the power source or hydraulic system and the associated regulating valves and controls.

OPA Test—Powertrain Function Check

This test checks for slippage of the friction components (clutches and band brakes) and the torque converter capacity.

1. To perform the test, start the engine and allow it to come to normal operating temperature or until the ATF temperature reaches 122–176°F (50–80°C). Apply both the parking and service brakes during the test.

TEST STEP		RESULT	ACTION TO TAKE
PPB1	SYSTEM INTEGRITY CHECK		
	• Visually inspect all wiring, wiring harness, connectors and components for evidence of overheating, insulation damage, looseness, shorting or other damage. • Is there any cause for concern?	Yes ► No ►	SERVICE as required. GO to PPB2.
PPB2	TP SIGNAL AT 4EAT		
	• Install 4EAT tester. • Key on. • Measure TP voltage between tester terminal 10 and ground. Key on .... 4–6v Key off .... below 1.5v Tester TP and ground terminals **Throttle** — **Approximate Voltage** Closed .... 0.5v 1/8 .... 1.0v 2/8 .... 1.5v 3/8 .... 2.0v 4/8 .... 2.5v 5/8 .... 3.0v 6/8 .... 3.5v 7/8 .... 4.0v Full .... above 4.3v • Are voltage readings OK?	Yes ► No ►	REPLACE 4EAT control module. GO to PPB3.

Service code 12 test information

TEST STEP		RESULT ►	ACTION TO TAKE
PPB3	TP SENSOR SIGNAL		
• Leave TP sensor connected. • Key on. • Measure voltage between TP sensor "LG/R" and "LG/Y." Key on .... 4–6v Key off .... below 1.5v TP sensor "LG/W" and "LG/Y" **Throttle** — **Voltage** Closed .... 0.5v 1/8 .... 1.0v 2/8 .... 1.5v 3/8 .... 2.0v 4/8 .... 2.5v 5/8 .... 3.0v 6/8 .... 3.5v 7/8 .... 4.0v Full .... above 4.3v • Are voltage readings OK?		Yes ► No ►	REPAIR wire(s) from 4EAT module to TP sensor. REPLACE TP sensor. NOTE: To prevent the replacement of a good TP sensor be aware that: – Idle speeds/throttle stop adjustments may need setting. – Binding shaft/linkage.

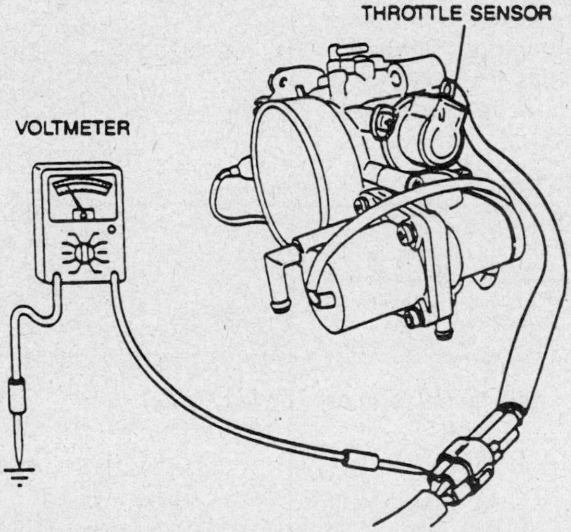
THROTTLE SENSOR
VOLTMETER

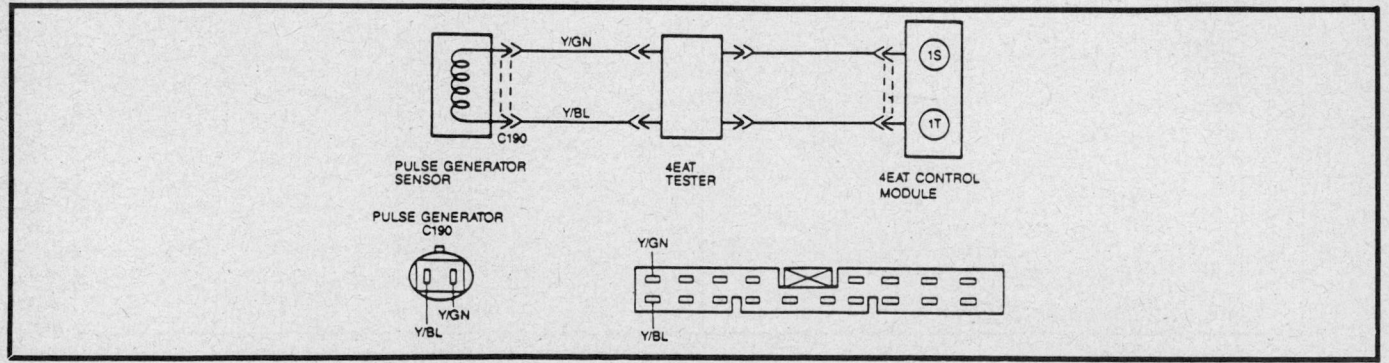

Service code 12 test information

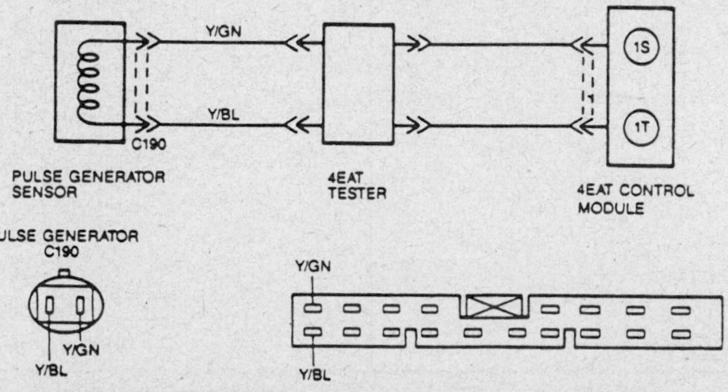
Y/GN
Y/BL
C190
PULSE GENERATOR SENSOR
4EAT TESTER
1S
1T
4EAT CONTROL MODULE
PULSE GENERATOR C190
Y/GN
Y/BL
Y/GN
Y/BL

Service code 55 – pinpoint test schematic

TEST STEP	RESULT ►	ACTION TO TAKE
PPC1 SYSTEM INTEGRITY CHECK		
• Visually inspect all wiring, wiring harness, connectors and components for evidence of overheating, insulation damage, looseness, shorting or other damage. • Is there any cause for concern?	Yes ► No ►	SERVICE as required. GO to PPC2 .
PPC2 PG SIGNAL AT 4EAT		
• Install 4EAT tester. • Drive vehicle. • Observe drum speed on tester and engine speed on tachometer. • Are speeds almost the same?	Yes ► No ►	REPLACE 4EAT control module. GO to PPC3 .
PPC3 RESISTANCE AT PG SENSOR		
• Key off. • VOM on 200 ohm scale. • Measure resistance between PG sensor terminals. • Is resistance 200–400 ohms?	Yes ► No ►	REPAIR wire(s) from PG sensor to 4EAT module. REPLACE PG sensor. NOTE: To prevent the replacement of a good PG sensor be aware that: – The rotor may be damaged – The rotor/sensor may not be installed properly.

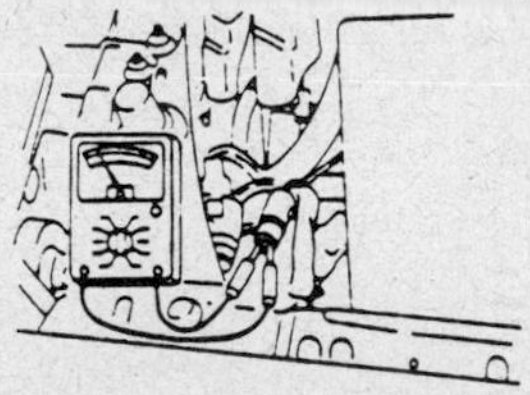

Service code 55 test information

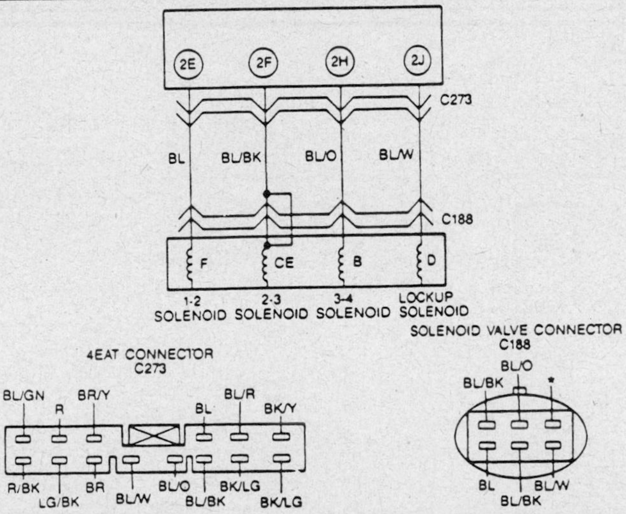
2E
2F
2H
2J
C273
BL
BL/BK
BL/O
BL/W
C188
F
CE
B
D
1-2 SOLENOID
2-3 SOLENOID
3-4 SOLENOID
LOCKUP SOLENOID
SOLENOID VALVE CONNECTOR C188
4EAT CONNECTOR C273
BL/GN
R
BR/Y
BL/R
BL
BK/Y
R/BK
LG/BK
BR
BL/W
BL/O
BL/BK
BK/LG
BK/LG
BL/O
BL/BK
BL
BL/W
BL/BK

Service code 60 – pinpoint test schematic

TEST STEP		RESULT ►	ACTION TO TAKE
PPD1	SYSTEM INTEGRITY CHECK		
	• Visually inspect all wiring, wiring harness, connectors and components for evidence of overheating, insulation damage, looseness, shorting or other damage. • Is there any cause for concern?	Yes ► No ►	SERVICE as required. GO to PPD2 .

TEST STEP		RESULT ►	ACTION TO TAKE
PPD2	SOLENOID RESISTANCE CHECK		
	• Disconnect solenoid valve (connector C188). • VOM on 200 ohm scale. • Measure resistance between connector C188 terminal F ("BL") and ground. • Is resistance between 13 to 27 ohms?	Yes ► No ►	GO to PPD3 . REPLACE solenoid. NOTE: To prevent the replacement of a good solenoid be aware that: – Mechanical functions in the transaxle must operate properly.

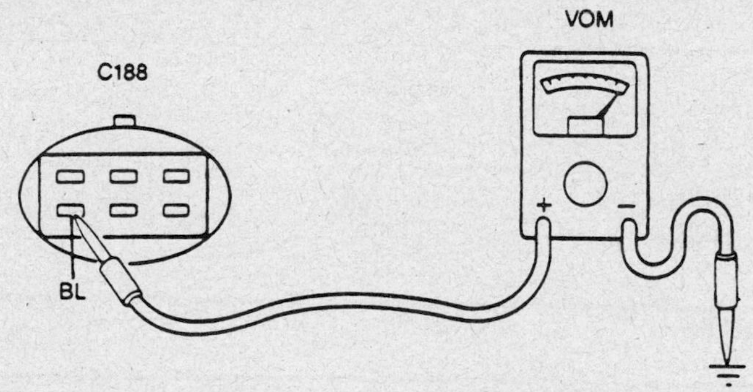
VOM
C188
BL
+
–

Service code 60 test information

TEST STEP		RESULT ►	ACTION TO TAKE
PPD3	CIRCUIT CONTINUITY CHECK		
• Disconnect 4EAT control module (connector C273). • Leave solenoid valve (connector C188) disconnected. • VOM on 200 ohm scale. • Measure resistance between connector C273 terminal 2E ("BL") and connector C188 terminal F ("BL"). • Is resistance less than 5 ohms?		Yes ► No ►	GO to PPD4 . REPAIR open in "BL" wire between 4EAT control module and solenoid valve.
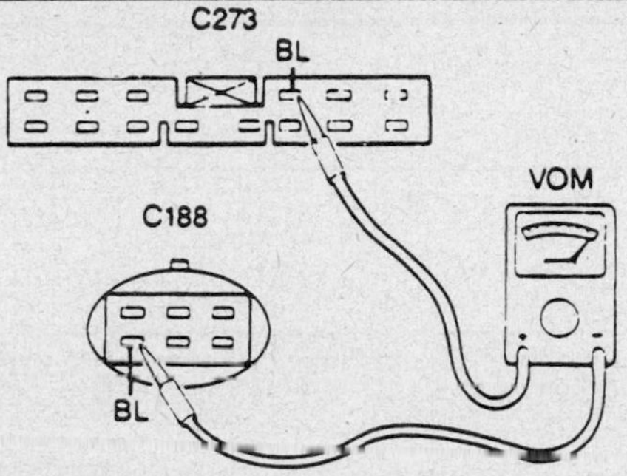C273 BL VOM C188 BL			
PPD4	SHORT TO VPWR CHECK		
• Leave 4EAT control module (connector C273) and solenoid valve (connector C188) disconnected. • Key on; engine off. • VOM on 20 volt scale. • Measure voltage between connector C273 terminal 2E ("BL") and ground. • Is voltage greater than 0 volts?		Yes ► No ►	REPAIR "BL" wire between 4EAT control module and solenoid valve for short to VPWR. GO to PPD5 .
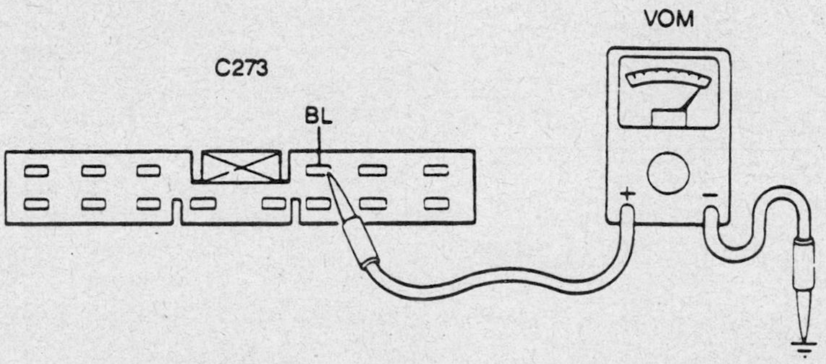VOM C273 BL + –			

Service code 60 test information

TEST STEP		RESULT ►	ACTION TO TAKE
PPD5	SHORT TO GROUND CHECK		
• Leave 4EAT control module (connector C273) and solenoid valve (connector C188) disconnected. • VOM on 200K ohm scale. • Measure resistance between connector C273 terminal 2E ("BL") and ground. • Is resistance greater than 10,000 ohms? See illustration in TEST STEP PPD4		Yes ► No ►	GO to 4EAT operational test OPS. REPAIR "BL" wire between 4EAT control module and solenoid valve for short to ground.

Service code 60 test information

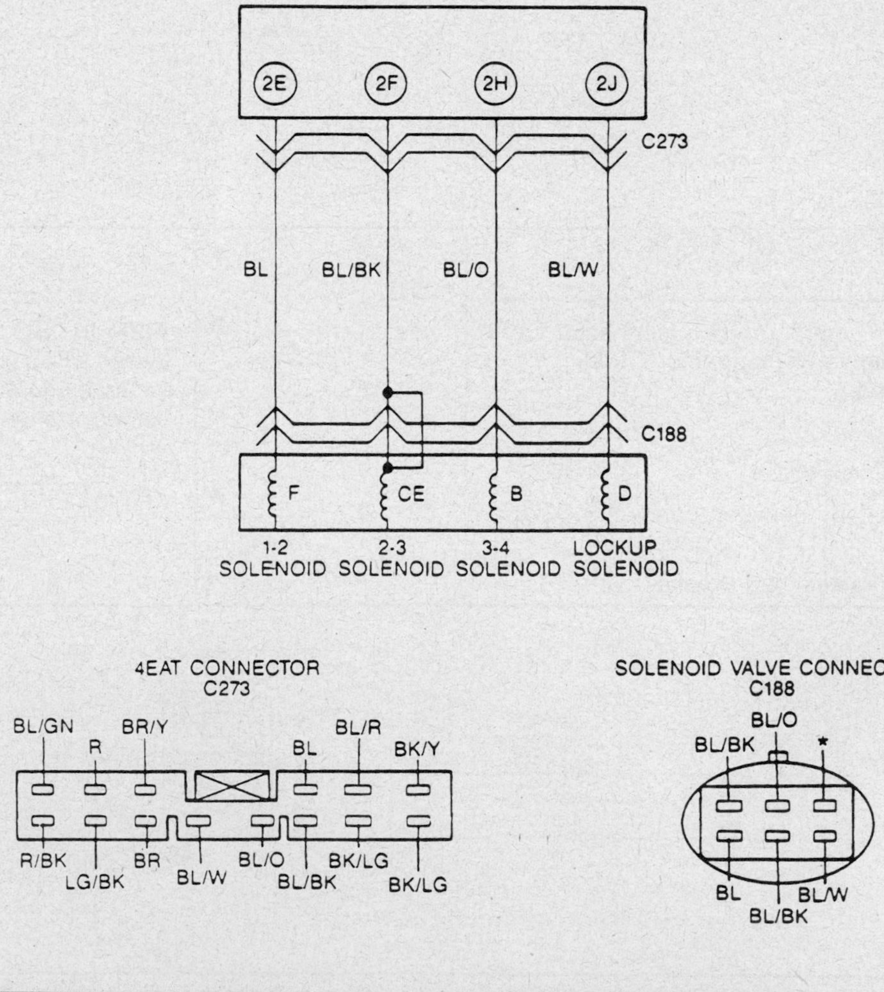
2E
2F
2H
2J
C273
BL
BL/BK
BL/O
BL/W
C188
F
CE
B
D
1-2 SOLENOID
2-3 SOLENOID
3-4 SOLENOID
LOCKUP SOLENOID
4EAT CONNECTOR C273
BL/GN
R
BR/Y
BL
BL/R
BK/Y
R/BK
LG/BK
BR
BL/W
BL/O
BL/BK
BK/LG
BK/LG
SOLENOID VALVE CONNECTOR C188
BL/O
BL/BK
*
BL
BL/BK
BL/W

Service code 61 – pinpoint test schematic

TEST STEP	RESULT ►	ACTION TO TAKE
PPE2 SOLENOID RESISTANCE CHECK		
• Disconnect solenoid valve (connector C188). • VOM on 200 ohm scale. • Measure resistance between connector C188 terminal C ("BL/BK") and ground, and between terminal E ("BL/BK") and ground. • Is resistance between 13 to 27 ohms?	Yes ► No ►	GO to **PPE3** . REPLACE solenoid. NOTE: To prevent the replacement of a good solenoid be aware that: – Mechanical functions in the transaxle must operate properly.

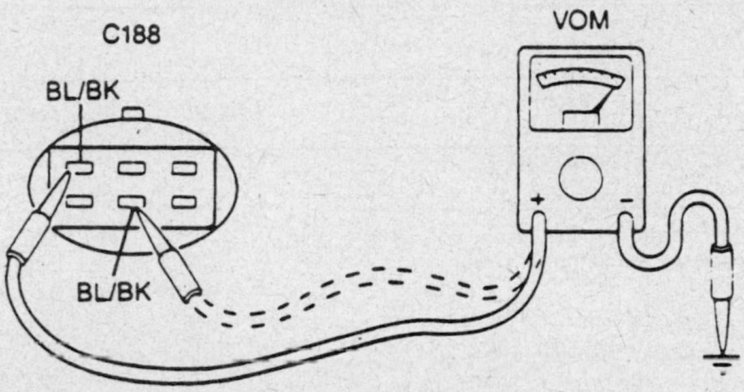
C188
BL/BK
VOM
BL/BK

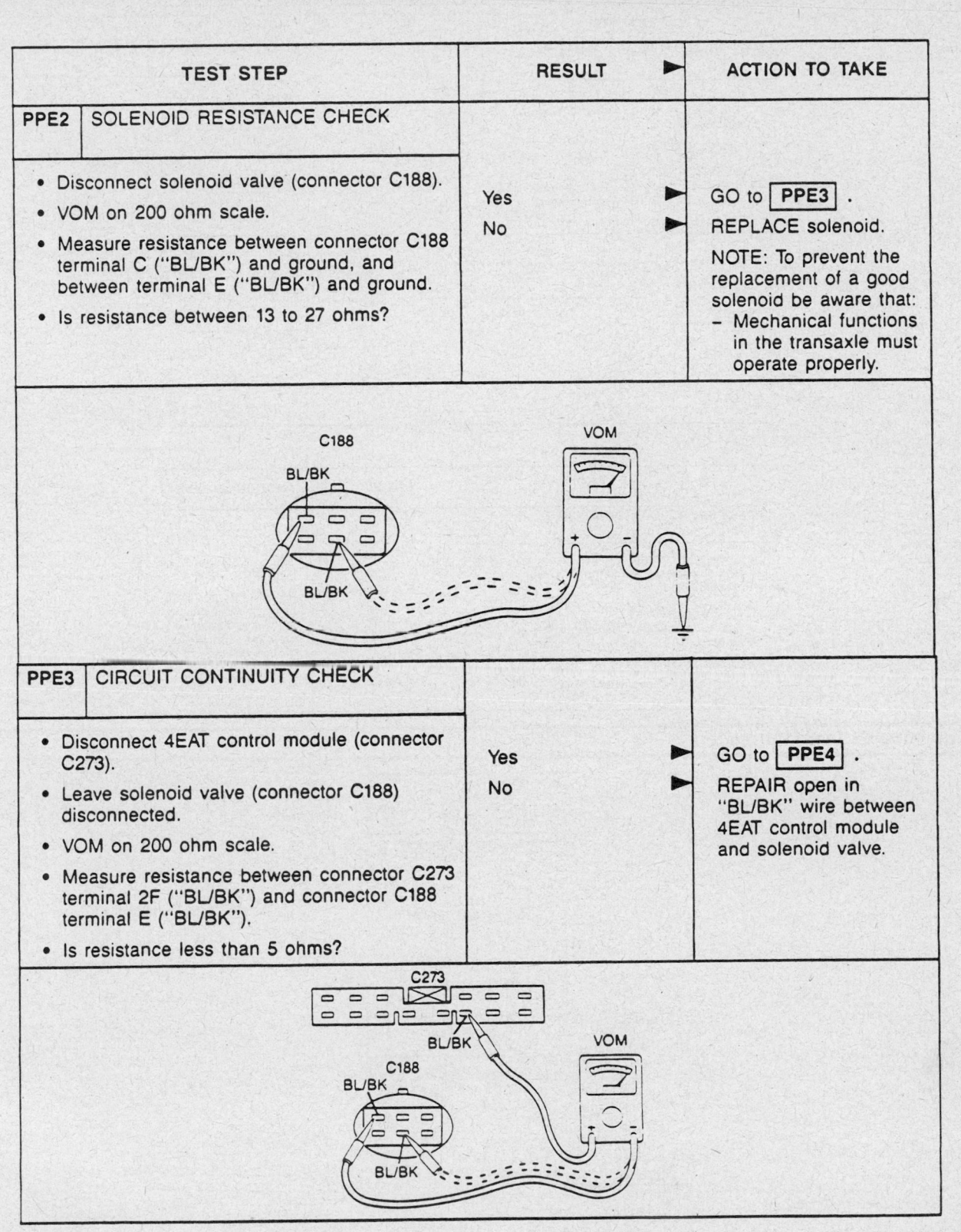

TEST STEP	RESULT ►	ACTION TO TAKE
PPE3 CIRCUIT CONTINUITY CHECK		
• Disconnect 4EAT control module (connector C273). • Leave solenoid valve (connector C188) disconnected. • VOM on 200 ohm scale. • Measure resistance between connector C273 terminal 2F ("BL/BK") and connector C188 terminal E ("BL/BK"). • Is resistance less than 5 ohms?	Yes ► No ►	GO to **PPE4** . REPAIR open in "BL/BK" wire between 4EAT control module and solenoid valve.

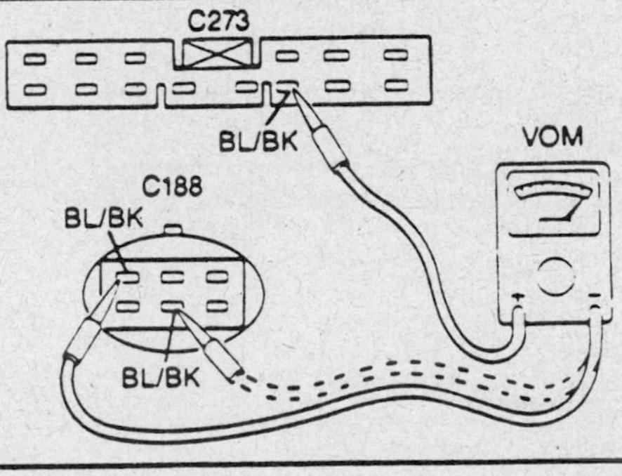
C273
BL/BK
VOM
C188
BL/BK
BL/BK

Service code 61 test information

TEST STEP		RESULT ►	ACTION TO TAKE
PPE4	SHORT TO VPWR CHECK		
• Leave 4EAT control module (connector C273) and solenoid valve (connector C188) disconnected. • Key on; engine off. • VOM on 20 volt scale. • Measure voltage between connector C273 terminal 2F ("BL/BK") and ground. • Is voltage greater than 0 volts?		Yes ► No ►	REPAIR "BL/BK" wire between 4EAT control module and solenoid valve for short to VPWR. GO to PPE5.

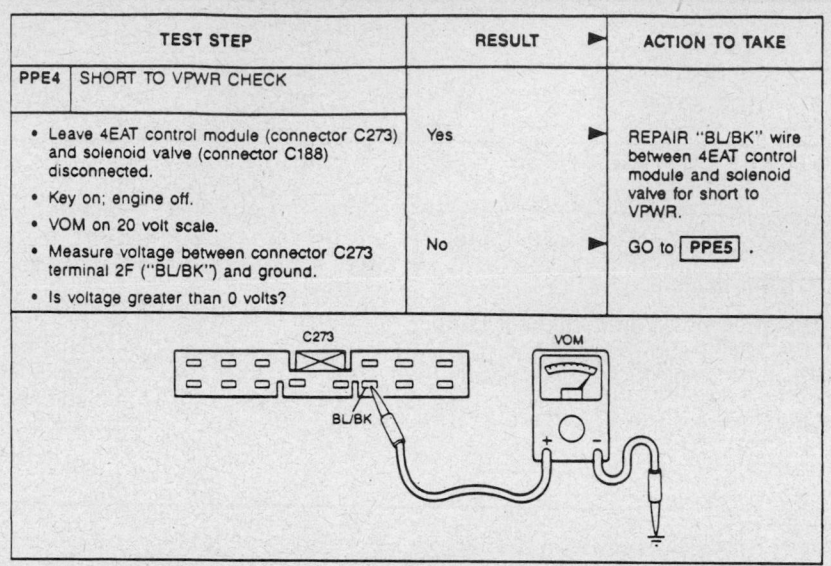

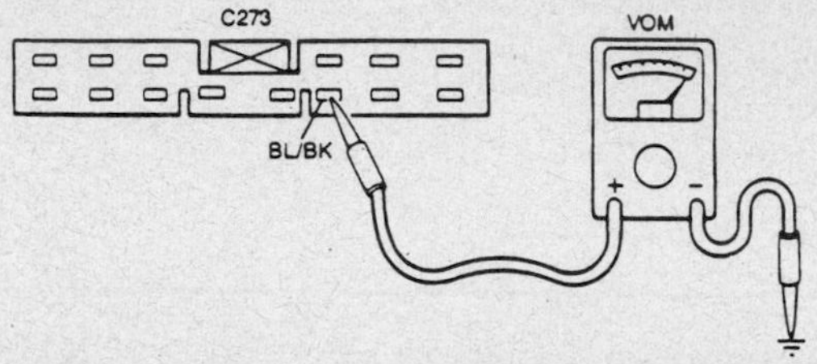
C273
BL/BK
VOM

TEST STEP		RESULT ►	ACTION TO TAKE
PPE5	SHORT TO GROUND CHECK		
• Leave 4EAT control module (connector C273) and solenoid valve (connector C188) disconnected. • VOM on 200K ohm scale. • Measure resistance between connector C273 terminal 2F ("BL/BK") and ground. • Is resistance greater than 10,000 ohms? See illustration in TEST STEP PPE4		Yes ► No ►	GO to 4EAT operational test OPS. REPAIR "BL/BK" wire between 4EAT control module and solenoid valve for short to ground.

Service code 61 test information

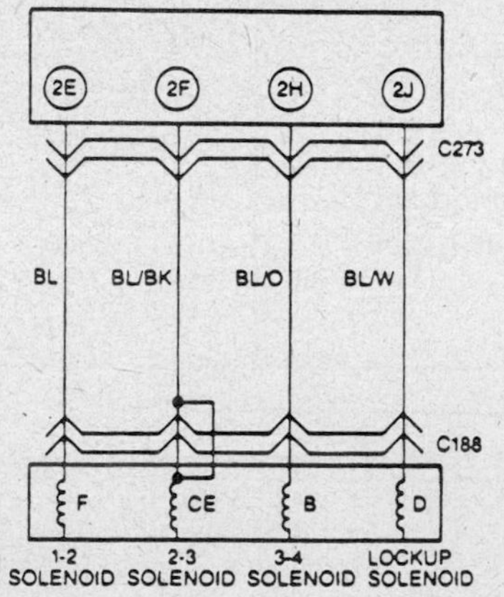
2E
2F
2H
2J
C273
BL
BL/BK
BL/O
BL/W
C188
F
CE
B
D
1-2 SOLENOID
2-3 SOLENOID
3-4 SOLENOID
LOCKUP SOLENOID

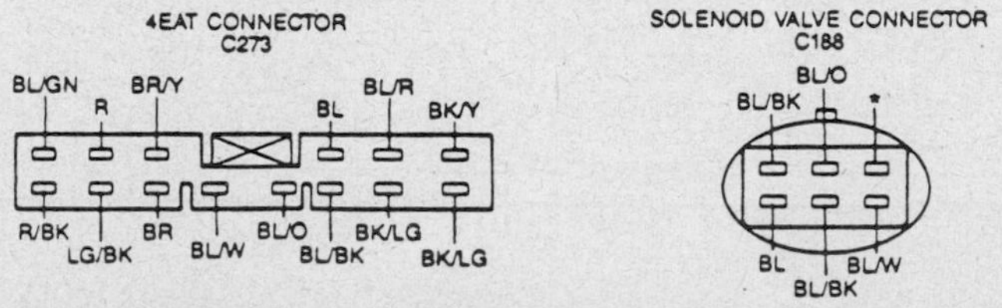
4EAT CONNECTOR C273
BL/GN
R
BR/Y
BL/R
BL
BK/Y
R/BK
LG/BK
BR
BL/W
BL/O
BL/BK
BK/LG
BK/LG
SOLENOID VALVE CONNECTOR C188
BL/O
BL/BK
*
BL
BL/W
BL/BK

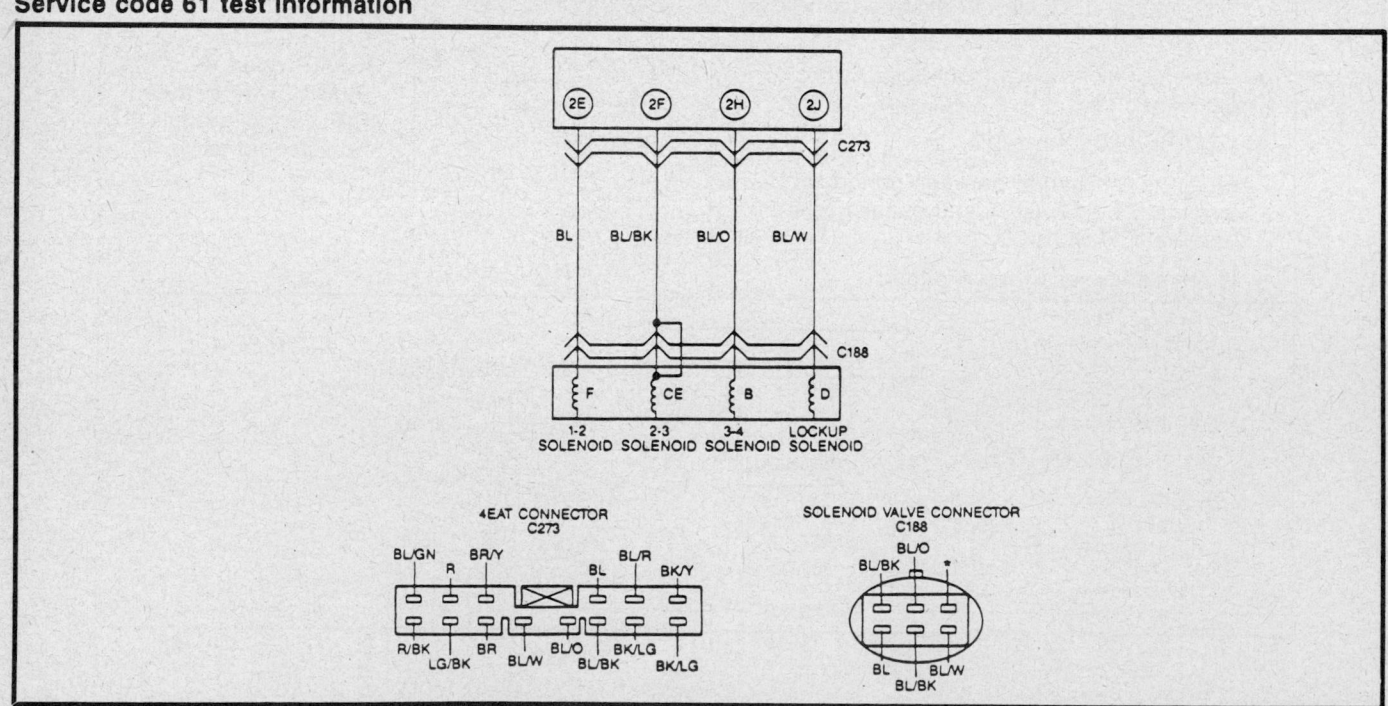

Service code 62 – pinpoint test schematic

TEST STEP	RESULT ►	ACTION TO TAKE
PPF1 SYSTEM INTEGRITY CHECK		
• Visually inspect all wiring, wiring harness, connectors and components for evidence of overheating, insulation damage, looseness, shorting or other damage. • Is there any cause for concern?	Yes ► No ►	SERVICE as required. GO to PPF2 .
PPF2 SOLENOID RESISTANCE CHECK		
• Disconnect solenoid valve (connector C188). • VOM on 200 ohm scale. • Measure resistance between connector C188 terminal B ("BL/O") and ground. • Is resistance between 13 to 27 ohms?	Yes ► No ►	GO to PPF3 . REPLACE solenoid. NOTE: To prevent the replacement of a good solenoid be aware that: – Mechanical functions in the transaxle must operate properly.

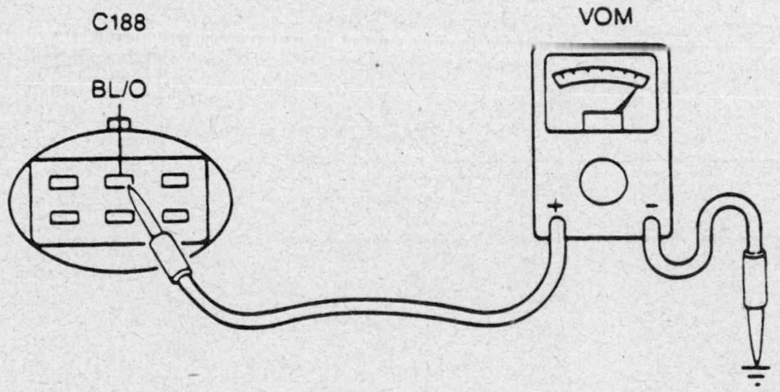
C188
BL/O
VOM
+
–

TEST STEP	RESULT ►	ACTION TO TAKE
PPF3 CIRCUIT CONTINUITY CHECK		
• Disconnect 4EAT control module (connector C273). • Leave solenoid valve (connector C188) disconnected. • VOM on 200 ohm scale. • Measure resistance between connector C273 terminal 2H ("BL/O") and connector C188 terminal B ("BL/O"). • Is resistance less than 5 ohms?	Yes ► No ►	GO to PPF4 . REPAIR open in "BL/O" wire between 4EAT control module and solenoid valve.

Service code 62 test information

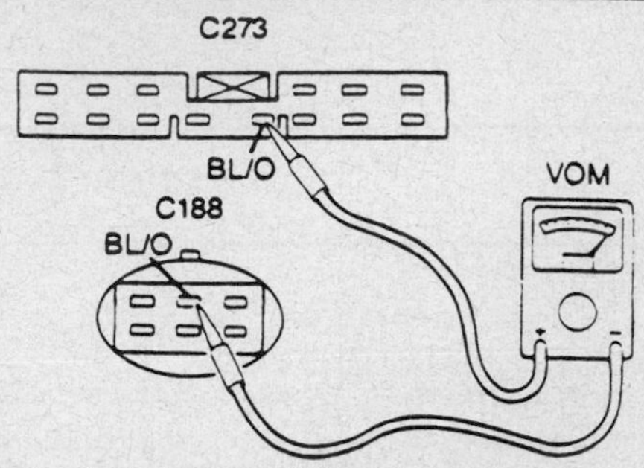
C273
BL/O
C188
BL/O
VOM

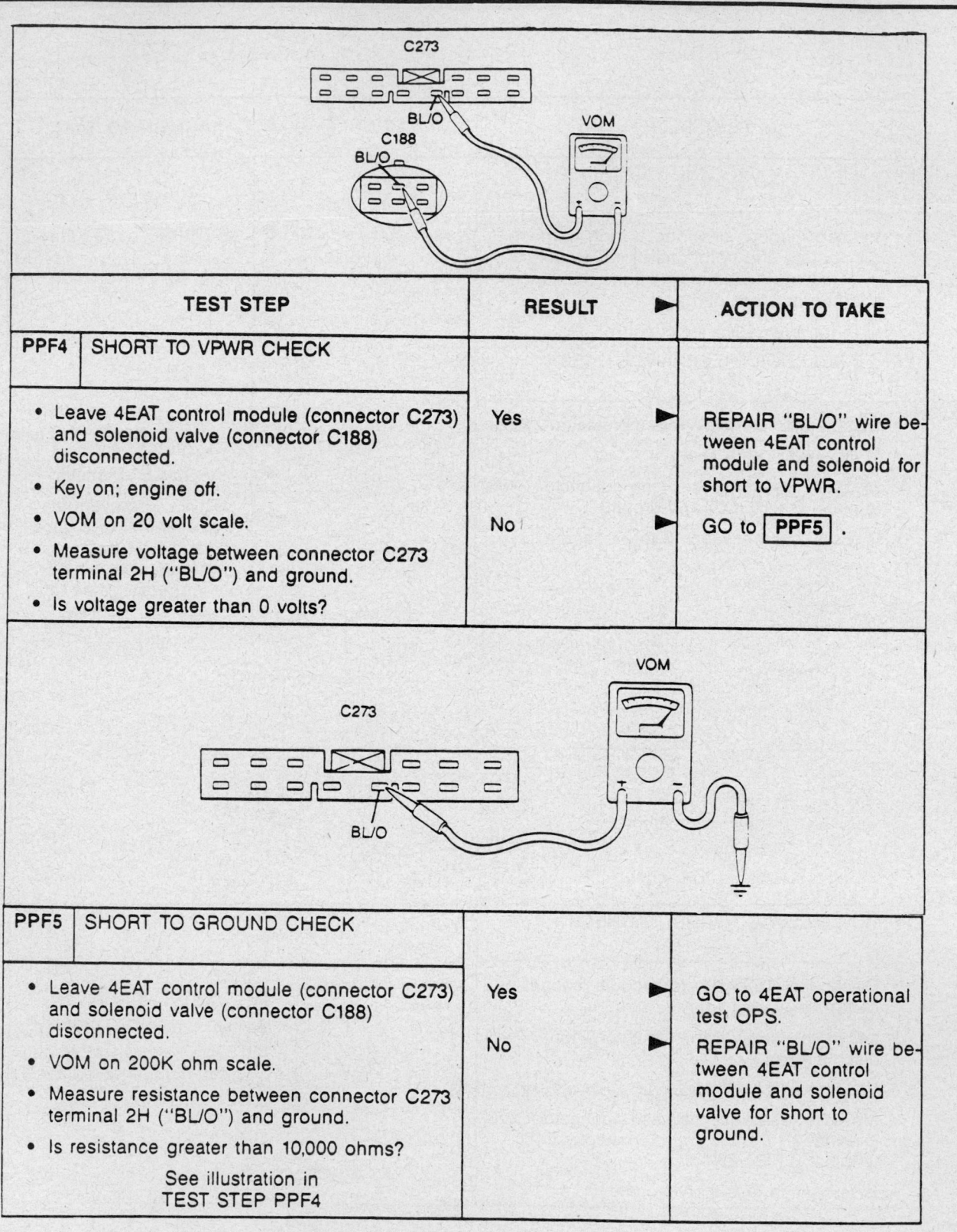

TEST STEP		RESULT ►	ACTION TO TAKE
PPF4	SHORT TO VPWR CHECK		
	• Leave 4EAT control module (connector C273) and solenoid valve (connector C188) disconnected. • Key on; engine off. • VOM on 20 volt scale. • Measure voltage between connector C273 terminal 2H ("BL/O") and ground. • Is voltage greater than 0 volts?	Yes ► No ►	REPAIR "BL/O" wire between 4EAT control module and solenoid for short to VPWR. GO to PPF5 .

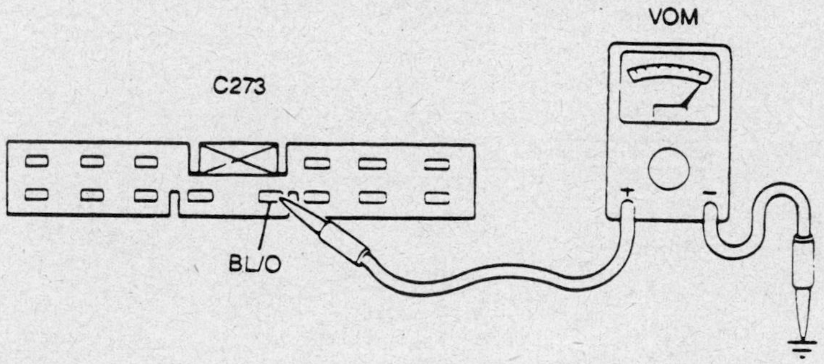
VOM
C273
BL/O

TEST STEP		RESULT ►	ACTION TO TAKE
PPF5	SHORT TO GROUND CHECK		
	• Leave 4EAT control module (connector C273) and solenoid valve (connector C188) disconnected. • VOM on 200K ohm scale. • Measure resistance between connector C273 terminal 2H ("BL/O") and ground. • Is resistance greater than 10,000 ohms? See illustration in TEST STEP PPF4	Yes ► No ►	GO to 4EAT operational test OPS. REPAIR "BL/O" wire between 4EAT control module and solenoid valve for short to ground.

Service code 62 test information

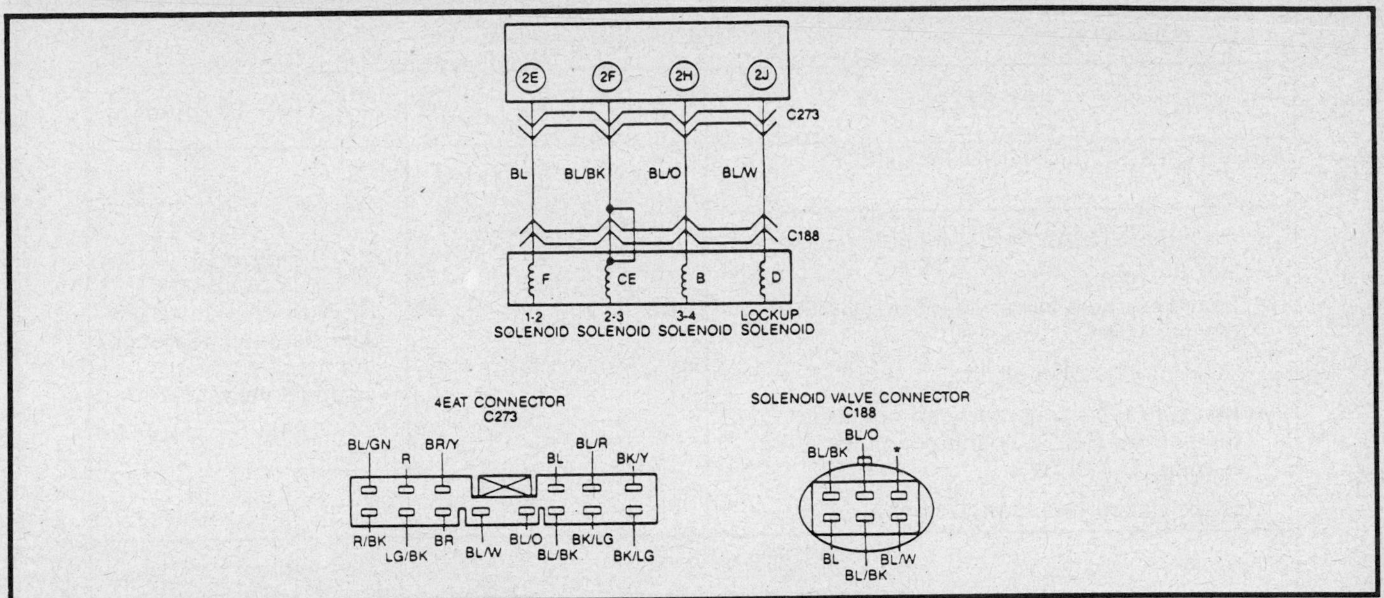
2E
2F
2H
2J
C273
BL
BL/BK
BL/O
BL/W
C188
F
CE
B
D
1-2 SOLENOID
2-3 SOLENOID
3-4 SOLENOID
LOCKUP SOLENOID
4EAT CONNECTOR C273
BL/GN
R
BR/Y
BL
BL/R
BK/Y
R/BK
LG/BK
BR
BL/W
BL/O
BL/BK
BK/LG
BK/LG
SOLENOID VALVE CONNECTOR C188
BL/O
BL/BK
*
BL
BL/W
BL/BK

Service code 63 – pinpoint test schematic

TEST STEP		RESULT	ACTION TO TAKE
PPG1	SYSTEM INTEGRITY CHECK		
• Visually inspect all wiring, wiring harness, connectors and components for evidence of overheating, insulation damage, looseness, shorting or other damage. • Is there any cause for concern?		Yes ► No ►	SERVICE as required. GO to PPG2 .
PPG2	SOLENOID RESISTANCE CHECK		
• Disconnect solenoid valve (connector C188). • VOM on 200 ohm scale. • Measure resistance between connector C188 terminal D ("BL/W") and ground. • Is resistance between 13 to 27 ohms?		Yes ► No ►	GO to PPG3 . REPLACE solenoid. NOTE: To prevent the replacement of a good solenoid be aware that: - Mechanical functions in the transaxle must operate properly.

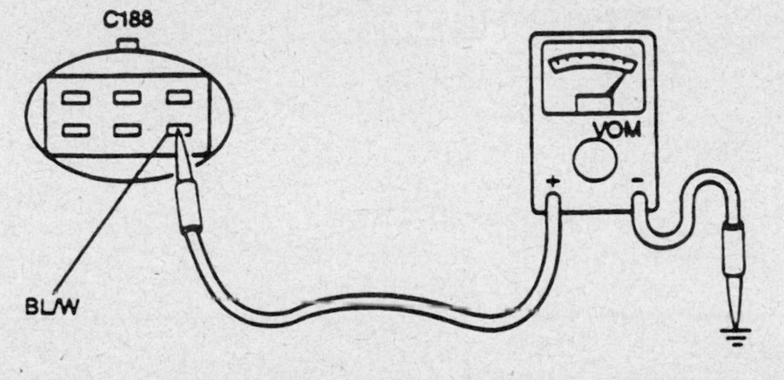
C188
VOM
+
-
BL/W

Service code 63 test information

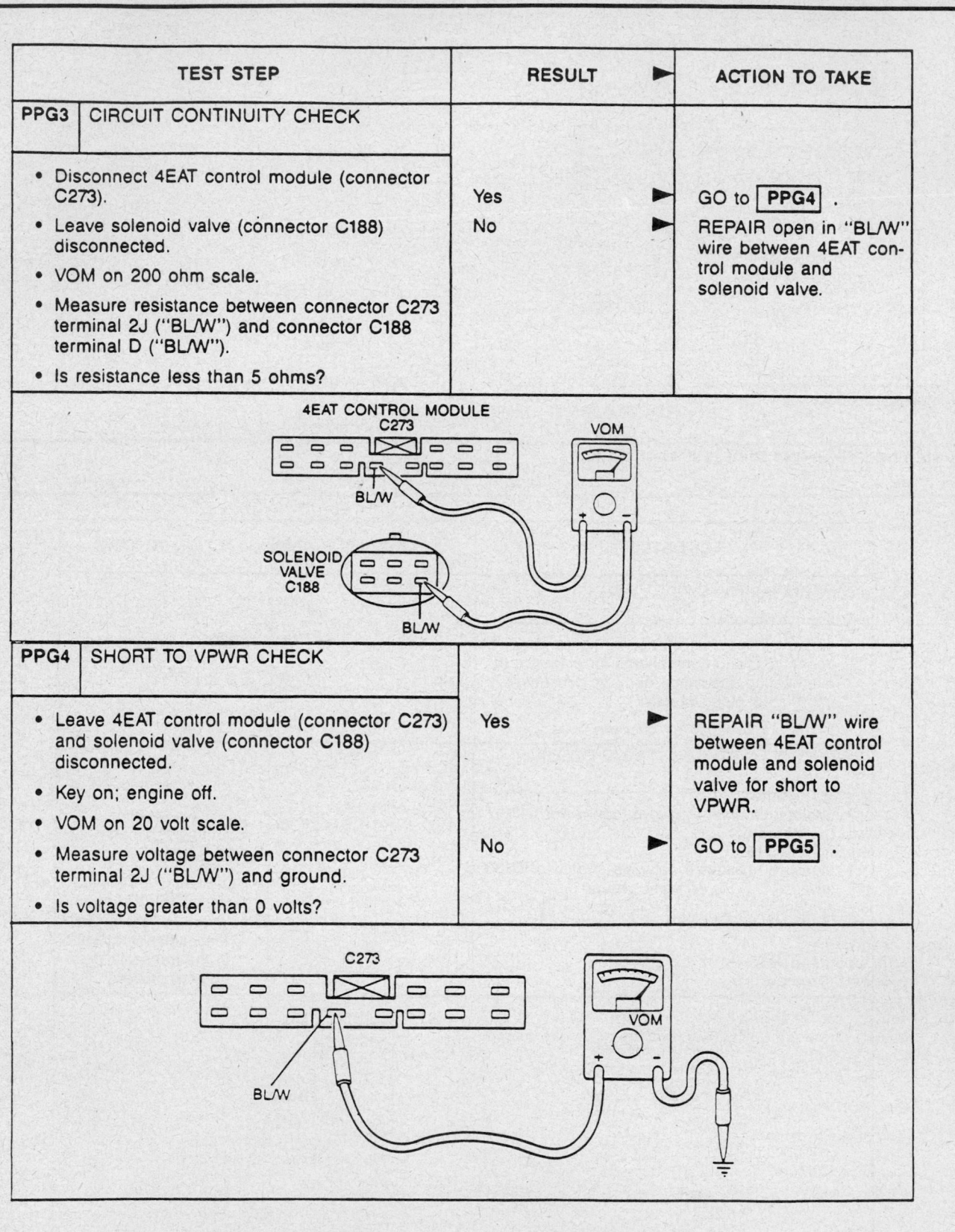

TEST STEP	RESULT ►	ACTION TO TAKE
PPG3 CIRCUIT CONTINUITY CHECK		
• Disconnect 4EAT control module (connector C273). • Leave solenoid valve (connector C188) disconnected. • VOM on 200 ohm scale. • Measure resistance between connector C273 terminal 2J ("BL/W") and connector C188 terminal D ("BL/W"). • Is resistance less than 5 ohms?	Yes ► No ►	GO to **PPG4**. REPAIR open in "BL/W" wire between 4EAT control module and solenoid valve.

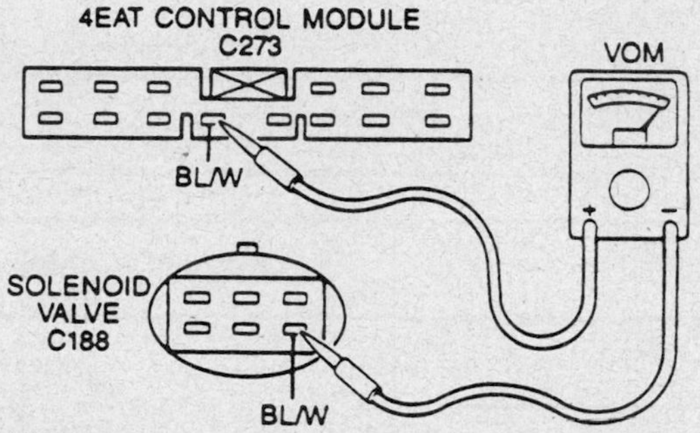
4EAT CONTROL MODULE
C273
VOM
BL/W
SOLENOID VALVE C188
BL/W

TEST STEP	RESULT ►	ACTION TO TAKE
PPG4 SHORT TO VPWR CHECK		
• Leave 4EAT control module (connector C273) and solenoid valve (connector C188) disconnected. • Key on; engine off. • VOM on 20 volt scale. • Measure voltage between connector C273 terminal 2J ("BL/W") and ground. • Is voltage greater than 0 volts?	Yes ► No ►	REPAIR "BL/W" wire between 4EAT control module and solenoid valve for short to VPWR. GO to **PPG5**.

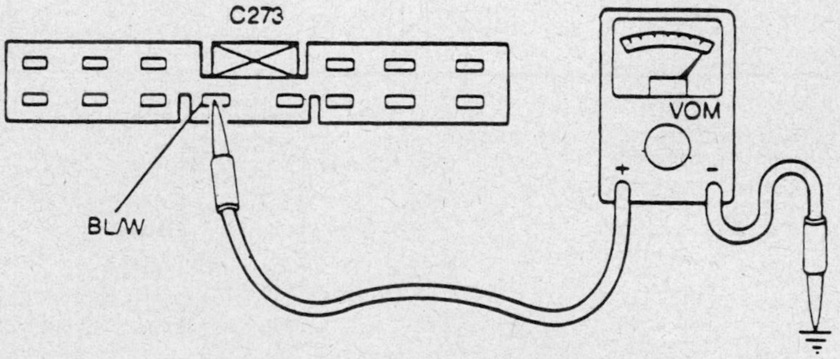
C273
VOM
BL/W

Service code 63 test information

TEST STEP		RESULT ►	ACTION TO TAKE
PPG5	SHORT TO GROUND CHECK		
• Leave 4EAT control module (connector C273) and solenoid valve (connector C188) disconnected. • VOM on 200K ohm scale. • Measure resistance between connector C273 terminal 2J ("BL/W") and ground. • Is resistance greater than 10,000 ohms? See illustration in TEST STEP PPG4		Yes ► No ►	GO to 4EAT operational test OPS. REPAIR "BL/W" wire between 4EAT control module and solenoid valve for short to ground.

Service code 63 test information

CAUTION

Do not allow any one to stand either in front of or behind the vehicle during the stall test. Personal injury could result.

2. Install a tachometer to the engine. Place the selector lever in the desired detent and depress the accelerator to the wide open throttle position, noting the total rpm achieved.

NOTE: Do not hold the throttle open for more than 5 seconds at a time during the test.

3. After the test, move the selector lever to the **N** position and let the engine idle for about a minute to cool the fluid before making a second or third test.
4. Use the Stall Test Evaluation chart to verify the correlation of observed test results with possible causes for the deviations from specifications.

OPB Test—Hydraulic Control System Time Lag Check

This test checks for the time lag between selector lever shift into gear and when a shock is felt, using a stopwatch.

1. To perform the test, start the engine and allow it to come to normal operating temperature or until the ATF temperature reaches 122–176°F (50–80°C). Apply the parking during the test.
2. With the engine idling in **P** at 725–775 rpm, shift from **N** to **OVERDRIVE** and note the elapsed time until a shock is felt, using the stopwatch.
3. Idle the engine in **N** for about a minute to cool the fluid.
4. Repeat test procedure for **N** to **OVERDRIVE** manual mode and **N** to **R**.
5. Repeat procedure 3 times and average the results.
6. Use the Time Lag Evaluation chart to verify the correlation of observed test results with possible causes for the deviations from specifications.

OPC Test—Oil Pressure and Control Check

This test checks the oil pump line pressure, line pressure control, throttle control pressure and oil leakage.

1. To perform the test, start the engine and allow it to come to normal operating temperature or until the ATF temperature reaches 122–176°F (50–80°C). Apply the parking during the test.
2. Connect a tachometer to the engine.
3. Connect a pressure tester with fittings at the line pressure inspection hole (square head plug marked **L**).
4. With the engine idling in **P** at 725–775 rpm, shift the selector lever to the **D** range, then read the line pressure at idle.
5. With the foot brake firmly applied, steadily increase the engine speed to its maximum quickly read the line pressure when the engine speed remains constant, then release the accelerator.

NOTE: This test must be completed within 5 seconds, followed by cooling the ATF in the N range idling for about a minute.

6. Repeat test for each range, making certain to cool the transaxle in between tests.
7. Use the Line Pressure Test Evaluation chart to verify the correlation of observed test results with possible causes for the deviations from specifications.

OPD Test—Throttle Pressure Test

This test checks the the line pressure for checking the hydraulic components and for improper throttle cable adjustments.

1. To perform the test, start the engine and allow it to come to normal operating temperature or until the ATF temperature reaches 122–176°F (50–80°C). Apply the parking during the test.
2. Connect a tachometer to the engine.
3. Connect a pressure tester with fittings at the throttle pressure inspection hole (square head plug marked **T**).
4. With the engine idling in **P** at 725–775 rpm, shift the selector lever to the **OD** range, then read the throttle pressure at idle.
5. With the foot brake firmly applied, steadily increase the engine speed to its maximum quickly read the throttle pressure when the engine speed remains constant, then release the accelerator.

NOTE: This test must be completed within 5 seconds, followed by cooling the ATF in the neutral range idling for about a minute.

6. Use the Throttle Pressure Test Evaluation chart to verify the correlation of observed test results with possible causes for the deviations from specifications.

ROAD TEST

Description

The Road Test is an evaluation of the 4EAT performance with

TEST STEP	RESULT ►	ACTION TO TAKE
PPM1 SYSTEM INTEGRITY CHECK		
• Visually inspect all wiring, wiring harness, connectors and components for evidence of overheating, insulation damage, looseness, shorting or other damage. • Is there any cause for concern?	Yes ► No ►	SERVICE as required. LOCATE malfunctioning switch in the table below and proceed to the appropriate Pinpoint Test Step as indicated in the "Action To Take" column.

SWITCH		ACTION TO TAKE
BRAKE LIGHT SWITCH		GO TO TEST STEP PPM2
IDLE SWITCH		GO TO TEST STEP PPM5
COOLANT TEMPERATURE SWITCH		GO TO TEST STEP PPM8
CHECK CONNECTOR		GO TO TEST STEP PPM12
NEUTRAL SAFETY SWITCH	L	GO TO TEST STEP PPM14
	D	GO TO TEST STEP PPM18
	(D)	GO TO TEST STEP PPM22
	N OR P	GO TO TEST STEP PPM26
MODE SWITCH		GO TO TEST STEP PPM28
MODE INDICATOR		GO TO TEST STEP PPM31
MANUAL SWITCH		GO TO TEST STEP PPM35
MANUAL INDICATOR		GO TO TEST STEP PPM39
NO LOAD SIGNAL		GO TO TEST STEP PPM41
THROTTLE POSITION		GO TO TEST STEP PPM44
AFT SWITCH		GO TO TEST STEP PPM45

System integrity check

TEST STEP	RESULT ►	ACTION TO TAKE
PPM2 BRAKE LIGHT SWITCH VOLTAGE CHECK		
• Disconnect 4EAT control module (connector C274). • Key on; engine off. • VOM on 20 volt scale. • Measure voltage at brake light switch connector C256 between "W/GN" wire and ground. Pedal Depressed — Above 10v Pedal Released — Below 1.5v • Are voltages OK?	Yes ► No ►	GO to **PPM3**. GO to **PPM4**.

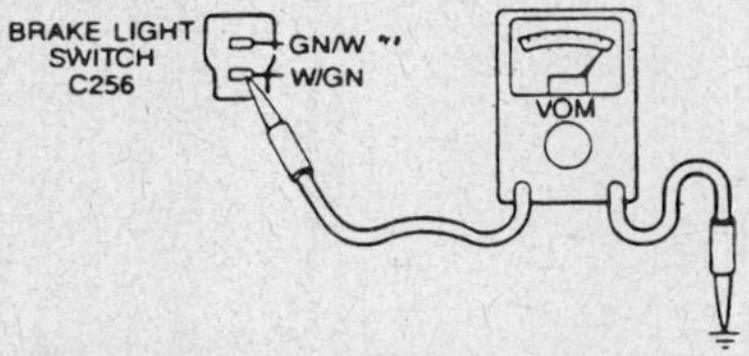
BRAKE LIGHT SWITCH C256
GN/W
W/GN
VOM

TEST STEP	RESULT ►	ACTION TO TAKE
PPM3 CHECK VOLTAGE AT 4EAT CONTROL UNIT		
• 4EAT control module (connector C273) disconnected. • Key on; engine off. • VOM on 20 volt scale. • Measure voltage at 4EAT control module connector C274 between terminal 1N ("W/GN") and ground. Pedal Depressed — Above 10v Pedal Released — 0.0v • Are voltages OK?	Yes ► No ►	REPLACE 4EAT control module. REPAIR "W/GN" wire between 4EAT control module and brake light switch.

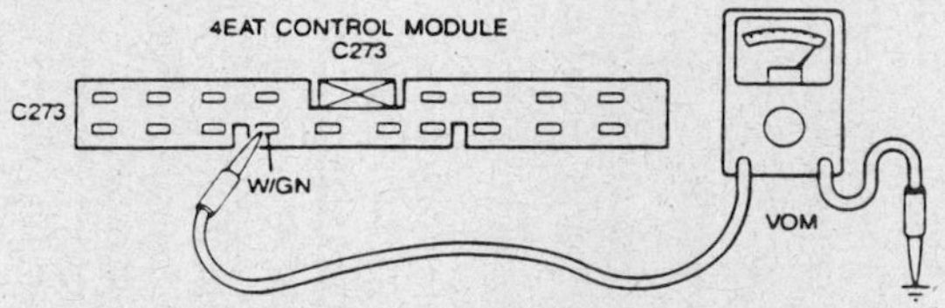
4EAT CONTROL MODULE C273
C273
W/GN
VOM

System integrity check

TEST STEP	RESULT ►	ACTION TO TAKE
PPM4 BRAKE LIGHT SWITCH POWER CHECK		
• Key on; engine off. • VOM on 20 volt scale. • Measure voltage at brake light switch (connector C256) between "GN/W" wire and ground. • Is voltage above 10 volts?	Yes ► No ►	REPAIR "W/GN" wire between brake light switch and 4EAT control module. If OK REPLACE brake light switch. REPAIR "GN/W" wire between brake light switch and fuse box.

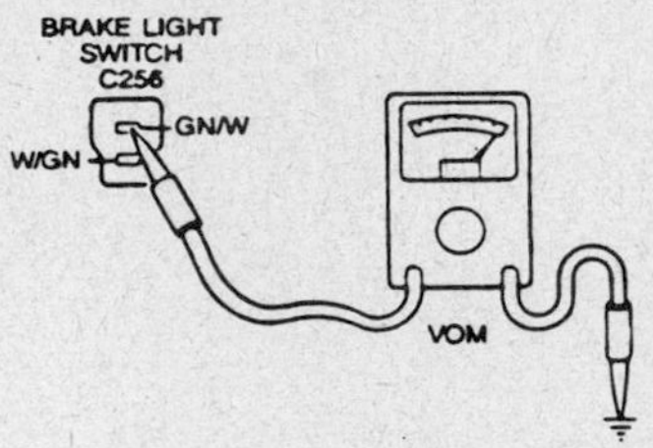
BRAKE LIGHT SWITCH C256
GN/W
W/GN
VOM

TEST STEP	RESULT ►	ACTION TO TAKE
PPM5 IDLE SWITCH FUNCTION CHECK		
• Disconnect idle switch (connector C132). • Measure continuity between connector C132 terminal and ground while exercising the idle switch. **Throttle Position** — **Continuity** Closed — Yes Open — No • Is the idle switch functioning OK?	Yes ► No ►	GO to **PPM6** . REPLACE idle switch.

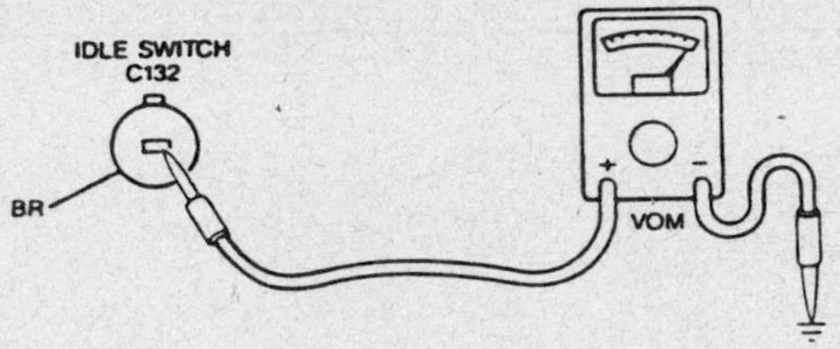
IDLE SWITCH C132
BR
VOM
+
−

System integrity check

TEST STEP	RESULT ►	ACTION TO TAKE
PPM6 CIRCUIT CONTINUITY CHECK		
• Idle switch (connector C132) disconnected. • Disconnect 4EAT control module (connector C274). • VOM on 200 ohm scale. • Measure resistance between connector C274 terminal 1L ("LG/BK") and connector C132 ("BR"). • Is resistance less than 5 ohms?	Yes ► No ►	GO to **PPM7** REPAIR wire between 4EAT control module and idle switch for open.

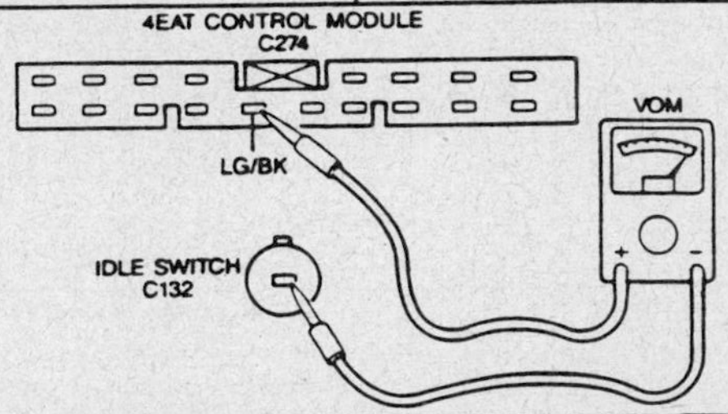
4EAT CONTROL MODULE C274
LG/BK
VOM
IDLE SWITCH C132
+
−

TEST STEP	RESULT ►	ACTION TO TAKE
PPM7 SHORT TO GROUND AND VPWR CHECK		
• Idle switch (connector C132) and 4EAT control module (connectors C273 and C274) disconnected. • Disconnect battery. • Key on. • VOM on 200K ohm scale. • Measure resistance between: Connector C274 terminal 1L ("LG/BK") and ground. Connector C274 terminal 1L ("LG/BK") and connector C273 terminal 2A ("BK/Y"). • Are resistances greater than 10,000 ohms?	Yes ► No ►	REPLACE 4EAT control module. REPAIR wire between 4EAT control module and idle switch for shorts.

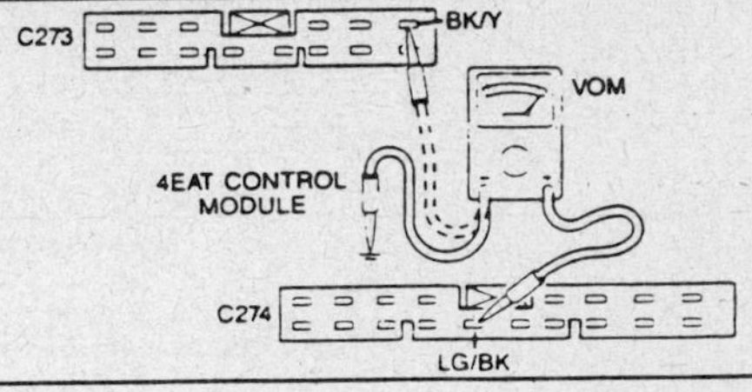
C273
BK/Y
VOM
4EAT CONTROL MODULE
C274
LG/BK

System integrity check

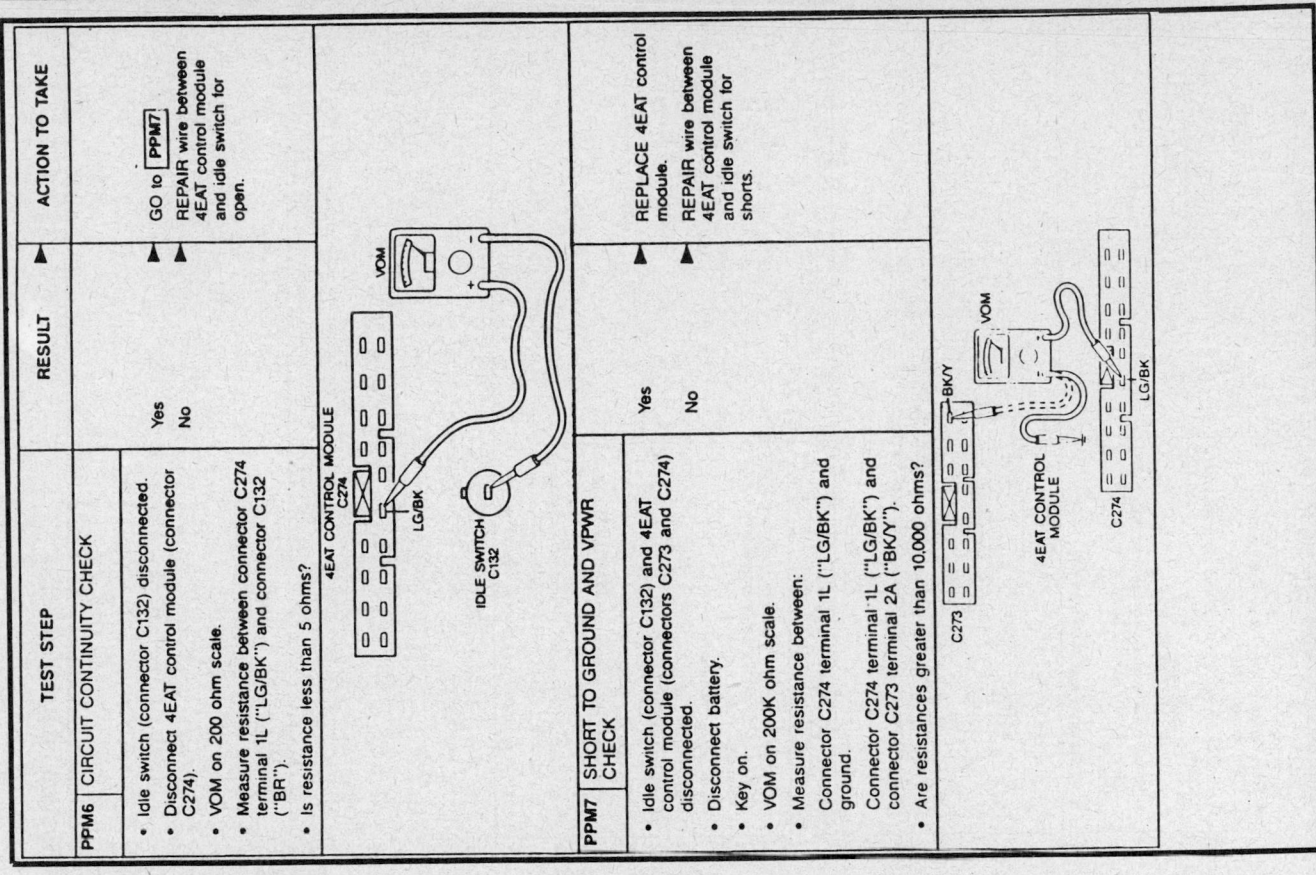

TEST STEP	RESULT ►	ACTION TO TAKE
PPM8 COOLANT TEMPERATURE SWITCH GROUND CHECK		
• Disconnect coolant temperature switch (connector C906). • VOM on 200 ohm scale. • Measure resistance between connector C906 "BK" wire and ground. • Is resistance less than 5 ohms?	Yes ► No ►	GO to PPM9. REPAIR "BK" wire between coolant temperature switch and ground.

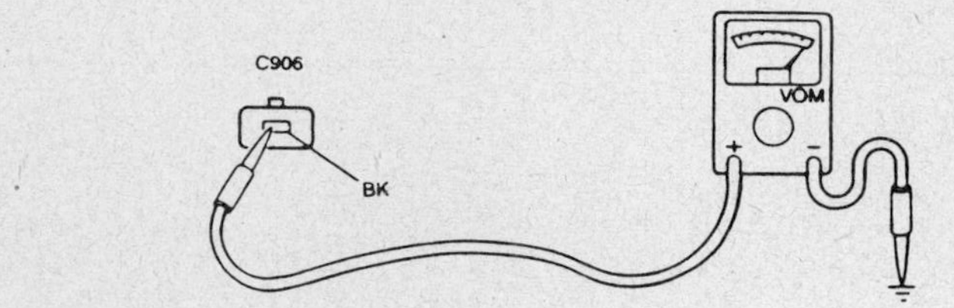
C906
BK
VOM
+
−

TEST STEP	RESULT ►	ACTION TO TAKE
PPM9 COOLANT TEMPERATURE SWITCH FUNCTION CHECK		
• Coolant temperature switch (connector C906) disconnected. • Check continuity between switch terminals as follows: **Coolant** — **Continuity** Below 65°C (149°F) — Yes Above 72°C (162°F) — No • Does the coolant temperature switch function OK?	Yes ► No ►	GO to PPM10. REPLACE coolant temperature switch.

System integrity check

TEST STEP	RESULT ►	ACTION TO TAKE
PPM10 CIRCUIT CONTINUITY CHECK		
• Coolant temperature switch (connector C191) disconnected. • Disconnect 4EAT control module (connector C274). • VOM on 200 ohm scale. • Measure resistance between connector C274 terminal 1G ("LG/R") and connector C191 ("LG/R"). • Is resistance less than 5 ohms?	Yes ► No ►	GO to PPM11. REPAIR "LG/R" wire between 4EAT control module and coolant temperature switch for open.

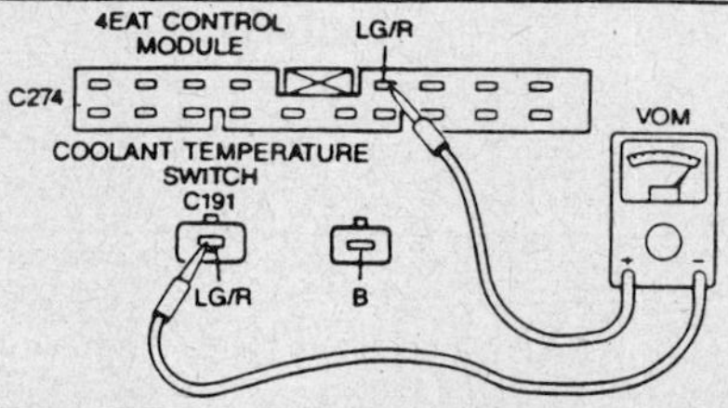
4EAT CONTROL MODULE
LG/R
C274
VOM
COOLANT TEMPERATURE SWITCH
C191
LG/R
B

TEST STEP	RESULT ►	ACTION TO TAKE
PPM11 SHORT TO GROUND AND VPWR CHECK		
• Coolant temperature switch (connector C191) and 4EAT control module (connectors C273 and C274) disconnected. • Disconnect battery. • Key on. • VOM on 200K ohm scale. • Measure resistance between: Connector C274 terminal 1G ("LG/R") and ground. Connector C274 terminal 1G ("LG/R") and connector C273 terminal 2A ("BK/Y"). • Are resistances greater than 10,000 ohms?	Yes ► No ►	REPLACE 4EAT control module. REPAIR "LG/R" wire between 4EAT control module and coolant temperature switch for shorts.

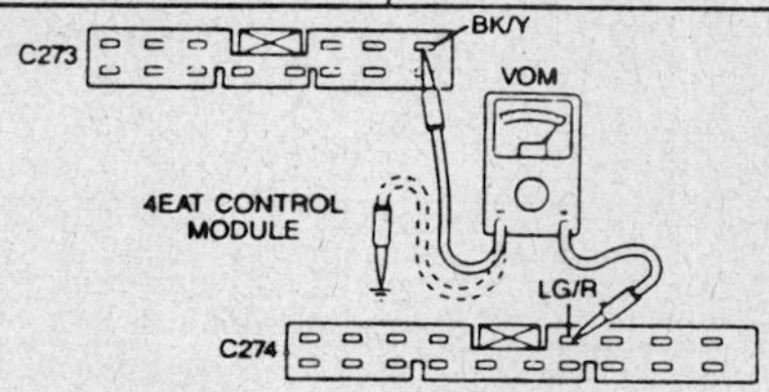
C273
BK/Y
VOM
4EAT CONTROL MODULE
LG/R
C274

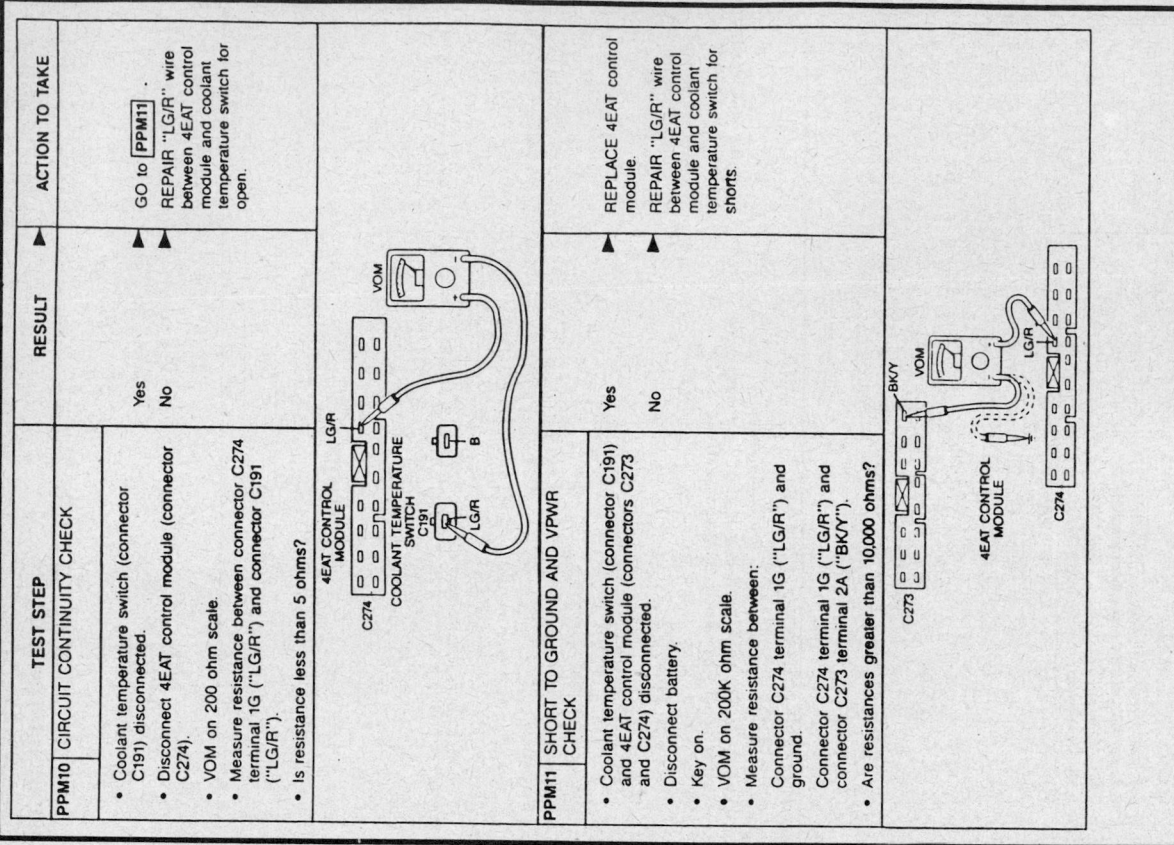

System integrity check

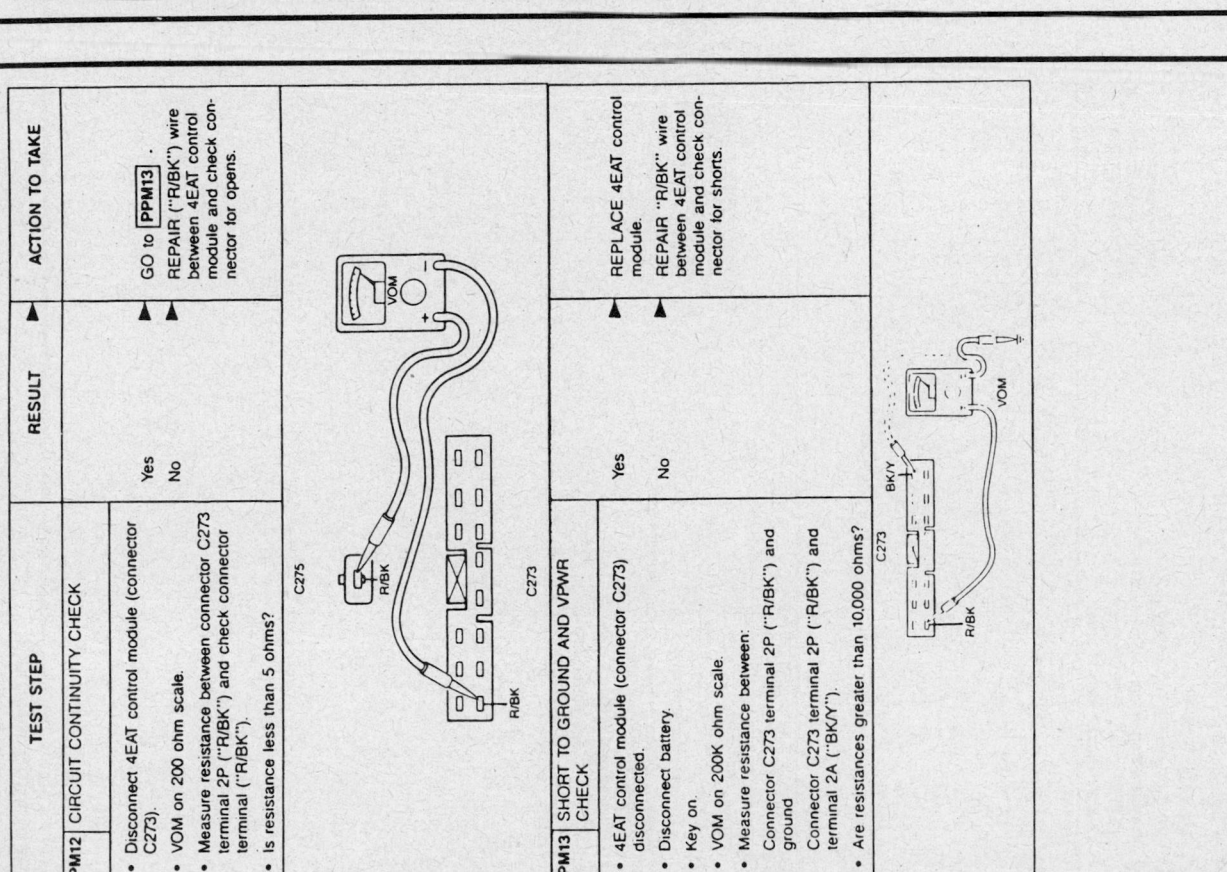

TEST STEP	RESULT ►	ACTION TO TAKE
PPM12 CIRCUIT CONTINUITY CHECK		
• Disconnect 4EAT control module (connector C273). • VOM on 200 ohm scale. • Measure resistance between connector C273 terminal 2P ("R/BK") and check connector terminal ("R/BK"). • Is resistance less than 5 ohms?	Yes ► No ►	GO to PPM13 . REPAIR ("R/BK") wire between 4EAT control module and check connector for opens.

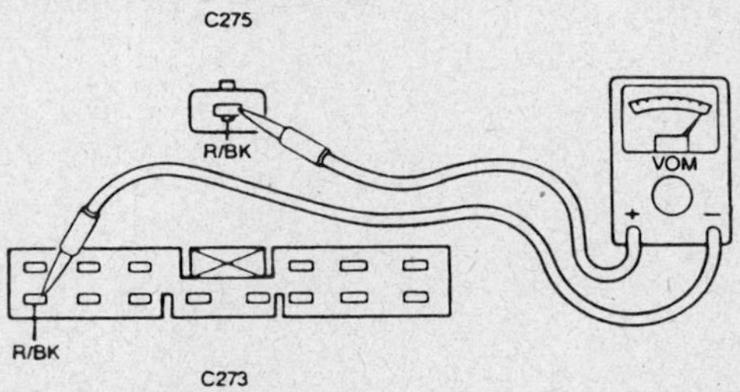
C275
R/BK
VOM
R/BK
C273

TEST STEP	RESULT ►	ACTION TO TAKE
PPM13 SHORT TO GROUND AND VPWR CHECK		
• 4EAT control module (connector C273) disconnected. • Disconnect battery. • Key on. • VOM on 200K ohm scale. • Measure resistance between: Connector C273 terminal 2P ("R/BK") and ground Connector C273 terminal 2P ("R/BK") and terminal 2A ("BK/Y"). • Are resistances greater than 10,000 ohms?	Yes ► No ►	REPLACE 4EAT control module. REPAIR "R/BK" wire between 4EAT control module and check connector for shorts.

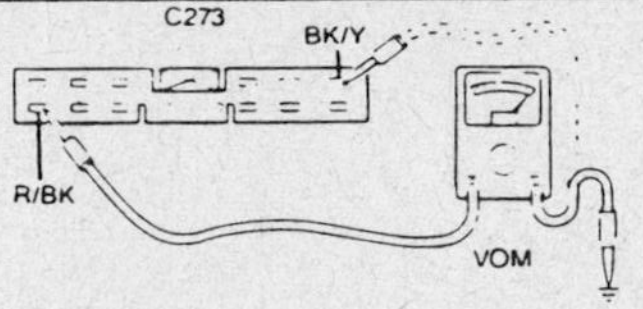
C273
BK/Y
R/BK
VOM

System integrity check

TEST STEP	RESULT ►	ACTION TO TAKE
PPM14 CHECK VOLTAGE AT NEUTRAL SAFETY SWITCH		
• Key on; engine off. • VOM on 20 volt scale. • Disconnect 4EAT control module. • Shift selector lever into "L" position. • Leave neutral safety switch connected. • Measure voltage at neutral safety switch "Y/W" terminal. • Is reading above 10 volts in "L" range and less than 2 volts in other ranges?	Yes ► No, always zero ► No, always above 10 volts ►	GO to PPM15 . GO to PPM16 . GO to PPM17 .

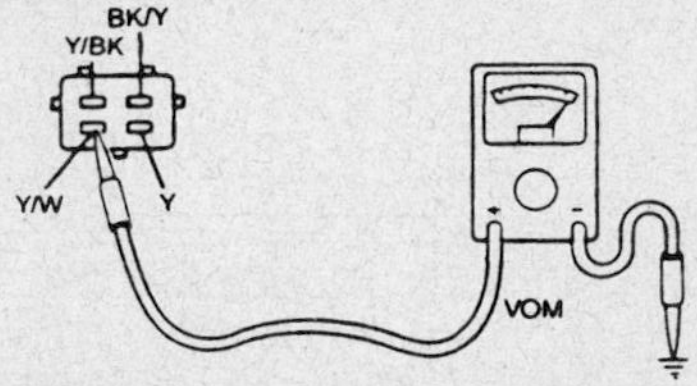
BK/Y
Y/BK
Y/W
Y
VOM

TEST STEP	RESULT ►	ACTION TO TAKE
PPM15 CHECK VOLTAGE AT CONTROL UNIT		
• Key on; engine off. • VOM on 20 volt scale. • Disconnect 4EAT control module. • Shift selector lever into "L" position. • Measure voltage at 4EAT control module 1C terminal. • Is reading above 10 volts in "L" range and less than 2 volts in other ranges?	Yes ► No ►	REPLACE 4EAT control module. REPAIR "Y/W" wire from neutral safety switch to control module.

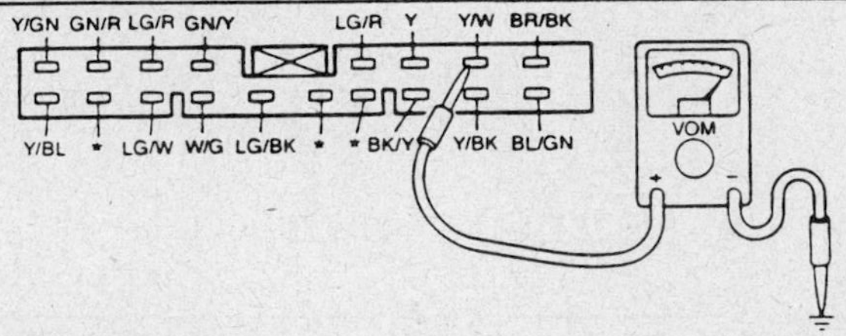
Y/GN GN/R LG/R GN/Y
LG/R Y Y/W BR/BK
Y/BL * LG/W W/G LG/BK * * BK/Y Y/BK BL/GN
VOM

System integrity check

TEST STEP	RESULT ►	ACTION TO TAKE
PPM16 CHECK VOLTAGE TO NEUTRAL SAFETY SWITCH		
• Key on; engine off. • VOM on 20 volt scale. • Disconnect neutral safety switch. • Measure voltage at neutral safety switch connector "BK/Y" terminal. • Is reading greater than 10 volts?	Yes ► No ►	REPLACE neutral safety switch. REPAIR "BK/Y" wire from neutral safety switch to fuse panel (15 amp meter fuse).

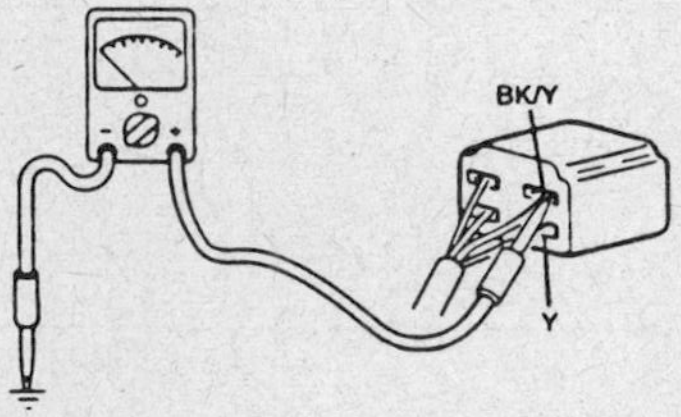
BK/Y
Y

TEST STEP	RESULT ►	ACTION TO TAKE
PPM17 CHECK FOR SHORT TO VPWR		
• Key on; engine off. • Disconnect 4EAT control module. • Disconnect neutral safety switch. • Measure voltage at neutral safety switch connector "Y/W" terminal. • Is reading greater than 4 volts? See illustration in TEST STEP PPM14	Yes ► No ►	REPAIR short in "Y/W" wire to VPWR or VREF. REPLACE neutral safety switch.

System Integrity check

TEST STEP	RESULT ►	ACTION TO TAKE
PPM18 CHECK VOLTAGE AT NEUTRAL SAFETY SWITCH		
• Key on; engine off. • VOM on 20 volt scale. • Disconnect 4EAT control module. • Shift selector lever into "D" position. • Leave neutral safety switch connected. • Measure voltage at neutral safety switch "Y/BK" terminal. • Is reading above 10 volts in "D" range and less than 2 volts in other ranges?	Yes ► No, always zero ► No, always above 10 volts ►	GO to PPM19 . GO to PPM20 . GO to PPM21 .

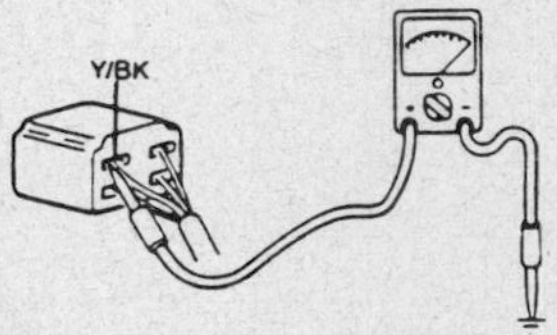
Y/BK

TEST STEP	RESULT ►	ACTION TO TAKE
PPM19 CHECK VOLTAGE AT CONTROL UNIT		
• Key on; engine off. • VOM on 20 volt scale. • Disconnect 4EAT control module. • Shift selector lever into "D" position. • Measure voltage at 4EAT control module 1D terminal. • Is reading above 10 volts in "D" range and less than 2 volts in other ranges?	Yes ► No ►	REPLACE 4EAT control module. REPAIR "Y/BK" wire from neutral safety switch to control module.

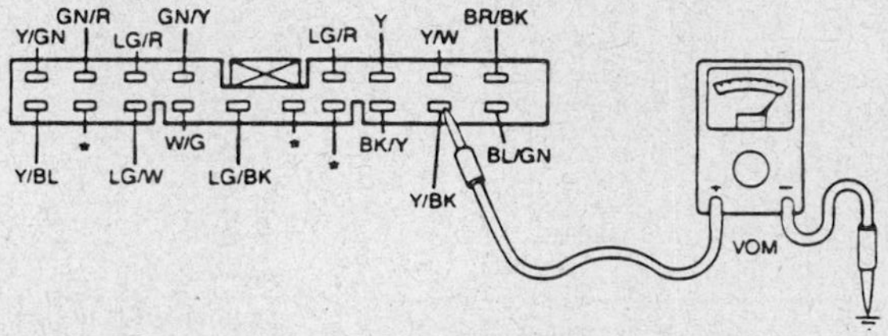
GN/R
Y/GN
LG/R
GN/Y
LG/R
Y
Y/W
BR/BK
Y/BL
LG/W
W/G
LG/BK
BK/Y
BL/GN
Y/BK
VOM

System Integrity check

TEST STEP	RESULT ►	ACTION TO TAKE
PPM20 CHECK VOLTAGE TO NEUTRAL SAFETY SWITCH		
• Key on; engine off. • VOM on 20 volt scale. • Disconnect neutral safety switch. • Measure voltage at neutral safety switch connector "BK/Y" terminal. • Is reading greater than 10 volts? See illustration in TEST STEP PPM16	Yes ► No ►	REPLACE neutral safety switch. REPAIR "BK/Y" wire from neutral safety switch to fuse panel (15 amp meter fuse).
PPM21 CHECK FOR SHORT TO VPWR		
• Key on; engine off. • Disconnect 4EAT control module. • Disconnect neutral safety switch. • Measure voltage at neutral safety switch connector "Y/BK" terminal. • Is reading greater than 4 volts? See illustration in TEST STEP PPM18	Yes ► No ►	REPAIR short in "Y/BK" wire to VPWR or VREF. REPLACE neutral safety switch.
PPM22 CHECK VOLTAGE AT NEUTRAL SAFETY SWITCH		
• Key on; engine off. • VOM on 20 volt scale. • Disconnect 4EAT control module. • Shift selector lever in "(D)" position. • Leave neutral safety switch connected. • Measure voltage at neutral safety switch "Y" terminal. • Is reading above 10 volts in "(D)" range and less than 2 volts in other ranges?	Yes ► No, always zero ► No, always above 10 volts ►	GO to PPM23. GO to PPM24. GO to PPM25.

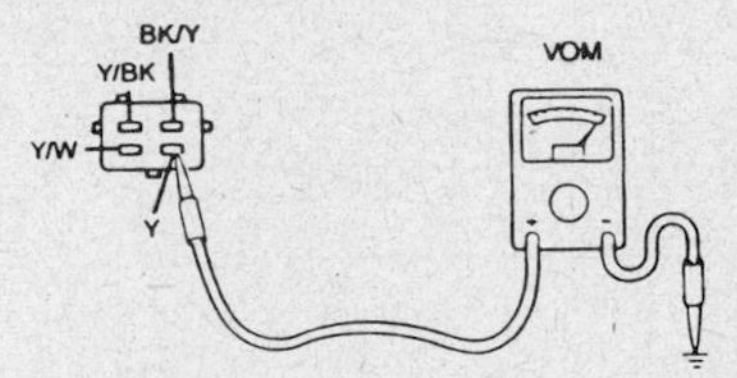
BK/Y
Y/BK
Y/W
Y
VOM

System Integrity check

TEST STEP	RESULT ►	ACTION TO TAKE
PPM23 CHECK VOLTAGE AT CONTROL UNIT		
• Key on; engine off. • VOM on 20 volt scale. • Disconnect 4EAT control module. • Shift selector lever into "(D)" position. • Measure voltage at 4EAT control module 1E terminal. • Is reading above 10 volts in "(D)" range and less than 2 volts in other ranges?	Yes ► No ►	REPLACE 4EAT control module. REPAIR "Y" wire from neutral safety switch to control module.

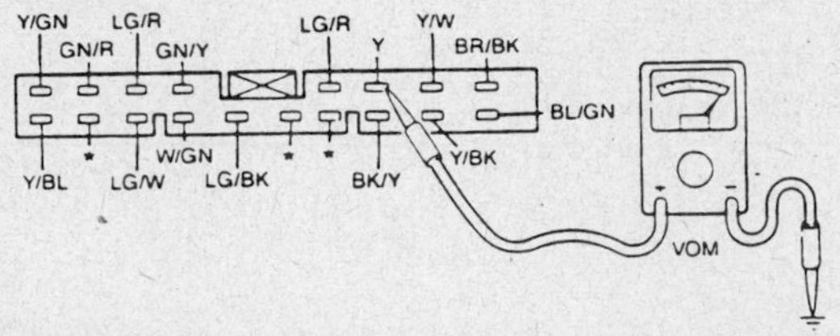
Y/GN
LG/R
GN/R
GN/Y
LG/R
Y
Y/W
BR/BK
BL/GN
Y/BL
LG/W
W/GN
LG/BK
BK/Y
Y/BK
VOM

TEST STEP	RESULT ►	ACTION TO TAKE
PPM24 CHECK VOLTAGE TO NEUTRAL SAFETY SWITCH		
• Key on; engine off. • VOM on 20 volt scale. • Disconnect neutral safety switch. • Measure voltage at neutral safety switch connector "BK/Y" terminal. • Is reading greater than 10 volts? See illustration in TEST STEP PPM16	Yes ► No ►	REPLACE neutral safety switch. REPAIR "BK/Y" wire from neutral safety switch to fuse panel (15 amp meter fuse).
PPM25 CHECK VOLTAGE TO NEUTRAL SAFETY SWITCH		
• Key on; engine off. • Disconnect 4EAT control module. • Disconnect neutral safety switch. • Measure voltage at neutral safety switch connector "Y" terminal. • Is reading greater than 4 volts? See illustration in TEST STEP PPM22	Yes ► No ►	REPAIR short in "Y" wire to VPWR or VREF. REPLACE neutral safety switch.

System Integrity check

TEST STEP	RESULT ►	ACTION TO TAKE
PPM26 CHECK VOLTAGE AT CONTROL MODULE		
• Key on; engine off. • VOM on 20 volt scale. • Selector lever in N or P position. • Measure voltage at 4EAT control module 1F terminal. • Is reading greater than 10 volts in N or P range and less than 2 volts in other ranges?	Yes ► No ►	GO to PPM27 . REPLACE 4EAT control module.

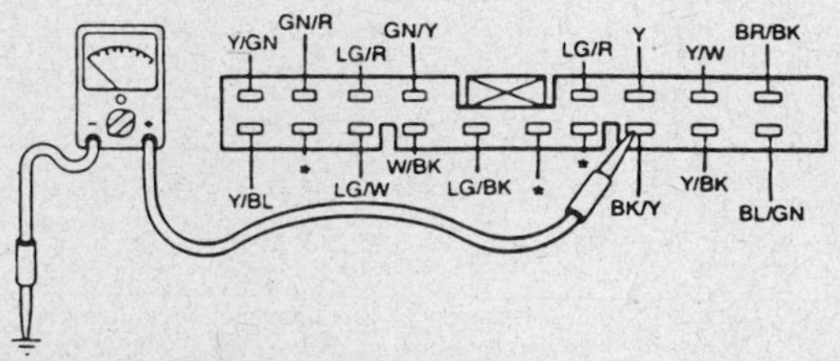
Y/GN
GN/R
LG/R
GN/Y
LG/R
Y
Y/W
BR/BK
Y/BL
LG/W
W/BK
LG/BK
BK/Y
Y/BK
BL/GN

TEST STEP	RESULT ►	ACTION TO TAKE
PPM27 CHECK VOLTAGE AT NEUTRAL SAFETY SWITCH		
• Key on; engine off. • VOM on 20 volt scale. • Selector lever in N or P position. • Measure voltage at neutral safety switch "BK/Y" terminal. • Is reading greater than 10 volts in N or P range or less than 2 volts in other ranges?	Yes ► No ►	REPLACE neutral safety switch. REPAIR "BK/Y" wire from neutral safety switch to 4EAT control module.

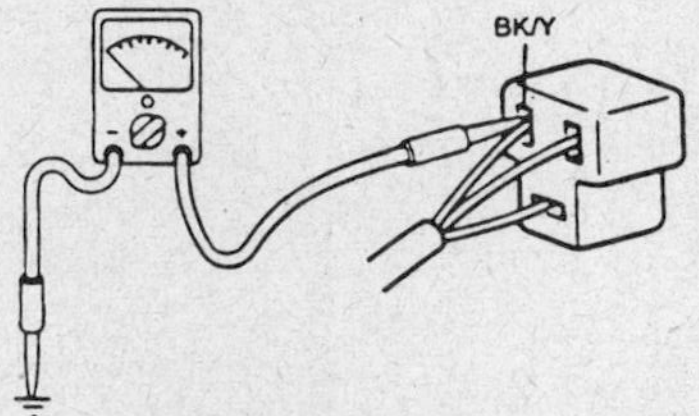
BK/Y

System integrity check

TEST STEP	RESULT ►	ACTION TO TAKE
PPM28 VOLTAGE AT 4EAT MODULE (1B)		
• Disconnect 4EAT module. • VOM on 20 volt scale. • Measure voltage between 4EAT module terminal 1B ("BL/BK" wire) and ground. Key off .... 0v Key on .... above 10v • Are voltage readings OK?	Yes ► No ►	GO to PPM29 . GO to PPM30 .

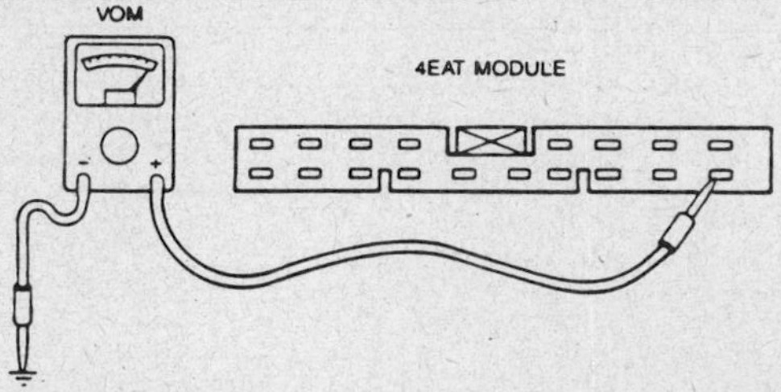
VOM
4EAT MODULE

TEST STEP	RESULT ►	ACTION TO TAKE
PPM29 OPERATION OF POWER LIGHT		
• Key on. • Operate 4EAT power/normal switch. • Observe "power" light. • Does "power" light operate properly?	Yes ► No ►	REPLACE 4EAT control module. GO to PPM31 .
PPM30 VOLTAGE AT LIGHT		
• Disconnect 4EAT module. • Connect power light "BL/BK" wire to ground with jumper wire. • Observe power light. Key off .... Light off Key on .... Light on • Does light operate properly?	Yes ► No ►	REPAIR "BL/BK" wire from light to 4EAT module. REPAIR Indicator Lights.

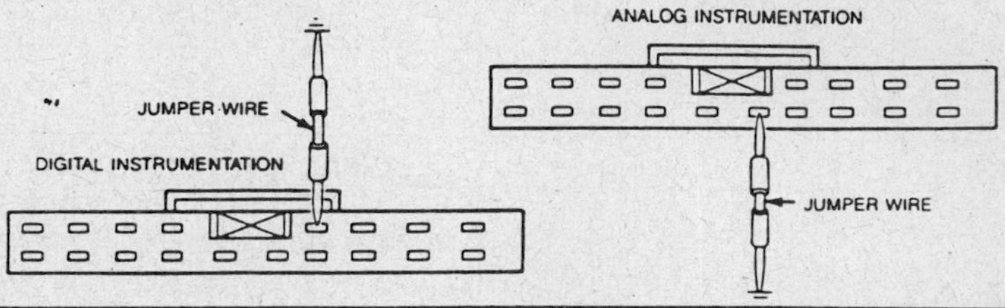
ANALOG INSTRUMENTATION
JUMPER WIRE
DIGITAL INSTRUMENTATION
JUMPER WIRE

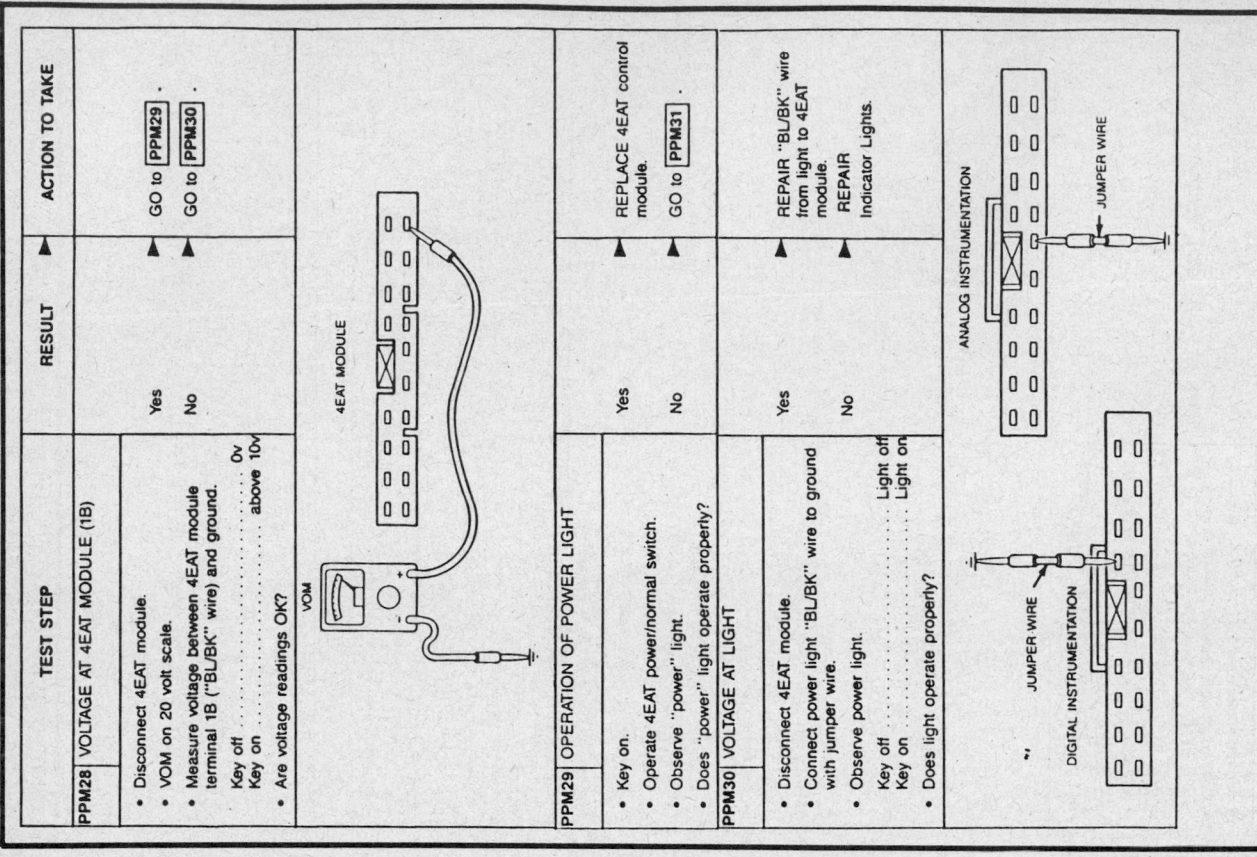

System integrity check

TEST STEP	RESULT	ACTION TO TAKE
PPM31 CHECK VOLTAGE AT 4EAT CONTROL MODULE		
• Disconnect 4EAT control module (connector C273). • Key on; engine off. • VOM on 20 volt scale. • Measure voltage between connector C273 terminal 2L ("BR") and ground. • Exercise the 4EAT switch between power and normal and verify the following: **4EAT Switch Position** — **Voltage** Power — 12v Normal — 0v • Are voltages OK?	Yes No	► REPLACE 4EAT control module. ► GO to PPM32 .

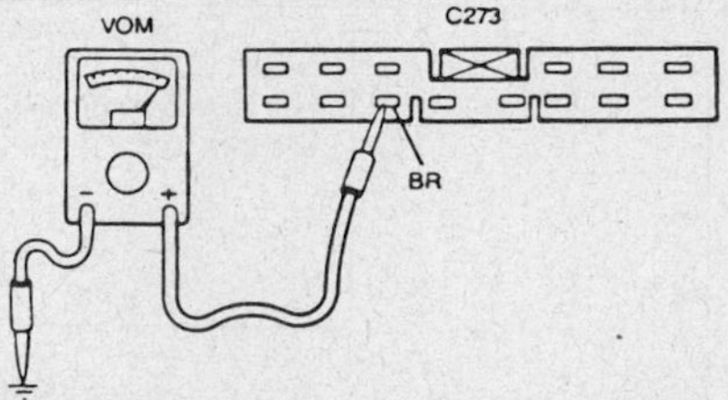
VOM
C273
BR

TEST STEP	RESULT	ACTION TO TAKE
PPM32 CHECK VOLTAGE AT 4EAT SWITCH		
• Key on; engine off. • VOM on 20 volt scale. • Measure voltage between connector C320 ("BR/BK" wire) and ground. • Exercise the 4EAT switch between power and normal and verify the following: **4EAT Switch Position** — **Voltage** Power — 12v Normal — 0v • Are voltages OK?	Yes No	► REPAIR "BR/BK" wire between 4EAT switch and 4EAT control module. ► GO to PPM33

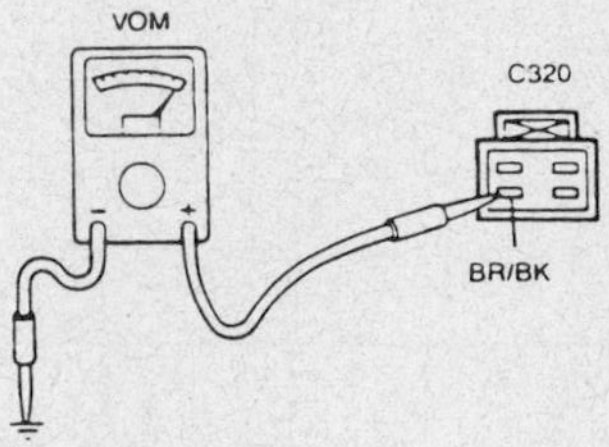
VOM
C320
BR/BK

System integrity check

TEST STEP	RESULT	ACTION TO TAKE
PPM33 4EAT SWITCH FUNCTION CHECK		
• Disconnect 4EAT switch (connector C320). • Check continuity between switch "BL/BK" and "BR/BK" terminals as follows: **4EAT Switch Position** — **Continuity** Power — Yes Normal — No • Is the 4EAT switch functioning OK?	Yes No	► GO to PPM34 . ► REPLACE 4EAT switch
PPM34 CHECK VOLTAGE AT INSTRUMENT PANEL		
• Key on; engine off. • VOM on 20 volt scale. • Measure voltage at instrument panel (connector C274) as follows: Analog Instrument Panel – measure voltage between terminal 2J ("BL/BK") and ground Digital Instrument Panel – measure voltage between terminal 2G ("BL/BK") and ground. • Is voltage above 10 volts?	**Yes** No	► REPAIR wire between instrument panel and 4EAT switch. ► REPAIR Warning and Indicator Lights.

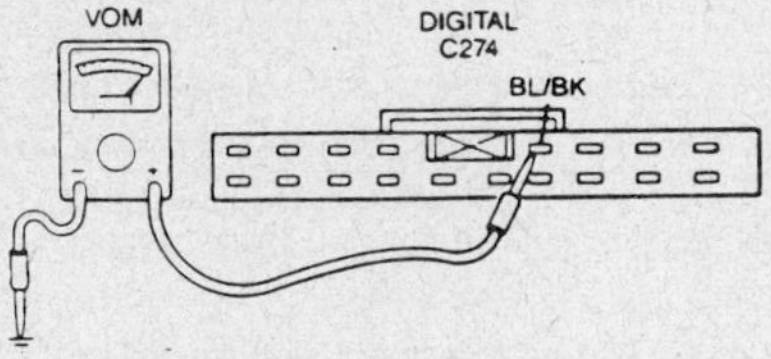
VOM
DIGITAL
C274
BL/BK

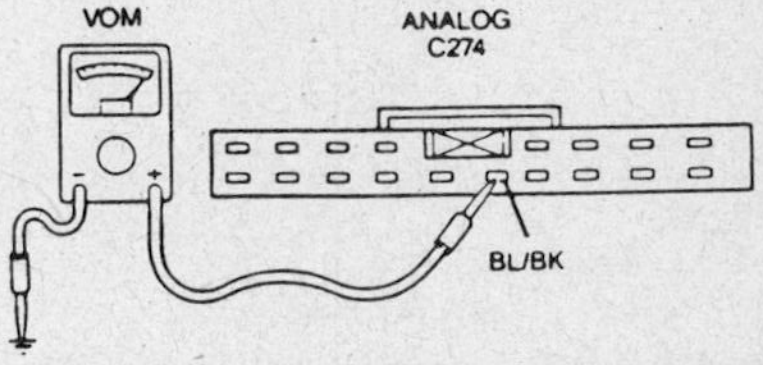
VOM
ANALOG
C274
BL/BK

TEST STEP	RESULT	ACTION TO TAKE
PPM35 MANUAL SWITCH OPERATION		
• Key off. • VOM on 200 ohm scale. • Measure resistance between manual switch terminals "BR/BK" and "BK." **Manual Switch** — **Resistance** Released .... above 10,000 ohms Depressed .... 0 ohms • Are resistance readings OK?	Yes No	► GO to PPM36 . ► REPLACE manual switch.

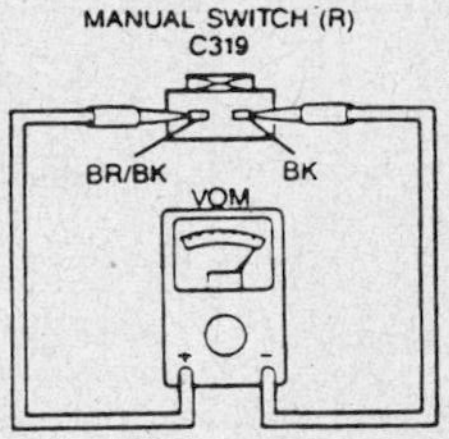
MANUAL SWITCH (R)
C319
BR/BK
BK
VOM

System integrity check

TEST STEP	RESULT ►	ACTION TO TAKE
PPM36 GROUND CHECK		
• Key off. • VOM on 200 ohm scale. • Measure resistance between manual switch "BK" and ground. • Is resistance below 5 ohms?	Yes ► No ►	GO to PPM37 . REPAIR "BK" wire to ground.
PPM37 CHECK FOR OPENS		
• Key off. • VOM on 200 ohm scale. • Measure resistance between 4EAT module terminal 1A ("BR/BK" wire) and manual switch ("BR/BK" wire). • Is resistance below 5 ohms?	Yes ► No ►	GO to PPM38 . REPAIR open in "BR/BK" wire.

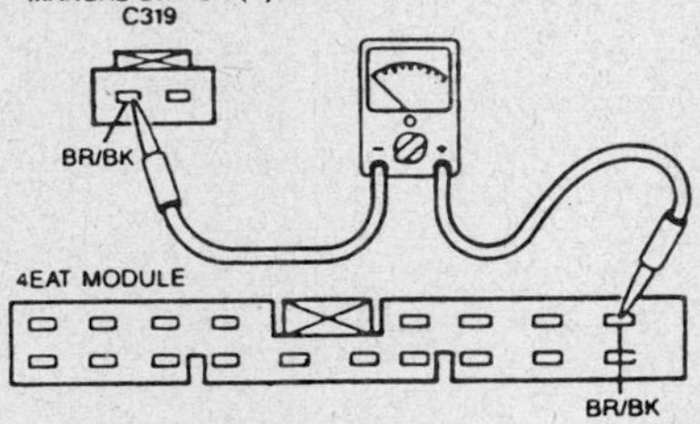
MANUAL SWITCH (R)
C319
BR/BK
4EAT MODULE
BR/BK

TEST STEP	RESULT ►	ACTION TO TAKE
PPM38 CHECK FOR SHORTS		
• Disconnect battery. • Key on. • VOM on 200 ohm scale. • Measure resistance between 4EAT module terminal 1A ("BR/BK" wire) and ground, battery power. • Is resistance above 10,000 ohms?	Yes ► No ►	REPLACE 4EAT control module. REPAIR short in "BR/BK" wire.

System integrity check

TEST STEP	RESULT ►	ACTION TO TAKE
PPM39 OPERATION AT 4EAT MODULE		
• Disconnect 4EAT module. • Key on. • Connect 4EAT module terminal 2K ("BR/Y" wire) to ground with jumper wire. • Observe manual shift light: **Jumper Wire** — **Light** With — On Without — Off • Does light operate properly?	Yes ► No ►	REPLACE 4EAT module. **NOTE:** Be certain manual switch operated properly in Quick Test Step 5.0. GO to PPM40

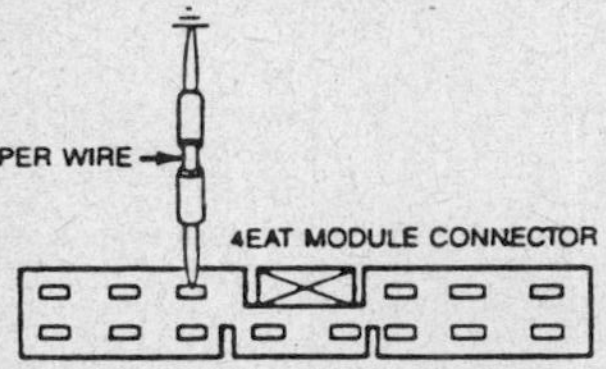
JUMPER WIRE
4EAT MODULE CONNECTOR

TEST STEP	RESULT ►	ACTION TO TAKE
PPM40 OPERATION AT LIGHT		
• Disconnect 4EAT module. • Key on. • Connect "BR/Y" wire at manual shift light to ground with jumper wire. • Observe manual shift light: **Jumper Wire** — **Light** With — On Without — Off • Does light operate properly?	Yes ► No ►	REPAIR "BR/Y" wire. REPAIR Indicator, Lights.

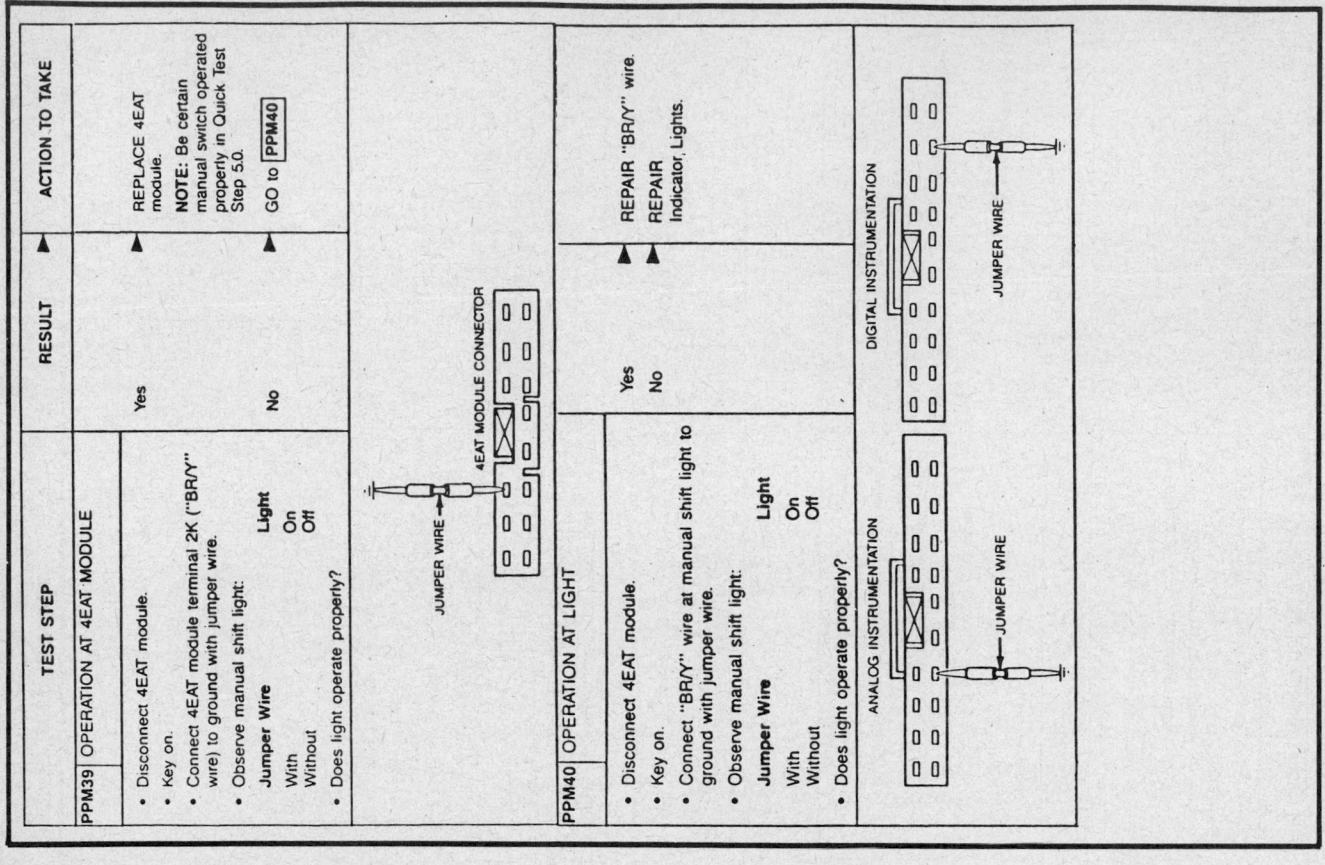

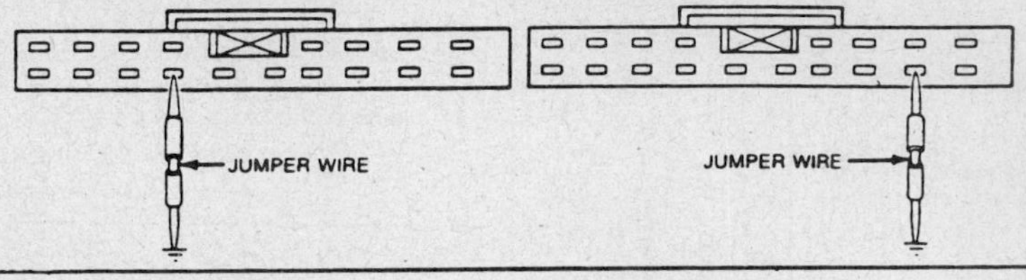
ANALOG INSTRUMENTATION
DIGITAL INSTRUMENTATION
JUMPER WIRE
JUMPER WIRE

System integrity check

TEST STEP	RESULT ►	ACTION TO TAKE
PPM41 NO LOAD OUTPUT		
• Disconnect the 4EAT module terminal 2N ("LG/BK" wire) only. • VOM on 20 volt scale. • Measure voltage between 4EAT module terminal 2N output and ground. **Drum Speed** / **Voltage** Below 80 rpm .... above 10v Above 640 rpm and in P or N range .... below 1.5v • Are voltage readings OK?	Yes ► No ►	GO to PPM42. REPLACE 4EAT control module.

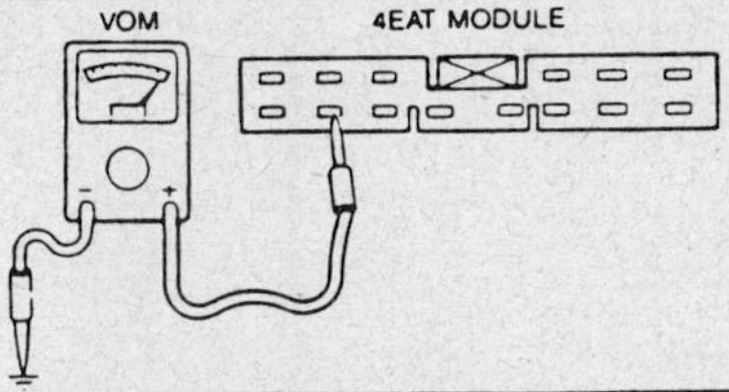
VOM
4EAT MODULE

TEST STEP	RESULT ►	ACTION TO TAKE
PPM42 CHECK FOR OPEN		
• Key off. • VOM on 200 ohm scale. • Measure resistance between 4EAT module 2N ("LG/BK" wire) and ECA 3D ("LG/BK" wire). • Is resistance below 5 ohms?	Yes ► No ►	GO to PPM43. REPAIR open in "LG/BK" wire.

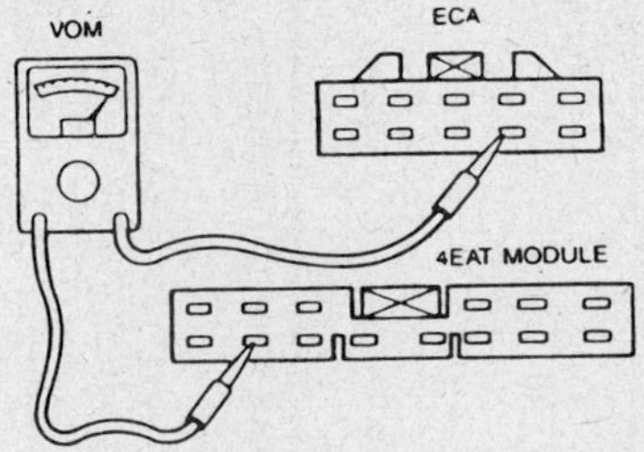
VOM
ECA
4EAT MODULE

System integrity check

TEST STEP	RESULT ►	ACTION TO TAKE
PPM43 CHECK FOR SHORTS		
• Disconnect battery, 4EAT module, and ECA. • VOM on 200,000 ohm scale. • Measure resistance between 4EAT module terminal 2N "LG/BK" wire and ground, battery power. • Are resistance readings above 10,000 ohms?	Yes ► No ►	REFER to engine emission diagnosis REPAIR short in "LG/BK" wire to ground or battery power.
PPM44 TP SIGNAL CHECK		
• Key on. • VOM on 20 volts scale. • Measure voltage between TP sensor "LG/W" wire and ground. **Throttle** / **Voltage** Closed .... 0.5v 1/8 .... 1.0v 2/8 .... 1.5v 3/8 .... 2.0v 4/8 .... 2.5v 5/8 .... 3.0v 6/8 .... 3.5v 7/8 .... 4.0v Full .... above 4.3v TP sensor "LG/R" wire and ground. Key off .... below 1.5v Key on .... 4–6v • Are voltage readings OK?	Yes ► No ►	REPAIR wire(s) in question. REPLACE throttle position sensor.
PPM45 AFT SWITCH GROUND CHECK		
• Disconnect AFT switch (connector C189). • VOM on 200 ohm scale. • Measure resistance between connector C189 "BK" wire and ground. • Is resistance less than 5 ohms?	Yes ► No ►	GO to PPM46. REPAIR "BK" wire between AFT switch and ground.

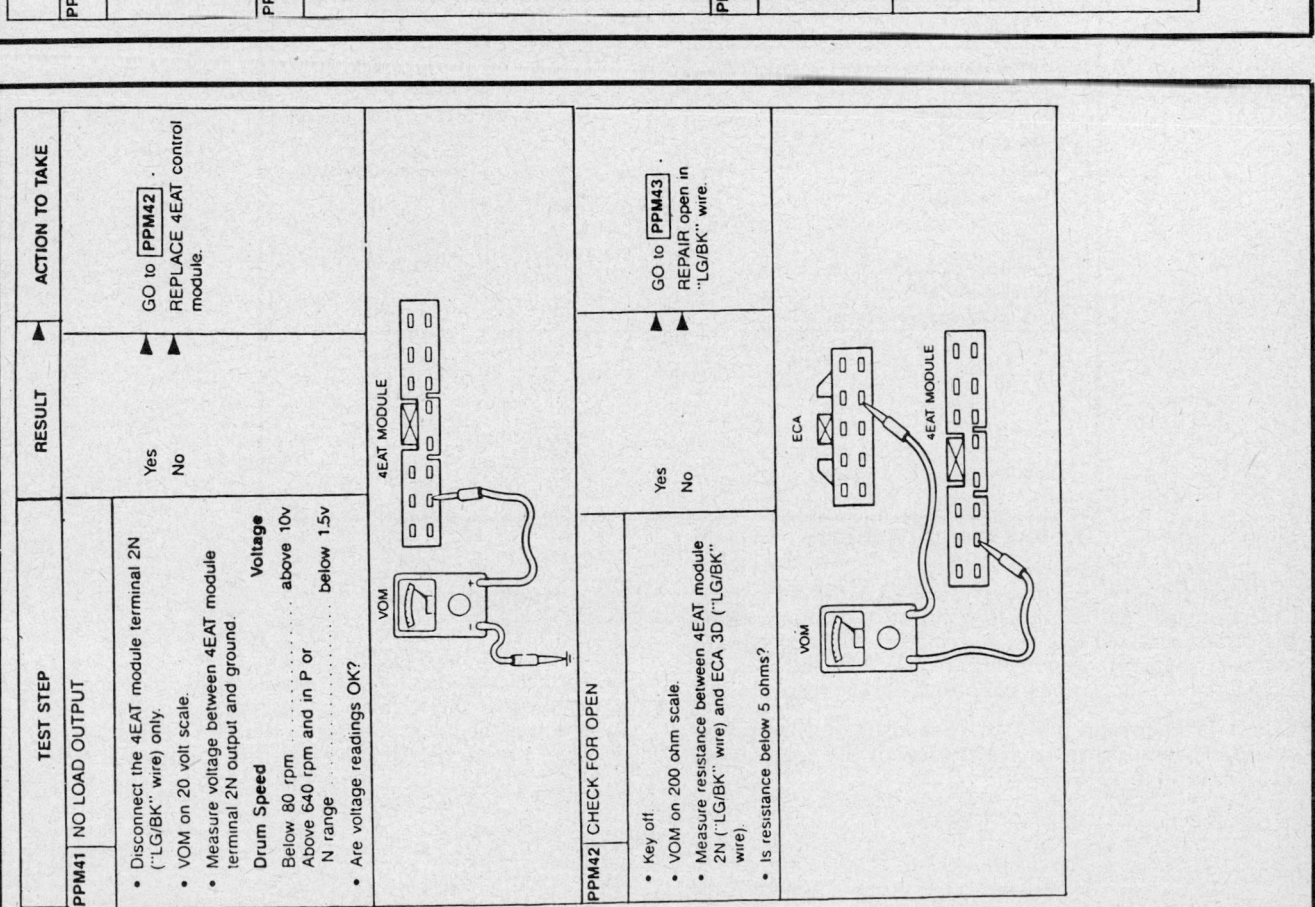

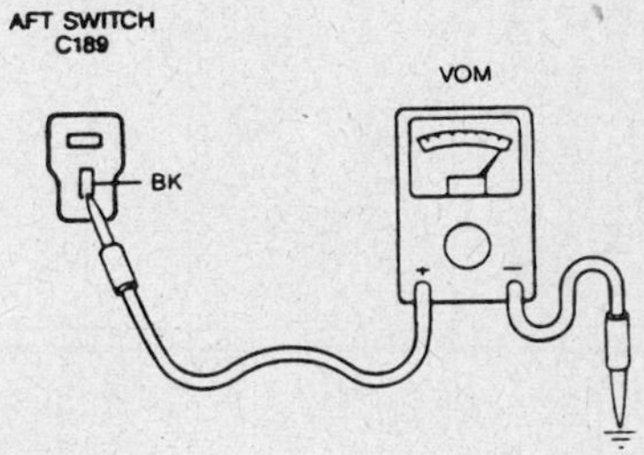
AFT SWITCH
C189
BK
VOM

System integrity check

TEST STEP	RESULT	► ACTION TO TAKE
PPM46 AFT SWITCH FUNCTION CHECK		
• AFT switch (connector C189) disconnected. • Check continuity between switch terminals as follows: AFT — Continuity Above 150ºC (302ºF) — Yes Below 143ºC (289ºF) — No • Does the AFT switch function OK?	Yes No	► GO to PPM47 . ► REPLACE AFT switch.
PPM47 CIRCUIT CONTINUITY CHECK		
• AFT switch. • Disconnect 4EAT control module (connector C274). • VOM On 200 ohm scale. • Measure resistance between connector C274 terminal 2O ("LG") and connector C189 ("LG"). • Is resistance less than 5 ohms?	Yes No	► GO to PPM48 . ► REPAIR "LG" wire between 4EAT control module and AFT switch for open.
PPM48 SHORT TO GROUND AND VPWR CHECK		
• AFT switch control module (connector C274) disconnected. • Disconnect battery. • Key on. • VOM on 200K ohm scale. • Measure resistance between: Connector C274 terminal 2O ("LG") and ground. Connector C274 terminal 2O ("LG") and terminal 2A ("BK/Y"). • Are resistances greater than 10,000 ohms?	Yes No	► REPLACE 4EAT control module. ► REPAIR "LG" wire between 4EAT control module and AFT temperature switch for shorts.

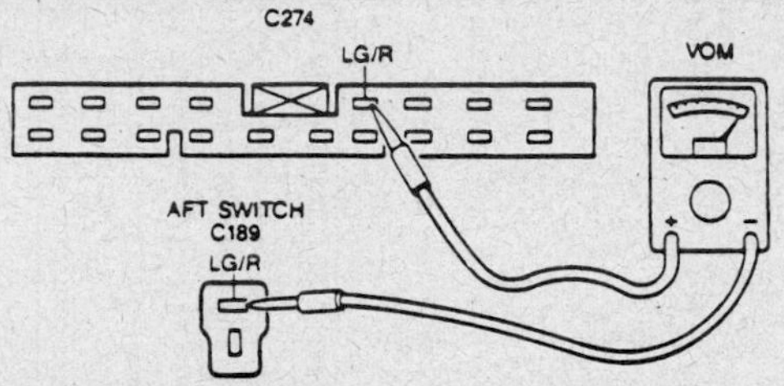
C274
LG/R
VOM
AFT SWITCH
C189
LG/R
+
-

System integrity check

the 4EAT tester in service. This test involves a driving evaluation of the transaxle shifting quality, ability and timing. Shift problems will be directed to a list of symptoms for appropriate action to take. The 3 sympton menus are given: Upshift, Downshift and Shift Feel for various symptoms encountered.

NOTE: It is recommended to test shift points more than once before making an evaluation of the transaxle.

Tester Hookup

1. Turn ignition key to the **OFF** position.
2. Disconnect the 4EAT control module.
3. Connect the 4EAT tester adapter cable between the control module and the harness connectors.
4. Connect the adapter cable to the 4EAT tester.
5. Refer to the SF1 Observe Solenoid Lamps test step.

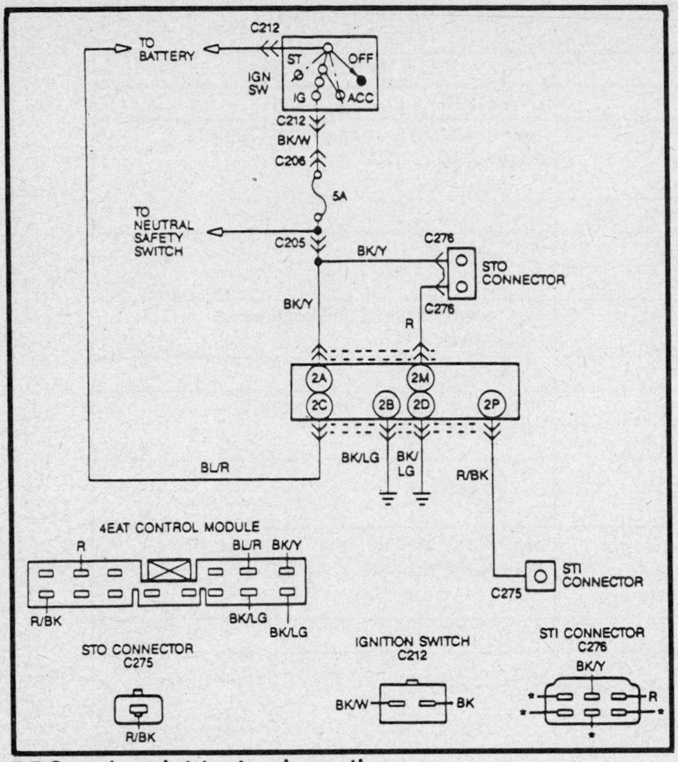
TO BATTERY
C212
IGN SW
ST
OFF
IG
ACC
BK/W
C206
5A
TO NEUTRAL SAFETY SWITCH
C205
BK/Y
C276
STO CONNECTOR
R
2A
2M
2C
2B
2D
2P
BL/R
BK/LG
BK/LG
R/BK
4EAT CONTROL MODULE
STI CONNECTOR
C275
STO CONNECTOR
IGNITION SWITCH
BK

PPQ—pinpoint test schematic

TEST STEP	RESULT ►	ACTION TO TAKE
PPQ1 INTEGRITY		
• Visually inspect all wiring, wiring harness, connectors and components for evidence of overheating, insulation damage looseness, shorting or other damage. • Is there any cause for concern?	Yes ► No ►	SERVICE as required. GO to PPQ2 .
PPQ2 POWER CHECK		
• VOM on 20 volt scale. • Measure voltage between: 4EAT control module terminal 2A ("BL/Y" wire) and ground. STO connector ("BK/Y" wire) and ground. STI connector ("R/BK" wire) and ground. Key off .... 0v Key on .... above 10v 4EAT control module terminal 2C ("BL/R" wire) and ground. Always above 10 volts. • Are voltage readings OK?	Yes ► No ►	GO to PPQ3 . REPAIR wire(s) to power in question.
PPQ3 GROUND CHECK		
• Key off. • VOM on 200 ohm scale. • Measure resistance between 4EAT control module terminals 2B ("BK/LG" wire), 2D ("BK/LG" wire) and ground. • Is resistance below 5 ohms?	Yes ► No ►	GO to PPQ4 . REPAIR "BK/LG" ground wire.
PPQ4 STO CONTINUITY		
• Key off. • VOM on 200 ohm scale. • Measure resistance between STO ("R" wire) connector "R" wire and 4EAT module terminal 2M. • Is resistance below 5 ohms?	Yes ► No ►	REPLACE 4EAT control module. REPAIR "R" wire from STO connector to 4EAT module.

PPQ test information

Test Result	Range	Possible Cause		Action To Take
Above specification*	In all ranges	Insufficient line pressure	Worn oil pump	Replace
			Oil leakage from oil pump, control valve, and/or transmission case	Tear down, inspect, and repair or replace as required.
			Stuck pressure regulator valve	
	In "Ⓓ" range	One-way clutch 2 slipping		Tear down, inspect, and repair or replace as required.
	In forward ranges	Forward clutch slipping One-way clutch 1 slipping		
	In " D " (Manual) and "L" (Manual) ranges	Coasting clutch slipping		
	In "Ⓓ" (Manual) and " D " (Manual) ranges	2-4 band slipping		Adjust and retest.
	In "R," "L" and "L" (Manual) ranges	Low and reverse slipping		Tear down, inspect, repair/replace as required.
	In "R" range	Low and reverse band slipping Reverse clutch slipping		Perform road test to determine whether problem is low and reverse band or reverse clutch a) Engine brake applied in 1st . . . Reverse clutch b) Engine brake not applied in 1st . . . Low and reverse band. Repair or replace as required.
Within specification*		All shift control elements within transmission are functioning normally.		Go to OPB
Below specification*		Engine out of tune		Tune engine before running Stall Test.
		One-way clutch slipping within torque converter		Tear down, inspect, repair or replace as required.

* Specification - Stall Speed
Ⓓ, D, L Ranges: 2120–2420 rpm
R Range: 2080–2380 rpm

Disassembly to step indicated if required after re-test.

Stall test evaluation

Shift	Result	Possible Cause	Action To Take
N - Ⓓ Normal Mode	More than specification*	Insufficient line pressure	Go to OPC.
		Forward clutch slipping One-way clutch 1 slipping One-way clutch 2 slipping	Tear down, inspect and repair, replace as required.
	Less than specification*	N–D accumulator not operating properly	
		Excessive line pressure	Go to OPC.
N - Ⓓ Manual Mode	More than specification*	Insufficient line pressure	
		Forward clutch slipping	Tear down, inspect, and replace as required.
		2-4 band slipping	Adjust and retest.
		One-way clutch 1 slipping	Tear down, inspect, and repair or replace as required.
	Less than specification*	1-2 accumulator not operating properly	
		Excessive line pressure	Go to OPC.
N-R	More than specification*	Insufficient line pressure	
		Low and reverse band slipping Reverse clutch slipping	Tear down, inspect, and repair or replace as required.
	Less than specification*	N–R accumulator not operating properly	
		Excessive line pressure	Go to OPC.

* Specified Time Lag:
N to D range 0.5–1.0 second
N to R range 0.5–1.0 second

\# Transaxle Disassembly Step required for access to component.

Time lag evaluation

LINE PRESSURE SPECIFICATIONS

	Line Pressure, kPa (psi)	
Range	Ⓓ, D, L	R
Idle	353–432 (51–63)	598–942 (87–137)
Stall Speed	873–1040 (127–151)	1668–2011 (242–292)

Pressure Test Result	Range	Possible Location of Problem	Action to Take
Low	All	Worn oil pump, fluid leaking from oil pump, control valve body or transaxle case. Pressure regulator valve sticking	Tear down, inspect, repair or replace as required the complete pump or valve assembly or components.
Low	Ⓓ, D	Fluid leaking from hydraulic circuit of forward clutch	Tear down, inspect, repair or replace components as required.
Low	R	Fluid leaking from hydraulic circuit of low and reverse band	Tear down, inspect, repair or replace components as required.
High	All	Throttle valve sticking. Throttle modulator valve sticking. Pressure regulator valve sticking.	Tear down, inspect, repair or replace components as required
Within Specified Limits	All	—	Go to OPD

Line pressure test evaluation

THROTTLE PRESSURE SPECIFICATIONS

	Throttle Pressure kPa (psi)
Idle	39–88 (6–13)
Stall Speed	471–589 (68–85)

Pressure Test Result	Position Location Of Problem	Action To Take
Not Within Specified Limits	Throttle valve sticking	Tear down, inspect, repair, clean, or replace the valve(s) as required.
	Improper adjustment of throttle cable	Remove, inspect for damage and freedom of movement, replace and adjust per shop manual as required.
Within Specified Limits	—	Go to Road Test.

Throttle pressure test evaluation

ON CAR SERVICE

Adjustments

LINE PRESSURE

1. Raise and support the vehicle safely. Remove the wheel and tire.
2. Remove the left front splash shield.
3. Remove the square head plug from the transaxle that is marked "L" and install the pressure gauge.
4. With the transaxle in **P**, start the engine. Warm up the engine to operating temperature and adjust the idle speed to 750–800 rpm for nonturbocharged vehicles or 725–775 rpm for turbocharged vehicles.
5. Adjust locknuts on the cable as follows to increase or decrease line pressure:
   a. Loosen the cable all the way so the locknuts are as far away from the throttle cam as possible.
   b. Turn the locknuts clockwise to increase or counterclockwise to decrease the line pressure to 63–66 psi.
6. Turn off the engine.
7. Remove the pressure gauge and install the square head plug in the transaxle.
8. Install the left front splash shield.
9. Install the wheel and tire.

2–4 BRAKE BAND

NOTE: The 2–4 brake band is also know as the "Busy Band" because it has a dual function in the transaxle.

1. Raise and support the vehicle safely. Remove the oil pan.
2. Loosen the locknut and tighten the piston stem to 78–95 inch lbs.
3. Loosen the piston stem 2 turns.
4. Tighten the locknut to 18–29 ft. lbs.
5. Install the oil pan. Refill the transaxle to the correct level.

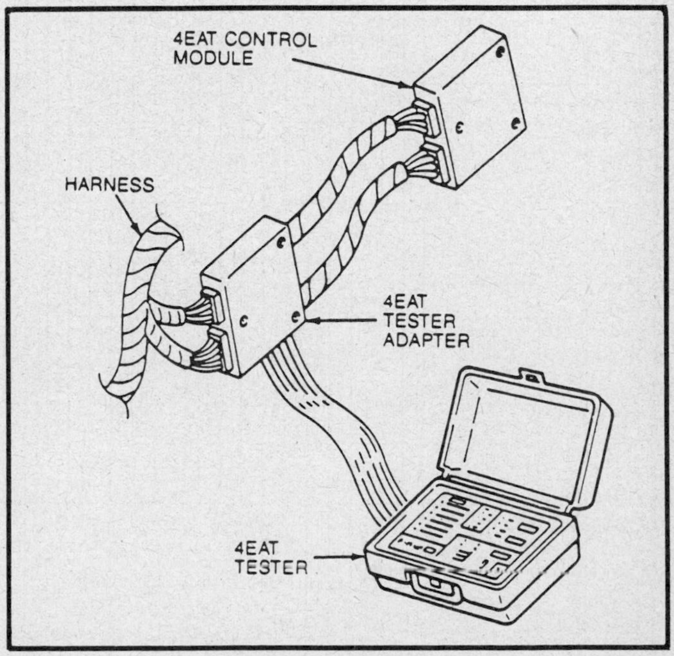
4EAT CONTROL MODULE
HARNESS
4EAT TESTER ADAPTER
4EAT TESTER

Road test—4EAT tester hookup

TEST STEP		RESULT ►	ACTION TO TAKE
SF1	OBSERVE SOLENOID LAMPS		
	• Warm engine at idle. • Drive vehicle through conditions on chart. • Observe lamps for solenoid functioning. • Do all lamps light at proper time?	Yes ► No ►	GO to OD1. GO to SF2.

SELECTOR LEVER POSITION	GEAR			SOLENOID VALVE LAMPS 1–2	2–3	3–4	Lockup
P	Non					ON	
R	Reverse			ON			
N	—	Below approx. 11 mph				ON	
		Above approx. 11 mph		ON			
Ⓓ	1st				ON	ON	
	2nd			ON	ON	ON	
	3rd	Below approx. 20–25 mph					
		Above approx. 25 mph	Lockup OFF	ON			
			Lockup ON	ON			ON
	OD	Lockup OFF		ON		ON	
		Lockup ON		ON		ON	ON
D	1st				ON	ON	
	2nd			ON	ON	ON	
	3rd	Below approx. 25 mph					
		Above approx. 25 mph		ON			
L	1st				ON	ON	
	2nd	Below approx. 68 mph		ON	ON		
		Above approx. 68 mph		ON	ON		
Manual Switch ON — D	2nd			ON	ON	ON	
	3rd	Below approx. 25 mph					
		Above approx. 25 mph		ON			
Manual Switch ON — D	2nd			ON	ON		
	2nd	Below approx. 25 mph		ON	ON		
		Above approx. 25 mph		ON	ON		
Manual Switch ON — L	1st				ON		
	1st	Below approx. 68 mph		ON	ON		
		Above approx. 68 mph		ON			

Road test diagnostic information

TEST STEP		RESULT ►	ACTION TO TAKE
SF2	CHECK SOLENOID VOLTAGES		
	• Warm engine at idle. • Drive vehicle through conditions on chart. • Check voltages at solenoid terminal(s) in question. • Are voltage readings OK?	Yes ► No ►	TESTER lamps are not working properly. REPLACE 4EAT control module.

SELECTOR LEVER POSITION	GEAR			SOLENOID VALVE VOLTAGE 1–2	2–3	3–4	Lockup
P	Non					10–14v	
R	Reverse			10–14v			
N	—	Below approx. 11 mph				10–14v	
		Above approx. 11 mph		10–14v			
Ⓓ	1st				10–14v	10–14v	
	2nd			10–14v	10–14v	10–14v	
	3rd	Below approx. 20 25 mph					
		Above approx. 25 mph	Lockup OFF	10–14v			
			Lockup ON	10–14v			10–14v
	OD	Lockup OFF		10–14v		10–14v	
		Lockup ON		10–14v		10–14v	10–14v
D	1st				10–14v	10–14v	
	2nd			10–14v	10–14v	10–14v	
	3rd	Below approx. 25 mph					
		Above approx. 25 mph		10–14v			
L	1st				10–14v	10–14v	
	2nd	Below approx. 68 mph		10–14v	10–14v		
		Above approx. 68 mph		10–14v	10–14v		
Manual Switch ON — D	2nd			10–14v	10–14v	10–14v	
	3rd	Below approx. 25 mph					
		Above approx. 25 mph		10–14v			
Manual Switch ON — D	2nd			10–14v	10–14v		
	2nd	Below approx. 25 mph		10–14v	10–14v		
		Above approx. 25 mph		10–14v	10–14v		
Manual Switch ON — L	1st				10–14v		
	1st	Below approx. 68 mph		10–14v	10–14v		
		Above approx. 68 mph		10–14v			

Road test diagnostic information

TEST STEP	RESULT ►	ACTION TO TAKE
OD1 SHIFT POINT CHECK		
• Warm engine to operating temperature (above 162°F). • Selector lever in (D) range. • Manual switch off. • Select the power mode. • Cruise control off. • Drive vehicle: Accelerator at 1/2 throttle Accelerator at full throttle Operate kickdown (sudden acceleration). • Compare shift point with chart. • Is shift point correct?	Yes ► No problem on upshift ► No problem on downshift ►	GO to OD2. GO to Upshift symptom menu. GO to Downshift symptom menu.

Throttle Position (Throttle Position Sensor Voltage)	Shifting (Gears)	Drum Speed (rpm)	Vehicle Speed (mph)
Fully opened (4.3 volts)	1 → 2	5000 - 5500	33 - 35
	2 → 3	5300 - 5700	65 - 70
	3 → OD	5400 - 5700	102 - 109
Half throttle (1.6-2.2 volts)	1 → 2	3500 - 4050	24 - 27
	2 → 3	3750 - 4250	47 - 53
	3 → OD	3600 - 4250	68 - 81
	Lockup ON (OD)	2500 - 3000	68 - 81
	Lockup OFF (OD)	2400 - 2850	64 - 77
	OD → 3	1950 - 2450	53 - 66
	3 → 2	1750 - 2300	33 - 43
Kickdown	OD → 3	3500 - 3700	95 - 101
	OD → 2	2150 - 2350	58 - 63
	OD → 1	950 - 1100	26 - 30
	3 → 2	3050 - 3350	40 - 63
	3 → 1	1350 - 1550	26 - 30
	2 → 1	2200 - 2400	26 - 30

Road test diagnostic information

TEST STEP	RESULT ►	ACTION TO TAKE
OD2 SHIFT POINT CHECK		
• Warm engine to operating temperature (above 162°F). • Selector lever in (D) range. • Select the economy mode. • Manual switch off. • Cruise control off. • Drive vehicle: Accelerate at 1/2 throttle Accelerate at full throttle Operate kickdown (sudden acceleration). • Watch 4EAT tester for shift point indication. • Compare shift point with chart. • Is shift point correct?	Yes ► No problem on upshift ► No problem on downshift ►	GO to OD3. GO to Upshift symptom menu. GO to Downshift symptom menu.

Throttle Position (Throttle Position Sensor Voltage)	Shifting (Gears)	Drum Speed (rpm)	Vehicle Speed (mph)
Fully opened (4.3 volts)	1 → 2	4900 - 5450	33 - 37
	2 → 3	5100 - 5500	63 - 68
	3 → OD	5400 - 5700	102 - 109
Half throttle (1.6-2.2 volts)	1 → 2	2800 - 3350	19 - 23
	2 → 3	3000 - 3400	37 - 42
	3 → OD	2900 - 3450	55 - 66
	Lockup ON (OD)	2050 - 2500	56 - 68
	Lockup OFF (OD)	1950 - 2350	53 - 64
	OD → 3	1600 - 1950	43 - 53
	3 → 2	1200 - 1550	24 - 30
Kickdown	OD → 3	3500 - 3700	95 - 101
	OD → 2	2050 - 2250	56 - 61
	OD → 1	950 - 1100	26 - 30
	3 → 2	2950 - 3200	56 - 61
	3 → 1	1350 - 1550	26 - 30
	2 → 1	2100 - 2400	26 - 30

Road test diagnostic information

TEST STEP	RESULT ►	ACTION TO TAKE
OD3 SHIFT FEEL CHECK		
• Warm engine to operating temperature (above 162°F). • Selector lever in (D) range. • Manual switch OFF. • Select both power and economy modes. • Cruise control OFF. • Drive vehicle from closed throttle to wide open throttle. • Does shift feel excessively harsh or slushy?	Yes ► No ►	GO to shift feel symptom menu. GO to [OD4].
OD4 MANUAL CHECK		
• Warm engine to operating temperature (above 162°F). • Selector lever in (D) range. • Select economy mode. • Manual switch ON. • Cruise control OFF. • Drive vehicle (accelerate/decelerate) • Watch 4EAT tester for shift point indication. • Check that the following conditions are met: 2nd–3rd upshift at 12 mph 3rd–2nd downshift at 6 mph No OD gear No 1st gear. • Are all conditions satisfied?	Yes ► No ►	GO to [OD5]. GO to road test symptom menu.
OD5 ENGINE BRAKING CHECK		
• Warm engine to operating temperature (above 162°F). • Selector lever in (D) range. • Select economy mode. • Manual switch OFF. • Cruise control OFF. • Drive vehicle until (D) gear is obtained. • Shift selecter into D range. • Is engine braking felt (in D3 only) immediately?	Yes ► No ►	GO to [D1]. GO to downshift symptom menu.

Road test diagnostic information

TEST STEP	RESULT ►	ACTION TO TAKE
D1 SHIFT POINT CHECK		
• Warm engine to operating temperature (above 162°F). • Cruise control off. • Selector lever in D range. • Manual switch off. • Power or normal mode. • Drive vehicle: Accelerate at 1/2 throttle Accelerate at full throttle. • Compare shift point with chart. • Is shift point correct?	Yes ► No problem on upshift ► No problem on downshift ►	GO to [D2]. GO to Upshift symptom menu. GO to Downshift symptom menu.

NOTE: Shift points in D range, Normal or Power mode are the same.

Throttle Position (Throttle Position Sensor Voltage)	Shifting (Gears)	Drum Speed (rpm)	Vehicle Speed (mph)
Fully opened (4.3 volts)	1 → 2	5000 – 5500	33 – 35
	2 → 3	5300 – 5700	65 – 70
	4 → 3	3750 – 4000	102 – 109
	3 → 2	3050 – 3350	40 – 63
	2 → 1	2200 – 2400	26 – 30
Half throttle (1.6–2.2 volts)	1 → 2	3500 – 4050	24 – 27
	2 → 3	3750 – 4250	47 – 53
	4 → 3	1950 – 2450	53 – 66
	3 → 2	1750 – 2300	33 – 43

TEST STEP	RESULT ►	ACTION TO TAKE
D2 MANUAL RANGE CHECK		
• Warm engine to operating temperature (above 162°F). • Cruise control OFF. • Selector lever in D range. • Manual switch OFF. • Power or Normal mode. • Drive vehicle until 3rd gear is obtained, then turn manual switch ON. • Decelerate vehicle. • Does 3–2 downshift occur at 66–70 mph?	Yes ► No ►	GO to [D3]. GO to road test symptom menu.

Road test diagnostic information

TEST STEP	RESULT ►	ACTION TO TAKE
D3 MANUAL RANGE CHECK		
• Warm engine to operating temperature (above 162ºF). • Cruise control OFF. • Selector lever in D range. • Manual switch ON. • Drive vehicle. • Is 2nd gear held?	Yes ► No ►	GO to D4. GO to road test symptom menu.
D4 ENGINE BRAKING CHECK		
• Warm engine to operating temperature (above 162ºF). • Cruise control OFF. • Selector lever in D range. • Manual switch OFF. • Drive vehicle until 3rd gear is obtained. • Shift selector lever into L range. • Is engine braking felt immediately?	Yes ► No ►	GO to L1. GO to downshift symptom menu.

TEST STEP	RESULT ►	ACTION TO TAKE
L1 SHIFT POINT CHECK		
• Warm engine to operating temperature (above 162ºF). • Cruise control off. • Selector lever in L range. • Power or Normal mode. • Drive vehicle: Accelerate at 1/2 throttle Accelerate at full throttle. • Compare shift point with chart. • Is shift point correct?	Yes ► No problem on upshift ► No problem on downshift ►	GO to L2. GO to Upshift symptom menu. GO to Downshift symptom menu.

NOTE: Shift points in L range, Normal or Power mode, are the same.

Throttle Position (Throttle Position Sensor Voltage)	Shifting (Gears)	Drum Speed (rpm)	Vehicle Speed (mph)
Fully opened (4.3 volts)	1 → 2	5000 - 5500	33 - 35
	2 → 1	2200 - 2400	26 - 30
Half throttle (1.6–2.2 volts)	1 → 2	3500 - 4050	24 - 27

Road test diagnostic information

TEST STEP	RESULT ►	ACTION TO TAKE
L2 MANUAL RANGE CHECK		
• Warm engine to operating temperature (above 162ºF). • Cruise control OFF. • Selector lever in L range. • Manual switch OFF. • Power or Normal mode. • Drive vehicle until 2nd gear is obtained, then turn manual switch ON. • Decelerate vehicle. • Does 2-1 downshift occur at 27–30 mph?	Yes ► No ►	GO to L3. GO to road test symptom menu.
L3 MANUAL RANGE CHECK		
• Warm engine to operating temperature (above 162ºF). • Cruise control. • Selector lever in L range. • Manual switch ON. • Drive vehicle. • Is 1st gear held?	Yes ► No ►	GO to L4. GO to road test symptom menu.
L4 ENGINE BRAKING CHECK		
• Warm engine to operating temperature (above 162ºF). • Cruise control OFF. • Selector lever in L range. • Manual switch ON. • Drive vehicle in 1st gear. • Decelerate vehicle. • Is engine braking felt?	Yes ► No ►	GO to P1. GO to Downshift symptom menu.
P1 VEHICLE STOPPING TEST		
• Drive vehicle on level surface. • Maximum speed of 2 mph. • Shift selector lever into P range. • Does vehicle stop?	Yes ► No ►	GO to S1. PERFORM parking pawl inspection.

Road test diagnostic information

TEST STEP		RESULT ►	ACTION TO TAKE
S1	CHECK SLIPPAGE		
	• Warm engine to operating temperature (above 165ºF). • Connect 4EAT tester. • Connect tachometer. • Drive vehicle. • Compare vehicle speed (and engine speed) to four indicated drum speeds. • Is vehicle speed (or engine speed) above or below indicated speed.	Yes ►	FOLLOW direction given in chart.
		Yes All speeds are incorrect. ►	INSPECT forward clutch.
		No ►	REFER to Chilton "3C" Diagnosis Chart

Driving condition		DRUM SPEED 1000	2000	3000	4000	
Gears	Other condition	VEHICLE SPEED (MPH)				ACTION TO TAKE
1st	L range, Manual mode	7	11	20	27	Inspect low and reverse clutch.
1st	(D) range, Normal mode	7	11	20	27	Inspect one-way clutch.
2nd	D range, Manual mode	20	25	37	50	Inspect 2–4 band.
3rd	(D) range, Manual mode	19	38	57	76	Inspect coasting clutch.
OD	(D) range, Normal mode	27	55	81	109	Inspect 3–4 clutch.
OD	(D) range, Normal mode, Lockup	1,000	2,000	3,000	4,000	Inspect Torque Converter.
		ENGINE SPEED (RPM)				

Road test diagnostic information

SHIFT FEEL SYMPTOM MENU		
CONDITION	POSSIBLE CAUSE	ACTION
• Shift shock in all ranges.	– Kickdown cable out of adjustment.	– Inspect cable adjustment.
	– Throttle valve sticking or damaged.	– Clean, service or replace.
	– Control valves.	– Check for clogging blockage, service as required.
	– Coasting clutch.	– Check for wear service or replace.
	– Low and reverse band.	– Check for adjustment, wear and damage, service as required.
	– Accumulators.	– Clean, service or replace.
	– 3–4 clutch.	– Inspect, service or replace.
	– CV joints or engine mounts.	– Service or replace.
	– 2–4 band and servo.	– Check adjustment.
	– Pressure regulator valve sticking or damaged.	– Clean, service or replace.
• Harsh 1–2 shift.	– Kickdown cable broken or out of adjustment.	– Check kickdown adjustment
• N–R shift shock.	– N–R accumulator sticking or damaged.	– Inspect and service or replace.
• 2–3 shift shock.	– 2–3 accumulator sticking or damaged.	– Inspect and service or replace.
	– 1–2 accumulator sticking or damaged.	– Inspect and service or replace.
	– Pulse generator not functioning.	– Check pickup and torque converter for damage.
• Erratic shifts.	– Kickdown cable broken or out of adjustment.	– Inspect cable adjustment.
	– Pulse generator not functioning.	– Inspect pickup and torque converter.
• Soft shift in all ranges.	– Kickdown cable broken or out of adjustment.	– Inspect cable adjustment.
	– Throttle valve sticking or damaged.	– Clean, service or replace.
	– Pressure regulator valve sticking or damaged.	– Clean, service or replace.
• 1–2 soft shift.	– Valve body.	– Inspect valve body, solenoid valves.
	– 2–4 band is too loose.	– Inspect adjustment.

Shift feel symptom diagnosis

SHIFT FEEL SYMPTOM MENU

CONDITION	POSSIBLE CAUSE	ACTION
• 2–3 soft shift.	– 2–3 accumulator sticking or damaged.	– Clean, service or replace.
	– Valve body.	– Inspect valve body, solenoid valves.
• N–R soft shift.	– N–R accumulator sticking or damaged.	– Clean, service or replace.
• No lockup	– Lockup valve sticking or damaged.	– Clean, service or replace.
• Drags in reverse link parking brake is applied.	– 2–4 band is too tight.	– Check adjustment
• Slow to engage in reverse.	– Reverse clutch.	– Inspect for damage or wear; service or replace.

Shift feel symptom diagnosis

UPSHIFT SYMPTOM MENU

CONDITION	POSSIBLE CAUSE	ACTION
• No 2–3 upshift.	– 3–4 clutch spring.	– Check clutch adjustment, damage.
	– Valve body.	– Inspect valve body, solenoid valves.
• No 2nd gear (transmission shifts 1–3).	– Valve body.	– Inspect valve body, solenoid valves.
	– Loose 2–4 band.	– Adjust.
• No lock up.	– Lockup solenoid not functioning.	– Inspect solenoid and related hydraulic circuit.
	– Torque converter.	– Inspect torque converter.
• Shift points incorrect.	– Valve body.	– Inspect valve body, solenoid valves.
	– 2–4 band out of adjustment.	– Check 2–4 band adjustments.
	– Damaged or worn forward clutch.	– Inspect and service or replace.
• Engine run away when upshifting.	– Neutral safety switch.	– Check adjustment and condition.
	– Valve body.	– Clean, service or replace.
	– One way clutch no. 1.	– Inspect, service, or replace.
	– 2–4 band and servo.	– Check adjustment and condition.
	– 3–4 clutch.	– Check condition, service.
	– Bypass valve sticking or damaged.	– Clean, service, or replace.
	– Forward clutch.	– Inspect, service or replace.
• No upshift into overdrive.	– One way clutch no. 1 stuck.	– Check clutch no. 1.
	– Valve body.	– Check orifices, solenoid valves, valve body.
	– Linkage.	
• Delayed 1–2 shift.	– Valve body.	– Inspect valve body, solenoid valves.

Upshift symptom diagnosis

DOWNSHIFT SYMPTOM MENU

CONDITION	POSSIBLE CAUSE	ACTION
• Engine has momentary run-away during 3–2 downshift.	– Coasting bypass valve sticking or damaged.	– Inspect, service or replace.
	– 2–4 band and servo.	– Inspect adjustment, service or replace.
• Hesitation in 3–2 shift.	– Valve body.	– Inspect valve body, solenoid valves.
• No engine braking Ⓓ to D.	– Fluid blockage to coasting clutch or failed coasting clutch.	– Check for blockage and coasting clutch condition.
	– Valve body.	– Inspect valve body, solenoid valves.
• No engine braking D to L.	– Fluid blockage to coasting clutch or failed coasting clutch.	– Inspect coasting for blockage or damage.
	– 2–4 band and servo.	– Check adjustment and inspect condition.
	– Valve body.	– Inspect valve body, solenoid valves.
	– Control valve.	– Inspect, clean or service.

Downshift symptom diagnosis

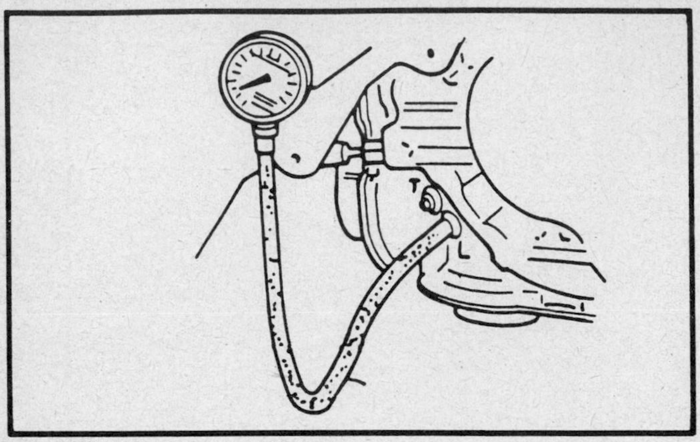

Installing pressure gauge

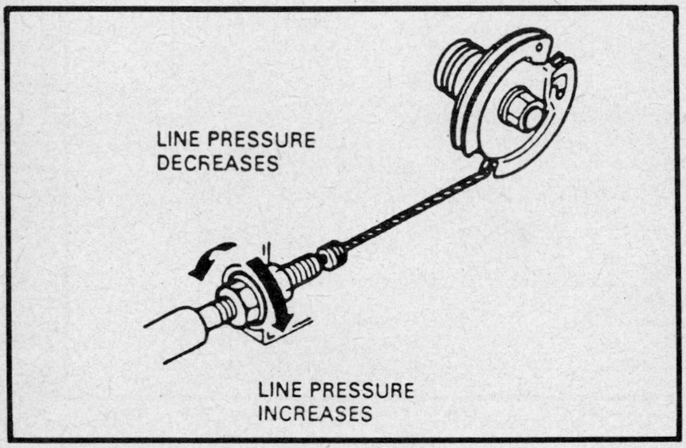
LINE PRESSURE
DECREASES
LINE PRESSURE
INCREASES

Line pressure adjustment

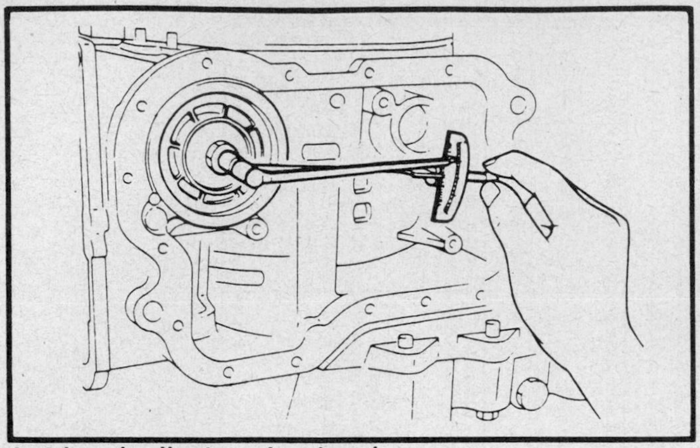

2–4 band adjustment—step 1

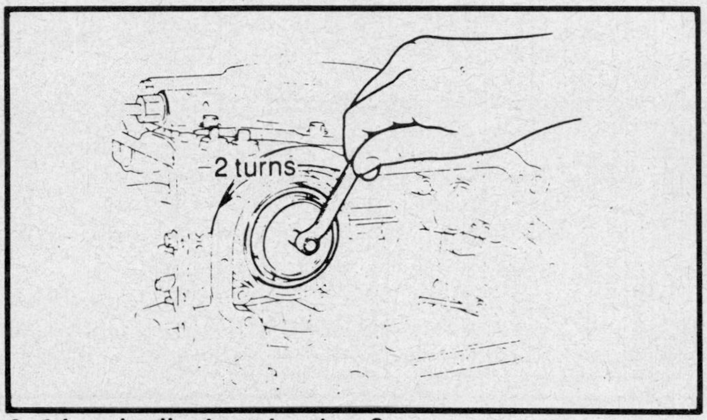
2 turns

2–4 band adjustment—step 2

SHIFT CONTROL CABLE

1. Remove the selector trim panel. Disconnect the wiring harness for the programmed ride control switch.
2. Remove 4 screws securing the transaxle selector bezel assembly.
3. Lift bezel assembly to gain access to the shift cable adjuster.
4. Loosen nuts A and B. Loosen bolt C.
5. Place transaxle in **P** by moving the transaxle mounted shift lever clockwise.
6. Place the selector lever in the **P** position.
7. Torque bolt C to 67–96 inch lbs.
8. Tighten nut A until nut touches the trunnion. Torque nut B to 67–96 inch lbs. Make sure nut B seats against spacer and not the spring.
9. Install transaxle selector bezel. Verify that there is a click at each range position.
10. Make sure the linkage adjustment has not affected operation of the neutral safety switch. With the parking brake and service brakes applied, try to start the engine in each gearshift position. The engine must crank only in the **N** and **P** positions. If the engine cranks in any other gear selector lever position, check the linkage adjustment and neutral safety switch operation.
11. Connect the programmed ride control switch wiring harness. Install the selector trim panel.

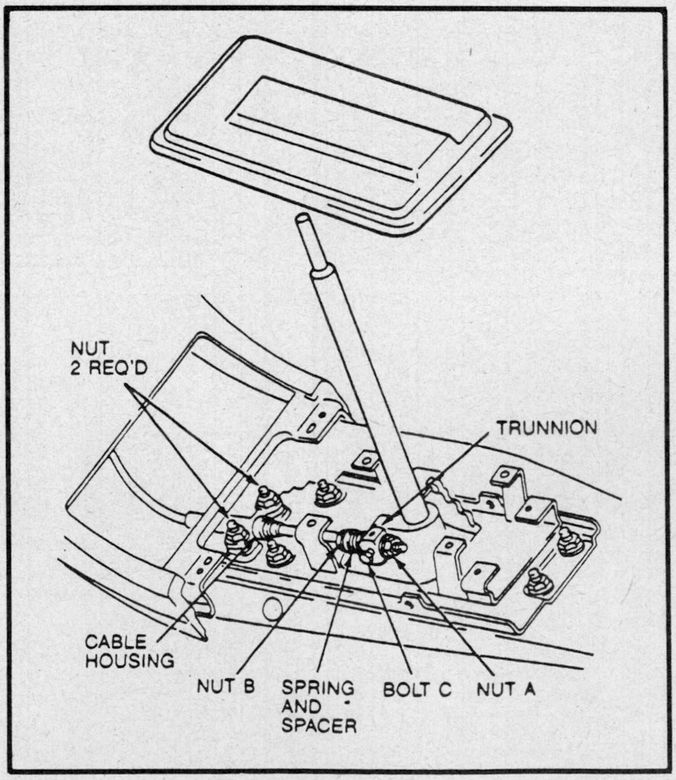
NUT
2 REQ'D
TRUNNION
CABLE
HOUSING
NUT B
SPRING
AND
SPACER
BOLT C
NUT A

Shift control cable adjustment

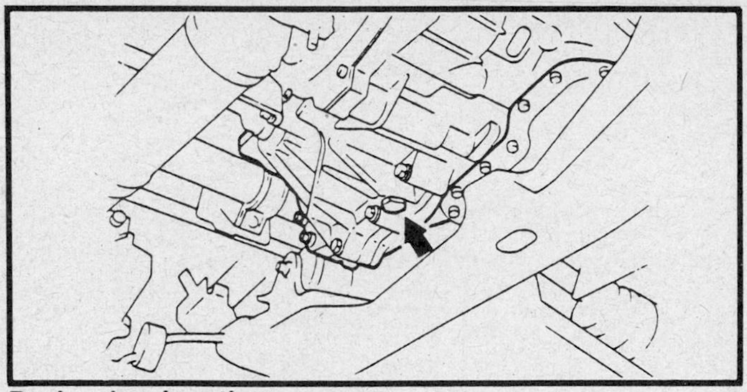

Drain plug location

Magnets in oil pan—correct location

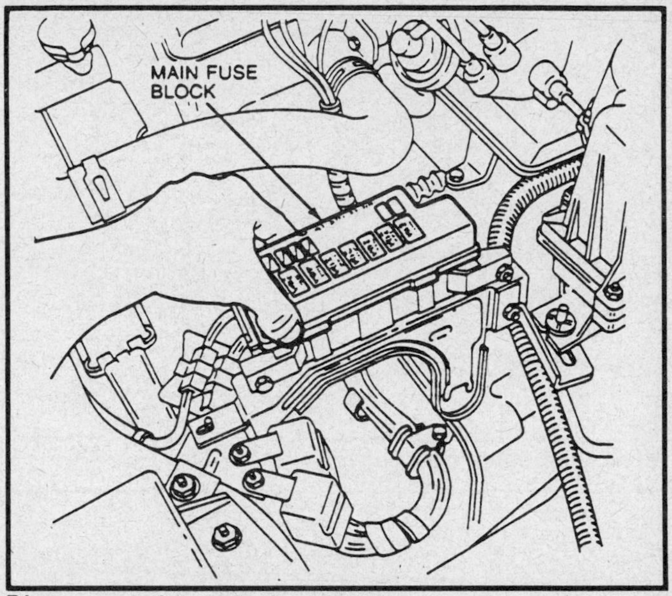
MAIN FUSE
BLOCK

Disconnect the main fuse block

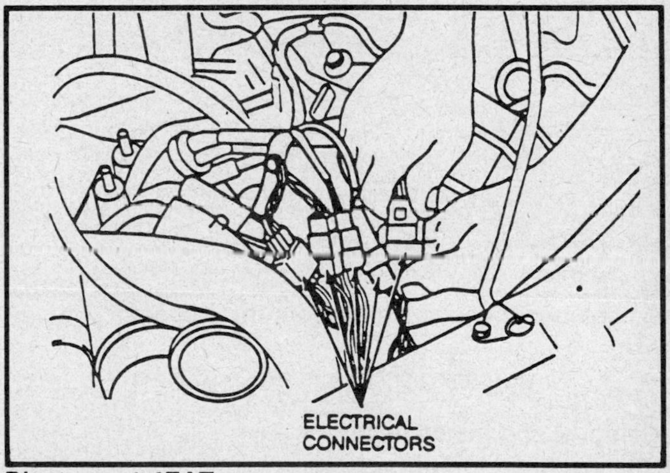
ELECTRICAL
CONNECTORS

Disconnect 4EAT connectors

Services

OIL PAN AND FILTER

Removal and Installation

1. Raise the vehicle and support it safely.
2. Drain the transmission fluid.
3. Remove the left side splash shield.
4. Remove the oil pan and gasket.
5. Remove the oil strainer.
6. Remove the O-ring from the oil strainer
7. To install reverse the removal procedures. Make sure to install the O-ring on the oil strainer and magnets in the correct position in the oil pan. Torque the oil strainer retaining bolts to 69–95 inch lbs. and the oil pan to case bolts to 69–95 inch lbs. Refill the transaxle.

VALVE BODY

Removal and Installation

NOTE: Before trying to service valve body make sure that a new separator plate gasket is available.

1. Remove the battery and battery carrier.
2. Disconnect the main fuse block.
3. Disconnect the 4EAT connectors and separate the 4EAT harness from the transaxle clips.
4. Raise and support the vehicle.
5. Drain the transaxle fluid.
6. Disconnect the oil cooler outlet and inlet hoses.
7. Remove the valve body cover and gasket.
8. Remove the kickdown cable from the throttle cam.
9. Disconnect the solenoid connector, pinch the tangs of the mating connector mounted on the transaxle case. Remove it by pushing inward.
10. Remove the attaching bolts from the valve body and carefully remove the valve body.

NOTE: Shift transaxle into reverse to place the manual plate in the correct position for installation.

11. Install the valve body, using a mirror if necessary to align the groove of the manual valve with the manual plate.
12. Tighten the valve body mounting bolts to 95–130 inch lbs.
13. Insert the solenoid connector into the transaxle case hole. Attach the mating connector.
14. Attach the kickdown cable to the throttle cam.
15. Install the valve body cover and new gasket. Tighten to 69–95 inch lbs.
16. Connect the oil cooler hoses.
17. Attach the 4EAT connectors and support the 4EAT harness on the transaxle clips.
18. Connect the main fuse block.
19. Install the battery carrier and battery.
20. Add the specified transaxle fluid and check for fluid leaks.

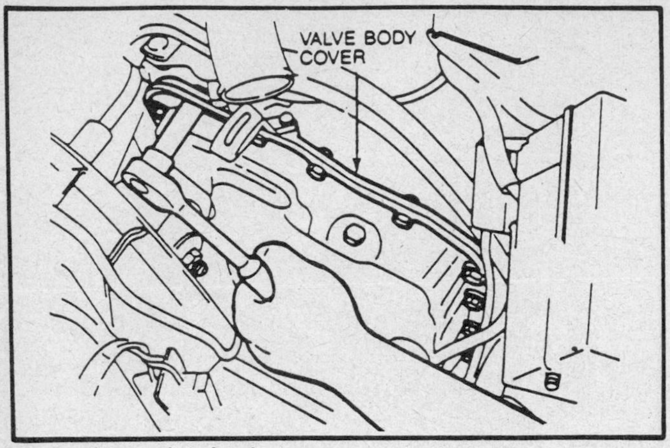
VALVE BODY
COVER

Removing valve body cover

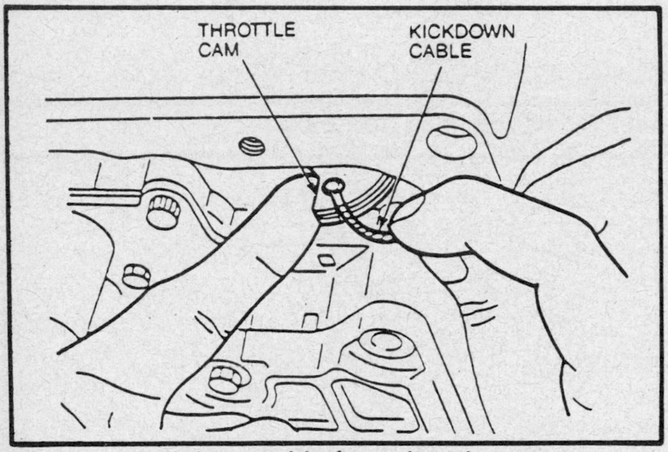
THROTTLE
CAM
KICKDOWN
CABLE

Removing kickdown cable from throttle cam

DIFFERENTIAL OIL SEALS

Removal and Installation

1. Raise and support the vehicle.
2. Remove the front wheels.
3. Remove the splash shields.
4. Drain the transaxle fluid.
5. Remove the tie rod nuts, cotter pins and disconnect the tie rod ends.
6. Remove the stabilizer link assemblies.
7. Remove the bolts and nuts from the lower arm ball joints.
8. Pull the lower arms to separate them from the knuckles.
9. Remove the right-hand joint shaft bracket.
10. Remove the halfshafts from the transaxle by prying with a bar inserted between the shaft and transaxle case. Support the halfshafts with wire.
11. Remove the differential oil seals with a flat-tip tool.
12. Tap in new differential oil seals using differential seal replacer tool T87C–77000–H or equivalent.
13. Replace the circlip located on the end of each halfshaft.
14. Install the halfshafts.
15. Attach the lower arm ball joints to the knuckles.
16. Install the tie rod ends and tighten the nuts to 22–33 ft. lbs. Install new cotter pins.
17. Install the bolts and nuts to the lower arm ball joints. Torque to 32–40 ft. lbs.
18. Install the stabilizer link assemblies. Turn the nuts on each asembly until 1.0 in. (25.4mm) of bolt thread can be measured from the upper nut. When this length is reached, secure

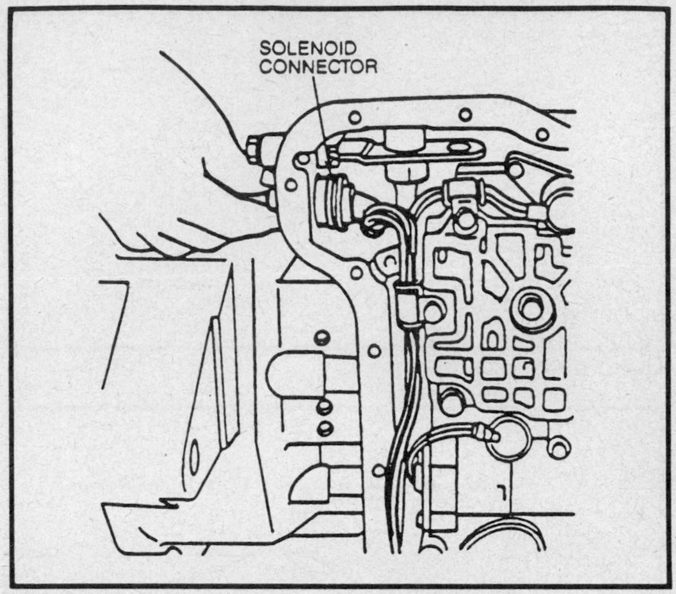
SOLENOID
CONNECTOR

Removing solenoid connector from case.

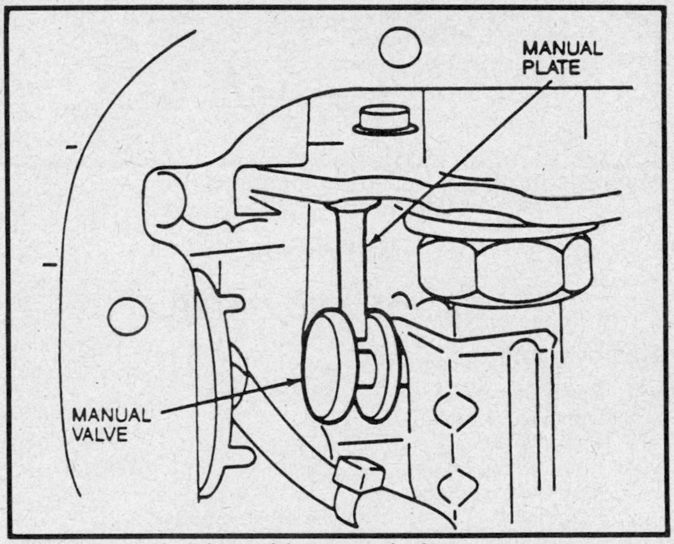
MANUAL
PLATE
MANUAL
VALVE

Align manual valve with manual plate

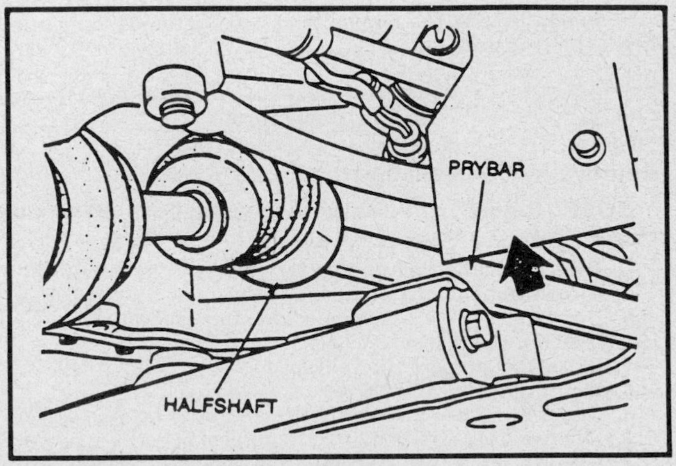
PRYBAR
HALFSHAFT

Removing halfshaft from transaxle

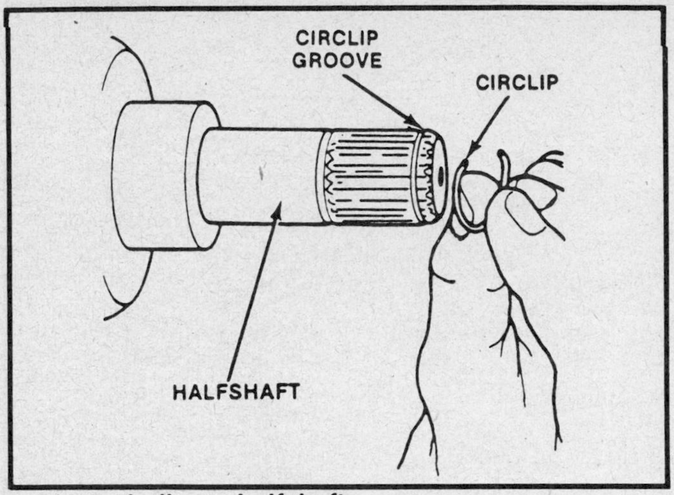
CIRCLIP
GROOVE
CIRCLIP
HALFSHAFT

Replace circlip on halfshaft

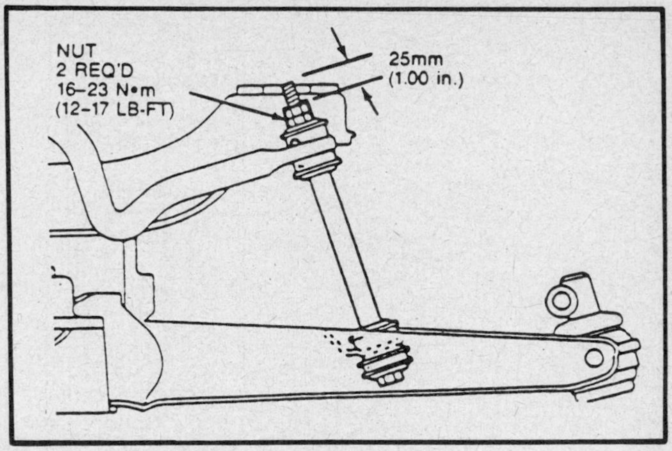
NUT
2 REQ'D
16–23 N•m
(12–17 LB-FT)
25mm
(1.00 in.)

Installing stabilizer link assemblies

the upper nut and back off the lower nut until a torque of 12–17 ft. lbs. is reached.

19. Install the splash shields. Install the front wheels and tighten the lug nuts to 65–87 inch lbs.
20. Add the specified transaxle fluid and check for leaks.

OIL COOLER

Removal and Installation

1. Disconnect the oil hoses.
2. Remove the oil cooler.
3. Straighten bent fins with a flat tool or equivalent.
4. To install reverse removal procedures.

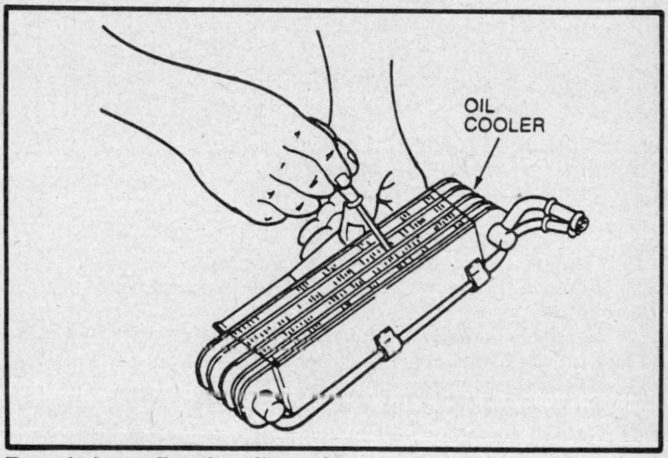
OIL
COOLER

Repair bent fins in oil cooler

REMOVAL AND INSTALLATION

TRANSAXLE REMOVAL

1. Remove the battery and battery carrier.
2. Disconnect the main fuse block.
3. Disconnect the lead from the center distributor terminal.
4. Disconnect the airflow meter connector and remove the air cleaner assembly.
5. Remove the resonance chamber and bracket.
6. Disconnect the speedometer cable (electromechanical cluster) or harnes (electronic cluster).
7. Disconnect the 4EAT electrical connectors and separate the 4EAT harness from the transaxle clips.
8. Disconnect the ground wires from the transaxle case.
9. Disconnect the range selector cable from the transaxle case.
10. Disconnect the kickdown cable.
11. Raise and support the vehicle safely.
12. Remove the front wheels.
13. Remove the splash shields.
14. Drain the transaxle fluid.
15. Disconnect the oil cooler outlet and inlet hoses. Insert plugs to prevent fluid leakage.

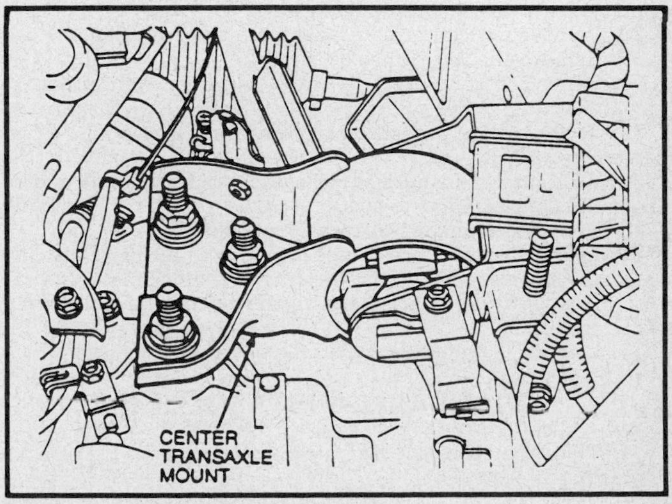
CENTER
TRANSAXLE
MOUNT

Remove center transaxle mount

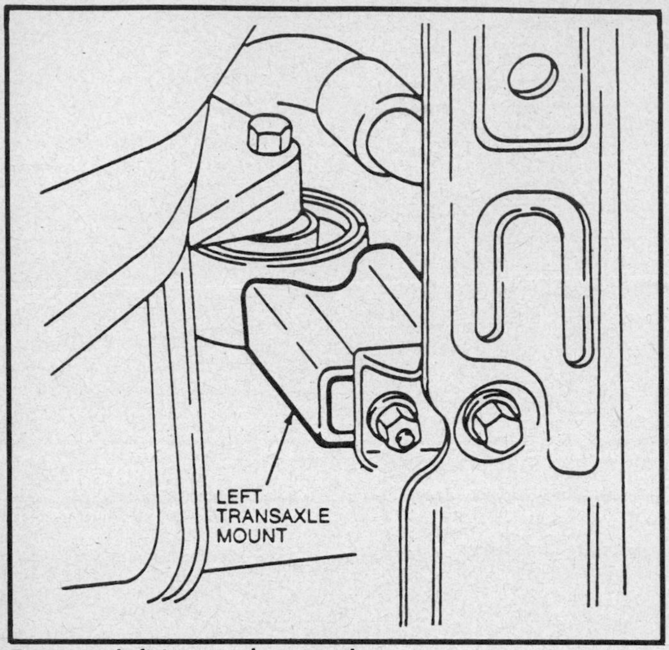
LEFT
TRANSAXLE
MOUNT
Remove left transaxle mount

16. Remove the stabilizer link assemblies.
17. Remove the tie rod nuts and cotter pins and disconnect the tie rod ends.
18. Remove the bolts and nuts from the lower arm ball joints.
19. Pull the lower arms to separate them from the knuckles.
20. Remove the right joint shaft bracket.
21. Remove the halfhafts from the transaxle by prying with a bar inserted between the shaft and transaxle case.
22. Install transaxle plugs T88C–7025–AH or equivalent, into the differential side gears.

NOTE: Failure to install the transaxle plugs may allow the differential side gears to become mispositioned.

23. Remove the gusset plate-to-transaxle bolts.
24. Remove the torque converter cover.
25. Remove the torque converter nuts.
26. Remove the starter motor and access brackets.
27. Mount an engine support bar, D79P–6000–B or equivalent and attach it to the engine hanger.
28. Remove the center transaxle mount and bracket.
29. Remove the left transaxle mount.
30. Remove the nut and bolt attaching the right transaxle mount to the frame.
31. Remove the crossmember and left lower arm as an assembly.
32. Position a transmission jack under the transaxle and secure the transaxle to the jack.
33. Remove the engine-to-transaxle bolts.
34. Before the transaxle can be lowered out of the vehicle, the torque converter studs must be clear of the flex plate. Insert a tool between the flex plate and converter and carefully disengage the studs.
35. Lower the transaxle out of the vehicle.

TRANSAXLE INSTALLATION

1. Place the transaxle on a transmission jack. Be sure the transaxle is secure.
2. Raise the transaxle to the proper height and mount the transaxle to the engine.

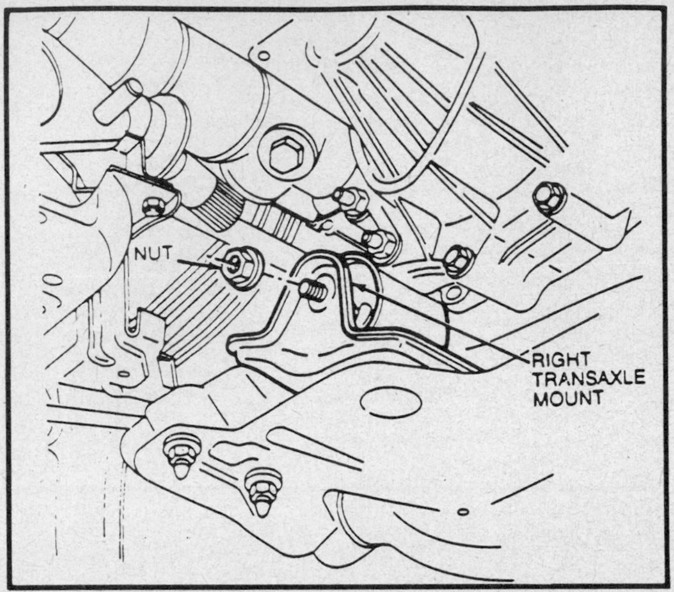
NUT
RIGHT
TRANSAXLE
MOUNT
Remove nut and bolt from right transaxle mount

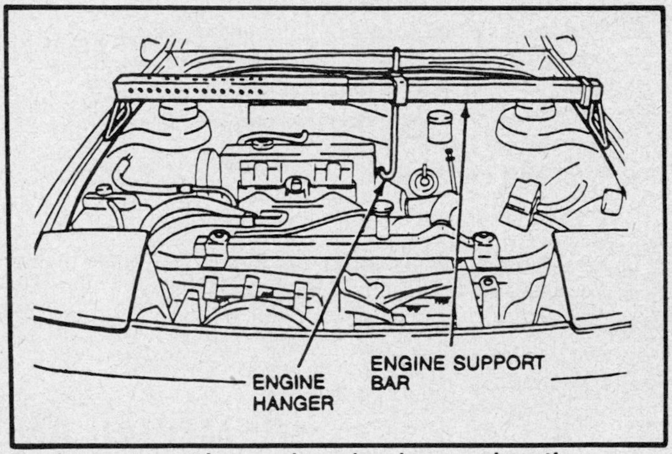
ENGINE
HANGER
ENGINE SUPPORT
BAR
Engine support bar and engine hanger location

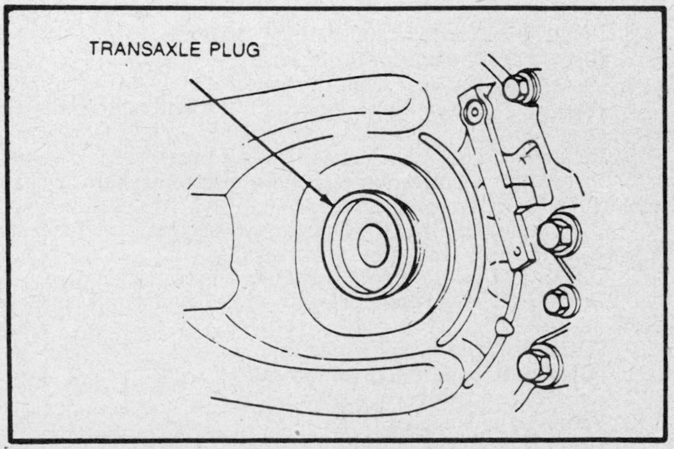
TRANSAXLE PLUG
Transaxle plugs—holds differential side gears in place

NOTE: Align the torque converter studs and flex plate holes.

3. Install the engine-to-transaxle bolts and tighten to 66–86 ft. lbs.
4. Install the center transaxle mount and bracket. Tighten the bolts to 27–40 ft. lbs. and the nuts to 47–66 ft. lbs.
5. Install the left transaxle mount. Tighten the transaxle-to-mount attaching nut to 63–86 ft. lbs. Tighten the mount-to-bracket bolt and nut to 49–69 ft. lbs.
6. Install the crossmember and left lower arm as an assembly. Tighten the bolts to 27–40 ft. lbs. and the nuts to 55–69 ft. lbs.
7. Install the right transaxle mount bolt and nut. Tighten to 63–86 ft. lbs.
8. Install the starter motor and access brackets.
9. Install the torque converter nuts and tighten to 32–45 ft. lbs.
10. Install the converter cover and tighten the bolts to 69–85 inch lbs.
11. Install the gusset plate-to-transaxle bolts and tighten to 27–38 ft. lbs.
12. Replace the circlip located on the end of each halfshaft.
13. Remove the transaxle plugs and install the halfshafts.
14. Attach the lower arm ball joints to the knuckles.
15. Install the tie rod ends and tighten the nuts to 22–33 ft. lbs. Install new cotter pins.
16. Install the bolts and nuts to the lower arm ball joints. Tighten to 32–40 ft. lbs.
17. Install the stabilizer link assemblies. Turn the nuts on each assembly until 1.0 in. (25.4mm) of bolt thread can be measured from the upper nut. When then length is reached, secure the upper nut and back off the lower nut until a torque of 12–17 ft. lbs. is reached.
18. Connect the oil cooler outlet and inlet hoses.
19. Install the splash shields.
20. Install the front wheels and tighten the lug nuts to 65–87 ft. lbs.
21. Connect the kickdown cable and adjust it while performing the oil pressure test.
22. Connect the range selector cable to the transaxle case and tighten the bolt to 22–29 ft. lbs.
23. Connect the ground wires to the transaxle case and tighten to 69–95 inch lbs.
24. Connect the 4EAT electrical connectors and attach the 4EAT harness to the transaxle clips.
25. Connect the speedometer cable (electromechanical cluster) or harness (electronic cluster).
26. Install the resonance chamber and bracket and tighten to 69–95 inch lbs.
27. Install the air cleaner assembly. Tighten the bolt to 23–30 ft. lbs. and the nuts to 69–95 inch lbs.

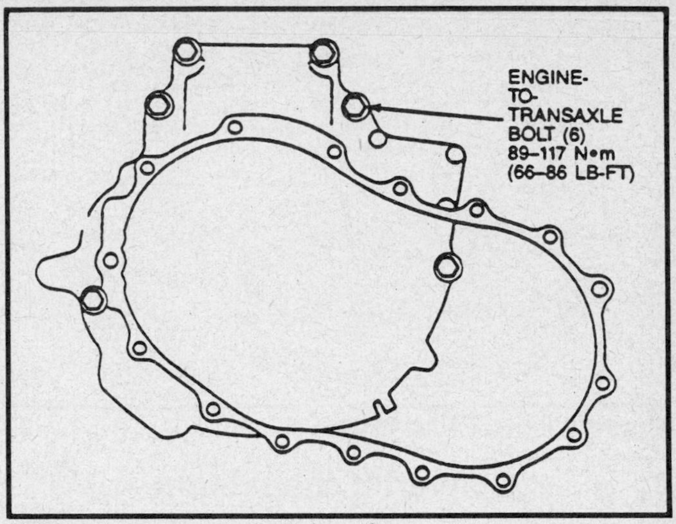
ENGINE-
TO-
TRANSAXLE
BOLT (6)
89–117 N•m
(66–86 LB-FT)

Engine to transaxle bolt location

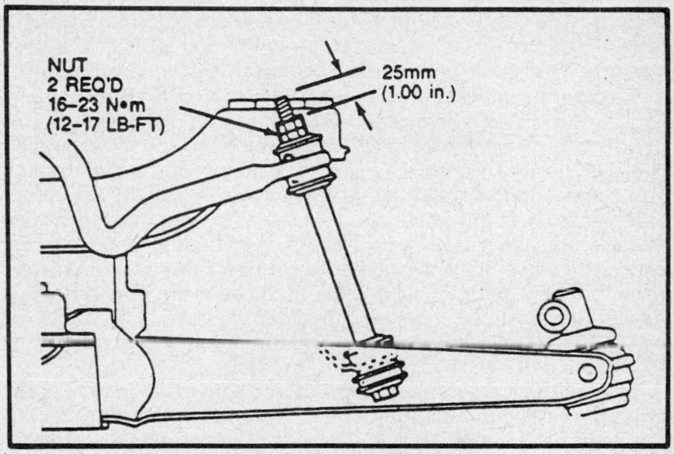
NUT
2 REQ'D
16–23 N•m
(12–17 LB-FT)
25mm
(1.00 in.)

Installing the stabilizer link assemblies

28. Connect the airflow meter connector.
29. Connect the center distributor terminal lead.
30. Connect the main fuse block and tighten to 69–95 inch lbs. Install the battery carrier and battery and tighten to 23–30 ft. lbs.
31. Remove the engine support bracket.
32. Add the specified transaxle fluid. Check for fluid leakage.
33. Road test vehicle for proper operation.

BENCH OVERHAUL

Before Disassembly

When servicing the unit, it is recommended that as each part is disassembled, it is cleaned in solvent and dried with compressed air. All oil passages should be blown out and checked for obstructions. Disassembly and reassembly of this unit and its parts must be done on a clean work bench. As is the case when repairing any hydraulically operated unit, cleanliness is of the utmost importance. Keep bench, tools, parts and hands clean at all times. Also, before installing bolts into aluminum parts, always dip the threads into clean transmission oil. Anti-seize compound can also be used to prevent bolts from galling the aluminum and seizing. Always use a torque wrench to keep from stripping the threads. Take care with the seals when installing them, especially the smaller O-rings. The slightest damage can cause leaks. Aluminum parts are very susceptible to damage so great care should be exercised when handling them. The internal snaprings should be expanded and the external snaprings compressed if they are to be re-used. This will help insure proper seating when installed. Be sure to replace any O-ring, gasket, or seal that is removed. Lubricate all parts with ATF when assembling.

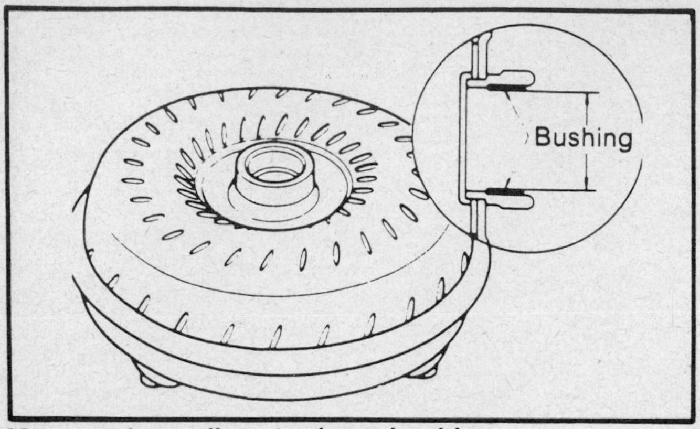
Bushing
Measure inner diameter boss bushing

Converter Inspection

The torque converter is welded together and cannot be disassembled. Check the torque converter for damage or cracks and replace, if necessary. Remove any rust from the pilot hub and boss of the converter. Measure the inner diameter of the boss bushing. If it exceeds 2.090 in., replace the torque converter.

When internal wear or damage has occurred in the transaxle, contaminants such as metal particles, clutch plate material, or band material may have been carried into the converter and oil cooler. These contaminants can be a major cause of recurring transaxle troubles and must be removed from the system before the transaxle is put back into service. Whenever the transaxle has been disassembled to replace worn or damaged parts or because the valve body sticks due to foreign material, the torque converter, oil and oil cooler lines must be cleaned and flushed using the Rotunda torque converter cleaner 014–00028 or equivalent. Under no circumstances should an attempt be made to clean converters by hand.

The lack of a drain plug in the 4EAT converter increases the amount of residual flushing solvent retained in the converter after cleaning. This retained solvent is not acceptable and a method of diluting is required. The following procedure is to be used after removal of the 4EAT torque converter from the cleaning equipment. Thoroughly drain the remaining solvent through the hub. Add about a ½ quart of clean transaxle fluid into the converter. Agitate by hand. Thoroughly drain the solution through the converter hub.

Transaxle Disassembly

NOTE: Whenever the transaxle is disassemblied, the bearing preload must be adjusted. The output gear and differential bearing preload are adjusted by selecting shim(s) to insert under the bearing cups.

1. Remove the torque converter.

CAUTION

The torque converter is heavy. Be careful not to drop it.

2. Remove the oil pump shaft.
3. Mount the transaxle on a bench mounted holding fixture or equivalent.
4. Remove the dipstick tube retaining bolts and pull the tube from its slot.
5. Remove the neutral safety switch.
6. Remove the fluid temperature switch.
7. Remove the pulse generator.
8. Disconnect the solenoid connector.
9. Remove the 4EAT wiring harness and harness clip.
10. Remove the oil pipes, oil hoses and switch box as an assembly.

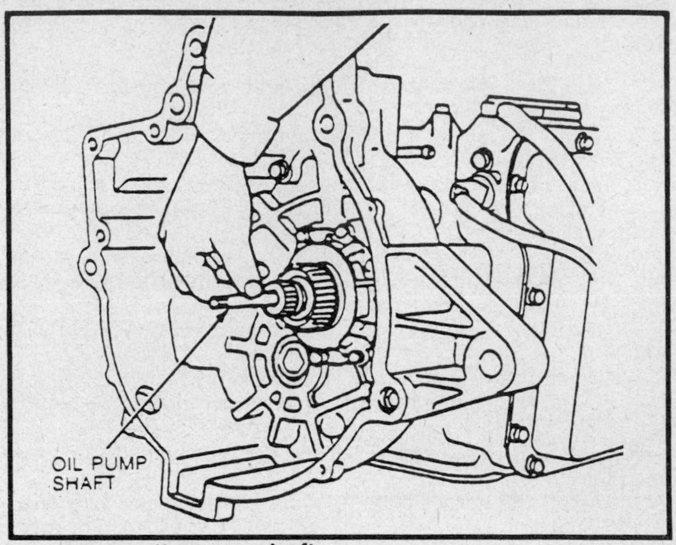
OIL PUMP
SHAFT
Removing oil pump shaft

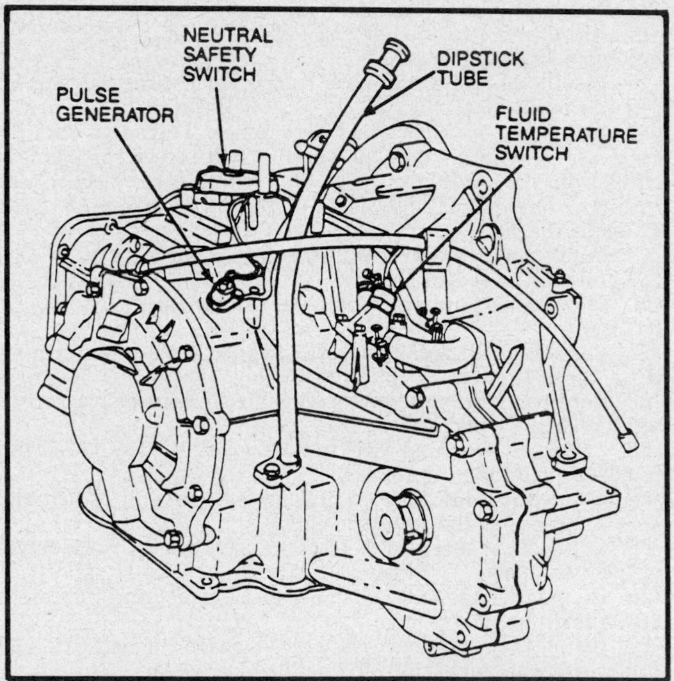
NEUTRAL
SAFETY
SWITCH
DIPSTICK
TUBE
PULSE
GENERATOR
FLUID
TEMPERATURE
SWITCH
Component location

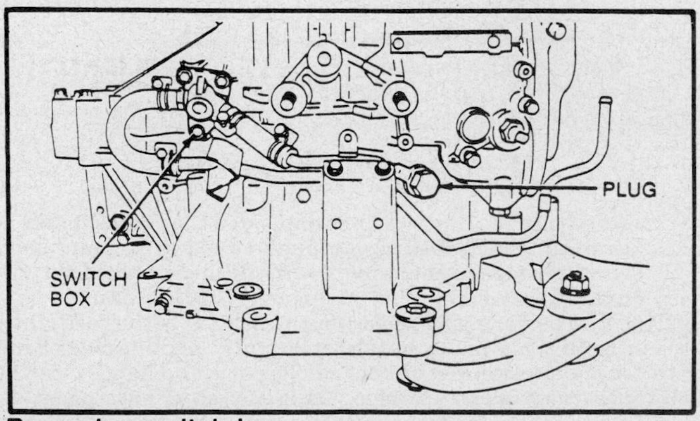
PLUG
SWITCH
BOX
Removing switch box

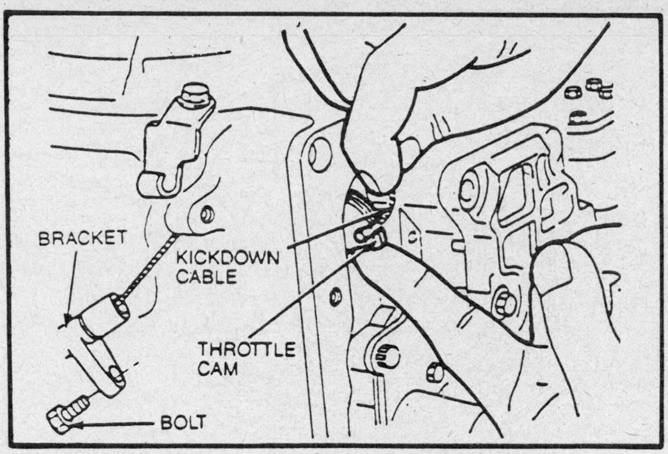
BRACKET
KICKDOWN CABLE
THROTTLE CAM
BOLT
Remove kickdown cable from throttle cam

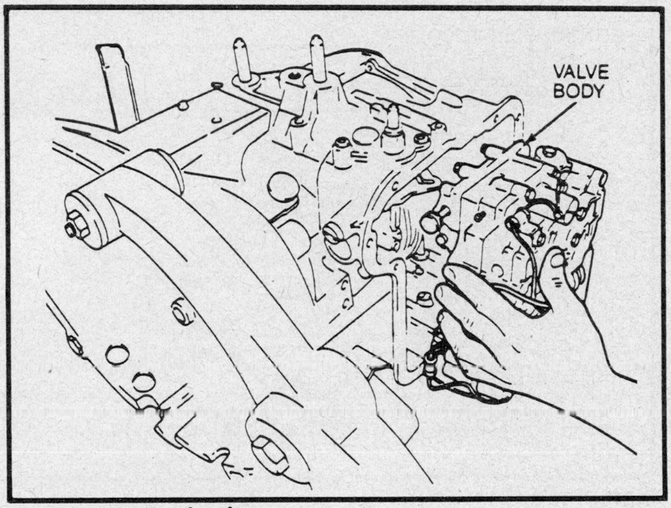
VALVE BODY
Remove valve body

NOTE: Use a magnet to remove the ball and spring from the plug hole.

11. Remove the oil pan and gasket.
12. Remove the oil strainer and O-ring.
13. Remove the valve body cover and gasket.
14. Remove the kickdown cable attaching bolt and bracket.
15. Remove the kickdown cable from the throttle cam.
16. Pinch the teeth of the solenoid connector mounted on the transaxle case. Remove it by pushing inward.
17. Remove the attaching bolts from the valve body and carefully remove the valve body.
18. Remove the oil pump and gasket.
19. Remove the piston stem from the servo.
20. Remove the turbine shaft snapring.
21. Remove the clutch assembly.
22. Remove the 2–4 band.

NOTE: Secure the 2–4 band with wire to prevent warping.

23. Remove the small sun gear and one-way clutch assembly.
24. Pull the anchor shaft while holding the strut, then remove the strut.
25. Use a C-clamp and socket to compress the servo. Remove the snapring, servo and spring.
26. Remove the one-way clutch snapring.
27. Remove the one-way clutch and carrier hub assembly.
28. Remove the low and reverse clutch snapring.

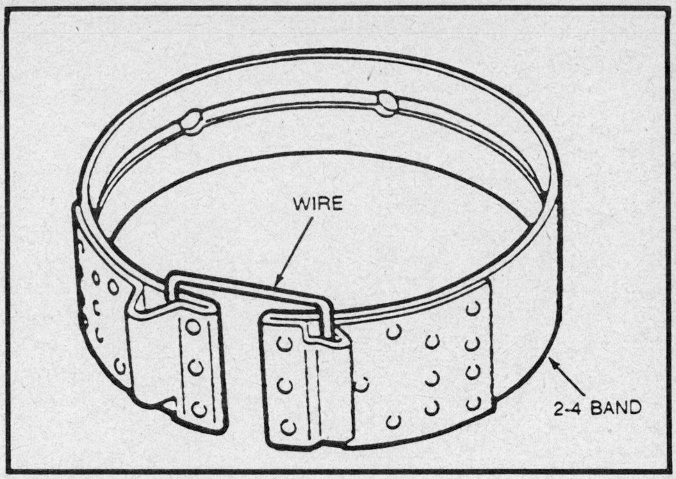
WIRE
2-4 BAND
Secure the 2–4 band

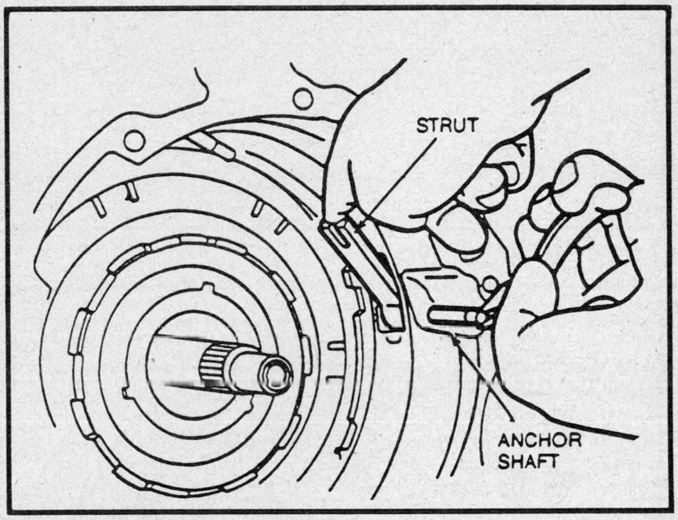
STRUT
ANCHOR SHAFT
Removing anchor shaft

29. Remove the low and reverse clutch retaining plate and drive and driven plates.
30. Remove the internal gear snapring.
31. Remove the internal gear.
32. Remove the O-ring located on the converter housing side of the turbine shaft.
33. Pull out the turbine shaft and remove the 3–4 clutch assembly.
34. Remove the transaxle case bolts and transaxle case from the converter housing. If necessary, tap lightly with a plastic hammer.
35. Remove the output shell from the output gear.
36. Compress the return spring and retainer using return spring compressor T88C–77000–AH and the plate from T87C–77000–B or equivalent.
37. Remove the retainer snapring, then the return spring and retainer.
38. Remove the return spring compressor.
39. Apply compressed air through the low and reverse clutch fluid passage to remove the low and reverse clutch piston.
40. Remove the plug, washer, spring and detent ball.
41. Remove the bracket.
42. Loosen the manual shaft nut and pull the manual shaft out.
43. Remove the nut, washer, spacer and manual plate.

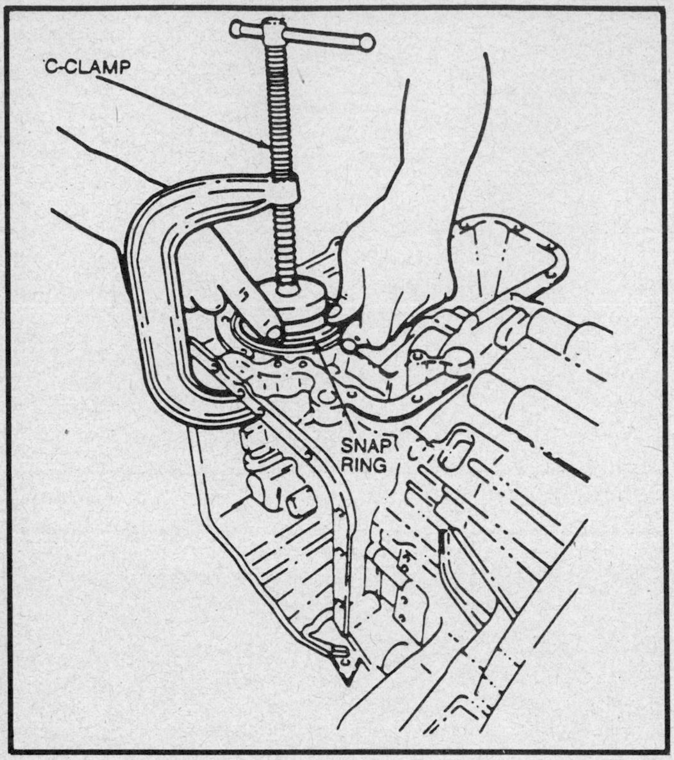
C-CLAMP
SNAP RING

Compressing the servo

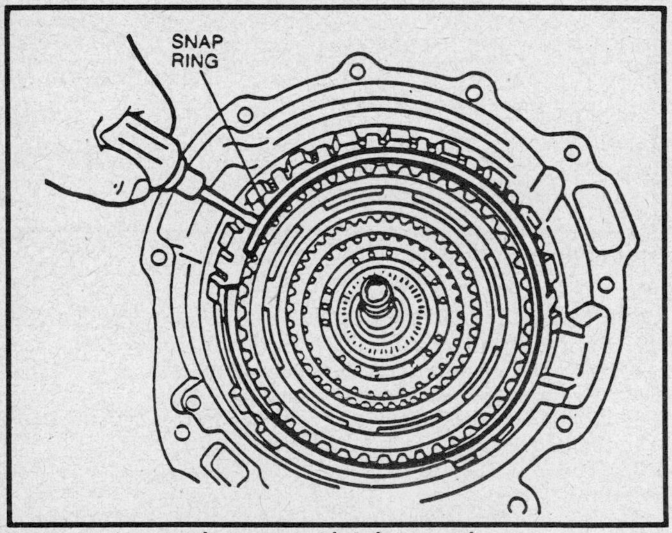
SNAP RING

Remove low and reverse clutch snapring

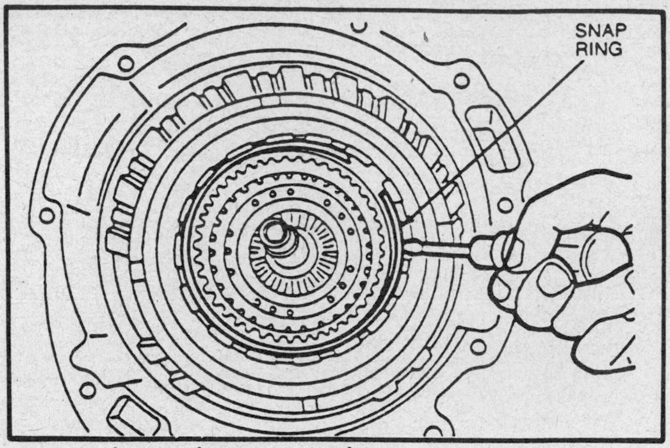
SNAP RING

Remove internal gear snapring

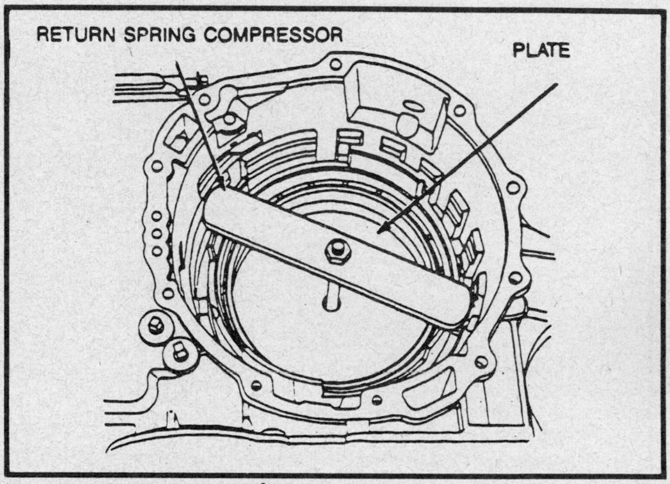
RETURN SPRING COMPRESSOR
PLATE

Compress return spring

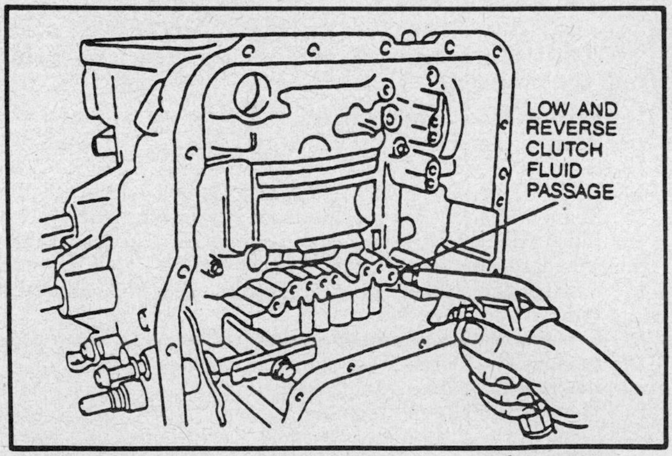
LOW AND REVERSE CLUTCH FLUID PASSAGE

Apply compressed air to remove low and reverse clutch piston

44. Remove the actuator support.
45. Remove the parking assist lever snapring.
46. Remove the parking assist lever.
47. Remove the parking pawl snapring.
48. Pull out the parking shaft, then remove the spring and parking pawl.
49. Remove the differential.
50. Remove the 2–3 accumulator.
51. Remove the bearing housing bolt to access the roll pin.
52. Remove the roll pin using a pin punch.
53. Remove the bearing housing. If necessary, tap lightly with a plastic hammer.
54. Use a socket or equivalent to tap out the idler and output gear assemblies from the torque converter housing.
55. Remove the converter seal from the bearing/stator support using puller tool–1175–AC and slide hammer T50T–100–A or equivalent.
56. Remove the converter housing from the holding fixture.
57. Remove the bearing/stator support bolts.
58. Press the bearing/stator support out of the torque converter housing using step plate D80L–630–10 or equivalent.

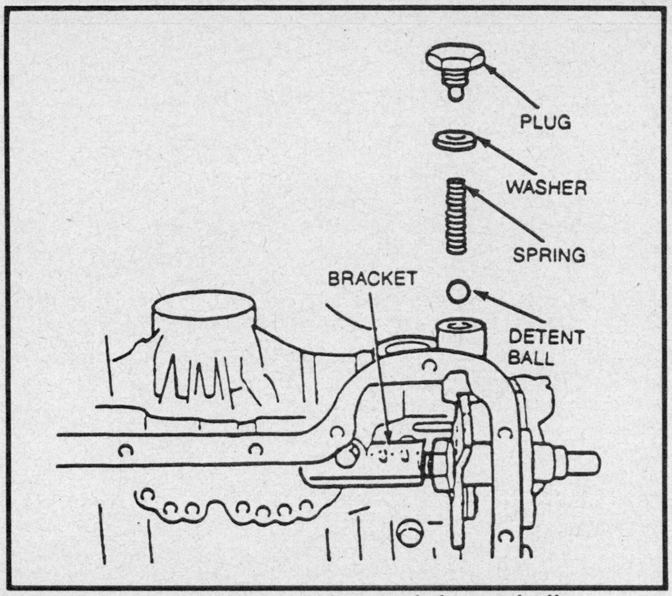
PLUG
WASHER
SPRING
BRACKET
DETENT
BALL
Remove plug, washer, spring and detent ball

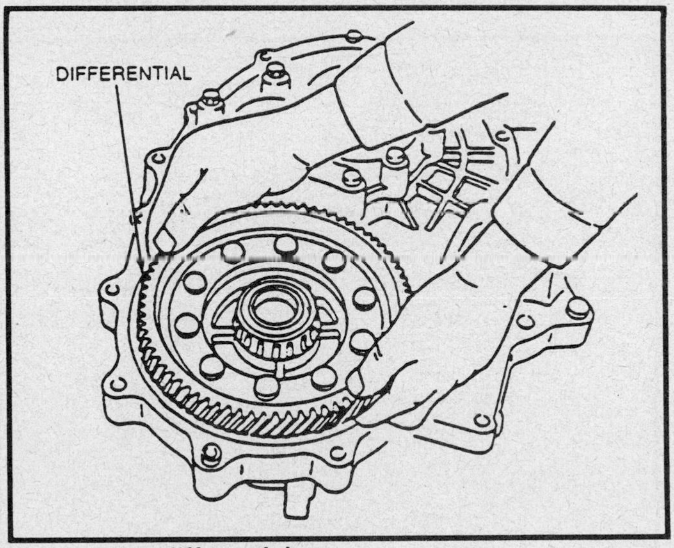
DIFFERENTIAL
Remove the differential

59. Remove the differential bearing cups using puller jaws T86P–70043–A, puller body T73L–2196–A and slide hammer T50T–100–A or equivalent. Remove the adjustment shim(s).

60. Remove the differential oil seals using puller T77F–1102–A and slide hammer T50T–100–A or equivalent.

DIFFERENTIAL BEARING PRELOAD SHIM SELECTION

1. Remove the rear bearing cup and shims from the transaxle case using puller jaw T86P–70043–A, puller body T73L–2196–A and slide hammer T50T–100–A or equivalent.
2. Install the front bearing cup into the converter housing using driver handle T80T–4000–W and differential bearing cup replacer T88C–77000–FH or equivalent.
3. Place the differential into the converter housing.
4. Place 6 collars (part of special tool T87C–77000–J) or equivalent on the converter housing.
5. Place the rear bearing cup over the differential bearing.

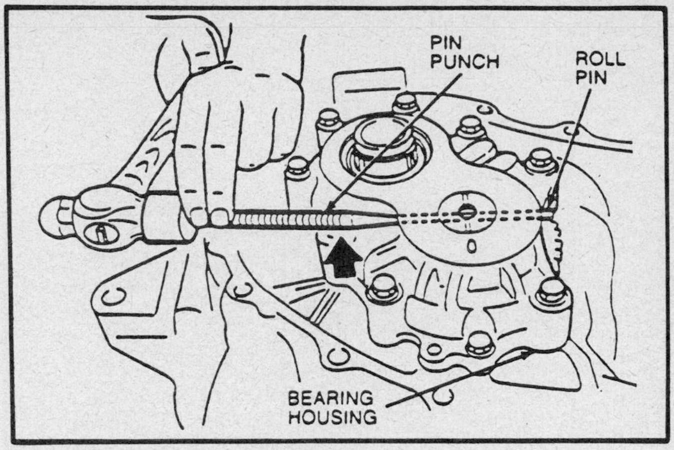
PIN
PUNCH
ROLL
PIN
BEARING
HOUSING
Remove roll pin from bearing housing

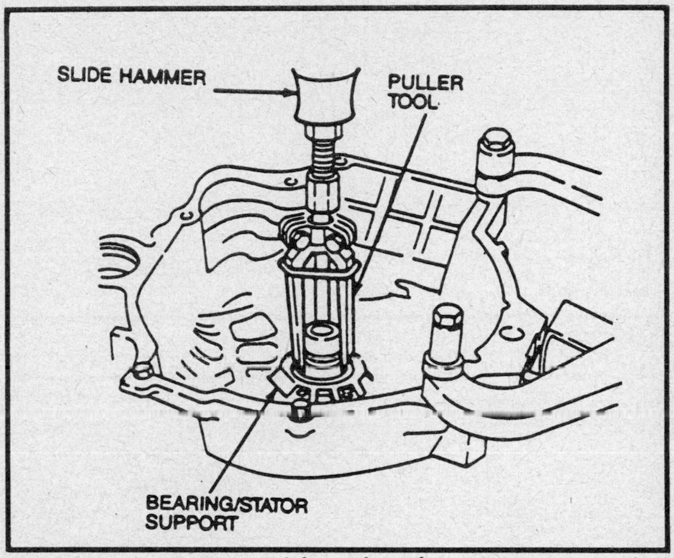
SLIDE HAMMER
PULLER
TOOL
BEARING/STATOR
SUPPORT
Remove converter seal from bearing stator support

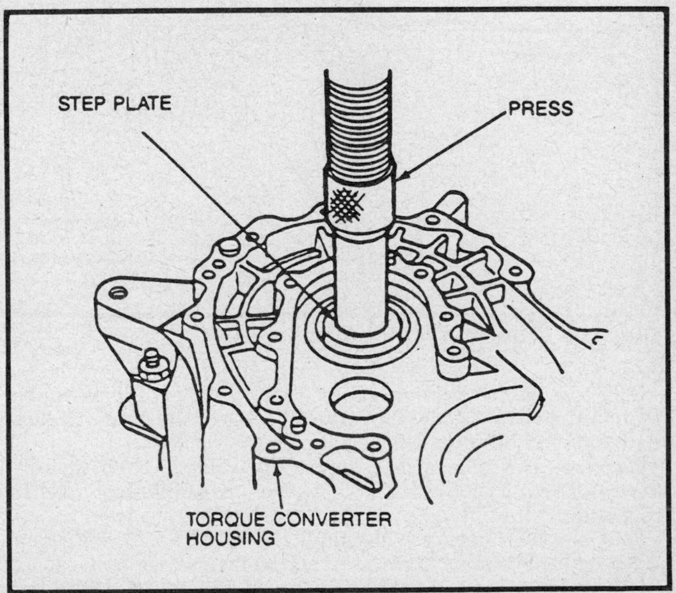
STEP PLATE
PRESS
TORQUE CONVERTER
HOUSING
Press the bearing stator out of converter housing

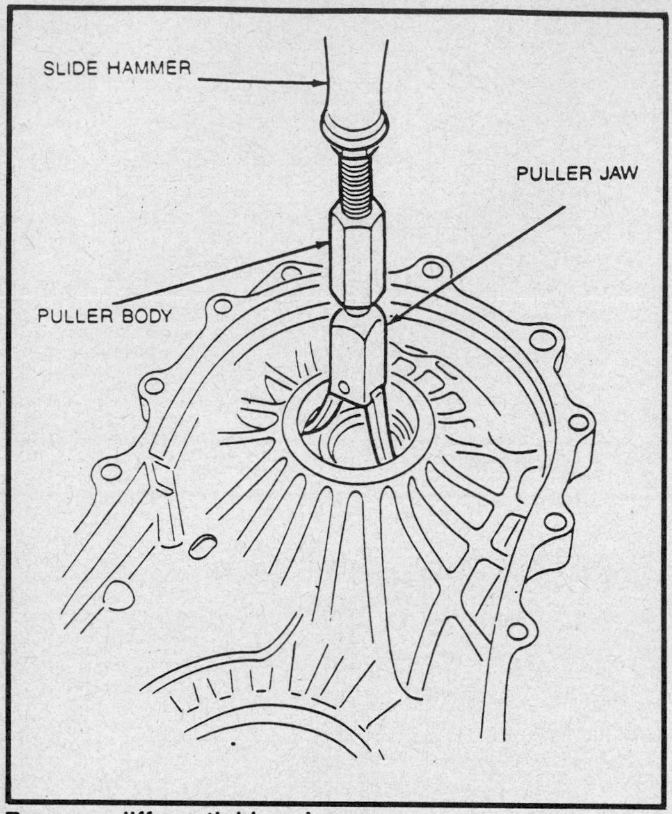
SLIDE HAMMER
PULLER JAW
PULLER BODY

Remove differential bearing cups

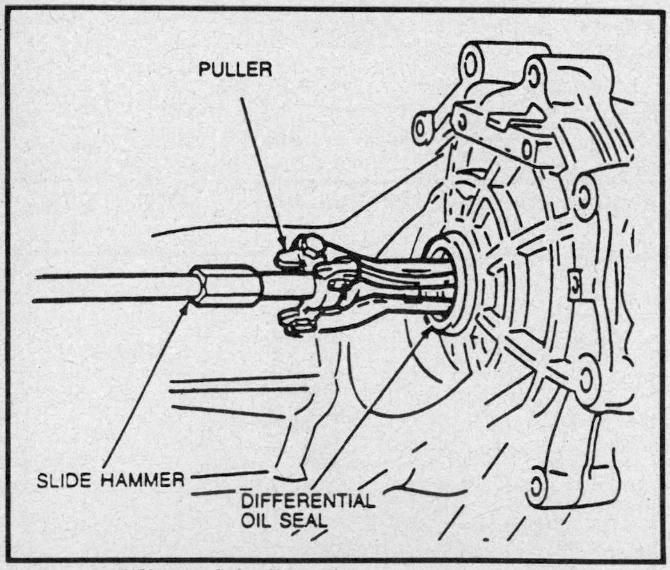
PULLER
SLIDE HAMMER
DIFFERENTIAL OIL SEAL

Remove differential oil seals

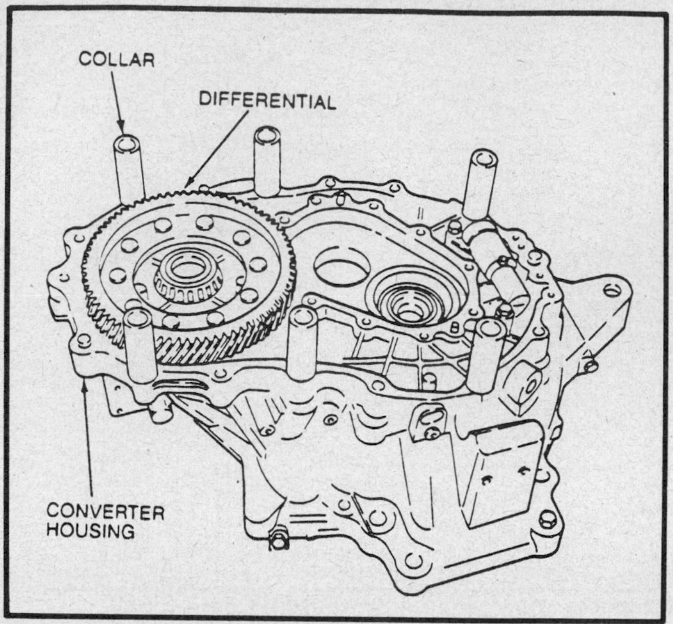
COLLAR
DIFFERENTIAL
CONVERTER HOUSING

Installing special tool on converter housing

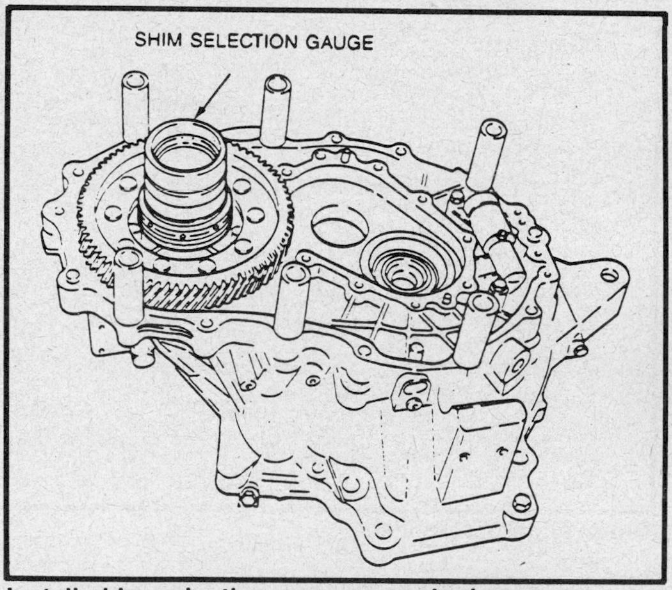
SHIM SELECTION GAUGE

Install shim selection gauge on output gear

6. Place shim selection gauge T88C–77000–CH1 or equivalent on the output gear. Turn the 2 halves of the gauge to eliminate any gap between them.

7. Place the transaxle case on the collars, then install 6 Screws T88C–77000–CH4 with washers or equivalent. Tighten to 27–38 ft. lbs.

8. Using the pins provided in T87C–77000–J or equivalent, unthread the gauge halves until all the freeplay is removed and the bearing cup is seated. Then thread the gauge halves back together.

9. Engage torque adapter T88C–77000–L or equivalent and attach a inch lbs. torque wrench to the adapter. Measure the drag on the differential bearing.

NOTE: Read the preload when the differential starts to turn.

10. Turn the gauge using the pins (part of T87C–77000–J) or equivalent until a reading of 4.3 inch lbs. is obtained on the torque wrench.

11. Use a feeler gauge to measure the gap between the 2 halves of the gauge. Measure the gap at 4 spots, at 90 degree intervals. Use the largest measurement.

12. Add 0.0079 in. to the largest measurement. Using the differential bearing preload shim selection chart, select the shim(s) closest (or slightly larger) to this final value. Use no more than 3 shims.

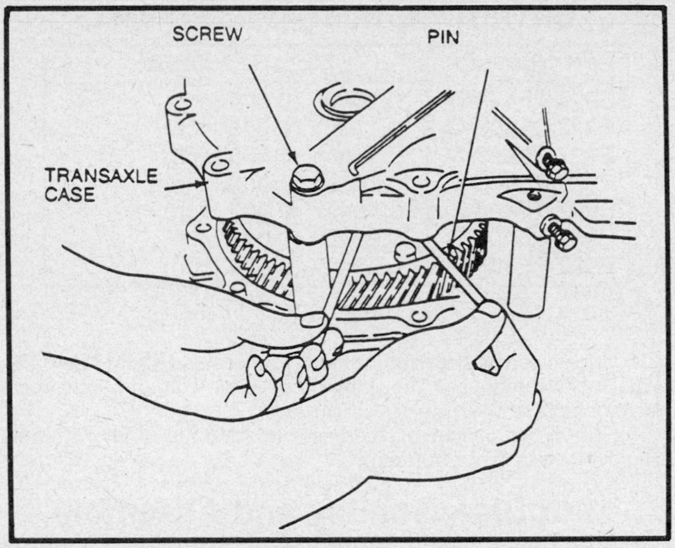
SCREW
PIN
TRANSAXLE CASE

Using special tool to seat bearing cup

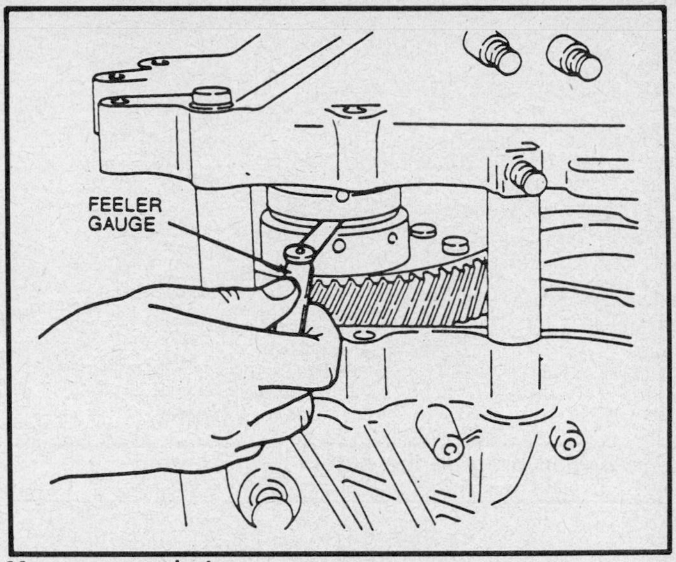
FEELER GAUGE

Measure gap between gauge

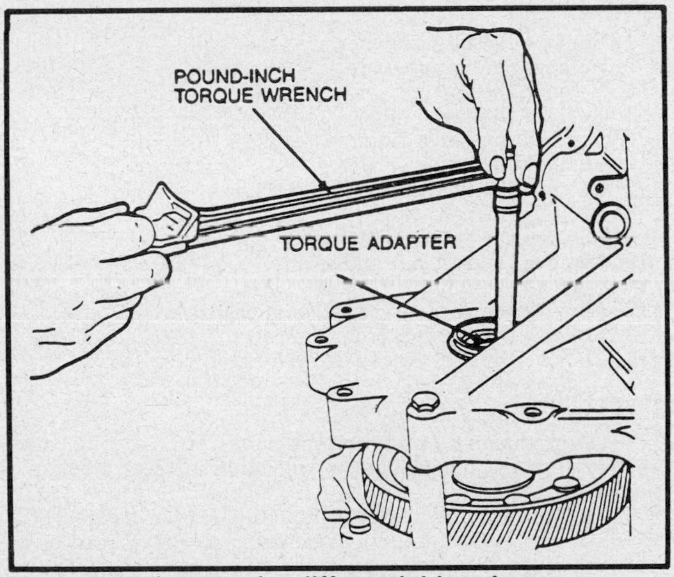
POUND-INCH TORQUE WRENCH
TORQUE ADAPTER

Measuring drag on the differential bearing

DIFFERENTIAL BEARING SHIM

Part Number	Shim Thickness
E92Z-4067-A	0.10mm (0.004 in.)
E92Z-4067-B	0.12mm (0.005 in.)
E92Z-4067-C	0.14mm (0.006 in.)
E92Z-4067-D	0.16mm (0.0063 in.)
E92Z-4067-E	0.18mm (0.007 in.)
E92Z-4067-F	0.20mm (0.008 in.)
E92Z-4067-G	0.25mm (0.010 in.)
E92Z-4067-H	0.30mm (0.012 in.)
E92Z-4067-J	0.35mm (0.014 in.)
E92Z-4067-K	0.40mm (0.016 in.)
E92Z-4067-L	0.45mm (0.018 in.)
E92Z-4067-N	0.50mm (0.020 in.)
E92Z-4067-P	0.55mm (0.022 in.)
E92Z-4067-Q	0.60mm (0.024 in.)
E92Z-4067-R	0.65mm (0.026 in.)
E92Z-4067-S	0.70mm (0.028 in.)
E92Z-4067-T	0.75mm (0.030 in.)
E92Z-4067-U	0.80mm (0.032 in.)
E92Z-4067-V	0.85mm (0.034 in.)
E92Z-4067-W	0.90mm (0.036 in.)
E92Z-4067-X	0.95mm (0.038 in.)
E92Z-4067-Y	1.00mm (0.040 in.)
E92Z-4067-Z	1.05mm (0.042 in.)
E92Z-4067-AA	1.10mm (0.044 in.)
E92Z-4067-AB	1.15mm (0.046 in.)
E92Z-4067-AC	1.20mm (0.048 in.)

13. Remove the screws, washers, transaxle case, gauge and bearing cup.
14. Install the selected shim(s) and bearing cup into the transaxle case using drive handle T80T–4000–W and differential bearing cup replacer T88C–77000–FH or equivalent.
15. Install the transaxle case. Tighten the retaining bolts to 27–38 ft. lbs.
16. Measure the bearing preload. The preload should be 26–35 inch lbs. Repeat the gauging process if the preload measurement is not within specification.
17. When the proper preload specification has been obtained, remove the transaxle case.

OUTPUT GEAR BEARING PRELOAD SHIM SELECTION

1. Align the bearing stator support using guide pins T80L–77100–A or equivalent then press the support into the converter housing using step plate D80L–630–6 or equivalent.
2. Remove the bearing cup and adjustment shim(s) from the bearing housing.
3. Place the output gear into the converter housing.
4. Place the bearing cup over the output gear bearing.
5. Place 4 collars (part of T87C–77000–J) or equivalent on the converter housing.
6. Place shim selection gauge T88C–77000–CH1 or equivalent on the output gear. Turn the 2 halves of the gauge to eliminate any gap between them.

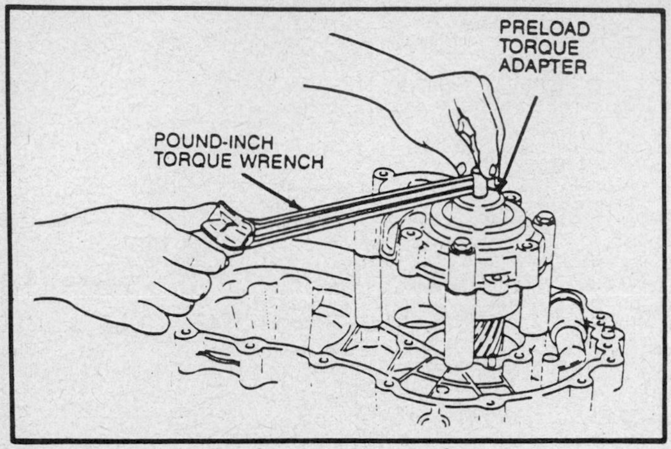
PRELOAD
TORQUE
ADAPTER
POUND-INCH
TORQUE WRENCH

Measuring drag on the output gear bearing

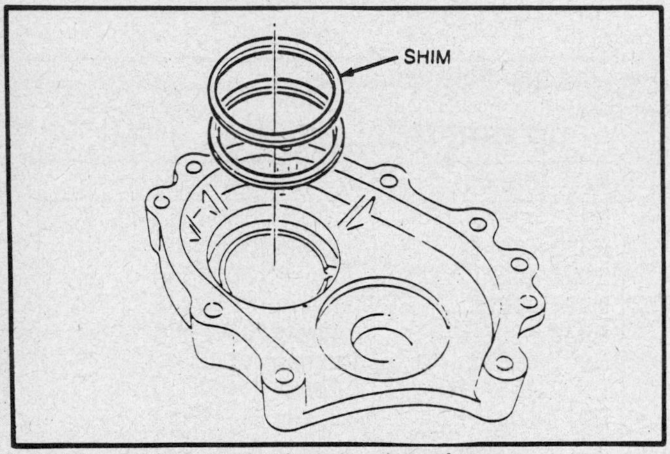
SHIM

Installing shims in the bearing housing

7. Place the bearing housing on the collars, install 4 Screws T88C–77000–CH4 with washers or equivalent. Tighten to 14–19 ft. lbs.
8. Place preload torque adapter T88C–77000–DH or equivalent on the output gear.
9. Using the pins provided in T87C–77000–J or equivalent, loosen the gauge halves until all of the freeplay is removed and the bearing cup is seated. Then thread the gauge halves back together.
10. Attach a inch pound torque wrench to the torque adapter. Measure the drag on the output gear bearing.

NOTE: Read the preload when the output gear starts to turn.

11. Turn the shim selection gauge using the pins (part of T87C–77000–J) or equivalent until a reading of 4.3–7.8 inch lbs. is obtained on the torque wrench.
12. Use a feeler gauge to measure the gap between the 2 halves of the gauge. Measure the gap at 4 spots, at 90 degree intervals. Use the largest measurement.
13. Using the output gear bearing preload shim selection chart, select the shim(s) that is closest (or slightly larger) to the measured value of the gauge gap. Use no more than 7 shims.
14. Remove the screws, washers, bearing housing, gauge and bearing cup.
15. Press the selected shim(s) and bearing cup into the bearing housing using bearing installer T60K–4616–A or equivalent.
16. Install the bearing housing. Tighten the retaining bolts to 14–19 ft. lbs.

OUTPUT GEAR BEARING SHIM

Part Number	Shim Thickness
E92Z-7F405-B	0.10mm (0.004 inch)
E92Z-7F405-C	0.12mm (0.005 inch)
E92Z-7F405-D	0.14mm (0.006 inch)
E92Z-7F405-E	0.16mm (0.0063 inch)
E92Z-7F405-F	0.18mm (0.007 inch)
E92Z-7F405-G	0.20mm (0.008 inch)
E92Z-7F405-A	0.50mm (0.020 inch)

17. Measure the bearing preload. The preload should be 0.26–7.81 inch lbs. Repeat the gauging process if the preload measurement is not within specification.
18. When the proper preload specification has been obtained, remove the bearing housing.

Unit Disassembly and Assembly

OIL PUMP

Disassembly

1. Remove the oil pump cover.
2. Remove the flange.
3. Remove the spring.
4. Remove the pivot roller.
5. Remove the guide ring.
6. Remove the guide spring.
7. Remove the vanes from the rotor.
8. Remove the cam ring.
9. Remove the rotor.
10. Remove the seal pin and spring.
11. Remove the plug, spring and valve.
12. Remove the thrust washer from the cover.
13. Remove the O-rings from the cover.
14. Remove the seal rings from the cover.

Inspection

1. Check the oil pump for a broken or worn seal ring, weakened springs and damaged or worn sliding surfaces. Replace as necessary.
2. Measure the following clearances using bar gauge T80L–77003–A or equivalent and a feeler gauge. If the clearances are not within specification, replace the oil pump as required..
   a. Seal pin to oil pump cover—The standard clearance should be 0.0002–0.0008 in. and the maximum allowable clearance is 0.002 in.
   b. Rotor to oil pump cover—The standard clearance should be 0.0002–0.0008 in. and the maximum allowable clearance 0.002 in.
   c. Cam ring to oil pump cover—The standard clearance should be 0.0002–0.0008 in. and the maximum allowable clearance 0.002 in.
   d. Vane to oil pump cover—The standard clearance should be 0.0006–0.0020 in. and the maximum allowable clearance 0.003 in.
   e. Vane to rotor groove—The standard clearance should be 0.0004–0.0018 in. and the maximum allowable clearance 0.0026 in.
3. Check each of the following parts for wear using the appropriate tool. If the wear limit is exceeded, replace the oil pump assembly.
   a. Sleeve to oil pump body—The standard outer diameter is 1.102 in.
   b. Rotor bushing—The standard inner diameter 1.102 in. and the maximum allowable inner diameter 1.104 in.

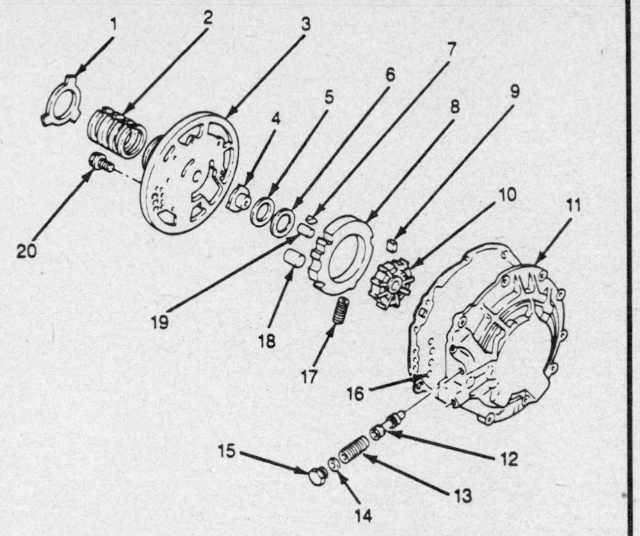
1
2
3
4
5
6
7
8
9
10
11
12
13
14
15
16
17
18
19
20

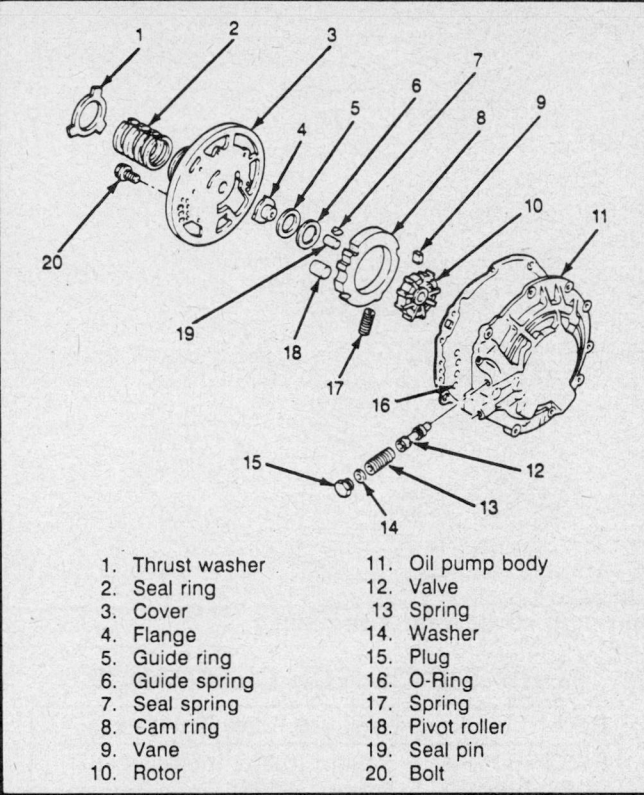

1. Thrust washer
2. Seal ring
3. Cover
4. Flange
5. Guide ring
6. Guide spring
7. Seal spring
8. Cam ring
9. Vane
10. Rotor
11. Oil pump body
12. Valve
13. Spring
14. Washer
15. Plug
16. O-Ring
17. Spring
18. Pivot roller
19. Seal pin
20. Bolt

Exploded view of oil pump

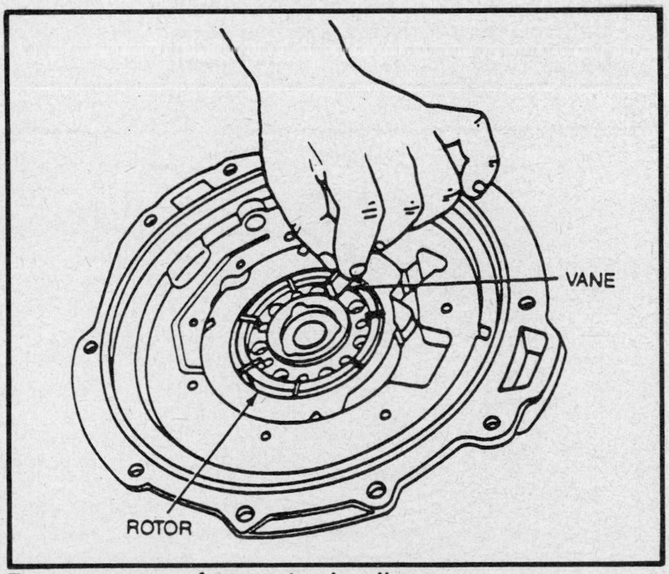
VANE
ROTOR

Remove vanes from rotor in oil pump

c. Guide ring—The standard outer diameter 0.278 in. and the minimum allowable outer diameter 0.272 in.

d. Valve—The standard outer diameter 0.472 in. and the minimum allowable outer diameter 0.467 in.

e. Seal pin—The standard outer diameter 0.236 in. and the minimum allowable outer diameter 0.232 in.

Assembly

1. Install the valve and spring into the oil pump body and check that the valve moves freely.

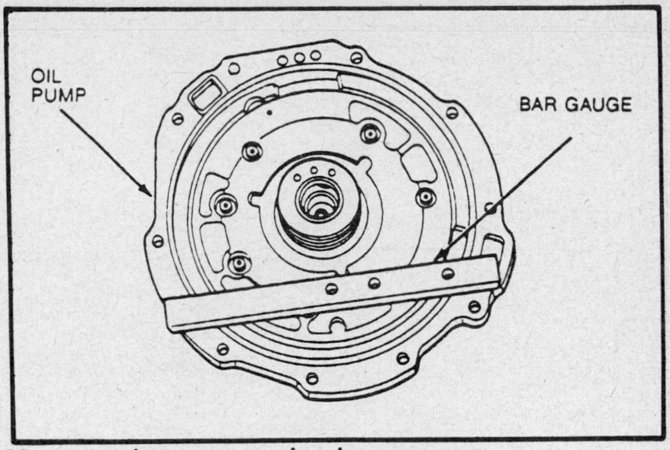
OIL PUMP
BAR GAUGE

Measure clearances using bar gauge

2. Install the plug and torque to 17–26 ft. lbs.
3. Install the cam ring and pivot roller.
4. Install the rotor.
5. Install the vanes into the rotor, with the flat edges and notches facing upwards.
6. Install the guide spring.
7. Install the guide ring.
8. Install the flange with the beveled edge down.
9. Install the spring.
10. Install new O-rings.
11. Install the seal pins and springs. Install the pins with the beveled edge down and the springs facing toward the cam ring.
12. Install the oil pump cover to the oil pump body.
13. Tighten the cover bolts in sequence. Torque to 71–97 inch lbs.
14. Install the oil pump shaft and check for smooth operation.
15. Install new seal rings.
16. Apply petroleum jelly to the thrust washer and install it on the oil pump cover. The outer diameter of the thrust washer should be 3.46 in.

FORWARD CLUTCH ASSEMBLY

Disassembly

1. Remove the needle bearing.
2. Remove the snapring.
3. Remove the pressure plate.
4. Remove the forward clutch pack.
5. Remove the dished plate.

Inspection

1. Check the drive and driven plates for damage or wear. The minimum thickness should be 0.055 in.
2. Check the clutch piston, clutch drum and seal contact areas for damage or wear. Check for broken or weakened springs. The free length of each spring should be 1.173 in. Replace as necessary.

Assembly

1. Install the dished plate with the beveled side facing upward.
2. Install the forward clutch pack, pressure plate and snapring.
3. Using a feeler gauge check the forward clutch clearance. Measure between the snapring and the pressure plate.
4. If the clearance is not within 0.040–0.047 in., adjust it by selecting an appropriate pressure.
5. Set the forward and reverse drum onto the oil pump. Check each clutch operation by applying a short burst of com-

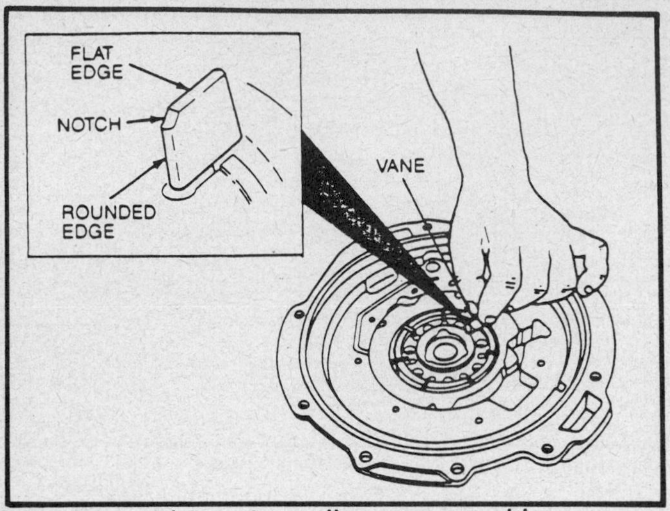
FLAT
EDGE
NOTCH
ROUNDED
EDGE
VANE
Install vanes into rotor—oil pump assembly

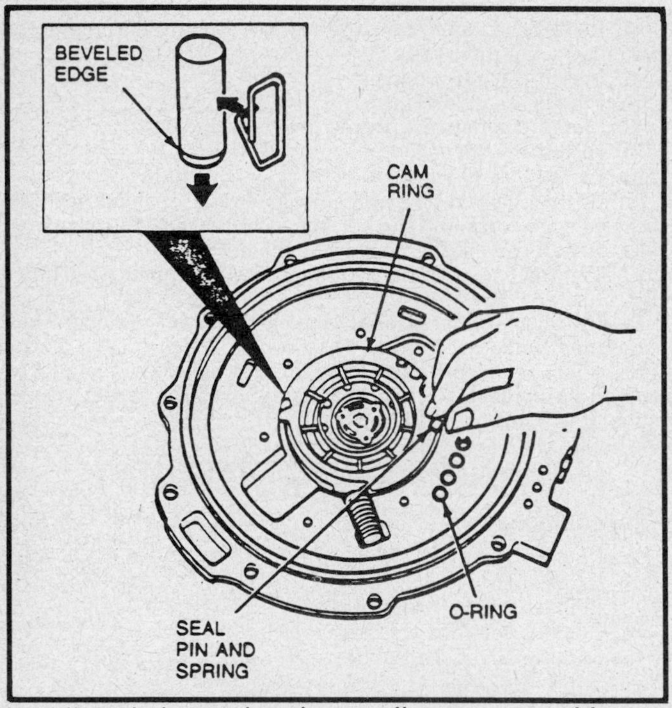
BEVELED
EDGE
CAM
RING
O-RING
SEAL
PIN AND
SPRING
Install seal pins and springs—oil pump assembly

pressed air through the fluid passages. As air pressure is applies, the clutch pack should compress. The pressure should not exceed 57 psi.

6. Pour the specified amount of transaxle fluid into a pan, until the reverse piston, coasting clutch drum and coasting piston are fully submerged in the fluid. Apply a short burst of compressed air through the fluid passages. Check that no bubbles come from between the piston and drum seal. The pressure should not exceed 57 psi.

7. Apply petroleum jelly to the needle bearings and install them on both sides of the clutch assembly. The outer diameter is 3.39 in. for the oil pump side and 2.21 in. for the one-way clutch side.

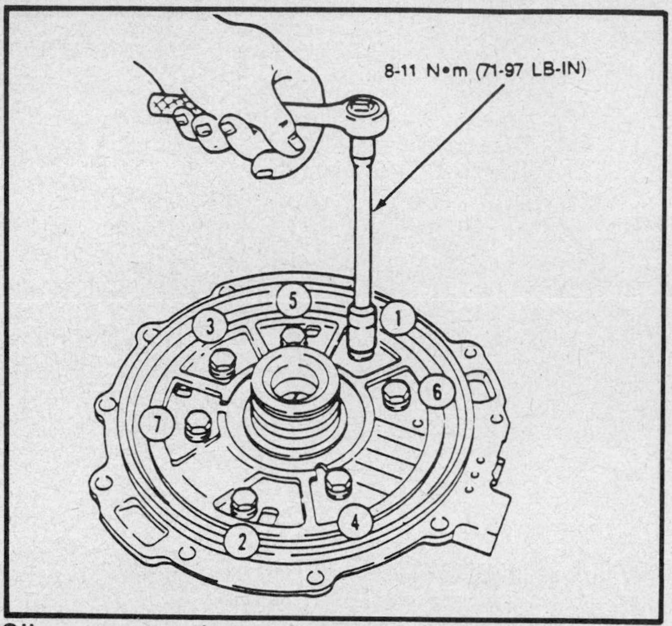
8-11 N•m (71-97 LB-IN)
1
2
3
4
5
6
7
Oil pump cover torque sequence

FORWARD CLUTCH CLEARANCE

Part Number	Pressure Plate Thickness
E92Z-7B066-A	5.9mm (0.232 inch)
E92Z-7B066-B	6.1mm (0.240 inch)
E92Z-7B066-C	6.3mm (0.248 inch)
E92Z-7B066-D	6.5mm (0.256 inch)
E92Z-7B066-E	6.7mm (0.264 inch)
E92Z-7B066-F	8.9mm (0.350 inch)

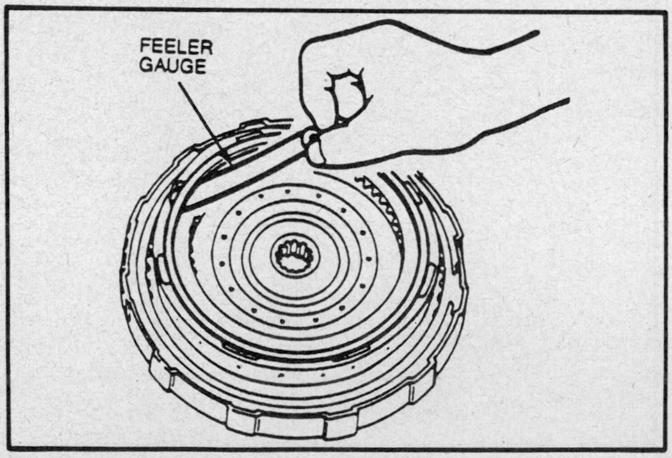
FEELER
GAUGE
Checking forward clutch clearance

COASTING CLUTCH ASSEMBLY

Disassembly

1. Remove the snapring and pressure plate.
2. Remove the coasting clutch pack.
3. Remove the dished plate.
4. Install spring compressor T65L–77505–A or equivalent and compress the return spring and retainer.
5. Remove the snapring.

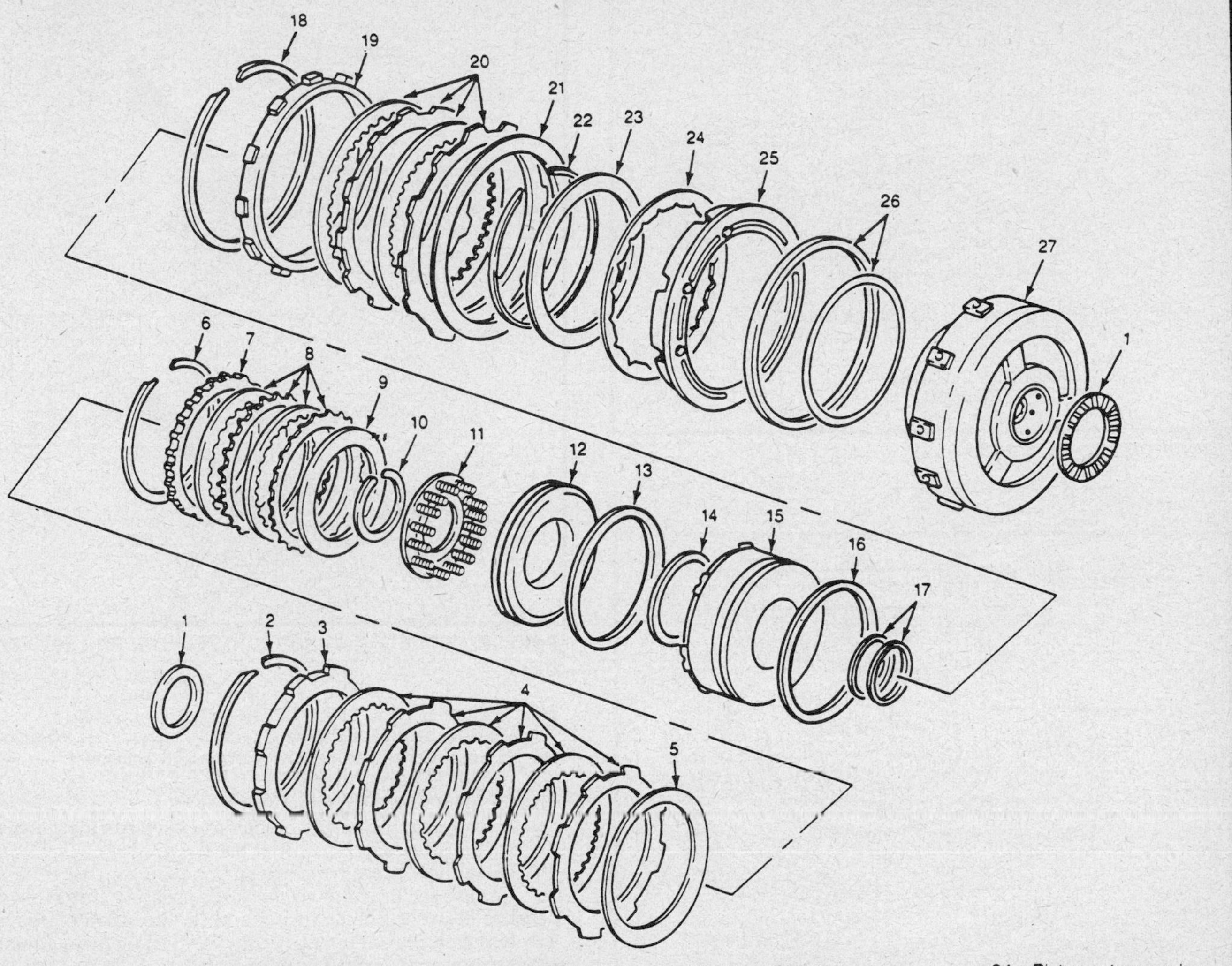

Forward Clutch
1. Needle bearing
2. Snapring
3. Pressure plate
4. Clutch pack
5. Dished plate

Coasting Clutch
6. Snapring
7. Pressure plate
8. Clutch pack
9. Dished plate
10. Snapring
11. Return spring and retainer
12. Coasting piston
13. Outer seal
14. Inner seal
15. Coasting clutch drum
16. Outer seal
17. Seal rings

Reverse Clutch
18. Snapring
19. Pressure plate
20. Clutch pack
21. Dished plate
22. Snapring
23. Return spring stopper
24. Piston return spring
25. Reverse piston
26. Seal rings
27. Reverse and forward drum

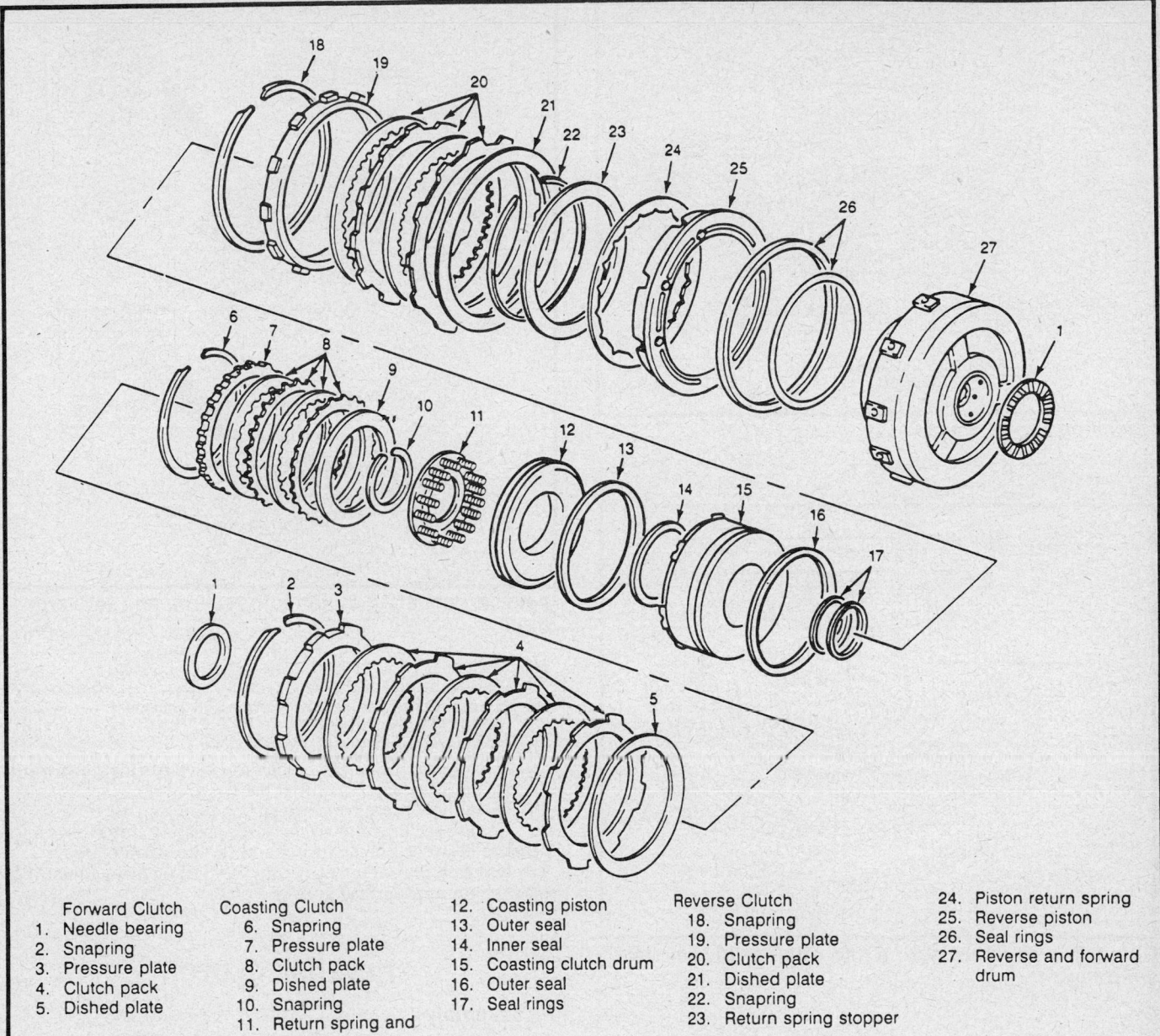

Exploded view of clutch assembly

6. Remove the spring compressor.
7. Remove the return spring and retainer.
8. Remove the coasting clutch drum from the clutch assembly by applying compressed air through the fluid passage.
9. Remove the coasting piston from the coasting clutch drum by applying low pressure compressed air through the fluid passage.

Inspection

1. Check the drive and driven plates for damage or wear. The minimum thickness should be 0.055 in.
2. Check the clutch piston, clutch drum and seal contact areas for damage or wear. Check for broken or weakened springs. The free length of each spring should be 1.173 in. Replace as necessary.

Assembly

1. Apply the specified transaxle fluid to the new seals and install them on the coasting piston.
2. Attach seal protector T88C–77000–HH or equivalent to the coasting piston and install the piston into the coasting clutch drum by pushing evenly around the circumference.
3. Apply the specified transaxle fluid to a new seal and install it on the coasting clutch drum.

NOTE: Roll the outer seal lip down to ease installation.

4. Install the coasting clutch drum into the reverse and forward drum.
5. Install the return spring and retainer.

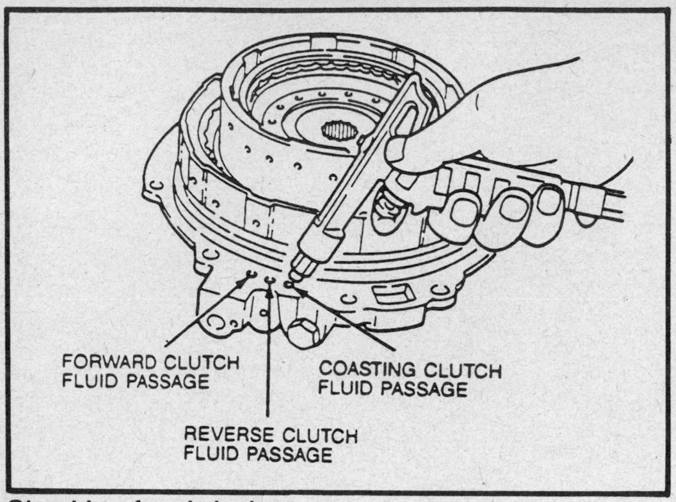
FORWARD CLUTCH
FLUID PASSAGE
COASTING CLUTCH
FLUID PASSAGE
REVERSE CLUTCH
FLUID PASSAGE

Checking for air leaks

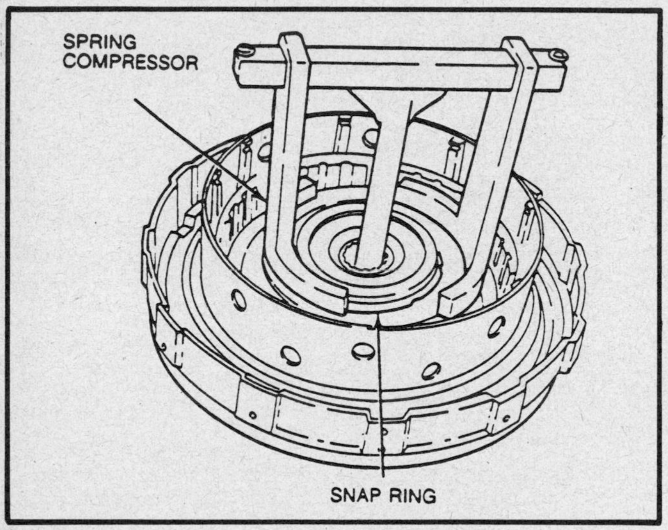
SPRING
COMPRESSOR
SNAP RING

Compressing return spring and retainer to remove snapring

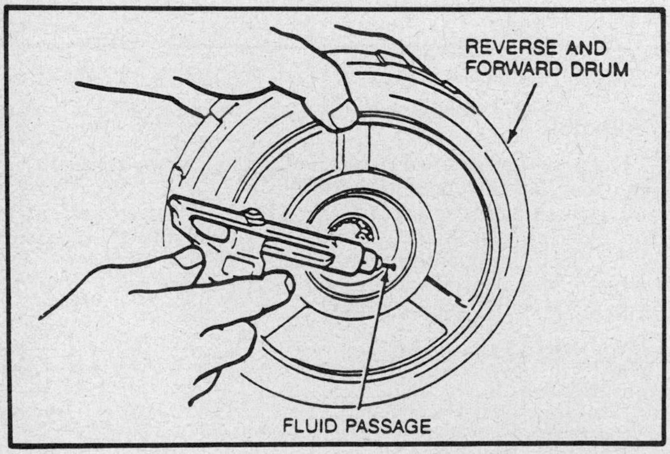
REVERSE AND
FORWARD DRUM
FLUID PASSAGE

Removing coasting clutch drum by air pressure

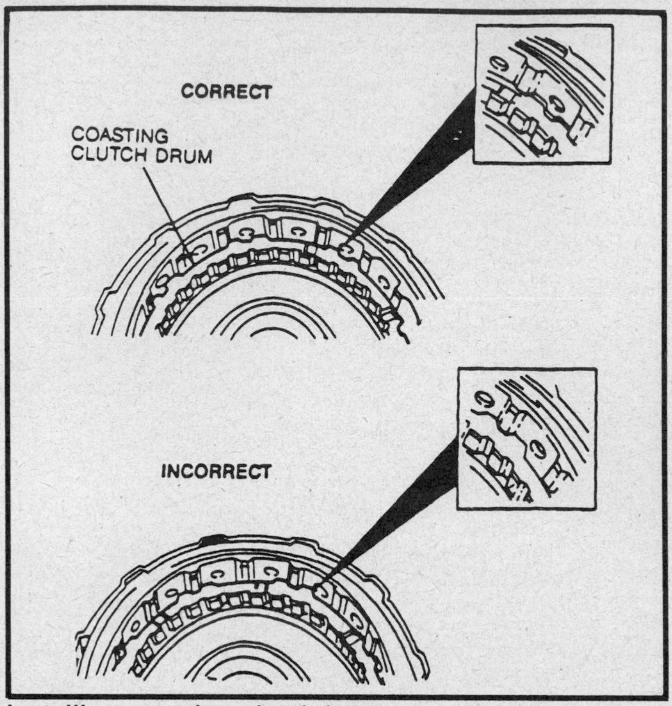
CORRECT
COASTING
CLUTCH DRUM
INCORRECT

Installing coasting clutch into reverse and forward drum

6. Install clutch spring compressor T65L–77515–A or equivalent and compress the return spring and retainer.
7. Install the snapring.
8. Remove the spring compressor.
9. Install the dished plate with the beveled side downward.
10. Install the coasting clutch pack.
11. Install the pressure plate and snapring.
12. Using a feeler gauge to check the coasting clutch clearance. Measure between the snapring and the pressure plate.
13. If the clearance is not within 0.040–0.047 in., adjust it by selecting an appropriate pressure plate.

REVERSE CLUTCH

Disassembly

1. Remove the snapring.
2. Remove the pressure plate.
3. Remove the reverse clutch pack.
4. Remove the dished plate.
5. Compress the piston return spring using return spring compressor T88C–77000–AH and the plate from T87C–77000–B or equivalent.
6. Remove one end of the snapring from the groove with snapring pliers. Once started, remove the snapring with a tool.
7. Remove the spring compressor.
8. Place the clutch assembly on the oil pump.
9. Apply low pressure compressed air through the fluid passage to remove the reverse piston.

Inspection

1. Check for damaged or worn drive and driven plates. The minimum allowable drive plate thickness is 0.055 in.
2. Check for a broken or worn piston or snapring.
3. Check for a broken or weakened spring. The free length of each spring should be 0.807 in. Replace as necessary.

COASTING CLUTCH CLEARANCE

Part Number	Pressure Plate Thickness
E92Z-7B066-M	4.6mm (0.181 inch)
E92Z-7B066-G	4.8mm (0.189 inch)
E92Z-7B066-H	5.0mm (0.197 inch)
E92Z-7B066-J	5.2mm (0.205 inch)
E92Z-7B066-K	5.4mm (0.213 inch)
E92Z-7B066-L	5.6mm (0.220 inch)

REVERSE CLUTCH CLEARANCE

Part Number	Pressure Plate Thickness
E92Z-7B066-N	6.6mm (0.260 inch)
E92Z-7B066-O	6.8mm (0.268 inch)
E92Z-7B066-P	7.0mm (0.276 inch)
E92Z-7B066-Q	7.2mm (0.283 inch)
E92Z-7B066-R	7.4mm (0.291 inch)
E92Z-7B066-S	7.6mm (0.299 inch)

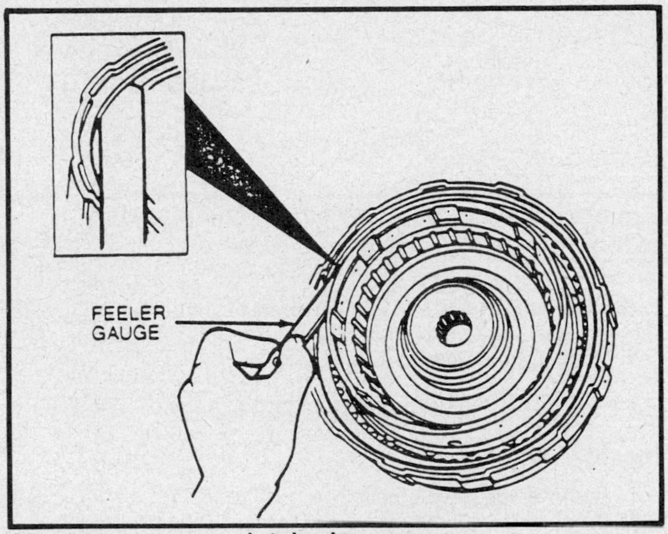
FEELER
GAUGE

Checking reverse clutch clearance

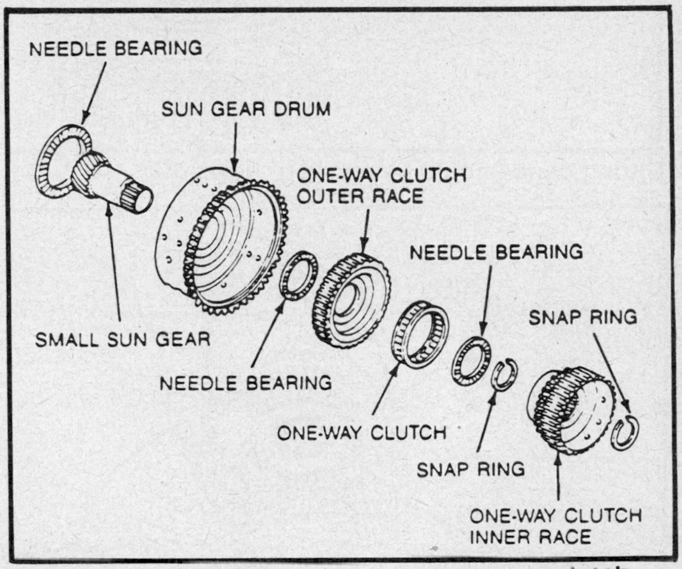
NEEDLE BEARING
SUN GEAR DRUM
ONE-WAY CLUTCH
OUTER RACE
NEEDLE BEARING
SNAP RING
SMALL SUN GEAR
NEEDLE BEARING
ONE-WAY CLUTCH
SNAP RING
ONE-WAY CLUTCH
INNER RACE

Exploded view small sun gear and one-way clutch

Assembly

1. Apply the specified transaxle fluid to the inner and outer faces of new seals and install them on the reverse piston.
2. Attach seal protector T88C–77000–GH or equivalent to the reverse piston. Install the reverse piston onto the reverse and forward drum by pushing evenly around the circumference. If necessary, use a tool to seat the piston. Remove the special tool seal protector.
3. Install the piston return spring with the tabs facing away from the reverse piston.
4. Install the return spring stopper with the step facing upwards.
5. Install the snapring half-way down the reverse and forward drum.
6. Compress the piston return spring using return spring compressor T88C–77000–AH and the plate from T87C–77000–B or equivalent.
7. Install the snapring with a tool.
8. Remove the spring compressor. Install the dished plate with the beveled side facing upward.
9. Install the reverse clutch pack.
10. Install the pressure plate with the step facing down.
11. Install the snapring.
12. Use a feeler gauge to check the reverse clutch clearance. Measure between the snapring and the pressure plate. if the clearance is not within 0.083–0.094 in., adjust it by selecting an appropriate pressure plate from the following chart.

SMALL SUN GEAR AND ONE-WAY CLUTCH

Disassembly

1. Remove the snapring.
2. Remove the one-way clutch inner and outer races.
3. Remove the snapring.
4. Remove the small sun gear from the sun gear drum.
5. Separate the one-way clutch inner race from the outer race.
6. Remove the one-way clutch.
7. Remove the needle bearing.

Inspection

1. Check the sun gear drum, small sun gear, bushing, clutch hub and inner and outer races for damage or wear.
2. Replace as necessary.

Assembly

1. Apply petroleum jelly to the needle bearing and install it to the one-way clutch inner race. The outer diameter is 2.44 in.
2. Install the one-way clutch into the one-way clutch outer race.

NOTE: Check that the spring cage faces toward the outer race.

3. Install the one-way clutch inner race into the one-way clutch outer race by turning the inner race counterclockwise. Make sure that the inner race turns only counterclockwise.
4. Install the small sun gear into the sun gear drum.
5. Install the snapring.
6. Install the one-way clutch inner and outer races to the sun gear drum.

NOTE: Align the splines of the one-way clutch inner race and small sun gear clutch hub.

7. Install the snapring.

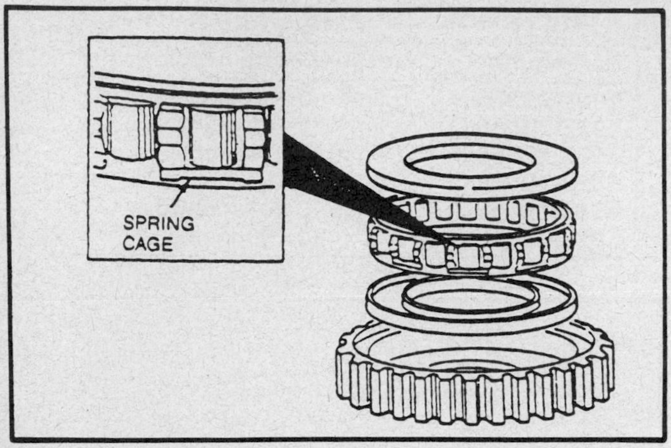
SPRING
CAGE

Spring cage should face toward outer race

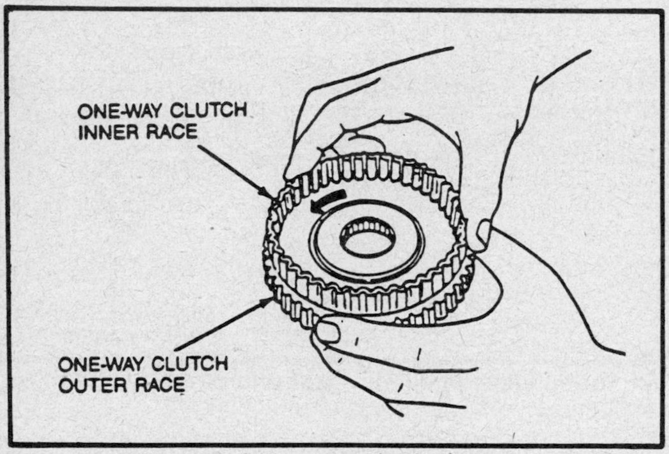
ONE-WAY CLUTCH
INNER RACE
ONE-WAY CLUTCH
OUTER RACE

Inner race turns only counterclockwise

8. Hold the small sun gear and make sure that the one-way clutch outer race turns smoothly and only clockwise.
9. Apply petroleum jelly to the needle bearing and install it to the sun gear drum. The outer diameter is 2.83 in.

ONE-WAY CLUTCH AND PLANETARY CARRIER ASSEMBLY

Disassembly

1. Remove the one-way clutch.
2. Remove the thrust washers.
3. Remove the snapring.
4. Remove the planetary carrier assembly from the inner race.
5. Place the one-way clutch on the inner race and make sure that the one-way clutch rotates smoothly and only clockwise.

Inspection

1. Check the inner race, thrust washers and gears for damage or wear.
2. Replace as necessary.

Assembly

1. Assemble the planetary carrier assembly to the inner race.
2. Install the snapring.
3. Apply petroleum jelly to the thrust washers and install them on the one-way clutch and planetary carrier assembly. The outer diameter of the sun gear drum side should be 2.83 in. for the sun gear drum side and 2.21 in. for the 3–4 clutch side.
4. Install the one-way clutch.

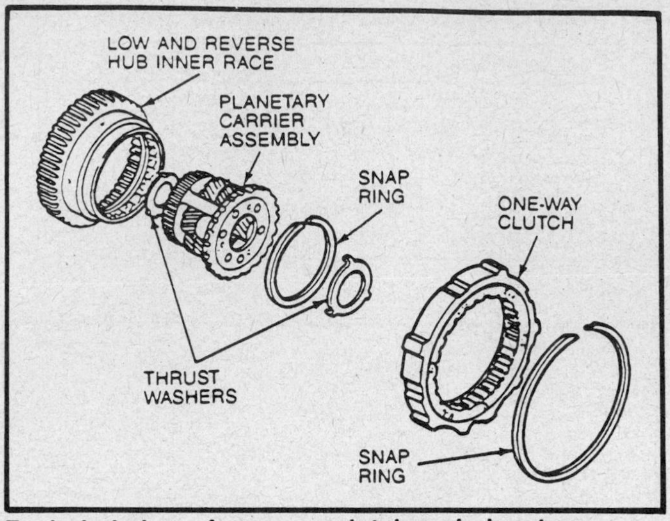
LOW AND REVERSE
HUB INNER RACE
PLANETARY
CARRIER
ASSEMBLY
SNAP
RING
ONE-WAY
CLUTCH
THRUST
WASHERS
SNAP
RING

Exploded view of one-way clutch and planetary carrier

3–4 CLUTCH

Disassembly

1. Remove the needle bearings.
2. Remove the snapring.
3. Remove the pressure plate.
4. Remove the 3–4 clutch pack.
5. Install clutch spring compressor T65L–77515–A or equivalent and compress the return spring and retainer assembly.
6. Remove the snapring.
7. Remove the spring compressor.
8. Remove the return spring and retainer assembly.
9. Remove the 3–4 clutch piston using compressed air applied through Leak check adapter T88C–77000–JH or equivalent.
10. Remove the inner and outer seals from the 3–4 clutch piston.

Inspection

1. Check the drive and driven plates for damage or wear. The minimum thickness should be 0.055 in.
2. Check the clutch piston and clutch drum and seal contact areas for damage.
3. Check for broken or worn springs. The free length of each spring should be 1.307 in. Replace as necessary.

Assembly

1. Apply the specified transaxle fluid to the inner and outer seals and install them onto the 3–4 clutch piston.
2. Install the 3–4 clutch piston by pushing evenly around the circumference.
3. Install the return spring and retainer assembly.
4. Install clutch spring compressor T65L–77515–A or equivalent and compress the return spring and retainer assembly.
5. Install the snapring.
6. Remove the clutch spring compressor.
7. Install the 3–4 clutch pack.
8. Install the pressure plate with the step facing upward.
9. Install the snapring.
10. Use a feeler gauge to check the 3–4 clutch clearance. Measure between the snapring and the pressure plate.

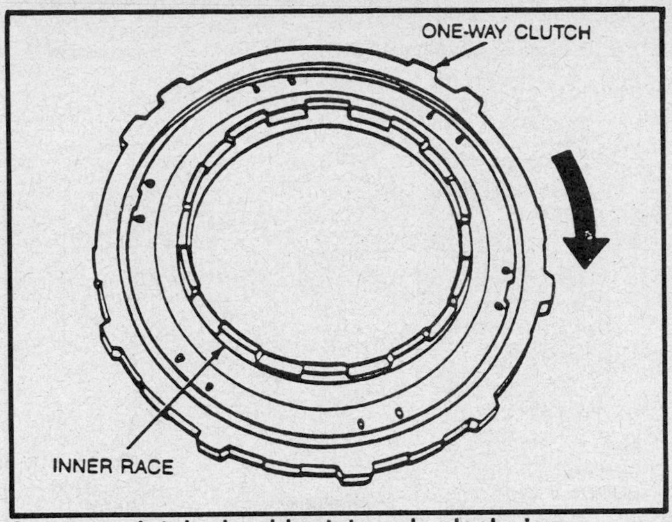
ONE-WAY CLUTCH
INNER RACE

One-way clutch should rotate only clockwise

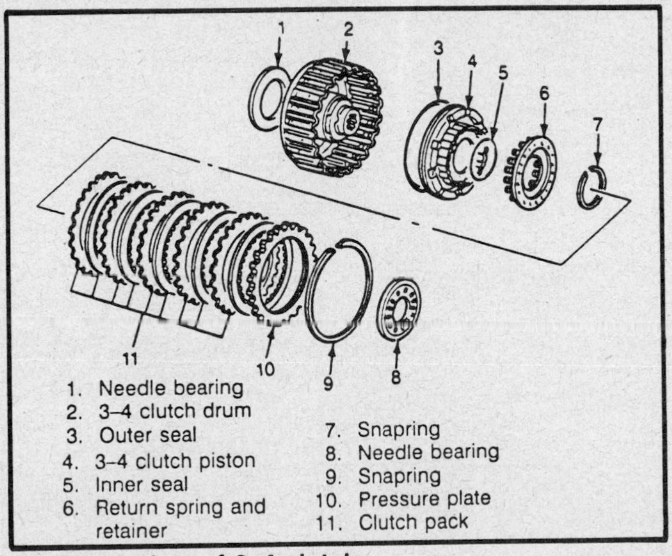
1
2
3
4
5
6
7
8
9
10
11
1. Needle bearing
2. 3–4 clutch drum
3. Outer seal
4. 3–4 clutch piston
5. Inner seal
6. Return spring and retainer
7. Snapring
8. Needle bearing
9. Snapring
10. Pressure plate
11. Clutch pack

Exploded view of 3–4 clutch

11. If the clearance is not within 0.051–0.059 in., adjust it by selecting a proper pressure plate.
12. Apply petroleum jelly to the needle bearings and install them on the 3–4 clutch. The outer diameter is 2.21 in. for the planetary carrier side and 2.84 in. for the output shell side.
13. Install Leak check adapter T88C–787000–JH or equivalent and apply compressed air to check operation.

NOTE: Do not apply over 57 psi of air pressure.

13. Pour the specified transaxle fluid into the clutch drum so the 3–4 clutch piston is fully submerged. Appy compressed air to check that no bubbles come from the clutch piston seal.

NOTE: Do no apply over 57 psi of air pressure. Do not apply the air pressure for more than few seconds.

2–3 ACCUMULATOR

Disassembly

1. Remove the snapring while holding in the stopper plug.
2. Remove the stopper plug.
3. Remove the spring.
4. Remove the piston.

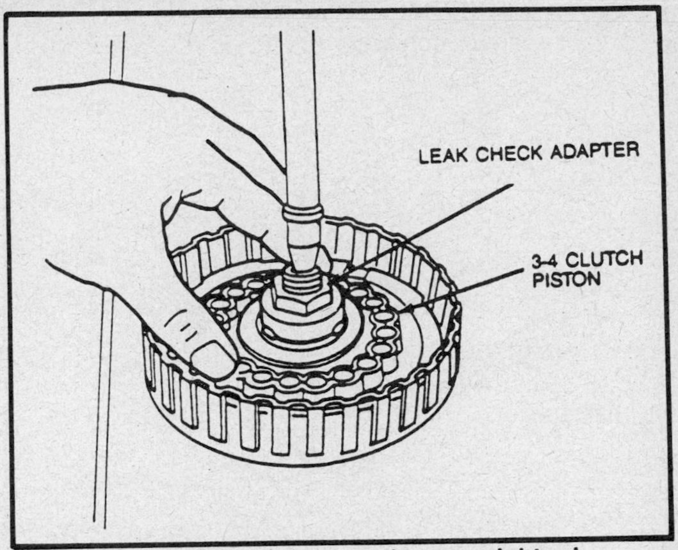
LEAK CHECK ADAPTER
3-4 CLUTCH PISTON

Removing 3–4 clutch piston using special tool

3–4 CLUTCH CLEARANCE

Part Number	Pressure Plate Thickness
E92Z-7B066-T	4.0mm (0.157 inch)
E92Z-7B066-U	4.2mm (0.165 inch)
E92Z-7B066-V	4.4mm (0.173 inch)
E92Z-7B066-W	4.6mm (0.181 inch)
E92Z-7B066-X	4.8mm (0.189 inch)

5. Remove the O-ring from the stopper plug.
6. Remove the seals from the piston.

Inspection

1. Check for a damaged or worn piston or stopper plug.
2. Check for a broken or worn spring. The spring free length for non-turbocharged vehicles should be 3.280 in. The spring free length for turbocharged vehicles should be 2.968 in.

Assembly

1. Apply the specified transaxle fluid to the seals and install them on the piston.
2. Apply the specified transaxle fluid to the O-ring and install it on the stopper plug.
3. Install the piston.
4. Install the spring.
5. Install the stopper plug.
6. Install the snapring while holding in the stopper plug.

VALVE BODY ASSEMBLY

Disassembly

NOTE: Each valve body bolt has a letter on the bolt head which matches the letter placed near the bolt hole. The bolts must be installed in the correct order.

1. Remove the 3–4 solenoid valve.
2. Remove the lockup solenoid valve.
3. Remove the 1–2 solenoid valve.
4. Remove the 2–3 solenoid valve.
5. Remove the brackets and wire harness.
6. Remove the fluid strainers.

NOTE: Do not turn the throttle valve adjusting screw in the main control body.

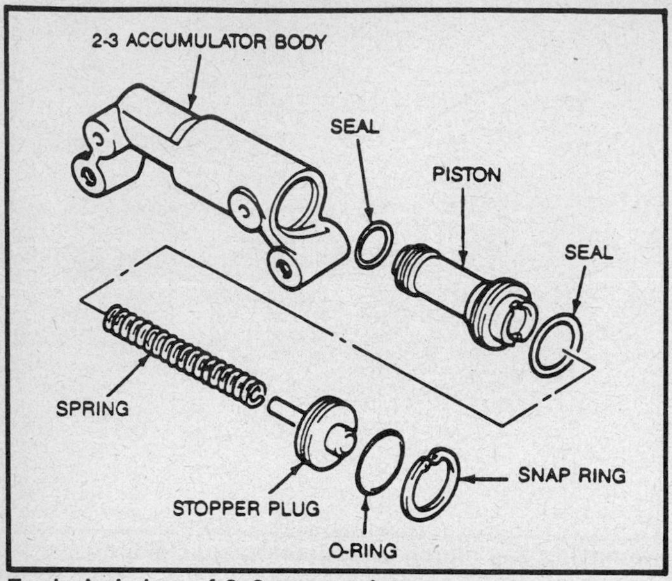
2-3 ACCUMULATOR BODY
SEAL
PISTON
SEAL
SPRING
SNAP RING
STOPPER PLUG
O-RING

Exploded view of 2–3 accumulator

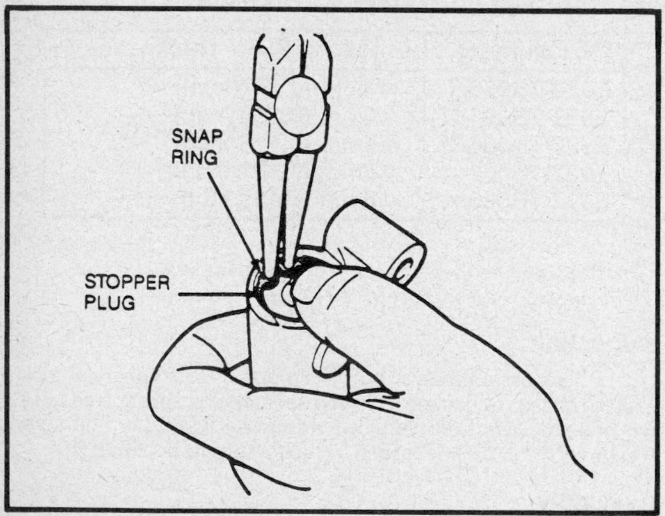
SNAP RING
STOPPER PLUG

Removing snapring from accumulator

7. Remove the O-rings.
8. Remove the front control body bolts.
9. Remove the front control body with the premain separator as a unit.
10. Remove the premain separator and front gasket from the front control body.
11. Remove the relief valve (0.031 in. orifice) and spring from the premain control body.
12. Remove the premain control body bolts, including the hexagonal head bolt.
13. Remove the premain control body and main separator as a unit.
14. Remove the premain rear gasket, main front gasket and main separator from the premain control body.
15. Remove the relief valves (0.079 in. orifice) and springs from the premain control body.
16. Remove the check ball and spring from the premain control body.
17. Remove the relief valve (0.031 in. orifice) and spring from the main control body.
18. Remove the check ball and spring from the main control body.

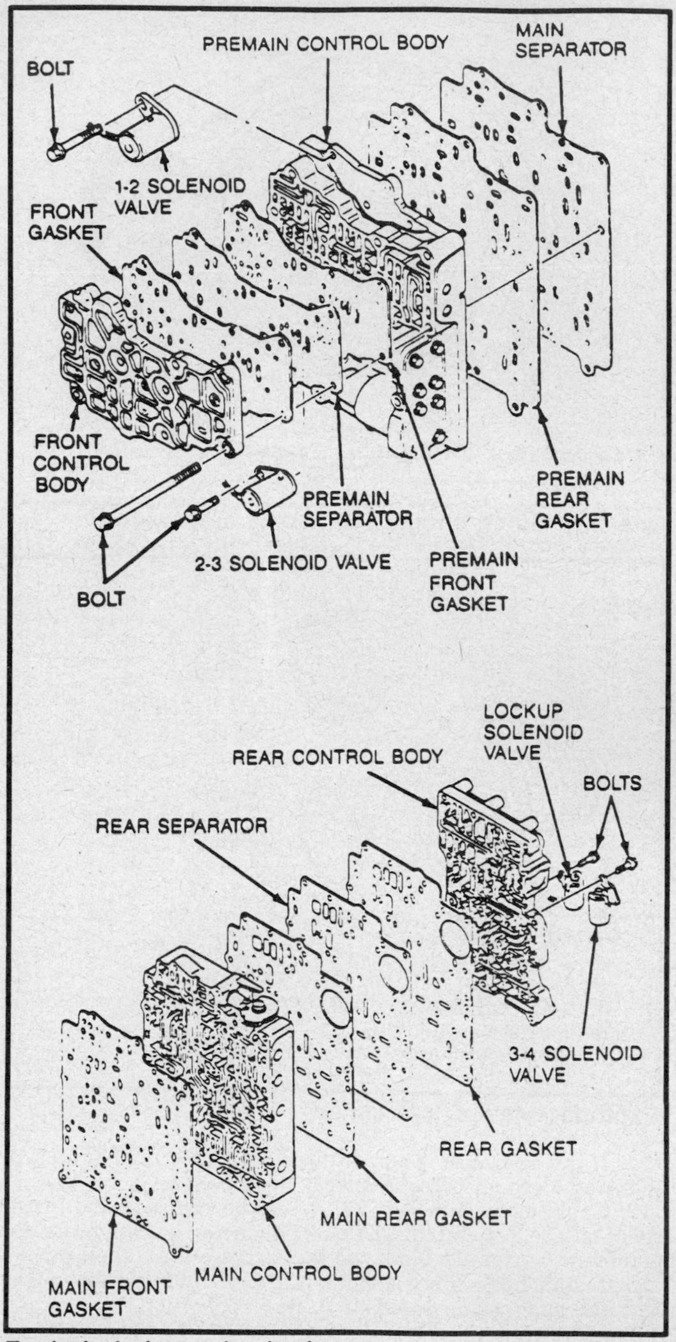
BOLT
PREMAIN CONTROL BODY
MAIN SEPARATOR
1-2 SOLENOID VALVE
FRONT GASKET
FRONT CONTROL BODY
PREMAIN SEPARATOR
PREMAIN REAR GASKET
2-3 SOLENOID VALVE
PREMAIN FRONT GASKET
BOLT
LOCKUP SOLENOID VALVE
REAR CONTROL BODY
BOLTS
REAR SEPARATOR
3-4 SOLENOID VALVE
REAR GASKET
MAIN REAR GASKET
MAIN CONTROL BODY
MAIN FRONT GASKET

Exploded view valve body

19. Turn the assembly over and remove rear control body bolts.
20. Remove the rear control body and rear separator as a unit.
21. Remove the main rear gasket, rear gasket and rear separator from the rear control body.
22. Remove the relief valves (0.039 in.; 0.059 in; 0.078 in. orifice) and springs from the rear control body.
23. Remove the relief valve (0.098 in. orifice) and spring from the main control body.
24. Remove the ruber ball from the main control body.

NOTE: The premain, main, rear control body individual valves and springs are removed by removing the retaining clips and bore plugs. Some valves are aluminum

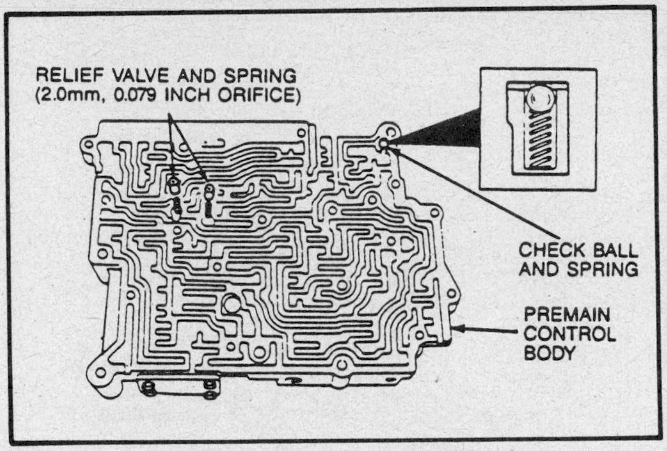
RELIEF VALVE AND SPRING
(2.0mm, 0.079 INCH ORIFICE)
CHECK BALL
AND SPRING
PREMAIN
CONTROL
BODY

Check ball location—premain control body

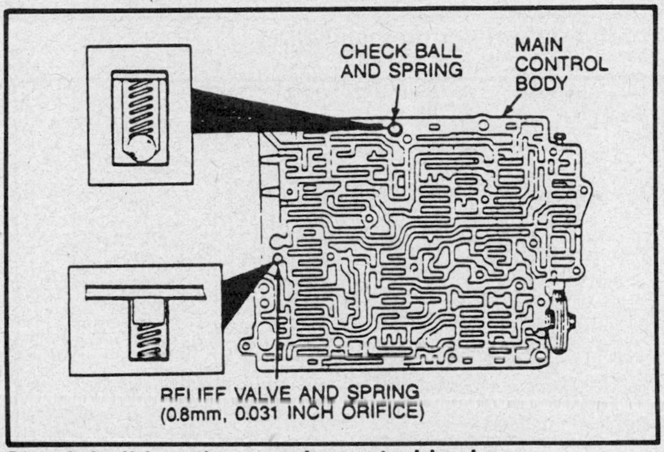
CHECK BALL
AND SPRING
MAIN
CONTROL
BODY
RELIEF VALVE AND SPRING
(0.8mm, 0.031 INCH ORIFICE)

Check ball location—main control body

and cannot be removed using a magnet. Remove these valves by tapping the valve body of the palm of the hand to slide the valve out of the bore. It may be necessary to remove the valves and springs using a pick. If so, use extreme care to prevent damaging valves or valve bores.

Inspection

1. Clean all parts thoroughly in clean solvent and blow dry with compressed air.
2. Inspect all valve and plug bores for scores. Check all fluid passages for obstructions. Inspect all mating surfaces for burrs and scores. If needed, use crocus cloth to polish valve and plugs. Avoid rounding the sharp edges of the valves and plugs with the crocus cloth.
3. Inspect all springs for distortion. Check all valves and plugs for free movement in their respective bores. Valve and plugs, when dry, must fall from their own weight into their respective bores.
4. Roll the manual valve on a flat surface to check for a bent condition. Replace any parts as necessary.

Assembly

MAIN CONTROL BODY

NOTE: When installing the throttle valve assembly, make sure that the groove is aligned with the bolt hole.

1. Install the throttle return spring on the throttle cam.
2. Tighten the throttle cam bolt to 69–95 inch lbs..

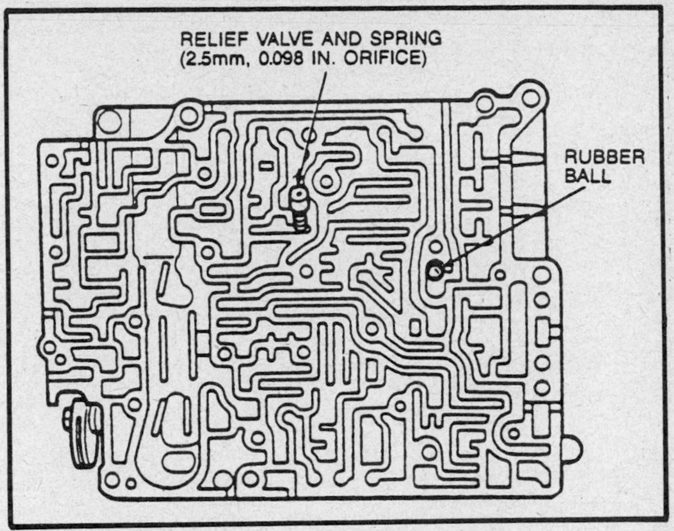
RELIEF VALVE AND SPRING
(2.5mm, 0.098 IN. ORIFICE)
RUBBER
BALL

Rubber ball location—main control body

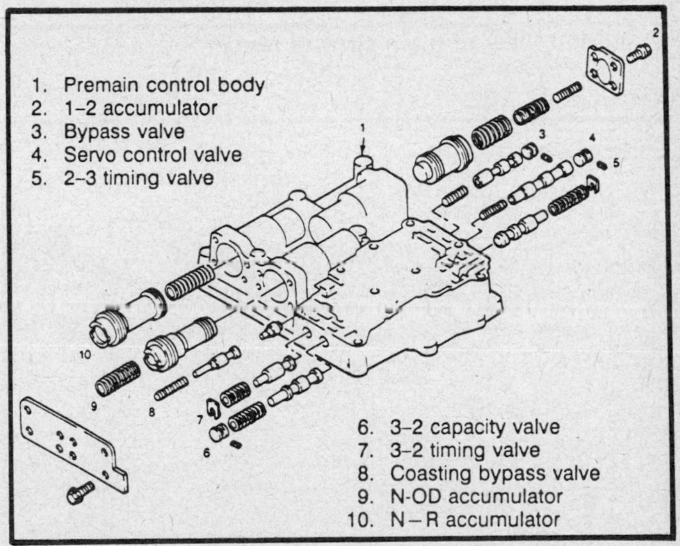
1. Premain control body
2. 1–2 accumulator
3. Bypass valve
4. Servo control valve
5. 2–3 timing valve
6. 3–2 capacity valve
7. 3–2 timing valve
8. Coasting bypass valve
9. N-OD accumulator
10. N–R accumulator

Exploded view of premain control body

PREMAIN CONTROL BODY

1. Tighten the N–R/N–OD accumulator plate to 57–69 inch lbs.
2. Tighten the 1–2 accumulator plate to 57–69 inch lbs.
3. Do not install the bolt which holds the harness bracket.

VALVE BODY

NOTE: Do not mix up the gaskets during assembly. Match the bolt head letter with the corresponding letter on the valve body.

1. Install the relief valves (0.039 in.; 0.059 in.; 0.078 in. orifice) and springs in the rear control body.
2. Install the gaskets on both sides of the rear separator, then install it onto the rear control body.

NOTE: The rear gasket and main rear gasket are not interchangeable.

3. Install the relief valve (0.098 in. orifice) and spring in the main control body.
4. Install the rubber ball in the main control body.

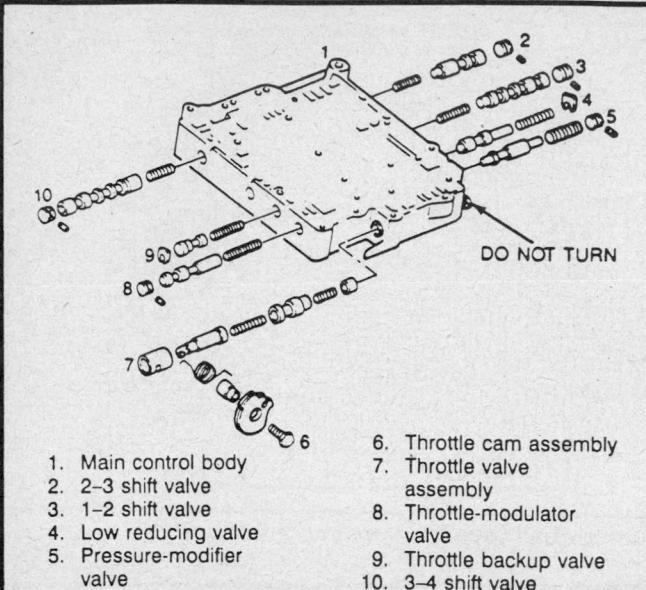

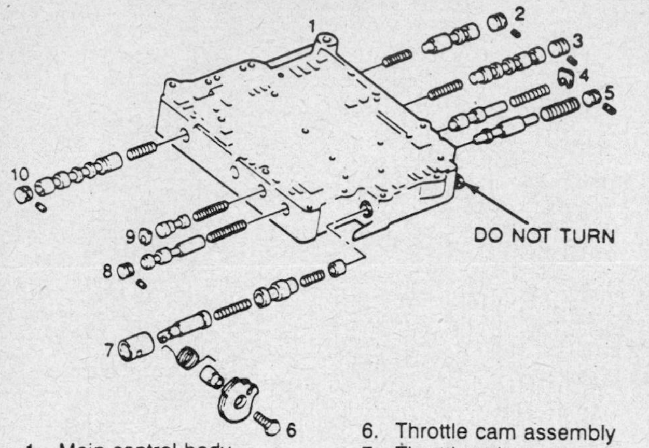
DO NOT TURN
1. Main control body
2. 2–3 shift valve
3. 1–2 shift valve
4. Low reducing valve
5. Pressure-modifier valve
6. Throttle cam assembly
7. Throttle valve assembly
8. Throttle-modulator valve
9. Throttle backup valve
10. 3–4 shift valve

Exploded view of main control body

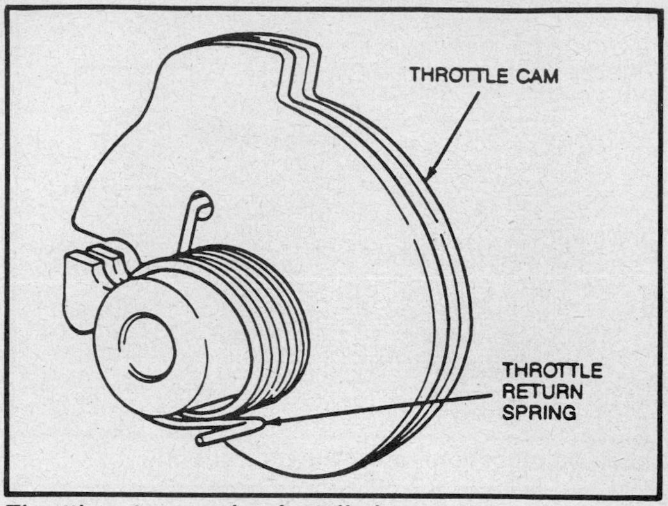
THROTTLE CAM
THROTTLE RETURN SPRING

Throttle return spring installation

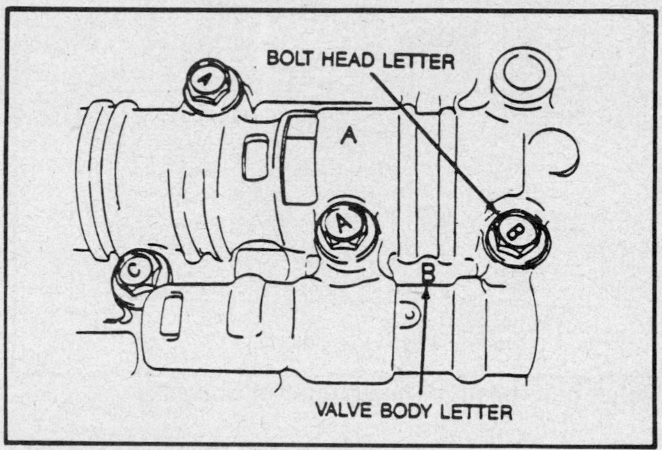
BOLT HEAD LETTER
VALVE BODY LETTER

Installing bolts in valve body

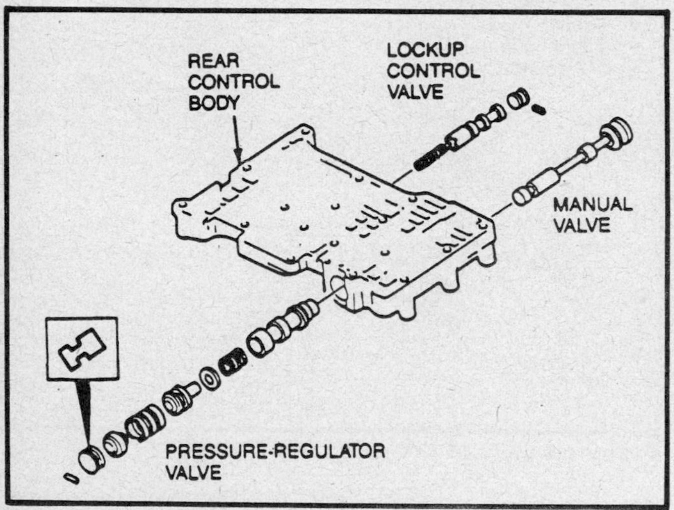
REAR CONTROL BODY
LOCKUP CONTROL VALVE
MANUAL VALVE
PRESSURE-REGULATOR VALVE

Exploded view of rear control body

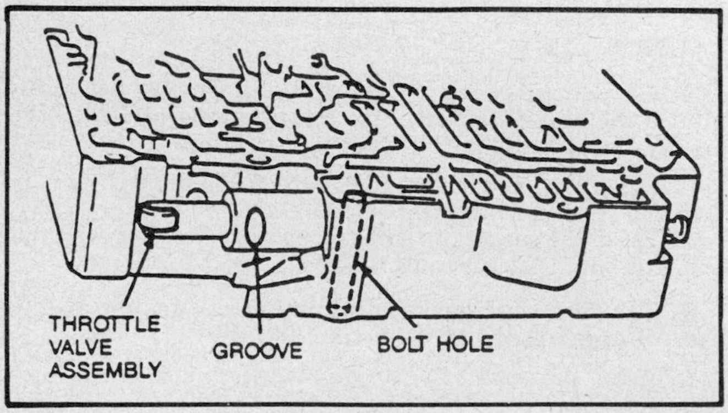
THROTTLE VALVE ASSEMBLY
GROOVE
BOLT HOLE

Installing throttle valve assembly

5. Install the rear control body to the main control body.
6. Loosely tighten the rear control body bolts.

NOTE: Match the bolt head letter with the letter on the valve body.

7. Turn the assembly over and install the relief valve (0.031 in. orifice) and spring in the main control body.
8. Install the check ball and spring in the main control body.
9. Install the relief valves (0.079 in. orifice) and springs into the premain control body.
10. Install the check ball and spring in the premain control body.
11. Install the gaskets on both sides of the main separator, then install it onto the premain control body.

NOTE: The premain rear gasket and main front gasket are not interchangeable.

12. Set the premain control body onto the main control body.
13. Loosely tighten the premain control body bolts, including the hexagonal head bolt.
14. Install the relief valve (0.031 in. orifice and spring into the premain control body.
15. Install the gaskets on both sides of the premain separator, then install it onto the front control body.

NOTE: The front gasket and premain front gasket are not interchangeable.

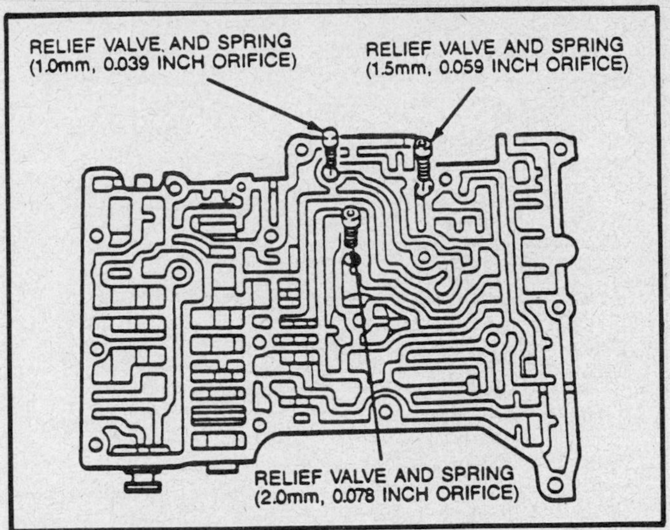
RELIEF VALVE AND SPRING
(1.0mm, 0.039 INCH ORIFICE)
RELIEF VALVE AND SPRING
(1.5mm, 0.059 INCH ORIFICE)
RELIEF VALVE AND SPRING
(2.0mm, 0.078 INCH ORIFICE)

Install relief valves in rear control body

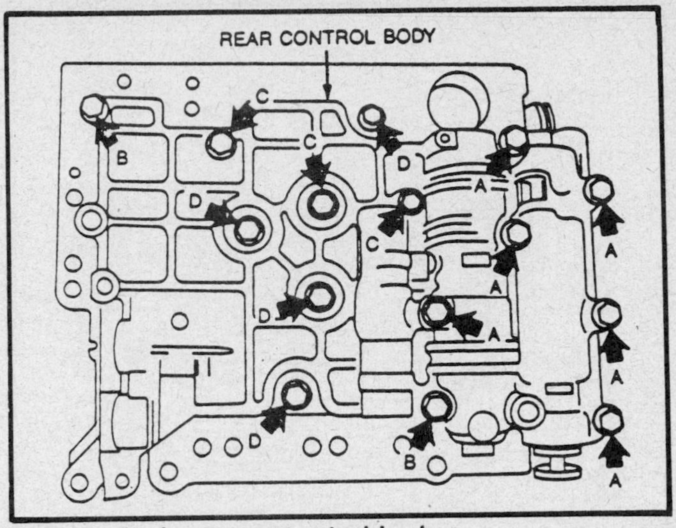
REAR CONTROL BODY
A
B
C
D

Bolt installation—rear control body

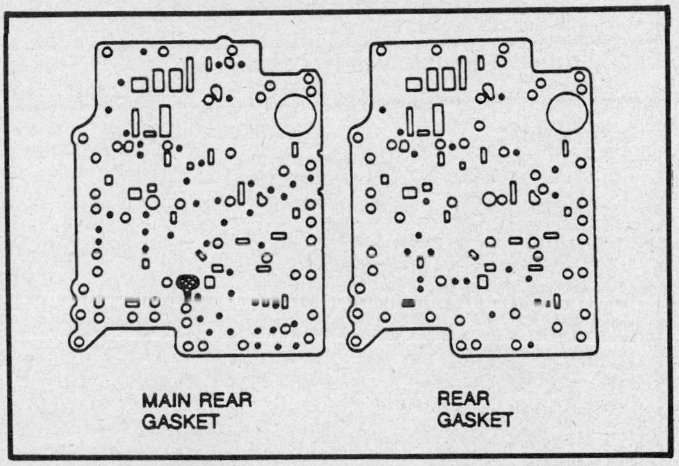
MAIN REAR
GASKET
REAR
GASKET

Valve body gasket

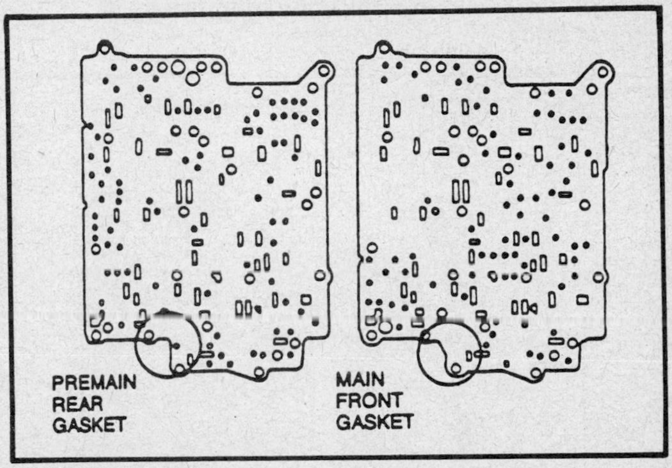
PREMAIN
REAR
GASKET
MAIN
FRONT
GASKET

Valve body gasket

16. Install the front control body on the premain control body.
17. Loosely tighten the front control body bolts.

NOTE: Match the bolt head letter with the letter on the valve body.

18. Install 2 valve body mounting bolts for alignment.
19. Tighten the bolts on the front face of the valve body to 57–69 inch lbs.
20. Tighten the bolts on the rear face of the valve body to 57–69 inch lbs.
21. Install new fluid strainers.
22. Install new O-rings on the solenoid valves.
23. Install the 3–4 solenoid valve.
24. Install the lockup solenoid valve.
25. Tighten the solenoid valve bolts to 57–69 inch lbs.
26. Install the 1–2 solenoid valve.
27. Install the 2–3 solenoid valve.
28. Tighten the solenoid valve bolts to 57–69 inch lbs.
29. Install the brackets and wire harness.

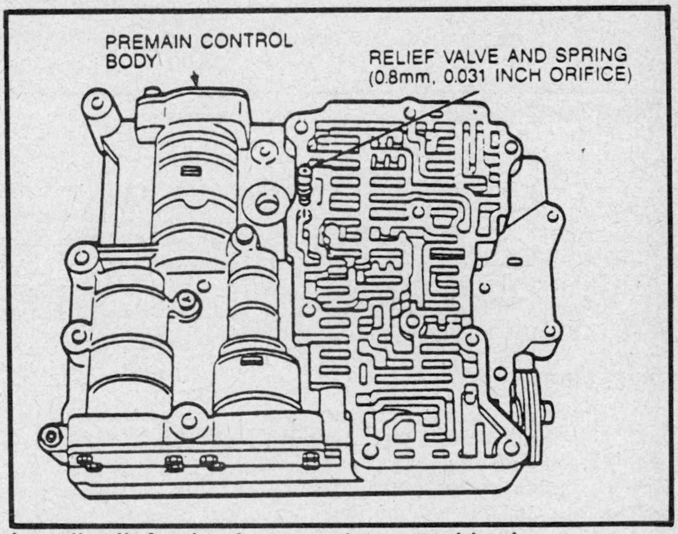
PREMAIN CONTROL
BODY
RELIEF VALVE AND SPRING
(0.8mm, 0.031 INCH ORIFICE)

Install relief valve in premain control body

DIFFERENTIAL

Disassembly

1. Remove the roll pin.
2. Remove the pinion shaft.
3. Remove the pinions and thrust washers by rotating them out of the gear case.
4. Remove the side gears and thrust washers.

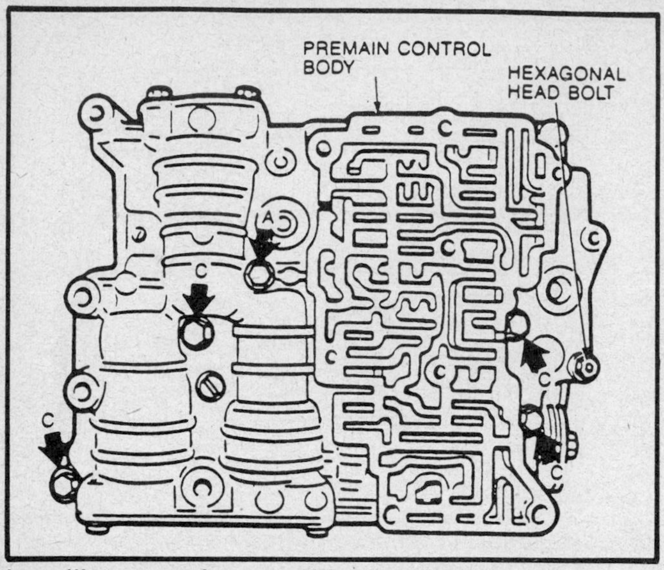
PREMAIN CONTROL BODY
HEXAGONAL HEAD BOLT
A
C
C
C
C
C
C

Installing premain control body onto the main control body

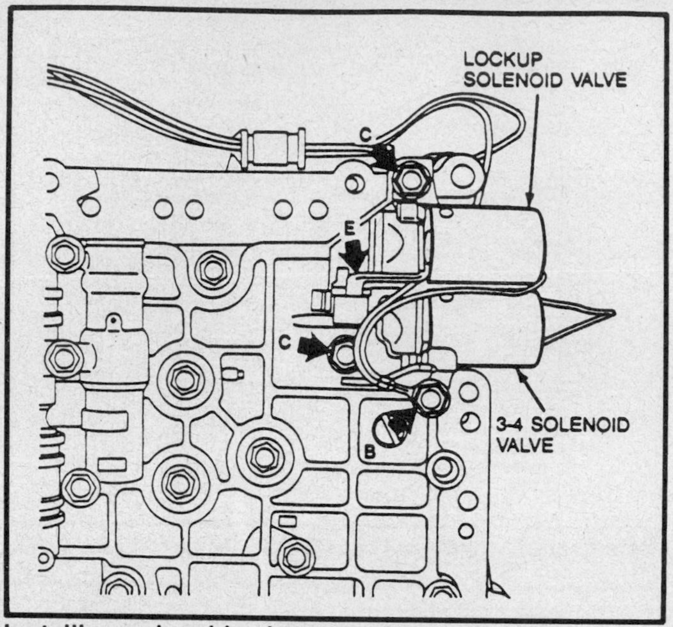
LOCKUP SOLENOID VALVE
C
E
C
B
3-4 SOLENOID VALVE

Installing solenoid valves to valve body

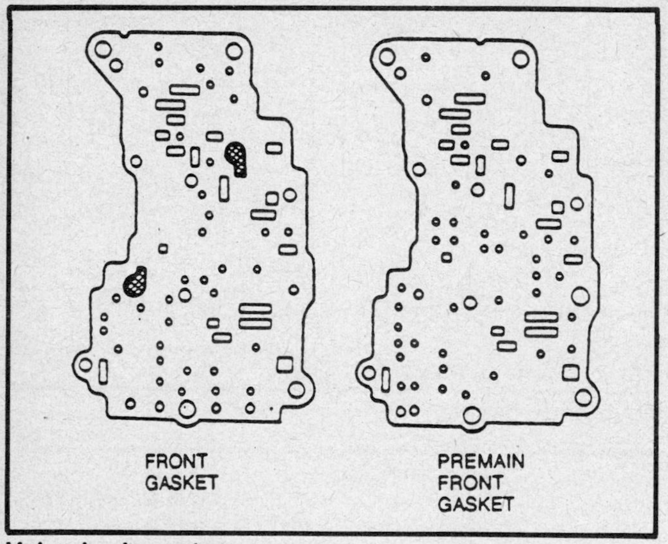
FRONT GASKET
PREMAIN FRONT GASKET

Valve body gasket

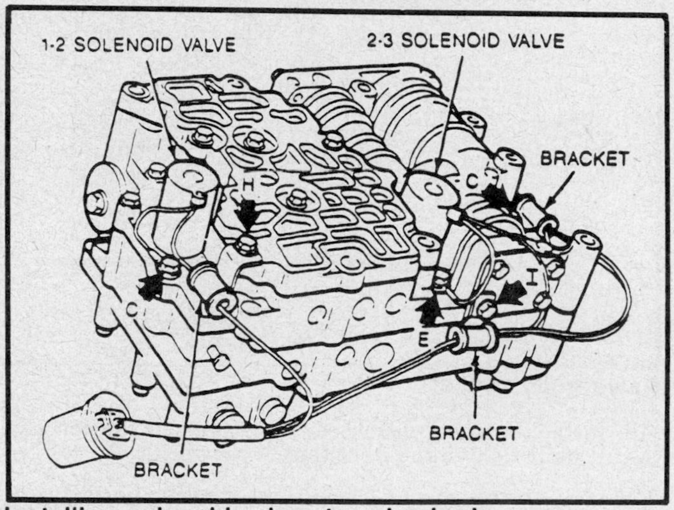
1-2 SOLENOID VALVE
2-3 SOLENOID VALVE
BRACKET
H
C
I
C
E
BRACKET
BRACKET

Installing solenoid valves to valve body

5. Remove the bearing core (speedometer drive gear end) using gear and pulley puller D80L–522–A, step plate D80L–630–3, bearing puller attachment D84–1123–A and the legs from push puller set D80L–927–A or equivalent.
6. Remove the speedometer drive gear.
7. Remove the bearing cone (ring gear end) using puller T77F–4220–B1 and step plate D80L–630–3 or equivalent.

Inspection

1. Check for damaged or worn gears.
2. Check for a cracked or damaged gear case.
3. Check side gear and pinion backlash.
   a. Install the left and right halfshafts into the differential.
   b. Support the halfshafts on V-blocks.
   c. Use dial indicator tool 4201–C with magnetic base/flex arm D78P–4201–C or equivalent to measure the backlash of both pinion gears. If the backlash is more then allowable, select a thrust washer with a different thickness. The backlash should be 0–0.004 in.

Assembly

NOTE: Whenever a bearing cone is removed, it must be replaced.

1. Install the speedomter drive gear and bearing cone using either driver handle T80T–4000–W or a press and differential bearing cone replacer T88C–77000–EH or equivalent.
2. Install the bearing cone (ring gear end) using either driver handle T80T–4000–W or a press and bearing cone replacer T88C–77000–EH or equivalent.
3. Install the thrust washers and pinions.
4. Install the pinion shaft.
5. Install the knock pin, then crimp it so that it cannot come out of the gear case.
6. Install the thrust washers and side gears.

OUTPUT GEAR

Disassembly

1. Remove the seal rings.

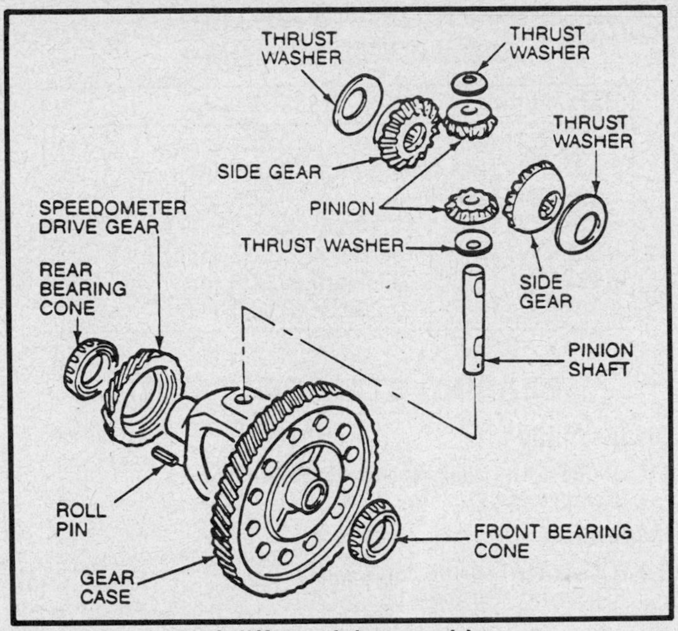
THRUST
WASHER
THRUST
WASHER
THRUST
WASHER
SIDE GEAR
SPEEDOMETER
DRIVE GEAR
PINION
THRUST WASHER
REAR
BEARING
CONE
SIDE
GEAR
PINION
SHAFT
ROLL
PIN
FRONT BEARING
CONE
GEAR
CASE

Exploded view of differential assembly

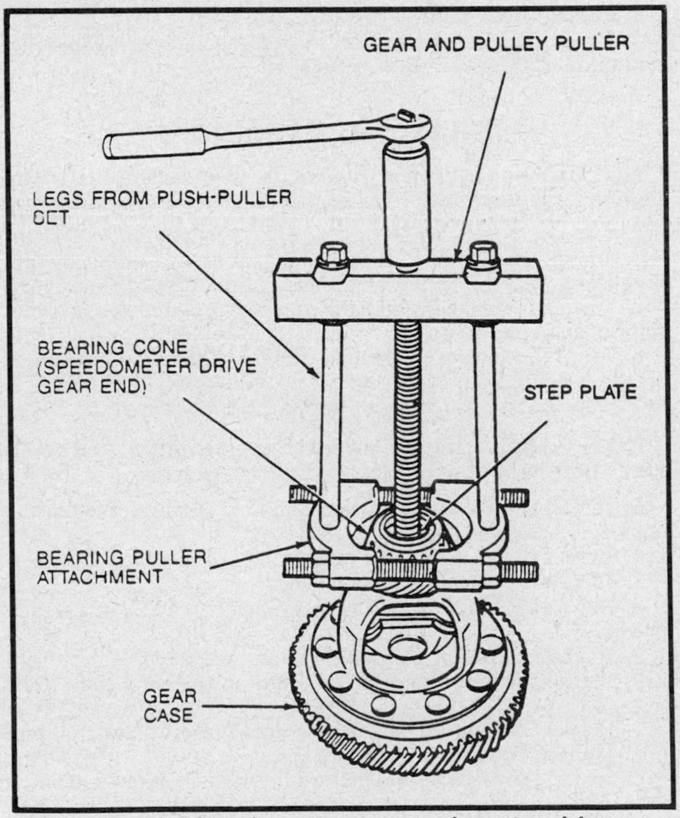
GEAR AND PULLEY PULLER
LEGS FROM PUSH-PULLER
SET
BEARING CONE
(SPEEDOMETER DRIVE
GEAR END)
STEP PLATE
BEARING PULLER
ATTACHMENT
GEAR
CASE

Removing the bearing cone—speedometer drive gear end

2. Press off the output gear bearings using step plate D80L–630–4 and puller D84L–1123–A or equivalent.

Inspection

1. Check for worn or damaged teeth or O-ring.
2. Check for worn or damaged seals.

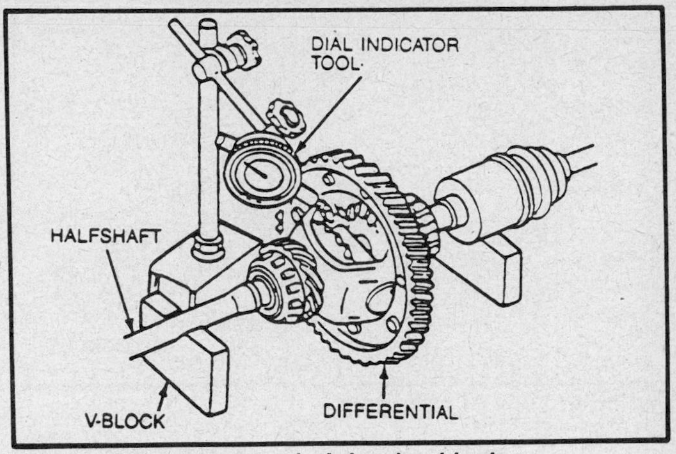
DIAL INDICATOR
TOOL
HALFSHAFT
V-BLOCK
DIFFERENTIAL

Checking side gear and pinion backlash

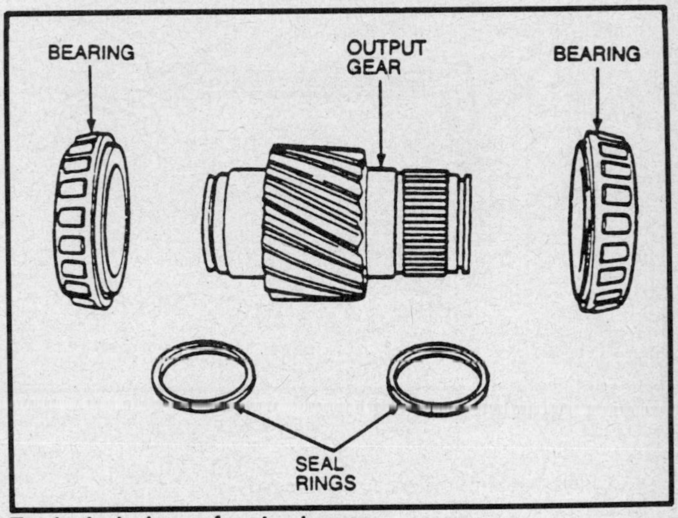
BEARING
OUTPUT
GEAR
BEARING
SEAL
RINGS

Exploded view of output gear

Assembly

1. Press on the output gear bearings using step plate D80L–630–4, bearing cone replacer T88T–7025–B and bearing installation plate T75L–1165–B or equivalent.
2. Install the seal rings.

IDLER GEAR

Disassembly

1. Secure the idler shaft in a vise using torque adapter T87C–77000–E or equivalent.
2. Remove the locknut using socket T88T–7025–A along with a 1⅝ inch socket or equivalent.
3. Remove the bearing.
4. Remove the spacer.
5. Remove the idler gear from the idler shaft.
6. Remove the adjustment shim.
7. Remove the other bearing.
8. Remove a bearing cup from the idler gear using puller D80L–943–A and slide hammer T50T–100–A or equivalent.
9. Press out the other bearing cup using step plate D80L–630–11 or equivalent.

Inspection

1. Check for worn or damaged gears.

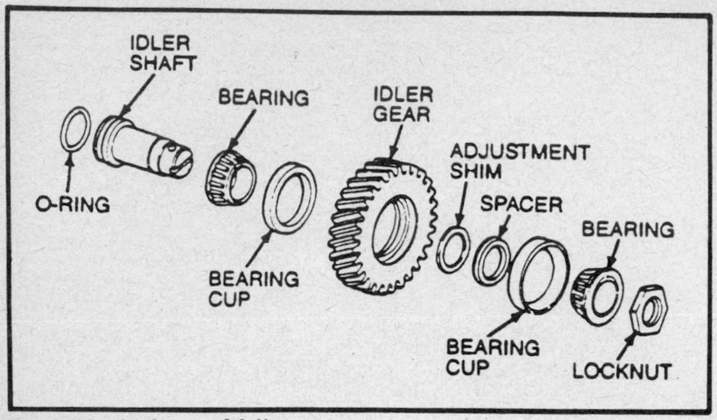
IDLER SHAFT
BEARING
IDLER GEAR
ADJUSTMENT SHIM
SPACER
BEARING
O-RING
BEARING CUP
BEARING CUP
LOCKNUT

Exploded view of idler gear

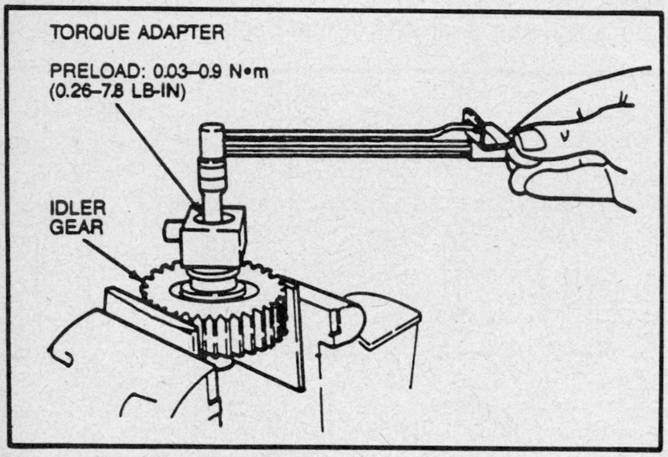
TORQUE ADAPTER
PRELOAD: 0.03–0.9 N•m
(0.26–7.8 LB-IN)
IDLER GEAR

Measuring idler gear preload

2. Check bearing rollers for damaged.
3. Check bearing cups wear.

Assembly

1. Press the bearing cups into the idler gear using bearing installation plate T80T–4000–E or equivalent.
2. Install the bearing onto the idler shaft.
3. Install the adjust shim.
4. Install the spacer.
5. Install the idler gear.
6. Install the other idler gear bearing.
7. Secure the idler shaft in a vise using torque adapter T87C–77000–E or equivalent.
8. Tighten locknut to 94 ft. lbs. using socket T88T–7025–A along with a 1⅝ in. socket or equivalent.
9. Turn the idler gear and adapter over and secure the gear in a vise. Use protective plates to prevent damage to the idler gear.
10. Attach a inch lbs. torque wrench and measure the preload while tightening the locknut to 94–130 ft. lbs. The preload should be 0.26–7.8 inch lbs.

NOTE: Read the preload when the idler shaft starts to turn.

11. If the specified preload is not reached within the specified tightening torque, select an appropriate adjustment shim(s).

NOTE: The preload can be reduced by increasing the thickness of the shims, or increased by reducing the thickness of the shims. Do not use more than 7 shims.

IDLER GEAR PRELOAD ADJUSTMENT SHIM

Part Number	Shim Thickness
E92Z-7N112-F	0.10mm (0.004 inch)
E92Z-7N112-A	0.12mm (0.005 inch)
E92Z-7N112-B	0.14mm (0.006 inch)
E92Z-7N112-C	0.16mm (0.0063 inch)
E92Z-7N112-G	0.18mm (0.007 inch)
E92Z-7N112-D	0.20mm (0.008 inch)
E92Z-7N112-E	0.50mm (0.020 inch)

BEARING STATOR SUPPORT

Disassembly

1. Remove the bearing cup with a pin punch.
2. Remove the O-ring.

Inspection

1. Check for bearing cup wear.
2. Check oil seal for damage.

Assembly

1. Install the O-ring.
2. Press the bearing cup into the cover using bearing installation plate T80T–4000–E or equivalent.

Transaxle Assembly

NOTE: Whenever the transaxle is disassembled, the bearing preload must be adjusted. Adjust the bearing preload by following the shim selection procedure.

1. Install the output gear and idler gear as an asembly by tapping them into the converter housing with a plastic hammer.
2. Install the bearing housing on the converter housing and tighten the bolts to 14–19 ft. lbs.
3. Align the groove on the idle shaft with the matching mark on the bearing housing.
4. Tap the roll pin with a pin punch and hammer.

NOTE: Apply the specified transaxle fluid to the O-rings before installing the 2–3 accumulator.

5. Install the 2–3 accumulator and new O-rings. Torque the bolts to 69–95 inch lbs.
6. Install the parking pawl and shaft.
7. Install the spring and snapring.
8. Move the shaft to check for proper operation of parking pawl.
9. Install the parking assist lever and snapring.
10. Install the actuator support. Tighten the bolts to 8–10 ft. lbs.
11. Install the manual shaft spacer, manual plate, washer and nut. Tighten the nut to 30–41 ft. lbs.
12. Install the bracket and bolt. Tighten the bolt to 69–95 inch lbs.
13. Install the detent ball, spring, washer and plug. Tighten plug to 8.7–13 ft. lbs.
14. Attach seal protector T88C–77000–GH or equivalent to the low and reverse clutch piston.
15. Install the low and reverse clutch piston by pushing evenly around the circumference. Remove the protector.

NOTE: Be careful not to damage the outer seal.

16. Install the return spring and retainer.
17. Compress the return spring and retainer using return

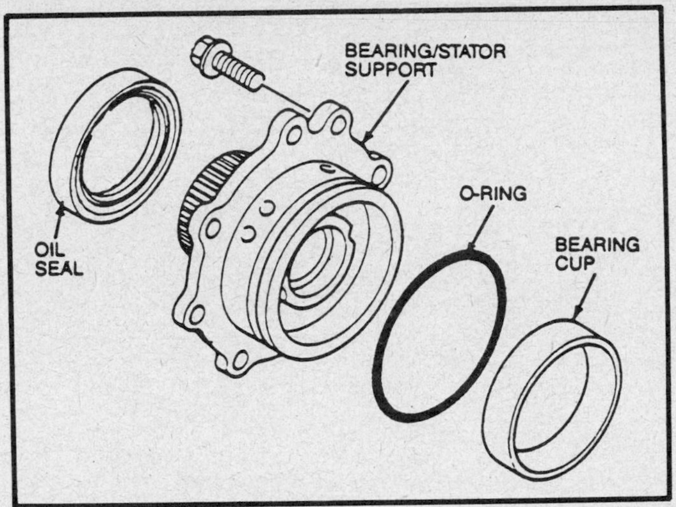
BEARING/STATOR SUPPORT
O-RING
OIL SEAL
BEARING CUP

Exploded view of bearing stator support

spring compressor T88C–77000–AH and the plate from T87C–77000–B or equivalent.
18. Install the snapring.
19. Remove the return spring compressor.
20. Pour the specified transaxle fluid over the low and reverse clutch piston until it is fully submerged. Check that no bubbles appear from between the piston and seals when applying compressed air through the fluid passage.

NOTE: The compressed air must be under 57 psi and not applied for more than a few seconds.

21. Install the output shell to the output gear.
22. Install the 2.83 in. thrust washer onto the output shell.
23. Apply a thin coat of silicon sealant to the contact surfaces of the converter housing and transaxle case.
24. Install new O-rings.
25. Install the transaxle case to the converter housing. Tighten the bolts to 27–38 ft. lbs.
26. Install transaxle plugs T88C–7025–AH or equivalent to the differential side gears.

NOTE: Failure to install the transaxle plugs may allow the differential side gears to become mispositioned.

27. Place the 3–4 clutch assembly over the turbine shaft.

NOTE: Be sure that the thrust washer and needle bearing are installed in the correct position.

28. Install the turbine shaft and 3–4 clutch assembly into the transaxle case.
29. Install turbine shaft holder T88C–77000–KH or equivalent and attach it to the turbine shaft.
30. Install the internal gear.
31. Install the internal gear snapring.

NOTE: Be sure the thrust washer and needle bearing are in the correct position before installing the carrier hub assembly.

32. Install the carrier hub assembly.
33. Install the low and reverse clutch pack, retaining plate and snapring.
34. Measure the clearance between the snapring and retaining plate. The clearance should be 0.083–0.094 in. If clearance is not with specification, adjust it by selecting a retaining plate with an appropriate thickness.
35. Install the one-way clutch.

NOTE: Turning the carrier hub assembly counterclockwise eases installation of the one-way clutch.

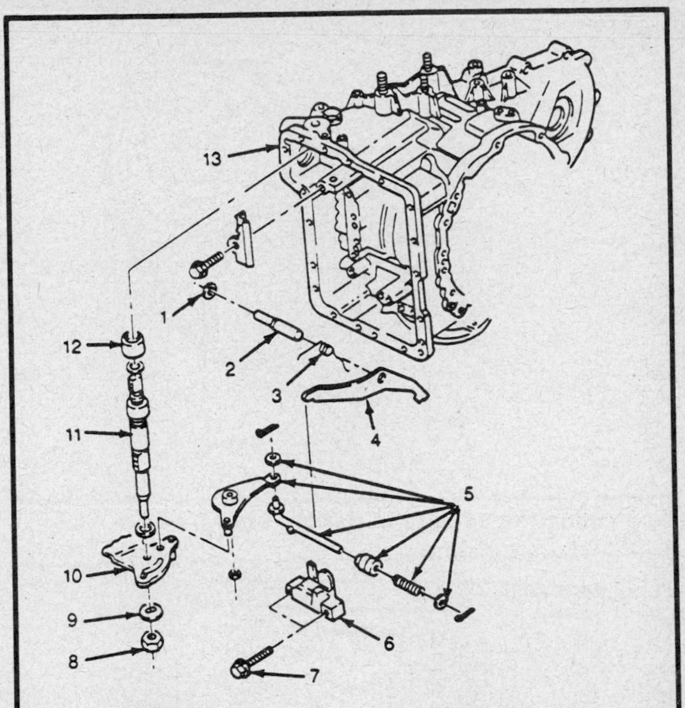
1
2
3
4
5
6
7
8
9
10
11
12
13

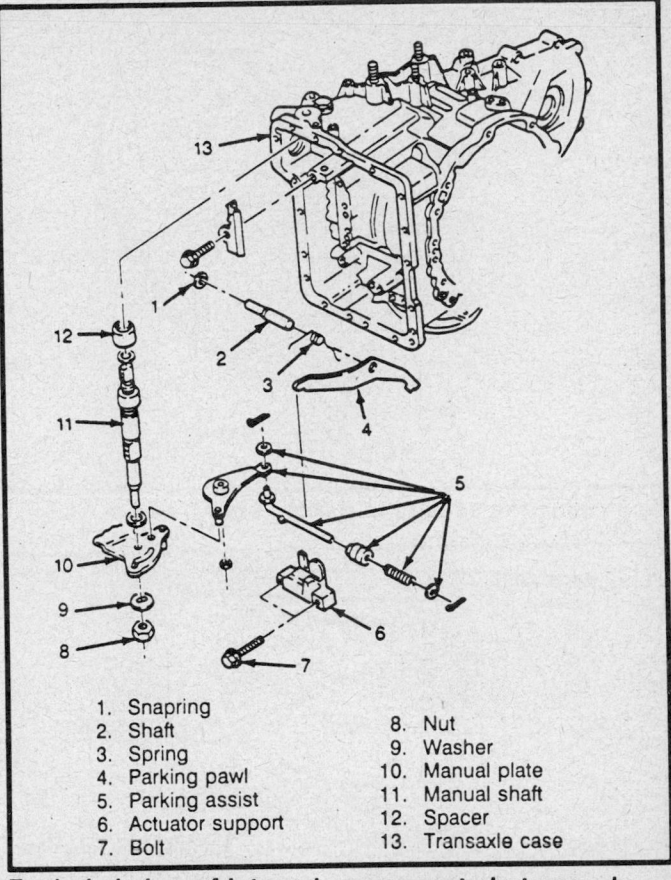

1. Snapring
2. Shaft
3. Spring
4. Parking pawl
5. Parking assist
6. Actuator support
7. Bolt
8. Nut
9. Washer
10. Manual plate
11. Manual shaft
12. Spacer
13. Transaxle case

Exploded view of internal components in transaxle

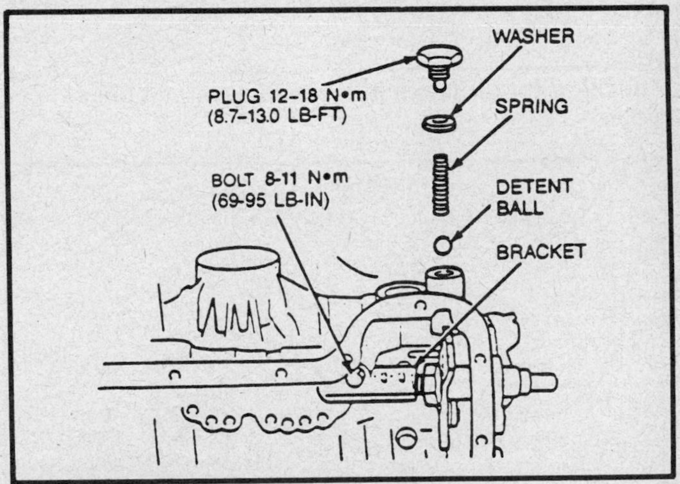
WASHER
PLUG 12–18 N•m
(8.7–13.0 LB-FT)
SPRING
BOLT 8-11 N•m
(69-95 LB-IN)
DETENT BALL
BRACKET

Exploded view of detent assembly

36. Install the one-way clutch snapring.
37. Install the servo spring and servo.
38. Compress the servo with a C-clamp.
39. Install the snapring, then remove the C-clamp.
40. Install the piston stem.
41. Install the anchor strut.
42. Install the 2–4 band in the transaxle case so it is fully expanded.

NOTE: Interlock the 2–4 band and anchor strut.

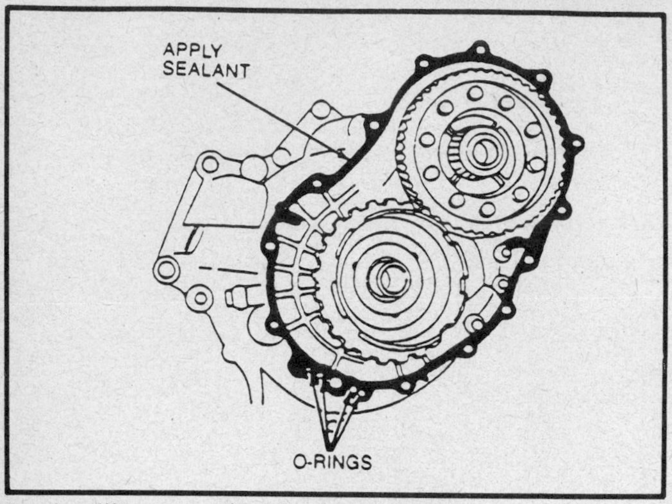
APPLY
SEALANT
O-RINGS

Apply silicone sealant in darken area

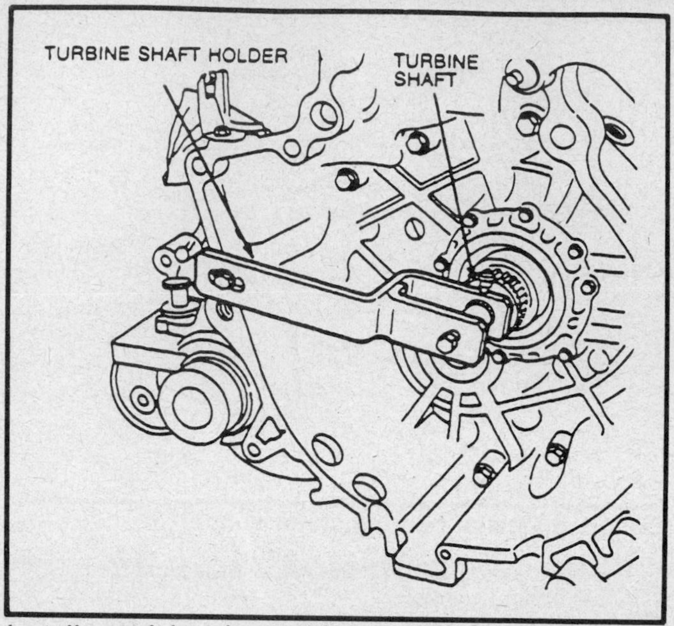
TURBINE SHAFT HOLDER
TURBINE
SHAFT

Install special tool to turbine shaft

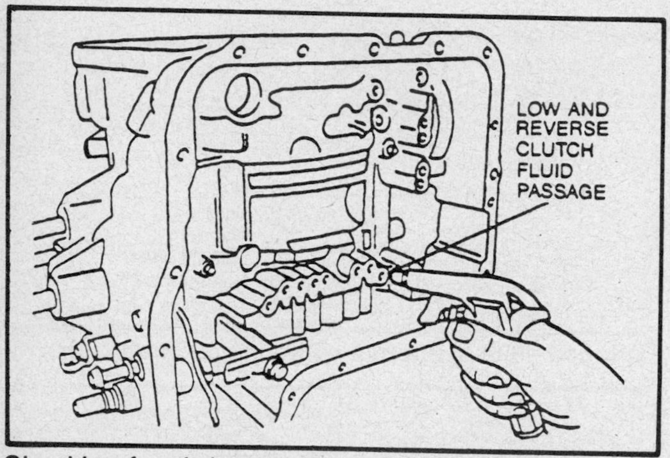
LOW AND
REVERSE
CLUTCH
FLUID
PASSAGE

Checking for air leaks in low and reverse clutch piston

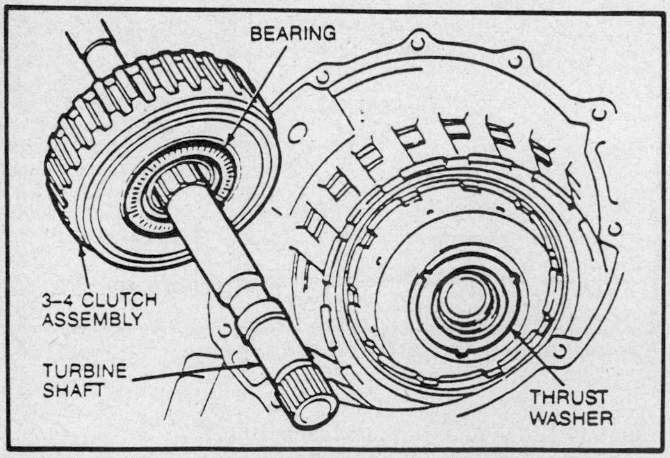
BEARING
3–4 CLUTCH
ASSEMBLY
TURBINE
SHAFT
THRUST
WASHER

Installing the turbine shaft and 3–4 clutch assembly

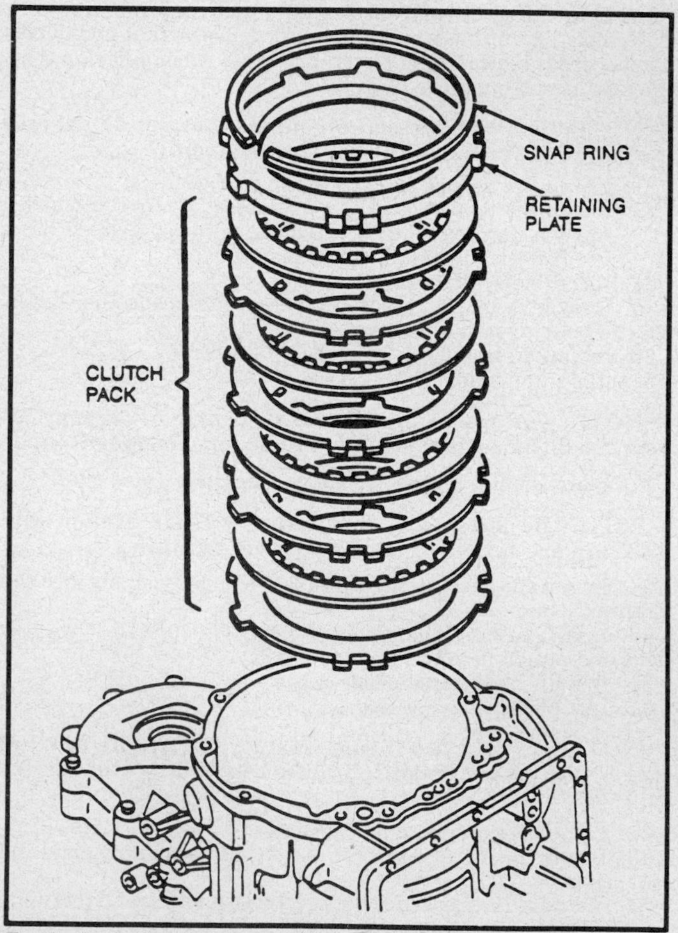
SNAP RING
RETAINING
PLATE
CLUTCH
PACK

Exploded view of low and reverse clutch pack

43. Install the small gun gear and one-way clutch by rotating it.

NOTE: Be sure the thrust washer and needle bearing are installed in the correct position.

44. Pull the 2–4 band with a tool and install the piston stem in the correct position. Loosely tighten the piston stem by hand.

NOTE: Be sure the needle bearing is in the correct position before installing the clutch assembly.

45. Install the clutch assembly by rotating it.

LOW AND REVERSE CLUTCH PACK CLEARANCE

Part Number	Pressure Plate Thickness
E92Z-7B066-AD	6.8mm (0.268 inch)
E92Z-7B066-Y	7.0mm (0.276 inch)
E92Z-7B066-Z	7.2mm (0.283 inch)
E92Z-7B066-AA	7.4mm (0.291 inch)
E92Z-7B066-AB	7.6mm (0.299 inch)
E92Z-7B066-AC	7.8mm (0.307 inch)

46. Measure the height difference between the reverse and forward drum and transaxle case. The height difference should be 0.032 in.
47. Place the needle bearing on the clutch assembly.
48. To adjust the total endplay, remove the previous thrust washer and gasket from the oil pump. Place a 0.087 in. thrust washer on the oil pump.
49. Set the oil pump onto the clutch assembly. Measure the clearance between the transaxle case and the oil pump. Select a suitable thrust washer from the chart.
50. Remove the oil pump.
51. Place the selected thrust washer and a new gasket on the oil pump.
52. Install the oil pump onto the clutch assembly. Tighten bolts to 14–19 ft. lbs.
53. Loosen the locknut and tighten the piston stem to 78–95 inch lbs.
54. Loosen the piston stem 2 turns.
55. Tighten the locknut to 18–29 ft. lbs.
56. Install the oil strainer with a new O-ring to the transaxle. Tighten the bolts to 69–95 inch lbs.

NOTE: Be sure the magnets in the oil pan are correctly positioned.

57. Install the oil pan with a new gasket. Tighten bolts to 69–95 inch lbs.
58. Align the manual valve with the pin on the manual plate and install the valve body into the transaxle case. Tighten the bolts to 95–130 inch lbs.
59. Install the solenoid connector with a new O-ring in the transaxle case.
60. Install a new O-ring on the bracket, then feed the kickdown cable through the transaxle case and connect it to the throttle cam.
61. Install the kickdown cable attaching bolt and bracket. Tighten the attaching bolt to 69–95 inch lbs. and the bracket bolt to 14–19 ft. lbs.
62. Install the valve body cover with a new gasket. Tighten to 69–95 inch lbs.
63. Install the oil pipes, oil hoses and switch box as as assembly. Tighten the switch box bolts to 12–17 ft. lbs.
64. Install the harness clip and tighten to 69–95 inch lbs.
65. Install the ball, spring, new washers and plug. Tighten plug to 23–35 ft. lbs.
66. Install the solenoid connector.
67. Install the pulse generator and fluid temperature switch. Tighten the pulse generator bolt to 69–95 inch lbs. Tighten the fluid temperature switch to 22–29 ft. lbs.
68. Install the dipstick tube with a new O-ring. Tighten the bolts to 61–87 inch lbs.
69. Turn the manual shaft to the **N** detent position.
70. Install the neutral safety switch and loosely tighten the bolts.
71. Remove the screw and insert a 0.079 in. pin. Move the neutral safety switch until the pin engages the switch alignment hole.

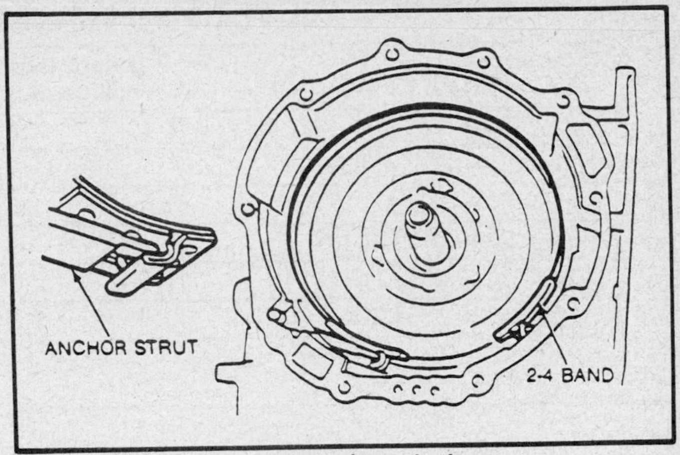
ANCHOR STRUT
2-4 BAND
Interlock 2–4 band and anchor strut

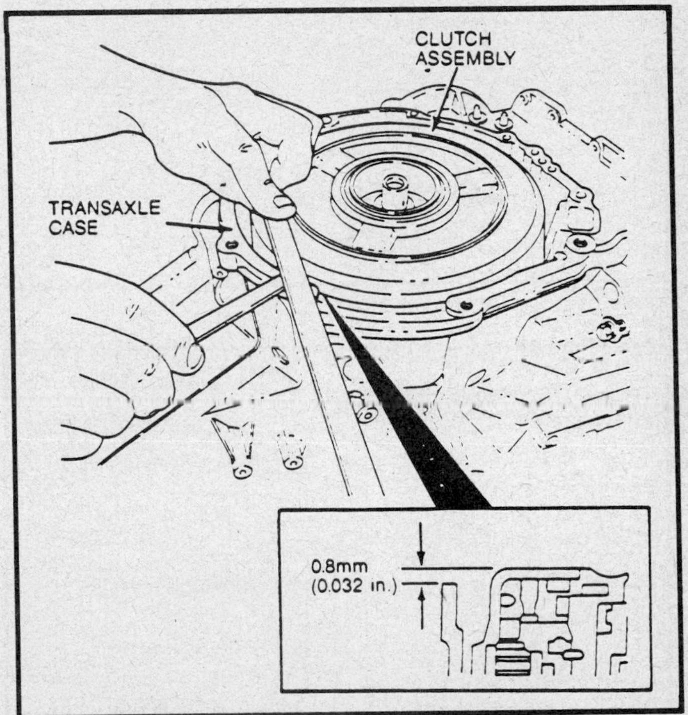
CLUTCH ASSEMBLY
TRANSAXLE CASE
0.8mm
(0.032 in.)
Measure the height between the reverse and forward drum and case

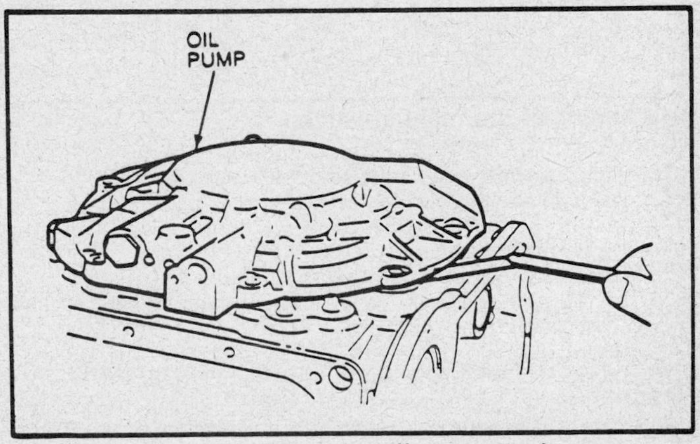
OIL PUMP
Measure clearance betwwen oil pump and case

OIL PUMP TO TRANSAXLE CASE CLEARANCE

Clearance mm (inch)	Select this Thrust Washer Thickness mm (inch)	Part Number
0.91–1.10 (0.036–0.043)	1.2 (0.047)	E92Z-7D014-E
0.71–0.90 (0.028–0.035)	1.4 (0.055)	E92Z-7D014-F
0.51–0.70 (0.020–0.027)	1.6 (0.063)	E92Z-7D014-A
0.31–0.50 (0.012–0.019)	1.8 (0.071)	E92Z-7D014-B
0.11–0.30 (0.004–0.011)	2.0 (0.078)	E92Z-7D014-C
0.00–0.10 (0.036–0.043)	2.2 (0.047)	E92Z-7D014-D

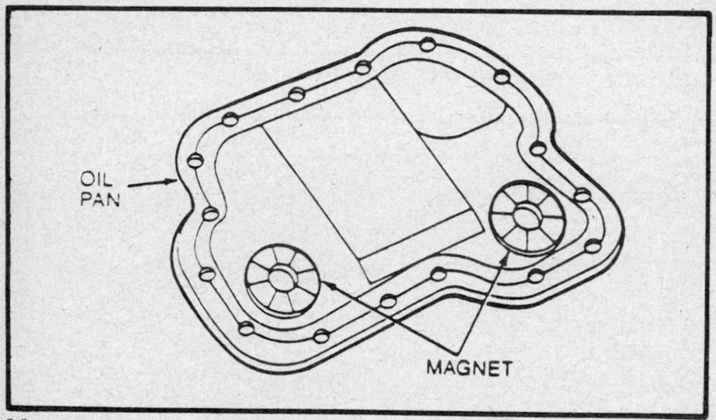
OIL PAN
MAGNET
Magnet position in oil pan

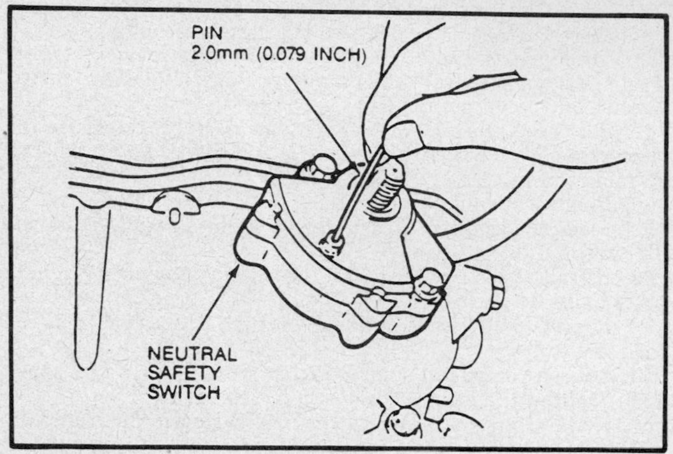
PIN
2.0mm (0.079 INCH)
NEUTRAL SAFETY SWITCH
Alignment of the neutral switch

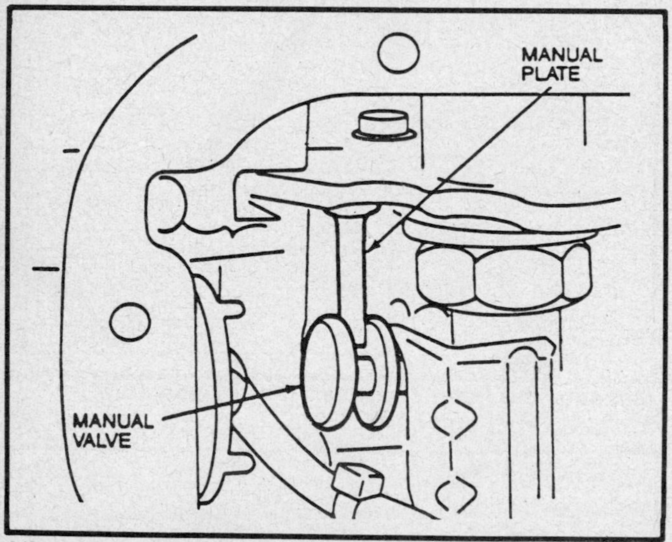
MANUAL PLATE
MANUAL VALVE
Alignment of the manual valve

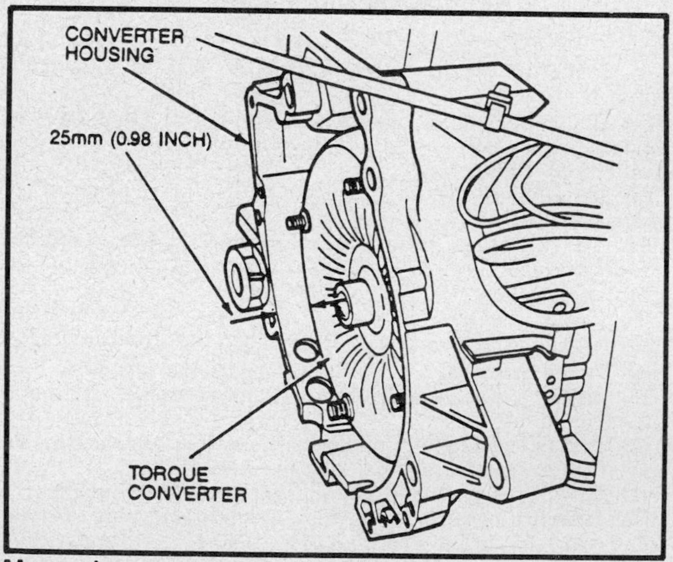
CONVERTER HOUSING
25mm (0.98 INCH)
TORQUE CONVERTER
Measuring converter to housing end

72. Tighten the switch bolts to 69–95 inch lbs.
73. Remove the pin and install the screw.
74. Install the harness with the remaining clip and tighten to 69–95 inch lbs.
75. Remove the transaxle from the holding fixture.
76. Install the converter seal using converter seal replacer T88C–77000–BH or equivalent.
77. Install the oil pump shaft.
78. Install a new O-ring onto the turbine shaft.
79. Fill the torque converter with specified transaxle fluid.
80.Install the torque converter in the converter housing while rotating it to align the splines.

NOTE: Do not try to force the torque converter in, install it carefully in the transaxle.

81. Measure the distance between the torque converter and the end of the converter housing. The distance should be 0.98 in.
82. Install the right transaxle mount and tighten the bolts to 43–49 ft. lbs.
83. Install the differential oil seals using differential seal replacer T87C–77000–H or equivalent.

SPECIFICATIONS

Torque converter stall torque ratio		1.700 – 1.900:1
Gear ratio	First	2.800:1
	Second	1.540:1
	Third	1.000:1
	Fourth (OD)	0.700:1
	Reverse	2.333:1
Final gear ratio		3.700
Number of drive plates/ driven plates	Forward clutch	3/3
	Coasting clutch	2/2
	3–4 clutch	5/5
	Reverse clutch	2/2
	Low and reverse brake	4/4
Servo diameter (Piston outer dia./retainer inner dia.) mm (in.)		78mm/40mm (3.07 in./1.57 in.)
Transaxle Fluid	Type	Motorcraft MERCON or equivalent
	Capacity liters (U.S. qt., Imp. qt.)	6.8 liters (7.2 U.S. qt., 6.0 Imp. qt.)

TORQUE SPECIFICATIONS

Description	Ft. Lbs.	Inch Lbs.
Line pressure plug	–	43-87
Bearing housing	14-19	–
Transaxle case to converter housing	27-38	–
Valve body	–	95-130
Transaxle to engine	66-86	–
Center transaxle mount bolts	27-40	–
Center transaxle mount nuts	47-66	–
Transaxle to left mount	63-86	–
Left mount to bracket	49-69	–
Crossmember bolts	27-40	–
Crossmember nuts	55-69	–
Right transaxle mount	63-86	–
Torque converter	32-45	–
Converter cover	–	69-95
Gusset plate to transaxle	27-38	–
Range sector to transaxle	22-29	–

Description	Ft. Lbs.	Inch Lbs.
2-3 accumulator	–	69-95
Actuator support	8-10	–
Manual plate	30-41	–
Oil pump	14-19	–
Oil strainer	–	69-95
Oil pan	–	69-95
Throttle cable bracket	14-19	–
Switch box	12-17	–
Oil line plug	23-35	–
Pulse generator	–	69-95
Fluid temperature switch	22-29	–
Dipstick tube	–	61-87
Neutral safety switch	–	69-95
Throttle cam	–	69-95
Drain plug	29-43	–

SPECIAL TOOLS

Tool Number	Description
T87C-77000-H	Differential seal replacer
T88C-7025-AH	Transaxle plug set
T88C-77000-AH	Return spring compressor
T88C-77000-BH	Converter seal replacer
T88C-77000-CH	Shim selection set
T88C-77000-DH	Preload torque adapter
T88C-77000-EH	Differential bearing cone replacer
T88C-77000-FH	Differential bearing cup replacer
T88C-77000-GH	Seal protector
T88C-77000-HH	Seal protector
T88C-77000-JH	Leak check adapter
T88C-77000-KH	Turbine shaft holder
T88C-77000-CH4	Screws
T88C-77000-CH5	Screws
D78P-4201-C	Magnetic base/flex arm
D80L-522-A	Puller
D80L-630-3	Step plate
D80L-630-4	Step plate
D80L-630-6	Step plate
D80L-630-10	Step plate
D80L-630-11	Step plate
D80L-927-A	Puller legs
T88C-77000-L	Torque adapter
D80L-943-A	Puller
D84L-1123-A	Puller
D87C-77000-A	Transmission test adapters
T87C-77000-J	Shim selection kit
D87L-6000-A	Engine support bar
T50T-100-A	Slide hammer

Model	Description
T57L-500-B	Bench mounting fixture
T57L-77820-A	Pressure gauge
T60K-4616-A	Bearing cup installer
T-65L-77515-A	Clutch spring compressor
T73L-2196-A	Puller body
T75L-1165-B	Bearing installation plate
T77F-1102-A	Puller
T80L-77003-A	Gauge bar
T80L-77100-A	Guide pins
T80T-4000-E	Bearing cup installation plate
T80T-4000-W	Driver handle
T86P-700043-A	Puller jaws
T87C-77000-E	Torque adapter
T88T-7025-A	Socket (55mm)
T88T-7025-B	Bearing cone replacer
TOOL-1175-AC	Puller
TOOL-4201-C	Dial indicator
T77F-4220-B1	Puller
007-00028	Super STAR II tester
059-00010	Inductive dwell-tach-volt-ohmmeter
007-00037	4EAT tester
055-00101	Tachometer
014-00737	Pressure tester
014-00456	Fittings
014-00210	Transmission jack
014-00028	Torque converter cleaner

Section 3

F3A Transaxle
Ford Motor Co.

APPLICATION

1989 Ford Festiva
1987–89 Mercury Tracer

GENERAL DESCRIPTION

The Tracer and Festiva use a Jatco F3A fully automatic transaxle, consisting of 3 forward speeds and a reverse speed. The internal components are similar to the 3N71B and R3A automatic transmissions.

The torque converter is located on the engine side and the oil pump is located on the other end of the transaxle. The front clutch, the rear clutch, front planetary and rear planetary gears are arranged in the respective order from the front, or oil pump end of the transaxle. During the section outline, the oil pump end will be referred to as the front and the converter end, or engine end, will be referred to as rear of the transaxle.

The control valve is located under the front clutch and the rear clutch assemblies. The governor is located on the outside of the case and responds to the speed of the output shaft to control operating oil pressure.

The low and reverse brake band is located on the outside of the rear planetary gears to shorten the total length of the transaxle.

The 3 shafts that are contained within the case are the oil pump driveshaft which transmit engine speed directly to the oil pump via a quill shaft inside the input shaft, the input shaft which transmits power from the torque converter turbine and drives the front clutch cover. The 3rd shaft is the output shaft which transmits power from the front planetary gear carrier and the rear planetary gear annulus, through the main drive idler gear to the differential drive gear.

Both the transaxle and the differential use a common sump with ATF fluid as the lubricant.

Transaxle and Converter Identification

TRANSAXLE

Identification tags are located on the front of the transaxle, under the oil cooler lines and identify the transaxle type and model. The Tracer transaxle part number for U.S. models is E7GZ–7000–B and for Canada models is E7GZ–7000–C. These part numbers are for complete transaxles including the torque converters. The Festiva complete assembly numbers were not available at the time of this publication. The Festiva and Tracer both use the F3A automatic transaxle and service procedures are the same. Many of the components are different and must be replaced with the correct components for the model being serviced.

CONVERTER

The torque converter is a welded unit and cannot be disassembled unless special tools are available for the purpose. A lockup torque converter is utilized, which consists of a lockup drive plate containing centrifugally operated shoe, bracket and spring assemblies and a one-way clutch.

Metric Fasteners

Metric bolt sizes and thread pitches are used for all fasteners on the Jatco transaxle. The metric fastener dimensions are close to the dimensions of the familiar inch system fasteners and for this reason, replacement fasteners must have the same measurement and strength as those removed. Do no attempt to interchange metric fasteners for inch system fasteners. Mismatched or incorrect fasteners can result in damage to the transaxle unit through malfunctions, breakage or possible personal injury. Care should be taken to reuse the fasteners in the same locations as removed whenever possible.

Capacities

The use of Dexron®II type automatic transaxle fluid or its equivalent, is recommended for use in the F3A automatic transaxle models.

The capacity of the F3A transaxle is 6.0 U.S. quarts (5.7 L).

Checking Fluid Level

With the engine/transaxle assemblies up to normal operating temperature, move the quadrant through all the selector positions and finish in the **P** position. The correct level is between the **F** and **L** marks on the dipstick. It is important to keep the level at, or slightly below, the **F** mark on the dipstick. Do not overfill the assembly.

Transaxle oil level should be checked, both visually and by smell, to determine that the fluid level is correct and to observe any foreign material in the fluid. Smelling the fluid will indicate if any of the bands or clutches have been burned through excessive slippage or overheating of the transaxle.

It is most important to locate the defect and its cause and to properly repair them to avoid having the same problem recur.

TRANSAXLE MODIFICATIONS

Low and Reverse Clutch Hub Snapring

Transaxles which have not been previously disassembled do not have a low and reverse clutch hub snapring. This clutch hub snapring will be install on all transaxles during the rebuilding process.

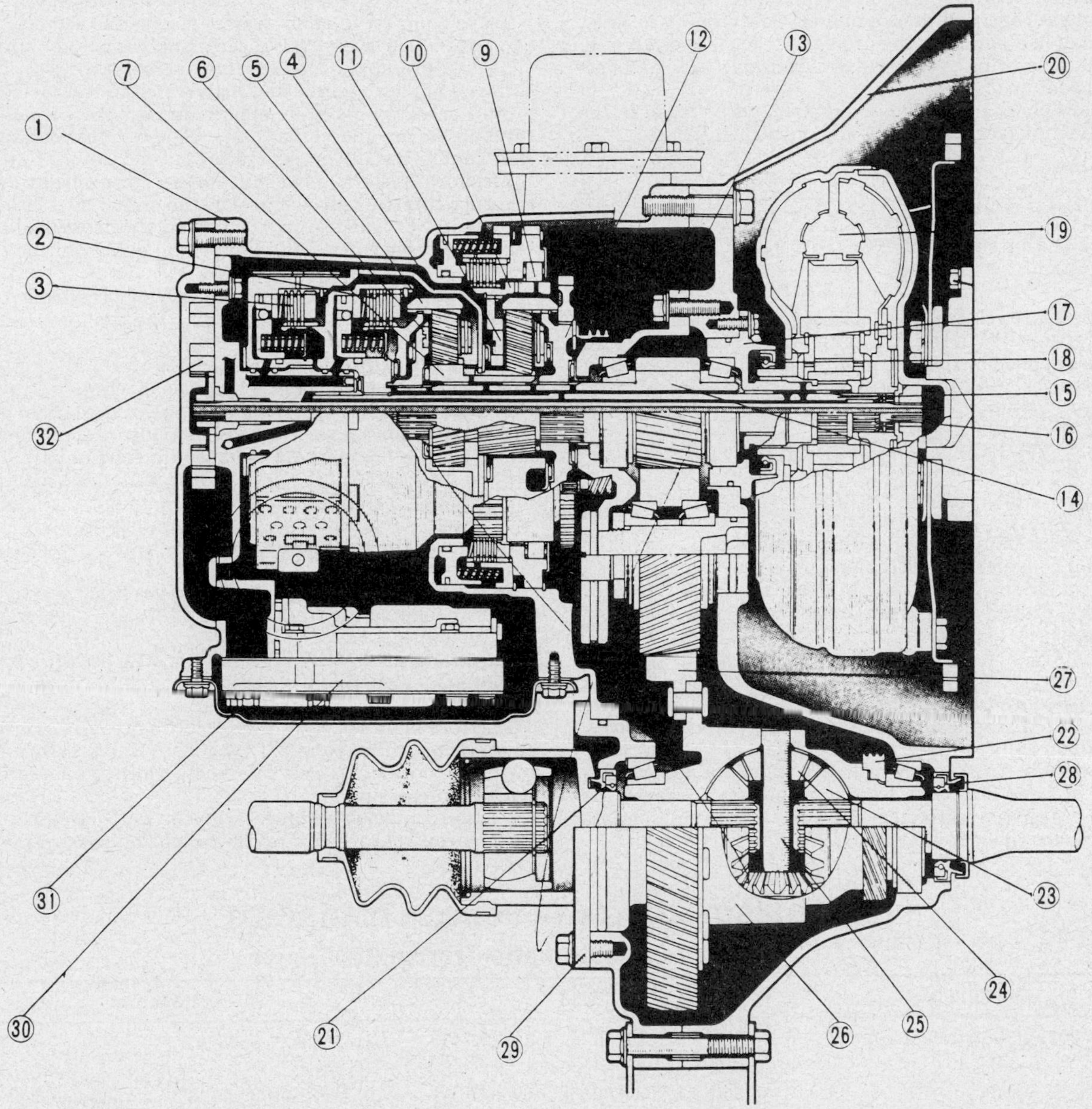
1
2
3
4
5
6
7
9
10
11
12
13
14
15
16
17
18
19
20
21
22
23
24
25
26
27
28
29
30
31
32

1. Transmission
2. Rear clutch
3. Front clutch
4. Connection shell
5. Rear clutch hub assembly
6. Planetary carrier
7. Sun gear
8. Low and reverse brake
9. One-way clutch
10. One-way clutch inner race
11. Planetary carrier
12. Drum hub assembly
13. Bearing housing
14. Output gear
15. Turbine shaft
16. Oil pump shaft
17. Bearing cover
18. Oil seal
19. Torque converter
20. Converter housing
21. Oil seal
22. Speedometer drive gear
23. Side gear
24. Pinion gear
25. Pinion shaft
26. Differential gear case
27. Ring gear
28. Oil seal
29. Side bearing housing
30. Control valve
31. Oil pan
32. Oil pump

Cross-section of the F3A transaxle

TROUBLE DIAGNOSIS

A logical and orderly diagnosis outline and charts are provided with clutch and band applications, shift speed and governor pressures, main control pressure and oil flow circuits to assist the repairman in diagnosing the problems, causes and extent of repairs needed to bring the automatic transaxle back to its acceptable level of operation.

Preliminary checks and adjustments should be made to the manual valve linkage, accelerator and downshift linkage.

Hydraulic Control System

NOTE: Please refer to Section 9 for all oil flow circuits.

Hydraulic pressure, clutch and band applications control the changing of gear ratios in the automatic transaxle. The clutches and bands are applied by the force of fluid pressure controlled by a system of valves and control mechanisms.

Major Components

The hydraulic control system consists of the following major components:

Main control pressure system which supplies pressure to the transaxle and converter when the engine is operating.

Converter and lubrication system which regulates converter fluid pressure, provides gear train lubrication and fluid cooling while the transaxle is operating.

Forward clutch pressure and governor pressure system applies the forward clutch, which is applied in all forward speeds and applies pressure to the governor valve. The governor valve supplies regulated pressure to the rear side of the shift valves, dependent upon the road speed of the vehicle.

Low and reverse brake apply system applies the low and reverse brake in **1** and **R** selector lever positions and locks out the 2nd and 3rd gears by directing pressure to the appropriate valves to prevent them from shifting.

First gear lock out system allows the transaxle to shift directly to 2nd speed and locks out the 1st and 3rd gears.

Brake band servo apply system applies the servo to hold the band to the surface of the reverse and high clutch cylinder.

Reverse pressure booster system increases control (line) pressure and applies reverse and high clutch in the reverse range.

Shift valve train system applies and exhausts the fluid pressures to servos and clutch assemblies for upshifts and downshifts automatically on demand.

Kickdown system (downshift) forces downshift by overriding governor/throttle valve control of the shift valves.

Governor provides a varying pressure proportional to engine vacuum to help control the timing and quality of the transaxle shifts.

Throttle T.V. system provides a varying pressure proportional to engine vacuum to help control the timing quality of the transaxle shifts.

Throttle backup system compensates for a lower rate of engine vacuum at ½ or more of throttle opening.

Pressure modifier system adjusts the control (line) pressures and 2-3 shift timing valve operation to insure smoother shifting under various engine load and vacuum conditions.

Fluid filter screens the fluid and cleans foreign material from the oil supply before entering the oil pump.

Oil pump supplies oil pressure to transaxle.

Converter pressure relief valve prevents converter pressure build up.

Transaxle fluid cooling system removes heat from torque converter by sending the transaxle fluid through a cooler in the engine cooling system.

Throttle control valve regulates throttle pressure in relation to the engine manifold vacuum through vacuum diaphragm (modulator).

Pressure modifier valve uses throttle pressure, controlled by governor pressure, to modify the main line pressure from the regulator valve. This prevents harsh shifting caused by excessive pump pressure.

Manual control valve moves with the shift selector and directs the line control pressure to the various oil passages.

CHILTON'S THREE C's TRANSAXLE DIAGNOSIS
Jatco F3A Automatic Transaxle

Condition	Cause	Correction
Engine does not start in any range	a) Neutral start/backup lamp switch wiring disconnected or damaged b) Neutral start/backup lamp switch sticking or failed	a) Replace, service b) Perform neutral start switch check
Engine does not start in P	a) Range selector and linkage	a) Service or adjust linkage
Engine starts in ranges other than P and N	a) Range selector linkage b) Neutral start/backup lamp switch loose c) Neutral start/backup lamp switch wiring short circuited	a) Perform linkage check b) Check and retighten c) Check for damage
Vehicle moves in P or parking gear not disengaged when P is disengaged	a) Range selector and linkage b) Parking linkage	a) Perform linkage check b) Check for proper operation
Vehicle moves in N	a) Range selector linkage b) Dirty or sticking valve body c) Rear clutch	a) Perform linkage check b) Clean, service or replace valve body c) Check for clutch not disengaging

CHILTON'S THREE C's TRANSAXLE DIAGNOSIS
Jatco F3A Automatic Transaxle

Condition	Cause	Correction
No drive in any gear	a) Valve body loose b) Sticky or dirty valve body c) Improper rear clutch application or damaged, worn clutch d) Low rear clutch application pressure e) Internal leakage f) Valve body loose g) Broken pump or turbine shaft	a) Tighten to specification b) Clean, service or repair valve body c) Service as required d) Perform line and pressure test e) Check pump seals f) Tighten to specification g) Perform stall test
Vehicle does not move in D (moves in 1, 2 and R)	a) Range selector linkage b) Oil pressure control system c) Dirty or sticking valve body d) One-way clutch	a) Perform linkage check b) Perform line and governor pressure tests c) Clean, service or repair valve body d) Service as required
Vehicle does not move in forward ranges, reverse OK	a) Dirty or sticking valve b) Improper rear clutch application or oil pressure control c) Damaged or worn rear clutch	a) Clean, service or repair valve body b) Check rear clutch for proper operation. Perform line pressure test c) Check and service as required
Vehicle does not move in reverse. Forward OK	a) Improper oil pressure b) Dirty or sticking valve body c) Damaged or worn low reverse clutch	a) Perform line pressure test b) Clean, service or replace valve body c) Check and service as required
Vehicle does not shift out of 1st gear in D	a) Dirty or sticking valve body b) Damaged or worn governor c) Improper oil pressure control	a) Check 1-2 shift valve operation. Clean, service or replace valve body b) Check governor valve for free movement. Service or replace governor c) Perform line pressure cut-back point and governor pressure test
Vehicle does not shift from 2 to 3 in D	a) Dirty or sticking valve body b) Governor valve c) Front clutch d) Improper oil pressure control	a) Check 2-3 shift valve operation. Clean, service or replave valve body b) Check governor valve for free movement. Service or replace governor c) Check for proper applicaton and for a worn clutch d) Perform line pressure, cut-back point and governor pressure tests
Shifts from 1 to 3 in D	a) Improper fluid level b) Dirty or sticking valve body c) Governor valve d) Band servo e) Polished or glazed band or drum	a) Perform fluid level check b) Check 1-2 shift valve for free movement. Clean, service or replace valve body c) Check governor valve for free movment. Clean, service or replace governor valve d) Check for seal leakage e) Service or replace as required
Engine overspeeds on 2-3 shift	a) Improper fluid level b) Vacuum diaphragm and piping c) Governor valve d) Improper front clutch application e) Damaged or worn front clutch f) Improper oil pressure	a) Perform fluid level check b) Service or replace c) Check governor valve for free movement. Clean, service or replace governor valve d) Check front clutch operation. Check fluid pressure line and governor e) Service as required f) Perform line pressure and cut-back point tests

CHILTON'S THREE C's TRANSAXLE DIAGNOSIS
Jatco F3A Automatic Transaxle

Condition	Cause	Correction
Practically no shift shock or slippage while 1-2 shifting	a) Improper fluid level b) Dirty or sticking valve body c) Oil pressure control d) Vacuum diaphragm and piping e) Band servo f) Polished or glazed band or drum	a) Perform fluid level check b) Clean, service or replace valve body c) Perform fluid pressure check line and governor d) Perform vacuum diaphragm test. Service a required e) Check for leaking seal, service as required f) Service or replace as required
Shift points incorrect	a) Kickdown switch, kickdown solenoid and wiring b) Vacuum diaphragm and piping c) Damaged or worn governor d) Improper clutch or band application or oil pressure control e) Damaged vacuum diaphragm	a) Check for loose connection, continuity and proper operation b) Check for proper operation and clogged or disconnected mline c) Perform governor pressure test. Check for free movement of governor valve or dirty governor d) Perform line pressure and cut-back point test. Check clutches and bands for proper engagement e) Perform vacuum diaphragm check
No forced downshifts in D	a) Improper band application or oil pressure control b) Dirty or sticking valve body c) Dirty or sticking governor valve d) Vacuum diaphragm and piping e) Kickdown solenoid kickdown switch and wiring	a) Perform line pressure test. Service or adjust as required b) Check for free movment of all valves. Clean, service or replace valve body c) Check governor valve for free movement. Service or replace governor valve d) Perform vacuum diaphragm test. Check for plugged vacuum line. Service or replace as required e) Perform kickdown switch and circuit test. Service or replace as required
Does not shift from 3-2 on D to 2 shift	a) Dirty or sticking valve b) Oil pressure control system c) Band servo d) Damaged or worn band, glazed or polished drum	a) Clean, service or replace valve body b) Perform line and governor pressure tests c) Check for proper operation. Service or replace as required d) Service or replace as required
Does not shift from 3 to 2 on D to 1 shift	a) Dirty or sticking valve body b) Oil pressure control system c) Band servo d) Damaged or worn band, glazed or polished drum	a) Clean, service or replace as required b) Perform line and governor pressure tests c) Check for proper operation. Service or replace as required d) Service or replace as required
Kickdown operates or engine overruns when depressing pedal in 3 beyond kickdown vehicle speed limit	a) Vacuum diaphragm and piping b) Dirty or sticking valve body c) Improper front clutch application or oil pressure control	a) Check for sticking vacuum diaphragm and throttle valve b) Clean, service or replace as required c) Check front clutch for proper application. Perform line and governor pressure test

CHILTON'S THREE C's TRANSAXLE DIAGNOSIS
Jatco F3A Automatic Transaxle

Condition	Cause	Correction
Runaway engine on 3-2 downshift	a) Improper fluid level b) Improper band application or oil pressure system c) Band servo d) Polished or glazed band drum	a) Perform fluid level check b) Check band for proper application. Perform line pressure, cut-back point and governor pressure test c) Check for proper operation and seal leak d) Replace or service as required
No engine braking in 1	a) Improper fluid level b) Damaged or improperly adjusted manual c) Oil pressure control system d) Dirty or sticking valve body e) Low reverse brake	a) Perform fluid level check b) Perform linkage check linkage c) Perform line pressure test d) Clean, service or replace as required e) Service as required
Slow initial engagement	a) Improper fluid level b) Contaminated fluid c) Dirty or sticking valve body d) Improper clutch application or oil control pressure	a) Perform fluid level check b) Check fluid for proper condition. Check for clogged filter c) Clean, service or replace valve body d) Check rear clutch for proper application Perform line and governor pressure test
Harsh initial engagement in either forward or reverse	a) High engine idle b) Looseness in halfshafts, CV joints or engine mounts c) Vacuum diaphragm and piping d) Improper rear clutch application or oil pressure control e) Sticking or dirty valve body	a) Adjust idle to specifiaction b) Service as required c) Service as required d) Check rear clutch for proper operation. Perform line and governor pressure test e) Clean, service or replace as required
Harsh 1-2 shift	a) Weak engine performance b) Dirty or sticking body c) Vacuum diaphragm and piping d) Improper brake band application or oil pressure control	a) Tune and adjust engine to specification b) Check for free movement of 1-2 shift valve. Clean, service or replace valve body c) Perform vacuum diaphragm test. Service or replace as required d) Check band for proper operation. Perform line and governor pressure tests
Harsh 2-3 shift	a) Dirty or sticking valve body b) Improper front clutch application or oil pressure control c) Band servo d) Brake band	a) Clean, service or replace valve body b) Perform line pressure, cut-back point and governor pressure tests c) Check for proper release d) Check for proper release
Vehicle braked when shifted from 1-2	a) Dirty or sticking valve body b) Improper front clutch application or oil pressure control c) Low reverse brake d) One-way clutch	a) Clean, service or replace valve body b) Check front clutch for proper engagement. Perform line and governor pressure tests c) Check for proper disengagement or dragging clutch d) Check for seized clutch
Vehicle braked when shifted from 2-3	a) Dirty or sticking valve body b) Brake band and servo	a) Clean, service or replace valve body b) Check for proper disengagement
Noise severe under acceleration or deceleration. OK in P or N or speed	a) Speedo cable grounding out b) Shift cable grounding out c) Engine mounts bound up	a) Install and route cable as specified b) Install and route cable as specified c) Neutralize engine mounts

CHILTON'S THREE C's TRANSAXLE DIAGNOSIS
Jatco F3A Automatic Transaxle

Condition	Cause	Correction
Noise in P or N. Does not stop in Drive	a) Loose flywheel to converter bolts b) Pump c) Torque converter	a) Torque to specification b) Examine, service pump c) Examine, service converter. Perform stall test
Noise in all gears, changes power to coast	a) Final drive gearset noisy	a) Examine, service final drive gearset
Noise in all gears, does not change power to coast	a) Defective speedo gears b) Bearings worn or damaged	a) Examine, replace speed drive or driven gear b) Examine, replace
Noise in Low	a) Planetary gearset noisy	a) Service planetary gearset
Transaxle noisy in D, 2, 1 & R	a) Improper fluid level b) Improper fluid pressure control c) Rear clutch d) Oil pump e) One-way clutch f) Planetary gears	a) Perform fluid level check b) Perform line and governor pressure tests c) Check and repair as necessary d) Check, repair or replace e) Check, repair or replace as necessary f) Check or replace as necessary
Transaxle noisy, (valve noise) NOTE: Gauges may aggravate any hydraulic noises. Remove gauge and check for noise level	a) Improper fluid level b) Improper band or clutch application or oil pressure control system c) Cooler line grounding d) Dirty or sticking valve body e) Internal leakage or pump cavitation	a) Perform fluid level check b) Perform line pressure test c) Free cooler lines d) Clean, service or replace valve body e) Service or replace as required
Transaxle overheats	a) Improper fluid level b) Incorrect engine performance c) Improper clutch or band application or oil pressure control d) Restriction in cooler lines e) Dirty or sticking valve body f) Seized converter one-way clutch	a) Perform fluid level check b) Adjust according to specifications c) Perform line and pressure governor pressure tests d) Check cooler lines for kinks and damage. Clean, service or replace cooler lines e) Clean, service or replace valve body f) Replace converter

Diagnosis Tests

OIL PRESSURE CIRCUITS

In order to more fully understand the Jatco automatic transmission and to diagnose possible defects more easily, the clutch and band applications charts and a general description of the hydraulic control system is given.

To utilize the oil flow charts for diagnosing transaxle problems, the repairman must have an understanding of the oil pressure circuits and how each circuit affects the operation of the transaxle by the use of controlled oil pressure.

Control (line) pressure is a regulated main line pressure, developed by the operation of the front pump. It is directed to the main regulator valve, where predetermined spring pressure automatically moves the regulator valve to control the pressure of the oil at a predetermined rate, by opening the valve and exhausting excessive pressured oil back into the sump and holding the valve closed to build up pressure when needed.

Therefore, it is most important during the diagnosis phase to test main line control pressure to determine if high or low pressure exits. Do not attempt to adjust a pressure regulator valve spring to obtain more or less control pressure. Internal transaxle damage may result.

The main valve is the controlling agent of the transaxle which directs oil pressure to 1 of 6 separate passages used to control the valve train. By assigning each passage a number, a better understanding of the oil circuits can be gained from the diagnosis oil flow schematics.

CONTROL PRESSURE SYSTEM TEST

Control pressure tests should be performed whenever slippage, delay or harshness is felt in the shifting of the transaxle. Throttle and modulator pressure changes can cause these problems also, but are generated from the control pressures and therefore reflect any problems arising from the control pressure system.

The control pressure is initially checked in all ranges without any throttle pressure input and then checked as the throttle pressure is increased by lowering the vacuum supply to the vacuum modulator with the use of the stall test.

CLUTCH AND BAND APPLICATION CHART
Jatco F3A Automatic Transaxle

Range		Front Clutch ①	Rear Clutch ②	Low & Reverse Brake Clutch	Brake Band Servo ③ Operation	Brake Band Servo ③ Release	One-Way Clutch	Parking Pawl
Park		—	—	On	—	—	—	On
Reverse		On	—	On	—	On	—	—
Neutral		—	—	—	—	—	—	—
Drive	Low D1	—	On	—	—	—	On	—
	Second D2	—	On	—	On	—	—	—
	Top D3	On	On	—	(On)	On	—	—
2	Second	—	On	—	On	—	—	—
1	Second 1_2	—	On	—	On	—	—	—
	Low 1_1	—	On	On	—	—	—	—

① Reverse and high clutch
② Forward clutch
③ Intermediate band

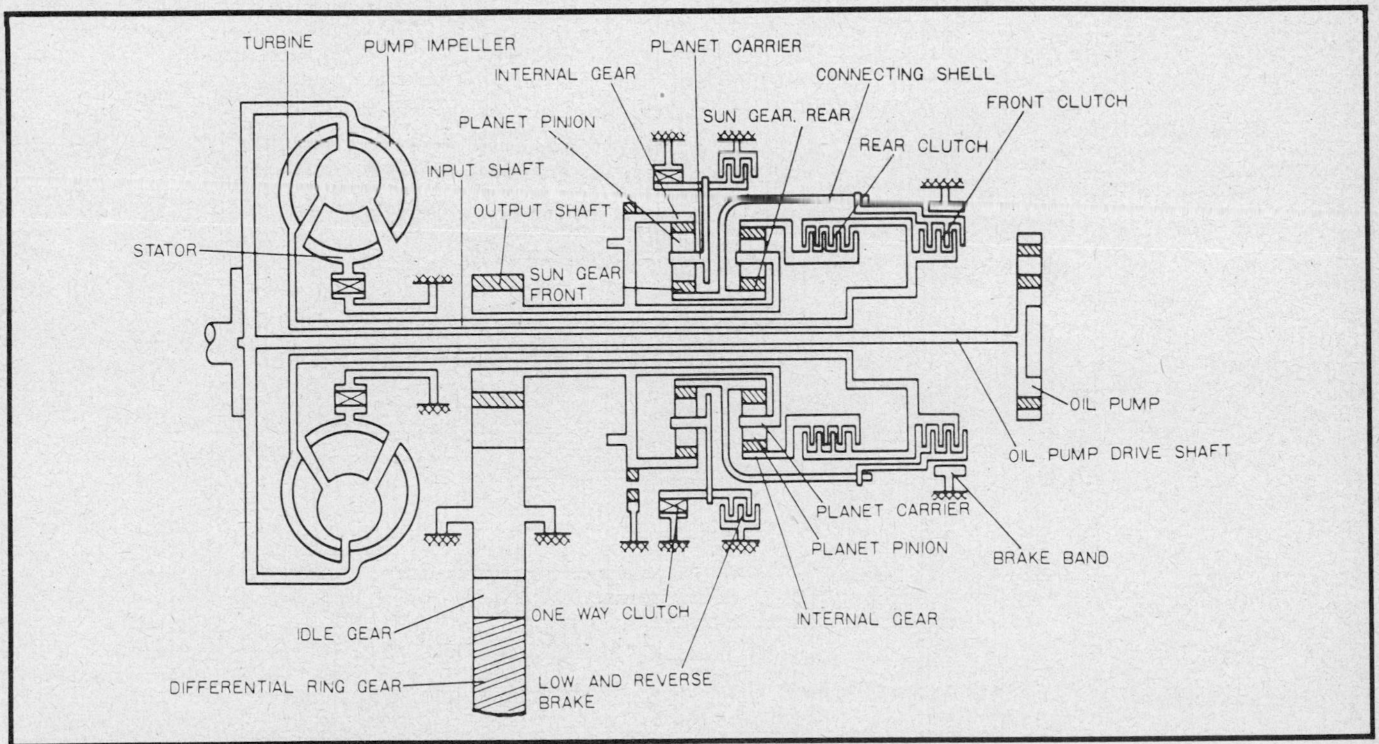
TURBINE
PUMP IMPELLER
PLANET CARRIER
INTERNAL GEAR
CONNECTING SHELL
FRONT CLUTCH
PLANET PINION
SUN GEAR, REAR
REAR CLUTCH
INPUT SHAFT
OUTPUT SHAFT
STATOR
SUN GEAR FRONT
OIL PUMP
OIL PUMP DRIVE SHAFT
PLANET CARRIER
PLANET PINION
BRAKE BAND
ONE WAY CLUTCH
INTERNAL GEAR
IDLE GEAR
DIFFERENTIAL RING GEAR
LOW AND REVERSE BRAKE

Power components of the F3A transaxle

The control pressure tests should define differences between mechanical or hydraulic failures of the transaxle.

Testing

1. Install a 0–400 psi pressure gauge to the main line control pressure tap. This may be marked ML on the side of the transaxle case.
2. Block wheels and apply both parking and service brakes.
3. Operate the engine/transaxle in the ranges on the following charts and at the manifold vacuum specified.
4. Record the actual pressure readings in each test and compare them to the given specifications.

Results

LOW PRESSURE AT IDLE IN ALL RANGES CAUSED BY:

1. EGR system, if equipped
2. Vacuum modulator
3. Manifold vacuum line
4. Throttle valve or control rod
5. Sticking regulator boost valve (pressure modifier valve)

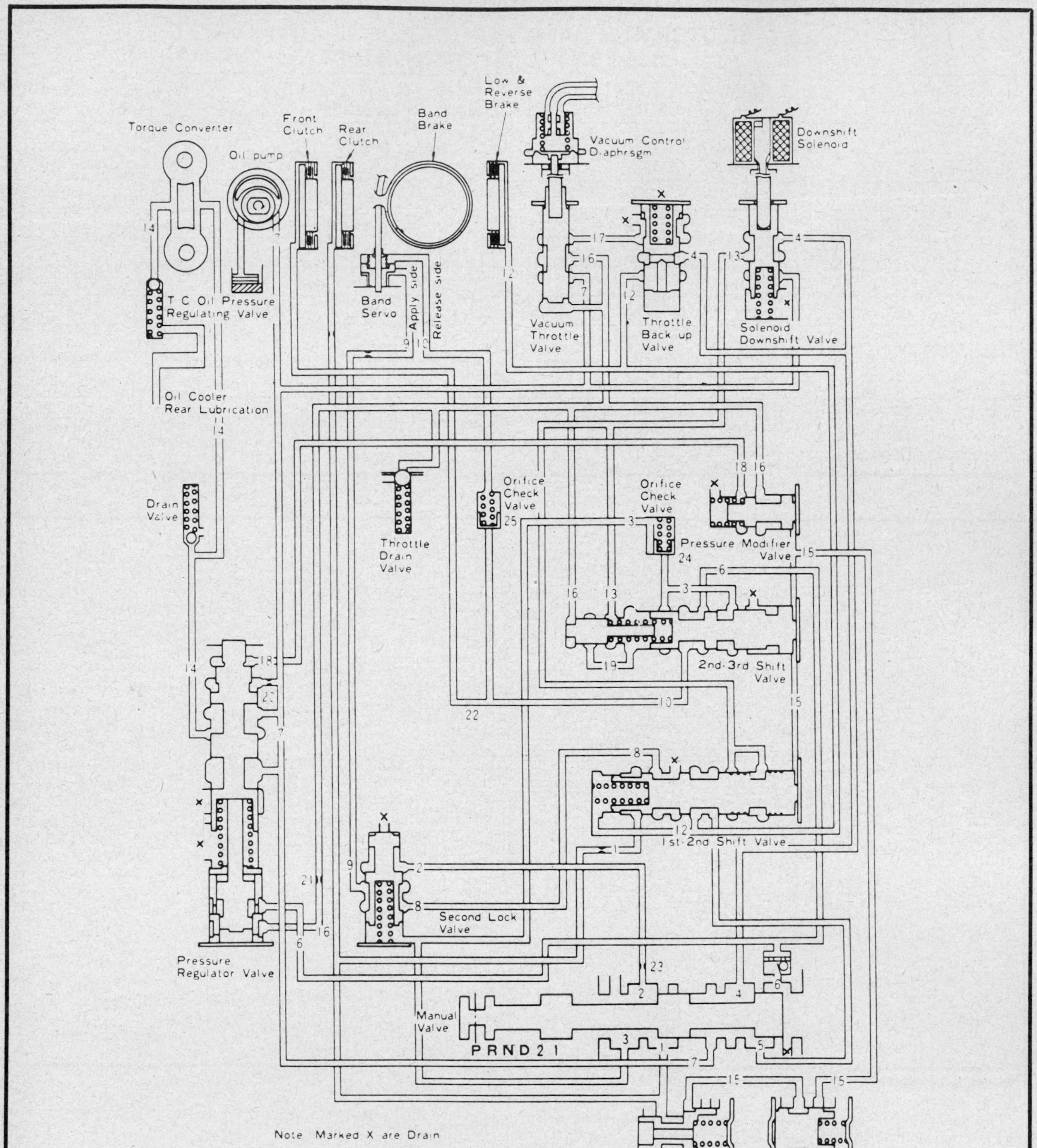
Torque Converter
Oil pump
Front Clutch
Rear Clutch
Band Brake
Low & Reverse Brake
Vacuum Control Diaphragm
Downshift Solenoid
Band Servo
Apply side
Release side
T C Oil Pressure Regulating Valve
Vacuum Throttle Valve
Throttle Back up Valve
Solenoid Downshift Valve
Oil Cooler Rear Lubrication
Drain Valve
Throttle Drain Valve
Orifice Check Valve
Orifice Check Valve
Pressure Modifier Valve
2nd-3rd Shift Valve
1st-2nd Shift Valve
Second Lock Valve
Pressure Regulator Valve
Manual Valve
P R N D 2 1
Note Marked X are Drain
Secondary Governor Valve
Primary Governor Valve

Hydraulic control schematic

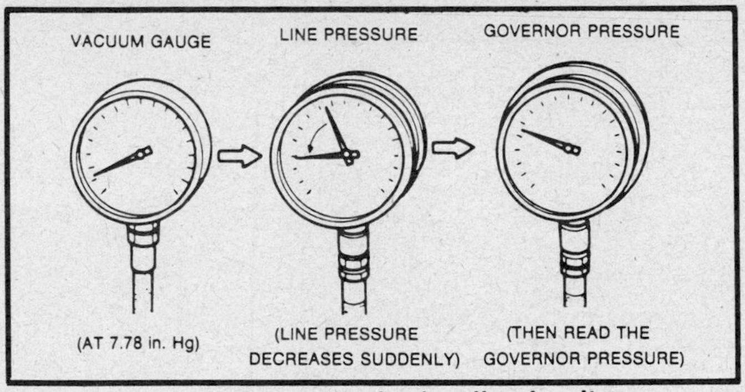
VACUUM GAUGE
LINE PRESSURE
GOVERNOR PRESSURE
(AT 7.78 in. Hg)
(LINE PRESSURE DECREASES SUDDENLY)
(THEN READ THE GOVERNOR PRESSURE)

Gauges needed to test the hydraulic circuits

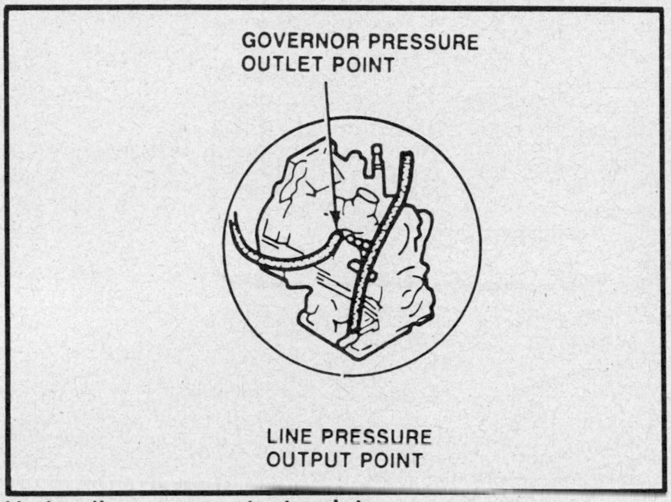
GOVERNOR PRESSURE OUTLET POINT
LINE PRESSURE OUTPUT POINT

Hydraulic pressure test points

OK AT IDLE IN ALL RANGES, BUT LOW AT 10 IN. OF VACUUM IS CAUSED BY:
1. Excessive leakage
2. Low pump capacity
3. Restricted oil pan screen or filter

PRESSURE LOW IN P RANGE IS CAUSED BY:
Valve body

PRESSURE LOW IN R IS CAUSED BY:
1. Front clutch
2. Low and reverse brake

PRESSURE LOW IN N RANGE IS CAUSED BY:
Valve body

PRESSURE LOW IN D RANGE IS CAUSED BY:
Rear clutch

PRESSURE LOW IN 2 RANGE IS CAUSED BY:
1. Rear clutch
2. Brake band servo

PRESSURE LOW IN 1 RANGE IS CAUSED BY:
1. Rear clutch
2. Low and reverse brake

HIGH OR LOW PRESSURE IN ALL TEST CONDITIONS IS CAUSED BY:
1. Modulator control rod broken or missing
2. Stuck throttle valve
3. Pressure modifier valve or regulator valve

POSSIBLE LOCATIONS OF PROBLEMS DUE TO LINE PRESSURE

Malfunctions

1. Low pressure when in **D**, **2**, or **R** positions could be the result of a worn oil pump, fluid leaking from the oil pump, control valve or transaxle case, or the pressure regulator valve sticking.
2. Low pressure when in **D** and **2** only could result from fluid leakage from the hydraulic circuit of the 2 ranges selected. Refer to the hydraulic fluid schematics.
3. Low fluid pressure when in the **R** position could result from a fluid leakage in the reverse fluid circuit. Refer to the hydraulic fluid schematic.
4. High pressure when idling could be the result of a broken or disconnected vacuum hose to the modulator or a defective vacuum modulator assembly.

Main Line Pressure Cut-Back Point Test

1. Connect the fluid pressure test gauge to the line pressure test port outlet of the transaxle case.
2. Connect a fluid pressure test gauge to the governor pressure test port on the transaxle case.
3. Position the gauges so that each can be seen from the driver's seat.
4. Disconnect the vacuum hose to the vacuum modulator and plug the hose.
5. Connect a vacuum pump to the vacuum modulator and position the pump so it can be operated from the driver's seat.
6. If the line pressure drops abruptly when the engine rpm is increased gradually while the selector lever is in the **D** position. Measure the governor pressure.
7. Measure the governor pressure when the vacuum is at 0 in. Hg. and at 7.9 in. Hg. The specifications are: 0.0 in. Hg—14–23 psi (98–157 Kpa) and 7.9 in. Hg (200mm–Hg)—6–14 psi (39–98 Kpa).
8. If the specifications are not met, check to see that the diaphragm rod has been installed or that it is more than standard. Check for a sticking valve inside the control valve assemble if the rod is correct.

Governor Pressure Test

1. Connect the fluid pressure gauge to the governor test port on the transaxle case. Position the gauge so that it is accessible to the operator.
2. Drive the vehicle with the selector lever in the **D** position.
3. Measure the governor pressure at the following speeds: The governor pressure should be 11.9–17.1 psi at 20 mph, 19.9–28.4 psi at 35 mph and 38.4–48.3 psi at 55 mph.
4. If the test results do not meet the specifications, the following should be checked:
   a. Fluid leakage from the line pressure hydraulic circuit.
   b. Fluid leakage from the governor pressure hydraulic circuit.
   c. Governor malfunctions.

AIR PRESSURE TEST

The control pressure test results and causes of abnormal pressure are to be used as a guide. Further testing or inspection could be necessary before repairs are made. If the pressures are found to be low in a clutch, servo or passageway, a verification can be accomplished by removing the valve body and performing an air pressure test. This test can be used to determine if a malfunction of a clutch or band is caused by fluid leakage in the system or is the result of a mechanical failure and also, to test the transaxle for internal fluid leakage during the rebuilding and before completing the assembly.
1. Obtain an air nozzle and adjust for 25 psi.
2. Apply air pressure (25 psi) to the passages.

Vacuum Modulator Test

The modulated throttle system, which adjusts throttle pressure for the control of the shift valves, is operated by engine manifold

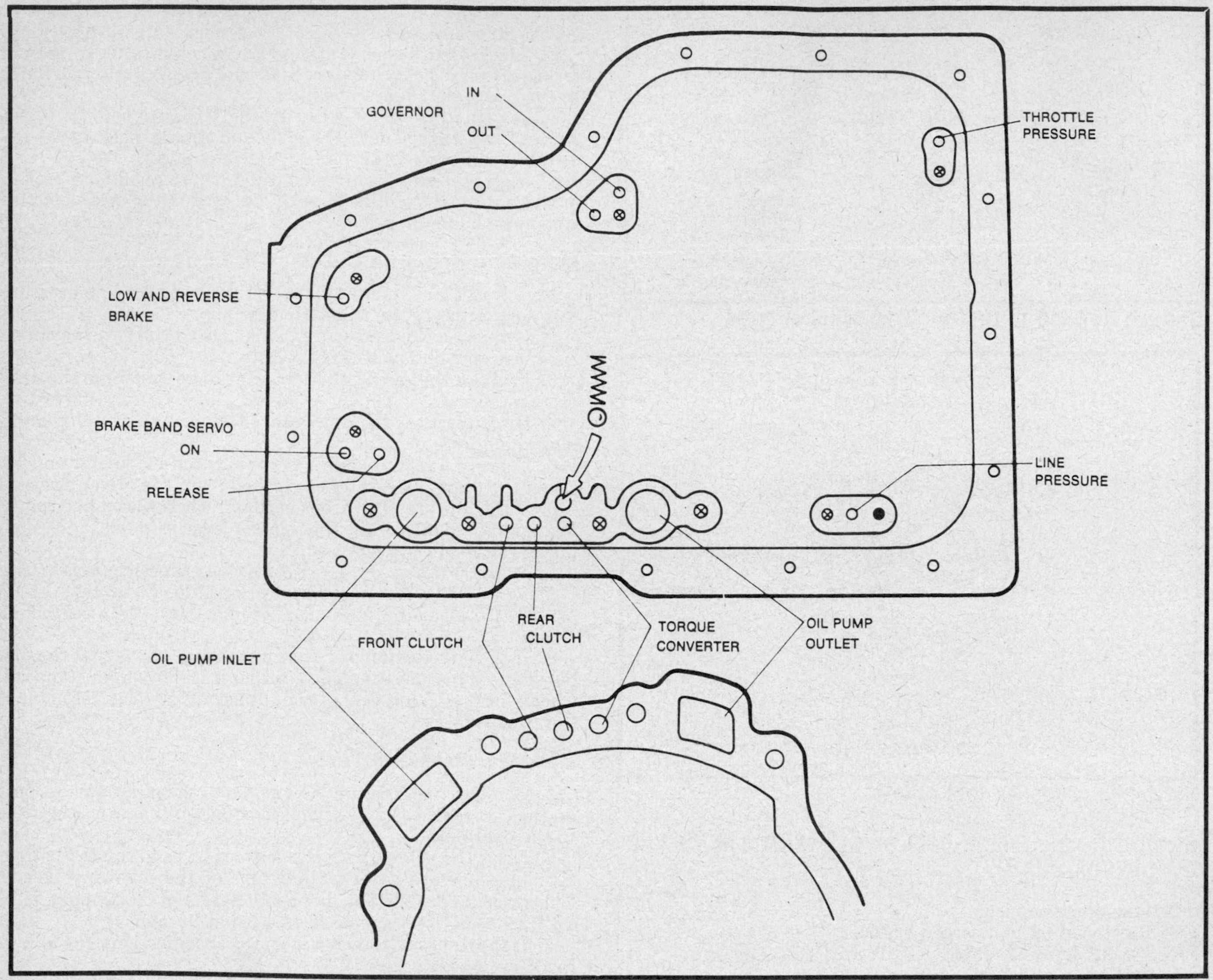
GOVERNOR
IN
OUT
THROTTLE
PRESSURE
LOW AND REVERSE
BRAKE
BRAKE BAND SERVO
ON
RELEASE
LINE
PRESSURE
OIL PUMP INLET
FRONT CLUTCH
REAR
CLUTCH
TORQUE
CONVERTER
OIL PUMP
OUTLET

Identification of the fluid passages in the transaxle case

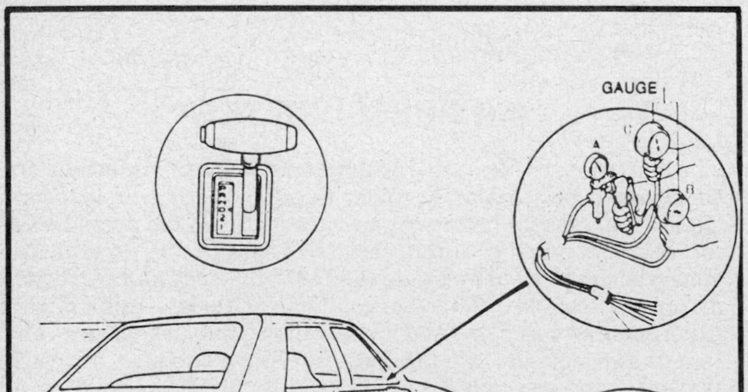
GAUGE

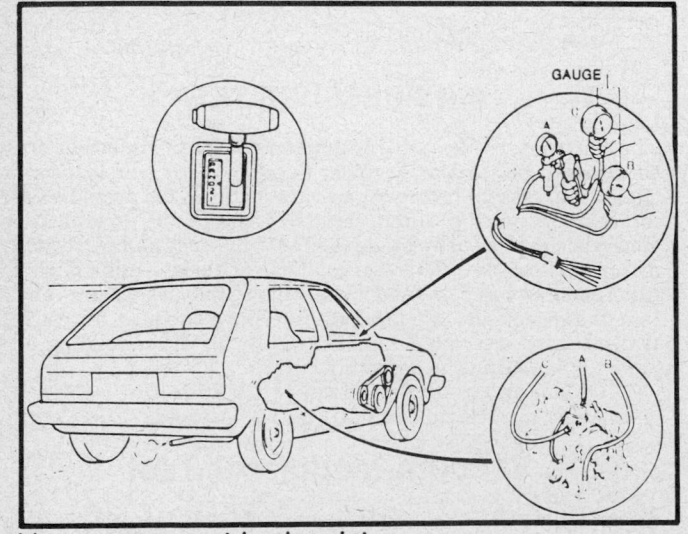

Line pressure cut back point

GAUGE

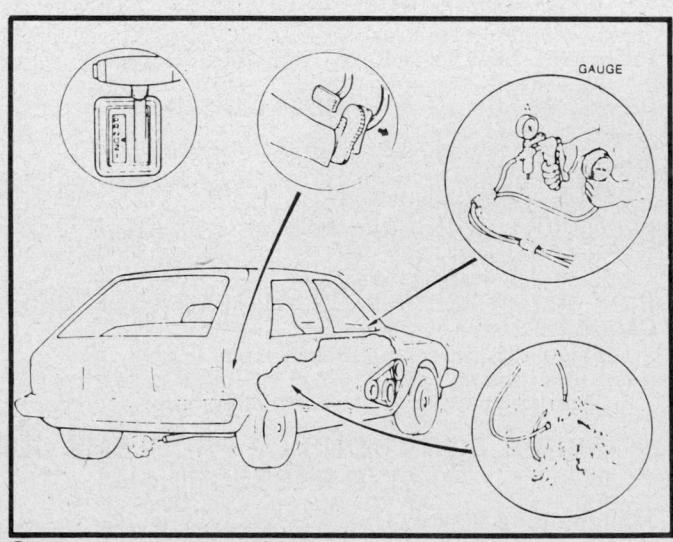

Governor pressure check

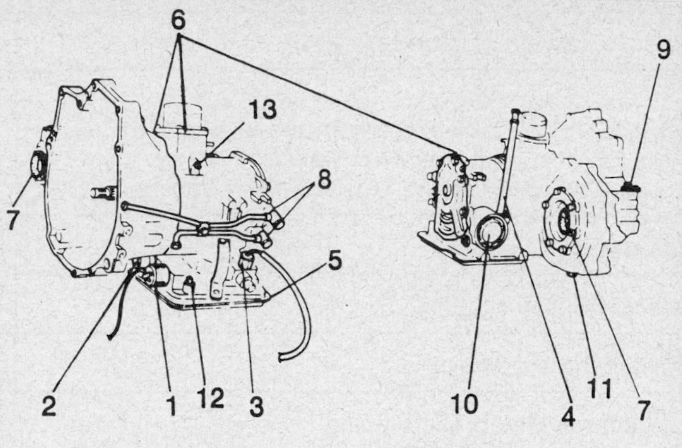
6
9
13
8
7
5
2
1
12
3
10
4
11
7

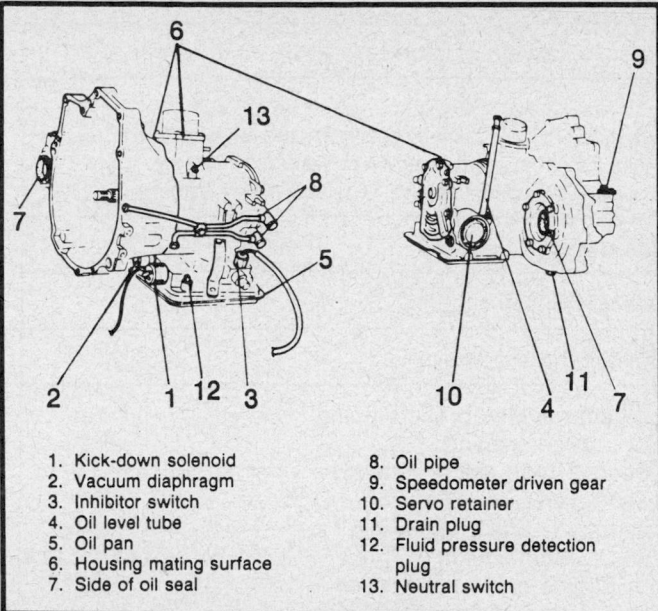

1. Kick-down solenoid
2. Vacuum diaphragm
3. Inhibitor switch
4. Oil level tube
5. Oil pan
6. Housing mating surface
7. Side of oil seal
8. Oil pipe
9. Speedometer driven gear
10. Servo retainer
11. Drain plug
12. Fluid pressure detection plug
13. Neutral switch

Possible fluid leakage locations

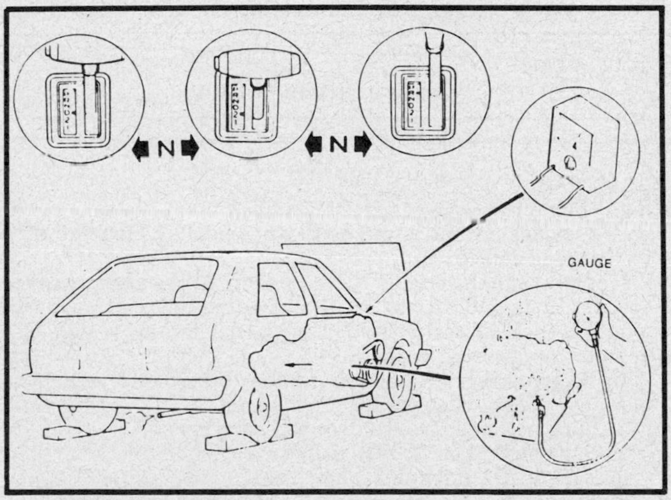
N
N
GAUGE

Line pressure test

vacuum through a vacuum diaphragm and must be inspected whenever a transaxle defect is apparent.

Before the vacuum modulator test is performed, check the engine vacuum supply and the condition and routing of the supply lines.

With the engine idling, remove the vacuum line at the vacuum modulator and install a vacuum gauge. There must be a steady, acceptable vacuum reading for the altitude at which the test is being performed.

If the vacuum is low, check for a vacuum leak or poor engine performance. If the vacuum is steady and acceptable, accelerate the engine sharply and observe the vacuum gauge reading. The vacuum should drop off rapidly at acceleration and return to the original reading immediately upon release of the accelerator.

If the vacuum reading does not change or changes slowly, check the vacuum supply lines for being plugged, restricted or connected to a vacuum reservoir supply. Repair the system as required.

MANIFOLD VACUUM CHECK

1. With the engine idling, remove the vacuum supply hose

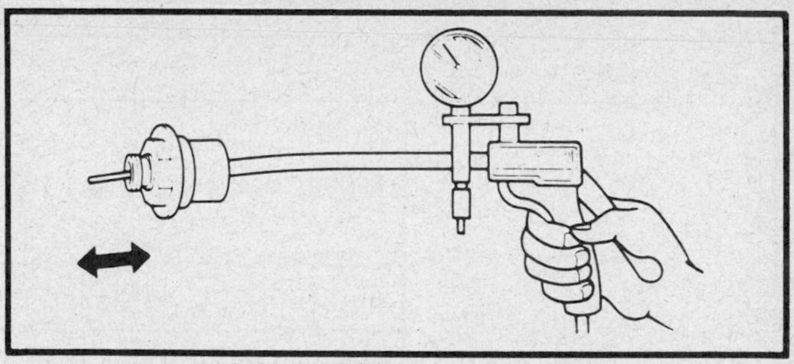

Testing the modulator with a hand vacuum pump

from the modulator nipple and check the hose end for the presence of engine vacuum with an appropriate gauge.

2. If vacuum is present, accelerate the engine and allow it to return to idle. A drop in vacuum should be noted during acceleration and a return to normal vacuum at idle.

3. If manifold vacuum is not present, check for breaks or restrictions in the vacuum lines and repair.

VACUUM MODULATOR CHECK

1. Apply at least 18 in. Hg. to the modulator vacuum nipple and observe the vacuum reading. The vacuum should hold.

2. If the vacuum does not hold, the diaphragm is leaking and the modulator assembly must be replaced.

NOTE: A leaking diaphragm causes harsh gear engagements and delayed or no up-shifts due to maximum throttle pressure developed.

On Vehicle Test

The vacuum modulator is tested on the vehicle with the aid of an outside vacuum source, which can be adjusted to maintain a certain amount of vacuum. Apply 18 inches Hg. to the vacuum modulator vacuum nipple, through a hose connected to the outside vacuum source. The vacuum should hold at the applied level without any leakdown. If the vacuum level drops off, the vacuum diaphragm is leaking and must be replaced.

Remove the vacuum hose and check transmission fluid in the hose. If the diaphragm has a leak, engine vacuum may draw transmission fluid through the hose and into the engine where it will be burned with the fuel.

Off Vehicle Test

With the vacuum modulator removed from the automatic transaxle, apply 18 in. Hg. to the modulator vacuum nipple.

The vacuum level should remain and not drop off. If the vacuum level drops, the diaphragm is leaking and the unit should be replaced.

Another test can be made with the modulator removed from the transaxle. Insert the control rod into the valve end of the diaphragm and apply vacuum to the nipple. Hold a finger over the control rod and release the vacuum supply hose. The control rod should be moved outward by the pressure of the internal return spring. If the control rod does not move outward, a broken return spring is indicated.

STALL TEST

The stall test is an application of engine torque, through the transaxle and drive train to lock up the rear wheels, which are held by the vehicle's brakes. The engine's speed is increased until the rpms are stabilized. Given ideal engine operating conditions and no slippage from transaxle clutches, bands or torque converter, the engine will stabilize at a specified test rpm.

Stall speed		Possible location of problem
Higher than standard	Higher at every position	Low line pressure a) Oil pump weak b) Oil leakage from the oil pump control valve body or transaxle case c) Pressure regulator valve sticking
	Higher in "D", "2", and "1"	Rear clutch slipping
	Higher only in "D"	One-way clutch slipping
	Higher only in "2"	Brake band slipping
	Higher only in "R"	Low and reverse brake slipping Front clutch slipping Perform road test to determine whether problem is low and reverse brake or front clutch a) Engine brake applied in 1st .....Front clutch b) Engine brake not applied in 1st .....Low and reverse brake
Within standard		Speed control elements in transaxle all normal
Lower then standard		Faulty engine One-way clutch in torque converter slipping

Stall speed trouble chart

Procedure

1. Check the engine oil level. Run the ngine until it reaches operating temperature.
2. Check the transaxle fluid level and correct as necessary. Attach a calibrated tachometer to the engine and a 0–400 psi oil pressure gauge to the transaxle control pressure tap on the right side of the case.
3. Mark the specified maximum engine rpm on the tachometer cover plate with a grease pencil to easily check if the stall speed is over or under specifications.
4. Apply the parking brake and block both front and rear wheels.

CAUTION

Do no allow anyone in front of the vehicle while performing the stall test. Secure vehicle with parking brake and blocking wheeling wheels or anchoring with chains.

5. While holding the brake pedal with the left foot, place the selector lever in **D** position and slowly depress the accelerator.
6. Read and record the engine rpm when the accelerator pedal is fully depressed and the engine rpm is stabilized. Read and record the oil pressure reading at the high engine rpm point. Stall speed – 2200–2450 rpm.

NOTE: The stall test must be made within 5 seconds.

7. Shift the selector lever into the **N** position and increase the engine speed to approximately 1000–1200 rpm. Hold this engine speed for 1–2 minutes to cool the transaxle and fluid.
8. Make similar tests in the **2**, **1** and **R** positions.

Results

HIGH ENGINE RPM

If a slipping condition occurs during the stall test, indicated by high engine rpm, the selector lever position at the time of slippage provides an indication as to what holding member of the transaxle is defective.

If at any time the engine rpm races above the maximum as per specifications, indications are that a clutch unit or band is slipping and the stall test should be stopped before more damage is done to the internal parts.

By determining the holding member involved, several possible causes of slippage can be diagnosed.

1. Slips in all ranges, control pressure low
2. Slips in **D**, **1** or **2**, rear clutch
3. Slips in **D1** only, one-way clutch
4. Slips in **R** only, front clutch or low and reverse brake

Perform a road test to confirm these conditions.

LOW ENGINE RPM

When low stall speed is indicated, the converter one-way clutch is not holding or the engine is in need of a major tune-up. To determine which is at fault, perform a road test and observe the operation of the transaxle and the engine. If the converter one-way clutch does not lock the stator, acceleration will be poor up to approximately 30 mph. Above mph the acceleration will be normal. With poor engine performance, acceleration will be poor at all speeds. When the one-way clutch is seized and locks the stator from turning either way, the stall test rpm will be normal. However, on a road test the vehicle will not go any faster than 50–55 mph because of the 2:1 reduction ratio in the converter.

If slippage was indicated by high engine rpm, the road test will help identify the problem area observing the transaxle operation during upshifts, both automatic and manual.

Road Test

The road test is used to confirm that malfunctions do exist within the transaxle unit, or that repairs have been accomplished

STALL TEST HOLDING MEMBER CHART

Selector Lever Position	Holding Member Applied
"D" 1st Gear	Rear clutch One-way clutch
"1" Manual	Rear clutch Low and reverse brake clutch
"2" Manual	Rear clutch Rear band
Reverse	Front clutch Low and reverse brake clutch

Line Pressure At Stall Speed

"D" Range	128 to 156 psi
"2" Range	114 to 171 psi
"R" Range	228 to 270 psi

Line Pressure Before Stall Test—At Idle

"D" Range	43 to 57 psi
"2" Range	114 to 171 psi
"R" Range	57 to 110 psi

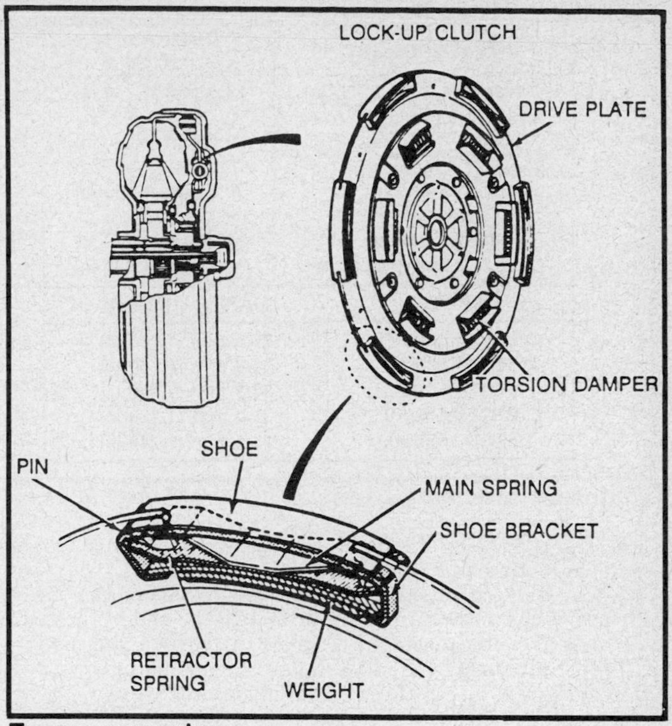
LOCK-UP CLUTCH
DRIVE PLATE
TORSION DAMPER
PIN
SHOE
MAIN SPRING
SHOE BRACKET
RETRACTOR SPRING
WEIGHT
Torque converter

and the transaxle unit is either operating properly or will require additional adjustments or repairs. The road test must be performed over a pre-determined drive course that has been used before to evaluate transaxle and/or transaxle operations.

Should malfunctions occur during the road test, the selector range and road speed should be noted, along with the particular gear and shift point. By applying the point of malfunction in the operation of the transaxle, to the Clutch and Band Application Chart and the Chilton's Three "C's"; diagnosis chart, the probable causes can be pinpointed.

Some of the points to be evaluated during the road test are as follows:

1. The shift point should be smooth and have a positive engagement.
2. The shifts speed are within specifications.
3. All shifts occur during the upshifts and downshifts when in the selector lever detents, as required.
4. All downshifts occur when a forced downshift is demanded.
5. No upshift to 3rd when the selector lever is in the **2** position and the transaxle is in the 2nd speed.
6. Only 1 upshift from the 1st speed when the selector lever is in the **1** position.
7. The vehicle is firmly locked when the lever is in the **P** position.

Converter Clutch Operation and Diagnosis

TORQUE CONVERTER CLUTCH

A lockup torque converter is utilized. This eliminates the slip which is inherent in conventional torque converters. The no slip characteristics are achieved by automatically locking the converter into direct mechanical drive at high engine speed.

The lockup converter consists of a lockup drive plate containing centrifugally operated shoe, bracket and spring assemblies and a one-way clutch.

The lockup drive plate assembly is attached to the splined turbine shaft and is located inside the torque converter between the turbine and the converter cover. Torque is transmitted when the centrifugal clutch linings contact the machined inner surface of the converter housing. Torsion dampers are provided in the drive plate to absorb shock when the clutch is engaged.

NOTE: Whenever a transaxle has been disassembled, the converter and oil cooler must be cleaned.

ON CAR SERVICES

Adjustments

VACUUM MODULATOR

The vacuum modulator has no adjustments other than the replacement of the diaphragm rod. The rods are available in varied lengths as follows.

1. Raise and support the vehicle safely. Remove the vacuum modulator from its mounting.
2. Insert the vacuum diaphragm rod gauge into the mounting hole, with the beveled side out, until the tool bottoms.
3. Place the rod through the opening of the vacuum diaphragm tool until the rod bottoms out against the valve.

MEASUREMENT	DIAPHRAGM ROD USED
Under 25.4 mm (1.000 in)	29.5 mm (1.160 in)
25.4 ~ 25.9 mm (1.000 ~ 1.020 in)	30.0 mm (1.180 in)
25.9 ~ 26.4 mm (1.020 ~ 1.039 in)	30.5 mm (1.200 in)
26.4 ~ 26.9 mm (1.039 ~ 1.059 in)	31.0 mm (1.220 in)
Over 26.9 mm (1.059 in)	31.5 mm (1.240 in)

PART NO.	DIAPHRAGM ROD
E7GZ-7A380-E	29.5 mm (1.160 in)
E7GZ-7A380-C	30.0 mm (1.180 in)
E7GZ-7A380-D	30.5 mm (1.200 in)
E7GZ-7A380-B	31.0 mm (1.220 in)
E7GZ-7A380-A	31.5 mm (1.240 in)

Modulator rod chart

4. Tighten the lock knob on the vacuum modulator tool and remove the tool and rod from the transaxle case.
5. Use a depth gauge to measure the distance from the flat surface of the vacuum modulator tool to the end of the rod.
6. Use this measurement to select the correct size rod.
7. Install the correct rod, lubricate the modulator O-ring with Dexron®II and install the vacuum modulator.

NOTE: The transaxle will have to be partially drained before the vacuum modulator is removed. Add the necessary fluid and correct the level as required.

KICKDOWN SWITCH

1. Move the ignition switch to the **ON** position.
2. Loosen the kickdown switch to engage when the accelerator pedal is between 7/8–15/16 in. of full travel. The downshift solenoid will click when the switch engages.
3. Tighten the attaching nut and check for proper operation.

NEUTRAL SAFETY SWITCH

No adjustment is possible on the neutral safety switch. If the engine will not start while the selector lever is in the **P** or **N** positions and the back-up lamps do not operate, check the shift control cable for proper adjustment. If shift cable adjustment is correct, the switch is defective and must be replaced.

Services

MANUAL SHIFT LINKAGE

REMOVAL AND INSTALLATION

1. Position the gear selector lever in the **N** position.
2. Remove the spring clip and pin attaching the shift cable trunnion to the transaxle shift lever.
3. Rotate the transaxle shift lever fully counterclockwise. This is the park position.
4. Rotate the transaxle shift lever clodkwise 2 detents. This is the neutral position. As the lever is rotated, position it between the ends of the shift cable trunnion.
5. If the hole in the shift lever aligns with the holes in the trunnion, the cable is properly adjusted. If the holes do not align proceed to the next step.
6. Remove the shift quadrant bezel. Lift the front of the bezel to disengage it from the console.
7. Lift and rotate the quadrant to provide access.
8. Loosen the adjuster nuts on the shift cable.

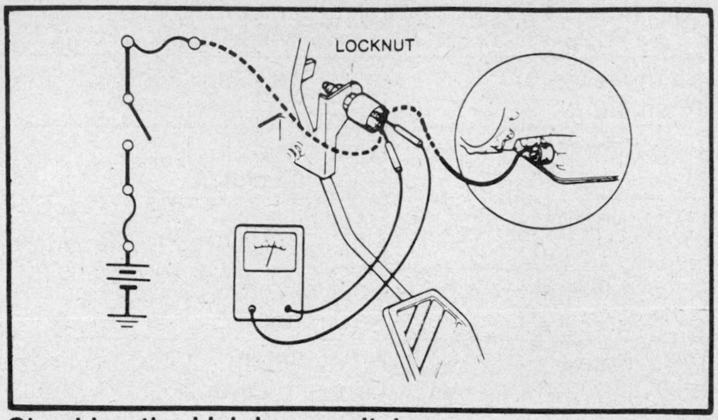
LOCKNUT
Checking the kickdown switch

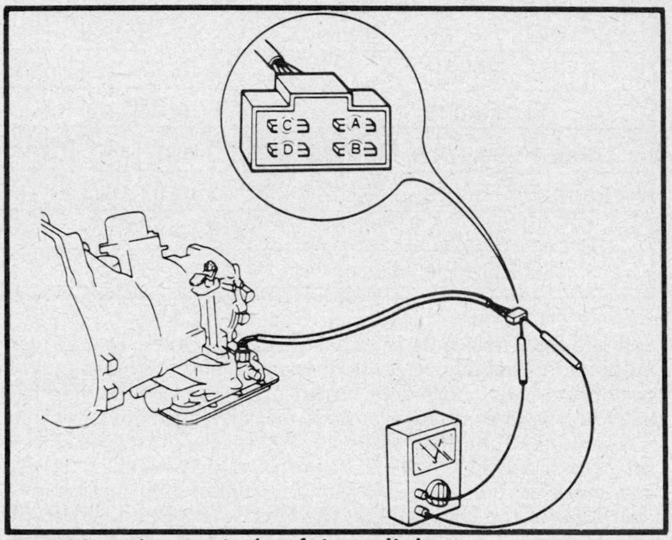
C
A
D
B
Checking the neutral safety switch

9. Position the gear selector lever in **P** position and inspect the position of the detent spring roller.
10. Loosen the attaching screws and move the detent spring forward or backward to center it in the detent.
11. Position the quadrant and install the attaching screws.
12. Position the selector lever in the **N** position.
13. Screw the adjuster nuts up or down the cable until the holes in the transaxle shift lever and the shift cable trunnion are aligned.
14. Torque the adjuster nut to 69–95 inch lbs. (8–11 Nm).
15. Check the alignment of the holes to make sure alignment was not disturbed.
16. Install the transaxle shift lever to shift cable attaching pin and retainer clip.
17. With an assistant, note the amount of freeplay when moving the shifter from **N** to **D** and compare to amount of freeplay between **N** and **R**. Adjust as necessary for equal amount of play in shifter and torque adjuster nut to 69–95 inch lbs. (8–11 Nm).

CAUTION

Make sure the linkage adjustment has not affected operation of the neutral safety switch. With the parking brake and service brake applied, try to start the engine in each gearshift position.

18. Position the shift quadrant bezel and install the attaching screws.

FLUID CHANGES

The Jatco transaxles do not have a specific or periodic fluid

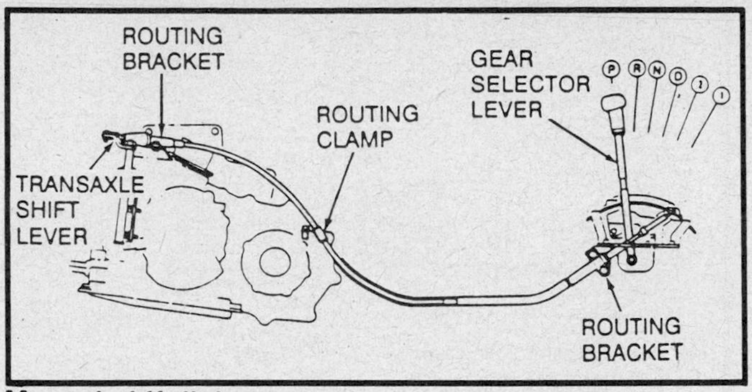
ROUTING
BRACKET
GEAR
SELECTOR
LEVER
ROUTING
CLAMP
TRANSAXLE
SHIFT
LEVER
ROUTING
BRACKET
Manual shift linkage

change interval for the normal maintenance of the units. However, at the time of any major repairs or when the fluid has been contaminated, the converter, cooler and lines must be flushed to remove any debris and contaminated fluid. If the vehicle is used in continuous service or driven under severe conditions (police or taxi type operations), the transaxle should be drained, flushed and refilled at mileage intervals of 18,000–24,000 or at time intervals of 18–24 months.

NOTE: The time or mileage intervals given are average. Each vehicle operated under severe conditions should be treated individually.

1. Raise and support vehicle safely.
2. Remove the undercover and side cover to gain access to the transaxle pan and drain plug.
3. Remove the drain plug at the bottom of the transaxle case.
4. Allow the fluid to drain completely and reinstall the drain plug. Torque drain plug to 29–40 ft. lbs. (39–54 Nm).
5. Add Dexron®II type fluid to the transaxle until the desired level is reached. Approximately 3 quarts if the transaxle, not including the torque converter, was drained.

VACUUM MODULATOR

Removal and Installation

NOTE: Drain the transaxle before removing the vacuum modulator.

1. Raise the vehicle and support safely. Disconnect the vacuum hose from the modulator unit.
2. Turn the threaded modulator unit to remove it from the transaxle case.
3. Pull the actuating pin and the throttle valve from the transaxle case.
4. Remove the O-ring from the assembly.
5. Install a new O-ring on the modulator unit.
6. Install the throttle valve, the actuating pin and the vacuum modulator tubes toward the transaxle case and install the assembly into the case.
7. Tighten the vacuum modulator unit securely.

OIL PAN

Removal and Installation

1. Raise and support vehicle safely.
2. Remove the undercover and side cover to gain access to the transaxle pan and drain plug.
3. Remove the drain plug at the bottom of the transaxle case.
4. Allow the fluid to drain completely.
5. Remove the pan from the transaxle case.
6. Thoroughly clean the oil pan and filter screen.
7. Install a new washer on the drain plug and torque to 29–40 ft. lbs. (39–54 Nm).

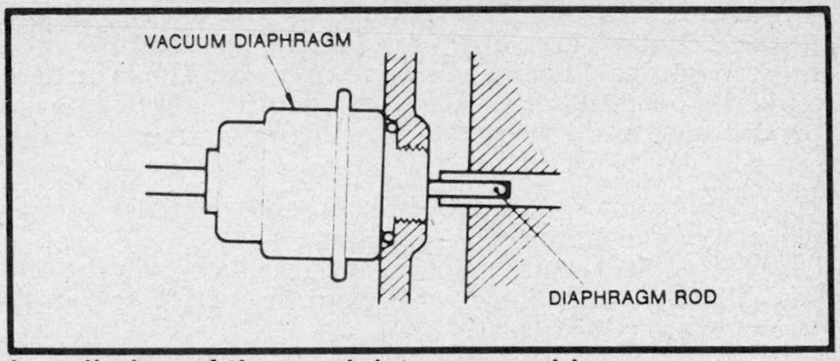
VACUUM DIAPHRAGM
DIAPHRAGM ROD
Installation of the modulator assembly

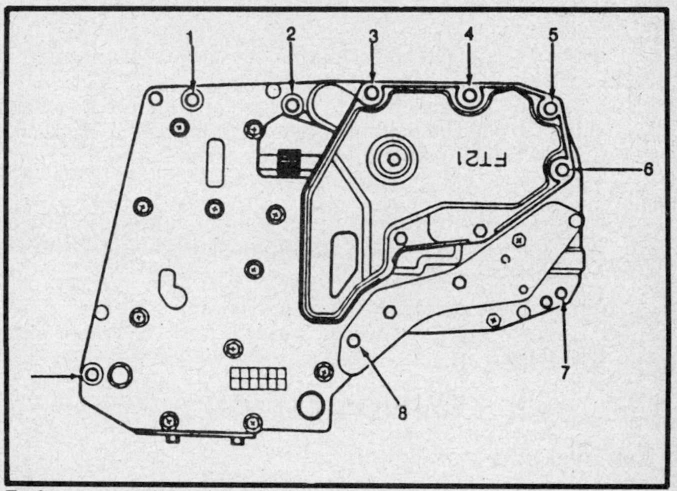
1
2
3
4
5
6
7
8
FT21
Bolt removal and torque sequence

8. Install a new pan gasket and replace pan. Torque pan bolts to 4–6 ft. lbs. (5–8 Nm).

NOTE: Do not overtighten bolts. Do not use any type of gasket sealer, RTV, etc., on the transaxle pan gasket. If necessary, soak the gasket in clean Dexron®II automatic transaxle fluid.

9. Install the undercover and side cover.
10. Remove the dipstick and add 3 quarts of Dexron®II transmission fluid.
11. Start and run engine until normal operating temperature is reached. Apply service brake and move selector through all the shift positions.

NOTE: Do no overspeed the engine during warm-up.

12. Place shift back in **P** and add fluid as necessary so reading is between the **F** and **L** marks on dipstick.

NOTE: Make certain fluid is just below the F mark. Do not overfill.

VALVE BODY

Removal and Installation

1. Disconnect the negative battery cable.
2. Raise and safely support the vehicle.
3. Remove the undercover and side cover.
4. Drain the transaxle fluid.
5. Remove the pan attaching bolts, pan and gasket.
6. Remove the valve body-to-case attaching bolts. Hold the manual valve to keep it from sliding out of the valve body and remove the valve body from the case.

NOTE: Failure to hold the manual valve while removing the control assembly could cause the manual valve to be dropped, causing the valve to become bent or damaged. Be careful not to loose the vacuum diaphragm rod or the ball and spring for the torque converter relief valve.

7. Thoroughly clean and remove all gasket material from the pan and pan mounting face of the case.
8. Install the vacuum diaphragm rod to its hole in the case. Install the check ball and spring into the slotted hole in the transaxle case.

NOTE: The ball is inserted first and then the spring. Use petroleum jelly to retain the spring and ball, if necessary.

9. Install the valve body, mating the groove of the manual valve with the driving pin of the shift rod.
10. Position the valve body to the case and install the attaching bolts. Torque the bolts to 70–95 inch lbs. (8–11 Nm).
11. Install the oil pan and torque the bolts to 43–69 inch lbs. (5–8 Nm).

NOTE: Do not use any type of gasket sealer or RTV on the pan gasket. If necessary, soak the gasket in clean transaxle fluid.

12. Install the undercover and side cover.
13. Lower vehicle. Refill the transaxle with the proper grade and type fluid.

SERVO ASSEMBLY

Removal and Installation

1. Raise and support the vehicle safely. Remove pan and valve body.
2. Remove left front wheel.
3. Remove the left lower ball joint bolt and separate the lower arm from the knuckle.
4. Separate the left drive shaft from the transaxle by prying with a bar inserted between the shaft and the case.

NOTE: A notch is provided in the side bearing housing to accommodate the bar. Do not insert the bar too far or damage to the lip of the oil seal may occur.

5. Support the halfshaft with a wire.
6. Loosen the anchor end-bolt and nut.
7. Remove the band strut.
8. Use a C-clamp and socket to compress the servo piston into the transaxle case.

CAUTION

Eye protection should be worn during servo removal.

9. Remove the servo snapring.
10. Remove the servo retainer, piston and spring by slowly loosening the C-clamp.
11. Lubricate the piston and spring with Dexron®II transmission fluid.
12. Replace the return spring. Replace the O-ring piston seal.
13. Use a C-clamp and socket to compress the assembly.
14. Install the snapring to the snapring groove.
15. Install the band strut to the band.
16. Install the anchor end-bolt to the band and torque to 8.7–10.8 ft. lbs. (12–15 Nm).
17. Back off the end-bolt 2 complete turns on carburetor equipped vehicles and 3 complete turns on EFI equipped vehicles.
18. Install the anchor end-bolt locknut and torque to 41–59 ft. lbs. (55–80 Nm).
19. Replace the clip at the end of the halfshaft with a new clip and install shaft with the clip gap at the top of the groove.

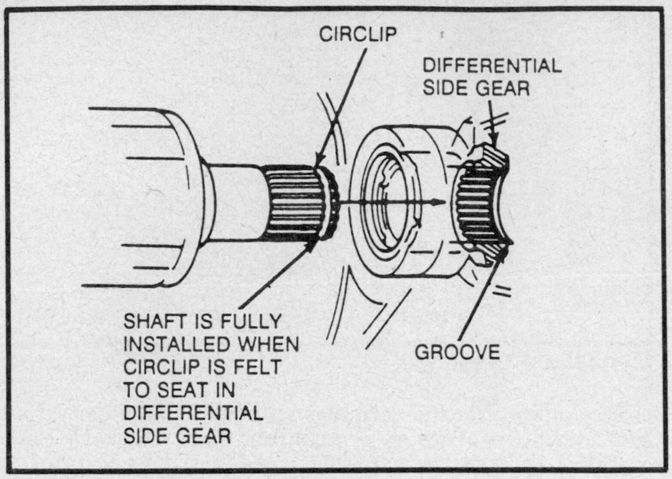
CIRCLIP
DIFFERENTIAL SIDE GEAR
SHAFT IS FULLY INSTALLED WHEN CIRCLIP IS FELT TO SEAT IN DIFFERENTIAL SIDE GEAR
GROOVE

Circlip location and installation

NOTE: Do not reuse the old clip. A new clip must be installed.

20. Slide the halfshaft horizontally into the transaxle differential, supporting it at the CV joint to prevent damage to the oil seal lip. Apply even pressure to the hub until the circlips are heard to engage.

NOTE: After installation, pull both front hubs outward to confirm that the drive shafts are retained by the circlips.

21. Install the lower arm ball joint to knuckle and torque nut to 32–40 ft. lbs. (43–54 Nm).
22. Install the underside covers.
23. Install the front wheel assembly and torque lugnuts to 65–87 ft. lbs. (90–120 Nm).
24. Refill the transaxle with the proper grade and type fluid.

GOVERNOR

Removal and Installation

1. Note the position of the governor, remove the 3 retaining bolts from the governor cover assembly. Lift the governor assembly from the transaxle case.
2. Remove the 2 governor retaining screws from the governor sleeve. Remove the governor valve body.
3. Disassemble the governor valve body as required.
4. Reassemble the governor valve body.
5. Install the governor valve body to the governor sleeve.
6. Mount the governor to the transaxle case in position noted during removal.
4. Install the 3 cover/governor retaining bolts and tighten to 69–95 inch lbs. (7.8–10.8 Nm).

DIFFERENTIAL OIL SEALS

The left and right axle seals can be installed with the axles removed. Conventional seal removing and installing tools can be used. Care must be exercised to prevent damage to the seals as the axles are reinstalled into the transaxle case.

Removal and Installation

1. Support vehicle safely, remove wheel assembly and underbody covers.
2. Remove the stabilizer bar to control arm bolts, washers and bushings.
3. Remove the lower ball joint bolt and separate the lower arm from the knuckle.

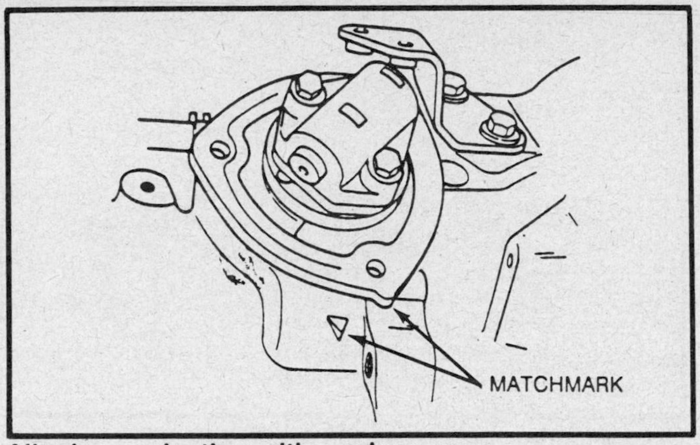
MATCHMARK

Aligning projection with mark on case

4. Partially drain the transaxle.
5. Separate the halfshaft from the transaxle by prying with a bar inserted between the shaft and the case. Tap the bar lightly to help loosen it from the differential gear.

NOTE: A notch is provided in the side bearing housing to accommodate the bar. Do not insert the bar too far or damage to the lip of the oil seal may occur.

6. Pull the halfshaft from the transaxle and support the it with a wire.
7. Pry the seal from the transaxle case using an appropriate tool.
8. Lubricate the new seal with transmission fluid and install using an appropriate tool.
9. Replace the clip at the end of the halfshaft with a new clip and install shaft with the clip gap at the top of the groove.

NOTE: Do not reuse the old clip. A new clip must be installed.

10. Slide the halfshaft horizontally into the transaxle differential, supporting it at the CV-joint to prevent damage to the oil seal lip. Apply even pressure to the hub until the circlips are heard to engage.

NOTE: After installation, pull both front hubs outward to confirm that the driveshafts are retained by the circlips.

11. Install the lower arm ball joint to knuckle and torque nut to 32–40 ft. lbs. (43–54 Nm).
12. Install the underside covers.
13. Install the front wheel assembly and torque lugnuts to 65–87 ft. lbs. (90–120 Nm).
14. Check and add Dexron®II transmission fluid as needed.

REMOVAL AND INSTALLATION

TRANSAXLE REMOVAL

The transaxle may be removed with the engine in place. The engine must be support from above, by using an engine support bar across the fenders. The procedures listed are for the Tracer. The technician should be aware that some of the procedures may vary slightly between the Tracer and Festiva, but the basic removal steps are the same.

1. Disconnect the negative cable from the battery.
2. Remove the air cleaner.
3. Loosen the front wheel lug nuts.
4. Remove the speedometer cable.
5. Disconnect the shift control cable from the transaxle by removing the clip and the bracket bolts.
6. Remove the engine ground wire from the cylinder head.
7. Remove the water pipe bracket.
8. Remove the secondary air pipe and EGR pipe bracket.
9. Remove the wire harness clip.
10. Disconnect the wiring to the inhibitor switch and kickdown solenoid.
11. Disconnect the body ground connector.
12. Remove the upper transaxle mounting bolts.
13. Disconnect the neutral switch connector at the transaxle.
14. Remove the vacuum hose from the vacuum modulator.
15. Remove and plug the cooler lines.
16. Mount the engine support bar across fenders and support the engine.
17. Raise and support the vehicle safely.
18. Remove the underbody covers.
19. Remove the tire assemblies.
20. Remove the stabilizer bar.
21. Remove the lower control arm ball joint clamp bolt and nut. Separate the ball joint from the steering knuckle.
22. Drain the transaxle.
23. Insert a pry bar between the transaxle case and halfshaft.

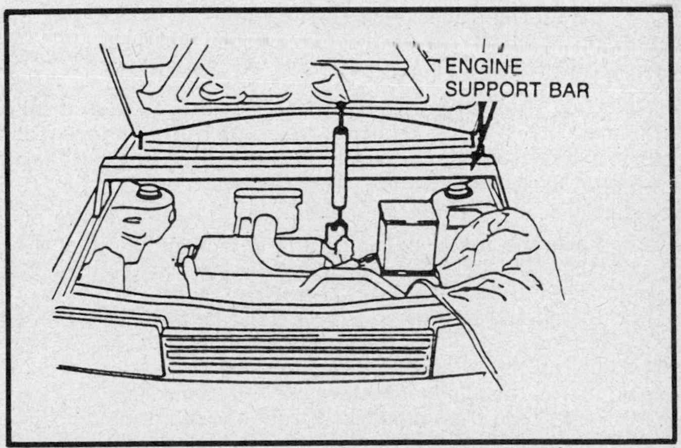
ENGINE
SUPPORT BAR

Supporting engine from above with support bar

Lightly tap on the end of bar until the halfshaft loosens from differential.

24. Pull the halfshaft from the transaxle and suspend under vehicle with wire.
25. Remove the crossmember to frame bolts.
26. Remove the nut attaching the positive cable to the starter and remove cable.
27. Remove the wire attach to the solenoid by grasping wire and pulling straight out.
28. Remove the starter.
29. Remove the bolts attaching the endplate to transaxle.
30. Remove the bolts attaching the torque converter to the flex plate.
31. Lean the engine toward the transaxle by loosening the engine support hook bolt.

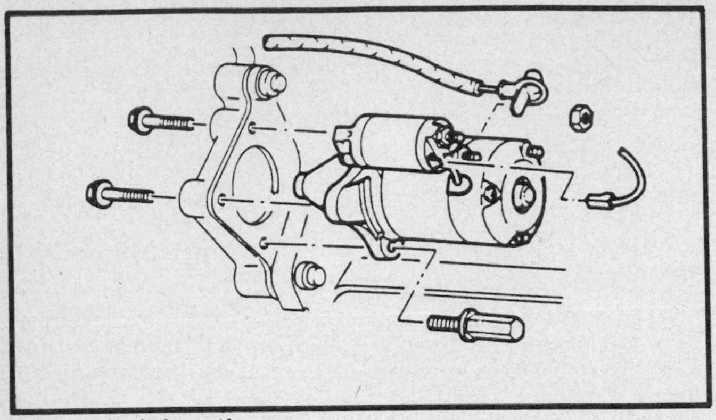

Starter bolt locations

32. Support the transaxle with a floor jack and block of wood.
33. Remove the nut and bolts that retain the engine mount bolts to the transaxle.
34. Remove the remaining transaxle mounting bolts and remove the transaxle, being careful not to allow the torque converter to separate from the transaxle.

TRANSAXLE INSTALLATION

1. Install the torque converter on the transaxle input shaft.

NOTE: Before installing the torque converter, pour a ½ qt. of Dexron®II transmission fluid into the torque converter.

2. To insure the converter is properly installed, measure the distance from converter housing to transaxle housing end. This measurement should be 0.790 in. (20mm).
3. Install the transaxle to the engine.
4. Install all but the top 2 retaining bolts and torque to 27–36 ft. lbs.(34–49Nm).

NOTE: If at all possible, it is advisable to install all of the retaining bolts, including the top bolts, before tightening. Tightening them all at the same time will avoid uneven stress on the transaxle case.

5. Raise the transaxle, using a floor jack and block of wood.
6. Align and install the torque converter to flex plate bolts and torque to 27–36 ft. lbs. (34–49Nm).
7. Install the starter and torque the bolts to 23–34 ft. lbs. (31–46 Nm).
8. Install the solenoid wire.
9. Install the positive battery cable to the solenoid.
10. Install the crossmember. Torque the motor mount nut to 21–34 ft. lbs. (28–46 Nm) and torque all other mounting bolts to 47–66 ft. lbs. (64–89 Nm).
11. Install a new clip on both halfshafts, making certain the gap in the clip is at the top of the groove when installing the shafts.

NOTE: Do not reuse the old clips. New clips must be installed.

12. Slide the halfshafts horizontally into the transaxle differential, supporting them at the CV-joint to prevent damage to the oil seal lip. Apply even pressure to the hub until the circlips are heard to engage.

NOTE: After installation, pull both front hubs outward to confirm that the driveshafts are retained by the circlips.

13. Install the lower arm ball joints to knuckle and torque nuts to 32–40 ft. lbs. (43–54 Nm).

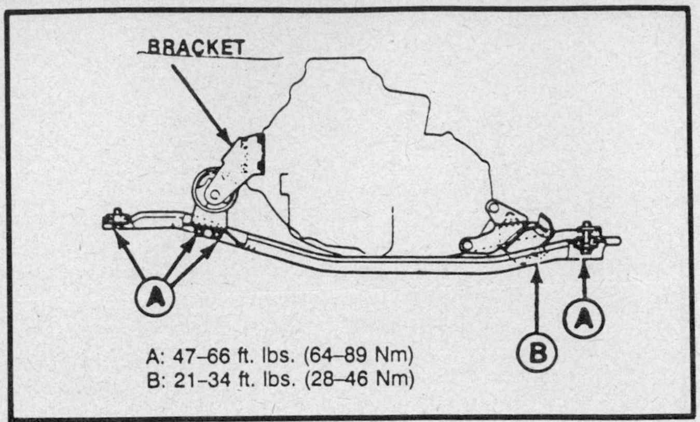
BRACKET
A
A
B
A: 47–66 ft. lbs. (64–89 Nm)
B: 21–34 ft. lbs. (28–46 Nm)

Crossmember bolt location and torque

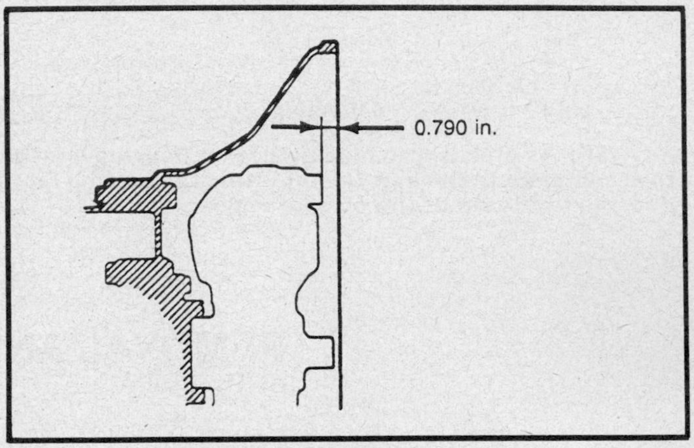
0.790 in.

Checking converter clearance

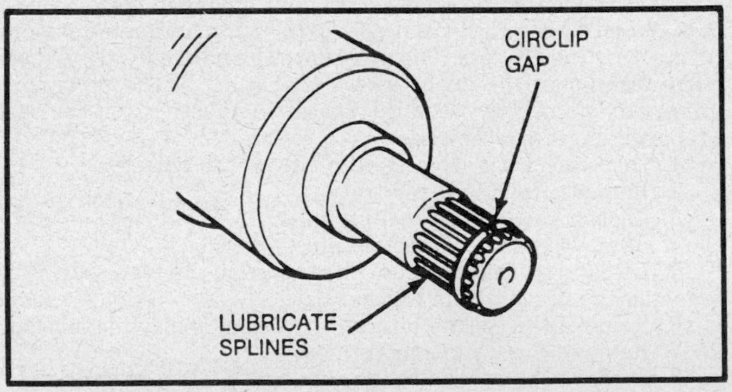
CIRCLIP
GAP
LUBRICATE
SPLINES

Circlip location and installation

14. Install the underside covers.
15. Install the front wheel assemblies and torque lug nuts to 65–87 ft. lbs. (90–120 Nm).
16. Lower vehicle, install upper 2 transaxle mounting bolts and torque to 47–66 ft. lbs. (64–89 Nm).
17. Remove the engine support bar.
18. Remove the plugs from the cooler lines and install cooler lines.
19. Install the vacuum line to modulator.
20. Install the neutral switch connector.
21. Install the body ground connector.
22. Connect the wiring to inhibitor switch and kickdown solenoid.
23. Install the wiring harness clips.
24. Install the secondary air pipe and EGR pipe bracket.

25. Install the engine ground wire.
26. Connect the change control cable to this transaxle. Install the mounting bracket bolts and tighten.
27. Install the speedometer cable and holddown.
28. Install the air cleaner.
29. Install all of the components that were necessary to remove the transaxle.
30. Refill the transaxle with the proper grade and type fluid.

BENCH OVERHAUL

Before Disassembly

Before removing any of the sub-assemblies, thoroughly clean the outside of the transaxle to prevent dirt from entering the mechanical parts during the repair operation.

CAUTION

Eye protection should be worn during procedures involving air pressure, when using spring compression tools or when removing snaprings.

During the repair of the subassemblies, certain general instructions which apply to all units of the transaxle must be followed. The instructions are given here to avoid unnecessary repetition.

Handle all transaxle parts carefully to avoid nicking or burring the bearing or mating surfaces.

Lubricate all internal parts of the transaxle before assembly with clean automatic transaxle fluid. Do not use any other lubricants except on gaskets and thrust washers which may be coated with petroleum jelly to facilitate assembly. Always install new gaskets when assembling the transaxle.

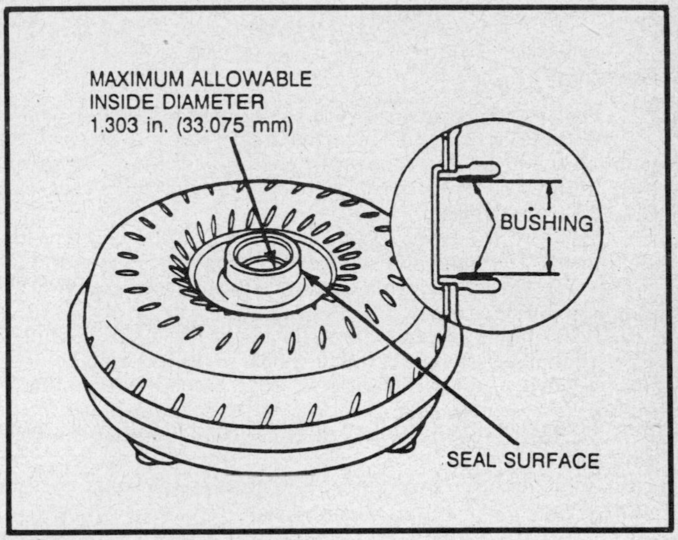
MAXIMUM ALLOWABLE
INSIDE DIAMETER
1.303 in. (33.075 mm)
BUSHING
SEAL SURFACE

Torque converter bushing

Converter Inspection

1. If the converter is to be reused, inspect the outer area of the converter for crack, inspect the bushing and seal surfaces for worn areas, scores, nicks or grooves.
2. The converter must be cleaned on the inside with cleaning solvent, dried, flushed with clean transmission fluid and drained until ready for installation.
3. Measure the converter bushing inside diameter.
4. Replace converter if bushing diameter is greater than 1.302 in. (33.075mm).

NOTE: Whenever a transaxle has been disassembled, the converter and oil cooler must be cleaned.

Transaxle Disassembly

CONVERTER AND OIL PAN

Removal

1. Remove the torque converter from transaxle by pulling it straight out of the housing.
2. If not already drained, remove the drain plug at the bottom of the transaxle case and allow the fluid to drain completely.
3. Remove the pan from the transaxle case.

VALVE BODY

Removal

1. Remove the vacuum modulator, taking care not to loosen vacuum diaphragm rod.
2. Remove the kickdown solenoid.
3. Remove the neutral safety switch.
4. Remove the valve body-to-case attaching bolts, note the position of each bolt for installation reference.

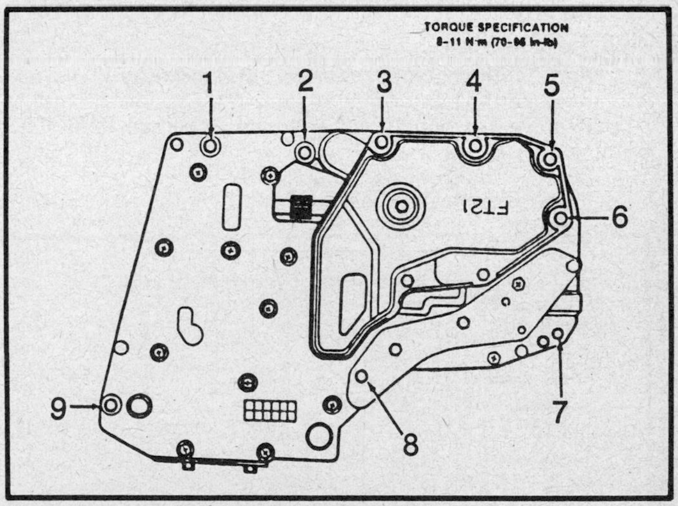
TORQUE SPECIFICATION
1
2
3
4
5
6
7
8
9
FT21

Bolt removal and torque sequence

NOTE: Be careful not to loose the ball and spring, located in the slotted hole.

5. Hold the manual valve to keep it from sliding out of the valve body and remove the valve body from the case.

NOTE: Failure to hold the manual valve while removing the control assembly could cause the manual valve to become bent or damaged. Be careful not to loose the vacuum diaphragm rod or the ball and spring for the torque converter relief valve.

6. If complete disassembly is to be done, remove the dipstick tube, speedometer driven gear, oil pump shaft and input shaft.

GOVERNOR

Removal

1. Note the position of the governor.
2. Remove the 3 retaining bolts from the governor cover assembly.
3. Lift the governor assembly from the transaxle case.
4. Remove the 2 governor retaining screws from the governor sleeve.
5. Remove the governor valve body.

OIL PUMP

Removal

1. Position the transaxle with the oil pump facing down.
2. With a flat blade tool inserted in the wide slot between the front clutch drum and connecting shell, pry down on the drum.
3. Rotate the drum 2 complete revolutions while repeating Step 2 several times.
4. Measure the front clutch drum endplay by checking the clearance of the small slot between the front clutch tabs and the connecting shell slots. This clearance is the front clutch drum endplay.
5. Record this measurement for reference upon assembly. The standard clearance is 0.020–0.031 in. (0.5–0.8mm).
6. Remove the oil cooler lines from the outside of the oil pump.
7. Secure the front clutch in place by tightening the band adjuster.
8. Remove the oil pump attaching bolts and note the position of each bolt for installation reference.

FRONT UNIT

Removal

1. Remove the band adjuster bolt and locknut.
2. Remove the band adjuster strut from inside the case.
3. Remove the band from the transaxle case from the oil pump side.
4. Remove the front clutch assembly.
5. Remove the needle bearing located on the rear clutch.

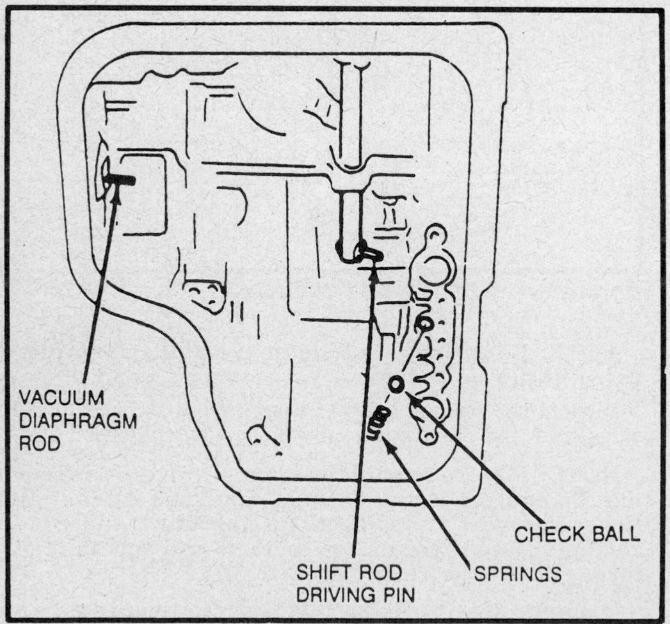
VACUUM DIAPHRAGM ROD
CHECK BALL
SHIFT ROD DRIVING PIN
SPRINGS

Transaxle and spring location

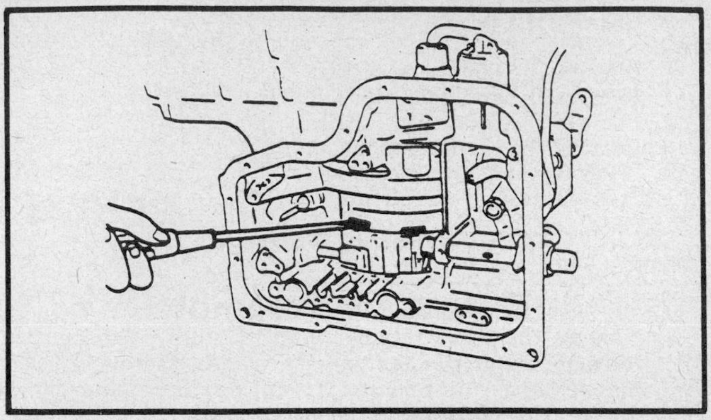

Prying clutch drum to check clearance

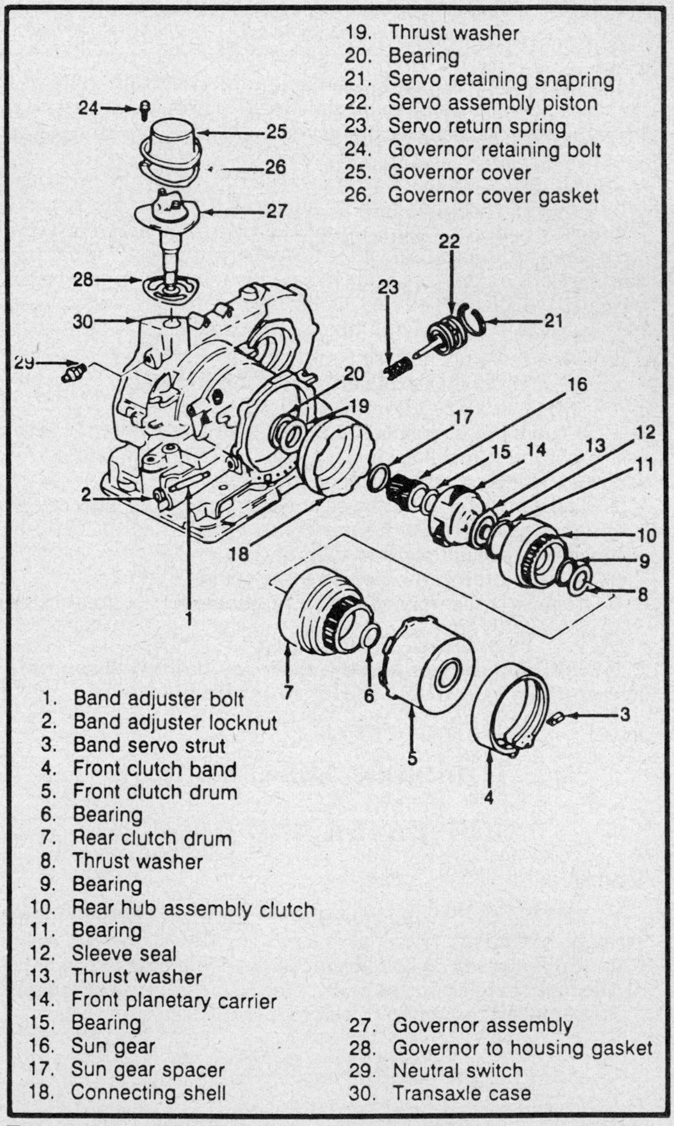
1. Band adjuster bolt
2. Band adjuster locknut
3. Band servo strut
4. Front clutch band
5. Front clutch drum
6. Bearing
7. Rear clutch drum
8. Thrust washer
9. Bearing
10. Rear hub assembly clutch
11. Bearing
12. Sleeve seal
13. Thrust washer
14. Front planetary carrier
15. Bearing
16. Sun gear
17. Sun gear spacer
18. Connecting shell
19. Thrust washer
20. Bearing
21. Servo retaining snapring
22. Servo assembly piston
23. Servo return spring
24. Governor retaining bolt
25. Governor cover
26. Governor cover gasket
27. Governor assembly
28. Governor to housing gasket
29. Neutral switch
30. Transaxle case

Transaxle exploded view—front section

6. Remove the rear clutch drum.
7. Remove the oil pump thrust washer, if washer did not stay on pump.
8. Remove the needle bearing located on the rear clutch hub assembly.

9. Remove the needle bearing and thrust washer from the planetary.
10. Remove the planetary carrier.
11. Remove the sun gear and spacer.
12. Remove the connecting shell.
13. Remove the thrust washer and needle bearing.
14. Use a C-clamp and socket to remove the servo piston into the transaxle case, if not already removed.

CAUTION

Eye protection should be worn during servo removal.

15. Remove the servo snapring.
16. Remove the servo retainer, piston and spring by slowly loosening the C-clamp.

INTERMEDIATE AND REAR UNIT

Removal

1. Remove the bolts attaching the transaxle case to the torque converter housing.
2. Carefully pry cases apart to separate the housings.
3. Pry oil lines away from case to remove.
4. Note the position of the parking pawl shaft and spring for installation reference. Remove pawl assembly by pulling the shaft straight out.
5. Remove the drum hub assembly from the case.
6. Remove the one-way clutch inner race assembly and planetary carrier from the case.
7. Remove the needle and thrust washer bearings from the planetary gear.
8. Before removing the one-way clutch, measure the clearance of the low and reverse clutch retaining plate and the one-way clutch, by using a feeler gauge. The clearance should be 0.032–0.041 in. (0.8–1.05mm).
9. Remove the snapring securing the one-way clutch and the retaining plate to the case.
10. Remove the clutch pack assembly.
11. Using a clutch compressor, place the recessed side of the large plate over the low and reverse clutch hub.
12. Place the small plate on the opposite side of the case. Insert the bolt through to clutch compressor plate and tighten the nut until the tension is relieved from the snapring.

NOTE: Transaxles which have not been previously disassembled do not have a low and reverse clutch hub snapring.

13. Remove the snapring from groove, if equipped.
14. Remove the clutch compressor tool.
15. Remove the low and reverse clutch hub.
16. Remove the springs from the low and reverse clutch piston.
17. To remove the low and reverse clutch piston, hold a wood block over the low and reverse clutch piston and apply a short burst of air to the piston application orifice in the transaxle case.

CAUTION

Wear eye protection and keep fingers out from between wood block and piston. Do not exceed 60 psi.

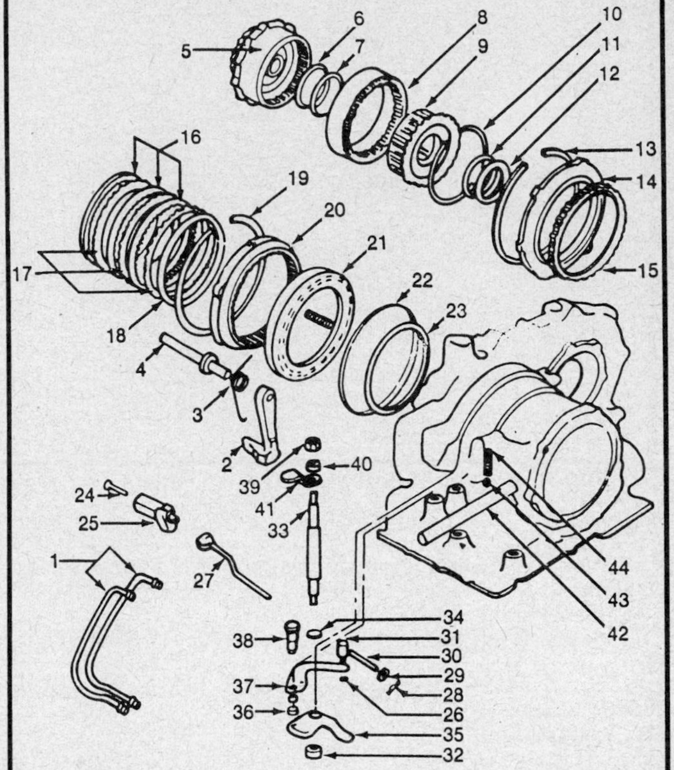
1
2
3
4
5
6
7
8
9
10
11
12
13
14
15
16
17
18
19
20
21
22
23
24
25
26
27
28
29
30
31
32
33
34
35
36
37
38
39
40
41
42
43
44

1. Oil lines
2. Parking pawl
3. Parking pawl return spring
4. Parking pawl shaft
5. Drum hub assembly
6. Needle bearing
7. Thrust washer
8. One-way clutch inner race
9. Planetary carrier
10. Retaining snapring
11. Needle bearing
12. Thrust washer
13. One-way clutch snapring
14. One-way clutch
15. One-way clutch retaining plate
16. Internal spline clutch plate
17. External spline clutch plate
18. Dished plate
19. Low-reverse retaining snapring
20. Low-reverse clutch hub
21. Low-reverse piston
22. Outer piston seal
23. Inner piston seal
24. Parking pawl actuator support retaining bolt
25. Parking pawl actuator support
26. Circlip parking pawl actuator ferrule
27. Parking pawl actuator rod
28. Clevis-pin
29. Flat washer
30. Spring
31. Ferrule
32. Lower manual shaft retaining nut
33. Manual shaft
34. O-ring
35. Control rod actuating lever
36. Circlip
37. Actuating lever
38. Pivot pin
39. Upper manual shaft retaining nut
40. Lock washer
41. Manual shaft arm
42. Control rod
43. Detent ball
44. Detent spring

Transaxle exploded view—Intermediate section

MANUAL LINKAGE

Removal

1. Remove the bolts attaching the parking pawl actuator guide to the case and remove the actuator guide.
2. Remove the lower nut at the end of the manual shaft assembly, hold the shaft with an open-end wrench.
3. Remove the 2 circlips from the manual shift linkage.
4. Remove the shift shaft linkage from the manual shaft.
5. Remove the bolts from the upper shaft support and slide the manual shaft out of the transaxle case.
6. Remove the parking pawl actuator rod from the transaxle case.
7. Position the transaxle with the oil pan opening up.
8. Remove the roll pin securing the control rod to the transaxle case by lightly tapping the roll pin with a $^{3}/_{32}$ in. pin punch and a hammer.

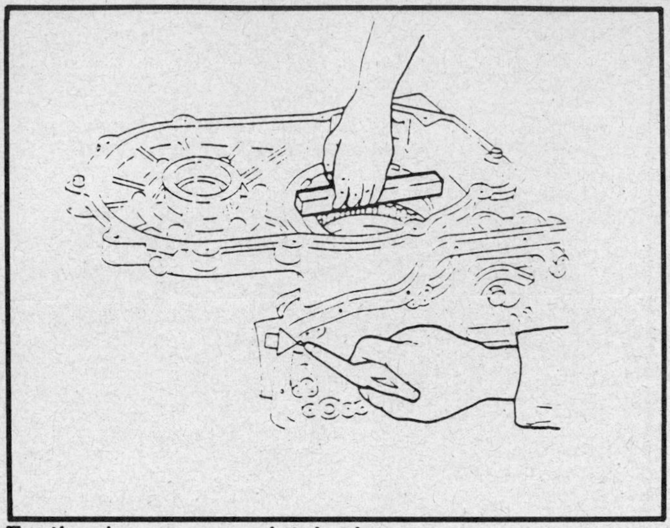

Testing low reverse clutch piston

9. Carefully slide the control rod out of the transaxle case, making sure not to lose the detent ball and spring.
10. Remove the detent ball.
11. Remove the spring.

FINAL DRIVE

Removal

1. Remove the differential assembly, by lifting it out of the case.
2. Remove the bolts attaching the output bearing and idler support to the converter housing.
3. Remove the support housing from the converter housing by lightly tapping the idler shaft with a brass drift and hammer.
4. Place the support assembly into a vise and remove the roll pin from the housing by tapping with a pin punch and hammer.
5. Carefully remove the bearing housing assembly from the vise.
6. Remove the idler gear assembly from the bearing cover.
7. Use the bearing remover to remove the bearing race from the support housing. Remove the adjustment shim from the housing. Separate or mark each shim for installation reference.
8. Using a puller and slide hammer remove the differential side bearing races from both the differential bearing housing and from the converter housing.
9. Remove the bolts and side bearing housing from the transaxle.
10. Remove the shim from the differential side bearing housing and drive out the oil seal.
11. Remove the O-ring from the differential side bearing housing.
12. Remove the oil seal from the output shaft bearing/stator support, using a puller and slide hammer.
13. Remove the output shaft bearing race from the bearing support, using a puller and slide hammer.
14. Remove the bolts and output shaft bearing/stator support by pressing it out of the case, using appropriate step plate tool.

Unit Disassembly and Assembly

NOTE: Allow all new clutch plates to soak in Dexron®II transmission fluid for a minimum of 2 hours before assembly.

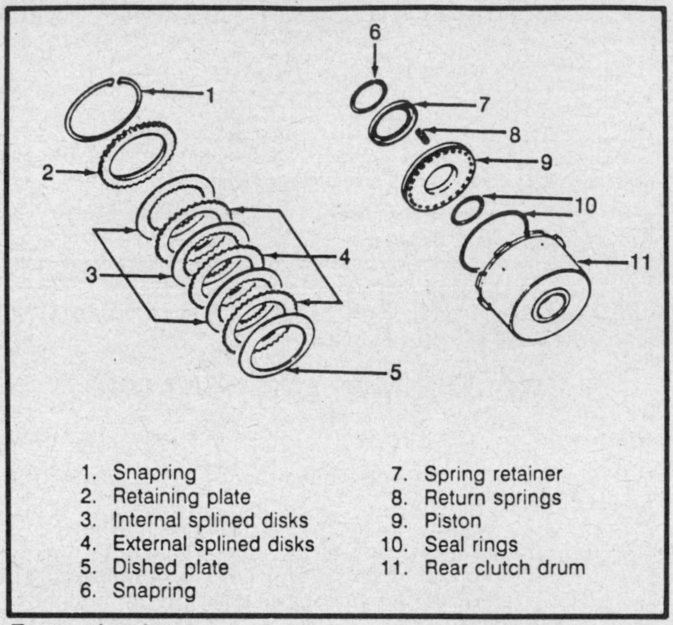
1
2
3
4
5
6
7
8
9
10
11
1. Snapring
2. Retaining plate
3. Internal splined disks
4. External splined disks
5. Dished plate
6. Snapring
7. Spring retainer
8. Return springs
9. Piston
10. Seal rings
11. Rear clutch drum

Front clutch

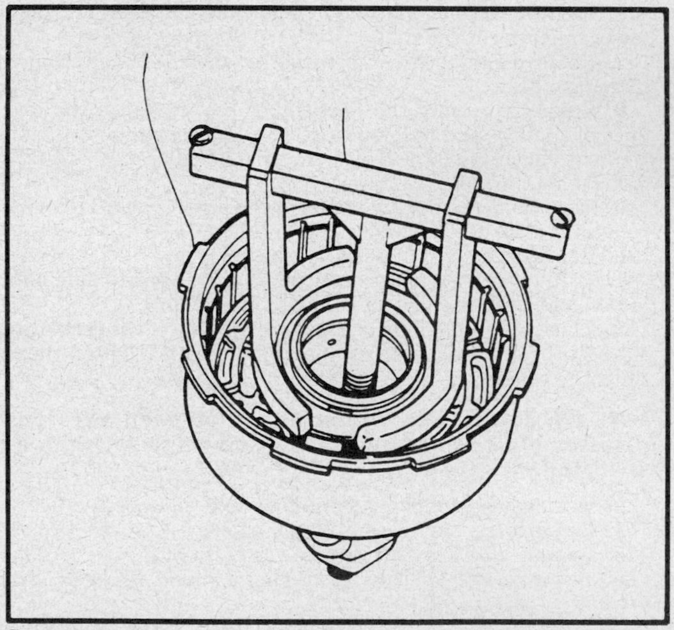

Compressing clutch plates

FRONT CLUTCH

Disassembly

1. Remove the retaining snapring from the drum.
2. Remove the retaining plate and the clutch plate assembly.
3. Remove the dished plate, noting the direction of the dish.
4. Remove the snapring from the drum hub with the use of a compressing tool.
5. Remove the piston from the drum by blowing compressed air into the apply hole in the drum. Remove the oil seals from the piston and drum hub.

Inspection

1. Inspect for damaged or worn drive plates, broken or worn

snaprings, deformed spring retainer, or weakened return springs.

NOTE: The free length of the return springs is 0.992–1.071 in. (25.2–27.2mm). If any spring is out of specification, replace all springs.

2. Inspect the drum bushing for being worn. Maximum inside diameter is 1.735 in. (44.075mm).

Assembly

1. It is good practice to install new clutch plates, both drive and driven, during the overhaul of the unit and not reuse the original plates.
2. Install the oil seals on the piston and the clutch drum hub. Lubricate the seals and grooves with vaseline or clean fluid.
3. Install the piston into the drum, being careful not to cut or damage the seals.
4. Install piston return springs to their mounting pegs on the clutch piston.
5. Place the spring retainer over the return spring.
6. Install the compressing tool and install the snapring holding the springs and spring retainer.
7. Install the dished plate with the protruding side facing the piston. Starting with a steel plate next to the dished plate, alternate with the lined plate and steel plate until 3 of each are installed.
8. Install the retaining plate and the retaining snapring.
9. Measure the front clutch clearance between the retaining plate and the snapring. The standard clearance is 0.063–0.071 in. (1.6–1.8mm). If the clearance is not correct, adjust it with the proper sized retaining plate.
10. After assembly, position the front clutch drum on the oil pump. Check clutch operation by applying air pressure to the application orifice.

REAR CLUTCH

Disassembly

1. Remove the clutch drum snapring, using an appropriate tool.
2. Remove the retaining plate, clutch plates, spacers and the dished plate from the clutch drum.
3. Place a "T" handled clutch spring compressor through the clutch drum. Install and tighten the nut of the compressor tool until the tension is relieved from the hub snapring.
4. Remove the clutch hub snapring, using an appropriate tool.
5. Remove the clutch compressor tool.
6. Place a wood block over the front of the front clutch drum. Remove the piston by applying compressed air into the fluid hole of the pump housing with the front and rear clutch assembly positioned on the oil pump extension.

CAUTION

Wear eye protection and keep fingers out from between wood block and piston. Do not exceed 60 psi.

7. Remove the inner and outer piston seal rings.

Inspection

Inspect for damaged or worn drive plates, broken or worn snaprings, deformed spring retainer, or weakened return springs.

NOTE: The free length of the return springs is 0.992–1.071 in. (25.2–27.2mm). If any spring is out of specification, replace all springs.

Assembly

1. Install the clutch piston into the clutch drum by pushing evenly, taking care not to damage the seals.

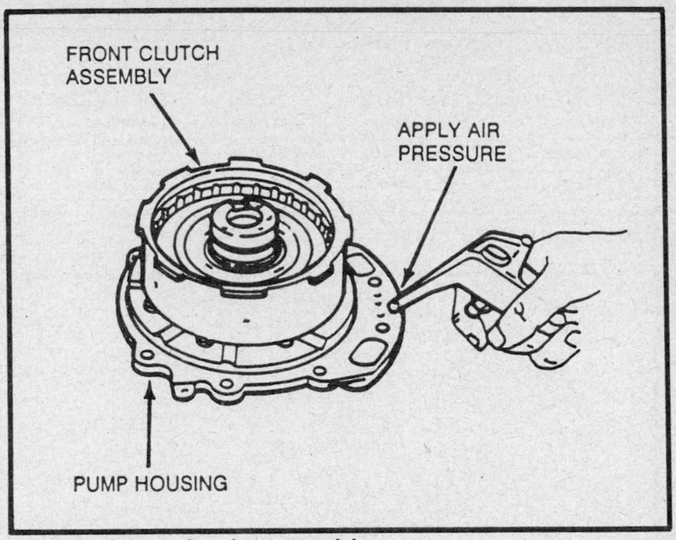
FRONT CLUTCH ASSEMBLY
APPLY AIR PRESSURE
PUMP HOUSING

Testing front clutch assembly

PART NUMBER	THICKNESS OF RETAINING PLATE	PART NUMBER	THICKNESS OF RETAINING PLATE
E7GZ-7B066-B	5.2 mm (0.205 in)	E7GZ-7B066-E	5.8 mm (0.228 in)
E7GZ-7B066-C	5.4 mm (0.213 in)	E7GZ-7B066-F	6.0 mm (0.236 in)
E7GZ-7B066-D	5.6 mm (0.221 in)	E7GZ-7B066-G	6.2 mm (0.244 in)

Retaining plate chart

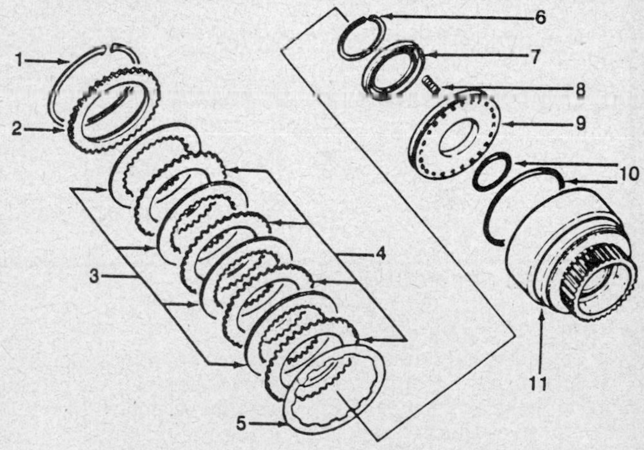
1
2
3
4
5
6
7
8
9
10
11

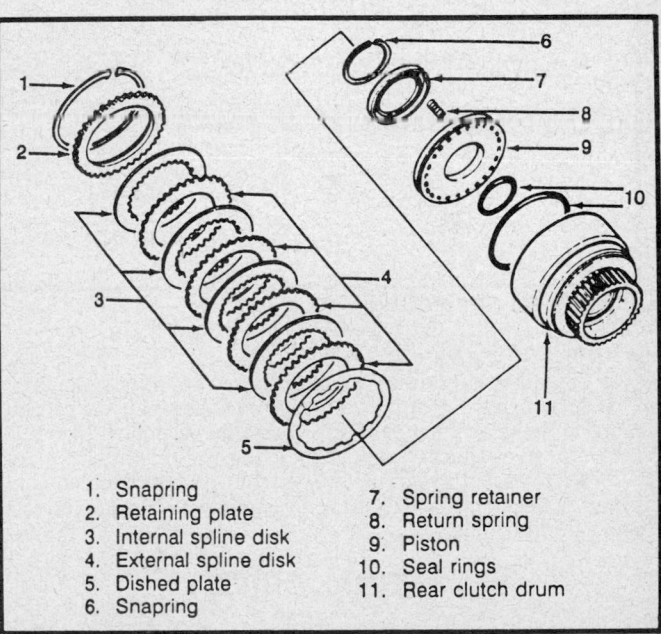

1. Snapring
2. Retaining plate
3. Internal spline disk
4. External spline disk
5. Dished plate
6. Snapring
7. Spring retainer
8. Return spring
9. Piston
10. Seal rings
11. Rear clutch drum

Rear clutch

2. Install the piston return springs to their mounting pegs on the clutch piston.
3. Place the spring retainer over the return springs.
4. Using the clutch spring compressor, press the clutch to gain access to the inner hub snapring groove.
5. Install the inner hub snapring, making certain it is fully seated in the groove. Remove compressor tool.
6. Install the dished plate with the protruding side facing the piston. Starting with a steel plate next to the dished plate, alternate with the lined plate and steel plate until 3 of each are installed.

PART NUMBER	THICKNESS OF RETAINING PLATE	PART NUMBER	THICKNESS OF RETAINING PLATE
E7GZ-7B066-A	5.0 mm (0.197 in)	E7GZ-7B066-E	5.8 mm (0.228 in)
E7GZ-7B066-B	5.2 mm (0.205 in)	E7GZ-7B066-F	6.0 mm (0.236 in)
E7GZ-7B066-C	5.4 mm (0.213 in)	E7GZ-7B066-G	6.2 mm (0.244 in)
E7GZ-7B066-D	5.6 mm (0.221 in)	E7GZ-7B066-H	4.8 mm (0.189 in)

Retaining plate chart

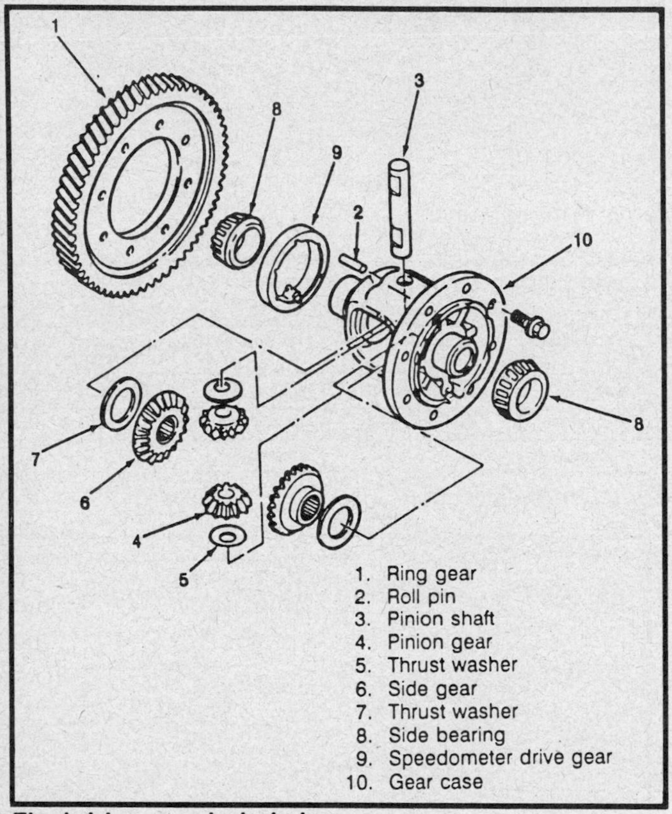
1
2
3
4
5
6
7
8
9
10
1. Ring gear
2. Roll pin
3. Pinion shaft
4. Pinion gear
5. Thrust washer
6. Side gear
7. Thrust washer
8. Side bearing
9. Speedometer drive gear
10. Gear case

Final drive—exploded view

7. Install the retaining plate and the retaining snapring.
8. Measure the front clutch clearance between the retaining plate and the snapring. If the clearance is not within 0.031–0.039 in. (0.8–1.0mm), adjust by using the proper size plate.
9. After assembly, position the front clutch drum on the oil pump. Check clutch operation by applying air pressure to the application orifice.

FINAL DRIVE

Disassembly

1. Install the halfshafts on differential and support unit in "V" blocks.
2. Measure and record the backlash of both pinion gears. Standard backlash is 0.000–0.004 in. (0.0–0.1mm).
3. Remove the ring gear.
4. Remove the pinion shaft roll pin, using a $^5/_{32}$ in. diameter punch or rod at least 3 in. long.
5. Press the front bearing from the differential case, using a puller tool.

NOTE: If differential bearing is removed, it must be replaced with a new bearing and race.

6. Remove the rear bearing from the differential case, using a puller.
7. Remove the pinion shaft by sliding it out of the gear case.
8. Remove the pinion gears and thrust washers by rotating them out of the gear case.
9. Remove the side gears from the case.
10. Remove the speedometer drive gear from the case.

Inspection

1. The inspection of the components consists of checking for broken teeth, worn gears and thrust washers, or a cracked carrier housing.
2. The roller bearing must be pressed from the carrier housing and new ones pressed back on. New races must be used with new bearings.
3. The backlash of the pinion gears is 0.0–0.039 in. (0.0–0.1mm).
4. If the backlash of the pinions are not correct, replace all the thrust washers with new and recheck. If excessive clearance still exists, check the carrier for wear.

Assembly

1. Install the speedometer drive gear, align the locating tang on the gear with the groove in the gear case.
2. Install the front and rear differential bearings to the gear case with a press.
3. Locate and record the identification number on each side gear thrust washer. This information may be used when setting the backlash of the side and pinion gears. If backlash was measured when disassembled, use proper thickness thrust washer to obtain necessary backlash.
4. Coat the side gear thrust washers with Dexron®II transmission fluid and install washers and gears into the case.
5. Coat the pinion gear thrust washers with Dexron®II transmission fluid and install pinion gear and washers into the case.
6. Align pinion shaft with gears and install pinion shaft, with flat on the shaft up and the roll pin hole entering case last.
7. Install the pin through the gear case and into the pinion shaft, using an appropriate tool, until $^1/_{16}$ in. below the surface of the gear case.
8. After installing the pin, stake the gear case to prevent the pin from coming out.
9. Install the ring gear to case, with the depression on the gear toward the case.
10. Align the holes in the gear with holes in case and install the bolts hand tight.
11. Tighten the bolts in a diagonal pattern in stages until torqued to 52–62 ft. lbs. (67–83 Nm).
12. Install the halfshafts on differential and support unit in "V" blocks.
13. Measure the backlash of both pinion gears. Standard backlash is 0.000–0.004 in. (0.0–0.1mm).
14. If the backlash is more than allowable, adjust by using different thickness thrust washers. Thrust washers should be the same thickness at each gear.

OUTPUT SHAFT

1. To disassemble, press off bearings.
2. Check for worn bearings or gears.
3. To reassemble, press bearing onto shaft.

IDLER GEAR

Disassembly

1. Insert hex torque adapter T87C–77000–E into the end of idler gear shaft and place the assembly into a vise.
2. Remove the locknut, using a 1½ in. socket.
3. Remove the idler gear and idler gear bearing.

Identification mark	Thickness
0	2.0 mm (0.079 in)
1	2.1 mm (0.083 in)
2	2.2 mm (0.087 in)

Pinion thrust washer identification

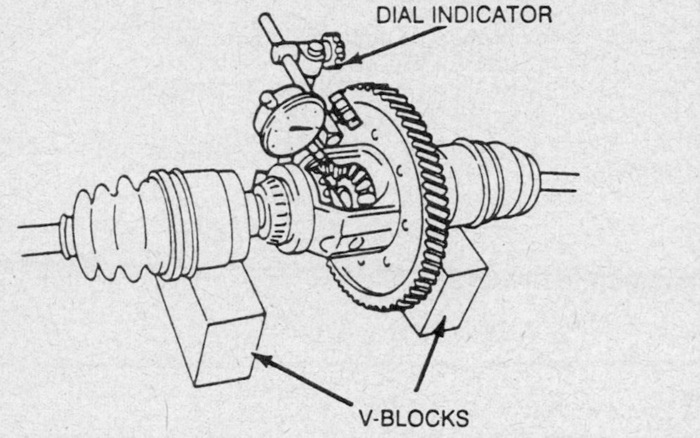
DIAL INDICATOR
V-BLOCKS

Checking backlash

4. If necessary, pull bearing race from gear, using appropriate puller.
5. Press other bearing race from gear, if necessary.

Inspection

Check for worn bearings or gears.

Assembly

1. Press the races into the idler gear, using a hydraulic press and appropriate adapters.
2. Assemble bearings, shims and locknut on the shaft and torque nut to 94 ft. lbs. (128 Nm).
3. Reposition the assembly in a vise, while protecting gear.
4. Using torque adapter tool and an inch lbs. wrench, measure the bearing preload. The correct amount of preload is 0.26–7.8 inch lbs. (0.3–9.0 Nm) while rotating tool.
5. Adjust shims as necessary to obtain proper preload.
6. When preload in correct, torque locknut to 94–130 ft. lbs. (128–177 Nm).

OIL PUMP

Disassembly

1. Remove the bolts that retain the pump cover to the pump housing.
2. Remove the pump cover, being careful not to allow the gears to fall out of the housing.
3. Mark the inner and outer gears with an indelible marker, prior to removing them.

NOTE: Do not mark the gears by pin-punching, or otherwise stressing the gear.

4. Remove the pump flange and inner and outer gears.

Inspection

1. Check the housing and cover for cracks or worn areas.

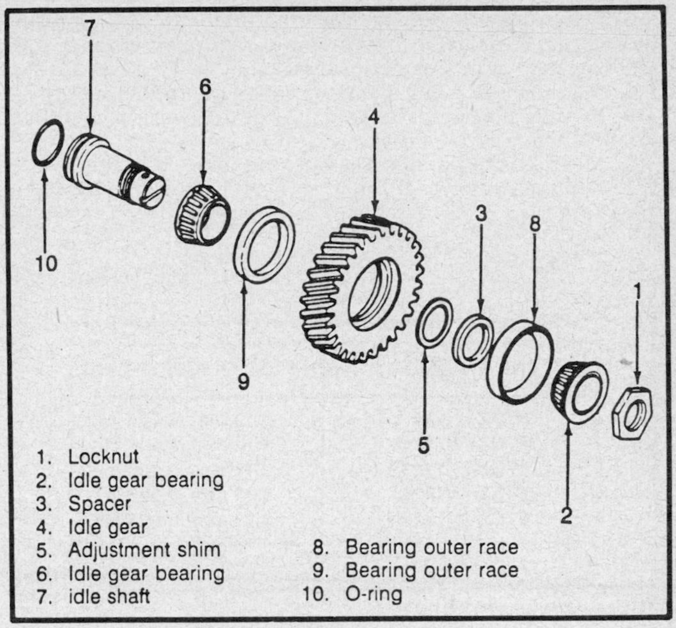
7
6
4
3
8
10
1
9
5
2
1. Locknut
2. Idle gear bearing
3. Spacer
4. Idle gear
5. Adjustment shim
6. Idle gear bearing
7. idle shaft
8. Bearing outer race
9. Bearing outer race
10. O-ring

Idler gear—exploded view

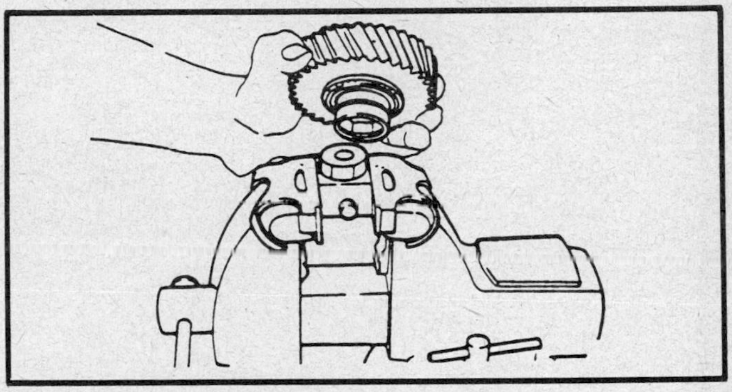

Using torque adapters

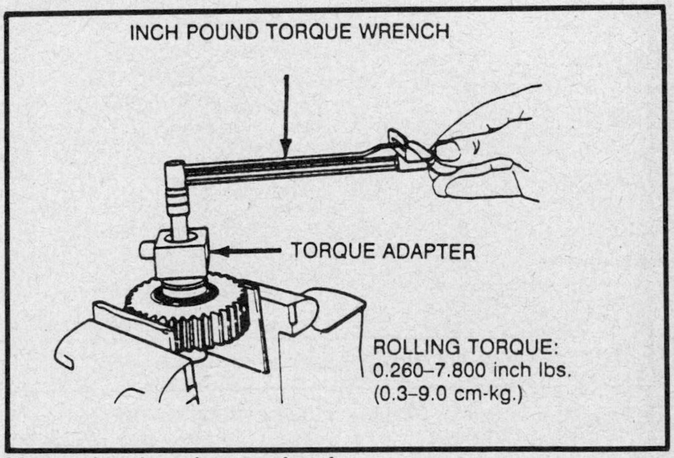
INCH POUND TORQUE WRENCH
TORQUE ADAPTER
ROLLING TORQUE:
0.260–7.800 inch lbs.
(0.3–9.0 cm-kg.)

Measuring bearing preload

2. Check the gears for wear, broken or damaged gear teeth.
3. Check the inner gear bushing of the pump housing sleeve for being worn or damaged.
4. Check the clearance of the inner gear to the pump cover and the outer gear to the pump cover.
5. Check the clearance of the outer gear teeth head to the crescent dam.

6. Check the clearance between the outer gear to the housing.
7. Check the clearance between the new seal rings and the seal ring groove in the pump cover hub.
8. Measure the outer diameter of the pump flange sleeve.
9. Replace the pump flange if the sleeve is worn beyond 1.492 in. (38.075mm).
10. Measure the pump cover bushing inside diameter.
11. Replace the cover if the bushing is worn beyond 1.499 in. (38.075mm).

PART NUMBER	THICKNESS OF SHIM
E7GZ-7N112-A	0.10 mm (0.004 in)
E7GZ-7N112-B	0.12 mm (0.005 in)
E7GZ-7N112-C	0.14 mm (0.006 in)
E7GZ-7N112-D	0.16 mm (0.007 in)
E7GZ-7N112-E	0.20 mm (0.008 in)

Shim thickness chart

Assembly

1. Assemble the gears to flange so the marks on the inner and outer gears are aligned and facing out.
2. Coat the gears with Dexron®II transmission fluid.
3. Install the pump cover to the pump housing.
4. Install the bolts and torque to 95–122 inch lbs. (11–14 Nm).
5. Install the oil pump shaft and make sure the gears turn easily.

SERVO

1. Discard the old seals.
2. Check for broken or damaged snapring.
3. Inspect piston for damage.
4. Check for weaken return spring. The free length of return spring should be 1.870 in. (47.5mm).
5. Coat new seals with Dexron®II transmission fluid before assembly.

ONE-WAY CLUTCH

Inspection

1. Check for worn parts.

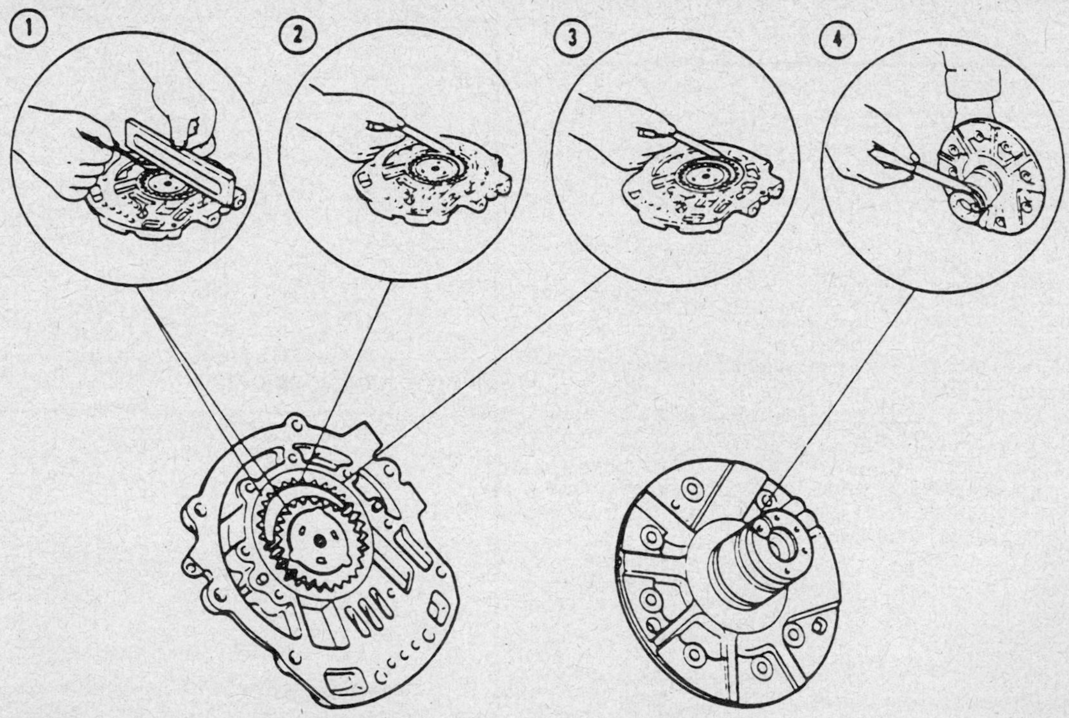
1
2
3
4

MEASURED LOCATION	STANDARD VALUE	LIMIT
1 INNER GEAR – PUMP COVER: OUTER GEAR – PUMP COVER	0.02 ~ 0.04 mm (0.001 ~ 0.002 in)	0.08 mm (0.003 in)
2 HEAD OF OUTER GEAR TEETH – CRESCENT DAM	0.14 ~ 0.21 mm (0.006 ~ 0.008 in)	0.25 mm (0.010 in)
3 OUTER GEAR – HOUSING	0.05 ~ 0.20 mm (0.002 ~ 0.008 in)	0.25 mm (0.010 in)
4 SEAL RING – SEAL RING GROOVE	0.04 ~ 0.16 mm (0.002 ~ 0.006 in)	0.40 mm (0.016 in)

Checking pump and gear clearance

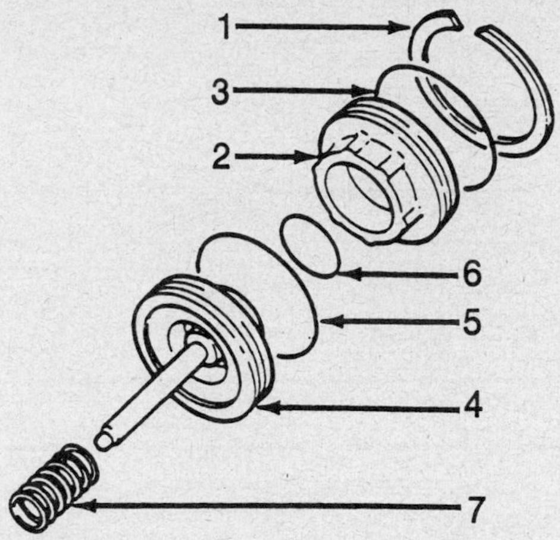
1
3
2
6
5
4
7

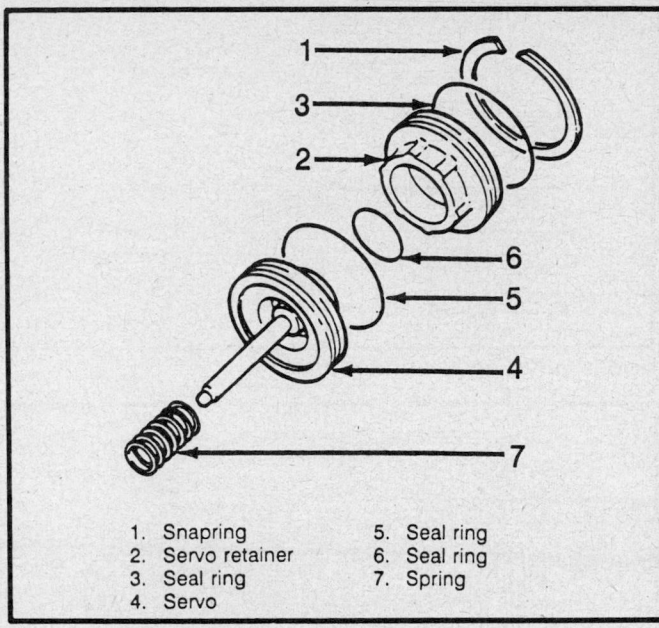

1. Snapring
2. Servo retainer
3. Seal ring
4. Servo
5. Seal ring
6. Seal ring
7. Spring

Servo—exploded view

2. Check for proper one-way operation.
3. Measure bushing, replace if greater than 5.121 in. (130.063mm).

LOW AND REVERSE CLUTCH

Inspection

1. Check for worn parts.
2. Check for weakened or damaged returned springs.
3. If any spring length is less than 1.07–1.11 in. (27.2–28.2mm), replace all springs.

DRUM HUB

Disassembly

1. Remove the parking gear spring.
2. Remove the parking gear by pushing the 2 pins which project from the drive hub.
3. Remove the snapring, the internal gear and the drive hub.

Inspection

Inspect the components for broken or worn snaprings, damaged or worn gears, or broken teeth.

Assembly

1. Assembly of the drum hub is in the reverse of its disassembly procedure.
2. Make certain the snapring and parking gear spring are in their proper positions.

PLANETARY UNITS

Disassembly

Remove snapring from one-way clutch inner race and separate planetary. The following procedures and measurements are the same for all planetary units.

Inspection

1. Inspect for worn snapring.
2. Inspect for binding, loose or rough rotation of gears.
3. Measure the clearance between the pinion washer and the planetary carrier, by using a feeler gauge. If the clearance exceeds 0.031 in. (0.8mm), replace the planetary unit.

Assembly

Clean and replace components in the reverse order of the removal procedure.

CONTROL VALVE BODY

The valve body is a high precision unit. It should be handled very carefully. Since many parts look alike, they should be kept in a well arranged order. If the clutches have been overheated or the band has been burn, make certain to disassemble, clean and inspect the valve body.

Disassembly

1. Remove the manual valve from the upper valve body.
2. Remove the bolts attaching the oil screen to the valve body and remove the oil screen.
3. Remove the 2-3 valve cover.
4. Remove the bolts attaching the upper and lower valve body, noting the position of each bolt.

NOTE: Keep the separator plate attached to the lower valve body to prevent losing the check ball, orifices and springs. If valve body is disassembled, note the locations of check ball, orifices and springs for reference during reassembly.

5. Carefully lift the lower body from the upper valve body, keeping the separator plate attached to the lower body.
6. Turn the assembly over, carefully remove the separator plate and remove the check ball and spring, 2 orifices and springs from the valve body, noting their locations.
7. Remove the orifice from the upper valve body with a magnet.
8. Note the location and position of each valve as the body is disassembled.
9. Carefully remove the side plates of the upper valve body to gain access to the valves.

NOTE: Keep the parts for each valve separated to prevent interchanging of springs that look alike.

Inspection

1. Check the valve body bores and valves for varnish or minor scoring.
2. Clean all parts, use carburetor cleaner to remove varnish.
3. If varnish or scoring is excessive, replace valve assembly.
4. Remove burrs from valves, using 600–800 grit finishing paper wet with Dexron®II transmission fluid.
5. Insert each valve separately into its bore. Do not lubricate at this time.
6. Check free movement of each valve, by tipping the valve body. Each valve should slide freely when valve body is shaken slightly.

Assembly

1. Lubricate all the components in Dexron®II transmission fluid.
2. Install the valves and springs in their correct bores.
3. Install the side plates at their correct positions and tighten the bolts to 22–30 inch lbs. (2.5–3.5 Nm).
4. Install the orifice in the upper half of the valve body.
5. Install the check ball and spring and the other 2 orifices and springs into their correct positions in the lower valve body.
6. Place the separator plate over the lower valve body. Hold-

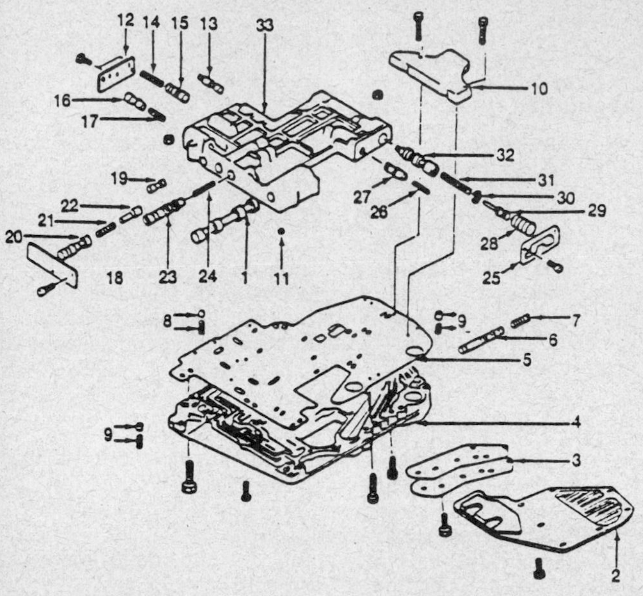
12 14 15 13 33
10
16
17
32
19
31
22
27
30
21
26
29
20
28
18 23 24 1 11
25
8
9
7
6
5
9
4
3
2

1. Manual valve
2. Oil strainer
3. 2–3 valve cover
4. Lower valve body
5. Separator plate
6. 3–2 timing valve
7. Spring
8. Check ball and spring
9. Orifice check valve and spring
10. Sub-body
11. Orifice check valve
12. Side plate
13. Vacuum throttle valve
14. Spring
15. Throttle backup valve
16. Downshift valve
17. Spring
18. Side plate
19. Modifier valve
20. 2–3 shift valve
21. Spring
22. 2–3 shift plug
23. 1–2 shift plug
24. Spring
25. Side plate
26. Spring
27. 2nd lock valve
28. Pressure regulator sleeve
29. Pressure regulator plug
30. Spring seat
31. Spring
32. Pressure regulator valve
33. Upper valve body

Valve assembly—exploded view

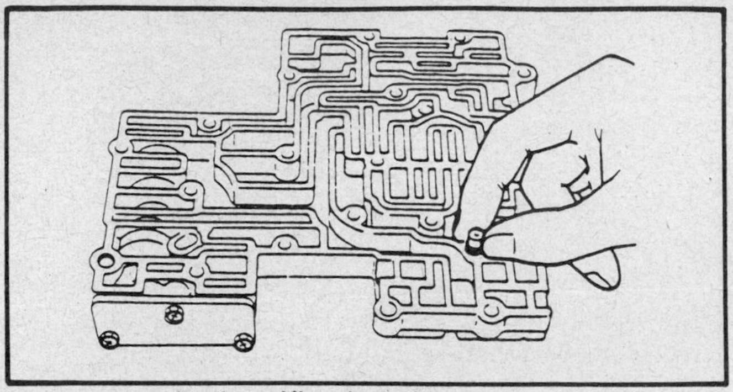

Upper valve body orifice location

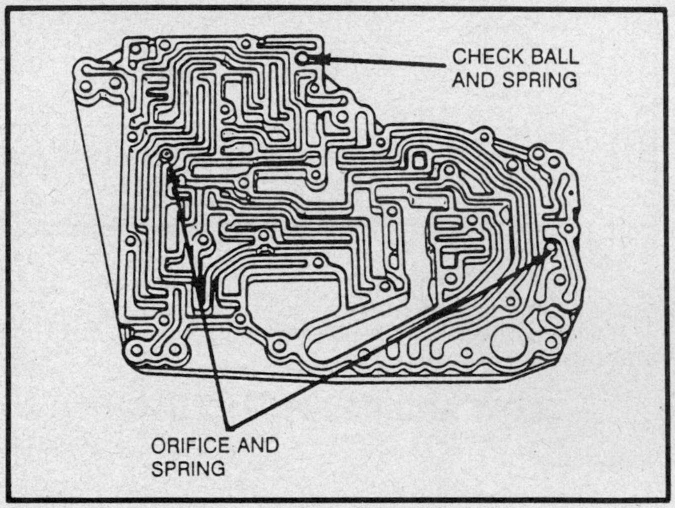
CHECK BALL
AND SPRING
ORIFICE AND
SPRING

Orifice, spring and check ball location

ing the separator plate and valve body together, turn the assembly over and place it onto the upper valve body.

7. Align the upper and lower valve body. Install attaching bolts and tighten to proper torque.
8. Install the 3-2 valve cover and torque the bolts to 26–35 inch lbs. (3–4 Nm).
9. Place valve body aside in a clean area, for installation at a later time.

GOVERNOR

Disassembly

1. Remove the governor body from the separator plate.
2. Remove the separator plate from the shaft.
3. Remove the filter from the separator plate.

NOTE: Cover the bore hole when removing the retainer plate to prevent losing the primary governor spring.

4. Remove the retainer plate for the primary governor spring by applying light pressure against the retainer plate and spring and sliding the plate out of the slot of the machined surface.
5. Remove the primary governor spring.

NOTE: Cover the bore hole when removing the retainer plate to prevent losing the secondary governor spring.

6. Remove the retainer plate for the secondary governor spring by applying light pressure against the retainer plate and spring and sliding the plate out of the slot of the machined surface.
7. Remove the secondary governor spring.
8. Remove the governor driven gear roll pin, using a pin punch and hammer.
9. Remove the governor driven gear and separate the governor shaft from the sleeve.
10. Remove the seal rings from the governor shaft.
11. Remove the bearing outer race from the governor shaft.
12. Remove the needle bearing from the sleeve.

Inspection

1. Inspect the valves for scoring or sticking.

NOTE: Minor scoring or varnish may be removed with fine 600–800 grit finishing paper wet with Dexron®II transmission fluid.

2. Inspect return springs.
3. Replace the primary spring, if the diameter is not 0.34–0.87 in. (8.7–9.3mm) or free length is not 0.66–0.70 in. (16.7–17.7mm).
4. Replace the secondary spring, if the diameter is not 0.35–0.38 in. (8.95–9.55mm) or free length is not 0.50–0.54 in. (12.7–13.7mm).
5. Inspect driven gear for damaged teeth.

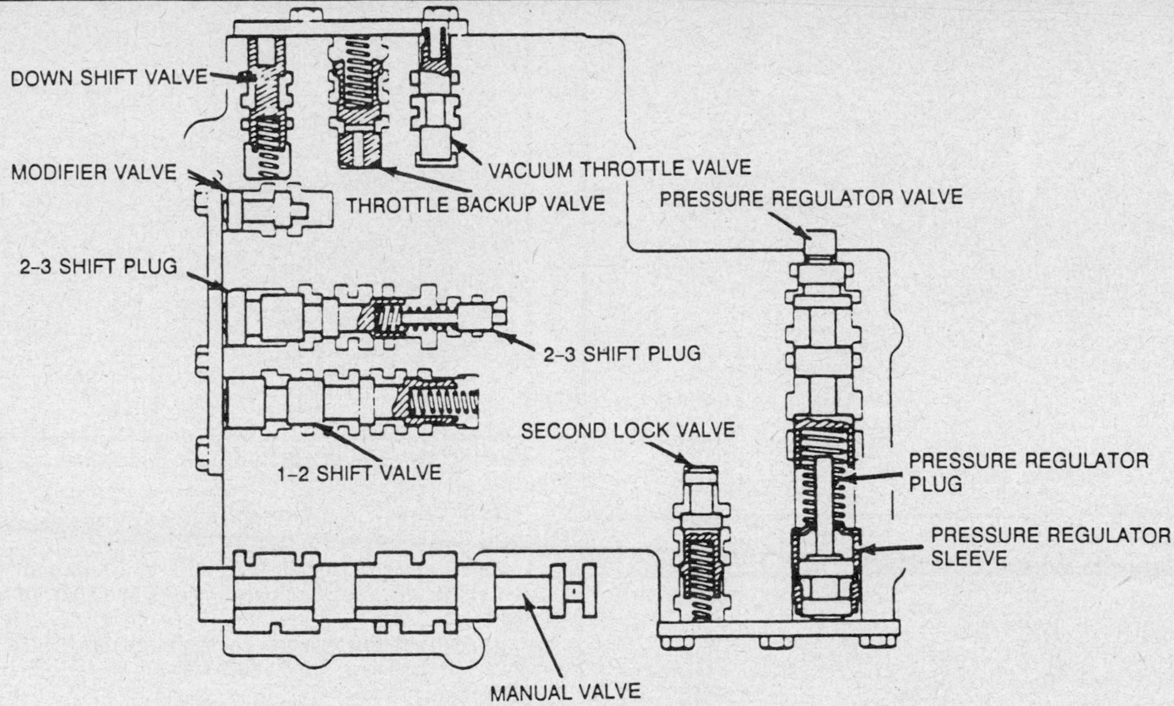
DOWN SHIFT VALVE
MODIFIER VALVE
VACUUM THROTTLE VALVE
THROTTLE BACKUP VALVE
PRESSURE REGULATOR VALVE
2-3 SHIFT PLUG
2-3 SHIFT PLUG
SECOND LOCK VALVE
1-2 SHIFT VALVE
PRESSURE REGULATOR PLUG
PRESSURE REGULATOR SLEEVE
MANUAL VALVE

Valve location

NAME	OUTER DIA. mm (in)	FREE LENGTH mm (in)	NO. OF COILS	WIRE DIA. mm (in)
THROTTLE BACK UP	7.3 (0.287)	36.0 (1.42)	16.0	0.8 (0.031)
DOWN SHIFT	5.5 (0.217)	21.9 (0.862)	14.0	0.55 (0.022)
2-3 SHIFT	6.4 (0.252)(carb.) 6.9 (0.272)(EFI)	39.2 (1.54)(carb.) 41.0 (1.61)(EFI)	20.0	0.7 (0.028)
1-2 SHIFT	6.55 (0.258)	32.0 (1.26)	18.7 (carb.) 18.0 (EFI)	0.55 (0.022)
SECOND LOCK	5.55 (0.219)	33.5 (1.32)	18.0	0.55 (0.022)
PRESSURE REGULATOR	11.7 (0.461)	43.0 (1.69)	15.0	1.2 (0.047)
THROTTLE RELIEF	7.0 (0.276)	11.2 (0.44)	6.0	0.9 (0.035)
ORIFICE CHECK	5.0 (0.197)	15.5 (0.61)	12.0	0.23 (0.009)
3-2 TIMING	7.5 (0.295)	22.1 (0.870)	13.0	0.8 (0.031)

Valve spring size chart

6. Inspect needle bearings and thrust washer for wear.
7. Inspect for clogged or torn filter.

Assembly

1. Coat all parts with Dexron®II transmission fluid.
2. Install the seals.
3. Install the needle bearing into sleeve with the exposed bearing facing up.
4. Install the thrust washer on the governor shaft, making certain the tangs of the washer are inserted into the holes of the governor shaft.
5. Install the governor shaft into the sleeve.

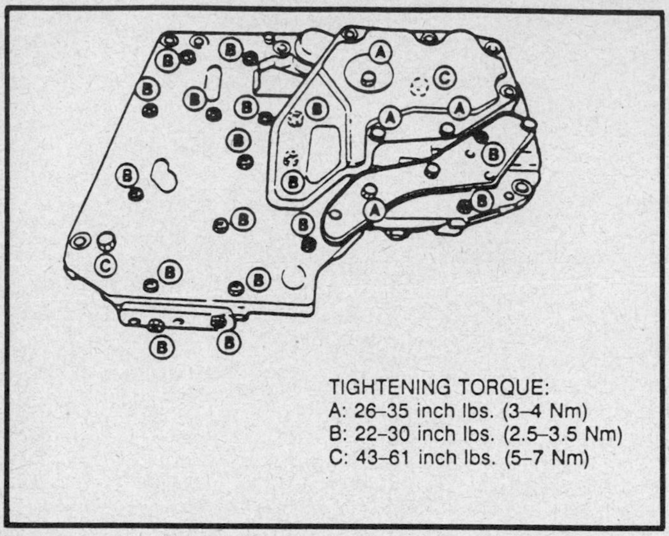
TIGHTENING TORQUE:
A: 26–35 inch lbs. (3–4 Nm)
B: 22–30 inch lbs. (2.5–3.5 Nm)
C: 43–61 inch lbs. (5–7 Nm)

Valve body torque chart

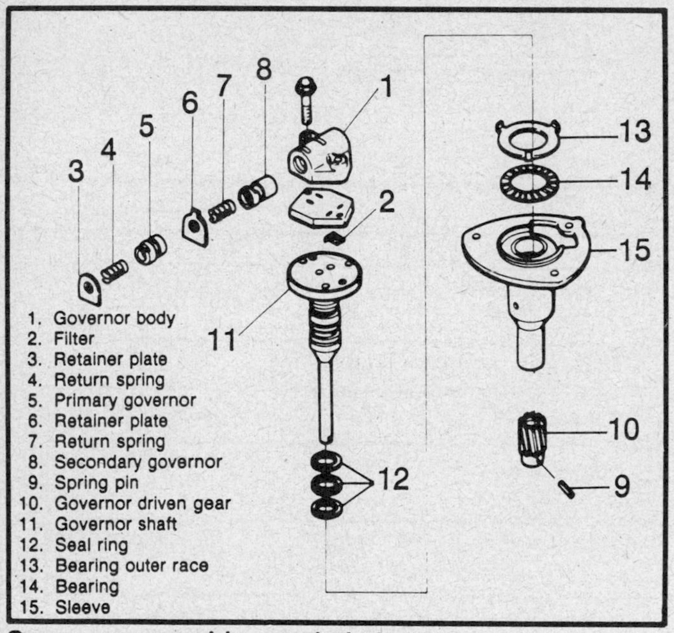
1. Governor body
2. Filter
3. Retainer plate
4. Return spring
5. Primary governor
6. Retainer plate
7. Return spring
8. Secondary governor
9. Spring pin
10. Governor driven gear
11. Governor shaft
12. Seal ring
13. Bearing outer race
14. Bearing
15. Sleeve

Governor assembly—exploded view

6. Install the governor driven gear to the governor shaft and install the roll pin, using a pin punch and hammer.
7. Install the secondary governor valve, narrow land first, into the large bore end of the governor body.

NOTE: The valve is fully seated when the narrow valve and extends out of the governor body case.

8. Install the secondary governor spring into the governor body until the spring end is seated in the recess of the secondary valve.
9. Compress the secondary spring. Install the retainer plate through the slot of the machined surface.
10. Install the primary governor valve, notched land last, into the large bore end of the governor body. The primary governor valve is fully seated when it contacts the retainer plate of the secondary valve.
11. Compress the primary spring. Install the retainer plate through the slot of the machined surface.
12. Install the filter onto the separator plate.

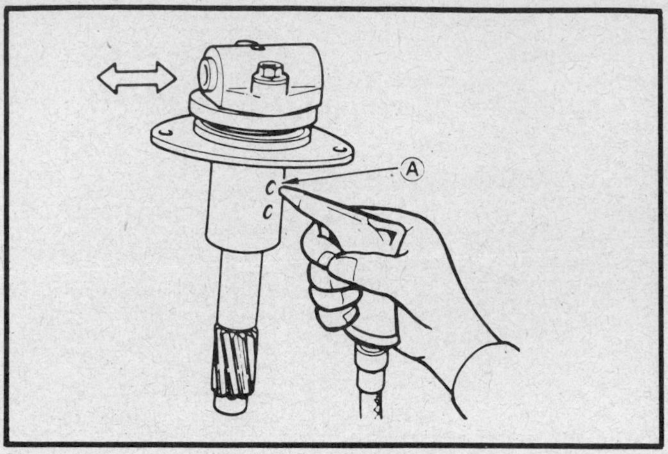
A

Testing the governor with air pressure

13. Install the separator plate, filter side down, to the governor body.
14. Install the governor body and separator plate on the governor shaft. Install the attaching bolts and tighten to 69–95 inch lbs. (8–11 Nm).
15. Apply compressed air through the upper hole in the side of governor housing. The valve should rattle when functioning properly.

Transaxle Assembly

CONVERTER HOUSING

Assembly

1. Install the differential output seal, using an appropriate tool.
2. Install the differential side bearing outer race in housing, using an appropriate tool.
3. Using guide pins, position output bearing/stator support in converter housing. Install and torque bolts to 8–10 ft. lbs. (11–14mm).
4. Using a driver and appropriate step plate press outer race for output shaft into bearing support.
5. Install the seal into stator support.
6. Using a new O-ring, install output bearing support in the transaxle case and torque bolts to 14–19 ft. lbs. (18–26 Nm).

DIFFERENTIAL BEARING PRELOAD

1. Position differential side bearing outer race in the recessed end of the gauge tool T87C–77000J. Screw the gauge tool on unit no clearance remains.
3. Place the differential in the converter housing. Place the gauge tool, with the outer race installed, over the differential side bearing.
4. Position the spacer collars in position and assemble the case halves. Torque bolts to 27–39 ft. lbs. (36–53 Nm).
5. Use the gauge wrench pins to unscrew the gauge tool and establish preload on the differential. Measure the preload using an inch lbs. torque wrench. Extend the gauge tool until the turning torque (drag) reads 4.3–6.9 inch lbs. (0.5–0.8 Nm).6. Measure the clearance at the separation of the gauge tool in order to determine the shim thickness.
7. Disassembly the gauge assembly. Install the necessary shims and bearing race in the transaxle housing and assemble the housing halves.

NOTE: Measure the clearance around the entire circumference and select shims equivalent to the maximum clearance. Do not use more than 5 shims.

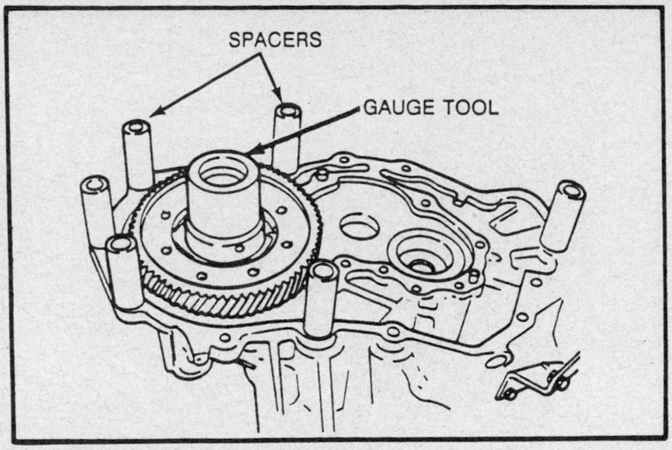
SPACERS
GAUGE TOOL
Installing bearing preload gauge

8. Test the bearing turning torque (drag) with housing assembled. If the turning torque is not 18–25 inch lbs. (2.1–2.9 Nm). repeat Steps 7 through 13.
9. Separate housing halves and install output shaft seals.

OUTPUT SHAFT BEARING PRELOAD

1.
Position the output gear and shaft into the converter housing.2.Position the 4 spacer collars.3.Insert the output shaft outer bearing race into the recessed end of the gauge tool and place the tool over the output gear shaft. Screw the halves of the gauge tool together so no clearance exists.4.Assemble the converter housing to the side bearing housing, using tool spacer bolts and torque bolts to 14–19 ft. lbs. (19–26 Nm).5.Unscrew the gauge tool, using gauge tool pins, until all the freeplay is removed and the bearing is seated.
6. Measure the drag on the output gear, by using the torque adapter tool.
7. Adjust drag to 0.36–0.65 inch lbs. (0.5–0.8 Nm), by adjusting shim thickness as necessary.
8. Disassemble the assembly tool and install selected shims and bearing race, using the appropriate tool and step plate.
9. Assembly output gear and bearing support. Torque bolts to 14–19 ft. lbs. (19–26 Nm).
10. Remeasure preload, if preload is not 0.26–7.81 inch lbs., repeat Steps 15 through 24.
11. After proper preload has been obtained, remove the bearing housing and install the idler gear assemble into the bearing housing. Replace the O-ring on the idler gear shaft.
12. Reinstall the bearing cone.
13. Torque the bearing housing bolts to 14–19 ft. lbs. (19–26 Nm).

MANUAL LINKAGE

Installation

1. Install the manual valve control rod, spring and ball detents.
2. Install the roll pin to retain the control rod.
3. Install the parking pawl actuator support and torque bolts to 8.7–11.6 ft. lbs. (12–16 Nm).
4. Install the actuating arm to the pawl rod and install circlips.
5. Install a new O-ring on the manual shaft.
6. Install the manual shaft to the transaxle case.
7. Install the manual shaft support to the transaxle case and torque the bolts to 8.7–11.6 ft. lbs. (12–16 Nm).
8. Install the shifter actuating arm to the manual shaft and align with the manual valve control rod.

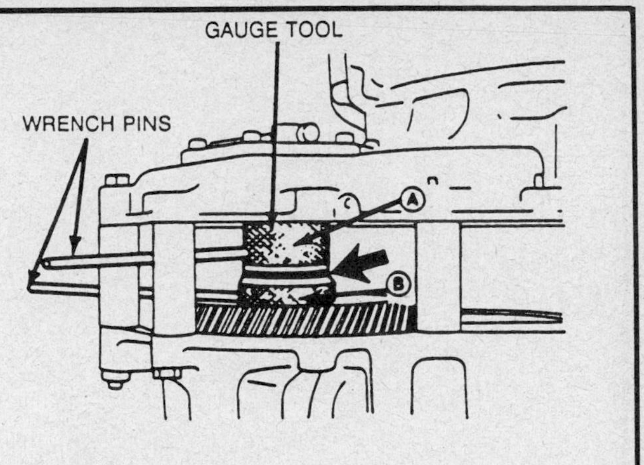
GAUGE TOOL
WRENCH PINS
A
B

PART NUMBER	THICKNESS OF SHIM
E7GZ4067A	0.10 mm (0.004 in)
E7GZ4067A	0.12 mm (0.005 in)
E7GZ4067B	0.14 mm (0.006 in)
E7GZ4067C	0.16 mm (0.007 in)
E7GZ4067B	0.20 mm (0.008 in)
E7GZ4067C	0.30 mm (0.012 in)
E7GZ4067D	0.40 mm (0.016 in)
E7GZ4067E	0.50 mm (0.020 in)
E7GZ4067F	0.60 mm (0.024 in)
E7GZ4067G	0.70 mm (0.028 in)
E7GZ4067H	0.80 mm (0.032 in)
E7GZ4067J	0.90 mm (0.036 in)

Adjusting bearing preload

PART NUMBER	THICKNESS OF SHIM
E7GZ-7F405-A	0.10 mm (0.004 in)
E7GZ-7F405-B	0.12 mm (0.005 in)
E7GZ-7F405-C	0.14 mm (0.006 in)
E7GZ-7F405-D	0.16 mm (0.007 in)
E7GZ-7F405-E	0.20 mm (0.008 in)
E7GZ-7F405-F	0.50 mm (0.020 in)

TEST PRELOAD: 0.5 ~ 0.9 N·m (0.05 ~ 0.09 m-kg, 0.36 ~ 0.65 ft-lb)

ASSEMBLED PRELOAD: 0.03 ~ 0.9 N·m (0.3 ~ 9.0 cm-kg, 0.26 ~ 7.81 in-lb)

CAUTION
A) MEASURE THE CLEARANCE AROUND THE ENTIRE CIRCUMFERENCE, AND SELECT SHIMS EQUIVALENT TO THE MAXIMUM CLEARANCE.
B) MAXIMUM ALLOWABLE NUMBER OF SHIMS: 7

Adjusting shim chart

9. Install the nut on manual shaft and torque to 22–29 ft. lbs. (29–39 Nm).

INTERMEDIATE UNIT

Installation

1. Install new seals in the low and reverse piston and install piston in bore taking care not to damage seal.

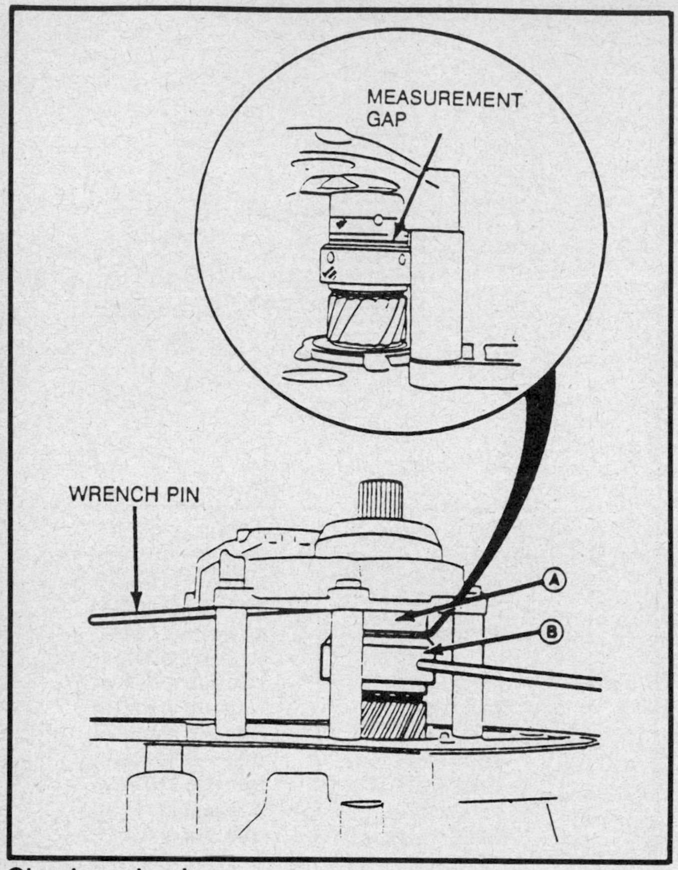
MEASUREMENT
GAP
WRENCH PIN
A
B

Check preload

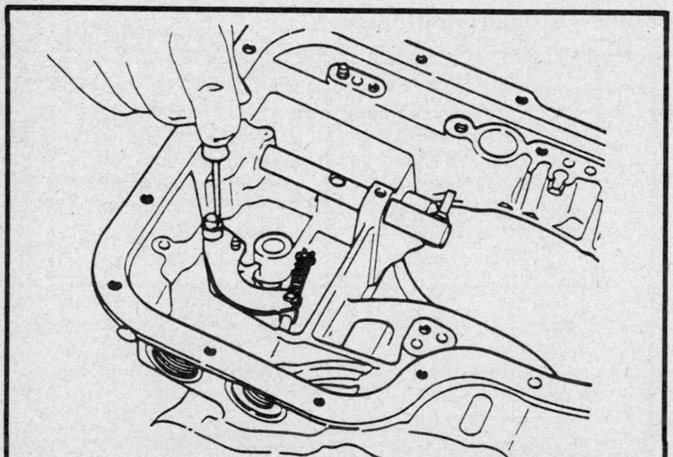

Install arm to pawl rod

2. Place the return spring and low and reverse clutch hub on top of the piston.
3. Install the clutch compressor and compress the assembly far enough to permit insertion of the retainer ring.
4. Install the dished plate to the low and reverse piston.
5. Install the clutch disc pack starting with a steel disc against the dished plate. Alternate internal and external tooth plates until all discs have been installed.
6. Install the retaining plate.
7. Install the one-way clutch with the bushing against the retaining plate. Compress the clutch assembly enough to install the retaining ring.

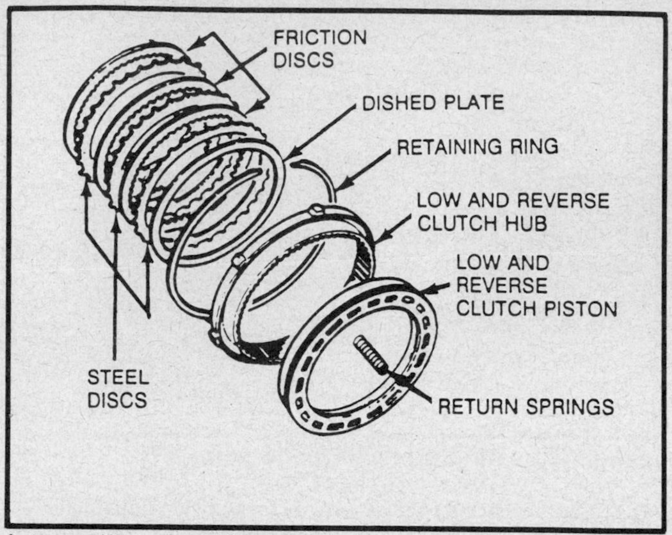
FRICTION
DISCS
DISHED PLATE
RETAINING RING
LOW AND REVERSE
CLUTCH HUB
LOW AND
REVERSE
CLUTCH PISTON
STEEL
DISCS
RETURN SPRINGS

Intermediate clutch pack

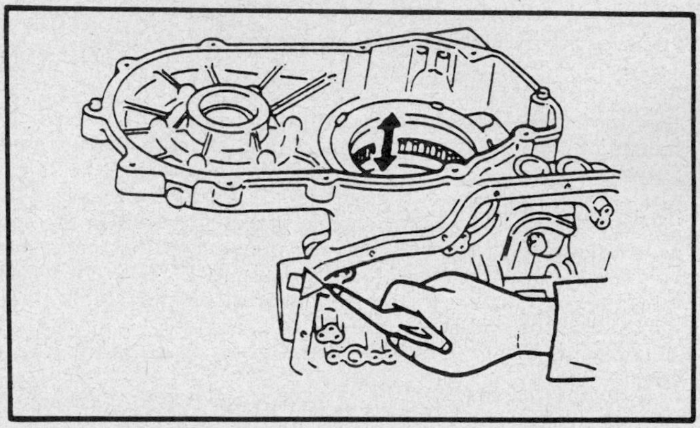

Test clutch pack

PART NUMBER	THICKNESS
E7GZ-7B066-J	7.8 mm (0.307 in)
E7GZ-7B066-K	8.0 mm (0.315 in)
E7GZ-7B066-L	8.2 mm (0.322 in)
E7GZ-7B066-M	8.4 mm (0.331 in)
E7GZ-7B066-N	8.6 mm (0.339 in)
E7GZ-7B066-P	8.8 mm (0.346 in)

Retaining plate thickness chart

8. Using a feeler gauge, measure the clearance between the one-way clutch and the retaining plate.
9. If the clearance is not 0.032–0.041 in. (0.81–1.05mm), adjust shim size as necessary.
10. Apply a burst of air pressure to the application port to test clutch plate action.

CAUTION

Do not allow air pressure to exceed 60 psi. Wear eye protection.

11. Install the parking pawl, spring and anchor pin to the case.
12. Position the converter housing to receive the differential. Install the differential, meshing ring gear with idler gear.
13. Install the thrust washer over output shaft.
14. Install the bearing so that the rollers contact the thrust washer.

15. Install the assembled drum hub onto the output shaft spline.
16. Install a bearing in the recess of the drum hub. Secure the opposing thrust washer to the planetary carrier with petroleum jelly.
17. Install the planetary carrier onto the one-way clutch inner race and secure it in place with the retainer ring.
18. Install the planetary carrier/inner race assembly into the drum hub.
19. Using a plastic mallet, install the governor oil transfer lines into the transfer case.
20. Apply a $^{1}/_{16}$ continuous bead of gasket eliminator E1FZ–19562–A (non-silicone) onto the converter housing mating surface.
21. Assemble the transaxle halves by rotating the one-way clutch inner race as the transaxle case is lowered onto the converter housing to engage the spline teeth of the inner race with the low and reverse disks.
22. Install the bolts and torque to 22–34 ft. lbs. (29–46 Nm).
23. After the cases are together, make certain all rotating parts rotate without resistance.

FRONT UNIT

Installation

1. Install the bearing to the rear planetary carrier.
2. Install the spacer over the small end of the sun gear and insert the sun gear into the connecting shell.
3. Place the thrust bearing over the end of the sun gear protruding from the connecting shell. Hold the washer in place with petroleum jelly.
4. Install the connecting shell/sun gear assembly into the rear planetary carrier.
5. Place the thrust bearing into the front planetary carrier using petroleum jelly. Face the rollers pointing out
6. Install the front planetary carrier into the connecting shell.
7. Install the thrust washer and matching bearing to the end of the front planetary carrier. Install the seal sleeve in the center of the front planetary carrier.
8. Install the rear clutch hub assembly over the front planetary carrier.
9. Install the thrust bearing, rollers up, in the rear clutch hub.
10. Coat the matching thrust washer with petroleum jelly. Index the tangs on the washer with the mating holes in the rear clutch and install the washer.
11. Install the rear clutch assembly to the rear clutch hub, while gently rotating the rear clutch.
12. Install the thrust bearing in the rear clutch hub, rollers up. The companion thrust washer will be installed to the end of the oil pump extension later.
13. Place the assembled front clutch over the splines of the rear clutch hub, while gently rotating the front clutch.
14. Install the intermediate band, servo, strut and adjuster bolt. Apply sealant E1FZ–19562–A (non-silicone) to threads and install bolts tight enough to hold components in place, but do not perform adjustments until pump has been installed.

GOVERNOR

Installation

1. Install the governor onto the transaxle case so that the sleeve projection is aligned with the mark on case.
2. Torque bolts to 69–95 ft. lbs. (7.8–10.8 Nm)

SERVO

Installation

1. Install the servo into the transaxle housing.
2. Use a C-clamp and deep socket to compress the servo return spring.
3. Install the retaining ring and remove C-clamp.

OIL PUMP

Installation and Measurement

TOTAL ENDPLAY

1. Remove the oil pump extension and place the housing with gears installed aside.
2. Install the oil pump extension into the front clutch housing, without the plastic adjusting washer.
3. Position a machinist straightedge over the oil pump cover in the transaxle case, be careful not to place the tool on the bolt holes.
4. Use a feeler gauge and measure the clearance between the pump and bar or transaxle and bar. The measurement should not be greater than 0.004 in. (0.10mm) with the pump below the transaxle surface, or 0.006 in. (0.15mm) with the pump cover above the case surface.
5. Replace thrust washer as necessary to obtain proper clearance.
6. Reassemble oil pump and torque the oil pump cover to housing bolts to 95–122 inch lbs. (11–14 Nm).
7. Lubricate pump with Dexron®II transmission fluid and check for free movement by inserting and turning shaft.

FRONT CLUTCH ENDPLAY

1. Use petroleum jelly to install the oil pump gasket to the oil pump.
2. Install the plastic adjusting washer to the oil pump cover using petroleum jelly.

PART NUMBER	THICKNESS OF BEARING RACE
E7GZ-7D014-A	1.2 mm (0.047 in)
E7GZ-7D014-B	1.4 mm (0.055 in)
E7GZ-7D014-C	1.6 mm (0.063 in)
E7GZ-7D014-D	1.8 mm (0.071 in)
E7GZ-7D014-E	2.0 mm (0.079 in)
E7GZ-7D014-F	2.2 mm (0.087 in)

Endplay race selection chart

3. Install the thrust washer onto the end of the pump extension using petroleum jelly.
4. Install the oil pump to the transaxle and torque bolts to 11–16 ft. lbs. (15–22 Nm).
5. Reposition the transaxle with the oil pump facing down.
6. While turning the connecting shell through 2 complete revolutions, push the front clutch down toward the oil pump, using an appropriate tool inserted into the tabs of the clutch drum.
7. Measure the clearance between the tabs of the clutch drum and the connecting shell.
8. If the endplay in not within 0.020–0.031 in (0.5–0.8mm), replace the plastic washer with an appropriate size washer.
9. Reassemble and install the oil pump.
10. Apply sealant E1FZ–19562–A (non-silicone) to threads and torque to 11–16 ft. lbs. (15–22 Nm).

VALVE BODY

Installation

1. Install the steel ball into the transaxle case.

PART NUMBER	THICKNESS OF SHIM
E7GZ-7F373-A	2.1 mm (0.083 in)
E7GZ-7F373-B	2.3 mm (0.091 in)
E7GZ-7F373-C	2.5 mm (0.098 in)
E7GZ-7F373-D	2.7 mm (0.106 in)
E7GZ-7F373-E	1.5 mm (0.059 in)
E7GZ-7F373-F	1.7 mm (0.067 in)
E7GZ-7F373-G	1.3 mm (0.051 in)

Plastic adjustment washer chart

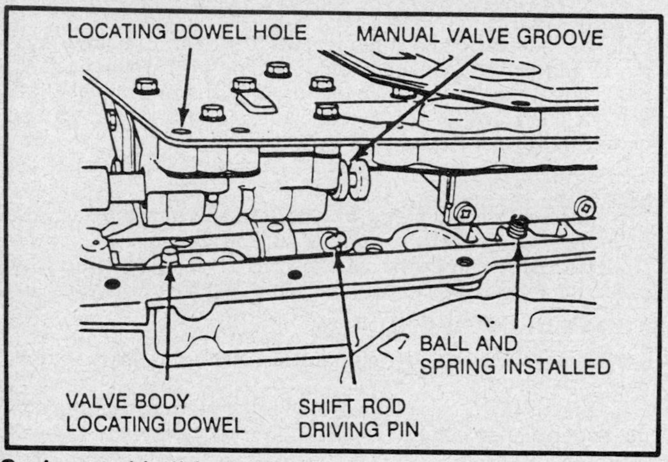
LOCATING DOWEL HOLE
MANUAL VALVE GROOVE
BALL AND
SPRING INSTALLED
VALVE BODY
LOCATING DOWEL
SHIFT ROD
DRIVING PIN

Spring and ball location

2. Install the manual valve into the valve body.

3. Align the manual valve land with the pin on the control rod. Mate the dowels in the transaxle case with the holes in the valve body.

4. Install the valve body retaining bolts and torque to 70–90 inch lbs. (8–11 Nm) in proper sequence.

INTERMEDIATE BAND ADJUSTMENT

1. Tighten adjuster bolt to 8.7–10.8 ft. lbs. (12–15 Nm).

2. Apply sealant E1FZ–19562–A (non-silicone) to adjuster screw threads and torque locknut to 41–59 ft. lbs. (55–80 Nm).

CONTROL VALVE DIAPHRAGM ROD

1. Insert the vacuum diaphragm rod gauge into the mounting hole, with the beveled side out, until the tool bottoms.

2. Place the rod through the opening of the vacuum diaphragm tool until the rod bottoms out against the valve.

3. Tighten the lock knob on the vacuum modulator tool and remove the tool and rod from the transaxle case.

4. Use a depth gauge to measure the distance from the flat surface of the vacuum modulator tool to the end of the rod.5. Use this measurement to select the correct size rod.6. Install the correct rod, lubricate the modulator O-ring with Dexron®II and install the vacuum modulator.

OIL PAN AND CONVERTER

Installation

1. Install the kickdown solenoid, coat O-ring with Dexron®II and threads with sealant E1FZ-19562 (non-silicone).

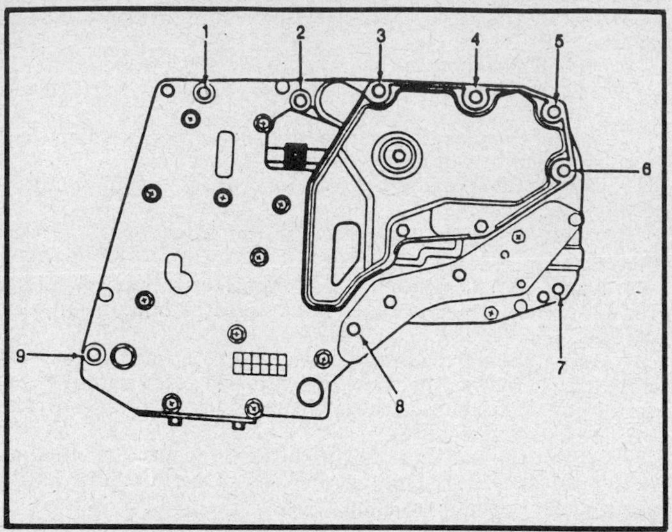
1
2
3
4
5
6
7
8
9

Bolt removal and torque sequence

MEASUREMENT	DIAPHRAGM ROD USED
Under 25.4 mm (1.000 in)	29.5 mm (1.160 in)
25.4 – 25.9 mm (1.000 – 1.020 in)	30.0 mm (1.180 in)
25.9 – 26.4 mm (1.020 – 1.039 in)	30.5 mm (1.200 in)
26.4 – 26.9 mm (1.039 – 1.059 in)	31.0 mm (1.220 in)
Over 26.9 mm (1.059 in)	31.5 mm (1.240 in)

PART NO.	DIAPHRAGM ROD
E7GZ-7A380-E	29.5 mm (1.160 in)
E7GZ-7A380-C	30.0 mm (1.180 in)
E7GZ-7A380-D	30.5 mm (1.200 in)
E7GZ-7A380-B	31.0 mm (1.220 in)
E7GZ-7A380-A	31.5 mm (1.240 in)

Vacuum diaphragm rod selection chart

2. Install the speedometer drive gear and torque bolts to 43–49 inch lbs. (5–8 Nm).

3. Lubricate oil filler tube end with Dexron®II and install. Torque retaining bolt to 43–69 inch lbs. (5–87 Nm).

4. Install a new pan gasket and oil pan. Torque pan bolts to 43–69 inch lbs. (5–87 Nm).

NOTE: Do not use any type of sealer or adhesive on the pan gasket. If necessary, soak the gasket in Dexron®II transmission fluid.

5. Coat the neutral safety switch threads with sealant E1FZ-19562 (non-silicone) and install.

NOTE: Switch must to torque to 14–19 ft.lbs. (19–26 Nm) for proper switch operation.

6. Install the turbine and oil pump shafts.

7. Lubricate the inside of torque converter with no more than ½ qt. of Dexron®II transmission fluid and install.

8. Measure the recess from converter to front housing mating surface. If distance is not 0.790 in. (20.0mm), converter is not seated properly.

SPECIFICATIONS

VALVE BODY

Item		EFI in. (mm)	Carburetor in. (mm)
Throttle back-up valve spring	Diameter	0.287 (7.3)	0.287 (7.3)
	Free length	1.417 (36.0)	1.420 (36.0)
Down shift valve spring	Diameter	0.219 (5.55)	0.217 (5.5)
	Free length	0.866 (22.0)	0.862 (21.9)
Throttle relief	Diameter	0.276 (7.0)	0.276 (7.0)
	Free length	0.440 (11.2)	0.440 (11.2)
2–3 shift valve spring	Diameter	0.272 (6.9)	0.252 (6.4)
	Free length	1.614 (41.0)	1.540 (39.2)
1–2 shift valve spring	Diameter	6.258 (6.55)	0.258 (6.55)
	Free length	1.260 (32.0)	1.260 (32.0)
Second lock valve spring	Diameter	0.219 (5.55)	0.219 (5.55)
	Free length	1.319 (33.5)	1.320 (33.5)
Pressure regulator valve spring	Diameter	0.461 (11.7)	0.461 (11.7)
	Free length	1.693 (43.0)	1.693 (43.0)
3-2 timing	Diameter	0.295 (7.5)	0.295 (7.5)
	Free length	0.870 (22.1)	0.870 (22.1)
Orifice check valve spring	Diameter	0.197 (5.0)	0.197 (5.0)
	Free length	—	0.610 (15.5)

SERVO

Item	Specification in. (mm)
Free length of return spring	1.87–1.93 (47.5–49.0)

GOVERNOR SPRINGS

Item		Specification in. (mm)
Primary spring	Outer diameter	0.343–0.366
	Free length	0.65–0.728 (16.5–18.5)
Secondary spring	Outer diameter	0.352–0.376 (8.95–9.55)
	Free length	0.488–0.567 (12.4–14.4)

GEAR ASSEMBLY

Item	Specification in. (mm)
Total endplay at pump	0.004–0.006 (0.1–0.15)
Endplay adjusting race	0.047 (1.2) 0.055 (1.4) 0.063 (1.6) 0.071 (1.8) 0.079 (2.0) 0.087 (2.2)
Idle gear bearing preload	0.3–7.8 (0.03–0.09)①
Preload adjusting shims	0.004 (0.10) 0.005 (0.12) 0.006 (0.14) 0.007 (0.16) 0.020 (0.50) 0.008 (0.20)
Output gear bearing preload	0.26–7.81 (0.03–0.9)①

① inch lbs. (Nm)

VACUUM DIAPHRAGM

Item	Specification in. (mm)
Available diaphragm rods	1.161 (29.5) 1.181 (30.0) 1.200 (30.5) 1.220 (31.0) 1.240 (31.5)

DRIVE AND DIFFERENTIAL

Item		Specification in. (mm)
Final gear	Type	Helical gear
	Reduction ratio	3.631
Side bearing prelooad		18–25 (2.1–2.9)①
Preload adjusting shims		0.004 (0.1) 0.008 (0.2) 0.012 (0.3) 0.016 (0.4) 0.020 (0.5) 0.024 (0.6) 0.028 (0.7) 0.031 (0.8) 0.035 (0.9) 0.047 (0.12) 0.055 (0.14) 0.063 (0.16)
Backlash of side gear and pinion		0–0.004 (0–0.1)
Backlash adjusting thrust washers		0.079 (2.0) 0.083 (2.1) 0.087 (2.2)

① inch lbs. (Nm)

TORQUE SPECIFICATIONS

Item	ft. lbs (Nm)
Drive plate to crankshaft	96–103 (71–76)
Drive plate to torque converter	25.3–36.2 (35–50)
Converter housing to engine	64–89 (47–66)
Converter halves	29–46 (22–34)
Converter housing to transaxle case	26.8–39.8 (37–55)
Bearing housing to converter housing	13.7–18.8 (19–26)
Side bearing housing to transaxle case	13.7–18.8 (19–26)
Bearing cover to transaxle case	8.0–10.1 (11–14)
Oil pump to transaxle case	11–16 (15–22)
Governor cover to transaxle case	7.8–10.8 (5.8–8.0)
Oil pan	3.6–5.8 (5–8)
Anchor end bolt (when adjusting band brake)	8.7–10.8 (12–15)
Anchor end bolt lock nut	41–59 (56–82)
Control valve body to transaxle case	5.8–8.0 (8–11)
Lower valve body to upper valve body	1.8–2.5 (2.5–3.5)
Side plate to control valve body	1.8–2.5 (2.5–3.5)
Reamer bolt of control valve body	3.6–5.1 (5–7)
Oil strainer of control valve	2.2–2.9 (3–4)
Governor valve body to governor shaft	5.8–8.0 (8–11)
Oil pump cover	8.0–10.1 (11–14)
Inhibitor switch	13.7–18.8 (19–26)
Manual shaft lock nut	21.7–29.0 (30–40)
Oil cooler pipe set bolt	11.6–17.4 (16–24)
Actuator for parking rod to transaxle case	8.7–11.6 (12–15)
Idle bear bearing lock nut	94–130 (130–180)

LINE PRESSURE SPECIFICATIONS

Gear	Condition	Specification psi (kPa)
R	Idling	57–100 (392–687)
	Stall	228–270 (1570–1864)
D	Idling	43–57 (294–392)
	Stall	128–156 (883–1079)
2	Idling	114–171 (785–1177)
	Stall	114–171 (785–1177)

CUT BACK POINT

Vacuum of Vacuum Pump in. Hg. (mm Hg.)	Governor Pressure psi (kPa)
0 (0)	14–23 (98–157)
7.87 (200)	6–14 (39–98)

GOVERNOR PRESSURE

mph	Governor Pressure psi (kPa)
20	13–21 (88–147)
35	27–36 (186–245)
55	58–70 (402–481)

SHIFT POINT SPEED

Throttle Condition		Shift Point mph (km/h)
CARBURETOR		
Wide open throttle	D^1–D^2	30–36 (48–58)
	D^2–D^3	60–68 (96–110)
	D^3–D^2	53–58 (85–93)
	D^2–D^1	24–26 (38–42)
Half throttle	D^1–D^2	10–19 (16–30)
	D^2–D^3	17–37 (28–60)
Fully closed throttle	D^3–D^1	6–9 (10–15)
	1_2–1_1	22–26 (35–42)
ELECTRONIC FUEL INJECTION		
Wide open throttle	D^1–D^2	30–36 (48–58)
	D^2–D^3	60–68 (96–110)
	D^3–D^2	53–58 (85–93)
	D^2–D^1	24–26 (38–42)
Half throttle	D^1–D^2	12–21 (20–34)
	D^2–D^3	37–48 (60–78)
Fully closed throttle	D^3–D^1	6–9 (10–15)
	1_2–1_1	22–26 (35–42)

TORQUE CONVERTER

Item	Specifications
Stall torque ratio	1.95–2.35
Stall torque rpm	2300–2500
Bushing diameter wear limit	1.302 in (33.075mm)

OIL PUMP

Clearance		Specification in. (mm)
Gear end float	Standard	0.0008–0.0016 (0.02–0.04)
	Limit	0.0031 (0.08)
Outer gear and crest	Standard	0.0055–0.0083 (0.14–0.21)
	Limit	0.0098 (0.25)
Outer gear and housing	Standard	0.002–0.0079 (0.05–0.20)
	Limit	0.0098 (0.25)
Oil seal ring and ring groove	Standard	0.0016–0.0063 (0.04–0.16)
	Limit	0.0157 (0.40)

LOW AND REVERSE BRAKE

Item	Specification in. (mm)
Number of friction and steel plates	3
Clearance between retaing plate and stopper	0.8–1.05 (0.032–0.041)
Clearance adjusting retaining plates	0.181 (4.6) 0.189 (4.8) 0.197 (5.0) 0.205 (5.2) 0.213 (5.4) 0.221 (5.6)
Free length of return spring	27.2–28.2 (1.07–1.11)

CLUTCH SPECIFICATIONS

Front Clutch	Specification in. (mm)
Number of driven & drive plates	3
Front clutch clearance	0.063–0.071 (1.6–1.8)
Clearance adjusting retaining plate	0.205 (5.2) 0.213 (5.4) 0.220 (5.6) 0.228 (5.8) 0.236 (6.0) 0.244 (6.2)
Return spring free length	0.992–1.071 (2.52–27.2)
Drum bushing inner diameter Standard Limit	 1.7322–1.7331 (44.0–44.025) 1.7354 (44.075)
Front clutch drum end prary (clearance between drum and connecting shell)	0.020–0.032 (0.5–0.8)

CLUTCH SPECIFICATIONS

Front Clutch	Specification in. (mm)
Endplay adjusting shim	0.051 (1.3) 0.059 (1.5) 0.067 (1.7) 0.075 (1.9) 0.083 (2.1) 0.091 (2.3) 0.098 (2.5) 0.106 (2.7)

Rear Clutch	Specification in. (mm)
Number of driven and drive plates	4
Rear clutch clearance	0.031–0.059 (0.8–1.5)
Return spring free length	0.992–1.071 (25.2–27.2)

SPECIAL TOOLS

Tool	Identification
Tool–1175–AC	Seal remover
Tool–4201–C	Dial indicator
D78P–4201–C	Magnetic base for dial indicator
D79P–6000–C	Engine support bar
D80L–630–A	Step plate adapter set
D80L–943–A	Puller
D83L–7059–A	Vacuum pump
D84L–1122–A	Bearing pulling attachment
T50T–100–A	Slide hammer
T57L–500–B	Bench mounted holding fixture
T57L–77820–A	Pressure gauge
T58L–101A	Puller
T65L–77515–A	Front clutch spring compressor
T77F–1217–B	Bearing cup puller
T77F–4220–B1	Differential bearing cap puller
T80T–4000–W	Handle
T80L–77003–A	Gauge bar

Tool	Identification
T86p–70043–A	Jaws (used with T58L–101–A)
T87C–7025–C	Differential plugs
T87C–77000–A	Vacuum diaphragm rod gauge
T87C–77000–B	Clutch compressor
T87C–77000–C	Bearing cone replacer
T87C–77000–D	Bearing cone replacer
T87c–77000–E	Torque adapter
T87C–77000–G	Converter seal replacer
T87C–77000–J	Shim selection tool
014–00456	Transmission tester (Tracer ATX adapter)
014–00028	Torque converter cleaner
D80L–630–3	Plate
D80L–630–7	Plate
D80L–943–A2	Jaw puller
D79P–6000–B	Engine support bar
T86P–70043–A2	Jaw puller
T87C–77000–H	Differential seal replacer

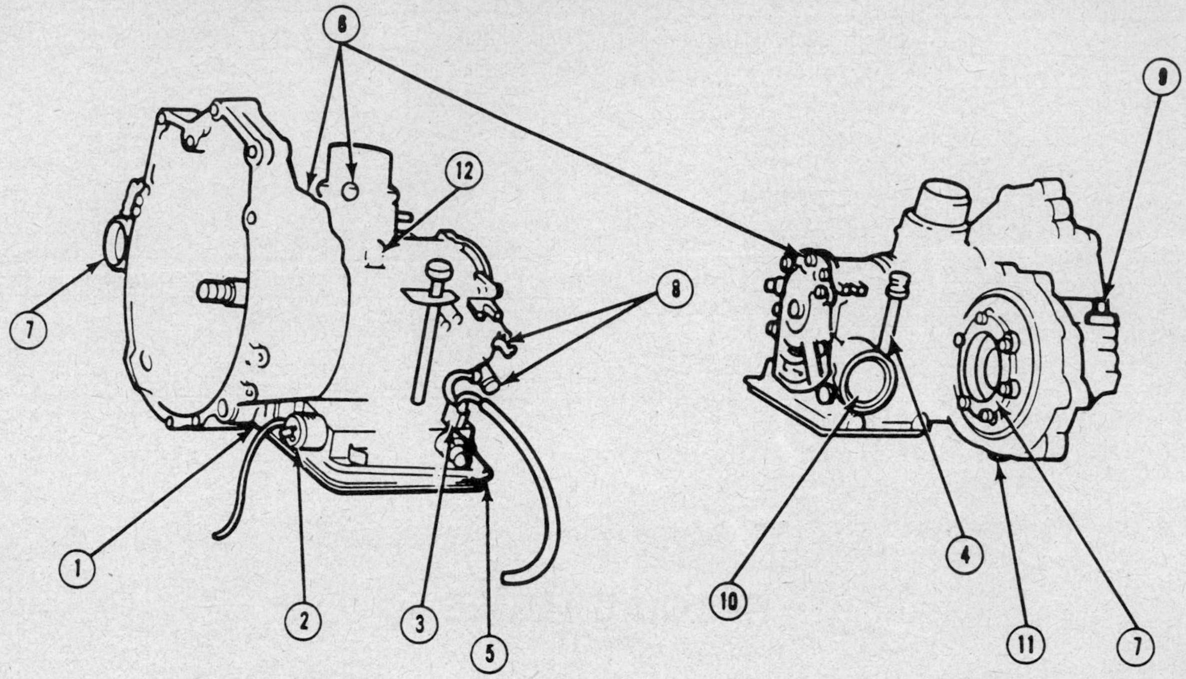
6
9
12
7
8
1
2
3
5
10
4
11
7

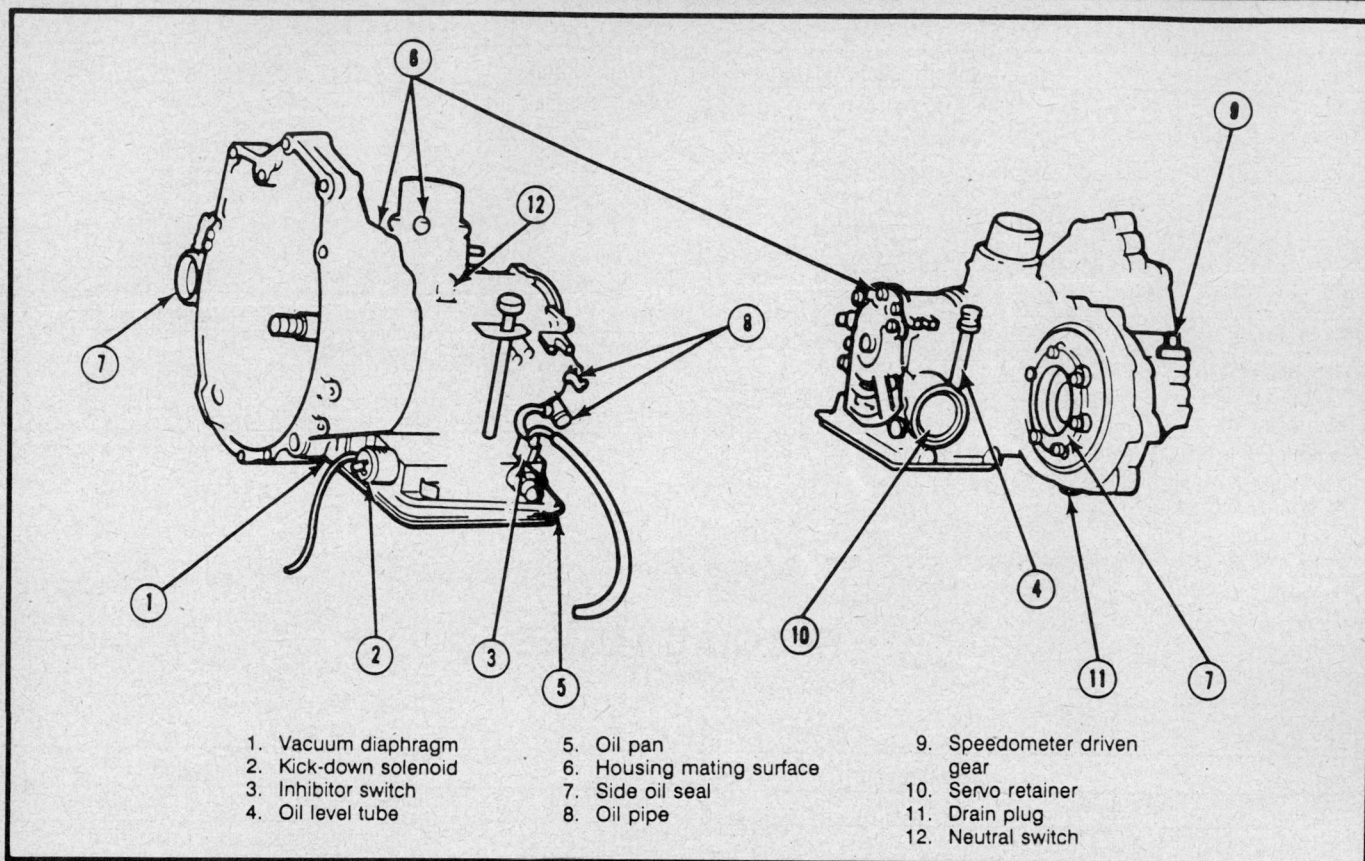

1. Vacuum diaphragm
2. Kick-down solenoid
3. Inhibitor switch
4. Oil level tube
5. Oil pan
6. Housing mating surface
7. Side oil seal
8. Oil pipe
9. Speedometer driven gear
10. Servo retainer
11. Drain plug
12. Neutral switch

Fluid leakage locations

Section 3

KF100 Transaxle
General Motors

APPLICATION

1985–88 Spectrum
1989 GEO Spectrum

GENERAL DESCRIPTION

The KF100 is a Jatco type automatic transaxle designed for front wheel drive vehicles. It is a compact type transaxle, consisting of a transmission and differential in 1 unit. The transaxle is designed so that the preload on the tapered roller bearings can be adjusted by the use of shims. The final gear in the transaxle is a helical type design requiring no tooth contact adjustment. The unit is designed so that the transmission portion and differential use Dexron® II automatic transmission fluid.

Transaxle and Converter Identification

TRANSAXLE

The transaxle identification is stamped on the flange at the front of the transaxle bell housing, on the top left hand side.

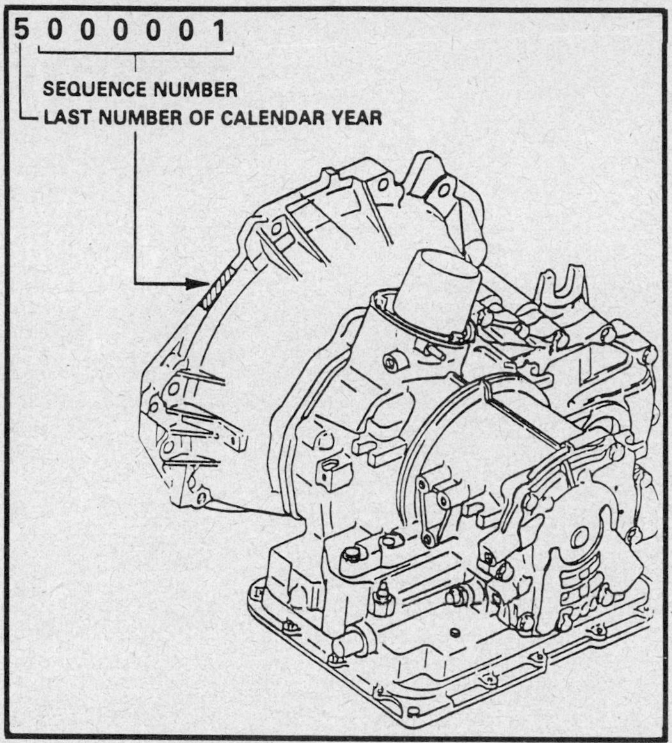
5 0 0 0 0 0 1
SEQUENCE NUMBER
LAST NUMBER OF CALENDAR YEAR

Transaxle identification

TORQUE CONVERTER

The torque converter is a welded unit and cannot be disassembled for repairs. Should this unit need to be replaced, both new and rebuilt units are available.

Metric Fasteners

This transaxle has metric bolts, nuts and threads. Do not replace metric size components with inch size equivalents. Do not use inch size tools to loosen, tighten, adjust or remove the metric size component parts.

Replacement fasteners must have the same measurement and strength as those removed. Mismatched or incorrect fasteners can result in damage to the transaxle unit through malfunctions, breakage or possible personal injury.

Capacities

The fluid capacity for the KF100 automatic transaxle is 6.1 qts. (5.8L) of Dexron® II automatic transmission fluid.

Checking Fluid Level

The vehicle must have been driven so that the engine and transaxle are at normal operating temperature (fluid temperature 158–176° F). If the fluid smells burnt or is black, replace it.

1. Park the vehicle on a level surface and set the parking brake.
2. With the engine idling, shift the selector into each gear from **P** range to **L** range and return to **P** range.
3. Pull out the transaxle dipstick and wipe it clean.
4. Push the dipstick back fully into the tube.
5. Pull out the dipstick and check that the fluid is in the **HOT** range.
6. If the level is low, add fluid. Do not overfill.

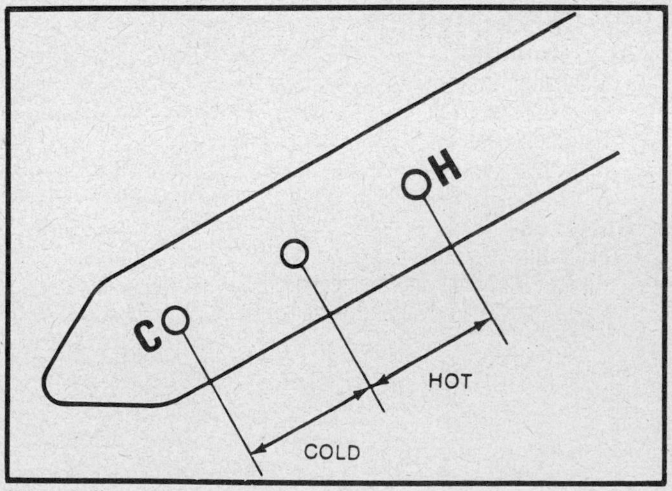
H
C
HOT
COLD

Dipstick

TRANSAXLE MODIFICATIONS

There are no transaxle modifications available at the time of publication.

TROUBLE DIAGNOSIS

CLUTCH AND BAND APPLICATION

Shift Position		Gear Ratio	Clutch Front	Clutch Rear	Low and Reverse Brake	Band Servo Operation	Band Servo Release	One-way Clutch
P		—	—	—	Applied	—	—	—
R		2.400	Applied	—	Applied	—	①	—
N		—	—	—	—	—	—	—
D	1st	2.841	—	Applied	—	—	—	Applied
	2nd	1.541	—	Applied	—	Applied	—	—
	3rd	1.000	Applied	Applied	—	①	①	—
2nd		1.541	—	Applied	—	Applied	—	—
1	2nd	1.541	—	Applied	—	Applied	—	—
	1st	2.841	—	Applied	Applied	—	—	—

① Part is operating under normal line pressure, but not transmitting power

CHILTON THREE "C" TRANSAXLE DIAGNOSIS

Condition	Cause	Correction
Fluid discolored or smells burnt	a) Fluid contaminated b) Torque converter faulty c) Transmission faulty	a) Replace fluid b) Replace torque converter c) Disassemble and inspect transmission
Vehicle does not move in any forward range or reverse	a) Transaxle manual control cable out of adjustment b) Valve body or primary regulator valve faulty c) Transmission faulty	a) Adjust control cable b) Inspect valve body c) Disassemble and inspect transmission
Vehicle does not move in any range	a) Parking lock pawl faulty b) Valve body or primary regulator valve faulty c) Torque converter faulty d) Converter drive plate broken e) Oil pump intake strainer blocked f) Transmission faulty	a) Inspect park pawl b) Inspect valve body c) Replace torque converter d) Replace torque converter e) Clean strainer f) Disassemble and inspect transmission
Shift lever position incorrect	a) Transaxle manual control cable out of adjustment b) Manual valve and lever faulty c) Transmission faulty	a) Adjust control cable b) Inspect valve body c) Disassemble and inspect transmission
Harsh engagement into any drive range	a) Throttle cable out of adjustment b) Valve body or primary regulator valve faulty c) Accumulator pistons faulty d) Transmission faulty	a) Adjust throttle cable b) Inspect valve body c) Inspect accumulator pistons d) Disassemble and inspect transmission
Delayed 1-2, 2-3 or 3-OD upshift, or downshifts from OD-3 or 3-2 then shifts back to OD or 3	a) Throttle cable out of adjustment b) Governor faulty c) Valve body faulty	a) Adjust throttle cable b) Inspect governor c) Inspect valve body
Slips on 1-2, 2-3 or 3-OD upshift; or slips or shudders on take-off	a) Transaxle manual control cable out of adjustment b) Throttle cable out of adjustment c) Valve body faulty d) Transmission faulty	a) Adjust control cable b) Adjust throttle cable c) Inspect valve body d) Disassemble and inspect transmission
Drag, binding or tie-up on 1-2, 2-3 or 3-OD upshift	a) Transaxle manual control cable out of adjustment b) Valve body faulty c) Transmission faulty	a) Adjust control cable b) Inspect valve body c) Disassemble and inspect transmission

CHILTON THREE "C" TRANSAXLE DIAGNOSIS

Condition	Cause	Correction
Harsh downshift	a) Throttle cable out of adjustment b) Accumulator pistons faulty c) Valve body faulty d) Transmission faulty	a) Adjust throttle cable b) Inspect accumulator pistons c) Inspect valve body d) Disassemble and inspect transmission
No downshift when coasting	a) Governor faulty b) Valve body faulty	a) Inspect governor b) Inspect valve body
Downshift occurs too quick or too late while coasting	a) Throttle cable out of adjustment b) Governor faulty c) Valve body faulty d) Transmission faulty	a) Adjust throttle cable b) Inspect governor c) Inspect valve body d) Disassemble and inspect transmission
No OD-3, 3-2 or 2-1 kick-down	a) Throttle cable out of adjustment b) Governor faulty c) Valve body faulty	a) Adjust throttle cable b) Inspect governor c) Inspect valve body
No engine braking in "2" range	a) Valve body faulty b) Transmission faulty	a) Inspect valve body b) Disassemble and inspect transmission
Vehicle does not hold in "P"	a) Transaxle manual control cable out of adjustment b) Parking lock pawl and rod	a) Adjust control cable b) Inspect lock pawl and rod

Hydraulic Control System

NOTE: Please refer to Section 9 for all oil flow circuits.

PUMP ASSEMBLY

The hydraulic pressure system requires a supply of transmission fluid and a pump to pressurize the fluid. The KF100 transaxle uses an internal-external gear type pump with its oil intake connected to a screen assembly.

The oil pump is designed to deliver fluid to the torque converter, lubricate the planetary gear unit and supply operating pressure to the hydraulic control system. The drive gear of the oil pump and the torque converter pump are driven by the engine. The pump has a sufficient capacity of oil to supply the necessary fluid pressure throughout all forward speed ranges and reverse.

MANUAL VALVE

The manual valve is linked to the gear shift lever and directs the fluid to the gear range circuit that the lever is positioned at.

PRIMARY REGULATOR VALVE

The primary regulator valve varies the hydraulic line pressure to each component in order to conform with engine power and operate all transaxle hydraulic systems.

SECONDARY REGULATOR VALVE

This valve regulates the converter pressure and lubrication pressure. Spring tension in the valve acts in an upward direction. Converter fluid pressure and lubrication pressure are determined by the spring tension.

THROTTLE VALVE

The throttle valve acts to produce throttle pressure in response to accelerator pedal modulation or engine output. When the accelerator pedal is depressed, the downshift plug is pushed upward by the throttle cable and throttle cam. The throttle valve also moves upward by means of the spring, opening the pressure passage for creation of throttle pressure.

THROTTLE MODULATOR VALVE

This valve produces throttle modulator pressure. It reduces throttle pressure when the throttle valve opening angle is high. It causes throttle modulator pressure to act on the primary regulator valve so that line pressure performance is close to engine power performance.

GOVERNOR VALVE

The governor valve is driven by the drive pinion worm gear and produces governor pressure in response to the vehicle speed. It balances the line pressure from the primary regulator valve and the centrifugal force of the governor weights to produce hydraulic pressure in proportion to vehicle speed.

ACCUMULATORS

The accumulators act to cushion the shifting shock. There are 3 accumulators: 1 each for the forward clutch (C_1), direct clutch (C_2) and the 2nd brake (B_2). The accumulators are located in the transaxle case. Accumulator control pressure is always acting on the back pressure side of the C_2 and B_2 pistons. This pressure along with spring tension, pushes down on the 2 pistons.

When line pressure is applied to the operating side, the pistons are pushed upward and shock is cushioned as the fluid pressure gradually rises. Operation of the C_1 piston is basically the same as that for C_2 and B_2. However the force pushing the piston downward is accomplished by spring tension only.

1–2 SHIFT VALVE

This valve automatically controls the 1–2 shift according to governor and throttle pressure. To improve the valve sliding characteristics, a 3 piece valve is used. When governor pressure is low and throttle pressure is high, the valve is pushed down by throttle pressure and because the 2nd brake circuit closes, the transaxle shifts into 1st gear.

When governor pressure is high and throttle pressure low, the valve is pushed up by governor pressure and the circuit to the 2nd brake piston opens so the transaxle will shift into 2nd gear. When the throttle pressure passage is closed, downshifting into

1st gear is dependant on spring tension and governor pressure only.

Unless the downshift plug actuates and allows the detent pressure to act on the 1–2 shift valve, downshifting into 1st gear will take place at a set vehicle speed. In the **L** range, there is no upshifting into 2nd gear because low modulator pressure is acting on the low coast shift valve.

2–3 SHIFT VALVE

This valve performs shifting between 2nd and 3rd gears. Control is accomplished by opposing throttle pressure and spring tension against governor pressure. When governor pressure is high, the valve is pushed up against the resistance of the throttle pressure and spring tension. This opens the passage to the direct clutch (C_2) piston to allow the shift into 3rd gear.

When governor pressure is low, the valve is pushed down by throttle pressure and spring tension to close the passage leading to the direct clutch piston, causing a downshift to 2nd gear. In the event of kickdown, the detent pressure acts on the 2–3 shift valve to permit a quicker downshift to 2nd gear. valve movement occurs due to the different size areas where pressure is applied. Since the area is larger for downshift than for upshift, manual **2** range, line pressure from the manual valve acts on the intermediate shift valve. The valve descends and shifting into 2nd gear is accomplished but there is no upshifting into 3rd gear. Line pressure passes through the 2nd modulator valve and 1–2 shift valve and acts on the 2nd coast brake to provide engine braking.

Diagnosis Tests

HYDRAULIC TESTS

1. Run the engine until it reaches normal operating temperature.
2. Raise the vehicle and support it safely.
3. Remove the transaxle case test plugs and install hydraulic pressure gauges.

LINE PRESSURE

1. Apply the parking brake.
2. Start the engine.
3. Apply the brake pedal while manipulating the accelerator pedal and measure the line pressure at the engine speeds as specified:
   D range—idling 43–57 psi
   D range—stall 128–156 psi
   2 range—idling 114–171 psi
   2 range—stall 114–171 psi
   R range—idling 57–110 psi
   R range—stall 228–270 psi
4. In the same manner, perform the test for the **R** range.
5. If the measured pressures are not up to specified values, recheck the throttle cable adjustment and retest.
6. If the measured values at all ranges are higher than specified, check the following:
   a. Throttle cable out of adjustment
   b. Throttle valve defective
   c. Regulator valve defective
7. If the measured values at all ranges are lower than specified, check the following:
   a. Throttle cable out of adjustment
   b. Throttle valve defective
   c. Regulator valve defective
   d. Oil pump defective
8. If the pressure is low in **D** and **2** range only, check the following:
   a. **D** and **2** range circuit fluid leakage
   b. Rear clutch governor defective
9. If the pressure is low in **R** range only, check the following:
   a. **R** range circuit fluid leakage
   b. First and reverse brake defective

GOVERNOR PRESSURE

1. Apply the parking brake.
2. Start the engine.
3. Shift into **D** range and measure the governor pressure at the speeds specified in the following chart:
   Vehicle speed 20 mph—gauge reading 17–21 psi
   Vehicle speed 35 mph—gauge reading 30–35 psi
   Vehicle speed 45 mph—gauge reading 58–64 psi
4. If the governor pressure is not to specification, check for the following:
   a. Line pressure not to specification
   b. Fluid leakage in the governor pressure cuircuit
   c. Improper governor valve operation

STALL SPEED TEST

The object of this test is to check the overall performance of the transaxle and engine by measuring the maximum engine speeds in the **D** and **R** ranges.

1. Run the engine until it reaches normal operating temperature.

NOTE: Do not continuously run this test longer than 5 seconds.

2. Apply the parking brake and block the front and rear wheels.
3. Connect an engine tachometer.
4. Shift into the **D** range.
5. While applying the brakes, step all the way down on the accelerator. Quickly read the highest rpm at this time. The stall speed should be 2050–2350 rpm and the main line pressure should be 128–156 psi.
6. Perform the same test in the **R** range.
7. If the engine speed is the same for all ranges, but lower than the specified value, check for the following:
   a. Engine output insufficient
   b. Stator one-way clutch not operating properly
8. If the engine speed is the same for all ranges and within the specified value, the speed control elements in the transaxle are all normal. Check for a faulty engine.
9. If the stall speed in **D** range is higher than specified, check for a slipping one-way clutch.
10. If the stall speed in **2** range is higher than specified, check for a slipping brake band.
11. If the stall speed in **D**, **2** and **1** is higher than the specified value, check for a slipping rear clutch.
12. If the stall speed in **R** range is higher than specified, check for the following:
    a. Line pressure too low
    b. Front clutch slipping
    c. Low and reverse brake slipping
13. If the stall speed is the same in all ranges, but higher than the specified value, check for the following:
    a. Line pressure too low
    b. Oil pump is weak
    c. Oil leaks from the oil pump control valve or the transaxle case.
    d. Pressure regulator valve is sticky.

ROAD TEST

NOTE: This test must be performed with the engine and transaxle at normal operating temperature.

DRIVE RANGE TEST 1

Shift into **D** range and while driving with the accelerator pedal

Check shift points under following conditions:

Throttle Valve Opening	Completely Closed	Speed Changing Point
Wide open throttle (Kickdown) (0 — 100 mm-Hg, 0 — 3.94 in-Hg)	$D_1 \rightarrow D_2$ $D_2 \rightarrow D_3$ $D_3 \rightarrow D_2$ $D_2 \rightarrow D_1$	51 — 60 km/h (32 — 37 mph) 101 — 110 km/h (63 — 68 mph) 90 — 99 km/h (56 — 62 mph) 40 — 49 km/h (25 — 30 mph)
Half throttle (200 ± 10 mm-Hg, 7.87 ± 0.39 in-Hg)	$D_1 \rightarrow D_2$ $D_2 \rightarrow D_3$ $D_3 \rightarrow D_2$ $D_2 \rightarrow D_1$	12 — 21 km/h (8 — 13 mph) 55 — 63 km/h (34 — 39 mph) 28 — 36 km/h (17 — 22 mph) 10 — 18 km/h (6 — 11 mph)
Fully closed throttle Manual "1"	$1_2 \rightarrow 1$	40 — 48 km/h (25 — 30 mph)

SHIFT SCHEDULE

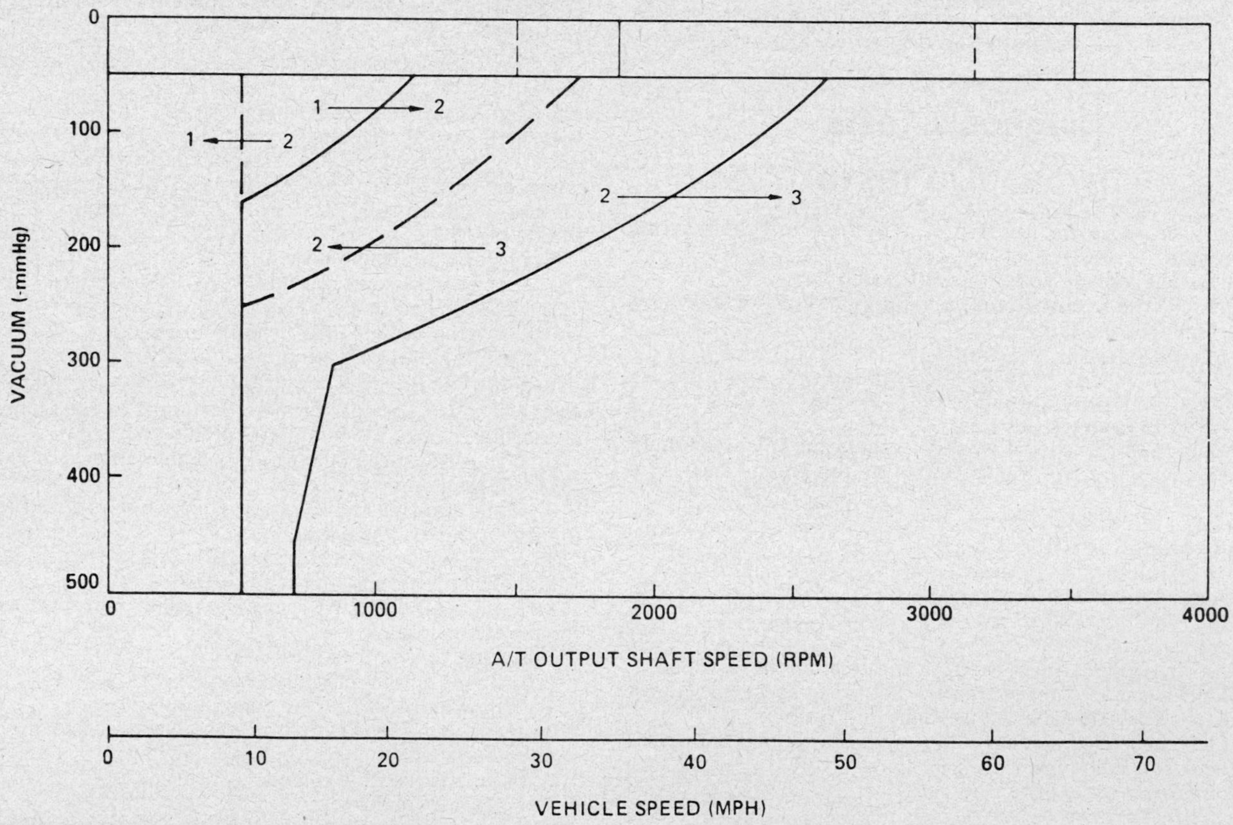
0
100
200
300
400
500
VACUUM (-mmHg)
1 → 2
1 ← 2
2 → 3
2 ← 3
0
1000
2000
3000
4000
A/T OUTPUT SHAFT SPEED (RPM)
0
10
20
30
40
50
60
70
VEHICLE SPEED (MPH)

Road test chart

held constant (throttle valve opening 50% and 100%) check the following points:

1. At each of the throttle openings, check that the 1–2 and 2–3 upshifts take place and that the shift points conform to those shown in the automatic shift diagram.

2. If there is no 1–2 upshift, the governor valve may be defective or the 1–2 shift valve stuck.

3. If there is no 2–3 upshift, the 2–3 shift valve is stuck.

4. If the shift point is defective, the throttle cable is out of adjustment or the throttle valve, 1–2 shift valve and the 2–3 shift valves are defective.

DRIVE RANGE TEST 2

Check for harsh shift or soft shift slip at 1–2 and 2–3 upshifts. If the shock is severe, the line pressure is too high, the accumulator is defective or the check ball is defective.

DRIVE RANGE TEST 3

1. While driving in **D** range, 3rd gear, check for abnormal noise and vibration. This condition could also be caused by unbalance in the differential, tires or the torque converter.
2. While driving in **D** range, 2nd and 3rd gears, check the kickdown vehicle speed limits for the 2–1, 3–1 and the 3–2 kickdowns conform to the automatic shift diagram.
3. While driving in **D** range, 3rd gear, shift to **2** and **L** ranges and check the engine braking effect at each of these ranges. If there is no engine braking effect at **2** range, the 2nd coast brake is defective. If there is no engine braking effect at **L** range, 1st and reverse is defective.

DRIVE RANGE TEST 4

While driving in **D** range, release the accelerator pedal and shift into **L** range. Check to see if the 3–2 and 2–1 downshift points conform to those given in the automatic shift diagram.

2 RANGE TEST

1. Shift to **2** range and drive with the throttle valve opening at 50% and 100% respectively. Check the 1–2 upshift points at each of the throttle valve openings to see that it conforms to those indicated in the automatic shift diagram.
2. While driving in **2** range, 2nd gear, release the accelerator pedal and check the engine braking effect.
3. Check the kickdown from **2** range. Check the 2–1 kickdown vehicle speed limit.
4. Check for abnormal noise at acceleration and deceleration and harshness at upshift and downshift.

L RANGE TEST

1. While driving in **L** range, check that there is no upshift to 2nd gear.
2. While driving in **L** range, release the accelerator pedal and check the engine braking effect.
3. Check for abnormal noise during acceleration and deceleration.

R RANGE TEST

Shift into **R** range and while starting at full throttle, check for slipping.

P RANGE TEST

Stop the vehicle on a slight hill and after shifting into **P** range, release the parking brake. Check to see that the parking lock pawl prevents the vehicle from moving.

ON CAR SERVICES

Adjustments

MANUAL LINKAGE

1. When shifting the select lever from **P** to **1**, a clicking should be felt at each shift position. Make sure that the gear corresponds to that of the position plate indicator.
2. Check that the lever can be shifted between **D** and **N** without depressing the push button.
3. Shifting from **D** to **R** cannot be done without depressing the push button. If the lever can be shifted from **D** to **R** without depressing the push button, or if the push button is loose, adjust it by unsrewing the locknut and twisting the selector lever knob.

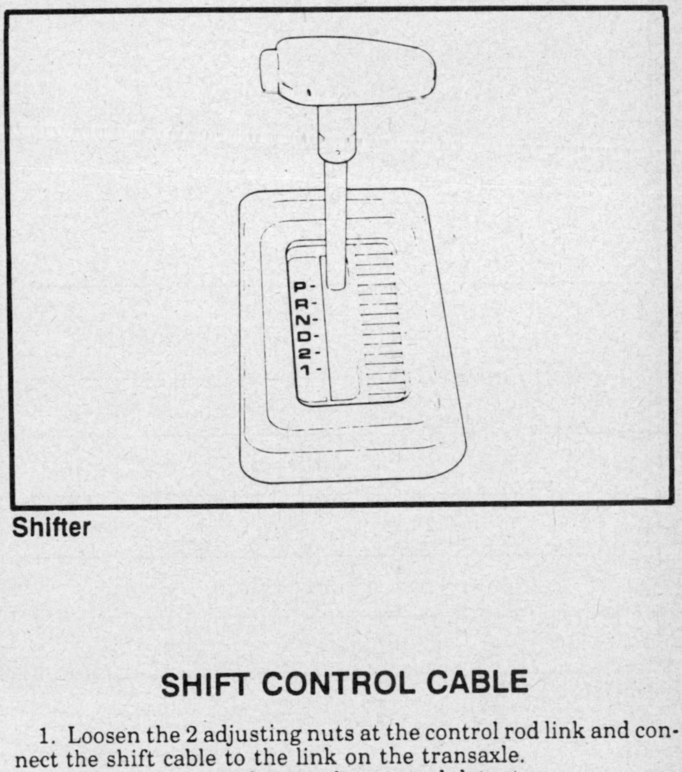
P
R
N
D
2
1
Shifter

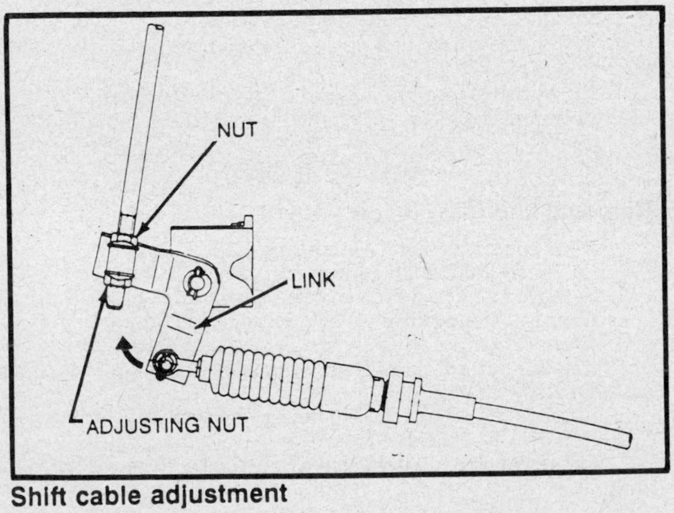
NUT
LINK
ADJUSTING NUT
Shift cable adjustment

SHIFT CONTROL CABLE

1. Loosen the 2 adjusting nuts at the control rod link and connect the shift cable to the link on the transaxle.
2. Shift the transaxle into the neutral detent.
3. Place the shifter lever in the **N** position.
4. Rotate the link assembly clockwise to remove the slack in the cable.
5. Tighten the rear adjusting nut until it makes contact with the link.
6. Tighten the front adjusting nut until it makes contact with the link and then tighten the nuts.

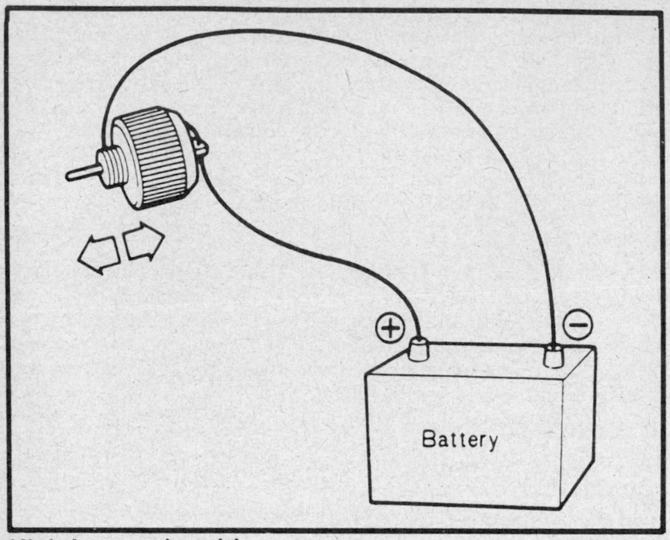
Battery

Kickdown solenoid

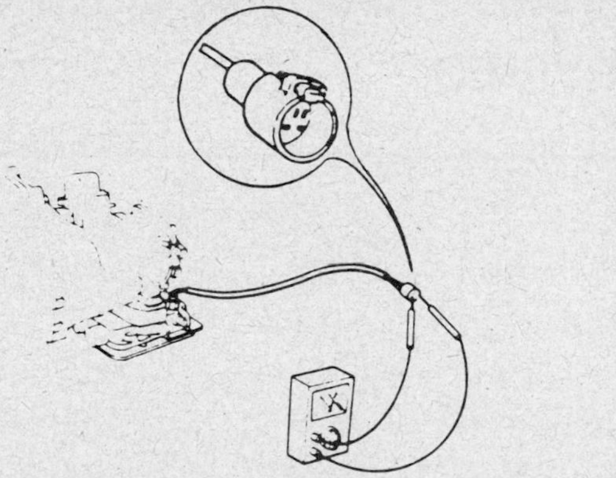

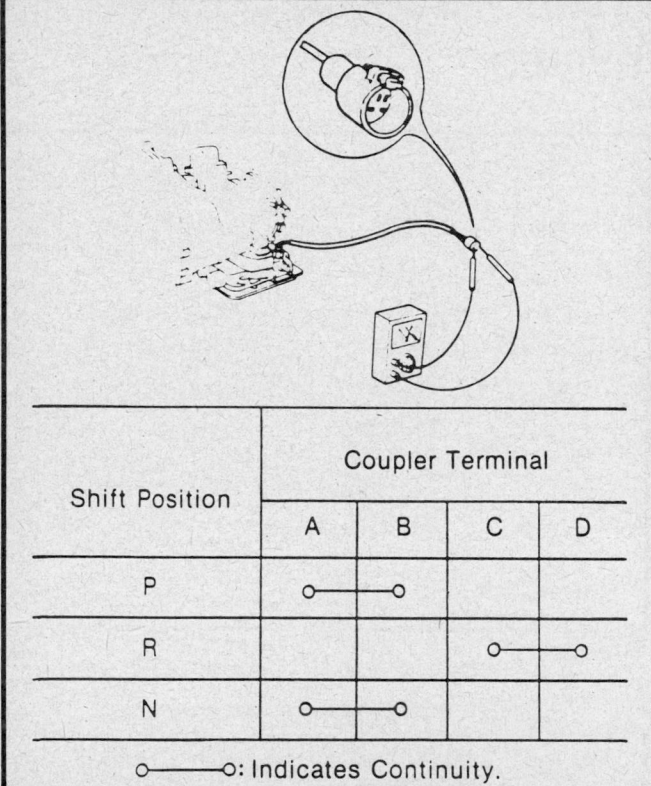

Shift Position	Coupler Terminal			
	A	B	C	D
P	o——	——o		
R			o——	——o
N	o——	——o		

o———o: Indicates Continuity.

Inhibitor switch

PARK LOCK CABLE

1. Place the ignition key in the **LOCK** position.
2. Place the shifter lever in the **P** position.
3. Pull the cable forward at the shifter bracket and tighten the forward nut until it makes contact with the bracket.
4. Tighten the rear nut until it makes contact with the bracket and then tighten the nuts.

KICKDOWN SOLENOID

1. Make sure the rod functions properly when 12 volts is applied to the kickdown solenoid.
2. Connect a circuit tester to the solenoid terminals. Check if continuity exists when depressing the pedal fully.
3. If continuity does not exist, adjust the kickdown switch by turning the switch so that continuity exists when depressing the pedal more than 7/8 of its stroke.

NEUTRAL SAFETY SWITCH

1. Check that the engine starts only in the **P** and **N** detents.
2. Check that the back-up light is **ON** when the selector lever is in the **R** detent.
3. If the neutral safety switch is faulty, disconnect the switch and check the continuity between each terminal.

Services

FLUID CHANGES

The conditions under which the vehicle is operated is the main consideration in determining how often the transaxle fluid should be changed. Different driving conditions result in different transaxle fluid temperatures. These temperatures effect the change intervals.

If the vehicle is driven under severe service conditions, change the fluid and filter every 15,000 miles. If the vehicle is not used under severe service conditions, change the fluid and replace the filter every 50,000 miles.

Do not overfill the transaxle. It takes 1 pint of fluid to change the level from **ADD** to **FULL** on the transaxle dipstick. Overfilling the unit can cause damage to the internal components of the automatic transaxle.

OIL PAN

Removal and Installation

1. Disconnect the negative battery cable.
2. Raise the vehicle and support it safely.
3. Remove the drain plug in the pan and drain the fluid.
4. Reinstall the drain plug and remove the oil pan. After removing the bolts, tap the pan lightly with a plastic hammer. Do not force the pan off by prying. This may cause damage to the gasket mating surface.
5. Remove the oil pan gasket material on the mating surface.
6. Remove the oil filter from the valve body.
7. Clean the inside of the oil pan before installation.
8. Install a new filter on the valve body.
9. Install the pan with a new gasket and torque the bolts to 5 ft. lbs. (7 Nm).
10. Fill with transmission fluid to the proper level.
11. Lower the vehicle and connect the battery cable.

VALVE BODY

Removal and Installation

1. Disconnect the negative battery cable.
2. Raise the vehicle and support it safely.
3. Remove the drain plug in the pan and drain the fluid.
4. Reinstall the drain plug and remove the oil pan. After removing the bolts, tap the pan lightly with a plastic hammer. Do not force the pan off by prying. This may cause damage to the gasket mating surface.
5. Remove the oil pan gasket material on the mating surface.
6. Remove the oil filter from the valve body.
7. Loosen and remove the valve body bolts. Remove the valve body.

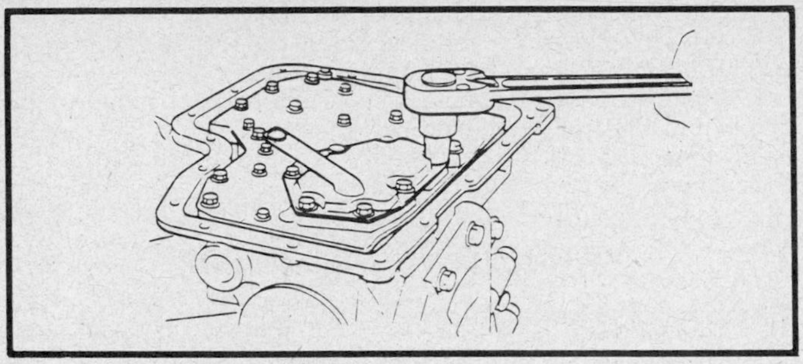

Valve body assembly

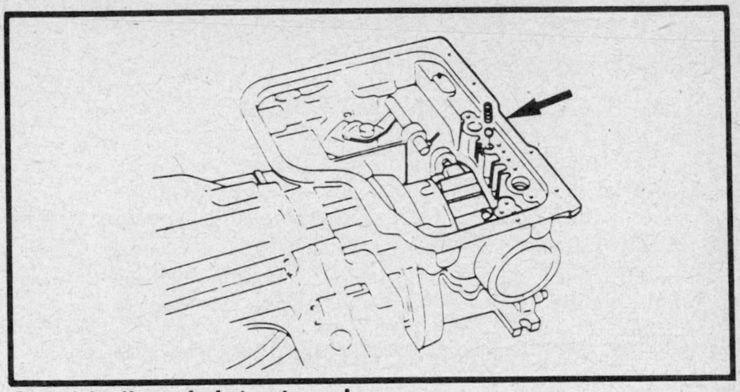

Steel ball and detent spring

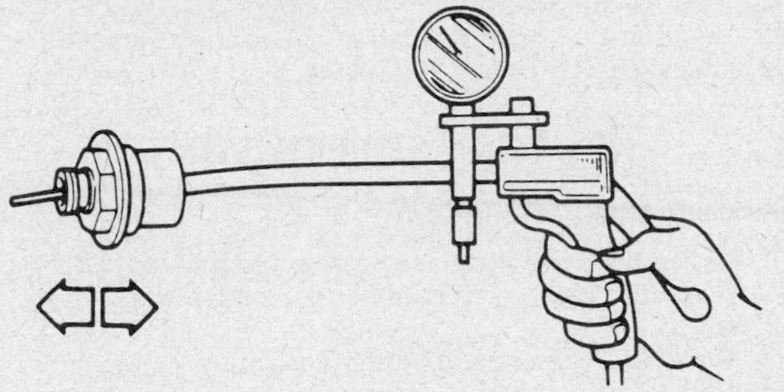

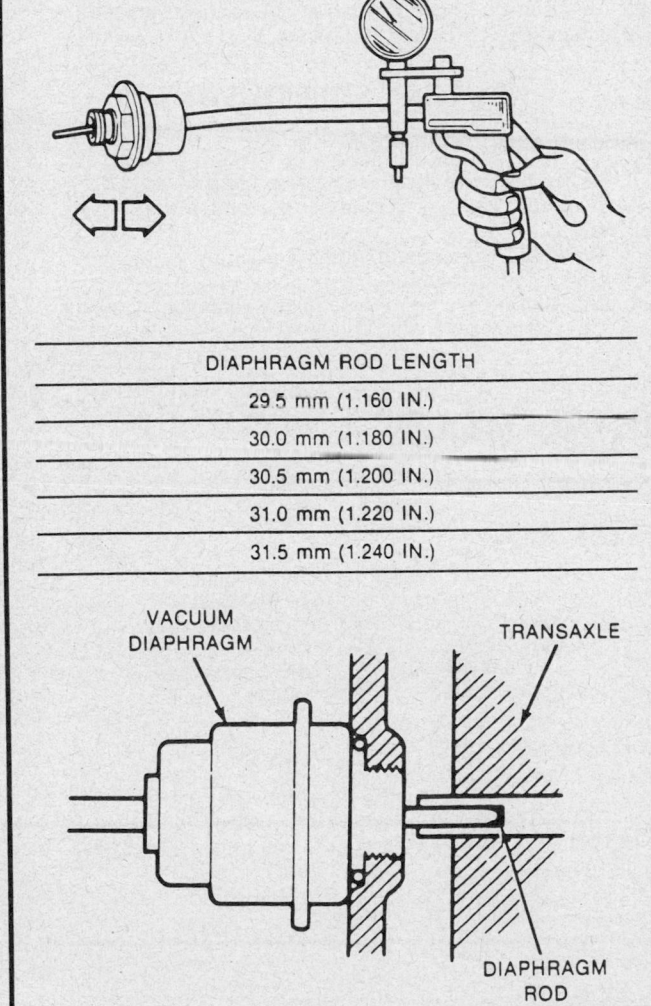

DIAPHRAGM ROD LENGTH
29.5 mm (1.160 IN.)
30.0 mm (1.180 IN.)
30.5 mm (1.200 IN.)
31.0 mm (1.220 IN.)
31.5 mm (1.240 IN.)

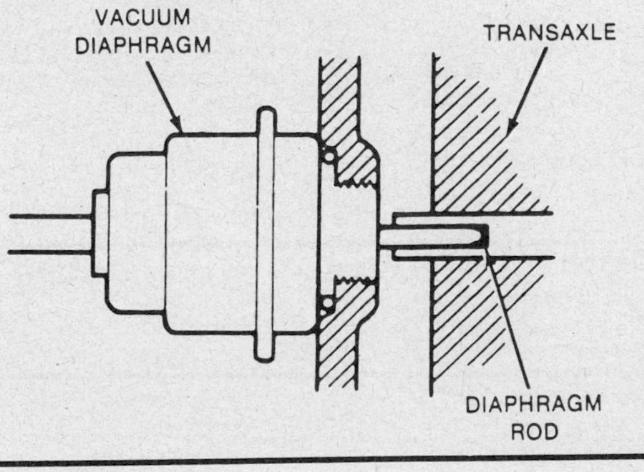
VACUUM DIAPHRAGM
TRANSAXLE
DIAPHRAGM ROD

Vacuum diaphragm

8. Remove the spring and steel ball from the oil passage.
9. On installation, install steel ball and spring in the oil passage.
10. Install the valve body with the bolts finger tight.
11. Align the manual valve with the shift lever arm and torque the bolts to 7.2 ft. lbs. (8 Nm).
12. Clean the inside of the oil pan before installation.
13. Install the oil pan with a new gasket and torque the bolts to 5 ft. lbs. (7 Nm).

SERVO ASSEMBLY

Removal and Installation

1. Disconnect the negative battery cable.
2. Raise the vehicle and support it safely.
3. Remove the left front wheel and tire assembly.
4. Remove the left lower control arm tension rod assembly.
5. Disconnect the tie rod from the left steering arm.
6. Remove the oil pan bolt nearest to the servo cover.
7. Install tool J-35278 and compress the servo cover. Remove the cover retaining ring and then the cover.
8. Remove tool J-35278 and the servo assembly from the transaxle.
9. Installation is the reverse of the removal procedure. Check the fluid level.

GOVERNOR

Removal and Installation

1. Disconnect the negative battery cable.
2. Raise the vehicle and support it safely.
3. Remove the governor cover retaining bolts.
4. Remove the governor assembly.
5. Installation is the reverse of the removal procedure.

SHIFT CONTROL CABLE

Removal and Installation

1. Disconnect the negative battery cable.
2. Remove the floor console.
3. Disconnect the cable at the shifter.
4. Pull the floor carpet rearward and remove the screws attaching the cable to the floor.
5. Raise the vehicle and support it safely.
6. Disconnect the cable from the transaxle.
7. Remove the cable from the vehicle.
8. Installation is the reverse of the removal procedure.

PARK LOCK CABLE

Removal and Installation

1. Disconnect the negative battery cable.
2. Remove the floor console.
3. Disconnect the cable and the shifter and remove the adjusting nuts.
4. Remove the lower trim cover from the steering column.
5. Pull the floor carpet rearward and remove the cable.
6. Installation is the reverse of the removal procedure.

SHIFTER CONTROL

Removal and Installation

1. Disconnect the negative battery cable.
2. Remove the floor console.
3. Disconnect the shift cable at the control.
4. Disconnect the park lock cable at the shifter.
5. Remove the bolts attaching the control to the floor and remove the control assembly.
6. Installation is reverse of the removal procedure.
7. Adjust the shift cable and the park lock cable.

SPEEDOMETER DRIVEN GEAR

Removal and Installation

1. Disconnect the negative battery cable.
2. Raise the vehicle and support it safely.
3. Disconnect the speedometer cable at the transaxle.
4. Remove the retainer bolt, retainer, speedometer driven gear and the O-ring seal.
5. Installation is the reverse of the removal procedure.
6. Use a new O-ring seal and check the fluid level.

VACUUM DIAPHRAGM

Removal and Installation

1. Disconnect the negative battery cable.
2. Disconnect the kickdown solenoid wire connector at the left fender.
3. Raise the vehicle and support it safely.
4. Remove the kickdown solenoid at the transaxle.
5. Remove the vacuum diaphragm.
6. Make sure that when vacuum is applied to the diaphragm, the diaphragm rod moves properly.
7. Installation is the reverse of the removal procedure.
8. Check the transaxle fluid level.

KICKDOWN SOLENOID

Removal and Installation

1. Disconnect the negative battery cable.
2. Disconnect the electrical connector for the solenoid at the fender.
3. Raise the vehicle and support it safely.
4. Remove the solenoid.
5. Installation is the reverse of the removal procedure.
6. Check the transaxle fluid level.

NEUTRAL SAFETY SWITCH

Removal and Installation

1. Disconnect the negative battery cable.
2. Disconnect the electrical connector for the switch at the left fender.
3. Raise the vehicle and support it safely.
4. Remove the switch.
5. Installation is the reverse of the removal procedure.
6. Check the transaxle fluid level.

REMOVAL AND INSTALLATION

TRANSAXLE REMOVAL

1. Disconnect the negative battery cable.
2. Remove the air intake hose from the air cleaner.
3. Disconnect the shift cable from the transaxle.
4. Disconnect the speedometer cable.
5. Disconnect the vacuum hose at the vacuum diaphragm.
6. Disconnect the engine wiring harness clamp at the transaxle.
7. Disconnect the neutral safety switch wire connector at the left fender.
8. Disconnect the transaxle cooler lines.
9. Remove the 3 upper transaxle to engine attaching bolts.
10. Raise the vehicle and support it safely.
11. Remove both front wheel and tire assemblies.
12. Remove the splash shield at the left front fender.
13. Disconnect both tie rod ends at the steering knuckles.
14. Remove both front tension rod brackets.
15. Disconnect both tension rods from the control arms.
16. Disengage both drive axle shafts from the transaxle.
17. Remove the flywheel dust cover.
18. Remove the converter to flywheel attaching bolts.
19. Remove the rear mount through bolts at the transaxle.
20. Remove the starter.
21. Support the transaxle.
22. Remove the lower transaxle to engine attaching bolts.
23. Remove the transaxle from the vehicle.

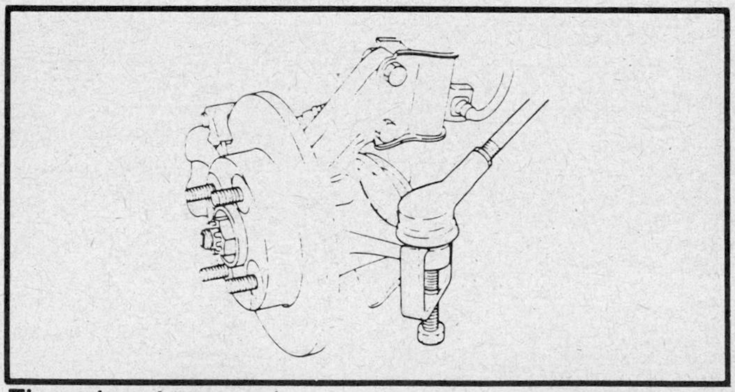

Tie rod end removal

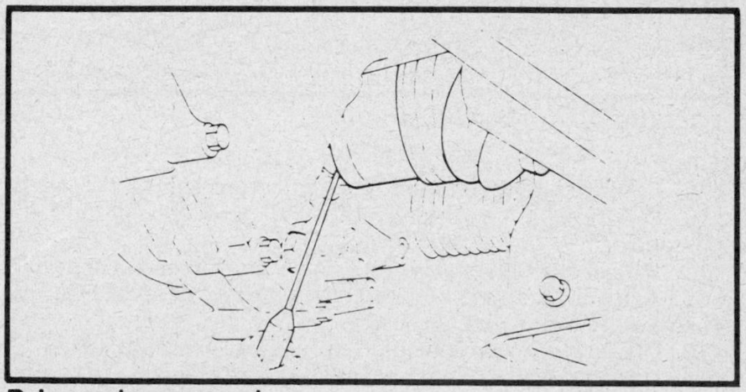

Drive axle removal

TRANSAXLE INSTALLATION

1. Fill the torque converter with Dexron® II automatic transmission fluid and install it on the transaxle mainshaft.
2. Position the transaxle on a suitable transmission jack.
3. Install the transaxle into the vehicle.

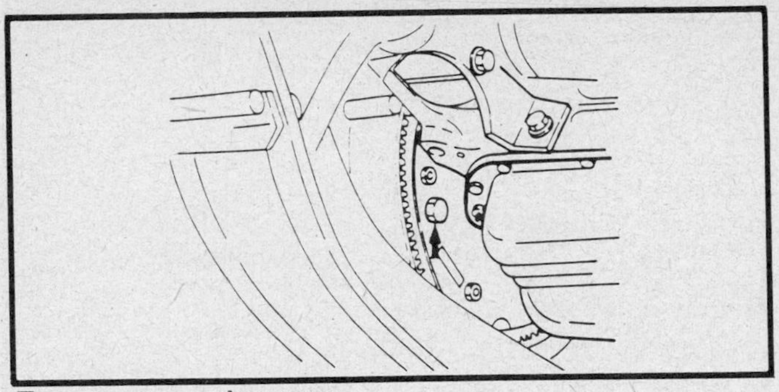

Torque converter cover

4. Install the lower transaxle to engine attaching bolts.
5. Install the starter.
6. Install the rear mount through bolts at the transaxle.
7. Install the converter to flywheel attaching bolts.
8. Install the flywheel dust cover.
9. Connect both drive axle shafts to the transaxle.
10. Connect both tension rods to the control arms.
11. Install both front tension rod brackets.
12. Connect both tie rod ends at the steering knuckles.
13. Install the splash shield at the left front fender.
14. Install both front wheel and tire assemblies.
15. Lower the vehicle.
16. Install the 3 upper transaxle to engine attaching bolts.
17. Connect the transaxle cooler lines.
18. Connect the neutral safety switch wire connector at the left fender.
19. Connect the engine wiring harness clamp at the transaxle.
20. Connect the vacuum hose at the vacuum diaphragm.
21. Connect the speedometer cable.
22. Connect the shift cable to the transaxle.
23. Install the air intake hose to the air cleaner.
24. Connect the negative battery cable.

BENCH OVERHAUL

Before Disassembly

Before opening up the transaxle, the outside of the unit should be thoroughly cleaned, preferably with high pressure cleaning equipment. Dirt entering the transaxle internal parts will negate all the effort and time spent on the overhaul. During inspection and reassembly, all parts should be thoroughly cleaned with solvent and then dried with compressed air. Cloths and rags should not be used to dry the parts since lint will find its way into the valve body passages.

Lube the seals with Dexron® II automatic transmission fluid and use unmedicated petroleum jelly to hold the thrust washers and ease the assembly of the seals. Do not use solvent on neoprene seals, friction plates or thrust washers. Be wary of nylon parts if the transaxle failure was due to a cooling system problem. Nylon parts exposed to antifreeze solutions can swell and distort, so they must be replaced. Before installing bolts into aluminum parts, dip the threads in clean oil.

Converter Inspection

Make certain that the transaxle is held securly. If the torque converter is equipped with a drain plug, open the plug and drain the fluid. If there is no drain plug, the converter must be drained through the hub after pulling the converter out of the transaxle. If the oil in the converter is discolored but does not contain metal bits or particles, the converter is not damaged. Color is no longer a good indicator of fluid condition.

If the oil in the converter contains metal particles, the converter is damaged internally and must be replaced. If the cause of the oil contamination was burned clutch plates or overheated oil, the converter is contaminated and should be replaced. If the pump gears or cover show signs of damage, the converter will contain metal particles and must be replaced.

Transaxle Disassembly

1. Position the transaxle in a suitable holding fixture. Remove the drain plug and drain the fluid from the transaxle.
2. Remove the neutral safety switch and the kickdown solenoid.
3. Remove the vacuum diaphragm and the rod.
4. Remove the dipstick and tube.
5. Remove the speedometer driven gear.

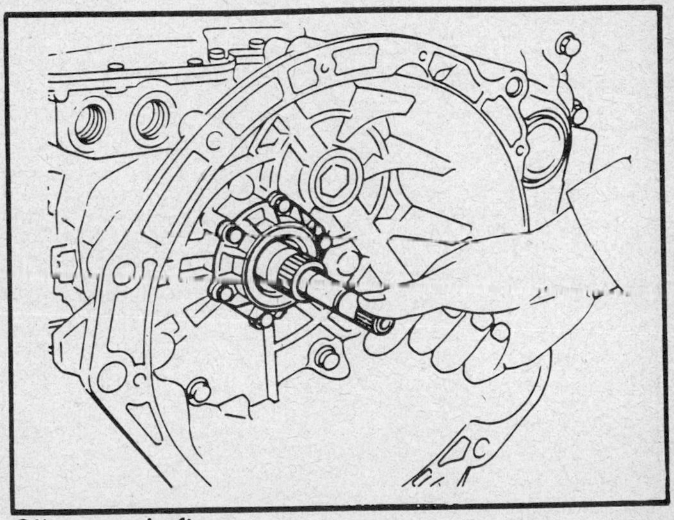

Oil pump shaft

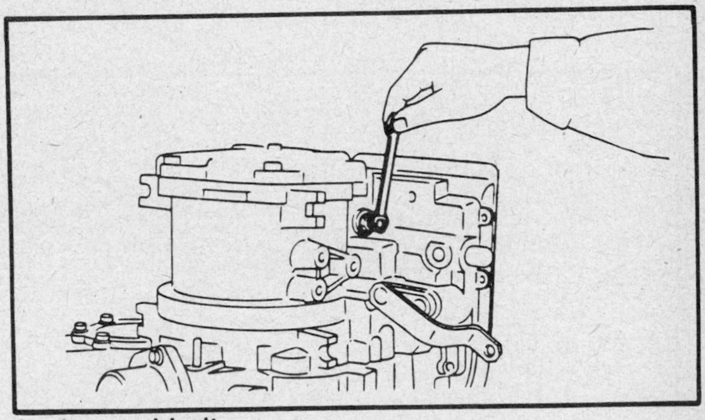

Anchor end bolt

6. Remove the oil pump shaft and then the turbine shaft by pulling outward.
7. Remove the oil pan.

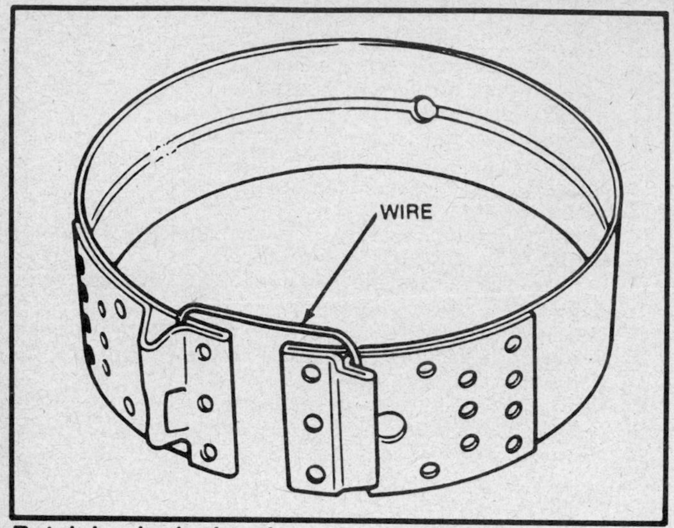
WIRE

Retaining brake band

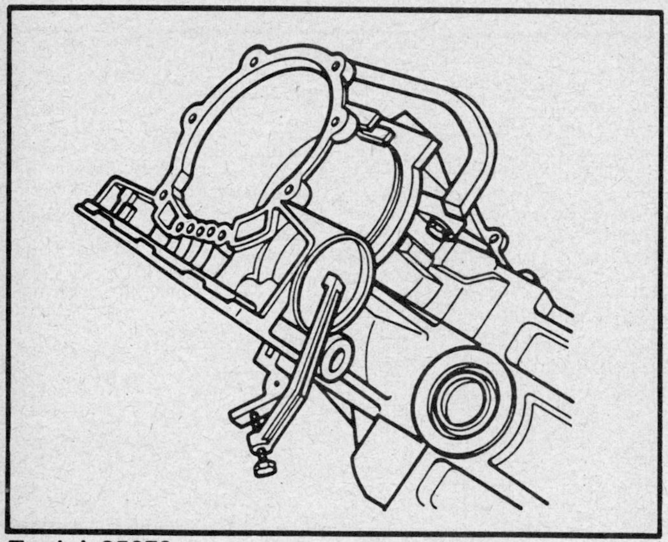

Tool J-35278

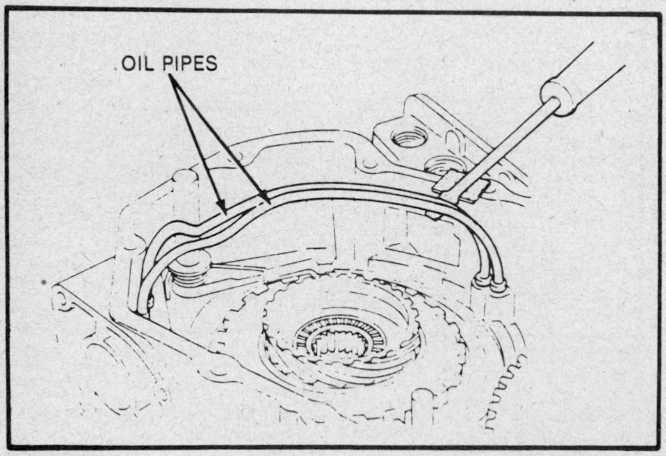
OIL PIPES

Oil pipes

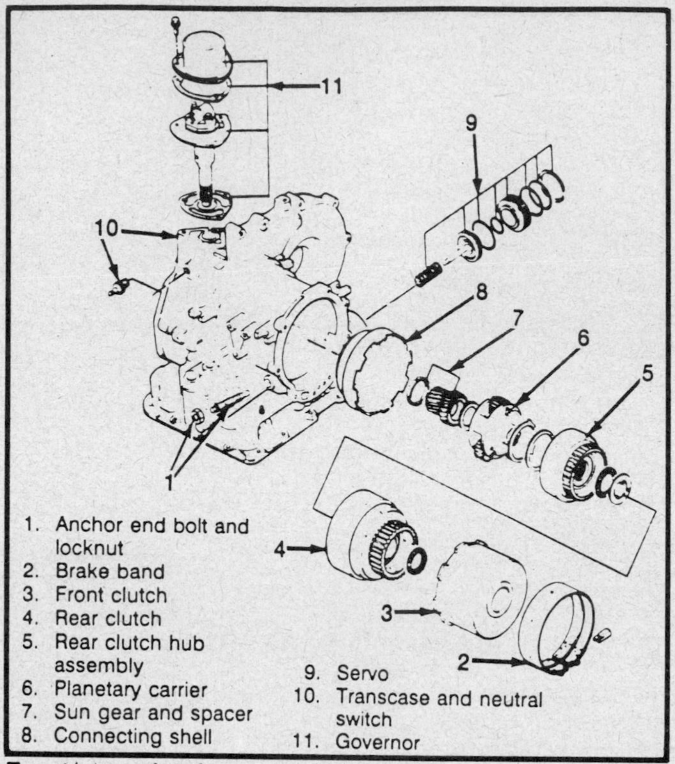
1. Anchor end bolt and locknut
2. Brake band
3. Front clutch
4. Rear clutch
5. Rear clutch hub assembly
6. Planetary carrier
7. Sun gear and spacer
8. Connecting shell
9. Servo
10. Transcase and neutral switch
11. Governor

Front/rear clutch, servo and governor assembly

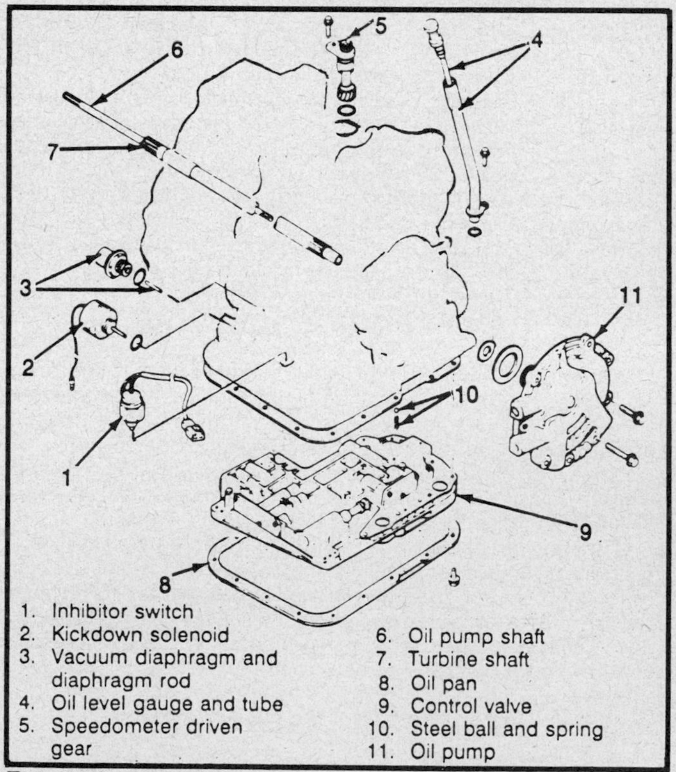
1. Inhibitor switch
2. Kickdown solenoid
3. Vacuum diaphragm and diaphragm rod
4. Oil level gauge and tube
5. Speedometer driven gear
6. Oil pump shaft
7. Turbine shaft
8. Oil pan
9. Control valve
10. Steel ball and spring
11. Oil pump

External controls

8. Remove the valve body assembly.
9. Remove the spring and steel check ball from the case.
10. Remove the oil pump assembly. If the oil pump is difficult to remove, tighten the anchor end bolt and lock the front clutch with the brake band.

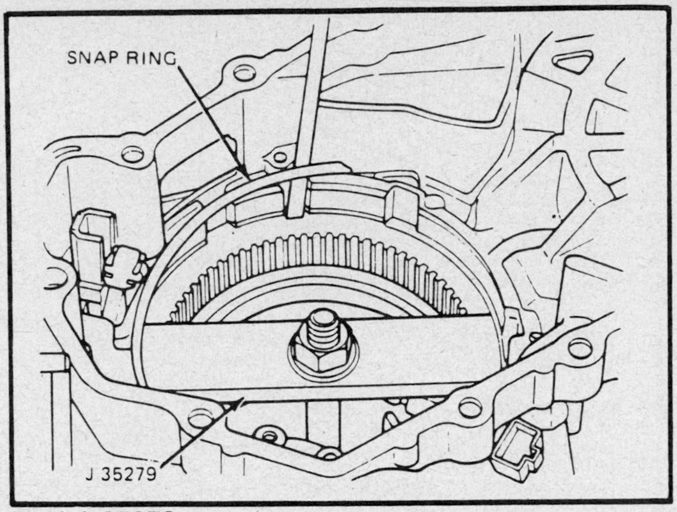
SNAP RING
J 35279
Tool J–35279

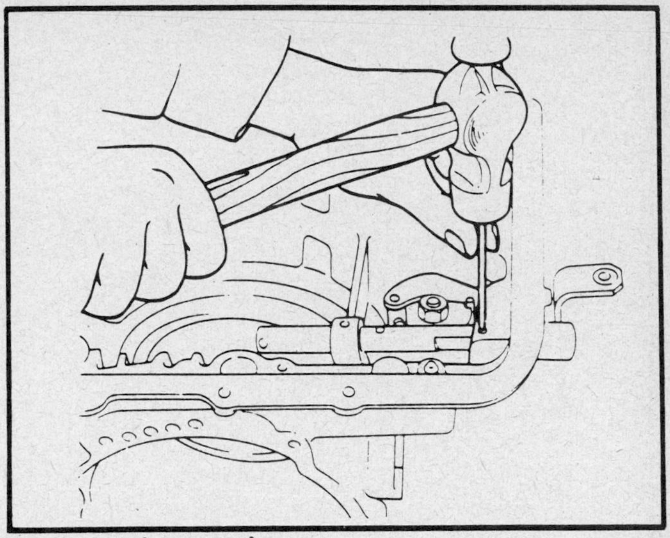
Control rod removal

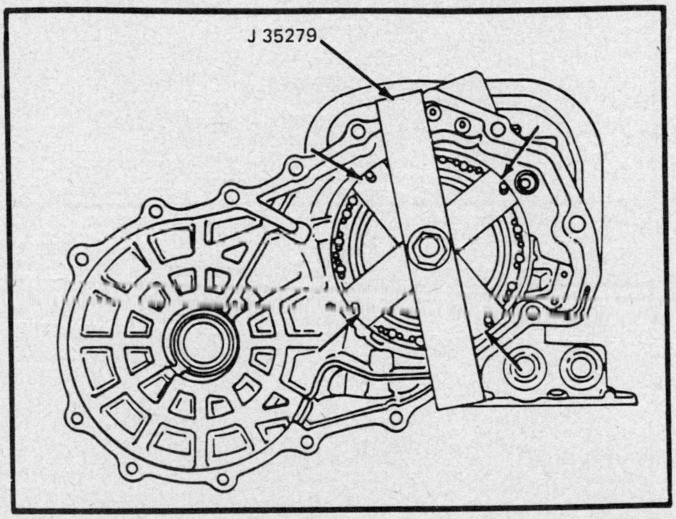
J 35279
Low/reverse piston removal

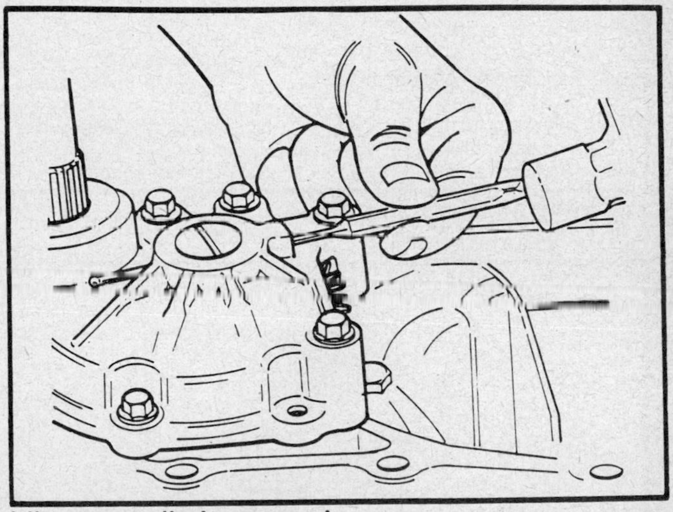
Idler gear roll pin removal

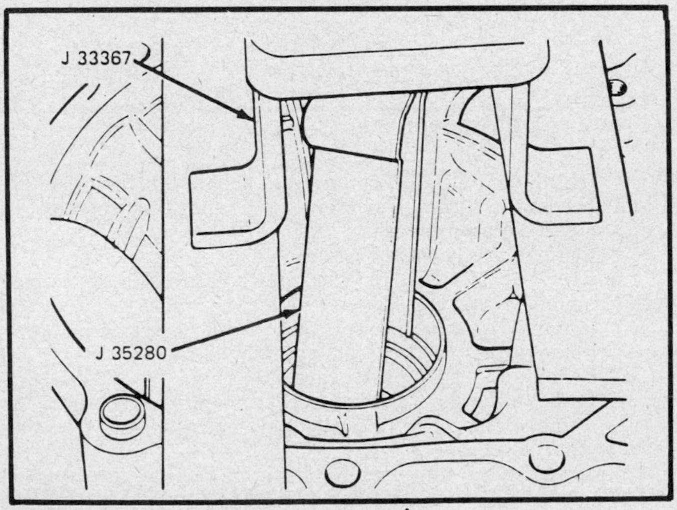
J 33367
J 35280
Differental bearing race removal

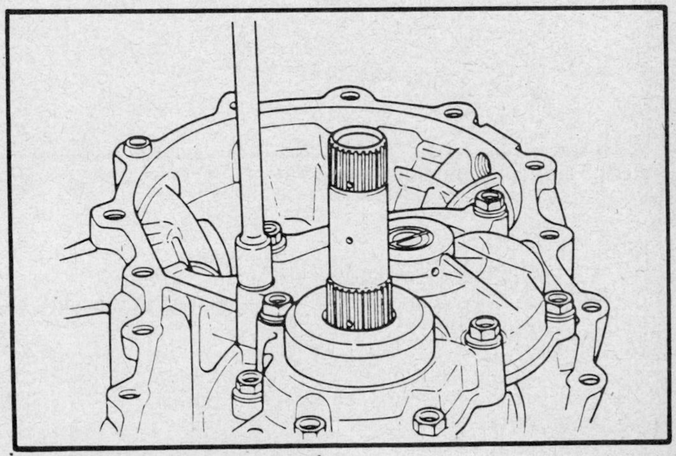
Bearing housing removal

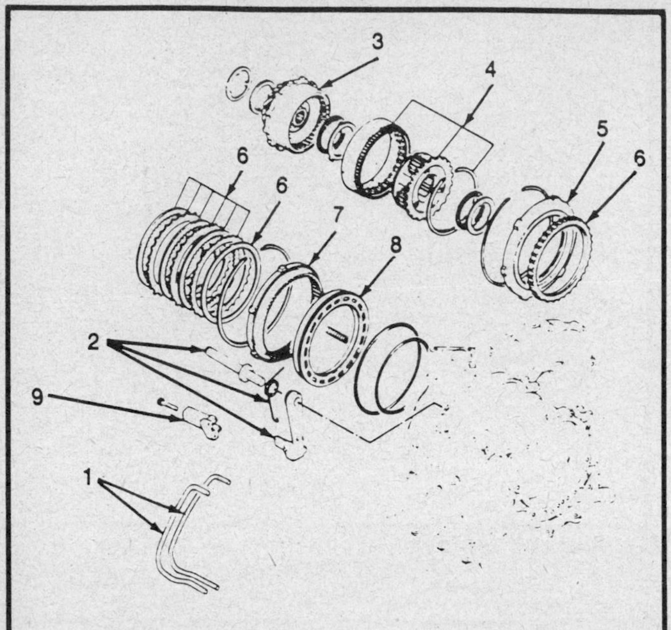
3
4
5
6
6
6
7
8
2
9
1

1. Oil pipe
2. Parking pawl assembly
3. Drum hub assembly
4. One-way clutch inner race assembly
5. One-way clutch
6. Low and reverse brake plate assembly
7. Low and reverse brake hub
8. Low and reverse brake piston
9. Actuator support

Low/reverse clutch assembly

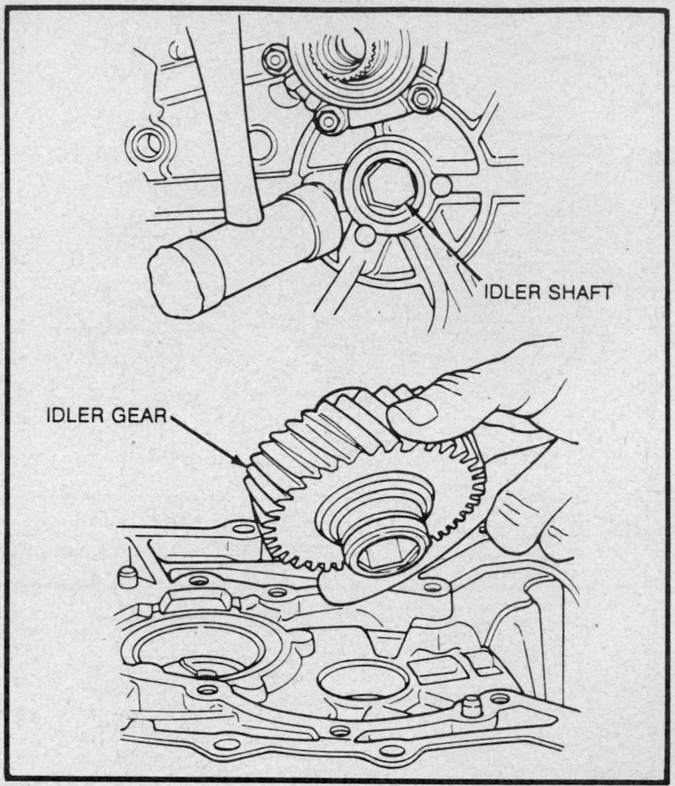
IDLER SHAFT
IDLER GEAR

Idler shaft removal

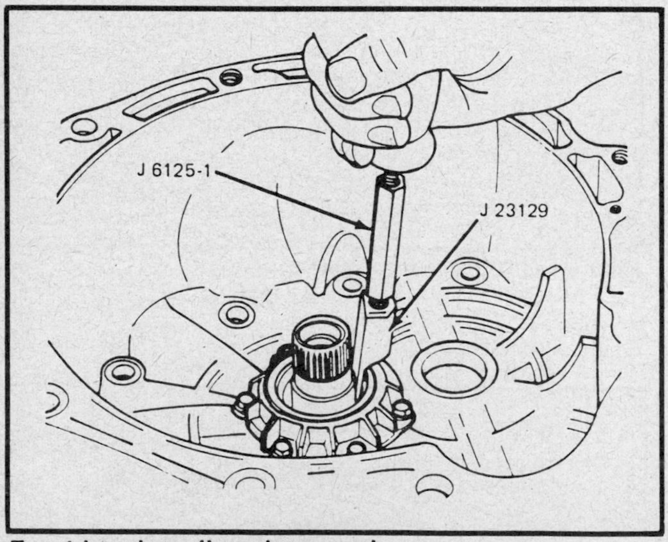
J 6125-1
J 23129

Front bearing oil seal removal

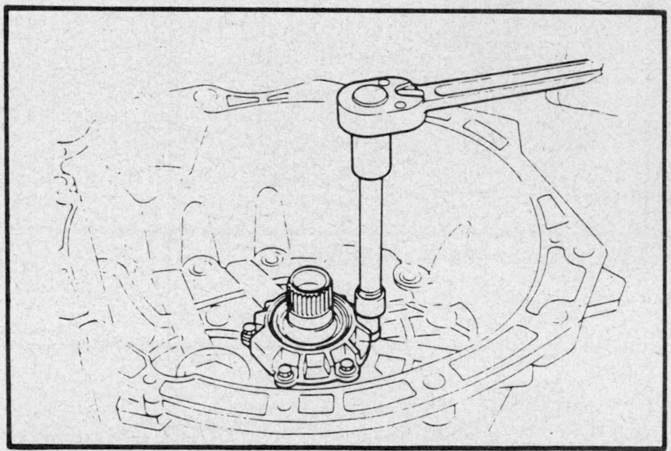

Front bearing cover

11. Remove the anchor end bolt and the locknut.
12. Remove the brake band. To avoid damage to the band, use a paper clip or piece of wire to prevent the band from fully expanding.
13. Remove the front/direct clutch assembly.
14. Remove the rear/forward clutch assembly.
15. Remove the rear clutch hub assembly.
16. Remove the planetary carrier, sun gear with the spacer and the connecting shell.
17. Install servo piston compressor tool J-35278 to the case and depress the servo cover. Remove the snapring, servo retainer and servo piston from the case.
18. Remove the governor assembly.
19. Remove the converter housing to case bolts and remove the housing.
20. Remove the oil pipes.
21. Remove the parking pawl assembly.
22. Remove the differential case assembly.
23. Remove the reaction internal gear and thrust washer.
24. Remove the drum/hub assembly and the one-way clutch inner race assembly.
25. Attach a dial indicator and measure the low/reverse clutch pack clearance. Use air to engage the clutch. The clearance should be 0.031–0.040 in.

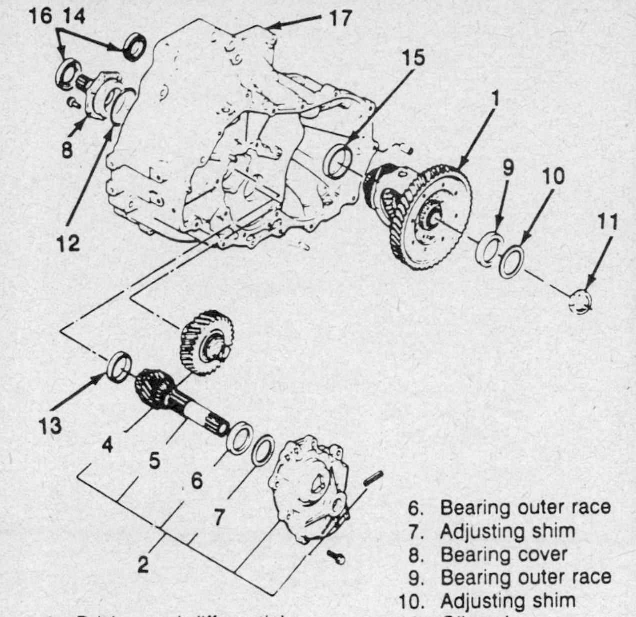
16 14
17
15
1
8
9
10
11
12
13
4
5
6
7
2

1. Driving and differential assembly
2. Bearing housing assembly
3. Spring pin
4. Idle gear assembly
5. Output gear assembly
6. Bearing outer race
7. Adjusting shim
8. Bearing cover
9. Bearing outer race
10. Adjusting shim
11. Oil seal
12. O-ring
13. Bearing outer race
14. Oil seal
15. Bearing outer race
16. Oil seal
17. Converter housing

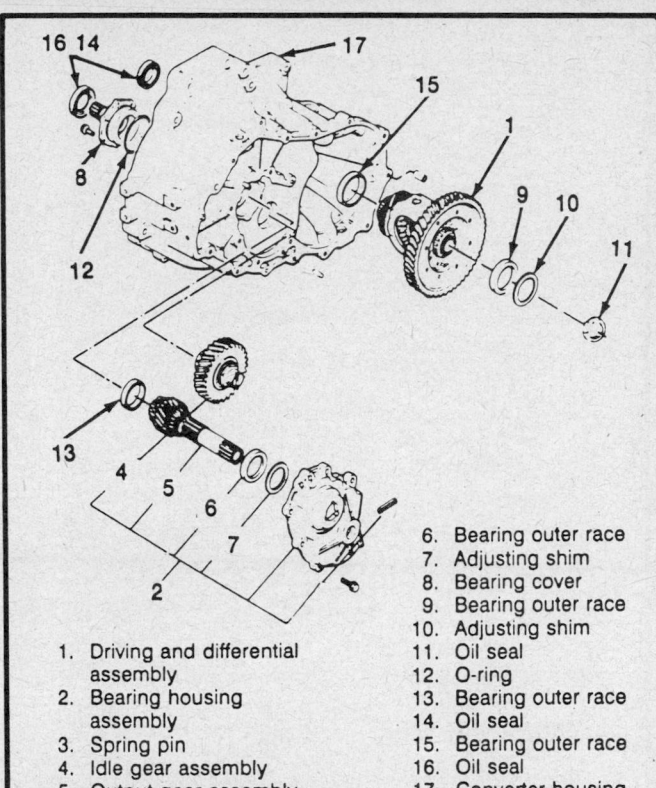

Final drive assembly

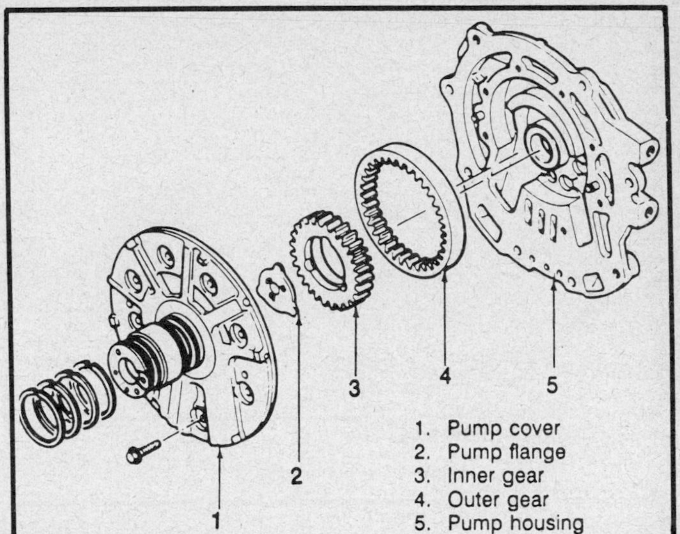
3
4
5
2
1

1. Pump cover
2. Pump flange
3. Inner gear
4. Outer gear
5. Pump housing

Oil pump exploded view

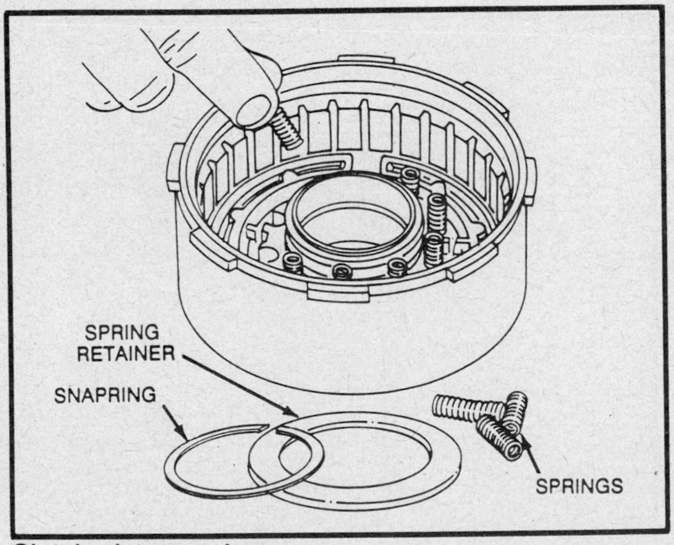
SPRING RETAINER
SNAPRING
SPRINGS

Clutch piston springs

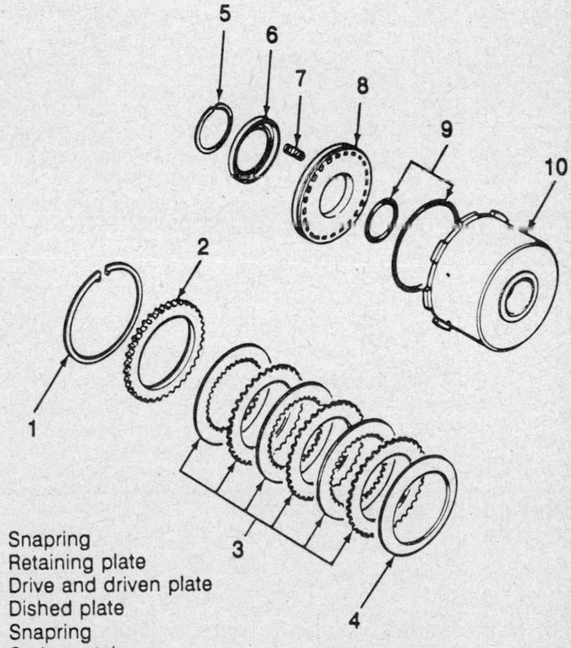
5
6
7
8
9
10
2
1
3
4

1. Snapring
2. Retaining plate
3. Drive and driven plate
4. Dished plate
5. Snapring
6. Spring retainer
7. Return spring
8. Piston
9. Seal ring
10. Front clutch drum

Front clutch exploded view

28. Install tool J–35279 and thread the 4 small bolts included through the bar and into the piston. Tighten the center bolt to pull the piston from the case.

26. Remove the snapring retaining the one-way clutch assembly and remove the one-way clutch from the case.

27. Install tool J–35279, or equivalent and tighten the bolt to compress the clutch return springs. Remove the snapring from the clutch return spring retainer. Remove tool J–35279 or equivalent, the spring retainer and the springs from the case.

29. Remove the roll pin retaining the manual shaft to the case and remove the manual shaft from the case.

30. Remove the differential side bearing outer race and shim from the case using puller J–33367 and J–35280 bearing race puller.

31. Using a drift and hammer, remove the roll pin retaining the idler gear to the bearing housing.

32. Remove the bolts retaining the bearing housing to the converter housing. Remove the bearing housing.

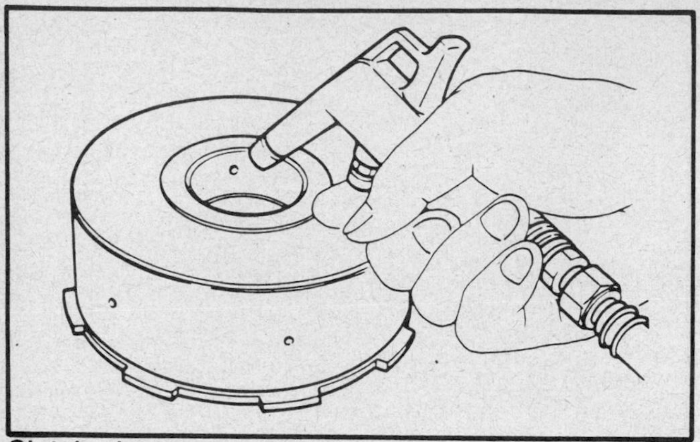

Clutch piston removal

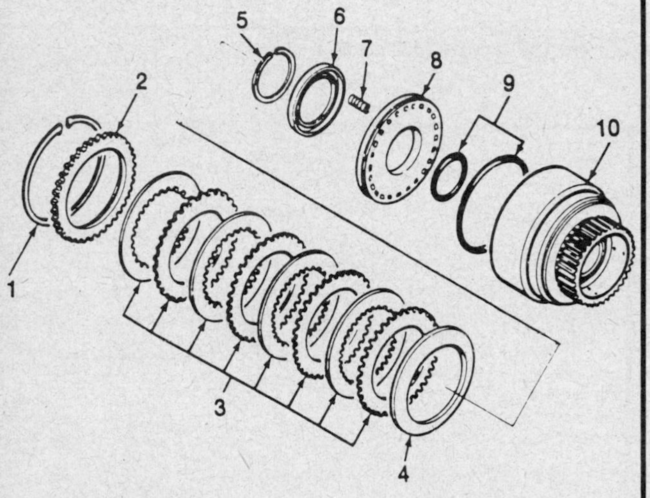
1
2
3
4
5
6
7
8
9
10

1. Snapring
2. Retaining plate
3. Drive and driven plate
4. Dished plate
5. Snapring
6. Spring retainer
7. Return spring
8. Piston
9. Seal ring
10. Rear clutch drum

Rear clutch exploded view

33. Remove the output gear from the housing.
34. Remove the idler gear from the housing. Tap the idler shaft from the converter side of the housing.
35. Remove the bearing outer race from the bearing housing.
36. Remove the oil seal from the front bearing cover using J-23129 seal remover and J-6125-1 slide hammer.
37. Remove the bearing race from the converter housing.
38. Remove the front bearing cover from the converter housing.

Unit Disassembly and Assembly

OIL PUMP

Disassembly

1. Remove the 7 bolts from the pump cover.
2. Separate the cover and hub from the pump body.

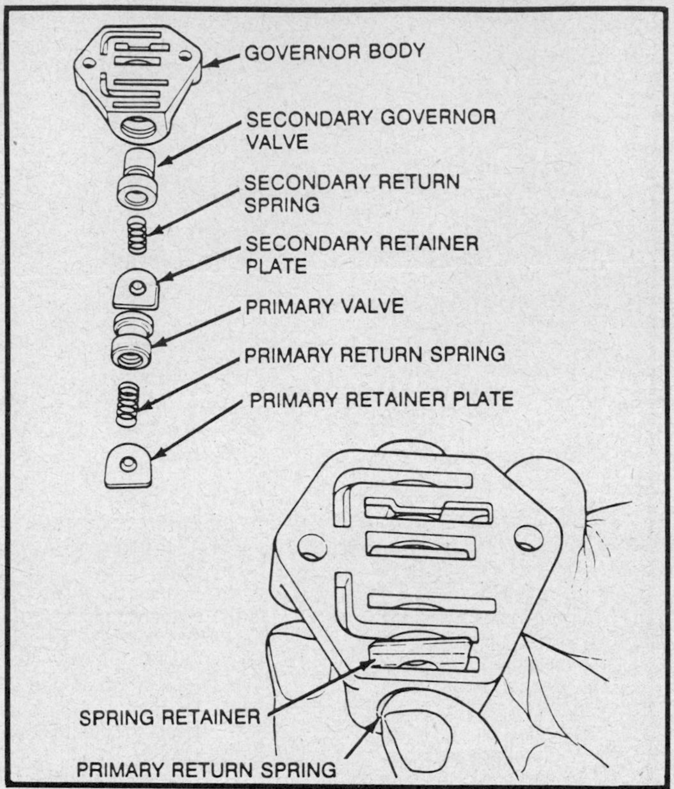
GOVERNOR BODY
SECONDARY GOVERNOR VALVE
SECONDARY RETURN SPRING
SECONDARY RETAINER PLATE
PRIMARY VALVE
PRIMARY RETURN SPRING
PRIMARY RETAINER PLATE
SPRING RETAINER
PRIMARY RETURN SPRING

Governor springs and retainers

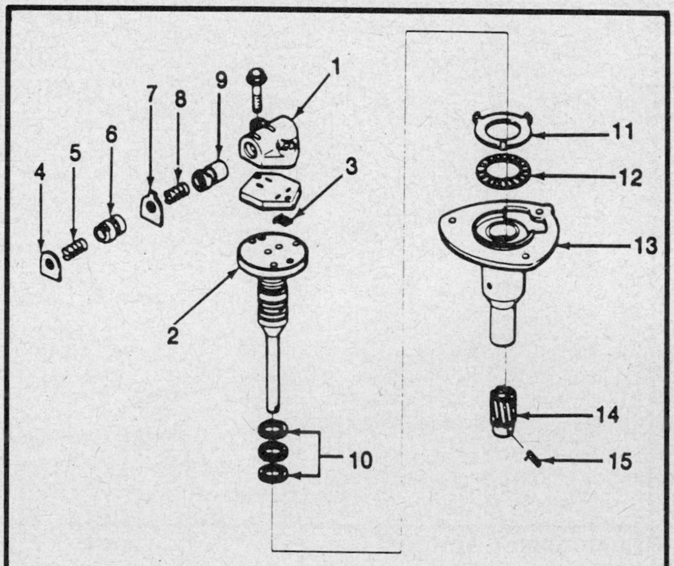
1
2
3
4
5
6
7
8
9
10
11
12
13
14
15

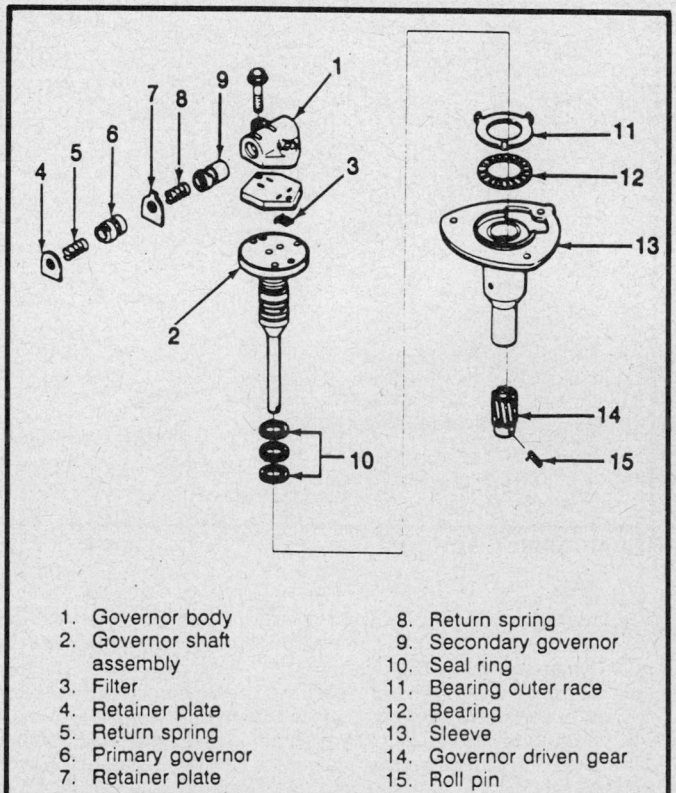

1. Governor body
2. Governor shaft assembly
3. Filter
4. Retainer plate
5. Return spring
6. Primary governor
7. Retainer plate
8. Return spring
9. Secondary governor
10. Seal ring
11. Bearing outer race
12. Bearing
13. Sleeve
14. Governor driven gear
15. Roll pin

Governor assembly

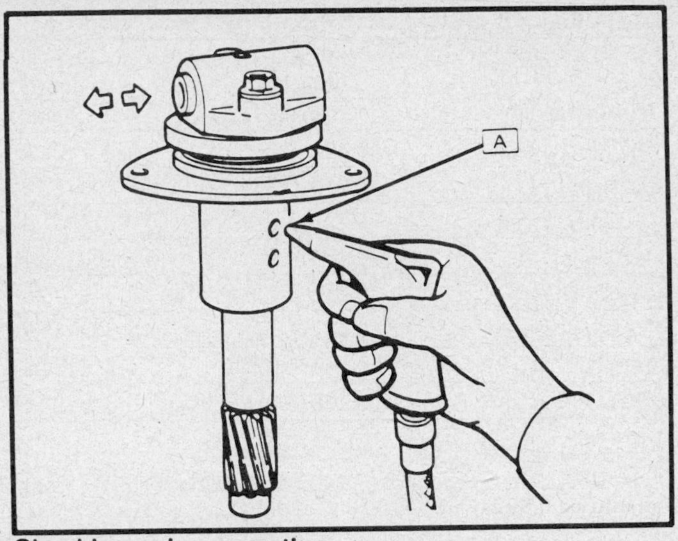
A
Checking valve operation

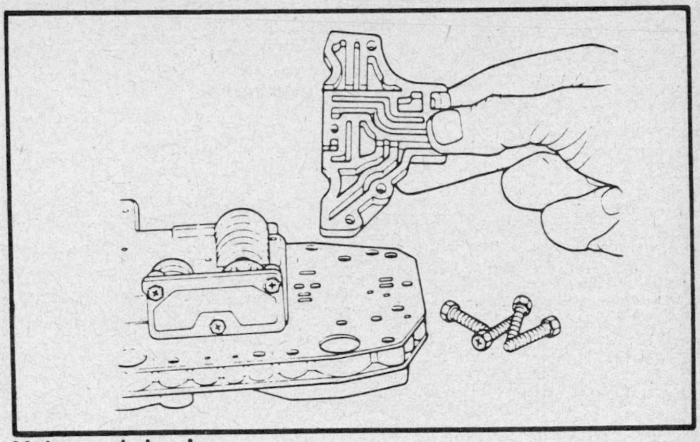
Valve sub-body

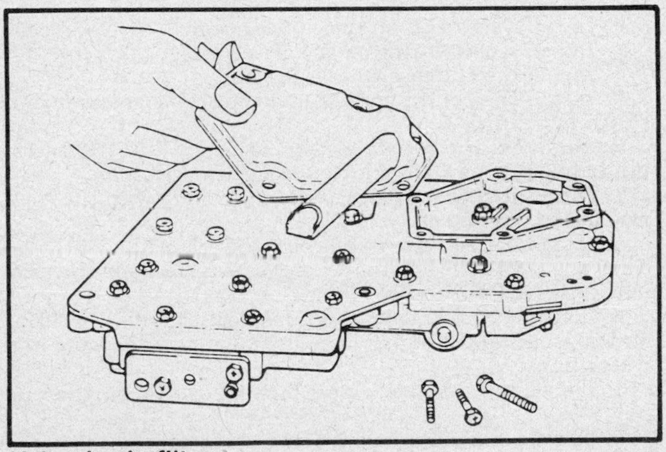
Valve body filter

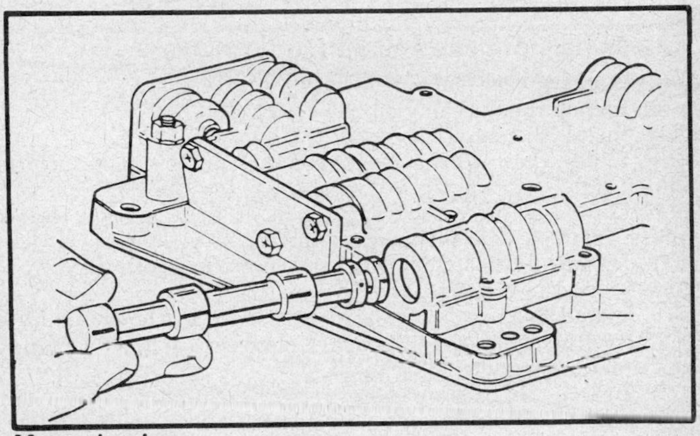
Manual valve

Inspection

1. Place a metal straight edge across the pump face. Use a feeler gauge to measure the inner gear to pump cover clearance. The clearance is 0.001–0.002 in. (0.02–0.04mm). The maximum clearance is 0.003 in. (0.08mm). Measure the outer gear to pump clearance in the same manner.
2. Use a feeler gauge to measure the head of the outer gear teeth to crescent clearance. The clearance is 0.006–0.008 in. (0.14–0.21mm). The maximum clearance is 0.010 in. (0.25mm).
3. Measure the outer gear to housing clearance. The clearance is 0.002–0.008 in. (0.05–0.20mm). The maximum clearance is 0.010 in. (0.25mm).
4. Use a feeler gauge to measure the pump hub seal ring to seal ring groove clearance. The clearance is 0.002–0.006 in. (0.04–0.16mm). The maximum clearance is 0.016 in. (0.40mm).
5. Use a micrometer to measure the pump housing sleeve outer diameter. The diameter should be 1.492 in. (37.900mm) minimum.
6. Use a micrometer to measure the inner gear bushing diameter. The diameter should be 1.499 in. (38.075mm) maximum.

Assembly

1. Reinstall the pump gears in the pump body.
2. Install the pump front cover and hub on the pump body.
3. Install the 7 bolts and tighten them to 17 ft. lbs.(23 Nm).

FRONT/DIRECT CLUTCH

Disassembly

1. Remove the large snapring from the front clutch pack.
2. Remove the multiple disc clutch pack including the dish plate from the hub.
3. Position the front/direct clutch in a press. Use spring compressor J-23327 and adapter J-25018-A to compress the piston return springs.
4. Remove the spring retainer and springs from the piston.
5. Apply air to the clutch hub to remove the piston.
6. Remove the O-ring from the clutch hub.

Inspection

1. Use a micrometer to measure the drum bushing inside diameter. The maximum inside diameter is 0.068 in. (1.735mm). If the bushing is out of specification, replace the drum.
2. Inspect the O-ring for wear. Replace as necessary.
3. Inspect the clutch plates, snaprings and spring retainer for wear.
4. Check the free length of the piston springs. The free length should be 0.992–1.071 in. (25.2–27.2mm).

Assembly

1. Apply Dexron® II automatic transmission fluid to the seals and seal surfaces.

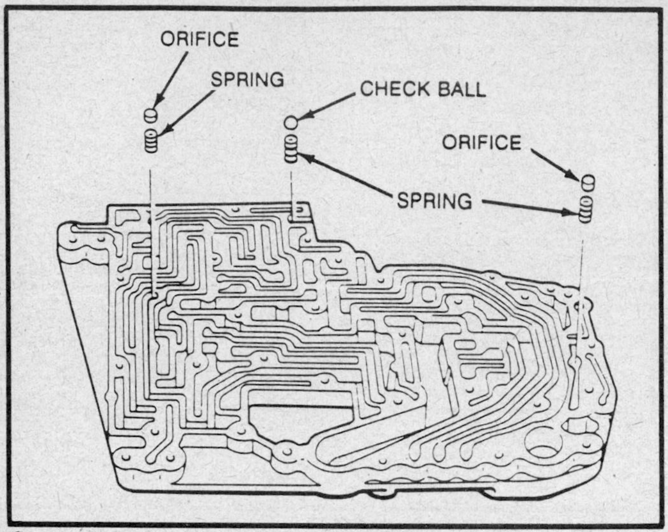
ORIFICE
SPRING
CHECK BALL
ORIFICE
SPRING

Check ball, orifices and spring locations

2. Install a new seal on the clutch piston.
3. Install a new O-ring on the clutch hub.
4. Install the clutch piston in the hub.
5. Install the return springs and the spring retainer in the clutch hub.
6. Compress the return springs using tool J-23327 with a press. Install a new snapring on the spring retainer.
7. Install the dished plate in the clutch hub with the convex side facing the piston.
8. Install 3 steel and 3 fiber clutch discs in the clutch hub. The 1st disc installed on the dished plate is steel and then alternate metal discs with fiber discs.
9. Install the backing plate onto the clutch disc with the smooth side down. Install a new snapring in the clutch hub.
10. Install the assembled front/direct clutch on the pump hub.
11. Install a dial indicator with the pin resting on the backing plate to measure the clutch pack travel.
12. Engage the clutch by applying air to the pump. The dial indicator measurement should be 0.063–0.071 in. (1.6–1.8mm).
13. If the dial indicator reading is not within specification, change the thickness of the retaining plate. The following is a list of the retaining plate sizes:
0.205 in. (5.2mm)
0.213 in. (5.4mm)
0.220 in. (5.6mm)
0.228 in. (5.8mm)
0.236 in. (6.0mm)
0.244 in. (6.2mm)

REAR/FORWARD CLUTCH

Disassembly

1. Remove the large snapring from the rear clutch pack.
2. Remove the clutch pack from the clutch hub.
3. Position the rear/forward clutch in a press. Use spring compressor J-23327 and adapter J-25018-A to compress the piston return springs.
4. Remove the spring retainer and springs from the piston.
5. Apply air to the clutch hub to remove the piston.
6. Remove the O-ring from the clutch hub.

Inspection

1. Inspect the O-ring for wear. Replace as necessary.
2. Inspect the clutch plates, snaprings and spring retainer for wear.

VALVE BODY SPRING DIMENSIONS

Name of spring	Outer Diameter in. (mm)	Free Length in. (mm)
Throttle backup	0.287 (7.3)	1.417 (36.0)
Downshift	0.218 (5.55)	0.866 (22.0)
2–3 shift	0.272 (6.9)	1.614 (41.0)
1–2 shift	0.258 (6.55)	1.260 (32.0)
Second lock	0.218 (5.55)	1.319 (33.5)
Pressure regulator	0.461 (11.7)	1.693 (43.0)
Steel ball	0.256 (6.5)	1.516 (26.8)
Orifice check	0.197 (5.0)	0.846 (21.5)

3. Check the free length of the piston springs. The free length should be 0.992–1.071 in. (25.2–27.2mm).

Assembly

1. Apply Dexron® II automatic transmission fluid to the seals and seal surfaces.
2. Install a new seal on the clutch piston.
3. Install a new O-ring on the clutch hub.
4. Install the clutch piston in the hub.
5. Install the return springs and the spring retainer in the clutch hub.
6. Compress the return springs using tool J-23327 with a press. Install a new snapring on the spring retainer.
7. Install the dished plate in the clutch hub with the convex side facing the piston.
8. Install 4 steel and 4 fiber clutch discs in the clutch hub. The 1st disc installed on the dished plate is steel and then alternate metal discs with fiber discs.
9. Install the backing plate onto the clutch disc with the beveled edge facing the clutch discs. Install a new snapring in the clutch hub.
10. Install the assembled rear/forward clutch on the pump hub.
11. Install a dial indicator with the pin resting on the backing plate to measure the clutch pack travel.
12. Engage the clutch by applying air to the pump. The dial indicator measurement should be 0.031–0.059 in. (0.8–1.5mm).
13. If the dial indicator reading is not within specification, replace the clutch discs.

GOVERNOR

Disassembly

1. Remove the bolts from the governor body.
2. Remove the body and spacer from the governor shaft assembly.
3. Remove the spring retainers from the governor body by pressing the primary return spring.
4. Remove the secondary retainer and return spring in the same manner.
5. Remove the roll pin from the governor driven gear and remove the gear from the shaft.
6. Remove the governor shaft, bearing and outer race from the sleeve.

Inspection

1. Inspect the valve and return springs for wear or damage.
2. Check the valve operation by applying air in the governor hole. The valve should vibrate and make a buzzing noise.
3. Check for a clogged filter.

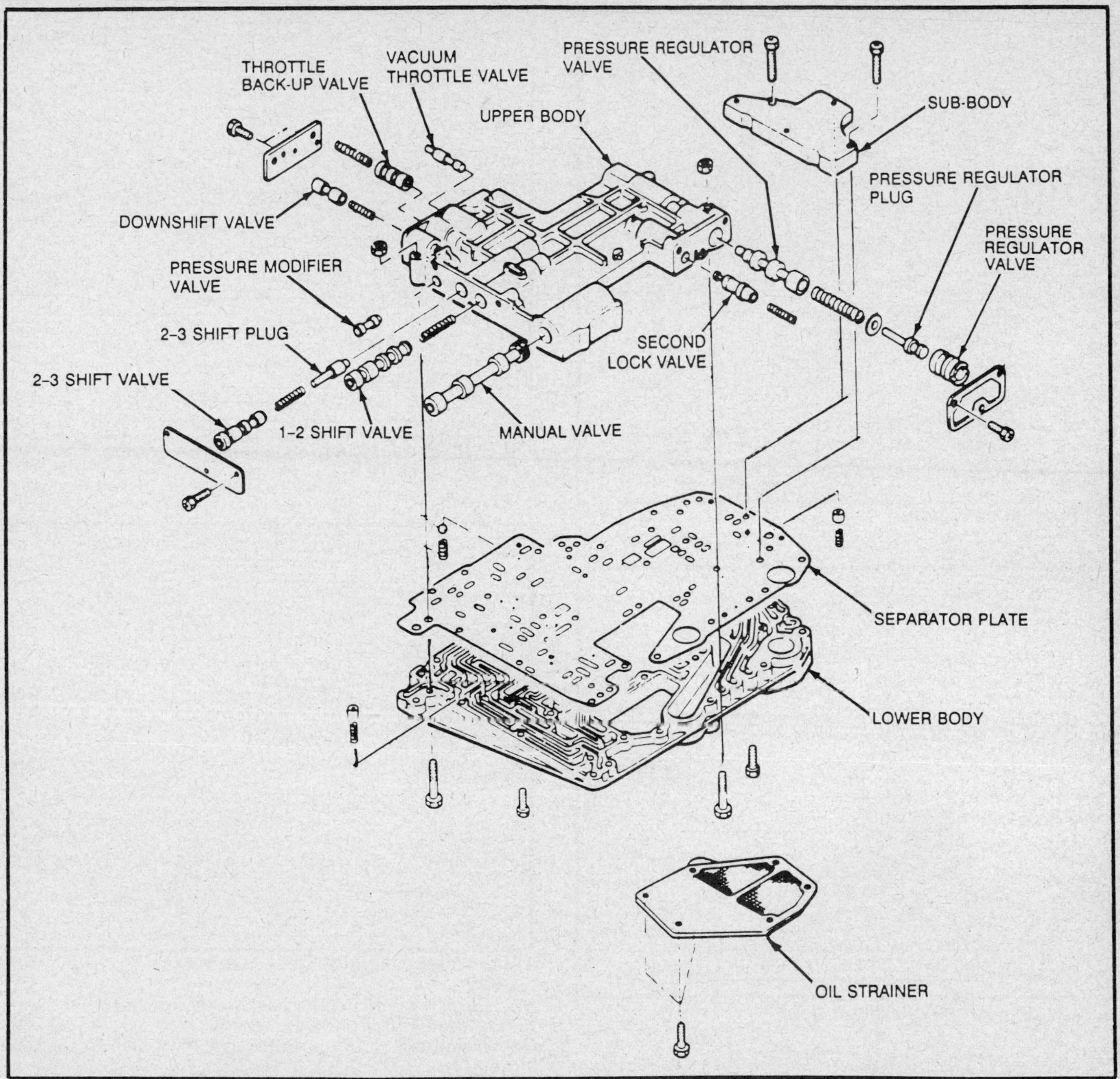
THROTTLE BACK-UP VALVE
VACUUM THROTTLE VALVE
PRESSURE REGULATOR VALVE
SUB-BODY
UPPER BODY
PRESSURE REGULATOR PLUG
DOWNSHIFT VALVE
PRESSURE REGULATOR VALVE
PRESSURE MODIFIER VALVE
2-3 SHIFT PLUG
SECOND LOCK VALVE
2-3 SHIFT VALVE
1-2 SHIFT VALVE
MANUAL VALVE
SEPARATOR PLATE
LOWER BODY
OIL STRAINER

Control valve exploded view

Assembly

1. Install the 3 seal rings on the shaft.
2. Install the bearing and the outer race on the sleeve.
3. Install the governor shaft in the sleeve.
4. Install the governor driven gear on the shaft and retain the gear with a roll pin.
5. Install the secondary valve, rerturn spring and retainer plate in the governor body.
6. Install the primary valve, return spring and retainer plate in the governor body.
7. Install a new filter, spacer and governor body on the shaft.
8. Install the retaining bolts and tighten to 5 ft. lbs. (7 Nm).

VALVE BODY

The control valve is one of the highest precision parts used in the automatic transaxle and it should be handled with the utmost care. If the clutch has been overheated or the brake band has been burnt, be sure to disassemble, clean and inspect the valve body.

Disassembly

1. Remove the 3 bolts retaining the filter to the valve body.
2. Remove the 4 bolts retaining the valve sub-body and remove the sub-body from the valve body.

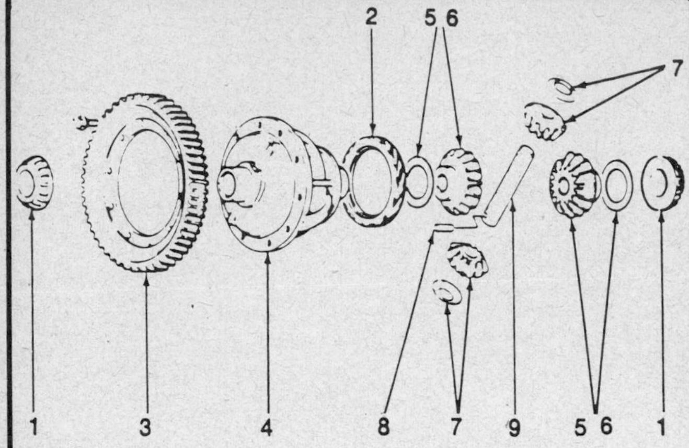
2
5 6
7
1
3
4
8
7
9
5 6
1

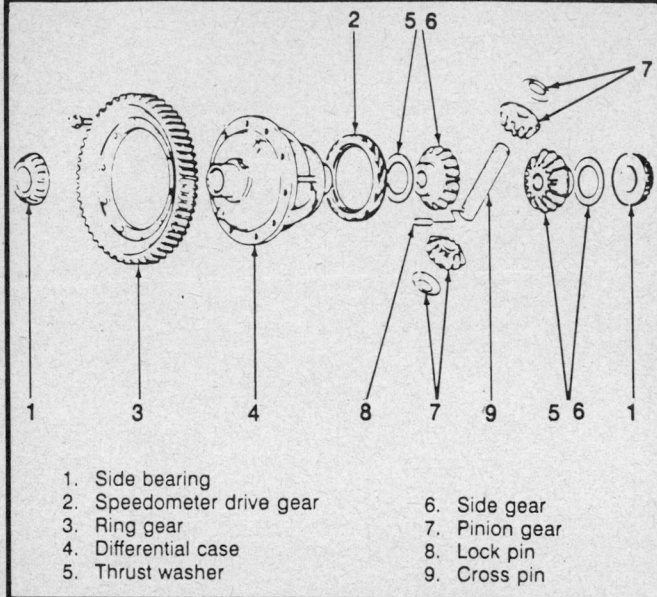

1. Side bearing
2. Speedometer drive gear
3. Ring gear
4. Differential case
5. Thrust washer
6. Side gear
7. Pinion gear
8. Lock pin
9. Cross pin

Differential assembly

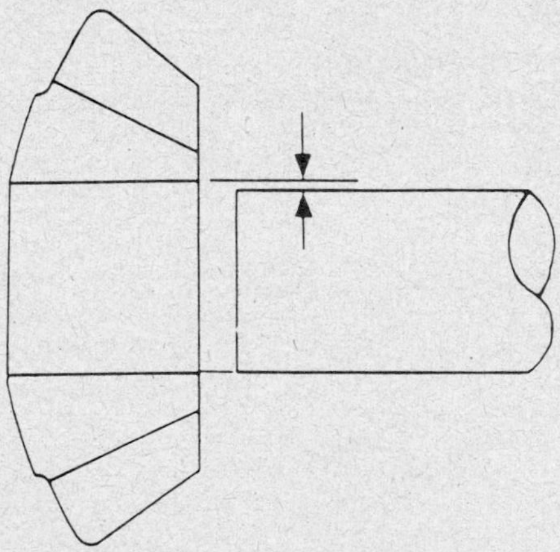

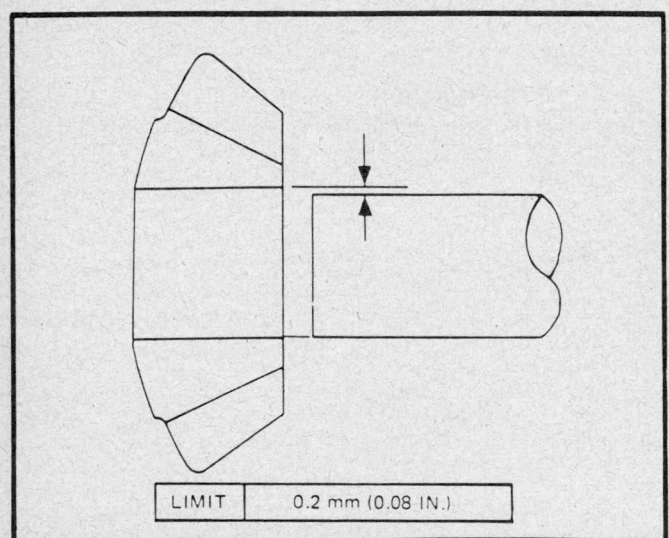

LIMIT	0.2 mm (0.08 IN.)

Pinion gear/cross pin clearance

3. Remove the manual valve from the valve body.
4. Remove the bolts retaining the upper and lower valve bodies. Separate the upper body from the lower body and remove the separator plate.
5. Remove the orifices, check ball and springs from the lower valve body.
6. Disassemble the upper valve body.

Inspection

1. Inspect each valve for damage or wear.
2. Check for damage in the oil passages.
3. Check each valve body for cracks.
4. Check the valve operations.
5. Check for spring fatigue.

Assembly

1. Assemble the upper valve body. When installing the side

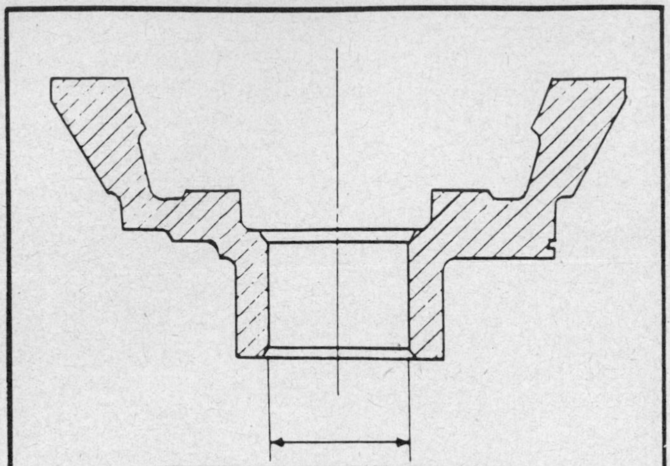

LIMIT	27.35 mm (9.358 IN.)

Differential case/drive axle shaft clearance

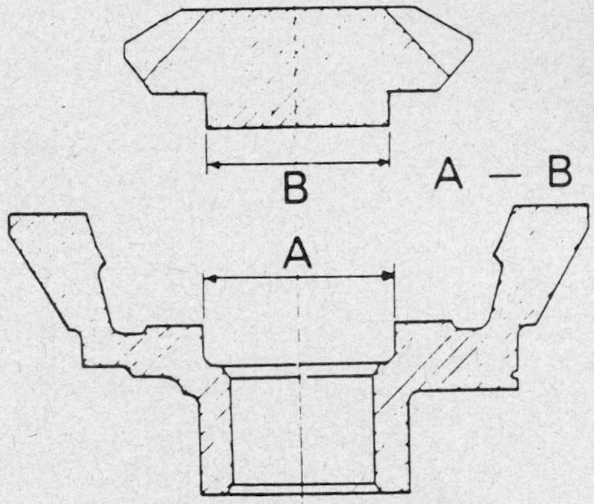
B
A — B
A

A-B	0.15 mm (0.059 IN.) OR LESS

Differential case/side gear clearance

plate, align the center of the hole that is arrowed with the center of the vacuum throttle valve.
2. Install the springs, orifices and check ball in the lower valve body.
3. Position the separator plate and the upper valve body on the lower valve body.
4. Install the manual valve in the upper valve body.
5. Position and align the valve sub-body on the separator plate and install the retaining bolts.
6. Install the valve body filter and the retaining bolts.

DIFFERENTIAL CASE

Disassembly

1. Remove the side bearings from the differential case using bearing puller J-22888 with J-35288 pilot.
2. Remove the speedometer drive gear. Heat the gear with a heat gun before pulling the gear off.
3. Remove the ring gear from the differential case. The ring gear bolts are not reusable.

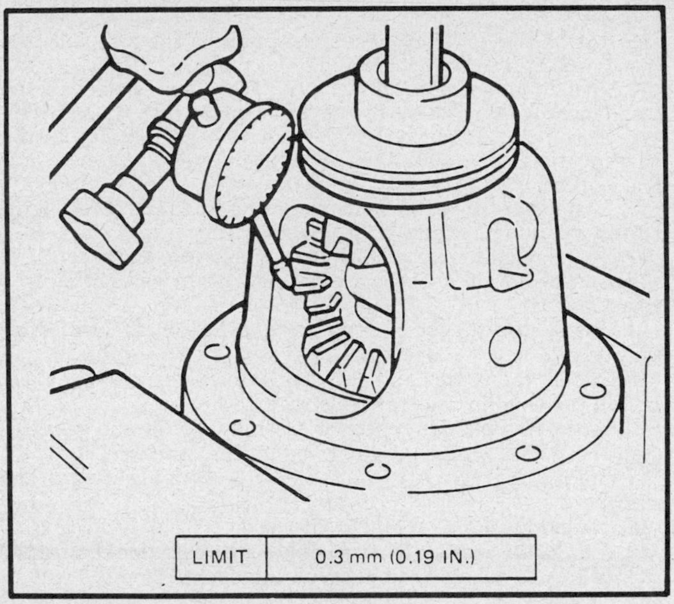
LIMIT
0.3 mm (0.19 IN.)
Side gear/pinion gear backlash

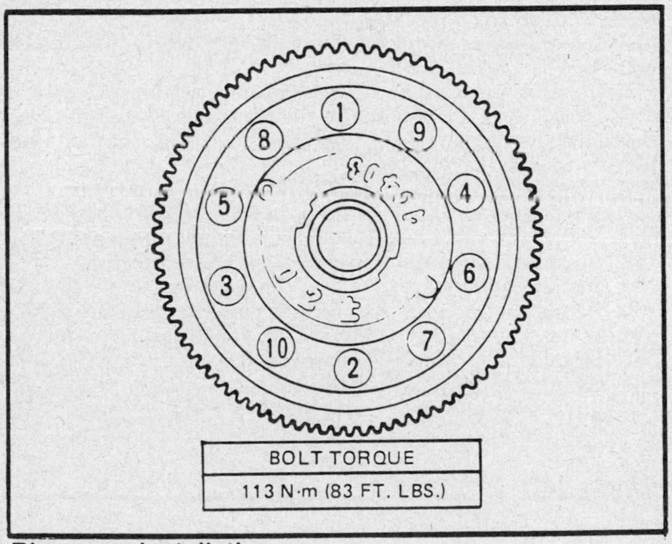
1
2
3
4
5
6
7
8
9
10
BOLT TORQUE
113 N·m (83 FT. LBS.)
Ring gear installation

4. Remove the lock pin used to retain the cross pin.
5. Remove the cross pin from the differential case.
6. Remove the pinion gears, side gears and thrust washers from the differential case.

Inspection

1. Measure the clearance between the pinion gear and the cross pin. The clearance is 0.08 in. (0.2mm).
2. Measure the clearance between the differential case and the side gear. The clearance is 0.059 in. (0.15mm) or less.
3. Measure the diameter in the differential case of the drive axle shaft. The clearance is 9.358 in. (27.35mm).
4. Measure the backlash between the side gear and the pinion gear. The limit is 0.19 in. (0.3mm). If the backlash is beyond the limit, install new thrust washers.

Assembly

1. Install the thrust washers in the case.
2. Install the side gears in the case.
3. Install the pinion gears in the case. Align the cross pin hole of the pinion gear with the cross pin hole of the differential case.
4. Install the cross pin. After installation of the pin, stake the edge of the lock pin hole in the case with a punch to prevent the loss of the lock pin.
5. Install a new speedometer drive gear. Heater the gear with a heat gun before installing. Do not use hot water to heat the gear.
6. Install both differential side bearings.
7. Install a new ring gear and tighten the bolts following the sequence to 83 ft. lbs. (113 Nm). Use new ring gear bolts if the ring has been removed.

Transaxle Assembly

1. Install the differential side bearing race in the converter housing.
2. Install the differential case assembly in the converter housing.
3. Set the shim selector gauge J-35284 bridge and leg assembly, with J-35284-4 gauge cylinder and J-35284-8 differential gear gauge pin on the transaxle case over the differential bearing housing.
4. Loosen the thumb screw allowing the gauge pin to rest on the bearing race seat. Tighten the thumb screw to remove the tool from the case.
5. Install the other side bearing race on the exposed side bearing. Set the shim selector gauge on the converter housing over the differential case.
6. Loosen the thumb screw allowing the gauge pin to rest on the bearing race and them tighten the thumb screw.
7. Select the appropriate side bearing shim according to the remaining gap in the gauge pin.
8. Install the selected shim into the side bearing race bore of the transaxle case. Install the bearing race.
9. Install a new O-ring on the bearing cover and install the cover on the converter housing. Tighten the bolts to 9 ft. lbs. (13 Nm).
10. Install a new oil seal in the bearing cover.
11. Install the bearing race in the bearing cover.
12. Set the shim selector gauge J-35284 bridge and leg assembly, with J-35284-4 gauge cylinder and J-35284-8 differential gear gauge pin on the bearing housing.
13. Loosen the thumb screw allowing the gauge pin to rest on the output shaft bearing race and shim seat. Tighten the thumb screw to remove the tool from the bearing housing.
14. Install the output gear in the converter housing with the bearing race on the exposed bearing.
15. Place the bridge assembly on the converter housing. Loosen the thumb screw so that the gauge pin rests on the output shaft bearing race. Tighten the thumb screw.
16. Select the appropriate output gear bearing shim according to the remaining gap in the gauge pin. The following is a list of available shim sizes:
    0.004 in. (0.10mm)
    0.005 in. (0.12mm)
    0.006 in. (0.14mm)
    0.007 in. (0.16mm)
    0.008 in. (0.20mm)
    0.020 in. (0.50mm)
17. Install the selected output gear shim into the bearing race bore of the bearing housing. Install the bearing race.
18. Install the idler gear assembly in the converter housing. Tap the idler shaft to seat the gear assembly.
19. Install the output gear assembly in the converter housing.

20. Install the bearing housing on the converter housing and tighten the bolts to 17 ft. lbs. (23 Nm).
21. Align the idler gear shaft roll pin hole with the bearing housing roll pin hole. Install the idler gear roll pin.
22. Install the low/reverse piston in the transaxle case.
23. Install the springs into the spring pockets on the low/reverse piston. Install the spring retainer plate on the springs. Install the snapring and make sure that it is seated in the groove.
24. Install the multiple disc clutch pack. Install the dish plate first with the concave side facing the piston, then alternate the clutch discs: 4 steel, 4 fibers starting with the steel disc first.
25. Install the backing plate on the clutch discs with the smooth flat side facing the discs. Install the one-way clutch on the backing plate with the machined surface facing the backing plate and retain it with a new snapring.
26. Measure the low/reverse clutch clearance by setting a dial indicator on the case with the gauge pin on the clutch plate.
27. Apply air through the oil passage to engage the clutch.
28. The clutch clearance is 0.031–0.041 in. (0.8–1.05mm). If the dial indicator reading is not within the clearance specification, change the thickness of the retaining plate. The following is a list of the available retaining plate sizes:
0.307 in. (7.8mm)
0.315 in. (8.0mm)
0.323 in. (8.2mm)
0.331 in. (8.4mm)
0.339 in. (8.6mm)
0.346 in. (8.8mm)
29. Install the low/reverse clutch pack spacer between the case and bottom disc.
30. Install the one-way clutch inner race assembly with the thrust washer.
31. Install the drum hub gear assembly with the thrust bearings on the one-way clutch inner race.
32. Install the parking rod assembly in the case.
33. Install the control rod in the case after the spring and detent ball has been installed.
34. Install the manual shaft with a new O-ring into the case. Install the manual plate lever on the end of the manual shaft and install the retaining nut.
35. Connect the manual plate lever to the parking rod assembly and install the retaining clip.
36. Install the actuator support in the transaxle case. Tighten the bolt to 10 ft. lbs. (14 Nm).
37. Install the parking pawl assembly.
38. Install the oil pipes in the transaxle case.
39. Install the differential case assembly in the case with the speedometer gear facing upward.
40. Install the governor assembly with a new gasket in the case, aligning thge tab on the governor plate with the mark on the case. Install the governor cover with a new gasket on the governor assembly. Install the bolts the bolts and tighten to 5 ft. lbs. (7 Nm).
41. Install the converter housing on the transaxle case. Tighten the retaining bolts to 30 ft. lbs. (40 Nm).
42. Install the servo and spring assembly into the case.
43. Install the servo cover using a new snapring to retain the cover.
44. Install the thrust bearing and washer on the planetary carrier.
45. Install the spacer and sun gear into the connecting shell. Install the shell in the transaxle case.
46. Install the planetary carrier assembly on the sun gear. Install the thrust washer and bearing on the carrier.
47. Install the rear clutch hub with the thrust bearing in the case.
48. Install the lube oil seal in the case.
49. Install the rear clutch assembly in the case. Make sure the tabbed thrust washer is in place on the back side of the clutch.
50. Install the front clutch assembly in the transaxle case.
51. Install the brake band with the strut in the case.
52. Install the anchor end bolt and tighten the bolt to 10 ft. lbs. (14 Nm) then loosen the bolt 2 full turns. Tighten the locknut to 50 ft. lbs. (68 Nm).
53. Install the oil pump assembly on the case and tighten the bolts to 17 ft. lbs. (23 Nm).
54. Install the spring and check ball in the transaxle case.
55. Install the control valve on the transaxle case aligning the manual valve with the shift lever arm. Install the bolts and tighten to 7 ft. lbs. (10 Nm).
56. Install the oil pan with a new gasket on the transaxle case. Install the bolts and tighten them to 5 ft. lbs. (7 Nm).
57. Install the turbin shaft and then the oil pump shaft.
58. Install the speedometer driven gear assembly.
59. Install the dipstick tube and the dipstick.
60. Install the rod and the vacuum diaphragm in the case.
61. Install the neutral safety switch in the case.
62. Install the kickdown solenoid in the case.
64. Install the drive axle shaft seals.

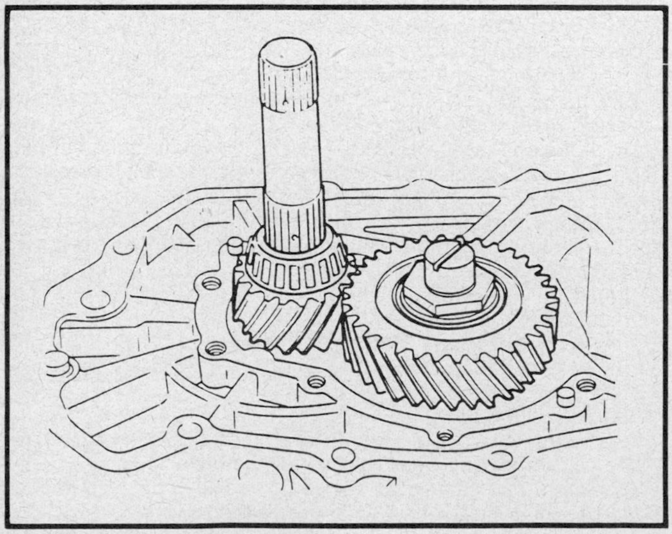

Installing the output gear

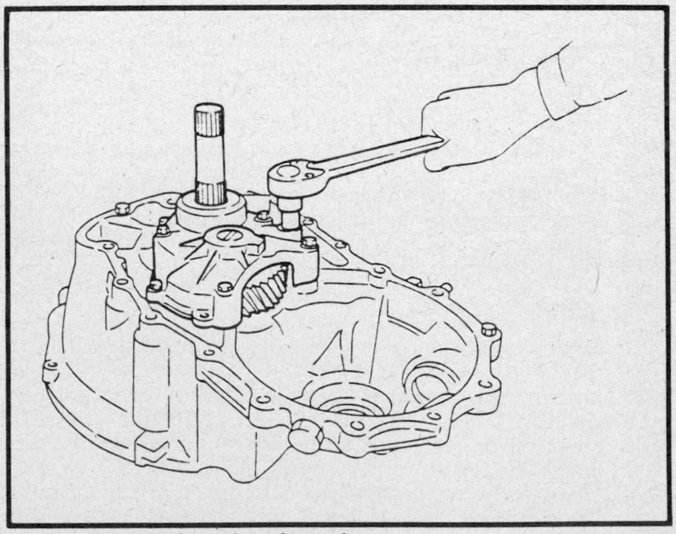

Installing the bearing housing

SPECIFICATIONS

GENERAL SPECIFICATIONS

Characteristics of torque converter	Stall capacity	3000 rpm
Planetary gear ratio	1st	2.841
	2nd	1.541
	3rd	1.000
	Reverse	2.400
	Number of front clutch plate	3
	Number of rear clutch plate	4
Hydraulic unit	Number of low and reverse brake plates	4
	Servo diameter	①
Speedometer gear ratio	—	0.857 (30/35)
Final gear ratio	—	3.526
Number of output gear teeth	—	19
Number of idler gear teeth	—	36
Number of ring gear teeth	—	67
Oil used	Type	ATF Dexron® II
	Capacity	5.8 liters (6.1 U.S. quarts)

① Piston outer diameter—2.520 in. (64 mm)
Retainer inner diameter—1.811 in. (46 mm)

SPECIAL TOOLS

Tool	Description
J 3289-20	Holding fixture base
J 35276	Holding fixture
J 35263	Output shaft bearing remover pilot
J 25695-10	Oil pressure gauge adapter
J 21867	Oil pressure gauge
J 35278	Servo piston compressor
J 35279	Low/reverse spring compressor
J 35280	Bearing outer race puller
J 26941	Bearing race puller
J 33367	Puller bridge
J 35281	Output gear bearing remover
J 29184	Front cover seal installer
J 35283	Output gear bearing installer
J 35286	Idler gear shaft holder
J 35287	Bearing outer race installer

Tool	Description
J 8092	Driver handle
J 35288	Differential side bearing puller pilot
J 22888	Differential side bearing puller
J 35290	Differential side bearing outer race installer
J 35291	Differential side bearing installer
J 35259	Bearing preload checker
J 544-01	Spring tension scale
J 35513	Low/reverse clutch pack support
J 35284	Shim selector
J 23327-A	Clutch spring compressor
J 25018-A	Clutch spring compressor adapter
J 29130	Axle seal installer
J 23129	Seal remover
J 6125-1	Slide hammer

SPECIAL TOOLS

ILLUSTRATION	NO.	NAME
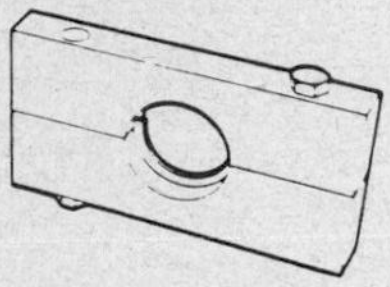	J-35281	Output gear bearing remover
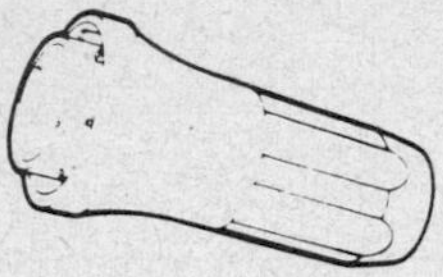	J-29184	Front cover seal installer
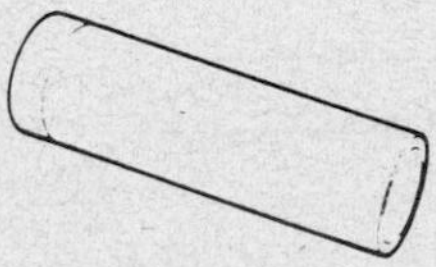	J-35283	Output gear bearing installer
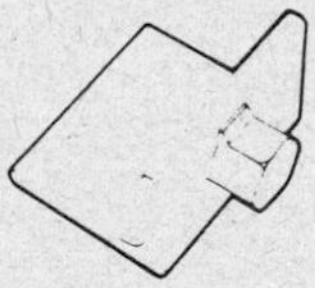	J-35286	Idle gear shaft holder
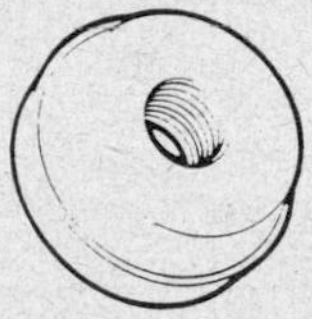	J-35287	Bearing outer race installer
	J-8092	Driver handle
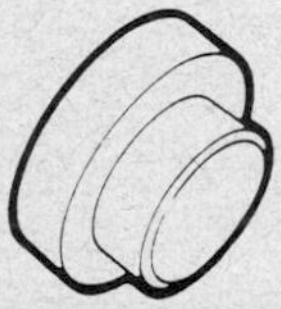	J-35288	Differential side bearing installer
	J-22888	Differential side bearing puller
	J-22888-30	Differential side bearing puller leg

SPECIAL TOOLS

ILLUSTRATION	NO.	NAME
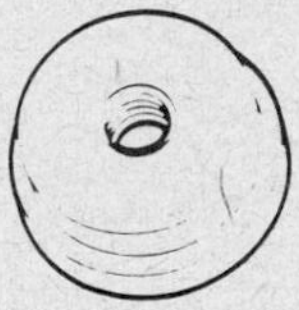	J-35290	Differential side bearing installer
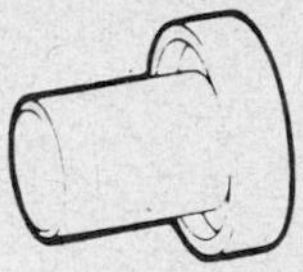	J-35291	Differential side bearing installer
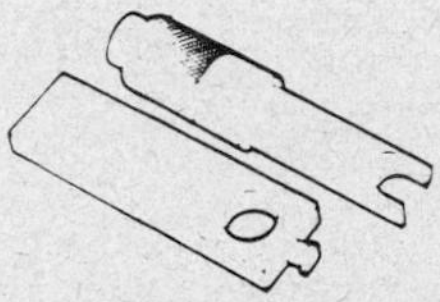	J-35259	Bearing preload checker
	J-35284	Shim selector
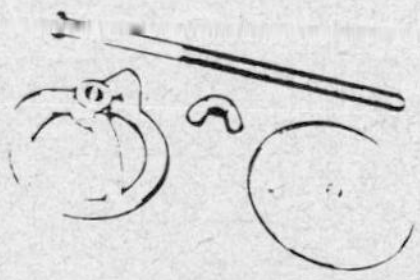	J-23327-A	Clutch spring compressor
	J-25018-A	Clutch spring compressor adapter
	J-29130	Axle seal installer
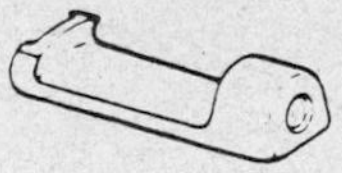	J-23129	Seal remover
	J-6125-1	Slide hammer

SPECIAL TOOLS

ILLUSTRATION	NO.	NAME
	J-3289-20	Holding fixture base
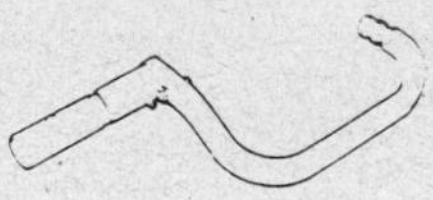	J-35276	Holding fixture
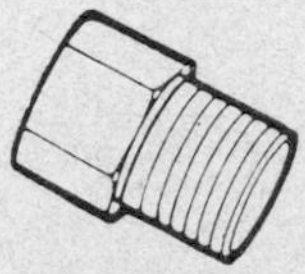	J-25695-10	Oil pressure gauge adapter
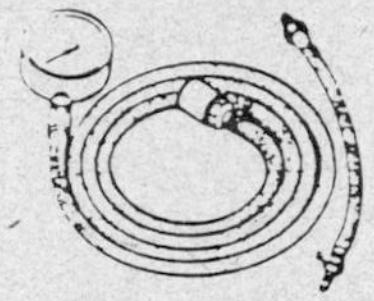	J-21867	Oil pressure gauge assembly
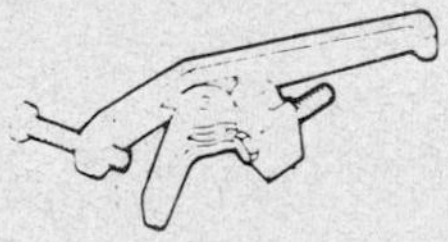	J-35278	Servo piston compressor
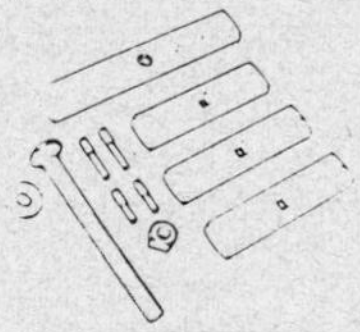	J-35279	Low reverse spring compressor
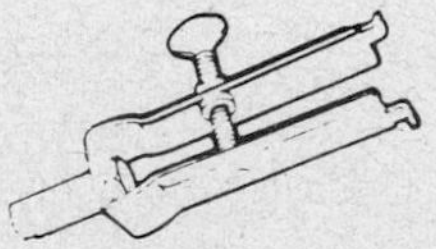	J-35280	Bearing outer race puller
	J-33367	Bearing puller bridge
	J-26941	Bearing outer race puller

Section 3

THM 125C Transaxle
General Motors

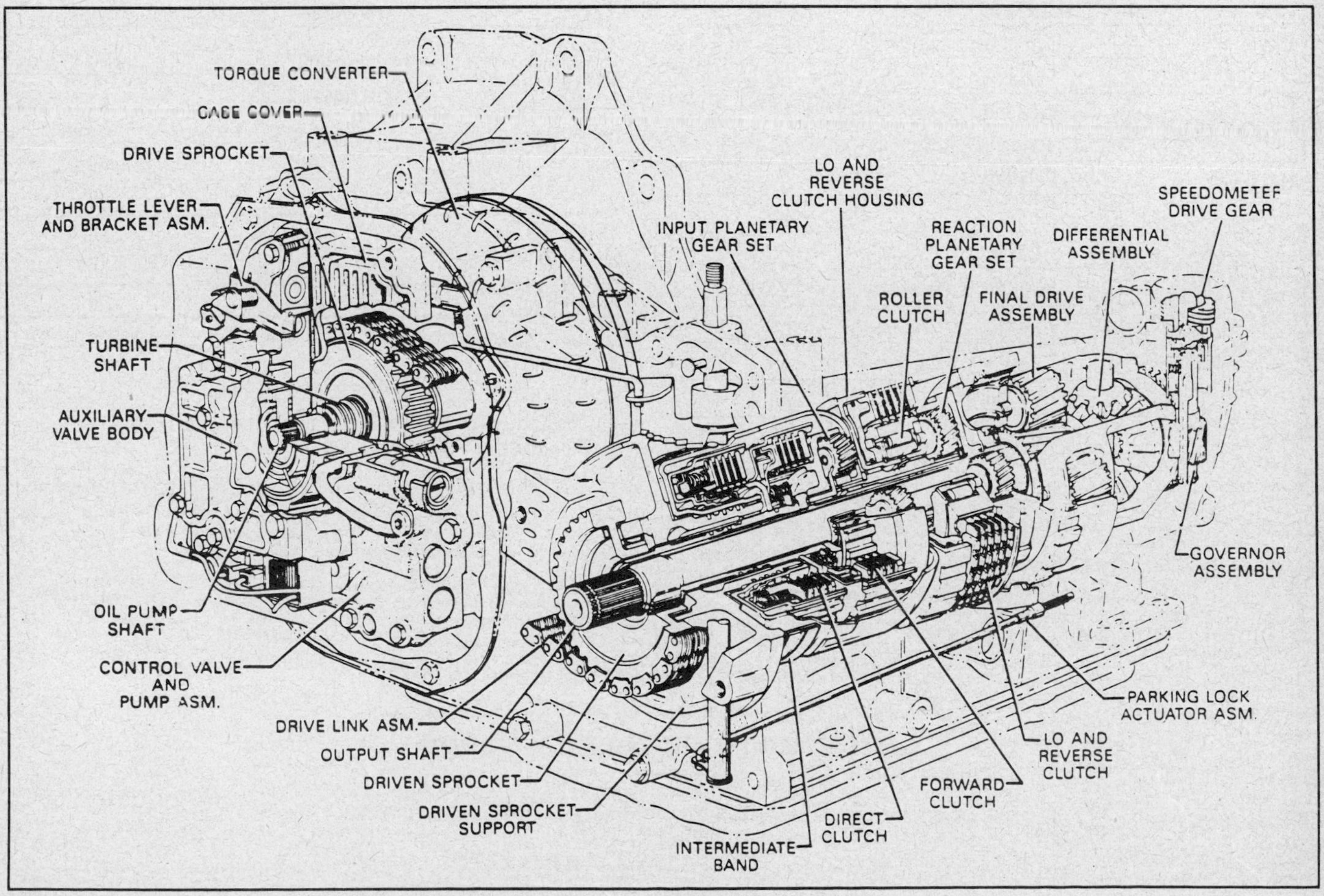

TORQUE CONVERTER
CASE COVER
DRIVE SPROCKET
THROTTLE LEVER AND BRACKET ASM.
TURBINE SHAFT
AUXILIARY VALVE BODY
OIL PUMP SHAFT
CONTROL VALVE AND PUMP ASM.
DRIVE LINK ASM.
OUTPUT SHAFT
DRIVEN SPROCKET
DRIVEN SPROCKET SUPPORT
INTERMEDIATE BAND
DIRECT CLUTCH
FORWARD CLUTCH
LO AND REVERSE CLUTCH
PARKING LOCK ACTUATOR ASM.
GOVERNOR ASSEMBLY
SPEEDOMETEF DRIVE GEAR
DIFFERENTIAL ASSEMBLY
FINAL DRIVE ASSEMBLY
REACTION PLANETARY GEAR SET
ROLLER CLUTCH
LO AND REVERSE CLUTCH HOUSING
INPUT PLANETARY GEAR SET

Cross section of General Motors THM 125C automatic transaxle

REFERENCES

APPLICATION
1980–84 Chilton's Professional Automatic Transmission Manual #7390
1984–89 Chilton's Professional Transmission Manual – Domestic Vehicles #7959

GENERAL DESCRIPTION
1980–84 Chilton's Professional Automatic Transmission Manual #7390

MODIFICATIONS
1980–84 Chilton's Professional Automatic Transmission Manual #7390
1984–89 Chilton's Professional Transmission Manual – Domestic Vehicles #7959

TROUBLE DIAGNOSIS
1980–84 Chilton's Professional Automatic Transmission Manual #7390

ON CAR SERVICES
1980–84 Chilton's Professional Automatic Transmission Manual #7390

REMOVAL AND INSTALLATION
1980–84 Chilton's Professional Automatic Transmission Manual #7390
1984–89 Chilton's Professional Transmission Manual – Domestic Vehicles #7959

BENCH OVERHAUL
1980–84 Chilton's Professional Automatic Transmission Manual #7390

SPECIFICATIONS
1980–84 Chilton's Professional Automatic Transmission Manual #7390
1984–89 Chilton's Professional Transmission Manual – Domestic Vehicles #7959

SPECIAL TOOLS
1980–84 Chilton's Professional Automatic Transmission Manual #7390

APPLICATION

THM 125C

Year	Vehicle	Engine
1984	Omega	All
	Phoenix	All
1984–85	Citation	All
	Skylark	All
1984–89	Celebrity	All
	Century	All①
	Ciera	All①
	A-6000	All③
	Cavalier	All
	Cimarron	All
	Firenza	All
	Skyhawk	All
1984–89	2000/Sunbird	All②
	Fiero	All
1985–89	Calais	All
	Grand Am	All
	Somerset Regal	All
1987–89	Beretta	All
	Corsica	All
1988–89	Lemans	All

① All engines except 3.8L V6 fuel injected and 1985 4.3L V6 diesel
② All engines except 2.0L L4 fuel injected
③ 1988–89 3.1L V6 engine is equipped with THM A-1 transmission

TRANSAXLE MODIFICATIONS

Transmission Name Change

By September 1, 1991, Hydra-matic will have changed the name designation of the THM 125C automatic transmission. The new designation for this transmission will be Hydra-matic 3T40. Transmissions built in 1989 and 1990 will serve as transitional years in which a dual system, made up of the old designation and the new designation will be in effect.

Service Package For Intermediate Band/Direct Clutch Housing Assembly

A new intermediate band and direct clutch housing service package has been assembled and released to repair all THM 125C automatic transaxles. This new service package, part number 8643941 consists of a direct clutch housing, drum assembly, and an intermediate band assembly. These parts must be used together as a complete unit. These items are no longer available individually and can only be obtained as a set.

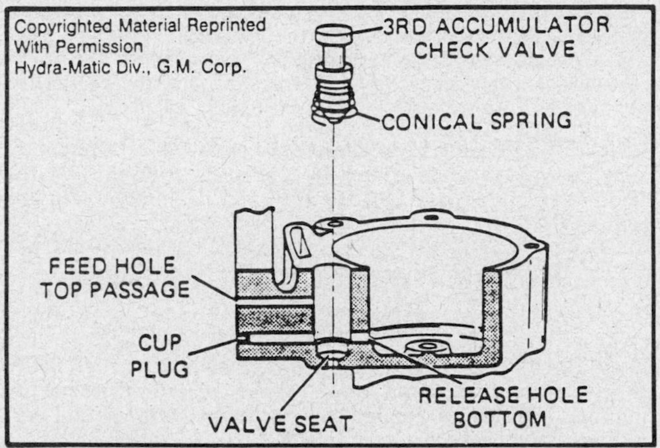
Copyrighted Material Reprinted
With Permission
Hydra-Matic Div., G.M. Corp.
3RD ACCUMULATOR CHECK VALVE
CONICAL SPRING
FEED HOLE TOP PASSAGE
CUP PLUG
RELEASE HOLE BOTTOM
VALVE SEAT

Installing the accumulator check valve into the case bore

Burnt Band and Direct Clutch Assembly Condition

Some THM 125C automatic transaxle equipped vehicles may experience a burnt band and direct clutch condition. A possible cause of the burnt band and direct drive condition might be the third accumulator check valve not seating properly. This condition allows the intermediate band to drag while the direct clutch is applied, causing excessive friction. If the third accumulator is found to be defective, order service package part number 8643964, which contains a new dual land third accumulator check valve and conical spring. Refer to the following procedure to replace the accumulator assembly.

1. Remove the intermediate servo cover and gasket.
2. Remove the third accumulator check valve and spring. Inspect the third accumulator valve bore for wear and damage to the valve seat.
3. Plug both the feed and exhaust holes in the bore using petroleum jelly.
4. Replace the third accumulator check valve with the new dual land check valve. Center the valve to be sure that it is seated properly.
5. Leak test the valve seat by pouring solvent into the accumulator check valve bore. Check for a leak on the inside of the case. A small amount of leakage is acceptable.
6. If the valve leaks, tap the assembly with a brass drift and rubber mallet in order to try and reseat the valve.
7. Repeat the leak test procedure. It may be necessary to replace the case should the valve continue to leak.
8. If the valve does not leak, remove the check valve and install the new conical valve spring onto the valve, with the small end first. Install the valve into the case bore.
9. Using a new gasket install the servo cover.

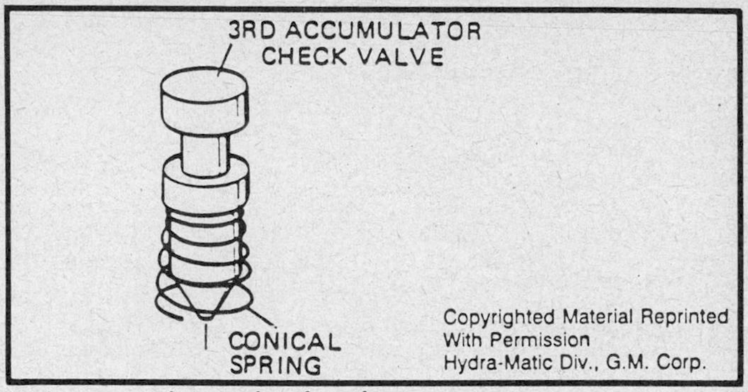
3RD ACCUMULATOR CHECK VALVE
CONICAL SPRING
Copyrighted Material Reprinted
With Permission
Hydra-Matic Div., G.M. Corp.

3rd accumulator check valve

Front Converter Seal Oil Leak

Some THM 125C transaxles may experience a front oil leak condition which is due to the front converter seal coming out of its seat in the case bore. This condition results from an oil pressure back-up behind the converter seal, which causes the converter seal to lose its seat in the case bore.

In servicing the transaxle for this condition, it is recommended that an oil drainback hole be drilled into the converter seal bore of the transaxle case. Proper location is important in drilling the hole. Use the following procedure:

1. With the converter removed, remove the valve body cover and gasket.
2. Remove the throttle lever and bracket assembly.
3. Disconnect the wire assembly from the electrical connector.
4. Remove the control valve assembly, spacer plate and gaskets.
5. Disconnect the manual valve from the rod and retainer assembly.
6. Remove the transaxle case cover.
7. Remove the drive link assembly.
8. Remove the converter oil seal.
9. Remove the drive sprocket support.
10. Drill, from the seal bore side, a $^{3}/_{16}$ in. (4.76mm) minimum diameter hole 0.080 in. (2.03mm) inside the radius of the con-

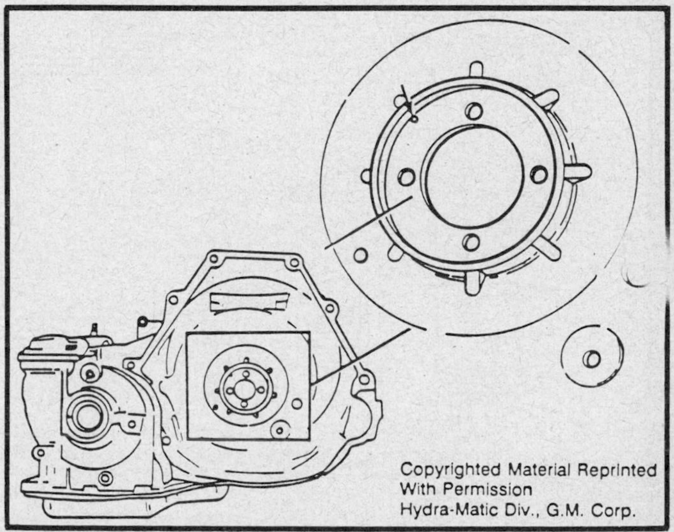
Copyrighted Material Reprinted
With Permission
Hydra-Matic Div., G.M. Corp.

Proper location for drilling the hole

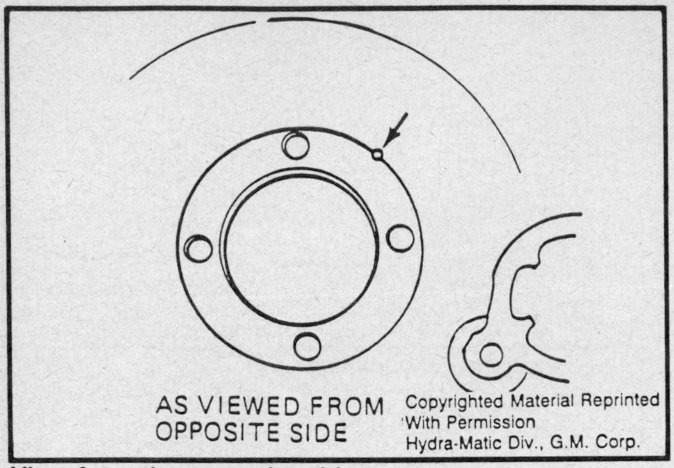
AS VIEWED FROM
OPPOSITE SIDE
Copyrighted Material Reprinted
With Permission
Hydra-Matic Div., G.M. Corp.

View from the opposite side

verter seal bore. The hole must be drilled through so that it exits the other side of the sprocket support mating face, partially outward from the machined surface.

11. Clean all aluminum chips from the case. The transaxle must be thoroughly cleaned and all burrs removed from the drilled hole before reassembly.
12. Install the drive socket support and torque the bolts to 18 ft. lbs. (24 Nm).
13. Install a new converter oil seal into the case using tool J–28540.
14. Install the drive link assembly.
15. Install the transaxle case cover. Torque all bolts to 18 ft. lbs. (24 Nm).
16. Connect the rod and retainer assembly to the manual valve.
17. Install the control valve assembly, spacer plate and gaskets, placing the check balls in their proper locations.
18. Connect the wire assembly to the electrical connector.
19. Install the throttle lever and bracket assembly.
20. Install the valve body cover and gasket. Torque all bolts to 12 ft. lbs. (16 Nm).

Transaxle Case

Beginning with 1984 production of the THM 125C transaxle, a new transaxle case went into production. The changes are as follows:

1. The manual shift hole was made larger in diameter to accommodate a larger manual shaft.
2. The manual shaft seal bore was made deeper to accommodate a double lip seal.
3. The transaxle case casting was changed to accommodate a neutral safety switch.
4. The low/reverse snapring groove was moved toward the case sprocket face to accomodate thicker low and reverse clutch plates.

NOTE: The new 1984 transaxle case cannot be used to service any models prior to the 1984 model. The past style transaxle case cannot be used to service the 1984 transaxle.

Rough Neutral to Drive Gear Engagement

Some 1984 THM 125C transaxles may exhibit a rough neutral to drive gear engagement. A new forward clutch wave plate, part number 8652126 is available. The model number and serial number are needed for application.

Oil Pump Driveshaft and Spacer, Pump Shaft Bearing and Seal, Control Valve and Pump

Starting in May, 1984, a new oil pump driveshaft and spacer assembly, pump shaft bearing and seal assembly and control valve

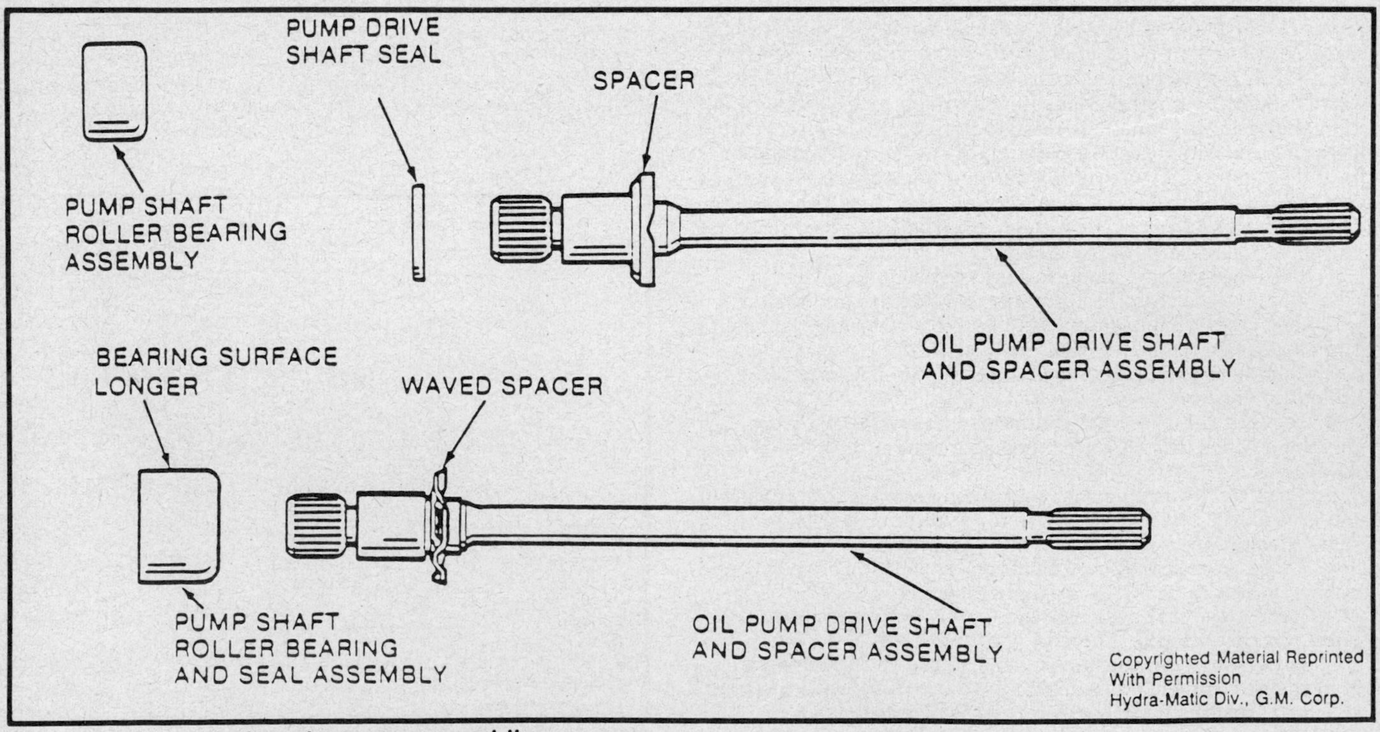
PUMP DRIVE
SHAFT SEAL
SPACER
PUMP SHAFT
ROLLER BEARING
ASSEMBLY
OIL PUMP DRIVE SHAFT
AND SPACER ASSEMBLY
BEARING SURFACE
LONGER
WAVED SPACER
PUMP SHAFT
ROLLER BEARING
AND SEAL ASSEMBLY
OIL PUMP DRIVE SHAFT
AND SPACER ASSEMBLY
Copyrighted Material Reprinted
With Permission
Hydra-Matic Div., G.M. Corp.

Oil pump drive shaft and spacer assemblies

and pump assembly went into production all THM 125C Transaxles built in the U.S.

The former parts are not interchangeable with the new parts for the following reasons:

1. The new oil pump driveshaft and spacer assembly has a thinner spacer and a longer bearing journal diameter to accommodate the new pump shaft bearing and seal assembly.
2. The new pump shaft bearing and seal assembly was also made longer in length to accommodate the wear as part of the bearing assembly.
3. The new control valve and pump assemblies were changed to accommodate the new oil pump driveshaft and roller pump shaft bearing and seal assemblies.

NOTE: All THM 125C transaxles built in Canada will have these new parts beginning start of production 1985. Some models built in Canada are also built in the U.S. and can be distinguished from each other by their nameplates. The U.S. models have serial numbers and Canadian models have julian dates.

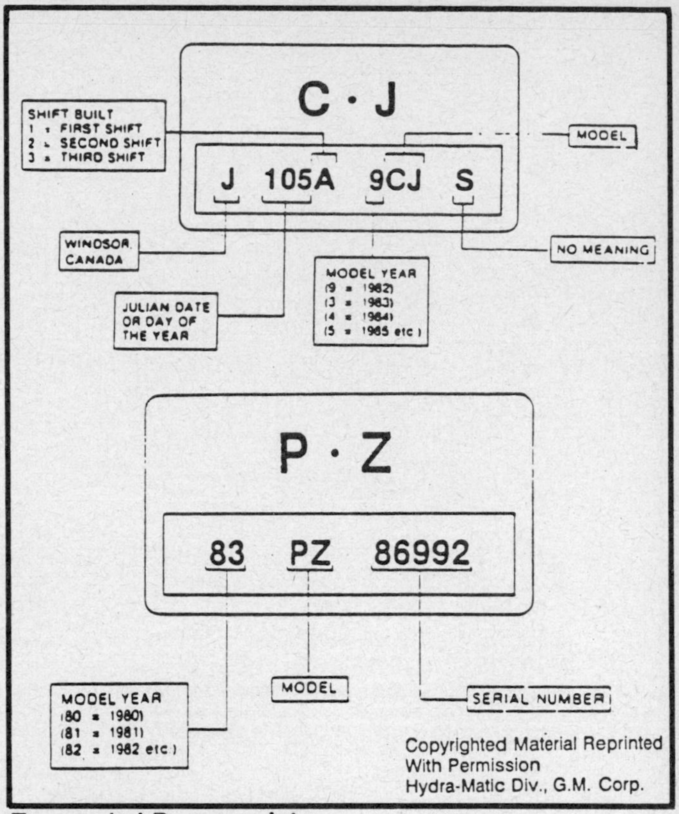
C · J
SHIFT BUILT
1 = FIRST SHIFT
2 = SECOND SHIFT
3 = THIRD SHIFT
MODEL
J 105A 9CJ S
WINDSOR CANADA
NO MEANING
MODEL YEAR
(9 = 1982)
(3 = 1983)
(4 = 1984)
(5 = 1985 etc)
JULIAN DATE OR DAY OF THE YEAR
P · Z
83 PZ 86992
MODEL
MODEL YEAR
(80 = 1980)
(81 = 1981)
(82 = 1982 etc)
SERIAL NUMBER
Copyrighted Material Reprinted With Permission Hydra-Matic Div., G.M. Corp.

Transaxle I.D. nameplates

Service Procedure for Checking Chain for Possible Wear

When disassembling any THM 125C, it is important to inspect the drive link assembly for possible wear. Use the following service procedure:

1. Midway between the sprockets and at right angles to the chain, push the slack strand of the chain down until finger tight and mark it on the opposite of the teeth.
2. Push up in the same manner and make a second mark, making sure that both marks are made from the same point on the chain.

NOTE: If the distance between the marks exceeds 27.4mm, the chain must be replaced.

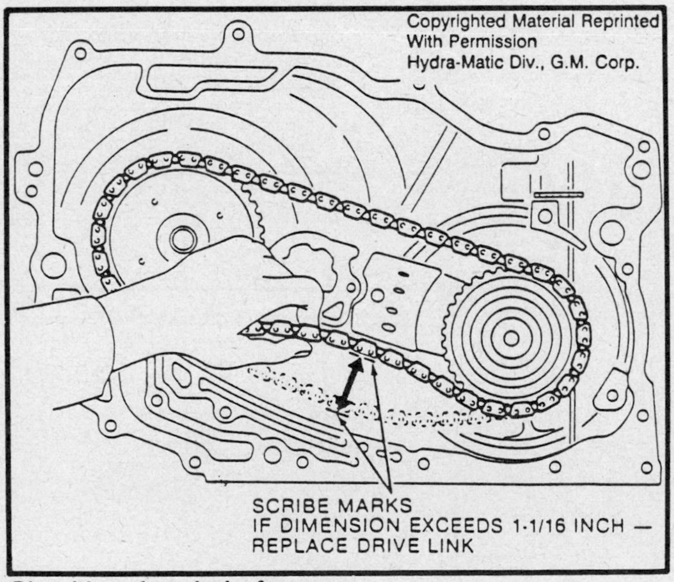
Copyrighted Material Reprinted With Permission Hydra-Matic Div., G.M. Corp.
SCRIBE MARKS
IF DIMENSION EXCEEDS 1-1/16 INCH — REPLACE DRIVE LINK

Checking the chain for wear

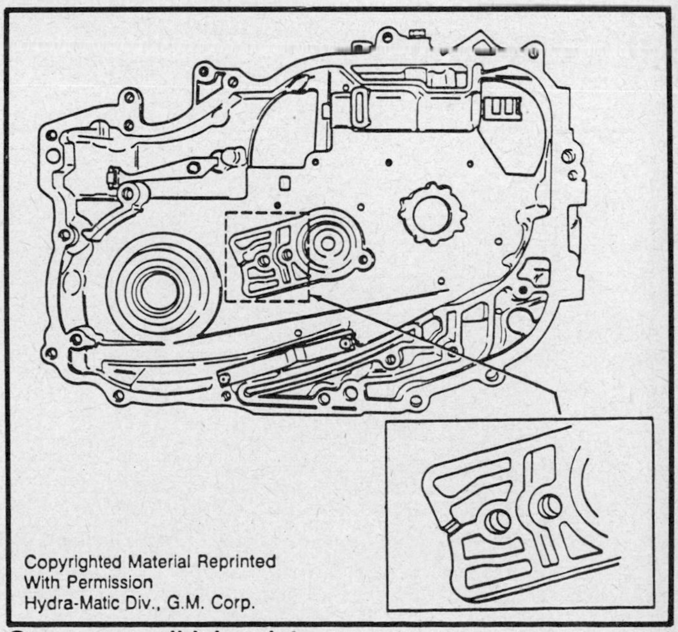
Copyrighted Material Reprinted With Permission Hydra-Matic Div., G.M. Corp.

Case cover oil lube slot

Case Cover Oil Slot

Beginning in April 1984, a slot was added to the case cover of the THM 125C transaxle. The slot was added to the drive feed oil circuit to allow transmission fluid to spray onto the drive chain to provide additional lubrication. The slot is a **V** shaped groove cut into the case cover. The slot is intentional and should not be considered as a damaged case cover.

Forward and Direct Clutch

At the start of production for 1985, all THM 125C transaxles are being built with a new forward and direct clutch assembly. The 1985 production forward and direct clutch assemblies are not interchangeable with the 1984 or prior model year forward and direct clutch assemblies.

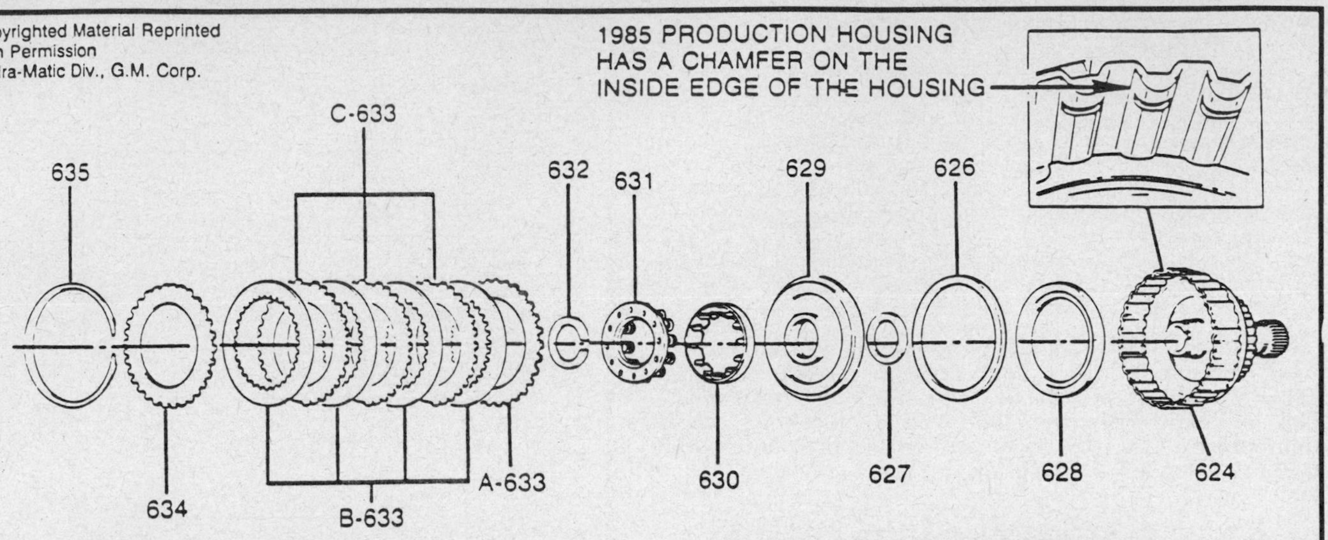
1985 PRODUCTION HOUSING
HAS A CHAMFER ON THE
INSIDE EDGE OF THE HOUSING
C-633
635
632
631
629
626
634
B-633
A-633
630
627
628
624

624 HOUSING ASSEMBLY, FORWARD CLUTCH
626 INSERT
627 SEAL, FORWARD CLUTCH PISTON INNER
628 SEAL, FORWARD CLUTCH PISTON OUTER
629 PISTON, FORWARD CLUTCH
630 CLUTCH RELEASE SPRING GUIDE
631 RETAINER, SPRING ASSEMBLY
A-633 WAVED STEEL PLATE
B-633 COMPOSITION FACED PLATE
C-633 STEEL CLUTCH PLATE
634 PLATE, CLUTCH BACKING, FORWARD
635 RING, SNAP

1985 Production forward clutch assembly

For the 1984 model, a new forward and direct clutch housing service assemblies are now available. The direct clutch housing service package, part number 8653978 contains the following:

1. Direct clutch housing assembly
2. Direct clutch piston assembly
3. Direct clutch piston inner, outer and center seals
4. Intermediate band assembly

The forward clutch housing service package, part number 8653977 contains the forward clutch housing assembly.

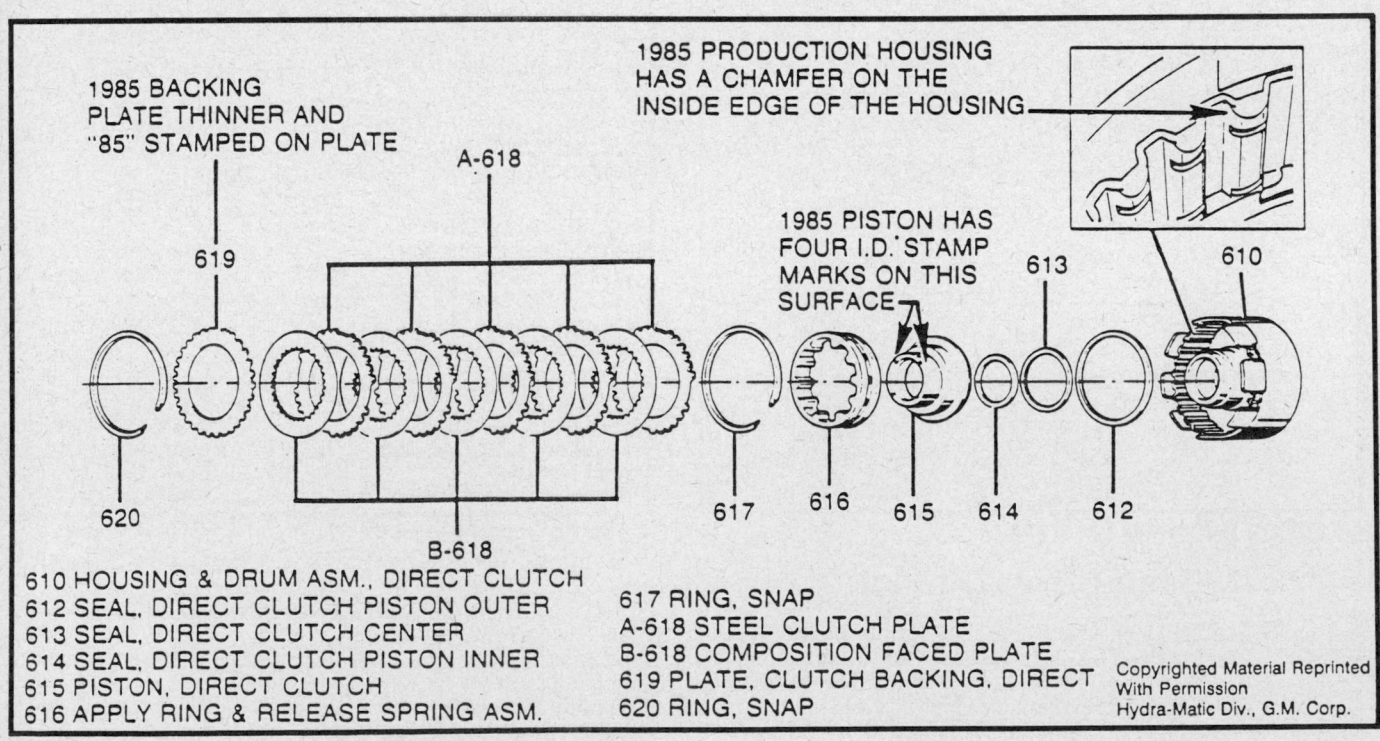

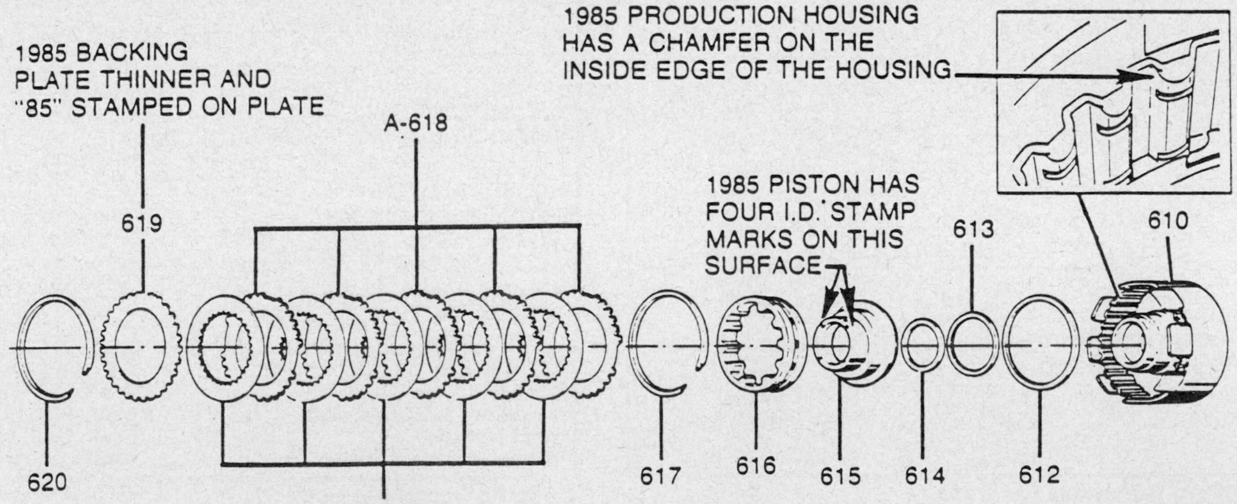
1985 PRODUCTION HOUSING
HAS A CHAMFER ON THE
INSIDE EDGE OF THE HOUSING
1985 BACKING
PLATE THINNER AND
"85" STAMPED ON PLATE
A-618
619
1985 PISTON HAS
FOUR I.D. STAMP
MARKS ON THIS
SURFACE
613
610
620
617
616
615
614
612
B-618

610 HOUSING & DRUM ASM., DIRECT CLUTCH
612 SEAL, DIRECT CLUTCH PISTON OUTER
613 SEAL, DIRECT CLUTCH CENTER
614 SEAL, DIRECT CLUTCH PISTON INNER
615 PISTON, DIRECT CLUTCH
616 APPLY RING & RELEASE SPRING ASM.
617 RING, SNAP
A-618 STEEL CLUTCH PLATE
B-618 COMPOSITION FACED PLATE
619 PLATE, CLUTCH BACKING, DIRECT
620 RING, SNAP

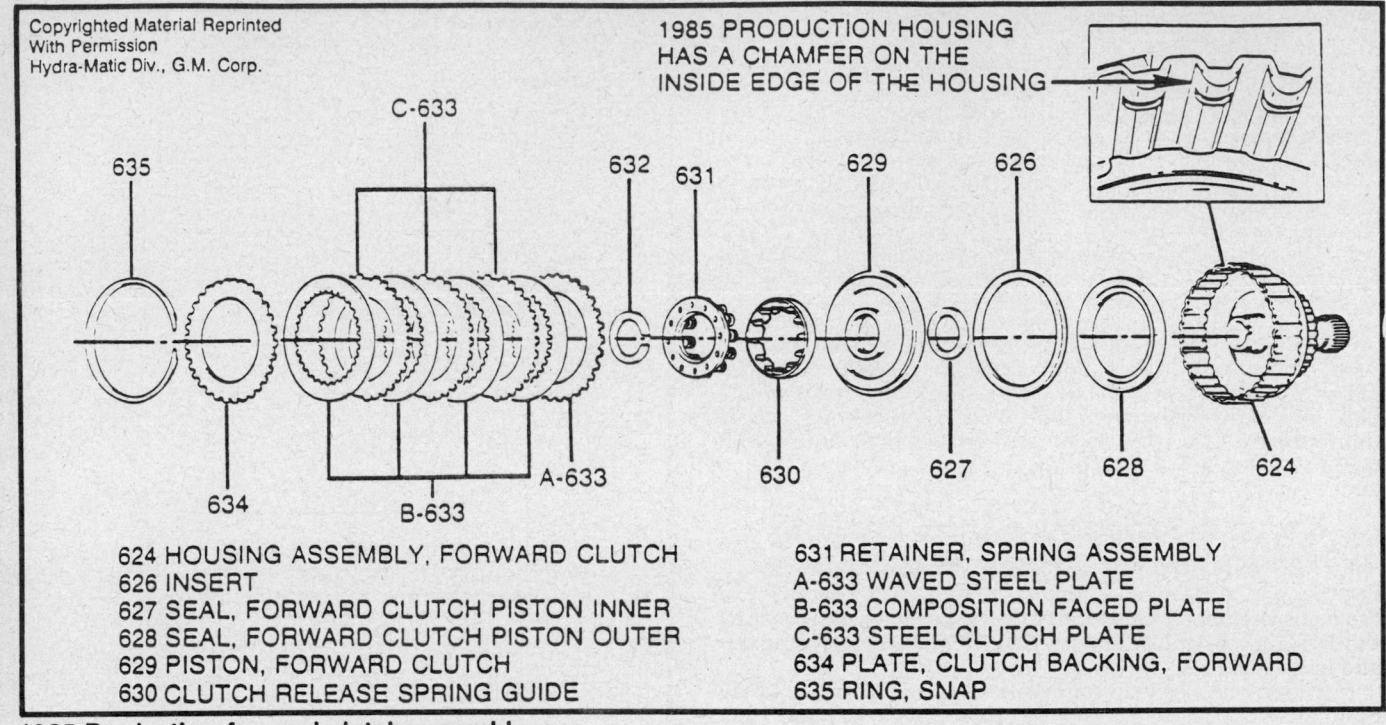

1985 Production direct clutch assembly

Throttle Valve Buzz

A 1985 or 1986 THM 125C transaxle built prior to October 29, 1985 (Julian Date 302) in Windsor, or prior to December 9, 1985 (Julian Date 343) in Ypsilanti, may exhibit a throttle valve buzz condition.

A new valve body to case cover spacer plate was introduced to eliminate the buzz by reducing the diameter of the 1–2 accumulator valve feedback orifice. To correct a throttle valve buzz condition for models built before these dates, replace the original valve body to case cover space plate with a new spacer plate, part number 8660569.

Intermediate Servo Orifice Eliminated

Beginning April 15, 1985 (Julian Date 105) for U.S. models, and April 11, 1985 (Julian Date 101) for Canadian models, 125C transaxles are being produced using a case with the intermediate servo bleed orifice eliminated. Either transaxle case is used for all applications in service.

New Gaskets

Beginning June 6, 1985 (Julian Date 157), all 1985 THM 125C transaxle are being produced using new gaskets for the oil pan and the valve body cover. The new gaskets improve sealing capability by better withstanding fastener torques.

When servicing 1984 and later THM 125C transaxles for oil pan and valve body cover fluid leaks, use the following procedure:

1. Diagnose the source of the fluid leak to be certain that the leak is from the transaxle oil pan and/or valve body cover.
2. If the leak is determined to be from the oil pan and/or valve body cover area, remove the pan and/or the cover.
3. Clean the case, oil pan and or valve body cover faces and visually inspect for damage or presence of gasket material.
4. Make sure all gasket material is removed from the case, oil pan and/or valve body faces.
5. Replace the oil pan and/or valve body cover gaskets with oil pan gasket part number 8660038 and valve body cover gasket part number 8660039.
6. When replacing the pan and/or cover, it is critical that a torque wrench is used to tighten the hex head screws. Torque all hex head screws to 9–12 ft. lbs. (12–16 Nm).

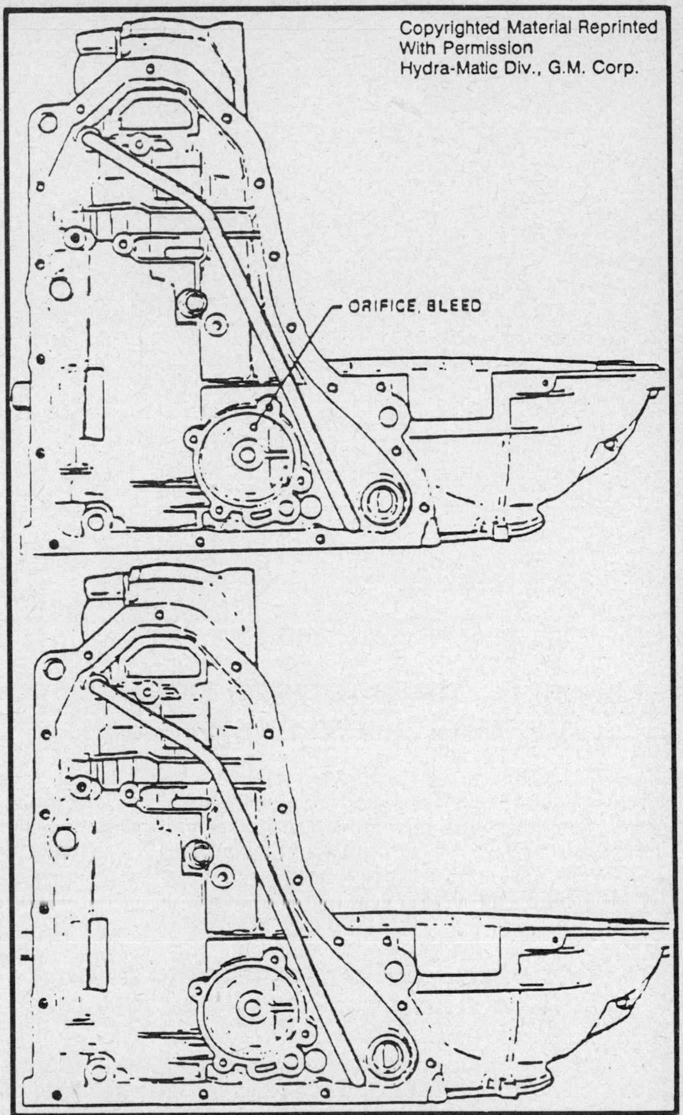
ORIFICE BLEED

Intermediate servo orifice eliminated

Slipping In All Ranges

Some THM 125C transaxles may exhibit a slipping in all ranges condition under hard acceleration and/or maneuvering operations, caused by oil pressure drop due to a cut or improperly seated oil strainer seal.

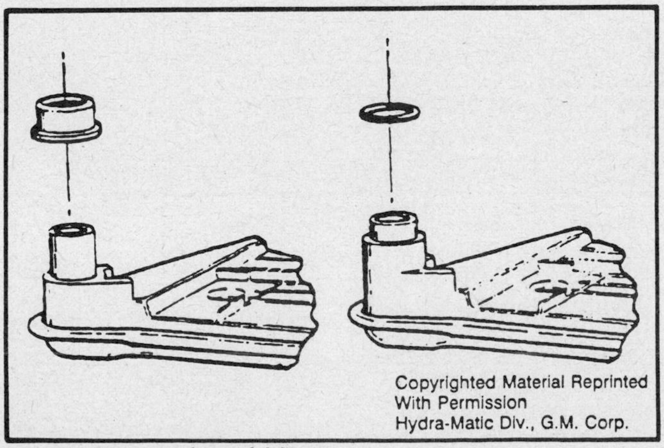

Oil strainer and seal kit

Use oil strainer seal kit part number 8652910 to replace oil strainer seal kit part number 8631951 on a transaxle that exhibits the slipping condition. Either design can be used to service all transaxles.

If replacement of the oil strainer during transaxle service is necessary, remove the original seal from the case and use the new seal provided in the kit. Always install the seal in the case first when installing the replacement oil strainer and seal assembly.

Soft Part-Throttle 1-2 Shift and Flare at 3-2 Shift Between 40–47 mph

A THM 125C transaxle model JSC built before December 11, 1985 (Julialn Date 345) may exhibit a soft-part throttle 1–2 shift and flare at the 3–2 shift between 40–47 mph. The JSC model is used only overseas applications.

This condition may be corrected by replacing the 3–2 control valve spring and 1–2 accumulator valve spring contasined in service package part number 8652920.

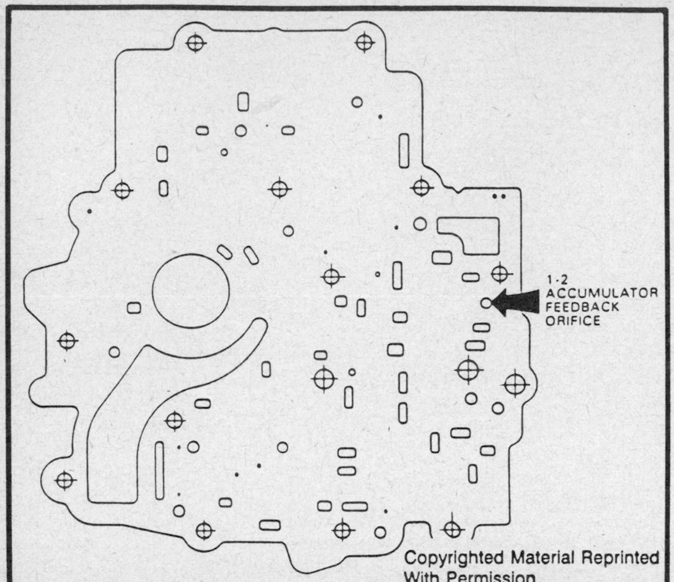
1-2
ACCUMULATOR
FEEDBACK
ORIFICE

1–2 Accumulator feedback orifice

Harsh 1-2 Upshift When Hot During Medium to Hard Acceleration

The THM 125C transaxle may experience a harsh 1-2 upshift when hot during medium to hard throttle acceleration. When servicing this condition, remove the spacer plate and measure the diameter of the 1-2 accumulator feedback orifice. If the hole is 0.040 in. (1.02mm), replace the spacer plate with a new spacer plate which has a larger 0.125 in. (3.18mm) 1-2 accumulator feedback orifice.

Forward Clutch Assembly and Low and Reverse Clutch Assembly

The neutral to drive package can be used to service a 1985 or 1986 (before Julian Date 184) THM 125C transaxle exhibiting a harsh drive engagement by replacing the forward clutch with a new forward clutch assembly.

The neutral to reverse package can be used to service a 1985 or 1986 (before Julian Date 189) THM 125C transaxle exhibiting a harsh reverse engagement by replacing the low and reverse clutch assembly with the new low and reverse housing assembly and low and reverse clutch pack assembly.

When servicing the forward and/or low and reverse clutch assemblies on a 1986 (after Julian Date 184) or 1987 THM 125C transaxle, it is important to select the correct backing plate. The backing plate selection effects piston travel and shift feel.

To check the forward clutch backing plate selection, apply 10 lbs. of pressure on the backing plate and check the gap between the backing plate and snapring. If the gap is larger than 1.49mm use the next thicker backing plate and if smaller than 0.67mm use the next thinner backing plate.

New Design Low and Reverse Pipe, Seal and Related Components

A new design low/reverse pipe, seal and components was implemented on all Windsor built 125C transaxles built between the dates of January 12, 1987 and April 6, 1987 (Julian Dates 012–096). This temporary change included a case with a machined boss, retainer bracket, reverse oil pipe and reverse oil seal.

NOTE: To assist in identification, the parts catalog refers to these changes as SECOND DESIGN.

Engine Stalls While Shifting Into Reverse or Drive

A new design torque converter clutch solenoid, which has an internal rubber seal that will prevent sediment from sticking to the electromagnetic portion of the solenoid, is in production in all 1987 THM 125C transaxles built after (Julian Date 160).

The new design solenoid will reduce the potential of the torque converter clutch solenoid sticking and the resulting condition of the engine stalling while shifting into reverse or drive.

Beginning June 9, 1987 (Julian Date 160) all 1987 THM 125C transaxles were produced with the new design solenoid. The new design solenoid can be used to service all past THM 125C transaxle models. Remove the side cover and replace the solenoid and conduit assembly.

Coastdown Whine Condition

A 1985 or 1986 J body vehicle with a 2.8L engine and THM 125C transaxle may exhibit an objectional coastdown whine from 50 mph to 20 mph. The noise may be most audible at 25–30 mph. To minimize this condition, replace the strut with a 1987 design engine strut assembly part, number 14106640.

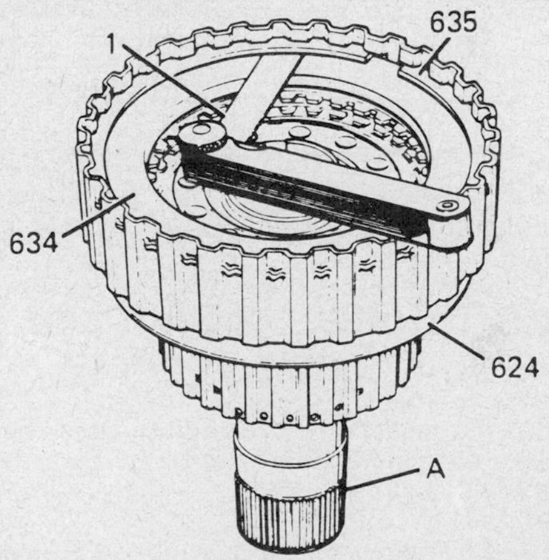
1
635
634
624
A

A INPUT SHAFT
1 FEELER GAGE .67-1.49mm (.026"-.059")
624 HOUSING ASSEMBLY, FORWARD CLUTCH
634 PLATE, FORWARD CL. BACKING (SELECTIVE)
635 RING, SNAP

BACKING PLATE THICKNESS		IDENTIFICATION CODE		PART NUMBER	
MM	Inches	Steel	Powdered Metal	Steel	Powdered Metal
5.0 - 4.9	.197 - .191	A	6	8664160(A)	8664156(6)
4.5 - 4.3	.175 - .170	B	7	8664161(B)	8664157(7)
3.9 - 3.8	.154 - .148	C	8	8664162(C)	8665158(8)
3.3 - 3.2	.132 - .126	D	9	8664163(D)	8664159(9)

Forward clutch backing plate selection

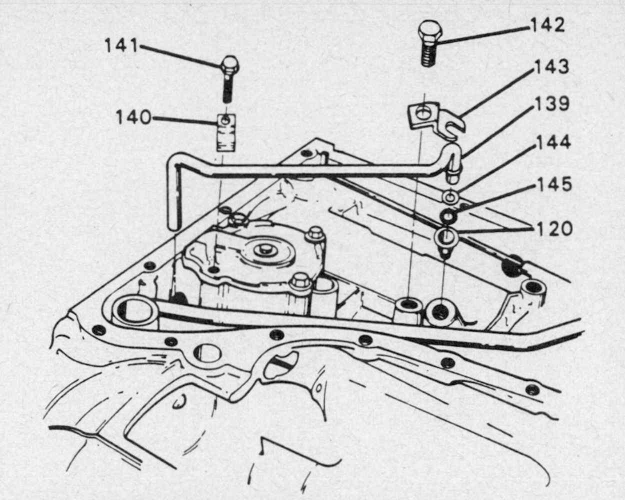
141
140
142
143
139
144
145
120

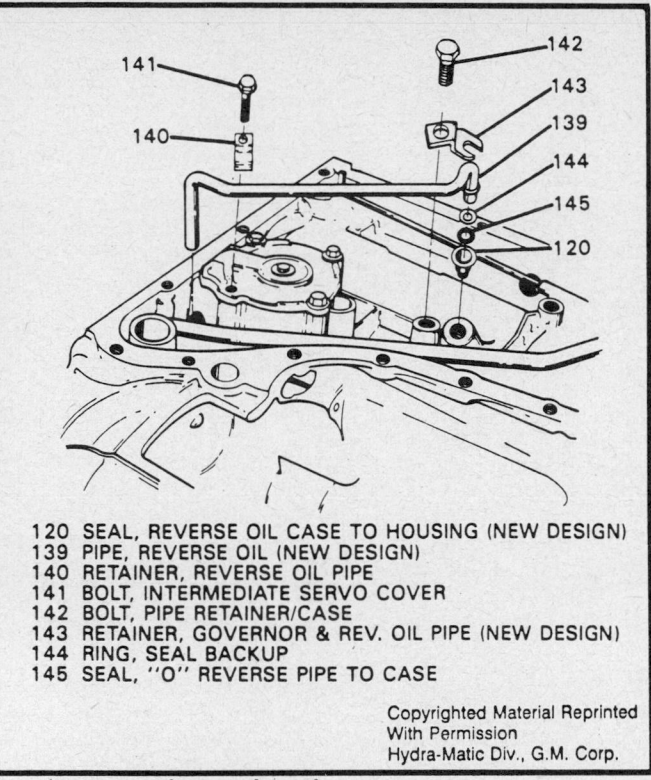

120 SEAL, REVERSE OIL CASE TO HOUSING (NEW DESIGN)
139 PIPE, REVERSE OIL (NEW DESIGN)
140 RETAINER, REVERSE OIL PIPE
141 BOLT, INTERMEDIATE SERVO COVER
142 BOLT, PIPE RETAINER/CASE
143 RETAINER, GOVERNOR & REV. OIL PIPE (NEW DESIGN)
144 RING, SEAL BACKUP
145 SEAL, "O" REVERSE PIPE TO CASE

Low/reverse pipe and seal

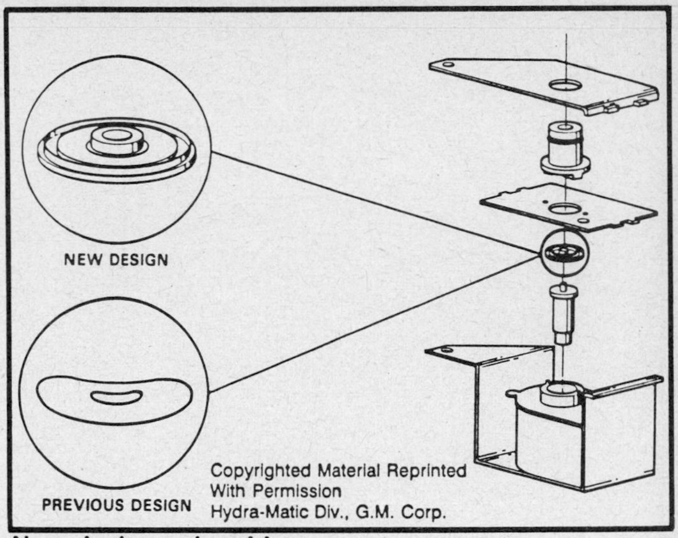
NEW DESIGN
PREVIOUS DESIGN

New design solenoid

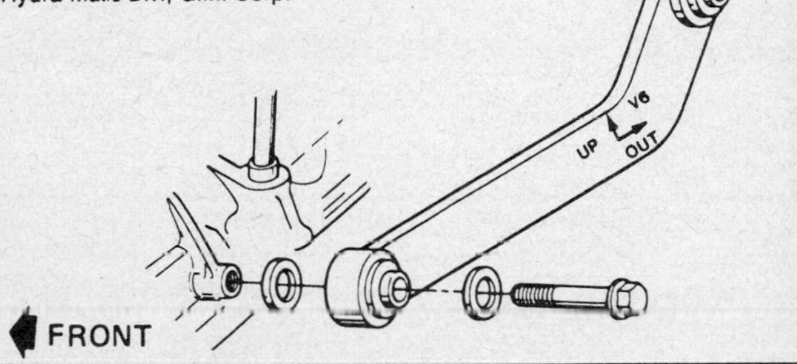
V6
UP
OUT
FRONT

2.8L engine strut

Harsh 1-2 Shift

On 1987 THM 125C transaxles built between the dates of April 13, 1987 through April 21, 1987 (Julian Dates 103–112) may have a rolled or misassembled 1–2 accumulator piston seal, resulting in a possible harsh 1–2 shift. These transaxles have a new design accumulator piston (flat inner surface) and a soft synthetic rubber 1–2 accumulator piston seal that may be inadvertently misassembled.

If a vehicle equipped with a 1987 THM 125C transaxle built between Julian Dates 103–112 experiences a harsh 1-2 shift, examine the 1-2 accumulator piston and seal for damage or misassembly. If necessary, replace the piston and seal with the original design accumulator piston (raised rib) and seal. These parts are not interchangeable.

Harsh 1-2 Shifts at WOT

A new control valve with a new 1–2 accumulator spring, valve and bushing went into production to reduce the potential for a harsh 1–2 upshift. The upshift may occur during wide open throttle in a 1988 N body vehicles equipped with a KDC model THM 125C transaxle and a 2.3L engine. This condition may be repaired by installing a 1–2 accumulator valve train.

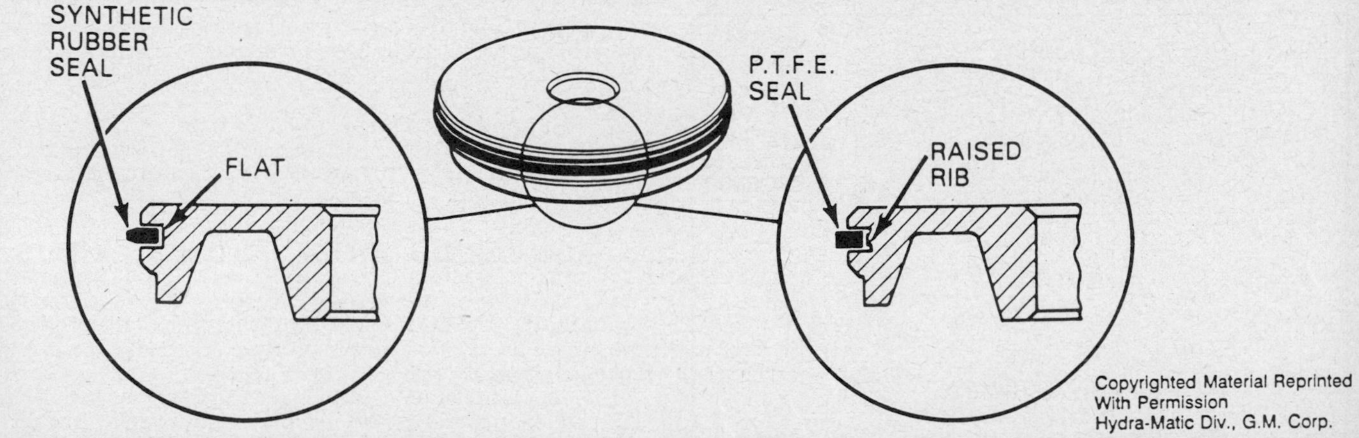
SYNTHETIC
RUBBER
SEAL
FLAT
P.T.F.E.
SEAL
RAISED
RIB

Accumulator piston seals

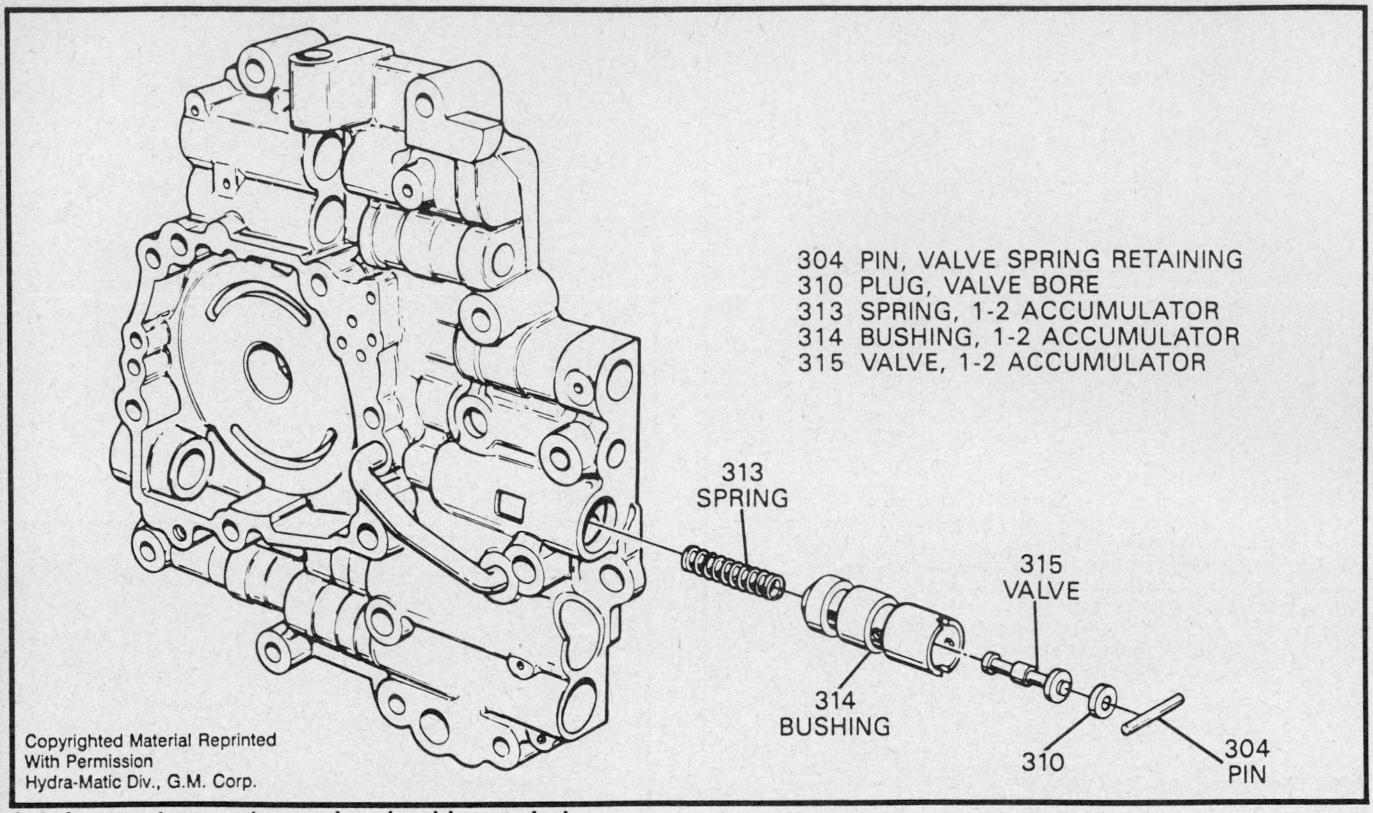
304 PIN, VALVE SPRING RETAINING
310 PLUG, VALVE BORE
313 SPRING, 1-2 ACCUMULATOR
314 BUSHING, 1-2 ACCUMULATOR
315 VALVE, 1-2 ACCUMULATOR
313 SPRING
315 VALVE
314 BUSHING
310
304 PIN

1–2 Accumulator valve spring, bushing and pin

Engine Stalls While Engaging In Reverse or Drive

A new design auxiliary valve body went into production March 9, 1987 (Julian Date 068). It has a new filter in the torque converter clutch feed passage. The new auxiliary valve body filter will reduce the potential of torque converter clutch solenoid sitcking and the resulting condition of the engine stalling while shifting into reverse or drive.

The new auxiliary valve body is not interchangeable with models built prior to March 9, 1987 (Julian Date 068). The past designed auxiliary valve bodies are not machined to allow for installation of the new filter. The auxiliary valve body gasket part number 8660789 is interchangeable with 1987 and 1988 THM 125C transaxles only. Gasket part number 8664847 can only be used on 1988 KDC models.

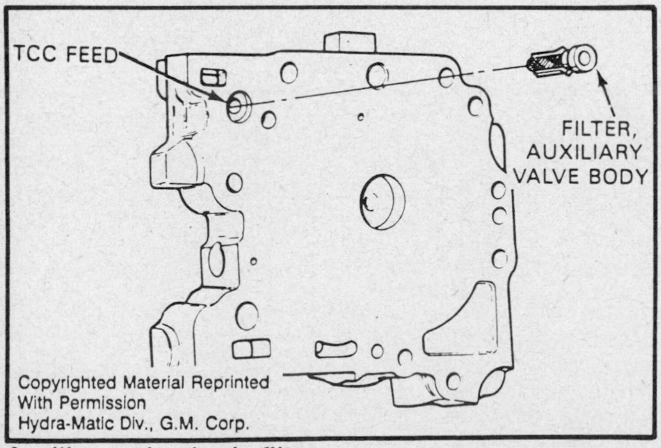
TCC FEED
FILTER, AUXILIARY VALVE BODY

Auxiliary valve body filter

Transaxle Side Cover or Bottom Pan Leaks

Whenever the side cover or the bottom pan has been removed on any THM 125C it is necessary to use Loctite 242® or equivalent to reduce the potential for fluid leaks on the bolts.

Clean the bolts with a suitable solvent to ensure that it is free of dirt and transmission fluid. Apply Loctite 242® or equivalent to each bolt and allow the compound to fully cure.

Slip Bump During 3-2 Downshift

A slip bump condition during a part throttle 3–2 downshift maneuver, may be caused by excess material left in the 3–2 control valve bore during machining of the control valve and oil pump assembly.

To repair this condition on a 1988 N body vehicle with a 2.3L or 3.0L engine and equipped with a THM 125C with a listed Julian Date lower than 315 install a new control valve and oil pump assembly.

New Input Carrier Thrust Washers

Input carriers with new Nibron coated pinion washers were built into THM 125C transaxles starting with (Julian Date 355) December 21, 1987. The pinion washers cannot be serviced individually.

The input carrier assembly should inspected for pinion damage, pinion tilt and pinion endplay. The endplay should be 0.009–0.027 in. (0.24–0.69mm). If the endplay is not within this range, a new input carrier should be installed.

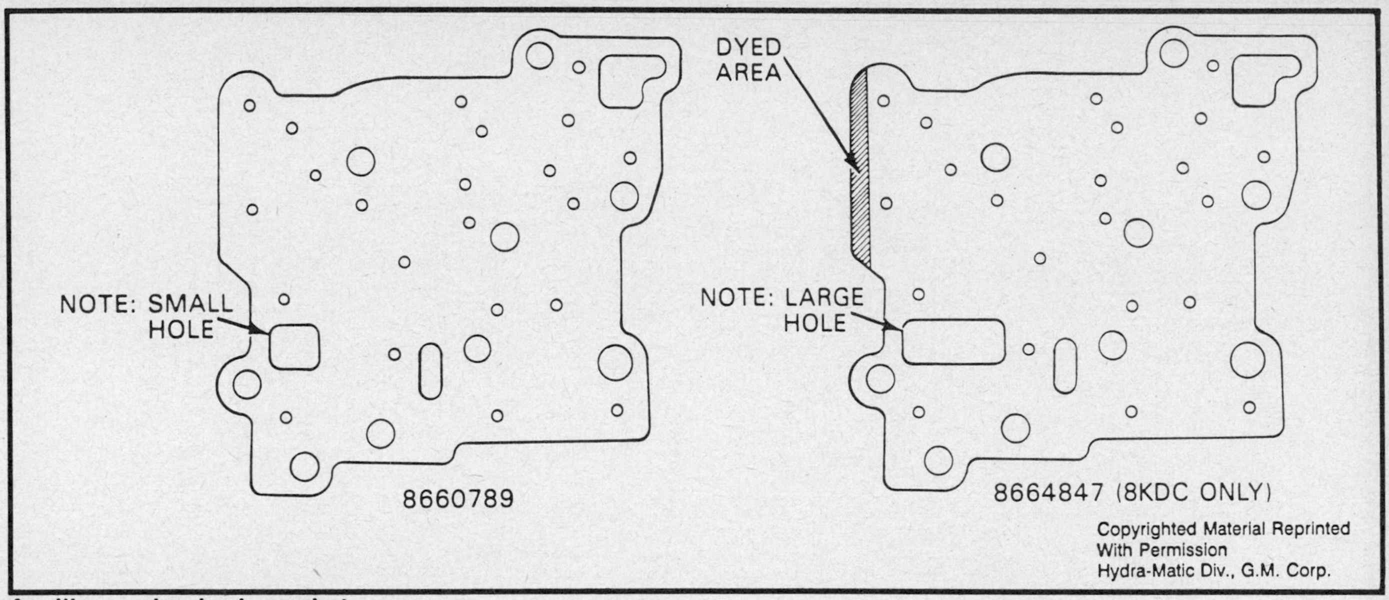
DYED
AREA
NOTE: SMALL
HOLE
NOTE: LARGE
HOLE
8660789
8664847 (8KDC ONLY)
Copyrighted Material Reprinted
With Permission
Hydra-Matic Div., G.M. Corp.

Auxiliary valve body gasket

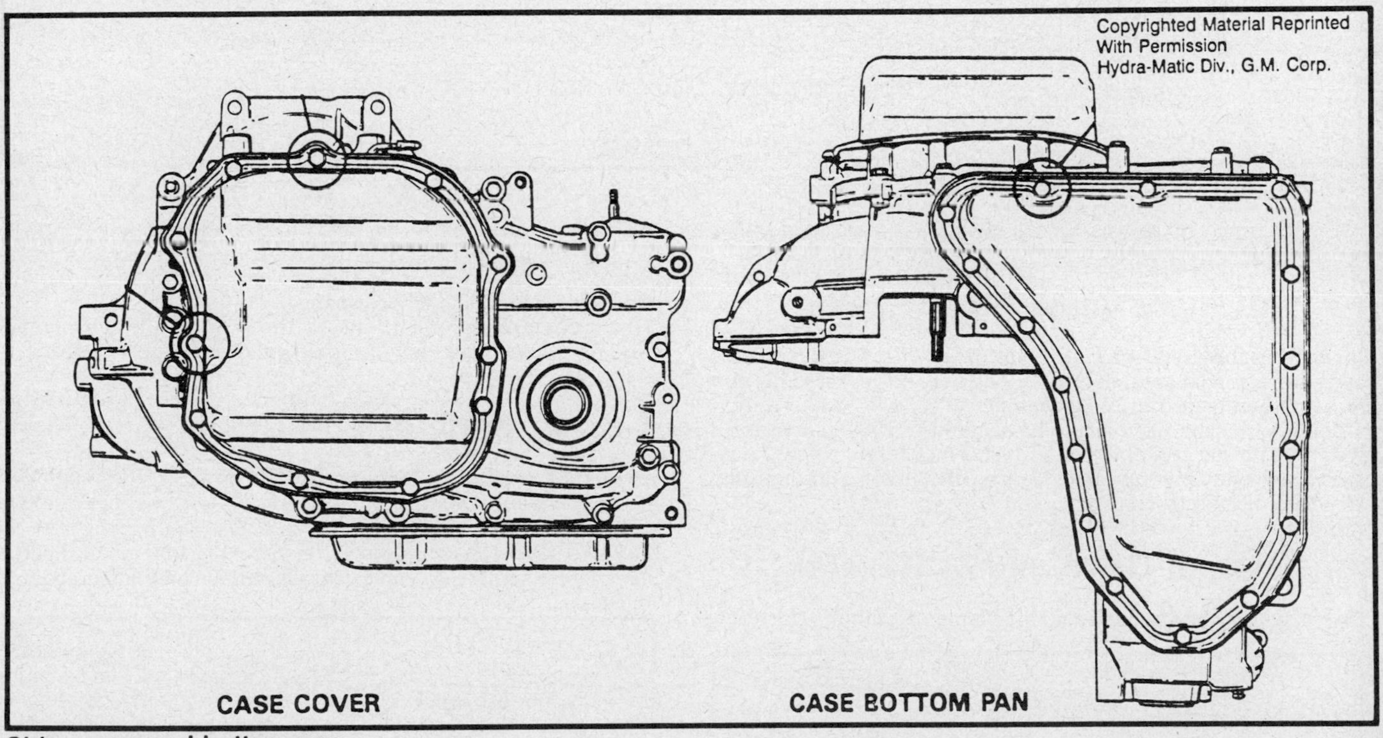
Copyrighted Material Reprinted
With Permission
Hydra-Matic Div., G.M. Corp.
CASE COVER
CASE BOTTOM PAN

Side cover and bottom pan

When servicing a THM 125C transaxle, if the inspection of the input carrier reveals any of the input carrier pinion washers are worn and/or missing, a new input carrier must be installed. This input carrier with new Nibron coating may be used on all past models.

Torque Control Clutch (TCC) Filter

A torque converter clutch filter was added to the torque converter clutch signal passage in the auxiliary valve body beginning in March 1987. This will reduce the possibility of sediment entering the solenoid.

If the solenoid gets contaminated with sediment, it could result in a complaint of engine stalling when the selector is placed in drive or reverse. The filter cannot be used in porevious models because of a machining process needed on the auxiliary valve body.

Governor Screen

The governor screen was added (1986 Julian Date 203) to the governor sleeve in the case to minimize the possibility of sediment entering the governor assembly and preventing the governor balls from seating properly.

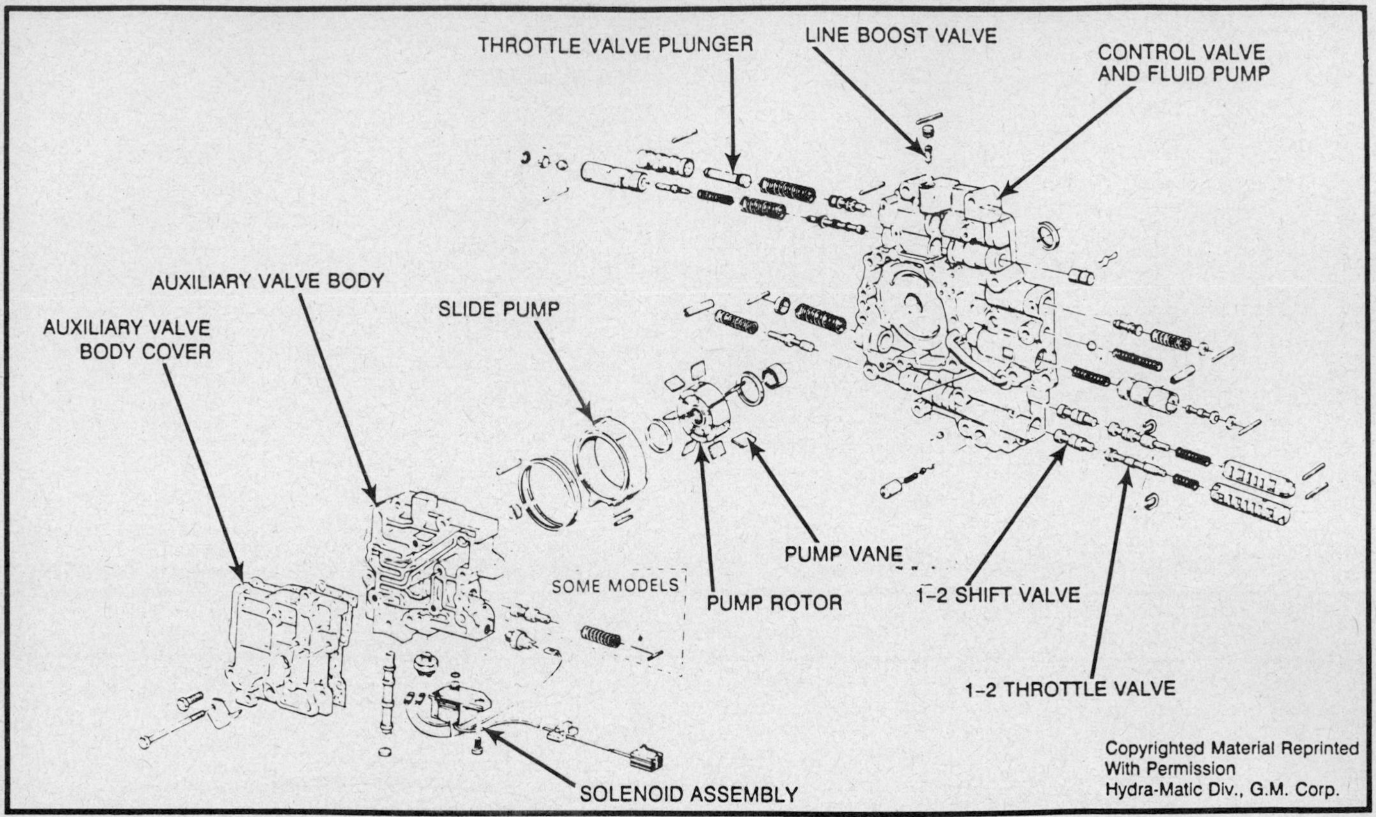
THROTTLE VALVE PLUNGER
LINE BOOST VALVE
CONTROL VALVE AND FLUID PUMP
AUXILIARY VALVE BODY
SLIDE PUMP
AUXILIARY VALVE BODY COVER
PUMP VANE
SOME MODELS
PUMP ROTOR
1-2 SHIFT VALVE
1-2 THROTTLE VALVE
SOLENOID ASSEMBLY
Copyrighted Material Reprinted With Permission Hydra-Matic Div., G.M. Corp.

Control valve and oil pump assembly

A machining change was needed in the case to make room for the new governor screen. It will not fit previous models.

Auxiliary Valve Body Gasket

For 1988, approximately 15 percent of the THM 125C transaxles require a new auxiliary valve body cover gasket. The new gasket is being used in N body vehicles.

The new gasket has a larger hole than the 1987 gasket and a dyed area on the gasket for identification. There are now 4 auxiliary valve body cover gaskets for the THM 125C transaxle and they cannot be interchanged.

Loss of Drive and/or Reverse

The cause may be a broken thermo-element retaining clip located inside the manual valve. This would allow the thermo-element to move forward, allowing drive and/or reverse oil to exhaust. A retaining clip that has broken and fallen out usually can be located in the bottom pan.

To correct the problem, install the previous design manual valve or replace the retaining clip and/or thermo-element as necessary.

Early or Soft Upshifts

Some 1985 Cimarrons equipped with the 2.8L V6 engine and automatic transaxle may exhibit early or soft upshifts due to a misadjusted throttle valve (T.V.) cable. Readjusting the T.V. cable will correct the condition. After repeated wide open throttle maneuvers, some T.V. cables can become misadjusted again.

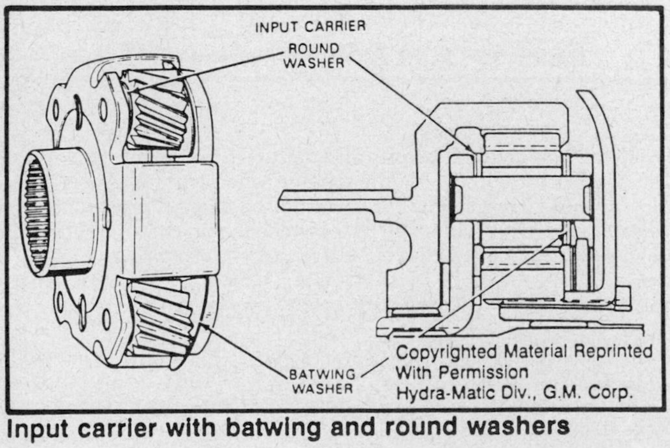
INPUT CARRIER
ROUND WASHER
BATWING WASHER
Copyrighted Material Reprinted With Permission Hydra-Matic Div., G.M. Corp.

Input carrier with batwing and round washers

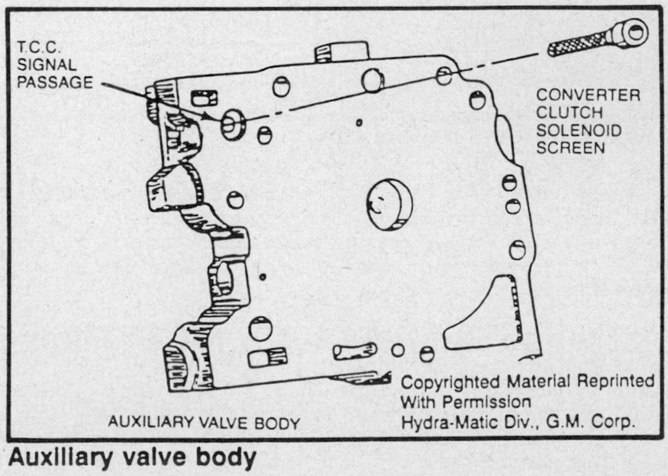
T.C.C. SIGNAL PASSAGE
CONVERTER CLUTCH SOLENOID SCREEN
AUXILIARY VALVE BODY
Copyrighted Material Reprinted With Permission Hydra-Matic Div., G.M. Corp.

Auxiliary valve body

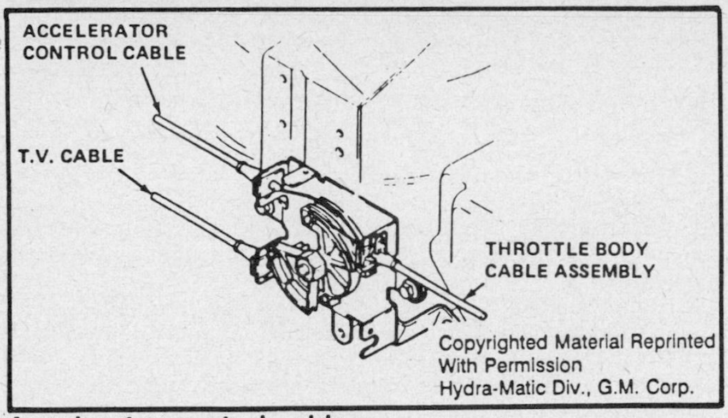
ACCELERATOR CONTROL CABLE
T.V. CABLE
THROTTLE BODY CABLE ASSEMBLY
Copyrighted Material Reprinted With Permission Hydra-Matic Div., G.M. Corp.

Accelerator control cable

This change in adjustment can occur if the T.V. cable readjust tab spring is not strong enough to maintain the proper cable setting during a wide open throttle maneuver.

Beginning with VIN broadcast number FJ511770, a new stronger spring was included in the T.V. cable assembly. When servicing a Cimarron produced prior to the above break point for this condition, replace the T.V. cable assembly with part number 14101470.

Intermittent No Upshift Due to Accelerator Cable Overtravel

Vehicles involved are 1988 Cutlass Calais with a Quad 4 engine built prior to VIN JM248915 and vehicles equipped with a THM 125C transaxle without cruise control. During vehicle operation, the accelerator control cable pulley/cam assembly may overtravel its closed throttle position if the accelerator pedal snaps back when the driver's foot is quickly removed from the pedal or if the pedal is pulled upward from the floor. This can allow the T.V. cable from the transaxle to retract too far into the transaxle allowing the line boost valve to seat which in turn will raise the T.V. pressure causing a delayed or no upshift.

T.V. pressure can be releived by shifting into **P** or **L**, or by stopping or restarting the engine. Upshifts will then be normal but the condition may return with another snap back of the accelerator. To correct this condition, install a new accelerator cable part number 22542664. The new cable has a stop to eliminate the overtravel.

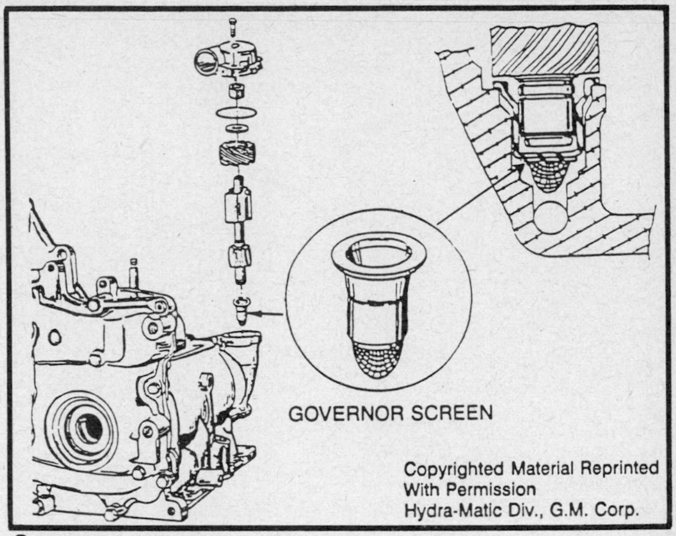
GOVERNOR SCREEN
Copyrighted Material Reprinted With Permission Hydra-Matic Div., G.M. Corp.

Governor screen

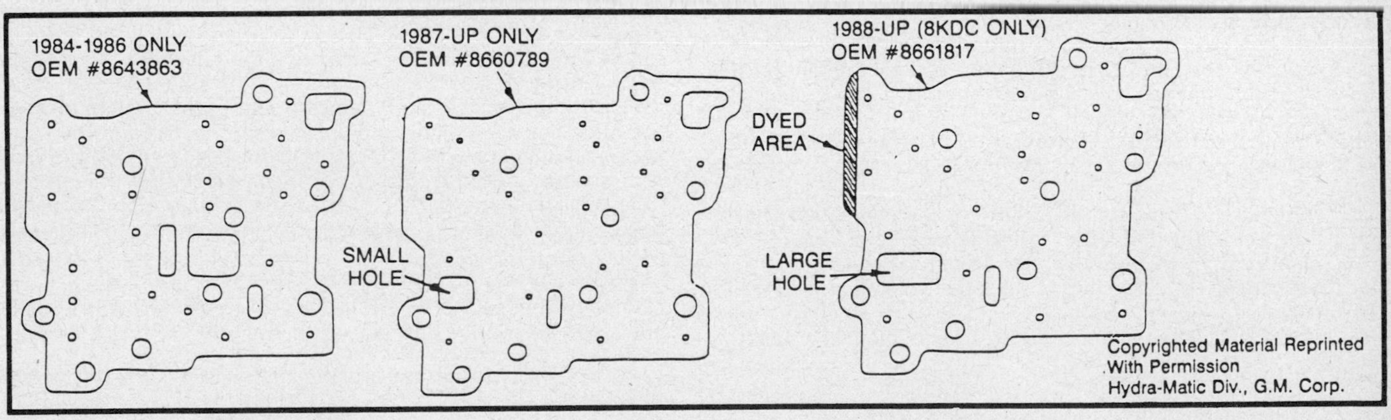
1984-1986 ONLY
OEM #8643863
1987-UP ONLY
OEM #8660789
SMALL HOLE
1988-UP (8KDC ONLY)
OEM #8661817
DYED AREA
LARGE HOLE
Copyrighted Material Reprinted With Permission Hydra-Matic Div., G.M. Corp.

Auxiliary valve body gasket

REMOVAL AND INSTALLATION

TRANSAXLE REMOVAL

THM 125C

CENTURY, SKYLARK, CITATION, CELEBRITY, OMEGA, CIERA, PHOENIX, 6000, REGAL, CALAIS, GRAND AM AND LEMANS

1. Disconnect the negative battery cable at the transaxle.
2. Remove:
   a. Air cleaner and disconnect the T.V. cable.
   b. T.V. cable lower attaching bolt and disconnect the cable from the transaxle.
   c. Strut shock bracket bolts from the transaxle.
   d. Oil cooler lines from the strut bracket.
   e. Transaxle-to-engine bolts, leaving the bolt near the starter installed loosely.
   f. Shift linkage retaining clip and washer at the transaxle.
   g. Shift linkage bracket bolts.
3. Disconnect the speedometer drive cable at the upper and lower couplings (at the transducer if equipped with cruise

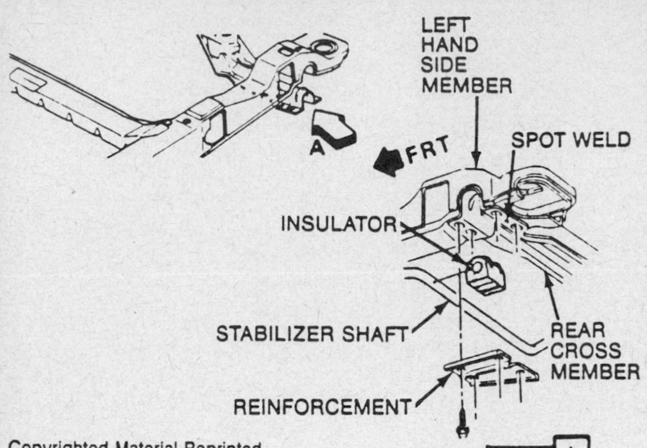
LEFT
HAND
SIDE
MEMBER
A
FRT
SPOT WELD
INSULATOR
STABILIZER SHAFT
REAR
CROSS
MEMBER
REINFORCEMENT
VIEW A

Spot weld location

control).

4. Disconnect the oil cooler lines at the transaxle.

5. Install the engine support fixture, locating it at the center of the cowl for 4 cylinder engines and on the strut towers for 6 cylinder engines.

NOTE: When installing a lift chain onto the aluminum cylinder head of the 4.3L V6 diesel engine, be sure that the bolt is tight or damage to the cylinder head may occur.

6. Rotate the steering wheel to position the steering gear stub shaft bolt in the upward position. Remove the bolt.

7. Raise the vehicle and place a jack under the engine to act as a support during removal and installation.

8. Remove the left front tire and wheel assembly.

9. Remove the power steering line brackets, remove the mounting bolts for the steering rack assembly, and support the assembly.

10. Disconnect the driveline vibration absorber, if equipped.

11. Disconnect the left side lower ball joint at the steering knuckle.

12. Remove the front stabilizer bar reinforcements and bushings from the right and left cradle side members.

13. Using a drill with a ½ in. bit, drill through the spot weld located between the rear holes of the left side front stabilizer bar mounting.

14. Disconnect the engine and transaxle mounts from the cradle.

15. Remove the sidemember-to-crossmember bolts.

16. Remove the bolts from the left side body mounts.

17. Remove the left side and front crossmember assembly. It may be necessary to carefully pry the crossmember loose.

18. Install axle shaft boot protectors, and using the appropriate special tools, pull the axle shaft cones out and away from the transaxle.

19. Pull the left axle shaft out of the transaxle.

20. Rotate the strut assembly so that the axle shaft is out of the way.

21. Remove:

a. Starter and converter shields.

b. Flywheel-to-converter bolts.

c. Two transaxle extension bolts from the engine-to-transaxle bracket.

d. Rear transaxle mount bracket assembly. It may be necessary to raise the transaxle assembly.

22. Securely attach a transaxle jack to the transaxle assembly.

23. Remove the 2 braces to the right end of the transaxle bolts.

24. Remove the remaining transaxle-to-engine bolt.

25. Remove the transaxle by moving it towards the driver's side, away from the engine.

SKYHAWK, CAVALIER, FIRENZA, CIMARRON AND 2000

1. Disconnect the negative battery cable, where it attaches to the transaxle.

2. Insert a ¼ × 2 in. bolt into the hole in the right front motor mount to prevent any mislocation during the transaxle removal.

3. Remove the air cleaner. Disconnect the T.V. cable.

4. Unscrew the bolt securing the T.V. cable to the transaxle. Pull up on the cable cover at the transaxle until the cable can be seen. Disconnect the cable from the transaxle rod.

5. Remove the wiring harness retaining bolt at the top of the transaxle.

6. Remove the hose from the air management valve and then pull the wiring harness up and out of the way.

7. Install and engine support bar. Raise the engine high enough to take the pressure off the motor mounts.

NOTE: The engine support bar must be located in the center of the cowl and the bolts must be tightened before attempting to support the engine.

8. Remove the transaxle mount and bracket assembly. It may be necessary to raise the engine slightly to aid in removal.

9. Disconnect the shift control linkage from the transaxle.

10. Remove the top transaxle-to-engine mounting bolts. Loosen, but do not remove, the transaxle-to-engine bolt nearest to the starter.

11. Unlock the steering column. Raise and support the front of the vehicle. Remove the front wheels.

12. Pull out the cotter pin and loosen the castellated ball joint nut until the ball joint separates from the control arm. Repeat on the other side of the vehicle.

13. Disconnect the stabilizer bar from the left lower control arm.

14. Remove the 6 bolts that secure the left front suspension support assembly.

15. Connect an axle shaft removal tool to a slide hammer.

16. Position the tool behind the axle shaft cones and then pull the cones out and away from the transaxle. Remove the axle shafts and plug the transaxle bores to reduce fluid leakage.

17. Remove the nut that secures the transaxle control cable bracket to the transaxle, then remove the engine-to-transaxle stud.

18. Disconnect the speedometer cable at the transaxle.

19. Disconnect the transaxle strut (stabilizer) at the transaxle.

20. Remove the 4 retaining screws and remove the torque converter shield.

21. Remove the 3 bolts securing the torque converter to the flex plate.

22. Disconnect and plug the oil cooler lines at the transaxle. Remove the starter.

23. Remove the screws that hold the brake and fuel line brackets to the left side of the underbody. This will allow the lines to be moved slightly for clearance during transaxle removal.

24. Remove the bolt that was loosened in Step 10.

25. Move the transaxle to the left and remove it from the vehicle.

BERETTA AND CORSICA

1. Disconnect the negative battery cable.

2. Remove the air cleaner assembly and the air intake duct.

3. Disconnect the T.V. cable at the throttle lever and the transaxle.

4. Remove the dipstick and the fill tube.

5. Install engine support fixture tool J28467 with adapter tool J35953.

6. Remove the nut securing the wiring harness to the transaxle.
7. Disconnect the wiring connectors at the speed sensor, torque converter clutch connector and the park/neutral and backup lamp switch.
8. Disconnect the shift linkage from the transaxle.
9. Remove the top 2 transaxle to engine bolts and the transaxle mount and bracket assembly.
10. Disconnect the rubber hose from the transaxle to the vent pipe.
11. Remove the remaining upper engine to transaxle bolts.
12. Raise the vehicle and support it safely. Remove both front wheels.
13. Disconnect the shift linkage and bracket from the transaxle.
14. Install the drive axle seal protectors.
15. Remove the left hand splash shield.
16. Disengage both drive axles.
17. Remove the transaxle strut.
18. Remove the left stabilizer shaft link pin bolt.
19. Remove the left stabilizer shaft bushing clamp nuts at the support.
20. Remove the left frame support bolts and swing aside.
21. Disconnect the speedometer to transaxle wire connector.
22. Remove the torque converter cover.
23. Remove the torque converter to flywheel bolts. Using a scribe, mark the flywheel to torque converter for proper reassembly.
24. Disconnect the transaxle cooler pipes and plug them to prevent leakage.
25. Remove the transaxle to engine support bracket. Position the transaxle jack.
26. Remove the remaining engine to transaxle bolts and remove the transaxle.

FIERO

1. Disconnect the negative battery cable.
2. Remove the air cleaner.
3. Remove the right and the left engine vent covers.
4. Disaconnect the throttle valve cable at the transaxle and at the carburetor.
5. Disconnect the shift cable at the transaxle bracket.
6. Disconnect the neutral safety switch electrical connection.
7. Disconnect the torque converter clutch electrical connection.
8. Disconnect the speedometer pickup electrical connection.
9. Disconnect the wire harness at the transaxle to engine retaining bolts.
10. Remove the cooler line support bracket.
11. Remove the transaxle to engine retaining bolts.
12. Remove the shift cable bracket to remove the neutral safety switch harness.
13. Install the engine fixture tool J28467–A.
14. Raise the vehicle and support it safely. Remove the rear wheels.
15. Install the rear axle boot protectors.
16. Remove the fixed adjusting link/lateral control arm through bolts.
17. Disconnect the trailing arms at the knuckles.
18. Disconnect the rear axle shafts from the transaxle. Support the axle shafts.
19. Remove the splash shields.
20. Disconnect the brake cables at the calipers.
21. Disconnect the brake control cable at the frame.
22. Disconnect the exhaust at the manifold.
23. Remove the engine mount and transaxle mount to cradle nuts and retaining bolts. Remove the cradle from the vehicle.
24. Remove the starter and the flywheel shield.
25. Remove the flywheel bolts.
26. Disconnect and plug the cooler lines.
27. Install the support jack.
28. Remove the transaxle support bracket at the back right.
29. Remove the remaining transaxle to engine retaining bolts including the ground wire.
30. Remove the transaxle from the vehicle.

THM A-1

6000

1. Disconnect the negative battery cable.
2. Install engine support fixture tool J28467–A.
3. Remove the air cleaner.
4. Disconnect the T.V. cable from the mounting clips or straps.
5. Remove the bolt securing the T.V. cable to the transaxle.
6. Disconnect the shift cable from the transaxle.
7. Disconnect the electrical connections from the speedometer gear, differential lock actuator, park/neutral safety and back up lamp switch.
8. Remove the oil cooler pipes from the transaxle bracket. Cap the open lines to prevent leakage.
9. Remove the shift linkage bracket.
10. Remove the engine to transaxle bolts leaving the bolt near the starter loosely insalled.
11. Remove the intermediate shaft to steering gear stub shaft retaining bolt.
12. Raise the vehicle and support it safely.
13. Disconnect the front and left sections of the front cradle.
14. Install drive axle seal protector tool J34754.
15. Disconnect the drive axle and plug the bore in the transaxle to minimize fluid loss.
16. Rotate the strut assembly so that the shaft is out of the way. Suspend the axle shaft with a wire.
17. Remove the splash shield from the left side of the vehicle.
18. Remove the starter and converter shields.
19. Remove the propshaft to transfer case attaching bolts. Suspend the propshaft with a wire.
20. Remove the flywheel to converter bolts.
21. Remove the rear transaxle mount bracket assembly.
22. Attach a transmission jack to the transaxle case.
23. Remove the 2 brace-to-right end of transaxle bolts.
24. Remove the remaining transaxle to engine bolt located near the starter.
25. Slide the transaxle toward the driver's side away from the engine.
26. Lower the transaxle out of the vehicle.

TRANSAXLE INSTALLATION

THM 125C

CENTURY, SKYLARK, CITATION, CELEBRITY, OMEGA, CIERA, PHOENIX, 6000, REGAL, CALAIS, GRAND AM AND LEMANS

1. Position the transaxle on a suitable jack.
2. Line up and insert the right hand axle shaft while the transaxle is being raised into place.
3. Install the transaxle to engine bolt near the starter finger tight.
4. Install the 2 braces to the right end of the transaxle bolts.
5. Remove the transaxle jack.
6. Install the rear mount bracket assembly.
7. Install the 2 transaxle extension bolts from the engine to the transaxle bracket.
8. Install the flywheel to converter bolts and torque to 46 ft. lbs. (62 Nm).
9. Install the starter and converter shields.
10. Install the left hand axle shaft.
11. Remove the axle shaft seal protectors.
12. Install the steering gear intermediate shaft and tighten the bolt to 40 ft. lbs. (54 Nm).

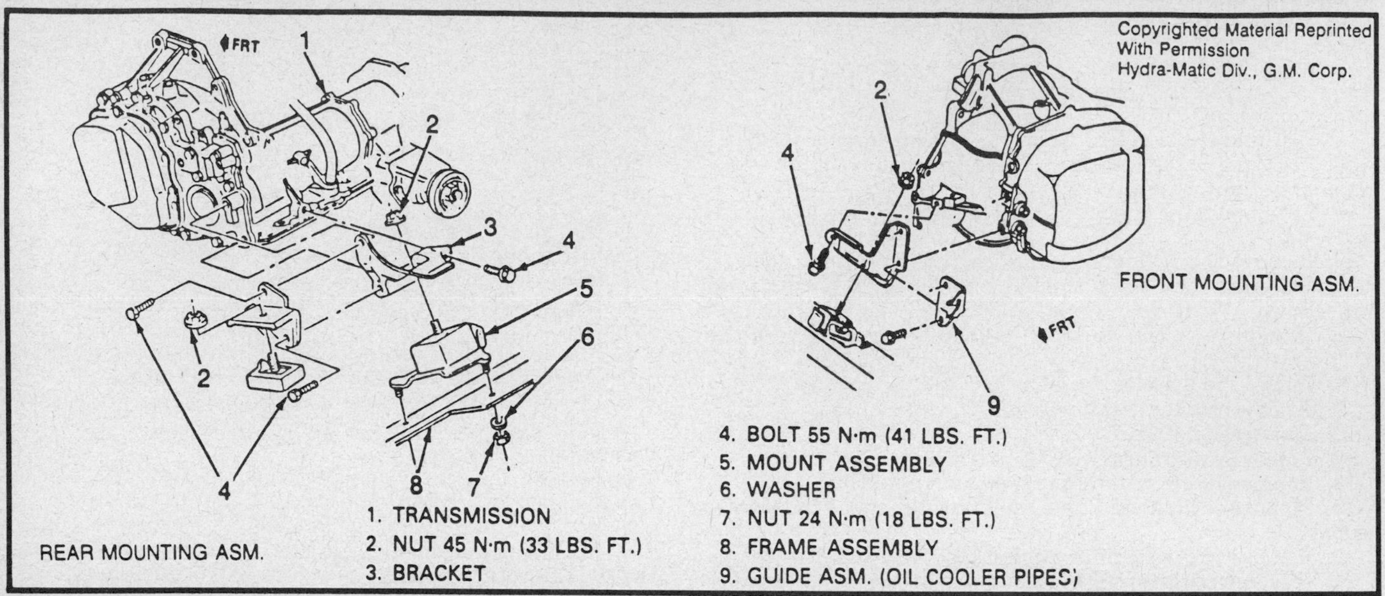
Copyrighted Material Reprinted With Permission Hydra-Matic Div., G.M. Corp.
FRT
REAR MOUNTING ASM.
FRONT MOUNTING ASM.
1. TRANSMISSION
2. NUT 45 N·m (33 LBS. FT.)
3. BRACKET
4. BOLT 55 N·m (41 LBS. FT.)
5. MOUNT ASSEMBLY
6. WASHER
7. NUT 24 N·m (18 LBS. FT.)
8. FRAME ASSEMBLY
9. GUIDE ASM. (OIL COOLER PIPES)

Transmission mounts

13. Install the front and left sections of the frame.
14. Connect the oil cooler pipes to the transaxle.
15. Install the shift linkage bracket and connect the shift linkage.
16. Connect the speedometer drive cable or speed sensor wiring connector.
17. Connect the torque converter clutch connector.
18. Install the remaining transaxle to engine bolts and tighten all bolts to 55 ft. lbs. (75 Nm).
19. Connect the oil cooler pipes to the strut bracket.
20. Connect the strut shock bracket bolts to the transaxle.
21. Lower the vehicle.
22. Connect the electrical connections for the park/neutral safety and backup lamp switch.
23. Connect the transaxle end of the cable to the T.V. link and secure it to the case with the bolt and washer assembly.
24. Install the cable routing clips or straps.
25. Install the air cleaner.
26. Connect the negative battery cable.
27. Adjust the shift linkage and T.V. cable. Check the fluid level.

SKYHAWK, CAVELIER, FIRENZA, CIMARRON AND 2000

1. Place a small amount of grease on the torque converter pilot hub.
2. Properly seat the torque converter in the oil pump.
3. Position the transaxle in the vehicle.
4. Install the lower engine to transaxle bolts and remove the jack.
5. Install the transaxle to engine support bracket.
6. Connect the cooler pipes.
7. Install the torque converter to flywheel bolts and torque to 46 ft. lbs. (62 Nm).
8. Install the torque converter cover.
9. Install the starter.
10. Connect the speedometer cable.
11. Connect the front exhaust manifold and pipe.
12. Install the left frame support assembly.
13. Install the left stabilizer shaft frame bushing nuts.
14. Install the stabilizer shaft link pin bolt.
15. Install the transaxle mounting strut.
16. Install the drive axles and seat them into the case.
17. Connect both ball joints to the control arms.
18. Remove the drive axle boot protectors.
19. Connect the shift linkage bracket to the transaxle.
20. Install both wheels. Lower the vehicle.
21. Install the upper engine to transaxle bolts.
22. Install the left side transaxle mount.
23. Connect the shift linkage to the transaxle.
24. Connect the wiring connectors at the speed sensor, torque converter clutch connector and at the neutral safety and backup lamp switch.

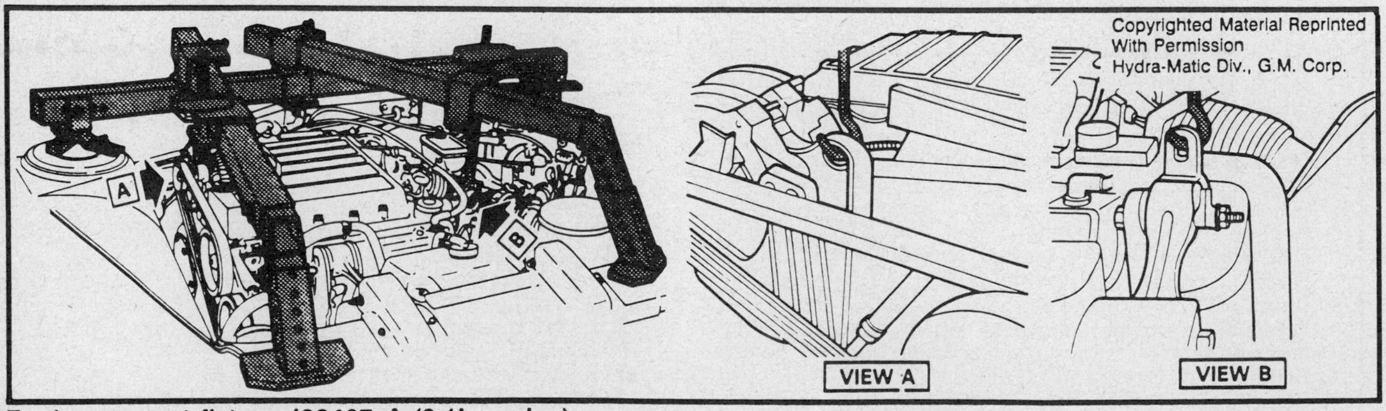
Copyrighted Material Reprinted With Permission Hydra-Matic Div., G.M. Corp.
A
B
VIEW A
VIEW B

Engine support fixture J28467-A (3.1L engine)

25. Install the nut securing the wiring harness to the transaxle.
26. Remove the alignment bolt from the right motor mount.
27. Remove engine support fixture tool J28467.
28. Install the dipstick and fill tube.
29. Install the T.V. cable at the transaxle and the throttle lever.
30. Connect the rubber hoses to the transaxle vent pipe.
31. Install the air cleaner.
32. Connect the negative battery cable.
33. Fill the transaxle to the proper level with Dexron® II transmission fluid.
34. Check for oil leaks. Adjust the T.V. cable and the shift linkage.

BERETTA AND CORSICA

1. Place a small amount of grease on the torque converter pilot hub.
2. Properly seat the torque converter in the oil pump.
3. Position the transaxle in the vehicle.
4. Install the lower engine to transaxle bolts and remove the jack.
5. Install the transaxle to engine support bracket.
6. Connect the cooler pipes.
7. Install the torque converter to flywheel bolts and torque to 46 ft. lbs. (62 Nm).
8. Install the torque converter cover.
9. Connect the speedometer wire connector.
10. Install the left hand drive axle into the transaxle.
11. Install the left frame support assembly.
12. Install the left stabilizer shaft frame bushing nuts.
13. Install the stabilizer shaft link pin bolt.
14. Install the transaxle mounting strut.
15. Install the drive axles and seat them into the case.
16. Remove the drive axle boot protectors.
17. Connect the shift linkage bracket to the transaxle.
18. Install both wheels. Lower the vehicle.
19. Install the upper engine to transaxle bolts.
20. Install the transaxle mount bolts.
21. Connect the shift linkage to the transaxle.
22. Connect the wiring connectors at the speed sensor, torque converter clutch connector and at the neutral safety and backup lamp switch.
23. Install the nut securing the wiring harness to the transaxle.
24. Remove engine support fixture.
25. Install the dipstick and fill tube.
26. Install the T.V. cable at the transaxle and the throttle lever.
27. Connect the rubber hoses to the transaxle vent pipe.
28. Install the air cleaner and air intake duct.
29. Connect the negative battery cable.
30. Fill the transaxle to the proper level with Dexron® II transmission fluid.
31. Check for oil leaks. Adjust the T.V. cable and the shift linkage.

FIERO

1. Position the transaxle to the engine.
2. Install the transaxle to engine retaining bolts including the ground wire.
3. Install the transaxle support bracket.
4. Remove the transaxle support jack.
5. Unplug and connect the cooler lines.
6. Install the flywheel bolts.
7. Install the starter and flywheel shield.
8. Connect the cradle to the vehicle.
9. Install the front and rear cradle retaining bolts.
10. Install the transaxle mounts to cradle nuts.
11. Install the engine mounts to the cradle nuts.
12. Connect the exhaust to the manifold.
13. Connect the brake control cable at the frame.
14. Connect the brake cables to the calipers.
15. Adjust the brake cables.
16. Install the splash shields.
17. Disconnect the rear axle shaft from the support.
18. Connect the rear axle shaft to the transaxle.
19. Connect the trailing arms to the knuckles and torque to 44 ft. lbs. (60 Nm) + 90 degrees.
20. Install the fixed adjusting link/lateral control arm through bolts and torque to 37 ft. lbs. (50 Nm) + 90 degrees.
21. Remove the rear axle boot protector.
22. Install the rear wheels. Lower the vehicle.
23. Remove the engine fixture tool J28467–A.
24. Install the neutral safety switch harness and shift cable bracket.
25. Install the transaxle to engine retaining bolts.
26. Install the transaxle cooler line support bracket.
27. Connect the wire harness at the transaxle to engine retaining bolts.

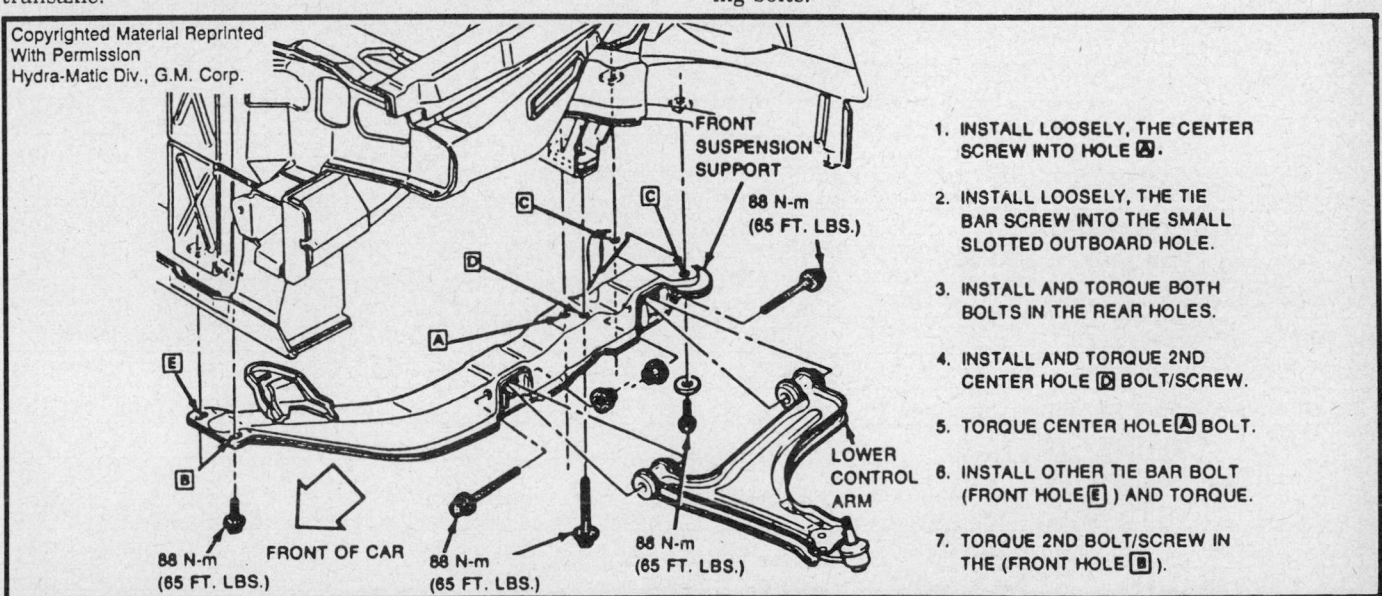

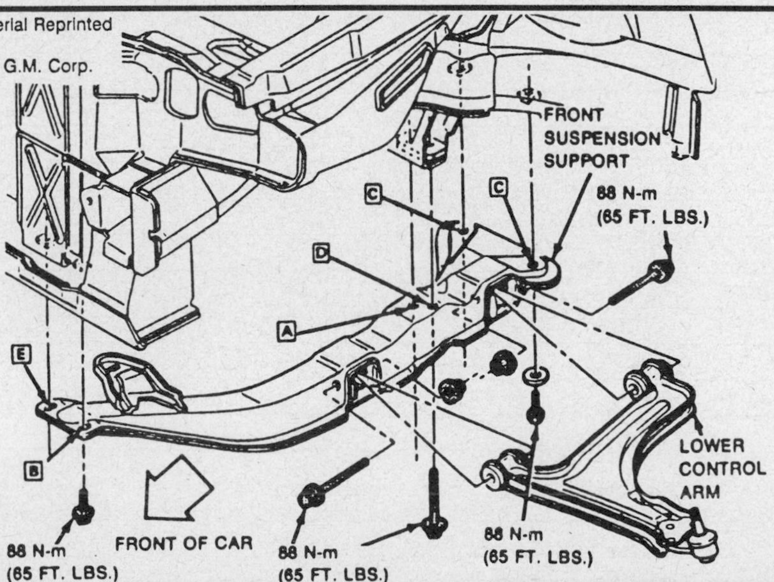
FRONT SUSPENSION SUPPORT
C
C
D
A
E
B
88 N-m (65 FT. LBS.)
LOWER CONTROL ARM
FRONT OF CAR
88 N-m (65 FT. LBS.)
88 N-m (65 FT. LBS.)
88 N-m (65 FT. LBS.)

1. INSTALL LOOSELY, THE CENTER SCREW INTO HOLE A.
2. INSTALL LOOSELY, THE TIE BAR SCREW INTO THE SMALL SLOTTED OUTBOARD HOLE.
3. INSTALL AND TORQUE BOTH BOLTS IN THE REAR HOLES.
4. INSTALL AND TORQUE 2ND CENTER HOLE D BOLT/SCREW.
5. TORQUE CENTER HOLE A BOLT.
6. INSTALL OTHER TIE BAR BOLT (FRONT HOLE E) AND TORQUE.
7. TORQUE 2ND BOLT/SCREW IN THE (FRONT HOLE B).

Front suspension support installation sequence—Skyhawk, Cavalier, Firenza, Cimarron and 2000

28. Connect the speedometer pickup electrical connection.
29. Connect the transaxle converter clutch electrical connection.
30. Connect the neutral safety switch electrical connection.
31. Connect the shift cable to the transmission bracket.
32. Connect the throttle valve cable to the transaxle and carburetor.
33. Install the left and right engine vent covers.
34. Install the air cleaner.
35. Connect the negative battery cable.

THM A-1
6000

1. Secure the transaxle to the transmission jack.
2. While the transaxle is being raised into place, line up and install the right hand axle shaft and differential lock actuator rod.
3. Install the transaxle to engine bolt near the starter finger tight.
4. Install the 2 brace-to-right end of transaxle bolts.
5. Remove the transmission jack.
6. Install the rear transaxle mount bracket assembly.
7. Install the flywheel to converter bolts and tighten to 46 ft. lbs. (62 Nm).
8. Connect the propshaft to the transfer case and tighten the bolts to 40 ft. lbs. (54 Nm).
9. Install the starter and converter shields.
10. Connect the left hand drive axle shaft.
11. Remove the axle shaft seal protectors.
12. Install the front and left sections of the frame.
13. Loosely install the remaining transaxle to engine bolts.
14. Lower the vehicle.
15. Install the intermediate shaft to steering gear stub shaft retaining bolt and tighten to 40 ft. lbs. (54 Nm).
16. Connect the oil cooler pipes to the transaxle.
17. Install the shift linkage bracket. Connect the shift linkage.
18. Connect the TCC connector.
19. Tighten the remaining transaxle to engine bolts to 55 ft. lbs. (75 Nm).
20. Connect the electrical connections at the park neutral safety and back up lamp switch, the differential lock actuator and the the speedometer gear.
21. Connect the transaxle end of the cable to the T.V. link and secure it to the transaxle case.
22. Install the cable routing clips or straps.
23. Install the air cleaner.
24. Remove the engine support fixture.
25. Connect the negative battery cable.
26. Adjust the shift linkage and the T.V. cable.
27. Bring to normal operating temperature and check the fluid level.

SPECIFICATIONS

TORQUE SPECIFICATION CHART

Description	Quantity	Torque Specification
Valve body to Case Cover	2	8 ft. lbs. (11 N•m)
Pump Cover to Case Cover	1	18 ft. lbs. (24 N•m)
Pump Cover to Valve Body	4	8 ft. lbs. (11 N•m)
Pump Cover to Valve Body	3	8 ft. lbs. (11 N•m)
Solenoid to Valve Body	1	8 ft. lbs. (11 N•m)
Valve Body to Case Cover	9	8 ft. lbs. (11 N•m)
Valve Body to Case	1	18 ft. lbs. (24 N•m)
Valve Body to Driven Sprocket Support	1	18 ft. lbs. (24 N•m)
Case Cover to Case	4	18 ft. lbs. (24 N•m)
Case Cover to Case	4	18 ft. lbs. (24 N•m)
Case Cover to Case	1	18 ft. lbs. (24 N•m)
Case Cover to Case	7	18 ft. lbs. (24 N•m)
Case Cover to Case	2	18 ft. lbs. (24 N•m)
Case to Drive Sprocket Support	4	18 ft. lbs. (24 N•m)
Oil Pan and Valve Body Cover	27	12 ft. lbs. (16 N•m)
Manual Detent Spring Assembly to Case	1	8 ft. lbs. (11 N•m)
Cooler Connector	2	23 ft. lbs. (38 N•m)
Line Pressure Take-Off	1	8 ft. lbs. (11 N•m)
Intermediate Servo Cover	4	8 ft. lbs. (11 N•m)
Parking Lock Bracket to Case	2	18 ft. lbs. (24 N•m)
Pipe Retainer to Case	2	18 ft. lbs. (24 N•m)
Governor Cover to Case	2	8 ft, lbs. (11 N•m)
Speedometer Driven Gear to Governor Cover	1	75 in. lbs. (9 N•m)
T.V. Cable to Case	1	75 in. lbs. (9 N•m)
Pressure Switch	2	8 ft. lbs. (11 N•m)

Section 3

AW131L Transaxle
General Motors

APPLICATION

1985–89 Nova

GENERAL DESCRIPTION

The A131L transaxle is a 3 speed automatic transaxle, consisting primarily of a 4 element hydraulic torque converter, 4 multiple disc clutches, 2 one-way clutches and a band which provides the friction elements required to obtain the desired function of the compound planetary gear set. The combination of the compound planetary gear set provides 3 forward ratios and 1 reverse.

Transaxle and Converter Identification

TRANSAXLE

The transaxle identification tag is located at the top front of the transaxle case.

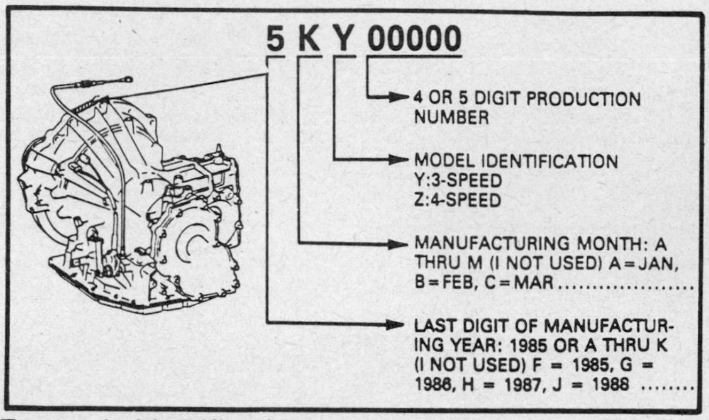
5 K Y 00000
4 OR 5 DIGIT PRODUCTION NUMBER
MODEL IDENTIFICATION Y:3-SPEED Z:4-SPEED
MANUFACTURING MONTH: A THRU M (I NOT USED) A=JAN, B=FEB, C=MAR ....
LAST DIGIT OF MANUFACTURING YEAR: 1985 OR A THRU K (I NOT USED) F = 1985, G = 1986, H = 1987, J = 1988 ....

Transaxle identification

CONVERTER

The torque converter is equipped with a built-in lock-up clutch. It smoothly couples the engine to the planetary gears with oil and hydraulically provides additional torque multiplication when required. Changing of the gear ratios is fully automatic in relation to vehicle speed and engine torque. Vehicle speed and engine torque signals are constantly fed to the transaxle to provide the proper gear ratio for maximum efficiency and performance at all throttle openings.

Metric Fasteners

The metric fastener dimensions are very close to the dimensions of the familiar inch system fasteners and for this reason, replacement fasteners must have the same measurement and strength as those removed.

Do not attempt to interchange metric fasteners for inch system fasteners. Mismatched or incorrect fasteners can result in damage to the transaxle unit through malfunctions, breakage or possible personal injury.

Capacities

The fluid used in both the transaxle and the differential is Dexron® II automatic transmission fluid. The fluid capacity for the transaxle is 6 qts. (5.6L). The fluid capacity for the differential is 1.5 qts. (1.4L).

Checking Fluid Level

The vehicle must have been driven so that the engine and transaxle are at normal operating temperature (fluid temperature 158–176° F). If the fluid smells burnt or is black, replace it.

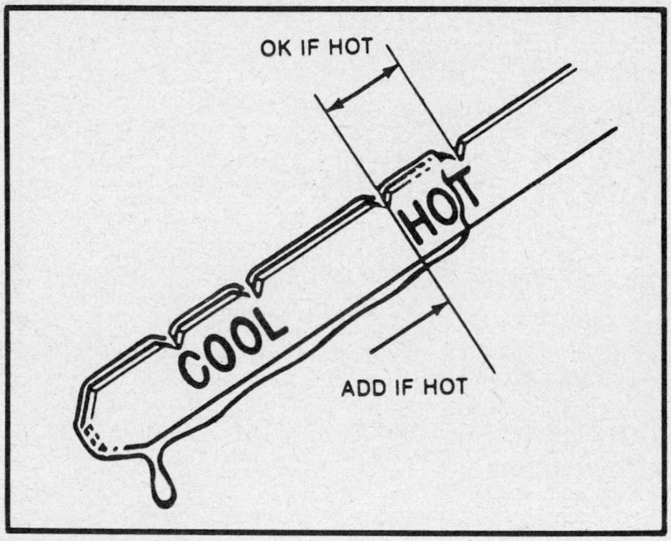
OK IF HOT
HOT
COOL
ADD IF HOT

Fluid level check

1. Park the vehicle on a level surface and set the parking brake.
2. With the engine idling, shift the selector into each gear from **P** range to **L** range and return to **P** range.
3. Pull out the transaxle dipstick and wipe it clean.
4. Push the dipstick back fully into the tube.
5. Pull out the dipstick and check that the fluid is in the **HOT** range.
6. If the level is low, add fluid. Do not overfill.

TRANSAXLE MODIFICATIONS

NOTE: There are no transaxle modifications available at the time of publication.

TROUBLE DIAGNOSIS

Hydraulic Control System

NOTE: Please refer to Section 9 for all oil flow circuits.

PUMP ASSEMBLY

The hydraulic pressure system requires a supply of transmission fluid and a pump to pressurize the fluid. The A131L transaxle uses an internal-external gear type pump with its oil intake connected to a screen assembly.

The oil pump is designed to deliver fluid to the torque converter, lubricate the planetary gear unit and supply operating pressure to the hydraulic control system. The drive gear of the oil pump and the torque converter pump are driven by the engine. The pump has a sufficient capacity of oil to supply the necessary fluid pressure throughout all forward speed ranges and reverse.

MANUAL VALVE

The manual valve is linked to the gear shift lever and directs the fluid to the gear range circuit that the lever is positioned at.

PRIMARY REGULATOR VALVE

The primary regulator valve varies the hydraulic line pressure to each component in order to conform with engine power and operate all transaxle hydraulic systems.

SECONDARY REGULATOR VALVE

This valve regulates the converter pressure and lubrication pressure. Spring tension in the valve acts in an upward direction. Converter fluid pressure and lubrication pressure are determined by the spring tension.

THROTTLE VALVE

The throttle valve acts to produce throttle pressure in response to accelerator pedal modulation or engine output. When the accelerator pedal is depressed, the downshift plug is pushed upward by the throttle cable and throttle cam. The throttle valve also moves upward by means of the spring, opening the pressure passage for creation of throttle pressure.

CUT-BACK VALVE

This valve regulates the cut-back pressure acting on the throttle valve and is activated by governor pressure and throttle pressure. By applying cut-back pressure to the throttle valve in this manner, the throttle pressure is lowered to prevent unnecessary power loss from the oil pump.

Governor pressure acts on the upper portion of the valve and as the valve is pushed downward, a passage from the throttle valve is opened and throttle pressure is applied. Because of the

CLUTCH AND BAND APPLICATION

Shift Lever Position	Gear Position	C_1	C_2	B_1	B_2	B_3	F_1	F_2
P	Parking	—	—	—	—	—	—	—
R	Reverse	—	Applied	—	—	Applied	—	—
N	Neutral	—	—	—	—	—	—	—
D	1st	Applied	—	—	—	—	—	Applied
	2nd	Applied	—	—	Applied	—	Applied	—
	3rd	Applied	Applied	—	Applied	—	—	—
2	1st	Applied	—	—	—	—	—	Applied
	2nd	Applied	—	Applied	Applied	—	Applied	—
L	1st	Applied	—	—	—	Applied	—	Applied
	2nd①	Applied	—	Applied	Applied	—	Applied	—

B_1—No. 1 brake, 2nd coast brake
B_2—No. 2 brake, 2nd brake
B_3—No. 3 brake, 1st and reverse brake
C_1—Front clutch, forward clutch
C_2—Rear clutch, direct clutch
F_1—No. 1 one-way clutch
F_2—No. 2 one-way clutch
① Downshift only in L range, 2nd gear. No upshift

CHILTON'S THREE C's TRANSAXLE DIAGNOSIS

Condition	Cause	Correction
Fluid discolored or smells burnt	a) Fluid contamination b) Torque converter faulty c) Transaxle faulty	a) Replace fluid b) Replace torque converter c) Disassemble and inspect transaxle
Vehicle does not move in any forward range or reverse	a) Control cable out of adjustment b) Valve body or primary regulator valve faulty c) Transaxle faulty	a) Adjust control cable b) Inspect valve body c) Disassemble and inspect transaxle
Vehicle does not move in any range	a) Parking lock pawl faulty b) Valve body or primary regulator valve faulty c) Torque converter faulty d) Converter drive plate broken e) Oil pump intake strainer blocked f) Transaxle faulty	a) Inspect parking pawl b) Inspect valve body c) Replace torque converter d) Replace torque converter e) Clean strainer f) Disassemble and inspect transaxle
Shift lever position incorrect	a) Control cable out of adjustment b) Manual valve and lever faulty c) Transaxle faulty	a) Adjust control cable b) Inspect valve body c) Disassemble and inspect transaxle
Harsh engagement into any drive range	a) Throttle cable out of adjustment b) Valve body or primary regulator valve faulty c) Accumulator pistons faulty d) Transaxle faulty	a) Adjust throttle cable b) Inspect valve body c) Inspect accumulator pistons d) Disassemble and inspect transaxle
Delayed 1–2, 2–3 or 3-OD upshift or downshifts from OD–3 or 3–2 then shifts back to OD or 3	a) Throttle cable out of adjustment b) Governor faulty c) Valve body faulty	a) Adjust throttle cable b) Inspect governor c) Inspect valve body
Slips on 1–2, 2–3 or 3–OD upshift or slips or shudders on take-off	a) Control cable out of adjustment b) Throttle cable out of adjustment c) Valve body faulty d) Transaxle faulty	a) Adjust control cable b) Adjust throttle cable c) Inspect valve body d) Disassemble and inspect transaxle
Drag, binding or tie-up on 1–2, 2–3 or 3–OD upshift	a) Control cable out of adjustment b) Valve body faulty c) Transaxle faulty	a) Adjust control cable b) Inspect valve body c) Disassemble and inspect transaxle
Harsh downshift	a) Throttle cable out of adjustment b) Accumulator pistons faulty c) Valve body faulty d) Transaxle faulty	a) Adjust throttle cable b) Inspect accumulator pistons a) Inspect valve body d) Disassemble and inspect transaxle
No downshift when coasting	a) Governor faulty b) Valve body faulty	a) Inspect governor b) Inspect valve body
Downshift occurs too quick or too late while coasting	a) Throttle cable out of adjustment b) Governor faulty c) Valve body faulty d) Transaxle faulty	a) Adjust throttle cable b) Inspect governor c) Inspect valve body d) Disassemble and inspect transaxle
No OD–3, 3–2 or 2–1 kickdown	a) Throttle cable out of adjustment b) Governor faulty c) Valve body faulty	a) Adjust throttle cable b) Inspect governor c) Inspect valve body
No engine braking in 2 range	a) Valve body faulty b) Transaxle faulty	a) Inspect valve body b) Disassemble and inspect transaxle
Vehicle does not hold in P	a) Control cable out of adjustment b) Parking lock pawl and rod	a) Adjust control cable b) Inspect lock pawl and rod

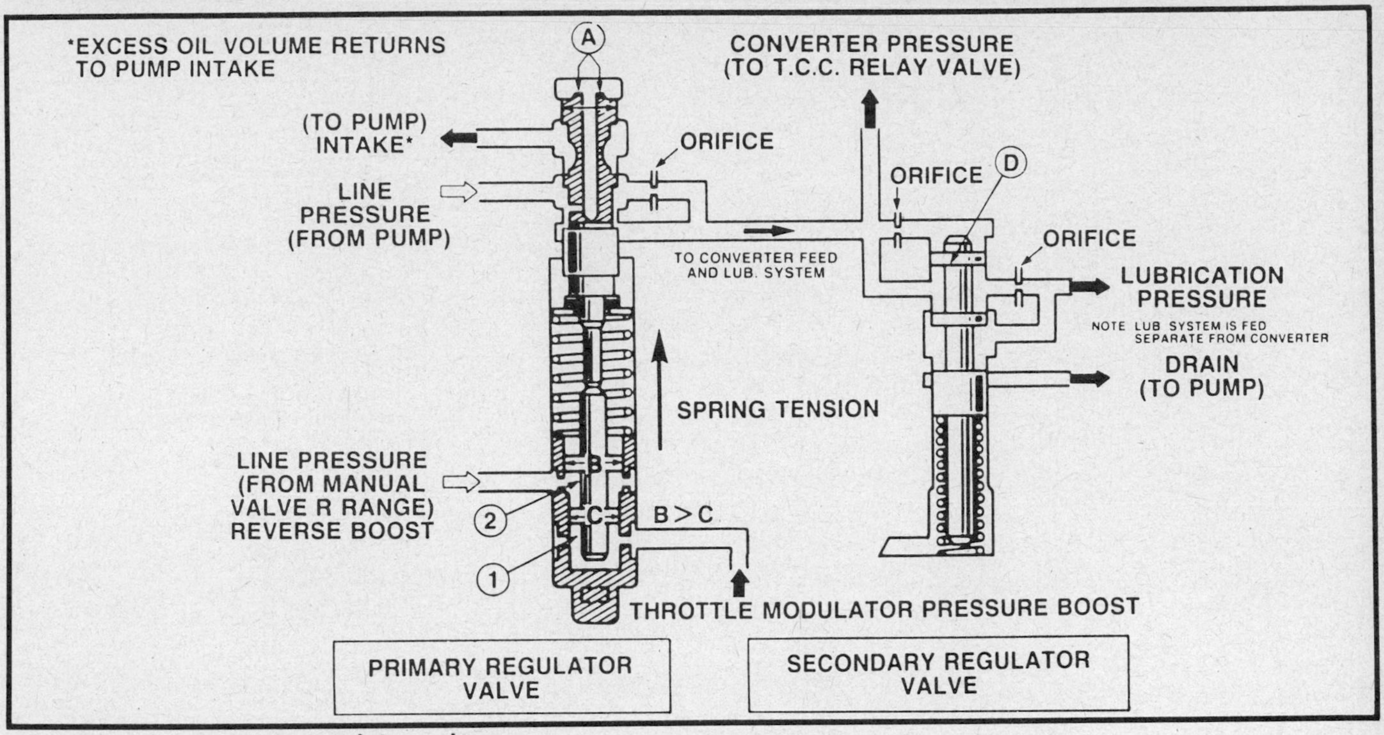
*EXCESS OIL VOLUME RETURNS
TO PUMP INTAKE
(TO PUMP)
INTAKE*
LINE
PRESSURE
(FROM PUMP)
A
CONVERTER PRESSURE
(TO T.C.C. RELAY VALVE)
ORIFICE
ORIFICE
D
ORIFICE
TO CONVERTER FEED
AND LUB. SYSTEM
LUBRICATION
PRESSURE
NOTE LUB SYSTEM IS FED
SEPARATE FROM CONVERTER
DRAIN
(TO PUMP)
SPRING TENSION
LINE PRESSURE
(FROM MANUAL
VALVE R RANGE)
REVERSE BOOST
B
C
B > C
2
1
THROTTLE MODULATOR PRESSURE BOOST
PRIMARY REGULATOR
VALVE
SECONDARY REGULATOR
VALVE
Primary and secondary regulator valves

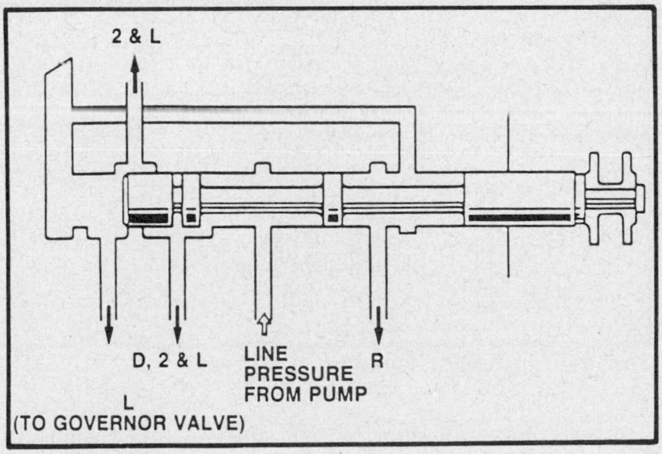
2 & L
D, 2 & L
LINE
PRESSURE
FROM PUMP
R
L
(TO GOVERNOR VALVE)
Manual valve

difference in the diameters of the valve pistons, the valve is pushed upward to balance the downward force of governor pressure and the throttle pressure becomes cut-back pressure. As the governor pressure rises, the valve is forced downward. Since the throttle pressure passage is open, the pressure becomes cut-back pressure.

THROTTLE MODULATOR VALVE

This valve produces throttle modulator pressure. It reduces throttle pressure when the throttle valve opening angle is high. It causes throttle modulator pressure to act on the primary regulator valve so that line pressure performance is close to engine power performance.

DOWNSHIFT PLUG

If the accelerator pedal is depressed to nearly the fully open position, the downshift plug opens the cut-back pressure passage very wide, then causes the detent regulator valve to operate and effect kickdown.

The cut-back pressure also acts on the downshift plug when the throttle valve opening angle is less than 85%. A power assist mechanism is provided to lighten spring tension in relation to the throttle cam only to the extent of the difference in the valve piston diameters.

GOVERNOR VALVE

The governor valve is driven by the drive pinion worm gear and produces governor pressure in response to the vehicle speed. It balances the line pressure from the primary regulator valve and the centrifugal force of the governor weights to produce hydraulic pressure in proportion to vehicle speed.

DETENT REGULATOR VALVE

During kickdown, the detent regulator valve stabilizes the hydraulic pressure acting on the 1-2 and 2-3 shift valves.

LOW MODULATOR VALVE

In the **L** range, the low modulator valve reduces the line pressure from the manual valve. Low modulator pressure pushes down the low coast shift valve and acts on the 1st and reverse brake to cushion the apply.

Torque Converter Clutch (TCC) SIGNAL VALVE

This valve detects governor pressure and determines the TCC apply point by controlling the pressure acting on the TCC relay valve. Over a certain governor pressure, the TCC signal valve is pushed upward and line pressure from the 2–3 shift valve acts on the end of the TCC relay valve. When not in 3rd gear with the

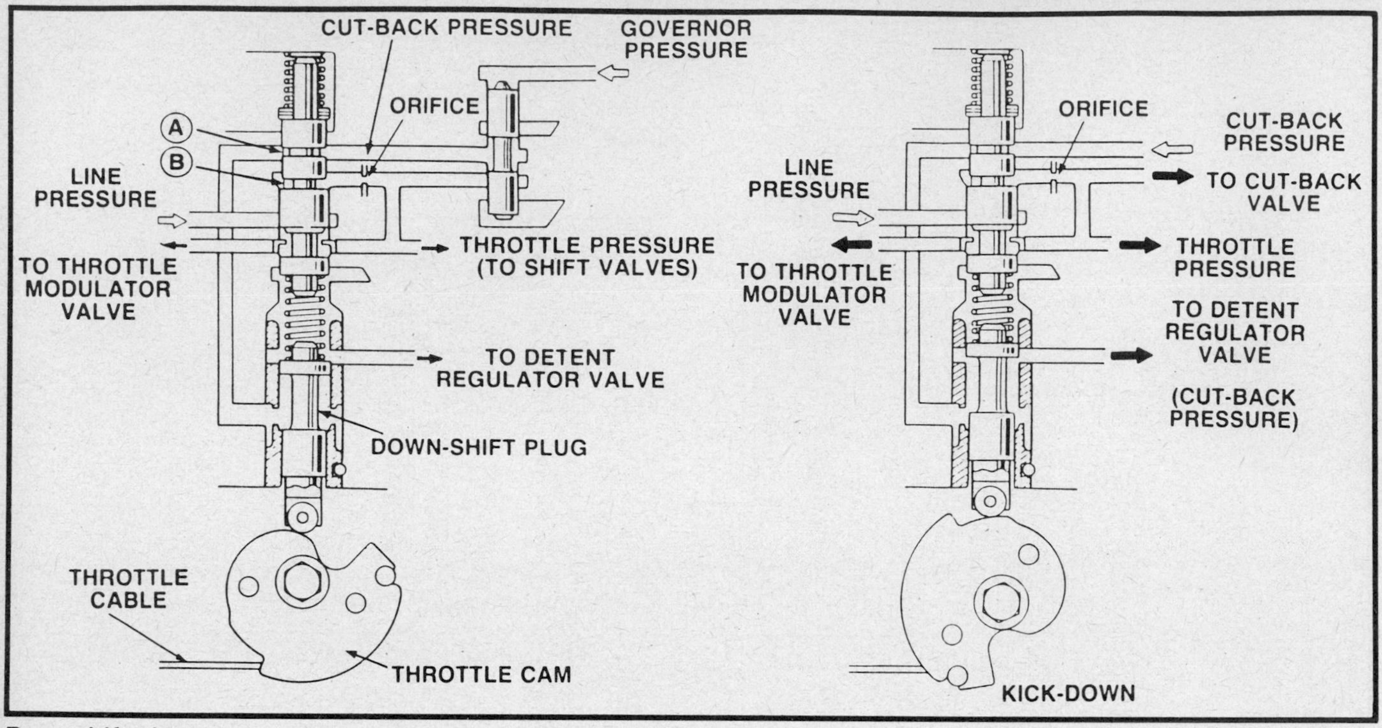
CUT-BACK PRESSURE
GOVERNOR PRESSURE
ORIFICE
A
B
LINE PRESSURE
TO THROTTLE MODULATOR VALVE
THROTTLE PRESSURE (TO SHIFT VALVES)
TO DETENT REGULATOR VALVE
DOWN-SHIFT PLUG
THROTTLE CABLE
THROTTLE CAM
ORIFICE
CUT-BACK PRESSURE
LINE PRESSURE
TO CUT-BACK VALVE
THROTTLE PRESSURE
TO THROTTLE MODULATOR VALVE
TO DETENT REGULATOR VALVE
(CUT-BACK PRESSURE)
KICK-DOWN

Downshift plug

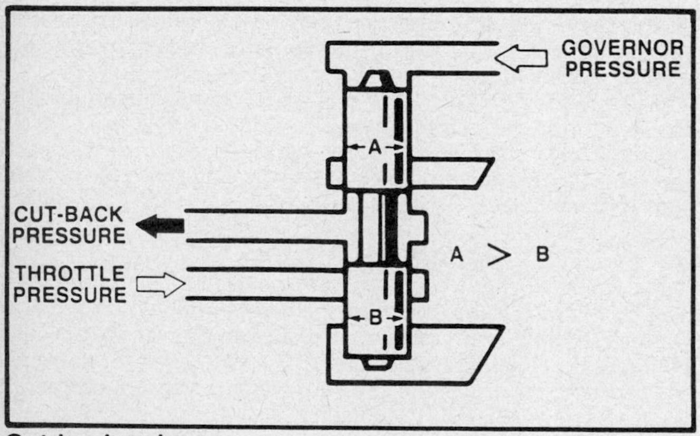
GOVERNOR PRESSURE
A
CUT-BACK PRESSURE
THROTTLE PRESSURE
A > B
B

Cut-back valve

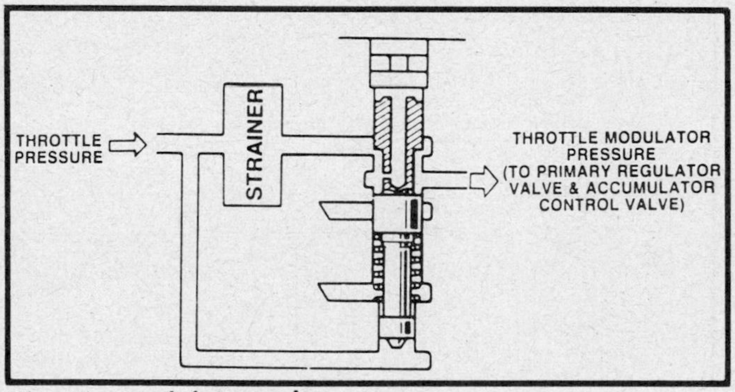
THROTTLE PRESSURE
STRAINER
THROTTLE MODULATOR PRESSURE (TO PRIMARY REGULATOR VALVE & ACCUMULATOR CONTROL VALVE)

Throttle modulator valve

shift lever in **D**, line pressure from the 2–3 shift valve is not applied to the TCC signal valve, so the TCC signal valve is pushed downward by spring force.

Torque Converter Clutch (TCC) RELAY VALVE

With a fluid pressure signal (C_2 pressure) from the TCC signal valve, the TCC relay valve reverses the fluid flow through the converter. When signal pressure is applied, the TCC apply relay valve is pushed upward and the TCC is actuated. When not in 3rd gear with the shift lever in **D**, the T.C.C. relay valve is in the downward position.

2ND MODULATOR VALVE

In the manual **2** range, this valve reduces line pressure from the intermediate shift valve. 2nd modulator pressure acts on the 2nd coast brake through the 1–2 shift valve to cushion the apply.

ACCUMULATORS

The accumulators act to cushion the shifting shock. There are 3 accumulators: 1 each for the forward clutch (C_1), direct clutch (C_2) and the 2nd brake (B_2). The accumulators are located in the transaxle case. Accumulator control pressure is always acting on the back pressure side of the C_2 and B_2 pistons. This pressure along with spring tension, pushes down on the 2 pistons.

When line pressure is applied to the operating side, the pistons are pushed upward and shock is cushioned as the fluid pressure gradually rises. Operation of the C_1 piston is basically the same as that for C_2 and B_2. However the force pushing the piston downward is accomplished by spring tension only.

ACCUMULATOR CONTROL VALVE

This valve cushions shifting shock by lowering the back pressure of the direct clutch (C_2) accumulator and 2nd brake (B_2) accumulator when the throttle opening angle is small.

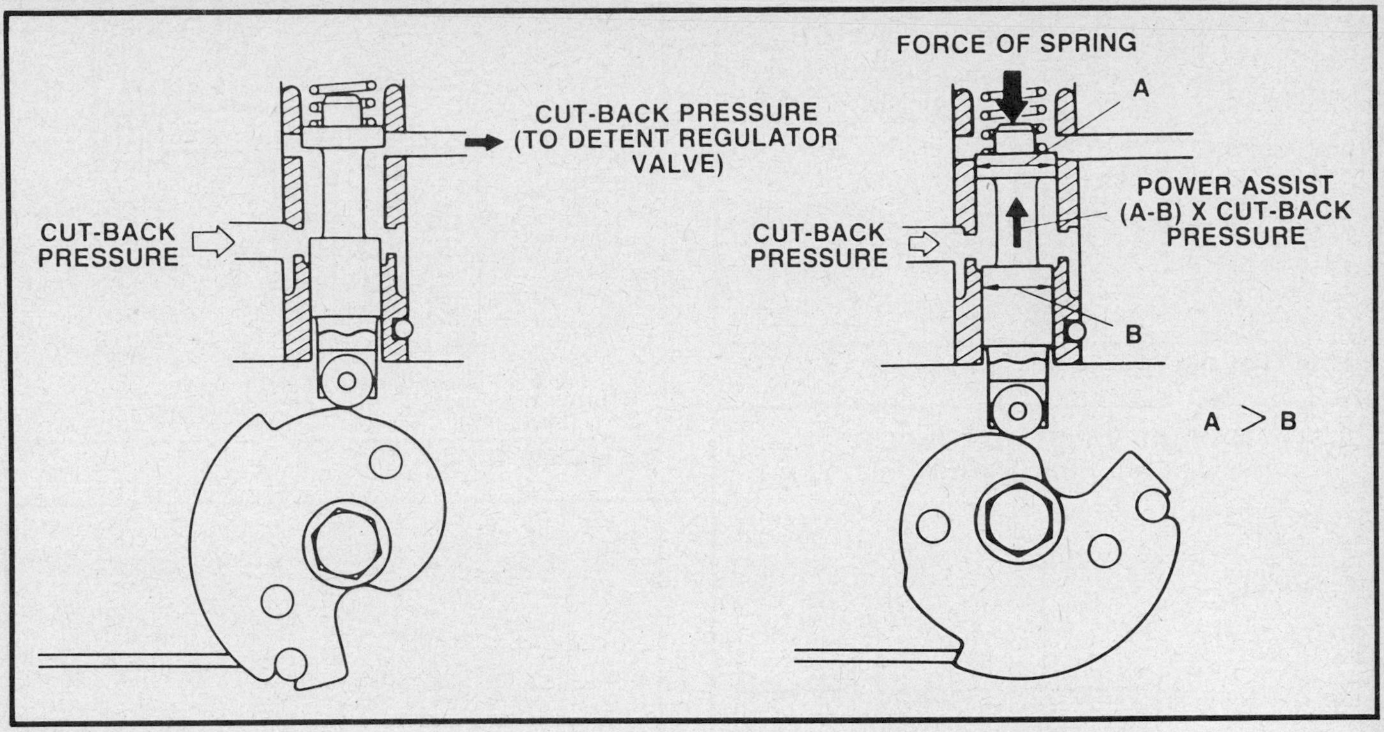
FORCE OF SPRING
A
CUT-BACK PRESSURE
(TO DETENT REGULATOR
VALVE)
POWER ASSIST
(A-B) X CUT-BACK
PRESSURE
CUT-BACK
PRESSURE
CUT-BACK
PRESSURE
B
A > B

Throttle valve

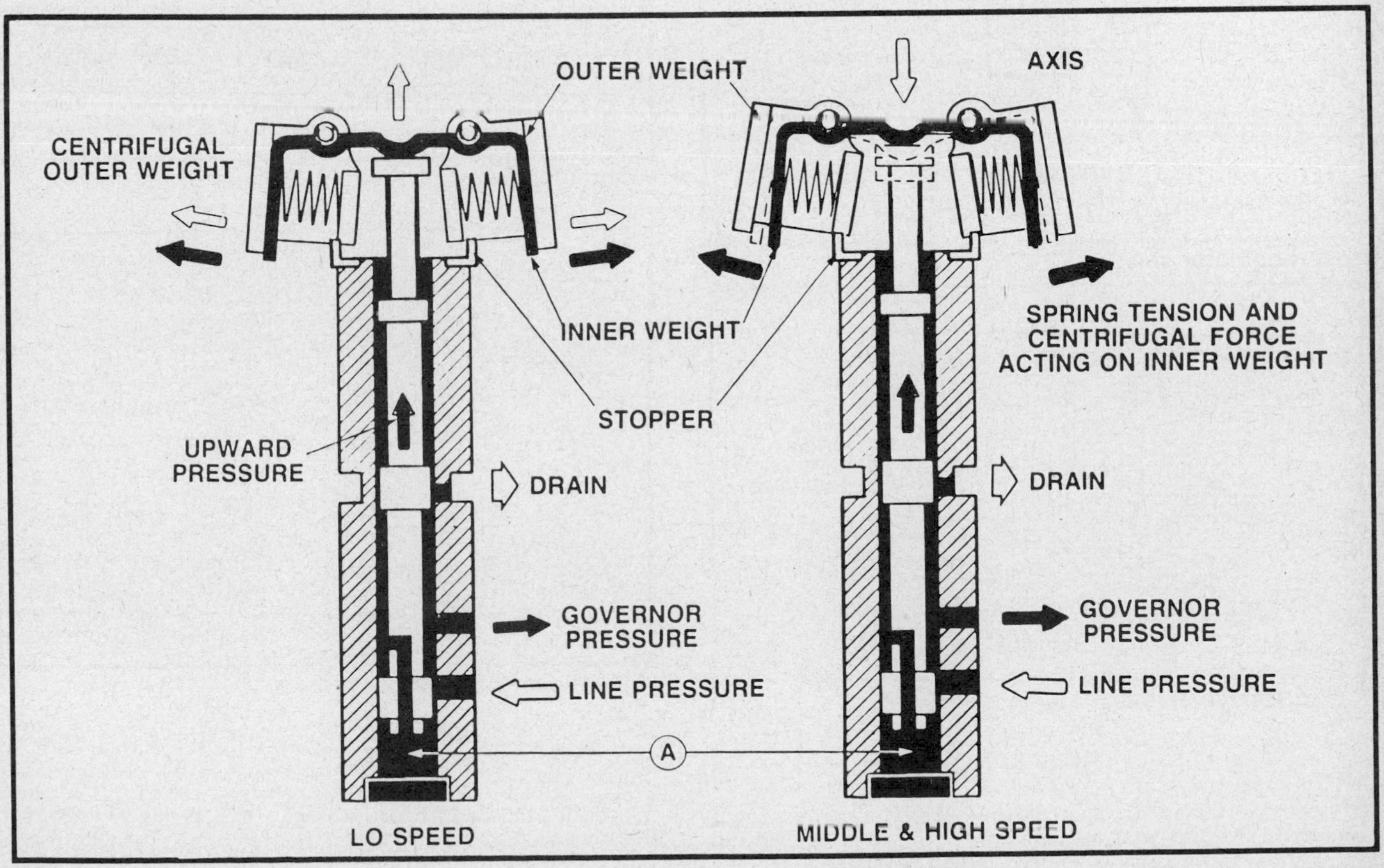
OUTER WEIGHT
AXIS
CENTRIFUGAL
OUTER WEIGHT
INNER WEIGHT
SPRING TENSION AND
CENTRIFUGAL FORCE
ACTING ON INNER WEIGHT
STOPPER
UPWARD
PRESSURE
DRAIN
DRAIN
GOVERNOR
PRESSURE
GOVERNOR
PRESSURE
LINE PRESSURE
LINE PRESSURE
A
LO SPEED
MIDDLE & HIGH SPEED

Governor valve

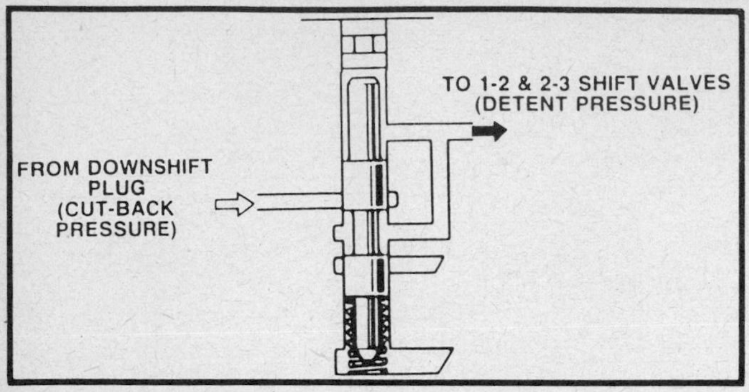
TO 1-2 & 2-3 SHIFT VALVES
(DETENT PRESSURE)
FROM DOWNSHIFT PLUG (CUT-BACK PRESSURE)

Detent regulator valve

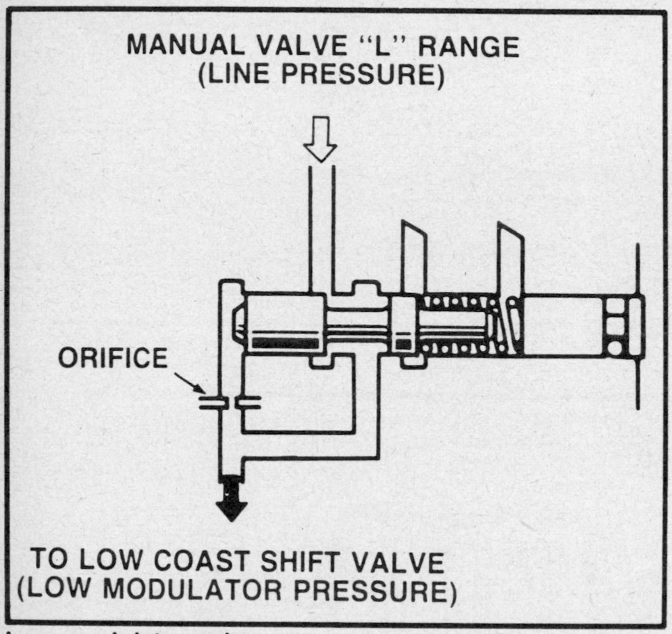
MANUAL VALVE "L" RANGE
(LINE PRESSURE)
ORIFICE
TO LOW COAST SHIFT VALVE
(LOW MODULATOR PRESSURE)

Low modulator valve

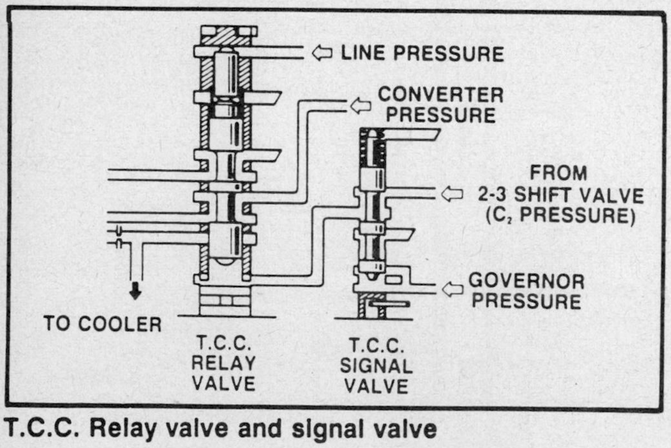
LINE PRESSURE
CONVERTER PRESSURE
FROM 2-3 SHIFT VALVE (C_2 PRESSURE)
GOVERNOR PRESSURE
TO COOLER
T.C.C. RELAY VALVE
T.C.C. SIGNAL VALVE

T.C.C. Relay valve and signal valve

1–2 SHIFT VALVE

This valve automatically controls the 1–2 shift according to governor and throttle pressure. To improve the valve sliding characteristics, a 3 piece valve is used. When governor pressure is low and throttle pressure is high, the valve is pushed down by throttle pressure and because the 2nd brake circuit closes, the transaxle shifts into 1st gear.

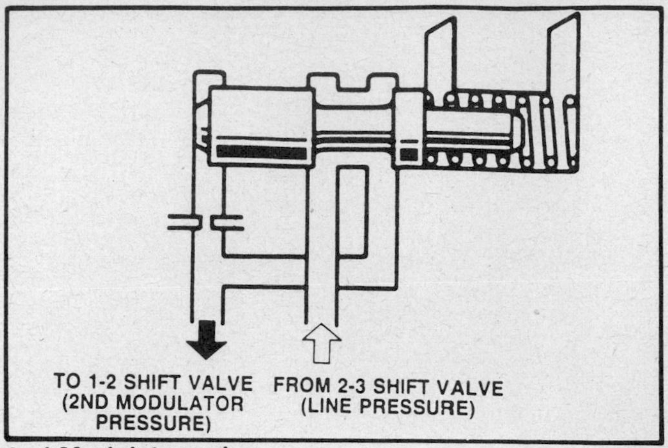
TO 1-2 SHIFT VALVE
(2ND MODULATOR PRESSURE)
FROM 2-3 SHIFT VALVE
(LINE PRESSURE)

2nd Modulator valve

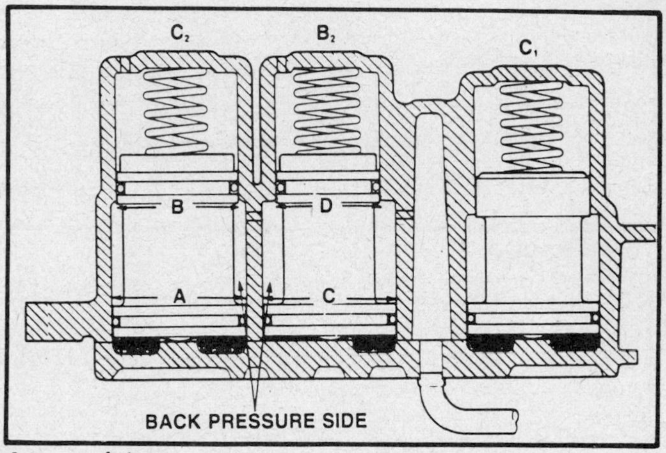
C_2
B_2
C_1
B
D
A
C
BACK PRESSURE SIDE

Accumulators

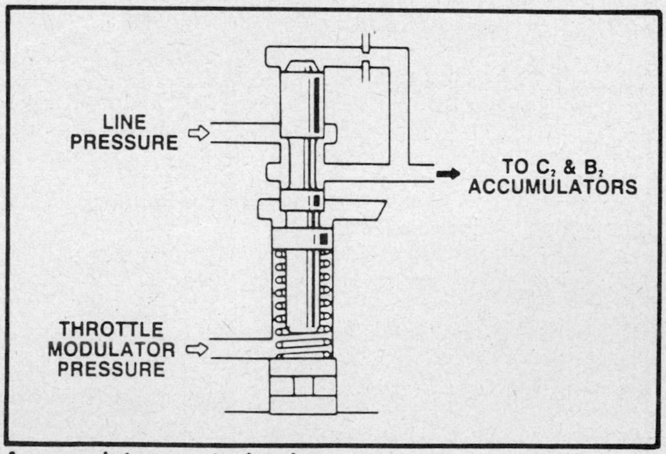
LINE PRESSURE
TO C_2 & B_2 ACCUMULATORS
THROTTLE MODULATOR PRESSURE

Accumulator control valve

When governor pressure is high and throttle pressure low, the valve is pushed up by governor pressure and the circuit to the 2nd brake piston opens so the transaxle will shift into 2nd gear. When the throttle pressure passage is closed, downshifting into 1st gear is dependant on spring tension and governor pressure only.

Unless the downshift plug actuates and allows the detent pressure to act on the 1–2 shift valve, downshifting into 1st gear will take place at a set vehicle speed. In the **L** range, there is no

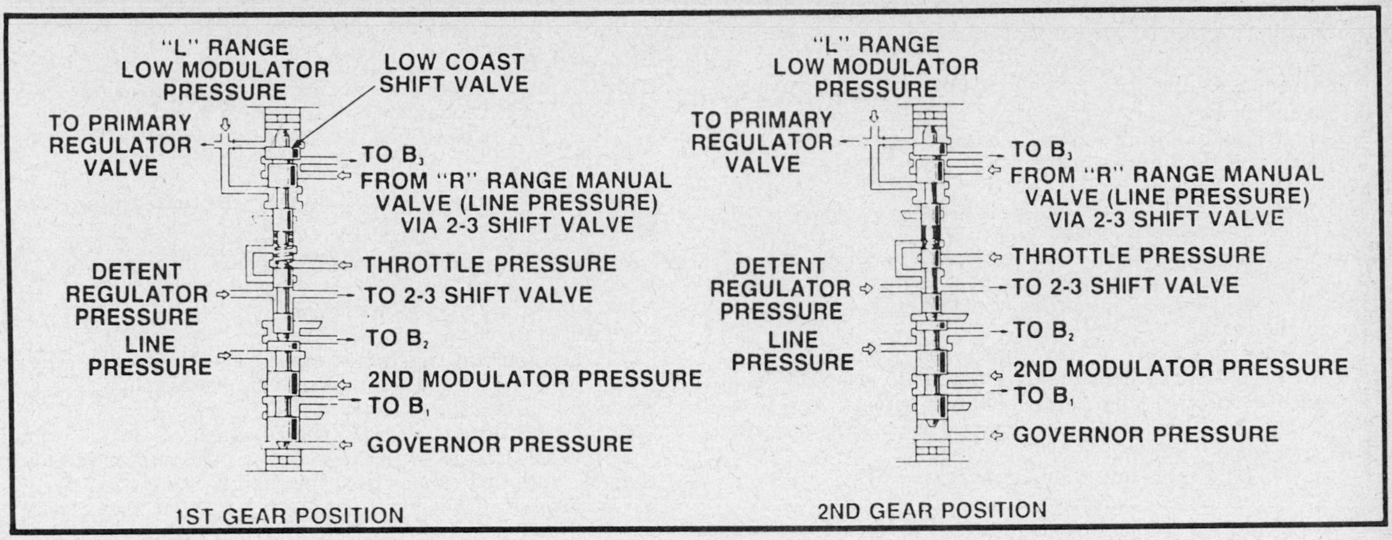
"L" RANGE
LOW MODULATOR
PRESSURE
LOW COAST
SHIFT VALVE
TO PRIMARY
REGULATOR
VALVE
TO B3
FROM "R" RANGE MANUAL
VALVE (LINE PRESSURE)
VIA 2-3 SHIFT VALVE
DETENT
REGULATOR
PRESSURE
THROTTLE PRESSURE
TO 2-3 SHIFT VALVE
LINE
PRESSURE
TO B2
2ND MODULATOR PRESSURE
TO B1
GOVERNOR PRESSURE
1ST GEAR POSITION
"L" RANGE
LOW MODULATOR
PRESSURE
TO PRIMARY
REGULATOR
VALVE
TO B3
FROM "R" RANGE MANUAL
VALVE (LINE PRESSURE)
VIA 2-3 SHIFT VALVE
DETENT
REGULATOR
PRESSURE
THROTTLE PRESSURE
TO 2-3 SHIFT VALVE
LINE
PRESSURE
TO B2
2ND MODULATOR PRESSURE
TO B1
GOVERNOR PRESSURE
2ND GEAR POSITION

1–2 Shift valve

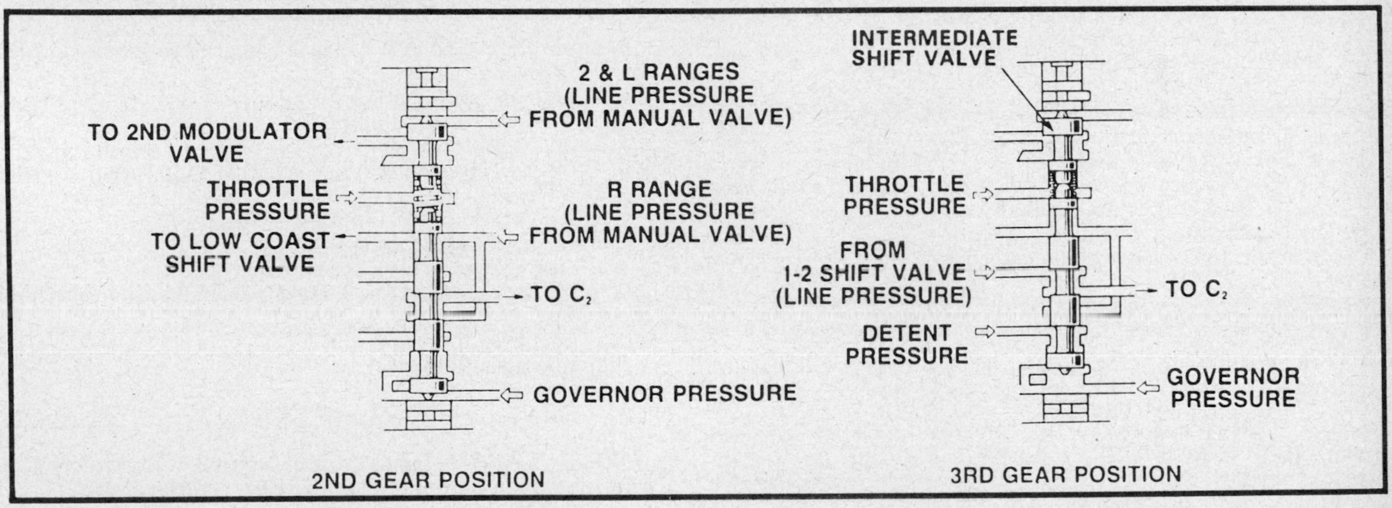
2 & L RANGES
(LINE PRESSURE
FROM MANUAL VALVE)
TO 2ND MODULATOR
VALVE
THROTTLE
PRESSURE
R RANGE
(LINE PRESSURE
FROM MANUAL VALVE)
TO LOW COAST
SHIFT VALVE
TO C2
GOVERNOR PRESSURE
2ND GEAR POSITION
INTERMEDIATE
SHIFT VALVE
THROTTLE
PRESSURE
FROM
1-2 SHIFT VALVE
(LINE PRESSURE)
TO C2
DETENT
PRESSURE
GOVERNOR
PRESSURE
3RD GEAR POSITION

2–3 Shift valve

upshifting into 2nd gear because low modulator pressure is acting on the low coast shift valve.

2–3 SHIFT VALVE

This valve performs shifting between 2nd and 3rd gears. Control is accomplished by opposing throttle pressure and spring tension against governor pressure. When governor pressure is high, the valve is pushed up against the resistance of the throttle pressure and spring tension. This opens the passage to the direct clutch (C_2) piston to allow the shift into 3rd gear.

When governor pressure is low, the valve is pushed down by throttle pressure and spring tension to close the passage leading to the direct clutch piston, causing a downshift to 2nd gear. In the event of kickdown, the detent pressure acts on the 2–3 shift valve to permit a quicker downshift to 2nd gear. valve movement occurs due to the different size areas where pressure is applied. Since the area is larger for downshift than for upshift, downshifting takes place at a lower vehicle speed.

In the manual **2** range, line pressure from the manual valve acts on the intermediate shift valve. The valve descends and shifting into 2nd gear is accomplished but there is no upshifting into 3rd gear. Line pressure passes through the 2nd modulator valve and 1–2 shift valve and acts on the 2nd coast brake to provide engine braking.

Diagnosis Tests

HYDRAULIC TESTS

1. Run the engine until it reaches normal operating temperature.
2. Raise the vehicle and support it safely.
3. Remove the transaxle case test plugs and mount hydraulic pressure gauges.

GOVERNOR PRESSURE

1. Apply the parking brake.
2. Start the engine.
3. Shift into **D** range and measure the governor pressure at the speeds specified in the chart.
4. If the governor pressure is not to specification, check for the following:
   a. Line pressure not to specification

b. Fluid leakage in the governor pressure cuircuit
c. Improper governor valve operation

Results

Vehicle speed 14 mph—gauge reading 10–21 psi
Vehicle speed 28 mph—gauge reading 19–29 psi
Vehicle speed 43 mph—gauge reading 33–44 psi

LINE PRESSURE

1. Apply the parking brake.
2. Start the engine.
3. Apply the brake pedal while manipulating the accelerator pedal and measure the line pressure at the engine speeds specified in the chart.
4. In the same manner, perform the test for the **R** range.
5. If the measured pressures are not up to specified values, recheck the throttle cable adjustment and retest.
6. If the measured values at all ranges are higher than specified, check the following:
   a. Throttle cable out of adjustment
   b. Throttle valve defective
   c. Regulator valve defective
7. If the measured values at all ranges are lower than specified, check the following:
   a. Throttle cable out of adjustment
   b. Throttle valve defective
   c. Regulator valve defective
   d. Oil pump defective
8. If the pressure is low in **D** range only, check the following:
   a. **D** range circuit fluid leakage
   b. Forward clutch defective
9. If the pressure is low in **R** range only, check the following:
   a. **R** range circuit fluid leakage
   b. Direct clutch defective
   c. First and reverse brake defective

Results

D range—idling 53-61 psi
D range—stall 131-152 psi
R range—idling 77-102 psi
R range—stall 205-239 psi

TIME LAG TEST

If the shift lever is shifted while the engine is idling, there will be a certain time elapse or lag before the shock can be felt. This is used for checking the condition of the forward clutch, direct clutch and 1st and reverse brake.

1. Run the engine until it reaches normal operating temperature.

NOTE: Allow 1 minute interval between tests. Make 3 measurements and take the average value.

2. Apply the parking brake.
3. Check the idle speed. With the cooling fan and A/C off and in **N** range, it should be 900 rpm with power steering and 800 rpm without power steering.
4. Move the shift lever from **N** to **D** range. Using a stop watch, measure the time it takes from shifting the lever until the shock is felt. The time lag should be less than 1.2 seconds.
5. Following Step 4, move the shift lever from **N** to **R** range. The time lag should be less than 1.5 seconds.
6. If the **N** time lag is longer than specified, check for the following:
   a. Line pressure too low
   b. Forward clutch worn
   c. Direct clutch worn
   d. First and reverse brake worn

STALL SPEED TEST

The object of this test is to check the overall performance of the transaxle and engine by measuring the maximum engine speeds in the **D** and **R** ranges.

1. Run the engine until it reaches normal operating temperature.

NOTE: Do not continuously run this test longer than 5 seconds.

2. Apply the parking brake and block the front and rear wheels.
3. Connect an engine tachometer.
4. Shift into the **D** range.
5. While applying the brakes, step all the way down on the accelerator. Quickly read the highest rpm at this time. The stall speed should be 2200–2350 rpm.
6. Perform the same test in the **R** range.
7. If the engine speed is the same for both ranges, but lower than the specified value, check for the following:
   a. Engine output insufficient
   b. Stator one-way clutch not operating properly

NOTE: If more than 600 rpm below the specified value, the torque converter could be at fault.

8. If the stall speed in **D** range is higher than specified, check for the following:
   a. Line pressure too low
   b. Forward clutch slipping
   c. One-way clutch No. 2 not operating properly
9. If the stall speed in **R** range is higher than specified, check for the following:
   a. Line pressure too low
   b. Direct clutch slipping
   c. First and reverse brake slipping
10. If the stall speed in **R** and **D** range is higher than specified, check for the following:
   a. Line pressure too low
   b. Improper fluid level

ROAD TEST

NOTE: This test must be performed with the engine and transaxle at normal operating temperature.

D RANGE TEST 1

Shift into **D** range and while driving with the accelerator pedal held constant (throttle valve opening 50% and 100%) check the following points:

1. At each of the throttle openings, check that the 1–2 and 2–3 upshifts take place and that the shift points conform to those shown in the automatic shift diagram.
2. If there is no 1–2 upshift, the governor valve may be defective or the 1–2 shift valve stuck.
3. If there is no 2–3 upshift, the 2–3 shift valve is stuck.
4. If the shift point is defective, the throttle cable is out of adjustment or the throttle valve, 1–2 shift valve and the 2-3 shift valves are defective.

D RANGE TEST 2

Check for harsh shift or soft shift slip at 1–2 and 2–3 upshifts. If the shock is severe, the line pressure is too high, the accumulator is defective or the check ball is defective.

D RANGE TEST 3

1. While driving in **D** range, 3rd gear, check for abnormal noise and vibration. This condition could also be caused by unbalance in the differential, tires or the torque converter.
2. While driving in **D** range, 2nd and 3rd gears, check the kickdown vehicle speed limits for the 2–1, 3–1 and the 3–2 kickdowns conform to the automatic shift diagram.

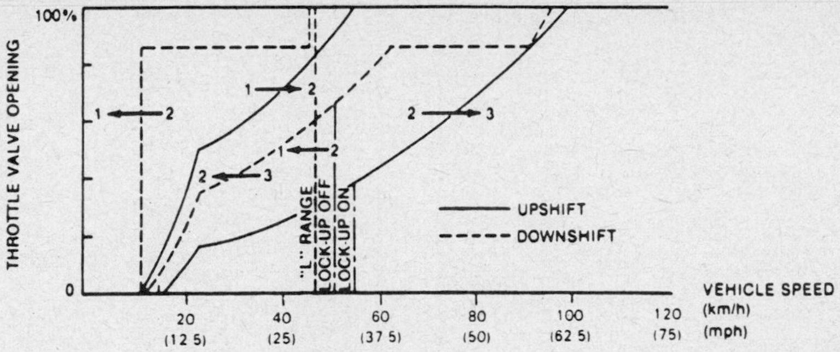
100%
THROTTLE VALVE OPENING
0
1→2
1←2
2→3
2←3
"L" RANGE
LOCK-UP OFF
LOCK-UP ON
UPSHIFT
DOWNSHIFT
VEHICLE SPEED
20 (12 5)
40 (25)
60 (37 5)
80 (50)
100 (62 5)
120 (75)
(km/h)
(mph)

km/h (mph)

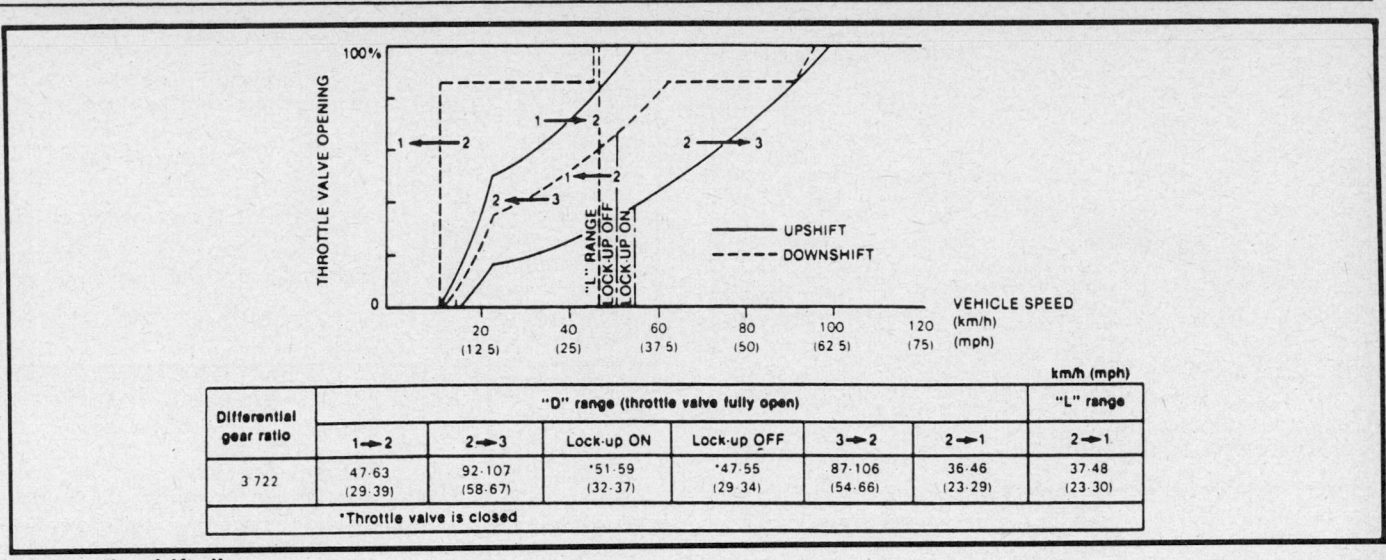

Differential gear ratio	"D" range (throttle valve fully open)						"L" range
	1→2	2→3	Lock-up ON	Lock-up OFF	3→2	2→1	2→1
3 722	47-63 (29-39)	92-107 (58-67)	*51-59 (32-37)	*47-55 (29-34)	87-106 (54-66)	36-46 (23-29)	37-48 (23-30)
	*Throttle valve is closed						

Automatic shift diagram

3. While driving in **D** range, 3rd gear, shift to **2** and **L** ranges and check the engine braking effect at each of these ranges. If there is no engine braking effect at **2** range, the 2nd coast brake is defective. If there is no engine braking effect at **L** range, 1st and reverse is defective.

D RANGE TEST 4

While driving in **D** range, release the accelerator pedal and shift into **L** range. Check to see if the 3–2 and 2–1 downshift points conform to those given in the automatic shift diagram.

LOCKUP MECHANISM

While driving in **D** range at a steady speed of about 34 mph (lock-up on), lightly depress the accelerator pedal and check that the engine speed does not change abruptly.

2 RANGE TEST

1. Shift to **2** range and drive with the throttle valve opening at 50% and 100% respectively. Check the 1–2 upshift points at each of the throttle valve openings to see that it conforms to those indicated in the automatic shift diagram.
2. While driving in **2** range, 2nd gear, release the accelerator pedal and check the engine braking effect.
3. Check the kickdown from **2** range. Check the 2–1 kickdown vehicle speed limit.
4. Check for abnormal noise at acceleration and deceleration and harshness at upshift and downshift.

L RANGE TEST

1. While driving in **L** range, check that there is no upshift to 2nd gear.
2. While driving in **L** range, release the accelerator pedal and check the engine braking effect.
3. Check for abnormal noise during acceleration and deceleration.

R RANGE TEST

Shift into **R** range and while starting at full throttle, check for slipping.

P RANGE TEST

Stop the vehicle on a slight hill and after shifting into **P** range, release the parking brake. Check to see that the parking lock pawl prevents the vehicle from moving.

ON CAR SERVICES

Adjustments

THROTTLE CABLE

1. Depress the accelerator pedal all the way and check that the throttle valve opens fully. If the throttle valve does not open fully, adjust the accelerator link.
2. Fully depress the accelerator.
3. Loosen the adjustment nuts.
4. Adjust the throttle cable housing so that the distance between the end of the boot and the stopper on the cable is correct. It should be 0–0.04 in. (0–1mm).
5. Tighten the adjusting nuts and recheck the adjustment.

TRANSAXLE CONTROL CABLE

1. Loosen the swivel nut on the lever.
2. Push the manual lever fully toward the right side of the vehicle.
3. Return the lever 2 notches toward the **N** position.
4. Set the shift lever in the **N** range.
5. While holding the lever lightly toward the **R** range side, tighten the swivel nut.

NEUTRAL SAFETY SWITCH

If the engine will start with the shift selector in any range other than **N** or **P** range, adjustment is required.

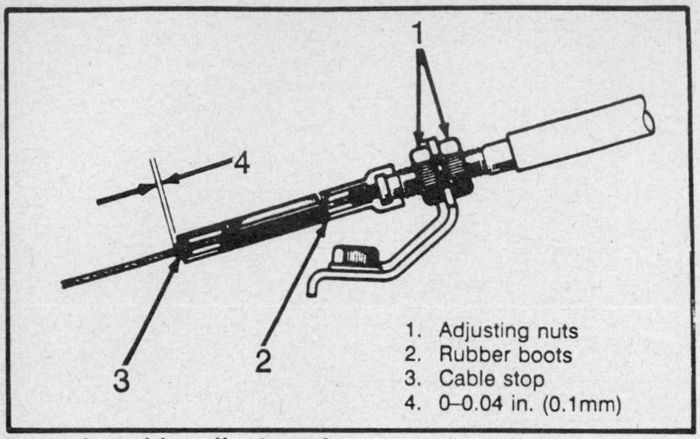
1
4
3
2
1. Adjusting nuts
2. Rubber boots
3. Cable stop
4. 0–0.04 in. (0.1mm)
Throttle cable adjustment

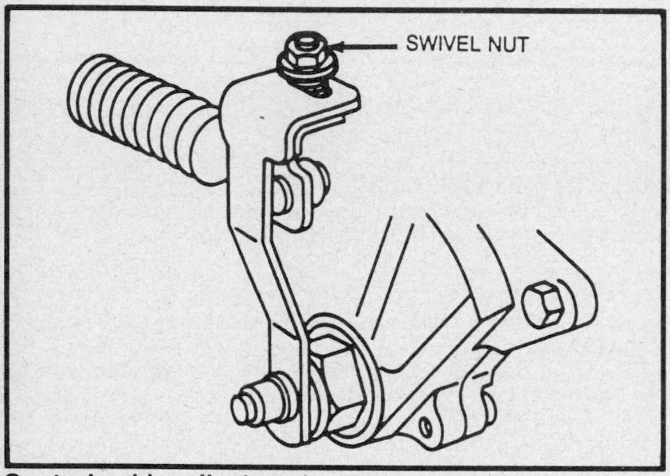
SWIVEL NUT
Control cable adjustment

1. Loosen the neutral safety switch bolts and set the shift selector in **N** range. Adjust the neutral safety switch.
2. Disconnect the neutral safety switch connector.
3. Connect an ohmmeter between the terminals.
4. Adjust the switch to the point where there is continuity between terminals.

Services

FLUID CHANGES

The conditions under which the vehicle is operated is the main consideration in determining how often the transaxle fluid should be changed. Different driving conditions result in different transaxle fluid temperatures. These temperatures effect the change intervals.

If the vehicle is driven under severe service conditions, change the fluid and filter every 15,000 miles. If the vehicle is not used under severe service conditions, change the fluid and replace the filter every 50,000 miles.

Do not overfill the transaxle. It takes 1 pint of fluid to change the level from **ADD** to **FULL** on the transaxle dipstick. Overfilling the unit can cause damage to the internal components of the automatic transaxle.

OIL PAN

Removal and Installation

1. Disconnect the negative battery cable.

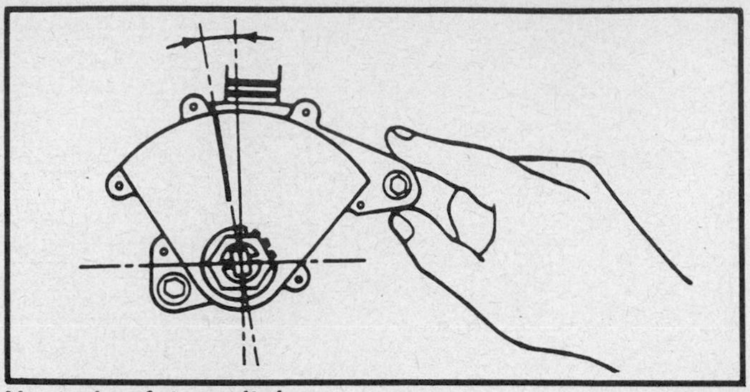
Neutral safety switch

2. Raise the vehicle and support it safely.
3. Remove the drain plug in the pan and drain the fluid.
4. Reinstall the drain plug and remove the oil pan. After removing the bolts, tap the pan lightly with a plastic hammer. Do not force the pan off by prying. This may cause damage to the gasket mating surface.
5. Remove the oil pan gasket material on the mating surface.
6. Remove the oil filter from the valve body.
7. Clean the inside of the oil pan before installation.
8. Clean the oil cleaner magnet and install it in the proper position.
9. Install a new filter on the valve body.
10. Install the pan with a new gasket and torque the bolts to 43 inch lbs. (4.9 Nm).
11. Fill with transmission fluid to the proper level.
12. Lower the vehicle and connect the battery cable.

VALVE BODY

Removal and Installation

1. Disconnect the negative battery cable.
2. Raise the vehicle and support it safely.
3. Remove the drain plug in the pan and drain the fluid.
4. Reinstall the drain plug and remove the oil pan. After removing the bolts, tap the pan lightly with a plastic hammer. Do not force the pan off by prying. This may cause damage to the gasket mating surface.
5. Remove the oil pan gasket material on the mating surface.
6. Remove the oil tube bracket and the oil strainer.
7. Pry out the 4 oil tubes and remove them.
8. Remove the detent spring.
9. Remove the 4 bolts and the manual valve body.
10. Loosen and remove the valve body bolts.
11. Lift the valve body assembly and disconnect the throttle cable from the cam.
12. Remove the valve body.
13. On installation, install the valve body on the case. Hold the cam down and insert the throttle cable into the slot.
14. Install the valve body with the bolts finger tight.
15. Align the manual valve with the manual lever and install the manual valve body. Torque the bolts to 84 inch lbs. (10 Nm).
16. Install the detent spring and tighten the bolts to 84 inch lbs. (10 Nm).
17. Make sure that the manual valve lever touches the center of the detent spring roller.
18. Tighten the valve body bolts to 84 inch lbs. (10 Nm).
19. Install the oil tubes, bracket and oil strainer.
20. Clean the inside of the oil pan before installation.
21. Clean the oil cleaner magnet and install it in the proper position.
22. Install the oil pan with a new gasket and torque the bolts to 43 inch lbs. (4.9 Nm).

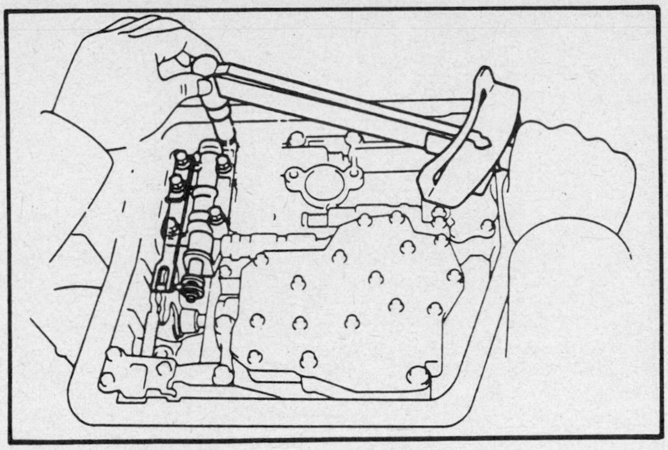

Manual valve body

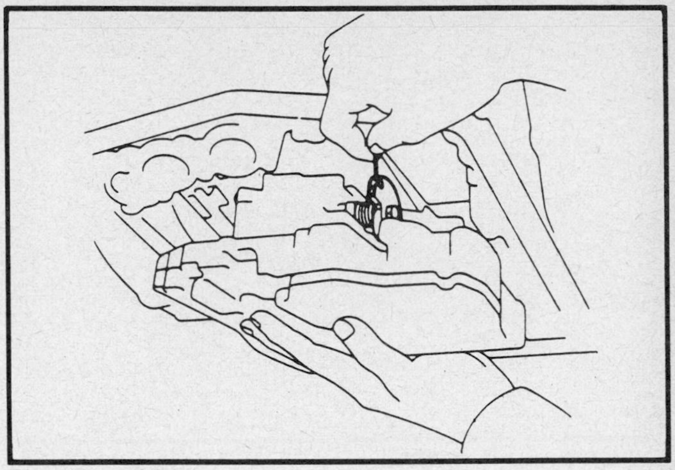

Throttle cable

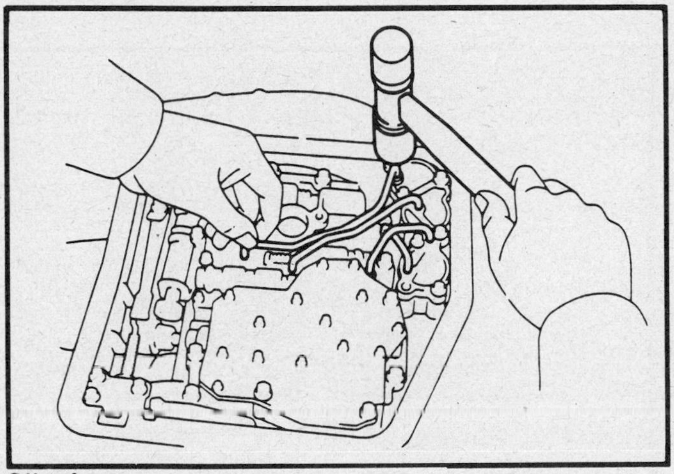

Oil tubes

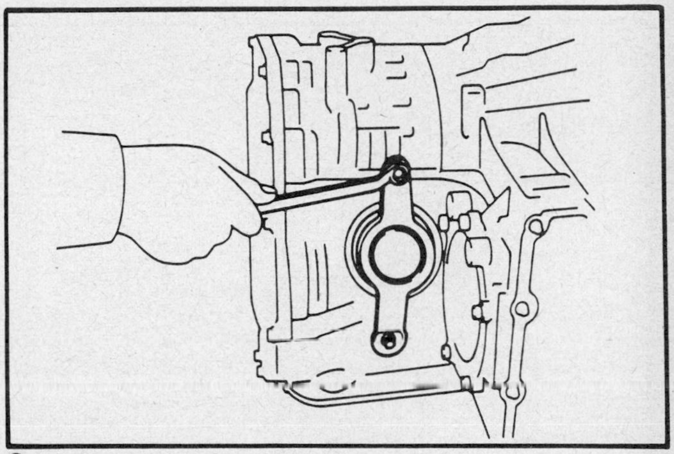

Governor cover

GOVERNOR

Removal and Installation

1. Disconnect the negative battery cable.
2. Raise the vehicle and support it safely.
3. Remove the transaxle dust cover.
4. Remnove the left hand driveshaft.
5. Remove the governor cover and O-ring.
6. Remove the governor body with the thrust washer.
7. Remove the washer.
8. Remove the governor body adapter.
9. For installation, install the governor body adapter.
10. Install the governor body with the thrust washer.
11. Install the governor cover with the O-ring.
12. Install the left hand driveshaft.
13. Install the transaxle dust cover.
14. Lower the vehicle and connect the battery cable.

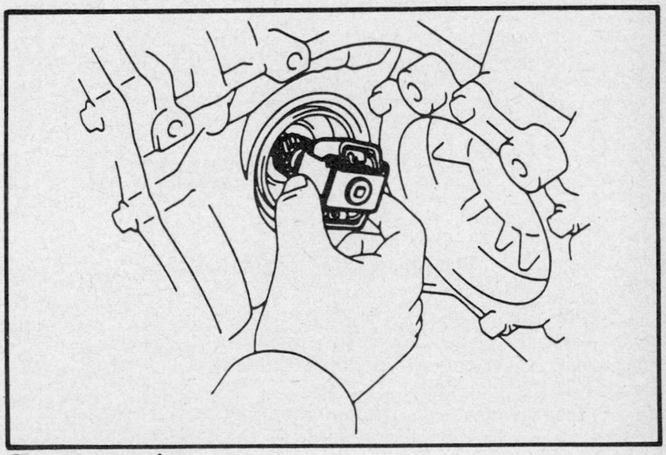

Governor valve

DIFFERENTIAL OIL SEAL

Removal and Installation

1. Disconnect the negative battery cable.
2. Raise the vehicle and support it safely.
3. Drain the fluid from the differential.
4. Remove the driveshaft from the transaxle and the steering knuckle.
5. Remove the differential oil seals using tools J–26941 and J–23907.
6. Using tool J–35553, drive the oil seal into the case until its surface is flush with the surface of the case.
7. Coat the top of the oil seal with grease.
8. Install the driveshaft to the transaxle and the steering knuckle.
9. Fill with transmission fluid and check the fluid level.

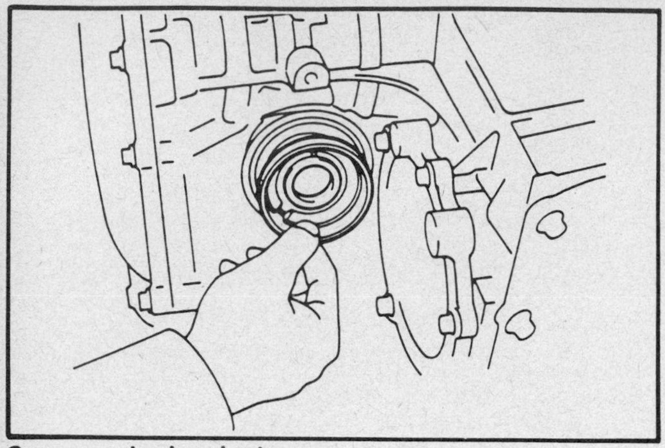

Governor body adapter

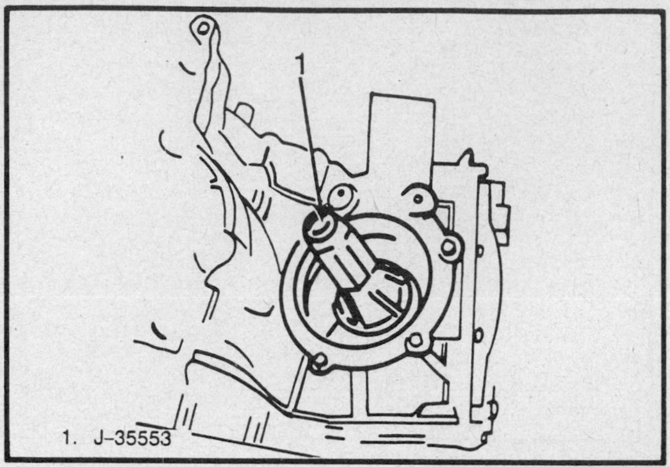
1
1. J-35553

Differential seal installation LH side

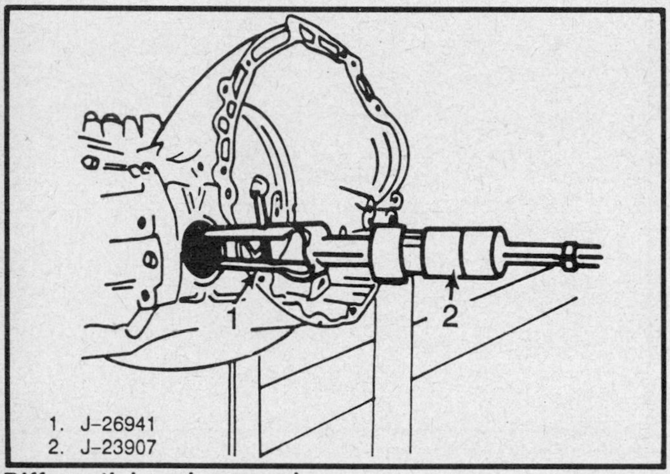
1
2
1. J-26941
2. J-23907

Differential seal removal

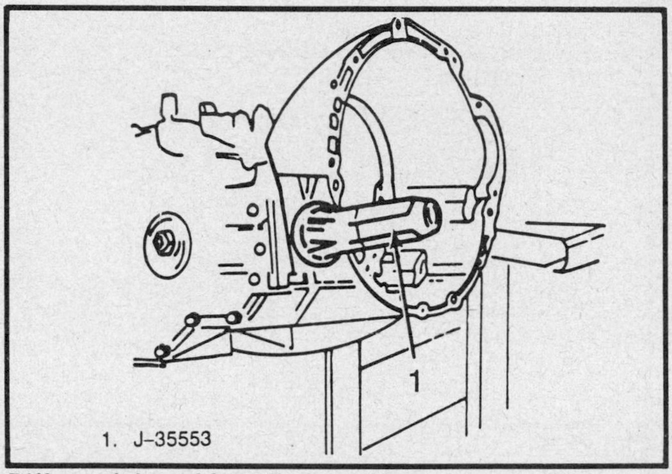
1
1. J-35553

Differential seal installation RH side

REMOVAL AND INSTALLATION

TRANSAXLE REMOVAL

1. Disconnect the negative battery cable.
2. Remove the air intake tube.
3. Disconnect the speedometer cable and the neutral safety switch at the transaxle.
4. Disconnect the thermostat housing at the transaxle.
5. Disconnect the ground cable at the transaxle.
6. Remove the upper mount to bracket bolt.
7. Disconnect the necessary electrical connections.
8. Disconnect the T.V. cable at the carburetor.
9. Remove the upper bell housing bolts.
10. Support the engine using an engine support tool.
11. Raise the vehicle and support it safely.
12. Remove the left wheel and tire assembly.
13. Remove the left, right and center splash shields.
14. Remove the center beam.
15. Disconnect the shift cable at the transaxle.
16. Remove the shift cable bracket.
17. Disconnect the cooler bracket.
18. Disconnect the cooler lines at the outlets.
19. Remove the inspection cover.
20. Remove the converter bolts.
21. Disconnect the left hand control arm at the ball joint.
22. Disconnect the right hand control arm at the ball joint.
23. Disconnect the right and left axle shafts at the tranxsaxle.
24. Remove the starter bolts. Remove the starter.
25. Remove the rear transaxle bolts.
26. Support the transaxle with a jack.
27. Remove the remaining mount bolts.
28. Remove the remaining bell housing bolts.
29. Remove the transaxle from the vehicle.

TRANSAXLE INSTALLATION

1. If the torque converter has been drained, refill it with 2.3 qts. (2.2L) Dexron® II automatic transmission fluid. Install the torque converter in the transaxle.

2. To be sure the torque converter is installed correctly, use a straight edge and measure from the installed surface to the front surface of the transaxle housing. The distance should be more than 0.79 in. (20mm).
3. Position the transaxle on a suitable transmission jack and align it in the vehicle.
4. Align the converter housing with the 2 dowel pins in the block and install a bolt.
5. Install the transaxle housing mounting bolts and torque the 12mm bolt to 47 ft. lbs. (64 Nm). Torque the 10mm bolts to 34 ft. lbs. (46 Nm).
6. Install the left hand engine mounting. Tighten the bolts to 38 ft. lbs. (52 Nm).
7. Install the torque converter bolts. Tighten the bolts evenly and torque to 13 ft. lbs. (18 Nm).
8. Install the engine rear end plate.
9. Install the starter.
10. Install the driveshafts.
11. Install the engine mounting center support. Tighten the bolts to 29 ft. lbs. (39 Nm).
12. Install the front and rear mounts. Tighten the bolts to 29 ft. lbs. (39 Nm).
13. Install the splash shields.
14. Install the engine under cover.
15. Lower the vehicle.
16. Install the thermostat housing.
17. Connect the oil cooler hose.
18. Connect and adjust the control cable.
19. Connect the speedometer cable.

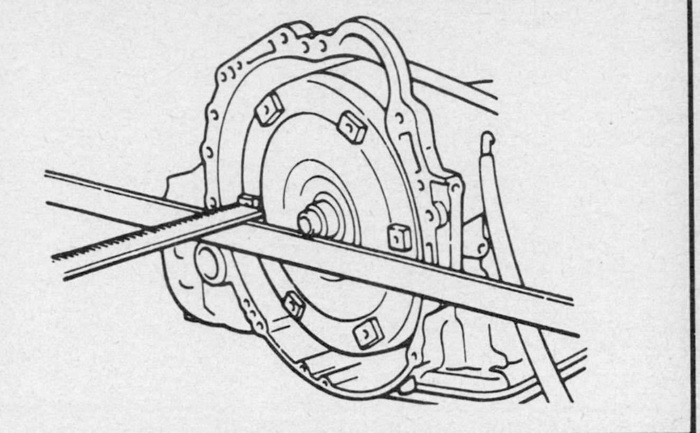
CORRECT DISTANCE: MORE THAN 20mm (0.79 IN.)
Checking the torque converter

20. Connect and adjust the throttle cable.
21. Connect the neutral safety switch connector.
22. Install the air cleaner.
23. Connect the negative battery cable.
24. Fill the transaxle with transmission fluid.

BENCH OVERHAUL

Before Disassembly

Before opening up the transaxle, the outside of the unit should be thoroughly cleaned, preferably with high pressure cleaning equipment. Dirt entering the transaxle internal parts will negate all the effort and time spent on the overhaul. During inspection and reassembly, all parts should be thoroughly cleaned with solvent and then dried with compressed air. Cloths and rags should not be used to dry the parts since lint will find its way into the valve body passages.

Lube the seals with Dexron® II automatic transmission fluid and use unmedicated petroleum jelly to hold the thrust washers and ease the assembly of the seals. Do not use solvent on neoprene seals, friction plates or thrust washers. Be wary of nylon parts if the transaxle failure was due to a cooling system problem. Nylon parts exposed to antifreeze solutions can swell and distort, so they must be replaced. Before installing bolts into aluminum parts, dip the threads in clean oil.

Converter Inspection

Make certain that the transaxle is held securey. If the torque converter is equipped with a drain plug, open the plug and drain the fluid. If there is no drain plug, the converter must be drained through the hub after pulling the converter out of the transaxle. If the oil in the converter is discolored but does not contain metal bits or particles, the converter is not damaged. Color is no longer a good indicator of fluid condition.

If the oil in the converter contains metal particles, the converter is damaged internally and must be replaced. If the cause of the oil contamination was burned clutch plates or overheated oil, the converter is contaminated and should be replaced. If the pump gears or cover show signs of damage, the converter will contain metal particles and must be replaced.

Transaxle Disassembly

1. Remove the oil cooler pipes.
2. Remove the manual shift lever.
3. Remove the neutral safety switch.
4. Remove the oil filler gauge and tube.
5. Remove the throttle cable retaining plate.
6. Remove the governor body using the following procedure:
   a. Remove the bolts and the cover bracket
   b. Remove the governor cover and O-ring
   c. Remove the thrust washer from the governor body
   d. Remove the governor body
   e. Remove the plate washer and the governor body adapter
7. Remove the pan and gasket.
8. Remove the magnet and check for chips and particles in the pan.
9. Remove the oil tube bracket and the oil strainer.
10. Remove the oil tubes.
11. Remove the manual detent spring.
12. Remove the manual valve and the manual valve body.
13. Disconnect the throttle cable from the cam and remove the valve body.
14. Remove the throttle cable from the case.
15. Remove the governor apply gasket.
16. Remove the governor oil strainer.
17. Remove the accumulator piston and springs. Loosen the bolts slowly until the spring tension is released.
18. Install the 2nd coast brake piston compressor J–35549 and remove the snapring.
19. Remove the bolts holding the oil pump to the transaxle case. Remove the pum using tools J–6125–B and J–35496 adapters.
20. Remove the direct clutch.
21. Remove the thrust washer.

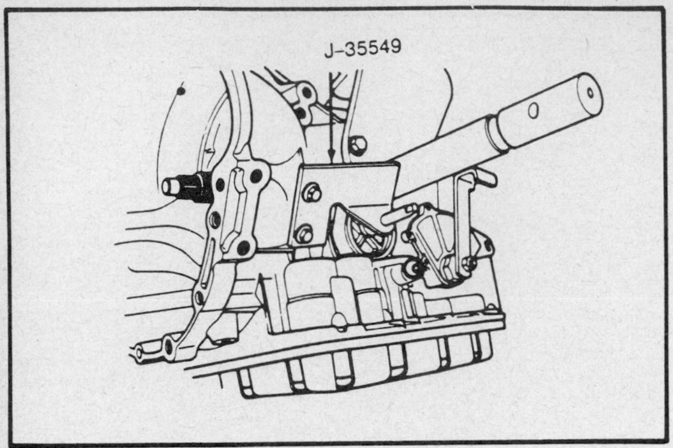
J-35549

2nd Coast brake piston compressor

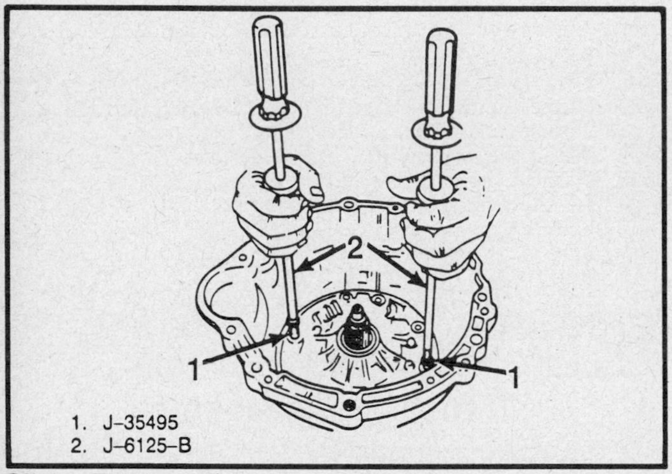
2
1
1
1. J-35495
2. J-6125-B

Oil pump removal

22. Remove the forward clutch drum.
23. Remove the 2nd coast brake band.
24. Remove the front planetary ring gear.
25. Remove the planetary gear.
26. Remove the sun gear, sun gear input drum, 2nd brake hub and No.1 one-way clutch.
27. Remove the 2nd coast brake band guide.
28. Remove the snapring holding the 2nd brake drum to the case.
29. Remove the 2nd brake drum and return springs.
30. Remove the 2nd brake drum gasket.
31. Remove the plates, discs and flange.
32. Blow out the piston with compressed air.
33. Remove the snapring holding the No. 2 one-way clutch outer race to the case.
34. Remove the No. 2 one-way clutch and rear planetary gear.
35. Remove the rear planetary ring gear and bearing.
36. Remove the snapring holding the flange to the case.
37. Remove the plates, discs and flanges.
38. Remove the bolts holding the rear cover to the transaxle case.
39. Remove the rear cover and intermediate shaft.
40. Remove the case gasket.
41. Remove the parking pawl bracket and parking lock rod.
42. Remove the parking lock pawl shaft, spring and parking lock pawl.

Unit Disassembly and Assembly

OIL PUMP

Disassembly

1. Remove the race from the stator shaft.
2. Remove the O-ring from the pump body.
3. Remove the oil seal rings from the back of the stator shaft.
4. Remove the thrust washer of the clutch drum from the stator shaft.
5. Remove the stator shaft. Keep the gears in assembly order.

Inspection

1. Check the body clearance of the driven gear. Push the driven gear to either side of the body. Using a feeler gauge, measure the clearance. If the clearance exceeds the limit, replace the drive gear, driven gear or pump body.
2. Check the tip clearance of both gears. Measure between the gear teeth and the crescent-shaped part of the pump body. If the clearance exceeds the limit, replace the drive gear, driven gear or pump body.
3. Check the side clearance of both gears. Using a steel straight edge and a feeler gauge, measure the side clearance of both gears. If the clearance exceeds the limit, replace the drive gear, driven gear or the pump body.
4. Inspect the front oil seal, check for wear, damage or cracks.
5. Replace the front oil seal. The seal end should be flush with the outer edge of the pump body.

Assembly

1. Install the driven gear and drive gear making sure the top of the gears are facing upward.
2. Install the stator shaft onto the pump body. Align the stator shaft with each bolt hole.
3. Tighten the bolts to 7 ft. lbs. (10 Nm).
4. Coat the thrust washer with petroleum jelly. Align the tab of the washer with the hollow of the pump body and install the thrust washer.
5. Install 2 oil seal rings on the oil pump. Do not spread the ring ends too far.
6. Check the pump drive gear rotation. Turn the drive gear and check that it rotates smoothly.
7. Install a new O-ring.
8. Install the race onto the stator shaft.

DIRECT CLUTCH

Disassembly

1. Remove the snapring from the clutch drum.
2. Remove the flange, discs and plates.
3. Compress the piston return springs and remove the snapring using tool J-23327.
4. Remove the spring retainer and springs.
5. Slide the direct clutch onto the oil pump. Apply compressed air to the oil pump to remove the piston. Remove the direct clutch from the oil pump.
6. Remove the clutch piston O-ring.

Inspection

Inspect the clutch piston. Check that the ball is free by shaking the piston. Check that the valve does not leak by applying low pressure compressed air. Inspect the discs.

Assembly

1. Coat the new O-rings with automatic transmission fluid and install them on the piston.

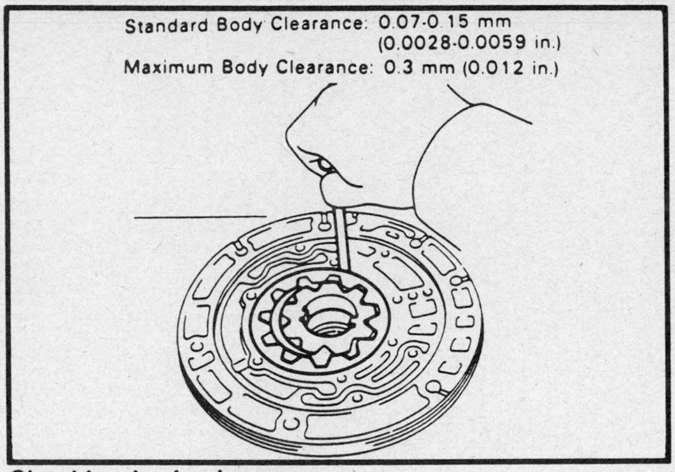
Standard Body Clearance: 0.07-0.15 mm
(0.0028-0.0059 in.)
Maximum Body Clearance: 0.3 mm (0.012 in.)

Checking body clearance

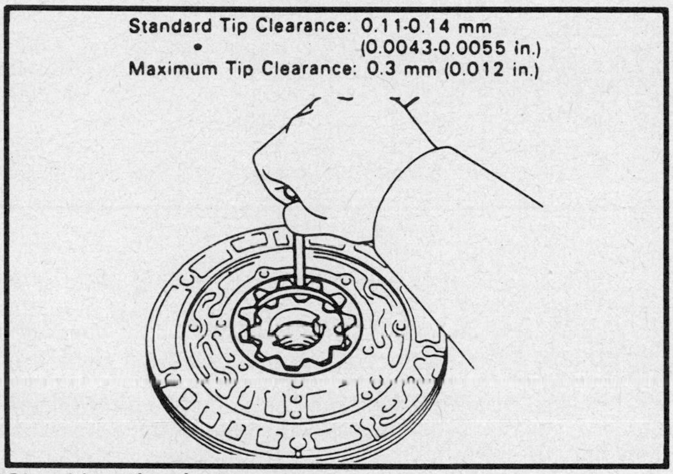
Standard Tip Clearance: 0.11-0.14 mm
(0.0043-0.0055 in.)
Maximum Tip Clearance: 0.3 mm (0.012 in.)

Checking tip clearance

2. Press the piston into the drum with the cup side up, being careful not to damage the O-ring.
3. Install the piston return springs and set the retainer and snapring in place.
4. Compress the return springs and install the snapring in the groove using tool J–23327. Install the snapring. Be sure the endgap of the snapring is not aligned with the spring retainer claw.
5. Install the plates, discs and flange. Install in order: plate-disc-plate-plate-disc. Install the flange, facing the flat end downward.
6. Install the outer snapring. Check that the endgap of the snapring is not aligned with a cutout.
7. Using compressed air, check the stroke of the direct clutch for only replacement of plate, discs and flange. If not within specification, select a proper flange. There are 2 different flange thickness.
8. Install the direct clutch onto the oil pump.
9. Apply compressed air into the passage of the oil pump body and be sure that the piston moves. If the piston does not move, disassemble and inspect.

FORWARD CLUTCH

Disassembly

1. Remove the thrust washer.
2. Remove the thrust bearings and races from both sides of the clutch.

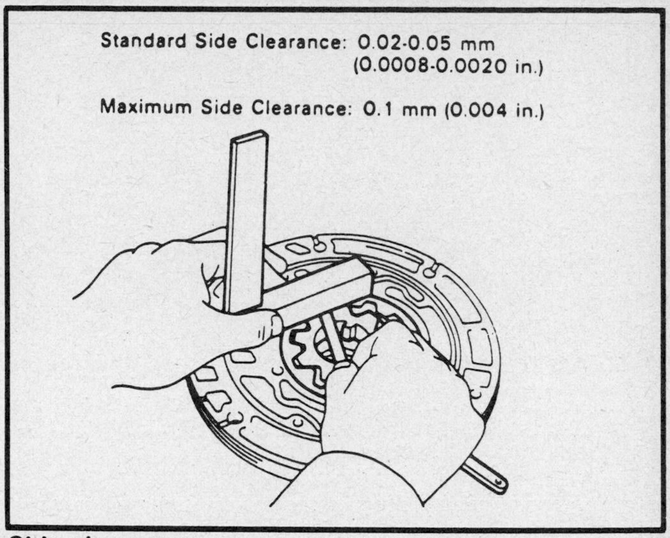
Standard Side Clearance: 0.02-0.05 mm
(0.0008-0.0020 in.)
Maximum Side Clearance: 0.1 mm (0.004 in.)

Side clearance

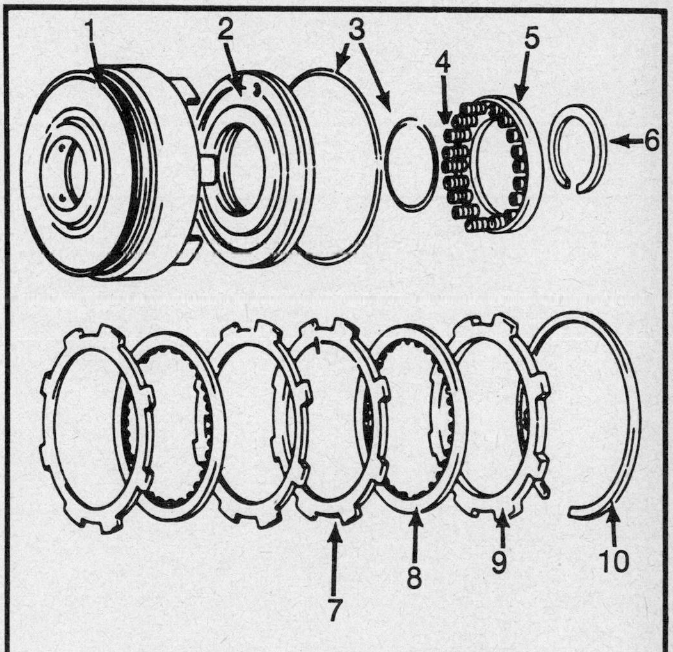
1
2
3
4
5
6
7
8
9
10

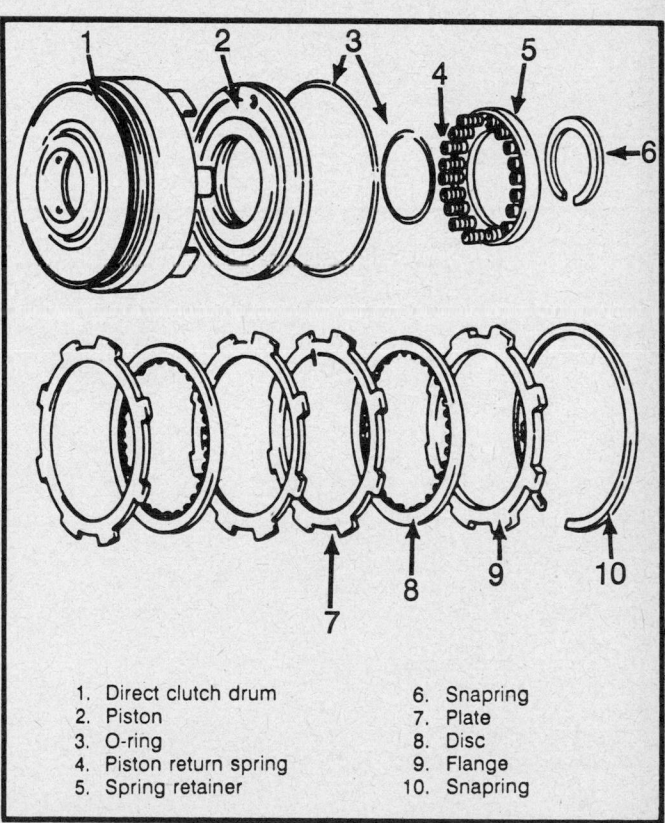

1. Direct clutch drum
2. Piston
3. O-ring
4. Piston return spring
5. Spring retainer
6. Snapring
7. Plate
8. Disc
9. Flange
10. Snapring

Direct clutch

3. Remove the snapring from the clutch drum.
4. Remove the flange, discs and plates.
5. Compress the piston return springs and remove the snapring using tools J–25018–A adapter and J–23327.
6. Remove the spring retainer and springs.
7. Apply compressed air into the oil passage to remove the piston.

Inspection

Make sure the check ball is free by shaking the piston. Check

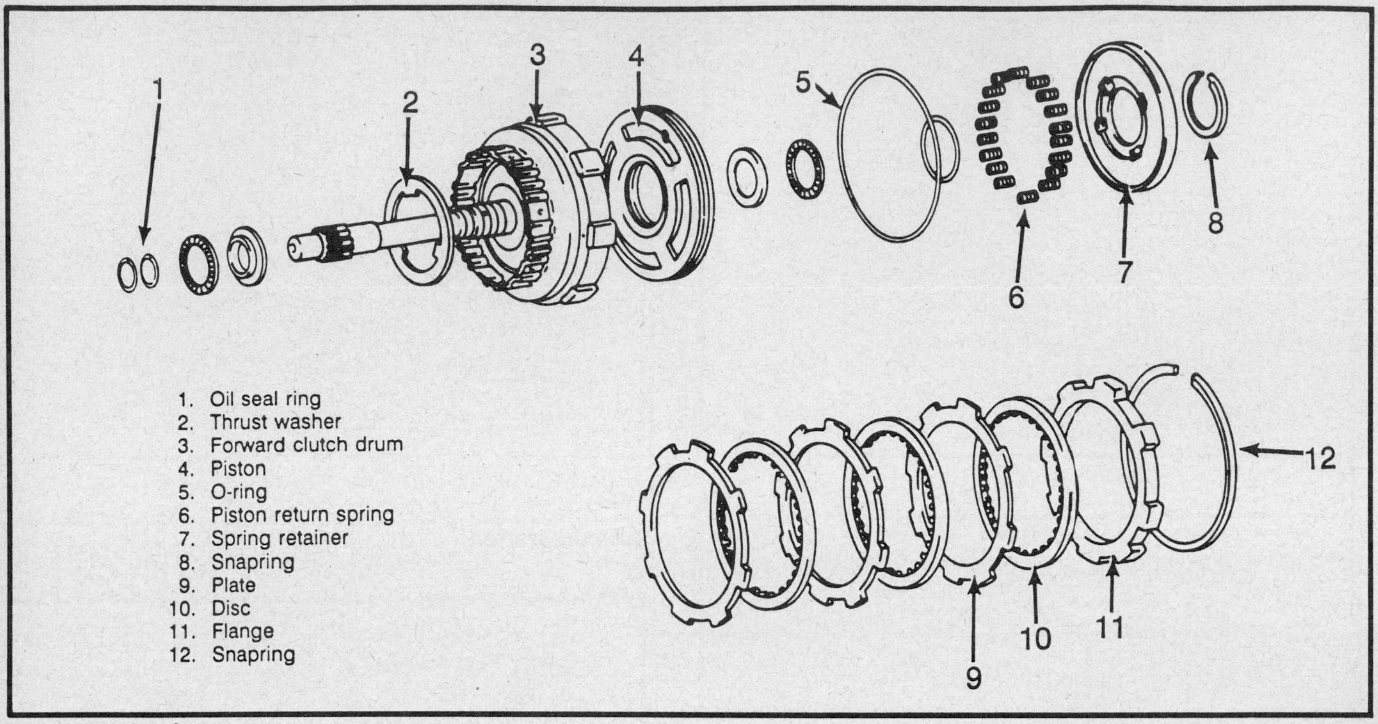
1
2
3
4
5
6
7
8
1. Oil seal ring
2. Thrust washer
3. Forward clutch drum
4. Piston
5. O-ring
6. Piston return spring
7. Spring retainer
8. Snapring
9. Plate
10. Disc
11. Flange
12. Snapring
9
10
11
12

Forward clutch

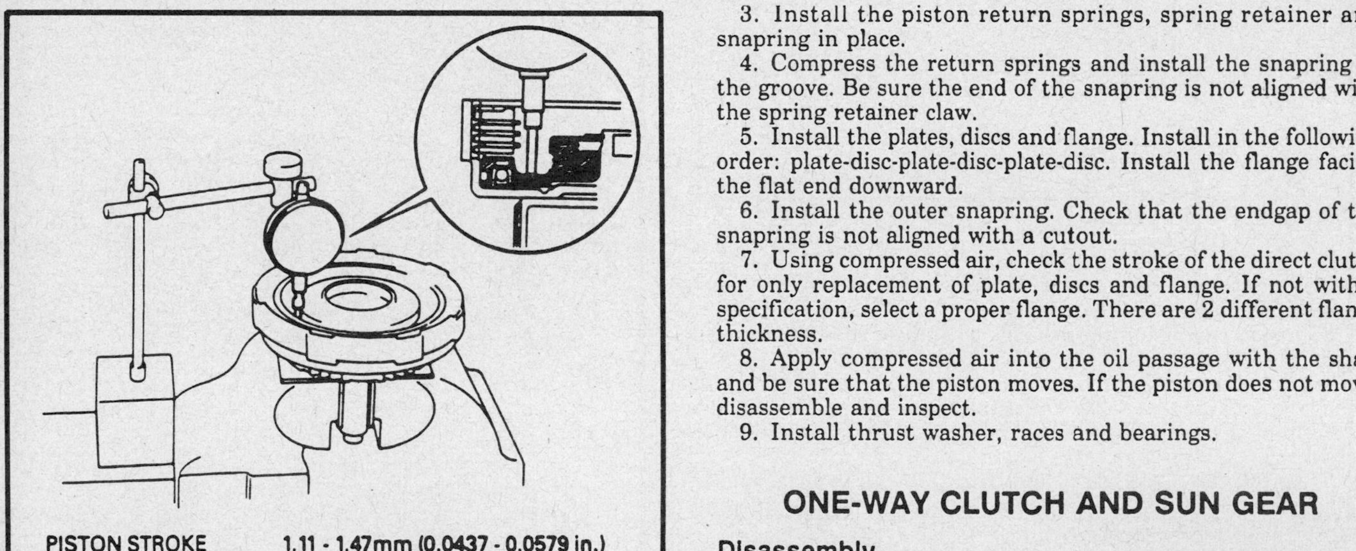
PISTON STROKE 1.11 - 1.47mm (0.0437 - 0.0579 in.)
FLANGE THICKNESS 3.00mm (0.1181 in.)
3.37 mm (0.1327 in.)

Snapring to flange stroke

that the valve does not leak by applying low pressure compressed air. Replace the oil seal rings. Slide the rings over top of the shaft and install them into the groove. Do not spread the ring ends.

Assembly

1. Install new O-rings on the piston, coat the O-rings with automatic transmission fluid.
2. Press the piston into the forward clutch drum with the cup side up, being careful not to damage the O-ring.
3. Install the piston return springs, spring retainer and snapring in place.
4. Compress the return springs and install the snapring in the groove. Be sure the end of the snapring is not aligned with the spring retainer claw.
5. Install the plates, discs and flange. Install in the following order: plate-disc-plate-disc-plate-disc. Install the flange facing the flat end downward.
6. Install the outer snapring. Check that the endgap of the snapring is not aligned with a cutout.
7. Using compressed air, check the stroke of the direct clutch for only replacement of plate, discs and flange. If not within specification, select a proper flange. There are 2 different flange thickness.
8. Apply compressed air into the oil passage with the shaft and be sure that the piston moves. If the piston does not move, disassemble and inspect.
9. Install thrust washer, races and bearings.

ONE-WAY CLUTCH AND SUN GEAR

Disassembly

1. Check the operation of the one-way clutch by holding the sun gear and turning the hub. The hub should turn freely clockwise and should lock counterclockwise.
2. Remove the 2nd brake hub and the one-way clutch from the sun gear.
3. Remove the No. 3 planetary carrier thrust washer from the sun gear input drum.
4. Remove the snapring and the sun gear input drum.

Inspection

If necessary, replace the one-way clutch.

1. Pry off the retainer.
2. Remove the one-way clutch.
3. Install the one-way clutch into the brake hub, facing the spring cage inward from the flanged side of the brake hub.

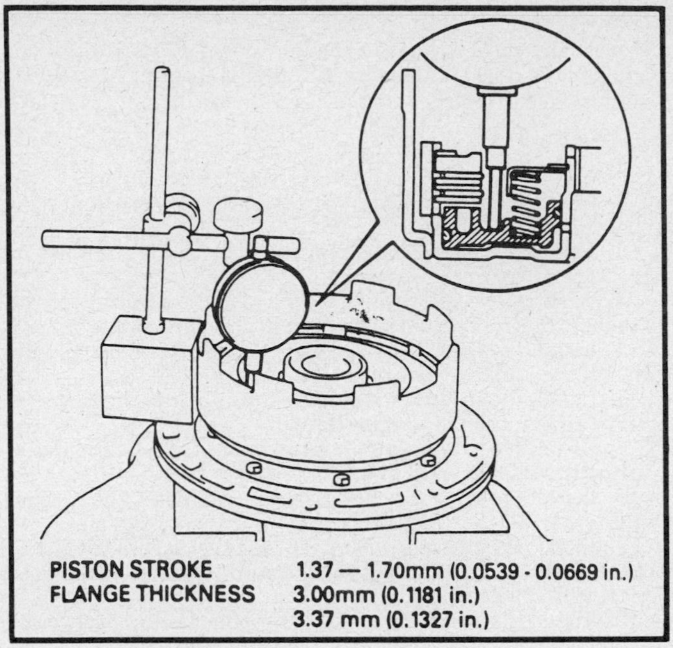
PISTON STROKE 1.37 — 1.70mm (0.0539 - 0.0669 in.)
FLANGE THICKNESS 3.00mm (0.1181 in.)
3.37 mm (0.1327 in.)

Forward clutch stroke

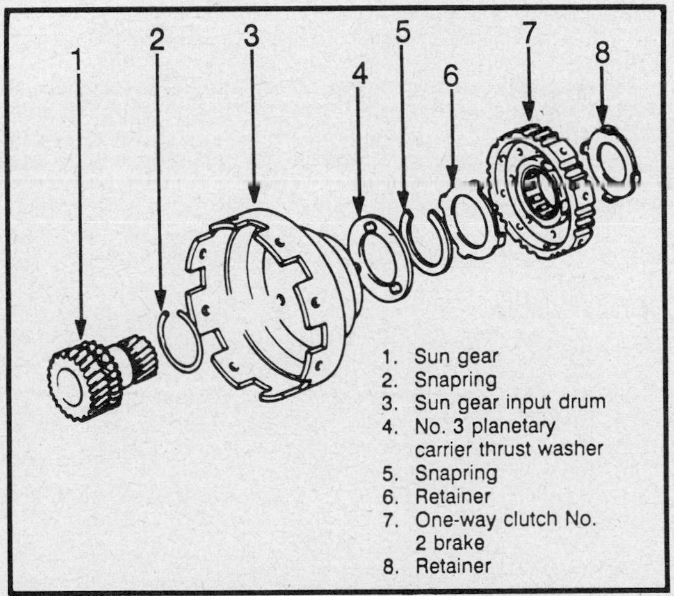
1
2
3
4
5
6
7
8
1. Sun gear
2. Snapring
3. Sun gear input drum
4. No. 3 planetary carrier thrust washer
5. Snapring
6. Retainer
7. One-way clutch No. 2 brake
8. Retainer

No. 1 one-way clutch and sun gear

4. Hold the brake hub in a vise and flatten the ears with a chisel.
5. Check the operation to make sure that the retainer is centered.

Assembly

1. Install the shaft snapring on the sun gear.
2. Install the sun gear input drum on the sun gear and install the shaft snapring.
3. Install the No. 3 planetary carrier thrust washer on the sun gear input drum.
4. Install the one-way clutch and the 2nd brake hub on the sun gear. While turning the hub clockwise, slide the one-way clutch into the inner race.

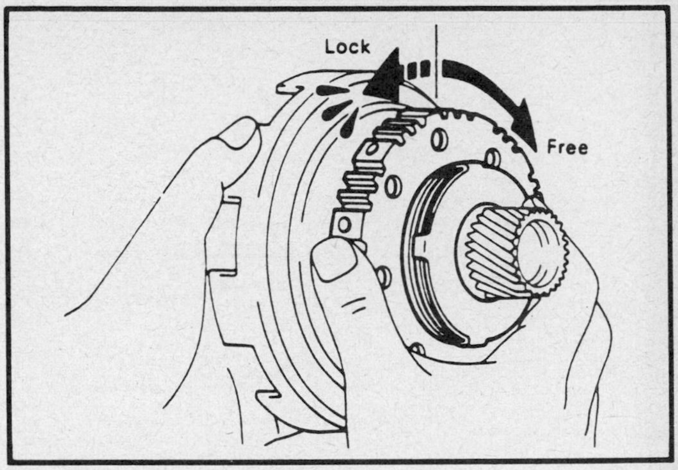
Lock
Free

One-way clutch operation

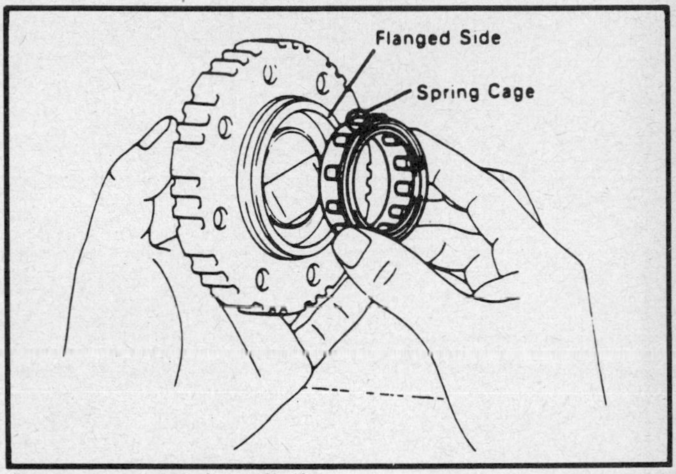
Flanged Side
Spring Cage

One-way clutch installation

VALVE BODY

Disassembly

1. Remove the 14 bolts and remove the remove the lower valve body cover.
2. Turn the assembly over and remove the 12 bolts from the upper valve body and upper valve body cover.
3. Remove the upper valve body cover, strainer gaskets and plate.
4. Turn the assembly over and remove the 3 bolts from the lower valve body.
5. Lift off the lower valve body and plate as a single unit. Hold the body plate to the lower valve body.

NOTE: Be careful that the check valve and ball do not fall out.

UPPER VALVE BODY

Disassembly

1. Remove the throttle valve retainer and check ball.
2. Remove the retainer for the plug with a magnetic finger and remove the plug.
3. Remove the lockup relay valve, control valve and spring.

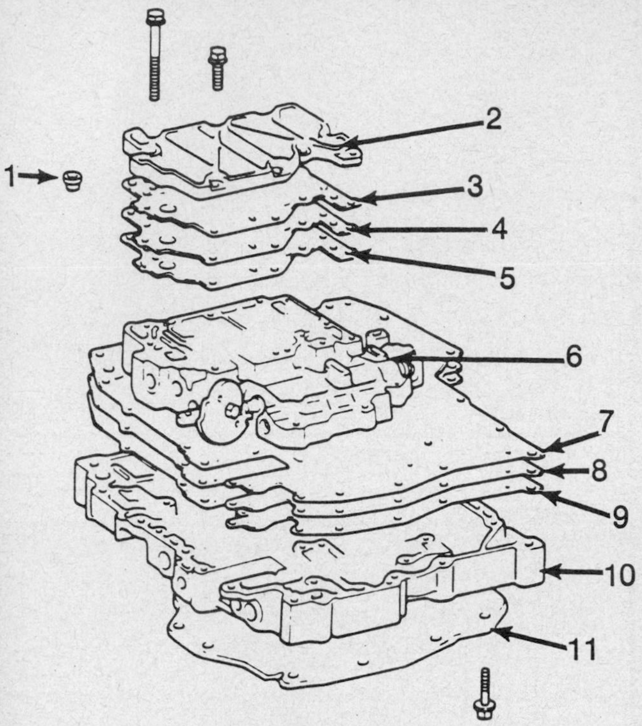
1
2
3
4
5
6
7
8
9
10
11

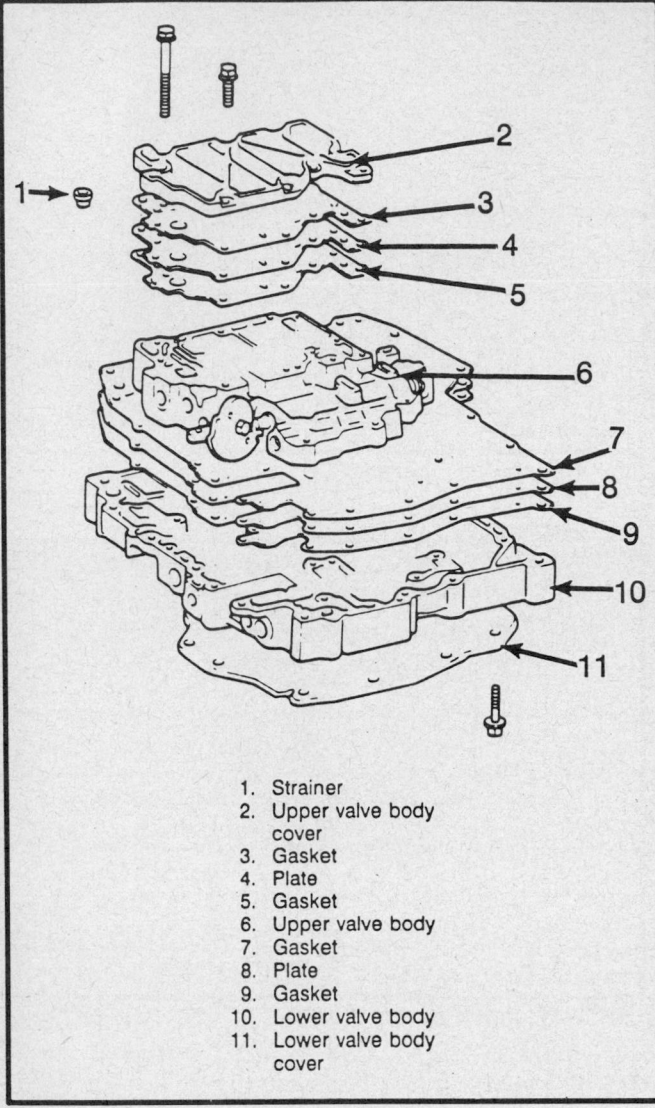

1. Strainer
2. Upper valve body cover
3. Gasket
4. Plate
5. Gasket
6. Upper valve body
7. Gasket
8. Plate
9. Gasket
10. Lower valve body
11. Lower valve body cover

Valve body

4. Remove the sleeve retainer with a magnetic finger and remove the sleeve.
5. Remove the cut-back valve by removing the retainer with a magnetic finger then remove the cut-back valve and plug.
6. Remove the throttle modulator valve by removing the retainer with a magnetic finger, then remove the plug, valve and spring.
7. Remove the accumulator control valve by removing the retainer with a magnetic finger and remove the plug, valve and spring.
8. Remove the low coast modulator valve by removing the pin with a magnetic finger then remove the plug, valve and spring.
9. Remove the 2nd coast modulator valve by removing the retainer with a magnetic finger then remove the spring and valve.
10. Remove the throttle cam. Loosen the bolt and remove the cam, spring and collar.
11. Remove the kickdown valve and spring by removing the pin with a magnetic finger, then remove the kickdown valve with the sleeve and spring.
12. Remove the throttle valve.
13. Remove the spring and adjusting rings.

Inspection

Inspect the valve springs. Check for damage, squareness, rust and collapsed coils. Measure the free length and replace any springs if necessary.

Assembly

1. Install the lockup relay valve sleeve into the bore.
2. Coat the retainer with petroleum jelly and install it into the end of the sleeve.
3. Install the control valve, spring and lockup relay valve into the bore in horizontal position.
4. Push in the relay valve by hand until the control valve touches the end of the sleeve. Install the plug and retainer.
5. Install the cut-back valve by installing the small end first. Install the plug and retainer.
6. Install the throttle modulator valve. Install the spring and valve. Install the plug and retainer.
7. Install the accumulator control valve. Install the valve and spring. Install the plug and retainer.
8. Install the low coast modulator valve. Install the valve and spring. Install the plug, thick end first. Install the pin.
9. Install the throttle valve and retainer. Install the throttle valve. Coat the retainer with petroleum jelly and install it into place in the valve body.
10. Install the adjusting rings and spring on the throttle valve shaft.
11. Install the spring into the throttle valve.
12. Install the kickdown valve and sleeve. Install the pin to hold the sleeve in place.
13. Install the 2nd coast modulator valve. Install the valve and spring.
14. Assemble the throttle cam. Insert the sleeve through a side of the cam. Install the spring with the hook through the hole in the cam.
15. Install the cam assembly on the upper valve body. Make sure that the cam moves on the roller of the kickdown valve.
16. Install the check ball.

LOWER VALVE BODY

Disassembly

1. Remove the lower valve body plate and gaskets.
2. Remove the cooler bypass valve and spring.
3. Remove the damping check valve and spring.
4. Remove the 3 check balls.
5. Remove the primary regulator valve.
6. Remove the secondary regulator valve.
7. Remove the 1–2 shift valve. Remove the 1–2 shift lower valve and spring.
8. Remove the low coast shift valve.
9. Remove the lockup control valve.
10. Remove the detent regulator valve.
11. Remove the 2–3 shift valve.
12. Remove the intermediate shift valve.
13. Remove the lockup signal valve.
14. Remove the 3–4 coast shift plug.
15. Remove the 3–4 shift plug.

Inspection

Inspect the valve springs. Check for damage, squareness, rust and collapsed coils. Measure the free length and replace any springs if necessary.

Assembly

1. Place the primary regulator valve into the bore in the horizontal position.
2. Push the valve into the bore until its tip bottoms in the bore.

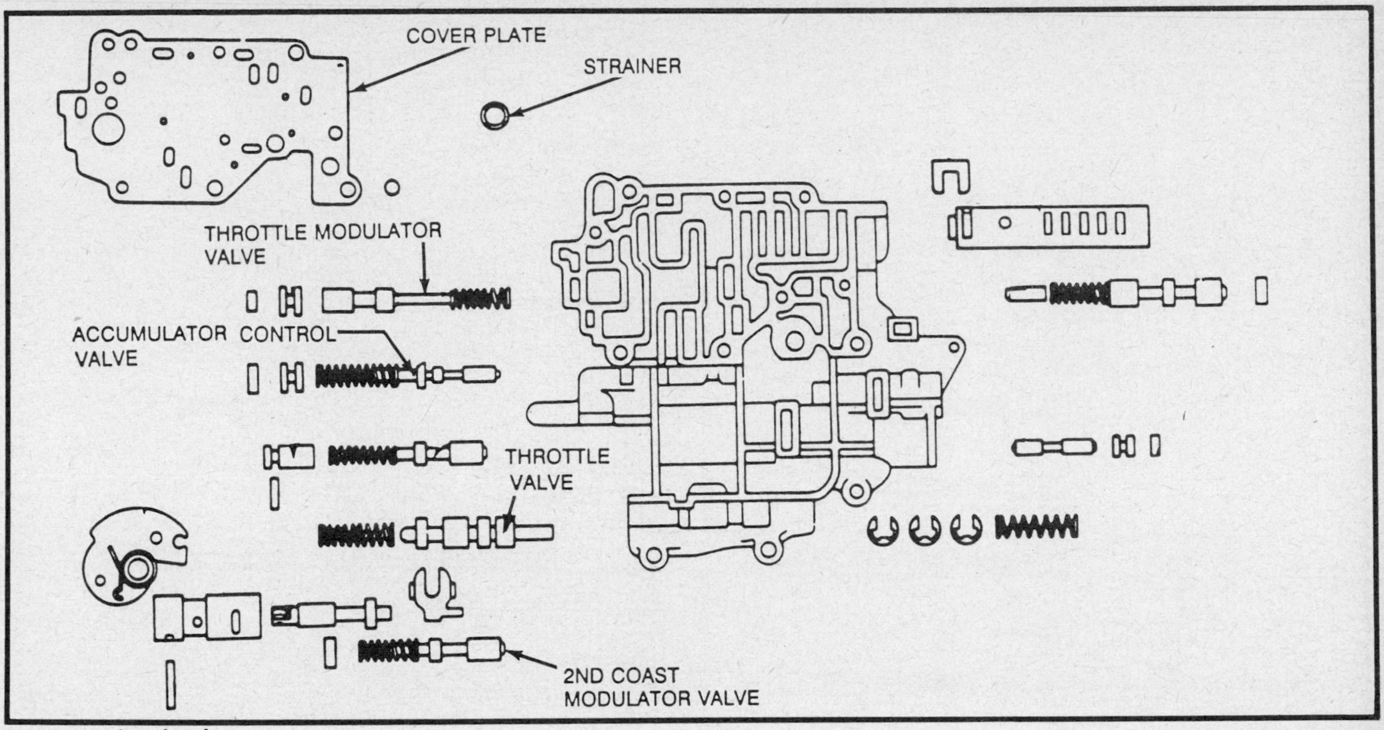
COVER PLATE
STRAINER
THROTTLE MODULATOR VALVE
ACCUMULATOR CONTROL VALVE
THROTTLE VALVE
2ND COAST MODULATOR VALVE

Upper valve body

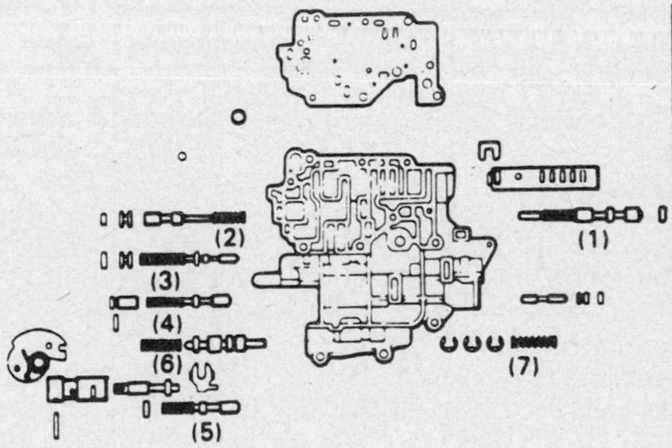
(2)
(3)
(4)
(6)
(5)
(1)
(7)

Spring	Free Length mm	Free Length (in.)	Color
(1) Lock-up relay valve	26.56	(1.0457)	White
(2) Throttle modulator valve	21.70	(0.8543)	None
(3) Accumulator control valve	33.20	(1.3071)	None
(4) Low coast modulator valve	23.40	(0.9213)	Red
(5) 2nd coast modulator valve	20.93	(0.8240)	Yellow-Green
(6) Kick-down valve	29.76	(1.1717)	White
(7) Throttle valve	30.70	(1.2087)	None

Valve spring chart

3. Install the valve spring.
4. Insert the plunger with the short end first. The plunger should be recessed inside the sleeve.
5. Install the sleeve with the plunger.
6. Install the secondary regulator valve.
7. Install the 1-2 shift valve. Install the spring and upper valve.
8. Install the low coast shift valve.
9. Install the lockup control valve.
10. Install the detent regulator valve.
11. Install the intermediate shift valve.
12. Install the 2-3 shift valve.
13. Install the lock-up signal valve.
14. Install the 3-4 coast shift plug.
15. Install the 3-4 shift plug.
16. Install the spring and cooler by-pass valve.
17. Install the spring and damping check valve.
18. Install the check balls.

VALVE BODY

Assembly

Install the lower valve body on the upper valve body together with the plate.

1. Position the new gaskets and the plate on the lower valve body. Assemble the gasket having the larger cooler by-pass hole to the lower valve body.
2. Install a new gasket and plate onto the lower valve body. Install another new gasket onto the plate. Align each bolt hole in the valve body with the 2 gaskets and plate.
3. Position the lower valve body and gaskets with the plate on

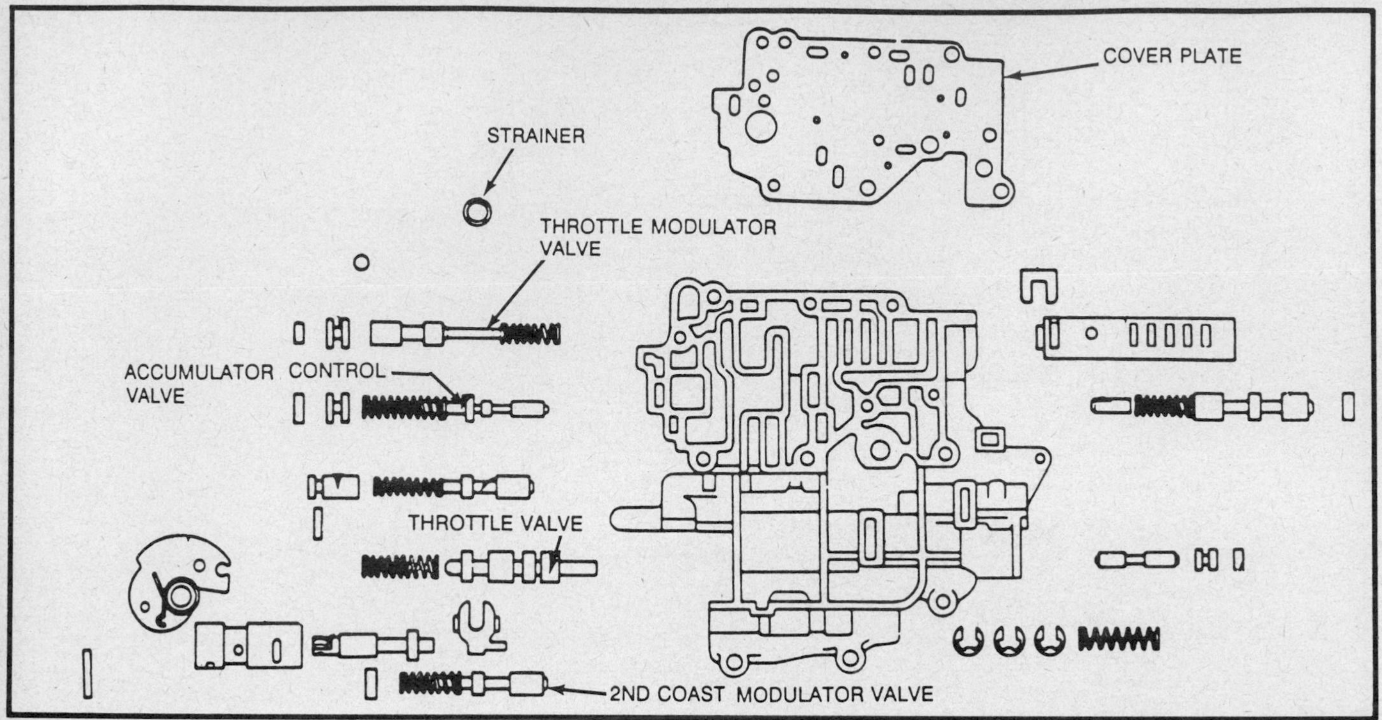
COVER PLATE
STRAINER
THROTTLE MODULATOR VALVE
ACCUMULATOR CONTROL VALVE
THROTTLE VALVE
2ND COAST MODULATOR VALVE

Upper valve body

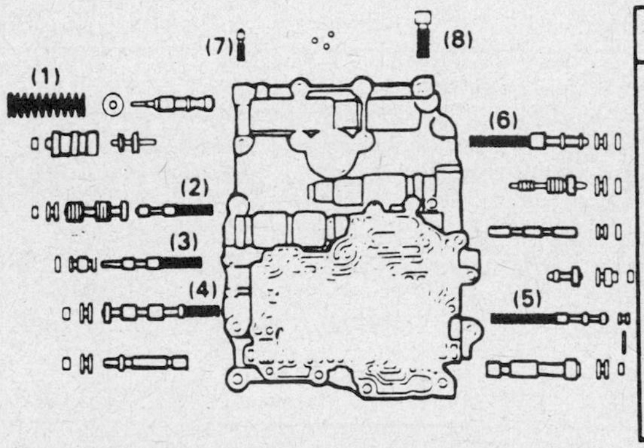
(1)
(2)
(3)
(4)
(5)
(6)
(7)
(8)

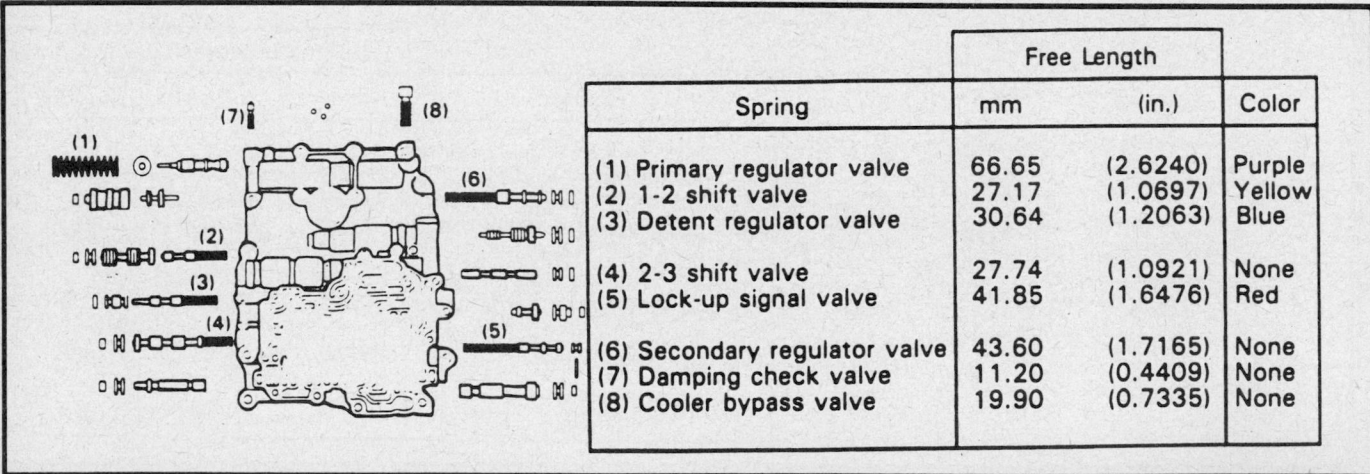

	Free Length		
Spring	mm	(in.)	Color
(1) Primary regulator valve	66.65	(2.6240)	Purple
(2) 1-2 shift valve	27.17	(1.0697)	Yellow
(3) Detent regulator valve	30.64	(1.2063)	Blue
(4) 2-3 shift valve	27.74	(1.0921)	None
(5) Lock-up signal valve	41.85	(1.6476)	Red
(6) Secondary regulator valve	43.60	(1.7165)	None
(7) Damping check valve	11.20	(0.4409)	None
(8) Cooler bypass valve	19.90	(0.7335)	None

Spring chart

top of the upper valve body. Align each bolt hole in the valve bodies with the gaskets and plate.

4. Install and finger tighten 3 bolts in the lower valve body to secure the upper valve body.

5. Turn the assembly over and finger tighten 3 bolts in the upper valve body.

6. Install the upper valve body cover new gaskets and strainer. Install the valve body cover and finger tighten the bolts.

7. Turn the assembly over and install a new gasket and the lower valve body cover.

8. Tighten the bolts of the upper and lower valve body to 48 inch lbs. (5.4 Nm).

Transaxle Assembly

Use the following guidelines before assembling the transaxle:

1. Before assembling new clutch discs, soak them in automatic transmission fluid for at least 2 hours.

2. Apply automatic transmission fluid to the sliding or rotating surfaces of parts before assembly.

3. Use petroleum jelly to keep small parts in their places.

4. Do not use adhesive cements on gaskets and similar parts.

5. When assembling the transaxle, be sure to use new gaskets and O-rings.

6. Dry all parts by blowing with compressed air. Never use the shop rags.

7. Be sure to install the thrust bearings and races in the correct direction and position.

Assembly

1. Install the parking pawl onto the case. Hook the spring ends to the case and pawl.

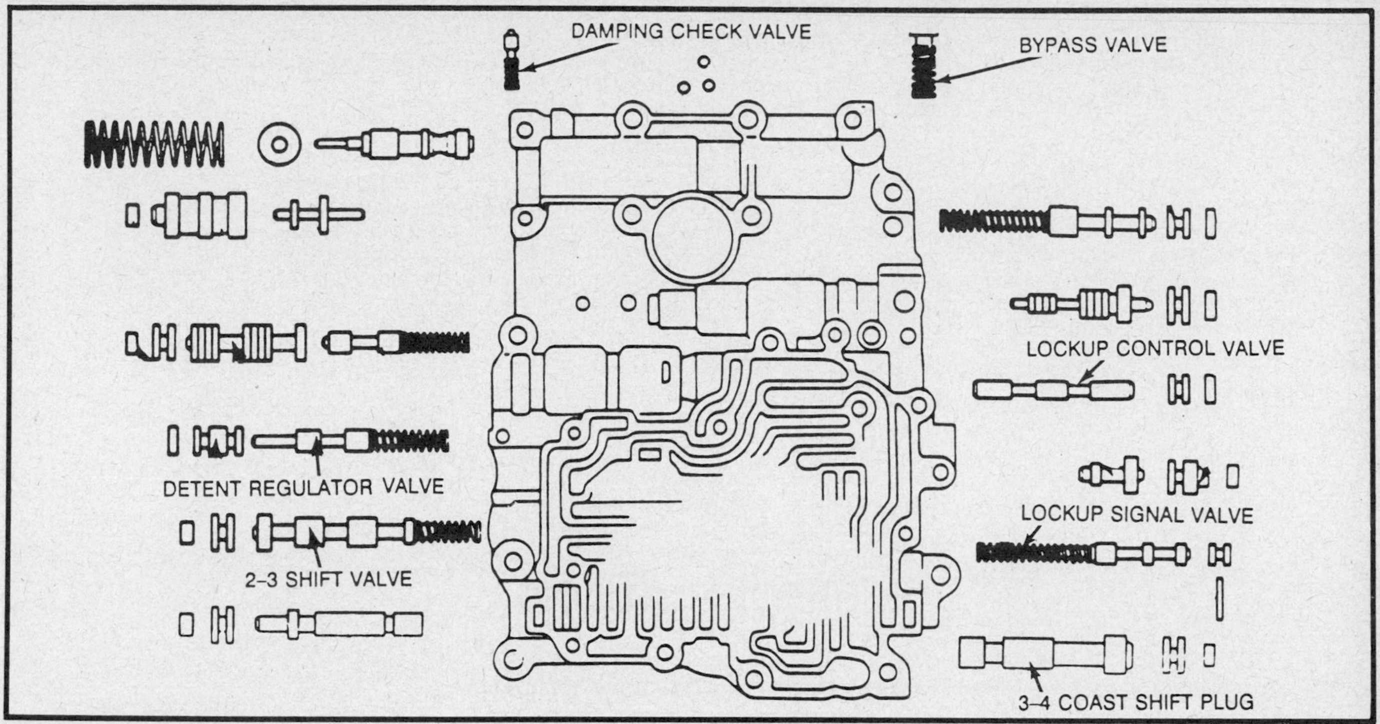
DAMPING CHECK VALVE
BYPASS VALVE
LOCKUP CONTROL VALVE
DETENT REGULATOR VALVE
LOCKUP SIGNAL VALVE
2–3 SHIFT VALVE
3–4 COAST SHIFT PLUG

Lower valve body

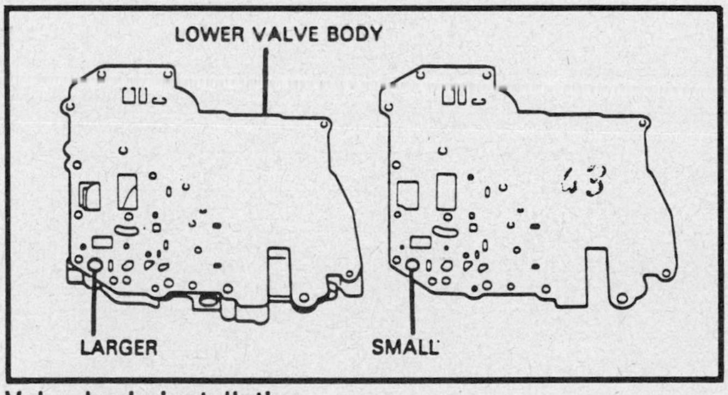
LOWER VALVE BODY
LARGER
SMALL

Valve body installation

2. Install the parking lock rod.
3. Install the parking pawl bracket.
4. Check the operation of the parking lock pawl. Make sure the counter driven gear is locked when the manual valve lever is in the **P** range.
5. Install the intermediate shaft.
6. Install the rear cover over a new gasket and torque the bolts to 18 ft. lbs. (25 Nm).
7. Check the intermediate shaft endplay. Make sure that the intermediate shaft turns smoothly.
8. Install the 1st and reverse brake in the case. Install the inner flange, facing the flat end toward the oil pump side. Install in order, 5 discs and 6 plates as follows: disc-plate-disc-plate-plate-disc-plate-plate-disc-plate-disc.
9. Install the outer flange, facing the flat end toward the piston side. Install the snapring with the endgap into the groove.
10. Check the operation of the 1st and reverse brake, apply compressed air into the oil passage in the case to check if the piston moves.
11. Install the No. 2 planetary carrier thrust washer.
12. Install the ring gear into the case.
13. Align the spline of the planetary carrier with flukes of the discs and install the planetary gear into the 1st and reverse brake discs.
14. Install the No. 2 one-way clutch into the case with the shiny side upward. Install the one-way clutch onto the inner race while turning the planetary gear clockwise.
15. Turn the planetary carrier. The carrier should turn freely clockwise and should lock counterclockwise.
16. Install the snapring endgap into the groove.
17. Install the 2nd coast brake band guide with its tip touching the case.
18. Install the 2nd brake into the case, install the flange, facing the flat end toward you.
19. Install in order: disc-plate-plate-disc-plate-plate.
20. Install the piston return spring assembly. Install each of the springs over the protrusions in the case.
21. Install the 2nd brake drum into the case. Align the groove of the drum with the bolt and place it into the case.
22. Install the snapring into the case so that the endgap is installed into the groove.
23. Check the operation of the 2nd brake. Apply compressed air into the oil passage of the case to be sure that the piston moves.
24. Install the 2nd brake drum gasket until it makes contact with the 2nd brake drum.
25. Install the No. 1 one-way clutch and 2nd brake hub. Align the flukes of the discs in the 2nd brake. Align the spine of the hub with the flukes of the discs and install the hub to the 2nd brake discs.
26. Install the sun gear and the sun gear input drum. While turning the sun gear clockwise, install it into the one-way clutch.
27. Install the front planetary gear and ring.
28. Install the planetary gear into the case.
29. Install the ring gear.
30. Install the 2nd coast brake band by installing the pin through the oil pump mounting bolt hole.
31. Install the forward clutch into the case.

32. Align the flukes of the disc in the forward clutch. Hold the sun gear input drum and rotate the input shaft to mesh the hub with the clutch discs of the forward clutch.
33. Align the center of the input shaft and intermediate shaft and while pushing on the input shaft, rotate it to mesh the hub and disc.
34. Install the direct clutch into the case.
35. Install the oil pump into the case. Install a new O-ring around the pump body. Torque the bolts to 16 ft. lbs. (22 Nm).
36. Make sure that the input shaft rotates freely.
37. Install the 2nd coast brake piston.
38. Install the accumulator pistons and springs.
39. Install the governor apply gasket and oil strainer.
40. Install the throttle cable in the case.
41. Install the valve body and torque the bolts to 7 ft. lbs. (10 Nm).
42. Install the manual valve body and torque the bolts to 7 ft. lbs. (10 Nm).
43. Install the detent spring.
44. Install the oil tubes, bracket and strainer.
45. Position the magnet in place.
46. Install the oil pan with a new gasket.
47. Install the governor body.
48. Install the throttle cable retaining plate.
49. Install the tube and filler gauge.
50. Install the neutral safety switch.
51. Install the manual shift lever.
52. Install the oil cooler pipes.

SPECIFICATIONS

VALVE BODY SPECIFICATIONS

Component	Free length		Coil outer diameter		No. coils	Wire diameter		Color
	in.	mm	in.	mm		in.	mm	
Upper valve body								
Throttle modulator valve	0.8543	21.70	0.3740	9.50	9.5	0.0354	0.90	None
Accumulator control valve	1.3071	33.20	0.4094	10.40	13.5	0.0394	1.00	None
Low coast modulator valve	0.9213	23.40	0.3110	7.90	11.5	0.0354	0.90	Red
Kickdown valve	1.1717	29.76	0.3437	8.73	13.5	0.0394	1.00	White
2nd coast modulator valve	0.8240	20.93	0.3346	8.50	10	0.0433	1.10	Light Green
Throttle valve	1.2087	30.70	0.3722	9.20	9.5	0.0276	0.70	None
Lock-up relay valve	1.0457	26.56	0.4016	10.20	11.5	0.0276	0.70	White
Lower valve body (USA)								
Primary regulator valve	2.6240	66.65	0.7323	18.60	12.5	0.0630	1.60	Purple
1-2 shift valve	1.0697	27.17	0.2516	6.39	15.5	0.0217	0.55	Yellow
Detent regulator valve	1.2063	30.64	0.3110	7.90	12.5	0.0354	0.90	Blue
2-3 shift valve	1.0921	27.74	0.3268	8.30	11	0.0236	0.60	None
Lockup signal valve	1.6476	41.85	0.3189	8.10	15.5	0.0276	0.70	Red
Secondary regulator valve	1.7165	43.60	0.4291	10.90	11.5	0.0551	1.40	None
Pressure relief valve	0.4409	11.20	0.2520	6.40	7.5	0.0354	0.90	None
Cooler by-pass valve	0.7835	19.90	0.4331	11.00	8.5	0.0394	1.00	None
Lower valve body (Canada)								
Primary regulator valve	2.6240	66.65	0.7323	18.60	12.5	0.0630	1.60	None
1-2 shift valve	1.0697	27.17	0.2516	6.39	15.5	0.0217	0.55	Yellow
Detent regulator valve	1.1701	29.72	0.3110	7.90	12.5	0.0354	0.90	Gray
2-3 shift valve	1.0921	27.74	0.3268	8.30	11	0.0236	0.60	Pink
Lockup signal valve	1.5858	40.28	0.3189	8.10	15.5	0.0276	0.70	Red
Secondary regulator valve	1.7165	43.60	0.4291	10.90	11.5	0.0551	1.40	None
Pressure relief valve	0.4409	11.20	0.2520	6.40	7.5	0.0354	0.90	None
Cooler by-pass valve	0.7835	19.90	0.4331	11.00	8.5	0.0394	1.00	None

BUSHING SPECIFICATIONS

Bushing		Finished bore in.	Finished bore mm	Bore Limit in.	Bore Limit mm
Stator support	Front	0.8465–0.8475	21.500–21.526	0.8494	21.576
	Rear	1.0630–1.0640	27.000–27.026	1.0660	27.076
Oil pump body		1.5005–1.5015	38.113–38.138	1.5035	38.188
Direct clutch drum		1.8504–1.8514	47.000–47.025	1.8533	47.075
Front planetary ring gear flange		0.7490–0.7500	19.025–19.050	0.7520	19.100
Input sun gear	Front and Rear	0.8671–0.8680	22.025–22.046	0.8699	22.096

ACCUMULATOR PISTON SPRING SPECIFICATIONS

Component	Free length in	Free length mm	Coil outer diameter in	Coil outer diameter mm	No. coils	Wire diameter in	Wire diameter mm	Color
B^2 (Center)	2.6252	66.68	0.6441	16.36	16.5	0.1024	2.60	Purple
C^1 (Transaxle rear cover side)	1.8898	48.00	0.5346	13.58	10.5	0.0906	2.30	Red
	3.1925	81.09	0.7323	18.60	17	0.0866	2.20	Yellow Green
C^2 (Torque converter side)	2.8417	72.18	0.6921	17.58	16.5	0.0906	2.30	Yellow

TORQUE SPECIFICATIONS

Part		ft. lbs.	Nm
Engine mounting		38	52
Transaxle case to Engine	12 mm	47	64
Transaxle case to Engine	10 mm	34	46
Drive plate to Crankshaft		47	64
Torque converter to Drive plate		20	27
Oil pump to Transaxle case		16	27
Oil pump body to Stator shaft		7	10
Second coast brake band guide		48 ①	5.4
Upper valve body to Lower valve body		48 ①	5.4
Valve body		7	10
Accumulator cover		7	10
Oil strainer		7	10
Oil pan		43 ①	4.9
Oil pan drain plug		36	49
Cooler pipe union nut		25	34

TORQUE SPECIFICATIONS

Part	ft. lbs.	Nm
Testing plug	65 ①	7.4
Parking lock pawl bracket	65 ①	7.4
Transaxle rear cover to Transaxle case	18	25
Neutral safety switch to Transaxle case (bolt)	48 ①	5.4
Neutral safety switch (nut)	61 ①	6.9

① Inch lbs.

OIL PUMP SPECIFICATIONS

		in.	mm
Side clearance	Std.	0.0008–0.0020	0.02–0.05
	Limit	0.004	0.1
Body clearance	Std.	0.0028–0.0059	0.07–0.15
	Limit	0.012	0.3
Tip clearance Driven gear	Std.	0.0043–0.0055	0.11–0.14
	Limit	0.012	0.3

SPECIAL TOOLS

Tool	Description
J-9617	Oil pump seal installer
J-29182	Pinion shaft bearing cup installer
J-35378	Bearing puller pilot
J-35399	Differential side bearing cup remover
J-35400	Differential side bearing housing support
J-35405	Differential preload wrench
J-35409	Differential side bearing installer
J-35455	Holding fixture
J-35467	One-way clutch tester
J-35495	Oil pump puller adapters
J-35549	Second coast brake piston compressor
J-35552	Differential side bearing cup installer
J-35553	Differential side bearing seal installer
J-35565	Intermediate shaft bearing installer
J-35661	Countergear bearing installer
J-35663	Countergear bearing cup installer
J-35664	Pinion shaft bearing installer

Tool	Description
J-35666	Pinion shaft bearing seal installer
J-35679	Band apply pin gauge
J-35683	First/reverse clutch spring compressor adapter
J-35752	Pressure gauge adapter
J-1859-03	Countergear puller
J-3289-20	Holding fixture mount
J-8001	Dial indicator set
J-8092	Driver handle
J-8614-01	Countergear holder
J-21867	Pressure gauge
J-22888	Right differential side bearing puller
J-22912-01	Bearing puller
J-23327	Clutch spring compressor
J-25018-A	Clutch spring compressor adapter
J-25025-1	Dial indicator mount
J-33411	Countergear installer
J-6125-B	Slide hammer set

Section 3

A240E Transaxle
General Motors

APPLICATION

1988–89 Nova

GENERAL DESCRIPTION

The A240E 4 speed transaxle, also called Electronic Control Transaxle (ECT), differs from the oil pressure control type transaxle in that it is controlled by a microcomputer. Trouble occurring in the ECT can stem from 1 of 3 sources: the engine, the ECT electronic control unit or the transaxle itself.

One of 2 driving modes (normal mode or power mode) can be selected in accordance with the driver's preference. If a malfunction occurs in the electrical circuit (solenoid valve or speed sensor system), this transaxle has a self diagnosis function which causes the **OD OFF** indicator light to flash on and off only when in the **OD** mode. Should the system malfunction, the Electronic Control Module (ECM) can control the transaxle to select as gear that will allow the vehicle to operate safely.

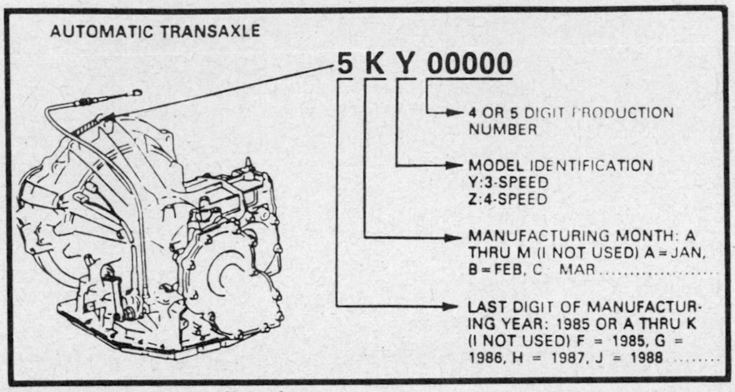
AUTOMATIC TRANSAXLE
5 K Y 00000
4 OR 5 DIGIT PRODUCTION NUMBER
MODEL IDENTIFICATION
Y:3-SPEED
Z:4-SPEED
MANUFACTURING MONTH: A THRU M (I NOT USED) A=JAN, B=FEB, C MAR
LAST DIGIT OF MANUFACTURING YEAR: 1985 OR A THRU K (I NOT USED) F = 1985, G = 1986, H = 1987, J = 1988

Transaxle identification

Transaxle and Converter Identification

TRANSAXLE

The transaxle identification tag is located at the top front of the transaxle case.

CONVERTER

The torque converter is a welded unit and cannot be disassembled for repairs. Should this unit need to be replaced, both new and rebuilt units are available.

Electronic Controls

The transaxle, being fully controlled by the computer system, is equipped with solenoids; it responds to road speed and engine torque demand.

The internal equipment, attached to the main control assembly, consists of the direct clutch and 2nd brake solenoids. A bulkhead connector/wiring assembly provides an electrical path to the computer system.

If diagnostic code Nos. 42, 61, 62 or 63 occur, the overdrive indicator light will begin to blink immediately to warn the driver. An impact or shock may cause the blinking to stop, but the code will still be retained in the ECT computer memory until cancelled out. There is no warning for diagnostic code No.64.

In the event of a simultaneous malfunction of both No.1 and No.2 speed sensors, no diagnostic code will appear and the failsafe system will not function. However, when driving in the **D** range, the transaxle will not upshift from the 1st gear, regardless of the vehicle speed.

The electronic control components include the following:

- ECT computer
- Torque converter clutch solenoid
- Neutral safety switch
- Throttle position sensor
- No. 2 speed sensor
- No. 1 speed sensor in combination meter
- No. 1 and No. 2 solenoids
- Pattern selection switch
- OD switch
- OD OFF indicator
- Stop light switch
- Lockup solenoid

The lockup clutch will turn on only infrequently during normal 2nd and 3rd gear operation. To trigger this action, press the accelerator pedal halfway down.

Reading Diagnostic Codes

1. Turn the ignition switch and the **OD** switch to **ON**.

NOTE: Do not start the engine. The warning and diagnostic codes can be read only when the overdrive switch on the shift lever is ON. If OFF, the overdrive light will light continuously and will not blink.

2. Short the service connector terminal to the body ground, using a service wire.
3. Read the diagnostic code as indicated by the number of times the **OD OFF** light flashes.
   a. If the system is operating normally, the light will blink for 0.25 seconds every 0.50 seconds.
   b. In the event of a malfunction, the light will blink once every 0.5 second. The number of blinks will equal the 1st number and after a 1.5 second pause, the second number of the 2 digit diagnostic code. If there are 2 or more codes, there will be a 2.5 second pause between each.
   c. In the event of several trouble codes occurring simultaneously, indication will begin from the smaller value and continue to the larger.
   d. If codes 62, 63 and 64 appear, there is an electrical malfunction in the solenoid.
   e. Causes due to mechanical failure, such as a stuck switch, will not appear.

Cancelling Diagnostic Codes

After repair in the trouble area, the diagnostic code retained in memory by the ECT computer must be canceled by removing the fuse marked **STOP** for 10 seconds or more, depending on the ambient temperature (the lower the temperature, the longer the fuse must be left out), with the ignition switch off.

Cancellation can also be done by removing the negative battery terminal, but in this case other memory systems (clock, radio, etc.) will also be canceled out. The diagnostic code can also be canceled out by disconnecting the ECT computer connector.

If the diagnostic code is not canceled out, it will be retained by the ECT computer and appear along with a new code in the event of future trouble.

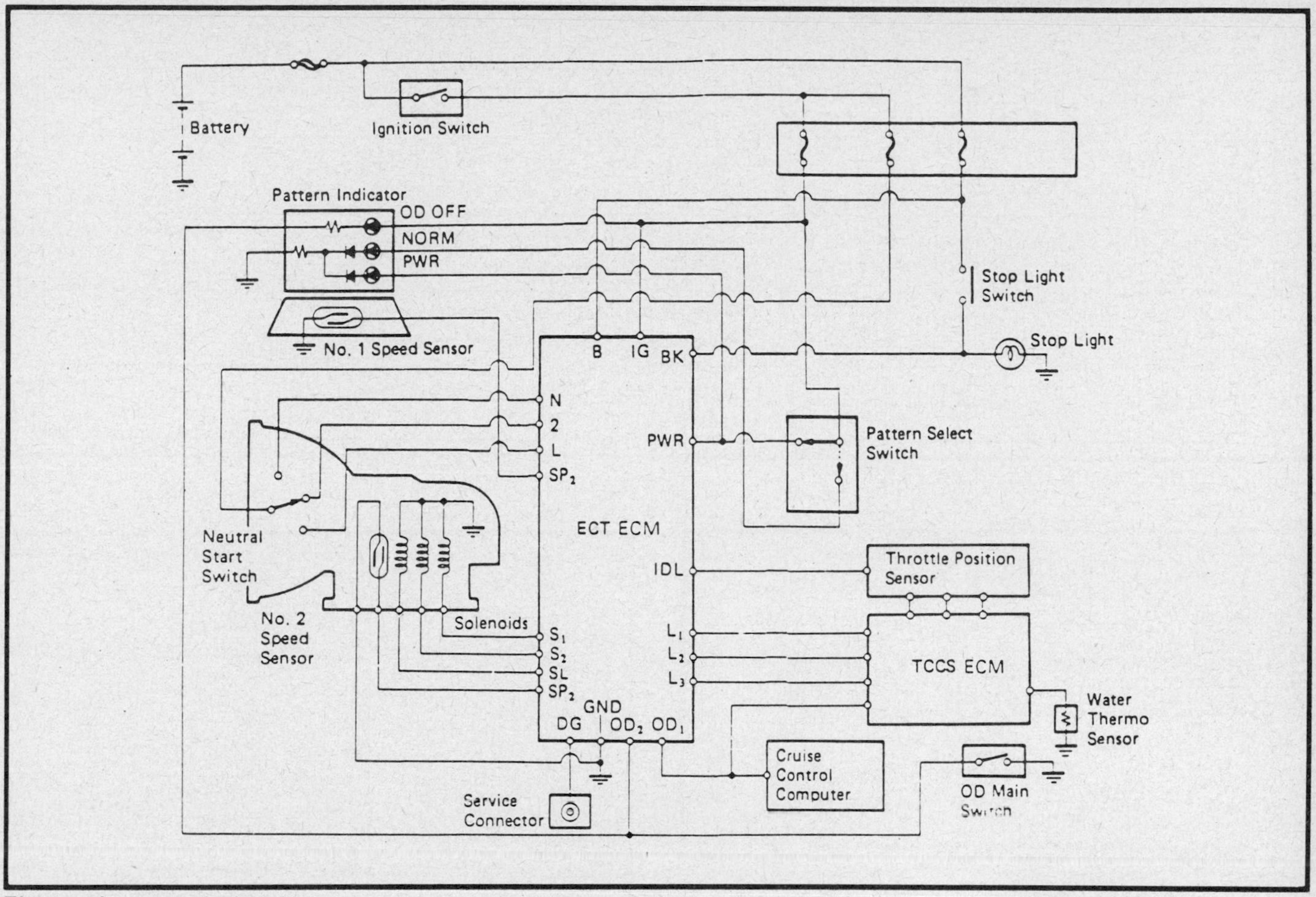
Battery
Ignition Switch
Pattern Indicator
OD OFF
NORM
PWR
No. 1 Speed Sensor
Stop Light Switch
Stop Light
B
IG
BK
N
2
L
SP2
PWR
Pattern Select Switch
Neutral Start Switch
ECT ECM
IDL
Throttle Position Sensor
No. 2 Speed Sensor
Solenoids
S1
S2
SL
SP2
L1
L2
L3
TCCS ECM
Water Thermo Sensor
GND
DG
OD2
OD1
Cruise Control Computer
OD Main Switch
Service Connector

Electronic control circuit

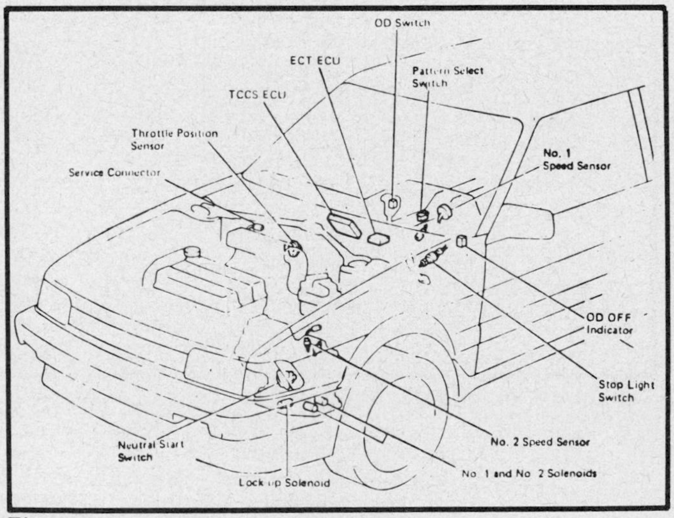
OD Switch
ECT ECU
Pattern Select Switch
TCCS ECU
Throttle Position Sensor
Service Connector
No. 1 Speed Sensor
OD OFF Indicator
Stop Light Switch
No. 2 Speed Sensor
Neutral Start Switch
Lock up Solenoid
No. 1 and No. 2 Solenoids

Electronic control components

ECT
NORMAL
POWER
O/D OFF

Reading diagnostic codes

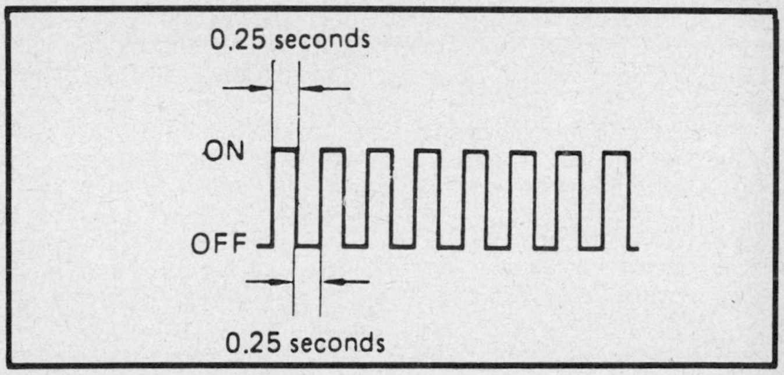
0.25 seconds
ON
OFF
0.25 seconds

System normal code

Metric Fasteners

The metric fastener dimensions are very close to the dimensions of the familiar inch system fasteners and for this reason, replacement fasteners must have the same measurement and strength as those removed.

Code No.	Light Pattern	Diagnosis System
42		Defective No. 1 speed sensor (in combination meter) – severed wire harness or short circuit
61		Defective No. 2 speed sensor (in ATM) – severed wire harness or short circuit
62		Severed No. 1 solenoid or short circuit – severed wire harness or short circuit
63	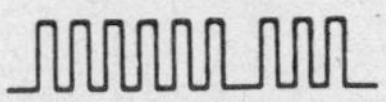	Severed No. 2 solenoid or short circuit – severed wire harness or short circuit
64		Severed No. 3 solenoid or short circuit – severed wire harness or short circuit

Identifying diagnostic code light patterns

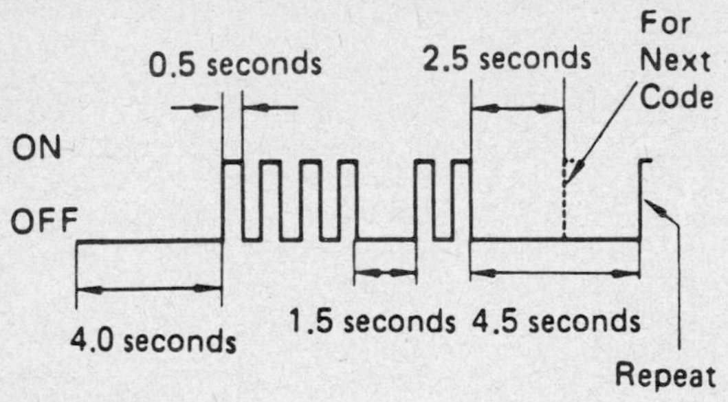
0.5 seconds
2.5 seconds
For
Next
Code
ON
OFF
4.0 seconds
1.5 seconds
4.5 seconds
Repeat

System malfunction code

Do not attempt to interchange metric fasteners for inch system fasteners. Mismatched or incorrect fasteners can result in damage to the transaxle unit through malfunctions, breakage or possible personal injury.

Capacities

The A240E automatic transaxle has a fluid capacity of 7.6 qts. (7.2 L) including the torque converter. The oil pan capacity for a drain and refill is 3.3 quarts (3.1 liters). This transaxle uses Dexron® II type automatic transmission fluid.

Checking Fluid Level

The vehicle must have been driven so that the engine and transaxle are at normal operating temperature (fluid temperature 158–176° F). If the fluid smells burnt or is black, replace it.

1. Park the vehicle on a level surface and set the parking brake.
2. With the engine idling, shift the selector into each gear from **P** range to **L** range and return to **P** range.
3. Pull out the transaxle dipstick and wipe it clean.

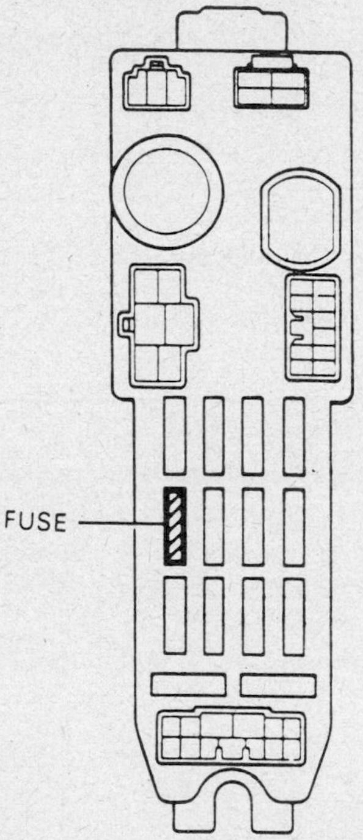
FUSE

Cancel the diagnostic code

4. Push the dipstick back fully into the tube.
5. Pull out the dipstick and check that the fluid is in the **HOT** range.
6. If the level is low, add fluid. Do not overfill.

TRANSAXLE MODIFICATIONS

There are no transaxle modifications available at the time of publication.

TROUBLE DIAGNOSIS

Hydraulic Control System

The hydraulic pressure system requires a supply of transmission fluid and a pump to pressurize the fluid. The A240E transaxle uses an internal-external gear type pump with its oil intake connected to a screen assembly.

The oil pump is designed to deliver fluid to the torque converter, lubricate the planetary gear unit and supply operating pressure to the hydraulic control system. The drive gear of the oil pump and the torque converter pump are driven by the engine. The pump has a sufficient capacity of oil to supply the necessary fluid pressure throughout all forward speed ranges and reverse.

MANUAL VALVE

The manual valve is linked to the gear shift lever and directs the fluid to the gear range circuit that the lever is positioned at.

PRIMARY REGULATOR VALVE

The primary regulator valve varies the hydraulic line pressure to each component in order to conform with engine power and operate all transaxle hydraulic systems.

SECONDARY REGULATOR VALVE

This valve regulates the converter pressure and lubrication pressure. Spring tension in the valve acts in an upward direction. Converter fluid pressure and lubrication pressure are determined by the spring tension.

THROTTLE VALVE

The throttle valve acts to produce throttle pressure in response to accelerator pedal modulation or engine output. When the accelerator pedal is depressed, the downshift plug is pushed upward by the throttle cable and throttle cam. The throttle valve also moves upward by means of the spring, opening the pressure passage for creation of throttle pressure.

CUT-BACK VALVE

This valve regulates the cut-back pressure acting on the throttle valve and is activated by governor pressure and throttle pressure. By applying cut-back pressure to the throttle valve in this manner, the throttle pressure is lowered to prevent unnecessary power loss from the oil pump.

Governor pressure acts on the upper portion of the valve and as the valve is pushed downward, a passage from the throttle valve is opened and throttle pressure is applied. Because of the difference in the diameters of the valve pistons, the valve is pushed upward to balance the downward force of governor pressure and the throttle pressure becomes cut-back pressure. As the governor pressure rises, the valve is forced downward. Since the throttle pressure passage is open, the pressure becomes cut-back pressure.

THROTTLE MODULATOR VALVE

This valve produces throttle modulator pressure. It reduces throttle pressure when the throttle valve opening angle is high. It causes throttle modulator pressure to act on the primary reg-

CLUTCH AND BAND APPLICATION

Gear position	C_1	C_2	C_3	B_1	B_2	B_3	B_4	F_1	F_2	F_3
P-Parking	—	—	—	—	—	—	Applied	—	—	—
R-Reverse	—	Applied	—	—	—	Applied	Applied	—	—	—
N-Neutral	—	—	—	—	—	—	Applied	—	—	—
0-1st	Applied	—	—	—	—	—	Applied	—	Applied	Applied
2nd	Applied	—	—	—	Applied	—	Applied	Applied	—	Applied
3rd	Applied	Applied	—	—	Applied	—	Applied	—	—	Applied
OD	Applied	Applied	Applied	—	Applied	—	—	—	—	—
2-1st	Applied	—	—	—	—	—	Applied	—	Applied	Applied
2nd	Applied	—	—	Applied	Applied	—	Applied	Applied	—	Applied
3rd	Applied	Applied	—	—	Applied	—	Applied	—	—	Applied
1-1st	Applied	—	—	—	—	Applied	Applied	—	Applied	Applied
2nd	Applied	—	—	Applied	Applied	—	Applied	Applied	—	Applied

CHILTON'S THREE C'S TRANSAXLE DIAGNOSIS

Condition	Cause	Correction
Fluid discolored or smells burnt	a) Fluid contaminated b) Torque converter faulty c) Transaxle faulty	a) Replace fluid b) Replace torque converter c) Disassemble and inspect
Vehicle does not move in any forward range or reverse	a) Control cable out of adjustment b) Valve body or primary regulator faulty c) Parking lock pawl faulty d) Torque converter faulty e) Oil pump intake screen blocked f) Transaxle faulty	a) Adjust control cable b) Inspect valve body c) Inspect parking lock pawl d) Replace torque converter e) Clean screen f) Disassemble and inspect transaxle
Shift lever position incorrect	a) Control cable out of adjustment b) Manual valve and lever faulty c) Transaxle faulty	a) Adjust control cable b) Inspect valve body c) Disassemble and inspect
Harsh engagement into any drive range	a) Throttle cable out of adjustment b) Valve body or primary regulator faulty c) Accumulator piston faulty d) Transaxle faulty	a) Adjust throttle cable b) Inspect valve body c) Inspect accumulator piston d) Disassemble and inspect
Delayed 1-2, 2-3 or 3-OD up-shift, or down-shifts from OD-3 or 3-2 and shifts back to 4 or 3 Slips on 1-2, 2-3 or 3-OD up-shift, or slips or shudders on acceleration	a) Throttle cable out of adjustment b) Valve body faulty c) Solenoid valve faulty d) Control cable out of adjustment e) Throttle cable out of adjustment f) Valve body faulty g) Solenoid valve faulty h) Transaxle faulty	a) Adjust throttle calbe b) Inspect valve body c) Inspect valve body d) Inspect governor e) Adjust control cable f) Adjust throttle cable g) Inspect valve body h) Inspect valve body
Vehicle does not hold in P range	a) Control cable out of adjustment b) Parking lock pawl cam and spring faulty	a) Adjust control cable b) Inspect cam and spring
Drag, binding or tie-up on 1-2, 2-3 or 3-OD up-shift	a) Control cable out of adjustment b) Valve body faulty c) Transaxle faulty	a) Adjust control cable b) Inspect valve body c) Disassemble and inspect
No lockup in 2nd, 3rd or OD	a) Electronic control faulty b) Valve body faulty c) Solenoid valve faulty d) Transaxle faulty	a) Inspect electronic control b) Inspect valve body c) Inspect valve body d) Disassemble and inspect transaxle
Harsh downshift	a) Throttle cable out of adjustment b) Valve body faulty c) Transaxle faulty	a) Adjust throttle cable b) Inspect valve body c) Disassemble and inspect transaxle
No downshift when coasting	a) Governor faulty b) Valve body faulty c) Solenoid valve faulty d) Electronic control faulty	a) Inspect governor b) Inspect valve body c) Inspect solenoid valve d) Inspect electronic control
Downshift occurs too quickly or too late while coasting	a) Throttle cable out of adjustment b) Valve body faulty c) Transaxle faulty d) Solenoid valve faulty e) Electronic control	a) Adjust throttle cable b) Inspect valve body c) Disassemble and inspect transaxle d) Inspect solenoid valve e) Inspect electronic control
No OD-3, 3-2 or 2-1 kickdown	a) Throttle cable out of adjustment b) Solenoid valve faulty c) Electronic control faulty d) Valve body faulty	a) Adjust throttle cable b) Inspect solenoid valve c) Inspect electronic control d) Inspect valve body
No engine braking in 2 or L range	a) Solenoid valve faulty b) Electronic control faulty c) Valve body faulty d) Transaxle faulty	a) Inspect solenoid valve b) Inspect electronic control c) Inspect valve body d) Disassemble and inspect transaxle

ECT ELECTRONIC CONTROL DIAGNOSIS

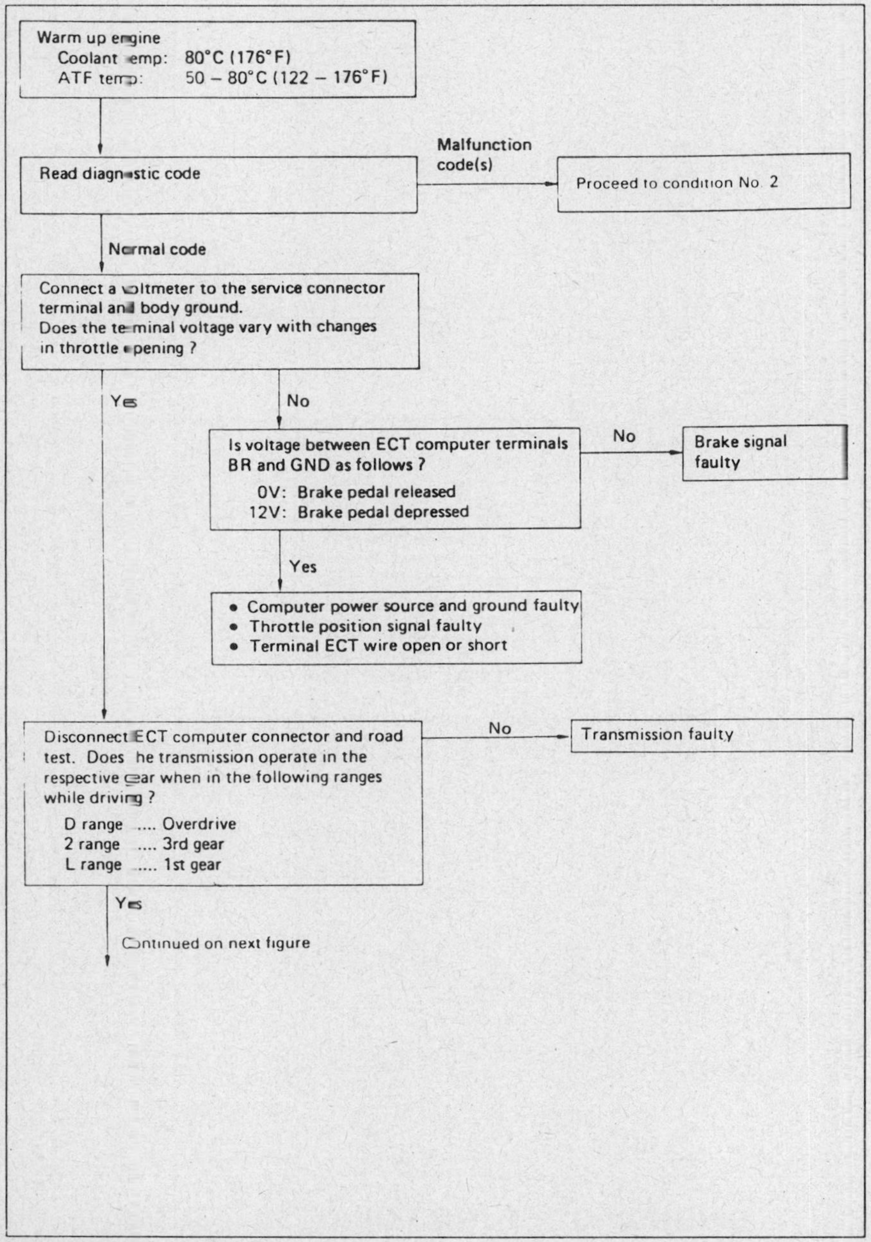
Warm up engine
Coolant temp: 80°C (176°F)
ATF temp: 50 – 80°C (122 – 176°F)
Read diagnostic code
Malfunction code(s)
Proceed to condition No. 2
Normal code
Connect a voltmeter to the service connector terminal and body ground. Does the terminal voltage vary with changes in throttle opening ?
Yes
No
Is voltage between ECT computer terminals BR and GND as follows ?
0V: Brake pedal released
12V: Brake pedal depressed
No
Brake signal faulty
Yes
• Computer power source and ground faulty
• Throttle position signal faulty
• Terminal ECT wire open or short
Disconnect ECT computer connector and road test. Does the transmission operate in the respective gear when in the following ranges while driving ?
D range .... Overdrive
2 range .... 3rd gear
L range .... 1st gear
No
Transmission faulty
Yes
Continued on next figure

Not shifting condition

ECT ELECTRONIC CONTROL DIAGNOSIS

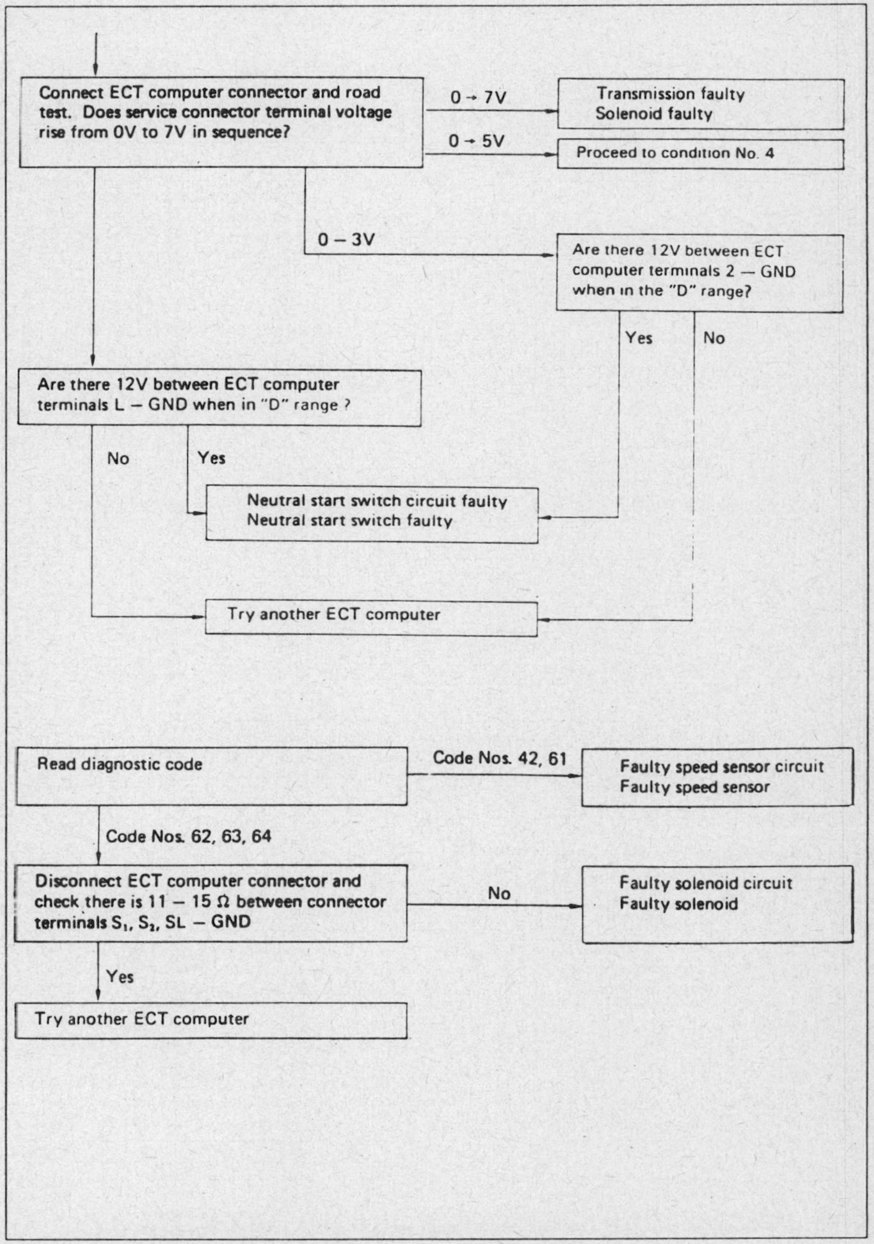
Connect ECT computer connector and road test. Does service connector terminal voltage rise from 0V to 7V in sequence?
0 → 7V
Transmission faulty
Solenoid faulty
0 → 5V
Proceed to condition No. 4
0 – 3V
Are there 12V between ECT computer terminals 2 — GND when in the "D" range?
Yes
No
Are there 12V between ECT computer terminals L – GND when in "D" range ?
No
Yes
Neutral start switch circuit faulty
Neutral start switch faulty
Try another ECT computer
Read diagnostic code
Code Nos. 42, 61
Faulty speed sensor circuit
Faulty speed sensor
Code Nos. 62, 63, 64
Disconnect ECT computer connector and check there is 11 – 15 Ω between connector terminals S1, S2, SL – GND
No
Faulty solenoid circuit
Faulty solenoid
Yes
Try another ECT computer

Blinking while driving, overdrive indicator lights

ECT ELECTRONIC CONTROL DIAGNOSIS

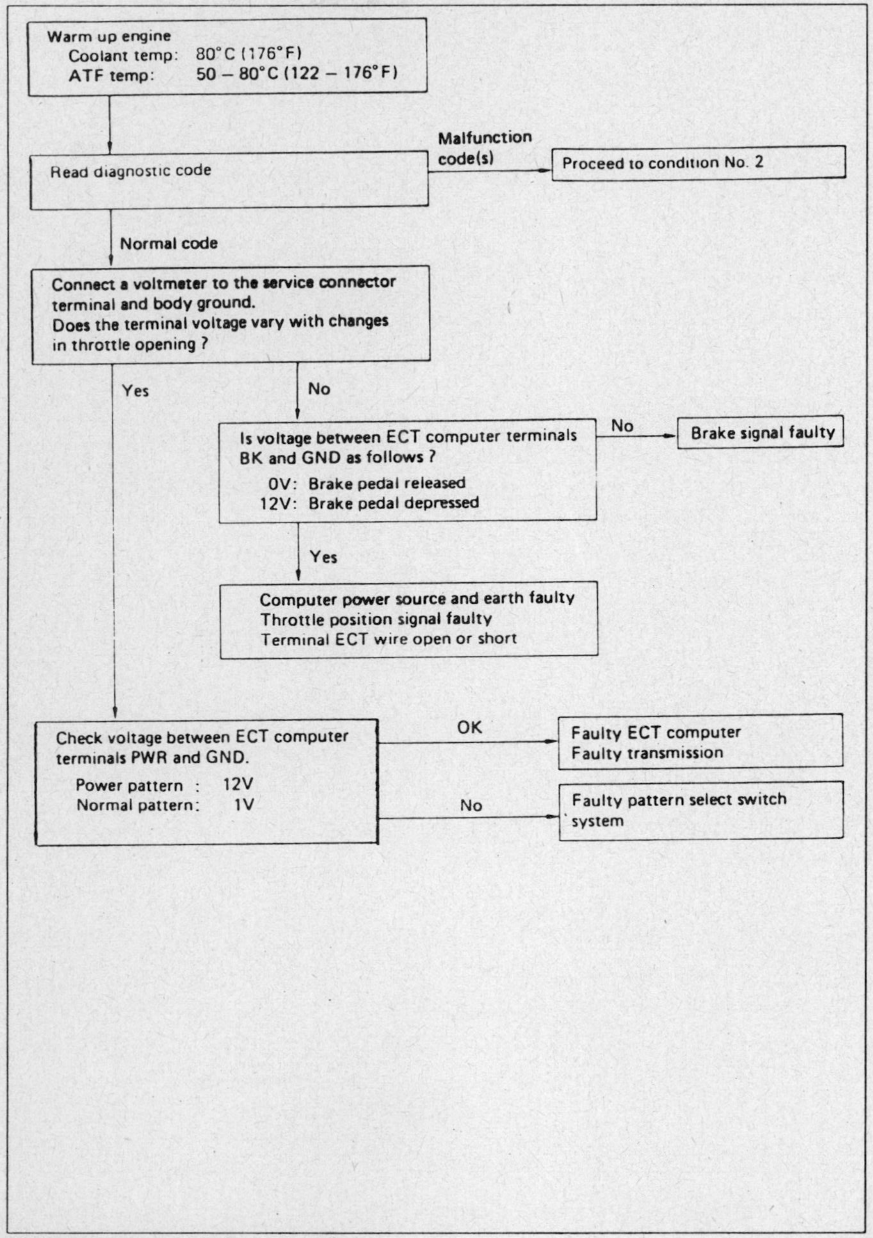
Warm up engine
Coolant temp: 80°C (176°F)
ATF temp: 50 – 80°C (122 – 176°F)
Read diagnostic code
Malfunction code(s)
Proceed to condition No. 2
Normal code
Connect a voltmeter to the service connector terminal and body ground.
Does the terminal voltage vary with changes in throttle opening ?
Yes
No
Is voltage between ECT computer terminals BK and GND as follows ?
0V: Brake pedal released
12V: Brake pedal depressed
No
Brake signal faulty
Yes
Computer power source and earth faulty
Throttle position signal faulty
Terminal ECT wire open or short
Check voltage between ECT computer terminals PWR and GND.
Power pattern : 12V
Normal pattern: 1V
OK
Faulty ECT computer
Faulty transmission
No
Faulty pattern select switch system

Shift point too high or too low

ECT ELECTRONIC CONTROL DIAGNOSIS

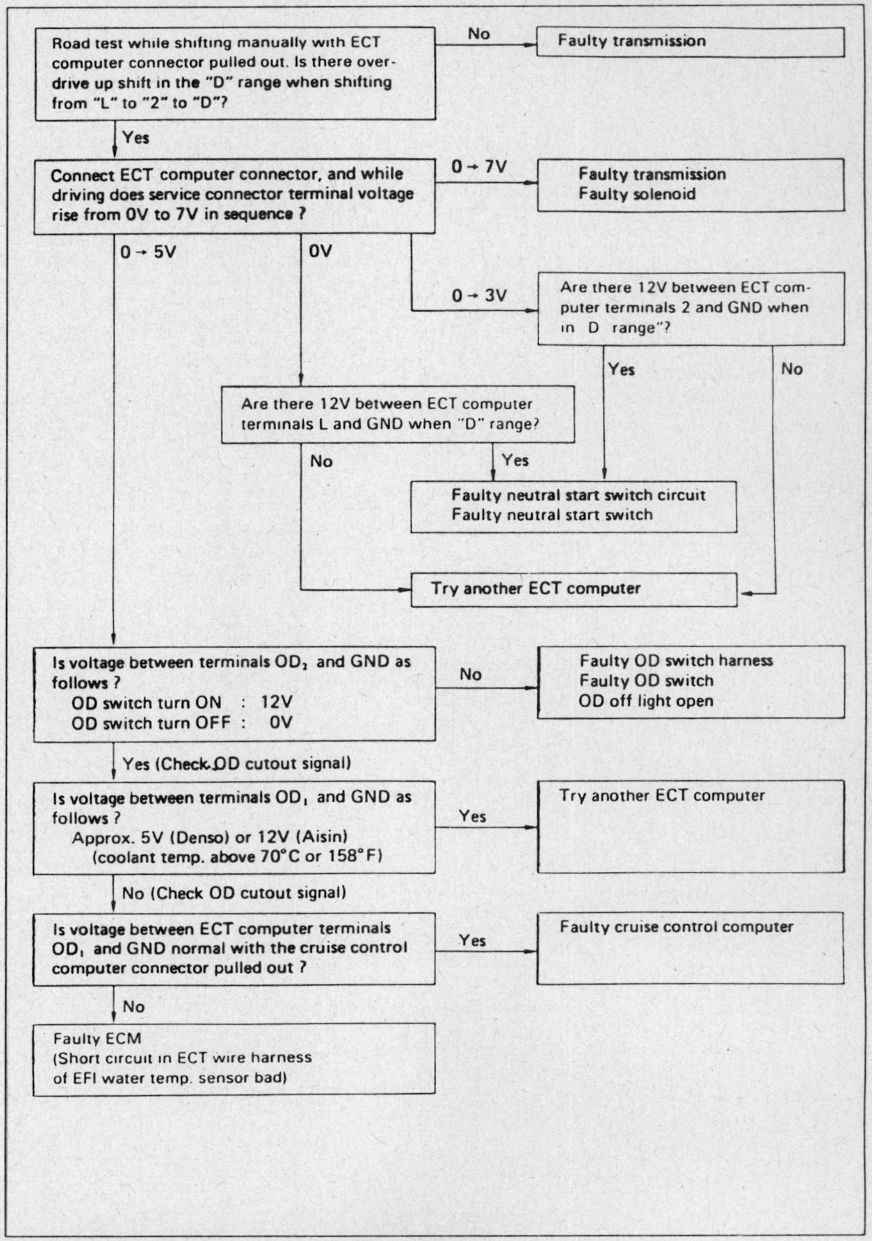
Road test while shifting manually with ECT computer connector pulled out. Is there over-drive up shift in the "D" range when shifting from "L" to "2" to "D"?
No
Faulty transmission
Yes
Connect ECT computer connector, and while driving does service connector terminal voltage rise from 0V to 7V in sequence ?
0 → 7V
Faulty transmission
Faulty solenoid
0 → 5V
0V
0 → 3V
Are there 12V between ECT computer terminals 2 and GND when in D range"?
Yes
No
Are there 12V between ECT computer terminals L and GND when "D" range?
No
Yes
Faulty neutral start switch circuit
Faulty neutral start switch
Try another ECT computer
Is voltage between terminals OD₂ and GND as follows ?
OD switch turn ON : 12V
OD switch turn OFF : 0V
No
Faulty OD switch harness
Faulty OD switch
OD off light open
Yes (Check OD cutout signal)
Is voltage between terminals OD₁ and GND as follows ?
Approx. 5V (Denso) or 12V (Aisin)
(coolant temp. above 70°C or 158°F)
Yes
Try another ECT computer
No (Check OD cutout signal)
Is voltage between ECT computer terminals OD₁ and GND normal with the cruise control computer connector pulled out ?
Yes
Faulty cruise control computer
No
Faulty ECM
(Short circuit in ECT wire harness of EFI water temp. sensor bad)

No upshift or overdrive after warm-up

ulator valve so that line pressure performance is close to engine power performance.

DOWNSHIFT PLUG

If the accelerator pedal is depressed to nearly the fully open position, the downshift plug opens the cut-back pressure passage very wide, then causes the detent regulator valve to operate and effect kickdown.

The cut-back pressure also acts on the downshift plug when the throttle valve opening angle is less than 85%. A power assist mechanism is provided to lighten spring tension.

GOVERNOR VALVE

The governor valve is driven by the drive pinion worm gear and produces governor pressure in response to the vehicle speed. It balances the line pressure from the primary regulator valve and the centrifugal force of the governor weights to produce hydraulic pressure in proportion to vehicle speed.

DETENT REGULATOR VALVE

During kickdown, the detent regulator valve stabilizes the hydraulic pressure acting on the 1-2 and 2-3 shift valves.

LOW MODULATOR VALVE

In the **L** range, the low modulator valve reduces the line pressure from the manual valve. Low modulator pressure pushes down the low coast shift valve and acts on the 1st and reverse brake to cushion the apply.

1–2 SHIFT VALVE

This valve automatically controls the 1–2 shift according to governor and throttle pressure. To improve the valve sliding characteristics, a 3 piece valve is used. When governor pressure is low and throttle pressure is high, the valve is pushed down by throttle pressure and because the 2nd brake circuit closes, the transaxle shifts into 1st gear.

When governor pressure is high and throttle pressure low, the valve is pushed up by governor pressure and the circuit to the 2nd brake piston opens so the transaxle will shift into 2nd gear. When the throttle pressure passage is closed, downshifting into 1st gear is dependant on spring tension and governor pressure only.

Unless the downshift plug actuates and allows the detent pressure to act on the 1–2 shift valve, downshifting into 1st gear will take place at a set vehicle speed. In the **L** range, there is no upshifting into 2nd gear because low modulator pressure is acting on the low coast shift valve.

2–3 SHIFT VALVE

This valve performs shifting between 2nd and 3rd gears. Control is accomplished by opposing throttle pressure and spring tension against governor pressure. When governor pressure is high, the valve is pushed up against the resistance of the throttle pressure and spring tension. This opens the passage to the direct clutch (C_2) piston to allow the shift into 3rd gear.

When governor pressure is low, the valve is pushed down by throttle pressure and spring tension to close the passage leading to the direct clutch piston, causing a downshift to 2nd gear. In the event of kickdown, the detent pressure acts on the 2–3 shift valve to permit a quicker downshift to 2nd gear. valve movement occurs due to the different size areas where pressure is applied.

Since the area is larger for downshift than for upshift, downshifting takes place at a lower vehicle speed. Line pressure from the manual valve acts on the intermediate shift valve. The valve descends and shifting into 2nd gear is accomplished but there is no upshifting into 3rd gear. Line pressure passes through the 2nd modulator valve and 1–2 shift valve and acts on the 2nd coast brake to provide engine braking.

Diagnosis Tests

LINE PRESSURE TEST

1. Run the engine until it reaches normal operating temperature.
2. Raise the vehicle and support it safely.
3. Remove the transaxle case test plugs and mount hydraulic pressure gauges.
4. Apply the parking brake.
5. Start the engine.
6. Apply the brake pedal while manipulating the accelerator pedal and measure the line pressure at the engine speeds specified in the chart.
7. In the same manner, perform the test for the **R** range.
8. If the measured pressure is not up to specified values, recheck the throttle cable adjustment and retest.
9. If the measured values at all ranges are higher than specified, check the following:
   a. Throttle cable out of adjustment
   b. Throttle valve defective
   c. Regulator valve defective
10. If the measured values at all ranges are lower than specified, check the following:
    a. Throttle cable out of adjustment
    b. Throttle valve defective
    c. Regulator valve defective
    d. Oil pump defective
    e. Underdrive brake defective
11. If the pressure is low in **D** range only, check the following:
    a. **D** range circuit fluid leakage
    b. Forward clutch defective
12. If the pressure is low in **R** range only, check the following:
    a. **R** range circuit fluid leakage
    b. Direct clutch defective
    c. First and reverse brake defective

The following is a list of line pressure specifications:

D range—idling 53–61 psi
D range—stall 131–152 psi
R range—idling 77–102 psi
R range—stall 205–239 psi

MANUAL SHIFTING TEST

With this test, it can be determined whether the trouble lies within the electrical circuit or is a mechanical problem in the transaxle.

1. With the engine **OFF**, remove the center cluster and disconnect the ECT computer connector.
2. If the **L**, **2** and **D** range gear positions are difficult to distinguish, do not perform the following test.
3. While driving, shift through **L**, **2** and **D** ranges and back up again. Check that the gear change corresponds to the gear position.
4. While driving, shift through **D**, **2** and **L** ranges and back down again. Check that the gear change corresponds to the gear position.
5. If any abnormality is found in the above test, do not perform the stall, time lag or road tests.

TIME LAG TEST

If the shift lever is shifted while the engine is idling, there will be a certain time elapse or lag before the shock can be felt. This is used for checking the condition of the overdrive clutch, forward clutch, direct clutch and 1st and reverse brake.

1. Run the engine until it reaches normal operating temperature.

NOTE: Allow 1 minute interval between tests. Make 3 measurements and take the average value.

2. Apply the parking brake.
3. Check the idle speed. It should be 800 rpm.
4. Move the shift lever from **N** to **D** range. Using a stop watch, measure the time it takes from shifting the lever until the shock is felt. The time lag should be less than 1.2 seconds.
5. Following Step 4, move the shift lever from **N** to **R** range. The time lag should be less than 1.5 seconds.
6. If the **N** to **D** time lag is longer than specified, check for the following:
   a. Line pressure too low
   b. Forward clutch worn
   c. No. 2 and UD one-way clutch not operating properly
7. If the **N** to **R** time lag is longer than specified, check for the following:
   a. Line pressure too low
   b. Direct clutch worn
   c. First and reverse brake worn
   d. UD brake worn

STALL SPEED

The object of this test is to check the overall performance of the transaxle and engine by measuring the maximum engine speeds in the **D** and **R** ranges.

1.Run the engine until it reaches normal operating temperature.

NOTE: Do not continuously run this test longer than 5 seconds.

2. Apply the parking brake and block the front and rear wheels.
3. Connect an engine tachometer.
4. Shift into the **D** range.
5. While applying the brakes, step all the way down on the accelerator. Quickly read the highest rpm at this time. The stall speed should be 2400–2550 rpm.
6. Perform the same test in the **R** range.
7. If the engine speed is the same for both ranges, but lower than the specified value, check for the following:
   a. Engine output insufficient
   b. Stator one-way clutch not operating properly
8. If the stall speed in **D** range is higher than specified, check for the following:
   a. Line pressure too low
   b. Forward clutch slipping
   c. One-way clutch No. 2 not operating properly
   d. UD one-way clutch not operating properly
9. If the stall speed in **R** range is higher than specified, check for the following:
   a. Line pressure too low
   b. Direct clutch slipping
   c. First and reverse brake slipping
   d. UD braking slipping
10. If the stall speed in **R** and **D** range is higher than specified, check for the following:
   a. Line pressure too low
   b. Improper fluid level
   c. UD braking slipping

ROAD TEST

NOTE: This test must be performed with the engine and transaxle at normal operating temperature.

DRIVE RANGE IN NORMAL AND POWER PATTERN RANGES

Shift into **D** range with the accelerator pedal held constant (throttle valve opening 50% and 100%). Push in one of the pattern selection buttons with the **OD** switch **ON** and check the following points:

1. 1–2, 2–3, 3–OD and lockup upshifts take place and that the shift points conform to those shown in the automatic shift diagram.
2. There is no OD upshift when the coolant temperature is below 122°F.
3. There is no lockup when vehicle speed is 6 mph less than set cruise control speed.
4. If there is no 1–2 upshift, check for the following:
   a. No. 2 solenoid stuck
   b. 1–2 shift valve stuck
5. If there is no 2–3 upshift, check for the following:
   a. No.1 solenoid stuck
   b. 2–3 shift valve stuck
6. If there is no 3–OD upshift, the 3–OD shift valve is stuck.
7. If the shift point is defective, the throttle valve, 1–2 shift valve, 2–3 shift valve and 3–OD shift valve are defective.
8. If the lockup is defective, check for the following:
   a. No. 3 solenoid stuck
   b. Lock-up relay valve stuck
9. While driving in the **D** range 2nd, 3rd gears and OD, check to see that possible kickdown vehicle speed limits for 2–1, 3–1, 3–2, OD–3 and OD–2 kickdowns conform to those in the automatic shift diagram.
10. While driving at 50 mph in **D** range OD gear, shift into **2** and **L** ranges and check the engine braking effect at each of these ranges.
11. If there is no engine braking effect at **2** range 3rd gear, the 2nd brake is defective.
12. If there is no engine braking effect at **2** range 2nd gear, the 2nd brake and 2nd coast brake are defective.
13. If there is no engine braking effect at **L** range 2nd gear, the 2nd brake and 2nd coast brake are defective.
14. If there is no engine braking effect at **L** range 1st gear, the 1st and reverse brake are defective.

LOCKUP MECHANISM

1. Connect a voltmeter to service connector terminal ECT.
2. Select normal pattern.
3. Drive at 37 mph to where 7, 5 or 3 volts appear on the voltmeter, this is the lockup range.
4. Depress the accelerator pedal and read the tachometer. If there is a big jump in the engine rpm, there is no lockup.

2 RANGE

1. Shift to **2** range and drive with the throttle valve opening at 50% and 100% respectively. Check the 1–2 upshift points at each of the throttle valve openings to see that it conforms to those indicated in the automatic shift diagram.
2. While driving in **2** range, 2nd gear, release the accelerator pedal and check the engine braking effect.
3. Check for abnormal noise at acceleration and deceleration and harshness at upshift and downshift.

L RANGE

1. While driving at 50 mph in **D** or **2** range, release the accelerator pedal and shift into **L** range and check that the 2–1 downshift points are at 32 mph.
2. While driving in **L** range, check that there is no upshift to 2nd gear.
3. While driving in **L** range, release the accelerator pedal and check the engine braking effect.
4. Check for abnormal noise during acceleration and deceleration.

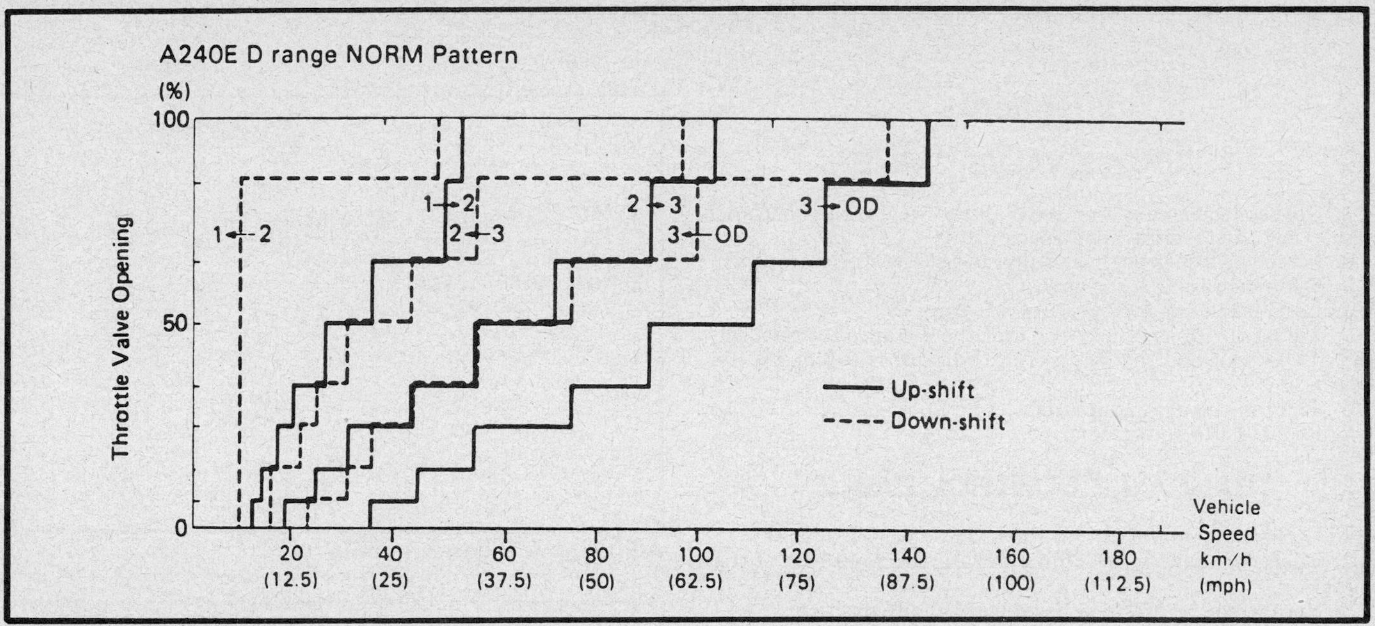
A240E D range NORM Pattern
(%)
100
50
0
Throttle Valve Opening
1←2
1→2
2←3
2→3
3←OD
3→OD
Up-shift
Down-shift
Vehicle
Speed
km/h
(mph)
20
(12.5)
40
(25)
60
(37.5)
80
(50)
100
(62.5)
120
(75)
140
(87.5)
160
(100)
180
(112.5)

Automatic shift diagram

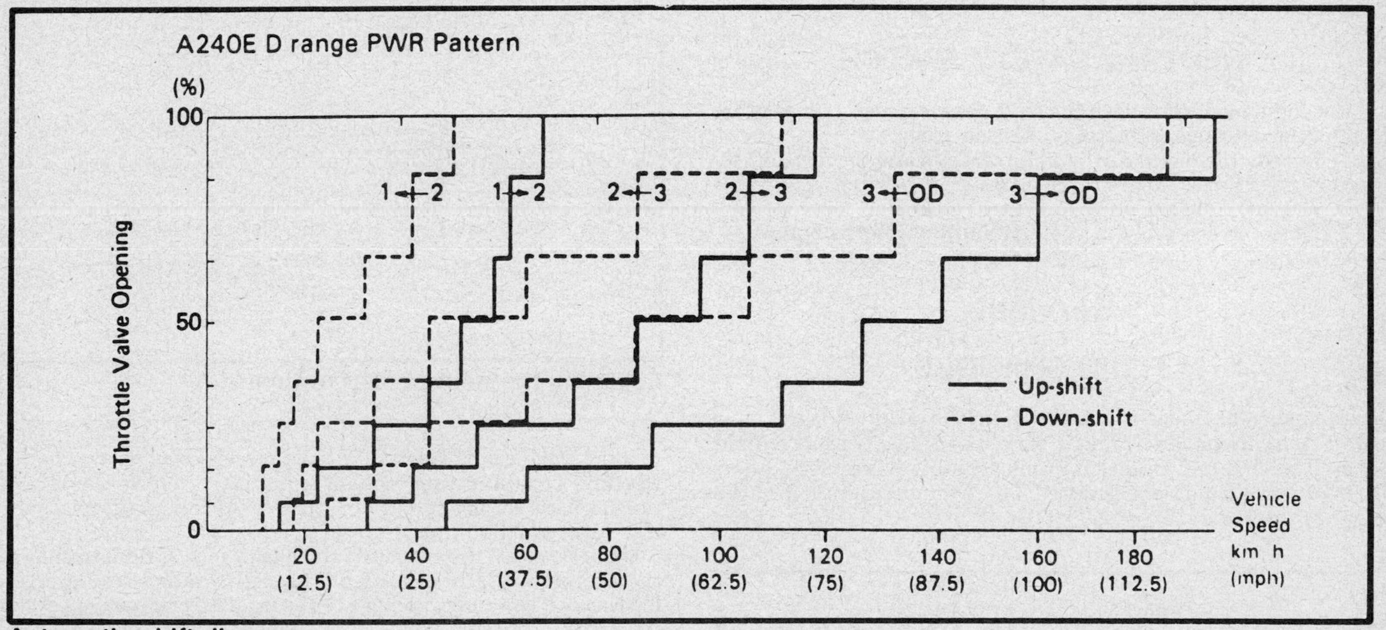
A240E D range PWR Pattern
(%)
100
50
0
Throttle Valve Opening
1←2
1→2
2←3
2→3
3←OD
3→OD
Up-shift
Down-shift
Vehicle
Speed
km/h
(mph)
20
(12.5)
40
(25)
60
(37.5)
80
(50)
100
(62.5)
120
(75)
140
(87.5)
160
(100)
180
(112.5)

Automatic shift diagram

R RANGE

Shift into **R** range and while starting at full throttle, check for slipping.

P RANGE

Stop the vehicle on a slight hill and after shifting into **P** range, release the parking brake. Check to see that the parking lock pawl prevents the vehicle from moving.

ON CAR SERVICES

Adjustments

THROTTLE VALVE CABLE

1. Depress the accelerator pedal all the way to the floor and check that the throttle valve opens fully.
2. If the valve does not open fully, adjust the accelerator link.
3. Fully depress the accelerator.
4. Loosen the adjustment nuts.
5. Adjust the cable housing so that the distance between the end of the boot and the stopper on the cable is 0–0.04 in. (0–1mm).
6. Tighten the adjusting nuts.
7. Recheck the adjustment.

TRANSAXLE CONTROL CABLE

1. Loosen the swivel nut on the lever.
2. Push the manual lever fully toward the right side of the vehicle.
3. Return the lever 2 notches toward the **N** position.
4. Set the shift lever in the **N** range.
5. While holding the lever lightly toward the **R** range side, tighten the swivel nut.

NEUTRAL SAFETY SWITCH

If the engine will start with the shift selector in any range other than **N** or **P** range, adjustment is required.

1. Loosen the neutral safety switch bolts and set the shift selector in **N** range. Adjust the neutral safety switch.
2. Align the groove and the neutral basic line.
3. Hold the assembly in position and tighten the bolts.
4. Adjust the idle speed in **N** range to 800 rpm.

Services

FLUID CHANGES

The conditions under which the vehicle is operated is the main consideration in determining how often the transaxle fluid should be changed. Different driving conditions result in different transaxle fluid temperatures. These temperatures effect the change intervals.

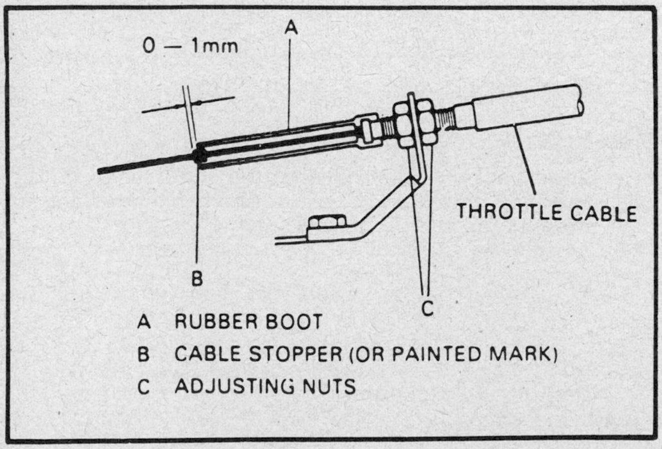
0 — 1mm
A
THROTTLE CABLE
B
C
A RUBBER BOOT
B CABLE STOPPER (OR PAINTED MARK)
C ADJUSTING NUTS

Adjusting the T.V. cable

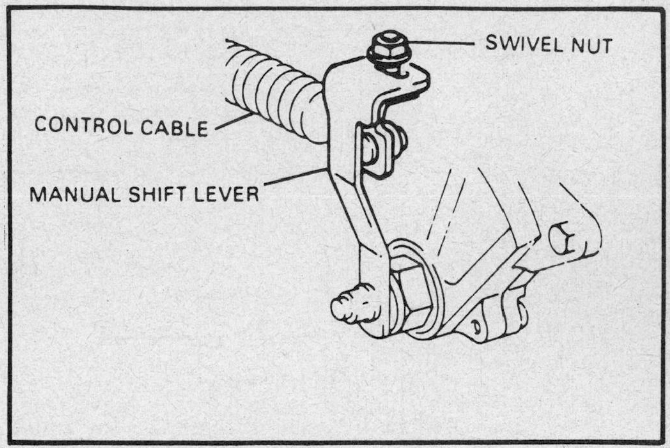
SWIVEL NUT
CONTROL CABLE
MANUAL SHIFT LEVER

Adjusting the control cable

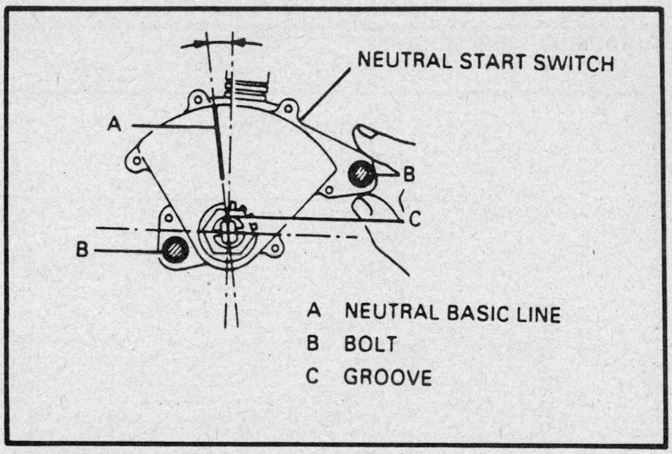
NEUTRAL START SWITCH
A
B
C
B
A NEUTRAL BASIC LINE
B BOLT
C GROOVE

Adjusting the neutral start switch

If the vehicle is driven under severe service conditions, change the fluid and filter every 15,000 miles. If the vehicle is not used under severe service conditions, change the fluid and replace the filter every 50,000 miles.

Do not overfill the transaxle. It takes 1 pint of fluid to change the level from **ADD** to **FULL** on the transaxle dipstick. Overfilling the unit can cause damage to the internal components of the automatic transaxle.

OIL PAN

Removal and Installation

1. Disconnect the negative battery cable.
2. Raise the vehicle and support it safely.
3. Remove the drain plug in the pan and drain the fluid.
4. Reinstall the drain plug and remove the oil pan. After removing the bolts, tap the pan lightly with a plastic hammer. Do not force the pan off by prying. This may cause damage to the gasket mating surface.
5. Remove the oil pan gasket material on the mating surface.
6. Remove the oil filter from the valve body.
7. Clean the inside of the oil pan before installation.
8. Clean the oil cleaner magnet and install it in the proper position.

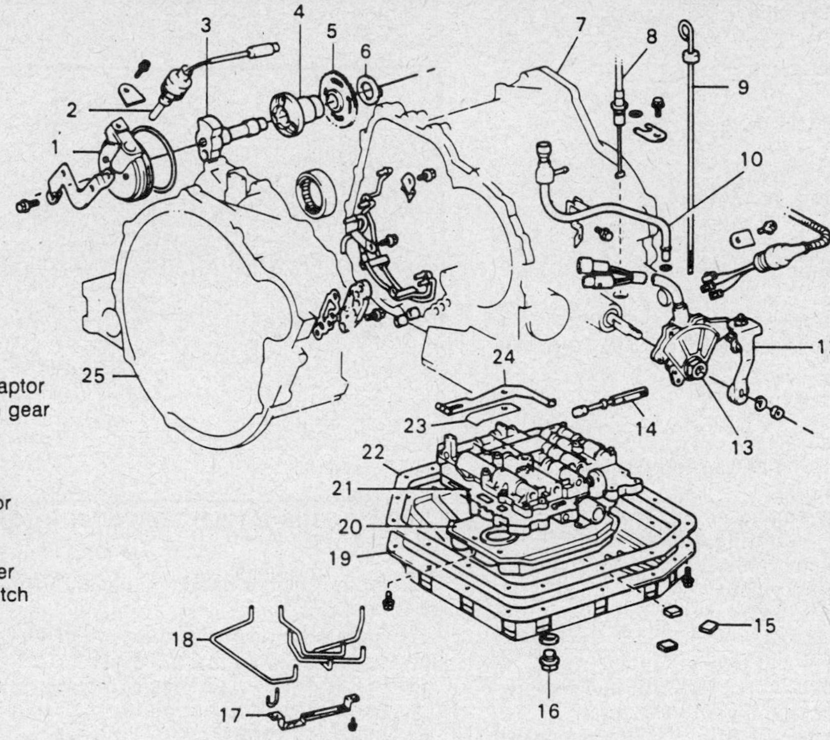

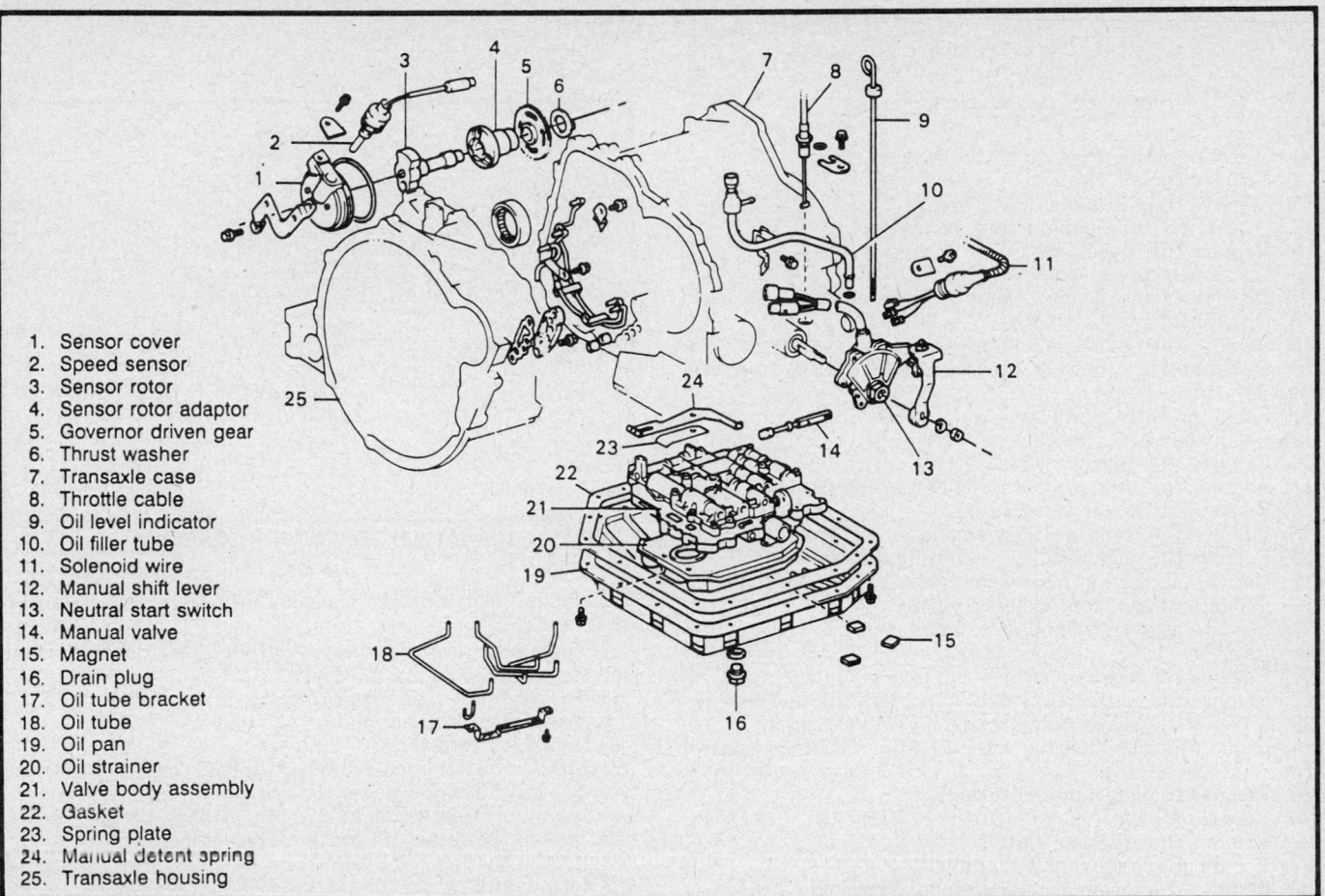

1. Sensor cover
2. Speed sensor
3. Sensor rotor
4. Sensor rotor adaptor
5. Governor driven gear
6. Thrust washer
7. Transaxle case
8. Throttle cable
9. Oil level indicator
10. Oil filler tube
11. Solenoid wire
12. Manual shift lever
13. Neutral start switch
14. Manual valve
15. Magnet
16. Drain plug
17. Oil tube bracket
18. Oil tube
19. Oil pan
20. Oil strainer
21. Valve body assembly
22. Gasket
23. Spring plate
24. Manual detent spring
25. Transaxle housing

Transaxle components

9. Install a new filter on the valve body.
10. Install the pan with a new gasket and torque the bolts to 43 inch lbs. (4.9 Nm).
11. Fill with transmission fluid to the proper level.
12. Lower the vehicle and connect the battery cable.

VALVE BODY

Removal and Installation

1. Disconnect the negative battery cable.
2. Raise the vehicle and support it safely.
3. Remove the drain plug in the pan and drain the fluid.
4. Reinstall the drain plug and remove the oil pan. After removing the bolts, tap the pan lightly with a plastic hammer. Do not force the pan off by prying. This may cause damage to the gasket mating surface.
5. Remove the oil pan gasket material on the mating surface.
6. Remove the oil tube bracket and the oil strainer.
7. Pry out the oil tubes and remove them.
8. Remove the detent spring.
9. Disconnect the 3 solenoid connectors.
10. Loosen and remove the valve body bolts.
11. Lift the valve body assembly and disconnect the throttle cable from the cam.
12. Remove the valve body.
13. Remove the 2nd brake apply gaskets.
14. On installation, install the valve body on the case. Hold the cam down and insert the throttle cable into the slot.
15. Install the valve body with the bolts finger tight.
16. Install the detent spring and tighten the bolts to 7 ft. lbs. (10 Nm).
17. Make sure that the manual valve lever touches the center of the detent spring roller.
18. Tighten the valve body bolts to 7 ft. lbs. (10 Nm).
19. Install the oil tubes, bracket and oil strainer.
20. Clean the inside of the oil pan before installation.
21. Clean the oil cleaner magnets and install it in the proper position.
22. Install the oil pan with a new gasket and torque the bolts to 43 inch lbs. (4.9 Nm).

REAR OIL SEAL

Removal and Installation

1. Disconnect the negative battery cable.
2. Raise the vehicle and support it safely.
3. Drain the fluid from the differential.
4. Remove the driveshaft from the transaxle and the steering knuckle.
5. Remove the differential oil seals using tools J–26941 or equivalent and J–23907 or equivalent.
6. Using tool J–35553 or equivalent, drive the oil seal into the case until its surface is flush with the surface of the case.
7. Coat the top of the oil seal with grease.
8. Install the driveshaft to the transaxle and the steering knuckle.
9. Fill with transmission fluid and check the fluid level.

REMOVAL AND INSTALLATION

TRANSAXLE REMOVAL

1. Disconnect the negative battery cable.
2. Remove the air cleaner.
3. Remove the neutral safety switch.
4. Disconnect the solenoid valve connector.
5. Remove the speed sensor.
6. Disconnect the speedometer cable.
7. Disconnect the throttle cable from the throttle linkage.
8. Disconnect the water inlet.
9. Remove the transaxle dipstick vent tube.
10. Disconnect the battery ground cable from the transaxle housing.
11. Raise the vehicle and support it safely.
12. Drain the transaxle fluid.
13. Remove the splash shields.
14. Remove the shift control cable and brackets.
15. Disconnect the oil cooler hoses.
16. Remove the front and rear engine mount bolts.
17. Remove the engine center mounting member.
18. Disconnect the oxygen sensor connector.
19. Disconnect the front exhaust pipe.
20. Remove the right and left hand driveshafts from the transaxle.
21. Remove the left hand wheel and tire assembly.
22. Remove the cotter pin, locknut cap, lock nut and washer.
23. Remove the brake caliper assembly and the rotor.
24. Disconnect the steering knuckle from the lower control arm.
25. Remove the left hand driveshaft.
26. Remove the starter.
27. Remove the stiffener plate.
28. Remove the engine rear end plate.
29. Remove the torque converter mounting bolts.
30. Lower the vehicle.
31. Install engine support fixture tool J-28467 or equivalent.
32. Remove the 3 bolts from the rear engine mount.
33. Raise the vehicle and support it safely.
34. Remove the transaxle from the vehicle.
35. Remove the torque converter from the transaxle.

TRANSAXLE INSTALLATION

1. Install the torque converter in the transaxle.
2. To be sure the torque converter is installed correctly, use a straight edge and measure from the installed surface to the front surface of the transaxle housing. The distance should be more than 0.79 in. (20mm).
3. Position the transaxle on a suitable transmission jack and align it in the vehicle.
4. Align the converter housing with the 2 dowel pins in the block and install a bolt.
5. Install the transaxle housing mounting bolts and torque the 12mm bolt to 47 ft. lbs. (64 Nm). Torque the 10mm bolts to 34 ft. lbs. (46 Nm).

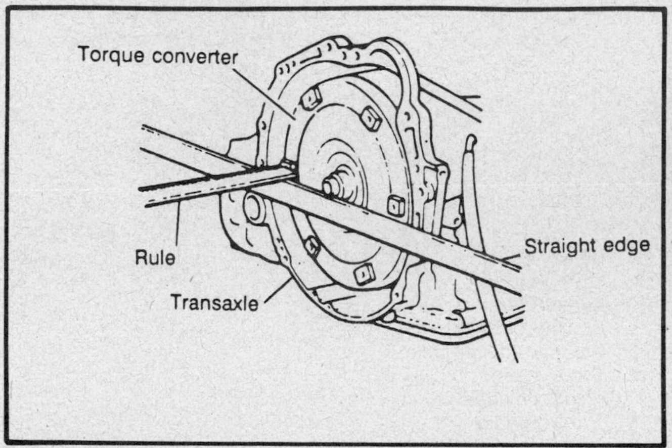
Torque converter
Rule
Straight edge
Transaxle

Checking the torque converter installation

6. Install the left hand engine mounting. Tighten the bolts to 38 ft. lbs. (52 Nm).
7. Install the torque converter bolts. Tighten the bolts evenly and torque to 20 ft. lbs. (27 Nm).
8. Install the engine rear end plate.
9. Install the stiffener plate.
10. Install the starter.
11. Install the left hand driveshaft. Install the cotter pin, locknut cap, locknut and washer.
12. Connect the steering knuckle to the lower control arm.
13. Install the rotor and brake caliper assembly.
14. Install the left hand wheel and tire assembly and tighten the nuts to 76 ft. lbs. (103 Nm).
15. Connect the left hand and right hand driveshafts to the transaxle.
16. Connect the front exhaust pipe.
17. Connect the oxygen sensor connector.
18. Install the engine center mounting member and tighten the bolts to 29 ft. lbs. (39 Nm).
19. Install the front and rear engine mount bolts and tighten them to 29 ft. lbs. (39 Nm).
20. Remove engine support fixture tool J-28467 or equivalent.
21. Connect the oil cooler hoses.
22. Install the shift control cable and brackets.
23. Install the splash shields.
24. Lower the vehicle.
25. Connect the battery ground cable to the transaxle housing.
26. Install the transaxle dipstick vent tube.
27. Install the water inlet.
28. Connect the throttle cable to the throttle linkage.
29. Connect the speedometer cable.
30. Install the speed sensor.
31. Connect the solenoid valve connector.
32. Install the neutral safety switch.
33. Install the air cleaner.
34. Connect the negative battery cable.

BENCH OVERHAUL

Before Disassembly

Before opening up the transaxle, the outside of the unit should be thoroughly cleaned, preferably with high pressure cleaning equipment. Dirt entering the transaxle internal parts will negate all the effort and time spent on the overhaul. During inspection and reassembly, all parts should be thoroughly cleaned with solvent and then dried with compressed air. Cloths and rags should not be used to dry the parts since lint will find its way into the valve body passages.

Lube the seals with Dexron® II automatic transmission fluid and use unmedicated petroleum jelly to hold the thrust washers and ease the assembly of the seals. Do not use solvent on neoprene seals, friction plates or thrust washers. Be wary of nylon parts if the transaxle failure was due to a cooling system problem. Nylon parts exposed to antifreeze solutions can swell and distort, so they must be replaced. Before installing bolts into aluminum parts, dip the threads in clean oil.

Converter Inspection

Make certain that the transaxle is held securey. If the torque converter is equipped with a drain plug, open the plug and drain the fluid. If there is no drain plug, the converter must be drained through the hub after pulling the converter out of the transaxle. If the oil in the converter is discolored but does not contain metal bits or particles, the converter is not damaged. Color is no longer a good indicator of fluid condition.

If the oil in the converter contains metal particles, the converter is damaged internally and must be replaced. If the cause of the oil contamination was burned clutch plates or overheated oil, the converter is contaminated and should be replaced. If the pump gears or cover show signs of damage, the converter will contain metal particles and must be replaced.

Transaxle Disassembly

1. Remove the oil cooler pipes.
2. Remove the transaxle dipstick and filler tube.
3. Remove the manual shift lever.
4. Remove the neutral safety switch.
5. Remove the throttle cable retaining plate.
6. Remove the solenoid and kickdown switch wire retaining plate.
7. Remove the speed sensor, sensor rotor and adapter.
8. Remove the pan and gasket.

NOTE: Do not turn the transaxle over, before pan removal, as it will contaminate the valve body with foreign material from the bottom of the pan.

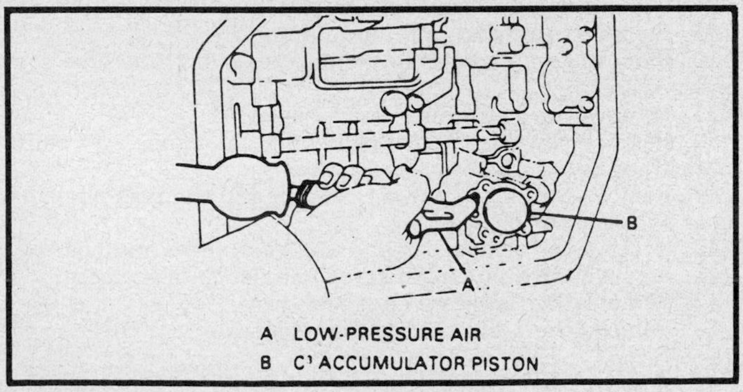
B
A
A LOW-PRESSURE AIR
B C' ACCUMULATOR PISTON

Removing the C_3 accumulator piston

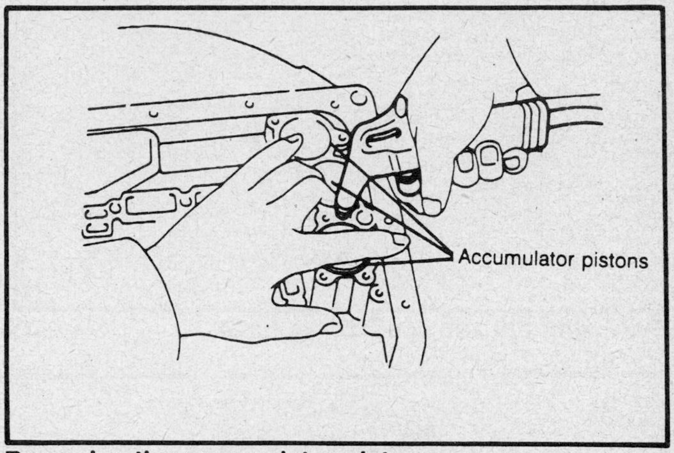
Accumulator pistons

Removing the accumulator pistons

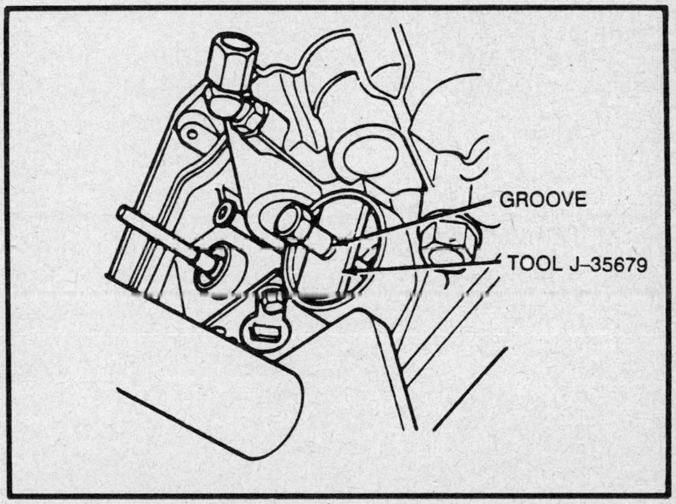
GROOVE
TOOL J-35679

Measuring the piston stroke

9. Turn the transaxle over and remove the oil tube bracket.
10. Remove the oil strainer.
11. Remove the oil tubes.
12. Remove the manual detent spring and the detent plate.
13. Disconnect the solenoid connectors.
14. Remove the valve body.
15. Remove the throttle cable from the case.
16. Disconnect the solenoid wire.
17. Remove the governor apply gasket.
18. Remove the C_3 accumulator piston and spring. Force low pressure compressed air into the hole to pop the piston out.
19. Remove the accumulator pistons and springs. Loosen the bolts 1 turn at a time until the spring tension is released. Remove the cover and gasket. It may be necessary to force low pressure compressed air into the hole to pop out pistons B_2 and C_2.
20. Measure the 2nd coast brake band piston stroke by applying a small amount of paint to the piston rod at the point where it meets the case. Using tool J-35679 or equivalent, measure the piston stroke by applying and releasing 57–114 psi (392–785 kPa) of compressed air to the hole. The piston stroke should be 0.059–0.118 in. (1.5–3.0mm).
21. If the piston stroke is more than specified, replace the pis-

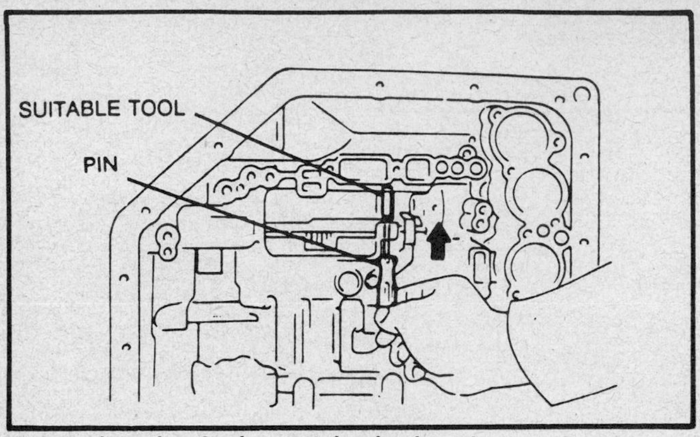
SUITABLE TOOL
PIN

Removing the 2nd coast brake band

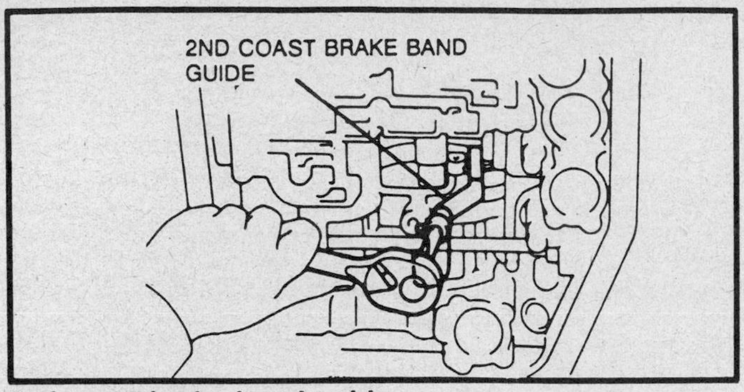
2ND COAST BRAKE BAND
GUIDE

2nd coast brake band guide

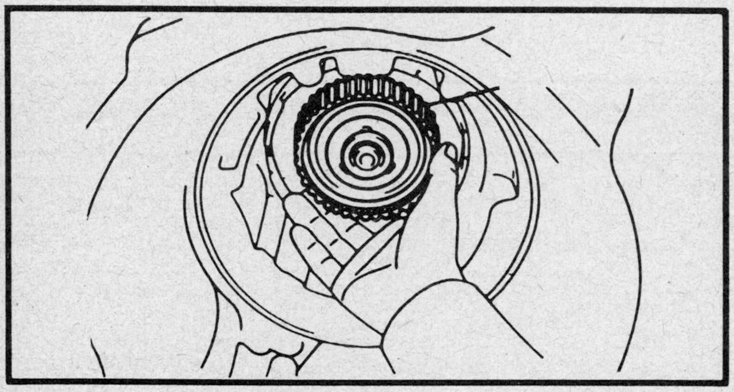

Front planetary ring gear

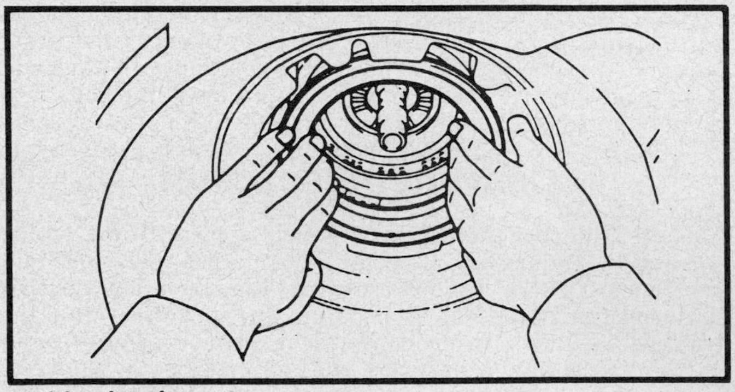

2nd brake drum

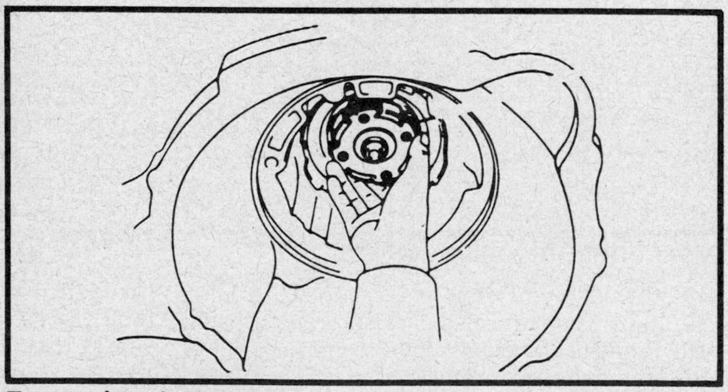

Front planetary gear

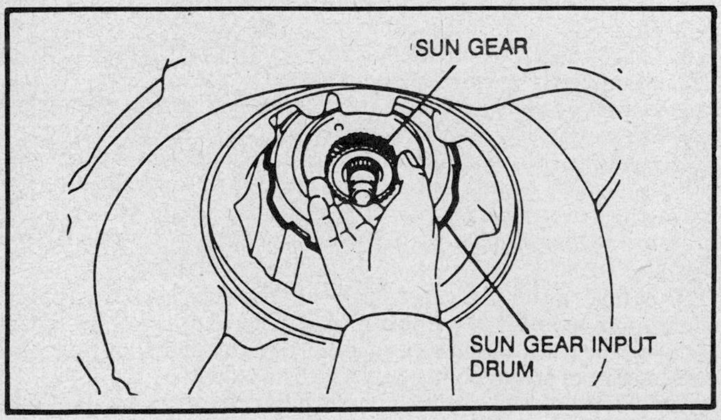
SUN GEAR
SUN GEAR INPUT
DRUM

Sun gear and input drum

ton rod with a longer rod. Piston rods are available in 2.870 in. (72.9mm) or 2.811 in. (71.4mm) lengths.

22. Re-measure the piston stroke. If it is still more than specified, replace the brake band with a new band.
23. Remove the 2nd coast brake band piston snapring, cover, piston and outer return spring.
24. Remove the oil pump using tools J–6125–B slide hammer and J–35495 pulley adapter.
25. Remove the direct clutch.
26. Remove the forward clutch.
27. Remove the 2nd coast brake band by removing the pin from the oil pump mounting bolt hole.
28. Remove the front planetary ring gear with the bearing and race.
29. Remove the front planetary gear with the race.
30. Remove the sun gear, sun gear input drum and thrust washer.
31. Remove the 2nd brake hub and the No. 1 one-way clutch.
32. Remove the second coast brake band guide.
33. Remove the snapring and the 2nd brake drum.
34. Remove the 2nd brake drum piston return spring.
35. Remove the plates, disc and flange.
36. Remove the 2nd brake piston using low pressure compressed air.
37. Remove the 2nd brake drum gasket.
38. Remove the snapring holding the No. 2 one-way clutch outer race to the case.
39. Remove the No. 2 one-way clutch and the rear planetary gear with the thrust washers.
40. Remove the rear planetary ring gear, races and bearing.
41. Remove the snapring holding the flange to the case.
42. Remove the flanges, plates and discs.
43. Remove the bolts holding the transaxle rear cover to the case.
44. Remove the transaxle rear cover and the intermediate shaft.

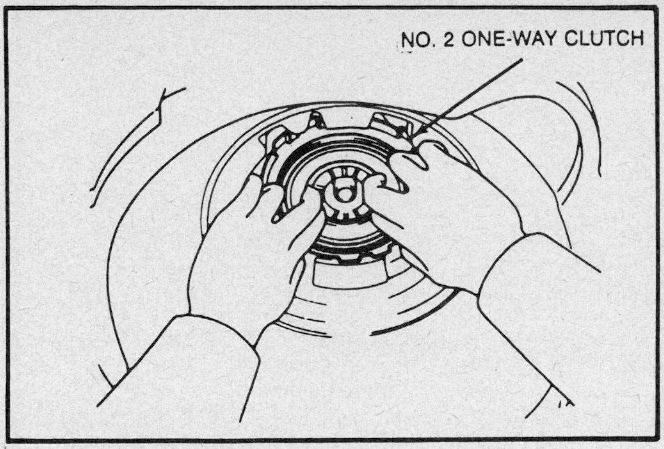
NO. 2 ONE-WAY CLUTCH
No. 2 one-way clutch and planetary gear

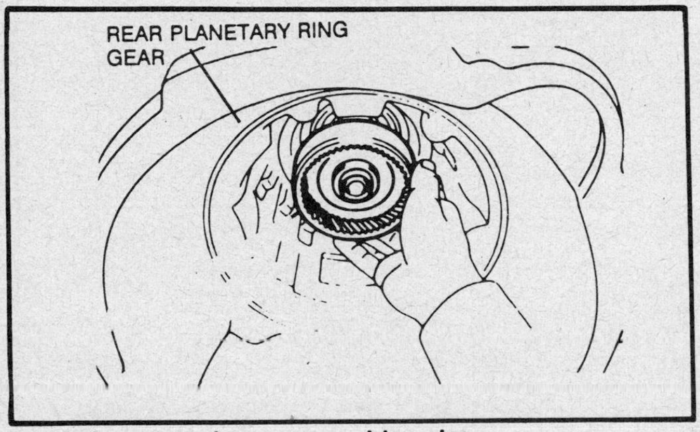
REAR PLANETARY RING GEAR
Rear planetary ring gear and bearing

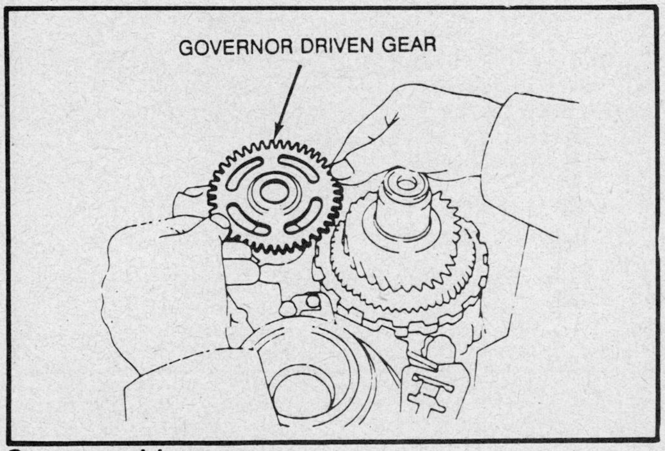
GOVERNOR DRIVEN GEAR
Governor driven gear

45. Remove the transaxle housing bolts and the transaxle housing.
46. Remove the differential.
47. Remove the governor driven gear and thrust washer.
48. Remove the oil seals.
49. Unstake the countershaft locknut on both sides. Attach tool J-8614-01 or equivalent to the counter driven gear and tool J-37271 or equivalent to the opposite locknut to remove the locknuts.
50. Remove the counter driven gear using tool J-1859-03 or equivalent.
51. Remove the thrust needle bearing.
52. Remove the countershaft assembly and the anti-rattle clip.
53. Remove the parking lock pawl stopper plate, torsion spring and spring guide.
54. Remove the pawl shaft clamp.
55. Remove the parking lock pawl shaft and the lock pawl.
56. Remove the parking lock sleeve.
57. Remove the cam guide bracket.
58. Remove the manual valve shaft sleeve.
59. Remove the retaining spring, manual valve shaft, manual valve lever and washer.
60. Replace the manual shaft oil seal, if necessary. Apply multipurpose grease to the oil seal lip.
61. Remove the oil seal rings.
62. Remove the oil galley cover and gasket.
63. Remove the B_4 accumulator piston and spring.
64. Remove the bearing from the case using tools J-23907 and J-26941 or equivalent.
65. Remove the 4 oil tube clamps and the oil tubes.
66. Remove the oil tube apply cover and gasket.
67. Remove the bearing from the housing using tools J-23907 and J-26941 or equivalent.

Cleaning

All disassembled parts should be washed clean, and the fluid passages and holes blown through with compressed air to make sure that they are not clogged. Dexron® II automatic transmission fluid or kerosene should be used for cleaning.

CAUTION

When using compressed air to dry parts, avoid spraying ATF or kerosene in your face.

Unit Disassembly and Assembly

OIL PUMP

Disassembly

1. Remove the race from the stator shaft.
2. Remove the O-ring from the pump body.
3. Remove the 2 oil seal rings from the back of the stator shaft.
4. Remove the thrust washer of the clutch drum from the stator shaft.
5. Remove the stator shaft.

Inspection

1. Check the body clearance of the driven gear. Push the driven gear to either side of the body. Using a feeler guage, measure the clearance. If the clearance exceeds the limit, replace the drive gear, driven gear or pump body. The standard body clearance is 0.0028–0.0059 in. (0.07–0.15mm). The maximum body clearance is 0.012 in. (0.3mm).
2. Check the tip clearance of both gears. Measure between the gear teeth and the crescent-shaped part of the pump body. If the clearance exceeds the limit, replace the drive gear, driven gear or pump body. The standard tip clearance is 0.0043–0.0055 in. (0.11–0.14mm). The maximum tip clearance is 0.012 in. (0.3mm).
3. Check the side clearance of both gears. Using a steel straight edge and a feeler gauge, measure the side clearance of both gears. If the clearance exceeds the limit, replace the drive gear, driven gear or the pump body. The standard side clearance is 0.0008–0.0020 in. (0.02–0.05mm). The maximum side clearance is 0.004 in. (0.1mm).
4. Inspect the front oil seal, check for wear, damage or cracks.

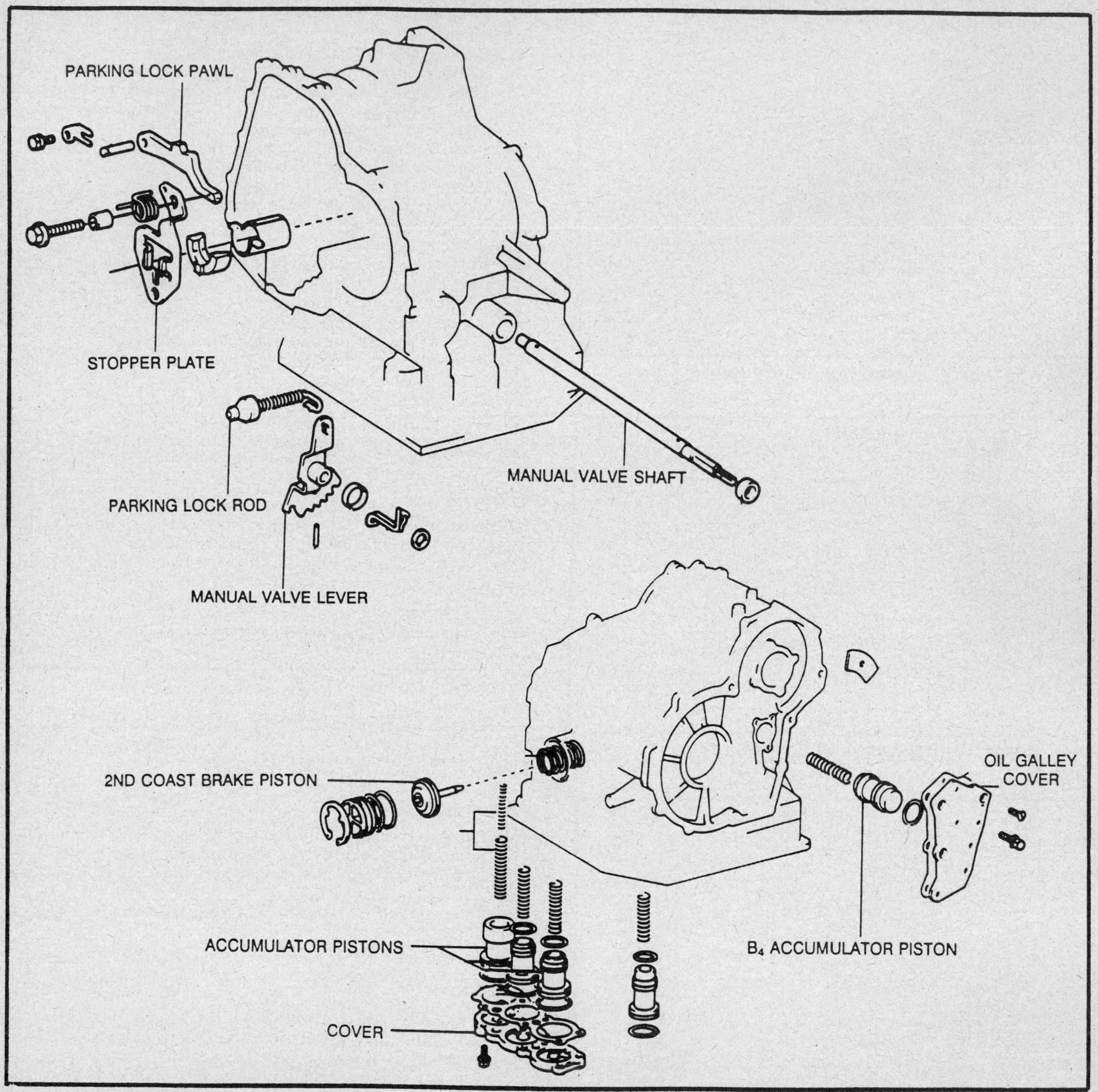
PARKING LOCK PAWL
STOPPER PLATE
PARKING LOCK ROD
MANUAL VALVE SHAFT
MANUAL VALVE LEVER
2ND COAST BRAKE PISTON
OIL GALLEY COVER
ACCUMULATOR PISTONS
B_4 ACCUMULATOR PISTON
COVER

Transaxle components

5. Replace the front oil seal. The seal end should be flush with the outer edge of the pump body.

Assembly

1. Install the driven gear and drive gear making sure the top of the gears are facing upward.
2. Install the stator shaft onto the pump body. Align the stator shaft with each bolt hole.
3. Tighten the bolts to 7 ft. lbs. (10 Nm).
4. Coat the thrust washer with petroleum jelly. Align the tab of the washer with the hollow of the pump body and install the thrust washer.
5. Install 2 oil seal rings on the oil pump. Do not spread the ring ends too far.
6. Check the pump drive gear rotation. Turn the drive gear and check that it rotates smoothly.
7. Install a new O-ring.
8. Install the race onto the stator shaft.

DIRECT CLUTCH

Disassembly

1. Remove the snapring from the clutch drum.

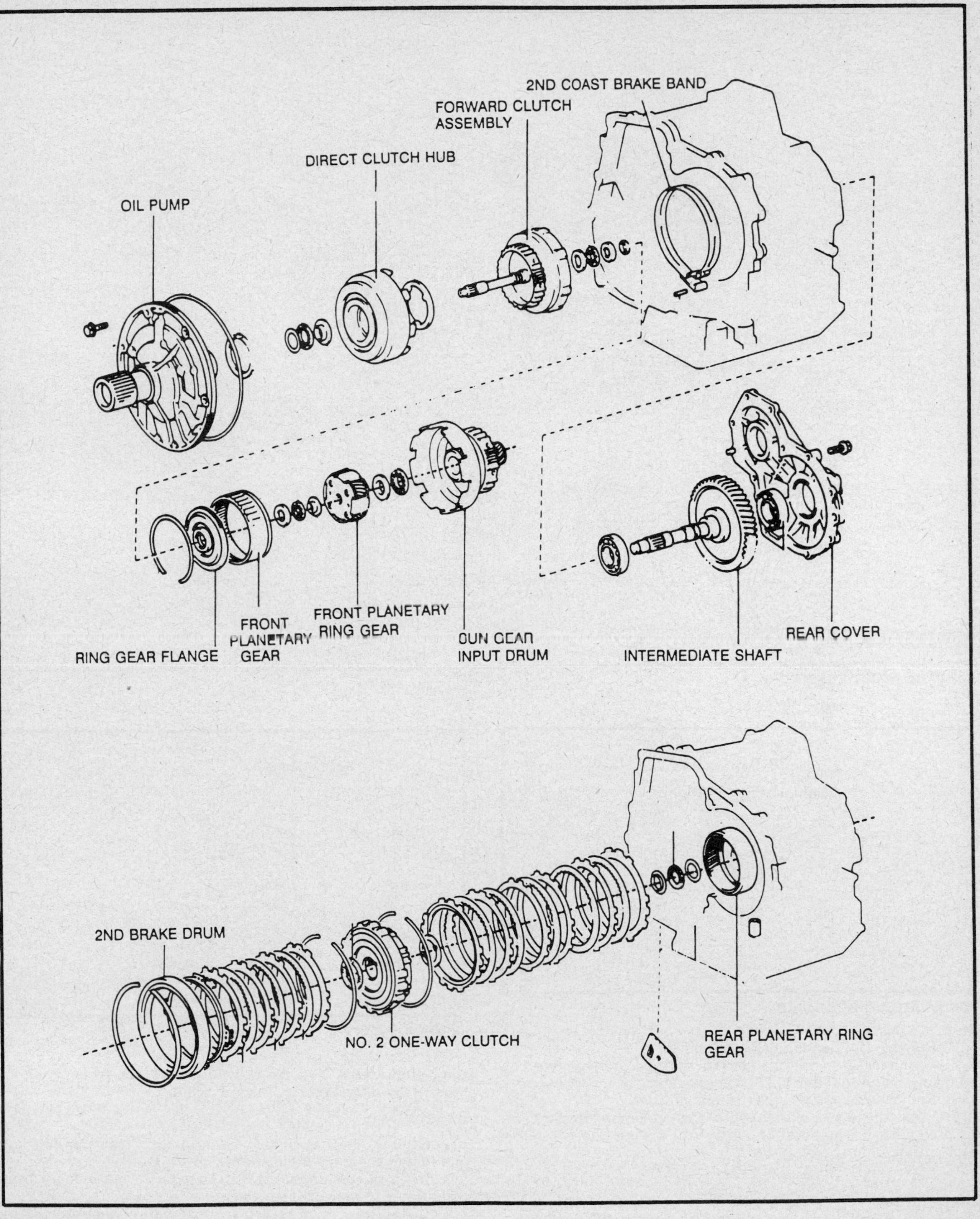
2ND COAST BRAKE BAND
FORWARD CLUTCH ASSEMBLY
DIRECT CLUTCH HUB
OIL PUMP
RING GEAR FLANGE
FRONT PLANETARY GEAR
FRONT PLANETARY RING GEAR
SUN GEAR INPUT DRUM
INTERMEDIATE SHAFT
REAR COVER
2ND BRAKE DRUM
NO. 2 ONE-WAY CLUTCH
REAR PLANETARY RING GEAR

Transaxle components

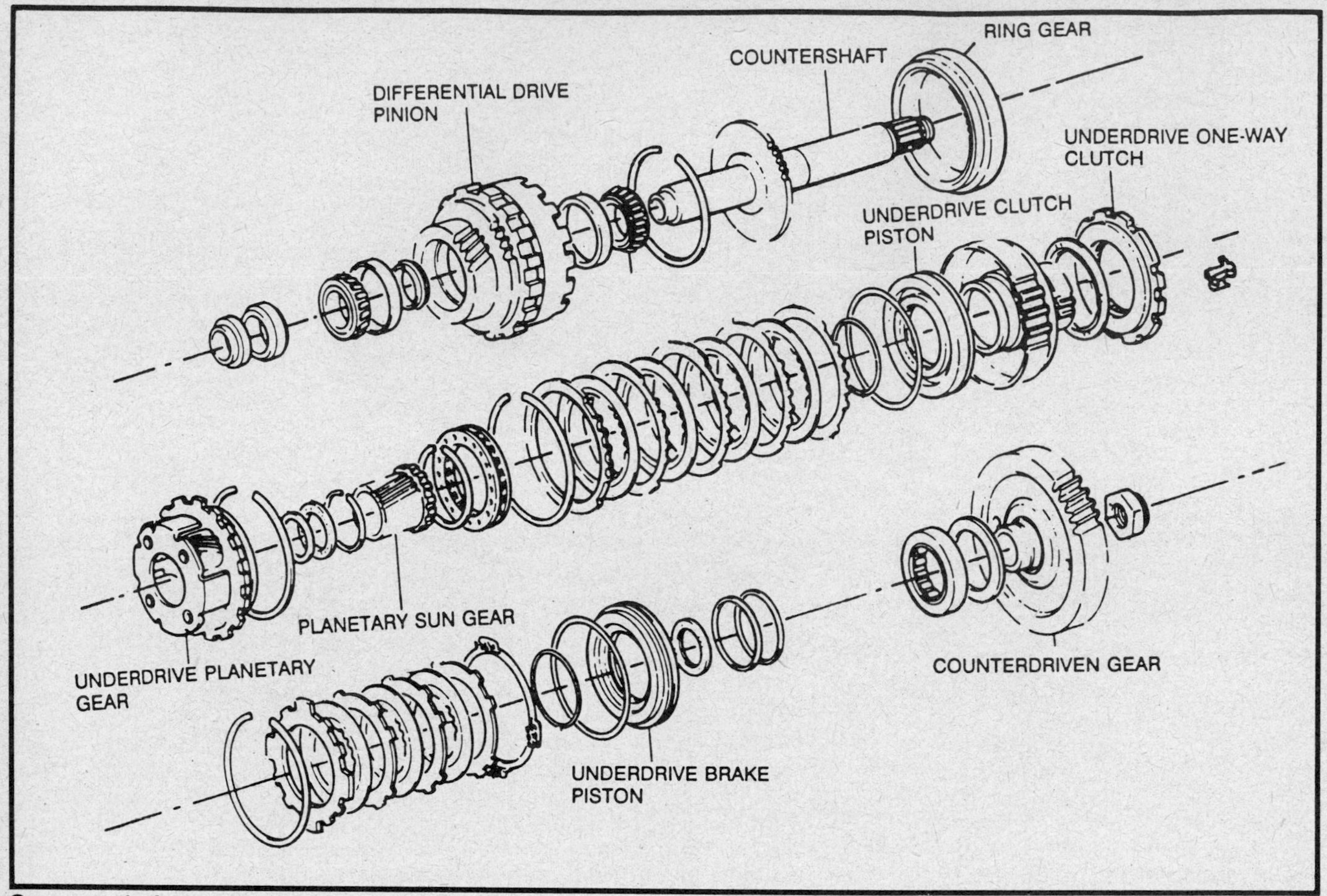
RING GEAR
COUNTERSHAFT
DIFFERENTIAL DRIVE PINION
UNDERDRIVE ONE-WAY CLUTCH
UNDERDRIVE CLUTCH PISTON
PLANETARY SUN GEAR
UNDERDRIVE PLANETARY GEAR
COUNTERDRIVEN GEAR
UNDERDRIVE BRAKE PISTON

Counter shaft assembly

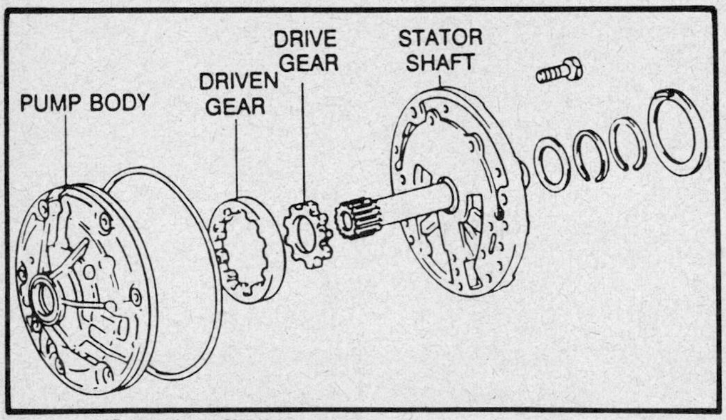
DRIVE GEAR
STATOR SHAFT
DRIVEN GEAR
PUMP BODY

Oil pump components

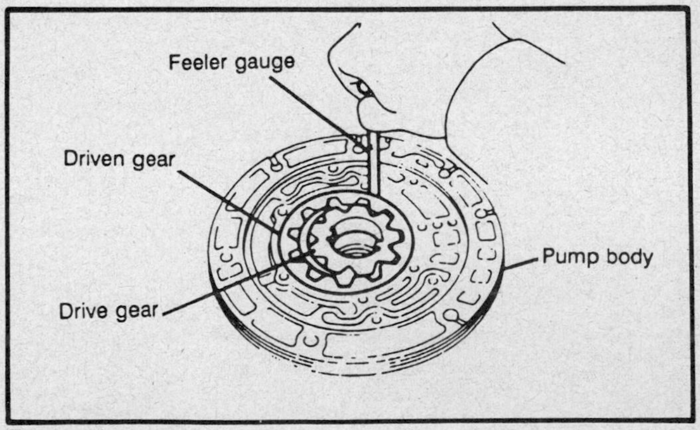
Feeler gauge
Driven gear
Pump body
Drive gear

Body clearance

2. Remove the flange, discs and plates.
3. Compress the piston return springs and remove the snapring using tool J–23327 or equivalent.
4. Remove the spring retainer and springs.
5. Slide the direct clutch onto the oil pump. Apply compressed air to the oil pump to remove the piston. Remove the direct clutch from the oil pump.
6. Remove the clutch piston O-ring.

Inspection

Inspect the clutch piston. Check that the ball is free by shaking the piston. Check that the valve does not leak by applying low pressure compressed air. Inspect the discs.

Assembly

1. Coat the new O-rings with automatic transmission fluid and install them on the piston.
2. Press the piston into the drum with the cup side up, being careful not to damage the O-ring.
3. Install the piston return springs and set the retainer and snapring in place.

4. Compress the return springs and install the snapring in the groove using tools J-23327-1 and J-37279 or equivalent. Install the snapring. Be sure the endgap of the snapring is not aligned with the spring retainer claw.

5. Install the plates, discs and flange. Install in order: plate-disc-plate-disc-plate-disc. Install the flange, facing the flat end downward.

6. Install the outer snapring. Check that the endgap of the snapring is not aligned with a cutout.

7. Using compressed air, check the stroke of the direct clutch piston for replacement of plate, discs or flange. If not within specification, select a proper flange. There are 2 different flange thickness. The piston stroke should be 0.0445–0.0591 in. (1.13–1.50mm). The flange thicknesses are 0.1024 in. (2.6mm) and 0.1181 in. (3.0mm).

FORWARD CLUTCH

Disassembly

1. Remove the thrust washer.
2. Remove the thrust bearings and races from both sides of the clutch.
3. Remove the snapring from the clutch drum.
4. Remove the flange, discs and plates.
5. Compress the piston return springs and remove the snapring using tools J-25018-A adapter and J-23327-1 or equivalent.
6. Remove the spring retainer and springs.
7. Apply compressed air into the oil passage to remove the piston.

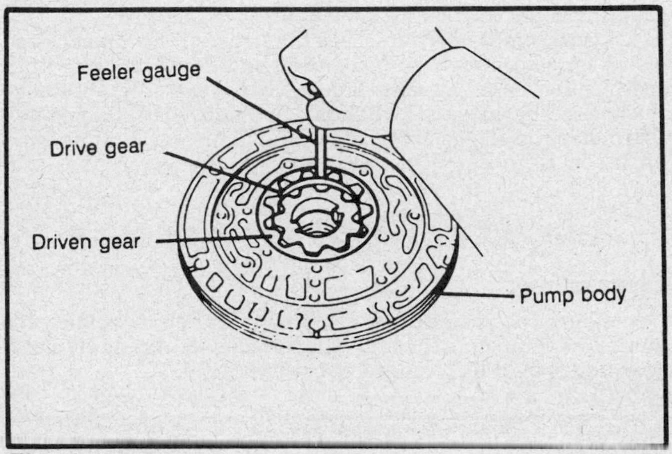
Feeler gauge
Drive gear
Driven gear
Pump body

Tip clearance

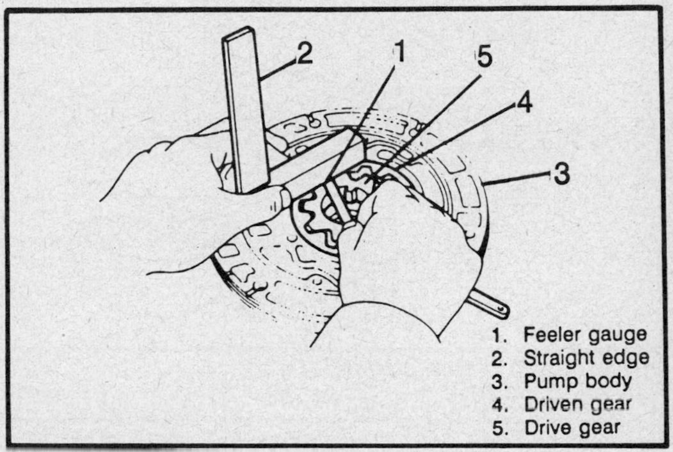
2
1
5
4
3
1. Feeler gauge
2. Straight edge
3. Pump body
4. Driven gear
5. Drive gear

Side clearance

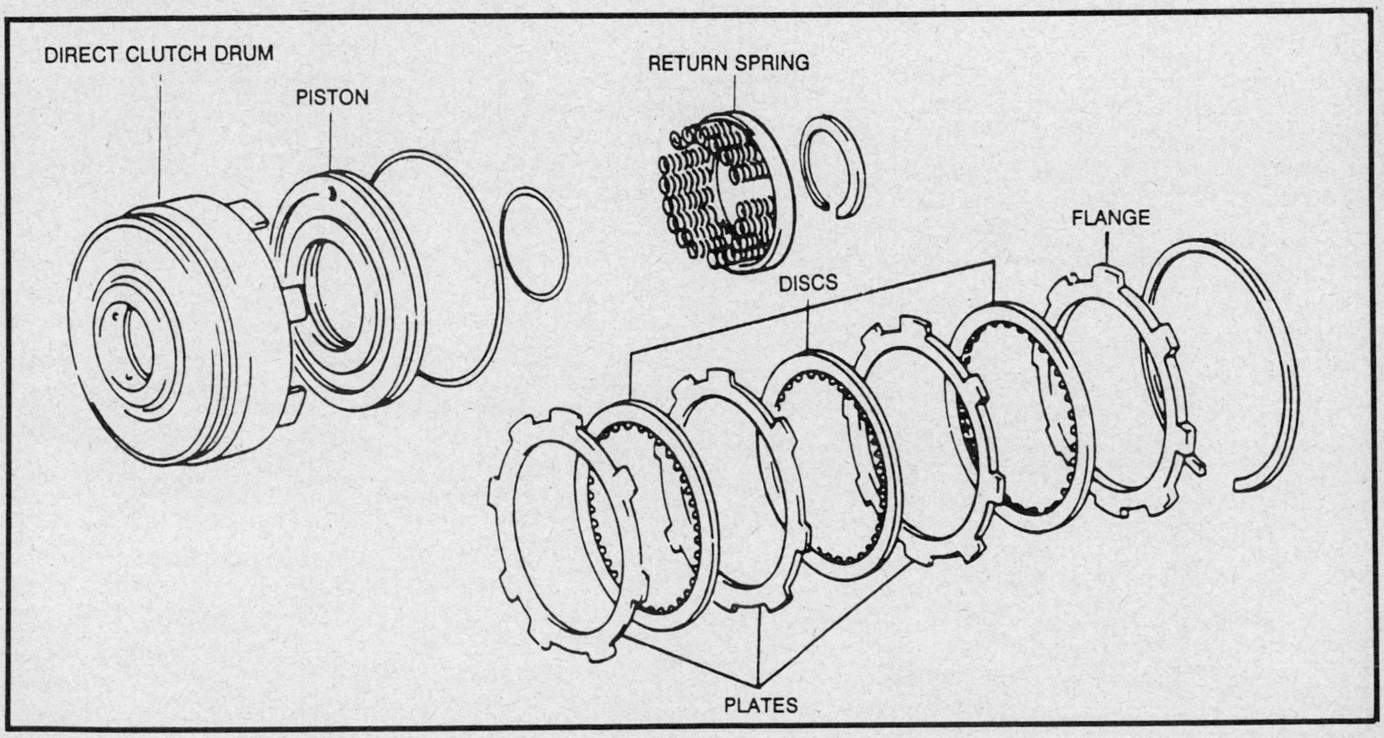
DIRECT CLUTCH DRUM
PISTON
RETURN SPRING
FLANGE
DISCS
PLATES

Direct clutch C_3 components

Inspection

Make sure the check ball is free by shaking the piston. Check that the valve does not leak by applying low pressure compressed air. Replace the oil seal rings. Slide the rings over top of the shaft and install them into the groove. Do not spread the ring ends.

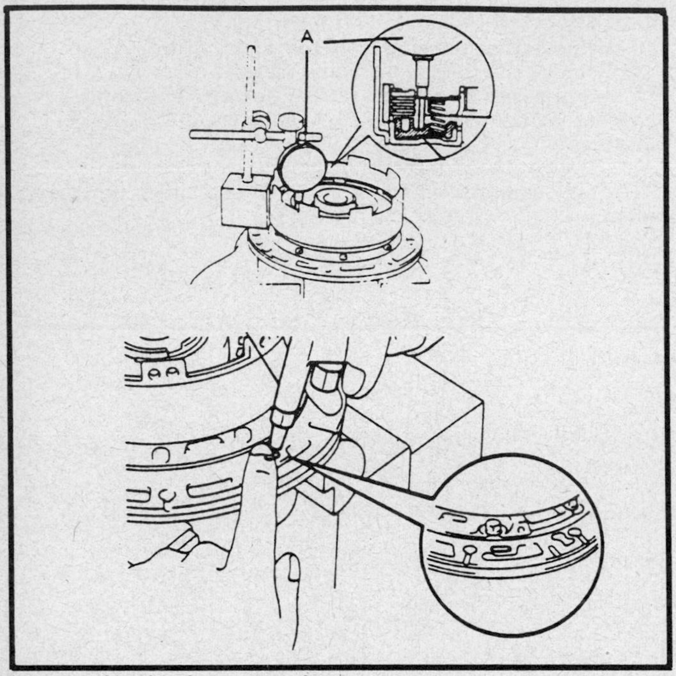
A

Direct clutch piston stroke

Assembly

1. Install new O-rings on the piston, coat the O-rings with automatic transmission fluid.
2. Press the piston into the forward clutch drum with the cup side up, being careful not to damage the O-ring.
3. Install the piston return springs, spring retainer and snapring in place.
4. Compress the return springs and install the snapring in the groove. Be sure the end of the snapring is not aligned with the spring retainer claw.
5. Install the plates, discs and flange. Install in the following order: plate-disc-plate-disc-plate-disc. Install the flange facing the flat end downward.
6. Install the outer snapring. Check that the endgap of the snapring is not aligned with a cutout.
7. Using compressed air, check the stroke of the direct clutch piston for replacement of plate, discs and flange. If not within specification, select a proper flange. There are 2 different flange thickness. The piston stroke should be 0.0559–0.0713 in. (1.42–1.81mm). The flange thicknesses are 0.1181 in. (3.0mm) and 0.1327 in. (3.37mm).

ONE-WAY CLUTCH AND SUN GEAR

Disassembly

1. Check the operation of the one-way clutch by holding the sun gear and turning the hub. The hub should turn freely clockwise and should lock counterclockwise.

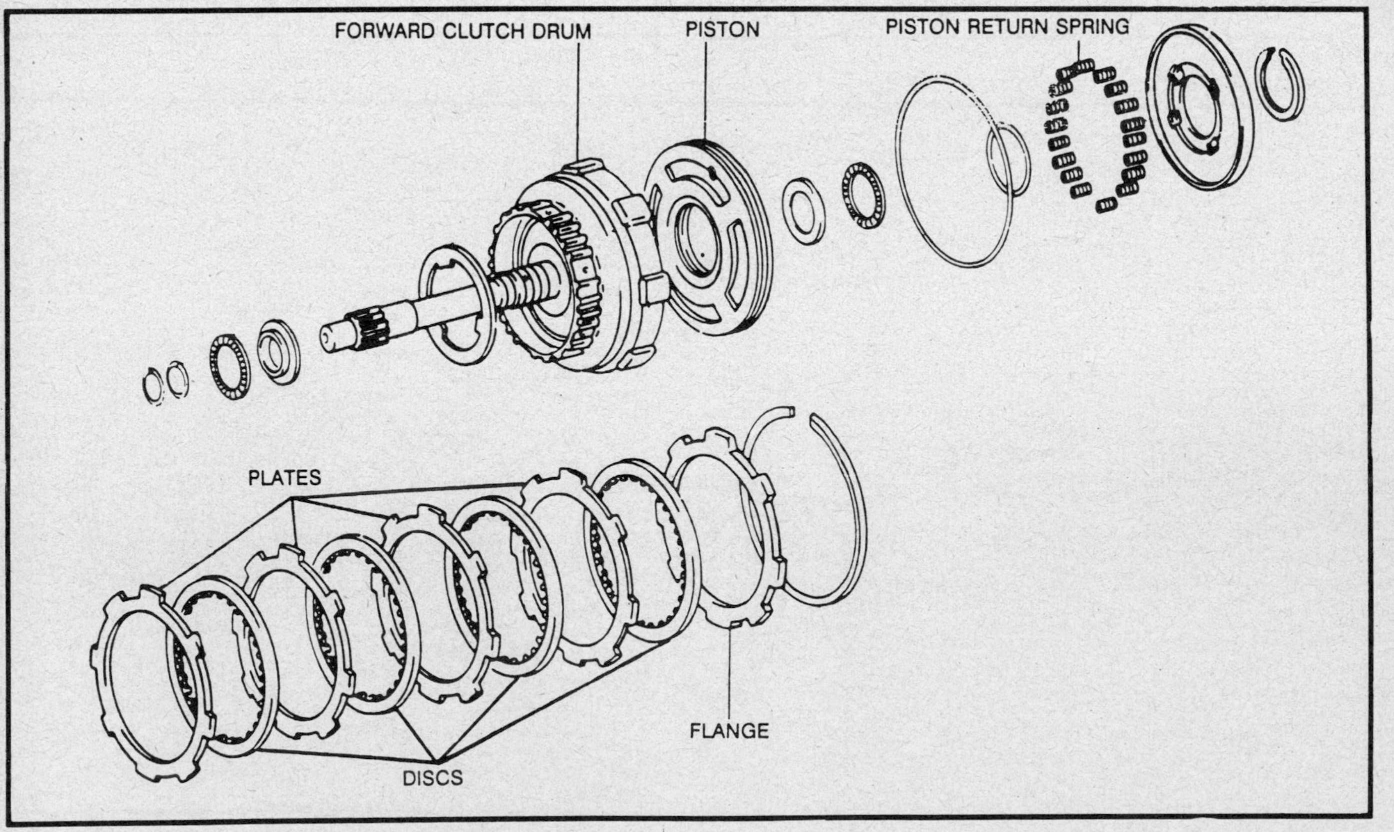
FORWARD CLUTCH DRUM
PISTON
PISTON RETURN SPRING
PLATES
FLANGE
DISCS

Forward clutch C_1 components

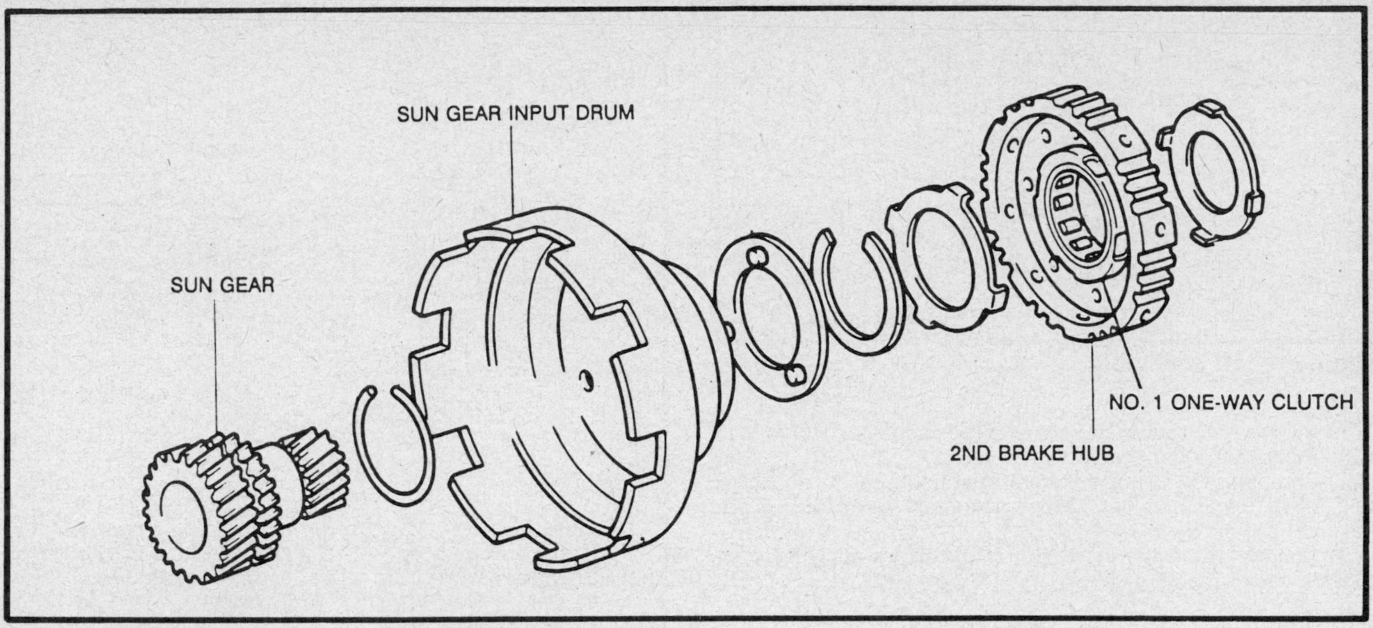
SUN GEAR INPUT DRUM
SUN GEAR
NO. 1 ONE-WAY CLUTCH
2ND BRAKE HUB

No. 1 one-way clutch and sun gear components

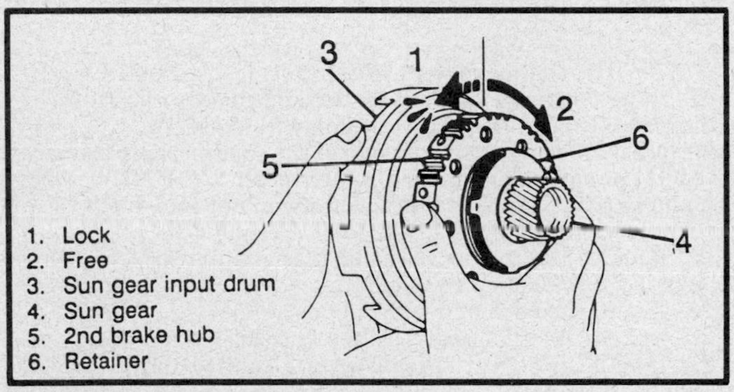
3
1
2
6
5
4
1. Lock
2. Free
3. Sun gear input drum
4. Sun gear
5. 2nd brake hub
6. Retainer

Checking No. 1 one-way clutch

2. Remove the 2nd brake hub and the one-way clutch from the sun gear.
3. Remove the No. 3 planetary carrier thrust washer from the sun gear input drum.
4. Remove the snapring and the sun gear input drum.

Inspection

If necessary, replace the one-way clutch.

1. Pry off the retainer.
2. Remove the one-way clutch.
3. Install the one-way clutch into the brake hub, facing the spring cage inward from the flanged side of the brake hub.
4. Hold the brake hub in a vise and flatten the ears with a chisel.
5. Check the operation to make sure that the retainer is centered.

Assembly

1. Install the shaft snapring on the sun gear.
2. Install the sun gear input drum on the sun gear and install the shaft snapring.
3. Install the No. 3 planetary carrier thrust washer on the sun gear input drum.
4. Install the one-way clutch and the 2nd brake hub on the sun gear. While turning the hub clockwise, slide the one-way clutch into the inner race.

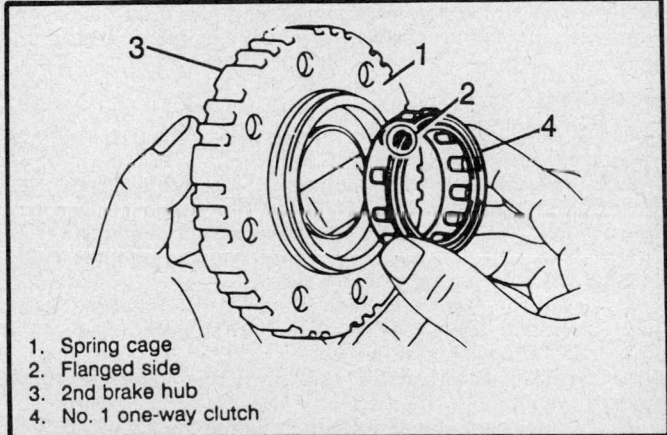

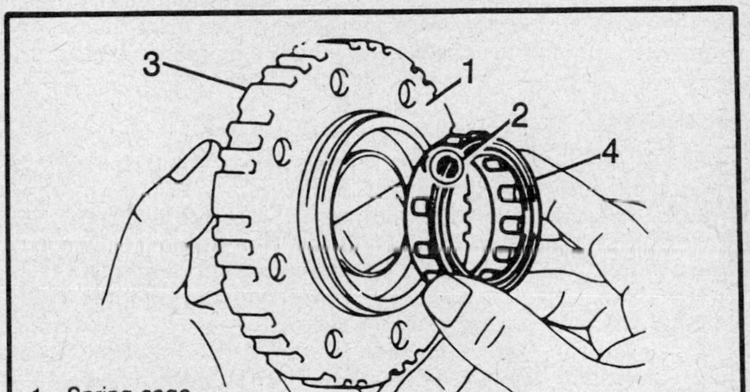
3
1
2
4
1. Spring cage
2. Flanged side
3. 2nd brake hub
4. No. 1 one-way clutch

Installing No. 1 one-way clutch

COUNTER SHAFT

Disassembly

1. Remove the bearing.
2. Remove the underdrive planetary sun gear.
3. Remove the snapring from the sun gear.
4. Remove the snapring from the countershaft assembly.
5. Remove the underdrive planetary gear.
6. Remove the thrust needle bearing and race.
7. Remove the drive pinion with the output flange, bearing, inner race and spacer. Press out the bearing using tool J–22912–01 or equivalent.
8. Remove the snapring and ring gear.
9. Remove the bearing using tools J–22912–01 and J–37273 or equivalent.
10. Remove the bearing outer race.
11. Remove the underdrive one-way clutch from the clutch drum.

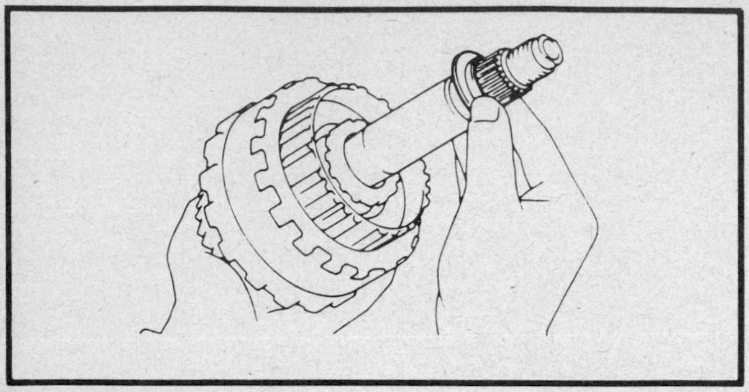

Countershaft assembly

12. Remove the thrust washer and the snapring.
13. Remove the flange, discs and plates.
14. Compress the piston return spring using tools J-23327-1 and J-37279 or equivalent. Remove the snapring and the return spring.
15. Remove the underdrive clutch piston by applying compressed air to the oil passage.

Inspection

Check that the underdrive clutch piston check ball is free by shaking the piston. Check that the valve does not leak by applying low pressure compressed air. Inspect the discs, plates and flange. If the sliding surfaces are burnt or worn, replace the part.

Assembly

1. Install the new O-rings on the piston. Coat the new O-rings with automatic transmission fluid.
2. Install the underdrive clutch in the clutch drum.
3. Compress the return spring on the clutch piston using tools J-32272-1 and J-37279 or equivalent and install the snapring in the groove. Be sure the snapring end gap is not aligned with the spring retainer claw.
4. Install the plates, disc and flange in the following order: plate-disc-plate-disc-plate-disc-plate-disc-flange.
5. Install the snapring.
6. If replacing the plates, disc or flange, first install the underdrive clutch in the transaxle case.
7. Measure the underdrive clutch piston stroke by applying and releasing low pressure compressed air and using a dial indicator. The piston stroke should be 0.0579–0.0744 in. (1.47–1.89mm).
8. If the piston stroke is less than the minimum, the parts may be misassembled.
9. If the piston stroke is not within specifications, select another flange. The flange thicknesses are 0.0803 in. (2.04mm) and 0.0945 in. (2.40mm).
10. Install the thrust washer in the clutch drum.
11. Install the underdrive one-way clutch.
12. Install the bearing outer races using tools J-8092, J-35287 and J-35663 or equivalent.
13. Install the bearing on the countershaft using tool J-37278 or equivalent.
14. Install the ring gear and snapring.
15. Install the new spacer and the drive pinion with the output flange.
16. Install the bearing inner race using a press and tool J-37278 or equivalent.
17. Install a new locknut.
18. Temporarily install the counter driven gear to the counter shaft using a press.
19. Install tool J-6814-01 or equivalent to the counter driven gear and secure the counter shaft in a vise.

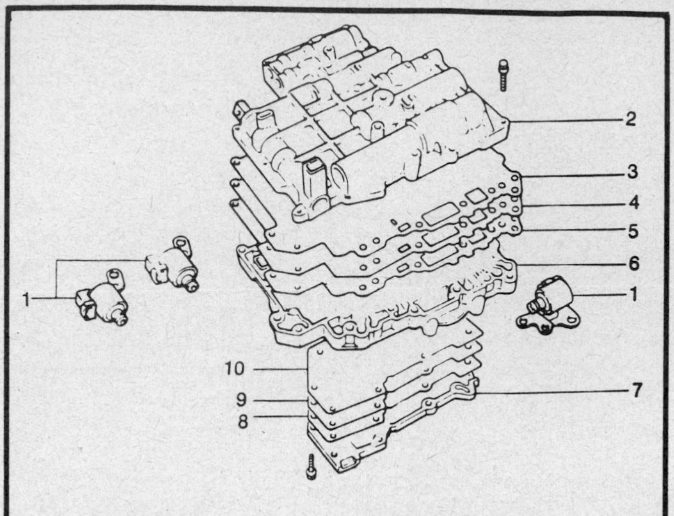
1
2
3
4
5
6
1
10
9
8
7

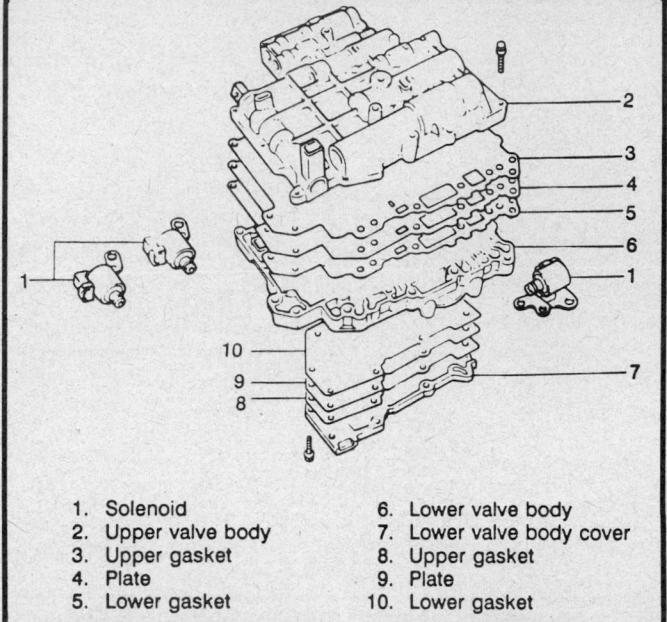

1. Solenoid
2. Upper valve body
3. Upper gasket
4. Plate
5. Lower gasket
6. Lower valve body
7. Lower valve body cover
8. Upper gasket
9. Plate
10. Lower gasket

Valve body

20. Tighten the locknut to 130–159 ft. lbs. (177–216 Nm).
21. Snug down the bearing by turning the countershaft.
22. Place the countershaft assembly in a vise.
23. Measure the starting torque of the countershaft using tool J-37271 or equivalent and a torque wrench. The starting torque should be 5.2–8.7 inch lbs. on the hexagon nut side. If the torque is exceeded, replace the spacer and retorque.
24. Remove the temporarily installed counter driven gear using tool J-1859-03 or equivalent.
25. Install the race and bearing.
26. Install the underdrive planetary gear and snapring.
27. Install the thrust bearing on the counter driven gear and stake the locknuts.

VALVE BODY

Disassembly

1. Remove the solenoid valves.
2. Remove the lower valve body cover, gaskets and plate.
3. Turn the assembly over and remove the 8 bolts from the upper valve body and upper valve body cover.
4. Turn the assembly over and remove the 9 bolts from the lower valve body.
5. Lift off the lower valve body and plate as a single unit. Hold the body plate to the lower valve body.

NOTE: Be careful of the bypass valve, pressure relief valve. Make sure the check balls, retainer, keys and pins do not fall out.

UPPER VALVE BODY

Disassembly

1. Remove the throttle valve retainer and check ball.
2. Remove the retainer for the plug with a magnetic finger and remove the plug.
3. Remove the lockup relay valve, control valve and spring.
4. Remove the sleeve retainer with a magnetic finger and remove the sleeve.

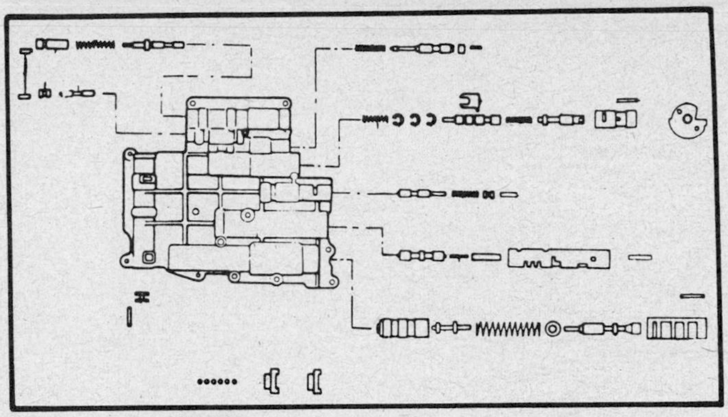

Upper valve body

UPPER VALVE BODY SPRING

Spring	Free Length in.	(mm)	Color
Primary regulator valve	2.6240	66.65	Purple
Lockup relay valve	0.7402	18.80	None
Low-coast modulator valve	1.0831	27.51	Yellow
Kickdown valve	1.1717	29.76	White
Throttle valve	1.1488	29.18	Yellow green
Throttle modulator valve	1.1772	29.90	Green
Accumulator control valve	1.5039	38.20	Yellow

5. Remove the cut-back valve by removing the retainer with a magnetic finger then remove the cut-back valve and plug.
6. Remove the throttle modulator valve by removing the retainer with a magnetic finger, then remove the plug, valve and spring.
7. Remove the accumulator control valve by removing the retainer with a magnetic finger and remove the plug, valve and spring.
8. Remove the low coast modulator valve by removing the pin with a magnetic finger then remove the plug, valve and spring.
9. Remove the 2nd coast modulator valve by removing the retainer with a magnetic finger then remove the spring and valve.
10. Remove the throttle cam. Loosen the bolt and remove the cam, spring and collar.
11. Remove the kickdown valve and spring by removing the pin with a magnetic finger, then remove the kickdown valve with the sleeve and spring.
12. Remove the throttle valve.
13. Remove the spring and adjusting rings.

Inspection

Inspect the valve springs. Check for damage, squareness, rust and collapsed coils. Measure the free length and replace any springs if necessary.

Assembly

1. Install the lockup relay valve sleeve into the bore.
2. Coat the retainer with petroleum jelly and install it into the end of the sleeve.
3. Install the control valve, spring and lockup relay valve into the bore in horizontal position.
4. Push in the relay valve by hand until the control valve touches the end of the sleeve. Install the plug and retainer.
5. Install the cut-back valve by installing the small end first. Install the plug and retainer.
6. Install the throttle modulator valve. Install the spring and valve. Install the plug and retainer.
7. Install the accumulator control valve. Install the valve and spring. Install the plug and retainer.
8. Install the low coast modulator valve. Install the valve and spring. Install the plug, thick end first. Install the pin.
9. Install the throttle valve and retainer. Install the throttle valve. Coat the retainer with petroleum jelly and install it into place in the valve body.
10. Install the adjusting rings and spring on the throttle valve shaft.
11. Install the spring into the throttle valve.
12. Install the kickdown valve and sleeve. Install the pin to hold the sleeve in place.
13. Install the 2nd coast modulator valve. Install the valve and spring.
14. Assemble the throttle cam. Insert the sleeve through a side of the cam. Install the spring with the hook through the hole in the cam.
15. Install the cam assembly on the upper valve body. Make sure that the cam moves on the roller of the kickdown valve.
16. Install the check ball.

LOWER VALVE BODY

Disassembly

1. Remove the lower valve body plate and gaskets.
2. Remove the cooler bypass valve and spring.
3. Remove the damping check valve and spring.
4. Remove the 3 check balls.
5. Remove the primary regulator valve.
6. Remove the secondary regulator valve.
7. Remove the 1–2 shift valve. Remove the 1–2 shift lower valve and spring.
8. Remove the low coast shift valve.
9. Remove the lockup control valve.
10. Remove the detent regulator valve.
11. Remove the 2–3 shift valve.
12. Remove the intermediate shift valve.
13. Remove the lockup signal valve.
14. Remove the 3–4 coast shift plug.
15. Remove the 3–4 shift plug.

Inspection

Inspect the valve springs. Check for damage, squareness, rust and collapsed coils. Measure the free length and replace any springs if necessary.

Assembly

1. Place the primary regulator valve into the bore in the horizontal position.
2. Push the valve into the bore until its tip bottoms in the bore.
3. Install the valve spring.
4. Insert the plunger with the short end first. The plunger should be recessed inside the sleeve.
5. Install the sleeve with the plunger.
6. Install the secondary regulator valve.
7. Install the 1–2 shift valve. Install the spring and upper valve.
8. Install the low coast shift valve.
9. Install the lockup control valve.
10. Install the detent regulator valve.
11. Install the intermediate shift valve.
12. Install the 2–3 shift valve.
13. Install the lockup signal valve.
14. Install the 3–4 coast shift plug.
15. Install the 3–4 shift plug.
16. Install the spring and cooler bypass valve.
17. Install the spring and damping check valve.
18. Install the check balls.

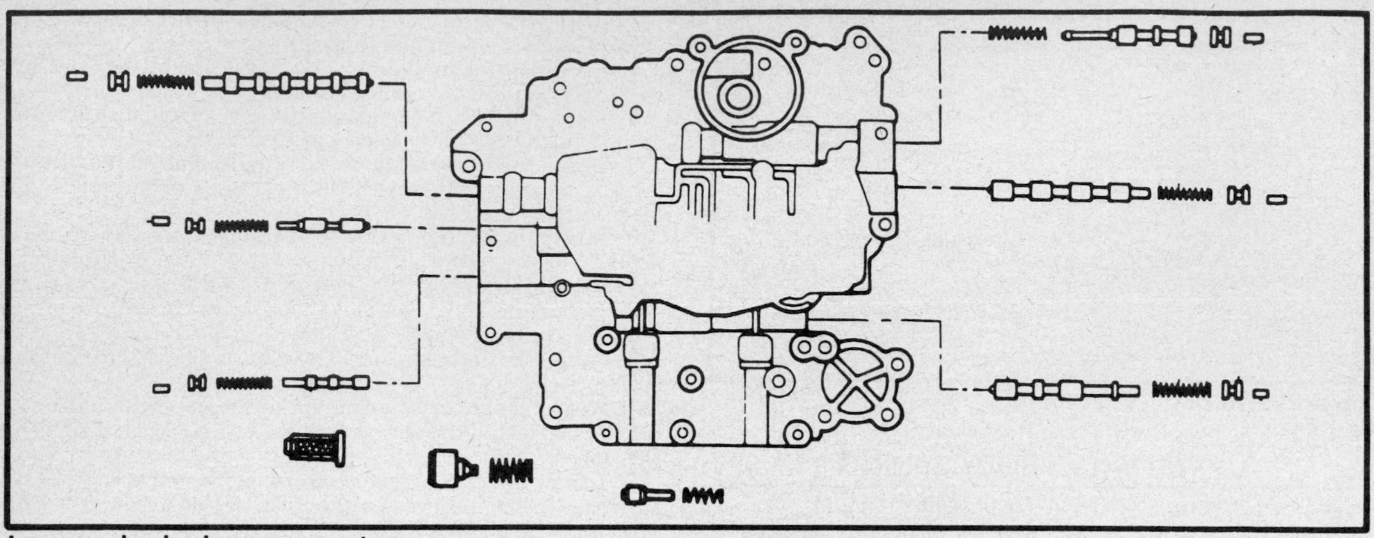

Lower valve body components

LOWER VALVE BODY SPRING

Spring	Free Length in.	(mm)	Color
Secondary regulator valve	1.2937	32.86	Orange
1-2 Shift valve	1.2114	30.77	Purple
3-4 Shift valve	1.2114	30.77	Purple
2-3 Shift valve	1.2114	30.77	Purple
2nd coast modulator valve	1.1665	29.63	Red
Lockup signal valve	1.1811	30.00	Orange

VALVE BODY

Assembly

Install the lower valve body on the upper valve body together with the plate.

1. Position the new gaskets and the plate on the lower valve body. Assemble the gasket having the larger cooler bypass hole to the lower valve body.
2. Install a new gasket and plate onto the lower valve body. Install another new gasket onto the plate. Align each bolt hole in the valve body with the 2 gaskets and plate.
3. Position the lower valve body and gaskets with the plate on top of the upper valve body. Align each bolt hole in the valve bodies with the gaskets and plate.
4. Install and finger tighten 3 bolts in the lower valve body to secure the upper valve body.
5. Turn the assembly over and finger tighten 3 bolts in the upper valve body.
6. Install the upper valve body cover new gaskets and strainer. Install the valve body cover and finger tighten the bolts.
7. Turn the assembly over and install a new gasket and the lower valve body cover.
8. Tighten the bolts of the upper and lower valve body to 48 inch lbs. (5.4 Nm).

Transaxle Assembly

1. Before assembling new clutch discs, soak them in automatic transmission fluid for at least 2 hours. Apply automatic transmission fluid to the sliding or rotating surfaces of parts before assembly.
2. Use petroleum jelly to keep small parts in their places. Do not use adhesive cements on gaskets and similar parts. When assembling the transaxle, be sure to use new gaskets and O-rings.
3. Install the bearing in the transaxle housing using tools J-8092 and J-37274 and a press. Install the bearing stopper.
4. Install the oil tube apply cover. Install the oil tubes. Install the bearing in the transaxle case with the lettering facing towards the case using tools J-8092 and J-37274 or equivalent and a press.
5. Install the B_4 accumulator piston and spring. Install the oil galley cover and gasket.
6. Install the oil seal rings in the transaxle case and check that they move smoothly.
7. Install the manual shaft with the washer, spacer, manual lever and retaining spring.
8. Install the pin and stake the spacer.
9. Install the cam guide bracket.
10. Install the parking lock sleeve with the protruding portion facing upward.
11. Install the parking lock pawl stopper plate, torsion spring and spring guide.
12. Install the parking lock pawl shaft, lock pawl and the pawl shaft clamp.
13. Install the underdrive brake piston and the return spring.
14. Install the plates, disc and flange in the following order: plate-disc-plate-disc-plate-disc. Install the flange facing the flat end upward.
15. Compress the return spring and install the snapring.
16. Install the underdrive one-way clutch. Install the anti-rattle clip.
17. Install the bearing and race. Install the snapring on the sun gear.
18. Install the sun gear in the transaxle case.
19. Install the countershaft assembly and the anti-rattle clip.
20. Install the thrust needle bearing.
21. Install the counter driven gear using tool J-1859-03 or equivalent.
22. Stake the countershaft locknut on both sides.
23. Install the intermediate shaft and the transaxle rear cover.
24. Install the governor driven gear and thrust washer.
25. Install the oil seals.
26. Install the 1st and reverse brake into the case.
27. Install the flange, plates and discs in the following order: flange-disc-plate-disc-plate-disc.

28. Install the snapring holding the flange to the case.
29. Install the No. 2 one-way clutch and the rear planetary gear with the thrust washers.
30. Install the rear planetary ring gear, races and bearing.
31. Install the snapring holding the No. 2 one-way clutch outer race to the case.
32. Install the second coast brake band guide.
33. Install the 2nd brake drum piston return spring.
34. Install the 2nd brake piston.
35. Install the snapring and the 2nd brake drum.
36. Install the 2nd brake drum gasket.
37. Install the No. 1 one-way clutch and the 2nd brake hub.
38. Install the sun gear, sun gear input drum and thrust washer.
39. Install the front planetary gear with the race.
40. Install the front planetary ring gear with the bearing and race.
41. Install the 2nd coast brake band.
42. Install the forward clutch.
43. Install the direct clutch.
44. Install the differential.
45. Install the transaxle housing to the case.
46. Install the oil pump.
47. Install the 2nd coast brake band piston.
48. Install the accumulator pistons and springs. Install the cover and gasket.
49. Install the 2nd brake apply gasket.
50. Install the throttle cable into the case.
51. Connect the solenoid wire.
52. Install the valve body.
53. Connect the solenoid connectors.
54. Install the manual detent spring and the detent plate.
55. Install the oil tubes.
56. Install the oil strainer.
57. Install the oil tube bracket.
58. Install the pan and gasket.
59. Install the speed sensor, sensor rotor and adapter.
60. Install the solenoid wire retaining plate.
61. Install the neutral start switch.
62. Install the manual shift lever.
63. Install the transaxle dipstick and filler tube.
64. Install the oil cooler pipes.

SPECIFICATIONS

VALVE BODY SPECIFICATIONS

Item	Free length		Coil outer diameter		No. coils	Wire diameter		Color
	in.	mm	in.	mm		in.	mm	
UPPER VALVE BODY								
Primary regulator valve	2.6240	66.65	0.732	18.6	12.5	0.063	1.6	Purple
Lockup relay valve	0.7402	18.80	0.201	5.1	14.5	0.020	0.5	None
Low coast modulator valve	1.0831	27.51	0.327	8.3	12.5	0.035	0.9	Yellow
Kickdown valve	1.1717	29.76	0.3437	8.73	13.5	0.039	1.0	White
Throttle valve	1.1488	29.18	0.362	9.2	9.5	0.028	0.7	Yellow green
Throttle modulator valve	1.1772	29.90	0.354	9.0	15.5	0.035	0.9	Green
Accumulator control valve	1.5039	38.20	0.394	10.0	11.5	0.035	0.9	Yellow
LOWER VALVE BODY								
Secondary regulator valve	1.2937	32.86	0.433	11.0	11.5	0.055	1.4	Orange
1-2 shift valve	1.2114	30.77	0.382	9.7	10.5	0.035	0.9	Purple
3-4 shift valve	1.2114	30.77	0.382	9.7	10.5	0.035	0.9	Purple
2-3 shift valve	1.2114	30.77	0.382	9.7	10.5	0.035	0.9	Purple
2nd coast modulator valve	1.1665	29.63	0.327	8.3	12.5	0.039	1.0	Red
Lockup signal valve	1.1811	30.00	0.323	8.2	11.5	0.028	0.7	Orange

ACCUMULATOR PISTON SPRING SPECIFICATIONS

Item		Free length		Coil outer diameter		No. coils	Wire diameter		Color
		in.	mm	in.	mm		in.	mm	
B_4 (Underdrive)		2.5756	65.42	0.7087	18.00	13.0	0.1024	2.60	None
B_2		2.6252	70.00	0.6976	17.72	15.5	0.1024	2.60	None
C_1	Inner	1.6732	42.50	0.6811	17.30	9.5	0.0906	2.30	None
	Outer	2.7992	71.10	0.9134	23.20	12.5	0.0906	2.30	White
C_2		2.4677	62.68	0.6890	17.50	18.5	0.0906	2.30	Black
C_3		2.4201	61.47	0.6193	15.73	14.5	0.0906	2.30	White

BUSHING SPECIFICATIONS

Bushing		Finished Bore in.	Finished Bore (mm)	Bore Limit in.	Bore Limit (mm)
Stator support	Front	0.8465–0.8475	21.500–21.526	0.8494	21.576
	Rear	1.0630–1.0640	27.000–27.026	1.0660	27.076
Oil pump body		1.5005–1.5015	38.138–38.138	1.5035	38.188
Direct clutch drum		1.8504–1.8514	47.000–47.025	1.8533	47.075
Front planetary ring gear flange		0.7490–0.7500	19.025–19.050	0.7520	19.100
Input sun gear (Front and Rear)		0.8671–0.8680	22.025–22.046	0.8699	22.096

OIL PUMP SPECIFICATIONS

Item		in.	mm
Side clearance	Std.	0.0008–0.0020	0.02–0.05
	Limit	0.004	0.1
Body clearance	Std.	0.0028–0.0059	0.07–0.15
	Limit	0.012	0.3
Tip clearance	Std.	0.0043–0.0055	0.11–0.14
Driven gear	Limit	0.012	0.3

CLUTCH PISTON STROKE SPECIFICATIONS

	in.	(mm)
Forward clutch (C_1)	0.0559–0.0713	1.42–1.81
Direct clutch (C_2)	0.0445–0.0591	1.13–1.50
Underdrive clutch (C_3)	0.0579–0.0744	1.47–1.89

TORQUE SPECIFICATIONS

Item	ft. lbs.	Nm
Engine mounting	38	52
Transaxle housing to Engine 12 mm	47	64
Transaxle housing to Engine 10 mm	34	46
Drive plate to Crankshaft	47	64
Torque converter to Drive plate	20	37
Oil pump to Transaxle case	18	25
Oil pump body to Stator shaft	7	10
Second coast brake band guide	48 ①	5.4
Upper valve body to Lower valve body	56 ①	6.4
Valve body	7	10
Accumulator cover	7	10
Oil strainer	7	10
Oil pan	43 ①	4.9
Oil pan drain plug	13	17
Cooler pipe union nut to Union elbow	25	34
Union elbow to Transaxle case	20	27
Testing plug	65 ①	7.4
Parking lock paw bracket	65 ①	7.4
Transaxle rear cover to Transaxle case	22	29
Neutral safety switch to Transaxle case	48 ①	5.4
Neutral safety switch	61 ①	6.9

① Inch lbs.

SPECIAL TOOLS

Tool	Description
J-9617	Oil pump seal installer
J-29182	Pinion shaft bearing cup installer
J-35378	Bearing puller pilot
J-35399	Differential side bearing cup remover
J-35400	Differential side bearing housing support
J-35405	Differential preload wrench
J-35409	Differential side bearing installer
J-35455	Holding fixture
J-35467	One-way clutch tester
J-35495	Oil pump puller adapters
J-35549	Second coast brake piston compressor
J-35552	Differential side bearing cup installer
J-35553	Differential side bearing seal installer
J-35565	Intermediate shaft bearing installer
J-35661	Countergear bearing installer
J-35663	Countergear bearing cup installer
J-35664	Pinion shaft bearing installer
J-35666	Pinion shaft bearing seal installer
J-35679	Band apply pin gauge
J-35683	First/reverse clutch spring compressor adapter
J-35752	Pressure gauge adapter
J-1859-03	Countergear puller
J-3289-20	Holding fixture mount
J-8001	Dial indicator set
J-8092	Driver handle
J-8614-01	Countergear holder
J-21867	Pressure gauge
J-22888	Right differential side bearing puller
J-22912-01	Bearing puller
J-23327	Clutch spring compressor
J-25018-A	Clutch spring compressor adapter
J-25025-1	Dial indicator mount
J-33411	Countergear installer
J-6125-B	Slide hammer set

Section 3

THM 325-4L Transaxle
General Motors

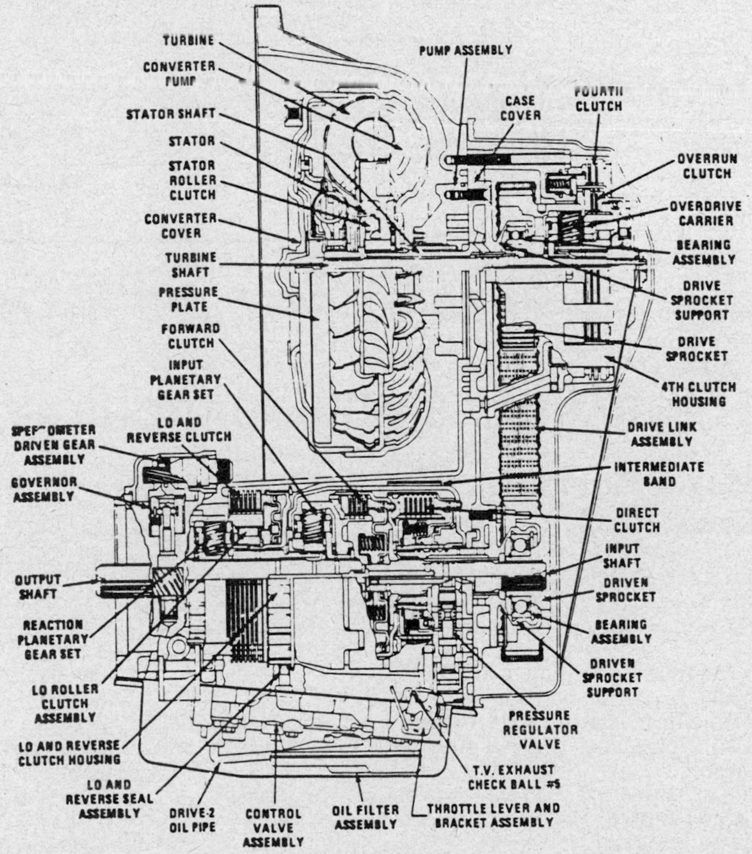

TURBINE
CONVERTER PUMP
STATOR SHAFT
STATOR
STATOR ROLLER CLUTCH
CONVERTER COVER
TURBINE SHAFT
PRESSURE PLATE
FORWARD CLUTCH
INPUT PLANETARY GEAR SET
LO AND REVERSE CLUTCH
DRIVEN GEAR ASSEMBLY
GOVERNOR ASSEMBLY
OUTPUT SHAFT
REACTION PLANETARY GEAR SET
LO ROLLER CLUTCH ASSEMBLY
LO AND REVERSE CLUTCH HOUSING
LO AND REVERSE SEAL ASSEMBLY
DRIVE-2 OIL PIPE
CONTROL VALVE ASSEMBLY
OIL FILTER ASSEMBLY
THROTTLE LEVER AND BRACKET ASSEMBLY
T.V. EXHAUST CHECK BALL #5
PRESSURE REGULATOR VALVE
DRIVEN SPROCKET SUPPORT
BEARING ASSEMBLY
DRIVEN SPROCKET
INPUT SHAFT
DIRECT CLUTCH
INTERMEDIATE BAND
DRIVE LINK ASSEMBLY
4TH CLUTCH HOUSING
DRIVE SPROCKET
DRIVE SPROCKET SUPPORT
BEARING ASSEMBLY
OVERDRIVE CARRIER
OVERRUN CLUTCH
FOURTH CLUTCH
CASE COVER
PUMP ASSEMBLY

Cut away view of the 325–4L

REFERENCES

APPLICATION

1980–84 Chilton's Professional Automatic Transmission Manual #7390
1984–89 Chilton's Professional Transmission Manual—Domestic Vehicles #7959

GENERAL DESCRIPTION

1980–84 Chilton's Professional Automatic Transmission Manual #7390

MODIFICATIONS

1980–84 Chilton's Professional Automatic Transmission Manual #7390
1984–89 Chilton's Professional Transmission Manual—Domestic Vehicles #7959

TROUBLE DIAGNOSIS

1980–84 Chilton's Professional Automatic Transmission Manual #7390

ON CAR SERVICES

1980–84 Chilton's Professional Automatic Transmission Manual #7390

REMOVAL AND INSTALLATION

1980–84 Chilton's Professional Automatic Transmission Manual #7390
1984–89 Chilton's Professional Transmission Manual—Domestic Vehicles #7959

BENCH OVERHAUL

1980–84 Chilton's Professional Automatic Transmission Manual #7390

SPECIFICATIONS

1980–84 Chilton's Professional Automatic Transmission Manual #7390
1984–89 Chilton's Professional Transmission Manual—Domestic Vehicles #7959

SPECIAL TOOLS

1980–84 Chilton's Professional Automatic Transmission Manual #7390

APPLICATION

THM 325-4L

Year	Vehicle	Engine
1984	Riviera	3.8L
		4.1L
		5.0L
		5.7L Diesel
	Seville/Eldorado	4.1L (HI 4100)
		5.7L Diesel
	Toronado	4.1L
		5.0L
		5.7L Diesel

Year	Vehicle	Engine
1985	Riviera	4.1L
		3.8L
		5.0L
	Seville/Eldorado	4.1L (HI 4100)
		5.7L Diesel
	Toronado	5.0L
		5.7L Diesel

TRANSAXLE MODIFICATIONS

Automatic Transaxle Fluid Leak in Final Drive Area

Some vehicles may experience a transaxle oil leak in the final drive-to-transaxle case mating area and/or an erratic upshift caused by a loose or missing governor cup plug in the transaxle case passage.

Automatic transaxles with the fluid leak problem are: model AJ serial number 58856, model AL serial number 8327 and model AM serial number 5587. These identification numbers are located on the transaxle identification tag, which is mounted on the left-side of the bell housing.

In order to correct an automatic transaxle with this problem the following procedure must be followed:

1. Remove the final drive assembly.
2. Remove the original governor cup plug.
3. Install a new governor cup plug, part number 8620318. Use Loctite® 240 or equivalent, to seal the cup plug to the transaxle case. Staking the plug cup into the transaxle case will not assure an adequate seal.
4. Using a new gasket, install the final drive assembly.

Hydraulic Noise in Park and/or Neutral

Some vehicles may experience a hydraulic buzzing sound in the **P** and/or **N** shift selector positions. This condition may be caused by an orifice cup plug missing from the pressure regulator valve.

In order to correct this condition, the following procedure must be performed:

1. Remove the automatic transaxle fluid pan and discard the pan gasket. Remove the pressure regulator valve.

SELECTIVE WASHER CHART

Identification Number	Color	Part Number	Thickness (inches)
1	----	8639291	.065-.070
2	----	8639292	.070-.075
3	Black	8639293	.076-.080
4	Light Green	8639294	.081-.085
5	Scarlet	8639295	.086-.090
6	Purple	8639296	.091-.095
7	Cocoa Brown	8639297	.096-.100
8	Orange	8639298	.101-.106
9	Yellow	8639299	.106-.111
10	Light Blue	8639300	.111-.116
11	Blue	8639301	.117-.121
12	----	8639302	.122-.126
13	Pink	8639303	.127-.131
14	Green	8639304	.132-.136
15	Gray	8639305	.137-.141

2. Inspect the valve for a missing orifice cup plug, which is located at the end of the valve.
3. If the cup plug is missing, inspect the pressure regulator valve bushing as well as the transaxle case for the original plug. If found, remove and discard.
4. Install the new pressure regulator valve, part number 8623422. Be sure the new valve is equipped with an orifice cup plug.
5. Using a new gasket, install the transaxle oil pan. Fill the transaxle as required with the proper grade and type automatic transmission fluid.

No Drive or Slipping

Some vehicles may experience a no drive or slipping condition. This condition, while not noticed in reverse, could be caused by a loose or missing solid cup plug in the input shaft of the forward clutch assembly.

When servicing any transaxles for this condition and the forward clutch assembly shows signs of burning, inspect the housing end of the input shaft which is part of the forward clutch assembly. If the solid cup plug is loose or missing, apply pressure to the forward clutch piston will be lost. This pressure loss will cause a no drive or slipping condition. If the solid cup plug is missing from the feed passage, it must be located before the transaxle is reassembled. It could possibly be in the open end of the output shaft. Tap the end of the output shaft on a table and remove the solid cup plug.

The solid cup plug, part number 8628145, is serviced separately. To install the cup plug, use a $^{5}/_{16}$ in. (7mm) punch. Apply Loctite® 290 or equivalent, to the plug before installation. Drive the plug into the larger of the 2 holes until it is 0.039 in. (1mm) below the surface.

Selective Washer for Input to Output Shaft

A selective washer is now available when servicing the input to output shaft assemblies. The selective washer is made from a different material and can be identified by a raised portion on the identification tab.

When servicing an automatic transaxle that necessitates the replacement of this selective washer due to wear or damage, the following procedure must be followed:

1. Inspect both shafts for wear or damage on the selective washer mating surfaces. If wear or damage is found, replace the components.
2. Use the proper selective washer from the selective washer chart.
3. Check the front endplay to verify the proper selective washer selection. Front endplay should be 0.022–0.051 in. (0.559–1.295mm).

Automatic Transaxle Bottom Oil Pan Leak

Some vehicles may exhibit a transaxle oil pan leak. This leak is caused by an interference between the 1–2 and 3–4 accumulator housing and the bottom oil pan.

Transaxles produced with a new design accumulator housing will correct the interference problem.

When servicing the transaxle for a bottom oil leak problem, be sure to inspect for interference with the accumulator housing when the oil pan is removed. If evidence of interference exists, remove the housing and grind off a portion of the casting boss. When performing this operation use caution to avoid damaging the cup plug.

4–2 Downshift Slip and Bump (Model AJ)

Some vehicles equipped with transaxle model AJ, may experience a 4–2 downshift as a 2 shift feel with a slight hesitation and a bump in between. This condition is most noticeable during a full throttle detent downshift.

All transaxles, beginning with transaxle serial number 83–AJ-30723, are built using a control valve assembly which correct the 2–4 downshift condition.

When servicing a model AJ transaxle for this condition, use service package 8635949, which contains the proper components to correct the problem.

No 4th Gear Engagement

Some transaxles may not engage into 4th gear (OD). If this con-

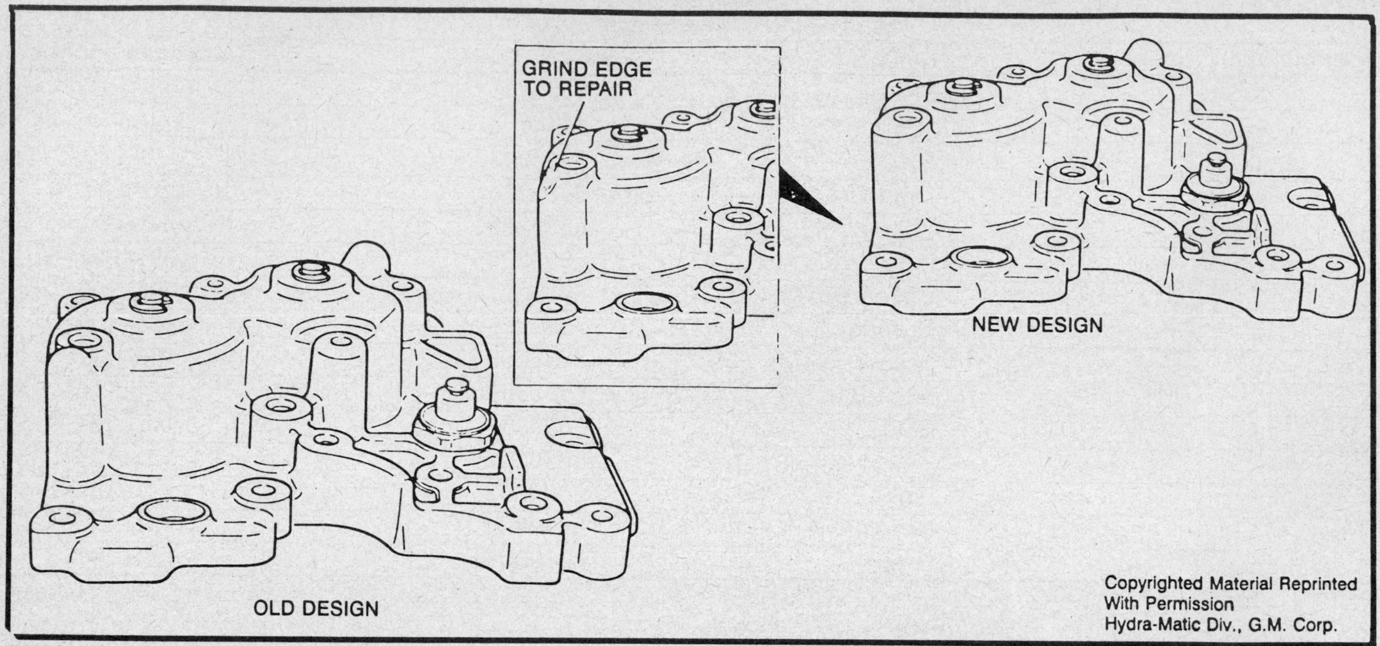
GRIND EDGE
TO REPAIR
NEW DESIGN
OLD DESIGN
Copyrighted Material Reprinted
With Permission
Hydra-Matic Div., G.M. Corp.

THM 325-4L accumulator housing

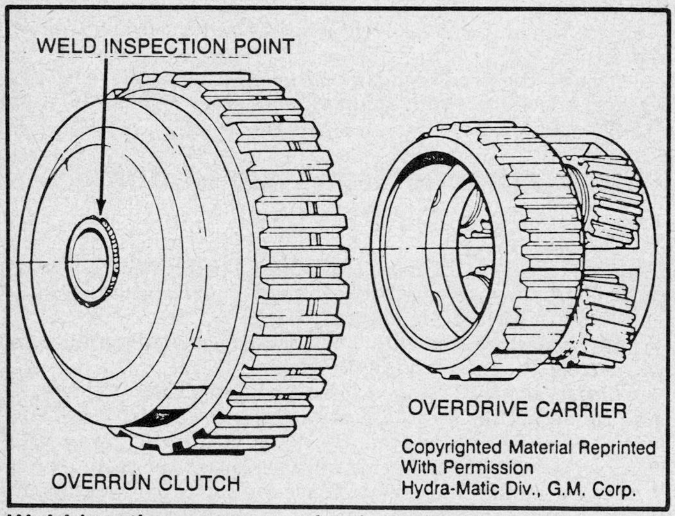
WELD INSPECTION POINT
OVERDRIVE CARRIER
OVERRUN CLUTCH
Copyrighted Material Reprinted
With Permission
Hydra-Matic Div., G.M. Corp.

Weld location – overrun clutch assembly

dition exists, the following diagnostic procedure will aid in correction of the problem:

Test drive the vehicle applying light throttle in **D** range (3rd gear) to about 30–35 mph (48–56 km/h). Move the selector lever into manual **2** and release the throttle. The automatic transaxle should downshift and coastdown overrun braking should occur.

If coastdown overrun braking does not occur, remove the transaxle from the vehicle. Inspect the overrun clutch housing for a break in the hub weld. If the hub weld is broken, the cause may be due to misaligned pinions in the overdrive carrier.

In all instances where the weld is broken on the overrun clutch housing, both the overrun clutch housing, part number 8635941 and the overdrive carrier, part number 8635036, must be replaced.

Shift Busyness, Cruise Control Engaged

Some vehicles, equipped with a model AM transaxle and the 4.1L engine, may experience a transaxle shift busyness at highway speeds with the cruise control engaged. Shift busyness refers to the cycling of the transaxle between 4th and 3rd gears; this condition is noticeable while driving up inclines. In order to correct this problem, install service package number 8635944 into the control valve assembly.

Delayed Engagement and/or Loss of Engine Power

Some transaxles may exhibit either delayed engagement when moving the selector lever from **P** to **R** or an apparent loss of engine power resulting from improper low and reverse clutch operation.

Delayed engagement (hot or cold) when moving the automatic transaxle selector lever from **P** to **R** after a complete overhaul can be caused by an insufficient number of flat steel plates installed in the low and reverse clutch assembly. The low and reverse clutch assembly should have 1 waved steel plate, 7 flat steel plates and 6 composition faced plates installed in it.

An apparent loss of engine power or automatic transaxle operation in 2 gears at the same time may be caused by the incorrect application of the low and reverse clutch assembly. The incorrect apply of the low and reverse clutch can be caused by a missing locator pin securing the low 1st detent valve bushing in the control valve assembly. This missing pin will permit the bushing to rotate in its control valve bore and apply the low and reverse clutch through the use of "detent" fluid. Whenever the throttle angle is above 60 degrees of travel, "detent" fluid is available.

To correct this condition, reroute the valve body, orient the bushing correctly in it's bore and install a locator pin, part number 8628400.

Check Ball Location Data

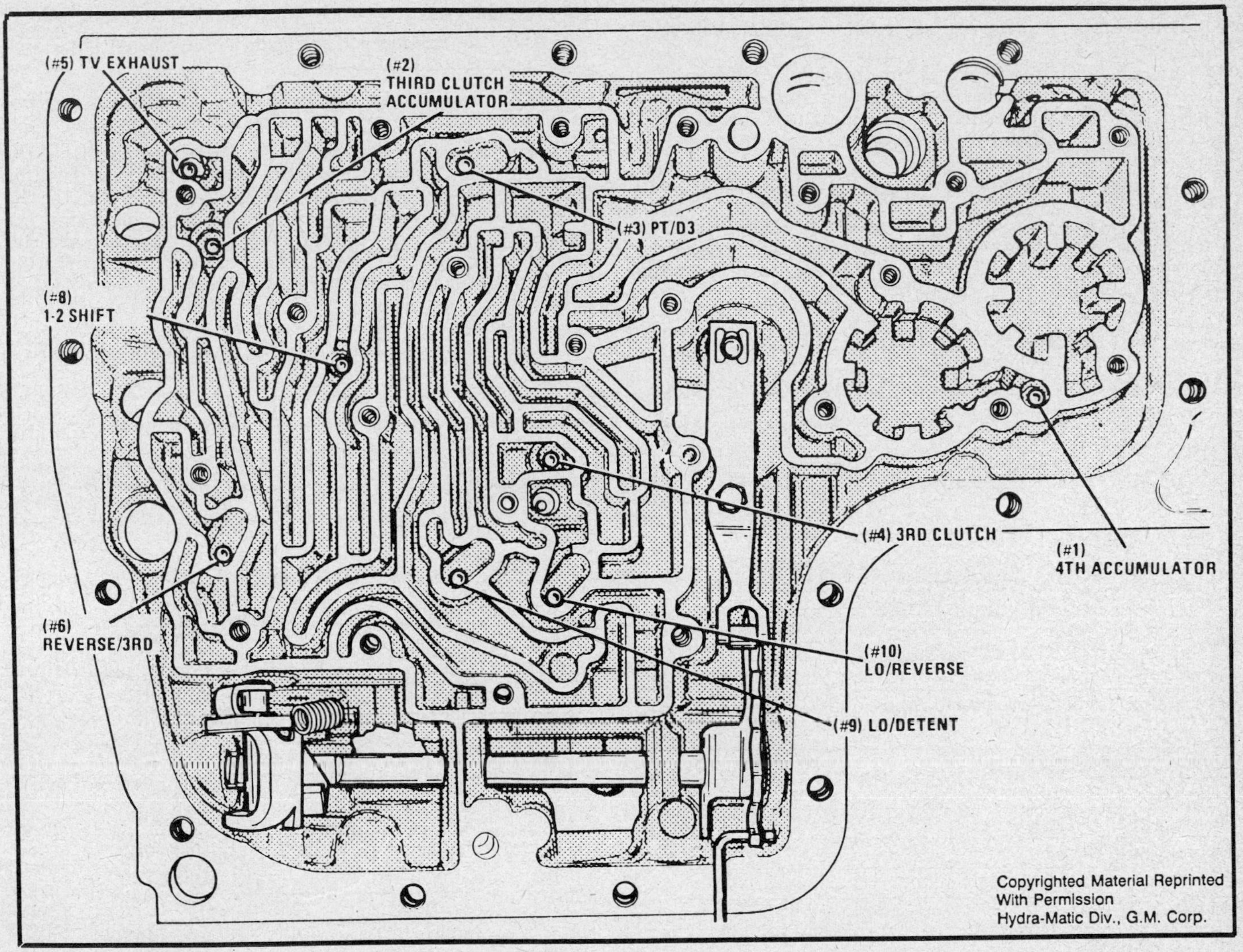
(#5) TV EXHAUST
(#2) THIRD CLUTCH ACCUMULATOR
(#3) PT/D3
(#8) 1-2 SHIFT
(#4) 3RD CLUTCH
(#1) 4TH ACCUMULATOR
(#6) REVERSE/3RD
(#10) LO/REVERSE
(#9) LO/DETENT
Copyrighted Material Reprinted With Permission Hydra-Matic Div., G.M. Corp.

Check ball location

REMOVAL AND INSTALLATION

TRANSAXLE REMOVAL

1. Open the hood and disconnect the negative battery cable (2 terminals on diesels).
2. Disconnect the speedometer cable from the transaxle. Remove the transaxle's oil dipstick tube.
3. Remove the air cleaner.
4. Disconnect the throttle valve (T.V./detent) cable from the transaxle at it's upper end. Disconnect the linkage by removing a nut from the shaft on the left side of the transaxle, if equipped.
5. Using an engine holding fixture tool between the cowl and the radiator support, safely support the engine.
6. Remove the top and 2 upper left final drive-to-transaxle bolts.
7. Remove the remaining accessible engine-to-transaxle bolts.
8. Raise and support the vehicle safely.
9. Remove the starter-to-transaxle bolts and the starter from the transaxle.
10. Disconnect the transaxle converter clutch connector.
11. Disconnect and plug the transaxle oil cooler lines.
12. Remove the flywheel inspection cover (loosen the top left bolt). Matchmark the flywheel-to-converter relationship for later assembly.
13. On the V8 engine, disconnect the exhaust Y-pipe connection to the left exhaust pipe. On all models, disconnect the right exhaust pipe at the manifold. On gasoline engines, disconnect both catalytic converter hanger bolts. On all models, lower the exhaust system about 5 in. (127mm) and support it.
14. Remove the crossmember-to-frame bolts.
15. Using a floor jack and a wooden block (to protect the transaxle case), position it under the transaxle case and raise the transaxle slightly.
16. Remove the remaining final drive-to-transaxle bolts.

17. Remove the torque converter-to-flywheel bolts.
18. Disconnect the shift linkage from the transaxle.
19. Remove the final drive support bracket bolt.
20. Remove the right transaxle mount (through bolt and bracket bolts).
21. Remove the left transaxle mount through bolt. Remove the lower bracket-to-transaxle bolt. Raise the transaxle assembly about 2 in. (51mm) for access to the remaining upper bracket-to-transaxle bolts. Remove the remaining transaxle-to-engine bolts.
22. Carefully lower the transaxle unit while disengaging the final drive.
23. Install a C-clamp or torque converter holding clamp in front of the torque converter (attached to the bell housing) to hold the converter in place. Remove the transaxle from the vehicle.

TRANSAXLE INSTALLATION

1. Using a new gasket, install the transaxle into the vehicle.

NOTE: Use care when engaging the final drive-to-transaxle splines and make sure the final drive-to-transaxle mounting faces are in alignment with each other.

2. After the splines are engaged, loosely install both final drive-to-transaxle lower bolts.

NOTE: Time can be saved here by installing 2 engine-to-transaxle bolts from above to aid engagement.

3. After the final drive and transaxle are aligned, install the attaching bolts. Before torquing the flywheel-to-converter bolts, make sure the weld nuts on the converter are flush with the flywheel and the converter rotates freely by hand. Hand start the bolts and tighten finger tight; this will insure proper converter alignment. Torque the transaxle-to-engine bolts to 30 ft. lbs. (40 Nm) and the final drive support bracket to final drive bolts to 35 ft. lbs. (48 Nm).
4. Adjust the T.V. cable.
5. Refill the transaxle to the proper level with Dexron®II.

SPECIFICATIONS

TORQUE SPECIFICATIONS

Description of Usage	Torque
Valve Body Assembly to Case	9-12 ft.-lbs.
Accumulator Housing to Case	9-12 ft.-lbs.
Manual Detent Spring Assembly to Case	9-12 ft.-lbs.
Governor Cover to Case	6-10 ft.-lbs.
Sprocket Cover to Case	9 ft.-lbs.
Case Cover to Pump Body (Flat Head)	15-20 ft.-lbs.
Driven Support to Case Cover	15-20 ft.-lbs.
Case Cover to Case	15-20 ft.-lbs.
Fourth Clutch Housing to Case	15-20 ft.-lbs.
Oil Pan to Case	7-10 ft.-lbs.
Cam to Manual Shaft	6-9 ft.-lbs.
Manual Shaft to Inside Detent Lever (Nut)	20-25 ft.-lbs.
Cooler Connector	26-30 ft.-lbs.
Line Pressure Take-Off	5-10 ft.-lbs.
Third Accumulator Take-Off	5-10 ft.-lbs.
Governor Pressure Take-Off	5-10 ft.-lbs.
Fourth Pressure Take-Off	5-10 ft.-lbs.
Pressure Switch	5-10 ft.-lbs.

INPUT UNIT ENDPLAY SELECTIVE WASHER CHART

Identification Number	Color	Thickness (inches)
One	—	.065-.070
Two	—	.070-.075
Three	Black	.076-.080
Four	Light Green	.081-.085
Five	Scarlet	.086-.090
Six	Purple	.091-.095
Seven	Cocoa Brown	.096-.100
Eight	Orange	.101-.206
Nine	Yellow	.106-.111
Ten	Light Blue	.111-.116
Eleven	—	.117-.121
Twelve	—	.122-.126
Thirteen	Pink	.127-.131
Fourteen	Green	.132-.136
Fifteen	Gray	.137-.141

DIRECT CLUTCH PLATE AND APPLY RING USAGE CHART

	Flat Steel Plate	Composition Faced Plate	Apply Ring
Number	Six	Six	—
Thickness	.91 Inch	—	—
Identification	—	—	Nine
Width	—	—	.492 Inch

LOW AND REVERSE CLUTCH PLATE AND APPLY RING USAGE CHART

	Wave Plate	Flat Steel Plate	Composition Faced Plate	Apply Ring
Number	One	Seven	Six	—
Thickness	.077 Inch	.077 Inch	—	—
Identification	—	—	—	Zero
Width	—	—	—	.516 Inch

Section 3

THM 440-4 and F7 Transaxles
General Motors

APPLICATION

THM 440-T4

Year	Body	Vehicle	Engine
1984	A	Century	2.8L 2bbl
	A	Century	3.0L 2bbl
	A	Century	3.8L MPI
	A	Celebrity	2.8L 2bbl
	A	Celebrity	4.3L Diesel
	A	Ciera	2.8L 2bbl
	A	Ciera	3.0L 2bbl
	A	Ciera	3.8L MPI
	A	6000	2.8L 2bbl
	A	6000	4.3L Diesel
1985	A	Century	2.8L 2bbl
	A	Century	3.0L 2bbl
	A	Century	3.8L MPI
	A	Century	4.3L Diesel
	C	Electra	3.0L 2bbl
	C	Electra	3.8L MPI
	C	Electra	4.3L Diesel
	C	Deville	4.1L DFI
	C	Fleetwood	4.1L DFI
	A	Celebrity	2.8L 2bbl
	A	Celebrity	2.8L MPI
	A	Celebrity	4.3L Diesel
	A	Ciera	2.8L 2bbl
	A	Ciera	3.8L MPI
	A	Ciera	4.3L Diesel
	C	Ninety-Eight	3.0L 2bbl
	C	Ninety-Eight	3.8L MPI
	C	Ninety-Eight	4.3L Diesel
	A	6000	2.8L 2bbl
	A	6000	2.8L MPI
	A	6000	4.3L Diesel
1986	A	Century	3.8L SFI
	A	Century	2.8L 2bbl
	C	Electra	3.8L SFI
	H	LeSabre	3.0L EFI
	H	LeSabre	3.8L SFI
	E	Riviera	3.8L SFI
	C	Deville	4.1L DFI
	E	Eldorado	4.1L DFI
	C	Fleetwood	4.1L DFI
	K	Seville	4.1L DFI
	A	Celebrity	2.8L 2bbl
	A	Celebrity	2.8L MPI

THM 440-T4

Year	Body	Vehicle	Engine
1986	A	Ciera	2.8L 2bbl
	A	Ciera	3.8L SFI
	C	Ninety-Eight	3.8L SFI
	H	Eighty-Eight	3.0L EFI
	H	Eighty-Eight	3.8L SFI
	E	Toronado	3.8L SFI
	A	6000	2.8L 2bbl
	A	6000	2.8L MPI
	A	6000	3.8L SFI
1987	A	Century	3.8L SFI
	A	Century	2.8L MFI
	C	Electra	3.8L SFI
	H	LeSabre	3.8L SFI
	E	Riviera	3.8L SFI
	V	Allante	4.1L DFI
	C	Deville	4.1L DFI
	E	Eldorado	4.1L DFI
	C	Fleetwood	4.1L DFI
	K	Seville	4.1L DFI
	A	Celebrity	2.8L MPI
	A	Ciera	2.8L MPI
	A	Ciera	3.8L SFI
	C	Ninety-Eight	3.8L SFI
	H	Eighty-Eight	3.8L SFI
	E	Toronado	3.8L SFI
	A	6000	2.8L MPI
	A	6000	3.8L SFI
	H	Bonneville	3.8L SFI
1988	A	Century	3.8L SFI
	A	Century	2.8L EFI
	C	Electra	3.8L SFI
	H	LeSabre	3.8L SFI
	E	Reatta	3.8L SFI
	E	Riviera	3.8L SFI
	W	Regal	2.8L MPI
	V	Allante	4.1L DFI
	C	Deville	4.5L DFI
	E	Eldorado	4.5L DFI
	C	Fleetwood	4.5L DFI
	K	Seville	4.5L DFI
	A	Celebrity	2.8L MPI
	A	Ciera	2.8L MPI
	A	Ciera	3.8L SFI

THM 440-T4

Year	Body	Vehicle	Engine
1988	C	Ninety-Eight	3.8L SFI
	H	Eighty-Eight	3.8L SFI
	E	Toronado	3.8L SFI
	W	Cutlass Supreme	2.8L MPI
	A	6000	2.8L MPI
	H	Bonneville	3.8L SFI
	W	Grand Prix	2.8L MPI
1989	A	Century	3.3L MFI
	A	Century	2.8L MFI
	C	Electra	3.8L SFI
	H	LeSabre	3.8L SFI
	E	Reatta	3.8L SFI
	E	Riviera	3.8L SFI
	W	Regal	2.8L MPI
	V	Allante	4.5L DFI
	C	Deville	4.5L DFI
	E	Eldorado	4.5L DFI
	C	Fleetwood	4.5L DFI
	K	Seville	4.5L DFI
	A	Celebrity	2.8L MPI
	A	Ciera	2.8L MPI
	A	Ciera	3.3L MFI
	C	Ninety-Eight	3.8L SFI
	H	Eighty-Eight	3.8L SFI
	E	Toronado	3.8L SFI

THM 440-T4

Year	Body	Vehicle	Engine
1989	W	Cutlass Supreme	2.8L MPI
	A	6000	2.8L MPI
	H	Bonneville	3.8L SFI
	W	Grand Prix	2.8L MPI

BBL—Barrel
DFI—Digital Fuel Injection
EFI—Electronic Fuel Injection
MFI—Multi-Port Fuel Injection
SFI—Sequential Fuel Injection

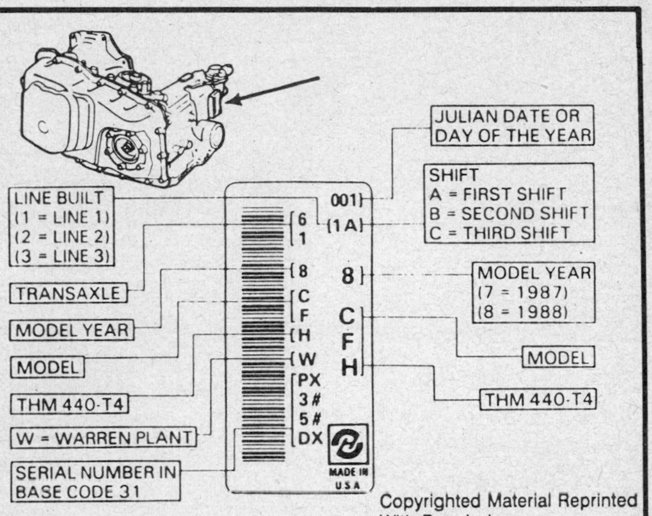
JULIAN DATE OR DAY OF THE YEAR
SHIFT
A = FIRST SHIFT
B = SECOND SHIFT
C = THIRD SHIFT
LINE BUILT
(1 = LINE 1)
(2 = LINE 2)
(3 = LINE 3)
TRANSAXLE
MODEL YEAR
MODEL
THM 440-T4
W = WARREN PLANT
SERIAL NUMBER IN BASE CODE 31
001
(1 A)
8
C
F
H
6
1
W
PX
3#
5#
DX
MADE IN U.S.A.
MODEL YEAR
(7 = 1987)
(8 = 1988)
MODEL
THM 440-T4

Transaxle identification data

GENERAL DESCRIPTION

Transaxle and Converter Identification

TRANSAXLE

The THM 440–T4 (ME9) and the F-7 automatic transaxles are fully automatic units. These transaxles consist of 4 multiple disc clutches, a roller clutch, a sprag and 2 bands, requiring hydraulic and mechanical applications to obtain the desired gear ratios from the compound planetary gears. The transaxle identification can be located on 1 of 3 areas of the unit. An identification plate on the side of the case, a stamped number on the governor housing or an ink stamp on the bell housing.

CONVERTER

Two types of converters are used in the varied vehicle applications. The first type is the 3 element torque converter combined with a lockup converter clutch. The second type is the 3 element torque converter combined with a viscous lockup converter clutch that has silicone fluid sealed between the cover and the body of the clutch assembly. Identification of the torque converters are either ink stamp marks or a stamped number on the shell of the converter.

Electronic Controls

The Torque Converter Clutch (TCC) system, or Cadillac's system, which uses a Viscous Converter Clutch (VCC) system, uses a solenoid operated valve in the automatic transaxle to couple the engine flywheel to the output shaft of the transaxle through the torque converter. This reduces the slippage losses in the converter, which increases fuel economy.

For the converter clutch to apply, 2 conditions must be met:

1. Internal transaxle fluid pressure must be correct.
2. The ECM grounds a switch internally to turn **ON** a solenoid in the transaxle. This moves a check ball, which will allow the converter clutch to apply, if the hydraulic pressure is correct.

The ECM controls the TCC apply solenoid by looking at several sensors:

1. Vehicle Speed Sensor (VSS)—Speed must be above a certain value before the clutch can apply.
2. Coolant Temperature Sensor (CTS)—Engine must be warmed up before clutch can apply.
3. Throttle Position Sensor (TPS)—After the converter clutch applies, the ECM uses the information from the TPS to

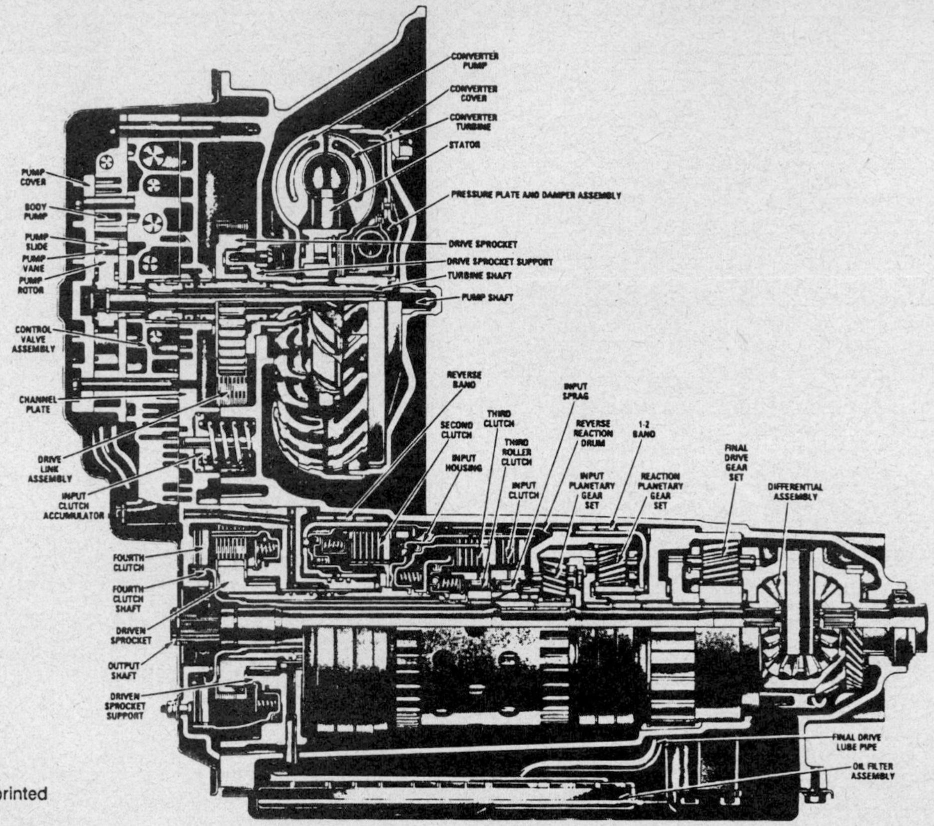
CONVERTER PUMP
CONVERTER COVER
CONVERTER TURBINE
STATOR
PRESSURE PLATE AND DAMPER ASSEMBLY
DRIVE SPROCKET
DRIVE SPROCKET SUPPORT
TURBINE SHAFT
PUMP SHAFT
PUMP COVER
BODY PUMP
PUMP SLIDE
PUMP VANE
PUMP ROTOR
CONTROL VALVE ASSEMBLY
CHANNEL PLATE
DRIVE LINK ASSEMBLY
INPUT CLUTCH ACCUMULATOR
REVERSE BAND
SECOND CLUTCH
THIRD CLUTCH
INPUT HOUSING
THIRD ROLLER CLUTCH
INPUT CLUTCH
INPUT SPRAG
REVERSE REACTION DRUM
INPUT PLANETARY GEAR SET
1-2 BAND
REACTION PLANETARY GEAR SET
FINAL DRIVE GEAR SET
DIFFERENTIAL ASSEMBLY
FOURTH CLUTCH
FOURTH CLUTCH SHAFT
DRIVEN SPROCKET
OUTPUT SHAFT
DRIVEN SPROCKET SUPPORT
FINAL DRIVE LUBE PIPE
OIL FILTER ASSEMBLY

Cross section of the THM 440–T4 automatic transaxle

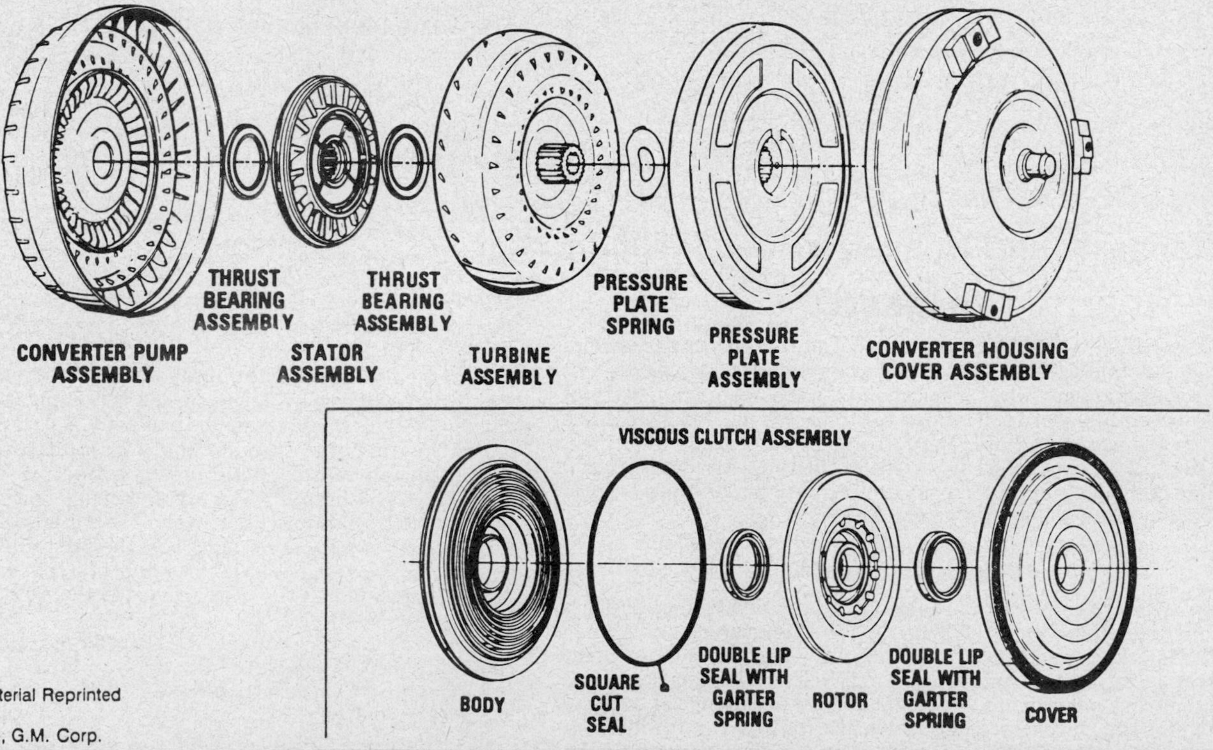
THRUST BEARING ASSEMBLY
THRUST BEARING ASSEMBLY
PRESSURE PLATE SPRING
CONVERTER PUMP ASSEMBLY
STATOR ASSEMBLY
TURBINE ASSEMBLY
PRESSURE PLATE ASSEMBLY
CONVERTER HOUSING COVER ASSEMBLY
VISCOUS CLUTCH ASSEMBLY
BODY
SQUARE CUT SEAL
DOUBLE LIP SEAL WITH GARTER SPRING
ROTOR
DOUBLE LIP SEAL WITH GARTER SPRING
COVER

Two types of converters used with the THM 440–T4 transaxle

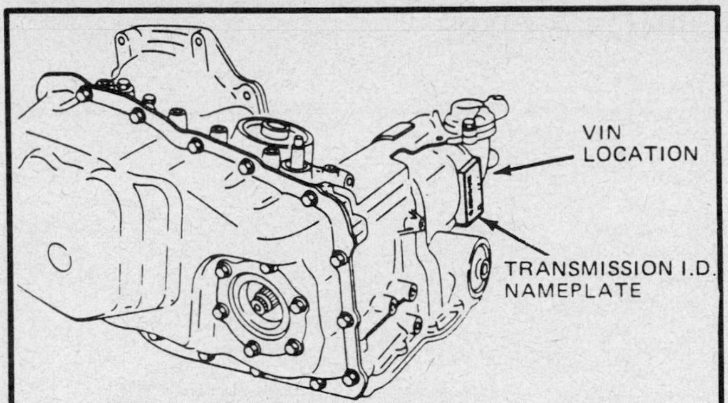
VIN
LOCATION
TRANSMISSION I.D.
NAMEPLATE

440-T4 TRANSMISSION I.D. NAMEPLATE LOCATION

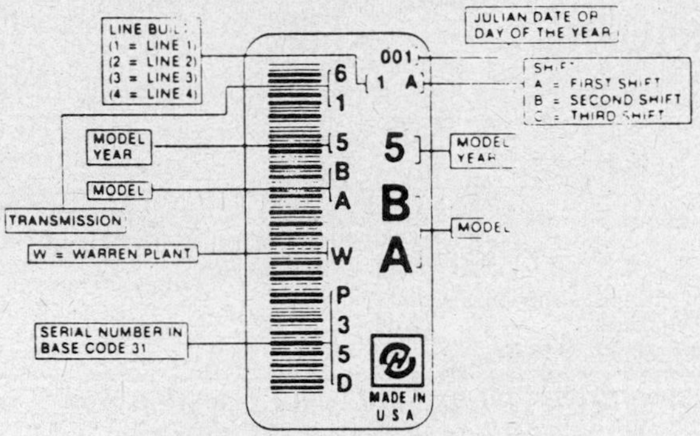
LINE BUILT
(1 = LINE 1)
(2 = LINE 2)
(3 = LINE 3)
(4 = LINE 4)
JULIAN DATE OR
DAY OF THE YEAR
001
A = FIRST SHIFT
B = SECOND SHIFT
C = THIRD SHIFT
6
1 A
1
MODEL
YEAR
5
5
MODEL
YEAR
B
MODEL
A
B
TRANSMISSION
MODEL
W = WARREN PLANT
W
A
P
3
SERIAL NUMBER IN
BASE CODE 31
5
D
MADE IN
U S A

THM 440-T4 TRANSMISSION I.D. NAMEPLATE
WARREN, MICHIGAN

Transaxle I.D. nameplate

release the clutch when the vehicle is accelerating or decelerating at a certain rate.

4. Brake switch—Another switch switch used in the TCC circuit is a brake switch which opens the 12 volt supply to the TCC solenoid when the brake is depressed.

5. 2nd, 3rd and 4th gear switches—The 440–T4 uses 3rd and 4th gear switches which are direct inputs to the ECM. The ECM uses the switch signals for TCC engage and disengage points. The 440–T4 with torque management uses 2nd, 3rd and 4th gear switches which are direct inputs to the ECM. The ECM uses the switch signals for TCC engage and disengage points.

On Allante vehicles, the transaxle has a shift control system. The ECM controls the shift into 3rd gear by grounding the 2–3 shift solenoid circuit. To control the 4th gear shift, the ECM grounds the 3–4 solenoid circuit. The shift solenoids control the movement of the 2–3 and 3–4 shift valves by blocking or opening oil passages to move shift valves. When the shift selector is moved to the **D** position, line oil is directed to areas of the shift valve. When the solenoid is de-energized, line oil flows through the solenoid pressurizing chamber.

When the solenoid is energized by the ECM, the control chamber is allowed to exhaust through the solenoid. The solenoid simultaneously blocks line oil from entering the solenoid (so as not to be constantly bleeding line pressure). The shift valve is moved to the upshift position by pressure. The 2–3 and 3–4 shift solenoids are controlled by the ECM for shift timing based upon other sensor input.

Metric Fasteners

The metric fastener dimensions are very close to the dimensions of the familiar inch system fasteners and for this reason, replacement fasteners must have the same measurement and strength as those removed.

Do not attempt to interchange metric fasteners for inch system fasteners. Mismatched or incorrect fasteners can result in damage to the transaxle unit through malfunctions, breakage or possible personal injury.

Care should be taken to re-use the fasteners in the same locations as removed.

Capacities

To refill the THM 440–T4 automatic transaxle after a complete overhaul, add 10 quarts of Dexron®II automatic transaxle fluid. To refill the 440–T4 automatic transaxle after the pan has been removed and replaced, add 6 quarts of Dexron®II automatic transaxle fluid. Recheck and correct the level after starting and warming the engine.

Checking Fluid Level

Transaxle Hot

1. Verify that the transaxle is at normal operating temperature. At this temperature, the end of the dipstick will be too hot to hold in the hand. Make sure that the vehicle is level.

2. With the selector in **P**, allow the engine to idle. Do not race engine. Move the selector through each range and back to **P**.

3. Immediately check the fluid level with the engine still running. Fluid level on the dipstick should be at the **FULL HOT** mark.

4. If the fluid level is low, add fluid as required, remembering that only 1 pint will bring the fluid level from **ADD** to **FULL**.

Transaxle Cold

Often it is necessary to check the fluid level when there is no time or opportunity to run the vehicle to warm the fluid to operating temperature. In this case, the fluid should be around room temperature (70°). The following Steps can be used.

1. Place the selector in **P** and start engine. Do not race engine. Move the selector through each range and back to **P**.

2. Immediately check the fluid level with the engine still running, off fast idle. Fluid level should be between the 2 dimples on the dipstick, approximately ¼ in. below the **ADD** mark on the dipstick.

3. If the fluid level is low, add fluid as required, to bring the fluid level to between the 2 dimples on the dipstick. Do not overfill. The reason for maintaining the low fluid level at room temperature is that the transaxle fluid level will rise as the unit heats up. If too much fluid is added when cold, then the fluid will rise to the point where it will be forced out of the vent and overheating can occur.

If the fluid level is correctly established at 70°F, it will appear at the **FULL** mark when the transaxle reaches operating temperature.

TRANSAXLE MODIFICATIONS

Transaxle Name Change

By September 1, 1991, Hydra-matic will have changed the name designation of the THM 440–T4 automatic transaxle. The new designation for this transaxle will be Hydra-matic 4L60. Transaxles built in 1989 and 1990 will serve as transitional years in which a dual system, made up of the old designation and the new designation will be in effect.

The name designation for the F-7 automatic transaxle will remain the same.

New Service Procedure for Checking Chain for Possible Wear

On all 1984 vehicles, when disassembling any THM 440–T4, it is important to inspect the drive link assembly (chain) for possible wear. Refer to the following service procedure:

1. Midway between the sprockets and at right angles to the chain, push the slack (bottom) strand of the chain down until finger tight and mark with crayon on opposite of teeth.
2. Push up in the same manner and put a 2nd mark, making sure that both marks are made from the same point on the chain.

NOTE: If the distance between the marks exceeds 1$^1/_{16}$ in. (27.4mm), the chain must be replaced.

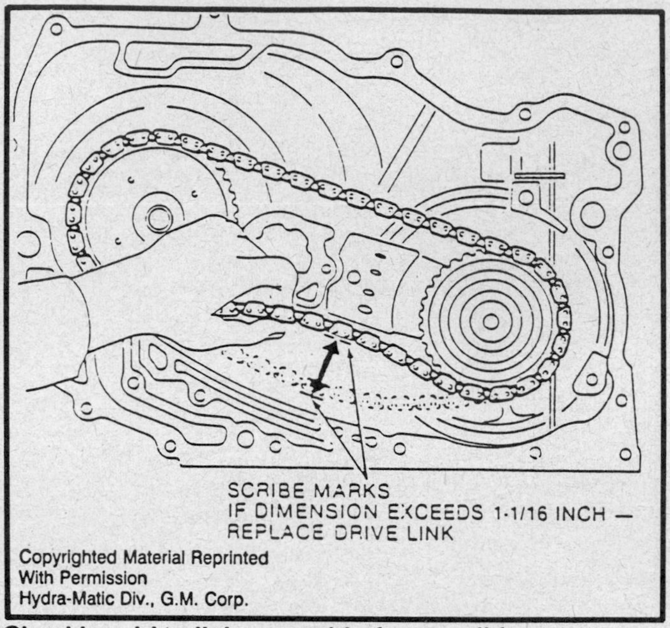
SCRIBE MARKS
IF DIMENSION EXCEEDS 1-1/16 INCH — REPLACE DRIVE LINK
Copyrighted Material Reprinted
With Permission
Hydra-Matic Div., G.M. Corp.

Checking drive link assembly for possible wear

Service Information on the Input and Fourth Clutch Apply Plates

On all 1984 vehicles, when disassembling THM 440–T4 transaxles, the word **UP** may be noted on the 4th clutch apply plate and the input clutch apply plate. During reassembly of the transaxle, follow these instructions:

1. Position the side of the 4th clutch apply plate that has the word **UP** stamped on it, against the 4th clutch piston.
2. Position the side of the input clutch apply plate that has the word **UP** stamped on it (stepped side) against the 3rd clutch backing plate snapring.

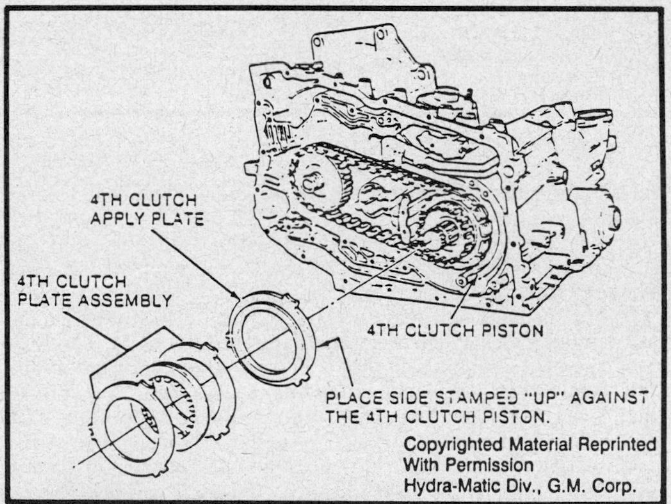
4TH CLUTCH APPLY PLATE
4TH CLUTCH PLATE ASSEMBLY
4TH CLUTCH PISTON
PLACE SIDE STAMPED "UP" AGAINST THE 4TH CLUTCH PISTON.
Copyrighted Material Reprinted
With Permission
Hydra-Matic Div., G.M. Corp.

4th clutch assembly replacement

Vacuum Related Shift Conditions

On all 1985 vehicles, this information is intended to assist the technician when diagnosing vacuum related shift problems. For proper operation, the THM 440–T4 requires 13–17 in. (44–57 kPa) Hg of engine vacuum at engine idle. A loss of engine vacuum can cause any of the following shift conditions:

Harsh park to reverse engagement
Harsh neutral to drive engagement
Harsh or firm light throttle upshifts
2nd speed stars
Harsh 3–2 coastdown shifts
Rough 4–3 and 2–3 manual downshifts
Slipping in drive or reverse

Low engine vacuum can be caused by a pinched, cut, plugged or disconnected vacuum line. It can also be the result of a poorly tuned engine or blocked exhaust. Also an incorrectly installed aspirator tee can give the appearance of low vacuum when the air conditioning is in operation. The aspirator tee has a "flag" with the words **MOD** and **MAN**. Make certain the modulator line is connected to the **MOD** nipple and the manifold line is connected to the **MAN** nipple as indicated by the arrows on the flag.

To check for proper vacuum, disconnect the vacuum line at the modulator and install a vacuum gauge to the line. If there is insufficient vacuum locate the cause and correct as required.

If there is sufficient engine vacuum available to the modulator, remove the modulator assembly and modulator valve. Inspect the valve for nicks or scoring. Connect a hand operated vacuum device (vacuum pump) to the modulator. Pump the device until 15–20 in. (51–68 kPa) Hg of vacuum is reached; at the same time, observe the modulator plunger, it should be drawn in as the vacuum pump is operated. After reaching 15–20 in (51–68 kPa) Hg the vacuum should not bleed down for at least 30 seconds.

If the modulator is found to be functioning properly and the valve is not damaged, the shift problem is not vacuum related. At this time, attach an oil pressure gauge to the pressure tap.

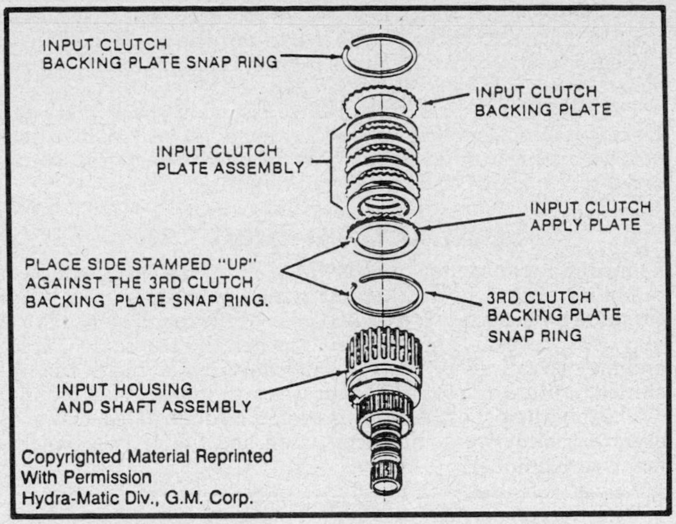
INPUT CLUTCH
BACKING PLATE SNAP RING
INPUT CLUTCH
BACKING PLATE
INPUT CLUTCH
PLATE ASSEMBLY
INPUT CLUTCH
APPLY PLATE
PLACE SIDE STAMPED "UP"
AGAINST THE 3RD CLUTCH
BACKING PLATE SNAP RING.
3RD CLUTCH
BACKING PLATE
SNAP RING
INPUT HOUSING
AND SHAFT ASSEMBLY
Copyrighted Material Reprinted
With Permission
Hydra-Matic Div., G.M. Corp.

Input clutch assembly

As a reminder, the T.V. does not control line pressure; it is used to control shift points (shift speed) only.

New Servo Spring Retainer Added

On all 1985 transaxles, models AY, BA and BS, to reduce the possibility of damage to the reverse servo seal during assembly, a servo spring retainer has been added. This retainer is identical to the one used for the 1–2 servo spring and the part number is 8656700.

Starting January 16, 1964 (Julian date, 016), all 1985 transaxles, models AY, BA and BS, were built with a new servo spring retainer.

The Julian date is on the transmission identification tag located on the side of the case by the governor.

Service Replacement Procedure for the Oil Pump Rotor, Slide and Vanes

On all 1984–85 vehicles, service parts are now available to service the oil pump rotor, slide and vanes in the pump body assembly on all THM 440–T4 transaxles. This service procedure on all THM 440–T4 transaxles must be followed to assure proper end clearance on the oil pump rotor and slide.

NOTE: Do not attempt to service the oil pump rotor, if either the pump body cover surfaces are scored. Servicing of the oil pump rotor, slides and vanes should be performed only if the selective pump rotor, slide or the vanes show wear.

1. Disassemble the oil pump.
2. Select the pump rotor, slide and vanes by the following:
   a. Use a micrometer to measure the original pump rotor, slide and vane thickness accurately. To obtain the most accurate reading, measure on flat, undamaged surfaces.
   b. Using the original part measurement, order replacement parts.
   c. Hone both sides of the replacement rotor, vanes and slide to remove any burrs and measure the replacement parts with a micrometer to assure proper selection. Incorrect rotor selection could result in a damaged oil pump assembly and/or low oil pressure. Incorrect slide could result in incorrect line pressure.

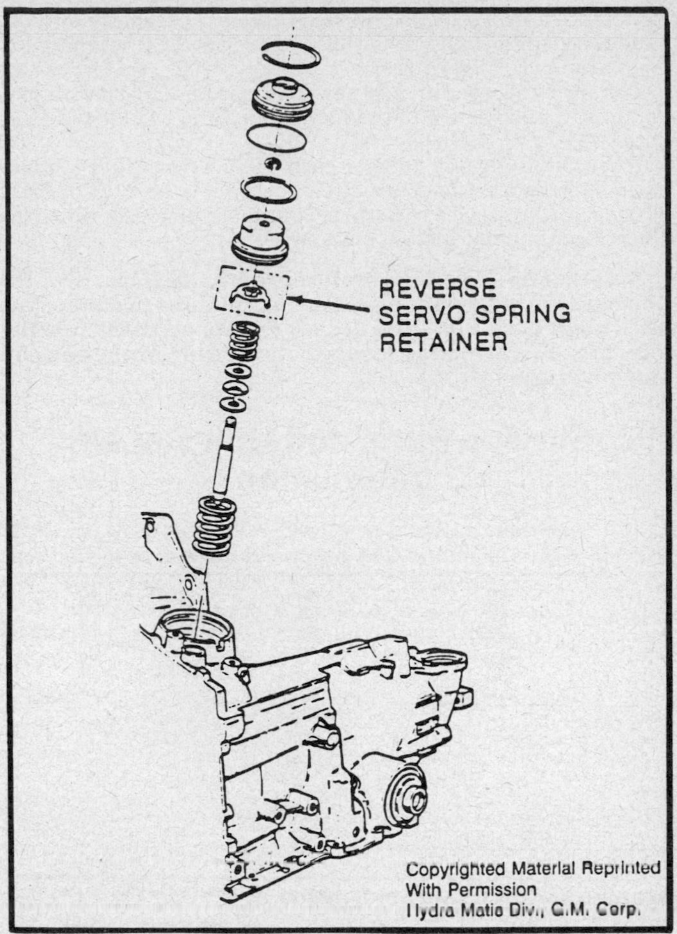
REVERSE
SERVO SPRING
RETAINER
Copyrighted Material Reprinted
With Permission
Hydra-Matic Div., G.M. Corp.

Reverse servo spring retainer

NOTE: The replacement part will provide the same clearance that the oil pump assembly was originally built with. The proper end clearance specification is 0.0025–0.0032 in. (0.0635–0.0813mm) for the rotor and vanes and 0.0013–0.0022 in. (0.0300–0.0559mm) for the slide.

Due to the factory selection of the oil seal ring, any field substitution of these parts will result in transaxle damage or could cause pump failure. These parts are serviceable only in the slide package.

The pump slide seal and the pump slide seal support can be serviced by ordering service package part number 8646916. These parts are not selective.

3. To assemble:
   a. Reassemble the oil pump assembly following.
   b. After assembling the oil pump, install the pump shaft and check for proper rotor and slide clearance by turning the pump shaft. The pump shaft should turn freely. If the pump shaft does not turn freely, recheck the pump rotor and slide selection.
   c. Install the transaxle into the vehicle.
   d. Before road testing the vehicle, install an oil pressure gauge to the transaxle and check oil pressure.

No Torque Converter Clutch (TCC) Apply or TCC Shudder

Some 1984 and some early 1985 THM 440–T4 transaxles may

experience a no Torque Converter Clutch (TCC) apply or TCC shudder condition, which may be due to a cut TCC accumulator piston seal.

To repair this condition, a new TCC accumulator piston pin, piston and seal went into production on June 24, 1984 (Julian date 176)

When Servicing any 1984 or early 1985 THM 440–T4 transaxles built before the above Julian date for a no TCC or TCC shudder, order service package part number 8646926, which includes accumulator piston, pin and seal.

NOTE: The TCC, accumulator piston seal can be purchased separately but only for TCC pistons produced after Julian date 176. (The Julian date is on the transmission identification tag located on the side of the case by the governor).

Service Parts and Unit Repair Information

Hydra-matic is releasing 8 new 1985½ THM 440–T4 transaxle to replace the 1985 units. The new units have a redesigned control valve assembly with a new 3–2 line control valve that makes the 3–2 downshift more responsive to throttle position. The oil passages in the control valve assembly, channel plate, spacer plate and gaskets have also changed.

The 1985½ transaxle control valve assembly, spacer plate and gaskets will not service the 1985 assembly. Until 1985½ part number listings are available, service replace parts can be ordered through GMWDD by description (part name).

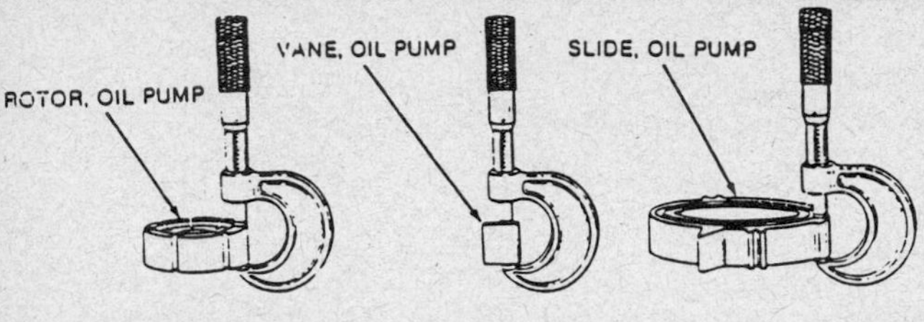
ROTOR, OIL PUMP
VANE, OIL PUMP
SLIDE, OIL PUMP

Measuring oil pump rotor, vane and slide for specification

Second Gear Start Out

Some 1985 transaxles, models CP, CW or HT, may exhibit a condition of 2nd gear start out with the selector in **D** after coastdown from 2nd, 3rd or 4th gear operation. The condition may occur several times (especially when the transaxle is cold) and may be caused by the 1–2 shift valve remaining in the upshifted position due to residual pressure in the governor circuit.

The following chart lists, by model, the Julian date a new governor assembly went into production and the service replacement part number.

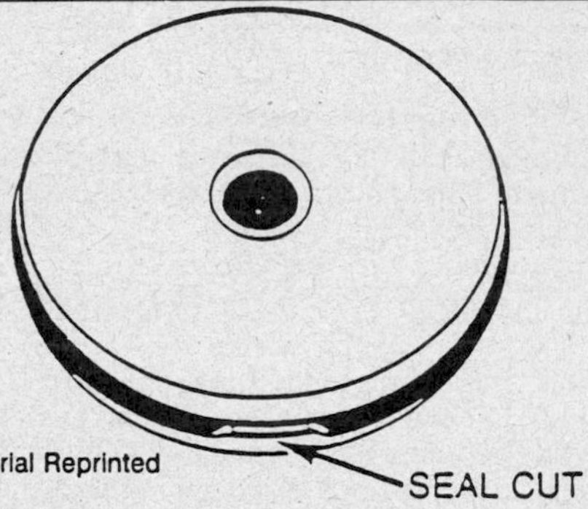
SEAL CUT

Torque converter clutch (TCC) accumulator piston and seal

PART SELECTION DATA

CHART A — Selective Pump Rotor

PART NUMBER	THICKNESS IN mm	THICKNESS IN INCHES
8656267	17.953 - 17.963	.7068 - .7072
8656268	17.963 - 17.973	.7072 - .7076
8656269	17.973 - 17.983	.7076 - .7080
8656270	17.983 - 17.993	.7080 - .7084
8656271	17.993 - 18.003	.7084 - .7088

CHART B — Selective Pump Slide, and Seals

PART NUMBER	THICKNESS IN mm	THICKNESS IN INCHES
8646911	17.983 - 17.993	.7080 - .7084
8646912	17.993 - 18.003	.7084 - .7088
8646913	18.003 - 18.013	.7088 - .7092
8646914	18.013 - 18.023	.7092 - .7096
8646915	18.023 - 18.033	.7096 - .7100

CHART C — Selective Pump Vane

PART NUMBER	THICKNESS IN mm	THICKNESS IN INCHES
8644661	17.943 - 17.961	.7064 - .7071
8644662	17.961 - 17.979	.7071 - .7078
8644663	17.979 - 17.997	.7078 - .7085

Model CP—Julian date 100—service part number 8b44961
Model CW—Julian date 81—service part number 8644963
Model HT—Julian date 82—service part number 8644963

To repair the 2nd gear start out on any 1985 THM 440–T4 model CP, CW, OR HT transaxle built before the listed Julian date, install the appropriate new governor assembly.

3–2 Downshift Bump

Some 1985 THM 440–T4 transaxle, models BA, BC, CM and CP may exhibit a 3–2 downshift bump. This condition may be most prevalent during a part-throttle downshift with the air conditioner operating. A slight transaxle tie-up condition which shakes or rattles the instrument panel during the downshift may also be experienced.

When servicing any 1985 transaxle, models CM and CP, for a 3–2 downshift bump, order and install service package part number 8646951.

Service package part number 8646951 contains a 3–2 control valve spring, a 1–2 servo boost valve and instruction sheet

When servicing any 1985 transaxle, models BA and BC for a 3–2 downshift bump, order and install service package part number 8646953.

Service package part number 8646593 contains a 3–2 control valve spring and instruction sheet.

Delayed Forward Engagement, Slips on Takeoff and/or 2–3 Upshift Bump or Shudder

A 1985 or 1986 THM 440–T4 transaxle may exhibit delayed forward engagement, slipping on take off from stops and/or a 2–3 upshift bump or shudder. The condition(s) may be caused by leakage in the 1–2 servo circuit, which controls apply and release of the 1–2 band.

Seals are included at both ends of the apply and release pipes and on the servo piston. A damaged seal in any position may affect operation of the servo. A leaking apply pipe circuit may cause delayed forward engagement and slipping on takeoff when in 1st or 2nd gear. A leak in the release pipe circuit may cause a 2–3 upshift bump or shudder. A 1–2 servo piston leak may cause any/all of the conditions(s).

When performing transaxle service, check transmission fluid for burned condition (burned fluid loses red color and has an acrid odor). Burned fluid may indicate major internal repair requirements.

If fluid condition is proper, inspect the servo pipe seals at both ends of each pipe and the 1–2 servo piston seals. The original pipes can be reused in field service. Replace any damaged seal when removing the seals, use care to ensure that damage to other transaxle components does not occur.

NOTE: When installing the servo pipes into the control valve assembly, it is critical that the pipes are correctly aligned with the assembly bore to ensure that seal damage does not occur.

Harsh Upshifts at High Altitudes

A 1985– A-body vehicles equipped with a 2.8L engine and a THM 440–T4 transaxle may exhibit harsh upshifts when operated at high altitudes. This condition may be caused by excessive line pressure resulting from insufficient engine vacuum.

A new vacuum modulator assembly and modulator valve have been released for service in a package, part number 8646956 for the harsh upshifts condition. The package consists of a transaxle vacuum modulator assembly, modulator valve, O-ring seal and instruction sheet

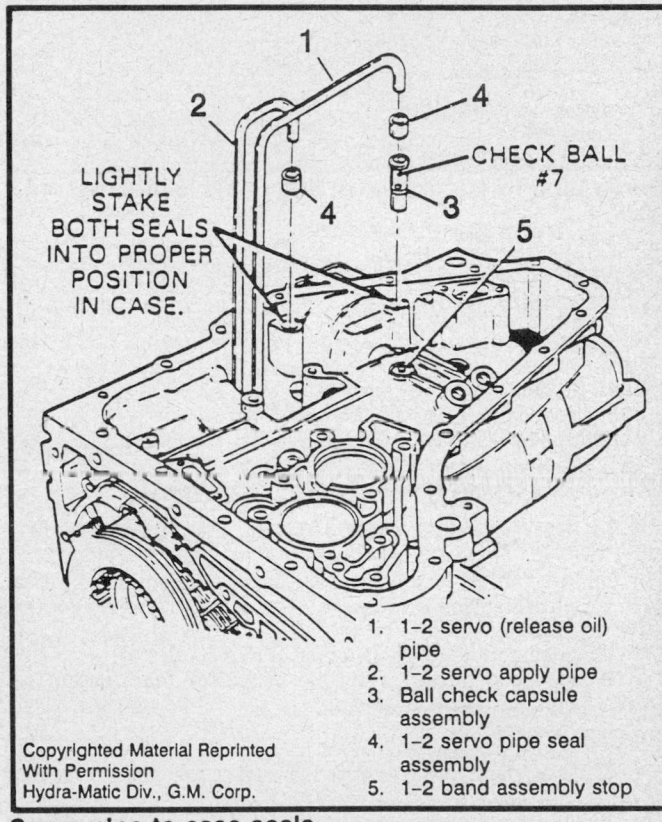

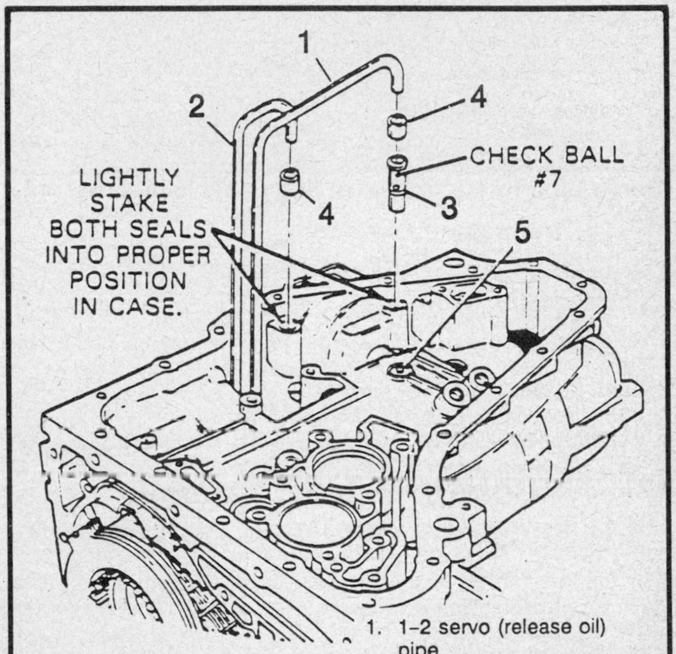
1
2
4
CHECK BALL #7
LIGHTLY STAKE BOTH SEALS INTO PROPER POSITION IN CASE.
4
3
5
1. 1–2 servo (release oil) pipe
2. 1–2 servo apply pipe
3. Ball check capsule assembly
4. 1–2 servo pipe seal assembly
5. 1–2 band assembly stop

Servo pipe to case seals

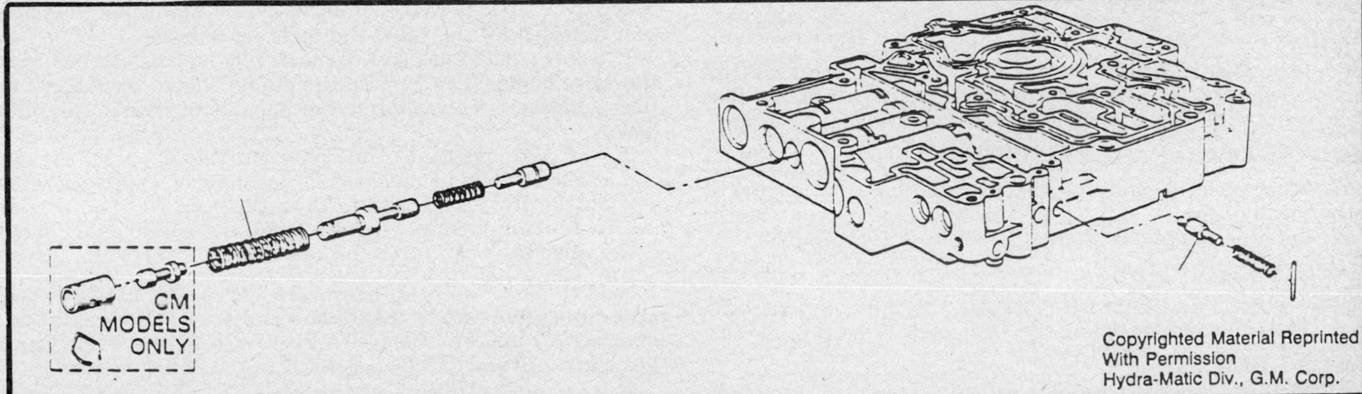
CM MODELS ONLY

3–2 control valve spring and 1–2 servo boost valve

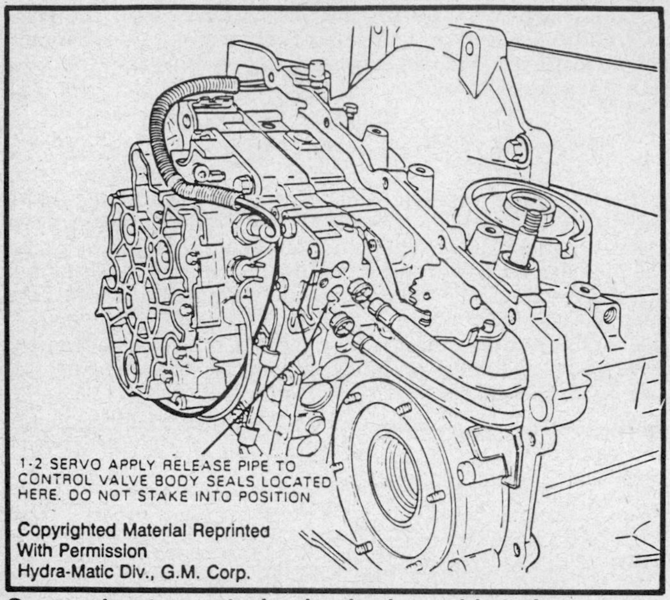
1-2 SERVO APPLY RELEASE PIPE TO CONTROL VALVE BODY SEALS LOCATED HERE. DO NOT STAKE INTO POSITION
Copyrighted Material Reprinted With Permission Hydra-Matic Div., G.M. Corp.

Servo pipe to control valve body seal locations

NOTE: A new engine vacuum line part number 14062624 must be installed with the service package.

The service package only applies to the following THM 440–T4 transaxles:
1985 Models—CW, CP and HT
1985½ Models—CM, CN, HJ and HA
1986 Models—CFH, CMH, CNH, HJH and HBH

Control Valve Assembly Cleaning and Servicing Procedures

The following information is provided as a recommended procedure for cleaning and servicing all automatic transaxle control valve assemblies when:

1. Diagnosis indicates a stuck or sticking valve(s).
2. A transaxle pump or torque converter malfunction requires a complete unit overhaul.

WORKBENCH

1. Must be clean and free of any foreign material (dirt, grease or other contaminants).
2. Work area should be large enough to allow for the disassembly of the control valve assembly and the layout of each valve train.

TOOLS

Should be cleaned with solvent before and during the disassembly procedures to ensure they are free of any grease or dirt that may contaminate the control valve assembly.

SOLVENT AND PARTS CLEANING TANK

1. Must be clean and free of contamination from engine or other components.
2. Must be clean and filtered if solvent is recirculated.

SAFETY EQUIPMENT

Safety glasses and rubber gloves are the minimum requirements to ensure personal safety.

TOOLS REQUIRED

a. Awl
b. Micro fine lapping compound
c. Pencil type magnet
d. Small flat blade screwdriver
e. Small non-abrasive parts cleaning brush
f. Tapered No.49 drill bit

Cleaning and Disassembly

1. Remove the control valve assembly from the transaxle.
2. Inspect the attaching bolts and bolt holes. Remove metal chips or foreign material that may be present.
3. Place the control valve assembly in a clean tank and with clean solvent thoroughly wash the entire outer surfaces of the valve body.
4. Remove the control valve assembly from the solvent and dry it using compressed air.
5. Inspect the casting and fluid passages for foreign materials that may have accumulated in pockets. Remove the foreign material with a small suitable tool.
6. As necessary, repeat the washing and drying procedures as described in Steps 3 and 4 until all foreign materials are removed.
7. While applying a slight pressure with your finger against spring force in the valve train, remove the coiled spring pin, sleeve or clip that retains the valve train.
8. Slowly release pressure on the valve train and remove it from the valve body. A small suitable tool may be used to remove the bushing from the bore.
9. Place the valve train on a clean surface in the exact sequence as it was removed from the valve body. Follow the procedures in Steps 7, 8 and 9 to remove all valves and bushings from the valve body. (Layout of the valves, springs and bushings on a clean, lint free towel will help to keep parts organized).
10. Remove pressure switches, pipe plugs and TCC solenoid, if applicable.
11. Clean one valve train at a time by washing the valves, springs and bushings in clean solvent and drying with compressed air.
12. Clean the valve body casting with clean solvent and dry with compressed air.

Inspection of Control Valve Assembly Components

1. Inspect the valve body casting for cracks, porosity, damaged machined surfaces, nicks or burrs in valve bores and/or flatness of valve body to case mating surface (using a straight edge or by inspecting the gaskets for uniform compression).
2. Inspect valves for burrs, nicks, scratches and/or scoring.
3. Inspect valve bushings for porosity, burrs, nicks, scratches and/or scoring.
4. Inspect springs for damaged and/or distorted coils.

Stuck Valve Servicing

If during disassembly of the control valve assembly a valve is found to be sticking in a bushing or bore, use the following procedure to service the valve and bushing or bore:

1. Apply a small amount of micro fine lapping compound to the valve lands. (The lapping compound should be 900 grit or finer.) Reinsert the valve into the bushing or bore in the valve body).
2. For the movement of the different valves:
   a. Steel valves using a pencil type magnet, rotate the valve while moving it back and forth in the bore.
   b. Aluminum valves using a small flat suitable tool, move the valve back and forth in the bore.

NOTE: Too much lapping of a valve will cause excessive clearance and increase the chance of a valve not operating. Clearance between a valve and it's bushing or bore are normally 0.001–0.0015 in. (0.028–0.04mm).

3. After lapping a steel valve with a magnet, check for magnetism in the valve by holding it near some steel fillings or chips. If

the valve can pick up the fillings, de-magnetize it by using a de-magnetizing tool (different types are available at most tool stores).

4. Again check the valve for magnetism and if necessary, again, de-magnetize it. Repeat this procedure until the magnetism has been removed.

5. Thoroughly clean the valve and bushing (or bore in the valve body) with solvent and dry using compressed air.

6. Place the valve in its bushing (or bore in the valve body) and check for freeness of movement by rocking the bushing (or valve body) back and forth. The valve should travel freely in its bore. If the valve still tends to stick, repeat the lapping procedure.

NOTE: The use of a honing stone, fine sandpaper or crocus cloth is not recommended for servicing stuck valves. All valve lands have sharply machined corners that are necessary for cleaning the bore. If these corners are rounded, foreign material could wedge between the valve and bore causing the valve to stick.

Assembly

1. Make sure the valve body casting, valves and bushings are completely dry and free of cleaning solvent.
2. Lubricate all springs, valves and bushings with clean transmission fluid.
3. Reassemble the valve trains in the valve body casting and check for freedom of movement using a small suitable tool.
4. Install the control valve assembly into the transaxle.
5. Tighten the control valve assembly to case attaching bolts by starting from the center of the control valve assembly and moving outward.

NOTE: Control valve assembly to case attaching bolts must be hand torqued to the specifications. Improper torque can cause a control valve assembly to not operate properly.

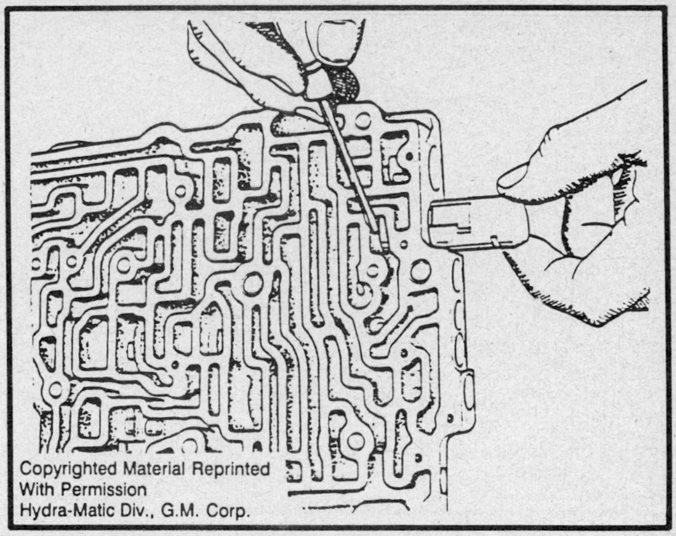

Control valve removal

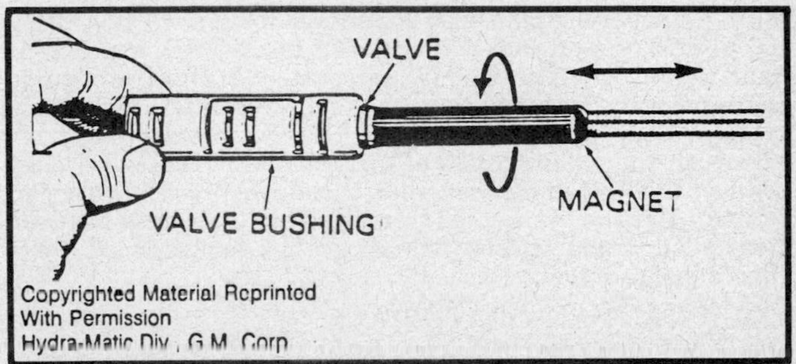
VALVE
MAGNET
VALVE BUSHING

Freeing a stuck steel type control valve with magnet

Torque Converter Clutch (TCC)

A 1986 THM 440–T4 transaxle built before January 27, 1986 (Julian date 027) may exhibit a torque converter clutch apply shudder at 45–55 mph. The shudder condition may be caused by the torque converter. If the TCC apply shudder condition is exhibited, replace the torque converter with a service converter.

3rd Clutch Plates Parts Change

Third clutch plates in 1985, 2.8L Chevrolet applications of the THM 440–T4 changed to single-sided 3rd clutch plates. At the start of 1986 production, all THM 440–T4 transaxles were built with single-sided clutch plates. When servicing 3rd clutch plates on any transaxle, use the single-sided clutch plates. service package part number 8646938.

Input Shaft Seal Eliminated

All 1986 transaxles were built without an input shaft to 4th clutch shaft seal. When servicing a THM 440–T4 built prior to 1986, it is not necessary to replace the input shaft to 4th clutch shaft seal.

Converter Clutch Regulator Valve Spring Eliminated

At the start of 1986 production, the converter clutch regulator valve spring was eliminated on the following models of the THM 440–T4 transaxle: AAH, ACH, AFH, AMH, ARH, BHH, CFH, CHH, CMH, CNH, HBH and JHJ, due to recalibration and fluid passage changes. Do not replace this spring on any listed model, as an increase of the apply feel of the transaxle converter clutch may result.

Loss of Drive and/or Reverse

Loss of drive and/or reverse on a 1984–1987 vehicles that have a THM 440–T4 transaxle with normal line pressures may be caused by the input sprag clutch assembly. This condition may be corrected by installing a new, wider input sprag clutch and related parts.

As of April 1, 1986 (Julian date 091), new input sprag clutch assembly used in later 1986 and all 1987 applications. This change is also on all Service Replacement Transmission Assemblys (SRTA) effective May 1, 1986 (Julian date 121).

Throttle Valve Lever Return Spring Eliminated

During 1985 transaxle production, the spring was removed from the throttle lever and bracket assembly to reduce throttle pedal effort. This change is also on all Service Replacement Transmission Assembly (SRTA) transaxles.

Do not install a throttle valve lever return spring on any THM 440–T4 that does not have one. When servicing a THM 440–T4, the spring can be eliminated on any 1984–85 transaxle to reduce throttle pedal effort.

Oil Cooler Flushing Procedure

On all transaxles, transmission oil cooler flushing must be per-

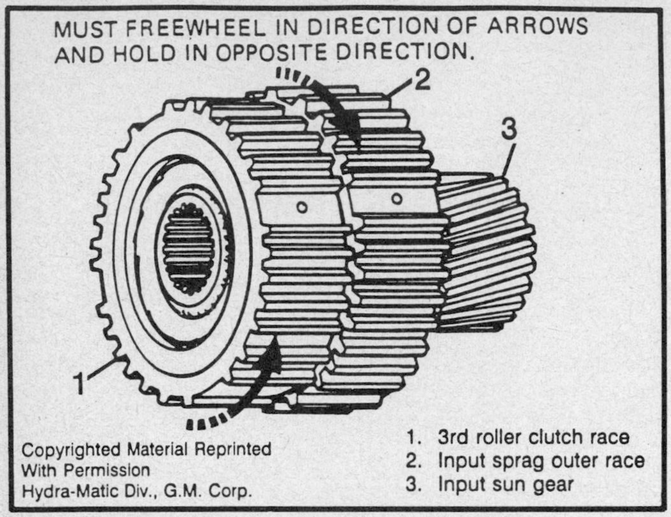
MUST FREEWHEEL IN DIRECTION OF ARROWS
AND HOLD IN OPPOSITE DIRECTION.
1
2
3
Copyrighted Material Reprinted
With Permission
Hydra-Matic Div., G.M. Corp.
1. 3rd roller clutch race
2. Input sprag outer race
3. Input sun gear

Checking the input sprag clutch assembly for proper operation

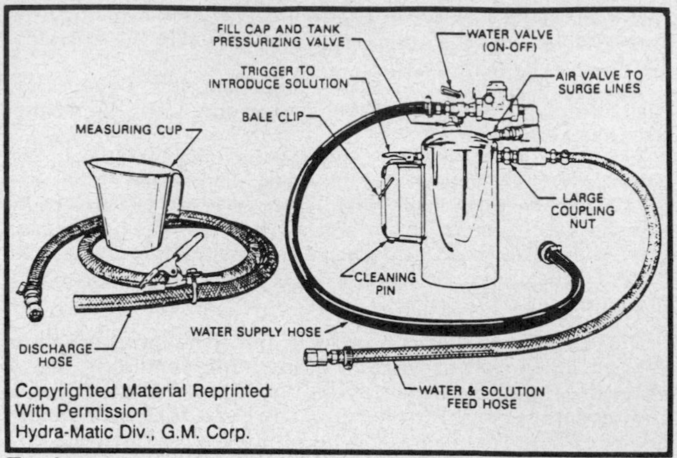
FILL CAP AND TANK
PRESSURIZING VALVE
WATER VALVE
(ON-OFF)
TRIGGER TO
INTRODUCE SOLUTION
AIR VALVE TO
SURGE LINES
BALE CLIP
MEASURING CUP
LARGE
COUPLING
NUT
CLEANING
PIN
WATER SUPPLY HOSE
DISCHARGE
HOSE
WATER & SOLUTION
FEED HOSE
Copyrighted Material Reprinted
With Permission
Hydra-Matic Div., G.M. Corp.

Equipment recommended for oil cooler flushing procedure

formed whenever a transaxle is removed for service. It is essential to flush the oil cooler after SRTA installation, major overhaul, if fluid contamination is suspected, or in any case of pump or torque converter replacement. A new, essential cooler flushing tool, Kent Moore tool J-35944, has been developed to properly flush the oil cooler. To ensure complete transaxle system service, it is recommended that the flush procedure be performed after the overhauled or replacement assembly has been reinstalled in the vehicle. The service procedure for oil cooler flushing is as follows:

EQUIPMENT AND MATERIALS

a. Kent Moore cooler flusher tool J-35944 (or equivalent)
b. Biodegradable flushing solution (Available by the gallon, J-35944-20, or case, J-35944-CSE)
c. Measuring cup (supplied with tool)
d. Funnel
e. Water supply (hot water recommended in all cases if available)
f. Water hose (at least ⅝ in. diameter)
g. Air supply (with water and oil filter)
h. Air chuck (with clip if available)
i. Air pressure gauge
j. Oil drain container
k. Five gallon pail (or larger with a lid)
l. Eye protection
m. Rubber gloves

Procedure

1. After overhaul or service replacement transaxle is reinstalled in vehicle, do not reconnect oil cooler pipes.
2. Remove fill cap on tool and fill can with 0.6 liter (20–21 oz.) of Kent Moore biodegradable flushing solution. Do not overfill or tool will need to be recharged with air before backflush. Follow manufacturer's suggested procedures for proper handling of solution.

CAUTION

Do not substitute with any other solution. The flushing tool is designed to use only this concentrate. Use of any other solution can result in damage to the tool, cooler components, or improper flushing of the cooler.

3. Secure fill cap and pressurize the flusher can with shop air to 80–100 psi (550–700 kPa).

CAUTION

Shop air supply must be equipped with a water/oil filter and not exceed 120 psi (825 kPa).

4. Connect the discharge hose to the transaxle end of the oil pipe that goes to the top fitting at the radiator.
5. Clip discharge hose onto the oil drain container.
6. Mount the flushing tool to undercarriage of vehicle with the hook provided and connect the hose from the flushing tool to the remaining oil cooler pipe.
7. With the water valve on the tool in the **OFF** position, connect the water hose from the water supply to the tool.
8. Turn **ON** the water supply at the faucet.
9. Initial flush—switch the water valve on the tool to the **ON** position and allow the water to flow through the oil cooler for 10 seconds to remove the supply of transaxle fluid in the system.

CAUTION

If water does not flow through the oil cooler (system is completely plugged) do not continue flushing procedure. Turn the water off immediately and inspect the pipes and cooler for restriction. Replace the oil pipe(s) and/or cooler.

10. Switch the water valve on the tool to the **OFF** position and clip the discharge hose onto a five gallon pail with a lid or position a shop towel over the end of the discharge hose to prevent splash. Discharge will foam vigorously when solution is introduced into water stream.
11. Switch the water valve on the tool to the **ON** position and depress the trigger to mix flushing solution into the water flow. Use the bale clip provided on the handle to hold the trigger down.
12. Flush oil cooler with water and solution for 2 minutes. During this flush, attach the air supply to the air valve located on plumbing of tool for 3–5 seconds at the end of every 15–20 second interval to create a surging action.

CAUTION

Shop air supply must be equipped with a water/oil filter and not exceed 120 psi (825 kPa).

13. Release the trigger and switch the water valve on the tool to the **OFF** position.
14. Disconnect both hoses from the oil cooler pipes.
15. Backflush—connect hoses to the oil cooler pipes opposite from the initial flush procedures to perform a backflush.
16. Repeat Steps 11 and 12.
17. Release the trigger and allow water only to rinse the oil cooler for one minute.
18. Switch the water valve on the tool to the **OFF** position and turn the water supply **OFF** at the faucet.
19. Attach air supply to the air valve located on plumbing of tool and dry the system out with air for at least 2 minutes, or

longer if moisture is visible exiting from the oil cooler line discharge hose. Use an air chuck clip, if available, to secure the air chuck onto the air valve for ease of operation.

NOTE: Excessive residual moisture can cause corrosion in the oil cooler or cooler pipes and can damage the transaxle. If Steps 20–23 cannot be completed at this time, rinse the oil cooler and cooler pipes with transaxle fluid. Complete Steps 20–23 after reinstallation of transaxle.

20. Connect the cooler feed pipe to the transaxle.
21. If not already connected, attach the discharge hose to the cooler return pipe and place into an appropriate drain container.
22. After filling the transaxle with automatic transaxle fluid, start the engine and run for 30 seconds. This will remove any residual moisture from the oil cooler and cooler pipes, protect all components from corrosion and check flow rate through the cooler. A minimum of 2 quarts must be obtained during this 30 second run. If fluid flow is insufficient, check the fluid flow out of the transaxle by disconnecting the oil cooler feed line at the radiator and restarting the engine. Do the following according to flow rate:
    a. Insufficient feed flow—inspect the transaxle for cause.
    b. Sufficient feed flow—inspect oil cooler pipes and fittings for restrictions or leaks and repeat the oil cooler flushing procedure. Repeat the check of fluid flow out the return line and if flow is still inhibited, replace the oil cooler.
23. Remove discharge hose, reconnect cooler return pipe to transaxle and refill unit to proper fluid level. Vertical (top) connector is cooler feed, horizontal (bottom) connector is cooler return.
24. Disconnect the water supply hose from the tool.
25. Bleed air pressure from the can, remove fill cap, return any unused solution to container and rinse the can out with water. Do not store tool with solution in tank.
26. Every 3rd tool cleaning, use loosen large coupling nut and remove plumbing from tank.
27. Remove screen from plumbing and wash with water.
28. Use the cleaning pin to remove any material in the solution orifice. Orifice is located in plumbing below screen.
29. Reconnect the plumbing and fill can half full with water, secure the fill cap and pressurize the can to 80–100 psi (550–700 kPa).
30. Aim tool into the 5 gallon pail or floor drain and depress the trigger to allow water from the can to flow through the solution orifice for 30 seconds to ensure proper cleaning.
31. Bleed air pressure from can, remove the fill cap and empty the can.
32. Reconnect fill cap flushing tool.

Long and/or Delayed 2–3 Shift

A long and/or delayed 2–3 shift on a THM 440–T4 transaxle may be caused by a leaking 3rd clutch inner seal.

As of February 12, 1987 (Julian date 043), the effected parts are a new 3rd clutch piston inner seal, input housing and shaft assembly and 3rd clutch roller assembly. This change is also on all Service Replacement Transmission Assembly (SRTA) transaxles effective November 15, 1986 (Julian date 319).

Remove the transaxle, disassemble, inspect and replace with parts included in the service package part number 8646991, which include an input housing and shaft assembly, 3rd clutch piston inner seal and a complete 3rd clutch roller assembly.

Harsh Shift Conditions and On Vehicle Service Information

NOTE: Before performing any service action, make sure to diagnose the specific shift condition and altitude that affects vehicle performance. Do not assume that every condition listed applies to the vehicle.

Harsh shift conditions and the specific service information that applies to each of these conditions. However, some other items that affect transaxle operation which may also contribute to these harsh shift conditions are:

a. Vacuum leaks (low vacuum supply to the vacuum modulator)—13–18 in. Hg required
b. Paint in the modulator tube restricting vacuum to the modulator causing a delayed response to throttle movement
c. Sticking modulator valve
d. Sticking 1–2 servo valve
e. Missing check balls (No. 2, No. 4, or No. 12)

Very Harsh Part Throttle 3–2 Shift at 20–40 mph

On all 1985–87 transaxles, a very harsh part throttle 3–2 shift at 20–40 mph may be caused by a missing or mislocated No. 2 control 3–2 check ball. Remove the valve body to inspect for proper check ball location or damage to the spacer plate. A mislocated check ball may have bent the spacer plate. Also inspect the No. 12 check ball (1–2 servo feed check ball) to ensure it is seating properly.

Harsh 3–2 and/or 4–2 Shifts

On all 1985–87 transaxles, harsh 3–2 and 4–2 shifts at any altitude may be caused by a missing or partial hole in the spacer plate. Remove the valve body and inspect the spacer plate for a missing or partial governor passage hole (No. 25). A partial hole can be drilled out using a No. 77 drill bit to meet the hole size requirement of 0.018 in. (0.46mm). If the hole is missing, use the spacer plate gasket and control valve body as a guide to locate and drill the hole, or replace the spacer plate.

Intermittent Harsh Part Throttle 3–2 Shift

On all 1985½ transaxles except transaxles with 2.8L or 3.0L engines and all 1986 transaxles except units with 2.8L or 3.0L engines and transaxles built prior to 7/21/87 (Julian date 202), an intermittent harsh 3–2 shift may be caused by a broken 3–2 line control retainer. Remove the transaxle side cover to access the 3–2 line control valve and, if necessary, replace a broken retainer. Also ensure that the 3–2 line control valve moves freely in its bore. Use service package part number 8658678.

Harsh 3–2 Shift (High Altitude Only)

On all 1985–87 vehicles except Seville and 1986 Seville, a harsh 3–2 shift condition at high altitude may be corrected by installing the appropriate service package that contains the following items:

a. 3–2 Control valve spring
b. Spacer plate gaskets
c. Instruction sheet

Before installing a Service Replacement Transaxle Assembly (SRTA) in areas designated as "high altitude," install the new 3–2 control valve spring.

Also ensure that the 3–2 control valve and the 1–2 servo control valve move freely in their bores.

Harsh 3–2 Shift (At Low Altitude)

On all 1986 vehicles with 2.8L, 3.0L and 3.8L engines and all 1987 vehicles with 3.8L engines, a harsh 3–2 shift at low altitude may be corrected by installing a service package that contains:

a. 3–2 control valve spring
b. Spacer plate gaskets
c. Instruction sheet

When servicing a transaxle for this condition, also ensure that the 3–2 control valve and 1–2 servo control valve move freely in their bores.

Harsh 3–2 Shift Maneuver At 17–20 mph

On all 1985½ vehicles except Seville and 1986 Seville, a 3–2

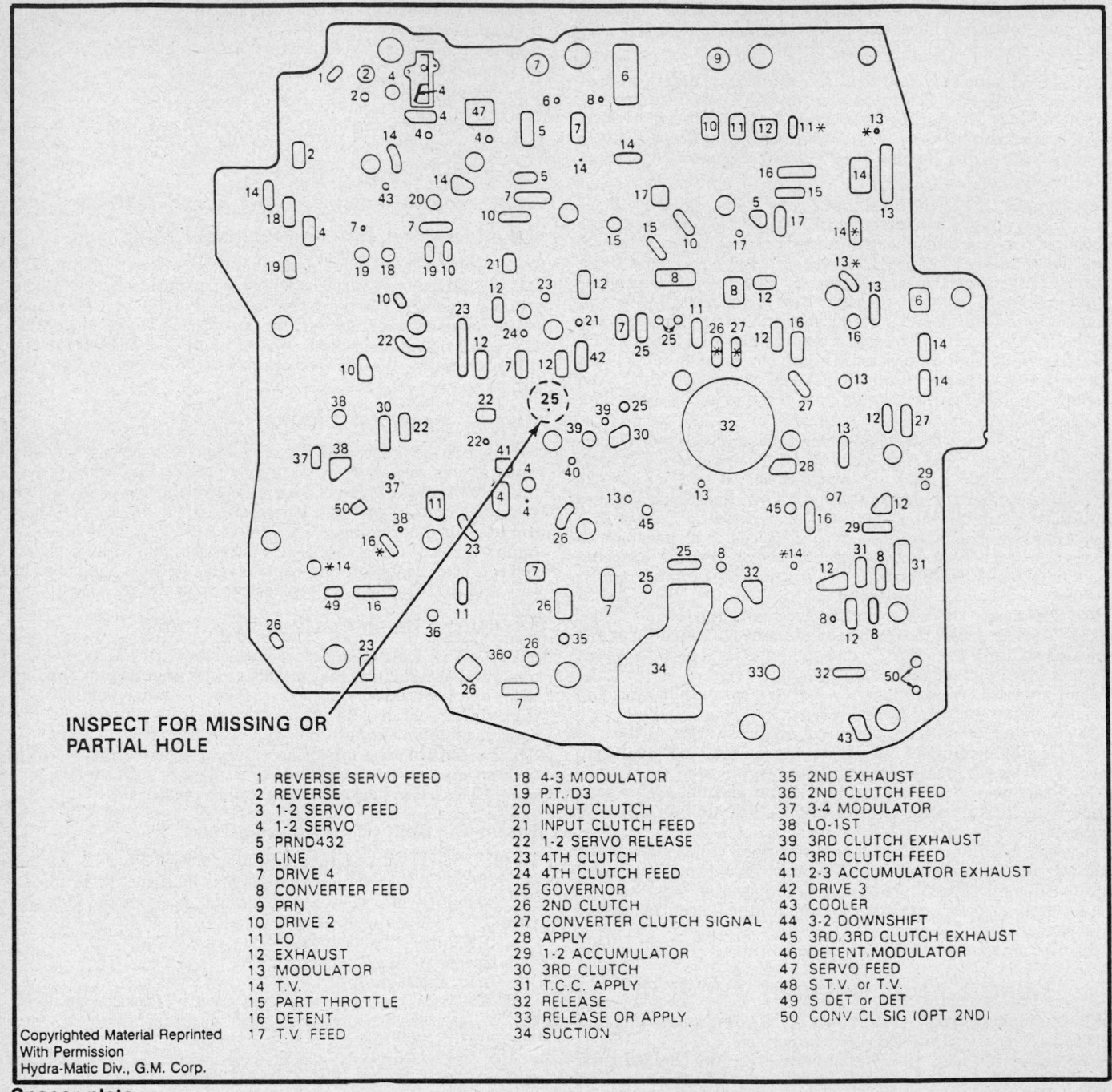

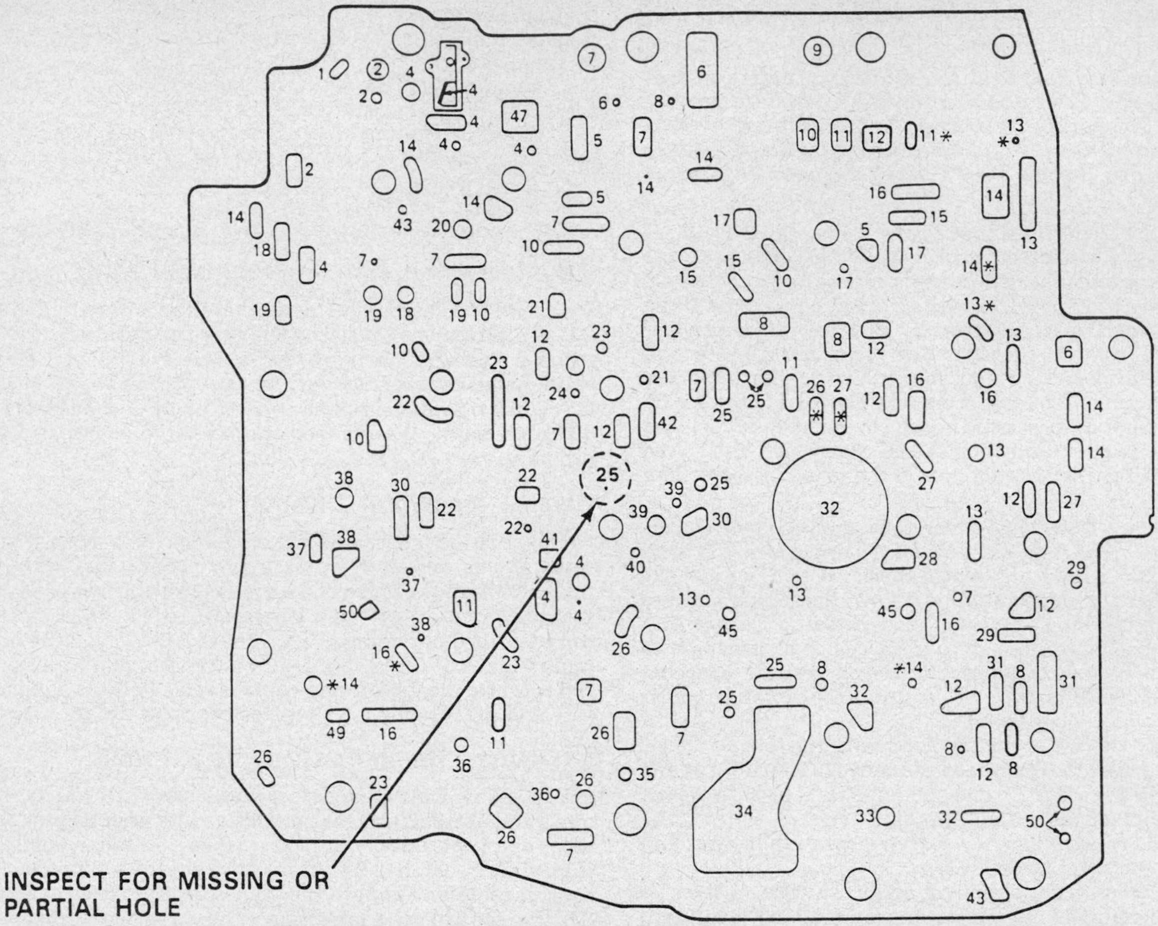
INSPECT FOR MISSING OR PARTIAL HOLE

1 REVERSE SERVO FEED
2 REVERSE
3 1-2 SERVO FEED
4 1-2 SERVO
5 PRND432
6 LINE
7 DRIVE 4
8 CONVERTER FEED
9 PRN
10 DRIVE 2
11 LO
12 EXHAUST
13 MODULATOR
14 T.V.
15 PART THROTTLE
16 DETENT
17 T.V. FEED
18 4-3 MODULATOR
19 P.T. D3
20 INPUT CLUTCH
21 INPUT CLUTCH FEED
22 1-2 SERVO RELEASE
23 4TH CLUTCH
24 4TH CLUTCH FEED
25 GOVERNOR
26 2ND CLUTCH
27 CONVERTER CLUTCH SIGNAL
28 APPLY
29 1-2 ACCUMULATOR
30 3RD CLUTCH
31 T.C.C. APPLY
32 RELEASE
33 RELEASE OR APPLY
34 SUCTION
35 2ND EXHAUST
36 2ND CLUTCH FEED
37 3-4 MODULATOR
38 LO-1ST
39 3RD CLUTCH EXHAUST
40 3RD CLUTCH FEED
41 2-3 ACCUMULATOR EXHAUST
42 DRIVE 3
43 COOLER
44 3-2 DOWNSHIFT
45 3RD, 3RD CLUTCH EXHAUST
46 DETENT MODULATOR
47 SERVO FEED
48 S T.V. or T.V.
49 S DET or DET
50 CONV CL SIG (OPT 2ND)

Spacer plate

Shift maneuver is obtained by allowing the vehicle to coastdown to 17–20 mph (with closed throttle) and then stepping into the throttle to obtain a 3–2 shift.

At start of production 1987, an orifice pipe was added between the 1–2 servo release cover plate and the 3–2 coastdown valve to allow a smoother 2nd gear apply. The 4 new parts installed as a set will service certain past transaxles with a harsh 3–2 shift maneuver condition.

If the new pump body assembly has a "1 terminal" 3rd clutch switch, remove it and replace with the "2 terminal" 3rd clutch switch from the original pump body assembly.

When servicing a transaxle for the above described condition in designated high altitude areas, also install the 3–2 control valve spring package.

Also ensure that the 3–2 control valve and 1–2 servo control valve move freely in their bores.

Harsh 3–2 Coastdown Bump 3.8L at 20–40 mph

On all 1985–87 vehicles with transaxles built prior to december 18,1986 (Julian date 352) and 3.8L engines, a harsh 3–2 coastdown bump may be caused by the 1–2 servo cushion

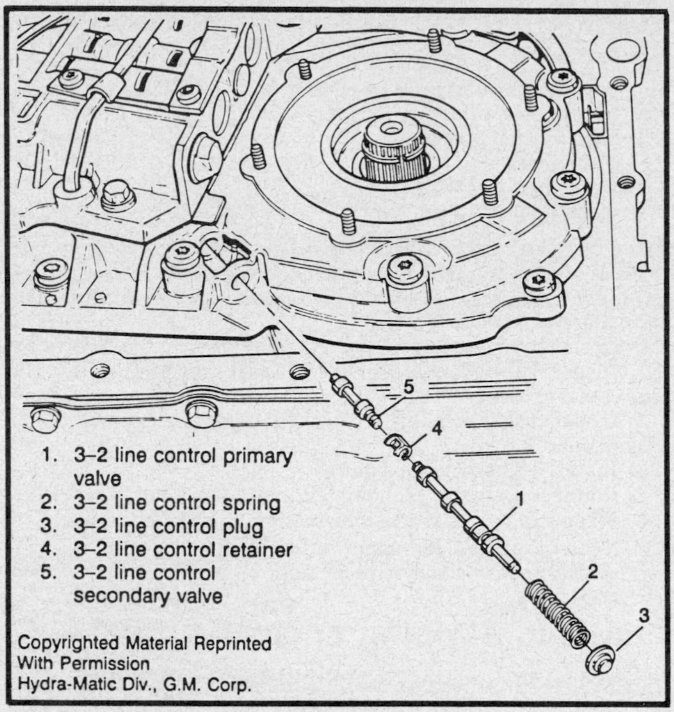
1. 3–2 line control primary valve
2. 3–2 line control spring
3. 3–2 line control plug
4. 3–2 line control retainer
5. 3–2 line control secondary valve
Copyrighted Material Reprinted With Permission Hydra-Matic Div., G.M. Corp.

3–2 line control valve assembly

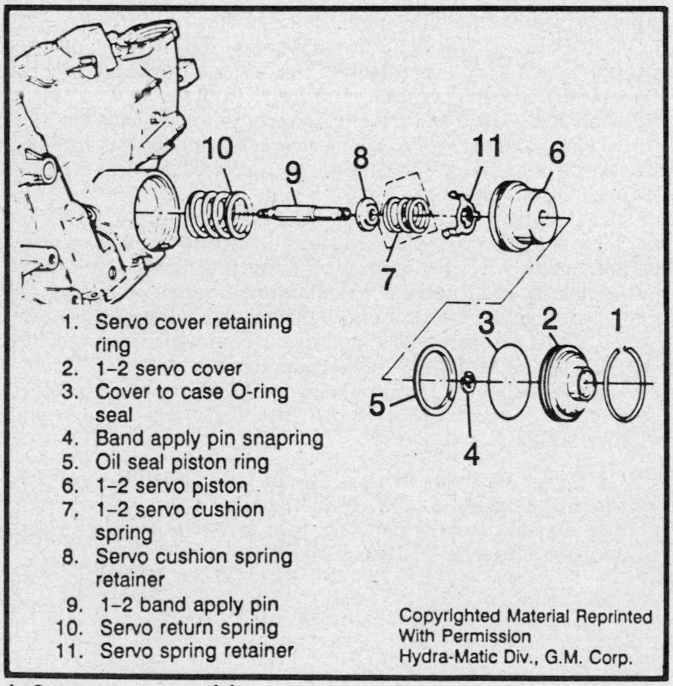
1. Servo cover retaining ring
2. 1–2 servo cover
3. Cover to case O-ring seal
4. Band apply pin snapring
5. Oil seal piston ring
6. 1–2 servo piston
7. 1–2 servo cushion spring
8. Servo cushion spring retainer
9. 1–2 band apply pin
10. Servo return spring
11. Servo spring retainer
Copyrighted Material Reprinted With Permission Hydra-Matic Div., G.M. Corp.

1–2 servo assembly

spring. This condition may be corrected by replacing the 1–2 servo cushion spring. Use service package part number 8646457.

If the above condition persists, make sure the 3–2 coastdown valve (located in the oil pump body) is not stuck and moves freely in its bore.

Harsh Upshifts (High Altitude Only)

On all 1985–86 A-body vehicles equipped with 2.8L engine, 1985–87 Cadillac C and E-body vehicles with 4.1L engine and 1986 Cadillac K-body vehicles with 4.1 engine, harsh upshifts at high altitudes may be caused by high lie pressure due to low engine vacuum. This condition may be corrected on certain transaxles, by installing the appropriate aneroid modulator service package.

Transaxle model identification 6CFH, 6CMH, 6CNH, 6HBH, 6HJH – service package number 8646956

Transaxle model identification 5CM, 5CN, 5CP, 5CW, 5HA, 5HJ, 5HT – service package number 8646956

Transaxle model identification 6AAH, 6ACH, 6AFH, 6AMH – service package number 8646966

Transaxle model identification 5AF, 5AM – service package number 8646966

Transaxle model identification 6ADH, 6ARH – service package number 8646967

A new vacuum line from the engine to the vacuum modulator must also be installed when using the service packages.

A-Body Vehicles – part number 14062624
C-Body Vehicles – part number 1637147
E/K-Body Vehicles – part number 1641441

Intermittent or Complete Loss of Drive and Reverse

Intermittent or complete loss of drive and reverse on a 1984–86 vehicle equipped with a THM 440–T4 transaxle having normal line pressures may be caused by the input sprag clutch assembly. This condition may be corrected by installing a new (wider) input sprag clutch assembly and related parts.

As of April 1, 1986 (Julian date 091), all 1986 THM 440–T4 transaxles were built using a new input sprag clutch assembly. Effective May 1, 1986 (Julian date 121), all Service Replacement Transmission Assembly (SRTA) transaxles were built using the new input sprag clutch assembly.

1. Remove the transaxle from the vehicle and disassemble it to the point of removing the input sprag clutch assembly. Complete disassembly of the transaxle is not always necessary.
2. Bench check the operation of the sprag and 3rd roller clutch assembly.
3. Inspect the input sprag clutch assembly for mispositioned sprag elements and broken ribbon tabs (between the sprag elements).
4. Disassemble the input sprag clutch assembly and inspect for cracked inner or outer races and indentation marks on the inner race that would indicate sprag element "rollover".
5. Replace the input sprag clutch assembly if any of the conditions described above are present. Use the appropriate service part(s).
   a. When replacing the input sprag clutch assembly on a transaxle built before April 1, 1986 (Julian date 091), use service part number 8646986.
   b. When replacing the input sprag clutch assembly on a transaxle built on or after April 1, 1986 (Julian date 091), use service part number 8646985.
   c. To update a transaxle built before April 1, 1986 (Julian date 0901) or a SRTA transaxle built before May 1, 1986 (Julian date 121), to use the new (wider) input sprag clutch assembly, the transaxle must contain part number 8658368 (lube dam), part number 8658407 (final drive sun shaft) and part number 8658408 (output shaft).

NOTE: These 3 correct design parts must be installed for proper lubrication of the 3rd roller clutch and new design sprag assemblies. After these parts have been assembled into the unit, the new input sprag clutch assembly and input carrier assembly, contained in the service package part number 8662906, may be used.

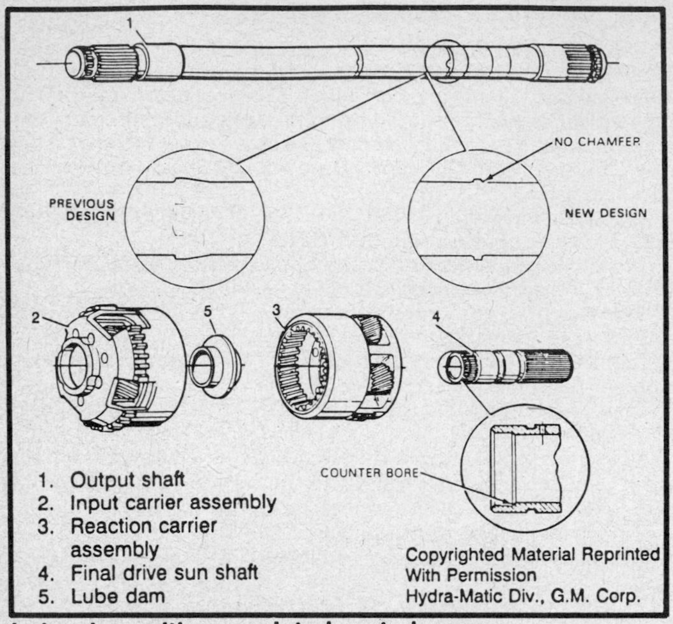
1
NO CHAMFER
PREVIOUS DESIGN
NEW DESIGN
2
5
3
4
COUNTER BORE
1. Output shaft
2. Input carrier assembly
3. Reaction carrier assembly
4. Final drive sun shaft
5. Lube dam
Copyrighted Material Reprinted With Permission Hydra-Matic Div., G.M. Corp.

Lube dam with associated part changes

Throttle Valve Conditions and Diagnosis

On all transaxles the throttle valve (T.V.) pressure controls the shift pattern and the amount of hydraulic line pressure used to apply clutches and bands.

NOTE: T.V. cable controls shift timing, not shift feel—line pressure is not directly affected by the T.V. cable setting.

If proper line pressure is not available, excess slippage may occur during the shift. If the T.V. cable is tailored in an effort to modify shift pattern/feel, this misadjustment can result in transaxle failure.

NOTE: Do not attempt to correct a condition by changing the T.V. cable setting from its proper adjustment. Apparent improvements from T.V. cable "tailoring" indicate a need for further diagnosis into the real cause of a condition (such as engine performance, the control valve, governor or servo assemblies, etc.).

Technicians and customers should know that certain types of heavy duty or high performance vehicles have transaxles calibrated to provide crisp, firm shifts for durability/performance. Once again, the T.V. cable must not be tailored in an effort to modify shift pattern/feel on these applications.

NOTE: Make sure engine fuel, electrical and mechanical systems are performing properly before attempting transaxle diagnosis. For example, a fouled fuel injector, cracked spark plug, damaged spark plug wire or vacuum leak can affect transaxle performance.

The T.V. system may cause one or more of the following conditions:

a. Delayed or harsh upshifts
b. Early and slipping upshifts
c. No upshifts
d. Chatters on take off
e. 1–2 shift at full throttle only
f. No full throttle or part throttle detent downshifts
g. Intermittent 2nd gear starts

The listed conditions may be caused by a T.V. cable that is:

a. Improperly adjusted
b. Kinked or binding
c. Disconnected
d. Broken
e. Wrong part

CAUTION

To avoid possible personal injury and/or damage to the vehicle, the following diagnosis procedure must be performed with the vehicle parking brake properly applied.

1. Inspect the T.V. cable for correct part number, listed in the parts catalog.
2. Inspect the T.V. cable for kinked, binding, disconnected or broken condition.
3. Install a line pressure gauge.
4. Connect a engine tachometer.
5. Warm up the engine to normal operating temperature.
6. Run engine at 1000 rpm.
7. With the gear selector in **P**, note oil pressure.

Soft, Slipping or No 1–2 Shift

On all 1984–86 transaxles, a soft, slipping or no 1–2 shift condition may result if the 2 cast iron seals between the driven sprocket support and 2nd clutch housing bind to the driven sprocket support and wear into the 2nd clutch housing.

Inspect the cast iron oil seal rings for wear (indicated by a shiny surface on the outer diameter) and the 2nd clutch housing for wear (grooves caused by the seal rings). The driven sprocket support may not appear affected, but must be replaced if these wear conditions are present.

For all 1984–86 transaxles, it is recommended that the new Vespel (TM) oil seal rings, 4-lobe ring seals and driven sprocket support be installed with other parts that indicate wear (such as the 2nd clutch housing).

Due to changes in the driven sprocket support for 1987, the new Vespel (TM) oil seal rings and 4-lobe ring seals are not interchangeable with the cast iron oil seal rings on transaxles built prior to 1987 start of production. Changes made to the driven sprocket support include deeper seal ring grooves and "cutouts" to accommodate the ring tangs which prevent the Vespel (TM) oil seal rings from rotating.

The new Vespel (TM) oil seal rings, 4-lobe ring seals and driven sprocket support are included in the service package part number 8662523.

NOTE: When installing a 4-lobe ring seal, make sure that it is not rolled or twisted. Lubricate the vespel rings with transaxle fluid or petroleum jelly for easier installation into the 2nd clutch housing.

Slip or No Drive

On all 1987 transaxles, a slip or no drive condition on a THM 440–T4 transaxle may be caused by low line pressure due to the pump slide binding on excessive casting flash when the slide is in the low pressure position. Modulator vacuum will not affect line pressures if this condition is present. This condition can be repaired by replacing the oil pump assembly.

THM 440–T4 transaxles built between November 26, 1986 (Julian date 330) and January 12, 1987 (Julian date 012) may exhibit this condition.

Remove the oil pump cover and pump priming springs and move the slide by hand. If excessive flash is present, the slide will bind.

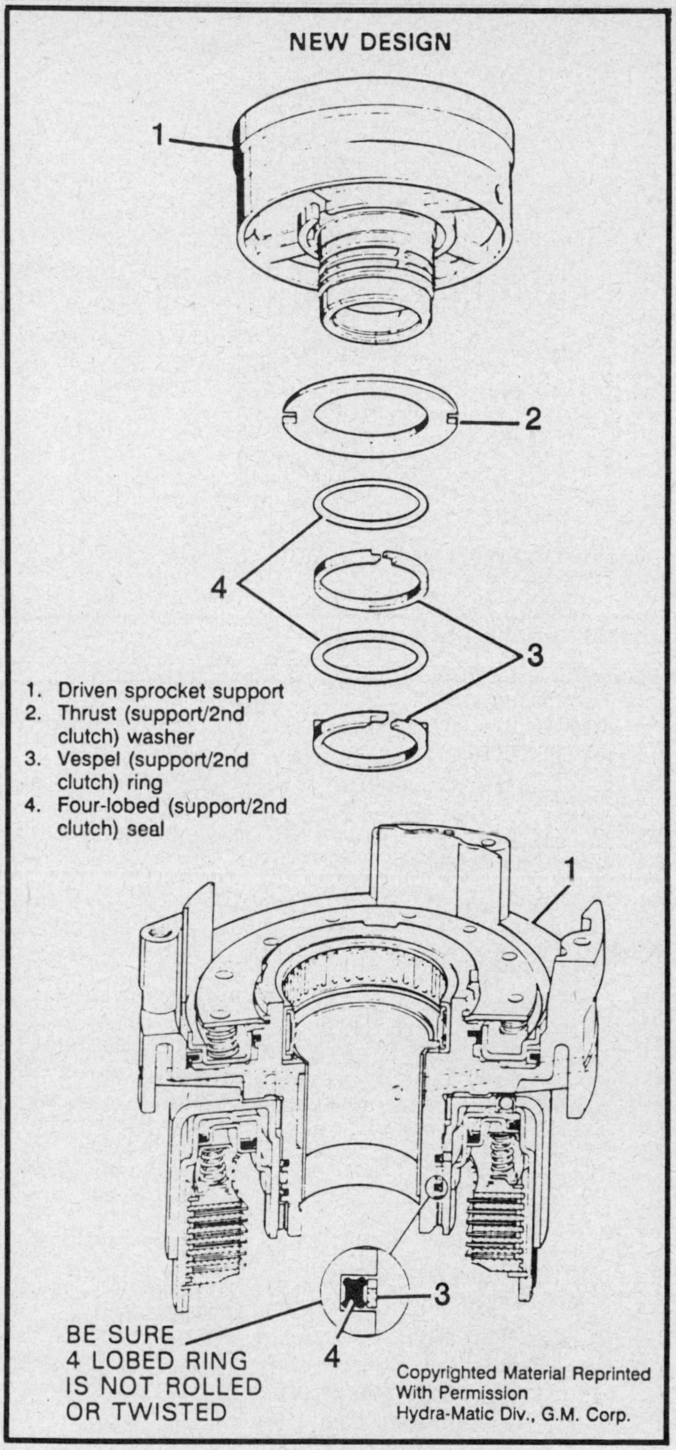
NEW DESIGN
1
2
4
3
1. Driven sprocket support
2. Thrust (support/2nd clutch) washer
3. Vespel (support/2nd clutch) ring
4. Four-lobed (support/2nd clutch) seal
1
3
4
BE SURE 4 LOBED RING IS NOT ROLLED OR TWISTED
Copyrighted Material Reprinted With Permission Hydra-Matic Div., G.M. Corp.

Redesigned driven sprocket support

Harsh Shifts

Harsh shifts in a 1987 2.8L fuel injected Pontiac 6000 or Pontiac 6000 LE may be caused by the installation of a 7CBH model transaxle (instead of a 7CAH model transaxle) at the vehicle assembly plant.

1987 vehicles built prior to December 1, 1986 may be equipped with a 7CBH transaxle which is calibrated for firmer shifts and used in the SE and STE models only.

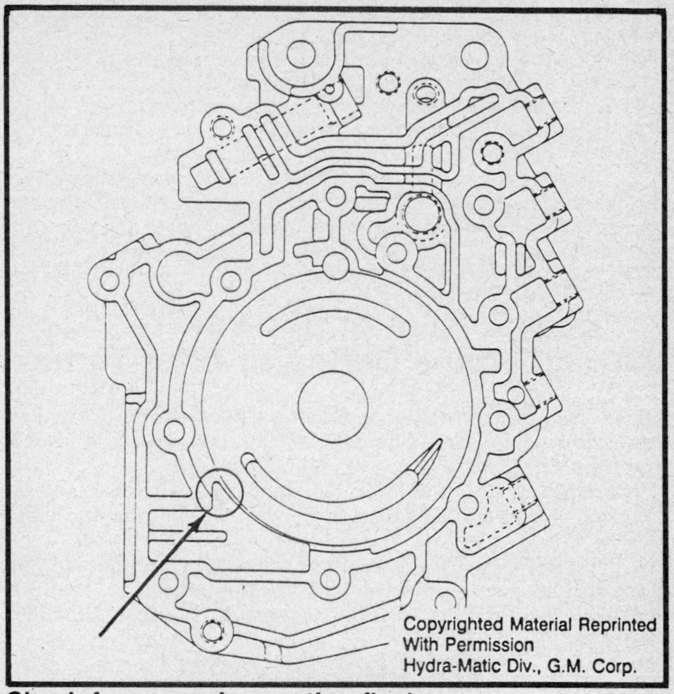
Copyrighted Material Reprinted With Permission Hydra-Matic Div., G.M. Corp.

Check for excessive casting flash

After verifying the vehicle and engine application, replace the 7CBH model transaxle with a 7CAH model transaxle.

Harsh Engagement, Double Bump or Engagement Shudder from Park or Neutral to Reverse

On all 1984–89 transaxles, a harsh engagement or engagement shudder from **P** or **N** to **R** in a THM 440–T4 transaxle may be caused by one of the items: oxidized fluid (the fluid may visually appear to be OK), or the reverse band assembly and 2nd clutch housing drum and bushing assembly.

Check engine idle speed and engine/transaxle mounts. If okay, drain and replace the transaxle fluid then road test vehicle for the above condition. If the condition still exists, replace the reverse band assembly and 2nd clutch housing drum and bushing assembly with the appropriate parts.

Reverse band assembly—1984–89, part number 8668012

2nd clutch housing drum and bushing assembly—1987–89, part number 8661912

2nd clutch housing drum and bushing assembly—1984–86, part number 8662936

Erratic Shifts/No Upshifts

On all 1984–87 transaxles, erratic shifts or no upshift in a THM 440–T4 transaxle may be caused by particles of hardened Imprex® material lodging in the governor pressure channel. (Imprex® is an "impregnation plastic" used in the manufacturing of cases to help seal the case and minimize porosity.)

1. Remove the governor assembly from transaxle and clean with solvent.
2. Inspect the governor for binding weights and/or mispositioned springs. To verify operation, turn the governor upside down and pour solvent into the governor shaft. Check for leakage past the checkballs. If leakage is noted, replace the governor assembly.

3. Reinstall the governor cover without the governor assembly.
4. Start the engine and move the selector lever to **D** position. Allow the engine to run for 15–20 seconds.
5. Move the selector lever to **P** and stop the engine.
6. Reinstall the governor assembly into the transaxle and road test the vehicle to verify transaxle operation.
7. If the erratic shifts or a no upshift condition continues, remove and clean the governor oil pipe retainer, orificed cup plug, governor feed and return pipes, accumulator cover and governor screen.

Engine Flare or Slips on Turns; Pump Whine During or After Turns

On all 1984–87 transaxles, engine flare, pump whine or transaxle slip during vehicle turns may be caused by low fluid level or the fluid filter.

Starting April 6, 1987 (Julian date 095), all THM 440–T4 transaxles were built using a filter with a new inlet and baffle design.

Check for proper fluid level and correct as necessary. If the flare still occurs, remove the bottom pan and replace the filter and seal with the filter and seal package part number 8646902.

1–2 Shift Shudder

On all 1987 transaxles, a 1–2 shift shudder on a 440–T4 transaxle may be caused by out of flat 2nd clutch plates. This condition may be repaired by replacing the 2nd clutch plates with the service package part number 8662914.

Production change occurred April 15, 1987 (Julian date 105) using the new 2nd clutch plates. This change is also on all Service Replacement Transmission Assembly (SRTA) transaxles built starting April 15, 1987 (Julian date 105).

When performing a complete overhaul or servicing a transaxle for 1–2 shift shudder, remove and discard the new fiber plates from the overhaul package and the original steel plates from the 2nd clutch assembly.

NOTE: 1987 steel plates are 0.075 in. thick, 1984–86 steel plates are 0.088 in. thick. Do not interchange these plates or improper clutch travel may result. There will be a package to service 1984–86 transaxles as soon as steel plates are available.

Modulator Valve Buzz Noise

On all 1986 Cadillac vehicles and all 1987 vehicles, modulator valve buzz noise can be caused by a missing orificed cup plug in the modulator circuit of the channel plate for transaxles built prior to February 1, 1987, (Julian date 032).

Remove the transaxle from the vehicle and remove the channel plate from the transaxle. Install the orificed cup lug from the service package, part number 8646987. The cup plug is properly installed at 0.020 in. (0.0–0.5mm) below the channel plate machined surface and must be staked securely.

Intermittent No Reverse or Locks In Reverse

A 1984 through 1987 vehicle equipped with a THM 440–T4 transaxle may have an intermittent no reverse or locking in reverse condition until shifted to a forward range prior to shifting into reverse ("unloading" the parking pawl). This condition may be caused by a "dent" on the parking pawl that does not allow the pawl to release until "unloaded".

Effective May 1, 1987 (Julian date 121) all 1987 THM 440–T4

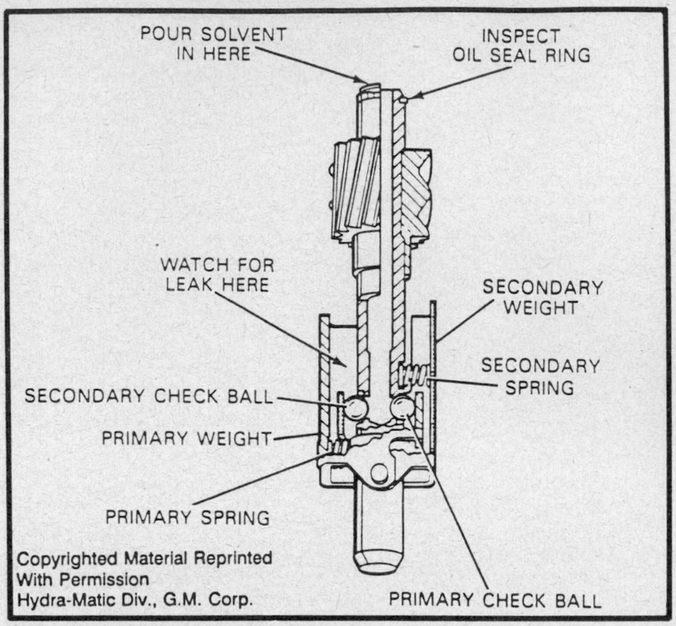
POUR SOLVENT IN HERE
INSPECT OIL SEAL RING
WATCH FOR LEAK HERE
SECONDARY WEIGHT
SECONDARY SPRING
SECONDARY CHECK BALL
PRIMARY WEIGHT
PRIMARY SPRING
PRIMARY CHECK BALL
Copyrighted Material Reprinted With Permission Hydra-Matic Div., G.M. Corp.

Checking governor assembly

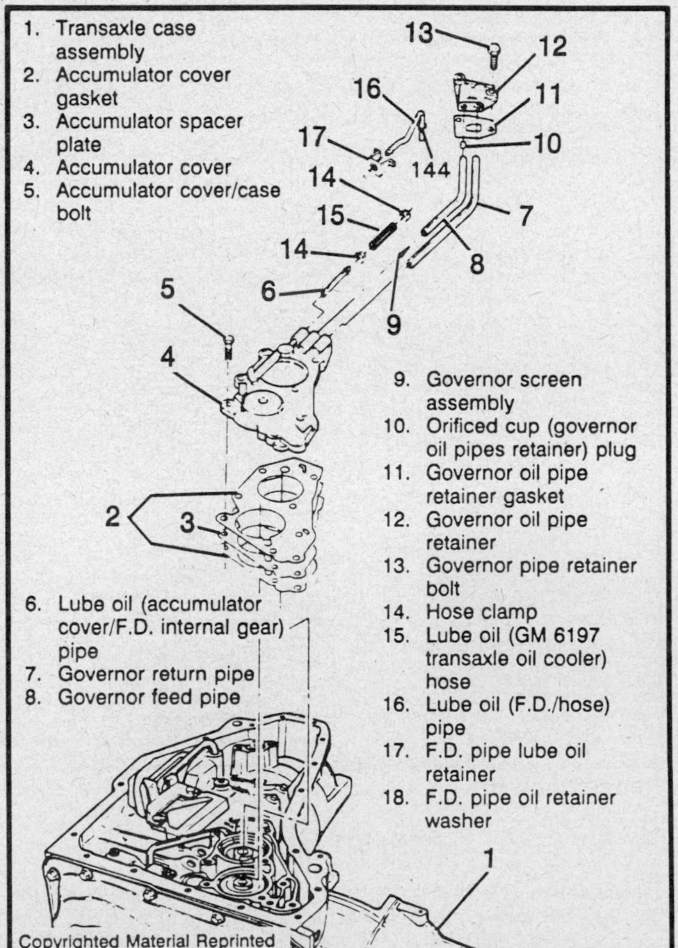
1. Transaxle case assembly
2. Accumulator cover gasket
3. Accumulator spacer plate
4. Accumulator cover
5. Accumulator cover/case bolt
6. Lube oil (accumulator cover/F.D. internal gear) pipe
7. Governor return pipe
8. Governor feed pipe
9. Governor screen assembly
10. Orificed cup (governor oil pipes retainer) plug
11. Governor oil pipe retainer gasket
12. Governor oil pipe retainer
13. Governor pipe retainer bolt
14. Hose clamp
15. Lube oil (GM 6197 transaxle oil cooler) hose
16. Lube oil (F.D./hose) pipe
17. F.D. pipe lube oil retainer
18. F.D. pipe oil retainer washer
144
Copyrighted Material Reprinted With Permission Hydra-Matic Div., G.M. Corp.

Governor oil pipe and accumulator assembly

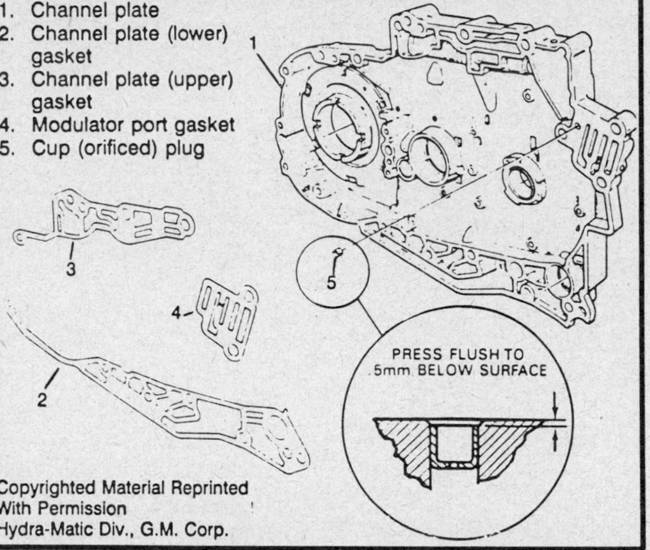
1
3
4
5
2
PRESS FLUSH TO .5mm BELOW SURFACE

1. Channel plate
2. Channel plate (lower) gasket
3. Channel plate (upper) gasket
4. Modulator port gasket
5. Cup (orificed) plug

Copyrighted Material Reprinted With Permission Hydra-Matic Div., G.M. Corp.

Checking orificed cup plug

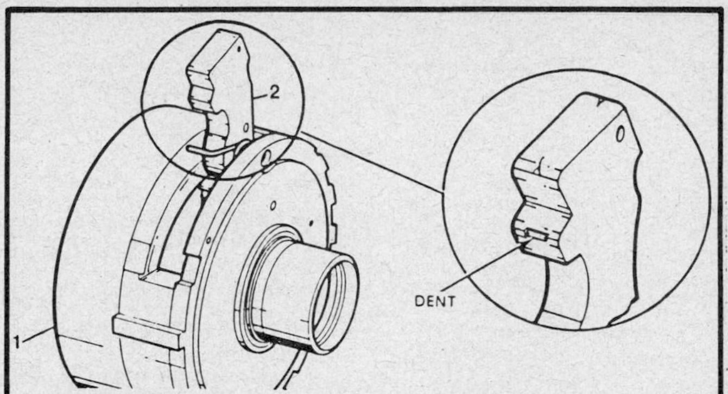
2
1
DENT

1. Final drive internal gear assembly
2. Parking lock pawl

Copyrighted Material Reprinted With Permission Hydra-Matic Div., G.M. Corp.

Checking parking lock pawl on final drive internal gear assembly

transaxles were produced with parking pawl assemblies not having a "dent".

Remove transaxle and replace the final drive internal gear/parking pawl assembly and actuator guide, plunger and spring assembly. Also inspect the parking gear for wear or damage and replace if necessary. Replacement parts are as follows:

Part number 8656372—final drive internal gear/parking pawl assembly

Part number 8656689—actuator guide, plunger and spring assembly

Part number 8644238—parking gear

Before installing the replacement parts, inspect the new final drive internal gear/parking pawl assembly to make sure it does not have a "dent".

Second Gear Starts

A 1984–87 vehicle equipped with a THM 440–T4 may start out in 2nd gear instead of 1st gear. This condition could occur intermittently or all the time.

High line pressure could be a cause by holding too much governor pressure and not allowing the 1–2 shift valve to downshift. Check line pressure. Governor pressure should be below 3.0 psi at 0 mph.

If governor pressure is okay, the 1–2 shift valve, may be stuck in the upshifted position. Remove the control valve assembly and check for a sticking 1–2 shift valve, 1–2 throttle valve, or the 1–2 throttle valve spring caused by the 1–2 throttle valve bushing retainer pushed in too far. The retainer should be installed flush with the machine surface of the valve body.

Proper Fluid Level Checking Procedure

On all 1987–88 transaxles, this bulletin outlines proper hot and cold fluid level checking procedures for the transaxles.

To obtain a proper cold fluid level check for transaxles:

1. Start vehicle and cycle gear range selector through **1, 2, D, OD,** and **R** for approximately 3 seconds in each range. Complete "cold" check preparation by letting vehicle idle in **P** for 3 minutes.
2. With vehicle level, accessories turned off and engine idling in **P**, check fluid level.
3. When cold, 80°F (27°C), a full THM 440–T4 transaxle will

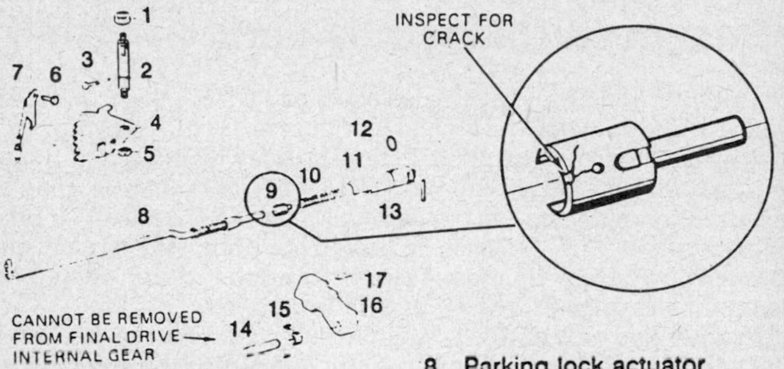
1
2
3
4
5
6
7
8
9
10
11
12
13
14
15
16
17
INSPECT FOR CRACK
CANNOT BE REMOVED FROM FINAL DRIVE INTERNAL GEAR

1. Manual shaft seal
2. Manual shaft
3. Manual shaft to case pin
4. Inside detent lever and pilot assembly
5. Hex nut
6. Manual detent spring bolt
7. Manual detent roller and spring assembly
8. Parking lock actuator assembly
9. Pawl lockout plunger assembly
10. Pawl lockout spring
11. Actuator guide
12. Bushing/actuator guide O-ring seal
13. Guide retaining pin
14. Parking lock pawl shaft
15. Parking pawl return spring
16. Parking lock pawl
17. Parking pawl lockout pin

Copyrighted Material Reprinted With Permission Hydra-Matic Div., G.M. Corp.

Inspecting pawl lockout plunger assembly for cracks

show the fluid level to be above the **FULL HOT** mark on the fluid level indicator. This is a result of the fluid that is stored in the bottom pan when the transaxle is cold. As the transaxle warms to normal operating temperature, fluid is captured in the side pan lowering the fluid level to within the cross-hatched area on the fluid level indicator.

NOTE: The cold level checking procedure cannot take the place of a "hot" level check. The cold level will let the technician know that there is enough fluid in the transaxle to perform a preliminary check procedure, an accurate road test and allow normal operating temperature to be obtained prior to the necessary hot check.

To obtain a proper hot fluid level check for all THM transaxles:

1. Drive vehicle in all ranges to attain a fluid temperature of 180–200°F (80–90°C).
2. Idle at normal idle speed in **P** for 3 minutes.
3. With vehicle level, accessories turned off and engine idling in **P**, check fluid level.

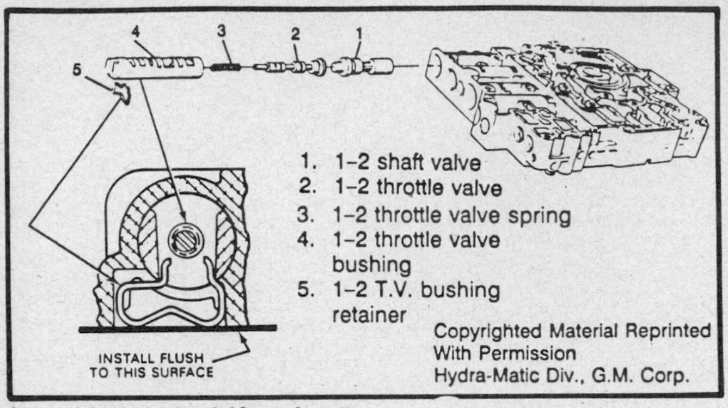
4
3
2
1
5
1. 1–2 shaft valve
2. 1–2 throttle valve
3. 1–2 throttle valve spring
4. 1–2 throttle valve bushing
5. 1–2 T.V. bushing retainer
INSTALL FLUSH TO THIS SURFACE
Copyrighted Material Reprinted With Permission Hydra-Matic Div., G.M. Corp.

Checking 1–2 shift valve

4. Fluid level should be within the cross-hatched area. (Read from and back of the fluid level indicator and use the lowest level.)

Torque Converter Clutch (TCC) Busyness Caused by Early 3–4 Shift

On 1986 C and H-body vehicles equipped with 3.0L or 3.8L engine, torque converter clutch (TCC) busyness caused by an early 3–4 shift condition in 1986 C and H-Body vehicles, equipped with a 3.0L or 3.8L engine and THM 440–T4 transaxle, may be repaired by replacing the 3–4 throttle valve (T.V.) spring. Under light throttle, TCC releases during the shift raising engine rpm momentarily (may be diagnosed as an engine flare), then TCC re-applies shortly after the shift is completed.

Inspect and adjust the T.V. cable and road test the vehicle. If the condition still exists, replace the 3–4 T.V. spring contained in service package part number 8662934.

Harsh 1–2 Shift

On all 1984–87 transaxles, a harsh 1–2 shift condition on a THM 440–T4 transaxle may be cause by a blocked or partially drilled 2nd clutch feed passage in the case. This condition may be repaired.

Before removing the transaxle from the vehicle, inspect the 2nd clutch feed passage for blockage or a partially drilled condition. To perform this inspection, remove the following items: transaxle bottom pan, filter, accumulator cover with governor pipes and governor pipe retainer, accumulator cover spacer plate and gaskets.

Inspect the 2nd clutch passage by measuring the hole depth which should be a minimum of 2.835 in. (72mm). If the 2nd clutch passage is properly drilled, refer to the Diagnosis Section for other possible causes of a harsh 1–2 shift condition.

If the 2nd clutch passage is not fully drilled:

1. Remove the transaxle and disassemble components up to and including, the channel plate.
2. Mark an 21.64 in. (8.5mm) drill bit at 2.835 in. (72mm) from the tip using tape or other acceptable method. From the bottom pan side of the case, finish drilling the 2nd clutch passage to the proper depth as marked on the drill bit.

NOTE: Avoid drilling too deep into the 2nd clutch passage because it will damage the case.

3. Remove metal chips caused by drilling and thoroughly flush the repaired case passages with clean solvent.
4. Reassemble the transaxle.

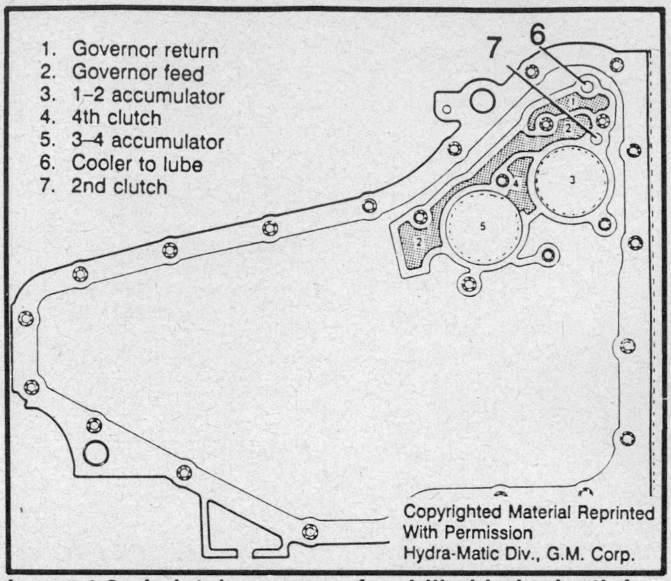
1. Governor return
2. Governor feed
3. 1–2 accumulator
4. 4th clutch
5. 3–4 accumulator
6. Cooler to lube
7. 2nd clutch
7
6
Copyrighted Material Reprinted With Permission Hydra-Matic Div., G.M. Corp.

Inspect 2nd clutch passage for drilled hole depth in case passage

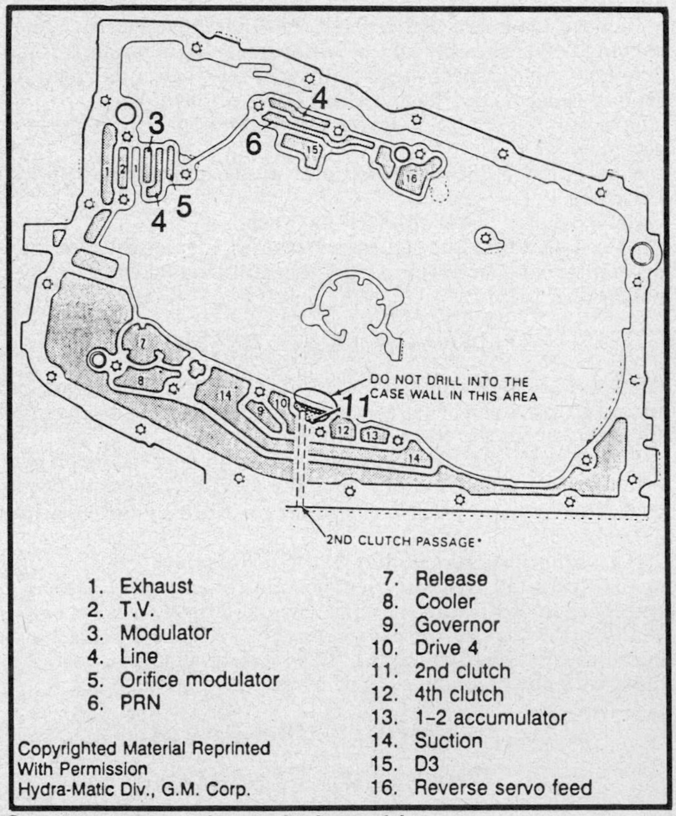
3
4
6
4
5
11
DO NOT DRILL INTO THE CASE WALL IN THIS AREA
2ND CLUTCH PASSAGE
1. Exhaust
2. T.V.
3. Modulator
4. Line
5. Orifice modulator
6. PRN
7. Release
8. Cooler
9. Governor
10. Drive 4
11. 2nd clutch
12. 4th clutch
13. 1–2 accumulator
14. Suction
15. D3
16. Reverse servo feed
Copyrighted Material Reprinted With Permission Hydra-Matic Div., G.M. Corp.

Case passage, channel plate side

Harsh Engagement Conditions

Harsh engagement conditions and specific service information for 1987–88 vehicles equipped with a THM 440–T4 transaxle follow. Other items which cause harsh engagement conditions are:

a. Vacuum leaks (low vacuum supply to the vacuum modulator) 13–18 in. Hg (33–46cm Hg) required
b. Stuck modulator valve
c. Missing No. 9 checkball

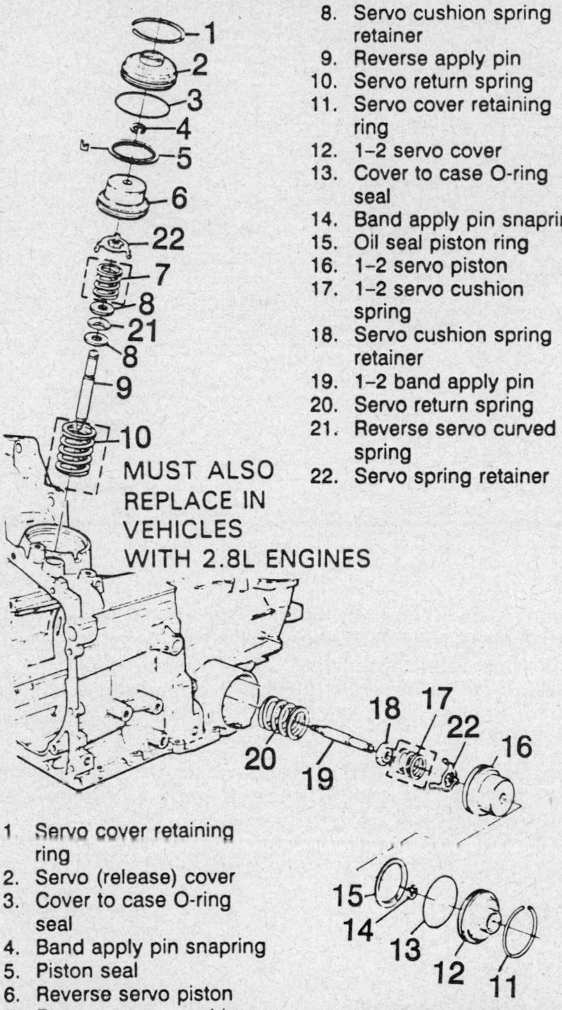
MUST ALSO REPLACE IN VEHICLES WITH 2.8L ENGINES

1. Servo cover retaining ring
2. Servo (release) cover
3. Cover to case O-ring seal
4. Band apply pin snapring
5. Piston seal
6. Reverse servo piston
7. Reverse servo cushion spring
8. Servo cushion spring retainer
9. Reverse apply pin
10. Servo return spring
11. Servo cover retaining ring
12. 1-2 servo cover
13. Cover to case O-ring seal
14. Band apply pin snapring
15. Oil seal piston ring
16. 1-2 servo piston
17. 1-2 servo cushion spring
18. Servo cushion spring retainer
19. 1-2 band apply pin
20. Servo return spring
21. Reverse servo curved spring
22. Servo spring retainer

Copyrighted Material Reprinted With Permission Hydra-Matic Div., G.M. Corp.

1-2 servo and reverse servo assemblies

Harsh Engagement From Park or Neutral To Drive or Neutral

1987-88 VEHICLES EXCEPT 2.8L ENGINE

1. Remove the 1-2 servo assembly and the reverse servo assembly from the transaxle.
2. Disassemble and replace the 1-2 servo cushion spring and the reverse servo cushion spring with part number 8668121 (reverse servo cushion spring) and part number 8668123 (1-2 servo cushion spring).
3. Install the 1-2 servo assembly and reverse servo assembly in the transaxle.

Harsh Engagement From Reverse To Drive, Park or Neutral To Drive or Reverse

1987-88 A-BODY AND 1988 W-BODY VEHICLES EQUIPPED WITH 2.8L ENGINE

1. Remove the 1-2 servo assembly and the reverse servo assembly from the transaxle.
2. Disassemble and replace the 1-2 servo cushion spring and the reverse servo cushion spring and the reverse servo return spring with part number 8668127 (reverse servo cushion spring) and part number 8668123 (1-2 servo cushion spring).

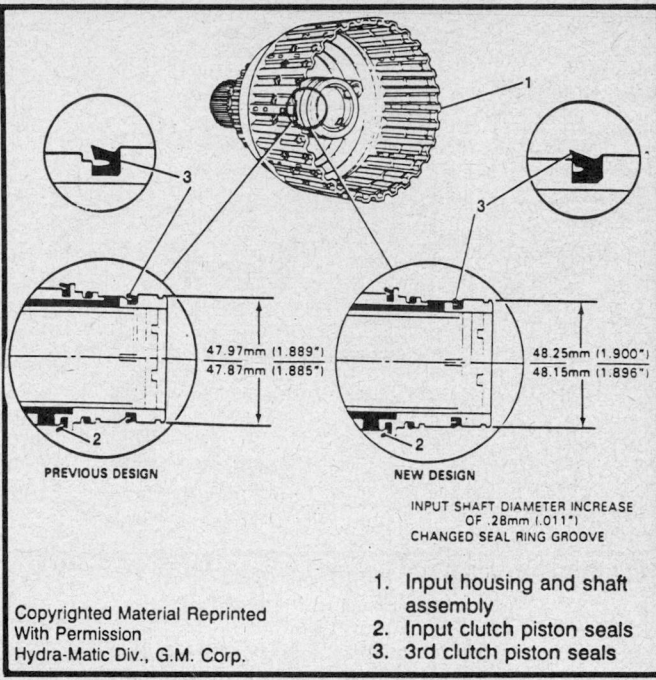

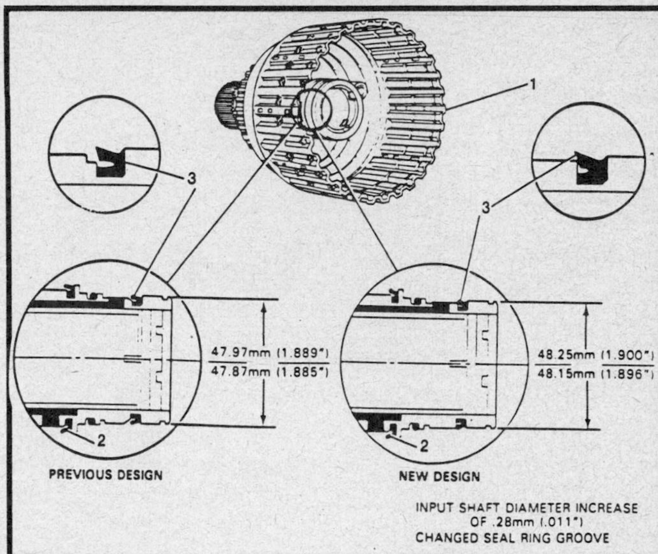
47.97mm (1.889")
47.87mm (1.885")
48.25mm (1.900")
48.15mm (1.896")
PREVIOUS DESIGN
NEW DESIGN
INPUT SHAFT DIAMETER INCREASE OF .28mm (.011")
CHANGED SEAL RING GROOVE

1. Input housing and shaft assembly
2. Input clutch piston seals
3. 3rd clutch piston seals

Copyrighted Material Reprinted With Permission Hydra-Matic Div., G.M. Corp.

Input clutch housing and shaft assembly

3. Install the 1-2 servo assembly and reverse servo assembly in the transaxle.

New Input Clutch and Third Clutch Seal Installers

New seal installation tools are available for the input clutch inner piston seal and the 3rd clutch inner piston seal. These tools are to be used when servicing the input clutch assembly on all 1984-88 vehicles equipped with a THM 440-T4 transaxle.

When servicing the input clutch housing and shaft assembly on a 1984-88 THM 440-T4 transaxle use the new seal installation tools, J-37361 and J-37362. The previous design tools, J-34091 and J-34092 will work on 1984-86 transaxles only.

Governor Screen Added to Transaxle Case

On some 1989 DeVilles, Eldorados, Sevilles and Allantes a production change has been made to add a screen to the governor shaft sleeve. The screen, which is identical to the one used in the THM 125C transaxle, was added to reduce the chance of sediment entering the governor assembly.

The production change occurred on some units beginning on November 30, 1988 (Julian date 335). The change occurred on all units by December 5, 1988 (Julian date 340).

The production change included dimensional changes in the case to provide clearance for assembly of the screen.

NOTE: Do not attempt to install the screen on earlier transaxles. Critical case dimensions will not allow any further operations.

On 1989½ vehicles—Governor shaft sleeve and screen assembly—part number 8660770.

On 1984-89½—Governor shaft sleeve—part number 8631328.

Transaxle Slips When Cold

On some 1989 DeVilles, Eldorados, Sevilles and Allantes may

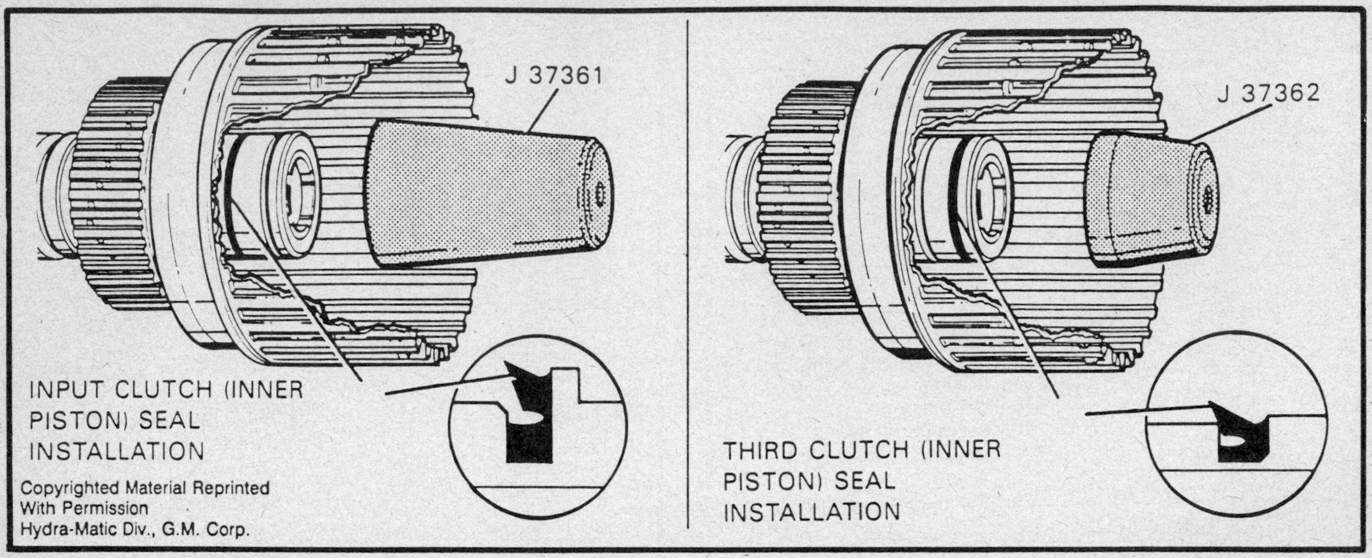
J 37361
INPUT CLUTCH (INNER PISTON) SEAL INSTALLATION
Copyrighted Material Reprinted
With Permission
Hydra-Matic Div., G.M. Corp.
J 37362
THIRD CLUTCH (INNER PISTON) SEAL INSTALLATION

Seal Installers J–37361 and J–37362

experience an intermittent transaxle slip in forward and/or reverse when cold. This may be due to 1 or more undersized input shaft oil rings that seal the input clutch and 3rd clutch feed holes. The slipping may occur less in reverse or manual low range (due to boosted line pressure) than in the other drive ranges.

To service the THM F–7 transaxle, replace the transaxle.

To service the THM 440–T4 transaxle, disassemble the transaxle, remove the input shaft and housing assembly and inspect the shaft oil seal rings in the grooves. Replace them if they do not occupy the full groove width. Use Overhaul seals and gaskets package, part number 8662959 or overhaul repair complete package for burned input and 3rd clutch plate assemblies, part number 8662960.

NOTE: Inspect the 3rd clutch and input clutch assemblies for damage or excess wear. Repair as necessary.

TROUBLE DIAGNOSIS

CLUTCH AND BAND APPLICATION CHART
THM 440-T4 Automatic Transaxle

Range		4th Clutch	Reverse Band	2nd Clutch	3rd Clutch	3rd Roller Clutch	Input Sprag	Input Clutch	1-2 Band
NEUTRAL PARK							①	①	
DRIVE	1						HOLD	ON	ON
	2			ON			OVER-RUNNING	①	ON
	3			ON	ON	HOLD			
	4	ON		ON	①	OVER-RUNNING			
MANUAL	3			ON	ON	HOLD	HOLD	ON	
	2			ON			OVER-RUNNING	①	ON
	1				ON	HOLD	HOLD	ON	ON
REVERSE		ON					HOLD	ON	

①APPLIED BUT NOT EFFECTIVE

CHILTON'S THREE "C's" DIAGNOSIS CHART
THM 440-T4 Automatic Transaxle

Condition	Cause	Correction
Oil Leakage	a) Side cover, bottom pan and gaskets, loose bolts	a) Repair or replace cover, gasket and torque bolts
	b) Damaged seal at T.V. cable, fill tube or electrical connector	b) Replace seal as required
	c) Damaged seal assembly on manual shaft	c) Replace seal assembly as required
	d) Leakage at governor cover, servo covers, modulator, parking plunger guide, speedometer driven gear sleeve	d) Replace damaged "O" ring seals as required
	e) Converter or converter seal leaking	e) Replace converter and seal as required
	f) Axle seals leaking	f) Remove axles and replace seals as required
	g) Pressure ports or cooler line fittings leaking	g) Tighten or repair stripped threads
Fluid foaming or blowing out the vent	a) Fluid level high	a) Correct fluid level
	b) Fluid foaming due to contaminates or overheating of fluid	b) Determine cause of contamination or overheating and repair
	c) Drive sprocket support has plugged drain back holes	c) Open drain back holes in sprocket support
	d) Thermo element not closing when hot	d) Replace thermo element
	e) Fluid filter "O" ring damaged	e) Replace fluid filter "O" ring
High or low fluid pressure, verified by pressure gauge	a) Fluid level high or low	a) Correct fluid level as required
	b) Vacuum modulator or hose leaking	b) Repair or replace hose or modulator
	c) Modulator valve, pressure regulator valve, pressure relief valve nicked, scored or damaged. Spring or ball checks missing or damaged	c) Repair or replace necessary components
	d) Oil pump or components damaged, parts missing	d) Repair or replace oil pump assembly
No drive in DRIVE range	a) Fluid level low	a) Correct fluid level
	b) Fluid pressure low	b) Refer to low fluid pressure causes
	c) Manual linkage misadjusted or disconnected	c) Repair or adjust manual linkage
	d) Torque converter loose on flex plate or internal converter damage	d) Verify malfunction and repair as required
	e) Oil pump or drive shaft damaged	e) Repair or replace oil pump and/or drive shaft
	f) Number 13 check ball misassembled or missing	f) Correct or install number 13 check ball in its proper location
	g) Damaged drive link chain, sprocket or bearings	g) Replace damaged components

CHILTON'S THREE "C's" DIAGNOSIS CHART
THM 440-T4 Automatic Transaxle

Condition	Cause	Correction
No drive in DRIVE range	h) Burned or missing clutch plates, damaged piston seals or piston, Housing check ball damaged, input shaft seals or feed passages blocked or damaged	h) Repair and/or replace damaged input clutch assembly components
	i) Input sprag and/or input sun gear assembly improperly assembled or sprag damaged	i) Correctly assemble or replace input sprag and input sun gear assembly
	j) Pinions, sun gear or internal gears damaged on input and reaction carrier assemblies	j) Repair or replace carrier assemblies as required
	k) 1-2 band or servo burned or damaged. Band apply pin incorrect in length	k) Repair or replace band and/or servo components as required
	l) 1-2 servo oil pipes leaking fluid	l) Correct oil tubes to prevent leakage
	m) Final drive assembly broken or damaged	m) Repair or replace necessary components of final drive
	n) Parking pawl spring broken	n) Replace parking pawl spring
	o) Output shaft damage, broken or misassembled	o) Repair, replace or re-assemble output shaft
First speed only, no 1-2 shift	a) Governor assembly defective	a) Repair or replace governor
	b) Number 14 check ball missing	b) Install number 14 check ball
	c) 1-2 shift valve sticking or binding	c) Repair or clean valve and bore
	d) Accumulator and/or pipes	d) Repair/renew components
	e) 2nd clutch assembly damaged	e) Repair/renew components
	f) Oil seal rings damaged on driven sprocket support	f) Replace oil seal rings
	g) Splines damaged or parts missing from reverse reaction drum	g) Repair or replace damaged components
Harsh or soft 1-2 shift	a) Fluid pressure	a) Check pressure and correct
	b) Defective accumulator assembly	b) Repair or replace accumulator assembly
	c) Accumulator valve stuck	c) Repair or clean valve and bore
	d) Missing or mislocated number 8 check ball	d) Install or re-locate number 8 check ball
High or low 1-2 shift speed	a) Disconnected or misadjusted T.V. cable	a) Connect and/or adjust T.V. cable
	b) Bent or damaged T.V. link, lever and bracket assembly	b) Repair or replace damaged components
	c) Stuck or binding T.V. valve and plunger	c) Correct binding condition or remove stuck valve and plunger
	d) Incorrect governor pressure	d) Correct governor pressure
No 2-3 upshift, 1st and 2nd speeds only	a) Defective 1-2 servo or components	a) Repair or replace 1-2 servo assembly
	b) Defective number 7 check ball and capsule assembly	b) Check, repair or replace check ball and capsule
	c) Number 11 check ball not seating	c) Check, repair or replace check ball
	d) 2-3 shift valve stuck in control valve assembly	d) Remove stuck 1-2 shift valve and repair

CHILTON'S THREE "C's" DIAGNOSIS CHART
THM 440-T4 Automatic Transaxle

Condition	Cause	Correction
No 2-3 upshift, 1st and 2nd speeds only	e) Seals damaged or blocked passages on input shaft f) Defective third clutch assembly g) Defective third roller clutch assembly h) Numbers 5, 6 and/or accumulator valve stuck	e) Replace seals and open passages f) Overhaul third clutch assembly g) Inspect, repair or replace necessary components h) Repair as required
Harsh or soft 2-3 shift	a) Fluid pressure b) Mislocated number 12 check ball	a) Test and correct fluid pressure b) Correct check ball location
High or low 2-3 shift speed	a) Disconnected or misadjusted T.V. cable b) Bent or damaged T.V. link, lever and bracket assembly c) Stuck or binding T.V. valve and plunger d) Incorrect governor pressure	a) Connect and/or adjust T.V. cable b) Repair or replace damaged components c) Correct binding condition or remove stuck valve and plunger d) Correct governor pressure
No 3-4 shift	a) Incorrect governor pressure b) 3-4 shift valve stuck in control valve assembly c) Defective 4th clutch assembly d) Spline damage to 4th clutch shaft	a) Correct governor pressure b) Free 3-4 shift valve and repair control valve assembly c) Overhaul 4th clutch assembly d) Replace 4th clutch shaft
Harsh or soft 3-4 shift	a) Fluid pressure b) Defective accumulator assembly c) Mislocated number 1 check ball	a) Test and correct fluid pressure b) Repair or replace accumulator assembly c) Correct check ball location
High or low 3-4 shift	a) Disconnected or misadjusted T.V. cable b) Bent or damaged T.V. link, lever and bracket assembly c) Stuck or binding T.V. valve and plunger d) Incorrect governor pressure	a) Connect and/or adjust T.V. cable b) Repair or replace damaged components c) Correct binding condition or remove stuck valve and plunger d) Correct governor pressure
No converter clutch apply (Vehicles equipped with E.C.M.)	a) Improper E.C.M. operation b) Electrical system of trans-axle malfunctioning c) Converter clutch apply valve stuck d) Number 10 check ball missing e) Converter clutch blow-off check ball not seating or damaged f) Seals damaged on turbine shaft g) Damaged seal on oil pump drive shaft	a) Verify proper E.C.M. operation b) Test and correct electrical malfunction c) Free converter clutch apply valve and repair d) Install missing check ball e) Inspect channel plate and check ball. Repair as required f) Replace seals and inspect shaft g) Replace seal on oil pump drive shaft
No converter clutch apply (vehicles not equipped with E.C.M.)	a) Electrical system of transaxle malfunctioning	a) Test and correct electrical malfunction

CHILTON'S THREE "C's" DIAGNOSIS CHART
THM 440-T4 Automatic Transaxle

Condition	Cause	Correction
No converter clutch apply (vehicles not equipped with E.C.M.)	b) Converter clutch shift and/or apply valves stuck	b) Free converter clutch shift and/or apply valves
	c) Number 10 check ball missing	c) Install missing check ball
	d) Converter clutch blow-off check ball not seated or damaged	d) Inspect channel plate and check ball. Repair as required
	e) Seals damaged on turbine shaft	e) Replace seals and inspect shaft
	f) Damaged oil seal on seal pump drive shaft	f) Replace seal on oil pump drive shaft
Converter clutch does not release	a) Converter clutch apply valve stuck in the apply position	a) Free apply valve for converter clutch and repair as required
Rough converter clutch apply	a) Converter clutch regulator valve stuck	a) Free converter clutch regulator valve and repair as required
	b) Converter clutch accumulator piston or seal damaged	b) Replace seal or piston as required Check accumulator spring
	c) Seals damaged on turbine shaft	c) Replace seals on turbine shaft
Harsh 4-3 downshift	a) Number 1 check ball missing in control valve assembly	a) Install number 1 check ball in control valve assembly
Harsh 3-2 downshift	a) 1-2 servo control valve stuck	a) Free 1-2 servo control valve
	b) Number 12 check ball missing	b) Install number 12 check ball
	c) 3-2 control valve stuck	c) Free 3-2 control valve
	d) Number 4 check ball missing	d) Install number 4 check ball
	e) 3-2 coast valve stuck	e) Free 3-2 coast valve
	f) Input clutch accumulator piston or seal damaged	f) Replace input clutch accumulator piston and/or seal
Harsh 2-1 downshift	a) Number 8 check ball missing	a) Install number 8 check ball
No reverse	a) Fluid pressure	a) Test and correct fluid pressure
	b) Defective oil pump	b) Test and correct oil pump malfunction
	c) Broken, stripped or defective drive link assembly	c) Repair, replace as required
	d) Reverse band burned or damaged	d) Replace rear band as required
	e) Defective input clutch	e) Repair, replace defective input clutch components
	f) Defective input sprag	f) Replace defective sprag
	g) Piston or seal damaged, pin selection incorrect for rear servo assembly	g) Repair rear servo as required and install correct pin if needed
	h) Defective input and reaction carriers	h) Replace input and reaction carriers as required
No park range	a) Parking pawl, spring or parking gear damaged	a) Repair as required
	b) Manual linkage broken or out of adjustment	b) Repair linkage or adjust as required
Harsh shift from Neutral to Drive or from Neutral to Reverse	a) Number 9 check ball missing	a) Install number 9 check ball
	b) Number 12 check ball missing	b) Install number 12 check ball
	c) Thermal elements not closing when warm	c) Replace thermal elements
No viscous clutch apply (Vehicles with E.C.M.)	a) Improper E.C.M. operation	a) Verify E.C.M. operation and repair as required
	b) Damaged thermister	b) Replace thermister
	c) Damaged temperature switch	c) Replace temperature switch

Hydraulic Control System

NOTE: Please refer to Section 9 for all oil flow circuits.

OPERATION PRINCIPLES

The torque converter smoothly couples the engine to the planetary gears and the overdrive unit through fluid and hydraulically/mechanically provides additional torque multiplication when required. The combination of the compound planetary gear set provides 4 forward gear ratios and 1 reverse. The changing of the gear ratios is fully automatic is relation to the vehicle speed and engine torque. Signals of vehicle speed and engine torque are constantly being directed to the transaxle control valve assembly to provide the proper gear ratio for maximum engine efficiency and performance at all throttle openings.

TORQUE CONVERTER

The torque converter assembly serves 3 primary functions. First, it acts as a fluid coupling to smoothly connect engine power through oil to the transaxle gear train. Second, it multiplies the torque or twisting effort from the engine when additional performance is desired. Thirdly, it provides direct drive through the torque converter.

The torque converter assembly consists of a 3-element torque converter combined with a friction clutch. The 3 elements are the pump (driving member), the turbine (driven or output member) and the stator (reaction member). The converter cover is welded to the pump to seal all 3 members in an oil filled housing. The converter cover is bolted to the engine flexplate which is bolted directly to the engine crankshaft. The converter pump is therefore mechanically connected to the engine and turns at engine speed whenever the engine is operating.

The stator is located between the pump and turbine and is mounted on a one-way roller clutch which allows it to rotate clockwise but not counterclockwise.

The purpose of the stator is to redirect the oil returning from the turbine and change its direction of rotation back to that of the pump member. The energy in the oil is then used to assist the engine in turning the pump. This increases the force of the oil driving the turbine; and as a result, mulitplies the torque or twisting force of the engine.

The force of the oil flowing from the turbine to the blades of the stator tends to rotate the stator counterclockwise, but the roller clutch prevents it from turning.

With the engine operating at full throttle, transaxle in gear and the vehicle standing still, the converter is capable of multiplying engine torque by approximately 2.0:1.

As turbine speed and vehicle speed increases, the direction of the oil leaving the turbine changes. The oil flows against the rear side of the stator vanes in a clockwise direction. Since the stator is now impeding the smooth flow of oil in the clockwise direction, its roller clutch automatically releases and the stator revolves freely on its shaft.

QUADRANT POSITION

The quadrant has 7 positions indicated in the following order: **P, R, N, OD, D, 2, 1.**

P—Park position enables the transaxle output shaft to be held, thus preventing the vehicle from rolling either forward or backward. (For safety reasons, the vehicle parking brake should be used in addition to the transaxle **P** position). Because the output shaft is mechanically locked, the **P** position should not be selected until the vehicle has come to a stop. The engine may be started in the **P** position.

R—Reverse enables the vehicle to be operated in a rearward direction.

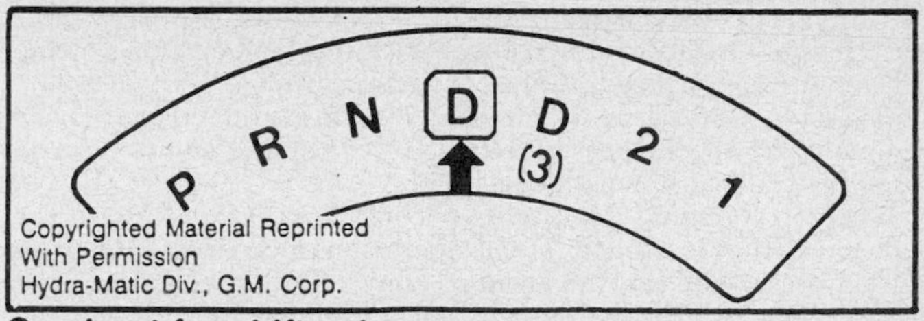
P
R
N
D
D
2
1
(3)

Quadrant for shift selector

N—Neutral position enables the engine to be started and operated without driving the vehicle. If necessary, this position must be selected if the engine has to be restarted while the vehicle is moving.

D—Drive is used for mostly highway driving conditions and maximum economy. Drive has 4 gear ratios, from the starting ratio, through direct drive to overdrive. Downshifts to a higher ratio are available for safe passing by depressing the accelerator.

D—Manual 3rd can be used for conditions where it is desired to use only 3 gears. This range is also useful for braking when descending slight grades. Upshifts and downshifts are the same as in **D** for 1st, 2nd and 3rd gears, but the transaxle will not shift to 4th gear.

2—Manual 2nd adds more performance. It has the same starting ratio as Manual 3rd range, but prevents the transaxle from shifting above 2nd gear, thus retaining 2nd gear for acceleration or engine braking as desired. Manual 2nd can be selected at any vehicle speed. If the transaxle is in 3rd or 4th gear it will immediately shift to 2nd Gear.

1—Manual low can be selected at any vehicle speed. The transaxle will shift to 2nd gear if it is in 3rd or 4th gear, until it slows below approximately 40 mph (64km/h), at which time it will downshift to 1st gear. This is particularly beneficial for maintaining maximum engine braking when descending steep grades.

DESCRIPTION OF HYDRAULIC COMPONENTS

Manual valve—Mechanically connected to the shift selector. It is fed by line pressure from the pump and directs pressure according to which range the driver has selected.

1–2 servo—A hydraulic piston and pin that mechanically applies the 1–2 band in 1st and 2nd gear. Also absorbs 3rd clutch oil to act as an accumulator for the 2–3 shift.

Reverse servo—A hydraulic piston and pin that mechanically applies the reverse band when **R** range is selected by the driver.

Modulator valve—Is controlled by the vacuum modulator assembly and regulates line pressure, into a modulator pressure that is proportional to engine vacuum (engine torque).

Modulator assembly—By sensing engine vacuum, it causes the modulator valve to regulate modulator pressure that is proportional to engine torque (inversely proportional to engine vacuum).

1–2 accumulator piston—Absorbs 2nd clutch oil to provide a cushion for the 2nd clutch apply. The firmness of the cushion is controlled by the 1–2 accumulator valve.

3–4 accumulator piston—Absorbs 4th clutch oil to provide a cushion for the 4th clutch apply. The firmness of the cushion is controlled by the 1–2 accumulator valve.

1–2 servo thermo elements—When cold, it opens another orifice to the servo, to provide less of a restriction for a quick servo apply. When warm, it blocks 1 of the 2 orifices to the servo apply. When warm, it blocks 1 of the 2 orifices to the servo and slow the flow of oil and provide a good neutral/drive shift feel.

Input clutch accumulator—Absorbs input clutch apply oil to cushion the input clutch apply.

Converter clutch accumulator—Cushions the converter clutch apply by absorbing converter clutch feed oil and slowing the amount of oil feed into the converter clutch apply passage.

3–2 downshift valve—Controlled by 2nd clutch oil that opens the valve when line pressure exceed 110 psi and allows T.V. oil to enter the 3–2 downshift passage.

Thermo element—Maintains a level of transaxle fluid in the side cover that is needed for the operation of the hydraulic pressure system. The thermo element allows fluid levels to increase or decrease with the increase or decrease of fluid temperature.

1–2 shift valve train—Shifts the transaxle from 1st to 2nd gear or 2nd to 1st gear, depending on governor, T.V., detent, or low oil pressures.

3–4 M.T.V. valve—Modulates T.V. pressure going to the 3–4 throttle valve to a lower pressure so that a light throttle 3–4 upshift will not be delayed.

2–3 accumulator valve—Receives line pressure from the manual valve and controlled by modulator pressure. The 2–3 accumulator valve, in 3rd gear and overdrive, varies 1–2 servo. Apply (2–3 accumulator) oil pressure in proportion to changes in modulator pressure (engine torque).

3–2 control valve—Controlled by governor oil, it controls the 3–2 downshift timing by regulating the rate at which the 3rd clutch releases and the 1–2 band applies.

2–3 shift valve train—Shifts the transaxle from 2nd to 3rd gear or 3rd to 1st gear, depending on governor T.V., detent or drive 2 oil pressures.

3–4 shift valve train—Shifts the transaxle from 3rd to 4th gear or 4th to 3rd gear, depending on governor, 3–4 M.T.V., 4–3 M.T.V., part throttle, or drive 3 oil pressures.

4–3 M.T.V. valve—Modulates T.V. pressure going to the 3–4 throttle valve to a lower pressure to prevent an early downshift at light to medium throttle.

Reverse servo boost valve—Under hard acceleration the higher line pressure will open the valve to provide a quick feed to the reverse servo and prevent the reverse band from slipping during application.

1–2 servo control valve—Closed by 2nd oil during a **D** range 3–2 downshift, it slows down the 1–2 servo apply.

1–2 servo boost valve—Under hard acceleration the higher line pressure will open the valve to provide a quick feed to the 1–2 servo and prevent the 1–2 band from slipping during application.

Converter clutch apply valve—Controlled by the converter clutch solenoid, it directs oil to either the release or the apply side of the converter clutch.

Converter clutch regulator valve—Controlled by T.V. pressure and fed by converter clutch feed pressure it regulates converter clutch apply pressure.

1–2 accumulator valve—Receives line pressure from the manual valve and controlled by modulator pressure. It varies 1–2 and 3–4 accumulator pressure in proportion to changes in modulator pressure (engine torque).

Pressure relief check ball—Prevents line pressure from exceeding 245–360 psi.

Converter clutch shift valve plug—Allows 2nd oil to feed into the converter clutch signal passage. The plug is used on models with vehicles equipped with computer command control.

Converter clutch shift valve train (non–C3 systems)—Sends signal oil to the converter clutch apply valve and together with the converter clutch solenoid determines whether the clutch should be released or applied. It is controlled by governor, T.V. and detent oil.

T.V. limit valve—Limits the line pressure fed to the throttle valve to 90 psi.

Throttle valve—A regulating valve that increases T.V. pressure as the accelerator pedal is depressed and is controlled by T.V. plunger movement.

T.V. plunger—Controlled by the throttle lever and bracket assembly and linked to the accelerator pedal. When accelerating, this valve compresses the throttle valve spring causing the throttle valve to increase T.V. pressure. It also controls the opening of the part throttle and detent ports.

Pressure regulator valve—Controls line pressure by regulating pump output and is controlled by the pressure regulator spring, the reverse boost valve and the line boost valve.

Pressure regulator valve with isolator—Same function as pressure regulator valve except isolator system assists in stabilizing the pressure regulator system.

Reverse boost valve—Boosts line pressure by pushing the pressure regulator valve up when acted on by Park, Reverse, Neutral (PRN) oil or low oil pressure.

Line boost valve—Boosts line pressure by pushing the pressure regulator valve up when acted on by modulator oil pressure.

Second clutch signal pipe—Directs 2nd clutch oil to apply or release the 1–2 control valve.

CLUTCH EXHAUST CHECK BALLS

To complete the exhaust of apply oil when the input, 2nd, or 3rd clutch is released, an exhaust check ball assembly is installed near the outer diameter of the clutch housings. Centrifugal force, resulting from the spinning clutch housings, working on the residual oil in the clutch piston cavity would give a partial apply of the clutch plates if it were not exhausted. The exhaust check ball assembly is designed to close the exhaust port by clutch apply pressure seating the check ball when the clutch is being applied.

When the clutch is released and clutch apply oil is being exhausted, centrifugal force on the check ball unseats it and opens the port to exhaust the residual oil from the clutch piston cavity.

CHECK BALLS

1. Fourth clutch check ball: Forces 4th clutch oil to feed through 1 orifice and exhaust through a different orifice.
2. 3–2 control check ball: Forces exhausting 1–2 servo release oil to either flow through an orifice or the regulating 3–2 control valve.
3. Part throttle and drive 3 check ball: Separates part throttle and drive 3 oil passages to the 3–4 shift valve.
4. Third clutch check ball: Forces 3rd clutch oil to feed through 1 orifice and exhaust through a different orifice.
5. 2–3 accumulator feed check ball: In 3rd gear and 4th gear forces D4 oil to be orificed into the 1–2 servo (2–3 accumulator).
6. 2–3 accumulator exhaust check ball: In 1st gear, allows the 2–3 accumulator exhaust passage to feed and apply the 1–2 servo unrestricted. In 3rd gear, forces exhausting 1–2 servo oil to either flow through an orifice or the regulating 2–3 accumulator valve.
7. Third clutch accumulator ball and spring: In 3rd gear, it closes the 1–2 servo release passage exhaust. On a 3–2 downshift after 1–2 servo release oil has dropped to a low pressure, the spring will unseat the check ball and allow the oil to exhaust completely.
8. Second clutch check ball: Forces 2nd clutch oil to feed through 1 orifice and exhaust through a different orifice.
9. Reverse servo feed check ball: Forces oil feeding the reverse servo to orifice, but allows the oil to exhaust freely.
10. Converter clutch release/apply check ball: Separates converter clutch release and converter clutch apply passages to the clutch blow-off ball.
11. Third/Lo-1st check ball: Separates the 3rd clutch and lo-1st passages to the 3rd clutch.
12. 1–2 servo feed check ball: Forces oil feeding the apply side of the 1–2 servo to orifice but allows the oil to exhaust unrestricted.
13. Input clutch/reverse check ball: Minimizes neutral—drive and neutral—reverse apply time by allowing the park, reverse, neutral (PRN) circuit to feed and apply the input clutch quicker.
14. Detent/modulator check ball: Allows detent oil to apply

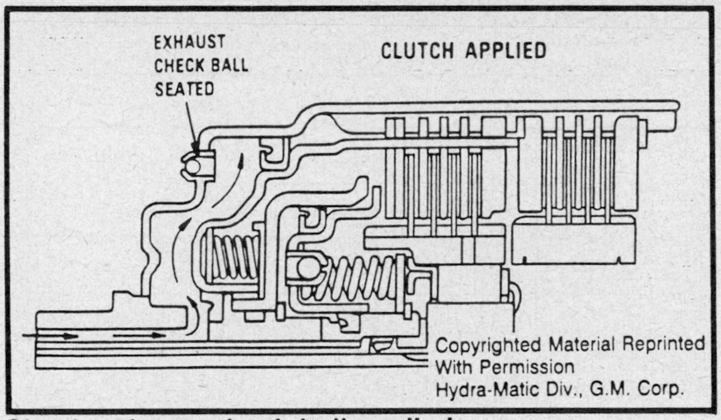
EXHAUST
CHECK BALL
SEATED
CLUTCH APPLIED
Copyrighted Material Reprinted
With Permission
Hydra-Matic Div., G.M. Corp.

Clutch exhaust check ball applied

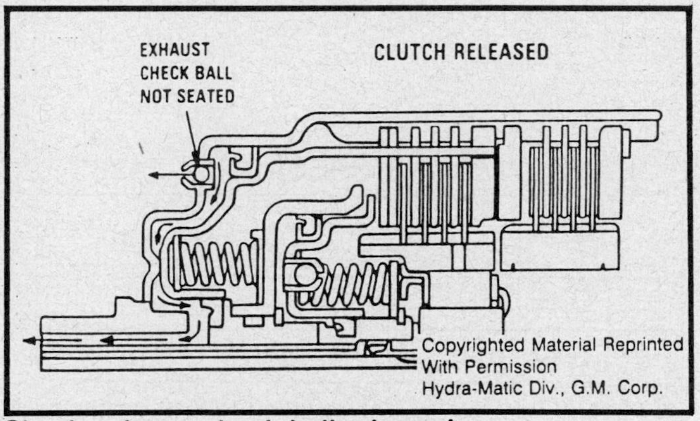
EXHAUST
CHECK BALL
NOT SEATED
CLUTCH RELEASED
Copyrighted Material Reprinted
With Permission
Hydra-Matic Div., G.M. Corp.

Clutch exhaust check ball released

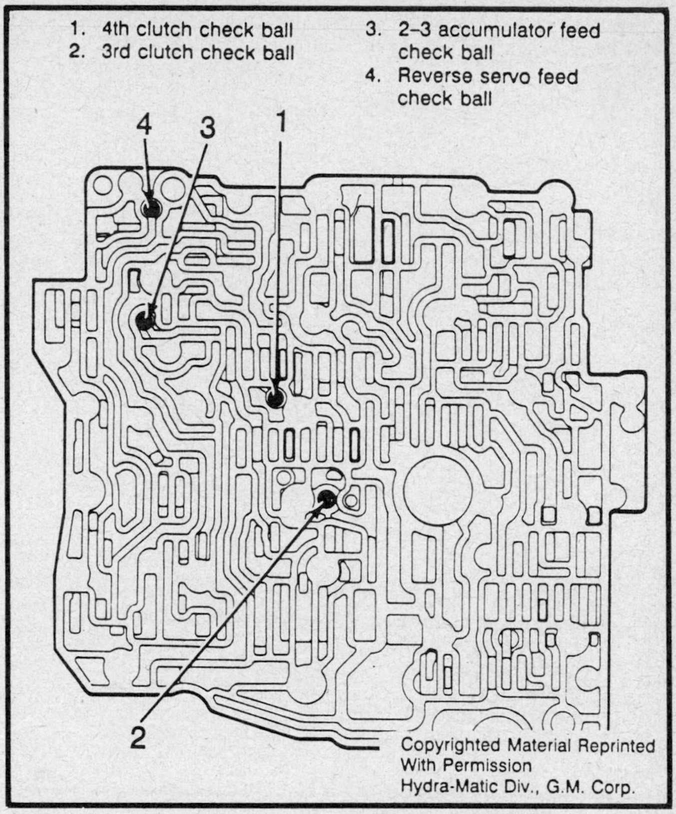
1. 4th clutch check ball
2. 3rd clutch check ball
3. 2–3 accumulator feed check ball
4. Reverse servo feed check ball
4
3
1
2
Copyrighted Material Reprinted
With Permission
Hydra-Matic Div., G.M. Corp.

Control valve check ball locations—1984–85

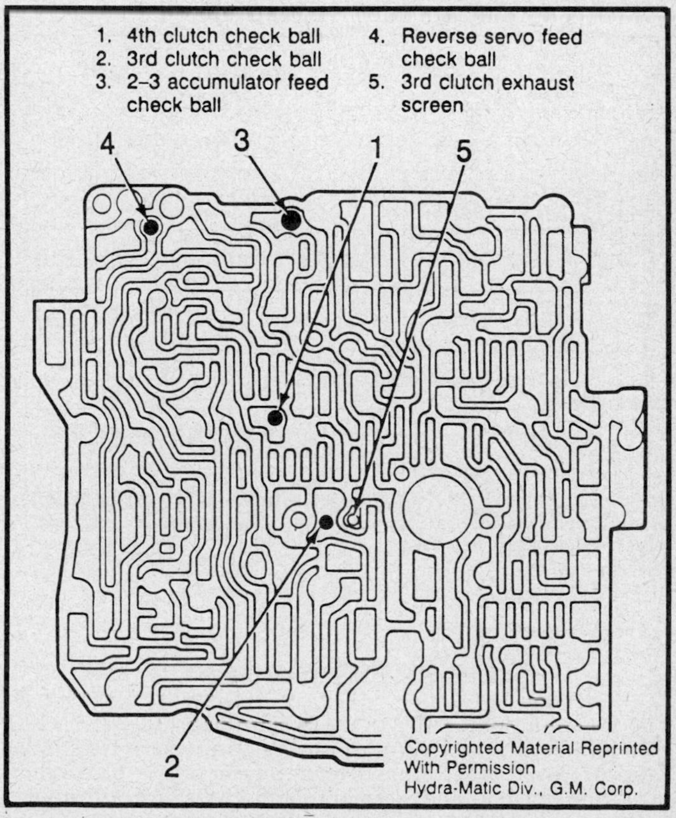
1. 4th clutch check ball
2. 3rd clutch check ball
3. 2–3 accumulator feed check ball
4. Reverse servo feed check ball
5. 3rd clutch exhaust screen
4
3
1
5
2
Copyrighted Material Reprinted
With Permission
Hydra-Matic Div., G.M. Corp.

Control valve check ball locations—1985½–89

force to the pressure regulator system during part or full throttle detent and when driving at high altitude.

15. Converter clutch blow off check ball: Prevents converter clutch release or apply pressure from exceeding 100 psi.

16. Low blow off check ball: Prevents low-1st pressure to the 3rd clutch from exceeding 70 psi in manual low.

17. Cooler check ball: When the engine is shut off the spring seats the ball to prevent converter drainback.

Diagnosis Tests

CONTROL PRESSURE TEST

Before proceeding, check the transaxle fluid to make sure that it is at the proper level. Note the fluid color. Burned fluid loses its red color and has an acrid odor.

NOTE: If the oil is burned and/or clutch plate material is found in the oil pan, overhaul may be required.

1. Check the following:
   a. T.V. cable adjustment
   b. Outside manual linkage and correct
   c. Engine tune
2. Connect an oil pressure gauge to the transaxle.
3. Connect a tachometer to the engine.
4. Apply the parking braking and chock the drive wheels.
5. Check the oil pressures with the brakes applied at all times. Take the line pressure readings in the ranges and at the engine rpm indicated in the chart.

NOTE: Total running time is not to exceed 2 minutes.

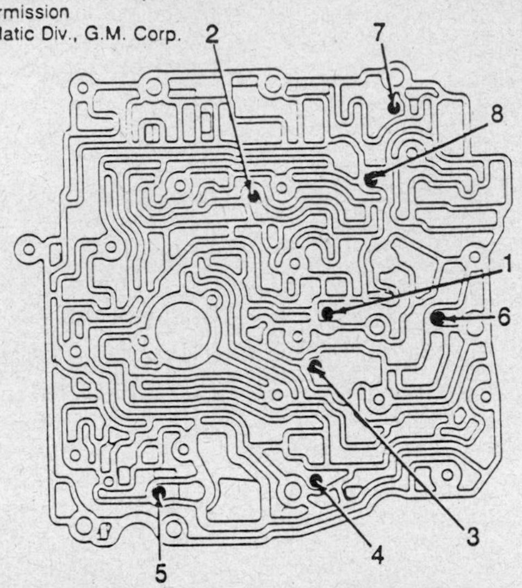
2
7
8
1
6
3
4
5

1. 3–2 control check ball
2. Part throttle and drive 3 check ball
3. 2–3 accumulator exhaust check ball
4. 2nd clutch check ball
5. Converter clutch release/apply check ball
6. 3rd/lo-1st check ball
7. 1–2 servo feed check ball
8. Input clutch/reverse check ball

Channel plate check ball locations

TRANSAXLE LINE PRESSURE—1984–85

MODEL			AY, BN, BS, BU, CW, HT		BA, BC		OB, OV	
		RANGE	kPa	PSI	kPa	PSI	kPa	PSI
TRANSMISSION LINE PRESSURE	ZERO T.V. (1000 R.P.M.)	D4,D3,D2	422-475	61-69	455-511	66-74	455-511	66-74
		D1	946-1324	137-192	968-1317	140-190	1074-1204	156-175
		Reverse	422-475	61-69	455-511	66-74	663-743	96-108
	FULL T.V. (1000 R.P.M.)	D4,D3,D2	1024-1260	149-183	1058-1297	153-188	1090-1189	158-172
		D1	946-1324	137-192	968-1317	140-190	1074-1204	156-175
		Reverse	1428-1757	207-255	1494-1831	217-266	1585-1729	230-250

AIR PRESSURE TEST

The positioning of the THM 440–T4 transaxle in the vehicle and the valve body location will cause the air pressure tests to be very difficult. It is advisable to make any air pressure tests during the disassembly and assembly of the transaxle to ascertain if a unit is operating.

TRANSAXLE LINE PRESSURE—1985½

MODEL			AF, AM, BV, CM, CN, HA, HJ		BR, BW	
		RANGE	kPa	PSI	kPa	PSI
TRANSMISSION LINE PRESSURE	MINIMUM LINE (1250 R.P.M.)	P,N, D4,D3,D2	422 - 475	61 - 69	455 - 511	66 - 74
		D1	946 - 1324	137 - 192	968 - 1317	140 - 191
		REV.	422 - 475	61 - 69	455 - 511	66 - 74
	FULL LINE (1250 R.P.M.)	N, D4, D3, D2	1030 - 1266	150 - 184	1064 - 1302	154 - 189
		D1	946 - 1324	137 - 192	968 - 1317	140 - 191
		REV.	1436 - 1764	209 - 257	1502 - 1838	218 - 267

TRANSAXLE LINE PRESSURE—1986

MODEL			AA, AU		AF, AM, AR, BA, BB, BC, BD, BH, BL, BM, BP, BT, CM, CN, HJ		BZ	
		RANGE	kPa	PSI	kPa	PSI	kPa	PSI
TRANSMISSION LINE PRESSURE	MINIMUM LINE (1250 R.P.M.)	P,N, D4,D3,D2	422-475	61-69	422-475	61-69	455-511	66-74
		D1	946-1324	137-192	998-1276	145-185	1112-1399	161-203
		REV.	422-475	61-69	422-475	61-69	455-511	66-74
	FULL LINE (1250 R.P.M.)	N, D4, D3, D2	1030-1266	150-184	1152-1393	167-202	1186-1429	172-207
		D1	946-1324	137-192	998-1276	145-185	1112-1399	161-203
		REV.	1436-1764	209-257	1573-1901	228-276	1619-1951	235-283

TRANSAXLE LINE PRESSURE—1987

MODEL			7ADH, 7AFH, 7AHH, 7ALH, 7ARH, 7CAH, 7CBH		7ACH, 7BBH, 7BCH, 7BJH, 7BKH, 7BNH, 7BRH, 7BSH, 7BTH, 7BUH, 7BZH, 7FBH, 7FCH, 7FJH, 7FKH, 7FLH, 7FNH, 7FRH, 7FSH, 7FTH, 7FUH, 7FZH, 7HAH, 7HCH	
		RANGE	kPa	PSI	kPa	PSI
TRANSMISSION LINE PRESSURE	MINIMUM LINE (1250 R.P.M.)	D4,D3,D2	422 - 475	61 - 69	422 - 475	61 - 69
		D1	946 - 1324	137 - 192	998 - 1276	145 - 185
		P,R,N	422 - 475	61 - 69	422 - 475	61 - 69
	FULL LINE (1250 R.P.M.)	D4,D3,D2	1030 - 1266	150 - 184	1152 - 1393	167 - 202
		D1	946 - 1324	137 - 192	998 - 1276	145 - 185
		P,R,N	1436 - 1764	209 - 257	1573 - 1901	228 - 276

TRANSAXLE LINE PRESSURE—1988–89 EXCEPT THM F7

		8BJH, 8BKH, 8BRH, 8BTH, 8BYH, 8FBH, 8FCH, 8FJH, 8FSH		8AAH, 8ABH, 8AFH, 8ANH, 8ATH, 8AWH		8CFH, 8CMH, 8CRH, 8CTH, 8CXH, 8CWH	
	RANGE	kPa	PSI	kPa	PSI	kPa	PSI
MINIMUM LINE @ 1250 R.P.M. (18 In. Hg. Vacuum At Modulator)	D4,D3,D2	422 - 475	61 - 69	544 - 1034	79 - 150	422 - 475	61 - 69
	D1	998 - 1276	145 - 185	998 - 1276	145 - 185	946 - 1324	137 - 192
	P,R,N	422 - 475	61 - 69	640 - 1216	93 - 176	422 - 475	61 - 69
FULL LINE @ 1250 R.P.M. (0 In. Hg. Vacuum At Modulator)	D4,D3,D2	1152 - 1393	167 - 202	1022 - 1536	148 - 223	1030 - 1266	149 - 184
	D1	998 - 1276	145 - 185	998 - 1276	145 - 185	946 - 1324	137 - 192
	P,R,N	1573 - 1901	228 - 275	1202 - 1807	174 - 262	1436 - 1764	208 - 256

TRANSXALE LINE PRESSURE 1988–89 THM F7

		8APZ	
	RANGE	kPa	PSI
MINIMUM LINE @ 1250 R.P.M. (18 In. Hg. Vacuum At Modulator)	D4,D3,D2	455 - 511	66 - 74
	D1	1112 - 1399	161 - 203
	P,R,N	455 - 511	66 - 74
FULL LINE @ 1250 R.P.M. (0 In. Hg. Vacuum At Modulator)	D4,D3,D2	1186 - 1429	172 - 207
	D1	1112 - 1399	161 - 203
	P,R,N	1619 - 1951	235 - 283

STALL SPEED TEST

General Motors does not recommend performing a stall test because of the excessive heat that is generated within the transaxle by the converter during the test.

Recommendations are to perform the control pressure test and road test to determine and localize any transaxle malfunctions.

ROAD TEST

Shift Check

1. Start the engine.
2. Depress the brake pedal.
3. Move the gear selector from **P** to **R** to **N** to **D**. The gear selections should be immediate and not harsh.

Upshifts and Torque Converter Clutch Apply

1. Position the gear selector in **OVERDRIVE D** range.
2. Accelerate using a steady increasing throttle pressure.
3. Note the shift speed point gear engagements for 2nd gear, 3rd gear and Overdrive.
4. Note the speed shift point for the torque converter clutch apply. This should occur while in 3rd or Overdrive. If the apply is not noticed, check the torque converter clutch.

NOTE: The torque converter clutch will not engage if the engine coolant has not reached a minimum operating temperature of 130°F (54°C).

Part Throttle Detent Downshift

At vehicle speeds of 40–55 mph (64–88 kph), quickly depress the accelerator to a half open position. The torque converter clutch should release and the transaxle should downshift to 3rd gear immediately.

Full Throttle Detent Downshift

At vehicle speeds of 40–55 mph (64–88 kph), quickly depress the accelerator to a wide open position. The torque converter clutch should release and the transaxle should downshift to 2nd gear immediately.

Manual Downshift

At vehicle speeds of 40–55 mph (64–88 kph), release the accelerator pedal while moving the gear selector lever to 3rd gear. The torque converter clutch should release and the transaxle should downshift to 3rd gear immediately. The engine should slow the vehicle down.

Move the gear selector to **OVERDRIVE D** and accelerate to 40–45 mph (64–72 kph). Release the accelerator pedal while moving the gear selector to 2nd gear. The torque converter clutch should be released and the transaxle should downshift to 2nd gear immediately. The engine should slow the vehicle down.

Move the gear selector to **OVERDRIVE D** and accelerate to 25 mph (40 kph). Release the accelerator pedal while moving the gear selector to 1st gear. The torque converter clutch should be released and the transaxle should downshift to 1st gear immediately. The engine should slow the vehicle down.

Coastdown Downshift

With the gear selector in **OVERDRIVE D**, accelerate the vehicle to 4th gear with the torque converter clutch applied. Release the accelerator pedal and lightly apply the brakes. The torque converter clutch should release and the transaxle should downshift.

Manual Gear Range Selection

MANUAL 3RD

With the vehicle stopped, position the gear selector in 3rd and accelerate. Note the 1st to 2nd gear shift point and the 2nd to 3rd gear shift point.

MANUAL 2ND

With the vehicle stopped, position the gear selector in 2nd and accelerate. Note the 1st to 2nd gear shift point. Accelerate to 25 mph (40 kph). A 2nd to 3rd gear shift should not occur and the torque converter clutch should not engage.

MANUAL 1ST

With the vehicle stopped, position the gear selector in 1st and accelerate to 15 mph (24 kph). No upshift should occur and the torque converter clutch should not engage.

REVERSE

With the vehicle stopped, position the gear selector in **R** and slowly accelerate and note the reverse gear operation.

CONVERTER STATOR OPERATION TEST

The torque converter stator assembly and its related roller clutch can possible have one of 2 different type malfunctions.

1. The stator assembly freewheels in both directions.
2. The stator assembly remains locked up at all times.

Malfunction Type One

If the stator roller clutch becomes ineffective, the stator assembly freewheels at all times in both directions. With this condition, the vehicle will tend to have poor acceleration from a standstill. At speeds above 30–35 mph (50–55 kph), the vehicle may act normal. If poor acceleration problems are noted, it should first be determined that the exhaust system is not blocked, the engine is in good tune and the transmission is in 1st gear when starting out.

If the engine will freely accelerate to high rpm in **N**, it can be assumed that the engine and exhaust system are normal. Driving the vehicle in **R** and checking for poor performance will help determine if the stator is freewheeling at all time.

Malfunction Type Two

If the stator assembly remains locked up at all times, the engine rpm and vehicle speed will tend to be limited or restricted at high speeds. The vehicle performance when accelerating from a standstill will be normal. Engine over-heating may be noted. Visual examination of the converter may reveal a blue color from the over-heating that will result.

Converter Clutch Operation and Diagnosis

TORQUE CONVERTER CLUTCH

The converter clutch mechanically connects the engine to the drive train and eliminates the hydraulic slip between the pump and turbine. When accelerating, the stator gives the converter the capability to multiply engine torque (maximum torque multiplication is 2.1:1) and improve vehicle acceleration. At this time the converter clutch is released. When the vehicle reaches cruising speed, the stator becomes inactive and there is no multiplication of engine torque (torque multiplication is 1.0:1). At this time the converter is a fluid coupling with the turbine at almost the same speed as the converter pump. The converter clutch can now be applied to eliminate the hydraulic slip between the pump and turbine and improve fuel economy.

The converter clutch cannot apply in **P**, **R**, **N** and **D** Range—1st gear.

On vehicles equipped with computer command control, converter clutch operation is controlled by the solenoid. The solenoid is controlled by the brake switch, 3rd clutch pressure switch and the computer command control system.

Converter Clutch Applied

The converter clutch is applied when oil pressure is exhausted between the converter pressure plate and the converter cover and pressure is applied to push the pressure plate against the converter cover.

When the solenoid is energized converter signal oil, from the converter clutch plug, pushes the converter clutch apply valve to the left. Converter clutch release oil then exhausts and converter clutch apply oil from the converter clutch regulator valve pushes the converter pressure plate against the converter to apply the converter clutch.

Converter Clutch Released

The converter clutch is released when oil pressure is applied between the converter cover and the converter pressure plate.

Converter feed pressure from the pressure regulator valve passes through the converter clutch apply valve into the release passage. The release oil feeds between the pump shaft and the turbine shaft to push the pressure plate away from the converter cover to release the converter clutch.

Converter Clutch Apply Feel

Converter clutch apply feel is controlled by the converter clutch regulator valve and the converter clutch accumulator.

The converter clutch regulator valve is controlled by T.V. and controls the pressure that applies the converter clutch. The converter clutch accumulator absorbs converter feed oil as converter release oil is exhausting. Less oil is then fed to the converter clutch regulator valve and the apply side of the converter clutch. This gives a cushion to the converter clutch apply.

VISCOUS CONVERTER CLUTCH

Viscous Converter Clutch Applied

The viscous converter clutch is capable of applying at approximately 25 mph providing that the transaxle is in 2nd gear and the transaxle oil temperature is below 200°F (93.3°C). (This temperature is monitored by the ECM through the thermistor). When transaxle oil temperatures are above 200°F (93.3°C) but below 315°F (157°C) the viscous clutch will not apply until approximately 38 mph. If transaxle oil temperature exceeds 315°F (157°C) a temperature switch located in the channel plate will open and release the viscous clutch to protect the transaxle from overheating.

TROUBLESHOOTING THE TORQUE CONVERTER CLUTCH

Before diagnosing the torque converter clutch system as being at fault in the case of rough shifting or other malfunctions,

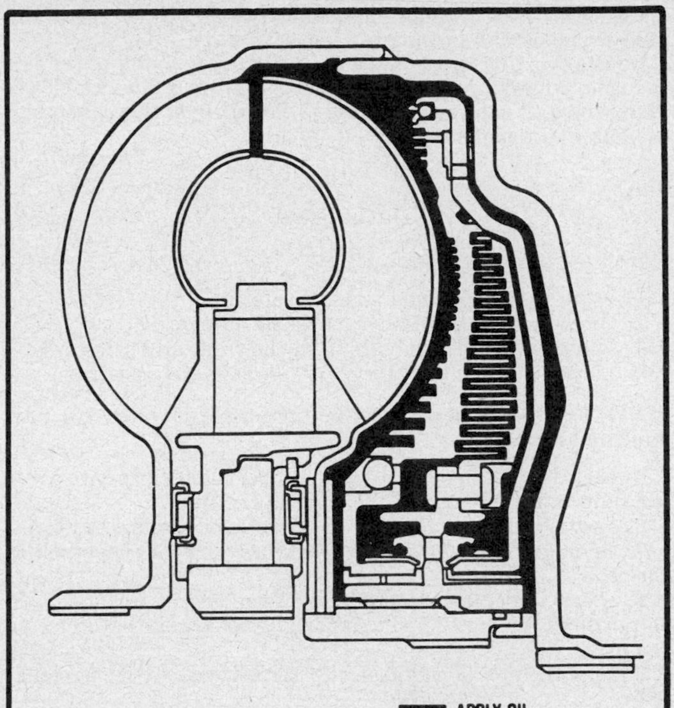
APPLY OIL
VISCOUS FLUID
SEALS
CLUTCH MATERIAL

Viscous converter clutch applied

make sure that the engine is in at least a reasonable state of tune. Also the following points should be checked:

1. Check the transmission fluid level and correct as necessary.
2. Check the manual linkage adjustment and correct as necessary.
3. Road test the vehicle to verify the complaint. Make sure that the vehicle is at normal operating temperature.

CAUTION

When inspecting the stator and turbine of the torque converter clutch unit, a slight drag is normal when turned in the direction of freewheel rotation because of the pressure exerted by the waved spring washer, located between the turbine and the pressure plate.

Torque Converter Clutch – Electrical Controls

Two types of electrical control systems are used to control the **APPLY** function of the torque converter clutch assembly. Both systems use the third clutch pressure switch and a solenoid. The difference in the 2 systems is that the vehicle speed sensing controls are not the same.

Computer Command Control System

Vehicles equipped with the computer command control system utilize the following components to accomplish the **APPLY** function of the torque converter clutch assembly.

1. Vacuum sensor – which sends engine vacuum information to the electronic control module.
2. Throttle position sensor – which sends throttle position information to the electronic control module.
3. Vehicle speed sensor – which sends vehicle speed information to the electronic control module.
4. Electronic control module – which energizes and grounds the transaxle electrical system.
5. Brake release switch – which avoids stalling the engine when braking. An time that the brakes are applied the torque converter clutch is released.
6. Third gear switch – which prevents operation until third gear speed is obtained.

ON CAR SERVICES

Adjustments

THROTTLE VALVE (T.V.) CABLE

2.5L AND 2.8L ENGINES

1. Check to see that the cable is in the full non-adjustment position.
2. Without twisting or kinking cable, insert cable slug into idler pulley (cam) slot.
3. Accelerator cable must be installed before adjustment.
4. Rotate the idler pulley (cam) in a counterclockwise direction to 65 inch lbs. (7.3 Nm).

3.8L ENGINE

1. With the engine stopped, depress and hold down the readjust tab at the T.V. cable adjuster.
2. Move the cable conduit until it stops against the fitting.
3. Release the readjustment tab.
4. Rotate the throttle lever by hand to its full travel position.
5. The slider must move (ratchet) toward the lever when the lever is rotated to its full travel position.

SHIFT CONTROL CABLE

COLUMN SHIFT CONTROL WITH CABLE ADJUSTER

1. Place the shift lever in **N**. Neutral can be found by rotating the selector shaft clockwise from **P** through **R** to **N**.
2. Place the shift control assembly in **N**.
3. Push the tab on the cable adjuster to adjust the cable in the cable mounting bracket.

COLUMN SHIFT CONTROL WITHOUT CABLE ADJUSTER

1. Position the selector lever in **N**.
2. Position the transaxle lever in **N**. Obtain **N** position by turning the transaxle lever clockwise from **P** through **R** into **N**.
3. Loosely assemble the pin part of the shift cable through the transaxle lever slotted hole.
4. Tighten the nut to 20 ft. lbs. (27 Nm). The lever must be held out of **P** when tightening the nut.

CONSOLE SHIFT CONTROL

1. Disconnect the negative battery cable at the battery.
2. Place the transaxle into **1st** gear position.

3. Loosen the shift cable attaching nuts at the transaxle lever.
4. Remove the console trim plate and slide shifter boot up shifter handle. Remove the console.
5. With the shift lever in **1st** gear position (pulled to the left and held against the stop) insert a yoke clip to hold the lever hard against the reverse lockout stop. Install af $^5/_{32}$ in. or No. 22 drill bit into the alignment hole at the side of the shifter assembly.
6. Remove the lash from the transaxle by rotating the lever in the direction away from the cable (do not force the cable) while tightening the nut.
7. Tighten the nut at the lever.
8. Remove the drill bit and yoke at the shifter assembly.
9. Install the console, shifter boot and the trim plate.
10. Connect the negative battery cable at the battery.
11. Road test the vehicle to check for a good neutral gate feel during shifting. It may be necessary to fine tune the adjustment after road testing.

PARK/NEUTRAL START AND BACKUP LIGHT SWITCH

1. Place the transaxle control shifter assembly in the **N** notch in the detent plate.
2. Loosen the switch attaching screws.
3. Rotate the switch on the sifter assembly to align the service adjustment hole with carrier tang hole. Insert a $^3/_{32}$ in. (2.34mm) maximum diameter gauge pin to a depth of $^5/_8$ in. (15mm).
4. Tighten the attaching screws.
5. Remove the gauge pin.

PARK LOCK CONTROL CABLE

1. Position the selector lever in **P**. Lock the steering column.
2. Position the transaxle lever into the **P** position.
3. Install the cable-to-shifter mounting bracket with the spring yoke.
4. Install the cable-to-park lock pin on the shifter. Install the lock pin. Push the lock button on the cable housing in to set the cable length.
5. Check the cable operation in the following manner:
   a. Turn the ignition key to the **LOCK** position.
   b. Press the detent release button in the shifter handle.
   c. Pull the shifter lever rearward.
   d. The shifter lock-hook must engage into the shifter base slot, within 2 degrees maximum movement of the shifter lever.
   e. Turn the ignition key to the **OFF** position.
   f. Repeat Steps b–c. The selector lever must be able to move rearward to the **L** (**1** on some vehicles) position.
   g. Repeat Steps a–d to assure that the adjustment nut has not slipped during check.
   h. Return the key to the **LOCK** position and check key removal.

Services

FLUID CHANGES

The conditions under which the vehicle is operated is the main consideration in determining how often the transaxle fluid should be changed. Different driving conditions result in different transaxle fluid temperatures. These temperatures affect change intervals.

If the vehicle is driven under severe service conditions, change the fluid and filter every 15,000 miles. If the vehicle is not used under severe service conditions, change the fluid and replace the filter every 50,000 miles.

Do not overfill the transaxle. It only takes one pint of fluid to change the level from add to full on the transaxle dipstick. Overfilling the unit can cause damage to the internal components of the automatic transaxle.

OIL PAN

Removal and Installation

1. Raise the vehicle and support safely.
2. Place a drain pan under transaxle oil pan.
3. Remove the oil pan bolts from the front and sides only.
4. Loosen the rear oil pan bolts approximately 4 turns.

NOTE: Do not damage the transaxle case or oil pan sealing surfaces.

5. Lightly tap the oil pan with a rubber mallet or pry to allow the fluid to drain. Inspect the color of the fluid.
6. Remove the remaining oil pan bolts, oil pan and gasket.
7. Remove the oil filter and O-ring. O-ring may be stuck in the case.
8. Clean the transaxle case and oil pan gasket surfaces with solvent and air dry. All traces of old gasket material must be removed.
9. Coat the new O-ring seal with a small amount of petroleum jelly.
10. Install the new O-ring onto the filter.
11. Install the new filter into the case.
12. Install a new gasket onto the oil pan and install the oil pan.
13. Install the oil pan bolts and tighten to 15 ft. lbs. (11 Nm).
14. Lower the vehicle.
15. Fill the transaxle to the proper level with Dexron®II fluid or equivalent.
16. Check the cold fluid level reading for initial fill. Do not overfill.
17. Check the oil pan gasket for leaks.

VALVE BODY

Removal and Installation

1. Disconnect the negative cable from the battery.
2. Raise and support the vehicle safely.
3. Remove the case side cover pan.
4. Remove the pump assembly.
5. Remove the 1–2 servo pipe clip from the valve body.
6. Remove the valve body-to-channel plate bolts.
7. Remove the valve body keeping the spacer plate with the transaxle.
8. Remove the spacer plate and gasket.
9. Retain the check balls in their proper locations with petroleum jelly.
10. Install the spacer plate and gasket.
11. Install the valve body and gasket.
12. Install the valve body-to-channel plate bolts. Torque to 10 ft. lbs. (14 Nm).

NOTE: Do not use an impact type tool on the valve body or pump assembly.

13. Install the 1–2 servo pipe clip to the valve body.
14. Install the pump assembly.
15. Install the case side cover pan.
16. Lower the vehicle.
17. Connect the negative battery cable to the battery.
18. Refill the transaxle to the proper level.
19. Adjust the T.V. cable. Check for leaks.

REVERSE SERVO ASSEMBLY

Removal and Installation

1. Raise and support the vehicle safely.
2. Disconnect the exhaust crossover pipe.
3. Depress the servo cover.
4. Remove the snapring, servo cover, servo piston, sealing ring, apply pin and servo spring.
5. Install the servo spring, apply pin, sealing ring, servo piston, servo cover and snapring.
6. Connect the exhaust crossover pipe.
7. Lower the vehicle.

1–2 SERVO ASSEMBLY

Removal and Installation

1. Raise and support the vehicle safely.
2. Disconnect the exhaust crossover pipe.
3. Depress the servo cover.
4. Remove the snapring, servo cover, servo piston, sealing ring, apply pin and servo spring.
5. Install the servo spring, apply pin, sealing ring, servo piston, servo cover and snapring.
6. Connect the exhaust crossover pipe.

GOVERNOR

Removal and Installation

1. Raise the vehicle and support safely.
2. Disconnect the speedometer cable.
3. Remove the governor cover attaching bolts.
4. Remove the governor cover and seal.
5. Remove the governor assembly (including the sleeve and the speedometer drive gear).
6. Install the governor assembly.
7. Install the governor cover and new seal.
8. Install the governor cover attaching bolts and torque to 97 inch lbs. (11 Nm).
9. Connect the speedometer cable.
10. Lower the vehicle.

REMOVAL AND INSTALLATION

TRANSAXLE REMOVAL

A-BODY VEHICLES

1. Disconnect the negative battery cable.
2. Remove the air cleaner and disconnect the T.V. cable at the throttle body.
3. Disconnect the shift linkage at the transaxle.
4. Install the engine support fixture. Tool J–28467 is recommended.
5. Disconnect all electrical connectors.
6. Remove the 3 bolts from the transaxle to the engine.
7. Disconnect the vacuum line at the modulator.
8. Raise the vehicle and suitably support it.
9. Remove the left front wheel and tire assembly.
10. Remove the left hand ball joint from the steering knuckle.
11. Disconnect the brake line bracket at the strut.

NOTE: A drive axle seal protector tool J–34754 should be modified and installed on any drive axle prior to service procedures on or near the drive axle. Failure to do so could result in seal damage or joint failure.

12. Remove the drive axles from the transaxle.
13. Disconnect the pinch bolt at the intermediate steering shaft. Failure to do so could cause damage to the steering gear.
14. Remove the frame to stabilizer bolts.
15. Remove the stabilizer bolts at the control arm.
16. Remove the left front frame assembly.
17. Disconnect the speedometer cable or wire connector from the transaxle.
18. Remove the extension housing to engine block support bracket.
19. Disconnect the cooler pipes.
20. Remove the right and left insulator attaching bolts.
21. Remove the flywheel splash shield.
22. Remove the converter cover and converter-to-flywheel bolts.
23. Remove all of the remaining transaxle to engine bolts except one.
24. Position a jack under the transaxle.
25. Remove the remaining transaxle to engine bolt and remove the transaxle.

BUICK AND OLDSMOBILE C-BODY VEHICLES

1. Disconnect the negative terminal from the battery. Disconnect the wire connector at the mass air flow sensor.
2. Remove the air intake duct and the mass air flow sensor as an assembly.
3. Disconnect the cruise control assembly and the the shift control linkage.
4. Label and disconnect the following:
   a. Park/Neutral switch
   b. Torque converter clutch
   c. Vehicle speed sensor
   d. Vacuum modulator hose at the modulator
5. Remove the 3 top transaxle-to-engine block bolts and install an engine support fixture.
6. Remove both front wheels and turn the steering wheel to the full left position.
7. Remove the right front ball joint nut and separate the control arm from the steering knuckle.
8. Remove the right halfshaft.

NOTE: Be careful not to allow the drive axle splines to contact any portion of the lip seal.

9. Using a medium pry bar, remove the left halfshaft; be careful not to damage the pan. Install drive axle boot seal protectors.
10. Remove 3 bolts at the transaxle and 3 nuts at the cradle member. Remove the left front transaxle mount.
11. Remove the right front mount-to-cradle nuts. Remove the left rear transaxle mount-to-transaxle bolts.
12. Remove the right rear transaxle mount. Remove the engine support bracket-to-transaxle case bolts.
13. Remove the flywheel cover, matchmark the flywheel-to-torque converter and remove the flywheel-to-converter bolts.

NOTE: Be sure to matchmark the flywheel-to-converter relationship for proper alignment upon reassembly.

14. Remove the rear cradle member-to-front cradle dog leg.
15. Remove the front left cradle-to-body bolt and the front cradle dog leg-to-right cradle member bolts.
16. Install a transaxle support fixture into position.

17. Remove the cradle assembly by swinging it aside and supporting it with jackstand.
18. Disconnect and plug the oil cooler lines at the transaxle.

NOTE: A bolt is located between the transaxle and the engine block; it is installed in the opposite direction.

19. Remove the remaining lower transaxle-to-engine bolts and lower the transaxle from the vehicle.

CADILLAC C-BODY VEHICLES

1. Disconnect the negative terminal from the battery. Remove the air cleaner and the T.V. cable.
2. Disconnect the shift linkage from the transaxle. Using an engine support fixture tool, connect it to and support the engine.
3. Label and disconnect the electrical connectors from the following items:
   a. Converter clutch
   b. Vehicle speed sensor
   c. Vacuum line at the modulator
4. Remove the upper bell housing-to-engine bolts and studs.
5. Raise and support the front of the vehicle. Remove both front wheels.
6. From the left side of the vehicle, disconnect the lower ball joint from steering knuckle. Remove both drive axles from the transaxle.
7. Remove the stabilizer bar-to-left control arm bolt.
8. Remove the left front cradle assembly.
9. Remove the extension housing-to-engine support bracket.
10. Disconnect and plug the oil cooler lines at the transaxle case.
11. Remove the right and left transaxle mount attachments.
12. Remove the flywheel splash shield. Matchmark the torque converter-to-flywheel and remove the converter-to-flywheel bolts.
13. Remove the lower bellhousing bolts except the lower rear on (No. 6).
14. Using a floor jack, position it under the transaxle and remove the last bell housing bolt.

NOTE: To reach the last bell housing bolt, use a 3 in. socket wrench extension through the right wheel arch opening.

15. Remove the transaxle assembly.

H-BODY VEHICLES

1. Disconnect the negative battery cable. Disconnect the wire connector at the mass air flow sensor.
2. Remove the air intake duct and the mass air flow sensor as an assembly.
3. Disconnect the cruise control assembly. Disconnect the shift control linkage.
4. Tag and disconnect the torque converter clutch, park/neutral switch, vehicle speed sensor and vacuum modulator hose at the modulator.
5. Remove the top transaxle to engine block bolts. Install an engine support fixture.
6. Remove both front wheels and turn the steering wheel to the full left position.
7. Remove the right front ball joint nut and separate the control arm from the steering knuckle.
8. Remove the right drive axle. Be careful not to allow the drive axle splines to contact any portion of the lip seal.
9. Remove the left drive axle using a suitable pry bar. Be careful not to damage the pan. Install drive axle boot seal protectors.
10. Remove bolts at the transaxle and nuts at the cradle member. Remove the left front transaxle mount.
11. Remove the right front mount to cradle nuts. Remove the left rear transaxle mount to transaxle bolts.
12. Remove the right rear transaxle mount as in Step 10. Remove the engine support bracket to transaxle case bolts.
13. Remove the flywheel cover. Remove the flywheel to converter bolts. Be sure to matchmark the flywheel to converter relationship for proper alignment upon reassembly.
14. Remove the bolts attaching the rear cradle member to the front cradle dog leg.
15. Remove the front left cradle to body bolt. Remove the front cradle dog leg to right cradle member bolts.
16. Install a transaxle support fixture into position.
17. Remove the cradle assembly by swinging it aside and supporting it with a suitable stand.
18. Disconnect and cap the oil cooler lines at the transaxle.

NOTE: A bolt is located between the transaxle and the engine block and is installed in the opposite direction.

19. Remove the remaining lower transaxle to engine bolts. Lower the transaxle assembly away from the vehicle.

1986–87 RIVIERA AND TORONADO

NOTE: To perform this procedure, secure an engine support tool J-28467 or equivalent and a drive axle remover tool J-33008 or equivalent.

1. Disconnect the negative battery cable. Install the engine support fixture.
2. Disconnect the vacuum line from the modulator; electrical connections involved with the transaxle; transaxle valve cable at the throttle body and at the transaxle; the cruise control servo.
3. Disconnect the shift selector bracket and cable from the transaxle. Disconnect the neutral safety switch.
4. Remove the top 3 transaxle mounting bolts.
5. Remove the bolts that fasten the wiring harness to the transaxle. Remove the driveline dampener bracket.
6. Raise and support the front of the vehicle.
7. Disconnect and drain the transaxle oil cooler lines at the transaxle.
8. Remove the torque converter cover. Scribe the relationship between the flexplate and the converter so the same relationship may be established on reinstallation for balance. Remove the converter-to-flexplate bolts, turning the crankshaft (as necessary).
9. Remove the left side transaxle mounting bolts. Remove the engine mounting nuts.
10. Disconnect the sway bar links. Disconnect the left side ball joint from the knuckle.
11. Disconnect the left side driveshaft from the transaxle using a special tool J-33008 or equivalent.
12. Disconnect the left side of the frame by removing the bolts.
13. Position a floor jack under the transaxle and support it securely.
14. Remove the 2 remaining engine-to-transaxle bolts.

NOTE: One of the bolts is located between the transaxle case and the block, it is installed in the direction opposite to the others.

15. Remove the engine-to-transaxle bracket.
16. Remove the right drive axle from the transaxle and hang it securely.
17. Remove the transaxle.

1986–89 ELDORADO, SEVILLE AND 1987–89 ALLANTE

1. Disconnect the negative battery cable. Remove the air cleaner assembly. Disconnect the transaxle throttle valve cable.
2. Remove the cruise control servo and bracket assembly. Disconnect the electrical connectors going to the distributor, oil pressure sending unit and transaxle.
3. Remove the bracket for the engine oil cooler lines.
4. Remove the shift linkage bracket from the transaxle and the manual shift lever from the manual shift shaft; leave the cable attached to the lever and bracket.

5. Remove the fuel line bracket and disconnect the neutral safety switch connector.
6. Remove the vacuum modulator.
7. Remove the throttle valve cable support bracket and engine oil cooler line bracket. Remove the bell housing bolts except the left and right side bolts; note the bolt lengths and positions.
8. Remove the air injection reactor crossover pipe fitting and reposition the pipe. Remove the radiator hose bracket and transaxle mount-to-bracket nuts.
9. Install an engine support fixture, noting the positions of the hooks.
10. Raise and support the front of the vehicle.
11. Remove both front wheels, the right and left stabilizer link bolts. Remove the ball joint cotter pins and nuts and press the ball joints from the steering knuckles.
12. Remove the air conditioner splash shield and the mount cover for the forwardmost cradle insulator.
13. Remove the hose connections from the ends of the air injection reactor pipes. Remove the vacuum hoses and the wire loom from the clips at the front of the cradle.
14. Remove the engine mount and dampener-to-cradle attachments. Remove the transaxle mount-to-cradle attachments. Remove the wire loom clip from the transaxle mount bracket and lower the vehicle.
15. Using the 2 left side support hooks on the engine support fixture to raise the transaxle 2 inches from its normal position. Raise and support the front of the vehicle.
16. Remove the right front and left rear transaxle-to-cradle bolts and the left stabilizer mount bolts. Remove the foremost cradle mount insulator bolt and the left cradle member, separate the right front corner first.
17. Remove the air injection reactor management valve/bracket assembly from the transaxle mount bracket and reposition the bracket to the transaxle stud bolts.
18. Lower the front of the vehicle. Lower the transaxle to its normal position to gain access to the transaxle mounting bracket. Remove the mounting bracket.
19. Raise and support the front of the vehicle. Remove the right rear transaxle mount-to-transaxle bracket. Remove the engine-to-transaxle brace bolts that pass into the transaxle VSS connector.
20. Mark the relationship between torque converter and flexplate for reassembly in the same position. Remove the flywheel covers, then, remove the torque converter bolts, rotating the crankshaft with a socket wrench as necessary to gain access. Position a jack under the transaxle to support it.
21. Remove the left and right bell housing bolts; note the bolt lengths and positions.

NOTE: Access may be gained through the right wheelhouse opening to remove the bolt on the right side; use a 3 foot long socket extension to reach it.

22. Disconnect the oil cooler lines at the transaxle, drain them and plug the openings. Then, install drive axle boot seal protectors and disconnect the driveshafts at the transaxle. Suspend the drive axles out of the way and remove the transaxle.

1988–89 REATTA, RIVIERA AND TORONADO

1. Disconnect the negative battery cable. Remove the air intake duct.
2. Disconnect the throttle valve cable from the transaxle and the throttle body. Disconnect the cruise control servo and cable.
3. Remove the exhaust pipe crossover.
4. Disconnect the shift control linkage lever from the manual shaft and the mounting bracket from the transaxle.
5. Disconnect the electrical harness connectors from the neutral start/backup light switch, the torque converter clutch (TCC) and the vehicle speed sensor (VSS).
6. From the vacuum modulator, disconnect the hose.
7. Remove the upper transaxle-to-engine bolts.
8. Using the engine support fixture tool J–28467 or equivalent, attach it to the engine, turn the wing nuts to relieve the tension on the engine cradle and mounts.
9. Turn the steering wheel to the full left position.
10. Raise and support the front of the vehicle. Remove both front wheel assemblies.
11. Using the drive axle seal protector tool J–34754 or equivalent, install one on each halfshaft. Remove both front ball joint-to-steering knuckle nuts and separate the control arms from the steering knuckles.
12. Using a medium pry bar, pry the halfshaft from the transaxle and support it on a wire; Do not remove the halfshaft from the steering knuckle.

NOTE: When removing the halfshaft, be careful not to damage the seal lips.

13. Remove the right rear transaxle-to-frame nuts, the left rear transaxle mount-to-transaxle bolts and the right rear transaxle mount.
14. From the left control arm, remove the stabilizer shaft.
15. Remove the flywheel cover bolts and the cover.
16. Matchmark the torque converter-to-flywheel bolts for reinstallation purposes. Remove the torque converter-to-flywheel bolts and push the torque converter back into the transaxle.
17. Remove the partial frame-to-main frame bolts, the partial frame-to-body bolts and the partial frame.
18. Disconnect and plug the oil cooler tubes from the transaxle.
19. Remove the lower transaxle-to-engine bolts.

NOTE: One bolt is located between the engine and the transaxle case and is positioned in the opposite direction.

20. Lower the transaxle from the vehicle; be careful not to damage the hoses, lines and wiring.

W-BODY VEHICLES

1. Disconnect the negative terminal from the battery. Remove the air cleaner, bracket, mass air flow (MAF) sensor and air tube as an assembly.
2. Disconnect the exhaust crossover from the right-side manifold and remove the left-side exhaust manifold, then, raise and support the manifold/crossover assembly.
3. Disconnect the T.V. cable from the throttle lever and the transaxle.
4. Remove the vent hose and the shift cable from the transaxle.
5. Remove the fluid level indicator and the filler tube.
6. Using the engine support fixture tool J–28467 or equivalent and the adapter tool J–35953 or equivalent, install them on the engine.
7. Remove the wiring harness-to-transaxle nut.
8. Label and disconnect the wires for the speed sensor, TCC connector and the neutral safety/backup light switch.
9. Remove the upper transaxle-to-engine bolts.
10. Remove the transaxle-to-mount through bolt, the transaxle mount bracket and the mount.
11. Raise and safely support the vehicle.
12. Remove the front wheel assemblies.
13. Disconnect the shift cable bracket from the transaxle.
14. Remove the left-side splash shield.
15. Using a modified drive axle seal protector tool J–34754 or equivalent, install one on each drive axle to protect the seal from damage and the joint from possible failure.
16. Using care not to damage the halfshaft boots, disconnect the halfshafts from the transaxle.
17. Remove the torsional and lateral strut from the transaxle. Remove the left-side stabilizer link pin bolt.
18. Remove the left frame support bolts and move it out of the way.
19. Disconnect the speedometer wire from the transaxle.

20. Remove the transaxle converter cover and matchmark the converter to the flywheel for assembly.
21. Disconnect and plug the transaxle cooler pipes.
22. Remove the transaxle-to-engine support.
23. Using a transmission jack, position and secure it to the transaxle and remove the remaining transaxle-to-engine bolts.
24. Make sure that the torque converter does not fall out and remove the transaxle from the vehicle.

NOTE: The transaxle cooler and lines should be flushed any time the transaxle is removed for overhaul, or to replace the pump, case or converter.

TRANSAXLE INSTALLATION

A-BODY VEHICLES

1. Install the transaxle into place.
2. Install the transaxle-to-engine bolts and torque to 55 ft. lbs. (75 Nm).
3. Install the converter-to-flywheel bolts and torque to 46 ft. lbs. (62 Nm). Retorque the first bolt after all 3 bolts have been tightened.
4. Install the flywheel splash shield.
5. Install the right and left insulator attaching bolts and torque to 40 ft. lbs. (55 Nm).
6. Connect the cooler pipes and torque tube nuts to 20 ft. lbs. (27 Nm).
7. Install the extension housing support bracket.
8. Connect the speedometer cable.
9. Install the frame assembly.
10. Install the stabilizer shaft.
11. Install the driveshafts.
12. Install the brake line bracket at the strut.
13. Install the pinch bolt and nut at the steering knuckle.
14. Install the wheel assemblies.
15. Lower the vehicle.
16. Connect the vacuum line at the modulator.
17. Tighten the remaining transaxle-to-engine bolts to 55 ft. lbs. (75 Nm).
18. Connect all electrical connectors.
19. Connect the shift linkage at the transaxle.
20. Connect the T.V. cable.
21. Install the air cleaner and the negative battery cable.
22. Fill the transaxle with transaxle fluid and check adjustments.

BUICK AND OLDSMOBILE C-BODY VEHICLES

1. Install the transaxle into the vehicle.
2. Install the lower transaxle-to-engine bolts and torque to 55 ft. lbs. (75 Nm).
3. Connect the cooler pipes at the transaxle and torque to 16 ft. lbs. (22 Nm).
4. Install the right and left drive axle splines into the case past the lip seal.
5. Attach the ball joint to the lower steering knuckle and torque to 48 ft. lbs. (65 Nm).
6. Remove the transaxle support fixture.
7. Install the cradle assembly.
8. Install the left frame-to-body attaching bolts.
9. Install the front frame strut to the right frame member bolts.
10. Install the front frame strut to the rear frame member bolts.
11. Install the flywheel-to-converter bolts and torque to 46 ft. lbs. (62 Nm). Retorque the first bolt after all 3 bolts have been tightened.
12. Install the flywheel cover and attaching bolts.
13. Connect the stabilizer link to the control arm bolt.
14. Install the engine support bracket-to-transaxle case bolts.
15. Install the right rear transaxle mount. Torque the 3 nuts to frame member to 30 ft. lbs. (41 Nm). Torque the 2 bolts to the transaxle case to 40 ft. lbs. (55 Nm).
16. Install the left rear transaxle mount to the transaxle bolts and torque to 30 ft. lbs. (41 Nm).
17. Install the right front transaxle mount to frame nuts.
18. Install the left front transaxle mount. Torque the 3 bolts at the transaxle to 40 ft. lbs. (55 Nm). Install the 3 nuts at the frame member.
19. Install the driveshafts.
20. Install the ball stud to the steering knuckle.
21. Install the wheel assemblies.
22. Lower the vehicle.
23. Remove the engine support fixture.
24. Install the remaining top transaxle to engine bolts and torque to 55 ft. lbs. (75 Nm).
25. Connect the vacuum modulator hose at the modulator.
26. Connect all electrical connectors.
27. Connect the shift control linkage.
28. Connect the cruise control accessories.
29. Connect the T.V. cable at the throttle body and at the transaxle.
30. Install the air intake duct and mass air flow sensor.
31. Connect the battery negative cable.
32. Check the fluid level and all adjustments.

CADILLAC C-BODY VEHICLES

1. Install the transaxle assembly.

NOTE: To reach the last bell housing bolt, use a 3 in. socket wrench extension through the right wheel arch opening.

2. Install the lower bellhousing bolts and torque to 55 ft. lbs. (75 Nm).
3. Install the flywheel splash shield.
4. Install the right and left transaxle mount attachments.
5. Connect the oil cooler lines at the transaxle case.
6. Install the extension housing-to-engine support bracket.
7. Install the left front cradle assembly.
8. Install the stabilizer bar-to-left control arm bolt.
9. Install both drive axles to the transaxle. From the left side of the vehicle, connect the lower ball joint to steering knuckle.
10. Install both front wheels.
11. Install the upper bell housing-to-engine bolts and studs.
12. Connect the electrical connectors to the following items:
    a. Converter clutch
    b. Vehicle speed sensor
    c. Vacuum line at the modulator
13. Connect the shift linkage to the transaxle. Remove engine support fixture tool.
14. Install the air cleaner and the TV cable. Connect the negative terminal to the battery.
15. Check the fluid level and all adjustments.

H-BODY VEHICLES

1. To install, position the transaxle into the vehicle.
2. Install the lower transaxle-to-engine bolts and torque to 55 ft. lbs. (75 Nm).
3. Connect the cooler pipes at the transaxle and torque to 16 ft. lbs. (22 Nm).
4. Install the right and left drive axle splines into the case past the lip seal.
5. Attach the ball joint to the lower steering knuckle and torque to 48 ft. lbs. (65 Nm).
6. Remove the transaxle support fixture.
7. Install the cradle assembly.
8. Install the left frame-to-body attaching bolts.
9. Install the front frame strut to the right frame member bolts.
10. Install the front frame strut to the rear frame member bolts.
11. Install the flywheel-to-converter bolts and torque to 46 ft. lbs. (62 Nm). Retorque the first bolt after all 3 bolts have been tightened.

12. Install the flywheel cover and attaching bolts.
13. Connect the stabilizer link to the control arm bolt.
14. Install the engine support bracket-to-transaxle case bolts.
15. Install the right rear transaxle mount. Torque the 3 nuts to frame member to 30 ft. lbs. (41 Nm). Torque the 2 bolts to the transaxle case to 40 ft. lbs. (55 Nm).
16. Install the left rear transaxle mount to the transaxle bolts and torque to 17 ft. lbs. (41 Nm).
17. Install the right front transaxle mount to frame nuts.
18. Install the left front transaxle mount. Torque the 3 bolts at the transaxle to 40 ft. lbs. (55 Nm). Install the 3 nuts at the frame member.
19. Install the driveshafts.
20. Install the ball stud to the steering knuckle.
21. Install the wheel assemblies.
22. Lower the vehicle.
23. Remove the engine support fixture.
24. Install the remaining top transaxle to engine bolts and torque to 55 ft. lbs. (75 Nm).
25. Connect the vacuum modulator hose at the modulator.
26. Connect all electrical connectors.
27. Connect the shift control linkage.
28. Connect the cruise control accessories.
29. Connect the T.V. cable at the throttle body and at the transaxle.
30. Install the air intake duct and mass air flow sensor.
31. Connect the battery negative cable.
32. Check the fluid level and all adjustments.

1986–87 RIVIERA AND TORONADO

1. Slide the transaxle into position and then install the 2 lower engine-to-transaxle bolts, torquing to 55 ft. lbs. (75 Nm).
2. Install the engine-to-transaxle bracket. Install the left side frame assembly bolts.
3. Install the engine mounting nuts. Install the left side transaxle mounting bolts.
4. Install the right driveshaft to the transaxle.
5. Install the engine-to-transaxle bracket.
6. Install the 2 remaining engine-to-transaxle bolts.
7. Connect the left side of the frame by removing the bolts.
8. Connect the left side driveshaft to the transaxle.
9. Connect the left side ball joint to the knuckle. Connect the sway bar links.
10. Install the engine mounting nuts. Install the left side transaxle mounting bolts.
11. Install the converter-to-flexplate bolts, turning the crankshaft (as necessary). Install the torque converter cover.
12. Connect the transaxle oil cooler lines at the transaxle.
13. Lower the front of the vehicle.
14. Install the driveline dampener bracket. Install the bolts that fasten the wiring harness to the transaxle.
15. Install the top 3 transaxle mounting bolts.
16. Connect the neutral safety switch. Connect the shift selector bracket and cable to the transaxle.
17. Connect the vacuum line to the modulator; electrical connections involved with the transaxle; transaxle valve cable at the throttle body and at the transaxle; the cruise control servo.
18. Remove the engine support fixture.
19. Connect the battery negative cable.
20. Check the fluid level and all adjustments.

1986–89 ELDORADO, SEVILLE AND 1987–89 ALLANTE

1. Install the transaxle into place.
2. Install the lower bell housing bolts and torque to 55 ft. lbs. (75 Nm).
3. Install the converter-to-flexplate bolts and torque to 46 ft. lbs. (63 Nm).
4. Install the flexplate splash shield.
5. Install the cooler lines to the case.
6. Install the driveshafts and remove the drive axle boot seal protectors.
7. Install the engine-to-transaxle brace bolts that pass into the transaxle VSS connector. Install the right rear transaxle mount-to-transaxle bracket. Lower the front of the vehicle.
8. Install the mounting bracket. Raise and safely support the vehicle.
9. Install the air injection reactor management valve/bracket assembly to the transaxle mount bracket.
10. Install the foremost cradle mount insulator bolt and the left cradle member. Install the right front and left rear transaxle-to-cradle bolts and the left stabilizer mount bolts.
11. Lower the front of the vehicle. Remove the engine support fixture.
12. Raise and support the vehicle. Install the wire loom clip to the transaxle mount bracket. Install the transaxle mount-to-cradle attachments. Install the engine mount and dampener-to-cradle attachments.
13. Install the vacuum hoses and the wire loom to the clips at the front of the cradle. Install the hose connections to the ends of the air injection reactor pipes.
14. Install the air conditioner splash shield and the mount cover for the forwardmost cradle insulator.
15. Install the ball joint cotter pins and nuts. Press the ball joints to the steering knuckles. Install the right and left stabilizer link bolts and both front wheels.
16. Lower the front of the vehicle.
17. Install the radiator hose bracket and transaxle mount-to-bracket nuts. Install the air injection reactor crossover pipe fitting.
18. Install the bell housing bolts. Install the throttle valve cable support bracket and engine oil cooler line bracket.
19. Install the vacuum modulator.
20. Install the fuel line bracket and connect the neutral safety switch connector.
21. Install the shift linkage bracket to the transaxle and the manual shift lever to the manual shift shaft.
22. Install the bracket for the engine oil cooler lines.
23. Install the cruise control servo and bracket assembly. Connect the electrical connectors going to the distributor, oil pressure sending unit and transaxle.
24. Connect the transaxle throttle valve cable. Install the air cleaner assembly. Connect the negative battery cable.
25. Adjust the transaxle valve cable and the shift linkage. Refill the transaxle to the proper level. Operate the engine until normal operating temperatures are reached. Adjust the level until it is correct.

1988–89 REATTA, RIVIERA AND TORONADO

1. Raise the transaxle to the vehicle; be careful not to damage the hoses, lines and wiring.
2. Install the lower transaxle-to-engine bolts.
3. Connect the oil cooler tubes to the transaxle.
4. Install the partial frame, the partial frame-to-main frame bolts and the partial frame-to-body bolts.
5. Install the torque converter-to-flywheel bolts.
6. Install the flywheel cover and bolts.
7. To the left control arm, install the stabilizer shaft.
8. Install the right rear transaxle mount, the left rear transaxle mount-to-transaxle bolts and the right rear transaxle-to-frame nuts.
9. Install the halfshaft to the transaxle.

NOTE: When installing the halfshaft, be careful not to damage the seal lips.

10. Connect the control arms to the steering knuckles. Install both front ball joint-to-steering knuckle nuts. Remove the drive axle seal protector tool.
11. Install both front wheel assemblies. Lower the front of the vehicle.
12. Install the upper transaxle-to-engine bolts.

13. Remove the engine support fixture tool.
14. Connect the hose to the vacuum modulator.
15. Connect the electrical harness connectors to the neutral start/backup light switch, the torque converter clutch (TCC) and the vehicle speed sensor (VSS).
16. Install the shift control linkage lever mounting bracket to the transaxle and connect the shift control linkage lever to the manual shaft.
17. Install the exhaust pipe crossover.
18. Connect the cruise control servo and cable. Connect the throttle valve (T.V.) cable to the transaxle and the throttle body.
19. Install the air intake duct. Connect the negative battery cable.
20. Check and/or adjust the T.V. and shift control cables. Check and/or refill the transaxle fluid. Road test the vehicle and check for leaks.

W-BODY VEHICLES

1. Put a small amount of grease on the pilot hub of the converter and make sure that the converter is properly engaged with the pump.
2. Raise the transaxle to the engine while guiding the right-side halfshaft into the transaxle.
3. Install the lower transaxle mounting bolts, tighten to 55 ft. lbs. and remove the jack.
4. Align the converter with the marks made previously on the flywheel and install the bolts hand tight.
5. Torque the converter bolts to 46 ft. lbs.; retorque the first bolt after the others.
6. Install the starter assembly. Install the left side halfshaft.
7. Install the converter cover, oil cooler lines and cover. Install the sub-frame assembly. Install the lower engine mount retaining bolts and the transaxle mount nuts.
8. Install the right and left ball joints. Install the power steering rack, heat shield and cooler lines to the frame.
9. Install the right and left inner fender splash shields. Install the tire assemblies.
10. Lower the vehicle. Connect all electrical leads. Install the upper transaxle mount bolts, tighten to 55 ft. lbs.
11. Attach the crossover pipe to the exhaust manifold. Connect the EGR tube to the crossover.
12. Connect the T.V. cable and the shift cable. Install the air cleaner and inlet tube.
13. Remove the engine support tool. Connect the negative battery cable.
14. Adjust the transaxle valve cable and the shift linkage. Refill the transaxle to the proper level. Operate the engine until normal operating temperatures are reached. Adjust the level until it is correct.

BENCH OVERHAUL

Before Disassembly

NOTE: Cleanliness is an important factor in the overhaul of the transaxle.

Before opening up the transaxle, the outside of the unit should be thoroughly cleaned, preferably with high-pressure cleaning equipment such as a car wash spray unit. Dirt entering the transaxle internal parts will negate all the effort and time spent on the overhaul. During inspection and reassembly, all parts should be thoroughly cleaned with solvent, then dried with compressed air. Wiping cloths and rags should not be used to dry parts since lint will find its way into valve body passages. Wheel bearing grease, long used to secure thrust washers and to lube parts should not be used. Lube seals with Dexron®II and use ordinary unmedicated petroleum jelly to hold thrust washers to ease assembly of seals, since it will not leave a harmful residue as grease often will. Do not use solvent on neoprene seals, friction plates or thrust washers. Be wary of nylon parts if the transaxle failure was due to a failure of the cooling system. Nylon parts exposed to antifreeze solutions can swell and distort and so must be replaced (Speedo gears, some thrust washers, etc.). Before installing bolts into aluminum parts, always dip the threads into clean oil. Anti-seize compound is also a good way to prevent the bolts from galling the aluminum and seizing. Always use a torque wrench to keep from stripping the threads. Take care of the seals when installing them, especially the smaller O-rings. The internal snaprings should be expanded and the external snaprings should be compressed, if they are to be reused. This will help insure proper seating when installed.

Converter Inspection

1. Inspect the converter for leaks as followed:
   a. Install pressurizing adapter tool J-21369-B or equivalent. Tighten the hex nut.
   b. Fill the converter with approximately 80 psi.
   c. Submerge in water and check for leaks.

CAUTION

After leak testing, bleed pressurized air from the converter before removing the tool.

   d. Remove the tool.
2. Inspect the converter hub surfaces for signs of scoring or wear.
3. Inspect the converter bushing for damage, cracks or scoring.
4. Inspect converter internal end clearance.
   a. Screw the upper and lower knurled knobs of tool J-35138 or equivalent, together. Be sure collet end is not expanded.
   b. Insert tool in the torque converter hub.
   c. Slide the tapered plug portion of the tool into the torque converter hub to ensure proper tool alignment.
   d. Hold the upper knurled knob and turn the wing nut clockwise until hand tight to expand the collet into the converter turbine hub. Then loosen the wing nut (counterclockwise) until the tool is just loose enough to allow up and down movement while not removable from the converter (generally 2½ turns or more).
   e. Hold the upper knurled knob and wing nut and turn the lower knurled knob clockwise until hand tight to clamp the tool in the converter turbine hub.
   f. Attach the magnetic base (tool J-26900-13 or equivalent) to the torque converter and set up the dial indicator (tool J-8001 or equivalent) so that the indicator tip is in the center of the tool wing nut.
   g. To make the end play check, push the upper knurled knob down, zero out dial indicator and pull the lower knurled knob up. Be sure the tapered plug stays seated in the converter hub. The end play is the difference between up and down position. Do not allow the wing nut to turn when measuring. Allowable clearance is 0–0.019 in. (0–0.50mm). Torque converter must be replaced if the clearance exceeds maximum limit.
   h. Remove the tool by holding the upper knurled counterclockwise until the collet can be removed from the torque converter turbine hub.

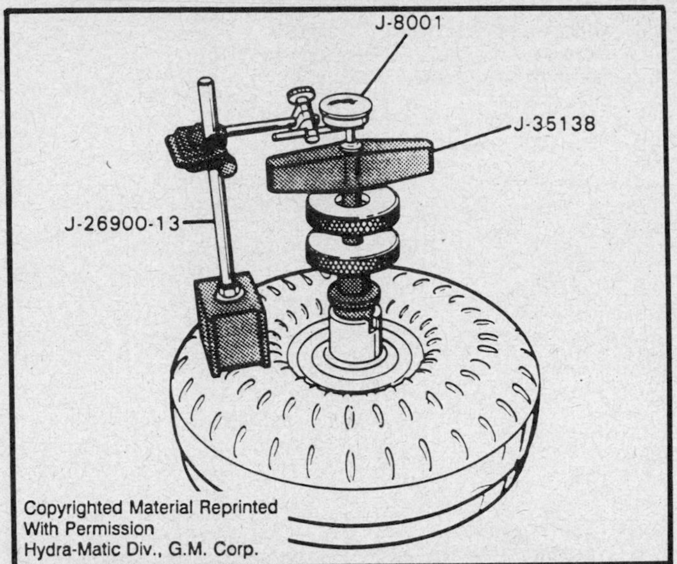
J-8001
J-35138
J-26900-13

Checking torque converter endplay

Transaxle Disassembly

TORQUE CONVERTER

Removal

1. Make certain that the transaxle is held securely.
2. Pull the converter straight out of the transaxle. Be careful since the converter contains a large amount of oil. There is no drain plug on the converter so the converter should be drained through the hub.

NOTE: The transaxle fluid that is drained from the converter can help diagnose transaxle problems.

3. If the oil in the converter is discolored but does not contain metal bits or particles, the converter is not damaged and need not be replaced. Remember that color is no longer a good indicator of transaxle fluid condition. In the past, dark color was associated with overheated transaxle fluid. It is not a positive sign of transaxle failure with the newer fluids like Dexron®II.
4. If the oil in the converter contains metal particles, the converter is damaged internally and must be replaced. The oil may have an aluminum pain appearance.
5. If the cause of oil contamination was due to burned clutch plates or overheated oil, the converter is contaminated and should be replaced.

OIL PAN

Removal

1. Remove the oil pan attaching bolts. Carefully bump the pan with a rubber mallet to free the pan. If the pan is pried loose instead, be very careful not to damage the gasket surfaces. Discard the pan gasket.
2. Inspect the bottom of the pan for debris that can give an indication of the nature of the transaxle failure.
3. Check the pan for distorted gasket flanges, especially around the bolt holes, since these are often dished-in due to overtorque. They can be straightened with a block of wood and a rubber mallet if necessary.

OIL FILTER

Removal

1. The oil filter is retained by a clip as well as the interface fit of the oil intake tube and O-ring. Move the clip out of the way and pull the filter from the case.
2. Discard the O-ring from the intake pipe. Often the O-ring will stick to its bore in the case.
3. The oil pressure regulator is in the bore next to the oil filter intake pipe opening. If it is to be removed, use snapring pliers to remove the retaining ring and then pull the pressure regulator bushing assembly from the case.

GOVERNOR

Removal

1. Since the speedometer drive gear is attached to the governor assembly, first remove the speedometer driven gear attaching bolt and the retaining clip.
2. Remove the speedometer driven gear from the governor cover. Remove the governor cover and discard the O-ring.
3. Remove the governor assembly along with the speedometer drive gear thrust bearing.
4. Remove the modulator retainer and lift out the modulator. Discard the O-ring. Lift out the modulator valve using a magnet.

REVERSE AND/OR 1–2 SERVO

Removal

1. To remove the reverse and/or 1–2 servo assembly, the cover will have to be depressed to relieve the spring pressure on the snapring that retains the cover. The factory type tool for this is similar to a clamp that hooks to the case and applies with a screw operated arm. The object is to release the spring holding pressure the cover has on the snapring.
2. Pliers can be used to grasp the servo cover to remove it. Discard the seals. Make sure that the cover seal ring is not stuck in the case groove where it might be overlooked.
3. Remove the servo assembly and the servo return springs. Remove the apply pin from the servo assembly.

NOTE: The servo assemblies are not interchangeable due to the reverse servo pin being longer than the 1–2 pin. Keep the servo assemblies separate.

OIL PUMP

Removal

1. Disconnect the side cover attaching nuts and bolts and remove the side cover. Discard the gaskets.
2. Detach the solenoid wiring harness from the case connector and pressure switch. Remove the throttle valve assembly from the valve body.
3. Remove the oil pump bolts and lift the oil pump assembly from the valve body.

VALVE BODY

Removal

1. Remove the valve body retaining bolts. Remove the valve body from the channel plate.
2. Remove the check balls from the spacer plate. Detach the spacer plate and discard the gaskets.
3. Remove the check balls from the channel plate.

OIL PUMP SHAFT AND CHANNEL PLATE

Removal

1. Remove the oil pump shaft by sliding it out of the channel

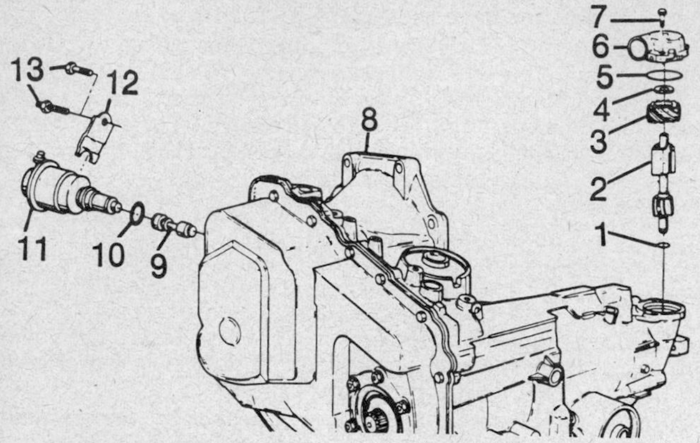
13
12
8
7
6
5
4
3
2
1
11
10
9

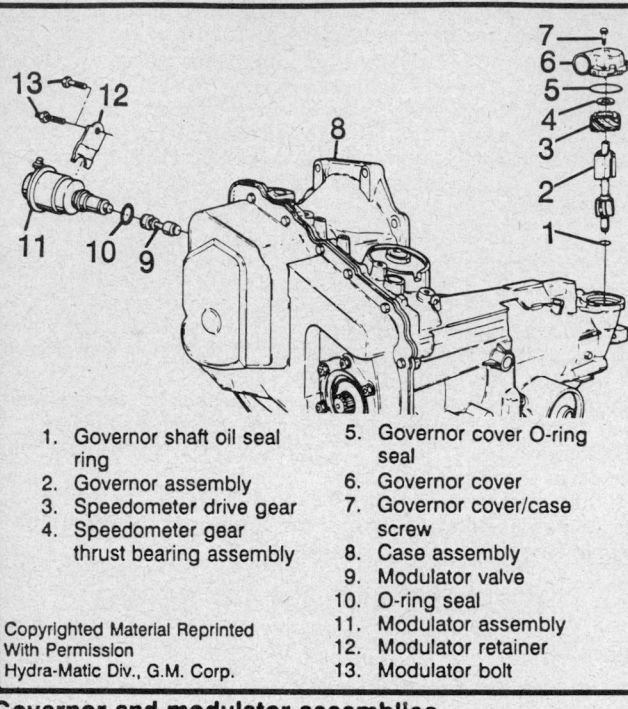

1. Governor shaft oil seal ring
2. Governor assembly
3. Speedometer drive gear
4. Speedometer gear thrust bearing assembly
5. Governor cover O-ring seal
6. Governor cover
7. Governor cover/case screw
8. Case assembly
9. Modulator valve
10. O-ring seal
11. Modulator assembly
12. Modulator retainer
13. Modulator bolt

Governor and modulator assemblies

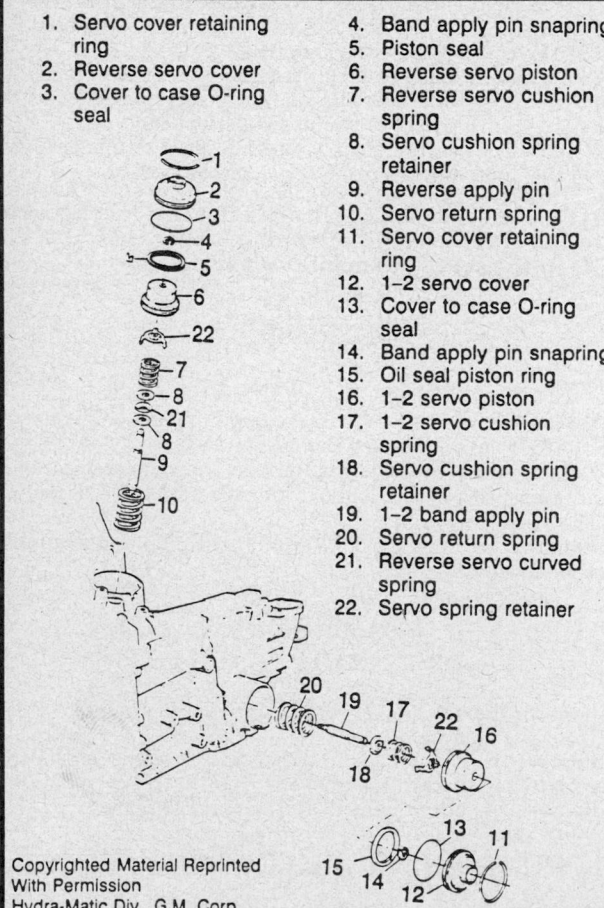

1. Servo cover retaining ring
2. Reverse servo cover
3. Cover to case O-ring seal
4. Band apply pin snapring
5. Piston seal
6. Reverse servo piston
7. Reverse servo cushion spring
8. Servo cushion spring retainer
9. Reverse apply pin
10. Servo return spring
11. Servo cover retaining ring
12. 1–2 servo cover
13. Cover to case O-ring seal
14. Band apply pin snapring
15. Oil seal piston ring
16. 1–2 servo piston
17. 1–2 servo cushion spring
18. Servo cushion spring retainer
19. 1–2 band apply pin
20. Servo return spring
21. Reverse servo curved spring
22. Servo spring retainer

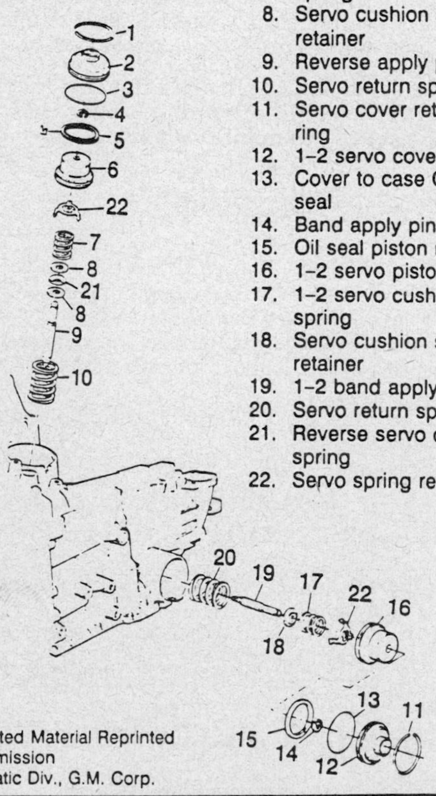
1
2
3
4
5
6
22
7
8
21
8
9
10
20
19
17
22
16
18
13
11
15
14
12

1-2 and reverse servo assemblies

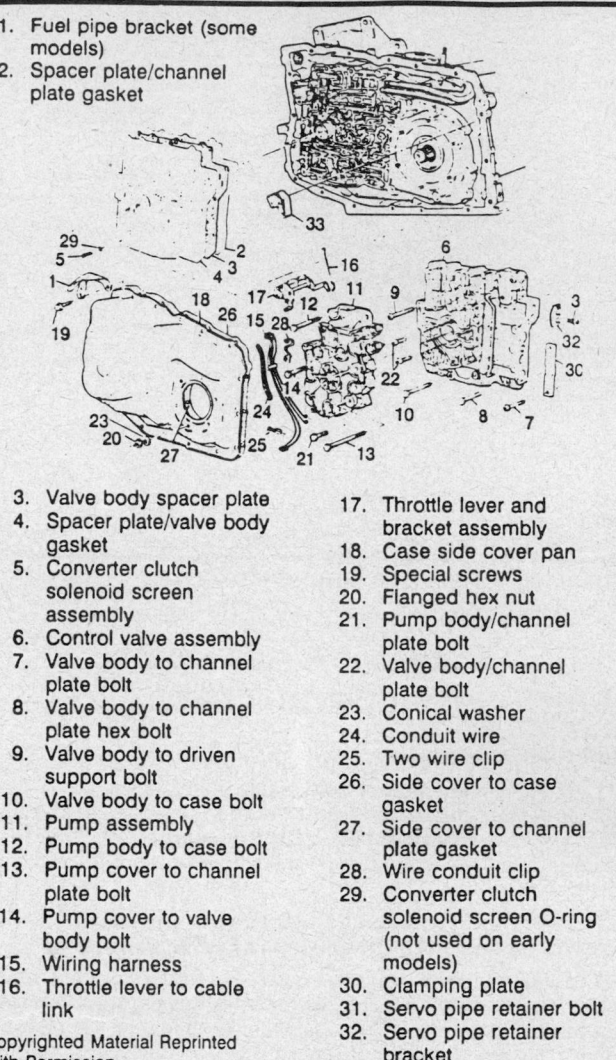

1. Fuel pipe bracket (some models)
2. Spacer plate/channel plate gasket

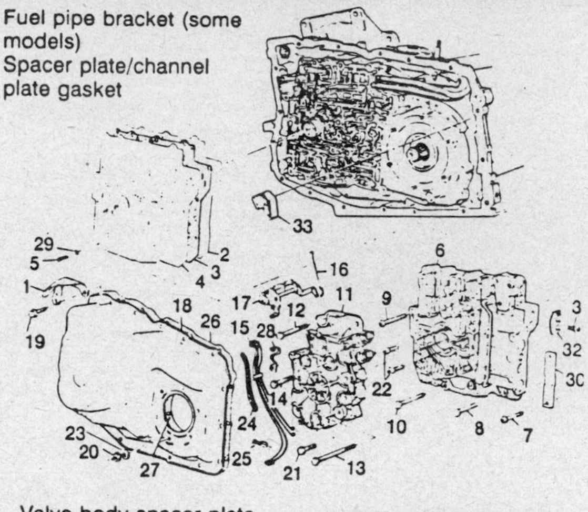

3. Valve body spacer plate
4. Spacer plate/valve body gasket
5. Converter clutch solenoid screen assembly
6. Control valve assembly
7. Valve body to channel plate bolt
8. Valve body to channel plate hex bolt
9. Valve body to driven support bolt
10. Valve body to case bolt
11. Pump assembly
12. Pump body to case bolt
13. Pump cover to channel plate bolt
14. Pump cover to valve body bolt
15. Wiring harness
16. Throttle lever to cable link
17. Throttle lever and bracket assembly
18. Case side cover pan
19. Special screws
20. Flanged hex nut
21. Pump body/channel plate bolt
22. Valve body/channel plate bolt
23. Conical washer
24. Conduit wire
25. Two wire clip
26. Side cover to case gasket
27. Side cover to channel plate gasket
28. Wire conduit clip
29. Converter clutch solenoid screen O-ring (not used on early models)
30. Clamping plate
31. Servo pipe retainer bolt
32. Servo pipe retainer bracket
33. Oil wier

Side cover, pump, valve body and parts

plate assembly. Place the detent lever in the **P** position and remove the manual valve clip.

2. Remove the channel plate attaching bolts and lift the channel plate from the transaxle case.

3. Remove the accumulator piston and the converter clutch accumulator piston and spring. Discard all gaskets.

FOURTH CLUTCH, SHAFT AND OUTPUT SHAFT

Removal

1. Remove the 3 clutch plates along with the apply plate. Remove the thrust bearing.

NOTE: The thrust bearing may still be on the case cover from earlier disassembly.

2. Remove the 4th clutch and shaft assembly. This will expose the output shaft.

3. Rotate the output shaft until the output shaft C-ring is visible. Remove the C-ring.

4. Pull the output shaft from the transaxle being careful not to damage it.

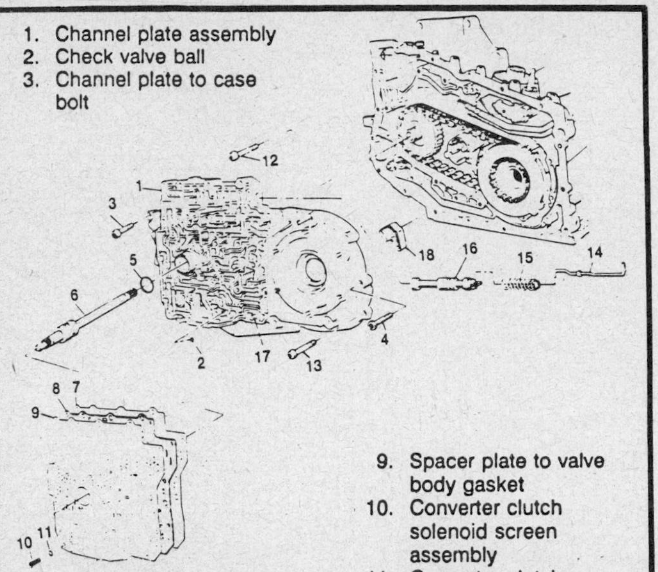
1
2
3
4
5
6
7
8
9
10
11
12
13
14
15
16
17
18

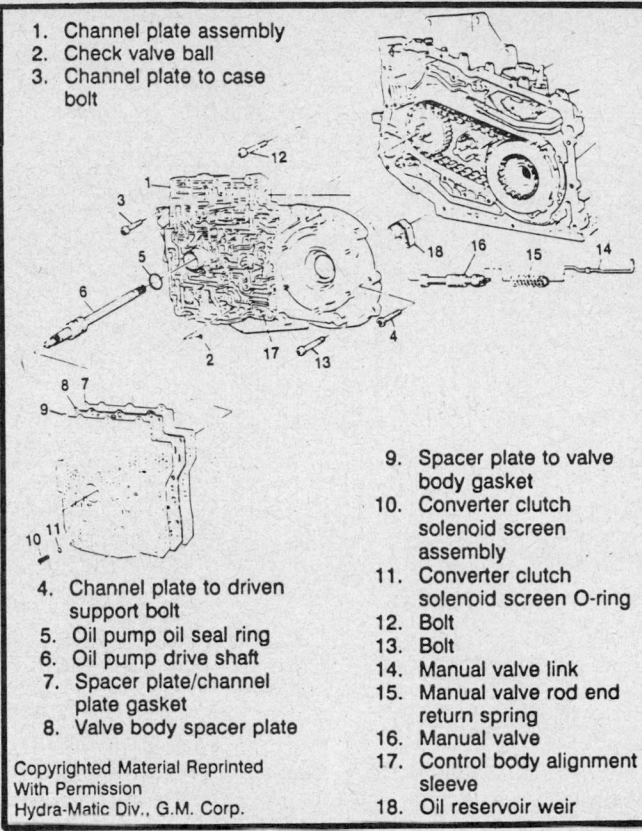

1. Channel plate assembly
2. Check valve ball
3. Channel plate to case bolt
4. Channel plate to driven support bolt
5. Oil pump oil seal ring
6. Oil pump drive shaft
7. Spacer plate/channel plate gasket
8. Valve body spacer plate
9. Spacer plate to valve body gasket
10. Converter clutch solenoid screen assembly
11. Converter clutch solenoid screen O-ring
12. Bolt
13. Bolt
14. Manual valve link
15. Manual valve rod end return spring
16. Manual valve
17. Control body alignment sleeve
18. Oil reservoir weir

Oil pump driveshaft and channel plate

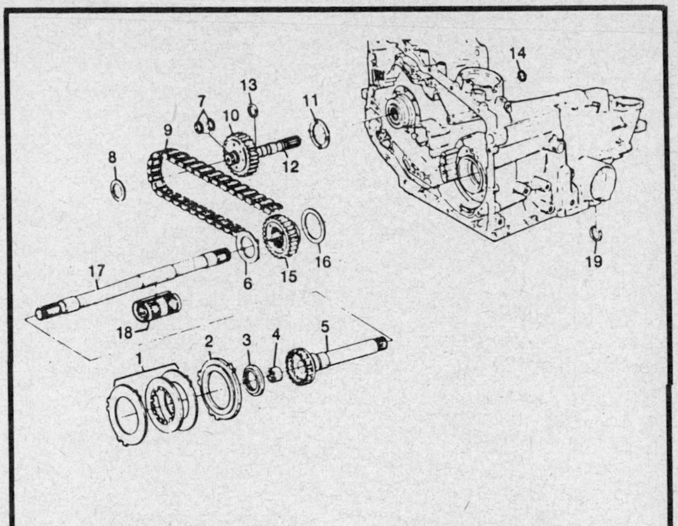
1
2
3
4
5
6
7
8
9
10
11
12
13
14
15
16
17
18
19

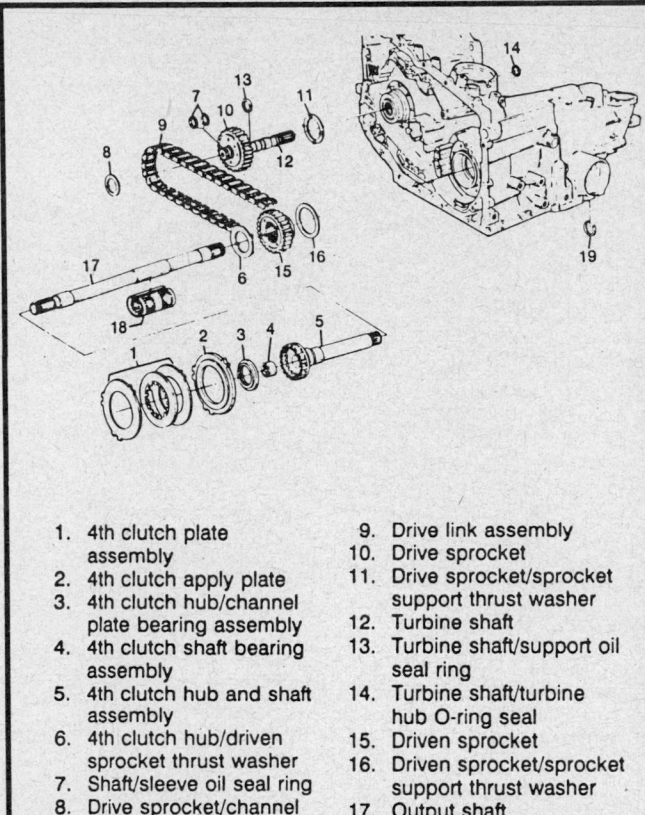

1. 4th clutch plate assembly
2. 4th clutch apply plate
3. 4th clutch hub/channel plate bearing assembly
4. 4th clutch shaft bearing assembly
5. 4th clutch hub and shaft assembly
6. 4th clutch hub/driven sprocket thrust washer
7. Shaft/sleeve oil seal ring
8. Drive sprocket/channel plate thrust washer
9. Drive link assembly
10. Drive sprocket
11. Drive sprocket/sprocket support thrust washer
12. Turbine shaft
13. Turbine shaft/support oil seal ring
14. Turbine shaft/turbine hub O-ring seal
15. Driven sprocket
16. Driven sprocket/sprocket support thrust washer
17. Output shaft
18. Input sun gear/output shaft bearing
19. Output shaft/differential snapring

Drive link assembly, output shaft, 4th clutch shaft and clutch assembly

DRIVE LINK ASSEMBLY

Removal

1. Remove the turbine shaft O-ring located at the front of the unit.
2. Reach through the access holes in the sprockets and slip the snaprings from their grooves.
3. Remove the sprockets and the chain as an assembly. It will required alternately pulling on the sprockets until the bearings come out of their support housings.

NOTE: If the sprockets are difficult to remove, use 2 small pieces of masonite or similar material to act both as wedges and as pads for a pry bar. Do not pry on the chain or the case.

4. After removing the drivelink assembly, take note as to the position of the colored link on the chain. It should be facing out. Remove the thrust washers.
5. Lift out the drive sprocket support and remove the thrust washer located on the casing.

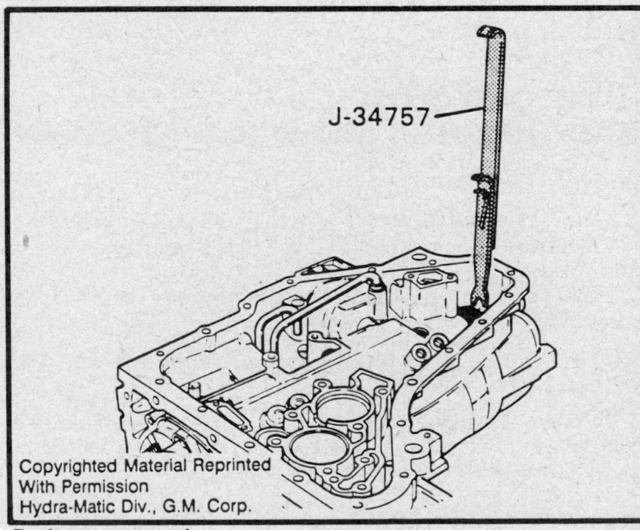
J-34757

C-ring removal

CHECK INPUT UNIT ENDPLAY

The factory tools for checking the endplay on this unit are somewhat elaborate and it is unlikely that every shop will be equipped in the same manner. However, the object is to pre-load the output shaft to remove the clearance. The factory tool mounts on the output end of the transaxle and a knob and screw arrangement can be tightened, forcing the output shaft upwards. A dial indicator is mounted on the end of the input shaft. By raising the input shaft with a suitable bar (again, the factory tool is special and grips the input shaft splines) the input unit endplay can be measured. Input shaft endplay should be 0.020–0.042 in.

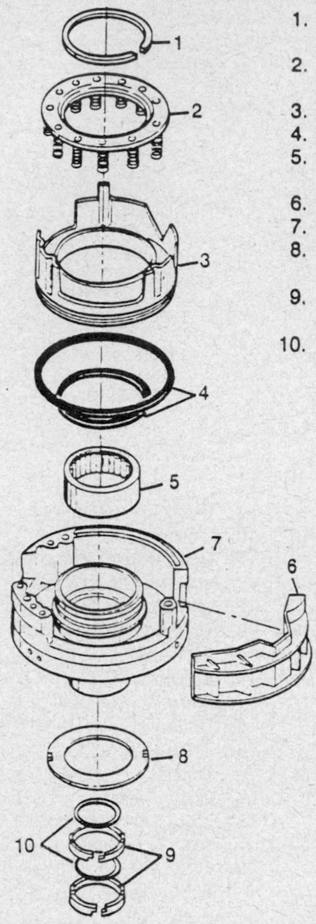
1
2
3
4
5
7
6
8
10
9

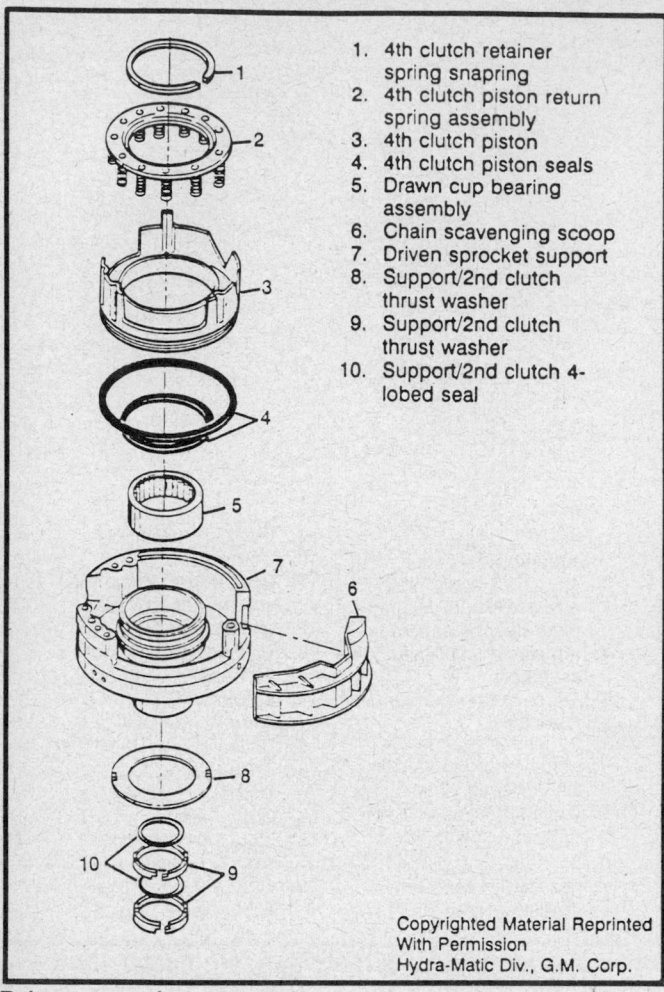

1. 4th clutch retainer spring snapring
2. 4th clutch piston return spring assembly
3. 4th clutch piston
4. 4th clutch piston seals
5. Drawn cup bearing assembly
6. Chain scavenging scoop
7. Driven sprocket support
8. Support/2nd clutch thrust washer
9. Support/2nd clutch thrust washer
10. Support/2nd clutch 4-lobed seal

Driven sprocket support

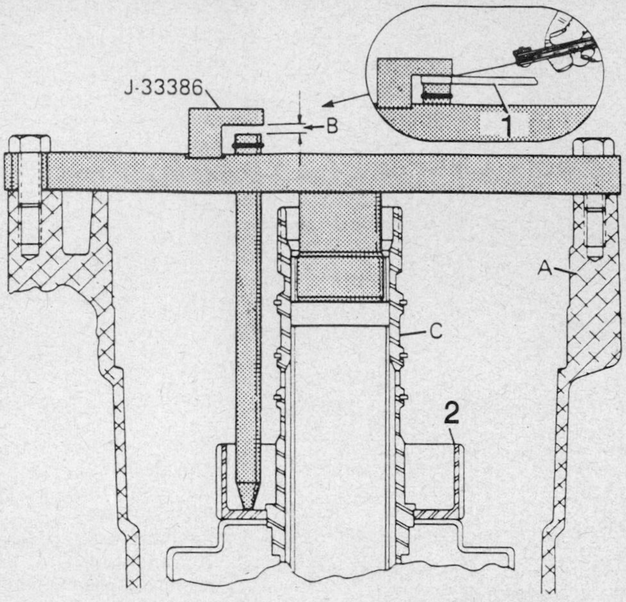
J-33386
B
1
A
C
2

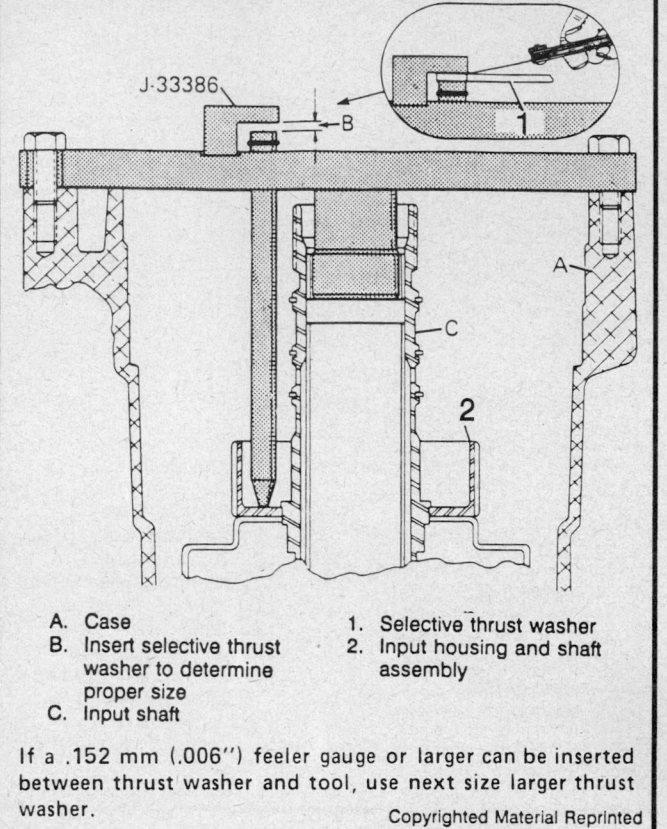

A. Case
B. Insert selective thrust washer to determine proper size
C. Input shaft

1. Selective thrust washer
2. Input housing and shaft assembly

If a .152 mm (.006") feeler gauge or larger can be inserted between thrust washer and tool, use next size larger thrust washer.

Checking input endplay and thrust washer selection chart

SECOND CLUTCH, INPUT CLUTCH, ROLLER CLUTCH AND SPRAG ASSEMBLY

Removal

1. Clamp the clutch and final drive tool to the 2nd clutch housing. Lift the 2nd clutch housing and the input shaft assembly out of the unit.
2. Remove the reverse band and the reverse reaction drum. Remove the input carrier assembly.

NOTE: The reverse band assembly may lift out with the 2nd clutch housing.

3. Remove the thrust bearing and the reaction carrier. The thrust bearing is located at one end of the reaction carrier.
4. Remove the thrust bearing which may be stuck to the reaction carrier.
5. Remove the reaction sun gear and the drum assembly. Remove the 1–2 band.

NOTE: The 1-2 band assembly should not be cleaned in cleaning solvent.

6. Remove the bearing ring and the sun gear shaft. A final drive internal bushing will be found on the sun gear shaft.

FINAL DRIVE ASSEMBLY

Removal

1. With a suitable tool, remove the snapring at the head of the internal final drive gear.
2. Clamp the clutch and final drive tool to the final drive internal gear and lift it out.
3. The final drive carrier will be removed with the final drive internal gear.

MANUAL SHAFT/DETENT LEVER AND ACTUATOR ROD

Removal

1. It is not necessary to disassembly the manual shaft, detent lever or the actuator rod unless replacement is needed.
2. Remove the manual shaft and detent lever retaining bolts and remove the shaft lever.
3. Remove the actuator rod assembly and check for wear or damage.
4. Remove the actuator guide assembly and check for damage.

MANUAL LINKAGE/ACTUATOR REPLACEMENT

Removal

1. Remove the manual shaft locknut and pin and lift out the manual shaft with the detent lever.

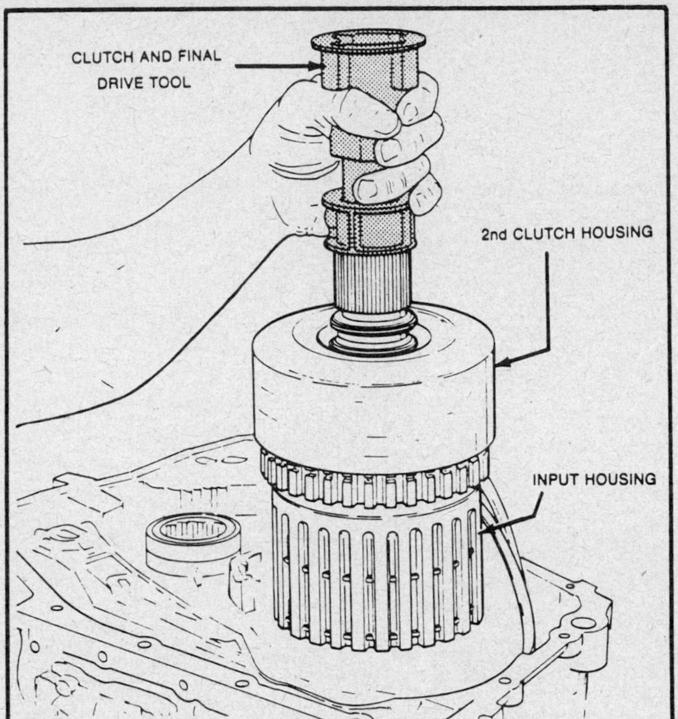
CLUTCH AND FINAL DRIVE TOOL
2nd CLUTCH HOUSING
INPUT HOUSING

Input housing and shaft removal

2. Remove the retaining pin from the case and detach the actuator guide assembly.
3. Remove the O-ring from the actuator guide and discard the O-ring.
4. The parking lock pawl assembly cannot be removed from the final drive internal gear.

NOTE: If the manual shaft seal is needed to be replaced, remove the axle oil seal along with the converter seal and then remove the manual shaft seal from its mounting in the transaxle case.

Unit Disassembly and Assembly

TRANSAXLE CASE

1. Clean the case and inspect carefully for cracks. Make certain that all passages are clean and that all bores and snapring grooves are clean and free from damage. Check for striped bolt holes. Check the case bushings for damage.
2. A new manual shaft seal can be installed at this time, along with a new axle oil seal and converter seal. Tap the seals into place with a suitable tool. The seal lips must face into the case.
3. Check the drive sprocket support assembly for damage. If it requires replacement, a slide hammer type puller can be used to pull the bearing from the sprocket support. Once the bearing is out, inspect the bore for wear or damage. The new bearing should be driven in straight and with care so as not to damage it.
4. If the parking pawl and related parts are to be removed, begin by turning the transaxle to the oil pan side up. Use a punch to remove the cup plug. Remove the parking pawl shaft retainer, then the shaft, pawl and return spring. Check the pawl for cracks.

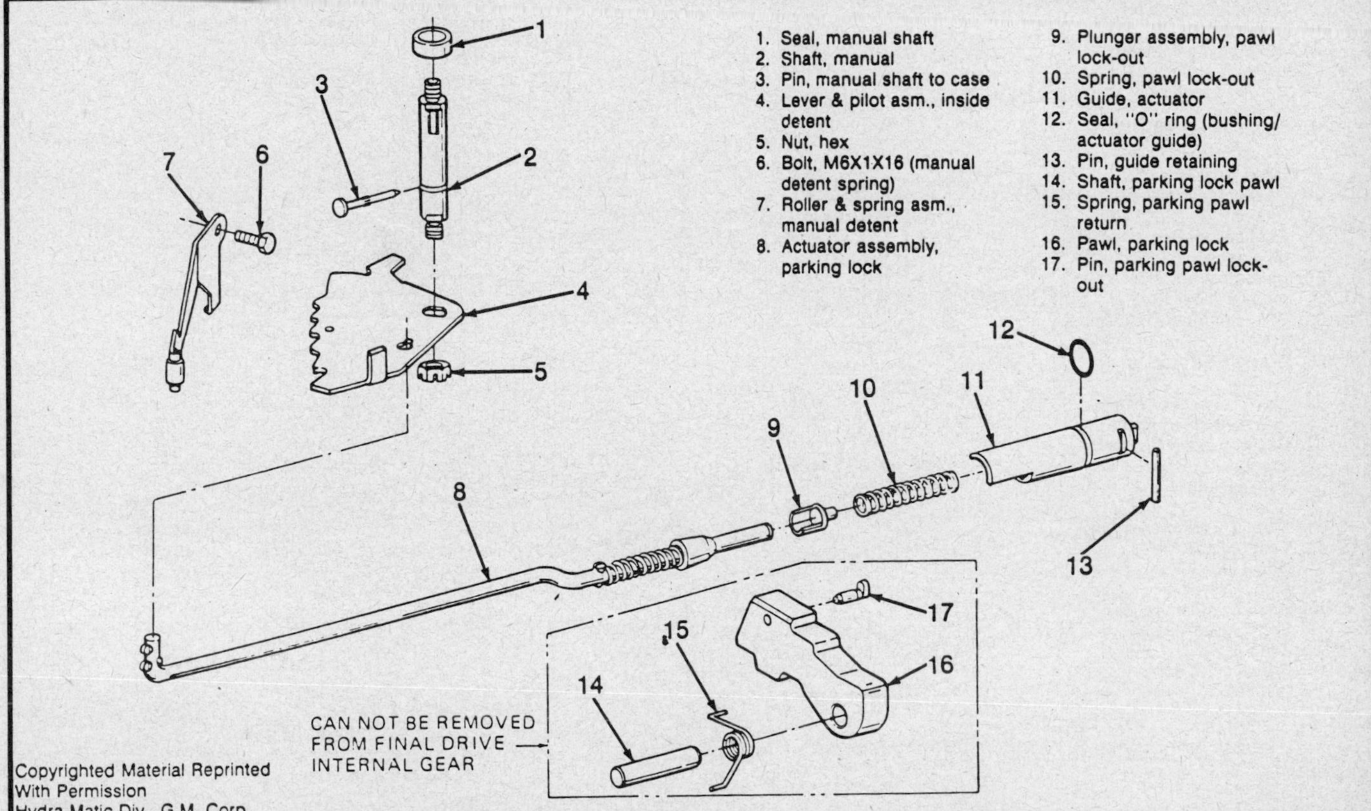
1. Seal, manual shaft
2. Shaft, manual
3. Pin, manual shaft to case
4. Lever & pilot asm., inside detent
5. Nut, hex
6. Bolt, M6X1X16 (manual detent spring)
7. Roller & spring asm., manual detent
8. Actuator assembly, parking lock
9. Plunger assembly, pawl lock-out
10. Spring, pawl lock-out
11. Guide, actuator
12. Seal, "O" ring (bushing/actuator guide)
13. Pin, guide retaining
14. Shaft, parking lock pawl
15. Spring, parking pawl return
16. Pawl, parking lock
17. Pin, parking pawl lock-out
CAN NOT BE REMOVED FROM FINAL DRIVE INTERNAL GEAR

Exploded view of manual linkage assembly

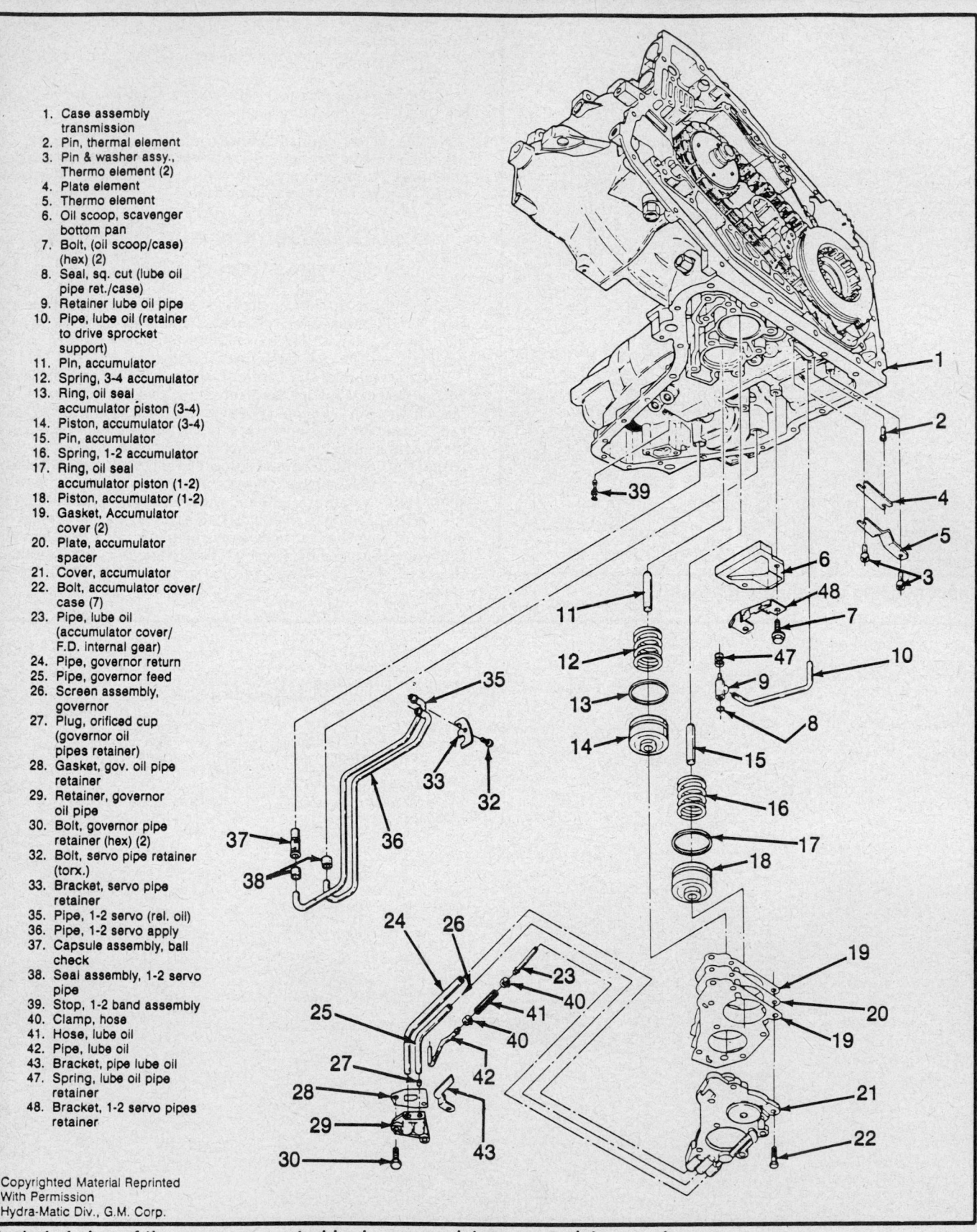

1. Case assembly transmission
2. Pin, thermal element
3. Pin & washer assy., Thermo element (2)
4. Plate element
5. Thermo element
6. Oil scoop, scavenger bottom pan
7. Bolt, (oil scoop/case) (hex) (2)
8. Seal, sq. cut (lube oil pipe ret./case)
9. Retainer lube oil pipe
10. Pipe, lube oil (retainer to drive sprocket support)
11. Pin, accumulator
12. Spring, 3-4 accumulator
13. Ring, oil seal accumulator piston (3-4)
14. Piston, accumulator (3-4)
15. Pin, accumulator
16. Spring, 1-2 accumulator
17. Ring, oil seal accumulator piston (1-2)
18. Piston, accumulator (1-2)
19. Gasket, Accumulator cover (2)
20. Plate, accumulator spacer
21. Cover, accumulator
22. Bolt, accumulator cover/ case (7)
23. Pipe, lube oil (accumulator cover/ F.D. internal gear)
24. Pipe, governor return
25. Pipe, governor feed
26. Screen assembly, governor
27. Plug, orificed cup (governor oil pipes retainer)
28. Gasket, gov. oil pipe retainer
29. Retainer, governor oil pipe
30. Bolt, governor pipe retainer (hex) (2)
32. Bolt, servo pipe retainer (torx.)
33. Bracket, servo pipe retainer
35. Pipe, 1-2 servo (rel. oil)
36. Pipe, 1-2 servo apply
37. Capsule assembly, ball check
38. Seal assembly, 1-2 servo pipe
39. Stop, 1-2 band assembly
40. Clamp, hose
41. Hose, lube oil
42. Pipe, lube oil
43. Bracket, pipe lube oil
47. Spring, lube oil pipe retainer
48. Bracket, 1-2 servo pipes retainer

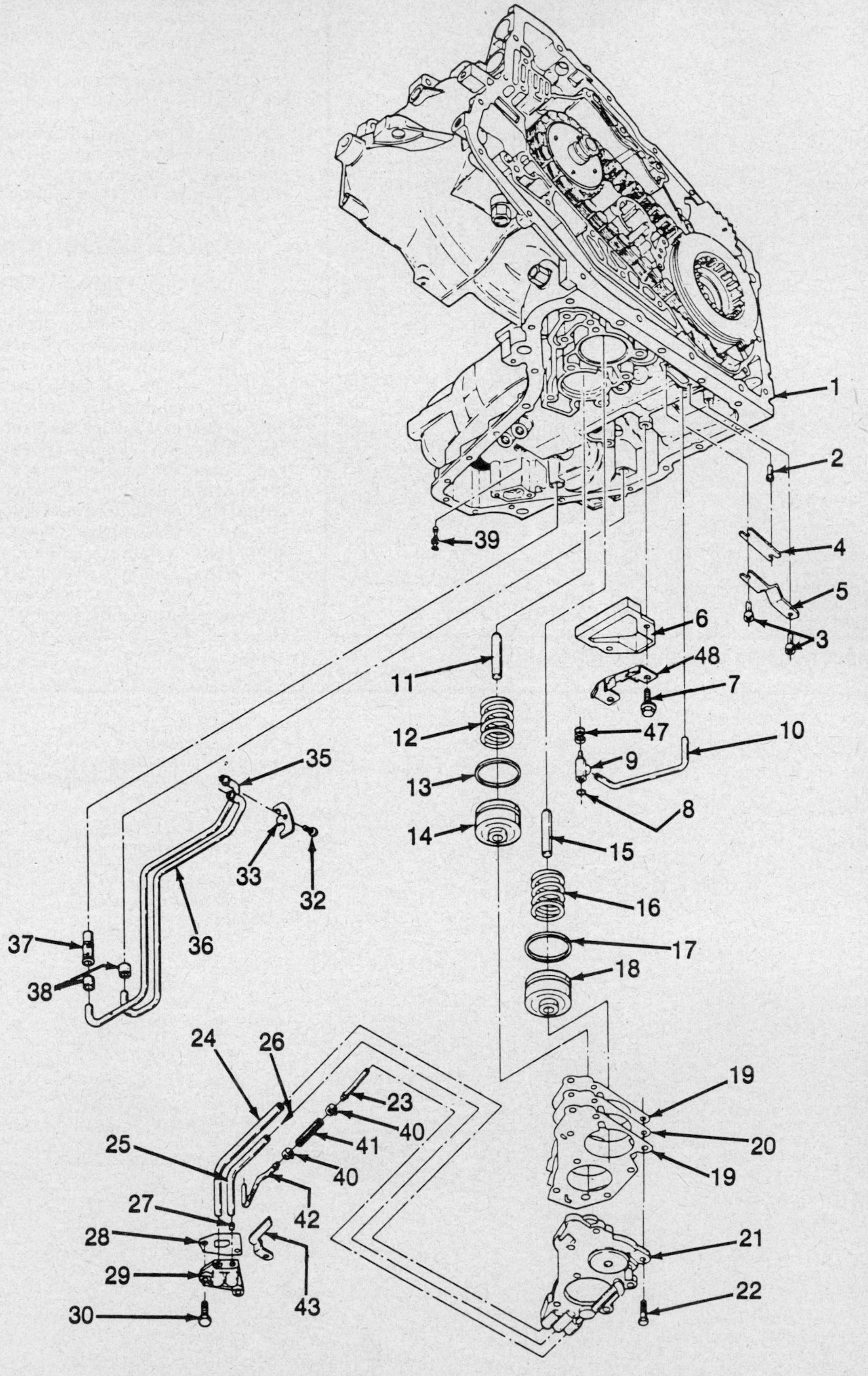
1
2
39
4
5
6
3
48
7
11
12
47
10
35
9
13
8
14
15
33
32
16
37
36
17
38
18
24
26
19
23
40
41
20
25
40
19
27
42
21
28
29
22
43
30

Exploded view of the governor control body, accumulator cover, pistons and component parts

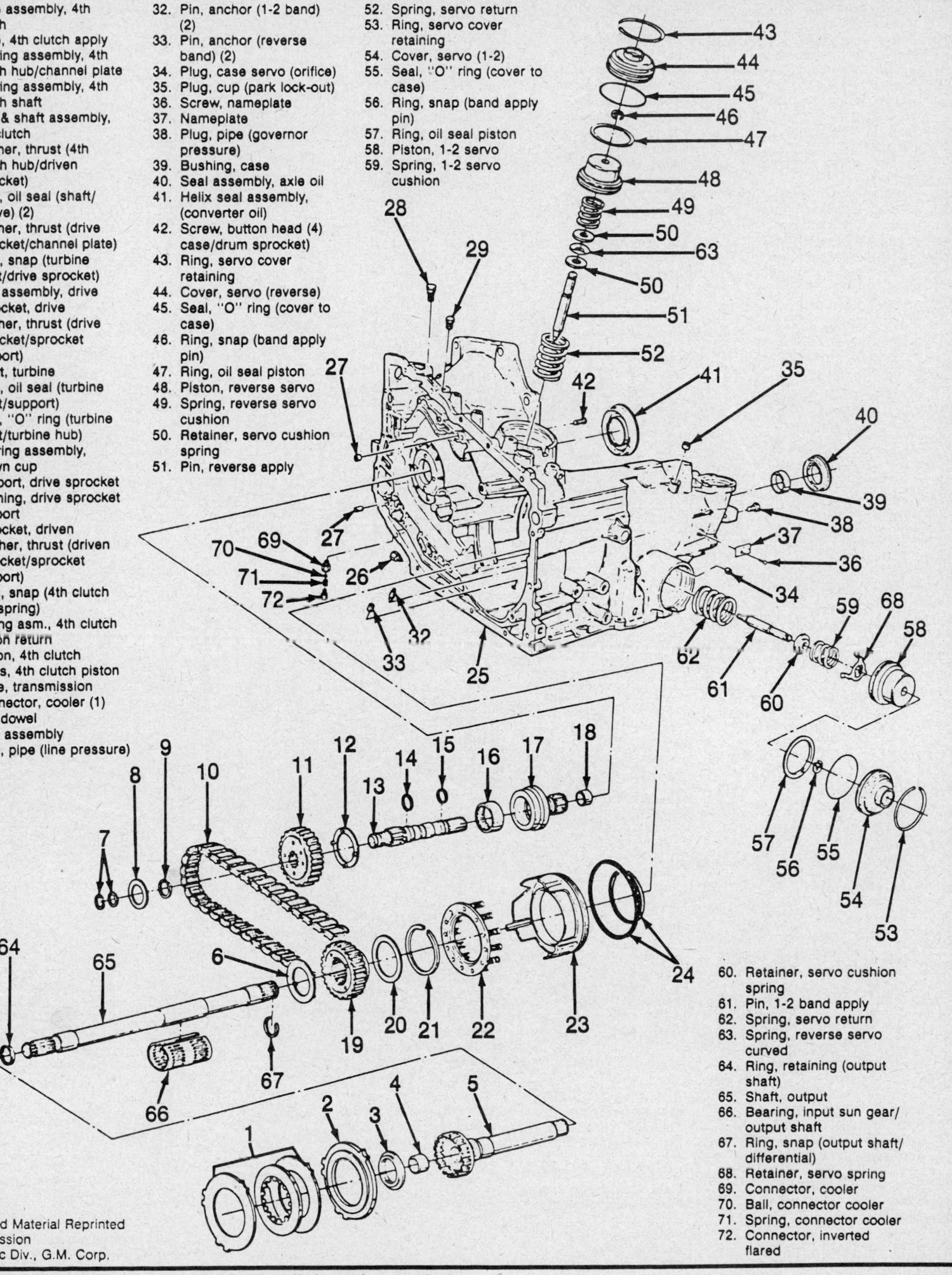

1. Plate assembly, 4th clutch
2. Plate, 4th clutch apply
3. Bearing assembly, 4th clutch hub/channel plate
4. Bearing assembly, 4th clutch shaft
5. Hub & shaft assembly, 4th clutch
6. Washer, thrust (4th clutch hub/driven sprocket)
7. Ring, oil seal (shaft/sleeve) (2)
8. Washer, thrust (drive sprocket/channel plate)
9. Ring, snap (turbine shaft/drive sprocket)
10. Link assembly, drive
11. Sprocket, drive
12. Washer, thrust (drive sprocket/sprocket support)
13. Shaft, turbine
14. Ring, oil seal (turbine shaft/support)
15. Seal, "O" ring (turbine shaft/turbine hub)
16. Bearing assembly, drawn cup
17. Support, drive sprocket
18. Bushing, drive sprocket support
19. Sprocket, driven
20. Washer, thrust (driven sprocket/sprocket support)
21. Ring, snap (4th clutch ret. spring)
22. Spring asm., 4th clutch piston return
23. Piston, 4th clutch
24. Seals, 4th clutch piston
25. Case, transmission
26. Connector, cooler (1)
27. Pin, dowel
28. Vent assembly
29. Plug, pipe (line pressure)
32. Pin, anchor (1-2 band) (2)
33. Pin, anchor (reverse band) (2)
34. Plug, case servo (orifice)
35. Plug, cup (park lock-out)
36. Screw, nameplate
37. Nameplate
38. Plug, pipe (governor pressure)
39. Bushing, case
40. Seal assembly, axle oil
41. Helix seal assembly, (converter oil)
42. Screw, button head (4) case/drum sprocket)
43. Ring, servo cover retaining
44. Cover, servo (reverse)
45. Seal, "O" ring (cover to case)
46. Ring, snap (band apply pin)
47. Ring, oil seal piston
48. Piston, reverse servo
49. Spring, reverse servo cushion
50. Retainer, servo cushion spring
51. Pin, reverse apply
52. Spring, servo return
53. Ring, servo cover retaining
54. Cover, servo (1-2)
55. Seal, "O" ring (cover to case)
56. Ring, snap (band apply pin)
57. Ring, oil seal piston
58. Piston, 1-2 servo
59. Spring, 1-2 servo cushion
60. Retainer, servo cushion spring
61. Pin, 1-2 band apply
62. Spring, servo return
63. Spring, reverse servo curved
64. Ring, retaining (output shaft)
65. Shaft, output
66. Bearing, input sun gear/output shaft
67. Ring, snap (output shaft/differential)
68. Retainer, servo spring
69. Connector, cooler
70. Ball, connector cooler
71. Spring, connector cooler
72. Connector, inverted flared

Exploded view of case assembly drive link and sprocket

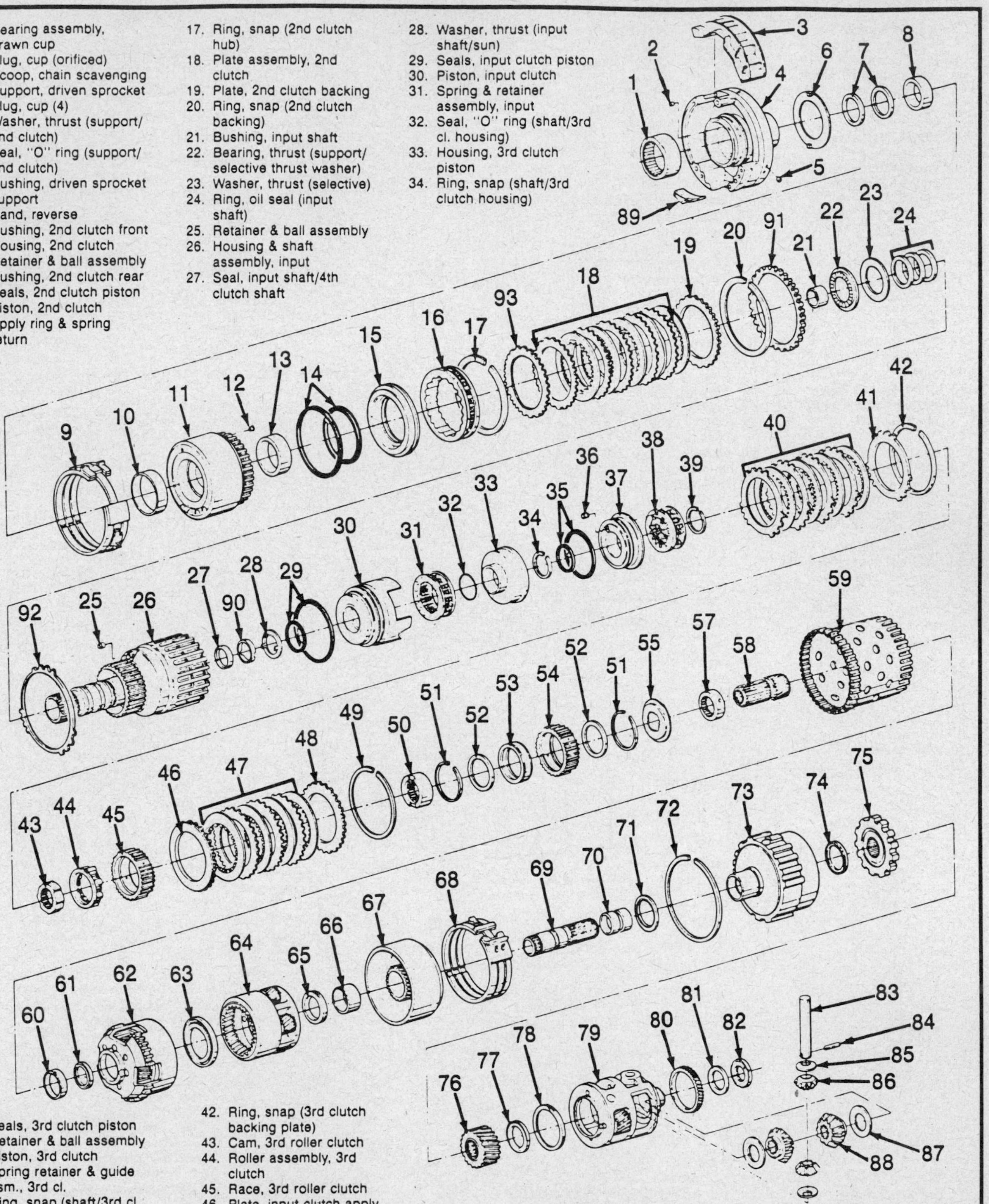

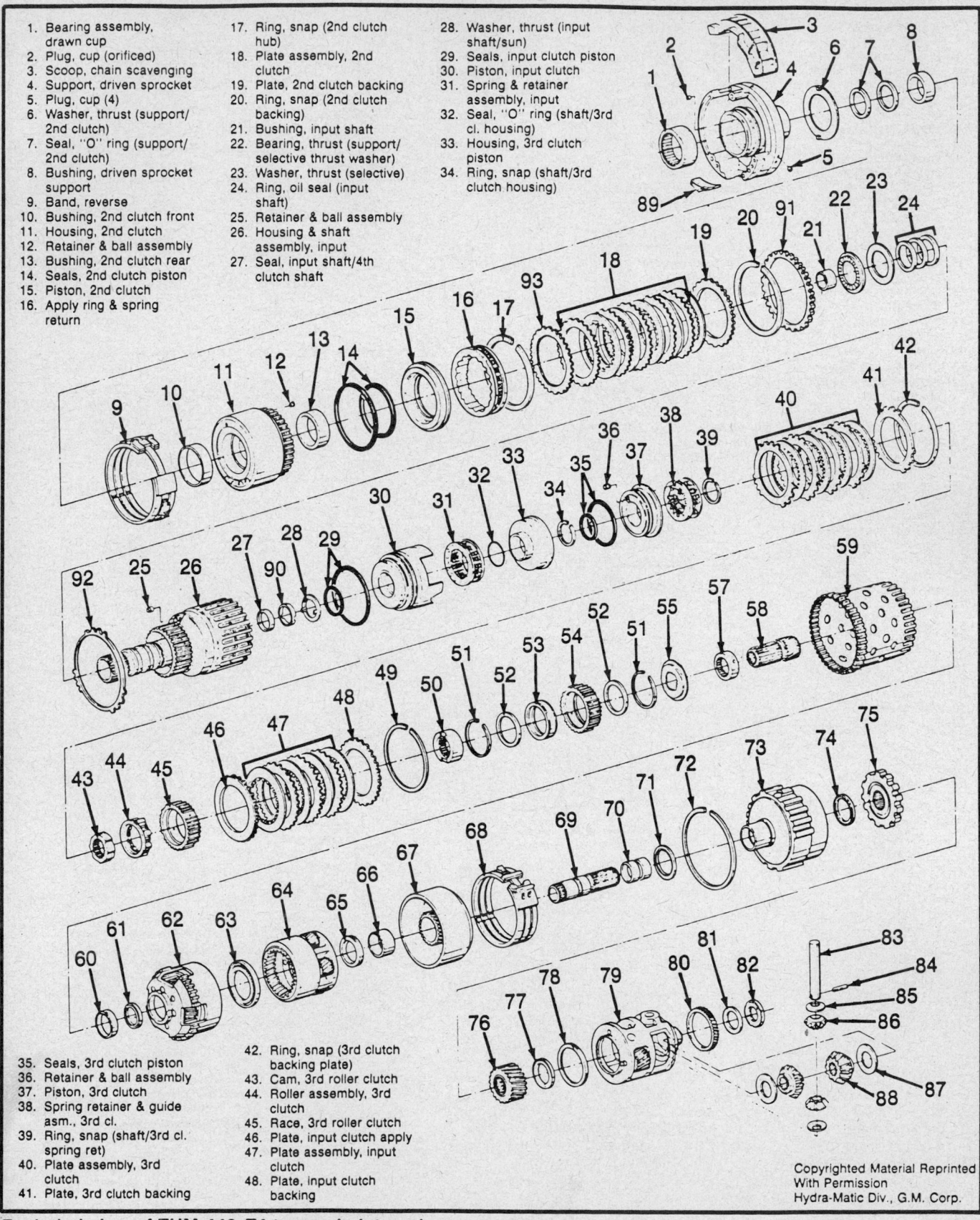

1. Bearing assembly, drawn cup
2. Plug, cup (orificed)
3. Scoop, chain scavenging
4. Support, driven sprocket
5. Plug, cup (4)
6. Washer, thrust (support/2nd clutch)
7. Seal, "O" ring (support/2nd clutch)
8. Bushing, driven sprocket support
9. Band, reverse
10. Bushing, 2nd clutch front
11. Housing, 2nd clutch
12. Retainer & ball assembly
13. Bushing, 2nd clutch rear
14. Seals, 2nd clutch piston
15. Piston, 2nd clutch
16. Apply ring & spring return
17. Ring, snap (2nd clutch hub)
18. Plate assembly, 2nd clutch
19. Plate, 2nd clutch backing
20. Ring, snap (2nd clutch backing)
21. Bushing, input shaft
22. Bearing, thrust (support/selective thrust washer)
23. Washer, thrust (selective)
24. Ring, oil seal (input shaft)
25. Retainer & ball assembly
26. Housing & shaft assembly, input
27. Seal, input shaft/4th clutch shaft
28. Washer, thrust (input shaft/sun)
29. Seals, input clutch piston
30. Piston, input clutch
31. Spring & retainer assembly, input
32. Seal, "O" ring (shaft/3rd cl. housing)
33. Housing, 3rd clutch piston
34. Ring, snap (shaft/3rd clutch housing)
35. Seals, 3rd clutch piston
36. Retainer & ball assembly
37. Piston, 3rd clutch
38. Spring retainer & guide asm., 3rd cl.
39. Ring, snap (shaft/3rd cl. spring ret)
40. Plate assembly, 3rd clutch
41. Plate, 3rd clutch backing
42. Ring, snap (3rd clutch backing plate)
43. Cam, 3rd roller clutch
44. Roller assembly, 3rd clutch
45. Race, 3rd roller clutch
46. Plate, input clutch apply
47. Plate assembly, input clutch
48. Plate, input clutch backing

Exploded view of THM 440–T4 transaxle internal components

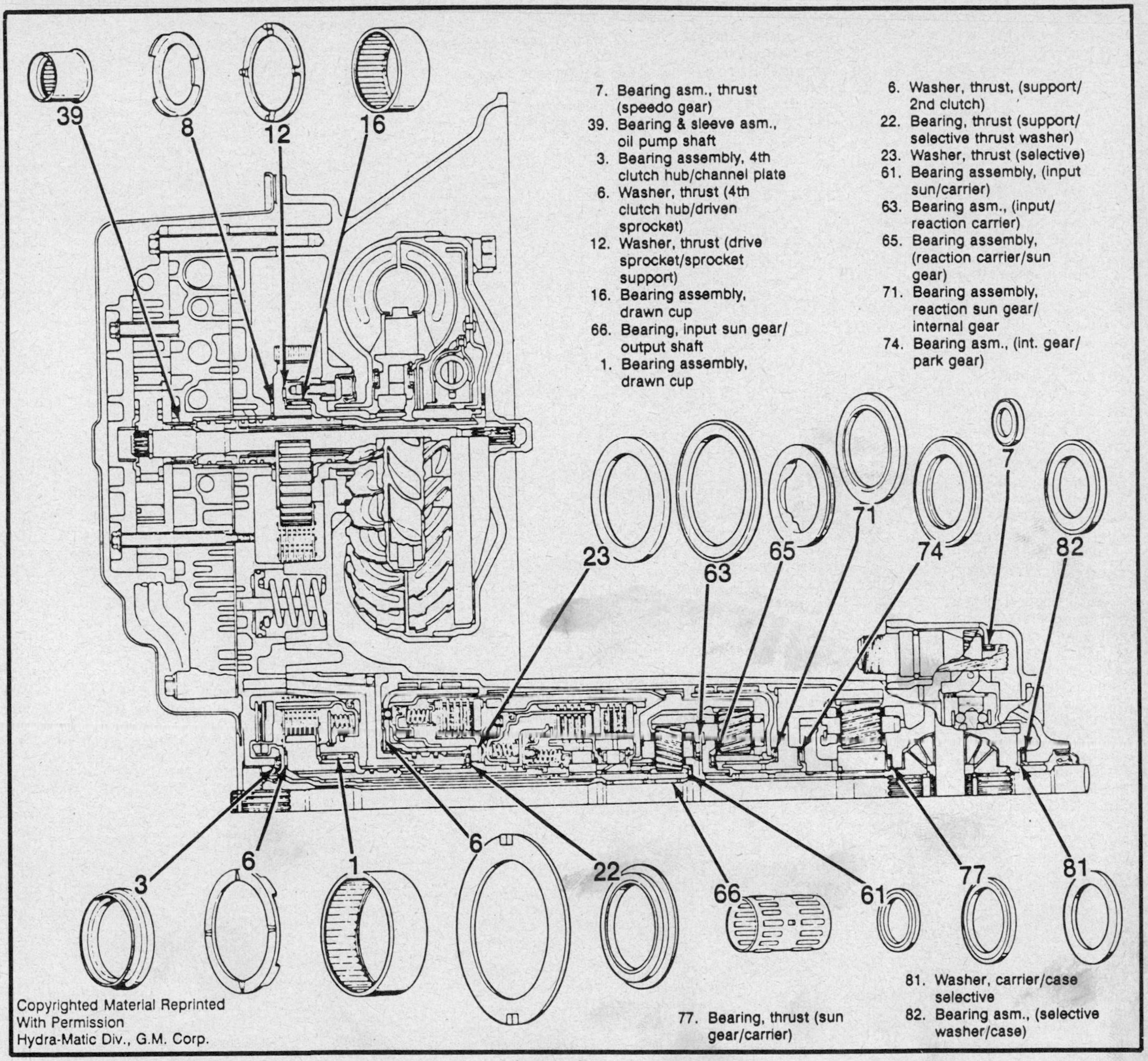
7. Bearing asm., thrust (speedo gear)
39. Bearing & sleeve asm., oil pump shaft
3. Bearing assembly, 4th clutch hub/channel plate
6. Washer, thrust (4th clutch hub/driven sprocket)
12. Washer, thrust (drive sprocket/sprocket support)
16. Bearing assembly, drawn cup
66. Bearing, input sun gear/output shaft
1. Bearing assembly, drawn cup
6. Washer, thrust, (support/2nd clutch)
22. Bearing, thrust (support/selective thrust washer)
23. Washer, thrust (selective)
61. Bearing assembly, (input sun/carrier)
63. Bearing asm., (input/reaction carrier)
65. Bearing assembly, (reaction carrier/sun gear)
71. Bearing assembly, reaction sun gear/internal gear
74. Bearing asm., (int. gear/park gear)
81. Washer, carrier/case selective
82. Bearing asm., (selective washer/case)
77. Bearing, thrust (sun gear/carrier)
39
8
12
16
23
63
65
71
74
7
82
3
6
1
6
22
66
61
77
81

Location of thrust bearings and washers in the drive train

FINAL DRIVE UNIT

Disassembly

1. Remove the final drive gear snapring and lift out the final drive gear unit. Lift out the bearing assembly and the parking gear.
2. Remove the final drive sun gear along with the final drive carrier and the governor drive gear.
3. Remove the carrier washer and the bearing assembly.
4. Place the final drive carrier on its side and remove the differential pinion shaft by tapping out the pinion shaft retaining pin.
5. Remove the pinion thrust washer and the differential pinion. Remove the 2 side gear thrust washers and the 2 differential side gears.

Inspection

1. Clean all parts well and check for damage or excessive wear. Check the gears for burrs and cracks.
2. Inspect the washers and replace any that appear damaged or warped.
3. Check the bearing assemblies for any mutilation and replace as needed.

Assembly

1. Install the differential side gears and washers into the final drive carrier.
2. Install the pinion thrust washers onto the differential pinions. A small amount of petroleum jelly can be used to hold the washers in place.
3. Place the pinions and washers into the final drive carrier.

1. Converter assembly
2. Bushing, converter pump
3. Sleeve, governor shaft
4. Ring, oil seal (governor shaft)
5. Governor assembly
6. Gear, speedometer drive
7. Bearing asm., thrust (speedo gear)
8. Seal, "O" ring (governor cover)
9. Cover, governor
10. Screw, governor cover/ case
11. Gear, speedo driven
12. Seal, "O" ring
13. Sleeve, speedo driven gear
14. Retainer, speedo driven gear
15. Bolt, speedo gear retaining
16. Case assembly
17. Valve, modulator
18. Seal, "O" ring
19. Modulator assembly
20. Retainer, modulator
21. Bolt, (modulator)
23. Electrical connector
24. Channel, plate assembly
25. Ball, check valve (7)
26. Bolt, channel plate to case (5)
27. Bolt, channel plate to driven support (6)
28. Ring, oil seal (oil pump)
29. Shaft, oil pump drive
30. Gasket, spacer plate/ channel plate
31. Plate, valve body spacer
32. Gasket, spacer plate/ valve body
33. Screen asm., conv. clutch solenoid
34. Ball, check valve (5)
35. Valve assembly, control
38. Screen assembly, oil pump pressure
39. Bearing & sleeve asm., oil pump shaft
40. Bolt, V.B. to C.P. (torque head) (6)
41. Bolt, valve body to C.P. (hex) (1)
42. Bolt, V.B. to driven support (torque) (2)
43. Bolt, valve body to case (hex) (3)
44. Pump assembly
45. Bolt, pump body to case (hex) (2)
46. Bolt, pump cover to C.P. (hex) (10)
47. Bolt, pump cover to valve body (hex) (1)
48. Harness, wiring
49. Link throttle lever to cable
50. Lever & bracket assembly, throttle
51. Pan, case side cover
52. Screw, special M8x1.25x16.0
53. Nut, flanged hex (M6x1.0)
54. Bolt, M6x1.0x35 LG. P.B./C.P. hex (1)
55. Bolt, M6x1.0x45 LG. V.B./C.P. (2)
56. Washer, conical
57. Seal assembly, oil filter
58. Filter assembly, oil
59. Pan, transmission oil
60. Screw, special M8x1.25x16.0)
61. Wire conduit
62. Clip, two wire
63. Gasket, transmission oil pan
64. Gasket, side cover to case
65. Gasket, side cover to channel plate
66. Magnet, chip collector

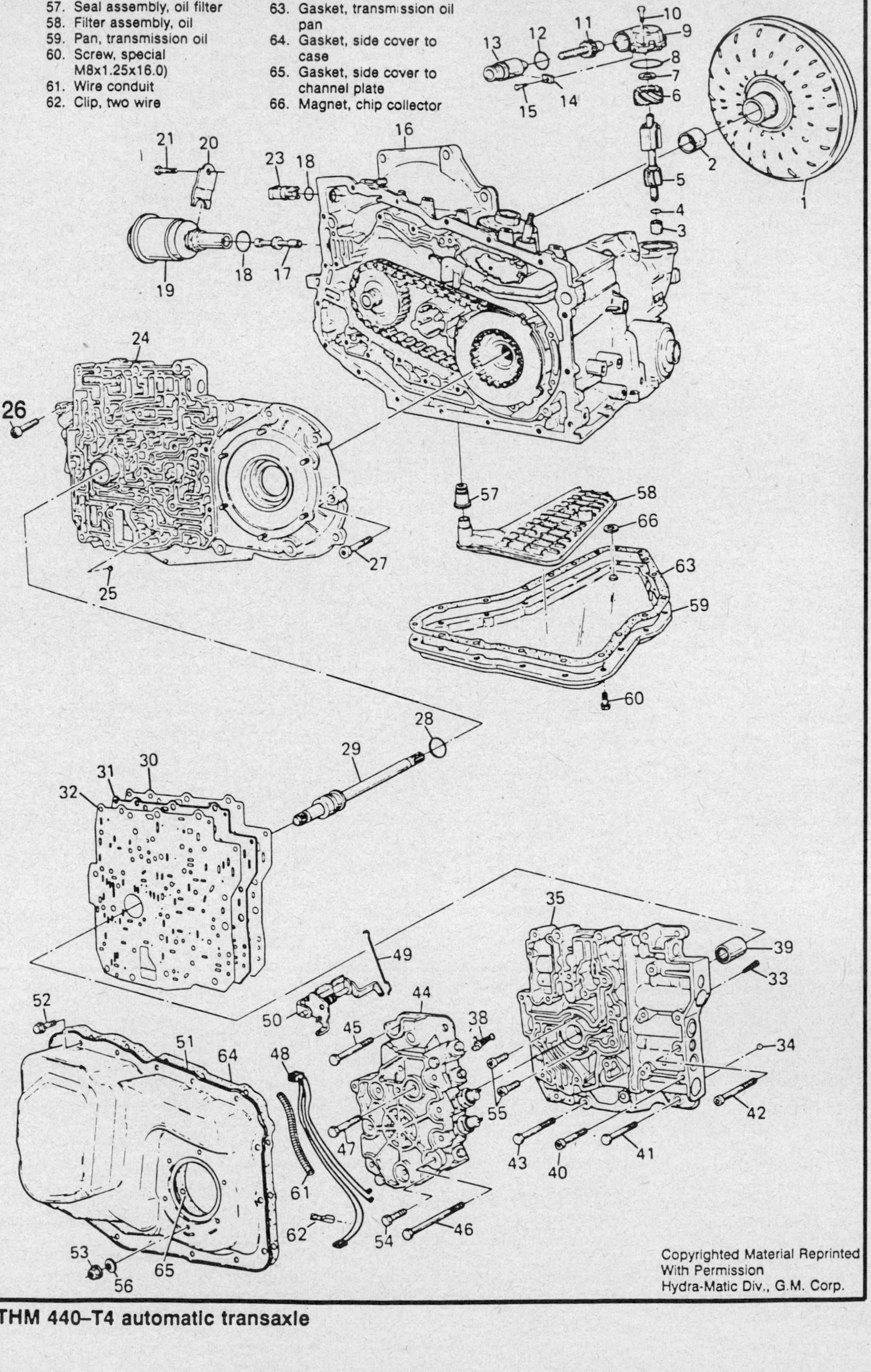

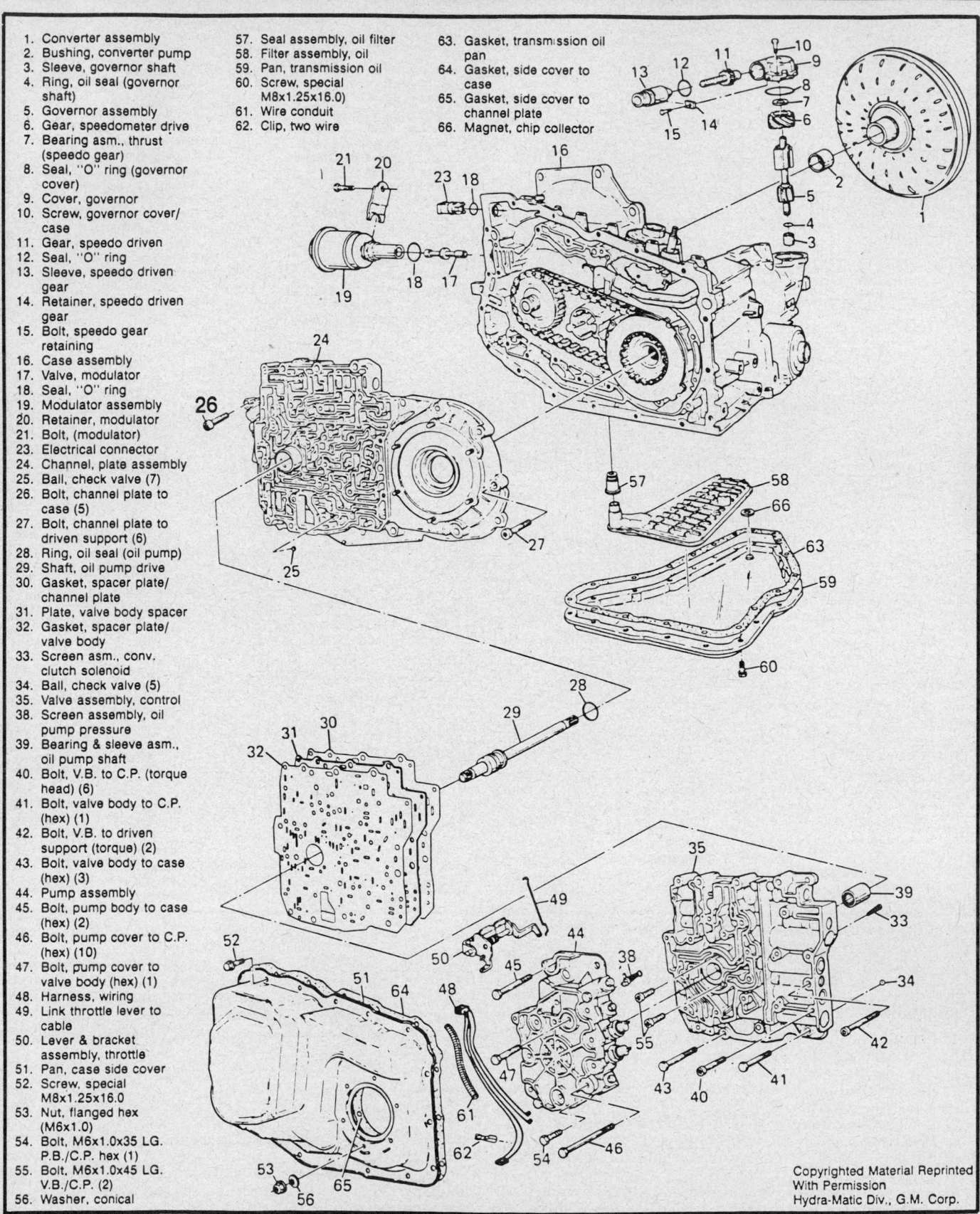

Exploded view of the THM 440–T4 automatic transaxle

4. Insert the differential pinion shaft into the final drive carrier to check alignment of the pinions, then remove.

5. If the pinions are out of alignment, correct and reinstall the differential pinion shaft. Tap the pinion shaft retaining pin into position.

6. Assemble the sun gear into the final drive carrier with the stopped side facing out. Install the parking gear onto the sun gear.

7. Install the thrust bearing assembly into the final drive internal gear and place the unit onto the final drive carrier.

8. Install the thrust washer and the thrust bearing onto the carrier hub. Install the snapring making sure it seats properly in its groove.

FINAL DRIVE SUN GEAR SHAFT

Disassembly

1. Lift off the reverse reaction drum and the input carrier assembly.

2. Remove the reaction carrier bearing, which may be stuck to the input carrier assembly.

3. Remove the internal gear bearing and the final drive sun gear.

4. Lift out the 1-2 band from the reaction sun gear and drum assembly.

5. Remove the bearing assembly from the reaction sun gear and drum assembly.

Inspection

1. Inspect the final drive sun gear shaft for damaged splines. Replace if necessary.

2. Inspect the 1-2 band assembly for damage from heat or excessive wear. Check the band assembly for lining separation or lining cracks.

NOTE: The 1-2 band assembly is presoaked in a friction solution and should not be washed in a cleaning solvent.

3. Inspect the sun gear/drum assembly for any scoring or damaged teeth.

4. Inspect the thrust bearings for damage. Replace as required.

5. Check the reaction carrier assembly for pinion endplay. Pinion endplay should be 0.23–0.61mm.

6. Check for pinion damage or internal gear damage. Replace as needed.

Assembly

1. Position the inside race of the thrust bearing against the final drive internal gear.

2. Install the reaction sun gear and drum assembly into the case.

3. Position the thrust bearing inside the reaction carrier and retain with petroleum jelly.

4. Install the reaction carrier and rotate until all the pinions engage with the sun gear.

5. Install the reverse reaction drum making sure that all the spline teeth engage with the input carrier.

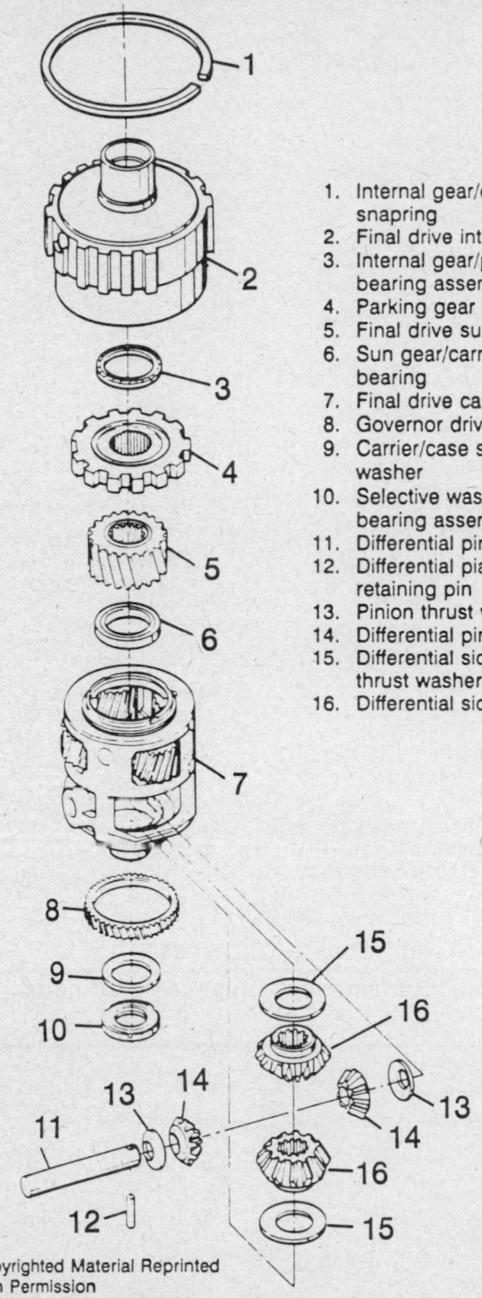
1
2
3
4
5
6
7
8
9
10
11
12
13
14
15
16
13
14
15
16

1. Internal gear/case snapring
2. Final drive internal gear
3. Internal gear/park gear bearing assembly
4. Parking gear
5. Final drive sun gear
6. Sun gear/carrier thrust bearing
7. Final drive carrier
8. Governor drive gear
9. Carrier/case selective washer
10. Selective washer/case bearing assembly
11. Differential pinion shaft
12. Differential pianion shaft retaining pin
13. Pinion thrust washer
14. Differential pinion
15. Differential side gear thrust washer
16. Differential side gear

Copyrighted Material Reprinted With Permission Hydra-Matic Div., G.M. Corp.

Final drive assembly

THIRD ROLLER CLUTCH AND INPUT SUN GEAR AND SPRAG

Disassembly

1. Remove the input sprag and 3rd roller clutch from the input sun gear.

2. Remove the input sun gear spacer and retainer from the sun gear.

3. Remove the snapring and lift off the input sprag wear plate and the 3rd roller clutch race and cam from the roller assembly.

4. Disassemble the input sprag assembly by removing the inner race from the sprag assembly.

5. Remove the snapring that holds the wear plate and lift out the wear plate and sprag assembly.

Inspection

1. Clean all parts in cleaning solvent and blow dry using compressed air.

2. Inspect the outer race and roller cam for any cracks or scoring. Replace as required.

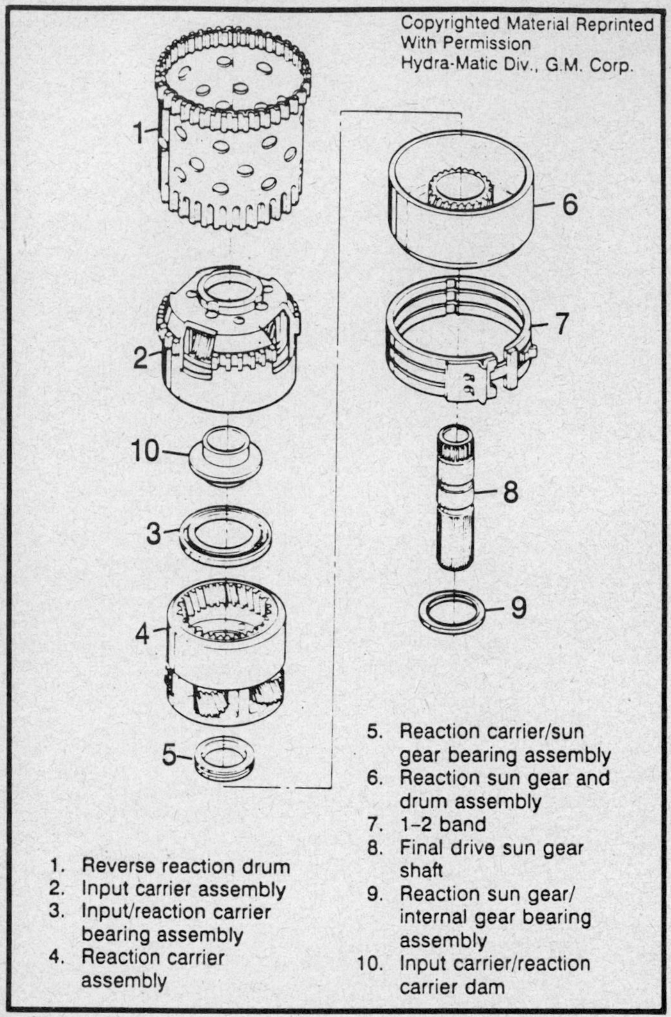
Copyrighted Material Reprinted With Permission Hydra-Matic Div., G.M. Corp.
1
2
10
3
4
5
6
7
8
9
1. Reverse reaction drum
2. Input carrier assembly
3. Input/reaction carrier bearing assembly
4. Reaction carrier assembly
5. Reaction carrier/sun gear bearing assembly
6. Reaction sun gear and drum assembly
7. 1-2 band
8. Final drive sun gear shaft
9. Reaction sun gear/internal gear bearing assembly
10. Input carrier/reaction carrier dam

Input and reaction carriers and associated parts

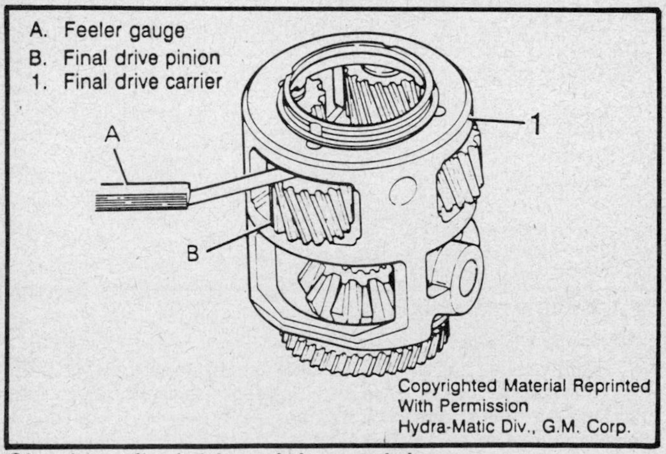
A. Feeler gauge
B. Final drive pinion
1. Final drive carrier
A
B
1
Copyrighted Material Reprinted With Permission Hydra-Matic Div., G.M. Corp.

Checking final drive pinion endplay

3. Inspect the roller assembly for damaged rollers and springs. Replace any loose rollers by depressing the spring and inserting the roller.
4. Inspect the sprag assembly for damaged sprags or cages and replace as required.
5. Inspect the inner race and wear plates for any scoring or damage. Replace as required.

Assembly

1. Position 1 wear plate against the snapring and hold in position with petroleum jelly.
2. Insert the wear plate with the snapring against the sprag assembly.
3. Install the spacer on the input sun gear and place the sprag retainer over the spacer.
4. Install the sprag snapring. Make sure the snapring seats properly.
5. Install the sprag assembly and the roller clutch onto the sun gear.

INPUT CLUTCH ASSEMBLY

Disassembly

1. Remove the input clutch snapring and remove the input clutch backing plate.
2. Remove the steel and composition clutch plates, along with the input clutch apply plate.
3. Remove the 3rd clutch snapring and remove the 3rd clutch backing plate.
4. Remove the steel and composition clutch plates, along with the 3rd clutch waved plate.
5. Remove the snapring from the spring retainer and lift out the 3rd clutch piston from its housing.
6. Remove the 3rd clutch piston inner seal from the shaft.
7. Compress the 3rd clutch piston housing and remove the snapring. Remove the 3rd clutch piston housing.
8. Remove the O-ring seal and take out the spring retainer. Remove the input clutch piston and inner seal.

Inspection

1. Wash all parts in cleaning solvent and blow dry using compressed air.
2. Inspect all parts for scoring, wear or damage.
3. Inspect the input clutch housing for damaged or worn bushings.
4. Check the 4th clutch shaft seal and replace if damaged or cut.
5. Repair or replace any parts found to be defective.

Assembly

1. Lubricate all parts with automatic transmission fluid prior to assembling.
2. Install the input clutch piston seal with the piston seal protector. Position the input piston in the housing.
3. Install the O-ring on the input shaft making sure it seats properly. Install the spring retainer in the piston.
4. Install the 3rd clutch piston housing into the input housing. Compress the 3rd clutch housing with a clutch spring compressor and install the snapring.
5. Install the 3rd clutch inner seal on the 3rd clutch piston and install the 3rd clutch piston.
6. Install the 3rd clutch spring retainer and compress the spring retainer using the clutch spring compressor and install the snapring.
7. Install the 3rd clutch plates and the 3rd clutch backing plate. install the snapring.

NOTE: When installing the 3rd clutch plates, start with a steel plate and alternate between composition and steel plates. When installing the 3rd clutch backing plate, make sure that the stepped side is facing up.

8. Install the input clutch apply plate with the notched side facing the snapring. Install the input clutch plates.

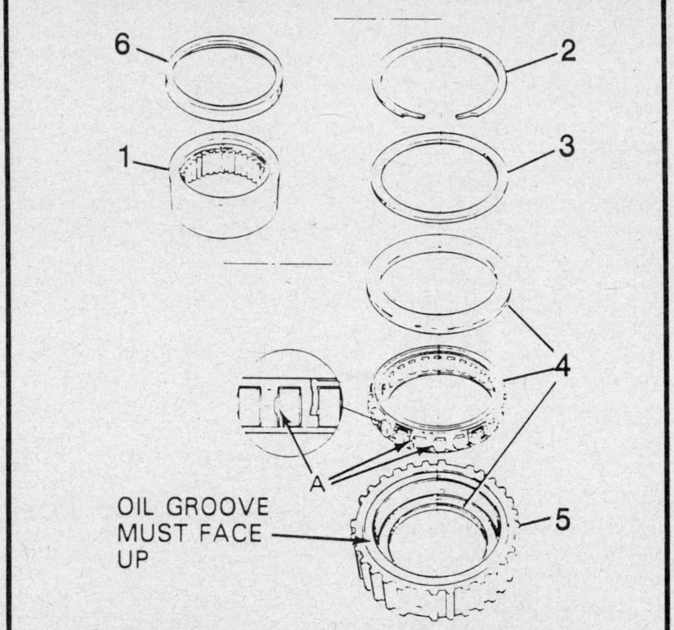
OIL GROOVE
MUST FACE
UP
A

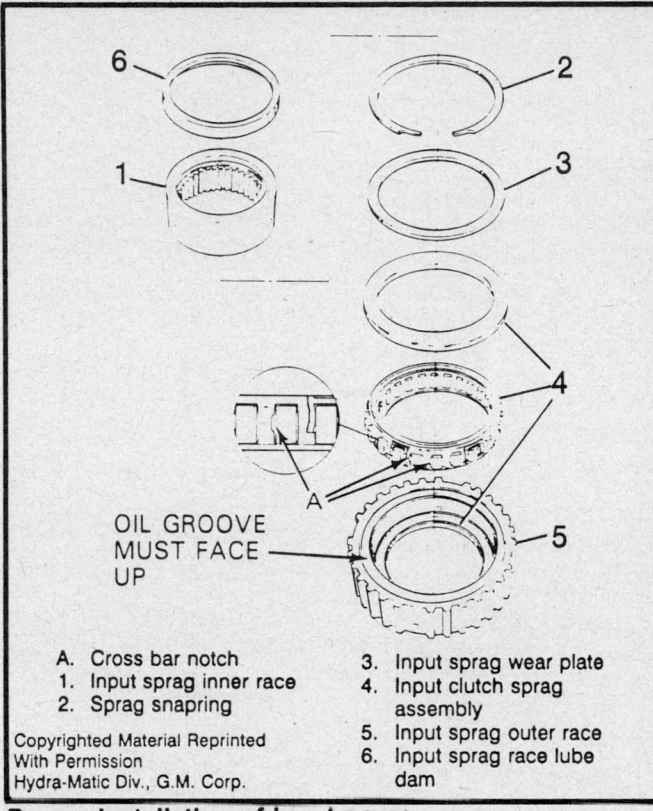

A. Cross bar notch
1. Input sprag inner race
2. Sprag snapring
3. Input sprag wear plate
4. Input clutch sprag assembly
5. Input sprag outer race
6. Input sprag race lube dam

Proper installation of input sprag

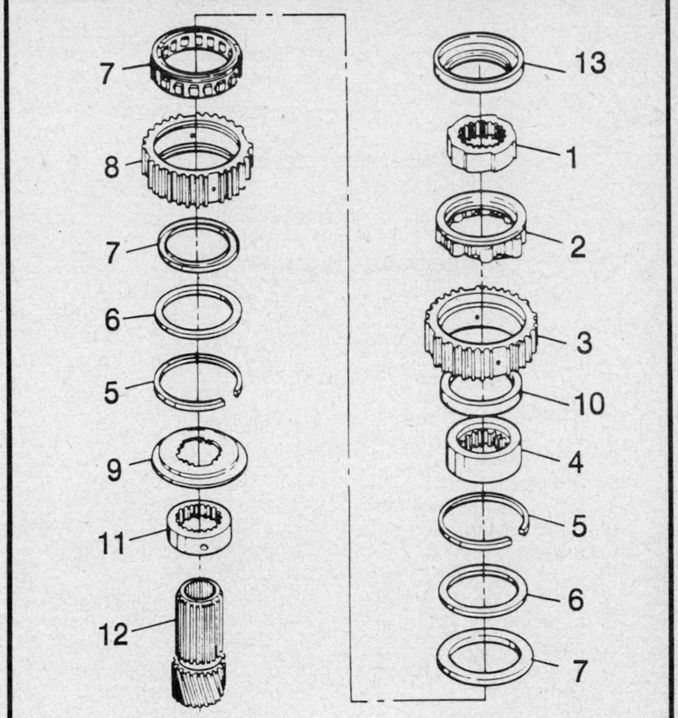

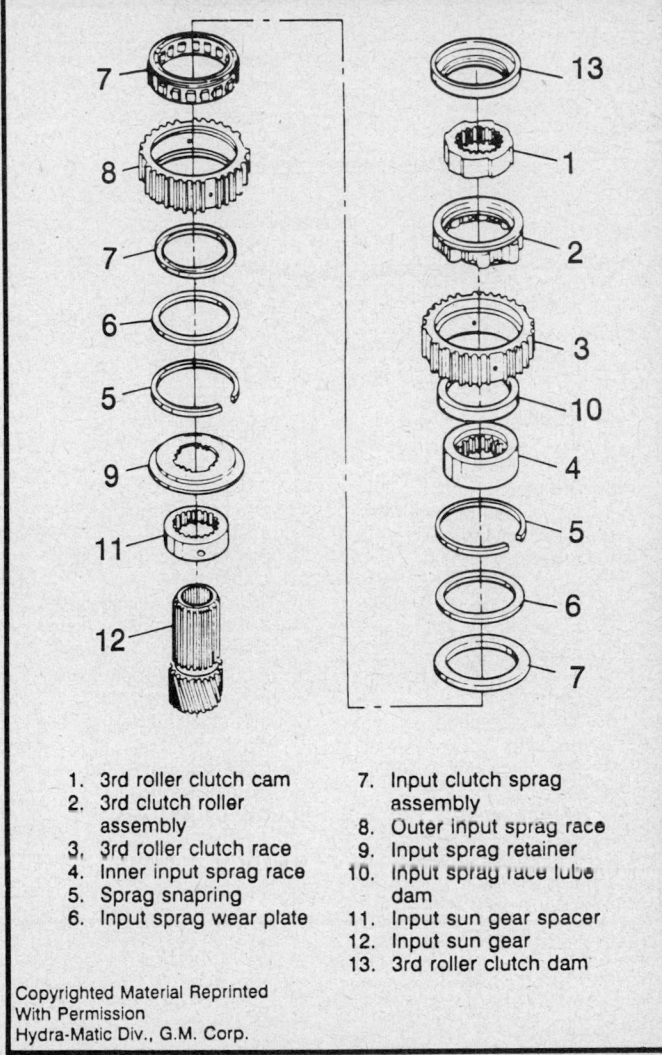

1. 3rd roller clutch cam
2. 3rd clutch roller assembly
3. 3rd roller clutch race
4. Inner input sprag race
5. Sprag snapring
6. Input sprag wear plate
7. Input clutch sprag assembly
8. Outer input sprag race
9. Input sprag retainer
10. Input sprag race lube dam
11. Input sun gear spacer
12. Input sun gear
13. 3rd roller clutch dam

Sprag and roller clutch assembly

NOTE: When installing the input clutch plates, start with a composition plate and alternate between steel and composition plates.

9. install the input clutch backing plate and secure with a snapring.

RETAINER AND BALL ASSEMBLY

Replacement (Optional)

1. Remove the retainer and ball assembly from the housing with a $^3/_8$ in. (9.5mm) drift.
2. Tap in a new retainer using a $^3/_8$ in. (9.5mm) drift.

PISTON SEAL

Replacement (Optional)

1. Remove the input clutch piston seal and the 3rd clutch piston seal.
2. Inspect the clutch piston for any remaining seals and remove.
3. Install a new input clutch piston seal and a new 3rd clutch piston seal. Lubricate with transmission fluid.

FOURTH CLUTCH SHAFT SEAL

Replacement (Optional)

1. Remove the lock up sleeve using a suitable tool and remove the oil seal.
2. Install the new oil seal into the input shaft making sure that the seal tab aligns with the slot in the shaft.
3. Install the lock up sleeve in the shaft using a bench press.

INPUT SHAFT SEAL

Replacement (Optional)

1. Remove the seal rings from the input shaft with a suitable tool.
2. Adjust the seal protector so that the bottom matches the seal ring groove.
3. Lubricate the oil seal ring and place it on the seal protector.
4. Slide the seal into position with the seal pusher over the seal protector.
5. Size the seal with a seal sizer and gently work the tool over the seal with a twisted motion.

SECOND CLUTCH HOUSING

Disassembly

1. Remove the 2nd clutch hub snapring and lift out the 2nd clutch wave plate.

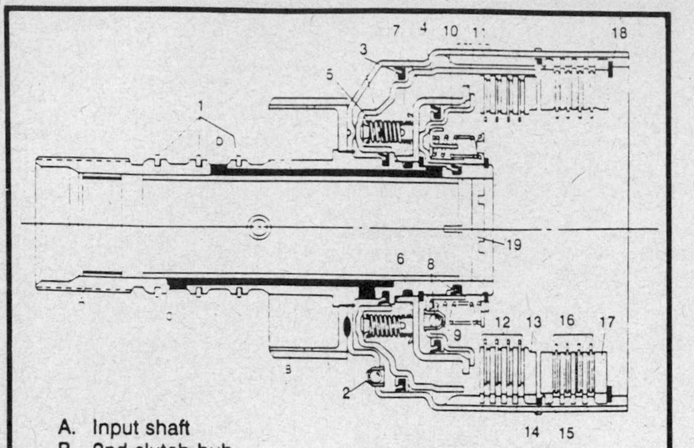

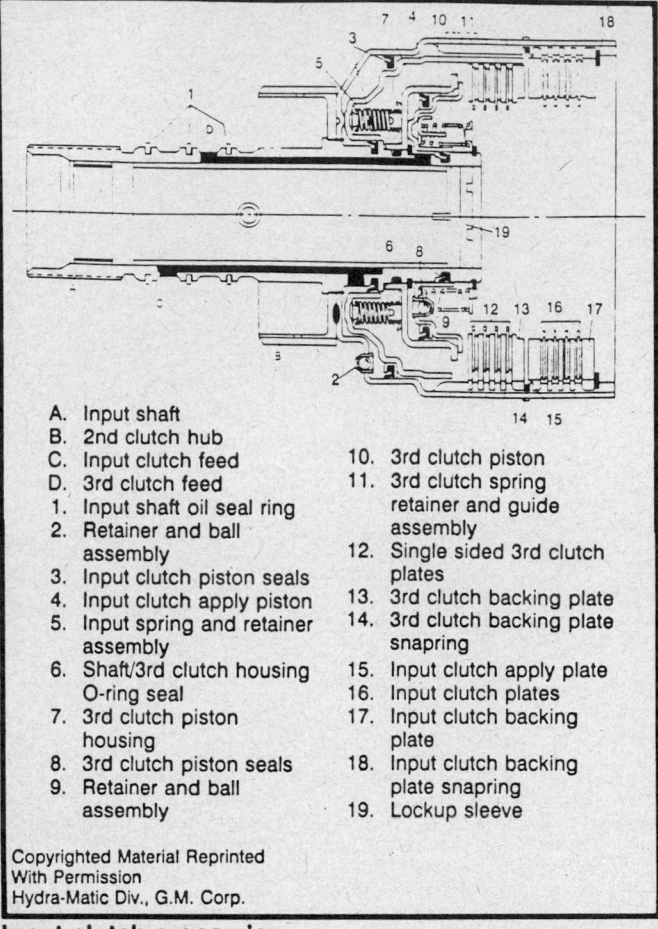

A. Input shaft
B. 2nd clutch hub
C. Input clutch feed
D. 3rd clutch feed
1. Input shaft oil seal ring
2. Retainer and ball assembly
3. Input clutch piston seals
4. Input clutch apply piston
5. Input spring and retainer assembly
6. Shaft/3rd clutch housing O-ring seal
7. 3rd clutch piston housing
8. 3rd clutch piston seals
9. Retainer and ball assembly
10. 3rd clutch piston
11. 3rd clutch spring retainer and guide assembly
12. Single sided 3rd clutch plates
13. 3rd clutch backing plate
14. 3rd clutch backing plate snapring
15. Input clutch apply plate
16. Input clutch plates
17. Input clutch backing plate
18. Input clutch backing plate snapring
19. Lockup sleeve

Input clutch cross view

2. Remove the 2nd clutch plate assembly and the 2nd clutch backing plate.
3. Remove the next snapring and remove the 2nd clutch housing support.

4. Remove the thrust bearing and thrust washer. Remove the reverse band.
5. Remove the 2nd clutch housing. Remove the 2nd clutch piston seals which may be stuck on the 2nd clutch housing.
6. Remove the 2nd clutch piston and the spring return apply ring.

Inspection

1. Wash all parts in cleaning solvent and blow dry using compressed air.
2. Inspect all parts for scoring, wear or damage.
3. Inspect the 2nd clutch piston and seal for damage or warping.
4. Repair or replace damaged parts as required.

Assembly

1. Lubricate a new piston seal with automatic transmission fluid and install in the 2nd clutch piston.
2. Install a new retainer and ball assembly into the 2nd clutch housing using a $^3/_8$ in. (9.5mm) drift.

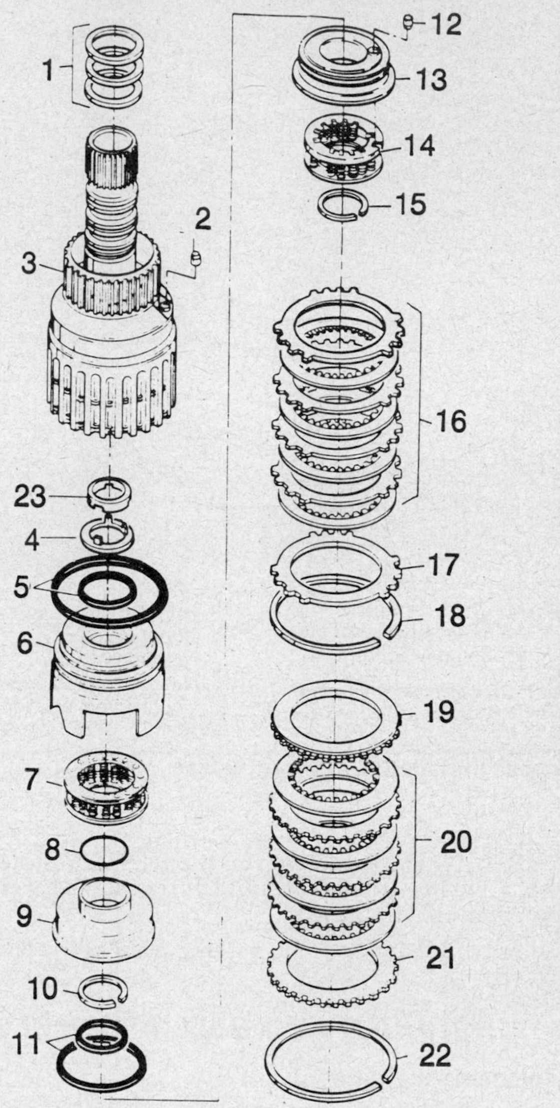

1. Input shaft oil seal ring
2. Retainer and ball assembly
3. Input housing and shaft assembly
4. Input shaft/sun thrust washer
5. Input clutch piston seals
6. Input clutch piston
7. Input spring and retainer assembly
8. Shaft/3rd clutch housing O-ring seal
9. 3rd clutch piston housing
10. Shaft/3rd clutch housing O-ring seal
11. 3rd clutch piston seals
12. Retainer and ball assembly
13. 3rd clutch piston
14. 3rd clutch spring retainer and guide assembly
15. Shaft/3rd clutch spring retainer snapring
16. 3rd clutch plate assembly
17. 3rd clutch backing plate
18. 3rd clutch backing plate snapring
19. Input clutch apply plate
20. Input clutch plate assembly
21. Input clutch backing plate
22. Input clutch backing plate snapring
23. Lockup sleeve

Input clutch assembly

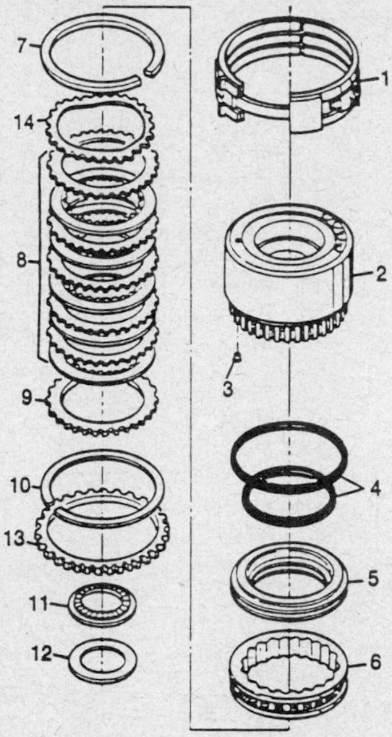

1. Reverse band
2. 2nd clutch housing
3. Retainer and ball assembly
4. 2nd clutch piston seals
5. 2nd clutch piston
6. Apply ring and spring return
7. 2nd clutch hub snapring
8. 2nd clutch plate assembly
9. 2nd clutch backing plate
10. 2nd clutch backing snapring
11. Support/selective thrust washer thrust bearing
12. Selective thrust washer
13. 2nd clutch housing support
14. 2nd clutch waved plate

2nd clutch assembly – 1984–86

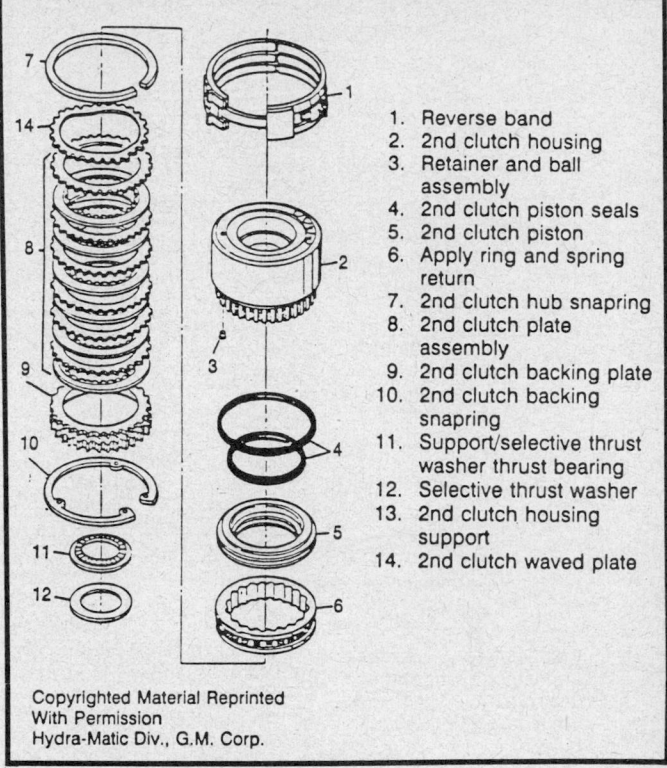

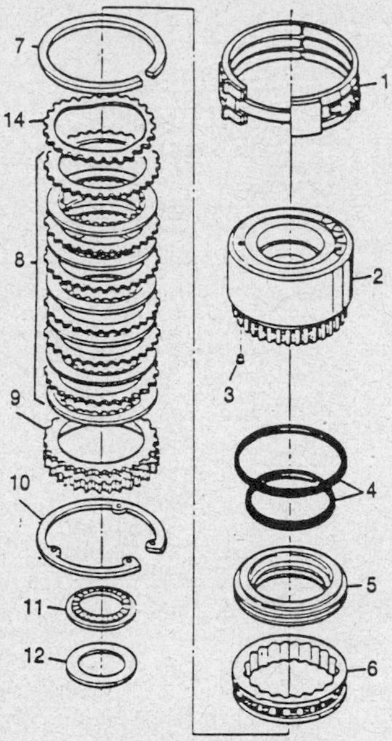

1. Reverse band
2. 2nd clutch housing
3. Retainer and ball assembly
4. 2nd clutch piston seals
5. 2nd clutch piston
6. Apply ring and spring return
7. 2nd clutch hub snapring
8. 2nd clutch plate assembly
9. 2nd clutch backing plate
10. 2nd clutch backing snapring
11. Support/selective thrust washer thrust bearing
12. Selective thrust washer
13. 2nd clutch housing support
14. 2nd clutch waved plate

2nd clutch assembly – 1987–89

3. Install the 2nd clutch piston into the 2nd clutch housing. Make sure the seals are not damaged.
4. Install the spring return apply ring into the 2nd clutch housing. Using a spring compressor, compress the spring return apply ring and insert the snapring.
5. Install the 2nd clutch plates starting with steel and alternating with composition plates.

NOTE: The 2nd clutch composition plates are presoaked and do not require soaking in solvent.

6. Install the reverse band and the thrust bearing and thrust washers.
7. Install the clutch housing support and secure with a snapring.
8. Install the backing plate and the 2nd clutch plates. Install the waved plate and the snapring.

DRIVEN SPROCKET SUPPORT

Disassembly

1. Using a suitable tool, compress the 4th clutch spring assembly and remove the snapring.
2. Remove the 4th clutch piston and the piston seals. Discard the piston seals.
3. Remove the bearing assembly from the driven sprocket support.
4. Remove the chain scavenging scoop and the oil reservoir wire.
5. Remove the thrust washer and the O-ring seals.

Inspection

1. Clean all parts thoroughly in solvent and blow dry using compressed air.
2. Inspect the driven sprocket support for cracks or damage.
3. Inspect the seals and pistons for damage. Replace as necessary.
4. Check the spring retainer for distorted or damaged springs.
5. Inspect the oil reservoir weir and the chain scavenging scoop for cracks or damage.
6. Repair or replace any damaged parts as required.

Assembly

1. Lubricate new O-rings and install them behind the thrust washer on the driven sprocket support.
2. Install the oil reservoir weir and the chain scavenging scoop on the driven sprocket support.
3. Install the bearing assembly on the driven sprocket support, between the oil reservoir weir and the chain scavenging scoop.
4. Install new seals on the 4th clutch piston and install the 4th clutch piston.
5. Install the 4th clutch spring assembly and using a suitable tool, compress the springs and insert the snapring.

OIL PUMP

Disassembly

1. Remove the pump bolts from the cover. Lift off the pump cover. Remove the pump cover sleeve and the vane ring.
2. Remove the pump rotor and the bottom vane ring. Remove the oil seal ring and the O-ring from the oil pump slide.
3. Remove the oil pump slide. Enclosed in the oil pump slide are the pump slide seal and support along with the inner and outer pump priming springs.

NOTE: The pump rotor, vane rings and oil pump slide are factory matched units. Therefore, if any parts need replacing, all parts must be replaced.

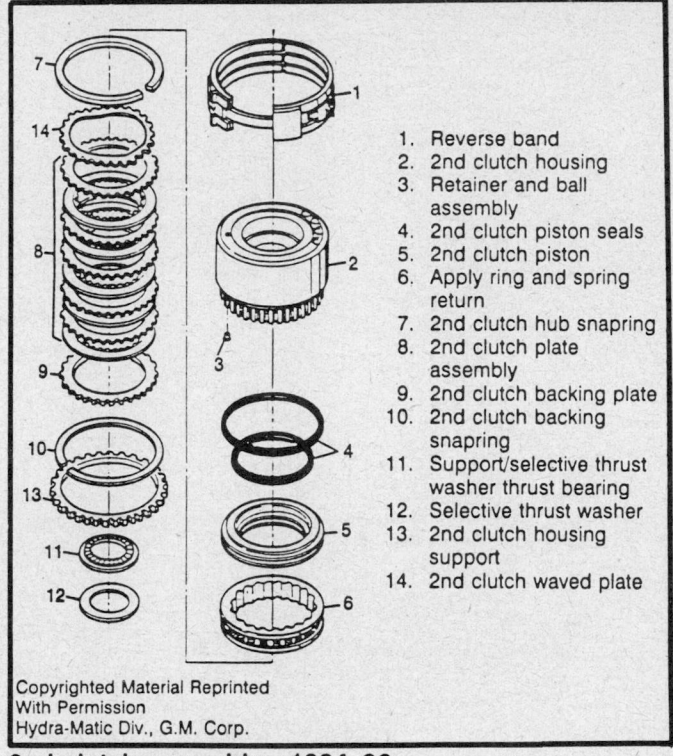

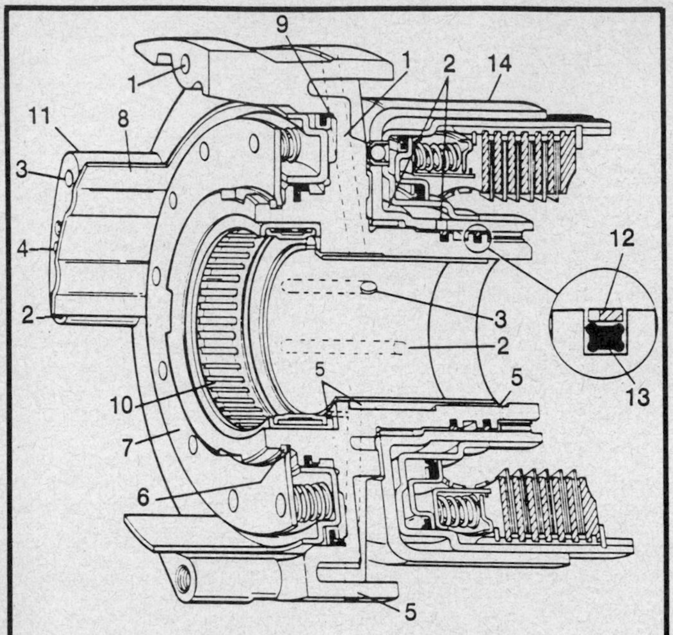

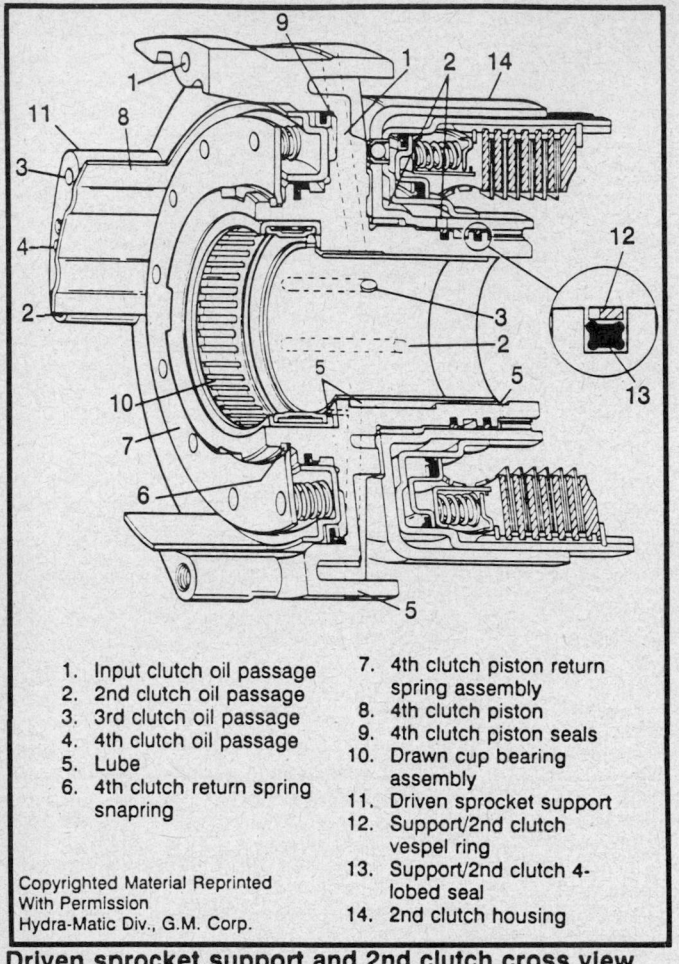

1. Input clutch oil passage
2. 2nd clutch oil passage
3. 3rd clutch oil passage
4. 4th clutch oil passage
5. Lube
6. 4th clutch return spring snapring
7. 4th clutch piston return spring assembly
8. 4th clutch piston
9. 4th clutch piston seals
10. Drawn cup bearing assembly
11. Driven sprocket support
12. Support/2nd clutch vespel ring
13. Support/2nd clutch 4-lobed seal
14. 2nd clutch housing

Driven sprocket support and 2nd clutch cross view

4. Remove the pivot pin and the roll pin from the oil pump body.

5. Remove the 3–2 coast down valve, spring and bore plug from the oil pump body. Remove the oil pressure switches.

Inspection

1. Clean all parts thoroughly in solvent and blow dry using compressed air.
2. Check the pump body for warping or cracks. Make sure that the oil passages are free of debris.
3. Check the oil pump slide and springs for excessive wear.
4. Check the rotor and vanes for any damage. Inspect the pump slide seal and support for any cracks.
5. Repair or replace damaged parts as needed.

Assembly

1. Install the 3–2 coast down valve along with the spring and bore plug into the oil pump body.
2. Install the lower vane ring into the pump pocket. Install the pump slide into the pump pocket being careful not to dislodge the lower vane ring.
3. Install the pump slide seal and support into the oil pump.
4. Insert the inner priming spring into the outer priming spring. Press the springs into the pump body.
5. Install the O-ring seal onto the oil pump slide. Install the oil seal ring onto the oil pump slide.
6. Install the oil pump rotor onto the oil pump body. Insert the pump vanes into the oil pump rotor. Install the upper vane ring onto the oil pump rotor.

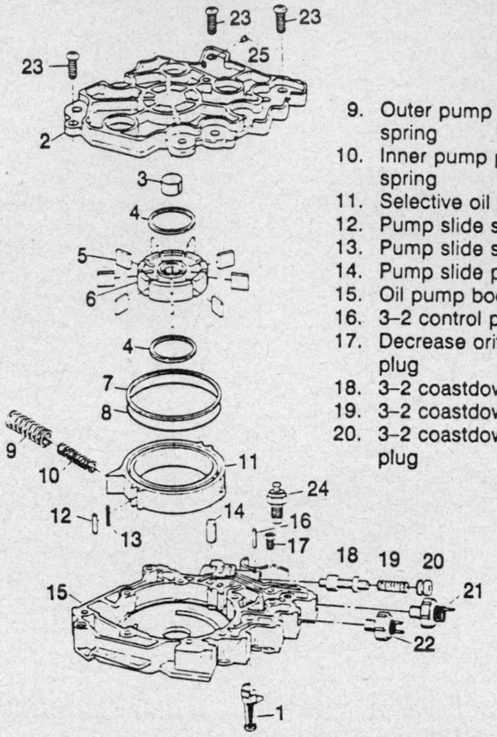

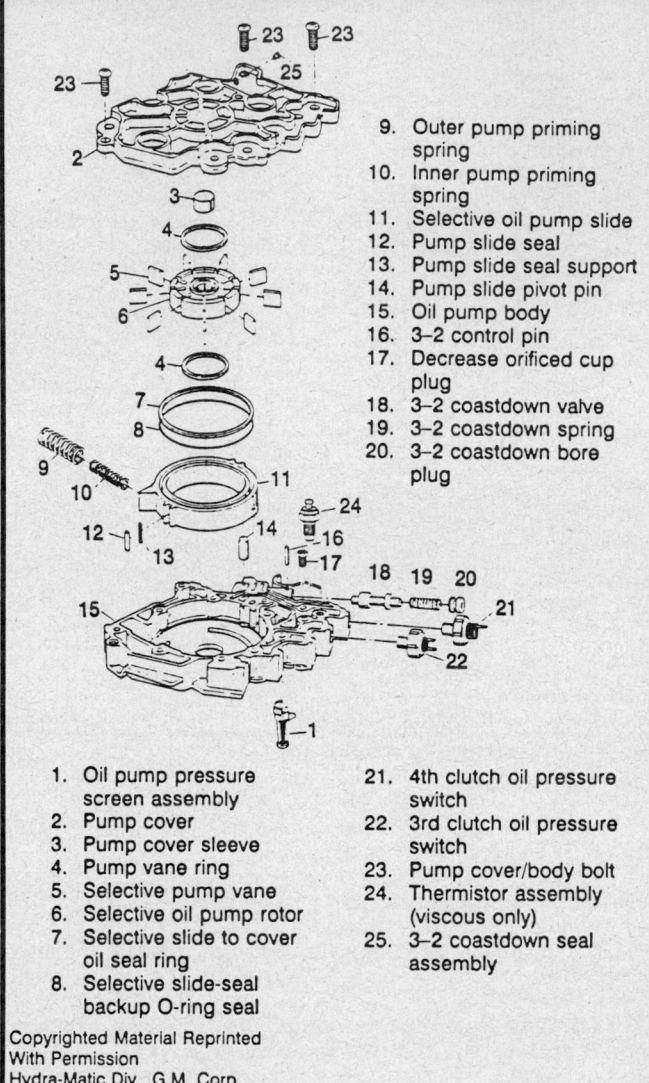

1. Oil pump pressure screen assembly
2. Pump cover
3. Pump cover sleeve
4. Pump vane ring
5. Selective pump vane
6. Selective oil pump rotor
7. Selective slide to cover oil seal ring
8. Selective slide-seal backup O-ring seal
9. Outer pump priming spring
10. Inner pump priming spring
11. Selective oil pump slide
12. Pump slide seal
13. Pump slide seal support
14. Pump slide pivot pin
15. Oil pump body
16. 3–2 control pin
17. Decrease orificed cup plug
18. 3–2 coastdown valve
19. 3–2 coastdown spring
20. 3–2 coastdown bore plug
21. 4th clutch oil pressure switch
22. 3rd clutch oil pressure switch
23. Pump cover/body bolt
24. Thermistor assembly (viscous only)
25. 3–2 coastdown seal assembly

Oil pump assembly

NOTE: The pump vanes must be installed flush with the top of the oil pump rotor.

7. Install the pump cover onto the oil pump body. Install the cover bolts.
8. Install the oil pressure switches into the oil pump body.

CONTROL VALVE

Disassembly

NOTE: As each part of the valve train is removed, place the pieces in order and in a position that is relative to the position on the valve body to lessen the chances for error in assembly. None of the valves, springs or bushings are interchangeable.

1. Lay the valve body on a clean work bench with the machined side up and the line boost valve at the top. The line boost valve should be checked for proper operation before its removal. If it is necessary to remove the boost valve, grind the end of a No. 49 drill to a taper (a small Allen wrench an sometimes be

substituted) and lightly tap the drill into the roll pin. Push the line boost valve out of the top of the valve body.

2. The throttle valve should be checked for proper operation before removing it, by pushing the valve against the spring. If it is necessary to remove the throttle valve, first remove the roll pin holding the T.V. plunger bushing and pull out the plunger and bushing. Remove the throttle valve spring. Remove the blind hole roll pin using the drill method.

3. On the same side of the valve body, move to the next bore down. Remove the straight pin and remove the reverse boost valve and bushing, as well as revere boost spring and the pressure regulator assembly.

NOTE: Transaxle on vehicles equipped with a diesel engine will not have a reverse boost spring or a pressure regulator modulator spring.

4. Move to the other side of the valve body and from the top bore remove the 1–2 throttle valve bushing retainer and lift out the throttle valve bushing and the 1–2 shift valve assembly.

5. Move to the next bore and remove the spring pin and lift out the 2–3 accumulator bushing assembly.

6. At the next bore down, remove the 3–2 control sleeve and the 3–2 control valve assembly.

7. Move to the next bore and remove the spring pin. Remove the 2–3 throttle valve bushing and its components. On Allante, remove the 2–3 shift solenoid assembly.

8. Move to the next bore and remove the coiled spring pin. Remove the 3–4 throttle valve assembly. On Allante, remove the 3–4 shift solenoid assembly.

9. The 2 remaining bores contain the 1–2 servo pipe lip seals, which are removed.

10. At the side of the valve body, are the servo assemblies. Remove the coiled spring pins and detach the servo valves and springs.

Inspection

1. Wash all bushings, springs and valves in solvent. Blow dry using compressed air.

2. Inspect all valves and bushings for any scoring or scratches.

3. Check the springs for collapsed coils and bore plugs for damage.

4. Repair or replace any defective parts as needed.

Assembly

1. Install the reverse servo boost valve and spring into its proper location in the valve body. Install the 1–2 servo control valve and the 1–2 servo boost valve into the valve body. Install the correct springs behind each valve and insert the coiled spring pins.

2. Install new servo pipe lip seals. Move to the bore next to the servo pipe lip seals and install the 3–4 throttle assembly. Install the coiled spring pin. On Allante, install the 3–4 shift solenoid assembly.

3. Move to the next bore and install the 2–3 throttle valve bushing and its components. Install the spring pin. On Allante, install the 2–3 shift solenoid assembly.

4. Move to the next bore and in the following order install 3–2 isolater valve and spring, 3–2 control valve and spring and the 3–2 control sleeve.

5. Move to the next bore and install the 2–3 accumulator bushing assembly. Install the spring pin.

6. Move to the last bore of the side and install in the following order the 1–2 shift valve and throttle valve, the throttle valve spring and the 1–2 throttle valve bushing. Insert the 1–2 throttle valve bushing retainer into the 1–2 throttle valve bushing.

7. Move to the other side of the valve body and install the pressure regulator valve assembly. Install the reverse boost valve and bushing. Install the straight pin.

NOTE: The reverse boost spring and a pressure regulator modulator spring will only be found on transaxles equipped with a gas engine.

8. Install in the next bore the throttle valve assembly. Secure the assembly with the spring pin.

9. Install the line boost valve assembly. Insert the retainer clip.

10. Move to the next bore and install throttle valve feed valve assembly. Make sure the valve stop plate is in its proper position.

11. Move to the next bore and install the converter clutch shift valve assembly securing it with the bushing and a coiled spring pin.

12. The next bore is the pump pressure valve and bushing locations. Secure these units with a spring pin.

13. Moving to the next bore, install the 1–2 accumulator assembly. Make sure that the valve bore plug is in place over the 1–2 accumulator valve.

14. Assemble the converter clutch valves into the next 2 bore holes.

15. Install the 2nd clutch pipes in the top of the valve body. Install the release cover plate gasket and cover.

CHANNEL PLATE

Disassembly

1. Remove the modulator port gasket and the upper channel plate gasket. Remove the lower channel plate gasket and discard all gaskets.

2. Remove the input clutch accumulator piston and spring. Discard the input clutch ring seal.

3. Remove the converter clutch accumulator piston and the converter clutch spring. Discard the converter clutch seal. Remove the axle oil seal.

4. Remove the manual valve assembly. Detach the manual valve clip. Remove the channel plate stud from the case side of the channel plate.

Inspection

1. Wash all parts in solvent and blow dry using compressed air.

2. Check all parts for excessive wear or damage. Check the channel plate for cracks or warping.

3. Repair or replace any defective parts as required.

Assembly

1. Install the channel plate stud to the channel plate. Attach the manual valve clip to the manual valve assembly. Install the manual valve assembly into the channel plate.

2. Press in a new axle oil seal using a suitable tool. Install a new converter clutch seal. Place the converter clutch piston on its shaft with a new seal. Install the converter clutch spring.

3. Install the input clutch piston with a new input clutch seal. Install the input clutch spring.

Transaxle Assembly

Before the assembly of the transaxle begins, make certain that the case and all other parts are clean that all parts are serviceable or have been overhauled. Inspect the case careful for cracks or any signs of porosity. Inspect the vents to make sure that they are open. Check the case lugs, the intermediate servo bore and snapring grooves for damage. Check the bearings that are in the case and replaced if they appear to be worn.

NOTE: If the bearings are replaced, they must be installed with the bearing identification facing up.

The converter seal should be replaced. After all these prelimi-

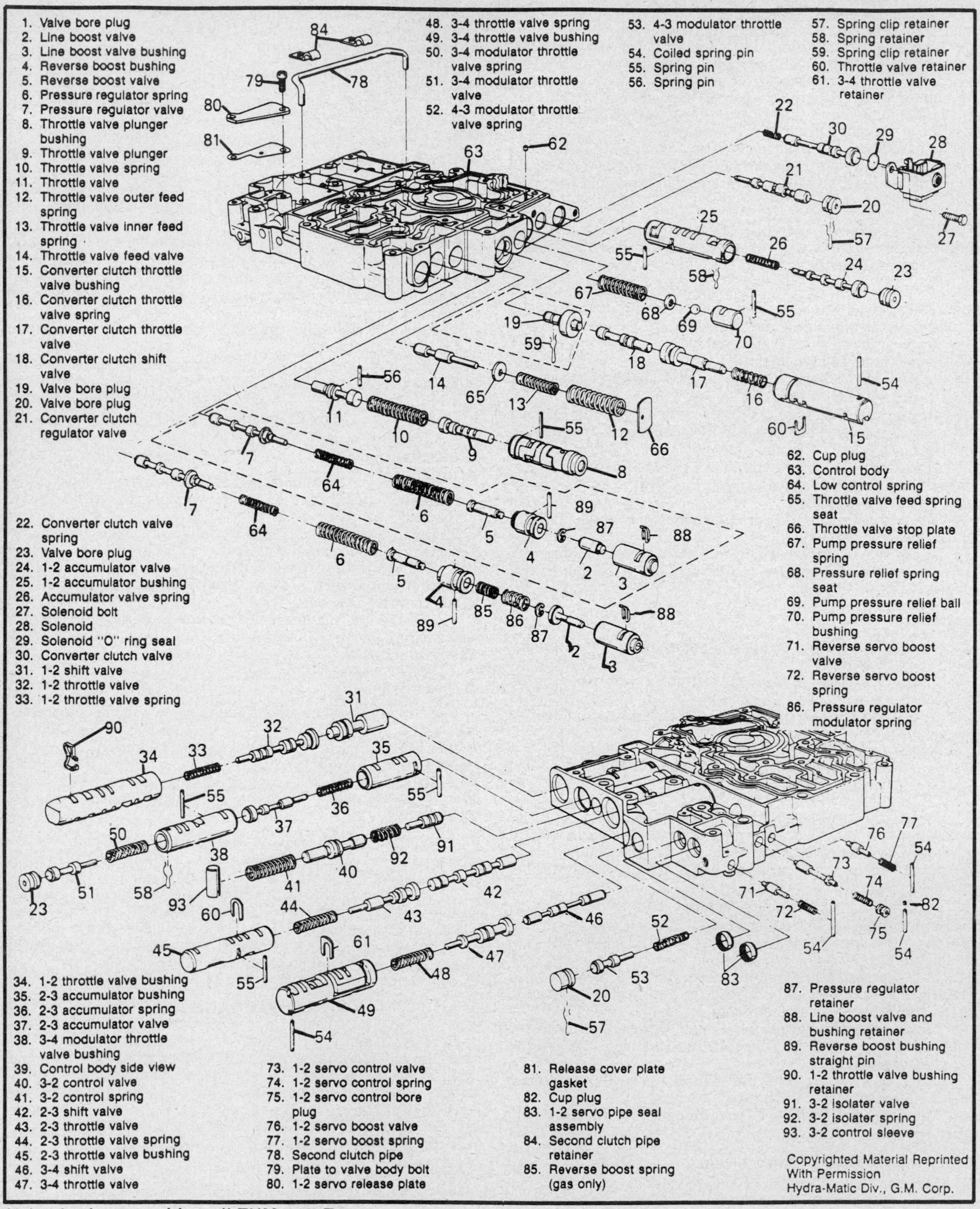

1. Valve bore plug
2. Line boost valve
3. Line boost valve bushing
4. Reverse boost bushing
5. Reverse boost valve
6. Pressure regulator spring
7. Pressure regulator valve
8. Throttle valve plunger bushing
9. Throttle valve plunger
10. Throttle valve spring
11. Throttle valve
12. Throttle valve outer feed spring
13. Throttle valve inner feed spring
14. Throttle valve feed valve
15. Converter clutch throttle valve bushing
16. Converter clutch throttle valve spring
17. Converter clutch throttle valve
18. Converter clutch shift valve
19. Valve bore plug
20. Valve bore plug
21. Converter clutch regulator valve
22. Converter clutch valve spring
23. Valve bore plug
24. 1-2 accumulator valve
25. 1-2 accumulator bushing
26. Accumulator valve spring
27. Solenoid bolt
28. Solenoid
29. Solenoid "O" ring seal
30. Converter clutch valve
31. 1-2 shift valve
32. 1-2 throttle valve
33. 1-2 throttle valve spring
34. 1-2 throttle valve bushing
35. 2-3 accumulator bushing
36. 2-3 accumulator spring
37. 2-3 accumulator valve
38. 3-4 modulator throttle valve bushing
39. Control body side view
40. 3-2 control valve
41. 3-2 control spring
42. 2-3 shift valve
43. 2-3 throttle valve
44. 2-3 throttle valve spring
45. 2-3 throttle valve bushing
46. 3-4 shift valve
47. 3-4 throttle valve
48. 3-4 throttle valve spring
49. 3-4 throttle valve bushing
50. 3-4 modulator throttle valve spring
51. 3-4 modulator throttle valve
52. 4-3 modulator throttle valve spring
53. 4-3 modulator throttle valve
54. Coiled spring pin
55. Spring pin
56. Spring pin
57. Spring clip retainer
58. Spring retainer
59. Spring clip retainer
60. Throttle valve retainer
61. 3-4 throttle valve retainer
62. Cup plug
63. Control body
64. Low control spring
65. Throttle valve feed spring seat
66. Throttle valve stop plate
67. Pump pressure relief spring
68. Pressure relief spring seat
69. Pump pressure relief ball
70. Pump pressure relief bushing
71. Reverse servo boost valve
72. Reverse servo boost spring
73. 1-2 servo control valve
74. 1-2 servo control spring
75. 1-2 servo control bore plug
76. 1-2 servo boost valve
77. 1-2 servo boost spring
78. Second clutch pipe
79. Plate to valve body bolt
80. 1-2 servo release plate
81. Release cover plate gasket
82. Cup plug
83. 1-2 servo pipe seal assembly
84. Second clutch pipe retainer
85. Reverse boost spring (gas only)
86. Pressure regulator modulator spring
87. Pressure regulator retainer
88. Line boost valve and bushing retainer
89. Reverse boost bushing straight pin
90. 1-2 throttle valve bushing retainer
91. 3-2 isolater valve
92. 3-2 isolater spring
93. 3-2 control sleeve

Valve body assembly—all THM 440–T4

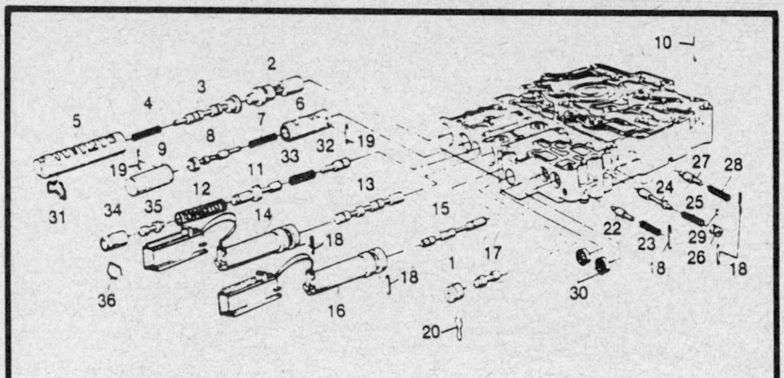

1. Valve bore plug
2. 1–2 shift valve
3. 1–2 throttle valve
4. 1–2 throttle valve spring
5. 1–2 throttle valve bushing
6. 2–3 accumulator bushing
7. 2–3 accumulator spring
8. 2–3 accumulator valve
9. 3–4 modulator plug
10. Spring pin
11. 3–2 control valve
12. 3–2 control spring
13. 2–3 shift valve
14. 2–3 shift solenoid assembly
15. 3–4 shift valve
16. 3–4 shift solenoid assembly
17. 4–3 manual throttle valve
18. Coiled spring pin
19. Spring pin
20. Spring clip retainer
21. Control body
22. Reverse servo boost valve
23. Reverse servo boost spring
24. 1–2 servo control valve
25. 1–2 servo control spring
26. 1–2 servo control bore plug
27. 1–2 servo boost valve
28. 1–2 servo boost spring
29. Cup plug
30. 1–2 servo pipe seal assembly
31. 1–2 throttle valve bushing retainer
32. 3–2 accumulator valve
33. 3–2 accumulator spring
34. 3–2 throttle valve bias bushing
35. 3–2 throttle valve bias valve
36. 3–2 throttle valve bias retainer

Valve body assembly—all THM F7

nary checks have been made, begin the reassembly of the transaxle. Make certain that the transaxle is held securely in the proper support fixture.

MANUAL LINKAGE/ACTUATOR REPLACEMENT

Installation

1. Install the O-ring to the actuator guide and install the actuator guide assembly.
2. Install the actuator guide retaining pin to the case.
3. Install the manual shaft with the detent lever. Install the locknut and torque to 25 ft. lbs. (34 Nm). Insert the retaining pin into the case.

MANUAL SHAFT/DETENT LEVER AND ACTUATOR ROD

Installation

1. Install the actuator guide assembly.
2. Install the actuator rod assembly.
3. Install the shaft lever and install the manual shaft and detent lever retaining bolts.

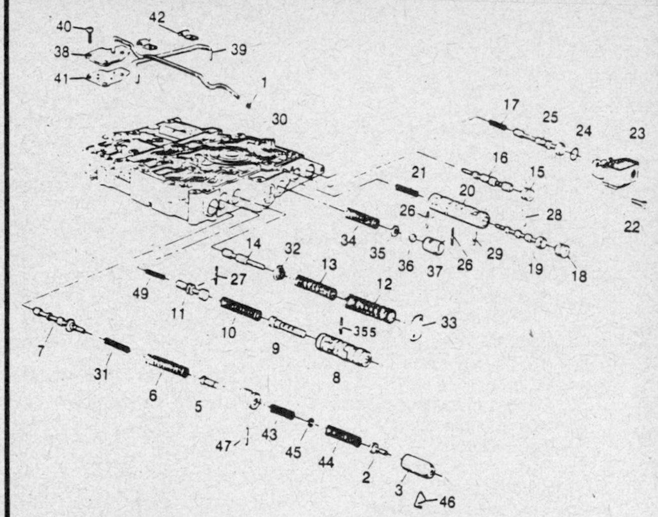

1. 3–2 coastdown seal assembly
2. Line boost valve
3. Line boost valve bushing
4. Reverse boost bushing
5. Reverse boost valve
6. Pressure regulator spring
7. Pressure regulator valve
8. Throttle valve plunger bushing
9. Throttle valve plunger
10. Throttle valve spring
11. Throttle valve
12. Throttle valve feed outer spring
13. Throttle valve feed inner spring
14. Throttle valve feed valve
15. Valve bore plug
16. Converter clutch regulator valve
17. Converter clutch valve spring
18. Valve bore plug
19. 1–2 accumulator valve
20. 1–2 accumulator bushing
21. Accumulator valve spring
22. Solenoid bolt
23. Solenoid
24. Converter clutch solenoid O-ring
25. Converter clutch valve
26. Spring pin
27. Spring pin
28. Spring clip retainer
29. Spring retainer
30. Control body
31. Pressure regulator isolator spring
32. Throttle valve feed spring seat
33. Throttle valve valve stop plate
34. Pump pressure relief spring
35. Pressure relief spring seat
36. Pump pressure relief ball
37. Pump pressure relief bushing
38. Plug and release cover assembly pipe
39. 2nd clutch-to-1-2 servo control valve pipe
40. Plate/valve body bolt
41. Release cover plate gasket
42. 2nd clutch pipe retainer
43. Reverse boost spring
44. Pressure regulator module boost spring
45. Pressure regulator retainer
46. Line boost valve and bushing retainer
47. Reverse boost bushing straight pin
48. Converter clutch regulator valve spring—some models
49. Throttle valve assist spring

Valve body assembly—all THM F7

FINAL DRIVE ASSEMBLY

Installation

1. The final drive carrier will be installed with the final drive internal gear.
2. Clamp the clutch and final drive tool to the final drive internal gear and install it.

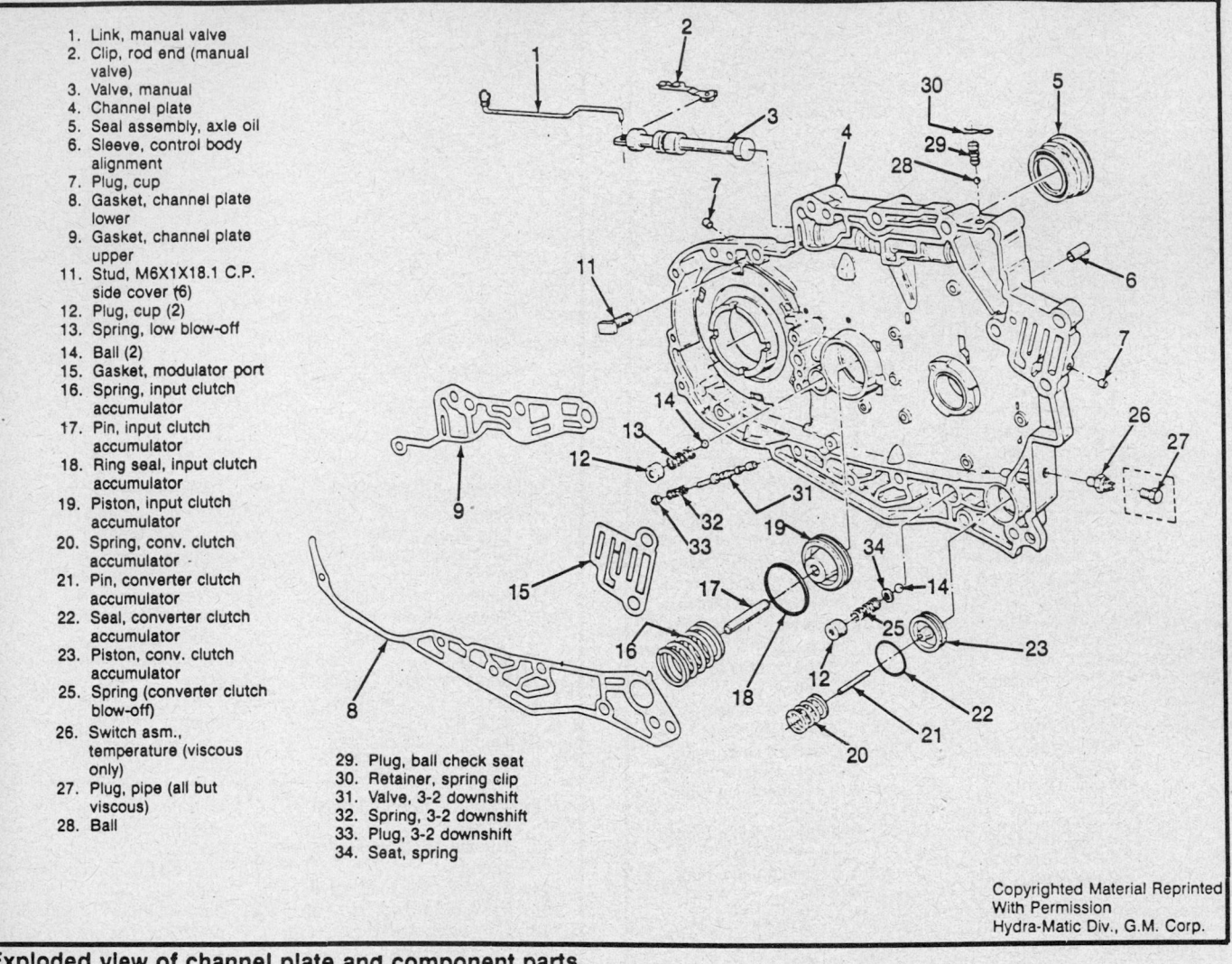

1. Link, manual valve
2. Clip, rod end (manual valve)
3. Valve, manual
4. Channel plate
5. Seal assembly, axle oil
6. Sleeve, control body alignment
7. Plug, cup
8. Gasket, channel plate lower
9. Gasket, channel plate upper
11. Stud, M6X1X18.1 C.P. side cover (6)
12. Plug, cup (2)
13. Spring, low blow-off
14. Ball (2)
15. Gasket, modulator port
16. Spring, input clutch accumulator
17. Pin, input clutch accumulator
18. Ring seal, input clutch accumulator
19. Piston, input clutch accumulator
20. Spring, conv. clutch accumulator
21. Pin, converter clutch accumulator
22. Seal, converter clutch accumulator
23. Piston, conv. clutch accumulator
25. Spring (converter clutch blow-off)
26. Switch asm., temperature (viscous only)
27. Plug, pipe (all but viscous)
28. Ball
29. Plug, ball check seat
30. Retainer, spring clip
31. Valve, 3-2 downshift
32. Spring, 3-2 downshift
33. Plug, 3-2 downshift
34. Seat, spring

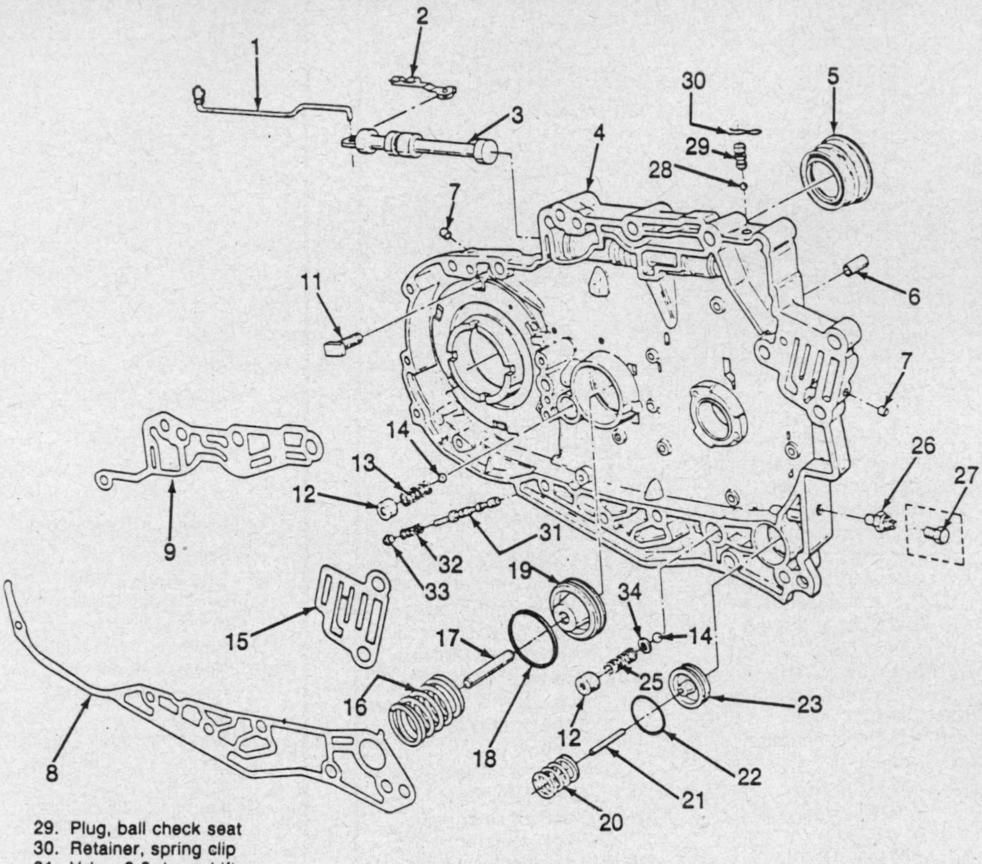

Exploded view of channel plate and component parts

3. With a suitable tool, install the snapring at the head of the internal final drive gear.

SECOND CLUTCH, INPUT CLUTCH, ROLLER CLUTCH AND SPRAG ASSEMBLY

Installation

1. Install the sun gear shaft and the bearing ring.
2. Install the 1–2 band. Install the reaction sun gear and the drum assembly.
3. Install the thrust bearing into the reaction carrier.
4. Install the reaction carrier.
5. Install the input carrier assembly. Install the reverse band and the reverse reaction drum.
6. Install the 2nd clutch housing and the input shaft assembly into the unit.

DRIVE LINK ASSEMBLY

Installation

1. Install the thrust washer. Install the drive sprocket support.
2. Position the drivelink assembly, onto the drive and driven sprockets. The drive link must be installed the same way as it was removed. It should be facing out.
3. Install the sprockets and the chain as an assembly. It will required alternately pushing on the sprockets until the bearings go onto of their support housings.
4. Reach through the access holes in the sprockets and slip the snaprings onto their grooves.
5. Install the turbine shaft O-ring located at the front of the unit.

FOURTH CLUTCH, SHAFT AND OUTPUT SHAFT

Installation

1. Install the output shaft into the transaxle being careful not to damage it.
2. Place the C-ring in the installation part of tool J-28583 and install the C-ring.
3. Install the 4th clutch shaft assembly.
4. Install the clutch apply plate with the stamped **DN** facing down. Install the steel clutch plate, then the composition plate, then the last steel clutch plate.

OIL PUMP SHAFT AND CHANNEL PLATE

Installation

1. Install the accumulator pistons and pins into the channel plate assembly. Install the 3 channel plate gaskets onto the channel plate and retain with petrolatum. Install the input clutch accumulator spring and the converter clutch spring into the case.
2. Install the channel plate into the transaxle case. Install the channel plate attaching bolts.
3. Align the manual detent roller and spring assembly to the detent lever and tighten the screws.

VALVE BODY

Installation

1. Install the 8 check balls into the channel plate.
2. Install the fluid reservoir weir.
3. Install the spacer plate/channel plate gasket. Install the spacer plate. Torque the bolts to 10 ft. lbs. (14 Nm). Install the spacer plate/valve body gasket.
4. Slide the fluid pump driveshaft through the drive sprocket hole.
5. Install the converter clutch solenoid screen.
6. Install the remaining check balls into the control valve assembly and retain with petrolatum.
7. Install the valve body onto the channel plate. Put the 1–2 servo pipes into the holes in the side of the valve body. Install the valve body retaining bolts. Torque the bolts to 10 ft. lbs. (14 Nm).

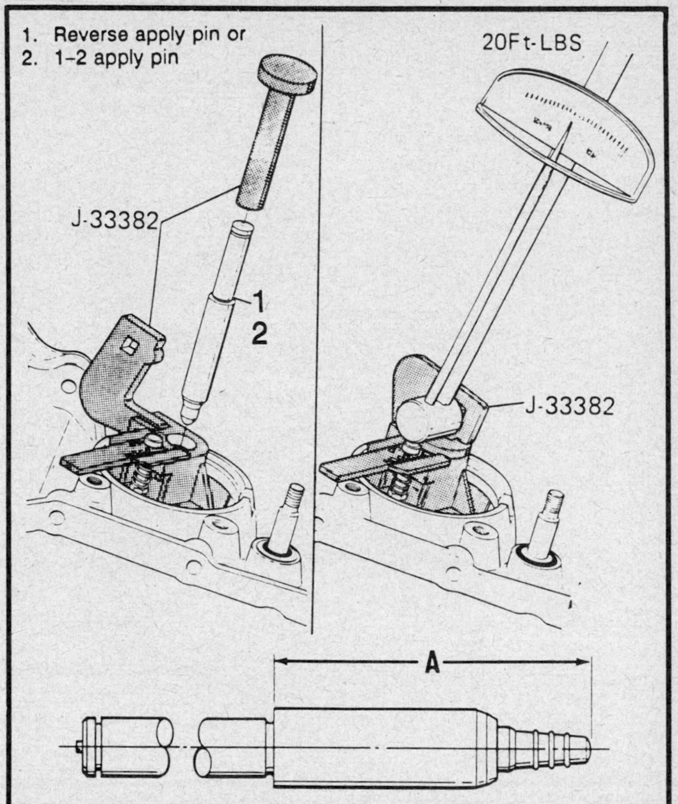
1. Reverse apply pin or
2. 1-2 apply pin
20Ft-LBS
J-33382
1
2
J-33382
A

Checking 1–2 apply pin or reverse apply pin

OIL PUMP

Installation

1. Position the oil pump assembly onto the valve body. Install the oil pump bolts. Torque to 10 ft. lbs. (14 Nm).
2. Install the solenoid wiring harness into the case connector and pressure switch.
3. Install the release pipe cover gasket and cover. Install the attaching screws.
4. Install the throttle valve assembly onto the valve body.
5. Install the side cover gasket on the case and position the side cover onto the case. Install the attaching screws, washer and nuts.

REVERSE AND/OR 1–2 SERVO

Installation

1. Install the accumulator springs and pistons. Install the accumulator pin through the piston hole.
2. To install the reverse and/or 1–2 servo assembly, install the piston and spring into the servo bore. Then install the cover. Depress the cover and install the snapring.

GOVERNOR

Installation

1. Install the speedometer drive gear onto the governor assembly.
2. Install the thrust bearing on the speedometer drive gear. The inside race goes against the gear or black side up.
3. Install the governor cover O-ring. Install the governor cover onto the case. Install the governor control body bolts and torque to 20 ft. lbs. (27 Nm).

4. Install the modulator valve into the case. Install the modulator O-ring onto the modulator and install in the case. Install the retainer and bolt. Torque to 20 ft. lbs. (27 Nm).

OIL FILTER

Installation

1. If the oil pressure regulator bushing assembly was removed, install it. Install the oil pressure regulator bushing assembly snapring.
2. Install the oil filter with a new O-ring. Attach the clips.

OIL PAN

Installation

1. Install the new oil pan gasket onto the oil pan.
2. Install the oil pan onto the case.
3. Install the oil pan attaching bolts. Torque to 10 ft. lbs. (13 Nm).

TORQUE CONVERTER

Installation

1. Install the converter assembly, making sure it is installed fully toward the rear of the transaxle.
2. The converter will be properly installed when the distance between the case mounting face and the front face of the converter lugs is a minimum of ½ in. (13mm).

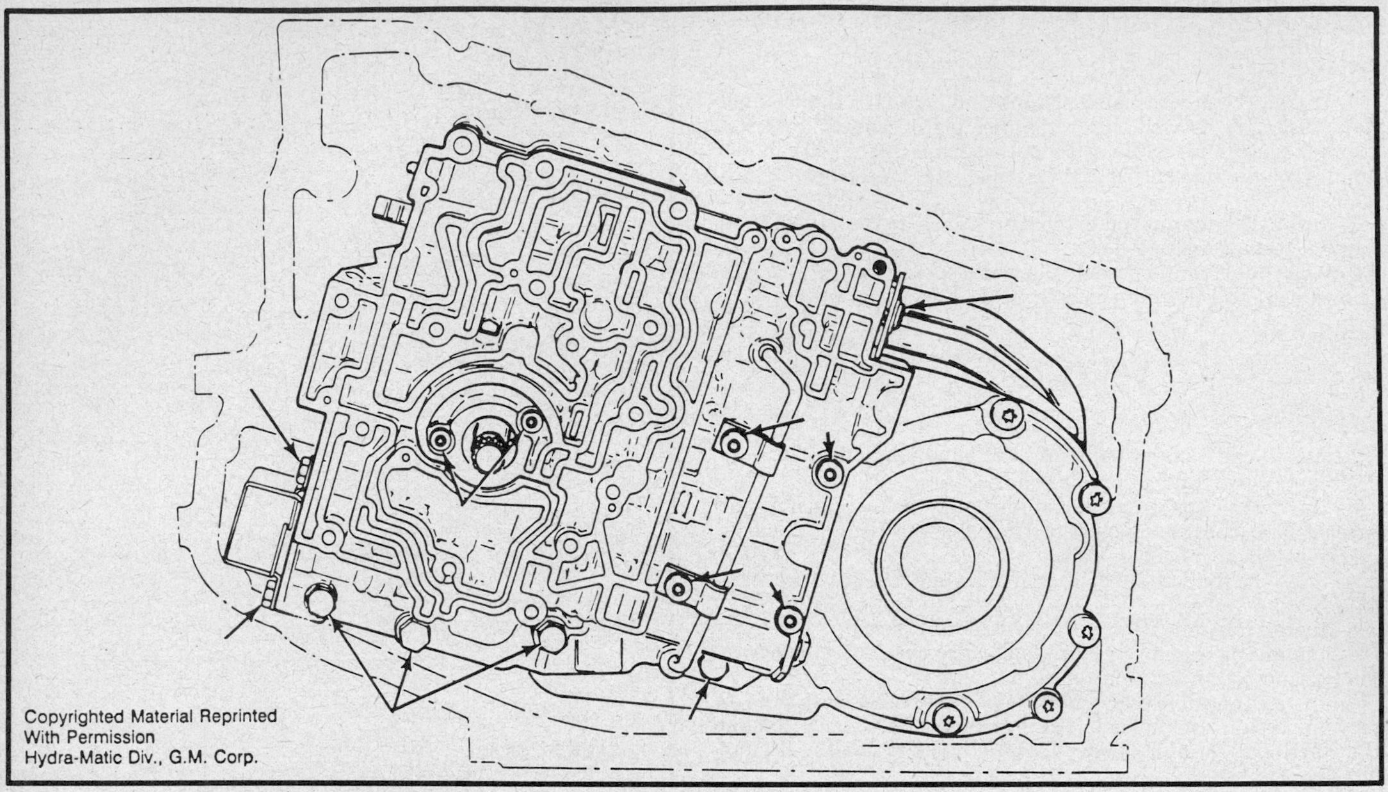
Copyrighted Material Reprinted
With Permission
Hydra-Matic Div., G.M. Corp.

Control valve assembly bolt locations

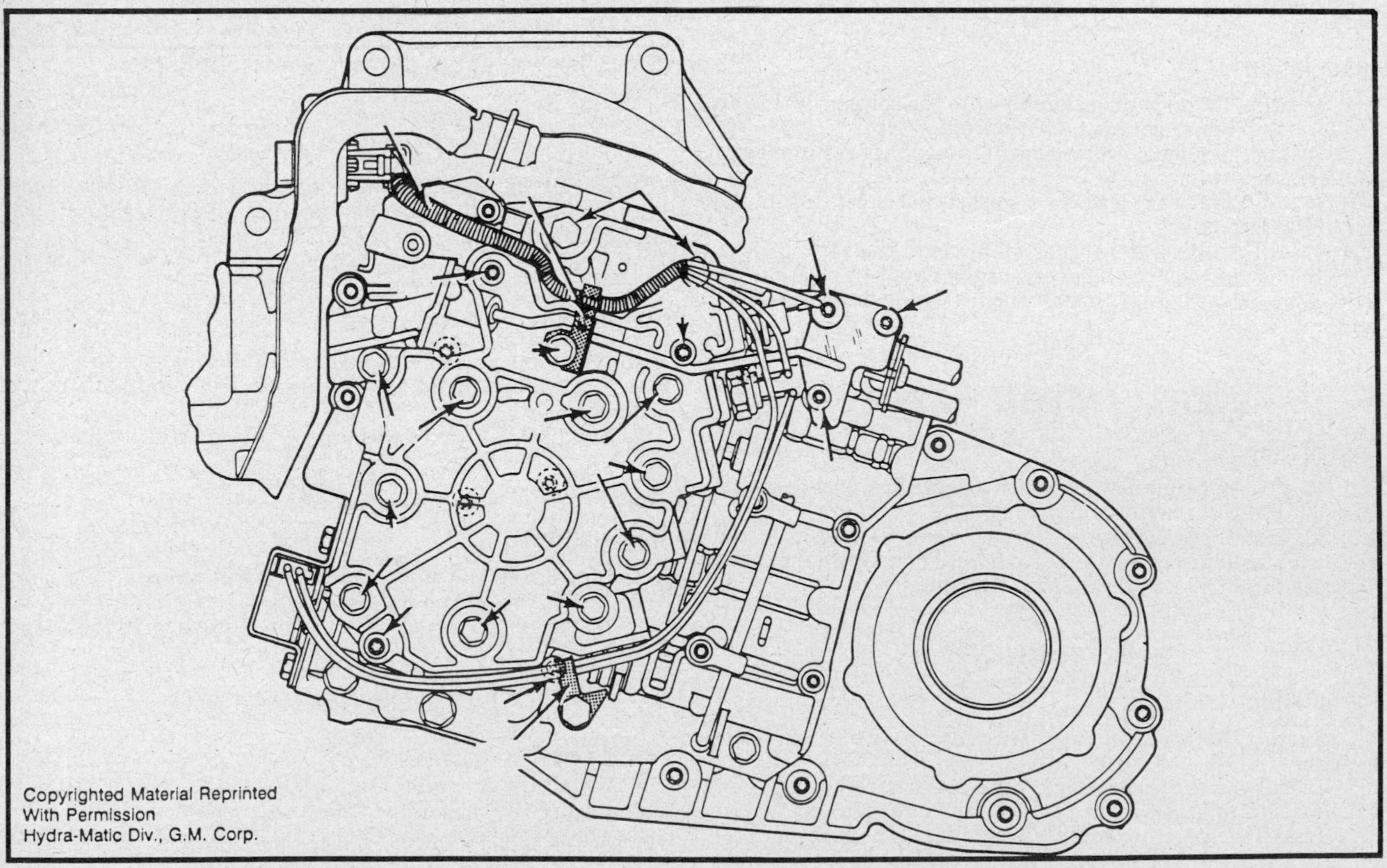
Copyrighted Material Reprinted
With Permission
Hydra-Matic Div., G.M. Corp.

Oil pump assembly bolt location

SPECIFICATIONS

TORQUE SPECIFICATIONS

Item	Foot Pounds	N•m
Cooler Fitting Connector	30	41
Modulator to Case	20	27
Pump Cover to Channel Plate	10	14
Pump Cover to Pump Body	20	27
Pump Cover to Pump Body (Torx Head)	20	27
Pipe Plug	10	14
Case to Drive Sprocket Support	20	27
Manifold to Valve Body	10	14
Governor to Case	20	27
Pressure Switch	10	14
Solenoid to Valve Body	10	14
Detent Spring to Valve Body	10	14
Case Side Cover to Channel Plate	10	14
Pump Cover to Valve Body	10	14
Pump Cover to Channel Plate	10	14
Valve Body to Case (Torx Head)	20	27
Valve Body to Case	20	27
Pump Body to Case	20	27
Valve Body to Channel Plate	10	14
Valve Body to Channel Plate (Torx Head)	10	14
Channel Plate to Case (Torx Head)	20	27
Channel Plate to Driven Sprocket Support (Torx Head)	20	27
Side Cover to Case	10	14
Accumulator Cover to Case	20	27
Oil Scoop to Case	10	14
Governor Control Body Retainer	20	27
Transmission Oil Pan to Case	10	14
Manual Shaft to Inside Detent Lever (Nut)	25	34

THRUST WASHER GUIDE

I.D. Number	Dimension (in.)	Color
1	2.90-3.00	Orange/Green
2	3.05-3.15	Orange/Black
3	3.20-3.30	Orange
4	3.35-3.45	White
5	3.50-3.60	Blue
6	3.65-3.75	Pink
7	3.80-3.90	Brown

THRUST WASHER GUIDE

I.D. Number	Dimension (in.)	Color
8	3.95-4.05	Green
9	4.10-4.20	Black
10	4.25-4.35	Purple
11	4.40-4.50	Purple/White
12	4.55-4.65	Purple/Blue
13	4.70-4.80	Purple/Pink
14	4.85-4.95	Purple/Brown
15	5.00-5.10	Purple/Green

FINAL DRIVE ENDPLAY

I.D. Number	Thickness
1	0.059-0.062 inches (1.50-1.60mm)
2	0.062-0.066 inches (1.60-1.70mm)
3	0.066-0.070 inches (1.70-1.80mm)
4	0.070-0.074 inches (1.80-1.90mm)
5	0.074-0.078 inches (1.90-2.00mm)
6	0.078-0.082 inches (2.00-2.10mm)

REVERSE BAND APPLY PIN

IDENTIFICATION	DIMENSION A
2 WIDE BANDS	70.86 - 71.01
3 GROOVES & WIDE BAND	71.91 - 72.06
2 GROOVES & WIDE BAND	72.96 - 73.11
1 GROOVE & WIDE BAND	74.01 - 74.16
NO GROOVE	75.03 - 75.18
1 GROOVE	76.08 - 76.23
2 GROOVE	77.13 - 77.28
3 GROOVE	78.18 - 78.33
4 GROOVE	79.20 - 79.35

1-2 BAND APPLY PIN

IDENTIFICATION	DIMENSION A
1 RING & WIDE BAND	56.24 - 56.39
1 RING	57.23 - 57.38
2 RINGS	58.27 - 58.42
3 RINGS	59.31 - 59.46
WIDE BAND	60.34 - 60.49
2 RINGS & WIDE BAND	61.34 - 61.49

SPECIAL TOOLS

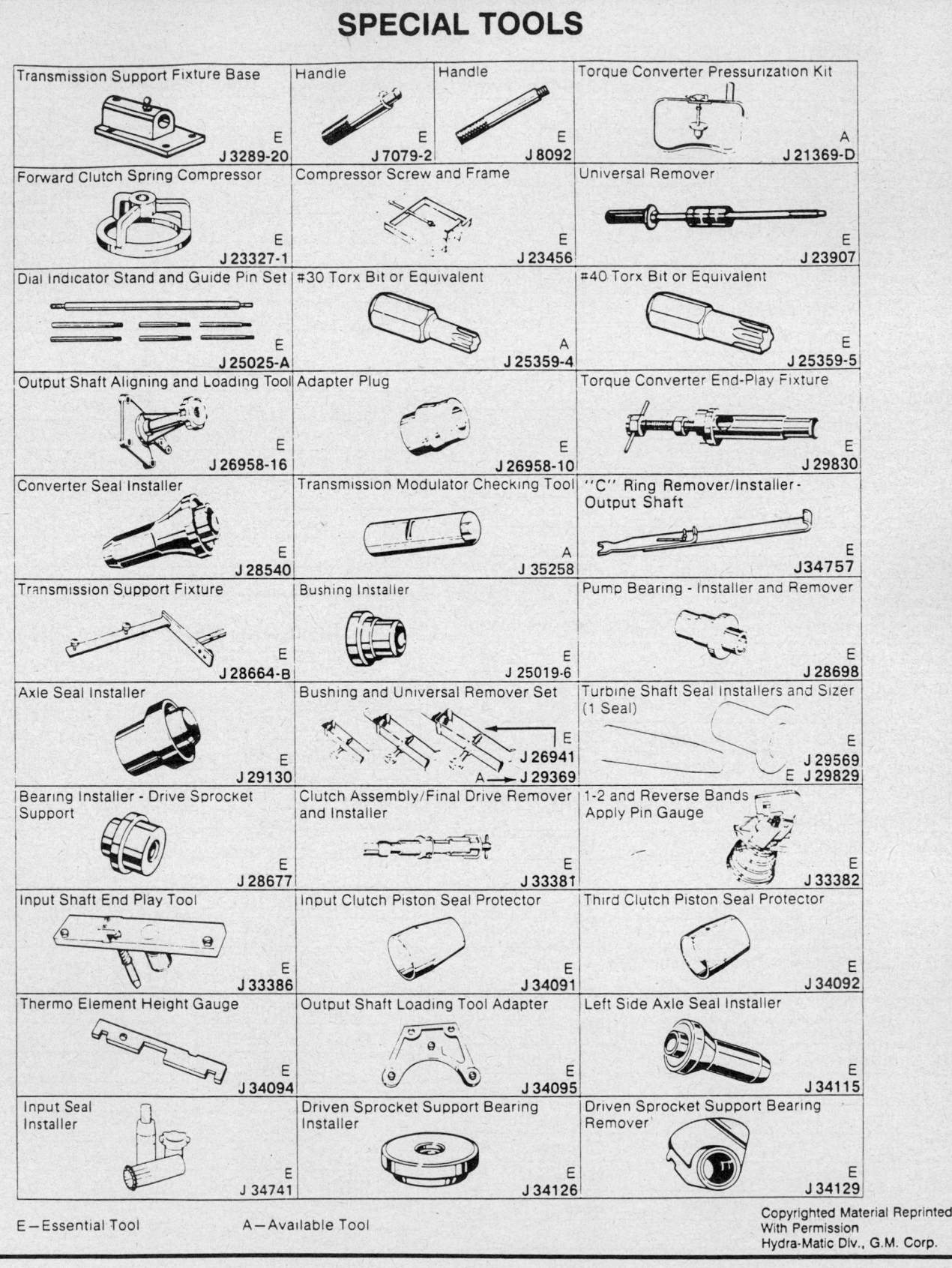

Transmission Support Fixture Base  E J 3289-20	Handle  E J 7079-2 Handle  E J 8092	Torque Converter Pressurization Kit 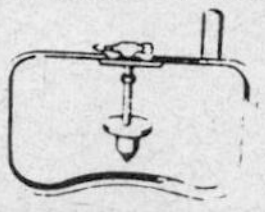 A J 21369-D
Forward Clutch Spring Compressor 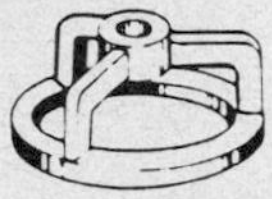 E J 23327-1	Compressor Screw and Frame 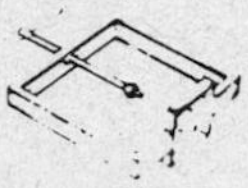 E J 23456	Universal Remover  E J 23907
Dial Indicator Stand and Guide Pin Set 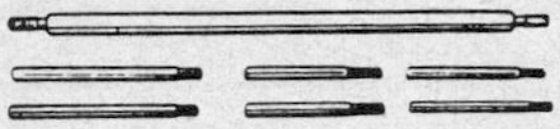 E J 25025-A	#30 Torx Bit or Equivalent 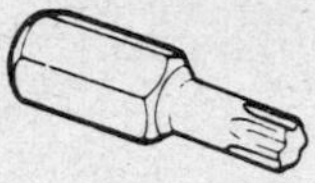 A J 25359-4	#40 Torx Bit or Equivalent 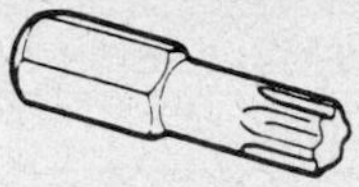 E J 25359-5
Output Shaft Aligning and Loading Tool 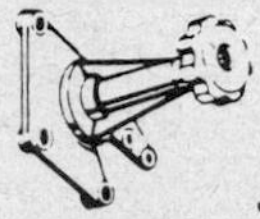 E J 26958-16	Adapter Plug  E J 26958-10	Torque Converter End-Play Fixture 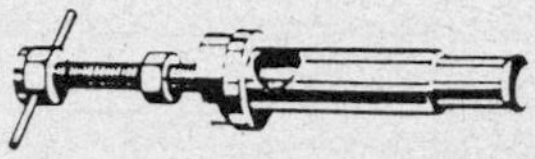 E J 29830
Converter Seal Installer 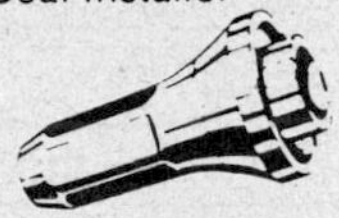 E J 28540	Transmission Modulator Checking Tool  A J 35258	"C" Ring Remover/Installer-Output Shaft 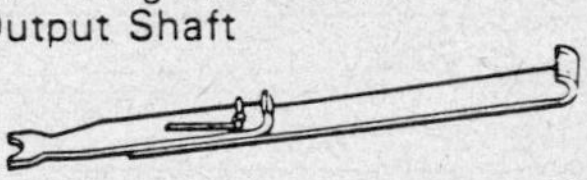 E J34757
Transmission Support Fixture 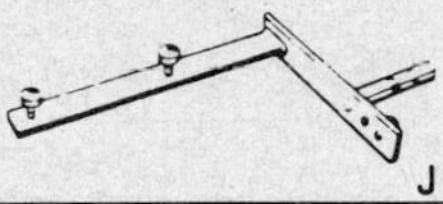 E J 28664-B	Bushing Installer  E J 25019-6	Pump Bearing - Installer and Remover  E J 28698
Axle Seal Installer  E J 29130	Bushing and Universal Remover Set  E J 26941 A J 29369	Turbine Shaft Seal Installers and Sizer (1 Seal)  E J 29569 E J 29829
Bearing Installer - Drive Sprocket Support  E J 28677	Clutch Assembly/Final Drive Remover and Installer  E J 33381	1-2 and Reverse Bands Apply Pin Gauge  E J 33382
Input Shaft End Play Tool  E J 33386	Input Clutch Piston Seal Protector 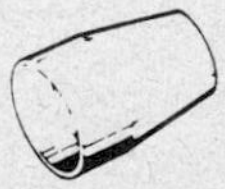 E J 34091	Third Clutch Piston Seal Protector  E J 34092
Thermo Element Height Gauge  E J 34094	Output Shaft Loading Tool Adapter  E J 34095	Left Side Axle Seal Installer 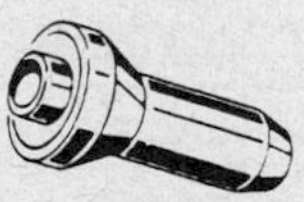 E J 34115
Input Seal Installer  E J 34741	Driven Sprocket Support Bearing Installer 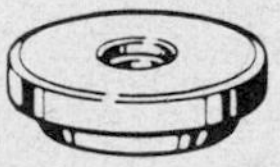 E J 34126	Driven Sprocket Support Bearing Remover  E J 34129

E—Essential Tool A—Available Tool

Section 3

Sprint and Metro Transaxle

General Motors

APPLICATION

1986–88 Sprint, 1989 Geo Metro

GENERAL DESCRIPTION

The transaxle is a 3 speed fully automatic overdrive transmission. The unit consists of 2 planetary gears, 2 disc clutches, a band brake, a disc brake and a one-way clutch. The fully automatic shift control which responds to the road speed and engine torque demand.

Transaxle and Converter Identification

TRANSAXLE

Code letters and numbers are stamped on the identification tag located on the left side (front) of the transaxle housing. The tag denotes the serial number and the date of manufacture. These numbers are important when ordering service replacement parts.

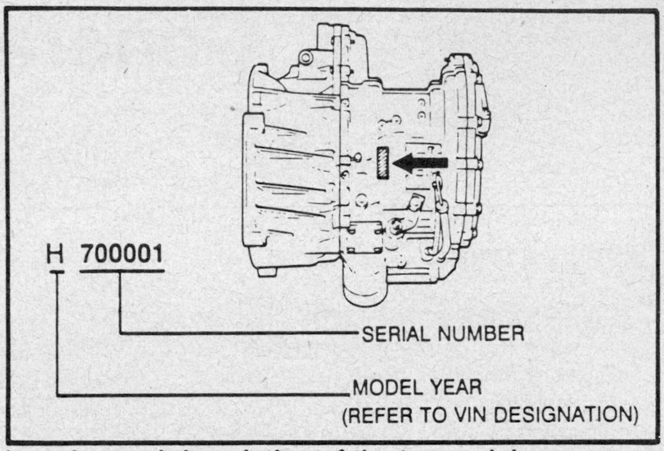
H 700001
SERIAL NUMBER
MODEL YEAR
(REFER TO VIN DESIGNATION)

Location and description of the transaxle's identification number

CONVERTER

The 3 element torque converter consists of a pump, turbine and stator; disassembly is not possible. The pump is mounted to the crankshaft, the turbine to the input shaft and stator to the transaxle case by way of a one-way clutch. Its torque increases when starting, accelerating and driving up hill; while driving at constant speed, it functions as a fluid clutch.

Electronic Controls

The transaxle, being fully controlled by the computer system, is equipped with 2 solenoids; it responds to road speed and engine torque demand.

The internal equipment, attached to the main control assembly, consists of the direct clutch and 2nd brake solenoids. A bulkhead connector/wiring assembly, attached to the middle, rear side of the chain cover assembly, provides an electrical path to the computer system.

Metric Fasteners

The transaxle is of a metric design; all bolt sizes and thread pitches are metric. Metric fastener dimensions are very close to the customary inch system fastener dimensions; replacement of the fasteners must be of the same measurement and strength as those removed.

Do not attempt to interchange metric fasteners with the customary inch system fasteners. Mismatched or incorrect fasteners can result in damage to the transaxle. Care should be taken to reuse the same fasteners in the location from which they were removed.

NOTE: Be sure to check the case holes for the quality of their threads. It is rather difficult to rethread bolt holes after the transaxle is installed in the vehicle.

Capacities

The fluid quantities are approximate and the correct fluid level should be determined by the dipstick indicator for the correct level. Dry fill is 4.7 qts. (4.5L)

Checking Fluid Level

Transaxle at Normal Operating Temperature

DIPSTICK COOL TO THE TOUCH

NOTE: The following procedure is used to check the fluid level after the transaxle has been overhauled, the fluid drained or the valve body serviced. This check is only temporary, be sure to perform the HOT check procedures before driving the vehicle.

1. With the vehicle on a level surface, place the transaxle in **P**, idle (do not race) the engine for 5 min. and apply the parking brake. Move the selector lever through each range; allow time in each range for the transaxle to engage.
2. Position the selector lever in **P**, fully apply the parking brake, block the drive wheels and allow the vehicle to idle.
3. Before removing the dipstick, wipe the dirt from the dipstick cap.
4. Remove the dipstick and wipe it clean. Insert it into the dipstick tube and be sure it fully seats.
5. Remove the dipstick again and observe the fluid level; it must be between the **FULL COLD** and **LOW COLD** indicators.
6. If necessary to add fluid, add it through the oil filler (dipstick) tube. Add enough fluid to raise the level. Do not overfill; overfilling will cause foaming, fluid loss and transaxle malfunction.

DIPSTICK HOT TO THE TOUCH

NOTE: When checking the fluid level, the vehicle must be at normal operating temperatures and positioned on a flat surface. The vehicle should not be driven if the fluid level is below the dipstick's DO NOT DRIVE hole.

1. If the oil is below 158–176°F (70–80°C), drive the vehicle until normal operating temperatures are reached.

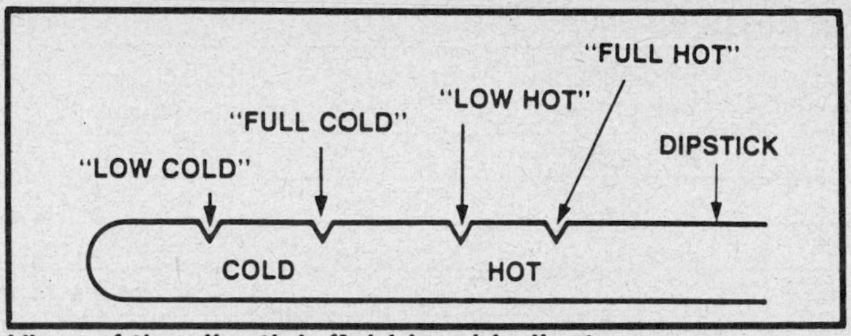
"FULL HOT"
"LOW HOT"
"FULL COLD"
"LOW COLD"
DIPSTICK
COLD
HOT

View of the dipstick fluid level indicators

NOTE: If the outside temperature is above 50°F (10°C), drive the vehicle 15–20 miles (24–32 km) of city driving. If the vehicle has been operated under extreme conditions in hot weather, turn the engine OFF and allow the fluid to cool for at least 30 minutes.

2. With the vehicle on a level surface, place the transaxle in **P**, idle (do not race) the engine and apply the parking brake. Move the selector lever through each range; allow time in each range for the transaxle to engage.
3. Position the selector lever in **P**, fully apply the parking brake, block the drive wheels and allow the vehicle to idle.
4. Before removing the dipstick, wipe the dirt from the dipstick cap.
5. Remove the dipstick and wipe it clean. Insert it into the dipstick tube and be sure it fully seats.
6. Remove the dipstick again and observe the fluid level; it must be between the **FULL HOT** and **LOW HOT** indicators.
7. If necessary to add fluid, add it through the oil filler (dipstick) tube. Add enough fluid to raise the level. Do not overfill; overfilling will cause foaming, fluid loss and transaxle malfunction.

TRANSAXLE MODIFICATIONS

Differential Side and Pinion Gear Wear

Some 1986–87 vehicles may experience operational problems due to wear of the differential side and pinion gears. The following serial numbers refer to the last 8 VIN digits and include all previous vehicles, HK740463 (2 Door) and HK740416 (4 Door).

Refer to the differential case assembly replacement procedures and replace the assembly with part number 90057455.

Direct Clutch Drum Bushing Wear

Some 1986–87 vehicles may experience operational problems due to wear of the differential side and pinion gears. The following serial numbers refer to the last 8 VIN digits and include all previous vehicles, HK740463 (2 Door) and HK740416 (4 Door).

Several sets of replacement parts are available from GM. Order the correct set according to the extent of repair or replacement that is required.

PART NUMBER INFORMATION

Previous Part No.	Qty.	New Part No.	Qty.	Part Name
35677–12010	3	35677–20020①	2	Disc, clutch (for 2nd brake)
35648–32010	3	Same①	4	Plate, clutch (for 2nd brake)
90501–25011	1	90501–26014①	1	Spring (B2 Accumulator piston)

① Parts are interchangable as a set only

TROUBLE DIAGNOSIS

NOTE: Before performing the wiring troubleshooting procedures, make sure the wiring is undamaged, the electrical connectors are firmly connected and the vacuum switch hoses are securely connected.

NOTE: Before performing the electrical checking procedures, make sure the ignition switch is turned OFF and the controller's electrical connector is disconnected.

NOTE: Before performing the electrical checking procedures on the accelerator and vacuum switches, firmly apply the parking brake, block the drive wheels, place the shift selector into P and the engine must be running.

Any checks with an asterisk mark (*) should be carried out within 5 seconds or damage may result. Release the accelerator after each continuity test, for long engine operation at a high speed can cause overheating of the transaxle fluid.

Hydraulic Control System

The main components of the hydraulic control system are: the valve body, oil pump, manual valve, throttle valve, primary regulator valve, secondary regulator valve, B2 control valve, cooler bypass valve, 1–2 shift valve, 2–3 shift valve, accumulator and direct clutch/2nd brake solenoids.

VALVE BODY

The valve body is located inside the oil pan. It contains the various valves which control oil pressure to the clutches and bands.

CLUTCH AND BAND APPLICATION

Range	Gear	Forward Clutch	Direct Clutch	Brake Band	Low-Reverse Brake Clutch	One-way Clutch	Parking Lock Pawl
P	Parking	—	—	—	Applied	—	Applied
R	Reverse	—	Applied	—	Applied	—	—
N	Neutral	—	—	—	—	—	—
D	1st	Applied	—	—	—	Applied	—
	2nd	Applied	—	Applied	—	—	—
	3rd	Applied	Applied	—	—	—	—
2	1st	Applied	—	—	—	Applied	—
	2nd	Applied	—	Applied	—	—	—
L	1st	Applied	—	—	Applied	Applied	—
	2nd①	Applied	—	Applied	—	—	—

① To prevent over-resolution of engine, this 2nd gear is operated only when selector lever is shifted to L range at the speed of more than 34 mph (55 km/h)

CHILTON'S THREE C's TRANSAXLE DIAGNOSIS

Condition	Cause	Correction
No 1–2 upshift	a) Sticking 1–2 shift valve b) 2nd gear solenoid stuck open c) Defective controller or poor electrical connection	a) Free or replace 1–2 shift valve b) Free or replace 2nd gear solenoid c) Replace controller or fix electrical connection
Harsh engagement during 1–2 upshift	a) Defective 2nd gear accumulator b) Worn brake band c) Sticking 2nd gear check ball	a) Replace 2nd gear accumulator b) Replace brake band c) Replace 2nd gear check ball
No 2–3 upshift	a) Sticking 2–3 shift valve b) Direct clutch solenoid stuck open c) Defectice controller or poor electrical connection	a) Free or replace 2–3 shift valve b) Free or replace direct clutch solenoid c) Replace controller or fix electrical connection
Harsh engagement during 2–3 upshift	a) Worn direct clutch b) Sticking direct clutch check ball	a) Replace direct clutch b) Replace direct clutch check ball
No upshift	a) Defective back drive solenoid b) Defective shift selector	a) Inspect wiring or replace back drive solenoid b) Adjust or replace shift selector
Harsh engagement during gear selection or gear change due to poor line pressure	a) Defective regulator valve b) Defective throttle valve c) Unadjusted accelerator cable	a) Replace regulator valve b) Replace throttle valve c) Adjust accelerator cable
Harsh engagement shifting from N to R	a) Worn low/reverse clutch b) Sticking low reverse clutch check ball	a) Replace low/reverse clutch b) Replace low/reverse check ball
Harsh engagement when shifting from N to D	a) Defective forward clutch accumulator b) Worn forward clutch c) Sticking forward clutch check ball	a) Replace forward clutch accumulator b) Replace forward clutch c) Replace forward clutch check ball
Slippage in any drive range with engine speed at 2000–2400 rpm	a) Transaxle failure a) Fluid below normal	a) Rebuild or replace transaxle b) Check and/or fill transaxle

CHILTON'S THREE C's TRANSAXLE DIAGNOSIS

Condition	Cause	Correction
Slippage in any drive range with engine speed below 2000–2400 rpm	a) Defective torque converter b) Poor engine power	a) Replace torque converter b) Tune or adjust engine
Slippage in any drive range with engine speed above 2000–2400 rpm and proper fluid pressure	a) Transaxle failure	a) Rebuild or replace transaxle
Slippage in any drive range with engine speed above 2000–2400 rpm and poor fluid pressure	a) Defective oil pump b) Defective regulator valve c) Defective throttle valve d) Misadjusted acceleartor cable	a) Replace oil pump b) Replace regulator valve c) Adjust or replace throttle valev d) Adjust accelerator cable
Slippage with shift selector in the D range only and engine speed is above 2000–2400 rpm	a) If the engine speed drops to the transaxle solenoid wire is disconnected a defective one-way clutch is indicated b) If the engine speed does not drop to 2000–2400 rpm when the transaxle solenoid wire is disconnected and the fluid pressure is OK a worn forward clutch is indicated c) If the engine speed does not drop to 2000–2400 rpm when the transaxle solenoid wire is disconnected and the fluid pressure is not OK an oil leak in the forward clutch or D range oil circuit is present	a) Replace one-way clutch b) Replace forward clutch c) Repair the forward clutch or D range circuit oil leak
Slippage with shift selector in the R range only and the engine speed is above the 2000–2400 range	a) If the engine speed drops to when the transaxle solenoid wire is disconnected a worn low reverse clutch is indicated b) If the engine speed does not drop to 2000–2400 rpm when the transaxle solenoid wire is disconnected and the fluid pressure is OK a worn direct clutch is indicated c) If the engine speed does not drop to 2000–2400 rpm when the transaxle solenoid wire is disconnected and the fluid pressure is not OK a worn direct clutch or R range oil circuit leak is present	a) Replace the low reverse clutch b) Replace the direct clutch c) Replace the direct clutch or R range circuits oil leak
No 3–2 or 2–1 downshift	a) Defective accelerator switch b) Defective controller or poor electrical connector	a) Replace accelerator switch b) Replace controller or fix electrical connector
No engine braking when shifted into 2nd range	a) Defective brake band	a) Replace brake band
No engine braking when shifted into low range	a) Defective low reverse clutch	a) Replace low reverse clutch

TROUBLESHOOTING SHIFT FAILURE IN D RANGE

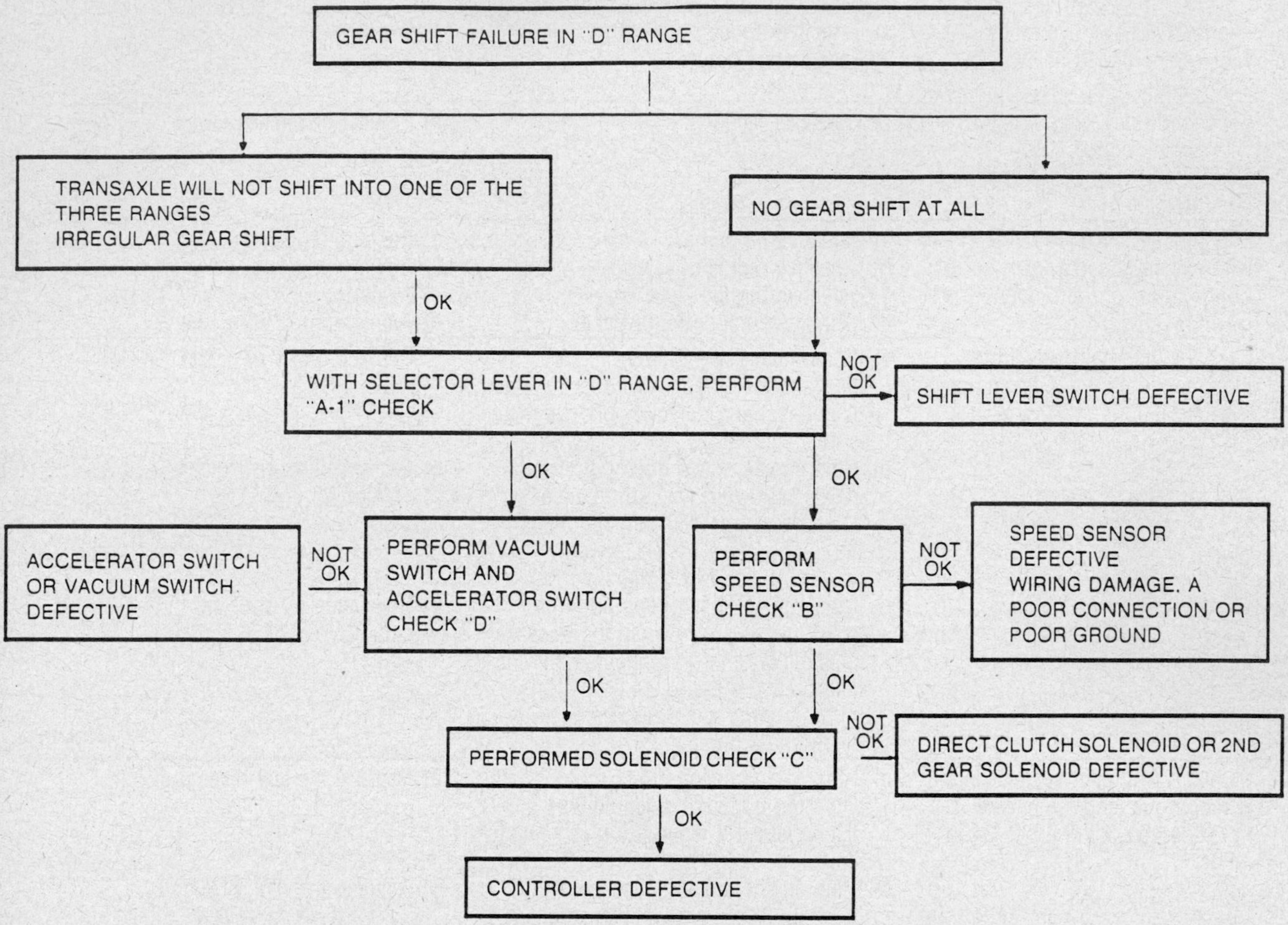
GEAR SHIFT FAILURE IN "D" RANGE
TRANSAXLE WILL NOT SHIFT INTO ONE OF THE THREE RANGES
IRREGULAR GEAR SHIFT
NO GEAR SHIFT AT ALL
OK
WITH SELECTOR LEVER IN "D" RANGE, PERFORM "A-1" CHECK
NOT OK
SHIFT LEVER SWITCH DEFECTIVE
OK
OK
ACCELERATOR SWITCH OR VACUUM SWITCH DEFECTIVE
NOT OK
PERFORM VACUUM SWITCH AND ACCELERATOR SWITCH CHECK "D"
PERFORM SPEED SENSOR CHECK "B"
NOT OK
SPEED SENSOR DEFECTIVE
WIRING DAMAGE, A POOR CONNECTION OR POOR GROUND
OK
OK
PERFORMED SOLENOID CHECK "C"
NOT OK
DIRECT CLUTCH SOLENOID OR 2ND GEAR SOLENOID DEFECTIVE
OK
CONTROLLER DEFECTIVE

OIL PUMP

The internal gear type oil pump, located directly behind the torque converter, is operated by a splined shaft to the torque converter hub. Its purpose is to feed oil to the torque converter, lubricate parts and deliver oil pressure to each clutch and brake solenoid.

MANUAL VALVE

The manual valve, located on the upper valve body, is directly connected the shift selector lever by a cable. It's purpose is to mechanically open and close the oil passage to the respective oil pressure circuit (range) according to the selector lever movement.

THROTTLE VALVE

The throttle valve, located in the upper valve body, is connected to the accelerator pedal by a cable and produces throttle pressure corresponding to the accelerator pedal movement.

Depressing the accelerator pedal, causes the throttle cam to push the shift plug, compressing 2 springs, to move the throttle valve; the line pressure passage opens and procedures throttle pressure. Throttle pressure is also applied to the rear of the throttle valve to push it backward. Throttle pressure is determined by the springs pressure and position of the shift plug, applied to the primary/secondary regulator valves to regulate the line pressure.

PRIMARY REGULATOR VALVE

The primary regulator valve, located in the upper valve body, regulates the oil pressure (produced by the oil pump) to correspond to each condition of use. It's operation is controlled throttle pressure, springs and line pressure (in reverse).

SECONDARY REGULATOR VALVE

The secondary regulator valve, located on the lower valve body, regulates the oil pressure to the torque converter and supplies lubrication to each part by means of the throttle pressure and spring.

B2 CONTROL VALVE

The B2 control valve, located in the lower valve body, operates

TROUBLESHOOTING
SHIFT FAILURE IN 2 RANGE

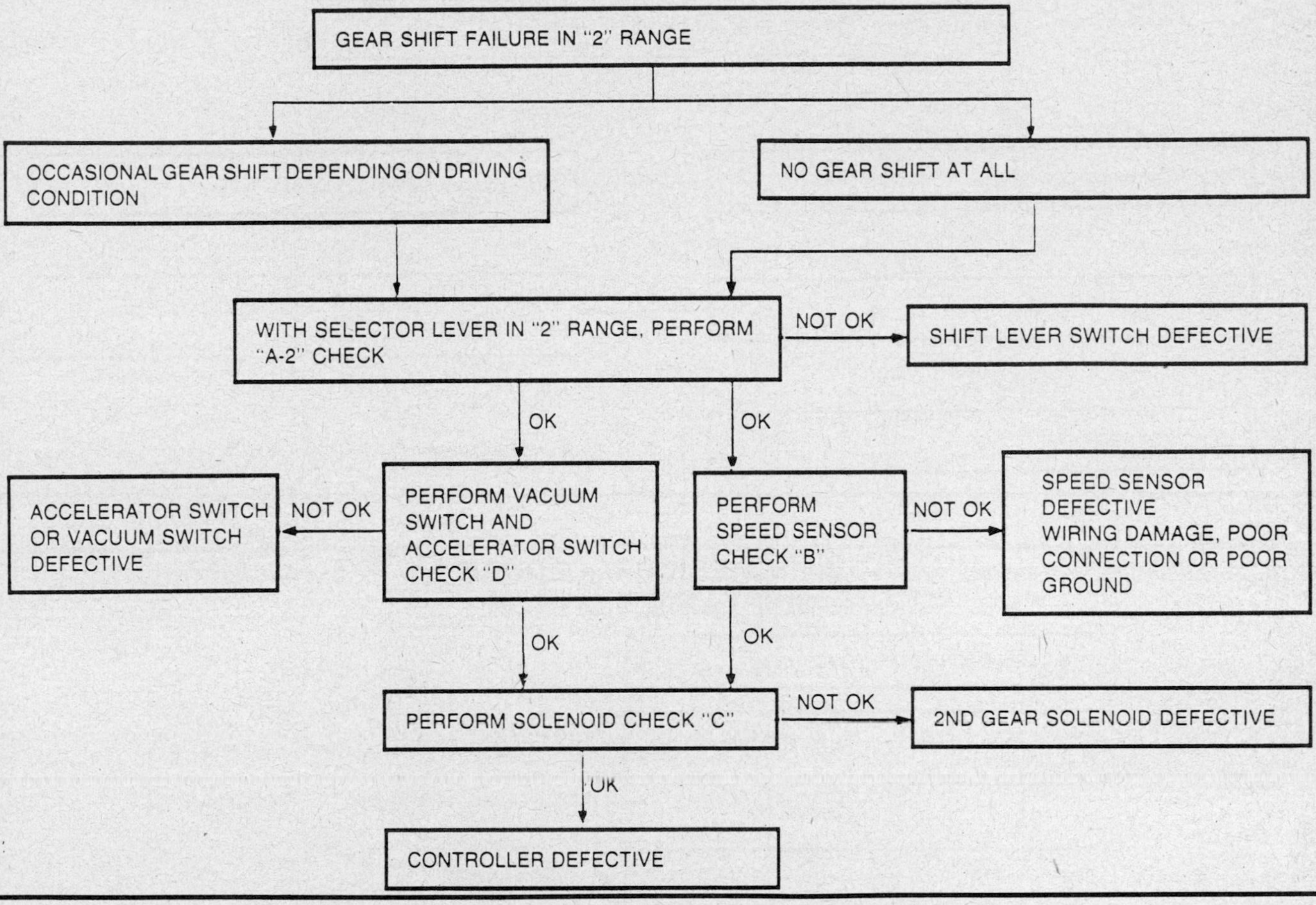
GEAR SHIFT FAILURE IN "2" RANGE
OCCASIONAL GEAR SHIFT DEPENDING ON DRIVING CONDITION
NO GEAR SHIFT AT ALL
WITH SELECTOR LEVER IN "2" RANGE, PERFORM "A-2" CHECK
NOT OK
SHIFT LEVER SWITCH DEFECTIVE
OK
OK
ACCELERATOR SWITCH OR VACUUM SWITCH DEFECTIVE
NOT OK
PERFORM VACUUM SWITCH AND ACCELERATOR SWITCH CHECK "D"
PERFORM SPEED SENSOR CHECK "B"
NOT OK
SPEED SENSOR DEFECTIVE
WIRING DAMAGE, POOR CONNECTION OR POOR GROUND
OK
OK
PERFORM SOLENOID CHECK "C"
NOT OK
2ND GEAR SOLENOID DEFECTIVE
OK
CONTROLLER DEFECTIVE

TROUBLESHOOTING
SHIFT FAILURE IN 2ND AND 3RD GEAR AND L RANGE

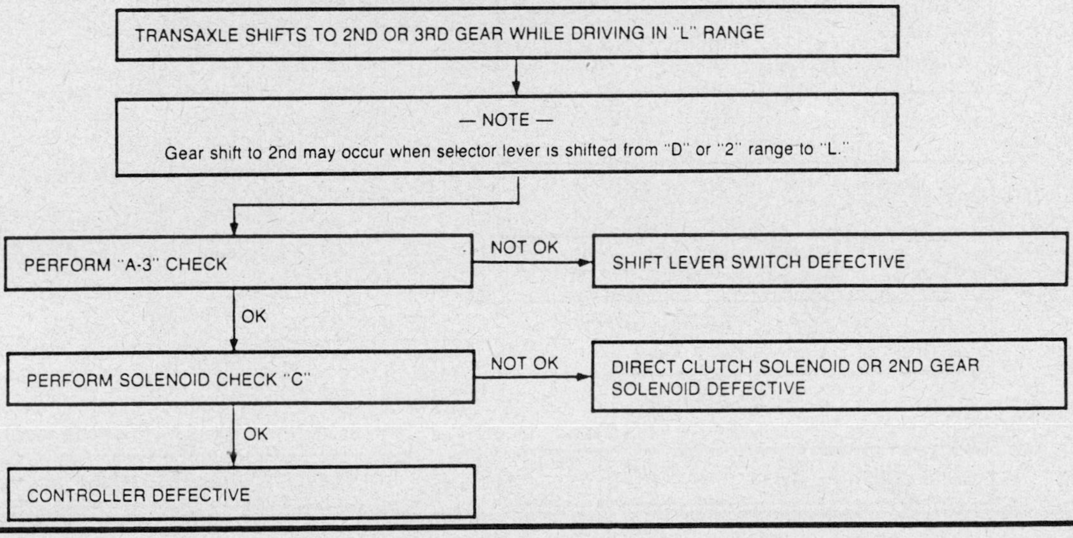
TRANSAXLE SHIFTS TO 2ND OR 3RD GEAR WHILE DRIVING IN "L" RANGE
— NOTE —
Gear shift to 2nd may occur when selector lever is shifted from "D" or "2" range to "L."
PERFORM "A-3" CHECK
NOT OK
SHIFT LEVER SWITCH DEFECTIVE
OK
PERFORM SOLENOID CHECK "C"
NOT OK
DIRECT CLUTCH SOLENOID OR 2ND GEAR SOLENOID DEFECTIVE
OK
CONTROLLER DEFECTIVE

TROUBLESHOOTING
NO REVERSE

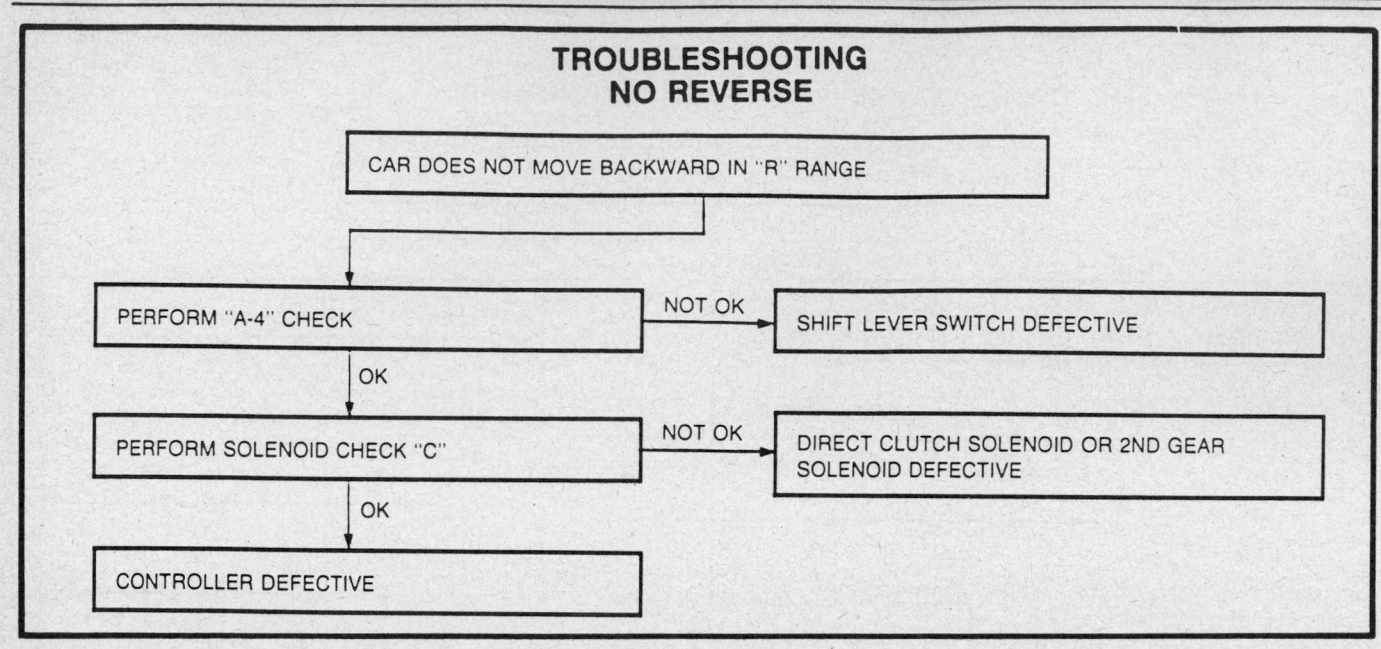

CAR DOES NOT MOVE BACKWARD IN "R" RANGE

↓

PERFORM "A-4" CHECK — NOT OK → SHIFT LEVER SWITCH DEFECTIVE

↓ OK

PERFORM SOLENOID CHECK "C" — NOT OK → DIRECT CLUTCH SOLENOID OR 2ND GEAR SOLENOID DEFECTIVE

↓ OK

CONTROLLER DEFECTIVE

A–1 CHECK PROCEDURE
SHIFT LEVER SWITCH

SHIFT SELECTOR LEVER TO "D" RANGE

↓

CHECK TERMINALS (12) - (4) ("D" RANGE) FOR CONTINUITY — OPEN → SHIFT LEVER SWITCH DEFECTIVE

↓ CONTINUITY

CHECK TERMINALS (12) - (16) ("P" AND "N" RANGES) FOR NO CONTINUITY — CONTINUITY → SHIFT LEVER SWITCH DEFECTIVE

↓ OPEN

TURN IGNITION SWITCH "ON" AND CHECK THE VOLTAGE BETWEEN TERMINALS (12) AND (5) ("R" RANGE)
TURN IGNITION SWITCH "OFF" — ABOUT 12V → SHIFT LEVER SWITCH DEFECTIVE

↓ OV

CHECK (12) - (17) ("2" RANGE) FOR NO CONTINUITY — CONTINUITY → SHIFT LEVER SWITCH DEFECTIVE

↓ OPEN

CHECK (12) - (8) ("L" RANGE) FOR NO CONTINUITY — CONTINUITY → SHIFT LEVER SWITCH DEFECTIVE

↓ OPEN

SHIFT LEVER SWITCH IN GOOD CONDITION

in the **L** range to reduce the line pressure shock acting on the 1st/reverse brake.

COOLER BYPASS VALVE

The cooler bypass valve, located in the lower valve body, is designed to keep the oil pressure in the torque converter constant.

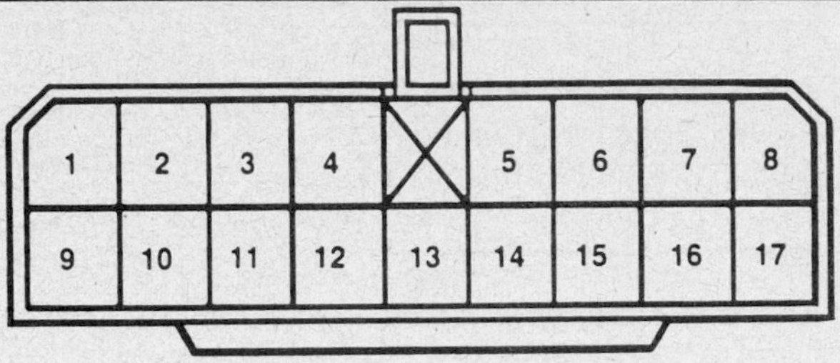
1
2
3
4
5
6
7
8
9
10
11
12
13
14
15
16
17

WIRE COLOR

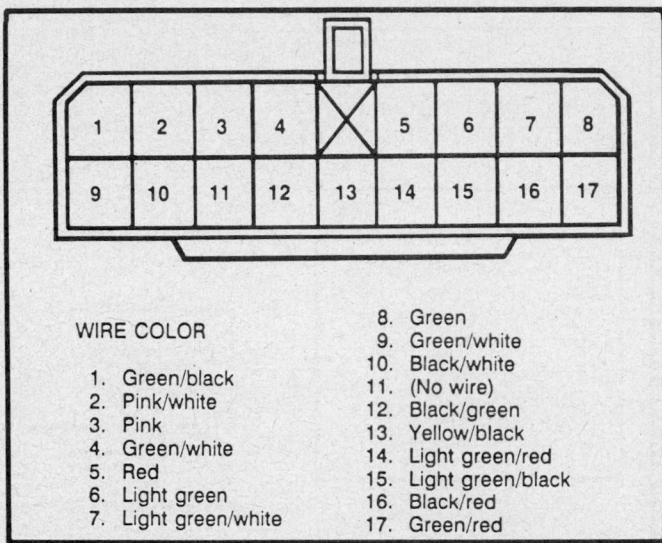

1. Green/black
2. Pink/white
3. Pink
4. Green/white
5. Red
6. Light green
7. Light green/white
8. Green
9. Green/white
10. Black/white
11. (No wire)
12. Black/green
13. Yellow/black
14. Light green/red
15. Light green/black
16. Black/red
17. Green/red

View of the controller's wiring harness connector

1–2 SHIFT VALVE

The 1–2 shift valve, located in the lower valve body, provides the gear shift between the 1st and 2nd gears.

When the controller operates the 2nd brake solenoid, line pressure is applied to the shift valve, the valve moves providing line pressure to the 2nd brake, thus, shifting from 1st-to-2nd gear. When the 2nd brake solenoid is de-energized, a spring forces the shift valve to return, thus, shifting from 2nd-to-1st gear.

NOTE: When the shift selector is placed in L or R range, the 2nd brake solenoid operates applying fluid pressure to the 1st/reverse brake.

2–3 SHIFT VALVE

The 2–3 shift valve, located in the lower valve body, provides the gear shift between the 2nd and 3rd gears.

When the controller operates the direct clutch solenoid, line pressure is applied to the shift valve, the valve moves providing line pressure to the direct clutch, thus, shifting from 2nd-to-3rd gear. When the direct clutch solenoid is de-energized, a spring forces the shift valve to return, thus, shifting from 3rd-to-2nd gear.

A–2 CHECK PROCEDURE
SHIFT LEVER SWITCH

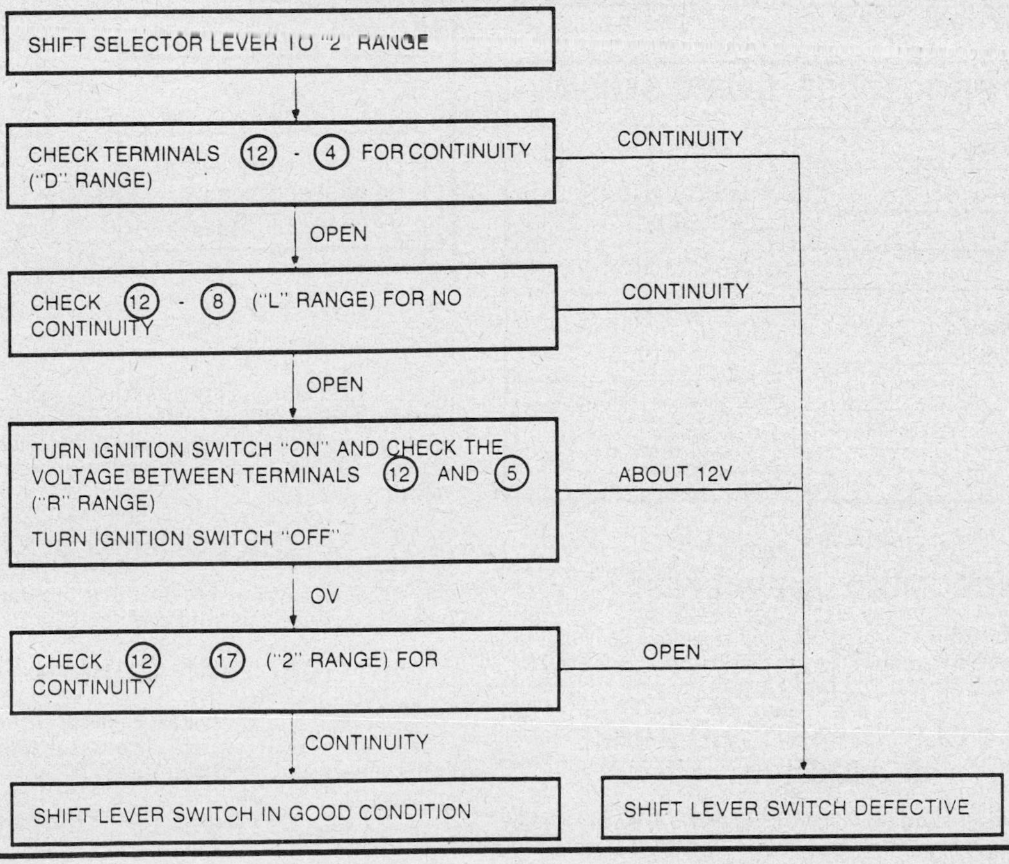
SHIFT SELECTOR LEVER TO "2" RANGE
CHECK TERMINALS (12) - (4) FOR CONTINUITY ("D" RANGE)
CONTINUITY
OPEN
CHECK (12) - (8) ("L" RANGE) FOR NO CONTINUITY
CONTINUITY
OPEN
TURN IGNITION SWITCH "ON" AND CHECK THE VOLTAGE BETWEEN TERMINALS (12) AND (5) ("R" RANGE)
TURN IGNITION SWITCH "OFF"
ABOUT 12V
OV
CHECK (12) (17) ("2" RANGE) FOR CONTINUITY
OPEN
CONTINUITY
SHIFT LEVER SWITCH IN GOOD CONDITION
SHIFT LEVER SWITCH DEFECTIVE

A-3 CHECK PROCEDURE SHIFT LEVER SWITCH

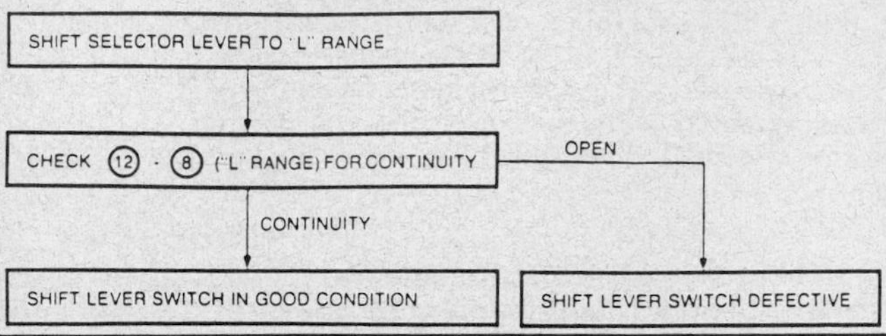
SHIFT SELECTOR LEVER TO "L" RANGE
CHECK (12) - (8) ("L" RANGE) FOR CONTINUITY
OPEN
CONTINUITY
SHIFT LEVER SWITCH IN GOOD CONDITION
SHIFT LEVER SWITCH DEFECTIVE

A-4 CHECK PROCEDURE SHIFT LEVER SWITCH

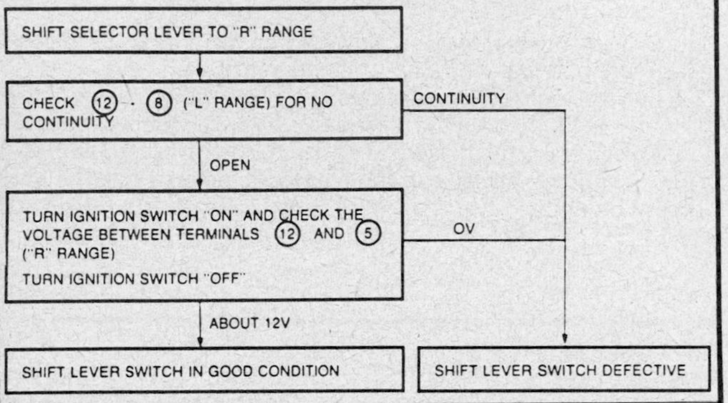
SHIFT SELECTOR LEVER TO "R" RANGE
CHECK (12) - (8) ("L" RANGE) FOR NO CONTINUITY
CONTINUITY
OPEN
TURN IGNITION SWITCH "ON" AND CHECK THE VOLTAGE BETWEEN TERMINALS (12) AND (5) ("R" RANGE)
TURN IGNITION SWITCH "OFF"
OV
ABOUT 12V
SHIFT LEVER SWITCH IN GOOD CONDITION
SHIFT LEVER SWITCH DEFECTIVE

B CHECK PROCEDURE, SPEED SENSOR

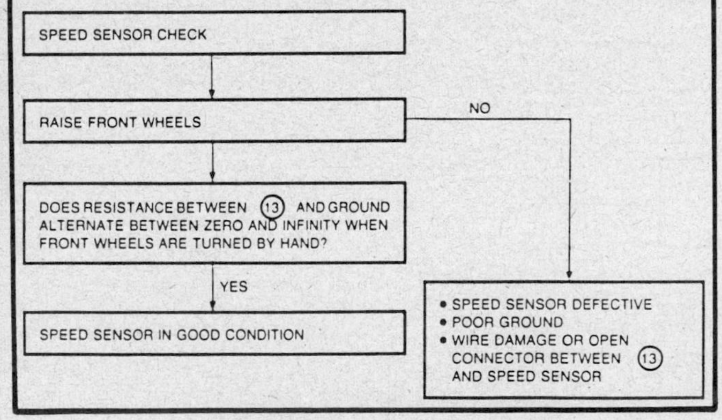
SPEED SENSOR CHECK
RAISE FRONT WHEELS
NO
DOES RESISTANCE BETWEEN (13) AND GROUND ALTERNATE BETWEEN ZERO AND INFINITY WHEN FRONT WHEELS ARE TURNED BY HAND?
YES
SPEED SENSOR IN GOOD CONDITION
• SPEED SENSOR DEFECTIVE
• POOR GROUND
• WIRE DAMAGE OR OPEN CONNECTOR BETWEEN (13) AND SPEED SENSOR

C CHECK PROCEDURE, SOLENOID

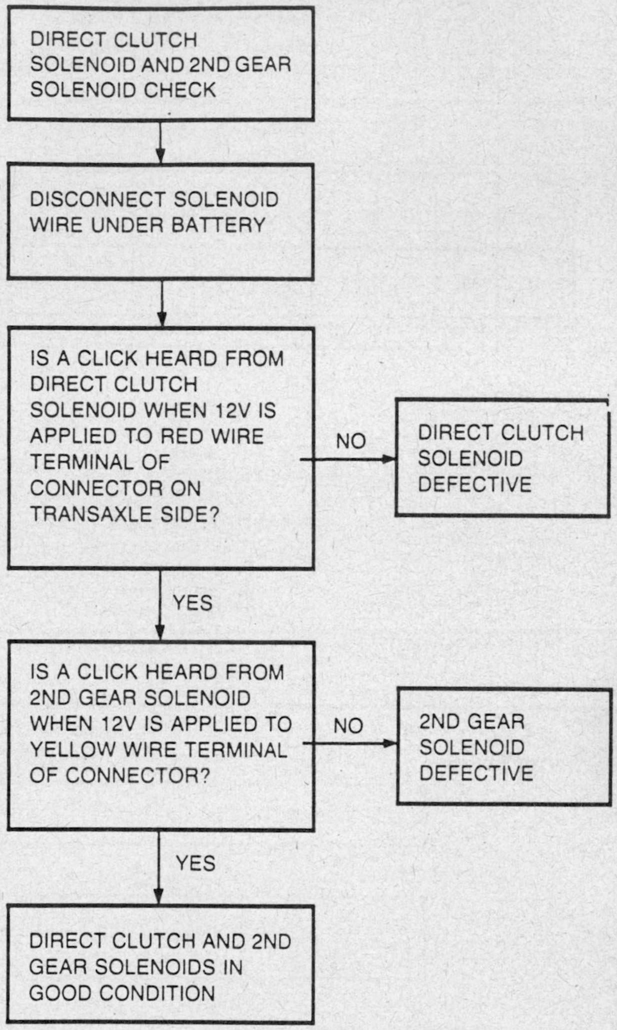
DIRECT CLUTCH SOLENOID AND 2ND GEAR SOLENOID CHECK
DISCONNECT SOLENOID WIRE UNDER BATTERY
IS A CLICK HEARD FROM DIRECT CLUTCH SOLENOID WHEN 12V IS APPLIED TO RED WIRE TERMINAL OF CONNECTOR ON TRANSAXLE SIDE?
NO
DIRECT CLUTCH SOLENOID DEFECTIVE
YES
IS A CLICK HEARD FROM 2ND GEAR SOLENOID WHEN 12V IS APPLIED TO YELLOW WIRE TERMINAL OF CONNECTOR?
NO
2ND GEAR SOLENOID DEFECTIVE
YES
DIRECT CLUTCH AND 2ND GEAR SOLENOIDS IN GOOD CONDITION

ACCUMULATOR VALVES

The accumulator valves, located between the valve body and the main housing, serve to reduce the gear shift shock; 1 for the forward clutch and 1 for the 2nd brake.

DIRECT CLUTCH AND 2ND BRAKE SOLENOIDS

The solenoids, located on the lower valve body, are activated by signals from the controller (computer) to control the gear shifting. The 2nd brake solenoid operates the 1–2 shift valve and the direct clutch solenoid operates the 2–3 shift valve. When the solenoid is turned **ON**, it's valve moves upward to relieve the line pressure on the shift valve; when the solenoid is turned **OFF**, it's valve moves downward applying line pressure to the shift valve.

OIL COOLING SYSTEM

The oil cooling system is a dual pipe type built into the lower radiator tank and is used to cool the transaxle fluid.

Gear Shift Control System

The gear shift control system consists of the controller, vacuum switch, shift lever switch, direct clutch/2nd brake solenoids, speed sensor and accelerator switch.

CONTROLLER (COMPUTER)

The controller, located inside the left corner of the main instru-

D CHECK PROCEDURE
ACCELERATOR AND VACUUM SWITCHES

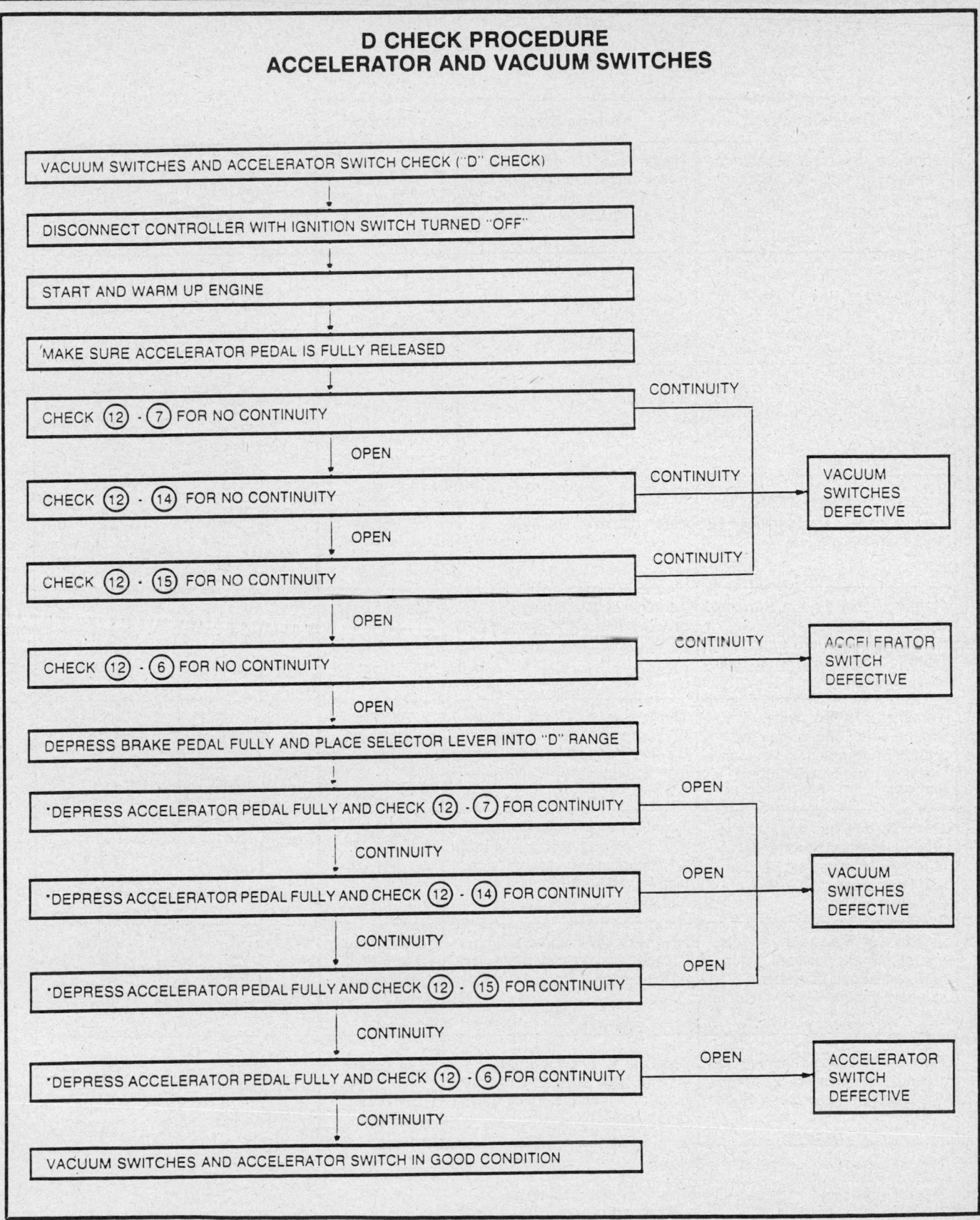

1. VACUUM SWITCHES AND ACCELERATOR SWITCH CHECK ("D" CHECK)
2. DISCONNECT CONTROLLER WITH IGNITION SWITCH TURNED "OFF"
3. START AND WARM UP ENGINE
4. MAKE SURE ACCELERATOR PEDAL IS FULLY RELEASED
5. CHECK (12) - (7) FOR NO CONTINUITY — CONTINUITY: VACUUM SWITCHES DEFECTIVE; OPEN: go to next step
6. CHECK (12) - (14) FOR NO CONTINUITY — CONTINUITY: VACUUM SWITCHES DEFECTIVE; OPEN: go to next step
7. CHECK (12) - (15) FOR NO CONTINUITY — CONTINUITY: VACUUM SWITCHES DEFECTIVE; OPEN: go to next step
8. CHECK (12) - (6) FOR NO CONTINUITY — CONTINUITY: ACCELERATOR SWITCH DEFECTIVE; OPEN: go to next step
9. DEPRESS BRAKE PEDAL FULLY AND PLACE SELECTOR LEVER INTO "D" RANGE
10. *DEPRESS ACCELERATOR PEDAL FULLY AND CHECK (12) - (7) FOR CONTINUITY — OPEN: VACUUM SWITCHES DEFECTIVE; CONTINUITY: go to next step
11. *DEPRESS ACCELERATOR PEDAL FULLY AND CHECK (12) - (14) FOR CONTINUITY — OPEN: VACUUM SWITCHES DEFECTIVE; CONTINUITY: go to next step
12. *DEPRESS ACCELERATOR PEDAL FULLY AND CHECK (12) - (15) FOR CONTINUITY — OPEN: VACUUM SWITCHES DEFECTIVE; CONTINUITY: go to next step
13. *DEPRESS ACCELERATOR PEDAL FULLY AND CHECK (12) - (6) FOR CONTINUITY — OPEN: ACCELERATOR SWITCH DEFECTIVE; CONTINUITY: go to next step
14. VACUUM SWITCHES AND ACCELERATOR SWITCH IN GOOD CONDITION

CONTROLLER CONNECTOR IDENTIFICATION

OHM Check	Voltage Check	Function		
Approximately 13 Ω resistance between terminal #9 and ground, such as terminal #12.	(Selector in "D" range.) Controller output: high voltage (12V) in 1st or 2nd gear operation, low voltage (0V) in 3rd gear operation.	Direct Clutch Solenoid Output	9 Green/White	1
No check.	Battery voltage, ignition on.	Switched Ignition	10 Black/White	2
			11 Not Used	3
Continuity between terminal #12 and alternate ground.	Zero volts at all times.	Ground	12 Black/Green	4
Referenced to ground, terminal #13 alternates between continuity and open circuit (zero Ω and infinity) when drive wheels turn.	Terminal #13 alternates between high (12V) and low (0V) voltage when drive wheels turn.	Speed Sensor Input	13 Yellow/Black	
Terminal #14 has continuity to ground when engine has less than 6.7" Hg manifold vacuum, and is an open circuit above 6.7" Hg.	Terminal #14 has low voltage (0V) when engine has less than 6.7" Hg manifold vacuum, and high voltage (12V) above 6.7" Hg.	Vacuum Switch #2 Input	14 Lt. Green/Red	5
Terminal #15 has continuity to ground when engine has less than 3.5" Hg manifold vacuum, and is an open circuit above 3.5" Hg.	Terminal #15 has low voltage (0V) when engine has less than 3.5" Hg manifold vacuum, and high voltage (12V) above 3.5" Hg.	Vacuum Switch #3 Input	15 Lt. Green/Black	6
Terminal #16 has continuity to ground in park position. (Will read resistance of starter solenoid circuit.) Open circuit is in gear.	No check. (Will read cranking voltage, ignition in start position, park position.)	Park Input From Shift Lever Switch	16 Black/Red	7
Terminal #17 has continuity to ground, shift lever in "2" range. Open circuit in any other range position.	Terminal #17 has low (0V) voltage, shift lever in "2" range, high (12V) voltage in any other range position.	"2" Range Input From Shift Lever Switch	17 Green/Red	8

CONTROLLER CONNECTOR IDENTIFICATION

		Function	Voltage Check	OHM Check
9	1 Green/Black	2nd Gear Solenoid Output	(Selector in "D" Range.) Controller output: high voltage (12V) in first gear operation, low voltage (0V) in 2nd or 3rd gear operation.	Approximately 13 Ω resistance between terminal #1 and ground such as terminal #12.
10	2 Pink/White	Output to ECM Speed Signal	Same as terminal #13.	Same as terminal #13.
11	3 Pink	Output to ECM Park Signal	Same as terminal #16.	Same as terminal #16.
12	4 Green/White	"D" Range Input From Shift Lever Switch	Terminal #4 has low voltage (0V), shift lever in "D" range, high voltage (12V) in any other range position.	Terminal #4 has continuity to ground, shift lever in "D" range, open circuit in any other range position.
13				
14	5 Red	Reverse Input From Shift Lever Switch	Terminal #5 has high voltage (12V), shift lever in "R" range, low voltage (0V) in any other range position.	No check. (Will read resistance of backup lamp circuit in any range position.)
15	6 Lt. Green	Wide-Open Throttle Input From Accelerator Switch	Terminal #6 has low voltage (0V) at more than 90% accelerator pedal travel, high voltage (12V) at less than 90% accelerator pedal travel.	Terminal #6 has continuity to ground at more than 90% accelerator pedal travel. Open circuit at less than 90% accelerator pedal travel.
16	7 Lt. Green/White	Vacuum Switch #1 Input	Terminal #7 has low voltage (0V) when engine has less than 11.8" Hg manifold vacuum, and high voltage (12V) above 11.8" Hg.	Terminal #7 has continuity when engine has less than 11.8" Hg manifold vacuum, and is an open circuit above 11.8" Hg.
17	8 Green	"L" Range Input From Shift Lever Switch	Terminal #8 has low voltage (0V), shift lever in "L" range, high voltage (12V) in any other range position.	Terminal #8 has continuity to ground, shift lever in "L" range, open circuit in any other range position.

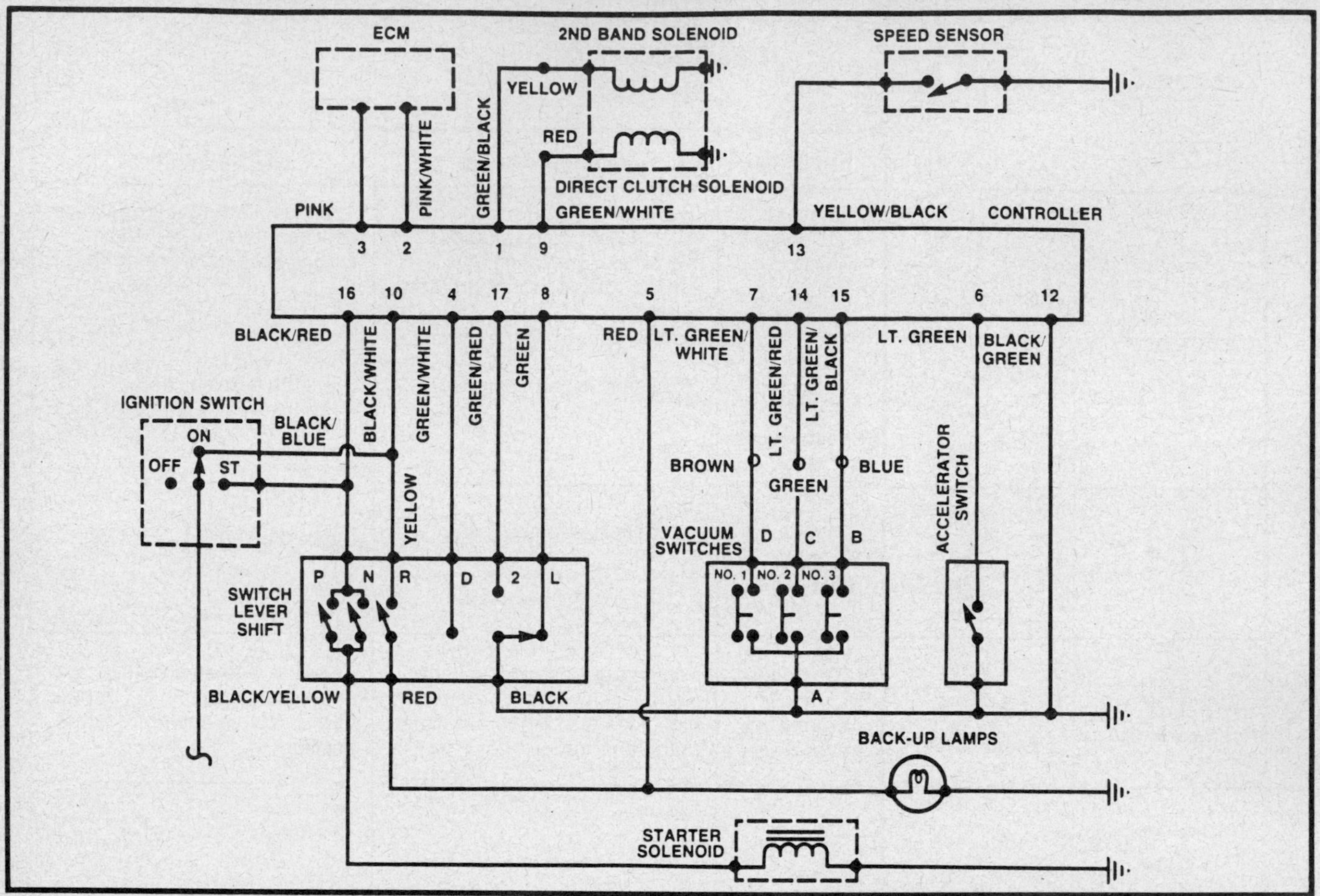
ECM
2ND BAND SOLENOID
SPEED SENSOR
YELLOW
RED
DIRECT CLUTCH SOLENOID
PINK
PINK/WHITE
GREEN/BLACK
GREEN/WHITE
YELLOW/BLACK
CONTROLLER
BLACK/RED
BLACK/WHITE
GREEN/WHITE
GREEN/RED
GREEN
RED
LT. GREEN/WHITE
LT. GREEN/RED
LT. GREEN/BLACK
LT. GREEN
BLACK/GREEN
IGNITION SWITCH
ON
OFF
ST
BLACK/BLUE
YELLOW
BROWN
GREEN
BLUE
VACUUM SWITCHES
ACCELERATOR SWITCH
SWITCH LEVER SHIFT
NO. 1
NO. 2
NO. 3
BLACK/YELLOW
RED
BLACK
BACK-UP LAMPS
STARTER SOLENOID

Schematic of the gear shift control system

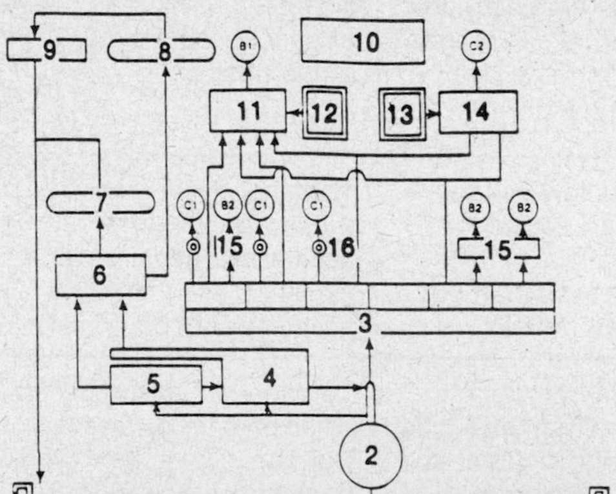

1. Oil pan
2. Oil pump
3. Manual valve
4. Primary regulator valve
5. Throttle valve
6. Secondary regulator valve
7. Lubrication
8. Torque converter
9. Oil cooler
10. Controller (computer)
11. 1-2 shift valve
12. 2nd brake solenoid
13. Direct clutch solenoid
14. 2-3 shift valve
15. B2 control valve
16. Accumulator
C1. Forward clutch
C2. Direct clutch
B1. 2nd brake
B2. 1st/reverse brake

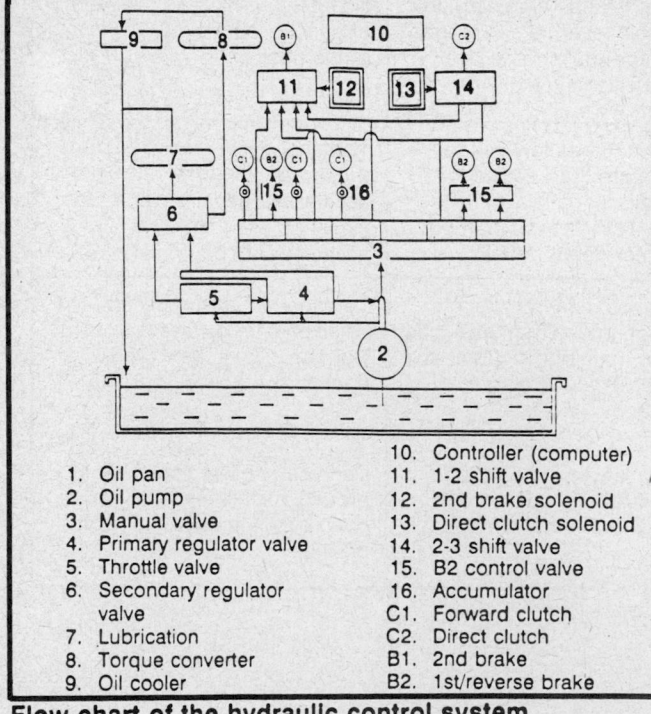

Flow chart of the hydraulic control system

ment panel, collects information from the accelerator switch, 3 vacuum switches, shift lever switch and speed sensor. After monitoring the information, it sends out signals to open/close the direct clutch and 2nd brake solenoids to control the transaxle's shifting.

VACUUM SWITCHES

Three vacuum switches, each having its own range of operation, monitor the throttle valve opening (engine load) and send the information to the controller.

SHIFT LEVER SWITCH

The shift lever switch, located on the gear shift selector, changes the shift lever positions into electrical signals and sends them to the controller. The switch's contact points (**P** and **N**) are connected to the starter motor circuit; the starter cannot be activated unless the shift selector is in the proper position.

DIRECT CLUTCH AND 2ND BRAKE SOLENOIDS

The solenoids, located on the lower valve body, are activated by signals from the controller (computer) to control the gear shifting. The 2nd brake solenoid operates the 1-2 shift valve and the direct clutch solenoid operates the 2-3 shift valve.

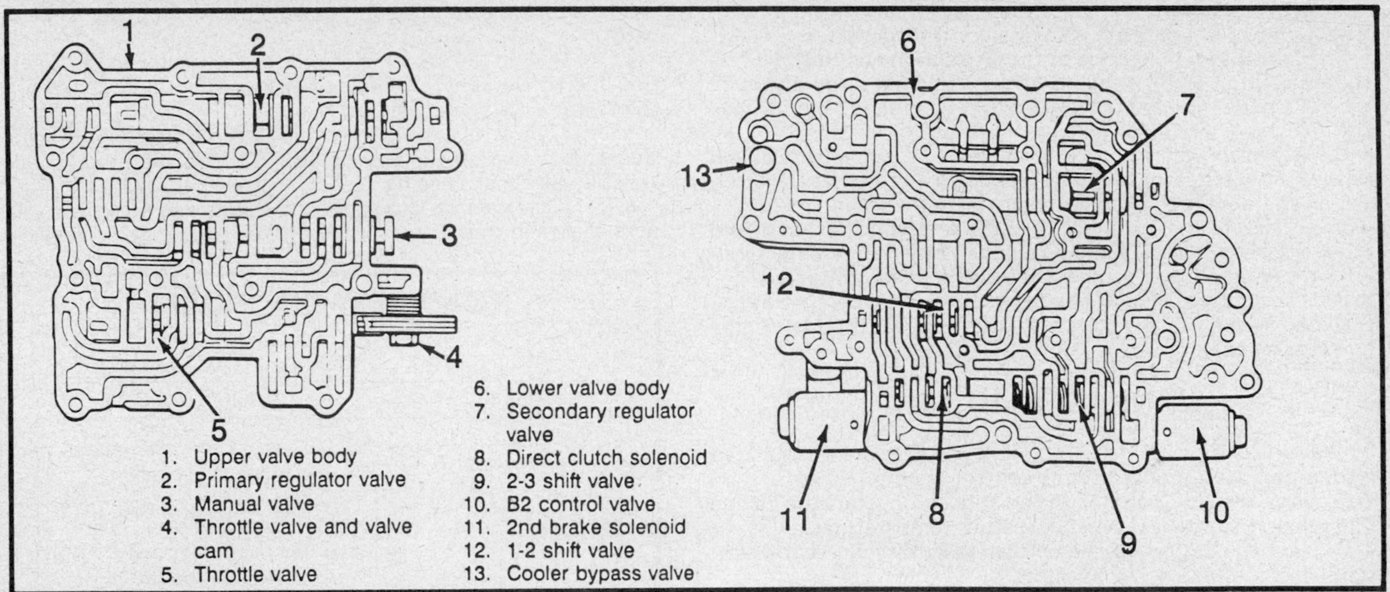
1
2
3
4
5
6
7
8
9
10
11
12
13
1. Upper valve body
2. Primary regulator valve
3. Manual valve
4. Throttle valve and valve cam
5. Throttle valve
6. Lower valve body
7. Secondary regulator valve
8. Direct clutch solenoid
9. 2-3 shift valve
10. B2 control valve
11. 2nd brake solenoid
12. 1-2 shift valve
13. Cooler bypass valve

Location of the valve body components

SPEED SENSOR

The speed sensor, built into the speedometer, consists of a lead switch and a magnet. As the magnet (attached to the speedometer cable) rotates, it's magnetic field causes the lead switch to turn **ON** and **OFF**. While the switching frequency increases or decreases in proportion with the vehicle speed, it is sent to the controller as pulse signals.

ACCELERATOR SWITCH

The accelerator switch, mounted on the accelerator pedal bracket, turns **ON** when the accelerator pedal is depressed 2.0–2.2 in. (50–56mm) and signals the controller of the throttle valve opening.

Diagnosis Tests

CONTROL PRESSURE TEST

This test is designed to check the oil pressure system by measuring the operating oil pressure.

1. Operate the vehicle until the transaxle's fluid is at normal operating temperatures. Make sure the oil level is between **FULL HOT** and **LOW HOT** and the transaxle shows no signs of oil leaks.
2. With the engine turned **OFF**, remove the oil plug (upper rear side of the oil pan) and install an oil pressure gauge in the threaded hole.
3. Install a tachometer onto the engine.
4. Firmly apply the parking brake and block the drive wheels.
5. Place the shift selector in the **P** position and start the engine.
6. Fully depress the brake pedal and place the shift selector in the **D**. Check the oil pressure at 700–800 rpm and at 2000–2400 rpm; the oil pressure should be 29–57 psi (200–400 kPa) and 57–85 psi (400–600 kPa) respectively.

NOTE: Do not operate the engine at high speed for more than 5 seconds for the oil temperature may rise excessively high.

7. Fully depress the brake pedal and place the shift selector in the **R** position. Check the oil pressure at 700–800 rpm and at 2000–2400 rpm; the oil pressure should be 29–57 psi (200–400 kPa) and 57–85 psi (400–600 kPa) respectively.

STALL SPEED TEST

This test is designed to check the overall performance.

NOTE: Do not operate the engine at high speed for more than 5 seconds for the oil temperature may rise excessively high.

1. Operate the vehicle until the transaxle's fluid is at normal operating temperatures. Make sure the oil level is between **FULL HOT** and **LOW HOT**.
2. Install a tachometer onto the engine.
3. Firmly apply the parking brake and block the drive wheels.
4. Place the shift selector in the **P** position and start the engine.
5. Fully depress the brake pedal. Place the shift selector in the **D** and fully depress the accelerator. While watching the tachometer, note when the engine speed becomes constant (stall speed). When the stall speed has been reached, immediately release the accelerator.
6. Fully depress the brake pedal and place the shift selector in the **R** position. While watching the tachometer, note when the engine speed becomes constant (stall speed). When the stall speed has been reached, immediately release the accelerator.
7. The stall speed should be within 2000–2400 rpm; if not, use the stall speed diagnosis chart to determine and correct the problem.

ROAD TEST

This test is designed to check the upshifts and downshifts at specific speeds.

1. Operate the engine until normal operating temperatures are reached.
2. Disconnect the vacuum switch(s) electrical harness connector.
3. Place the shift selector in the **D** position, depress the accelerator ½ stroke and accelerate the vehicle. Note, if the upshifts occur from 1st-to-2nd at 15 mph (25 km/h) and from 2nd-to-3rd at 38 mph (62 km/h).

4. Stop the vehicle and place the shift selector in the **P** position. Connect the vacuum switch electrical connector.

5. Place the shift selector in the **D** position and fully depress the accelerator. Note, it the upshifts occur from 1st-to-2nd at 32 mph (52 km/h) and from 2nd-to-3rd at 60 mph (97 km/h). Stop the vehicle.

6. Accelerate and operate the vehicle at 18 mph (30 km/h) and release the accelerator (for 1–2 seconds). Fully depress the accelerator and note if the vehicle downshifts from 2nd-to-1st.

7. Accelerate and operate the vehicle at 47 mph (75 km/h) and release the accelerator (for 1–2 seconds). Fully depress the accelerator and note if the vehicle downshifts from 3rd-to-2nd.

8. If the correct upshifts or downshifts do not occur at each specified speed, use the road test diagnosis chart to determine and correct the problem.

MANUAL ROAD TEST

This test is designed to check the **L**, **2** and **D** gears when driven with a non-operating gear shift control system.

1. Place the shift selector in the **P** position, start the engine and allow it to reach normal operating temperatures.

2. Near the bottom of the battery tray, disconnect the solenoid electrical harness connector.

3. Place the shift selector in the **L** position and accelerate the vehicle to 18 mph (30 km/h); make sure the 1st gear is being used.

4. At 18 mph (30 km/h), move the shift selector to the **2** position and accelerate the vehicle to 37 mph (60 km/h); make sure the 2nd gear is being used.

5. At 37 mph (60 km/h), move the shift selector to the **D** position and accelerate the vehicle over 37 mph (60 km/h); make sure the 3rd gear is being used.

6. Stop the vehicle, turn **OFF** the engine and reconnect solenoid electrical connector.

Engine rpm	Line pressure	
	"D" range	"R" range (changed)
700 - 800 rpm	2 - 4 kg/cm² 28.5 - 56.8 psi 200 - 400 kPa	5.5 - 8 kg/cm² 78.3 - 113.7 psi 550 - 800 kPa
2,000 - 2,400 rpm (Stall speed)	4 - 6 kg/cm² 56.9 - 85.3 psi 400 - 600 kPa	8.5 - 12 kg/cm² 120.9 - 170.6 psi 850 - 1,200 kPa

Control pressure test chart

ROAD TEST DIAGNOSIS

Condition	Possible causes
No upshift from 1st to 2nd	1 – 2 shift valve defective 2nd brake solenoid defective Controller defective, or disconnection or poor connection in controller electric circuit
No upshift from 2nd to 3rd	2 – 3 shift valve defective Direct clutch solenoid defective Controller defective, or disconnection or poor connection in controller electric circuit
No downshift from 2nd to 1st or 3rd to 2nd	Accelerator switch defective Controller defective, or disconnection or poor connection in controller electric circuit

CONTROL PRESSURE DIAGNOSIS

Line pressure measured	Possible cause
Higher than specification in "D" & "R" ranges	Regulator valve defective Throttle valve defective Accelerator cable and oil pressure control cable maladjusted
Lower than specification in "D" & "R" ranges	Oil pump defective Regulator valve defective Throttle valve defective Accelerator cable and oil pressure control cable maladjusted
Lower than specification only in "D" range	Forward clutch oil pressure system oil leakage "D" range oil pressure system oil leakage
Lower than specification only in "R" range	Direct clutch oil pressure system oil leakage 1st – reverse brake oil pressure system oil leakage "R" range oil pressure system oil leakage

STALL SPEED DIAGNOSIS

Stall speed measured	Possible causes
Lower than specification	Engine output insufficient Torque converter defective
Higher than specification in "D" range	Forward clutch slippage One-way clutch defective
Higher than specification in "R" range	Direct clutch slippage 1st – reverse brake slippage

ENGINE BRAKE TEST

1. While driving the vehicle in the **D** position and in 3rd gear, move the shift selector to the **2** position and note if the engine brake operates. If the engine fails to brake, the 2nd brake is defective.
2. While driving the vehicle in the **D** position and in 3rd gear, move the shift selector to the **L** position and note if the engine brake operates. If the engine fails to brake, the 1st/reverse brake is defective.

PARK TEST

1. Stop the vehicle on a slope, move the shift selector to the **P** position and firmly apply the parking brake.
2. Turn the engine **OFF**, release the parking brake and note if the vehicle remains stationary.

ON CAR SERVICES

Adjustments

OIL PRESSURE CONTROL CABLE

The oil pressure control cable operates the throttle valve cam.

1. Inspect and/or adjust the accelerator cable play by performing the following procedures:
   a. At the carburetor, check the amount of play in the accelerator cable; it should be 0.40–0.59 in. (10–15mm) cold or 0.12–0.19 in. (3–5mm) warm.
   b. If necessary to adjust, loosen the locknut and turn the adjustment nut until the correct specifications are met.
   c. After adjustment, tighten the locknut.
2. Operate the engine until normal operating temperatures are reached and allow the engine to idle; make sure the carburetor is not on the fast idle step.
3. From near transaxle's dipstick tube, remove the oil pressure control cable cover. Using a feeler gauge, check that the boot-to-inner cable stopper clearance is 0–0.02 in. (0–0.5mm).
4. If the clearance is not within specifications, perform the following procedures:
   a. Loosen the **A** adjusting nuts (engine side of bracket) and turn them to adjust the clearance.
   b. If the adjustment (engine side) fails to establish the clearance, tighten the nuts and move the other side (dipstick side) of the bracket.
   c. Loosen the **B** adjusting nuts (dipstick side of bracket) and turn them to adjust the clearance.
   d. After adjustment is complete, tighten the adjusting nuts, recheck adjustment and install the cover.

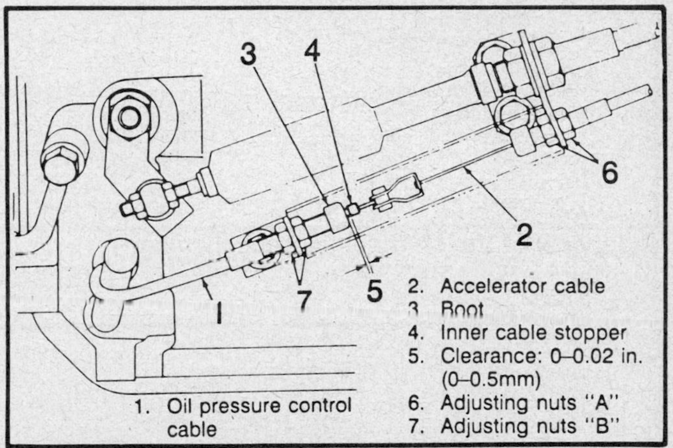
3
4
6
2
1
7
5
2. Accelerator cable
3. Boot
4. Inner cable stopper
5. Clearance: 0–0.02 in. (0–0.5mm)
6. Adjusting nuts "A"
7. Adjusting nuts "B"
1. Oil pressure control cable

Checking and adjusting the oil pressure control cable

BACK DRIVE SOLENOID

1. Remove the console cover and loosen the solenoid mounting screws.
2. Move the shift selector into the **P** position.
3. Using grease, lubricate the upper/lower edges of the lock plate.
4. Adjust the solenoid so the following situations exist:
   a. When the ignition switch is turned **OFF**, the solenoid is inoperative; when the ignition switch is turned **ON**, the solenoid is operative.
   b. There is to be no clearance between the lock plate and the guide plate.
5. After adjusting the solenoid, tighten the solenoid's mounting screws.
6. After tightening the solenoid screws, the following situations must exist:
   a. When the ignition switch is turned **OFF**, the shift selector should be locked in the **P** position and cannot be moved to any other position.
   b. When the ignition switch is turned **ON**, the shift selector can be moved from **P** to any other position.
   c. If the manual release knob is pulled and the ignition switch is turned **OFF**, the shift selector can be shifted from **P** to any other position.
7. Install the console cover.

BACK DRIVE CABLE

1. Place the shift selector into the **P** position.
2. Loosen both (**A** and **B**) back drive cable nuts.
3. Pull the outer wire forward so there is no deflection on the inner wire and tighten both nuts (hand tight); tighten **A** first and **B** second. Tighten both nuts securely.
4. After tightening the nuts, check the following situations:
   a. With the shift selector in the **P** position, the ignition key can be turned from **ACC** to **LOCK** position and removed from the ignition switch.

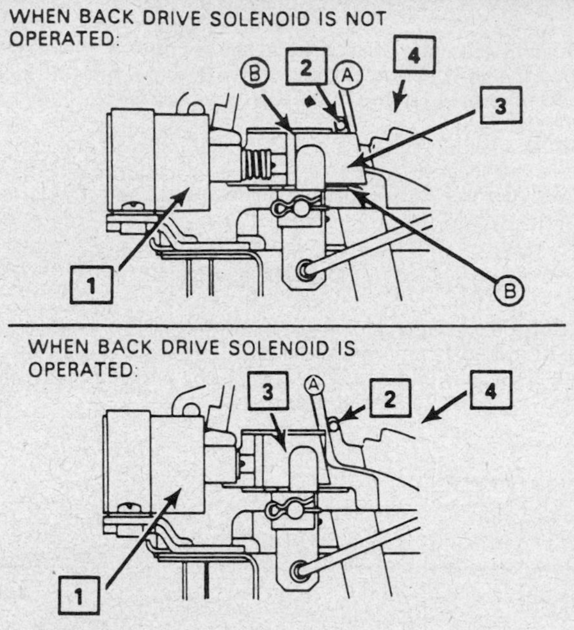
WHEN BACK DRIVE SOLENOID IS NOT OPERATED:
WHEN BACK DRIVE SOLENOID IS OPERATED:

When back drive solenoid is not operated
When back drive solenoid is operated
1. Back drive solenoid
2. Detent pin
3. Lock plate
4. Detent plate
A. More than 0.04 in. (1mm)
B. Apply grease here

Side view of the shift selector mechanism

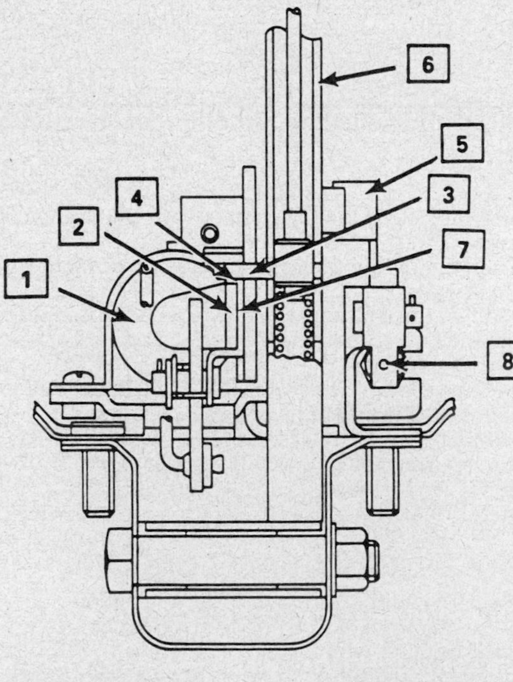

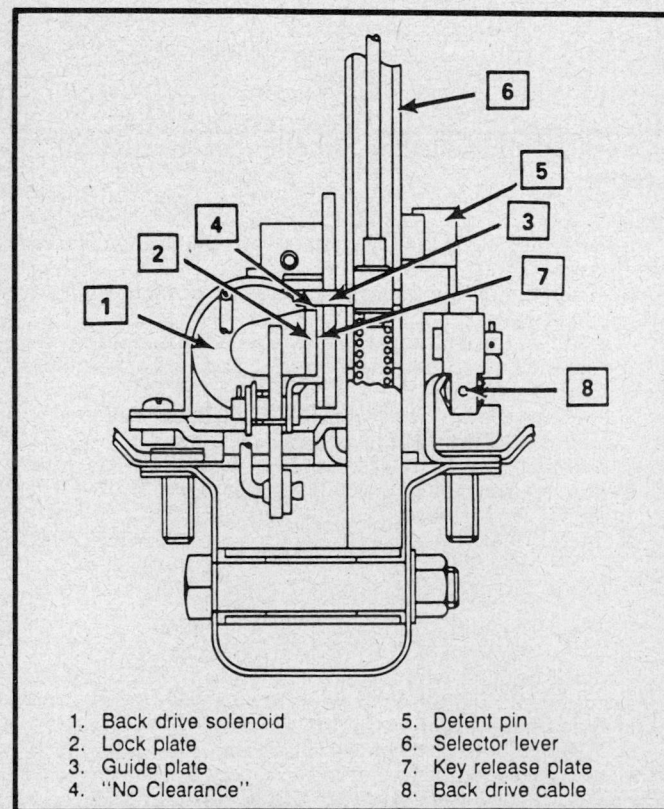

1. Back drive solenoid
2. Lock plate
3. Guide plate
4. "No Clearance"
5. Detent pin
6. Selector lever
7. Key release plate
8. Back drive cable

Front view of the shift selector mechanism

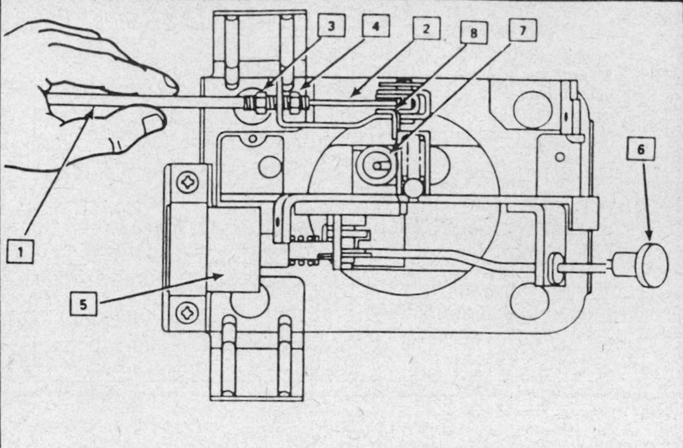

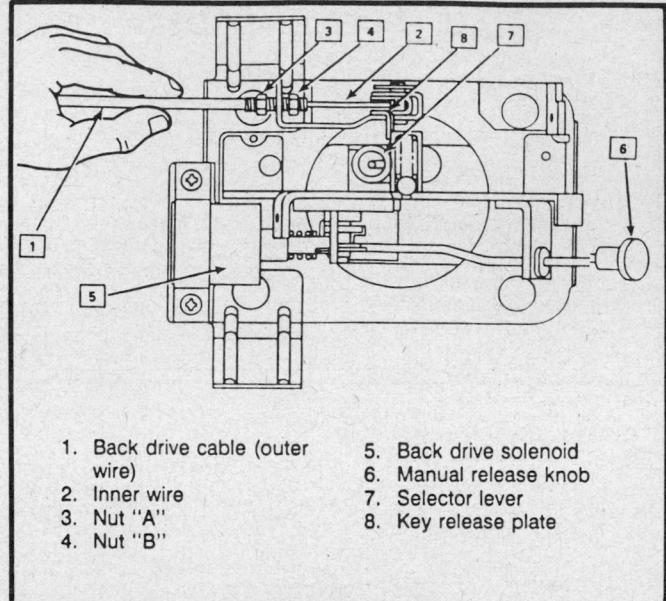

1. Back drive cable (outer wire)
2. Inner wire
3. Nut "A"
4. Nut "B"
5. Back drive solenoid
6. Manual release knob
7. Selector lever
8. Key release plate

Adjusting the back drive cable

b. With the shift selector in any position, other than **P**, the ignition key cannot be turned from **ACC** to **LOCK** position.

SHIFT SELECTOR CABLE

1. At the shift lever on the transaxle, loosen the shift selector cable adjusting nuts.

NOTE: It may be necessary to remove the shift lever switch.

2. Move the shift lever into the **N** position. From inside the vehicle, move the shift selector lever into the **N** position.
3. At the shift selector cable, turn adjusting nut **A** (hand tight) until it contacts the manual shift cable joint. Using a wrench, tighten adjusting nut **B**.
4. After the cable is adjusted, check for the following situations:
   a. Move the shift selector lever into the **P** position and start the engine; the vehicle should not move.
   b. Move the shift selector lever into the **N** position; the vehicle should not move under power.
   c. Move the shift selector lever into the **D**, **2** and **L** positions; the vehicle should move under power.
   d. Move the shift selector lever into the **R** position; the vehicle should move rearward under power.

SHIFT LEVER SWITCH

1. Remove the shift lever switch from the transaxle.
2. Using a flat blade prybar, turn the shift lever switch joint to align it with the shift lever shaft; stop at the position where a click is heard from the joint.
3. Install the switch and torque the bolt to 10–16 ft. lbs. (13–23 Nm). Install the couplers and the clamp.
4. After the switch is installed, check for the following situations:
   a. Firmly apply the parking brake and block the drive wheels.
   b. Move the shift selector lever to the **P** position. Turn the ignition switch **ON** and verify that the starter motor operates.

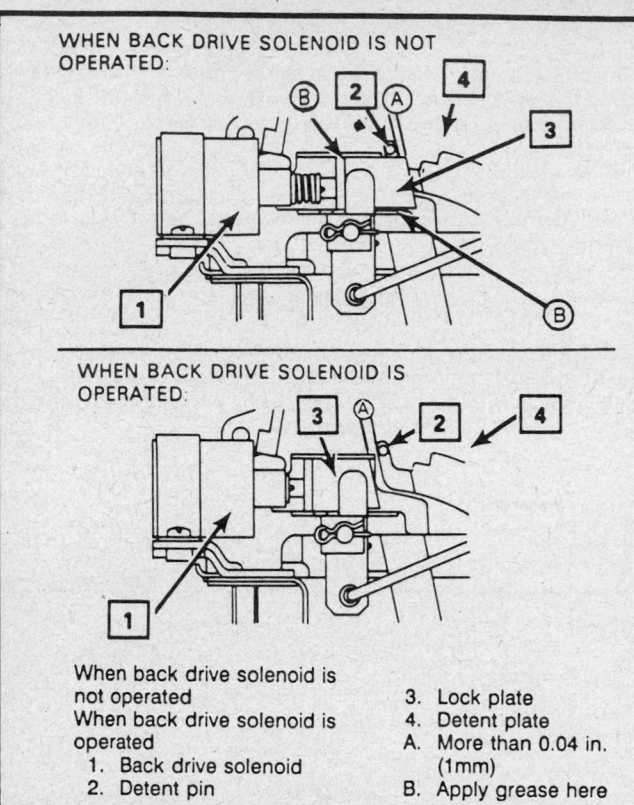

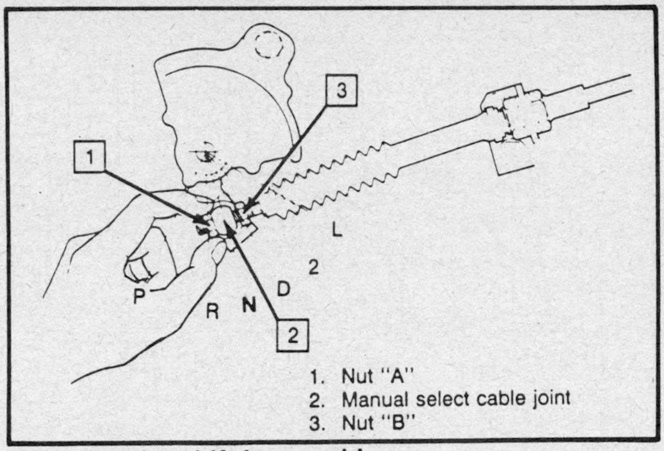
3
1
L
2
P
R
N
D
2
1. Nut "A"
2. Manual select cable joint
3. Nut "B"

Adjusting the shift lever cable

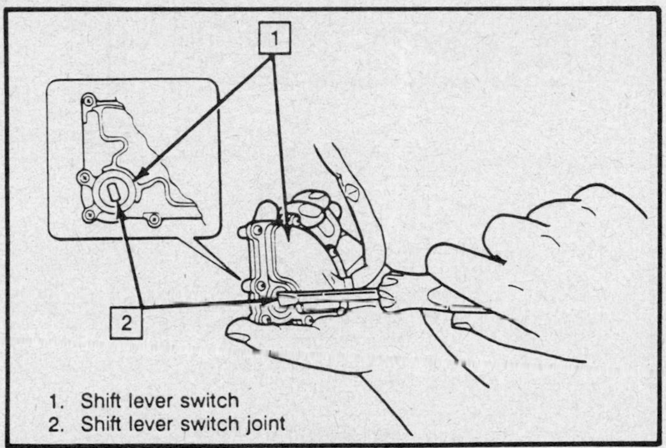
1
2
1. Shift lever switch
2. Shift lever switch joint

Adjusting the shift lever switch

c. Move the shift selector lever to the **N** position. Turn the ignition switch **ON** and verify that the starter motor operates.

d. Move the shift selector lever to the **L** position and back to the **N** position. Turn the ignition switch **ON** and verify that the starter motor operates.

e. Move the shift selector lever to the **P** position. Turn the ignition switch **ON** and verify that the starter motor operates.

f. Move the shift selector lever to any position, except the **P** or **N** positions. Turn the ignition switch **ON** and verify that the starter motor operates.

g. Turn the ignition switch **ON**, do not start the engine, move the shift selector lever into the **R** position and check that the backup lights turn **ON**.

h. If any of these situations do not exist, remove the shift lever switch and reperform the adjustment procedures.

Services

FLUID CHANGES

Under normal driving conditions, the fluid should be changed every 100,000 miles (160,000 km). Under severe driving conditions, the fluid should be changed every 15,000 miles (25,000 km). The following conditions are considered severe:

a. Heavy city traffic where the outside temperature reaches 90°F (32°C)

b. Very hilly or mountainous areas

c. Commercial use, such as, taxi, police or delivery service

Do not overfill the transaxle. It only takes 1 pint of fluid to change the level from add to full on the dipstick. Overfilling can cause damage to the internal components of the automatic transaxle.

Procedure

1. Raise and support the vehicle safely. If the transaxle is hot, allow it to cool.
2. Remove the drain plug and drain the fluid into a drain pan.
3. After draining, install the drain plug and gasket into the pan; torque the plug to 13–16 ft. lbs. (18–23 Nm).
4. Remove the dipstick and install a funnel in it's place. Using 3.16 qts. (1.5L) of Dexron®II automatic transmission fluid, install it through the dipstick tube.

NOTE: If refilling the transaxle after an overhaul (torque converter reused), use 3.7 qts. (3.5L) of fluid. If refilling the transaxle after an overhaul (torque converter dry), use 4.7 qts. (4.5L) of fluid.

5. Check the fluid level at both room temperature and operating temperature.

OIL PAN

Removal and Installation

1. Raise and support the vehicle safely. Drain the transaxle; if it is hot, allow it to cool.
2. Remove the stabilizer bar-to-chassis bolts and the bar.
3. Using a floor jack and a block of wood, support the transaxle by not blocking the oil pan.
4. Remove the transaxle mounting member bolts and the member.
5. Remove the oil pan bolts. Using a plastic hammer, tap around the oil pan to remove it; do not use a prybar.
6. Using a gasket scraper or a putty knife, clean the gasket from the oil pan and transaxle.
7. Using solvent, clean the oil pan; be sure to place the magnet directly below the oil strainer.
8. Using a new gasket, sealant (if necessary), install the oil pan. Using sealant on the cross grooved (on head) bolts, torque the bolts to 3–4 ft. lbs. (4–6 Nm).
9. Install the transaxle mounting member. Torque the member-to-transaxle nuts to 29–36 ft. lbs. (40–50 Nm) and the member-to-chassis bolts to 36–43 ft. lbs. (50–60 Nm).
10. Install the stabilizer bar. Torque the stabilizer bar-to-chassis bolts to 22–40 ft. lbs. (30–55 Nm).
11. Refill the transaxle with Dexron®II transmission fluid.

DIRECT CLUTCH AND 2ND BRAKE SOLENOIDS

Removal and Installation

1. Raise and support the vehicle safely.
2. Drain the transaxle and remove the oil pan.
3. Disconnect the electrical connectors from the direct clutch and 2nd brake solenoids.
4. Remove the solenoids.

NOTE: If removing the wiring harness, remove it with the grommet from the upper side of the transaxle.

5. When installing the wiring harness, use a new grommet (O-ring) seal if the old one is damaged.
6. Install the solenoids into the case; if the O-ring seals are damaged, replace them.
7. Install the electrical connectors to the solenoids.
8. Install the oil pan and refill with clean fluid.

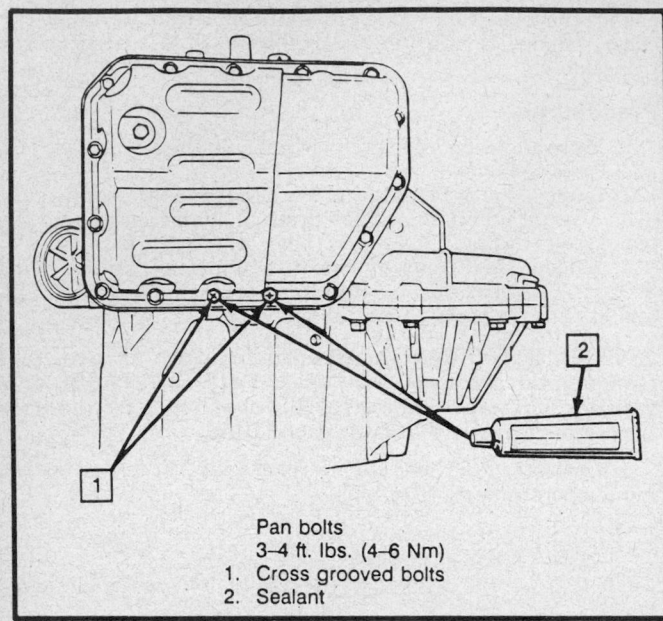

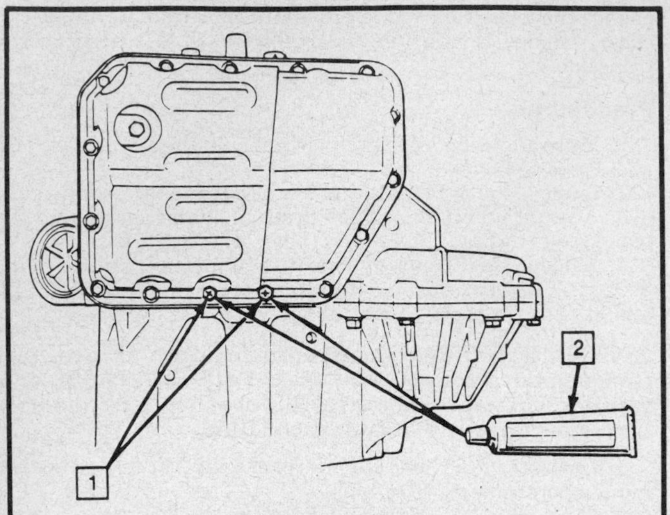
1
2
Pan bolts
3–4 ft. lbs. (4–6 Nm)
1. Cross grooved bolts
2. Sealant

Installing sealant on the cross grooved oil pan bolts

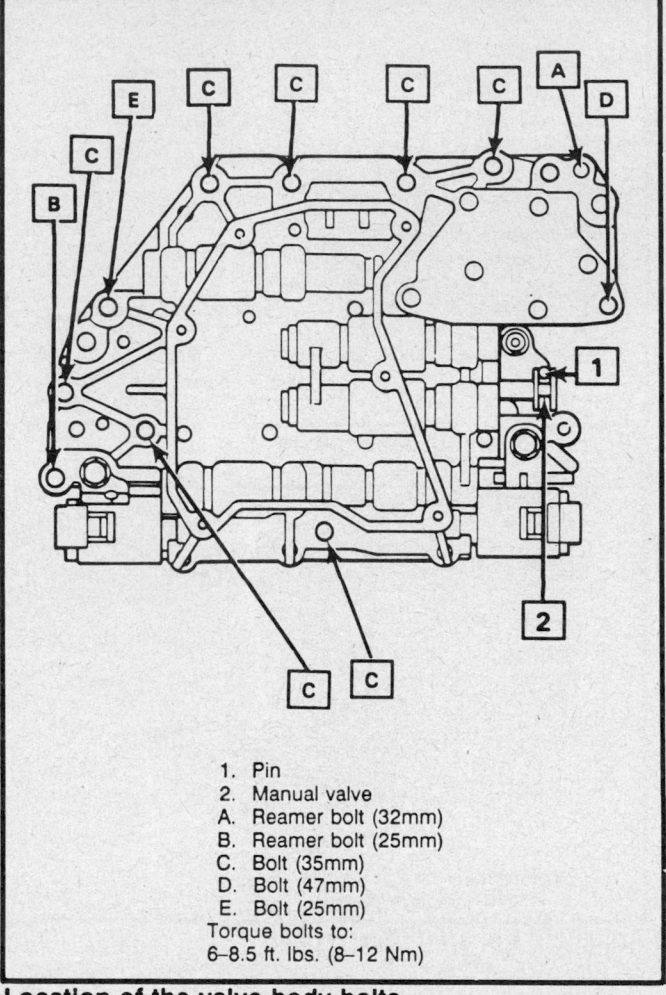

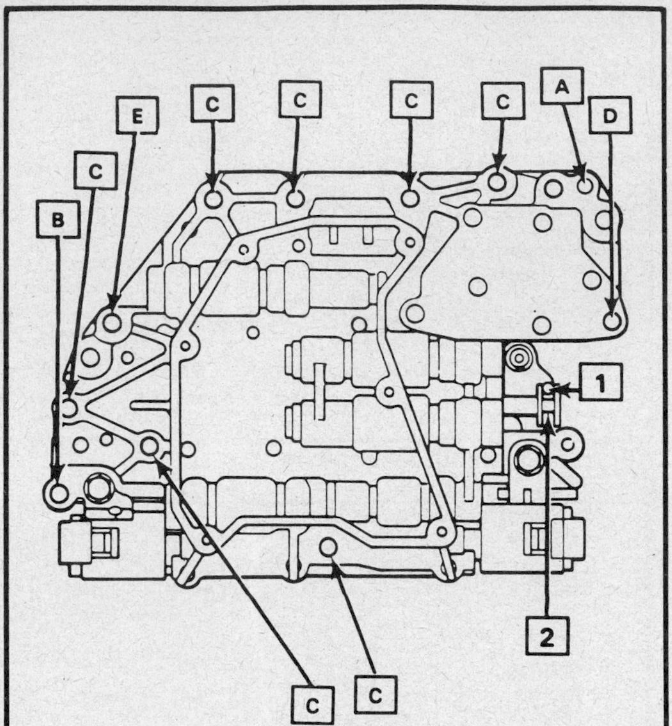
E
C
C
C
C
A
D
C
B
1
2
C
C
1. Pin
2. Manual valve
A. Reamer bolt (32mm)
B. Reamer bolt (25mm)
C. Bolt (35mm)
D. Bolt (47mm)
E. Bolt (25mm)
Torque bolts to:
6–8.5 ft. lbs. (8–12 Nm)

Location of the valve body bolts

VALVE BODY

Removal and Installation

1. Raise and support the vehicle safely.
2. Drain the fluid and remove the oil pan.
3. Remove the direct clutch and 2nd brake solenoids.
4. Using a small prybar, pry both oil tubes from the lower valve body.
5. Disconnect the oil pressure control cable from the throttle valve cam and remove the cable.
6. Remove the oil strainer-to-valve body bolts and the strainer.
7. Remove the valve body-to-transaxle bolts and the valve body.
8. To install the valve body, position it into the case by aligning the manual valve with the manual shift lever pin; make sure that the accumulator spring is seated. First, torque both reamer (A and B) bolts to 6–8.5 ft. lbs. (8–12 Nm), secondly, torque all remaining bolts (in diagonal order) to 6–8.5 ft. lbs. (8–12 Nm).
9. Connect the oil pressure cable to the throttle valve cam, by holding the cam down and sliding the cable end into the slot.
10. Using a soft hammer, carefully tap the oil tubes into the valve body; make sure to insert them into the flange position securely.
11. Install the solenoids and connect the electrical connectors to them.
12. Install the oil strainer-to-valve body and torque the bolts to 3–4 ft. lbs. (4–6 Nm); be sure to install the solenoid wire clamp.
13. Using solvent, clean the oil pan; be sure to place the magnet directly below the oil strainer.
14. Using a new gasket, sealant (if necessary), install the oil pan. Using sealant on the cross grooved (on head) bolts, torque the bolts to 3–4 ft. lbs. (4–6 Nm).

REMOVAL AND INSTALLATION

TRANSAXLE REMOVAL

1. From the air cleaner, remove the air suction guide.
2. Disconnect both cables from the battery and the negative (−) cable from the transaxle. Remove the battery and the battery tray.
3. From the transaxle, disconnect the solenoid coupler, the shift lever switch coupler and the wiring harness.
4. Separate the oil pressure control cable from the accelerator cable. From the transaxle, disconnect the accelerator cable and the shift selector cable.
5. Remove the starter motor. Place a catch pan under the transaxle and drain the fluid.
6. Disconnect and plug the oil cooler tubes at the transaxle.
7. Raise and support the vehicle safely. Remove the exhaust pipe and the lower clutch housing plate.

NOTE: Before removing the torque converter-to-drive plate bolts, make alignment marks on the torque converter and drive plate for assembly purposes.

8. Using a prybar, insert it through the notch (underside of transaxle) to lock the drive plate gear. Remove the torque converter-to-drive plate bolts.

9. To remove the left halfshaft, perform the following procedures:

a. From the wheel hub, remove the center cap, the split pin and the driveshaft nut.

b. Remove the lug nuts and the front wheels.

c. Using a prybar, position it between the differential case and the halfshaft's inboard joint, pry the joint until the snapring disconnects from the side gear.

d. Remove both stabilizer bar-to-chassis brackets and the ball stud-to-steering knuckle bolt. Pull the stabilizer bar downward to disconnect the ball joint from the steering knuckle.

e. Carefully remove the halfshaft from the differential case and the steering knuckle to prevent tearing the boots.

10. Using a prybar, disconnect the right halfshaft from the differential case.

11. Remove the transaxle mounting member bolts and the member. Using a floor jack and a piece of wood, support the transaxle.

12. Remove the left transaxle mount.

13. Remove the transaxle-to-engine bolts. Slide the transaxle from the engine (to prevent damaging the crankshaft, drive plate or torque converter) and lower it from the vehicle.

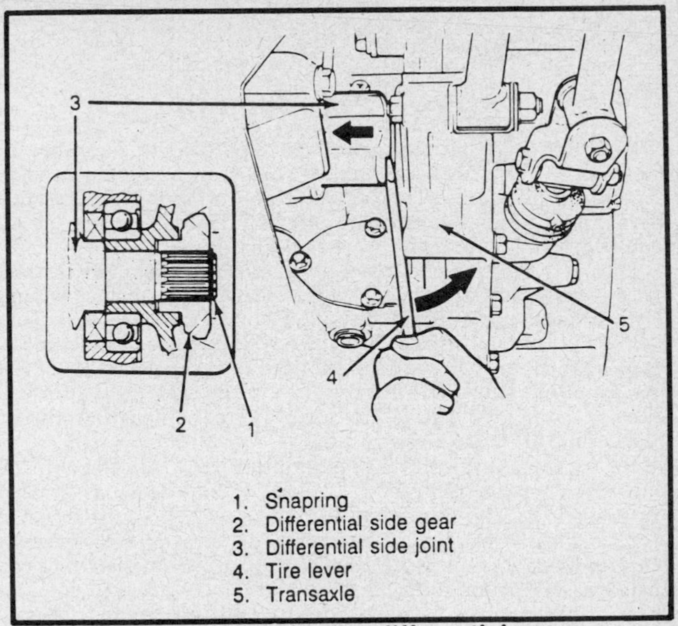
3
4
5
2
1
1. Snapring
2. Differential side gear
3. Differential side joint
4. Tire lever
5. Transaxle

Prying the halfshaft from the differential case

TRANSAXLE INSTALLATION

1. Using grease, lubricate the cup around the center of the torque converter.

2. Measure the distance **A** between the torque converter and the transaxle housing; it should be at least 0.85 in. (21.4mm). If the distance is less than specified, the torque converter is improperly installed; remove and reinstall it.

3. When installing the transaxle, guide the right halfshaft into the differential case; make sure the snapring seats in the differential gear.

4. To install the left halfshaft, perform the following procedures:

a. Clean and lubricate the halfshaft splines with grease.

b. Carefully install the halfshaft into the steering knuckle and the differential case to prevent tearing the boots; make sure the snapring seats in the differential gear.

c. Install the ball joint stud into the steering knuckle and torque the bolt to 36–50 ft. lbs. (50–70 Nm).

d. Install the stabilizer bar-to-chassis brackets and torque the bolts to 22–39 ft. lbs. (30–55 Nm).

e. Torque the halfshaft hub nut to 108–195 ft. lbs. (150–270 Nm) and install the split pin (to the shaft) and the center cap.

f. Torque the lug nuts to 29–50 ft. lbs. (40–70 Nm).

5. Torque the transaxle housing-to-engine bolts to 12–16.5 ft. lbs. (16–23 Nm), the mounting member-to-chassis bolts to 40 ft. lbs. (55 Nm), the mounting member-to-transaxle nuts to 33 ft. lbs. (45 Nm) and the transaxle-to-mount bolts to 40 ft. lbs. (55 Nm).

6. Using a prybar, insert it through the notch (underside of transaxle) to lock the drive plate gear and torque the torque converter-to-drive plate bolts to 13–14 ft. lbs. (18–19 Nm).

7. Install the oil cooler lines and the starter.

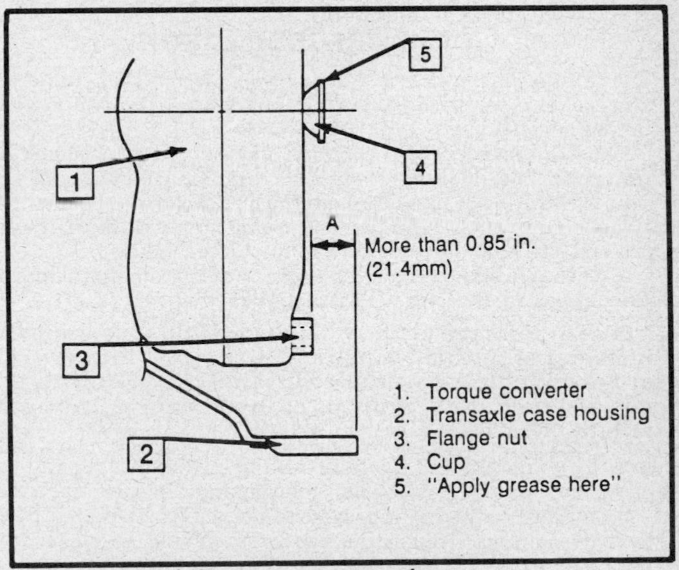
5
4
1
A
More than 0.85 in. (21.4mm)
3
2
1. Torque converter
2. Transaxle case housing
3. Flange nut
4. Cup
5. "Apply grease here"

Installation of the torque converter

8. After connecting the oil pressure control cable to the accelerator cable, check and/or adjust the cable play.

9. Connect the wiring harness, the shift lever switch coupler and the solenoid coupler to the transaxle.

10. Install and adjust the select cable and shift switch.

11. Install the battery tray and the battery. Connect the battery cables to the battery and the negative (−) battery cable to the transaxle.

12. Install the air suction guide to the air cleaner.

13. Refill and check the fluid level.

BENCH OVERHAUL

Before Disassembly

When servicing the transaxle it is important to be aware of cleanliness. Before disassembling the transaxle, the outside should be throughly cleaned, preferably with a high-pressure spray cleaning equipment. Dirt entering the unit may negate all the effort and time spent on the overhaul.

During inspection and reassembly, all parts should be cleaned with solvent and dried with compressed air; do not use wiping rags or cloths for lint may find its way into the valve body passages. Lubricate the seals with Dexron®II and use petroleum jelly to hold the thrust washers; this will ease the assembly of the seals and not leave harmful residues in the system. Do not use solvent on neoprene seals, friction plates or composition thrust washers, if they are to be reused.

Before installing bolts into aluminum parts, dip the threads into clean transmission fluid. Anti-seize compound may be used to prevent galling the aluminum or seizing. Be sure to use a torque wrench to prevent stripping the threads. Be especially careful when installing the seals (O-rings), the smallest nick can cause a leak. Aluminum parts are very susceptible to damage; great care should be used when handling them. Reusing snaprings is not recommended but should they be: compress the internal ones and compress the external ones.

Converter Inspection

The torque converter is a sealed, welded design that cannot be disassembled for service or repair; there are a few checks that can be made.

The torque converter contains approximately 1 quart of transmission fluid. Since there is no drain plug on the unit, the fluid can only be drained through the hub. To drain the converter, invert it over a catch pan and drain the fluid. Fluid that is drained can help diagnosis the converter's condition.

1. If the fluid is discolored but does not contain metal bits or particles, the converter is usable and need not be replaced.

NOTE: Remember the fluid color is no longer a good indicator of the fluid condition. In the past, the dark color would indicate overheated transaxle fluid; with the newer fluids, this is not a positive sign of transaxle failure.

2. Metal particles in the fluid, having an aluminum paint appearance, indicating converter damage and replacement.
3. If fluid contamination is due to burned clutch plates, overheated oil or antifreeze, the converter should be cleaned or replaced.

The converter should be checked carefully for damage, especially around the seal area; remove any sharp edges or burrs from the seal surface. Do not expose the seal to any type of solvent. If the converter is to be washed with solvent, the seal must be removed.

Transaxle Disassembly

1. Pull the torque converter from the front of the transaxle.
2. Remove the bolts from the rear cover and the solenoid wire harness clamps. Using a holding fixture tool, mount the transaxle to it.
3. Remove the dipstick, the dipstick tube and the fluid cooling lines.
4. Place an oil catch pan under the transaxle and drain the fluid.

NOTE: Before draining the fluid from the transaxle, do not turn it over for the foreign matter in the pan will contaminate the valve body.

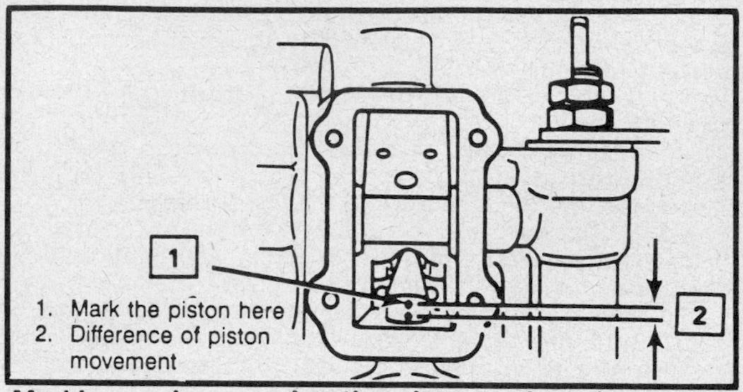
1
2
1. Mark the piston here
2. Difference of piston movement
Marking and measuring the piston rod stroke

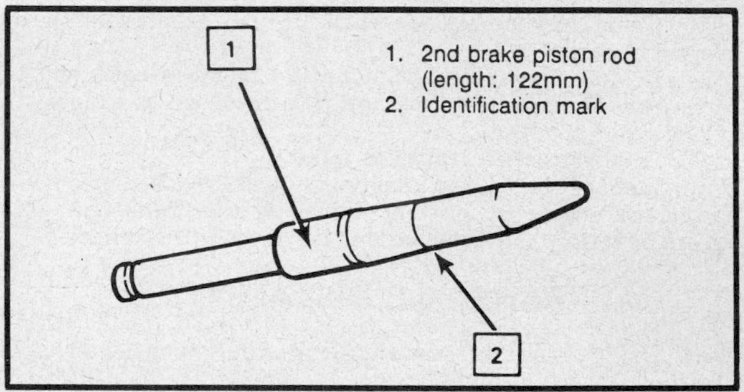
1
2
1. 2nd brake piston rod (length: 122mm)
2. Identification mark
View of the 2nd brake band piston

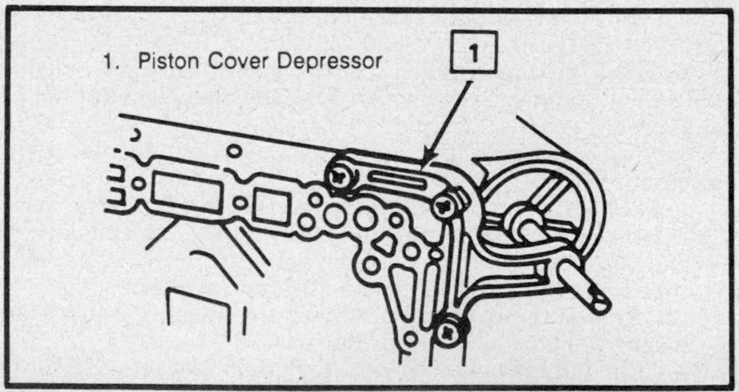
1
1. Piston Cover Depressor
Using the piston cover depressing tool to compress the 2nd brake piston

5. Remove the oil pan bolts. Using a soft mallet, tap the oil pan from the transaxle and remove the gasket.
6. From the bottom of the transaxle, disconnect the electrical connectors from the direct clutch and 2nd brake solenoids.
7. Using a prybar, pry the oil tubes from the valve body.
8. Using a pair of pliers, disconnect the oil control cable from the throttle valve cam and remove the cable from the transaxle.
9. Remove the oil strainer-to-valve body bolts, the strainer, the valve body-to-transaxle case bolts and the valve body.
10. Using a rag and 15 psi (100 kPa) of compressed air, place the rag over the accumulator pistons, force the compressed air into the piston holes, pop them out and catch the them with the rag.
11. Remove the 2nd brake band cover and gasket. To check the 2nd brake piston stroke, perform the following procedures:
    a. Scribe a mark on the piston rod.

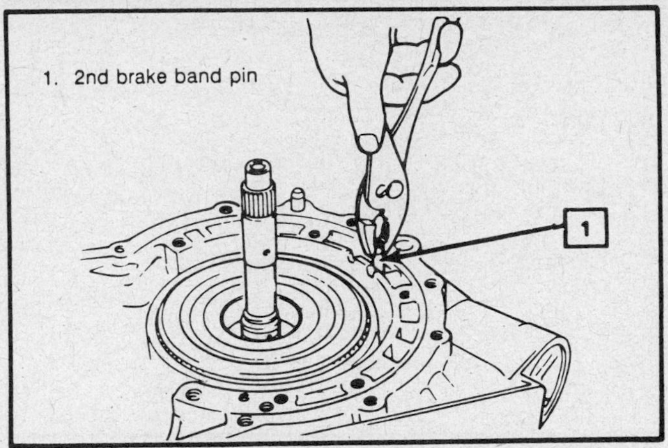
1. 2nd brake band pin
1

Removing the 2nd brake band pin

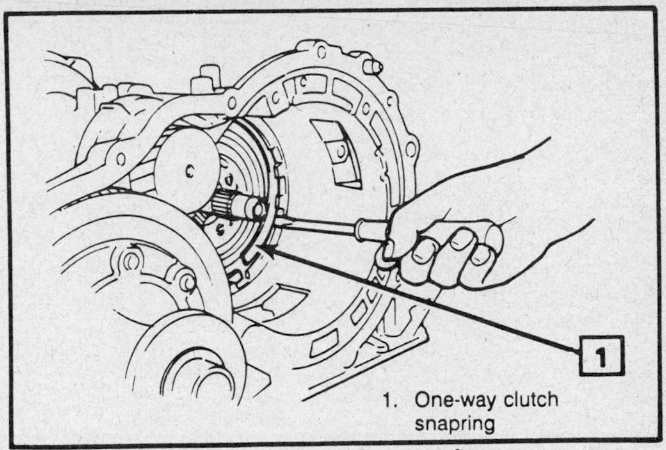
1
1. One-way clutch snapring

Removing the one-way clutch snapring

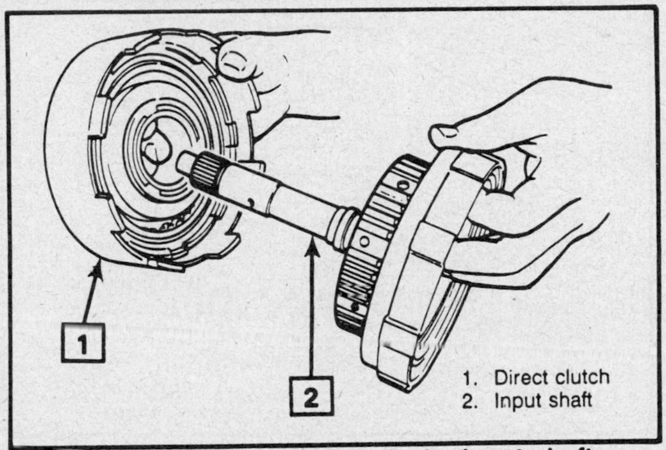
1
2
1. Direct clutch
2. Input shaft

Removing the direct clutch from the input shaft

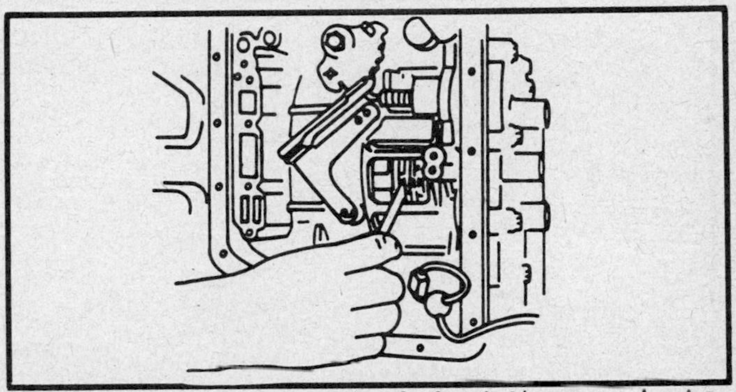

Using a feeler gauge to check the 1st/reverse brake clearance

b. Apply low pressurized air into the oil hole and measure the 2nd brake piston rod stroke; it should be 0.06–0.11 in. (1.5–3.0mm).

c. If out of specifications, replace the 2nd brake band or the piston rod (with a different length); the piston rod is available in 2 lengths.

12. To remove the 2nd brake piston, install a piston cover depressor tool or equivalent, onto the transaxle case and tighten the bolt to compress the piston spring. Using a small prybar, remove the snapring. Remove the depressor tool and the 2nd brake piston.

13. At the solenoid wire harness, remove the wire holding the plate retaining nut and pull out the solenoid wire.

14. To remove the oil pump, perform the following procedures:

a. Remove the oil pump bolts.

b. Using a pair of slide hammers, equipped with adapters, pull the oil pump from the transaxle.

15. Remove the bell housing-to-transaxle case bolts, tap around with housing (with a plastic hammer) and remove the bell housing.

16. Using a pair of pliers, pull the 2nd brake band pin from the transaxle housing.

17. Grasp the input shaft and remove the direct/forward clutch assembly; be careful, for the ring gear bearing race and bearing, which may be sticking to the clutch assembly, may drop out.

18. From the housing, remove the 2nd brake band, the front planetary ring gear, the ring gear bearing, the planetary sun gear and the front planetary gear bearing.

19. Using a small prybar, remove the one-way clutch snapring from the housing. Remove the one-way clutch, the rear planetary gear, the rear planetary ring gear, the ring gear bearing and the washers.

20. Using a feeler gauge, place it between the snapring and the 1st/reverse brake; the clearance should be 0.023–0.075 in. (0.58–1.92mm). If the measurement is out of specifications, replace the 1st/reverse brake plate or disc with a new one.

21. Using a small prybar, remove both 1st/reverse snaprings. Remove the 1st/reverse brake flange, discs, plates, damper plate and the differential gear assembly.

22. From the rear of the transaxle, remove the rear cover bolts/nuts and the cover; it may be necessary to use a plastic hammer to lightly tap the rear cover from the transaxle.

23. To remove the reduction driven gear, perform the following procedures:

a. Using a hammer and a chisel, remove the stake mark from the retaining nut.

b. Move the shift selector lever to the **P** position; make sure the output shaft cannot turn.

c. Using a wrench, carefully loosen and remove the retaining nut; do not use a hammer, for the parking lock pawl and output shaft may be damaged.

d. Remove the reduction driven gear.

24. Using a plastic hammer, drive the counter shaft from the transaxle.

25. Using the output shaft remover tool or equivalent, position it (the tool's legs in the case notches) on the outer race of the internal output shaft bearing (inside the transaxle) and drive the output shaft from the transaxle.

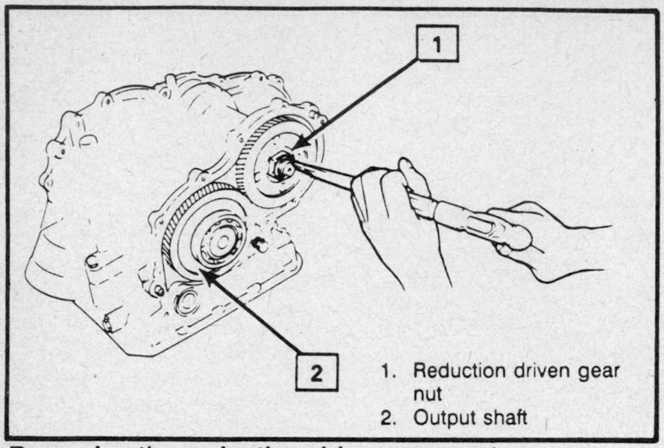
1
2
1. Reduction driven gear nut
2. Output shaft

Removing the reduction driven gear nut

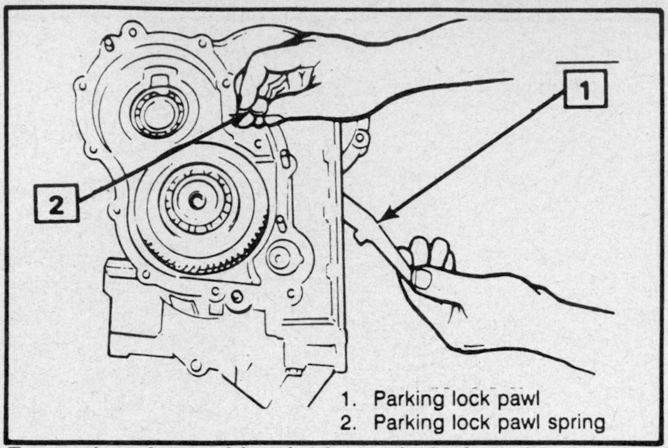
1
2
1. Parking lock pawl
2. Parking lock pawl spring

Removing the parking lock pawl shaft and parking lock pawl

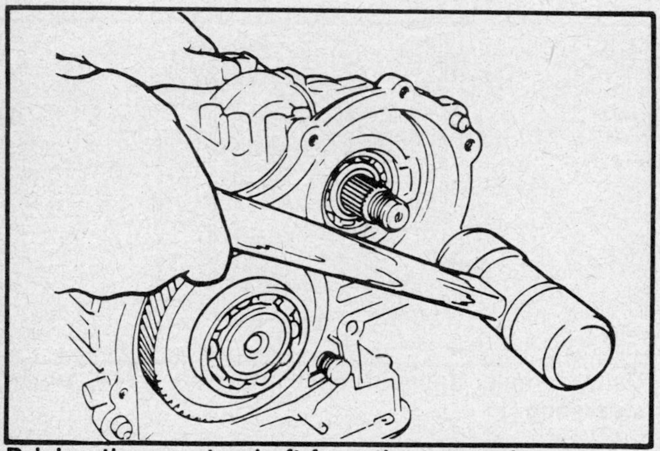

Driving the countershaft from the transaxle

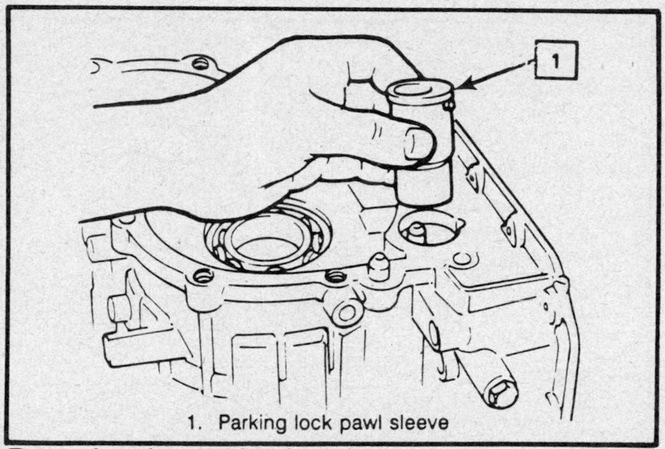
1
1. Parking lock pawl sleeve

Removing the parking lock pawl sleeve

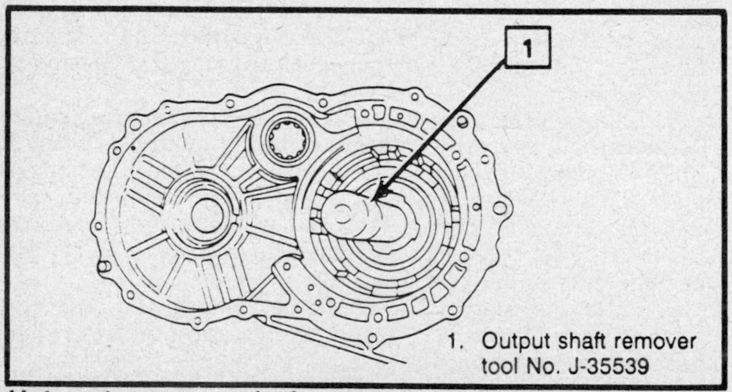
1
1. Output shaft remover tool No. J-35539

Using the output shaft remover tool to drive the output shaft from the transaxle

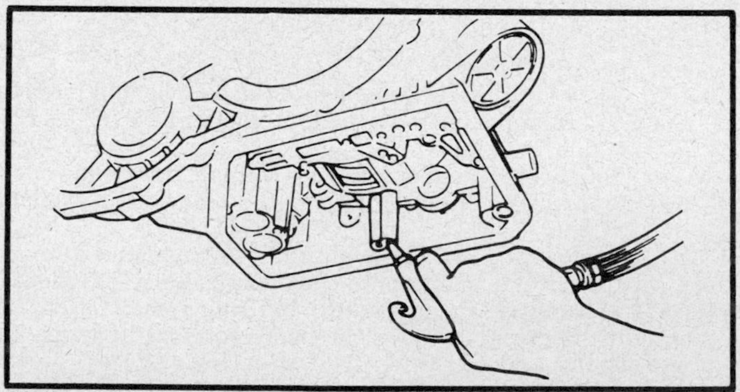

Using air pressure to remove the 1sr/reverse brake piston from the case

26. To remove the parking lock pawl, perform the following procedures:
   a. Pull out the parking lock pawl shaft and the spring.
   b. Remove the parking lock pawl.
   c. Pull out the parking lock pawl sleeve.
   d. Remove the manual detent spring assembly and the manual shift shaft.
27. To remove the 1st/reverse brake piston, perform the following procedures:
   a. Using a small prybar, push the return spring assembly downward and pry the snapring from the flange.
   b. Lift the return spring assembly from the case.
   c. Using low pressurized air, apply it to the oil hole and force the 1st/reverse piston from the case.

Unit Disassembly and Assembly

DIRECT CLUTCH

Disassembly

1. Using a Vernier® scale, measure the height between the snapring and the clutch flange; the height should be 0.098–0.120 in. (2.49–3.06mm). If the height is within specifications,

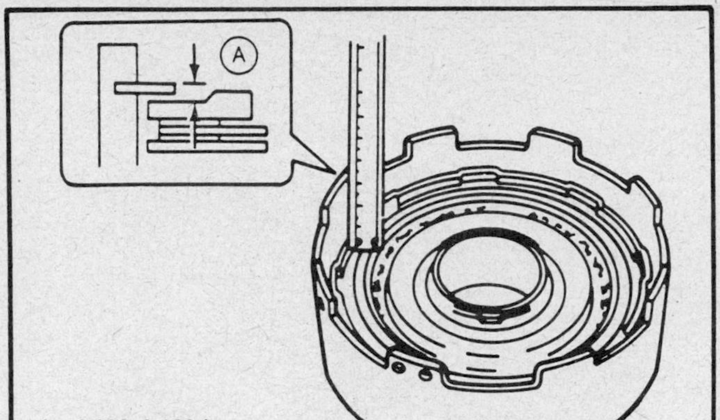
A
A. 0.098–0.120 in.
(2.49–3.06mm)

Measuring the direct clutch height

the clutch is OK; if the height is not within specifications, replace the clutch discs or plates.

2. Using a prybar, remove the large clutch plate snapring.
3. Remove the clutch flange, discs and plates.
4. Using a clutch spring compressor tool or equivalent, and a shop press, compress the piston return springs and remove the small snapring with a small prybar.
5. Remove the spring seat and the return spring assembly.
6. To remove the direct clutch piston, apply low pressurized air through oil hole in the drum; the piston should pop out. If the piston does not come out, use a pair of needle-nose pliers and lift it out.
7. Remove the inner seal from the drum and the outer seal from the piston.

Inspection

1. Check the piston for free movement of the check valve (steel ball).
2. Using low pressurized air, check for leakage of the check valve; if faulty, replace the piston.

Assembly

1. Using Dexron®II transmission fluid, lubricate the new O-rings seals; fit the inner seal into the drum and the outer seal onto the piston.
2. When installing the piston into the drum, be careful the O-ring does not get twisted or caught.
3. Install the clutch return spring assembly and the spring seat.
4. Using a clutch spring compressor tool or equivalent, and a shop press, compress the piston return springs and install the small snapring with a small prybar.

NOTE: Make sure the snapring is fully seated in the 4 spring seat projections. Do not compress the return spring more than necessary.

5. Install the discs, plates and flange in the following order: plate, disc, plate, plate, disc and flange.

NOTE: If installing new clutch discs, soak them in Dexron®II fluid for at least 2 hours before assembly.

6. Install the large snapring. After assembly, measure the height between the clutch flange and the snapring; it should be 0.098–0.120 in. (2.49–3.06mm). If the height is not within specifications, with new clutch plates and discs installed, install the different clutch flange. Clutch flanges are available in 2 thicknesses: 0.118 in. (3.00mm) and 0.132 in. (3.37mm).
7. Apply low pressurized air to the oil hole in the drum to check the piston movement.

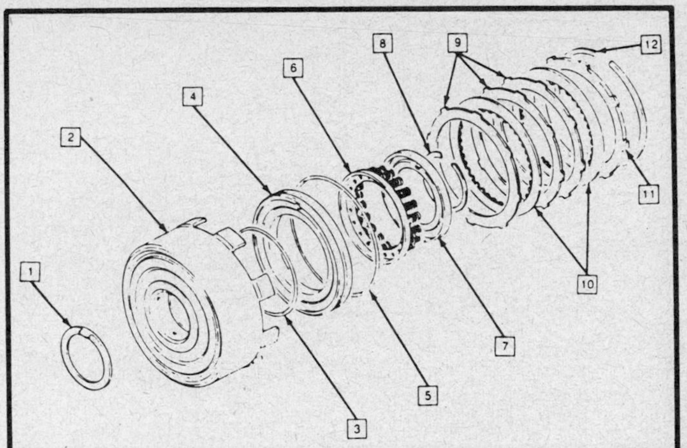
1
2
3
4
5
6
7
8
9
10
11
12

1. Direct clutch washer
2. Direct clutch drum
3. Inner seal
4. Direct clutch piston
5. Outer seal
6. Return spring assembly
7. Return spring seat
8. Clutch plate snapring
9. Clutch plate
10. Clutch disc
11. Clutch flange
12. Clutch plate snapring

Exploded view of the direct clutch assembly

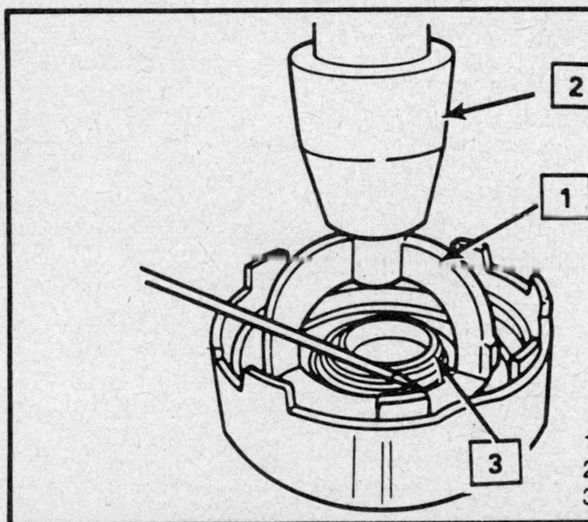
1
2
3

1. J-23327
2. Press
3. Snapring

Removing the clutch plate snapring

FORWARD CLUTCH

Disassembly

1. Using a Vernier® scale, measure the height between the snapring and the clutch flange; the height should be 0.079–0.105 in. (2.01–2.68mm). If the height is within specifications, the clutch is OK; if the height is not within specifications, replace the clutch discs or plates.
2. Using a prybar, remove the large clutch plate snapring.
3. Remove the clutch flange, discs and plates.
4. Using a clutch spring compressor tool or equivalent, an adapter tool and a shop press, compress the return springs and remove the small snapring with a small prybar.

NOTE: When compressing the return spring, be careful not to compress them more than necessary.

5. Remove the spring seat and the return springs.
6. To remove the forward clutch piston, apply low pressurized air through oil hole in the input shaft; the piston should pop out. If the piston does not come out, use a pair of needle-nose pliers and lift it out.

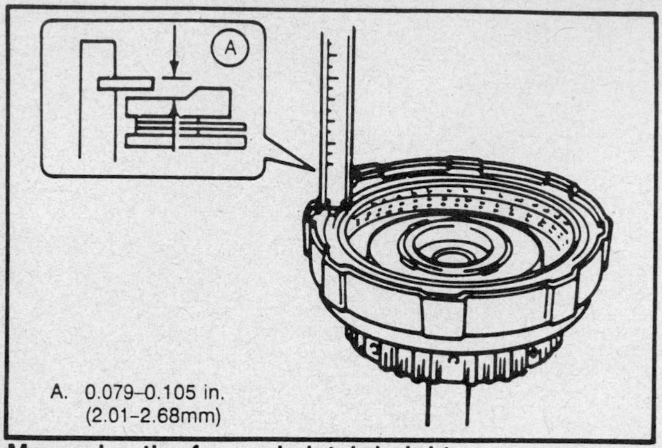
A
A. 0.079–0.105 in.
(2.01–2.68mm)

Measuring the forward clutch height

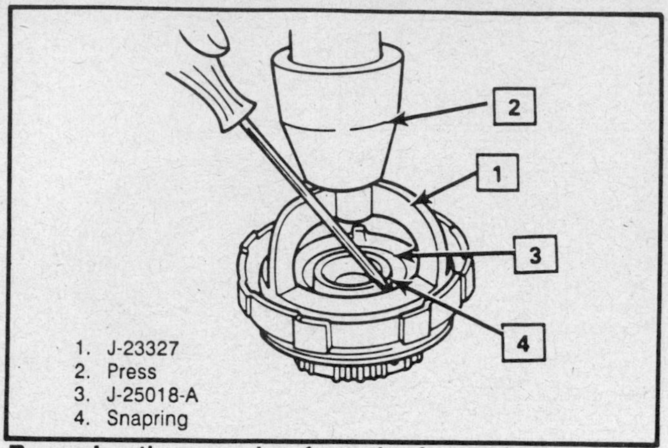
2
1
3
4
1. J-23327
2. Press
3. J-25018-A
4. Snapring

Removing the snapring from the forward clutch assembly

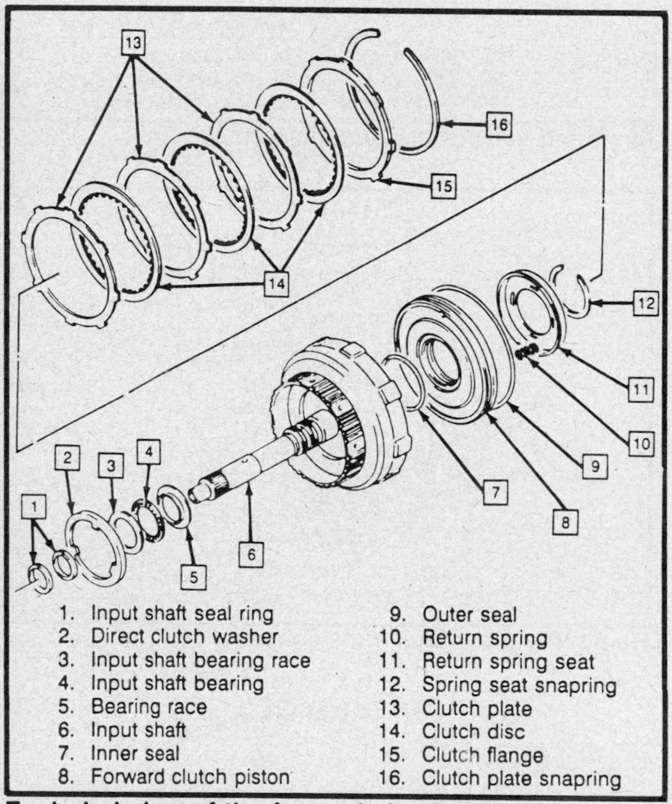
13
16
15
14
12
11
10
9
7
8
6
5
4
3
2
1
1. Input shaft seal ring
2. Direct clutch washer
3. Input shaft bearing race
4. Input shaft bearing
5. Bearing race
6. Input shaft
7. Inner seal
8. Forward clutch piston
9. Outer seal
10. Return spring
11. Return spring seat
12. Spring seat snapring
13. Clutch plate
14. Clutch disc
15. Clutch flange
16. Clutch plate snapring

Exploded view of the forward clutch assembly

7. Remove the inner and outer seals (O-rings) from the piston.

Inspection

1. Check the clutch piston for free movement of the check valve (ball).
2. Using low pressurized air, check for leakage of the check valve; if faulty, replace the clutch piston.

Assembly

1. Using Dexron®II transmission fluid, lubricate the new O-rings seals; fit them onto the piston.
2. When installing the piston into the input shaft drum, be careful the O-ring seals do not get twisted or caught.
3. Install the piston return springs and the spring seat.
4. Using a clutch spring compressor tool or equivalent, an adapter tool and a shop press, compress the return springs and install the small snapring.

NOTE: Make sure the snapring is fully seated in the 4 spring seat projections. Do not compress the return spring more than necessary.

5. Install the discs, plates and flange in the following order: plate, disc, plate, disc, plate, disc and flange.

NOTE: If installing new clutch discs, soak them in Dexron®II fluid for at least 2 hours before assembly.

6. Install the large snapring. After assembly, measure the height between the clutch flange and the snapring; it should be 0.079–0.105 in. (2.01–2.68mm). If the height is not within specifications, with new clutch plates and discs installed, install the different clutch flange. Clutch flanges are available in 2 thicknesses: 0.118 in. (3.00mm) and 0.132 in. (3.37mm).
7. Apply low pressurized air to the oil hole in the input shaft to check the clutch piston movement.

DIFFERENTIAL CASE

Disassembly

1. Using a wheel puller tool and a plug tool, pull the side bearings and the speedometer driven gear from the differential case.
2. Using a hammer and a pin punch, drive the pinion shaft roll pin from the differential case.
3. Remove the pinion shaft, pinion gears and side gears from the differential case.
4. Remove the ring gear-to-differential case bolts and the ring gear.

Inspection

1. Using Dexron®II fluid or kerosene, wash the parts throughly.
2. Inspect the gear teeth for excessive signs of wear; minor nicks or scratches may be removed with an oil stone. If necessary, replace the damaged or worn parts.
3. Inspect the thrust washers for signs of excessive wear, distortion or damage; if necessary, replace them.

Assembly

1. Using Dexron®II fluid, lubricate the parts before installation.
2. Install the side washers and the side gears.
3. Install the pinion washers, the pinion gears and the pinion shaft.

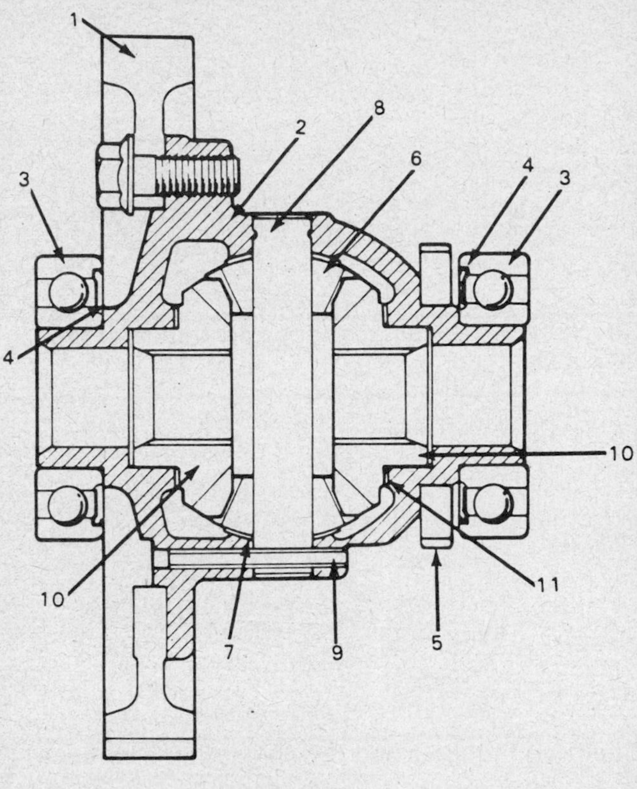
1
2
3
4
5
6
7
8
9
10
11

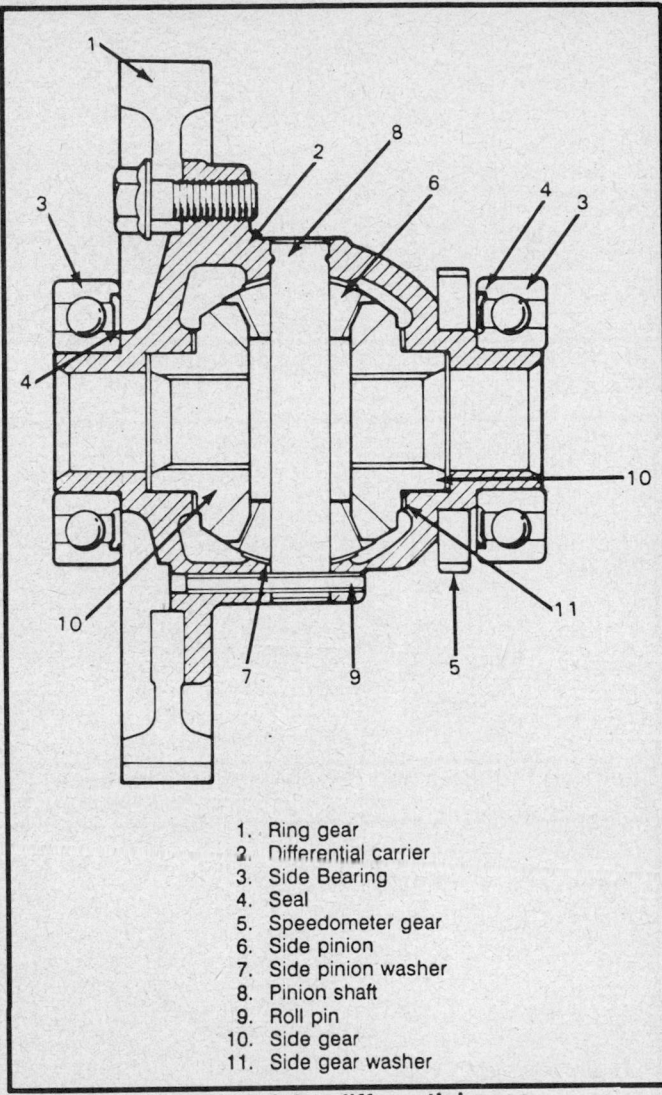

1. Ring gear
2. Differential carrier
3. Side Bearing
4. Seal
5. Speedometer gear
6. Side pinion
7. Side pinion washer
8. Pinion shaft
9. Roll pin
10. Side gear
11. Side gear washer

Crossectional view of the differential case

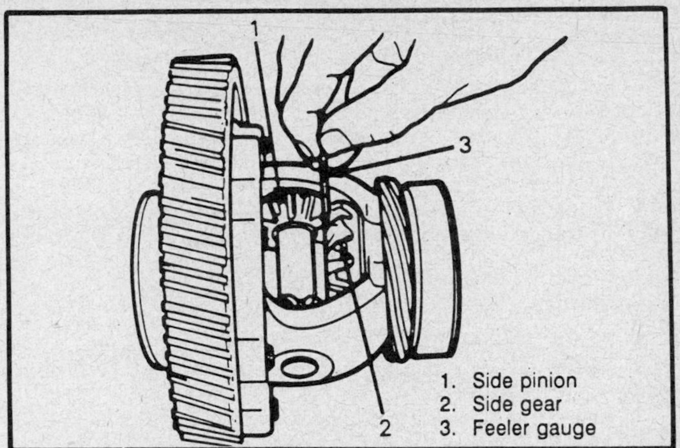
1
2
3

1. Side pinion
2. Side gear
3. Feeler gauge

Measuring the backlash of the pinion and side gears

	mm	IN.
Available side gear	0.90	0.035
	0.95	0.037
Washer sizes	1.00	0.039
	1.05	0.041
	1,10	0.043
	1.15	0.045
	1.20	0.047

List of the side gear washer sizes

4. Using a feeler gauge, measure the backlash of the side pinions and side gears; it should be 0–0.009 in. (0.01–0.25mm). If the backlash is not within specifications, adjust it by changing the thickness of the side washers.
5. Align the pinion shaft hole with the differential case hole and install the roll pin.
6. Install the ring gear and torque the bolts to 58–72 ft. lbs. (80–100 Nm).
7. Using a shop press, press the speedometer gear onto the differential case.
8. Using a bearing installation tool or equivalent, and a hammer, drive both side bearings (seal side facing case) onto the differential case.

OIL PUMP

Disassembly

1. Remove both cover seal rings and the O-ring.
2. Remove the cover-to-body bolts and the front cover.

Inspection

1. Check the oil pump seal for wear, cracks or damage; if necessary, replace the seal.

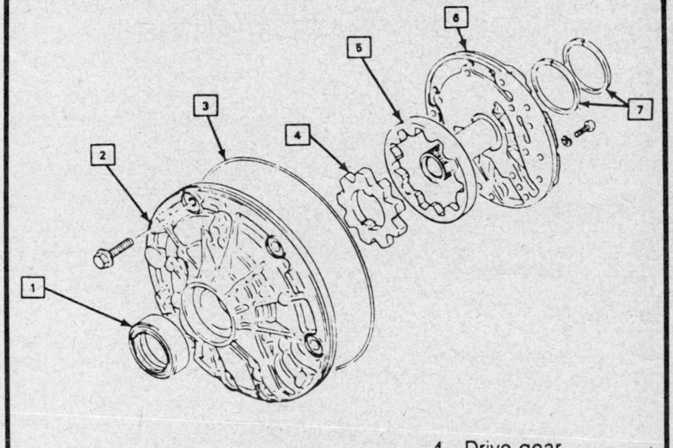
1
2
3
4
5
6
7

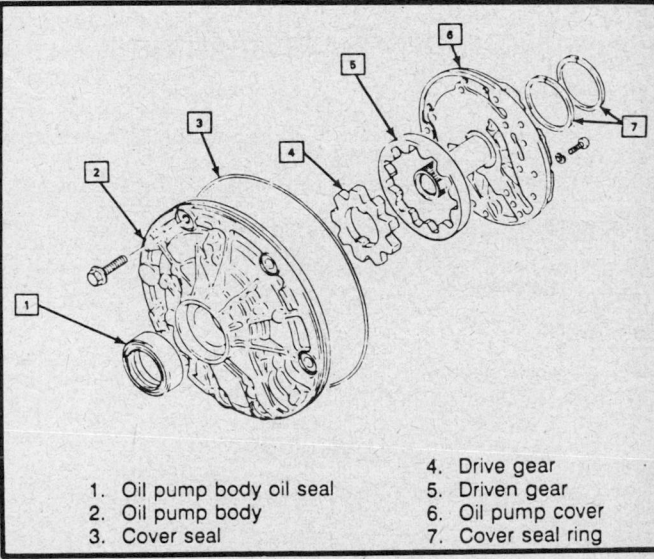

1. Oil pump body oil seal
2. Oil pump body
3. Cover seal
4. Drive gear
5. Driven gear
6. Oil pump cover
7. Cover seal ring

Exploded view of the oil pump assembly

2. To check the driven gear clearance, perform the following procedures:

a. Push the driven gear to one side of the pump body.

b. Using a feeler gauge, measure the driven gear-to-body clearance; it should be 0.0028–0.0059 in. (0.07–0.15mm). If the clearance exceeds the limits, replace the driven gear.

3. Using a feeler gauge, check the gear tooth-to-cresent clearance between the drive and driven gears.

4. Using a steel straight edge and a feeler gauge, check the side clearance between the drive/driven gears and the pump body.

Assembly

1. Using Dexron®II fluid, lubricate the parts.
2. Install the driven gear and drive gear into the oil pump body.
3. Install the cover onto the body and torque the bolts to 6–8.5 ft. lbs. (8–12 Nm).
4. Lubricate both oil pump seal rings and install them onto the pump.
5. Install the O-ring onto the cover; make sure it is not twisted and fully seated into the groove.
6. Rotate the drive gear and check it for smooth rotation.

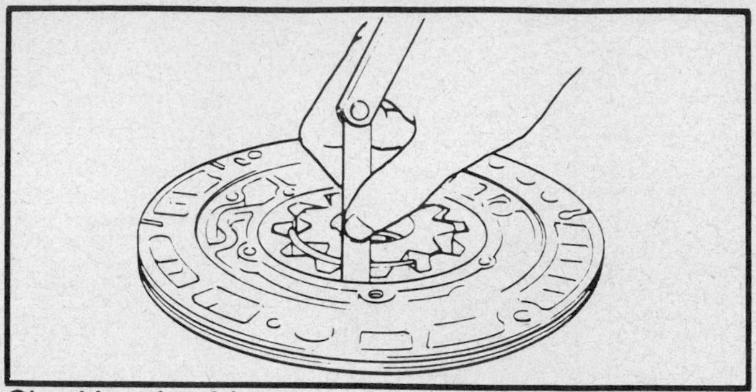

Checking the driven gear body clearance of the oil pump

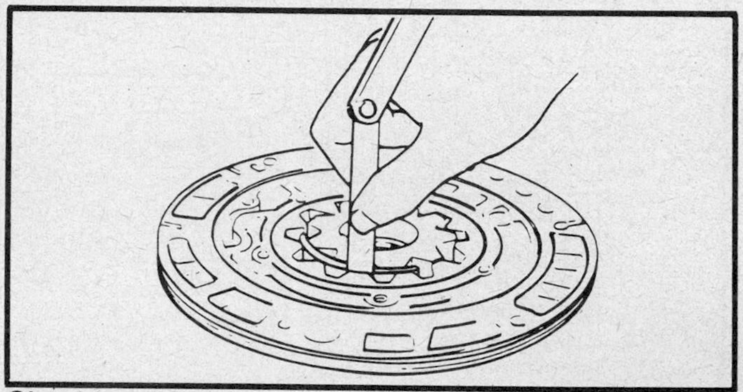

Checking the drive and driven gear tip clearance

Checking the drive and driven gear side clearance

COUNTERSHAFT BEARINGS

Disassembly

1. Using a prybar inside the case, remove the countershaft bearing snaprings.
2. Remove the rear cover (backing plate).
3. Using a bearing remover tool and a slide hammer, remove both countershaft bearings.

Inspection

Check the bearings for excessive wear and damage; if necessary, replace the bearings.

Assembly

1. Insert the countershaft spacer and roller bearing into the case. Using the bearing installation tool and a hammer, drive the bearing into the case until it is seated. Install the snapring.
2. Install the ball bearing into the case. Using the bearing installation tool and a hammer, drive the bearing into the case until it is seated.
3. Install the bearing backing plate and the snapring.

OUTPUT SHAFT BEARINGS

Disassembly

1. Using a bearing puller tool and a shop press (bearing puller), remove the bearings from the output shaft.
2. After the bearings have been removed, discard them.

Inspection

Check the bearings for excessive wear and damage; if necessary, replace the bearings.

Assembly

Using a bearing installation tool and a shop press, press the new bearings onto the output shaft.

VALVE BODY

The valve body consists of an upper and lower portion.

Disassembly

1. Remove the upper valve body-to-lower valve body bolts; when separating the valve bodies, be careful not drop the check balls.
2. From the upper valve body, remove the throttle valve cam bolt, cam and spring.
3. From the lower valve body, remove the 2nd brake solenoid, the direct clutch solenoid, the lower valve body cover bolts, cover and gasket (discard it).
4. Using a needle-nose pliers, remove the valve plug keys.

NOTE: Even though the valve components appear to be similar, they are different; be sure to keep them separated.

5. When removing the valves (from each body) and keep the corresponding spring together with each valve.
6. Using Dexron®II fluid, throughly wash each part. Using compressed air, blow out the fluid passages and holes.

Inspection

1. Inspect the check balls for damage or a sticking condition.
2. Check the valves for free movement, damage or scoring.
3. Check the springs for binding.

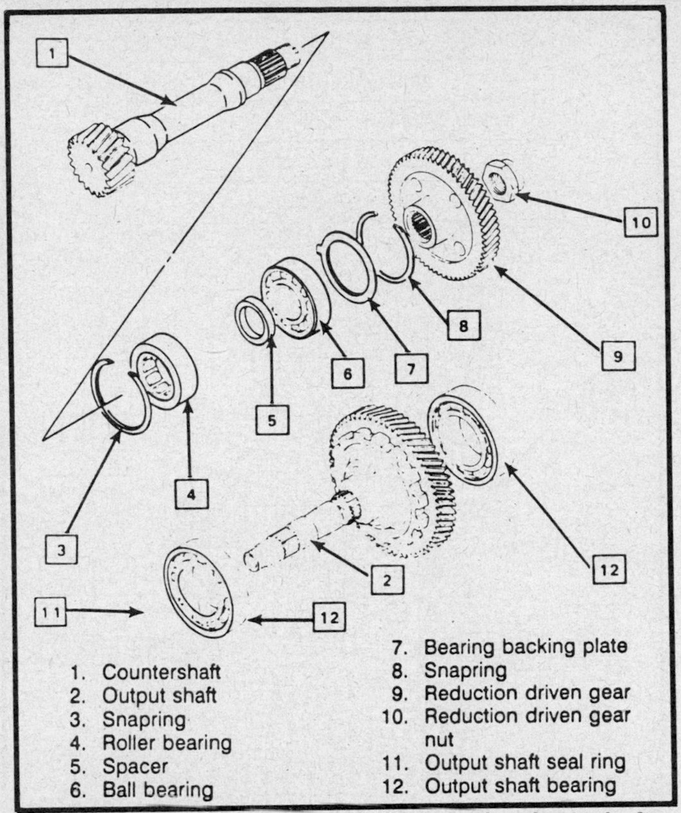
1. Countershaft
2. Output shaft
3. Snapring
4. Roller bearing
5. Spacer
6. Ball bearing
7. Bearing backing plate
8. Snapring
9. Reduction driven gear
10. Reduction driven gear nut
11. Output shaft seal ring
12. Output shaft bearing

Exploded view of the countershaft and output shaft assemblies

4. Check the seals for nicks, cracks or deterioration; replace them, if necessary.

Assembly

1. When assembling the lower valve body, use a new gasket, install the lower valve body cover and torque the bolts to 3–4 ft. lbs. (4–6 Nm).
2. Assemble the throttle valve cam to the upper valve body and torque the bolt to 4.5–6.5 ft. lbs. (6–9 Nm).
3. Position the check balls in the correct positions in the lower valve body.
4. When assembling the valve bodies, use a new gasket, install the bolts and perform the following procedure:
   a. Install all of the bolts, in their correct locations, finger tight.
   b. Torque the No. 2 bolts to 3.7–4.3 ft. lbs. (5–6 Nm).
   c. Torque the No. 3 bolts to 3.7–4.3 ft. lbs. (5–6 Nm).
   d. Torque the No. 4 bolts to 3.7–4.3 ft. lbs. (5–6 Nm).
   e. Torque the No. 1 bolts to 3.7–4.3 ft. lbs. (5–6 Nm).

Transaxle Assembly

1. To install the manual shift shaft and parking lock pawl into the transaxle case, perform the following procedures:
   a. Install the lower washer and parking lock rod to the manual shift shaft.

NOTE: When installing the manual shift shaft, be careful not to damage the oil seal lip.

   b. Install the manual shift shaft into the case, followed by the manual detent spring. Torque the manual detent spring nut/bolt to 6.0–8.5 ft. lbs. (8–12 Nm).
   c. Install the upper washer and manual shift lever to the manual shift shaft. Torque the upper/lower nuts to 20–23.5 ft. lbs. (27–33 Nm).
   d. After torquing the nuts, check the manual shift shaft for rotation smoothness.
   e. Assemble the restrictor pin and snapring to the parking lock pawl sleeve and install the assembly into the case.
   f. Shift the manual shift lever into **P**, install the parking lock pawl, the lock pawl shaft and the lock pawl spring. Move the manual shift lever to make sure the parking lock pawl moves smoothly.
2. To install the 1st/reverse brake piston, perform the following procedures:
   a. Using Dexron®II fluid, lubricate the piston's inner/outer seals and fit them to the piston.
   b. Install the piston into the case so the spring holes are facing upward; be sure the seals are not twisted or caught.
   c. Position the return spring assembly on the piston; make sure the springs (of the assembly) are fitted securely in the piston holes.
   d. Using prybars, press the return spring assembly downward and install the snapring.
3. Using a driver handle tool or equivalent, a countershaft in-

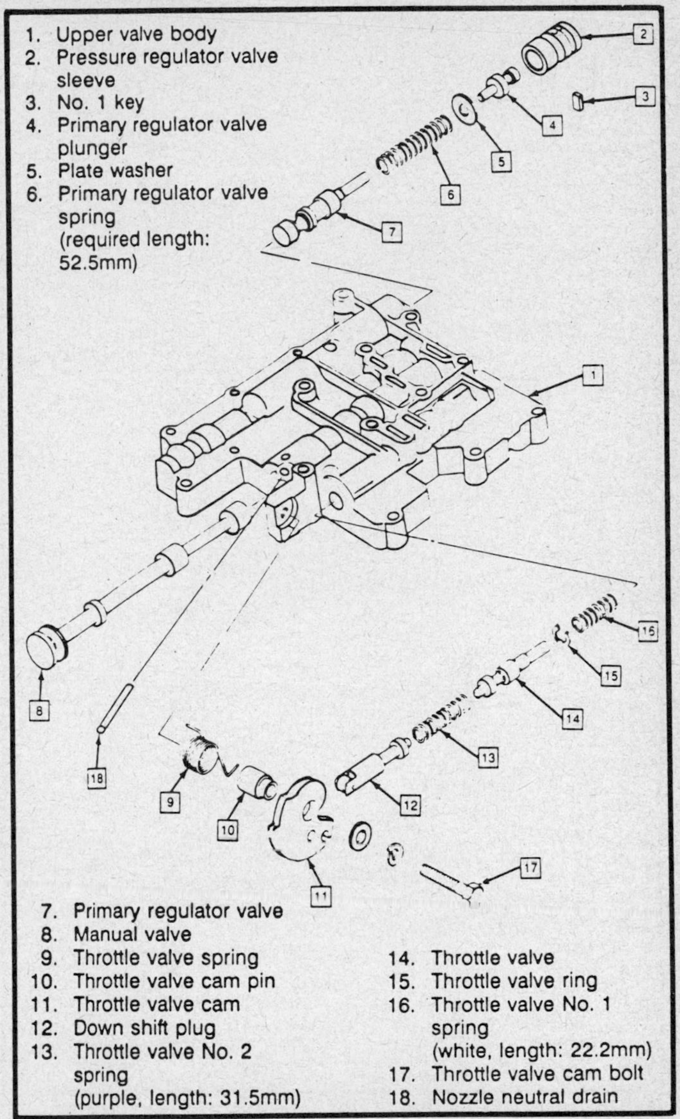
1. Upper valve body
2. Pressure regulator valve sleeve
3. No. 1 key
4. Primary regulator valve plunger
5. Plate washer
6. Primary regulator valve spring (required length: 52.5mm)
7. Primary regulator valve
8. Manual valve
9. Throttle valve spring
10. Throttle valve cam pin
11. Throttle valve cam
12. Down shift plug
13. Throttle valve No. 2 spring (purple, length: 31.5mm)
14. Throttle valve
15. Throttle valve ring
16. Throttle valve No. 1 spring (white, length: 22.2mm)
17. Throttle valve cam bolt
18. Nozzle neutral drain

Exploded view of the upper valve body components

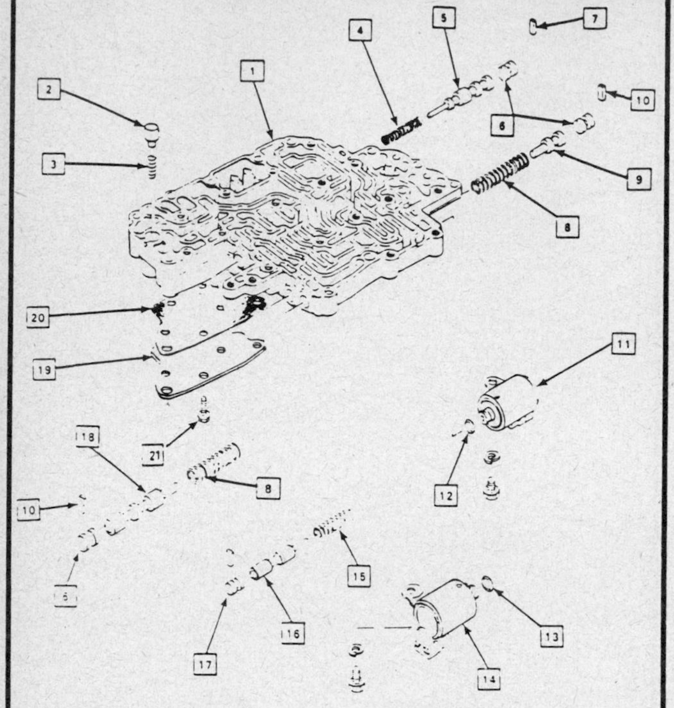

1. Lower valve body
2. Cooler bypass valve
3. Spring (orange, length: 19.9mm)
4. Secondary regulator valve spring (yellow, length: 31.4mm)
5. Secondary regulator valve
6. Plug
7. No. 1 key
8. Shift valve spring (pink, length: 39.6mm)
9. 2-3 shift valve
10. No. 2 key
11. Direct clutch solenoid
12. Seal
13. Seal
14. 2nd brake solenoid
15. B2 Control valve spring (blue, length: 28.1mm)
16. B2 control valve
17. B2 control valve plug
18. 1-2 shift valve
19. Lower valve body cover
20. Gasket
21. Lower valve body cover bolt

Exploded view of the lower valve body components

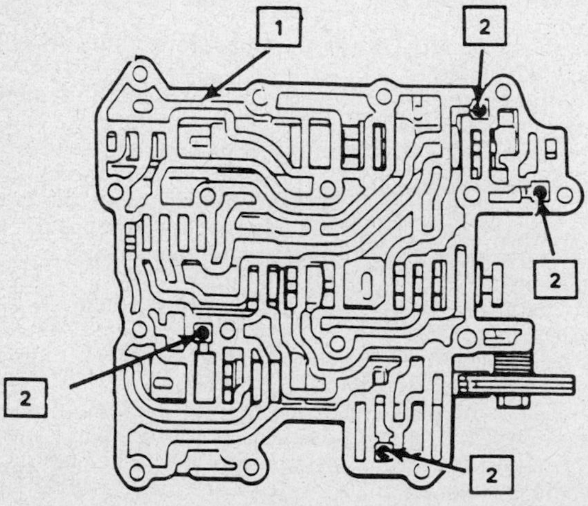

1. Upper valve body
2. Steel ball

Location of the check balls in the upper valve body

BOLT	LENGTH	PIECES
A	29.5 mm (1.16 IN.)	6
B	38 mm (1.49 IN.)	6
C	44 mm (1.73 IN.)	2
D	REAMER BOLT	2

SPECIFIED TORQUE: 5-6 N·m (3.7-4.3 lb. ft.)

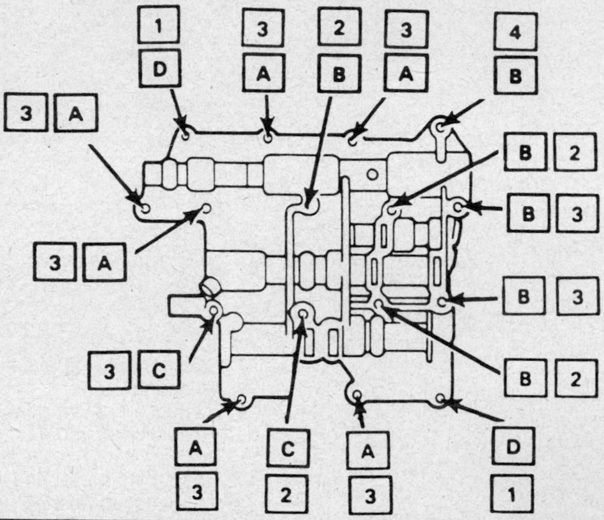

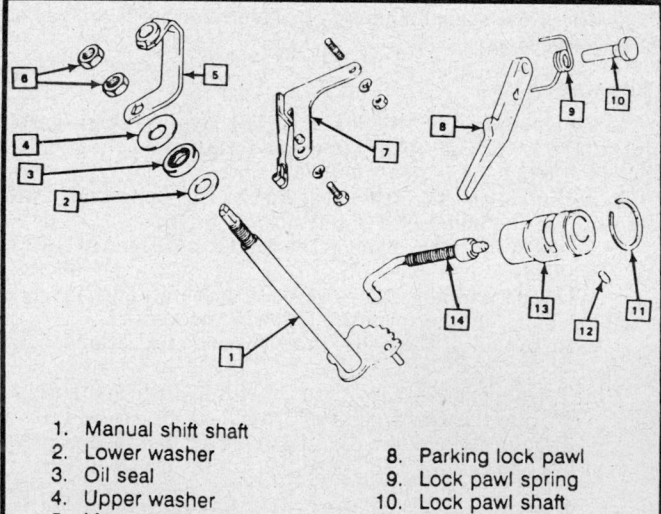

Description and location of the valve body bolts

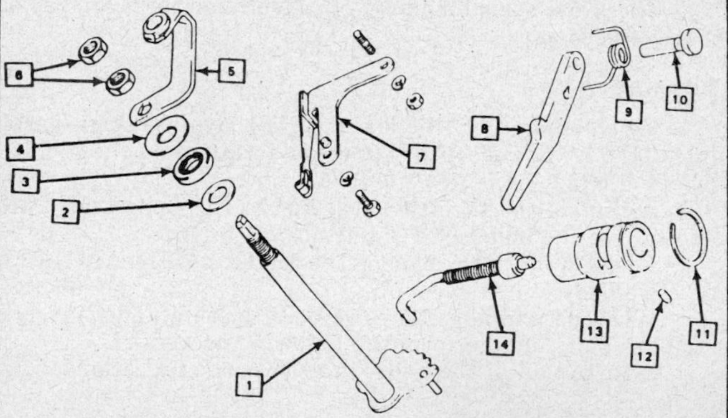

1. Manual shift shaft
2. Lower washer
3. Oil seal
4. Upper washer
5. Manual shift shaft
6. Nut
7. Manual detent spring assembly
8. Parking lock pawl
9. Lock pawl spring
10. Lock pawl shaft
11. Sleeve snapring
12. Pin
13. Parking lock pawl sleeve
14. Parking lock rod

Exploded view of the manual shift shaft, parking lock pawl and related parts

stallation tool or equivalent, and a hammer, drive the countershaft into the case until it is seated in the bearings.

NOTE: When driving the countershaft, make sure the spacer is in position and do not drive the shaft excessively hard, for the snapring may become damaged.

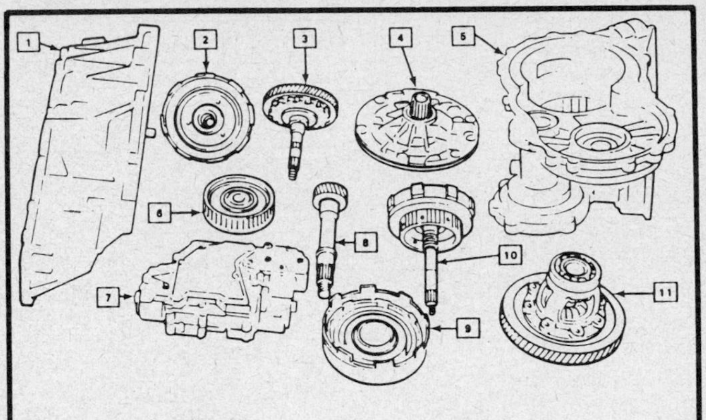
1
2
3
4
5
6
7
8
9
10
11

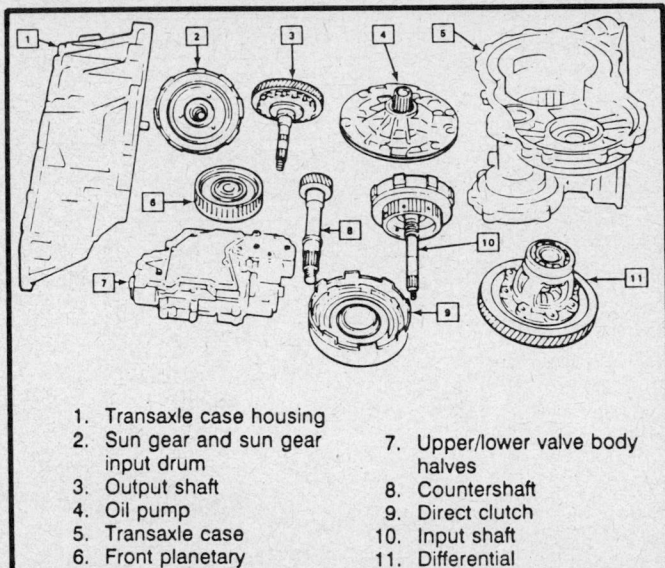

1. Transaxle case housing
2. Sun gear and sun gear input drum
3. Output shaft
4. Oil pump
5. Transaxle case
6. Front planetary
7. Upper/lower valve body halves
8. Countershaft
9. Direct clutch
10. Input shaft
11. Differential

View of the transaxle components

4. To install the output shaft, perform the following procedures:

a. Move the manual shift lever into a position other than **P**.

b. Using a bearing installation tool or equivalent, and a hammer, drive the output shaft into the case until it seats.

5. To install the reduction driven gear, perform the following procedures:

a. Position the reduction driven gear on the countershaft, engaging the gear teeth with the output shaft gear.

b. Move the manual shift lever into the **P** position; make sure the output shaft is locked and cannot move.

c. Install the driven gear nut and torque it to 86–108 ft. lbs. (110–150 Nm).

d. Using a chisel and a hammer, stake the gear nut in 2 places.

6. To install the rear transaxle cover, perform the following procedures:

a. Using a new gasket, install the rear cover. Make sure the output shaft bearing enters the rear cover bearing hole smoothly. Rotate the output shaft and check for abnormal gear sounds.

b. Install the rear cover bolts/nuts and torque the bolts to 12–16 ft. lbs. (16–23 Nm) and the nuts to 8–10 ft. lbs. (11–15 Nm).

NOTE: If the holding fixture tool is being used, only ½ the rear cover bolts can be installed; install the remaining bolts after the tool is removed.

7. To seat the output shaft, perform the following procedures:

a. Using the output shaft remover/installer tool or equivalent, align 4 of the tool's projections in 4 case notches.

b. Using a hammer, lightly tap the tool to seat the bearing and output shaft against the rear cover; do not hammer on the shaft directly.

8. Install the differential gear assembly by engaging the final gear teeth with the countershaft teeth; be careful not to damage the gear teeth surfaces.

9. To install the 1st/reverse brake parts, perform the following procedures:

a. Install the damper plate (convex side facing upward) onto the return spring assembly; do not reverse the direction of the plate.

b. Install the discs, plates and flange in the following order: plate, disc, plate, disc, plate, disc, plate, disc and flange (flat side facing downward).

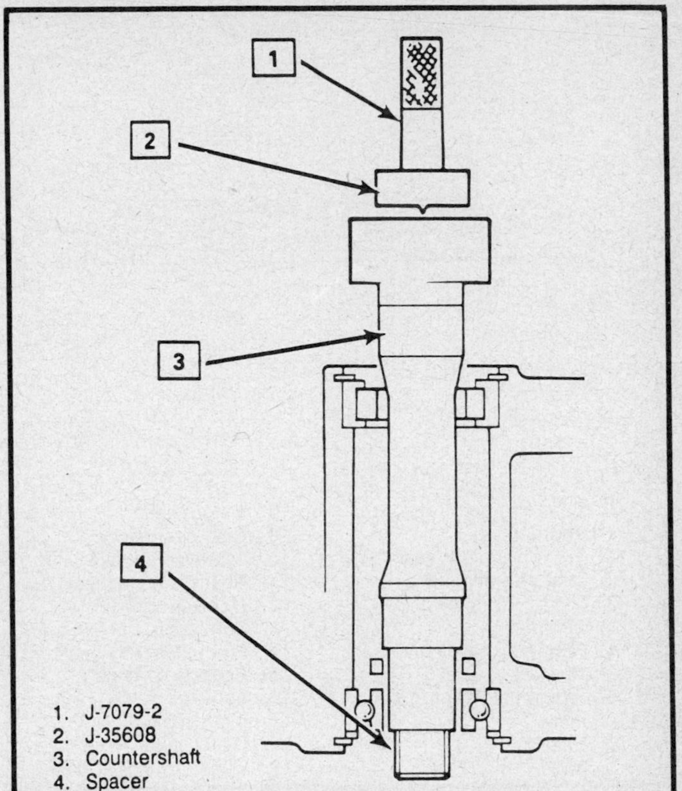
1
2
3
4

1. J-7079-2
2. J-35608
3. Countershaft
4. Spacer

Installing the countershaft into the case

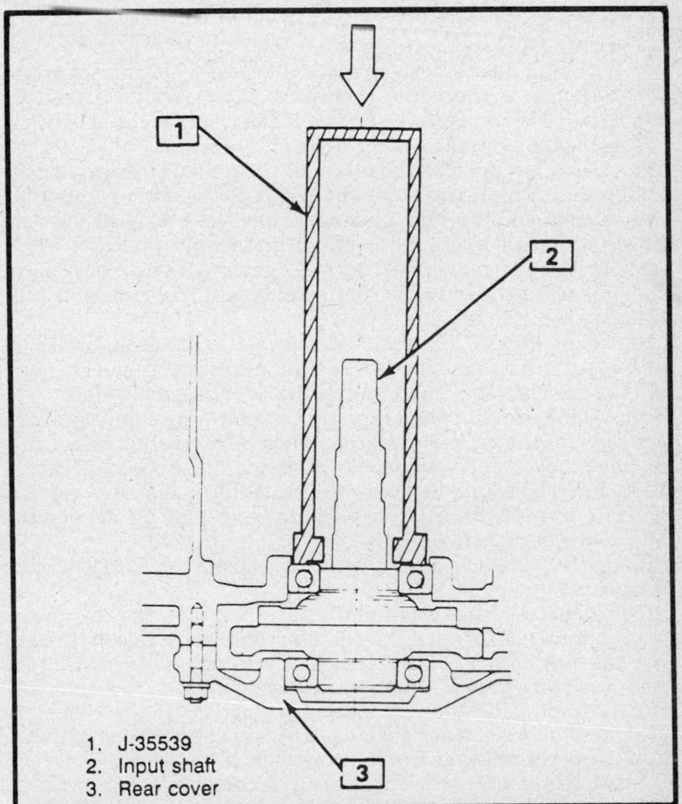
1
2
3

1. J-35539
2. Input shaft
3. Rear cover

Seating the output shaft and bearing against the rear cover

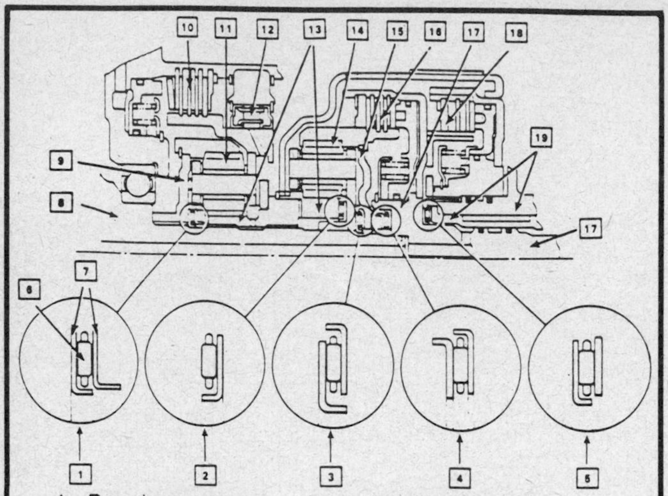
10
11
12
13
14
15
16
17
18
19
9
8
17
7
6
1
2
3
4
5

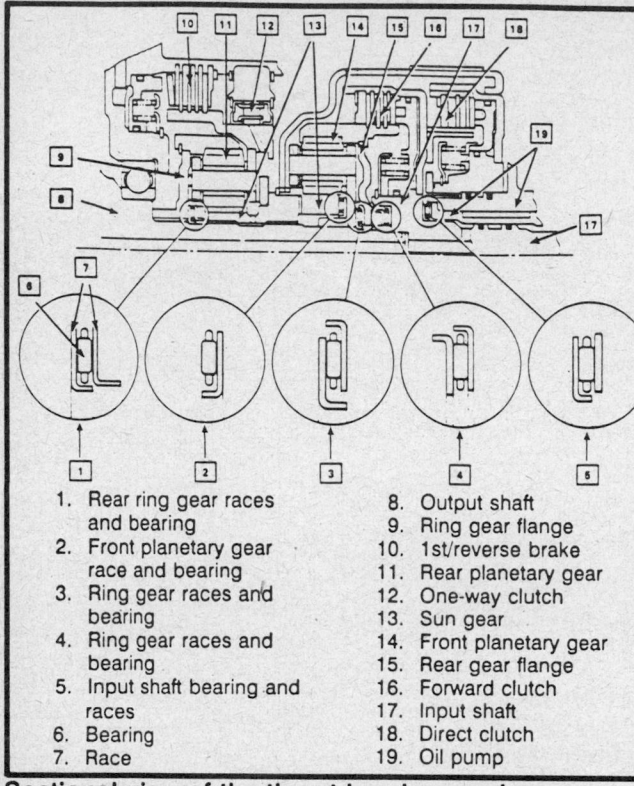

1. Rear ring gear races and bearing
2. Front planetary gear race and bearing
3. Ring gear races and bearing
4. Ring gear races and bearing
5. Input shaft bearing and races
6. Bearing
7. Race
8. Output shaft
9. Ring gear flange
10. 1st/reverse brake
11. Rear planetary gear
12. One-way clutch
13. Sun gear
14. Front planetary gear
15. Rear gear flange
16. Forward clutch
17. Input shaft
18. Direct clutch
19. Oil pump

Sectional view of the thrust bearings and races

NOTE: If installing new discs, soak them in Dexron®II fluid for at least 2 hours before installation.

c. Install the snapring.

d. Using a feeler gauge, place it between the snapring and the flange to check the clearance; the clearance should be 0.023–0.075 in. (0.58–1.92mm). If the clearance is not within specifications, replace the flange.

10. Using low pressurized air, apply it to the 1st/reverse brake piston oil hole (oil pan side) and check the piston operation.

11. To engage the rear planetary ring gear, engage the ring gear and output shaft spline and insert it into the case.

12. Install the rear planetary ring gear races and bearing in the following order: race (flange side up), bearing and race (flange side up).

13. Using grease, lubricate the rear planetary thrust washers and place 1 on each side of the rear planetary gear assembly; make sure the lug shapes match the gear assembly slots.

14. Install the rear planetary gear assembly by aligning it with the teeth of the 1st/reverse brake discs. After installation, lightly move the assembly up and down; a clear "click" sound should be heard. The sound indicates the assembly is installed correctly; if no sound is heard, the washers/races may be out of place and requires reinstallation.

15. Install the one-way clutch race snapring into the case groove.

16. To install the one-way clutch, position it on the rear planetary gear assembly, turn the planetary assembly clockwise until the one-way clutch falls into place. After installation, rotate the rear planetary gear assembly clockwise, it should turn smoothly (freely) indicating the installation is correct. If the planetary gear assembly does not turn clockwise but does turn counterclockwise, the one-way clutch is installed in the wrong direction; it must be removed and reinstalled in the opposite direction.

17. Push the one-way clutch snapring by hand; make sure it is fully seated and the snapring ring ends are between the lugs.

18. Using grease, apply it to the sun gear thrust washer. Install the thrust washer and pin to the sun gear. Make sure the pin is fitted in the thrust washer notch.

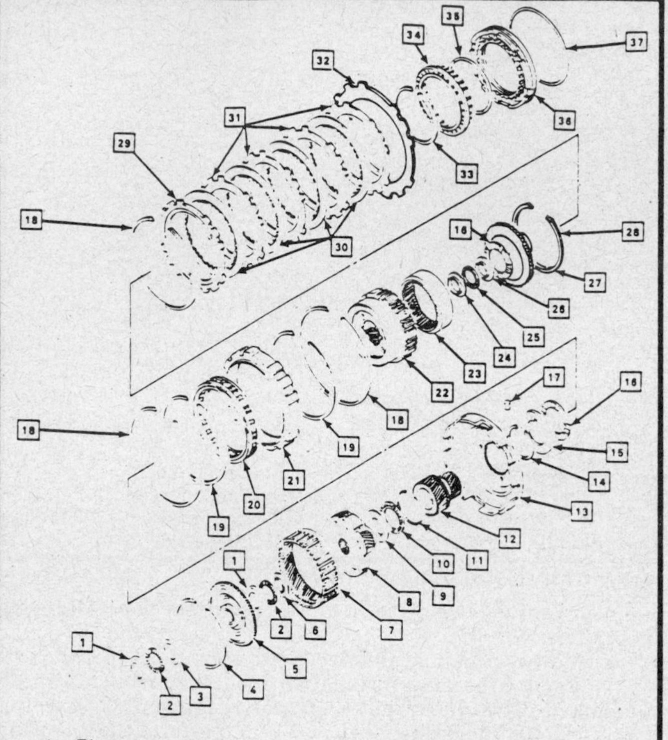
35
34
37
32
36
31
29
33
18
30
16
28
27
26
25
24
23
22
18
17
16
15
14
13
12
11
10
19
21
20
19
18
9
8
7
1
2
6
5
1
4
2
3

1. Ring gear race
2. Ring gear bearing
3. Ring gear race
4. Snapring
5. Ring gear flange
6. Ring gear race
7. Front planetary ring gear
8. Front planetary gear assembly
9. Front planetary gear race
10. Front planetary gear bearing
11. Input drum snapring
12. Sun gear
13. Sun gear input drum
14. Snapring
15. Planetary thrust washer
16. Rear planetary thrust washer
17. Sun gear pin
18. One-way clutch race snapring
19. One-way clutch snapring
20. One-way clutch
21. One-way clutch race
22. Rear planetary gear assembly
23. Rear planetary ring gear
24. Rear ring gear race
25. Ring gear bearing
26. Rear ring gear race
27. Ring gear flange
28. Ring gear snapring
29. 1st/reverse brake flange
30. 1st/reverse brake disc
31. 1st/reverse brake plate
32. 1st/reverse brake damper plate
33. Return spring snapring
34. 1st/reverse brake return spring
35. Piston inner seal
36. 1st/reverse brake piston
37. Piston outer seal

Exploded view of the planetary gears and related parts

19. Install the sun gear assembly by engaging it with the rear planetary gear; be careful not to damage the bushing (inside the sun gear). After installation, lightly move the sun gear up and down; it should make a clear "click" sound. If no sound is heard, the washers may be out of place; remove and reinstall the sun gear.

20. Install the front planetary gear bearing and race (flange side down) on the sun gear.

21. To install the front planetary gear assembly, turn it in either direction until it engages with the sun gear. After installation, lightly move the assembly up and down; it should make a clear "click" sound, indicating it is properly installed. If no sound is heard, the bearing and race may be out of position; remove, check and reinstall the assembly.

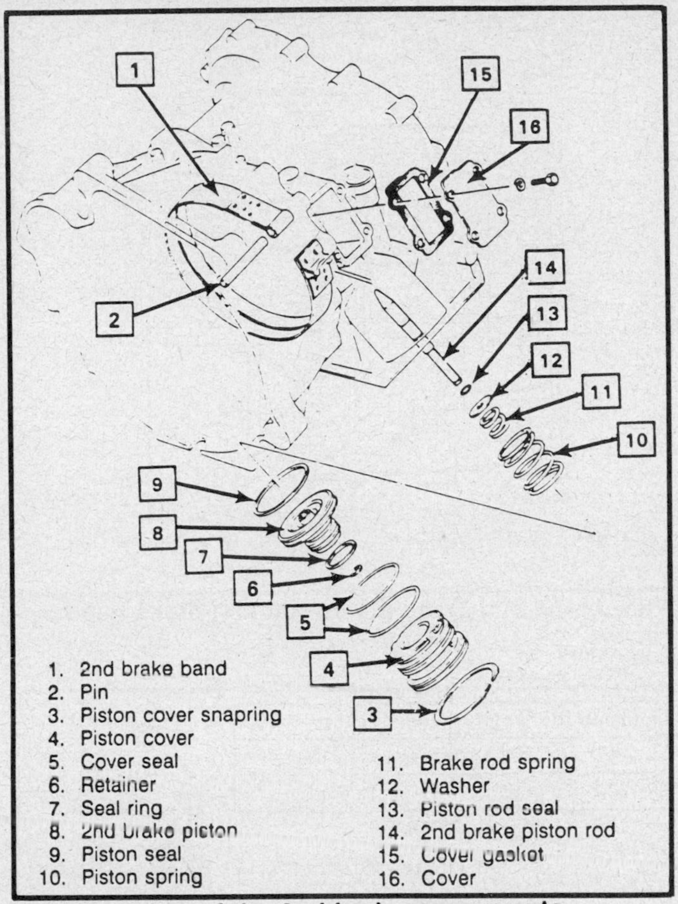
1
2
3
4
5
6
7
8
9
10
11
12
13
14
15
16
1. 2nd brake band
2. Pin
3. Piston cover snapring
4. Piston cover
5. Cover seal
6. Retainer
7. Seal ring
8. 2nd brake piston
9. Piston seal
10. Piston spring
11. Brake rod spring
12. Washer
13. Piston rod seal
14. 2nd brake piston rod
15. Cover gasket
16. Cover

Exploded view of the 2nd brake components

22. On the front planetary gear assembly, install these items in the following order: the ring gear race (flange side up), the ring gear bearing and ring rear race (flange side down).

23. Install the front planetary ring gear assembly. After installation, lightly move the assembly up and down; it should make a clear "click" sound, indicating it is properly installed. If no sound is heard, the bearing and races may be out of position; remove, check and reinstall the assembly.

24. Install the 2nd brake band in the correct direction in the case; do not bend it too much, for it can become damaged.

25. To install the 2nd brake band pin, dip it in Dexron®II fluid and insert it in the case hole aligned with the 2nd brake band hole.

26. To install the 2nd brake piston, perform the following procedures:

a. Using Dexron®II fluid, lubricate the piston rod, seal and seal ring. Install the piston spring and piston assembly into the case.

b. Using Dexron®II fluid, lubricate both cover seals and install the piston cover on the case.

c. Using the piston cover depressor tool or equivalent, compress the piston cover and install the snapring.

27. Inspect the output shaft seal ring for wear or damage; replace it, if necessary. Install the seal on the output shaft; be careful not to expand it too much.

28. Using grease, lubricate the input shaft and install the input shaft seal rings on the shaft; be careful not to expand it too much.

29. Using grease, apply it to the direct clutch washer and position it (grooved face outward) on the direct clutch; the grease will hold the washer in place.

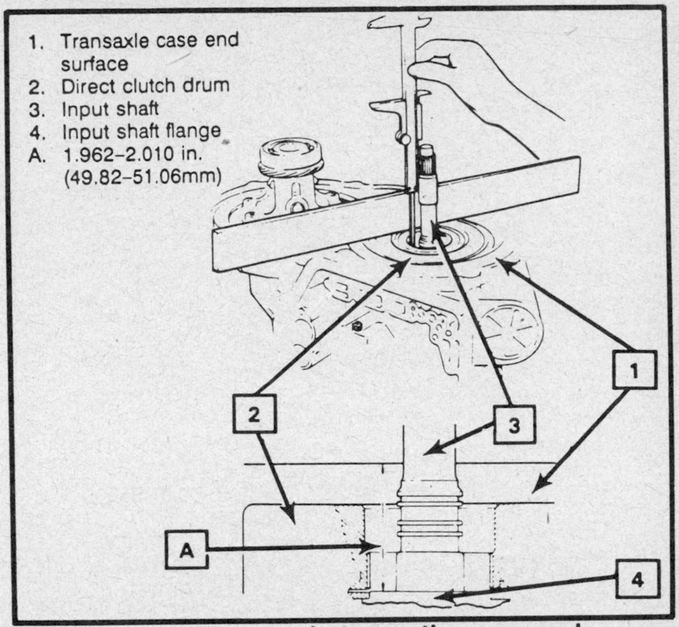
1. Transaxle case end surface
2. Direct clutch drum
3. Input shaft
4. Input shaft flange
A. 1.962–2.010 in. (49.82–51.06mm)
1
2
3
4
A

Measuring the distance between the case end surface and input shaft flange.

30. To install the direct clutch, align the discs teeth and place it on the input shaft. After installation, lightly move the assembly up and down; it should make a clear "click" sound, indicating it is properly installed. If no sound is heard, the assembly is not installed correctly; remove and reinstall it.

31. Using grease, apply it to the ring gear races and bearing. Install these items in the following order: the 1.41 in. (35.8mm) ring gear race (flange side down), the bearing and the 1.49 in. (37.9mm) ring gear race on the ring gear.

32. To install the input shaft and forward/direct clutch assembly, support the input shaft (direct clutch installed), align the forward clutch discs and lower it into the case by turning it back and forth; be careful the bearing and race do not fall off. After installation, lightly move the assembly up and down; it should make a clear "click" sound, indicating it is properly installed. If no sound is heard, the assembly is not installed correctly; remove and reinstall it.

33. To check the components for correct installation, perform the following procedures:

a. Using a steel straight-edge, position it across the transaxle case.

b. Using a depth micrometer or vernier scale, measure the distance from the input shaft flange to the upper straight-edge surface (A). Subtract the straight-edge's width from measurement (A); the final calculation is the input shaft flange-to-case distance, it should be 1.962–2.010 in. (49.82–51.06mm).

c. If the measurement is not within specifications, disassemble the component parts and reinstall them properly.

34. To install the transaxle case housing, perform the following procedures:

a. Using a new gasket, install the transaxle case housing; be sure the gasket does not protrude inside the housing.

b. When installing the case bolts, notice that 3 have star-shaped grooves in their heads, these bolts require sealant on their threads; do not apply sealant on the other bolts. Torque the bolts to 12–16.5 ft. lbs. (16–23 Nm).

35. Using grease, apply it to the input shaft bearing race and install it (flange side outward), with the bearing, on the forward clutch; make sure the bearing does not get positioned on the race flange. Using grease, apply it to the other input shaft bearing race and install it onto the oil pump.

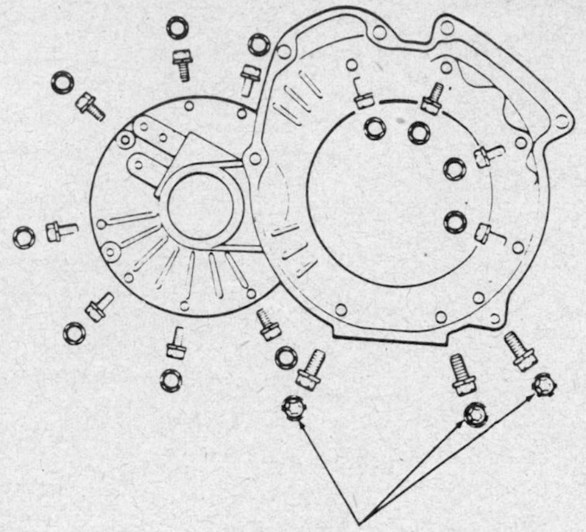
APPLY SEALANT TO THESE
BOLT THREADS
BOLT TORQUE
12–16 FT. LBS.
(16–23 NM)

Location of the case housing bolts

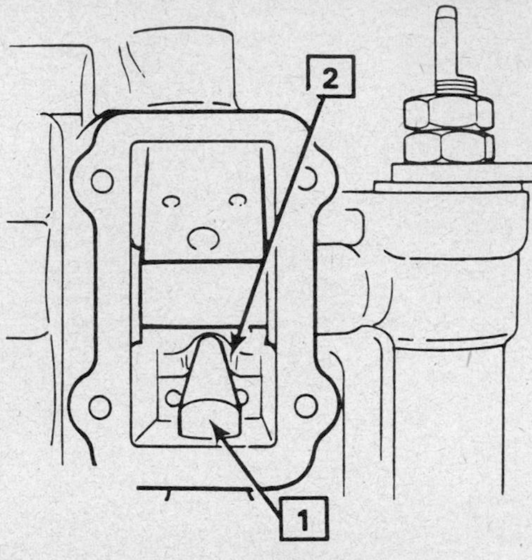
2
1
1. Piston rod
2. Recess

View of the 2nd brake piston rod and brake band recess

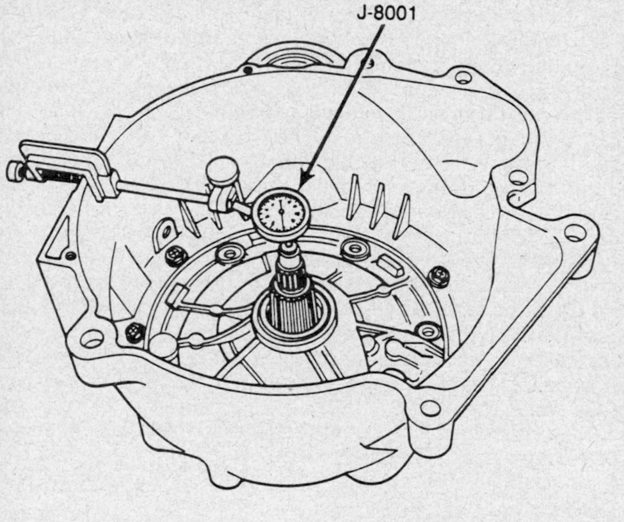
J-8001
INPUT SHAFT ENDPLAY
0.012–0.035 IN. (0.3–0.9MM)
BEARING RACE THICKNESS
0.031 IN. (0.8MM)
0.055 IN. (1.4MM)

Measuring the endplay of the input shaft

Piston rod length	Identification mark
121.3 mm (4.77 in)	Unmarked
122.7 mm (4.83 in)	Marked

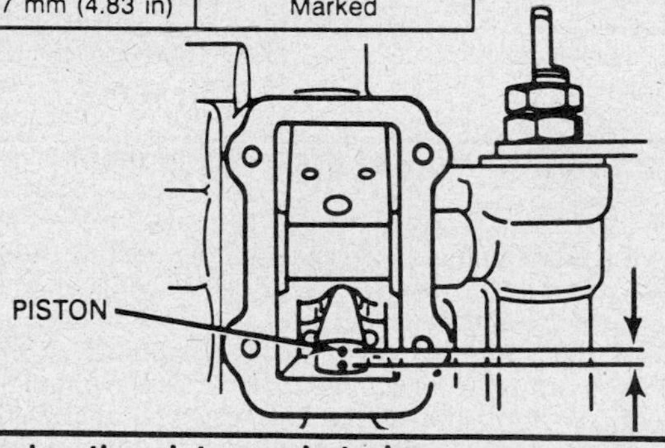
PISTON

Measuring the piston rod stroke

36. Using grease, apply it to the direct clutch washer and install it on the oil pump; fit the washer's flange into notch of the oil pump body.

37. Using grease, lubricate a new oil pump cover seal (O-ring) and install it the outer groove of the oil pump; make sure the seal is not twisted or extruded.

38. To install the oil pump, position the pump in the case, align the bolt holes and install the bolts. Make sure the direct clutch washer does not fall off and the input shaft seal rings or pump cover seal rings do not come off or get damaged. Torque the bolts to 13.5–19.5 ft. lbs. (18–27 Nm).

39. To check the input shaft endplay, perform the following procedures:

a. Using a dial indicator, position it on the tip of the input shaft and zero it.

b. Lift the input shaft and measure the amount of movement; it should be 0.012–0.035 in. (0.3–0.9mm). If the endplay is not within specifications, remove the oil pump and replace the input shaft bearing race (oil pump side).

40. To install the solenoid wiring harness, perform the following procedures:

a. Insert the solenoid wire holding plate in the groove of the solenoid wire grommet and the solenoid wire to the stud bolt.

b. Using a lock washer and a nut, secure the holding plate.

c. To the rear cover, install both solenoid wire clamps and secure them with the rear cover bolts.

41. To check the 2nd brake band for correct installation, look through the 2nd brake band cover hole to make sure the 2nd brake piston rod end is aligned with the center of the brake band recess. If the rod end contacts outside the brake band recess, insert a thin wire in the brake band fitting and pull the 2nd brake band up so it's recess aligns properly with the rod end.

42. To check the 2nd brake piston stroke, perform the following procedures:

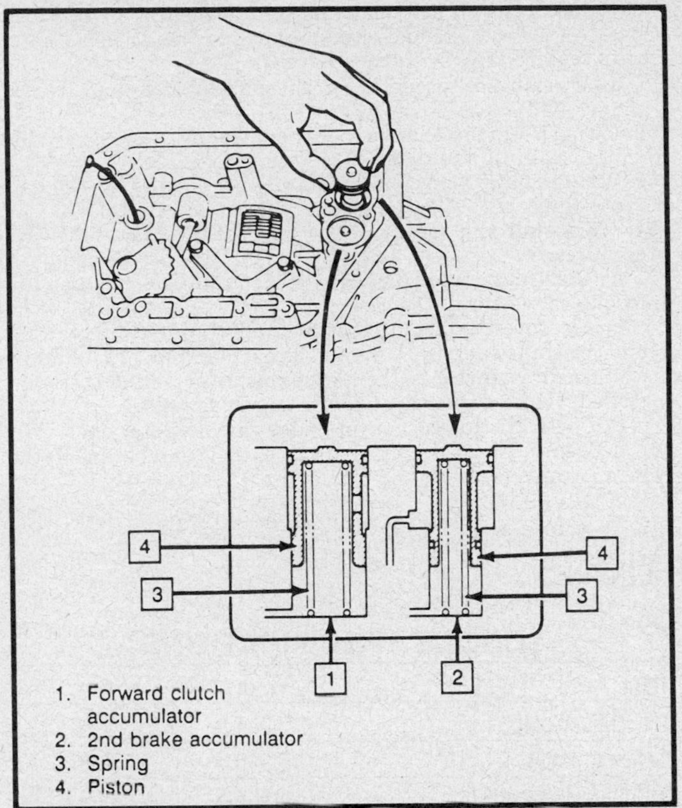
4
4
3
3
1
2
1. Forward clutch accumulator
2. 2nd brake accumulator
3. Spring
4. Piston

Installing the accumulator pistons and spring

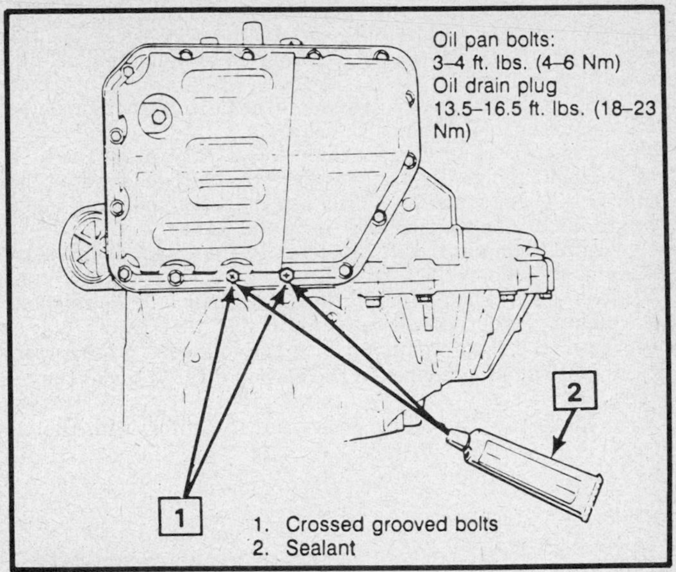
Oil pan bolts:
3–4 ft. lbs. (4–6 Nm)
Oil drain plug
13.5–16.5 ft. lbs. (18–23 Nm)
2
1
1. Crossed grooved bolts
2. Sealant

Installing the oil pan bolts

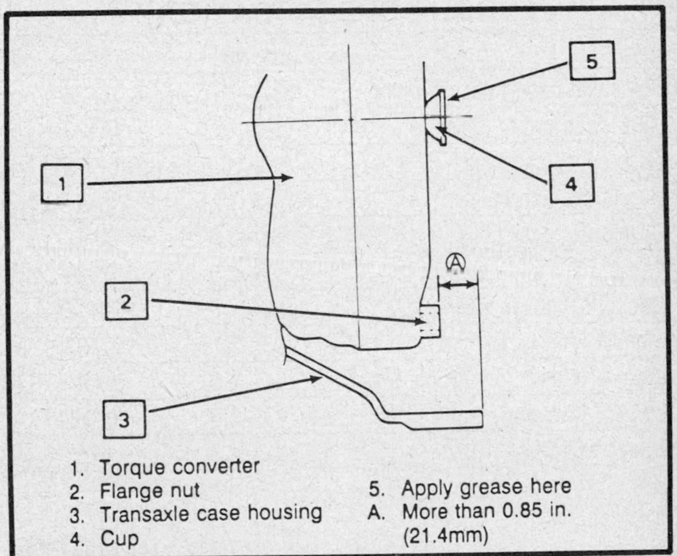
5
1
4
Ⓐ
2
3
1. Torque converter
2. Flange nut
3. Transaxle case housing
4. Cup
5. Apply grease here
A. More than 0.85 in. (21.4mm)

Installation of the torque converter

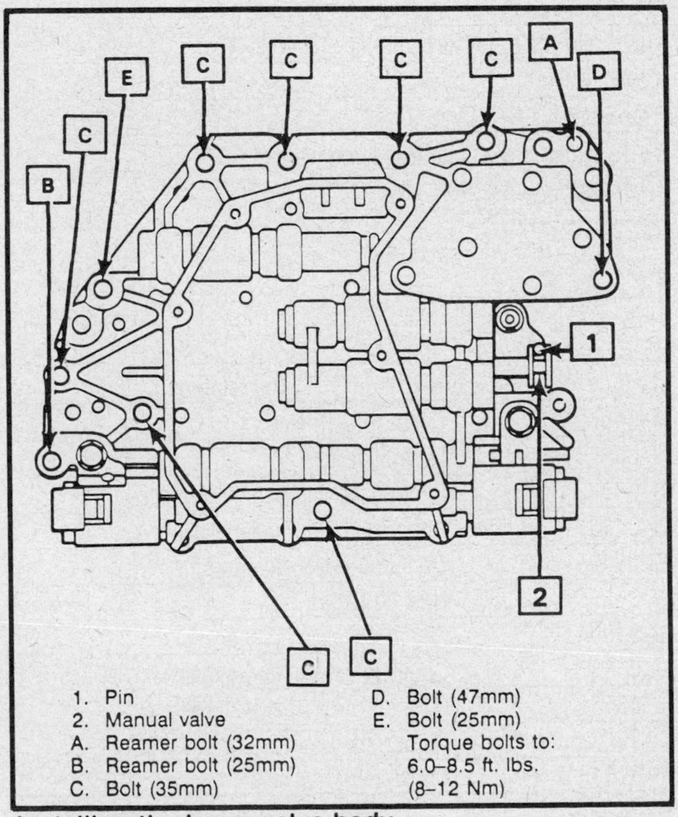
A
C
C
C
C
E
D
C
B
1
2
C
C
1. Pin
2. Manual valve
A. Reamer bolt (32mm)
B. Reamer bolt (25mm)
C. Bolt (35mm)
D. Bolt (47mm)
E. Bolt (25mm)
Torque bolts to:
6.0–8.5 ft. lbs.
(8–12 Nm)

Installing the lower valve body

a. Scribe a mark on the piston rod.

b. Apply low pressurized air into the oil hole and measure the 2nd brake piston rod stroke; it should be 0.06–0.11 in. (1.5–3.0mm).

c. If out of specifications, replace the 2nd brake band or the piston rod (with a different length); the piston rod is available in 2 lengths.

43. Using a new gasket, install the 2nd brake band cover and torque the bolts to 5.5–6.5 ft. lbs. (7–9 Nm).

44. Install the oil pressure control cable into the case.

45. To install the accumulator pistons, perform the following procedures:

a. Using Dexron®II fluid, dip the new seal rings and install them onto the pistons.

b. Install the pistons in the case; be careful the seal rings do not fall off.

c. Insert the spring into the 2nd brake accumulator piston.

46. To install the lower valve body into the case, align the manual valve with the manual shift lever pin and lower the valve

body into place; be careful the accumulator spring does not incline. Torque the valve body-to-case bolts to 6.0–8.5 ft. lbs. (8–12 Nm); be sure to tighten both reamer bolts first, then, all others in diagonal order.

47. To install the oil pressure control cable, hold the cam down and slide the cable end into the slot.

48. To install the oil tubes into the valve body, insert the tube end (without flange) into the lower valve body followed by the other end. Using a mallet, gently tap them into position; make sure to install the flanged end up to the flange.

49. Connect an electrical connector to the direct clutch solenoid and the 2nd brake solenoid.

50. To the valve body, install the oil strainer and the solenoid wire clamp. Torque the bolts to 3–4 ft. lbs. (4–6 Nm).

51. To install the oil pan, perform the following procedures:

a. Position the magnet in the oil pan directly under the oil strainer.

b. Make sure the oil tubes are not in contact with the oil pan.

c. Install the oil pan bolts and torque them to 3–4 ft. lbs. (4–6 Nm). Two bolts have cross grooves in their heads; these bolts require sealant on their threads.

d. If removed, torque the drain plug to 13.5–16.5 ft. lbs. (18–23 Nm).

52. Using a seal installation tool or equivalent, drive a new differential seal into the case.

53. Using a new O-ring, install the oil filler tube as far as possible into the case.

54. To install the torque converter, perform the following procedures:

a. Using care not to damage the oil pump seal, position the torque converter in the case.

b. Measure the distance between the torque converter lug and the transaxle case; it should be over 0.85 in. (21.4mm). If the specification is less, the torque converter is improperly installed; remove and reinstall the torque converter.

c. Check the torque converter for smooth rotation.

d. Using grease, apply it around the center cup of the torque converter.

SPECIFICATIONS

TORQUE SPECIFICATIONS

Item	ft. lbs.	Nm
Transaxle case housing bolts	12.0–16.5	16–23
Rear case cover bolts	12.0–16.5	16–23
Rear cover case nuts	8.0–10.5	11–15
Oil pump case bolts	13.5–19.5	18–27
Oil pump cover bolts	6.0–8.5	8–12
2nd brake band cover bolts	5.5–6.5	7–9
Reduction driven gear nut	80.0–108.0	110–150
Manual shift lever nuts	20.0–23.5	27–33
Manual detent spring bolt & nut	6.0–8.5	8–12
Lower valve body bolts	6.0–8.5	8–12
Upper valve body bolts	3.7–4.3	5–6

TORQUE SPECIFICATIONS

Item	ft. lbs.	Nm
Oil strainer bolts	3.0–4.0	4-6
Solenoid bolt	5.5–6.5	7–9
Throttle valve cam bolt	4.5–6.5	6–9
Oil pan bolts	4–6	–
Oil drain plug	13.5–16.5	18–23
Drive plate bolts (8mm)①	13.0–14.0	18–19
Drive rear plate bolts (10mm)②	40.5–45.0	56–62
Lower valve body cover bolt	3.0–4.0	4–6
Throttle valve cam bolt	4.5–6.5	6–9

① To install drive plate torque converter
② To install drive plate and

SPECIAL TOOLS

Tool	Description
J–6125–B	Oil pump puller slide hammers
J–7079–2	Driver handle
J–8001	Dial indicator set
J–8092	Driver handle
J–9617	Oil pump seal installer
J–22888	Output shaft bearing installer
J–23327	Clutch spring compressor
J–23907	Slide hammer
J–25018–A	Clutch spring compressor adapter
J–29369	Differential bearing puller
J–29369–1	Countershaft ball bearing remover
J–29369–2	Countershaft roller bearing remover

Tool	Description
J–34672	Depth micrometer
J–34673	Flat gauge bar
J–34842	Differential side bearing installer
J–34851	Differential side bearing pilot
J–35495	Oil pump puller adapters
J–35525	Holding fixture
J–35534	Piston cover depressor
J–35537	Countershaft bearing installer
J–35538	Differential oil seal installer
J–35539	Output shaft remover/installer
J–35608	Countershaft installer

Section 4

AW4 Transmission
AMC/Jeep-Eagle

APPLICATION

1986–89 Cherokee, Wagoneer and Comanche

GENERAL DESCRIPTION

The AW-4 automatic transmission is a 4-speed, electronically controlled transmission. This transmission consists of a lockup torque converter, oil pump, 3 planetary gear sets, clutch and brake units, hydraulic accumulators, a valve body which is controlled by electronic solenoids and a transmission computer unit.

The valve body solenoids are activated by electrical signals generated from the transmission computer unit. Signal sequence is determined by throttle position and vehicle speed.

Transmission and Converter Identification

TRANSMISSION

The AW-4 automatic transmission identification plate is attached to the right rear side of the automatic transmission case. This data plate provides the transmission part number, build date, model code and transmission serial number. Also included on this identification tag are the year and month that the transmission was built.

CONVERTER

The AW-4 automatic transmission uses a lockup torque converter. The lockup mechanism consists of a sliding clutch piston, coil springs and clutch friction material. All of the lockup clutch components are incorporated inside of the torque converter assembly. The clutch friction material is attached to the torque converter front cover. The clutch piston and torsion springs are attached to the turbine hub. The function of the torsion springs is to dampen engine firing impulses and loads that occur during the initial phase of the torque converter lockup function.

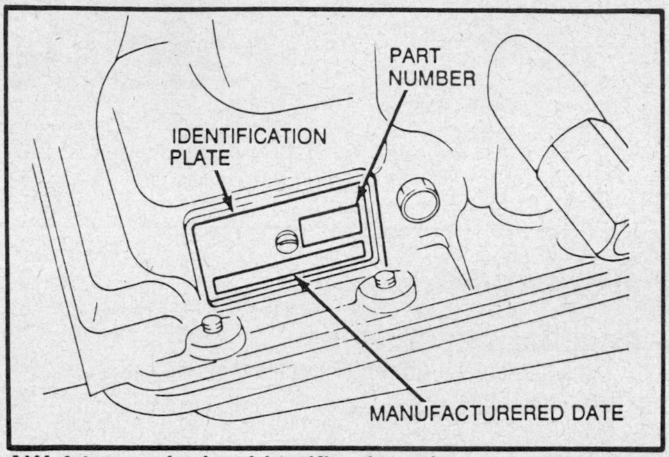
PART NUMBER
IDENTIFICATION PLATE
MANUFACTURERED DATE
AW-4 transmission identification plate

Torque converter lockup is controlled by valve body solenoid number 3 and by the lockup relay valve. When lockup speed is reached, the solenoid routes line pressure to the lockup clutch through the relay valve. Lockup operation is provided in 2nd, 3rd, and 4th gear ranges only.

Electronic Controls

The AW-4 automatic transmission is electronically controlled in the forward gear ranges. The controls consist of the transmission computer unit, valve body solenoids, and sensors. These sensors monitor vehicle speed, throttle opening, shift lever position and brake pedal application.

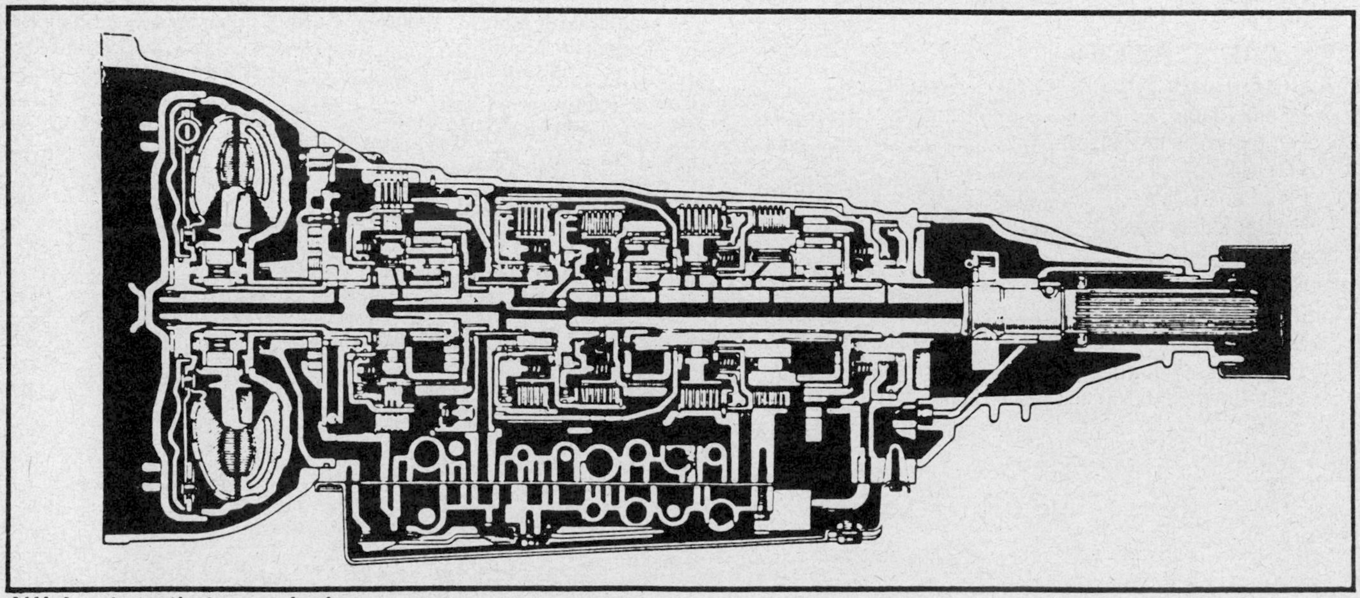
AW-4 automatic transmission

TRANSMISSION COMPUTER UNIT

The transmission computer unit determines shift and torque converter lockup timing based on signals received from the sensors. The transmission computer unit also activates and deactivates the valve body solenoids.

Two separate shift modes are programmed into the transmission computer unit. The first (comfort) mode, provides normal shift speeds and points. The second (power) mode, provides higher engine speeds and shift points when extra acceleration and torque are needed. Both of these shift modes are activated by a switch, which is located on the instrument panel.

The transmission computer unit incorporates a built in diagnostic program. Problems in the transmission circuitry or components can be diagnosed with the use of the MS1700 tester or the DRB II tester. Once the problem is noted it is stored in the transmission computer unit memory. The problem is stored until it is fixed and the computer is cleared by the technican. To erase the stored problem, disconnect and reconnect the **TRANS** fuse in the transmission computer unit wiring harness.

The transmission computer unit is located under the instrument panel on the passenger side of the vehicle.

VALVE BODY SOLENOIDS

The solenoids are mounted on the valve body and operated by the transmission control unit. These solenoids control operation of the torque converter lockup and shift valves in response to input signals from the transmission computer unit.

SENSORS

The sensors include the throttle position sensor, the speed sensor, the neutral safety switch and the brake pedal application switch.

The throttle position sensor is located on the throttle body. It electronically determines throttle position and relays this data to the transmission computer unit in order to control shift points and torque converter lockup.

The speed sensor consists of a rotor and a magnet. It is located on the transmission output shaft. A switch is mounted on the transmission extension housing or adapter. The sensor switch is activated each time the rotor and the magnet complete a revolution. The speed sensor signals are transmitted to the transmission computer unit through a wiring harness.

The neutral safety switch is located on the valve body manual shaft. The switch sends a signal from the shift linkage and manual valve position to the transmission computer unit, through an interconnecting wiring harness. The switch also prevents the engine from starting in any selectlor lever detent position other than **P** or **N**.

The brake application switch releases the lockup clutch in the torque converter whenever the vehicle brakes are applied. The switch is located on the brake pedal bracket and signals the transmission computer unit when the brake pedal is depressed or released.

Metric Fasteners

Metric tools will be required to service this transmission. Due to the large number of alloy parts used in this transmission, torque specifications should be strictly observed. Before installing capscrews into aluminum parts, dip the bolts into clean transmission fluid as this will prevent the screws from galling the aluminum threads, thus causing damage.

Metric fastener dimensions are very close to the dimensions of the familiar inch system fasteners. For this reason replacement fasteners must have the same measurement and strength as the original fastener.

Do not attempt to interchange meteric fasteners for inch system fasteners. Mismatched or incorrect fasteners can cause damage to the automatic transmission unit and possible personal injury. Care should be taken to reuse fasteners in their original locations.

Capacities

The fluid capacities are approximate and the correct fluid level should be determined by the dipstick indicator. After complete transmission overhaul the AW-4 transmission should hold approximately 8½ quarts of automatic transmission fluid. After pan and filter service, the AW-4 transmission should hold approximately 5 quarts of automatic transmission fluid. Transmissions used in 1986–88 vehicles should use Jeep automatic transmission fluid or Dexron® II. Transmissions used in 1989 vehicles should use Mercon® automatic transmission fluid.

Checking Fluid Level

The correct transmission fluid level is to the **FULL** mark on the dipstick indicator stick. Before checking the transmission fluid level be sure that the transmission is at normal operating temperature. Before checking the fluid level shift the transmission selector lever from the **P** detent to the **1–2** detent and then back into the **P** detent. Check the fluid level with the transmission selector lever in the **P** detent and the engine at curb idle speed.

The transmission fluid should be clear and free of foreign material or particles. If the transmission fluid is dark brown or black in color and smells burnt, it has been overheated and should be changed.

The transmission operation chould be checked if the unit contains large quanities of metal particles or clutch disc friction material. A small quanity of friction material or metal particles in the fluid pan is normal. These particles are usually generated during the brake in-period and indicate normal seating of the various transmission components.

The automatic transmission fluid should be changed every 30,000 miles or 30 months under normal driving conditions.

1. Run the engine until normal operating temperature is reached.
2. Position the vehicle on a flat surface. Depress the parking brake.
3. Check the idle speed and adjust, as required.
4. Move the transmission selector lever through all the shift detents and then position the selector lever in the **P** detent.
5. Remove the dipstick indicator level and wipe it clean. Insert the dipstick indicator level until it seats.
6. Remove the dipstick indicator level and check the fluid reading. It should be between the **ADD** and **FULL** marks on the dipstick. Correct the fluid level, as required.
7. Do not overfill the transmission, it takes about a pint of transmission fluid to raise the level from the **ADD** to the **FULL** mark on the dipstick indicator.

TRANSMISSION MODIFICATIONS

Information regarding any modifications to the AW-4 automatic transmission is not available at the time of publication.

TROUBLE DIAGNOSIS

CLUTCH AND BAND APPLICATION
AW–4 Automatic Transmission

Shift Lever Position	Gear	Valve Body Solenoid No.1	Valve Body Solenoid No.2	Over-Drive Clutch	Forward Clutch	Direct Clutch	Over-Drive Brake	Second Coast Brake	Second Brake	First/ Reverse Brake	Over-Drive One-way Clutch	No. 1 One-way Clutch	No. 2 One-way Clutch
P	Park	On	Off	Applied	–	–	–	–	–	–	–	–	
R	Reverse	On	Off	Applied	–	Applied	–	–	–	Applied	Applied	–	–
N	Neutral	On	Off	Applied	–	–	–	–	–	–	–	–	–
D	First	On	Off	Applied	Applied	–	–	–	–	–	Applied	–	Applied
	Second	On	On	Applied	Applied	–	–	–	Applied		–	Applied	Applied
	Third	Off	On	Applied	Applied	Applied	–	–	Applied	–	Applied	–	–
	OD	Off	Off	–	Applied	Applied	Applied	–	Applied	–	–	–	–
3	First	On	Off	Applied	Applied	–	–	–	–	–	Applied	–	Applied
	Second	On	On	Applied	Applied	–	–	Applied	Applied	–	Applied	Applied	–
	Third	Off	On	Applied	Applied	Applied	–	–	Applied	–	Applied	–	–
1–2	First	On	Off	Applied	Applied	–	–	–	–	Applied	Applied	–	Applied
	Second	On	On	Applied	Applied	–	–	Applied	Applied	–	Applied	Applied	–

CHILTON'S THREE C's TRANSMISSION DIAGNOSIS
AW–4 Transmission

Condition	Cause	Correction
Fluid discolored or smells burnt	a) Fluid contaminated b) Torque converter faulty c) Transmission faulty	a) Replace fluid b) Replace torque converter c) Disassemble and repair transmission
Vehicle does not move in any forward range or reverse	a) Shift linkage out of adjustment b) Valve body or primary regulator faulty c) Park lock pawl faulty d) Torque converter faulty e) Converter drive plate broken f) Oil pump intake screen blocked g) Transmission faulty	a) Adjust linkage b) Inspect/repair valve body c) Repair park pawl d) Replace torque converter e) Replace drive plate f) Clean screen g) Disassemble and repair transmission
Shift lever position incorrect	a) Shift linkage out of adjustment b) Manual valve and lever faulty	a) Adjust linkage b) Repair valve body
Harsh engagement (all ranges)	a) Throttle cable out of adjustment b) Valve body or primary regulator faulty c) Accumulator pistons faulty d) Transmission faulty	a) Adjust throttle cable b) Repair valve body c) Repair pistons d) Disassemble and repair

CHILTON'S THREE C's TRANSMISSION DIAGNOSIS
AW-4 Automatic Transmission

Condition	Cause	Correction
Delayed 1–2, 2–3 or 3–OD upshift or downshifts from 4–3 or 3–2 and shifts back to 4 or 3	a) Electronic control problem b) Valve body faulty c) Solenoid faulty	a) Find faulty part with MS 1700 tester or DRB II tester b) Repair valve body c) Repair solenoid
Slips on 1–2, 2–3 or 3–OD upshift or slips or shudders on take-off	a) Shift linkage out of adjustment b) LP cable out of adjustment c) Valve body faulty d) Solenoid faulty e) Transmission faulty	a) Adjust linkage b) Adjust cable c) Repair valve body d) Replace solenoid e) Disassemble and repair transmission
Drag or bind in 1–2, 2–3 or 3–OD upshift	a) Shift linkage out of adjustment b) Valve body faulty c) Transmission faulty	a) Adjust shift linkage b) Repair valve body c) Disassemble and repair transmission
No lockup in 2nd, 3rd or OD	a) Electronic control problem b) Valve body faulty c) Solenoid faulty d) Transmission faulty	a) Repair MS 1700 tester or DRB II tester b) Repair valve body c) Replace solenoid d) Disassemble and repair transmission
Harsh downshift	a) Throttle cable out of adjustment b) Throttle cable and cam faulty c) Accumulator pistons faulty d) Valve body faulty e) Transmission faulty	a) Adjust cable b) Replace cable and cam c) Repair pistons d) Repair valve body e) Transmission faulty
No downshift when coasting	a) Valve body faulty b) Solenoid faulty c) Electronic control problem	a) Repair valve body b) Replace solenoid c) Locate problem with MS 1700 tester or DRB II tester
Downshift late or early during coast	a) Throttle cable faulty b) Valve body faulty c) Transmission faulty d) Solenoid faulty e) Electronic control problem	a) Replace cable b) Repair valve body c) Disassemble and repair transmission d) Replace solenoid e) Locate problem with MS 1700 tester or DRB II tester
No OD–3, 3–2 or 2–1 kickdown	a) Solenoid faulty b) Electronic control problem c) Valve body faulty	a) Replace solenoid b) Locate problem with MS 1700 tester or DRB II tester c) Repair valve body
No engine braking in 1–2 position	a) Solenoid faulty b) Electronic control problem c) Valve body faulty d) Transmission faulty	a) Replace solenoid b) Loacte problem with MS 1700 tester or DRB II tester c) Repair valve body d) Disassemble and repair transmission
Vehicle does not hold in Park	a) Shift linkage out of adjustment b) Parking lock pawl cam and spring faulty	a) Adjust shift linkage b) Replace cam and spring

Hydraulic Control System

The hydraulic system consists of the transmission oil pump, the valve body and solenoids and 4 hydraulic accumulators.

The transmission fluid pump provides the required system lubrication and operating pressure.

The valve body controls the application of the clutches, brakes, second coast band and torque converter lockup clutch. The valve body solenoids control the sequencing of the 1–2, 2–3 and 3–4 shift valves within the valve body. The solenoids are activated by signals from the transmission computer unit.

The accumulators are used in the clutch and brake feed circuits to control initial apply pressure. Spring loaded accumulator pistons modulate the initial surge of apply pressure for smooth engagement.

TRANSMISSION OIL PUMP COMPONENTS

A rotor type oil pump is used in this transmission. The pump gears are mounted in the oil pump body. The drive gear is operated by the torque converter hub. Drive tangs on the hub engage in the drive slots of the drive gear.

VALVE BODY COMPONENTS

Automatic transmission working pressure is supplied to the clutch and brake apply circuits through the valve body assembly. The valve body consists of an upper body assembly, lower body assembly, separator plate and upper and lower gaskets. Various spool valves, sleeves, plugs and springs are also located within both valve body assembly sections. The manual valve, 1-2 shift valve, primary regulator valve, accumulator control valve, check balls, solenoids and oil strainers are located in the lower valve body assembly. The remaining control and shift valves plus check balls and an additional oil strainer are located in the upper valve body assembly.

The manual valve is operated by the gearshift linkage. The valve diverts fluid to the apply circuits according to shift lever position

The primary regulator valve modulates line pressure to the clutches and brakes according to engine load. The valve is actuated by throttle valve pressure. During high load operation, the valve increases line pressure to maintain positive clutch and

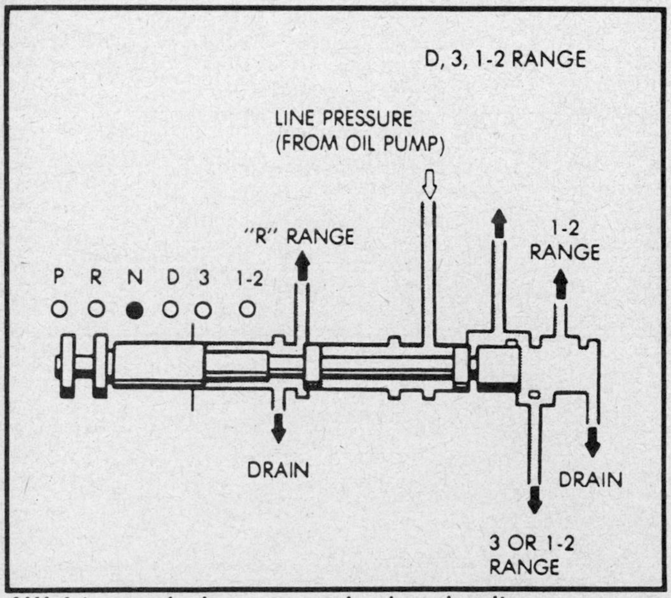
D, 3, 1-2 RANGE
LINE PRESSURE
(FROM OIL PUMP)
"R" RANGE
1-2
RANGE
P R N D 3 1-2
DRAIN
DRAIN
3 OR 1-2
RANGE

AW-4 transmission—manual valve circuitry

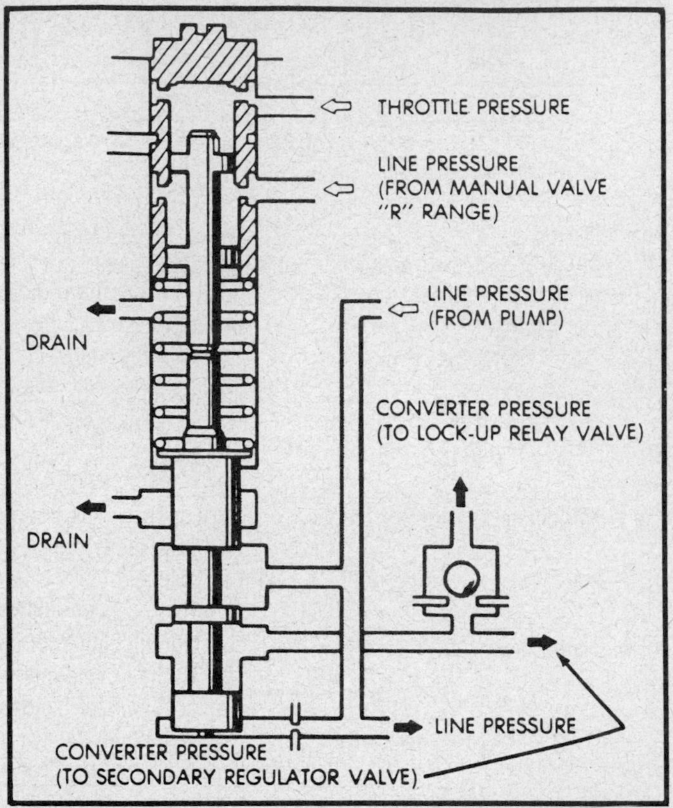
THROTTLE PRESSURE
LINE PRESSURE
(FROM MANUAL VALVE
"R" RANGE)
LINE PRESSURE
(FROM PUMP)
DRAIN
CONVERTER PRESSURE
(TO LOCK-UP RELAY VALVE)
DRAIN
LINE PRESSURE
CONVERTER PRESSURE
(TO SECONDARY REGULATOR VALVE)

AW-4 transmission—primary regulator valve circuitry

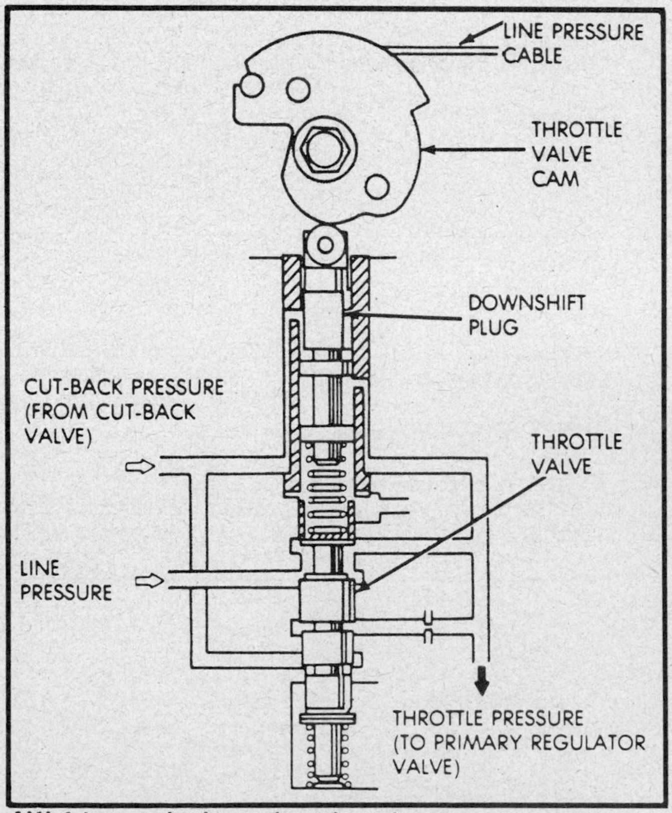
LINE PRESSURE
CABLE
THROTTLE
VALVE
CAM
DOWNSHIFT
PLUG
CUT-BACK PRESSURE
(FROM CUT-BACK
VALVE)
THROTTLE
VALVE
LINE
PRESSURE
THROTTLE PRESSURE
(TO PRIMARY REGULATOR
VALVE)

AW-4 transmission—throttle valve and downshift plug circuitry

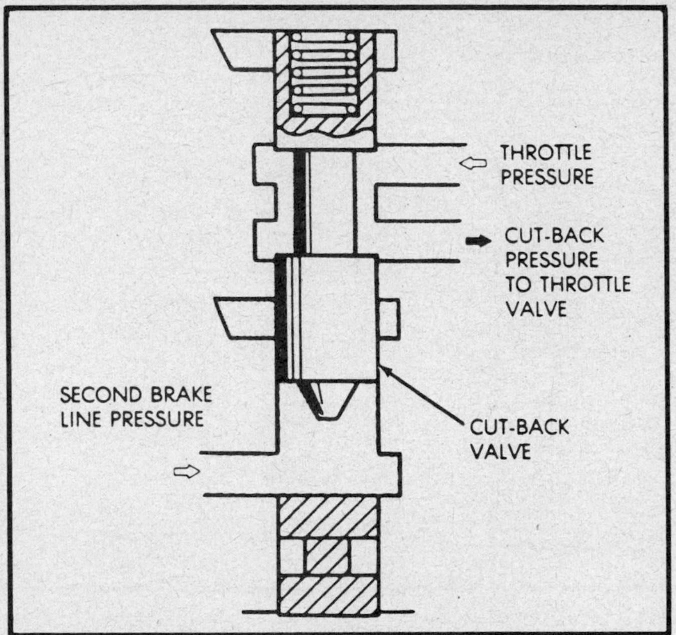
THROTTLE
PRESSURE
CUT-BACK
PRESSURE
TO THROTTLE
VALVE
SECOND BRAKE
LINE PRESSURE
CUT-BACK
VALVE

AW-4 transmission–cut back valve circuitry

brake engagement. At light load, the valve decreases line pressure just enough to maintain somooth engagemant.

The throttle valve and downshift plug control throttle pressure to the primary regulator valve. These valves are operated by the throttle valve cam line pressure cable in response to engine throttle position. Throttle valve pressure is also modulated by the cut back valve in 2nd, 3rd and 4th gear ranges.

The cut back valve helps prevent excessive pump pressure buildup in 2nd, 3rd and 4th gear. This valve is actuated by throttle pressure and by line pressure from the second brake. This valve also aids in regulating line pressure by controlling the amount of cut back pressure to the throttle valve.

The secondary regulator valve regulates converter lockup clutch and transmission lubrication pressure. When primary regulator valve pressure exceeds the specification for converter lockup clutch engagement or transmission lubrication, the secondary regulator valve is moved upward, exposing the drain port. Excess pressure than bleeds off, as required. As pressure drops, spring tension moves the valve downward closing the drain port.

The lockup relay valve controls the fluid flow to the torque converter lockup clutch. The valve is operated by line pressure from the 1–2 shift valve and is controlled by the third solenoid valve.

The 1–2 shift valve controls 1–2 upshifts and downshifts. The valve is operated by the second valve body solenoid and line pressure from the manual valve, second coast modulator valve and the 2–3 shift valve. When the transmission computer unit deactivates the solenoid, line pressure at the top of the valve moves the valve down, thus closing the second brake accumulator feed port. As the solenoid is activated and the drain port opens, spring force moves the valve upward and exposes the second brake feed port for the shift into 2nd gear.

The 2–3 shift valve controls 2–3 upshifts and downshifts. The valve is acutated by the first valve body solenoid and by line pressure from the manual valve and primary regulator valve. When the transmission computer unit activates the first valve body solenoid, line pressure at the top of the 2–3 valve is released through the solenoid drain port. Spring tension than moves the valve upward to hold the valve in the 2nd gear position. As the solenoid is deactivated, line pressure than moves

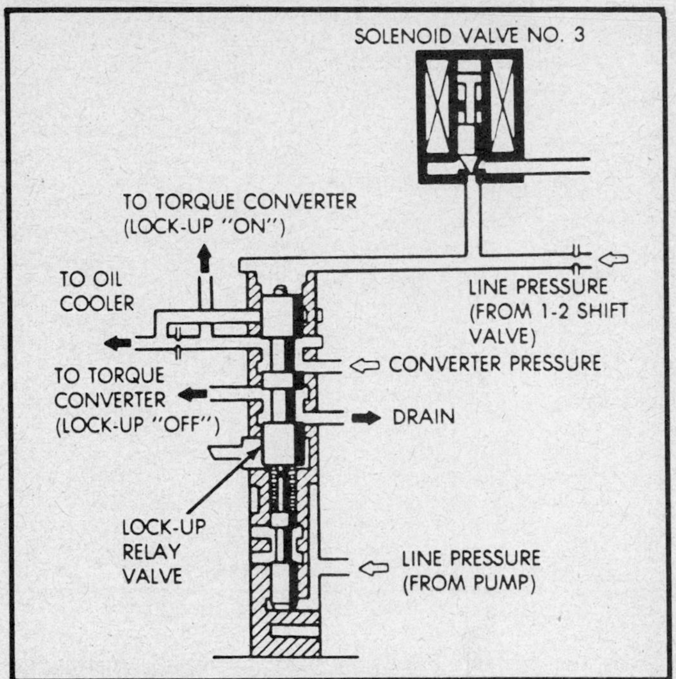
SOLENOID VALVE NO. 3
TO TORQUE CONVERTER
(LOCK-UP "ON")
TO OIL
COOLER
LINE PRESSURE
(FROM 1-2 SHIFT
VALVE)
CONVERTER PRESSURE
TO TORQUE
CONVERTER
(LOCK-UP "OFF")
DRAIN
LOCK-UP
RELAY
VALVE
LINE PRESSURE
(FROM PUMP)

AW-4 transmission–lockup relay valve circuitry

the valve down exposing the direct clutch feed port for the shift to 3rd gear.

The 3–4 shift valve is operated by the second valve body solenoid and by line pressure from the manual valve, the 2–3 valve and the primary regulator valve. As the transmission computer unit activates the second valve body solenoid, line pressure at the top of the 3–4 valve is released through the solenoid valve drain port. Spring tension than moves the valve upward, exposing the overdrive clutch accumulator feed port in order to apply the clutch. When the solenoid is deactivated and the drain port

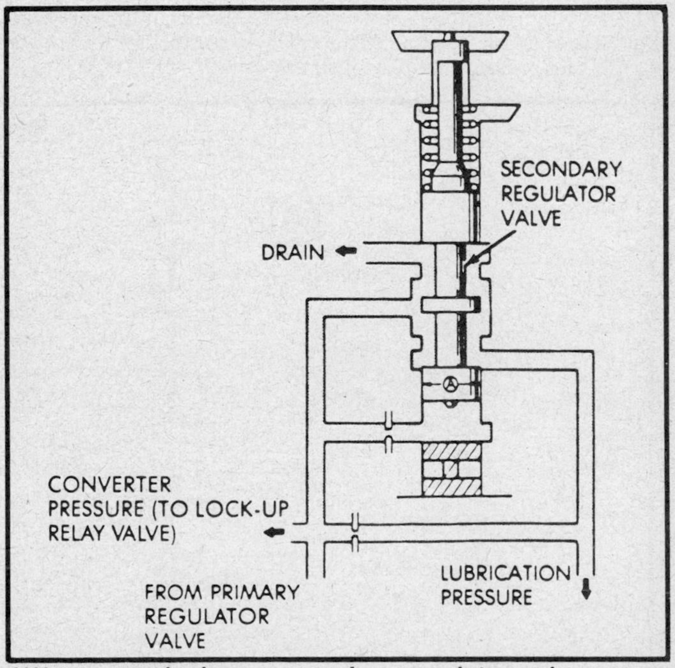
SECONDARY
REGULATOR
VALVE
DRAIN
CONVERTER
PRESSURE (TO LOCK-UP
RELAY VALVE)
LUBRICATION
PRESSURE
FROM PRIMARY
REGULATOR
VALVE

AW-4 transmission–secondary regulator valve circuitry

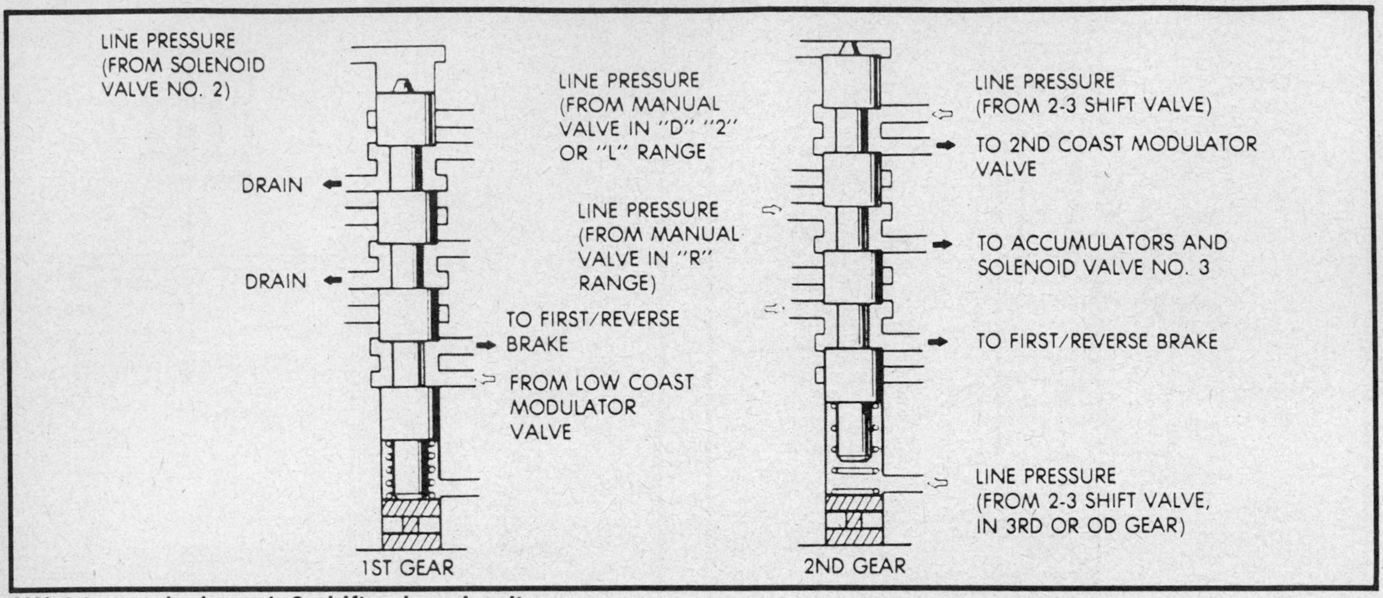
LINE PRESSURE
(FROM SOLENOID
VALVE NO. 2)
LINE PRESSURE
(FROM MANUAL
VALVE IN "D" "2"
OR "L" RANGE
DRAIN
DRAIN
LINE PRESSURE
(FROM MANUAL
VALVE IN "R"
RANGE)
TO FIRST/REVERSE
BRAKE
FROM LOW COAST
MODULATOR
VALVE
1ST GEAR
LINE PRESSURE
(FROM 2-3 SHIFT VALVE)
TO 2ND COAST MODULATOR
VALVE
TO ACCUMULATORS AND
SOLENOID VALVE NO. 3
TO FIRST/REVERSE BRAKE
LINE PRESSURE
(FROM 2-3 SHIFT VALVE,
IN 3RD OR OD GEAR)
2ND GEAR

AW-4 transmission–1–2 shift valve circuitry

closes, line pressure moves the valve downward, exposing the overdrive brake accumulator feed port for the shift into 4th gear. In the **1–2** or **3** gearshift lever detents, line pressure from the 2–3 shift valve is applied to the lower end of the 3–4 valve. This holds the valve upward thus closing off the overdrive brake feed port to prevent a shift into 4th gear.

The second coast modulator valve momentarily reduces line pressure from the 1–2 shift valve in order to cushion application of the second coast brake. The valve is operative only when the shift lever and manual valve are in the 3rd gear detent.

The low coast modulator valve momentarily reduces line pressure from the 2–3 shift valve to cushion the application of the first/reverse brake. The valve is operative only when the shift lever and the manual valve are in the 1–2 gear detent.

The accumulator control valve cushions clutch and brake application by reducing back pressure to the accumulators when throttle opening is small. The valve is controlled by both oil pump line pressure and throttle pressure.

This transmission uses 3 valve body solenoids. The first and second valve body solenoids control shift valve operation by applying or releasing line pressure as indicated by the transmission computer unit signal. The third valve body solenoid controls operation of the torque converter lockup clutch in response to signals from the transmission computer unit. When the first and second valve body solenoids are activated, the solenoid plunger is moved off of its seat thus opening the drain port to release line pressure. When either solenoid is deactivated, the plunger closes the drain port. The third valve body solenoid operates in reverse. When the solenoid is deactivated, the solenoid plunger is moved from its seat thus opening the drain port in order to release line pressure. When the solenoid is activated, the plunger closes the drain port.

ACCUMULATORS

Four accumulators are used to cushion the application of the clutches and brakes. The accumulators consist of spring loaded

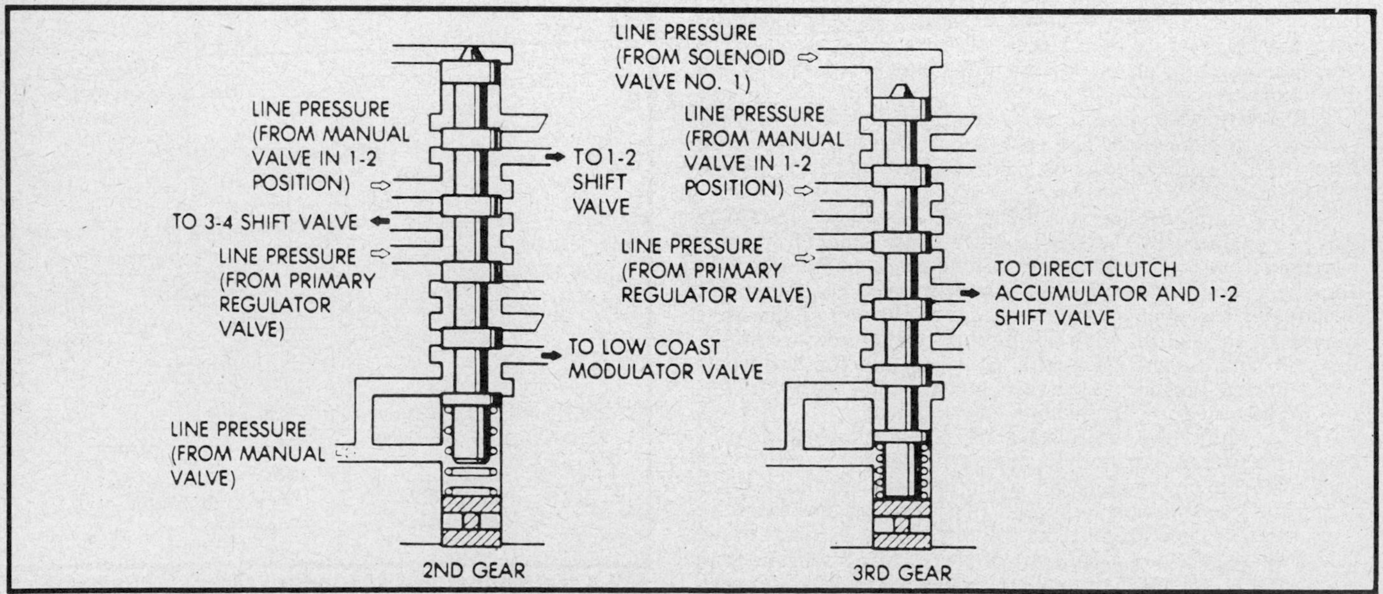
LINE PRESSURE
(FROM MANUAL
VALVE IN 1-2
POSITION)
TO 1-2
SHIFT
VALVE
TO 3-4 SHIFT VALVE
LINE PRESSURE
(FROM PRIMARY
REGULATOR
VALVE)
TO LOW COAST
MODULATOR VALVE
LINE PRESSURE
(FROM MANUAL
VALVE)
2ND GEAR
LINE PRESSURE
(FROM SOLENOID
VALVE NO. 1)
LINE PRESSURE
(FROM MANUAL
VALVE IN 1-2
POSITION)
LINE PRESSURE
(FROM PRIMARY
REGULATOR VALVE)
TO DIRECT CLUTCH
ACCUMULATOR AND 1-2
SHIFT VALVE
3RD GEAR

AW-4 transmission–2–3 shift valve circuitry

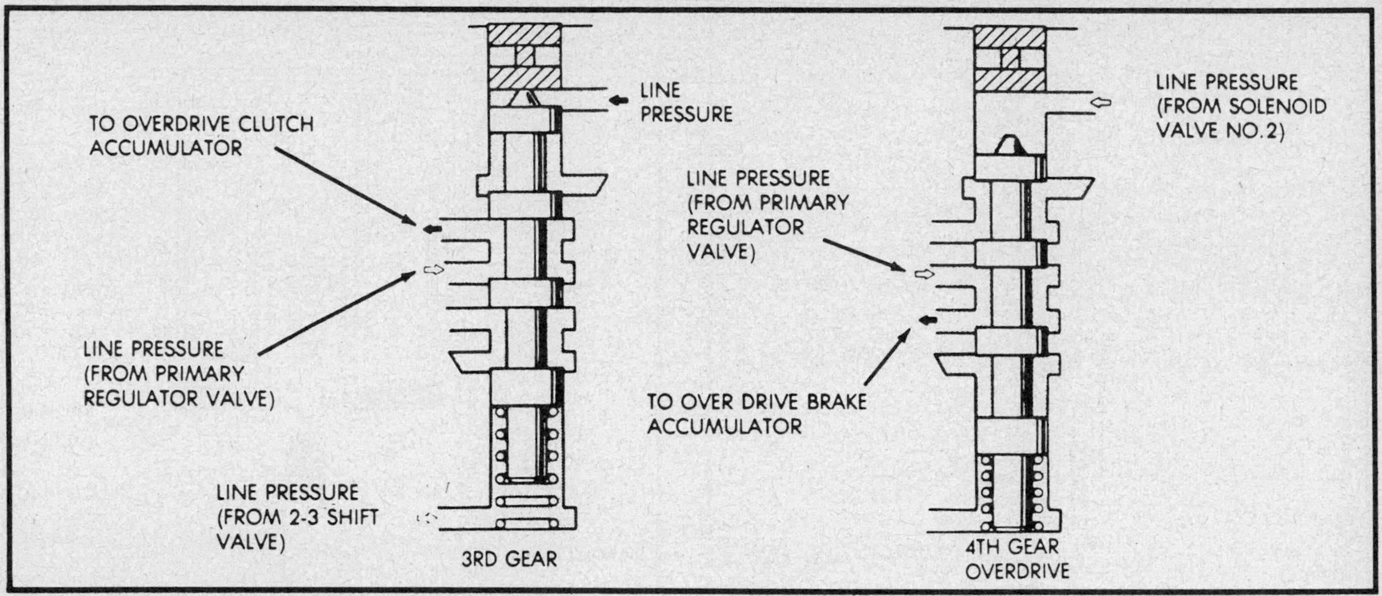
TO OVERDRIVE CLUTCH ACCUMULATOR
LINE PRESSURE
LINE PRESSURE (FROM PRIMARY REGULATOR VALVE)
LINE PRESSURE (FROM 2-3 SHIFT VALVE)
3RD GEAR
LINE PRESSURE (FROM SOLENOID VALVE NO. 2)
LINE PRESSURE (FROM PRIMARY REGULATOR VALVE)
TO OVER DRIVE BRAKE ACCUMULATOR
4TH GEAR OVERDRIVE

AW-4 transmission—3–4 shift valve circuitry

pistons which dampen the initial surge of apply pressure in order to provide smooth engagement during transmission gear shifts. Control pressure from the accumulator control valve is continously applied to the back pressure side of the accumulator pistons. This pressure, plus spring tension holds the pistons down. As line pressure from the shift valves enters the opposite end of the piston bore, control pressure and spring tension momentarily delay application of full line pressure in order to cushion engagement. All of the accumulators are located inside the transmission case assembly.

Diagnosis Tests

The AW-4 automatic transmission is an electronically controlled transmission. Shift points and sequence in the forward gear ranges are controlled by the transmission computer unit. Before attempting transmission repair, it will be necessary to determine if the transmission is at fault and than whether the fault is mechanical or electrical.

The transmission computer unit that is used with the AW-4 transmission has a self diagnostic program. The program is compatible with the MS1700 tester or the DRB II tester. The tester will identify faults in the electrical control system. The road test, control pressure test, stall speed test and time lag test will identify faults in the mechanical functions of the transmission.

PRE TEST DATA

1. Check and adjust the shift linkage.
2. Check line pressure cable operation. Correct, as required.
3. Check engine throttle operation. Correct, as required.

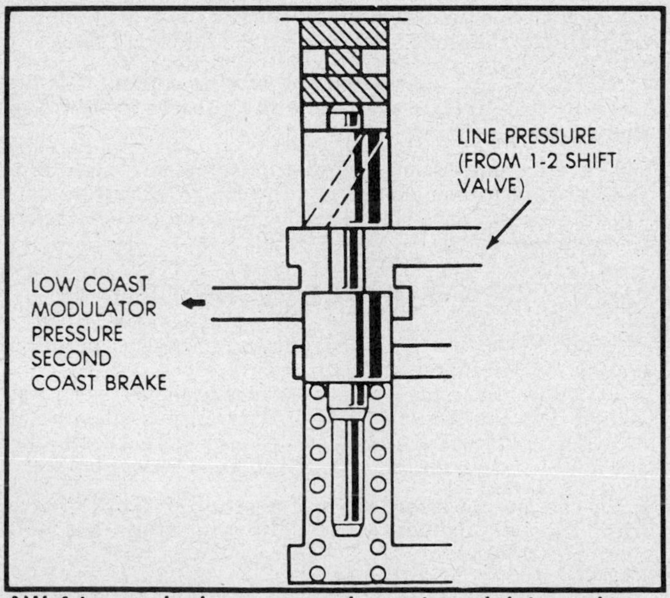
LINE PRESSURE (FROM 1-2 SHIFT VALVE)
LOW COAST MODULATOR PRESSURE SECOND COAST BRAKE

AW-4 transmission—second coast modulator valve circuitry

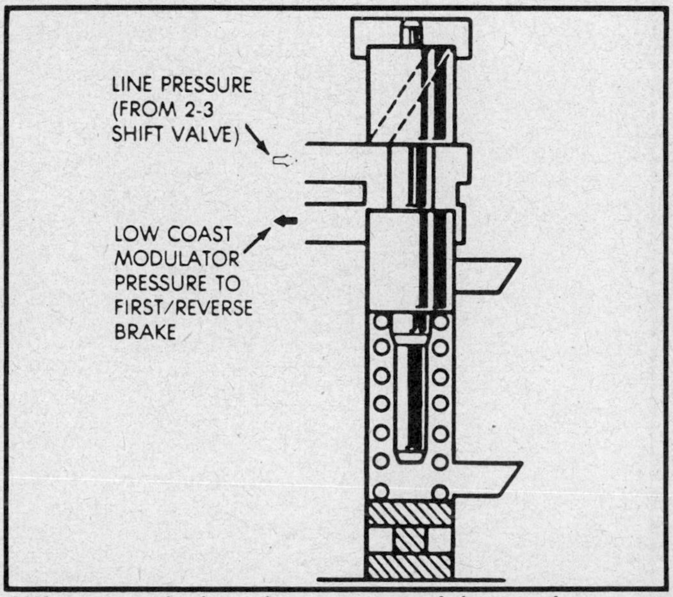
LINE PRESSURE (FROM 2-3 SHIFT VALVE)
LOW COAST MODULATOR PRESSURE TO FIRST/REVERSE BRAKE

AW-4 transmission—low coast modulator valve circuitry

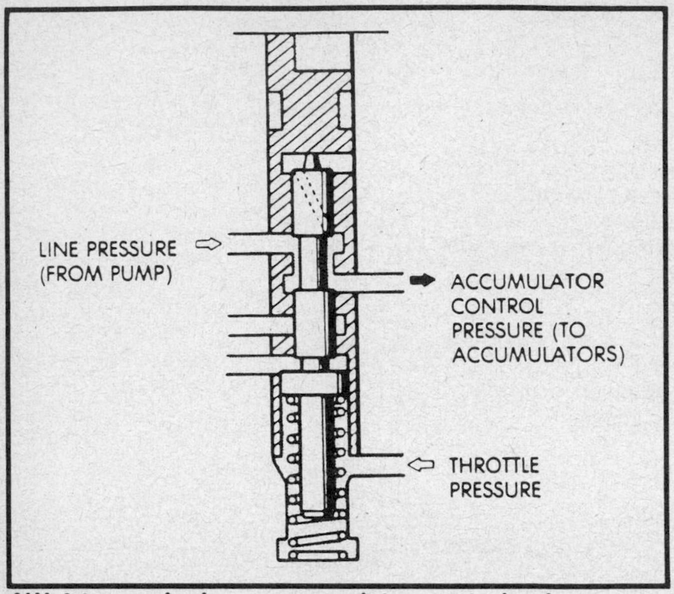
LINE PRESSURE
(FROM PUMP)
ACCUMULATOR
CONTROL
PRESSURE (TO
ACCUMULATORS)
THROTTLE
PRESSURE

AW-4 transmission—accumulator control valve circuitry

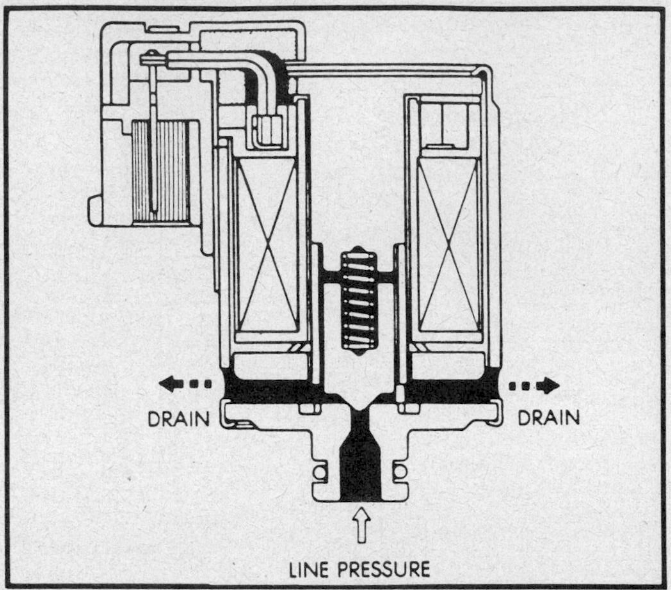
DRAIN
DRAIN
LINE PRESSURE

AW-4 transmission—valve body solenoids

4. Check transmission fluid level. Correct, as required.
5. Check the neutral safety switch. Adjust, as required.
6. Check TPS adjustment and operation. Adjust or replace the sensor, as required.

MANUAL SHIFTING TEST

1. Stop the engine. Disconnect the transmission computer unit or the transmission computer unit fuse.

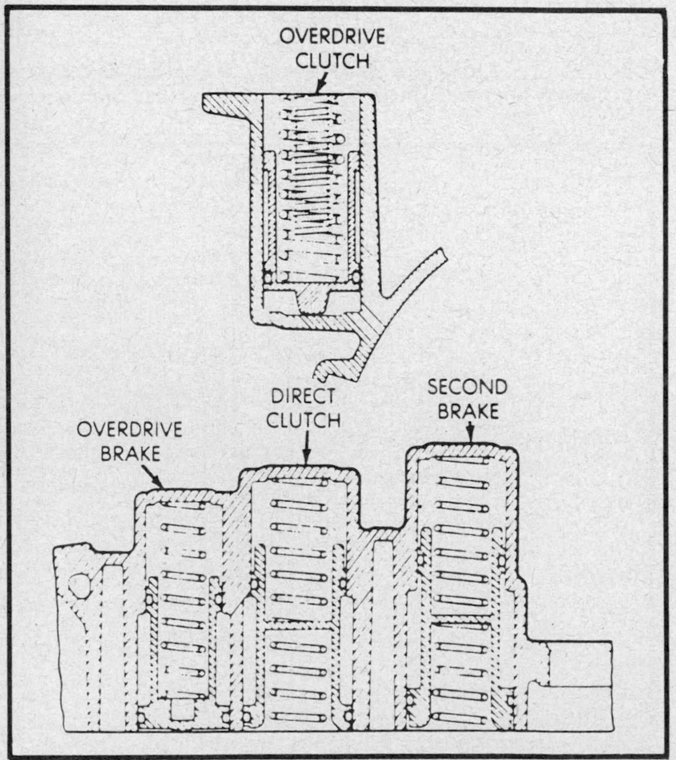
OVERDRIVE
CLUTCH
DIRECT
CLUTCH
SECOND
BRAKE
OVERDRIVE
BRAKE

AW-4 transmission—accumulator assemblies

2. Road test the vehicle. Shift the transmission selector lever into each gear range.
3. The transmission should lock in the **P** detent. The transmission should back up in the **R** detent. The transmsision should not move in the **N** detent.
4. The transmission should have 1st gear only when the selector lever is positioned in the **1-2** detent. The transmission should have 3rd gear only when the selector lever is positioned in the **3** detent. The transmission should have overdrive (4th) gear only when the selector lever is positioned in the **D** detent.
5. If the transmission does not operate as indicated, refer to the diagnosis charts. Do not perform the stall speed test or the timelag test.
6. If the transmission does perform as indicated, and all forward gear ranges were not difficult to distinguish, continue the road test.
7. Manually downshift the transmission from the **D** detent to the **3** detent and than from the **3** detent to the **1–2** detent. Manually upshift the transmission through all forward ranges.

NOTE: Do not overspeed the engine during this test. Ease off the throttle and allow the vehicle to slow down before downshifting.

8. If the transmission operation is acceptable, perform the stall speed test, time lag test and the control pressure test.
9. If a transmission shifting problem is encountered, refer to the diagnosis charts.

CONTROL PRESSURE TEST

1. Run the engine until the transmission reaches normal operating temperature.
2. Connect the control pressure gauge to the test port, which is located on the passenger side of the transmission.
3. Depress the parking brake. Block the drive wheels. Check and adjust the engine idle speed. Apply the service brakes.

NOTE: Do not allow anyone to stand in front of or behind the vehicle while the following steps are being performed.

4. Shift the transmission into the **D** detent. Record the line pressure with the engine at idle speed. The control pressure specification should be 53–61 psi.

5. With the selector lever in the **D** detent, depress the accelerator to the wide open throttle position. Record the line pressure. The control pressure specification should be 161–196 psi. Do not maintain wide open throttle for more than a few seconds at a time.

6. Shift the transmission into the **R** detent. Record the line pressure with the engine at idle speed. The control pressure specification should be 73–87 psi.

7. With the selector lever in the **R** detent, depress the accelerator to the wide open throttle position. Record the line pressure. The control pressure specification should be 223–273 psi. Do not maintain wide open throttle for more than a few seconds at a time.

8. If the control line pressure is not within specification, adjust the line pressure cable and repeat the control pressure test.

9. If the line pressures are higher than specification, check for a defective line pressure cable, or a worn, sticking or damaged throttle valve, downshift plug, throttle cam or primary regulator valve.

10. If the line pressures are lower than specification, check for a defective line pressure cable, or a worn, sticking or damaged throttle valve, downshift plug, throttle cam or primary regulator valve. Also check for defective oil pump gears and housing and a worn overdrive clutch assembly.

11. If the line pressure is low in the **D** detent, check the forward clutch assembly for wear and damage. Also check for fluid leakage in the drive circuit.

12. If the line pressure is low in the **R** detent, check the shift linkage and the manual valve for proper adjustment. Check the direct clutch assembly and the first/reverse brake for wear and damage. Also check for fluid leakage in the reverse circuit.

AIR PRESSURE TEST

The air pressure test can be used to determine if cross passages are present within the transmission case or valve body assembly. This test is also used to determine if the clutch passages are open.

STALL SPEED TEST

1. Run the engine until the transmission reaches normal operating temperature.

2. Connect a tachometer to the engine, position it so that it can be viewed from the drivers seat.

3. Depress the parking brake. Block the drive wheels. Check and adjust the engine idle speed. Apply the service brakes.

4. If equipped with 4WD, shift the transfer case into 2WD **HIGH** position. Start the engine.

NOTE: Do not allow anyone to stand in front of or behind the vehicle while the following steps are being performed.

5. Shift the transmission selector lever into the **D** detent. Depress the accelerator to the wide open throttle position and record the maximum rpm reading. Stall speed should be 2100–2400 rpm. Do not maintain wide open throttle for more than a few seconds at a time.

6. Release the throttle and shift the transmission into the **N** detent. Allow the transmission fluid to cool for about 15–20 seconds.

7. Shift the transmission selector lever into the **R** detent. Depress the accelerator to the wide open throttle position and record the maximum rpm reading. Stall speed should be 2100–2400 rpm. Do not maintain wide open throttle for more than a few seconds at a time.

8. If the engine rpm specification is lower than specification, check engine performance. Check the stator clutch in the torque converter, as it may not be holding if the engine rpm speed was less than 1500 rpm.

9. If the engine rpm specification is higher than specification, check for low fluid level, low line pressure or the overdrive one-way clutch not holding.

10. If the engine rpm specification is higher than specified in the **D** range, check for low line pressure, forward clutch slippage, the number 2 one-way clutch not holding or the overdrive one-way clutch not holding.

11. If the engine rpm specification is higher than specified in the **R** range, check for low line pressure, direct clutch slippage, first/reverse brake slippage or the overdrive one-way clutch not holding.

ROAD TEST

When road testing the vehicle, be sure that the transmission is at normal operating temperature. Operate the transmission in each shift detent to check for slipping or any variation in the shifting patern. Note whether the shifts are harsh or spongy. By utilizing the clutch and band application chart with any malfunctions noted, the defective unit or circuit can be found.

TIME LAG TEST

The time lag test checks the general condition of the overdrive clutch, forward clutch, rear clutch and first/reverse brake. Condition is indicated by the amount of time required for clutch/brake engagement with the engine at normal idle speed. Engagement time is measured for the **D** and **R** detents.

1. Check and adjust the transmission fluid level. Run the engine until the transmission reaches normal operating temperature.

2. Depress the parking brake. Turn the air conditioning off. If equipped with 4WD, shift the transfer case into the 2WD **HIGH** position.

3. Start the engine and check the idle speed, correct as required.

4. Position the transmission selector lever in the **N** detent. Set the stop watch.

5. During the following tests, start the stop watch as soon as the selector lever reaches the **D** and **R** detents.

6. Shift the transmission selector lever into the **D** detent and record the time it takes for engagement. Engagement time should be 1.2 seconds maximum. Repeat the test at least twice.

7. Reset the stop watch. Shift the transmission selector lever into the **N** detent.

8. Shift the transmission selector lever into the **r** detent and record the time it takes for engagement. Engagement time should be 1.5 seconds maximum. Repeat the test at least twice.

9. If engagement is longer than specification with the selector lever in the **D** detent, check for misadjusted shift linkage, low line pressure, worn forward clutch assembly or a worn and damaged overdrive clutch.

10. If engagement is longer than specification with the selector lever in the **R** detent, check for misadjusted shift linkage, low line pressure, worn direct clutch assembly, worn first/reverse brake or a worn and damaged overdrive clutch.

ON CAR SERVICES

Adjustments

THROTTLE CABLE

1. Turn the ignition switch to the **OFF** position.
2. Fully retract the cable plunger. Press the cable button all the way down. Push the cable button inward.

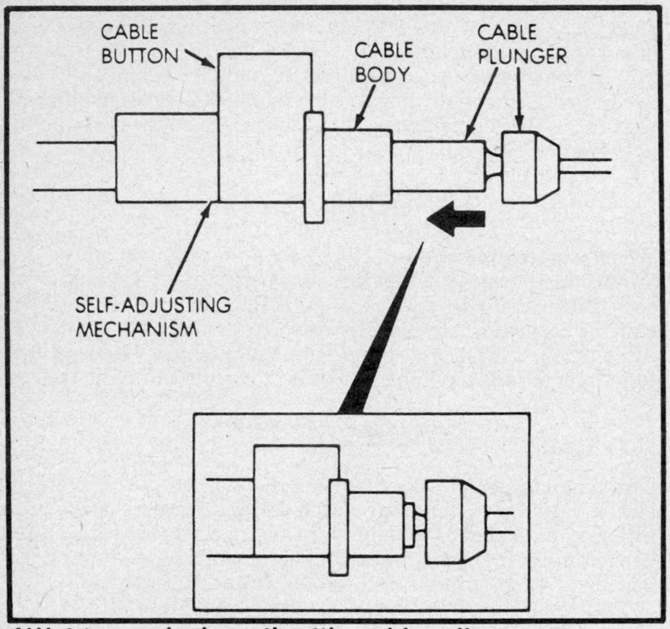
CABLE
BUTTON
CABLE
BODY
CABLE
PLUNGER
SELF-ADJUSTING
MECHANISM

AW-4 transmission—throttle cable adjustment

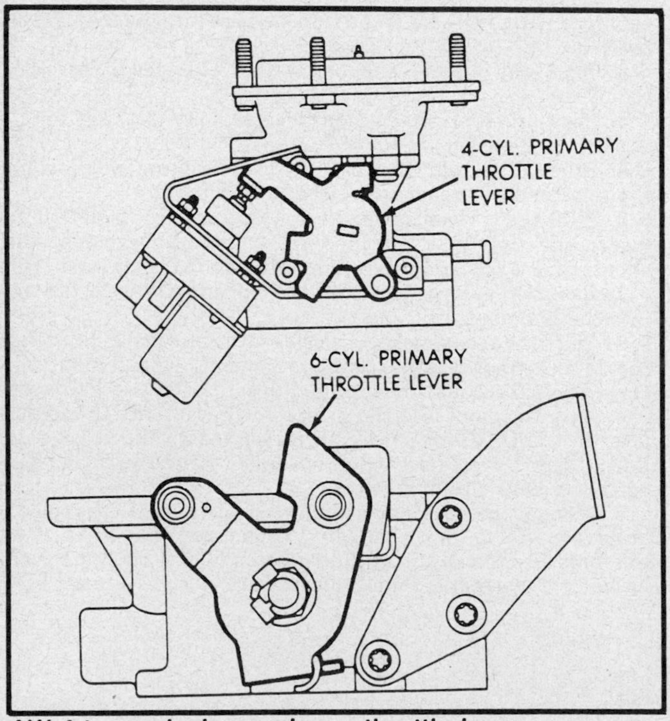
4-CYL. PRIMARY
THROTTLE
LEVER
6-CYL. PRIMARY
THROTTLE LEVER

AW-4 transmission—primary throttle lever positioning

3. Rotate the primary throttle lever to the wide open throttle lever position and let the cable plunger extend.
4. Release the lever when the plunger is fully extended. The cable is now adjusted.

THROTTLE POSITION SENSOR

1. Loosen the TPS adjusting screws. Partially retighten a retaining screw in order to secure the sensor for adjustment.
2. If a voltmeter is used for adjustment, connect the positive lead to terminal **B** and the negative lead to terminal **D**.
3. If the MS1700 tester or the DRB II tester is being used, connect it according to the manufacturers instructions.
4. Observe the equipment reading and rotate the TPS until output voltage is 82% of input voltage, or about 4.2 volts.
5. Tighten the TPS retaining screws. Recheck the adjustment.
6. If required output voltage cannot be obtained or input voltage is considerably less than 5.0 volts, replace the TPS assembly.

SHIFT CABLE ADJUSTMENT

1. Be sure that the transmission selector lever is in the **P** detent.
2. Raise and support the vehicle safely.
3. Release the cable adjuster clamp in order to unlock the cable. Unsnap the cable from the bracket.
4. Move the transmission selector lever all the way rear ward into the **P** detent. The lever is located on the manual valve shaft at the left side of the transmission case.

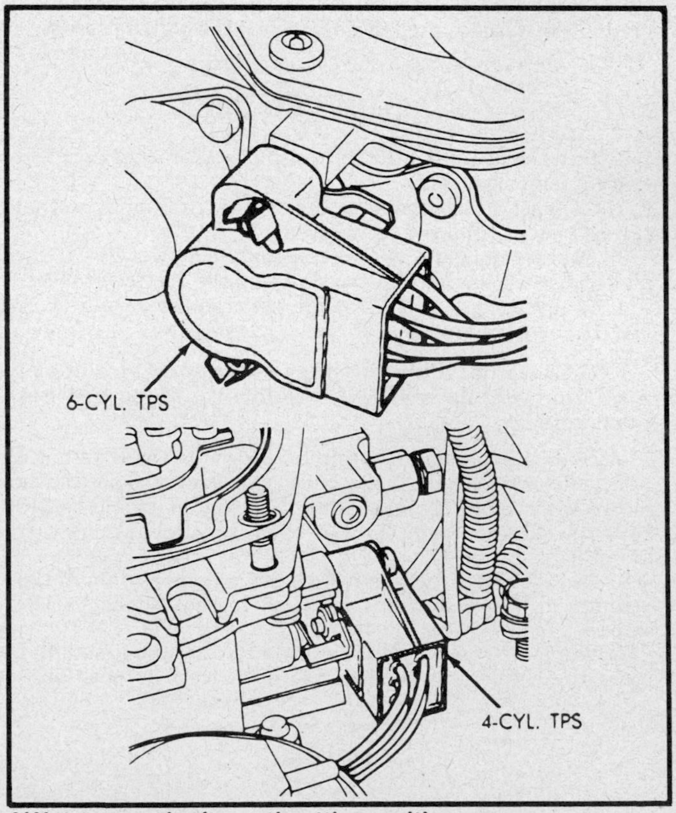
6-CYL. TPS
4-CYL. TPS

AW-4 transmission—throttle position sensor identification

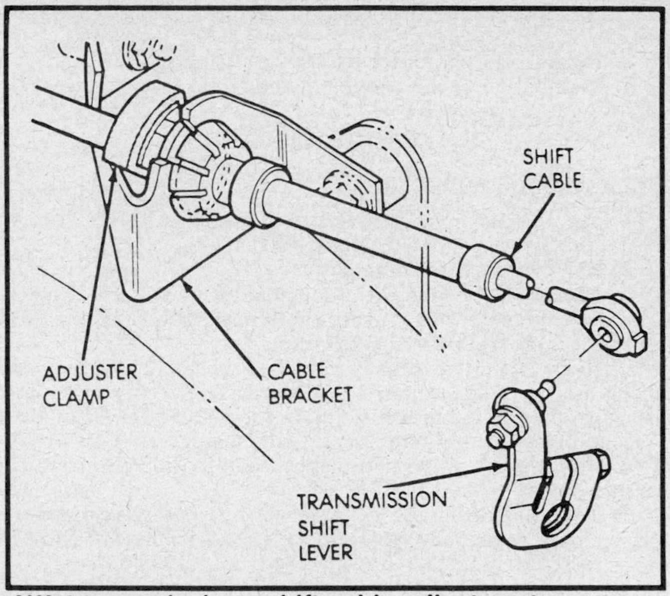
SHIFT CABLE
ADJUSTER CLAMP
CABLE BRACKET
TRANSMISSION SHIFT LEVER

AW-4 transmission—shift cable adjustment

5. Verify that the transmission is in the **P** position by attempting to rotate the driveshaft, it should not turn.
6. Snap the cable into the cable bracket. Lock the cable by pressing the cable adjuster clamp down until it snaps into place.
7. The engine should only start in the **P** or **N** detent.

PARK LOCK CABLE ADJUSTMENT

1. Be sure that the transmission selector lever is in the **P** detent.
2. Be sure that the ignition switch is in the **LOCK** position.
3. Remove the shift lever bezel and console screws. Raise the bezel and the console to gain access to the cable.
4. Pull the cable lock button upward in order to release the cable.
5. Pull the cable forward. Release the cable and press the cable lock button down until it snaps into place.
6. Check the movement of the release shift handle button on vehicles equipped with a floor shifter, it should not press inward.
7. Check the movement of the release lever on vehicles equipped with a column shifter, the selector lever should not move.
8. Turn the ignition switch to the **ON** position.
9. If equipped with a floor shifter, press the shifter release button. If equipped with a column shifter, move the selector lever.
10. Shift the selector lever into the **N** detent. If the cable adjustment is correct, the ignition switch cannot be turned to the **LOCK** position.
11. Shift the selector lever into the **D** detent. If the cable adjustment is correct, the ignition switch cannot be turned to the **LOCK** position.
12. Position the selector lever into the **P** detent. Check the ignition switch operation. It is correct if the ignition switch can be turned to the **LOCK** position and the release button, on floor shifter equipped vehicles or the selector lever, on column shifter equipped vehicles does not move.

NEUTRAL SAFETY SWITCH

1. Position the transmission selector lever in the **N** detent.
2. Raise and support the vehicle safely.

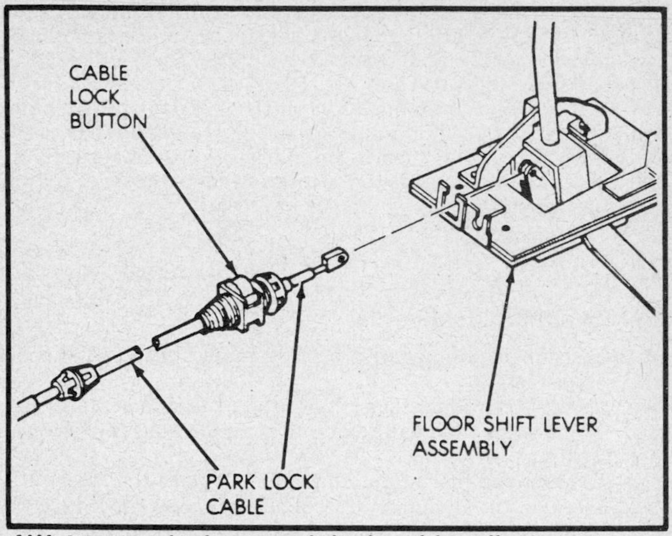
CABLE LOCK BUTTON
FLOOR SHIFT LEVER ASSEMBLY
PARK LOCK CABLE

AW-4 transmission—park lock cable adjustment

3. Loosen the switch retaining bolt. Rotate the switch in order to align the neutral standard line with the vertical groove on the manual valve shaft.
4. Align the switch standard line with the groove or flat on the manual valve shaft.
5. Torque the switch adjusting bolt to 9 ft. lbs. (13 Nm). Bend the washer lock tabs over the switch retaining nut in order to secure it in place.
6. Lower the vehicle and check the switch operation.

Services

FLUID CHANGES

The conditions under which the vehicle is operated is the main consideration in determining how often the transmission fluid should be changed. Different driving conditions result in different transmission fluid temperatures. These temperatures affect change intervals.

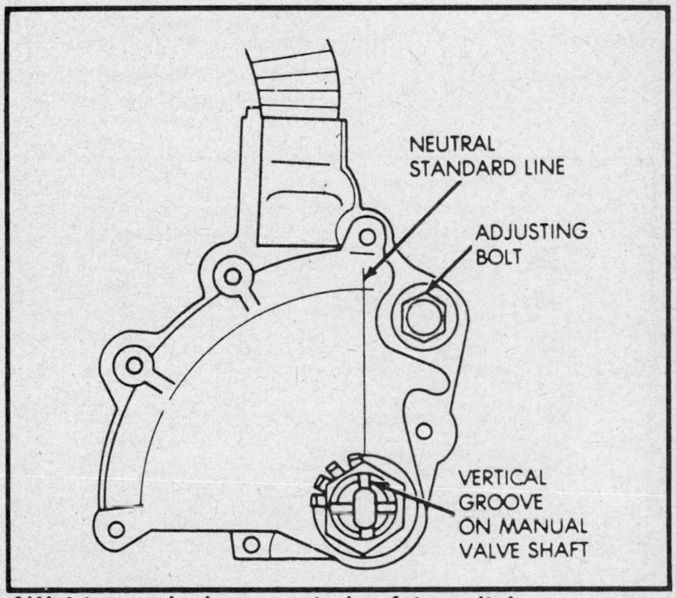
NEUTRAL STANDARD LINE
ADJUSTING BOLT
VERTICAL GROOVE ON MANUAL VALVE SHAFT

AW-4 transmission—neutral safety switch adjustment

If the vehicle is driven under severe conditions, change the transmission fluid every 15,000 miles. If the vehicle is not used under severe conditions, change the fluid and replace the filter every 30,000 miles.

Do not overfill the transmission. It takes about a pint of automatic transmission fluid to raise the level from the **ADD** to the **FULL** mark on the transmission indicator dipstick. Overfilling the unit can cause damage to the internal components of the automatic transmission.

OIL PAN

Removal and Installation

1. Disconnect the negative battery cable. Raise and support the vehicle safely.
2. Drain the transmission fluid into a suitable container.
3. Remove the fluid pan retaining bolts. Remove the fluid pan and gasket.
4. Remove the filter retaining bolts. Remove the filter.
5. Installation is the reverse of the removal procedure. Be sure to use a new gasket or RTV sealant, as required.
6. Fill the transmission to specification with the proper grade and type automatic transmission fluid.
7. Run the engine until the transmission reaches normal operating temperature. Recheck the fluid level and correct, as required.

NEUTRAL SAFETY SWITCH

Removal and Installation

1. Disconnect the negative battery cable. Raise and support the vehicle safely.
2. Disconnect the electrical connector from the switch assembly.
3. Pry the washer lock tabs upward. Remove the switch retaining nut and the tabbed washer.
4. Remove the switch adjusting bolt. Slide the neutral safety switch off of the manual valve shaft.
5. Installation is the reverse of the removal procedure. Adjust the neutral safety switch, as required.

TRANSMISSION COMPUTER UNIT

Removal and Installation

1. Disconnect the negative battery cable. The unit is located under the instrument panel on the left side of the vehicle.

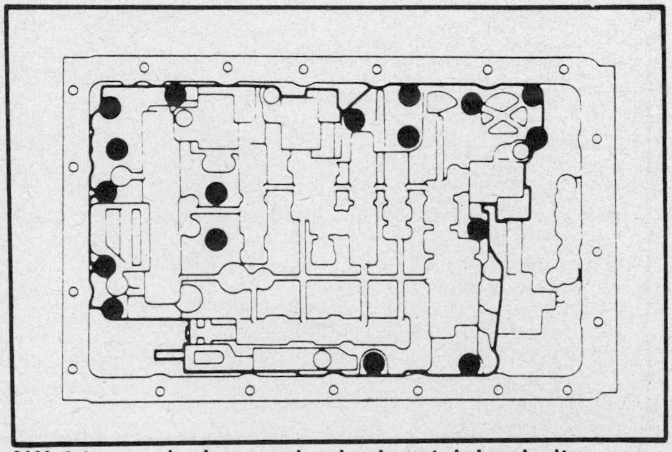

AW-4 transmission—valve body retaining bolt location

2. Disconnect the electrical connector from the computer unit.
3. Remove the computer unit from its mounting.
4. Installation is the reverse of the removal procedure.

VALVE BODY

Removal and Installation

1. Raise and support the vehicle safely.
2. Remove the transmission fluid pan.
3. Remove the filter assembly.
4. Disconnect and mark the solenoid electrical wires. Remove the valve body oil tubes, using the proper tool. Disconnect the throttle cable from the throttle cam.
5. Remove the valve body retaining bolts. Lower the valve body assembly and remove the overdrive clutch accumulator springs, direct clutch accumulator spring and spacer and the second brake accumulator spring and spacer.
6. Remove the valve body, check ball and spring from the transmission.
7. Installation is the reverse of the removal procedure. Torque the valve body retaining bolts to 7 ft. lbs. (10 Nm). Be sure to properly install the valve body oil tubes.

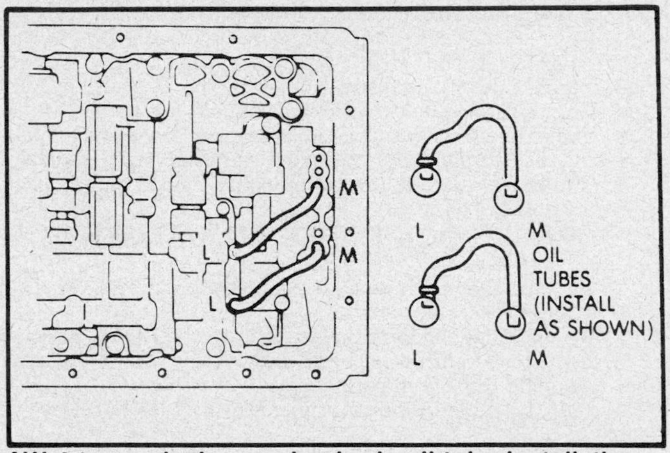
M
M
L
L
L
M
OIL TUBES (INSTALL AS SHOWN)
L
M

AW-4 transmission—valve body oil tube installation

8. Be sure to use a new gasket or RTV sealant, as required. Fill the transmission to specification with the proper grade and type automatic transmission fluid.
9. Run the engine until the transmission reaches normal operating temperature. Recheck the fluid level and correct, as required.

VALVE BODY SOLENOIDS

Removal and Installation

1. Raise and support the vehicle safely.
2. Remove the transmission fluid pan. Remove the filter assembly.
3. Disconnect and mark the solenoid electrical wires. Remove the solenoid retaining bolt. Remove the solenoid from the valve body.

NOTE: Be sure that no other components fall out of the valve body when the solenoids are removed from the valve body.

4. To test the solenoid connect the test leads of an ohmmeter to the solenoid mounting bracket and to the solenoid wire terminal. Resistance should be 11–15 ohms. Replace the solenoid, as required.
5. Installation is the reverse of the removal procedure.

Torque the solenoid retaining bolts to 7 ft. lbs. (10 Nm).

6. Be sure to use a new gasket or RTV sealant, as required. Fill the transmission to specification with the proper grade and type automatic transmission fluid.

7. Run the engine until the transmission reaches normal operating temperature. Recheck the fluid level and correct, as required.

MANUAL VALVE SHAFT SEAL

Removal and Installation

1. Disconnect the negative battery cable. Raise and support the vehicle safely.
2. Remove the neutral safety switch. Remove the transmission fluid pan. Remove the valve body assembly.
3. Remove the bolts that retain the park rod bracket to the transmission case. Remove the park rod from the shift sector.
4. Cut the spacer sleeve, using a chisel. Remove it from the manual valve shaft.
5. Remove the pin from the shaft and sector, using a punch. Remove the shaft and the sector from the transmission case.
6. Pry the shaft seals out of the transmission case.
7. Inspect the components for wear and damage. Repair or replace defective components as required.
8. Installation is the reverse of the removal procedure. Be sure to coat the new seals with petroleum jelly prior to installation.
9. Be sure to use a new gasket or RTV sealant, as required. Fill the transmission to specification with the proper grade and type automatic transmission fluid.
10. Run the engine until the transmission reaches normal operating temperature. Recheck the fluid level and correct, as required.

THROTTLE CABLE

Removal and Installation

1. Disconnect the negative battery cable.
2. Disconnect the throttle cable from the throttle linkage. Compress the cable mounting ears and remove the cable from the linkage bracket.
3. Raise and support the vehicle safely. Remove the transmission fluid pan.
4. Disengage the cable from the throttle valve cam. Remove the cable bracket bolt. Remove the cable and the bracket from the transmission.
5. Remove and discard the cable seal.
6. Installation is the reverse of the removal procedure. Be sure to lubricate the new seal with clean transmission fluid prior to installation.
7. Be sure to use a new gasket or RTV sealant, as required. Fill the transmission to specification with the proper grade and type automatic transmission fluid.
8. Run the engine until the transmission reaches normal operating temperature. Recheck the fluid level and correct, as required.

ACCUMULATOR PISTONS AND SPRINGS

Removal and Installation

1. Disconnect the negative battery cable. Raise and support the vehicle safely.
2. Drain the transmission fluid. Remove the transmission fluid pan and filter assembly. Remove the valve body assembly.
3. Using compressed air, remove the accumulator pistons. To accomplish this, apply air through the small feed hole next to each accumulator piston bore. Catch the accumulator piston assembly as it exits the bore.

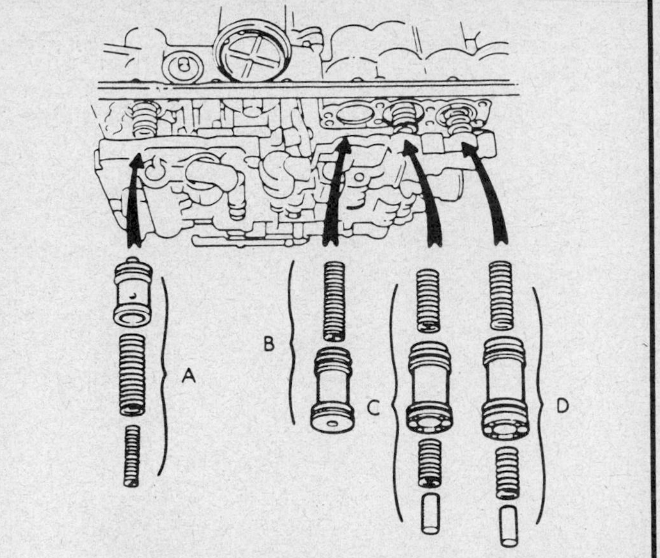
A
B
C
D

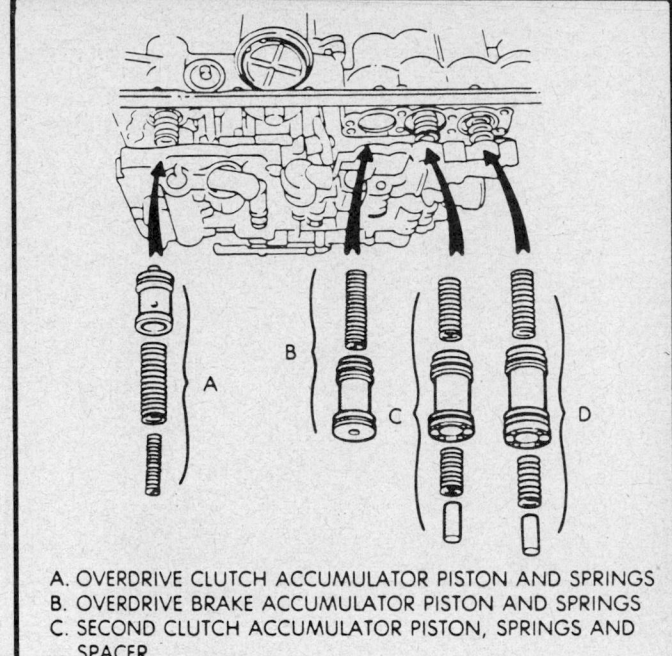

A. OVERDRIVE CLUTCH ACCUMULATOR PISTON AND SPRINGS
B. OVERDRIVE BRAKE ACCUMULATOR PISTON AND SPRINGS
C. SECOND CLUTCH ACCUMULATOR PISTON, SPRINGS AND SPACER
D. SECOND CLUTCH ACCUMULATOR PISTON, SPRINGS AND SPACER

AW-4 transmission—accumulator piston assembly

NOTE: Use only enough air pressure to ease each accumulator piston out of the bore. Mark each accumulator piston assembly for reinstallation. Do not intermix the accumulator piston assemblies.

4. Remove and discard the accumulator piston O-ring seals. Check the accumulator piston springs, replace as required. Check the accumulator piston assemblies, repair or replace defective components as required.
5. Installation is the reverse of the removal procedure. Be sure to lubricate the O-rings with clean transmission fluid prior to installation.
6. Be sure to use a new gasket or RTV sealant, as required. Fill the transmission to specification with the proper grade and type automatic transmission fluid.
7. Run the engine until the transmission reaches normal operating temperature. Recheck the fluid level and correct, as required.

SECOND COAST BRAKE SERVO

Removal and Installation

1. Disconnect the negative battery cable. Raise and support the vehicle safely.
2. Drain the transmission fluid. Remove the transmission fluid pan and filter assembly. Remove the valve body assembly.
3. Using compressed air, remove the servo piston and cover. To accomplish this, apply air through the oil hole in the servo boss to ease the piston out of the bore.
4. Remove and discard the seal and O-rings from the cover and piston. Repair or replace defective components, as required.
5. Installation is the reverse of the removal procedure. Be sure to lubricate the O-rings with clean transmission fluid prior to installation.
6. Be sure that the servo piston rod is properly engaged in the second coast brake band.
7. Be sure to use a new gasket or RTV sealant, as required.

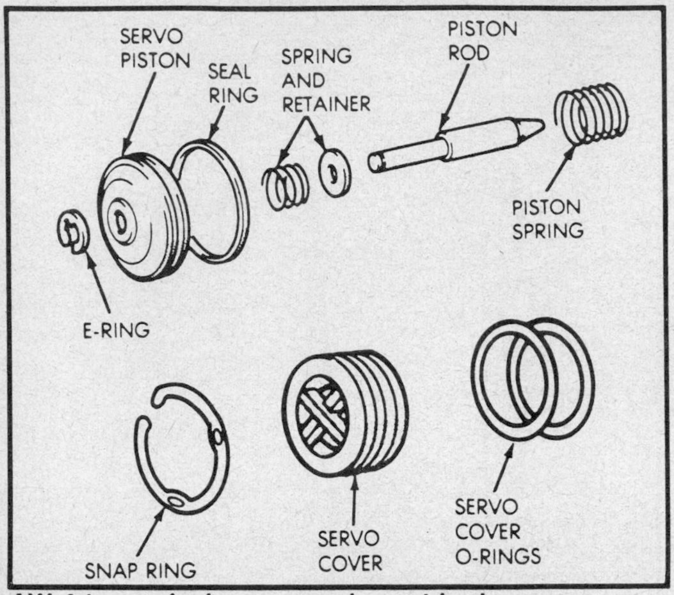
SERVO
PISTON
SEAL
RING
SPRING
AND
RETAINER
PISTON
ROD
PISTON
SPRING
E-RING
SNAP RING
SERVO
COVER
SERVO
COVER
O-RINGS

AW-4 transmission – second coast brake servo assembly

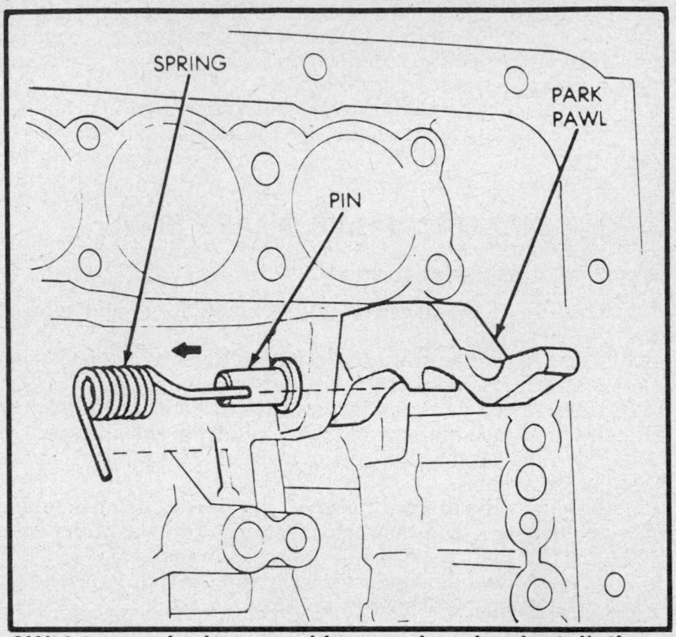
SPRING
PARK
PAWL
PIN

AW-4 transmission – parking pawl spring installation

Fill the transmission to specification with the proper grade and type automatic transmission fluid.
8. Run the engine until the transmission reaches normal operating temperature. Recheck the fluid level and correct, as required.

PARK ROD AND PAWL

Removal and Installation

1. Disconnect the negative battery cable. Raise and support the vehicle safely.
2. Drain the transmission fluid. Remove the transmission fluid pan and filter assembly. Remove the valve body assembly.
3. Remove the bolts that retain the park lock bracket to the transmission case. Remove the park lock rod from the manual valve shaft sector. Remove the park rod.
4. Remove the park pawl, pin and spring. Repair or replace defective components as required.
5. Installation is the reverse of the removal procedure. Be sure that the spring is properly positioned.
6. Be sure to use a new gasket or RTV sealant, as required. Fill the transmission to specification with the proper grade and type automatic transmission fluid.
7. Run the engine until the transmission reaches normal operating temperature. Recheck the fluid level and correct, as required.

REAR OIL SEAL

Removal and Installation

1. Disconnect the negative battery cable. Raise and support the vehicle safely.
2. As necessary, disconnect or remove the driveshaft, crossmember, shift linkage, electrical connectors, hoses, transfer case and exhaust system in order to gain access to the rear oil seal.
3. On 4WD vehicles, remove the seal from the adaptor housing, using the proper tool.
4. On 2WD vehicles, remove the dust shield and remove the seal from the extension housing, using the proper tool.
5. Installation is the reverse of the removal procedure. Fill the transmission to specification with the proper grade and type automatic transmission fluid.
6. Run the engine until the transmission reaches normal operating temperature. Recheck the fluid level and correct, as required.

REMOVAL AND INSTALLATION

TRANSMISSION REMOVAL

1. Disconnect the negative battery cable. Remove the transmission dipstick indicator.
2. Raise and support the vehicle safely. Drain the transmission fluid.
3. Remove the upper half of the transmission dipstick tube. Disconnect and plug the fluid cooler lines.
4. Properly support the engine and the transmission assembly, using the proper equipment.
5. Disconnect the transmission linkage. Disconnect the transfer case linkage, if equipped with 4WD.
6. Disconnect all the required electrical and vacuum connections. Disconnect the speedometer cable.
7. Remove the required exhaust system components. Remove the driveshaft. If equipped wth 4WD, remove the front driveshaft.
8. Remove the rear crossmember. Disconnect the transmission throttle cable.
9. Remove the starter. Remove the torque converter to flex plate retaining bolts.
10. Remove the torque converter housing to engine retaining bolts.
11. Properly secure the transmission and the transfer case, if equipped to a transmission jack.

12. Remove the assembly from the vehicle. Remove the torque converter from the transmission. Separate the transmission from the transfer case, if equipped.

TRANSMISSION INSTALLATION

1. Position and secure the transmission on the transmission jack. Install the torque converter. If equipped, install the transfer case.
2. Install the assembly. Install the torque converter housing to engine retaining bolts. Install the torque converter to flex plate retaining bolts.
3. Install the starter. Connect the transfer case linkage, vacuum lines and electrical connections, if equipped.
4. Install the required exhaust components. Install the rear crossmember.
5. Install the driveshaft. Install the front driveshaft, if equipped.
6. Connect the speedometer cable. Connect the transmission electrical harness. Connect the fluid lines.
7. Connect the throttle valve cable. Install the upper dipstick tube into the lower dipstick tube, using a new O-ring.
8. Lower the vehicle. Fill the transmission to specification with the proper grade and type automatic transmission fluid.
9. Run the engine until the transmission reaches normal operating temperature. Recheck the fluid level and correct, as required.

BENCH OVERHAUL

Before Dissassembly

Cleanliness is an important factor in the overhaul of the AW-4 automatic transmission. Before opening up this unit, the entire outside of the transmission assembly should be cleaned, preferable with a high pressure washer such as a car wash spray unit. Dirt entering the transmission internal parts will negate all the time and effort spent on the overhaul. During inspection and reassembly all parts should be thoroughly cleaned with solvent then dried with compressed air. Wiping cloths and rags should not be used to dry parts since lint will find its way into the valve body passages.

Wheel bearing grease, long used to hold thrust washers and lube parts, should not be used. Lube seals with clean transmission fluid and use ordinary unmedicated petroleum jelly to hold the thrust washers and to ease the assembly of seals, since it will not leave a harmful residue as grease often will. Do not use solvent on neoprene seals, friction plates if they are to be reused, or thrust washers. Be wary of nylon parts if the transmission failure was due to failure of the cooling system. Nylon parts exposed to water or antifreeze solutions can swell and distort and must be replaced.

Before installing bolts into aluminum parts, always dip the threads into clean transmission fluid. Antiseize compound can also be used to prevent bolts from galling the aluminum and seizing. Always use a torque wrench to keep from stripping the threads. Take care when installing new O-rings, especially the smaller O-rings. The internal snaprings should be expanded and the external rings should be compressed, if they are to be reused. This will help insure proper seating when installed.

Converter Inspection

After the torque converter is removed from the transmission, the stator roller clutch can be checked by inserting a finger into the splined inner race of the roller clutch and trying to turn the race in both directions. The inner race should turn freely in the clockwise direction, but not turn in the counterclockwise direction. The inner race may tend to turn in the counterclockwise direction, but with great difficulty, this is to be considered normal. Do not use such items as the driven sprocket support or the shafts to turn the race, as the results may be misleading. Inspect the outer hub lip and the inner bushing for burrs or jagged edges to avoid injury to your fingers when testing the torque converter.

Transmission Disassembly

1. Position the transmission assembly in a suitable holding fixture.
2. The converter pulls out of the transmission. Be careful since the converter contains a large amount of fluid.
3. Remove both halves of the dipstick indicator tube. Remove the clamp that attaches the wire harness and throttle pressure cable to the transmission.
4. Remove the shift lever from the manual valve shaft at the left side of the transmission.
5. Remove the neutral safety switch. If equipped, remove the speedometer driven gear. Remove the speed sensor.
6. Remove the torque converter housing retaining bolts. Separate the torque converter housing from the transmission assembly.
7. On 2WD vehicles, remove the extension housing retaining bolts. Remove the extension housing from the transmission case.
8. On 4WD vehicles, remove the adaptor housing retaining bolts. Remove the adaptor housing from the transmission case.
9. On 2WD vehicles, use an inside micrometer and measure the inside diameter of the extension housing bushing. The diameter should be 1.4996 in. (38.09mm) or less. If not within specification, replace the extension housing.
10. Remove the speedometer drive gear snapring. Remove the gear and the gear spacer, if equipped.
11. Remove the speed sensor rotor and key. Position a block of wood between the rotor and the transmission in order to remove the rotor.
12. Remove the transmission fluid pan. Remove the filter.
13. Remove the valve body oil feed tubes. Disconnect the solenoid electrical wires.
14. Remove the harness bracket bolt. Remove the harness and the bracket.
15. Remove the valve body retaining bolts. Disconnect the throttle cable from the throttle cam. Remove the valve body from the transmission.
16. Remove the accumulator springs, spacers and check ball and spring.
17. Using compressed air, remove the second brake and clutch accumulator pistons. Apply air pressure through the feed port and ease the pistons out of the bore.
18. Using compressed air, remove the overdrive brake accumulator piston and the overdrive clutch accumulator piston. Remove the throttle cable.
19. Remove the oil pump retaining bolts. Using a suitable puller remove the oil pump assembly from the transmission. Remove the race from the oil pump.
20. Pull out the 4th gear overdrive planetary gear and overdrive direct clutch assembly. Remove the race from the 4th gear overdrive planetary.
21. Remove the thrust bearing, race and overdrive planetary ring gear.

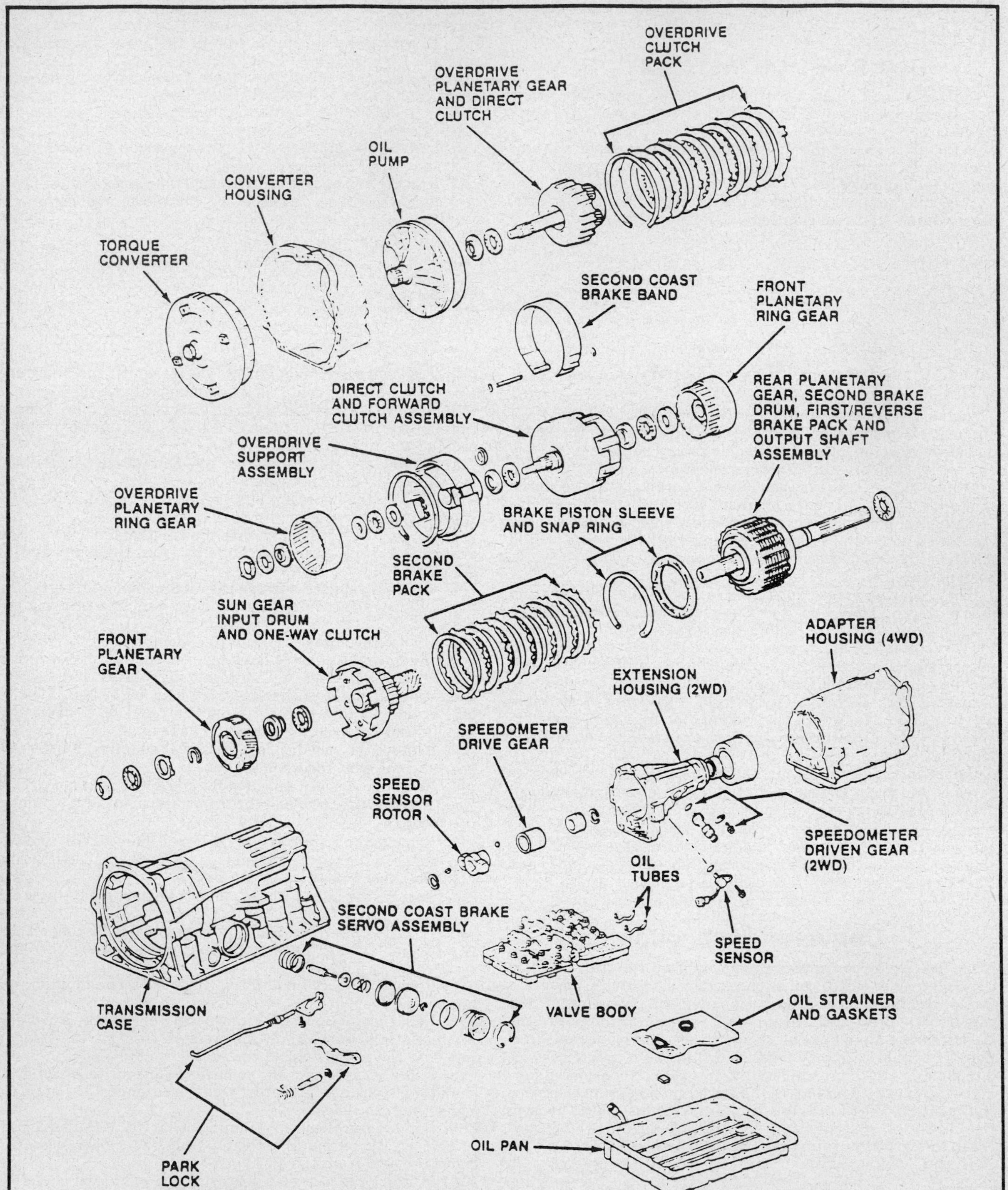
OVERDRIVE CLUTCH PACK
OVERDRIVE PLANETARY GEAR AND DIRECT CLUTCH
OIL PUMP
CONVERTER HOUSING
TORQUE CONVERTER
SECOND COAST BRAKE BAND
FRONT PLANETARY RING GEAR
REAR PLANETARY GEAR, SECOND BRAKE DRUM, FIRST/REVERSE BRAKE PACK AND OUTPUT SHAFT ASSEMBLY
DIRECT CLUTCH AND FORWARD CLUTCH ASSEMBLY
OVERDRIVE SUPPORT ASSEMBLY
OVERDRIVE PLANETARY RING GEAR
BRAKE PISTON SLEEVE AND SNAP RING
SECOND BRAKE PACK
SUN GEAR INPUT DRUM AND ONE-WAY CLUTCH
FRONT PLANETARY GEAR
ADAPTER HOUSING (4WD)
EXTENSION HOUSING (2WD)
SPEEDOMETER DRIVE GEAR
SPEED SENSOR ROTOR
SPEEDOMETER DRIVEN GEAR (2WD)
OIL TUBES
SECOND COAST BRAKE SERVO ASSEMBLY
SPEED SENSOR
TRANSMISSION CASE
VALVE BODY
OIL STRAINER AND GASKETS
OIL PAN
PARK LOCK COMPONENTS

AW-4 transmission—exploded view of major assemblies

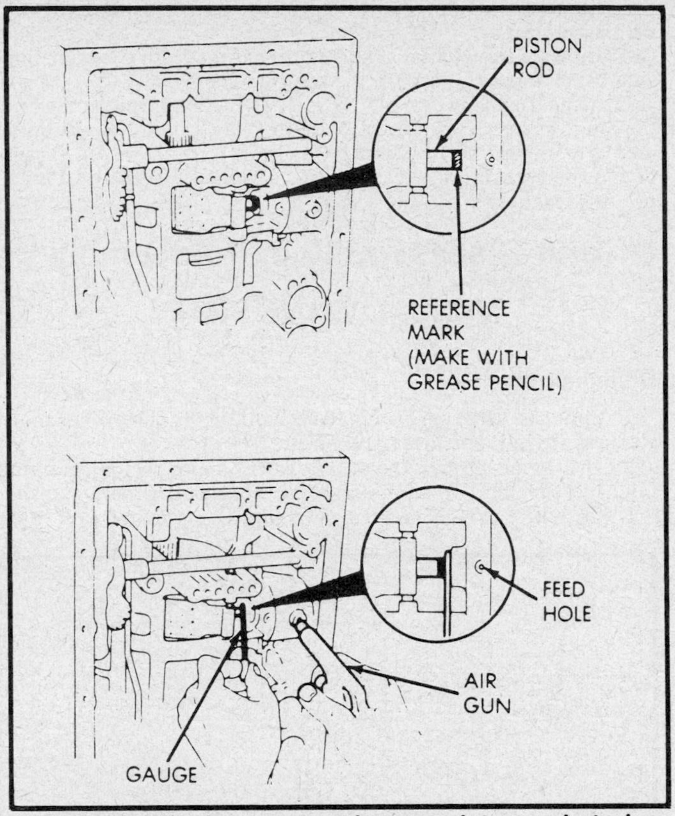
PISTON ROD
REFERENCE MARK (MAKE WITH GREASE PENCIL)
FEED HOLE
AIR GUN
GAUGE

AW-4 transmission – second coast piston rod stroke measurement

22. To measure the stroke length of the overdrive brake piston, install a dial indicator gauge on the transmission case.
23. Mount the gauge tool so that it contacts the piston. Apply 57–114 psi air pressure through the piston apply port.
24. Note the piston stroke reading on the dial indicator gauge, it should be 0.055–0.669 in. (1.40–1.70mm) for vehicles equipped with a 6 cylinder engine and 0.052–0.0638 in. (1.32–1.62mm) for vehicles equipped with a 4 cylinder engine. If not within specification, replace the brake pack retainer.

OVERDRIVE BRAKE RETAINER SELECTION

Retainer No.	Thickness in.	Thickness mm
26	0.130	3.3
25	0.138	3.5
12	0.124	3.6
24	0.146	3.7
11	0.150	3.8
23	0.154	3.9
Not Marked	0.157	4.0

25. Remove the overdrive brake snapring. Remove the overdrive support lower race. Remove the upper bearing and race assembly.
26. Remove the overdrive support bolts. Remove the overdrive support snapring, using the proper tool.
27. Using a bridge type puller, remove the overdrive support from the transmission case.
28. Remove the race from the hub of the overdrive support. Remove the overdrive brake discs and plates.
29. Measure the disc thickness using a micrometer. Minimum disc thickness specification should be 0.0724 in. (1.84mm). Replace the discs if not within specification.
30. To measure the stroke length of the second coast brake piston, make a reference mark on the piston rod.
31. Apply 57–114 psi of air pressure through the piston feed hole and check the stroke length using gauge tool BVIFM-40/41 or equivalent.
32. The stroke length specification should be 0.059–0.118 in (1.5–3.0mm). If not within specification, install a new piston rod and recheck the stroke length.

NOTE: Replacement piston rods are available in 2.811 in (71.4mm) length or 2.870 in. (72.9mm) length.

33. If the stroke length is still not correct, replace the second coast brake band.
34. Using tool BVIFM-29 or equivalent, remove the second coast brake piston snapring. Remove the piston cover and piston assembly by applying compressed air through the piston feed hole.
35. Disassemble the second coast brake piston, as required.
36. Remove the direct and the forward clutch assembly. Remove the thrust bearing and the race from the clutch hub.
37. Remove the second coast brake band E-clip from the band pin. Remove the brake band.
38. Remove the front planetary ring gear front bearing race. Remove the front planetary ring gear.
39. Remove the thrust bearing and the rear race from the ring gear. Remove the planetary thrust race.
40. To relieve the load on the planetary snapring, loosen the transmission holding fixture. Turn the transmission over and allow the output shaft to support the transmission weight. Position wood blocks under the shaft in order to protect the splines.
41. Remove the planetary snapring. Remove the planetary gear assembly.
42. Remove the sun gear, input drum and one-way clutch as an assembly.
43. Measure the second brake clutch pack clearance. It should be 0.0244–0.0780 in (0.62–1.98mm) for vehicles equipped with a 6 cylinder engine and 0.0350–0.0846 in. (0.89–2.15mm) for vehicles equipped with a 4 cylinder engine. Replace the discs, if not within specification.
44. Remove the second brake clutch pack snapring. Remove the second brake clutch pack. Measure the disc thickness, using a micrometer. Minimum disc thickness should be 0.0724 in. (1.84mm). Replace the discs, if not within specification.

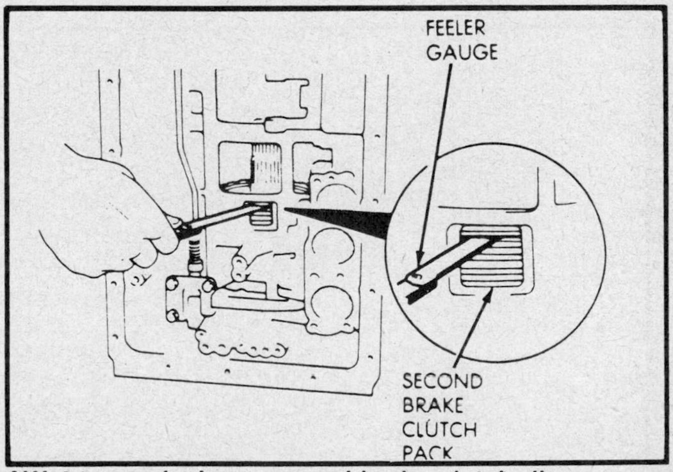
FEELER GAUGE
SECOND BRAKE CLUTCH PACK

AW-4 transmission – second brake clutch disc thickness check

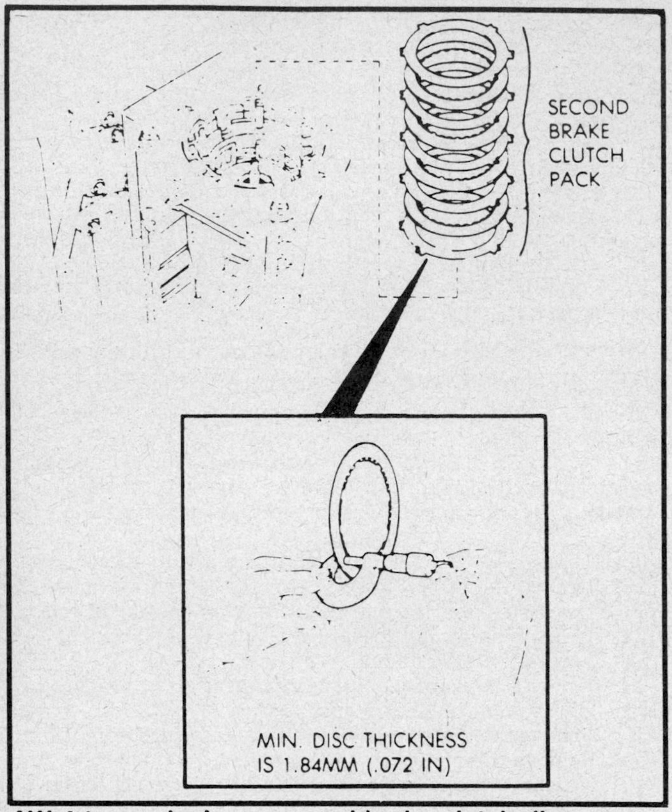
SECOND
BRAKE
CLUTCH
PACK
MIN. DISC THICKNESS
IS 1.84MM (.072 IN)

AW-4 transmission – second brake clutch disc thickness check

45. Remove the park rod bracket retaining bolts. Disconnect the park rod from the manual shaft lever. Remove the park rod and bracket. Remove the park pawl spring, pin and pawl.
46. Measure the clearance of the 1st/reverse brake clutch pack. Clearance should be 0.0276–0.787 in (0.70–2.00mm) for vehicles equipped with a 6 cylinder engine and 0.0236–0.0685 in. (0.60–1.74mm) for vehicles equipped with a 4 cylinder engine. Replace the discs, if not within specification.
47. Remove the second brake piston sleeve. To avoid damaging the transmission case, cover the removal tool with tape.
48. Remove the rear planetary gear, second brake drum and output shaft as a complete assembly.

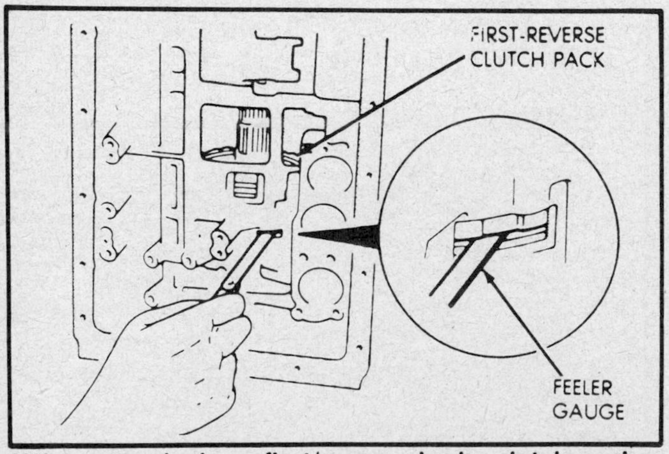
FIRST-REVERSE
CLUTCH PACK
FEELER
GAUGE

AW-4 transmission – first/reverse brake clutch pack clearance check

49. Remove the planetary and brake drum thrust bearing and race assembly.
50. Remove the second brake drum gasket from the transmission case, using the proper gasket removal tool.
51. Measure the inside diameter of the transmission case rear bushing, using an inside micrometer. The maximum allowable diameter should be 1.5031 in. (38.18mm). If not within specification the transmission case must be replaced as the bushing is not serviceable.

Unit Disassembly and Assembly

MANUAL VALVE SHAFT

Disassembly

1. Split the shaft spacer sleeve in half, using the proper tools. Remove it from the lever and shaft.
2. Using the proper tool, remove the shift sector retaining pin. Pull the shaft from the case and remove the manual lever.
3. Carefully pry the shaft seals from the transmission case.

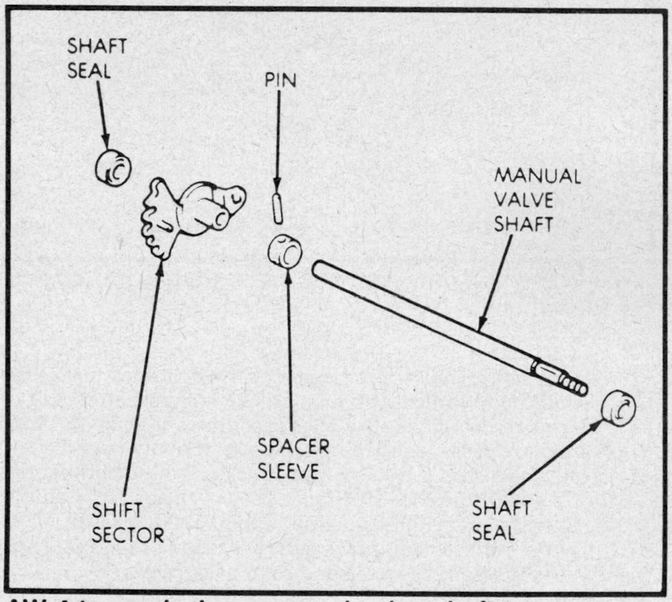
SHAFT
SEAL
PIN
MANUAL
VALVE
SHAFT
SPACER
SLEEVE
SHIFT
SECTOR
SHAFT
SEAL

AW-4 transmission – manual valve shaft components

Inspection

1. Inspect all components. Repair or replace defective components as required.
2. Replace all seals, prior to installation coat them with petroleum jelly.

Assembly

1. Install a new spacer sleeve on the shift sector. Install the sector and sleeve on the shaft. Install the shaft into the transmission case.
2. Align the sector and sleeve. Install a new retaining pin.
3. Align the notch in the sleeve with the depression in the sector. Stake the sleeve in 2 positions.
4. Check that the lever and the shaft rotate freely.

OIL PUMP

Disassembly

1. Remove the pump body O-ring. Remove the pump seal rings.

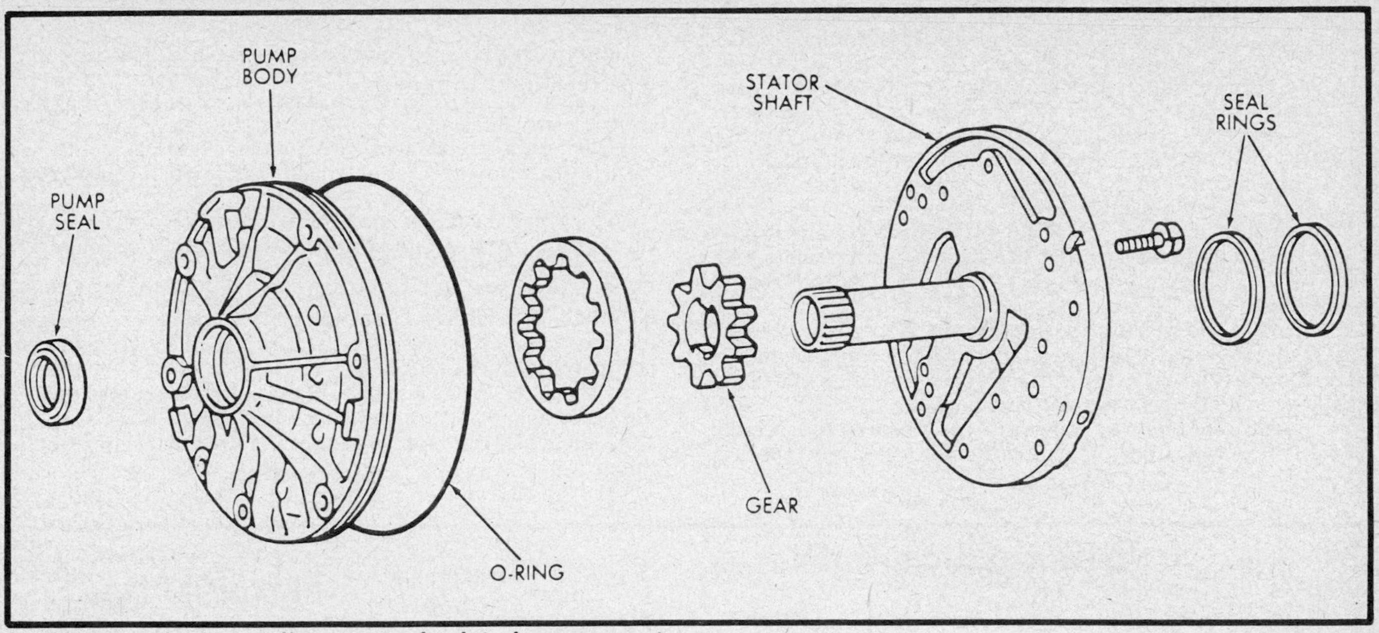
PUMP BODY
STATOR SHAFT
SEAL RINGS
PUMP SEAL
GEAR
O-RING

AW-4 transmission—oil pump and related components

2. Remove the bolts that retain the stator shaft to the pump body. Separate the components.
3. Remove the drive gear and driven gear and driven gear from the pump body.
4. Measure the inside diameter of the pump body, using a bore gauge. The specification should be 1.5035 in. (38.19mm) or less. If not within specification, replace the pump body.
5. Measure the inside diameter of the stator shaft bushing. Take these measurements at the front and rear of the bushing. The front diameter specification should be 0.08496 in. (21.58mm) or less. The rear diameter specification should be 1.0661 in. (27.08mm) or less. If not within specification, replace the stator shaft.
6. Measure the oil pump clearances. Clearance between the oil pump driven gear and the oil pump body should be 0.012 in. (0.3mm) or less. Clearance between the tips of the oil pump gear teeth should be 0.012 in. (0.3mm) or less. Clearance between the rear surface of the oil pump housing and the oil pump gears should be 0.004 in. (0.1mm) or less. Replace the oil pump body and gears if not within specification.
7. Remove the oil pump seal, using the proper removal tool.

Inspection

1. Inspect all components. Repair or replace defective components as required.
2. Replace all seals, prior to installation coat them with clean transmission fluid.
3. Lubricate the oil pump gears with clean transmission fluid, prior to installation.

Assembly

1. Using tool BVIFM-38 or equivalent lubricate and install a new oil pump seal. Lubricate and install the pump gears into the pump body.
2. Assemble the stator shaft and pump body. Torque the retaining bolts to 7 ft. lbs. (10 Nm).
3. Install a new O-ring on the pump body and new seal rings on the stator shaft.
4. Install the oil pump into the torque converter to check pump gear rotation. The gears must rotate smoothly when turned clockwise and counterclockwise.

OVERDRIVE PLANETARY GEAR AND CLUTCH

Disassembly

1. To check the operation of the one-way clutch in the clutch drum, hold the drum and turn the planetary shaft clockwise and counterclockwise.
2. The planetary shaft should turn counterclockwise freely, but lock when turned counterclockwise. Replace the assembly if it does not function properly.
3. Remove the overdrive clutch from the planetary gear. Remove the thrust bearing and race assembly from the clutch drum.
4. To measure the stroke length of the clutch piston, mount the oil pump on the torque converter, than mount the the clutch on the oil pump.
5. Install a dial indicator gauge on the clutch and position the indicator stylus on the clutch piston.
6. Apply compressed air through the feed hole in the oil

CLUTCH/BRAKE PACK SELECTION

Component	Engine	Discs Required	Plates Required	Retainer Required
Overdrive brake	6 cyl.	4	3	2
	4 cyl.	3	2	2
Second brake	6 cyl.	5	4	1
	4 cyl.	4	3	1
Overdrive direct clutch	6 cyl.	2	2	1
	4 cyl.	2	2	1
Direct clutch	6 cyl.	4	4	1
	4 cyl.	3	3	1
Forward clutch	6 cyl.	6	6	1
	4 cyl.	5	5	1
1st/reverse brake	6 cyl.	7	7	1
	4 cyl.	6	6	1

pump. Record the piston stroke length. It should be 0.0728–0.0846 in. (1.85–2.15mm). If not within specification, replace the clutch pack retainer.

7. Remove the clutch pack snapring. Remove the clutch pack. Using tool BVIFM-27 or equivalent compress the piston return spring. With the aid of a shop press remove the piston snapring.

8. Remove the compressor tool and the piston return springs.

9. Install the oil pump assembly onto the torque converter. Install the clutch assembly onto the oil pump.

10. Hold the clutch piston and apply compressed air through the oil pump feed hole, in order to ease the piston from its bore. Use only enough compressed air to remove the piston from its mounting.

11. Remove the bearing and race from the ring gear. Remove the snapring from the ring gear. Remove the ring gear hub.

12. Remove the race from the planetary gear. Remove the snapring and remove the retaining plate.

13. Remove the one-way clutch and outer race as an assembly. Separate the race from the clutch. Remove the thrust washer.

Inspection

1. Inspect all components. Repair or replace defective components as required.

2. Replace all seals, prior to installation coat them with clean transmission fluid.

3. Using the proper tool, measure the clutch disc thickness. Minimum allowable specification should be 0.0724 in. (1.84mm).

4. Measure the free length of the piston return springs, with the springs in the retainer. Proper specification should be 0.0661 in. (16.8mm).

5. Check the clutch piston check ball. Shake the piston to see if the ball moves freely. Check the ball sealing function by applying low volume compressed air to the ball inlet. Air should not leak past the check ball.

6. Check the inside diameter of the clutch drum bushings, using a bore gauge. The inside diameter specification should be 1.0673 in. (27.11mm) or less. If not within specification, replace the drum.

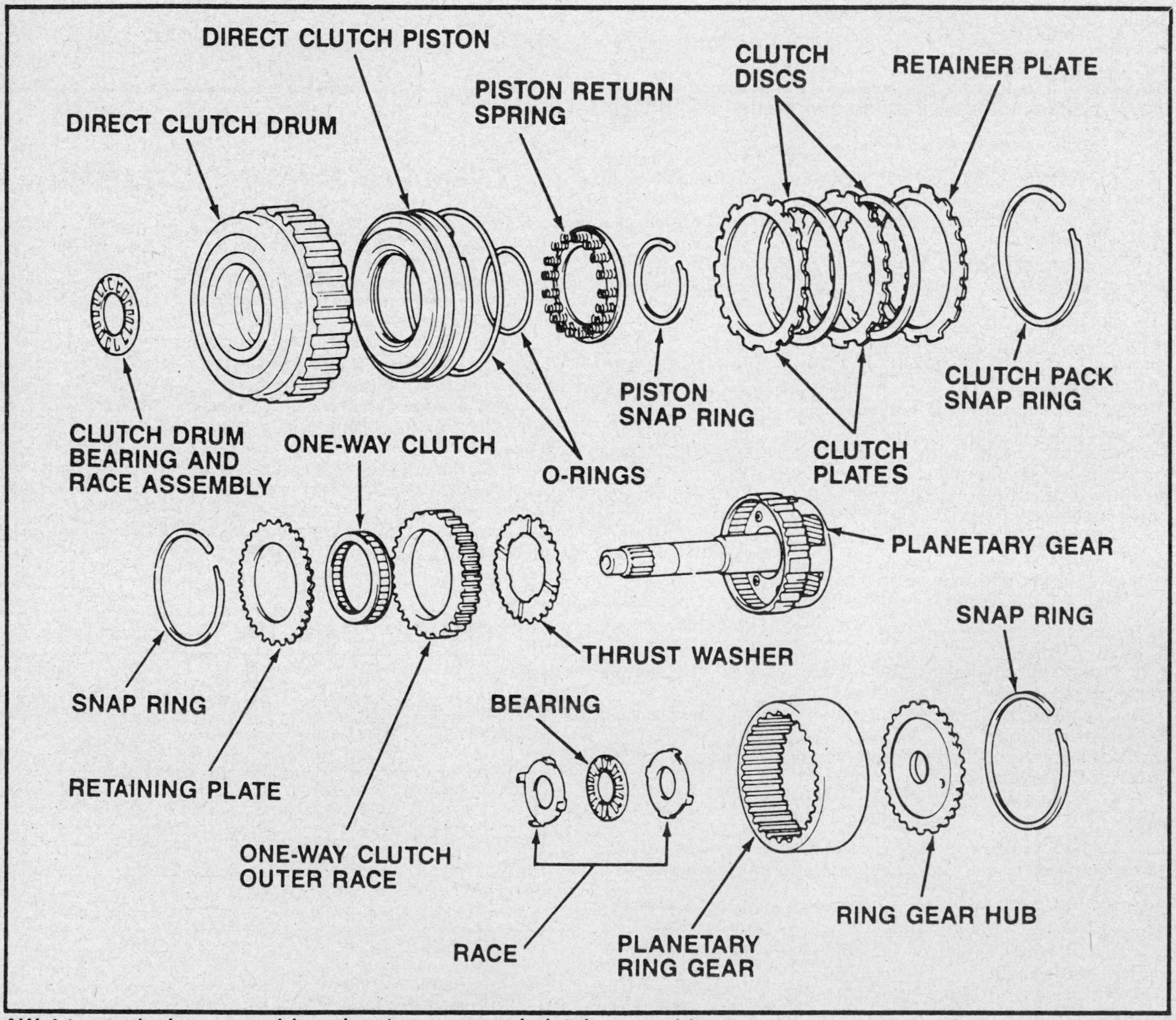
DIRECT CLUTCH PISTON
CLUTCH DISCS
RETAINER PLATE
PISTON RETURN SPRING
DIRECT CLUTCH DRUM
PISTON SNAP RING
CLUTCH PACK SNAP RING
CLUTCH DRUM BEARING AND RACE ASSEMBLY
ONE-WAY CLUTCH
O-RINGS
CLUTCH PLATES
PLANETARY GEAR
SNAP RING
THRUST WASHER
SNAP RING
BEARING
RETAINING PLATE
ONE-WAY CLUTCH OUTER RACE
RING GEAR HUB
RACE
PLANETARY RING GEAR

AW-4 transmission—overdrive planetary gear and clutch assembly

7. Check the inside diameter of the planetary gear bushing. The inside diameter specification should be 0.04437 in. (11.27mm) or less. If not within specification, replace the planetary gear.

Assembly

1. Install the thrust washer in the planetary gear. Note that the grooved side of the washer faces up and toward the front.
2. Install the one-way clutch in the race. The flanged side of the clutch must face upward.
3. Install the assembled one-way clutch and outer race in the planetary gear. Be sure that the flanged side of the clutch is facing upward.
4. Install the clutch pack retaining plate and snapring in the planetary gear.
5. Coat the planetary race with petroleum jelly and install it on the planetary gear. Check the outside diameter of the race, it should be 1.646 in. (41.8mm). Check the inside diameter of the race, it should be 1.067 in. (27.1mm).
6. Install the hub in the planetary ring gear. Install the snapring. Coat the race and the bearing with petroleum jelly. Install the planetary ring gear.
7. Check the race size. The outside diameter of the race should be 1.882 in. (47.8mm). The inside diameter of the race should be 0.953 in. (24.2mm).
8. Check the bearing size. The outside diameter of the bearing should be 1.843 in. (46.8mm). The inside diameter of the race should be 1.024 in. (26mm).
9. Lubricate and install new O-rings on the clutch piston. Install the piston in the clutch drum.
10. Install the piston return springs in the clutch piston. Install the piston snapring. Compress the piston return springs using the spring compressor tool and the shop press.
11. Install the clutch pack in the drum. Install the steel plate first, then a disc. Continue this sequence until all plates and discs are installed.
12. Install the clutch pack retainer with the flat side side facing downward. Install the retainer snapring. Compress the springs, using tool BVIFM-27 or equivalent.
13. Recheck the clutch piston stroke length. If not within specification, install new clutch discs or a select fit retainer.
14. Install the clutch drum bearing and race assembly. Be sure that the bearing rollers face upward. The outside diameter of the assembled bearing and race should be 1.976 in. (50.2mm). The inside dioameter of the component should be 1.138 in. (28.9mm).

PISTON STROKE LENGTH SELECTION

		Specification	
Location	Engine	in.	mm
Direct clutch	All	0.0539–0.0640	1.37–1.60
Overdrive brake	6 cyl.	0.0551–0.0669	1.40-1.70
	4 cyl.	0.0200–0.0638	1.32–1.62
Second coast brake	All	0.0590–0.1180	1.50–3.00
Forward clutch	6 cyl.	0.1469–0.1807	3.73–4.59
	4 cyl.	0.1346–0.1655	3.42–4.23
Overdrive direct clutch	All	0.0278–0.0846	1.82–2.15

15. Install the clutch on the planetary gear.
16. To check the one-way clutch operation, hold the drum and turn the planetary shaft clockwise and counterclockwise. The shaft clockwise, it should turn freely. Turn the shaft counterclockwise, it should lock.

OVERDRIVE SUPPORT

Disassembly

1. To check the brake piston operation, mount the overdrive support on the clutch assembly.
2. Apply compressed air through the support feed hole and observe the brake piston movement.
3. The piston should move smoothly and not bind or stick. If the piston does not perform properly, replace the piston and the support.
4. Remove the thrust bearing front race, thrust bearing and rear race.
5. Turn the overdrive support over and remove the bearing race and the clutch drum thrust washer.
6. Using tool BVIFM-26 or equivalent, compress the piston return spring and remove the piston snapring.
7. Mount the overdrive support in the direct clutch and remove the piston, using compressed air. Use the same feed hole as used when checking piston operation.
8. Remove and discard the overdrive support O-rings. Remove the overdrive support seal rings.

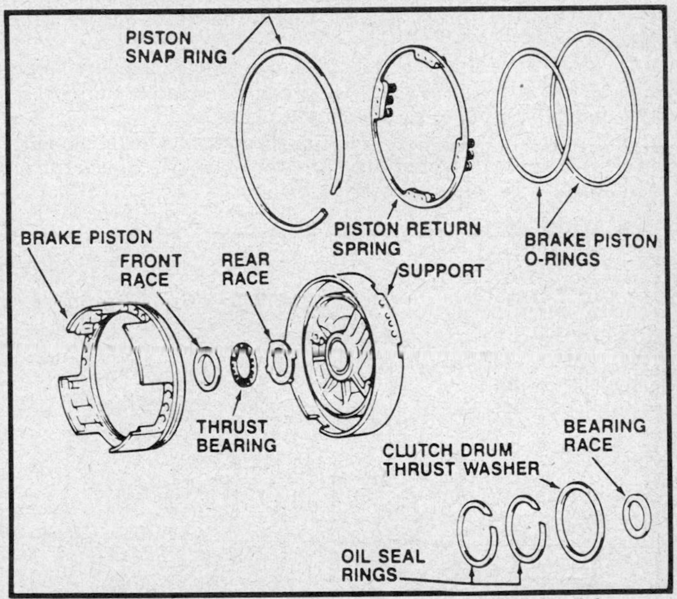
PISTON SNAP RING
PISTON RETURN SPRING
BRAKE PISTON O-RINGS
BRAKE PISTON
FRONT RACE
REAR RACE
SUPPORT
THRUST BEARING
CLUTCH DRUM THRUST WASHER
BEARING RACE
OIL SEAL RINGS

AW-4 transmission—overdrive support assembly

Inspection

1. Measure the free length of the piston return springs with the springs mounted in the retainer. Spring length should be 0.733 in. (18.61mm).
2. Clean the overdrive support components and dry them using compressed air.
3. Inspect all components. Repair or replace defective components as required.
4. Replace all seals, prior to installation coat them with clean transmission fluid.
5. Inspect the overdrive support and brake piston. Replace the overdrive support and piston if either part is worn or damaged.

Assembly

1. Lubricate the overdrive support seal rings. Compress the rings and install them on the overdrive support.
2. Lubricate and install new O-rings on the brake piston. Carefully seat the piston in the overdrive support.

3. Install the return springs on the brake piston. Using the proper tool, compress the return springs and install the snapring.
4. Install the support bearing race and clutch drum thrust washer.
5. Install the thrust bearing and the front and rear bearing races. The thrust bearing rollers should face upward.
6. Check the thrust race sizes.
The front race outer diameter should be 1.882 in. (47.8mm).
The front race inside diameter should be 1.209 in. (30.7mm).
The rear race outer diameter should be 1.882 in. (47.8mm).
The rear race inside diameter should be 1.350 in. (47.7mm).
7. Check the thrust bearing sizes. The outer bearing diameter should be 1.878 in. (47.7mm). The inside bearing diameter should be 1.287 in. (32.7mm).
8. Check the brake piston operation. The piston should move smoothly and not bind or stick.

DIRECT CLUTCH

Disassembly

1. Remove the direct clutch from the forward clutch. Remove the clutch drum thrust washer. Check the clutch piston stroke length.
2. Mount the direct clutch on the overdrive support assembly. Mount the dial indicator gauge on the clutch and position the indicator plunger on the clutch piston.
3. Apply 57–114 psi air pressure through the feed hole in the overdrive support. Note the piston stroke length. Check the piston stroke a couple of times.

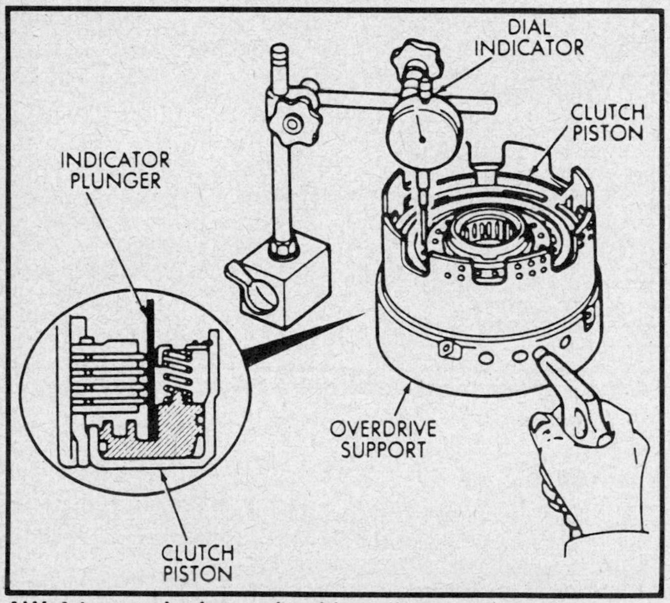
DIAL INDICATOR
CLUTCH PISTON
INDICATOR PLUNGER
OVERDRIVE SUPPORT
CLUTCH PISTON
AW-4 transmission—checking direct clutch piston stroke length

4. Piston stroke length should be 0.0539–0.0642 in. (1.37–1.60mm). If the specification is not within the limits either the clutch pack retainer or the clutch discs must be replaced.
5. Using tool BVIFM-27 or equivalent, remove the clutch pack snapring. Remove the retainer and the clutch pack from the drum.
6. Compress the clutch piston return springs using tool BVIFM-27. Remove the clutch piston snapring.
7. Remove the compressor tool and the return spring. Remove the clutch piston.
8. Remount the clutch on the overdrive support. Apply compressed air through the piston feed hole in the overdrive support in order to remove the piston. Use only enough compressed air to remove the piston.
9. Remove and discard the clutch piston O-rings.

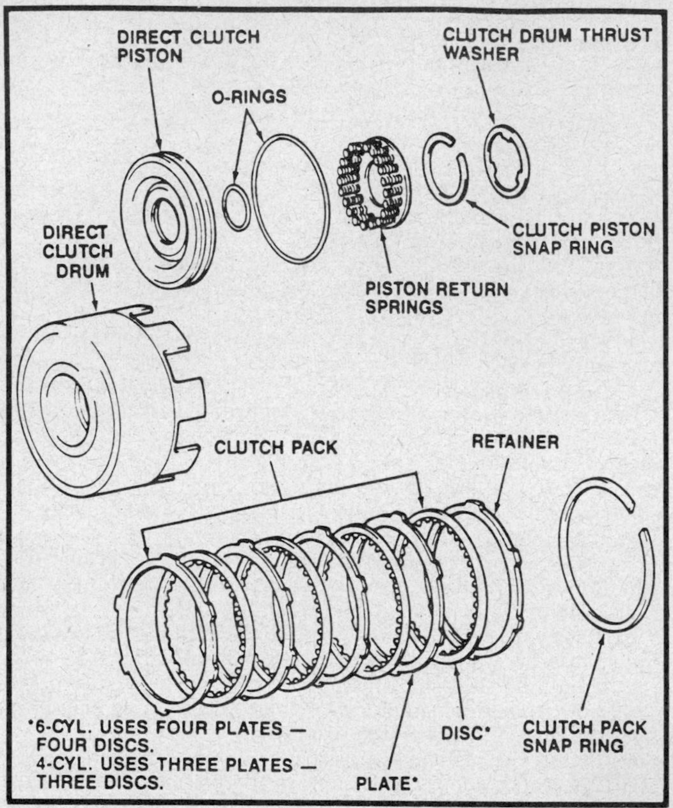
DIRECT CLUTCH PISTON
CLUTCH DRUM THRUST WASHER
O-RINGS
DIRECT CLUTCH DRUM
CLUTCH PISTON SNAP RING
PISTON RETURN SPRINGS
CLUTCH PACK
RETAINER
*6-CYL. USES FOUR PLATES — FOUR DISCS.
4-CYL. USES THREE PLATES — THREE DISCS.
DISC*
CLUTCH PACK SNAP RING
PLATE*
AW-4 transmission—direct clutch assembly

Inspection

1. Inspect all components. Repair or replace defective components as required.
2. Replace all seals, prior to installation coat them with clean transmission fluid.
3. Measure the clutch disc thickness. The minimum allowable thickness is 0.0724 in. (1.84mm). If not within specification, replace the clutch discs.
4. Measure the free length of the piston return springs with the springs in the retainer. The free length specification should

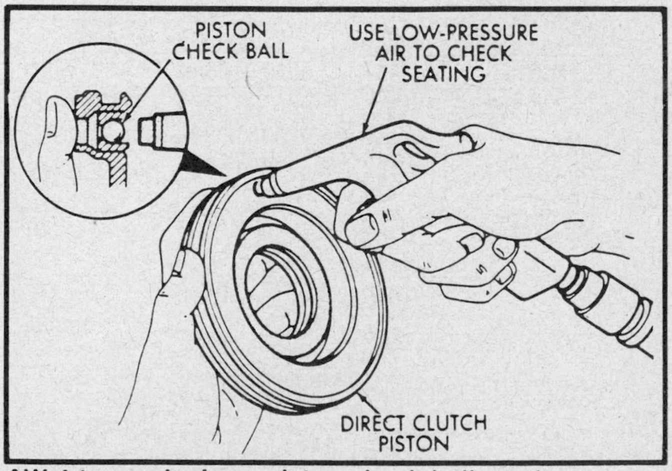
PISTON CHECK BALL
USE LOW-PRESSURE AIR TO CHECK SEATING
DIRECT CLUTCH PISTON
AW-4 transmission—piston check ball seating

be 0.0839 in. (2.32mm). If not within specification, replace the return springs.

5. Check the clutch piston check ball. Shake the piston to see if the ball moves freely. Check the ball seat by applying low volume compressed air to the ball inlet.

6. Measure the inside diameter of the clutch drum bushing. The inside diameter should be less than 2.1248 in. (53.97mm). If not within specification, replace the drum.

Assembly

1. Lubricate and install the replacement O-rings on the clutch piston. Install the clutch piston in the drum. Install the return springs on the piston.

2. Compress the piston return springs and install the snapring. Be sure that the snapring end gap is not aligned with the spring retainer tab.

3. Install the clutch discs and plates. Install the plate and then the disc, until all the plates and discs are installed.

4. Install the clutch pack retainer in the drum. Install the clutch pack snapring.

5. Check the snapring position. As required, shift the snapring until the end gap is not aligned with any notches in the clutch drum.

6. Recheck the clutch piston stroke length. If correct continue.

7. If the clutch piston stroke length is incorrect, replace the clutch discs or use a different thickness clutch pack retainer.

8. Lubricate the clutch drum thrust washer with pertoleum jelly and install it in the drum.

9. Mount the direct clutch assembly on the forward clutch assembly. Check the assembled height it should be 2.767–2.815 in. (70.3–71.5mm).

10. If the assembled height is incorrect, the clutches are not seated properly.

11. If the assembled height is correct, remove the direct clutch from the forward clutch.

FORWARD CLUTCH

Disassembly

1. To check the forward clutch piston stroke, position the overdrive support assembly on wood blocks. Mount the forward clutch drum on the support.

2. Remove the bearing and the race form the forward clutch drum.

3. Install a dial indicator gauge on the clutch drum. Position the dial indicator plunger against the clutch piston.

4. Apply compressed air through the right side feed hole in the support. Note the piston stroke length on the dial indicator.

5. Stroke length should be 0.1469–0.1807 in. (3.73–4.59mm) for transmissions used in vehicles equipped with a 6 cylinder engine. Stroke length should be 0.1346–0.1665 in. (3.42–4.23mm) for transmissions used in vehicles equipped with a 4 cylinder engine.

6. Replace the clutch discs if the stroke length is not within specification.

7. Remove the clutch pack snapring. Remove the retainer and the clutch pack. Remove the clutch pack cushion plate.

8. Using tool BVIFM-27, compress the clutch springs and remove the piston snapring. Remove the tool and the piston return springs.

9. Remount the forward clutch drum on the overdrive support. Apply compressed air through the feed hole in the support in order to remove the piston.

10. Use only enough compressed air to ease the piston out of the drum. Remove and discard the clutch piston O-rings.

11. Remove the clutch drum O-ring from the rear hub of the drum. Remove the seal rings from the clutch drum shaft. Remove the thrust bearing and the race assembly from the clutch drum.

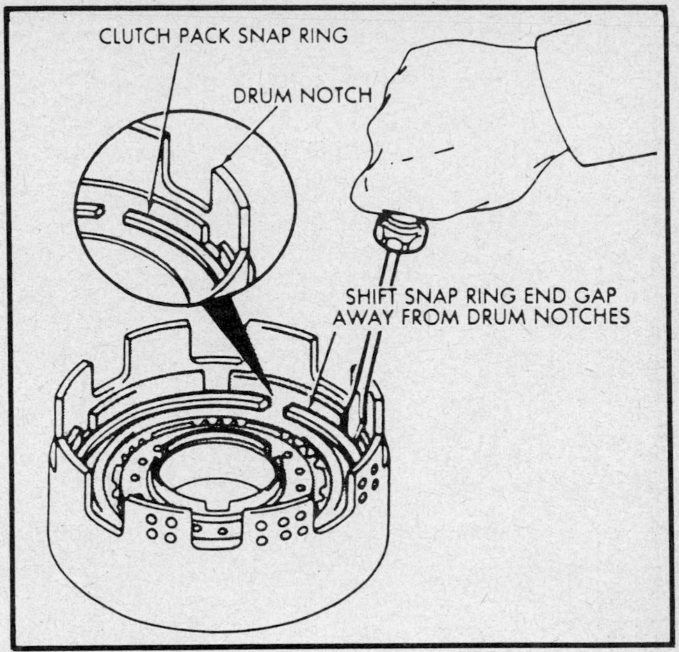
CLUTCH PACK SNAP RING
DRUM NOTCH
SHIFT SNAP RING END GAP
AWAY FROM DRUM NOTCHES

AW-4 transmission—clutch pack snapring positioning

Inspection

1. Inspect all components. Repair or replace defective components as required.

2. Replace all seals, prior to installation coat them with clean transmission fluid.

3. Measure the clutch disc thickness, using a micrometer. The minimum allowable thickness should be 0.0595 in. (1.51mm) for transmissions used in vehicles equipped with a 6 cylinder engine and 0.0724 in. (1.84mm) for transmissions used in vehicles equipped with a 4 cylinder engine.

4. Measure the free length of the piston return springs with the springs in the retainer. The free length specification should be 0.767 in. (19.47mm). If not within specification, replace the return springs and the retainer.

5. Check the clutch piston check ball. The ball should move freely. Check the ball seat by applying low volume compressed air to the ball inlet. The ball should seat firmly and not leak air.

6. Measure the inside diameter of the clutch drum hub. The inside diameter should be less than 0.9480 in. (24.08mm). If not within specification, replace the clutch drum.

Assembly

1. Lubricate the bearing and the race assembly with petroleum jelly and install it in the clutch drum. Be sure that the race side of the assembly faces downward and toward the drum and that the bearing rollers face up.

2. Coat the new clutch drum shaft seal rings with petroleum jelly. Before installing the drum shaft seal rings, squeeze each ring so that the ring ends overlap. This tightens the ring and makes clutch installation easier.

3. Install the seal rings on the shaft. Keep the rings closed as tightly as possible during installation. Avoid overspreading them.

4. Mount the clutch drum on the overdrive support. Lubricate and install a new O-ring on the clutch drum hub.

5. Lubricate and install new O-rings on the clutch piston. Install the piston in the drum. Install the piston return springs.

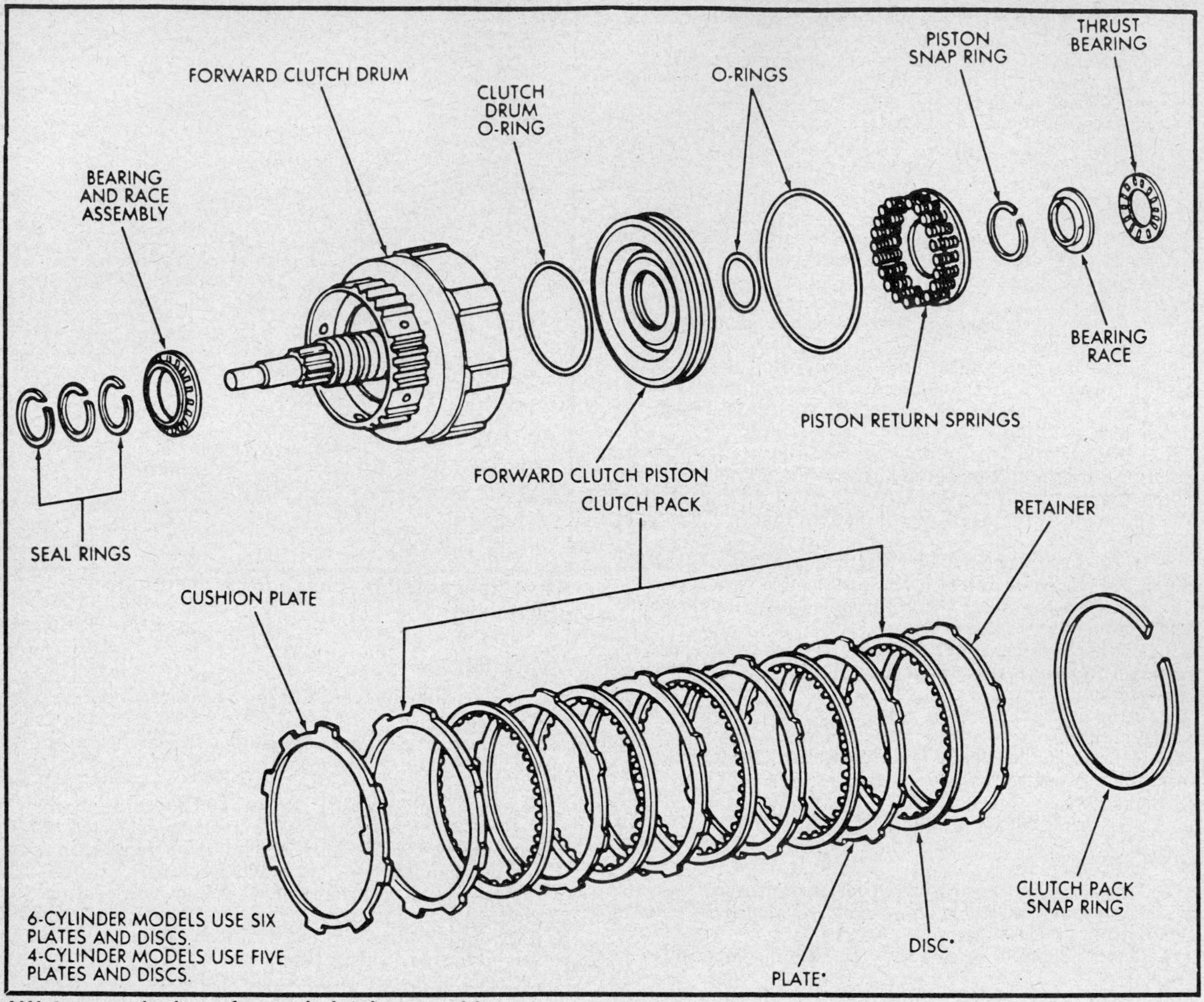
FORWARD CLUTCH DRUM
CLUTCH DRUM O-RING
O-RINGS
PISTON SNAP RING
THRUST BEARING
BEARING AND RACE ASSEMBLY
BEARING RACE
PISTON RETURN SPRINGS
FORWARD CLUTCH PISTON
SEAL RINGS
CLUTCH PACK
RETAINER
CUSHION PLATE
CLUTCH PACK SNAP RING
6-CYLINDER MODELS USE SIX PLATES AND DISCS.
4-CYLINDER MODELS USE FIVE PLATES AND DISCS.
DISC*
PLATE*

AW-4 transmission—forward clutch assembly

6. Using tool BVIFM-27 and a shop press, install the piston snapring. Be sure that the snapring end gap is not aligned with any notches in the drum.

7. Install the cushion plate in the drum. Be sure that the concave side of the plate faces down.

8. Install the clutch discs, plates and retainer. Install the tabbed plate followed by a disc, until all the plates and discs are installed. Install the clutch pack snapring.

9. Recheck the clutch piston stroke length. If not within specification, replace the clutch discs.

10. Lubricate the race and bearing with petroleum jelly. Install them in the clutch drum. Be sure that the bearing rollers face upward and the race lip is seated in the drum.

11. Check the bearing and the race size. The outer diameter of the bearing should be 1.839 in. (46.7mm). The outer diameter of the race should be 1.925 in. (48.9mm). The inner diameter of the bearing and the race should be 1.024 in. (26.0mm).

12. Mount the forward clutch on the direct clutch. Check the assembled height. It should be 2.767–2.815 in. (70.3–71.5mm).

FRONT PLANETARY GEAR

Disassembly

1. Remove the ring gear from the planetary gear. Remove the front bearing and both races from the ring gear.

2. Remove the tabbed thrust race from the planetary gear. Remove the snapring that retains the planetary gear to the shaft. Remove the gear.

3. Remove the rear bearing and the race from the planetary gear.

4. Measure the inside diameter of the ring gear bushing. The maximum allowable diameter is 0.9489 in. (24.08mm). Replace the ring gear if the bushing inside diameter is greater than specification.

Inspection

1. Inspect all components. Repair or replace defective components as required.

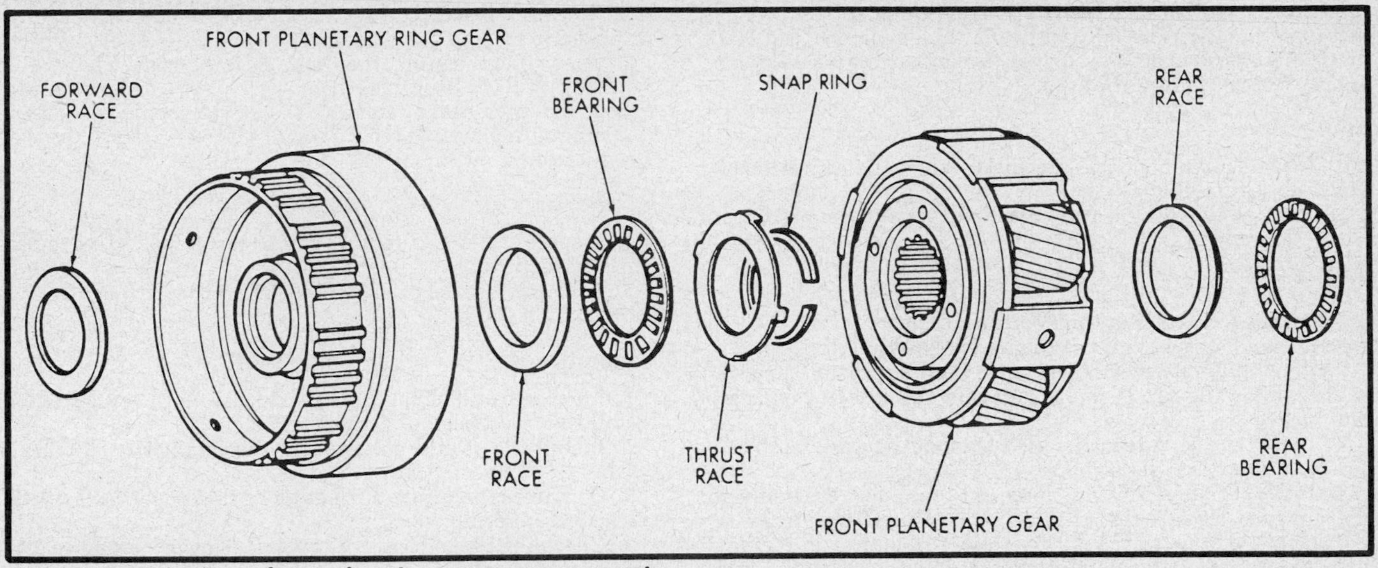
FRONT PLANETARY RING GEAR
FORWARD RACE
FRONT BEARING
SNAP RING
REAR RACE
FRONT RACE
THRUST RACE
REAR BEARING
FRONT PLANETARY GEAR

AW-4 transmission–front planetary gear components

2. Replace all seals, prior to installation coat them with clean transmission fluid.

Assembly

1. Lubricate the planetary and ring gear bearings and their races with petroleum jelly.
2. Identify the planetary bearings and races before installation.

The outer diameter of rear bearing is 1.878 in. (47.7mm).
The inner diameter of the rear bearing is 1.398 in. (35.5mm).
The outer diameter of the rear race 1.874 in. (47.6mm).
The inner diameter of the rear race is 1.327 in. (33.7mm).
The outer diameter of the front race is 2.110 in. (53.6mm).
The inner diameter of the front race is 1.201 in. (30.5mm).
The outer diameter of the front bearings is 1.878 in. (47.7mm).
The inner diameter of the front bearings is 1.283 in. (32.6).
The outer diameter of the forward race is 1.850 in. (47.0mm).
The inner diameter of the forward race is 1.043 in. (26.5mm).

3. Install the rear race and bearing in the gear. Turn the planetary assembly over and install the thrust race.
4. Install the front race and bearing and the forward race in the ring gear.

SUN GEAR AND NO. 1 ONE-WAY CLUTCH

Disassembly

1. Hold the sun gear and turn the second brake hub clockwise and than counterclockwise. The hub should rotate freely clockwise, but lock when turned counterclockwise. Replace the one-way clutch and hub if it does not operate properly.
2. Remove the one-way clutch/second brake hub assembly from the drum. Remove the thrust washer from the drum. Remove both seal rings from the sun gear.
3. Support the sun gear on a wood block. Remove the first sun gear snapring and separate the drum from the gear.
4. Remove the remaining snapring from the sun gear.

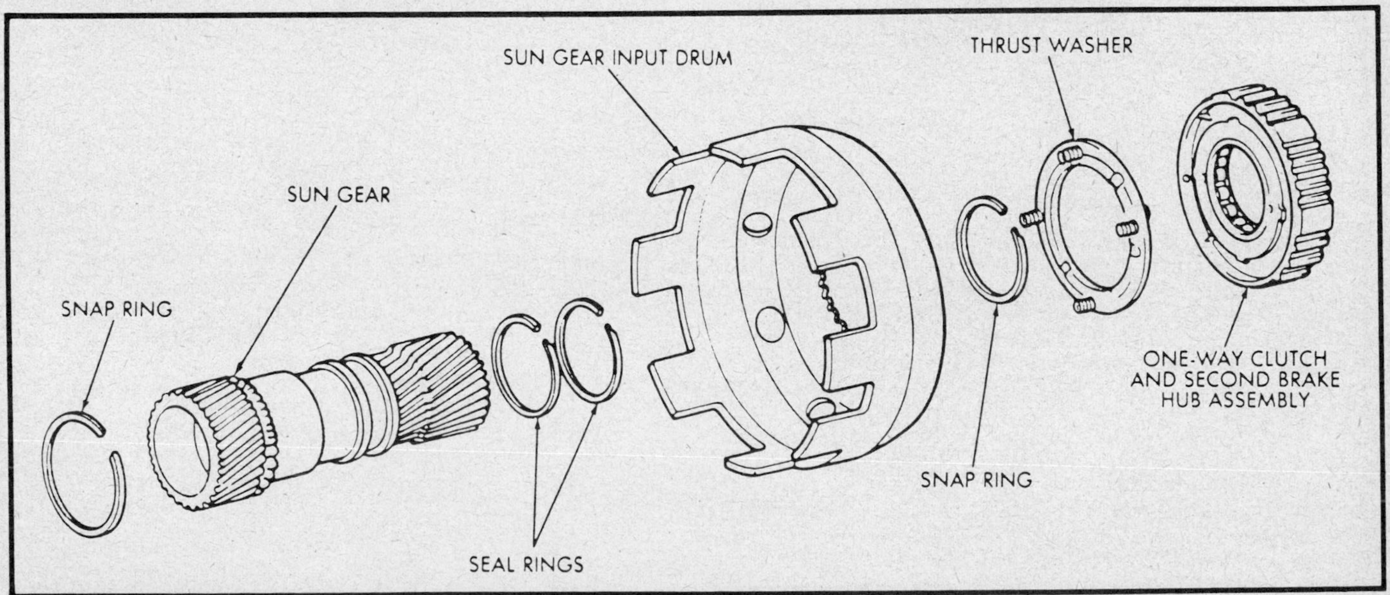
SUN GEAR INPUT DRUM
THRUST WASHER
SUN GEAR
SNAP RING
ONE-WAY CLUTCH AND SECOND BRAKE HUB ASSEMBLY
SNAP RING
SEAL RINGS

AW-4 transmission–sun gear and one-way clutch assembly

5. Measure the inside diameter of the sun gear bushings using a micrometer. The maximum allowable diameter is 1.0661 in. (27.08mm). Replace the sun gear if the bushing inside diameter is greater than specification.

Inspection

1. Inspect all components. Repair or replace defective components as required.
2. Replace all seals, prior to installation coat them with clean transmission fluid.

Assembly

1. Install the first snapring on the sun gear. Install the sun gear in the drum. Install the remaining snapring.
2. Coat the replacement seal rings with petroleum jelly. Install them on the sun gear. Be sure that the seal ring ends are interlocked.
3. Install the thrust washer. Be sure that the washer tabs are seated in the drum slots.
4. Install the one-way clutch/second brake hub assembly on the sun gear. Be sure that the deep side of hub flange faces up.
5. Check the one-way clutch operation. Hold the sun gear and turn the second brake hub clockwise and than counterclockwise. The hub should turn clockwise freely, but lock when turned counterclockwise.

SECOND BRAKE

Disassembly

1. Remove the second brake drum from the output shaft. Set the output shaft assembly aside.
2. Remove the thrust washer from the second brake drum.
3. Using tool BVIFM-27 or equivalent and a shop press, compress the piston return springs.
4. Remove the compressor tool. Remove the spring retainer and the return springs.
5. Remove the second brake piston and sleeve from the drum, using low pressure compressed air. Be sure to use only enough air pressure to ease the piston out of the drum.
6. Remove and discard the brake piston O-rings.
7. Measure the free length of piston return springs with springs mounted in the retainer. The length should be 0.632 in. (16.05mm). Replace the return springs if the length is less than specifiication.

Inspection

1. Inspect all components. Repair or replace defective components as required.
2. Replace all seals, prior to installation coat them with clean transmission fluid.

Assembly

1. Lubricate and install the new O-rings on the brake piston. Install the brake piston in the drum.
2. Install the return springs and the retainer on the brake piston.
3. Compress the return springs using the proper tools. Install the piston snapring.
4. Check the brake piston operation, using low pressure compressed air. Apply air pressure through the feed hole in the drum. The piston should move smoothly when applying and releasing air pressure.
5. Coat the thrust washer with petroleum jelly and install it in the drum. Be sure that the washer notches are aligned with the tabs on the spring retainer.

REAR PLANETARY, NO. 2 ONE-WAY CLUTCH AND OUTPUT SHAFT

Disassembly

1. Remove the output shaft from the gear assembly. Remove and discard the shaft seal ring. Remove the brake pack from the planetary gear.
2. Measure the thickness of each brake pack disc. The mini-

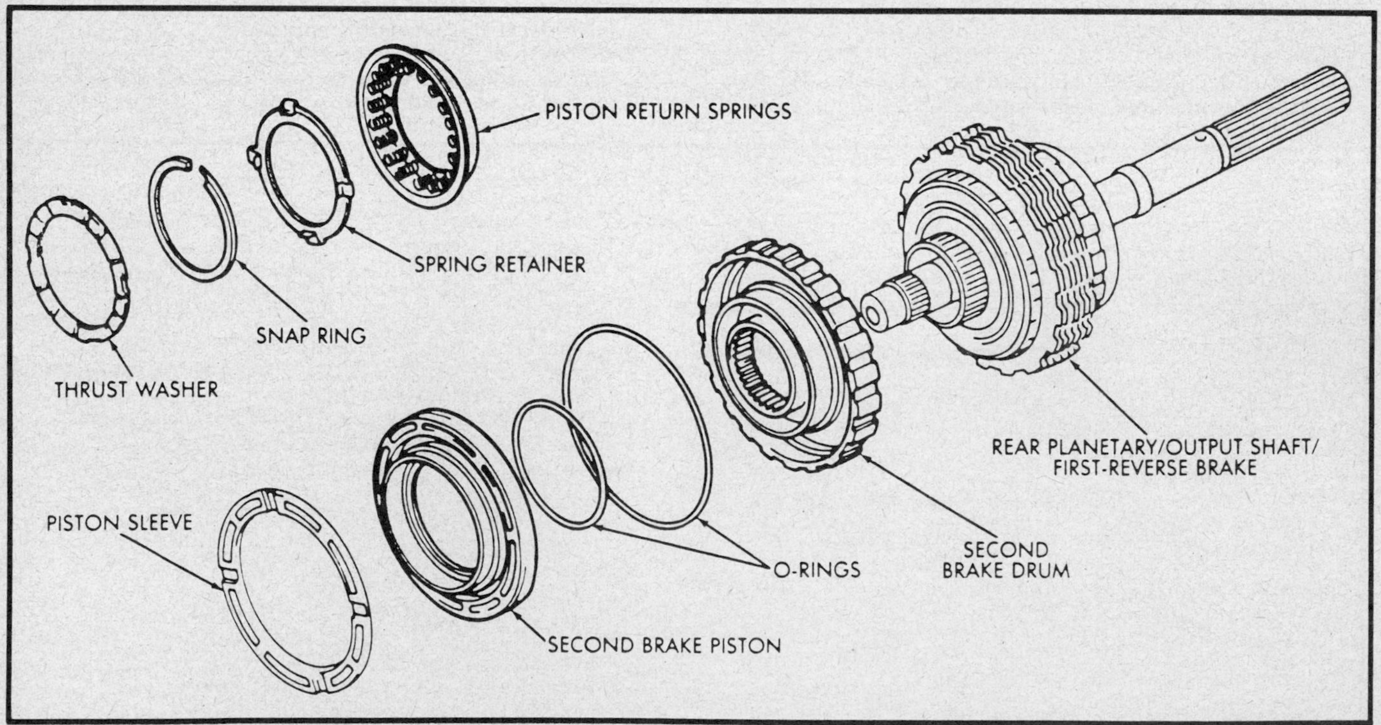
PISTON RETURN SPRINGS
SPRING RETAINER
SNAP RING
THRUST WASHER
REAR PLANETARY/OUTPUT SHAFT/
FIRST-REVERSE BRAKE
PISTON SLEEVE
O-RINGS
SECOND
BRAKE DRUM
SECOND BRAKE PISTON

AW-4 transmission—second brake assembly

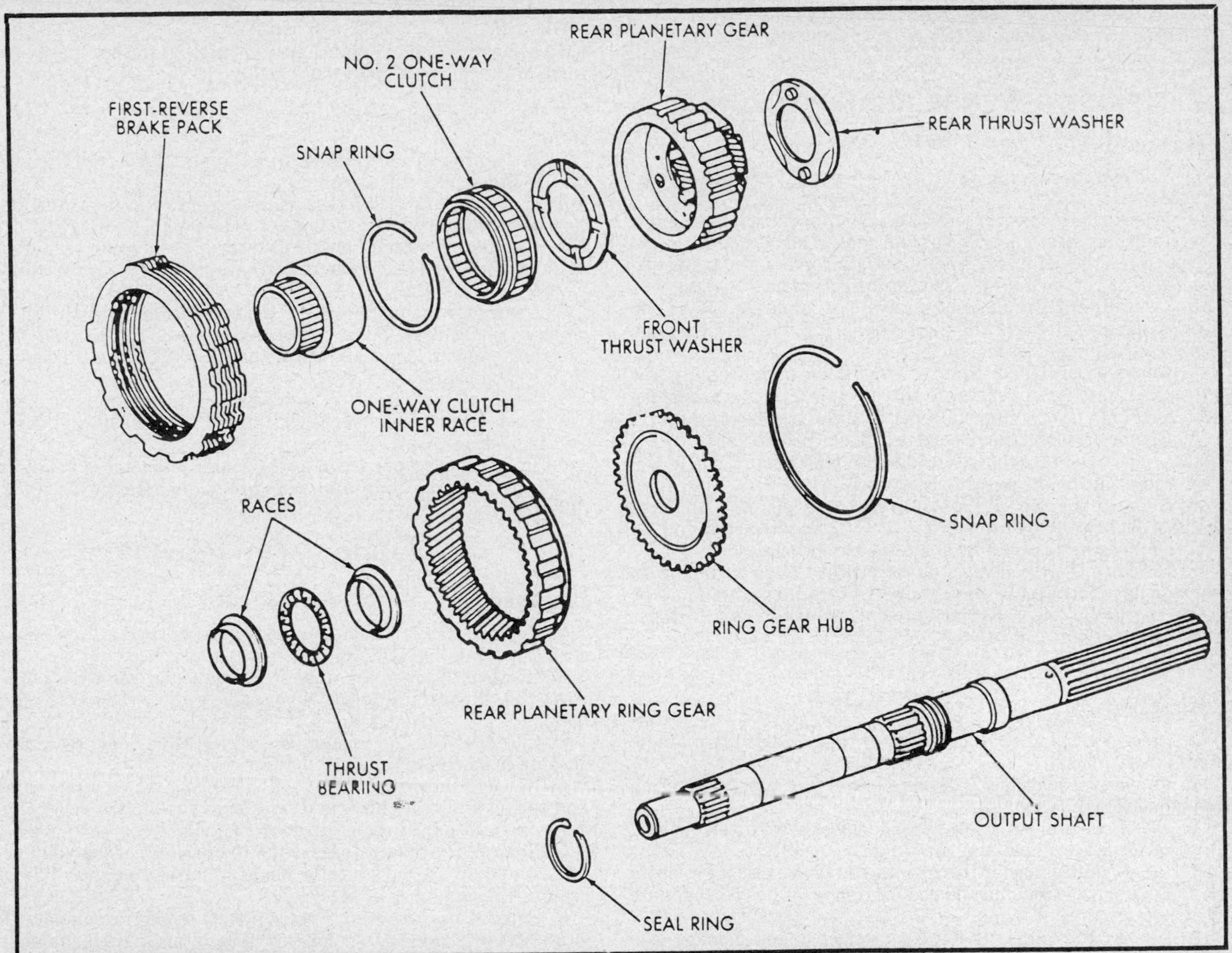
REAR PLANETARY GEAR
NO. 2 ONE-WAY CLUTCH
FIRST-REVERSE BRAKE PACK
SNAP RING
REAR THRUST WASHER
FRONT THRUST WASHER
ONE-WAY CLUTCH INNER RACE
RACES
SNAP RING
RING GEAR HUB
REAR PLANETARY RING GEAR
THRUST BEARING
OUTPUT SHAFT
SEAL RING

AW-4 transmission—rear planetary, brake pack, clutch and output shaft assembly

mum thickness is 0.0594 in. (1.51mm). Replace all of the discs if any disc is thinner than specification. Remove the planetary gear from the ring gear.

3. Check the No. 2 one-way clutch. Hold the planetary gear and turn the clutch inner race in both directions. The race should turn freely counterclockwise, but lock when turned clockwise. Replace the one-way clutch if it does not perform properly.

4. Remove the clutch inner race from the planetary gear. Remove the clutch snapring. Remove the No. 2 one-way clutch from the planetary gear assembly.

5. Remove the front and rear thrust washers from the planetary gear assembly. Remove the thrust bearing and washers from the ring gear.

6. Remove the ring gear snapring. Remove the ring gear hub.

Inspection

1. Inspect all components. Repair or replace defective components as required.

2. Replace all seals, prior to installation coat them with clean transmission fluid.

Assembly

1. Install the hub and the snapring in the ring gear.

2. Identify the ring gear thrust bearing and races.
The outer diameter of the bottom race is 1.764 in. (44.8mm).
The inner diameter of the bottom race is 1.087 in. (27.6mm).
The outer diameter of the bearing is 1.760 in. (44.7mm).
The inner diameter of the bearing is 1.185 in. (30.1mm).
The outer diameter of the upper race is 1.764 in. (44.8mm).
The inner diameter of the upper race is 1.134 in. (28.8mm).

3. Lubricate the ring gear thrust bearing and races with petroleum jelly. install them in ring gear.

4. Coat the planetary thrust washers with petroleum jelly. Install them in gear.

5. Install the No. 2 one-way clutch in the planetary gear. Be sure that the flanged side of clutch faces up.

6. Install the clutch retaining snapring. Install the clutch inner race. Turn the race counterclockwise to ease installation.

7. Verify that the one-way clutch is operating properly. Hold the gear and turn the inner race in both directions. The race should turn freely counterclockwise, but lock when turned clockwise.

8. Install the planetary gear in the ring gear.

9. Assemble the clutch discs and the clutch plates. The installation sequence is disc first, then a plate. Use 7 discs and plates in a transmission that is used in a vehicle equipped with a 6 cylinder engine. Use 6 discs and plates in a transmission that is used in a vehicle equipped with a 4 cylinder engine.

10. Install the brake pack on the planetary gear. Install a new seal ring on the output shaft. Be sure that the ring ends are interlocked.

FIRST/REVERSE BRAKE PISTON

Disassembly

1. Remove the bearing and the race assembly from the transmission case.
2. Check the first/reverse brake piston operation, using low pressure compressed air. The piston should move smoothly and not bind or stick. If piston operation is incorrect, the transmission case or the piston may require replacement.
3. Using tool BVIFM-28 or equivalent compress the piston return springs. Remove the piston snapring. Remove the tool. Remove the piston return springs.
4. Remove the No. 2 first/reverse brake piston, using low pressure compressed air. Apply air through the same transmission feed hole that is used for checking the piston operation.
5. Remove the reaction sleeve using tool BVIFM-31 or equivalent. Insert the tool flanges under the sleeve and lift the tool and sleeve out of the transmission case.
6. Remove the No. 1 first/reverse brake piston using tool BVIFM-32 or equivalent. Slip the tool under the piston and lift the tool and the piston out of the transmission case.
7. Measure the free length of the piston return springs with the springs mounted in the retainer. Spring length should be 0.724 in. (18.382mm). Replace the springs if the length is less than specification.

Inspection

1. Inspect all components. Repair or replace defective components as required.
2. Replace all seals, prior to installation coat them with clean transmission fluid.
3. Clean the transmission case thoroughly with solvent and dry it with compressed air.
4. Blow compressed air through the oil feed passages in order to remove solvent residue and ensure that the fluid passages are clear.
5. Inspect the transmission case for wear and damage. Replace the transmission case, as required.

Assembly

1. Lubricate and install the new O-rings on the No. 1 first/reverse brake piston and on the reaction sleeve. Install the piston in the sleeve.
2. Lubricate and install the new O-ring on the No. 2 brake piston.
3. Install the assembled No. 1 piston and the reaction sleeve on the No. 2 piston.
4. Lubricate and install the piston assembly in the transmission case. Align the piston and the case slots.
5. Press the piston assembly into the transmission case with hand pressure. Position the piston return springs on the No. 2 piston.
6. Compress the piston return springs, using tool BVIFM-28 or equivalent. Install the piston snapring. Be sure that the snapring end gap is not aligned with any of the tangs on the return spring retainer.
7. Verify piston operation using low pressure compressed air.
8. Coat the bearing and the race assembly with petroleum jelly. Install it in piston assembly.
9. The bearing and race assembly outer diameter is 2.272 in. (57.7mm). The bearing and race inner diameter is 1.543 in. (39.2mm).

LOWER VALVE BODY

Disassembly

1. Remove the 2 piece detent spring. Note the position of the spring sections for reassembly.
2. Remove the manual valve from the lower valve body. Remove the bolts retaining the upper valve body to the lower valve body.
3. Carefully lift and remove the upper valve body, plate and gaskets from the lower valve body.
4. Remove the check valve and spring, pressure relief valve and spring and ball check and seat from the lower valve body. Note the location of each valve for reassembly.
5. Remove the oil strainers. Note the position of the valve retainers and pressure reducing plug clip for reassembly. Do not remove the retainers at this time.
6. Remove the valve body solenoids. Discard the solenoid O-rings. Remove the release control valve retainer with the magnet. Remove the release control valve and plug.

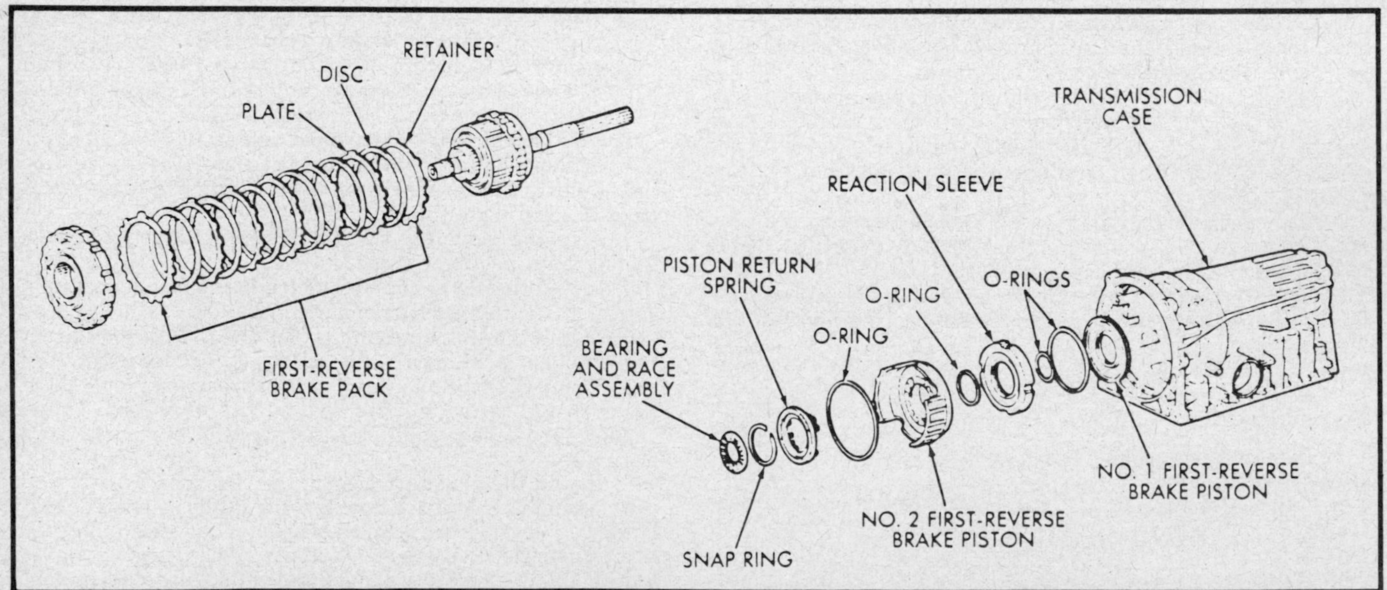
RETAINER
DISC
PLATE
TRANSMISSION CASE
REACTION SLEEVE
PISTON RETURN SPRING
O-RINGS
O-RING
O-RING
BEARING AND RACE ASSEMBLY
FIRST-REVERSE BRAKE PACK
NO. 1 FIRST-REVERSE BRAKE PISTON
NO. 2 FIRST-REVERSE BRAKE PISTON
SNAP RING

AW-4 transmission—first/reverse brake piston assembly

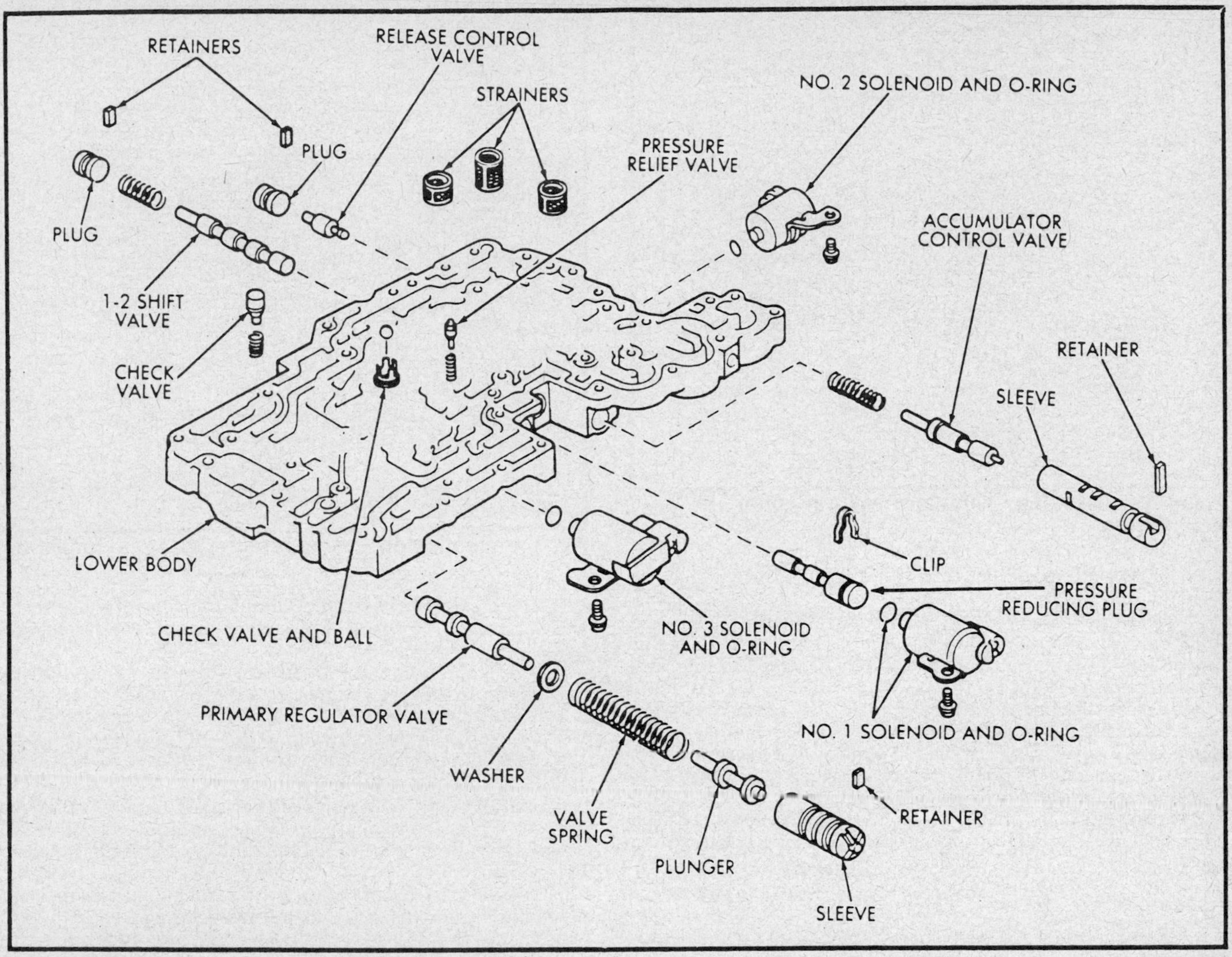
RETAINERS
RELEASE CONTROL VALVE
STRAINERS
NO. 2 SOLENOID AND O-RING
PLUG
PRESSURE RELIEF VALVE
PLUG
ACCUMULATOR CONTROL VALVE
1-2 SHIFT VALVE
CHECK VALVE
RETAINER
SLEEVE
LOWER BODY
CLIP
PRESSURE REDUCING PLUG
CHECK VALVE AND BALL
NO. 3 SOLENOID AND O-RING
PRIMARY REGULATOR VALVE
NO. 1 SOLENOID AND O-RING
WASHER
VALVE SPRING
RETAINER
PLUNGER
SLEEVE

AW-4 transmission—lower valve body assembly

7. Remove the 1–2 shift valve retainer. Remove the valve plug, valve spring and valve.

NOTE: The primary regulator valve sleeve and plunger are under tension from the valve spring. Be sure to exert counterpressure on the spring while removing the valve retainer in order to prevent components from flying out.

8. To remove the primary regulator valve, note the position of the valve retainer for reassembly. Press the valve sleeve inward and remove the retainer, using a magnet.
9. Slowly release the pressure on the sleeve and remove the sleeve, spring and washer and the valve. Use of a magnet to remove the valve is necessary.
10. Remove the regulator valve and plunger from the sleeve. Remove the retaining clip. Using the proper tool, remove the pressure reducing plug.
11. Remove the accumulator control valve retainer. Remove the control valve assembly. Remove the spring and the control valve from the valve sleeve.

Inspection

1. Inspect all components. Repair or replace defective components as required.

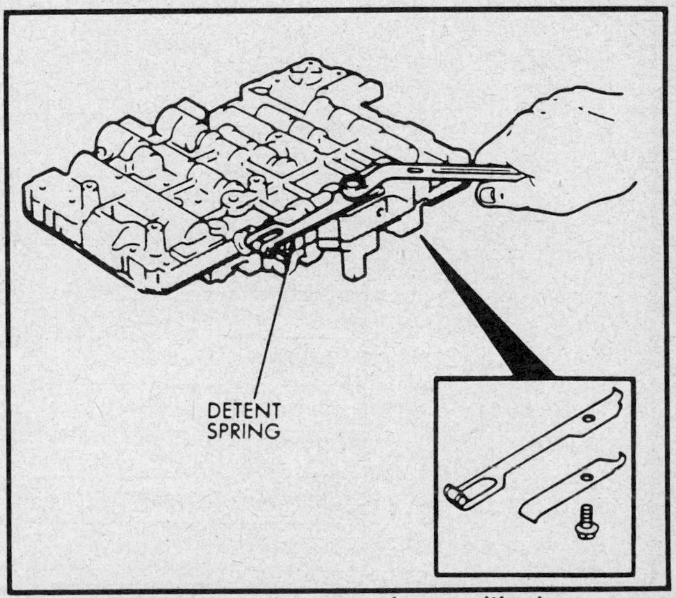
DETENT SPRING

AW-4 transmission—detent spring positioning

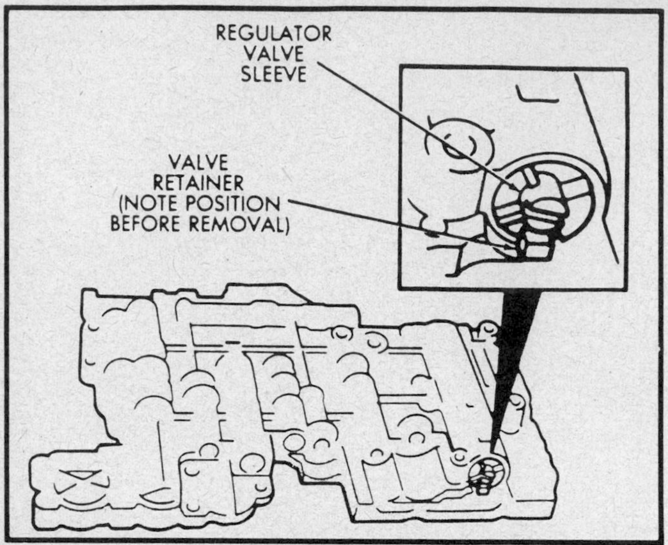
REGULATOR
VALVE
SLEEVE
VALVE
RETAINER
(NOTE POSITION
BEFORE REMOVAL)

AW-4 transmission—regulator valve retainer positioning

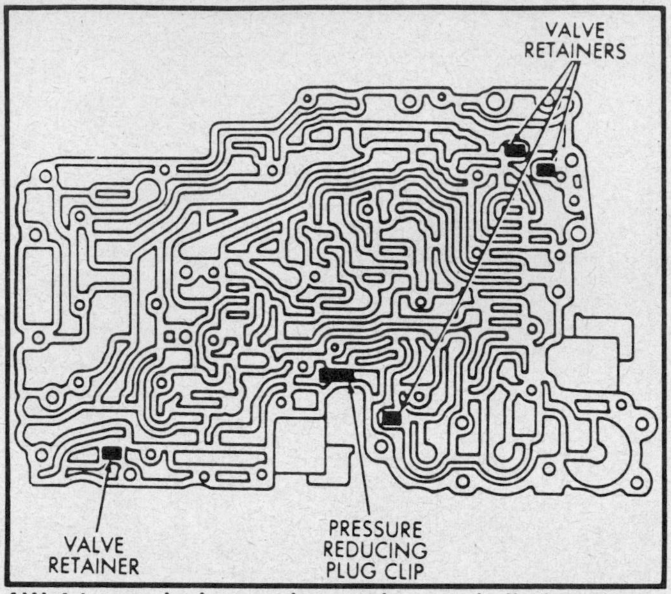
VALVE
RETAINERS
VALVE
RETAINER
PRESSURE
REDUCING
PLUG CLIP

AW-4 transmission—valve retainer and clip location

2. Replace all seals, prior to installation coat them with clean transmission fluid.
3. Clean the lower body valve components with solvent and dry them with compressed air. Do not use shop towels or rags, as lint or foreign material from the towels or rags can interfere with valve operation.
4. Inspect the condition of lower valve body components. Replace the lower valve body if any bores are scored or corroded. Replace any valves, plugs or sleeves that are scored or worn. Replace the oil strainers if cut, torn or damaged in any way.
5. Inspect the valve body springs. Replace any springs having rusted, distorted, or collapsed coils. Measure the length of each valve body spring. Replace any spring if free length is less than specification.

Assembly

1. Lubricate the lower valve body components with clean automatic transmission fluid.
2. Install the spring and accumulator control valve in the sleeve. Install the assembled components in the lower valve body.
3. Press the accumulator control valve assembly into the valve bore. Install the retainer.
4. Install the pressure reducing plug in the plug bore. Secure the plug with the retaining clip.
5. Install the washer on the primary regulator valve plunger. Install the primary regulator valve plunger in the valve sleeve.
6. Install the valve spring and the regulator valve sleeve and plunger. Press the regulator valve sleeve into the bore. Install the retainer. Be sure that the retainer is positioned in the sleeve lugs properly.
7. Install the 1–2 shift valve, spring and plug. Press the valve assembly into the bore and install the retainer.
8. Install the release control valve and plug in the bore. Install the valve retainer.

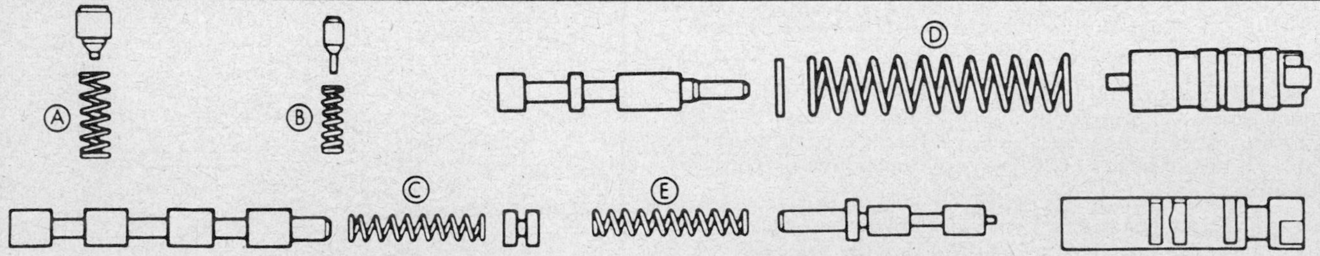
A
B
C
D
E

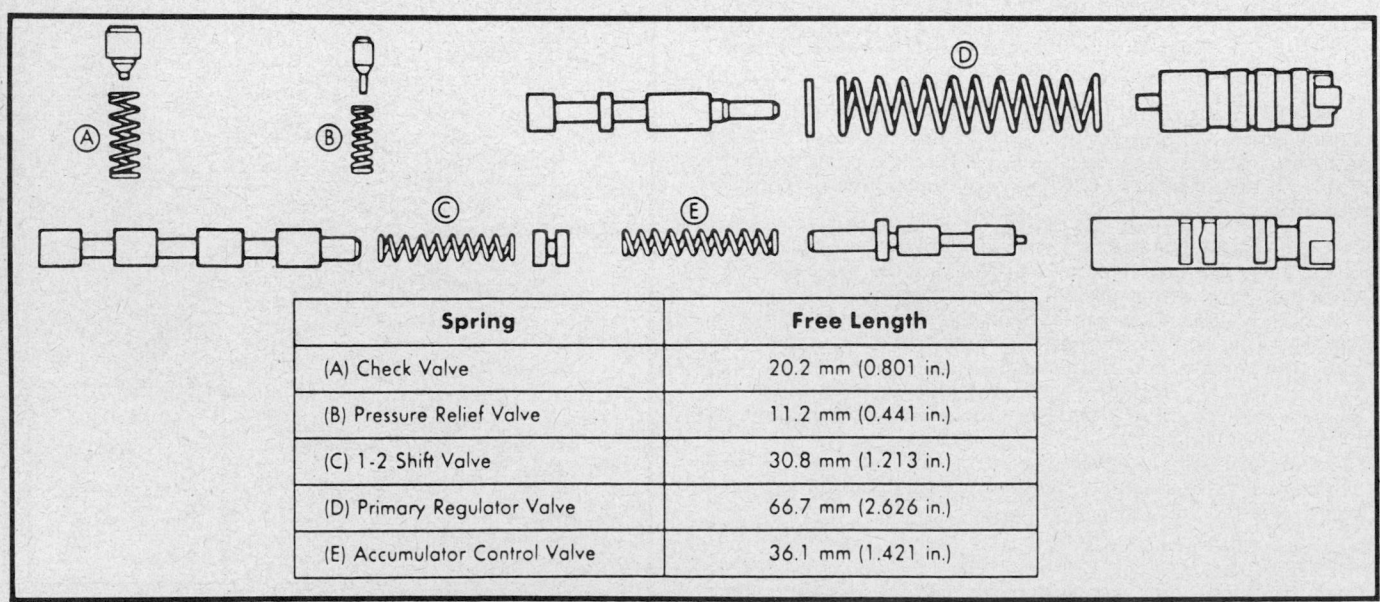

Spring	Free Length
(A) Check Valve	20.2 mm (0.801 in.)
(B) Pressure Relief Valve	11.2 mm (0.441 in.)
(C) 1-2 Shift Valve	30.8 mm (1.213 in.)
(D) Primary Regulator Valve	66.7 mm (2.626 in.)
(E) Accumulator Control Valve	36.1 mm (1.421 in.)

AW-4 transmission—lower valve body spring dimension data

9. Install the replacement O-rings on the solenoids. Install the solenoids on the valve body. Torque the solenoid retaining bolts to 7 ft. lbs. (10 Nm).

10. Install the oil strainers. Be sure to identify the strainers prior to installation. The strainers are all the same diameter but are different lengths. Two strainers are 0.443 in. (11.0mm) long and the other strainer is 0.76 in. (19.5mm) long.

11. Install the check valves and springs seats.

UPPER VALVE BODY

1. Remove the 2 piece detent spring. Note the position of the spring sections for reassembly.
2. Remove the manual valve from the lower valve body. Remove the bolts retaining the upper valve body to the lower valve body.
3. Carefully lift and remove the upper valve body, plate and gaskets from the lower valve body.
4. Remove the upper valve body plate and gaskets. Discard the gaskets.
5. Remove the strainer and the 9 check balls. Note the check ball and strainer position for reassembly.
6. Remove the valve stop and the throttle cam. Remove the throttle valve pin using a magnet. Remove the downshift plug, valve spring and throttle valve.
7. Turn the upper valve body over and remove the throttle valve adjusting rings and spring. Note the number of adjusting rings, if the valve is equipped with them.
8. Remove the 3–4 shift valve retainer, using a magnet. Remove the valve plug, spring and the 3–4 shift valve.
9. Remove the second coast modulator valve retainer. Remove the valve plug, spring and valve.
10. Remove the lock-up relay valve retainer. Remove the relay valve and sleeve assembly. Remove the lock-up relay valve and the spring and plunger from the valve sleeve.
11. Remove the secondary pressure regulator valve retainer. Remove the plug, regulator valve and spring.
12. Remove the cut-back valve retainer. Remove the plug, cut-back valve and spring.
13. Remove the 2–3 shift valve retainer. Remove the plug, spring and the 2–3 shift valve.
14. Remove the low coast modulator valve retainer. Remove the valve plug, spring and the low coast modulator valve.

Inspection

1. Inspect all components. Repair or replace defective components as required.
2. Replace all seals, prior to installation coat them with clean transmission fluid.
3. Clean the upper body components with solvent and dry them with compressed air. Do not use shop towels or rags as lint or foreign material from the towels or rags can interfere with valve operation.
4. Inspect the condition of the upper valve body components. Replace the upper valve body if any of the bores are scored or corroded. Replace any valves, plugs or sleeves if scored or worn. Replace the oil strainer if cut, torn or damaged in any way.
5. Inspect the valve body springs. Replace any spring having rusted, distorted, or collapsed coils. Measure the length of each spring. Replace any spring if the free length is less than specification.

Assembly

1. Lubricate the valves, springs, plugs, sleeves and the valve bores in the upper valve body with clean automatic transmission fluid. Note position of the valve retainers and stop for reassembly reference.
2. Install the low coast modulator valve, spring and plug in the valve bore. Press valve plug inward and install the retainer.

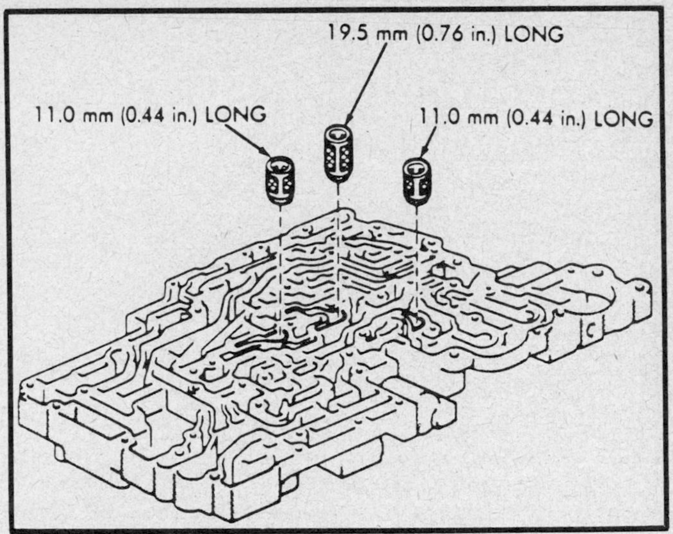
19.5 mm (0.76 in.) LONG
11.0 mm (0.44 in.) LONG
11.0 mm (0.44 in.) LONG

AW-4 transmission—oil strainer identification and location

3. Install the 2–3 shift valve, spring and plug in the valve bore. Press the plug inward and install retainer.
4. Install the cut-back valve spring, valve and plug. Press the plug inward and install retainer.
5. Install the secondary regulator valve spring, valve and plug in the valve bore. Press the plug inward and install the retainer.
6. Assemble the lock-up relay valve. Install the spring and plunger in the valve sleeve. Install the assembled valve in the sleeve.
7. Install the assembled lock-up relay valve in the valve bore. Install the retainer.

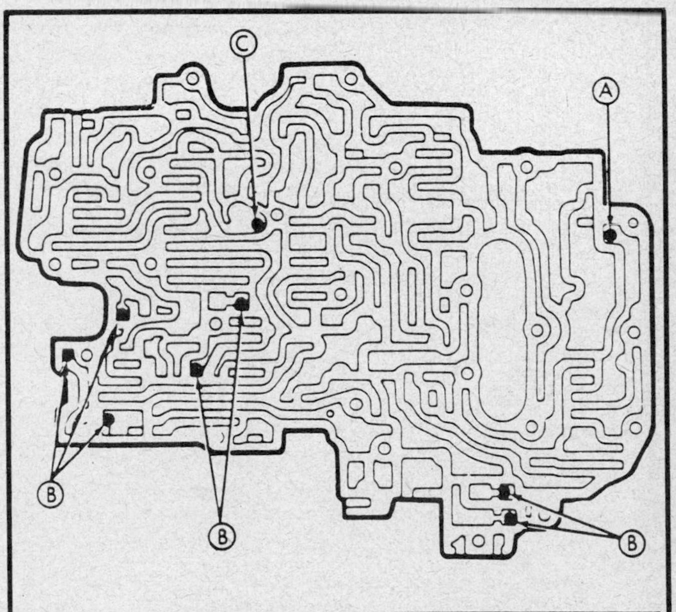
C
A
B
B
B

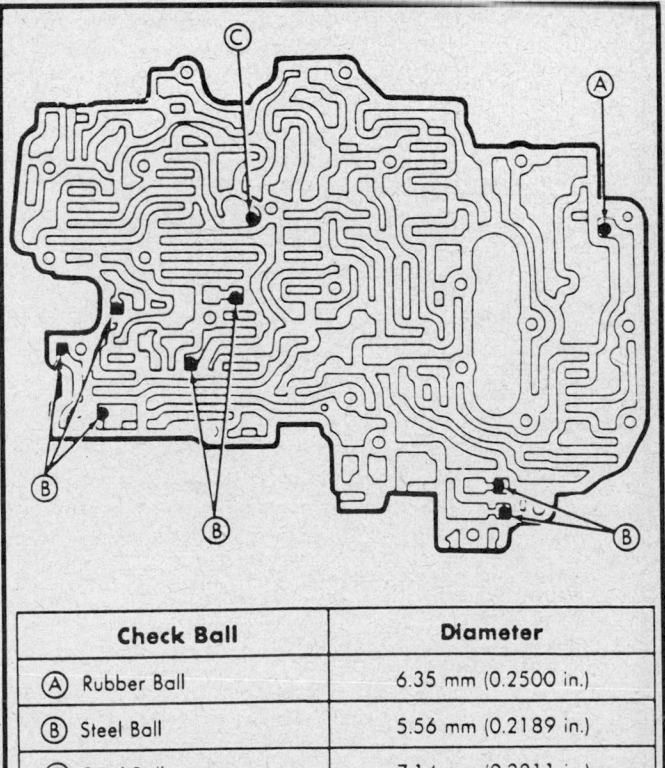

Check Ball	Diameter
Ⓐ Rubber Ball	6.35 mm (0.2500 in.)
Ⓑ Steel Ball	5.56 mm (0.2189 in.)
Ⓒ Steel Ball	7.14 mm (0.2811 in.)

AW-4 transmission—check ball and strainer location

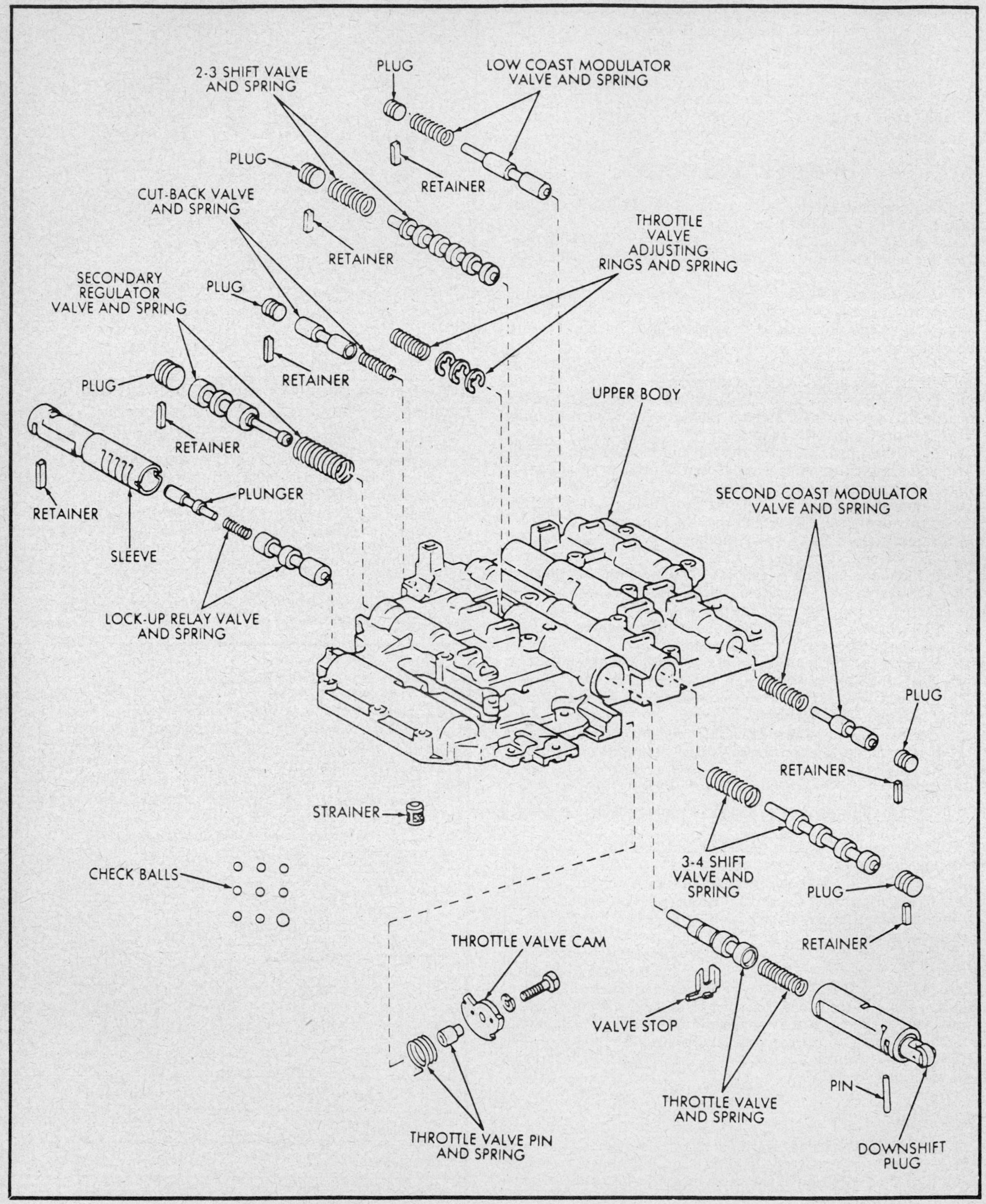
2-3 SHIFT VALVE AND SPRING
PLUG
LOW COAST MODULATOR VALVE AND SPRING
PLUG
RETAINER
CUT-BACK VALVE AND SPRING
RETAINER
THROTTLE VALVE ADJUSTING RINGS AND SPRING
SECONDARY REGULATOR VALVE AND SPRING
PLUG
RETAINER
PLUG
UPPER BODY
RETAINER
PLUNGER
RETAINER
SLEEVE
SECOND COAST MODULATOR VALVE AND SPRING
LOCK-UP RELAY VALVE AND SPRING
PLUG
RETAINER
STRAINER
CHECK BALLS
3-4 SHIFT VALVE AND SPRING
PLUG
RETAINER
THROTTLE VALVE CAM
VALVE STOP
PIN
THROTTLE VALVE AND SPRING
THROTTLE VALVE PIN AND SPRING
DOWNSHIFT PLUG

AW-4 transmission—upper valve body assembly

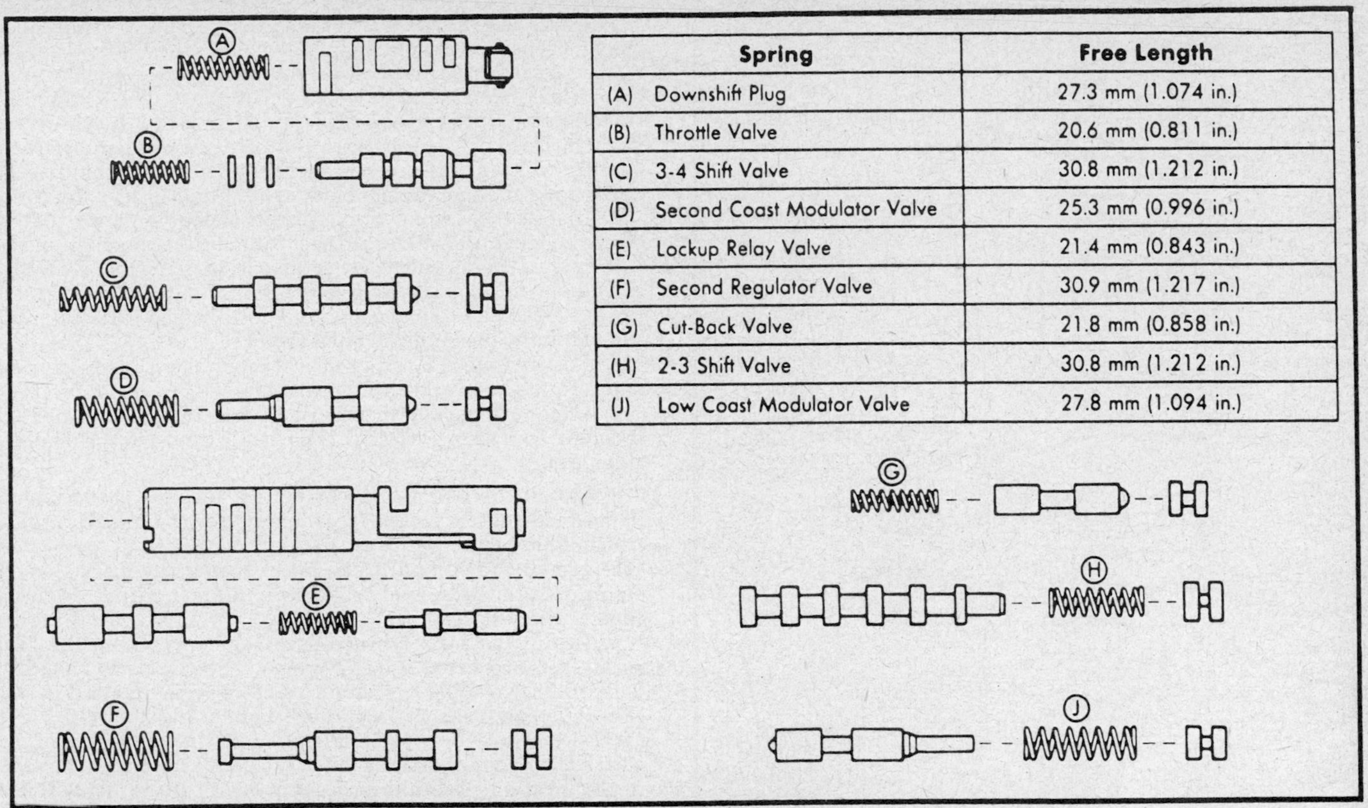
A
B
C
D
E
F
G
H
J

Spring	Free Length
(A) Downshift Plug	27.3 mm (1.074 in.)
(B) Throttle Valve	20.6 mm (0.811 in.)
(C) 3-4 Shift Valve	30.8 mm (1.212 in.)
(D) Second Coast Modulator Valve	25.3 mm (0.996 in.)
(E) Lockup Relay Valve	21.4 mm (0.843 in.)
(F) Second Regulator Valve	30.9 mm (1.217 in.)
(G) Cut-Back Valve	21.8 mm (0.858 in.)
(H) 2-3 Shift Valve	30.8 mm (1.212 in.)
(J) Low Coast Modulator Valve	27.8 mm (1.094 in.)

AW-4 transmission—upper valve body and spring dimension data

8. Install the second coast modulator valve, spring and plug in the valve bore. Press the plug inward and install the retainer.

9. Install the 3–4 shift valve, spring and plug in the bore. Press the plug inward and install the retainer.

10. Install the throttle valve in the valve bore. Push the valve into place and install the valve stop.

11. On transmissions that use adjusting rings, turn the upper valve body over and install the adjusting rings. Be sure to install the same number of rings as removed.

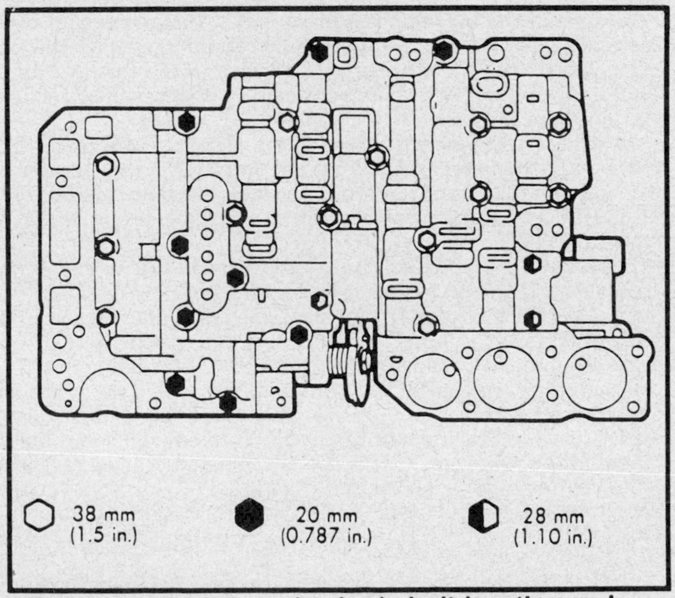
38 mm
(1.5 in.)
20 mm
(0.787 in.)
28 mm
(1.10 in.)

AW-4 transmission—valve body bolt location and identification

12. Install the throttle valve adjusting spring in the bore and onto the end of the throttle valve.

13. Install the downshift spring and plug in the throttle valve bore. Press the plug inward against the throttle valve and spring. Install the retainer pin.

14. Install the sleeve in the throttle cam. Install the spring on the cam. Hook the curved end of the spring through the hole in the cam.

15. Mount the cam on the upper body. Install the cam retaining bolt and spacer. Torque the bolt to 7 ft. lbs. (10 Nm).

16. Be sure that the straight end of the spring is seated in the upper valve body slot.

17. Install the check balls in the upper valve body. Refer to illustration for check ball identification and location. Install the oil strainer.

18. To install the upper valve body to the lower valve body, position a new No. 1 gasket on the upper valve body.

19. Positon the valve body plate on the No. 1 gasket. Positon a new No. 2 gasket on the valve body plate and align the gaskets and plate, usie the bolt holes as guides.

20. Install the valve body bolts. Different length bolts are used. Torque the valve body bolts to 56 inch lbs. (6.4 Nm).

21. Install manual valve. Install the detent spring. Torque the spring retaining bolt to 7 ft. lbs. (10 Nm).

Transmission Assembly

NOTE: During the assembly of the transmission, be sure to lubricate components with clean transmission fluid or petroleum jelly, as required.

1. If any of the tranmission components are still assembled after the overhaul checking procedures have been done, disassemble these components as necessary in preparation for transmission reassembly.

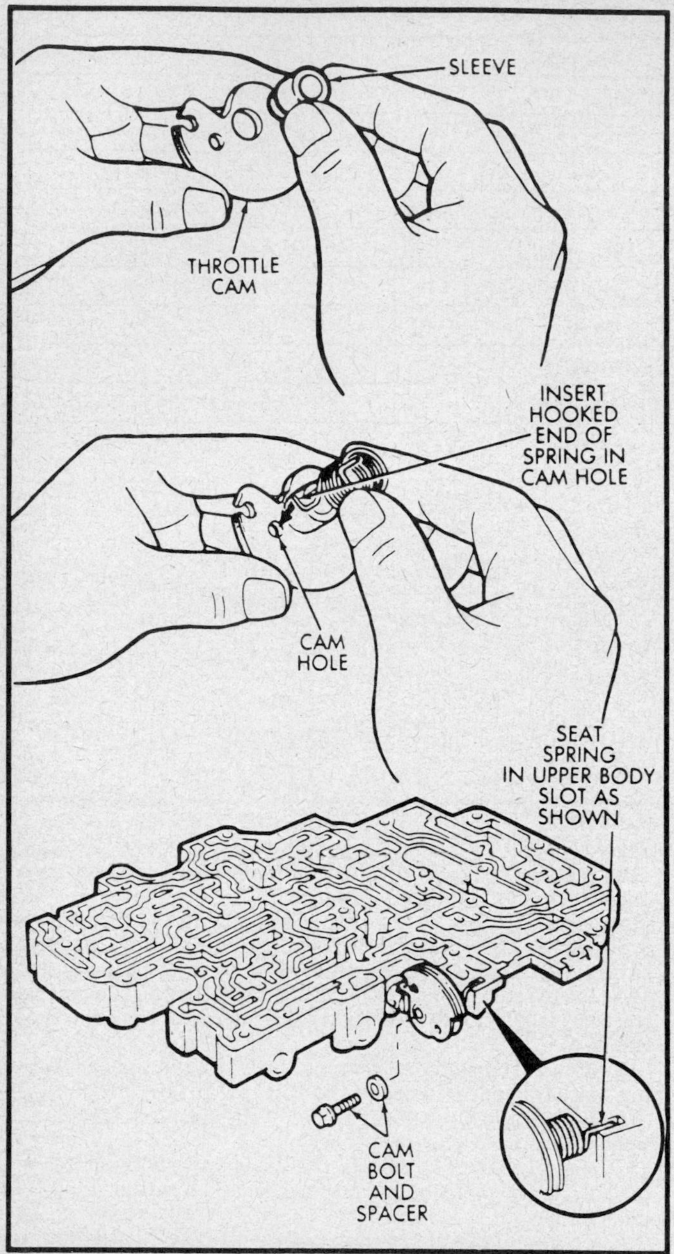
SLEEVE
THROTTLE CAM
INSERT HOOKED END OF SPRING IN CAM HOLE
CAM HOLE
SEAT SPRING IN UPPER BODY SLOT AS SHOWN
CAM BOLT AND SPACER

AW-4 transmission—throttle cam positioning

2. Verify correct thrust bearing and race installation during reassembly.
3. To install the rear planetary gear, second brake drum and output shaft, verify the No. 10 thrust bearing and race specification. The bearing and race outer diameter is 2.272 in. (57.7mm). The bearing and race inside diameter is 1.543 in. (39.2mm).
4. Coat the thrust bearing and race assembly with petroleum jelly and install it in the transmission case. Be sure that the race faces down and the bearing rollers face up.
5. Align the teeth of the second brake drum and the clutch pack. Align the rear planetary output shaft assembly teeth with the tranmsission case slots and install assembly in the tranmsission case.
6. Install the rear planetary snapring using tool BVIFM-29, or equivalent. The chamfered side of the snapring must face up and toward the front of the transmission case.
7. Check the first/reverse brake pack clearance, using a feeler gauge. Clearance should be, 0.024–0.069 in (0.60–1.74mm) for transmissions used in vehicles equipped with a 4 cylinder engine. Clearance should be 0.028–0.079 in. (0.070–2.00mm) for transmissions used in vehicles equipped with a 6 cylinder engine. If clearance is incorrect the planetary assembly, thrust bearing or snapring is not properly seated inside the case. Remove and reinstall components, as required.
8. Install the second brake piston sleeve. Be sure that the sleeve lip faces up and toward the front of the transmission case.
9. Install the second brake drum gasket using tool BVIFM-33 or equivalent. The gasket depth is 1.720 in. (43.7mm). Install the park lock pawl, spring and pin. Connect the park lock rod to the manual valve and the shift sector.
10. Position the park lock rod bracket on the transmission case and torque the bracket retaining bolts to 7 ft. lbs. (10 Nm).
11. Check the park lock operation. Move the shift sector to the **P** detent. The park pawl should be firmly locked in the planetary ring gear.
12. Install the No. 1 one-way clutch. Be sure that the short flanged side of the clutch faces up and toward the front of the transmission case.
13. Install the second brake pack. Install the discs then the plates. Install the second brake pack retainer with the rounded edge of the retainer facing the disc.
14. Install the second brake pack snapring. Check the brake pack clearance using a feeler gauge. The clearance should be, 0.035–0.084 in (0.89–2.15mm) for transmissions used in vehicles equipped with a 4 cylinder engine and 0.024–0.078 in. (.062–1.98mm) for transmissions used in vehicles equipped with a 6 cylinder engine. If the brake pack clearance is not within specification the brake pack components are not seated. Reassemble the brake pack, as required.
15. Install the planetary sun gear and the input drum. Be sure that the drum thrust washer tabs are seated in the drum. Use petroleum jelly to hold the thrust washer in position, as necessary.
16. Install the front planetary gear on the sun gear. Support the output shaft with wood blocks. Install the planetary snapring on the sun gear, using tool BVIFM-30 or equivalent.
17. Install the tabbed thrust race on the front planetary gear. The washer tabs face down and toward the gear. The race outer diameter is 1.882 in. (47.8mm). the race inner diameter is 1.350 in. (34.3mm).
18. Install the second coast brake band. Install the pin in the second coast brake band. Install the retaining ring on the pin.
19. Install the thrust bearing and race in the forward direct clutch. Coat the bearing and bearing race with petroleum jelly to retain them in place.
20. Check the forward direct clutch thrust bearing size. The race outer diameter is 1.925 in. (48.9mm). The race innner diameter is 1.024 in. (26.0mm). the bearing outer diameter is 1.839 in. (46.7mm). the bearing inner diameter is 1.024 in. (26.0mm).
21. Coat the front of the planetary ring gear race with petroluem jelly and install it in the ring gear.
22. Check the ring gear race size. The outer diameter is 1.850 in. (47.0mm). The inner diameter is 1.045 in. (26.5mm).
23. Align the forward direct clutch disc splines, using the proper tool. Align and install the front planetary ring gear in the forward direct clutch.
24. Coat the bearing and race with petroleum jelly and install them in the ring gear. Check the bearing and bearing race size. The bearing outer diameter is 1.878 in. (47.7mm). The bearing inner diameter is 1.283 in. (32.6mm). The race outer diameter is 2.110 in. (53.6mm). The race inner diameter is 1.205 in. (30.6mm).
25. Rotate the front of the transmission case downward and install the assembled planetary gear forward direct clutch assembly.

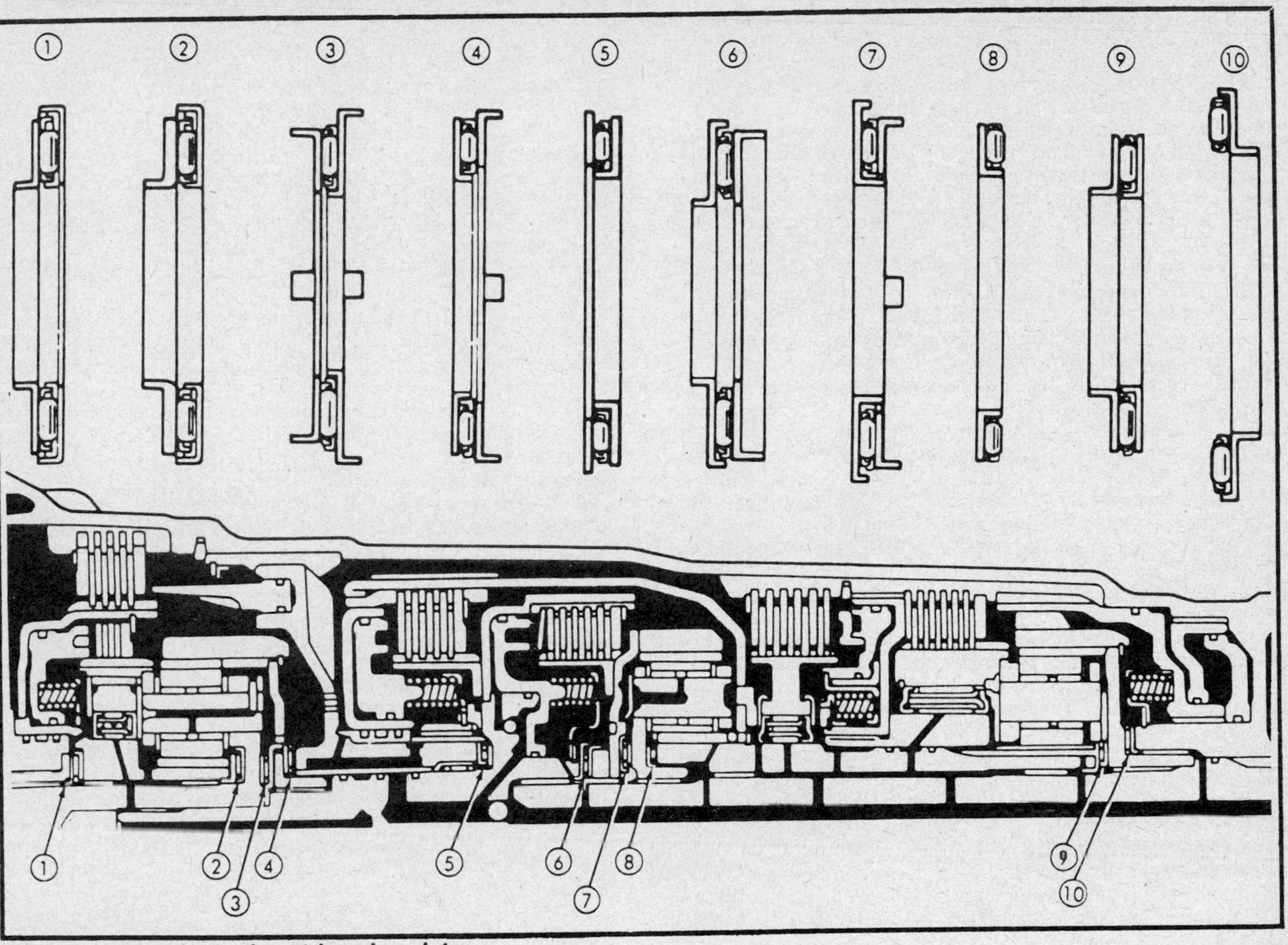
1
2
3
4
5
6
7
8
9
10
1
2
3
4
5
6
7
8
9
10

AW-4 transmission—thrust bearing data

26. Check the clearance between the sun gear input drum and the direct clutch drum. The clearance should be 0.3860–0.4654 in. (9.8–11.8mm). If the clearance is incorrect the planetary gear forward direct clutch assembly is not seated or it is improperly assembled. Remove the assembly and correct, as required.

27. Coat the thrust bearing and the race assembly with petroleum jelly and install it on the clutch shaft. Be sure that the bearing faces up and toward the front of the transmission case. Check the bearing and bearing race size. The bearing and race outer diameter is 1.882 in. (47.8mm). The bearing and race inner diameter is 1.301 in. (33.6mm).

28. Assemble the second coast brake piston components. Install the assembled second coast brake piston in the transmission case.

29. Install the replacement seals on the second coast brake piston cover. Install the cover in the transmission case.

30. Install the second coast brake piston snapring, using tool BVIFM-29 or equivalent.

31. To check the second coast brake piston stroke, make a reference mark on the brake piston rod. Apply 57–114 psi of compressed air through the feed hole. Alternately apply and release the air pressure in order to operate the piston.

32. Check the stroke, using the proper gauge tool. Use gauge tool BVIFM40 or equivalent for transmissions used in vehicles wquipped with a 4 cylinder engine and use gauge tool BVIFM-41 or equivalent for transmissions used in vehicles equipped with a 6 cylinder engine.

33. If the stroke length is incorrect, the piston, cover or snapring is not seated properly. Reassemble these components and recheck the stroke, as required.

34. Coat the thrust race and tabbed washer with petroleum jelly and install them on overdrive support. Check the race size. the race outer diameter is 2.004 in. (50.9mm). The race inner diameter is 1.426 in. (36.2mm).

35. Install the overdrive support in the transmission case. Use both of the long bolts to help align and guide the support into position.

36. Install the overdrive support snapring using tool BVIFM-29 or equivalent. Be sure that the chamfered side of the snapring faces up and toward the front of the transmission case. The snapring ends must be aligned with the case opening and with the ring ends approximately 0.94 in. from the centerline of the case opening.

37. Install and torque the overdrive support bolts to 19 ft. lbs. (25 Nm). Using a dial indicator gauge check the output shaft endplay. The endplay should be 0.0106–0.0339 in. (0.27–0.86mm).

38. If the output shaft endplay is incorrect, one or more of the installed components is not seated properly. Reassemble the components, as required and recheck the endplay.

39. Install the overdrive clutch pack. Be sure to install the thickest clutch plate first and check that the rounded edge of the plate faces up. Install the first disc followed by a plate until the correct number of discs and plates are used. Install 4 discs and 3

plates in transmissions that are used in vehicles equipped with a 6 cylinder engine. Install 3 discs and 2 plates in transmissions that are used in vehicles equipped with a 4 cylinder engine.

40. Install the stepped ring retainer plate with the flat side facing the disc. Install the brake pack snapring.

41. To check the overdrive brake piston stroke, install tool BVIFM-35 or equivalent, in the dial indicator and position the gauge tool against the overdrive brake piston.

42. Using compressed air, apply and release the brake piston. Apply air pressure through the feed hole in the transmission case.

43. The piston stroke length should be 0.052–0.063 in. (1.32–1.62mm) for transmissions used in vehicles equipped with a 4 cylinder engine and 0.55–0.66 in. (1.40–1.70mm) for transmissions used in vehicles equipped with a 6 cylinder engine.

44. If the stroke is incorrect, the brake pack or piston is installed incorrectly. Recheck and correct as necessary and measure piston stroke again.

45. Remove the dial indicator and gauge tool. Remove the overdrive brake piston snapring and remove overdrive clutch pack components.

46. Coat the overdrive lower race, thrust bearing and upper race with petroleum jelly and install them in overdrive support. Be sure that the races and bearing are assembled and installed properly.

47. Check the bearing and bearing race sizes. The outer diameter of the lower race is 1.882 in. (47.8mm). the inner diameter of the lower race is 1.350 in. (34.3mm). The outer diameter of the bearing is 1.878 in. (47.7mm). The inner diameter of the bearing is 1.287 in. (32.7mm). The outer diameter of the upper race is 1.882 in. (47.8mm). The inner diameter of the upper race is 1.209 in. (30.7mm).

48. Install the overdrive planetary ring gear in the support. Coat the ring gear thrust race and the thrust bearing assembly with petorleum jelly and install them in the gear.

49. Check the bearing and the bearing race size. The outer diameter of the ring gear race bearing is 1.882 in. (47.8mm). The inner diameter of the ring gear race bearing is 0.953 in. (24.2mm). The outer diameter of the bearing is 1.844 in. (46.8mm). The inner diameter of the bearing is 1.024 in. (26.0mm).

ACCUMULATOR COMPONENT SELECTION

Item	Component	Diameter in. (mm)	Length in. (mm)
Second brake accumulator	Pin A	0.472 (12.0)	1.386 (35.2)
	Spring B, 4 cyl.	0.748 (19.0)	1.653 (42.0)
	Spring B, 6 cyl.	0.764 (19.4)	1.496 (38.0)
	Piston C	1.453 (36.9)	2.697 (68.5)
	Spring D, 4 cyl.	0.772 (19.6)	2.106 (53.5)
	Spring D, 6 cyl.	0.775 (19.7)	2.106 (53.5)
Direct clutch accumulator	Pin E	0.539 (13.7)	1.307 (33.2)
	Spring F, 4 cyl.	0.819 (20.8)	1.535 (39.0)
	Spring F, 6 cyl.	0.831 (21.1)	1.433 (36.4)
	Piston G	1.453 (36.9)	2.465 (62.6)
	Spring H	0.799 (20.3)	1.893 (48.1)
Overdrive brake accumulator	Piston I	1.256 (31.9)	2.047 (52.0)
	Spring J	0.626 (15.9)	2.598 (66.0)
Overdrive clutch accumulator	Spring K	0.551 (14.0)	1.811 (46.0)
	Spring L	0.799 (20.3)	2.937 (74.6)
	Piston M	1.177 (29.9)	1.929 (49.0)

50. Coat the tabbed thrust race with petroleum jelly and install it on the planetary gear. The race outer diameter is 1.646 in. (41.8mm). The race inner diameter is 1.067 in. (27.1mm).

51. Install the assembled overdrive planetary gear and clutch.

52. Coat the thrust bearing and the race assembly with petroleum jelly. Install it on the clutch input shaft. The bearing and race outer diameter is 1.976 in. (50.2mm). The bearing and race inner diameter is 1.138 in. (28.9mm).

53. To install the overdrive brake pack, install the 0.157 in. (4.0mm) thick plate first. Be sure that the rounded edge of the plate faces upward.

54. Install a disc followed by a plate until the required number of discs and plates are installed. Be sure to install the stepped plate last with the plate side of the plate facing the disc.

55. Install 4 discs and 3 plates in vehicle transmissions equipped with a 6 cylinder engine. Install 3 discs and 2 plates in vehicle transmissions equipped with a 4 cylinder engine.

56. Install the clutch pack snapring. Coat the thrust bearing race with petroleum jelly and install it in the oil pump. The bearing race outer diameter is 1.858 in. (47.2mm). The bearing race inner diameter is 1.106 in. (28.1mm).

57. Lubricate and install the replacement O-ring on the oil pump body. Install the oil pump in the transmission case. Align the pump and the case bolt holes and carefully ease the pump into position.

NOTE: Do not use force to seat the pump. The seal rings on the stator shaft could be damaged if they bind or stick to the direct clucth drum.

58. Torque the oil pump bolts to 16 ft. lbs. (22 Nm). Check the input shaft rotation. The shaft should rotate smoothly and not bind.

59. Lubricate and install a new O-ring on the throttle cable adapter and install the cable in the transmission case.

60. Check the clutch and brake operation. Operate the clutches and brakes with compressed air applied through the feed holes in the transmission case. Listen for clutch and brake application. If application is not heard, diassemble the transmission and repair the fault.

NOTE: It is necessary to block the overdrive clutch accumulator feed hole No. 8 in order to check the direct clutch operation.

61. Lubricate and install new O-rings on the accumulator pistons. Assemble and install the accumulator piston components.

62. Install a new check ball body and spring. Position the valve body assembly on the transmission case. Install the detent spring.

63. Align the manual valve, detent spring and shift sector.

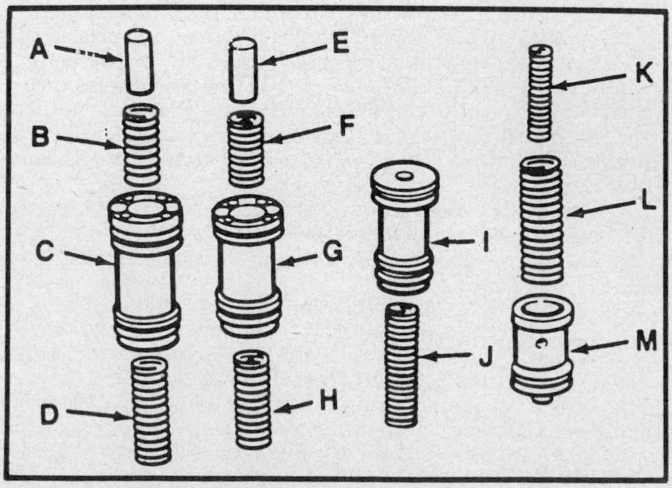
A
B
C
D
E
F
G
H
I
J
K
L
M

Accumulator pistion identification

Connect the throttle cable to the throttle valve cam. Install and torque the valve body to transmission case bolts to 7 ft. lbs. (10 Nm).

64. Connect the valve body solenoid wires to the solenoids. Install a new O-ring on the solenoid harness adapter and secure the adapter to the transmission case.

65. Install the valve body oil tubes. Tap the tubes into place using a plastic mallet. Be sure that the flanged tube ends and straight tubes ends are installed correctly..

66. Install new gaskets on the oil screen and install the screen on the valve body. Torque the screen bolts to 7 ft. lbs. (10 Nm).

67. Install a magnet in the oil pan. Be sure that the magnet does not interfere with valve body oil tubes.

68. Apply RTV sealer, to the gasket surface of the oil pan. The sealer bead should be 0.04 in. (1mm) wide. Install a new gasket on the oil pan. Install the oil pan. Torque the pan bolts to 65 inch lbs. (7.4 Nm).

69. Install the speed sensor rotor and key on the output shaft. Install the spacer and speedometer drive gear on the output shaft. Install the retaining snapring.

70. Install the spacer and the speedometer drive gear on the output shaft. Install the retaining snapring.

71. Apply a bead of sealer to the sealing surface at rear of the transmission case. Install the extension housing (2WD vehicles) or adapter housing (4WD vehicles). Torque the retaining bolts to 25 ft. lbs. (34 Nm).

72. Install the speed sensor. Torque the sensor bolt to 65 inch lbs. (7.4 Nm) Connect the sensor wire harness connector.

73. Install the speedometer driven gear. Torque the gear retaining bolt to 175 inch lbs. (19 Nm).

74. Install the torque converter housing. Torque the 12mm diameter housing bolts to 42 ft. lbs. (57 Nm). Torque the 10mm diameter housing bolts to 25 ft. lbs. (34 Nm).

75. Install the transmission shift control lever on the manual valve shaft. Do not install the lever retaining nut at this time.

76. Move the shift control lever all the way to the rear. Then move it 2 detent positions forward.

77. Install the neutral saftey switch on the manual valve shaft and tighten switch adjusting bolt just enough to keep the switch from moving.

78. Install the neutral safety switch tabbed washer and retaining nut. Torque the nut to 61 inch lbs. (6.9 Nm), but do not bend any of the washer tabs against the nut.

79. Align the neutral safety switch standard line with the groove or flat on the manual shaft. Torque the neutral safety switch adjusting bolt to 9 ft. lbs. (13 Nm)

80. Install the shift control lever on the manual valve shaft. Torque the lever retaining nut to 12 ft. lbs. (16 Nm).

81. Install the retaining clamp for the wire harness and the throttle cable. Install the torque converter.

82. Check that the converter is seated by measuring the distance between the torque converter housing flange and one of the torque converter mounting pads.

83. Use a straightedge and vernier calipers to measure this distance. On vehicles equipped with a 4 cylinder engine, the distance should be 0.689 in. (17.5 mm). On vehicles equipped with a 6 cylinder engine, the distance should be 0.650 in. (16.5mm).

84. Install the lower half of the transmission fill tube. Install the upper half of the transmission fill tube after the transmission is installed in the vehicle.

SPECIFICATIONS

TORQUE SPECIFICATIONS

Component	Service Set-to-torque ft. lbs.	Service Set-to-torque Nm	Service Recheck Torque ft. lbs.	Service Recheck Torque Nm
Converter housing bolt				
10mm	25	34	23–27	32–36
12mm	42	57	40–43	55–59
Extension housing bolt	25	34	23–27	32–36
Speed sensor bolt	65①	7.4	57–75①	6.4–8.4
Speedometer housing bolt	175①	19	160–185①	18–20
Shift lever nut	12	16	11–13	15–17
Neutral safety switch bolt	9	13	8–10	12–14
Neutral safety switch nut	61①	6.9	53–70	5.9–7.9
Solenoid harness bolt	65①	7.4	57–75	6.4–8.4
Oil pan bolts	65①	7.4	57–75	6.4–8.4
Oil pan drain plugs	15	20	14–16	19–21
Oil screen bolt	88①	10	80–96①	9–11
Valve body bolt (to case)	88①	10	80–96	9–11
Valve body bolt (to valve body)	56①	6.4	54–58①	6–6.8
Detent spring bolt	88①	10	80–96①	9–11
Oil pump bolt (to case)	17	22	16–18	21–23

TORQUE SPECIFICATIONS

Component	Service Set-to-torque ft. lbs.	Service Set-to-torque Nm	Service Recheck Torque ft. lbs.	Service Recheck Torque Nm
Oil pump bolt (to stator shaft)	88①	10	80–96①	9–11
OD support bolt (to case)	19	25	18–20	23–27
Park pawl bracket	88①	10	80–96①	9–11

① Measurement in inch lbs.

BUSHING SPECIFICATIONS

Bushing Location	Maximum Allowance Inside Diameter in.	Maximum Allowance Inside Diameter mm
Extension housing	1.4996	38.09
Direct clutch drum	2.1248	53.97
Overdrive planetary gear	0.4437	11.27
Overdrive direct clutch drum	1.0673	21.11
Front startor shaft	0.8496	21.58
Rear startor shaft	1.0661	27.08
Oil pump body	1.5035	38.19
Transmission case	1.5031	38.18

ENDPLAY AND CLEARANCE SPECIFICATIONS

Component	Engine	Specification in.	Specification mm
Output shaft endplay		0.0106–0.0339	0.27–0.86
1st/Reverse brake pack clearance	6 cyl.	0.0280–0.0790	0.70–2.00
	4 cyl.	0.0240–0.0690	0.60–1.74
Second brake pack clearance	6 cyl.	0.0350–0.0840	0.89–2.15
	4 cyl.	0.0240–0.0780	0.62–1.98
Clutch discs All except 1st/reverse and forward		0.00724	1.84

Component	Engine	Specification in.	Specification mm
Forward clutch disc	6 cyl.	0.0594	1.51
	4 cyl.	0.0724	1.84
6 cyl. direct clutch plates	Thin (1)	0.905	2.3
	Thick (3)	0.118	3.0
4 cyl. Direct clutch plates	Thin (1)	0.118	3.0
	Thick (2)	0.1574	4.0
Forward clutch plate	6 cyl.	0.070	1.8
	4 cyl.	0.078	2.0
1st/reverse brake disc		0.0594	1.51

SPECIAL TOOLS

TOOL REF. NUMBER	DESCRIPTION
B.Vi. FM 25	PUMP PULLER
B.Vi. FM 26	OVERDRIVE SPRING COMPRESSOR
B.Vi. FM 27	PISTON SPRING COMPRESSOR
B.Vi. FM 28	PISTON SPRING COMPRESSOR
B.Vi. FM 29	SNAP RING PLIERS
B.Vi. FM 30	SNAP RING PLIERS
B.Vi. FM 31	BRAKE SLEEVE PULLER
B.Vi. FM 32	PISTON PULLER
B.Vi. FM 33	SEAL INSTALLER
B.Vi. FM 34	SEAL INSTALLER
B.Vi. FM 35	GAUGE
B.Vi. FM 36	CLUTCH TEST TOOL (CONVERTER)
B.Vi. FM 37	CLUTCH TEST TOOL (CONVERTER)
B.Vi. FM 38	SEAL INSTALLER
B.Vi. FM 39	SEAL REMOVER
B.Vi. FM 40	1.5 mm WIRE GAUGE
B.Vi. FM 41	3.0 mm WIRE GAUGE
B.Vi. KM 01	SEAL INSTALLER 2WD
J-29184	SEAL INSTALLER 4WD

B.Vi. FM 35

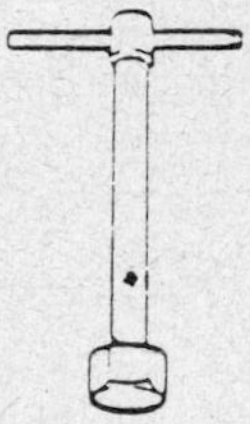
B.Vi. FM 36

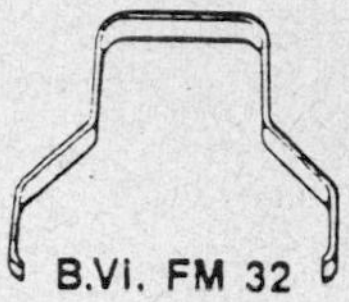
B.Vi. FM 32

B.Vi. FM 37

B.Vi. FM 40

B.Vi. FM 41

B.Vi. FM 31

B.Vi. FM 27

B.Vi. FM 38

B.Vi. KM.01 OR J-29184

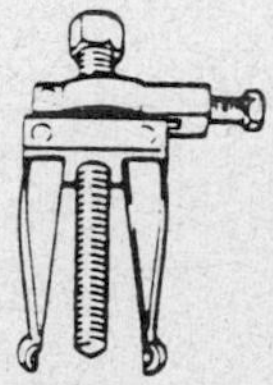
B.Vi. FM 39

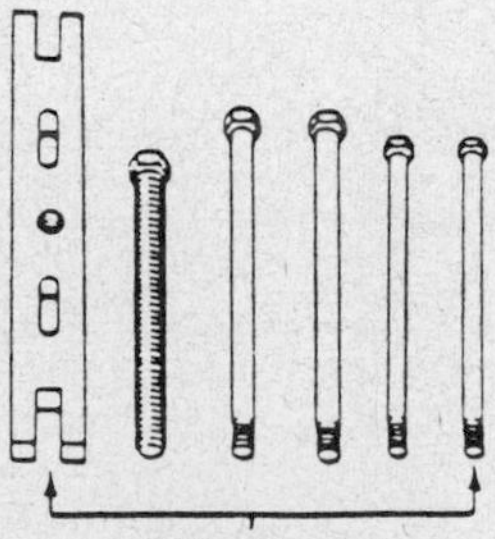
B.Vi. FM 25

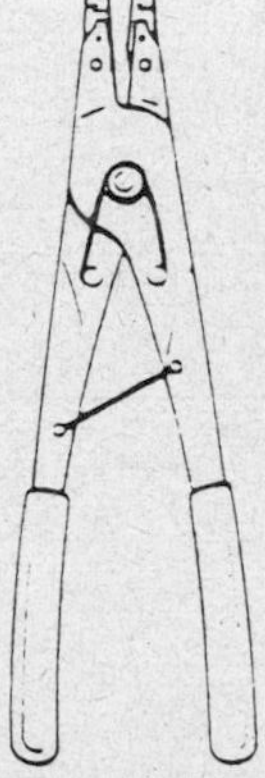
B.Vi. FM 29

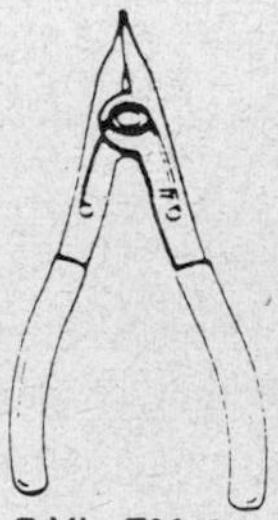
B.Vi. FM 30

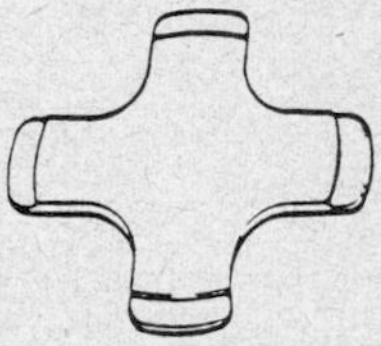
B.Vi. FM 26

B.Vi. FM 33

B.Vi. FM 34

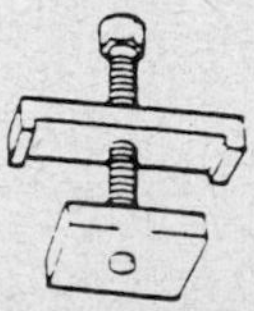
B.Vi. FM 28

Section 4

A904, A999 and A727 Transmissions Chrysler Corp.

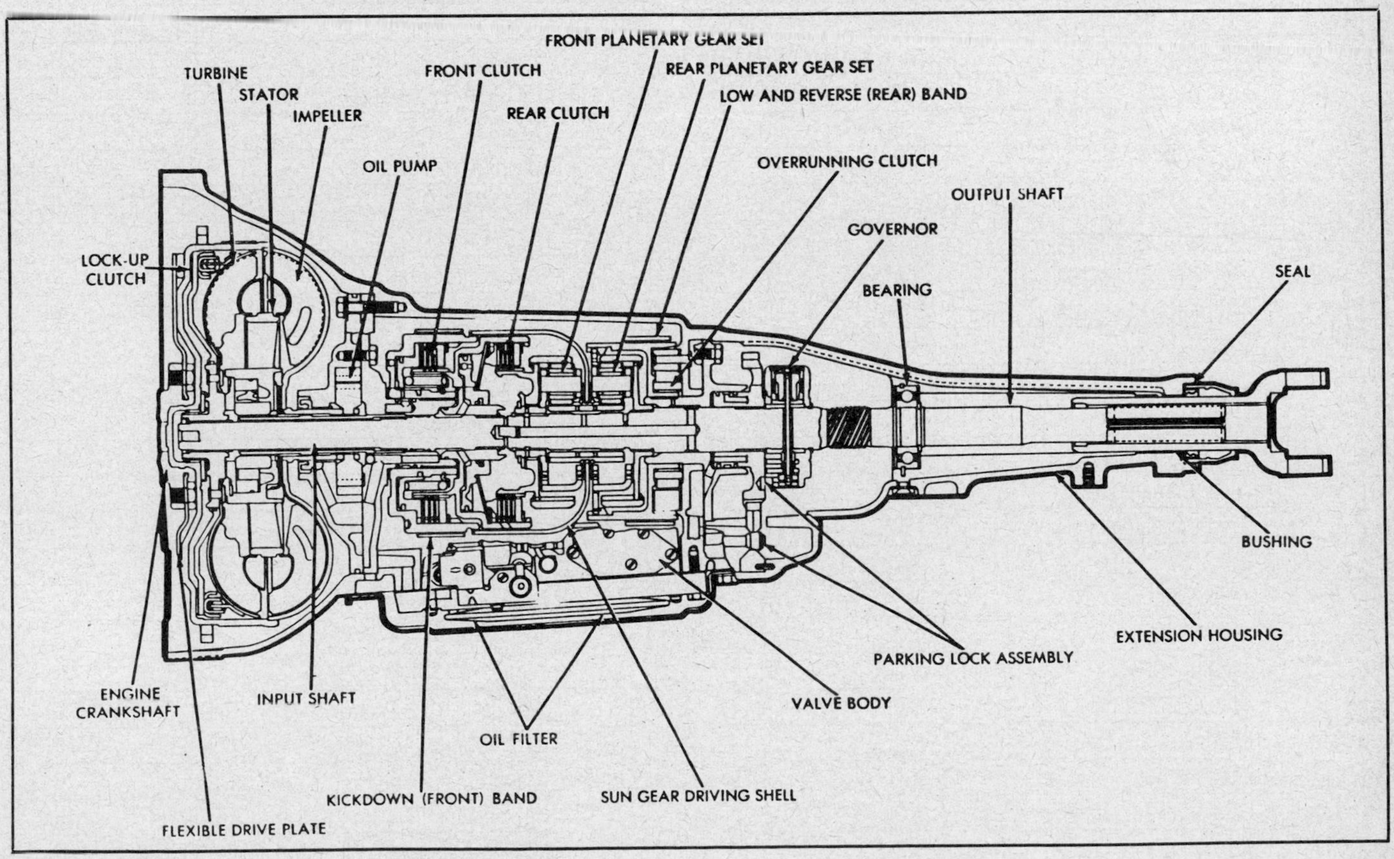

FRONT PLANETARY GEAR SET
TURBINE
FRONT CLUTCH
REAR PLANETARY GEAR SET
STATOR
LOW AND REVERSE (REAR) BAND
IMPELLER
REAR CLUTCH
OIL PUMP
OVERRUNNING CLUTCH
OUTPUT SHAFT
GOVERNOR
LOCK-UP CLUTCH
SEAL
BEARING
BUSHING
EXTENSION HOUSING
PARKING LOCK ASSEMBLY
ENGINE CRANKSHAFT
INPUT SHAFT
VALVE BODY
OIL FILTER
KICKDOWN (FRONT) BAND
SUN GEAR DRIVING SHELL
FLEXIBLE DRIVE PLATE

Cut away view of the automatic transmission

REFERENCES

APPLICATION

1974–80 Chilton's Professional Automatic Transmission Manual #6927
1984–89 Chilton's Professional Transmission Manual—Domestic Vehicles #7959

GENERAL DESCRIPTION

1974–80 Chilton's Professional Automatic Transmission Manual #6927

MODIFICATIONS

1974–80 Chilton's Professional Automatic Transmission Manual #6927
1984–89 Chilton's Professional Transmission Manual—Domestic Vehicles #7959

TROUBLE DIAGNOSIS

1974–80 Chilton's Professional Automatic Transmission Manual #6927

ON CAR SERVICES

1974–80 Chilton's Professional Automatic Transmission Manual #6927

REMOVAL AND INSTALLATION

1974–80 Chilton's Professional Automatic Transmission Manual #6927
1984–89 Chilton's Professional Transmission Manual—Domestic Vehicles #7959

BENCH OVERHAUL

1974–80 Chilton's Professional Automatic Transmission Manual #6927

SPECIFICATIONS

1974–80 Chilton's Professional Automatic Transmission Manual #6927
1984–89 Chilton's Professional Transmission Manual—Domestic Vehicles #7959

SPECIAL TOOLS

1974–80 Chilton's Professional Automatic Transmission Manual #6927

APPLICATION

A-904, A-999 AND A-727

Year	Vehicle	Engine	Vin Code
1984	Diplomat	5.2L	G
	Fifth Avenue	5.2L	F
	Grand Fury	5.2L	B
	Newport	5.2L	F
	150-350 Ramcharger	3.7L	H
	150-350 Ramcharger	5.2L	T
	150-350 Ramcharger	5.9L	W
	150-350 Ramcharger	5.9L	I
	Jeep/CJ-7	4.2L	C
	Jeep/Cherokee	2.5L	B
	Jeep/Cherokee	2.8L	W
	Jeep/Comanche	2.5L	B
	Jeep/Comanche	2.8L	
	Jeep/Scrambler	4.2L	C
	Jeep/Wagoneer	4.2L	C
	Jeep/Wagoneer	5.9L	N
	Jeep/J-10 Truck	4.2L	C
	Jeep/J-10 Truck	5.9L	N
	Jeep/J-20 Truck	5.9L	N
1985	Diplomat	5.2L	G
	Fifth Avenue	5.2L	F
	Grand Fury	5.2L	B
	Newport	5.2L	F
1985	150-350 Ramcharger	3.7L	H
	150-350 Ramcharger	5.2L	T
	150-350 Ramcharger	5.9L	W
	150-350 Ramcharger	5.9L	I ①
	Jeep/CJ-7	4.2L	C
	Jeep/Cherokee	2.5L	U ③
	Jeep/Cherokee	2.5L	H ④
	Jeep/Cherokee	2.8L	W
	Jeep/Comanche	2.1L	B
	Jeep/Comanche	2.5L	U ③
	Jeep/Comanche	2.5L	H ④
	Jeep/Comanche	2.8L	W
	Jeep/Scrambler	4.2L	C
	Jeep/Wagoneer	4.2L	C
	Jeep/Wagoneer	5.9L	N
	Jeep/J-10 Truck	4.2L	C
	Jeep/J-10 Truck	5.9L	N
	Jeep/J-20 Truck	5.9L	N
1986	Diplomat	5.2L	G
	Fifth Avenue	5.2L	F
	Grand Fury	5.2L	B
	Newport	5.2L	F
	150-350 Ram Van/Wagon	3.7L	H

A-904, A-999 AND A-727

Year	Vehicle	Engine	Vin Code
1986	150-350 Ram Van/Wagon	5.2L	T
	150-350 Ram Van/Wagon	5.9L	V ②
	150-350 Ram Van/Wagon	5.9L	W
	150-350 Ram Van/Wagon	5.9L	I ①
	Jeep/CJ-7	4.2	C
	Jeep/Cherokee	2.5	U ③
	Jeep/Cherokee	2.5L	H ④
	Jeep/Cherokee	2.8L	W
	Jeep/Comanche	2.5L	U ③
	Jeep/Comanche	2.5L	H ④
	Jeep/Comanche	2.8L	W
	Jeep/Scrambler	4.2L	C
	Jeep/Wagoneer	4.2L	C
	Jeep/Wagoneer	5.9L	N
	Jeep/J-10 Truck	4.2L	C
	Jeep/J-10 Truck	5.9L	N
	Jeep/J-20 Truck	5.9L	N
1987	Diplomat	5.2L	G
	Fifth Avenue	5.2L	F
	Grand Fury	5.2L	B
	Newport	5.2L	F
	150-350 Ram Van/Wagon	3.7L	H
	150-350 Ram Van/Wagon	5.2L	T
	150-350 Ram Van Wagon	5.9L	V ②
	150-350 Ram Van/Wagon	5.9L	W
	150-350 Ram Van/Wagon	5.9L	I ①
	Jeep/Cherokee	2.5L	U ③
	Jeep/Cherokee	2.5L	H ④
	Jeep/Cherokee	2.8L	W
	Jeep/Comanche	2.5L	U ③
	Jeep/Comanche	2.5L	H ④
	Jeep/Comanche	4.0	M
	Jeep/Wagoneer	4.2L	C
	Jeep/Wagoneer	5.9L	N
	Jeep/Wrangler	4.2L	C
	Jeep/J-10 Truck	4.2L	C
	Jeep/J-10 Truck	5.9L	N
	Jeep/J-20 Truck	5.9L	N
1988	Diplomat	5.2L	G
	Fifth Avenue	5.2L	F
1988	Grand Fury	5.2L	B
	Newport	5.2L	F
	150-350 Ram Van/Wagon	3.9L	X
	150-350 Ram Van/Wagon	5.2L	Y
	150-350 Ram Van/Wagon	5.9L	W
	Jeep/Cherokee	2.5L	U ③
	Jeep/Cherokee	2.8L	H ④
	Jeep/Cherokee	2.8L	W
	Jeep/Comanche	2.5L	U ③
	Jeep/Comanche	2.5L	H ④
	Jeep/Comanche	4.0L	M
	Jeep/Wagoneer	4.2L	C
	Jeep/Wagoneer	5.9L	N
	Jeep/Wrangler	4.2L	C
	Jeep/J-10 Truck	4.2L	C
	Jeep/J-10 Truck	5.9L	N
	Jeep/J-20 Truck	5.9L	N
1989	Diplomat	5.2L	G
	Fifth Avenue	5.2L	F
	Grand Fury	5.2L	B
	Newport	5.2L	F
	150-350 Ram Charger	3.9L	X
	150-350 Ram Charger	5.2L	Y
	150-350 Ram Charger	5.9L	W
	150-350 Ram Charger	5.9L Diesel	8
	Jeep/Cherokee	2.5L	3
	Jeep/Cherokee	2.5L	H ④
	Jeep/Cherokee	2.5L	W
	Jeep/Comanche	2.5L	U ③
	Jeep/Comanche	2.5L	H ④
	Jeep/Comanche	4.0L	M
	Jeep/Wagoneer	4.2L	C
	Jeep/Wagoneer	5.9L	N
	Jeep/Wrangler	4.2L	C
	Jeep/J-10 Truck	4.2L	C
	Jeep/J-10 Truck	5.9L	N
	Jeep/J-20 Truck	5.9L	N

① California
② 2bbl
③ Carbureted
④ Throttle Body Injection (TBI)

TRANSMISSION MODIFICATIONS

Part Throttle Unlock

Beginning 1986, the part throttle unlock feature was incorporated in all 5.2L engines equipped with the A-999 transmission and all 3.7L engines equipped with the A-904 truck transmission with lockup feature. Performance is improved by unlocking the torque converter with application of heavy throttle speeds up to 50 mph. Lockup and unlock points are computer controlled by sensing road speed, engine vacuum and engine temperature.

Addition of a solenoid in the valve body assembly requires changes to the valve body, steel plate, transfer plate, throttle pressure adjusting bracket, case and reverse band. The following is a description of the parts that have been changed.

Case – a threaded hole was added in the rear face of the solenoid connector. Casting revisions in accumulator bore area to provide clearance for solenoid. The new cases can be used to service prior models by plugging the connector hole with a plug (part number 6028873). The new case number for the 318/A-999 combination is 4202767 (old number was 4130977). The new case number for the 225/A-904 combination is 4202768 (old number was 4202047).

Transfer plate – The hydraulic passages revised and bore added for the solenoid. Tapped hole for attaching the solenoid also added. New part used only with part throttle unlock assemblies.

Steel plates – Orifice added and provisions added for part throttle unlock solenoid. Used only with part throttle unlock.

Throttle Position Adjusting Bracket – Redesigned for added strength. Can be used in prior model valve body assemblies except those used in MMC transmissions. The new part number is 434882 and the old part number 3743840.

Valve body – Hydraulic passages revised to provide control circuit to part throttle unlock solenoid. New valve body is not interchangeable with old part. However, in conjunction with new steel plates, it will be used in regular lockup and no lockup valve bodies for production and service.

New Extension Housing

A new extension housing was released for use in the long wheelbase trucks on the A-904 and A-999 transmissions. The bearing access hole was rotated 90 degrees, flange thickness increased and ribs added to increase strength. The new heavy duty part can be used in place of the regular extension in trucks. The new part number is 4329883 and the old part number is 3681600.

New Transmission Information

As the A-727 transmission will not be used with the 3.7L truck engine starting in 1986, a new non lockup transmission was released for California usage. The valve body assembly requires a new steel plate part number 4348702 and uses lockup body (without valves) part number 4348863 as a stiffener instead of flat plate number 3410380.

Transmission Case Modification

The A-904 case, used in the XJ Jeep, was modified by adding a cast slot and 2 reamed holes for attaching the Renix timing probe. The new case will be used to service prior models. The new case part number is 4348603 and the old case part number is 4269738.

A-727 Transmission Case Design Change

The depth of 3 of the 14 $^{7}/_{16}$ holes at the bottom of the A engine bell has been increased to permit use of a longer strut bolts. The new case will service the prior models. The new case part number is 4377912 and the old case part number is 4130190.

Lockup Speed Change

The A-999 transmission lockup speed decreased in units used with the 2.2 axle ratio and the 5.2L engine. New valve body assemblies part number 4412488 are to replace part number 4295888 for this application. The transmission and valve bodies are not interchangeable.

REMOVAL AND INSTALLATION

TRANSMISSION REMOVAL

CHRYSLER CORPORATION

1. The transmission and torque converter must be removed as an assembly.

CAUTION

The drive plate is not meant to support the load of holding the converter if disengaged from the transmission and serious damage will result. Remove transmission and converter together.

2. For safety, remove the ground (negative) cable from the battery. Raise and support the vehicle safely. On vehicles equipped with a transfer case, the transfer case must be removed first.
3. It may be necessary to lower the exhaust system for easy transmission removal, lower the exhaust system by removing the bolts and nuts attaching the exhaust pipe(s) to the manifold. Remove and exhaust system support straps or braces and lower the exhaust system.
4. Remove the engine to transmission struts, if so equipped. Remove and plug the cooler lines at the transmission.
5. Remove the starter motor and cooler line bracket.
6. Remove the torque converter access cover. Loosen the oil pan bolts, tap the pan to break it loose and let the fluid drain into a suitable drain pan.
7. Reinstall the pan temporarily with a few of the pan bolts. Mark the torque converter and drive plate to aid in reassembly. The crankshaft flange bolt circle, inner and outer circle of the holes in the drive plate and the 4 tapped holes in the front face of the torque converter all have 1 hole offset so these parts will be installed in the original position. This maintains the balance of the engine and torque converter.
8. Rotate the engine clockwise with the socket wrench on the crankshaft bolt to position the bolts attaching the torque converter to the drive plate and remove the bolts.

9. Mark the parts for reassembley, then disconnect the driveshaft at the rear universal joint Carefully pull the shaft assembly out of the extension housing.
10. Disconnect the wire connector from the back-up lamp and neutral starting switch and lockup solenoid wiring connector. Disconnect the speedometer cable and remove the oil fill tube.
11. Disconnect the gear shift rod and torque shaft assembly from the transmission.

NOTE: When it is necessary to disassemble the linkage rods from levers that use a plastic grommet as retainers, the grommets should be replaced with new ones.

12. Use a suitable tool to force the rod from the grommet in the lever and cut away the old grommet. Use pliers to snap the new grommet into the lever and rod into the grommet.
13. Disconnect the throttle rod from the lever at the left side of the transmission. Remove the linkage bell crank from the transmission, if so equipped.
14. Install the engine support fixture tool C–3487–A or equivalent with frame hooks, so as to support the rear of the engine.
15. Raise the transmission slightly with a suitable transmission jack to relieve the load on the supports.
16. Remove the bolts supporting the transmission mount to the crossmember and crossmember to frame and remove the crossmember.
17. Remove all bell housing bolts. Carefully work the transmission and torque converter assembly rearward off the engine block dowels and disengage the torque converter hub from the end of the crankshaft.

NOTE: Attach a small C-clamp to the edge of the bell housing to hold the torque converter in place during transmission removal.

18. Lower the transmission and remove the assembly from under the vehicle. To remove the torque converter assembly, remove the C-clamp from the edge of the bell housing and carefully slide the assembly out the transmission.

CJ, SCRAMBLER, GRAND WAGONEER AND TRUCK

1. Disconnect the negative battery cable. Disconnect the fan shroud, if equipped. Disconnect the transmission oil fill tube top bracket.
2. Raise the vehicle and support it safely. Remove the inspection cover from the lower part of the converter housing.
3. Remove the oil filler tube and dipstick. Remove the starter.
4. Mark the driveshafts and yokes for position. Disconnect the driveshafts from the transfer case yokes. Secure the shafts to the frame with wire so they are out of the way.
5. On V8 engines, disconnect the exhaust pipes at the exhaust manifolds and remove them as required, to gain working clearance. Drain the transfer case lubricant and transmission fluid.
6. Disconnect the speedometer cable, gearshift linkage, throttle linkage and the wires to the neutral safety switch.
7. Mark the converter drive plate and converter for location reference. Remove the bolts that attach the converter to the drive plate.
8. Properly support the transmission and transfer case assembly on a suitable jack. Be sure the transmission assembly is firmly chained or secured for removal.
9. Remove the rear crossmember. Lower the transmission slightly and disconnect the oil cooler lines. Remove the bolts that mount the transmission to the engine.
10. Move the transmission and converter back and away from the engine. Make sure the converter breaks loose from the drive plate and is firmly mounted on the transmission.
11. Hold the converter in position and lower the transmission assembly until the converter housing clears the engine. Remove the transmission and transfer case assembly from the vehicle. Remove the transfer case from the transmission.

2WD CHEROKEE, WAGONEER AND COMANCHE

1. Disconnect the negative battery cable. Raise and support the vehicle safely.
2. Mark the rear propeller shaft and yoke for reassembly. Disconnect and remove the rear propeller shaft.
3. Remove the torque converter inspection cover. Mark the converter drive plate and converter assembly for reassembly.
4. Remove the bolts attaching the torque converter to the flex plate. Properly support the transmission assembly.
5. Remove the bolts attaching the rear crossmember to the transmission side rail. Disconnect the exhaust pipe at the catalytic converter.
6. Lower the transmission slightly in order to disconnect the fluid cooler lines.
7. Disconnect the backup light switch wire and the speedometer cable. Disconnect the transmission linkage.
8. Remove the bolts attaching the transmission assembly to the engine. Move the transmission assembly and the torque converter rearward to clear the crankshaft.
9. Carefully lower the transmission assembly from the vehicle.

4WD CHEROKEE, WAGONEER AND COMANCHE

1. Disconnect the negative battery cable. Raise and support the vehicle safely.
2. Mark the rear propeller shaft and yoke for reassembly. Disconnect and remove the rear propeller shaft.
3. Remove the torque converter inspection cover. Mark the converter drive plate and converter assembly for reassembly.
4. Remove the bolts attaching the torque converter to the flex plate. Properly support the transmission assembly.

NOTE: If the vehicle is equipped with a diesel engine, remove the left motor mount and starter in order to gain access to the torque converter drive plate bolts through the starter opening.

5. Remove the bolts attaching the rear crossmember to the transmission side rail. Disconnect the exhaust pipe at the catalytic converter.
6. Lower the transmission slightly in order to disconnect the fluid cooler lines. Mark the front propeller shaft assembly for reinstallation. Disconnect the propeller shaft at the transfer case and secure the assembly out of the way.
7. Disconnect the backup light switch wire and the speedometer cable. Disconnect the transfer case and the transmission linkage. Disconnect the vacuum lines and the vent hose.
8. Remove the bolts attaching the transmission assembly to the engine. Move the transmission assembly and the torque converter rearward to clear the crankshaft.
9. Carefully lower the transmission assembly from the vehicle. Remove the transfer case retaining bolts from the transmission assembly.

WRANGLER

1. Disconnect the negative battery cable. Remove the fan shroud retaining screws. Disconnect the transmission fill tube upper bracket.
2. Raise and support the vehicle safely. Remove the torque converter cover. Remove the transmission fill tube. As required, remove the starter.
3. Matchmark the front and rear driveshafts and remove them. Disconnect the transmission gearshift and throttle linkage.
4. Disconnect the transfer case vent and vacuum hoses and the neutral safety switch wires.
5. Matchmark the torque converter to drive plate. Remove the torque converter retaining bolts.
6. Support the transmission using the proper equipment. Remove the rear crossmember retaining bolts. Remove the crossmember from the vehicle.

7. Drain the transfer case lubricant. Disconnect the speedometer cable. Disconnect and plug the fluid lines at the transmission assembly. Disconnect the transfer case shift linkage.

8. Properly support the transmission and transfer case assembly. Properly support the engine under the rear main.

9. Remove the transmission to engine block retaining bolts. Carefully remove the transmission and the transfer case from the vehicle.

10. Separate the transmission from the transfer case.

11. Installation is the reverse of the removal procedure. Be sure to fill the transmission and the transfer case with the proper grade and type lubricant.

TRANSMISSION INSTALLATION

CHRYSLER CORPORATION

If the transmission was removed because of malfunction caused by sludge, accumulated friction material or metal particles, the oil cooler and lines must be flushed thoroughly. Also, once the transmission is out, the engine block can be inspected for leaking core hole plugs or oil gallery plugs that would otherwise be inaccessable. The transmission and converter must be installed as an assembly; otherwise, the converter drive plate, pump bushing and oil seal will be damaged.

1. The pump rotors must be aligned so that the converter will fully seat. Slide converter over the input and reaction shaft. Make sure the converter hub slots are engaging the pump inner rotor lugs as the converter is slid inward. Test for full engagement by placing a straightedge on the face of the case. Measure from the straightedge to the lug (square piece with tapped hole for drive plate) on the front cover. The dimension should be at least ½ in. when the converter is pushed all the way into the transmission. This is very important.

2. Attach a small C-clamp to the edge of the converter housing to hold the converter in place during installation. Check the converter drive plate for cracks or distortion and if necessary, replace it. Drive plate to crankshaft bolts should be torqued to 55 ft. lbs. and the torque should be checked even if bolts were not removed.

3. Coat the hole in the crankshaft that will receive the converter hub with Multi-Purpose grease. With transmission and converter assembly held securely on the jack, align as necessary with the engine.

4. Rotate the converter so that the mark that was made on the converter during removal will align with the mark on the drive plate. The offset holes in the plate are located next to the ⅛ in. hole in the inner circle of the plate. Carefully work transmission forward onto the block dowels with converter hub entering the crankshaft opening. Remove the C-clamp before pushing the assembly fully home.

NOTE: If the converter or drive plate has been replaced, or the marks made at removal are lost, examine crankshaft flange to see how the plate and converter must line up because of the offset holes. Then mark the plate and converter (paint, chalk) for assembly.

5. When the transmission is in position, install the converter housing bolts and torque then to 30 ft. lbs. (41 Nm) install the vibration dampener removed from the driveline tunnel, if so equipped.

6. Install the crossmember to the frame and lower the transmission to install the mount on the extension to the crossmember, tighten all bolts. Remove engine support fixture.

7. Using alignment marks made at disassembly, align speedometer adapter and install. Install oil fill tube.

8. Connect the throttle rod to the transmission lever. Connect the gearshift rod and torque the shaft assembly to the transmission lever and frame.

9. When it is necessary to disassemble the linkage rods from levers that use a plastic grommet as retainers, the grommets should be replaced with new ones. Use a suitable tool to force the rod from the grommet in the lever and cut away the old grommet. Use pliers to snap the new grommet into the lever and rod into the grommet.

10. Place the wire connector on the combination back-up and neutral/park starting switch. Connect the wiring to the lockup solenoid wiring connector at the rear of the transmission case.

11. Using a socket on the engine vibration damper, rotate engine clockwise as needed to install converter to drive plate bolts, making sure the marks made previously match up. Torque these bolts to 22 ft. lbs.

12. Install converter access cover, starter and hook up cooler lines.

13. Install the engine to transmission struts, if so equipped. Tighten the bolts holding the strut to transmission before the strut to engine bolts.

14. Reinstall the exhaust system if it was disturbed.

15. Carefully guide the sliding yoke into the extension housing and on the outout shaft splines. Align the marks made at removal and connect the driveshaft to the axle pinion shaft yoke.

16. Adjust the gearshift and throttle linkage.

17. Refill transmission with Mopar® ATF. Road test the vehicle.

CJ, SCRAMBLER, GRAND WAGONEER AND TRUCK

1. Hold the converter in position and raise the transmission and transfer case assembly until the converter housing meets the engine.

2. Install the bolts that mount the transmission to the engine. Install the oil cooler lines. Install the rear crossmember.

3. Install the bolts that attach the torque converter to the drive plate.

4. Reconnect the speedometer cable, gearshift linkage, throttle linkage and the wires to the neutral safety switch.

5. On V8 engines, reconnect the exhaust pipes at the exhaust manifolds.

6. Reconnect the driveshafts from the transfer case yokes.

7. Install the oil filler tube and dipstick. Install the starter.

8. Install the inspection cover to the lower part of the converter housing. Lower the vehicle.

9. Reconnect the negative battery cable. Reconnect the fan shroud, if equipped. Reconnect the transmission oil fill tube top bracket.

10. Refill the transmision with the proper automatic transmission fluid and refill the transfer case. Check all operations of the transmission, check for leaks and road test the vehicle.

2WD CHEROKEE, WAGONEER AND COMANCHE

1. Carefully raise the transmission assembly into the vehicle.

2. Install the bolts attaching the transmission assembly to the engine.

3. Reconnect the backup light switch wire and the speedometer cable. Reconnect the transmission linkage.

4. Install the transmission fluid cooler lines.

5. Install the bolts attaching the rear crossmember to the transmission side rail. Reconnect the exhaust pipe at the catalytic converter.

6. Install the bolts attaching the torque converter to the flex plate.

7. Install the torque converter inspection cover.

8. Install the rear propeller shaft.

9. Reconnect the negative battery cable. Lower the vehicle.

10. Refill the transmision with the proper automatic transmission fluid. Check all operations of the transmission, check for leaks and road test the vehicle.

4WD CHEROKEE, WAGONEER AND COMANCHE

1. Hold the converter in position and raise the transmission and transfer case assembly until the converter housing meets the engine.

2. Install the bolts that mount the transmission to the engine.

3. Reconnect the backup light switch wire and the speedometer cable. Reconnect the transfer case and the transmission linkage. Reconnect the vacuum lines and the vent hose.
4. Install the fluid cooler lines. Install the propeller shaft at the transfer case and secure the assembly with the retaining bolts.
5. Install the bolts attaching the rear crossmember to the transmission side rail. Reconnect the exhaust pipe at the catalytic converter.
7. Install the bolts attaching the torque converter to the flex plate.

NOTE: If the vehicle is equipped with a diesel engine, install the left motor mount and starter motor.

8. Install the torque converter inspection cover.
9. Install the rear propeller shaft. Reconnect the negative battery cable. lower the vehicle.
10. Refill the transmision with the proper automatic transmission fluid and refill the transfer case. Check all operations of the transmission, check for leaks and road test the vehicle.

WRANGLER

1. Hold the converter in position and raise the transmission and transfer case assembly until the converter housing meets the engine.
2. Install the bolts that mount the transmission to the engine.
3. Reconnect the speedometer cable. Reconnect the fluid lines at the transmission assembly. Reconnect the transfer case shift linkage.
4. Install the rear crossmember and the rear crossmember retaining bolts.
5. Install the torque converter retaining bolts.
6. Reconnect the transfer case vent and vacuum hoses and the neutral safety switch wires.
7. Install front and rear driveshafts. Reconnect the transmission gearshift and throttle linkage.
8. Install the torque converter cover. Install the transmission fill tube. Install the starter, if removed. Lower the vehicle.
9. Reconnect the negative battery cable. Install the fan shroud retaining screws. Install the transmission fill tube upper bracket.
10. Refill the transmision with the proper automatic transmission fluid and refill the transfer case. Check all operations of the transmission, check for leaks and road test the vehicle.

SPECIFICATIONS

TORQUE SPECIFICATIONS
904 Transmission

Component	Torque ft. lbs.	Torque Nm
Oil cooler line radiator fitting	110 ①	12
Oil cooler line flared fittings nut	175 ①	20
Converter drive plate to crankshaft bolts 4 cylinder	58	79
Converter drive plate to crankshaft bolts 6 cylinder	50	68
Converter drive plate to torque converter bolts 4 cylinder	40	54
Converter drive plate to torque converter bolts 6 cylinder	40	54
Adapter housing-to-transmission case bolt	24	33
Governor body bolt	100 ①	11
Front band adjusting screw locknut	35	47
Kickdown lever shaft plug	150 ①	17
Rear band adjusting screw locknut	35	47
Neutral starter switch	24	33
Oil filler tube bracket bolt	150 ①	17
Oil pan bolt	150 ①	17
Oil pump housing-to-transmission case bolt	175 ①	20
Output shaft support bolt	150 ①	17
Overrunning clutch cam setscrew	40 ①	4
Pressure test port plug	110 ①	12
Reaction shaft support to oil pump bolt	160 ①	18
Transmission-to-engine bolt	28	38

TORQUE SPECIFICATIONS
904 Transmission

Component	Torque ft. lbs.	Torque Nm
Valve body screw	35 ①	4
Valve body-to-transmission case screw	100 ①	11

① inch lbs.

TORQUE SPECIFICATIONS
999 Transmission

Component	ft. lbs.	Nm
Cooler line fitting	155 ①	18
Cooler line nut	85 ①	10
Converter drive plate to crankshaft bolt	55	75
Converter drive plate to torque converter bolt	270 ①	31
Extension housing to transmission case bolt	32	43
Extension housing to insulator mounting bolt	50	68
Governor body to support bolt	95 ①	11
Kickdown band adjusting screw lock nut	30	41
Kickdown lever shaft plug	150 ①	17
Lockup solenoid wiring connector	150 ①	17
Neutral starter switch	25	34
Oil pan bolt	150 ①	17
Oil pump housing to transmission case bolt	175 ①	20
Output shaft support bolt	150 ①	17
Overrunning clutch cam set screw	40 ①	5

TORQUE SPECIFICATIONS
999 Transmission

Component	ft. lbs.	Nm
Pressure test take-off plug	120 ①	14
Reaction shaft support to oil pump bolt	175 ①	20
Reverse band adjusting screw lock nut	25	34
Speedometer drive clamp screw	100 ①	11
Transmission to engine bolt	30	41
Valve body screw	35 ①	4
Valve body to transmission case bolt	105 ①	12

① Inch lbs.

TORQUE SPECIFICATIONS
727 Transmission

Component	inch lbs.	Nm
Cooler line fitting	160	18
Cooler line nut	150	17
Converter drive plate-to-crankshaft bolts	105 ①	142
Converter drive plate-to-torque converter bolts	26 ①	35
Adapter housing-to-transmission case bolt	24 ①	33
Governor body bolt	100	11
Front band adjusting screw locknut	35 ①	47
Kickdown lever shaft plug	150	17
Rear band adjusting screw locknut	35 ①	47
Neutral switch	24 ①	33
Oil filler tube bracket bolt	150	17
Oil pan bolt	150	17
Oil pump housing-to-transmission case bolt	175	20
Ouptut shaft support bolt	150	17
Overrunning clutch cam setscrew	40	4
Pressure test port plug	110	12
Reaction shaft support-to-oil pump bolt	160	18
Transmission-to-engine bolt	160	38
Valve body screw	35	4
Valve body-to-transmission case screw	100	11

① Ft. lbs.

THRUST WASHER SPECIFICATIONS

Item	(in.)	
A904/A999		
Engine 3.7L/5.2L		
Reaction shaft support thrust washer No. 1	.061–.063	
Rear clutch retainer thrust washer No. 2	.061–.063	
Input shaft thrust plate	.024–.026	
Output shaft thrust washer No. 3 selective	.052–.054	(Tin)
	.068–.070	(Red)
	.083–.085	(Green)
Output shaft thrust plate	—	
Front annulus thrust washer No. 4	.121–.125	
Front carrier (to annulus) thrust washer No. 5	.048–.050	
Drive shell (to front annulus) thrust washer	—	
Front carrier (to drive shell) thrust washer No. 6	.048–.050	
Sun gear drive shell thrust plate No. 7	.050–.052	
Sun gear drive shell thrust plate No. 8	.050–.052	
Rear carrier (to drive shell) thrust washer No. 9	.048–.050	
Rear carrier (to annulus) thrust plate	—	
Rear carrier (to annulus) thrust washer No. 10	.048–.050	
A727		
Engine—All		
Reaction shaft support thrust washer No. 1	.061–.063	(Natural)
	.084–.086	(Red)
	.102–.104	(Yellow)
Rear clutch retainer thrust washer No. 2	.061–.063	(Natural)
Output shaft thrust washer No. 3	.062–.064	
Output shaft thrust plate	.030–.032	
Front carrier (to annulus) thrust washer No. 4	.059–.062	
Drive shell (to front annulus) thrust washer No. 5	.059–.062	
Sun gear drive shell thrust plate No. 6	.034–.036	
Rear carrier (to drive shell) thrust No. 7	.059–.062	
Rear carrier (to annulus) thrust plate No. 8	.034–.036	

Section 4

A500 Transmission
Chrysler Corp.

APPLICATION

1988–89 Dodge Dakota

GENERAL DESCRIPTION

The front portion of the A-500 4 speed overdrive automatic transmission is a modified version of the A-998 3 speed Loadflite automatic transmission. The rear unit, or overdrive unit, replaces the extension housing and provides a overdrive gear with a gear ratio of 0.69–1.

The first 3 gear ratios of the A-500 have the same gear ratios and torque capacity that was offered in the 3 speed Loadflite transmission. The addition of the 4th gear overdrive to the Loadflite gives the added features of increased fuel economy , prolonged engine life and less engine noise at cruising speed.

The overdrive unit is designed to withstand up to 400 lbs. of torque. Because of the high loads and long periods of time that can be spent in overdrive, all the thrust bearings are needle bearings. The overdrive has added approximately 50 lbs. of weight and 6½ in. of length to the A-998 transmission.

The 4th gear (overdrive) is electronically controlled and hydraulically activated. A variety of sensor inputs are fed to the Single Module Engine Controller (SMEC) which controls a solenoid mounted on the valve body. The solenoid will energize and close a vent, allowing a 3–4 upshift. The SMEC also controls the operation of the lockup torque converter using many of the same sensor inputs.

A lockup torque converter, which is electronically controlled and hydraulically activated, will also be used with this transmission. 4th gear (overdrive) and lockup will only occur during certain conditions determined by the SMEC.

When the vehicle is traveling in 3rd gear over 25 mph, the SMEC uses the following information to allow the transmission to shift. The SMEC checks the coolant sensor signal for a 60° fahrenheit minimum temperature. It also checks the engine sensor speed, the vehicle speed sensor, the throttle position sensor and the MAP sensor.

The steering column shift selector remains at 6 positions. Overdrive will be engaged automatically in **D**. A separate overdrive **OFF** switch will be located on the instrument panel. This switch will override the SMEC and shift out of overdrive and prevent further shifts into overdrive. If the overdrive **OFF** switch is activated again the automatic operation is restored. The switch has an indicator light when the overdrive is turned off. The switch also resets on key-off so that the automatic overdrive feature is restored.

The use of fault codes aid in diagnosing the electronic components used to operate the overdrive and lockup torque converter. Other features in conjunction with this unit include.

1. The output shaft of the 3 speed section is now an intermediate shaft.
2. The 3 speed section rear drum is retained on the support with a snapring.
3. The governor and speedometer drive have been relocated to the rear of the output shaft.
4. The overdrive case now contains 2 output shaft bearings.
5. There are no rotating seal rings or pressurized oil for the overdrive and direct clutches in the overdrive housing. The governor is the only component receiving pressurized oil through the slip fit tubes. Pressurized oil for the overdrive lubrication circuit is supplied through the intermediate shaft.
6. Governor pressure and overdrive pressure taps are provided in the rear of the transmission case for in-vehicle transmission pressure testing.
7. The valve body is modified by adding several new valves. There is an overdrive solenoid, a 3–4 shift valve, a 3–4 timing valve, a 3–4 accumulator and a 3–4 shuttle valve. Once in 4th (overdrive) gear, the lockup solenoid, lockup valve and lockup timing valve accomplish the hydraulics to lock the converter turbine to the torque converter housing.
8. The direct drive and overdrive gear ratios are supplied by a 3rd planetary gear set, a direct clutch, an overdrive clutch and an overrunning clutch. A very strong spring, rated at up to 800 lbs. (5516 kPa), holds the sun gear to the annulus for direct drive. For coasting or reverse gear, power flows only through the direct clutch.
9. The lockup timing valve releases the torque converter to normal operation prior to the 4–3 downshift.
10. All closed throttle 3–4 upshifts will occur at 25–28 mph., regardless of the axle ratio.
11. All closed throttle 4–3 downshifts will occur at 25 mph., regardless of the axle ratio.
12. No 3–4 upshifts can be achieved, regardless of the vehicle speed, if the throttle opening is greater than 70% approximately.

Transmission and Converter Identification

TRANSMISSION

The 7 digit transmission part number is usually stamped on the left side of the transmission case just above the oil pan mating surface. This number is followed by a 4 digit code number which indicates the date of manufacture. The final 4 digit number group stamped on the transmission case represents the transmission serial number.

CONVERTER

Because the lockup converter is completely enclosed within the converter and cannot be seen, look for lockup converters to have an identifying decal attached to the front cover. The decal is circular in shape and states converter type and stall ratio such as **LOCKUP** and **LS** (Low Stall) or **HS** (High Stall).

NOTE: The torque converters no longer come equipped with drain plugs and therefore cannot be flushed if contaminated, only replaced.

Electronic Controls

The 4th gear (overdrive) is electronically controlled and hydraulically activated. A variety of sensor inputs are fed to the Single Module Engine Controller (SMEC) which controls a solenoid mounted on the valve body. The solenoid will energize and close a vent, allowing a 3–4 upshift. The SMEC also controls the operation of the lockup torque converter using many of the same sensor inputs.

A lockup torque converter, which is electronically controlled and hydraulically activated, will also be used with this transmission. 4th gear (overdrive) and lockup will only occur during certain conditions determined by the SMEC.

When the vehicle is traveling in 3rd gear over 25 mph, the SMEC uses the following information to allow the transmission to shift. The SMEC checks the coolant sensor signal for a 65°F minimum temperature. It also checks the engine sensor speed, the vehicle speed sensor, the throttle position sensor and the MAP sensor.

Th 25 mph speed limit is actually the lowest speed the transmission will stay in overdrive, during a coast down. The upshift into overdrive will be somewhat higher than 25 mph.

The engine speed is compared to the vehicle speed so that the SMEC can determine if the transmission is in 3rd gear . The SMEC must know this before it will allow the 3–4 shift to occur. This also prevents the possibility of a 2–4 downshift occuring. This signal comes from the ignition distributor.

The throttle position signal is compared to the engine speed signal to determine when to engage or disengage the overdrive unit. This signal comes from the throttle position sensor.

The SMEC must see all of the criteria stated above before it will energize the overdrive solenoid and go into overdrive.

The steering column shift selector remains at 6 positions. Overdrive will be engaged automatically in **D**. A separate overdrive **OFF** switch will be located on the instrument panel. This switch will override the SMEC and shift out of overdrive and prevent further shifts into overdrive. If the overdrive **OFF** switch is activated again the automatic operation is restored. The switch has an indicator light when the overdrive is turned off. The switch also resets on key-off so that the automatic overdrive feature is restored.

The use of fault codes help to diagnose the electronic components used to operate the overdrive and lockup torque converter.

Metric Fasteners

The metric fastener dimensions are very close to the dimensions of the familiar inch system fasteners and for this reason, replacement fasteners must have the same measurement and strength as those removed.

Do not attempt to interchange metric fasteners for inch system fasteners. Mismatched or incorrect fasteners can result in damage to the transmission unit through malfunctions, breakage or possible personal injury.

Care should be taken to reuse the fasteners in the same locations as removed.

Common metric fastener strength property classes are 9.8 and 12.9 with class identification embossed on the head of each bolt. The inch strength classes range from grade 2–8 with the line identification embossed on each bolt head. Markings on the bolt head correspond to 2 lines less than actual grade (for example grade 8 bolt will exhibit 6 embossed lines on the bolt head). Some metric nuts will be marked with a single digit strength identification numbers on the nut face.

Capacities

Chrysler products using lockup torque converters and models equipped with extra transmission coolers will all vary slightly in their capacity. Therefore, check fluid level carefully. The complete refill oil capacity of the A-998 transmission is 8.5 quarts (8.1 liters) and 10.5 quarts (9.6 liters) for the A-500 transmission.

Checking Automatic Transmission Fluid Level

The Chrysler type automatic transmission are designed to operate with fluid level at the **FULL** mark on the dipstick indicator. The fluid level can be checked with the transmission at normal operating temperature or with the transmission at room temperature. Do not overfill the transmission.

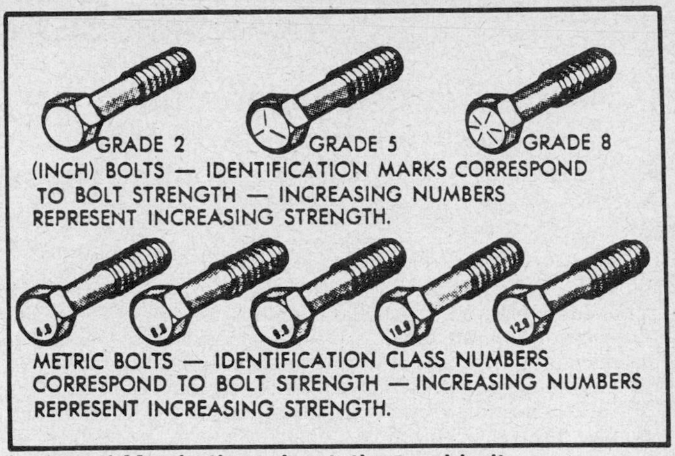
GRADE 2
GRADE 5
GRADE 8
(INCH) BOLTS — IDENTIFICATION MARKS CORRESPOND TO BOLT STRENGTH — INCREASING NUMBERS REPRESENT INCREASING STRENGTH.
METRIC BOLTS — IDENTIFICATION CLASS NUMBERS CORRESPOND TO BOLT STRENGTH — INCREASING NUMBERS REPRESENT INCREASING STRENGTH.

Inch and Metric thread notation and bolt identification

NOTE: When the automatic transmission fluid is replaced or topped off, MOPAR® ATF should be used. DEXRON® II type fluid should be used, only if the recommended fluid is not available.

TRANSMISSION AT NORMAL OPERATING TEMPERATURE

Approximately 180°F — Dipstick Hot To The Touch

1. With the vehicle on a level surface, engine idling, wheels blocked and the parking brake applied, move the selector lever through all the gear positions and return to the **N** position. Allow the engine to idle.
2. Clean the dipstick area of dirt and remove the dipstick from the filler tube. Wipe the dipstick indicator clean and reinsert the dipstick back into the filler tube and seat firmly.
3. Remove the dipstick from the filler tube again and check the fluid level as indicated on the dipstick indicator. The fluid level should be at the **FULL** mark.
4. If necessary, add fluid through the filler tube to bring the fluid level to its proper height.
5. When the fluid level is correct, fully seat the dipstick in the filler tube to avoid entrance of dirt or other foreign matter.

TRANSMISSION AT ROOM TEMPERATURE

Approximately 70°F — Dipstick Cool To The Touch

1. With the vehicle on a level surface, engine idling, wheels blocked and the parking brake applied, move the selector lever through all the gear positions and return to the **N** position. Allow the engine to idle.
2. Clean the dipstick area of dirt and remove the dipstick from the filler tube. Wipe the dipstick indicator clean and reinsert it back into the filler tube and seat firmly.
3. Remove the dipstick from the filler tube again and check the fluid level as indicated on the dipstick indicator. The fluid level should be at the **ADD ONE PINT** mark on the dipstick.
4. If necessary, add fluid through the filler tube to bring the fluid level to its proper height.
5. Reset the dipstick into the filler tube to avoid the entrance of dirt or foreign materials.
6. Upon bringing the transmission to its normal operating temperature, the fluid level should be at the **FULL** mark on the dipstick, due to the expansion of the fluid through heat.

TRANSMISSION MODIFICATIONS

Driveshaft Shudder During Acceleration From 0–10 MPH And Or A Vibration And Booming Noise At 35–50 MPH

The following modification is to be performed on 1987–89 trucks equipped with the A-998 transmission. Since the A-500 transmission is a modified A-998 transmission, this modification may apply to the A-500 transmission as well as the A-998 transmission. The vehicle may experience a driveshaft shudder during acceleration from 0–10 mph and or a vibration and booming noise at 35–50 mph. Repair or adjust the problem as follows:

1. Attach a suitable tachometer to the engine. Road test the vehicle to verify the condition.
2. The diagnosis procedure for acceleration shudder is performed from a standing start. Accelerate with medium to heavy acceleration from 0–15 mph while watching for the shudder. If the shudder is evident, note the engine rpm.
3. In the case of vibration and/or booming noise, road test the vehicle in the 35–55 mph range. If a vibration and/or booming noise is evident, note the engine rpm where the problem is most perceptible.
4. With the vehicle stopped and transmission in the **N** position, run the engine up through the same rpm ranges and note if the same condition is present.
5. If neither of these problems are present while running up the engine and either or both of the above conditions are have been verified, it will be necessary to remove the 2 piece driveshaft and install a single piece driveshaft as follows:
   a. Remove the front and rear driveshaft assembly.
   b. Unbolt and remove the driveshaft center bearing support cross member.
   c. Install a new 1 piece driveshaft and torque the U-joint strap bolts to 14 ft. lbs. (68 Nm).
   d. Road test the vehicle to verify repairs. The driveshaft required are as follows.
   e. A wheel base of 124 in. with a 7¼ rear axle requires driveshaft part number 4384625.
   f. A wheel base of 124 in. with a 8¼ rear axle requires driveshaft part number 4384605.

TROUBLE DIAGNOSIS

CLUTCH AND BAND APPLICATION
A-500 Overdrive Transmission

	A500			Transmission						Overdrive		
				Clutches				Bands		Clutches		
Lever Position	Over-Drive	Start Safety	Parking Sprag	Front	Rear	Running	Lockup	KD Front	Reverse/Rear	OD	Running	Direct
P-Park	—	Applied	Applied	—	—	—	—	—	—	—	—	—
R-Reverse	2.21	—	—	Applied	—	—	—	—	Applied	—	—	Applied
0-Drive												
First	2.74	—	—	—	Applied	Applied	—	—	—	—	Applied	Applied
Second	1.54	—	—	—	Applied			Applied	—	—	Applied	Applied
Third	1.00	—	—	Applied	Applied	—	—	—	—	—	Applied	Applied
2-Second												
First	2.74	—	—	—	Applied	Applied	—	—	—	—	Applied	Applied
Second	1.54	—	—	—	Applied	—	—	Applied	—	—	Applied	Applied
1-Low	2.74	—	—	—	Applied	Applied	—	—	Applied	—	Applied	Applied

CHILTON'S THREE C'S TRANSMISSION DIAGNOSIS PROCEDURE
A-500 Overdrive Transmission

Condition	Cause	Correction
Harsh engagement from N to D or R	a) Engine idle speed too high b) Valve body malfunction c) Hydraulic pressure too high d) Worn or faulty rear clutch	a) Adjust to specification b) Clean or overhaul c) Adjust to specification d) Overhaul rear clutch

CHILTON'S THREE C'S TRANSMISSION DIAGNOSIS PROCEDURE
A-500 Overdrive Transmission

Condition	Cause	Correction
Delayed engagement from N to D or R	a) Hydraulic pressure too low	a) Adjust to specification
	b) Valve body malfunction	b) Clean or overhaul
	c) Malfunction in low/reverse servo, band or linkage	c) Overhaul
	d) Low fluid level	d) Add as required
	e) Manual linkage adjustment	e) Adjust as required
	f) Oil filter clogged	f) Change filter and fluid
	g) Faulty oil pump	g) Overhaul pump
	h) Bad input shaft seals	h) Replace seal rings
	i) Idle speed too low	i) Adjust to specifications
	j) Bad reaction shaft support seals	j) Replace seal rings
	k) Bad front clutch	k) Overhaul
	l) Bad rear clutch	l) Overhaul
Runaway upshift	a) Hydraulic pressure too low	a) Adjust to specifications
	b) Valve body malfunction	b) Clean or overhaul
	c) Low fluid level	c) Add as required
	d) Oil filter clogged	d) Change filter and fluid
	e) Aerated fluid	e) Check for overfilling
	f) Manual linkage ajustment	f) Adjust as required
	g) Bad reaction shaft support seals	g) Replace seal rings
	h) Malfunction in kickdown servo, band or linkage	h) Overhaul
	i) Bad front clutch	i) Repair as needed
No upshift	a) Hydraulic pressure too low	a) Adjust to specifications
	b) Valve body malfunction	b) Clean or overhaul
	c) Low fluid level	c) Add as required
	d) Manual linkage adjustment	d) Adjust as required
	e) Incorrect throttle linkage adjustment	e) Adjust as required
	f) Bad seals on governor support	f) Replace seals
	g) Bad reaction shaft support seals	g) Replace seal rings
	h) Governor malfunction	h) Service or replace unit
	i) Malfunction in kickdown servo, band or linkage	i) Overhaul
	j) Bad front clutch	j) Overhaul
3-2 Kickdown runaway	a) Hydraulic pressure too low	a) Adjust to specifications
	b) Valve body malfunction	b) Clean or overhaul
	c) Low fluid level	c) Add as required
	d) Aerated fluid	d) Check for overfilling
	e) Incorrect throttle linkage adjustment	e) Adjust as reqiured
	f) Kickdown band out of adjustment	f) Adjust to specifications
	g) Bad reaction shaft support seals	g) Replace seal rings
	h) Malfunction in kickdown servo, band or linkage	h) Overhaul
	i) Bad front clutch	i) Overhaul
No kickdown or normal downshift	a) Valve body malfunction	a) Clean or overhaul
	b) Incorrect throttle linkage adjustment	b) Adjust as required
	c) Governor malfunction	c) Service or replace unit
	d) Malfunction in kickdown servo, band or linkage	d) Overhaul

CHILTON'S THREE C'S TRANSMISSION DIAGNOSIS PROCEDURE
A-500 Overdrive Transmission

Condition	Cause	Correction
Shifts erratic	a) Hydraulic pressure too low	a) Adjust to specifications
	b) Valve body malfunction	b) Clean or overhaul
	c) Low fluid level	c) Add as required
	d) Manual linkage adjustment	d) Adjust as required
	e) Oil filter clogged	e) Change filter and fluid
	f) Faulty oil pump	f) Overhaul oil pump
	g) Aerated fluid	g) Check for overfilling
	h) Incorrect throttle linkage adjustment	h) Adjust as required
	i) Bad seals on governor support	i) Replace seals
	j) Bad reaction shaft support seals	j) Replace seal rings
	k) Governor malfunction	k) Service or replace unit
	l) Malfunction in kickdown servo, band or linkage	l) Overhaul
	m) Bad front clutch	m) Overhaul
Slips in forward drive positions	a) Hydraulic pressure too low	a) Adjust to specifications
	b) Valve body malfunction	b) Clean or overhaul
	c) Low fluid level	c) Add as required
	d) Manual linkage adjustment	d) Adjust as required
	e) Oil filter clogged	e) Change filter and fluid
	f) Faulty oil pump	f) Overhaul pump
	g) Bad input shaft seals	g) Replace seal rings
	h) Aerated fluid	h) Check for overfilling
	i) Incorrect throttle linkage adjustment	i) Adjust as required
	j) Overrunning clutch not holding	j) Overhaul or replace
	k) Bad rear clutch	k) Overhaul
Slips in R only	a) Hydraulic pressure too low	a) Adjust as required
	b) Low/reverse band out of adjustment	b) Adjust to specifications
	c) Valve body malfunction	c) Clean or overhaul
	d) Malfunction in low/reverse servo, band or linkage	d) Overhaul
	e) Low fluid level	e) Add as required
	f) Manual linkage adjustment	f) Adjust as required
	g) Faulty oil pump	g) Overhaul pump
	h) Aerated fluid	h) Check for overfilling
	i) Bad reaction shaft support seals	i) Replace seal rings
	j) Bad front clutch	j) Overhaul
Slips in all positions	a) Hydraulic pressure too low	a) Adjust as required
	b) Valve body malfunction	b) Clean or overhaul
	c) Low fluid level	c) Add as required
	d) Oil filter clogged	d) Change fluid and filter
	e) Faulty oil pump	e) Overhaul pump
	f) Bad input shaft seals	f) Replace seal rings
	g) Aerated fluid	g) Check for overfilling
No drive in any position	a) Hydraulic pressure too low	a) Adjust to specifications
	b) Valve body malfunction	b) Clean or overhaul
	c) Low fluid level	c) Add as required
	d) Oil filter clogged	d) Change filter and fluid
	e) Faulty oil pump	e) Overhaul pump
	f) Planetary gear sets broken or seized	f) Replace affected parts
No drive in forward drive positions	a) Hydraulic pressure too low	a) Adjust to specifications
	b) Valve body malfunction	b) Clean or overhaul
	c) Low fluid level	c) Add as required
	d) Bad input shaft seals	d) Replace seal rings
	e) Overrunning clutch not holding	e) Overhaul or replace
	f) Bad rear clutch	f) Overhaul
	g) Planetary gear sets broken or seized	g) Replace affected parts

CHILTON'S THREE C'S TRANSMISSION DIAGNOSIS PROCEDURE
A-500 Overdrive Transmission

Condition	Cause	Correction
No drive in R	a) Hydraulic pressure too low b) Low/reverse band out of adjustment c) Valve body malfunction d) Malfunction in low/reverse servo, band or linkage e) Manual linkage adjustment f) Bad input shaft seals g) Bad front clutch h) Bad rear clutch i) Planetary gear sets broken or seized	a) Adjust to specifications b) Adjust to specifications c) Clean or overhaul d) Overhaul e) Adjust as required f) Replace seal rings g) Overhaul h) Overhaul i) Replace affected parts
Drives in N	a) Valve body malfunction b) Manual linkage adjustment c) Insufficient clutch plate clearance d) Bad rear clutch e) Rear clutch dragging	a) Clean or overhaul b) Adjust as required c) Overhaul clutch pack d) Overhaul e) Overhaul
Drags or locks	a) Stuck lock-up valve b) Low/reverse band out of adjustment c) Kickdown band adjustment too tight d) Planetary gear sets broken or seized e) Overrunning clutch broken or seized	a) Clean or overhaul b) Adjust to specifications c) Adjust to specifications d) Replace affected parts e) Overhaul or replace
Grating, scraping or growling noise	a) Low/reverse band out of adjustment b) Kickdown band out of adjustment c) Output shaft bearing or bushing bad d) Planetary gear sets broken or seized e) Overrunning clutch broken or seized	a) Adjust to specifications b) Adjust to specifications c) Replace d) Replace affected parts e) Overhaul or replace
Buzzing noise	a) Valve body malfunction b) Low fluid level c) Aerated fluid d) Overrunning clutch inner race damaged	a) Clean or overhaul b) Add as required c) Check for overfilling d) Overhaul or replace
Hard to fill, oil blows out filler tube	a) Oil filter clogged b) Aerated fluid c) High fluid level d) Breather clogged	a) Change filter and fluid b) Check for overfilling c) Bad converter check valve d) Clean, change fluid
Transmission overheats	a) Engine idle speed too high b) Hydraulic pressure too low c) Low fluid level d) Manual linkage adjustment e) Faulty oil pump f) Kickdown band adjustment too tight g) Faulty cooling system h) Insufficient clutch plate clearance	a) Adjust to specifications b) Adjust to specifications c) Add as required d) Adjust as required e) Overhaul pump f) Adjust to specifications g) Service vehicle's cooling system h) Overhaul clutch pack
Harch upshift	a) Hydraulic pressure too low b) Incorrect throttle linkage adjustment c) Kickdown band out of adjustment d) Hydraulic pressure too high	a) Adjust to specifications b) Adjust as required c) Adjust to specifications d) Adjust to specifications
Delayed upshift	a) Incorrect throttle linkage adjustment b) Kickdown band out of adjustment c) Bad seals on governor, support d) Bad reaction shaft support seals e) Governor malfunction f) Malfunction in kickdown servo, band or linkage g) Bad front clutch	a) Adjust as required b) Adjust as required c) Replace seals d) Replace seal rings e) Service or replace unit f) Overhaul g) Overhaul

CHILTON THREE C's TRANSMISSION DIAGNOSIS
A-500 Overdrive Transmission

Condition	Cause	Correction
No reverse or slips in reverse	a) Failed direct clutch b) Overdrive spring lost c) Wrong overdrive piston selected proper spacer	a) Replace the clutch b) Replace the spring load c) Replace with the bearing spacer
No overdrive shift	a) Blown fuse b) Faulty overdrive c) Faulty wiring or connectors d) Faulty overdrive switch e) Faulty SMEC f) Failed overdrive clutch g) Wrong overdrive piston bearing spacer selected h) Valve body malfunction	a) Replace the fuse b) Replace the solenoid solenoid c) Repair or replace as necessary d) Repair or replace as necessary e) Replace SMEC f) Replace the clutch g) Replace with the proper spacer h) Repair or replace the internal components as necessary
Runaway overdrive shift	a) Failed overdrive overrunning clutch	a) Replace the overrunning clutch
Overdrive shift occurs immediately after the 2-3 shift	a) Faulty overdrive solenoid (not venting) b) Lower valve body malfunction c) Faulty wiring or connectors d) Faulty SMEC	a) Replace overdrive solenoid b) Repair or replace the valve body or its internal components as necessary c) Repair or replace as necessary d) Replace SMEC
Excessively delayed overdrive shift	a) Wrong overdrive piston bearing b) Faulty sensor	a) Replace with the proper spacer b) Replace the sensor
No 4-3 downshift	a) Faulty lockup solenoid b) Lower valve body malfunction c) Faulty wiring or connectors d) Faulty SMEC	a) Replace lockup solenoid b) Repair or replace the valve body or its internal components as necessary c) Repair or replace as necessary d) Replace SMEC
No 4-3 downshift with overdrive off switch	a) Faulty overdrive b) Faulty lockup solenoid c) Faulty wiring or connectors d) Faulty SMEC	a) Replace overdrive switch a) Replace lockup solenoid c) Repair or replace as necessary d) Replace SMEC
Torque converter locks up in 2nd and 3rd gears	a) Faulty lockup solenoid (not venting)	a) Replace or repair solenoid valve
Harsh shift 1–2, 2–3 & 3–2	a) Faulty lockup solenoid (not venting)	a) Replace lockup solenoid
Low governor pressure	a) Leaking governor tubes. Bent, loose fit or governor seal rings broken or worn	a) Repair or replace as necessary
Noisy	a) Failed overdrive piston bearing b) Failed gear train needle thrust bearing c) Failed overdrive planetary d) Failed overdrive overrunning clutch	a) Replace bearing b) Replace thrust bearing c) Replace planetary assembly d) Replace the clutch

CHILTON THREE C's LOCKUP TORQUE CONVERTER DIAGNOSIS
A–500 Overdrive Transmission

Condition	Cause	Correction
No lockup	a) Faulty oil pump b) Sticking governor valve c) Valve body malfunction. Stuck switch valve, lock valve or fail-safe valve d) Faulty torque converter e) Failed locking clutch f) Leaking turbine hub seal g) Faulty input shaft or seal ring	a) Replace the oil pump b) Repair or replace as necessary c) Repair or replace valve body or its internal components d) Replace torque converter e) Replace torque converter f) Replace torque converter g) Repair or replace as necessary
Will not unlock	a) Sticking governor valve b) Valve body malfunction. Stuck switch valve, lockup valve or fail-safe valve	a) Repair or replace as necessary b) Repair or replace the valve body or its internal components as necessary
Stays locked up at too a speed in Direct	a) Sticking governor valve b) Valve body malfunction. Stuck switch valve, lockup valve or fail-safe valve	a) Repair or replace as necessary b) Repair or replace the valve body or its internal components as necessary
Locks up or drags in low or second	a) Faulty oil pump b) Valve body malfunction. Stuck switch valve, lockup valve or fail-safe valve	a) Replace the oil pump b) Repair or replace the valve body or its internal components as necessary
Sluggish or stalls in reverse	a) Faulty oil pump b) Plugged oil cooler, cooler lines c) Valve body malfunction. Stuck switch valve, lockup valve or fail-safe valve	a) Replace oil pump b) Flush or replace cooler and flush line and fittings c) Repair or replace the valve body or its internal components as necessary
Loud chatter during lockup engagement (cold)	a) Faulty torque converter b) Failed locking clutch c) Leaking turbine hub seal	a) Replace torque converter b) Replace torque converter c) Replace torque converter
Vibration or shudder during lockup Engagement	a) Faulty oil pump b) Valve body malfunction c) Faulty torque converter d) Engine needs tune-up	a) Replace oil pump b) Repair or replace the valve body or its internal components as necessary c) Replace torque converter d) Tune-up engine
Vibration after lockup engagement	a) Faulty torque converter b) Exhaust system strikes the underbody c) Engine needs tune-up d) Throttle linkage misadjusted	a) Replace torque converter b) Align the exhaust system c) Tune-up engine d) Adjust throttle linkage
Vibration when revved in neutral	a) Torque converter out of balance	a) Replace torque converter
Overheating. Oil blows out of the dipstick tube or pump seal	a) Plugged cooler, cooler lines or or fittings b) Stuck switch valve	a) Flush or replace cooler and flush lines b) Repair switch valve in the valve body or replace valve body
Shudder after lockup engagement	a) Plugged cooler, cooler lines fittings b) Faulty oil pump c) Valve body malfunction d) Faulty torque converter e) Engine needs tune-up f) Exhaust system strikes the underbody g) Failed locking clutch	a) Flush or replace cooler and flush lines b) Replace oil pump c) Repair or replace the valve body or its internal components as necessary d) Replace torque converter e) Tune-up engine f) Align the exhaust system g) Replace torque converter
Torque converter locks up in 2nd and 3rd gears	a) Faulty lockup solenoid not venting	a) Replace or repair

Hydraulic Control System

NOTE: Please refer to Section 9 for all oil flow circuits.

The hydraulic control system has 4 important functions to perform in order to make the transmission fully automatic. These functions are, the pressure supply system, the pressure regulating system, the flow control valve system and the clutches, band servos and accumulator systems.
An explanation of each system follows to assist the technician in understanding the oil flow circuitry within the torqueflite automatic transmissions.

THE PRESSURE SUPPLY SYSTEM

The pressure supply system consists of the oil pump driven by the engine through the torque converter. The oil pump furnishes the fluid pressure for all the hydraulic and lubrication requirements.

THE PRESSURE REGULATING SYSTEM

The pressure regulating system consists of the pressure regulator valve which controls line pressure at the value dependent upon throttle opening. The governor valve transmits regulated pressure to the valve body to control upshifts and downshifts, in conjunction with the vehicle speed.

The throttle valve transmits regulated pressure to the transmission to control the upshifts, downshifts and lockup speeds when the transmission is equipped with a lockup converter assembly. The throttle pressure operates in conjunction with the throttle opening.

THE FLOW CONTROL VALVE SYSTEM

1. The manual valve provides the different transmission drive ranges as selected by the vehicle operator.
2. The 1–2 shift valve automatically shifts the transmissionfrom 1st speed to 2nd or from 2nd speed to 1st speed, depending upon the vehicle road speed.
3. The 2–3 shift valve automatically shifts the transmission from 2nd to 3rd speed or from 3rd speed to 2nd speed, depending upon the vehicle speed.
4. The kickdown valve makes possible a forced downshift from 3rd to 2nd, 2nd to 1st, or 3rd to 1st speeds, depending upon the vehicle speed, by depressing the accelerator past the detent feel near the wide open throttle (WOT).
5. The throttle pressure plug, at the end of the 2–3 shift valve, provides a 3–2 downshift with varying throttle openings, again depending upon the vehicle speed.
6. The 1–2 shift control valve transmits 1–2 shift control pressure to the transmission accumulator piston to control the kickdown band capacity on the 1–2 upshift and 3–2 downshifts. The limit valve is used to determine the maximum speed at which a 3–2 throttle downshift can be made.
7. The shuttle valve has 2 functions and they are accomplished independently of each other. The first function is to provide a fast release of the kickdown band and to provide a smooth front clutch engagement when the vehicle operator makes a closed throttle upshift from 2nd to 3rd speed. The second function of the shuttle valve is to regulate the application of the kickdown servo and band when making 3rd to 2nd kickdown.
8. On automatic transmissions equipped with the lockup converter assembly, the lockup valve automatically applies the torque converter lockup clutch if the vehicle is above a predetermined speed when in the 3rd (direct) gear ratio.
9. The failsafe valve restricts feed to the lockup clutch if the front clutch apply pressure drops. This valve permits lockup only in the 3rd (direct) gear and provides a fast lockup release during a kickdown.
10. The switch valve directs fluid to apply the lockup clutch in 1 position and release it in the other and also directs fluid to the cooling and lubrication systems.
11. Transmissions not equipped with the lockup converter use a control valve to limit the pressure to the lubrication and cooling system and to direct main line pressure to the converter.

THE CLUTCHES, BAND SERVOS AND ACCUMULATOR SYSTEM

The front and rear clutch pistons and both servo pistons are moved hydraulically to engage the clutches and apply the bands. The pistons are released by spring tension when the hydraulic pressure is released. On the 2–3 upshift, the kickdown servo piston is released by spring tension and hydraulic pressure.
The accumulator controls the hydraulic pressure on the apply side of the kickdown servo during the 1–2 shift, thereby cushioning the kickdown band application at any throttle opening.

Diagnostic Tests

HYDRAULIC PRESSURE TEST

Before starting the pressure test, make sure the fluid level is correct and linkage is adjusted properly. A good pressure test is an important part of diagnosing transmission problems. Oil flow charts are supplied for hydraulic circuitry details.

1. Check fluid level and linkage adjustments. Remember that fluid must be at operating temperature (150–200°F) during tests.
2. Install engine tachometer and route wires so that it can be read under the car.
3. Raise vehicle on hoist so that rear wheels can turn. It may be helpful to disconnect throttle valve and shift rod from transmission levers so that they can be shifted from under the vehicle.
4. Two size gauges are needed: one 100 psi and the other a 300 or 400 psi. The higher pressure gauge is required for the "reverse" test.

Test 1 (Selector in 1)

The purpose of this test is to check pump output, the pressure regulation and also to check on the condition of the rear clutch and rear servo hydraulic circuits.

1. Hook up gauges to line and rear servo ports.
2. Adjust engine speed to 1000 rpm.
3. Shift into **1** position. (Selector lever on transmission all the way forward.)
4. Read pressures on both gauges as the throttle lever on transmission is moved from the full forward position to full rearward position.
5. Line pressure should read from 54–60 psi with the throttle lever forward and it should gradually increase as the lever is moved rearward, to 90–96 psi.
6. The rear servo pressure should read about the same as the line pressure, to within 3 psi.

Test 2 (Selector in 2)

The purpose of this test is to check pump output, the pressure regulation and also to check on the condition of the rear clutch and lubrication hydraulic circuits.

1. Attach gauge to the line pressure port on the right hand side of the unit and install a T-fitting into the rear cooler line so that lubrication pressure can be recorded.
2. Adjust engine speed to 1000 rpm.
3. Shift into **2** position (this is 1 detent rearward from the full forward position of the selector lever).
4. Read pressures on both gauges as the throttle lever on the transmission is moved from the full forward position to full rearward position.

5. Line pressure should read from 54–60 psi with the throttle lever forward and it should gradually increase as the lever is moved rearward, to 90–96 psi.

6. Lubrication pressure should be between 5–15 psi with the throttle lever forward and between 10–30 psi with lever rearward.

Test 3 (Selector In D)

The purpose of this test is to check pump output, the pressure regulation and also to check on the condition of the rear clutch and front clutch and hydraulic circuits.

1. Attach gauges to the line pressure port and also to the front servo release port, on the right hand side of the unit.
2. Adjust engine speed to 1600 rpm.
3. Shift into **D** position. (This is 2 detents rearward from the full forward position of the selector lever).
4. Read pressures as the throttle lever on the transmission is moved from the full forward position to full rearward position.
5. Line pressure should read from 54–60 psi with the throttle lever forward and it should gradually increase as the lever is moved rearward.
6. The front servo release is pressurized only in direct drive and should be the same as the line pressure reading within 3 psi, up to the downshift point.

TEST 4 (Selector In Reverse)

The purpose of this test is to check pump output, pressure regulation and also to check on the condition of the front clutch and rear servo hydraulic circuits. Also, at this time, a check can be made for leakage into the rear servo, due to case porosity, cracks, valve body or case warpage which can cause reverse band burn out.

1. Attach the 300 psi (or higher) gauge to the rear servo apply port at the right hand rear of the unit.
2. Adjust engine speed to 1600 rpm.
3. Shift into **R** position. (This is 4 detents rearward from the full forward position of the selector lever.)
4. Read pressure on gauge. It should be between 230–260 psi.
5. By moving the selector lever back to the **D** position, the pressure at the rear servo should drop to zero. This checks for leakage into the rear servo, which would result in the reverse band burning out.

Analyzing the Pressure Test

1. If the proper line pressure from the minimum to maximum is found in any one test, the pump and pressure regulator are working properly.
2. If there is low pressure in **D**, **1** and **2**, but there is a correct pressure reading in **R**, then there is leakage in the rear clutch circuit.
3. if there is low pressure in **D** and **R**, but there is correct pressure in **1**, then there is leakage in the front clutch circuit.
4. If there is low pressure in **R** and **1**, but there is correct pressure in **2**, then there is leakage in the rear servo circuit.
5. If there is low line pressure in all positions then there could be a defective pump, a clogged filter, or a stuck pressure regulator valve.

GOVERNOR PRESSURE

The governor pressure only needs to be tested if the transmission shifts at the wrong vehicle speeds, when the throttle rod is correctly adjusted.

1. Connect a 100 psi gauge to the governor pressure opening. This is on the left hand side, near the bottom, near the extension housing mounting flange.
2. Pressures are recorded by operating in 3rd gear.
3. If the pressures are wrong at a given speed, the governor valve or the weights are sticking. The pressure should respond smoothly to any changes in rpm and should drop to 0–1½ psi when the vehicle is stopped. If there is high pressure at stand still (more than 2 psi) then the transmission will be prevented from downshifting.

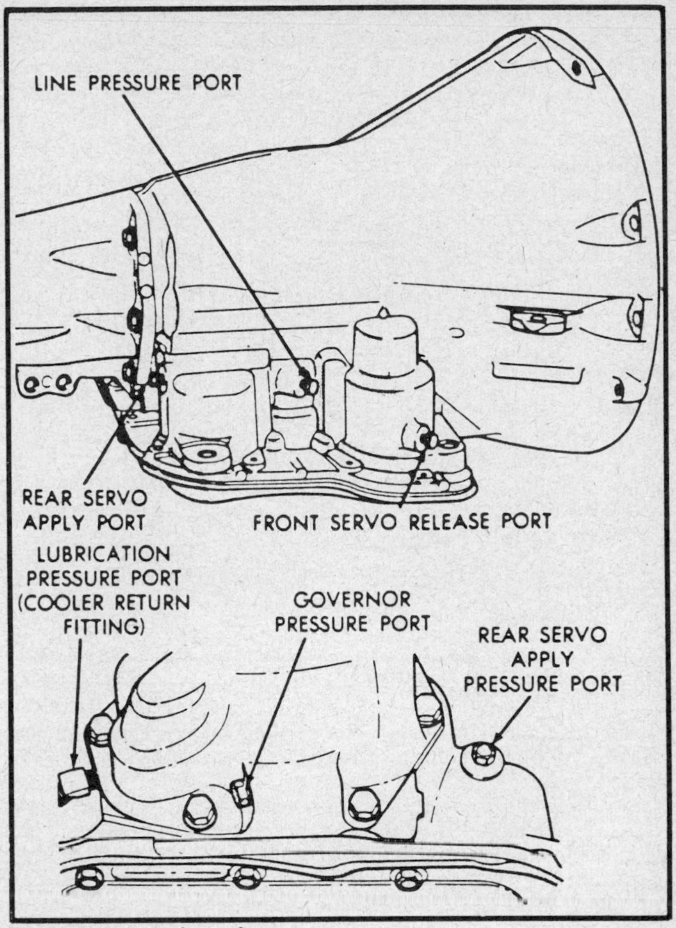
LINE PRESSURE PORT
REAR SERVO APPLY PORT
FRONT SERVO RELEASE PORT
LUBRICATION PRESSURE PORT (COOLER RETURN FITTING)
GOVERNOR PRESSURE PORT
REAR SERVO APPLY PRESSURE PORT

Pressure test locations

THROTTLE PRESSURE

This transmission has no provision for testing throttle pressure with a gauge. The only time incorrect throttle pressure should be suspected is if the part throttle up-shift speeds are either too slow in coming or occur too early in relation to vehicle speeds. Engine runaway on either up shifts or down shifts can also be an indicator of incorrect (low) throttle pressure setting. The throttle pressure really should not be adjusted until the throttle linkage is checked and adjustment has been verified to be right.

CLUTCH AND SERVO AIR PRESSURE TESTS

Even though all fluid pressures are correct and the hydraulic pressure test checks, out, it is still possible to have a no drive condition, due to inoperative clutches or bands. The inoperative units can be located by using tests, with air pressure instead of hydraulic pressure. To make an air pressure test the following points should be observed.

1. Compressed air should be set to about 30 psi and must be free of dirt and moisture.
2. After the vehicle is safely supported or on a hoist, remove pan and carefully remove the valve body assembly. (If necessary,

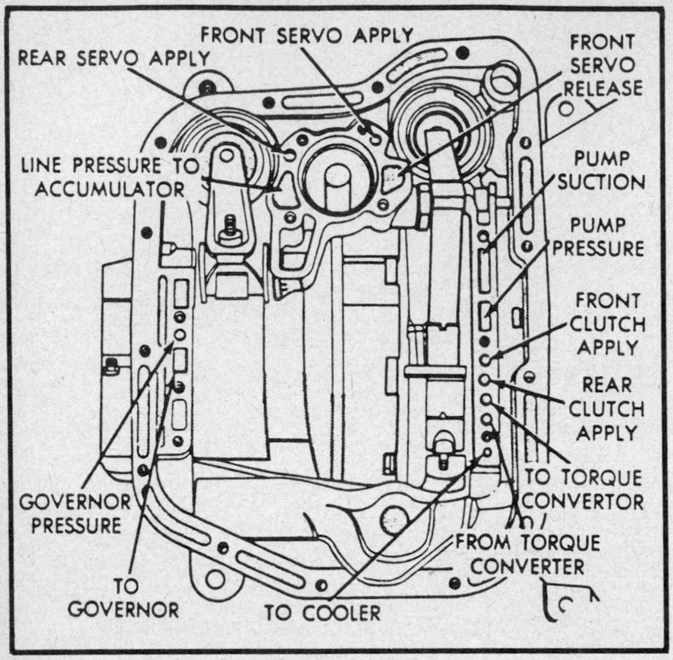
FRONT SERVO APPLY
REAR SERVO APPLY
FRONT SERVO RELEASE
LINE PRESSURE TO ACCUMULATOR
PUMP SUCTION
PUMP PRESSURE
FRONT CLUTCH APPLY
REAR CLUTCH APPLY
TO TORQUE CONVERTOR
GOVERNOR PRESSURE
FROM TORQUE CONVERTER
TO GOVERNOR
TO COOLER

Air pressure test locations

rotate the driveshaft to align the parking gear and sprag to permit the knob on the end of the parking control rod to pass the sprag.)

3. Locate the front clutch apply passage and apply air pressure. Listen for a full thud and/or place fingertips on front clutch housing to feel a slight movement of the piston. This indicates that the front clutch is operating. Hold the air pressure on for a few seconds and inspect the system for excessive oil leaks.

4. Locate the rear clutch apply passage air pressure. Listen for a dull thud and/or place fingertips on rear clutch housing to feel a slight movement of the piston. This indicates that the rear clutch is operating. Hold the air pressure on for a few seconds and inspect the system for excessive oil leaks.

5. Locate the front servo apply passage and apply air pressure. This tests the kickdown servo and operation of the servo is indicated by the front band tightening. The spring tension on the servo piston should release the band.

6. Locate the rear servo apply passage and apply air pressure. This tests the Low and Reverse servo and operation of the servo is indicated by the rear band tightening. The spring tension on the servo piston should release the band.

7. If, after the air pressure tests, correct operation of the clutches and servos are confirmed and the complaint was no upshift or erratic shifts, then the problem is in the valve body.

STALL SPEED TEST

Stall speed testing involves determining the maximum engine rpm obtainable at full throttle with the rear wheels locked and the transmission in **D** position.

CAUTION

Never allow anyone to stand in front of the car when performing a stall test. In addition, always block the front wheels and have both the parking and service brakes fully applied during the test.

CAUTION

Do not hold the throttle open any longer than necessary and never longer than 5 seconds at a time. If more than 1 stall test is required, operate the engine at 1,000 RPM in neutral for at least 20 seconds to cool the transmission fluid between runs.

1. Connect tachometer to engine.
2. Check and adjust transmission fluid level as necessary.
3. Operate engine until transmission fluid reaches operating temperature (approx. 175° F).
4. Block front wheels.
5. Fully apply parking brakes.
6. Fully apply service brakes.
7. Open throttle completely and record maximum engine rpm registered on tachometer, in **D**.
8. If engine speed exceeds maximum stall speed, release the accelerator immediately. The indicates that transmission slippage is occurring.
   a. Torque converter diameter – 10 ¾ in.
   b. Engine stall speed rpm – 1900–2100.
9. Shift transmission into **N**, operate engine for 20 seconds, stop engine, shift into **P** and release brakes.
10. If the values read on the tachometer do not agree with the information in the stall speed specifications chart, refer to the problem diagnosis below.

Stall Speed Too High

If the stall speed exceeds the maximum specified in the chart by more than 200 rpm, transmission clutch slippage is indicated. Refer to the hydraulic pressure test and air pressure test procedures to find the cause of slippage.

Stall Speed Too Low

Low stall speeds with a properly tuned engine indicate a torque converter stator clutch problem. The condition should be confirmed by road testing prior to converter replacement. If stall speeds are 250–350 rpm below the minimum specified in the cart and the car operates properly at highway speeds but has poor low speed acceleration, the stator overrunning clutch is slipping and the torque converter should be replaced.

Stall Speed Normal

If stall speeds are normal but road testing shows that abnormally high throttle opening is required to maintain highway speeds even though low speed acceleration is normal, the stator overrunning clutch is seized and the torque converter must be replaced.

Converter Clutch Operation and Diagnosis

TORQUE CONVERTER CLUTCH

The 4th gear (overdrive) is electronically controlled and hydraulically activated. A variety of sensor inputs are fed to the Single Module Engine Controller (SMEC) which controls a solenoid mounted on the valve body. The solenoid will energize and close a vent, allowing a 3–4 upshift. The SMEC also controls the operation of the lockup torque converter using many of the same sensor inputs.

A lockup torque converter, which is electronically controlled and hydraulically activated, will also be used with this transmission. 4th gear (overdrive) and lockup will only occur during certain conditions determined by the SMEC.

The transmission locked or unlocked is controlled by a solenoid. First the coolant temperature is sensed. If the actual temperature is below 150°F, the solenoid will remain de-energized (unlocked). Once this reference temperature is reached, mph, vacuum and the throttle position sensor determine the locked or unlocked of the transmission.

If the throttle position sensor is closed, the transmission will always be unlocked. The transmission will remain unlocked for a programmed period after the throttle is opened. If the actual mph is above the low reference value entry into the lockup region occurs. If the actual mph is below the low reference value, entry into the lockup region will not occur (solenoid de-energized).

The state of the transmission locked or unlocked, is determined by a set of vacuum verus mph curves. Two hysteresis bands are present. Once in a lockup region the vacuum must fall below or rise above the unlock vacuum reference points in order to enter the unlock (lockup) region. Lockup (solenoid energized) of the transmission is delayed a programmable period of time after entry into the lockup region occurs. There will be a drop in rpm when lockup takes place.

TROUBLESHOOTING THE LOCKUP TORQUE CONVERTER

The Single Module Engine Controller (SMEC) has been programmed to monitor several different circuits of the fuel injection system. This monitoring is called On Board Diagnosis. If a problem is sensed with a monitored circuit, often enough to indicate an actual problem, its fault code is stored in the SMEC module for eventual display to the service technician. If the problem is repaired or ceases to exist, the SMEC cancels the fault code after 50–100 ignition key on/off cycles.

Fault Codes

Fault codes are 2 digit numbers that identify which circuit is bad. In most cases, they do not identify which component is bad in a circuit. When a fault code appears (either by flashes of the power loss/limited (check engine) lamp or by watching the diagnostic read out tool (DRB-II), or equivalent, it indicates that the SMEC has recognized an abnormal signal in the system. Fault codes indicate the results of a failure but do not always identify the failed component directly.

Accessing Trouble Code Memory

There are 2 methods used in accessing trouble codes. The first method is the use of a diagnostic readout box tool (DRB—II) or equivalent, the second method is observing the power loss/limited (check engine) lamp.

The diagnostic readout box (DRB—II) is used to put the system into a Diagnostic Test Mode, Circuit Actuation Test Mode, Switch Test Mode, Engine Running Test Mode and Sensor Test Mode. Four of these modes of testing are called for at certain points of the driveability test procedure. A fifth test mode is available with the engine running. The following is a description of each test mode:

1. Diagnostic Test Mode—This mode is used to see if there are any fault codes stored in the on-board diagnostic system memory.
2. Circuit Actuation Test Mode (CTM Test)—This mode is used to turn a specific circuit on and off in order to check it. CTM test codes are used in this mode.
3. Switch Test Mode—This mode is used to determine if specific switch inputs are being received by the logic module.
4. Sensor Test Mode—This mode is used to see the output signals of certain sensors as received by the logic module.
5. Engine Running Test Mode—This mode is used to determine if the oxygen feedback system is switching from rich to lean and lean to rich.

DIAGNOSIS USING READOUT BOX

The diagnostic readout box (DRB—II) is used to put the on-board diagnostic system in 4 different modes of testing as called for in the driveability test procedure.

1. Connect tool diagnostic read out box (DRB-II) or equivalent, to the mating connector located in the wiring harness in the engine compartment near the SMEC.
2. Start the engine if possible, cycle the transmission selector and the A/C switch if applicable. Shut off the engine.

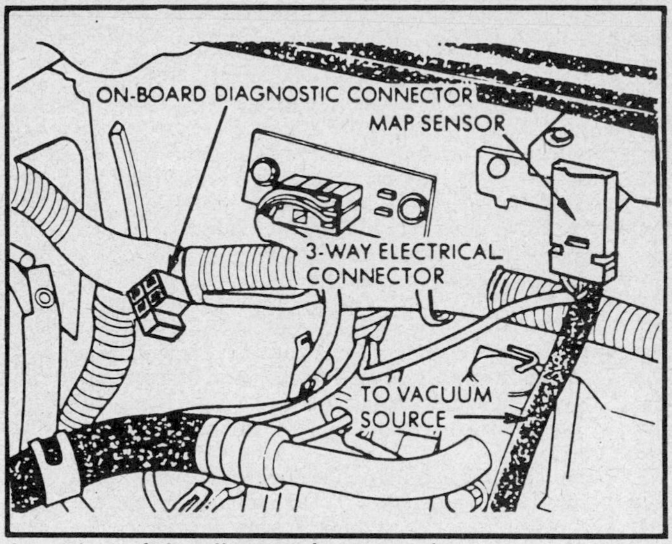
ON-BOARD DIAGNOSTIC CONNECTOR
MAP SENSOR
3-WAY ELECTRICAL CONNECTOR
TO VACUUM SOURCE

Location of the diagnostic connector

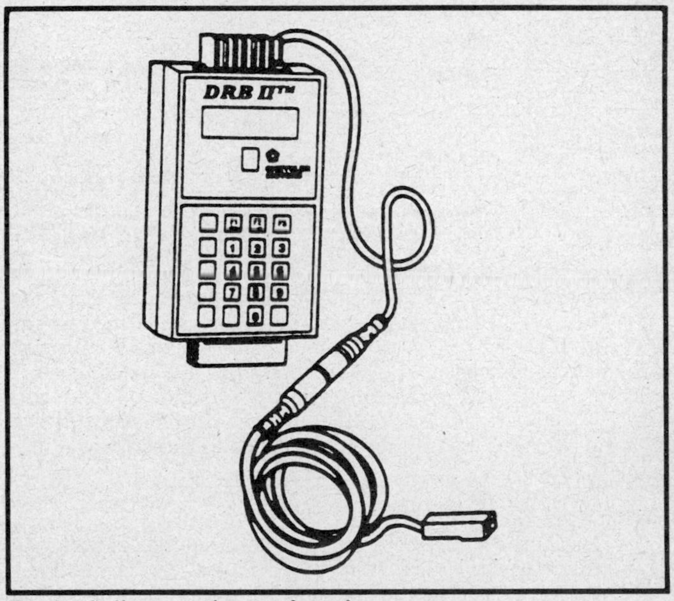
DRB II™

Typical diagnostic readout box

3. Place the read/hold switch on the readout box (DRB—II) in the read position. Turn the ignition key **ON-OFF-ON-OFF-ON** within 5 seconds. Record all the fault codes shown on the diagnostic readout box (DRB—II). Observe the check engine lamp on the instrument panel the lamp should light for 3 seconds then go out (bulb check).

NOTE: The display of codes can be stopped by moving the read/hold switch to the hold position. Returning to the read position will continue the displaying of codes. To erase the fault codes, access the erase fault code data on the diagnostic readout box (DRB—II).

DIAGNOSIS USING POWER LOSS/LIMIT (CHECK ENGINE) LAMP

The power loss/limit (check engine) lamp has 2 modes of operation. If for some reason the diagnostic readout box (DRB—II) is not available, the logic module can show fault codes by means of flashing the power loss (check engine) lamp on the instrument panel

NOTE: In some cases proper diagnostic cannot be determined by only using fault codes. Remember they are only the result, not necessarily the reason for the problem. The driveability test procedure is designed for use with the Diagnostic Readout Box (DBR—II) not the flashing power loss (check engine) lamp.

To activate this function, turn the ignition key **ON-OFF-ON-OFF-ON** within 5 seconds. The power loss (check engine) lamp will then come on for 2 seconds as a bulb check. Immediately following this it will display a fault code by flashing on and off. There is a short pause between flashes and a longer pause between digits. All codes displayed are two digit numbers with a four second pause between codes. Use the following for an example.

1. Lamp on for 2 seconds, then turns off.
2. Lamp flashes 5 pauses and then flashes once.
3. Lamp pauses for 4 seconds, flashes 5 times, pauses and then flashes 5 times.

The two codes are 51 and 55 . Any number of codes can be displayed as long as they are stored in the computer memory. The lamp will flash until all of them are displayed.

NOTE: In some cases proper diagnosis cannot be determined by only using fault codes. Remember they are only the results, not necessarily the reason for the problem.

4. Unlike the diagnostic readout box (DRB—II), the power loss lamp (check engine) cannot do the following.
   a. Once the lamp begins to display fault codes, it can not be stopped. If the code count is forgotten or lost, it is necessary to start over.
   b. The lamp can not perform the actuation test, sensor read test, switch test mode and engine running test modes.

Fault Code Description

The only fault codes that actually pertains to the transmission itself, are fault codes 37 and 45. Code 37 is used one both the Dakota trucks equipped with single point and dual point injection systems. This code is the PTU Solenoid Circuit. Code 37 is stating that an open or shorted condition is detected in the torque converter part throttle unlock solenoid circuit. Code 45 is used on the Dakota trucks equipped with only the dual point injection system. This code is for the Overdrive Circuit. Code 45 is stating an open or shorted condition detected in the overdrive solenoid circuit.

ON CAR SERVICES

Adjustments

THROTTLE LINKAGE

Throttle linkage adjustment is important to proper operation of this transmission. This adjustment positions a valve which controls shift speed, shift quality and downshift sensitivity at part throttle. If the setting is too short, early shifts and slippage between shifts may occur. If the linkage setting is too long, shifts may be delayed and part throttle downshifts may be very sensitive. This adjustment is so critical that the use of a throttle lever holding spring is advised to remove slack in the linkage during adjustment.

1. Raise and safely support the vehicle safely. Make the adjustment at the throttle lever.
2. Loosen the adjustment swivel lock screw. To insure proper adjustment, the swivel must be free to slide along the flat end of the throttle rod so that the preload spring action is not restricted. Disassemble and clean or repair the parts to assure free action, if necessary.
3. Hold the transmission lever firmly forward against its internal stop and tighten the swivel lock screw to 100 inch lbs.
4. The adjustment is finished and linkage backlash was automatically removed by the preload spring.
5. Lower the vehicle. Test the linkage freedom of operation by moving the throttle rod rearward, slowly releasing it to confirm it will return fully forward.

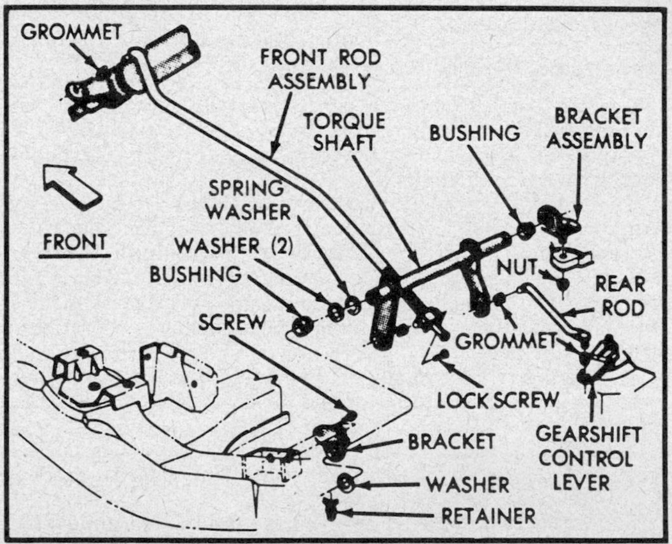
GROMMET
FRONT ROD ASSEMBLY
TORQUE SHAFT
BUSHING
BRACKET ASSEMBLY
SPRING WASHER
FRONT
WASHER (2)
BUSHING
NUT
REAR ROD
SCREW
GROMMET
LOCK SCREW
BRACKET
GEARSHIFT CONTROL LEVER
WASHER
RETAINER
Exploded view of the gearshift linkage

GEAR SHIFT LINKAGE

The gear shift linkage adjustment is important because the linkage positions the manual valve in the valve body. Incorrect adjustment will result in the vehicle creeping in **N**, premature clutch wear, delayed engagement in any gear or a no-start in the **P** or **N** condition.

Proper operation of the neutral start switch will provide a quick check of linkage adjustment as follows:

1. Turn key to **ON** to unlock column and shift lever.
2. Move shift lever slowly until it clicks into the **P** detent. Try to start engine. If starter does operate, **P** position is correct.
3. Stop the engine. Repeat, only this time moving lever to **N**. Try to start engine. If starter does operate, the **N** position is correct and linkage is properly adjusted.
4. Adjust the linkage as follows:
   a. Use a suitable tool to force the rod from the grommet in the lever (pry only where the grommet and rod attach, not the rod itself) then cut away the old grommet. Use suitable pliers to snap the new grommet into the lever and rod into the grommet.
   b. To insure the proper adjustment, make sure the adjustable swivel block is free to turn on the shift rod. Disassemble and clean or repair parts to assure free action , if necessary.

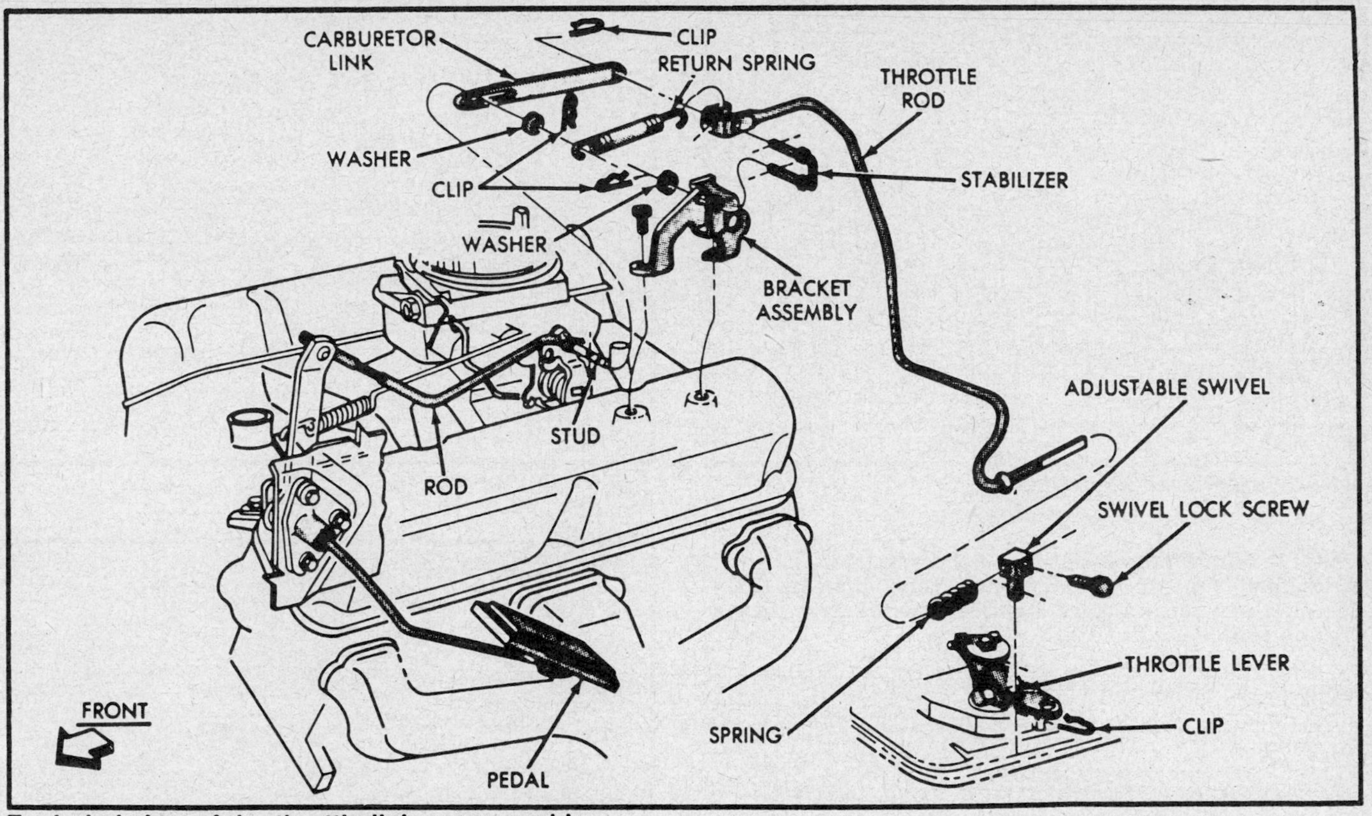
CARBURETOR LINK
CLIP
RETURN SPRING
THROTTLE ROD
WASHER
CLIP
STABILIZER
WASHER
BRACKET ASSEMBLY
ADJUSTABLE SWIVEL
STUD
ROD
SWIVEL LOCK SCREW
THROTTLE LEVER
FRONT
SPRING
CLIP
PEDAL

Exploded view of the throttle linkage assembly

c. Place the gearshift in the **P** position. With all the linkage assembled and the adjustable swivel lockbolt loose, move the shift lever on the transmission all the way to the rear detent (park) position.

d. Tighten the adjustment swivel lockbolt to 90 inch lbs. (10 Nm).

e. Check the detent position for **N** and **D**. They should be within limits of hand lever gate stops. The key start must occur only when the shift lever is in the **P** or **N** positions.

KICKDOWN (FRONT) BAND ADJUSTMENTS

The kickdown (front) band adjustment is done from the outside of the transmission. While somewhat difficult to reach, it can nevertheless be adjusted while still in the vehicle. The front band adjusting screw is located on the left side of the transmission case just above the manual lever and throttle lever.

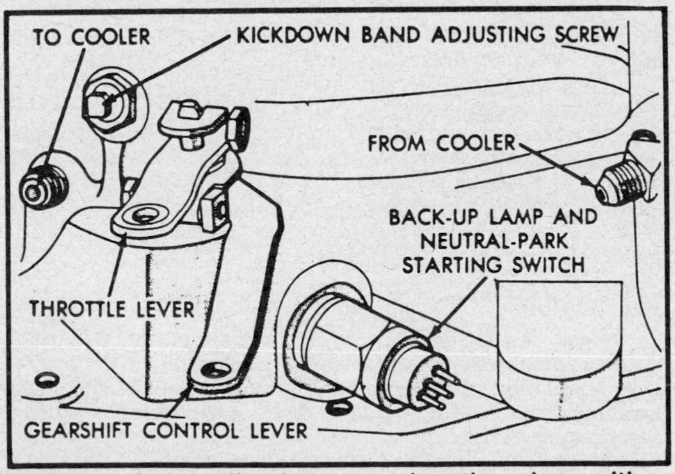
TO COOLER
KICKDOWN BAND ADJUSTING SCREW
FROM COOLER
BACK-UP LAMP AND NEUTRAL-PARK STARTING SWITCH
THROTTLE LEVER
GEARSHIFT CONTROL LEVER

Kickdown band adjusting screw location along with other external controls

1. Raise and support the vehicle safely.
2. Loosen adjusting screw locknut and back off locknut 5 turns.
3. Check that adjusting screw turns freely. Using special tool wrench C-3380-A with adapter C-3705 or equivalent. Tighten the band adjusting screw to 47–50 inch lbs. (5 Nm). If the adapter is not being used, tighten adjusting screw to 72 inch lbs. (8 Nm) torque, using a small torque wrench and a $^5/_{16}$ square socket.
4. Back off adjusting screw 2 turns.
5. Hold the adjusting screw in this position and tighten the screw locknut to 30 ft. lbs. (41 Nm) making sure that adjuster screw setting does not change.

REAR BAND ADJUSTMENT

The rear band adjustment is an inside adjustment so the pan must be removed.

1. Raise and support the vehicle safely.
2. Remove oil pan and drain fluid.
3. Look carefully at fluid, filter and pan bottom for a heavy accumulation of friction material or metal particles. A little accumulation can be considered normal, but a heavy concentration indicates damaged or worn parts.
4. Adjust the band by loosening locknut approximately 5 turns, then tightening the band adjusting screw to 72 inch lbs. (8 Nm) using a small torque wrench and a ¼ hex head socket or tool C-3380-A.
5. Back off adjusting screw approximately 4 turns. Hold the adjusting screw in this position and tighten the locknut to 25 ft. lbs.

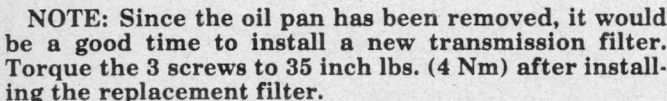

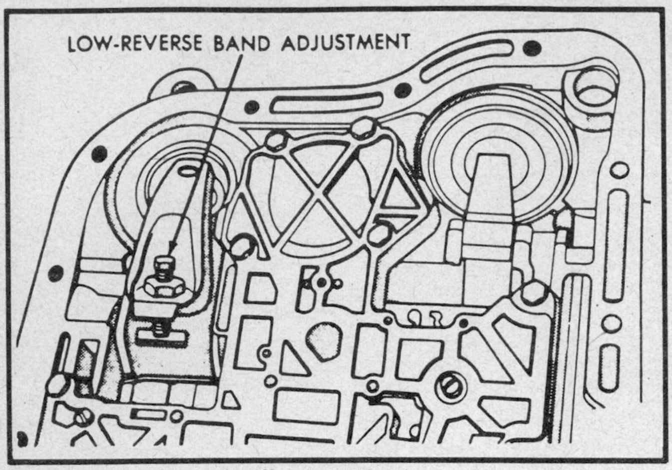
LOW-REVERSE BAND ADJUSTMENT
Low reverse band adjustment

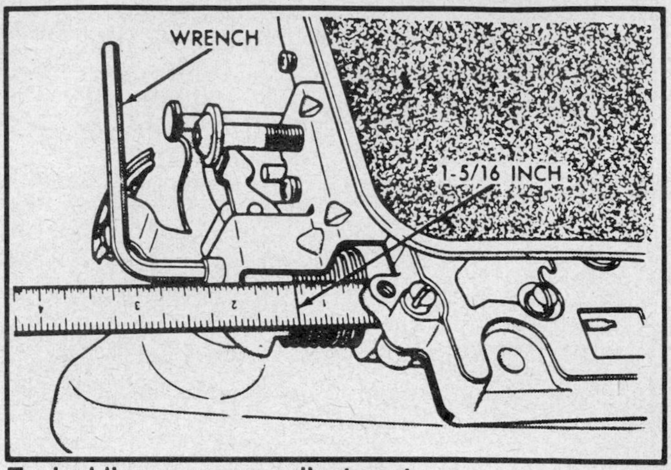
WRENCH
1-5/16 INCH
Typical line pressure adjustment

NOTE: Since the oil pan has been removed, it would be a good time to install a new transmission filter. Torque the 3 screws to 35 inch lbs. (4 Nm) after installing the replacement filter.

6. Using a new gasket on the pan, install and torque bolts evenly to 150 inch lbs. (17 Nm).
7. Lower the vehicle and fill transmission with specified amount of Mopar® ATF.

HYDRAULIC CONTROL PRESSURE ADJUSTMENT

There are 2 pressure adjustments that can be performed on the valve body. They are line pressure and throttle pressure both the line pressure and throttle pressure are independent of each other since each affects the shift quality and timing. Both adjustments must be performed properly and in the correct sequence. The line pressure adjustment must be performed first. To adjust the line pressure, remove the valve body, then follow this procedure:

1. Measure the distance from the valve body to the inner edge of the adjusting screw, using an accurate steel scale. The distance should be $1^5/_{16}$ in. (3.3cm).
2. If adjustment is needed, turn the screw in or out to obtain the $1^5/_{16}$ in. (3.3cm) setting.

NOTE: Because of manufacturing tolerances, the above dimension is an approximate setting and it may be necessary to vary this setting to get the proper pressure. The adjusting screw may be turned with an allen wrench. One complete turn of the adjusting screw changes the line pressure about $1^2/_3$ psi. To increase the pressure, turn the adjusting screw counterclockwise, while clockwise decreases the pressure.

3. Adjust the throttle pressure as follows:
   a. Since a special tool C-3763 is recommended by the factory, it will require that a tool be made to insert between the throttle lever cam and kickdown valve.
   b. Push the gauge tool (¼ in. block, 7 in.long x 0.629 in. wide) inward to compress the kickdown valve against its spring and to bottom the throttle valve in the valve body.
   c. Maintain the pressure against the kickdown valve spring and turn the throttle lever stop screw until the screw head touches the throttle lever tang and throttle lever cam touches the gauge tool along the 0.629 in. width. Be sure the adjustment is made with the spring fully compressed and the valve bottomed in the valve body.

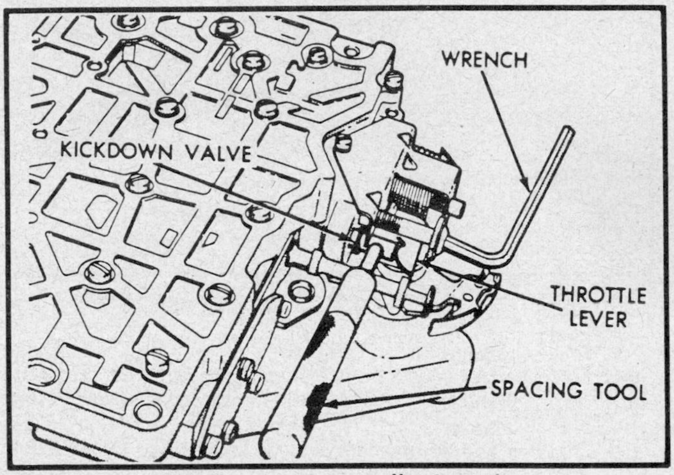
WRENCH
KICKDOWN VALVE
THROTTLE LEVER
SPACING TOOL
Making the throttle pressure adjustment

Services

FLUID CHANGES

The automatic transmission fluid and filter should be changed and the bands adjusted as follows:

a. Normal usage — every 37,500 miles.
b. Severe usage — every 12,000 miles.

Severe usage would be any prolonged operation with heavy loading especially in hot weather. Off-highway operations (city driving) and towing a trailer.

When refilling the transmission, make sure only MOPAR© ATF is used. A band adjustment and filter change should be made at the time of the oil change.

Drain and Refill

1. Raise and safely support vehicle. Place a container with a large opening under the transmission oil pan.
2. Loosen, but do not remove all of the pan bolts and gently pull one corner down so fluid will drain. If transmission is hot be careful of spilling oil.
3. Remove oil pan bolt and pan.
4. Perform any necessary band adjustments and/or filter change.

5. Check pan carefully for distortion, straightening the edges with a straight block of wood and a rubber mallet if necessary. Clean pan well.

NOTE: If there is evidence of contamination or if trouble shooting indicates a problem in the converter, it must be replaced.

6. Install new gasket on pan and torque the bolts to 150 inch lbs. (17 Nm).
7. Lower the vehicle, pour 4 quarts of MOPAR© ATF through the filler tube.
8. Start the engine and allow it to idle for at least 2 minutes. Apply the parking brake and block the drive wheels. Move the selector lever momentarily to each position, ending in the **N** position.
9. Recheck the ATF fluid level, if necessary, add sufficient fluid to bring level to the **ADD ONE PINT** mark.
10. Recheck fluid level after transmission is at normal operating temperature. The normal level should be between the **FULL** mark and **ADD ONE PINT** mark.
11. Do not overfill. Make sure the dipstick is properly seated to seal against dirt and water entering system.

OIL PAN

Removal and Installation

1. Raise and safely support the vehicle.
2. Loosen oil pan bolts and gently pull one corner down so fluid will drain. If transmission is hot, be careful of spilling oil.
3. Remove oil pan bolts and pan.
4. Carefully inspect filter and pan bottom for a heavy accumulation of friction material or metal particles. A little accumulation can be considered normal, but a heavy concentration indicates damaged or worn parts.

NOTE: Filter replacement is recommended when a pan is removed. Also, rear band can be adjusted if necessary.

5. Check pan carefully for distortion, straightening the edges with a straight block of wood and a rubber mallet if necessary. Clean pan well.
6. Install new gasket on pan and install. Torque bolts to 150 inch lbs. (17 Nm).
7. Lower the vehicle, pour 4 quarts of MOPAR© ATF through the filler tube.
8. Start the engine and allow it to idle for at least 2 minutes. Apply the parking brake and block the drive wheels. Move the selector lever momentarily to each position, ending in the **N** position.
9. Recheck the ATF fluid level, if necessary, add sufficient fluid to bring level to the **ADD ONE PINT** mark.
10. Recheck fluid level after transmission is at normal operating temperature. The normal level should be between the **FULL** mark and **ADD ONE PINT** mark.
11. Do not overfill. Make sure the dipstick is properly seated to seal against dirt and water entering system.

VALVE BODY AND ACCUMULATOR PISTON

Removal and Installation

1. Raise and safely support vehicle.
2. Remove pan and drain fluid.
3. Disconnect throttle and shift levers from transmission.
4. Unplug neutral switch harness and unscrew switch. Disconnect the electronic lockup solenoid wire from inside of the wiring connector at the rear of the transmission case.
5. Place large opening drain pan under transmission, then remove the 10 bolts holding the valve body to the transmission. Hold the valve body in position while the bolts are being removed.
6. Gently lower valve body out of the transmission and pull it forward, out of the case. It may be necessary to rotate the driveshaft so that the parking gear and sprag align to permit the knob on the end of the parking control rod to pass the sprag.
7. Remove accumulator piston and spring from transmission case and inspect for wear, nicks or cracks. Check spring for cracks or distortion. Clean for inspection or replacement of parts as necessary.

NOTE: If seal for valve body manual shaft is leaking, it is not necessary to remove valve body to replace seal. Simply drive the seal out of the case with a small punch and drive in a new seal using a $^{15}/_{16}$ in. socket as a driver. Be careful not to scratch the manual lever shaft or the manual lever shaft seal.

CAUTION

Do not clamp any portion of the valve body or transfer plate in a vise. Even the slightest distortion will result in stuck valves and leakage or fluid cross leakage. Handle the valve body with care at all times.

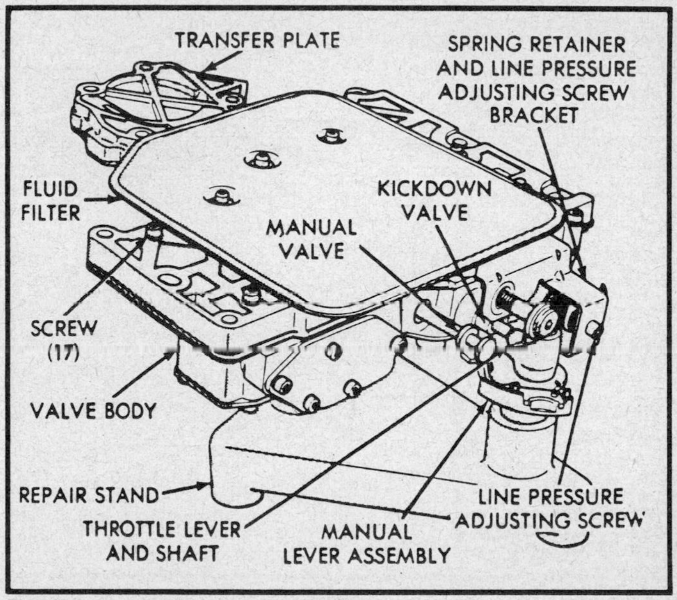
TRANSFER PLATE
SPRING RETAINER AND LINE PRESSURE ADJUSTING SCREW BRACKET
FLUID FILTER
KICKDOWN VALVE
MANUAL VALVE
SCREW (17)
VALVE BODY
REPAIR STAND
LINE PRESSURE ADJUSTING SCREW
THROTTLE LEVER AND SHAFT
MANUAL LEVER ASSEMBLY

Typical valve body assembly

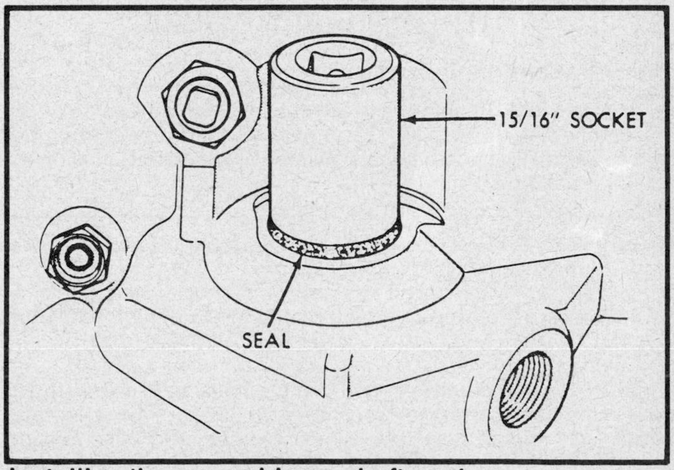
15/16" SOCKET
SEAL

Installing the manual lever shaft seal

EXTENSION HOUSING YOKE SEAL

Removal and Installation

1. Raise and support the vehicle safely. Mark the parts for reassembly and disconnect the driveshaft at the rear universal joint, Carefully pull the shaft yoke out of the transmission extension housing.

CAUTION

Be careful not to scratch or nick the ground surface on the sliding spline yoke during removal and installation of the shaft assembly.

2. Using seal removal tool C-3985 or equivalent, remove the old seal.
3. To install a new seal, position the seal in the opening of the extension housing and drive it into the housing with a suitable seal driver.
4. Carefully guide the front universal joint yoke into the extension housing and on the mainshaft splines. Align the marks made at removal and connect the driveshaft to the rear axle pinion shaft yoke.

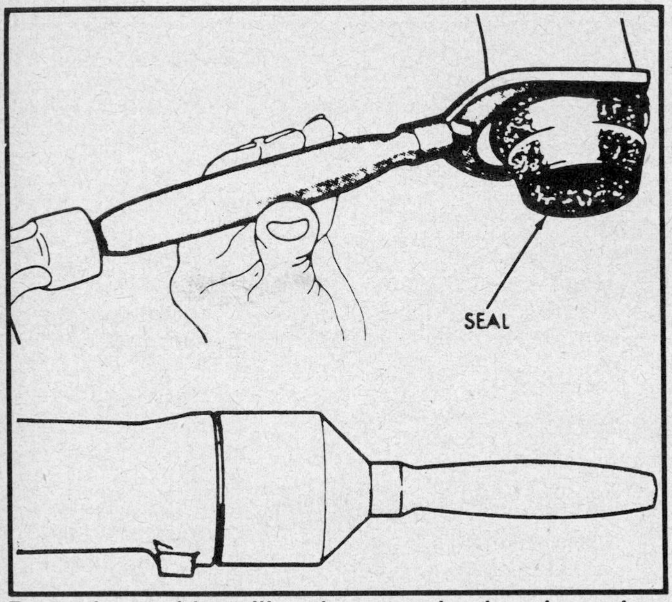
SEAL

Removing and installing the extension housing yoke seal

EXTENSION HOUSING BUSHING AND OUTPUT SHAFT BEARING

Removal and Installation

1. Raise and support the vehicle safely. Mark the parts for reassembly and disconnect the driveshaft at the rear universal joint, Carefully pull the shaft yoke out of the transmission extension housing.

CAUTION

Be careful not to scratch or nick the ground surface on the sliding spline yoke during removal and installation of the shaft assembly.

2. Remove the speedometer pinion and adapter assembly. Drain approximately 2 quarts of ATF fluid from the transmission.
3. Remove the bolts securing the extension housing to the crossmember. Raise the transmission slightly with a suitable floor jack and remove the center crossmember and support assembly.
4. Remove the extension housing to transmission bolts.

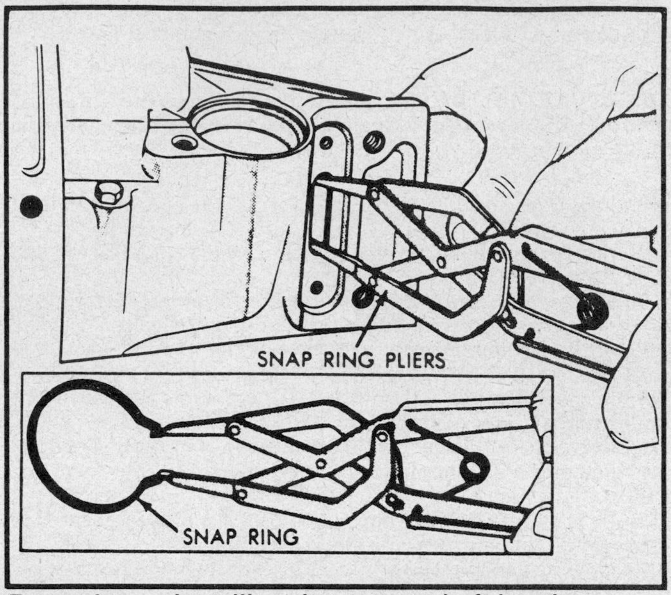
SNAP RING PLIERS
SNAP RING

Removing or installing the output shaft bearing snapring

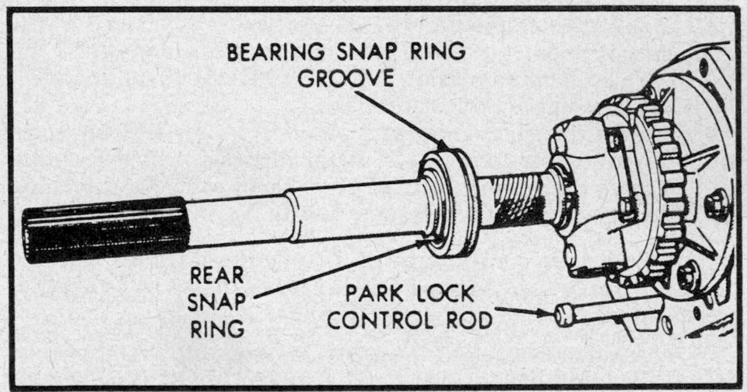
BEARING SNAP RING GROOVE
REAR SNAP RING
PARK LOCK CONTROL ROD

Installing a new oil seal into the housing

NOTE: When removing or installing the extension housing, the gearshift lever must be in the 1 (low) position. This positions the parking lock control rod rearward so it can be disengaged or engaged with the parking lock sprag.

5. Remove the screws, plate and gasket from the bottom of the extension housing mounting pad. Spread the snapring from the output shaft bearing
6. With the snapring spread as far as possible, carefully tap the extension housing rearward to allow the parking lock control knob to clear the parking sprag and remove the housing.
7. Using a heavy duty pair of snapring pliers, remove the output shaft bearing snapring and remove the bearing from the shaft.
8. Install a new bearing onto the shaft with the outer race ring groove toward the front and install the heavy duty snapring.
9. Remove the oil seal with a seal removal tool C-3985 or equivalent. Press or drive out the bushing with tool C-3396 or equivalent.
10. Slide a new bushing on the installing end of bushing tool C-3396 or equivalent. Align the oil hole in the bushing with the oil slot in the housing, press or drive the bushing into place.
11. Drive a new seal into the housing with tool C-3995 or equivalent.
12. Install the extension housing with a new gasket and torque

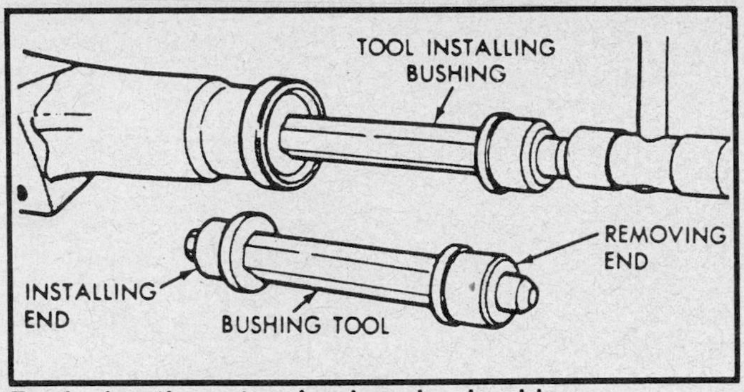
TOOL INSTALLING
BUSHING
REMOVING
END
INSTALLING
END
BUSHING TOOL

Replacing the extension housing bushing

the retaining bolts to 32 ft. lbs. (43 Nm). Torque the housing to support bolt to 50 ft. lbs. (68 Nm). Add the proper ATF fluid to the transmission to bring it up to the proper level.

GOVERNOR AND PARKING GEAR

Removal and Installation

1. Raise and support the vehicle safely. Remove the extension housing and output shaft bearing.
2. Carefully remove the snapring from the weight end of the governor valve shaft. Slide the valve and shaft assembly out of the governor body.
3. Remove the large snapring from the weight end of the governor body, lift out the governor weight assembly.
4. Remove the snapring from inside the governor weight, remove the inner weight and spring from the outer weight.
5. Remove the snapring from behind the governor body. Slide the governor and support assembly off the output shaft.
6. Remove the bolts and separate the governor body and screen from the parking gear.
7. Inspect all parts for burrs and wear. Inspect all moving parts for free movement, replace or repair as necessary.
8. Install by assembling the governor body and screen to support and tighten bolts finger tight. Be sure that the oil passage of the governor body aligns with the passage in the support.
9. Position the support and governor assembly on the output shaft. Align the assembly so the valve shaft hole in the governor body aligns with the hole in output shaft, slide the assembly into place. Install the snapring behind the governor body. Use new valve body support bolts and tighten the valve body to support bolts to 95 inch lbs. (11 Nm). The support bolts have a self locking nylon patch and cannot be reused.
10. Assemble the governor weights and spring and secure with a snapring inside of the larger governor weight. Place the weight assembly in the governor body and install the snapring.
4. Place the governor valve onto the valve shaft, insert the assembly into the body through governor weights. Install the valve shaft retaining snapring. Inspect the valve and weight assembly for free movement after installation.
5. Install the output shaft bearing and extension housing.

PARK LOCK

Removal and Installation

1. Raise and support the vehicle safely. Remove the extension housing.
2. Slide the shaft out of the extension housing to remove the parking sprag and spring.
3. Remove the snapring and slide the curved reaction plug and pin assembly out of the housing.
4. To remove the parking lock control rod, it is necessary to remove the valve body.
5. Installation is the reverse order of the removal procedure.

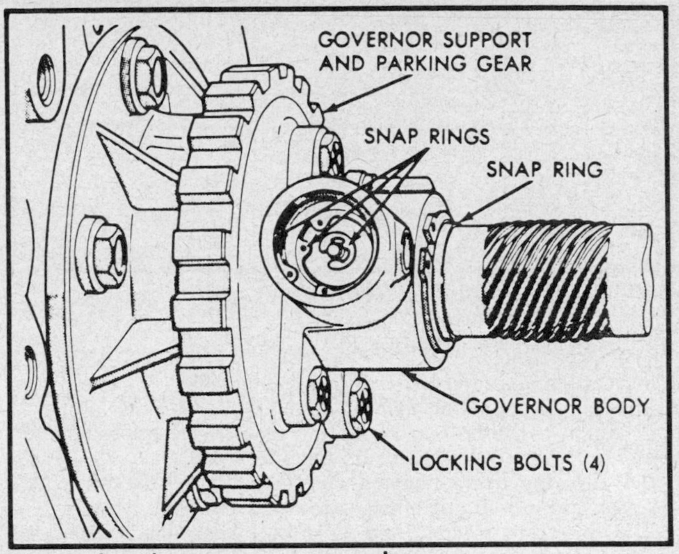
GOVERNOR SUPPORT
AND PARKING GEAR
SNAP RINGS
SNAP RING
GOVERNOR BODY
LOCKING BOLTS (4)

Removing the governor snapring

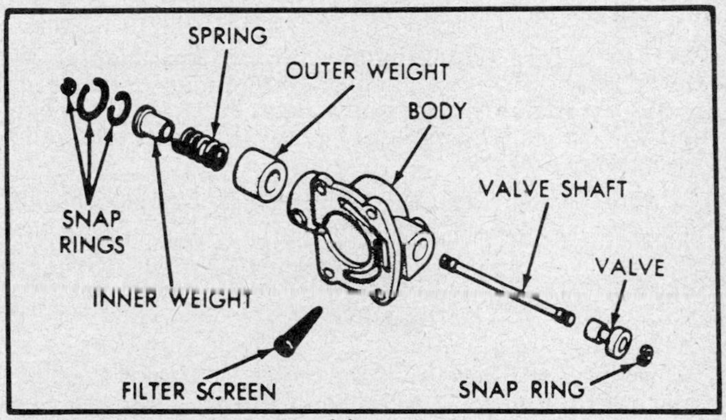
SPRING
OUTER WEIGHT
BODY
VALVE SHAFT
SNAP
RINGS
VALVE
INNER WEIGHT
FILTER SCREEN
SNAP RING

Typical governor assembly

SPEEDOMETER PINION GEAR

Any time the speedometer pinion adapter is removed, a new O-ring (black in color) must be installed on the outside diameter of the adapter.

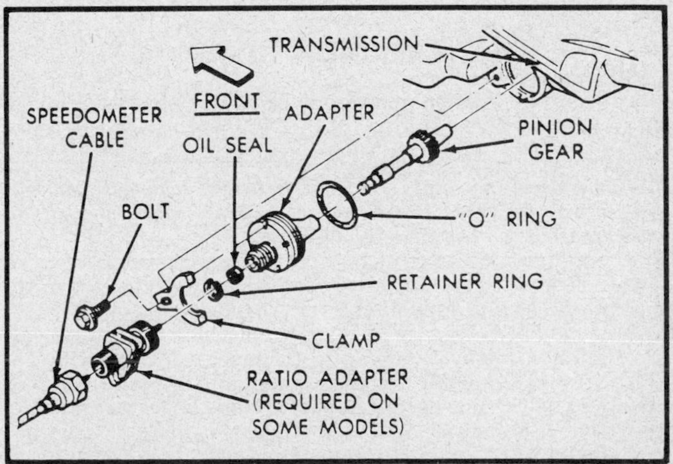
TRANSMISSION
FRONT
SPEEDOMETER
CABLE
ADAPTER
OIL SEAL
PINION
GEAR
BOLT
"O" RING
RETAINER RING
CLAMP
RATIO ADAPTER
(REQUIRED ON
SOME MODELS)

Exploded view of the speedometer drive assembly

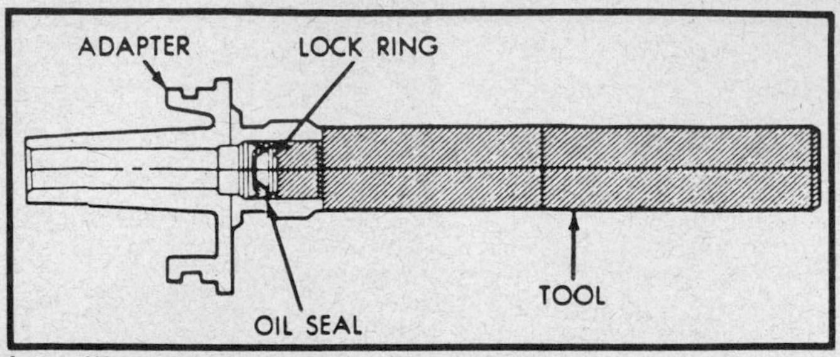
ADAPTER
LOCK RING
TOOL
OIL SEAL
Installing the speedometer pinion seal

Removal and Installation

1. Raise and support the vehicle safely. Place the drain pan under the speedometer adapter.
2. Remove the bolt and retainer securing the speedometer pinion adapter in the extension housing.
3. With the cable housing connected, carefully work the adapter and pinion out of the extension housing.
4. If the transmission fluid is found in the cable housing, replace the seal in the adapter.
5. Start the seal and retainer ring in the adapter and push them into the adapter with seal installer tool C-4004 or equivalent until the tool bottoms.

NOTE: Before installing the pinion and adapter assembly, make sure the adapter flange and mating area on the extension housing are perfectly clean. Dirt or sand will cause misalignment resulting in speedometer pinion gear noise.

6. Note the number of gear teeth and install the speedometer pinion gear into the adapter.

NEUTRAL STARTING AND BACK-UP LAMP SWITCH

The neutral starting switch is the center terminal of the 3 terminal switch. It provides a ground for the starter solenoid circuit through the selector lever in only the **P** and **N** positions. Proper operation of the neutral start switch can be checked as follows:

1. Turn key to **ON** to unlock column and shift lever.
2. Move shift lever slowly until it clicks into the **P** detent. Try to start engine. If starter does operate, **P** position is correct.

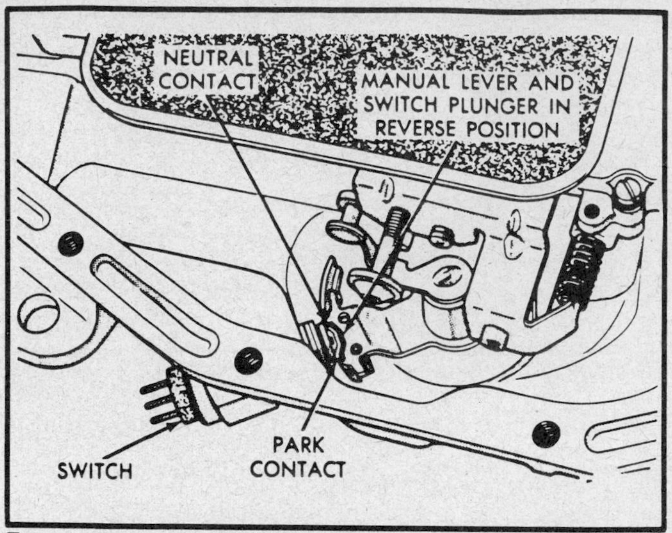
NEUTRAL CONTACT
MANUAL LEVER AND SWITCH PLUNGER IN REVERSE POSITION
SWITCH
PARK CONTACT
Removing or installing the neutral/back-up lamp switch

3. Stop the engine. Repeat, only this time moving lever to the **N** position. Try to start engine. If starter does operate, **N** position is correct and linkage is properly adjusted.
4. Remove the switch as follows:
   a. Unscrew the switch from the transmission case allowing the fluid to drain into a suitable container.
   b. Move the selector lever to **P** and then to **N** positions and inspect to see that the switch operating lever fingers are centered in the switch opening in the case.
   c. Screw the switch and new seal into the transmission case and torque the switch to 25 ft. lbs. (34 Nm). Retest the switch with a suitable test lamp.
   d. Add the proper ATF fluid to the transmission to bring it up to the proper level.
   e. The back-up lamp switch circuit is through the 2 outside terminal of the 3 terminal switch.
   f. To test the switch, remove the wiring connector from the switch and test for continuity between the 2 outside pins.
   g. Continuity should exist only with the transmission in the **R** position.
   h. No continuity should exist from either pin to the transmission case.

REMOVAL AND INSTALLATION

TRANSMISSION REMOVAL

1. The transmission and torque converter must be removed as an assembly.

CAUTION

The drive plate is not meant to support the load of holding the converter if disengaged from the transmission and serious damage will result. Remove transmission and converter together.

2. For safety, remove the ground (negative) cable from the battery. Raise and support the vehicle safely.
3. It may be necessary to lower the exhaust system for easy transmission removal, lower the exhaust system by removing the bolts and nuts attaching the exhaust pipe(s) to the manifold. Remove and exhaust system support straps or braces and lower the exhaust system.
4. Remove the engine to transmission struts, if so equipped. Remove and plug the cooler lines at the transmission.
5. Remove the starter motor and cooler line bracket.
6. Remove the torque converter access cover. Loosen the oil pan bolts, tap the pan to break it loose and let the fluid drain into a suitable drain pan.
7. Reinstall the pan temporarily with a few of the pan bolts. Mark the torque converter and drive plate to aid in reassembly. The crankshaft flange bolt circle, inner and outer circle of the holes in the drive plate and the 4 tapped holes in the front face of the torque converter all have 1 hole offset so these parts will be installed in the original position. This maintains the balance of the engine and torque converter.
8. Rotate the engine clockwise with the socket wrench on the crankshaft bolt to position the bolts attaching the torque converter to the drive plate and remove the bolts.
9. Mark the parts for reassemble then disconnect the driveshaft at the rear universal joint Carefully pull the shaft assembly out of the extension housing.
10. Disconnect the wire connector from the back-up lamp and neutral starting switch and lockup solenoid wiring connector.

Disconnect the spedometer cable and remove the oil fill tube.
11. Disconnect the gear shift rod and torque shaft assembly from the transmission.

NOTE: When it is necessary to disassemble the linkage rods from levers that use a plastic grommet as retainers, the grommets should be replaced with new ones.

12. Use a suitable tool to force the rod from the grommet in the lever and cut away the old grommet. Use pliers to snap the new grommet into the lever and rod into the grommet.
13. Disconnect the throttle rod from the lever at the left side of the transmission. Remove the linkage bell crank from the transmission, if so equipped.
14. Install the engine support fixture tool C-3487-A or equivalent with frame hooks, so as to support the rear of the engine.
15. Raise the transmission slightly with a suitable transmission jack to relieve the load on the supports.
16. Remove the bolts supporting the transmission mount to the crossmember and crossmember to frame and remove the crossmember.
17. Remove all bell housing bolts. Carefully work the transmission and torque converter assembly rearward off the engine block dowels and disengage the torque converter hub from the end of the crankshaft.

NOTE: Attach a small C-clamp to the edge of the bell housing to hold the torque converter in place during transmission removal.

18. Lower the transmission and remove the assembly from under the vehicle. To remove the torque converter assembly, remove the C-clamp from the edge of the bell housing and carefully slide the assembly out the transmission.
19. Remove the overdrive unit retaining bolts. Very carefully pull the overdrive unit off of the intermediate shaft. A bearing and select spacer may be either on the overdrive piston on the rear of the transmission case, sliding hub or intermediate shaft.

NOTE: Once the overdrive unit is pulled back approximately 1 in., it is free to fall, if unsupported.

TRANSMISSION INSTALLATION

If the transmission was removed because of malfunction caused by sludge, accumulated friction material or metal particles, the oil cooler and lines must be flushed thoroughly. Also, once the transmission is out, the engine block can be inspected for leaking core hole plugs or oil gallery plugs that would otherwise be inaccessable. The transmission and converter must be installed as an assembly; otherwise, the converter drive plate, pump bushing and oil seal will be damaged.
1. Reinstall the overdrive unit, if it had been removed and torque the retaining bolts to 25 ft. lbs.
2. The pump rotors must be aligned so that the converter will fully seat. Slide converter over the input and reaction shaft. Make sure the converter hub slots are engaging the pump inner rotor lugs as the converter is slid inward. Test for full engagement by placing a straightedge on the face of the case. Measure from the straightedge to the lug (square piece with tapped hole for drive plate) on the front cover. The dimension should be at least ½ in. when the converter is pushed all the way into the transmission. This is very important.
3. Attach a small C-clamp to the edge of the converter housing to hold the converter in place during installation. Check the converter drive plate for cracks or distortion and if necessary, replace it. Drive plate to crankshaft bolts should be torqued to 55 ft. lbs. and the torque should be checked even if bolts were not removed.
4. Coat the hole in the crankshaft that will receive the converter hub with Multi-Purpose grease. With transmission and

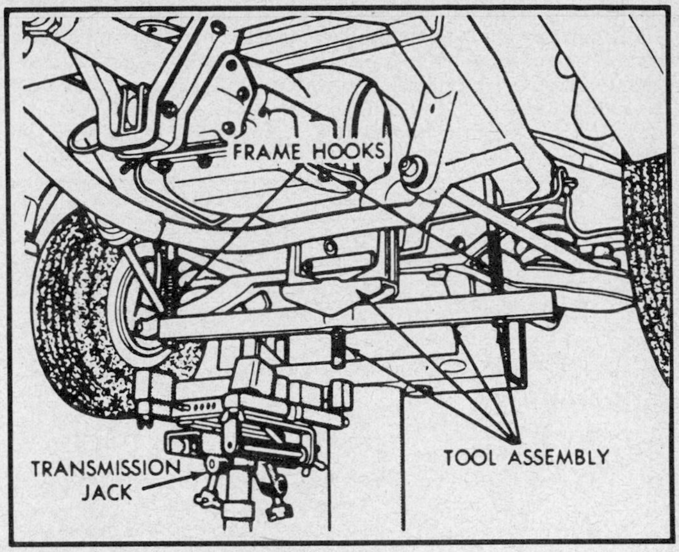
FRAME HOOKS
TRANSMISSION JACK
TOOL ASSEMBLY
Typical engine support fixture

converter assembly held securely on the jack, align as necessary with the engine.
5. Rotate the converter so that the mark that was made on the converter during removal will align with the mark on the drive plate. The offset holes in the plate are located next to the ⅛ in. hole in the inner circle of the plate. Carefully work transmission forward onto the block dowels with converter hub entering the crankshaft opening. Remove the C-clamp before pushing the assembly fully home.

NOTE: If the converter or drive plate has been replaced, or the marks made at removal are lost, examine crankshaft flange to see how the plate and converter must line up because of the offset holes. Then mark the plate and converter (paint, chalk) for assembly.

6. When the transmission is in position, install the converter housing bolts and torque then to 30 ft. lbs. (41 Nm) install the vibration dampener removed from the driveline tunnel, if so equipped.
7. Install the crossmember to the frame and lower the transmission to install the mount on the extension to the crossmember, tighten all bolts. Remove engine support fixture.
8. Using alignment marks made at disassembly, align speedometer adapter and install. Install oil fill tube.
9. Connect the throttle rod to the transmission lever. Connect the gearshift rod and torque the shaft assembly to the transmission lever and frame.
10. When it is necessary to disassemble the linkage rods from levers that use a plastic grommet as retainers, the grommets should be replaced with new ones. Use a suitable tool to force the rod from the grommet in the lever and cut away the old grommet. Use pliers to snap the new grommet into the lever and rod into the grommet.
11. Place the wire connector on the combination back-up and neutral/park starting switch. Connect the wiring to the lockup solenoid wiring connector at the rear of the transmission case.
12. Using a socket on the engine vibration damper, rotate engine clockwise as needed to install converter to drive plate bolts, making sure the marks made previously match up. Torque these bolts to 22 ft. lbs.
13. Install converter access cover, starter and hook up cooler lines.
14. Install the engine to transmission struts, if so equipped. Tighten the bolts holding the strut to transmission before the strut to engine bolts.

15. Reinstall the exhaust system if it was disturbed.
16. Carefully guide the sliding yoke into the extension housing and on the outout shaft splines. Align the marks made at removal and connect the driveshaft to the axle pinion shaft yoke.
17. Adjust the gearshift and throttle linkage.
18. Refill transmission with Mopar® ATF. Road test the vehicle.

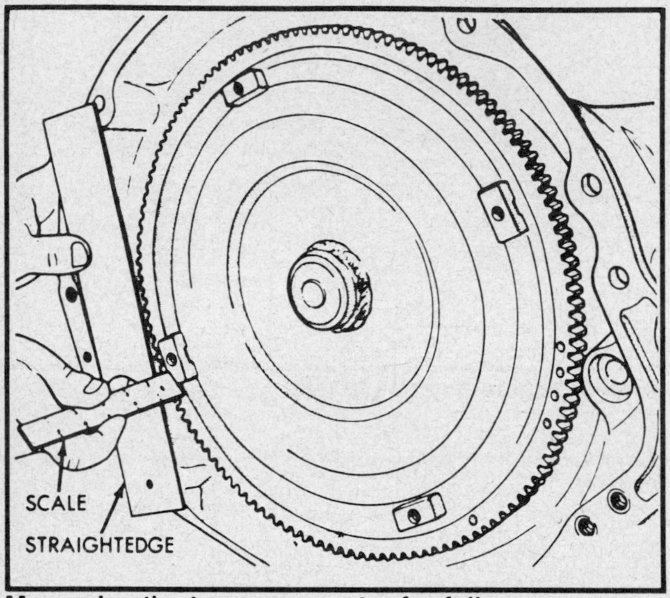
SCALE
STRAIGHTEDGE
Measuring the torque converter for full engagement into transmission

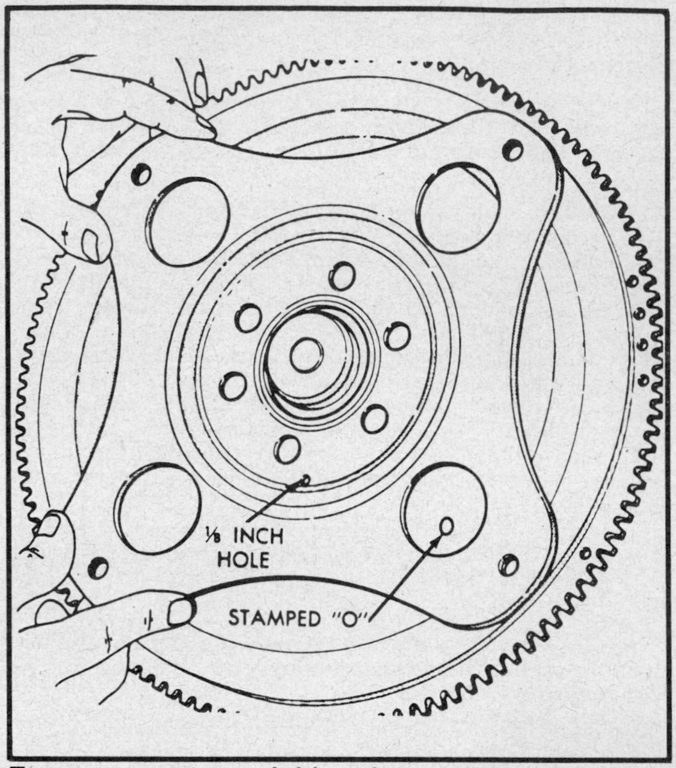
⅛ INCH HOLE
STAMPED "O"
Torque converter and drive plate marking

BENCH OVERHAUL

Before Disassembly

Cleanliness during disassembly and assembly is necessary to avoid further transmission trouble after assembly. Before removing any of the transmission subassemblies, plug all the openings and clean the outside of the of the transmission thoroughly. Steam cleaning or car wash type high pressure equipment is preferable. During disassembly, clean all parts in a suitable solvent and dry each part. Do not use cloth or paper towels to dry parts. Use compressed air only.

ENDPLAY MEASUREMENT

Measuring the endplay before disassembling the unit will indicate whether a thrust washer change is required and will save a considerable amount of time at assembly

1. Mount transmission in a holding fixture or otherwise secure and rig up a dial indicator so that the stylus is against the forward end of the input shaft.
2. Move input shaft to the back as far as it will go and zero the indicator.
3. Pull the input shaft forward to obtain the endplay reading.
4. Mark this figure down for future reference and remove indicator.

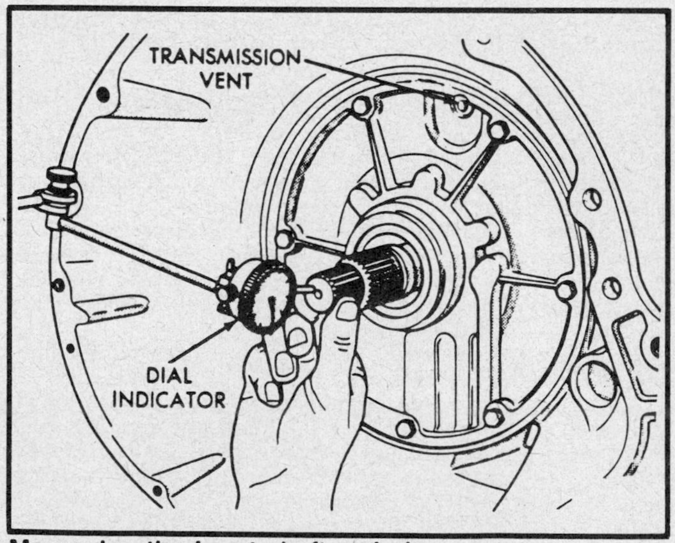
TRANSMISSION VENT
DIAL INDICATOR
Measuring the input shaft end play

Converter Inspection

The torque converter is removed by simply sliding the unit out of the transmission off the input and reaction shaft. If the converter is to be reused, set aside so it will not be damaged. Since the units are welded and have no drain plugs, converters subjected to burnt fluid or other contamination must be replaced.

Transmission Disassembly

OIL PAN AND FILTER

Removal

1. Make sure transmission is held securely either in a stand or fixture.

2. Remove the pan to case bolts and gently tap pan loose. Do not insert a tool between pan and case as a prying tool or case damage may result.

3. Check pan carefully for distortion, straightening the edges of the pan with a straight block of wood and a rubber mallet if necessary. A power driven wire wheel is useful in removing glued on gaskets.

4. There are only 3 screws used to hold the filter to the valve body. A new filter must be installed whenever the pan is removed for transmission service. Remove screws, discard filter. The filter retaining screws may be a special Torx® drive screw head which requires the proper Torx® drive bit to remove the screws.

5. When the pan is eventually reinstalled torque the bolts evenly to 150 inch lbs.

VALVE BODY

Removal

1. Loosen the clamp bolts and remove the gearshift and throttle levers from the transmission.

2. Unscrew the neutral/back-up light switch. Disconnect the lockup solenoid wire from the wiring connector at the rear of the transmission case.

3. Remove the 10 bolts holding the valve body to the transmission. Remove the **E** clip that holds the parking lock rod to the valve body manual lever.

4. Lift valve body, freeing the lock rod from the manual lever. Remove the valve body assembly from its mating surface and place it on a shop bench.

5. At this time, remove the accumulator piston spring and piston, keeping them separate from the servo pistons and springs remove next.

EXTENSION HOUSING

Removal

1. Remove the speedometer pinion and adapter assembly.
2. Remove the extension housing to transmission bolts.

NOTE: When removing or installing the extension housing, the gearshift lever must be in the 1 (low) position. This positions the parking lock control rod rearward so it can be disengaged or engaged with the parking lock sprag.

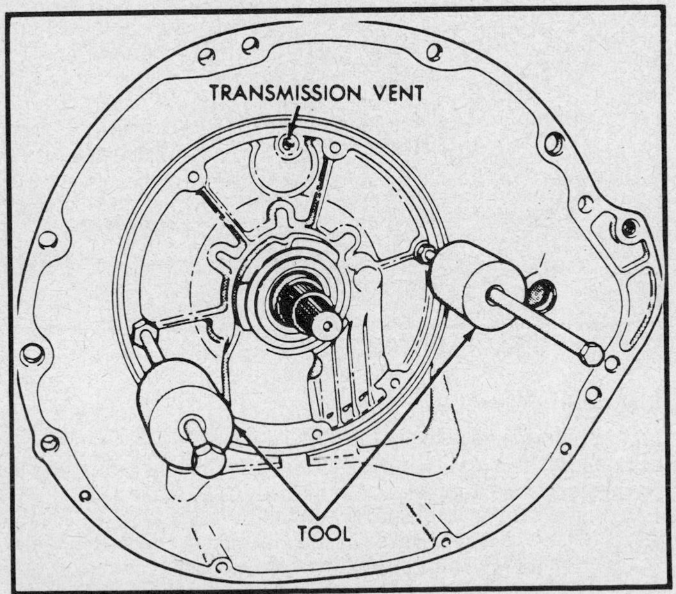
TRANSMISSION VENT
TOOL
Removing the oil pump assembly

3. Remove the screws, plate and gasket from the bottom of the extension housing mounting pad. Spread the snapring from the output shaft bearing

4. With the snapring spread as far as possible, carefully tap the extension housing rearward to allow the parking lock control knob to clear the parking sprag and remove the housing.

5. Using a heavy duty pair of snapring pliers, remove the output shaft bearing snapring and remove the bearing from the shaft.

GOVERNOR AND SUPPORT

Removal

1. Carefully remove the snapring from the weight end of the governor valve shaft (smaller side of governor) slide valve and shaft assembly out of the governor body.

2. Carefully remove the snapring from behind the governor body and slide the governor body and support assembly off of the output shaft.

OIL PUMP AND REACTION SHAFT SUPPORT

Removal

1. Tighten the front band adjusting screw until the band is tight on the front clutch retainer. This prevents the front clutch retainer from coming out with the pump which might cause unnecessary damage to the clutches.
2. Remove the the oil pump housing retaining bolts.
3. Using 2 slide hammers, bump the unit outward evenly.

FRONT BAND AND FRONT CLUTCH

Removal

1. Loosen the front band adjuster screws and remove the band strut. Slide the band out of the case.
2. Slide front clutch assembly out of the case.

INPUT SHAFT AND REAR CLUTCH

Removal

1. Grasp input shaft and slide input shaft and rear clutch assembly out of the case.

2. Be careful not to lose the thrust washer located between the rear end of the input shaft and the forward end of the output shaft.

PLANETARY GEAR ASSEMBLIES, SUN GEAR AND DRIVING SHELL

Removal

1. While supporting the output shaft and driving shell, carefully slide the assembly forward and out of the case, being very careful of the output shaft's surface.

NOTE: While it is possible to remove the planetaries and driving shell at one time, it may be easier to remove the front planetary/annulus first.

REAR BAND AND LOW-REVERSE DRUM

Removal

1. Loosen the band adjusting screw and remove the band strut and link.

2. Remove the low reverse drum from the case.

OVERRUNNING CLUTCH

Removal

Due to the possibility of the overrunning clutch breaking loose from the case on high mileage vehicles, the clutch should be carefully inspected and the springs replaced. The rollers tend to apply unevenly due to decreasing spring pressure. This forces the inner race, or hub to one side moving the rest of the gear train with it. With long-nose pliers, pull out 1 roller and spring and inspect for wear. If badly worn, unit replacement is recommended.

1. Note the position of the overrunning clutch rollers and springs before disassembly, to assist in reassembly.
2. Carefully slide out the clutch hub and remove the rollers and springs.

KICKDOWN (FRONT) SERVO

Removal

1. Compress the kickdown servo piston rod guide until it bottoms in case bore. This can be done by pushing with a suitable tool or by compressing with a C-clamp on a suitable adapter, like a socket, or by using an engine valve spring compressor type tool. (In this case it is not necessary to remove port plug).
2. If not using a C-clamp or compressor, then insert a suitable tool through the port to hold the servo rod guide down far enough to remove the snapring.
3. After snapring removal, press guide back down to release the tool and slowly release rod guide.
4. Remove rod guide, springs and piston rod from the case.

CAUTION

Do not attempt to remove the rod using pliers. If the rod sticks in the case, work it gently to release it. Keep spring and parts separate. Do not allow front and rear servo parts to get mixed.

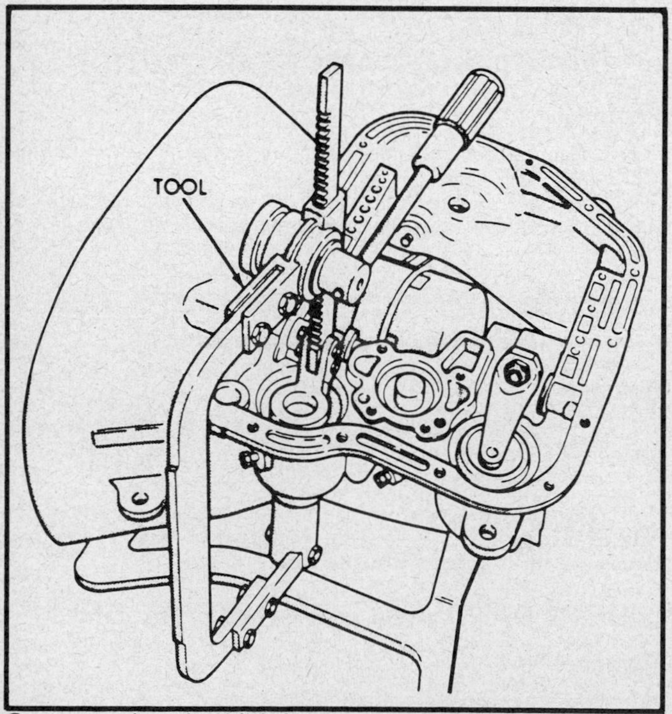
TOOL
Compressing the kickdown servo spring

LOW-REVERSE (REAR) SERVO

Removal

1. Compress the low and reverse servo piston spring by using a engine valve spring compressor tool and remove the snapring.
2. Remove the spring retainer, spring and servo piston and plug assembly from the case.

OVERDRIVE UNIT

Removal

1. Remove the overdrive unit retaining bolts.
2. Very carefully pull the overdrive unit off of the intermediate shaft.
3. A bearing and select spacer may be either on the overdrive piston on the rear of the transmission case, sliding hub or intermediate shaft.

NOTE: Once the overdrive unit is pulled back approximately 1 in., it is free to fall, if it is unsupported.

Unit Disassembly and Assembly

FRONT CLUTCH

Disassembly

1. Remove large snapring and lift off pressure plate and clutch plates.
2. With a spring compressor, squeeze piston spring retainer down far enough to remove snapring. Remove spring and retainer.
3. Turn clutch retainer assembly over and tap on a block of wood to remove piston. Remove and discard seals from hub and piston.

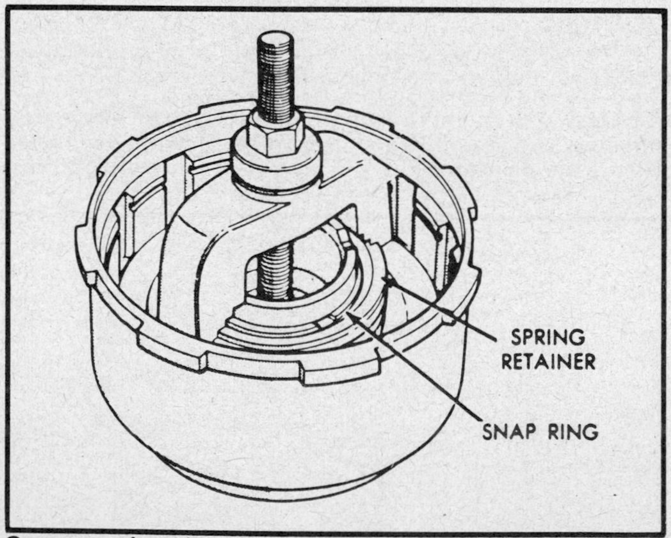
SPRING RETAINER
SNAP RING
Compressing the piston spring

Inspection

1. If plates show any sign of deterioration or wear, they must be replaced. The plates and disc should be flat, they must not be warped or coned shaped.
2. Inspect the facing material on all driving discs. Replace the discs that are charred, glazed or heavily pitted. Discs should also be replaced if they show any evidence of material flaking off or if the facing material can be scrapped off easily. Inspect the driving disc splines for wear or other damage.

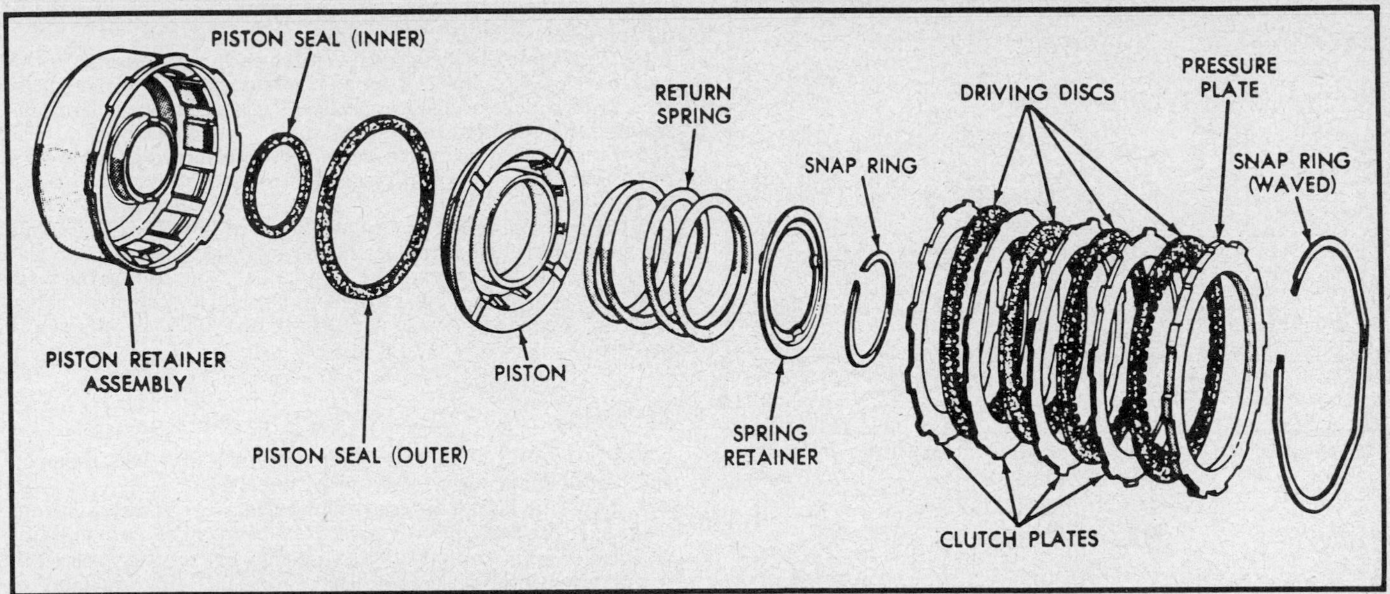
PISTON SEAL (INNER)
PRESSURE PLATE
RETURN SPRING
DRIVING DISCS
SNAP RING
SNAP RING (WAVED)
PISTON RETAINER ASSEMBLY
PISTON
SPRING RETAINER
PISTON SEAL (OUTER)
CLUTCH PLATES

Exploded view of the front clutch assembly

3. Inspect the steel plate and pressure plate surface for burning, scoring or damaged driving lugs. Replace if necessary.
4. Inspect the steel plate grooves in the clutch retainer for smooth surfaces, plates must travel freely in the grooves. Inspect the band contacting surface on the clutch retainer for scores, the contact surface should be protected from damage during disassembly and handling.
5. Note the ball check in the clutch retainer, make sure the ball moves freely. Inspect the piston seal surfaces in the clutch retainer for nicks or deep scratches. Light scratches will not interfere with sealing of the seals.
6. Inspect retainer assembly bushing for wear or scores. If bushing is to be replaced, use the following procedure:
   a. Turn the clutch retainer over (open end down) on a clean smooth surface and place removing head tool SP-3627 or equivalent in the bushing. Install the handle tool C-4171 or equivalent in removing head.
   b. Drive the bushing straight down and out of the clutch retainer bore. Be careful not to cock the tool in the bore.
   c. Turn the clutch retainer over (open end up) on a clean smooth surface. Slide a new bushing on the head tool SP-3626 or equivalent and start them in the clutch retainer bore.
   d. Drive the new bushing into the clutch retainer until it bottoms out. Thoroughly clean the clutch retainer before assembly and installation.
7. Inspect the inside bore of the piston for score marks, if light, remove with a suitable crocus cloth. Inspect the seals for deterioration, wear and hardness. Inspect the piston spring, retainer and snapring for distortion.
8. Clean all parts well, then install seal on retainer hub.

Assembly

1. Lubricate and install the inner seal on the hub of the clutch retainer. Make sure the lip of the seal faces down and properly seated in the groove.
2. Install the outer seal on the clutch piston, with the lip of the seal toward the bottom of the clutch retainer. Apply a coat of ATF fluid to the outer edge of the seals and press the seal to the bottom of its groove around the piston diameter for easier installation of the piston assembly.
3. Place the piston assembly into the retainer and carefully seat the piston in the bottom of the retainer.
4. Install spring, retainer and with spring compressor, install snapring into the hub groove. Remove the spring compressor.
5. Soak all plates in suitable automatic transmission fluid and install 1 steel plate followed by a lined plate (disc) making sure the correct number of plates are used, since it may vary from unit to unit and application to application.
6. Install pressure plate and snapring, be sure that snapring is properly seated.
7. Insert a feeler gauge between the pressure plate and the snapring, under the raised waved part to measure the maximum clearance. The clutch plate clearance should be with 5 discs — 0.075–0.152 in.

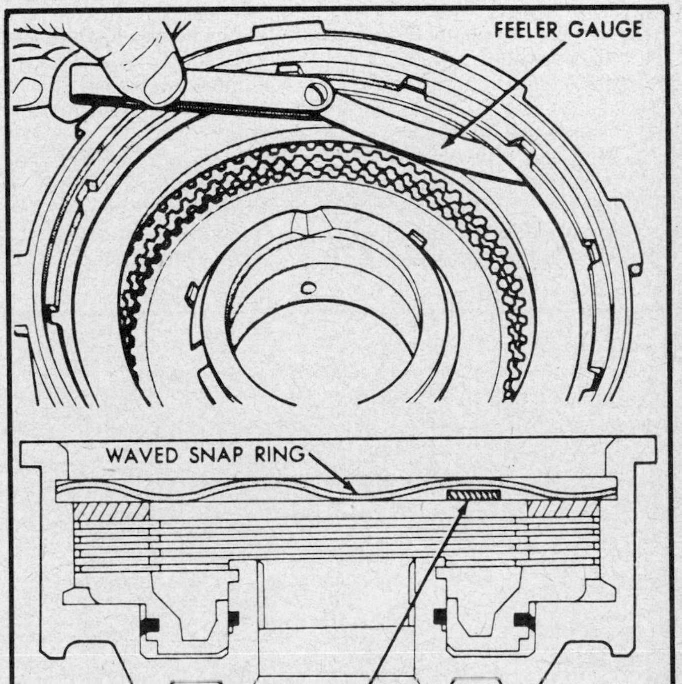
FEELER GAUGE
WAVED SNAP RING
FEELER GAUGE

Measure the front clutch plate clearance

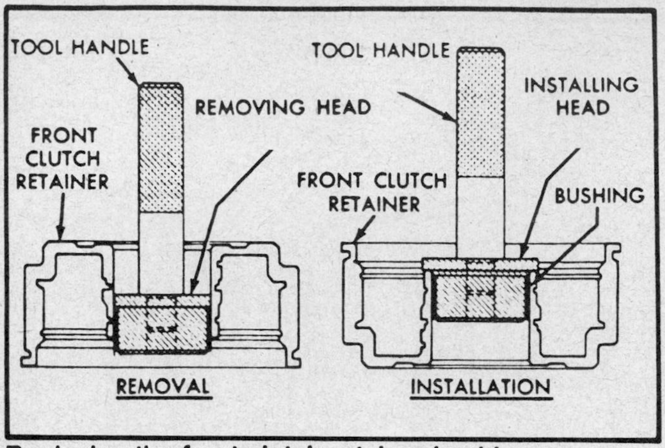
TOOL HANDLE
TOOL HANDLE
REMOVING HEAD
INSTALLING HEAD
FRONT CLUTCH RETAINER
FRONT CLUTCH RETAINER
BUSHING
REMOVAL
INSTALLATION

Replacing the front clutch retainer bushing

REAR CLUTCH

Disassembly

1. Remove the large snapring retaining the pressure plate. This is a selective fit snapring; handle it carefully. Take out pressure plate, clutches and inner pressure plate.
2. Pry one end of the wave spring out of its groove in the clutch retainer and remove along with clutch piston spring.
3. Turn retainer assembly over and tap on a block of wood to remove piston. Remove and discard seals from piston.
4. Remove snapring and press out input shaft if required.

Inspection

1. If plates show any sign of deterioration or wear, they must be replaced. The plates and disc should be flat, they must not be warped or coned shaped.
2. Inspect the facing material on all driving discs. Replace the discs that are charred, glazed or heavily pitted. Discs should also be replaced if they show any evidence of material flaking off or if the facing material can be scrapped off easily. Inspect the driving disc splines for wear or other damage.
3. Inspect the steel plate and pressure plate surface for burning, scoring or damaged driving lugs. Replace if necessary.
4. Inspect the steel plate grooves in the clutch retainer for smooth surfaces, plates must travel freely in the grooves. Inspect the band contacting surface on the clutch retainer for scores, the contact surface should be protected from damage during disassembly and handling.
5. Note the ball check in the clutch retainer, make sure the ball moves freely. Inspect the piston seal surfaces in the clutch retainer for nicks or deep scratches. Light scratches will not interfere with sealing of the seals.
6. Inspect the neoprene seals for deterioration, wear and hardness. Inspect the piston spring and wave spring for distortion.
7. Inspect the teflon or cast iron seal rings on the input shaft for wear. If required, replacement rings will cast iron hooked-joint type. Do not remove the rings unless the conditions warrant. Inspect the rear clutch to front clutch No. 2 thrust washer for wear. Washer thickness should be 0.061–0.063 in., replace as necessary.

Assembly

1. Press input shaft into piston retainer, if it was removed and install snapring.
2. Lube and install inner and outer seals on the clutch piston. Make sure the seal lips face toward the head of the clutch retainer (head of input shaft) and see that they are properly seated in the piston in bottom of retainer.
3. Install the piston assembly into the retainer and with a twisting motion, seat the piston in the bottom of the retainer. Place the clutch piston spring on top of the piston in the retainer. Start one end of the wave spring in its retainer groove, tapping it progressively in place until it is fully seated in the groove.
4. Install inner pressure plate, with the raised portion of the plate resting on the spring.
5. Soak all plates in automatic transmission fluid and install making sure correct No. of plates is used, since it may vary from unit to unit and application to application. Make sure a lined plate is installed first, then a steel, until all are installed. Install the outer pressure plate and selective snapring.
6. Measure rear clutch plate clearance by pressing down firmly on the outer plate (it may require an assistant) then using a feeler gauge between the plate and snapring. The clutch plate clearance should be with 4 discs – 0.032–0.055 in.

NOTE: Rear clutch pack clearance is very important in obtaining proper clutch engagement and shaft quality. Clearance can be adjusted by the use of various thickness outer snaprings. Snaprings are available in 0.060, 0.076 and 0.098 in. thickness.

7. Grease the cupped side of the No. 3 thrust plate and install the cupped side over the input shaft.

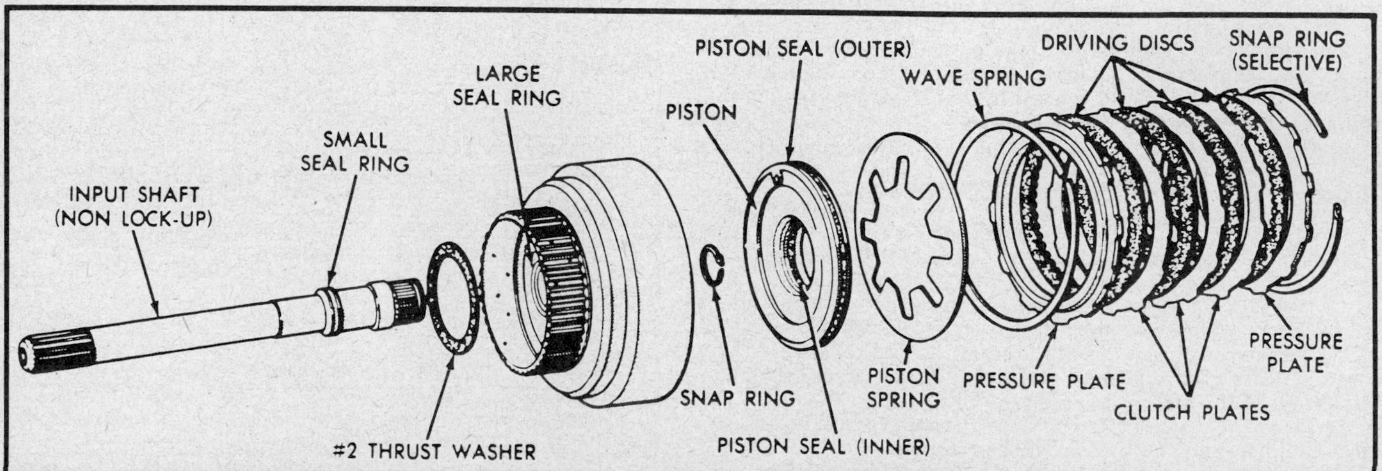
PISTON SEAL (OUTER)
DRIVING DISCS
SNAP RING (SELECTIVE)
LARGE SEAL RING
WAVE SPRING
PISTON
SMALL SEAL RING
INPUT SHAFT (NON LOCK-UP)
PRESSURE PLATE
PISTON SPRING
PRESSURE PLATE
SNAP RING
CLUTCH PLATES
#2 THRUST WASHER
PISTON SEAL (INNER)

Exploded view of the rear clutch

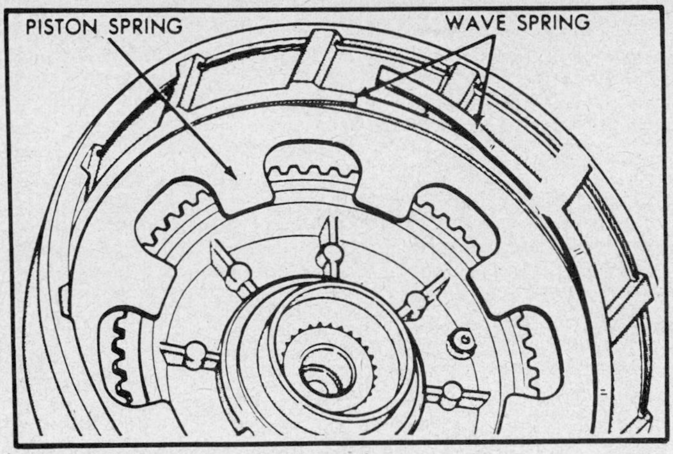
PISTON SPRING
WAVE SPRING

Installing the rear clutch spring

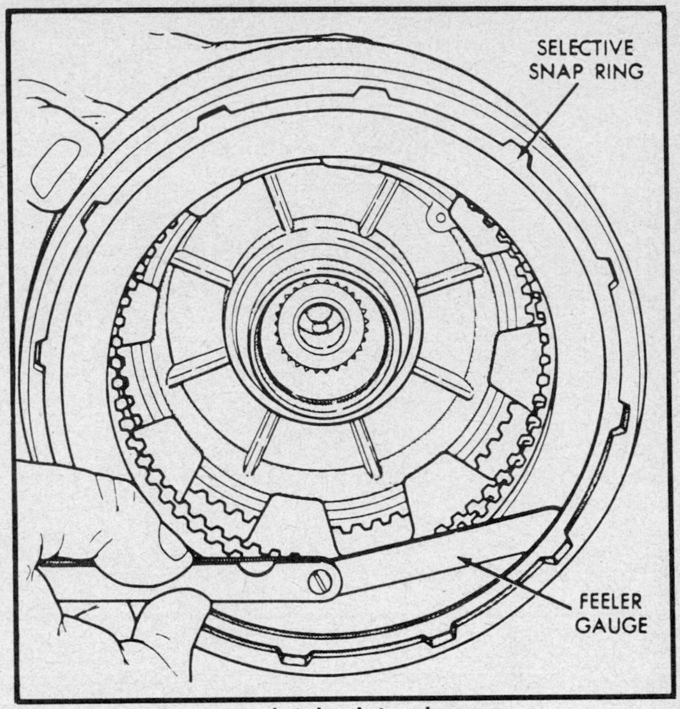
SELECTIVE
SNAP RING
FEELER
GAUGE

Measuring the rear clutch plate clearance

PLANETARY GEAR TRAIN

Endplay

Before disassembly, endplay of the planetary gear assemblies, sun gear and driving shell should be measured. Stand the gear train upright with the forward end of the output shaft on a block of wood. This causes all the parts to slide forward against the selective snapring on the front of the shaft. Insert feeler gauges between the rear annulus gear support hub and the shoulder on the output shaft. The clearance should be between 0.005–0.048 in. If clearance is not within these sizes, replace thrust washers, any worn parts and selective thickness snaprings at assembly.

Disassembly

1. Remove the selective No. 3 thrust washer from the tip of the output shaft.
2. Remove the selective snapring and slide the front planetary assembly from the shaft.
3. Remove the snapring and No. 4 thrust washer from the hub of the planetary; slide front annulus gear and support off the planetary gear set. Remove the No. 5 thrust washer from the front side of the planetary gear assembly. Remove the No. 6 thrust washer from the rear side of the planetary gear assembly. If necessary, remove the snapring from the front of the annulus gear to separate the support from the annulus gear.
4. Slide the sun gear, driving shell and the rear planetary off of the output shaft.
5. Lift the sun gear and the driving shell off of the rear planetary assembly. Remove the snapring and the No. 8 thrust plate (steel) from the sun gear. Slide the sun gear out of the driving shell and remove the snapring and the No. 7 thrust plate from the opposite end of the sun gear, if necessary.
6. Remove the No. 9 thrust washer from the forward side of the rear planetary and separate the planetary from the rear annulus.
7. Remove the No. 10 thrust washer from the rear side of the planetary assembly and if necessary, remove the snapring from the rear annulus gear to separate the support from the annulus gear.

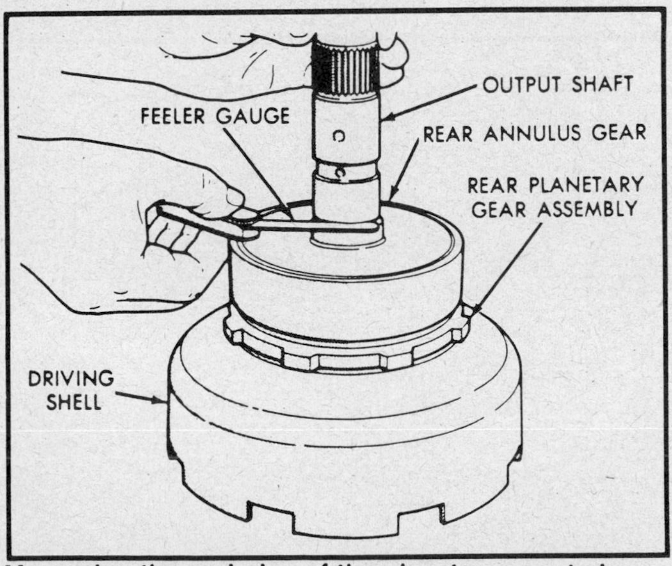
OUTPUT SHAFT
FEELER GAUGE
REAR ANNULUS GEAR
REAR PLANETARY
GEAR ASSEMBLY
DRIVING
SHELL

Measuring the end play of the planetary gear train

Inspection

1. Inspect all oil passages in the shaft and make sure they are open and clean.
2. Inspect the bearing surfaces on the output shaft for nicks, burrs, scores or other damage. Light scratches , small nicks or burrs can be removed with crocus cloth or a fine stone.
3. Inspect the speedometer drive gear for any nicks or burrs and remove with a sharp edged stone.
4. Inspect the bushing in the sun gear for wear or scores, replace the sun gear assembly if the bushings are damaged.
5. Inspect all thrust washers for wear and scores, replace if damaged or worn below specifications.
6. Inspect the thrust faces of the planetary gear carriers for wear, scores or other damage, replace as required. Inspect the planetary gear carrier for cracks and pinions for broken or worn gear teeth and for broken pinion shaft welds.
7. Inspect the annulus gear and driving gear teeth for damage. Replace distorted lock rings. Clean all parts well, blow dry.

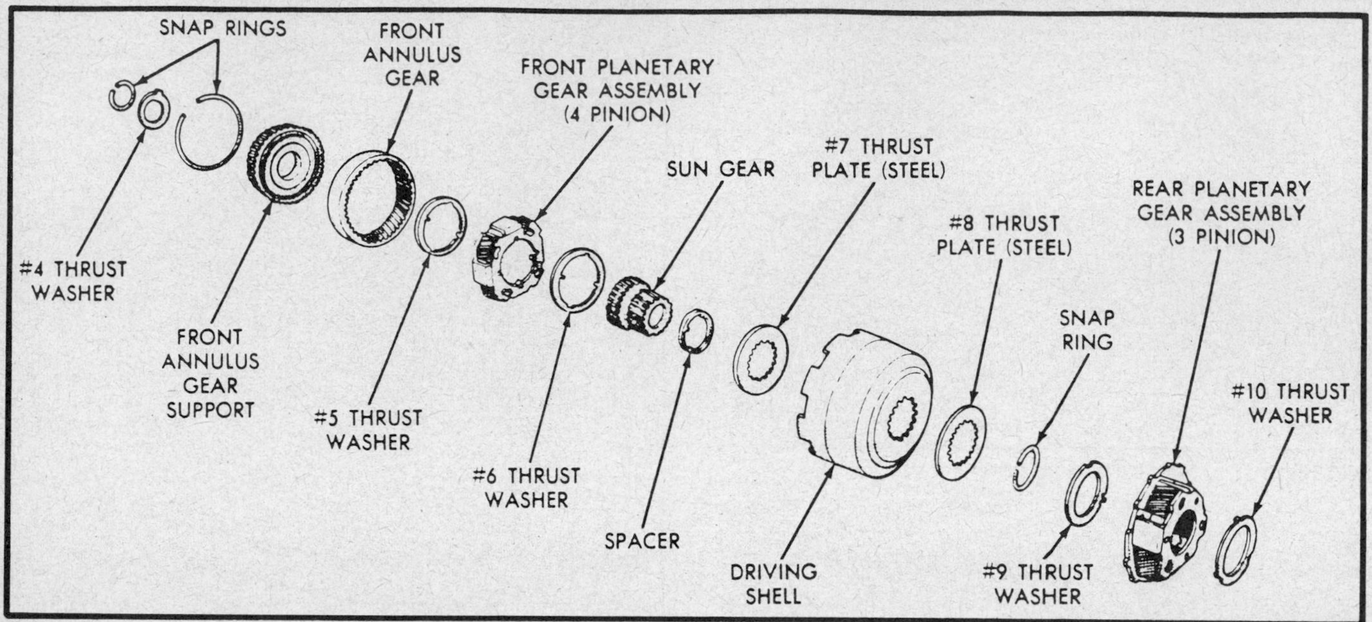
SNAP RINGS
FRONT ANNULUS GEAR
FRONT PLANETARY GEAR ASSEMBLY (4 PINION)
#7 THRUST PLATE (STEEL)
SUN GEAR
REAR PLANETARY GEAR ASSEMBLY (3 PINION)
#8 THRUST PLATE (STEEL)
#4 THRUST WASHER
SNAP RING
FRONT ANNULUS GEAR SUPPORT
#10 THRUST WASHER
#5 THRUST WASHER
#6 THRUST WASHER
SPACER
DRIVING SHELL
#9 THRUST WASHER
Exploded view of the planetary gear train

Assembly

1. To assemble, place rear annulus gear support in annulus gear and install snapring.
2. Put the No. 10 thrust washer on the back of the rear planetary, insert into the rear annulus and follow with the No. 9 thrust washer on the front side of the planetary.
3. From the back of the rear annulus gear, insert the output shaft, slowly working it through the planetary until splines on the shaft are meshed with the splines of annulus gear support.
4. Install the No. 7 thrust plate and snapring on one end of the sun gear. Insert the sun gear through the front side of the driving shell, install No. 8 thrust plate and snapring.
5. Carefully slide the driving shell and sun gear assembly onto the output shaft, meshing the sun gear teeth with the rear planetary pinion teeth.
6. Place the front annulus gear support in the annulus gear and install the snapring.
7. Put the No. 5 thrust washer on the front side of the front planetary gear assembly. Put the front planetary in front of the front annulus gear, install the No. 4 thrust washer over the planetary hub and install the snapring. Put the No. 6 thrust washer on the rear side of the planetary gear assembly.
8. Carefully work the front planetary and annulus gear assembly onto the shaft meshing the planetary pinions with the sun gear.
9. With all components in place, install the selective snapring on the front of the output shaft. Remeasure the endplay. The clearance can be adjusted by the use of various thickness snaprings. Snaprings are available in 0.042, 0.064 and 0.084 in. thickness.

OVERRUNNING CLUTCH

Inspection

Check rollers for signs of wear or out-of-round. They must be free from flat spots and chipped edges. Check the cam for wear and the springs for distortion. If the overrunning clutch cam or spring retainer is worn, or damaged, the service replacement contains a cam with tapped holes and retaining bolts.

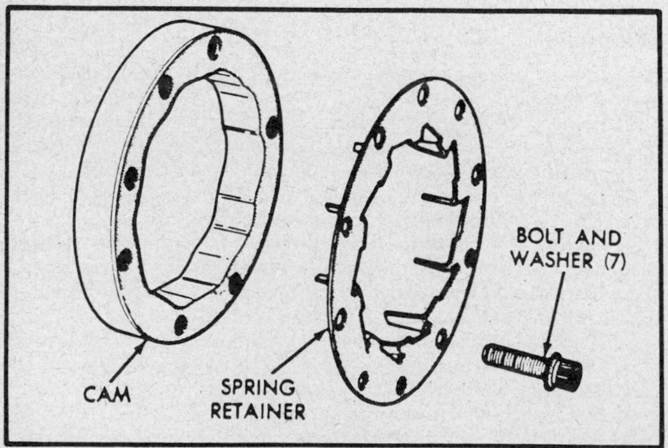
BOLT AND WASHER (7)
CAM
SPRING RETAINER
Overrunning clutch replacement

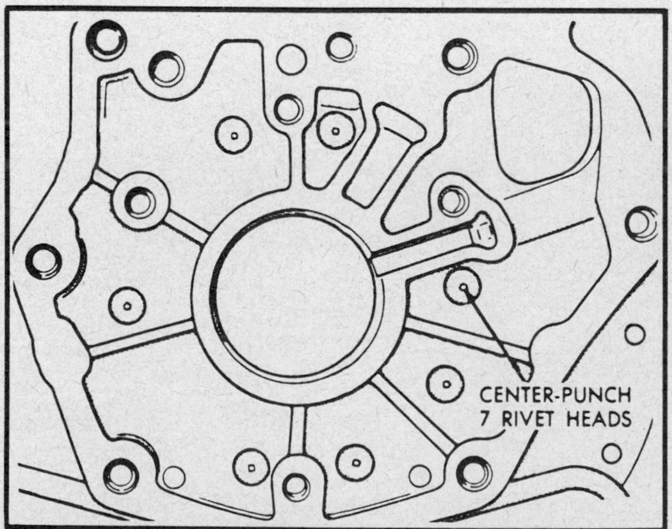
CENTER-PUNCH 7 RIVET HEADS
Center punch rivet heads

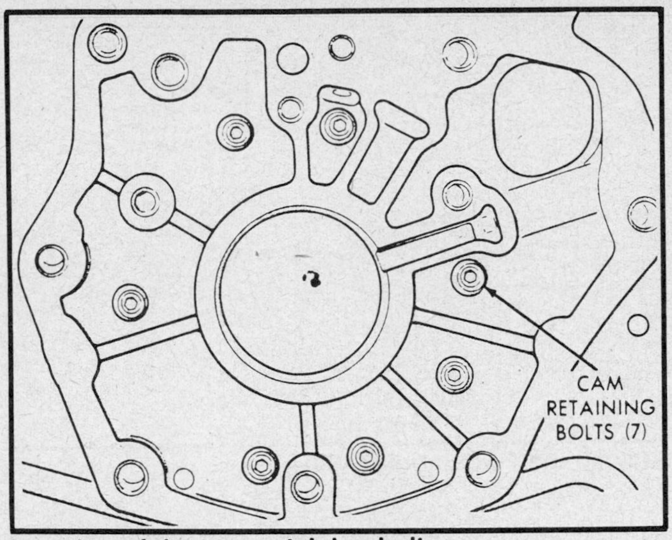
CAM
RETAINING
BOLTS (7)

Location of the cam retaining bolts

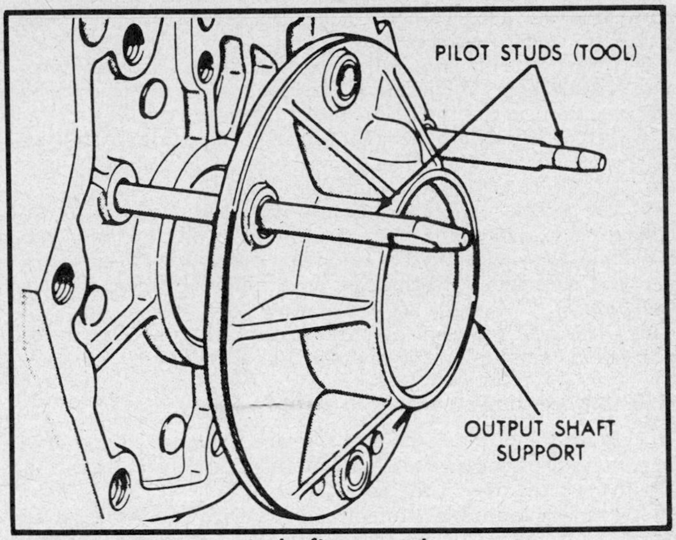
PILOT STUDS (TOOL)
OUTPUT SHAFT
SUPPORT

Installing the output shaft support

Overrunning Clutch Cam Replacement

The service parts are retained in the case with bolts instead of rivets. To install proceed as follows:

1. Remove the 4 bolts holding the output shaft support to the transmission case and remove, tapping with a soft hammer if required.
2. Center punch the rivets exactly in the center of each rivet head.
3. Using a 3/8 in. drill bit, drill through each rivet head being careful not to drill into the transmission case. Remove remainder of rivet head and drive out rivets with a punch.
4. Enlarge carefully the holes with a 17/64 in. drill bit. Clean case so that no clips remain and check for burrs around drilled holes.
5. Drive cam from case.
6. Position new cam and spring retainer in case, align bolt holes and install, but do not tighten bolts. Note that the cone shaped washers on the bolts must be installed to that the inner diameter is coned toward the bolt head.
7. If necessary, tap the cam into the case and tighten bolt evenly to 100 inch lbs.
8. Install output shaft support. It may require using 2 bolts with heads cut off, as pilot studs to help align the support. Tap firmly into case with soft hammer.
9. Tighten bolts to 150 inch lbs. (17 Nm).

OIL PUMP

Disassembly

Because this unit is equipped with a lockup torque converter, it is important that the oil pump be within the specified clearance limits. In any overhaul, the oil pump bushing should be replaced.

1. Remove the 6 bolts attaching the pump to the reaction shaft support and remove the support. Remove and discard O-ring and oil seal, which may need to be driven out with a blunt punch.
2. Apply a small amount of machinist's bluing on one spot on the inner and outer rotors and scribe a mark across the rotors for correct tooth alignment at reassembly.

Inspection

1. Inspect reaction shaft support for cracks or burrs on the sealing ring grooves; seal rings will have to be removed to allow clearance for the No. 1 thrust washer to be removed. Inspect pump rotors for scoring or pitting and clean well.
2. Inspect front clutch piston retainer to reaction shaft sup-

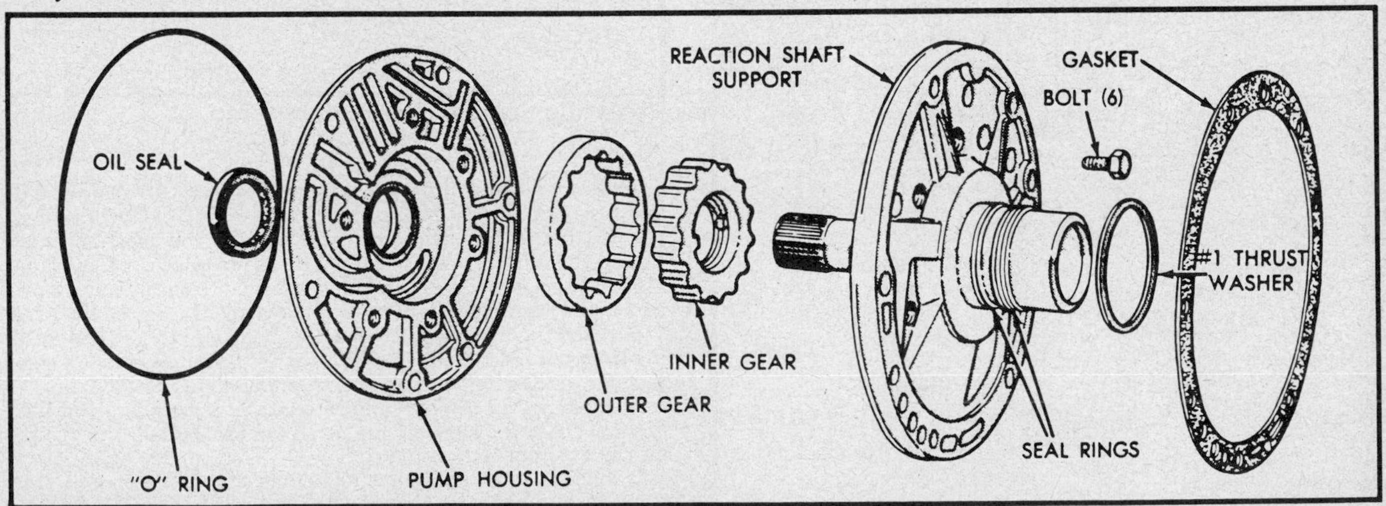
OIL SEAL
REACTION SHAFT
SUPPORT
GASKET
BOLT (6)
#1 THRUST
WASHER
INNER GEAR
OUTER GEAR
SEAL RINGS
"O" RING
PUMP HOUSING

Exploded view of the oil pump and reaction shaft support

port thrust washer for wear. Washer thickness should be 0.061–0.063 in., replace if necessary.

3. Inspect the machined surfaces on the pump body and reaction shaft support for nicks and burrs. Inspect the pump body and reaction shaft support bushings for wear or scores.

4. Inspect the pump gears for scoring and pitting. With gears cleaned and installed in the pump body, place a straight edge across the face of the gears and pump body.

5. Use a feeler gauge to measure the clearance between the straightedge and face of the gears. Clearance limits from 0.001–0.0025 in. Also, measure the rotor tip clearance between the inner and outer rotor teeth. Clearance limits are from 0.0045–0.0095 in.

6. Clearance between the outer gears and its bore in the oil pump body should be 0.0035–0.0075 in.

Bushing Replacement

1. Place the pump body (seal side down) on bench and drive bushing out of bore with a suitable bushing driver. Be careful not to cock the tool in the bore.

2. Using a suitable bushing installer, drive in a new bushing until it bottoms in the oil pump cavity. Be careful not to cock the tool during installation. When the bushing is in place, stake it in several places with a blunt punch. A gentle tap will do.

3. Using a suitable tool, clean off the burrs and high spots caused by the stake. Don't use a file or any tool that would remove too much metal. Rinse pump body in solvent.

4. Install the pump rotors in the pump body. Place a straightedge across the rotor faces and pump body. Using a feeler gauge measure the clearance between the straightedge and feeler gauge. Clearance limits are 0.001–0.003 in. (0.254–0.762mm).

5. Turn the rotors so that the center of a tooth on each rotor is aligned. Measure the clearance between the tips of the teeth. Measure 4 times, turning the inner rotor ¼ turn between measurements. Rotor tip clearance should be 0.005–0.010 in. (0.1270–0.2540mm). Measure the clearance between the outer surface of the outer rotor and the pump bore. The clearance should be 0.004 to 0.008 in. (0.1016 to 0.2043mm).

6. Inspect reaction shaft support bushing. If the bushing has failed or is badly worn, then inspect support assembly. If the input shaft or rear clutch retainer seal ring lands are worn or badly grooved, the entire support assembly should be replaced. If just the bushing is to be replaced, use the following procedure.

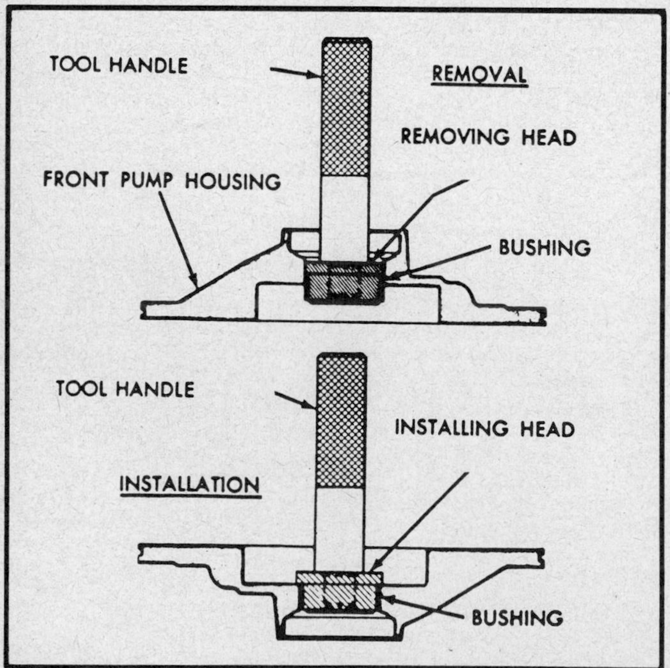
TOOL HANDLE
REMOVAL
REMOVING HEAD
FRONT PUMP HOUSING
BUSHING
TOOL HANDLE
INSTALLING HEAD
INSTALLATION
BUSHING
Replacing the oil pump bushing

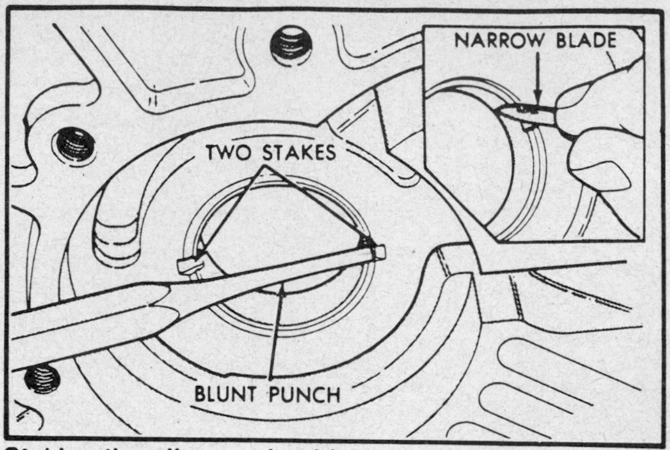
NARROW BLADE
TWO STAKES
BLUNT PUNCH
Staking the oil pump bushing

Reaction Shaft Bushing Replacement

Do not clamp any part of the reaction shaft or support in a vise. Special type removal tools grab the bushing from the inside which is then bumped out with the tool slide hammer. Do not damage support assembly.

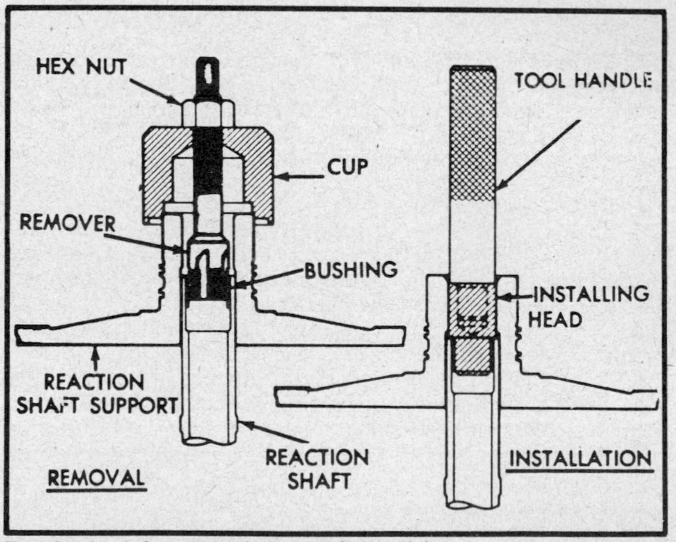
HEX NUT
CUP
REMOVER
BUSHING
REACTION SHAFT SUPPORT
REACTION SHAFT
REMOVAL
TOOL HANDLE
INSTALLING HEAD
INSTALLATION
Replacing the reaction shaft bushing

1. Assemble bushing remover tool SP-5324, cup tool SP-3633 and hex tool SP-1191 or equivalents.

2. With the cup held firmly in against the reaction shaft, thread the remover into the bushing as far as possible by hand.

3. Using a wrench, screw the remover into the bushing 3 or 4 additional turns to firmly engage the threads into the bushing.

4. Turn the hex head down against the cup to pull the bushing from the reaction shaft. Thoroughly clean the reaction shaft to remove the chips made by the remover threads.

5. Lightly grip the bushing in a vise or with pliers and back the tool out of the bushing. Be careful not to damage the threads on the bushing remover.

6. Slide a new bushing on the bushing installer and start them into the bore of the reaction shaft.

7. Support the reaction shaft upright on a clean smooth sur-

face and drive the bushing into the shaft until the tool bottoms out.

8. Thoroughly clean the reaction shaft support assembly before installation.

Assembly

1. Place the reaction shaft support in assembling tool C-3759, with the hub of the support tool resting on a smooth flat surface bench. Screw 2 pilot studs into the threaded holes of the reaction shaft support flange.

2. Assemble and place the pump gears in the center of the support.

3. Lower the pump body over the pilot studs, insert tool C-3756 or equivalent through the pump body and engage the pump inner gear.

4. Rotate the pump gears with the tool to center the gears in the pump body, then with the pump body firm against the reaction shaft support, tighten the clamping tool securely.

5. Invert the pump and reaction shaft support assembly with clamping tool intact. Install the support to pump body bolts and tighten to 175 inch lbs. (20 Nm). Remove the clamping tool, pilot studs and gear alignment tool.

NOTE: If difficulty was encountered removing the pump at first, it may be well to expand the case with a heat lamp before attempting to install pump.

6. Place a new oil seal in the opening of the pump housing (lip of the seal facing inward), drive a seal into the housing until the tool bottoms out.

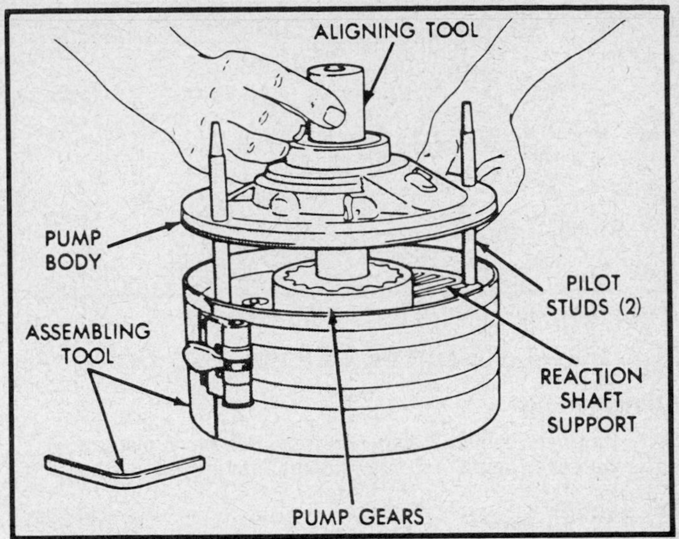
ALIGNING TOOL
PUMP BODY
PILOT STUDS (2)
ASSEMBLING TOOL
REACTION SHAFT SUPPORT
PUMP GEARS
Assembling the oil pump and reaction shaft support

OIL PUMP SEAL

Removal and Installation

The pump oil seal can be replaced without removing the pump and reaction shaft support assembly from the transmission case.

1. Screw the seal remover tool C-3981 or equivalent into the seal, tighten the screw portion of the tool to remove the seal.

2. To install a new seal, place the seal in the opening of the pump housing (lid side facing inward).

3. Using tool C-4193 and handle tool C-4171 or equivalent, drive the new seal into the housing until the tool bottoms out.

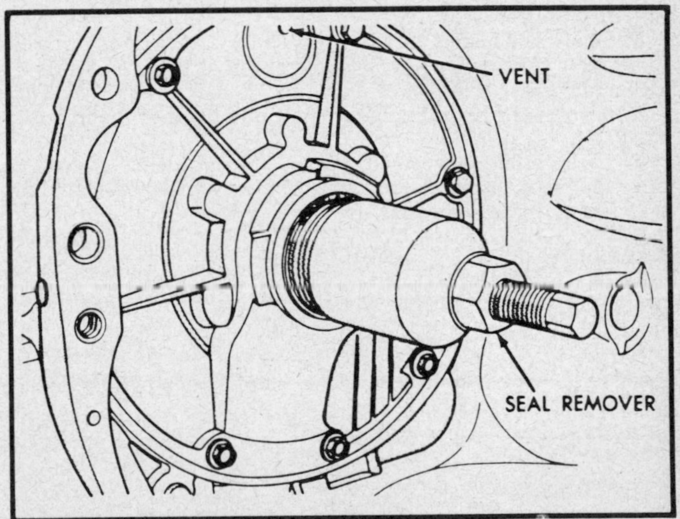
VENT
SEAL REMOVER
Removing the oil pump seal

KICKDOWN SERVO AND BAND

Disassembly

1. Remove the small snapring from the servo piston.

2. Remove the washer, spring and piston rod assembly from the servo piston.

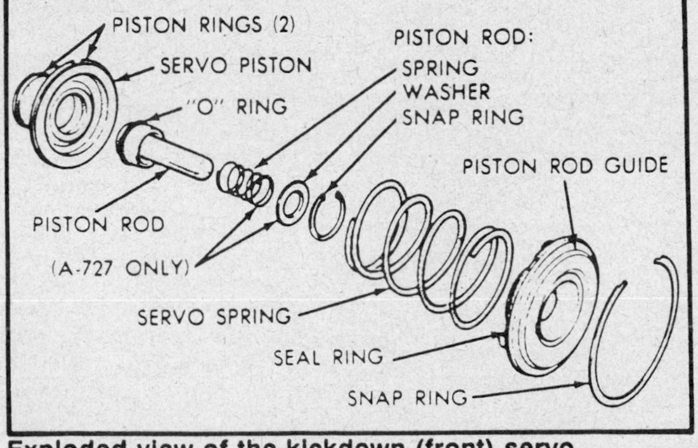
PISTON RINGS (2)
SERVO PISTON
"O" RING
PISTON ROD:
SPRING
WASHER
SNAP RING
PISTON ROD GUIDE
PISTON ROD
(A-727 ONLY)
SERVO SPRING
SEAL RING
SNAP RING
Exploded view of the kickdown (front) servo

Inspection

1. Inspect carefully the piston and guide seal rings for wear and make sure they turn freely. It usually is not necessary to remove the seal rings unless they show signs of wear or deterioration.

2. Check the piston bore in the case for scores or any marks. Make sure springs are not distorted and that guide slides freely on piston rod.

3. Inspect the band lining for wear and bond of the lining to the band. Inspect the lining for black burn marks, glazing, non-uniform wear pattern and flaking.

4. If the lining is worn so grooves are not visible at ends or any portion of the bands, replace the band. Inspect the band for distortion or cracked ends.

Assembly

1. Grease the O-ring and install the piston rod.

2. Install the piston rod into the servo piston.

3. Install the spring, flat washer and snapring to complete the assembly.

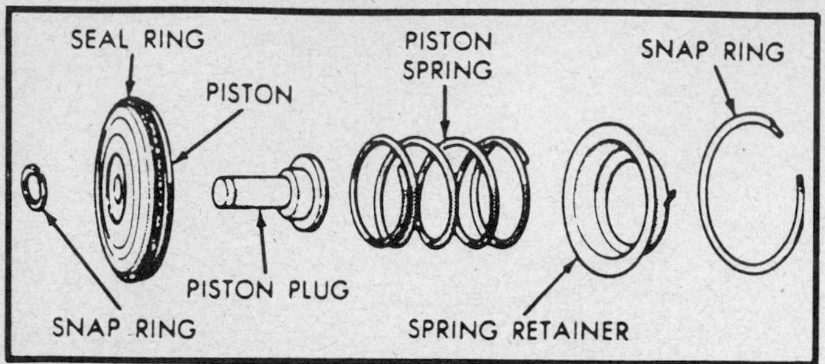
SEAL RING
PISTON
PISTON SPRING
SNAP RING
PISTON PLUG
SNAP RING
SPRING RETAINER
Exploded view of the low-reverse servo

LOW—REVERSE SERVO AND BAND

Disassembly

1. Remove the small snapring from the servo piston.
2. Remove the piston plug and cushion spring.

Inspection

1. Inspect carefully the piston and guide seal rings for wear and make sure they turn freely. It usually is not necessary to remove the seal rings unless they show signs of wear or deterioration.
2. Check the piston bore in the case for scores or any marks. Make sure springs are not distorted and that guide slides freely on piston rod.
3. Inspect the band lining for wear and bond of the lining to the band.
4. If the lining is worn so grooves are not visible at ends or any portion of the bands, replace the band. Inspect the band for distortion or cracked ends.

Assembly

1. Lubricate and install the piston plug and cushion spring in the piston.
2. Secure it with a snapring.

VALVE BODY

Disassembly

CAUTION

Do not clamp any part of the valve body or plate in a vise, for this will cause sticking valves or excessive leakage or both. When removing and installing valves or plugs, slide them in or out very carefully. Do not use force to remove or install the valves.

NOTE: When disassembling the valve body identify all valve springs with a tag for assembly reference later. Also, oil filter screws are longer than transfer plate screws. Do not mix them.

1. Remove the E-clip and park control rod from the manual lever.
2. Place the valve body on a suitable clean work bench. Remove the screws from the oil filter and remove the filter.
3. Remove the top and bottom screws from the spring retainer and adjustment screw bracket.
4. Hold the spring retainer firmly against the spring force while removing the last retaining screw from the side of the valve body. Do not alter the settings of the throttle pressure adjusting screws, while removing the line pressure adjusting screw assembly and switch valve regulator springs and valves from their bores.
5. Slide the switch valve and regulator valve out of their bores.
6. Remove the screws from the lockup module and carefully remove the tube (note the long end of the tube for assembly) and lockup module. Disassemble lockup module, tagging the springs.
7. Remove the transfer plate retaining screws and lift off the transfer plate and separator plate assembly.
8. Remove the lockup solenoid retaining screw and pull the solenoid from its bore in the transfer plate. Remove the screws from the separator plate and separate parts for cleaning.
9. Remove the rear clutch check ball, reverse servo ball check

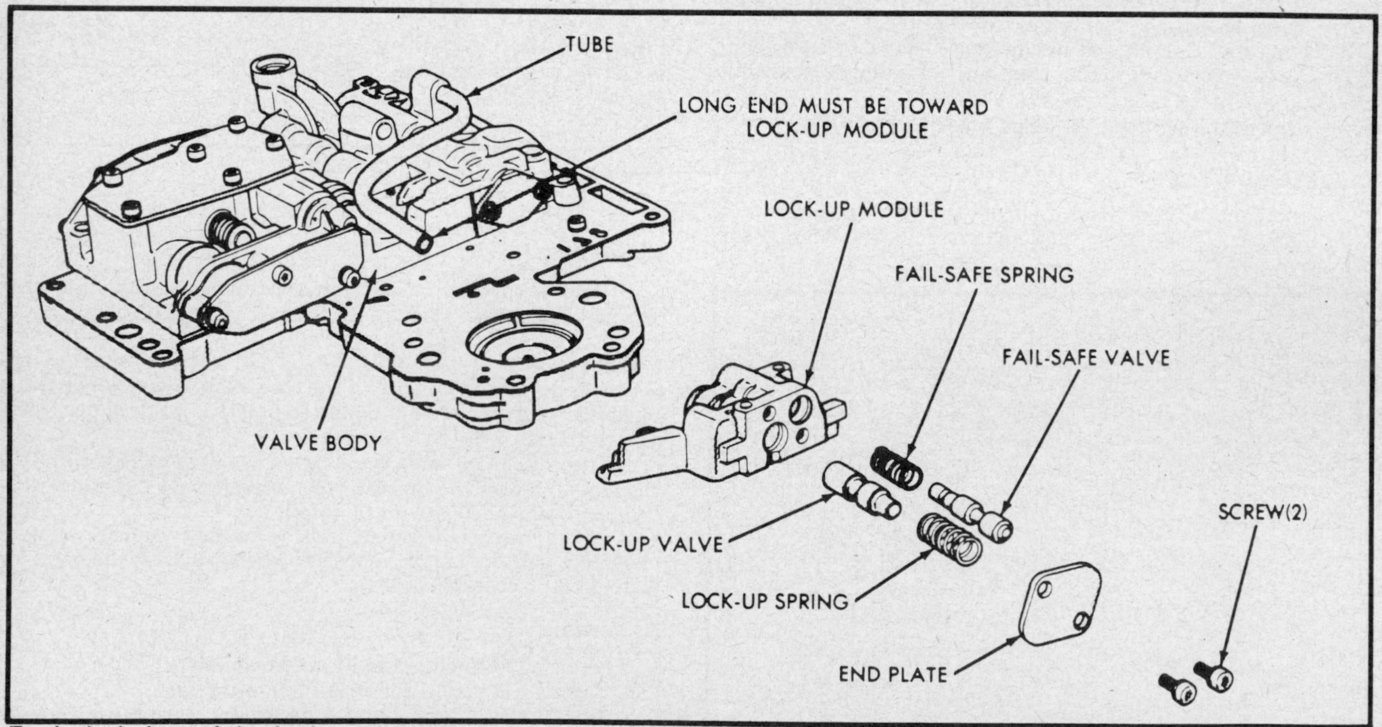
TUBE
LONG END MUST BE TOWARD LOCK-UP MODULE
LOCK-UP MODULE
FAIL-SAFE SPRING
FAIL-SAFE VALVE
VALVE BODY
SCREW(2)
LOCK-UP VALVE
LOCK-UP SPRING
END PLATE
Exploded view of the lockup module

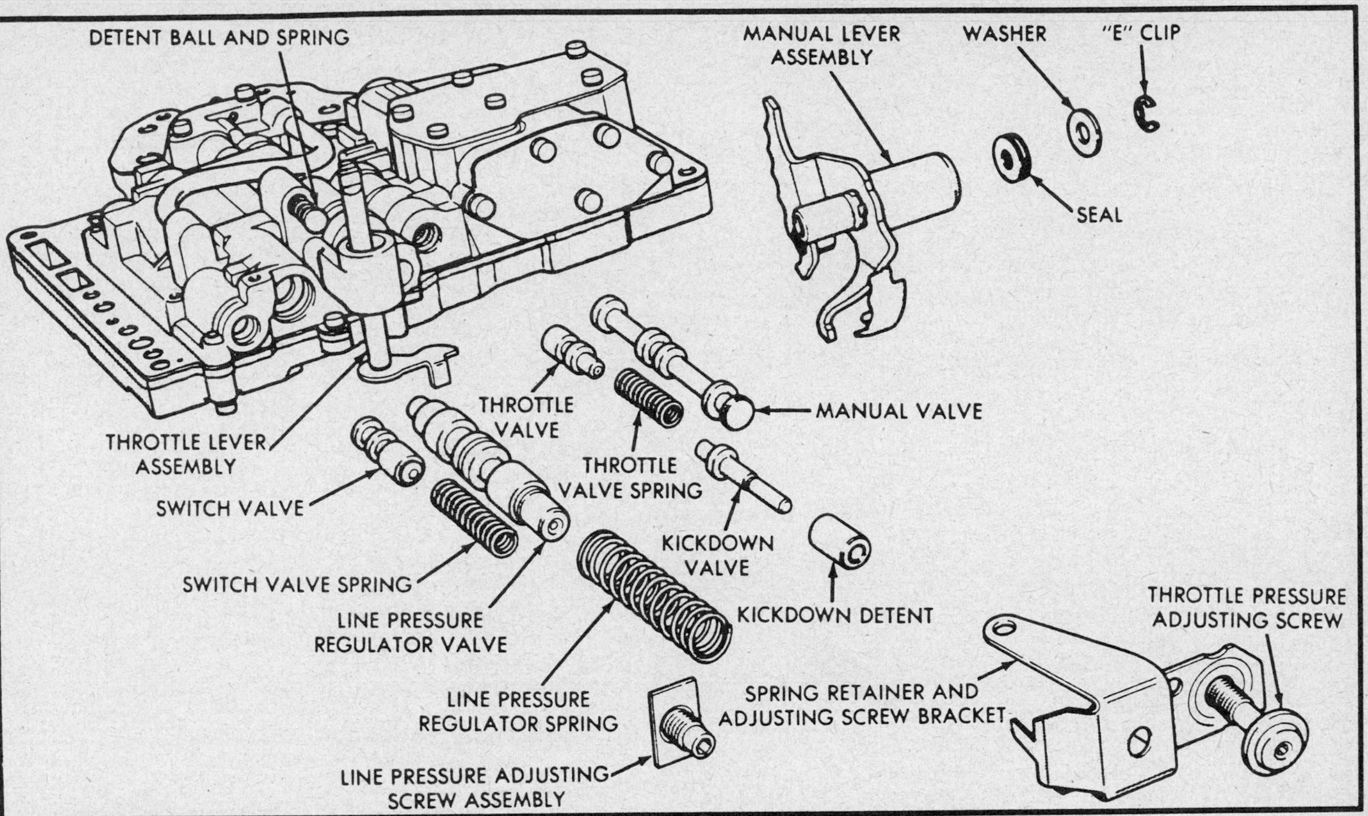
DETENT BALL AND SPRING
MANUAL LEVER ASSEMBLY
WASHER
"E" CLIP
SEAL
THROTTLE VALVE
MANUAL VALVE
THROTTLE LEVER ASSEMBLY
THROTTLE VALVE SPRING
SWITCH VALVE
KICKDOWN VALVE
SWITCH VALVE SPRING
KICKDOWN DETENT
THROTTLE PRESSURE ADJUSTING SCREW
LINE PRESSURE REGULATOR VALVE
LINE PRESSURE REGULATOR SPRING
SPRING RETAINER AND ADJUSTING SCREW BRACKET
LINE PRESSURE ADJUSTING SCREW ASSEMBLY

Exploded view of the pressure regulators and manual controls

and line pressure regulator valve screen from the separator plate, clean all parts.

10. Remove the 7 balls from the valve body.
11. Disasemble the lockup module as follows:
    a. Remove the end cover.
    b. Remove the lockup spring and valve.
    c. Remove the fail-safe valve and spring.
    d. Be sure to tag these springs as they are removed, for reassembly identification.

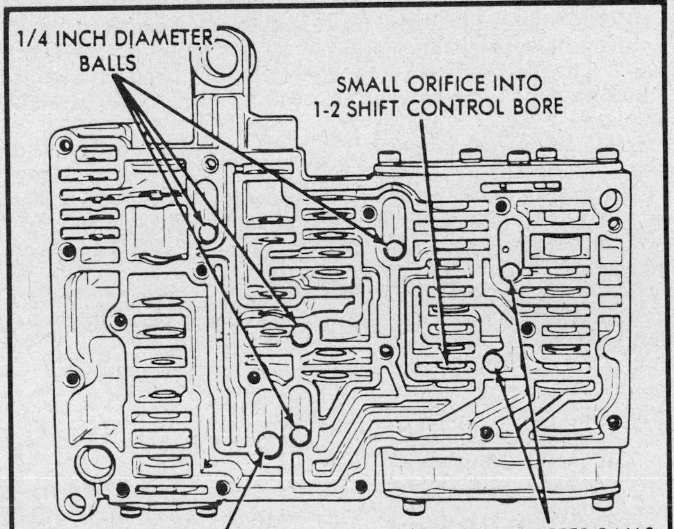
1/4 INCH DIAMETER BALLS
SMALL ORIFICE INTO 1-2 SHIFT CONTROL BORE
1/4 INCH DIAMETER BALLS
11/32 INCH DIAMETER BALL

Location of the seven steel check balls in the valve body

12. Turn the valve body over and remove the shuttle valve cover plate. Remove the governor plug end plate and slide out the shuttle valve throttle plug and spring, the 1–2 shift valve governor plug and the 2–3 shift valve governor plug.
13. To remove the shuttle valve, remove the E-clip and slide the shuttle valve out of its bore. Also remove the secondary spring and guides which were held in by the E-clip.
14. To remove the manual lever and throttle lever, first remove the E-clip and washer. The manual lever detent ball is

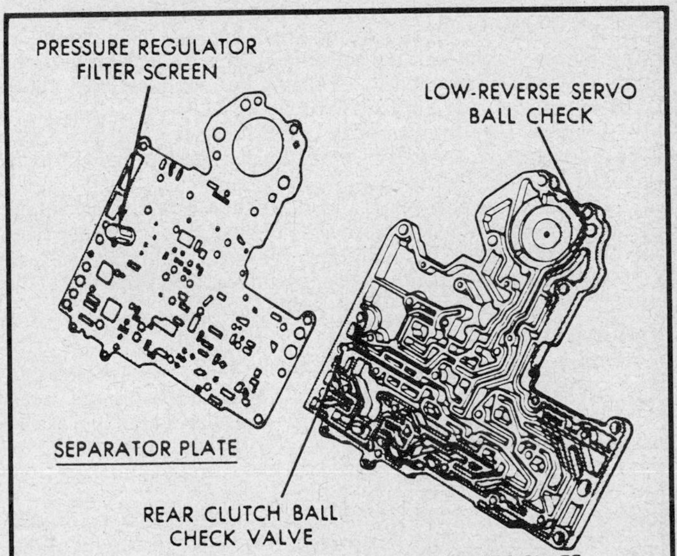
PRESSURE REGULATOR FILTER SCREEN
LOW-REVERSE SERVO BALL CHECK
SEPARATOR PLATE
REAR CLUTCH BALL CHECK VALVE
TRANSFER PLATE

Typical transfer plate and separator plate

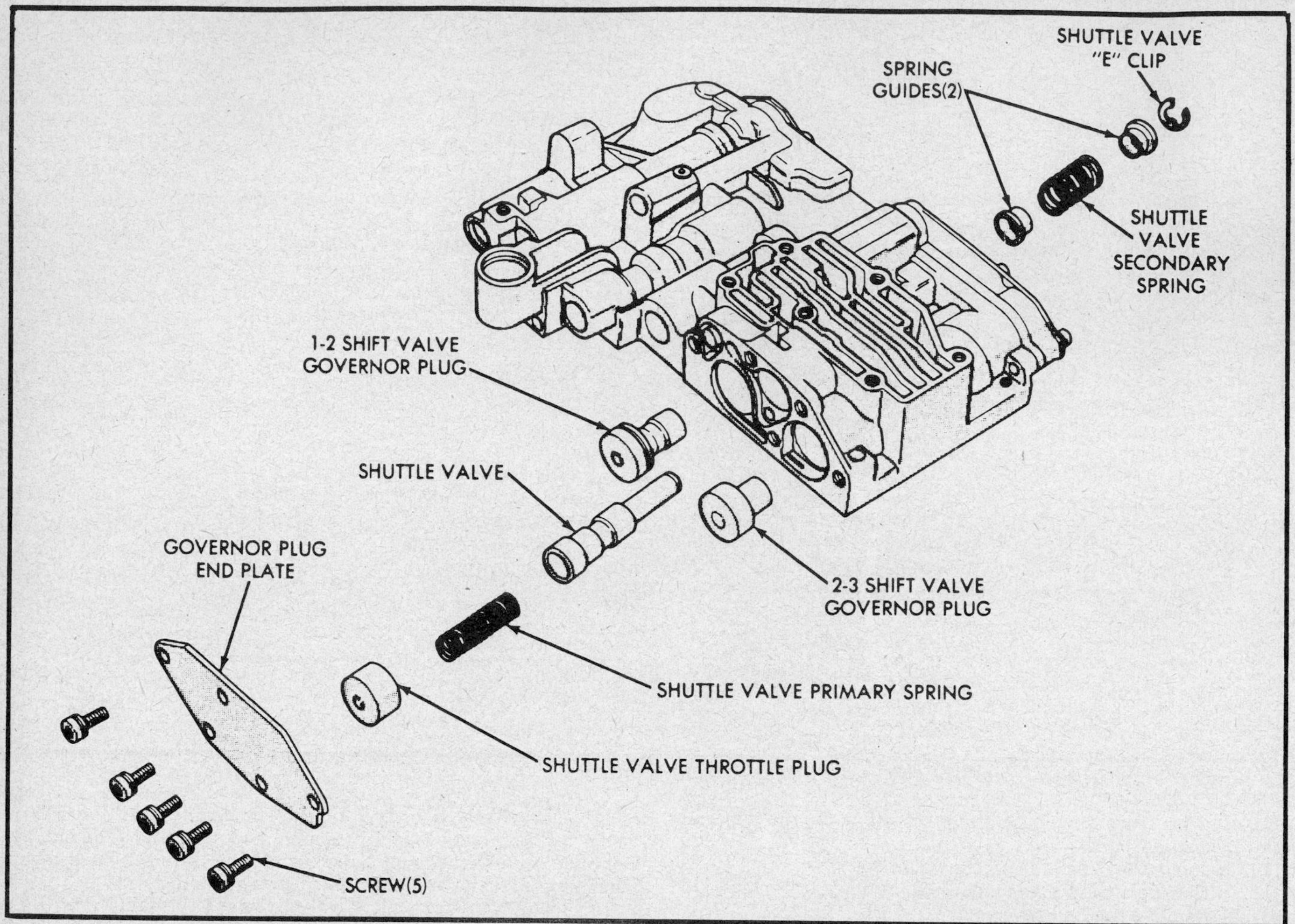
SHUTTLE VALVE "E" CLIP
SPRING GUIDES(2)
SHUTTLE VALVE SECONDARY SPRING
1-2 SHIFT VALVE GOVERNOR PLUG
SHUTTLE VALVE
GOVERNOR PLUG END PLATE
2-3 SHIFT VALVE GOVERNOR PLUG
SHUTTLE VALVE PRIMARY SPRING
SHUTTLE VALVE THROTTLE PLUG
SCREW(5)

Exploded view of the shuttle valve and governor plugs

spring loaded and if the lever is removed the ball will jump out and may cause injury or be lost. Hold detent ball with a suitable tool or other object while sliding the lever off the shaft.

15. While detent ball and spring are being held, remove the lever, then carefully pick up ball and spring, then slide out the manual valve. Slide out the kickdown detent, kickdown valve, throttle valve spring and the throttle valve.

16. Remove the line pressure regulator valve end plate and slide out the regulator valve sleeve, line pressure plug and throttle pressure plug and spring.

17. Remove the end plate and downshift housing assembly. Remove the throttle plug from the housing. Slide retainer from the other end of the housing and remove the limit valve and spring. Remove the springs and shift valves from the valve body.

Inspection

1. Thoroughly wash and blow dry all parts. Be sure all passages are clean and free from dirt or other obstructions.

2. Check the manual and throttle levers and shaft for looseness or being bent or worn excessively. If bent, replace the assembly. Check all parts for burrs or nicks. Slight imperfections can be removed with crocus cloth.

3. Using a straightedge, inspect all mating surfaces for warpage or distortion. Slight distortion can be corrected by abrading the mating surfaces on a sheet of crocus cloth on a flat piece of glass, using very light pressure. Be sure all metering holes are open in both the valve body and separator plate.

4. Use a penlight to inspect valve body bores for scratches, burrs, pits or scores. Inspect valve springs for distortion. Check valves and plugs for burrs, nicks and scores. Remove slight irregularities with crocus cloth, but do not round off the sharp edges. The sharpness of these edges is vitally important because it prevents foreign matter from lodging between the valve and the bore.

5. When valves, plugs and bores are clean and dry, they should fall freely in the bores.

6. Inspect the lockup solenoid assembly for a cut and or broken wire, melted or disorted coil, cut or nicked O-rings, etc.

7. Shake the solenoid to verify that plunger is free to travel. Replace the solenoid if the plunger is stuck.

8. Check the orifice in the solenoid nozzle and drilled crosshole at the solenoid bore in the transfer plate for dirt or forgien material.

9. To check the solenoid operation, hold the solenoid with the nozzle pointing up and apply 12 volts between the solenoid wire and solenoid frame at the screw hole.

10. The plunger should travel up and down as the 12 volts is applied intermittently.

11. Make sure the small orifice in the 1–2 shift control bore is open by inserting a $^1/_{32}$ in. diameter drill through it into the 1–2 shift control valve bore.

Assembly

All screws used in the valve body are tightened to the same torque, 35 inch lbs. (4 Nm).

1. After all cleaning and checking, slide shift valves and springs into their proper valve body bores.

2. Assemble the 3–2 limit valve housing by inserting the limit valve and spring into the downshift assembly housing and slide the retainer into its groove. Install the throttle plug in the housing then position the assembly against the shift valve springs in the valve body.

3. Install end plate and tighten screws.

4. Install regulator valve throttle pressure plug spring, plug, sleeve and line pressure plug then fasten end plate to the valve body.

5. To assembly the shuttle valve and governor plugs, plate the 1–2 and 2–3 shift valve governor plugs in their respective bores. Install the shuttle valve and hold it in the bore while putting on the secondary spring and guides and the E-clip. Install the primary shuttle valve primary spring and shuttle valve throttle plug. Install the governor plug end plate and tighten the retaining screws. Install the shuttle valve cover plate and tighten the retaining screws.

6. Install the throttle valve, throttle valve spring, kickdown valve and kickdown detent.

7. Slide the manual valve into its bore, then install the throttle lever and shaft on the valve body. Insert the detent spring and ball in its bore in the valve body. Depress ball with a piece of tubing cut on an angle and slide manual valve over the throttle shaft so that it engages the manual valve and detent ball. Install seal, retaining washer and E-clip.

8. Install the 7 balls in the transfer plate. Install rear clutch ball and low-reverse servo ball check into the transfer plate and regulator valve screen into the separator plate.

9. Install the screws and place the transfer plate on the valve body, being careful to align the filter screen. Install the 17 screws finger tight, (the 3 longer screws are for the oil filter) starting at the center and working outward, tighten the screws to 35 inch lbs. (4 Nm).

NOTE: To help insure against cross-leakage due to a warped case, some rebuilders do not tighten the valve body assembly screws until they have torqued the valve body itself to the case.

10. Slide the switch valve and line pressure valves and their spring into their respective bores.

11. Install the pressure adjusting screw and bracket assembly on the springs and fasten the screw that goes into the side of the valve body. Start the top and bottom screws, then torque the side screw to 35 inch lbs. (4 Nm), then, the top and bottom screws. Install the oil filter and torque the screws to 35 inch lbs. (4 Nm).

12. Install the lockup valve and spring. Install the fail-safe spring and valve into the lockup module. Install the lockup module to the transfer and separator plate assembly with the retaining screws.

13. Insert the lockup solenoid nozzle (with O-ring) into the bore in the transfer plate and install the retaining screw. Route the lockup solenoid wire between the solenoid and limit valve housing cover and underneath the edge of the oil filter.

NOTE: The correct wire routing is very important. The wire must be routed away from the low-reverse band lever.

14. After the valve body has been serviced and completely assembled, measure the throttle and line pressure. If the diagnostic test, done before the transmission was disassembled, were satisfactory, use the original settings.

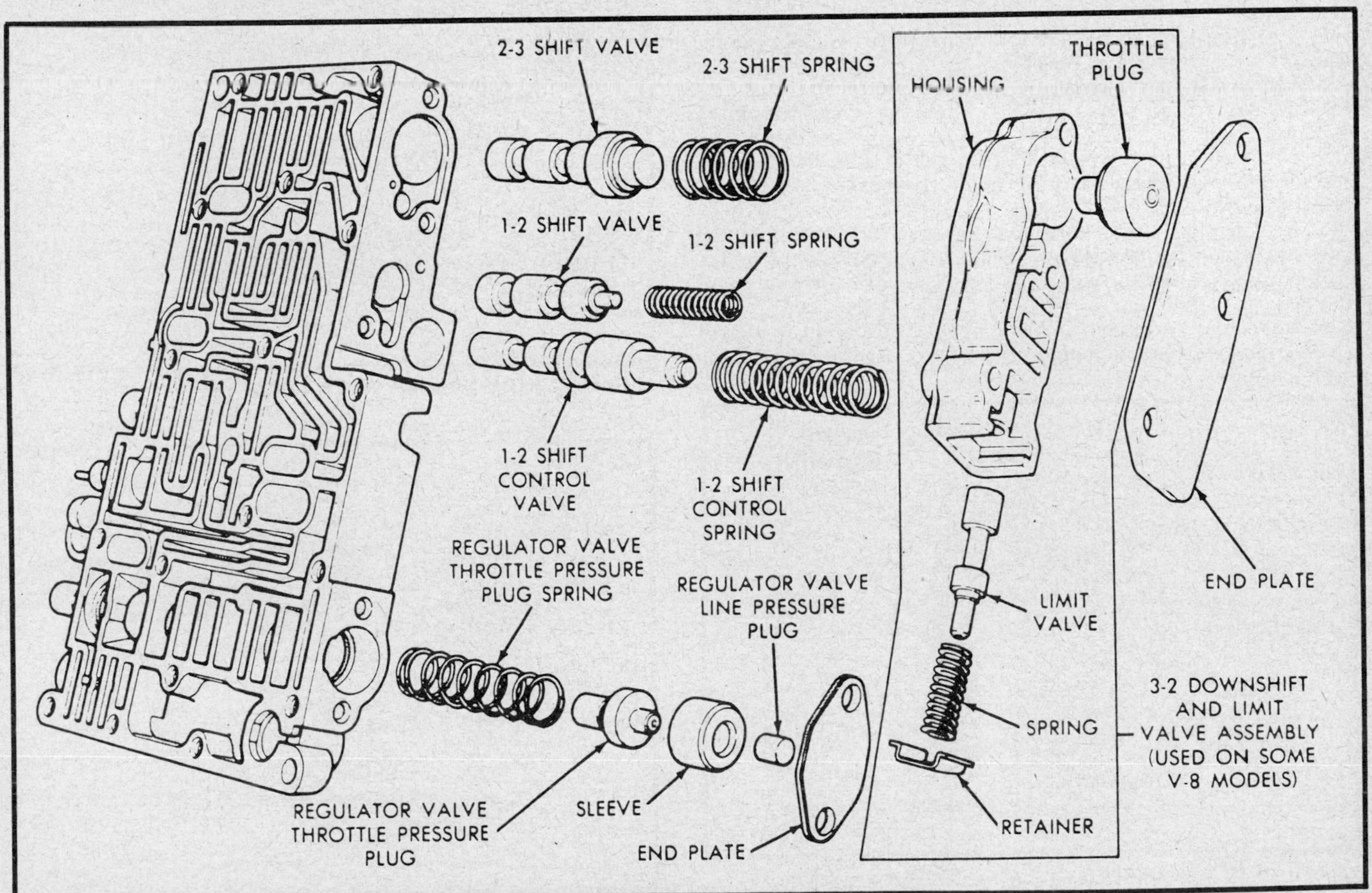
2-3 SHIFT VALVE
2-3 SHIFT SPRING
HOUSING
THROTTLE PLUG
1-2 SHIFT VALVE
1-2 SHIFT SPRING
1-2 SHIFT CONTROL VALVE
1-2 SHIFT CONTROL SPRING
REGULATOR VALVE THROTTLE PRESSURE PLUG SPRING
REGULATOR VALVE LINE PRESSURE PLUG
LIMIT VALVE
END PLATE
SPRING
3-2 DOWNSHIFT AND LIMIT VALVE ASSEMBLY (USED ON SOME V-8 MODELS)
REGULATOR VALVE THROTTLE PRESSURE PLUG
SLEEVE
END PLATE
RETAINER

Exploded view of the shift valves and pressure regulator valve plugs

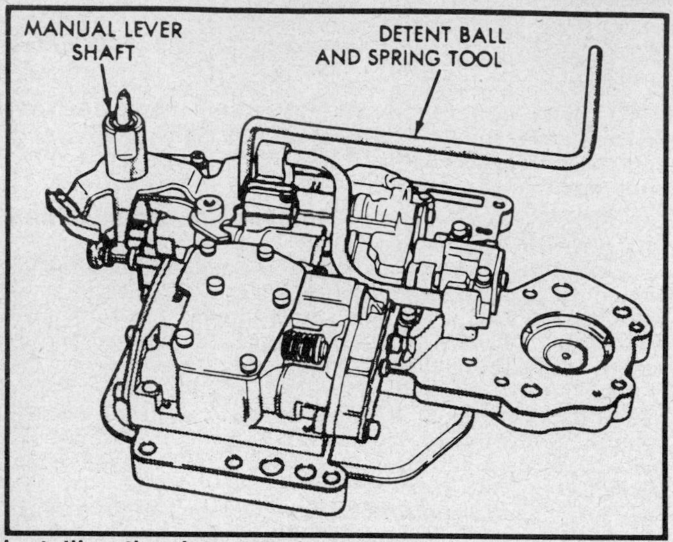
MANUAL LEVER
SHAFT
DETENT BALL
AND SPRING TOOL

Installing the detent spring and ball

15. Install the parking lock rod and E-clip retainer to the manual lever.

OVERDRIVE UNIT

Disassembly

1. Remove the overdrive clutch wire retaining rings and pull out the alternating plates and discs. Note that the heaviest metal plate is placed in the front of the clutch pack. This is the pressure plate.
2. Take a close look at each overdrive clutch component for signs of wear. Replace as necessary.
3. Remove the wave snapring. This is a special ring that acts as a cushion to absorb the shock when the overdrive clutch engages.
4. In the same groove remove the large flat snapring. Remove the access cover. The snapring holds the output shaft front bearing in place.
5. Because the entire case must be inverted to remove the gear train, insert alignment tool C-6277-2 or equivalent into the sun gear. After seating it, invert the case onto the alignment tool.
6. Use expanding type snapring pliers to expand the output shaft front bearing snapring and carefully lift the case off the gear train.

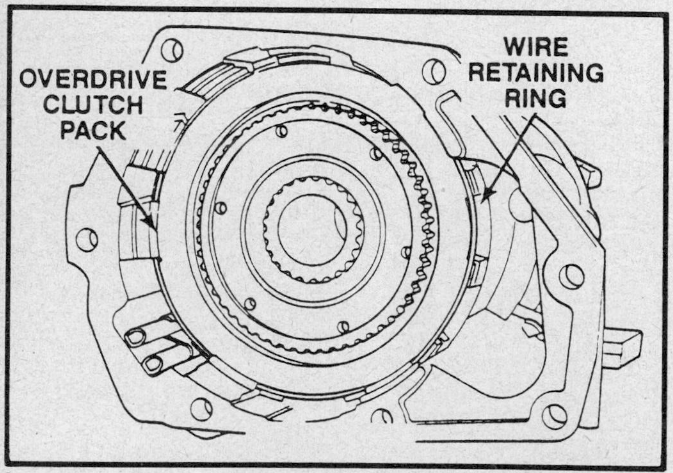
OVERDRIVE
CLUTCH
PACK
WIRE
RETAINING
RING

Location of the overdrive clutch pack and wire retaining ring

7. Remove the governor retaining snapring. Remove the governor and shaft key. This will prevent damaging the governor when the direct clutch spring is compressed in an arbor press. Set the gear train aside and continue disassembly of the case components.
8. Remove the output shaft front bearing snapring and remove the governor support snapring. Take the governor support with the slip-fit (pressure) tubes out of the case.

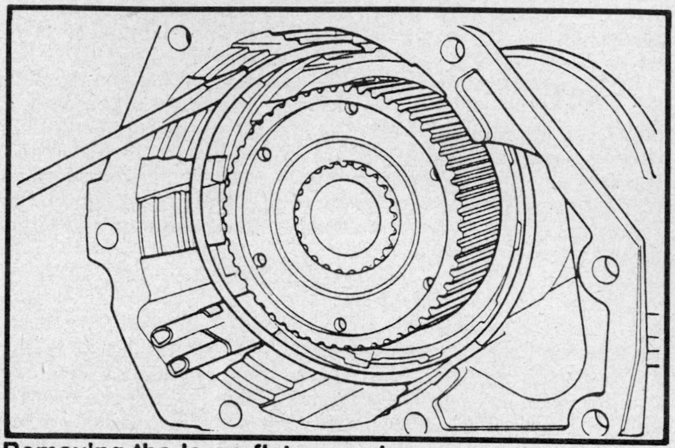

Removing the large flat snap ring

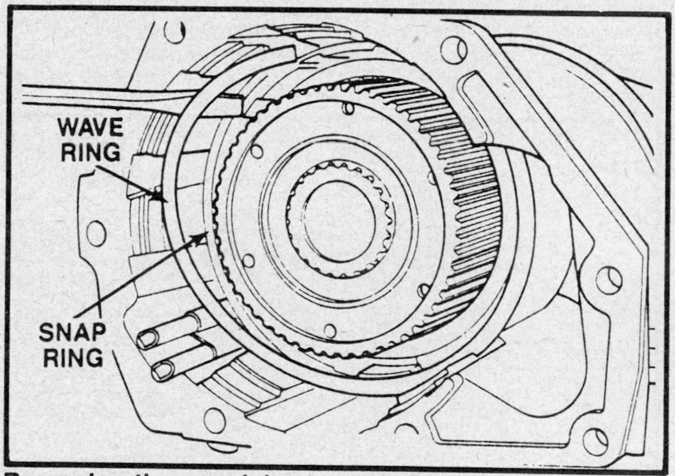
WAVE
RING
SNAP
RING

Removing the special wave snapring

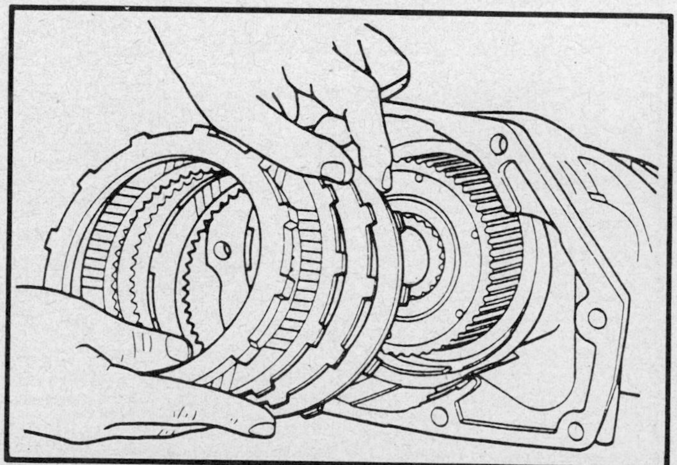

Inspecting the clutch components

9. Using locking snapring pliers, remove the output shaft rear bearing snapring. Tap the overdrive case downward on the bench to remove the rear bearing.

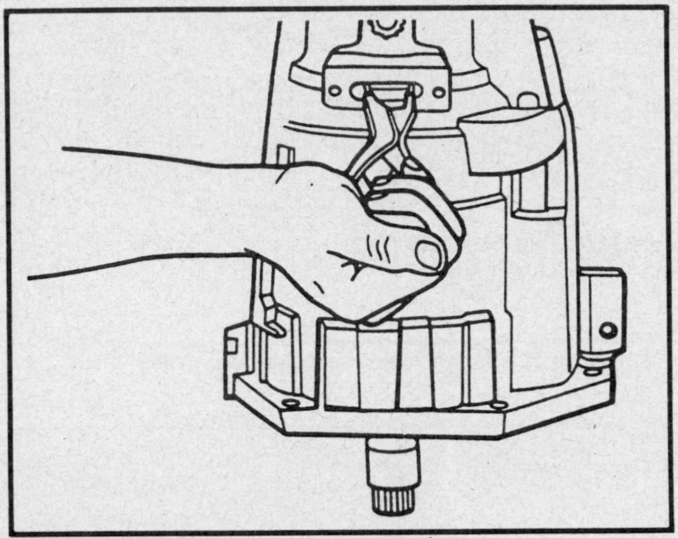
Removing the front bearing snapring

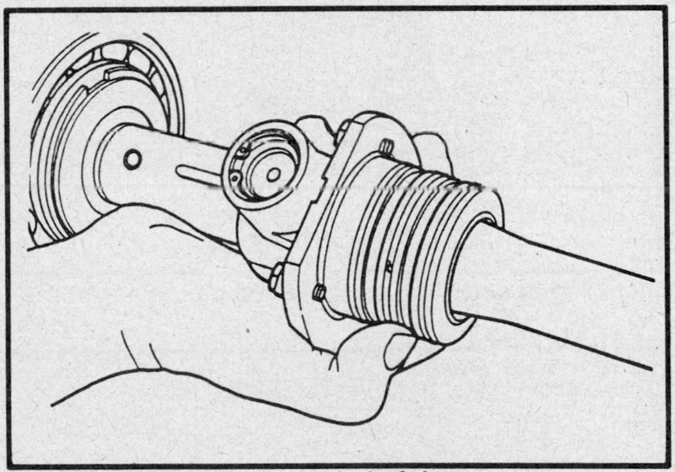
Removing the governor and shaft key

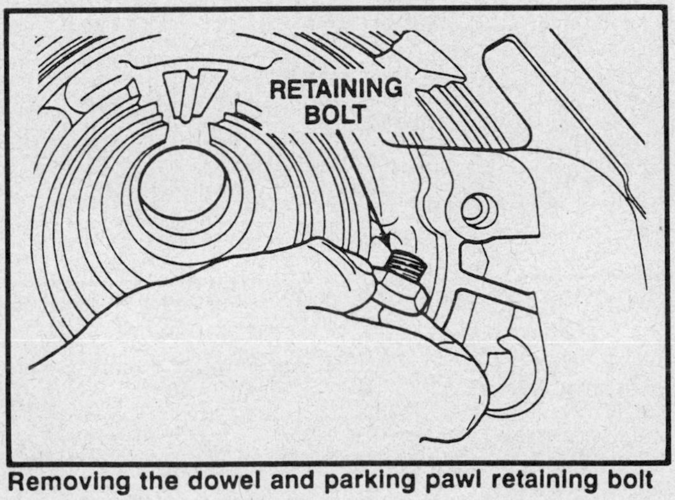
RETAINING
BOLT
Removing the dowel and parking pawl retaining bolt

10. To remove the parking mechanism, first remove the reaction plug snapring. Compress the snapring just enough to allow its removal.

11. Unscrew the bolt securing the dowel and parking pawl, another light tap to the case on the bench will cause the dowel and the parking pawl components to drop out on the bench.

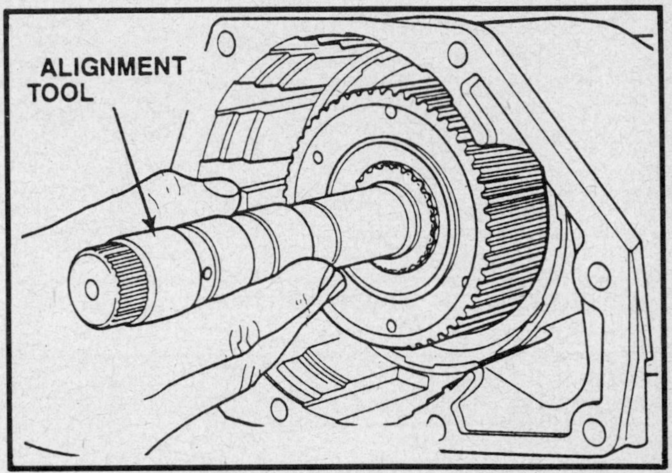
ALIGNMENT
TOOL
Installing the alignment tool in the overdrive case

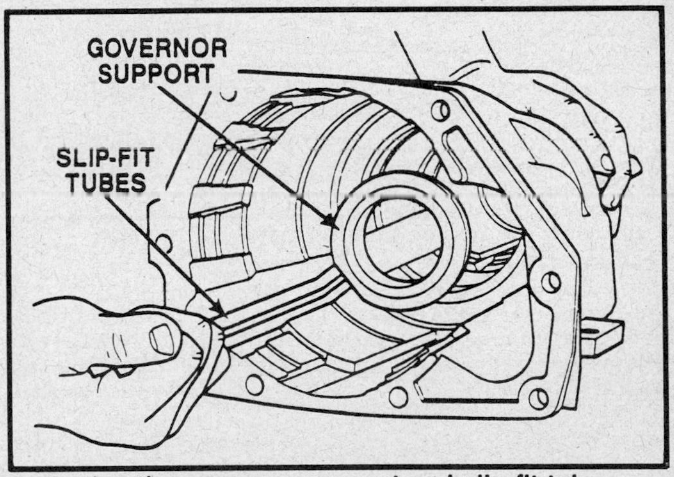
GOVERNOR
SUPPORT
SLIP-FIT
TUBES
Removing the governor support and slip-fit tubes

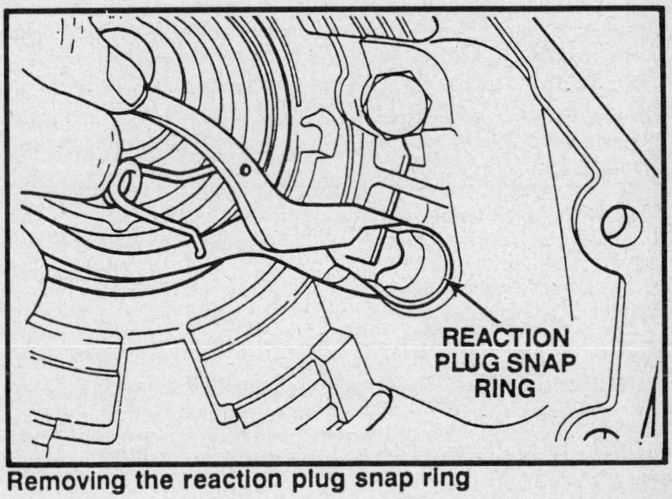
REACTION
PLUG SNAP
RING
Removing the reaction plug snap ring

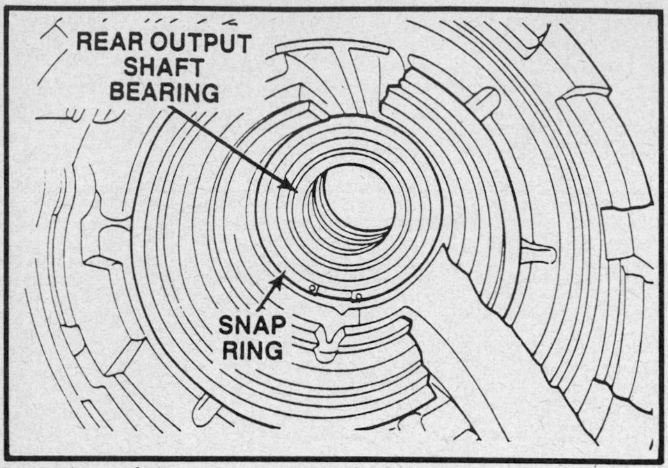
REAR OUTPUT SHAFT BEARING
SNAP RING
Location of the rear output shaft bearing

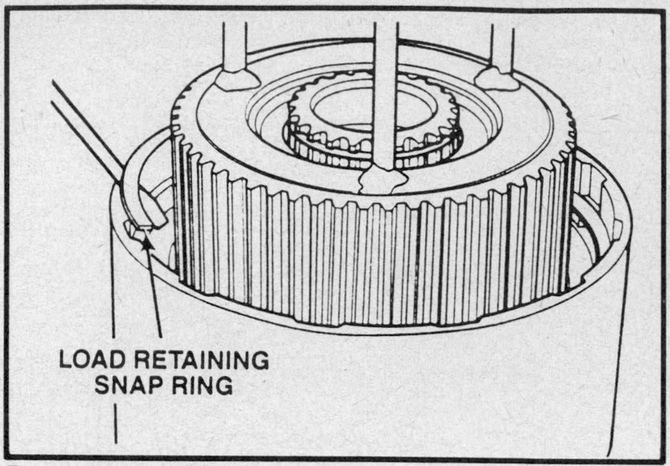
LOAD RETAINING SNAP RING
Removing the large load retaining snapring

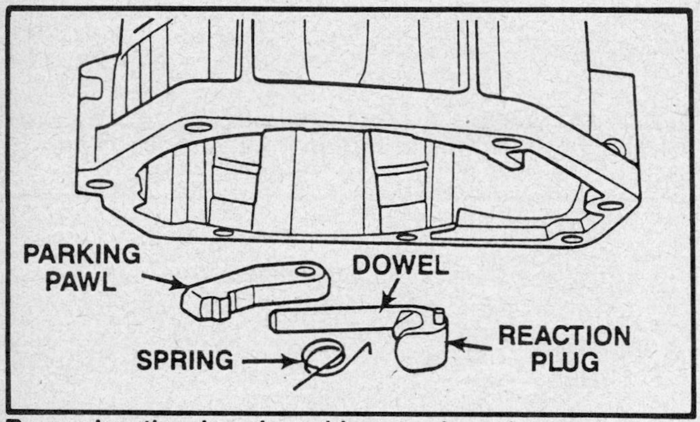
PARKING PAWL
DOWEL
REACTION PLUG
SPRING
Removing the dowel, parking pawl, spring and reaction plug

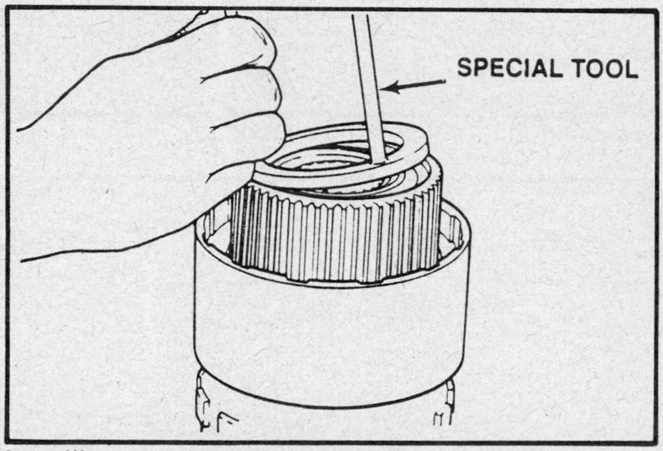
SPECIAL TOOL
Installing special tool on the direct clutch assembly

12. At the rear of the transmission, a standard rear transmission bushing and seal are used at the output shaft. Now the case is disassembled you may continue the disassembly of the gear train if necessary.

OVERDRIVE SUBASSEMBLY RECONDITION

Disassembly

It is of the upmost importance to use of a press capable of 3 in. of travel to compress the direct clutch spring. A press must have the 3 in. travel required to perform this critical step safely.

CAUTION

The direct clutch spring exerts over 800 lbs. of force on the sliding hub. Spring tension must be released slowly and completely to avoid personal injury.

1. Place the output shaft in a suitable fixture that will support the output shaft flange. With the assembly properly supported in the press, place tool C-6227-1 or equivalent into place on the direct clutch and operate the press (an assistant may be required) to compress the direct clutch spring.
2. When the hub is compressed, the large load retaining ring and the small load retaining ring can be removed safely. When unloading the press, the direct clutch spring tension is relieved. The rest of the unit can now be disassembled.
3. Remove the sliding hub with the direct clutch on it. Remove the components of the clutch from the hub and inspect them 1 at a time.

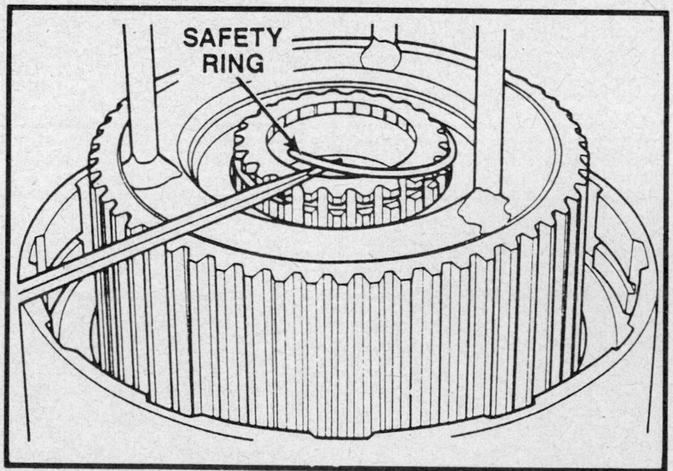
SAFETY RING
Removing the small load retaining safety snapring

4. Remove the direct clutch spring, sun gear, needle bearing and planet carrier.
5. To remove the overrunning clutch, invert the assembly and with expanding snapring pliers, reach into the inner splines of the clutch. Remove the overrunning clutch intact with a quick counterclockwise twist. Also, remove the needle bearing.

6. Mark the direct clutch drum and annulus for reassembly. Two wire retaining rings secure the direct clutch drum to the annulus. Remove the inner one first, then the one behind the rear of the drum. Slide the drum from the annulus.

7. Mark the annulus and output shaft for exact reassembly. To remove the annulus gear (a snapring secures it to the output shaft), a light tap with a soft mallet will pop it off the shaft. Also remove the output shaft front bearing.

Assembly

Before assembling, clean all parts and dry them with compressed air. Do not clean or dry parts with shop towels as lint deposits could plug the oil filter.

1. To assemble the overdrive unit, align the mating marks and insert the shaft through the back of the annulus and secure it with a snapring.

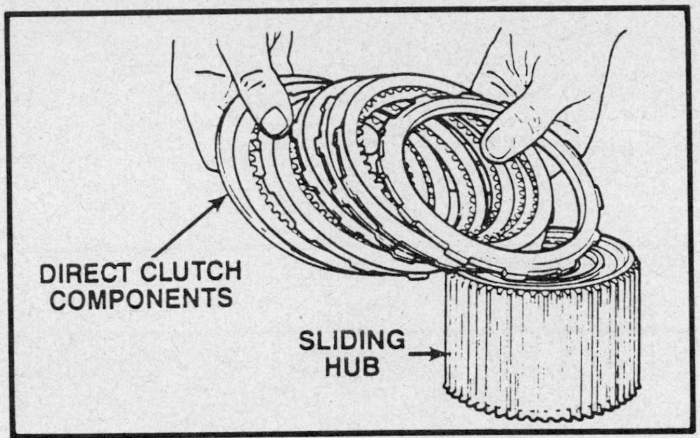
DIRECT CLUTCH
COMPONENTS
SLIDING
HUB

Inspecting the direct clutch components

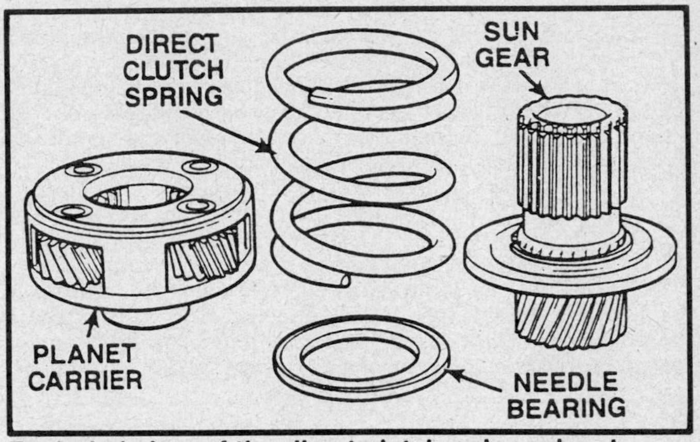
DIRECT
CLUTCH
SPRING
SUN
GEAR
PLANET
CARRIER
NEEDLE
BEARING

Exploded view of the direct clutchspring, planet carrier, sun gear and needle bearing

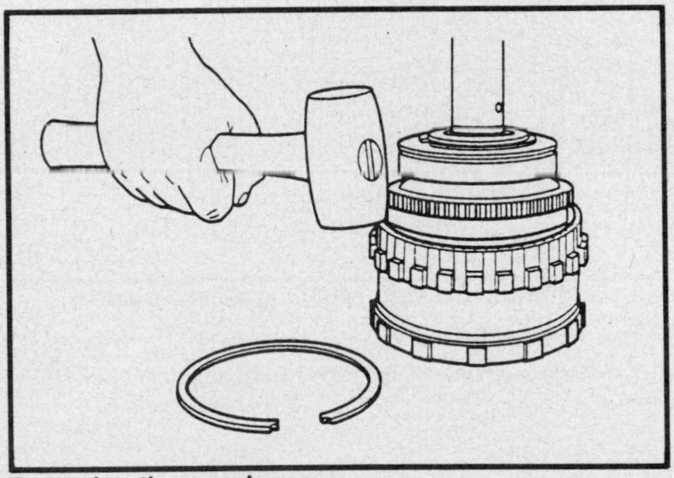

Removing the annulus

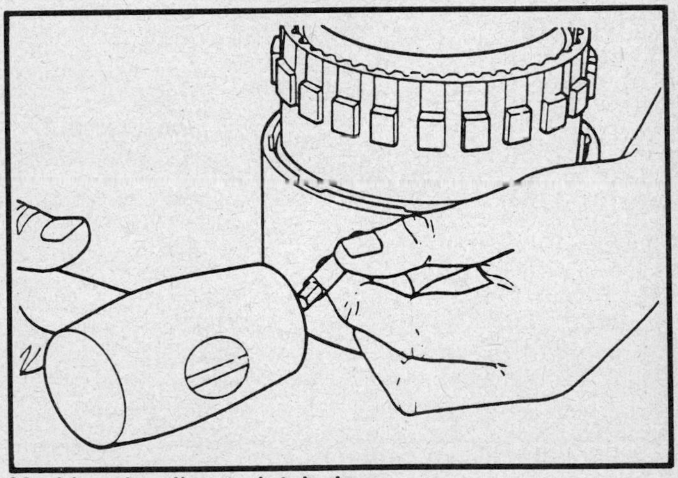

Marking the direct clutch drum

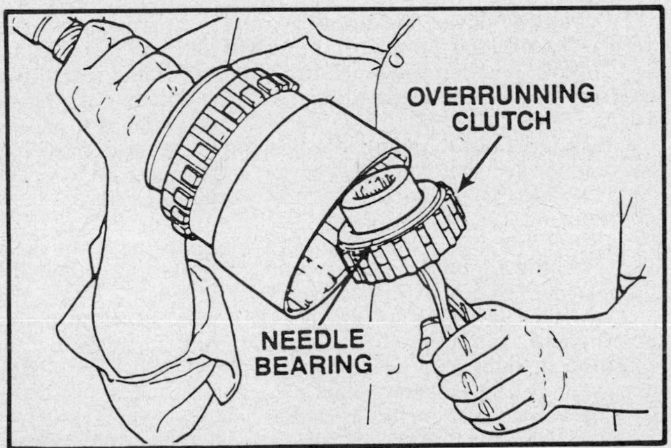
OVERRUNNING
CLUTCH
NEEDLE
BEARING

Removing the overrunning clutch

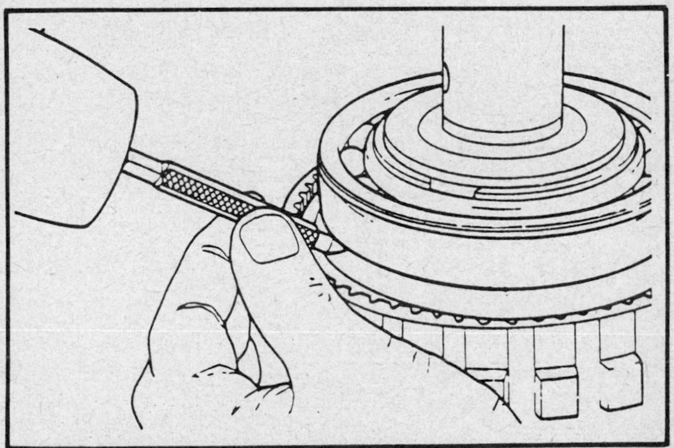

Marking the annulus and output shaft

2. Set the direct clutch drum face down and align the mating marks. Insert the annulus lugs into the slots inside the drum.

3. Install the rear retaining ring first. Invert the assembly. Slide the drum forward to expose the retaining ring groove. Install the front wire retaining ring to secure the drum.

4. Hold the overrunning clutch upside down with expanding snapring pliers. place the needle bearing against the back face of the clutch. Hold the shaft assembly upside down and with an upward, counterclockwise twisting motion, install the overrunning clutch.

5. Carefully set the carrier assembly into the annulus and align the splines using tool C-6227-2 or equivalent.

6. Set the needle bearing in place, remove the tool and install the sun gear. Set the spring on the sun gear and place the sliding hub on the spring. Reinstall the alignment tool and install the direct clutch plates and discs, one at a time on the hub.

7. With the aid of an arbor press, set tool C-6227-1 or equivalent on the hub and compress the direct clutch spring (an assistant may be required). As the spring is compressed, slide the clutch plates into their grooves and down on the hub.

8. Install the the large load retaining snapring, seating it firmly with a suitable tool. Install the safety ring. Slowly release the press, ensuring that the load-retaining rings are properly seated. Install the governor and snapring.

CAUTION

The direct clutch spring exerts over 800 lbs. of force on the sliding hub. Spring tension must be released slowly to avoid personnal injury.

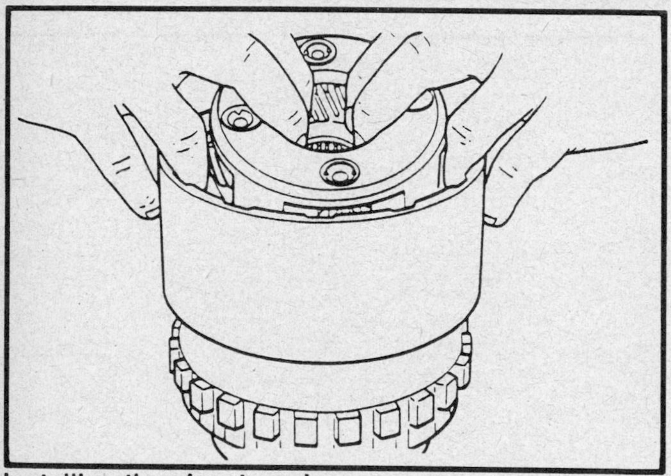
Installing the planet carrier

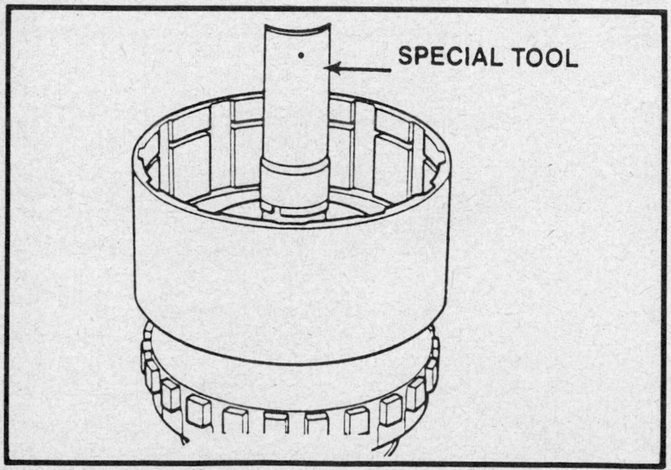
SPECIAL TOOL
Aligning the splines

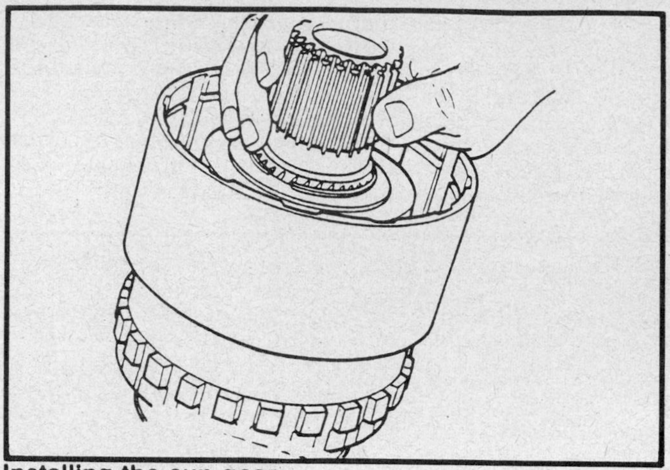
Installing the sun gear

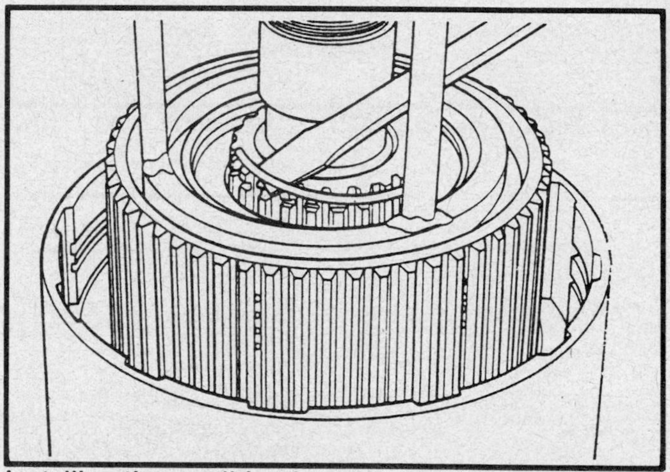
Installing the small load retaining safety snapring

9. Set the gear train aside, with special alignment tool still in place.

OVERDRIVE UNIT

Assembly

1. Install the dowel, parking pawl and spring back into the overdrive case. Reinstall the retaining bolt and torque it to 20 ft. lbs. (27 Nm). Position the reaction plug in place and install the snapring. Use care to squeeze the snapring only enough to install it.

2. Install the output shaft rear bearing in the case. Snapring groove is toward the front of the case.

3. Install the governor support and install the governor support snapring.

4. Install the output shaft front bearing snapring in the case. Make sure the alignment tool is installed in the gear train. Invert the gear train and slip the case over the shaft.

5. Expand the snapring and slip the case down until the snapring locks in the bearing groove. Release the snapring.

6. After installing the access plate and gasket, install the flat snapring. Use a suitable tool to be sure that it is properly seated. Reinstall the wave snapring using tha same seating technique.

7. One by one, install the overdrive clutch plates and discs, be certain to put the thichest plate in last.

8. Position the overdrive unit vertically in a large vise. To determine the proper intermediate shaft spacer thickness, insert tool C-6312 (depth gauge) or equivalent through the sun gear. Be sure that the tool bottoms out against the carrier spline shoulder. Position tool C-6311 across the overdrive case face. Using a dial caliper tool C-6311 Measure the distance to the top of the tool C-6312.

9. Use the measurements that follow as a guide to select the proper thickness spacer.

a. A measurement of 0.7336–0.7505 in. will require a spacer thickness of 0.159–0.158 in., the spacer part number will be 4431916.

b. A measurement of 0.7506–0.7675 in. will require a spacer thickness of 0.176–0.175 in., the spacer part number will be 4431917.

c. A measurement of 0.7676–0.7855 in. will require a spacer thickness of 0.194–0.193 in., the spacer part number will be 4431918.

d. A measurement of 0.7856–0.8011 in. will require a spacer thickness of 0.212–0.211 in., the spacer part number will be 4431919.

10. To determine the proper shim thickness for the overdrive piston, position tool C-6311 or equivalent across the overdrive case face. Using a suitable dial caliper tool C-4962 position it over tool C-6311 (straight edge) or equivalent, measure the distance to the sliding hub bearing seat.

11. This measurement should be taken at 4 locations 90° apart. Add all measurements together and divide by 4. Use the measurements that follow as a guide to select the proper thickness spacer.

a. A measurement of 1.7500–1.7649 in. will require a spacer thickness of 0.108–0.110 in., the spacer part number will be 4431730.

b. A measurement of 1.7650–1.7799 in. will require a spacer thickness of 0.123–0.125 in., the spacer part number will be 4431585.

c. A measurement of 1.7800–1.7949 in. will require a spacer thickness of 0.138–0.140 in., the spacer part number will be 4431731.

d. A measurement of 1.7950–1.8099 in. will require a spacer thickness of 0.153–0.155 in., the spacer part number will be 4431586.

e. A measurement of 1.8100–1.8249 in. will require a spacer thickness of 0.168–0.170 in., the spacer part number will be 4431732.

f. A measurement of 1.8250–1.8399 in. will require a spacer thickness of 0.183–0.185 in., the spacer part number will be 4431587.

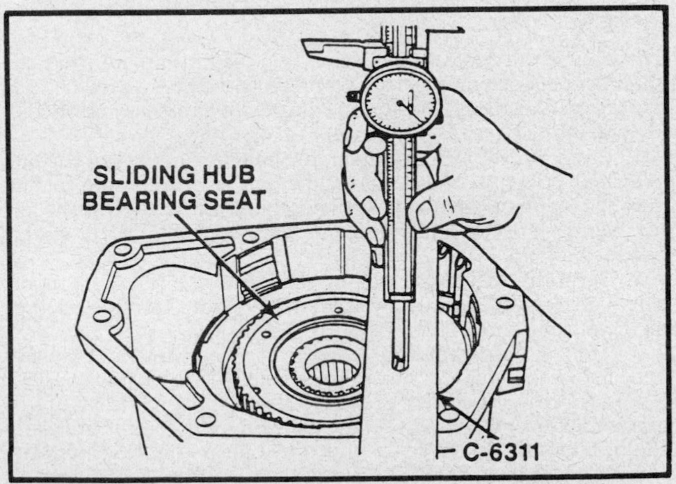
SLIDING HUB
BEARING SEAT
C-6311
Making the measurements necessary to determine the proper overdrive piston shim thickness

g. A measurement of 1.8400–1.8549 in. will require a spacer thickness of 0.198–0.200 in., the spacer part number will be 4431733.

h. A measurement of 1.8550–1.8699 in. will require a spacer thickness of 0.213–0.215 in., the spacer part number will be 4431588.

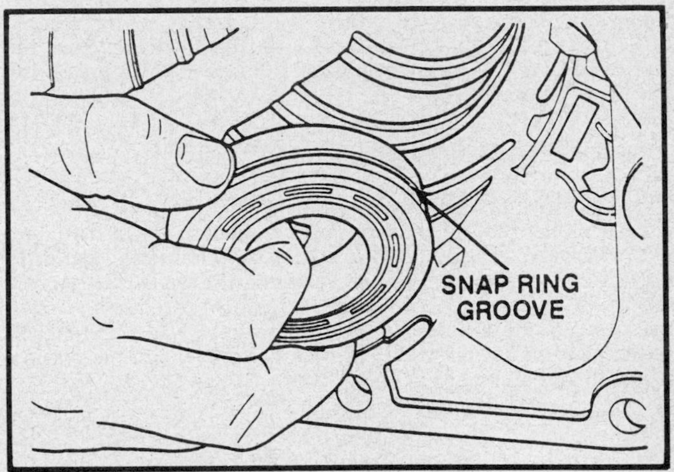
SNAP RING
GROOVE
Installing the rear output shaft bearing

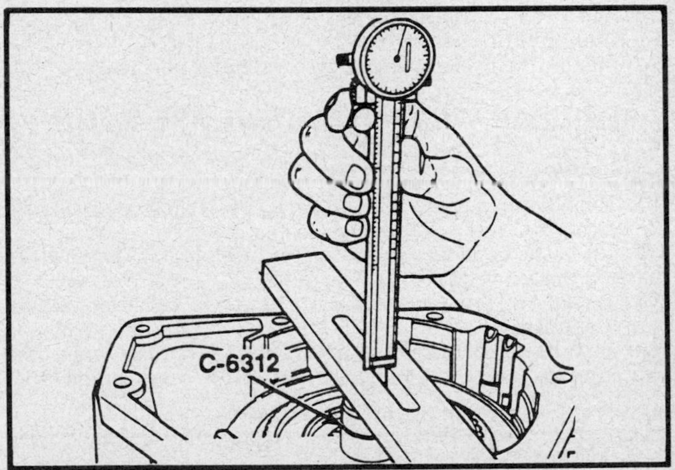
C-6312
Making the measurements necessary to determine the proper intermediate shaft spacer thickness

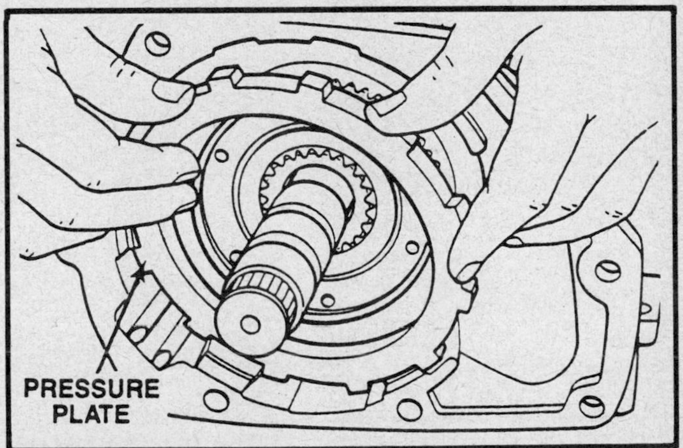
PRESSURE
PLATE
Installing the overdrive clutch plates

i. A measurement of 1.8700–1.8849 in. will require a spacer thickness of 0.228–0.230 in., the spacer part number will be 4431734.

j. A measurement of 1.8850–1.8999 in. will require a spacer thickness of 0.243–0.245 in., the spacer part number will be 4431590.

NOTE: Before installing the overdrive unit, it will be necessary to cut out the old gasket. Using a sharp tool, cut out the old gasket around the piston. Place the old gasket down on the new gasket for a template and trim the new gasket to fit.

Transmission Assembly

The assembly procedures given here include installation of subassemblies in the transmission case and adjusting input shaft endplay. Do not use force to assemble mating parts. If parts so not assemble freely, invesitgate the cause and correct the trouble before proceeding with assembly procedures. Always use new gaskets during assembly operations. Use only Mopar® automatic transmission fluid to lubricate the transmission parts during assembly.

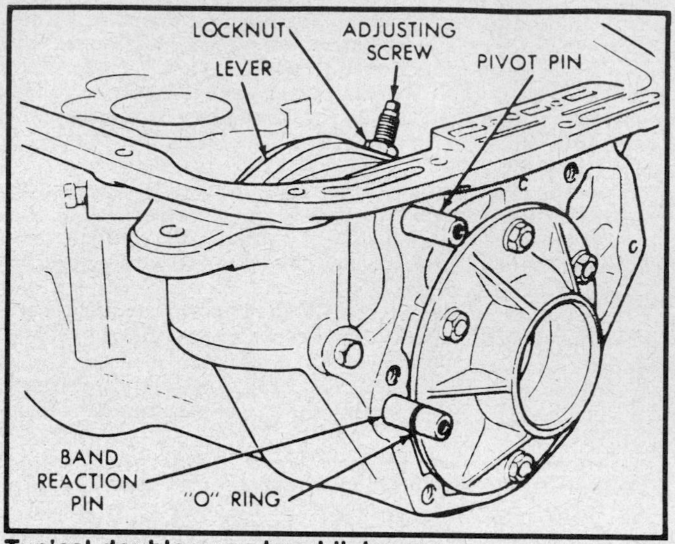
LOCKNUT
ADJUSTING SCREW
LEVER
PIVOT PIN
BAND REACTION PIN
"O" RING
Typical double wrap band linkage

OVERRUNNING CLUTCH

Installation

1. With the transmision case in an upright position, insert the clutch hub inside the cam.
2. Install the overrunning clutch rollers and springs.

DOUBLE WRAP LOW-REVERSE BAND

Installation

1. Push the band reaction pin (with new O-ring) into the case flush with the gasket surface.
2. Place the band into the case resting the 2 lugs against the reaction pin.
3. Install the low-reverse band drum into the overrunning clutch and band.
4. Install the operating lever with pivot pin flush in the case and adjusting screw touching the center lug on the band.

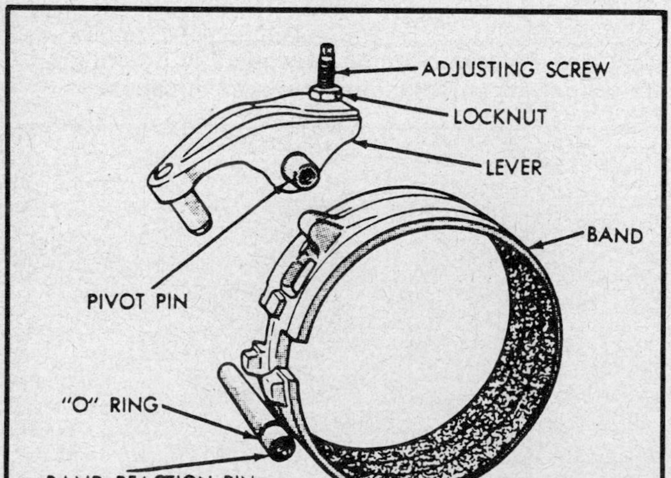
ADJUSTING SCREW
LOCKNUT
LEVER
BAND
PIVOT PIN
"O" RING
BAND REACTION PIN
Exploded view of the double wrap band assembly

KICKDOWN SERVO

Installation

1. Carefully push the servo piston assembly into the case bore. Install the spring, guide and snapring.
2. Compress the kickdown servo spring by using the engine valve spring compressor tool C-3422-A or equivalent and install the snapring.

PLANETARY GEAR ASSEMBLIES, SUN GEAR AND DRIVING SHELL

Installation

1. While supporting the assembly in the case, insert the output shaft through the rear support.
2. Carefully work the assembly rearward engaging the rear planetary carrier lugs into the low-reverse drum slots.
3. Be careful not to damage the ground surfaces on the output shaft during installation.

FRONT AND REAR CLUTCH ASSEMBLIES

Installation

The front and rear clutches, front band, oil pump and reaction shaft support are more easily installed with the transmission in the upright position. Step 1 and 2 is to used as an easy method of supporting the transmission.

1. Cut a 3½ in. (89mm) diameter hole in a bench, in the end of a small oil drum or a large wooden box strong enough to support the transmission. Cut or file notches at the edge of the 3½ in. (89mm) hole so that the output shaft support will fit and lay flat in the hole.
2. Carefully insert the output shaft into the hole to support the transmission upright, with its weight resting on the flange of the output shaft support.
3. Apply a light coat of grease to the selective thrust washer and install the washer on the front end of the output shaft.
4. Apply a light coat of grease to the input shaft thrust plate and install over the input shaft. Apply a light coat of grease to the input shaft thrust plate and install over the input shaft.
5. If the input shaft endplay is not within specifications, (0.022–0.091 in.), when tested before disassembly, replace the thrust washer with one of proper thickness.

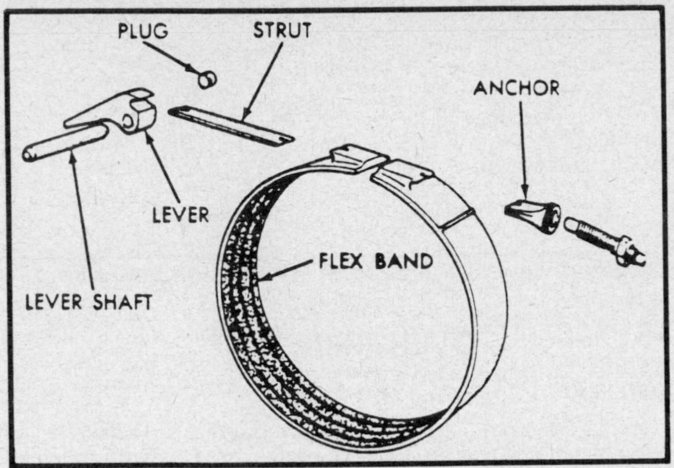
PLUG
STRUT
ANCHOR
LEVER
FLEX BAND
LEVER SHAFT

Exploded view of the kickdown band assembly

6. Align the front clutch plate inner splines and place the assembly in position on the rear clutch. Be sure the front clutch plate splines are fully engaged on the rear clutch splines.

7. Align the rear clutch plate inner splines, grasp the input shaft and lower the clutch assemblies into the transmission case.

8. Carefully work the clutch assemblies in a circular motion to engage the rear clutch splines over the splines of the front annulus gear. Be sure the front clutch drive lugs are fully engaged in the slots in the driving shell.

KICKDOWN BAND

Installation

1. Slide the kickdown band over the front clutch assembly.
2. Install the kickdown band strut, screw in the adjuster just enough to hold the strut and anchor in place.

OIL PUMP AND REACTION SHAFT SUPPORT

Installation

If difficulty was encountered removing the pump, it may be well to expand the case with a heat lamp before attempting to install pump.

1. Install the No. 1 thrust washer on the reaction shaft support hub.
2. Screw the pilot studs tool C-3288-B or equivalent into the opening of the pump case. Install a new gasket over the pilot studs.
3. Place a new rubber seal ring in the groove on the outer flange of the pump housing. Make sure the seal ring is not twisted. Coat the seal ring with grease for easy installation.
4. Install the pump assembly into the case, tap it lightly with a soft mallet, if necesary. Remove the pilot studs and install the bolts and snug down evenly. Rotate the input and output shafts to see if any binding exists and torque the bolts to 175 inch lbs. (20 Nm). Check the shafts again for free rotation.
5. Adjust both bands.

GOVERNOR AND SUPPORT

Installation

1. Position the support and governor body assembly on the output shaft. Align the assembly so that the governor valve shaft hole in the governor body aligns with the hole in the output shaft and slide the assembly into place.

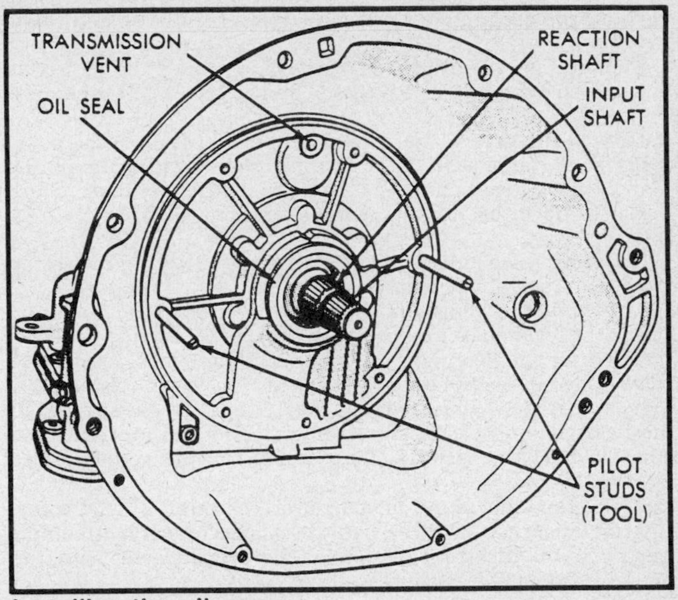
TRANSMISSION VENT
OIL SEAL
REACTION SHAFT
INPUT SHAFT
PILOT STUDS (TOOL)

Installing the oil pump

2. Install the snapring behind the governor body. Torque the body to support self-locking bolts to 95 inch lbs. (11 Nm).
3. Place the governor valve on the valve shaft, insert the assembly into the body and through the governor weights. Install the valve shaft retaining snapring.

OUTPUT SHAFT BEARING AND EXTENSION (OR ADAPTER ON 4WD VEHICLES)

Installation

1. Install the bearing on the shaft with its outer race ring groove toward the front. Press or tap the bearing tight against the shoulder and install the rear snapring.
2. On 4WD vehicles, use ATF lubricant on the output shaft bearing O-ring. Drive the bearing into the adapter using tool C-4203 or equivalent inverter with handle C-4171 or equivalent.
3. Place a new extension housing gasket on the transmission case. Position the output shaft bearing retaining snapring as far as possible then carefully tap the extension housing (or adapter) into place. Make sure that the snapring is fully seated in the bearing groove.
4. Install and tighten the extension housing (or adapter) bolts to 32 ft. lbs.
5. Install the gasket, plate and screws on the bottom of the extension housing mounting pad.
6. Be sure to re-measure the input shaft endplay and adjust as necessary.

VALVE BODY AND ACCUMULATOR PISTON INSTALLATION

Installation

1. Make sure that the combination back-up lamp/neutral start switch is not installed in the transmission case.
2. Place the valve body manual lever in the **MANUAL LOW** position to move the parking rod to the rear position.
3. Using a suitable tool to push the park sprag into the engagement with the parking gear, turning the output shaft to verify engagement. This will allow the knob of the end of the parking rod to move past the sprag as the valve body is installed.
4. Install the accumulator piston in the transmission case.

Position the accumulator spring between the piston and valve body.

5. Place the valve body in position, working the park rod through the opening and past the sprag. Install the retaining bolts finger tight.

6. With the neutral starting switch installed, place the manual lever in the **N** position. Shift the valve body if necessary, to center the neutral finger over the neutral switch plunger. Snug the bolts down evenly then torgue them to 105 inch lbs. (12 Nm).

7. Connect the lockup solenoid wire to the wiring connector pin at the rear of the transmission case.

8. Install the gearshift lever and tighten the clampbolt. Check the lever shaft for binding in the case by moving the lever through all the detent positions. If binding exists, loosen the valve body bolts and realign.

9. Make sure that the throttle shaft seal is in place and install the flat washer with the throttle lever and tighten the clamp bolt. Connect the throttle and gearshift linkage and adjust as required.

10. Position the round magnet over the bump in the front, right hand corner of the oil pan. Install the oil pan with a new gasket. Refill the transmission to the proper level, with the proper automatic transmission fluid.

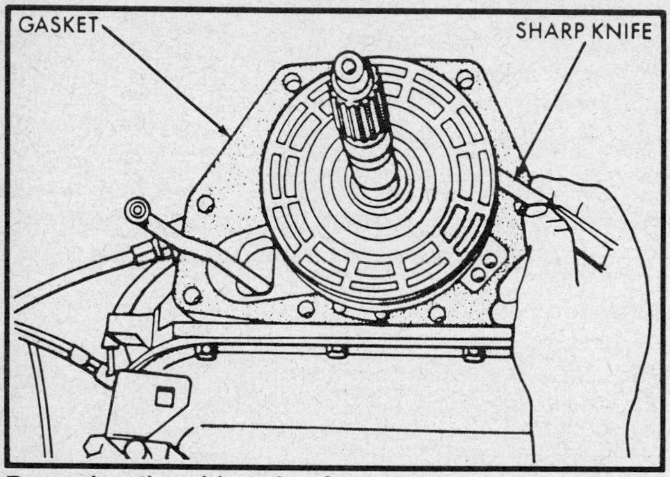
GASKET
SHARP KNIFE
Removing the old gasket from the overdrive unit

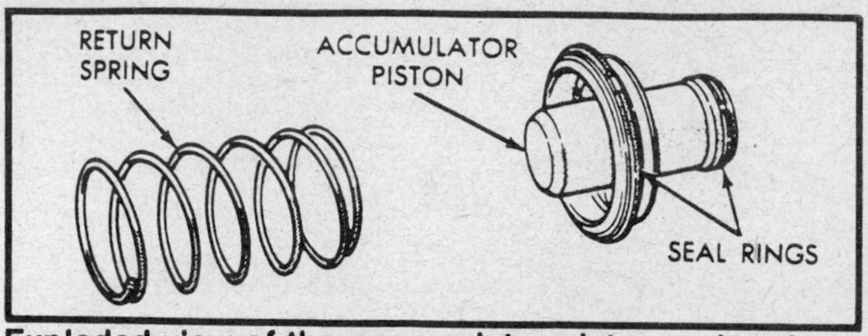
RETURN SPRING
ACCUMULATOR PISTON
SEAL RINGS
Exploded view of the accumulator piston and spring

OVERDRIVE UNIT

Installation

1. Place the new gasket in place on the rear of the overdrive case. Install the spacer, selected earlier in the overdrive reassembly, on the intermediate shaft.

2. Place the spacer, selected earlier in the overdrive reassembly, in position over the piston on the main portion of the transmission. Install the sliding hub bearing over the intermediate shaft against the sliding hub.

CAUTION

The shoulder on the inside diameter of the bearing must face forward. A small amount of petroleum jelly or equivalent should be used to hold the shim and bearing into position.

3. Carefully lift the overdrive unit and slide it onto the intermediate shaft. Insert the parking rod into the reaction plug.

CAUTION

Extreme caution must be used not to tilt the unit as this could cause the carrier and overrunning clutch splines to rotate out of alignment. If this happens, it will be necessary to remove the overdrive unit and align them with tool C-6227-2 or equivalent.

4. Align the slip-fit governor tubes and push the unit forward until it touches the transmission case. Install the attaching bolts and torque them in a crisscross pattern to 25 ft. lbs. (34 Nm).

5. Install the crossmember, speedometer cable and driveshaft using the marks made at disassembly. Refill to the proper level with Mopar® ATF.

SPECIFICATIONS

TORQUE SPECIFICATIONS

A500 Automatic (LoadFlite)	Ft. Lbs.	N·m
Cooler Line Fitting	155*	18
Converter Drive Plate to Crankshaft Bolt	55	75
Converter Drive Plate to Torque Converter Bolt	270*	31
Extension Housing to Transmission Case Bolt	32	43
Extension Housing to Insulator Mounting Bolt	50	68
Governor Body to Support Bolt	95	11
Kickdown Band Adjusting Screw Locknut	30	41
Kickdown Lever Shaft Plug	150*	17
Lock-up Solenoid Wiring Connector	150*	17
Neutral Starter Switch	25	34
Oil Pan Bolt	150*	17
Oil Pump Housing to Transmission Case Bolt	175	20
Output Shaft Support Bolt	150*	17
Pressure Test Take-Off Plug	120*	14
Reaction Shaft Support to Oil Pump Bolt	175*	20
Reverse Band Adjusting Screw Locknut	25	34
Speedometer Drive Clamp Screw	100*	11
Transmission to Engine Bolt	30	41
Valve Body Screw	35*	4
Valve Body to Transmission Case Bolt	105*	12

*Inch Pounds

GENERAL TRANSMISSION DATA

Transmission Model:	A-998		A-500		
TYPE	Automatic 3-Speed		Automatic 4-Speed Overdrive		
TORQUE CONVERTER DIAMETER (Standard)	10-3/4 inches		10-3/4 inches		
	U.S.A. Measure	Metric Measure	U.S.A. Measure	Metric Measure	
OIL CAPACITY—TRANSMISSION AND TORQUE CONVERTER	17.1 pts.	8.1 Liter	20.4 pts.	9.6 Liter	
Use "MOPAR ATF PLUS" (Automatic Transmission Fluid) Type 7176					
COOLING METHOD	Water-Heat Exchanger				
LUBRICATION	Pump (Gear Type)				
GEAR RATIOS:	First 2.74	Second 1.54	Third 1 to 1	Reverse 2.21	Overdrive 0.69 to 1
PUMP CLEARANCES:					
Outer Gear to Case Bore	.0035 to .0075 inch				
End Clearance—Gears	.0004 to .0025 inch				
GEAR TRAIN END PLAY	.005 to .048 inch				
INPUT SHAFT END PLAY	.022 to .091 inch				
SNAP RINGS:					
Rear Clutch Snap Ring (Selective)	.060 to .062 inch .068 to .070 inch .076 to .078 inch .098 to .100 inch				
Output Shaft (Forward End)	.040 to .044 inch .062 to .066 inch .082 to .086 inch				
CLUTCH PLATE CLEARANCE:					
Front Clutch	5 Disc. .075 to .152 inch				
Rear Clutch	4 Disc .032 to .055 inch				
CLUTCHES:					
Number of Front Clutch Discs	5		5		
Number of Rear Clutch Discs	4		4		
Number of Direct Clutch Discs	—		5		
Number of Overdrive Clutch Discs	—		3		
BAND ADJUSTMENTS:					
Kickdown (Front) Turns*	2-1/2		2-1/2		
Low-Reverse (Internal) Turns*	4		4		

*Backed off from 72 inch-pounds (8 N·m).

THRUST WASHERS:		A-998/A-500
Reaction Shaft Support Thrust Washer	#1	.061 to .063 inch
Rear Clutch Retainer Thrust Washer	#2	.061 to .063 inch
Input Shaft Thrust Plate		.024 to .026 inch
Output Shaft Thrust Washer	#3	Selective .052 to .054 inch (Tin) .068 to .070 inch (Red) .083 to .086 inch (Green)
Front Annulus Thrust Washer	#4	.121 to .125 inch
Front Carrier (To Annulus) Thrust Washer	#5	.048 to .050 inch
Front Carrier (To Drive Shell) Thrust Washer	#6	.048 to .050 inch
Sun Gear Drive Shell Thrust Plate	#7	.050 to .052 inch
	#8	.050 to .052 inch
Rear Carrier (To Drive Shell) Thrust Washer	#9	.048 to .050 inch
Rear Carrier (To Annulus) Thrust Washer	#10	.048 to .050 inch

SPECIAL TOOLS

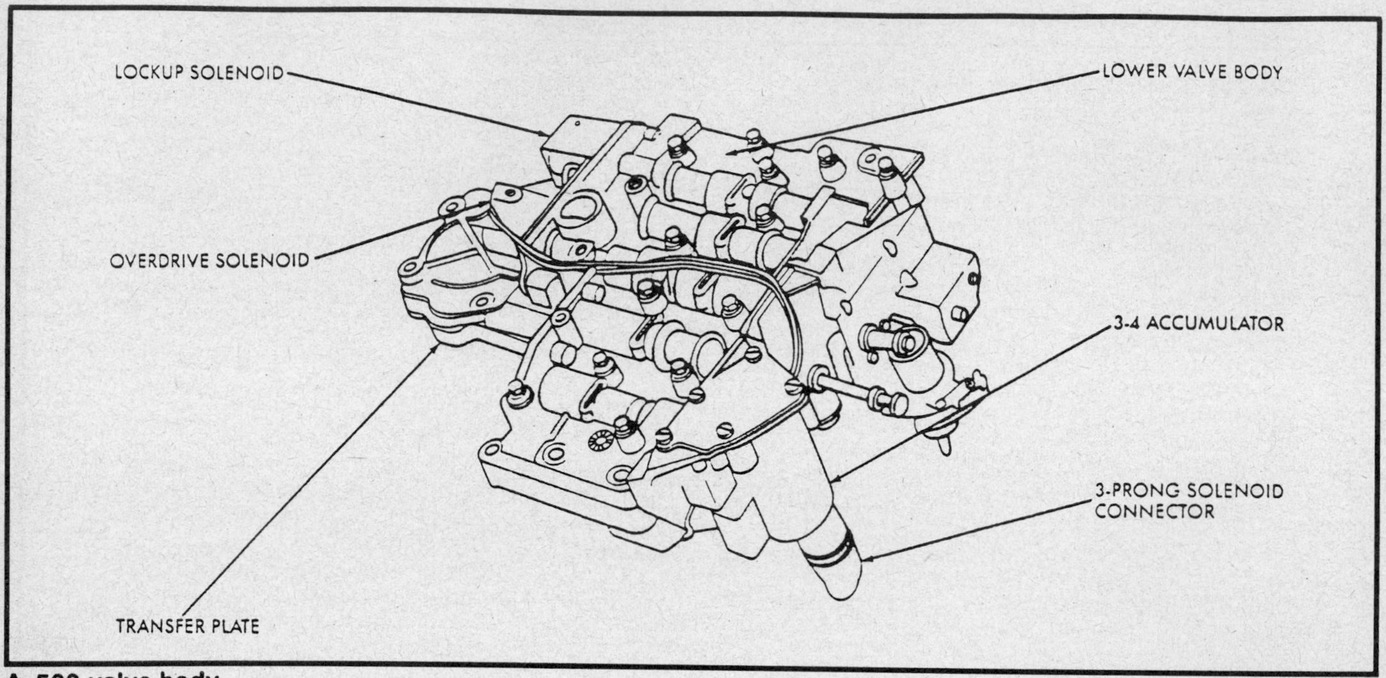
LOCKUP SOLENOID
LOWER VALVE BODY
OVERDRIVE SOLENOID
3-4 ACCUMULATOR
3-PRONG SOLENOID CONNECTOR
TRANSFER PLATE

A–500 valve body

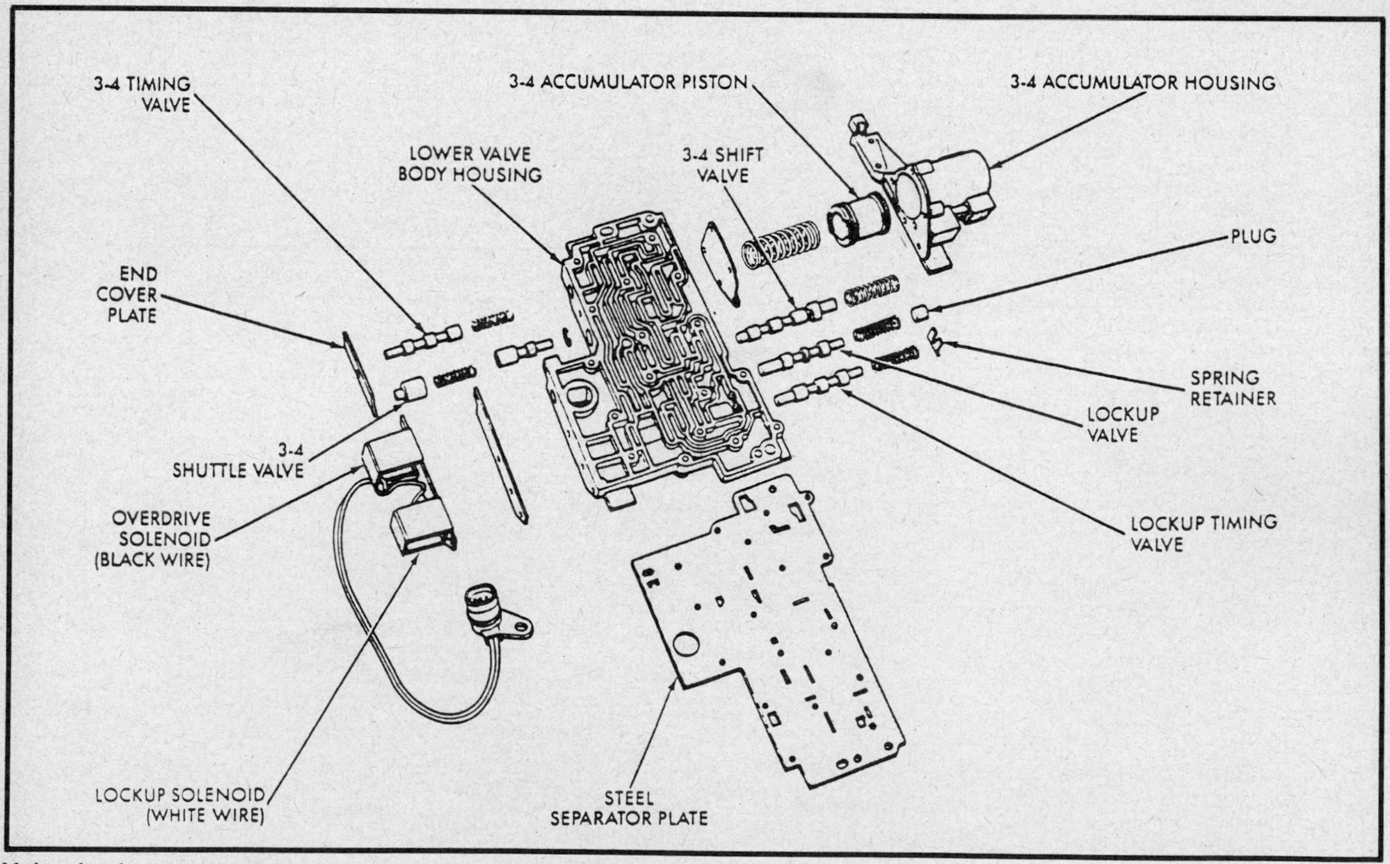
3-4 TIMING VALVE
3-4 ACCUMULATOR PISTON
3-4 ACCUMULATOR HOUSING
LOWER VALVE BODY HOUSING
3-4 SHIFT VALVE
PLUG
END COVER PLATE
SPRING RETAINER
LOCKUP VALVE
3-4 SHUTTLE VALVE
OVERDRIVE SOLENOID (BLACK WIRE)
LOCKUP TIMING VALVE
LOCKUP SOLENOID (WHITE WIRE)
STEEL SEPARATOR PLATE

Valve body components

Section 4

KM148, AW372 and AW372L Transmissions

Chrysler Corp.

APPLICATION

KM148

Year	Vehicle	Engine
1987–89	Dodge Ram 50	2.6L
	Dodge Raider	2.6L

AW372 AND AW372L

Year	Vehicle	Engine
1987–89	Dodge Ram	2.0L

GENERAL DESCRIPTION

The automatic transmissions used are the AW372, AW372L and KM148. The AW372 transmission is designed for use in vehicles equipped with 2WD and the 2.0L or 2.6L engines. The AW372L transmission is designed for use in vehicles equipped with 2WD and the 2.4L engine. The KM148 transmission is designed for use in vehicles equipped with 4WD and the 2.6L engine.

Basically these transmissions are the same with a few slight differences: The AW372 and KM148 are identical, except the AW372 does not incorporate a transfer case, the AW372L, like the AW372 does not use a transfer case but utilizes a hydraulically controlled lockup type torque converter clutch. The AW372 and KM148 use a one-way clutch (sprag type) torque converter.

All of these transmissions consists of 3 multiple disc clutches, 3 one-way clutches, 4 multiple disc brakes and 2 planetary gear sets to provide 4 forward ratios and a reverse ratio.

The transfer case on the KM148 has a high/low and a 2WD/4WD position. By operating the transfer control lever, running at 2WD high (2H), 4WD high (4H) of 4WD low (4L) can be selected freely.

Transmission and Converter Identification

The transmission can be identified by the third line on the Vehicle information code plate, which is located on the top section of the cowl outer panel, in the engine compartment. The plate shows model code, engine model, transmission model and body color code. Additional information such as, the serial number, fixed number, classification, manufacturing year and month can all be found on the left side of the transmission directly above the pan.

The KM148 and AW372 automatic transmissions utilizes a torque converter that consists of an impeller turbine stator, a one-way clutch and front and rear covers. It is a non serviceable, sealed constructed unit, in which the surfaces of the outer and inner covers are welded together.

The AW372L automatic transmission utilizes a torque converter that has a built in hydraulically controlled lockup clutch, which prevents torque converter loss or slippage in medium and high speed ranges, as well as economizing fuel consumption.

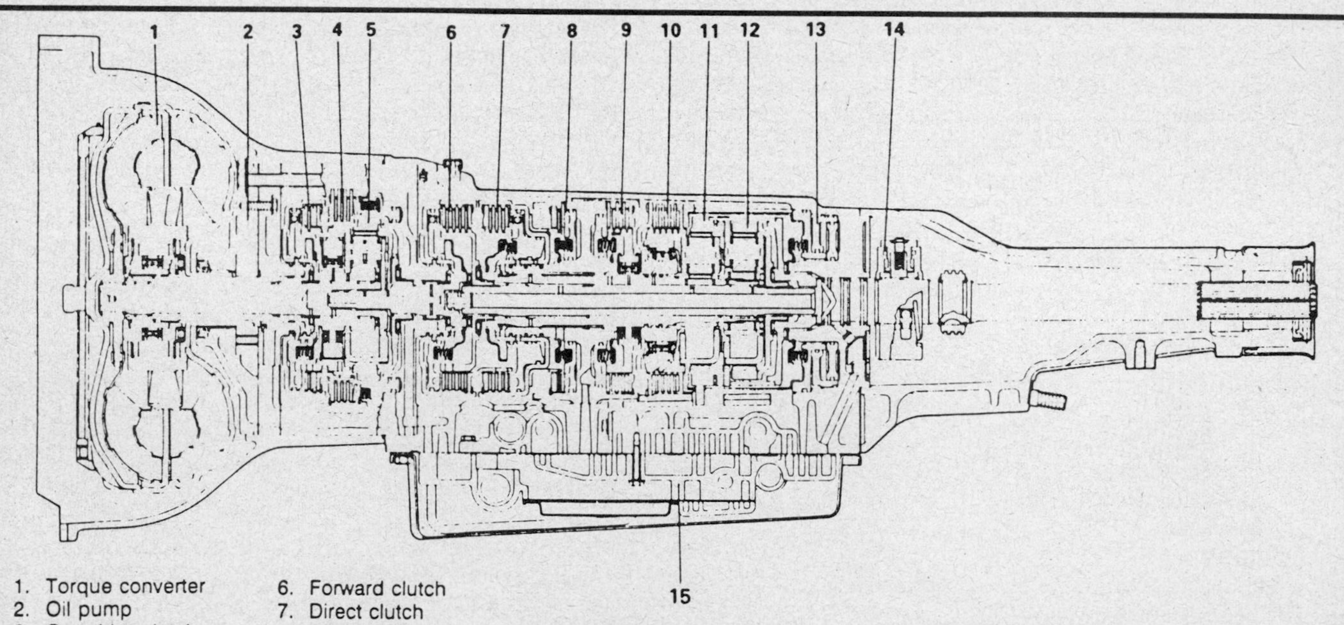
1
2
3
4
5
6
7
8
9
10
11
12
13
14
15

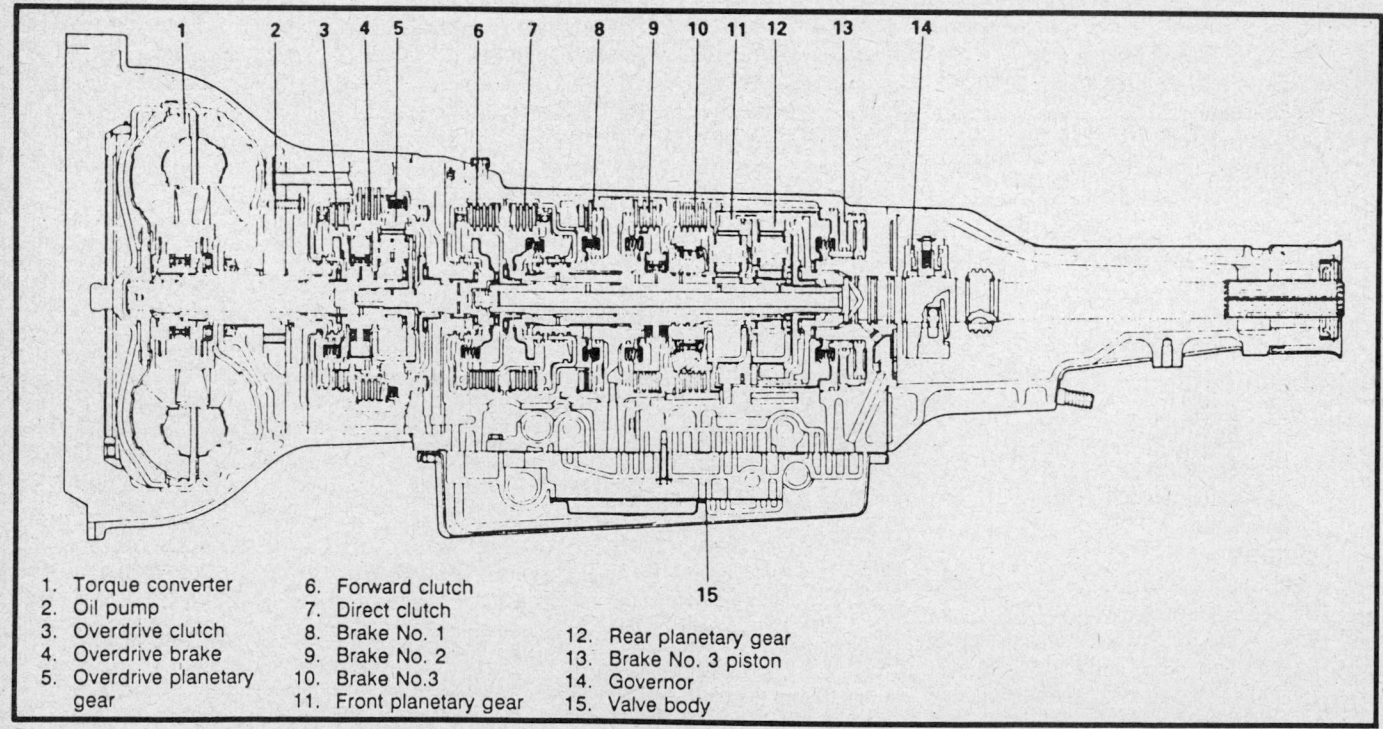

1. Torque converter
2. Oil pump
3. Overdrive clutch
4. Overdrive brake
5. Overdrive planetary gear
6. Forward clutch
7. Direct clutch
8. Brake No. 1
9. Brake No. 2
10. Brake No.3
11. Front planetary gear
12. Rear planetary gear
13. Brake No. 3 piston
14. Governor
15. Valve body

Cross-sectional view of the AW372 automatic transmission

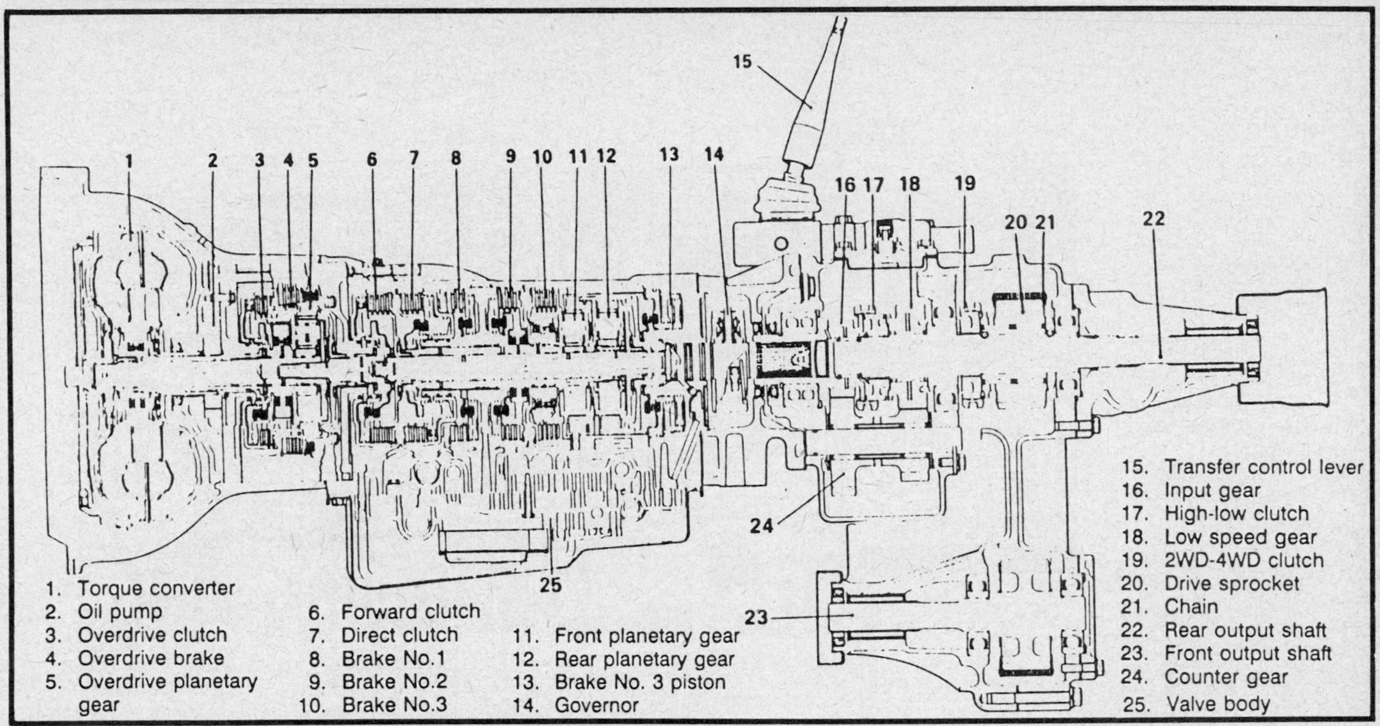
15
1
2
3
4
5
6
7
8
9
10
11
12
13
14
16
17
18
19
20
21
22
24
25
23
1. Torque converter
2. Oil pump
3. Overdrive clutch
4. Overdrive brake
5. Overdrive planetary gear
6. Forward clutch
7. Direct clutch
8. Brake No.1
9. Brake No.2
10. Brake No.3
11. Front planetary gear
12. Rear planetary gear
13. Brake No. 3 piston
14. Governor
15. Transfer control lever
16. Input gear
17. High-low clutch
18. Low speed gear
19. 2WD-4WD clutch
20. Drive sprocket
21. Chain
22. Rear output shaft
23. Front output shaft
24. Counter gear
25. Valve body
Cross-sectional view of the KM148 automatic transmission

Engaging and disengaging the lockup clutch is controlled by the change in direction of hydraulic fluid flow in the torque converter. This unit is also non serviceable.

Torque converters are coded by the manufacturer and sold through varied parts networks. Specific part numbers are used to identify the converters and to allow proper match up to the transmission. When replacing a converter, verify the replacement unit is the same as the unit originally used with the engine/transmission combination.

Metric Fasteners

Metric bolts and fasteners are used in attaching the transmission to the engine and also in attaching the transmission to the chassis crossmember mount.

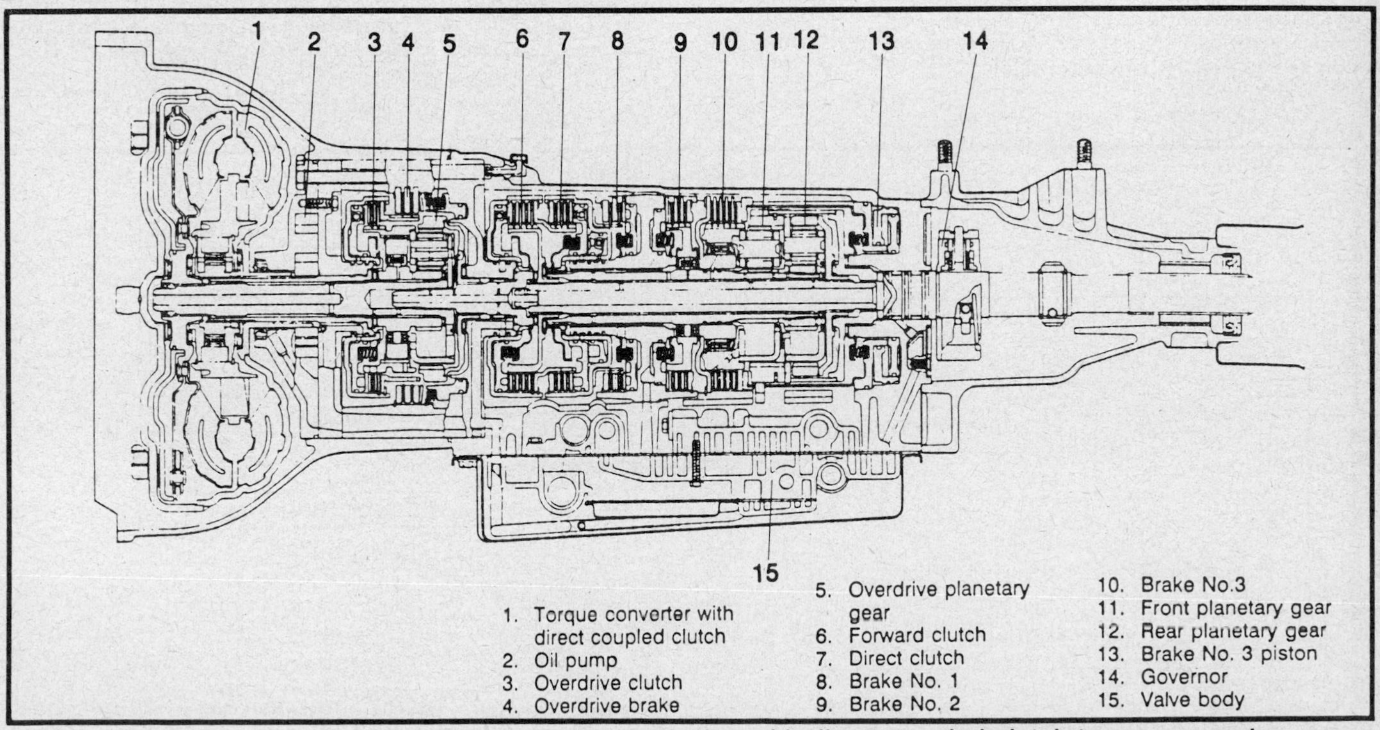
1
2
3
4
5
6
7
8
9
10
11
12
13
14
15
1. Torque converter with direct coupled clutch
2. Oil pump
3. Overdrive clutch
4. Overdrive brake
5. Overdrive planetary gear
6. Forward clutch
7. Direct clutch
8. Brake No. 1
9. Brake No. 2
10. Brake No.3
11. Front planetary gear
12. Rear planetary gear
13. Brake No. 3 piston
14. Governor
15. Valve body
Cross-sectional view of the AW372L automatic transmission with direct coupled clutch torque converter

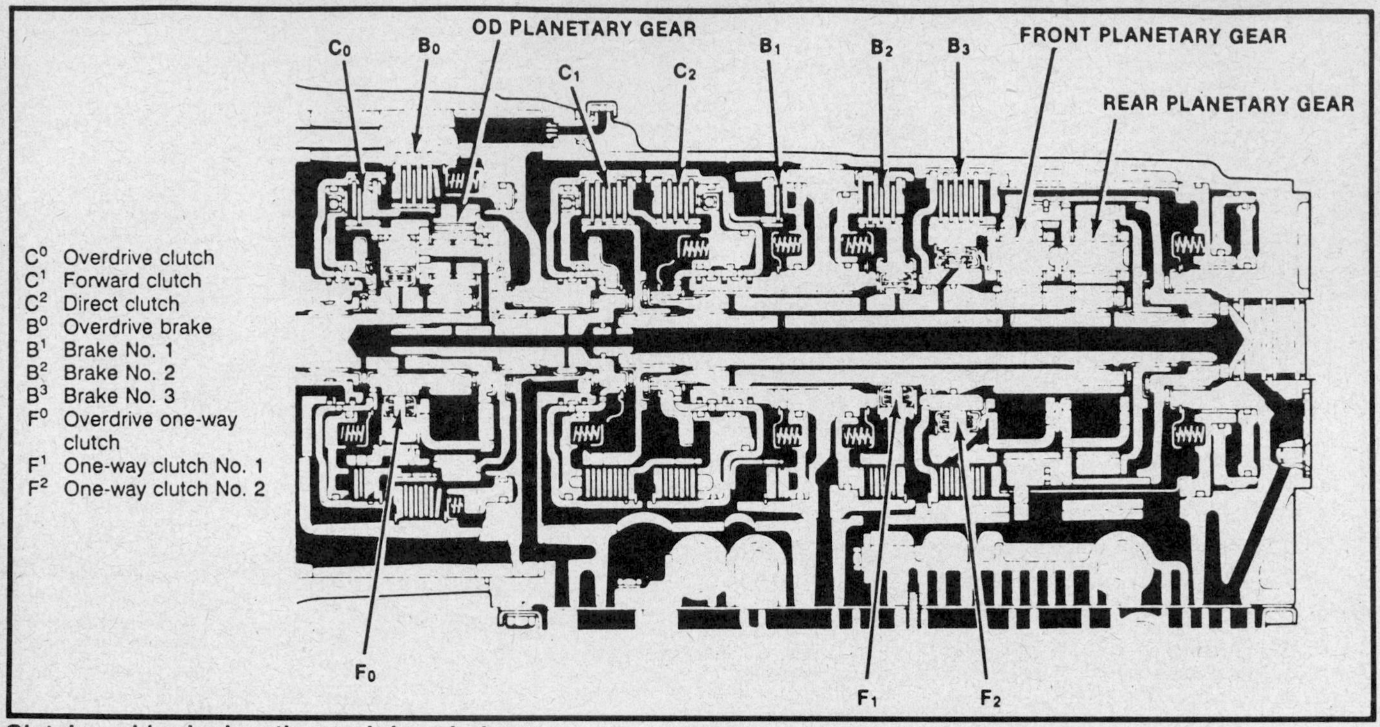
OD PLANETARY GEAR
FRONT PLANETARY GEAR
REAR PLANETARY GEAR
C_0
B_0
C_1
C_2
B_1
B_2
B_3
F_0
F_1
F_2
C^0 Overdrive clutch
C^1 Forward clutch
C^2 Direct clutch
B^0 Overdrive brake
B^1 Brake No. 1
B^2 Brake No. 2
B^3 Brake No. 3
F^0 Overdrive one-way clutch
F^1 One-way clutch No. 1
F^2 One-way clutch No. 2

Clutch and brake location and description

The metric fastener dimensions are very close to the dimensions of the familiar inch system fasteners and for this reason replacement fasteners must have the same measurement and strength as those removed.

Do not attempt to interchange metric fasteners for inch system fasteners. Mismatched or incorrect fasteners can result in damage to the transmission unit through malfunctions, breakage or possible personal injury.

Care should be taken to reuse the fasteners in the same location as removed, whenever possible.

Capacities

The fluid capacities are approximate and the correct fluid level should be determined by the dipstick indicator. After complete transmission overhaul the AW372, AW372L and KM148 transmissions should hold approximately 7.6 quarts of automatic transmission fluid. After pan and filter service the AW372, AW372L and KM148 transmissions should hold approximately 3 quarts of automatic transmission fluid type Dexron® II.

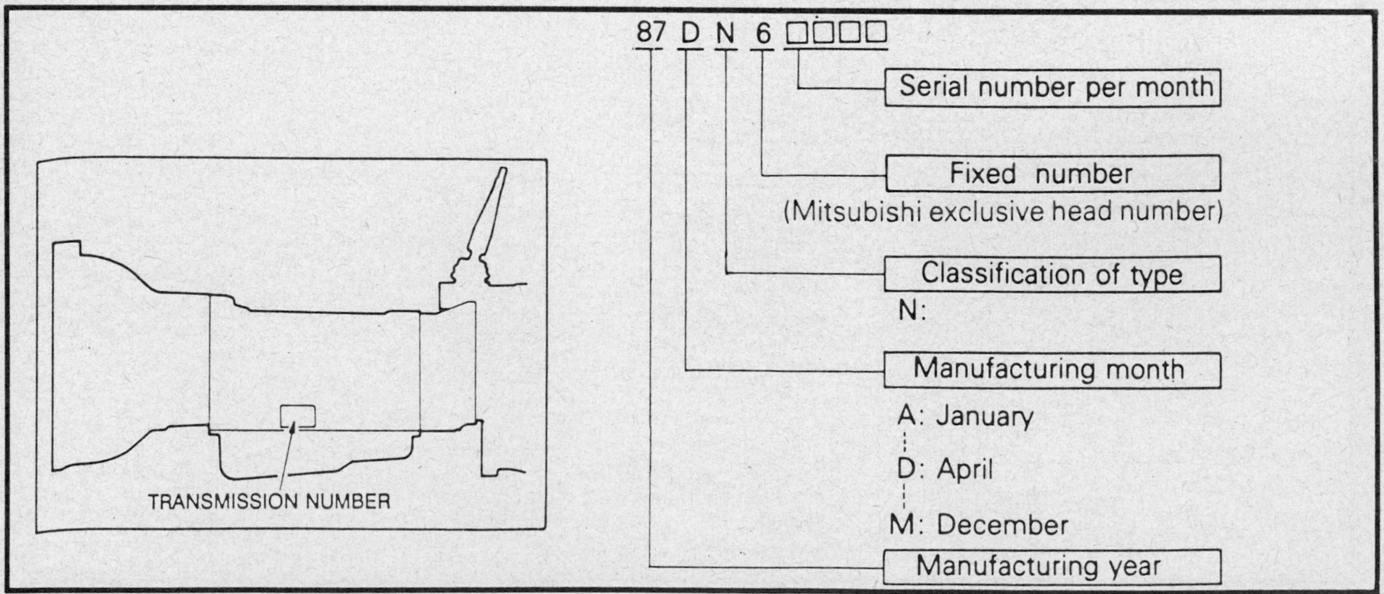
87 D N 6 □□□□
Serial number per month
Fixed number
(Mitsubishi exclusive head number)
Classification of type
N:
Manufacturing month
A: January
D: April
M: December
Manufacturing year
TRANSMISSION NUMBER

Transmission identification information plate

FLUID CAPACITY
All capacities given in quarts

Year	Vehicle	Transmission	Fluid Type	Pan Capacity①	Overhaul Capacity
1987–89	Dodge Raider	KM148	Dexron®II	3.2	7.6
	Dodge Ram 50	AW372	Dexron®II	2.9	7.2
	Dodge Ram 50	KM148	Dexron®II	3.2	7.6

① Approximate specification

Checking Fluid Level

1. Start the engine and allow to idle for at least a couple of minutes. With the parking brake applied, move the selector lever through all of its positions. Finally, place the selector lever in the **N** position.

2. After the transmission has warmed up to the normal operating temperature, recheck the fluid level. The level must be between the cold upper limit and hot lower limit marks on the dipstick. Insert the dipstick fully to prevent dirt from entering the transmission.

NOTE: A fully cooled transmission will show an oil level below the bottom the of the dipstick mark, even when correctly filed with oil.

3. If the oil level is too low, the oil pump will suck up air which can be clearly heard. The oil will foam and provide incorrect dipstick level results during an oil level checkup. Wait a few minutes until the oil foam has resided, add oil and recheck the oil level.

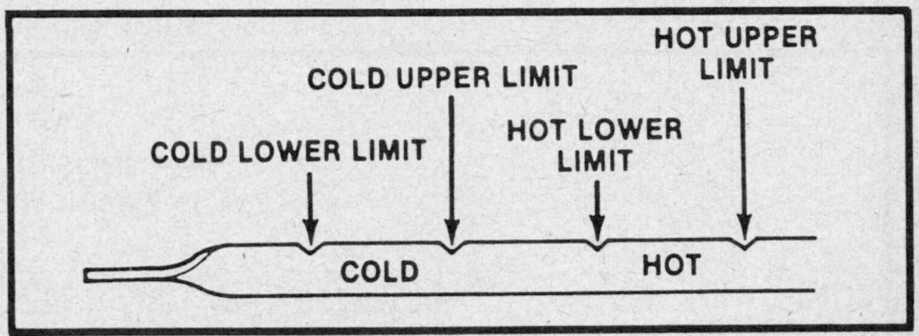
HOT UPPER LIMIT
COLD UPPER LIMIT
HOT LOWER LIMIT
COLD LOWER LIMIT
COLD
HOT
Transmission dipstick levels

4. Excessive transmission fluid must be drained, otherwise the transmission gears would be splashing in oil. The temperature will increase unnecessarily, resulting in foaming oil being ejected through the breather. Continuous operation under such circumstances can lead to transmission damage.

TRANSMISSION MODIFICATIONS

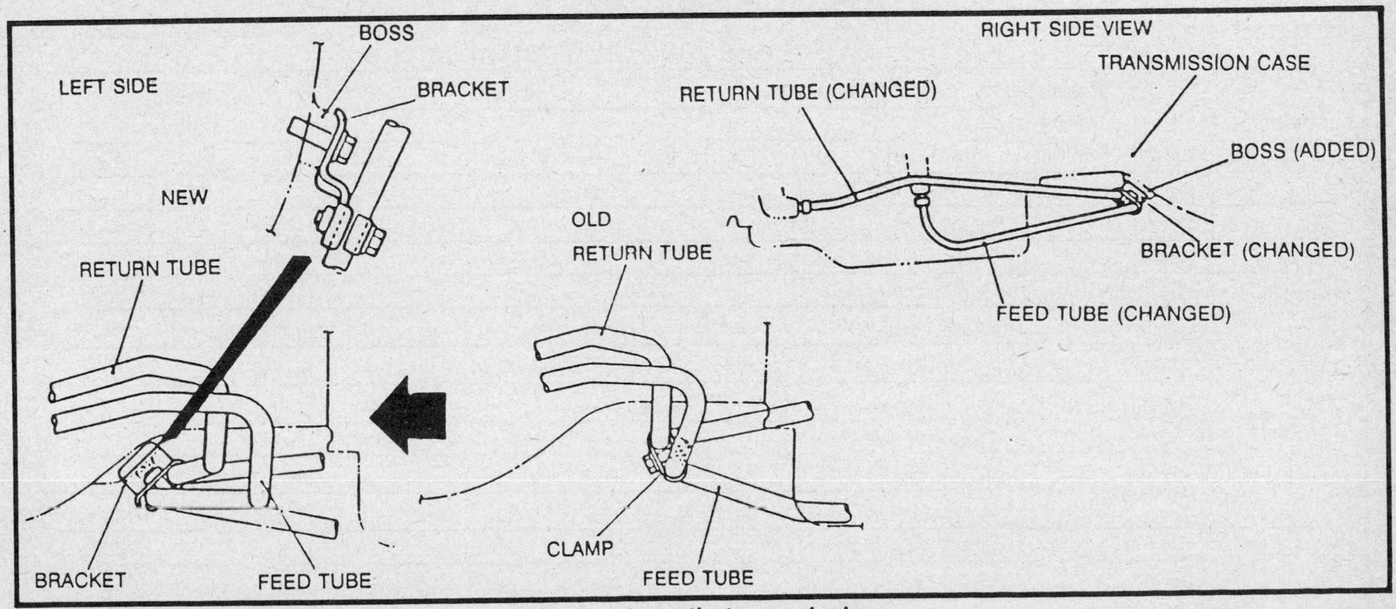
BOSS
LEFT SIDE
BRACKET
NEW
RETURN TUBE
BRACKET
FEED TUBE
OLD
RETURN TUBE
CLAMP
FEED TUBE
RIGHT SIDE VIEW
TRANSMISSION CASE
RETURN TUBE (CHANGED)
BOSS (ADDED)
BRACKET (CHANGED)
FEED TUBE (CHANGED)
Transmission cooler line modification—KM148 automatic transmission

Check Ball Positioning and Cooler Line Changes

Corrections to the positioning of the valve body check balls and transmission cooler lines on the KM148 require that changes be made on the 1987 Mitsubishi Montero, Chrysler Ram 50 and Ram Raider.

RAM 50

On the 2WD vehicles, the shape of the transmission lines have been changed.

On both the 2WD and 4WD vehicles, the cooler line clamp has been changed to a bracket.

On both 2WD and 4WD vehicles, a threaded boss has been added to the transmission case to accommodate the new bracket mounting bolt.

RAM RAIDER AND MONTERO

The cooler line clamp has been eliminated and a threaded boss has been added to the transmission case to accommodate the new bracket mounting bolt.

RAM 50, RAM RAIDER AND MONTERO

Install the steel check ball in its designated position. The other 3 rubber check balls are identical and may be installed in any other position.

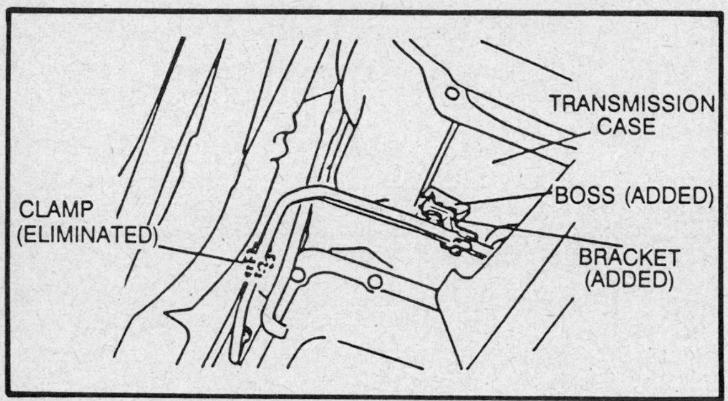
TRANSMISSION CASE
BOSS (ADDED)
CLAMP (ELIMINATED)
BRACKET (ADDED)

Transmission cooler line clamp and threaded boss modification—KM148 automatic transmission

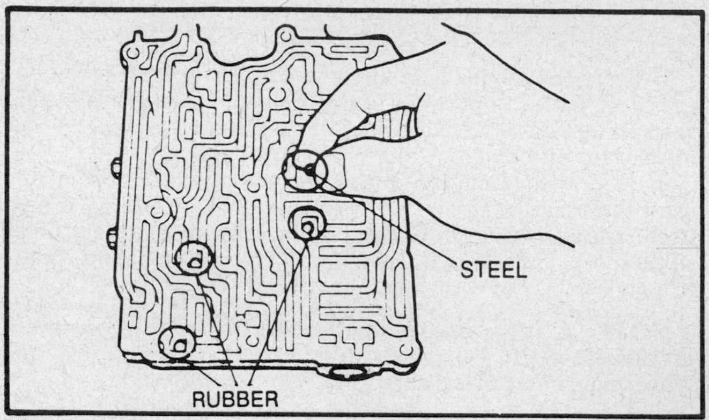
STEEL
RUBBER

Valve body check ball positions

TROUBLE DIAGNOSIS

CLUTCH AND BAND APPLICATION
KM148, AW372 and AW372L Transmissions

Selector Position	Gear Position	Parking Mechanism	Clutch C_0	Clutch C_1	Clutch C_2 IP	Clutch C_2 OP	Brake B_0	Brake B_1	Brake B_2	Brake B_3 IP	Brake B_3 OP	One-Way Clutch F_0	One-Way Clutch F_1	One-Way Clutch F_2
P	Neutral	Holding	Applied	–	–	–	–	–	–	–	Applied	–	–	–
R	Reverse	–	Applied	–	Applied	Applied	–	–	–	Applied	Applied	Holding	–	–
N	Neutral		–	Applied	–	–	–	–	–	–	–	–	–	–
D–On	First	–	Applied	Applied	–	–	–	–	–	–	–	Holding	–	Holding
	Second	–	Applied	Applied	–	–	–	–	Applied	–	–	Holding	Holding	–
	Third	–	Applied	Applied	–	Applied	–	–	Applied	–	–	Holding	–	–
	Fourth	–	–	Applied	–	Applied	Applied	–	Applied	–	–	–	–	–
D–Off	First	–	Applied	Applied	–	–	–	–	–	–	–	Holding	–	Holding
	Second	–	Applied	Applied	–	–	–	–	Applied	–	–	Holding	Holding	–
	Third	–	Applied	Applied	–	–	–	–	Applied	–	–	Holding	–	–
2	First	–	Applied	Applied	–	–	–	–	–	–	–	Holding	–	Holding
	Second	–	Applied	Applied	–	–	–	Applied	Applied	–	–	Holding	Holding	–
L	First	–	Applied	Applied	–	–	–	–	–	Applied	Applied	Holding	–	Holding

CHILTON'S THREE C's TRANSMISSION DIAGNOSIS
KM148, AW372 and AW372L Transmissions

Condition	Cause	Correction
No starter action in P or N) Starter motor action in all other selector positions) Back-up lights inoperative	a) Neutral safety switch out of adjustment	a) Replace or adjust switch as needed
Excessive thump into D, 1st or R	a) Engine idle too high b) Throttle cable out of adjustment c) Valves sticking in control valve body	a) Adjust idle speed b) Adjust or replace throttle cable c) Clean control valve body or replace as necessary
Vehicle moves with selector lever in N	a) Manual linkage out of adjustment b) Fault in front clutch support housing c) Fault in stator supportshaft bearing d) Fault in forward sun gear shaft seals	a) Adjust manual likage b) Replace front clutch support housing c) Replace stator support shaft bearing d) Replace forward sun gear shaft seals
Stall speed in D and R range is equal to each other but lower than nominal value	a) Engine output is low b) Stator one-way clutch is faulty c) Faulty torque converter is suspected if it is lower than nominal by more than 600 rpm	a) Check engine performance and repair condition b) Replace stator one-way clutch c) Replace torque converter
Stall speed in D range is higher than nominal	a) OD clutch slipping b) OD one-way clutch faulty c) Forward clutch slipping d) One-way clutch No. 2 faulty e) Low line pressure	a) Replace OD clutch b) Replace one-way clutch c) Replace forward clutch d) Replace No. 2 one-way clutch e) See line pressure diagnosis
Stall speed in R range is higher than nominal	a) OD clutch slipping b) OD one-way clutch faulty c) Direct clutch slipping d) Brake No. 3 slipping e) Low line pressure	a) Replace OD clutch b) Replace one-way clutch c) Replace direct clutch d) Replace brake No. 3 e) See line pressure diagnosis
Hydraulic pressure higher than nominal in all ranges	a) Regulator valve faulty b) Throttle valve faulty c) Throttle control cable incorrectly adjusted	a) Replace regulator valve b) Replace throttle valve c) Adjust throttle valve cable
Hydraulic pressure lower than nominal in D range	a) Large fluid leaks in D range hydraulic circuit b) Forward clutch faulty c) OD clutch faulty	a) Eliminate leakage b) Replace forward clutch c) Replace OD clutch
Hydraulic pressure lower than nominal in R range	a) Large fluid leaks in R range hydraulic circuit b) Brake No. 3 faulty c) Direct clutch faulty d) OD clutch faulty	a) Eliminate leakage b) Replace brake No. 3 c) Replace direct clutch d) Replace OD clutch

CONVETER CLUTCH DIAGNOSIS
KM148, AW372 and AW372L Transmissions

Condition	Cause	Correction
No drive at any position due to clutch torque converter engaged	a) Abnormal signal slippage in damper clutch torque converter system b) Malfunctioning sealing in solenoid valve torque converter	a) Replace b) Replace

CONVETER CLUTCH DIAGNOSIS
KM148, AW372 and AW372L Transmissions

Condition	Cause	Correction
Excessive vibration	a) Decreased signal slippage from CPU (Computer Processing Unit)	a) Replace
Inoperative damper clutch torque	a) No signal lock-up from CPU b) Lock-up line pressure low c) Opened or shorted circuit of solenoid valve	a) Replace b) Replace c) Replace
Increased fuel consumption	a) Damper clutch torque converter does not engage because of a stuck valve	a) Clean up
Lock-up torque converter does not release	a) Decreased driving effort in facing of clutch plate b) Burn out clutch disc c) Damper clutch torque converter system solenoid valve stuck open	a) Replace b) Replace c) Replace
Increased vibration due to no control of slipping ratio	a) Sticking shaft in throttle opening sensor	a) Replace
No drive at any position	a) Seized or stuck thrust bearing in torque converter b) Deformed crankshaft bushing in torque converter c) Broken or cracked drive plate d) Low oil level	a) Replace b) Replace c) Replace d) Refill with fluid
Increased noise due to inoperative	a) Deformed or worn locking-ring in torque converter	a) Replace
Excessive slips when starting	a) Low oil level b) Worn over running clutch in torque converter	a) Refill with fluid b) Replace
Selector lever operation is stiff	a) Incorrect adjustment of pushbutton b) Incorrect adjustment of control cable c) Excessive wear of detent plate d) Excessive wear of pin at end of selector e) Worn contact surfaces of bushbutton and sleeve	a) Adjust b) Adjust c) Replace d) Replace e) Replace
Starter motor does not operate with the selector lever in the N or P position	a) Malfunction in inhibitor switch b) Incorrect adjustment of control cable	a) Replace b) Adjust
Vehicle does not move	a) Low automatic transmission fluid level b) Broken planetary gear carrier	a) Correct b) Replace

Hydraulic Control System

NOTE: Please refer to Section 9 for all oil flow circuits.

The hydraulic control system mainly consists of the oil pump to generate hydraulic pressure, the governor to detect vehicle speed and the valve body assembly to control the clutches and brakes.

The oil pump is of the gear type. It generates hydraulic pressure suppling fluid to the torque converter, to activate the hydraulic control system and provides lubrication to all frictional parts.

The valve body contains valves to distribute the hydraulic pressure from the oil pump to each component and to control the hydraulic pressure. The throttle valve opening and vehicle speed determine hydraulic pressure. The primary regulator valve automatically controls hydraulic pressure, preventing excessive pressure build up by the pump.

The secondary regulator valve controls the converter pressure, the lubrication oil pressure and the cooler pressure. The pressure relief valve regulates maximum hydraulic pressure generated by the oil pump to maintain the safety of the hydraulic pressure circuit.

The torque converter fluid, which has been heated up during operation is channeled into the cooler on the engine side, here the fluid is cooled down. The check valve (for cooler bypass), reduces the fluid pressure for the cooler in order to ensure safe cooler operation.

The governor is installed on the output shaft and changes line pressure from the primary regulator valve to governor pressure. Governor pressure increases as the vehicle speed increases.

The manual valve is is mechanically connected by the manual shift rod or cable, which in turn is linked to the selector lever in the passenger compartment. This valve changes the fluid passages for each gear selector position (P, R, N, D, 2 and L).

Downshift is achieved by movement of the kickdown valve. At ¾ throttle position, the kickdown valve moves until the detent regulator valve port is opened. This allows detent pressure to flow to the 1–2, 2–3 and 3–4 shift valves to downshift. The detent regulator valve regulates the hydraulic detent pressure applied to the 1–2, 2–3 and 3–4 shift valves through the kickdown valve to limit the speed during a downshift.

The intermediate modulator valve regulates line pressure and supplies it to brake No. 1 to reduce the shock when the engine brake is applied in the 2nd speed in **D2** range. The low coast modulator, reduces line pressure to low coast modulator pressure when the selector lever is in the **L** range operation.

The 1-2 shift valve automatically controls the shift from 1st gear to 2nd gear or vice versa depending on governor pressure and throttle pressure. The 2–3 shift valve automatically controls a 2–3 or 3–4 shift in accordance with the governor pressure and throttle pressure.

The 3–4 shift valve automatically controls a 3–4 or 4–3 shift in accordance with the governor pressure and throttle pressure. The 3–4 shift valve receives upward force from throttle pressure at the top end, line pressure at the middle portion and spring force. It also receives upward force from the governor pressure at its lower piston.

The cutback valve, operated by governor pressure and throttle pressure, controls cutback pressure which is applied to the throttle valve. Applying cutback pressure to the throttle valve lowers the throttle pressure. Consequently, line pressure is lowered by applying the lowered throttle pressure to the primary regulator valve, thus preventing unnecessary loss of power by the oil pump.

The reverse clutch sequence valve reduces shock when a shift is made into **R** range. It is controlled by line pressure acting on the inner piston of the direct clutch C2.

When the shift lever is changed from **D** position to **D2** position while in 4th gear (OD), the transmission shifts to 3nd gear and to 2nd gear. Thus the D–2 down timing valve allows a smooth transition from 4th gear to 2nd gear.

Accumulators are provided for C1, C2 and B2, respectively to dampen shocks when each is actuated. The area of the pressure receiving side of the accumulator piston is made greater than that of the back pressure side. The line pressure is always applied to the back pressure side, keeping the piston in the up position. When the circuit to the pressure receiving side opens causing the line pressure to be applied to the piston, the piston is slowly pushed downward, thus dampening shocks when each device is operated.

The 4th (OD) control system utilizes a solenoid valve to electrically select the hydraulic circuit to shift into 4th speed. When the OD-off switch on the select lever is turned ON, the OD solenoid is energized to select a circuit to set the 4th speed. When the OD-off switch is turned off, the OD-off indicator light, on the meter panel comes on.

The OD-off switch is located on the select lever on the driver's side seat. The OD solenoid valve is installed to the left of the transmission case.

THROTTLE VALVE SYSTEM

The throttle valve controls the throttle pressure with the power demand of the driver.

When the accelerator pedal is depressed, the throttle cable, which is linked to the pedal, is pulled to turn the throttle cam. This in turn pushes the kickdown valve to the right. Consequently, the throttle valve is moved to the right by a spring, thus opening a line pressure passage. On the other hand, throttle pressure is applied to the back of throttle valve land No. 2, pushing the valve to the left. Cut back pressure acts on land No. 3 to the position where all forces balance. In this condition, the throttle valve closes the line pressure port.

The throttle pressure is applied to the 1–2, 2–3 and 3–4 shift valve and works against governor pressure. This pressure is applied to the primary and secondary regulator valves, thus controlling line pressure by balancing throttle valve pressure against governor pressure.

Diagnosis Tests

GENERAL DIAGNOSTIC SEQUENCE

Before starting any test procedures, a selected sequence should be followed in the diagnosis of any automatic transmission malfunction. A suggested sequence is as follows:

1. Inspect the fluid level and correct as required.
2. Check the freedom of movement of the downshift linkage and adjust as required.
3. Check the manual linkage synchronization and adjust as required.
4. Install a 400 psi oil pressure gauge to the main line pressure port on the transmission case. Should this arrangement be used on a road test, route and secure the hose so as not to drag or be caught during the test.
5. Perform the road test over a predetermined route to verify shift speeds and engine performance. Refer to the pressure gauge during all shifts for irregularities in the pressure readings. With the aid of a helper, record all readings for reference.
6. During the road test, governor operation can be noted and the shift speeds recorded as the throttle valves are moved through various positions. Should further testing of the governor system be needed, this can be accomplished when the vehicle is returned to the service center, by a shift test.
7. Should a verification of engine performance or initial gear engagement be needed, a stall test can be used to aid in pinpointing a malfunction.
8. Perform a case air pressure test, should the malfunction indicated internal transmission pressure leakage.

CONTROL PRESSURE TEST

Control pressure tests should be performed whenever slippage, delay or harshness is felt in the shifting of the transmission. Throttle pressure changes can cause these problems also, but are generated from the control pressures and therefore reflect any problems arising from the control pressure system.

The control pressure is first checked in all ranges without any throttle pressure. The control pressure tests should define differences between mechanical or hydraulic failures of the transmission.

CONTROL PRESSURE DIAGNOSIS

Condition	Cause	Correction
Hydraulic pressure higher than nominal in all ranges	a) Regulator valve faulty b) Throttle valve faulty c) Throttle control cable incorrectly adjusted	a) Replace regulator valve b) Replace throttle valve c) Adjust throttle valve

CONTROL PRESSURE DIAGNOSIS

Condition	Cause	Correction
Hydraulic pressure lower than nominal in D range	a) Large fluid leaks in D range hydraulic circuit	a) Eliminate leakage
	b) Forward clutch faulty	c) Replace forward clutch
	c) OD clutch faulty	c) Replace OD clutch
Hydraulic pressure lower than nominal in R range	a) Large fluid leaks in R range hydraulic circuit	a) Eliminate leakage
	b) Brake No. 3 faulty	b) Replace brake No. 3
	c) Direct clutch faulty	c) Replace direct clutch
	d) OD clutch faulty	d) Replace OD clutch
Hydraulic pressure lower than nominal in all ranges	a) Oil pump faulty	a) Replace oil pump
	b) Regulator valve faulty	b) Replace regulator valve
	c) Throttle valve faulty	c) Replace throttle valve
	d) Throttle control cable incorrectly adjusted	d) Check and adjust throttle cable
	e) OD clutch faulty	e) Replace OD clutch

AIR PRESSURE

Air pressure testing is helpful in locating leak points during diassembly and in verifying that the fluid circuits are not leaking during build-up. If the road test discloses which clutch or servo is not holding, that is the circuit to be air tested. Use air pressure regulated to about 25 psi and check for air escaping to detect leakage. If the pressures are found to be low in a clutch, servo or passage-way, a verification can be accomplished by removing the valve body and performing an air pressure test.

1. To determine if a malfunction of a clutch or brake is caused by fluid leakage in the system or is the result of a mechanical failure.
2. To test the transmission for internal fluid leakage during the rebuilding and before completing the assembly.

Procedure

1. Raise and support the vehicle safely. Remove the oil pan.
2. Remove the valve body and place it in and clean area.
3. Obtain an air nozzle and adjust for 25 psi.
4. Apply low compressed air pressure (25 psi.) to the each passage and test the operation of the clutches and brakes.

NOTE: When appling low pressure compressed air to each passage, a dull thud will normally be heard, if a hissing or no thud is heard the transmission must be removed, disassembled and checked further.

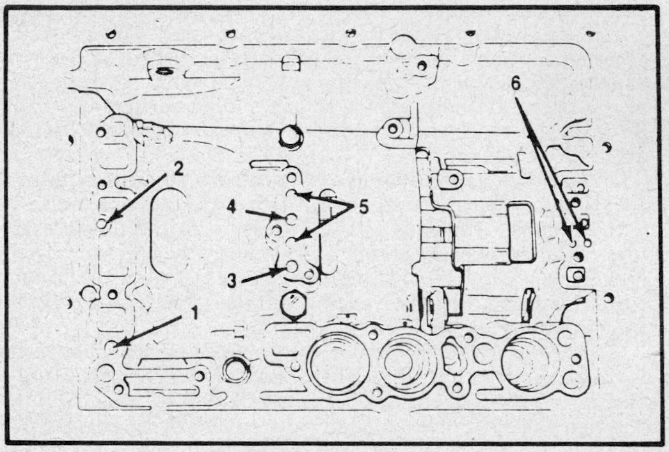
1
2
3
4
5
6
Air pressure test ports

STALL TEST

The stall test is used to check the maximum engine rpm (no more increase in engine rpm at wide open throttle) with the selector lever in the **D, 2, 1, R** positions and to determine if any slippage is occurring from the clutches, bands or torque converter. The engine operation is noted and a determination can be made as to its performance.

Before performing pressure test, perform basic checks and adjustment including fluid level and condition check and throttle cable adjustment. Prior to pressure test, the engine and transmission must have been warmed up enough, the engine coolant temperature to 180–195° F (80–90°C) and the transfer must be in 2WD high.

1. Check the engine oil level, start the engine and bring to normal operating temperature.
2. Check the transmission fluid level and correct as necessary. Attach a calibrated tachometer to the engine and a 0–400 psi oil pressure gauge to the transmission control pressure tap on the left side of the case.
3. Mark the specified maximum engine rpm on the tachometer cover plate with a grease pencil to immediately check if the stall speed is over or under specifications.
4. Apply the parking brake and block both front and rear wheels.

CAUTION

Do not allow anyone in front or rear of the vehicle while performing the stall test.

5. While holding the brake pedal with the left foot, place the selector lever in the **D** position and slowly depress the accelerator.
6. Read and record the engine rpm when the accelerator pedal is fully depressed and the engine rpm is stabilized. Read and record the oil pressure reading at the high engine rpm point.

CAUTION

The stall test must be made within 8 seconds.

STALL SPEED SPECIFICATIONS

Vehicle Application	Engine (liters)	Transmission	Stall Speed rpm Minimum	Maximum
Mitsubishi Montero	2.6	KM148	2100	2400
Mitsubishi Truck	2.0	AW372	1800	2100
Mitsubishi Truck	2.6	KM148	2100	2400
Dodge Raider	2.6	KM148	2100	2400
Dodge Ram 50	2.0	AW372	1800	2100
Dodge Ram 50	2.6	KM148	2100	2400

7. Shift the selector lever into the **N** position and increase the engine rpm to approximately 1000–1200. Hold the engine speed for 1–2 minutes to cool the converter, transmission and fluid.
8. Make similar tests in the **R** position.

CAUTION

If at any time the engine rpm exceeds the maximum as per the specifications, indications are that a clutch unit or brake is slipping and the stall test should be stopped before more damage is done to the internal parts.

STALL SPEED TEST RESULTS

Condition	Cause
Stall speed in D and R range is equal to each other but lower than nominal value	a) Engine output is low b) Stator one-way clutch is faulty. Faulty torque converter is suspected if it is lower than nominal by more than 600 rpm
Stall speed in D range is higher than nominal	a) OD clutch slipping b) OD one-way clutch faulty c) Forward clutch slipping d) One-way clutch No. 2 faulty e) Low line pressure
Stall speed in R range is higher than nominal	a) OD clutch slipping b) OD one-way clutch faulty c) Direct clutch slipping d) Brake No. 3 slipping e) Low line pressure

GOVERNOR PRESSURE TEST

Procedure

1. Raise and support the vehicle safely. Remove the plug from the governor pressure gauge using a suitable adapter tool.
2. Position the gauge inside the vehicle for easy access.
3. Appy the parking brake and start the engine.
4. Release the parking brake.
5. Place the selector lever in the **D** range and measure the governor pressure.

NOTE: If the governor pressure is not within specification, incorrect line pressure, oil leakes from the governor pressure circuit or faulty governor is suspected.

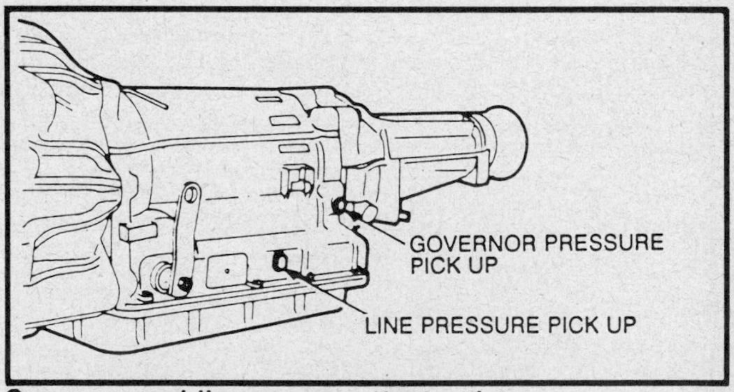
GOVERNOR PRESSURE PICK UP
LINE PRESSURE PICK UP

Governor and line pressure test points

GOVERNOR PRESSURE TEST

Output Shaft Speed rpm	Governor Pressure 2.0L psi (kPa)	2.4L psi (kPa)	2.6L psi (kPa)
1000	19–22 (128)–150)	18–23 (130–160)	20–24 (138–166)
2000	33–38 (226)–264)	33–38 (230–270)	36–41 (246–284)
3200	54–62 (373)–431)	54–63 (380–440)	59–66 (402–460)

LINE PRESSURE TEST

	Line Pressure D Range psi (kPa)	R Range psi (kPa)
At idle	66–76 (452–529)	100–116 (687–804)
At stall	144–169 (991–1166)	214–270 (1471–1863)

ROAD TEST

Prior to performing the road test, be sure to make basic checks including check and adjustment of fluid level and condition and adjustment of the throttle cable. For road testing, the transfer case must be placed in the 2H (2WD-high) position.

In the road test, various changes during transmission operation must be observed, slippage conditions are checked at each shift position.

D Range 3rd Speed Test

1. Place the selector in **D** position range 3rd speed and listen for abnormal noise and vibration.
2. Check carefully as abnormal noise and vibration are sometimes caused by unbalanced driveshaft, differential, tires, torque converter, engine, etc.

2 Range Test

1. Shift to **2** position and start with the throttle valve at 50% full open. Check to see if the 1st and 2nd speed upshift point at these throttle openings, meet the shift pattern.
2. Kickdown at 2nd speed in **2** range, to check that 2nd to 1st kickdown speed limits meet the shift pattern.
3. Check for abnormal noise during acceleration and deceleration, check for shock during upshift and downshift.

L Range Test

1. Shift the selector into the **L** range and check that upshift to 2nd speed does not take place.
2. Check for abnormal noise during acceleration and deceleration.

R Range Test

1. Shift to the **R** range and start forward with full throttle to check for slippage.
2. While the vehicle is running, lightly depress the accelerator pedal to check for slippage.

P Range Test

With the vehicle parked on a slope (about 5 degrees or steeper), shift to the **P** range and release the parking brake to check to see that the parking brake system functions to keep the vehicle stationary.

Converter Clutch Operation and Diagnosis

A direct-coupled clutch is incorporated in the torque converter; it corresponds to a manual transmission clutch.

As a result, torque converter loss or slippage in the medium and high speed ranges is virtually eliminated, thus improving fuel consumption.

The direct-coupled clutch is installed to the turbine's hub, and is engaged to or disengaged from the front cover by hydraulic control.

ENGAGEMENT

When the vehicle is driven in overdrive 4th gear, the line pressure circuit acting on the 2 part of the signal valve is obstructed by the 3–4 shift valve, with the result that the line pressure from the 3–4 shift valve is then applied to the 1 part of the signal valve.

When the speed of the vehicle equals a certain specified speed (approximately 34 mph (55 km/h) or higher, the spring force pressing the signal valve is overcome by the governor pressure, with the result that the signal valve is pressed upward, and line pressure is applied to the lower part of the relay valve. Line pressure is applied to the upper part of the relay valve at all times, and this plus the spring force press the relay valve downward, but, because the (B) diameter is greater than the diameter of the (A) part, the (B-A) X line pressure) force overcomes the force of the spring, thus pressing the relay valve upward. As a result, the converter pressure acts upon the right side of the direct-coupled clutch, thus causing it to engage with the front cover.

DISENGAGEMENT

When the vehicle is driven in a gear other than overdrive (4th gear), the line pressure from the 3–4 shift valve does not act upon the (1) part of the signal valve. The line pressure from the 3–4 shift valve, however, does act upon the (2) part of the signal valve, and this plus the force of the spring cause the signal valve to be forced upward.

When the vehicle speed becomes the certain designated speed or less, the governor pressure decreases, and the signal valve is pressed upward by the spring force. As a result, because the line pressure does not act upon the lower part of the relay valve, the relay valve remains in the pushed down condition due to the spring force and the line pressure acting on the upper part of the relay valve. Consequently, the converter pressure acts upon the left side of the direct-coupled clutch, and the clutch is disengaged from the front cover.

Testing the Converter Clutch

1. Perform a converter stall speed test to determine if trouble is in the converter clutch or transmission.
2. If the stall speed in **D** and **R** range is equal to each other but lower than the nominal valve:
   a. The engine output is low.
   b. The stator one-way clutch is faulty.

NOTE: A faulty torque converter is suspected if the stall speed is lower than nominal by more than 600 rpm.

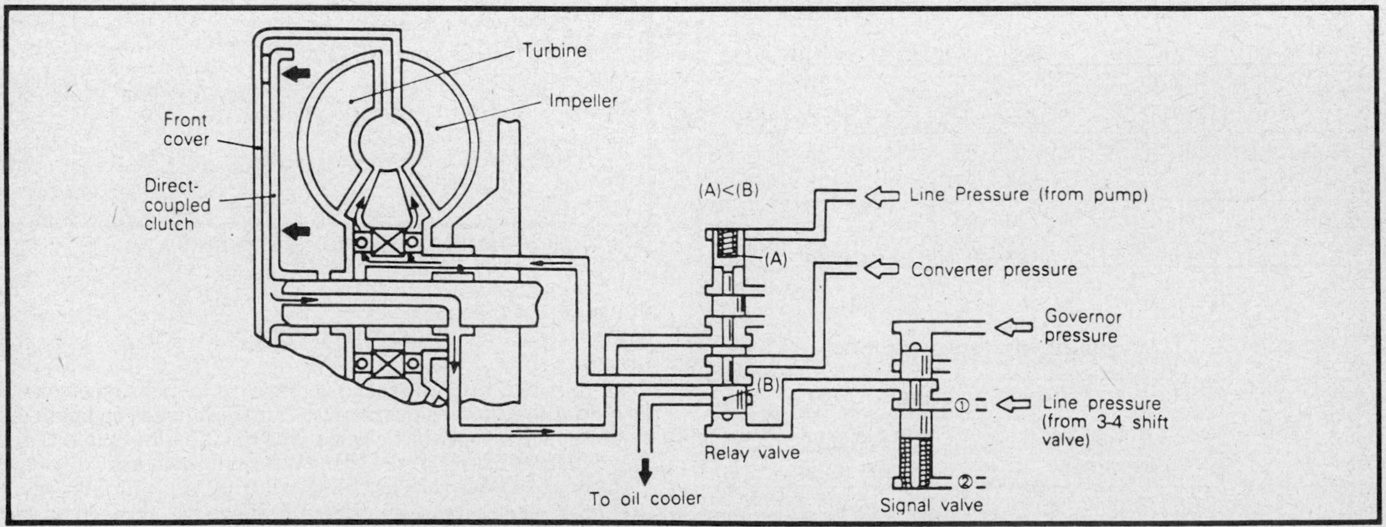
Turbine
Impeller
Front cover
Direct-coupled clutch
(A)<(B)
Line Pressure (from pump)
(A)
Converter pressure
Governor pressure
(B)
Line pressure (from 3-4 shift valve)
①
②
Relay valve
To oil cooler
Signal valve

Torque converter direct-coupled engagement—AW372L automatic transmission

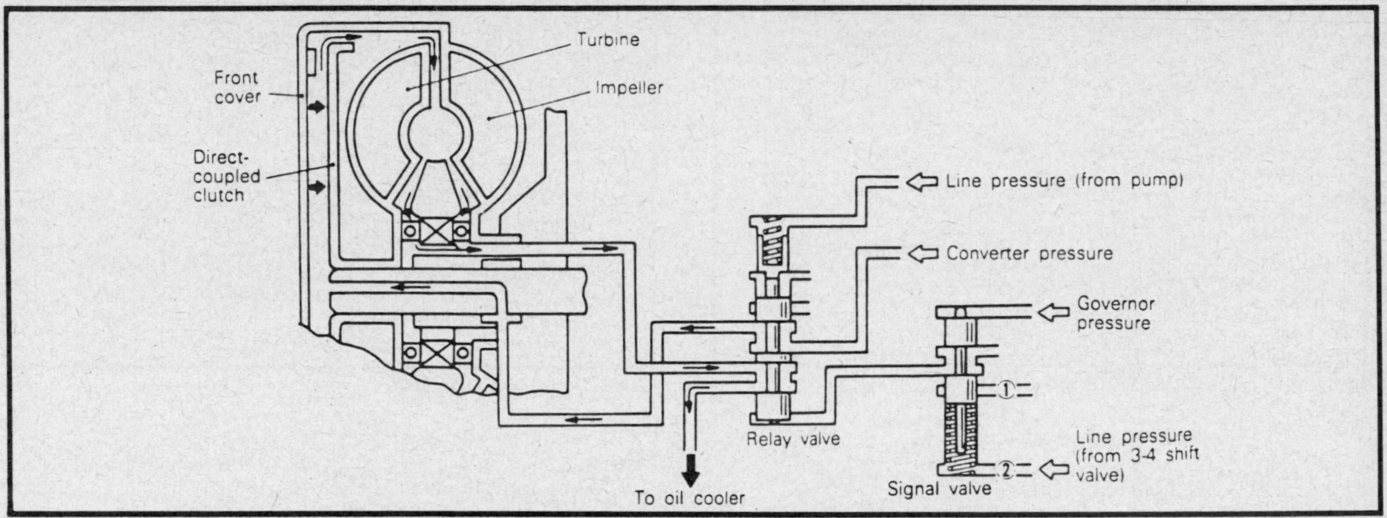
Turbine
Front cover
Impeller
Direct-coupled clutch
Line pressure (from pump)
Converter pressure
Governor pressure
①
Relay valve
Line pressure (from 3-4 shift valve)
②
Signal valve
To oil cooler
Torque converter direct-coupled disengagement—AW372L automatic transmission

ON CAR SERVICES

Adjustments

THROTTLE CABLE

1. Check the engine idle adjustment. If necessary, readjust.

CAUTION

When engine idle adjustment has been performed, always adjust the throttle control cable.

2. Make sure that no bending or deformation exists on the carburetor throttle lever and throttle cable bracket.
3. Measure the length between the inner cable stopper and the cover end with the carburetor throttle valve full open. If it is not within the standard value, adjust the inner cable bracket moving upward or downward. Standard value: 2.047–2.086 in. (52–53mm).

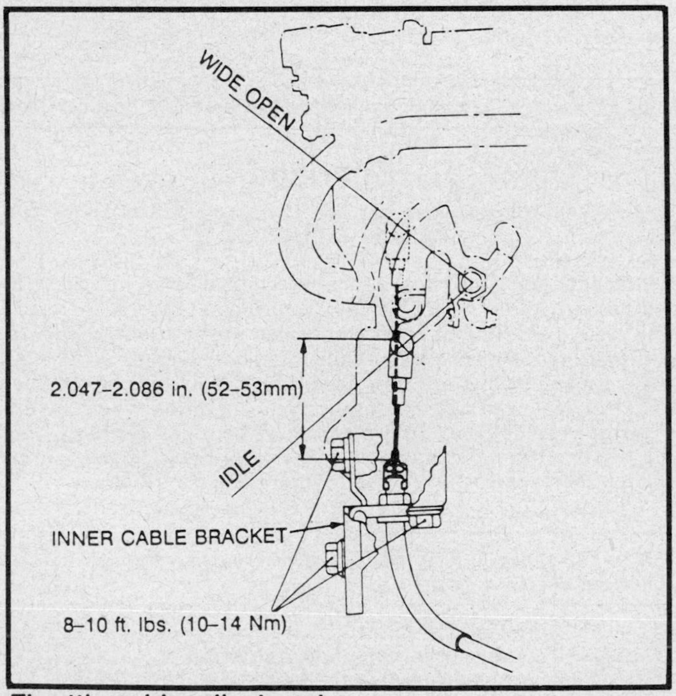
WIDE OPEN
2.047–2.086 in. (52–53mm)
IDLE
INNER CABLE BRACKET
8–10 ft. lbs. (10–14 Nm)
Throttle cable adjustment

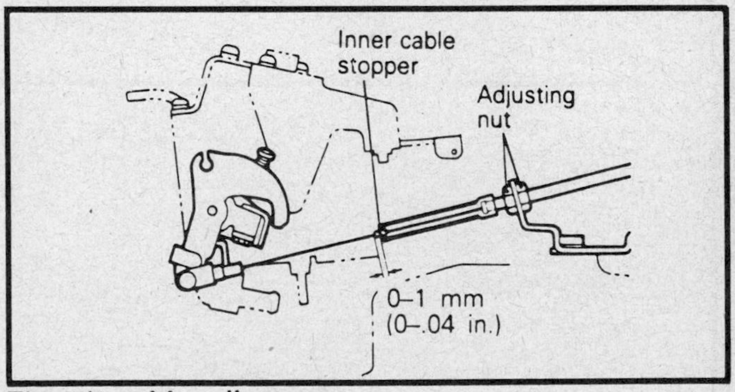
Inner cable stopper
Adjusting nut
0–1 mm
(0–.04 in.)
Throttle cable adjustment

CONTROL ROD

1. Move the transmission and shift lever to the **N** position and install the transmission control arm, rod and nut about center. Tighten the nut.
2. Check while driving, to be sure that the transmission is set to each range when the selector lever is shifted to each position.
3. Check while driving, to be sure that the overdrive is activated and canceled correctly when the overdrive switch is used.

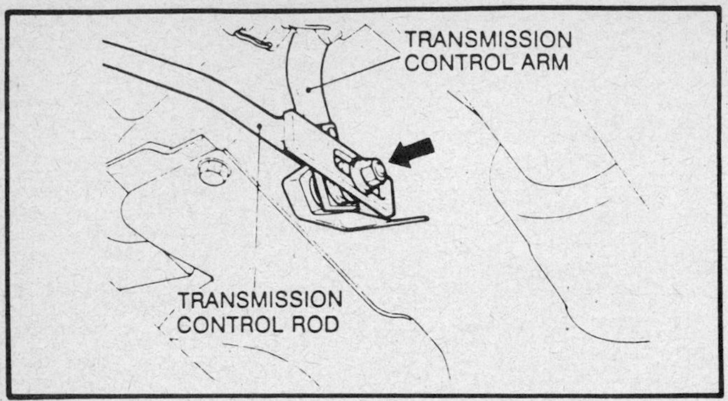
TRANSMISSION
CONTROL ARM
TRANSMISSION
CONTROL ROD
Control rod adjustment position

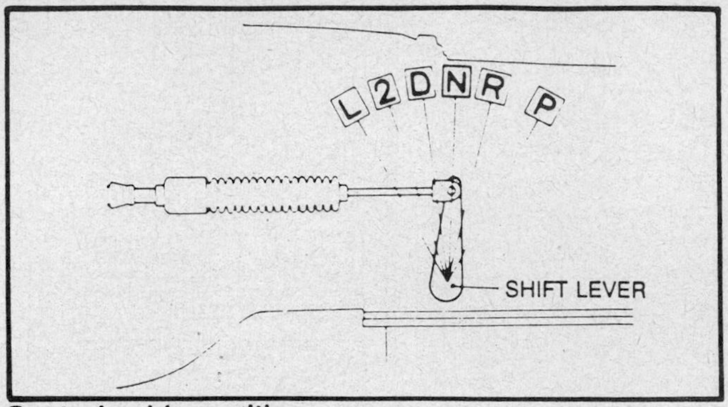
L 2 D N R P
SHIFT LEVER
Control cable position

CONTROL CABLE

No adjustment is necessary to the control cable. Visually inspect the cable for wear and smooth operation.

NOTE: When handling the control cable be careful not to bend or kink it.

NEUTRAL SAFETY SWITCH

ALL VEHICLES

1. Be sure that the engine only starts with selector lever in **N** or **P** and not in any other range.
2. Be sure that the backup lamp lights up only with selector lever in **R** and not in any other range. Be sure that all indicators light in **P-L**.

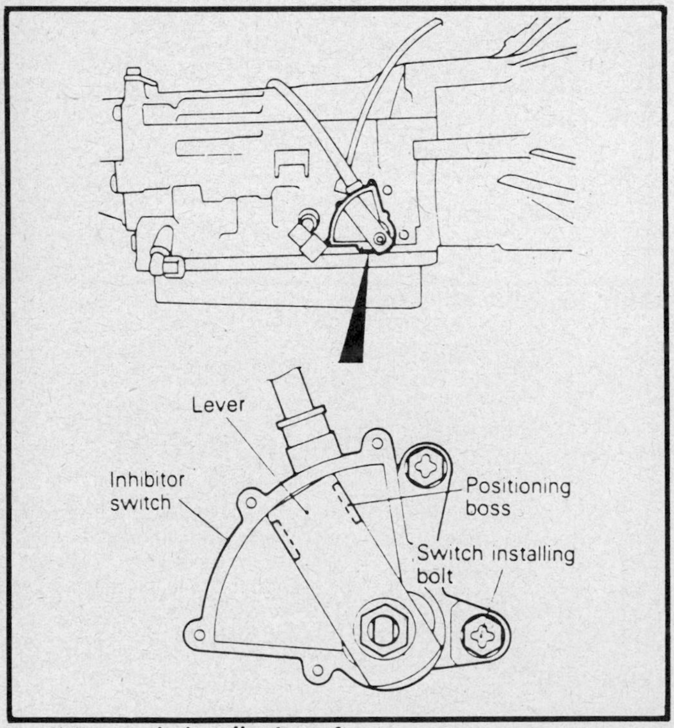
Lever
Inhibitor
switch
Positioning
boss
Switch installing
bolt
Inhibitor switch adjustment

3. When faulty, adjust as follows:
   a. Adjust the control cable.
   c. Shift the lever on the transmission side to **N** range.
   b. Loosen the switch installing bolt.
   c. Adjust the switch by aligning the shift lever with the positioning boss located on the switch.
   d. Tighten the switch mounting bolt to 3–5 ft. lbs. (4–7 Nm).
   e. Recheck for correct inhibitor switch operation.

MANUAL LINKAGE

Before starting the engine, move the shift lever through each gear range, feeling the detents in the transmission. The detents and the shift selector should be synchronized. The neutral safety switch is installed on the selector lever. After checking normal operation of this switch, place the selector lever in the the **N** position. If the notch of the selector lever on the transmission side faces directly down, the linkage has been adjusted correctly. If not an adjustment is required.

CAUTION

Do not roadtest the vehicle until the adjustments have been completed.

Services

FLUID CHANGES

Transmission fluid changes are determined on an individual basis as per operating conditions. As a rule of tumb, If the vehicle is used in continuous service or driven under severe conditions, the automatic transmission fluid must be changed every 22,500. Severe conditions are described as extensive idling, frequent short trips of 10 miles or less, vehicle operation when the temperature remains at 10°F for 60 days or more, sustained high speed operation during hot weather of 90°F, towing a trailer for a long distance or driving in severe dusty conditions.

CAUTION

If an ATF change is required due to damage to the transmission be sure to clean the cooler system.

1. Raise the vehicle and support it safely. Place a drain container with a large opening under the drain plug.
2. Remove the drain plug to allow the fluid to drain.
3. Tighten the drain plug to 13–17 ft. lbs. (18–23 Nm).
4. Pour clean ATF through the oil level gauge hole until its level reaches the cold lower limit of the level gauge.
5. Start the engine and allow it to idle for at least 2 minutes.

With the parking brake and service brake applied, move the selector lever through all positions and finally place it in the **N** or **P** position.

6. Add fluid to bring the level to the **Cold** mark on the dipstick. With the transmission at normal operating temperature, recheck the fluid level. The fluid level must be between the **Hot** marks.

7. Insert the oil level gauge securely to avoid water and dust from entering.

NOTE: By smelling the fluid, a burnt odor indicates a major transmission failure with overhaul required. Should burned flakes, solid residue or varnish in the fluid or on the dipstick be evident, overhaul of the unit is indicated.

OIL PAN

Removal and Installation

1. Raise and support the vehicle safely, remove the pan drain plug and allow the transmission fluid to drain.
2. Remove the transmission pan retaining bolts.
3. Remove the transmission oil pan and gasket.
4. Examine metallic particles in the oil pan. Remove the magnet and use it to collect any steel chips. Look carefully at the chips and particles in the oil pan and on the magnet to determine where the wear occurs in the transmission.
5. Clean the pan throughly, examine for distortion and repair or replace as needed.

NOTE: Steel metalic particles indicate: bearing wear or gear and clutch plate wear. Brass non-magnetic particles indicate bushing wear.

6. Remove all gasket material on the mating surfaces of the pan and transmission case.
7. Install the magnet in the oil pan and install the oil pan with a new gasket. Torque the retaining bolts to 3.3 ft. lbs. (4.5 Nm).

CAUTION

Make sure that the magnet does not interfere with the oil pipes.

8. Install the drain plug with a new gasket. Torque to 5.4 ft. lbs. (7.5 Nm).

VALVE BODY

Removal and Installation

1. Raise and support the vehicle safely. Remove the pan drain plug and allow the transmission fluid to drain out.
2. Remove the transmission pan retaining bolts.
3. Remove the transmission oil pan and gasket.
4. Remove the tubes by prying up both ends of each tube with a with a suitable tool.
5. Remove the strainer.
6. Remove the valve body retaining bolts.
7. Lift the valve body and disconnect the throttle cable from the cam. Remove the valve body.
8. Prior to installing the valve body, make sure the accumulator pistons are pressed fully into the bore.
9. Align the manual valve with the pin on the manual valve lever and position the valve body into place.
10. Prior to installing the retaining bolts, attach the throttle cable and make sure that the lower spring is installed on the B2 or C2 piston.
11. Install the retaining bolts in the valve body and tighten the bolts uniformly to 7.2 ft. lbs. (10 Nm).
12. Install the detent spring.
13. Install the oil strainer and tighten the bolts to 3.3 ft. lbs. (4.5 Nm).

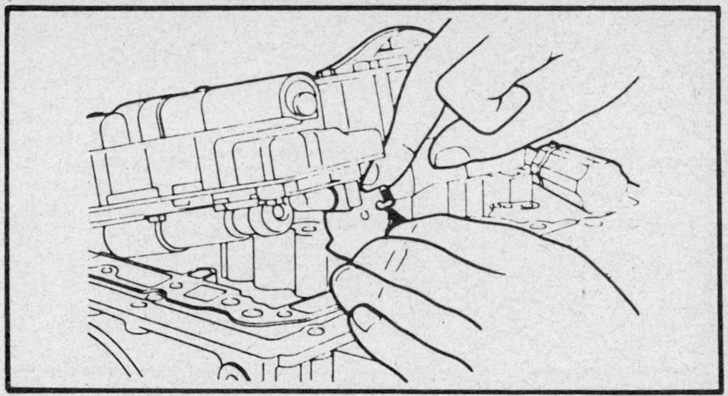

Removing the throttle cable from the valve body

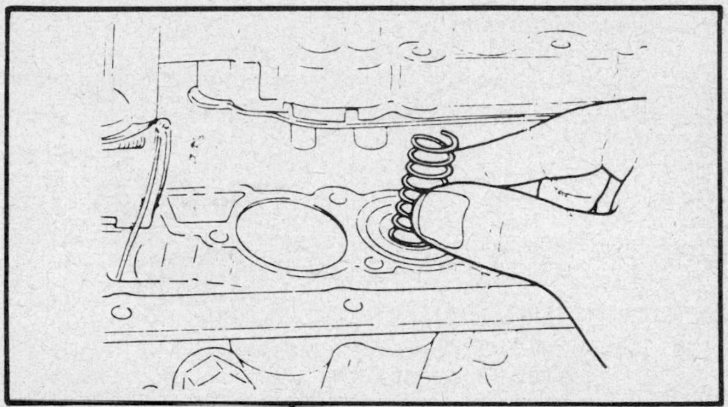

Installing the B^2 or C^2 piston lower spring

14. Using a plastic hammer, install the oil pipes into the position.

CAUTION

Be careful not to bend or damage the pipes.

15. Install the magnet in the oil pan and install the oil pan with a new gasket.

CAUTION

Make sure that the magnet does not interfere with the oil pipes.

16. Install the drain plug with a new gasket. Torque to 5.4 ft. lbs. (7.5 Nm).

GOVERNOR

Removal and Installation

4WD VEHICLES

1. Raise the vehicle and support it safely.
2. Drain the transmission and transfer case fluids.
3. Remove both front and rear driveshafts from the vehicle.
4. Position an adjustable stand or transmission jack to support the transmission.
5. Remove the transfer case assembly.
6. Remove the governor body lock bolt.
7. While lifting the retaining clip with a suitable tool, slide the governor off the output shaft.
8. Installation is the reverse order of the removal procedure.

2WD VEHICLES

1. Raise the vehicle and support it safely.
2. Matchmark the driveshaft and remove from the vehicle.

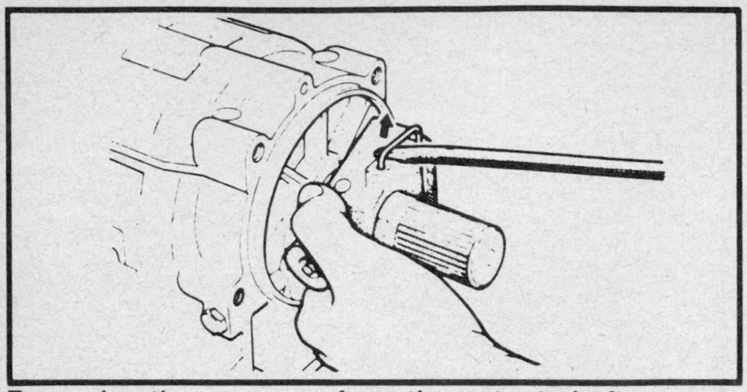
Removing the governor from the output shaft

3. Position and adjustable stand or transmission jack to support the transmission.
4. Remove the speedometer driven gear.
5. Remove the extension housing retaining bolts and remove the extension housing and gasket.
6. Remove the speedometer drive gear.
7. Loosen the staked part of the governor lock plate.
8. Remove the governor body lock bolt.
9. While lifting the retaining clip with a suitable tool, slide the governor off the output shaft.
10. Remove the governor strainer lock plate screws, plate and remove the strainer.
11. Installation is the reverse order of the removal procedure. Torque the extension housing retaining bolts to 25 ft. lbs. (34.5 Nm) and the driveshaft retaining bolts to 36–43 ft. lbs. (50 Nm).

REAR OIL SEAL

Removal and Installation

ALL VEHICLES

1. Raise and support the vehicle safely.
2. Remove the driveshaft. Use a special puller tool to remove the dust shield and oil seal from the extension housing.
3. When installing the oil seal, apply grease to the lip.
4. When installing the dust seal, soak it with fluid.
5. Install the oil seal and dust shield using a special installer tool.

REMOVAL AND INSTALLATION

TRANSMISSION REMOVAL

KM148 TRANSMISSION

1. Remove the transfer shift lever knob.
2. Remove the floor console.
3. Raise the vehicle and support it safely.
4. Remove the skid plate and undercover.
5. Drain the transmission and transfer case fluids.
6. Remove the throttle cable snap pin and clevis pin, position the throttle cable to the side.
7. Remove the transfer case protector.
8. Remove the shift control cable snap pin and disconnect the cable from the shaft.
9. Disconnect the speedometer cable.
10. Disconnect the neutral safety switch connection from the inhibitor switch.
11. Remove the rear driveshaft.
12. Remove the front driveshaft.
13. Remove the starter motor mounting bolts and remove the starter motor.
14. Remove exhaust pipe bracket retaining bolts.
15. Remove the torque converter housing inspection cover retaining bolts.
16. Remove the torque converter to flywheel retaining bolts.

CAUTION

The torque converter to flywheel retaining bolts are special bolts and should not be mixed in or exchanged with bolts not specifically designed for the converter.

17. Disconnect the oil cooler return lines.
18. Remove the oil fill tube.
19. Position a transmission jack to support the transmission.
20. Disconnect the engine support insulator and crossmember.
21. Disconnect the transfer case mounting bracket.
22. Lower the crossmember from the vehicle.
23. Disconnect the transmission and transfer case assembly from the engine by pulling it slowly toward the rear of the vehicle.
24. When lowering the transmission and transfer case assembly, tilt the front of the transmission downward and slowly lower the assembly forward until it is clear.
25. Place the transmission in a suitable holding fixture.

AW372 TRANSMISSION

1. Raise the vehicle and support it safely.
2. Remove the under cover.
3. Drain the transmission fluid.

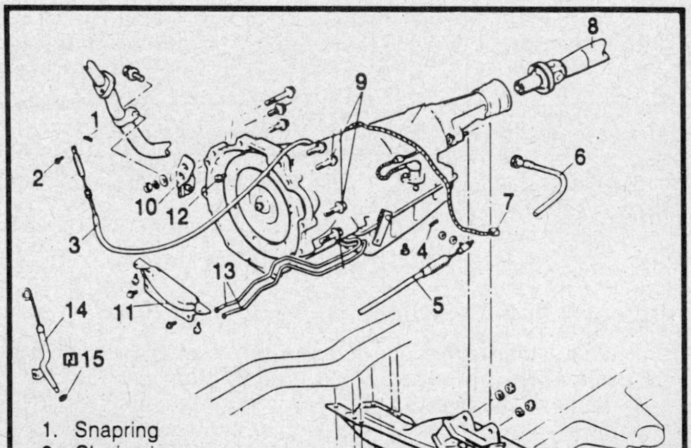
1
2
3
4
5
6
7
8
9
10
11
12
13
14
15
16

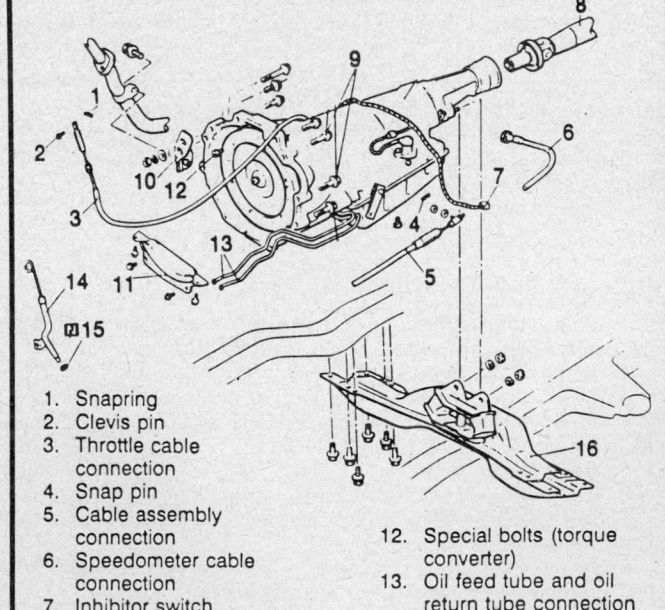

1. Snapring
2. Clevis pin
3. Throttle cable connection
4. Snap pin
5. Cable assembly connection
6. Speedometer cable connection
7. Inhibitor switch connection
8. Driveshaft
9. Starter motor retaining bolts
10. Exhaust pipe mounting bracket
11. Bell housing cover
12. Special bolts (torque converter)
13. Oil feed tube and oil return tube connection
14. Filler tube
15. O-ring
16. Rear engine support insulator and crossmember

Exploded view of the AW372 automatic transmission and related components

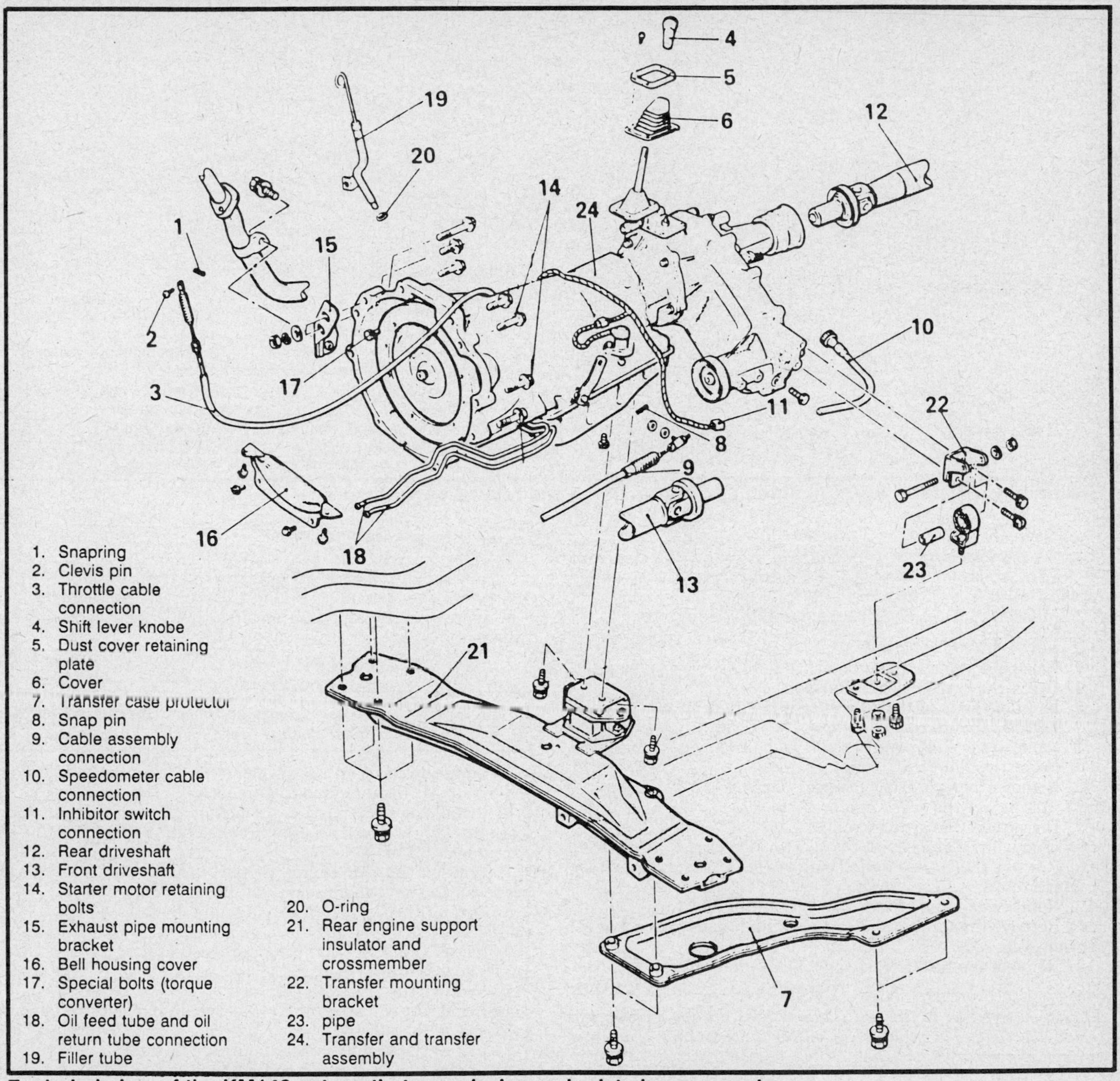

1. Snapring
2. Clevis pin
3. Throttle cable connection
4. Shift lever knobe
5. Dust cover retaining plate
6. Cover
7. Transfer case protector
8. Snap pin
9. Cable assembly connection
10. Speedometer cable connection
11. Inhibitor switch connection
12. Rear driveshaft
13. Front driveshaft
14. Starter motor retaining bolts
15. Exhaust pipe mounting bracket
16. Bell housing cover
17. Special bolts (torque converter)
18. Oil feed tube and oil return tube connection
19. Filler tube
20. O-ring
21. Rear engine support insulator and crossmember
22. Transfer mounting bracket
23. pipe
24. Transfer and transfer assembly

Exploded view of the KM148 automatic transmission and related components

4. Remove the throttle cable snap pin and clevis pin, position the throttle cable to the side.
5. Remove the shift control cable snap pin and disconnect the cable from the shaft.
6. Disconnect the speedometer cable.
7. Disconnect the neutral safety switch connection from the inhibitor switch.
8. Remove the driveshaft.
9. Remove the starter motor mounting bolts and remove the starter motor.
10. Remove exhaust pipe bracket retaining bolts.
11. Remove the torque converter housing inspection cover retaining bolts.
12. Remove the torque converter to flywheel retaining bolts.

CAUTION

The torque converter to flywheel retaining bolts are special bolts and should not be mixed in or exchanged with bolts not specifically designed for the converter.

13. Disconnect the oil cooler return lines.
14. Remove the oil fill tube.
15. Position a transmission jack to support the transmission.
16. Disconnect the rear engine support insulator and crossmember.
17. Lower the crossmember from the vehicle.
18. Disconnect the transmission assembly from the engine by pulling it slowly toward the rear of the vehicle.

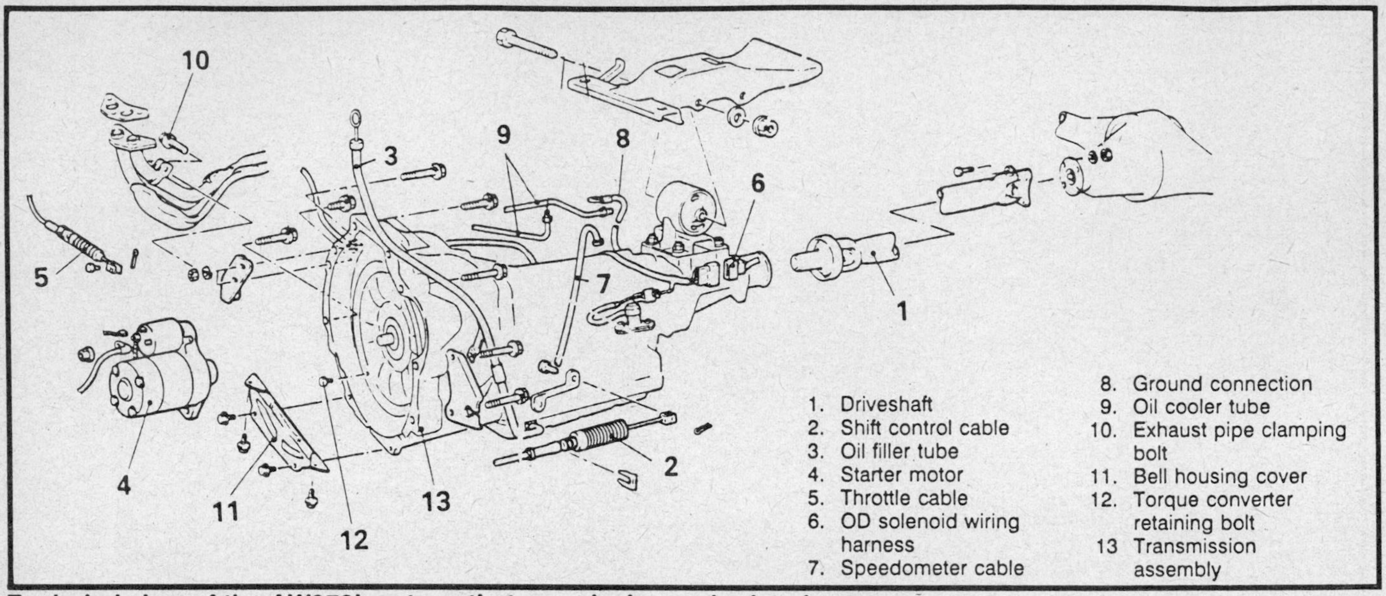
10
9
8
3
6
5
7
1
4
2
11
13
12
1. Driveshaft
2. Shift control cable
3. Oil filler tube
4. Starter motor
5. Throttle cable
6. OD solenoid wiring harness
7. Speedometer cable
8. Ground connection
9. Oil cooler tube
10. Exhaust pipe clamping bolt
11. Bell housing cover
12. Torque converter retaining bolt
13 Transmission assembly

Exploded view of the AW372L automatic transmission and related components

19. When lowering the transmission assembly, tilt the front of the transmission downward and slowly lower the assembly forward until it is clear.

20. Place the transmission in a suitable holding fixture.

AW372L TRANSMISSION

1. Raise the vehicle and support it safely.
2. Drain the transmission fluid.
3. Remove the throttle cable snap pin and clevis pin, position the throttle cable to the side.
4. Remove the shift control cable snap pin and disconnect the cable from the shaft.
5. Remove the overdrive solenoid harness.
6. Disconnect the speedometer cable.
7. Disconnect the ground cable.
8. Remove the driveshaft.
9. Remove the starter motor mounting bolts and remove the starter motor.
10. Remove exhaust pipe bracket retaining bolts.
11. Remove the torque converter housing inspection cover retaining bolts.
12. Remove the torque converter to flywheel retaining bolts.

CAUTION

The torque converter to flywheel retaining bolts are special bolts and should not be mixed in or exchanged with bolts not specifically designed for the converter.

13. Disconnect the oil cooler return lines.
14. Remove the oil fill tube.
15. Position a transmission jack to support the transmission.
16. Disconnect the rear engine support insulator and crossmember.
17. Disconnect the transmission assembly from the engine by pulling it slowly toward the rear of the vehicle.
18. When lowering the transmission assembly, tilt the front of the transmission downward and slowly lower the assembly forward until it is clear.
19. Place the transmission in a suitable holding fixture.

TRANSMISSION INSTALLATION

KM148 TRANSMISSION

1. Raise the vehicle and support it safely.
2. Prior to installing the transmission:
   a. Assemble the transfer case to the transmission.
   b. Fill the transmission and transfer case with fluid.
3. Place the transmission and transfer case assembly on a transmission jack.
4. Place the torque converter in position on the input shaft.

CAUTION

Make sure that the converter is completly seated, by turning and pushing the converter until a slight clunk is heard.

5. Carefully position the transmission and transfer case assembly up to the engine in alignment with the dowel pins.
6. Install the transmission to engine bolts and torgue to 31–40 ft. lbs. (43–55 Nm), for 10mm bolts and 14–20 ft. lbs. (20–27 Nm), for 8mm bolts.
7. Position the rear engine support insulator and crossmember in up to the transmission and frame.
8. Torque the crossmember to frame bolts to 40–54 ft. lbs. (55–75 Nm).
9. Install the rear engine support insulator bolts. Torque to 13–18 ft. lbs. (18–25 Nm).
10. Remove the transmission jack.
11. Install the converter to flywheel retaining bolts and torque to 25–30 ft. lbs. (35–42 Nm).
12. Install the transfer case mounting bracket.
13. Connect the oil cooler return lines.
14. Install the special torque converter to flywheel retaining bolts.

CAUTION

The torque converter to flywheel retaining bolts are special bolts and should not be mixed in or exchanged with bolts not specifically designed for the converter.

15. Install the torque converter housing inspection cover retaining bolts.
16. Install the exhaust pipe bracket retaining bolts.
17. Install the starter motor mounting bolts and install the starter motor.
18. Install the front driveshaft.
19. Install the rear driveshaft.
20. Connect the neutral safety switch connection to the neutral safety switch.

21. Fill the transmission and transfer case with fluid.
22. Start the engine and test transmission operation in all ranges.

AW372 TRANSMISSION

1. Raise the vehicle and support it safely.
2. Place the transmission on a transmission jack.
3. Place the torque converter in position on the input shaft.

CAUTION

Make sure that the converter is completely seated, by turning and pushing the converter until a slight clunk is heard.

4. Carefully position the transmission assembly up the the engine in alignment with the dowel pins.
5. Install the transmission to engine bolts and torgue to 31–40 ft. lbs. (43–55 Nm), for 10mm bolts and 14–20 ft. lbs. (20–27 Nm), for 8mm bolts.
6. Position the rear engine support insulator and crossmember in up to the transmission and frame.
7. Torque the crossmember to frame bolts to 29–36ft. lbs. (40–50 Nm).
8. Install the rear engine support insulator nuts. Torque to 14–17 ft. lbs. (20–24 Nm).
9. Remove the transmission jack.
10. Install the converter to flywheel retaining bolts and torque to 25–30 ft. lbs. (35–42 Nm).
11. Connect the oil cooler return lines.
12. Install the torque converter to flywheel retaining bolts.

CAUTION

The torque converter to flywheel retaining bolts are special bolts and should not be mixed in or exchanged with bolts not specifically designed for the converter.

13. Install the torque converter housing inspection cover retaining bolts.
14. Install the exhaust pipe bracket retaining bolts and bracket.
15. Install the starter motor mounting bolts and install the starter motor.
16. Install the driveshaft.
17. Connect the neutral safety switch connection to the neutral safety switch.
18. Connect the speedometer cable.
19. Install the shift control cable snap pin and connect the cable to the shaft.
20. Install the throttle cable snap pin and clevis pin and connect the throttle cable.
21. Fill the transmission with fluid.
22. Lower the vehicle to the ground and test the transmission operation in all ranges.

AW372L TRANSMISSION

1. Raise the vehicle and support it safely.
2. Place the transmission on a transmission jack.
3. Place the torque converter in position on the input shaft.

CAUTION

Make sure that the converter is completely seated, by turning and pushing the converter until a slight clunk is heard.

4. Carefully position the transmission assembly up to the engine in alignment with the dowel pins.
5. Install the transmission to engine bolts and torgue to 31–40 ft. lbs. (43–55 Nm), for 10mm bolts and 14–20 ft. lbs. (20–27 Nm), for 8mm bolts.
6. Install the rear engine support mount bolt and nut. Torque to 50–65 ft. lbs. (70–90 Nm).
7. Remove the transmission jack.
8. Install the converter to flywheel retaining bolts and torque to 25–30 ft. lbs. (35–42 Nm).
9. Install the oil fill tube.
10. Connect the oil cooler return lines.
11. Install the torque converter to flywheel retaining bolts.

CAUTION

The torque converter to flywheel retaining bolts are special bolts and should not be mixed in or exchanged with bolts not specifically designed for the converter.

12. Install the torque converter housing inspection cover and retaining bolts.
13. Install the exhaust pipe bracket and retaining bolts.
14. Install the starter motor mounting bolts and install the starter motor.
15. Install the driveshaft.
16. Connect the ground cable.
17. Connect the speedometer cable.
18. Install the overdrive solenoid harness.
19. Install the shift control cable snap pin and disconnect the cable from the shaft.
20. Install the throttle cable snap pin and clevis pin and install the throttle cable.
21. Lower the vehicle to the floor. Fill the transmission with fluid and test the transmission operation in all ranges.

BENCH OVERHAUL

Before Disassembly

All disassembled parts should be washed clean and the fluid passages and holes blown through with compressed air to make sure that they are not clogged.

CAUTION

When using compressed air to dry parts, avoid spraying ATF or cleaning solvents, as personal injury may occur.

After cleaning, all parts should be arranged in proper order to allow a thorough inspection.

When disassembling the valve body, be sure to keep each valve together with its own spring.

New brake and clutch discs that are to be used for replacement must be soaked in ATF for at least 2 hours before assembly.

Converter Inspection

1. Remove the torque from the transmission, if not already done.
2. Insert a special wrench tool into the inner race of the one-way clutch of the torque converter.
3. Insert a special stopper tool so that it installs in the notch of the converter hub and the other race of the one-way clutch.
4. Test the one-way clutch: The clutch should lock when turned counterclockwise and should rotate freely and smoothly clockwise. Less than 22 inch lbs. (2.5 Nm) of torque should be required to rotate the clutch clockwise. If necessary, clean the converter and retest the clutch. Replace the converter if the clutch still fails the test.

Transmission Disassembly

TORQUE CONVERTER

Removal

With the transmission secured on a work bench or a suitable transmission holding fixture, grasp the torque converter firmly and pull the assembly straight out of the transmission.

NOTE: The torque converter is a heavy unit and care must be exercised to handle the weight.

The torque converter is a sealed unit and is non-serviceable. If failure occurs, it must be replaced.

OIL PAN

Removal

1. Removal the pan retaining bolts and remove the pan and gasket.
2. Examine metallic particles in the oil pan. Remove the magnet and use it to collect any steel chips. Look carefully at the chips and particles in the oil pan and on the magnet to determine where the wear occurs in the transmission.

VALVE BODY

Removal

1. With the transmission mounted in a suitable holding fixture remove the oil pan.
2. Remove the tubes by prying up both ends of each tube with a with a suitable tool.
3. Remove the strainer.
4. Remove the valve body retaining bolts.
5. Lift slightly the valve body and disconnect the throttle cable from the cam, and remove the valve body.

GOVERNOR

Removal

1. On the AW372 and AW372L, remove the extension housing.
2. Remove the speedometer drive gear.
3. Remove the governor body lock bolt.
4. While lifting the retaining clip with a suitable tool, slide the governor off the output shaft.

OIL PUMP

Removal

1. Remove the oil pump retaining bolts.

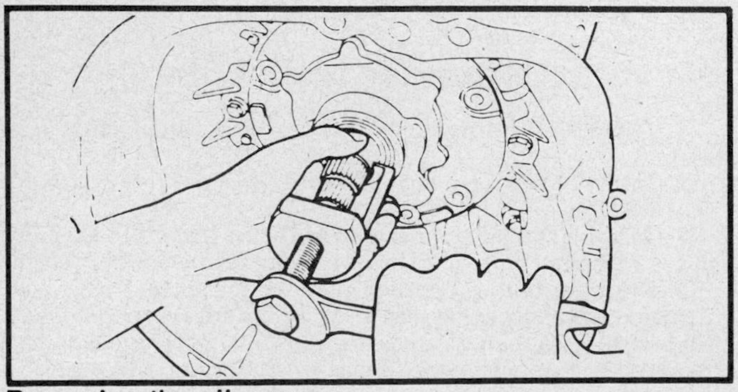

Removing the oil pump

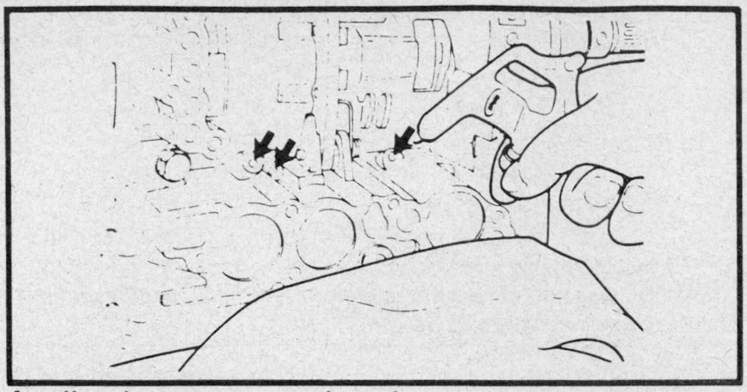

Appling low pressure air to force pistons out the case

2. Position a suitable pulling tool on the shaft in back of the spline.
3. Turn bolt of the pulling tool to free the pump.

CAUTION

Be careful not to damage the shaft bushing surface.

4. Grasp the front pump stator shaft and pull the pump from the case.
5. Remove the bearing and race behind the oil pump.

OVERDRIVE CLUTCH ASSEMBLY

Removal

1. Remove the converter housing.
2. Using a 10mm socket, push the plastic throttle cable retainer out of the transmission case to remove the cable with the retainer.
3. Position a rag to catch each piston. Blow low pressure compressed air of about 14.5 psi into each of the holes. Remove the pistons and springs.

CAUTION

Wear protective eye wear and keep face away to avoid injury. Do not use high pressure air.

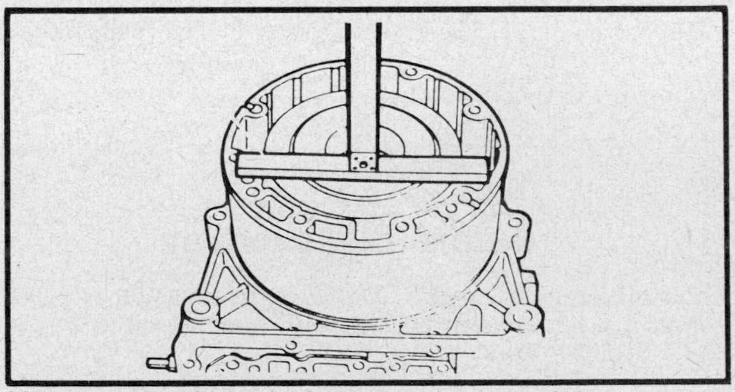

Measure the distance between the top of the overdrive case and clutch cylinder

4. Remove the parking lock linkage.
5. Remove the cam plate.
6. Remove the parking lock rod.
7. Remove the spring, pivot pin and parking lock pawl.
8. Remove the manual lever and shaft.
9. Using a hammer and punch, drive out the pin retaining the manual lever shaft.

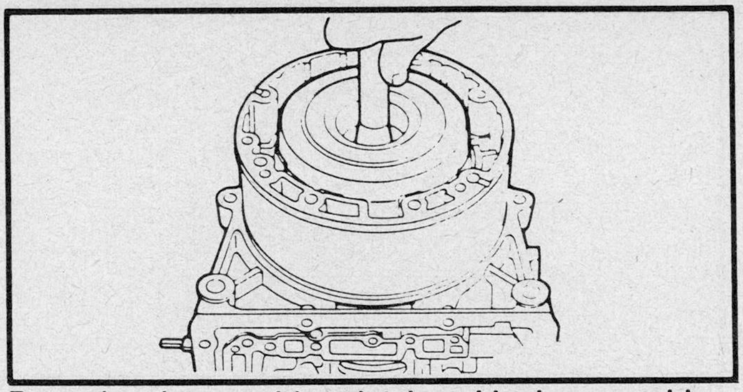

Removing the overdrive clutch and brake assembly

10. Slide the shaft out of the case and remove the detent plate.
11. Place the transmission on a suitable holding stand for more efficient work.
12. Position a suitable measuring tool on the overdrive case and measure the distance between the top of overdrive case and clutch cylinder. Make a note of the distance for reassembly.
13. Grasp the shaft and pull out the overdrive clutch assembly. Watch for bearings and races on both sides of the assembly.
14. Remove the overdrive case and brake as follows: Hold both sides of the overdrive case and pull it out from the transmission case. Watch for bearings and races on both sides of the assembly.
15. Position a suitable measuring tool in the case. Measure the distance between the top of case flange and the clutch drum. Make a note of the finding for reassembly.

FORWARD CLUTCH

Removal

1. After the overdrive unit has been removed, the forward clutch may be removed.
2. Grasp the shaft and pull out the forward clutch assembly. Remove the bearings and races on both sides of the assembly.

DIRECT CLUTCH

Removal

1. After the forward clutch has been removed the direct clutch may be removed.
2. Remove the direct clutch by grasping the clutch hub and pulling it out from the case.

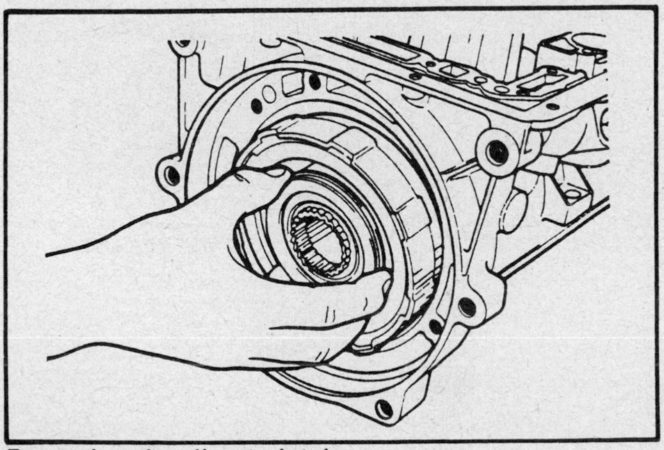

Removing the direct clutch

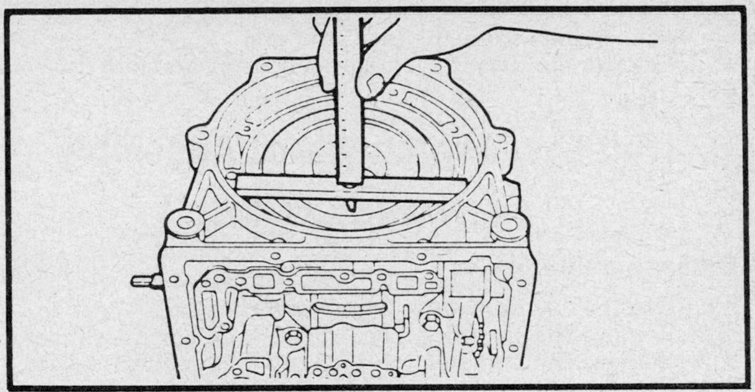

Measure the distance between the top of the case and the clutch drum

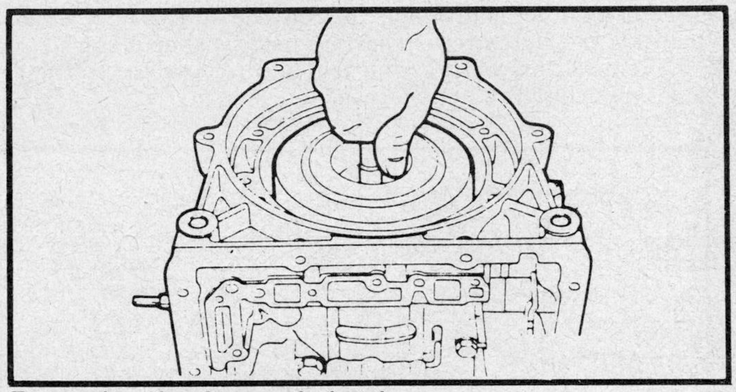

Removing the forward clutch

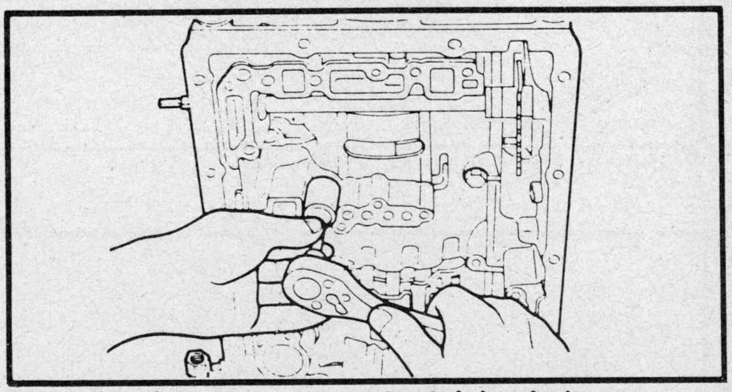

Removing the center support retaining bolts

CENTER SUPPORT

Removal

1. Remove the center support retaining bolts.
2. Grasp the center support assembly and pull out the center support with sun gear from the case.

NO. 3 BRAKE AND PLANETARY CARRIER

Removal

1. Remove the snapring in front of the planetary carrier.
2. Remove the reaction plate retaining ring using a suitable tool.
3. Remove the No. 3 brake and planetary carrier assembly by pulling out the intermediate shaft. If the brake apply tube and

rear thrust bearing and races do not come out with the assembly, remove them from the case.

4. Remove the output shaft thrust bearing and race from the case.

Unit Disassembly and Assembly

FORWARD CLUCTH

Disassembly

1. Use the extension housing as a work stand.
2. Remove the snapring from the forward clutch cylinder.
3. Remove the direct clutch hub and forward hub.
4. Remove the thrust bearing No. 12 and the races No. 11 and No. 13.
5. Remove the clutch disc.
6. Remove the snapring.
7. Remove the remaining clutch plates and discs.
8. Position a compressing tool on the spring retainer and compress the clutch return springs using a shop press.
9. Remove the compressing tool and remove the spring retainer and clutch return springs.

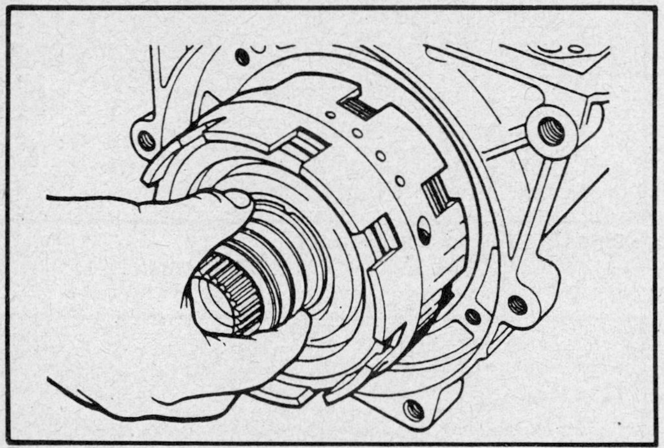
Removing the center support

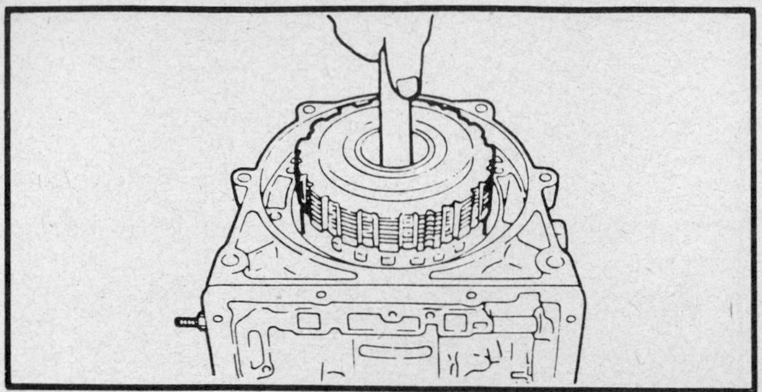
Removing the No. 3 brake assembly

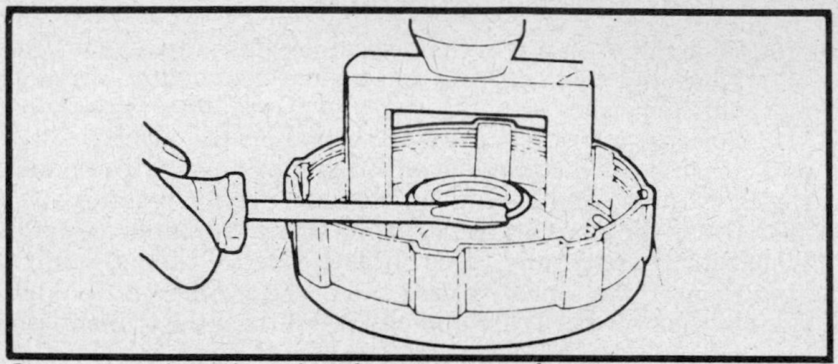
Compressing the clutch return springs

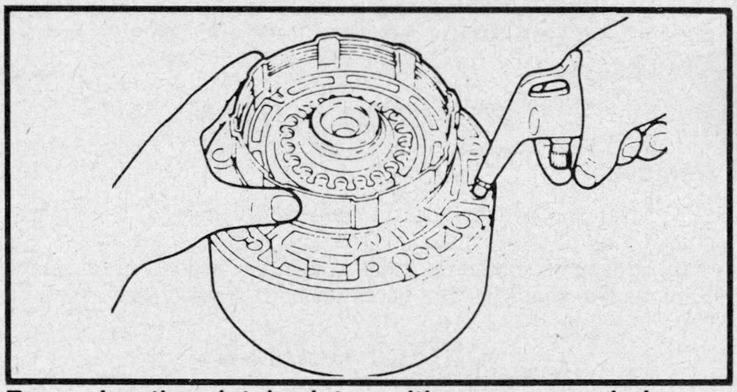
Removing the clutch piston with compressed air

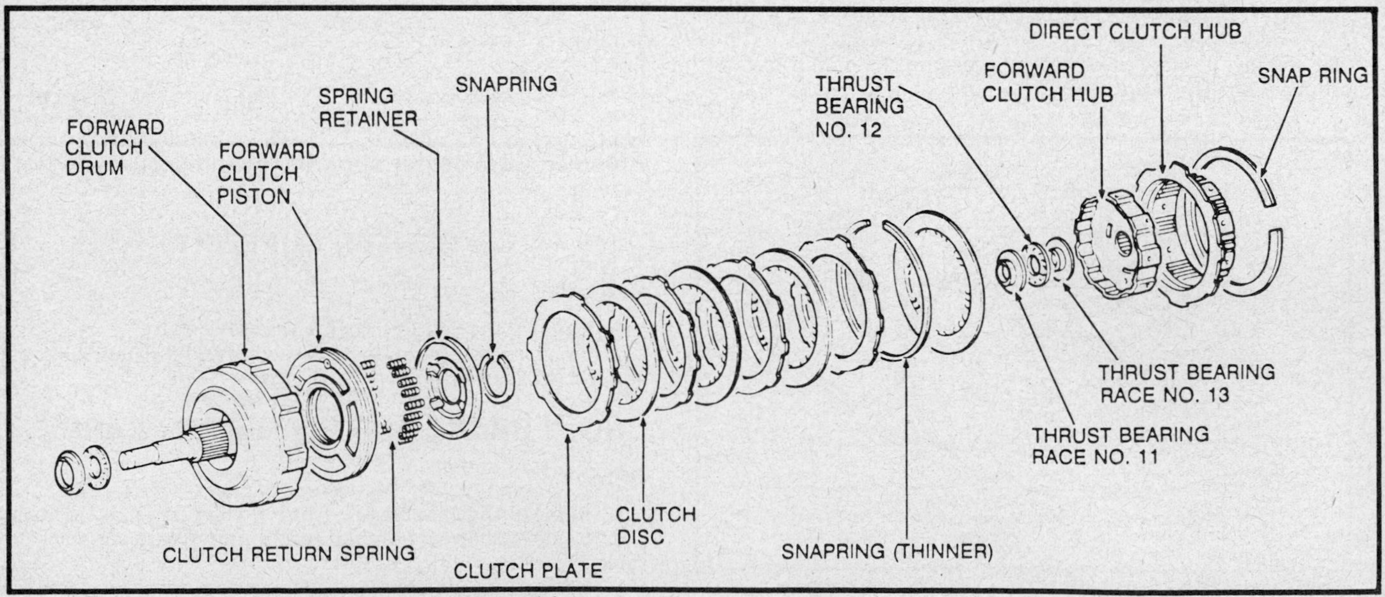
DIRECT CLUTCH HUB
FORWARD CLUTCH HUB
SNAP RING
THRUST BEARING NO. 12
SNAPRING
SPRING RETAINER
FORWARD CLUTCH DRUM
FORWARD CLUTCH PISTON
THRUST BEARING RACE NO. 13
THRUST BEARING RACE NO. 11
CLUTCH DISC
CLUTCH RETURN SPRING
SNAPRING (THINNER)
CLUTCH PLATE
Exploded view of the forward clutch assembly

10. Assemble the forward clutch cylinder and piston on the overdrive case and force out the piston by compressed air.

a. Slide the forward clutch cylinder and piston onto the overdrive case.

b. Apply compressed air to the overdrive case to remove the piston.

c. Remove the forward clutch cylinder from the overdrive case.

11. Remove the O-rings from the forward clutch cylinder and piston.

Inspection

Wash the removed parts and dry with compressed air. Check the the following and replace faulty parts.

1. Check the input shaft and clutch cylinder for excessive wear and binding of thrust bearing contact surfaces. Check for damage of splines and for wear of the OD case seal ring contact surface.
2. Check the forward clutch cylinder for wear and damage of clutch drum teeth and for wear or binding of the piston sliding surface. Also, check for damage and binding of the thrust bearing seating surface.
3. Check the front clutch for abnormal wear and damage of teeth, splines and hub thrust surface.
4. Check the clutch disc and clutch plate for wear and binding of friction surfaces and from wear and damage from engagement with the cylinder and hub.
5. Check the clutch return spring for damage and cracks. Check for spring outside wear and deterioration.
6. Check the clutch piston for wear and damage of the surface in contact with the cylinders. Shake the piston to check if the check ball is binding or free. Also apply low pressure compressed air to the piston for air leaks.
7. Check the direct clutch for abnormal wear or damage.

NOTE: Prepare the new discs by soaking them at least 2 hours in clean ATF.

Assembly

1. Install new O-rings on the forward clutch piston and coat with clean ATF.
2. Press the forward clutch piston into forward clutch cylinder with the cup side up (check ball down). Be careful not to damage the O-rings.
3. Install the clutch return springs, spring retainer and snapring in place.
4. Compress the clutch return springs and install the spapring in the groove.

a. Position the compressing tool on the spring retainer and compress the springs using a shop press.

b. Install the snapring in place. Be sure the end gap of the snapring is not aligned with the spring retainer claw.

5. Install the clutch discs and plates without assembling the snapring.

NOTE: New clutch discs should be soaked in automatic transmission fluid for at least 2 hours before installation.

6. Using low pressure compressed air, blow all excess fluid from the discs.

CAUTION

High pressure air will damage discs.

7. Install the clutch plates and discs alternately. Do not install the snapring (thinner) at this time.
8. Check the piston stroke of the forward clutch.

a. Install the direct clutch hub and snapring.

b. Install the forward clutch cylinder assembly onto the overdrive case. Using a dial indicator, measure the stroke by applying and releasing the compressed air of 58–116 psi. (400–800 kPa). Standard stroke is 0.563–0.1154 in. (1.43–2.93mm). If the stroke exceeds the limit, the clutch discs and plates are probably worn. If the stroke is less than the limit, parts may be misassembled or there may be excess automatic transmission fluid on the discs.

Checking the piston stroke with a dial indicator

c. After the check, remove the snapring and direct clutch hub.

9. Compress and lower the snapring into the groove by hand. Check that the ends of the snapring are not aligned with any of the cutouts.
10. Install the clutch disc.
11. Install the thrust bearing No. 12 and the races No. 11 and No. 13, coat with petroleum jelly.
12. Install the forward clutch hub while aligning the disc lugs with the hub teeth. Make sure the hub meshes with all the discs and is fully inserted.
13. Install the direct clutch hub and snapring. Check that the snapring ends are not aligned with any of the cutouts.

DIRECT CLUTCH

Disassembly

1. Remove the snapring from the direct clutch cylinder.
2. Remove the flange, clutch discs and plates.

NOTE: Do not allow the clutch discs to dry out.

3. Position a compressing tool on the spring retainer and compress the piston return springs with a shop press. Remove the snapring.
4. Remove the spring retainer and piston return springs.
5. Assemble the direct clutch cylinder and piston set, on the center support and force out the piston by compressed air.
6. Side the direct clutch cylinder and piston set onto the center support.
7. Appy compressed air to the center support to remove the piston.
8. Remove the direct clutch from the center support.
9. Remove the O-rings from the direct clutch piston.

Inspection

Wash the removed parts and dry with compressed air. Check the the following and replace faulty parts.

1. Check the clutch cylinder for wear and damage to the grooves and piston sliding surfaces, thrust bearing surface and the seal ring sliding surface.
2. For inspection of the clutch disc, plate, piston and spring.
3. Check the clutch disc and clutch plate for wear and binding of friction surfaces and from wear and damage from engagement with the cylinder and hub.
4. Check the clutch return spring for damage and cracks. Also check for spring outside wear and deterioration.

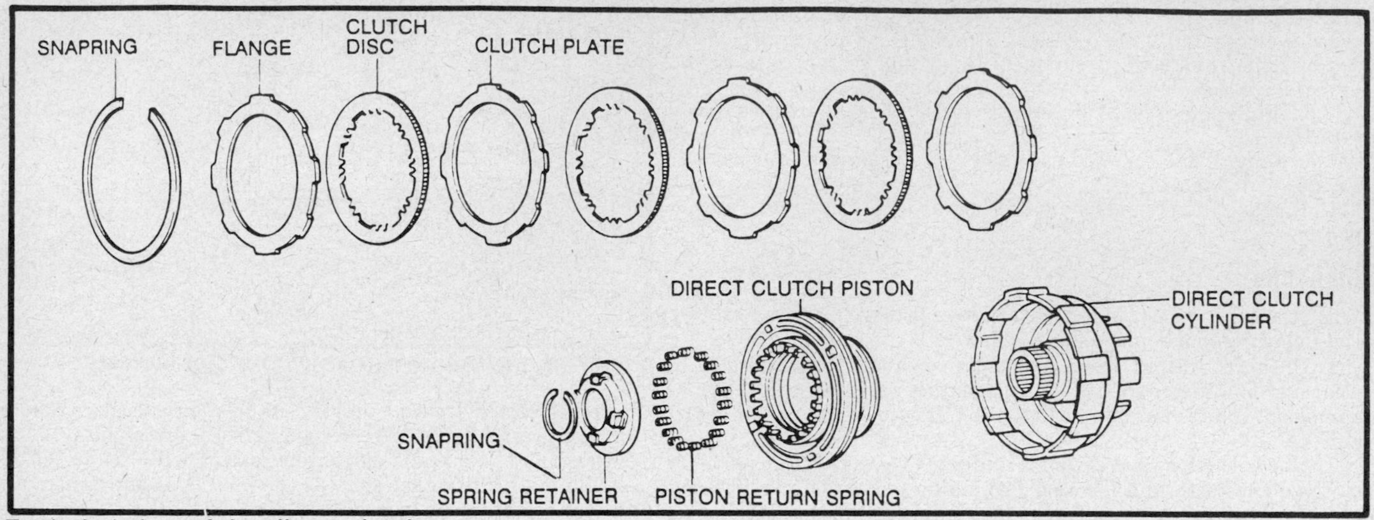
SNAPRING
FLANGE
CLUTCH DISC
CLUTCH PLATE
DIRECT CLUTCH PISTON
DIRECT CLUTCH CYLINDER
SNAPRING
SPRING RETAINER
PISTON RETURN SPRING

Exploded view of the direct clutch

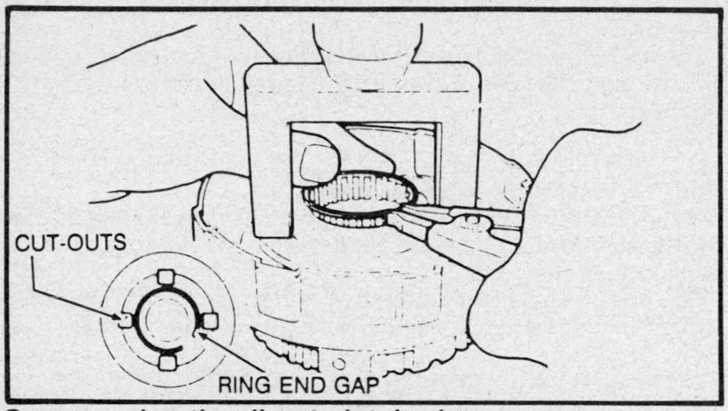
CUT-OUTS
RING END GAP

Compressing the direct clutch piston

5. Check the clutch piston for wear and damage of the surface in contact with the cylinders. Shake the piston to check if the check ball is binding or free. Also apply low pressure compressed air to the piston for air leaks.
6. Check the direct clutch for abnormal wear or damage.

Assembly

1. Install new O-rings on the direct clutch piston. Coat the O-rings with automatic transmission fluid.
2. Install the direct clutch piston in the direct clutch cylinder. Press the direct clutch piston into the cylinder with the cup side facing up, being careful not to damage the O-rings.
3. Install the piston return springs and set the retainer with the snapring in place.
4. Compress the piston return springs and install the snapring.
5. Position the compressing tool on the spring retainer and compress the springs using a shop press. Install the snapring. Be sure the end gap of the snapring is not aligned with the spring retainer claw.
6. Install the clutch discs, plates and flange.

NOTE: New clutch discs should be soaked in automatic transmission fluid for at least 2 hours before installation.

7. Using low pressure compressed air, blow all excess fluid from the clutch discs.

CAUTION

High pressure air will damage the discs.

8. Install the parts in the following order: clutch plate, clutch disc, clutch plate, clutch disc, flange (flat end facing down).
9. Install the snapring. Check that the snapring ends are not aligned with any of the cutouts.
10. Check the piston stroke of the direct clutch. Install the direct clutch onto the center support. Using a dial indicator, measure the stroke by applying and releasing the compressed air of 58–116 psi. (400–800 kPa). Standard stroke is 0.0358–0.0783 in. (0.91–1.99mm). If the stroke exceeds the limit, the clutch discs and/or plates are probably worn. If the stroke is less than the limit, parts may be misassembled or there may be excess automatic transmission fluid on the discs.

CENTER SUPPORT

Disassembly

1. Remove the snapring from the end of the planetary sun gear shaft.
2. Pull the center support assembly from the planetary sun gear.
3. Remove the snapring from the front of the center support assembly (No. 1 brake).
4. Remove the flange, clutch disc and plate No. 1 brake.
5. Position a compressing tool on the spring retainer and compress the springs with a shop press. Remove the snapring.
6. Remove the spring retainer and the brake return springs.
7. Blow compressed air through the center support oil hole to remove the No. 1 brake piston.
8. Remove the No. 1 brake piston O-rings.

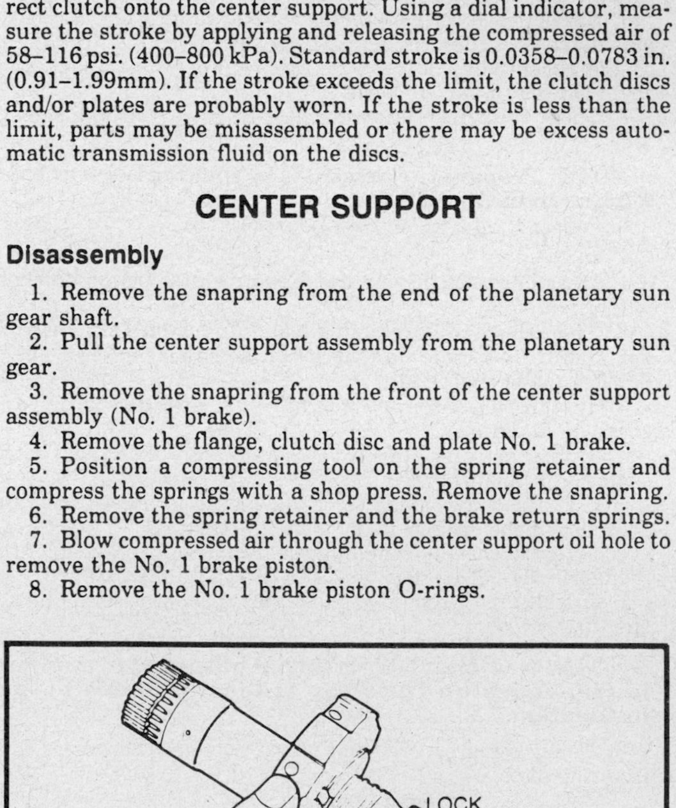
LOCK
FREE

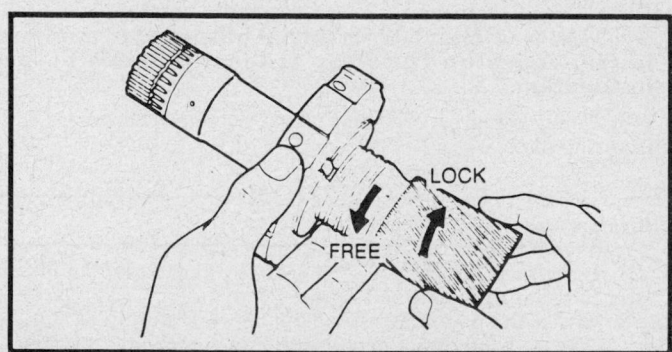

Checking the one-way clutch operation

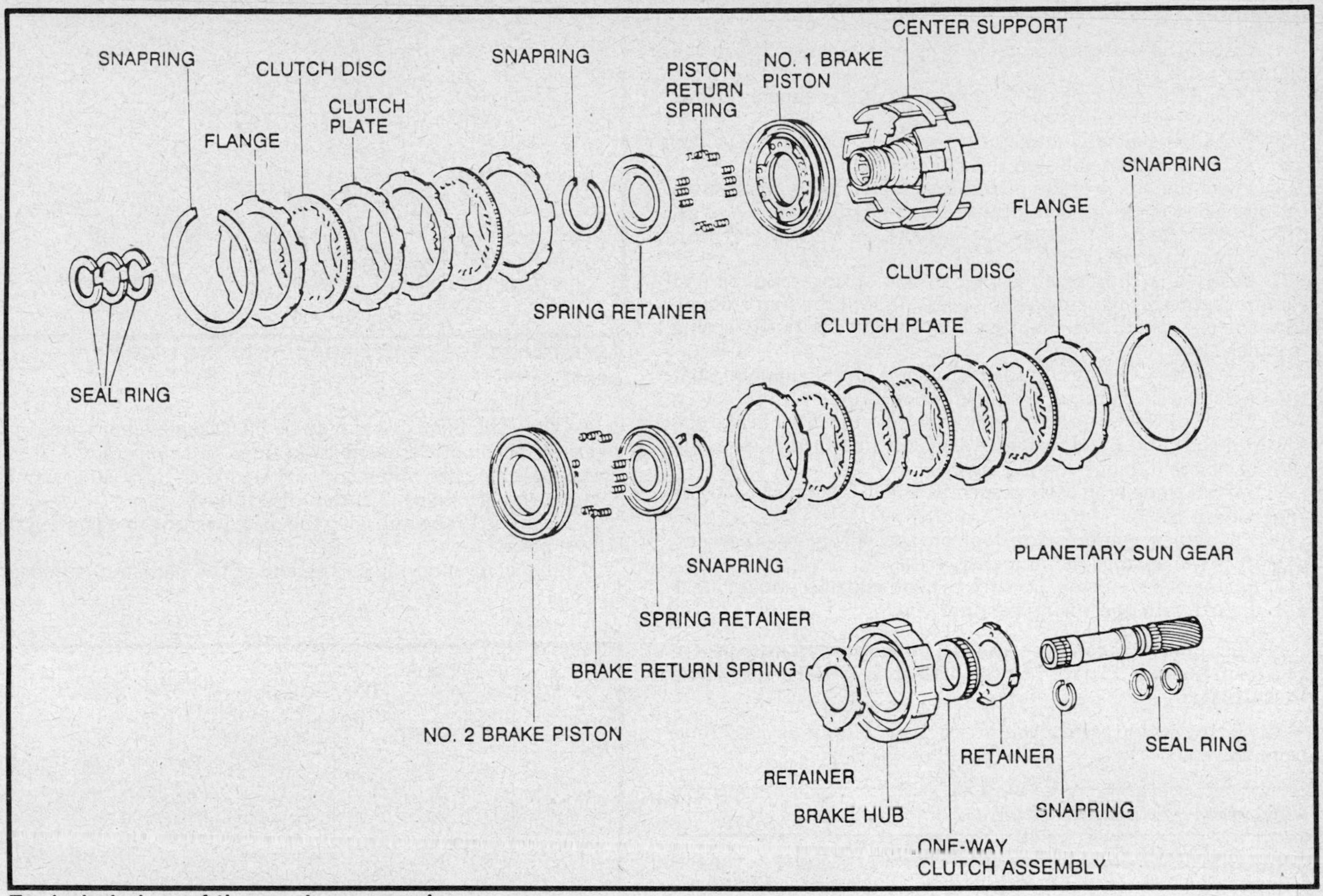
CENTER SUPPORT
SNAPRING
CLUTCH DISC
SNAPRING
PISTON RETURN SPRING
NO. 1 BRAKE PISTON
CLUTCH PLATE
FLANGE
SNAPRING
FLANGE
CLUTCH DISC
SPRING RETAINER
CLUTCH PLATE
SEAL RING
SNAPRING
PLANETARY SUN GEAR
SPRING RETAINER
BRAKE RETURN SPRING
NO. 2 BRAKE PISTON
SEAL RING
RETAINER
RETAINER
BRAKE HUB
SNAPRING
ONE-WAY CLUTCH ASSEMBLY

Exploded view of the center support

9. Turn the center support assembly over and remove the rear snapring (No. 2 brake).
10. Position the compressing tool on the spring retainer and compress the springs with a shop press. Remove the snapring.
11. Remove the spring retainer and the brake return springs.
12. Blow compressed through the center support oil hole to remove the No. 2 brake piston.
13. Remove the No. 2 brake piston O-rings.
14. Remove the oil seal rings from the center support.
15. Remove the one-way clutch assembly and oil seal rings from the planetary sun gear.
16. Inspect the one-way clutch assembly by holding the No. 2 brake hub and turn the planetary sun gear. The sun gear should turn freely counterclockwise and should lock clockwise. If the one-way clutch does not operate properly, replace it.

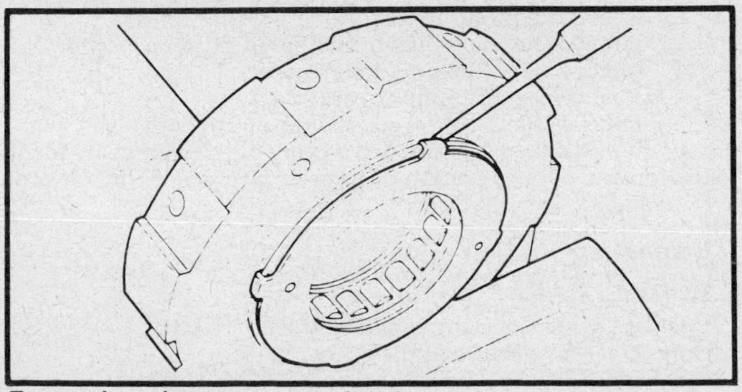

Removing the one-way clutch

17. If it is necessary to replace the one-way, bend the tabs back with a tapered punch.
18. Pry off the retainer with a suitable tool. Leave the other retainer on the hub.
19. Remove the one-way clutch.
20. Install the one-way clutch into the brake hub facing the spring cage toward the front.
21. Hold the brake hub in a soft jawed vice and bend the tabs with a suitable tool.
22. Check to make sure that the retainer is centered.

Inspection

Wash the removed parts and dry with compressed air. Check the the following and replace faulty parts.

1. Check the center support for damage and deterioration of the seal rings, for abnormal wear and binding of bushin and ofr wear of clutch plate slots.
2. Check the brake piston for damage of its outside surface in contact with the center support cylinder.
3. Check the clutch disc and clutch plate for wear and binding of friction surfaces and from wear and damage from engagement with the cylinder and hub.
4. Check the clutch return spring for damage and cracks. Also check for spring outside wear and deterioration.
5. Check the clutch piston for wear and damage of the surface in contact with the cylinders. Shake the piston to check if the check ball is binding or free. Also apply low pressure compressed air to the piston for air leaks.
6.Check the direct clutch for abnormal wear or damage.

Assembly

1. Install the seal rings and one-way clutch assembly, on the planetary sun gear.
2. Install the seal rings in the groove of center support. Hook both ends of the ring by hand.
3. Install the new O-rings on the piston. Coat the O-rings with automatic transmission fluid.
4. Press the No. 1 brake piston into the center support with the cup side facing up, being careful no to damage the O-rings.
5. Install the piston return springs and set the retainer with the snapring in place.
6. Position a compressing tool on the spring retainer and compress the spring using a shop press. Install the snapring. Be sure the end gap of the snapring is not aligned with the spring retainer claw.
7. Install new O-rings on the piston and center support. Coat the O-rings with automatic transmission fluid.
8. Turn the center support over and press the No. 2 brake piston into the center support with the cup side facing up, being careful not to damage the O-rings.
9. Install the piston return springs and set retainer with the snapring in place.
10. Position a compressing tool on the spring retainer and compress the springs using a shop press.
11. Install the snapring. Be sure the end gap of the snapring is not aligned with the spring retainer claw.

NOTE: New clutch disc should be soaked in automatic transmission fluid for at least 2 hours berfore installation.

12. Using low pressure compressed air, blow all excess fluid from the disc.

CAUTION

High pressure air will damage the disc.

13. Install the parts in this order: Install the parts in the following order: clutch plate, clutch disc, clutch plates (2 pieces), clutch disc, flange (rounded end facing down).
14. Install the snapring in the center support. Check that the snapring ends are not aligned with any of the cutouts.
15. Check the piston stroke of the No. 1 brake. Using a dial indicator, measure the stroke by applying and releasing the compressed air of 58–116 psi. (400–800 kPa). Standard stroke is 0.0315–0.0681 in. (0.80–1.73mm). If the stroke exceeds the limit, the clutch discs and/or plates are probably worn. If the stroke is less than the limit, parts may be misassembled or there may be excess automatic transmission fluid on the discs.
16. Turn the center support over and install the No. 2 brake, clutch plates, discs and flange.

NOTE: New clutch discs should be soaked in automatic transmission fluid for at least 2 hours before installation.

17. Using low pressure compressed air blow the excess fluid from the discs.

CAUTION

High presure air will damage the the disc.

18. Install the parts in the following order: Install the parts in the following order: clutch plate, clutch disc, clutch plate, clutch disc, clutch plate, clutch disc, flange.
19. Install the snapring in the center support. Check that the snapring ends are not aligned with any of the cutouts.
20. Check the piston stroke of the No. 2 brake. Using a dial indicator, measure the stroke by applying and releasing the compressed air of 58–116 psi. (400–800 kPa). Standard stroke is 0.0398–0.0896 in. (1.01–2.25mm). If the stroke exceeds the limit, the clutch discs and/or plates are probably worn. If the stroke

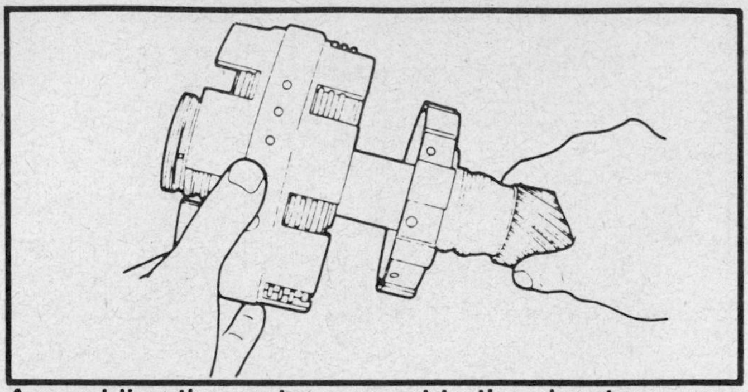

Assembling the center support to the planetary sun gear

is less than the limit, parts may be misassembled or there may be excess automatic transmission fluid on the discs.
21. Assemble the center support and planetary sun gear by aligning the brake No. 2 clutch disc flukes.
22. Mesh the brake hub with the disc, twisting and jiggling the hub in place.
23. Install the snapring on the end of the planetary sun gear.

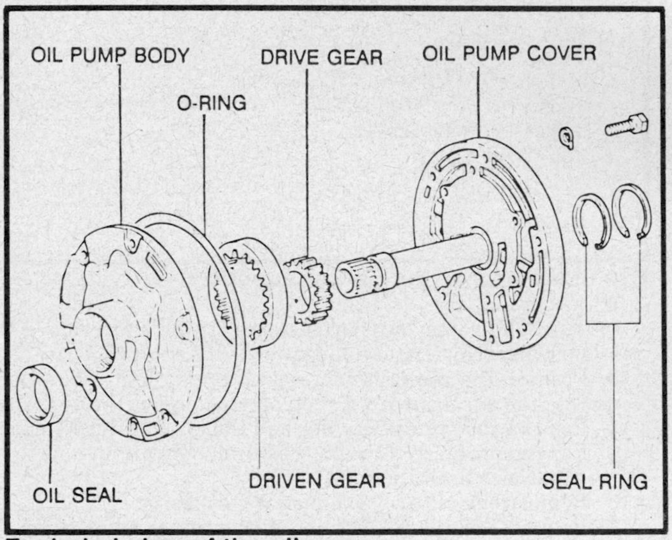
OIL PUMP BODY
DRIVE GEAR
OIL PUMP COVER
O-RING
OIL SEAL
DRIVEN GEAR
SEAL RING

Exploded view of the oil pump

OIL PUMP

Disassembly

1. Secure the pump in a suitable holding fixture.
2. Remove the seal rings from the oil pump cover.
3. Remove the attaching bolts.
4. Remove the oil pump cover.
5. Remove the O-ring from the oil pump body.
6. Remove the oil pump drive gear and driven gear. Identify the top and bottom section by appropriately marking for proper reassembly.

Inspection

BODY CLEARANCE

Push the driven gear to the side of the pump body. Using a feeler gauge, measure the clearance.

Standard value: 0.003–0.006 in. (0.07–0.15mm).
Limit: 0.012 in. (0.3mm).

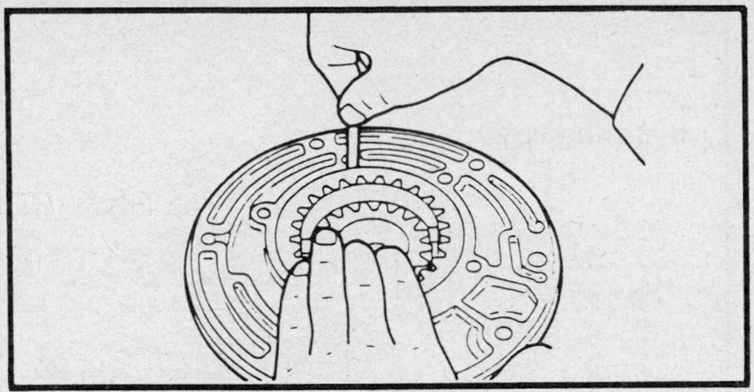
Pump body clearance check

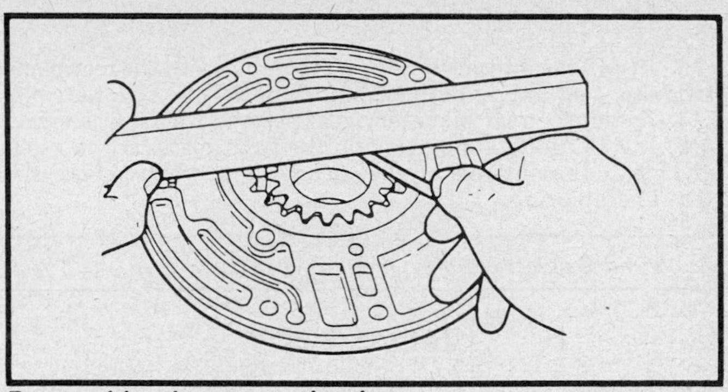
Pump side clearance check

TIP CLEARANCE

Measure the gap between the drive and driven gear teeth and the cresent-shaped part of pump body.
Standard value: 0.0043–0.0055 in. (0.11–0.14mm).
Limit: 0.012 in. (0.3mm).

SIDE CLEARANCE

Using a steel straightedge and a feeler gauge, measure the side clearance of drive and driven gears.
Standard value: 0.0008–0.0020 in. (0.02–0.05mm).
Limit: 0.004 in. (0.2 mm).

FRONT OIL SEAL

Check for wear, damage or cracks. If necessary, replace the oil seal by the following steps.

a. Pry the oil seal off using a suitable tool.
b. Install a new oil seal.

NOTE: The seal end should be flush with outer edge of pump body.

Assembly

1. Secure the pump in a suitable holding fixture.
2. Install the driven and drive gears on the oil pump body in correct directions according to the marks put during disassembly.
3. Install the oil pump cover on the body.
4. Align the bolt holes in cover with those in body. Install the attaching bolts with wave washers finger tight.
5. Install a pointed tool to align the body and cover.
6. Tighten the pump cover bolts to 4.5–6.5 ft. lbs. (6–8.5 Nm).
7. Remove the tool.
8. Install the seal rings on the pump cover by spreading apart and sliding them into the groove. Hook both ends by hand.
9. Install a new O-ring on the pump. Make sure the O-ring is not twisted and is fully seated in the groove.

REAR PLANAETARY UNIT

Disassembly

1. Loosen the retaining ring and remove the intermediate shaft (front planetary ring gear and rear planetary gear) from the output shaft assembly.
2. Remove the front planetary ring gear, thrust washer and rear planetary gear from the intermediate shaft.
3. Remove retaining ring from the rear of the intermediate shaft and remove rear planetary ring gear and thrust bearing.

Inspection

Wash the removed parts and dry with air. Check the following and replace faulty parts.

1. Check the front planetary ring gear for wear and damage of internal gear teeth and parking pawl teeth.
2. Check the intermediate shaft for wear and damage of the splines and bushing seating surfaces and for clogging of oil holes in the shaft.
3. Check the rear planetary gear for wear of the carrier thrust surface and for play in the thrust direction of the pinion.
4. Check the rear planetary ring gear for wear and damage of the internal gear teeth and internal splines.
5. Check the output shaft for wear and damage of flange thrust bearing surface and shaft bushing seating surface and for clogging of shaft oil hole and governor oil way.
6. Check the thrust washer and thrust race for wear and binding of the bearing surface.
7. Check the seal ring for wear and damage. Also check the groove.

Reassembly

1. Install a thrust bearing on the intermediate shaft and install the rear planetary ring gear and hold it in place with the retaining ring.

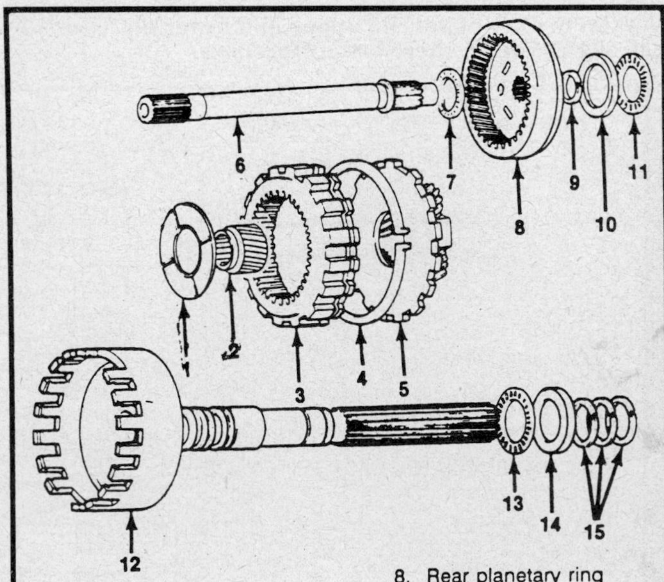
6
7
8
9
10
11
1
2
3
4
5
12
13
14
15

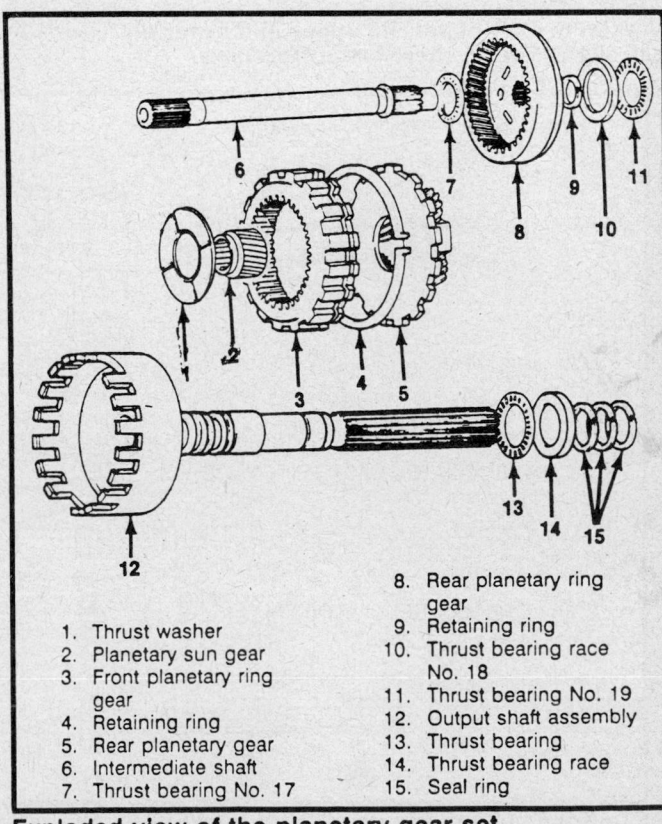

1. Thrust washer
2. Planetary sun gear
3. Front planetary ring gear
4. Retaining ring
5. Rear planetary gear
6. Intermediate shaft
7. Thrust bearing No. 17
8. Rear planetary ring gear
9. Retaining ring
10. Thrust bearing race No. 18
11. Thrust bearing No. 19
12. Output shaft assembly
13. Thrust bearing
14. Thrust bearing race
15. Seal ring

Exploded view of the planetary gear set

2. Install a thrust bearing and bearing race on the output shaft.
3. Install the intermediate shaft together with the rear planetary ring gear on the output shaft flange.
4. Install the rear planetary gear set and thrust washer.
5. Install the retaining ring on the front planetary ring gear and install the ring gear on the output flange while compressing the retaining ring.

CAUTION

Pay attention to location of retaining ring ends.

VALVE BODY

Disassembly

NOTE: Keep the disassembled parts orderly for efficient reassembly operation. Attach tags to springs for identification. When disassembling the valve, do not attempt to remove the valve with undue force. The valve and valve bore could be damaged or burred, leading to faulty valve operation. When removing the front upper and rear valve bodies from the lower valve body, use care not to lose check balls and springs.

1. Place the valve body assembly on a clean work bench.
2. Remove the manual valve.
3. Remove the lower valve body retaining bolts.
4. Turn the valve body upside down and remove the upper valve body retaining bolts.
5. Separate the lower valve body from the upper front valve body.
6. Remove the detent plate.
7. Remove the lower valve body from the rear upper valve body.
8. Remove the separator attaching bolts and remove the separator plate.

NOTE: Do not get the upper and lower plates mix up, altough similar, there are differences.

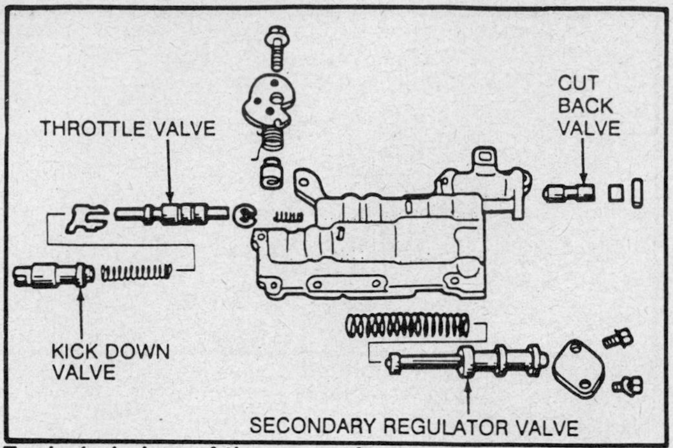
THROTTLE VALVE
CUT BACK VALVE
KICK DOWN VALVE
SECONDARY REGULATOR VALVE

Exploded view of the upper front valve body and related components

9. Remove the lower valve body gasket from the lower valve body.
10. Remove the rubber check ball at this time and take note to its location.
11. Remove the cutback valve plug and remove the cutback valve and retainer.
12. Remove the throttle cam spring and cam from the upper front valve body. Take note to the spring location prior to removal.
13. Remove the kickdown valve, throttle valve primary spring, remove the key plate and E-rings, throttle secondary spring and the throttle valve. Note order of removal for correct installation later.
14. Remove front valve end cover and remove the secondary regulator valve spring and secondary valve from the front upper valve body.
15. Remove the lower valve body cover and gasket.

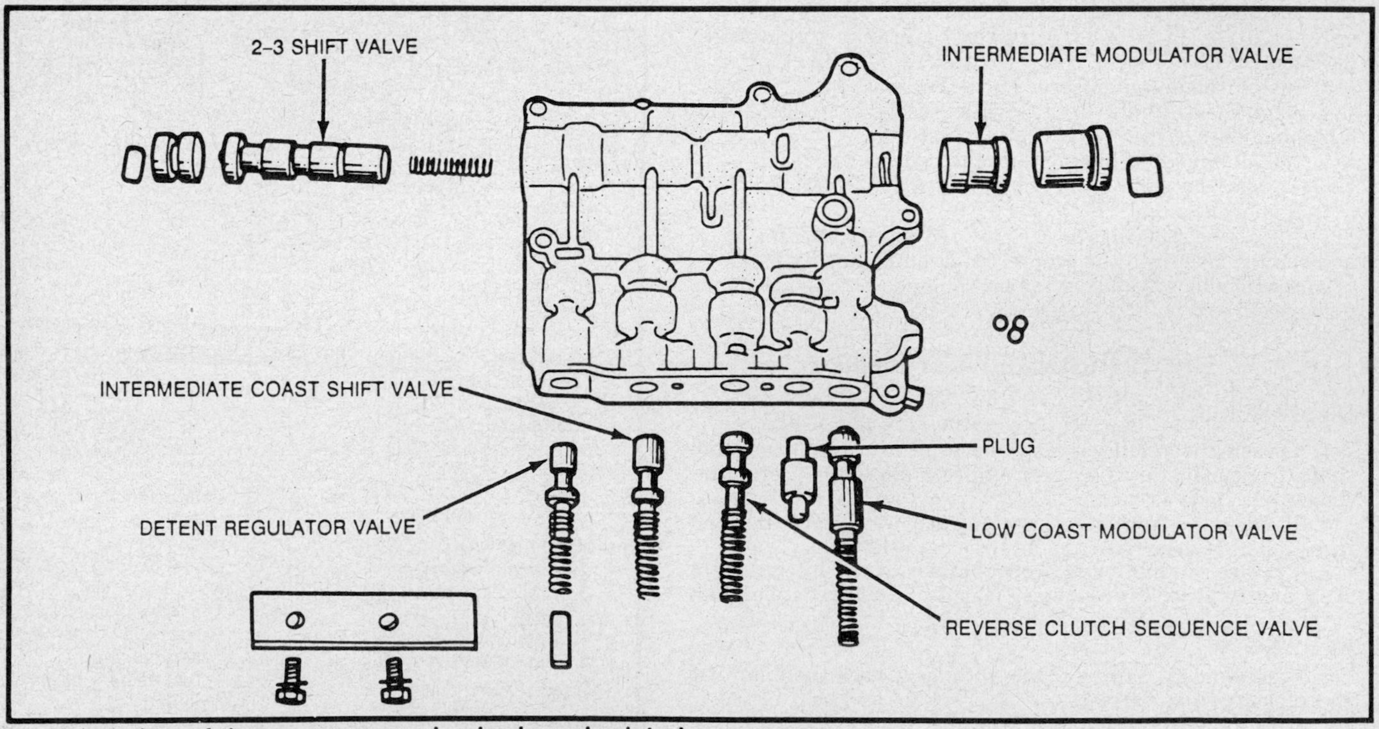
2-3 SHIFT VALVE
INTERMEDIATE MODULATOR VALVE
INTERMEDIATE COAST SHIFT VALVE
PLUG
DETENT REGULATOR VALVE
LOW COAST MODULATOR VALVE
REVERSE CLUTCH SEQUENCE VALVE

Exploded view of the upper rear valve body and related components

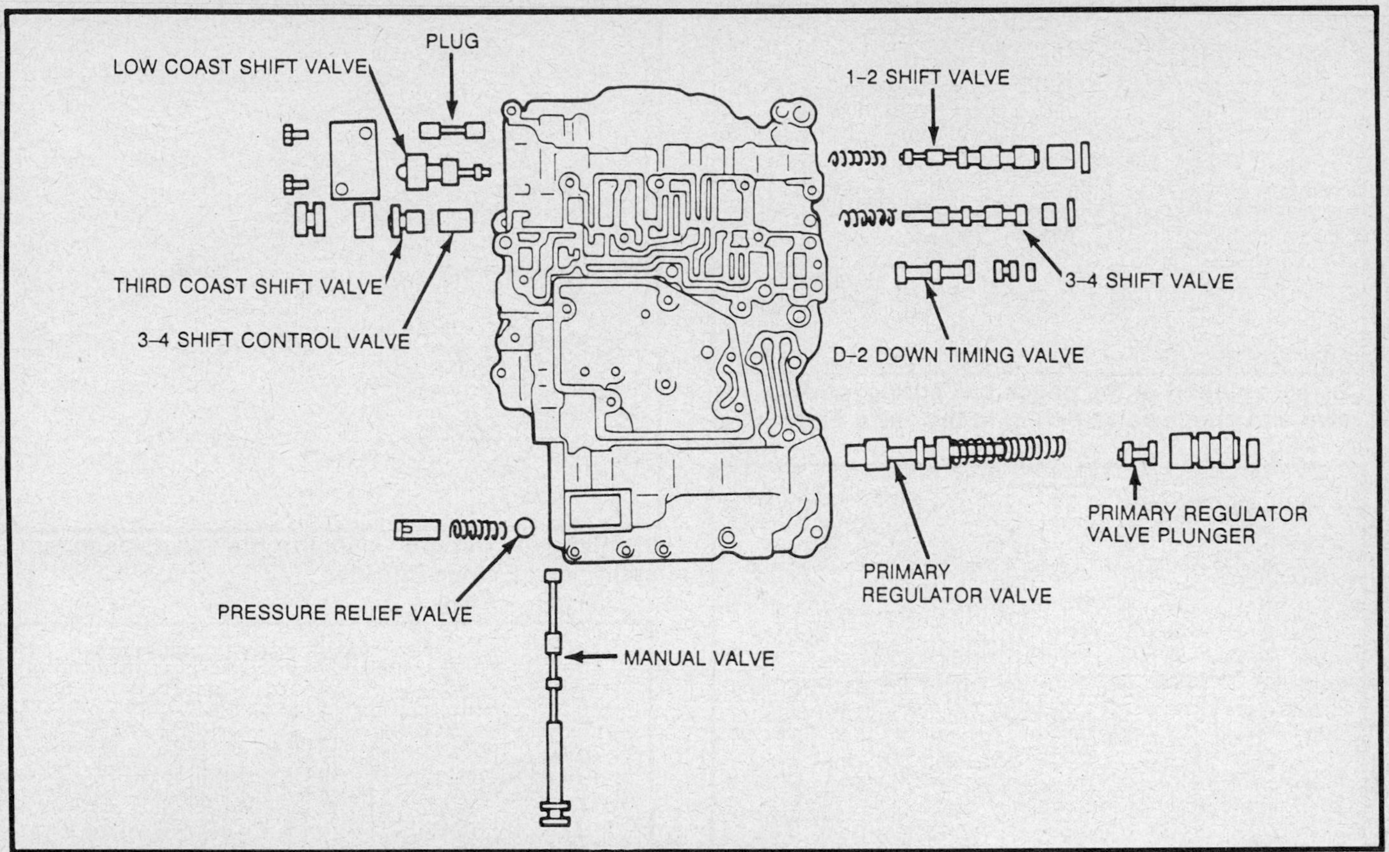
PLUG
LOW COAST SHIFT VALVE
1–2 SHIFT VALVE
THIRD COAST SHIFT VALVE
3–4 SHIFT VALVE
3–4 SHIFT CONTROL VALVE
D–2 DOWN TIMING VALVE
PRIMARY REGULATOR
VALVE PLUNGER
PRIMARY
REGULATOR VALVE
PRESSURE RELIEF VALVE
MANUAL VALVE

Exploded view of the lower valve body and related components

NOTE: Take particular notice in comparing the upper and lower gaskets at this time, the lower gasket has oval holes to accommodate the check balls. Do not get the upper and lower gaskets mix up.

16. Remove the rubber check balls at this time paying particular attention to the location of each ball.
17. Remove the rear valve cover and remove the valve springs, intermediate coast modulator valve, reverse brake sequence valve, plug and low coast modulator.
18. Remove the detent regulalor valve, spring and retainer.
19. Remove the intermediate coast shift valve, plug and retainer.
20. Remove the 2–3 shift valve, spring, 2–3 shift valve plug and retainer from the upper rear valve body.
21. Remove theD–2 down timing valve, plug and seat.
22. Remove the 3–4 shift valve, spring, plug and locating pin.
23. Remove the check valve and check valve spring. Note location.
24. Remove the pressure relief valve, spring and retainer.
25. Remove the 1–2 shift valve spring, low coast shift valve, plug and low coast shift valve cover.
26. Remove the 1–2 shft valve, valve plug and insert retainer.
27. Remove the primary regulator valve retainer, primary regulator valve, spring, pluger and sleeve from the lower valve body.

INSPECTION

Wash the removed parts and dry with air. Then make the following checks.

CAUTION

When making checks, use care not to damage the valve land outside and valve body bores.

1. Check the valves for damage and wear.
2. Insert the valves in the valve body and check for smooth rotation and sliding.
3. Check for damage and wear of valve bores and for clogging of oil passages.
4. Check for damage or wear of the valve body plate and check balls.
5. Check for clogging of the oil strainer.
6. Check the springs and replace if broken or excessively deteriorated.

ASSEMBLY

CAUTION

Before reassembly, wash the parts in a clean detergent and dry with air. Do not wipe dry with rags. Entry of dust could cause faulty valve operation.

1. Install primary regulator valve, spring, plunger and sleeve in the lower valve body in order. Insert retainer to hold the valve in the valve body.
2. Install 1–2 shift valve, valve plug and insert retainer.
3. Install 1–2 shift valve spring, low coast shift valve, plug, and low coast shift valve cover.
4. Install pressure relief valve, spring and retainer.
5. Install spring, check valve and check valve spring at illustrated locations.
6. Install the spring, 3–4 shift valve, plug, and locating pin.
7. Install the D–2 down timing valve, plug, and seat.
8. Install 2–3 shift valve spring, 2–3 shift valve, 2–3 shift valve plug, and retainer in the upper rear valve body.
9. Install intermediate coast shift valve, plug and retainer.

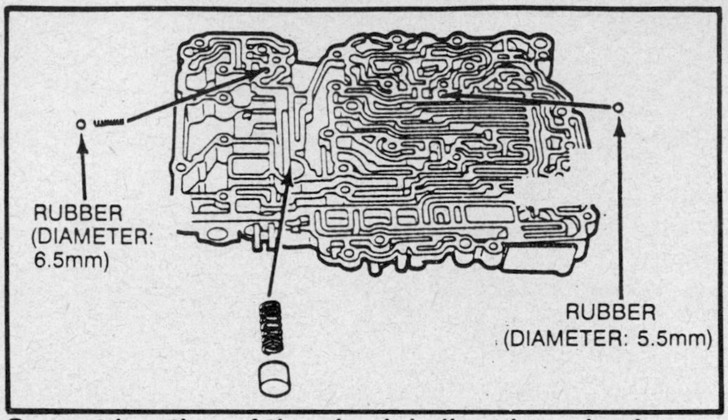
RUBBER
(DIAMETER:
6.5mm)
RUBBER
(DIAMETER: 5.5mm)

Correct location of the check ball spring, check valve and check valve spring in the valve body

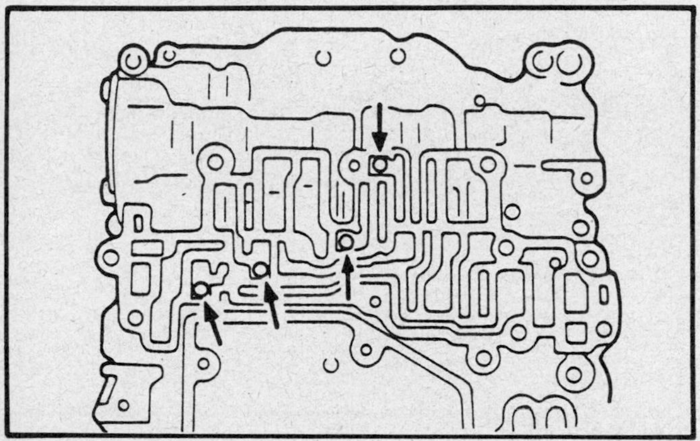

Correct location of the rubber check balls in the lower valve boby

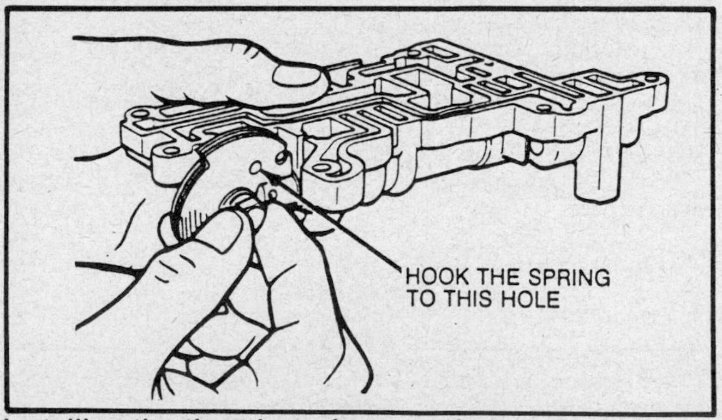
HOOK THE SPRING
TO THIS HOLE

Installing the throttle and cam spring

10. Install detent regulator valve, spring and retainer in that order.
11. Install low coast modulator valve, plug, reverse brake sequence valve and intermediate coast modulator valve.
12. Install valve springs.
13. Install rear valve cover.
14. Place the rubber check balls in the lower valve body at proper locations.
15. When installing the lower valve body cover, use the correct gasket. The lower valve body gasket is equipped with oval holes for the check balls.
16. Insert secondary regulator valve spring and the secondary regulator valve in the front upper valve body. Install front valve end cover.

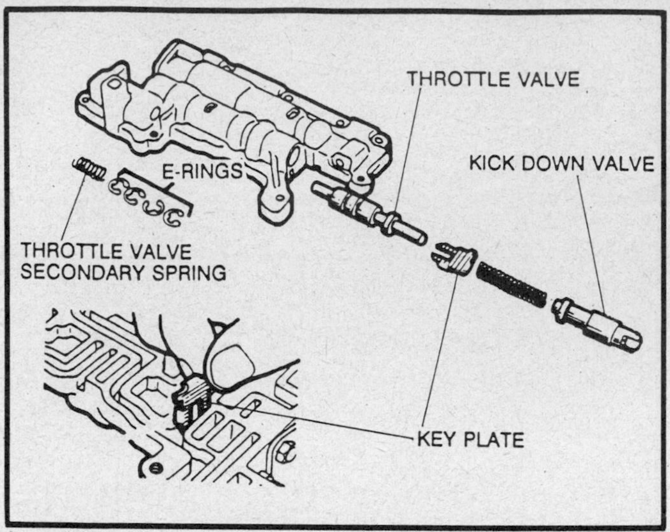
THROTTLE VALVE
E-RINGS
KICK DOWN VALVE
THROTTLE VALVE
SECONDARY SPRING
KEY PLATE

Installing the throttle valve, throttle valve secondary spring and E-rings in order

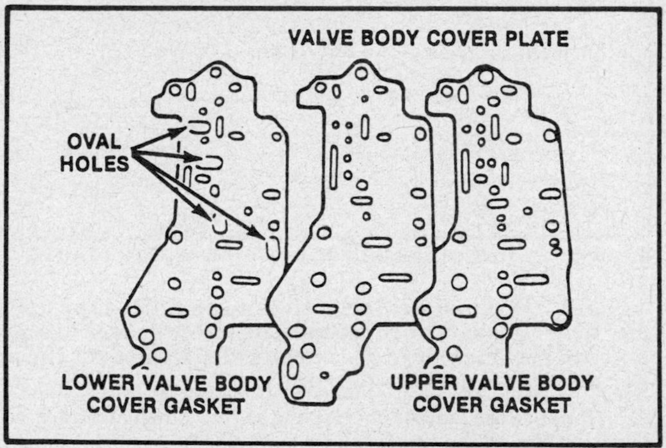
VALVE BODY COVER PLATE
OVAL
HOLES
LOWER VALVE BODY
COVER GASKET
UPPER VALVE BODY
COVER GASKET

Oval hole location for check balls—lower valve body

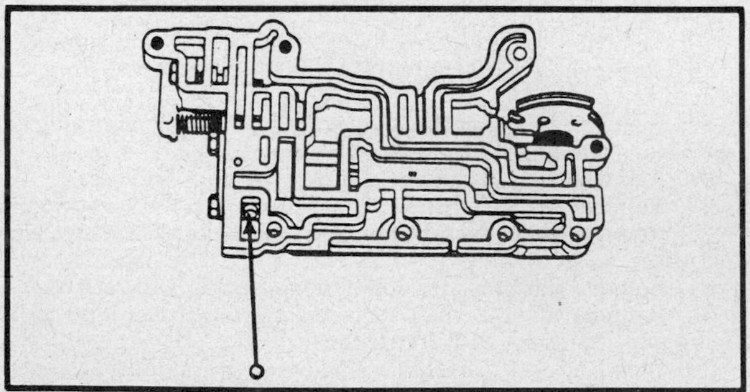

Install the rubber check ball in this location—lower front valve body

17. Insert throttle valve, throttle valve secondary spring, E-rings, and key plate in the proper location. Install the throttle valve primary and kickdown valve.

CAUTION

Install the same number of E-rings as before disassembly for the correct throttle valve adjustment. Insertion of the throttle valve key plate at the incorrect location could cause faulty valve operation.

18. Install the throttle cam and spring on the upper front valve body and tighten the bolts temporarily. When installing, note the location of the spring end of the body side. Hook the other end of the spring to the cam and bolt the cam to the valve body. After installation, check that the throttle cam turns through a full stroke smoothly.

CAUTION

Hook the spring to the correct hole.

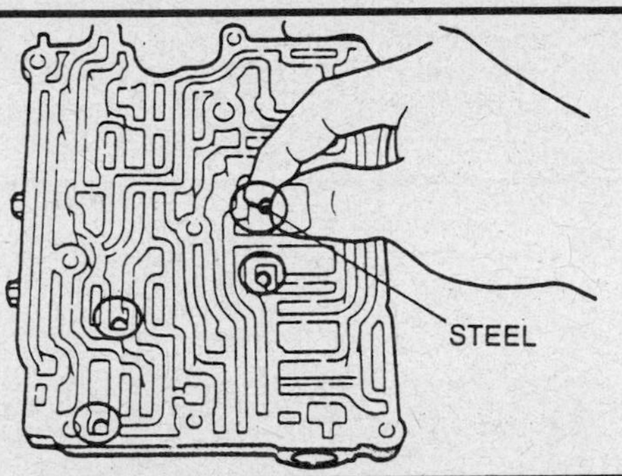
STEEL

Steel check ball must be place in this location

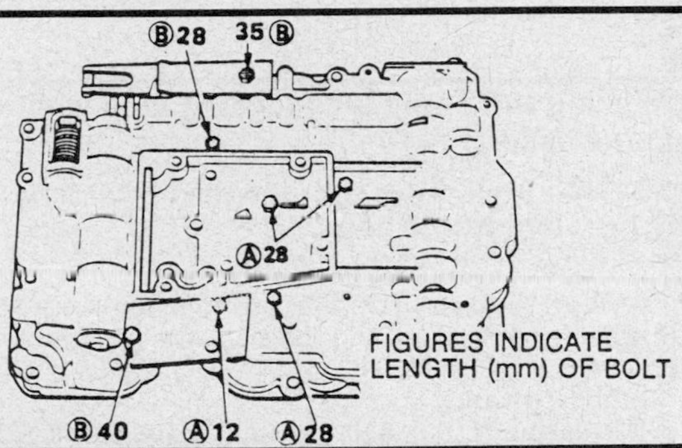
Ⓑ28
35Ⓑ
Ⓐ28
FIGURES INDICATE
LENGTH (mm) OF BOLT
Ⓑ40
Ⓐ12
Ⓐ28

Tighting bolts indicated by A or B—lower valve body side

19. Install cutback valve, valve plug and retainer. Install the cutback plug with the larger land end facing out.
20. Install the rubber check ball in its location.
21. Place a new lower valve body gasket on the lower valve body.

CAUTION

Do not mix the upper and lower gaskets up, although similar, there are differences.

22. Install the separator plate and temporarily tighten the bolts.
23. Place a gasket for upper valve body, aligning with separator plate.
24. Install the steel check ball in its designated location. The 3 rubber check balls are identical and may be installed in any other location.
25. Install the lower valve body onto the rear upper valve body and temporarily tighten the bolts indicated by A. Be careful not to disturb the check valve position on the rear upper valve body.
26. Remove the 2 bolts previously tighten.
27. Install the lower valve body onto the rear upper valve body and temporarily tighten the bolts indicated by B.
28. Install the detent plate.
29. Install the lower valve body onto the upper front valve body and temporarily tighten the bolts indicated at the left of the lower valve body side.

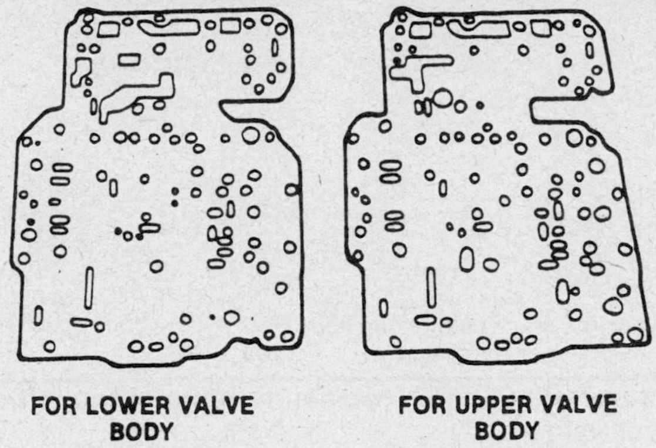
FOR LOWER VALVE BODY
FOR UPPER VALVE BODY

Upper and lower valve body gaskets

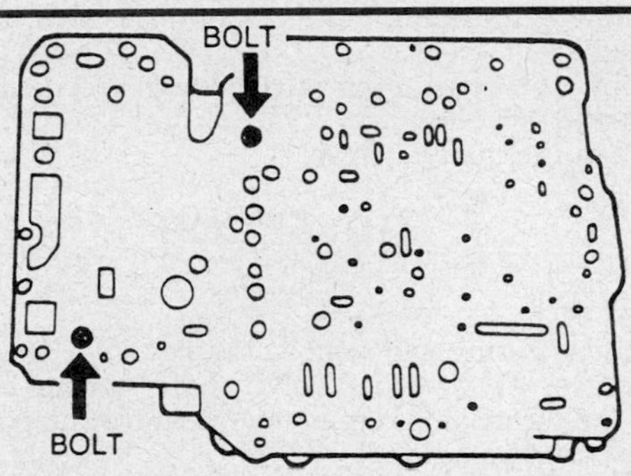
BOLT
BOLT

Separator plate

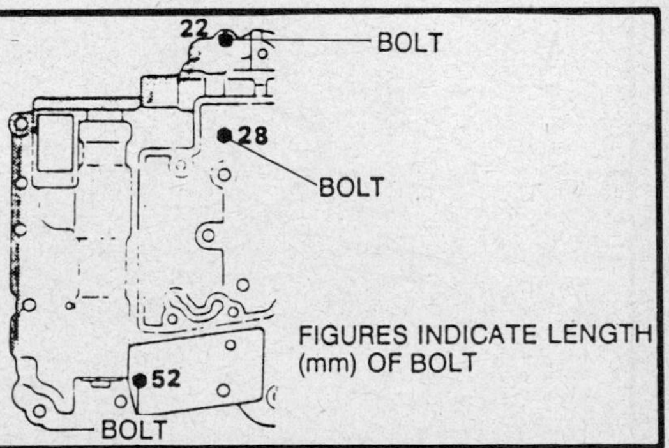
22
BOLT
28
BOLT
FIGURES INDICATE LENGTH
(mm) OF BOLT
52
BOLT

Tighten bolts indicated—upper valve body side

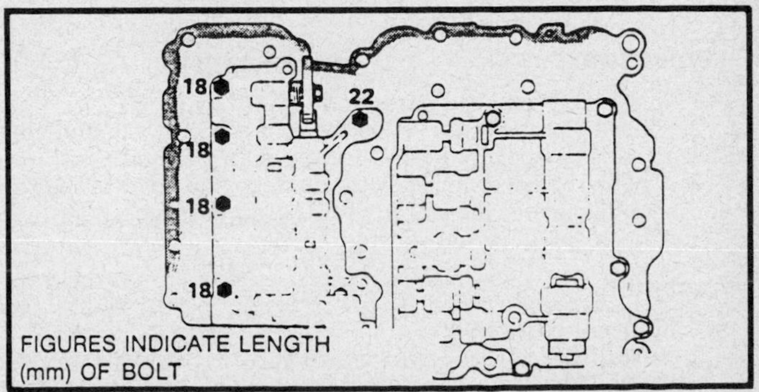
18
18
18
18
22
FIGURES INDICATE LENGTH
(mm) OF BOLT

Tighten bolts indicated—upper valve body side

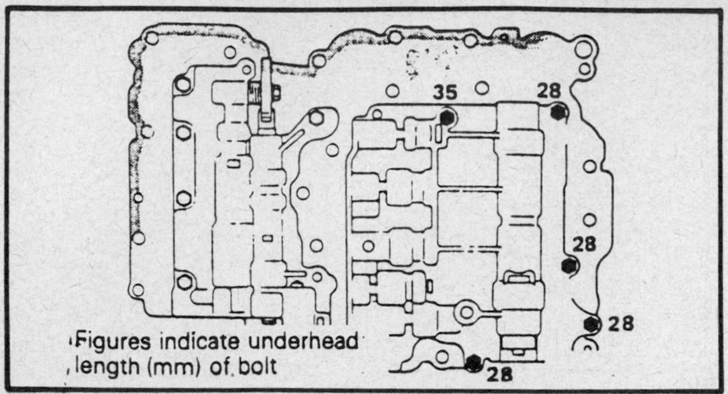
35
28
28
28
28
Figures indicate underhead length (mm) of bolt

Temporarilly tighten remaining bolts at the left of the lower valve body

30. Turn the valve body upside down and temporarily tighten the indicated bolts from the upper valve body side.
31. Temporarily tighten the remaining valve body bolts as illustrated.
32. Tighten all bolts of the valve body to specified torque. 3.6–4.3 ft. lbs. (5–5.5 Nm).
33. Install the manual valve.

GOVERNOR

Disassembly

1. Remove E-ring and remove the governor weight.
2. Remove the governor valve shaft, spring and governor valve in the direction of the arrow. Remove the governor valve from the output shaft hole.

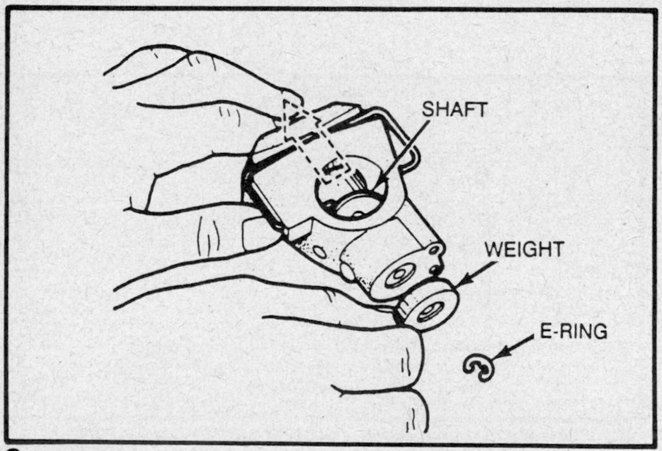
SHAFT
WEIGHT
E-RING

Governor

Inspection

1. Wash the removed parts and dry with air.
2. Check the governor valve for damage or wear and check that it slides smoothly while rotating it in the body.
3. Check the governor body for damage and wear of the valve sliding surface and for clogging of the oil passages.
4. Check the governor spring.

Assembly

1. Coat all parts in ATF.
2. Install the governor valve into the output shaft hole.
3. Install the spring and shaft.
4. Install the weight and E-ring.

Transmission Assembly

ASSEMBLY

NOTE: Before assembly, make sure that all component assemblies are assembled correctly.

1. Place the transmission in a suitable holding fixture for more efficient work. Place shock absorbing material between the case and holding fixture to prevent damage to the case.
2. Install the thrust bearing No. 20 and then the race No. 21 facing the cup side downward.
3. Install the brake apply tube onto the case, aligning the tube's locking tab part A with part B of the case.

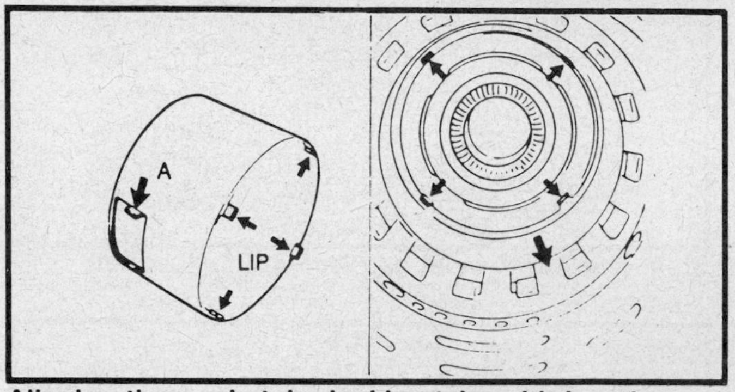
A
LIP

Aligning the apply tube locking tabs with locations in the case

NOTE: Make sure that the tab of the tube is completely installed in the case.

4. Insert the output shaft assembly into the case.
5. With the case in upright position, make sure that the No. 3 brake is lower than the ledge below the retaining snapring groove.
   Standard value:
   Dimension A: 0.024–0.104 in. (0.61–2.64mm).

NOTE: If the the No. 3 brake is not lower than the ledge, components may be misassembled or there may be excess ATF between the disc and plate.

6. Install the reaction plate positioning the notched tooth of the reaction plate toward the valve body side of the case. Push it into place.

NOTE: The reaction plate is correctly installed if the retaining snapring groove is fully visible.

7. Using a suitable tool, install the retaining snapring. Compress the snapring and push into place by hand. Work around

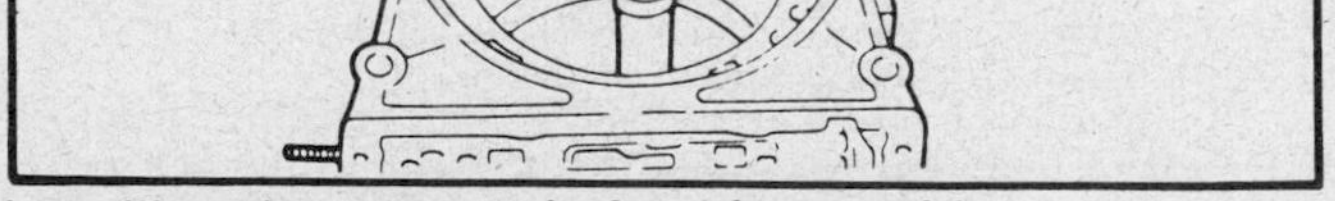

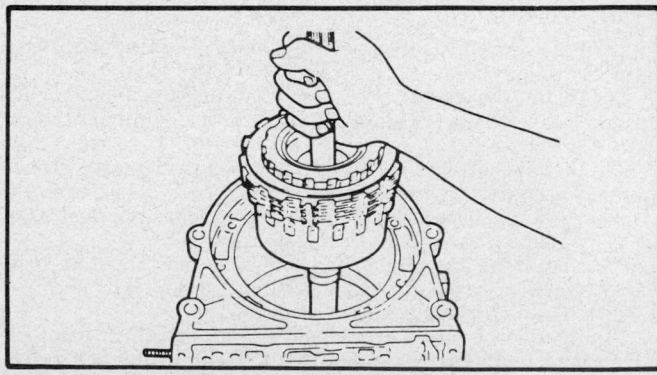

Installing the output shaft and assembly

the case. Visually check to make sure that the ring is fully seated. Make sure that the ends of the snapring are between the lugs.

8. Push the center support assembly into the case while aligning the oil hole and bolt hole of the center support with those of the body side.

9. Install the center support bolts with wave washers. Finger tighten the bolts.

10. Install the direct clutch in the case while turning the clutch to mesh its hub with the center support.

11. Check for correct installation of the direct clutch. If the direct clutch is fully meshed with the center support, the splined center of the clutch will be flush with the end of the planetary sun gear shaft.

12. After being coated with petroleum jelly, install the thrust bearing race No. 16 over the splined end of the direct clutch in case with its lip toward the direct clutch.

13. After being coated with petroleum jelly, install the thrust bearing No. 15 and race No. 14 on the forward clutch, with the race's lip outward.

14. Install the forward clutch assembly in the case by aligning the flukes of the direct clutch discs and mesh them with the forward clutch hub. Push the forward clutch assembly into the case.

CAUTION

Be careful not to allow the thrust bearing to drop out.

15. Check for proper installation of the forward clutch as follows:

a. Position a suitable measuring tool on the transmission case.

b. Measure the distance between the top surface of the tool and forward clutch assembly. If the distance corresponds to that during disassembly, the forward clutch is installed correctly.

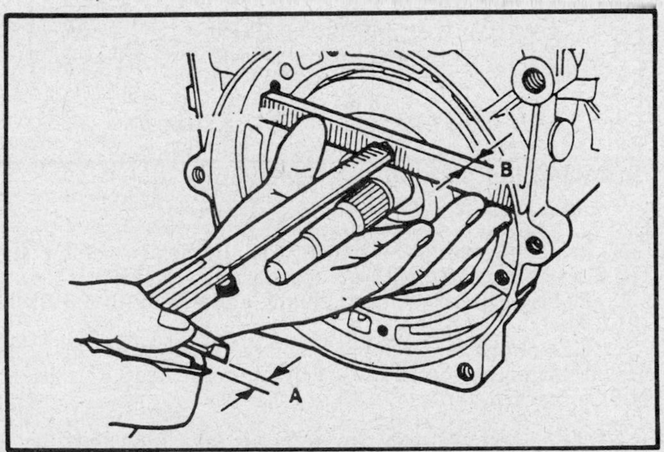
B
A

Checking the forward clutch installation height

16. Remove the measuring tool from the case.

17. After being coated with petroleum jelly, install the thrust bearing No. 10 on the forward clutch.

18. After being coated with petroleum jelly, install the thrust race No. 9 on the overdrive case end with its lip toward the overdrive case.

19. Insert the overdrive case gently into the transmission case through the guide pins with the part indicated by arrow facing in the direction shown.

20. Coat the thrust washers with petroleum jelly. Install the washers on the overdrive planetary gear.

NOTE: The washer lugs should be inserted in the holes.

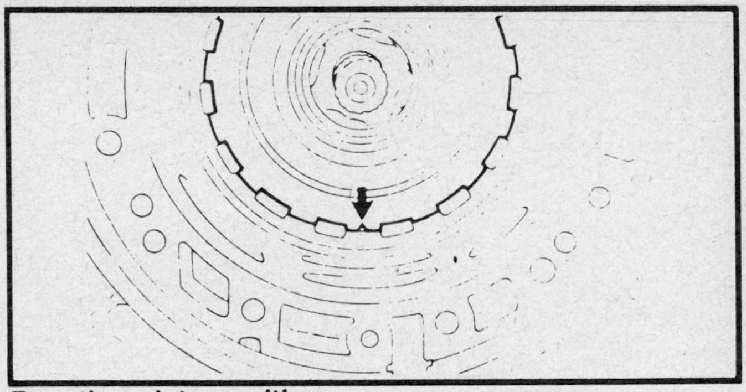

Reaction plate position

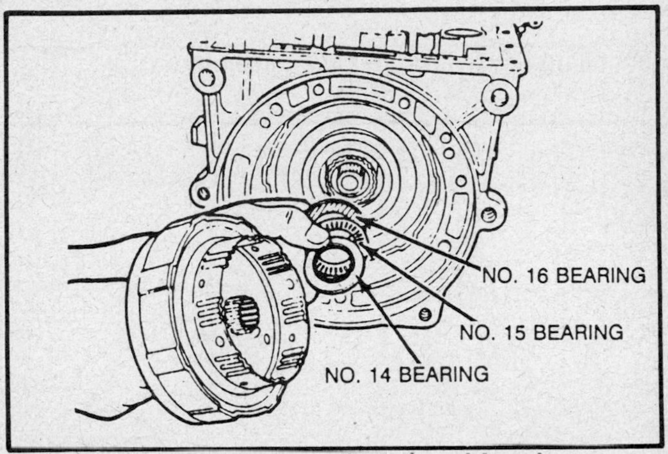
NO. 16 BEARING
NO. 15 BEARING
NO. 14 BEARING

Installing the No. 14, 15 and 16 thrust bearing

21. Install the overdrive clutch in the case by aligning the disc flukes in the overdrive case. Align the flukes with the slots of the overdrive clutch and press the overdrive clutch into the overdrive case.

CAUTION

Be careful not to let the thrust washer drop.

22. Check for correct installation of the overdrive clutch as follows:

a. Position a suitable measuring tool on the overdrive case.

b. Measure the distance between the top surface of the tool and the overdrive clutch. If the distance corresponds to that during disassembly, the overdrive clutch is installed correctly.

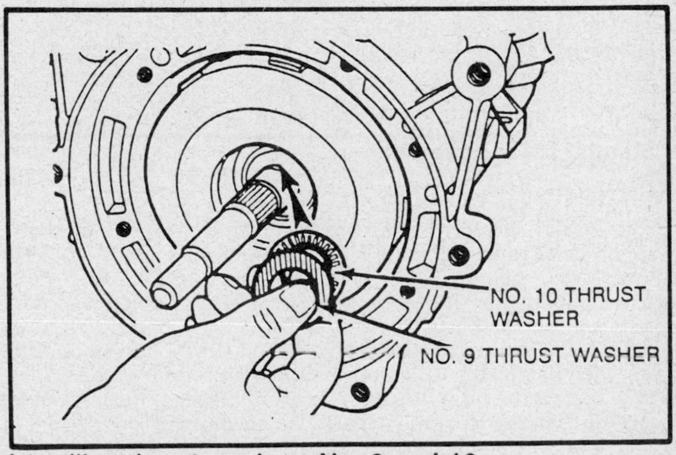
NO. 10 THRUST WASHER
NO. 9 THRUST WASHER

Installing thrust washers No. 9 and 10

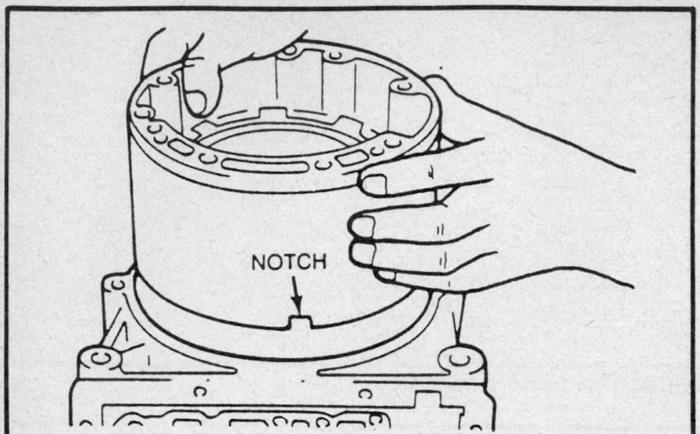
NOTCH

Installing the overdrive case assemby

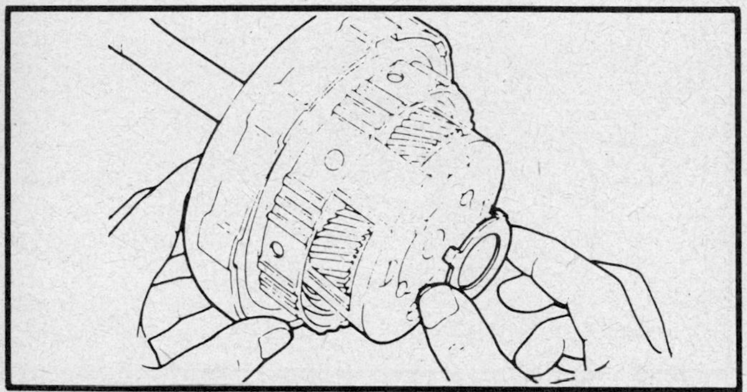

Installing the thrush washer on the overdrive planetary gear

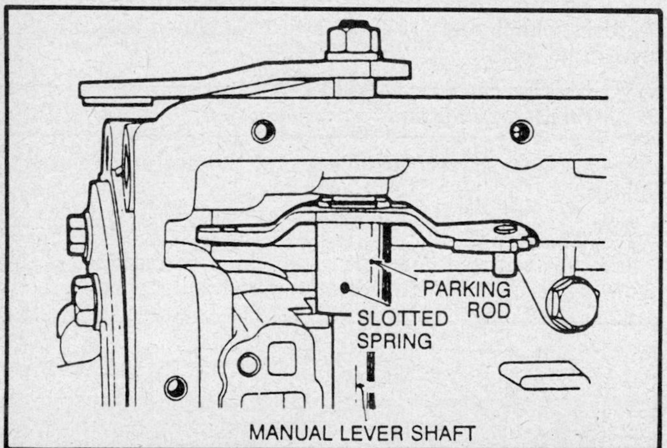
PARKING
ROD
SLOTTED
SPRING
MANUAL LEVER SHAFT

Manual lever assembly

23. Install the O-ring on the overdrive case.
24. Install the torque converter housing. Tighten the bolts: 12mm bolts to 42 ft.lbs (58 Nm), 10mm bolts to 25 ft. lbs. (34.5 Nm).
25. Coat the thrust bearing race No. 3 with petroleum jelly and install in on the overdrive clutch.
26. Coat the thrust bearing No. 2 and the No. 1 combination racewith petroleum jelly and install it on the oil pump.
27. Install the oil pump gently through the guide bolts, being careful that the thrust washer does not fall out.
28. Coat the pump retaining with sealant and finger tighten them.

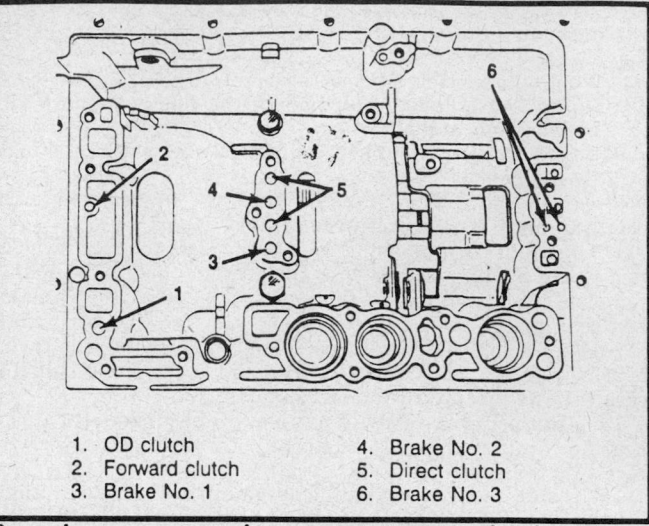

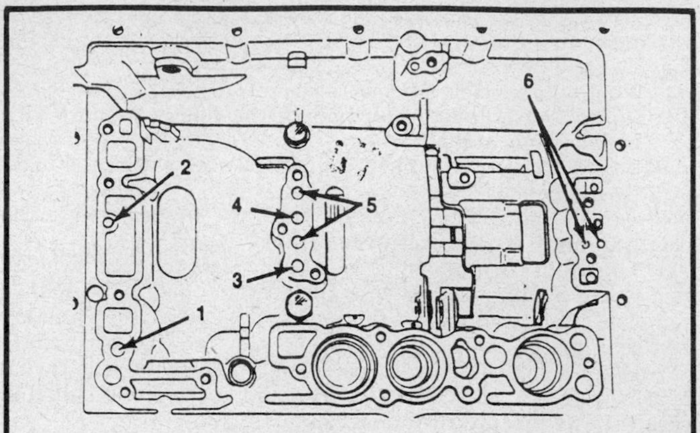
1
2
3
4
5
6

1. OD clutch
2. Forward clutch
3. Brake No. 1
4. Brake No. 2
5. Direct clutch
6. Brake No. 3

Appy low pressure air pressure to test clutch and brake operation

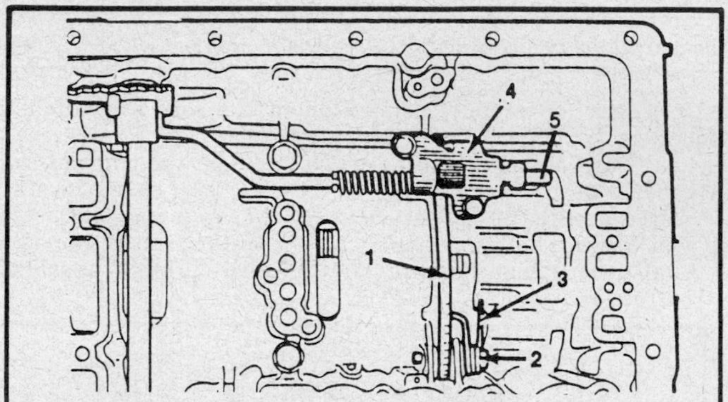
1
2
3
4
5

1. Parking pawl
2. Pivot pin
3. Spring
4. Cam plate
5. Rod

Installing the parking pawl assembly

29. Remove a suitable measuring tool. In the place of it, install the 2 set bolts coated with an appropriate sealant.
30. Tighten the set bolts gradually and evenly to 5.6 ft. lbs. (21.5 Nm).
31. Tighten the center support bolts alternately in 5.16 ft. lbs. (7 Nm) increments until the specified torque of 18.8 ft. lbs. (26 Nm) is reached.

NOTE: First tighten the accumulator side bolt.

32. Check the operation of the pistons by appling low pressure compressed air into the passages indicated in the figure and listen for noise from piston movement.
   a. Overdrive clutch
   b. Overdrive brake
   c. Forward clutch
   d. Direct clutch
   e. Brake No. 1
   f. Brake No. 2
   g. Brake No. 3

If the pistons do not move, disassemble and inspect them.
33. Check the input shaft and output shaft:
   a. Make sure that the input shaft has play in axial direction and that it turns.
   b. Make sure that the output shaft has an appropriate endplay of: 0.012–0.035 in. (0.3–0.9mm).

34. Install the parking rod assembly on the manual valve lever and insert the manual shaft in transmission case.
35. Drive in a new slotted spring pin with the slot at a right angle to the shaft.
36. Install the parking pawl, pivot pin and spring in the case.
37. Install the cam plate on the case with the attaching bolts. Make sure that the parking rod protrudes from the cam plate. Tighten the bolts to 4.5–6.5 ft. lbs. (6–8.5 Nm). Make sure the pawl moves freely.

CAUTION

Be careful, as it is possible for the cam plate to be installed too far forward, where it will bind the pawl.

38. Check the operation of the parking lock pawl. The planetary gear output shaft must be locked when the manual valve lever is in the **P** range.
39. Install a new O-ring on the throttle cable.
40. Install the throttle cable in the case by pushing the cable through the case, being careful not to damage the O-ring. Check for full seating.
41. Install the accumulator piston and springs.
42. Place the valve body on the transmission as follows: Make sure the accumulator pistons are pressed fully into the bore. Align the manual valve with the pin on the manual valve lever, and lower valve body into place.
43. Lift a side of the valve body and attach the throttle cable.
44. Make sure that the lower spring is installed on the B^2 or C^2 piston.

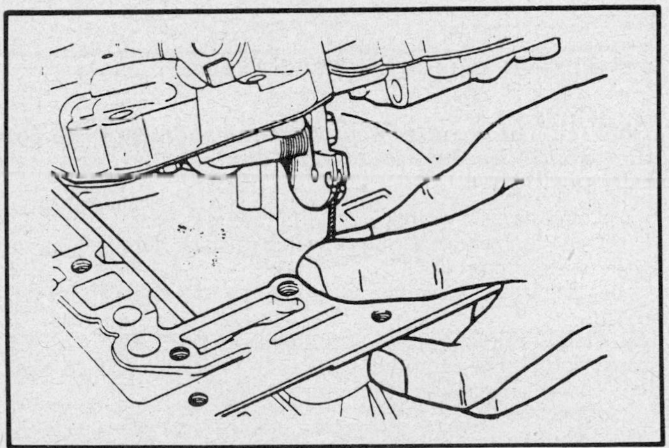

Connecting the throttle cable to the valve body

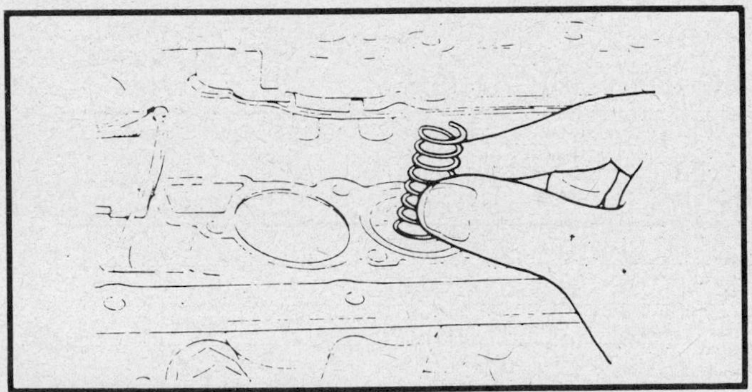

Installing the lower spring

45. Install the bolts in the valve body and tighten the bolts to 6–85 ft. lbs. (8–11 Nm).
46. Install the detent spring.
47. Install the oil strainer and tighten the bolts to 3.6–4.3 ft. lbs. (5–5.5 Nm).

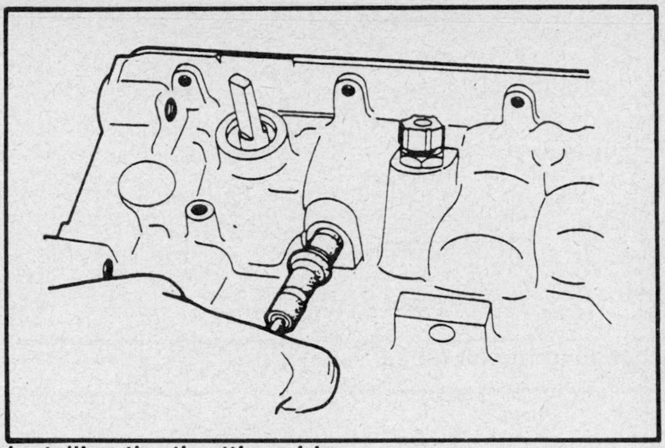

Installing the throttle cable

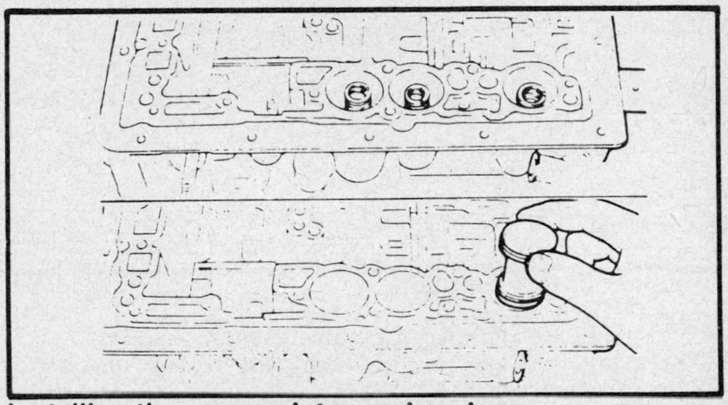

Installing the accumulator and springs

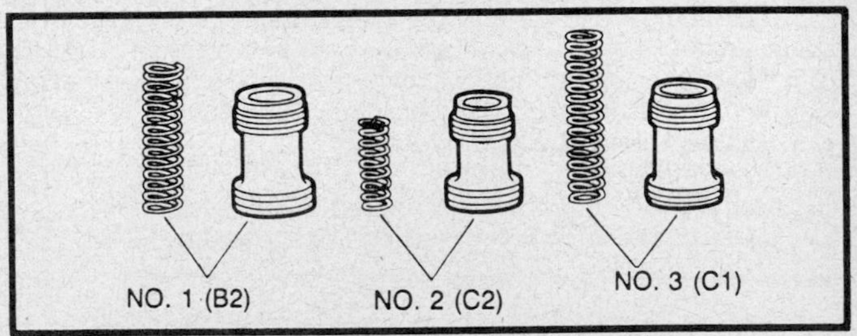
NO. 1 (B2)
NO. 2 (C2)
NO. 3 (C1)

Accumlator piston and spring identification

48. Using a plastic hammer, install the oil pipes into position.

CAUTION

Be careful not to bend or damage the pipes.

49. Install the magnet in the oil pan and install the oil pan with a new gasket. Torque pan retaining bolts to 3–3.5 ft. lbs. (4–4.5 Nm).

CAUTION

Make sure that the magnet does not interfere with the oil pipes.

50. Install the drain plug with a new gasket. Torque to 13–16 ft. lbs. (18–22 Nm).
51. Install the governor line strainer on the transmission case and then install the plate.
52. Install the governor and speedometer drive gear on the output shaft.
53. Install the lock plate and bolt and stake the lock plate.
54. Install the snapring and lock ball.
55. Slide the speedometer gear onto the shaft.

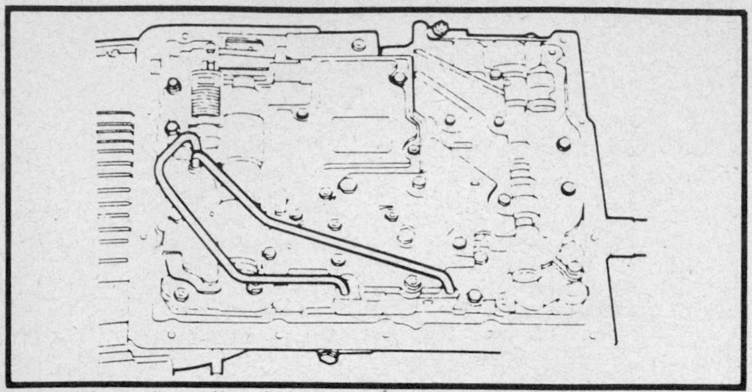

Installing the oil tubes

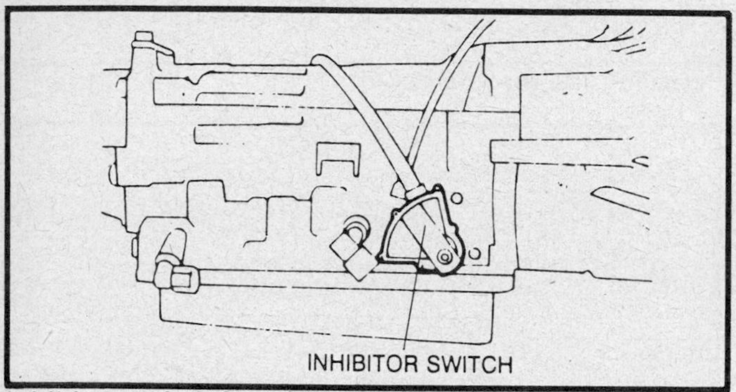
INHIBITOR SWITCH

Inhibitor switch location

56. Using snapring pliers, install the outer snapring.
57. Install the rear extension housing with a new gasket. Torque to 25 ft. lbs. (34.5 Nm).

NOTE: Do not use sealant on gasket.

54. Install new O-rings and the speedometer driven gear to the shaft sleeve.
58. Install the speedometer driven gear assembly in the extension housing and install the lock plate and bolt.
59. Install the neutral safety switch.
60. Install the shift handle.
61. Adjust the neutral safety switch as follows:
   a. Turn the shift lever to the **N** range position.

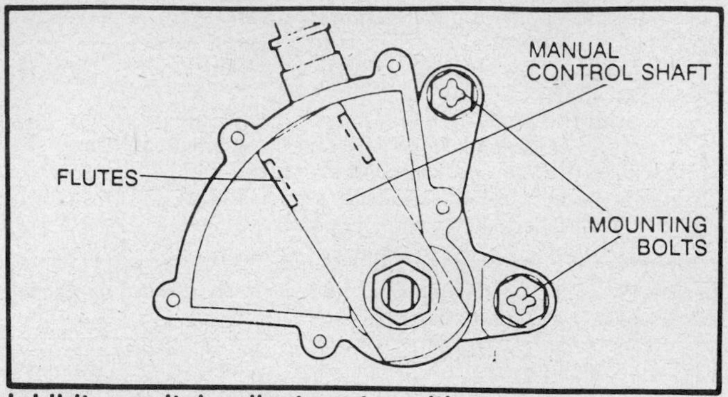
MANUAL
CONTROL SHAFT
FLUTES
MOUNTING
BOLTS

Inhibitor switch adjustment position

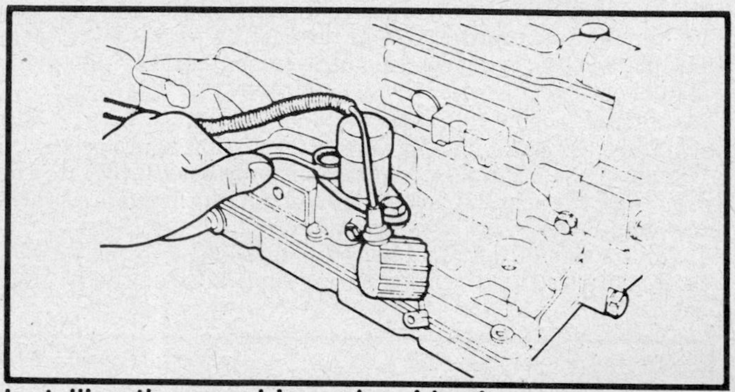

Installing the overdrive solenoid valve

NOTE: The N range position corresponds to the position where the flats on manual control shaft are vertically positioned.

   b. Loosen the switch mounting bolts.
   c. Move the neutral safety switch until the lever aligns with the positioning boss.
   d. Tighten the switch mounting bolt to 2.9–5.1 ft. lbs. (4–7 Nm).
62. Install the overdrive solenoid valve on the case with 2 O-rings. Torque the retaining bolts to 8–11 ft. lbs. (10–15 Nm).
63. Install the transmission in the vehicle.

SPECIFICATIONS

ACCUMULATOR SPRING IDENTIFICATION

Spring	Engine (liters)	Free Length in. (mm)	OD in. (mm)	Wire Diameter in. (mm)	Color
B_2, No. 1	2.0	2.626 (66.7)	0.705 (17.9)	0.102 (2.6)	–
B_2, No. 1	2.6	2.626 (66.7)	0.681 (17.3)	0.110 (2.8)	Red
B_2	2.4	2.6252 (66.68	0.627 (17.34)	0.1102 (2.8)	Red
C_2, No. 2	2.0	2.409 (61.2)	0.650 (16.5)	0.098 (2.5)	Yellow
C_2, No. 2	2.6	2.173 (55.2)	0.646 (16.4)	0.09 (2.3)	–
C_2, No. 1	2.4	1.2886 (32.73)	0.5827 (14.80)	0.512 (1.3)	Green
C_2, No. 2	2.4	1.7016 (43.22)	0.5449 (13.84)	0.0787 (2.0)	Red
C_1, No. 3	2.0, 2.6	2.547 (64.7)	0.689 (17.5)	0.079 (2.0)	–
C_1	2.4	2.5465 (64.88)	0.6890 (17.50)	0.787 (2.0)	–

ACCUMULATOR PISTON IDENTIFICATION

Piston	OD in. (mm)	Length in. (mm)
B_2, No. 1	1370 (34.8)	1909 (48.5)
C_2, No. 2	1252 (31.8)	1772 (45)
C_1, No. 3	1252 (31.8)	1949 (49.5)

TORQUE SPECIFICATIONS
AW372L Transmission

Items	ft. lbs.	Nm
12mm torque converter housing	42.0	58
10mm torque converter housing	25.0	34.5
Extension housing	25.0	34.5
Oil pump	15.6	21.5
Center pump	18.8	26
Upper valve body to lower valve body	4.0	5.5
Valve body	7.2	10
Oil strainer	4.0	5.5
Oil pan	3.3	4.5
Oil drain plug	13–17	18–23
Oil pump cover bolt	5.4	7.5
Cooler pipe union nut	18.1	25
Testing plug	5.4	7.5
Parking lock pawl bracket	5.4	7.5
Inhibitor switch bolt	2.9–5.1	4–7
Selector handle installing screw	1.4 or more	2 or more
Shift lever shaft	13–17	18–24
Oil cooler tube flare nut	29–36	40–50
Oil cooler tube clamp	2–4	3–5
Oil cooler tube bracket	7–10	9–14
Propeller shaft installing bolt	36–43	50–60
Rear engine mount installing bolt	50–65	70–95
Exhaust pipe clamp bolt	14–22	20–30
Starter motor installing bolt	20–25	27–34
Torque converter installing bolt	25–30	35–42
Engine and transmission tightening bolt		
Bolts with 0.39 (10mm) outside dia.	31–40	43–55
Bolts with 0.31 (8mm) outside dia.	14–20	20–27

TORQUE SPECIFICATIONS
KM148 and AW372 Transmissions

Items	ft. lbs.	Nm
Converter housing attaching bolt		
10mm diameter bolt	20–30	27–41
12mm diameter bolt	35–49	47–66
Oil pump assembly attaching bolt	13–18	18–24
Oil pump body and cover tightening bolt	4.5–6.5	6–8.5
Center support attaching bolt	18–20	24–27
Adapter attaching bolt	20–30	27–41
Cover plate attaching screw	4.5–6.5	6–8.5
All bolts for valve body	3.6–4.3	5–5.5
Throttle cam attaching bolts	4.5–6.5	6–8.5
Valve body assembly attaching bolt	6–8.5	8–11
Oil screen attaching bolt	3.6–4.3	5–5.5
Parking cam plate attaching bolt	4.5–6.5	6–8.5
Oil pan attaching bolt	3–3.5	4–4.5
Union	15–21	20–29
Elbow connector	15–21	20–29
Plug for hydraulic test	4.5–6.5	6–8.5
Oil pan drain plug	13–16	18–22
OD solenoid valve attaching bolt	8–11	10–15
Plug	8–11	10–15
Manual lever attaching nut	11–13	14–17
Transmission mounting bolt	31–40	43–55
Exhaust pipe mounting bracket mounting bolt	15–20	20–27
Special bolt	25–30	35–42
Rear engine support insulator to transmission		
2WD	14–17	20–24
4WD	13–18	18–25
Bell housing cover mounting bolt	6–7	8–10
Exhaust clamp mounting bolt	14–21	20–30
Starter motor mounting bolt	16–23	22–32
No. 2 crossmenber mounting bolt		
2WD	29–36	40–50
4WD	40–54	55–75
Steering wheel lock nut	25–33	35–45

GENERAL SPECIFICATIONS

Items	AW372	KM148	AW372L
Torque converter type	3 elements, 1 stage, 2 phase	3 elements, 1 stage, 2 phase	3 element, 1 stage, 2 phase with a lock-up clutch
Stall torque ratio	1.96	1.96	2.26
One-way clutch	Sprag type	Sprag type	Sprag type
Transmission Type	Forward 4 stages Reverse 1 stage Single row planetary gear and Simpson planetary gear type	Forward 4 stages Reverse 1 stage Single row planetary gear and Simpson planetary gear type	Forward 4 stages Reverse 1 stage Single row planetary gear and Simpson planetary gear type

GENERAL SPECIFICATIONS

Items	AW372	KM148	AW372L
Control element			
Clutch	Multi disc type, 3 sets	Multi disc type, 3 sets	Mutli disc type, 3 sets
Brake	Multi disc type, 4 sets	Multi disc type, 4 sets	Multi disc type, 4 sets
One-way clutch	Sprag type, 3 sets	Sprag type, 3 sets	Sprag type, 3 sets
Gearbox ratio			
1st	2.826	2.826	2.828
2nd	1.493	1.493	1.493
3rd	1.000	1.000	1.000
4th	0.688	0.688	0.688
Reverse	2.703	2.703	2.703
Shift control method	Column shift type	Column shift type	Floor shift type
Select pattern	PRND2L and overdrive switch	PRND2L and overdrive switch	PRND2L and overdrive switch
Oil pump type	Gear type	Gear type	Gear type
Driving method	Directly connected to engine via torque converter	Directly connected to engine via torque converter	Directly connected to engine via torque converter
Hydraulic control method	Detection of throttle opening and vehicle speed	Detection of throttle opening and vehicle speed	Detection of throttle opening and vehicle speed
Oil cooling method	Water cooling type	Air and water cooling type (dual system)	Water cooling type
Transfer type	—	Constant mesh type	—
Shift control method	—	Single lever, floor shift type	—
Speed change ratio			
low	—	1.944	—
high	—	1.000	—
Speed meter gear ratio	—	22/8	—

COMPONENT SPECIFICATIONS

Items		Standard in.	Standard mm	Limit in.	Limit mm
Transmission control	Sleeve and selector lever end dimension	0.559–0.587	14.2–14.9	—	—
Oil pump	Side clearance	0.0008–0.0020	0.02–0.50	0.004	0.1
	Body clearance	0.0028–0.0058	0.07–0.15	0.012	0.3
	Tip clearance (driven gear)	0.0043–0.0055	0.11–0.14	0.012	0.3
Clutch and brake piston stroke	Overdrive clutch (C^0)	0.0614–0.0996	1.56–2.53	—	—
	Forward clutch (C^1)	0.0563–0.1154	1.43–2.93	—	—
	Direct clutch (C^2)	0.0358–0.783	0.91–1.99	—	—
	No. 1 brake (B^1)	0.0135–0.0681	0.80–1.73	—	—
	No. 2 brake (B^2)	0.0398–0.0886	1.01–2.25	—	—
Brake clearance	Overdrive brake (B^0)	0.0256–0.0870	0.65–2.21	—	—
	No. 3 brake (B^3)	0.0240–0.1309	0.61–2.64	—	—
Bushing bore	Stator support—front	0.8465–0.8475	21.501–21.527	0.8495	21.577
	Stator support—rear	0.9065–0.9075	23.025–23.501	0.9095	23.138
	Oil pump body	1.5005–1.5015	38.113–38.138	1.5035	38.188

COMPONENT SPECIFICATIONS

Items		Standard in.	Standard mm	Limit in.	Limit mm
Bushing bore	Overdrive sun gear (front and rear)	0.9080–0.9090	23.062–23.088	0.9109	23.138
	Overdrive input shaft	0.4409–0.4418	11.200–11.221	0.4437	11.271
	Sun gear (front and rear)	0.8465–0.8475	21.501–21.527	0.8495	21.577
	Center support	1.4325–1.4335	36.386–36.411	1.4355	36.461
	Transmission case	1.5005–1.5015	38.113–38.138	1.5035	38.188
	Output shaft	0.7087–0.7097	18.001–18.026	0.7117	18.076
	Extension housing	1.5605–1.5615	39.636–39.661	1.5634	39.711
Thermo switch	Continuity temperature	122°F	50°C	—	—
	Input shaft endplay	0.012–0.035	0.3–0.9	—	—
	Overdrive brake clearance	0.014–0.062	0.35–1.6	—	—
Overdrive clutch return spring	Free length	0.587	14.9	—	
Overdrive brake return spring	Free length	0.634	16.1	—	—

VALVE BODY VALVE SPRING IDENTIFICATION

Valve Body	Springs	Outer Diameter in. (mm)	Free Length in. (mm)	Number of Turns	Wire Diameter in. (mm)	Identification color
Upper front	Spring for throttle valve	0.338 (8.58)	0.757 (19.24)	8	0.028 (0.71)	—
	Spring for kickdown valve	0.428 (10.87)	1.710 (43.44)	15.5	0.047 (1.20)	Orange
	Spring for secondary regulator valve	0.686 (17.43)	2.806 (71.21)	15	0.076 (1.93)	Green
Upper rear	Spring for intermediate modulator valve	0.356 (9.04)	1.073 (27.26)	9.5	0.043 (1.10)	Green
	Spring for sequence valve	0.367 (9.32)	1.327 (33.72)	13	0.052 (1.32	Yellow
	Spring for low coast modulator valve	0.364 (9.24)	1.667 (42.35)	15	0.033 (0.84)	—
	Spring for 2–3 shift valve	0.353 (0.96)	1.382 (35.10)	12.5	0.030 (0.76)	White
	Spring for detent regulator valve	0.350 (8.90)	1.198 (30.43)	13	0.035 (0.90)	Green
Lower	Spring for 1–2 shift valve	0.298 (7.56)	1.363 (34.62)	13	0.022 (0.56)	—
	Spring for 3–4 shift valve	0.417 (10.60)	1.385 (35.18)	14.5	0.043 (1.10)	Green
	Spring for pressure relief valve	0.517 (13.14)	1.265 (32.14)	9	0.080 (2.03)	—
	Spring for check valve	0.544 (13.82)	1.312 (33.32)	7	0.052 (1.32)	Yellow
	Spring for primary regulator valve	0.677 (17.20)	2.409 (61.20)	13	0.071 (1.80)	White
	Spring for primary regulator valve damping	0.196 (4.97)	0.787 (20.00)	16	0.016 (0.40)	—

SPECIAL TOOLS

Tool (Number and name)	Use	Tool (Number and name)	Use
MD998218 Wrench 	Inspection of torque converter	MD998212 Oil pump puller 	Removal of oil pump
MD998219 Stopper 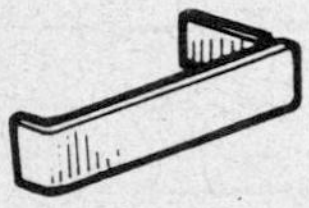	Inspection of torque converter	MD998412 Guide 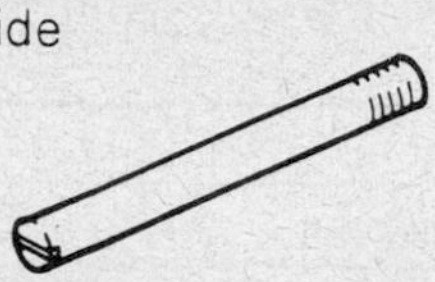	Installation of oil pump
MD999563 (includes MD998331) Oil pressure gage (1000 kPa) (142 psi) 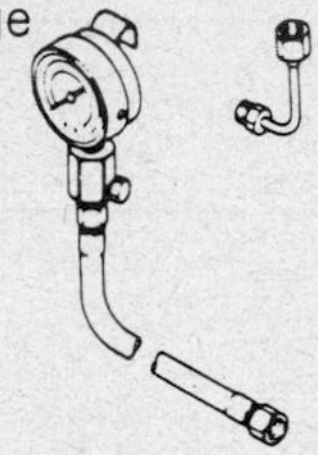	Measurement of oil pressure	MD998330 (includes MD998331) Oil pressure gage (3000 kPa) (427 psi) 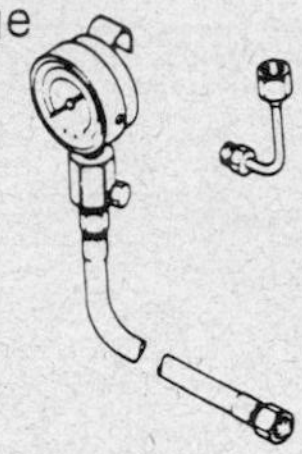	Measurement of oil pressure
MD998206 Adapter 	Connection of oil pressure gage	MD998903 Spring compressor 	Disassembly and assembly of clutch and brake
MD998335 Oil pump band 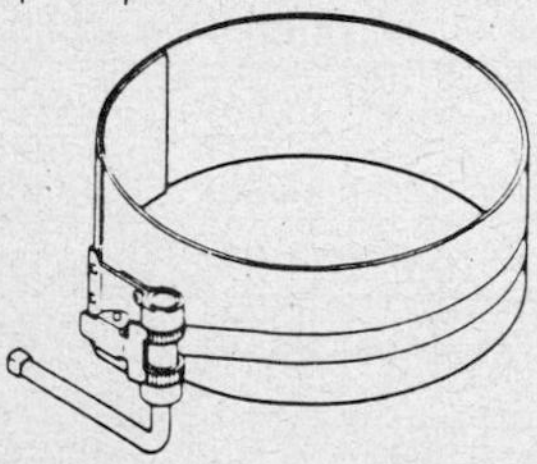	Assembly of oil pump	MD998353 Torque driver set 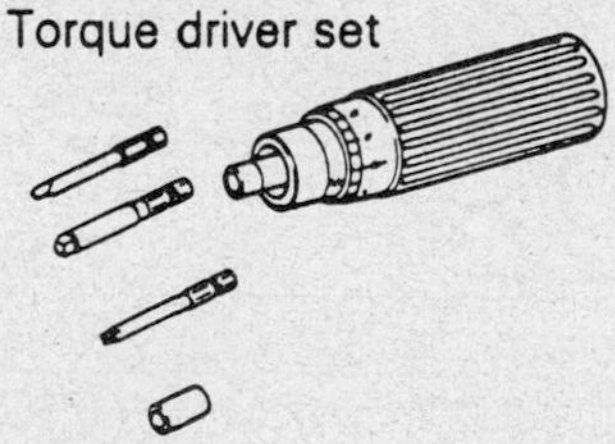	Tightening of valve body screw
MD998210 Bolt 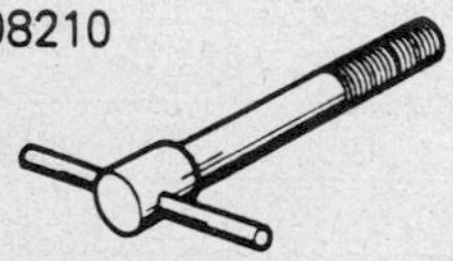	Disassembly and assembly of No.3 brake spring	MD998217 Gage 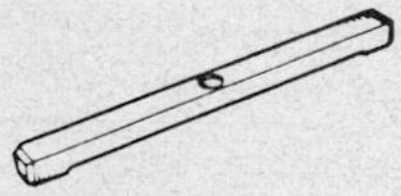	Check of quality of assembly condition
MD998211 Retainer 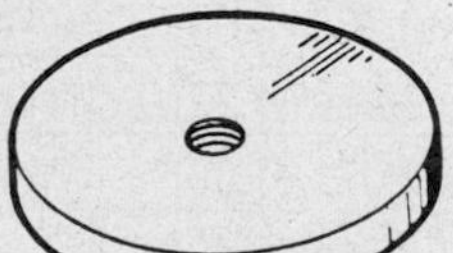	Disassembly and assembly of No.3 brake spring	DT-1001-A Steering wheel puller 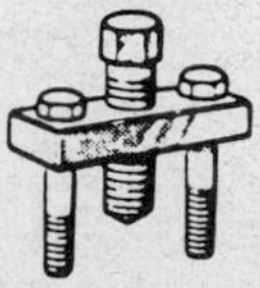	Removal of the steering wheel

Section 4

C3 Transmission
Ford Motor Co.

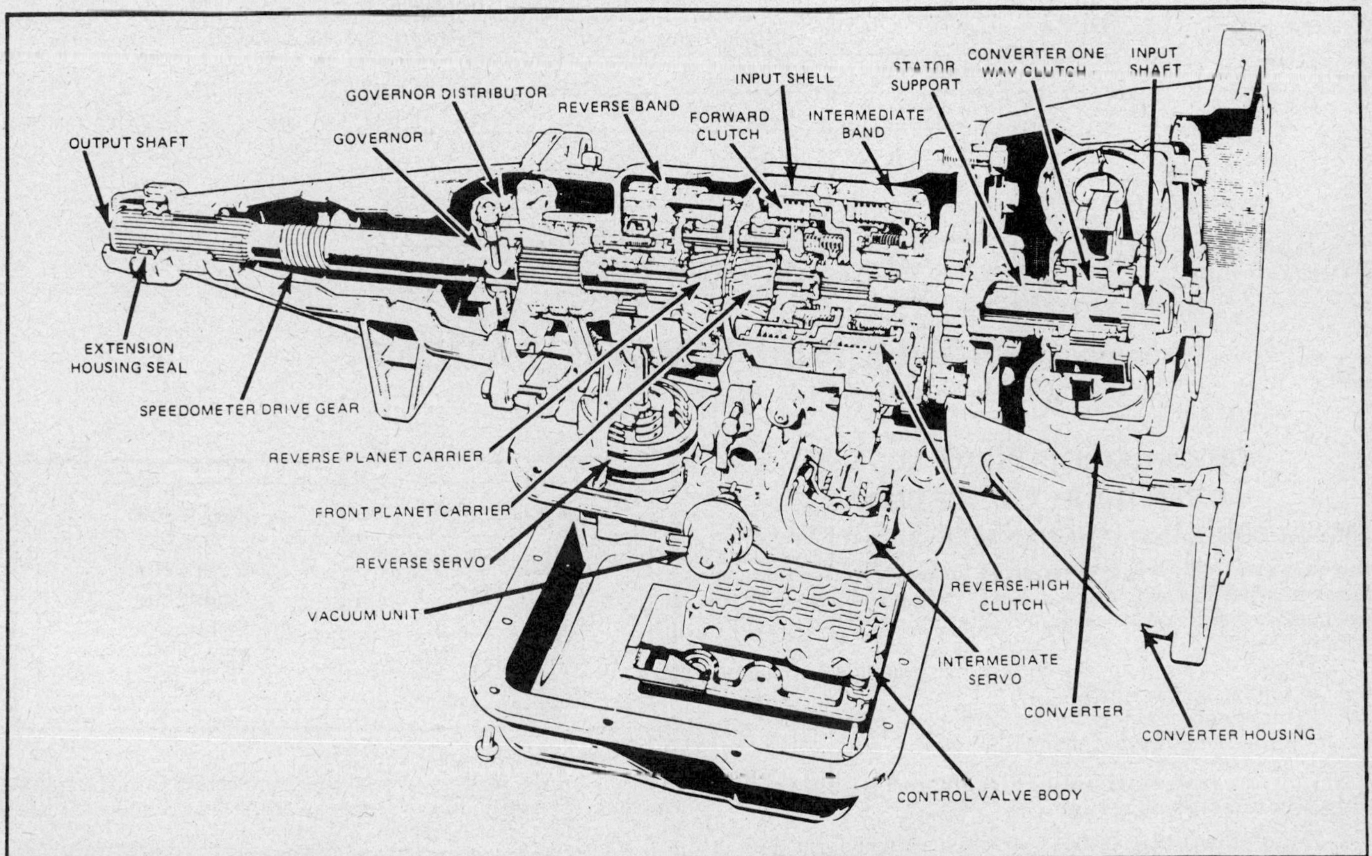

OUTPUT SHAFT
GOVERNOR DISTRIBUTOR
GOVERNOR
REVERSE BAND
INPUT SHELL
FORWARD CLUTCH
INTERMEDIATE BAND
STATOR SUPPORT
CONVERTER ONE WAY CLUTCH
INPUT SHAFT
EXTENSION HOUSING SEAL
SPEEDOMETER DRIVE GEAR
REVERSE PLANET CARRIER
FRONT PLANET CARRIER
REVERSE SERVO
VACUUM UNIT
REVERSE-HIGH CLUTCH
INTERMEDIATE SERVO
CONVERTER
CONVERTER HOUSING
CONTROL VALVE BODY

Cross section of the C3 transmission

REFERENCES

APPLICATION

1980–84 Chilton's Professional Automatic Transmission Manual #7390
1984–89 Chilton's Professional Transmission Manual—Domestic Vehicles #7959

GENERAL DESCRIPTION

1980–84 Chilton's Professional Automatic Transmission Manual #7390

MODIFICATIONS

1980–84 Chilton's Professional Automatic Transmission Manual #7390
1984–89 Chilton's Professional Transmission Manual—Domestic Vehicles #7959

TROUBLE DIAGNOSIS

1980–84 Chilton's Professional Automatic Transmission Manual #7390

ON CAR SERVICES

1980–84 Chilton's Professional Automatic Transmission Manual #7390

REMOVAL AND INSTALLATION

1980–84 Chilton's Professional Automatic Transmission Manual #7390
1984–89 Chilton's Professional Transmission Manual—Domestic Vehicles #7959

BENCH OVERHAUL

1980–84 Chilton's Professional Automatic Transmission Manual #7390

SPECIFICATIONS

1980–84 Chilton's Professional Automatic Transmission Manual #7390
1984–89 Chilton's Professional Transmission Manual—Domestic Vehicles #7959

SPECIAL TOOLS

1980–84 Chilton's Professional Automatic Transmission Manual #7390

APPLICATION

C-3

Year	Vehicle	Engine
1984–86	Mustang/Capri	2.3L OHC
1984–86	LTD/Marquis	2.3L OHC
1984–86	Thunderbird/Cougar	2.3L Turbo
1984	Ranger 4×2 Pick-up	2.3L
1986	Merkur	2.3L Turbo
1987–89	Merkur XR4Ti	2.3L Turbo

TRANSMISSION MODIFICATIONS

Transmission Fluid Leak at Cooler Line Connections

1985–87 FORD AND LINCOLN/MERCURY

In the event that service is required for a disconnected transmission cooler line, the quick-connect fitting should not be reused because it may not reseal properly and can result in repeat repair.

To correct the problem, the cooler lines must be modified using the following procedure:

1. Remove and throw away the quick-connect fitting connector from the radiator or transmission case.

NOTE: Rework the connectors in sets, do not rework only 1 connector at a time.

2. Cut off the tubing as close as possible to the formed bead, using a tubing cutter.
3. Slide a $^5/_{16}$ × ½–20 thread tube nut on the tube.

Part Number	Part Name
EQUZ-7D273-A	Connector
E2TZ-7D273-A	Connector
E7TZ-7D273-B	Connector
E1FZ-7D273-A	Connector
87944-S8	Nut

4. Make a double flare tube end.
5. Install the proper size and type connectors into the transmission case or radiator. Use thread sealer on the male threads of the connector.
6. Install thread tube nuts onto connectors. Torque to 12–18 ft. lbs.

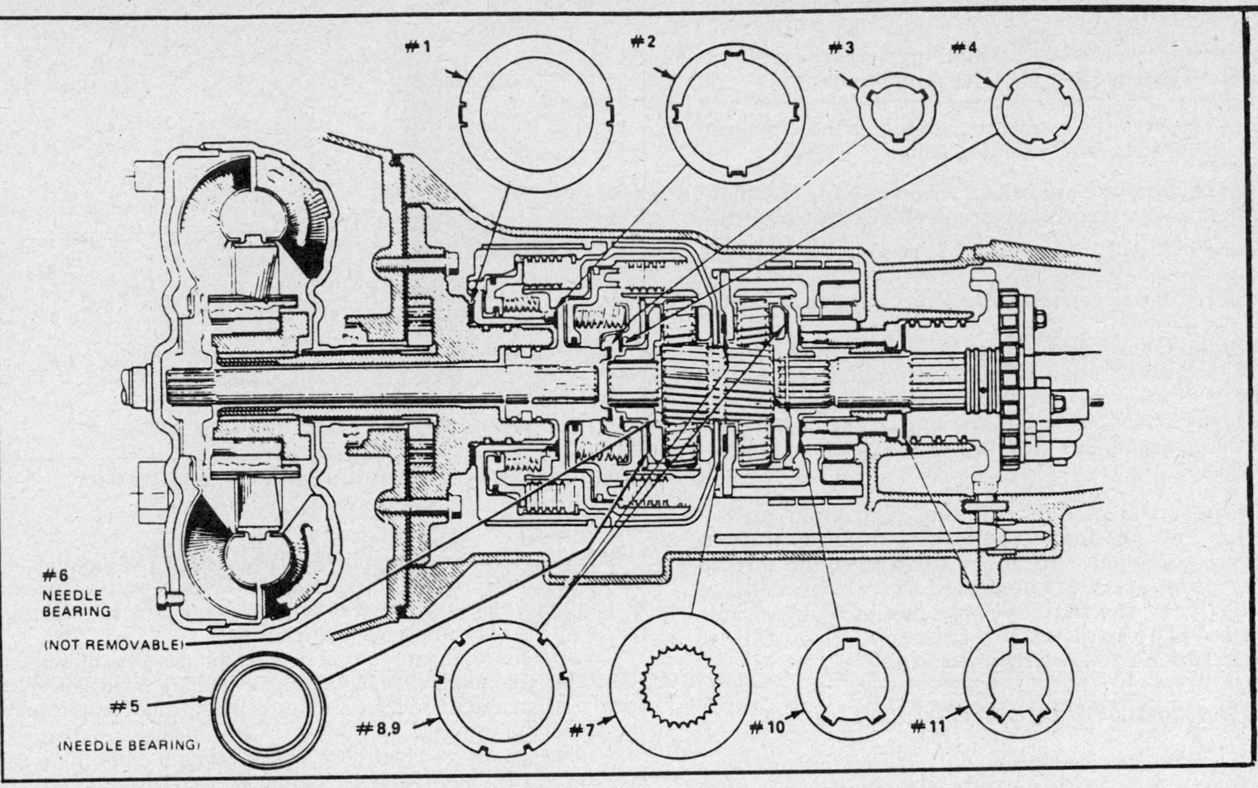
#1
#2
#3
#4
#6
NEEDLE
BEARING
(NOT REMOVABLE)
#5
(NEEDLE BEARING)
#8,9
#7
#10
#11

REMOVAL AND INSTALLATION

TRANSMISSION REMOVAL

1. Disconnect the battery ground cable.
2. Raise the vehicle and support safely. Position a drain pan under the transmission oil pan. Starting at the rear of the pan, loosen, but do not remove the pan bolts, allowing the fluid to drain gradually. When the fluid has stopped draining, remove all the pan bolts except 2 at the front of the oil pan and allow the fluid to drain from the pan. After all the fluid has drained, install 2 pan bolts loosely at the rear of the pan to hold the pan in place.
3. Remove the torque converter access cover and adapter plate bolts from the lower end of the converter housing.
4. Remove the 3 converter-to-flywheel retaining bolts, Rotate the engine in the normal direction of rotation to gain access to the bolts, by using a wrench on the crankshaft pulley bolt.

NOTE: On belt-driven overhead cam engines, do not turn the engine backwards as this may cause the camshaft drive belt to slip, throwing the engine valve timing out of adjustment and causing mechanical damage to the engine.

5. Rotate the engine until the converter drain plug is accessible. Place a drain pan under the plug area and remove the plug from the converter. When all the fluid has drained from the converter, replace the drain plug.
6. Matchmark and remove the driveshaft. Plug the seal end of the rear extension housing to prevent fluid leakage.
7. Disconnect the speedometer cable from the rear extension housing.
8. Disconnect the shift rod at the transmission manual lever. Disconnect the downshift rod at the transmission downshift lever.

NOTE: On Merkur models disconnect the oxygen sensor harness and unclip the wires from the clip. Remove the stabilizer bar U-brackets and the body stiffener rod, then position a block of wood between the stabilizer bar and the body side rail.

9. Remove the bolts holding the starter to the converter housing and position the starter out of the way.
10. Disconnect the neutral start switch wires at the switch.
11. Disconnect the vacuum line at the transmission vacuum diaphragm.
12. Place a transmission jack under the transmission and raise it slightly.
13. Remove the nuts retaining the engine rear support to the crossmember. Remove the crossmember side support bolts to the frame and remove the crossmember.

NOTE: On 4 cylinder turbocharged engines remove the nuts attaching the catalytic converter inlet pipe to the turbocharger. Remove the support bracket attaching bolt and then the catalytic converter and inlet pipe as an assembly.

14. Disconnect the exhaust pipe from its supports and disconnect the pipe flange at the exhaust manifold.

15. Lower the transmission jack and allow the transmission to hand free.

CAUTION

For safety reasons, keep the jack under the transmission.

16. Raise the front of the engine to gain access to the 2 upper converter housing-to-engine retaining bolts.

NOTE: Care must me exercised when raising the front of the engine, not to stretch or break the engine mounts.

17. Disconnect and plug the 2 oil cooler lines at the transmission.
18. Remove the lower converter housing-to-engine retaining bolts.
19. Remove the transmission filler tube and dipstick.
20. Secure the transmission to the transmission jack with a safety chain.
21. Remove the 2 upper converter housing-to-engine retaining bolts. Pull the transmission assembly to the rear and lower it from the vehicle.

NOTE: Unless the vehicle remains stationary on the lifting device, do not leave the engine support jack in place. Severe damages can occur to the engine and to the engine compartment components, if the vehicle should drop. Keep the transmission assembly level during the removal to prevent the converter from falling out. Lock in place with a wire or a bracket.

TRANSMISSION INSTALLATION

1. Position the converter to the transmission, making sure that the converter hub is fully engaged in the pump gear. With the converter properly installed, the distance between the converter hub to the front of the converter housing should be ⅜ in. (dimension "A"). Block the converter in place to maintain the required distance until installation.
2. Place the transmission on a transmission jack and secure with a safety chain.
3. Rotate the converter so that the bolt drive studs and drain plug are in alignment with their corresponding holes in the flywheel.
4. Reinstall the jack at the front of the engine, if removed. Position the transmission assembly and transmission jack under the vehicle and raise into position. Remove the converter retaining wire or bracket. Keep the transmission assembly level to prevent the converter from disengaging from the pump gear. Move the transmission assembly forward to mate the converter squarely with the flywheel to avoid the converter pilot from binding in the engine crankshaft.
5. Install the 2 upper converter housing-to-engine retaining bolts and torque to specifications.
6. Remove the safety chain from the transmission.
7. Install the filler tube and dipstick into the transmission stub tube and secure to the engine.

NOTE: Replace the stub tube if loosened or dislodged.

8. Connect the oil cooler lines to the transmission.

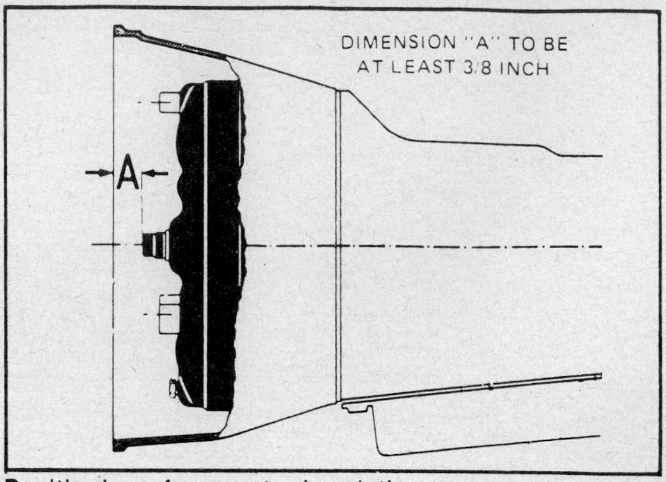
DIMENSION "A" TO BE
AT LEAST 3/8 INCH
A

Positioning of converter in relation to converter housing

9. Remove the jack supporting the front of the engine.
10. Position the exhaust pipe support bracket to the converter housing and install the lower converter housing to engine retaining bolts. Torque to specifications.
11. Raise the transmission and position the crossmember to the frame side supports. Install the attaching bolts, lower the transmission and install the rear support nuts and bolt. Remove the transmission jack.
12. Connect the vacuum hose to the vacuum diaphragm unit and connect the neutral start switch wires on the switch.
13. Install the starter motor and tighten the retaining bolts.
14. To tighten the converter to the flywheel, install one of the flywheel-to-converter bolts through the pilot hole and torque to specifications. Install the other 2 bolts and torque them to the same specification as the first bolt. This method insures proper alignment of the flywheel holes and the mating of the converter weld nuts.
15. Install the converter drain plug access cover and adapter plate bolts.
16. Connect the exhaust pipe to the exhaust manifold, using a new gasket as required.
17. Connect the transmission shift rod to the manual control lever.
18. Connect the downshift rod to the downshift lever.
19. Connect the speedometer to the rear extension housing.
20. Install the driveshaft to its original position.
21. Adjust the manual and downshift linkage as required.
22. Lower the vehicle and connect the battery ground cable.
23. Install 5 quarts of the specified transmission fluid and start the engine. Gradually add 3 more quarts of the specified fluid to the transmission.
24. Check that level of fluid is maintained exactly at the full mark with the engine idling in **P** or **N** and on a level surface.
25. Inspect the transmission, converter assembly and oil cooler lines for leakage of fluid. Correct as necessary.
26. Roadtest vehicle for proper operation.

SPECIFICATIONS

LINE PRESSURE

Year	Transmission Model	Range	Idle Manifold Vacuum 16 in. Hg and Above	15 in. Hg and Above	10 in.	WOT Stall Thru Detent ③
1984	83DT-AAB/AAB/ACB/ADB	D, 2, 1 ①	—	50-60	74–94	165–195
		R	—	66–78	129–148	278–316
		P, N	—	50–60	—	—
	83DT-ALB/AMB/AGB/AHB	D, 2, 1 ①	—	50–70	90–113	167–195
		R	—	75–109	158–178	282-316
		P,N	—	50–70	—	—
	83DT-ACB/ADB	D, 2, 1 ②	—	50-60	50–75	144–177
		R	—	66–78	94–116	244–282
		P, N	—	50-60	—	—
	83DT-AMB/AHB	D, 2, 1 ②	—	50–60	70–93	144–177
		R	—	66-78	122–145	247-282
		P, N	—	50-60	—	—
1985	85DT-AAA/BAA/DAA/JAA	D, 2, 1 ①	—	50-60	74–94	165–195
		R	—	66–78	129–148	278–316
		P, N	—	50–60	—	—
	85DT-DAA/JAA	D, 2, 1 ②	—	50–60	53–75	144–177
		R	—	66–78	94–116	244–282
		P, N	—	50–60	—	—
	85DT-KAA/LAA	D, 2, 1 ①	51-61	—	70–90	180–222
		R	67-82	—	123–143	302–330
		P, N	51-61	—	—	—
	85DT-LAA	D, 2, 1 ②	51-61	—	62–82	171–195
		R	67-82	—	109–129	292–311
		P, N	51-61	—	—	—
1986	86DT-AAA/ABA/ACA/ADA	D, 2, 1 ①	—	50-60	74–94	165–195
		R	—	66–78	129–148	278–316
		P, N	—	50–60	—	—
	86DT-ACA/ADA	D, 2, 1 ②	—	50–60	53–75	144–177
		R	—	66–78	94–116	244–282
		P, N	—	50–60	—	—
	86DT-AEA/AFA/AGA/AHA	D, 2, 1 ①	51-61	—	70-90	180–222
		R	67-82	—	123–143	302–330
		P, N	51-61	—	—	—
	86DT-AFA/AHA	D, 2, 1 ②	51-61	—	62-82	171–195
		R	67-82	—	109–129	292–311
		P, N	51-61	—	—	—

NOTE: The use of vacuum bleed valve to regulate manifold vacuum is necessary
① Absolute barometric pressure (ABP) 29.0–30.0
② Absolute barometric pressure (ABP) 24.0–25.0
③ Wide open throttle

STALL SPEED SPECIFICATIONS

Year	Vehicle	Engine	Converter Size Inches	Converter ID	Stall Speed (rpm) Min.	Stall Speed (rpm) Max.
1984	Mustang/Capri LTD/Marquis	2.3L	10¼	HA	2256	2654
	Thunderbird/Cougar	2.3L Turbo	10¼	HA	2963	3439
	Ranger 4×2	2.3L	10¼	HA	2381	2760
1985	Mustang/Capri LTD/Marquis	2.3L	10¼	HA	2256	2654
	Thunderbird/Cougar	2.3LT Turbo	10¼	HA	2963	3439
1986	Mustang/Capri LTD/Marquis	2.3L OHC	10¼	—	2442	2827
	Thunderbird/Cougar	2.3L Turbo	10¼	—	2701	3222
	Merkur	2.3L Turbo	10¼	—	3000	3300
1987–89	Merkur XR4Ti	2.3L Turbo	10¼	—	3000	3300

SELECTIVE THRUST WASHERS

Year	Endplay	Description	Part Numbers	Thickness (mm)	Identification Number
1984	.001–.025 (Less Gasket)	No. 1 Thrust Washer Front Pump Support (Selective)	74DT-7D014-EA	0.1091–0.1110	5
			76DT-7D014-DA	0.0929–0.0949	4
			76DT-7D014-GA	0.0768–0.0787	3
			76DT-7D014-BA	0.0610–0.0630	2
			76DT-7D014-AA	0.0488–0.0507	1
1985	.001–.025 (Less Gasket)	No. 1 Thrust Washer Front Pump Support (Selective)	74DT-7D014-EA	0.1091–0.1110	5
			74DT-7D014-DA	0.0929–0.0949	4
			74DT-7D014-CA	0.0768–0.0787	3
			74DT-7D014-BA	0.0610–0.0630	2
			74DT-7D014-AA	0.0488–0.0507	1
1986	.001–.025 (Less Gasket)	No. 1 Thrust Washer Front Pump Support (Selective)	84DT-7D014-AA	1.35–1.40(.053)	1
			84DT-7D014-BA	1.55–1.60(.060)	2
			84DT-7D014-CA	1.75–1.80(.068)	3
			84DT-7D014-DA	1.95–2.00(.076)	4
			84DT-7D014-EA	2.15–2.20(.084)	5
			84DT-7D014-FA	2.35–2.40(.092)	6

TORQUE SPECIFICATIONS

Item	ft. lbs.
Converter housing to case	27–39
Extension housing to case	27–39
Oil pump to converter housing	7–10
Flywheel to converter housing	27–49
Main control to case	7–9
Plate to valve body	7–9
Servo cover to case	7–10
Oil pan to case	12–17
Governor to collector body	7–10
Converter housing to engine	28–38

Item	ft. lbs.
Nut-downshift lever-outer	7–11
Nut-manual lever-inner	30–40
Neutral switch to case	7–10
Front band adjusting locknut	35–45
Vacuum diaphragm retaining clip to case	80–106 ①
Oil cooler line or by-pass tube to connector	7–10
Connector to case	10–15
Drain plug-converter	20–30
Flywheel to crankshaft	48–53
Filler tube to engine clip	28–38

① In.-Lbs.

Section 4

C5 Transmission
Ford Motor Co.

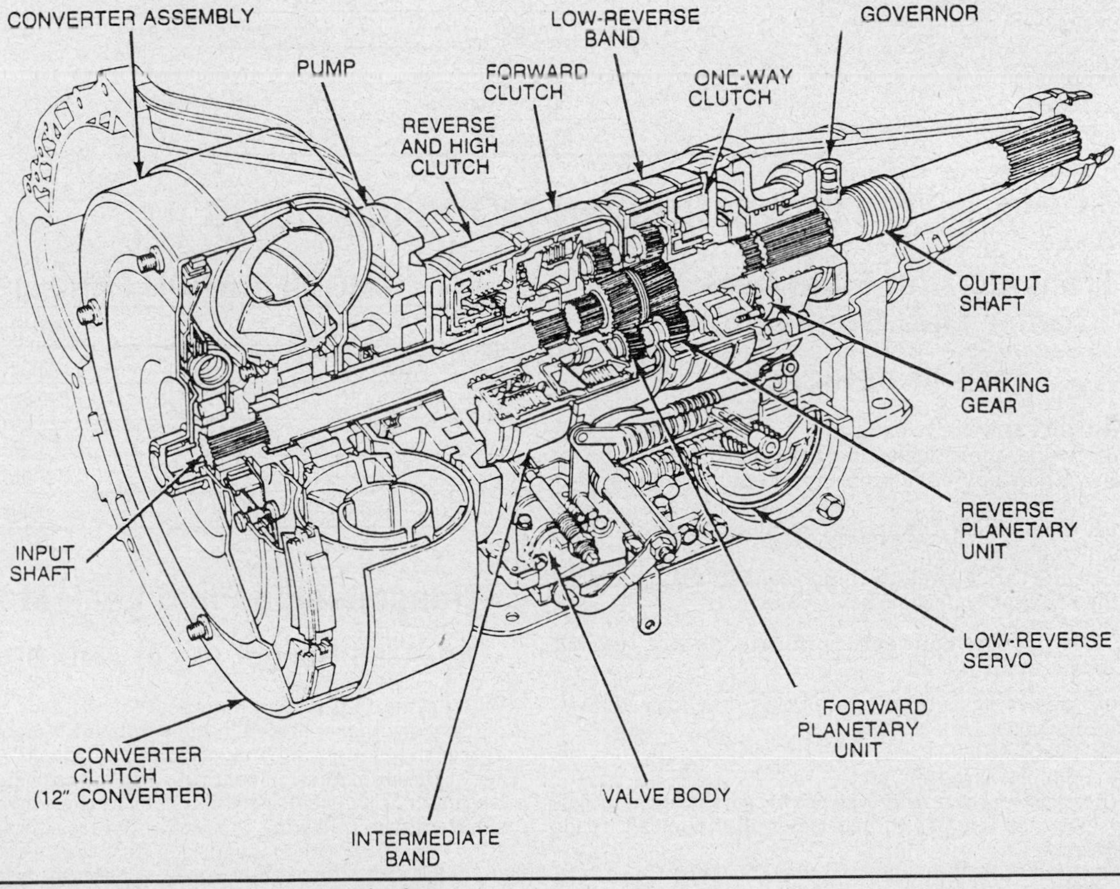

CONVERTER ASSEMBLY
LOW-REVERSE BAND
GOVERNOR
PUMP
FORWARD CLUTCH
ONE-WAY CLUTCH
REVERSE AND HIGH CLUTCH
OUTPUT SHAFT
PARKING GEAR
REVERSE PLANETARY UNIT
INPUT SHAFT
LOW-REVERSE SERVO
FORWARD PLANETARY UNIT
CONVERTER CLUTCH (12″ CONVERTER)
VALVE BODY
INTERMEDIATE BAND

Cut away view of the C5 automatic transmission

REFERENCES

APPLICATION

1980–84 Chilton's Professional Automatic Transmission Manual #7390
1984–89 Chilton's Professional Transmission Manual—Domestic Vehicles #7959

GENERAL DESCRIPTION

1980–84 Chilton's Professional Automatic Transmission Manual #7390

MODIFICATIONS

1980–84 Chilton's Professional Automatic Transmission Manual #7390
1984–89 Chilton's Professional Transmission Manual—Domestic Vehicles #7959

TROUBLE DIAGNOSIS

1980–84 Chilton's Professional Automatic Transmission Manual #7390

ON CAR SERVICES

1980–84 Chilton's Professional Automatic Transmission Manual #7390

REMOVAL AND INSTALLATION

1980–84 Chilton's Professional Automatic Transmission Manual #7390
1984–89 Chilton's Professional Transmission Manual—Domestic Vehicles #7959

BENCH OVERHAUL

1980–84 Chilton's Professional Automatic Transmission Manual #7390

SPECIFICATIONS

1980–84 Chilton's Professional Automatic Transmission Manual #7390
1984–89 Chilton's Professional Transmission Manual—Domestic Vehicles #7959

SPECIAL TOOLS

1980–84 Chilton's Professional Automatic Transmission Manual #7390

APPLICATION

C–5

Year	Vehicle	Engine
1984–87	LTD/Marquis	3.8L
	Cougar	3.8L
	Thunderbird	3.8L
	F150 Pick-up	4.9L
	Bronco II	3.8L

TRANSMISSION MODIFICATIONS

Transmission Fluid Leak at Cooler Line Connections

1985–87 VEHICLES

In the event that service is required for a disconnected transmission cooler line, the quick-connect fitting should not be reused because it may not reseal properly and can result in repeat repair.

To correct the problem, the cooler lines must be modified using the following procedure:

1. Remove and throw away the quick-connect fitting connector from the radiator or transmission case.

NOTE: Rework the connectors in sets, do not rework only 1 connector at a time.

2. Cut off the tubing as close as possible to the formed bead, using a tubing cutter.
3. Slide a $^5/_{16}$ × ½–20 thread tube nut on the tube.
4. Make a double flare tube end.
5. Install the proper size and type connectors into the transmission case or radiator. Use thread sealer on the male threads of the connector.
6. Install thread tube nuts onto connectors. Torque to 12–18 ft. lbs.

MODIFICATION PART LIST

Part Number	Part Name
EQUZ–7D273–A	Connector
E2TZ–7D273–A	Connector
E7TZ–7D273–B	Connector
E1FZ–7D273–B	Connector
87944–S8	Nut

Transmission Fluid Leak at the Vent

1984–85 Ranger (4×4) AND BRONCO II

When some C–5 transmissions are overfilled, subjected to extending driving, operated in high ambient temperatures or used in trailer towing, they may experience transmission fluid leakage at the transmission vent tube rubber hose due to swelling or looseness. If the vehicle exhibits any of the above and leakage is apparent, the following service procedure should be followed:

1. Verify that the transmission fluid leakage is originating at the left hand rear of the transmission case, in the vent fitting area.

2. Check the transmission vent tube connecting hose to the vent fitting for swelling, looseness of any evidence of transmission fluid on the hose.

3. In those instances where the hose exhibits swelling, looseness, or is contaminated with transmission fluid, remove and discard the vent hose. If during servicing of the transmission for other repairs it is evident that the vent tube hose has been contaminated with transmission fluid, the hose also should be replaced.

4. Replace the affected hose with a modified vent tube hose part No. E3TZ–7A246–F. Cut the hose to 2.5 in.

REMOVAL AND INSTALLATION

TRANSMISSION REMOVAL

EXCEPT TRUCKS

1. Protect fender areas and disconnect the battery cables.
2. Remove the air cleaner assembly on vehicles equipped with the 3.8L engine.
3. After removing the attaching bolts, position the fan shroud back over the fan assembly.
4. Disconnect the thermactor air injection hose at the catalytic converter check valve, were applicable.

NOTE: The check valve is located on the right side of the engine compartment, near the firewall.

5. Remove the 2 upper transmission to engine attaching bolts, accessible from the engine compartment, on vehicles equipped with the 3.8L engine.
6. Raise the vehicle and support safely.
7. Matchmark and remove the driveshaft.
8. While supporting the exhaust system, disconnect the muffler inlet pipe from the catalytic converter outlet pipe. Wire the exhaust system assembly to the vehicle's undercarriage.
9. Remove the exhaust pipe(s) from the manifold(s) and by pulling back on the converters, release the converter hangers from their mounting brackets. Lower the pipe assemblies and set aside.
10. Remove the speedometer driven gear from the extension housing.
11. Disconnect the neutral safety switch wiring from the neutral safety switch. Remove vacuum hose from modulator nipple.
12. Disconnect the downshift rod at the transmission manual lever. On floor mounted shift equipped vehicles, remove the shift cable routing bracket and disconnect the cable from the transmission manual lever.
13. Remove the converter housing bust shield and remove the converter to drive plate attaching nuts.

NOTE: The crankshaft must be turned to gain access to all the converter to drive plate attaching nuts.

14. Remove starter cable, remove starter attaching bolts and lower starter from the engine.
15. Loosen the attaching nuts from the rear support to the No. 3 crosmember.
16. Position a transmission jack under the transmission and secure the transmission to the jack with a safety chain.
17. Remove the through bolts securing the No. 3 crossmember to the body brackets.
18. Lower the transmission enough to gain working room and disconnect the oil cooler lines.
19. Remove the remaining transmission to engine attaching bolts. On all other models, remove the remaining bolts.
20. Pull the transmission rearward to clear the converter studs from the drive plate and lower the transmission.

TRUCKS

1. Disconnect the battery cables. Raise the vehicle and support safely.
2. Place a drain pan under the transmission oil pan, loosen the oil pan bolts and allow the fluid to drain into the container. Carefully, lower the oil pan and allow the remainder of the fluid to drain from the pan. Reinstall the pan with a few bolts to hold the pan in place.

NOTE: Certain oil pans will have the filler tube installed. Remove to drain.

3. Remove the converter drain plug access cover from the bottom of the converter housing.
4. Remove the converter to flywheel attaching nuts. Turn the engine crankshaft to locate each nut.
5. Turn the converter to place the converter drain plug in the bottom position. Place a drain pan under the converter and remove the drain plug. Reinstall the plug after the converter has drained.
6. Matchmark the driveshaft and disconnect from the rear yoke. Pull the shaft from the rear of the transmission.
7. Disconnect the starter cable and remove the starter from the engine.
8. Disconnect the neutral safety switch wire at the connector.
9. Position a transmission jack under the transmission and remove the rear mount-to-crossmember insulator attaching nuts and the crossmember-to-frame attaching bolts. Remove the right and left crossmember gussets.
10. Remove the rear insulator-to-extension housing attaching bolts.
11. Disconnect the downshift and selector linkage from the transmission manual lever.
12. Remove the bellcrank bracket from the converter housing.
13. Raise the transmission assembly enough to gain clearance fro removal of the crossmember. Remove the rear mount from the crossmember and remove the crossmember from the side supports.
14. Lower the transmission to gain access to the oil cooler lines and disconnect.
15. Disconnect the speedometer cable and remove the speedometer driven gear from the rear extension housing.
16. Remove the bolt holding the transmission filler tube to the engine block.
17. Be sure the safety chain is securing the transmission to the transmission jack. Remove the converter housing-to-engine bolts.
18. Pull the transmission to the rear while lowering the unit. Remove the transmission from under the vehicle.

TRANSMISSION INSTALLATION

EXCEPT TRUCKS

1. With the vehicle in the air and supported safely, raise the transmission on a transmission jack and into position to mate the converter studs with the drive plate hoses and the transmission dowels on the rear of the engine to the transmission bell housing.

NOTE: It will be necessary to rotate the converter to align the converter studs and the converter drain plug to the holes in the drive plate. Be sure during the installation that the converter studs are in the drive plate holes before bolting the transmission bell housing to the engine.

2. Install the transmission to engine bolts. Tighten the attaching bolts to 40–50 ft. lbs.
3. Connect the cooler lines to the transmission.
4. Raise the transmission by the jack mechanism and install the No. 3 crossmember through bolts. Install the attaching nuts and tighten to 20–30 ft. lbs.
5. Remove the safety chain and the transmission jack from under the vehicle.
6. Tighten the rear support attaching bolt nuts to 30–50 ft. lbs.
7. Install the starter assembly, tighten the bolts to 15–20 ft. lbs., and install the starter cable.
8. Install the converter to drive plate attaching nuts and torque to 20–30 ft. lbs. Install the dust shield and if previously remove, position the linkage bellcrank bracket and install the attaching bolts.
9. Connect the shift linkage to the transmission manual lever. If equipped with floor mounted shift, connect the cable to the manual lever and install the routing bracket with the attaching bolt.
10. Connect the downshift rod to the transmission lever.
11. Connect the neutral safety switch wiring harness to the switch. Install vacuum hose to modulator nipple.
12. Install the speedometer driven gear assembly in the rear extension housing. Tighten the clamp bolt to 36–54 inch lbs.
13. Install the catalytic converters into their hanger brackets and install the exhaust pipe(s) to the exhaust manifold. Install the attaching nuts, but do not tighten.

NOTE: If the exhaust pipe(s) were disconnected from the converters, install new gaskets or seal before connecting.

14. Disconnect the wire holding the exhaust system to the body undercarriage and connect the pipe to the converter outlet. Do not tighten the attaching nuts.
15. Align the exhaust system and tighten the manifold and converter outlet attaching nuts.
16. Install the driveshaft in the vehicle, aligning the previously made matchmarks.
17. Check and if necessary, adjust the shift linkage.
18. Recheck undercarriage assembly and lower the vehicle.
19. Install the transmission to engine attaching bolts from the engine compartment.
20. Connect the thermactor air injection hose to the converter check valve.
21. Position the fan shroud and install the retaining bolts.
22. Install the air cleaner or vehicles equipped with the 3.8L engine.
23. Connect the battery cables.
24. Add the correct type and amount of fluid to the transmission assembly as required.
25. Start the engine, being sure the starter will only operate in **P** or **N** positions.
26. Check and correct the transmission fluid level.
27. Raise the vehicle and inspect for leakage and correctness of assembly. Lower and road test as required.

TRUCKS

1. Raise the vehicle and supported safely.
2. Position the transmission the transmission jack and secure with the safety chain.
3. Move the transmission and jack assembly under the vehicle and position the transmission in-line with the engine block, having rotated the converter to align the attaching bolt holes in the flywheel with the attaching bolts of the converter. Push the assembly forward, mating the dowel pins on the engine with the holes in the converter housing.
4. Install the converter housing-to-engine attaching bolts. Torque to 40–50 ft. lbs.

CAUTION

Be sure drain plug is in place, tightened and positioned properly in the flywheel.

5. Install the bolt holding the filler tube to the engine block.
6. Install the speedometer driven gear assembly and speedometer cable to the rear extension housing.
7. Raise or lower the transmission to connect the oil cooler lines to the transmission case.
8. Raise the transmission assembly enough to install the crossmember to the side supports. Install the rear support to the crossmember.
9. Install the downshift and selector linkage to the transmission manual levers, while installing the bellcrank bracket to the converter housing.
10. Install the rear insulator-to-extension housing attaching bolts.
11. Install the rear mount-to-crossmember insulator attaching nuts and install the left and right crossmember gussets. Remove the transmission jack.
12. Install the neutral safety switch wiring connector and properly route the wiring through the retaining clips.
13. Install the starter assembly to the engine and install the starter cable.
14. Install the driveshaft into the rear of the transmission and align the matchmarks at the yoke and secure.
15. Install the converter attaching nuts and tighten to 20–34 ft. lbs. Install the converter drain plug access cover.
16. Install the oil filler tube to the oil pan, if equipped. Otherwise, be sure a new gasket is on the oil pan and the retaining bolts are properly tightened.
17. Lower the vehicle, install the proper type and quantity of fluid into the transmission. Install the battery cables and start the engine. Recheck the fluid level and correct as required.
18. Raise the vehicle and check for leakage. Inspect the assembly for correct installation. Road test as required when lowered. Make any further adjustments.

SPECIFICATIONS

CLUTCH PLATES

	Forward Clutch			Reverse Clutch		
Model	External Spline (Steel)	Internal Spline (Composition)	Free Pack Clear	External Spline (Steel)	Internal Spline (Composition)	Free Pack Clear
Thunderbird/Cougar, LTD/Marquis, Mustang/Capri	4	5	0.025–0.050 in. 0.635–1.27mm	3	3	0.025–0.050 in. (0.635–1.27mm)
F150, Ranger①	4	5	0.025–0.050 in.	4	4	0.025–0.050 in.

① With 4.9L engine

TORQUE SPECIFICATIONS

Description	N•m	in. lbs.
End Plates To Valve Body	2.82-4.51	25-40
Separator Plate To Timing Valve Body	2.82-4.51	25-40
Lower Body To Upper Body (10-24)	4.51-6.77	40-60
Screen To Timing Valve Body	2.82-4.51	25-40
Governor To Governor Oil Collector Body	9.03-12.55	80-120
Pump Assembly To Case	2.25-3.95	20-38
Main Control To Case	9.03-13.55	80-120
Neutral Switch To Case	6.21-8.47	55-75
Upper Body To Lower Body (Long) (¼-20)	9.03-12.55	80-120
Upper Body To Lower Body (Short) (10-24)	4.51-6.77	40-60
3-2 Timing Valve Body To Upper Body (10-24)	4.51-6.77	40-60
3-2 Timing Valve Body To Lower Body (¼-20)	5.9-8.1	52-72
Detent Spring and Lower Body To Upper Body	4.51-6.77	40-60
Detent Spring and Main Control To Case	9.03-12.55	80-120
3-2 Timing Valve Body To Lower Body (10-24)	4.51-6.77	50-60
Speedometer Clamp Bolt	4-6	36-54
Overrunning Clutch Race To Case	18-27	13-20
Push Connector To Transmission Case	24-31	18-23
Oil Pan To Case	16-22	12-16

Description	N•m	in. lbs.
Stator Support To Pump	17-27	12-20
Converter to Flywheel	27-46	20-34
Converter Housing Cover To Converter Housing	17-21	12-16
Converter Housing To Case	38-55	28-40
Engine Rear Cover Plate To Transmission	17-21	12-16
Rear Servo Cover To Case	17-27	12-20
Intermediate Servo Cover To Case	22-30	16-22
Oil Distributor Sleeve To Case	16-27	12-20
Extension Housing To Case	38-54	28-40
Pump and Converter Housing To Case	38-51	28-38
Engine To Transmission (3.8L)	38-51	28-38
Transmission To Engine (3.3L, 4.2L)	55-67	40-50
Outer Throttle Lever To Shaft	17-21	12-16
Band Adjusting Screws To Case	13.5	10
Inner Manual Lever To Shaft	41-54	30-40
Pump Pressure Plug To Case	9-16	6-12
Intermediate Band and Reverse Band Adjusting Screw Locknut	47-61	35-45
Drain Plug To Converter Cover	20-24	15-18

VACUUM DIAPHRAGM ASSEMBLY SPECIFICATIONS

Diaphragm Type	Part No.	Identification	Throttle Valve Rod①② Length in.	Throttle Valve Rod①② Identification
S-HAD	D7DZ-7A377-A	1 white stripe	1.5925–1.5876 1.6075–1.6025	Green daub Blue daub
SAD	E2TZ-7A377-A	No identification	1.6225–1.6175 1.6375–1.6325	Orange daub Black daub
S-SAD	D5AZ-7A377-B	1 green stripe	1.6585–1.6535	Pink/white daub
HAD	D7BZ-7A377-B	Blue cover	—	—
SAD	D2AZ-7A377-A	2 yellow stripes	1.5925–1.5876 1.6075–1.6025 1.6225–1.6175 1.6375–1.6325 1.6585–1.6535	Green daub Blue daub Orange daub Black daub Pink/white daub

SAD Single Area Diaphragm
HAD High Altitude Diaphragm
S-SAD Super Single Area Diaphragm
S-HAD Super High Altitude Diaphragm
① Selective fit rods
② Part of diaphragm kit

FLUID CAPACITY

All capacities given in quarts

Year	Vehicle	Transmission	Fluid Type	Pan Capacity	Overhaul Capacity
1984–87	LTD/Marquis	C-5	Type H	8.8	11.0
	Mustang/Capri	C-5	Type H	8.8	11.0
	Thunderbird/Cougar	C-5	Type H	8.8	11.0
	F150 Truck	C-5	Type H	9.2	11.5
	Ranger	C-5	Type H	9.2	11.5

CONTROL PRESSURE SPECIFICATIONS

		Idle		
Transmission Model	Range①	15 in. Hg. & Above (psi)	10 in. Hg. (psi)	WOT② Stall thru Detent psi
PEP-Z, AC,	D	53–70	93–107	162–182
AD, AE, AL,	2, 1	107–121	100–112	162–182
AM, AN, AP	R	88–116	156–178	271–303
	P, N	53–70	93–107	162–182
PEP-AF	D	64–68	86–97	160–176
	2, 1	105–115	100–110	160–176
	R	73–95	144–162	267–294
	P, N	64–68	86–97	160–176

① At sea level baromteric pressure is 29.5
② Wide open throttle

Section 4

C6 Transmission
Ford Motor Co.

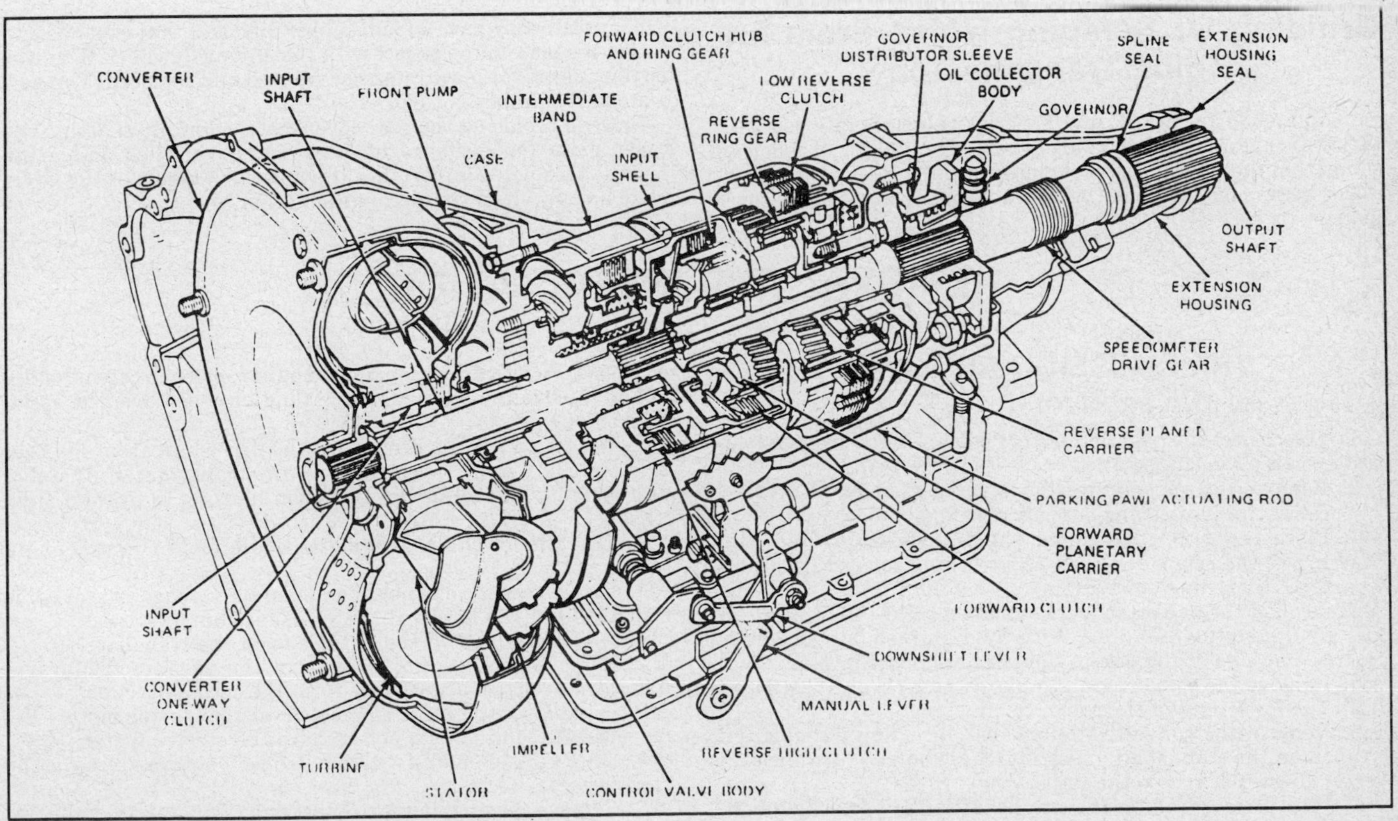

FORWARD CLUTCH HUB AND RING GEAR
GOVERNOR DISTRIBUTOR SLEEVE
SPLINE SEAL
EXTENSION HOUSING SEAL
CONVERTER
INPUT SHAFT
FRONT PUMP
INTERMEDIATE BAND
LOW REVERSE CLUTCH
OIL COLLECTOR BODY
GOVERNOR
REVERSE RING GEAR
CASE
INPUT SHELL
OUTPUT SHAFT
EXTENSION HOUSING
SPEEDOMETER DRIVE GEAR
REVERSE PLANET CARRIER
PARKING PAWL ACTUATING ROD
FORWARD PLANETARY CARRIER
FORWARD CLUTCH
INPUT SHAFT
DOWNSHIFT LEVER
CONVERTER ONE-WAY CLUTCH
MANUAL LEVER
IMPELLER
REVERSE HIGH CLUTCH
TURBINE
STATOR
CONTROL VALVE BODY

Cut away view of the C6 automatic transmission

REFERENCES

APPLICATION

1974–80 Chilton's Professional Automatic Transmission Manual #6927
1984–89 Chilton's Professional Transmission Manual – Domestic Vehicles #7959

GENERAL DESCRIPTION

1974–80 Chilton's Professional Automatic Transmission Manual #6927

MODIFICATIONS

1974–80 Chilton's Professional Automatic Transmission Manual #6927
1984–89 Chilton's Professional Transmission Manual – Domestic Vehicles #7959

TROUBLE DIAGNOSIS

1974–80 Chilton's Professional Automatic Transmission Manual #6927

ON CAR SERVICES

1974–80 Chilton's Professional Automatic Transmission Manual #6927

REMOVAL AND INSTALLATION

1974–80 Chilton's Professional Automatic Transmission Manual #6927
1984–89 Chilton's Professional Transmission Manual – Domestic Vehicles #7959

BENCH OVERHAUL

1974–80 Chilton's Professional Automatic Transmission Manual #6927

SPECIFICATIONS

1974–80 Chilton's Professional Automatic Transmission Manual #6927
1984–89 Chilton's Professional Transmission Manual – Domestic Vehicles #7959

SPECIAL TOOLS

1974–80 Chilton's Professional Automatic Transmission Manual #6927

APPLICATION

1984–86 Ford E and F Series and Bronco

TRANSMISSION MODIFICATIONS

Grinding Or Scraping Noise From Engine/Transmission

On some 1986 Bronco, E and F Series vehicles equipped with the 5.8L engine a loud grinding or scraping noise that seems to come from the transmission/engine area during high speed driving or heavy load load, may be coming from an engine rear cover plate.

This plate may have a foam rubber dust seal that expands and forces the plate into contact with the engine flywheel. If so, the engine rear cover plate must be removed and replaced. The new plate part number is C8DZ–7007–A

Prior to replacing the rear cover plate, carefully examine the cover place for evidence of loose metal particles and grind marks. This will confirm that the noise is a result of the cover plate and flywheel contacting each other.

REMOVAL AND INSTALLATION

TRANSMISSION REMOVAL

F–150, F–350 AND BRONCO

1. Disconnect the negative battery cable. Disconnect the neutral switch plug connector.
2. Remove the upper converter housing-to-engine bolts.
3. Raise the vehicle and support it safely.
4. Place the drain pan under the transmission fluid pan. Starting at the rear of the pan and working toward the front, loosen the attaching bolts and allow the fluid to drain. Finally remove all of the pan attaching bolts execpt 2 at the front, to allow the fluid to further drain.With fluid drained, install 2 bolts on the rear side of the pan to temporaily hold it in place.
5. Remove the converter drain plug access cover, if equipped from the lower end of the converter housing.
6. Remove the converter-to-flywheel attaching nuts.Place a wrench on the crankshaft pulley attaching bolt to turn the converter to gain access to the nuts.
7. With the wrench on the crankshaft pulley attaching bolt, turn the converter to gain access to the converter drain plug, if equipped. Place a drain pan under the converter to catch the fluid and remove the plug. After the fluid has been drained, re-install the plug.
8. On 4×2 vehicles, disconnect the driveshaft from the rear axle and slide shaft rearward from the transmission. Install a seal installation tool in the extension housing to prevent fluid leakage.
9. Disconnect the speedometer cable from the extension housing.
10. Disconnect the downshift and manual linkage rods or cable controls from the levers at the transmission.
11. Disconnect the oil cooler lines from the transmission.
12. Remove the vacuum hose from the vacuum diaphragm unit. Remove the vacuum line from the retaining clip.
13. Disconnect the cable connection at the starter motor. Remove the attaching bolts and remove the starter motor.
14. On F–150, F–350 4×4 and Bronco vehicles, remove the transfer case.
15. Remove the engine rear suport and insulator assembly attaching bolts.

16. Remove the engine rear support and insulator assembly-to-extension housing attaching bolts.
17. Remove the bolts securing the No. 2 crossmember to the frame side rails.
18. Raise the transmission with a transmission jack and remove the crossmember.
19. Secure the transmission to the jack with the safety chain.
20. Remove the remaining converter housing-to-engine attaching bolts.
21. Move the transmission away from the engine. Lower the jack and remove the converter and transmission assembly from under the vehicle.

E–150, E–250 AND E–350

1. Disconnect the negative battery cable. Working from inside the vehicle, remove the engine compartment cover.
2. Disconnect the neutral safety switch wires at the plug connector.
3. If the vehicle is equipped with a V8 engine, remove the flexhose from the air cleaner heat tube.
4. Remove the upper converter housing-to-engine attaching bolts.
5. Remove the bolt securing the filter tube to the engine.
6. Raise the vehicle and support it safely.
7. Place the drain pan under the transmission fluid pan. Starting at the rear of the pan and working toward the front, loosen the attaching bolts and allow the fluid to drain. Finally remove all of the pan attaching bolt except 2 at the front, to allow the fluid to further drain. With fluid drained, install 2 bolts on the rear side of the pan to temporarily hold it in place.
8. Remove the converter drain plug access cover from the lower end of the converter housing, if equipped.
9. Remove the converter-to-flywheel attaching nuts. Turn the engine crankshaft pulley to gain access to the nuts.
10. Turn the crankshaft pulley once again to gain access to the converter drain plug, if equipped. Place a drain pan under the converter to catch the fluid. Remove the plug. With fluid drained, re-install the plug.
11. Matchmark and remove the driveshaft.
12. Remove fluid filler tube.
13. Disconnect the starter cable at the starter. Remove the starter-to converter housing attaching bolts and remove the starter.
14. Position an appropriate engine support bar to the frame and engine oil pan flanges.
15. Disconnect the cooler lines from the transmission. Disconnect the vacuum line from the vacuum diaphragm unit.Remove the vacuum line from the retaining clip at the transmission.
16. Remove the speedometer driven gear from the extension housing.
17. Disconnect the manual and downshift linkage rods or cable controls from the transmission control levers.
18. Position a transmission jack to support the transmission. Install the safety chain to hold the transmission.
19. Remove the bolts and nuts securing the rear support and insulator assembly to the crossmember. Remove the bolts retaining the crossmember to the side rails and remove the 2 support gussets. Raise the transmission with the jack and remove the crossmember.
20. Remove the remaining converter housing-to-engine attaching bolts. Lower the jack and remove the converter and transmission assembly from under the vehicle.
21. Remove the converter and mount the transmission in a holding fixture.

TRANSMISSION INSTALLATION

F–150, F–350 AND BRONCO

1. Tighten the converter drain plug to 18–28 ft. lbs. (11–37 Nm), if equipped.
2. Position the converter on the transmission making sure the converter drive flats are fully engaged in the pump gear.
3. With the converter properly installed, place the transmission on the jack. Secure the transmission to the jack with the chain.
4. Rotate the converter until the studs and drain plug are in alignment with their holes in the flywheel.
5. Move the converter and transmission assembly forward and into position, using care not to damage the flywheel and the converter pilot. The converter must rest squarely against the flywheel. This indicates that the converter pilot is not binding in the engine crankshaft. Do not allow converter drive flats to disengage from pump gear.
6. Install and tighten the converter housing-to-engine attaching bolts to 40–50 ft. lbs. (55–67 Nm) on gasoline engines and 50–65 ft. lbs. (67–87 Nm) on diesel engines.
7. Remove the transmission jack safety chain from around the transmission.
8. Position the No.2 crossmember to the frame side rails. Install and tighten the attaching bolts to specifications.
9. Install transfer case on F–150, F–250 4×4 and Bronco.
10. Position the engine rear support and insulator assembly above the crossmember. Install the rear support and insulator assembly-to-extension housing mounting bolts and tighten the bolts to.
11. Lower the transmission and remove the jack.
12. Secure the engine rear support and insulator assembly to the crossmember with the attaching bolts and tighten them to specifications.
13. Connect the vacuum line to the vacuum diaphragm making sure that the line is in the retaining clip.
14. Connect the oil cooler lines to the transmission.
15. Connect the downshift and manual linkage rods or cable controls to their respective levers on the transmission.
16. Connect the speedometer cable to the extension housing.
17. Secure the starter motor in place with the attaching bolts. Connect the cable to the terminal on the starter.
18. Install a new O-ring on the lower end of the transmission filler tube and insert the tube in the case.
19. Secure the converter-to-flywheel attaching nuts and tighten them to 20–30 ft. lbs (28–40 Nm).
20. Install the converter housing access cover and secure it with the attaching bolts.
21. Connect the driveshaft.
22. Adjust the shift linkage as required.
23. Lower the vehicle. Install the upper converter housing-to-engine bolts and tighten to 40–50 ft. lbs. (55–67 Nm) on gasoline engines and 50–65 ft. lbs. (67–87) Nm on diesel engines.
24. Connect neutral switch wire to plug connector.
25. Make sure the pan is securely attached, and fill the transmission to the correct level with the specified fluid.

E–150, E–250 AND E–350

1. Tighten the converter drain plug to specifications, if equipped.
2. Position the converter on the transmission making sure the converter drive flats are fully engaged in the pump gear.
3. With the converter properly installed, place the transmission on the jack. Secure the transmission to the jack with the safety chain.
4. Rotate the converter until the studs and drain plug are in alignment with their holes in the flywheel, if equipped.
5. Move the converter and transmission assembly forward into positon, using care not to damage the flywheel and the converter pilot.

NOTE: The converter must rest squarely against the flywheel. This indicates that the converter pilot is not binding in the engine crankshaft. Do not allow converter drive flats to disengage from the pump gear.

6. Install the lower converter housing-to-engine attaching bolts. Tighten the bolts to 40–50 ft. lbs. (55–67 Nm) on gasoline

engines and 50–65 ft. lbs. (67–87 Nm) on diesel engines. Install the converter-to-flywheel attaching nuts. Tighten the nuts to 20–30 ft. lbs. (28–40 Nm).

7. Install the crossmember. Install the rear support and insulator assembly-to-crossmember attaching bolts and nuts.
8. Remove the safety chain and remove the jack from under the vehicle. Remove the engine support bar.
9. Install a new O-ring on the lower end of the transmission filler tube and insert the tube and dipstick in the case.
10. Connect the vacuum line to the vacuum diaphragm making sure the line is secured in the retaining clip.
11. Connect the cooler lines to the transmission.
12. Install the speedometer driven gear into the extension housing.
13. Connect the transmission linkage rods to the transmission control levers. When making transmission control attachments new retaining rings and grommets should always be used .
14. Attach the shift rod to the steering column shift lever.
15. Align the flats of the adjusting stud with the flats of the rod slot and insert the stud through the rod.
16. Assemble the adjusting stud nut and washer to a loose fit.
17. Perform a linkage adjustment.
18. Install the converter housing access cover and tighten the attaching bolts to 12-16 ft. lbs. (17-21 Nm).
19. Position the starter into the converter housing and install the attaching bolts. Tighten the bolts to 40–50 ft. lbs. (55–67 Nm) on gasoline engines and 50–65 ft. lbs. (67–87 Nm) on diesel engines. Install the starter cable.
20. Install driveshaft.
21. Lower the vehicle to the ground.
22. Install the upper converter housing-to-engine attaching bolts.
23. On V8 engines, install the flex hose to the air cleaner heat tube. Install the bolt that retains the filler tube to the cylinder block.
24. Connect the neutral safety switch wires at the plug connector.
25. Fill the transmission to the proper level with the specified fluid.
26. Raise the vehicle and check for transmission fluid leakage. Make sure the transmission fluid pan is securely attached. Lower the vehicle and adjust the downshift and manual linkage.
27. Install the engine compartment cover.
28. Road test the vehicle and check transmission operation in all ranges.

SPECIFICATIONS

CONTROL PRESSURE SPECIFICATIONS

All specifications are given in psi

Transmission		Idle 15 in. Hg. & Above		10 in. Hg. Vacuum		WOT Stall	
Model	Range	Altitude	Non-Altitude	Altitude	Non-Altitude	Altitude	Non-Altitude
PGD-EV-EY-FD DW-FG	D, 2, 1	42-61 ② 53-81 ①	52–76	68–95 ② 86–113 ①	88–111	134–159 ② 150–185 ①	150–185
	R	66–95 ② 81–126 ①	81–119	106–148 ② 135–177 ①	137–73	209–249 ② 235–285 ①	245–275
	P, N	42-61 ② 53-81 ①	52–76	—	—	—	—
PGD-AW-EG-FE-FF PJE-B-C	D, 2, 1	42-61 ② 53-81 ①	42–63	68–95 ② 86–113 ①	75–110	134–159 ② 150–185 ①	155–180
	R	66–95 ② 81–126 ①	66–69	106–148 ② 135–177 ①	117–157	209–249 ② 235–285 ①	245–275
	P, N	42-61 ② 53-81 ①	42–63	—	—	—	—
PGD-EK-FB-FC PJE-A	D, 2, 1	42–63		75–110		115–180	
	R	66–99		117–157		245–275	
	P, N	42–63		—	—	—	—
PJD-BA-BB-BC	D, 2, 1	67–91		99–119		155–180	
	R	94–142		155–186		245–275	
	P, N	67–91		—	—	—	—

① At sea level bar = 29.5
② At 5000 ft bar = 24.5

SELECTIVE SNAPRING AVAILABILITY

Part Number	Thickness in.	Thickness MM	Forward Clutch	Reverse High Clutch
377434	0.060–0.056	1.52–1.42	Used	Used
377126	0.069–0.064	1.75–1.62	Not Used	Used
377127	0.078–0.074	1.98–1.87	Used	Used
377128	0.087–0.083	2.20–2.10	Not Used	Used
377444	0.096–0.092	2.43–2.33	Used	Used
386841	0.114–0.110	2.89–2.79	Used	Not Used
386842	0.132–0.128	3.35–3.25	Used	Not Used

TORQUE SPECIFICATIONS

Item	ft. lbs.	Nm
Converter to flywheel	20–34	28–45
Front pump to transmission case	16–30	22–40
Overrunning clutch race to case	18–25	25–33
Oil pan to case	8–12	11–16
Stator support to pump	12–16	17–21
Converter cover to converter housing	12–16	17–21
Guide plate to case	12–16	17–21
Intermediate servo cover to case	14–20	19–27
Diaphragm assembly to case	12–16	17–21
Distributor sleeve to case	12–16	17–21
Extension assembly to transmission case	25–35	34–47
Plug—case front pump or line pressure	6–12	8.5–16
Pressure gauge tap	6–12	8.5–16
Band adjust screw locknut to case	35–45	48–61
End plates to body	20–45 ①	2.5–5
End plates to body	20–40 ①	2.5–4.5
Inner downshift lever stop	20–45 ①	2.5–5
Reinforcement plate to body	20–45 ①	2.5–5
Screen and lower to upper valve body	40–55 ①	5–6.2
Shift valve plate to upper body	20–45 ①	2.5–5
Upper to lower body	40–55 ①	5–6.2
VRV to fuel injector pump	75–90 ①	8–10.5
Converter drain plug	8–28	11–37
Manual valve inner lever to shaft	30–40	41–54
Downshift lever to shaft	12–16	17–21
Filler tube to engine (econoline—5.0L/5.8L/7.5L)	40–50	54–67
Filler tube to engine (econoline 4.9L)	33–42	44–56
Filler tube to engine (econoline 6.9L)	24–35	32–47
Transmission to engine (diesel only)	50–65	67–87
Transmission to engine (all gasoline engines)	40–50	55–67
Rear engine support to transmission	60–80	80–107
Plug case—throttle pressure	6–12	8.5–16
5/16 in. fitting—cooler line connector to case—front and rear (case fitting)	18–23	25–32
5/16 in. tube nut—cooler line to transmission case fitting	12–18	17–24
Reinforcing right side plate to lower body	20–45	2.5–5
Converter housing cover to converter housing (7.5L)	30–60	3.5–6.5
Control assembly to case	95–125	11–14
Governor body to collector body	90–120	10.5–13.5
Detent spring to case	80–120	9.5–13.5
Rear engine support to frame	40–60	5–6.5
Neutral switch to case	55–75	6.5–8

① inch lbs.

THRUST WASHER SELECTION AND APPLICATION

Identification Color	Thickness in.	Thickness mm
Blue	0.056–0.060	1.42–1.52
Natural (White)	0.073–0.077	1.85–1.95
Red	0.088–0.092	2.23–2.33

VACUUM DIAPHRAGM ASSEMBLY IDENTIFICATION

Year	Type	Part Number	Identification	Selective Fit Rods Length (in.)	Selective Fit Rods Color
1984–86	HAD	D7AP-7A337-AA	Part No. Stamped	1.677–1.667	No color
	SAD	D7OP-7A377-BA	1 green stripe	1.727-1.717	Purple
				1.611–1.601	Yellow
	SAD	D4TP-7A377-BA	1 black strip	1.644–1.634	Blue
	SAD	D5AP-7A377-AA	1 purple stripe	1.660-1.650	Green
				1.710–1.700	White
				1.694–1.684	Brown

SAD—Single Area Diaphragm
S-SAD—Super Single Area Diaphragm
HAD—High Attitude Diaphragm

CONVERTER STALL SPEEDS

Year	Application	Engine Displacement	Converter Size (in.)	Stall Speed (rpm)
1984–86	F-150/250/350 E-150/250/350 Bronco	4.9L	12	1616–1871
	F-150/250 E-150/250 Bronco	5.0L	12	1616–1871
	F-150/250/350 E-150/250/350 Bronco	5.8L	12	1569–1729
	F-250/350 E-250/350	6.9L	12	1715–1966
	F-250/350 E-250/350	7.5L	12	1610–1891

SNAPRING CLEARANCE CHART

Year		Clearance in.	Clearance mm
1984–86	Forward clutch	0.021–0.046	0.533–1.168
	High clutch	0.022–0.036	0.558–0.914
	Reverse clutch	0.030–0.056 ①	0.762–1.422
		0.040–0.075 ②	1.016–1.905

① 4.9L Engine
② 5.0L EF1, 5.0L HO, 5.8L Engine

CLUTCH PLATE IDENTIFICATION

Year	Clutch Type	Transmission Model	Steel Plates	Friction Plates
1984–86	Forward	PGD, PJD	4 ①	4
	High	PGD, PJD	3	3
	Reverse	PJD	5 ②	5
		PGD	4 ②	4

① Plus a waved plate (7E457) next to inner pressure plate
② Plus a waved plate next to the piston

Section 4

AOD Transmission
Ford Motor Co.

APPLICATION

AOD

Year	Engine	Vehicle
1984	5.0L (CFI)	Mustang/Capri
	3.8L (CFI)/5.0L (CFI)	LTD/Marquis
	3.8L (CFI)/5.0L (CFI)	Thunderbird/Cougar
	5.0L (CFI)	Mark VII/Continental
	5.0L (CFI)/5.8L (Carb.)	Crown Victoria/Grand Marquis
	5.0L (CFI)	Town Car
	5.0L (Carb.)	E-150/250
	5.0L (Carb.)	F-150/250 Pickup
1985–86	5.0L (CFI)/5/0L (EFI)	Mustang/Capri
	3.8L (CFI)/5.0L (CFI)	LTD/Marquis
	3.8L (CFI)/5.0L (EFI)	Thunderbird/Cougar
	5.0L (EFI)	Mark VII/Continental
	5.0L (EFI)/5.8L (Carb.)	Crown Victoria/Grand Marquis
1985–86	5.0L (CFI)/5.0L (EFI)	Town Car
	5.0L (Carb.)	E-150/250
	5.0L (Carb.)	F-150/250 Pickup
1987	5.0L HO (EFI)	Mustang
	3.8L (CFI)/5.0L (EFI)	Thunderbird/Cougar
	5.0L (EFI)	Mark VII/Continental
	5.0L (EFI)/5.8L (Carb.)	Crown Victoria/Grand Marquis
	5.0L (EFI)	Town Car
1988–89	5.0L HO (EFI)	Mustang
	3.8L (EFI)/5.0L (EFI)	Thunderbird/Cougar
	5.0L (EFI)	Mark VII
	5.0L (EFI)/5.8L (Carb.)	Crown Victoria/Grand Marquis
	5.0L (EFI)	Town Car

GENERAL DESCRIPTION

The automatic overdrive transmission (AOD) differs from conventional 3 speed transmissions in that the planetary gear set operates in 4th gear. The AOD provides fully automatic operation in either **D** or **OD** positions.

OD—this is the normal driving position for the AOD transmission. In this position the transmission starts in 1st gear and as the vehicle accelerates, it automatically upshifts to 2nd, 3rd and 4th gear. The transmission will automatically downshift as vehicle speed decreases.

NOTE: The transmission will not shift into or remain in overdrive when the accelerator is pushed to the floor.

D—in this position the transmission operates as in **OD** except that there will be no shift into the overdrive gear. This position may be used when driving up or down mountainous roads to provide better perfomance and engine breaking than the **OD** position. The transmission may be shifted from **D** to **OD** or **OD** to **D** at any time and at any speed, while in motion.

1 (LOW)—this position can be used when maximum engine braking is desired. To help brake the vehicle on hilly roads where **D** does not provide enough braking, shift the selector to **1** (LOW). At speeds above 25 mph, the transmission will shift to 2nd gear and remain in 2nd gear. When the vehicle speed drops below 25 mph, the transmission will shift to 1st gear and remain in 1st gear.

Forced downshifts—at vehicle speeds from 55–25 mph in **OD** or **D**, the transmission will downshift to 2nd gear when the accelerator is pushed to the floor. At vehicle speeds below 25 mph, the transmission will downshift to 1st gear when the accelerator is pushed to the floor. At most vehicle speeds in **OD**, the transmission will downshift from 4th to 3rd gear when the accelerator is pushed for moderate to heavy acceleration.

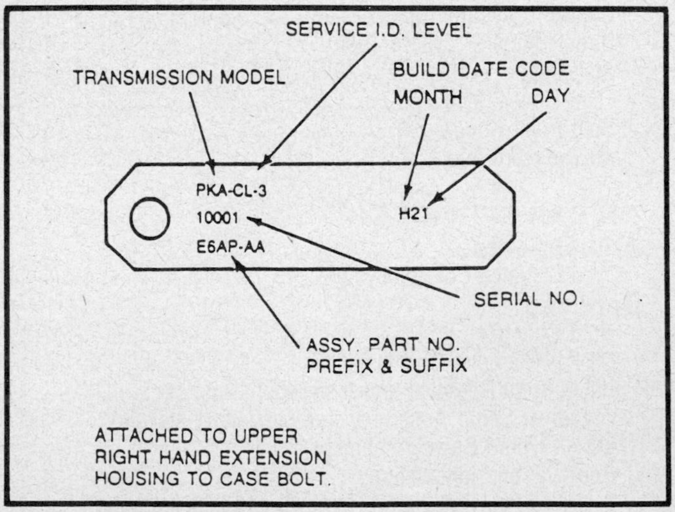
SERVICE I.D. LEVEL
TRANSMISSION MODEL
BUILD DATE CODE
MONTH
DAY
PKA-CL-3
10001
E6AP-AA
H21
SERIAL NO.
ASSY. PART NO.
PREFIX & SUFFIX
ATTACHED TO UPPER
RIGHT HAND EXTENSION
HOUSING TO CASE BOLT.

AOD identification tag

Transmission and Converter Identification

TRANSMISSION

The AOD automatic transmission is identified by the letter **T** stamped on the vehicle certification label, mounted on the left

TORQUE CONVERTER APPLICATION

Vehicle Application	Engine Disp.	Trans. Type	Converter Size	Stall Speed (RPM) Min.	Max.
Thunderbird/Cougar	3.8L CFI 2V	AOD	12"	2082	2409
Mustang	5.0L (HO) SEFI	AOD	12"	2066	2457
Ford Crown Victoria/ Mercury Grand Marquis	5.0L SEFI	AOD	12"	2062	2400
Lincoln Town Car	5.0L SEFI	AOD	12"	2061	2399
Continental/Mark	5.0L SEFI	AOD	12"	2032	2446
Thunderbird/Cougar/ Mark VII LSC	5.0L SEFI	AOD	12"	2032	2346
Ford Crown Victoria/ Mercury Grand Marquis	5.8L (HO)	AOD	12"	1543	1857

driver's door body pillar post. The identification tag is located on one of the extension housing bolts. The identification tag indicates the model code, the part number prefix, build code and transmission serial number.

CONVERTER

The AOD torque converter is identified by either a reference or part number stamped on the converter body and is matched to a specific engine. The torque converter is a welded unit and is not repairable. If internal problems exists, the torque converter must be replaced.

Metric Fasteners

Metric bolts and nuts are used in the construction of the transmission, along with the familiar inch system fasteners. The dimensions of both systems are very close and for this reason, replacement fastener must have the same measurement and strength as those removed. Do not attempt to interchange metric fasteners for inch system fasteners. Mismatched or incorrect fasteners can result in damage to the transmission unit through malfunctions, breakage or personal injury. Care should be exercised to replace the fasteners in the same locations as removed.

Capacities

Only Motorcraft Dexron®II or equivalent fluid, meeting Ford Motor Co.'s specifications ESP-M2C1-166HP, should be used in the AOD transmission. Failure to use the proper fluid could result in internal transmission damage. The pan capacity is approximately 4 quarts. The total overhaul capacity is 12 quarts.

Checking Fluid Level

The AOD transmission is designed to operate with the fluid level between the **ADD** and **FULL** mark of the dipstick, with the transmission unit at normal operating temperature of 155–170°F. If the fluid level is at or near the bottom indicator on the dipstick, either cold or hot, do not drive the vehicle until fluid has been added.

Transmission at Room Temperature

70–95°F, DIPSTICK COOL TO THE TOUCH

1. With the vehicle on a level surface, engine idling, wheels blocked and foot brake applied, move the selector lever through the gear positions to engage each gear and to fill the oil passages with fluid.
2. Place the selector lever in the **P** position and apply the parking brakes, allowing the engine to idle.

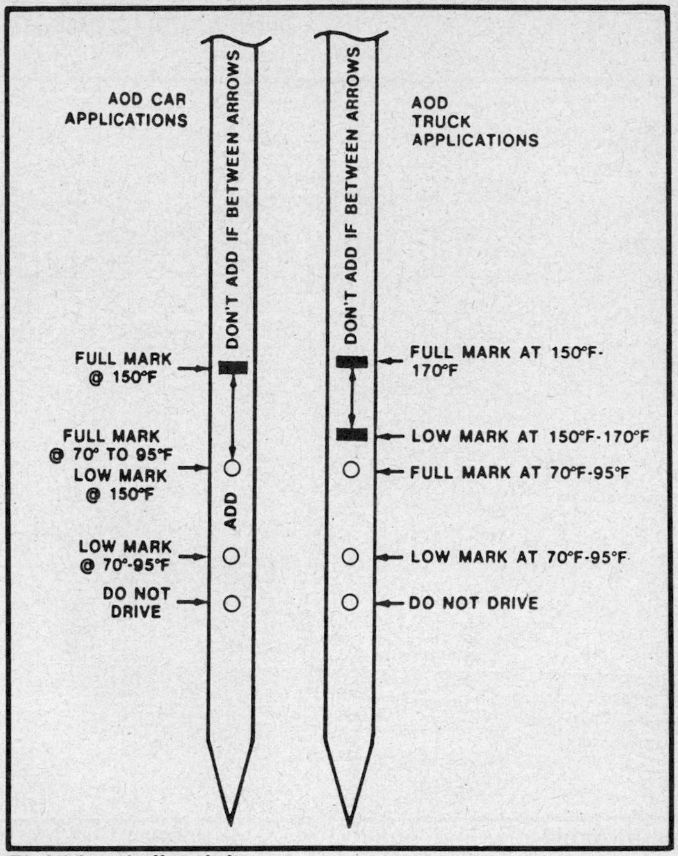
AOD CAR APPLICATIONS
DON'T ADD IF BETWEEN ARROWS
DON'T ADD IF BETWEEN ARROWS
AOD TRUCK APPLICATIONS
FULL MARK @ 150°F
FULL MARK AT 150°F-170°F
FULL MARK @ 70° TO 95°F
LOW MARK @ 150°F
LOW MARK AT 150°F-170°F
ADD
FULL MARK AT 70°F-95°F
LOW MARK @ 70°-95°F
LOW MARK AT 70°F-95°F
DO NOT DRIVE
DO NOT DRIVE

Fluid level dipstick

3. Clean the dipstick area of dirt and remove the dipstick from the filler tube. Wipe the dipstick clean and re-insert it back into the filler tube and seat it firmly.
4. Remove the dipstick from the filler tube again and check the fluid level as indicated on the dipstick. The level should be between the cold low mark and the cold full mark on the dipstick indicator.
5. If necessary, add enough fluid to bring the level to its cold full mark. Re-install the dipstick and seat it firmly in the filler tube.
6. When the transmission reaches normal operating temperature of 155–170°F, re-check the fluid level and correct as required to bring the fluid to its hot level mark.

Transmission at Normal Operating Temperature

155–170°F, DIPSTICK HOT TO THE TOUCH

1. With the vehicle on a level surface, engine idling, wheels blocked and foot brake applied, move the selector lever through the gear positions to engage each gear and to fill the passageways with fluid.
2. Place the selector lever in the **P** position and apply the parking brake, allowing the engine to idle.
3. Clean the dipstick area of dirt and remove the dipstick from the filler tube. Wipe the dipstick clean, re-insert the dipstick into the filler tube and seat firmly.
4. Again remove the dipstick from the filler tube and check the fluid level as indicated on the dipstick. The level should be between the **ADD** and **FULL** marks. If necessary, add enough fluid to bring the fluid level to the **FULL** mark.
5. When the fluid level is correct, fully seat the dipstick in the filler tube.

TRANSMISSION MODIFICATIONS

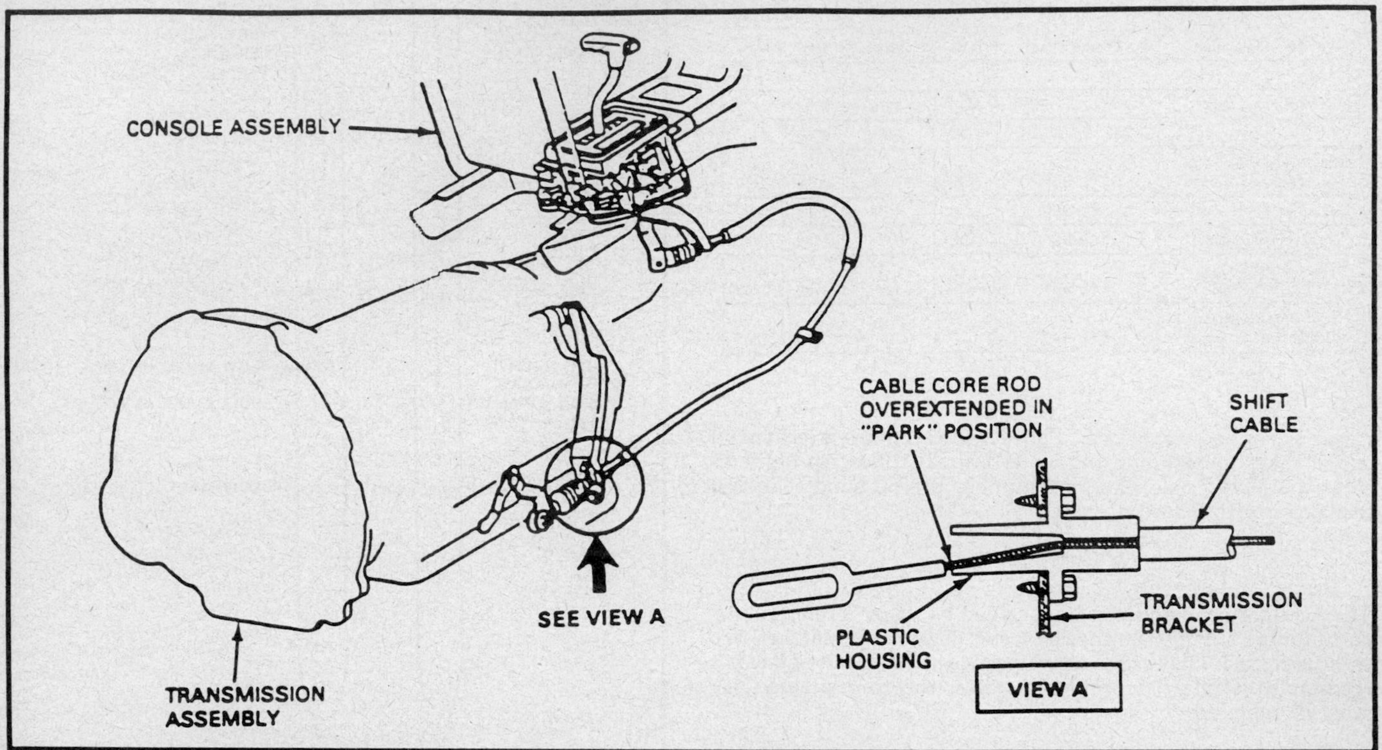
CONSOLE ASSEMBLY
CABLE CORE ROD OVEREXTENDED IN "PARK" POSITION
SHIFT CABLE
SEE VIEW A
TRANSMISSION BRACKET
PLASTIC HOUSING
VIEW A
TRANSMISSION ASSEMBLY

Shift cable binding repair

Transmission Fluid Leak at Cooler Line Connections

1985–87 FORD AND LINCOLN/MERCURY

In the event that service is required for a disconnected transmission cooler line, the quick-connect fitting should not be reused because it may not reseal properly and can result in repeat repair.

To correct the problem, the cooler lines must be modified using the following procedure:

1. Remove and discard the quick-connect fitting connector from the radiator or transmission case.

NOTE: Rework the connectors in sets, do not rework only one connector at a time.

2. Cut off the tubing as close as possible to the formed bead, using a tubing cutter.
3. Slide a $^5/_{16}$ × ½–20 thread tube nut on the tube.
4. Make a double flare tube end.
5. Install the proper size and type connectors into the transmission case or radiator. Use thread sealer on the male threads of the connector.
6. Install thread tube nuts onto connectors. Torque to 12–18 ft. lbs.

Shift cable Lockup in Park

Transmissions that lockup in the **P** mode on some 1987 Thunderbird/Cougar vehicles with floor shift gear selectors may be caused by a shift cable bracket that is not positioned correctly on the gear selector. This allows the shift cable to catch on the plastic cable housing when the gear selector is positioned in **P**.

To correct this, install a new design gear selector and shift cable bracket using the following service procedure:

1. Remove the boot that seals the shift cable onto the transmission.
2. Check if shift cable core rod extends past the plastic cable housing when gear selector is positioned in **P**.
3. If the cable core rod extends past the housing, replace the gear selector with part number E8SZ–7210 and shift cable bracket with part number E7SZ–7B229–B.

TROUBLE DIAGNOSIS

CHILTON THREE C's TRANSMISSION DIAGNOSIS

Ford AOD

Condition	Cause	Correction
No / delayed forward engagement. Reverse OK	a) Improper fluid level	a) Check fluid level
	b) Manual linkage misadjusted or damaged	b) Check and adjust or repair as required
	c) Low line pressure	c) Control pressure test, note results
	d) Valve body bolts too loose or tight	d) Torque to specification 80–100 inch lbs. (9–11 Nm)
	e) Valve body dirty or sticking valves	e) Determine source of contamination. Repair as required
	f) Misassembled or leaking oil filter	f) Visually check oil filter for leaks at crimps. Verify proper assembly of oil filter to valve body gasket. Inspect rubber sealing grommet for nicks or cuts. Torque attaching bolts to specification
	g) 2–3 accumulator piston seal (small) cut/worn-piston bore in case damaged; piston damaged	g) Determine cause of failure and repair as required
	h) Forward clutch assembly burnt or damaged. Piston seals worn or cut. Cylinder ball check not seating. Stator support seal rings or ring grooves damaged or worn	h) Determine cause of failure and repair as required
No forward engagement. OD and 3	a) Planetary (low) one-way clutch	a) Repair as required
No / delayed reverse engagement. Forward OK	a) Improper fluid level	a) Check fluid level
	b) Low line pressure in reverse	b) Control pressure test, note results
	c) Manual linkage misadjusted or damaged	c) Check and adjust or repair as required. Manual linkage check
	d) Valve body bolts too loose or tight	d) Torque to specification 80–100 inch lbs. (9–11 Nm)
	e) Misassembled or leaking oil filter	e) Visually check oil filter for leaks at crimps. Verify proper assembly of oil filter to valve body gasket. Inspect rubber sealing grommet for nicks or cuts. Torque attaching bolts to specification
	f) Valve body dirty or sticking valves	f) Determine source of contamination then repair as required
	g) Reverse clutch assembly burnt or worn. Piston seals worn or cut. Piston ball check not seating. Stator support seal rings or ring grooves worn or damaged. Abuse	g) Determine cause of failure then repair as required
No / delayed engagement forward and reverse	a) Pump gear damaged (no engagement)	a) Replace and repair as required
	b) Output shaft broken (no engagement only)	b) Repair as required
	c) Turbine shaft broken (no engagement only)	c) Repair as required
No / delayed reverse engagement and no engine breaking in manual low (1)	a) Low reverse band or servo piston burnt or worn. Servo seal worn or cut. Servo bore damaged. Servo piston sticking in bore. Low line pressure	a) Determine cause of failure, then repair as required
Harsh engagements or initial engagement clunk (warm engine)	a) Improper fluid level	a) Check fluid level
	b) Throttle linkage misadjusted (long), disconnected, sticking, damaged or return spring disconnected	b) Adjust throttle linkage
	c) Engine curb idle too high	c) Check engine curb idle

Condition	Cause	Correction
Harsh engagements or initial engagement clunk (warm engine)—Continued	d) Valve body bolts—loose/too tight	d) Torque to specification 80–100 inch lbs. (9–11 Nm)
	e) Misassembled/leaking oil filter	e) Visually check oil filter for leaks at crimps. Verify proper assembly of oil filter to valve body gasket. Inspect rubber sealing grommet for nicks or cuts. Torque attaching bolts to specification
	f) Valve body dirty or sticking valves. Throttle valve (forward and reverse). 2–3 backout valve (forward only)	f) Determine source of contamination. Repair as required
	g) Engine rpm above specification	g) Adjust engine rpm to specification
	h) Throttle valve linkage misadjusted	h) Adjust throttle linkage
	i) Worn, damaged or loose U-joint (front/rear), slip yoke, rear axle, rear suspension	i) Repair as necessary
	j) Excessive transmission endplay	j) Check transmission endplay. Replace selective thrust washer if necessary
Forward engagement slips, shudders or chatters	a) Improper fluid level	a) Check fluid level
	b) Low throttle pressure or throttle valve rod misadjusted (short)	b) Adjust throttle valve rod
	c) Manual linkage misadjusted or damaged	c) Check and adjust or repair as required. Manual linkage check
	d) Low line pressure	d) Control pressure test
	e) Valve body bolts too loose or tight	e) Torque to specification 80–100 inch lbs. (9–11 Nm)
	f) Misassembled or leaking oil filter	f) Visually check oil filter for leaks at crimps. Verify proper assembly of oil filter to valve body gasket. Inspect rubber sealing grommet for nicks or cuts. Torque attaching bolts to specification
	g) Valve body dirty or sticking valves	g) Determine source of contamination. Repair as required
	h) Forward clutch piston ball check not seating	h) Replace forward clutch cylinder. Repair transmission as required
	i) Forward clutch piston seal cut or worn	i) Replace seal and repair clutch as required
	j) Contamination blocking forward clutch feed hold	j) Determine source of contamination, then repair as required
	k) Low one-way clutch (planetary) damaged	k) Repair as required. Determine cause of failure
Reverse shudder/chatters/slips	a) Improper fluid level	a) Check fluid level
	b) Low line pressure in reverse	b) Control pressure test
	c) Reverse servo or servo bore damaged	c) Determine cause of failure. Repair as required
	d) Misassembled or leaking oil filter	d) Visually check oil filter for leaks at crimps. Verify proper assembly of oil filter to valve body gasket. Inspect rubber sealing grommet for nicks or cuts. Torque attaching bolts to specification
	e) Reverse clutch drum bushing damaged	e) Determine cause of failure. Repair as required
	f) Reverse clutch—short one friction plate	f) Rebuild with proper number of plates
	g) Reverse clutch stator support seal rings/ring grooves worn/damaged	g) Determine cause of failure. Repair as required
	h) Reverse clutch piston seal cut or worn	h) Determine cause of failure then repair as required

Condition	Cause	Correction
Poor vehicle acceleration at higher speed 1st and 2nd gears. All engine speeds in 3rd and 4th. (Stall test will check OK)	a) Torque converter one-way clutch locked up	a) Replace torque converter
All upshifts harsh or delayed or no upshifts	a) Improper fluid level	a) Check fluid level
	b) Throttle linkage misadjusted (long), disconnected, sticking, or damaged or return spring disconnected	b) Adjust throttle linkage. Repair as required
	c) Manual linkage misadjusted or damaged	c) Check and adjust or repair as required. Manual linkage check
	d) Governor sticking	d) Perform governor test. Repair as required
	e) Line pressure too high or low	e) Control pressure test. Repair as required
	f) Valve body bolts too loose or tight	f) Torque to specification 80–100 inch lbs. (9–11 Nm)
	g) Misassembled or leaking oil filter	g) Visually check oil filter for leaks at crimps. Verify proper assembly of oil filter to valve body gasket. Inspect rubber sealing grommet for nicks or cuts. Torque attaching bolts to specification
	h) Valve body dirty or sticking valves. Throttle valve or throttle pressure limit valve	h) Determine source of contamination. Repair as required
Mushy, early or pile up—all upshifts	a) Improper fluid level	a) Check fluid level
	b) Throttle linkage misadjusted (short), sticking or damaged	b) Adjust throttle linkage. Repair as required
	c) Low line pressure	c) Control pressure test
	d) Valve body bolts too loose or tight	d) Torque to specification 80–100 inch lbs. (9–11 Nm)
	e) Misassembled or leaking oil filter	e) Visually check oil filter for leaks at crimps. Verify proper assembly of oil filter to valve body gasket. Inspect rubber sealing grommet for nicks or cuts. Torque attaching bolts to specification
	f) Valve body valve sticking. Throttle valve or throttle pressure limit valve	f) Determine source of contamination. Repair as required
	g) Governor valve sticking	g) Perform governor test. Repair as required
No 1–2 upshifts	a) Improper fluid level	a) Check fluid level
	b) Throttle linkage misadjusted (long), disconnected or sticking	b) Adjust throttle linkage
	c) Manual linkage misadjusted or damaged	c) Check and adjust or repair as required. Manual linkage check
	d) Low line pressure to intermediate friction clutch	d) Control pressure test. Note results
	e) Governor valve sticking	e) Perform governor test. Repair as required
	f) Valve body bolts too loose or tight	f) Torque to specification 80–100 inch lbs. (9–11 Nm)
	g) Misassembled or leaking oil filter	g) Visually check oil filter for leaks at crimps. Verify proper assembly of oil filter to valve body gasket. Inspect rubber sealing grommet for nicks or cuts. Torque attaching bolts to specification
	h) Valve body dirty or sticking valves. 1–2 shift valve or throttle valve	h) Determine source of contamination. Repair as required
Upshifts 1–3	a) Intermediate clutch assembly burnt. Piston seals worn or cut. Piston not positioned properly. Improper stack up, low line pressure.	a) Determine cause of failure then repair as required
	b) Intermediate one-way clutch damaged	b) Repair as required

Condition	Cause	Correction
Rough or harsh delayed 1–2 upshift	a) Improper fluid level	a) Check fluid level
	b) Poor engine performance	b) Tune-up engine
	c) Throttle linkage misadjusted (long), damaged or disconnected	c) Adjust throttle linkage. Repair as required
	d) Line pressure too high or low	d) Control pressure test. Note results
	e) Valve body bolts too loose or tight	e) Torque to specification 80–100 inch lbs. (9–11 Nm)
	f) Misassembled or leaking oil filter	f) Visually check oil filter for leaks at crimps. Verify proper assembly of oil filter to valve body gasket. Inspect rubber sealing grommet for nicks or cuts. Torque attaching bolts to specification.
	g) Valve body dirty or sticking valves. Rough shift—1–2 capacity modulator, 1–2 accumulator, 1–2 accumulator exhaust port blocked. Delayed shift—2–3 throttle modulator	g) Determine source of contamination, then repair as required
	h) Governor valve sticking	h) Perform governor test. Repair as required
Mushy, early, soft or slipping 1–2 upshift	a) Improper fluid level	a) Check fluid level
	b) Throttle linkage misadjusted (short) or damaged	b) Adjust throttle linkage. Repair as required
	c) Low line pressure	c) Control pressure test. Note results
	d) Valve body bolts loose/too tight	d) Torque to specification 80–100 inch lbs. (9–11 Nm)
	e) Misassembled/leaking oil filter	e) Visually check oil filter for leaks at crimps. Verify proper assembly of oil filter to valve body gasket. Inspect rubber sealing grommet for nicks or cuts. Torque attaching bolts to specification
	f) Valve body dirty or sticking valves. Throttle valve, throttle pressure limit valve, 2–3 throttle pressure limit (early) or 1–2 capacity modulator	f) Determine source of contamination. Repair as required
	g) Intermediate friction clutch burnt or worn. Clutch piston seals cut or worn. Clutch piston not aligned with pump body. Excessive clutch pack clearance	g) Determine cause of failure then repair as required
	h) Clutch piston upside down	h) Install properly. Bleedhole to be at 12 o'clock
	i) Governor valve sticking	i) Perform governor test. Repair as required
No 2–3 upshift	a) Improper fluid level	a) Check fluid level
	b) Throttle linkage misadjusted (long), sticking or damaged	b) Adjust throttle linkage. Repair as required
	c) Low line pressure to direct clutch	c) Control pressure test. Note results
	d) Valve body bolts too loose or tight	d) Torque to specification 80–100 inch lbs. (9–11 Nm)
	e) Misassembled or leaking oil filter	e) Visually check oil filter for leaks at crimps. Verify proper assembly of oil filter to valve body gasket. Inspect rubber sealing grommet for nicks or cuts. Torque attaching bolts to specification
	f) Valve body dirty/sticking valves. 2–3 shift valve	f) Determine source of contamination, then repair as required
	g) Direct clutch assembly burnt or worn. Piston seals cut/worn. Piston ball check stuck (not seating). Output shaft seal rings (small, Teflon) worn/cut. Collector body seal rings (large cast iron) worn	g) Perform stall test. Determine cause of failure then repair as required

Condition	Cause	Correction
No 2–3 upshift—Continued	h) Converter damper hub weld broken	h) Perform Converter Damper Hub Weld Check. Replace torque converter if required
Harsh / delayed 2–3 upshift	a) Improper fluid level	a) Check fluid level
	b) Poor engine performance	b) Tune-up engine
	c) Throttle linkage misadjusted (long), sticking or damaged	c) Adjust throttle linkage. Repair as required
	d) Valve body bolts too loose or tight	d) Torque to specification 80–100 inch lbs. (9–11 Nm)
	e) Misassembled or leaking oil filter	e) Visually check oil filter for leaks at crimps. Verify proper assembly of oil filter to valve body gasket. Inspect rubber sealing grommet for nicks or cuts. Torque attaching bolts to specification
	f) Valve body dirty/sticking valves. Throttle valve, throttle pressure limit valve, 2-3 backout valve (harsh backout shift), 2–3 capacity modulator (rough), or 2–3 throttle pressure modulator (delayed)	f) Determine source of failure. Repair as required
	g) 2–3 accumulator piston drain hole plugged/omitted	g) Remove 2–3 accumulator piston and visually inspect for plugging condition or omission
	h) 2–3 accumulator piston seals cut/worn	h) Replace seals, determine cause of failure, repair as required
Soft, early or mushy 2–3 upshift	a) Improper fluid level	a) Check fluid level
	b) Throttle linkage misadjusted (short) or bent	b) Adjust throttle linkage. Repair as required
	c) Valve body bolts too loose or tight	c) Torque to specification 80–100 inch lbs. (9–11 Nm)
	d) Misassembled or leaking oil filter	d) Visually check oil filter for leaks at crimps. Verify proper assembly of oil filter to valve body gasket. Inspect rubber sealing grommet for nicks or cuts. Torque attaching bolts to specification
	e) Valve body dirty or sticking valves. Throttle valve, throttle pressure limit valve, 2–3 throttle pressure limit valve (early), 2–3 capacity modulator valve, or 2–3 backout valve (mushy)	e) Determine source of contamination then repair as required
	f) Direct clutch assembly burnt/worn. Clutch piston seals cut/worn. Clutch piston ball check stuck/not seating. Output shaft seal rings (small Teflon) worn. Collector body seal rings (large cast) worn on output shaft	f) Perform stall test. Determine cause of failure. Repair as required
No 3–4 upshift	a) Improper fluid level	a) Check fluid level
	b) Throttle linkage—misadjusted (long)/ bent/sticking	b) Adjust throttle linkage. Repair as required
	c) Low pressure to overdrive band servo	c) Control pressure test. Note results
	d) Valve body bolts too loose or tight	d) Torque to specification 80–100 inch lbs. (9–11 Nm)
	e) Misassembled or leaking oil filter	e) Visually check oil filter for leaks at crimps. Verify proper assembly of oil filter to valve body gasket. Inspect rubber sealing grommet for nicks or cuts. Torque attaching bolts to specification
	f) Valve body dirty/sticking valves. 3–4 shift valve. Throttle valve. 3–4 shuttle valve	f) Determine source of contamination, then repair as required

Condition	Cause	Correction
No 3–4 upshift—Continued	g) Overdrive band assembly burnt or worn. OD band not seated to case. OD servo not seated to band end seat. OD servo apply blocked. OD servo seals worn/cut. OD servo bore damaged	g) Determine cause of failure, repair as required
	h) 3–4 accumulator piston seals worn or cut. 3–4 accumulator piston drain passage blocked	h) Determine cause of failure. Repair as required
Harsh/delayed 3–4 upshift	a) Improper fluid level	a) Check fluid level
	b) Throttle linkage misadjusted (long) or bent	b) Adjust throttle linkage. Repair as required
	c) Valve body bolts too loose or tight	c) Torque to specification 80–100 inch lbs. (9–11 Nm)
	d) Misassembled or leaking oil filter	d) Visually check oil filter for leaks at crimps. Verify proper assembly of oil filter to valve body gasket. Inspect rubber sealing grommet for nicks or cuts. Torque attaching bolts to specification
	e) Valve body dirty or sticking valves. 3–4 shift valve, throttle valve, throttle pressure limit valve, 3–4 shuttle valve (rough), 3–4 backout valve (rough backout shifts) or 3–4 throttle pressure modulator valve	e) Determine source of contamination then repair as required
Slips, shudder soft or early 3–4 upshift	a) Improper fluid level	a) Check fluid level
	b) Throttle linkage misadjusted (short), bent or sticking	b) Adjust throttle linkage. Repair as required
	c) Low pressure to overdrive band servo	c) Control pressure test. Note results
	d) Valve body bolts too loose or tight	d) Torque to specification 80–100 inch lbs. (9–11 Nm)
	e) Misassembled or leaking oil filter	e) Visually check oil filter for leaks at crimps. Verify proper assembly of oil filter to valve body gasket. Inspect rubber sealing grommet for nicks or cuts. Torque attaching bolts to specification
	f) Valve body dirty or sticking valves. 3–4 backout valve (soft), throttle valve, 3–4 shuttle valve (soft), or 3–4 throttle pressure modulator valve (early)	f) Determine source of contamination, then repair as required
	g) Overdrive band assembly burnt/worn. Overdrive band not seated to case. Overdrive servo not seated to band end seat. Overdrive servo apply blocked. Overdrive servo seals worn/cut. Overdrive servo bore damaged. 3–4 accumulator piston seals worn or cut. 3–4 accumulator piston drain passage blocked	g) Determine cause of failure. Repair as required
Erratic shifts	a) Improper fluid level	a) Check fluid level
	b) Throttle linkage binding or sticking	b) Inspect throttle linkage. Repair as required
	c) Valve body bolts—loose/too tight	c) Torque to specification 80–100 inch lbs. (9–11 Nm)
	d) Misassembled/leaking oil filter	d) Visually check oil filter for leaks at crimps. Verify proper assembly of oil filter to valve body gasket. Inspect rubber sealing grommet for nicks or cuts. Torque attaching bolts to specification

Condition	Cause	Correction
	e) Valve body dirty or sticking valves	e) Perform line pressure test, note results and source of contamination, then repair as required
	f) Governor valve sticking	f) Perform governor test. Repair as required
	g) Output shaft collector body seal rings (large cast iron) worn/cut	g) Repair as required
Shift hunting 3–4, 4–3	a) Poor engine performance. EGR solenoid defective	a) Tune-up engine. Replace solenoid
	b) Throttle linkage misadjusted	b) Inspect and adjust throttle linkage as required
	c) Manual linkage misadjusted	c) Check and adjust or repair as required. Manual linkage check adjustment
No engine breaking—manual low "pull-in"	a) Low reverse band is not holding	a) Refer to condition "No reverse engagement. Forward OK"
Shift efforts high	a) Manual shift linkage damaged/misadjusted	a) Check and adjust or repair as required. Manual linkage check adjustment
	b) Inner manual lever nut loose	b) Torque nut to specification
	c) Manual lever retainer pin damaged	c) Adjust manual linkage and install new pin
No kickdown or hard to get kickdown	a) Misadjusted throttle linkage (short). Carburetor throttle lever sticking. Carburetor misadjusted	a) Inspect and adjust throttle linkage. Repair as required. Adjust to specification
Transmission leaks	a) Torque converter Seam weld leaks. Drain plug loose/cross threaded. Impeller hub pitted/worn	a) Replace or repair as required
	b) Oil pump Impeller hub seal cut or worn. Overdrive seal cut/pinched. Gasket misaligned. Bolts loose	b) Replace or repair as required
	c) Lever seals in case Manual lever seal cut/worn. Throttle lever seal cut/worn	c) Replace or repair as required
	d) Case fittings Cooler line fittings loose or cross threaded. Pressure plugs loose or cross threaded	d) Replace or repair as required
	e) Case Porosity or cracked	e) Replace or repair as required
	f) Oil pan Gasket misaligned or damaged. Bolts loose or cross threaded. Pan damaged	f) Replace or repair as required
	g) Extension housing Gasket misaligned or damaged. Bolts loose. Seal cut/worn. Bushing worn or damaged. Speedometer cable connection	g) Replace or repair as required
	h) Filler tube Out top of tube, at bottom of tube. Cut or worn seal. Case porosity. Damaged tube	h) Check converter drainback valve. Replace seal. Replace case. Replace tube
	i) Case breather vent	i) Repair as required
	j) Engine-power steering leak	j) Correctly identify fluid; isolate leakage area; repair leak
Harsh coasting downshift clunk	a) Anti-clunk spring not seated properly	a) Reposition anti-clunk spring properly
	b) Throttle linkage misadjusted (long)	b) Adjust throttle linkage, repair as required

Condition	Cause	Correction
Improper shift timing—4-3 Backout shift 4-3 Shift on full back out (foot off accelerator Pedal) accompanied by 3–4 shift when accelerator depressed	a) Loose governor—output shaft snap ring not seated b) Worn or broken output shaft seal rings-large diameter c) Worn seal ring grooves d) Worn collector bore e) Output shaft holes blocked	a) Repair as required b) Repair as required c) Repair as required d) Repair as required e) Repair as required
No start in P	a) Manual linkage misadjusted b) Plug connector for the neutral start switch does not fit properly	a) Adjust manual linkage b) Check plug connection
No start in P and N	a) Plug connector for the neutral start switch does not fit properly	a) Check plug connection

CLUTCH AND BAND APPLICATION

	Interm. Friction Clutch	Interm. One-Way Clutch	Overdrive Band	Reverse Clutch	Forward Clutch	Planetary One-Way Clutch	Low-Reverse Band	Direct Clutch
1st Gear Manual Low					Applied	Holding	Applied	
2nd Gear Manual Low	Applied	Holding	Applied		Applied			
1st Gear — Ⓓ(OVERDRIVE) or D(3)					Applied	Holding		
2nd Gear — Ⓓ(OVERDRIVE) or D(3)	Applied	Holding			Applied			
3rd Gear — Ⓓ(OVERDRIVE) or D(3)	Applied				Applied			Applied
4th Gear — Ⓓ(OVERDRIVE)	Applied		Applied					Applied
Reverse (R)				Applied			Applied	

Hydraulic Control System

Before the transmission can transfer the input power from the engine to the drive wheels, hydraulic pressures must be developed and routed to the varied components to cause them to operate, through numerous internal systems and passages. A basic understanding of the components, fluid routings and systems, will aid in the trouble diagnosis of the transmission.

OIL PUMP

The oil pump is a positive displacement pump, meaning that as long as the pump is turning and fluid is supplied to the inlet, the pump will deliver fluid in a volume proportionate to the input drive speed. The pump is in operation whenever the engine is operating, delivering more fluid than the transmission needs, with the excess being bled off by the pressure regulator valve and routed to the sump. It should be remembered, the oil pump is driven by a shaft which is splined into the converter cover and through a drive gear insert. The gears, in turn, are installed in a body, which is bolted to the pump support at the rear of the transmission case.

Should the oil pump fail, fluid would not be supplied to the transmission to keep the converter filled, to lubricate the internal working parts and to operate the hydraulic controls.

PRESSURE REGULATOR SYSTEM

The pressure regulator system controls the main line pressure at pre-determined levels during the vehicle operation. The main oil pressure regulator valve and spring determines the psi of the main line control pressure. The main line control pressure is regulated by balancing pressure at the end of the inner valve land, against the valve spring. When the pump begins fluid delivery and fills the passages and transmission components, the spring holds the valve closed and there is no regulation. As the pressure rises, the pressure regulator valve is moved against the spring tension, opening a passage to the torque converter. Fluid then flows into the converter, the cooler system and back to the lubrication system. As the pressure continues its rise, the pressure regulator valve is moved further against the spring tension and at a predetermined psi level and spring tension rate, the valve is moved further to open a passage, allowing excess pressurized fluid to return to the sump. The valve then opens and closes in a vibrating type action, dependent upon the fluid requirements of the transmission. A main line oil pressure booster valve is used to increase the line pressure to meet the needs of higher pressure, required when the transmission torque load increases, to operate the clutches and band servo.

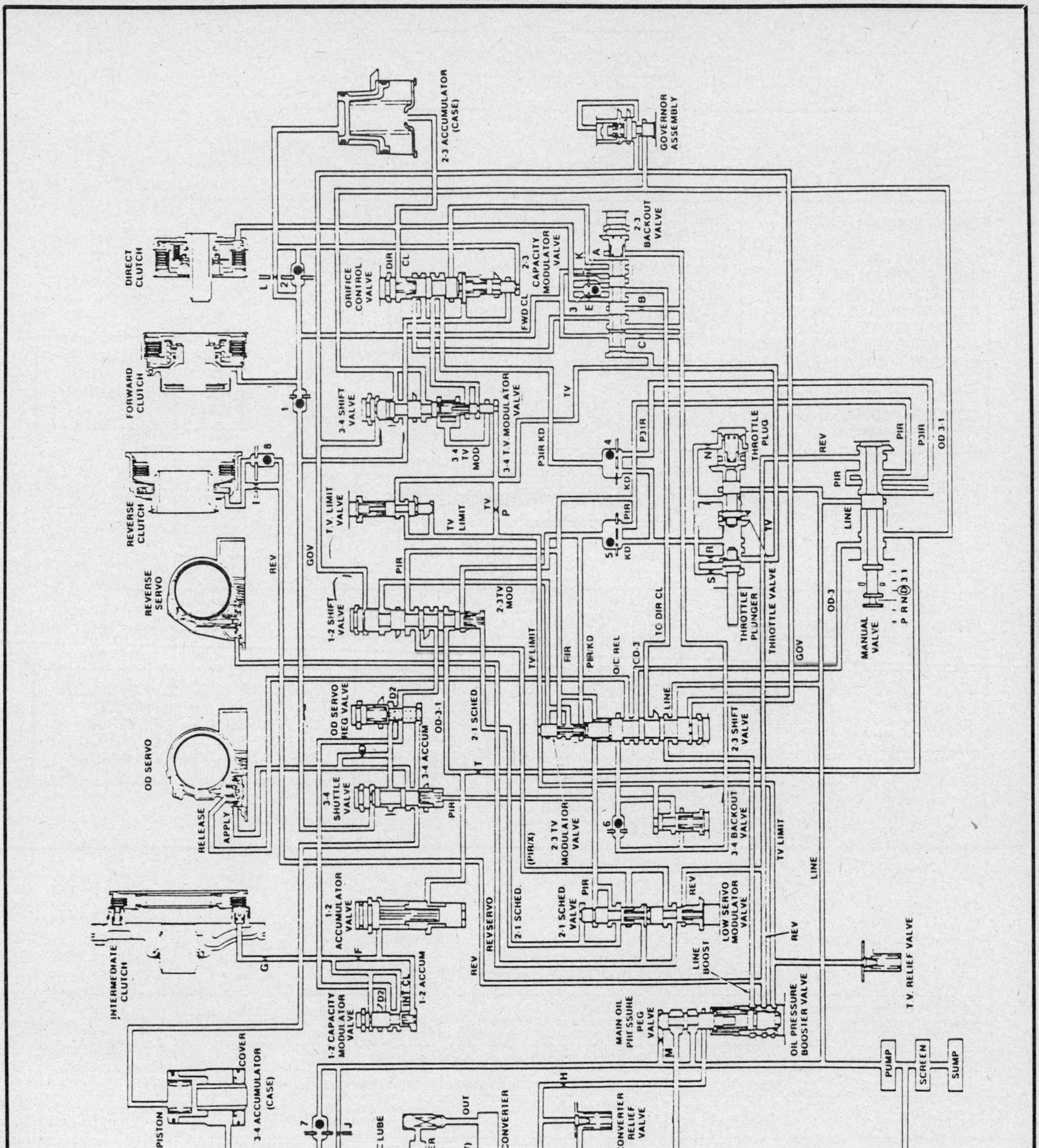

2-3 ACCUMULATOR (CASE)
GOVERNOR ASSEMBLY
DIRECT CLUTCH
ORIFICE CONTROL VALVE
2-3 CAPACITY MODULATOR VALVE
2-3 BACKOUT VALVE
FORWARD CLUTCH
3-4 SHIFT VALVE
3-4 T.V. MODULATOR VALVE
THROTTLE PLUG
REVERSE CLUTCH
T.V. LIMIT VALVE
REVERSE SERVO
1-2 SHIFT VALVE
2-3 TV MOD
THROTTLE PLUNGER
THROTTLE VALVE
MANUAL VALVE
P R N D 3 1
OD SERVO
OD SERVO REG VALVE
2-3 SHIFT VALVE
RELEASE
APPLY
3-4 SHUTTLE VALVE
2-3 TV MODULATOR VALVE
3-4 BACKOUT VALVE
INTERMEDIATE CLUTCH
1-2 ACCUMULATOR VALVE
2-1 SCHED VALVE
LOW SERVO MODULATOR VALVE
T.V. RELIEF VALVE
LINE BOOST
1-2 CAPACITY MODULATOR VALVE
MAIN OIL PRESSURE REG VALVE
OIL PRESSURE BOOSTER VALVE
PUMP
SCREEN
SUMP
COVER
PISTON
3-4 ACCUMULATOR (CASE)
LUBE
OUT
CONVERTER
IN
COOLER CHECK VALVE (PUMP ASSY)
CONVERTER RELIEF VALVE

AOD hydraulic system schematic

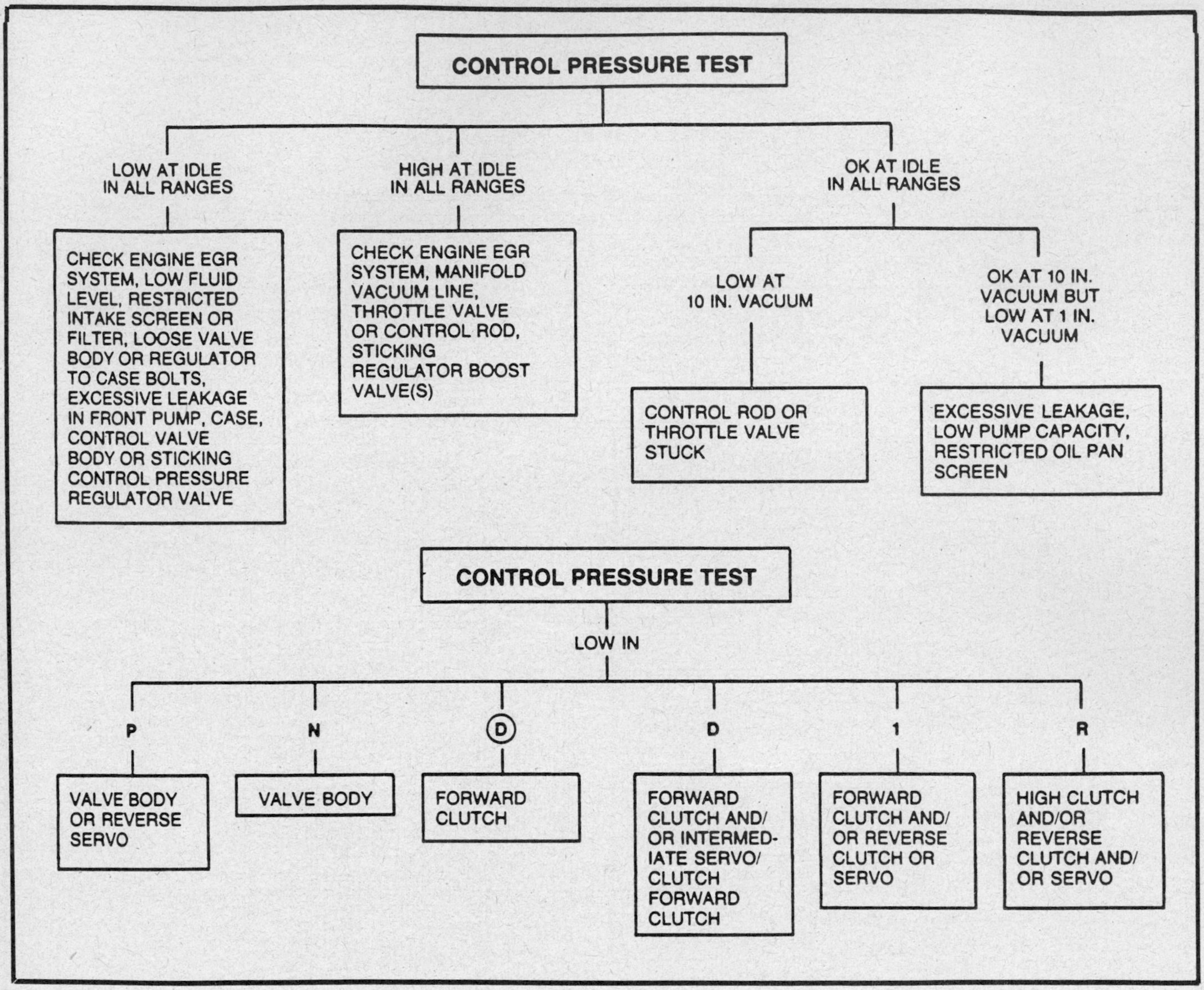
CONTROL PRESSURE TEST
LOW AT IDLE IN ALL RANGES
HIGH AT IDLE IN ALL RANGES
OK AT IDLE IN ALL RANGES
CHECK ENGINE EGR SYSTEM, LOW FLUID LEVEL, RESTRICTED INTAKE SCREEN OR FILTER, LOOSE VALVE BODY OR REGULATOR TO CASE BOLTS, EXCESSIVE LEAKAGE IN FRONT PUMP, CASE, CONTROL VALVE BODY OR STICKING CONTROL PRESSURE REGULATOR VALVE
CHECK ENGINE EGR SYSTEM, MANIFOLD VACUUM LINE, THROTTLE VALVE OR CONTROL ROD, STICKING REGULATOR BOOST VALVE(S)
LOW AT 10 IN. VACUUM
OK AT 10 IN. VACUUM BUT LOW AT 1 IN. VACUUM
CONTROL ROD OR THROTTLE VALVE STUCK
EXCESSIVE LEAKAGE, LOW PUMP CAPACITY, RESTRICTED OIL PAN SCREEN
CONTROL PRESSURE TEST
LOW IN
P
N
Ⓓ
D
1
R
VALVE BODY OR REVERSE SERVO
VALVE BODY
FORWARD CLUTCH
FORWARD CLUTCH AND/ OR INTERMED-IATE SERVO/ CLUTCH FORWARD CLUTCH
FORWARD CLUTCH AND/ OR REVERSE CLUTCH OR SERVO
HIGH CLUTCH AND/OR REVERSE CLUTCH AND/ OR SERVO

CONTROL PRESSURE DIAGNOSIS

Control Pressure Condition	Possible Cause(s)
Low in P	Valve body bolts loose, main oil regulator valve sticking, low reverse servo leakage.
Low in R	Reverse clutch leakage, low reverse servo leakage.
Low in N	Valve body bolts loose, main oil regulator valve sticking.
Low in Ⓓ	Forward clutch leakage, overdrive servo leakage, valve body bolts loose, main oil regulator valve sticking.
Low in D	Forward clutch leakage.
Low in 1	Forward clutch leakage, low. Reverse servo leakage.
Low at idle in all ranges	Low fluid level, restricted oil filter, loose valve body bolts, pump leakage, case leakage, valve body leakage, excessively low engine idle, fluid too hot, main oil regulator valve sticking.
High at idle in all ranges	T.V. linkage, valve body (throttle valve or main oil regulator valve sticking).
OK at idle but low at WOT	Internal leakage, pump leakage, restricted inlet screen, T.V. linkage, valve body (T.V. or T.V. limit valve sticking, main oil regulator valve sticking).

LINE PRESSURE SPECIFICATIONS

Engine	Operation Range	Idle		W.O.T. Stall	
		Throttle Pressure	Line Pressure	Throttle Pressure	Line Pressure
5.0L SEFI,	P,N, Ⓓ, D,1	0	55-65	79-91	180-215
5.0L HO SEFI, 5.8L	R	0	75-90	79-91	250-290
3.8L	P,N, Ⓓ, D,1	0	55-65	74-86	176-204
	R	0	75-90	74-86	241-279

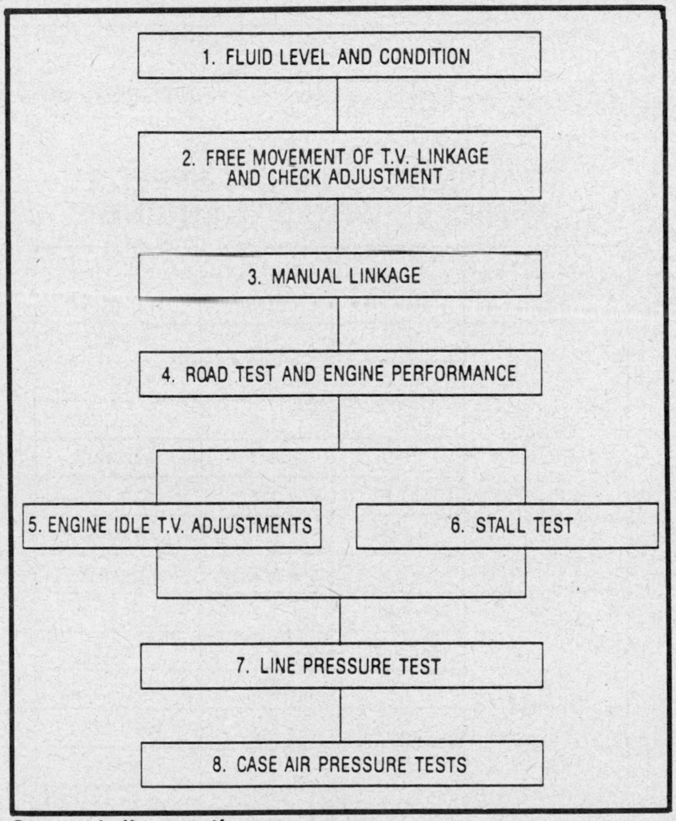
1. FLUID LEVEL AND CONDITION
2. FREE MOVEMENT OF T.V. LINKAGE AND CHECK ADJUSTMENT
3. MANUAL LINKAGE
4. ROAD TEST AND ENGINE PERFORMANCE
5. ENGINE IDLE T.V. ADJUSTMENTS
6. STALL TEST
7. LINE PRESSURE TEST
8. CASE AIR PRESSURE TESTS

General diagnostic sequence

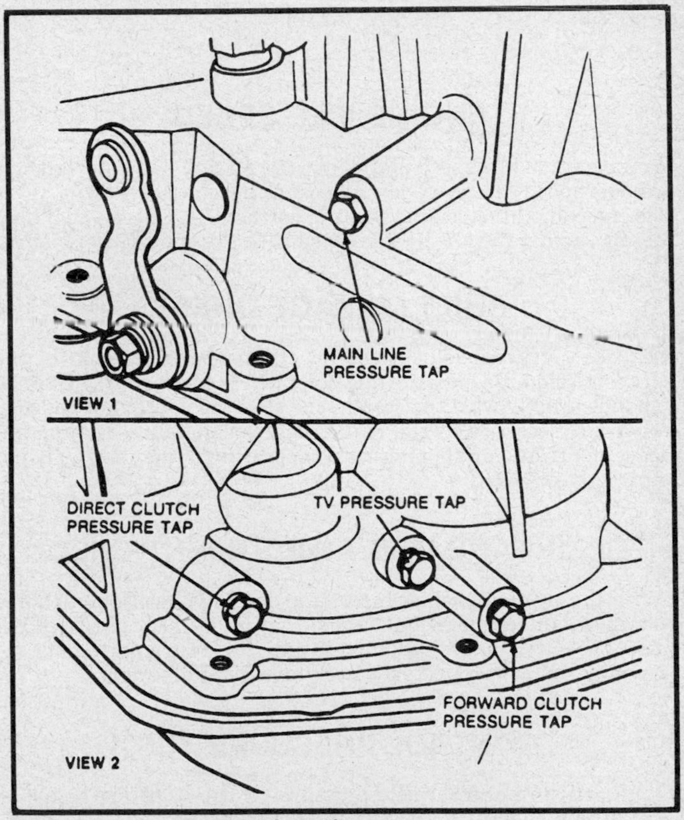
MAIN LINE PRESSURE TAP
VIEW 1
DIRECT CLUTCH PRESSURE TAP
TV PRESSURE TAP
FORWARD CLUTCH PRESSURE TAP
VIEW 2

Control pressure tap locations

MANUAL CONTROL VALVE

Main line control pressure is always present at the manual control valve. Other than required passages, such as to the converter fill, the lubricating system and certain valve assemblies, the manual valve must be moved to allow the pressurized fluid to flow to the desired components or to charge certain passages in order to engage the transmission components in their applicable gear ratios.

GOVERNOR ASSEMBLY

The governor assembly reacts to vehicle road speed and provides a pressure signal to the control valves. This pressure signal causes automatic upshifts to occur as the road speed increases and permits downshifts as the road speed decreases. The governor has 3 hydraulic passages, an exhaust governor pressure out and line pressure in, controlled by springs and weights, with the weight position determined by centrifugal force as the governor assembly rotates.

Throttle Valve (T.V.) Control Systems

The AOD transmission uses three T.V. control systems. On all vehicles equipped with 3.8L CFI engines and 5.8L carbureted engines, a T.V. control rod linkage is used. On vehicles equipped with 3.8L EFI engines, a manual locking T.V. cable control system is used. All vehicles equipped with 5.0L SEFI and 5.8L HO SEFI engines, use a self locking T.V. cable system.

Diagnostic Tests

A general diagnosis sequence should be followed to determine in what area of the transmission a malfunction exists. The following sequence is suggested by the manufacturer to diagnose and test the operation of the AOD transmission.

1. Inspect fluid level and condition.
2. Freedom of movement of the T.V. linkage and verify adjustment.
3. Correct positioning of the manual linkage.
4. Road test and engine performance.
5. Engine idle and T.V. adjustment.
6. Stall test.
7. Main line pressure test.
8. Air pressure test of case.

ENGINE IDLE SPEED

If the engine idle speed is too low, the engine will run roughly. An idle speed that is too low will cause harsh engagement, vehicle creep and harsh closed throttle downshifts. Check and if necessary, adjust the engine idle speed with the throttle positioner.

SHIFT LINKAGE CHECK

This is a critical adjustment for the proper operation of the transmission. Be sure the **OD** detent in the transmission corresponds exactly with the stop in the steering column or console. Hydraulic leakage at the manual valve can cause delay in engagements and/or slipping while operating if the linkage is not correctly adjusted.

TV LINKAGE ROD ADJUSTMENT CHECK

With the transmission selector lever in the **N** position, carburetor de-cammed, there should be no gap at TV linkage. If gap exists, check for binding grommets and TV return spring not returning TV at carburetor. Check linkage adjustment.

CONTROL PRESSURE TEST

Line pressure and throttle pressure on the AOD are tested in the idle position (0 T.V.) and the wide-open throttle (WOT) position. In each of the modes the reverse reading will be higher than the others.

1. Check the T.V. linkage for correct aadjustment.
2. Connect a 0–300 psi. gauge to the line pressure port of the transmission case. The gauge should have enough hose to be read while running the engine.
3. Connect a 0–100 psi. pressure gauge to the T.V. port on the right side of the transmission case.

NOTE: WOT readings should be taken at full stall. Run the engine at fast idle in NEUTRAL between tests for cooling.

4. Run the engine until it reaches normal operating temperature.

TRANSMISSION SHIFT SPEEDS VEHICLES WITH 5.8L ENGINE

Throttle	Range	Shift	OPS (RPM)	Column Number
				1
Closed Throttle (See Note)	Ⓓ,D	1-2	370-530	11-15
	Ⓓ,D	2-3	720-890	21-26
	Ⓓ,	3-4	1460-1850	42-53
	Ⓓ,	4-3	1580-1200	46-35
	Ⓓ,	3-2	870-700	25-20
	Ⓓ,D	2-1	470-320	14-9
	1	3-1, 2-1	730-810	36-23
Part Throttle (See Note)	Ⓓ,D	1-2	670-1050	19-30
	Ⓓ,D	2-3	1350-1780	39-51
	Ⓓ,	3-4	1790-2420	52-70
	Ⓓ,	4-3	1760-1340	51-39
	Ⓓ,D	3-2	1390-830	40-24
	Ⓓ,D	2-1	750-500	22-14
Wide Open	Ⓓ,D	1-2	1400-1940	40-56
	Ⓓ,D	2-3	2650-3020	77-87
	Ⓓ,D	3-2	2530-2170	73-63
	Ⓓ,D	2-1	1560-1060	45-31

Note: Part throttle shift speeds cannot be checked unless a TV pressure gauge is installed.

TRANSMISSION SHIFT SPEEDS VEHICLES WITH 3.8L ENGINE

Throttle	Range	Shift	OPS (RPM)
Closed Throttle	Ⓓ,D	1-2	370-530
	Ⓓ,D	2-3	740-910
	Ⓓ	3-4	1680-2060
	Ⓓ	4-3	1840-1460
	Ⓓ,D	3-2	870-720
	Ⓓ,D	2-1	470-320
	1	3-1, 2-1	1230-810
Part Throttle	Ⓓ,D	1-2	730-1140
	Ⓓ,D	2-3	1500-1890
	Ⓓ	3-4	1970-2590
	Ⓓ	4-3	1990-1780
	Ⓓ,D	3-2	1580-1170
	Ⓓ,D	2-1	770-540
Wide Open	Ⓓ,D	1-2	1430-1900
	Ⓓ,D	2-3	2630-3000
	Ⓓ,D	3-2	2570-2220
	Ⓓ,D	2-1	1520-1060

5. Apply the parking and service brakes. Shift the transmission through all of the ranges and record the line pressure and throttle pressure for each position. Compare the readings with the specifications.

STALL SPEED DIAGNOSIS

Selector Position	Stall Speeds High	Stall Speeds Low
Overdrive & D	Planetary One-Way Clutch	
Overdrive, D & 1	Forward Clutch	
Overdrive, D, 1 & R	General Problems Pressure Test. Check TV Cable Adjustment if Not Done Prior to Test.	Converter Stator One-Way Clutch or Engine Performance
R	High and/or Reverse Clutch or Low Reverse Band or Servo	

TRANSMISSION SHIFT SPEEDS VEHICLES WITH 5.0L ENGINE

Throttle	Range	Shift	OPS (RPM)
Closed Throttle (See Note)	Ⓓ,D	1-2	310-460
	Ⓓ,D	2-3	680-830
	Ⓓ	3-4	1300-1680
	Ⓓ	4-3	1440-1060
	Ⓓ,D	3-2	800-660
	Ⓓ,D	2-1	410-240
	1	3-1, 2-1	1100-730
Part Throttle (See Note)	Ⓓ,D	1-2	670-1020
	Ⓓ,D	2-3	1330-1710
	Ⓓ	3-4	1600-2210
	Ⓓ	4-3	1600-1200
	Ⓓ,D	3-2	1440-1020
	Ⓓ,D	2-1	690-480
Wide Open	Ⓓ,D	1-2	1370-1800
	Ⓓ,D	2-3	2470-2810
	Ⓓ,D	3-2	2420-2080
	Ⓓ,D	2-1	1450-1020

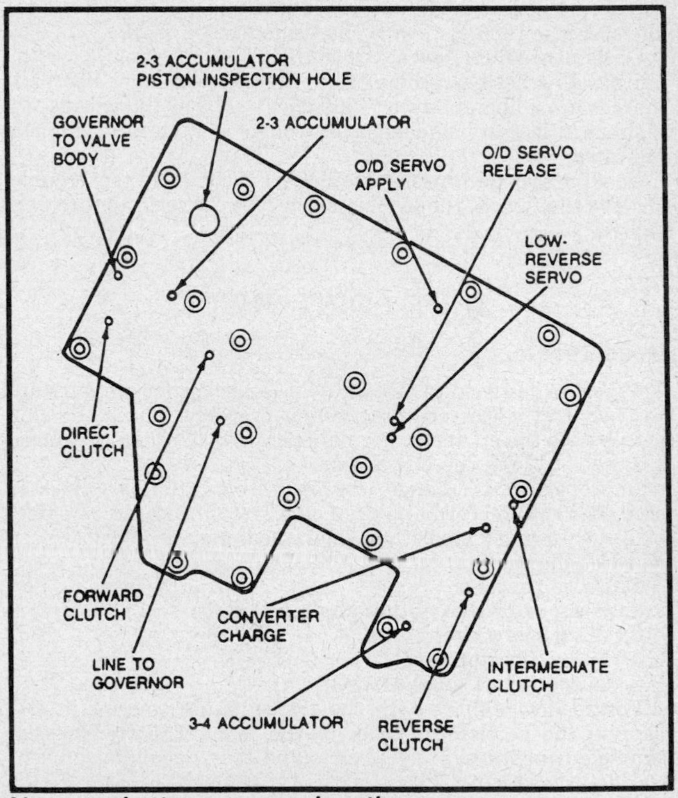
2-3 ACCUMULATOR PISTON INSPECTION HOLE
GOVERNOR TO VALVE BODY
2-3 ACCUMULATOR
O/D SERVO APPLY
O/D SERVO RELEASE
LOW-REVERSE SERVO
DIRECT CLUTCH
FORWARD CLUTCH
CONVERTER CHARGE
LINE TO GOVERNOR
INTERMEDIATE CLUTCH
3-4 ACCUMULATOR
REVERSE CLUTCH

Air test adapter passage locations

NOTE: Clutch and servo leakage may or may not show up on the control pressure test. This is because the pump has a high output volume and the leak may not be severe enough to cause a pressure drop. Pressure loss caused by a major leak is more likely to show up at idle than WOT, where the pump is delivering full volume.

DIRECT CLUTCH PRESSURE TEST

The direct clutch pressure test is designed to to diagnose a low-pressure condition or leakage in the direct clutch circuit. A difference of 15 psi. or more between direct clutch pressure and line pressure (read at the forward clutch pressure tap) will prevent a normal 3–4 shift.

1. Attach 0–300 psi. pressure gauges to the forward clutch pressure tap and to the direct clutch pressure tap.

NOTE: Gauges must be accurate enough to distinguish a 15 psi. difference and have enough hose to be read while operating the vehicle.

2. Drive the vehicle. When pressure is applied to the direct clutch , note the difference between the line pressure and direct clutch pressure.
3. If the difference in the gauge readings is more than 15 psi. there could be a leak in the direct clutch pressure circuit.

STALL TEST

The stall test checks converter clutch operation and installation, the holding ability of the forward clutch, reverse clutch, the low/reverse bands, the planetary one-way clutch and engine performance.

1. Run the engine until it reaches normal operating temperature. Check the T.V. for proper adjustment.
2. Connect a tachometer to the engine. Apply the parking and service brakes.
3. In each of the gearshift positions, press the accelerator to the floor and hold it long enough to let the engine get to full rpm. While making the test, do not hold the throttle open for more than 5 seconds at a time.

4. Note the results of the test in each gear range. After checking each range, run the engine at 1000 rpm for 15 seconds to cool the converter before the next test.

TRANSMISSION FLUID FLOW CHECK

The linkage, fluid and control pressure must be within specifications before performing this flow check.

Remove the transmission dipstick from the filler tube. Place a funnel in the transmission filler tube. Raise the vehicle, remove the cooler return line from its fitting in the case. Attach a hose to the cooler return line and fasten the free end of the hose in the funnel installed in the filler tube. Start the engine and set idle speed at 1000 rpm with the transmission in **N**.

Observe the fluid flow at the funnel. When the flow is "solid" (air bleeding has been completed), the flow should be liberal. If there is not a liberal flow at 1000 rpm in **N**, low pump capacity, main circuit system leakage or cooler system restriction is indicated.

To separate transmission trouble from cooler system trouble, observe the flow at the transmission case converter-out fitting.

SHIFT POINT CHECK

Road Test

This test is designed to determine if the governor pressure and shift control valves are functioning properly. Check the shift points with the engine at normal operating temperature. Operate the vehicle in the **OD** detent.

Improper Shift Timing–improper shift timing 4–3 backout shift (4–3 shift on full backout of accelerator pedal, accompanied by 3–4 shift when accelerator depressed) may be diagnosed/serviced by the verification of the following:

1. Loose governor
2. Worn or broken output shaft seal rings
3. Worn seal ring grooves
4. Worn collector bore
5. Output shaft holes blocked

Forced downshifts–with the transmission selector in **OD**, depress the accelerator pedal to the floor. The transmission should downshift to 3rd gear or to 2nd gear, depending on vehicle road speed.

Closed throttle downshifts–closed throttle downshifts should be extremely difficult to detect. It may be necessary to attach 0–100 psi. pressure gauges to the forward and direct clutch taps in order to detect 4–3 and 2–3 coast down shifts. A 4–3 coast (closed throttle) downshift is signified by the application of the forward clutch (the pressure reading from the forward clutch pressure apply will indicate an increase in pressure from 0–60 psi). A 3–2 coast downshift is signified by the release of the direct clutch. A 2–1 downshift should be imperceptible (the pressure reading from the gauge on the direct clutch pressure tap will indicate a decrease in pressure from 60–0 psi.

In Shop Check

A shift test can be performed in the shop to check for 1–2, 2–3 and 3–4 upshifts.

1. Raise and safely support the vehicle so that the rear wheels are off the ground.

CAUTION

Never exceed 60 mph speedometer speed with the vehicle raised.

2. Place the transmission lever in **OD**and make a minimum throttle 1–2, 2–3 and 3–4 upshift.
3. When the shifts occur, the speedometer needle will make a momentary surge and the driveline will bump.
4. Check the shift points against the specifications. If the shift points are within specification, the 1–2, 2–3 and 3–4 shift valves and governor are ok.
5. If the shift points are not within specification, perform a governor check to isolate the problem.

AIR PRESSURE TEST

A no drive condition can exist, even with the correct transmission fluid pressure, because of inoperative clutches or band. Erratic shifts could be caused by a stuck governor valve. The inoperative units can be located through a series of checks by substituting air pressure for the fluid pressure to determine the location of the malfunction.

A no drive condition in **D** and **2** may be caused by an inoperative band or one-way clutch. When there is no drive in **1**, the difficulty could be caused by improper functioning of the direct clutch or band and the one-way clutch. Failure to drive in **R** range could be caused by a malfunction of the reverse clutch or one-way clutch.

When a slip problem is evident, whether it is in the valve body or in the hydraulic system beyond the valve body, the air pressure tests can be very valuable.

To properly air test the transmission, a main control to case gasket and the following special service tools or equivalent will be required.

a. Adapter plate tool T82L–7006–A
b. Adapter plate attaching screws tool T82P–7006–C
c. Air nozzle tool 7000–DE
d. Air nozzle rubber tip tool 7000–DD

With the main control body removed, position the adapter plate and gasket on the transmission. Install the adapter plate attaching screws and tighten the screws to 80–100 inch lbs. (9–11 Nm) torque. Note that each passage is identified on the plate. Using the air nozzle equipped with the rubber tip, apply air pressure to each passage in the following order:

Overdrive Servo

Apply air pressure to the OD servo appy passage in the service plate tool. Operation of the band is indicated by the tightening of the band around the reverse clutch drum. The OD servo will return to the release position as a result of spring force from the release spring. Also, when the servo returns to the release position, a thud can be felt on the OD servo cover. The band will then relax.

Direct Clutch

Apply air pressure to the forward clutch apply passage in the service tool plate. A dull thud can be heard or movement of the piston can be felt on the case as the clutch piston is applied. If the clutch seal(s) are leaking a hissing sound will be heard.

Intermediate Clutch

Apply air pressure to the intermediate clutch apply passage in the service tool plate. A dull thud can be heard or movement of the piston can be felt on the case as the clutch piston is applied. If the clutch seal(s) are leaking a hissing sound will be heard.

Reverse Clutch

Apply air pressure to the reverse clutch apply passage in the service tool plate. A dull thud can be heard or movement of the piston can be felt on the case as the clutch piston is applied. If the clutch seal(s) are leaking a hissing sound will be heard.

Forward Clutch

Apply air pressure to the forward clutch apply passage in the service plate. A dull thud can be heard or movement of the piston can be felt on the case as the clutch piston is applied.

3–4 Accumulator

Apply air pressure to the 3–4 accumulator apply passage, the accumulator should unseat.

Low/Reverse Servo

Apply air pressure to the low/reverse servo apply passage in the service plate. A dull thud can be heard when the low/everse servo band tightens around the planetarý assembly drum surface. Also, movement of the ring gear can be detected.

2–3 Accumulator

Apply air pressure to the 2–3 accumulator apply passage in the service plate. The accumulator piston should unseat. This can be detected by inserting a metal rod into the 2–3 piston inspection hole. When the piston unseats, the rod will move. Also a dull thud can be heard when the piston is applied.

Governor

In order to air pressure test the line to governor passage and the governor to valve body passage, the driveshaft, crossmember and extension housing must be removed.

Apply air pressure to the line to governor passage in the service plate while checking the governor valve. If air is escaping from the governor valve then the passage is unobstructed.

To air pressure check the governor to valve body passage, remove the governor. Apply air pressure to the passage while checking the holes in the output shaft. If air escapes any of the holes, the passage is unobstructed.

ON CAR SERVICES

Adjustments

THROTTLE VALVE CONTROL SYSTEM

5.8L Carbureted Engine

The T.V. control linkage is set to its proper length during initial assembly using the sliding trunnion block at the transmission end of the T.V. control rod. Under normal circumstances, it should not be necessary to to alter this adjustment. Any required adjustment of the T.V. control linkage can normally be accomplished using the adjustment screw on the linkage lever at the carburetor.

Major linkage adjustment (sliding trunnion on rod) may only be required after maintenance involving the removal and/or replacement of the carburetor, T.V. control rod assembly or the transmission. Minor linkage adjustment (adjustment screw on the linkage lever) may be required after installing a new main control assembly, or after idle speed adjustments greater than 50 rpm and to correct poor transmission shift quality.

MINIMUM IDLE STOP

When the linkage is correctly adjusted, the T.V. control lever on the transmission will be at its internal idle stop position (lever up as far as it will travel) when the carburetor throttle lever is at its minimum idle stop. There will be a light contact force between the the throttle lever and end of the linkage lever adjustment screw. Due to flexibility in the linkage system, the linkage lever adjustment screw would have to be backed out approximately 3 turns before a gap between the screw and throttle lever could be detected.

At WOT, the T.V. control lever on the transmission may or may not be at its wide open stop. The WOT throttle position must not be used as the reference point in adjusting the linkage.

LINKAGE ADJUSTMENT AT CARBURETOR

1. Set carburetor at minimum idle stop. Place shift lever in **N** and set parking brake.
2. Back out linkage lever adjusting screw all the way (screw end is flush with lever face).
3. Turn in adjusting screw until a clearance of 0.005 in. is reached (check with feeler gauge).

NOTE: To eliminate the effect of friction, push the linkage lever forward and release before checking clearance between end of screw and throttle.

4. Turn in screw 3 turns (3 turns are preferred. If travel is limited, 1 turn is sufficient).
5. If it is not possible to turn in adjusting screw at least 1 turn, if there was insufficient screw adjusting capacity to obtain an initial gap, adjustment at the transmission is necessary.

LINKAGE ADJUSTMENT AT TRANSMISSION

The linkage lever adjustment screw has limited adjustment capability. If it is not possible to adjust the T.V. linkage using this screw, the length of the T.V. control rod assembly must be readjusted using the following procedure. This procedure must also be followed whenever a new control rod assembly is installed.

1. Set carburetor at its minimum idle stop. Place the shift lever in **N** and set the parking brake.
2. Set the linkage adjustment screw at approximately midrange.
3. If a new T.V. control rod assembly is being installed, connect the rod to the linkage lever at the carburetor.
4. Raise and safley support the vehicle.
5. Loosen the bolt on the sliding trunnion block on the T.V. control rod assembly. Remove any corrosion from the control rod and free the trunnion block so that it slides freely on the control rod.
6. Push up on the lower end of the control rod to ensure that the linkage lever at carburetor is firmly against the throttle lever. Release the tension on the rod, rod must stay up.
7. Push the T.V. control lever on the transmission up against its internal stop firmly. Tighten the bolt on the trunnion block, do not release the force on the block until the bolt is tightened.
8. Lower the vehicle and verify that the throttle lever is still against the minimum idle stop or throttle solenoid positioner stop. If not, repeat the adjustment procedure.

3.8L EFI and 5.0L HO/SEFI Engines

The T.V. control system is set and locked to its proper length during initial assembly by pushing down on the locking lever at the throttle body end of the cable assembly. When the lever is unlocked, the cable is released for adjustment. The take up spring at this end of the cable automatically tensions the cable when released. With the slack taken up and the locking lever pushed, the take up spring plays no part in the operation of the system.

Under normal circumstances, it should not be necessary to alter or readjust, the initial setting of the T.V. control cable. Situations requiring the readjustment of the T.V. control cable include maintenance involving the removal and/or replacement of the throttle body, transmission, T.V. cable assembly or main control assembly.

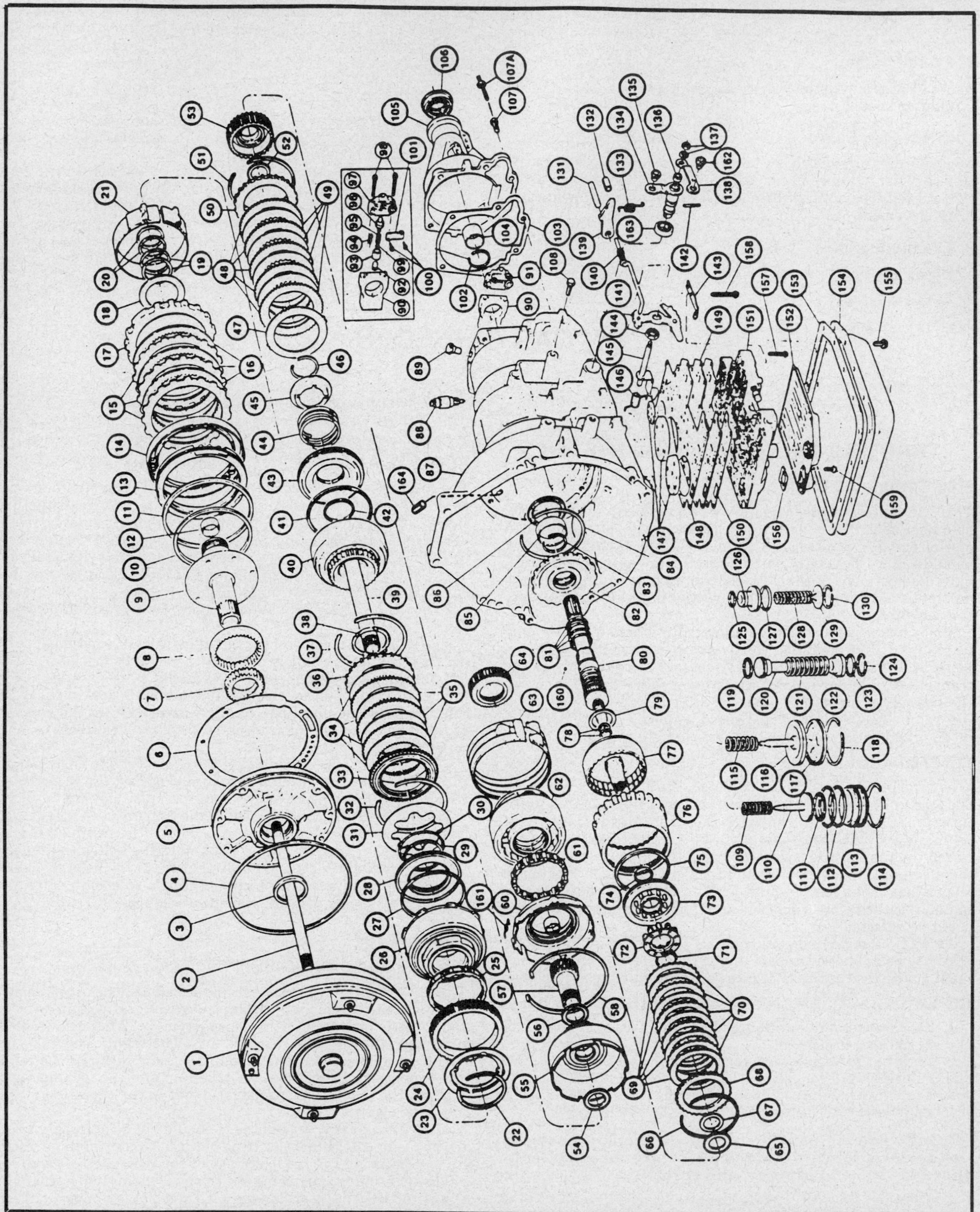

Exploded view of the AOD transmission

1. Torque converter
2. Direct drive shaft
3. Front pump seal
4. Front pump O-ring
5. Front pump body
6. Front pump gasket
7. Front pump drive gear
8. Front pump driven gear
9. Front pump stator support
10. Intermediate piston inner lip seal
11. Intermediate piston outer lip seal
12. Front pump bushing
13. Intermediate clutch piston
14. Intermediate clutch piston return spring
15. Intermediate clutch external spline steel plates
16. Intermediate clutch internal spline friction plates
17. Intermediate clutch pressure plates
18. Front pump No. 1 thrust washer
19. Stator support seal rings – reverse clutch
20. Stator support seal rings – forward clutch
21. Overdrive band
22. Intermediate one-way clutch retaining snapring
23. Intermediate one-way clutch retaining plate
24. Intermediate one-way clutch outer race
25. Intermediate one-way clutch assembly
26. Reverse clutch drum
27. Reverse clutch piston seal
28. Reverse clutch piston
29. Reverse clutch inner piston seal
30. Thrust ring
31. Reverse clutch piston return spring
32. Retaining snapring
33. Reverse clutch front pressure plate
34. Forward and reverse clutch internal spline friction plate
35. Reverse clutch external spline steel plate
36. Forward and reverse clutch rear pressure plate
37. Reverse clutch retaining ring
38. No. 2 thrust washer
39. Turbine shaft
40. Forward clutch cylinder and turbine shaft
41. Forward clutch outer piston seal
42. Forward clutch inner piston seal
43. Forward clutch piston
44. Forward clutch piston return spring
45. Return spring retainer
46. Retaining snapring
47. Waved plate
48. Forward clutch external spline steel plate
49. Forward and reverse clutch internal spline friction plate
50. Forward and reverse clutch pressure plate
51. Retaining snapring
52. No. 3 needle bearing – forward clutch
53. Forward clutch hub
54. No.4 needle bearing
55. Reverse sun gear and drive shell
56. No. 5 needle bearing
57. Forward sun gear
58. Center support retaining ring
59. Center support
60. Center support planetary
61. Planetary one-way clutch and cage spring and roller
62. Planetary assembly
63. Reverse band
64. Direct clutch hub
65. No. 7 needle bearing
66. Retaining snapring
67. Thrust spacer
68. Direct clutch pressure plate
69. Direct clutch internal splined plate
70. Direct clutch external splined plate
71. Retaining snapring
72. Return spring and retainer
73. Direct clutch piston
74. Direct clutch piston inner seal
75. Direct clutch piston outer seal
76. Ring gear and park gear
77. Direct cylinder
78. Output shaft steel seal rings
79. No. 8 needle bearing
80. Output shaft
81. Large output shaft seal
82. Output shaft seal
83. Retaining snapring
84. Retaining snapring
85. Rear case bushing
86. No. 9 needle bearing
87. Case assembly
88. Neutral start switch
89. Vent cap
90. Governor counterweight
91. Governor body assembly
92. Governor plug
93. Governor sleeve
94. Governor oil screen assembly
95. Governor valve spring
96. Governor valve
97. Governor body
98. Bolt
99. Clip
100. Bolt
101. Governor valve cover
102. Snapring
103. Extension housing bracket
104. Extension housing bushing
105. Extension housing
106. Extension housing seal
107. Extension housing bolt
107A. Extension housing stud
108. Pipe plug
109. Overdrive servo piston return spring
110. Overdrive servo piston
111. Overdrive servo piston seal
112. Overdrive servo cover seals
113. Overdrve servo cover
114. Retaining snapring
115. Reverse servo piston return spring
116. Reverse servo piston
117. Reverse servo cover
118. Snapring
119. 3–4 accumulator valve seal
120. 3–4 accumulator valve
121. 3–4 accumulator valve return spring
122. 3–4 accumulator cover
123. 3–4 accumulator cover seal
124. Snapring
125. 2–3 accumulator valve small seal
126. 2–3 accumulator valve
127. 2–3 accumulator valve large seal
128. 2–3 accumulator valve return spring
129. 2–3 accumulator cover
130. Snapring
131. Park pawl shaft
132. Guide cup
133. Park pawl return spring
134. Manual lever
135. Grommet
136. Throttle lever oil seal
137. Attaching nut and lock washer
138. Throttle lever
139. Park pawl
140. Park pawl actuating rod
141. Inner manual lever
142. Roll pin
143. Detent spring
144. Attaching nut
145. Inner throttle lever
146. Throttle torsion spring
147. Valve body reinforcement plate
148. Separator plate gasket
149. Separator plate
150. Separator plate gasket
151. Valve body
152. Oil filter and grommet
153. Oil pan gasket
154. Oil pan
155. Bolt
156. Oil filter gasket
157. Bolt
158. Bolt
159. Bolt
160. Governor drive ball
161. Anti clunk spring
162. Grommet
163. Oil seal assembly
164. Connector assembly
165. Bolt

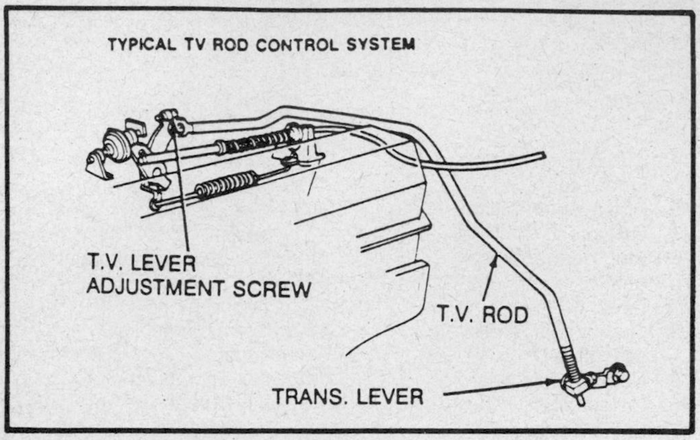
TYPICAL TV ROD CONTROL SYSTEM
T.V. LEVER ADJUSTMENT SCREW
T.V. ROD
TRANS. LEVER

T.V rod system components

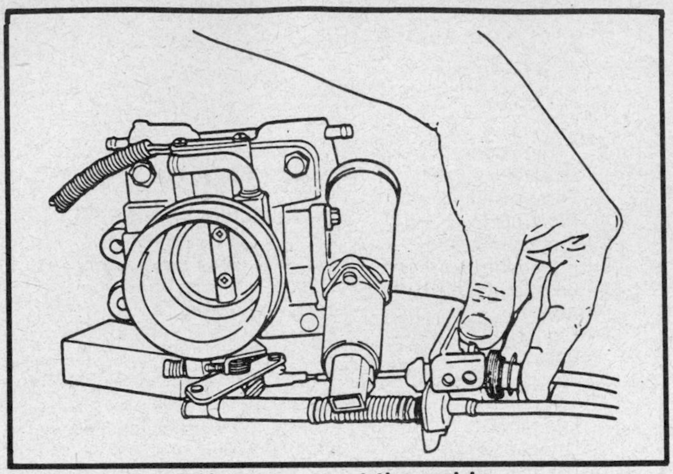

Adjusting the T.V. pressure at the cable

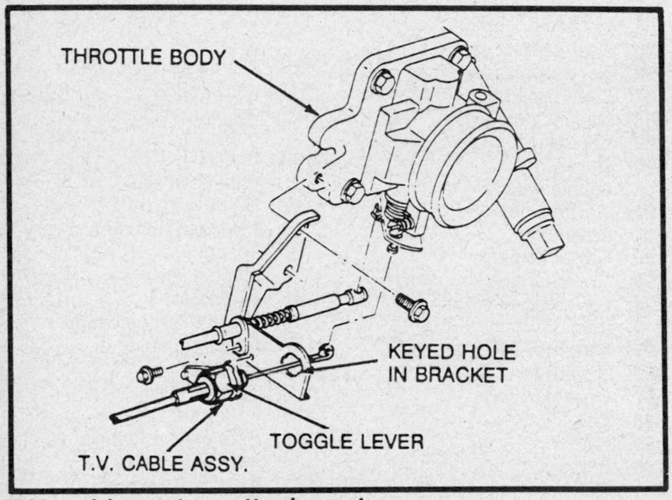
THROTTLE BODY
KEYED HOLE IN BRACKET
TOGGLE LEVER
T.V. CABLE ASSY.

T.V. cable system attachment

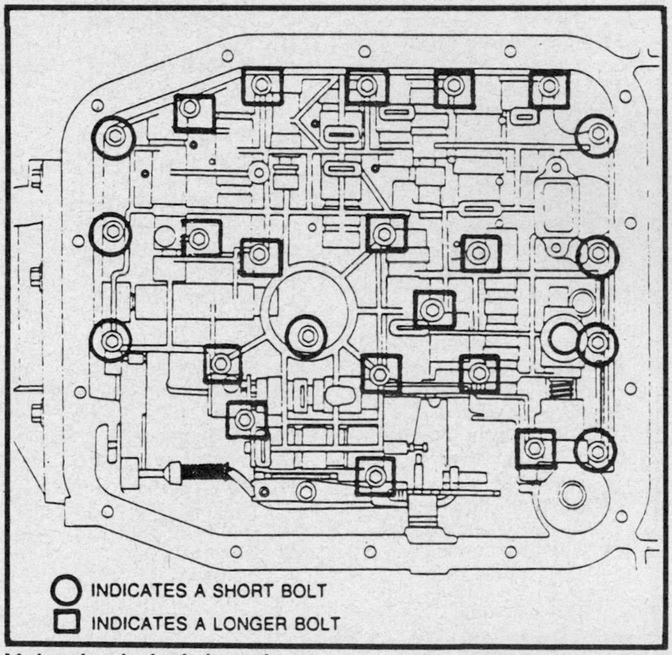
INDICATES A SHORT BOLT
INDICATES A LONGER BOLT

Valve body bolt locations

NOTE: The EFI engine uses an air bypass valve (ISC, that does not affect throttle position. Therefore, idle automatic setting does not affect T.V. cable adjustment.

LINKAGE ADJUSTMENT USING T.V. CONTROL PRESSURE

To correctly perform this procedure, the use of a T.V. pressure gauge with hose (0–60 psi) T86L–70002–A or equivalent is required. Also, T.V. cable gauge tool T86L–70332–A or equivalent is needed for correct adjustment.

1. Attach the T.V. pressure gauge to the T.V. port on the transmission.
2. Insert the tapered end of the cable gauge tool between the crimped slug end of the cable and plastic cable fitting that attaches to the throttle lever. Push gauge as far in as it will go.
3. Run the engine until it reaches normal operating temperature. The transmission fluid temperature should be approximatelt 100–150°F.
4. Set the parking brake and place the shift lever in **N**, the T.V. pressure should be between 30–40 psi. For best transmission operaton, set T.V. pressure as close as possible to 33 psi.

NOTE: T.V. pressure must be set with the transmission in N.

5. Using a small pry bar, pry up the white toggle lever on the cable adjuster located behind the throttle body cable mounting bracket. The adjuster pre-load spring should cause the the adjusting slider to move away from the throttle body and T.V. pressure should increase.
6. Push on the slider form behind bracket until T.V. pressure is 33 psi. While still holding slider, push down on toggle lever as far as it will go, locking the slider in position. Toggle lever must be completely down for proper locking on cable.
7. Remove gauge tool, allowing cable spring to return to its normal idling position. With the engine still idling, T.V. pressure must be at or near 0 psi (less than 5 psi.). If the T.V. pressure is not in this range, repeat the adjustment procedure but lower the setting to 30 psi. Recheck the pressure.
8. Remove the T.V. gauge and check transmission shift operation.

Services

FLUID CHANGES

Normal maintenance and lubrication requirements do not necessitate periodic transmission fluid changes. When used under

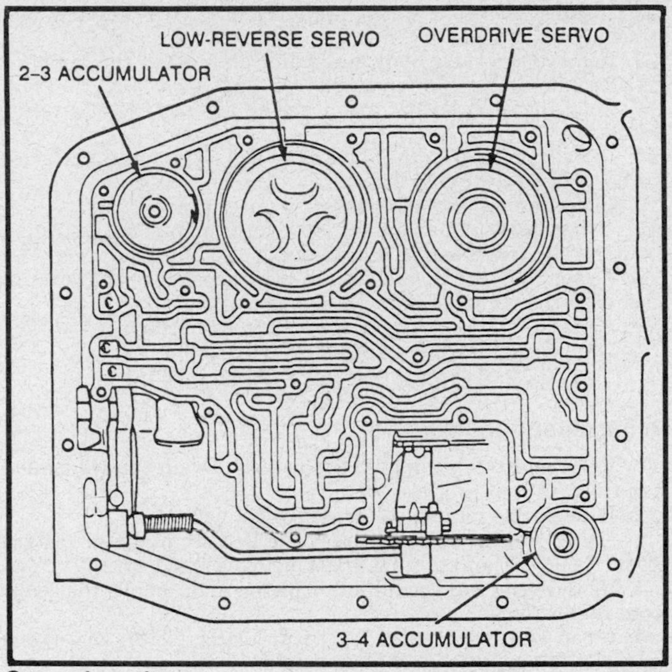
LOW-REVERSE SERVO
OVERDRIVE SERVO
2-3 ACCUMULATOR
3-4 ACCUMULATOR

Servo locations

continuous or severe conditions the fluid should be changed more frequently than when used in normal circumstances.

The specified fluid to use in the AOD transmission is Mercon® fluid or an equivalent fluid that meets Ford specification ESP-M2C-185-C. Use of fluid that does not meet this specification could result in transmission failure.

The fluid capacity of the AOD for an complete service, including converter drain and fill, is 12.3 qts. The drain and refill service capacity is approximately 3 qts.

OIL PAN

Removal and Installation

1. Raise the vehicle and support safely. Place a drain pan under the oil pan and loosen the pan attaching bolts.
2. Allow the fluid to drain from the oil pan, to the level of the pan flange. Remove the pan attaching bolts in such a manner as to allow the oil pan to drop slowly, draining more of the fluid from the oil pan.
3. Remove the oil pan and drain the remaining fluid into the container. Discard the old gasket and clean the oil pan.

NOTE: The torque converter on the AOD transmission can be drained also. This can be accomplished by removing the dust cover and then removing the drain plug from the converter. The drain plug is accessible by turning the converter.

4. Clean or replace the oil filter screen. Install a new gasket on the oil pan and install the pan on the transmission. Tighten the oil pan attaching bolt to 6–10 ft. lbs.
5. Lower the vehicle and fill the transmission to the correct level with the specified fluid. Re-check the level as required, using the room temperature checking procedure.

VALVE BODY

Removal and Installation

1. Raise and safely support the vehicle.
2. Starting at the rear and working toward the front, loosen the oil pan attaching bolts and drain the fluid from the transmission.
3. Remove the transmission pan attaching bolts, the pan and gasket.
4. Remove the filter-to-valve body attaching bolts and remove the filter, grommet and gasket.
5. Remove the detent spring attaching bolt and spring.
6. Remove the 24 valve body attaching bolts and remove the valve body. Discard the gasket.
7. Clean and inspect the valve body.
8. Using the valve body guide pins, T80L-77100-A or equivalent, position the valve body to the case. Make sure the inner manual lever and inner T.V. levers are engaged.
9. Install the 24 valve body-to-case bolts and tighten to 80–100 inch lbs.
10. Install the detent spring and attaching bolt. Tighten the attaching bolt to 80–100 inch lbs.
11. Remove the valve body guide pins and install the remaining attaching bolts, tighten to 80–100 inch lbs.
12. Load the T.V. lever torsion spring against the separator plate.
13. Install the filter to the valve body and tighten the 3 attaching bolts to 80–100 inch lbs.
14. Clean the oil pan and gasket surfaces thoroughly.
15. Using a new oil pan gasket, attach the oil pan to the transmission. Tighten the attaching bolts to 12–16 ft. lbs.
16. Lower the vehicle and fill the transmission to the correct level with the correct fluid. Check the fluid level.
17. Adjust the linkage rod or cable as needed.

OVERDRIVE SERVO ASSEMBLY

Removal and Installation

1. Raise and safely support the vehicle. Drain the transmission fluid from the vehicle.
2. Remove the oil pan, filter and valve body.
3. Depress the overdrve servo piston cover with a blunt object and remove the snapring retaining the piston.
4. Using servo piston remover, T80L-77030-B or equivalent, apply low air pressure to the servo piston release passage in order to remove the overdrive servo piston cover and spring.
5. Remove the piston from the cover. Remove the rubber seals from the piston and cover.
6. Clean and inspect the servo piston for nicks or burrs. Clean and inspect the servo piston pocket in the case for burrs.
7. Install new seals on the servo piston and cover. Lubricate the seals with clean transmission fluid. Install the servo piston into the cover and lubricate with clean transmission fluid.
8. Assemble the return spring to the servo piston. Install the overdrive piston cover and spring into the overdrive servo pocket and case. Make sure the servo rod contacts the overdrive band apply pocket.
9. Using a blunt object, depress the overdrive servo and install the retaining snapring.
7. Using the valve body guide pins, T80L-77100-A or equivalent, position the valve body to the case. Make sure the inner manual lever and inner T.V. levers are engaged.
8. Install the 24 valve body-to-case bolts and tighten to 80–100 inch lbs.
9. Install the detent spring and attaching bolt. Tighten the attaching bolt to 80–100 inch lbs.
10. Remove the valve body guide pins and install the remaining attaching bolts, tighten to 80–100 inch lbs.
11. Load the T.V. lever torsion spring against the separator plate.
12. Install the filter to the valve body and tighten the 3 attaching bolts to 80–100 inch lbs.
13. Clean the oil pan and gasket surfaces thoroughly.
14. Using a new oil pan gasket, attach the oil pan to the trans-

mission. Tighten the attaching bolts to 12–16 ft. lbs.

15. Lower the vehicle and fill the transmission to the correct level with the correct fluid. Check the fluid level.

16. Adjust the linkage rod or cable as needed.

LOW/REVERSE SERVO ASSEMBLY

Removal and Installation

1. Raise and safely support the vehicle. Drain the transmission fluid.
2. Remove the oil pan, filter and valve body.
3. Depress the reverse servo piston cover with a blunt object and remove the retaining snapring.
4. To remove the reverse servo piston and spring, apply low air pressure to the servo piston release passage using tool T80L–77030–B or equivalent.
5. Cover the servo piston pocket to prevent the piston from falling out of the case when the air pressure is applied.
6. Clean and inspect the reverse servo piston, cover and spring.
7. Assemble the return spring to the servo piston. Install the servo piston and spring into the cover and into the case.
8. Using a blunt object, depress the reverse servo piston and install the retaining snapring.
9. Using the valve body guide pins, T80L–77100–A or equivalent, position the valve body to the case. Make sure the inner manual lever and inner T.V. levers are engaged.
10. Install the 24 valve body-to-case bolts and tighten to 80–100 inch lbs.
11. Install the detent spring and attaching bolt. Tighten the attaching bolt to 80–100 inch lbs.
12. Remove the valve body guide pins and install the remaining attaching bolts, tighten to 80–100 inch lbs.
13. Load the T.V. lever torsion spring against the separator plate.
14. Install the filter to the valve body and tighten the 3 attaching bolts to 80–100 inch lbs.
15. Clean the oil pan and gasket surfaces thoroughly.
16. Using a new oil pan gasket, attach the oil pan to the transmission. Tighten the attaching bolts to 12–16 ft. lbs.
17. Lower the vehicle and fill the transmission to the correct level with the correct fluid. Check the fluid level.
18. Adjust the linkage rod or cable as needed.

3–4 ACCUMULATOR PISTON

Removal

1. Raise and safely support the vehicle. Drain the transmission fluid.
2. Remove the oil pan, filter and valve body.
3. Depress the 3–4 accumulator piston cover with a blunt object and remove the retaining snapring.
4. Slowly release the tension on the 3–4 accumulator cover and remove the piston cover and piston. On some models a spring is also used.
5. Remove the seals from the accumulator piston and cover.
6. Clean and inspect the accumulator piston and cover.
7. Install new seals on the piston and cover.
8. Lubricate the seals with clean transmission fluid. Install the 3–4 accumulator piston and spring into the case. Install the cover.
9. Depress the cover into the piston pocket and install the retaining snapring.
10. Using the valve body guide pins, T80L–77100–A or equivalent, position the valve body to the case. Make sure the inner manual lever and inner T.V. levers are engaged.
11. Install the 24 valve body-to-case bolts and tighten to 80–100 inch lbs.
12. Install the detent spring and attaching bolt. Tighten the attaching bolt to 80–100 inch lbs.
13. Remove the valve body guide pins and install the remaining attaching bolts, tighten to 80–100 inch lbs.
14. Load the T.V. lever torsion spring against the separator plate.
15. Install the filter to the valve body and tighten the 3 attaching bolts to 80–100 inch lbs.
16. Clean the oil pan and gasket surfaces thoroughly.
17. Using a new oil pan gasket, attach the oil pan to the transmission. Tighten the attaching bolts to 12–16 ft. lbs.
18. Lower the vehicle and fill the transmission to the correct level with the correct fluid. Check the fluid level.
19. Adjust the T.V linkage rod or cable as needed.

2–3 ACCUMULATOR PISTON

Removal and Installation

1. Raise and safely support the vehicle. Drain the transmission fluid.
2. Remove the oil pan, filter and valve body.
3. Using a blunt object depress the 2–3 accumulator piston cover. Remove the retaining snapring, cover and spring.
4. Remove the 2–3 accumulator piston and remove the seals from the piston.
5. Clean and inspect the 2–3 accumulator piston and cover for nicks or burrs. Replace if damaged.
6. Install new seals on the 2–3 accumulator piston and install the piston into the case.
7. Install the return spring and cover. Depress the 2–3 accumulator cover and install the retaining snapring.
8. Using the valve body guide pins, T80L–77100–A or equivalent, position the valve body to the case. Make sure the inner manual lever and inner T.V. levers are engaged.
9. Install the 24 valve body-to-case bolts and tighten to 80–100 inch lbs.
10. Install the detent spring and attaching bolt. Tighten the attaching bolt to 80–100 inch lbs.
11. Remove the valve body guide pins and install the remaining attaching bolts, tighten to 80–100 inch lbs.
12. Load the T.V. lever torsion spring against the separator plate.
13. Install the filter to the valve body and tighten the 3 attaching bolts to 80–100 inch lbs.
14. Clean the oil pan and gasket surfaces thoroughly.
15. Using a new oil pan gasket, attach the oil pan to the transmission. Tighten the attaching bolts to 12–16 ft. lbs.
16. Lower the vehicle and fill the transmission to the correct level with the correct fluid. Check the fluid level.
17. Adjust the T.V linkage rod or cable as needed.

EXTENSION HOUSING BUSHING AND REAR SEAL

Removal and Installation

1. Raise and safely support the vehicle.
2. Disconnect the driveshaft at the transmission. To keep the driveline balance, mark the driveshaft yoke and axle flange so that the driveshaft can be installed in its original position.
3. Remove the rear seal from the transmission using a seal remover, T74P–77248–A or equivalent.
4. After removing the seal, the bushing can be removed using a bushing remover, T77L–7697–A or equivalent.
5. Inspect the seal bore for nicks or burrs, any burrs can be removed with crocus cloth.
6. Install new bushing using a bushing installer, T80L–77034–A or equivalent.
7. Install a new seal into the extension housing using a seal installer, T61L–7657–A or equivalent. The seal should be firmly

seated in the bore. Coat the inside diameter of the seal with a long life multi-purpose lubricant.

8. Install the driveshaft aligning the marks made during removal.

EXTENSION HOUSING

Removal and Installation

1. Raise and safely support the vehicle.
2. Disconnect the parking brake cable from the equalizer.
3. Disconnect the driveshaft at the transmission. To keep the driveline balance, mark the driveshaft yoke and axle flange so that the driveshaft can be installed in its original position.
4. Disconnect the speedometer cable from the extension housing.
5. Remove the engine rear suport-to-extension housing bolts. Place a suitable lifting device under the transmission and raise it enough to remove the weight from the engine support.
6. Remove the bolt that secures the rear engine support to the crossmember and remove the support.
7. Place a drain pan under the rear of the transmission case. Lower the transmission and remove the extension housing attaching bolts. Slide the extension housing off the output shaft and allow the fluid to drain.
8. Remove and discard the extension housing gasket.
9. Clean the extension housing and transmission mounting surfaces. Remove any sealant from bolts and bolt holes. Position a new gasket on the transmission.
10. Coat all of the attaching bolts with thread sealer. Position the extension housing and install the attaching bolts, tightening to 16–20 ft. lbs.
11. Using a suitable lifting device, raise the transmission high enough to position the rear engine support on the crossmember.
12. Install the support-to-crossmember bolt and tighten to 25–35 ft. lbs.
13. Lower the transmission and install the support-to-extension housing bolts, tighten to 50–70 ft. lbs.
14. Connect the speedometer cable to the extension housing. Connect the parking brake cable to the equalizer.
15. Install the driveshaft in the transmission, aligning the matchmarks made during removal.
16. Lower the vehicle and fill the transmission to the correct level with the appropriate fluid. Run the engine and recheck the fluid level.

GOVERNOR

Removal and Installation

1. Raise and safely support the vehicle.
2. Remove the extension housing. Remove the governor-to-output shaft retaining snapring.
3. Using a soft faced hammer, tap the governor assembly off the output shaft. Remove the governor drive ball.
4. Remove the governor-to-counterweight attaching screws and lift the governor from the counterweight.
5. Lubricate the governor valve parts with clean transmission fluid. Make certain that the valve moves freely in the valve body bore.
6. Position the governor valve body on the counterweight with the cover facing towards the front of the vehicle. Install the attaching screw and tighten to 50–60 inch lbs.
7. Position the governor drive ball into the pocket of the output shaft.
8. Align the keyway in the counterweight to the governor drive ball. Slide the governor assembly onto the output shaft. If necessary, gently tap the governor into position with a soft faced hammer.
9. Reinstall the governor-to-output shaft retaining snapring. Clean the mounting surface on the transmission and on the extension housing. Position a new gasket on the transmission.
10. Install the extension housing.

NEUTRAL SAFETY SWITCH

Removal and Installation

TOWN CAR, 1984–87 CONTINENAL, CROWN VICTORIA/GRAND MARQUIS

1. Set the parking brake and palce the transmission shift lever in the 1 position.
2. Open the hood and remove the air cleaner assembly. Disconnect the negative battery cable.
3. Disconnect the neutral safety switch connector by lifting the harness off the switch without using a side to side motion.
4. Remove the neutral safety switch and O-ring seal. The switch can be removed by using a 24 in. long extension and universal adapter, the access path is along the left side of the firewall.
5. Install the neutral safety switch and O-ring into the transmission, tightening to 8–11 ft. lbs.
6. Connect the neutral safety switch electrical lead.
7. Connect the negative battery cable. Check the operation of the switch in each gear position.
8. Install the air cleaner assembly.

MARK VII, THUNDERBIRD/COUGAR, MUSTANG/CAPRI

1. Place the transmission shift lever in the **1** position. Disconnect the negative battery cable.
2. Raise and safely support the vehicle.
3. Disconnect the neutral safety switch electrical connector.
4. Remove the neutral safety switch and seal O-ring.
5. Install the neutral safety switch and O-ring into the transmission, tightening to 8–11 ft. lbs.
6. Connect the neutral safety switch electrical lead.
7. Lower the vehicle.
8. Connect the negative battery cable. Check the operation of the switch in each gear position.

REMOVAL AND INSTALLATION

TRANSMISSION REMOVAL

1. Disconnect the negative battery cable. Raise and safely support the vehicle.
2. Place a drain pan under the transmission and drain the fluid. Once the fluid is drained, install 2 bolts in the oil pan to keep it in place.
3. Remove the converter drain plug access cover. Remove the converter to flywheel attaching nuts.
4. Place a drain pan under the converter. Turn the converter to gain access to the converter drain plug and remove the plug. When the fluid is drained, reinstall the drain plug.
5. Disconnect the dirveshaft from the rear axle and slide the shaft back to remove it form the transmission. To maintain driveline balance matchmark the driveshaft flange and the rear axle yoke.
6. Disconnect the cable from the terminal on the starter. Remove the attaching bolts and remove the starter.
7. Disconnect the neutral safety switch wire from the neutral safety switch.

8. Remove the rear mount-to-crossmember attaching bolts and the crossmember-to-frame attaching bolts. Remove the engine rear support-to-extension housing attaching bolts.

9. Disconnect the T.V. linkage from the T.V. lever ball stud. On Thunderbird/Cougar, disconnect the T.V. linkage from the bellcrank lever stud and remove the bracket bolt.

10. Disconnect the manual rod from the transmission manual lever. Remove the bolts securing the bellcrank bracket to the case.

11. Using a suitable lifting device, raise the transmission enough to allow removal of the rear mount. Remove the rear mount from the crossmember and remove the crossmember from the side supports.

12. Disconnct the oil cooler lines from the fittings on the transmission. Disconnect the speedometer cable from the extension housing.

13. Remove the bolts securing the transmission filler tube to the cylinder block. Remove the filler tube from the transmission.

14. Secure the transmission to the lifting device and remove the engine-to-transmission bolts.

15. Carefully move the transmission and converter assembly away from the engine and lower the assembly from under the vehicle.

TRANSMISSION INSTALLATION

1. Position the converter on the transmission, make sure the converter is fully seated on the pump gear by rotating the converter.

2. With the converter properly installed, place the transmission on a lifting device. Lift the transmission into position under the vehicle.

3. Align the converter with the flywheel. Move the transmission assembly forward into position, use care not to damage the flywheel and converter pilot.

4. Install the transmission-to-engine bolts and tighten to 40–50 ft. lbs.

5. Install a new O-ring on the transmission filler tube and insert the filler tube in the transmission case. Install the attaching bolt.

6. Connect the speedometer cable to the extension housing. Connect the cooler lines and tighten to 18–23 ft. lbs.

7. Position the crossmember on the side rails and install the rear nount on the crossmember. Install the attaching bolt and nut.

8. Secure the engine support to the extension housing. Secure the crosmember to the side supports tightening the bolts to 70–100 ft. lbs.

9. Install the bellcrank assembly onto the converter housing and tighten the bolts to 20–30 ft. lbs.

10. Connect the T.V. linkage rod to the to the T.V. lever ball stud. Connect the manual linkage rod to the manual lever.

11. Install the converter to flywheel attaching nuts, tighten to 20–34 ft. lbs.

12. Install the converter housing access cover, tighten the bolts to 12–16 ft. lbs.

13. Install the starter motor and connect the cable to the terminal. Connect the neutral safety switch wires.

14. Install the driveshaft, aligning the matchmarks made during disassembly.

15. Lower the vehicle and attach the negative battery cable. Fill the transmission to the correct level with the specified fluid.

16. Adjust the T.V. linkage, idle and manual linkage as necessary.

BENCH OVERHAUL

Before Disassembly

When servicing this unit, it is recommended that as each part is disassembled, it is cleaned in solvent and dried with compressed air. All oil passages should be blown out and checked for obstructions. Disassembly and reassembly of this unit and its parts must be done on a clean work bench. As is the case when repairing any hydraulically operated unit, cleanliness is of the utmost importance. Keep work area, tools, parts and hands clean at all times. Also, before installing bolts into aluminum parts, always dip the threads into clean transmission oil. Antiseize compound can also be used to prevent bolts from galling the aluminum and seizing. Always use a torque wrench to keep from stripping the threads. Take care with the seals when installing them, especially the smaller O-rings. The slightest damage can cause leaks. Aluminum parts are very susceptible to damage so great care should be exercised when handling them. The internal snaprings should be expanded and the external snaprings compressed if they are to be re-used. This will help insure proper seating when installed. Be sure to replace any O-ring, gasket, or seal that is removed, although often the Teflon seal rings, when used, will not need to be removed unless damaged. Lubricate all parts with Dexron® II when assembling.

Inspection

CONVERTER ENDPLAY CHECK

1. Insert endplay checking tool, T80L–7902–A or equivalent, into the converter pump drive hub until it bottoms.

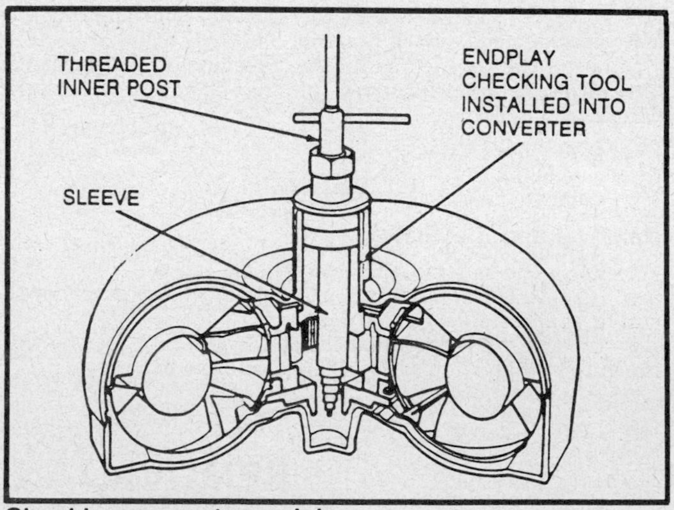
THREADED
INNER POST
ENDPLAY
CHECKING TOOL
INSTALLED INTO
CONVERTER
SLEEVE
Checking converter endplay

2. Expand the sleeve in the turbine spline by tightening the threaded inner post until the tool is securely locked into the spline.

3. Attach dial indicator tool, 4201–C or equivalent, to the tool. Position the indicator button on the converter pump drive hub and set the dial face at 0.

4. Lift the tool upward as far as it will go and note the indicator reading. The indicator is the total endplay which the turbine

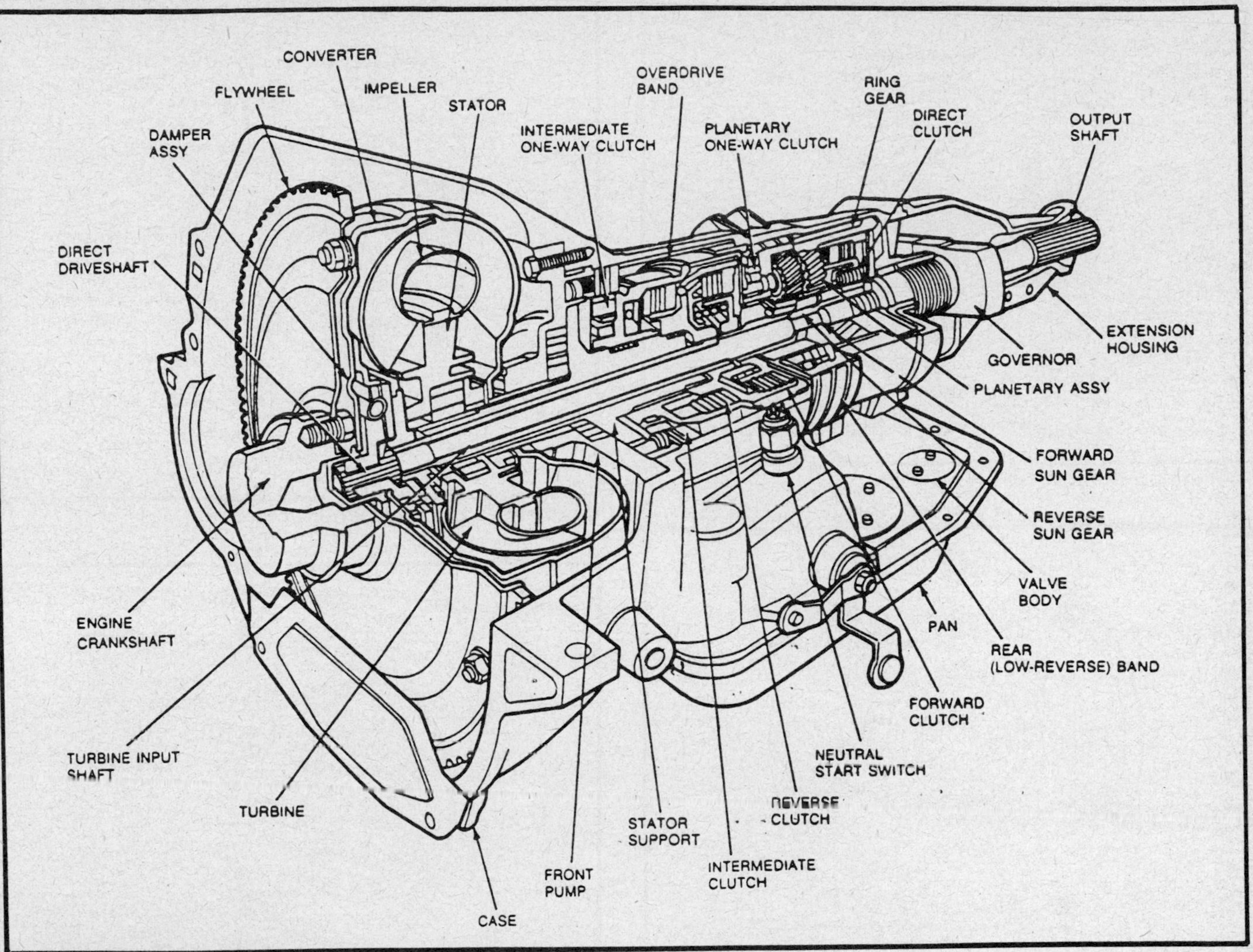
CONVERTER
FLYWHEEL
IMPELLER
STATOR
OVERDRIVE BAND
RING GEAR
DIRECT CLUTCH
OUTPUT SHAFT
DAMPER ASSY
INTERMEDIATE ONE-WAY CLUTCH
PLANETARY ONE-WAY CLUTCH
DIRECT DRIVESHAFT
EXTENSION HOUSING
GOVERNOR
PLANETARY ASSY
FORWARD SUN GEAR
REVERSE SUN GEAR
VALVE BODY
PAN
REAR (LOW-REVERSE) BAND
ENGINE CRANKSHAFT
FORWARD CLUTCH
NEUTRAL START SWITCH
TURBINE INPUT SHAFT
TURBINE
REVERSE CLUTCH
STATOR SUPPORT
INTERMEDIATE CLUTCH
FRONT PUMP
CASE

AOD component locations

and stator share. Replace the converter unit if the total endplay exceeds 0.50 in.

5. Remove the tools from the converter.

CONVERTER DAMPER/HUB ASSEMBLY WELD CHECK

1. Position a suitable converter holding fixture, T83L–7902–A3 or equivalent, in a vise.
2. Place the converter on top of the holding fixture, aligning the pilot hub and 1 stud in the appropriate holes.
3. Insert a rod torque adapter tool, T83L–7902–A1 or equivalent, into the converter making sure the splines engage the damper assembly.
4. Install the pilot guide T83L–7902–A2 or equivalent, over the rod torque adapter turning tool and onto impeller hub.
5. Hold the converter snug to the holding fixture while tightening.
6. Turn the shaft clockwise and counterclockwise applying approximately 50 ft. lbs. of pressure with a ¾ in. drive socket and torque wrench.
7. The shaft should not turn more than 2 degrees.
8. If there is a grinding noise and/or if the shaft turns more than 2 degrees, the converter damper assembly, welds, rivets or reaction hub are broken. Replace the torque converter.

REACTOR ONE-WAY CLUTCH CHECK

1. Align the slot in the thrust washer with the slot in the holding lug.

NOTE: To align the slots, use tool, T81P–7902–B or its equivalent to turn the reactor.

2. Position the holding wire, T77L–7902–A or its equivalent, in the holding lug.
3. While holding the wire in position in the lug, install the one-way clutch torquing tool, T76L–7902–C or equivalent, in the reactor spline.
4. Continue holding the wire and turn the torquing tool counterclockwise with a torque wrench. The converter one-way clutch should lock-up and hold at 10 ft. lbs. of torque. The converter one-way clutch should rotate freely in clockwise direction.
5. If the clutch fails to lock-up and hold at 10 ft. lbs, replace the converter.

Transmission Disassembly

1. Remove the transmission from the vehicle and mount it in a suitable holding fixture.

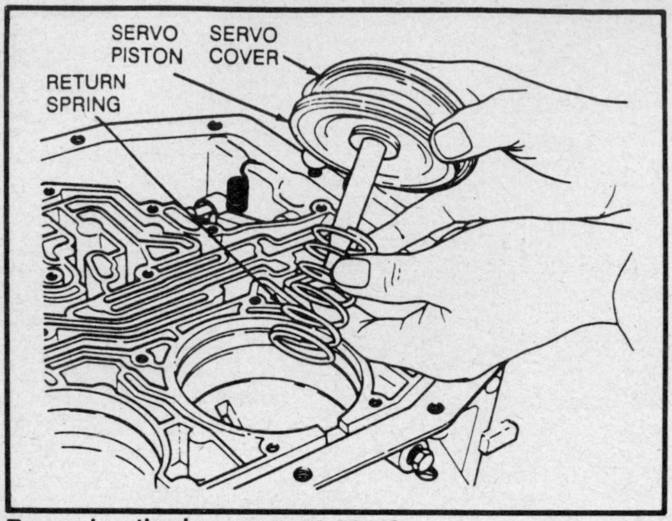
SERVO PISTON
SERVO COVER
RETURN SPRING
Removing the low-reverse servo

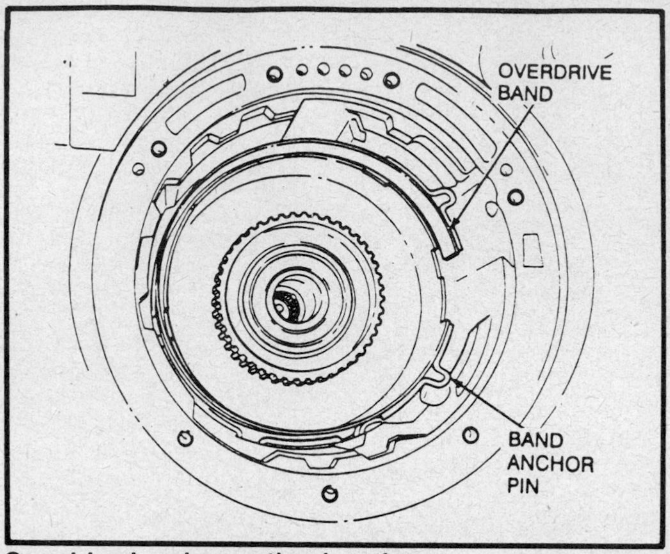
OVERDRIVE BAND
BAND ANCHOR PIN
Overdrive band mounting location

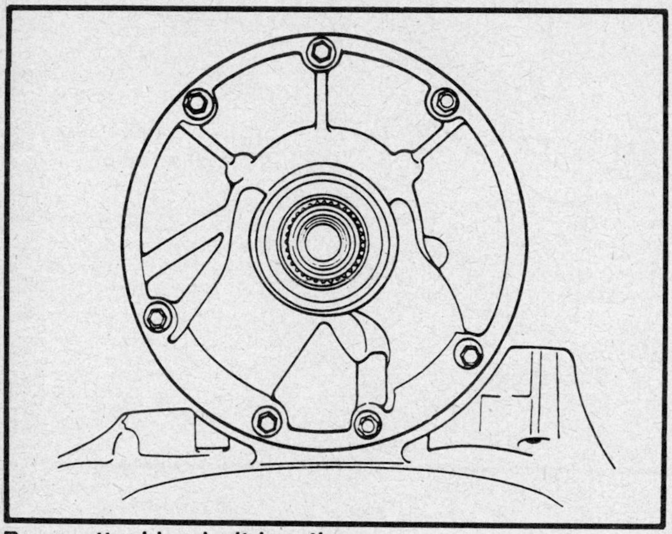
Pump attaching bolt locations

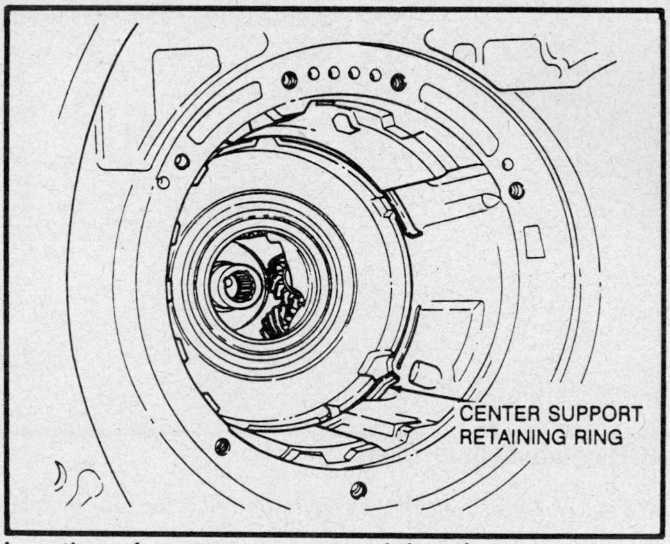
CENTER SUPPORT RETAINING RING
Location of center support retaining ring

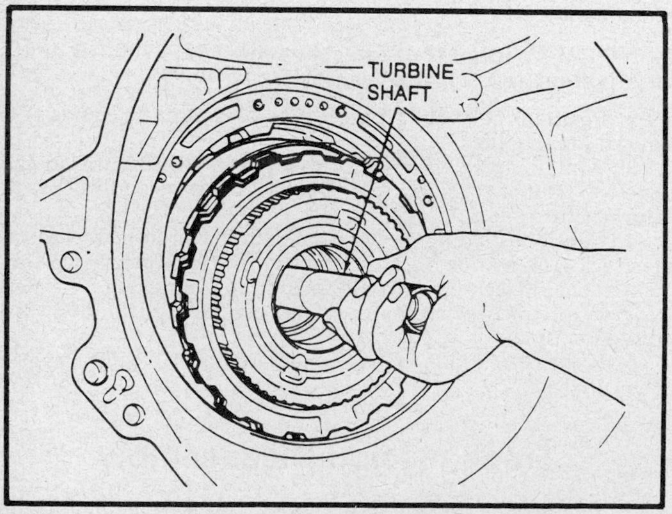
TURBINE SHAFT
Removing the turbine shaft and clutch packs

2. Remove the torque converter from the inside the bell housing by pulling it straight out.

3. Remove the bolts attaching the oil pan and remove the oil pan. Discard the gasket. Remove the bolts retaining the oil filter and remove the filter, grommet and gasket.

4. Remove the manual lever detent spring and roller assembly. Remove the remaining valve body retaining bolts and remove the valve body.

5. Remove the 3–4 accumulator cover and piston, the overdrive servo cover and piston, the low reverse servo cover and piston and the 2–3 accumulator.

6. Place the transmission in a vertical position. Remove the pump body attaching bolts and remove the pump assembly from the case, using slide hammer tool T59L–100–B or equivalent.

7. Place the transmission in a horizontal position. Grasp the turbine shaft and pull it from the case.

NOTE: When the turbine shaft is removed, the intermediate clutch pack, reverse and forward clutch packs and the intermediate one-way clutch will also come out. Care should be taken not to damage the overdrive band friction material when removing these components.

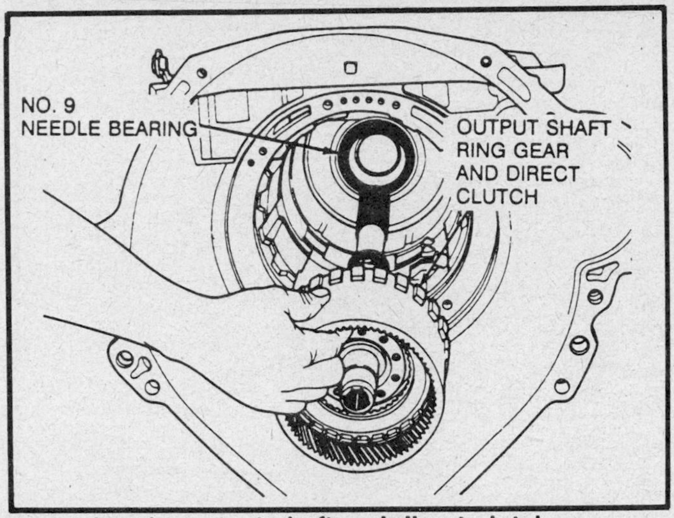
NO. 9
NEEDLE BEARING
OUTPUT SHAFT
RING GEAR
AND DIRECT
CLUTCH

Removing the output shaft and direct clutch assembly

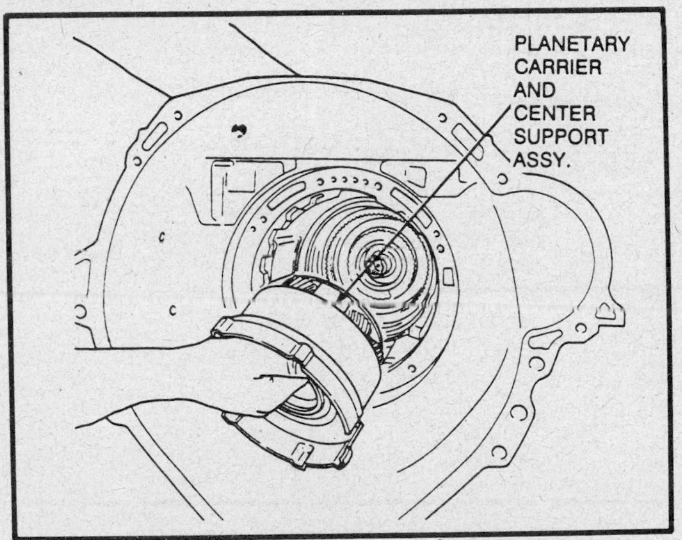
PLANETARY
CARRIER
AND
CENTER
SUPPORT
ASSY.

Removing the planetary carrier and center support

8. Disengage the overdrive band from the anchor pins and remove it from the case.
9. Remove the forward clutch hub and the No. 3 needle bearing from the case as an assembly.
10. Remove the forward sun gear, No. 5 needle bearing, reverse sun gear, drive shell and the No. 4 needle bearing as an assembly.
11. Remove the center support retaining ring, note the position of the tabs for reassembly.
12. Using a small pry bar, pry the anti-clunk spring from between the center support and the case. Note the location of the spring for reassembly.
13. Remove the center support and planetary carrier as an assembly. Remove the reverse band.
14. If the direct clutch hub did not come out with the planetary, lift it out of the direct clutch.
15. Rotate the transmission so that the extension housing is in an upright position. Remove the extension housing mounting bolts and remove the extension housing.
16. Rotate the transmission into a horizontal position. Remove the governor retaining ring and the governor from the output shaft.
17. Remove the governor drive ball from the output shaft. Remove the output shaft, the ring gear and the direct clutch through the front of the case, as an assembly.
18. Remove the output shaft No. 9 needle bearing from the rear of the case.
19. Remove the intermediate clutch pack from the intermediate one-way clutch. Remove the reverse clutch assembly from the forward clutch assembly.

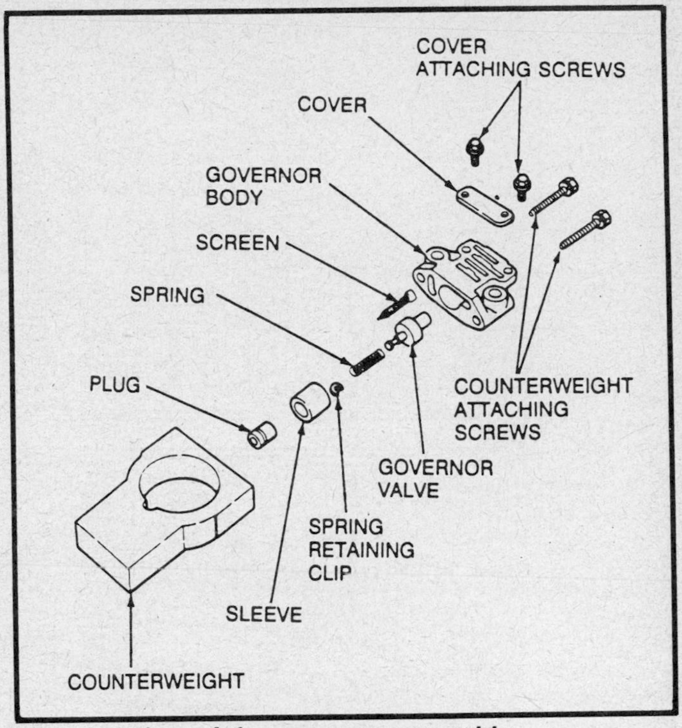
COVER
ATTACHING SCREWS
COVER
GOVERNOR
BODY
SCREEN
SPRING
PLUG
COUNTERWEIGHT
ATTACHING
SCREWS
GOVERNOR
VALVE
SPRING
RETAINING
CLIP
SLEEVE
COUNTERWEIGHT

Exploded view of the governor assembly

Unit Disassembly and Assembly

GOVERNOR

Disassembly

1. Remove the screws attaching the counterweight to the governor body and remove the governor cover screws.
2. Remove the governor cover. Remove the plug, sleeve and governor valve from the governor body.
3. Remove the screen from the governor body.

Inspection

1. Inspect the governor valves and bores for scores. Minor scores may be removed from the valves with crocus cloth. Replace the governor if the valves or body are deeply scored.
2. Inspect the governor screen for obstructions. The screen must be free of foreign material. If contaminated, clean thoroughly in a suitable solvent and blow dry.
3. Check for free movement of the valves in their bores. The valves should slide freely in the bores when dry. Inspect the fluid passages in the governor body and counterweight for obstructions.
4. Check the mating surfaces of the governor valve and the counterweight for burrs or scratches.

Assembly

1. Install the clip and sleeve on the governor valve. Install the governor valve in the governor body.

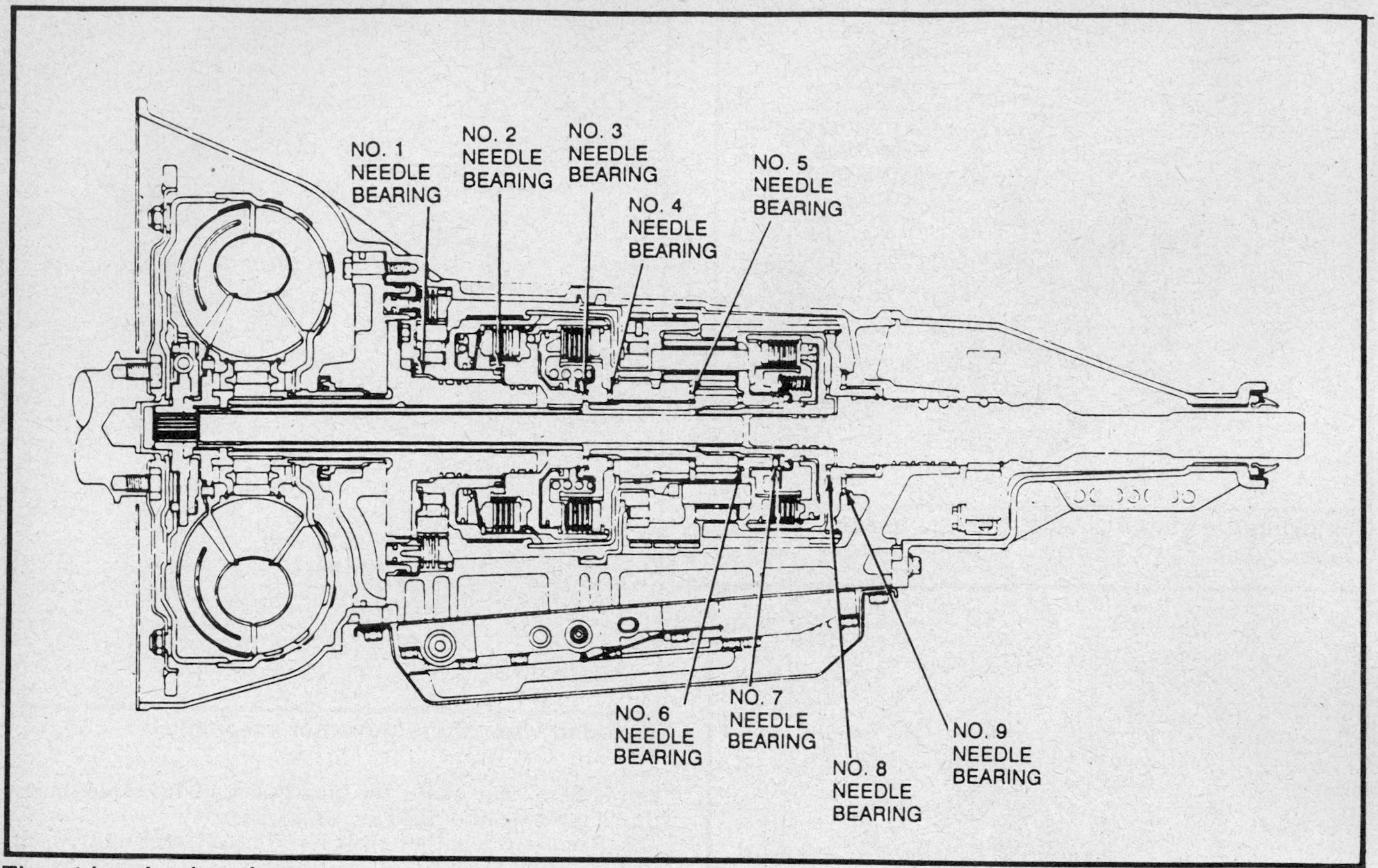
NO. 1
NEEDLE
BEARING
NO. 2
NEEDLE
BEARING
NO. 3
NEEDLE
BEARING
NO. 4
NEEDLE
BEARING
NO. 5
NEEDLE
BEARING
NO. 6
NEEDLE
BEARING
NO. 7
NEEDLE
BEARING
NO. 8
NEEDLE
BEARING
NO. 9
NEEDLE
BEARING

Thrust bearing locations

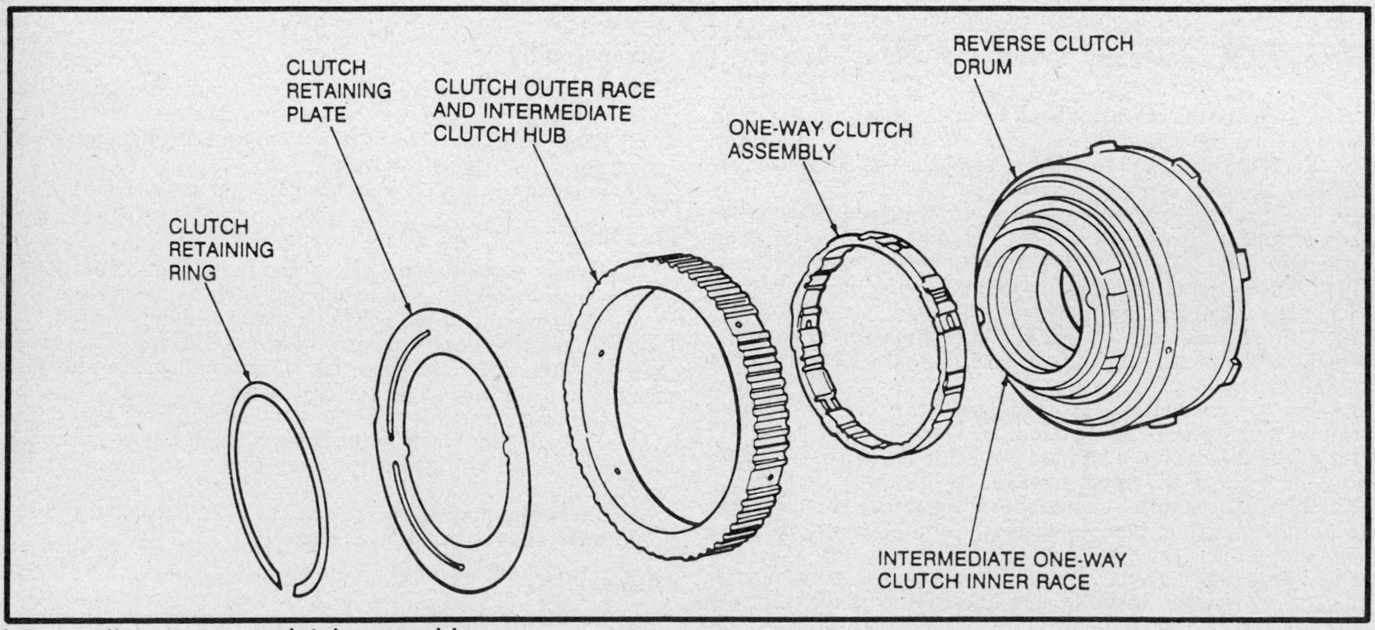
CLUTCH
RETAINING
PLATE
CLUTCH OUTER RACE
AND INTERMEDIATE
CLUTCH HUB
REVERSE CLUTCH
DRUM
ONE-WAY CLUTCH
ASSEMBLY
CLUTCH
RETAINING
RING
INTERMEDIATE ONE-WAY
CLUTCH INNER RACE

Intermediate one-way clutch assembly

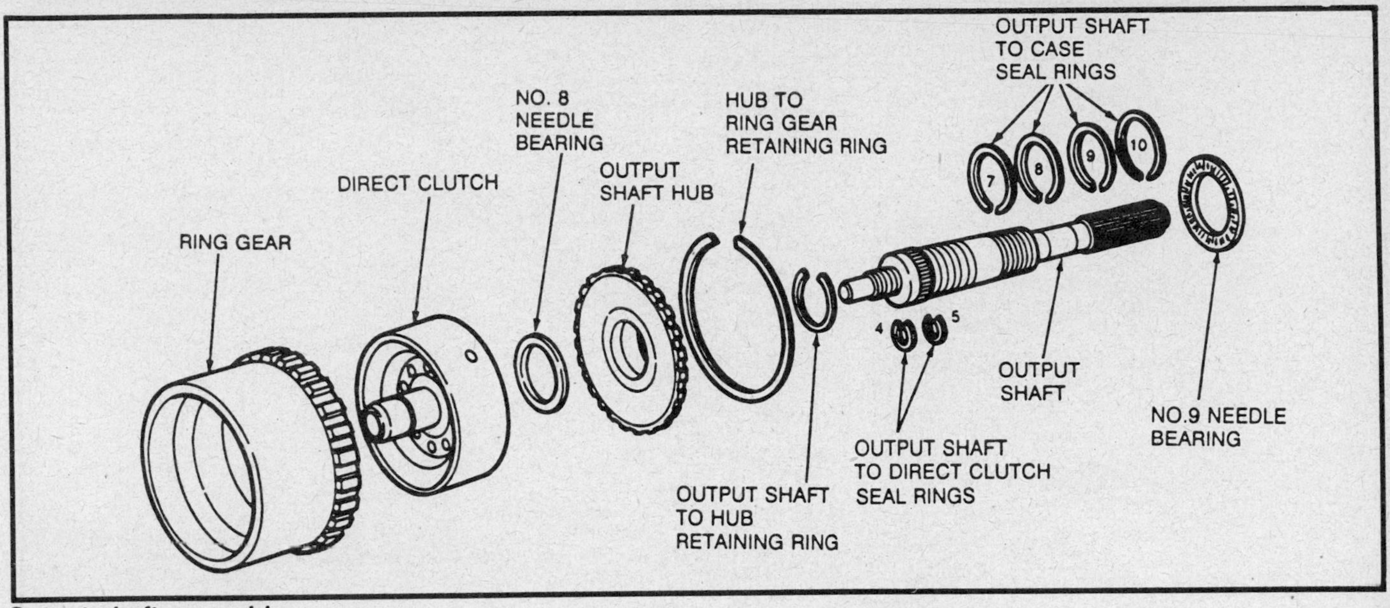
OUTPUT SHAFT
TO CASE
SEAL RINGS
NO. 8
NEEDLE
BEARING
HUB TO
RING GEAR
RETAINING RING
OUTPUT
SHAFT HUB
DIRECT CLUTCH
RING GEAR
OUTPUT
SHAFT
NO.9 NEEDLE
BEARING
OUTPUT SHAFT
TO DIRECT CLUTCH
SEAL RINGS
OUTPUT SHAFT
TO HUB
RETAINING RING

Output shaft assembly

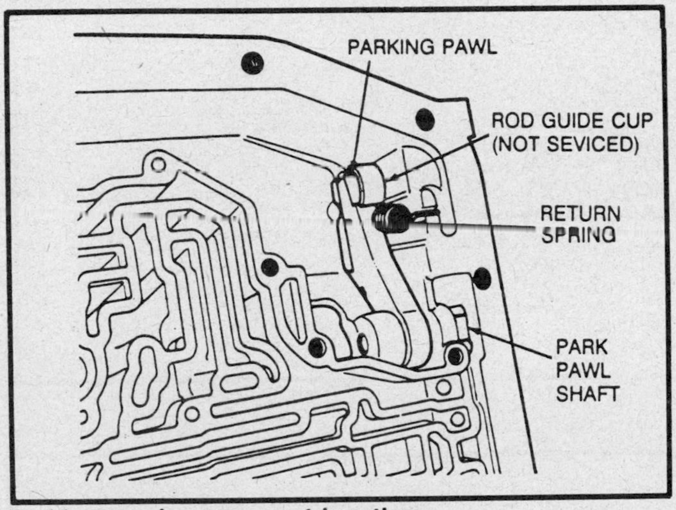
PARKING PAWL
ROD GUIDE CUP
(NOT SEVICED)
RETURN
SPRING
PARK
PAWL
SHAFT

Parking pawl component locations

2. Install the sleeve in the governor body with the points outward. Install the plug in the sleeve with the knurled face inward.
3. Position the cover and install the attaching screws, tightening to 20–30 inch lbs.
4. Install the screen in the body with the steel band facing inward and the top of the screen facing out.
5. Position the governor body on the counterweight and install the retaining screws. Tighten the retaining screws to 50–60 inch lbs.

PARKING PAWL

Disassembly

1. Slide the park pawl shaft out of the rear of the case and remove the parking pawl.
2. Remove the return spring.

Assembly

1. Hook the squared end of the spring into the notch on the park pawl.
2. Hold the pawl and spring in place and hook the curved end of the spring into the recess in the case.
3. Install the park pawl shaft by sliding it into the case.

INTERMEDIATE/ONE-WAY CLUTCH

Disassembly

1. Using expanding type snapring pliers, remove the snapring that retains the clutch.
2. Remove the clutch retaining plate. Remove the clutch outer race by lifting on the race while turning counterclockwise.
3. Carefully lift the one-way clutch from the inner race.

Inspection

1. Inspect the clutch outer race for burrs or spline damage.
2. Inspect the clutch inner race for burrs.
3. Check the clutch hub splines for wear. The clutch hub is attached to the reverse clutch drum.

Assembly

1. Install the one-way clutch over the inner race.
2. Install the clutch outer race by placing it over the one-way clutch and turning it counterclockwise.
3. Install the clutch retaining plate and install the retaining snapring.

OUTPUT SHAFT

Disassembly

1. Remove the ring retaining the output shaft hub to the ring gear.
2. Separate the output shaft and hub assembly from the ring gear.
3. Remove the direct clutch from the ring gear and the No. 8 needle bearing from the back of the direct clutch.
4. Remove the 4 output shaft seal rings and the hub-to-output shaft retaining ring. Separate the hub from the output shaft.
5. Remove the 2 direct clutch seal rings from the output shaft.

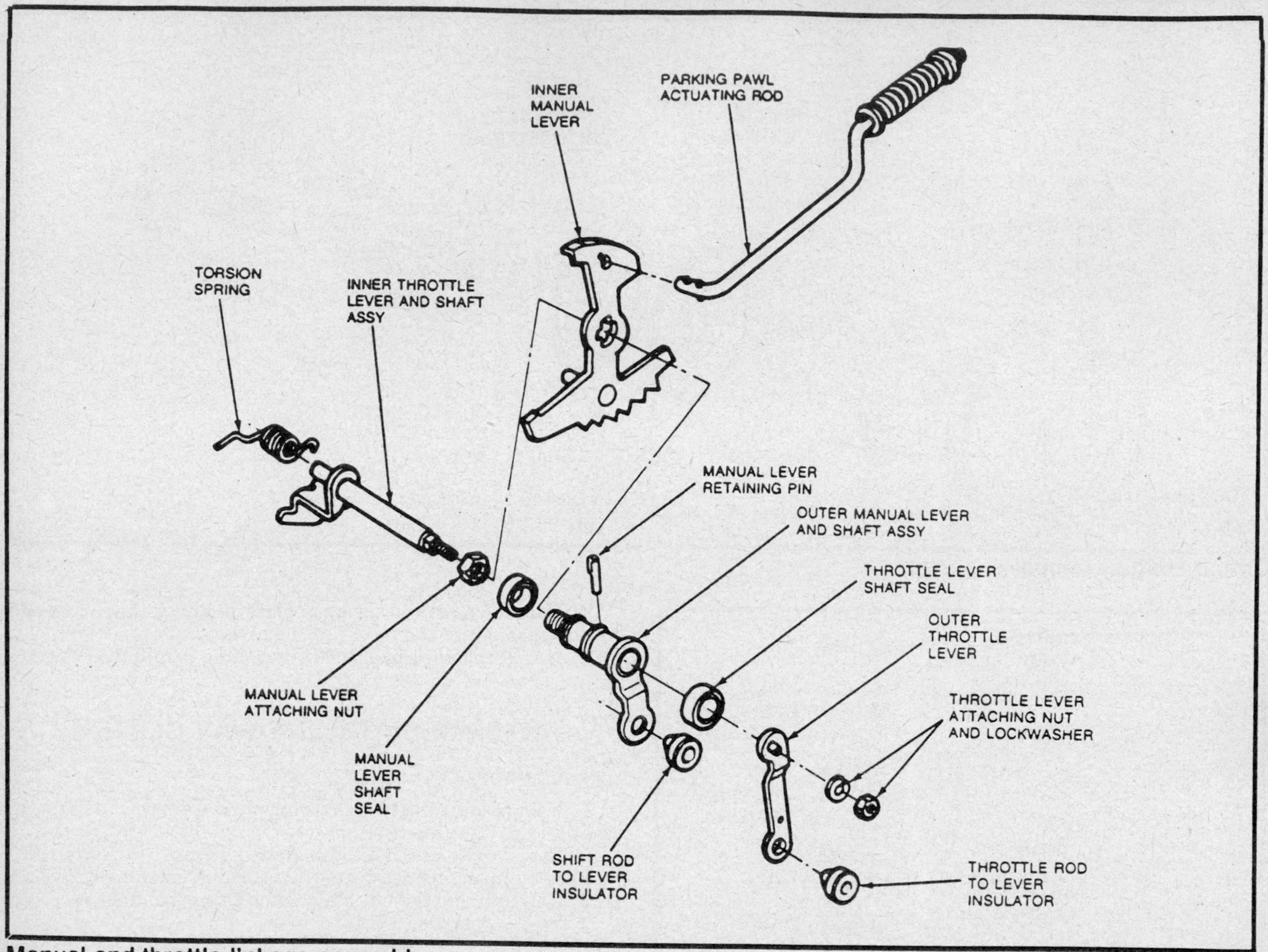
INNER
MANUAL
LEVER
PARKING PAWL
ACTUATING ROD
TORSION
SPRING
INNER THROTTLE
LEVER AND SHAFT
ASSY
MANUAL LEVER
RETAINING PIN
OUTER MANUAL LEVER
AND SHAFT ASSY
THROTTLE LEVER
SHAFT SEAL
OUTER
THROTTLE
LEVER
MANUAL LEVER
ATTACHING NUT
THROTTLE LEVER
ATTACHING NUT
AND LOCKWASHER
MANUAL
LEVER
SHAFT
SEAL
SHIFT ROD
TO LEVER
INSULATOR
THROTTLE ROD
TO LEVER
INSULATOR

Manual and throttle linkage assembly

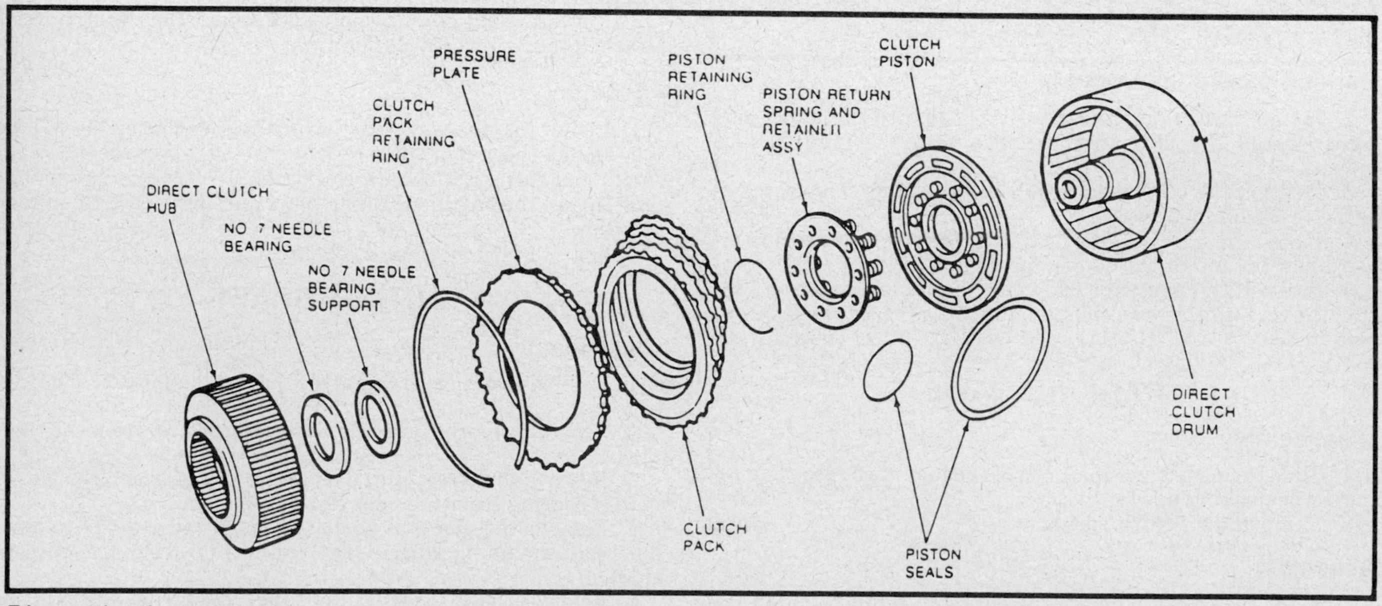
PRESSURE
PLATE
PISTON
RETAINING
RING
CLUTCH
PISTON
PISTON RETURN
SPRING AND
RETAINER
ASSY
CLUTCH
PACK
RETAINING
RING
DIRECT CLUTCH
HUB
NO. 7 NEEDLE
BEARING
NO. 7 NEEDLE
BEARING
SUPPORT
DIRECT
CLUTCH
DRUM
CLUTCH
PACK
PISTON
SEALS

Direct clutch assembly

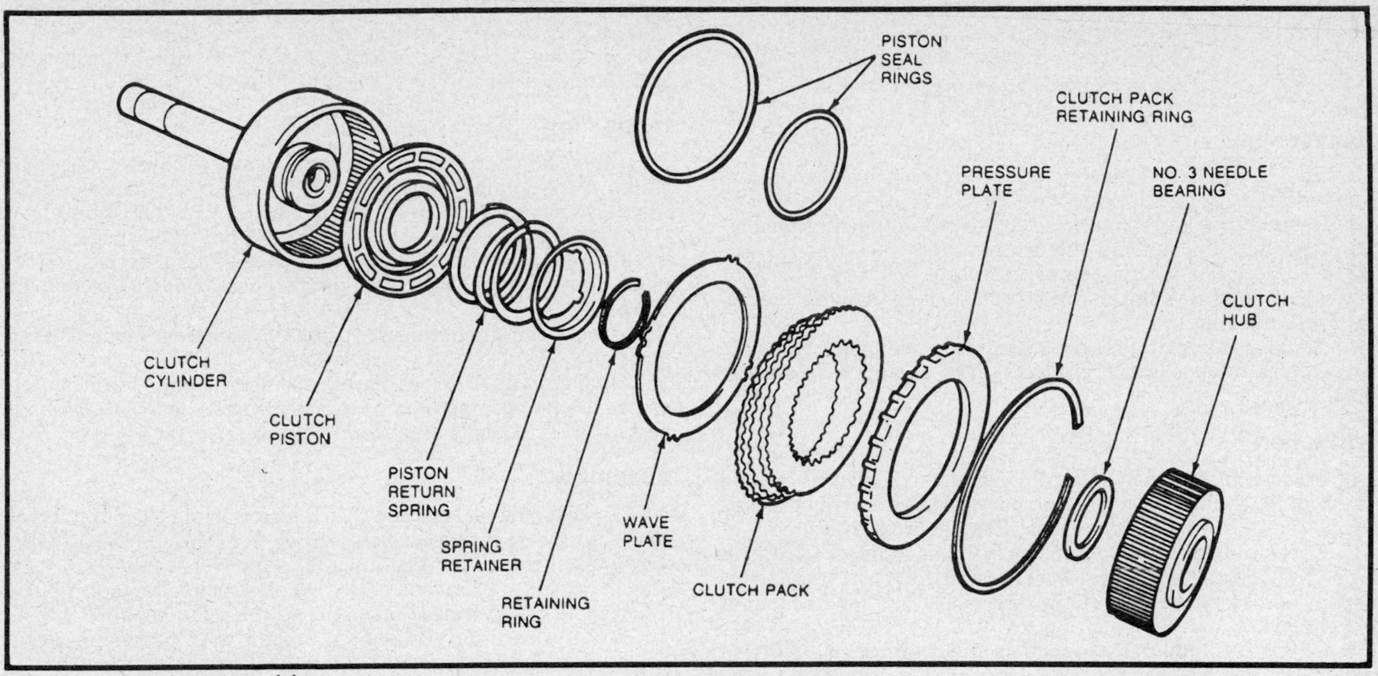
PISTON
SEAL
RINGS
CLUTCH PACK
RETAINING RING
PRESSURE
PLATE
NO. 3 NEEDLE
BEARING
CLUTCH
HUB
CLUTCH
CYLINDER
CLUTCH
PISTON
PISTON
RETURN
SPRING
SPRING
RETAINER
WAVE
PLATE
RETAINING
RING
CLUTCH PACK

Forward clutch assembly

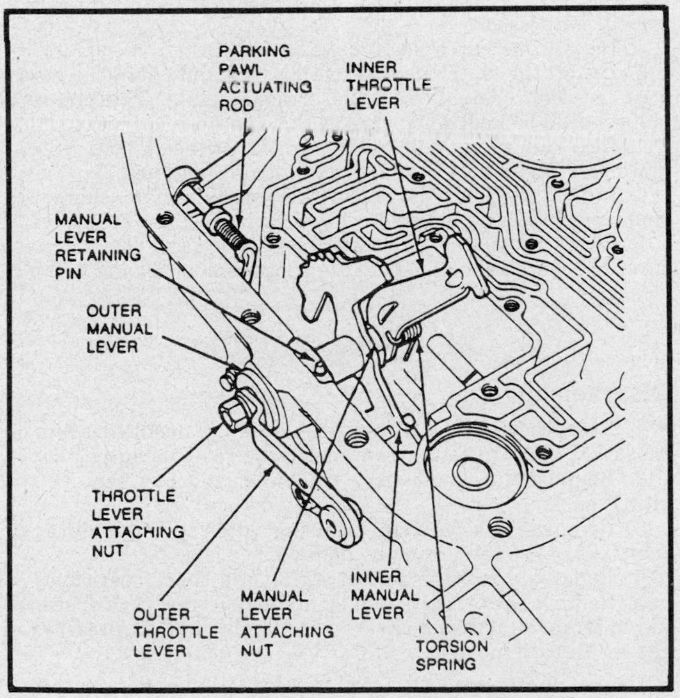
PARKING
PAWL
ACTUATING
ROD
INNER
THROTTLE
LEVER
MANUAL
LEVER
RETAINING
PIN
OUTER
MANUAL
LEVER
THROTTLE
LEVER
ATTACHING
NUT
INNER
MANUAL
LEVER
OUTER
THROTTLE
LEVER
MANUAL
LEVER
ATTACHING
NUT
TORSION
SPRING

Manual and throttle linkage component locations

Inspection

1. Check the output shaft splines for damage or wear. Replace the shaft if the splines are excessively worn.
2. Inspect the ring gear for damaged or chipped teeth.
3. Check the needle bearing for freedom of movement and damaged bearings.
4. Replace any damaged or worn components. Replace all seal rings.

Assembly

1. Install new direct clutch O-rings on the output shaft.
2. Install 4 outshaft seal rings and the hub. Install the hub-to output shaft retaining ring.
3. Install the direct clutch assembly and the needle bearing onto the direct clutch.
4. Install the output shaft and hub assembly onto the ring gear.
5. Install the output shaft retaining ring.

MANUAL AND THROTTLE LINKAGE

Disassembly

1. Loosen the attaching nut on the throttle lever. Hold the lever firmly while loosening to prevent damage to the valve body and remove the throttle lever.
2. Using an awl, remove the lever shaft seal from the case.
3. Using a small prybar, remove the manual shaft retaining pin from the case.
4. Using a 21mm wrench, remove the manual lever attaching nut while securely holding the inner manual lever. Thread the nut off the shaft and remove the lever.
5. Remove the inner throttle lever and T.V. lever torsion spring.
6. Remove the inner manual lever and parking pawl actuating rod as an assembly.
7. Remove the manual lever shaft seal from the case using an slide hammer type, seal removal tool.

Assembly

1. Install a new manual lever seal into the case.
2. Install the inner manual lever and park pawl actuating rod as an assembly.
3. Install the inner throttle lever and T.V. torsion spring. Install the outer manual lever into the case and install the inner manual lever retaining nut. Tighten to 19–27 ft. lbs.
4. Install the manual lever shaft retaining pin. Install a new outer manual lever seal.

5. Install the outer throttle lever, lockwasher and the attaching nut. Tighten the attaching nut to 12–16 ft. lbs.

DIRECT CLUTCH

Disassembly

1. Remove the No. 7 direct clutch hub inner needle bearing. Remove the bearing support.
2. Remove the clutch pack selective retaining ring and remove the clutch pack from the drum.
3. Using clutch spring compressor tool, T65L–77515–A or equivalent, compress the piston return springs. Remove the piston retaining ring.
4. Remove the spring retainer assembly and piston from the drum. Note the position and direction of the lip seals and remove them.

Inspection

1. Inspect the clutch cylinder thrust surfaces, piston bore and clutch plate splines for scores or burrs. Minor burrs or scores can be removed with crocus cloth. Replace if badly scored.
2. Check the fluid passages in the cylinder for blockage. Clean out all fluid passages. Inspect the clutch piston for scores and replace if necessary. Inspect the piston check ball for freedom of movement.
3. Check the clutch release springs and retainer for distortion and cracks, replace if any are evident.
4. Inspect the friction plates, springs, retainer and clutch pressure plate for worn or scored bearing surfaces. Replace all parts that are deeply scored or polished.

Assembly

1. Install the inner piston seal on the clutch drum as follows:
   a. Position a direct clutch lip seal protector tool, T80L–77234–A or equivalent, over the clutch drum hub.
   b. Lubricate the seal and protector with petroleum jelly.
   c. Position the seal over the installer tool with the sealing lip facing down.
   d. Push the seal down until it snaps off the end of the protector onto the clutch hub.
   e. Remove the seal protector from the hub.
   f. Slide the seal up until it seats on the seal groove.
2. Install the outer clutch piston seal. Note the direction of the sealing lip, the lip points away from the spring posts.
3. Install the clutch apply piston, coat the piston, seals and the clutch drum sealing area with petroleum jelly.
4. Install the piston spring, retainer and retaining ring using clutch spring compressor tool, T65L–77515–A.
5. Install the clutch pack. Install the clutch pack selective retaining ring and check the clearance between the ring and the pressure plate using a feeler gauge. The clearance should be; 0.050–0.067 in. for V8 engines and 0.040–0.057 in. for V6 engines.
6. If the clearance is not within specification, there are snaprings available in various sizes, to correct the clearance.
7. Check the clutch for proper operation using low air pressure. The clutch should be heard to apply smoothly and without leakage.
8. Install the No. 7 needle bearing support. Install the No. 7 needle bearing.

FORWARD CLUTCH

Disassembly

1. Remove the clutch hub and the No. 3 needle bearing.
2. Remove the clutch pack selective retaining ring and remove the clutch pack.
3. Compress the piston return spring using clutch spring compressor tool, T65L–77515–A or equivalent. Remove the retaining ring and spring retainer.
4. Remove the clutch piston, note the position of the lip seals and remove the seals.

Inspection

1. Inspect the clutch cylinder thrust surfaces, piston bore and clutch plate splines for scores or burrs. Minor burrs or scores can be removed with crocus cloth. Replace if badly scored.
2. Check the fluid passages in the cylinder for blockage. Clean out all fluid passages. Inspect the clutch piston for scores and replace if necessary. Inspect the piston check ball for freedom of movement.
3. Check the clutch release springs and retainer for distortion and cracks, replace if any are evident.
4. Inspect the friction plates, springs, retainer and clutch pressure plate for worn or scored bearing surfaces. Replace all parts that are deeply scored or polished.

Assembly

1. Install new seals on the clutch piston. Note the direction of the sealing before installation.
2. Install the clutch piston as follows:
   a. Position a forward clutch lip seal protector tool, T80L–77140–A or equivalent, over the clutch cylinder hub.
   b. Lubricate the seals and protector with petroleum jelly.
   c. Position the seal and protector over the clutch drum.
   d. Push the piston to the bottom of the drum.
3. Install the piston return spring, spring retainer and retaining ring. Compress the piston return spring using the clutch spring compressor.
4. Install the clutch pack, wave spring first.
5. Install the clutch pack retaining ring and check the clearance between the ring and the pressure plate. The pressure plate should be held downward as the clearance is checked. The clearance should be; 0.050–0.089 in. for vehicles with V8 engines and 0.040–0.071 for vehicles with V6 engines.
6. If the clearance is not within specification, there are snaprings available in various sizes, to correct the clearance.
7. Check the clutch for proper operation using low air pressure. The clutch should be heard to apply smoothly and without leakage.

REVERSE CLUTCH

Disassembly

1. Remove the No. 2 thrust washer (No. 2 needle bearing on Mark VII and Mustang) from inside the reverse clutch.
2. Remove the clutch pack retaining ring and remove the clutch pack.
3. Remove the wave snapring using clutch spring compressor tool, T65L–77515–A or equivalent.
4. Remove the piston return spring and the thrust ring.
5. Remove the clutch piston from the reverse clutch drum. To aid in removal, it may be necessary to apply low air pressure to the drum to remove it.

Inspection

1. Inspect the drum band surface, the bushing and thrust surfaces for scoring. Minor scoring can be removed with crocus cloth.
2. Inspect the clutch piston bore and the piston inner and outer bearing surfaces for scoring. Check the air bleed ball valve in the clutch piston for free movement. Check the orifice to make sure it is not plugged.
3. Check the fluid passages for obstructions. All fluid passages must be clean and free of obstruction.
4. Inspect the clutch plates for wear, scoring and fit on the clutch splines. Replace all plates that are badly heat distressed,

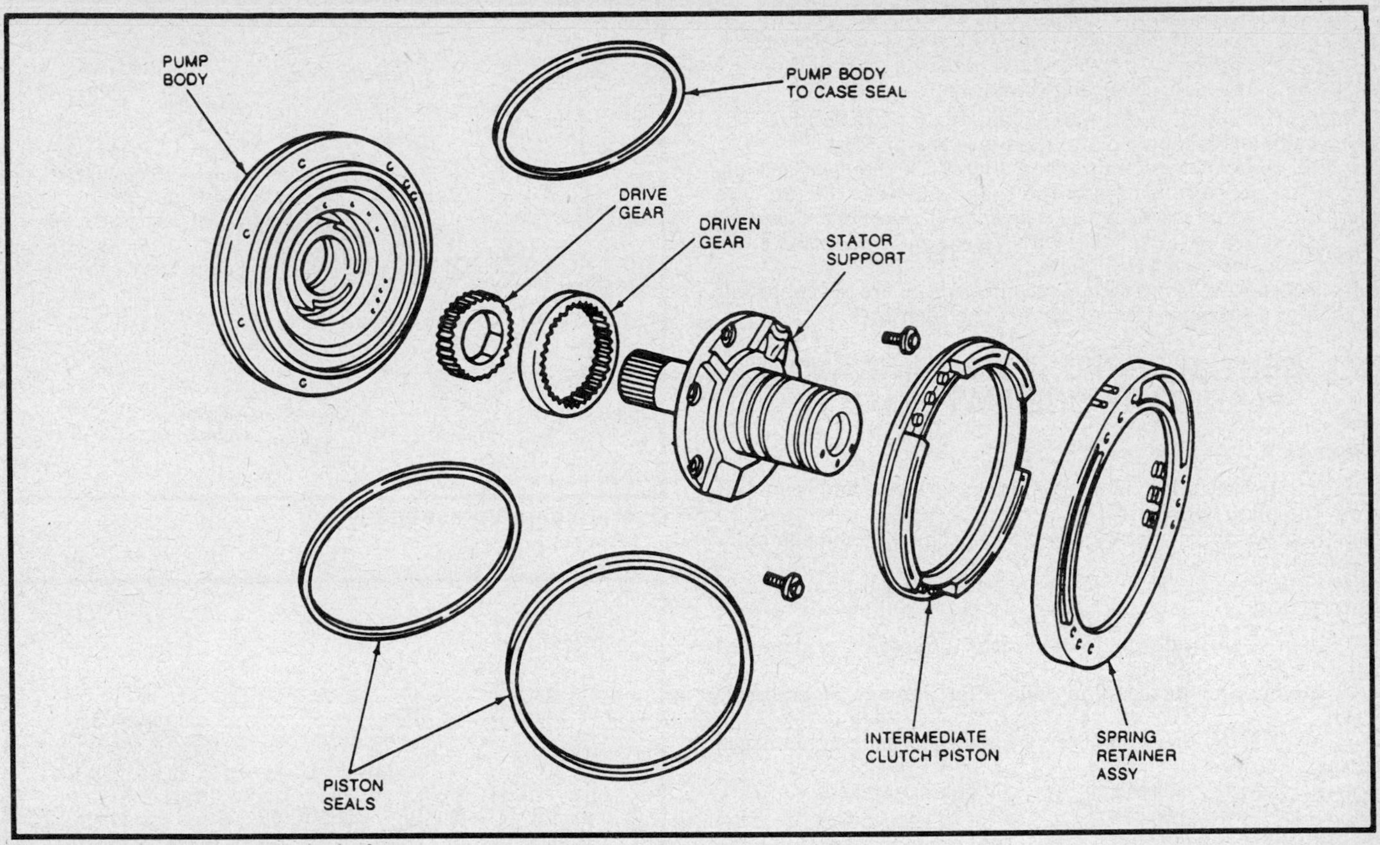
PUMP BODY
PUMP BODY TO CASE SEAL
DRIVE GEAR
DRIVEN GEAR
STATOR SUPPORT
INTERMEDIATE CLUTCH PISTON
SPRING RETAINER ASSY
PISTON SEALS

Pump assembly–exploded view

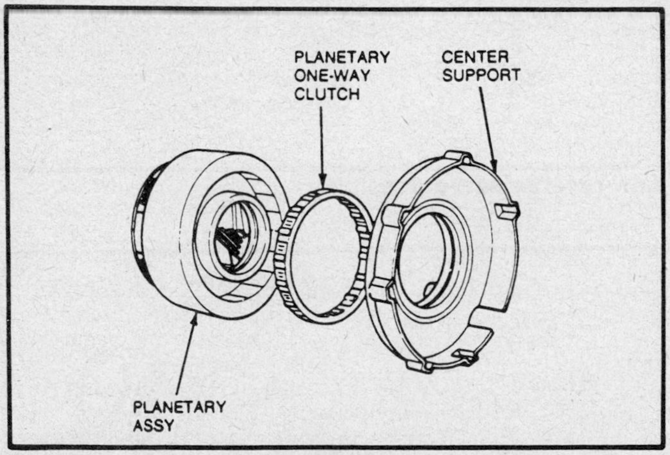
PLANETARY ONE-WAY CLUTCH
CENTER SUPPORT
PLANETARY ASSY

Center support and planetary one-way clutch assembly

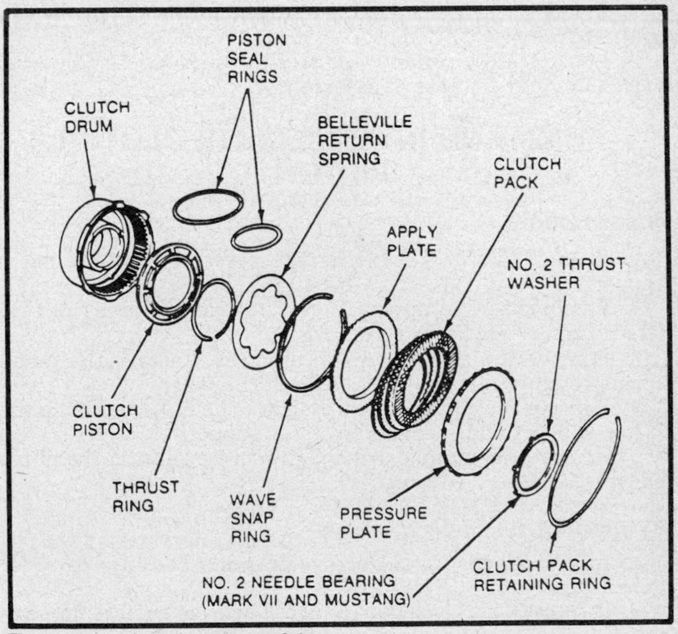
PISTON SEAL RINGS
CLUTCH DRUM
BELLEVILLE RETURN SPRING
CLUTCH PACK
APPLY PLATE
NO. 2 THRUST WASHER
CLUTCH PISTON
THRUST RING
WAVE SNAP RING
PRESSURE PLATE
NO. 2 NEEDLE BEARING (MARK VII AND MUSTANG)
CLUTCH PACK RETAINING RING

Reverse clutch assembly

worn or do not move freely in the hub.

5. Inspect the clutch pressure plate for scores on the clutch plate bearing surface. Check the clutch release springs for distortion.

Assembly

1. Install new seals on the clutch piston.

NOTE: Because the seals used on the reverse clutch piston are cut square, the direction of the seal during installation is not important.

2. Install the clutch piston into the reverse clutch drum in the following sequence:

a. Coat the piston seals and the inside of the clutch drum with petroleum jelly.

b. Install inner seal protector, T80L–7403–B or equivalent and outer seal protector, T80L–77403–A or equivalent, into the clutch drum.

c. Install the piston and push it down into the drum until it is snug against the bottom. Remove the seal protector tools.

3. Install the piston spring and the thrust ring. Install the wave snapring using a clutch spring compressor.

4. Install the clutch pack, putting the apply plate in first. The dished side of the apply plate must face the piston.

5. Install the clutch retaining ring and check the clearance between the pressure plate and the snapring. Hold the pressure plate down while checking the clearance. The correct clearance is; 0.040–0.075 in for vehicles with V8 engines and 0.030–0.056 in. for vehicles with V6 engines.

6. If the clearance is not within specification, there are snaprings available in various sizes, to correct the clearance.

CENTER SUPPORT BEARING AND PLANETARY ONE-WAY CLUTCH

Disassembly

1. Rotate the center support counterclockwise and remove it from the planetary carrier.
2. Remove the planetary one-way clutch from the planetary assembly.

Inspection

1. Inspect the inner and outer races for scoring or damaged surface areas.
2. Inspect the springs and rollers for damage or excessive wear.
3. Inspect the spring and roller case for bent or damaged spring retainers.
4. Replace any excessively worn or damaged parts.

Assembly

1. Install the one-way clutch into the planetary carrier.
2. Install the center support into the one-way clutch, using a counterclockwise motion.
3. Lubricate the clutch races and the clutch assembly with petroleum jelly for ease of assembly.

PUMP AND INTERMEDIATE CLUTCH PISTON

Disassembly

1. Remove the No. 1 selective thrust washer from the stator support.
2. Remove the 4 pump seal rings. The reverse clutch rings are larger than the forward clutch rings.
3. Remove the pump body-to-case seal. Remove the pump spring retainer assembly by releasing the tabs.
4. Remove the clutch piston. Remove the bolts retaining the stator support and remove the stator support.
5. Remove the pump drive and driven gears from the pump body.

Inspection

1. Inspect the mating surfaces of the pump body and cover for burrs.
2. Inspect the drive and driven gear bearing surface for scoring. Check the gear teeth for burrs.
3. Check fluid passages for obstructions.
4. If any parts are excessively worn or damaged, replace the pump as a unit.
5. Check the large seal ring groove on the pump body for damage.

Assembly

1. Install the pump gears into the pump body. The chamfers on the gears must face towards the pump.
2. Install the stator support to the pump body, tightening the attaching bolts to 12–16 ft. lbs.

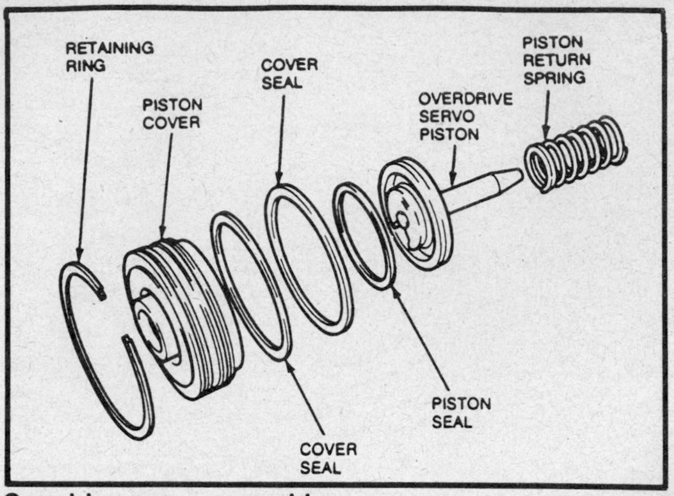
RETAINING RING
PISTON COVER
COVER SEAL
OVERDRIVE SERVO PISTON
PISTON RETURN SPRING
PISTON SEAL
COVER SEAL
Overdrive servo assembly

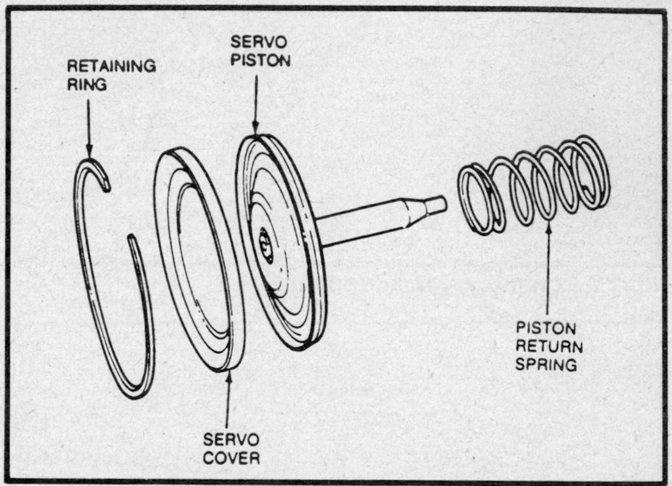
RETAINING RING
SERVO PISTON
PISTON RETURN SPRING
SERVO COVER
Low-reverse servo assembly

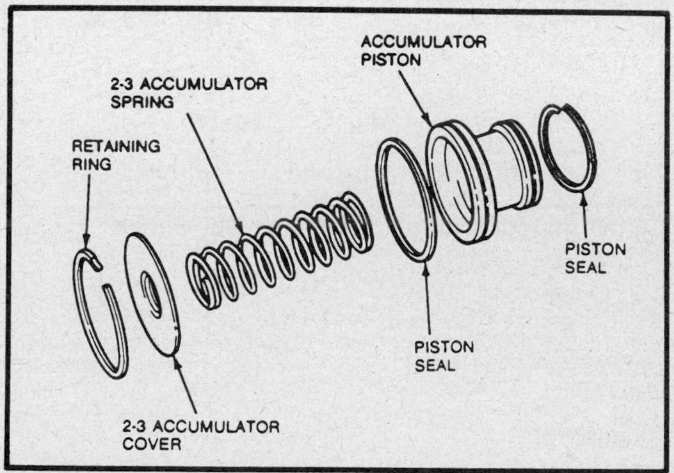
ACCUMULATOR PISTON
2-3 ACCUMULATOR SPRING
RETAINING RING
PISTON SEAL
PISTON SEAL
2-3 ACCUMULATOR COVER
2–3 accumulator assembly

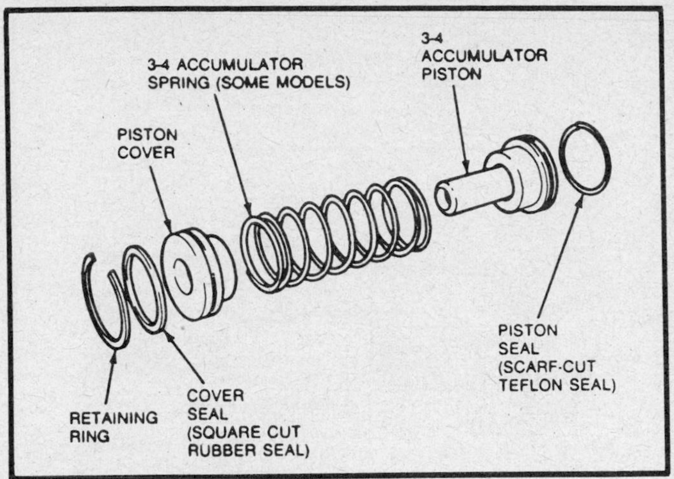
3-4 ACCUMULATOR SPRING (SOME MODELS)
3-4 ACCUMULATOR PISTON
PISTON COVER
PISTON SEAL (SCARF-CUT TEFLON SEAL)
COVER SEAL (SQUARE CUT RUBBER SEAL)
RETAINING RING

3–4 accumulator assembly

3. Install a new pump to case seal. Install new seal on the piston.

4. Install the clutch piston by first coating the seals with petroleum jelly and then installing the piston into the pump body. Push it all the way into the bore.

NOTE: The piston bleed hole must be located at the 12 o'clock position.

5. Snap the spring retainer assembly into position on the pump.

6. Install new pump seal rings. Stator support seal rings are are the largest seal rings and are for the reverse clutch. These rings are closest to the pump. Stator support seal rings are for the forward clutch and are the furthest from the pump.

ACCUMULATORS AND SERVOS

Disassembly and Assembly

3–4 AND 2–3 ACCUMULATORS

Install new seals on the accumulator piston and cover, be sure the diagonal cuts on the piston seal are aligned properly. On some transmission applications, the 3–4 accumulator may be built with a spring.

LOW/REVERSE SERVO

Inspect the sealing edge on both the servo cover and the apply piston. Replace the cover or piston if any damage is evident. The length of the rod that is attached to the piston may vary from transmission to transmission. There are 3 possible lengths; a single groove on the piston indicates the shortest possible length and 2 or 3 grooves on the rod indicate longer lengths. Do not interchange rods when assembling the transmission, use only the length rod that was removed.

OVERDRIVE SERVO

Pull the overdrive servo from the piston cover and inspect it. Install new seals on the piston and cover before reassembly. To aid in assembly, the piston seal should be lubricated with petroleum jelly.

VALVE BODY

Disassembly

1. Remove and discard the valve body gasket. Remove the bolts from the reinforcement plates and detent spring guide bolt from the separator plate.

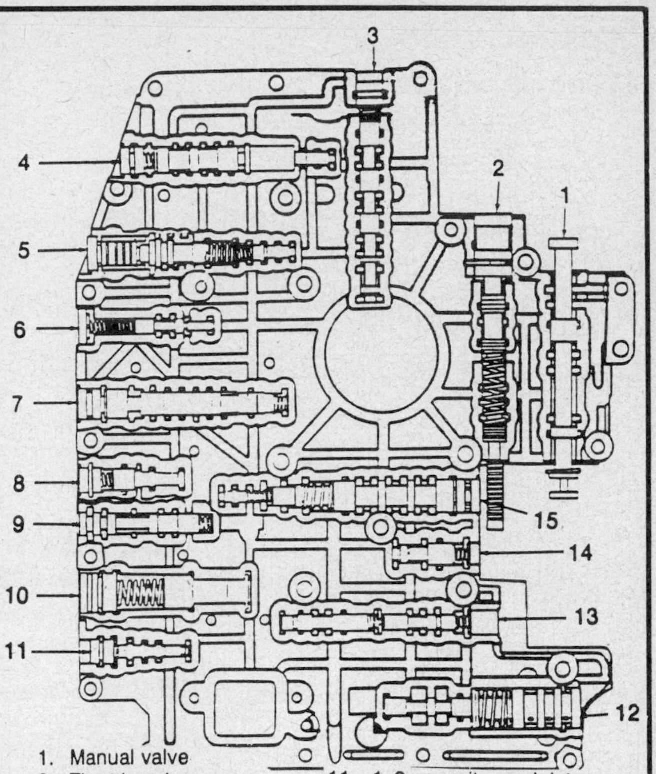
1
2
3
4
5
6
7
8
9
10
11
12
13
14
15

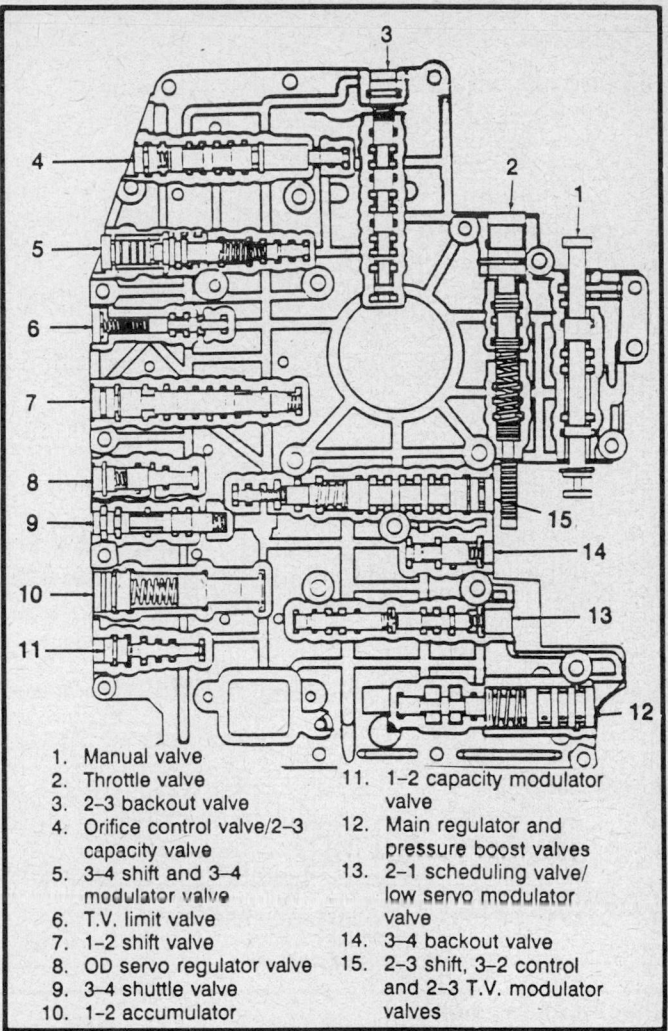

1. Manual valve
2. Throttle valve
3. 2–3 backout valve
4. Orifice control valve/2–3 capacity valve
5. 3–4 shift and 3–4 modulator valve
6. T.V. limit valve
7. 1–2 shift valve
8. OD servo regulator valve
9. 3–4 shuttle valve
10. 1–2 accumulator
11. 1–2 capacity modulator valve
12. Main regulator and pressure boost valves
13. 2–1 scheduling valve/ low servo modulator valve
14. 3–4 backout valve
15. 2–3 shift, 3–2 control and 2–3 T.V. modulator valves

Valve body—valve locations

2. Remove the separator plate, reinforcement plates and separator plate gasket.

3. Remove the 2 relief valves and the 7 check balls from the valve body. Note the location of the orange check ball, it is not interchangeable with the black check balls.

NOTE: The check balls are numbered 1 through 8, check ball number 7 was eliminated.

4. Slide the manual valve out of its bore.

5. To remove the throttle control valve, remove the retaining clip and slide the throttle sleeve out of the bore. Remove the preload spring, throttle plug, throttle control valve and throttle plunger.

6. To remove the 2–3 back out valve, remove the retaining clip and slide the valve and spring from the bore.

7. To remove the 2–3 capacity modulator valve, remove the spring retainer plate and slide the valve bore plug out of the bore. Remove the orifice control valve and spring, the second spring retainer plate and the 2–3 capacity modulator valve and sring.

8. To remove the 3–4 modulator/shift valves, remove the clip retaining the valves and slide the sleeve and plug from the bore. Remove the 3–4 shift valve and spring, 3–4 T.V. modulator valve and spring from the bore.

9. To remove the T.V. limit valve, remove the spring retainer plate and slide the limit valve and spring from the bore.

10. To remove the 1–2 shift valve, remove the clip retaining

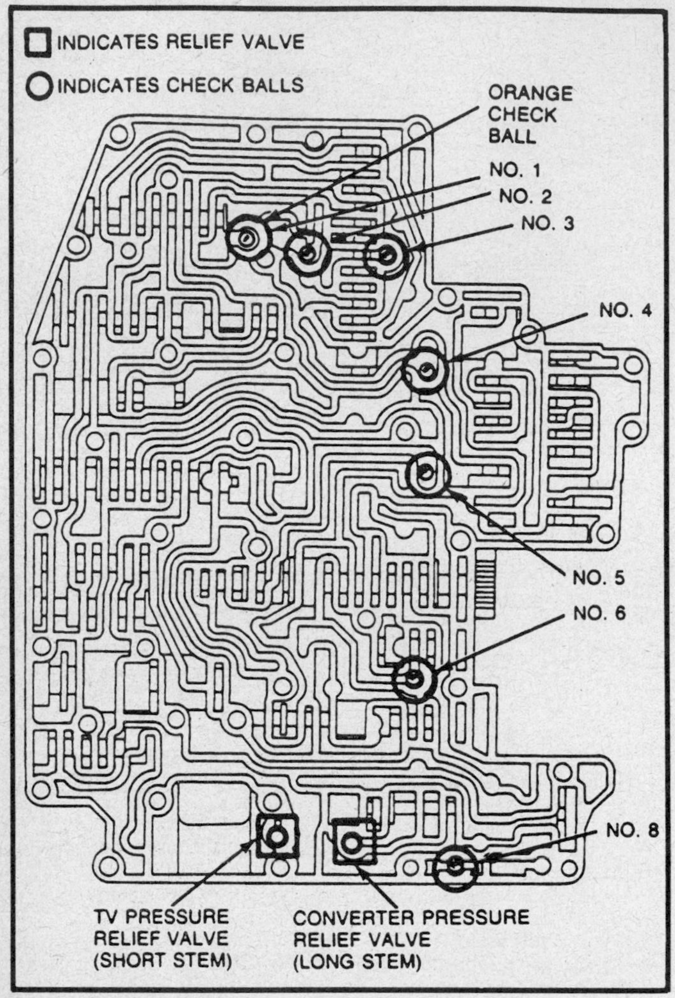
INDICATES RELIEF VALVE
INDICATES CHECK BALLS
ORANGE CHECK BALL
NO. 1
NO. 2
NO. 3
NO. 4
NO. 5
NO. 6
NO. 8
TV PRESSURE RELIEF VALVE (SHORT STEM)
CONVERTER PRESSURE RELIEF VALVE (LONG STEM)

Check ball locations

the bore plug and slide the bore plug, valve and spring from the bore.

11. To remove the OD servo regulator valve, remove the spring retainer plate and slide the bore plug, valve and spring from the valve body.

12. To remove the 3–4 shuttle valve, remove the spring retainer plate and slide the bore plug, valve and spring from the valve body.

13. To remove the 1–2 accumulator valve, remove the clip retaining the bore plug and remove it. Remove the O-ring seal and slide the 1–2 accumulator valve and spring from the bore.

14. To remove the 1–2 capacity modulator valve, remove the clip retaining the bore plug and remove it. Remove the the 1–2 accumulator valve and spring from the bore. Remove the 1–2 capacity modulator valve and spring.

15. To remove the main pressure regulator valve, remove the clip retaining the boost sleeve. Remove the boost valve and spring, main regulator valve spring and seat. Slide the main pressure regulator valve from the bore.

16. To remove the low servo/2–1 scheduling valves, remove the spring retaining plate, slide the low servo modulator valve and spring. Remove the second spring retainer and slide the 2–1 scheduling valve from the bore.

17. To remove the 3–4 back out valve, remove the spring retainer plate and slide the valve from the bore.

18. To remove the 2–3 shift/2–3 moulator valves – remove the clip retaining the bore plug and remove the bore plug. Slide the

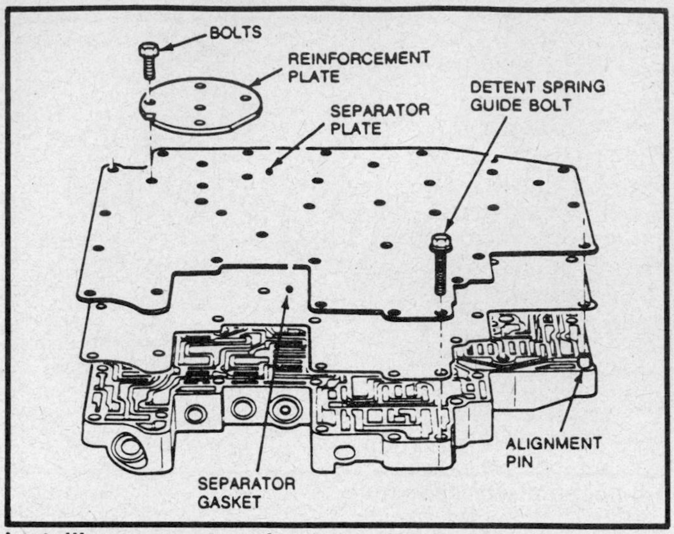
BOLTS
REINFORCEMENT PLATE
SEPARATOR PLATE
DETENT SPRING GUIDE BOLT
ALIGNMENT PIN
SEPARATOR GASKET

Installing separator plate on valve body

2–3 shift valve and spring, 3–2 control valve and 2–3 modulator valve and spring from the bore.

Inspection

1. Clean all parts thoroughly in solvent and blow dry with free compressed air.
2. Inspect all valve and plug bores for scoring. Check all fluid passages for obstructions. Inspect the check valves for freedom of movement and all mating surfaces for burrs or distortion.
3. Inspect all springs for distortion. Check all valves and plugs for freedom of movement in their bores. Valves and plugs should fall freely from their bores when dry.
4. Roll the manual valve on a flat surface to make sure it is not bent.
5. All burrs or scoring can be removed with crocus cloth. Do not round out the flat edges of valves with the cloth.

Assembly

1. Assemble each of the valves in their respective bores making sure that they move freely.
2. Lay the valve body on a flat surface and install the 7 check balls in their correct location. Make sure the orange ball is in the correct location, it is not interchangeable with the black check balls.
3. Install the 2 pressure relief valves in the valve body.
4. Install 2 alignment pins, T80L–77100–A or equivalent, into the valve body. Using a new separator plate gasket, install the separator plate over the alignment pins. Position the 3 reinforcement plates and loosely install the valve body bolts. Loosely install the detent spring guide bolt.
5. Starting at the center reinforcement plate and working outward, tighten the 11 attaching bolts to 80–100 inch lbs. Tighten the detent spring guide bolt to 80–100 inch lbs. Remove the alignment pins.

Transmission Assembly

1. Install the No. 9 needle bearing in the transmission case. Install the No. 7 bearing support and direct clutch hub in the direct clutch.
2. Install the output shaft, direct clutch and ring gear as an assembly. Install the governor drive ball and the governor assembly. Install the governor retaining ring, the face of the governor should be almost flush with the counterweight.
3. Install the low/reverse band into the front of the case. Make sure the band is seated on the anchor pin.

NOTE: When the band is properly installed, the center of the band actuating rod seat can be seen through the servo piston bore.

4. Install the center support and planetary assembly into the case. Align the planet carrier splines with the direct clutch hub splines.

NOTE: The planet carrier and center support assembly cannot be installed unless the notch in the center support is aligned with the overdrive band anchor pin.

5. Install the center support anti-clunk spring, the tabs on the spring must face outward. Install the center support retaining ring.

6. Before installing the low/reverse servo piston, the piston rod length must be determined. Use the following procedure to determine the length of piston rod needed:

a. Lubricate the low/reverse piston seal to ease assembly.

b. Install the low/reverse servo piston and return spring. Do not install the cover or retaining ring.

c. Install servo piston selection tool, T80L–77030–A or equivalent. Tighten the band apply bolt on the servo piston select tool to 50 inch lbs.

d. Attach dial indicator tool, TOOL–4201–C or equivalent and position the indicator stem on the flat portion of the piston. Zero the dial indicator.

e. Back the bolt out of the selector tool until the piston stops against the bottom of the tool.

f. Read the amount of piston travel on the dial indicator. If the travel is within 0.112–0.237 in., the travel is within specification. If the travel is not within specification, there are 3 piston lengths available; 2.936 in. (1 groove), 2.989 in. (2 grooves) and 3.043 in. (3 grooves). Select the proper rod to bring the travel within specification. Remove the service tools.

g. Lubricate the low/reverse servo piston cover seal and install the cover and retaining ring.

7. Install the reverse clutch on the forward clutch assembly. Be sure the No. 2 thrust washer is in position.

8. Install the No. 3 needle bearing and forward clutch hub in the forward clutch. Position the No. 4 needle bearing on the forward clutch hub and install the drive shell.

9. Install the No. 5 needle bearing and forward sun gear on the drive shell. Install the drive shell, forward clutch and reverse clutch as an assembly. Rotate the output shaft as necessary, to aid in engaging the sun gear with the planetary gears.

10. Install the overdrive band, make sure the band anchor is properly positioned on the anchor pin.

11. Lubricate the overdrive servo cover seals to ease assembly and install the servo in the case. With the overdrive servo installed, inspect the apply pin and band for proper position and engagement.

12. Install the intermediate clutch pack components, by first installing the pressure plate then the clutch pack and finally the selective steel plate.

13. Measure the intermediate clutch clearance using a depth micrometer D80P–4201–A or equivalent and endplay gauge bar T80L–77003–A or equivalent. Set the endplay tool across the pump case mounting. Locate the micrometer end play gauge bar and read the depth. The depth at the intermediate clutch separator plate should be; 1.634–1.636 in. for vehicles with V8 engines and 1.629–1.640 in. for vehicles with V6 engines.

NOTE: Maintain downward pressure on the clutch pack while measuring depth.

14. If the depth is not within tolerance, there are 4 different sized separator plates available to correct the depth; 0.071–0.067 in., 0.081–0.077 in., 0.091–0.087 in. and 0.101–0.097 in. Install the corrective plate, if necessary and recheck the clearance.

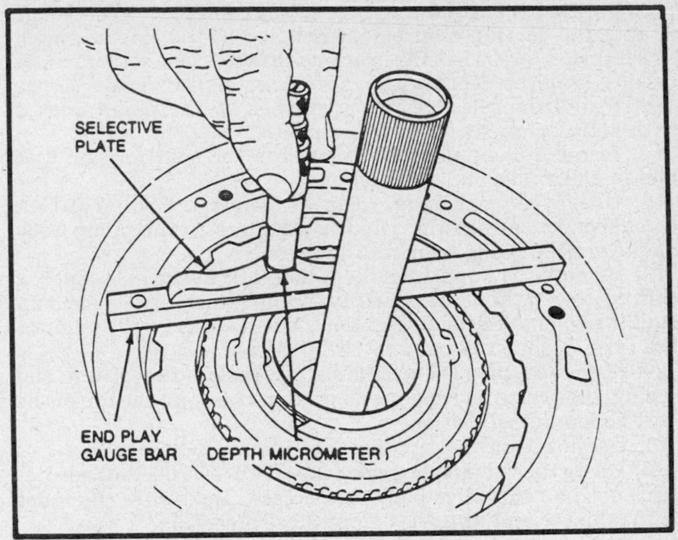
SELECTIVE PLATE
END PLAY GAUGE BAR
DEPTH MICROMETER

Checking intermediate clutch clearance

SELECTIVE THRUST WASHER ENDPLAY

Depth	Washer No.	Washer Size	Washer Color
37.668-38.113mm (1.483-1.500 inch)	1	0.050-0.054 inch	Green
38.114-38.540mm (1.501-1.517 inch)	2	0.068-0.072 inch	Yellow
38.541-38.970mm (1.518-1.534 inch)	3	0.085-0.089 inch	Natural
38.971-39.408mm (1.535-1.551 inch)	4	0.102-0.106 inch	Red
39.409-39.827mm (1.552-1.568 inch)	5	0.119-0.123 inch	Blue

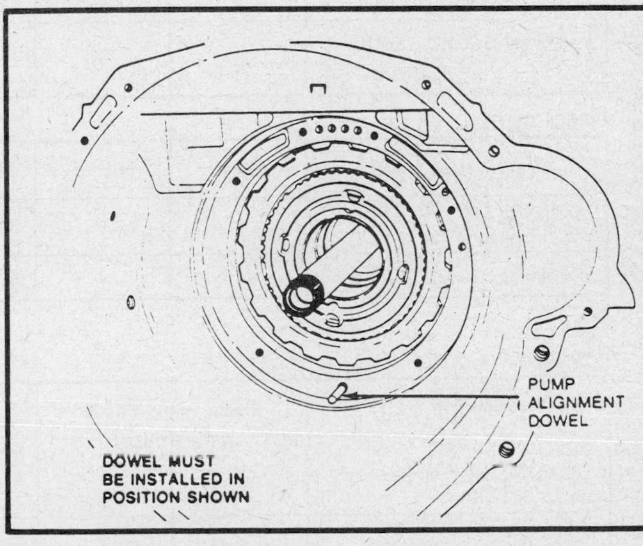
PUMP ALIGNMENT DOWEL
DOWEL MUST BE INSTALLED IN POSITION SHOWN

Installing the pump alignment dowel

15. Position the depth micrometer and end play depth gauge so that the depth measurement is taken at the reverse clutch drum thrust face. There are various thrust washers that can be used to keep the depth within specification.
16. Install the selective thrust washer on the pump, coat it with petroleum jelly to hold it in place.
17. Install a pump alignment dowel in the transmission case and install a new pump gasket.
18. Using the front pump remover adapters, T80L–77103–A or equivalent, attach the removal adapters to the pump body and lower the pump body into the case.
19. Remove the pump alignment dowel from the case. Apply a thin coating of thread sealer to the pump attaching bolts and install the bolts. Alternately tighten the bolts to draw the pump to the case. Torque the bolts to 16–20 ft. lbs.
20. Assemble the 3–4 accumulator. Install the piston and spring, lubricate the rubber seal on the cover and the top of the bore to aid in installation.
21. Assemble the 2–3 accumulator piston and cover.
22. Using the valve body guide pins, T80L–77100–A or equivalent, position the valve body to the case. Make sure the inner manual lever and inner T.V. levers are engaged.
23. Install the 24 valve body-to-case bolts and tighten to 80–100 inch lbs.
24. Install the detent spring and attaching bolt. Tighten the attaching bolt to 80–100 inch lbs.
25. Remove the valve body guide pins and install the remaining attaching bolts, tighten to 80–100 inch lbs.
26. Load the T.V. lever torsion spring against the separator plate.
27. Install the filter to the valve body and tighten the 3 attaching bolts to 80–100 inch lbs.
28. Clean the oil pan and gasket surfaces thoroughly.
29. Using a new oil pan gasket, attach the oil pan to the transmission. Tighten the attaching bolts to 12–16 ft. lbs.
30. Clean the mounting surface on the transmission and on the extension housing. Remove any sealant from the bolts and the case bolt holes. Install a new gasket on the transmission, coat the extension housing bolts with thread sealant and install them. Tighten the mounting bolts to 16–20 ft. lbs.
31. Install the direct driveshaft by sliding it into the pump. Install the torque converter, be sure the converter is fully seated in the pump.

SPECIFICATIONS

CLUTCH SPECIFICATIONS

FORWARD CLUTCH

Transmission Model	Steel	Friction	Clearance	Selective Snap Rings-Thickness
Models with 3.8L (232 CID)	4*	4	1.02-1.80mm (0.040-0.071 inch)	0.060-0.064 0.074-0.078 0.088-0.092 0.102-0.106
Models with 5.0L HO S.E.F.I., 5.0L S.E.F.I. or 5.8L (302 or 351 CID)	5*	5	1.27-2.26mm (0.050-0.089 inch)	

*Plus a waved plate (installed next to piston).

REVERSE CLUTCH

Transmission Model	Steel	Friction	Clearance	Selective Snap Rings-Thickness
Models with 3.8L (232 CID)	2	3	0.76-1.42mm (0.030-0.056 inch)	0.060-0.064 0.074-0.078 0.088-0.092 0.102-0.106
Models with 5.0L HO S.E.F.I., 5.0L S.E.F.I. or 5.8L (351 CID)	3	4	1.02-1.91mm (0.040-0.075 inch)	

DIRECT CLUTCH

Transmission Model	Steel	Friction	Clearance	Selective Snap Rings-Thickness
Models with 5.0L HO S.E.F.I., 5.0L S.E.F.I. or 5.8L	5	5	1.3-1.77mm (0.050-0.067 inch)	0.050-0.054 0.064-0.068 0.078-0.082 0.092-0.096
Models with 3.8L (232 CID)	4	4	1.02-1.44mm (0.040-0.057 inch)	

INTERMEDIATE CLUTCH

Transmission Model	Steel	Friction	Gauge Dim.	Selective Steel Plates-Thickness
Models with 5.8L or 5.0L HO S.E.F.I., 5.0L S.E.F.I., (351 or 302 CID)	3	3	41.504-41.808mm (1.634-1.646 inch)	0.067-0.071 0.077-0.081 0.087-0.091 0.097-0.101
Models with 3.8L (232 CID)	2	2	41.4-41.7mm (1.628-1.640 inch)	

TORQUE CONVERTER ENDPLAY DATA

Transmission	New or Rebuilt	Used
All	.023" Max.	.050" Max.

SELECTIVE SERVO PISTON

Rod Length*	I.D.	Rod Length*	I.D.	Rod Length*	I.D.
2.936	1 Groove	2.989	2 Groove	3.043	3 Groove

*Measured from the piston surface to the end of the rod.

SELECTIVE THRUST WASHER DATA

Depth	Thickness	Color Code	Depth	Thickness	Color Code
37.668-38.113mm (1.483-1.500 inch)	0.050-0.054	Green	38.971-39.408mm (1.535-1.551 inch)	.102-.106	Red
38.114-38.540mm (1.501-1.517 inch)	0.068-0.072	Yellow	39.409-39.827mm (1.552-1.568 inch)	.119-.123	Blue
38.541-38.970mm (1.518-1.534 inch)	0.085-0.089	Natural			

*The thrust washer is located on the stator support which is attached to the back of the pump housing.

TORQUE SPECIFICATIONS

Description	N·m	Lb-Ft	Description	N·m	Lb-Ft
Stator Support to Pump Body	16-22	12-16	Radiator Connectors Push Connector to Tank	16-24	12-18
Front Pump to Case	22-27	16-20	Threaded Connector to Tank	24-31	18-23
Reinforcing Plate to Valve Body	9-14	80-120 lb-in	Tube Nut to Threaded Connector	16-24	12-18
Separator Plate to Valve Body	9-11	80-100 lb-in	Converter to Flywheel	27-46	20-34
Valve Body to Case	9-11	80-100 lb-in	Converter Housing Access Cover to Converter Housing	16-22	12-16
Filter to Valve Body	9-14	80-120 lb-in	Detent Spring Attaching Bolt	9-14	80-120 lb-in
Oil Pan to Case	8-13.5	6-10	Inner Manual Lever to Shaft	26-37	19-27
Extension to Case	22-27	16-20	Outer Throttle Lever to Shaft	16-22	12-16
Governor Body to Counterweight	6-7	50-60	Push Connect Fitting to Case	24-31	18-23
Governor Body Cover to Governor Body	2.3-3.4	20-30	Converter Plug to Converter	11-38	8-28
Transmission Connectors Push Connector to Case	24-31	18-23	Neutral Start Switch to Case	11-15	8-11
Tube Nut to Connector	16-24	12-18	Pressure Plug to Case	8-16	6-12
			Transmission to Engine	55-68	40-50

SPECIAL TOOLS

Tool Number	Description	Tool Number	Description
T50T-100-A	Slide Hammer	T83L-7902-A	Converter Checking Tool
T59L-100-B	Impact Slide Hammer	T83L-7902-A1	Rod Torque Adapter
T58L-101-A	Shift Shaft Seal Remover	T83L-7902-A2	Pilot Guide
TOOL-1175-AC	Front Pump Seal and Rear Case Bushing Remover	T83L-7902-A3	Holding Fixture
T71P-19703-C	O-Ring Pick	T82L-9500-AH	Cooler Line Disconnect Tool
D80P-4201-A	Depth Micrometer (Also Commercially Available)	T80L-77030-B	Servo Piston Remover (Also used for air pressure checks)
TOOL-4201-C	Dial Indicator	T80L-77034-A	Extension Housing Bushing Replacer
T57L-500-B	Bench Mounted Holding Fixture	T80L-77100-A	Valve Body Guide Pins
T73L-6600-A	Pressure Gauge	T80L-77103-A	Front Pump Remover Adapter
TOOL-7000-DE	Air Nozzle	T80L-77110-A	Rear Case Bushing Replacer
T82L-7006-A	Air Test Adapter Plate	T80L-77140-A	Forward Clutch Lip Seal Protector (Inner)
T82P-7006-C	Air Test Adapter Plate Screws	T80L-77234-A	Direct Clutch Lip Seal Protector (Inner)
T84P-7341-A	Shift Linkage Grommet Removal Tool	T74P-77247-A	Neutral Start Switch Socket
T61L-7657-A	Extension Housing Seal Replacer	T74P-77248-A	Extension Housing Seal Remover
T77L-7697-A	Extension Housing Bushing Remover	T80L-77254-A	Lip Seal Protector
D84P-70332-A	Rod TV Control Pressure Gauge Block	T80L-77268-A	Front Pump Bushing Replacer
T86L-70332-A	Cable TV Control Pressure Gauge Tool	T80L-77268-B	Front Pump Bushing Remover
T86L-70002-A	TV Pressure Gauge with Hose	T80L-77403-A	Reverse Clutch Seal Protector (Outer)
D80L-77001-A	Adapter Fitting	T80L-77403-B	Reverse Clutch Seal Protector (Inner)
T80L-77003-A	End Play Gauge Bar	T80L-77405-A	Reverse Clutch Spring Compressor Plate
T80L-77005-A	Intermediate Clutch Lip Seal Protector (Inner and Outer)	T74P-77498-A	Shift Shaft Seal Replacer
T80L-77030-A	Servo Piston Selection Tool	T65L-77515-A	Clutch Spring Compressor
TOOL-7000-DD	Rubber Tip for Air Nozzle	T80L-77515-A	Forward Clutch Spring Compressor Adapter
T80L-7902-A	End Play Checking Tool	T63L-77837-A	Front Pump Seal Replacer
T76L-7902-C	Converter Clutch Torquing Tool	T68P-7D158-A	Forward Clutch Lip Seal Protector (Outer)
T77L-7902-A	Converter Clutch Holding Tool	T86P-77265-AH	Cooler Line Disconnect Tool (Push Connect Fittings)

Section 4

A4LD Transmission
Ford Motor Co.

APPLICATION

A4LD

Year	Vehicle	Engine
1985	Bronco II	All
	Ranger	All
1986	Aerostar	All
	Bronco II	All
	Ranger	All
1987	Mustang	2.3L, 5.0L
	Thunderbird	2.3L
	Aerostar	All
	Bronco II	All
	Ranger	All
1988–89	Mustang	2.3L, 5.0L
	Scorpio	2.9L
	Thunderbird	2.3L
	Aerostar	All
	Bronco II	All
	Ranger	All

GENERAL DESCRIPTION

Transmission and Converter Identification

TRANSMISSION

The A4LD is a 4 speed overdrive automatic transmission with a lockup torque converter. It is similar to the C-3 (3 speed) automatic transmission. It is the first Ford Motor Company production automatic transmission to use electronic controls integrated in the on-board computer system. The hydraulic lockup and unlock function of the torque converter is electronically controlled by the on-board computer EEC-IV system.

All vehicles are equipped with a safety standard certification Label on the driver's side door lock post. Refer to the stamped code in the space marked **Trans.** for proper transmission identification. The transmission is also identified by a tag on the transmission body, attached to the lower left hand extension attaching bolt.

CONVERTER

The converter is a welded unit and is not to be repaired. If a transmission malfunction is traced to the torque converter, replacement of the converter is recommended.

Electronic Controls

The torque converter incorporates electrical and hydraulic lockup components. The EEC-IV system controls a converter clutch solenoid in the main control which hydraulically operates the piston/plate clutch in the converter to provide a solid drive transmission function.

Metric Fasteners

Metric bolts may be used in attaching the transmission to the engine and also in attaching the transmission to the crossmember mount. The metric fastener dimensions are very close to the dimensions of the familiar inch system fasteners and for this

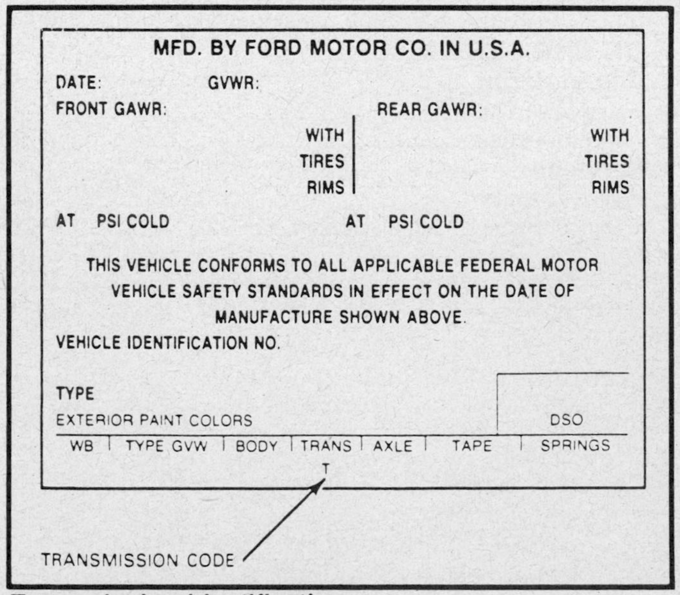
MFD. BY FORD MOTOR CO. IN U.S.A.
DATE:
GVWR:
FRONT GAWR:
REAR GAWR:
WITH
TIRES
RIMS
WITH
TIRES
RIMS
AT PSI COLD
AT PSI COLD
THIS VEHICLE CONFORMS TO ALL APPLICABLE FEDERAL MOTOR VEHICLE SAFETY STANDARDS IN EFFECT ON THE DATE OF MANUFACTURE SHOWN ABOVE.
VEHICLE IDENTIFICATION NO.
TYPE
EXTERIOR PAINT COLORS
DSO
WB
TYPE GVW
BODY
TRANS
AXLE
TAPE
SPRINGS
T
TRANSMISSION CODE

Transmission Identification

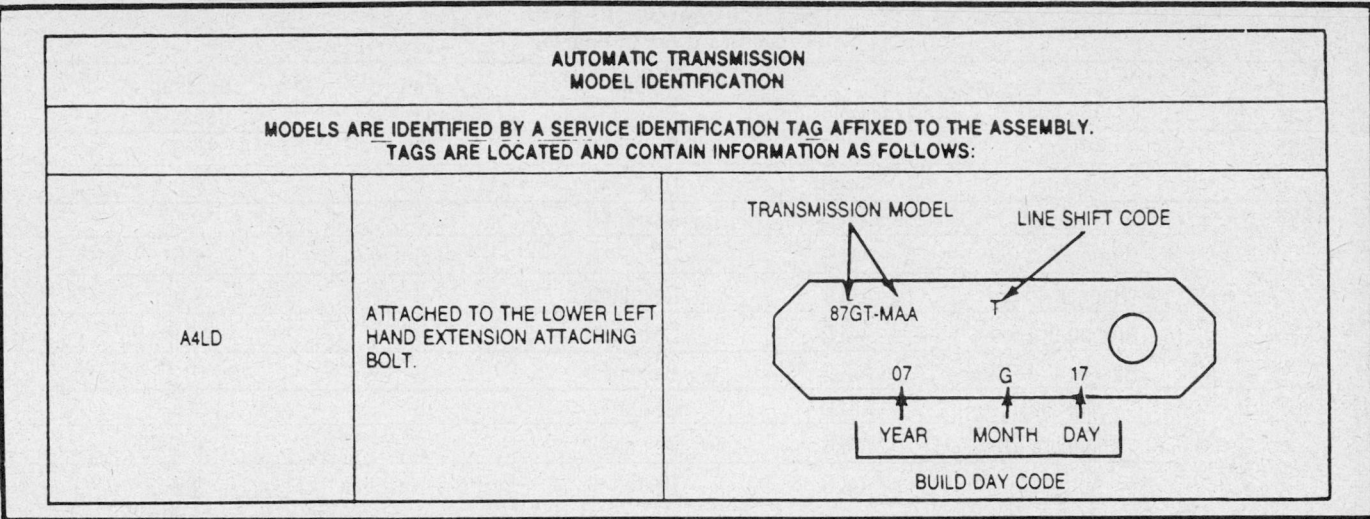

AUTOMATIC TRANSMISSION MODEL IDENTIFICATION		
MODELS ARE IDENTIFIED BY A SERVICE IDENTIFICATION TAG AFFIXED TO THE ASSEMBLY. TAGS ARE LOCATED AND CONTAIN INFORMATION AS FOLLOWS:		
A4LD	ATTACHED TO THE LOWER LEFT HAND EXTENSION ATTACHING BOLT.	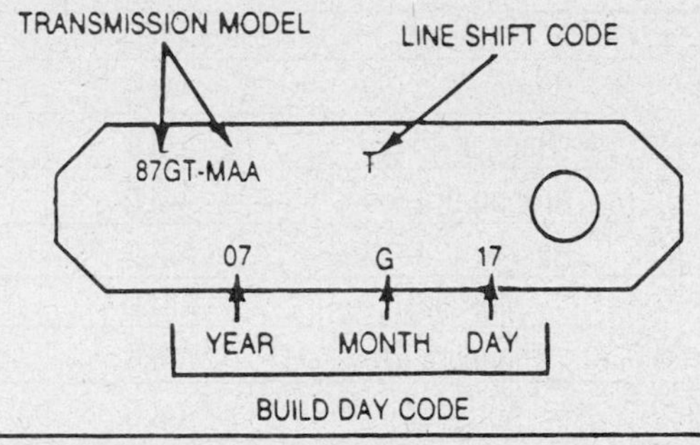 TRANSMISSION MODEL LINE SHIFT CODE 87GT-MAA 07 G 17 YEAR MONTH DAY BUILD DAY CODE

Transmission tag

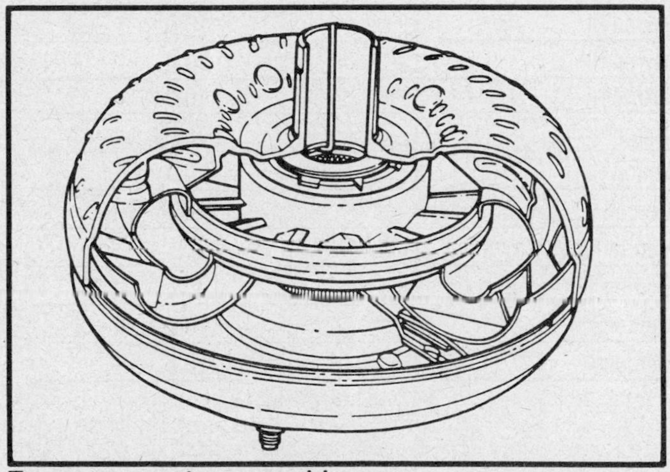

Torque converter assembly

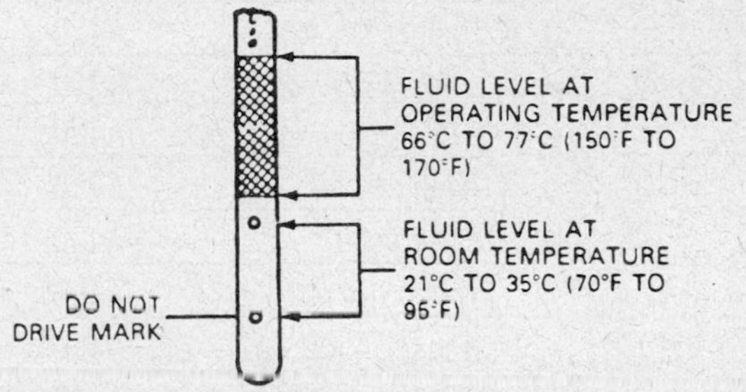
FLUID LEVEL AT
OPERATING TEMPERATURE
66°C TO 77°C (150°F TO
170°F)
FLUID LEVEL AT
ROOM TEMPERATURE
21°C TO 35°C (70°F TO
95°F)
DO NOT
DRIVE MARK

Dipstick reading

reason, replacement fasteners must have the same measurement and strength as those removed.

CAUTION

Do not attempt to interchange metric fasteners for inch system fasteners. Mismatched or incorrect fasteners can result in damage to the transmission unit through malfunctions or breakage and cause personal injury.

Capacities

The A4LD automatic transmission has a fluid capacity of approximately 8–10 quarts including the torque converter if it has been overhauled or flushed. If the torque converter has not been drained, the fluid capacity is 4 quarts. Use Dexron® II or Motorcraft Mercon® automatic transmission fluid. Bring the transmission to normal operating temperature and recheck the fluid level. Do not overfill the unit.

Checking Fluid Level

OPERATING TEMPERATURE

The automatic transmission should be checked at an operating temperature of 150–170° F (66–77° C). The dipstick should be hot to the touch. The dipstick reading should be within the cross-hatched area.

ROOM TEMPERATURE

If the transmission is not at operating temperature and it becomes necessary to check the fluid level, it may be checked at a room temperature of 70–95° F (21–35° C). The dipstick should be cool to the touch. The dipstick reading should be between the holes.

1. With the transmission in **P**, engine at idle rpm, foot brakes applied and vehicle on level surface, move the selector lever through each range allowing time to engage the transmission.
2. Return the selector lever to **P**, applying the parking brake fully and block the wheels. Do not turn off the engine during the fluid level check.
3. Pull the dipstick out of the filler tube, wipe it clean and push it all the way back into the tube. Make sure that it is fully seated.
4. Pull the dipstick out of the filler tube again and check the fluid level. Before adding fluid, check for the correct type to use. It is usually stamped on the dipstick.
5. If necessary, add enough fluid through the filler tube to raise the level to the correct position. Over filling the transmission will result in foaming, loss of fluid through the vent and possible transmission malfunction.
6. Install the dipstick checking that it is fully seated in the tube.

FLUID CAPACITY

All capacities given in quarts

Year	Vehicle	Transmission	Fluid Type	Pan Capacity	Overhaul Capacity
1985	Bronco II	A4LD	Dexron®II	3	9.0
	Ranger	A4LD	Dexron®II	3	9.0
1986	Aerostar	A4LD	Dexron®II	3	9.5
	Bronco II 4×2	A4LD	Dexron®II	3	9.5
	Ranger 4×2	A4LD	Dexron®II	3	9.5
	Bronco II 4×4	A4LD	Dexron®II	3	10.3
	Ranger II 4×4	A4LD	Dexron®II	3	10.3
1987	Mustang	A4LD	Dexron®II	3	9.5
	Thunderbird	A4LD	Dexron®II	3	9.5
	Aerostar	A4LD	Dexron®II	3	9.5
	Bronco II 4×2	A4LD	Dexron®II	3	9.5
	Ranger 4×2	A4LD	Dexron®II	3	9.5
	Bronco II 4×4	A4LD	Dexron®II	3	10.0
	Ranger 4×4	A4LD	Dexron®II	3	10.0
1988–89	Mustang	A4LD	Dexron®II	3	10.0
	Scorpio	A4LD	Dexron®II	3	9.5
	Thunderbird	A4LD	Dexron®II	3	10.0
	Aerostar	A4LD	Mercon®	3	9.5
	Bronco II 4×2	A4LD	Mercon®	3	9.5
	Ranger 4×2	A4LD	Mercon®	3	9.5
	Bronco II 4×4	A4LD	Mercon®	3	10.0
	Ranger 4×4	A4LD	Mercon®	3	10.0

TRANSMISSION MODIFICATIONS

Front Pump Installation

A new pump bolt (part number E800512-S72) silver color, has been released with increased torque specifications, 9.6–11 ft. lbs. (13–15 Nm). Whenever overhauling the A4LD transmission, replace all old design pump bolts (black colored) with the new bolts and increase the torque. Do not increase the torque specification with the old design bolts.

When installing the housing and pump to case, always install new aluminum washers (part number E830124-S) under the bolt. Use front pump alignment tools (kit no. T74P-77103-X) for installation.

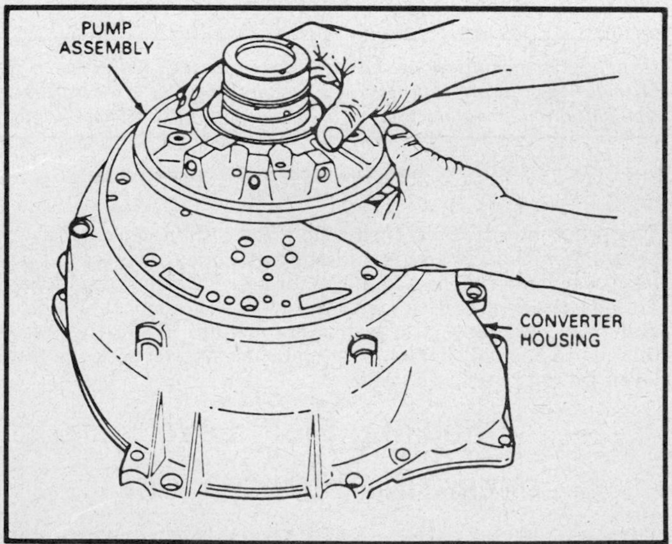
PUMP
ASSEMBLY
CONVERTER
HOUSING

Front oil pump installation

Front Pump Seal Blow Out

Whenever this condition is found in 1985–87 vehicles, a new housing/seal assembly has been released and must be used for service to prevent repeat repairs. If new tools are not available to remove, replace and restake the seal, then seal repairs on housings with staked seals will require replacement of the housing assembly.

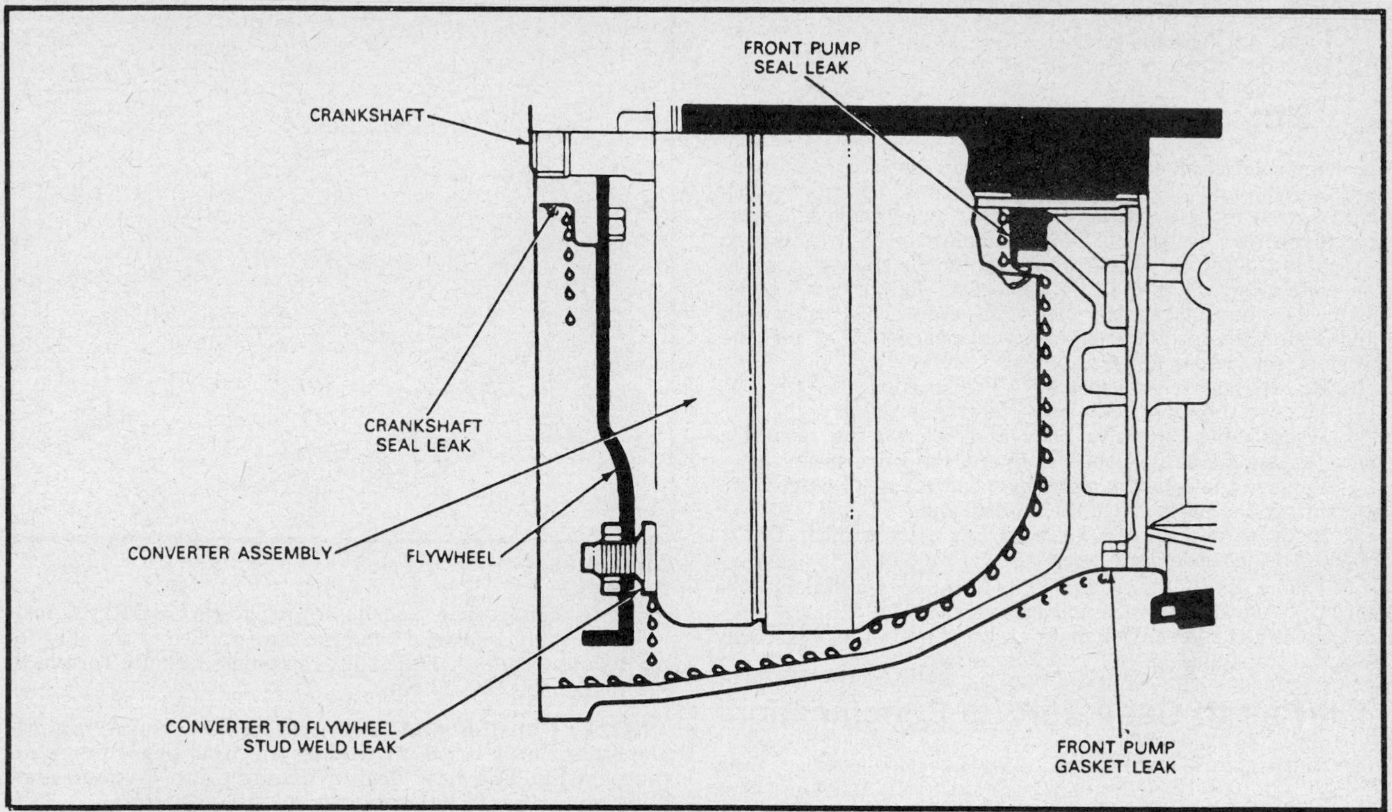
FRONT PUMP
SEAL LEAK
CRANKSHAFT
CRANKSHAFT
SEAL LEAK
CONVERTER ASSEMBLY
FLYWHEEL
CONVERTER TO FLYWHEEL
STUD WELD LEAK
FRONT PUMP
GASKET LEAK

Fluid leak points

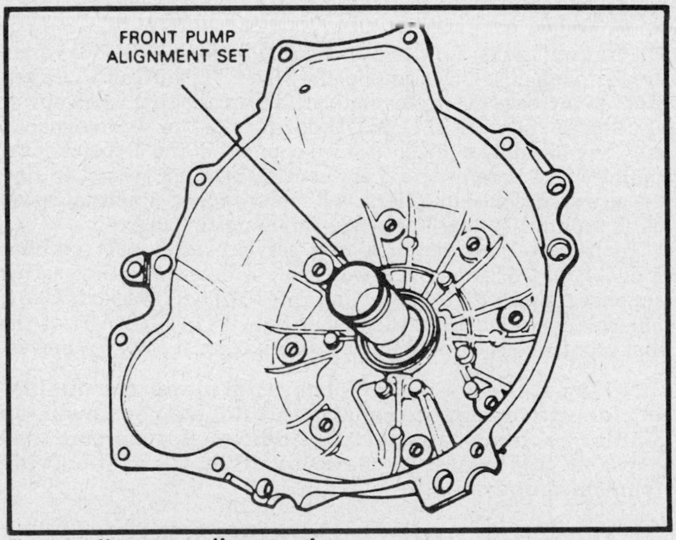
FRONT PUMP
ALIGNMENT SET

Front oil pump alignment

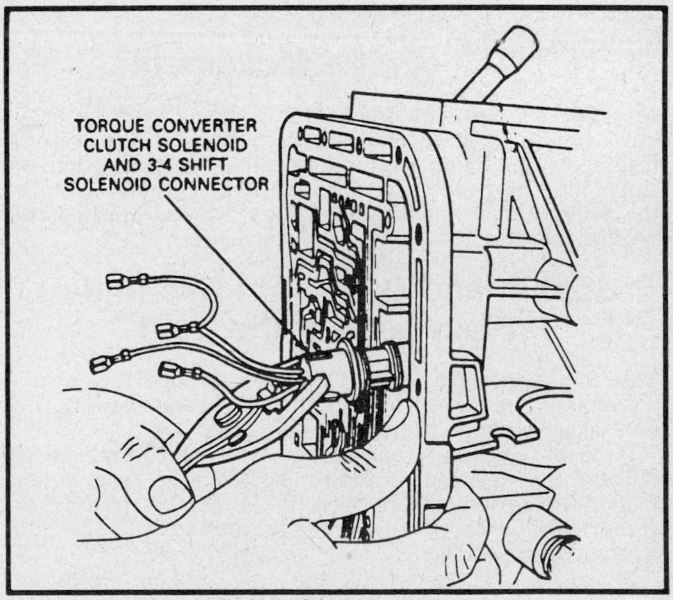
TORQUE CONVERTER
CLUTCH SOLENOID
AND 3-4 SHIFT
SOLENOID CONNECTOR

Converter clutch solenoid connector

Converter Override Clutch Connector Leak

There may be a leak at the converter clutch override connector and transmission case. This condition may be caused by sump fluid migrating up the override connector wires and depositing at the terminals at the top of the connector. This condition may be corrected by replacing the switch (part number E5TZ-7E449-A).

Burnt Fluid After Towing

Some Bronco II/Ranger vehicles, equipped with 4WD and an electronic transfer case, may sustain transmission damage while being towed by another vehicle. The electronic transfer case does not have a neutral position, so it cannot be shifted into **N** as can be done with the manually operated transfer case. If towing is necessary, make sure of the following:

1. Transmission is placed in **N**.
2. Front hubs are disengaged.
3. Rear driveshaft is removed.

Engine Stalls On Engagement

On some vehicles an engine stall condition that occurs after engine start up and when shifting the transmission into any forward gear or reverse gear may be caused by a broken converter clutch shuttle valve spring. This will allow the shuttle valve to remain in the bottom of its bore, continuously applying the piston plate clutch in the torque converter. This provides a mechanical connection between the engine and wheels, resulting in engine stall when the transmission is engaged. To correct the problem, proceed as follows:

1. Be sure that engine speed is set to specification.
2. Remove the valve body from the transmission.
3. Disassemble the valve body and remove the separator plate. Locate the shuttle valve bore and the valve spring.
4. Remove the retainer plate, override solenoid, plug, valve and spring. Do not dislodge the shuttle balls.
5. Install a new shuttle valve spring (part number E5TZ-7L490-A) color code dark green.
6. Reinstall the other components in the reverse order. Make sure that the shuttle valve moves freely.
7. Install the separator plate and assemble the valve body into the transmission.

Oil Pan to Case Torque Specification

The oil pan to case fastener torque on the A4LD is revised from 5–10 ft. lbs. (6.8–13.6 Nm) to 8–10 ft. lbs. (10.8–13.6 Nm). This new torque specification applies to all A4LD transmissions regardless of case type and will ensure a proper oil pan to case seal when service is performed.

No Forward Movement

On some 1985 and later Bronco II/Ranger and Aerostar, no forward movement at normal throttle position in the overdrive may be caused by an improperly installed overdrive one-way clutch into the overdrive center shaft. When properly installed, the center shaft should turn freely in a clockwise direction when holding the overdrive planet assembly.

Upshift and Downshift Slip and/or Fluid Leak at the Vent

This condition my be caused by governor seal ring wear or transmission case wear in the counterbore area, allowing internal leakage past the transmission governor.

If the seal rings and transmission case need to be replaced because of this wear, new Viton rubber seal rings (part number E7TZ-7D011-A) should be used. If the original transmission case is used, the cast iron seal rings (part number D4ZZ-7D011-B) should be installed.

NOTE: Use of Viton rubber seals in a worn or grooved transmission case will result in a recurrence of leakage past the governor.

Selector Lever Vibration

On some 1987 Mustangs, a rattling noise from the transmission shifter housing and lever assembly while traveling on rough roads may be caused by the spring tension of the transmission shifter assembly being too weak.

To correct this, install a new design transmission housing and lever assembly (part number E7ZZ-7210-C) which has higher spring tension. It must be adjusted in the **OVERDRIVE** position with slight rearward hand pressure applied to the shift lever during torquing of the shift rod adjusting nut. Torque to 9.5–18.5 ft. lbs. (13–25 Nm).

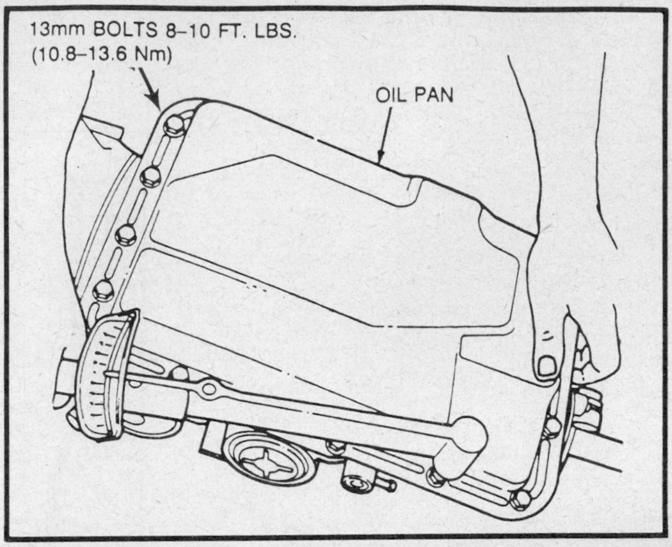
13mm BOLTS 8–10 FT. LBS.
(10.8–13.6 Nm)
OIL PAN
Oil pan assembly

NOTE: Housing and lever assemblies requiring replacement may be identified by a yellow paint mark on the casting. The new design housing and lever assemblies have purple paint marks on the castings.

Engine RPM Increases on Downshift

On transmissions built between 07/30/86 and 10/01/86, which are installed in vehicles equipped with a 2.3L, 2.9L and 3.0L engine, an increase of approximately 1000 engine rpm when downshifting to 2nd gear in **D** or **OD** during a 3–2 or 4–2 downshift may be caused by a mispositioned channel in the 3–2 kickdown timing valve bore of the main control casting. This condition usually occurs within a narrow 3 mph range at a normal speed of 38 mph depending upon axle ratio and tire size.

To correct this, install a new service valve (part number E7TZ-7D054-A). Remove the oil pan to determine the casting supplier code (1-7010-D), casting date (07/30/86 through 10/01/86), cavity No.1 and (part number P86GT-7A101-AAR). If the main control has the matching identification, it can be removed.

NOTE: If the main control does not match the qualifying identification, the repair must not be made even if a similar problem is reported. Additional problems may occur if this repair is made to other than qualifying transmissions.

Upshift and Downshift Slipping

Some 1986–87 Aerostars may be equipped with a transmission that slips during upshifts and downshifts. This problem may be caused by the manual linkage not properly adjusted to specification. To correct this, adjust the manual linkage using the following service procedure:

1. From inside the vehicle, place the shift lever in the **OD** position.
2. From below the vehicle, loosen the adjustment screw on the shift cable and remove the end fitting from the manual lever ball stud.
3. Position the manual lever in the **OD** position by moving the lever all the way rearward, and then moving it 3 detents forward. Hold the shift lever against the **OD** rear stop.

4. Connect the cable end fitting to the manual lever. Tighten the adjustment screw to 45–60 inch lbs. (5–7 Nm).

5. After adjustment, check for **P** engagement. The control lever must move to the right when engaged in the **P** detent. Check the transmission control lever in all detent positions with the engine running to make sure of correct detent action.

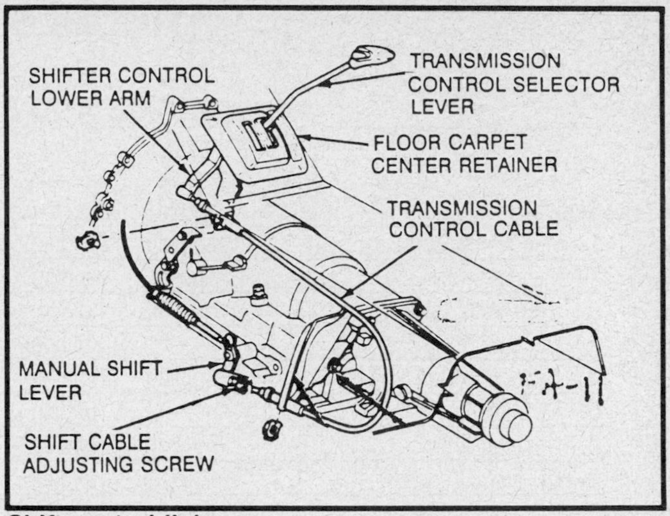
SHIFTER CONTROL LOWER ARM
TRANSMISSION CONTROL SELECTOR LEVER
FLOOR CARPET CENTER RETAINER
TRANSMISSION CONTROL CABLE
MANUAL SHIFT LEVER
SHIFT CABLE ADJUSTING SCREW
Shift control linkage

Erratic or No 3–4 Upshift or 4–3 Downshift

Some 1987–88 vehicles may experience and erattic or no 3–4 upshift or 4–3 downshift condition. This condition may be caused by normal levels of contamination in the transmission fluid. This causes the 3–4 shift solenoid to stick in the open or closed position. The 3–4 shift solenoid is external of the transmission filter which makes it sensitive to normal levels of contamination.

To correct this, install a new 3–4 shift solenoid and sleeve/screen assembly. The sleeve/screen assembly will protect the new 3–4 shift solenoid from contamination. Both components are contained in service kit (part number E8TZ-7M107-A).

Front Oil Pump to Converter Housing Bolt

On the 1987–88 Mustang and Thunderbird, 1986–88 Aerostar and 1985–88 Bronco II/Ranger, the front oil pump to converter housing bolts for the 1988 transmission cannot be used on prior model year transmissions. Two types of front oil pump to converter housing bolts are currently available for service. The M–6 bolts are used to service 1987 and prior transmission. The M–8 bolts are used to service the 1988 transmission only. The M–8 bolt size is M–8 × 1 with fine threads and has a torque range of 17–20 ft. lbs. (23–27 Nm).

NOTE: The 1987 and prior model year M–6 bolts, front oil pump support, oil pump adapter plate and converter housing assemblies are not to be used to service 1988 A4LD transmissions.

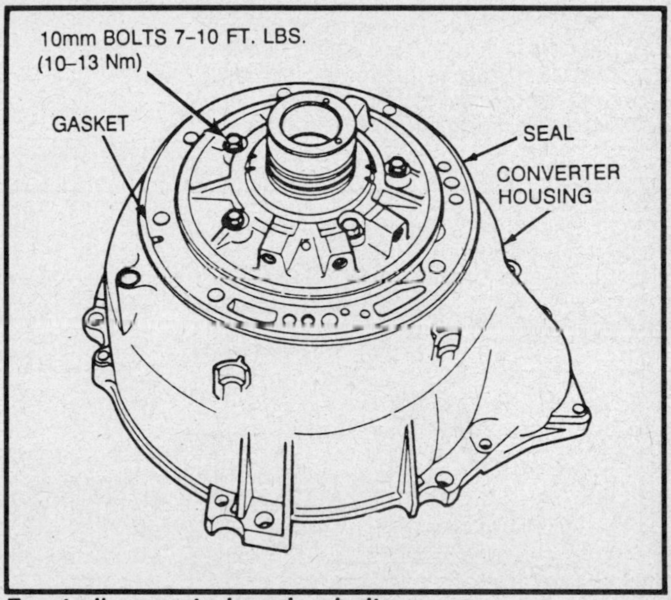
10mm BOLTS 7–10 FT. LBS. (10–13 Nm)
GASKET
SEAL
CONVERTER HOUSING
Front oil pump to housing bolts

Snapring and Output Shaft

Aerostar extended vans require the use of a 2.0mm snapring and matching output shaft. During the 1989 model year some A4LD transmissions used in other vehicle and light truck applications, are built with either the 1.2mm snapring and matching output shaft or the 2.0mm snapring and matching output shaft. The snaprings and output shafts are not interchangeable because of the difference in size.

If service is required on the snapring or output shaft, it will be necessary to determine which size snapring is needed. The 1.2mm and 2.0mm snaprings are packaged together for service.

NOTE: The 2.0mm snapring and matching output shaft can be used to service prior model year A4LD transmissions.

Transmission Fluid Leak

On some 1988 Scoropios, transmission fluid leakage at the converter housing to transmission case may be caused by the existing converter housing to transmission case attaching bolts. The bolts may not provide adequate sealing qualities.

To correct this, install new design converter housing to transmission case attaching bolts (part number E804594–S72) that have an integral O-ring. Torque the new bolts to 28–38 ft. lbs. (38–52 Nm).

NOTE: If the transmission requires disassembly, always install the new design converter housing to transmission case attaching bolts at the time of reassembly.

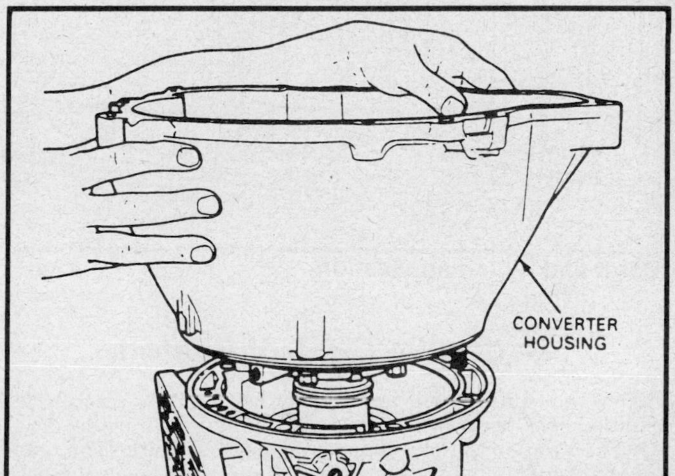
CONVERTER HOUSING
Converter housing to transmission case

TROUBLE DIAGNOSIS

CLUTCH AND BAND APPLICATION

Gear	Over-drive B and A	Over-drive Clutch B	Drive One-way Clutch C	Inter-mediate Band D	Reverse and High Clutch E	Forward Clutch F	Low and Reverse Band G	One-way Clutch H	Gear Ratio
1—Manual 1st (low)	—	Applied	Holding	—	—	Applied	Applied	Holding	2.47:1
2—Manual 2nd	—	Applied	Holding	Applied	—	Applied	—	—	1.47:1
D—Drive auto 1st	—	Applied	Holding	—	—	Applied	—	Holding	2.47:1
D—OD auto 1st	—	—	Holding	—	—	Applied	—	Holding	2.47:1
D—Drive auto 2nd	—	Applied	Holding	Applied	—	Applied	—	—	1.47:1
D—OD auto 2nd	—	—	Holding	Applied	—	Applied	—	—	1.47:1
D—Drive auto 3rd	—	Applied	Holding	—	Applied	Applied	—	—	1.00:1
D—OD auto 3rd	—	—	Holding	—	Applied	Applied	—	—	1.00:1
D—OD automatic 4th	Applied	—	—	—	Applied	Applied	—	—	0.75:1
Reverse	—	Applied	Holding	—	Applied	—	Applied	—	2.10:1

NOTE: Lock-up converter may apply in 3rd or 4th depending on both engine and vehicle speed as determined by transmission hydraulics and on-board Computer Electronic Controls (EEC IV)

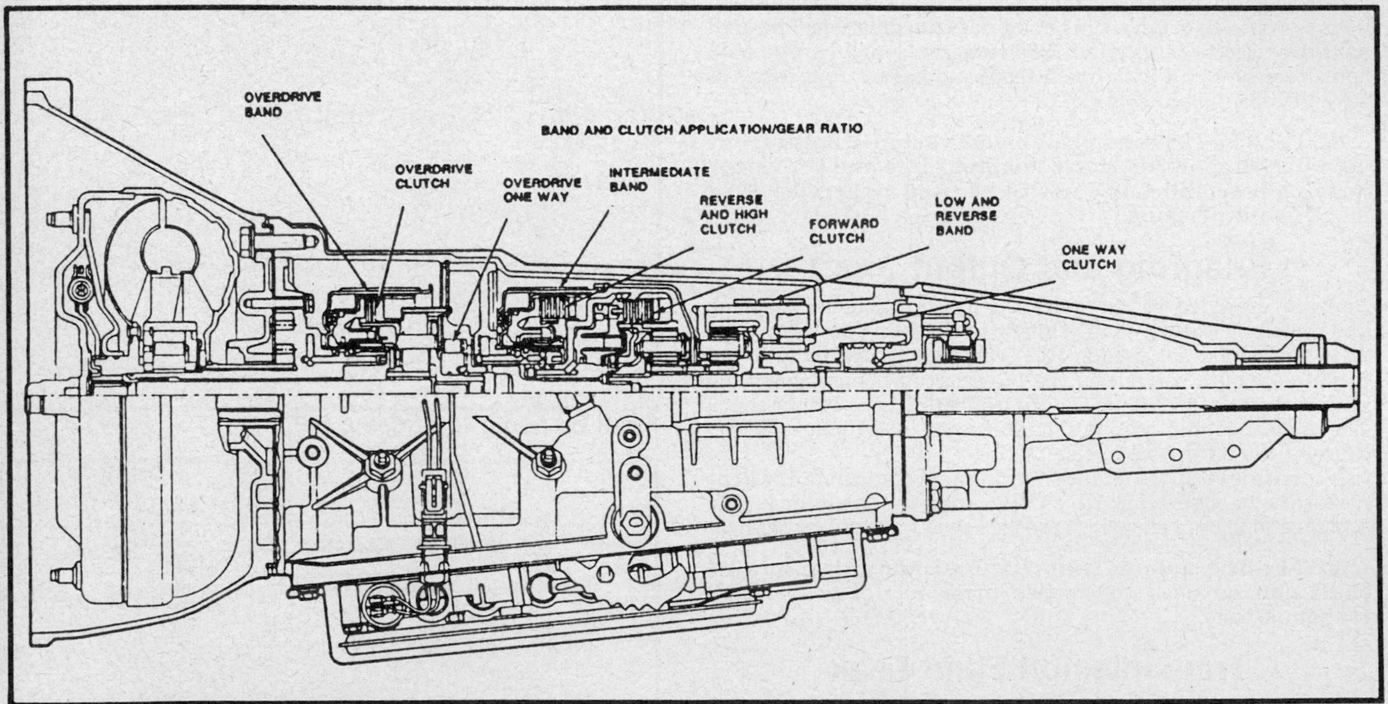
BAND AND CLUTCH APPLICATION/GEAR RATIO
OVERDRIVE BAND
OVERDRIVE CLUTCH
OVERDRIVE ONE WAY
INTERMEDIATE BAND
REVERSE AND HIGH CLUTCH
FORWARD CLUTCH
LOW AND REVERSE BAND
ONE WAY CLUTCH

Band and clutch application

Hydraulic Control System

Before the transmission can transfer the input power from the engine to the drive wheels, hydraulic pressures must be developed and routed to the varied components to cause them to operate, through numerous internal systems and passages. A basic understanding of the components, fluid routings and systems, will aid in the trouble diagnosis of the transmission.

OIL PUMP

The oil pump is a positive displacement pump, meaning that as long as the pump is turning and fluid is supplied to the inlet, the pump will deliver fluid in a volume proportionate to the input drive speed. The pump is in operation whenever the engine is operating, delivering more fluid than the transmission needs, with the excess being bled off by the pressure regulator valve

CHILTON'S THREE C's TRANSMISSION DIAGNOSIS
A4LD Automatic Transmission

Condition	Cause	Correction
Converter clutch does not engage	a) Converter clutch solenoid is not being energized electrically. Wires to solenoid shorted or an open circuit. Transmission case connector not sealed. Open or short circuit inside of solenoid. Malfunctioning engine coolant temperature sensor, throttle position sensor or manifold absolute pressure sensor. Vacuum line disconnected from the MAP sensor	a) Perform EEC-IV diagnostic check key on, engine off
	b) Brake switch malfunction	b) Perform EEC-IV diagnostic check, engine running
	c) Malfunctioning EEC-IV processor	c) Run diagnostic check on processor
	d) Converter clutch solenoid is being energized electronically but foreign material on solenoid valve is preventing valve closure	d) Remove transmission oil pan. Remove valve body. Remove solenoid and check operation of solenoid. Clean if necessary
	e) Converter clutch shuttle valve stuck in unlock position (against plug) or too high a load spring	e) Remove valve body. Check operation of converter clutch shuttle valve. Remove any contamination. Spring load should be approximateley 2 lbs @ 0.512 in.
	f) Converter clutch shift valve stuck in downshift position	f) Remove valve body. Check operation of converter clutch shift valve. Remove any contamination. Be sure valve moves freely
	g) Torque converter internal malfunction preventing lock-up piston application	g) Remove transmission. Replace converter
Converter clutch always engaged even at zero road speed. (Symptom: vehicle will move only when the engine is accelerated to a high rpm and transmission selector level is into D	a) Converter clutch shift valve stuck in lock position	a) Remove tranmission valve body. Check to see that converter clutch shift valve moves freely
	b) Converter clutch shuttle valve stuck in locked position	b) Remove valve body. Check converter clutch shuttle valve for ease of movement
	c) Lock-up piston in torque converter	c) Remove transmission. Replace converter
Converter clutch will not disengage on coastdown	a) malfunctioning throttle position sensor (should unlock at closed throttle)	a) Perform EEC-IV diagnpostic check, key on, engine off
	b) Converter clutch solenoid sticking	b) Remove valve body. Check operation of of solenoid. Replace if required
Slow initial engagement	a) Improper fluid level	a) Perform fluid level check
	b) Damaged or improperly adjusted manual linkage	b) Service or adjust manual linkage
	c) Contaminated fluid	c) Perform fluid condition check
	d) Improper clutch and band application or low main control pressure	d) Perform control pressure test
Rough initial engagement in either forward or reverse	a) Imprpoer fluid level	a) Perform fluid level check
	b) High engine idel	b) Adjust to specification
	c) Automatic choke on (warm temperature)	c) Service as required
	d) Looseness in the driveshaft U-joints or engine mounts	d) Service as required

CHILTON'S THREE C's TRANSMISSION DIAGNOSIS
A4LD Automatic Transmission

Condition	Cause	Correction
Rough initial engagement in either forward or reverse	e) Improper clutch band application or oil control pressure	e) Perform control pressure test
	f) Sticking or dirty valve body	f) Clean, service or replace valve body
	g) Converter clutch not disengaging	g) Check converter clutch engagement/ disengagement
Harsh engagements (warm engine)	a) Improper fluid level	a) Perform fluid level check
	b) Engine curb idle speed too high	b) Check engine curb idle speed
	c) Valve body bolts too loose or tight	c) Tighten to specification
	d) Valve body dirty, sticking valves	d) Determine source of contamination. Service as required
No or delayed forward engagement. Reverse OK	a) Improper fluid level	a) Perform fluid level check
	b) Manual linkage misadjusted or damaged	b) Check and adjust or service as required
	c) Low main control pressure (leakage). Forward clutch center support seal rings leaking	c) Control pressure test. Note results
	d) Forward clutch assembly burnt, damaged, leaking. Check ball in cylinder, leaking piston seal rings	d) Perform air pressure test
	e) Valve body bolts too loose ot tight	e) Tighten to specification
	f) Valve body dirty, sticking valves	f) Determine source of contamination. Service as required
	g) Transmission filter plugged	g) Replace filter
	h) Pump damaged or leaking	h) Visually inspect pump gear. Replace pump if necessary
No or delayed reverse engagement	a) Improper fluid level	a) Perform fluid level check
	b) Manual linkage misadjusted or damaged	b) Check and adjust or service as required
	c) Low main control pressure in reverse	c) Control pressure test
	d) Reverse clutch assembly burnt, worn or leaking; check ball in piston, leaking piston seal rings	d) Perform air pressure test
	e) Valve body bolts too loose or tight	e) Tighten to specification
	f) Valve body dirty, sticking valves	f) Determine source of contamination. Service as required
	g) Transmission filter plugged	g) Replace filter
	h) Pump damaged	h) Visually inspect pump gears. Replace pump if necessary
	i) Low/reverse servo piston seal cut or leaking	i) Perform air pressure test. Check and replace piston seal. Check and replace low/reverse band
No engagement in drive or forward (any position) or reverse	a) Improrper fluid level	a) Perform fluid level check
	b) Low main control pressure	b) Perform control pressure test
	c) Mechanical damage	c) Check splines on turbine, input shaft, and OD carrier, OD one-way clutch, center shaft, forward clutch, forward carrier and output shaft. Replace if necessary
No engagement or drive in D (2 & 1 OK)	a) Manual linkage misadjusted	a) Adjust manual linkage
	b) Rear one-way clutch damaged	b) Replace rear one-way clutch
	c) Dirty or contaminaited transmission fluid	c) Clean transmission and valve body
	d) Overdrive one-way clutch damaged	d) Repair or replace
Vehicle creeping in neutral	a) Forward clutch failing to disengage	a) Clean transmission

CHILTON'S THREE C's TRANSMISSION DIAGNOSIS
A4LD Automatic Transmission

Condition	Cause	Correction
No or delayed reverse engagement and/or no engine braking in manual low (1)	a) Improper fluid level	a) Perform fluid level check
	b) Linkage out of adjustment	b) Service or adjust linkage
	c) Low reverse servo piston seal leaking	c) Check and replace piston seal
	d) Low reverse band burnt or worn	d) Perform air pressure test
	e) Overdrive clutch, overdrive one-way clutch damaged	e) Replace as required
	f) Polished, glazed low/reverse band or drum	f) Service or replace as require4d
	g) Rear one-way clutch damaged	g) Replace
No engine braking in manual 2nd gear	a) Intermediate band out of adjustment	a) Adjust intermediate band
	b) Improper band or clutch application or oil pressure control system	b) Perform control pressure test
	c) Intermediate servo leaking	c) Perform air pressure test of intermediate servo for leakage. Service as required
	d) Overdrive clutch, OD one-way clutch damaged	d) Replace as required
	e) Glazed band	e) Service or replace as required
Forward engagement slips, shudders, chatters	a) Improper fluid level	a) Perform fluid level check
	b) Manual linkage misadjusted, damaged	b) Check and adjust or service as required
	c) Low main control pressure	c) Perform control pressure test
	d) Valve body bolts too loose or tight	d) Tighten to specification
	e) Valve body dirty, sticking valves	e) Determine source of contamination. Service as required
	f) Forward clutch piston ball check not sealing, leaking	f) Replace forward clutch piston. Service transmission as required
	g) Forward clutch piston seals cut or worn	g) Replace seal and service clutch as required
	h) OD one-way clutch damaged	h) Replace as required
	i) Rear one-way clutch damaged	i) Determine cause of condition. Service as required
Reverse shudders, chatters, slips	a) Improper fluid level	a) Perform fluid level check
	b) Low main control pressure in reverse	b) Perform control pressure test
	c) Low/reverse servo leaking	c) Air pressure test. Visually inspect seal rings and piston bore
	d) OD and/or rear one-way clutch damaged	d) Determine cause of condition. Service as required
	e) OD and/or rear revers/high clutch drum bushing damaged	e) Determine cause of condition. Service as required
	f) OD and/or rear/high clutch center support seal rings, ring grooves worn or damaged	f) Detetermine condition. Service as mt,f) required
	g) OD and/or rear reverse/high clutch piston seals cut or worn	g) Determine cause of condition. Service mt,g) as required
	h) Low/reverse servo piston damaged or worn	h) Service as required
	i) Low/reverse band out of adjustment or damaged	i) Adjust and inspect low/reverse band
	j) Looseness in the driveshaft, U-joints or engine mounts	j) Service as required
	k) Low/reverse servo piston, or bores damaged	k) Perform air pressure check

CHILTON'S THREE C's TRANSMISSION DIAGNOSIS
A4LD Automatic Transmission

Condition	Cause	Correction
No drive, slips or chatters in 1st gear in D. All other gears normal	a) Damaged or worn rear one-way clutch	a) Service or replace rear one-way clutch
No drive, slips or chatters in 2nd gear	a) Intermediate band out of adjustment b) Improper band or clutch application or control pressure c) Damaged or worn intermediate servo d) Dirty or sticking valve body e) Polished, glazed intermediate band	a) Adjustment intermediate band b) Perform control pressure test c) Perform air pressure test d) Clean, service or replace valve body e) Replace or service as required
Starts up in 2nd or 3rd	a) Improper band and/or clutch application or oil pressure control system b) Damaged, worn or sticking governor c) Valve body loose d) Dirty or sticking valve body e) Cross leaks between valve body and case mating surfce	a) Perform control pressure test b) Perform governor check. Replace or service governor, clean screen c) Tighten to specification d) Clean, service or replace valve body e) Service or replace valve body and or case as required
Shift points incorrect	a) Improper fluid level b) Vacuum line damaged, clogged or leaks c) Improper operation of EGR system d) Improper speedometer gear installed e) Improper clutch or band application or oil pressure control system f) Damaged or worn governor g) Vacuum diaphragm bent, sticks or leaks h) Dirty or sticking valve body	a) Perform fluid level check b) Perform vacuum supply test c) Service or replace as required d) Replace gear e) Perform shift test and control f) Service or replace governor, clean screen g) Service or replace as required h) Clean, service or replace valve body
All upshifts harsh, delayed or no upshifts	a) Improper fluid level b) Manual linkage misadjusted or damaged c) Governor sticking d) Main control pressure too high e) Valve body bolts too loose or tight f) Valve body dirty or sticking valves g) Vacuum leak to diaphragm unit h) Vacuum diaphragm bent, sticking, leaks	a) Perform fluid level check b) Check and adjust or service as required c) Perform governor test. Service as required d) Perform control test. Service as required e) Tighten to specification f) Determine source of contamination. Service as required g) Perform vacuum supply and diaphragm test. Check vacuum lines to diaphragm unit. Service as required h) Check diaphragm unit. Service as required
All upshifts mushy or early	a) Low main control pressure b) Valve body bolts too loose or tight c) Valve body or throttle control valve sticking d) Governor valve sticking e) Kickdown linkage misadjusted, sticking or damaged	a) Perform control pressure. Note results b) Tighten to specification c) Determine source of contamination. Service as required d) Perform governor test. Repair as required e) Adjust linkage, service as required

CHILTON'S THREE C's TRANSMISSION DIAGNOSIS
A4LD Automatic Transmission

Condition	Cause	Correction
No 1–2 upshift	a) Improper fluid level	a) Peform fluid level check
	b) Kickdown system damaged	b) Replace damaged parts
	c) Manual linkage misadjusted or damaged	c) Check and adjust or service as required
	d) Governor valve sticking	d) Perform governor test. Service as required
	e) Intermediate band out of adjustment	e) Adjust intermediate band
	f) Vacuum leak to diaphragm unit	f) Check vacuum lines to diaphragm. Service as required
	g) Vacuum diaphragm bent, sticking, leaks	g) Check diapraghm unit. Service as required
	h) Valve body bolts too loose or tight	h) Valve body dirty or sticking valves
	i) Intermediate band and/or servo assembly burnt	i) Perform air pressure test
Rough, harsh or delayed 1–2 upshift	a) Improper fluid level	a) Perform fluid level check
	b) Poor engine performance	b) Tune engine
	c) Kickdown linkage misadjusted	c) Adjust linkage
	d) Intermediate band out of adjustment	d) Adjust intermediate band
	e) Main control pressure too high	e) Perform control pressure test
	f) Governor valve sticking	f) Perform governor test. Service as required
	g) Damaged intermediate servo	g) Perform air pressure test on intermediate servo
	h) Engine vacuum leak	h) Check engine vacuum lines. Check vacuum diaphragm unit. Perform vacuum supply and diaphragm test. Service as required
	i) Valve body too loose or tight	i) Tighten to specifications
	j) Valve body dirty or sticking valves	j) Determine source of contaminiation. Service as required
	k) Vacuum leak to diaphragm unit	k) Check vacuum lines to diaphragm unit. service as required
	l) Vacuum diaphragm bent, sticking or leaks	l) Check diaphragm unit. Service as required
Mushy, early, soft or slipping	a) Improper fluid level	a) Perform fluid level check
	b) Main regulator or throttle valve sticking	b) Service as required
	c) Incorrect engine performance	c) Tune engine as required
	d) Intermediate band out of adjustment	d) Adjust intermeidate band
	e) Low main control pressure	e) Perform control pressure test
	f) Valve body bolts too loose or tight	f) Tighten to specification
	g) Valve body dirty or sticking valves	g) Determine source of contamination. Service as required
	h) Governor valve sticking	h) Perform governor test. Service as required
	i) Damaged intermediate servo or band	i) Perform air pressure test. Service as required
	j) Polished or glazed intermediate band or drum	j) Service or replace as required
No 2–3 upshift	a) Low fluid level	a) Perform fluid level check
	b) Kickdown system damaged	b) Replace damaged parts
	c) Low main control pressure to reverse high clutch	c) Perform control pressure test. Note results
	d) Valve body bolts too loose or tight	d) Tighten to specification

CHILTON'S THREE C's TRANSMISSION DIAGNOSIS
A4LD Automatic Transmission

Condition	Cause	Correction
No 2–3 upshift	e) Valve body dirty or sticking valves	e) Determine source of contamination. Service as required
	f) Reverse/high clutch assembly burnt or worn	f) Determine cause of condition. Service as required
Harsh or delayed 2–3 upshift	a) Incorrect engine performance	a) Check engine tune-up
	b) Engine vacuum leak	b) Check engine vacuum lines. Check vacuum diaphragm unit. Perform vacuum supply and diaphragm test. Service as required
	c) Kickdown system damaged	c) Replace damaged parts
	d) Damaged or worn intermediate servo release and reverse/high clutch piston check ball	d) Air pressure test the intermediate servo. Apply and release the reverse/high clutch piston check ball
	e) Valve body bolts too loose or tight	e) Tighten to specification
	f) Valve body dirty or sticking valves	f) Determine source of condition. Service as required
	g) Vacuum diaphragm bent, sticking or leaks	g) Check diaphragm. Replace as required
	h) Throttle valve stuck	h) Service as required
Soft, early or mushy 2–3 upshift	a) Kickdown system damaged	a) Replace damaged parts
	b) Valve body bolts too loose or tight	b) Tighten to specification
	c) Valve body dirty or sticking valves	c) Determine source of contamination. Service as required
	d) Vacuum diaphragm or T.V. control rod bent, sticking, leaks	d) Check diaphragm and rod. Replace as required
	e) Throttle valve stuck	e) Service as required
Erratic shifts	a) Poor engine performance	a) Check engine tune-up
	b) Vacuum line damaged	b) Service as required
	c) Valve body bolts too loose or tight	c) Tighten to specification
	d) Valve body dirty or sticking valves	d) Perform air pressure test. Note results Determine source of contamination. Service as required
	e) Governor valve stuck	e) Perform governor test. Service as required
	f) Output shaft collector body seal rings damaged	f) Service as required
Shifts 1–3 in OD or D	a) Intermediate band out of adjustment	a) Adjust band
	b) Damaged intermediate servo and/or internal leaks	b) Perform air pressure test. Service front servo and/or internal leaks
	c) Improper band or clutch application or oil pressure control system	c) Perform control pressure test
	d) Polished or glazed band or drum	d) Service or replace band or drum
	e) Dirty or sticking valve body or governor	e) Clean, service or replace valve body or governor
	f) Governor valve stuck	f) Perform governor test. Service as required
	g) Kickdown system out of adjustment	g) Adjust kickdown system
Engine overspeeds on 2–3 shift	a) Kickdown system damaged	a) Replace damaged parts
	b) Improper band or clutch application or oil pressure control system	b) Perform control pressure test
	c) Damaged or worn reverse/high clutch and/or intermediate servo piston	c) Perform air pressure test. Service as required

CHILTON'S THREE C's TRANSMISSION DIAGNOSIS
A4LD Automatic Transmission

Condition	Cause	Correction
Engine overspeeds on 2–3 shift	d) Intermediate servo piston seals cut or leaks e) Dirty or sticking valve body f) Throttle valev stuck g) Damaged vacuum diaphragm	d) Replace seals. Check for leaks e) Clean, service or replace valve body f) Service as required g) Replace vacuum diaphragm
Rough or shudder 3–2 shift at closed throttle in D	a) Incorrect engine idle or performance b) Improper kickdown linkage adjustment c) Improper clutch or band application or oil pressure control system d) Improper governor operation e) Dirty or sticking valve body	a) Tune and adjust engine idle b) Service or adjust kickdown linkage c) Perform control pressure test d) Perform governor test. Service as required e) Clean, service or replace valve body
No 3–4 upshift	a) Kickdown system damaged b) Vacuum line damaged c) Vacuum diaphragm damaged d) Throttle valve sticking e) OD servo damaged or leaking f) Polished or glazed OD band or drum g) Dirty or sticking valve body h) Dirty or sticking 3–4 solenoid	a) Replace damaged parts b) Repair or replace as required c) Repair or replace as required d) Repair or replace as required e) Check and replace OD piston seal if required f) Service or replace OD band or drum g) Clean, service or replace valve body. Check 3–4 shift valve for freedom of movement h) Clean, service or replace 3–4 solenoid and filter sleeve assembly
Slipping 4th gear	a) OD servo damaged or leaking b) Polished or glazed OD band or drum	a) Check and replace OD piston seal b) Service or replace OD band on drum
Engine stall speed exceeded in OD, D or R	a) Vacuum system b) Low main control pressure	a) Check and service vacuum system b) Control pressure test. Check and clean valve body. Replace valve body gasket. Check or service pump
Engine stall speed exceeded in R	a) Low/reverse servo or band damaged b) Reverse and high clutch damaged	a) Check engine braking in 1. If not OK check, service or replace if required the low/reverse servo and band b) If low/reverse servo OK, check and repair reverse and high clutch
Engine stall speed exceeded in OD or D. OK in R	a) OD one-way clutch or rear one-way clutch damaged	a) Check engine stall speeds in 2 and 1 If OK, repair OD or rear one-way clutches. Clean transmission
1–2 upshift is above 40 mph (64 km/h)	a) Vacuum system b) Main control pressure c) Governor damaged or worn d) Dirty or sticking valve body	a) Check and service hoses and vacuum diaphragm if required b) Perform control pressure test c) Perform governor check. Replace or service governor d) Clean, service or replace valve body
Kickdown shift speeds too early	a) Kickdown system damaged b) Main control pressure c) Governor damaged or worn	a) Replace damaged parts b) Perform control pressure test c) Perform governor check. Replace or service governor

CHILTON'S THREE C's TRANSMISSION DIAGNOSIS
A4LD Automatic Transmission

Condition	Cause	Correction
No kickdown into 2nd gear between 40–60 mph (64–100 km/h) on OD or D	a) Kickdown system damaged b) Main control pressure c) Dirty or sticking valve body d) Kickdown cable overadjusted	a) Replace damaged parts b) Perform control pressure test c) Ckeck kickdown valve. Clean or replace valve body d) Adjust kickdown cable
No shift into 2nd gear with accelerator 3–4 depressed at 25 mph (40 km/h) in OD or D	a) Main control pressure b) Governor damaged or worn c) Dirty or sticking valve body	a) Perform control pressure test b) Check governor c) Clean or replace valve body
When moving selector from OD or D to manual 1 at 55 mph (86 km/h) with accelerator released, no braking felt from downshift to 2nd gear	a) Main control pressure b) Intermediate band out of adjustment c) Overdrive clutch damaged	a) Perform control pressure test b) Adjust band. Check intermediate servo c) Repair or replace overdrive clutch
When moving selector from OD or D to manual 1 at 55 mph (86 km/h) with accelerator released, shift into 1st gear occurs over 45 mph (72 km/h)	a) Main control pressure b) Dirty or sticking valve body c) Governor damaged or worn d) Kickdown linkage misadjusted or stuck	a) Perform main control pressure test b) Clean or replace valve body c) Perform governor check. Replace or service governor d) Adjust or repair kickdown linkage
When moving selector from OD or D to manual 1 at 55 mph (86 km/h) with accelerator released, 1st gear shift occurs under 15 mph (24 km/h)	a) Main control pressure b) Dirty or sticking valve body c) Low/reverse servo damaged d) Governor damaged or worn service governor e) Overdrive clutch damaged	a) Perform main control pressure test b) Clean or replace valve body c) Check and service as required d) Perform governor check. e) Repair or replace as required
No forced downshifts	a) Kickdown cable damaged b) Kickdown cable overadjusted c) Damaged internal kickdown linkage d) Improper clutch or band application or oil pressure control system e) Dirty or sticking governor f) Dirty or sticking valve governor	a) Replace damaged parts b) Adjust kickdown cable c) Service internal kickdown linkage d) Perform control pressure test e) Service or replace governor, clean screen f) Clean, service or replace valve body
Engine overspeeds on 3–2 downshift	a) Linkage out of adjustment b) Intermediate band out of adjustment c) Improper band or clutch application and one-way clutch or oil pressure control system d) Damaged or worn intermediate servo e) Polished or glazed band or drum f) Dirty or sticking valve body	a) Service or adjust linkage b) Adjust intermediate band c) Perform control pressure test. Service clutch d) Perform air pressure test. Check the intermediate servo. Service servo and/or seals e) Service or replace as required f) Clean, service or replace valve body
Shift efforts high	a) Manual shaft linkage damaged or misadjusted b) Inner manual lever nut loose misadjusted c) Manual lever retainer pin damaged	a) Check and adjust or service as required b) Tighten nut to specification c) Adjust linkage and install pin
Transmission overheats	a) Improper fluid level b) Incorrect engine idle or performance c) Improper clutch or band application or oil pressure control system	a) Perform fluid level check b) Tune or adjust engine idle c) Perform control pressure test

CHILTON'S THREE C's TRANSMISSION DIAGNOSIS
A4LD Automatic Transmission

Condition	Cause	Correction
Transmission overheats	d) Restriction in cooler lines e) Seized converter one-way clutch f) Dirty or sticking valve body	d) Service rectriction e) Replace one-way clutch f) Clean, service or replace valve body
Transmission leaks	a) Case breather vent b) Leakage at gasket, seal, etc.	a) Check the vent for free breathing. Repair as required b) Remove all traces of lube on exposed surfaces of transmission. Check the vent for free breathing. Operate transmission at normal temperatures and perform fluid leakage check. Service as required
Poor vehicle acceleration	a) Poor engine performance b) Torque converter one-way clutch slipping	a) Check engine tune-up b) Replace torque converter
Transmission noisy. Valve resonance b) Service or adjust linkage NOTE: Gauges may aggravate any hydraulic resonance. Remove gauge and check for resonance level	a) Improper fluid level c) Improper band or clutch application or oil pressure control system d) Cooler lines grounding e) Dirty or sticking valve body f) Internal leakage or pump cavitation	a) Perform fluid level check b) Linkage out of adjustment c) Perform control pressure test d) Free up cooler lines e) Clean, service or replace valve body f) Service as required
Engine stalls when shifting into forward or reverse	a) Low engine idle b) Broken converter clutch shuttle valve spring	a) Verify that engine idle speeds are set b) Replace converter clutch shuttle valve spring

and routed to the sump. It should be remembered, the oil pump is driven by a shaft which is splined into the converter cover and through a drive gear insert. The gears, in turn, are installed in a body, which is bolted to the pump support at the rear of the transmission case.

Should the oil pump fail, fluid would not be supplied to the transmission to keep the converter filled, to lubricate the internal working parts and to operate the hydraulic controls.

PRESSURE REGULATOR SYSTEM

The pressure regulator system controls the main line pressure at pre-determined levels during vehicle operation. The main oil pressure regulator valve and spring determines the psi of the main line control pressure. The main line control pressure is regulated by balancing pressure at the end of the inner valve land, against the valve spring. When the pump begins fluid delivery and fills the passages and transmission components, the spring holds the valve closed and there is no regulation. As the pressure rises, the pressure regulator valve is moved against the spring tension, opening a passage to the torque converter. Fluid then flows into the converter, the cooler system and back to the lubrication system. As the pressure continues its rise, the pressure regulator valve is moved further against the spring tension and at a predetermined psi level and spring tension rate, the valve is moved further to open a passage, allowing excess pressurized fluid to return to the sump. The valve then opens and closes in a vibrating type action, dependent upon the fluid requirements of the transmission. A main line oil pressure booster valve is used to increase the line pressure to meet the needs of higher pressure, required when the transmission torque load increases, to operate the clutches and band servo.

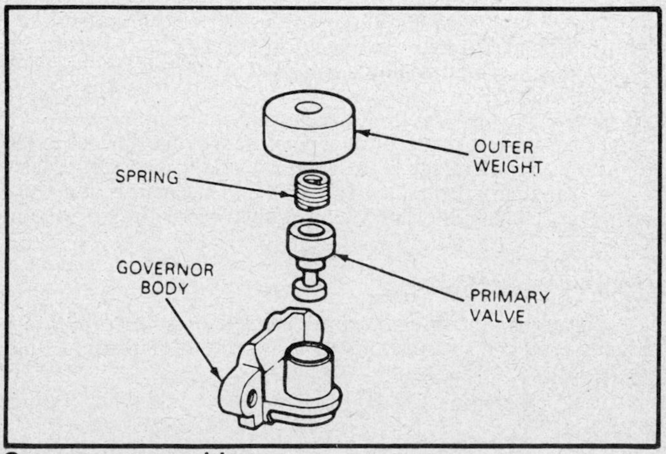
OUTER WEIGHT
SPRING
GOVERNOR BODY
PRIMARY VALVE
Governor assembly

MANUAL CONTROL VALVE

Main line control pressure is always present at the manual control valve. Other than required passages, such as to the converter fill, the lubricating system and to certain valve assemblies, the manual valve must be moved to allow the pressurized fluid to flow to the desired components or to charge certain passages in order to engage the transmission components in their applicable gear ratios.

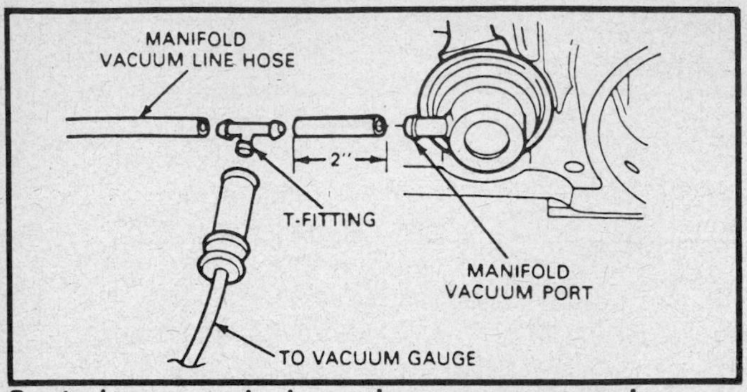
MANIFOLD
VACUUM LINE HOSE
2"
T-FITTING
MANIFOLD
VACUUM PORT
TO VACUUM GAUGE
Control pressure test—engine vacuum procedure

GOVERNOR ASSEMBLY

The governor assembly reacts to vehicle road speed and provides a pressure signal to the control valves. This pressure signal causes automatic upshifts to occur as the road speed increases and permits downshifts as the road speed decreases. The governor has 3 hydraulic passages, an exhaust governor pressure out and line pressure in, controlled by springs and weights, with the weight position determined by centrifugal force as the governor assembly rotates.

Diagnosis Tests

CONTROL PRESSURE TEST

There are 2 methods of performing the control pressure test. The first is to perform the test using the engine vacuum. The other is to use a remote vacuum source, such as a hand operated vacuum pump.

Engine Vacuum Method

1. Attach a tachometer to the engine. Connect vacuum gauge, Rotunda Number 059-00008 or equivalent to the transmission vacuum line at the manifold vacuum port.
2. Attach a pressure gauge to the control pressure outlet at the transmission.
3. Apply the parking brake and start the engine.
4. Check the throttle and downshift linkage for a binding condition. If the linkage is satisfactory, check for vacuum leaks at the transmission diaphragm unit and its connecting tubes and hoses. Check all other vacuum operated units for vacuum leaks.

Vacuum Pump Method

1. Install an adjustable vacuum source. Disconnect and temporarily plug the vacuum line at the vacuum diaphragm unit.
2. Apply both the parking and service brakes.
3. Start the engine and the vacuum pump. Set the vacuum at 15 in. Hg.
4. Read and record the control pressure in all selector positions.
5. Run the engine up to 1000 rpm and reduce the vacuum to 10 in. Hg.
6. Read and record the pressure in all of the forward drive ranges.
7. Keep the engine at 1000 and reduce the vacuum to 1 in. Hg.
8. Read and record the pressure in all forward drive ranges and reverse.

AIR PRESSURE TEST

A no drive condition can exist, even with the correct transmission fluid pressure, because of inoperative clutches or bands. The inoperative units can be located through a series of checks by substituting air pressure for fluid pressure to determine the locations of the malfunction.

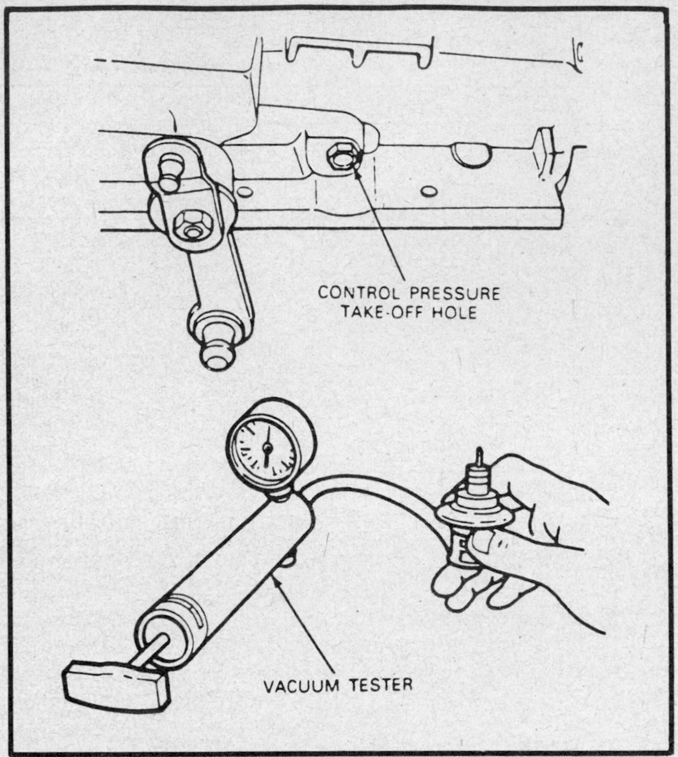
CONTROL PRESSURE
TAKE-OFF HOLE
VACUUM TESTER
Control pressure test—vacuum pump method

To make the air pressure checks, loosen the oil pan bolts and lower the edge to drain the transmission fluid. Remove the oil pan and the control valve body assembly. The inoperative clutches or bands can be located by introducing low pressure compressed air into the various transmission case passages. If the servos do not operate, disassemble, clean and inspect them to locate the source of the trouble. If air pressure applied to either of the clutch passages fails to operate a clutch or operates both clutches at once, remove and with air pressure check the fluid passages in the case and front pump to detect obstructions.

FORWARD CLUTCH

Apply low pressure compressed air to the transmission case forward clutch passages. A dull thud can be heard when the clutch piston is applied. If no noise is heard, place the finger tips on the input shell and again apply air pressure to the forward clutch passage. Movement of the piston can be felt as the clutch is applied.

GOVERNOR

Apply low pressure compressed air pressure to the forward clutch feed to governor passage and listen for a sharp clicking or whistling noise. The noise indicates governor valve movement.

OVERDRIVE SERVO

Hold the air nozzle in the overdrive servo apply passage. Operation of the servo is indicated by a tightening of the overdrive band around the overdrive drum. Continue to apply low pressure compressed air to the servo apply passage and introduce air pressure into the overdrive servo release passage. The overdrive servo should stroke off releasing the overdrive band.

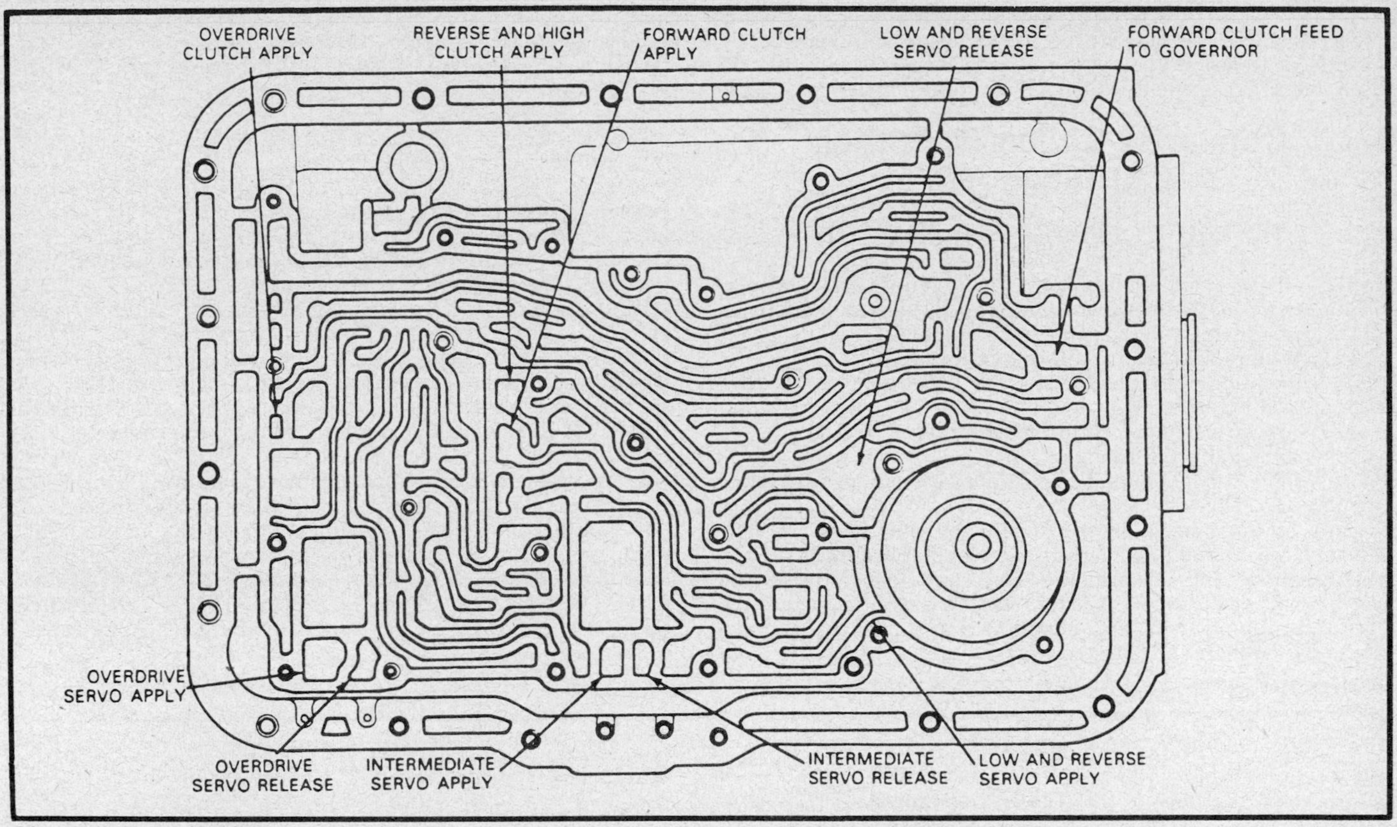
OVERDRIVE CLUTCH APPLY
REVERSE AND HIGH CLUTCH APPLY
FORWARD CLUTCH APPLY
LOW AND REVERSE SERVO RELEASE
FORWARD CLUTCH FEED TO GOVERNOR
OVERDRIVE SERVO APPLY
OVERDRIVE SERVO RELEASE
INTERMEDIATE SERVO APPLY
INTERMEDIATE SERVO RELEASE
LOW AND REVERSE SERVO APPLY

Air pressure checks

OVERDRIVE CLUTCH

Applied in **D, 2, 1** and **R** ranges. Apply low pressure compresed air to the overdrive clutch feed passage. A dull thud indicates that the overdrive clutch piston has moved to the applied position.

REVERSE/HIGH CLUTCH

Apply low pressure compressed air to the reverse/high clutch. A dull thud indicates that the piston has moves to the applied position. If no noise is heard, place the finger tips on the clutch drum and again apply air pressure to detect movement of the piston.

INTERMEDIATE SERVO

Hold the air nozzle in the intermediate servo apply passages. Operation of the servo is indicated by a tightening of the intermediate band around the drum. Continue to apply low pressure compressed air to the servo apply passage and introduce air pressure into the intermediate servo release passage. The intermediate servo should release the band against the apply pressure.

LOW-REVERSE SERVO

Apply low pressure compressed air to the low-reverse servo. The low-reverse band should tighten around the drum if the servo is operating properly.

STALL SPEED TEST

The stall test checks converter one-way clutch operation and installation, the holding ability of the forward clutch, reverse clutch, the low-reverse bands, the planetary one-way clutch and engine performance.

The test should be done only with the engine coolant and transmission fluid at the proper levels and at operating temperature. Apply the service and the parking brakes for each stall test.

1. Mark the specified stall rpm for the vehicle on the tachometer.
2. Connect the tachometer to the engine.
3. In each of the drive and reverse range, press the accelerator to the floor and hold it just long enough to let the engine get to full rpm. Do not hold the throttle wide open for more than a few seconds at a time.
4. After each range, move the selector lever to **N** and run the engine at 1000 rpm for 15 seconds to cool the converter before making the next test.

NOTE: If the engine speed recorded by the tachometer exceeds the maximum limit, releases the accelerator immediately, because clutch or band slippage is indicated.

ROAD TEST

This test will determine if the electronics, governor and shift control valves are functioning properly. Check the throttle upshifts with the selector in the **OD** range.

The transmission should start in 1st gear and shift through 2nd, 3rd and 4th **OD** gears and then lock the converter clutch. A wide open throttle lockup and 3–4 shift is not possible. A 4–3 wide open throttle kickdown can always be obtained regardless of road speed.

When the selector lever is in the **D** position, the transmission will make all automatic upshifts except the 3–4. When the selector lever is at **2**, the transmission can operate only in second gear.

With the transmission in 3rd gear and the road speed over 45 mph, the transmission should shift to second gear when the selector lever is moved from **D** to **2** to **1**.

If the vehicle is traveling at 45 mph and the selector lever is moved from **OD** or **D** to **1**, the transmission will immediately downshift to second gear. As the road speed drops below 30 mph, the transmission will downshift to first gear.

Converter Clutch Operation and Diagnosis

TORQUE CONVERTER CLUTCH

In the A4LD transmission, converter clutch upshifts and downshifts are scheduled electronically. The converter clutch does not engage during the following driving modes.

1. Engine coolant below 128° F or above 240° F.
2. Application of brakes.
3. Closed throttle.
4. Heavy or wide open throttle acceleration.
5. Quick tip ins and tip outs.
7. When actual engine speed is below a certain value at lower vacuums.

In any of the named driving modes, current does not flow through the solenoid. When the converter clutch shuttle valve is resting on the plug, line pressure is directed through the shuttle valve and to the torque converter in a flow path that pushes the lockup piston off (converter clutch unlocked). When line pressure on the spring end of the converter clutch shuttle valve is exhausted, line pressure on the plug end of the valve forces the valve to move and compress the spring. Line pressure is now directed through the shuttle valve to the converter in a flow path that pushes the piston on (converter clutch locked).

In the **CONVERTER CLUTCH LOCKED** position, lockup is permitted electronically because the vehicle is not operating in any of the named driving modes.

In the **CONVERTER CLUTCH UNLOCKED** position, governor pressure acting on the converter clutch shift valve has not yet moved the valve to upshifted postion. Line pressure is therefore acting on the spring end of the converter clutch shuttle valve. The torque converter is unlocked.

As the vehicle speed increases, governor pressure increases and the converter clutch shift valve moves to the upshifted position. Oil on the spring end of the converter clutch shuttle valve now drains to exhaust at the converter clutch shift valve. The shuttle valve takes the position as shown in the **CONVERTER CLUTCH LOCKED** schematic and the torque converter locks up.

If the brakes are now applied, current will not flow through the solenoid. With no current to the solenoid, line pressure can flow through the solenoid valving and enter the lockup inhibition passage. Line pressure in the inhibition passage forces the shuttle ball to take the position as shown in the **CONVERTER CLUTCH UNLOCKED ELECTRONICALLY INHIBITED** schematic. The shuttle valve moves up against the plug and the converter unlocks.

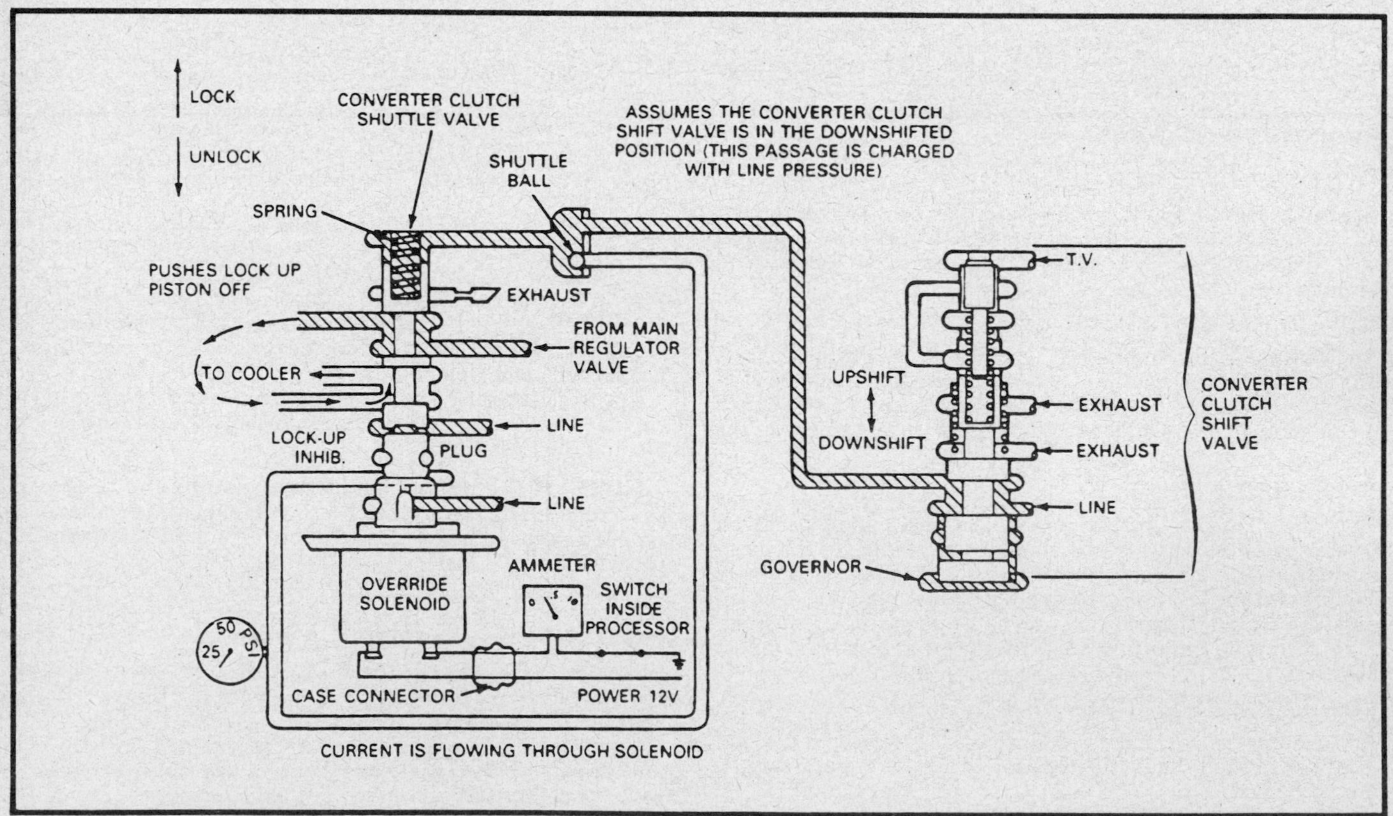
LOCK
UNLOCK
CONVERTER CLUTCH SHUTTLE VALVE
ASSUMES THE CONVERTER CLUTCH SHIFT VALVE IS IN THE DOWNSHIFTED POSITION (THIS PASSAGE IS CHARGED WITH LINE PRESSURE)
SHUTTLE BALL
SPRING
PUSHES LOCK UP PISTON OFF
EXHAUST
FROM MAIN REGULATOR VALVE
TO COOLER
LINE
LOCK-UP INHIB.
PLUG
LINE
T.V.
UPSHIFT
DOWNSHIFT
EXHAUST
EXHAUST
LINE
CONVERTER CLUTCH SHIFT VALVE
GOVERNOR
OVERRIDE SOLENOID
AMMETER
SWITCH INSIDE PROCESSOR
50 PSI
25
CASE CONNECTOR
POWER 12V
CURRENT IS FLOWING THROUGH SOLENOID

Converter clutch system

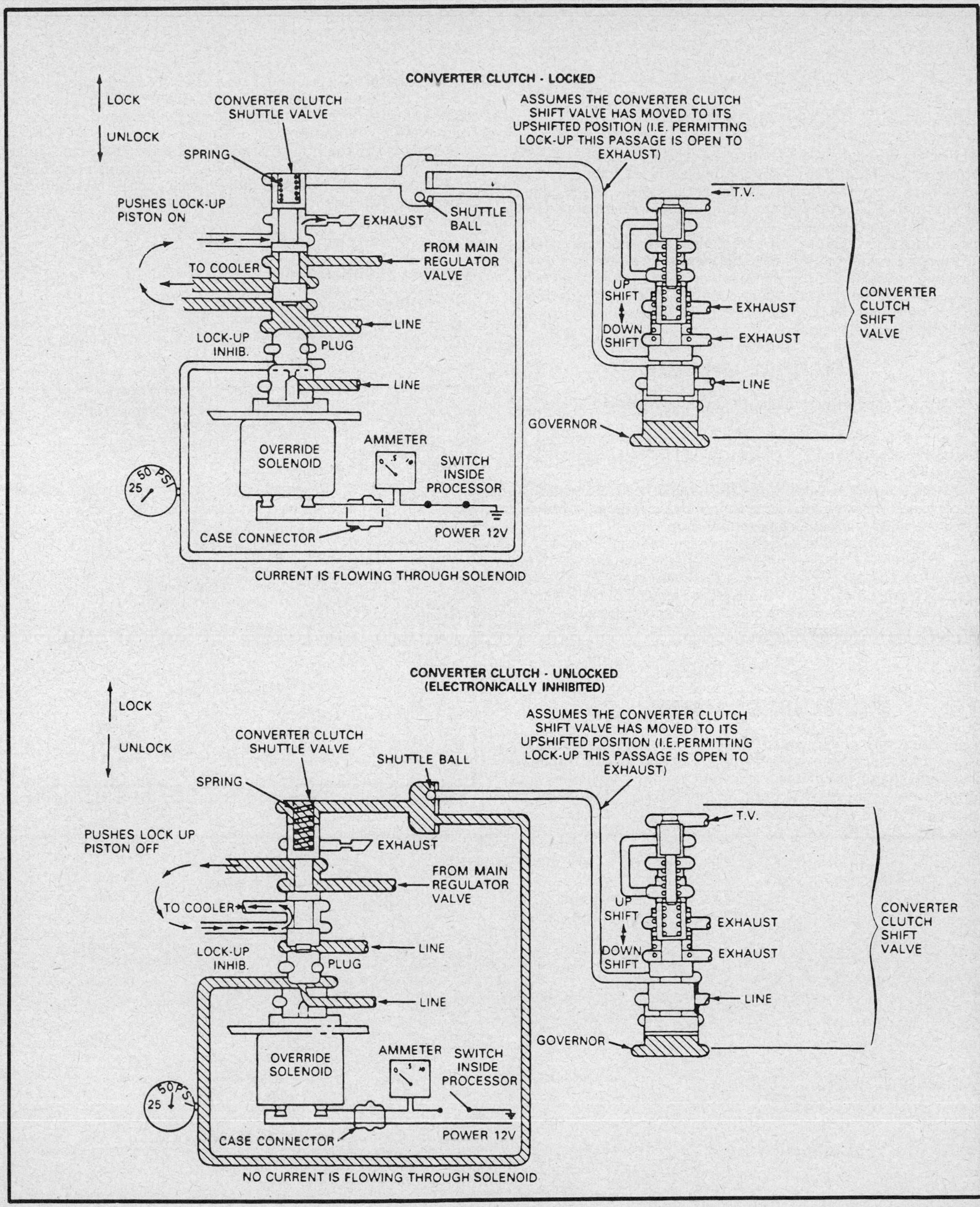
CONVERTER CLUTCH - LOCKED
LOCK
UNLOCK
CONVERTER CLUTCH SHUTTLE VALVE
ASSUMES THE CONVERTER CLUTCH SHIFT VALVE HAS MOVED TO ITS UPSHIFTED POSITION (I.E. PERMITTING LOCK-UP THIS PASSAGE IS OPEN TO EXHAUST)
SPRING
PUSHES LOCK-UP PISTON ON
EXHAUST
SHUTTLE BALL
T.V.
TO COOLER
FROM MAIN REGULATOR VALVE
UP SHIFT
DOWN SHIFT
EXHAUST
EXHAUST
CONVERTER CLUTCH SHIFT VALVE
LINE
LOCK-UP INHIB.
PLUG
LINE
LINE
GOVERNOR
AMMETER
OVERRIDE SOLENOID
SWITCH INSIDE PROCESSOR
25
50 PSI
CASE CONNECTOR
POWER 12V
CURRENT IS FLOWING THROUGH SOLENOID
CONVERTER CLUTCH - UNLOCKED (ELECTRONICALLY INHIBITED)
LOCK
UNLOCK
CONVERTER CLUTCH SHUTTLE VALVE
SHUTTLE BALL
ASSUMES THE CONVERTER CLUTCH SHIFT VALVE HAS MOVED TO ITS UPSHIFTED POSITION (I.E.PERMITTING LOCK-UP THIS PASSAGE IS OPEN TO EXHAUST)
SPRING
T.V.
PUSHES LOCK UP PISTON OFF
EXHAUST
FROM MAIN REGULATOR VALVE
TO COOLER
UP SHIFT
DOWN SHIFT
EXHAUST
EXHAUST
CONVERTER CLUTCH SHIFT VALVE
LOCK-UP INHIB.
PLUG
LINE
LINE
LINE
GOVERNOR
OVERRIDE SOLENOID
AMMETER
SWITCH INSIDE PROCESSOR
25
50 PSI
CASE CONNECTOR
POWER 12V
NO CURRENT IS FLOWING THROUGH SOLENOID

Converter clutch system

ON CAR SERVICES

Adjustments

KICKDOWN CABLE

The kickdown cable is attached to the accelerator pedal near the accelerator cable. The kickdown cable is routed from the transmission through the dash to the accelerator pedal. A self adjuster mechanism is located in the engine compartment at the inlet for the cable on the dash.

The kickdown cable is self-adjusting over a tolerance range of 1 in. If the cable requires readjustment, reset the cable by depressing the semi-circular metal tab on the self-adjuster mechanism and pulling the cable forward to the **ZERO** position setting. The cable will then automatically readjust to the proper length when kicked down.

MANUAL LINKAGE

1. From inside the vehicle, place the shift lever in the **OD** position.
2. From below the vehicle, loosen the adjustment screw on the shift cable and remove the end fitting from the manual lever ball stud.
3. Position the manual lever in the **OD** position by moving the lever all the way rearward, then move it 3 detents forward. Hold the shift lever against the rear stop.
4. Connect the cable end fitting to the manual lever.
5. Tighten the adjustment screw to 45–60 inch lbs. (5–7 Nm).
6. After the adjustment, check for **P** engagement. The control lever must move to the right when engaged in the **P** detent.
7. Check the transmission control lever in all detent positions with the engine running.

Services

FLUID CHANGES

The conditions under which the transmission is operated are the main consideration in determining how often the transmission fluid should be changed. Different driving conditions result in different transmission fluid temperatures. These temperatures affect change intervals.

If the vehicle is driven under severe service conditions, change the fluid and filter every 15,000 miles. If the vehicle is not used under severe service conditions, change the fluid and replace the filter every 50,000 miles.

Do not overfill the transmission. It only takes 1 pint of fluid to change the level from **ADD** to **FULL** on the transmission dipstick. Overfilling the unit can cause damage to the internal components of the automatic transmission.

OIL PAN

Removal and Installation

1. Disconnect the negative battery cable.
2. Raise the vehicle and support it safely.
3. Position a drain pan under the transmission pan.

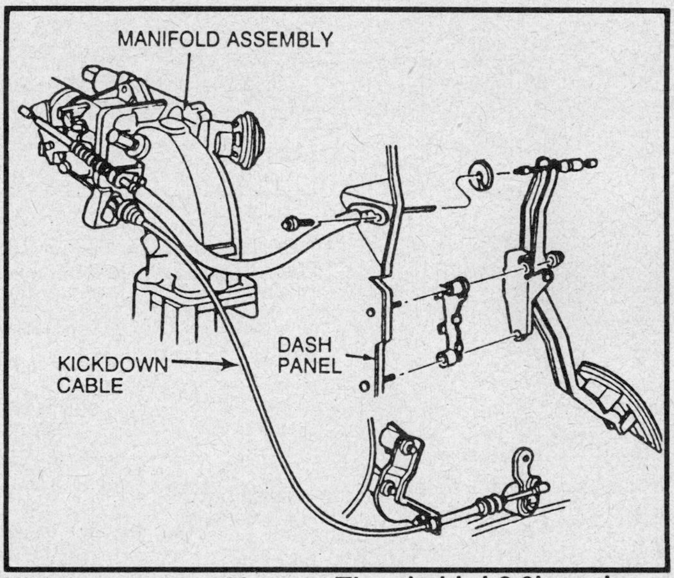
MANIFOLD ASSEMBLY
KICKDOWN CABLE
DASH PANEL

Kickdown cable—Mustang/Thunderbird 2.3L engine

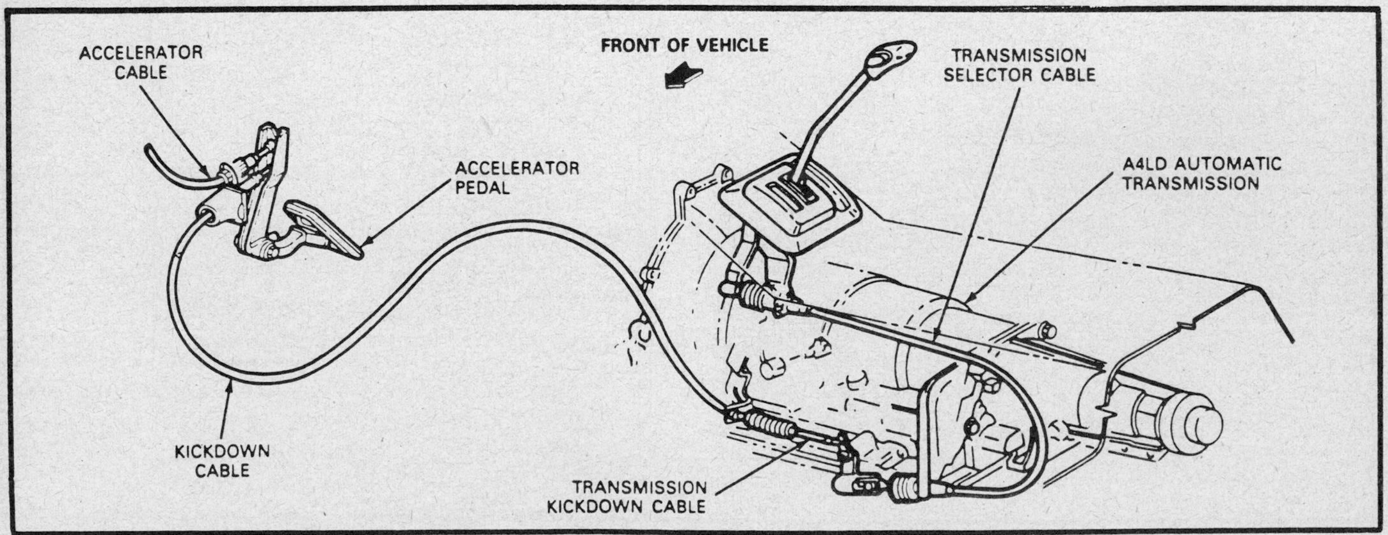
ACCELERATOR CABLE
FRONT OF VEHICLE
TRANSMISSION SELECTOR CABLE
ACCELERATOR PEDAL
A4LD AUTOMATIC TRANSMISSION
KICKDOWN CABLE
TRANSMISSION KICKDOWN CABLE

Aerostar kickdown cable

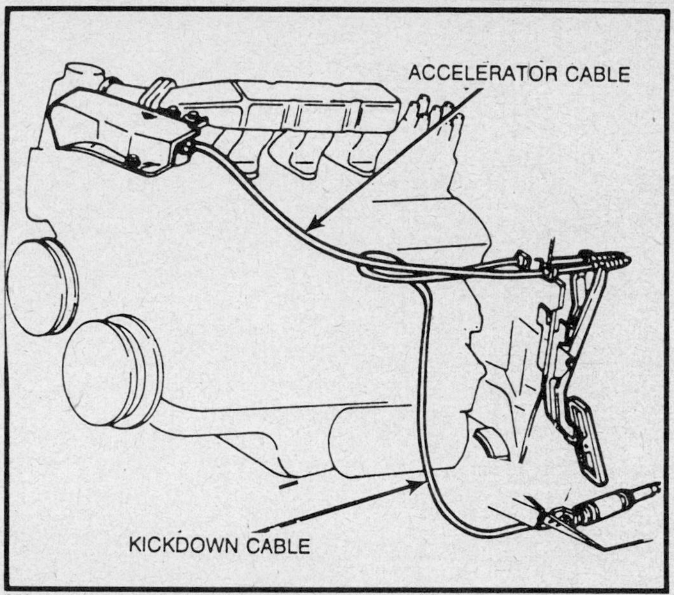
ACCELERATOR CABLE
KICKDOWN CABLE

Kickdown cable—Ranger/Bronco II

4. Starting at the rear, loosen, but do not remove the pan bolts.
5. Loosen the pan from the transmission and allow the fluid to drain gradually.
6. Remove all of the pan bolts except 2 at the front or rear and allow the fluid to continue draining.
7. Remove the pan. Clean the old gasket from the pan and the transmission case.
8. Install a new gasket on the pan. Install the pan on the transmission case.
9. Install the pan bolts and torque them to 8–10 ft. lbs. (11–13 Nm).
10. Install 3 quarts of Dexron® II or Mercon® type transmission fluid into the filler tube (converter not drained). When refilling a dry transmission and converter, install 5 quarts of fluid into the filler tube.
11. Start the engine and run it until it reaches normal operating temperature.
12. Check the fluid level after moving the gear selector through all ranges. Correct the fluid level as necessary.

VALVE BODY

Removal and Installation

1. Disconnect the negative battery cable.
2. Raise the vehicle and support it safely.
3. Remove the oil pan.
4. Remove the filter screen and gasket.
5. Remove the low/reverse servo cover, piston, spring and gasket.
6. Disconnect the 2 wires at the converter clutch solenoid and the 2 wires at the 3–4 shift solenoid.
7. Remove the bolts and while easing the valve body out of the transmission, unlock and detach the selector lever connecting link. Remove the valve body and the gasket.
8. Attach and lock the selector lever connecting rod to the manual valve and ease the valve body into the case.
9. Tighten all bolts except the filter screen bolt in the correct sequence to 71–97 inch lbs. (8.0–11.0 Nm).
10. Install the low-reverse servo cover, piston, spring and gasket.
11. Install the converter clutch solenoid wires.
12. Install the filter screen and gasket.

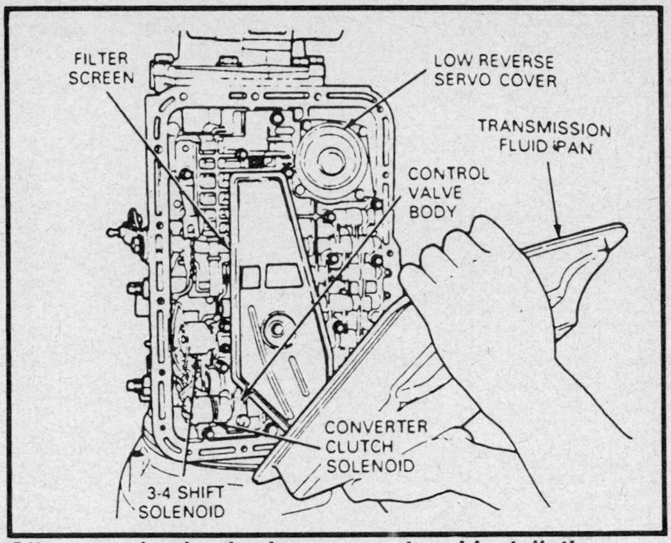
FILTER SCREEN
LOW REVERSE SERVO COVER
TRANSMISSION FLUID PAN
CONTROL VALVE BODY
CONVERTER CLUTCH SOLENOID
3-4 SHIFT SOLENOID

Oil pan and valve body—removal and installation

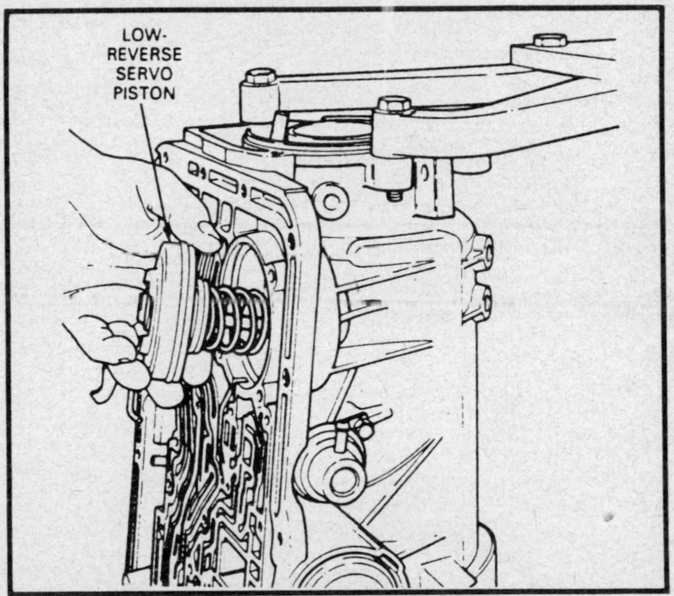
LOW-REVERSE SERVO PISTON

Low/Reverse servo assembly

13. Install the oil pan and fill with specified transmission fluid to the correct level.

LOW/REVERSE SERVO ASSEMBLY

Removal and Installation

1. Disconnect the negative battery cable.
2. Raise the vehicle and support it safely.
3. Remove the oil pan.
4. Remove the filter screen and gasket.
5. Remove the low/reverse servo cover, piston, spring and gasket.
6. Installation is the reverse of the removal procedure. Refill the transmission.

GOVERNOR

Removal and Installation

1. Disconnect the negative battery cable.

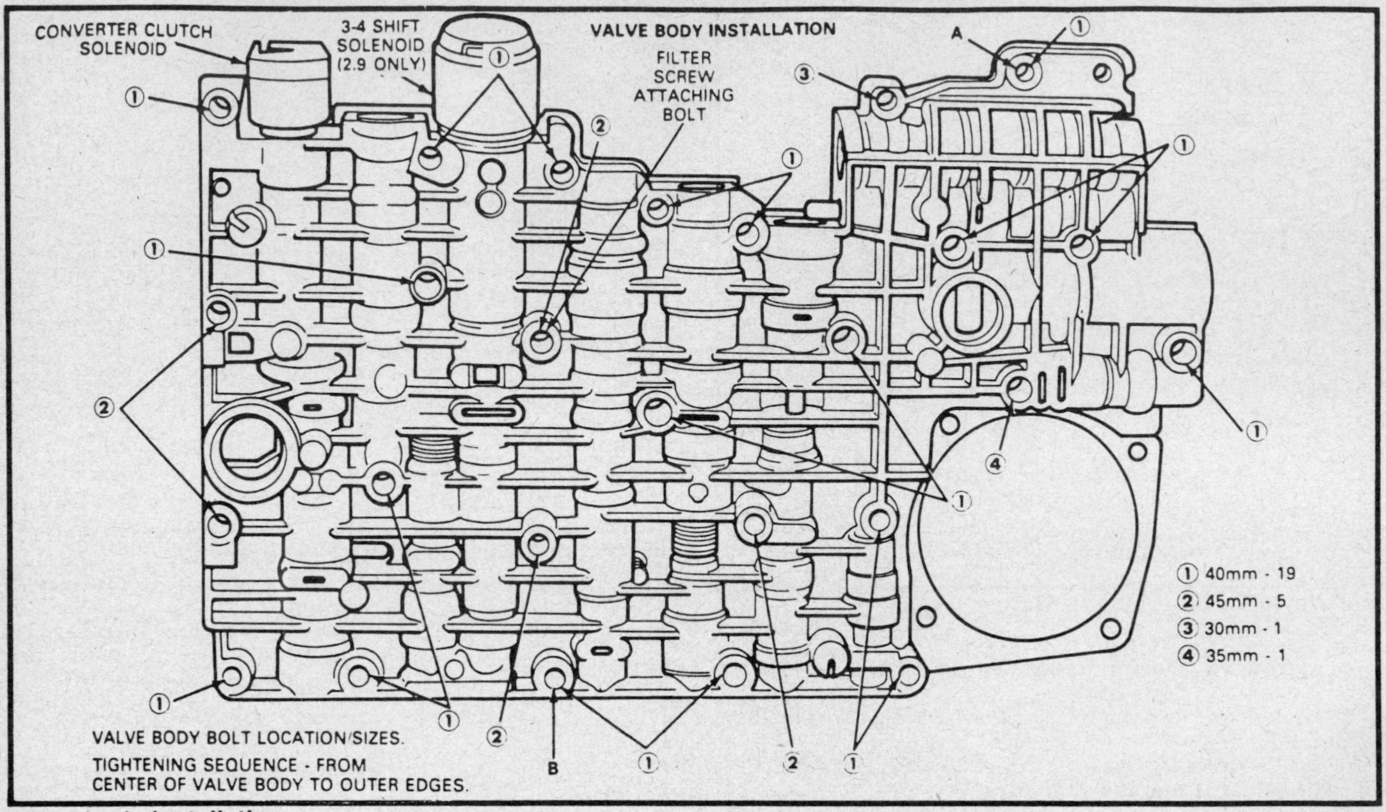
CONVERTER CLUTCH SOLENOID
3-4 SHIFT SOLENOID (2.9 ONLY)
VALVE BODY INSTALLATION
FILTER SCREW ATTACHING BOLT
A
B
① 40mm - 19
② 45mm - 5
③ 30mm - 1
④ 35mm - 1
VALVE BODY BOLT LOCATION/SIZES.
TIGHTENING SEQUENCE - FROM CENTER OF VALVE BODY TO OUTER EDGES.

Valve body installation

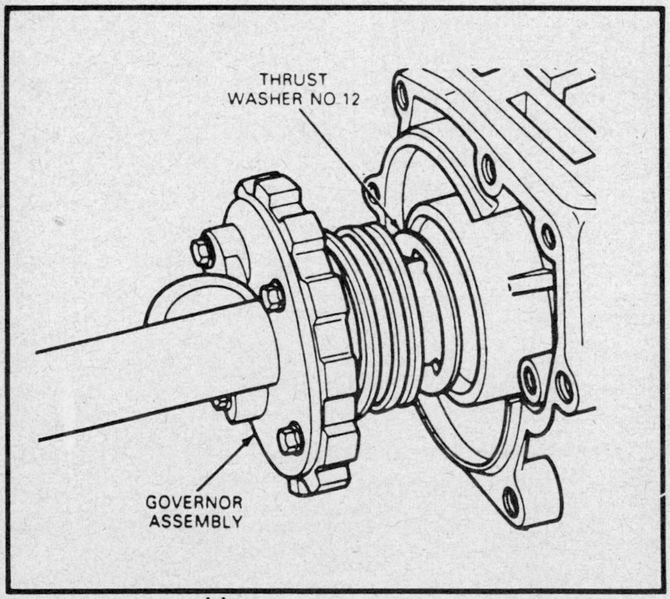
THRUST WASHER NO 12
GOVERNOR ASSEMBLY

Governor assembly

2. Raise the vehicle and support it safely.
3. Remove the extension housing.
4. Remove the governor body to oil collector body attaching bolts.
5. Remove the governor body, valve, spring and weight from the collector body.

NOTE: The components are not retained once the governor body to oil collector body attaching bolts have been removed. It is necessary to hold the governor body and components while removing or installing.

6. Assemble the governor body and components.
7. Position the governor body over the oil feed holes of the oil collector body.
8. Install the governor body to oil collector body attaching bolts and tighten to 84–120 inch lbs. (9.5–13.6 Nm).

REAR OIL SEAL

Removal and Installation

1. Raise the vehicle and support it safely.
2. Remove the driveshaft. Make scribe marks on the driveshaft end yoke and rear axle companion flange to assure proper positioning of the driveshaft during assembly.
3. Remove the extension housing seal using tool T71P-7657-A or equivalent.
4. Before installing a new seal, inspect the sealing surface of the universal joint yoke for scores. Inspect the counterbore of the housing for burrs. Remove any burrs using crocus cloth.
5. Install the new seal using tool T74P-77052-A or equivalent. Coat the inside diameter at the end of the rubber boot portion of the seal and the front universal joint spline with Multi-Purpose Long-Life Lubricant.

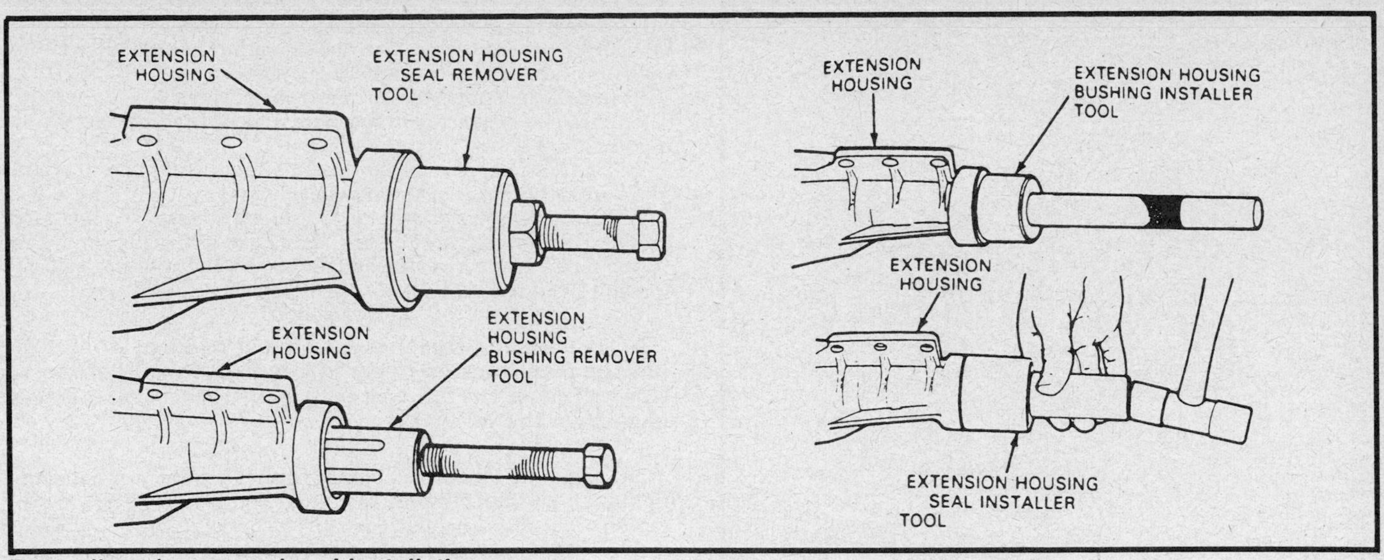
EXTENSION HOUSING
EXTENSION HOUSING SEAL REMOVER TOOL
EXTENSION HOUSING
EXTENSION HOUSING BUSHING REMOVER TOOL
EXTENSION HOUSING
EXTENSION HOUSING BUSHING INSTALLER TOOL
EXTENSION HOUSING
EXTENSION HOUSING SEAL INSTALLER TOOL

Rear oil seal—removal and installation

REMOVAL AND INSTALLATION

TRANSMISSION REMOVAL

1. Disconnect the negative battery cable.
2. Raise the vehicle and support it safely.
3. Position a drain pan under the transmission pan.
4. Starting at the rear, loosen, but do not remove the pan bolts.
5. Loosen the pan from the transmission and allow the fluid to drain gradually.
6. Remove all of the pan bolts except 2 at the front or rear and allow the fluid to continue draining.
7. Remove the converter access cover from the lower right side of the converter housing on the 3.0L engine. Remove the cover from the bottom of the engine oil pan on the 2.3L engine. Remove a bolt on the access cover of the 2.9L engine and swing the cover open. Remove the access cover and adapter plate bolts from the lower left side of the converter housing on all other applications.
8. Remove the flywheel to converter attaching nuts. Use a socket and breaker bar on the crankshaft pulley attaching bolt. Rotate the pulley clockwise as viewed from the front to gain access to each of the nuts.

NOTE: On belt driven overhead cam engines, never rotate the pulley in a counterclockwise direction as viewed from the front.

9. Scribe a mark indexing the driveshaft to the rear axle flange. Remove the driveshaft.
10. Remove the speedometer cable from the extension housing.
11. Disconnect the shift rod or cable at the transmission manual lever and retainer bracket.
12. Disconnect the downshift cable from the downshift lever. Depress the tab on the retainer and remove the kickdown cable from the bracket.
13. Disconnect the neutral start switch wires, converter clutch solenoid and the 3–4 shift soleniod connector.
14. Remove the starter mounting bolts and the ground cable. Remove the starter.
15. Remove the vacuum line from the transmission vacuum modulator.
16. Remove the filler tube from the transmission.
17. Position a transmission jack under the transmission and raise it slightly.
18. Remove the engine rear support to crossmember bolts.

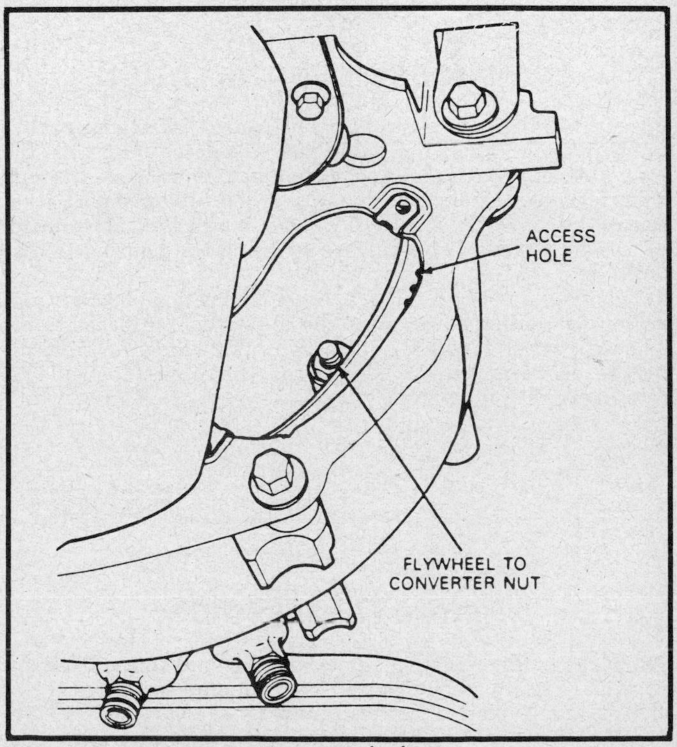
ACCESS HOLE
FLYWHEEL TO CONVERTER NUT

Torque converter nut access hole

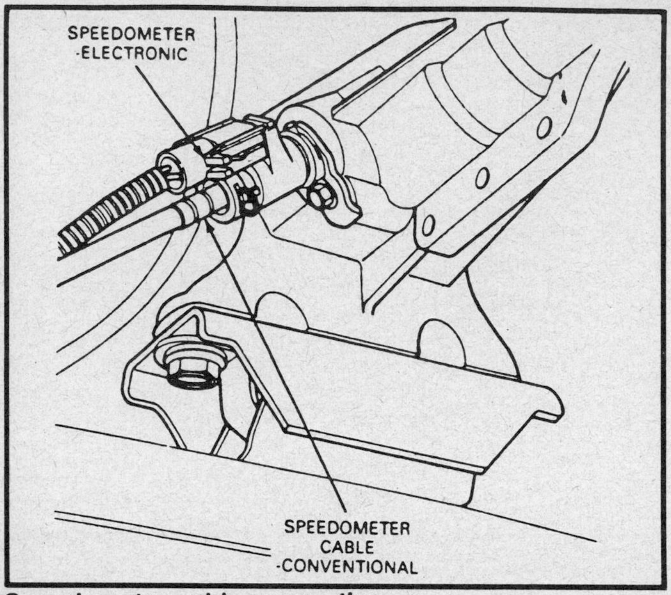
SPEEDOMETER
-ELECTRONIC
SPEEDOMETER
CABLE
-CONVENTIONAL

Speedometer cable connection

19. Remove the crossmember to frame side support attaching nuts and bolts. Remove the crossmember.
20. Remove the converter housing to engine bolts.
21. Slightly lower the jack to gain access to the oil cooler lines. Disconnect the oil cooler lines at the transmission. Plug all openings to keep dirt and contamination out.
22. Move the transmission to the rear so it disengages from the dowel pins and the converter is disengaged from the flywheel. Lower the transmission from the vehicle.
23. Remove the torque converter from the transmission.

NOTE: If the transmission is to be removed for a period of time, support the engine with a safety stand and wood block.

TRANSMISSION INSTALLATION

Proper installation of the converter requires full engagement of the converter hub in the pump gear. To accomplish this, the converter must be pushed and at the same time rotated through what feels like 2 notches or bumps. When fully installed, rotation of the converter will usually result in a clicking noise heard, caused by the converter surface touching the housing to case bolts.

This should not be a concern, but an indication of proper converter installation since, when the converter is attached to the engine flywheel, it will be pulled slightly forward away from the bolt heads. Besides the clicking sound, the converter should rotate freely with no binding.

For reference, a properly installed converter will have a distance from the converter pilot nose from face to converter housing outer face of $^7/_{16}$–$^9/_{16}$ in. (10.23–14.43mm).

1. Install the converter on the transmission.
2. With the converter properly installed, position the transmission on the jack.
3. Rotate the converter so that the drive studs are in alignment with the holes in the flywheel.
4. Move the converter and transmission assembly forward into position, being careful not to damage the flywheel and converter pilot. The converter housing is piloted into position by the dowels in the rear of the engine block.

NOTE: During this move, to avoid damage, do not allow the transmission to get into a nose down position as this will cause the converter to move forward and disengage from the pump gear.

5. Install the converter housing to engine attaching bolts and tighten to 28–38 ft. lbs. (38–51 Nm). The 2 longer bolts are located at the dowel holes.
6. Remove the jack supporting the engine.
7. Raise the transmission. Position the crossmember to the frame side supports. Install the attaching bolts and tighten to 20–30 ft. lbs. (27–41 Nm).
8. Lower the transmission and install the rear engine to crossmember nut and tighten to 60–80 ft. lbs. (82–108 Nm). Remove the transmission jack.
9. Install the filler tube in the transmission.
10. Install the oil cooler lines in the retaining clip at the cylinder block. Connect the lines to the transmission case.
11. Install the vacuum hose on the transmission vacuum unit. Install the vacuum line into the retaining clip.
12. Connect the neutral start switch plug to the neutral start switch. Connect the converter clutch solenoid wires and the 3–4 shift solenoid wires.
13. Install the starter and tighten the bolts to 15–20 ft. lbs. (20–27 Nm).
14. Install the flywheel to converter attaching nuts and tighten to 20–34 ft. lbs. (27–46 Nm).
15. Connect the muffler inlet pipe to the exhaust manifold.
16. Connect the transmission shift rod or cable to the manual lever.
17. Connect the downshift cable to the downshift lever.
18. Install the speedometer cable or sensor.
19. Install the driveshaft making sure to line up the scribe marks made during removal on the driveshaft and axle flange. Tighten the companion flange U-bolt attaching nuts to 70–95 ft. lbs. (95–130 Nm).
20. Adjust the manual and downshift linkages.
21. Lower the vehicle. Connect the negative battery cable.
22. Fill the transmission to the proper level with the specified fluid.
23. Check the transmission, converter and oil cooler lines for leaks.

BENCH OVERHAUL

Before Disassembly

If the transmission is being removed for major overhaul, it is important to completely clean all transmission components, including the converter, cooler, cooler lines, main control valve body, governor, all clutches and all check balls after any transmission servicing that generates contamination. These contaminants are a major cause for recurring transmission troubles and must be removed from the system before the transmission is put back into service.

Thorough cleaning of the transmission exterior will reduce the possibility that damaging contaminants might enter the sub-assemblies during disassembly and assembly.

When building up sub-assemblies, each component part should be lubricated with clean transmission fluid. Also lubri-

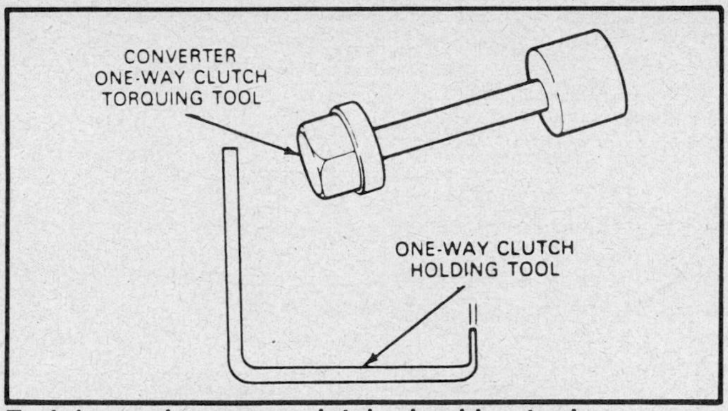
CONVERTER
ONE-WAY CLUTCH
TORQUING TOOL
ONE-WAY CLUTCH
HOLDING TOOL

Endplay and one-way clutch checking tools

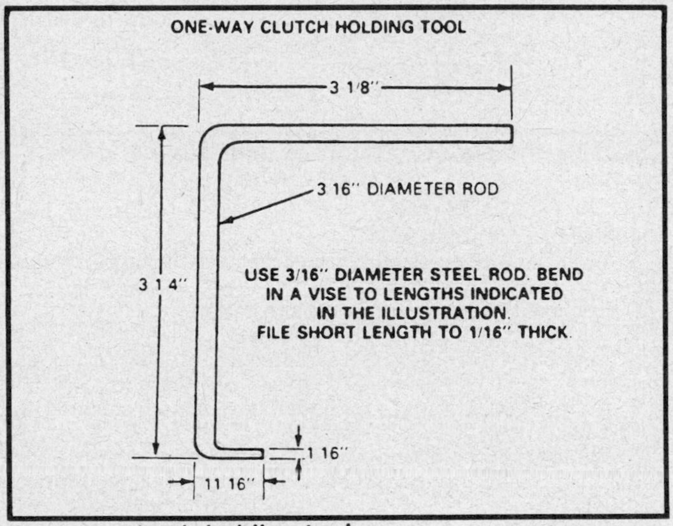
ONE-WAY CLUTCH HOLDING TOOL
3 1/8"
3/16" DIAMETER ROD
3 1/4"
USE 3/16" DIAMETER STEEL ROD. BEND
IN A VISE TO LENGTHS INDICATED
IN THE ILLUSTRATION.
FILE SHORT LENGTH TO 1/16" THICK.
1/16"
11/16"

One-way clutch holding tool

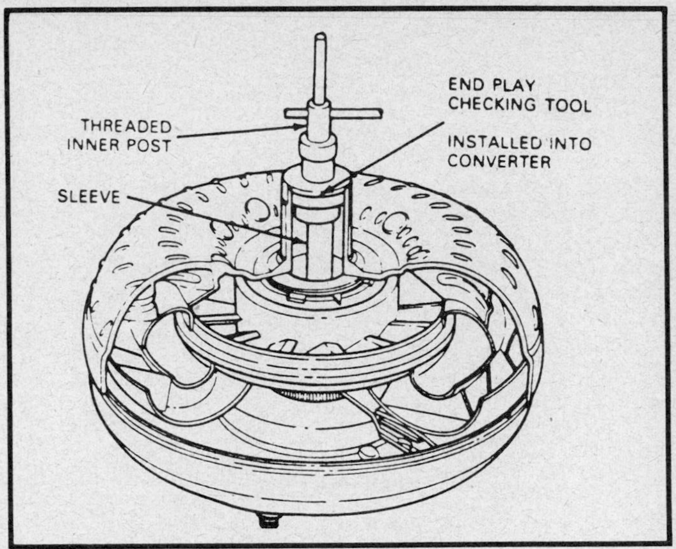
END PLAY
CHECKING TOOL
THREADED
INNER POST
INSTALLED INTO
CONVERTER
SLEEVE

Endplay tool installed into the converter

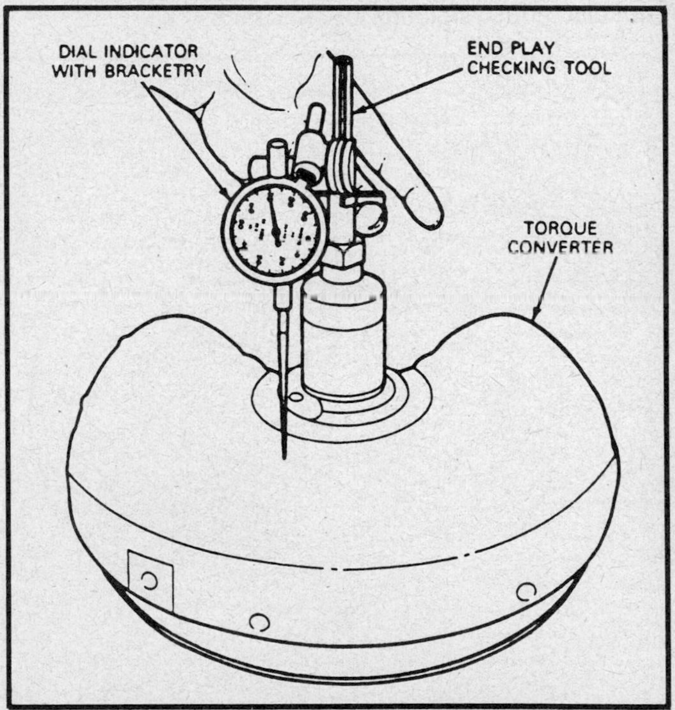
DIAL INDICATOR
WITH BRACKETRY
END PLAY
CHECKING TOOL
TORQUE
CONVERTER

Dial indicator attached to the checking tool

cate the sub-assemblies as they are installed in the case. Needle bearings, thrust washers and seals should be lightly coated with petroleum jelly during transmission assembly.

Converter Inspection

Checking Converter Endplay

1. Insert endplay checking tool T80L-7902-A or equivalent into the converter impeller hub until it bottoms.
2. Expand the sleeve in the turbine spline by tightening the threaded inner post until the tool is securely locked in the spline.
3. Attach a dial indicator tool 4201-C or equivalent to the endplay checking tool. Position the indicator button on the converter impeller housing and set the dial face at 0.
4. Lift the tool upward as far as it will go and note the indicator reading. The indicator reading is the total endplay which the turbine and stator share. Replace the converter unit if the total endplay exceeds the limits.
5. Loosen the threaded inner post to free the tool and then remove the tool from the converter.

Checking Converter One-Way Clutch

1. Use one-way clutch holding tool D84L-7902-A or equivalent. Insert the tool in 1 of the grooves in the stator thrust washer.
2. Insert the converter one-way clutch torquing tool T77L-7902-B or equivalent in the converter impeller hub so as to engage the one-way clutch inner race.
3. Attach a torque wrench to the one-way clutch torquing tool. With the one-way clutch holding tool held stationary, turn the torque wrench counterclockwise. The clutch should lock up and hold a 10 ft. lbs. (14 Nm) torque. The clutch should rotate freely in a clockwise direction.
4. If the clutch fails to lock up and hold at 10 ft. lbs. (14 Nm) torque, replace the torque converter.

Checking Stator to Impeller Interference

1. Position the front pump assembly on a bench with the spline end of the stator support pointing up.
2. Mount a converter on the pump with the splines of the one-way clutch inner race engaging the mating splines of the stator support. The impeller hub will then engage the pump drive gear.
3. Hold the pump stationary and try to rotate the torque con-

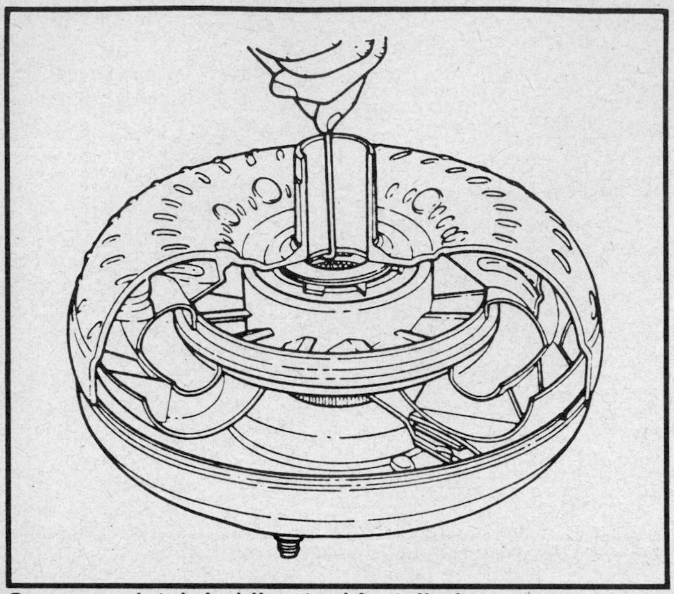

One-way clutch holding tool installed

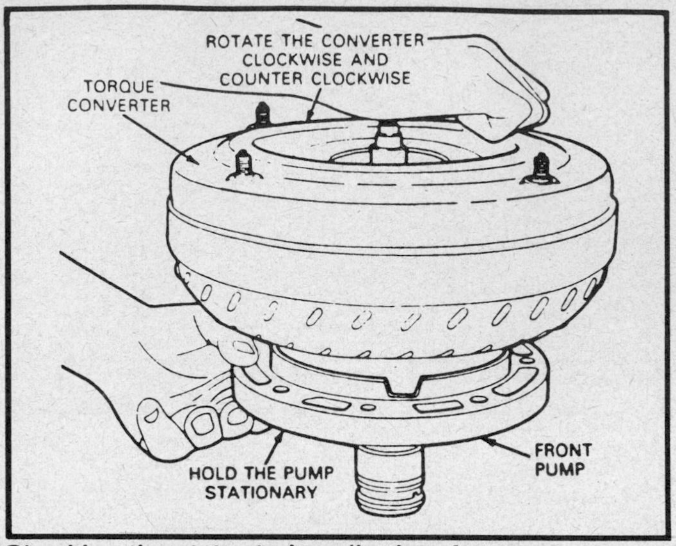
ROTATE THE CONVERTER
CLOCKWISE AND
COUNTER CLOCKWISE
TORQUE
CONVERTER
HOLD THE PUMP
STATIONARY
FRONT
PUMP

Checking the stator to impeller interference

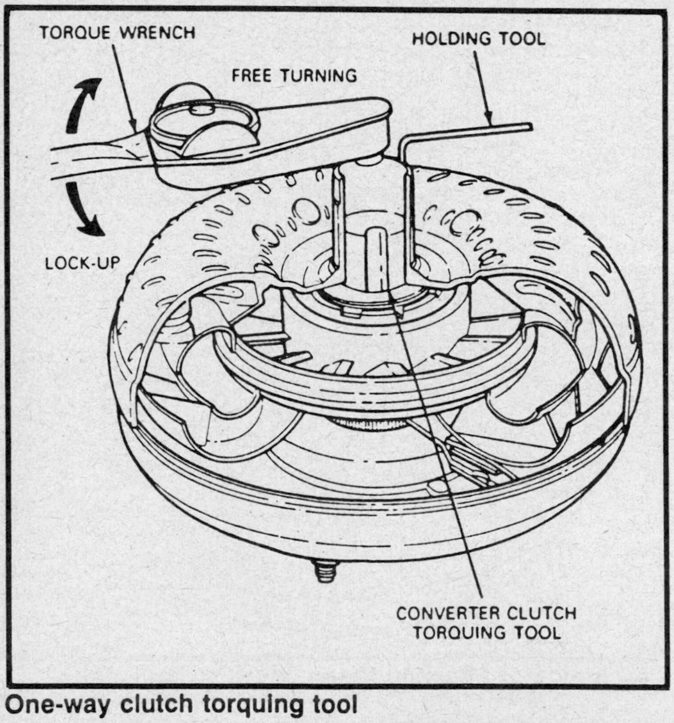
TORQUE WRENCH
HOLDING TOOL
FREE TURNING
LOCK-UP
CONVERTER CLUTCH
TORQUING TOOL

One-way clutch torquing tool

verter both clockwise and counterclockwise. The converter should rotate freely without any signs of interference or scraping within the converter assembly.

4. If there is an indication of scraping, the trailing edges of the stator blades may be interfering with the leading edges of the impeller blades. In such cases, replace the converter.

Transmission Disassembly

1. If the converter is equipped with a drain plug, remove the plug and drain the fluid from the converter.
2. Pull the converter off of the mainshaft and if necessary, drain the fluid out of the hub opening.
3. Mount the transmission in a suitable holding fixture.
4. Remove the input shaft.
5. With the transmission inverted, remove the oil pan retaining bolts.
6. Remove the pan and discard the gasket.
7. Remove the filter screen and gasket.
8. Remove the low/reverse servo cover, piston, spring and gasket.
9. Disconnect both wires at the converter clutch solenoid and both wires at the 3–4 shift solenoid.
10. Remove the bolts and while easing the valve body out of the transmission, unlock and detach the selector lever connecting link.
11. Remove the valve body and the gasket.
12. Remove the retaining bolt that holds the center support.
13. Remove the extension housing.
14. Remove the parking pawl and the return spring.
15. Remove the governor body to oil collector body attaching bolts.
16. Remove the governor body, valve, spring and weight from the collector body.

NOTE: The components are not retained once the governor body to oil collector body attaching bolts have been removed. It is necessary to hold the governor body and components while removing or installing.

17. Remove the converter housing and the oil pump as an assembly. Rotate and lift so that the clutches stay in place.
18. Loosen the overdrive band locknut and back off the adjusting screw.
19. Remove the anchor and apply struts.
20. Lift out the overdrive clutch assembly and band.
21. Remove the hydraulic pump oil seal.
22. Remove the hydraulic pump from the converter housing and remove the steel plate with the O-ring.
23. Lift out the overdrive one-way clutch and planetary assembly.
24. Remove the center support retaining snapring.
25. Remove the overdrive apply lever and shaft.
26. Remove the overdrive control bracket from the valve body side of the case.

NOTE: The overdrive apply lever does not have a boss on the shaft hole as compared to the intermediate apply lever. The overdrive apply lever shaft is longer as compared to the intermediate apply lever shaft.

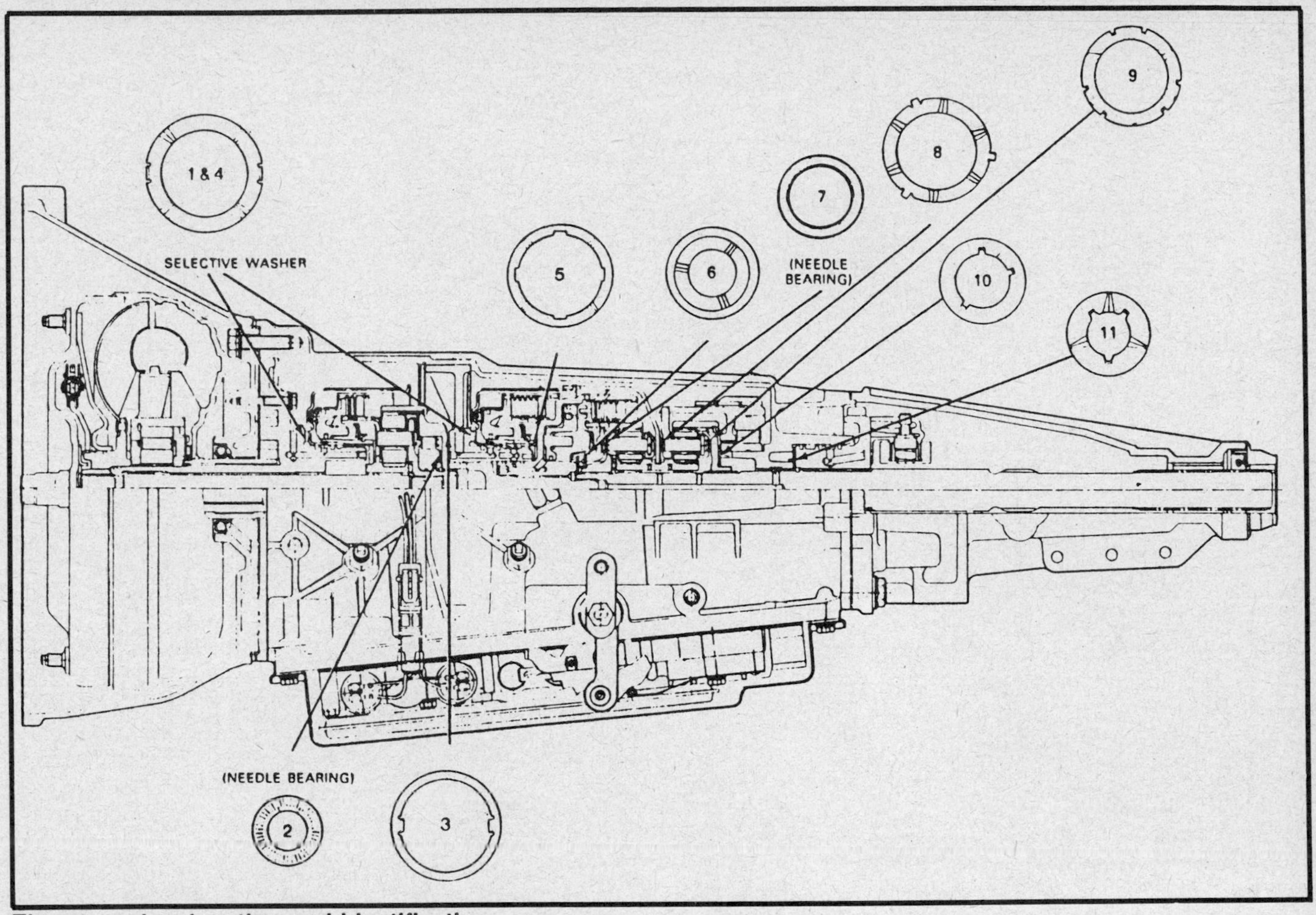
1 & 4
SELECTIVE WASHER
5
6
(NEEDLE BEARING)
7
8
9
10
11
(NEEDLE BEARING)
2
3
Thrust washer location and identification

27. Remove the thrust washer on top of the center support. Identify the thrust washer for reassembly.
28. Remove the center support being careful to pry upward evenly.
29. Remove the thrust washer below the center support. Identify the thrust washer for reassembly.
30. Loosen the intermediate band locknut and back off the adjusting screw.
31. Remove the anchor and apply struts.
32. Remove the reverse/high and forward clutch assembly.
33. Remove the intermediate band. Identify as intermediate and identify which end is the apply side or the anchor side for reinstallation.
34. Remove the forward planet assembly.
35. Remove the sun gear shell.
36. Remove the reverse planet assembly. Note and identify the thrust washers on both sides. They are identical.
37. Remove the snapring and the output shaft ring gear.
38. Remove the low/reverse drum and one-way clutch assembly.
39. Remove the low/reverse servo from the valve body side of the case. Remove the low/reverse band.
40. Remove the intermediate apply lever and shaft. This apply lever has a boss on the shaft hole and the shaft is shorter than the overdrive shaft.
41. Remove the output shaft.
42. Remove the park gear/collector body assembly from the rear of the case.
43. Remove the vacuum diaphragm and the throttle valve actuator rod.
44. Remove the throttle valve from the bore after verifying that it moves freely in the bore.
45. Remove the intermediate servo cover snapring. The transmission case is notched out to permit easy snapring removal.
46. Remove the intermediate servo cover, piston and spring and tag them for identification.
47. Remove the overdrive servo cover, piston and spring and tag the parts for identification.

CAUTION

Covers can pop off due to spring pressure behind the piston.

48. Remove the neutral start switch.
49. Remove the kickdown lever nut and O-ring seal.
50. Remove the linkage centering pin taking precautions not to damage the case flange.
51. Remove the manual lever, internal kickdown lever and park pawl rod and detent plate assembly.
52. Remove the lever shaft oil seal.
53. Remove the torque converter clutch solenoid connector. A tab on the outside of the case on the backside of the connector must be depressed while pulling with pliers.

Unit Disassembly and Assembly

VALVE BODY

Disassembly

1. Remove the screws that retain the separator plate and gasket to the valve body.

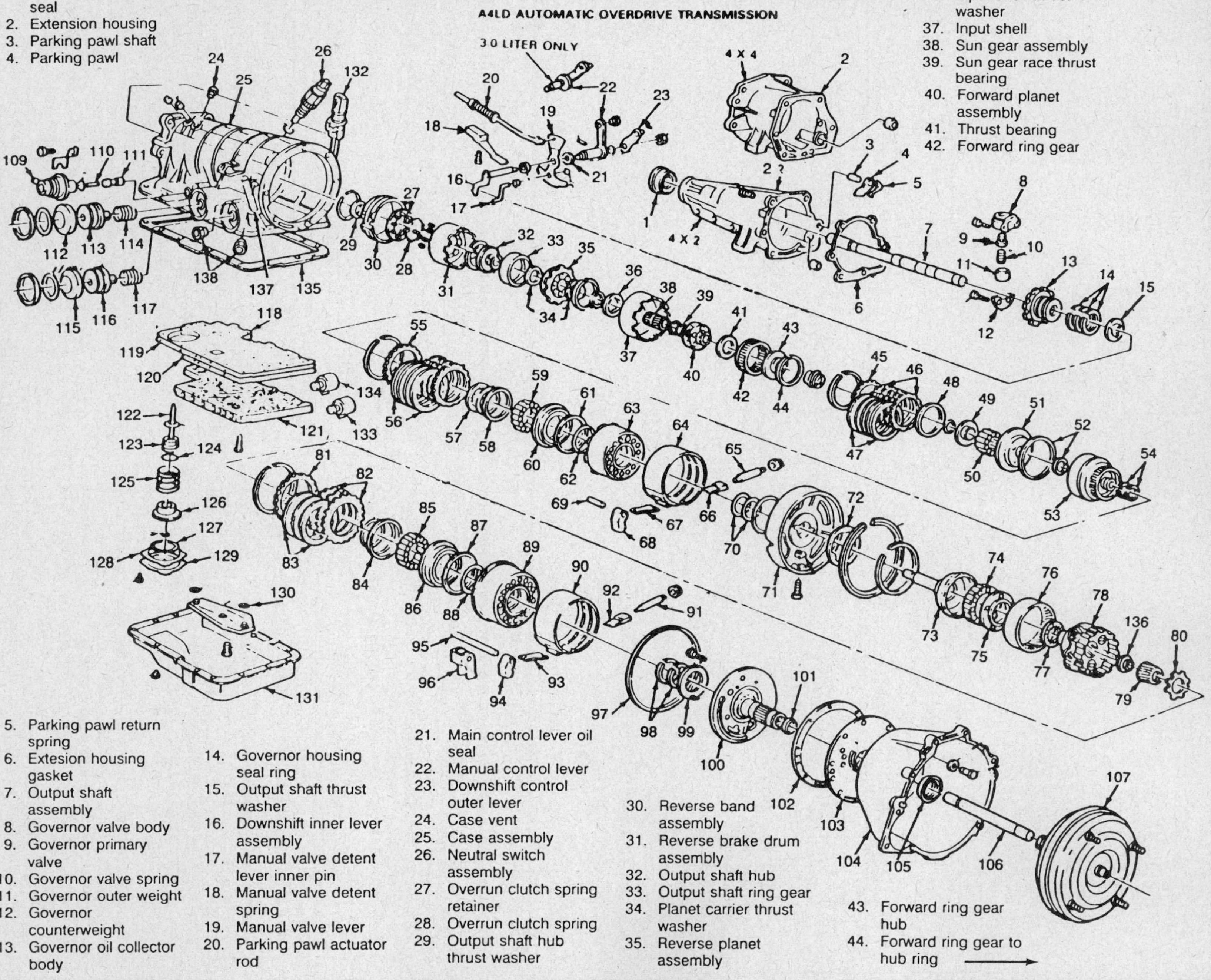
A4LD AUTOMATIC OVERDRIVE TRANSMISSION
3.0 LITER ONLY
4 X 4
4 X 2

1. Extension housing oil seal
2. Extension housing
3. Parking pawl shaft
4. Parking pawl
5. Parking pawl return spring
6. Extesion housing gasket
7. Output shaft assembly
8. Governor valve body
9. Governor primary valve
10. Governor valve spring
11. Governor outer weight
12. Governor counterweight
13. Governor oil collector body
14. Governor housing seal ring
15. Output shaft thrust washer
16. Downshift inner lever assembly
17. Manual valve detent lever inner pin
18. Manual valve detent spring
19. Manual valve lever
20. Parking pawl actuator rod
21. Main control lever oil seal
22. Manual control lever
23. Downshift control outer lever
24. Case vent
25. Case assembly
26. Neutral switch assembly
27. Overrun clutch spring retainer
28. Overrun clutch spring
29. Output shaft hub thrust washer
30. Reverse band assembly
31. Reverse brake drum assembly
32. Output shaft hub
33. Output shaft ring gear
34. Planet carrier thrust washer
35. Reverse planet assembly
36. Input shell thrust washer
37. Input shell
38. Sun gear assembly
39. Sun gear race thrust bearing
40. Forward planet assembly
41. Thrust bearing
42. Forward ring gear
43. Forward ring gear hub
44. Forward ring gear to hub ring →

Exploded view of the transmission assembly

45. Forward clutch pressure plate
46. Forward clutch plate
47. Forward clutch plate
48. Forward clutch cushion spring
49. Forward clutch cushion spring retainer
50. Forward clutch piston spring
51. Forward clutch piston
52. Clutch piston oil seal
53. Forward clutch cylinder
54. Forward clutch cylinder seal
55. Clutch pressure reverse plate
56. Clutch high plate
57. Brake drum thrust washer
58. Reverse clutch piston spring retainer
59. Reverse clutch piston spring
60. Reverse clutch piston
61. Clutch piston oil seal
62. High clutch piston inner seal
63. Intermediate brake drum
64. Intermediate servo band
65. Reverse band adjusting screw
66. Intermediate brake band anchor strut
67. Intermediate brake band apply strut
68. Intermediate band servo lever
69. Intermediate band actuator lever shaft
70. High clutch seal ring
71. Center overdrive support
72. Center suport thrust washer
73. Center overdrive shaft
74. Overdrive overrun clutch
75. Overdrive clutch washer
76. Overdrive ring gear
77. Overdrive inner race bearing
78. Overdrive planet gear carrier
79. Overdrive sun gear
80. Overdrive clutch adapter
81. Overdrive clutch pressure plate
82. Overdrive clutch plate
83. Overdrive clutch spline plate
84. Spring retainer
85. Overdrive clutch piston spring
86. Overdrive clutch piston
87. Overdrive clutch piston outer seal
88. Overdrive clutch piston inner seal
89. Overdrive drum assembly
90. Overdrive band assembly
91. Overdrive band adjusting screw
92. Overdrive brake drum anchor strut
93. Overdrive brake drum apply strut
94. Overdrive band servo lever
95. Overdrive band adjusting lever shaft
96. Overdrive bracket
97. Front oil pump seal
98. Intermediate brake drum seal
99. Input thrust weasher
100. Front pump support and gear assembly
101. Front pump support seal
102. Oil pump gasket
103. Oil pump adapter plate
104. Converter housing
105. Front oil pump seal
106. Input shaft
107. Converter assembly
108. TV Control diaphragm clamp
109. TV Control diaphragm assembly
110. TV Control rod
111. Throttle control valve
112. Intermediate band servo cover and seal
113. Intermediate piston and rod assembly
114. Intermediate band servo piston spring
115. Overdrive band servo cover and seal
116. Overdrive piston and rod assembly
117. Overdrive band servo piston spring
118. Valve body separater gasket
119. Valve body separating plate
120. Valve body separating gasket
121. Main control assembly
122. Reverse band servo piston rod
123. Reverse servo spring
124. Reverse band servo piston oil seal
125. Reverse servo piston spring
126. Reverse servo piston and rod assembly
127. Reverse band servo retainer oil seal
128. Reverse servo separator plate cover gasket
129. Reverse band servo piston cover
130. Oil pan screen assembly
131. Oil pan
132. Converter clutch override connecter
133. Overdrive shift solenoid connector
134. 3-4 Shift solenoid connector
135. Oil pan gasket
136. Sun gear thrust bearing race
137. Lube oil inlet tube
138. Integral thrust washer

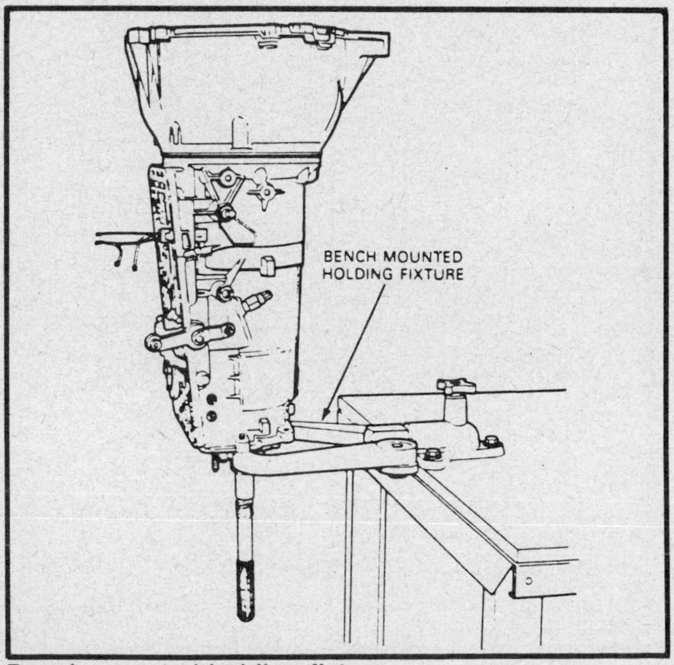

BENCH MOUNTED
HOLDING FIXTURE

Bench mounted holding fixture

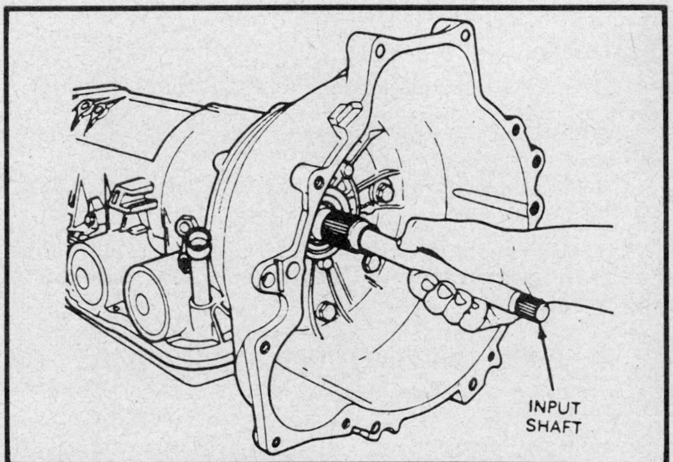

INPUT
SHAFT

Removing the input shaft

2. With the separator plate and gasket removed, the following can be removed from the valve body:
   a. Converter pressure relief valve and spring
   b. T.V. pressure relief valve and spring
   c. Shuttle balls
   d. Accumulator check valve

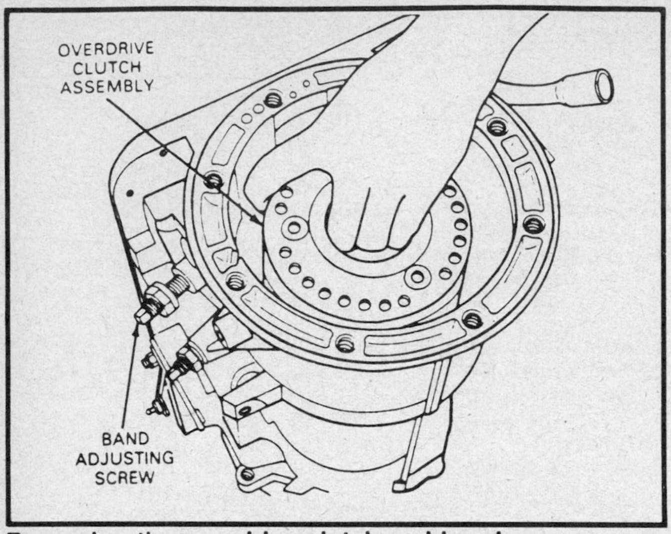
OVERDRIVE CLUTCH ASSEMBLY
BAND ADJUSTING SCREW

Removing the overdrive clutch and band

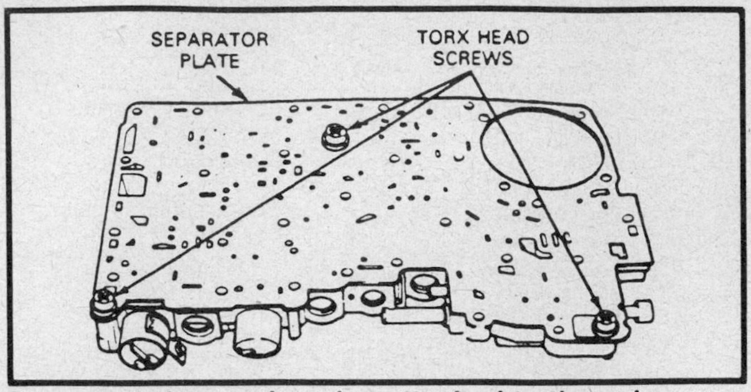
SEPARATOR PLATE
TORX HEAD SCREWS

Separator plate and gasket attached to the valve body

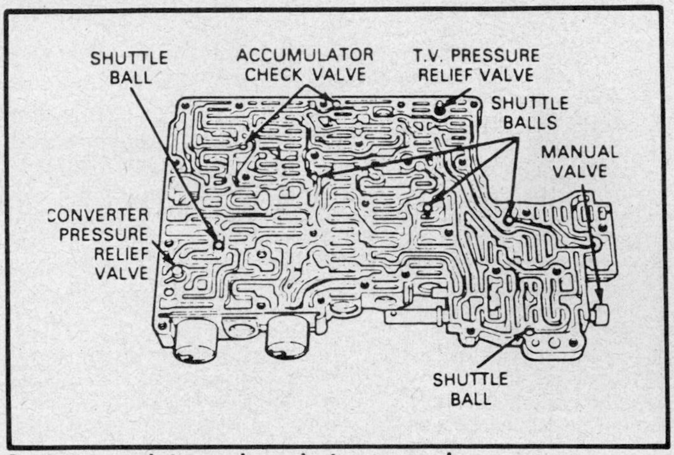
SHUTTLE BALL
ACCUMULATOR CHECK VALVE
T.V. PRESSURE RELIEF VALVE
SHUTTLE BALLS
MANUAL VALVE
CONVERTER PRESSURE RELIEF VALVE
SHUTTLE BALL

Separator plate and gasket removed

e. Filter that is retained by the separator plate at the converter clutch solenoid location

Inspection

1. Clean all parts thoroughly in clean solvent and blow dry with moisture free compressed air.
2. Inspect all valve and plug bores for scores.
3. Check all fluid passages for obstructions.
4. Inspect the check valve for free movement.
5. Inspect all mating surfaces for burrs or distortion.
6. Inspect all plugs and valves for burrs or scores.
7. Use crocus cloth to polish valves and plugs. Avoid rounding the sharp edges of the valves and plugs with the cloth.
8. Inspect all springs for distortion.
9. Check all valves and plugs for free movement in their respective bores. Valves and plugs, when dry, must fall from their own weight in their respective bores.
10. Roll the manual valve on a flat surface to check for bent condition.

Assembly

1. Install all valves, springs and check balls that were removed.
2. Align the valve body to the separator plate and gasket using tapered punches.
3. Install the bolts and tighten to 84–107 inch lbs. (9.5–12.1 Nm).

NOTE: Petroleum jelly must be used to keep the gasket in the proper location on the separator plate during assembly.

OVERDRIVE ONE-WAY CLUTCH

Disassembly

After the overdrive one-way clutch has been removed from the transmission case, it is very easily disassembled by removing the retaining ring.

1. Remove the retaining ring.
2. Lift out the cage with the springs and bearing rollers as a unit.

Assembly

1. Clean all parts and check for wear.
2. Install the cage with the springs.

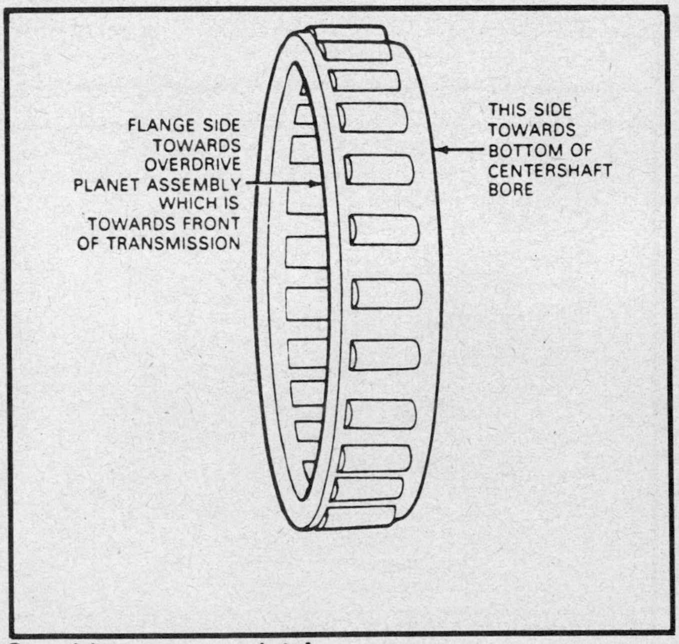
FLANGE SIDE TOWARDS OVERDRIVE PLANET ASSEMBLY WHICH IS TOWARDS FRONT OF TRANSMISSION
THIS SIDE TOWARDS BOTTOM OF CENTERSHAFT BORE

Overdrive one-way clutch

3. Insert the bearing rollers one by one and install the retaining ring.
4. Install the one-way clutch assembly into the center shaft so that the flanges of the inner and outer cages are toward the

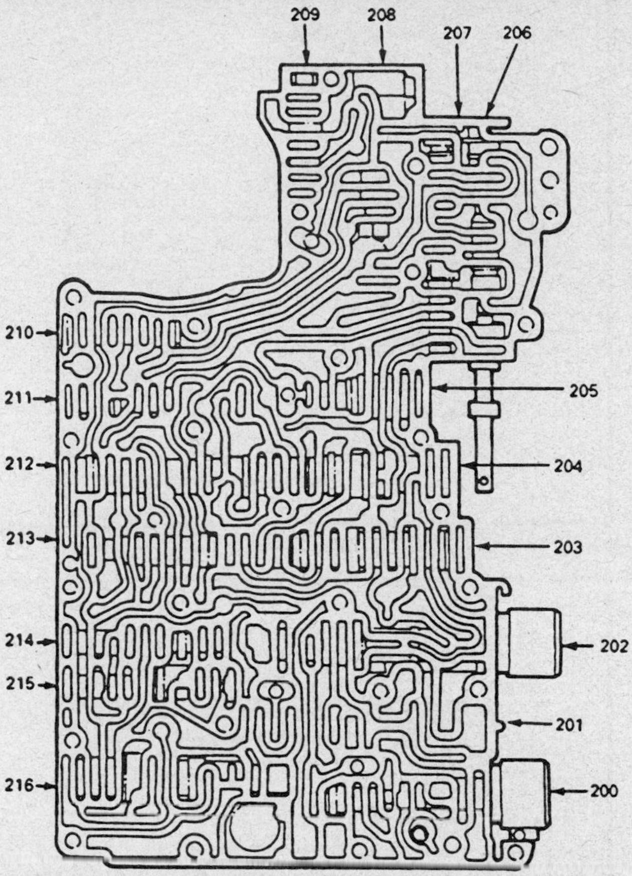
209
208
207
206
210
211
205
212
204
213
203
214
202
215
201
216
200

200. CONVERTER CLUTCH OVERRIDE SOLENOID, CONVERTER CLUTCH CONTROL VALVE
201. 3-4 SHIFT VALVE, 3-4 T.V. MODULATOR VALVE
202. 3-4 SHIFT SOLENOID, 3-4 SHIFT VALVE, OIL INLET SCREEN, CONVERTER CLUTCH SHIFT VALVE
203. 2-3 SHIFT VALVE, 2-3 T.V. MODULATOR VALVE
204. 1-2 SHIFT VALVE, D2 SHIFT VALVE
205. GOVERNOR COAST BOOST VALVE, LINE PRESURE COAST BOOST VALVE (2 AND LOW)
206. MANUAL VALVE
207. THROTTLE DOWNSHIFT VALVE (KICKDOWN)
208. MAIN OIL PRESSURE BOOSTER VALVE, MAIN OIL PRESSURE REGULATOR VALVE
209. CUTBACK VALVE
210. TORQUE DEMAND CONTROL VALVE
211. 1-2 TRANSITION VALVE, 2-3 BACKOUT VALVE, ENGAGEMENT CONTROL VALVE
212. T.V. PRESSURE BOOST VALVE
213. 3-2 COAST CONTROL VALVE
214. 3-2 KICKDOWN TIMING
215. CLUTCH RELEASE VALVE, 3-2 INTERMEDIATE SERVO RELEASE CONTROL VALVE
216. INTERMEDIATE SERVO ACCUMULATOR VALVE, OVERDRIVE SERVO ACCUMULATOR VALVE, 3-4 BACKOUT VALVE

Valve body valve identification

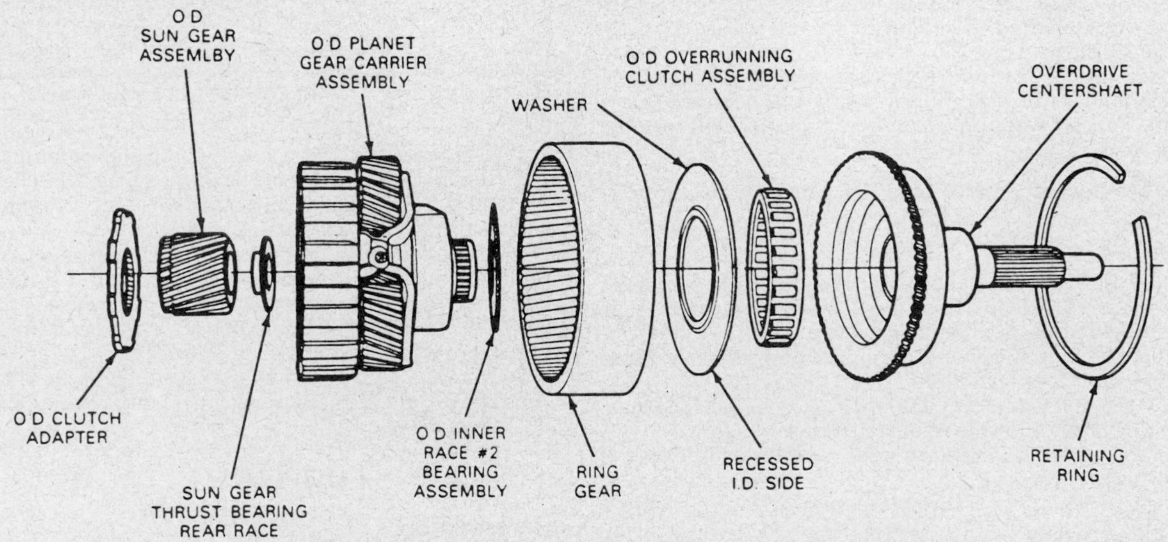
O D SUN GEAR ASSEMLBY
O D PLANET GEAR CARRIER ASSEMBLY
WASHER
O D OVERRUNNING CLUTCH ASSEMBLY
OVERDRIVE CENTERSHAFT
O D CLUTCH ADAPTER
SUN GEAR THRUST BEARING REAR RACE
O D INNER RACE #2 BEARING ASSEMBLY
RING GEAR
RECESSED I.D. SIDE
RETAINING RING

One-way clutch and planetary assembly

overdrive planet assembly which is toward the front of the transmission.

5. Position the overdrive clutch washer between the overdrive planet carrier and centershaft. It must be installed so that the recessed I.D. faces toward not against the sprag clutch.

6. After assembly, perform a build-up check. The center shaft should turn clockwise when holding the overdrive planet assembly.

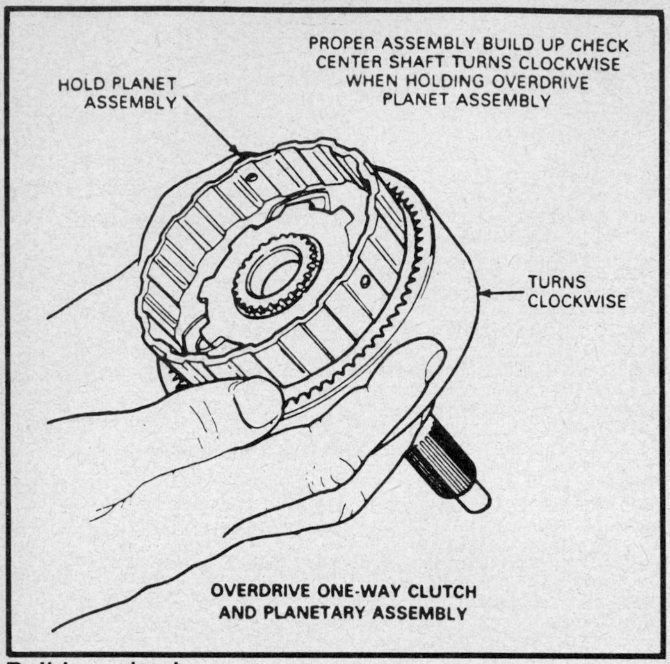
PROPER ASSEMBLY BUILD UP CHECK CENTER SHAFT TURNS CLOCKWISE WHEN HOLDING OVERDRIVE PLANET ASSEMBLY
HOLD PLANET ASSEMBLY
TURNS CLOCKWISE
OVERDRIVE ONE-WAY CLUTCH AND PLANETARY ASSEMBLY

Build-up check

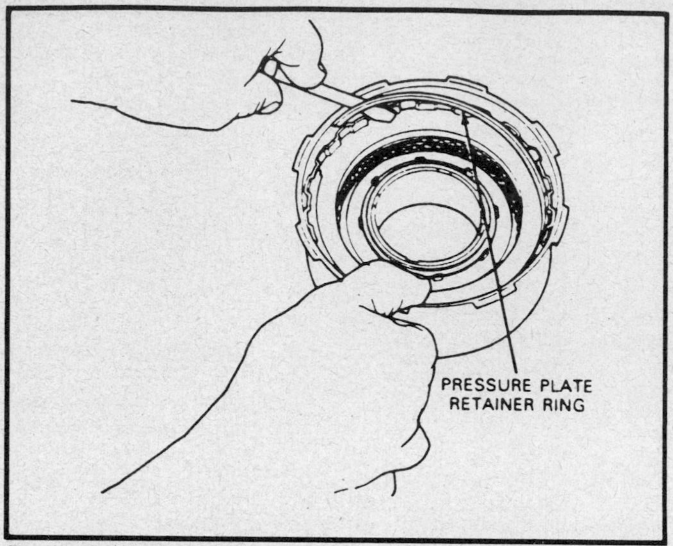
PRESSURE PLATE RETAINER RING

Reverse/High clutch assembly

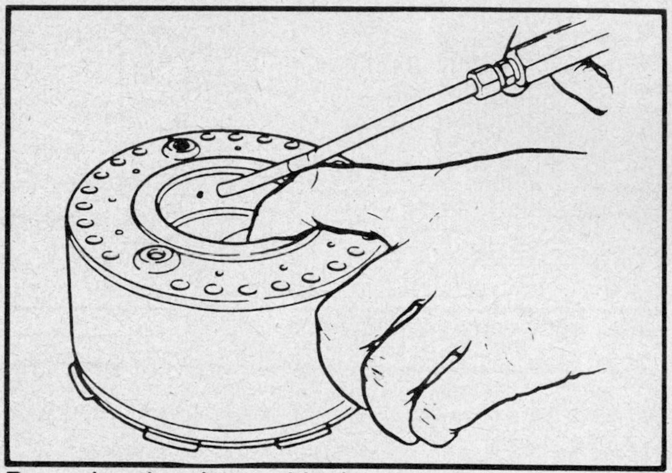

Removing the piston with air pressure

REVERSE/HIGH CLUTCH

Disassembly

1. Remove the pressure plate retainer ring and remove the plate neck.
2. Inspect the steel clutch plates and clutch lining plates for wear, damage or effects of overheating. Replace the entire set if necessary.
3. If new plates are to be used, immerse them in transmission fluid for 30 minutes before assembly.
4. Compress the compression springs using tool T65L-77515-A or equivalent. Remove the retaining ring and release the pressure on the spring.
5. Remove the spring and compression spring retainer.
6. The piston is removed by air pressure. Use a finger to close off the opposite hole and apply air pressure to blow out the clutch piston.

NOTE: Do not exceed 20 psi (137 kPa) air pressure.

Inspection

1. Inspect the drum band surface, the bushing and the thrust surfaces for scores. Minor scores may be removed with crocus cloth. Badly scored parts must be replaced.
2. Inspect the clutch piston bore and the piston inner and outer bearing surfaces for scores.
3. Check the fluid passages for obstructions. All fluid passages must be clean and free of obstructions.
4. Inspect the clutch plates for wear, scoring and fit on the clutch hub serrations.
5. Inspect the clutch pressure plate for scores on the clutch plate bearing surface.
6. Check the clutch release spring for distortion.
7. The clutch cylinders have check balls. Inspect the check balls for freedom of movement and proper seating.

Assembly

1. Install new seal rings on the clutch piston.
2. Install the clutch piston into the clutch body.
3. Install the compression spring and the spring retainer.
4. Compress the springs and install the retaining ring.
5. Install the clutch plates beginning with a steel plate, then a friction plate alternately in that order, then the pressure plate and secure it with the retaining ring.
6. Use a feeler gauge to check the clearance between the retaining ring and the pressure plate. Push downward on the plates while making this check. If the clearance is not between 0.051–0.079 in. (1.3–2.0mm), install a different suitable retaining ring.
7. Air test blocking the hole with a finger to prevent air leakage. The piston must apply when pressurized and released when air is removed.

FORWARD CLUTCH

Disassembly

1. Remove the pressure plate retainer ring and remove the plate neck.
2. Inspect the steel clutch plates and clutch lining plates for wear, damage or effects of overheating. Replace the entire set if necessary.
3. If new plates are to be used, immerse them in transmission fluid for 30 minutes before assembly.
4. Compress the compression springs using tool T65L-77515-

A or equivalent. Remove the retaining ring and release the pressure on the spring.

5. Remove the spring and compression spring retainer.

6. The piston is removed by air pressure. Use a finger to close off the opposite hole and apply air pressure to blow out the clutch piston.

NOTE: Do not exceed 20 psi (137 kPa) air pressure.

Inspection

1. Inspect the clutch cylinder thrust surfaces, piston bore, and clutch plate serrations for scores or burrs. Minor scores or burrs may be removed with crocus cloth. Replace the clutch cylinder if it is badly scored or damaged.
2. Check the fluid pressure in the clutch cylinder for obstructions. Clean out all fluid passages.
3. Inspect the clutch piston for scores and replace if necessary. Inspect the piston check ball for freedom of movement and proper seating.
4. Check the clutch release springs for distortion and cracks.
5. Inspect the composition clutch plates, steel clutch plates and clutch pressure plate for worn or scored bearing surface.
6. Check the clutch plates for flatness and fit on the clutch hub serrations.
7. Check the clutch hub thrust surfaces for scores and the clutch hub splines for wear.
8. Check the input shaft for damaged or worn splines.
9. Inspect the bushing in the stator support for scores.

Assembly

1. Install new seal rings on the clutch piston.
2. Install the clutch piston into the clutch body.
3. Install the compression spring and the spring retainer.
4. Compress the springs and install the retaining ring.
5. Install the clutch plates beginning with a steel plate, then a friction plate alternately in that order, then the pressure plate and secure it with the retaining ring.
6. Use a feeler gauge to check the clearance between the retaining ring and the pressure plate. Push downward on the plates while making this check. If the clearance is not between 0.055–0.083 in. (1.4–2.1mm), install a different suitable retaining ring.
7. Air test blocking the hole with a finger to prevent air leakage. The piston must apply when pressurized and released when air is removed.

OVERDRIVE CLUTCH

Disassembly

1. Remove the pressure plate retainer ring and remove the plate neck.
2. Inspect the steel clutch plates and clutch lining plates for wear, damage or effects of overheating. Replace the entire set if necessary.
3. If new plates are to be used, immerse them in transmission fluid for 30 minutes before assembly.
4. Compress the compression springs using tool T65L-77515-A or equivalent. Remove the retaining ring and release the pressure on the spring.
5. Remove the spring and compression spring retainer.
6. The piston is removed by air pressure. Use a finger to close off the opposite hole and apply air pressure to blow out the clutch piston.

NOTE: Do not exceed 20 psi (137 kPa) air pressure.

Inspection

1. Inspect the outer and inner races for scores or damaged surface areas where rollers contact the races.

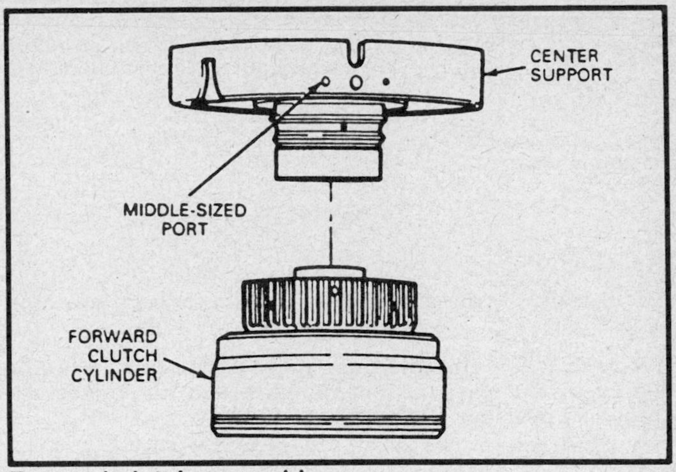
CENTER SUPPORT
MIDDLE-SIZED PORT
FORWARD CLUTCH CYLINDER

Forward clutch assembly

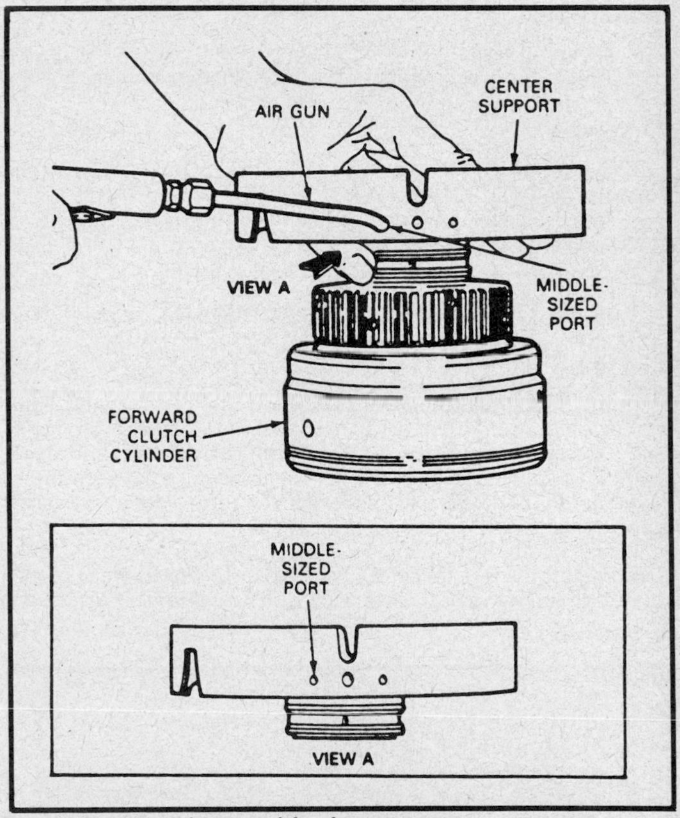
AIR GUN
CENTER SUPPORT
VIEW A
MIDDLE-SIZED PORT
FORWARD CLUTCH CYLINDER
MIDDLE-SIZED PORT
VIEW A

Applying the piston with air pressure

2. Inspect the rollers and springs for excessive wear or damage.
3. Inspect the spring and roller cage for bent or damaged spring retainers.

Assembly

1. Install new seal rings on the clutch piston.
2. Install the clutch piston into the clutch body.
3. Install the compression spring and the spring retainer.
4. Compress the springs and install the retaining ring.
5. Install the clutch plates beginning with a steel plate, then a friction plate alternately in that order, then the pressure plate and secure it with the retaining ring.
6. Use a feeler gauge to check the clearance between the re-

taining ring and the pressure plate. Push downward on the plates while making this check. If the clearance is not between 0.039–0.067 in. (1.0–1.7mm), install a different suitable retaining ring.

7. Air test blocking the hole with a finger to prevent air leakage. The piston must apply when pressurized and released when air is removed.

GOVERNOR

Disassembly

1. Remove the governor body to oil collector body attaching bolts.

NOTE: When the governor body attaching bolts are removed, the governor comonents are no longer retained in position to the governor body.

2. Remove the governor components from the governor body.
3. Remove the counterwieght.

Inspection

1. Clean all parts.
2. Replace any components that are worn or damaged.

Assembly

1. Assemble the outer weight spring and primary valve in the governor body.
2. Assemble the governor body and counterweight to the oil collector body.

FORWARD GEARTRAIN

Assembly

1. Assemble the forward clutch to the reverse and high clutch, positioning the No. 5 thrust washer between them.
2. Assemble the forward planet gear carrier to the internal gear, with needle bearing thrust washer No. 8 in between them.
3. Install a new thrust washer on the forward planet carrier hub.
4. Install the front planet assembly into the forward clutch.
5. Install the needle bearing washer into the forward planet gear carrier and install the clutch hub and sun gear.

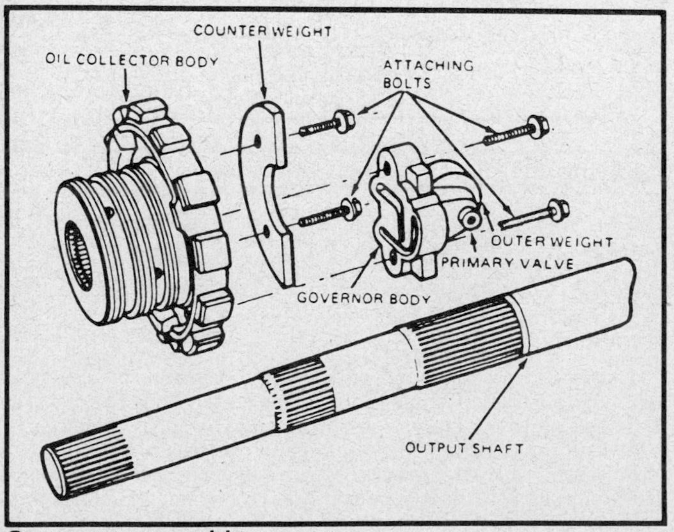
COUNTER WEIGHT
OIL COLLECTOR BODY
ATTACHING BOLTS
OUTER WEIGHT
PRIMARY VALVE
GOVERNOR BODY
OUTPUT SHAFT

Governor assembly

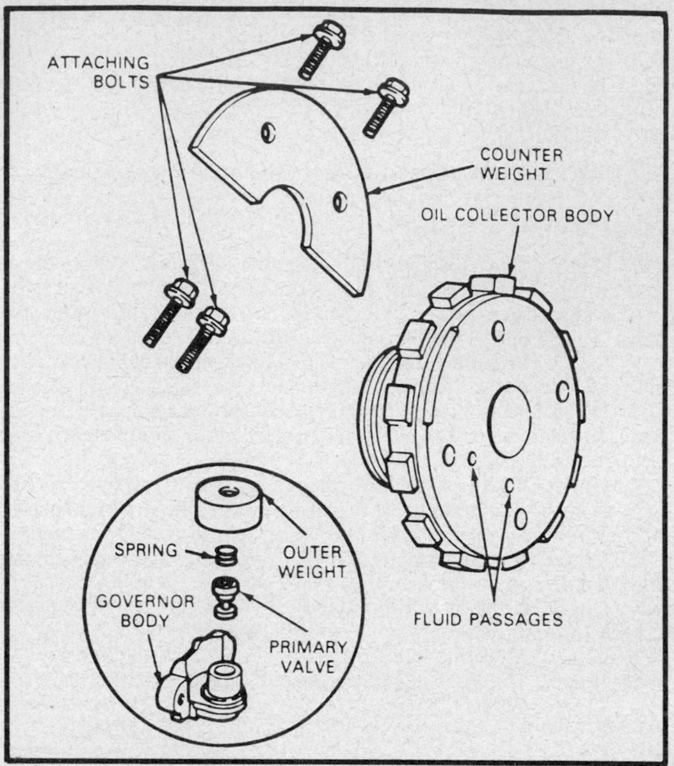
ATTACHING BOLTS
COUNTER WEIGHT
OIL COLLECTOR BODY
SPRING
OUTER WEIGHT
GOVERNOR BODY
PRIMARY VALVE
FLUID PASSAGES

Governor components

Transmission Assembly

Before beginning assembly of the transmission, the following high clutch seal sizing must be performed:

1. Install the new high clutch seals on the support hub.
2. Apply a liberal amount of petroleum jelly to the center support hub and seals.
3. Use the overdrive brake drum as a sizing tool. Rotate the center support while inserting it into the brake drum.
4. Observe the seals as they enter the cavity to see that they do not roll over or get cut.
5. Seat the center support fully into the overdrive drum and allow to stand for several minutes so that the seals seat in the grooves.

NOTE: If this is not done, the seals can be cut or rolled over when entering the intermediate brake drum cavity.

After the seal sizing procedure is completed, set the assembly aside until it is required for reassembly.

1. Install thrust washer No. 12 into the back of the transmission case.
2. Install the collector body in the rear of the case.
3. Install the output shaft.
4. Install the governor on the collector body and tighten to 84–120 inch lbs. (9–14 Nm).
5. Install thrust washer No. 11 into the case from the front.
6. Install the low/reverse drum using the overrunning clutch replacement guide tool T74P-77193-A or equivalent.
7. Install the output shaft ring gear and snapring onto the output shaft.
8. Install thrust washer No. 10, reverse planet assembly and thrust washer No. 9.
9. Use petroleum jelly to hold the thrust washers in position on the planet assembly.
10. Install the snapring in the drum to hold the planet assembly in place.

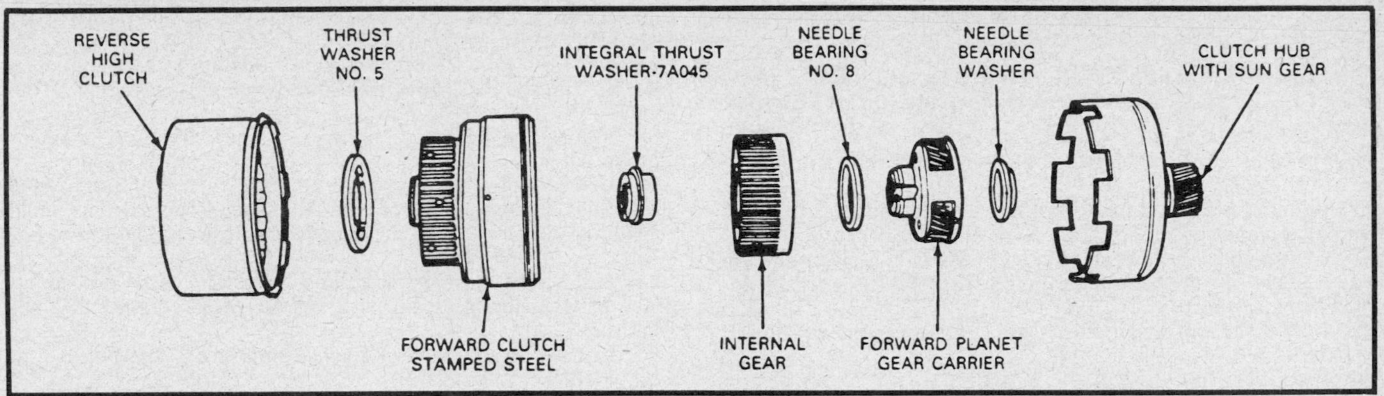
REVERSE HIGH CLUTCH
THRUST WASHER NO. 5
INTEGRAL THRUST WASHER-7A045
NEEDLE BEARING NO. 8
NEEDLE BEARING WASHER
CLUTCH HUB WITH SUN GEAR
FORWARD CLUTCH STAMPED STEEL
INTERNAL GEAR
FORWARD PLANET GEAR CARRIER

Forward geartrain assembly

11. Install the low/reverse band.
12. Replace the servo piston or O-ring, if necessary.
13. Install the low/reverse servo piston to hold the band in position.
14. Replace the piston or O-ring, if necessary.
15. Install the intermediate servo spring, piston, cover and snapring.
16. Replace the piston or O-ring, if necessary.
17. Install the overdrive servo spring, piston, cover and snapring.
18. Locate and identify the intermediate servo apply lever and shaft. The intermediate servo apply lever is the lever that has the boss on the shaft hole and the shaft is shorter than the overdrive shaft.
19. Install the intermediate apply lever and shaft into the case.
20. Install the complete forward clutch and reverse and high clutch assemblies.
21. Install the intermediate band and apply strut.
22. Install the intermediate band anchor strut and proceed to the transmission rear endplay check.

The transmission rear endplay check determines the amount of space existing between the thrust washer surfaces of the overdrive center support and the intermediate brake drum. It also determines the thickness of the No. 4 thrust washer that is required to obtain an endplay of 0.012–0.022 in. (0.30–0.54mm).

To perform the endplay check, fabricate a depth gauge fixture from an overdrive center support. An ⅛ in. hole must be drilled through the thrust washer surface of the center support. This allows depth micrometer D80P-4201-A or equivalent access to the area between the thrust surfaces of the support and the intermediate brake drum. Remove the rubber seals from the center support to allow easy insertion into the intermediate brake drum.

23. Place the depth micrometer over the drilled hole in the fabricated depth gauge fixture. Extend the micrometer probe until it is flush with the thrust washer surface of the fixture. Record the micrometer reading. This is reading **A**.
24. Install the depth gauge and input shaft fixture into the intermediate brake drum and make sure it is fully seated in the transmission case. Allow the center support fixture to slide into the intermediate brake drum using its own weight. The fixture axially locates the drum in its proper position.
25. Position the depth micrometer over the drilled hole in the fixture.
26. Continue extending the micrometer probe until it contacts the thrust washer surface of the intermediate brake drum. This is reading **B**.
27. Subtract reading **A** from reading **B**. The difference between these readings is dimension **A**. This is the space between the thrust surfaces.
28. Remove and rotate the fixture 180° and repeat Steps 23 through 27.

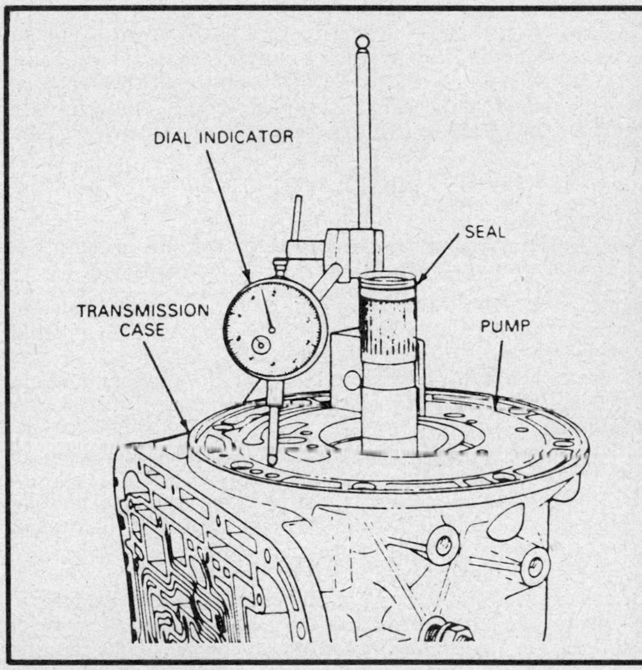
DIAL INDICATOR
SEAL
TRANSMISSION CASE
PUMP

Checking endplay

29. Average the 2 dimension A readings to obtain a final dimension A reading.
30. Select the proper thrust washer required to obtain the specified endplay of 0.012–0.022 in. (0.30–0.54mm). If dimension **A** is outside the specified limits, this indicates improper asssembly, missing parts or parts out of specification.
31. Remove the depth gauge and overdrive fixture from the overdrive drum.
32. Position the correct No. 4 selective washer on the rear of the center support using petroleum jelly.
33. Insert the input shaft through the center support and into the splines in the forward clutch cylinder.
34. Install the center support into the case checking to make sure that it is square with the case and the 5mm allen bolt retainer nut is oriented with the bolt hole in the case.
35. Do not apply any pressure to the center support. Allow the center support to slide into the intermediate brake drum using its own weight. When the support is fully seated, remove the input shaft.
36. Position the No. 3 thrust washer on top of the center support.
37. Install the large snapring to retain the center support in

position with the taper of the snapring towards the front of the transmission.

NOTE: The 2 ends of the snapring should be positioned in the wide shallow cavity located in the 5 o'clock position.

38. Install the 5mm allen head bolt that retains the center support to the case.
39. Install the sun gear and overdrive clutch adapter into the overdrive planet assembly and one-way clutch.
40. Center the needle bearing race inside of the planetary. Be sure it stays centered and positioned with the extruded lip toward the sun gear.
41. Install the overdrive planet assembly and one-way clutch into the case.
42. Install the overdrive drum assembly.
43. Install the overdrive bracket, apply lever and shaft.
44. Install the overdrive band and apply strut. Install the anchor strut.
45. Check that the needle bearing race in the overdrive planetary is centered and the overdrive clutch is seated down fully.
46. Install the No. 1 selective washer on top of the overdrive clutch drum and temporarily install the pump assembly into the case. The pump body must be below the level of the case gasket surface.
47. Check the front pump seal for damage. Replace if necessary.

NOTE: The rough casting portion of the crescent is not a flaw and the pump should not be replaced.

48. Mount dial indicator D78P-4201-G or equivalent block on the pump with the plunger resting on the transmission housing. Set the dial indicator to 0.
49. Swing the indicator around so that the plunger contacts the pump. Check the dial reading. This reading is the amount of the endplay.
50. Move the dial indicator block to the opposite side of the pump and repeat Steps 48 and 49.
51. Find the average of the 2 readings. This reading of endplay should be from 0.001–0.025 in. (0.025–0.64mm). If the reading exceeds the limits, change the No. 1 selective washer.
52. Press in a new hydraulic pump oil seal.
53. Install the separator plate on the converter housing.
54. Install the pump gears into the pump housing. The inside edge of the small gear has a chamfer on the side. This chamfer must be positioned toward the front of the transmission. The larger gear has a dimple on the side which must be positioned toward the front of the transmission.
55. Install the pump assembly into the separator plate and converter housing and install the bolts finger tight.
56. Align the pump in the converter housing using tool T74P-77103-X or equivalent. This tool must be used in order to prevent seal leakage, pump breakage or bushing failure.
57. Before removing the alignment tool, tighten the bolts to 7–10 ft. lbs. (10–13 Nm).
58. Install the input shaft into the pump and install the converter into the pump gears. Rotate the converter to check for free movement, then remove the converter and the input shaft.
59. Coat the converter housing gasket with petroleum jelly and position it on the housing. Install the seal on the converter housing.
60. Install the No. 1 selective washer on the rear of the pump using petroleum jelly.
61. Align the converter housing and the pump to the transmission. Install the bolts with new aluminum washers and tighten to 27–38 ft. lbs. (37–52 Nm).
62. Adjust the overdrive band using tool T71P-77370-A or equivalent.
63. Install a new locknut on the adjusting screw and tighten to 10 ft. lbs. (14 Nm). Back off the adjusting screw 2 turns. Hold the adjusting screw from turning and tighten the locknut to 35–45 ft. lbs. (48–61 Nm).
64. Adjust the intermediate band using the same procedure. Back off the adjusting screw 2½ turns before tightening the locknut.
65. Install the shift lever oil seal using tool T74P-77498-A or equivalent.
66. Install the internal shift linkage including the external manual control lever and centering pin. Tighten to 30–40 ft. lbs. (41–54 Nm).
67. Install the O-ring, kickdown lever and 13mm nut. Tighten to 7–10 ft. lbs. (10–14 Nm).
68. Install the neutral start switch and tighten to 84–120 inch lbs. (9.5–13.6 Nm).
69. Install the converter clutch solenoid connector.
70. Install the throttle valve, vacuum diaphragm, retaining clamp and bolt.
71. Align the valve body to separator plate and gasket using tapered punches. Install the bolts and tighten to 84–107 inch lbs. (9.5–12.1 Nm).
72. Attach and lock the selector lever connecting rod to the manual valve and ease the control body into the case.
73. Tighten all bolts except the filter screen bolt in the correct sequence to 71–97 inch lbs. (8.0–11.0 Nm).
74. Install the low-reverse servo cover, piston, spring and gasket.
75. Install the converter clutch solenoid wires.
76. Install a new servo cover gasket and tool T74P-77190-A or equivalent and tighten with the attaching bolts. Tighten the servo tool attaching screw to 35 inch lbs. (4 Nm).
77. Install a dial indicator on the transmission case and position the indicator on the piston pad. Set the dial indicator to 0.
78. Back out the servo tool adjusting screw until the piston bottoms out on the tool. Record the distance the servo piston travelled. If the piston travel is between 0.120–0.220 in. (3–5.6mm), it is within specification. If the piston travel is greater than 0.220 in. (5.6mm), use the next longer piston and rod.
79. Remove the servo adjusting tool and reverse servo piston checking spring.
80. Install the servo piston assembly, accumulator spring, gasket and cover. Tighten the bolts to 7–10 ft. lbs. (10–13 Nm).
81. Install new O-rings on the screen and lubricate with petroleum jelly. Install the filter screen and tighten the bolt to 71–97 inch lbs. (8–11 Nm).
82. Install the oil pan and gasket to the case. Tighten the bolts to 5–10 ft. lbs. (7–14 Nm).
83. Install the parking pawl and its return spring in the extension housing and preload.
84. Using a new gasket, install the extension housing. correctly seat the operating parking rod in the extension guide cup. Tighten the bolts and studs to 27–38 ft. lbs. (37–52 Nm).
85. Replace the extension housing seal and bushing.

SPECIFICATIONS

TORQUE SPECIFICATIONS

Description	ft. lbs.	Nm
Transmission to engine	28–38	38–51.5
Converter housing lower cover to converter housing	12–16	16.3–21.7
Converter housing and pump to case	27–39	36.6–52.9
Oil pump to converter housing	9.6–11	13–15
Center support (OD) to case	80–115①	9.0–13.0
Extension housing to case	27–39	36.6–52.9
Oil pan to case	8–10	11–13.5
Main control to case	71–97①	8–11
Separator plate to valve body	54–72①	6.1–8.1
Detent spring to valve body	80–107①	9–12.1
Neutral start switch to case	84–120①	9.5–13.6
Reverse servo to case	80–115①	9–13
Vacuum diaphragm retainer clip to case	80–106①	9–12
Governor assembly to oil collector body	84–120①	9.5–13.6
Outer downshift lever to inner lever shaft nut	7–11	9.5–15
Manual lever nut	30–40	40.7–54.2
Overdrive band adjusting screw locknut to case	35–45	47.5–61
Intermediate band adjusting screw locknut to case	35–45	47.5-61
Converter to flywheel attaching nut	20–34	27.1–46.1
Cooler line to case connector	18–23	24.4–31.2
Push connect cooler to line fitting case	18–23	24.4–31.2
Pressure plug to case	7–11	9.5–14.9

① inch lbs.

TORQUE CONVERTER ENDPLAY

New or rebuilt converter	Used converter
0.023 in. (Max.) 0.58mm (Max.)	0.050 in. 1.27mm (Max.)

SELECTIVE SNAPRINGS

Part Number	Thickness in.	Thickness mm	Diameter in.	Diameter mm
OVERDRIVE AND REVERSE/HIGH CLUTCH				
E 860126–S	0.0539	1.37	5.122	130.1
E 860127–S	0.0681	1.73	5.122	130.1
E 860128–S	0.0819	2.08	5.122	130.1
E 860129–S	0.0961	2.44	5.122	130.1
FORWARD CLUTCH				
E 860115–S	0.0539	1.37	4.925	125.1
E 860116–S	0.0681	1.73	4.925	125.1
E 860117–S	0.0819	2.08	4.925	125.1
E 860118–S	0.0961	2.44	4.925	125.1

CLUTCH PLATES

Engine (liter)	Steel	Friction	Clearance in.
FORWARD CLUTCH			
2.3 EFI	5	5	0.055–0.083
2.3 turbo	5	5	0.055–0.083
3.0 EFI	5	5	0.055–0.083
OVERDRIVE CLUTCH			
2.3 EFI	3	3	0.039–0.067
2.3 EFI turbo	3	3	0.039–0.067
3.0 EFI	3	3	0.039–0.067
REVERSE/HIGH CLUTCH			
2.3 EFI	4	4	0.051–0.079
2.3 turbo	5	5	0.051–0.079
3.0 EFI	5	5	0.051–0.079

CHECKS AND ADJUSTMENTS

Operation	Specification
Transmission endplay (front)	0.001–0.025 in. 0.025–0.685mm Less gasket
Transmission endplay (rear)	0.012–0.022 in. 0.30–0.54mm
Overdrive and intermediate band	Remove and discard locknut. Install new lockout. Tighten adjusting screw 10 turns. Back off 2 turns for overdrive or 2½ turns for intermediate band. Hold screw and tighten locknut

SELECTIVE THRUST WASHERS

Location	Transmission Endplay	Part Number	Thickness in.	Thickness (mm)	Number Stamped on Washer
No. 1 thrust washer front pump support	Front 0.001–0.0025 in. 0.0025–0.625mm without gasket	84DT–7D014–AA	0.053–0.055	1.35–1.40	1
		84DT–7D014–BA	0.060–0.062	1.55-1.60	2
		84DT–7D014–CA	0.068–0.070	1.75–1.80	3
		84DT–7D014–DA	0.076–0.078	1.95–2.00	4
		84DT–7D014–EA	0.084–0.086	2.15–2.20	5
		84DT–7D014–FA	0.092–0.094	2.35–2.40	6
No. 4 thrust washer OD center support	Rear 0.012–0.022 in. 0.30–0.54mm	84DT–7D014–FA	0.092–0.094	2.35–2.40	6

SPECIAL TOOLS

Tool Number	Description
T50T–100–A	Impact slide hammer
T57L–500–B	Bench mounted holding fixture
TOOL–4201–C	Dial indicator with bracket
D78P–4201–G	Dial indicator
TOOL–7000–DD	Rubber tip for air nozzle
TOOL–7000–DE	Air nozzle assembly
T67P–7341–A	Shift linkage tool
T84P–7341–A	Shift linkage tool
T84P–7341–B	Shift linkage tool
T71P–7657–A	Extension housing seal remover
T77L–7697–E	Extension housing bushing remover
T77L–7697–F	Extension housing bushing replacer
T80L–7902–A	Torque converter endplay checking tool
T74P–77000–A	C–3 service set
T74P–77001–A	Transmission mounting adapter
T74P–77028–A	Front servo cover compressor
D80P–4201–A	Depth micrometer
T74P–77052–A	Extension housing seal replacer
T74P–77103–X	Front pump alignment set
T74P–77190–A	Servo rod selecting guide

Tool Number	Description
T74P–77193–A	Overrunning clutch replacing guide
T74P–77247–A	Neutral start switch socket
T74P–77248–A	Seal remover
T74P–77248–B	Front pump seal replacer
T71P–77370–A	Band adjustment torque wrench set
T74P–77404–A	Lip seal protector
T74P–77498–A	Seal replacer
T65L–77515–A	Clutch spring compressor
T74P–77548–B	Lip seal protector
T57L–77820–A	Pressure gauge 0–700 psi
T77L–7902–B	Converter one-way clutch torquing tool
D84L–7902–A	One-way clutch holding tool
T82L–9500–AH	Cooler line disconnect tool
T87L–77248–AH	Front pump seal replacer
T87L–77248–BH	Seal staking tool
ROTUNDA EQUIPMENT	
014–00737	Automatic transmission tester
014–00028	Torque converter cleaner
021–00047	Torque converter leak tester

Section 4

E40D Transmission
Ford Motor Co.

APPLICATION

1989 E–250, E–350 over 8500 lbs. GVW except 4.9L engine
1989 F–250, F–350 over 8500 lbs. GVW except 4.9L engine
1989 F-Super Duty series vehicles

GENERAL DESCRIPTION

The E40D transmission is a fully automatic electronically controlled, 4 speed unit with a 3 element locking torque converter. The main operating components of the E40D transmission include a converter clutch, 6 multiple disc friction clutches, a band, 2 sprag one-way clutches and a roller one-way clutch which provide for the desired function of 3 planetary gear sets.

In the **OVERDRIVE** range, automatic operation of all 4 gears is possible. The overdrive cancel switch, located on the vehicle's dashboard, disables overdrive operation and enables automatic operation through the first 3 gears.

Manual gear selection is available in the **1** and **2** range. The 2nd gear is commanded when the gear selector is in the **2** range and when downshifted into the **1** range at speeds above approximately 35 mph for gasoline engines and 30 mph for diesel engines. The 1st gear is commanded in the **1** range at startups and when downshifted into **1** range below approximately 35 mph for gasoline engines and 30 mph for diesel engines.

Transmission and Converter Identification

TRANSMISSION

The transmission identification tag is located on the left hand side of the transmission case to the rear of the manual level position sensor.

CONVERTER

The torque converter is identified by either a reference or part number stamped on the converter body and is matched to a specific engine. The torque converter is a welded unit and is not repairable. If internal problems exists, the torque converter must be replaced.

Electronic Controls

On gasoline engine equipped vehicles, with gear selection in the **OVERDRIVE** range, the converter clutch operation is controlled by the EEC–IV control system. Operating conditions are relayed to EEC–IV by various sensors throughout the vehicle. The EEC–IV compares these conditions with electronically stored parameters and logically determines the state that the transmission should operate at.

For diesel engine applications, the E40D transmission utilizes a computerized electronic transmission control module which will process various engine, transmission and vehicle inputs but without the engine control function.

Metric Fasteners

Metric bolts and fasteners may be used in attaching the transmission to the engine and also in attaching the transmission to the chassis crossmember mount. The metric fastener dimensions are very close to the dimensions of the familiar inch system fasteners, and for this reason, replacement fasteners must have the same measurement and strength as those removed.

CAUTION

Do not attempt to interchange metric fasteners for inch system fasteners. Mismatched or incorrect fasteners can result in damage to the transmission unit through malfunctions or breakage and possible personal injury.

Capacities

The E40D automatic transmission use Motorcraft Mercon® automatic transmission fluid. The capacity of the E40D automatic transmission is 16.4 qts. or 15.5L for 2WD drive transmission types and 16.9 qts. or 16L for 4WD drive transmission.

Checking Fluid Level

The automatic transmission is designed to operate with the fluid level between the **ADD** and **FULL** mark on the dipstick at an operating temperature of 150°F–170°F (65°C–77°C). Fluid level

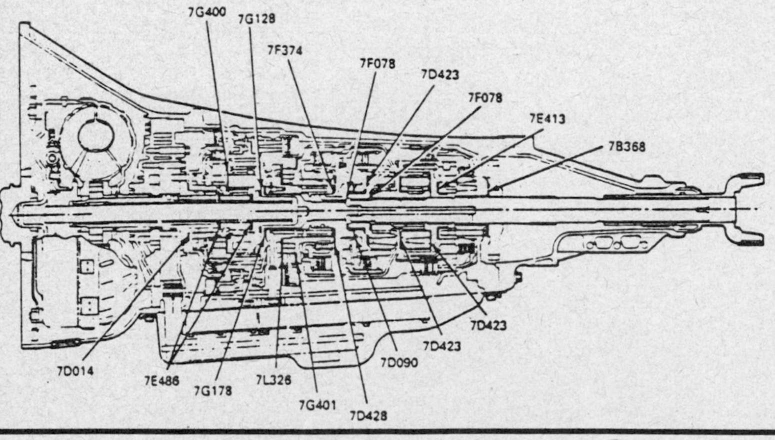
7G400
7G128
7F374
7F078
7D423
7F078
7E413
7B368
7D423
7D423
7D090
7D014
7E486
7G178
7L326
7G401
7D428

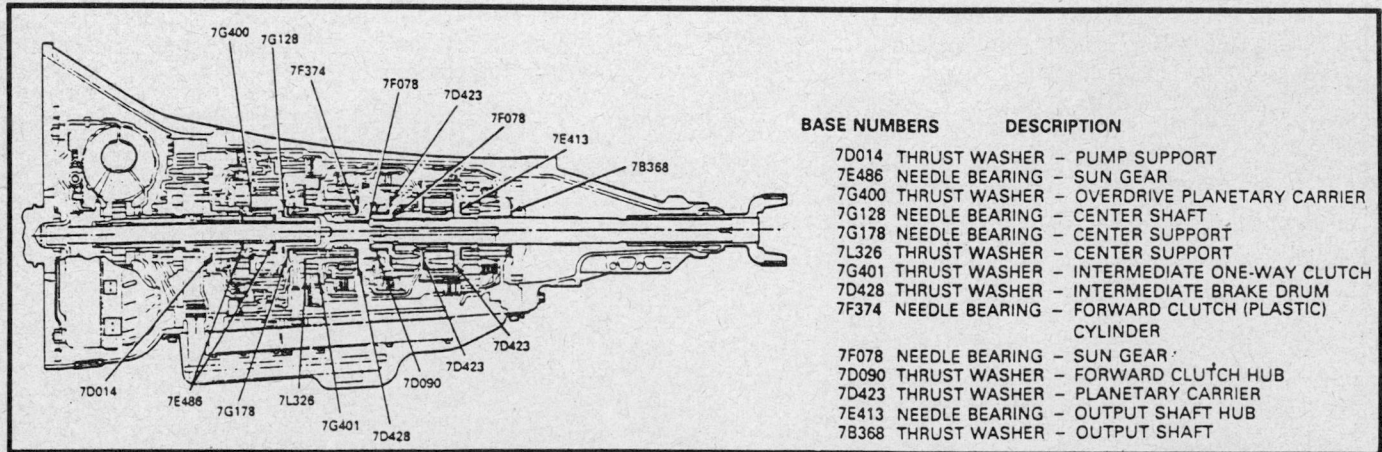

BASE NUMBERS	DESCRIPTION
7D014	THRUST WASHER – PUMP SUPPORT
7E486	NEEDLE BEARING – SUN GEAR
7G400	THRUST WASHER – OVERDRIVE PLANETARY CARRIER
7G128	NEEDLE BEARING – CENTER SHAFT
7G178	NEEDLE BEARING – CENTER SUPPORT
7L326	THRUST WASHER – CENTER SUPPORT
7G401	THRUST WASHER – INTERMEDIATE ONE-WAY CLUTCH
7D428	THRUST WASHER – INTERMEDIATE BRAKE DRUM
7F374	NEEDLE BEARING – FORWARD CLUTCH (PLASTIC) CYLINDER
7F078	NEEDLE BEARING – SUN GEAR
7D090	THRUST WASHER – FORWARD CLUTCH HUB
7D423	THRUST WASHER – PLANETARY CARRIER
7E413	NEEDLE BEARING – OUTPUT SHAFT HUB
7B368	THRUST WASHER – OUTPUT SHAFT

Thrust washer and neddle bearing location

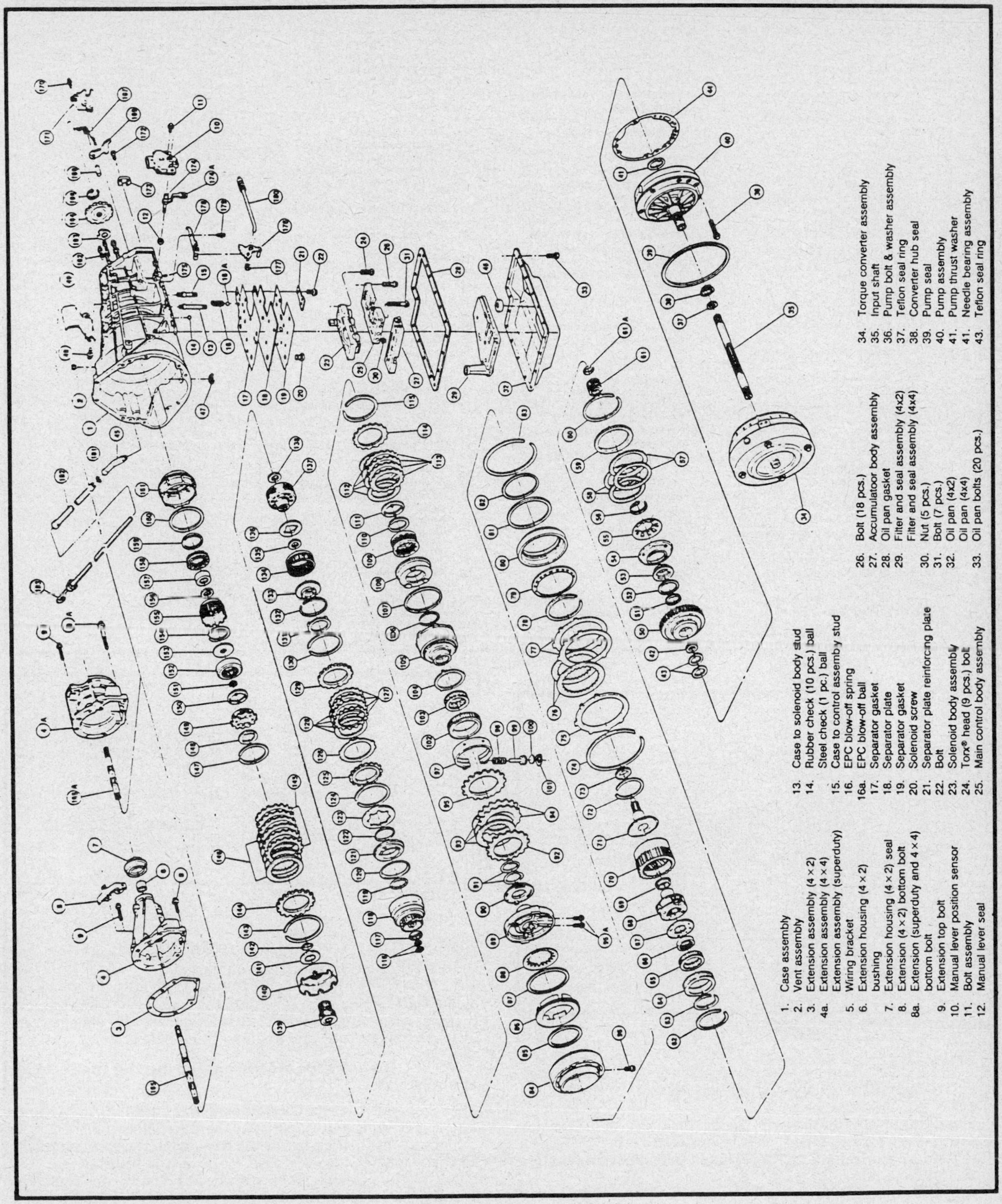
1. Case assembly
2. Vent assembly
3. Extension assembly (4 × 2)
4. Extension assembly (4 × 4)
4a. Extension assembly (superduty)
5. Wiring bracket
6. Extension housing (4 × 2) bushing
7. Extension housing (4 × 2) seal
8. Extension (4 × 2) bottom bolt
8a. Extension (superduty and 4 × 4) bottom bolt
9. Extension top bolt
10. Manual lever position sensor
11. Bolt assembly
12. Manual lever seal
13. Case to solenoid body stud
14. Rubber check (10 pcs.) ball
Steel check (1 pc.) ball
15. Case to control assembly stud
16. EPC blow-off spring
16a. EPC blow-off ball
17. Separator gasket
18. Separator plate
19. Separator gasket
20. Solenoid screw
21. Separator plate reinforcing plate
22. Bolt
23. Solenoid body assembly
24. Torx® head (9 pcs.) bolt
25. Main control body assembly
26. Bolt (18 pcs.)
27. Accumulatoor body assembly
28. Oil pan gasket
29. Filter and seal assembly (4x2)
Filter and seal assembly (4x4)
30. Nut (5 pcs.)
31. Bolt (7 pcs.)
32. Oil pan (4x2)
Oil pan (4x4)
33. Oil pan bolts (20 pcs.)
34. Torque converter assembly
35. Input shaft
36. Pump bolt & washer assembly
37. Teflon seal ring
38. Converter hub seal
39. Pump seal
40. Pump assembly
41. Pump thrust washer
42. Needle bearing assembly
43. Teflon seal ring

Exploded view E40D transmission

44. Pump gasket
45. Stub tube
46. Pan magnet
47. Converter access plug
48. Heat shield bolt
49. Solenoid body connector heat shield
50. Coast clutch cylinder assembly
51. Inner seal
52. Outer seal
53. Piston
54. Piston apply ring
55. Piston return spring
56. Retaining ring
57. Coast clutch external spline plate
58. Coast clutch internal spline plate
59. Coast clutch pressure plate
60. Retaining selective fit ring
61. Overdrive sun gear
61a. Retaining ring
62. Retaining (outer race to overdrive ring gear) ring
63. Retaining (overdrive owc to outer race) ring
64. Overdrive one-way clutch outer race
65. Overdrive one-way clutch assembly
66. Overdrive one-way clutch inner race
67. Thrust washer
68. Overdrive planetary carrier assembly
69. Needle bearing assembly
70. Overdrive ring gear
71. Center shaft
72. Retaining (center shaft to overdrive ring gear) ring
73. Needle bearing assembly
74. Overdrive retaining ring
75. Overdrive clutch pressure plate
76. Overdrive clutch internal spline plate
77. Overdrive clutch external spline plate
78. Return spring retaining ring
79. Overdrive return spring
80. Overdrive piston
81. Overdrive outer seal
82. Overdrive inner seal
83. Intermediate/overdrive cylinder retaining ring
84. Intermediate/overdrive cylinder
85. Intermediate inner seal
86. Intermediate piston
87. Intermediate outer seal
88. Intermediate return spring
89. Center support assembly
90. Thrust washer
91. Direct clutch cast iron (2 pcs.) seal
92. Intermediate clutch apply plate
93. Intermediate clutch internal spline plate
94. Intermediate clutch external spline plate
95. Intermediate clutch pressure plate
96. Cylinder hydraulic feed bolt
96a. Center support hydraulic feed bolt
97. Band assembly
98. Servo return spring
99. Servo piston assembly
100. Servo cover plate
101. Servo retaining ring
102. Intermediate one way clutch outer race
103. Intermediate one way clutch assembly
104. Thrust washer
105. Intermediate brake drum assembly
106. Inner seal
107. Outer seal
108. Piston assembly
109. Piston return spring
110. Spring retaining spring
111. Thrust washer
112. Direct clutch internal spline plate
113. Direct clutch external spline plate
114. Direct clutch pressure plate
115. Retaining selective fit ring
116. Teflon seal ring
117. Needle bearing assembly
118. Forward clutch assembly cylinder
119. Inner seal
120. Outer seal
121. Piston assembly
122. Piston apply ring
123. Piston return spring
124. Retaining (for return spring) ring
125. Forward clutch pressure plate
126. Cushion spring
127. Forward clutch external spline plate
128. Forward clutch internal spline plate
129. Forward clutch pressure plate
130. Retaining selective fit ring
131. Plastic thrust washer
132. Retaining ring
133. Forward hub
134. Forward ring gear
135. Needle bearing assembly
136. Thrust washer
137. Forward planetary carrier assembly
138. Needle bearing assembly
139. Forward/reverse sun assembly gear
140. Input shell
141. Thrust washer
142. Retaining ring
143. Retaining ring
144. Reverse clutch pressure plate
145. Reverse clutch external spline plate
146. Reverse clutch internal spline plate
147. Retaining ring
148. Thrust washer
149. Reverse planetary carrier assembly
150. Thrust washer
151. Retaining (for output shaft) ring
152. Reverse ring gear
153. Output shaft hub
154. Retaining ring
155. Reverse hub and clutch assembly (4x2)
Reverse hub and clutch assembly (4x4)
156. Needle bearing assembly
157 Low/reverse one way clutch inner race
158. Piston return spring
159. Inner seal
160. Outer seal
161. Piston
162. One way clutch to case bolts
163. Thrust washer
164. Parking gear
165. Output shaft assembly (4x2)
165a.
Output shaft assembly (4x4)
166. Retaining ring
167. Parking pawl return spring
168. Parking pawl pin
169. Parking pawl
170. Bolt and washer assembly
171. Parking rod guide plate
172. Bolt
173. Parking pawl actuating abutment
174. Manual control lever assembly
174a
Insulator
175. Manual lever retaining pin
176. Inner detent lever
177. Inner detent lever nut
178. Manual valve detent spring assembly
179. Hex flange head bolt
180. Parking pawl actuating rod assembly
181. O-ring filler tube
182. Oil filler tube assy.
183. Oil level indicator assy.

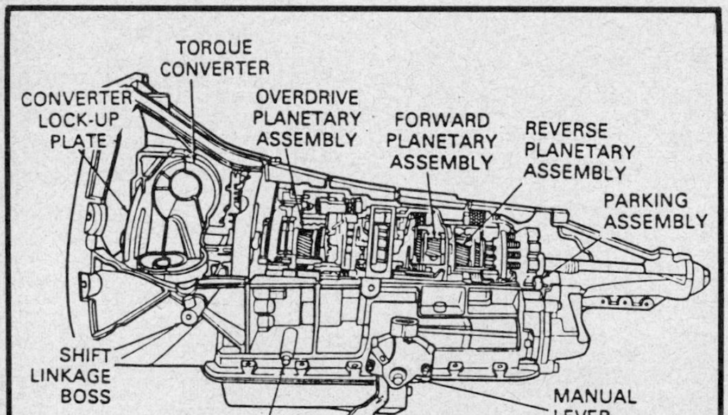
TORQUE CONVERTER
CONVERTER LOCK-UP PLATE
OVERDRIVE PLANETARY ASSEMBLY
FORWARD PLANETARY ASSEMBLY
REVERSE PLANETARY ASSEMBLY
PARKING ASSEMBLY
SHIFT LINKAGE BOSS
LINE PRESSURE TAP
MANUAL LEVER
MANUAL LEVER POSITION SENSOR

Main operation components

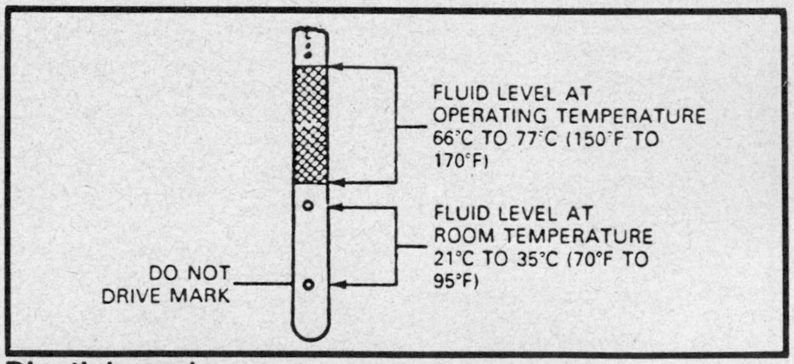
FLUID LEVEL AT OPERATING TEMPERATURE 66°C TO 77°C (150°F TO 170°F)
FLUID LEVEL AT ROOM TEMPERATURE 21°C TO 35°C (70°F TO 95°F)
DO NOT DRIVE MARK

Dipstick marks

should be checked under these temperatures. Obtain the operating temperature by driving 15–20 miles in city traffic if necessary.

The fluid level indication on the dipstick will be different at operating and room temperature. Using a fluid other than specified fluid or equivalent, could result in transmission malfunction and or failure.

A fluid level that is too high will cause the fluid to become aerated. Aerated fluid will cause low control pressure and the aerated fluid may be forced out the vent. A fluid that is too low will affect transmission operation. Low level may indicate fluid leaks that could cause transmission damage.

1. With transmission in **P**, engine at curb idle rpm, foot brake applied and vehicle on level surface, move the selector lever through each range. Allow time in each range to engage transmission, return to **P** and apply parking brake.

NOTE: Do not turn off the engine during the fluid level check.

2. Clean all dirt from the transmission fluid dipstick cap before removing the dipstick from the filler tube.
3. Pull the dipstick out of the tube, wipe it clean and push it all the way back into the tube. Be sure it is fully seated.
4. Pull the dipstick out of the tube again and check fluid level. The fluid level should be between the **ADD** and **FULL** mark.
5. Correct the fluid level, as required. Install the dipstick, making sure it is fully seated in the tube.

TRANSMISSION MODIFICATIONS

There have been no modifications to the E40D transmission at the time of this printing.

TROUBLE DIAGNOSIS

A logical and orderly diagnosis outline and charts are provide to assist the repairman in diagnosing the problems, causes and the extent of repairs needed to bring the automatic transmission back to its acceptable level of operation.

Preliminary checks and adjustments should be made to all related components. Transmission oil level should be checked, both visually and by smell, to determine whether the fluid level is correct and to observe any foreign material in the fluid, if present. Smelling the fluid will indicate if any of the bands or clutches have been burned through excessive slippage or overheating of the transaxle.

It is most important to locate the defect and its cause and to properly repair them to avoid having the same problem reoccur.

In order to more fully understand the E40D automatic transmission and to diagnose possible defects more easily, the clutch and band application chart and a general description of the hydraulic and electrical control systems are given.

Hydraulic Control System

TORQUE CONVERTER

The torque converter consists of: converter clutch (piston plate clutch and damper assembly), converter cover, turbine, impeller, reactor (sometimes called a stator) and a impeller hub. The torque converter couples the engine to the input shaft. Provides torque multiplication and absorbs engine shock caused by gear shifting.

OIL PUMP

The E40D uses a positive displacement gerotor type pump. The pump provides transmission fluid to the torque converter, lube circuit and the remaining hydraulic circuits. The pump assembly is primarily made up of 3 parts, the pump body, the control body and the reactor support. The pump body contains the pocket for the gerotor gearset. The control body contains the active ports for the pump and the bores for 3 of the hydraulic control valves: the main regulator valve, converter clutch regulator valve and coverter clutch control valve. The reactor support is pressed into the control body and supports the coast clutch cylinder, torque coverter and input shaft.

The gerotor pump consists of 2 elements. These include the inner rotor which is driven at engine speed and the outer rotor that is driven by the inner rotor. Because the inner and outer rotors have 1 tooth difference, the outer rotor runs offset from the inner and at a slower speed. This action causes the rotors to separate, causing a void in the pump cavity. This void is filled by fluid caused by atmospheric pressure on the sump fluid.

As the pump rotates, a separate chamber is created which moves

CLUTCH AND BAND APPLICATION

	Friction Elements							One-way Clutches		
									Drive	
Gear	Coast	Inter-mediate	Direct	Forward	Reverse	Over-drive	Band	O/D OWC	Inter-mediate OWC	Low Reverse OWC
OD-1st	①	—	—	Applied	—	—	—	Holding	—	Holding
D1	Coast	Coast	Coast	Coast	Coast	Coast	Coast	OR ①	—	OR
OD-2nd	①	Applied	—	Applied	—	—	—	Holding	Holding	OR
D2	Coast	Coast	Coast	Coast	Coast	Coast	Coast	OR ①	OR	OR
OD-3rd	①	Applied	Applied	Applied	—	—	—	Holding	OR	OR
D3	Coast	Coast	Coast	Coast	Coast	Coast	Coast	OR ①	OR	OR
OD-4th	—	Applied	Applied	Applied	—	Applied	—	OR	OR	OR
D4	Coast	Coast	Coast	Coast	Coast	Coast	Coast	OR	OR	OR
1	Applied	—	—	Applied	Applied	—	—	Ineff	—	Ineff
2	Applied	Applied	—	Applied	—	—	Applied	—	Ineff	OR
R	Applied	—	Applied	—	Applied	—	—	Ineff	—	Ineff

OD—Overdrive
OWC—One-way clutch
OR—Overrunning
① In D Range with overdrive cancel switch pressed, the coast clutch is applied and the OD one-way is bypassed

Ineff—Indicates that a related component renders the one-way inoperative (without load) unless the component slips

CHILTON'S THREE C'S DIAGNOSIS

Condition	Cause	Correction
Fluid venting or foaming	a) Check fluid level b) Inspect transmission fluid for contamination with anti-freeze or engine overheating c) Inspect transmission fluid filter for damaged seal or misassembly to pump	a) Drain transmission to proper level b) Determine source of leak. Service as required c) Replace filter seals or reassemble fluid filter
Stalls when stopping	a) Poor engine performance b) Check fluid level c) Check electronic engine control operation d) Test converter clutch. Converter clutch does not release	a) Check engine and service as required b) Drain or fill transmission to proper level c) Perform Quick Test d) Refer to service procedure
Shift efforts high	a) Inspect manual shift linkage for damage or misadjustment b) Inspect manual lever retainer pin for damage c) Check detent spring d) Inspect inner manual lever nut	a) Service as required b) Adjust linkage and install new pin c) Replace detent spring d) Tighten nut to specifications
Poor vehicle performance	a) Poor engine performance b) Test converter clutch. Converter clutch does not release c) Inspect torque converter one-way clutch. One-way clutch locked up	a) Perform Quick Test b) Refer to service procedure c) Replace converter
Vehicle will not start	a) Inspect ignition switch. Misadjusted or defective b) Check electronic engine control operation	a) Adjust or replace as required b) Perform Quick Test
Transmission overheats	a) Excessive tow loads b) Check fluid level c) Check electronic engine control operation d) Inspect transmission cooler and cooler lines for restriction e) Test converter clutch. Converter clutch does not apply f) Inspect valve body. Dirty or sticky valves g) Inspect torque converter one-way clutch. One-way clutch locked up	a) Check owner's manual for tow restriction b) Drain or fill transmission to proper level c) Perform Quick Test d) Service as required e) Refer to service procedure f) Clean, service or replace valve body g) Replace converter
No 1st gear, starts in higher gear	a) Check line pressure. Low line pressure b) Check solenoid operation c) Inspect D2 valve, 2-3 shift valve and 3-4 shift valve for missing springs or sticky valves	a) Perform line pressure test b) Refer to electrical diagnosis. Service as required c) Service as required
No 1-2 upshift	a) Check fluid level b) Check manual linkage c) Test line pressure. Low to intermediate friction clutch d) Check solenoid operation (S2 solenoid suspected) e) Inspect valve body bolts for correct torque f) Inspect valve body. Dirty/sticky valves g) Inspect 1-2 shift valve for damage h) Inspect D2 valve for missing spring or sticky valve i) Inspect intermediate clutch accumulator regulator valve for damage	a) Drain or fill transmission to the proper level b) Service as required c) Perform line pressure test d) Service as required e) Tighten bolts to specification f) Service as required g) Service as required h) Service as required i) Service as required

CHILTON'S THREE C'S DIAGNOSIS

Condition	Cause	Correction
No 1-2 upshift	j) Inspect intermediate clutch accumulator. Plunger stuck or damaged. Springs missing or damaged	j) Service as required
	k) Inspect intermediate clutch assembly. Clutch plates damaged/missing. Piston or seals damaged. Ball check stuck/missing. Feedbolt loose/missing/sealant leak. Clutch hub damaged	k) Service as required
	l) Inspect intermediate one-way clutch assembly for damaged cage/sprags or misassembled on inner race	l) Disassemble and inspect. Service as required
1-2 Shift harsh or soft	a) Check for high or low line pressure	a) Perform line pressure test
	b) Service line modulator pressure high or low	b) Refer to service procedure
	c) Inspect valve body bolts for correct torque	c) Tighten bolts to specification
	d) Inspect intermediate clutch accumulator regulator valve. Valve stuck, nicked or damaged. Spring missing or tangled	d) Service as required
	e) Inspect valve body. Dirty or sticky valves	e) Service as required
	f) Inspect intermediate clutch accumulator. Plunger stuck or damaged. Springs missing or tangled	f) Service as required
2-3 Shift harsh or soft	a) Check for high or low line pressure	a) Perform line pressure test
	b) Sevice line modulator pressure high or low	b) Refer to service procedure
	c) Inspect valve body bolts for correct torque	c) Tighten bolts to specification
	d) Inspect intermediate clutch accumulator regulator valve. Valve stuck, nicked or damaged. Spring missing or tangled	d) Service as required
	e) Inspect intermediate clutch accumulator. Plunger stuck or damaged. Springs missing or tangled	e) Service as required
	f) Inspect valve body. Dirty or sticky valves	f) Service as required
	g) Inspect pump air bleed check valve for leak or damage	g) Service as required
	h) Inspect intermediate clutch assembly. Clutch plates damaged/missing. Piston or seals damaged. Ball check stuck or missing. Feedbolt loose/missing sealant leak. Clutch hub damaged	h) Service as required
No 2-3 upshift	a) Check fluid level. Fluid level high or low	a) Drain or fill transmission to the proper level
	b) Check line pressure. Low to direct clutch	b) Perform line pressure test
	c) Check solenoid operation (S1 solenoid)	c) Refer to electrical diagnosis
	d) Inspect valve body bolts for correct torque	d) Tighten bolts to specification
	e) Inspect valve body. Dirty or sticky valves	e) Service as required
	f) Inspect 2-3 shift valve for being stuck, nicked or damaged	f) Service as required
	g) BS5 check ball missing. Plate seat damaged	g) Replace BS5 check ball and plate
	h) Inspect direct clutch assembly. Clutch plates damaged/missing. Piston or seals damaged. Ball check assembly stuck or missing	h) Service as required
	i) Inspect direct clutch cylinder. Seals damaged or missing or holes blocked	i) Service as required

CHILTON'S THREE C'S DIAGNOSIS

Condition	Cause	Correction
No 2-3 upshift	j) Inspect center support. Damaged, feedbolts loose or missing. Center support O.D. or case bore damaged/leaking. Teflon seal damaged	j) Service as required
2-3 Shift harsh or soft	a) Check line pressure. High or low line pressure	a) Perform line pressure test
	b) Service line modulator pressure high or low	b) Refer to service procedure
	c) Inspect valve body bolts for correct torque	c) Tighten bolts to specification
	d) Inspect valve body. Dirty or sticky valves	d) Service as required
	e) Inspect direct clutch accumulator regulator valve. Valve stuck, nicked or damaged. Spring missing or tangled	e) Service as required
	f) Inspect direct clutch accumulator. Springs missing or tangled. Plunger nicked or damaged	f) Service as required
	g) Inspect direct clutch assembly. Clutch plates damaged/missing. Piston or seals damaged. Ball check assembly stuck or missing	g) Service as required
	h) Inspect direct clutch cylinder. Seals damaged, missing or holes blocked	h) Service as required
	i) Inspect center support. Damaged, feedbolts loose or missing. Center support O.D. or case bore damaged/leaking. Teflon seal damaged	i) Service as required
No 3-4 upshift	a) Check fluid level. Fluid level high or low	a) Drain or fill transmission to the proper level
	b) Check for high or low line pressure	b) Perform line pressure test
	c) Check S2 solenoid operation	c) Refer to electrical diagnosis procedure. Service as required
	d) Inspect valve body bolts for correct torque	d) Tighten bolts to specification
	e) Inspect valve body. Dirty or sticky valves	e) Service as required
	f) Inspect 3-4 shift valve. Valve stuck, nicked or damaged. Springs missing or tangled	f) Service as required
	g) Inspect overdrive accumulator regulator valve. Valve stuck, nicked or damaged. Spring missing or tangled	g) Service as required
	h) Inspect overdrive clutch assembly. Clutch plates burnt or worn. Overdrive clutch cylinder damaged/feedbolt loose or missing/sealant leaking. Cylinder ball check assembly stuck or missing	h) Service as required
3-4 Shift harsh or soft	a) Check for line pressure high or low	a) Perform line pressure test
	b) Service line modulator pressure high or low	b) Refer to service procedure
	c) Inspect valve body bolts for correct torque	c) Tighten bolts to specification
	d) Inspect valve body. Dirty or sticky valves	d) Service as required
	e) Inspect overdrive accumulator regulator valve. Valve stuck, nicked/damaged. Spring missing or tangled	e) Service as required
	f) Inspect overdrive accumulator. Accumulator plunger stuck or damaged. Springs missing or tangled	f) Service as required

CHILTON'S THREE C'S DIAGNOSIS

Condition	Cause	Correction
3-4 Shift harsh or soft	g) Inspect overdrive clutch assembly. Clutch plates burnt or worn. Overdrive clutch cylinder damaged or feedbolt loose or missing. Cylinder ball check assembly stuck or missing	g) Service as required
Shifts 1-3	a) Check for fluid level high or low	a) Drain or fill transmission to the proper level
	b) Check S1 solenoid operation	b) Refer to electrical service procedure. Service as required
	c) Inspect D2 shift valve. Dirty or sticky. Spring missing or damaged	c) Service as required
	d) Inspect intermediate clutch accumulator regulator valve. Valve sticky or dirty	d) Service as required
	e) Inspect intermediate friction clutch. May be burnt or worn	e) Replace intermediate friction clutch
	f) Inspect intermediate one-way clutch assembly. Damaged cage/sprags. Misassembled on inner race	f) Disassemble and inspect. Service as required
Shift speed high or low	a) Check for fluid level high or low	a) Drain or fill transmission to the proper level
	b) Check electronic engine control operation	b) Perform Quick Test
	c) Inspect vehicle speed sensor. Wrong gear/damaged gear	c) Repair or replace as necessary
4-3 Downshift harsh	a) CB7 check ball missing. Plate seat damaged	a) Replace CB7 check ball and plate
3-2 Downshift harsh	a) CB6 check ball missing. Plate seat damaged	a) Replace CB6 check ball and plate
2-1 Downshift harsh	a) CB14 check ball missing. Plate seat damaged	a) Replace CB14 check ball and plate
No drive in drive range	a) Check for low fluid level	a) Fill transmission to the proper level
	b) Check for low line pressure	b) Perform line pressure test
	c) Inspect manual linkage (internal and external). Misadjusted, disconnected, damaged, broken or bent	c) Service as required
	d) Check transmission filter inside oil pan	d) Replace filter if plugged
	e) Inspect valve body and pump control body bolts for correct torque	e) Tighten bolts to specification
	f) Inspect pump control body and valve body. Dirty or sticky valves	f) Service as required
	g) Inspect overdrive one-way clutch. Improperly assembled/damaged. Damaged sprags or races	g) Service as required
	h) Inspect forward clutch assembly. Burnt or missing clutch plates. Damaged piston or seals. Forward clutch ball check assembly missing or damaged. Center support seals damaged or missing/holes blocked/feedbolt loose or missing. Forward clutch hub damaged	h) Service as required
	i) Inspect reverse one-way clutch. Improper installation or damaged rollers	i) Service as required
	j) Inspect front sun gear/shell for damage	j) Replace front gear/shell
	k) Inspect front and rear carrier. Damaged pinions/lugs to rear ring gear	k) Determine source of damage. Service as required
	l) Inspect reverse ring gear for damaged gears/lugs to forward carrier	l) Service as required
	m) Inspect output shaft for damage splines	m) Replace output shaft

CHILTON'S THREE C'S DIAGNOSIS

Condition	Cause	Correction
No reverse	a) Check for low fluid level	a) Fill transmission to the proper level
	b) Inspect manual linkage. Misadjusted, disconnected, damaged, broken or bent	b) Service as required
	c) Check for low line pressure	c) Perform line pressure test
	d) Check transmission filter inside oil pan	d) Replace filter if plugged
	e) Inspect valve body and pump control body bolts for correct torque	e) Tighten bolts to specification
	f) Inspect pump control body and valve body. Dirty or sticky valves	f) Service as required
	g) Inspect direct clutch accumulator regulator valve. Valve stuck, nicked or damaged. Spring missing or tangled	g) Service as required
	h) BS5 check ball missing. Plate seat damaged	h) Replace BS5 check ball and plate
	i) Inspect direct clutch assembly (if 3rd gear inoperative). Damaged piston or seals. Burnt or missing clutch plates. Direct clutch ball check assembly missing or damaged. Center support seals damaged or missing or holes blocked. Direct clutch hub damaged	i) Service as required
	j) Inspect coast clutch assembly for leakage	j) Service as required
	k) Inspect reverse clutch. Burnt or missing clutch plates. Damaged piston or seals	k) Service as required
	l) Inspect front and rear carrier. Damaged pinions/lugs to rear ring gear	l) Determine source of damage. Service as required
No park range	a) Inspect manual shift linkage for damage or misadjustment	a) Service as required
	b) Damaged park mechanism. Chipped or broken parking pawl or parking gear. Broken park pawl return spring. Bent or broken actuating rod	b) Determine source of damage. Service as required
Harsh neutral to drive or neutral to reverse engagements	a) Check for low fluid level	a) Fill transmission to the proper level
	b) Check electronic engine control operation	b) Perform Quick Test
	c) Worn/damaged/loose U-joint, slip yoke, rear axle or rear suspension	c) Service as required
	d) Inspect valve body bolts for correct torque	d) Tighten bolts to specification
	e) Engagement control valve. Valve stuck, nicked or damaged	e) Service as required
	f) CB13 check ball missing. Plate seat damaged	f) Replace CB13 check ball and plate
	g) Inspect direct clutch accumulator regulator valves. Valve sticking or dirty. Spring missing or tangled	g) Service as required
	h) Inspect direct clutch accumulator. Accumulator plunger stuck. Accumulator seal damaged or missing. Springs missing or tangled	h) Service as required
	i) Inspect forward clutch assembly. Burnt or missing clutch plates. Damaged piston or seals. Forward clutch ball check assembly missing or damaged. Center support seals damaged or missing/holes blocked/feedbolt loose or missing. Forward clutch hub damaged	i) Service as required
	j) Inspect reverse clutch for leakage	j) Identify source of leakage. Service as required
	k) Excessive transmission end play	k) Check transmission end play. Replace selective thrust washer

CHILTON'S THREE C'S DIAGNOSIS

Condition	Cause	Correction
No forced downshifts	a) Check for fluid level high or low	a) Drain or fill transmission to the proper level
	b) Check electronic engine control operation	b) Perform Quick Test
	c) Inspect valve body bolts for correct torque	c) Tighten bolts to specification
	d) Inspect valve body. Dirty or sticky valves	d) Service as required
No engine braking in manual one	a) Check for low fluid level	a) Fill transmission to the proper level
	b) Check for low line pressure	b) Perform line pressure test
	c) Check S1 solenoid operation	c) Refer to electrical diagnosis procedure
	d) Inspect for dirty or sticky valves. Reverse clutch modulator, D2 4-3-2 timing or 2-3 or coast clutch shift valves	d) Service as required
	e) Check ball missing. BS1, BS3 or CB1. Plate seat damaged	e) Replace check balls and plate
	f) Inspect coast clutch. Worn or burnt. Piston or seals damaged. Stator support damaged or holes blocked. Coast clutch hub damaged or holes blocked	f) Service as required
	g) Inspect reverse clutch. Worn or burnt. Piston or seals damaged	g) Service as required
No engine braking in manual second	a) Check for low fluid level	a) Fill transmission to the proper level
	b) Check for low line pressure	b) Perform line pressure test
	c) Inspect for dirty or sticky valves. 4-3-2 timing, D2, 2-3 or coast clutch shift valve	c) Service as required
	d) Check ball missing, BS1, BS3 or CB1. Plate seat damaged	d) Replace check balls and plate
	e) Check intermediate servo	e) Perform air pressure test of servo for leakage. Service as required
	f) Inspect intermediate band or drum, may be worn or burnt	f) Service as required
	g) Inspect coast clutch, may be worn or burnt. Piston or seals damaged. Stator support damaged or holes blocked. Coast clutch hub damaged or holes blocked	g) Service as required
Erratic shifts	a) Check for high or low fluid level	a) Drain or fill transmission to the proper level
	b) Check electronic engine control operation	b) Perform Quick Test
	c) Inspect vehicle speed sensor. Damaged or defective	c) Service as required
	d) Inspect valve body bolts for correct torque	d) Tighten bolts to specification
	e) Inspect valve body. Dirty or sticky valves	e) Service as required
Shift hunting	a) Check for high or low fluid level	a) Drain or fill transmission to the proper level
	b) Check electronic engine control operation	b) Perform Quick Test
High or low line pressure	a) Check for high or low fluid level	a) Drain or fill transmission to the proper level
	b) Electronic pressure control solenoid malfunction	b) Refer to electrical diagnosis procedure. Service as required
	c) Main regulator valve or spring. Dirty or sticky valve. Damaged spring	c) Determine source of damage or contamination. Service as required
	d) Pump assembly. Gears damaged, broken or worn	d) Determine source of damage. Service as required
No convertor clutch apply	a) Check for high or low fluid level	a) Drain or fill transmission to the proper level

CHILTON'S THREE C'S DIAGNOSIS

Condition	Cause	Correction
No convertor clutch apply	b) Electrical system or electronic engine control. No lock-up signal. S3 solenoid malfunction. Bulkhead connector damaged. Pinched wires c) Inspect stator shaft Teflon seal for damage d) Converter clutch control valve dirty or sticky	b) Refer to electrical diagnosis procedure. Service as required c) Replace stator shaft seal d) Service as required
Converter clutch does not release	a) Check for high or low fluid level b) Electrical system or electronic engine control. No unlock signal. S3 solenoid malfunction. Bulkhead connector damaged. Pinched wires c) Converter clutch control valve dirty or stuck valve	a) Drain or fill transmission to the proper level b) Refer to electrical diagnosis procedure. Service as required c) Service as required
Line modulator pressure high or low	a) Check for high or low line pressure b) Inspect line pressure modulator valve. Valve stuck or damaged. Plunger or sleeve stuck or damaged	a) Perform line pressure test b) Service as required

fluid from the void to the compression area. The rotors squeeze the fluid out to the hydraulic system which creates the flow necessary for control line pressure from the pressure regulator.

CONTROL VALVES

MAIN REGULATOR BOOSTER VALVE

This valve reacts to pressure from the manual 1 circuit and the electronic pressure control solenoid to force, directly or through a spring force, the main regulator valve back toward its rest position resulting in increased line pressure.

MAIN REGULATOR VALVE

This valve regulates line pressure by allowing excess pump output to force the valve from its rest position. As the valve moves, the excess flow is exhausted into the converter charge passage or the sump, thereby relieving excess pressure. The regulated line pressure is variable and dependent on the spring force on the valve and the magnitude of the manual 1 circuit and electronic pressure control force against the main regulator booster valve.

NOTE: This valve boosts control line pressure during reverse or manual gear operation.

CONVERTER CLUTCH REGULATOR VALVE

This valve regulates line pressure to the converter. The maximum pressure to the torque converter is 113 psi.

CONVERTER CLUTCH CONTROL VALVE

The valve position is solenoid controlled and directs flow to the converter, which either applies or releases the converter clutch.

LINE MODULATOR VALVE

The line modulator valve provides a controlled pressure to an accumulator valve train which controls the pressure applied to the friction clutches. Line modulator pressure acts on an accumulator plunger which sets the output of the accumulator regulator valve during a shift.

MANUAL VALVE

The manual valve motion is controlled by the position of the transmission manual lever. This is connected mechanically to the transmission shift selector. The manual valve directs line pressure input from the main regulator valve to 1 of 4 circuits. The circuits are the overdrive–2–1 circuit, 2 circuit, 1 circuit and the reverse circuit.

SOLENOID REGULATOR VALVE

This valve regulates pressure to 50 psi for use by the 4 shift solenoids. The output is also used for 4–3–2 shift timing.

SHIFT VALVES

LOW/REVERSE MODULATOR VALVE

This valve regulates pressure into the low and reverse clutch assembly.

3–4 SHIFT VALVE

This valve directs flow into the overdrive clutch when in rest position (during 4th gear) and directs flow into coast clutch shift valve when shifted.

2–3 SHIFT VALVE

This valve supplies fluid to the direct clutch in 3rd and 4th gears. It directs the flow in 1st and 2nd gear drive range to shift 3–4 shift valve. It also supplies fluid to the 1–2 shift valve in manual 1 and manual 2.

D2 VALVE

This valve shifts 1–2 shift valve during reverse operation.

1–2 SHIFT VALVE

This valve directs fluid to the intermediate clutch in 2nd, 3rd and 4th gears. It stops the flow in 1st gear. In manual gears it also sends flow to the band in 2 and to the reverse clutch in 1.

SHIFT TIMER PLUNGER

In manual pull in, it delays the shifting of the 4–3–2 shift timer valve thereby delaying the application of the band or the reverse clutch.

4–3–2 SHIFT TIMER VALVE

During manual pull in, it directs the flow into the 2–3 shift valve.

ENGAGEMENT CONTROL VALVE

During higher electronic pressure control this valve increases pressure to the forward clutch during forward engagement and direct clutch during reverse engagement.

1-2 MANUAL TRANSITION VALVE

This valve prevents the band from coming on during manual 1 and 2 upshift until the low/reverse clutch is off.

COAST CLUTCH SHIFT VALVE

This valve when shifting during manual pull in or when overdrive range is cancelled, directs fluid to the coast clutch, otherwise it stops oil flow.

OVERDRIVE CLUTCH ACCUMLATOR PLUNGER AND VALVE

This valve regulates line pressure supplied to the overdrive friction clutch by using the valve to constrict the flow. Valve regulated pressure is determined by the force of line modulator pressure and the inner and outer springs acting against the accumulator plunger.

DIRECT CLUTCH ACCUMULATOR PLUNGER AND VALVE

This valve regulates line pressure supplied to the direct friction clutch by using the valve to constrict the flow. Valve regulated pressure is determined by the force of line modulator pressure and the inner and outer springs acting against the accumulator plunger.

INTERMEDIATE CLUTCH ACCUMULATOR

This valve regulates line pressure supplied to the intermediate friction clutch by using the valve to constrict the flow. Valve regulated pressure is determined by the force of line modulator pressure and the inner and outer springs acting against the accumulator plunger.

LINE PRESSURE MODULATOR PLUNGER AND VALVE

This valve provides a modulated pressure dependent upon the electronic pressure control, which acts in the accumulator body to resist accumulator valve movement and thereby increase clutch application pressure.

Electronic Controls

GASOLINE ENGINE

On vehicles equipped with gasoline engines, the operation of E40D automatic transmission is controlled by the EEC–IV system.

THROTTLE POSITION SENSOR (TP S)

The throttle position sensor is a potentiometer mounted on the throttle body. It consists of a lever positioned between the throttle valve and a variable resistor. The throttle position sensor detects the opening of the throttle plate and sends this information to the ECA as a varying voltage signal.

MANIFOLD ABSOLUTE PRESSURE SENSOR (MAP)

The manifold absolute pressure sensor uses pressure to produce an electrical voltage signal. The frequency of this voltage signal varies with intake manifold pressure. This sensor sends the signal to the ECA, which determines altitude from manifold pressure. The ECA can then adjust the transmission shift schedule for different altitudes.

PROFILE IGNITION PICKUP SIGNAL (PIP)

The profile ignition pickup signal is produced by a Hall Effect device in the distributor. It tells the ECA the engine rpm and the crankshaft position.

BRAKE ON/OFF SWITCH (BOO)

The brake on/off switch tells the ECA whether the brakes are applied or not. The switch is closed when the brakes are applied and open when they are not.

MANUAL LEVER POSITION SENSOR (MLPS)

This sensor tells the ECA which position the shift lever is in. It is located on the outside of the transmission, at the manual lever.

VEHICLE SPEED SENSOR (VSS)

The vehicle speed sensor is a magnetic pickup that sends an signal to the ECA. This signal is proportional to the transmission output shaft rpm and tells the ECA what the vehicle speed is.

TRANSMISSION OIL TEMPERATURE SENSOR (TOT)

The transmission oil temperature sensor is a temperature sensitive device called a thermistor. It sends a voltage signal that varies with the transmission oil temperature to the ECA. The ECA uses this signal to determine whether a cold start shift schedule is necessary. The cold start shift schedule lowers shift speeds to allow for increased viscosity of the cold transmission fluid. This sensor is located on the solenoid body in the transmission sump.

OVERDRIVE CANCEL SWITCH AND INDICATOR LIGHT

When the overdrive cancel switch is pressed, the indicator light comes on and a signal is sent to the ECA. The ECA then energizes solenoid 4, applying the coast clutch cancelling 4th gear operation.

NOTE: The overdrive cancel switch indicator light will come on if the variable force solenoid (VFS) fails.

SOLENOIDS

The ECA controls the E40D transmission operation through 4 on/off solenoids and 1 variable force solenoid. These solenoids are housed in the transmission valve body assembly. Solenoids 1 and 2 provide gear selection of the 1st through 4th gears by controlling the pressure to the 3 shift valves. Solenoid 3 provides converter clutch control by shifting the converter clutch control valve. Solenoid 4 provides coast clutch control for overdrive lockout by shifting the coast clutch shift valve. This solenoid can be activated either by pressing the overdrive cancel switch or by selecting the **R**, **1** or **2** range with the transmission selector lever.

The variable force solenoid (VFS) is an electrohydraulic actuator combining a solenoid and a regulating valve. It produces electronic pressure control which regulates transmission line pressure by producing resisting forces to the main regulator valve and the line modulator valve. These 2 modified pressures control the clutch application pressures.

DIESEL ENGINE

On vehicles equipped with diesel engines, the operation of the E40D transmission is controlled by the transmission control system (Transmission electronic control assembly).

FUEL INJECTION PUMP LEVER SENSOR (FIPL)

The fuel injection pump lever sensor is a potentiometer similar to the TP sensor used on gasoline engines. This sensor is attached to the fuel injection pump and is operated by the throttle lever. It sends a varying voltage signal to the transmission electronic control assembly module, telling the module how much fuel is being delivered to the engine.

If a malfunction occurs in the sensor circuit, the transmission control system will recognize that the sensor signal is out of specification. The transmission electronic control assembly module will then operate the transmission in a high capacity mode to prevent transmission damage. This high capacity mode causes harsh upshifts and engagements, a sign that transmission diagnosis is required.

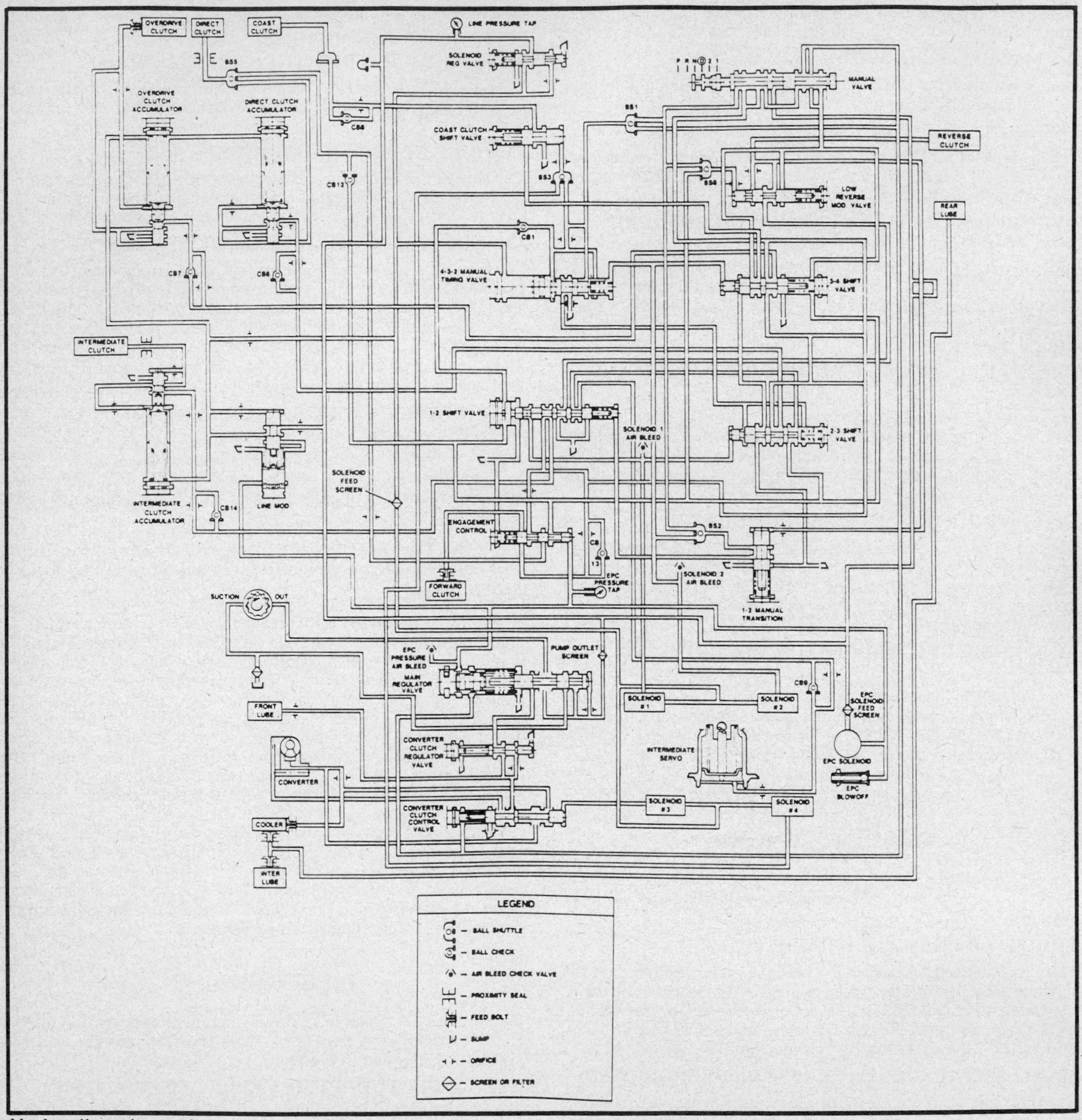
OVERDRIVE CLUTCH
DIRECT CLUTCH
COAST CLUTCH
LINE PRESSURE TAP
SOLENOID REG VALVE
MANUAL VALVE
REVERSE CLUTCH
REAR LUBE
OVERDRIVE CLUTCH ACCUMULATOR
DIRECT CLUTCH ACCUMULATOR
COAST CLUTCH SHIFT VALVE
LOW REVERSE MOD VALVE
4-3-2 MANUAL TIMING VALVE
3-4 SHIFT VALVE
INTERMEDIATE CLUTCH
1-2 SHIFT VALVE
2-3 SHIFT VALVE
SOLENOID 1 AIR BLEED
SOLENOID FEED SCREEN
INTERMEDIATE CLUTCH ACCUMULATOR
LINE MOD
ENGAGEMENT CONTROL
FORWARD CLUTCH
EPC PRESSURE TAP
SOLENOID 2 AIR BLEED
1-2 MANUAL TRANSITION
SUCTION
OUT
EPC PRESSURE AIR BLEED
MAIN REGULATOR VALVE
PUMP OUTLET SCREEN
FRONT LUBE
SOLENOID #1
SOLENOID #2
EPC SOLENOID FEED SCREEN
CONVERTER CLUTCH REGULATOR VALVE
INTERMEDIATE SERVO
EPC SOLENOID
EPC BLOWOFF
CONVERTER
CONVERTER CLUTCH CONTROL VALVE
SOLENOID #3
SOLENOID #4
COOLER
INTER LUBE
LEGEND
BALL SHUTTLE
BALL CHECK
AIR BLEED CHECK VALVE
PROXIMITY SEAL
FEED BOLT
SUMP
ORIFICE
SCREEN OR FILTER

Hydraulic schematic

BAROMETRIC PRESSURE SENSOR (BP)

The barometric pressure sensor operates in the same as the MAP sensor, except that its measures barometric pressure instead of intake manifold pressure. The transmission electronic control assembly module uses the signal from the BP sensor to determine the altitude at which the vehicle operates then adjust the shift schedule for that altitude.

ENGINE RPM SENSOR

The engine RPM sensor indicates the engine speed with information from the fuel injection pump gear.

BRAKE ON/OFF SWITCH (BOO)

The brake on/off Switch tells the transmission electronic control

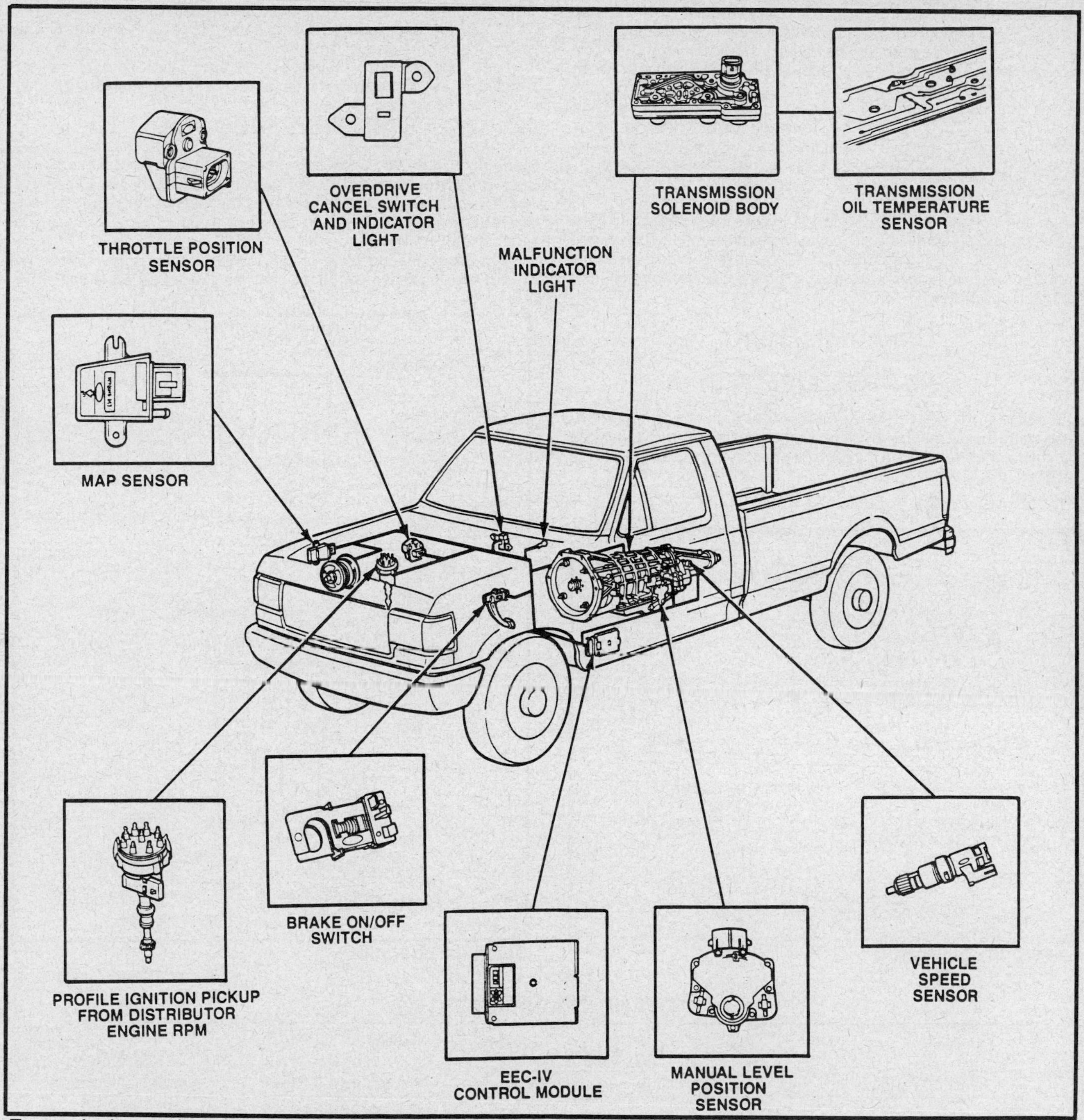
THROTTLE POSITION SENSOR
OVERDRIVE CANCEL SWITCH AND INDICATOR LIGHT
MALFUNCTION INDICATOR LIGHT
TRANSMISSION SOLENOID BODY
TRANSMISSION OIL TEMPERATURE SENSOR
MAP SENSOR
PROFILE IGNITION PICKUP FROM DISTRIBUTOR ENGINE RPM
BRAKE ON/OFF SWITCH
EEC-IV CONTROL MODULE
MANUAL LEVEL POSITION SENSOR
VEHICLE SPEED SENSOR

Transmission and engine sensor locations

assembly whether the brakes are applied or not. The switch is closed when the brakes are applied and open when they are not.

MANUAL LEVER POSITION SENSOR (MLPS)

This sensor tells the transmission electronic control assembly which position the shift lever is in. It is located on the outside of the transmission, at the manual lever.

VEHICLE SPEED SENSOR (VSS)

The vehicle speed sensor is a magnetic pickup that sends an signal to the transmission electronic control assembly. This signal is proportional to the transmission output shaft rpm and tells to the transmission electronic control assembly vehicle speed.

TRANSMISSION OIL TEMPERATURE SENSOR (TOT)

The transmission oil temperature sensor is a temperature sensitive device called a thermistor. It sends a voltage signal that varies with the transmission oil temperature to the transmission electronic control assembly. The transmission electronic control assembly uses this signal to determine whether a cold start shift schedule is necessary. The cold start shift schedule lowers shift speeds to allow for increased viscosity of the cold transmission fluid. This sensor is located on the solenoid body in the transmission sump.

OVERDRIVE CANCEL SWITCH AND INDICATOR LIGHT

When the overdrive cancel switch is pressed, the indicator light comes on and a signal is sent to the transmission electronic control assembly. The transmission electronic control assembly then energizes solenoid 4, applying the coast clutch cancelling 4th gear operation.

Diagnosis Tests

LINE PRESSURE TEST

NOTE: Line pressure does not vary directly with engine rpm. Instead the ECA or the transmission ECA calls for an increase in line pressure just before the clutch is applied.

1. Raise and safely support the vehicle.

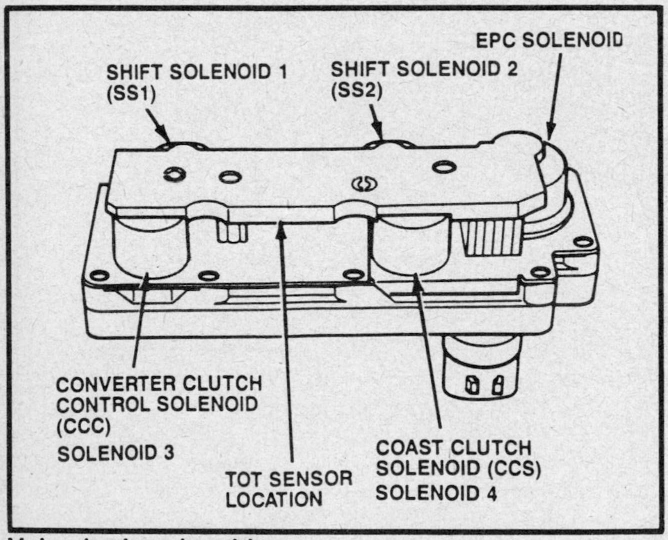
EPC SOLENOID
SHIFT SOLENOID 1
(SS1)
SHIFT SOLENOID 2
(SS2)
CONVERTER CLUTCH
CONTROL SOLENOID
(CCC)
SOLENOID 3
TOT SENSOR
LOCATION
COAST CLUTCH
SOLENOID (CCS)
SOLENOID 4

Valve body solenoids

2. Connect pressure gauge to line pressure tap.
3. Start engine and check line pressures. Determine if line pressure is within specification.
4. If line pressure is not within specification, perform air pressure test and service the main control system as necessary.

AIR PRESSURE TEST

A no drive condition can exist, even with the correct transmission fluid pressure, because of inoperative clutches or bands. An erratic shift can be located through a series of checks by substituting air pressure for fluid pressure to determine the location of the malfunction.

When the selector lever is in a forward gear range (overdrive, 2 and 1) a no drive condition may be caused by an inoperative

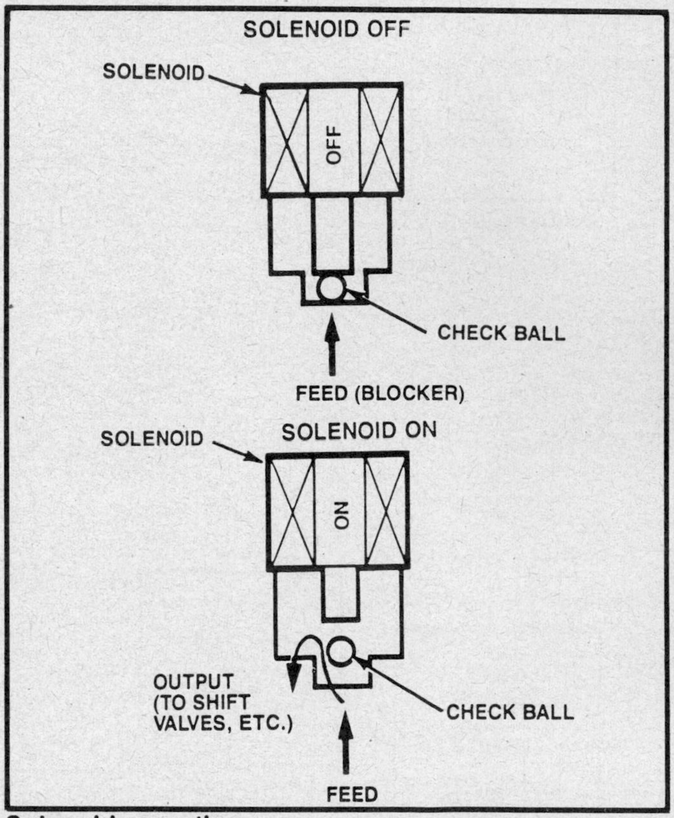
SOLENOID OFF
SOLENOID
OFF
CHECK BALL
FEED (BLOCKER)
SOLENOID
SOLENOID ON
ON
OUTPUT
(TO SHIFT
VALVES, ETC.)
CHECK BALL
FEED

Solenoid operation

SOLENOID APPLICATIONS

Gear Selector Position	Gear	Shift Control Solenoid 1	Shift Control Solenoid 2	Converter Clutch Control Solenoid 3	Coast Clutch Control Solenoid 4
OD	4	Off	Off	On/Off	Off
	3	Off	On	On/Off	Off
	2	On	On	On/Off	Off
	1	On	Off	On/Off	Off
OD	3	Off	On	Based on EEC-IV Strategy	On
Overdrive Cancel	2	On	On	Based on EEC-IV Strategy	On
Switch Pressed	1	On	Off	Based on EEC-IV Strategy	On
2	2	Off	Off	Based on EEC-IV Strategy	Off
1	1	On	Off	Off	Off

LINE PRESSURE SPECIFICATIONS CHART

Engine	Range	Idle		Stall	
		psi	kPa	psi	kPa
5.8L	P, N	55–65	379–448	—	—
	R	85–110	586–758	240–265	1655–1827
	OD, 2	55–65	379–448	156–174	1076–1200
	1	74–99	510-682	157–182	1082–1255
7.3L Diesel	P,N	55–65	379–448	—	—
	R	105–130	723–896	240–265	1655–1827
	OD, 2	55–65	379–448	156–174	1076–1200
	1	74–99	510–682	161–186	1110–1282
7.5L	P, N	55–65	379–448	—	—
	R	90–115	621–792	240–265	1655–1827
	OD, 2	55–65	379–448	156–174	1076–1200
	1	74–99	510–682	157–182	1082–1255

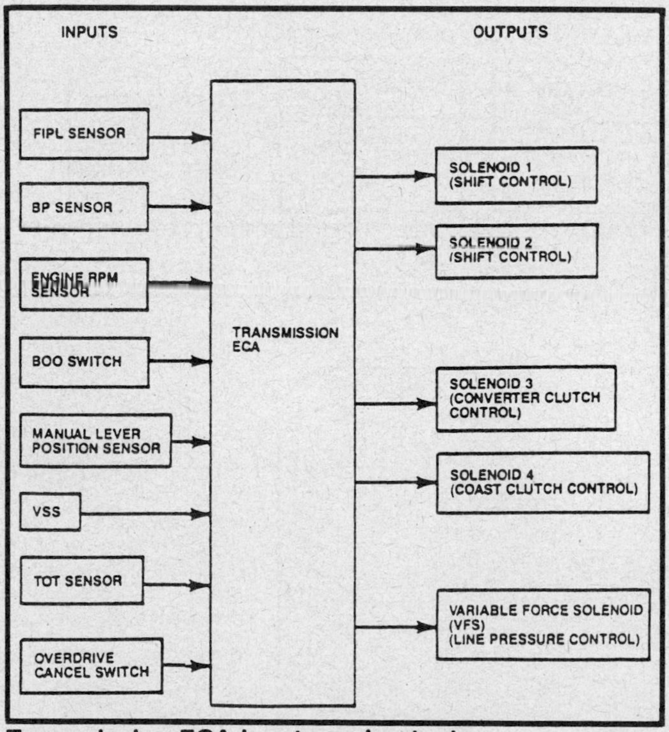
INPUTS
OUTPUTS
FIPL SENSOR
BP SENSOR
ENGINE RPM SENSOR
BOO SWITCH
MANUAL LEVER POSITION SENSOR
VSS
TOT SENSOR
OVERDRIVE CANCEL SWITCH
TRANSMISSION ECA
SOLENOID 1 (SHIFT CONTROL)
SOLENOID 2 (SHIFT CONTROL)
SOLENOID 3 (CONVERTER CLUTCH CONTROL)
SOLENOID 4 (COAST CLUTCH CONTROL)
VARIABLE FORCE SOLENOID (VFS) (LINE PRESSURE CONTROL)

Transmission ECA inputs and outputs

forward clutch, overdrive one-way clutch or low/reverse one-way clutch. No manual low (1) coast could be caused by an inoperative coast clutch or the reverse clutch. Failure to move in reverse gear could be caused by a problem in the reverse clutch, overdrive one-way clutch or the direct clutch.

1. Raise the vehicle and safely support.
2. Drain the transmission fluid and remove the oil pan.
3. Remove the filter and the seal asembly, solenoid body and the main control assemblies.
4. The inoperative clutches can be located by introducting air pressure into the various test passages.

NOTE: A dull thud can be heard, or movement of the piston felt when clutch piston is applied. If clutch seal(s) are leaking, a hissing sound will be heard.

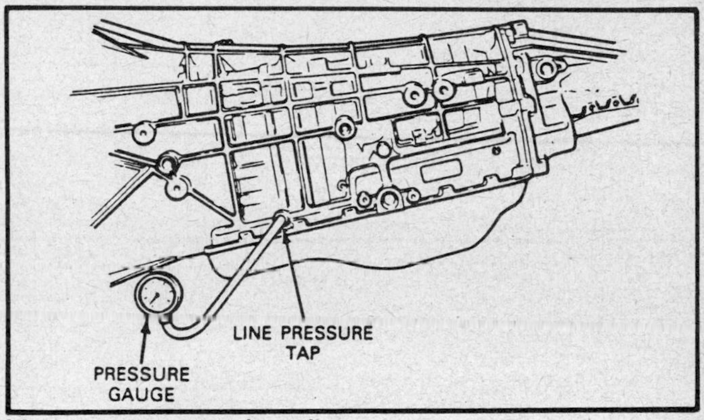
LINE PRESSURE TAP
PRESSURE GAUGE

Line pressure tap location

STALL TEST

The stall test checks the operation of the following items: converter one-way clutch, forward clutch, low/reverse one-way clutch, reverse clutch, overdrive one-way clutch, direct clutch, coast clutch and the engine performance.

CAUTION

Do not allow any one to stand either in front of or behind the vehicle during the stall test. Personal injury could result.

1. Apply the brakes. Brock the drive wheels. Run engine to normal operating temperature then connect the tachometer to the engine.
2. Press the accelerator pedal to floor (WOT) in each transmission range. Record the rpm reached in each range. The stall speeds should be in the appropriate range. On the 5.8L engine the stall speed should be between 2100–2460 rpm. On the 7.3L Diesel engine the stall speed should be between 1650–1910 rpm and on the 7.5L engine it should be betwwen 1840–2155 rpm.
3. After testing each of the following ranges overdrive, 2, 1 and reverse, move selector lever to **N** and run engine for about 15 seconds to allow converter to cool before testing next range.

NOTE: Do not maintain wide open throttle (WOT) in any gear range for more than 5 seconds.

4. If engine rpm recorded by the tachometer exceeds maxi-

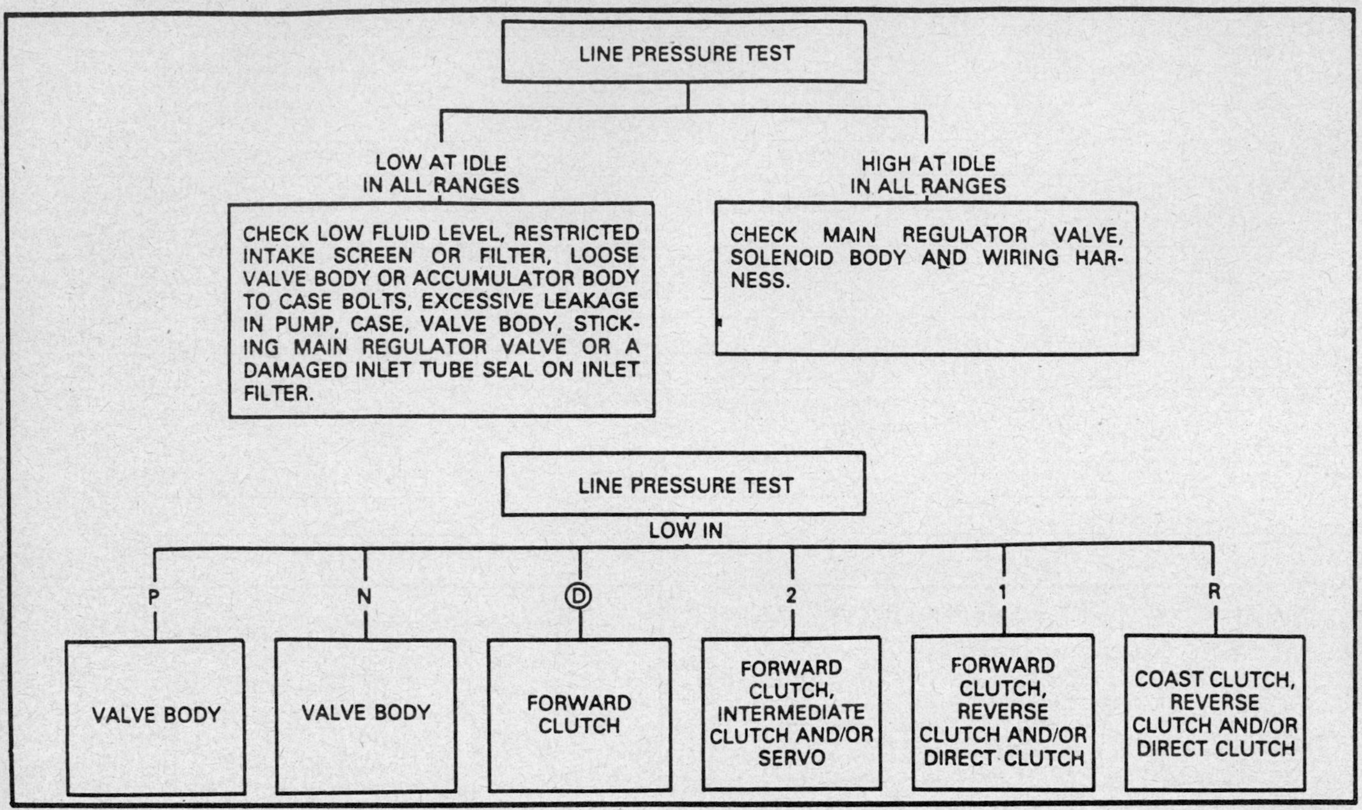
LINE PRESSURE TEST
LOW AT IDLE IN ALL RANGES
CHECK LOW FLUID LEVEL, RESTRICTED INTAKE SCREEN OR FILTER, LOOSE VALVE BODY OR ACCUMULATOR BODY TO CASE BOLTS, EXCESSIVE LEAKAGE IN PUMP, CASE, VALVE BODY, STICKING MAIN REGULATOR VALVE OR A DAMAGED INLET TUBE SEAL ON INLET FILTER.
HIGH AT IDLE IN ALL RANGES
CHECK MAIN REGULATOR VALVE, SOLENOID BODY AND WIRING HARNESS.
LINE PRESSURE TEST
LOW IN
P
N
D
2
1
R
VALVE BODY
VALVE BODY
FORWARD CLUTCH
FORWARD CLUTCH, INTERMEDIATE CLUTCH AND/OR SERVO
FORWARD CLUTCH, REVERSE CLUTCH AND/OR DIRECT CLUTCH
COAST CLUTCH, REVERSE CLUTCH AND/OR DIRECT CLUTCH

Line pressure test chart

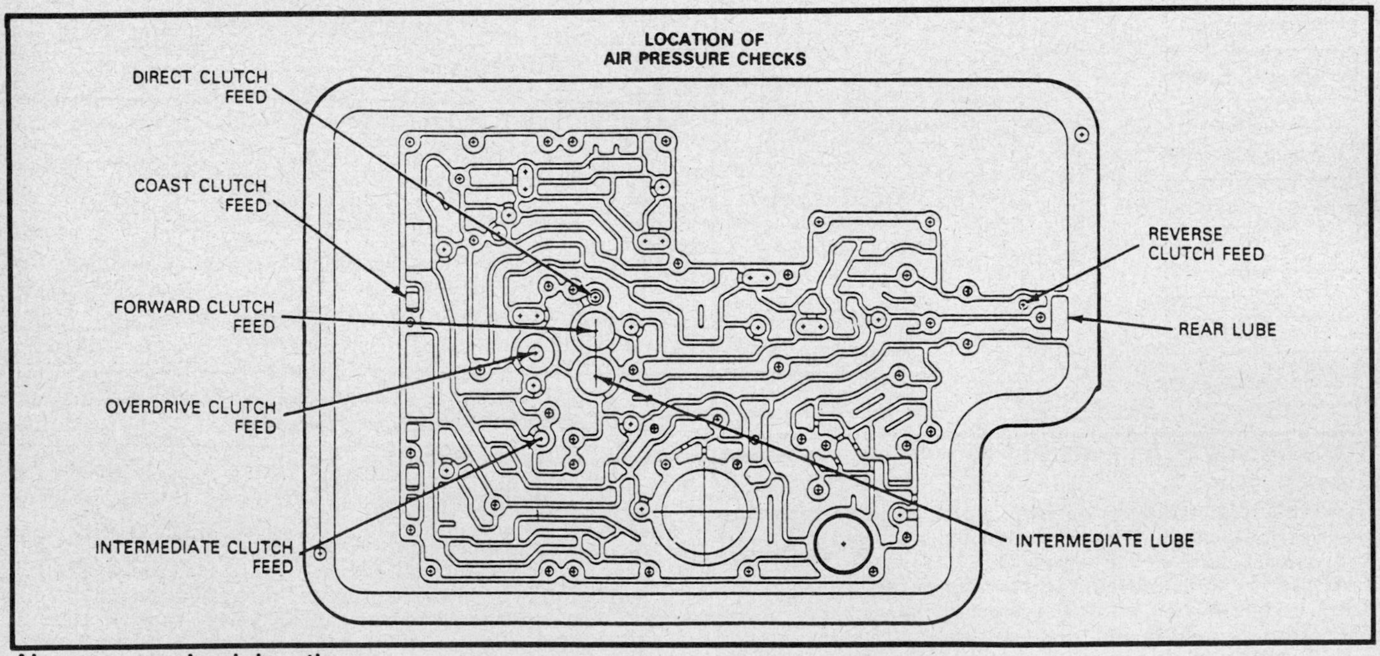
LOCATION OF AIR PRESSURE CHECKS
DIRECT CLUTCH FEED
COAST CLUTCH FEED
FORWARD CLUTCH FEED
OVERDRIVE CLUTCH FEED
INTERMEDIATE CLUTCH FEED
REVERSE CLUTCH FEED
REAR LUBE
INTERMEDIATE LUBE

Air pressure check locations

mum specified rpm, then release the accelerator pedal immediately. Clutch or band slippage is indicated by this test result.

5. If the stall speeds were too high, refer to the stall speed diagnosis chart. If the stall speeds were too low, check engine performance. If engine performance is correct, remove the torque converter and check the torque converter reactor one-way clutch for slippage.

Range	Possible Source
Ⓓ	Forward Clutch Overdrive One-Way Clutch Low/Reverse One-Way Clutch
R	Direct Clutch Overdrive One-Way Clutch Reverse Clutch
2	Forward Clutch Overdrive One-Way Clutch Intermediate Clutch Intermediate One-Way Clutch
1	Forward Clutch Reverse Clutch and Low/Reverse One-Way Clutch Coast Clutch and Overdrive One-Way Clutch

Stall speed diagnosis chart

TRANSMISSION FLUID COOLER FLOW TEST

The transmission linkage adjustment, fluid level and line pressure must be within specifications before performing this test.

1. Remove the dipstick from filler tube and place a funnel in the filler tube.
2. Raise the vehicle and support safely.
3. Remove the cooler return line (rear fitting) from fitting on transmission case and connect a hose to the cooler return line. Insert other end of hose into the funnel in the dipstick tube.
4. Start the engine and run at idle with transmission in **N** range.
5. Observe the fluid flow at the funnel. When fluid flow is solid, the flow should be liberal and the test is completed.
6. If the flow is not liberal, stop the engine. Disconnect the hose from the cooler return line and connect it to the converter out line fitting (front fitting) on the transmission case.
7. Start engine and observe the fluid flow. If fluid flow is not liberal, check the transmission oil cooler for blockage and or the transmission oil pump assembly.

SHIFT POINT TEST

This test verifies that the shift control system is operating properly.

Road Test

1. Bring engine transmission up to normal operating temperature.
2. Operate the vehicle in the overdrive range.
3. Apply minimum throttle pressure and observe the upshift speeds and speeds at which the converter and clutch apply.
4. With vehicle in overdrive (4th gear), depress overdrive cancel switch. The transmission should downshift into 3rd gear.
5. Depress the accelerator pedal to the floor (WOT). The transmission should shift from 3rd to 2nd, or 3rd to 1st depending on vehicle speed and the converter clutch should release and then reapply.
6. With vehicle in the overdrive range above 50 mph and less than half throttle, move transmission selector from overdrive range to the 2nd range and release accelerator pedal. The transmission should immediately downshift into 2nd gear. With vehicle remaining in 2nd gear range, move transmission selector into 1st gear range, and release accelerator pedal. Transmission should downshift into 1st gear at speeds below 30–35 mph.
7. If transmission fails to upshift or downshift, adjustment and or repair is necessary.

In-Shop Test

1. Raise and support the vehicle safely. Run the engine to the normal operating temperature is reached.

NOTE: Do not exceed 60 mph indicated speedometer speed. Do not exceed recommended tire speed rating.

2. To check shift valves, place the transmission in the overdrive range. Apply throttle pressure and observe the upshift speeds.
3. At the shift points, the speedometer needle will make a momentary surge, a slight drive line bump may be felt and the engine speed will drop without releasing the accelerator pedal.
4. If transmission fails to upshift or downshift adjustment and or repair is necessary.

ELECTRICAL TESTS FOR GASOLINE ENGINES

When performing trouble diagnosis to E40D transmission, perform the Electronic Engine Control (EEC–IV) Quick Test. This will determine if any service codes for the transmission exist. These service codes may appear during the EEC–IV Quick Test. Service these codes first and repeat the EEC–IV Quick Test before continuing with the transmission diagnosis.

EEC–IV Quick Test Service Codes

CODE 26
TOT SENSOR OUT OF SELF-TEST RANGE

The transmission oil temperature (TOT) sensor registers a temperature not in the allowable range of testing. The test should be repeated with the transmission heated to the correct testing temperature.

CODE 47
4WD SWITCH CLOSED

The transmission transfer case is activated into 4WD drive. Release the 4WD drive and repeat the test.

CODE 65
OVERDRIVE CANCEL SWITCH NOT CHANGING STATE

The operation of the overdrive cancel switch was not recorded during the Engine On Quick Test. Service the switch and or wiring connections

CODE 67
MANUAL LEVER POSITION SENSOR OUT OF RANGE/AC ON

If the AC clutch is on during the test, this code will appear. Shut off the AC or defrost and repeat the test. If the AC unit was off during the test, go to Transmission Quick Test Service Code 67.

Transmission Quick Test Service Codes

NOTE: If any of the following service codes appear during the EEC–IV Quick Test, perform the Drive Cycle Test for continuous codes.

CODE 49
1–2 SHIFT ERROR

The engine speed drop during the 1 to 2 shift. It does not fall within tolerance limits.

CODE 56
TRANSMISSION OIL TEMPERATURE SENSOR INDICATED -40°F SENSOR CIRCUIT OPEN

The voltage drop across the transmission oil temperature sensor exceeds the scale set for the temperature of -40°F.

CODE 59
2–3 SHIFT ERROR

The engine speed drop during the 2 to 3 shift. It does not fall within tolerance limits.

CODE 62
CONVERTER CLUTCH FAILURE

The EEC–IV module picks up excessive amount of converter slip while converter is scheduled to be locked up.

CODE 66
TRANSMISSION OIL TERMPERATURE OIL SENSOR INDICATED 315°F SENSOR CIRCUIT GROUNDED

The voltage drop across the transmission oil temperature sensor does not reach the scale set for the temperature of 315°F.

CODE 67
MANUAL LEVER POSITION SENSOR OUT OF RANGE/ AC ON

The indicated voltage drop across the MLPS (manual lever position sensor) exceeds the limits established for each position. With AC or defrost on the fault results from the AC clutch being on during Quick Test.

CODE 69
3–4 SHIFT ERROR

The engine speed drop during the 3 to 4 shift. It does not fall within tolerance limit.

CODE 91
SHIFT SOLENOID 1 CIRCUIT FAILURE

The solenoid 1 circuit fails to provide voltage drop across the solenoid.The circuit is open or shorted, or an EEC driver failure exist.

CODE 92
SHIFT SOLENOID 2 CIRCUIT FAILURE

The solenoid 2 circuit fails to provide voltage drop across the solenoid. The circuit is open or shorted, or an EEC driver failure exist.

CODE 93
CCS SOLENOID CIRCUIT FAILURE

Solenoid 4 (coast clutch solenoid) fails to provide voltage drop across the solenoid. The circuit is open or shorted, or an EEC driver failure exist.

CODE 94
CCS SOLENOID CIRCUIT FAILURE

Solenoid 3 (converter clutch solenoid) fails to provide voltage

Error Codes	Pinpoint Test
49	AA
56	BB
59	AA
62	CC
66	BB
67	EE
69	AA
91	GG
92	GG
93	GG
94	GG
98	HH
99	HH

Electrical diagnosis chart index

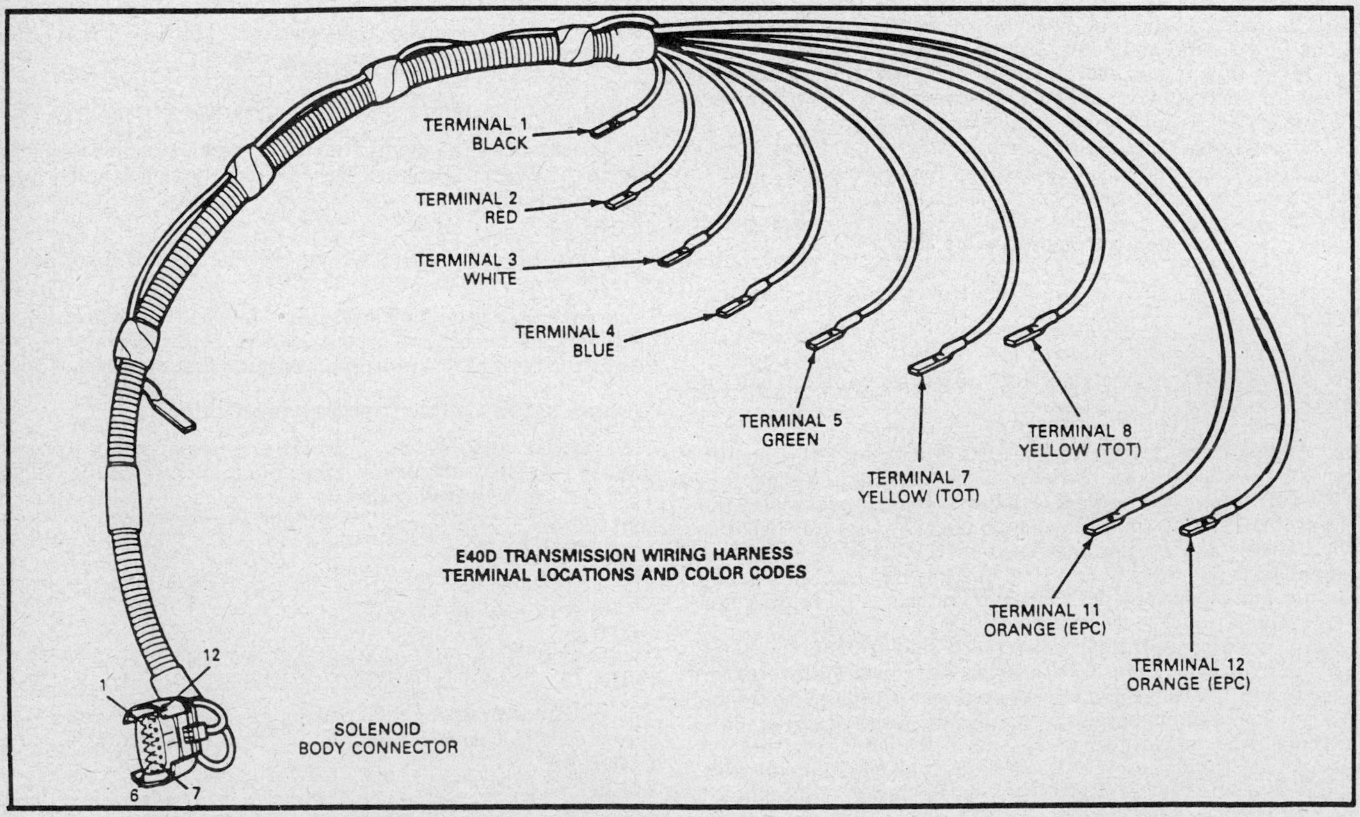
TERMINAL 1
BLACK
TERMINAL 2
RED
TERMINAL 3
WHITE
TERMINAL 4
BLUE
TERMINAL 5
GREEN
TERMINAL 7
YELLOW (TOT)
TERMINAL 8
YELLOW (TOT)
TERMINAL 11
ORANGE (EPC)
TERMINAL 12
ORANGE (EPC)
E40D TRANSMISSION WIRING HARNESS
TERMINAL LOCATIONS AND COLOR CODES
1
12
6
7
SOLENOID
BODY CONNECTOR

E40D transmission test harness

drop across the solenoid. The Circuit open or shorted, or EEC driver failure exist.

CODE 98
FAILURE MODE AND EFFECTS MAGAGEMENT FAILURE/FAILED EPC OUTPUT DRIVER

During the Quick Test, the voltage through the EPC (electronic pressure control) solenoid is checked and compared to a voltage through the solenoid after a time delay. An error will be noted if the change tolerance is exceeded.

CODE 99
EPC SOLENOID CIRCUIT FAILURE/SHORT

The voltage measured across the electronic pressure control (EPC) solenoid is less than a calculated minimum voltage.

Drive Cycle Test

After performing the EEC–IV Quick Test, the following drive cyle test for checking the E40D transmission continuous codes should be performed. Faults have to appear 4 times consecutively for continuous codes 49, 59 and 69 to be set in memory and 5 times consecutively for continuous code 62 to be set in memory.

1. Record the EEC–IV Quick Test codes.
2. Verify that the transmission fluid level is correct and engine is at operating temperature.
3. With transmission in **OVERDRIVE** range, press the overdrive cancel switch (light should illuminate) and moderately accelerate from stop to 40 mph. This will allow the transmission to shift into 3rd gear. Hold speed and throttle opening steady for a minimum of 15 seconds (30 seconds above 4000 feet altitude).
4. Depress the overdrive cancel switch (light should turn off) and accelerate from 40 mph to 50 mph). This will allow the transmission to shift into 4th gear. Hold speed and the throttle position steady for a minimum of 15 seconds.
5. With transmission in 4th gear, maintaining steady speed and throttle opening, lightly apply and release the brake (to operate brake lights). Then hold speed and throttle steady for an additional 15 seconds minimum.
6. Bring vehicle to a stop and remain stopped for a minimum of 20 seconds with the transmission in **OVERDRIVE** range.
7. Perform test at least 5 times then perform EEC–IV Quick Test and record continuous codes. If the codes appear, refer to the Electrical Diagnosis Chart Index for appropriate Pinpoint Test.

NOTE: If any other service codes appear, service those codes first as they could affect the electrical operation of the transmission. After the servicing of any error codes perform the Quick Test again to recheck for the fault.

SERVICE CODES: 49, 59 AND 69 — PINPOINT TESTS AA

TEST STEPS		RESULTS ▶	ACTION TO TAKE
AA1	CHECK HARNESS CONNECTIONS		
	Check that the vehicle harness connector is fully engaged on the transmission bulkhead connector. Check that the vehicle harness connector terminals are fully engaged in the connector.	OK ▶	GO to **AA2**.
		NOT OK ▶	SERVICE or REPLACE as required. REPEAT QUICK TEST
AA2	CHECK RESISTANCE OF SOLENOID		
	NOTE: Refer to the E4OD Transmission Wiring Harness Terminal Locations and Color Codes preceding these Pinpoint Tests. Install service jumper harness to the transmission bulkhead connector. (Do not pry vehicle harness connector off with a screwdriver.)* Connect ohmmeter negative lead to the black wire on the service harness and the positive lead to the white wire on the service harness. This is to test solenoid 1. Record the resistance. Resistance should be between 20-30 ohms. Connect ohmmeter negative lead to the black wire on the service harness and the positive lead to the red wire on the service harness. This is to test solenoid 2. Record the resistance. Resistance should be between 20-30 ohms.	20-30 ohms ▶	GO to **AA3**.
		High resistance ▶	REPLACE solenoid body and REPEAT QUICK TEST

SERVICE CODES: 49, 59 AND 69 — PINPOINT TESTS AA (Continued)

TEST STEPS		RESULTS ▶	ACTION TO TAKE
AA3	CHECK SOLENOID FOR SHORT TO GROUND		
	Install service jumper harness to transmission bulkhead connector. (Do not pry vehicle harness connector off with a screwdriver.)* Check for continuity between an engine ground and appropriate wire with an ohmmeter or other low current tester (less than 200 milliamps). Solenoid 1 — Wire White; Solenoid 2 — Wire Red Connection should show no continuity (infinite resistance).	Continuity ▶	REPLACE Solenoid Body. REPEAT QUICK TEST.
		No continuity ▶	GO to **AA4**.
AA4	CHECK SOLENOID REGULATOR VALVE		
	Tear down to solenoid regulator valve. Inspect solenoid regulator valve for damage or contamination. Check for stuck or missing spring.	OK ▶	CLEAR errors and REPEAT QUICK TEST
		NOT OK ▶	Service as required.

Solenoid	Wire
1	White
2	Red

*Remove solenoid body connector by pushing on the center tab and pulling on the wiring harness. Do not attempt to pry tab with a screwdriver. Remove heat shield from transmission before removing connector.

SERVICE CODES: 56 AND 66 — PINPOINT TESTS BB

TEST STEPS		RESULTS ▶	ACTION TO TAKE
BB1	CHECK HARNESS CONNECTIONS		
	Check that the vehicle harness connector is fully engaged on the transmission bulkhead connector. Check that the vehicle harness connector terminals are fully engaged in the connector.	OK ▶	GO to **BB2**.
		NOT OK ▶	SERVICE or REPLACE as required. REPEAT QUICK TEST
BB2	CHECK TOT SENSOR RESISTANCE		
	NOTE: Refer to the E4OD Transmission Wiring Harness Terminal locations and Color Codes preceding these Pinpoint Tests. Install service jumper harness to the transmission bulkhead connector. (Do not pry vehicle harness connector off with a screwdriver.)*	Resistance in range ▶	GO to **BB3**.
		Resistance greater than 100K ▶	REPLACE solenoid body and REPEAT QUICK TEST

SERVICE CODES: 56 AND 66 — PINPOINT TESTS BB

TEST STEPS	RESULTS	ACTION TO TAKE
BB2 CHECK TOT SENSOR RESISTANCE		
Carefully touch the transmission oil pan on the driver's side, away from the exhaust system, to approximate the temperature. After running the Quick Test, the transmission oil pan should be warm to the touch. (As a guide, warm to the touch is about 41-70 degrees C [105-158 degrees F]). Connect ohmmeter negative lead and the positive lead to the yellow wires on the service harness. Record the resistance. Resistance should be approximately in the following ranges. (See table below.) If the resistance was not the appropriate temperature range but was between 0.8K and 100K ohms, perform the following test. If the transmission is cold, run the transmission to heat it up. If the transmission is warm, allow the transmission to cool. Check TOT sensor resistance again. Compare the resistance with the initial resistance. Resistance should decrease if transmission was heated and should increase if transmission was allowed to cool. If the correct change in resistance occurs, REPEAT QUICK TEST.	Resistance out of range	▶ PERFORM SECOND TEST listed in this step. REPEAT QUICK TEST
BB3 CHECK TOT SENSOR FOR SHORT TO GROUND		
Install service jumper harness to transmission bulkhead connector. (Do not pry vehicle harness connector off with a screwdriver.)* Check tor continuity between engine ground and one yellow wire with an ohmmeter or other low current tester (less than 200 milliamps). Repeat the continuity check with the other yellow wire. Connection should show no continuity (infinite resistance).	Continuity	▶ REPLACE solenoid body and REPEAT QUICK TEST.
	No continuity	▶ If code was a continuous code, inspect transmission fluid to determine if fluid is burnt. If burnt, teardown transmission and inspect for damage. SERVICE as required and REPEAT QUICK TEST

TRANSMISSION FLUID TEMPERATURE

Degrees C	(Degrees F)	Resistance (Ohms)
0- 20	(32- 58)	37K- 100K
21- 40	(59-104)	16K- 37K
41- 70	(105-158)	5K- 16K
71- 90	(159-194)	2.7K- 5K
91-110	(195-230)	1.5K- 2.7K
111-130	(231-266)	0.8K- 1.5K

*Remove solenoid body connector by pushing on the center tab and pulling on the wiring harness. Do not attempt to pry tab with a screwdriver. Remove heat shield from transmission before removing connector.

SERVICE CODE: 62 — PINPOINT TEST CC

TEST STEPS		RESULTS ▶	ACTION TO TAKE
CC1	CHECK HARNESS CONNECTIONS		
	• Check that the vehicle harness connector is fully engaged on the transmission bulkhead connector. • Check that the vehicle harness connector terminals are fully engaged in the connector.	OK ▶	GO to **CC2**.
		NOT OK ▶	SERVICE or REPLACE as required. REPEAT QUICK TEST
CC2	CHECK RESISTANCE OF SOLENOID		
	NOTE: Refer to the E4OD Transmission Wiring Harness Terminal locations and Color Codes preceding these Pinpoint Tests. • Install service jumper harness to the transmission bulkhead connector. (Do not pry vehicle harness connector off with a screwdriver.)* • Connect ohmmeter negative lead to the black wire on the service harness and the positive lead to the green wire on the service harness. This is to test converter clutch solenoid. • Record the resistance. • Resistance should be between 20-30 ohms.	20-30 ohms ▶	GO to **CC3**.
		High resistance ▶	REPLACE solenoid body and REPEAT QUICK TEST
CC3	CHECK SOLENOID FOR SHORT TO GROUND		
	• Install service jumper harness to transmission bulkhead connector. (Do not pry vehicle harness connector off with a screwdriver.)* • Check for continuity between engine ground and green wire with an ohmmeter or other low current tester (less than 200 millilamps).	No continuity ▶	GO to **CC4**.
		Continuity ▶	Replace Solenoid Body and REPEAT QUICK TEST
CC4	CHECK CONVERTER CLUTCH REGULATOR VALVE AND CONVERTER CLUTCH CONTROL VALVE		
	• Tear down to converter clutch regulator valve and converter clutch control valve. • Inspect valves for damage or contamination. • Check for struck or missing spring.	OK ▶	CLEAR errors and REPEAT continuous drive tests.
		NOT OK ▶	SERVICE as required.

*Remove solenoid body connector by pushing on the center tab and pulling on the wiring harness. Do not attempt to pry tab with a screwdriver. Remove heat shield from transmission before removing connector.

SERVICE CODE: 67 — PINPOINT TEST EE

TEST STEPS		RESULTS ▶	ACTION TO TAKE
EE1	ADJUST MANUAL LEVER POSITION SENSOR		
	Apply the parking brake. Place transmission in Neutral position.	OK ▶	GO to **EE2**.
		NOT OK ▶	ADJUST sensor according to adjustment procedures
EE2	CHECK OPERATION OF MANUAL LEVER POSITION SENSOR		
	Insert Manual Lever Position Sensor test harness into the Manual Lever Position Sensor connector. Plug test box into power supply With transmission in Park, press the buttons on the box. The light should light only when the P button is pushed. Repeat the test for R, N, Ⓓ, 2 and 1.	OK ▶	REPEAT QUICK TEST
		NOT OK ▶	REPLACE Manual Lever Position Sensor and REPEAT QUICK TEST

SERVICE CODES: 91, 92, 93 AND 94 — PINPOINT TEST GG

TEST STEPS		RESULTS ▶	ACTION TO TAKE
GG1	CHECK HARNESS CONNECTIONS		
	Check that the vehicle harness connector is fully engaged on the transmission bulkhead connector. Check that the vehicle harness connector terminals are fully engaged in the connector.	OK ▶	GO to **GG2**.
		NOT OK ▶	SERVICE or REPLACE as required. REPEAT QUICK TEST
GG2	CHECK RESISTANCE OF SOLENOID		
	NOTE: Refer to the E4OD Transmission Wiring Harness Terminal locations and Color Codes preceding these Pinpoint Tests Install service jumper harness to the transmission bulkhead connector. (Do not pry vehicle harness connector off with a screwdriver.)* Connect ohmmeter negative lead to the black wire on the service harness and the positive lead to the appropriate wire on the service harness. Error Code / Wire: 91 White; 92 Red; 93 Green; 94 Blue Record the resistance. Resistance should be between 20-30 ohms.	20-30 ohms ▶	GO to **GG3**.
		High resistance ▶	REPLACE solenoid body and REPEAT QUICK TEST

Error Code	Wire
91	White
92	Red
93	Green
94	Blue

SERVICE CODES: 91, 92, 93 AND 94 — PINPOINT TEST GG

TEST STEPS	RESULTS ▶	ACTION TO TAKE
GG3 CHECK SOLENOID FOR SHORT TO GROUND		
Install service jumper harness to transmission bulkhead connector. (Do not pry vehicle harness connector off with a screwdriver.)* Check for continuity between engine ground and appropriate wire with an ohmmeter or other low current tester (less than 200 milliamps). Error Code / Wire: 91 White; 92 Red; 93 Green; 94 Blue Connection should show no continuity (infinite resistance).	Continuity ▶	REPLACE solenoid body and REPEAT QUICK TEST
	No continuity ▶	REPEAT QUICK TEST Problem should not reoccur if the solenoid body passed previous tests.

Error Code	Wire
91	White
92	Red
93	Green
94	Blue

*Remove solenoid body connector by pushing on the center tab and pulling on the wiring harness. Do not attempt to pry tab with a screwdriver. Remove heat shield from transmission before removing connector.

SERVICE CODES: 98 AND 99 — PINPOINT TEST HH

TEST STEPS	RESULTS ▶	ACTION TO TAKE
HH1 CHECK HARNESS CONNECTIONS		
Check that the vehicle harness connector is fully engaged on the transmission bulkhead connector. Check that the vehicle harness connector terminals are fully engaged in the connector.	OK ▶	GO to **HH2**.
	Not OK ▶	SERVICE or REPLACE as required. REPEAT QUICK TEST
HH2 CHECK RESISTANCE OF SOLENOID		
NOTE: Refer to the E4OD Transmission Wiring Harness Terminal locations and Color Codes preceding these Pinpoint Tests. Install service jumper harness to the transmission bulkhead connector. (Do not pry vehicle harness connector off with a screwdriver.)* Connect ohmmeter negative lead and positive lead to the orange wires on the service harness. Record the resistance. Resistance should be between 4.0-6.5 ohms.	4.0-6.5 ohms ▶	GO to **HH3**.
	High resistance ▶	REPLACE solenoid body and REPEAT QUICK TEST
HH3 CHECK SOLENOID FOR SHORT TO GROUND		
Install service jumper harness to transmission bulkhead connector. (Do not pry vehicle harness connector off with a screwdriver.)* Check for continuity between engine ground and one of the orange wires with an ohmmeter or other low current tester (less than 200 milliamps). Repeat the continuity check with the other orange wire. Connection should show no continuity (infinite resistance).	Continuity ▶	REPLACE solenoid body and REPEAT QUICK TEST
	No continuity ▶	REPEAT QUICK TEST Problem should not reoccur if the solenoid body passed previous tests.

*Remove solenoid body connector by pushing on the center tab and pulling on the wiring harness. Do not attempt to pry tab with a screwdriver. Remove heat shield from transmission before removing connector.

ON CAR SERVICES

Services

FLUID CHANGE

The conditions under which the vehicle is operated is the main consideration in determining how often the transmission fluid should be changed. Different driving conditions result in different transmission fluid temperatures. These temperatures affect change intervals.

If the vehicle is driven under severe service conditions, change the fluid and filter every 40,000 miles. If the vehicle is not used under severe service conditions, change the fluid and replace the filter every 100,000 miles.

OIL PAN

Removal and Installation

1. Raise the vehicle and support safely.
2. Place a drain pan under the transmission.
3. Loosen the oil pan attaching bolts and drain the fluid from the transmission.
4. When fluid has drained to level of pan flange, remove the rest of the pan bolts working from the right hand side and allow it to drop and drain slowly.
5. When all fluid has drained from the transmission, remove and thoroughly clean the pan. Discard the gasket.
6. Place a new gasket on the pan and install the pan on the transmission.
7. Lower the vehicle refill the transmission. Start the engine and check the fluid level.

EXTENSION HOUSING GASKET

Removal and Installation

1. Raise and support the vehicle safely.
2. Remove the front driveshaft, if equipped. Remove the rear driveshaft.
3. Remove transmission mounting pad nuts and bolts.
4. On 4WD models only, remove shift linkage from transfer case shift lever.
5. On 4WD models only, remove 4WD drive switch connector from transfer case, use care not to overextend tabs.
6. Disconnect wire harness locators from extension housing wire bracket.
7. On 4WD models only, remove wire harness locators from left hand side of the crossmember.
8. Remove the speedometer cable.
9. On 4WD models only, remove transfer case vent hose from the detent place.
10. On 4WD models only, place transmission stand fixture on universal high lift transmission jack or equivalent and position under transfer case.
11. Remove the 9 extension housing bolts using a 13mm box wrench or equivalent.
12. Slide the transfer case rearward and downward to remove, and discard extension housing gasket from housing and transfer case mating surfaces.
13. Position the extension housing gasket on the extension housing.
14. Raise the extension housing into position and install the 9 extension housing bolts. Tighten to 20–29 ft. lbs.

NOTE: On 4WD models only, attach transfer case vent hose to detent plate and install the 4WD drive connector.

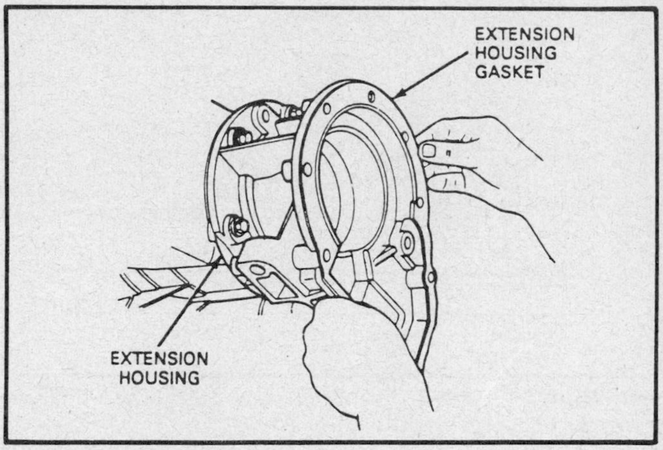
EXTENSION
HOUSING
GASKET
EXTENSION
HOUSING
Installing extension housing gasket

15. Install the wire harness locator into extension housing wire bracket.
16. Install the transmission mounting pad bolts and nuts and torque to 60–80 ft. lbs. for all F Series or 50–70 ft. lbs. for all E Series.
17. Remove the universal high lift transmission jack or equivalent. Install the speedometer cable.
18. Install the front driveshaft, if equipped. Install the rear driveshaft.
19. Start the engine and check the fluid level. Fill the transmission to the proper level with the specified fluid.

PARKING MECHANISM

Removal and Installation

1. Raise and support the vehicle safely. Remove the extension housing from the transmission.
2. Remove the 2 bolts using a 13mm socket or equivalent from the park rod guide plate.
3. Remove the parking pawl return spring, pin and parking pawl from the transmission case.
4. Remove the Torx® head bolt (40A bit) and the parking pawl abutment.
5. Install the parking pawl, pin and the return spring. The spring end rests on the inside surface of case.
6. Install the parking pawl abutment with Torx® head bolt (40A bit). Torque to 16–20 ft. lbs.
7. Using a 13mm socket or equivalent, attach the park rod guide plate with 2 bolts and washers. Tighten to 16–20 ft. lbs. Check that the plate dimple is facing inward.
8. Install the extension housing and refill the transmission. Start engine and check the fluid level.

VALVE BODY AND INTERMEDIATE BAND SERVO

Removal and Installation

1. Raise and support the vehicle safely. Remove the solenoid body connector heat shield, loosen both bolts using an 8mm socket or equivalent.
2. Remove the slotted heat shield and the solenoid body connector by pushing on the center tab and pulling on the wire harness.

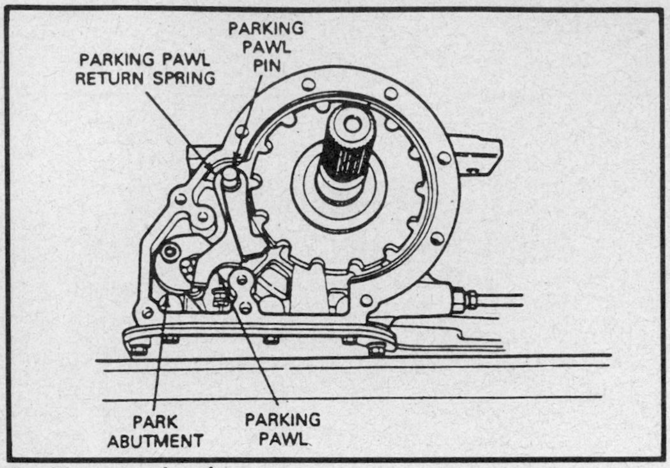
PARKING PAWL RETURN SPRING
PARKING PAWL PIN
PARK ABUTMENT
PARKING PAWL
Parking mechanism

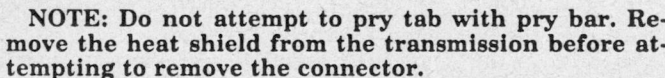

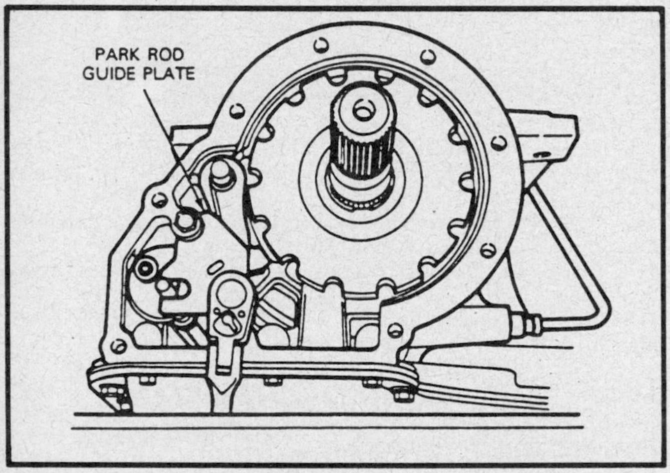
PARK ROD GUIDE PLATE
Removing park rod guide plate

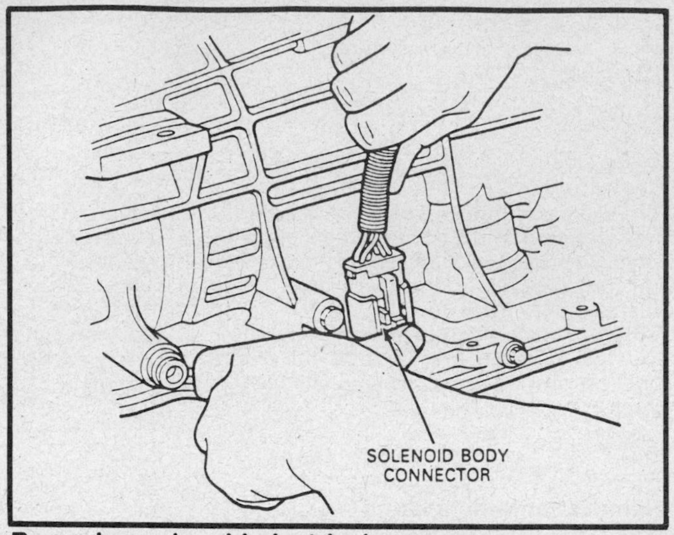
SOLENOID BODY CONNECTOR
Removing solenoid electrical connector

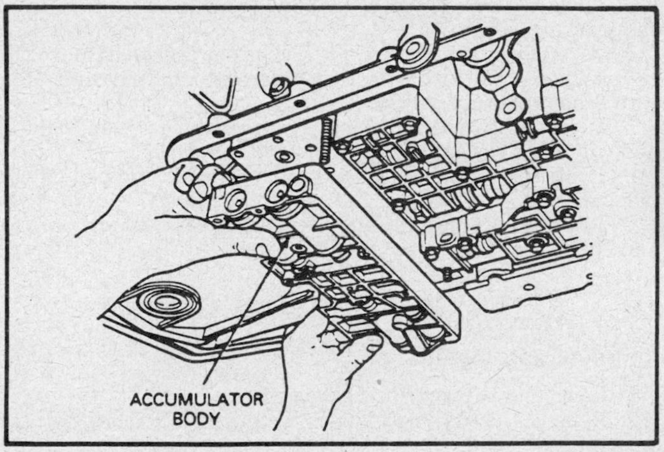
ACCUMULATOR BODY
Removing accumulator body

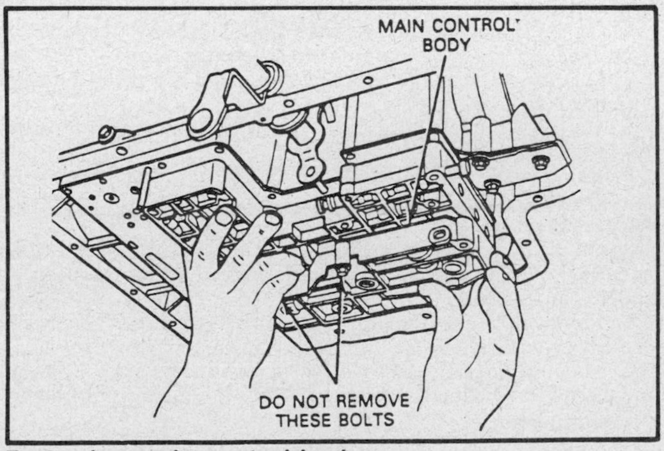
MAIN CONTROL BODY
DO NOT REMOVE THESE BOLTS
Removing main control body

NOTE: Do not attempt to pry tab with pry bar. Remove the heat shield from the transmission before attempting to remove the connector.

3. Check the electrical connectors for terminal conditions, corrosion and contamination. Service as required.
4. Remove the 15 back and side pan bolts using a 10mm socket or equivalent.
5. Place a drain pan or equivalent under the transmission pan. Loosen transmission pan bolts and remove the transmission pan.
6. Remove the filter and seal assembly by carefully pulling and rotating the filter as necessary. If seal remains in bore, carefully remove using O-ring tool T71P–19703–C or equivalent. Discard the filter and seal.
7. Remove the 11 accumulator body bolts using an 8mm socket or equivalent and 2 nuts using a 10mm socket or equivalent. Remove the accumulator body.
8. Remove the 14 main control body bolts using an 8mm socket or equivalent and 2 nuts using a 10mm socket or rquivalent. Remove the main control body.

NOTE: Do not remove the 2 center bolts on the main control body.

9. Remove the 9 solenoid body bolts using a Torx® (30A bit) and 1 nut using a 10mm socket or equivalent. Push down on the solenoid body receptacle to remove solenoid body.
10. Remove the solenoid screen by turning counterclockwise and pulling out. Remove the 3 reinforcing plate bolts using an 8mm socket or equivalent. Remove the plate.

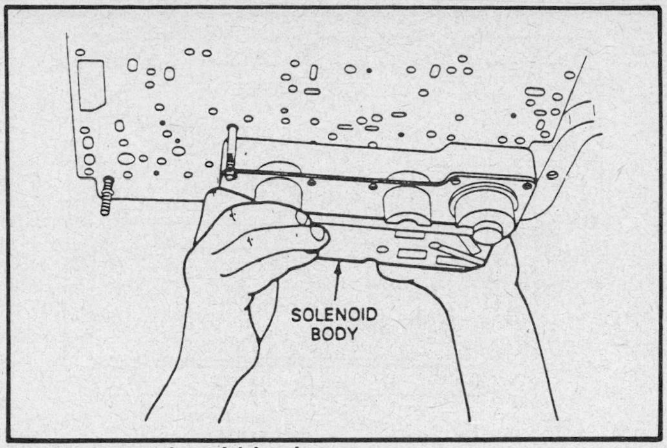
SOLENOID
BODY
Removing solenoid body

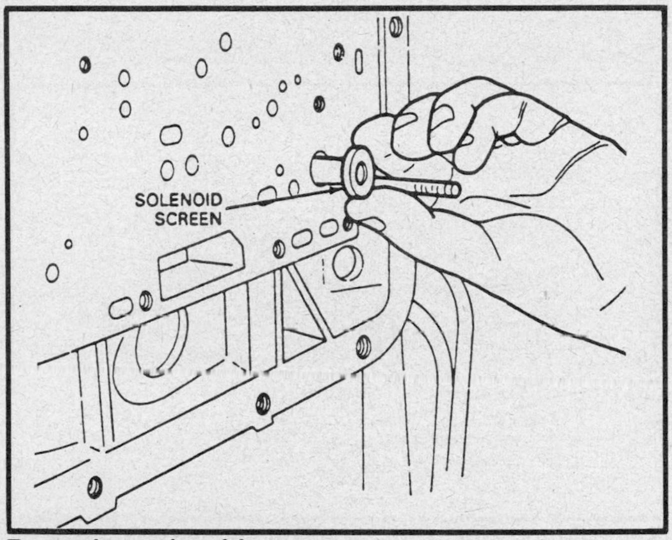
SOLENOID
SCREEN
Removing solenoid screen

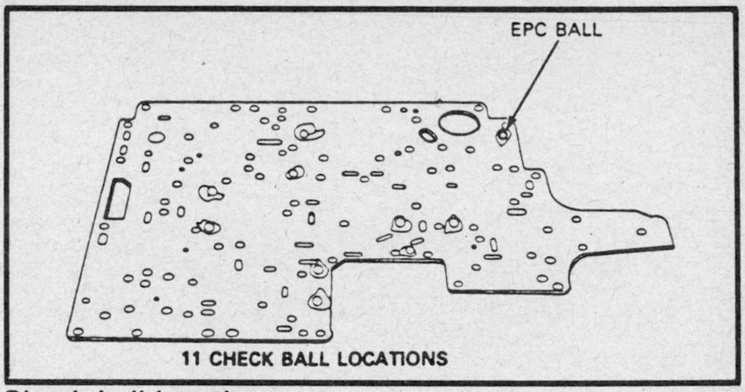
EPC BALL
11 CHECK BALL LOCATIONS
Check ball locations

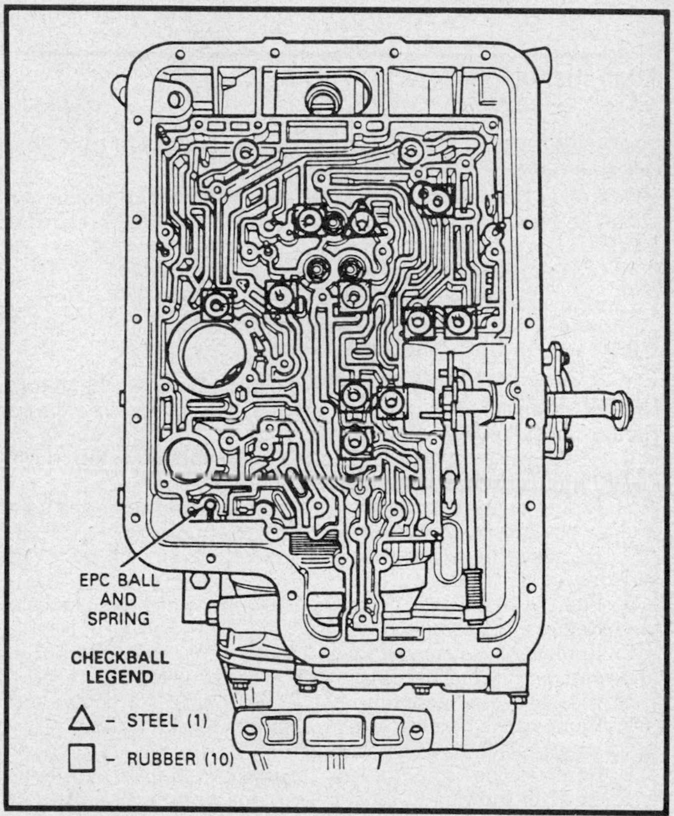
EPC BALL
AND
SPRING
CHECKBALL
LEGEND
△ - STEEL (1)
□ - RUBBER (10)
Check ball locations

11. Carefully lower the separator plate and gasket so that check balls, EPC ball and spring are retained.
12. Remove the servo snapring and retaining plate, piston and rod assembly and servo spring.
13. Install the servo spring, servo piston and rod assembly.
14. Install the servo retaining plate and snapring. Grease the separator plate with petroleum jelly to hold new separator to control gasket.
15. Position the new separator to case gasket on separator plate.
16. Lubricate the separator case gasket with petroleum jelly.
17. Grease the valve body pockets with petroleum jelly. Put 11 check balls (10 rubber and 1 steel), EPC spring and greased (petroleum jelly) EPC ball into position.
18. Install the separator plate and gaskets. Install the 3 reinforcing plate bolts using an 8mm socket or equivalent. Tighten to 80–100 inch lbs.

NOTE: Check the location of check balls and EPC ball at this point of the installation. The stamped marking UP on the reinforcing plate must be visible.

19. Install the solenoid screen and lock in place by turning clockwise. Install the main control body over the studs. Align the manual valve with the manual lever.
20. Attach the valve body with 2 nuts and 14 bolts. Torque to 80–100 inch lbs.
21. Install the accumulator body over the studs and attach with 2 nuts and 11 bolts. Torque to 80–100 inch lbs.
22. Install the solenoid body over the stud and attach with 9 Torx® (30A bit) bolts and 1 nut. Torque to 80–100 inch lbs.

NOTE: Prior to installing the solenoid body assembly, coat the case connector bore with M1C172-A grease or equivalent.

23. Install a new filter and a seal assembly by lubricating the seal with transmission fluid and pressing the filter into place. Position the gasket onto pan and check the condition and placement of pan magnet.
24. Install the 20 pan bolts. Tighten to 10–12 ft. lbs.
25. Completely seat the solenoid body connector into the solenoid valve body receptacle. An audible click sound indicates proper installation of this connector.

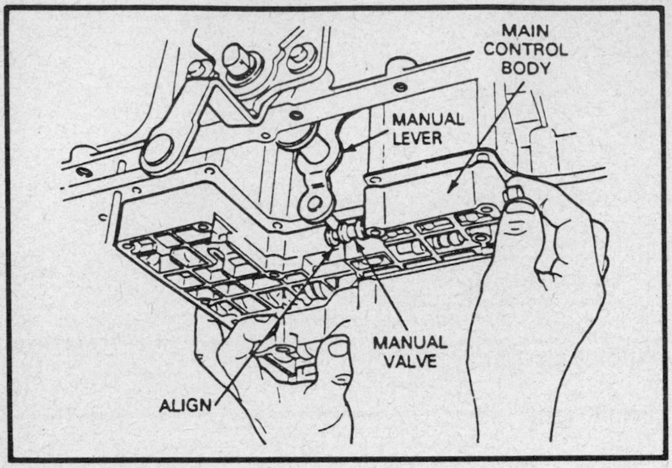
MAIN
CONTROL
BODY
MANUAL
LEVER
MANUAL
VALVE
ALIGN
Align manual valve with manual lever

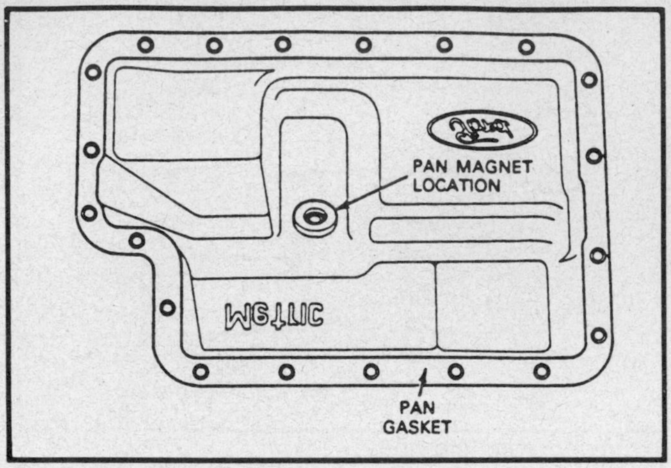
PAN MAGNET
LOCATION
METRIC
PAN
GASKET
Pan magnet location

26. Install the solenoid body connector heat shield with offset bending inward. Tighten to 6–9 ft. lbs.
27. Lower the vehicle. Fill the transmission to the proper level with the specified fluid. Start the engine and check the fluid level.

MANUAL LEVER SEAL

Removal and Installation

1. Raise and support the vehicle safely. Remove the manual lever position sensor connector by squeezing connector tabs and pulling on the connector harness.
2. Check the electrical connectors for terminal condition, corrosion and contamination. Service as requirerd.
3. Remove the lever control rod from the manual lever using a large pry bar or equivalent.
4. Remove the 2 manual lever position sensor bolts and sensor.
5. Place a drain pan or rquivalent under the transmission pan. Remove the transmission pan.
6. Remove the filter and seal assembly by carefully pulling and rotating filter as necessary. If seal remains in bore, carefully remove using O-ring using tool T71P–19703–C or equivalent.
7. Remove the manual lever roll pin using the locknut pin remover tool T78P–3504–N or equivalent.
8. Remove the inner detent lever nut using a 21mm box wrench while holding lever with crescent wrench or equivalent.
9. Remove the inner detent lever and park actuating rod assembly from manual lever. Remove manual the lever.
10. Remove the manual lever seal using a pry tool, being careful not to score the transmission bore.
11. Clean bore opening with cleaner and install a new seal using shift lever seal replacer tool T74P–77498–A or equivalent.
12. Install the manual lever, inner detent lever, park actuating rod assembly and nut. The inner detent lever must be sealed on flats of the shaft and rod assembly must be through guide plate. The inner lever pin must be aligned with manual lever.
13. Torque the inner detent lever nut using a 21mm crows foot tool while holding lever with a crescent wrench or equivalent. Torque to 30–40 ft. lbs.
14. Install the manual lever roll pin and manual lever position sensor with 2 bolts and washers. Do not tighten bolts at this time. Align manual lever position sensor for **N** gear position using gear position sensor adjuster tool T89T–70010–J or equivalent. Tighten bolts to 55–75 inch lbs.
15. Install the manual lever position sensor connector. An audible click sound indicates proper installation.
16. Install the shift linkage.

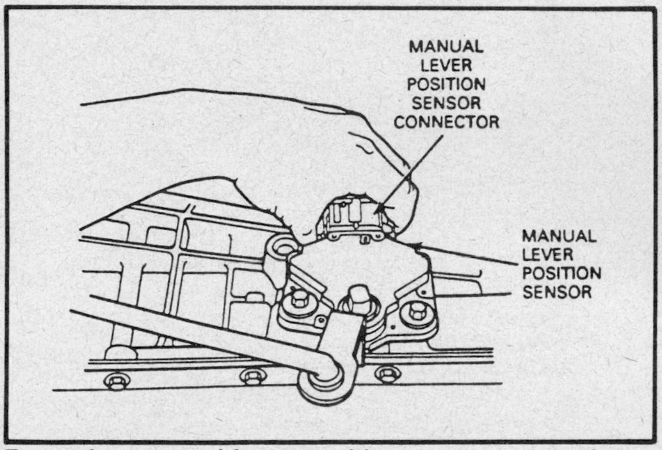
MANUAL
LEVER
POSITION
SENSOR
CONNECTOR
MANUAL
LEVER
POSITION
SENSOR
Removing manual lever position sensor connector

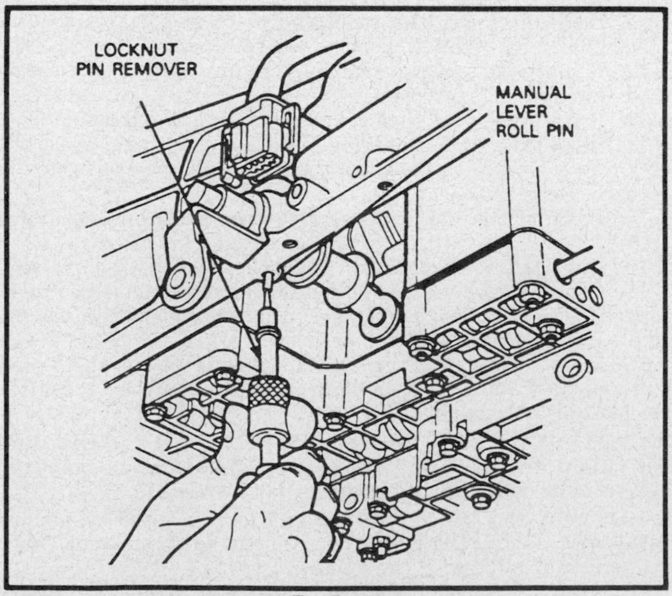
LOCKNUT
PIN REMOVER
MANUAL
LEVER
ROLL PIN
Removing manual lever roll pin

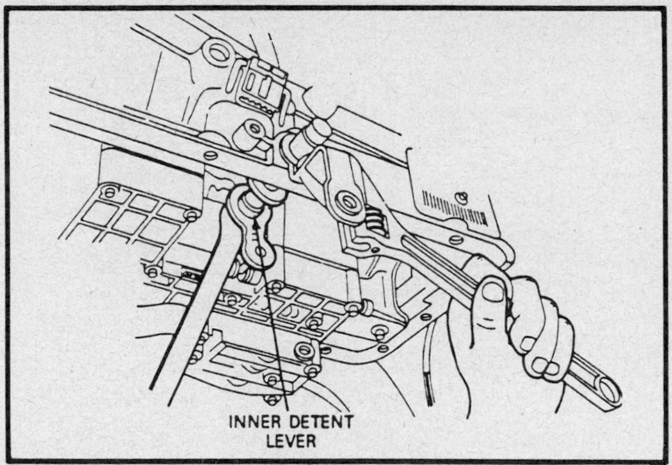
INNER DETENT
LEVER
Removing inner detent lever nut

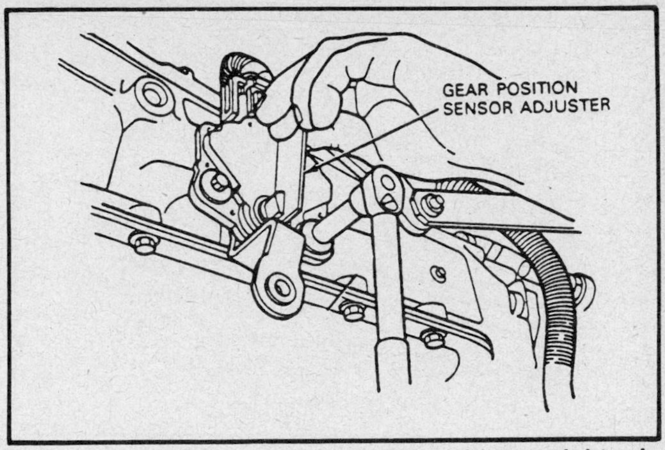
GEAR POSITION
SENSOR ADJUSTER
Align manual lever position sensor with special tool

17. Install a new filter and seal assembly by lubricating the seal with transmission fluid and pressing the filter into place. Using petroleum jelly to hold new pan gasket, position gasket onto pan. Check condition and placement of pan magnet.

18. Install the transmission pan. Torque to 10–12 ft. lbs.
19. Lower the vehicle and refill the transmission. Start the engine and check the fluid level.

REMOVAL AND INSTALLATION

TRANSMISSION REMOVAL

1. Disconnect negative battery cable at the battery. Remove the transmission dipstick.
2. Place transmission selector in **N** position. Raise and safely support the vehicle.
3. On 4WD models only, remove the front driveshaft. Remove the rear driveshaft. On F-Super Duty vehicles, remove the transmission mounted parking brake.
4. Disconnect the shift linkage. On 4WD models only, remove the shift linkage from transfer case shift lever.
5. Remove the manual lever position sensor connector by squeezing connector tabs and pulling on connector.

NOTE: Do not attempt to pry tab with pry bar. Remove the heat shield from the transmission before attempting to remove the connector.

6. Remove the solenoid body connector heat shield.
7. Remove the solenoid body connector by pushing on the center tab and pulling on the wire harness.
8. On 4WD models only, remove the 4WD drive switch connector from the transfer case. Use care not to overextend tabs.
9. Remove the wire harness locator from the extension housing wire bracket. On 4WD models only, remove the wire harness locators from left hand side of the crossmember.
10. Remove the speedometer cable and the lower converter bolts.
11. Remove the rear engine cover plate bolts. Remove the starter.
12. Using a $^{15}/_{16}$ socket or equivalent, rotate the crankshaft bolt to gain access to converter nuts. Remove the 4 converter mounting nuts and discard the nuts.
13. Place a transmission stand fixture tool 014–00763 or equivalent on a universal transmission jack and position under the transmission.

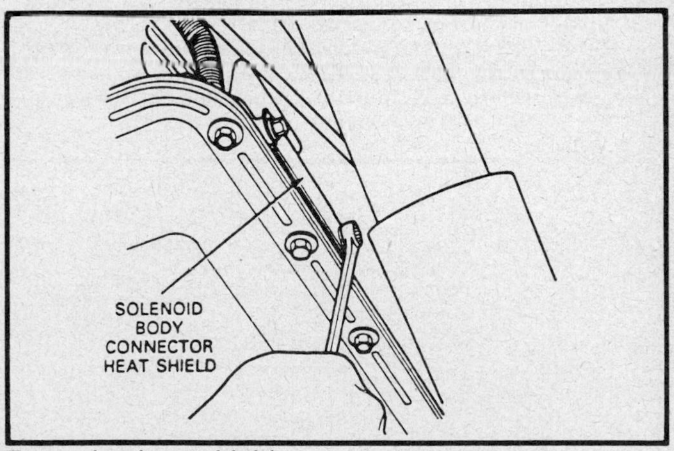
SOLENOID
BODY
CONNECTOR
HEAT SHIELD
Removing heat shield

CAUTION

Use a safety strap to secure the transmission to the transmission stand fixture.

14. Loosen the 2 rear transmission mounting pad nuts. Remove the retaining bolts and remove the crossmember from the transmission.
15. Remove the transmission cooling lines from the case. Cap cooling lines and plug fittings at transmission.
16. Remove the 6 bell housing bolts. Back out the converter pilot from the flywheel and gently lower the transmission while observing for obstructions.
17. Install torque converter handles, T81P–8902–C or equivalent on the converter with handles in the 6 and 12 o'clock positions.

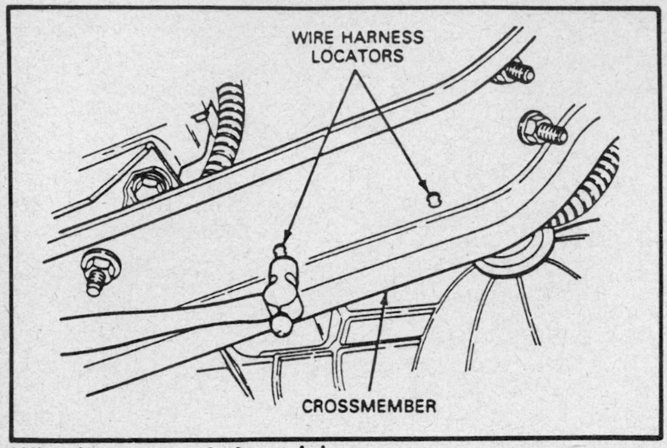
WIRE HARNESS
LOCATORS
CROSSMEMBER

Wire harness—4x4 models

18. Remove the transmission filler tube. On 4WD models only, remove the transfer case vent hose form detent bracket and the transfer case from the transmission. On F-Super Duty models, remove the transmission mounted parking brake.

TRANSMISSION INSTALLATION

1. Place the transmission onto a transmission stand fixture tool 014–4–763 or equivalent.
2. On 4WD models only, install the transfer case to transmission. On F-Super Duty models, install transmission mounted parking brake.
3. Install the torque converter using torque converter handles T81P–7902–C or the equivalent. Carry the converter with the handles in the 6 and 12 o'clock positions. Push and rotate the converter onto the pump until it bottoms out.

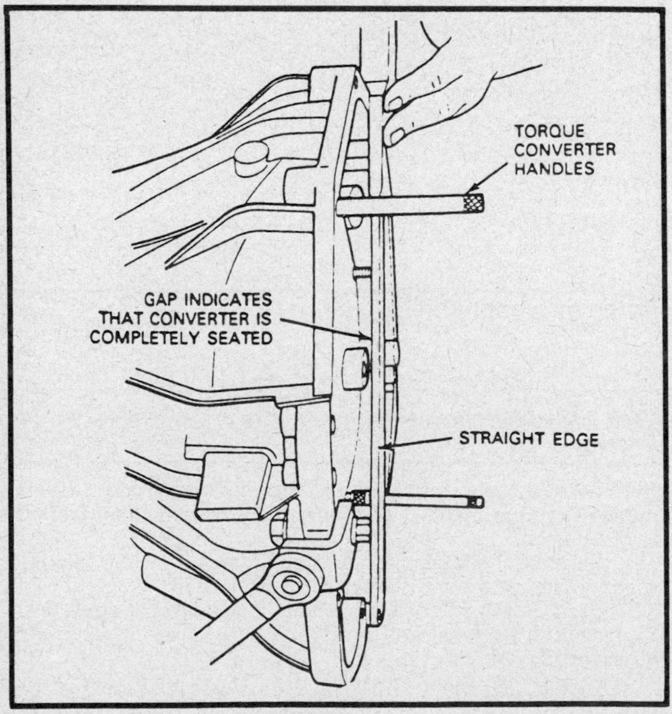
TORQUE
CONVERTER
HANDLES
GAP INDICATES
THAT CONVERTER IS
COMPLETELY SEATED
STRAIGHT EDGE

Checking converter installation

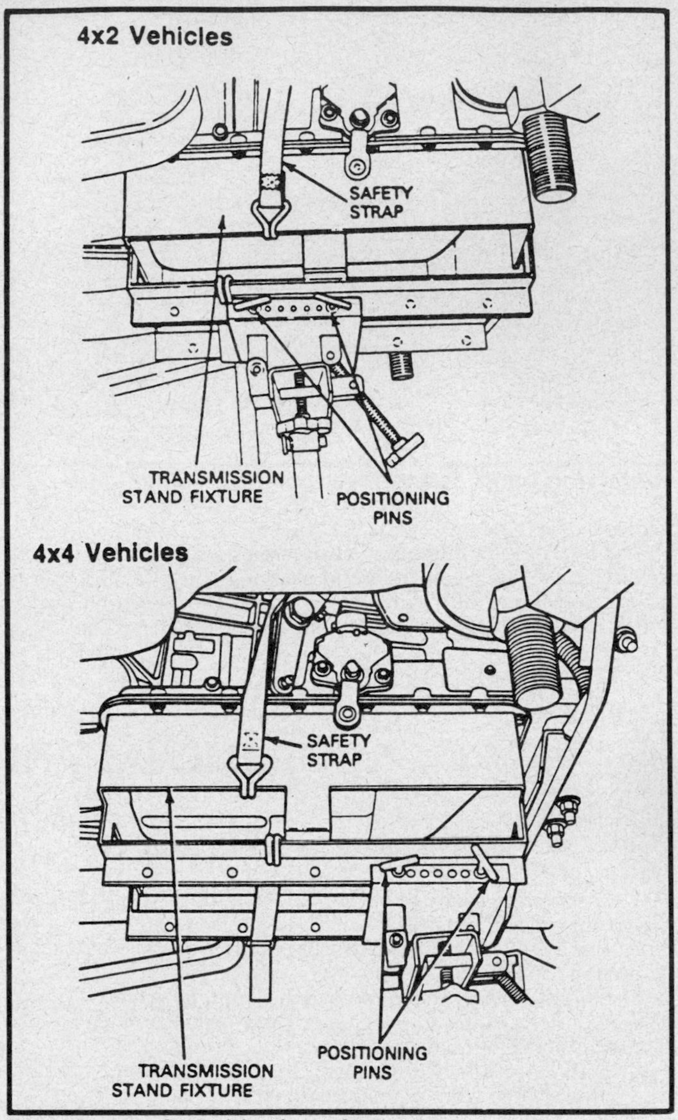
4x2 Vehicles
SAFETY
STRAP
TRANSMISSION
STAND FIXTURE
POSITIONING
PINS
4x4 Vehicles
SAFETY
STRAP
TRANSMISSION
STAND FIXTURE
POSITIONING
PINS

Placement of positioning pins

NOTE: Check the seating of the converter by placing a straightedge across the bell housing. There must be a gap between the converter pilot face and the straightedge.

4. Remove the converter handles. Check the condition of filler tube O-ring, if damaged or worn replace the O-ring. Install the filler tube.
5. Rotate the converter studs to align with flywheel mounting holes. Raise transmission into position while observing for any obstructions. Do not allow converter drive flats to disengage from pump gear. Use rubber converter drain plug cover or equivalent to aid in the alignment of the converter studs.

NOTE: Use care not to damage the flywheel and convert pilot. The converter must rest squarely against the flywheel. This indicates that the converter pilot is not binding in the engine crankshaft.

6. Alternately snug up the bell housing bolts and final torque to 40–50 ft. lbs.
7. Install the rubber converter drain plug cover and transmission cooling lines. Tighten lines to 18–23 ft. lbs.

8. Install the crossmember and the transmission retaining bolts. Remove the safety strap and the universal high lift transmission jack.
9. Rotate the crankshaft to gain access to converter studs. Install new stud nuts and torque to 20–30 ft. lbs.
10. Install the starter motor, rear engine plate cover and lower dust cover. Tighten to 12–16 ft. lbs.
11. Install the speedometer cable.
12. Completely seat the solenoid body connector into solenoid valve body recepticle. An audible click sound indicates proper installation.
13. Install the solenoid body connector heat shield with offset bending inward. Tighten to 6–9 ft. lbs.
14. On 4WD models only, install wire harness locators into crossmember.
15. Install the wire harness locator into extension housing wire bracket.
16. On 4WD models only, install the 4WD drive switch connector and connect the transfer case shift linkage.
17. Install the manual lever position sensor connector. An audible click sound indicates proper installation.
18. Install the shift linkage. On 4WD models only, install the shift rod to transfer case shift lever.
19. Install the rear driveshaft and install the front driveshaft on 4WD models.
20. Lower the vehicle, connect the negative battery cable and refill the transmission. Start the engine and check the fluid level.
21. Road test vehicle for proper operation and correct shift patterns.

BENCH OVERHAUL

Before Disassembly

When servicing the unit, it is recommended that as each part is disassembled, it is cleaned in solvent and dried with compressed air. All oil passages should be blown out and checked for obstructions. Disassembly and reassembly of this unit and its parts must be done on a clean work bench. As is the case when repairing any hydraulically operated unit, cleanliness is of the utmost importance. Keep bench, tools, parts and hands clean at all times. Also, before installing bolts into aluminum parts, always dip the threads into clean transmission oil. Anti-seize compound can also be used to prevent bolts from galling the aluminum and seizing. Always use a torque wrench to keep from stripping the threads. Take care with the seals when installing them, especially the smaller O-rings. The slightest damage can cause leaks. Aluminum parts are very susceptible to damage so great care should be exercised when handling them. The internal snaprings should be expanded and the external snaprings compressed if they are to be reused. This will help insure proper seating when installed. Be sure to replace any O-ring, gasket, or seal that is removed. Lubricate all parts with the specified transmission fluid when assembling.

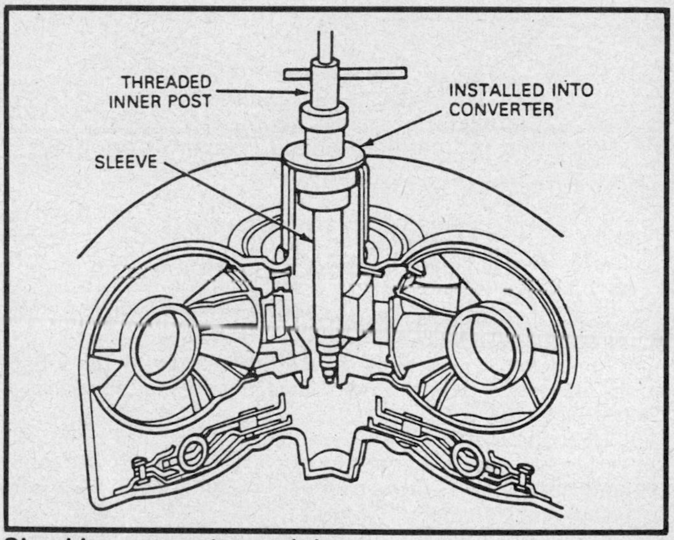
THREADED
INNER POST
INSTALLED INTO
CONVERTER
SLEEVE
Checking converter endplay

Converter Inspection

The torque converter is welded together and cannot be disassembled. Check the torque converter for damage or cracks and replace, if necessary. Remove any rust from the pilot hub and boss of the converter.

When internal wear or damage has occurred in the transmission, metal particles, clutch plate material, or band material may have been carried into the converter. These contaminants are a major cause or recurring transmission troubles and must be removed from the system before the transmission is put back into service. The converter must be cleaned by using the torque converter cleaner tool 014–00028 or equivalent. Under no circumstances should an attempt be made to clean converter by hand agitation with solvent.

CONVERTER ENDPLAY CHECK

1. Insert tool T80L–7902–A or equivalent into the converter pump drive hub until it bottoms out.
2. Expand the sleeve in the turbine spline by tightening the threaded inner post until the tool is securely locked into the spline.
3. Attach a dial indicator with bracketry tool–4201–C or equivalent to the tool. Position the indicator button on the converter pump drive hub and set the dial face at zero.
4. Lift the tool upward as far as it will go and note the indicator reading. The indicator reading is the total endplay which the turbine and stator share. Replace the converter unit if the total endplay exceeds the limits. The torque converter endplay on a new or rebuilt converter should be 0.021 in. (0.533mm) maximum. The torque converter endplay on a used converter should be 0.50 in. (1.25mm) maximum.
5. Loosen the threaded inner post to free the tool, and then remove the tool from the converter.

TORQUE CONVERTER ONE-WAY CLUTCH INSPECTION

In order to test the converter one-way clutch, insert fingers into the torque converter. Reaching the first splined segment, attempt to spin it. The segment should rotate freely clockwise and should not turn counterclockwise without the converter turning with it. If the segment rotates freely counterclockwise or does not rotate freely clockwise, the one-way clutch has failed and the torque converter should be replaced.

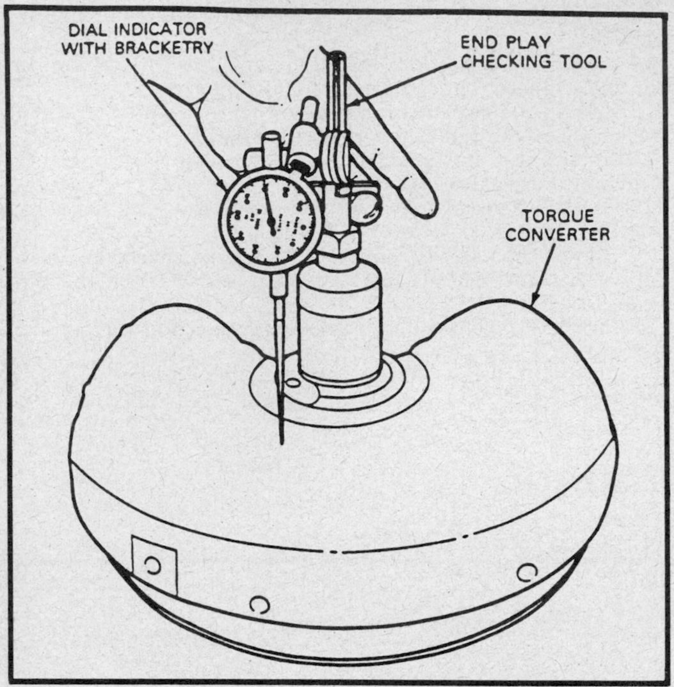
DIAL INDICATOR
WITH BRACKETRY
END PLAY
CHECKING TOOL
TORQUE
CONVERTER
Installing dial indicator to check converter endplay

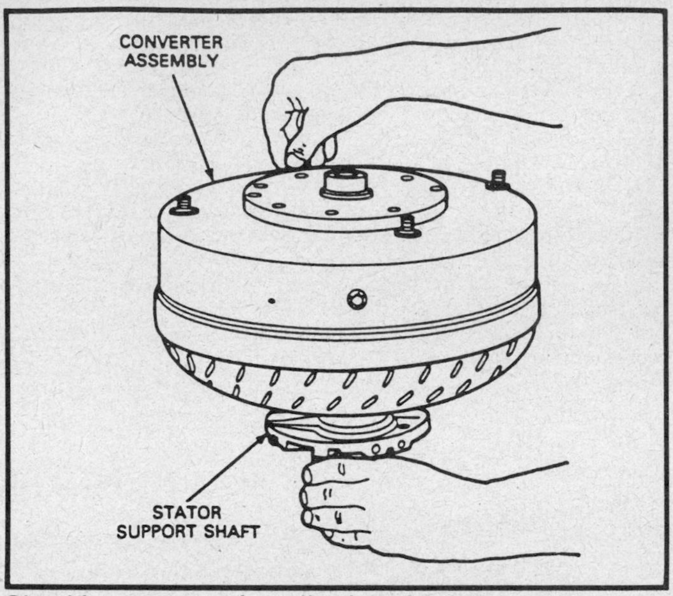
CONVERTER
ASSEMBLY
STATOR
SUPPORT SHAFT
Checking stator to impeller interference

STATOR TO IMPELLER INTERFERENCE CHECK

1. Position the stator support on a work bench with the spline end of the shaft pointing up.
2. Mount the converter on the stator support with the splines on the one-way clutch inner race engaging the mating splines of the stator support.
3. Hold the stator support stationary and try to rotate the converter counterclockwise. The converter should rotate freely without any signs of interference or scraping within the converter assembly.
4. If there is an indication of scraping, the trailing edges of the stator blades may be interfering with the leading edges of the impeller blades. In such cases, replace the converter. The stator support may remain in pump assembly during this test.

STATOR TO TURBINE INTERFERENCE CHECK

1. Position the converter on the work bench front side down.
2. Install a stator support to engage the mating splines of the stator support shaft.
3. Install the input shaft, engaging the splines with the turbine hub.
4. Hold the stator shaft stationary and attempt to rotate the turbine with the input shaft. The turbine should rotate freely in both directions without any signs of interference or scraping noise.
5. If interference exists, the stator front thrust washer may be worn, allowing the stator to hit the turbine. In such cases, the converter must be replaced.
6. Check the converter crankshaft pilot for nicks or damaged surfaces that could cause interference when installing the converter into the crankshaft.
7. Check the converter impeller hub for nicks or sharp edges that would damage the pump seal. The stator support may remain in pump assembly during this test.

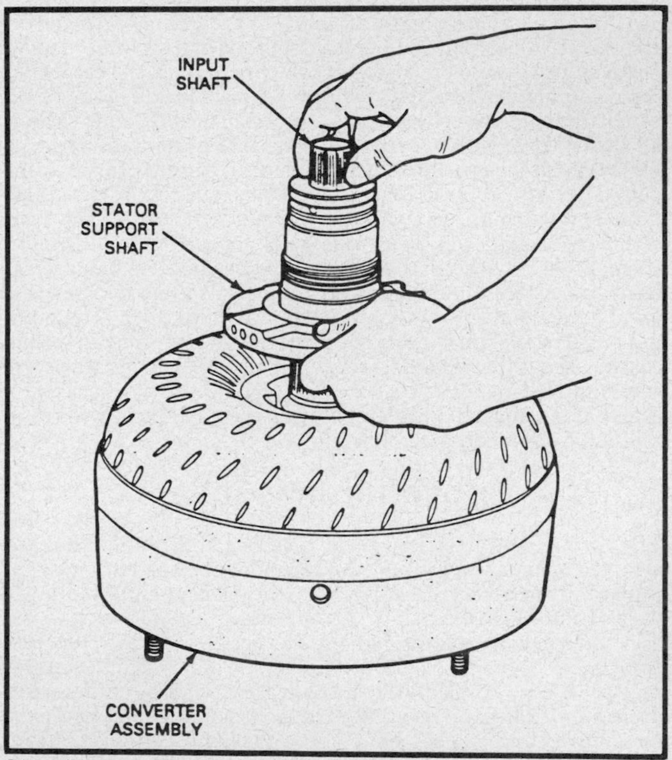
INPUT
SHAFT
STATOR
SUPPORT
SHAFT
CONVERTER
ASSEMBLY
Checking stator to turbine interference

Transmission Disassembly

1. Remove the input shaft from transmission and mount the transmission in a suitable holding fixture.
2. Thoroughly clean the solenoid body connector area to avoid contamination. Rotate the transmission so that pan is facing up. Remove the pan and gasket, discard the gasket.
3. Remove the filter and seal assembly by carefully pulling and rotating the filter as necessary. If seal remains in bore, carefully remove using O-ring tool T71P–19703–C or equivalent.

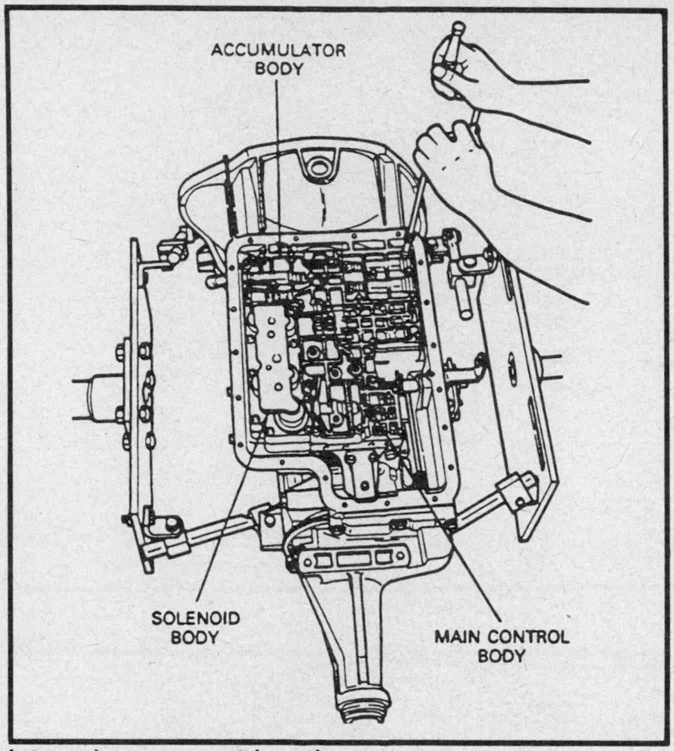
ACCUMULATOR BODY
SOLENOID BODY
MAIN CONTROL BODY

Internal component locations

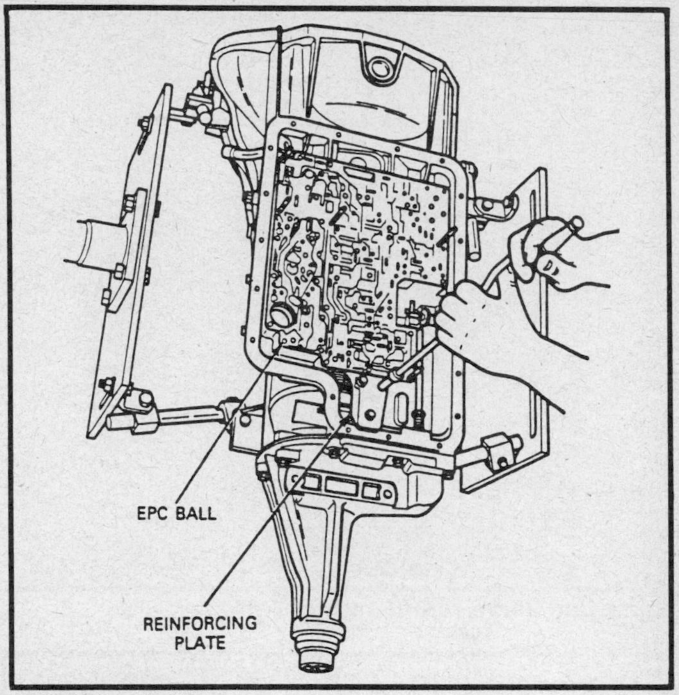
EPC BALL
REINFORCING PLATE

EPC ball location

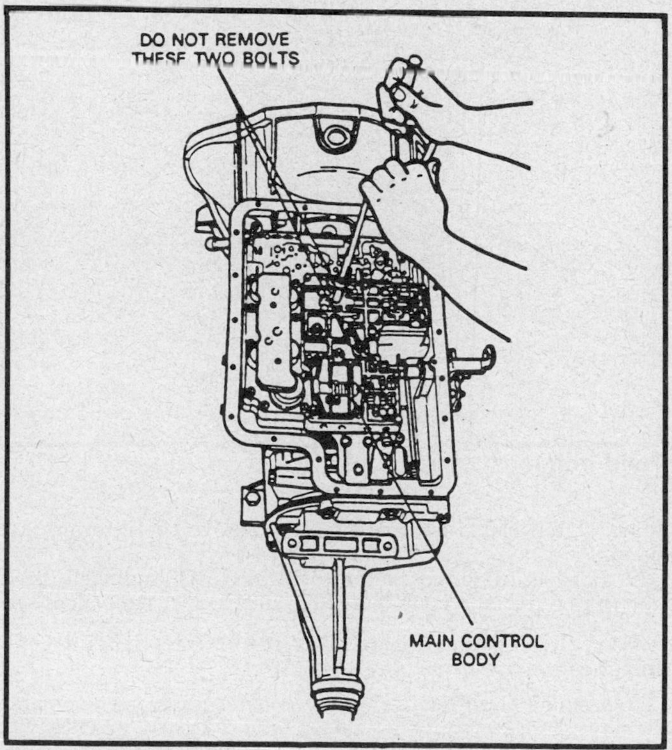
DO NOT REMOVE THESE TWO BOLTS
MAIN CONTROL BODY

Main control body locations

4. Remove the accumulator body and the main control body

NOTE: Do not remove the 2 center bolts.

5. Remove the 9 solenoid body bolts using a Torx® (30A bit) and 1 nut.

6. Push up on the solenoid body connector while removing solenoid body. Remove the solenoid screen, by turning counterclockwise and pull out.

7. Remove the 3 reinforcing plate bolts and remove the plate.

NOTE: The EPC ball is spring loaded under the separator plate.

8. Remove the separator plate, 2 gaskets, EPC ball and blowoff spring, discard gaskets. Remove 1 steel and 10 rubber check balls from transmission, using a small tool. Do not damage the rubber check balls.

9. Remove the servo snapring, retaining plate, piston/rod assembly and servo spring. Apply slight downward pressure to plate while remove snapring.

10. Remove the 3 feed bolts. Discard the feed bolts.

11. Rotate the transmission so that bell housing is facing up. Remove 9 pump bolts. Discard the pump bolt washers.

12. Use 2 threaded holes in pump and install pump puller adapter T89T–70010–A or equivalent. Install a slide hammer T59L–100–B or equivalent into adapter and remove the pump.

13. Remove the pump gasket and No. 7D014 thrust washer. Discard the gasket.

14. Lift out the coast clutch assembly. Remove the needle bearing assembly No. 7E486 between the front pump and the sun gear.

15. Remove the large snapring using a large removal tool.

16. Remove the overdrive pressure plate and clutch pack and mark for reassembly. Remove the overdrive ring gear and center shaft assembly and needle bearing assembly No. 7G178.

17. Install the clutch spring compressor T89T–70010–F or equivalent into the case. Tighten center bolt to 65 inch lbs.

18. Remove the large snapring with a large removal tool. Loosen the spring compressor center bolt and remove the compressor tool.

19. Remove the intermediate/overdrive cylinder assembly. Remove the intermediate return spring.

20. Remove the center support and thrust washer No. 7L326.

21. Remove the intermediate pressure plate and clutch plates.

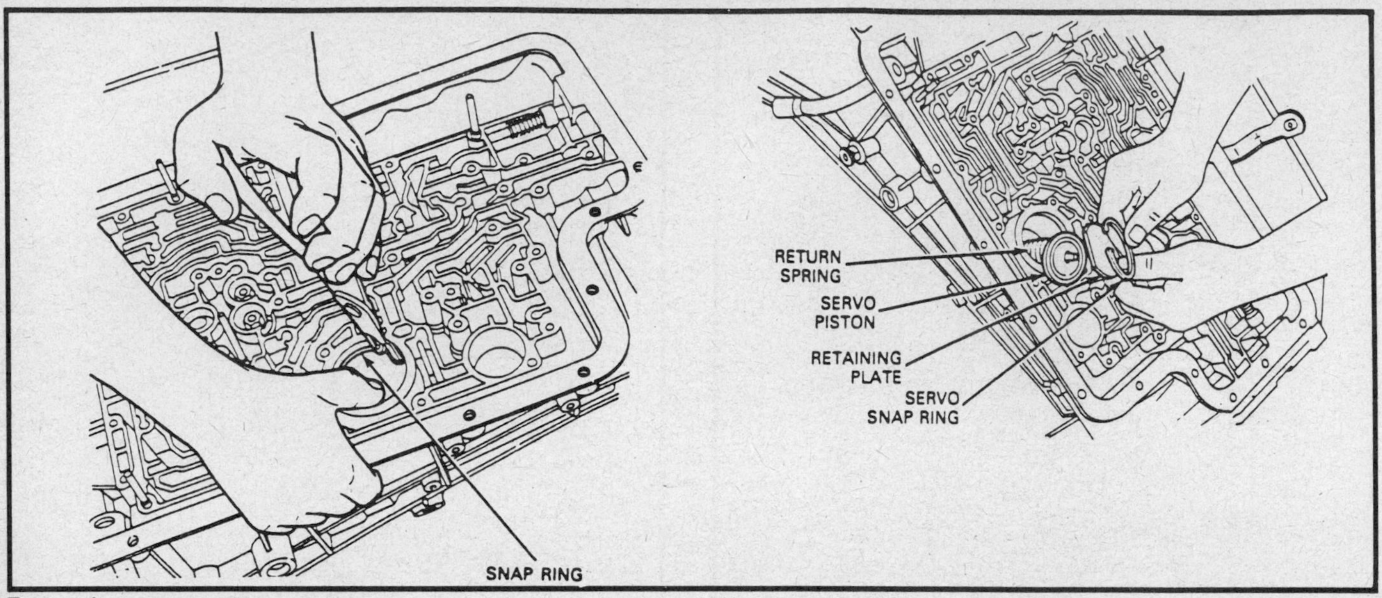
RETURN SPRING
SERVO PISTON
RETAINING PLATE
SERVO SNAP RING
SNAP RING

Removing servo return spring

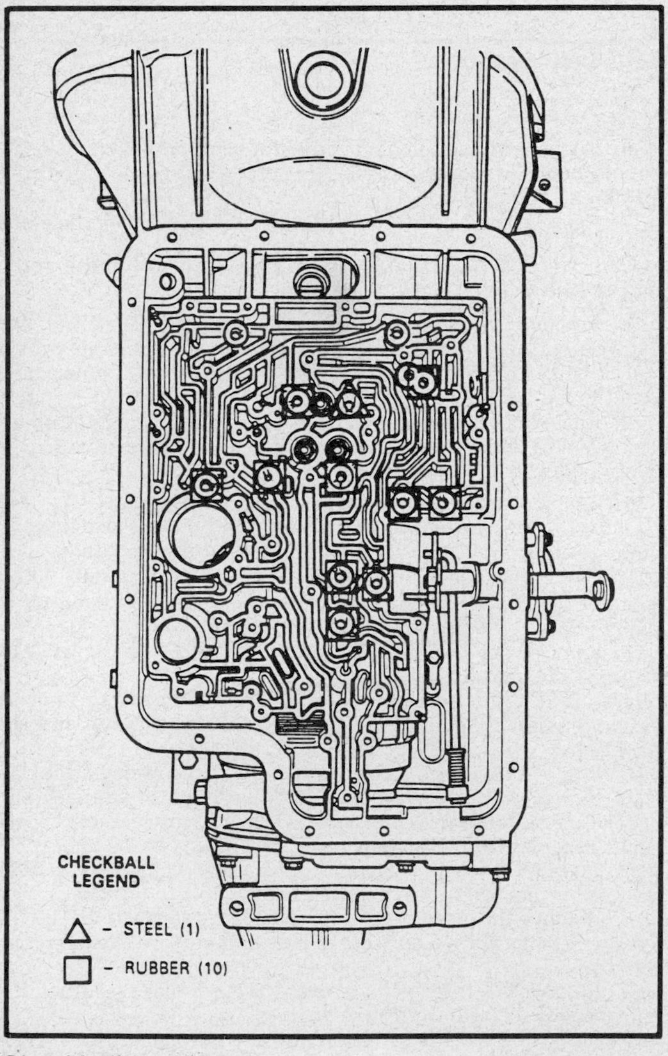
CHECKBALL LEGEND
△ - STEEL (1)
□ - RUBBER (10)

Checkball location

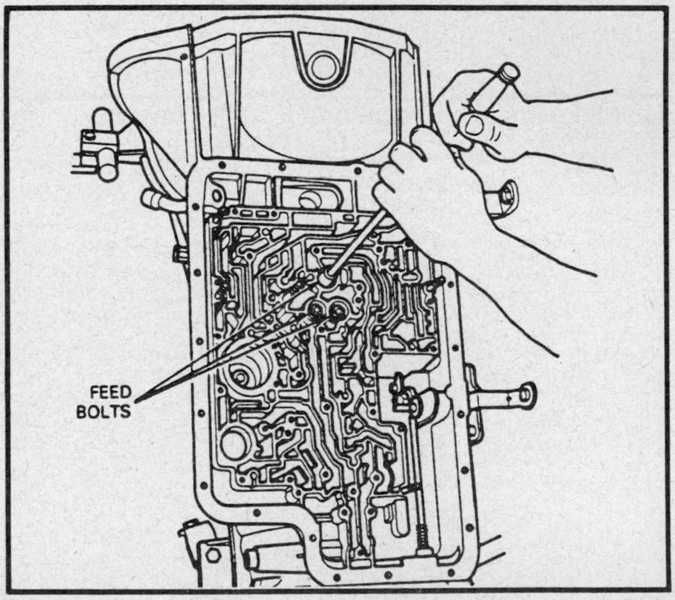
FEED BOLTS

Feed bolt locations

Mark clutch plates for reassembly. Remove the intermediate band.

22. Remove the direct clutch forward clutch and shell using clutch remover and installer T89T–70010–E or equivalent.

NOTE: Hooks on crossbar must be rotated into notches on the input shell.

23. Using a large removal tool, remove the reverse snapring. Remove the reverse planet carrier and 2 thrust washers No. 7D423.

24. Remove the output shaft snapring. Discard the snapring.

25. Remove the ring gear and hub assembly, and needle bearing assembly No. 7E413.

26. Remove the reverse hub and the one-way clutch assembly.

27. Using the proper tool remove the reverse clutch snapring. Remove the reverse pressure plate and clutch pack. Mark for reassembly.

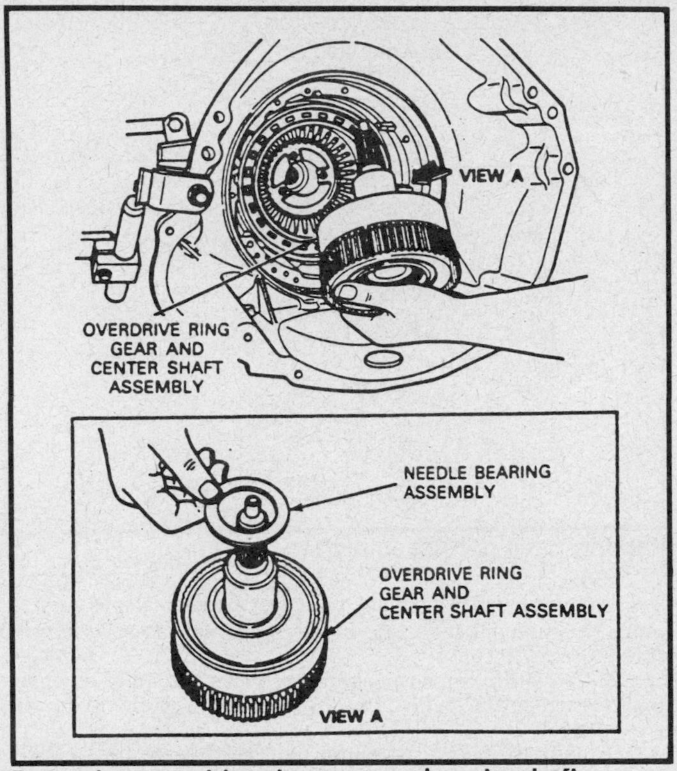
VIEW A
OVERDRIVE RING GEAR AND CENTER SHAFT ASSEMBLY
NEEDLE BEARING ASSEMBLY
OVERDRIVE RING GEAR AND CENTER SHAFT ASSEMBLY
VIEW A

Removing overdrive ring gear and centershaft assembly

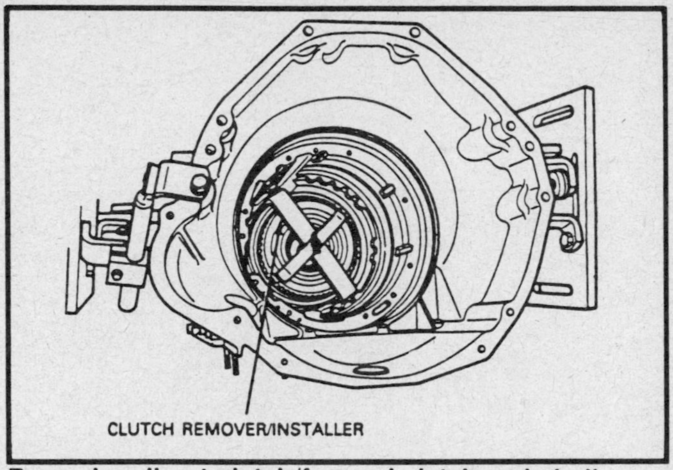
CLUTCH REMOVER/INSTALLER

Removing direct clutch/forward clutch and shell from case

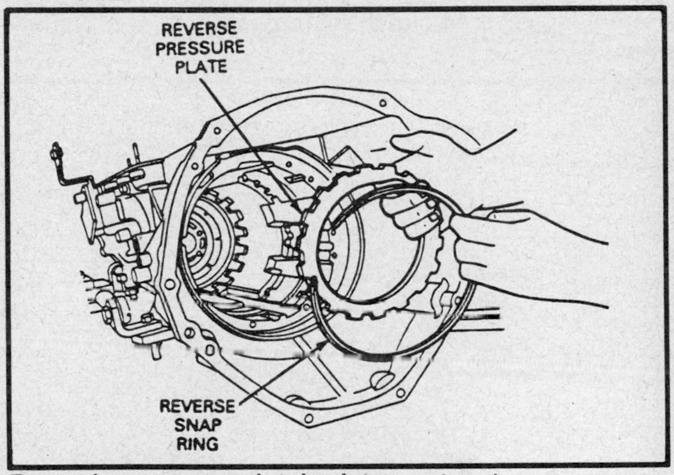
REVERSE PRESSURE PLATE
REVERSE SNAP RING

Removing reverse clutch piston—step 1

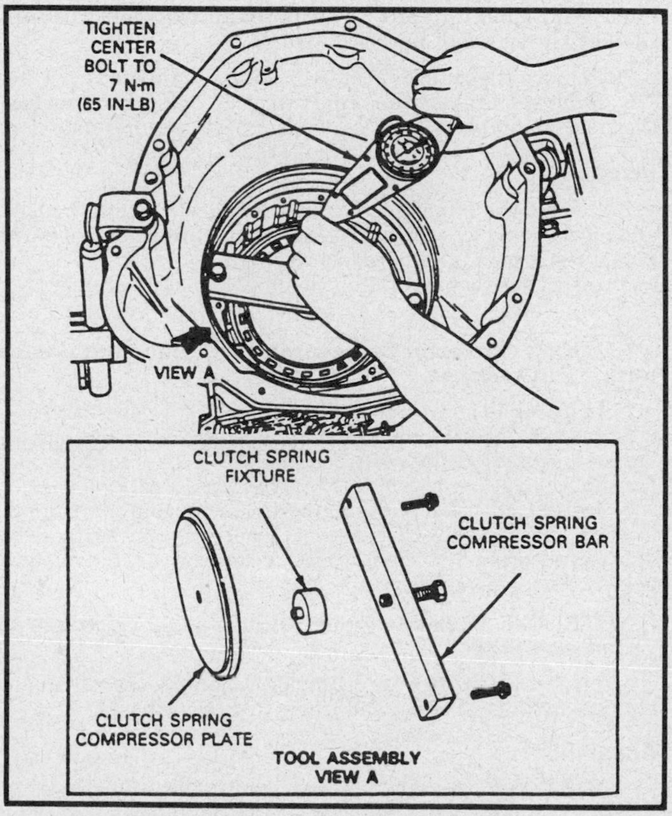
TIGHTEN CENTER BOLT TO 7 N·m (65 IN-LB)
VIEW A
CLUTCH SPRING FIXTURE
CLUTCH SPRING COMPRESSOR BAR
CLUTCH SPRING COMPRESSOR PLATE
TOOL ASSEMBLY VIEW A

Install clutch spring compressor into case

28. Rotate the transmission so that pan surface is facing up. Remove the 9 extension housing bolts. Remove the wiring bracket, extension housing and gasket, discard gasket.
29. Remove the output shaft, park gear and thrust washer No. 7B368.
30. Remove the 5 bolts from the low/reverse one-way clutch inner race. Remove the reverse clutch, return spring and inner race. Install the reverse clutch pressure plate and snapring, to hold reverse clutch piston during removal.
31. Blow compressed air into the reverse clutch feed port. This will blow out the reverse clutch piston against the pressure plate.
32. Remove the snapring, reverse clutch pressure plate and piston from case.
33. Rotate the transmission so that pan surface is facing down. Remove the park pawl return spring, pin and parkwing pawl from the case.
34. Remove the 2 bolts from the parking rod guide plate.
35. Remove the Torx® head bolt (40A bit) and parking pawl abutment.
36. Using side cutters or locknut pin remover T78P-3504-N or equivalent remove manual lever roll pin from the case.
37. Remove the inner detent lever nut, while holding the lever with crescent wrench or equivalent.
38. Remove the inner detent lever and parking pawl actuating rod assembly from the manual lever.
39. Remove the 2 bolts and the manual lever position sensor.

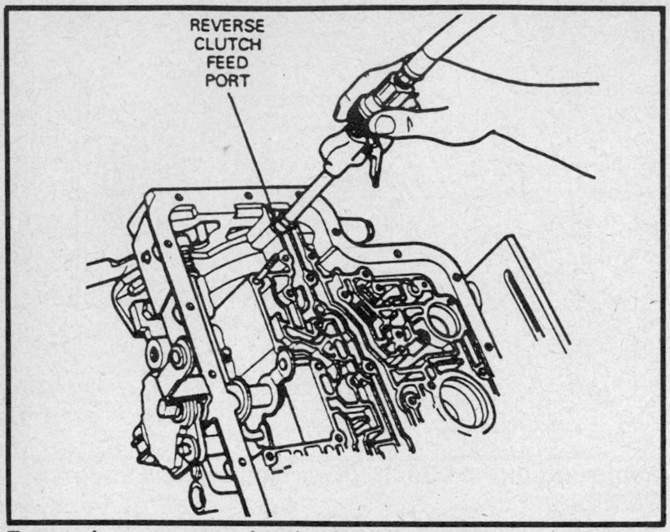
REVERSE
CLUTCH
FEED
PORT
Removing reverse clutch piston—step 2

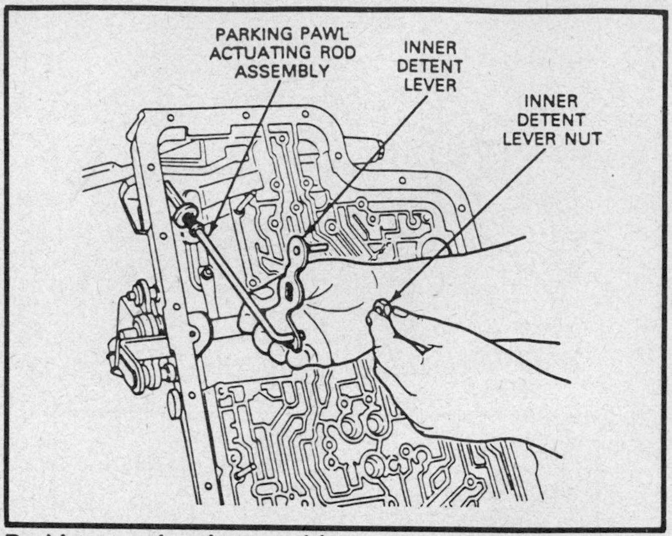
PARKING PAWL
ACTUATING ROD
ASSEMBLY
INNER
DETENT
LEVER
INNER
DETENT
LEVER NUT
Parking pawl rod assembly

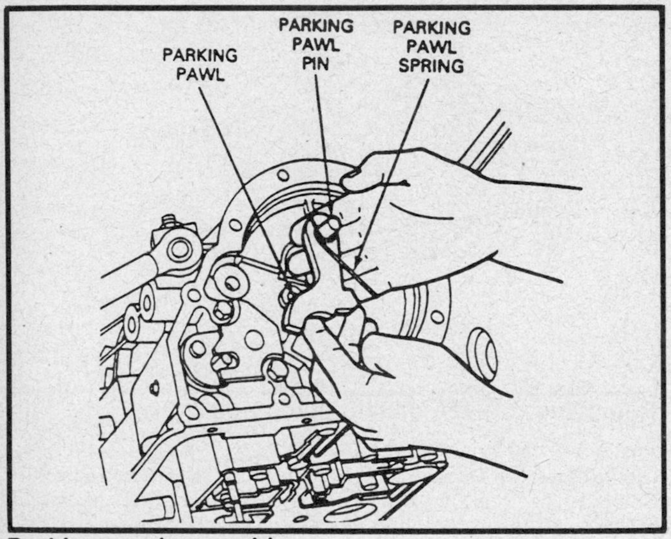
PARKING
PAWL
PARKING
PAWL
PIN
PARKING
PAWL
SPRING
Parking pawl assembly

40. Remove the manual lever and the manual lever seal using seal remover T74P–77248–A and slide hammer T50T–100–A or equivalents.
41. If necessary, remove the stub tube using channel lock pliers or equivalent.

Unit Disassembly and Assembly

OIL PUMP

Disassembly

1. Remove the 2 Teflon® coast clutch seals from the stator support. Remove the converter clutch seal from the front of the stator support. Remove the pump outer diameter square cut seal.
2. Obtain a banding tool prior to removing pump body bolts. This tool is needed to align the pump with the control body assembly during reassembly.
3. Remove the 11 bolts and separate the pump control body from the pump body.
4. Apply pressure to main regulator booster sleeve and remove the internal snapring. Remove the main regulator valve train.
5. Remove the converter regulator valve assembly by applying pressure to the end plug and removing the retainer clip with a small tool.
6. Remove the converter clutch shift valve assembly by applying pressure to the end plug and removing retainer clip with a small tool.

NOTE: Do not remove any of the cup plugs unless they are damaged or leaking. Do not remove the stator support from the control body as this may distort the surface of the control body.

7. Remove the gerotor gearset from the pump body.
8. Remove the gasket material from the control body surface. Clean all pump parts in solvent and dry with compressed air.

Inspection

1. Inspect the pump gears, faces, gear teeth, pump housing and mating surfaces for damage or scoring. Replace the entire pump if any part is damaged or worn.
2. Inspect the converter hub bushing. Replace if scored or excessively worn.

NOTE: If converter hub bushing is replaced, stake the bushing at notches.

3. Inspect the converter hub seal. Replace if necessary.
4. Inspect the stator input shaft bushings. If bushings are worn or scored, replace the complete oil pump control body assembly.
5. Inspect all valve and plug bores for scoring, or damage. Check all passages for obstructions. Inspect mating surfaces for burrs and scoring. If necessary, use crocus cloth or equivalent to polish components.

NOTE: Use care to avoid rounding sharp edges of valves and plugs.

6. Inspect all springs for distortion. Check all the valves and plugs that when dry they fall freely from their bores.

Assembly

1. Install the main regulator valve assembly. Apply pressure to main regulator booster sleeve and install internal snapring. Make sure that the snapring is properly seated.

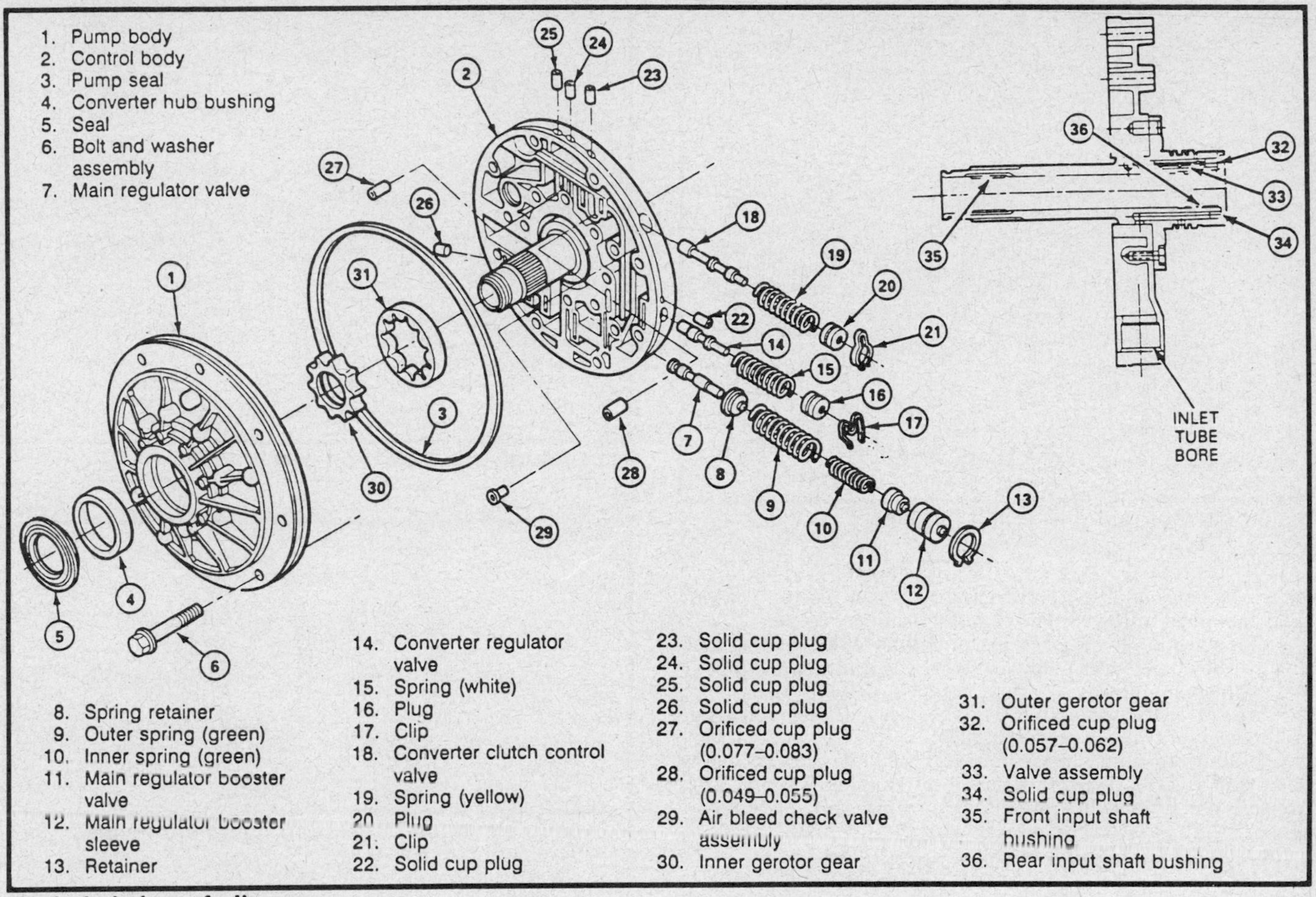
1. Pump body
2. Control body
3. Pump seal
4. Converter hub bushing
5. Seal
6. Bolt and washer assembly
7. Main regulator valve
8. Spring retainer
9. Outer spring (green)
10. Inner spring (green)
11. Main regulator booster valve
12. Main regulator booster sleeve
13. Retainer
14. Converter regulator valve
15. Spring (white)
16. Plug
17. Clip
18. Converter clutch control valve
19. Spring (yellow)
20. Plug
21. Clip
22. Solid cup plug
23. Solid cup plug
24. Solid cup plug
25. Solid cup plug
26. Solid cup plug
27. Orificed cup plug (0.077–0.083)
28. Orificed cup plug (0.049–0.055)
29. Air bleed check valve assembly
30. Inner gerotor gear
31. Outer gerotor gear
32. Orificed cup plug (0.057–0.062)
33. Valve assembly
34. Solid cup plug
35. Front input shaft bushing
36. Rear input shaft bushing
INLET TUBE BORE

Exploded view of oil pump

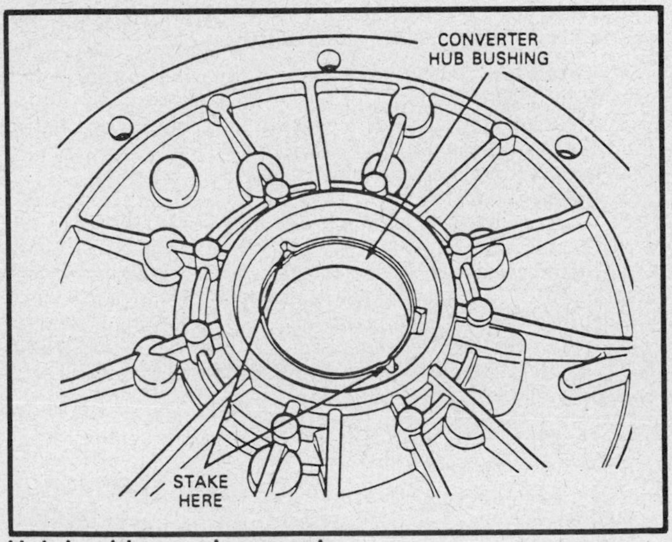
CONVERTER HUB BUSHING
STAKE HERE

Hub bushing replacement

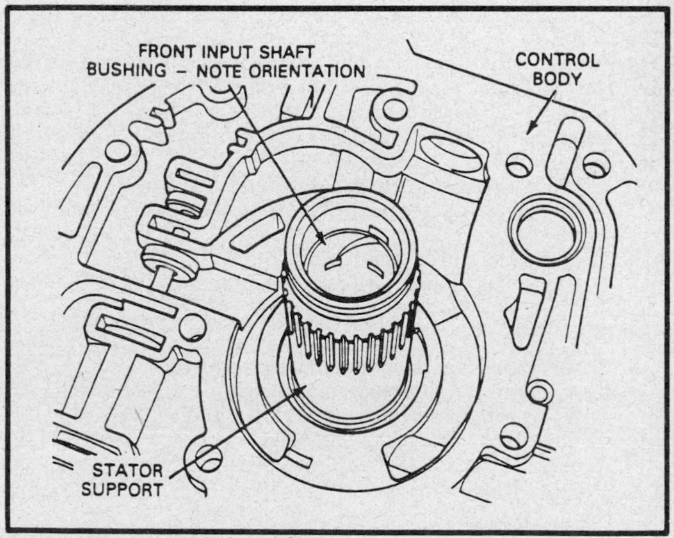
FRONT INPUT SHAFT BUSHING – NOTE ORIENTATION
CONTROL BODY
STATOR SUPPORT

Stator input shaft bushing

2. Install the converter shuttle valve assembly.

3. Lightly coat the gerotor gears with petroleum jelly and install in the pump housing.

NOTE: The dot on the inner gerotor gear must face the control body assembly.

4. Lower the control body and stator assembly onto the pump body, aligning the 28mm round hole in the control body with the 28mm hole in the pump body.

5. Loosely install bolts into the pump body. Install banding tool D89L–77000 or equivalent with clamp by filter inlet. Align

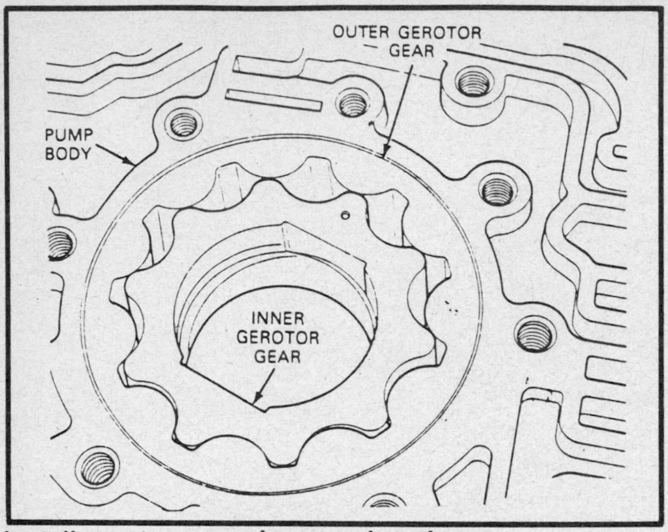
OUTER GEROTOR GEAR
PUMP BODY
INNER GEROTOR GEAR
Install gerotor gears in pump housing

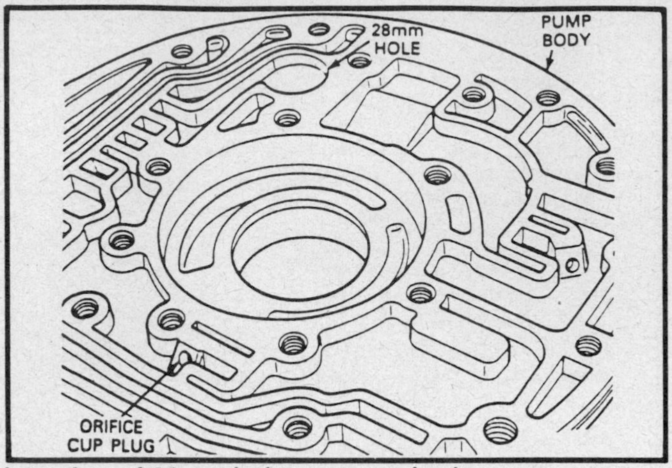
28mm HOLE
PUMP BODY
ORIFICE CUP PLUG
Location of 28mm hole on pump body

outer bolt holes and tighten banding tool. This aligns the input shaft bushings to the converter hub bushings.

6. Torque bolts to 18–23 ft. lbs. and remove the banding tool. Ensure the outer edges of the control body and the pump body are completely aligned.

7. Install the coast clutch Teflon® seals. Install converter lockup seal on front of stator support.

8. Install the pump outer diameter seal. Be sure groove is clean and free of burrs. Lubricate outer diameter seal with transmission fluid before installing pump into transmissin case.

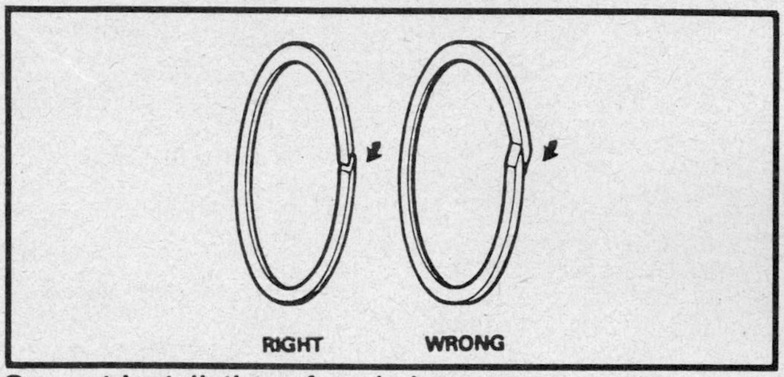
RIGHT
WRONG
Correct installation of seals in pump assembly

COAST CLUTCH CYLINDER ASSEMBLY

Disassembly

1. Remove the sun gear assembly from the coast clutch cylinder.
2. Remove the snapring and discard. Remove the pressure and clutch plates from the cylinder. Mark for reasembly.
3. Using the clutch spring compressor T65L–77515–A or equivalent remove the return spring retaining ring. Remove the compressor tool.
4. Remove the piston return spring.
5. Remove the piston apply plate and piston from the cylinder.
6. Remove the outer seal from the piston.
7. Remove the inner seal from the cylinder.

Inspection

1. Inspect the clutch cylinder thrust surfaces, piston bore, and clutch plate serrations for scores or burrs. Minor scores or burrs may be removed with crocus cloth. Replace the clutch cylinder if it is badly scored or damaged.
2. Check the fluid passage in the clutch cylinder for blockage. Clean out all fluid passages. Inspect the clutch piston for scores and replace if necessary.
3. Check the clutch release spring for distortion and cracks.
4. Inspect the composition clutch plates, steel clutch plates, and clutch pressure plate for worn or scored bearing surfaces. Replace all parts that are deeply scored or burred.
5. Check the clutch plates for flatness and fit on the clutch hub serrations. Discard any plate that does not slide freely on the serrations or that is not flat.
6. Check the clutch hub thrust surfaces for scores and the clutch hub splines for wear.

Assembly

NOTE: Soak all the friction plates in clean transmission fluid for 15 minutes and lube all O-ring seals with transmission fluid before installing.

1. Install the inner seal so that lip is facing toward bottom (down) into the cylinder.
2. Install the outer seal so that lip is facing toward bottom (down) onto the piston.
3. Install the piston into the cylinder.
4. Install the piston apply plate and piston return spring.
5. Using a clutch spring compressor T65L–77151–A or equivalent, install the snapring.
6. Install the clutch pack plates, alternately starting with the steel plate.
7. Install the pressure plate.
8. Install the selective snapring and check stack-up using a feeler gauge. If not within specification, install correct selective snapring and recheck.
   Specification
   0.045–0.025 in. (1.14–0.62mm)
   Selective snaprings
   0.057–0.053 in. (1.45–1.35mm)
   0.072–0.068 in. (1.85–1.75mm)
   0.088–0.084 in. (2.25–2.15mm)
9. Install the overdrive sun gear with the short end of gear down into the coast clutch cylinder.

OVERDRIVE RING GEAR AND CENTER SHAFT ASSEMBLY

Disassembly

1. Remove the inner race.

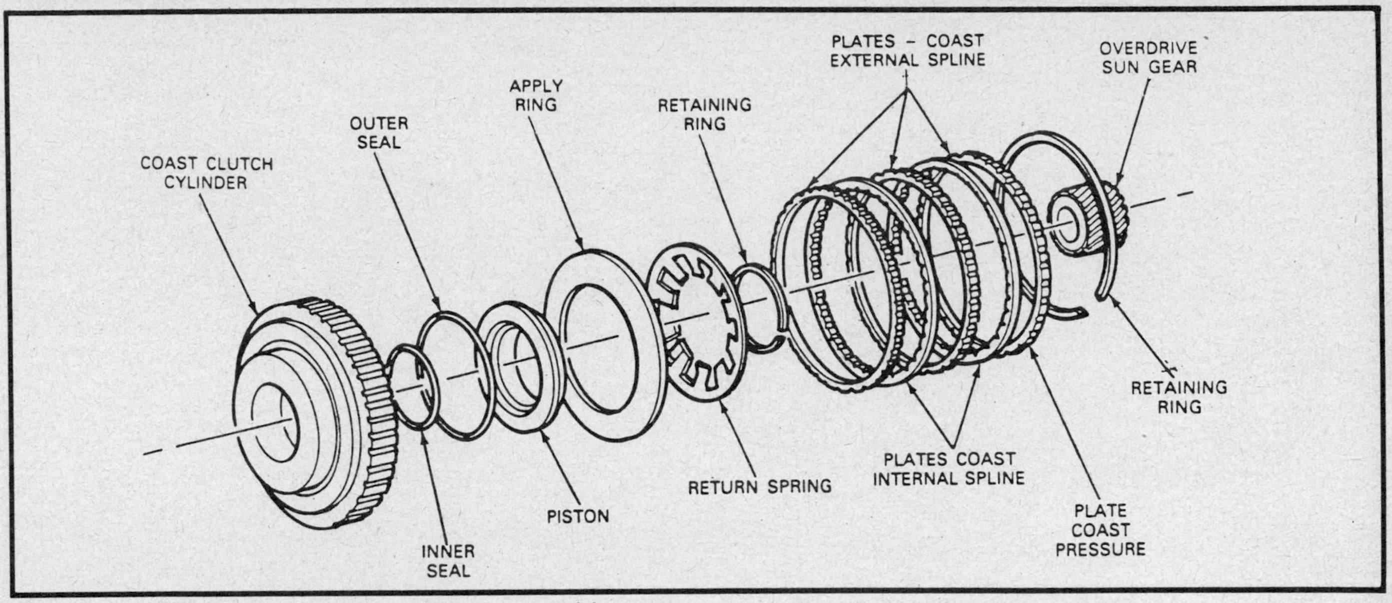
PLATES - COAST
EXTERNAL SPLINE
OVERDRIVE
SUN GEAR
APPLY
RING
RETAINING
RING
OUTER
SEAL
COAST CLUTCH
CYLINDER
RETAINING
RING
PLATES COAST
INTERNAL SPLINE
RETURN SPRING
PISTON
PLATE
COAST
PRESSURE
INNER
SEAL

Exploded view of coast clutch cylinder assembly

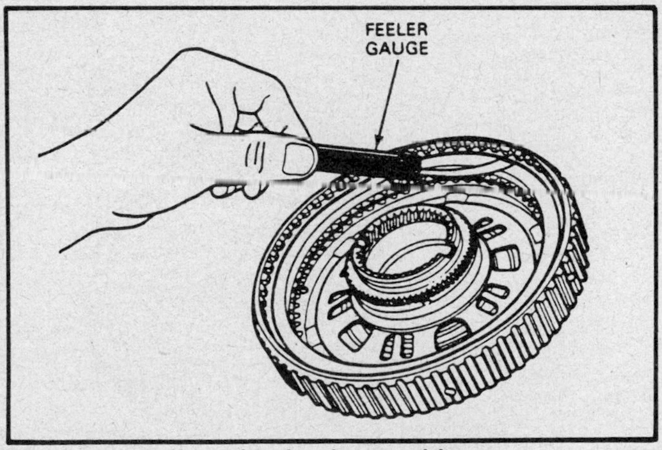
FEELER
GAUGE

Checking stack-up in clutch asembly

2. Remove the small (inner) snapring and one-way clutch assembly.
3. Remove the large snapring and the outer race assembly from ring gear.
4. Remove the thrust washer No. 7G400 from front of the carrier.
5. Remove the overdrive carrier from ring gear assembly.
6. Remove the needle bearing assembly No. 7G128 from rear face of carrier using the appropriate tool.
7. Remove the center shaft to ring gear wave type snapring. Remove the center shaft from ring gear.

Inspection

1. Inspect the outer and inner races for scores or damaged surface areas.
2. Inspect the rollers for excessive wear or damage.

NOTE: Individual parts of the planet carrier are not serviceable.

3. The pins and shafts in the planet assembly should be checked for loose fit and or complete disengagement.
4. Inspect the pinion gears for damaged or excessively worn teeth.
5. Check for free rotation of the pinon gears.

Assembly

1. Install the center shaft into the overdrive ring gear with retaining ring.
2. Install the needle bearing assembly No. 7G128 on rear face of carrier.
3. Install the overdrive carrier into center shaft and ring gear assembly.
4. Install the thrust washer No. 7G400 on front of the carrier.
5. Install the outer race assembly into the ring gear with snapring groove facing up and attach ring gear with snapring.

NOTE: The overdrive one-way clutch end caps must be installed correctly. The end cap with the scallops on the inner diameter must be toward the front of the transmission for proper lubrication.

6. Place the top (thick) end cap onto the one-way clutch. Place the thin end cap onto the bottom of the one-way clutch.
7. Install the one-way clutch assembly.The date code on the outside of thick end cap must be visible. Secure in place with snapring.
8. Install the inner race and make sure the inner race rotates counterclockwise.

INTERMEDIATE/OVERDRIVE CYLINDER ASSEMBLY

Disassembly

1. Using a suitable tool compress the overdrive return spring.
2. Remove the snapring and compressor tool assembly.
3. Remove the return spring and the overdrive piston.
4. Remove the outer and inner seals, using O-ring tool T71P–19703–C or equivalent.
5. Remove the intermediate piston.
6. Remove the intermediate/overdrive inner seal from cylinder bore, using O-ring tool T71P–19703–C or equivalent.
7. Remove the outer seal from the intermediate piston.

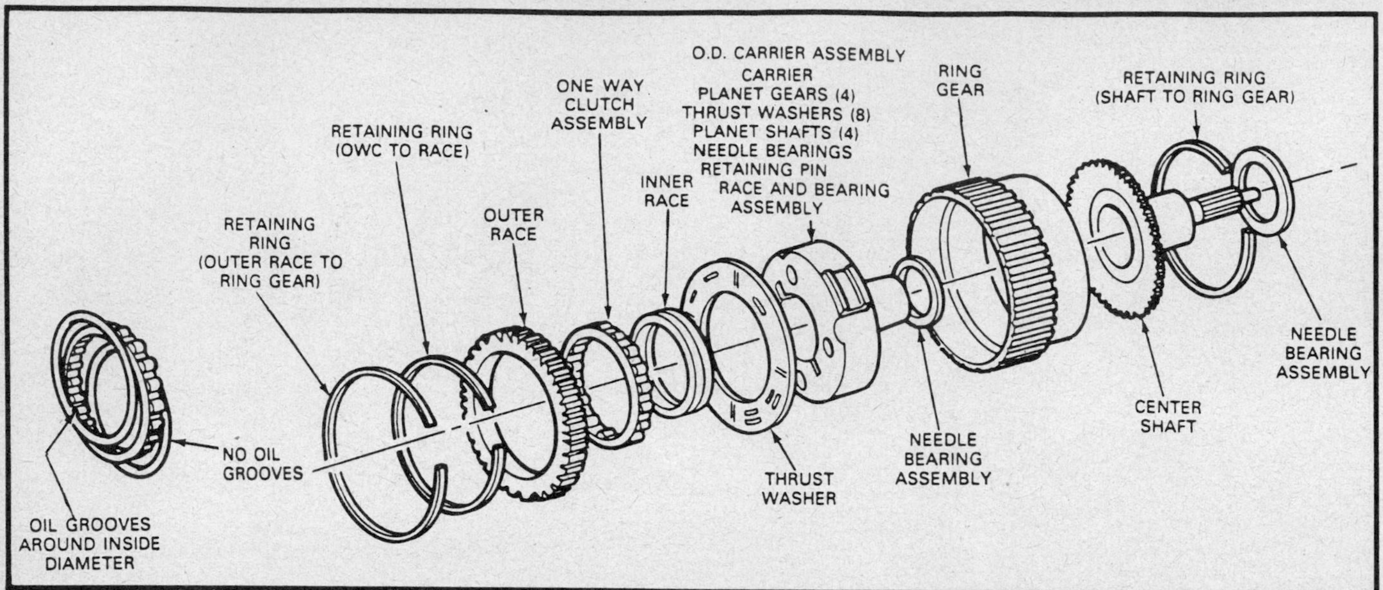
O.D. CARRIER ASSEMBLY
CARRIER
PLANET GEARS (4)
THRUST WASHERS (8)
PLANET SHAFTS (4)
NEEDLE BEARINGS
RETAINING PIN
RACE AND BEARING ASSEMBLY
RING GEAR
RETAINING RING (SHAFT TO RING GEAR)
ONE WAY CLUTCH ASSEMBLY
RETAINING RING (OWC TO RACE)
INNER RACE
OUTER RACE
RETAINING RING (OUTER RACE TO RING GEAR)
NEEDLE BEARING ASSEMBLY
CENTER SHAFT
NO OIL GROOVES
NEEDLE BEARING ASSEMBLY
THRUST WASHER
OIL GROOVES AROUND INSIDE DIAMETER

Exploded view of overdrive ring gear and center shaft assembly

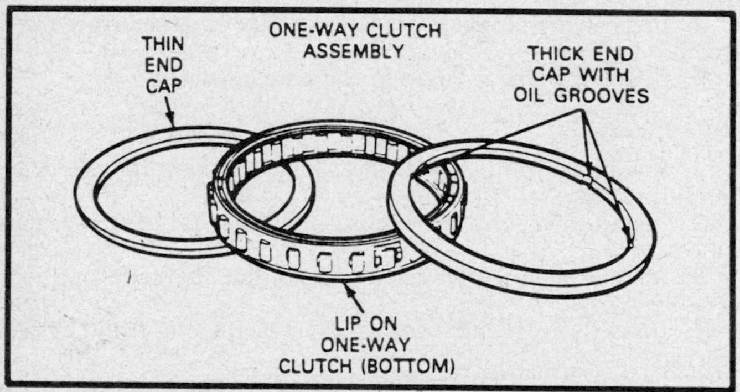
THIN END CAP
ONE-WAY CLUTCH ASSEMBLY
THICK END CAP WITH OIL GROOVES
LIP ON ONE-WAY CLUTCH (BOTTOM)

Correct installation of one-way clutch end caps

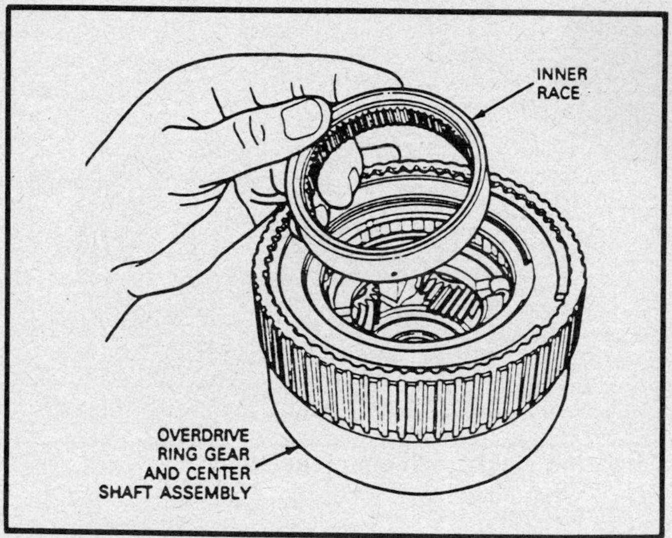
INNER RACE
OVERDRIVE RING GEAR AND CENTER SHAFT ASSEMBLY

Installation of inner race

Inspection

1. Inspect the clutch cylinder thrust surfaces, piston bore, and clutch plate serrations for scores or burrs. Minor scores or burrs may be removed with crocus cloth. Replace the clutch cylinder if it is badly scored or damaged.
2. Check the fluid passage in the clutch cylinder for blockage. Clean out all fluid passages. Inspect the clutch piston for scores and replace if necessary.
3. Check the clutch release spring for distortion and cracks.
4. Inspect the composition clutch plates, steel clutch plates, and clutch pressure plate for worn or scored bearing surfaces. Replace all parts that are deeply scored or burred.
5. Check the clutch plates for flatness and fit on the clutch hub serrations.

NOTE: Replace any plate that does not slide freely on the serrations or that is not flat.

6. Check the clutch hub thrust surfaces for scores and the clutch hub splines for wear.

Assembly

1. Install the outer seal onto intermediate piston with lip seal facing down towards cylinder.
2. Install the intermediate/overdrive inner seal onto cylinder bore with lip seal facing down towards cylinder.
3. Install the intermediate piston.
4. Install the overdrive outer and inner seals with lip seal facing down towards the cylinder.
5. Install the overdrive piston and return spring. Make note that the spring fingers are facing up.
6. Using a suitable tool compress the return spring and install the snapring. Remove the tool assembly.

CENTER SUPPORT

Disassembly

Remove the 2 cast iron outer seal rings using a suitable tool.

Assembly

Install the 2 cast iron seal rings on the center support.

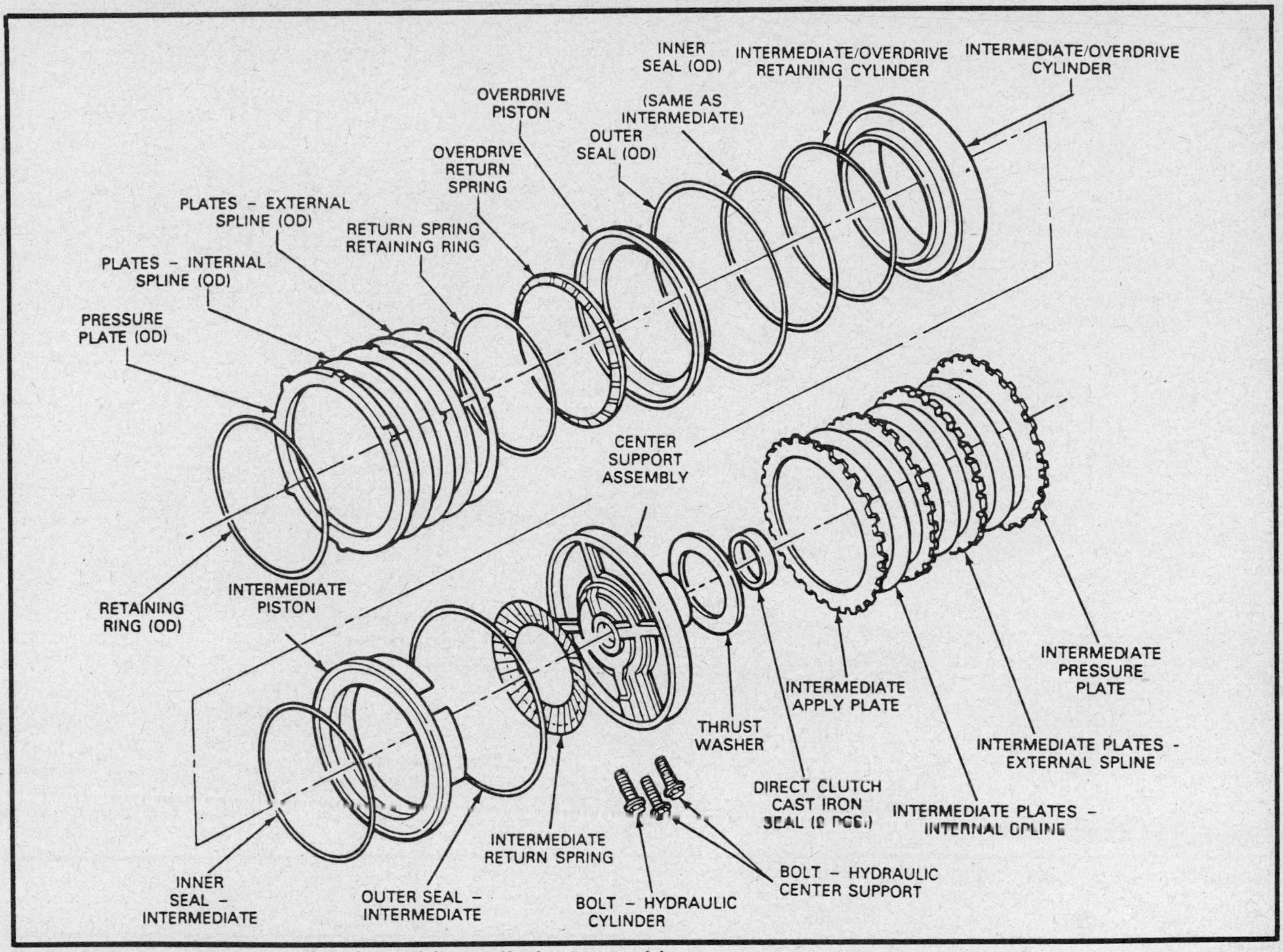
INNER SEAL (OD)
INTERMEDIATE/OVERDRIVE RETAINING CYLINDER
INTERMEDIATE/OVERDRIVE CYLINDER
OVERDRIVE PISTON
(SAME AS INTERMEDIATE)
OUTER SEAL (OD)
OVERDRIVE RETURN SPRING
PLATES - EXTERNAL SPLINE (OD)
RETURN SPRING RETAINING RING
PLATES - INTERNAL SPLINE (OD)
PRESSURE PLATE (OD)
CENTER SUPPORT ASSEMBLY
RETAINING RING (OD)
INTERMEDIATE PISTON
INTERMEDIATE PRESSURE PLATE
INTERMEDIATE APPLY PLATE
THRUST WASHER
INTERMEDIATE PLATES - EXTERNAL SPLINE
DIRECT CLUTCH CAST IRON SEAL (2 PCS.)
INTERMEDIATE PLATES - INTERNAL SPLINE
INTERMEDIATE RETURN SPRING
BOLT - HYDRAULIC CENTER SUPPORT
INNER SEAL - INTERMEDIATE
OUTER SEAL - INTERMEDIATE
BOLT - HYDRAULIC CYLINDER

Exploded view of intermediate/overdrive cylinder assembly

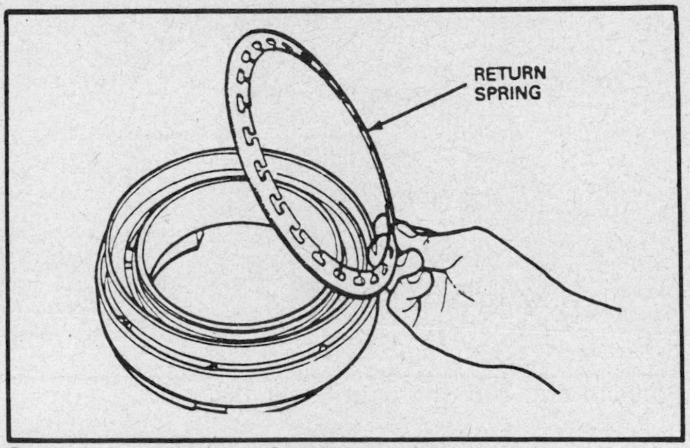
RETURN SPRING

Correct installation of return spring

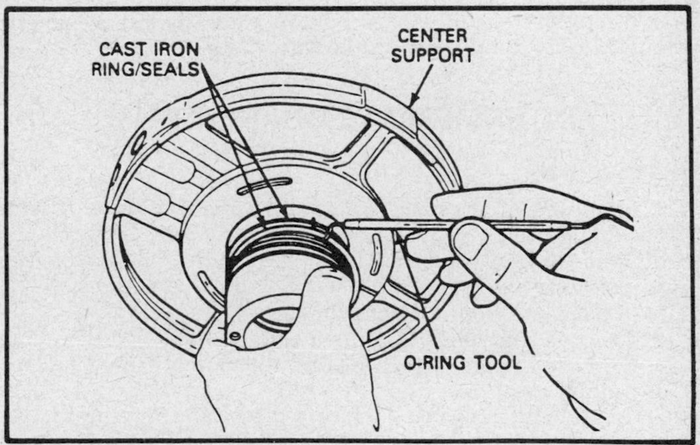
CAST IRON RING/SEALS
CENTER SUPPORT
O-RING TOOL

Center support assembly

FORWARD HUB AND RING GEAR

Disassembly

1. Remove the plastic thrust washer No. 7D090 from the front face of hub.
2. Remove the snapring using suitable tool.
3. Remove the forward hub from the ring gear.

Inspection

Check the clutch hub thrust surfaces for scores and the clutch hub splines for wear or damage.

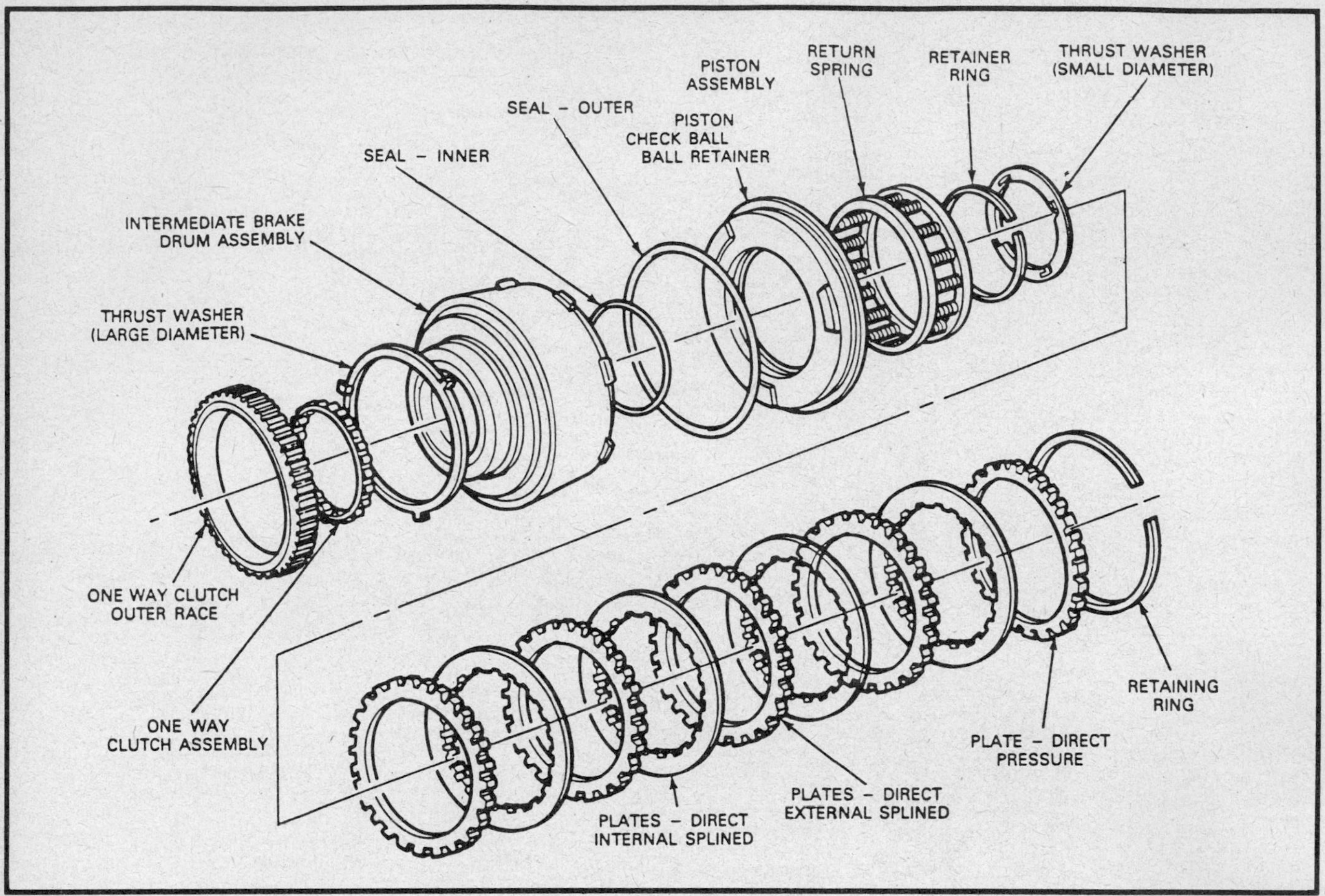
PISTON ASSEMBLY
RETURN SPRING
RETAINER RING
THRUST WASHER (SMALL DIAMETER)
SEAL - OUTER
PISTON CHECK BALL BALL RETAINER
SEAL - INNER
INTERMEDIATE BRAKE DRUM ASSEMBLY
THRUST WASHER (LARGE DIAMETER)
ONE WAY CLUTCH OUTER RACE
ONE WAY CLUTCH ASSEMBLY
RETAINING RING
PLATE - DIRECT PRESSURE
PLATES - DIRECT EXTERNAL SPLINED
PLATES - DIRECT INTERNAL SPLINED

Exploded view of intermediate brake drum

Assembly

1. Install the forward hub into the gear.
2. Install the snapring.
3. Using petroleum jelly to hold in place, install the plastic thrust washer No. 7D090 on the hub.

INTERMEDIATE BRAKE DRUM

Disassembly

1. Remove the outer race. Remove the one-way clutch assembly and end caps.
2. Remove the large brass thrust washer No. 7G401 from the rear face of cylinder and the small brass thrust washer No. 7D428 from the front face of cylinder.
3. Remove the snapring using a suitable tool.
4. Remove the pressure plate and the clutch pack. Mark for reassembly.
5. Install clutch spring compressor tool and remove the return spring snapring.
6. Remove the return spring assembly.
7. Remove the piston from the intermediate brake drum.
8. Remove inner and outer seals from the drum using O-ring tool T71P–19703–C or equivalent.

Inspection

1. Inspect the clutch cylinder thrust surfaces, piston bore, and clutch plate serrations for scores or burrs. Minor scores or burrs may be removed with crocus cloth. Replace the clutch cylinder if it is badly scored or damaged.
2. Check the fluid passage in the clutch cylinder for blockage. Clean out all fluid passages. Inspect the clutch piston for scores and replace if necessary. Inspect the check balls for freedom of movement and proper seating.
3. Check the clutch release spring for distortion and cracks.
4. Inspect the composition clutch plates, steel clutch plates, and clutch pressure plate for worn or scored bearing surfaces.

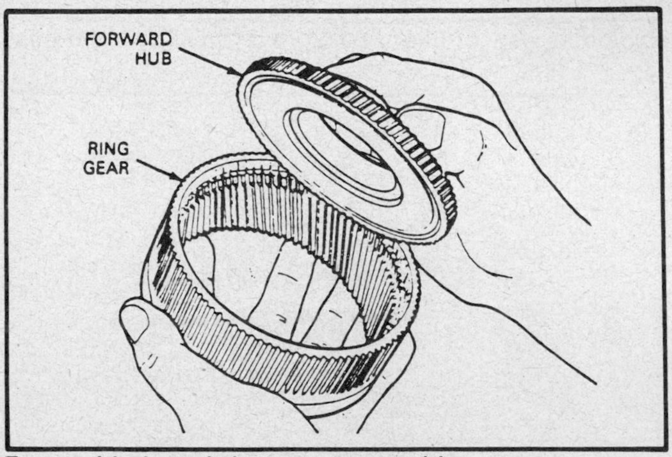
FORWARD HUB
RING GEAR

Forward hub and ring gear assembly

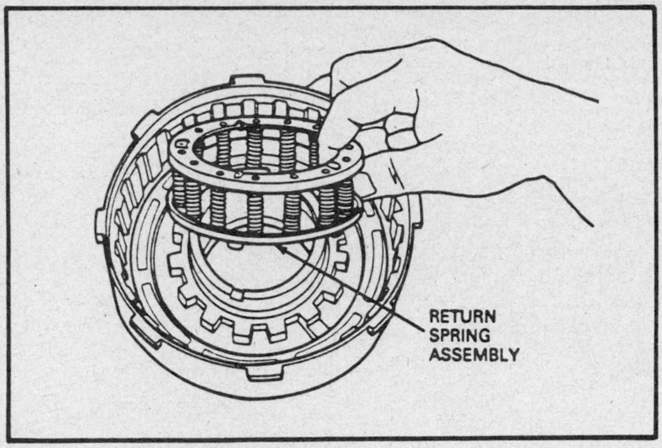
RETURN
SPRING
ASSEMBLY

Return spring assembly

5. Check the clutch plates for flatness and fit on the clutch hub serrations. Discard any plate that does not slide freely on the serrations or that is not flat.

6. Check the clutch hub thrust surfaces for scores and the clutch hub splines for wear.

Assembly

1. Install the inner seal into the cylinder with seal groove facing down.
2. Install the outer seal into the intermediate brake drum with seal groove facing down.
3. Inspect the piston check ball for freedom of movement.
4. Install piston into drum and return spring assembly. Using clutch spring compressor tool T65l–77515–A or equivalent, install the snapring. Ensure that the protrusions on the spring retainer are properly engaged with the lugs on the clutch piston.
5. Install the 4 plate clutch pack, starting with the steel plate. Install the pressure plate.

NOTE: Soak the clutch plates with clean transmission for 15 minutes before installing.

6. Install the selective snapring. Check the stack-up using a feeler gauge. If not with specification, install correct snapring and recheck.
   Specification:
   0.060–0.045 in. (1.52–1.15mm)
   Selective snaprings:
   0.065–0.069 in. (1.65–1.75mm)
   0.074–0.078 in. (1.88–1.98mm)
   0.083–0.087 in. (2.10–2.20mm)
7. Install the small brass thrust washer No. 7D428 on face of the cylinder.
8. Install the large brass thrust washer No. 7G401 on face of cylinder.
9. Install the intermediate one-way clutch end cap, one-way clutch assembly, outer race, and outer end cap over inner race.

NOTE: The lip is up on one-way clutch.

10. Install outer race so that the race turns counterclockwise.

FORWARD CLUTCH ASSEMBLY

Disassembly

1. Remove the needle bearing assembly from inner face of the cylinder.
2. Remove the needle bearing assembly No. 7F374.
3. Remove both Teflon® seal rings from the grooves.
4. Remove the snapring and rear pressure plate.

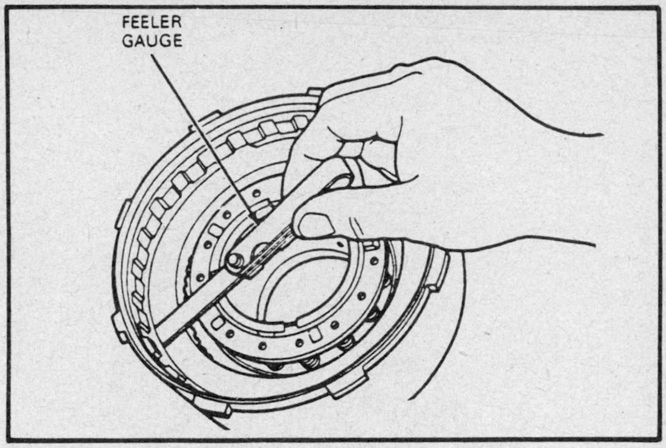
FEELER
GAUGE

Checking stack-up using feeler gauge

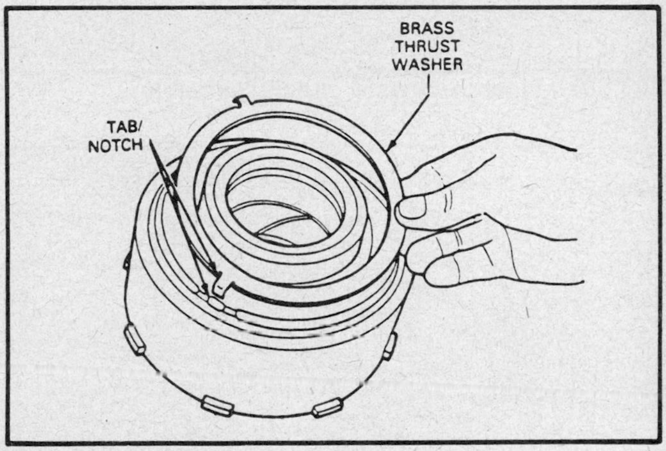
BRASS
THRUST
WASHER
TAB/
NOTCH

Correct installation of large brass washer on assembly

5. Remove the 4 plate clutch pack, cushion spring and front pressure plate. Mark for reassembly.
6. Remove the return spring snapring and return spring.
7. Remove the steel ring from the piston groove.
8. Remove the piston from the cylinder by applying compressed air to the piston.
9. Remove the outer seal from piston.
10. Remove the inner seal from the cylinder.

Inspection

1. Inspect the clutch cylinder thrust surfaces, piston bore, and clutch plate serrations for scores or burrs. Minor scores or burrs may be removed with crocus cloth. Replace the clutch cylinder if it is badly scored or damaged.
2. Check the fluid passage in the clutch cylinder for blockage. Clean out all fluid passages. Inspect the clutch piston for scores and replace if necessary. Inspect the check balls for freedom of movement and proper seating.
3. Check the clutch release spring for distortion and cracks.
4. Inspect the composition clutch plates, steel clutch plates, and clutch pressure plate for worn or scored bearing surfaces.
5. Check the clutch plates for flatness and fit on the clutch hub serrations. Discard any plate that does not slide freely on the serrations or that is not flat.
6. Check the clutch hub thrust surfaces for scores and the clutch hub splines for wear.

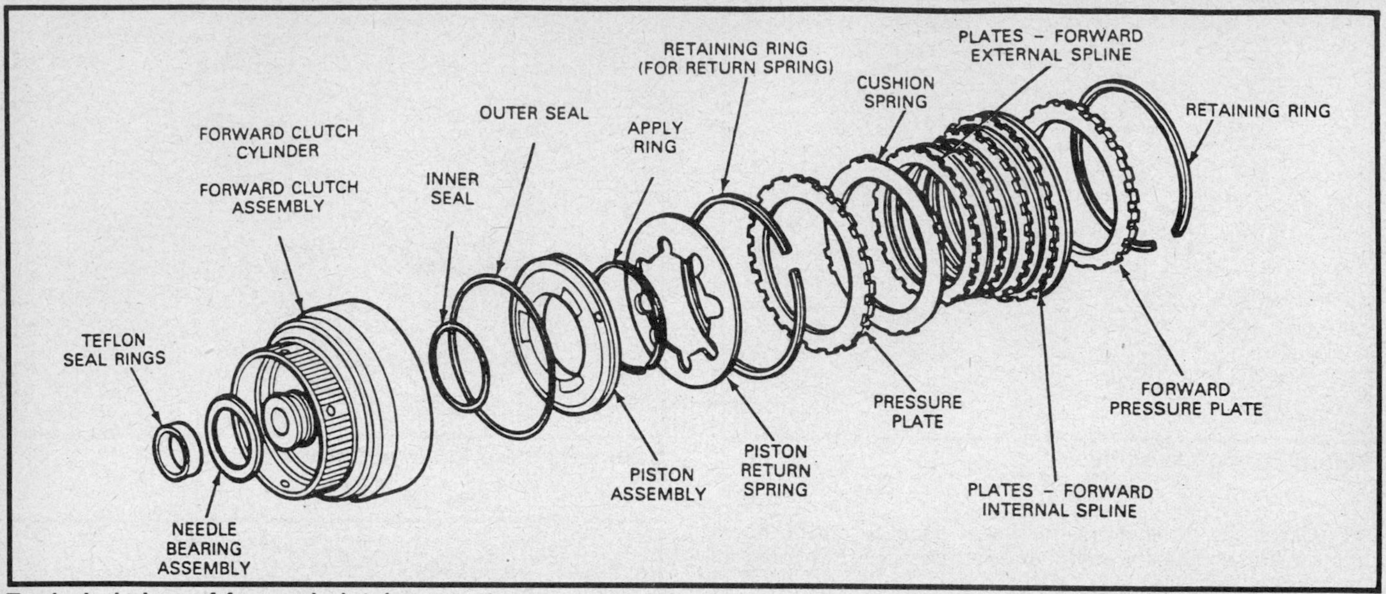
RETAINING RING
(FOR RETURN SPRING)
PLATES – FORWARD
EXTERNAL SPLINE
CUSHION
SPRING
RETAINING RING
OUTER SEAL
FORWARD CLUTCH
CYLINDER
APPLY
RING
FORWARD CLUTCH
ASSEMBLY
INNER
SEAL
TEFLON
SEAL RINGS
PRESSURE
PLATE
FORWARD
PRESSURE PLATE
PISTON
RETURN
SPRING
PISTON
ASSEMBLY
PLATES – FORWARD
INTERNAL SPLINE
NEEDLE
BEARING
ASSEMBLY

Exploded view of forward clutch assembly

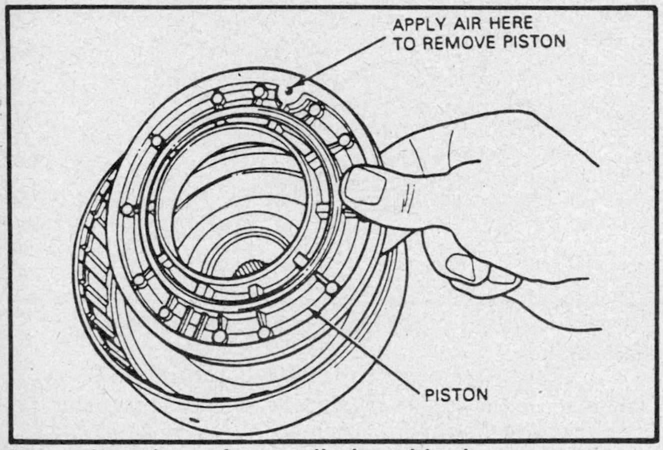
APPLY AIR HERE
TO REMOVE PISTON
PISTON

Removing piston from cylinder with air pressure

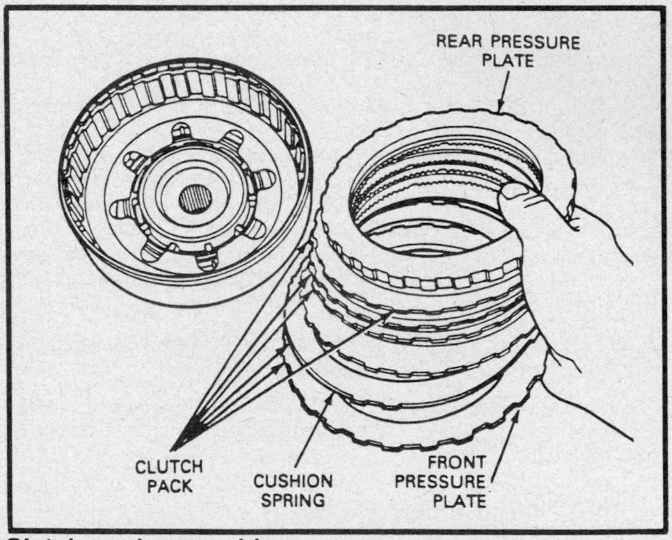
REAR PRESSURE
PLATE
CLUTCH
PACK
CUSHION
SPRING
FRONT
PRESSURE
PLATE

Clutch pack assembly

Assembly

1. Install the inner seal in the cylinder.
2. Install the outer seal on the piston.
3. Inspect the piston check ball for freedom of movement.
4. Install piston into cylinder using lip seal protector T77L–77548–A or equivalent.
5. Install the steel ring into the groove on the piston.
6. Install the retrun spring with the return spring fingers against the piston steel ring.
7. Install the snapring and the front pressure plate.
8. Install cushion spring.
9. Install the 4 steel plates and 4 friction plates alternately, starting with a steel plate.

NOTE: Soak the clutch plates with clean transmission fluid for 15 minutes before installing.

10. Install the rear pressure plate and the selective snapring.
11. Check the stack-up clearance, using a feeler gauge. If not with specification install correct snapring and recheck.
    Specification:
    0.055–0.030 in. (1.40–0.76mm)
    Selective snaprings:
    0.056–0.060 in. (1.42–1.52mm)
    0.074–0.078 in. (1.88–1.98mm)
    0.092–0.096 in. (2.34–2.44mm)
    0.110–0.114 in. (2.79–2.90mm)
    0.128–0.132 in. (3.25–3.35mm)
12. Install the Teflon® seal rings in the grooves.
13. Install the needle bearing assembly No. 75374 over Teflon® seal snout.
14. Install the needle bearing assembly on inner face of cylinder, with notched inner race facing outward.

FORWARD CARRIER

Disassembly

1. Remove the needle bearing assembly No. 7F078 from the carrier.
2. Remove the thrust washer No. 7D423 from the front side of the carrier.

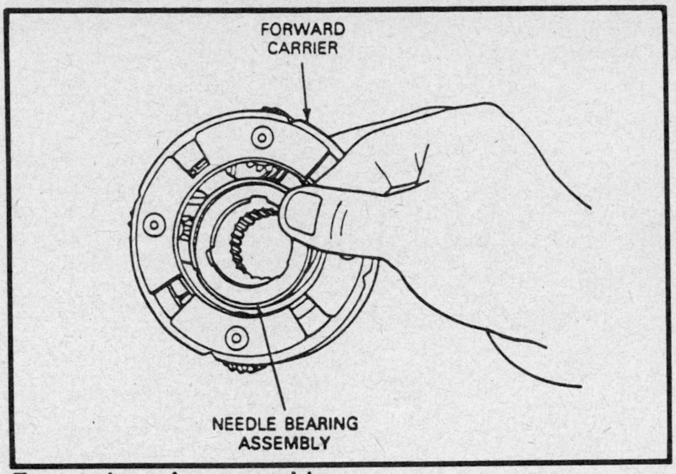
FORWARD CARRIER
NEEDLE BEARING ASSEMBLY
Forward carrier assembly

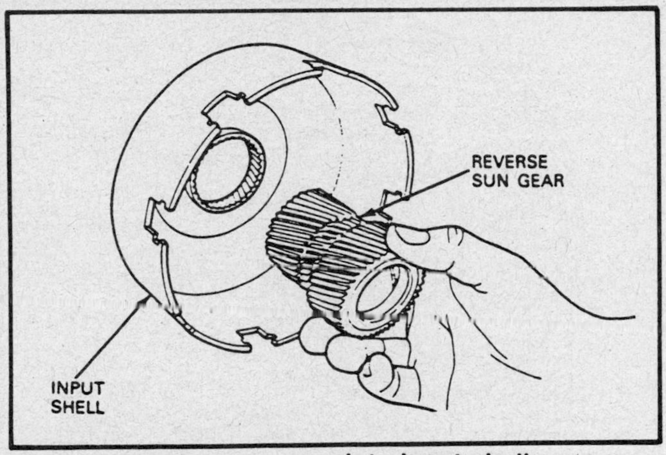
REVERSE SUN GEAR
INPUT SHELL
Installing reverse sun gear into input shell

Inspection

1. The pins and shafts in the planet assembly should be checked for loose fit and or complete disengagement.

NOTE: Individual parts of the planet carrier are not serviceable.

2. Inspect the pinion gears for damaged or excessively worn teeth.
3. Check for free rotation of the pinon gears.

Assembly

1. Place thrust washer No. 7D423 on front side of carrier, using petroleum jelly to hold in place.The thrust washer tabs go into the carrier.
2. Install the needle bearing assembly on the inner face of the carrier. The notched inner race should face outward.

INPUT SHELL

Disassembly

1. Remove the snapring from the reverse sun gear using snapring pliers.
2. Remove the thrust washer No. 7D066 from the input shell.
3. Remove the reverse sun gear from input shell.

Inspection

1. Check the input shell for cracks or damage.
2. Check the reverse sun gear for wear.

Assembly

1. Install the reverse sun gear into the input shell so that lube hole in the sun gear is between the stand off pads on the shell.
2. Install the input shell thrust washer No. 7D066 onto the reverse sun gear.
3. Install the snapring onto the reverse sun gear using snapring pliers.

REVERSE PLANET CARRIER

Disassembly

1. Remove the rear thrust washer No. 7D423.
2. Remove the front thrust washer No. 7D423.

Inspection

1. The pins and shafts in the planet assembly should be checked for loose fit and or complete disengagement.

NOTE: Individual parts of the planet carrier are not serviceable.

2. Inspect the pinion gears for damaged or excessively worn teeth.
3. Check for free rotation of the pinon gears.

Assembly

1. Install the front thrust washer No. 7D423. Hold the thrust washer in place by using petroleum jelly.
2. Install rear thrust washer No. 7D423. Hold the thrust washer in place by using petroleum jelly.

REVERSE ONE-WAY CLUTCH

Disassembly

1. Remove the reverse hub and needle bearing assembly No. 7E413 from the reverse one-way clutch hub.
2. Remove the snapring from the reverse one-way clutch hub.
3. Remove the brass thrust washer and rollers from the reverse clutch hub.

Inspection

1. Inspect the outer and inner races for scores or damaged surface areas where the rollers or sprags contact the races.
2. Inspect the rollers, sprags and springs for excessive wear or damage.
3. Inspect the spring and cage for bent or damaged spring retainers.

Assembly

1. Install the one-way clutch rollers and the brass thrust washer No. 7E194.
2. Install the snapring onto the one-way clutch hub.
3. Install the needle bearing assembly No. 7E413 with the smooth race surface facing up. Lightly grease the thrust washer with petroleum jelly to hold in place upon installation.

REVERSE CLUTCH PISTON

Disassembly

1. Remove the outer piston seal.
2. Remove the inner piston seal.

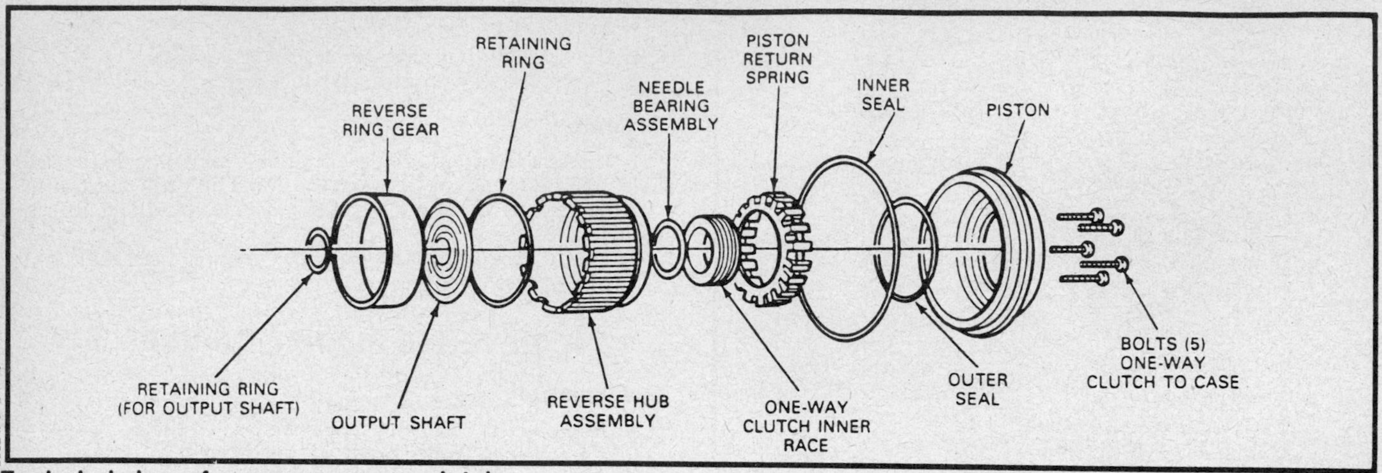
RETAINING RING
PISTON RETURN SPRING
NEEDLE BEARING ASSEMBLY
INNER SEAL
REVERSE RING GEAR
PISTON
BOLTS (5) ONE-WAY CLUTCH TO CASE
OUTER SEAL
RETAINING RING (FOR OUTPUT SHAFT)
OUTPUT SHAFT
REVERSE HUB ASSEMBLY
ONE-WAY CLUTCH INNER RACE
Exploded view of reverse one-way clutch

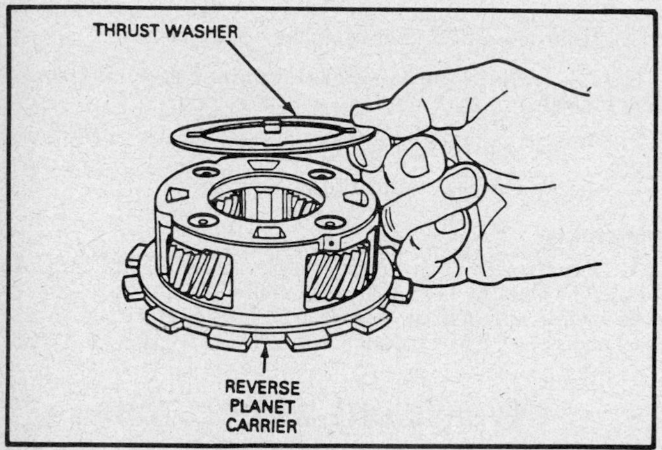
THRUST WASHER
REVERSE PLANET CARRIER
Reverse planet carrier

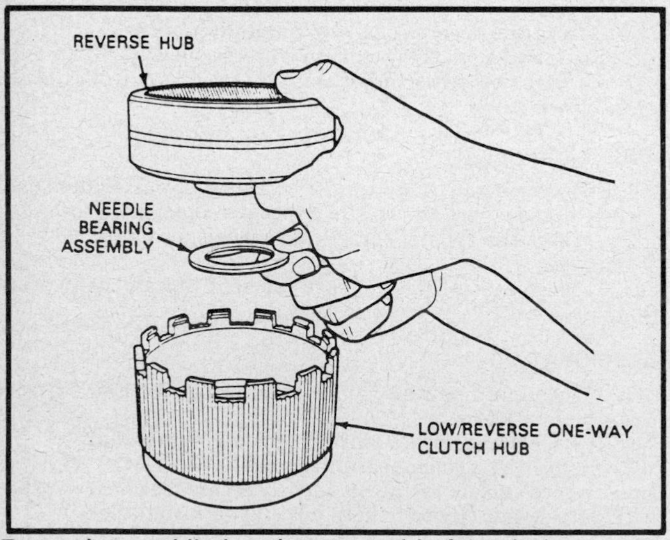
REVERSE HUB
NEEDLE BEARING ASSEMBLY
LOW/REVERSE ONE-WAY CLUTCH HUB
Removing neddle bearing assembly from hub

Inspection

Check piston surface for cracks or damage. Always check piston groves for distortion so that the piston seal will seat properly upon installation.

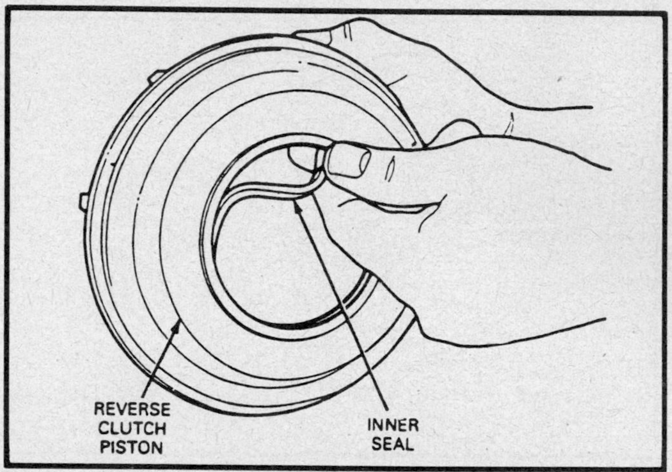
REVERSE CLUTCH PISTON
INNER SEAL
Removing inner seal from reverse clutch piston

Assembly

1. Install the inner piston seal.
2. Install the outer piston seal.

EXTENSION HOUSING

Disassembly

1. Using a suitable tool remove the extension housing seal.
2. On 2WD models, check the extension housing bushing for wear or damage. If wear is visible replace the bushing as the new extension housing seal will fail under these conditions.

Assembly

1. Install extension housing seal, using a suitable tool.
2. Install the extension housing bushing in the tailshaft by carefuly driving the bushing in place using a suitable tool.

ASSEMBLY OF SUBASSEMBLIES

Assembly

1. Place the thrust washer No. 7D428 onto the intermediate brake drum. Lightly grease thrust washer with petroleum jelly to hold in correct location.
2. Install the forward clutch onto the intermediate brake drum.

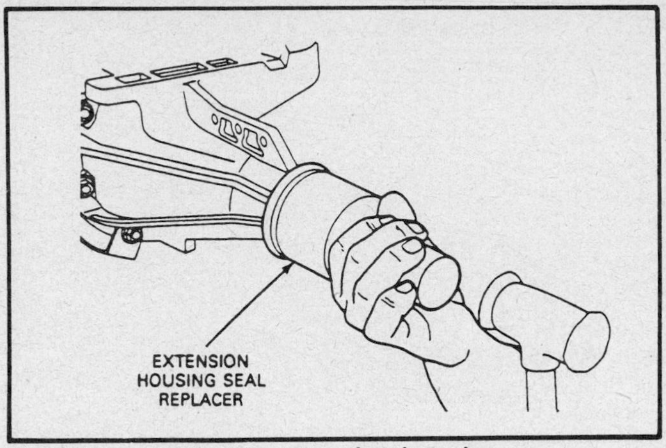
EXTENSION
HOUSING SEAL
REPLACER

Installing rear seal in extension housing

3. Install the needle bearing assembly No. 7F078 onto the intermediate brake drum and forward clutch assembly.

NOTE: Lightly grease with petroleum jelly the needle bearing assembly. Notched inner race must be facing outward (up).

4. Grease with petroleum jelly the plastic thrust washer No. 7D090 and place onto the forward clutch hub. Place forward clutch hub into the intermediate brake drum and forward clutch assemblies.
5. Grease with petroleum jelly the thrust washer No. 7D423 and place onto the forward planet carrier. Place carrier into assembly.
6. Install the needle bearing assembly No. 7F078 into forward carrier assembly.

NOTE: Lightly grease with petroleum jelly the needle bearing to hold in place. The notched inner race surface should face up.

7. Align the input shell notches with intermediate brake drum.
8. Install the input shell onto the assembly.
9. Install the needle bearing assembly No. 7F374 into the front end of forward clutch assembly.
10. Install the intermediate brake drum, forward clutch and input shell remover/install T89T–70010–E or equivalent.

Transmission Assembly

NOTE.: Soak all the friction clutch plates in clean transmission for 15 minutes. Lightly lubricate all the O-ring seals before installing using transmission fluid and grease all thrust washers with petroleum jelly, to hole in place during assembly.

1. Mount transmission in suitable holding fixture. Rotate the transmission so that bell housing is facing up.
2. Install the inner and outer seals on the reverse clutch piston.
3. Install the reverse clutch piston using a suitable tool. Remove tool after installing piston.
4. Install the reverse piston return spring assembly and one-way clutch inner race.
5. Attach to case with 5 (11mm) bolts and torque to 18–25 ft. lbs.
6. Install a 6 internal spline plate reverse clutch pack starting with an external spline plate. Alternate external spline plates with internal spline plates. Install snapring.

NOTE: No stack-up clearance measurement is required.

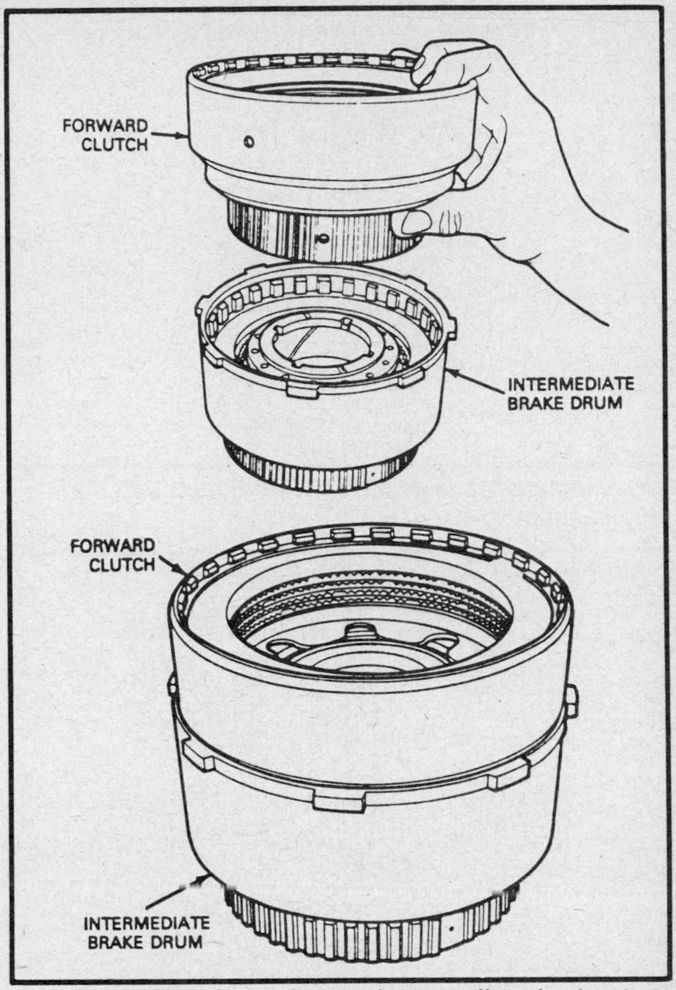
FORWARD
CLUTCH
INTERMEDIATE
BRAKE DRUM
FORWARD
CLUTCH
INTERMEDIATE
BRAKE DRUM

Installing forward clutch onto intermediate brake drum

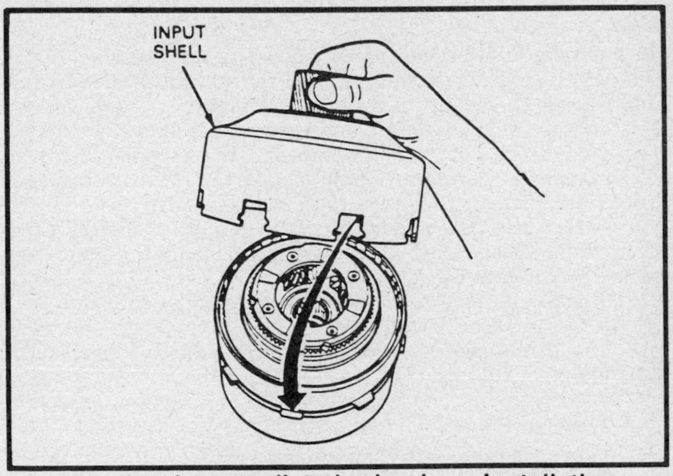
INPUT
SHELL

Input shell to intermediate brake drum installation

7. Rotate transmission to horizontal position. Grease with petroleum jelly the steel side of the thrust washer No. 7B368 and place on the rear of case so that bronze side is facing outward.
8. Install the snapring onto the output shaft. Slide park gear

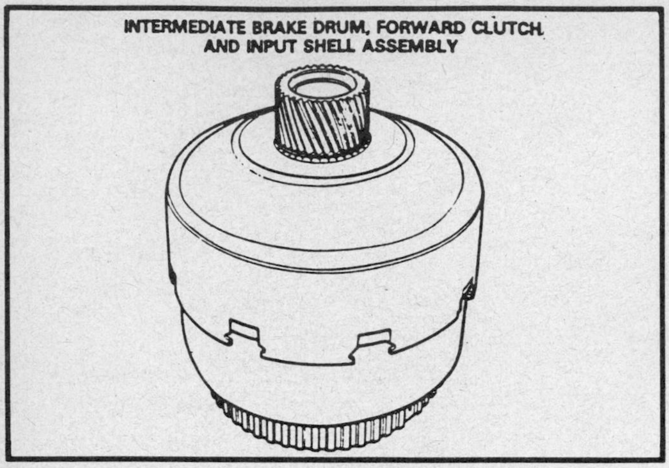
INTERMEDIATE BRAKE DRUM, FORWARD CLUTCH AND INPUT SHELL ASSEMBLY

Intermediate brake drum, forward clutch and input shell assembly

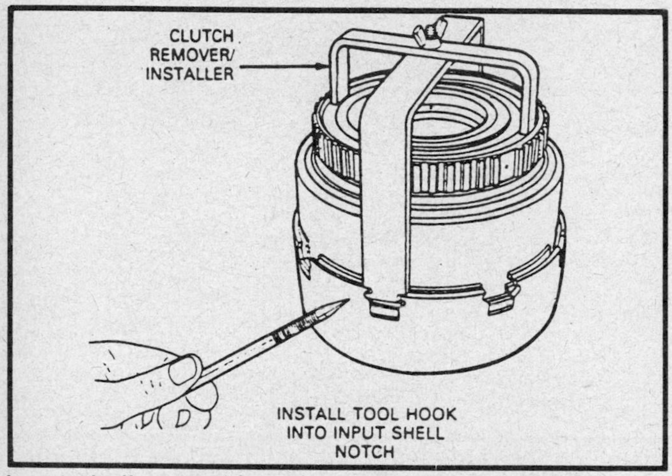
CLUTCH REMOVER/ INSTALLER
INSTALL TOOL HOOK INTO INPUT SHELL NOTCH

Input shell assembly with special tool installed

onto the shaft with thrust surface opposite the snapring. Install the output shaft. Do not overextend the snapring when installing. Ensure the snapring is securely seated in the groove.

9. Install the reverse hub and low reverse one-way clutch.
10. Install the output shaft hub and the reverse ring gear, placing needle bearing assembly No. 7E413 on rear surface of hub. Hold bearing in place with petroleum jelly.
11. Install the new snapring onto the output shaft. Do not overextend the snapring when installing. Ensure the snapring is securely seated in the groove.
12. Install the reverse plate carrier assembly into hub with thrust washer No. 7D423.
13. Install the snapring. Rotate the transmission to that bell housing is facing up.
14. Attach clutch remover/installer T89T–70010–E or equivalent, onto the input shell and lower the entire assembly (intermediate brake drum assembly, forward clutch assembly and input shell assembly) into case.

NOTE: It may be necessary to rotate output shaft to seat reverse sun gear.

15. Install the intermediate band so that 1 ear is resting on the reaction pin.
16. Install the servo snapring, retaining plate, piston and rod assembly and servo spring. Apply slight downward pressure to plate while removing the snapring.

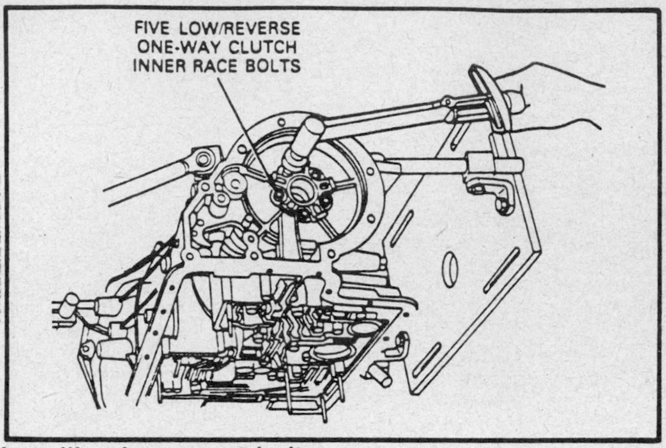
FIVE LOW/REVERSE ONE-WAY CLUTCH INNER RACE BOLTS

Installing inner race bolts

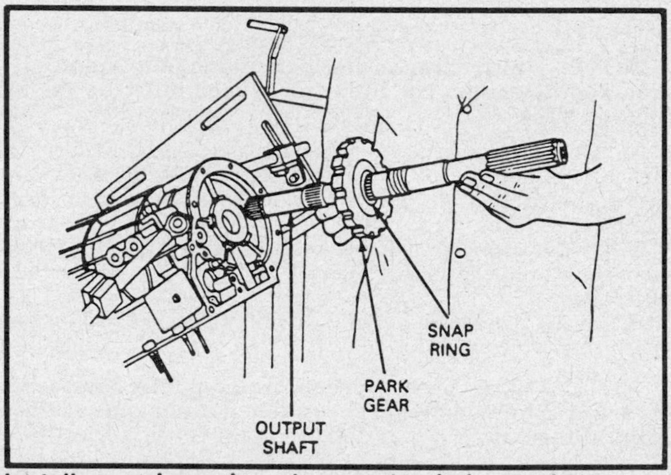
SNAP RING
PARK GEAR
OUTPUT SHAFT

Install snapring – do not overextended snapring

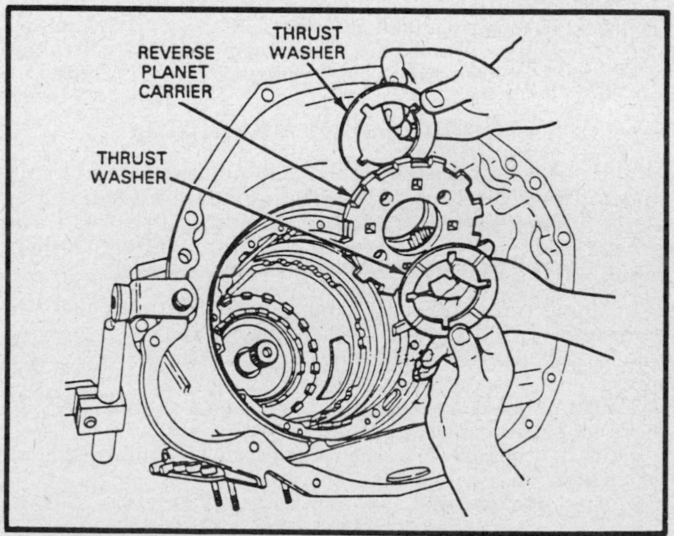
THRUST WASHER
REVERSE PLANET CARRIER
THRUST WASHER

Installing reverse plate carrier assembly

17. Install the intermediate pressure plate. Install the cluch pack starting with internal spline plate. Install the apply plate.
18. Determine the endplay. The transmission rear endplay check determines the amount of space existing between the

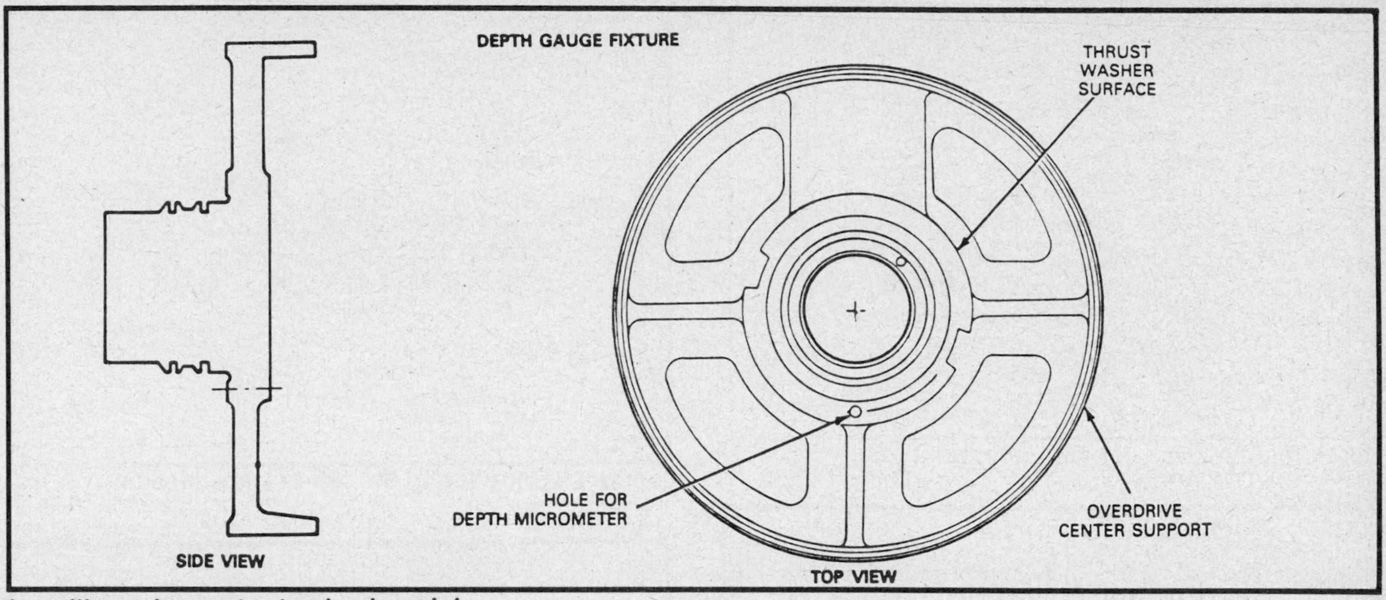
DEPTH GAUGE FIXTURE
THRUST
WASHER
SURFACE
HOLE FOR
DEPTH MICROMETER
OVERDRIVE
CENTER SUPPORT
SIDE VIEW
TOP VIEW

Installing micrometer to check endplay

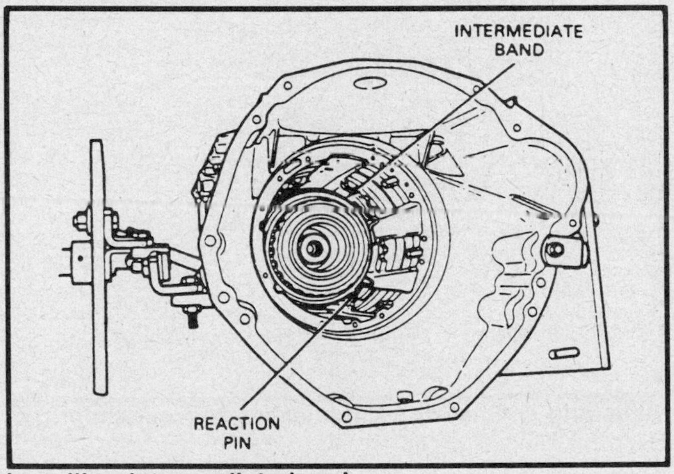
INTERMEDIATE
BAND
REACTION
PIN

Installing intermediate band

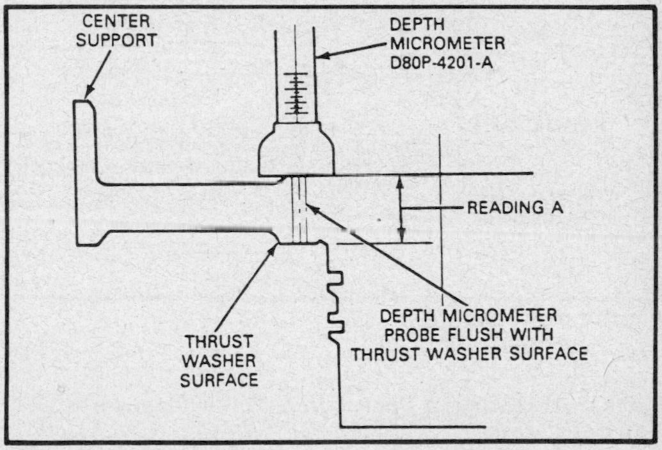
CENTER
SUPPORT
DEPTH
MICROMETER
D80P-4201-A
READING A
DEPTH MICROMETER
PROBE FLUSH WITH
THRUST WASHER SURFACE
THRUST
WASHER
SURFACE

Micrometer reading A

thrust washer surfaces of the center support and the intermediate brake drum 0.081–0.032 in. (2.06–8.1mm).

19. Use depth micrometer D80P–4201–A or equivalent to access the area between the thrust surfaces of the support and the intermeidate brake drum. Remove the cast iron seals from the center support to allow easy insertion into the intermediate brake drum.

20. Place depth micrometer D80P–4201–A or equivalent over the drilled hole in center support fixture. Extend the micrometer probe until it is flush with thrust washer surface. Record micrometer reading. This is reading A.

21. Install the center support into intermediate brake drum and gently wiggle the input shaft to allow center support fixture to slide into the intermediate brake drum using its own weight. Ensure it is fully seated in transmission case.

22. Place the depth micrometer over the drilled hole in the center support.

23. Continue the extending micrometer probe until it contacts thrust washer surface of intermediate brake drum. This is reading B.

24. Subtract reading A from reading B. The difference be-

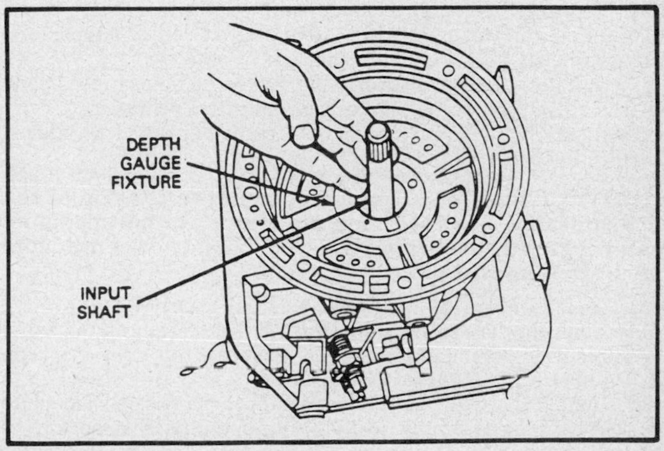
DEPTH
GAUGE
FIXTURE
INPUT
SHAFT

Installing center support

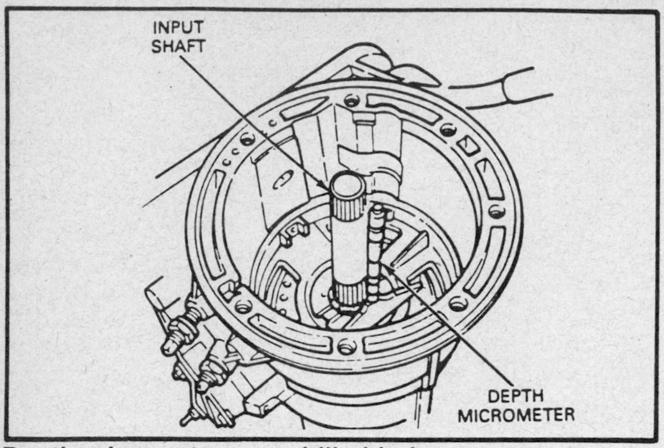
INPUT SHAFT
DEPTH MICROMETER
Depth micrometer over drilled hole

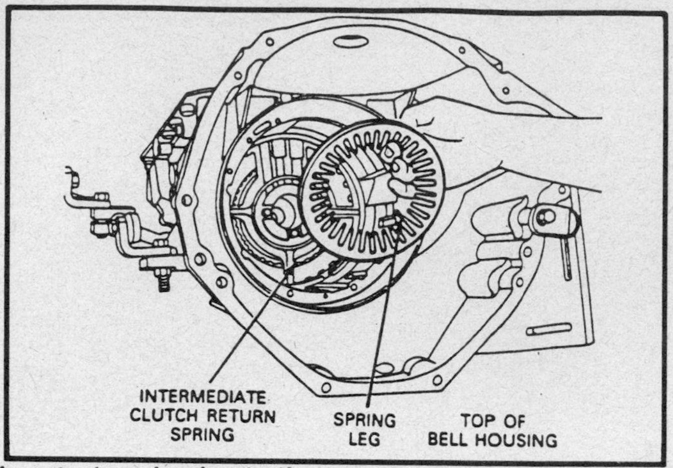
INTERMEDIATE CLUTCH RETURN SPRING
SPRING LEG
TOP OF BELL HOUSING
Locate 1 spring leg to the top of transmission

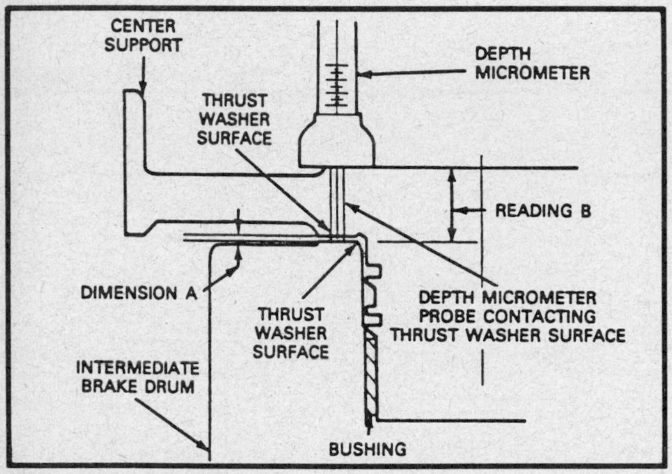
CENTER SUPPORT
DEPTH MICROMETER
THRUST WASHER SURFACE
READING B
DIMENSION A
THRUST WASHER SURFACE
DEPTH MICROMETER PROBE CONTACTING THRUST WASHER SURFACE
INTERMEDIATE BRAKE DRUM
BUSHING
Micrometer reading B

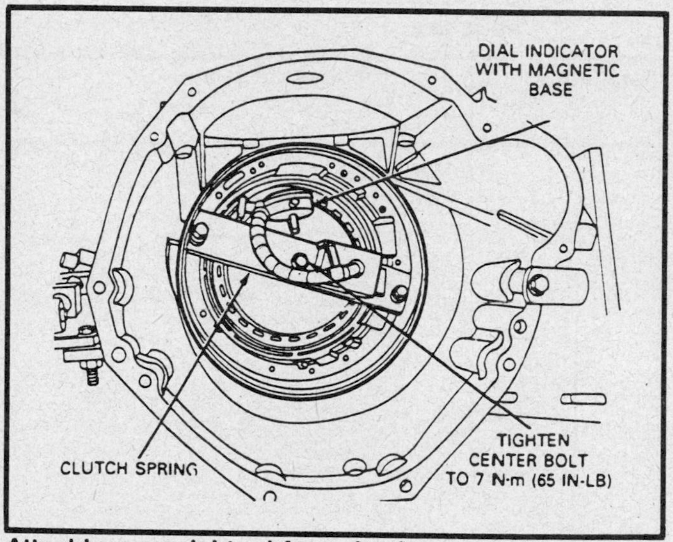
DIAL INDICATOR WITH MAGNETIC BASE
CLUTCH SPRING
TIGHTEN CENTER BOLT TO 7 N·m (65 IN-LB)
Attaching special tool for selecting correct snapring

tween these readings is Dimension A. This is the space between thrust surfaces.

25. Subtract the thrust washer thickness from Dimension A to determine final endplay. Specification is 0.081–0.032 in. (2.06–0.81mm). If the final dimension is outside specified limits, this indicates improper assembly, missing parts or parts out of specification. This requires rebuilding the unit again.

26. Remove the center support from the intermediate brake drum. Position washer No. 7L326 on rear of center support using petroleum jelly to hold in place.

27. Install the center support, align with the holes in feed port. Install the 2 feed bolts. Do not tighten at this time.

28. Install the intermediate clutch return spring with dished surface inward.

NOTE: Locate 1 spring leg pointing to the top of the transmission. The intermediate clutch return spring locator legs must be properly located inside of center support circular cast rib.

29. Install intermediate overdrive cylinder assembly into case. Align the cylinder assembly locator tab with center support and feel hole with hole in case. Install 1 feed bolt but do not tighten it. Do not cock cylinder when installing.

30. Install the trail selective snapring over the intermediate clutch cylinder assembly so that ring opening is at bottom of case for proper oil drainback. Place clutch spring compressor plate T89T–70010–F and intermediate clutch spring fixture T89T–70010–C or equivalents onto the intermediate clutch cylinder assembly.

31. Tighten the center bolt to 65 inch lbs. Seat the selective snapring into the case ring groove.

32. Attach a dial indicator with magnetic base D78P–4201–B or equivalent or to bar. Place stylus onto the compressor plate T89T–70010–F or equivalent and zero the dial.

33. Release the torque on center bolt and record the reading on the indicator. If reading is not within specifications, repeat procedure using the correct selective snapring.

Specifications:
0.054–0.026 in. (1.37–0.67mm)
Selective snaprings:
0.061–0.057 in. (1.55–1.45mm)
0.080–0.076 in. (2.05–1.95mm)
0.100–0.098 in. (2.60–2.50mm)

34. Remove the clutch spring tool. Tighten the 3 (13mm) feed bolts into the intermediate overdrive cylinder assembly and the center support. Tighten the front feed bolt to 6–10 ft. lbs. and both rear feed bolts to 8–12 ft. lbs.

35. Using petroleum jelly to hold in place, postion needle bearing assembly No. 7G178 on the rear face of the center shaft. Install the center shaft, overdrive ring gear, overdrive planetary gearset and coast clutch cylinder as an assembly.

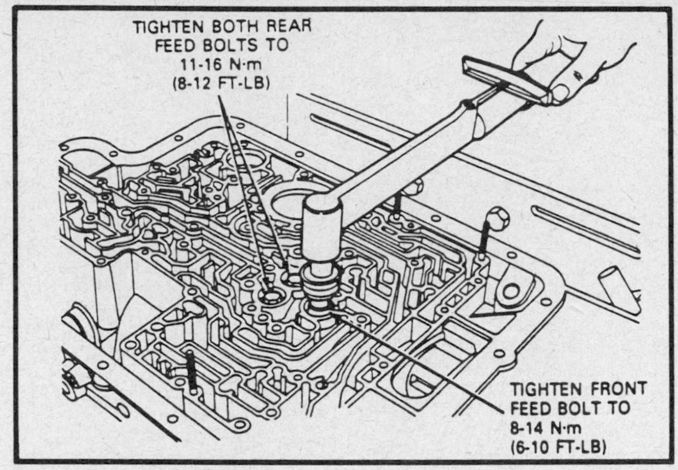
TIGHTEN BOTH REAR
FEED BOLTS TO
11-16 N·m
(8-12 FT-LB)
TIGHTEN FRONT
FEED BOLT TO
8-14 N·m
(6-10 FT-LB)
Tighten feed bolts

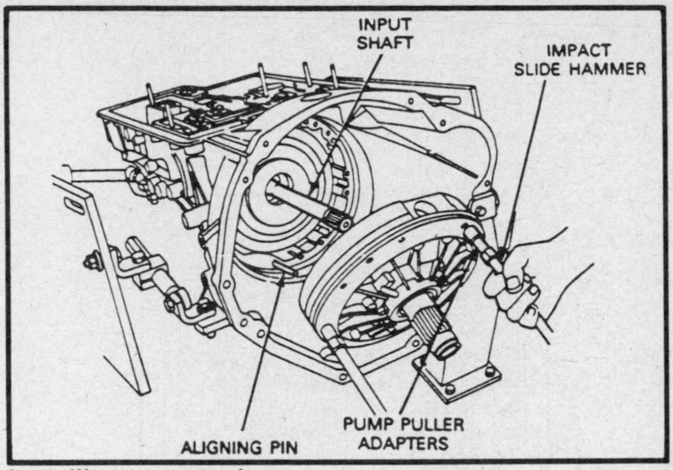
INPUT
SHAFT
IMPACT
SLIDE HAMMER
ALIGNING PIN
PUMP PULLER
ADAPTERS
Installing pump onto case

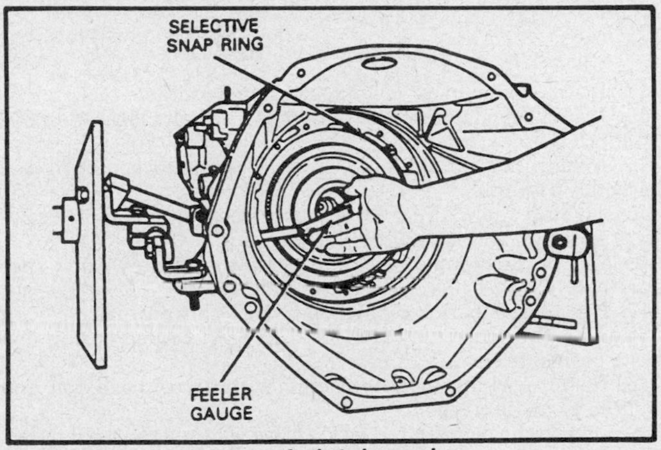
SELECTIVE
SNAP RING
FEELER
GAUGE
Measuring clearance of clutch pack

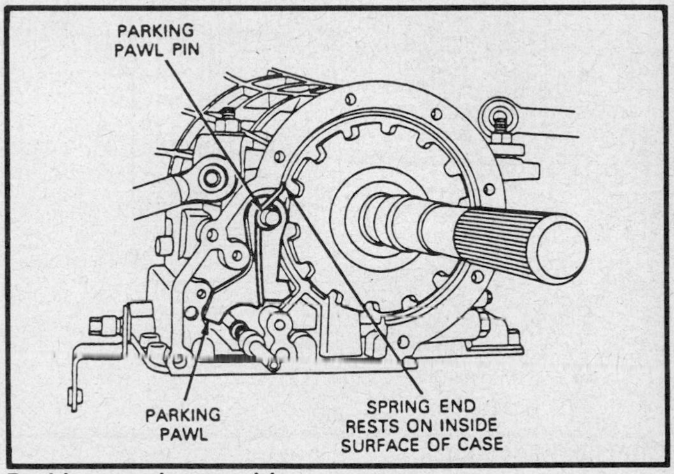
PARKING
PAWL PIN
PARKING
PAWL
SPRING END
RESTS ON INSIDE
SURFACE OF CASE
Parking pawl assembly

36. Install the overdrive clutch pack starting with a steel plate. Install pressure plate with dot facing outward and toward the top of the transmission. Install the trial selective snapring with opening at bottom of case.

37. Check the stack-up clearance using a feeler gauge. If not with specification, install correct selective snapring and recheck.

Specification:
0.047–0.022 in. (1.20–0.55mm)
Selective Snaprings
0.061–0.057 in. (1.55–1.45mm)
0.80–0.076 in. (2.05–1.95mm)
0.10–0.098 in. (2.60–2.50mm)
0.12–0.118 in. (3.10–3.00mm)
0.14–0.137 in. (3.60–3.50mm)

38. Install the pump gasket into the case. Screw the pump puller adapters T89T–70010–A or equivalent into pump threaded holes. Screw on impact slide hammers T59L–100–B or equivalent.

39. Install the thrust washer No. 7D014 and needle bearing assembly 7E486 onto pump. Use petroleum jelly to hold in place.

40. Install the input shaft (long splines end 1st) and alignment pin T89T–70010–B or equivalent into the case. Install the pump into the case. Position the filter inlet tube bore towards the valve body mounting surfaces.

41. Remove the old rubber coated washers from the 9 pump to case bolts. Install the new pump bolt washers. Remove aligning pin T89T–70010–B or equivalent. Install the pump using 9 bolts. Torque to 18–23 ft. lbs.

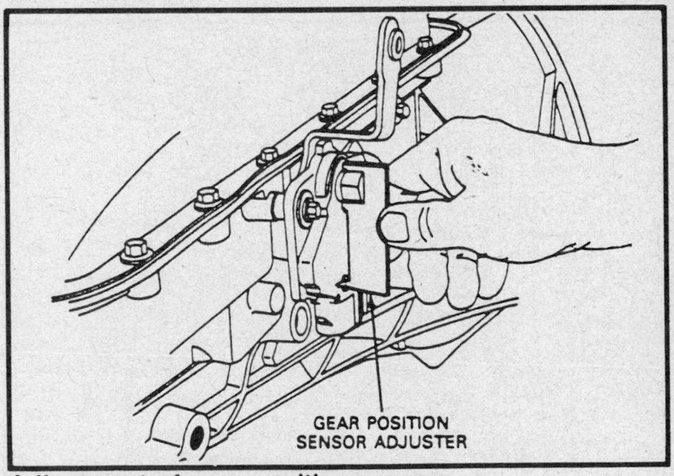
GEAR POSITION
SENSOR ADJUSTER
Adjustment of gear position sensor

NOTE: Draw the pump into the case assembly evenly, by tighten all bolts evenly, to avoid seal damage. Always remove the input shaft.

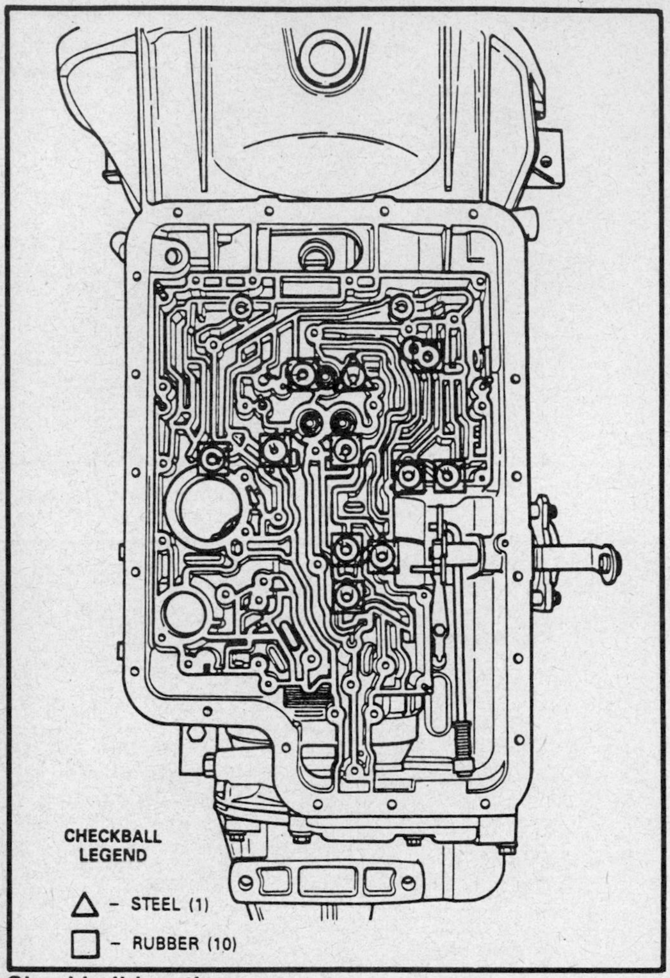
CHECKBALL
LEGEND
△ - STEEL (1)
□ - RUBBER (10)
Checkball location

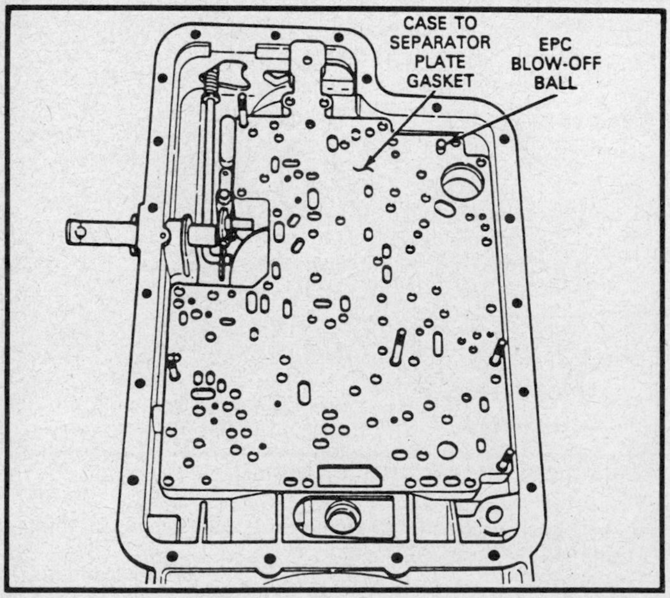
CASE TO
SEPARATOR
PLATE
GASKET
EPC
BLOW-OFF
BALL
EPC ball location

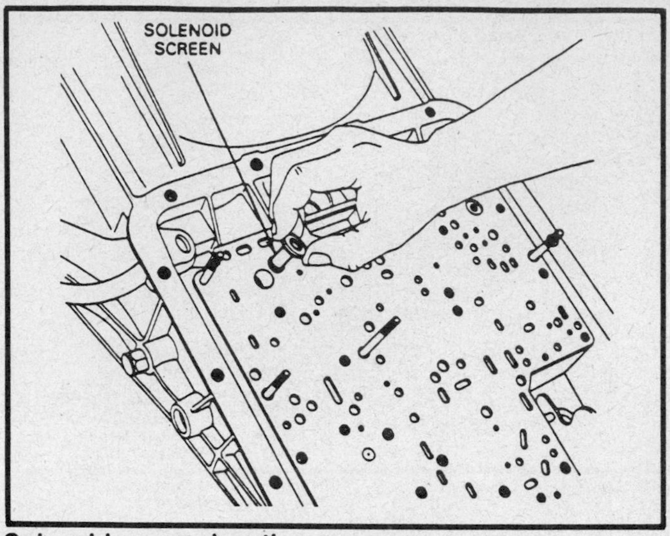
SOLENOID
SCREEN
Solenoid screen location

42. Install the manual lever seal using a suitable tool.
43. Install the manual lever detent spring bolt. Tighten to 80–100 inch lbs.
44. Install the manual lever, inner lever, park actuating rod assembly and nut.
45. Tighten the manual lever nut using suitable tool (crows foot) to 30–40 ft. lbs).
46. Install the manual lever roll so that pin is just below case surface.
47. Install the parking pawl, pin and parking pawl return spring on the rear face. The parking pawl return spring end rests on inside surface of case.
48. Install parking pawl abutment with Torx® head bolt and tighten to 16–20 ft. lbs).
49. Attach the parking rod guide plate. Torque to 16–20 ft. lbs. Ensure the plate dimple is facing inward.
50. Install the manual lever position sensor. Do not tighten bolts at this time. Align manual lever position sensor for **N** gear position using gear position sensor adjuster T89T–70010–J or equivalent.
51. Tighten the (8mm) bolts to 55–75 inch lbs.
52. Using petroleum jelly to hold in place position gasket on the extension housing.
53. Install the extension housing and wiring bracket on rear of case. Tighten bolts to 20–29 ft lbs.

NOTE: The 2 bottom bolts are longer on 4WD vehicles.

54. Rotate the transmission so that pan is facing up. Install 1 steel and 10 rubber check balls, EPC blow off spring and ball into the case pockets. Make sure you install the check balls in the correct location.
55. Install the case to separator plate gasket.
56. Install the separator plate.
57. Attach the reinforcing plate with 3 (8mm) bolts with marking stamped **UP** facing up. Tighten to 80–100 in lbs. Check the placement of EPC blow off ball.
58. Install the new separator to the control gasket.
59. Install the solenoid screen into the separator plate. Turn and lock solenoid screen.
60. Install the accumulator body over studs and attach with 2 nuts and 11 bolt. Tighten to 80–100 inch lbs.
61. Lower the main control body over studs. Align manual valve with manual lever.
62. Attach the valve body with 2 nuts and 14 bolts. Tighten to 80–100 inch lbs.

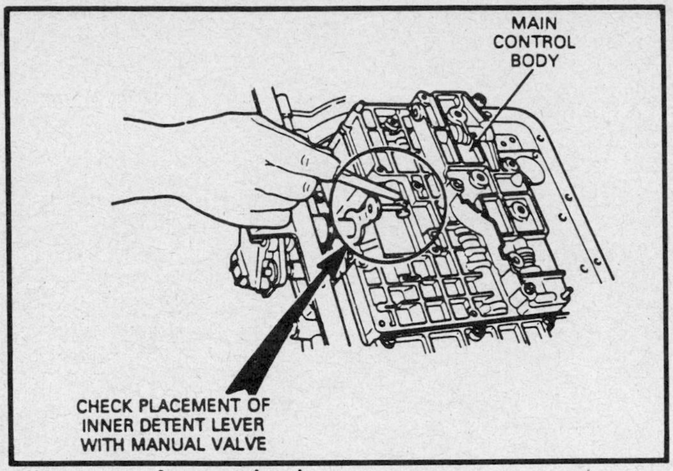
MAIN CONTROL BODY
CHECK PLACEMENT OF INNER DETENT LEVER WITH MANUAL VALVE
Alignment of manual valve

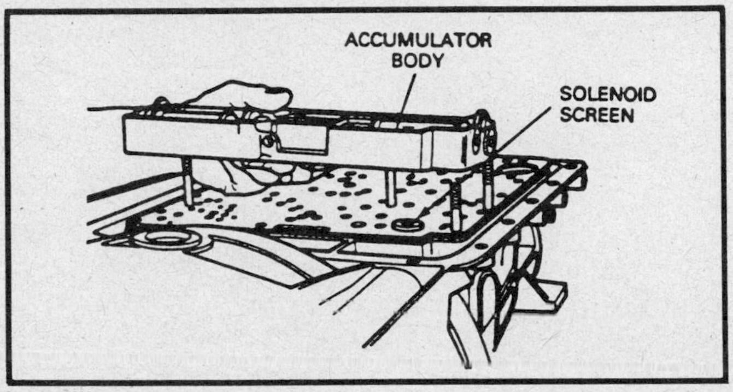
ACCUMULATOR BODY
SOLENOID SCREEN
Installing accumlator body

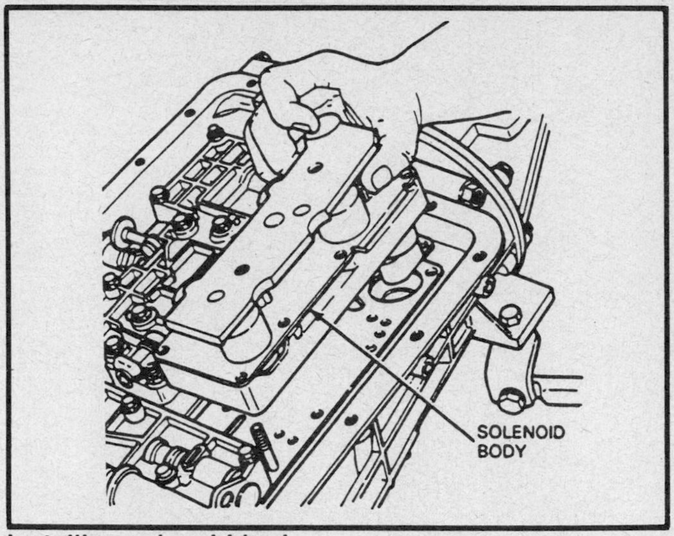
SOLENOID BODY
Installing solenoid body

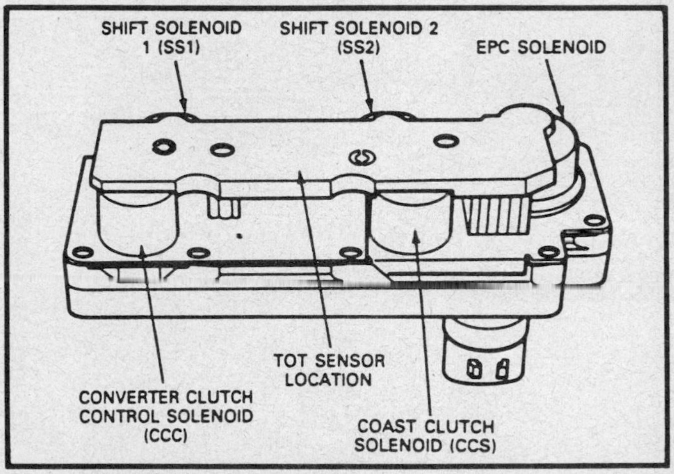
SHIFT SOLENOID 1 (SS1)
SHIFT SOLENOID 2 (SS2)
EPC SOLENOID
TOT SENSOR LOCATION
CONVERTER CLUTCH CONTROL SOLENOID (CCC)
COAST CLUTCH SOLENOID (CCS)
Shift solenoid body assembly

63. Install the solenoid body over stud and attach with 9 Torx® bolts and 1 nut. Tighten to 80–100 inch lbs. Prior to installing the solenoid body assembly, coast the case connector bore with M1C172-A grease or equivalent.
64. Install a new filter and seal assembly by lubricating the seal with transmission fluid and pressing the filter into place.
65. Place pan magnet on dimple in bottom of pan. Install new pan gasket.
66. Attach pan with 20 bolts. Tighten bolts to 10–12 ft. lbs.
67. If necessary, install the stub tube using suitable tool. Use the stripe on the side of tube for alignment. The stripe should be farthest outboard when installed.
68. Reinstall the input shaft, long splined end installed first.

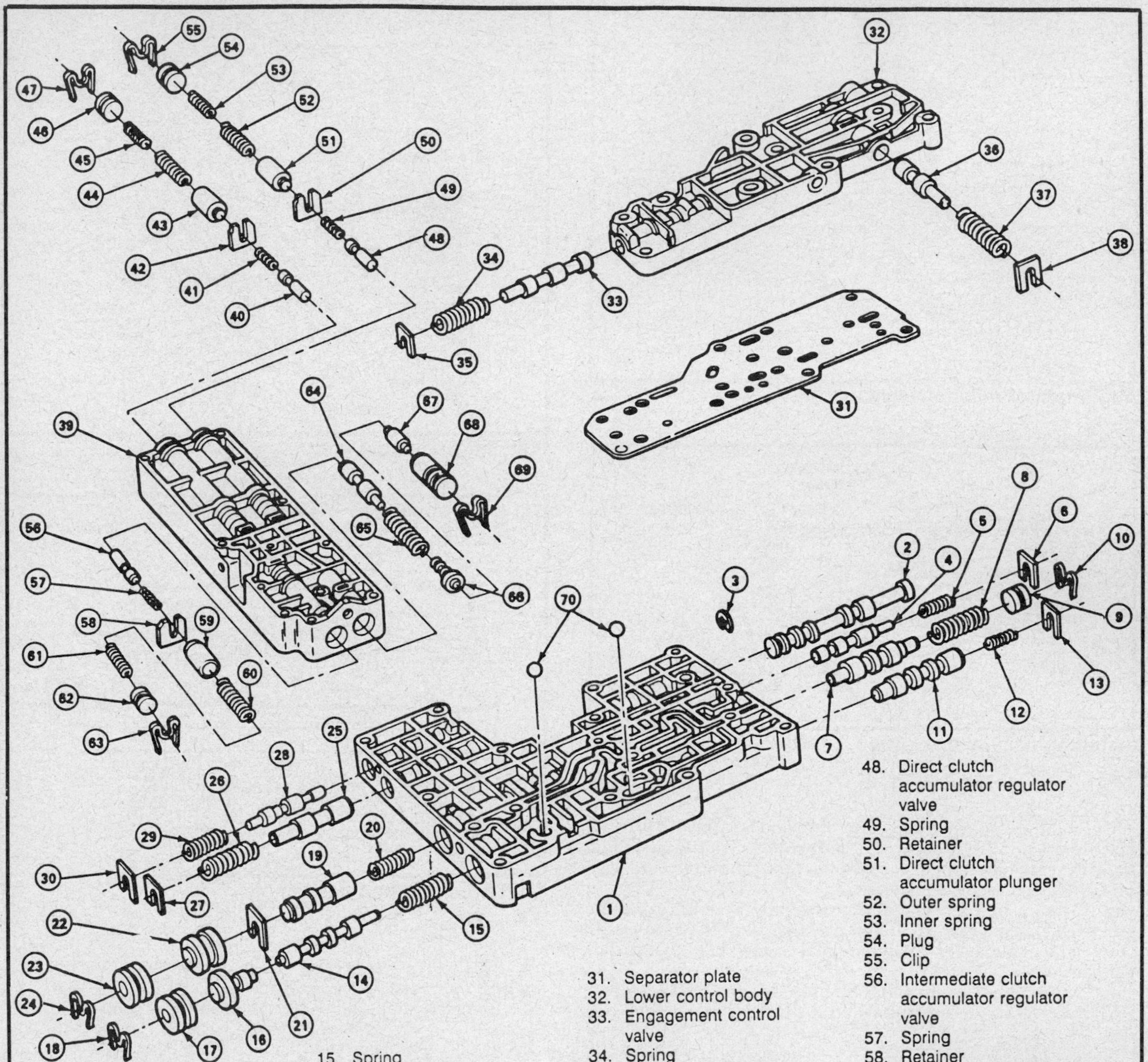

1. Main control body
2. Manual valve
3. Retainer
4. Low/reverse modulator valve
5. Spring
6. Retainer
7. 3-4 shift valve
8. Spring
9. Plug
10. Clip
11. 2-3 shift valve
12. Spring
13. Retainer
14. D2 shift valve
15. Spring
16. 1-2 shift valve
17. Plug
18. Clip
19. 4-3-2 manual timing valve
20. Spring
21. Retainer
22. 4-3-2 manual timing plunger
23. Plug
24. Clip
25. Coast clutch shift valve
26. Spring
27. Retainer
28. Solenoid regulator valve
29. Spring
30. Retainer
31. Separator plate
32. Lower control body
33. Engagement control valve
34. Spring
35. Retainer
36. 1-2 manual transition valve
37. Spring
38. Reatiner
39. Accumulator body
40. Overdrive clutch accumulaotr regulaotr valve
41. Spring
42. Retainer
43. Overdrive clutch accumulaotr plunger
44. Outer spring
45. Inner spring
46. Plug
47. Clip
48. Direct clutch accumulator regulator valve
49. Spring
50. Retainer
51. Direct clutch accumulator plunger
52. Outer spring
53. Inner spring
54. Plug
55. Clip
56. Intermediate clutch accumulator regulator valve
57. Spring
58. Retainer
59. Intermediate clutch accumulator plunger
60. Outer spring
61. Inner spring
62. Plug
63. Clip
64. Line pressure modulator valve
65. Outer spring
66. Spring and retainer assembly
67. Line pressure modulaotr plunger valve
68. Line pressure modulator sleeve
69. Clip
70. Check ball

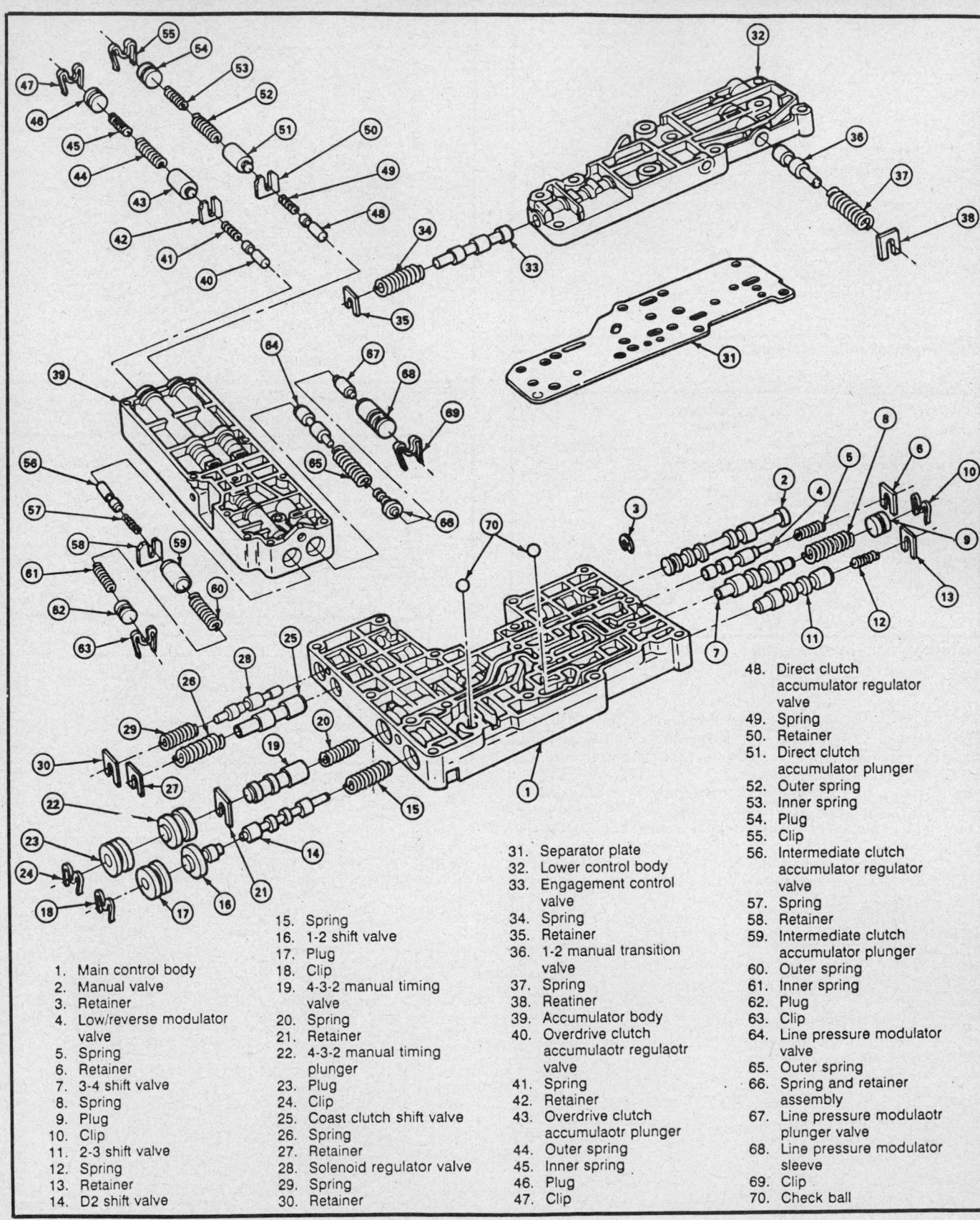

Exploded view of valve body

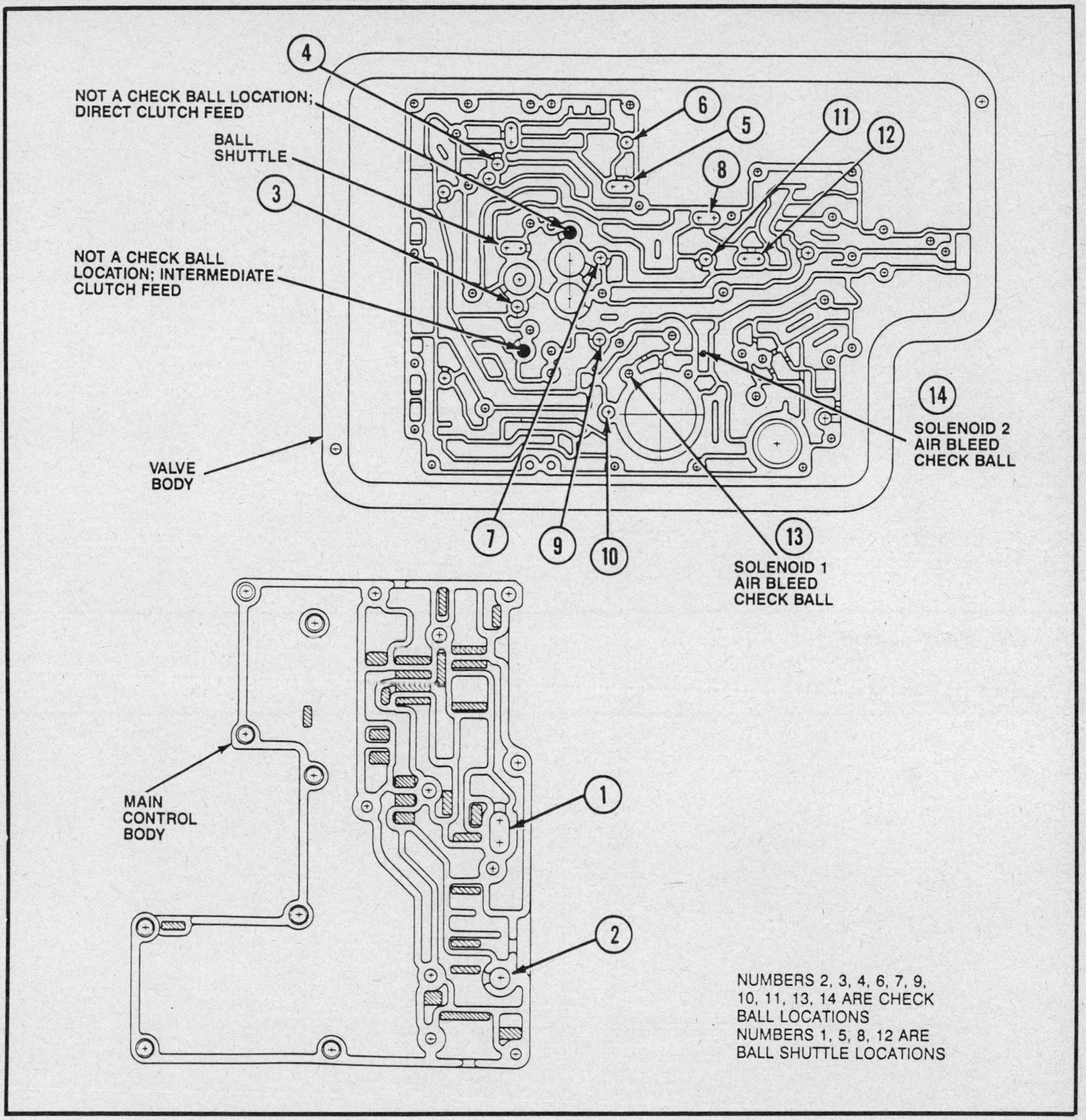
4
NOT A CHECK BALL LOCATION; DIRECT CLUTCH FEED
BALL SHUTTLE
3
NOT A CHECK BALL LOCATION; INTERMEDIATE CLUTCH FEED
VALVE BODY
6
5
8
11
12
14
SOLENOID 2 AIR BLEED CHECK BALL
7
9
10
13
SOLENOID 1 AIR BLEED CHECK BALL
MAIN CONTROL BODY
1
2
NUMBERS 2, 3, 4, 6, 7, 9, 10, 11, 13, 14 ARE CHECK BALL LOCATIONS
NUMBERS 1, 5, 8, 12 ARE BALL SHUTTLE LOCATIONS

Check ball and ball shuttle locations

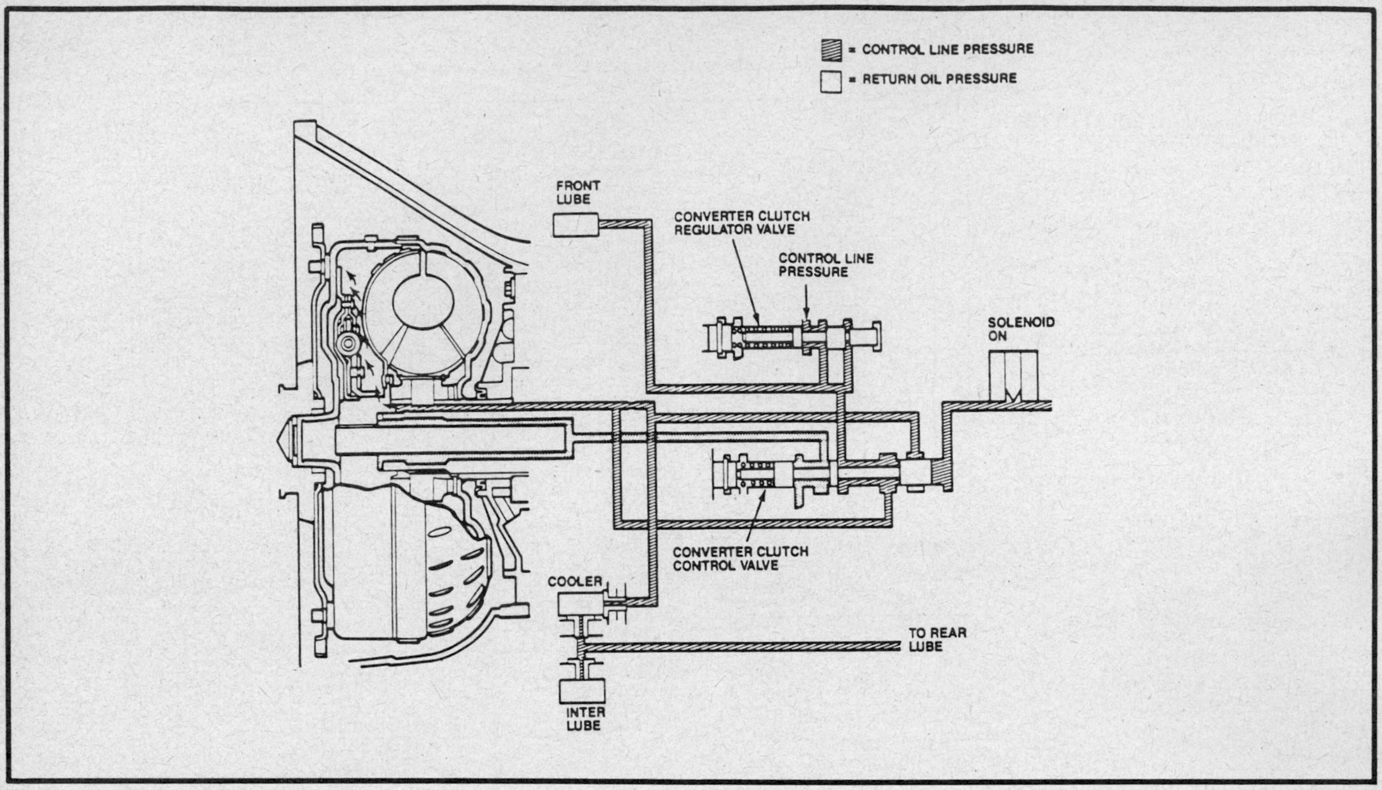
= CONTROL LINE PRESSURE
= RETURN OIL PRESSURE
FRONT LUBE
CONVERTER CLUTCH REGULATOR VALVE
CONTROL LINE PRESSURE
SOLENOID ON
CONVERTER CLUTCH CONTROL VALVE
COOLER
TO REAR LUBE
INTER LUBE

Converter clutch applied

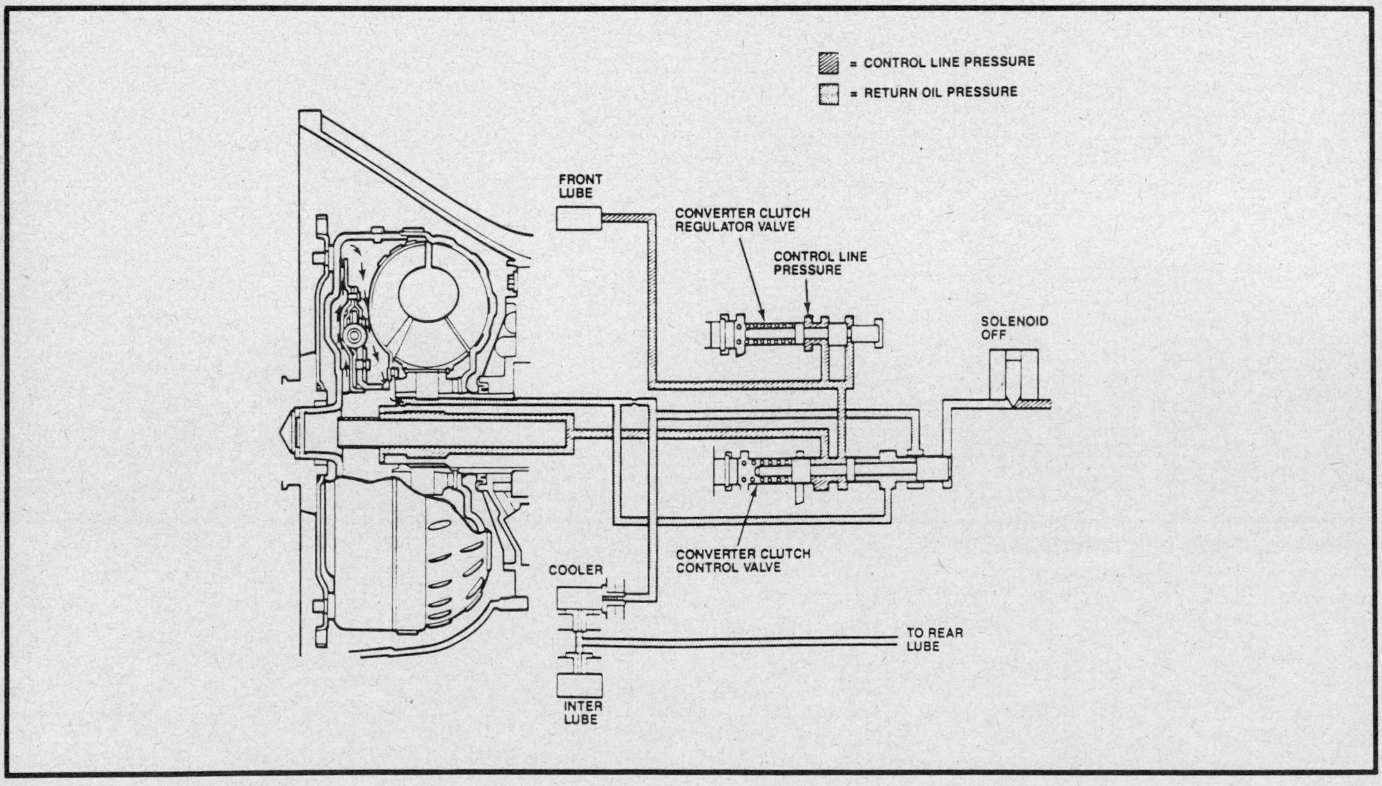
= CONTROL LINE PRESSURE
= RETURN OIL PRESSURE
FRONT LUBE
CONVERTER CLUTCH REGULATOR VALVE
CONTROL LINE PRESSURE
SOLENOID OFF
CONVERTER CLUTCH CONTROL VALVE
COOLER
TO REAR LUBE
INTER LUBE

Converter clutch released

SPECIFICATIONS

TORQUE SPECIFICATIONS

Description	ft. lbs.	Nm
Inner OWC race to case	18–25	24–34
Connector to case (fluid) cooler line	18–23	24–31
Plug line pressure case	6–12	8–16
Throttle pressure case plug	6–12	8–16
Inner & outer lever to manual control shaft	30–40	40–54
Positive detent spring to case	80–100 ①	9–11
Parking rod guide plate to case	16–20	22–27
Neutral switch assembly to case	55–75 ①	6–8
Center support to hub	80–120 ①	9–14
Center support fluid feed	8–12	11–16
Extension housing to case	20–29	27–39
Extension housing to case (4x2)	20–29	27–39
Extension housing to case (4x4)	20–29	27–39
Stator support to pump body	80–100 ①	9–11
Oil pump body to case	18–23	24–31
Reinforcing plate to case	80–100 ①	9–11
Main accumulator and solenoid body to case	80–100 ①	9–11
Main and lower body to case	80–100 ①	9–11
Lower body to main body	80–100 ①	9–11
Solenoid body to case	80–100 ①	9–11
Park rod abutment to case	16–20	22–27
Control assembly to pump	18–23	24–31
Oil pan to case	10–12	14–16
Converter drain plug	18–20	24–27
Overdrive cylinder fluid feed	6–10	8–14
Stud—valve body to case short	80–100 ①	9–11
Stud—valve body to case long	80–100 ①	9–11
Nut—valve body to case	80–100 ①	9–11
Nut—manual detent lever	30–40	41–54

① inch lbs.

CLUTCH PLATE USAGE AND CLEARANCE SPECIFICATIONS

	Steel	Friction	Selective Snapring Clearance in. (mm)	Selective Snapring Thickness in. (mm)
Forward	4	4	(0.030–0.055) 0.76–1.40	(0.128–0.132) 3.25–3.35
				(0.110–0.114) 2.79–2.90
				(0.092–0.096) 2.34–2.44
				(0.074–0.078) 1.88–1.98
				(0.056–0.06) 1.42–1.52
Direct	4	4	(0.039–0.062) 0.99–1.57	(0.083–0.087) 2.11–2.21
				(0.074–0.078) 1.88–1.98
	3	3	(0.034–0.044) 0.86–1.12	(0.065–0.069) 1.65–1.75
Intermediate	3	3	(0.030–0.06) 0.76–1.52	(0.098–0.102) 2.49–2.59
				(0.077–0.081) 1.95–2.05
	2	2	(0.020–0.040) 0.51–1.02	(0.057–0.061) 1.45–1.55
Overdrive	3	3	(0.030–0.060) 0.76–1.52	(0.138–0.142) 3.5–3.6
				(0.118–0.122) 3.00–3.1
				(0.098–0.102) 2.49–2.59
				(0.077–0.081) 1.95–2.05
	2	2	(0.02–0.04) 0.51–1.02	(0.057–0.061) 1.45–1.55
Coast	2	2	(0.025–0.045) 0.635–1.140	(0.085–0.089) 2.15–2.26
				(0.069–0.073) 1.75–1.85
				(0.053–0.057) 1.35–1.45

SPECIAL TOOLS

Tool Number	Description
T50T-100-A	Impact slide hammer (use w/T89T-70010-A)
T59L-100-B	Impact slide hammer (use w/T89T-70010-A)
T58L-101-B	Puller
T77F-1102-A	Puller
T00L-1175-AC	Seal remover
T77F-1176-A	Clutch spring compressor
D79P-2100-T30	Torx® bit (T-30)
D79P-2100-T40	Torx® bit (T-40)
T78P-3504-N	Roll pin remover
T80T-4000-W	Handle (use w/PS88B800-10)
D78P-4201-B	Dial indicator with magnetic base
T00L-4201-C	Dial indicator with bracketry
T67P-7341-A	Shift linkage insulation tool
T84P-7341-A/B	Shift linkage grommet remover/replacer
T61L-7657-B	Extension housing seal replacer
T77L-7697-C	Extension housing bushing replacer
T77L-7697-D	Extension housing bushing remover
D89L-77000-A	Banding tool
T57L-77820-A	Pressure gauge
T80L-7902-A	Endplay checking tool
T81P-7902-C	Torque converter handles
T71P-19703-C	O-ring tool
T89T-70010-A	Pump puller adaptors

Tool Number	Description
T89T-70010-B	Aligning pin
T89T-70010-C	Clutch spring fixture
T89T-70010-E	Clutch remover/installer
T89T-70010-F	Clutch spring compressor plate
T89T-70010-G	Stub tube installer
T89T-70010-J	Gear position sensor adjuster
T88C-77000-AH2	Clutch spring compressor bar
T74P-77248-A	Seal remover
T80L-77405-A	Clutch spring compressor
T74P-77498-A	Shift lever seal replacer
T65L-77515-A	Clutch spring compressor
T77L-77548-A	Lip seal protector
T63L-77837-A	Pump seal replacer
T89T-70100-A	E40D test harness
Not available at time of printing	Manual lever position sensor tester
	ROTUNDA EQUIPMENT
014-00028	Torque converter and oil cooler cleaner
014-00104	C-3 transmission adapter kit
014-00106	Rotunda twin post engine stand
021-00054	Torque converter leak test kit
014-00763	Transmission stand fixture

Section 4

ZF-4HP-22 Transmission
Ford Motor Co.

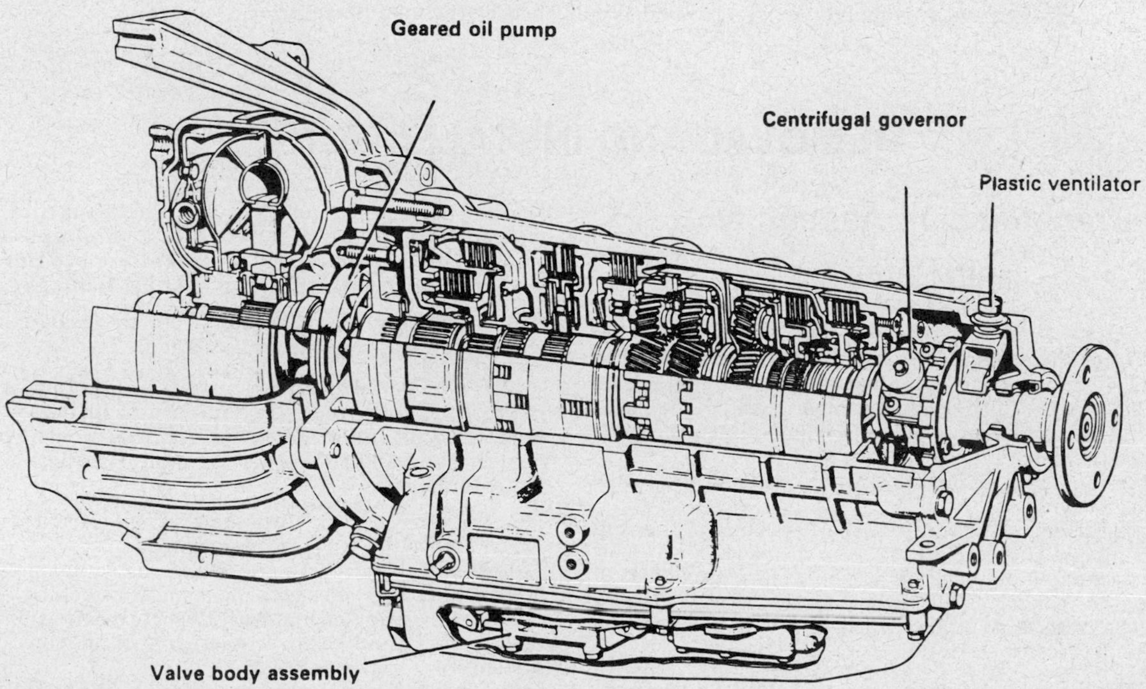

Geared oil pump
Centrifugal governor
Plastic ventilator
Valve body assembly

Cross section of the automatic transmission

REFERENCES

APPLICATION

1980–84 Chilton's Professional Automatic Transmission Manual #7390
1984–89 Chilton's Professional Transmission Manual—Domestic Vehicles #7959

GENERAL DESCRIPTION

1980–84 Chilton's Professional Automatic Transmission Manual #7390

MODIFICATIONS

1980–84 Chilton's Professional Automatic Transmission Manual #7390
1984–89 Chilton's Professional Transmission Manual—Domestic Vehicles #7959

TROUBLE DIAGNOSIS

1980–84 Chilton's Professional Automatic Transmission Manual #7390

ON CAR SERVICES

1980–84 Chilton's Professional Automatic Transmission Manual #7390

REMOVAL AND INSTALLATION

1980–84 Chilton's Professional Automatic Transmission Manual #7390
1984–89 Chilton's Professional Transmission Manual—Domestic Vehicles #7959

BENCH OVERHAUL

1980–84 Chilton's Professional Automatic Transmission Manual #7390

SPECIFICATIONS

1980–84 Chilton's Professional Automatic Transmission Manual #7390
1984–89 Chilton's Professional Transmission Manual—Domestic Vehicles #7959

SPECIAL TOOLS

1980–84 Chilton's Professional Automatic Transmission Manual #7390

APPLICATION

1984 Lincoln Continential and Mark VII with 2.4L turbocharged engine
1984 Volvo 760GLE with D24T engine

TRANSMISSION MODIFICATIONS

No modifications have been published by the manufacturer at the time of this publication.

REMOVAL AND INSTALLATION

Transmission Assembly

REMOVAL

FORD

1. Disconnect the negative battery cable. Remove the kickdown (T.V.) cable and insert from the injector pump side lever and cable bracket to the engine compartment.
2. Place the transmission selector lever in **N**. Raise the vehicle and support safely.
3. Remove the outer manual lever and nut from the transmission selector shaft.
4. Remove the engine brace from the lower end of the converter housing and engine block.
5. Place a transmission jack under the transmission.
6. Remove the converter-to-flywheel attaching nuts. Place a wrench on the crankshaft pulley attaching bolt to turn the converter to gain access to the nuts.

NOTE: The converter studs are installed in the converter with LocTite®. During disassembly the nuts may override the LocTite® and the nut and stud come out as a "bolt". This poses no concern. The stud and converter threads are to be cleaned, LocTite® applied and the "bolt" can be reinstalled and tightened to normal specifications without removing the nut from the stud.

7. Disconnect the driveshaft from the rear axle and slide shaft rearward from the transmission.

NOTE: To maintain driveshaft balance, mark the rear driveshaft yoke and axle companion flange so the driveshaft can be installed in its original position. Install a seal installation tool in the extension housing to prevent fluid leakage.

8. Disconnect the neutral start switch electrical connector.
9. Remove the extension housing damper.
10. Remove the rear mount-to-crossmember attaching nuts and the 2 crossmember-to-side support attaching bolts.
11. Remove the 2 engine rear support-to-extension housing attaching bolts and remove the rear mount from the exhaust system.
12. On Continental models with column shift, remove the bolts securing the bellcrank bracket to the engine to transmission brace.

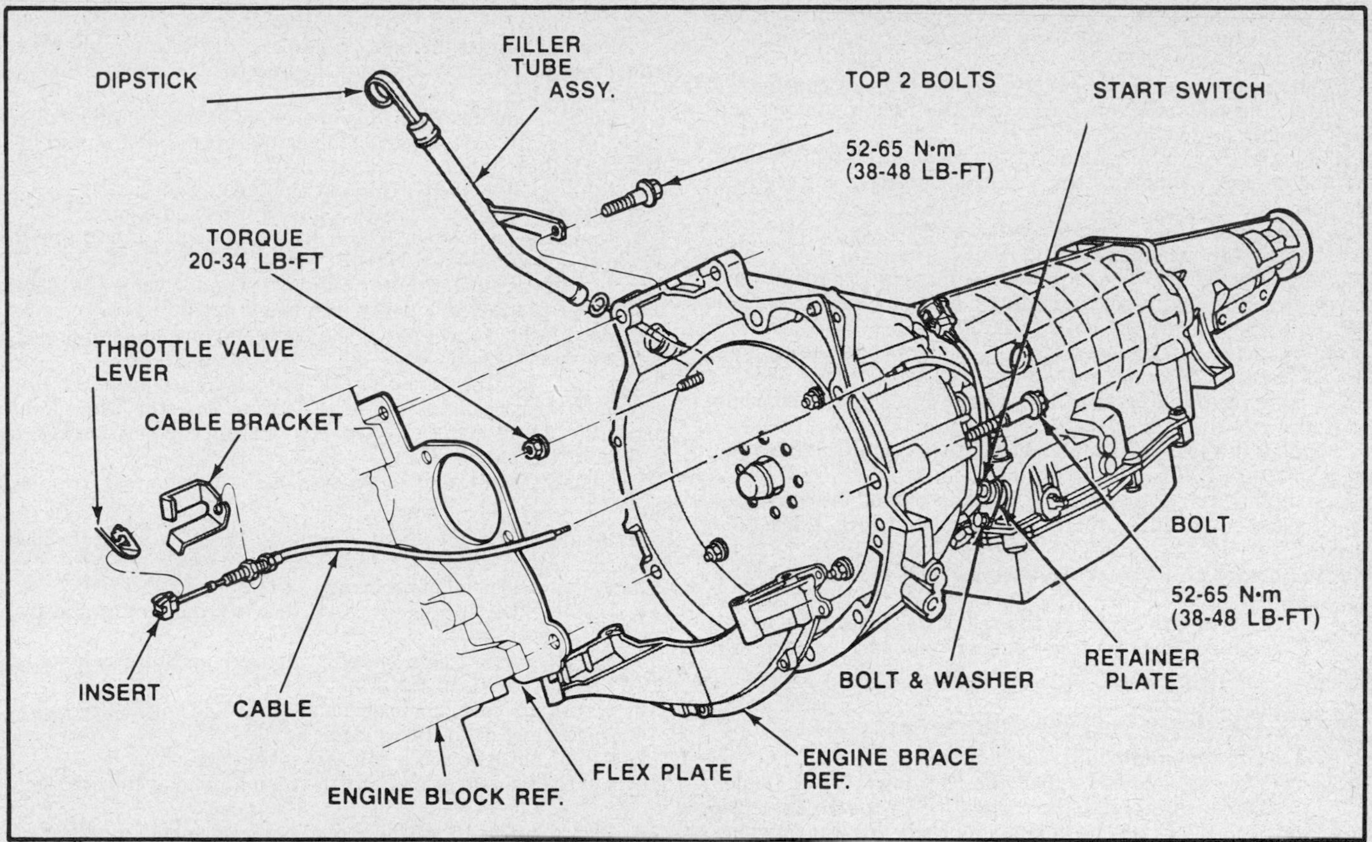
DIPSTICK
FILLER TUBE ASSY.
TOP 2 BOLTS
START SWITCH
52-65 N•m (38-48 LB-FT)
TORQUE 20-34 LB-FT
THROTTLE VALVE LEVER
CABLE BRACKET
BOLT
52-65 N•m (38-48 LB-FT)
RETAINER PLATE
BOLT & WASHER
INSERT
CABLE
FLEX PLATE
ENGINE BRACE REF.
ENGINE BLOCK REF.

Removal or installation of transmission and converter assembly

NOTE: Some exhaust system hardware may have to be removed to facilitate removal of crossmember and transmission.

13. Disconnect each oil line from the fittings on the transmission.
14. Disconnect the speedometer cable from the extension housing.
15. Remove the converter housing to starter motor bolts.
16. Secure the transmission to the jack with a chain and lower it slightly.
17. Remove the converter housing-to-cylinder block attaching bolts.
18. Remove the filler tube and dipstick.
19. Carefully move the transmission and converter assembly away from the engine and, at the same time, lower the jack to clear the underside of the vehicle.

VOLVO

1. Remove air cleaner.
2. Disconnect throttle cable at pulley and cable sheath at bracket.
3. Remove the two upper converter housing to engine bolts.
4. Disconnect transmission oil filler pipe from the engine.
5. Raise the vehicle and support safely. Disconnect oil filler pipe from oil pan and drain transmission oil.
6. Remove the retaining bolts and take off the splash guard.
7. Pry off the rubber suspension rings from the front muffler.
8. Mark the flanges and disconnect the drive shaft at the rear flange. Remove drive shaft.
9. Remove the exhaust pipe clamps.
10. Remove the bolts securing the transmission support member. Pull support member back, twist and lift out.
11. Remove the rear engine mount securing bolts. Remove attachment and bracket.
12. Disconnect speedometer cable at transmission extension.
13. Remove transmission oil cooler pipes.
14. Remove transmission neutral safety switch. On later models the switch is located at and directly controlled by the gearshift control lever.
15. Disconnect gearshift control rod.
16. Remove cover plate between engine and transmission, remove starter motor blind cover and remove starter motor.
17. Remove bolts attaching converter to drive plate.
18. Position a transmission fixture under transmission and remove the lower retaining bolts and separate the converter from the drive plate.
19. Lower the transmission assembly and slide it out from under vehicle.

Installation

FORD

1. Place the transmission on a jack. Secure the transmission to the jack with a chain.
2. Rotate the converter until the studs are in alignment with the holes in the flywheel and flexplate.
3. Move the converter and transmission assembly forward into position, using care not to damage the flywheel, flexplate and the converter pilot. The converter face must rest squarely against the flexplate. This indicates that the converter pilot is not binding in the engine crankshaft.
4. Install the filler tube and dipstick, position bracket over the upper right housing to engine bolt hole.
5. Install and tighten the converter housing-to-engine attaching bolts to 38–48 ft. lbs.(52–65 Nm).
6. Remove the safety chain from around the transmission.
7. Connect the oil cooler lines by pushing them into the fitting on the transmission (located on the intermediate plate).

8. Connect the speedometer cable to the extension housing.
9. Install the extension housing damper. Torque the bolts to 18–25 ft. lbs. (24–43 Nm).
10. Secure the crossmember on the side support and install the attaching bolts and nuts. Position the rear mount on the crossmember and tighten nuts.
11. Install the rear mount on to the exhaust system. Secure the engine rear support to the extension housing and tighten the bolts.
12. If removed, install exhaust system hardware.
13. Lower the transmission and remove the jack.
14. On Continental models with column shift, position the bellcrank to the engine to transmission brace and install the attaching bolts. Torque the bolts to 10–20 ft. lbs. (14–27 Nm).
15. Guide the Kickdown (T.V.) Cable up into the engine compartment.
16. Install the outer manual lever on the transmission selector shaft. Torque the nut to 10–20 ft. lbs. (14–27 Nm).
17. Install the converter to flywheel attaching nuts and torque them to 20–34 ft. lbs. (27–46 Nm).
18. Install the engine brace on the lower end of the converter housing and engine block. Torque the bolts to 15–18 ft. lbs. (20–24 Nm).
19. Connect the neutral start switch harness at the transmission.
20. Connect the driveshaft to the rear axle. Install the driveshaft so the index marks, made during removal, are correctly aligned.

NOTE: Lubricate the yoke splines.

21. Adjust the manual shift linkage.
22. Lower the vehicle and adjust the Kickdown (T.V.) Cable.
23. Fill the transmission to the correct level with the specified fluid (Dexron® II). Start the engine and shift the transmission through all ranges, then recheck the fluid level.

VOLVO

1. Position the transmission and converter assembly on a transmission fixture. Raise and position the transmission behind the engine.
2. Line up and install the lower transmission retaining bolts, adjust the plate between the starter motor and casing and install the starter motor.
3. Connect the oil filler pipe at the lower end.
4. Install the upper transmission to the engine bolts.
5. Install the converter to the drive plate bolts and torque to 30–36 ft. lbs. (40.7–47.6 Nm).
6. Install the starter motor blind plate and lower cover plate.
7. Move gear selector lever into the **2** position.
8. Attach the control rod at the front end and adjustable clevis to the gear selector lever.
9. Check control adjustment. The clearance from **D** stop should be approximately the same as from **2** to stop. Move lever to position **1** and then to **P**. Recheck clearance in position **D** and **2**. Readjust if necessary.
10. Install starter neutral safety switch. Torque to 4–7 ft. lbs. (5.4–9.4 Nm).
11. Install oil cooler pipes. Torque to 14–22 ft. lbs. (19–29.8 Nm).
12. Install driveshaft and attach at rear flange.
13. Install exhaust pipe brackets, rear engine mount and the speedometer cable.
14. Install transmission support member and torque bolts to 30–37 ft. lbs. (40.6–50.1 Nm).
15. Install exhaust pipe clamps and muffler suspender.
16. Install engine splash guard.
17. Attach throttle cable to bracket and adjust cable.
18. Fill the transmission with the recommended transmission fluid.
19. Install air cleaner. Connect the negative battery cable.
20. Road test vehicle and recheck the fluid level.

SPECIFICATIONS

TORQUE SPECIFICATIONS

Description	Torque ft. lbs. (Nm)
Cylinder F—Countersunk Bolts	7 (10)
Park Assembly Bolts	7 (10)
Pump Assembly Bolts	7 (10)
Intermediate Plate Plugs	36 (50)
Bell Housing Bolts	33 (46)
Governor Housing and Hub Bolts	7 (10)
Extension Housing Bolts	16 (23)
Output Flange Collar Nut	73 (100)
Valve Body Bolts	4 (6)
Oil Pan Drain Plug	11 (15)
Oil Pan Cap Nut	14 (20)
Oil Pan Attaching Bolts	6 (8)
Oil Screen Attaching Bolts	6 (8)

Section 4

C3 Transmission
Ford Motor Co.

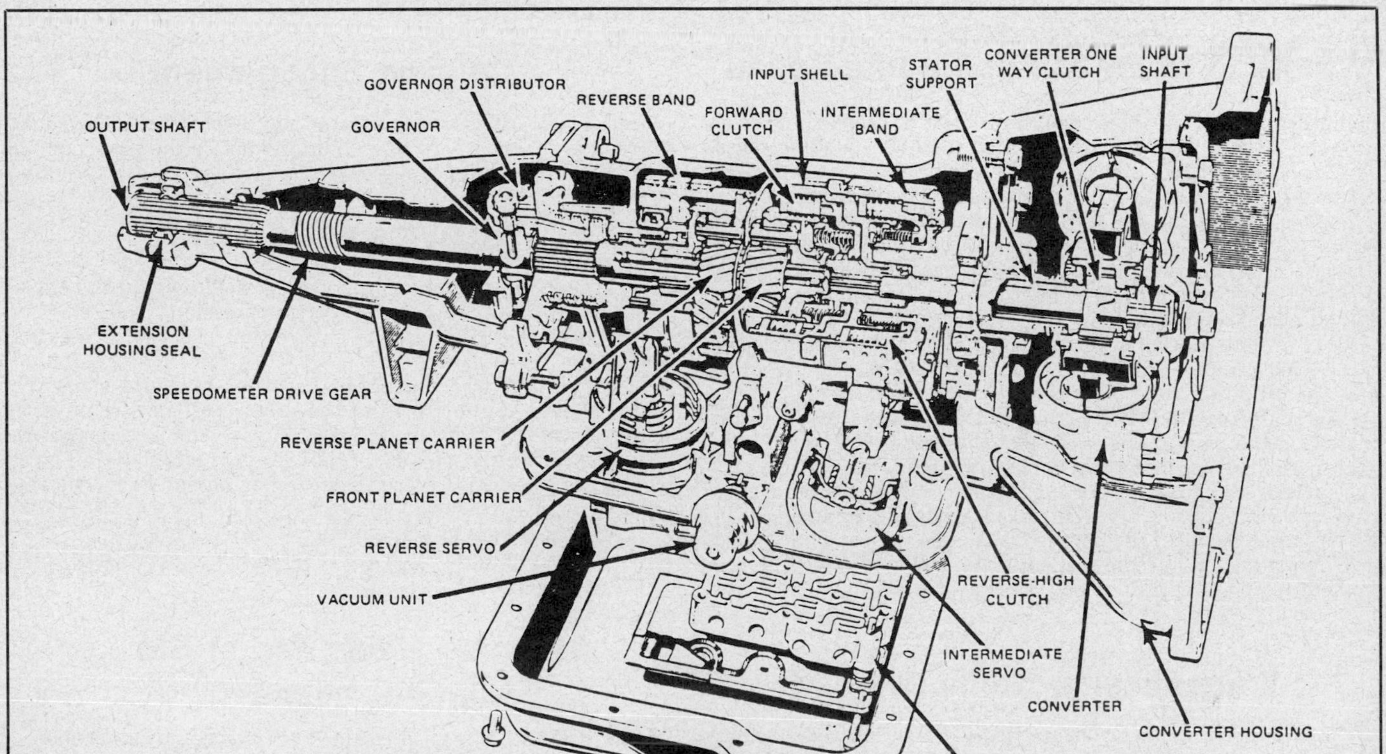

OUTPUT SHAFT
GOVERNOR DISTRIBUTOR
GOVERNOR
REVERSE BAND
INPUT SHELL
FORWARD CLUTCH
INTERMEDIATE BAND
STATOR SUPPORT
CONVERTER ONE WAY CLUTCH
INPUT SHAFT
EXTENSION HOUSING SEAL
SPEEDOMETER DRIVE GEAR
REVERSE PLANET CARRIER
FRONT PLANET CARRIER
REVERSE SERVO
VACUUM UNIT
REVERSE-HIGH CLUTCH
INTERMEDIATE SERVO
CONVERTER
CONVERTER HOUSING
CONTROL VALVE BODY

Cross section of the C3 transmission

REFERENCES

APPLICATION

1980–84 Chilton's Professional Automatic Transmission Manual #7390
1984–89 Chilton's Professional Transmission Manual—Domestic Vehicles #7959

GENERAL DESCRIPTION

1980–84 Chilton's Professional Automatic Transmission Manual #7390

MODIFICATIONS

1980–84 Chilton's Professional Automatic Transmission Manual #7390
1984–89 Chilton's Professional Transmission Manual—Domestic Vehicles #7959

TROUBLE DIAGNOSIS

1980–84 Chilton's Professional Automatic Transmission Manual #7390

ON CAR SERVICES

1980–84 Chilton's Professional Automatic Transmission Manual #7390

REMOVAL AND INSTALLATION

1980–84 Chilton's Professional Automatic Transmission Manual #7390
1984–89 Chilton's Professional Transmission Manual—Domestic Vehicles #7959

BENCH OVERHAUL

1980–84 Chilton's Professional Automatic Transmission Manual #7390

SPECIFICATIONS

1980–84 Chilton's Professional Automatic Transmission Manual #7390
1984–89 Chilton's Professional Transmission Manual—Domestic Vehicles #7959

SPECIAL TOOLS

1980–84 Chilton's Professional Automatic Transmission Manual #7390

APPLICATION

1984–87 Chevrolet Chevette and Pontiac 1000
1989 Chevrolet Geo Tracker

TRANSMISSION MODIFICATIONS

Whistling Noise at Engine Idle Speed

Some THM 180C transmissions (built prior to breakpoint serial number 2649550) may exhibit a high pitched whistling noise which occurs in **R**, **D**, and **L**, with the engine warm to hot. The noise, which is only heard at engine idle speeds, sounds like a whistling tea kettle. Do not confuse this whistle noise with pump whine or cavitation, which can be caused by either a plugged oil filter or leaking oil filter gasket. This whistling noise, if found, is coming from the converter feed hole, located in the oil pump wear plate.

To eliminate this whistling noise, obtain a new oil pump wear plate, part number 96013212, from W.D.D.G.M. and install it.

If the oil pump wear plate is obtained from dealership stock rather than from W.D.D.G.M., check the surface finish prior to installing the part. If the surface of the part is bright or shiny with wire brush or grind marks running in one direction across the surface—do not use the part. It may result in a re-occurance of the noise condition. If the surface is dull with small marks in an irregular pattern, the part is OK to use.

When reassembling the transmission, follow the steps in the procedure and be certain to use the converter-to-oil pump alignment tool J-23082-01.

Transmission Name Change

By September 1, 1991, Hydra-matic will have changed the name designation of the THM 180C automatic transmission. The new designation for this transmission will be Hydra-matic 3L30. Transmissions built in 1989 and 1990 will serve as transitional years in which a dual system, made up of the old designation and the new designation will be in effect.

Delayed Engagement

Some THM 180C transmissions may experience delayed engagement in 1st gear. This condition may be the governor hub seal rings seized in their grooves.

SERVICE PROCEDURE

1. Check the engine vacuum using a vacuum gauge to ensure there is sufficient vacuum to the transmission.
2. Check the modulator to ensure that it is the correct part number.
3. Check for freeness of the governor hub seal rings. The rings should rotate freely in the grooves. If not, remove the rings, clean the grooves with cleaning solvent, and inspect the governor hub ring grooves for nicks or burrs. Remove all burrs, if any are present. Replace the governor hub and governor hub seal rings, if necessary.
4. Check the band adjustment.
5. Check the 3rd clutch input sprag for wear. If the sprag is worn, replace it.

Whinning Noise and/or Slow or Slipping Engagement

Some THM 180C transmissions built before serial number TP 2928976 may exhibit a whinning noise and/or slow or slipping engagement in all ranges. The noise increases when the engine rpm is increased. This whinning noise is similar to the noise caused from the steering wheel being turned to its maximum travel.

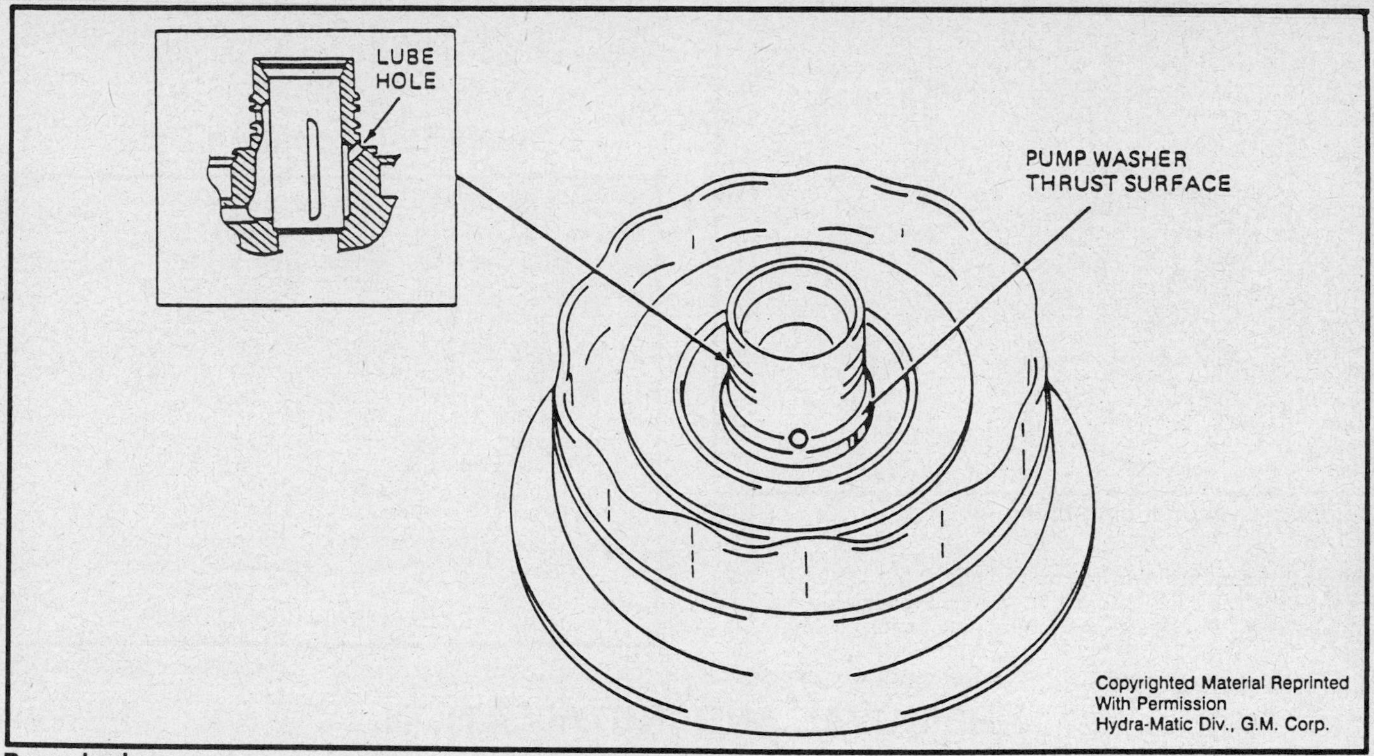
LUBE HOLE
PUMP WASHER THRUST SURFACE
Copyrighted Material Reprinted With Permission Hydra-Matic Div., G.M. Corp.

Pump body

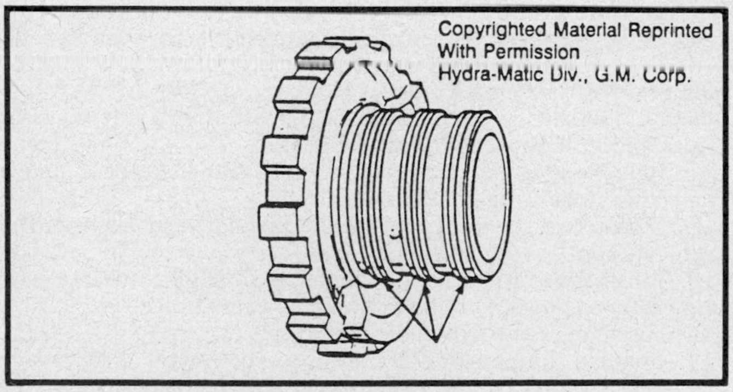
Copyrighted Material Reprinted With Permission Hydra-Matic Div., G.M. Corp.

Governor hub seal rings locations

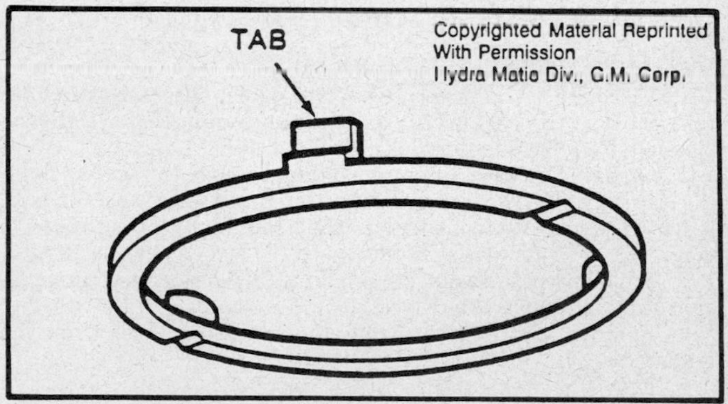
TAB
Copyrighted Material Reprinted With Permission Hydra-Matic Div., G.M. Corp.

Endplay washer

SERVICE PROCEDURE

1. Remove the oil filter and visually inspect for restriction.
2. If the oil filter is restricted, remove the transmission from the vehicle, and remove the pump.
3. After removing the pump, inspect the selective endplay washer for wear, especially on the inside diameter of the washer. If the washer is worn, replace the oil pump assembly and selective endplay washer, using service package 96040245. This package contains the pump assembly and selective endplay washer. The new endplay washer is not interchangeable with the past oil pump assembly.

NOTE: Starting March 22, 1984, serial number 2928976, a new oil pump assembly and selective endplay washer went into production on all THM 180C transmissions. The new oil pump assembly has a casting change to accommodate the new selective endplay washer. The new endplay washer has 3 raised tabs to hold the washer stationary on the pump.

4. If the oil filter is not restricted.

Slipping or No Upshift Burned 2nd Clutch

Some THM 180C transmissions may exhibit a slipping or no upshift condition due to a burned 2nd clutch. The burned 2nd clutch may be due to distortion of the valve body spacer plate which results in loss of 2nd clutch apply pressure. The valve body spacer plate may become distorted when the vacuum pipe fitting becomes blocked by carbon at the engine intake manifold. As the fitting is blocked off, the 2nd clutch check ball vibrates back and forth against the spacer plate causing the distortion.

SERVICE PROCEDURE

To service a THM 180C transmission for slipping or no upshift condition, inspect and replace as necessary, the 2nd clutch

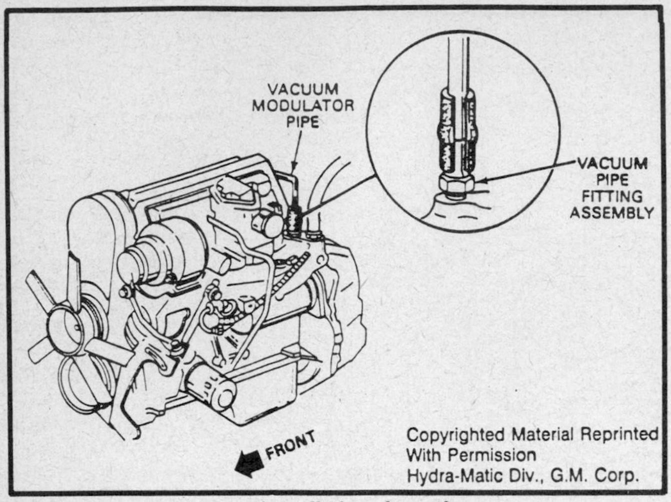
VACUUM
MODULATOR
PIPE
VACUUM
PIPE
FITTING
ASSEMBLY
FRONT
Copyrighted Material Reprinted
With Permission
Hydra-Matic Div., G.M. Corp.

Modulator vacuum pipe fitting location

plates, 3rd clutch plates, the valve body spacer plate and gaskets. Check the vacuum modulator, using modulator checking tool J-24466; and clean the vacuum pipe fitting assembly.

NOTE: To prevent a vacuum leak when reinstalling the fitting, use pipe sealant, service part number 1052080 or equivalent, around the threads.

Check Ball Location Data

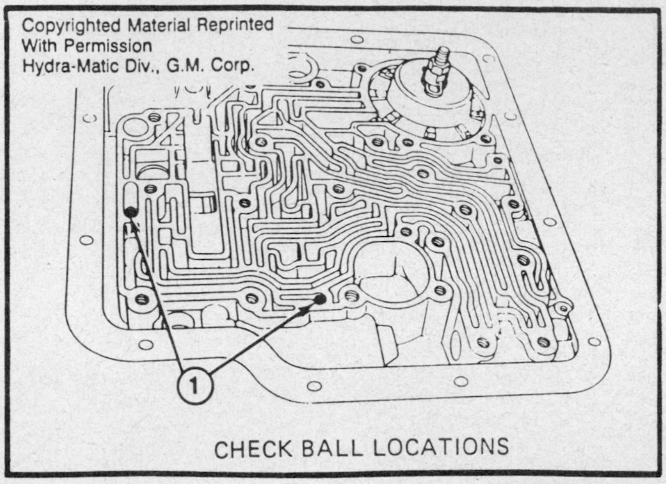
Copyrighted Material Reprinted
With Permission
Hydra-Matic Div., G.M. Corp.
1
CHECK BALL LOCATIONS

REMOVAL AND INSTALLATION

TRANSMISSION REMOVAL

1984–87 CHEVROLET CHEVETTE AND PONTIAC 1000

1. Disconnect the negative battery cable. Disconnect the detent cable at the carburetor or injection pump. Remove the air cleaner. Remove the dipstick.
2. On vehicles with air conditioning, remove the screws holding the heater core cover, disconnect the wire connector and, leaving the hoses attached, place the heater core cover out of the way.
3. Raise the vehicle and support it safely. Drain the transmission fluid. Remove the driveshaft.
4. Remove speedometer cable and the cooler lines from the transmission.
5. Disconnect the shift control linkage.
6. Raise the transmission slightly with a jack and remove the support bolts and converter bracket.
7. Disconnect the exhaust pipe at the rear of the catalytic converter and at the manifold and remove exhaust as an assembly.
8. Remove the pan under the torque converter and remove the flexplate bolts. Mark the converter and flexplate so that they can be reassembled in the same manner.
9. Lower the transmission until there is access to the engine mounting bolts and remove them.
10. Raise the transmission to its normal position and support the engine with a jack. Slide transmission back and down.

NOTE: Hold on to the converter and/or keep the rear of the transmission lower than the front so it won't fall from the bell housing.

1989 CHEVROLET GEO TRACKER

1. Disconnect the negative battery cable. Remove the dipstick.
2. Remove the transfer case shift control lever knob. Remove the breather hose at the rear end of the cylinder head.
3. Remove the wiring harness clamp at the rear end of the intake manifold to free the wiring harness. Disconnect the wiring harness connectors.
4. Remove the starter attaching bolts and remove the starter.
5. Raise the vehicle and support it safely. Drain the transmission and transfer case fluids.
6. Remove the flange bolts of the driveshaft shafts' universal joints and pull out the driveshaft shafts.
7. Remove speedometer cable from the transfer case. Remove the cooler lines from the transmission.
8. Disconnect the shift control and kickdown cables from the transmission.
9. Raise the transmission/transfer case slightly with a jack and remove the support bolts and converter bracket.
10. Disconnect the exhaust center pipe.
11. Remove the pan under the torque converter and remove the flexplate bolts. Mark the converter and flexplate so that they can be reassembled in the same manner.
12. Lower the transmission/transfer case until there is access to the engine mounting bolts and remove them.
13. Raise the transmission/transfer case to its normal position and support the engine with a jack. Slide transmission/transfer case back and down.

NOTE: Hold on to the converter and/or keep the rear of the transmission/transfer case lower than the front so it won't fall from the bell housing.

14. Separate the transfer case from the transmission.

TRANSMISSION INSTALLATION

1984–87 CHEVROLET CHEVETTE AND PONTIAC 1000

1. Transmission installation is the reverse of the removal procedure.
2. Make sure that the converter pump hub keyway is seated into the oil pump drive lugs. The distance from the face of the bell housing to the end of the hub should be 0.200–0.280 in.

(0.508–0.711mm). Check that the converter has free movement.

3. Start all bolts in the converter finger tight so that the flexplate will not be distorted, and align the marks made at disassembly. Then torque the bolts to 20–30 ft. lbs. (27–41 Nm). The transmission support-to-extension should be torqued to 29–36 ft. lbs. (40–49 Nm) and the exhaust system parts to 35 ft. lbs. (48 Nm).

4. Lower the vehicle. Fill the transmission with the proper grade and type automatic transmission fluid.

5. Make linkage and cable adjustments as required. Road test the vehicle as required.

1989 CHEVROLET GEO TRACKER

1. Transmission/transfer case installation is the reverse of the removal procedure.

2. Make sure that the converter pump hub keyway is seated into the oil pump drive lugs. The distance from the face of the bell housing to the end of the hub should be 0.200–0.280 in. (0.508–0.711mm). Check that the converter has free movement.

3. Start all bolts in the converter finger tight so that the flexplate will not be distorted, and align the marks made at disassembly. Then torque the bolts to 35 ft. lbs. (48 Nm).

4. Lower the vehicle. Fill the transmission with the proper grade and type automatic transmission fluid. Fill the transfer case with the proper grade gear oil.

5. Make cable adjustments as required. Road test the vehicle as required.

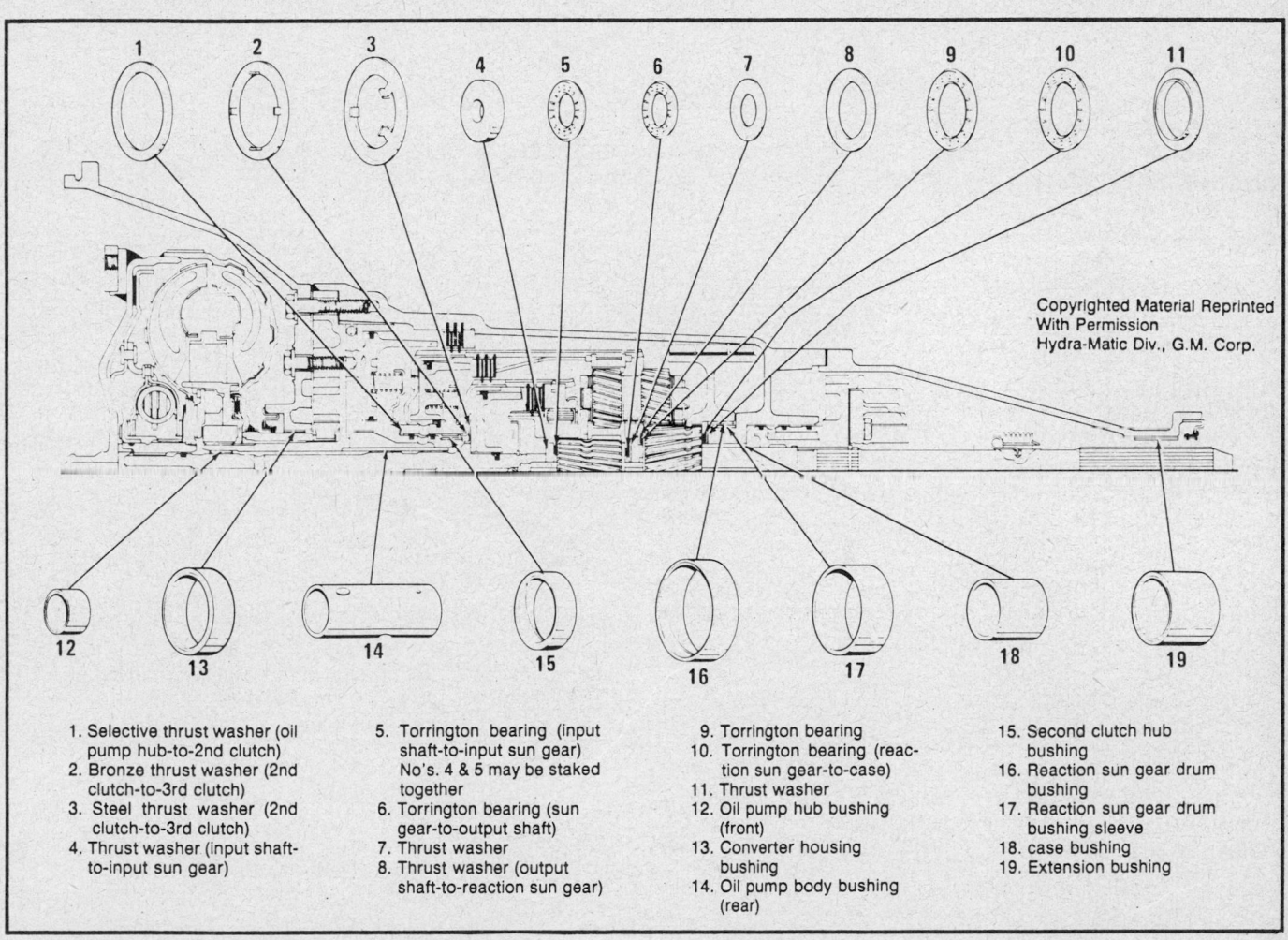

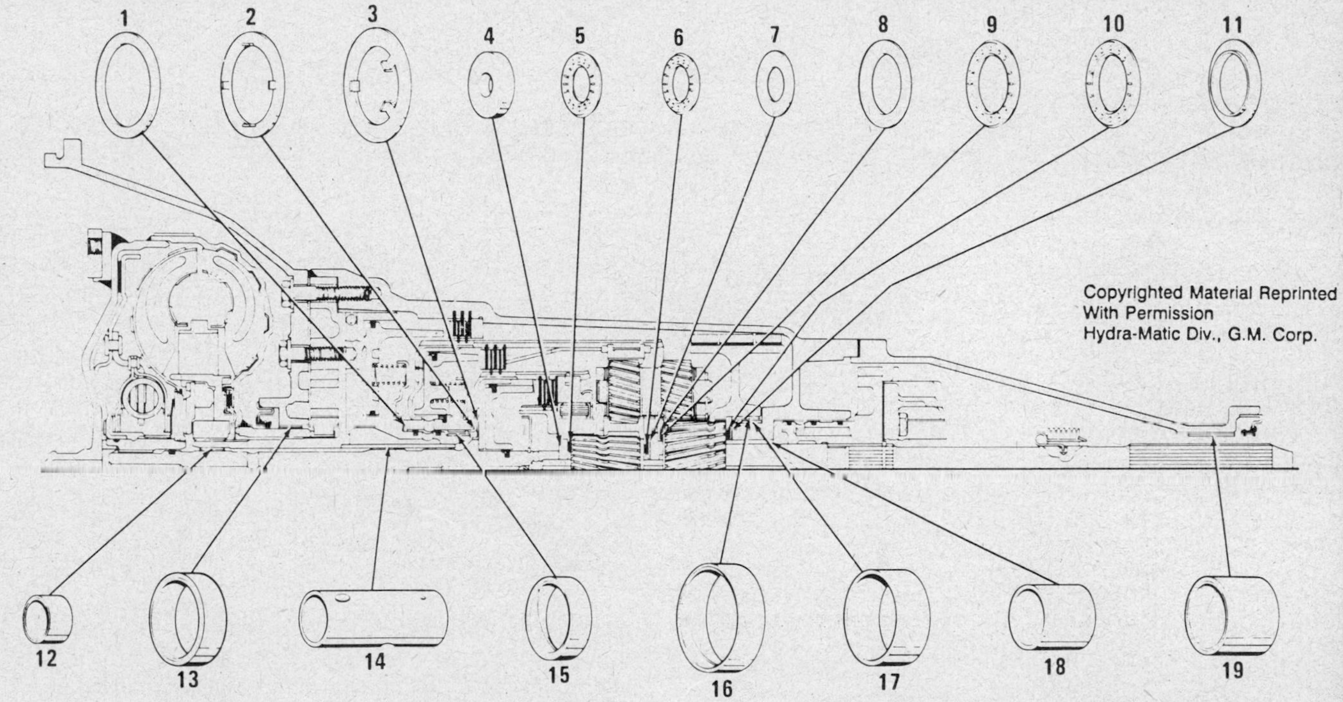
1
2
3
4
5
6
7
8
9
10
11
12
13
14
15
16
17
18
19
Copyrighted Material Reprinted With Permission Hydra-Matic Div., G.M. Corp.

1. Selective thrust washer (oil pump hub-to-2nd clutch)
2. Bronze thrust washer (2nd clutch-to-3rd clutch)
3. Steel thrust washer (2nd clutch-to-3rd clutch)
4. Thrust washer (input shaft-to-input sun gear)
5. Torrington bearing (input shaft-to-input sun gear) No's. 4 & 5 may be staked together
6. Torrington bearing (sun gear-to-output shaft)
7. Thrust washer
8. Thrust washer (output shaft-to-reaction sun gear)
9. Torrington bearing
10. Torrington bearing (reaction sun gear-to-case)
11. Thrust washer
12. Oil pump hub bushing (front)
13. Converter housing bushing
14. Oil pump body bushing (rear)
15. Second clutch hub bushing
16. Reaction sun gear drum bushing
17. Reaction sun gear drum bushing sleeve
18. case bushing
19. Extension bushing

Thrust washer and bearing location guide

SPECIAL TOOLS

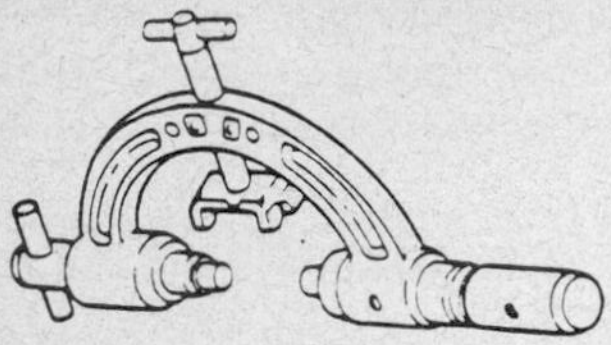
TRANSMISSION HOLDING FIXTURE

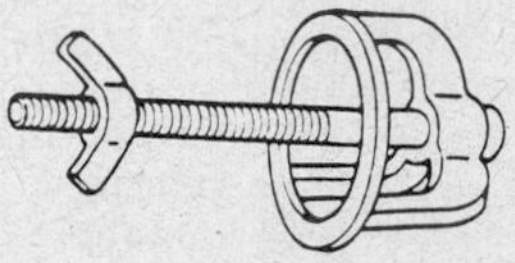
CLUTCH PISTON COMPRESSOR

CONVERTER HOUSING BUSHING REMOVER/ INSTALLER

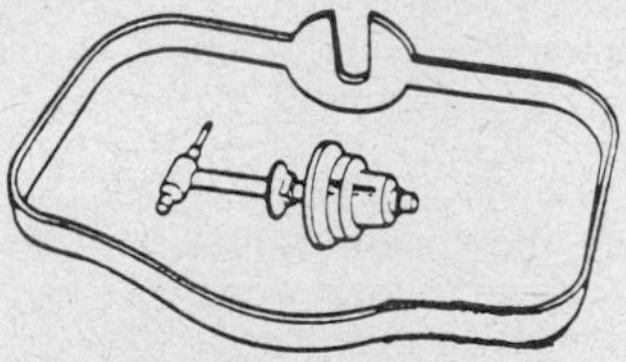
CONVERTER LEAK TEST FIXTURE

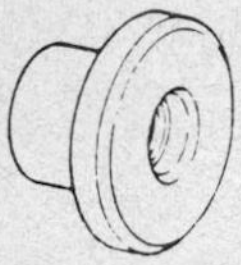
REACTION SUN GEAR DRUM BUSHING SLEEVE INSTALLER

2ND CLUTCH PISTON SEAL INSTALLER

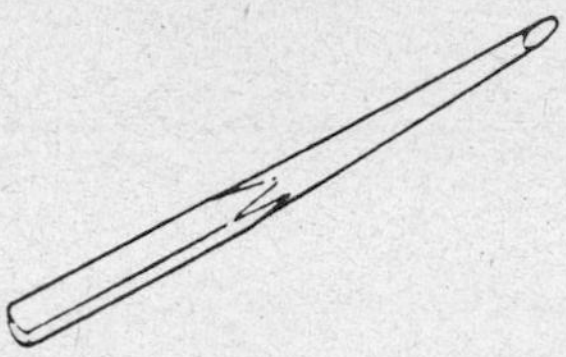
CAPE CHISEL

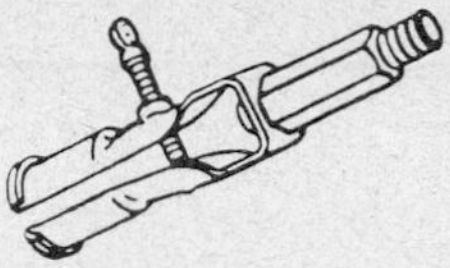
BUSHING REMOVER

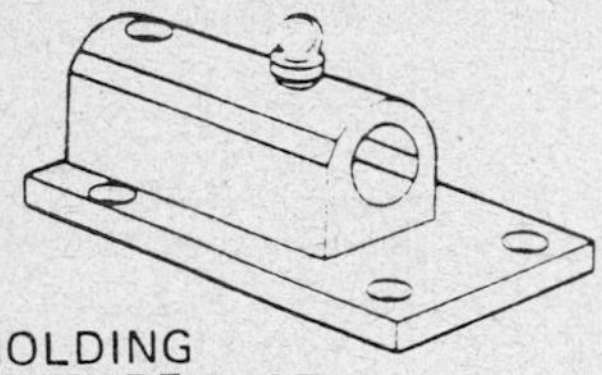
HOLDING FIXTURE BASE

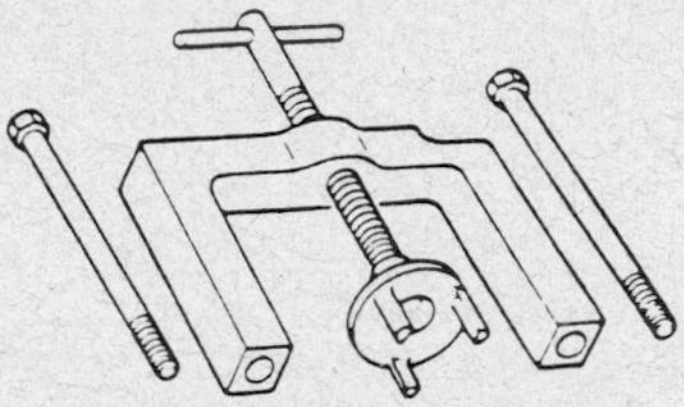
SERVO/3RD CLUTCH PISTON SPRING COMPRESSOR

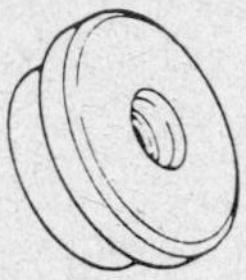
REACTION SUN GEAR DRUM BUSHING INSTALLER

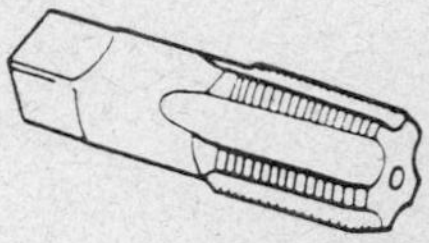
OIL PUMP BUSHING REMOVER (3/4 - 14 NPT)

REAR CASE BUSHING REMOVER/INSTALLER

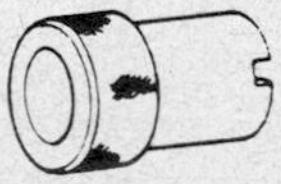
CONVERTER-TO-OIL PUMP ALIGNMENT TOOL

DRIVER HANDLE

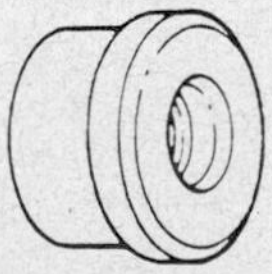
2ND CLUTCH DRUM BUSHING REMOVER/ INSTALLER

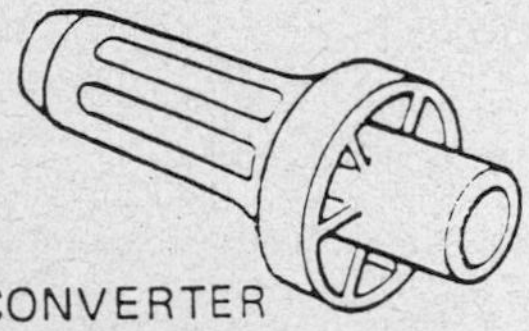
CONVERTER HOUSING OIL SEAL INSTALLER

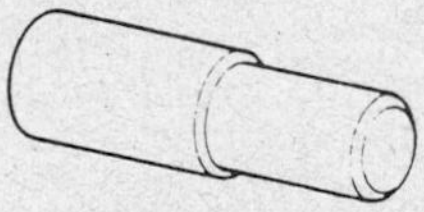
OIL PUMP BUSHING INSTALLER

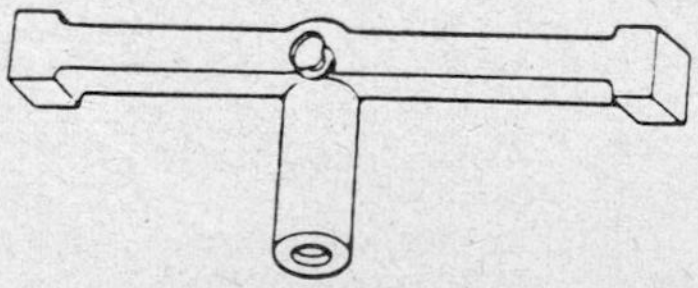
OIL PUMP-TO-2ND CLUTCH DRUM GAGING TOOL

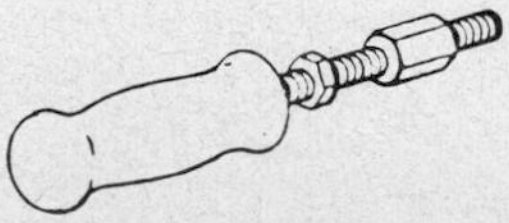
SLIDE HAMMER

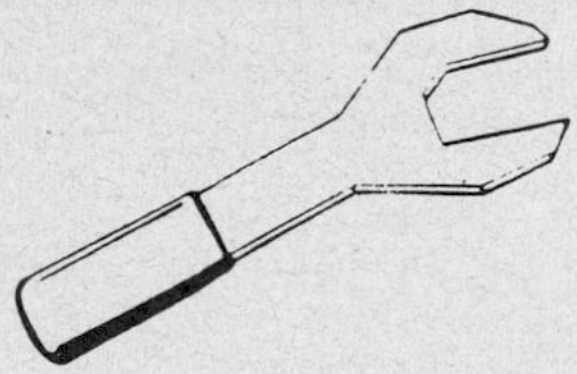
VACUUM MODULATOR WRENCH

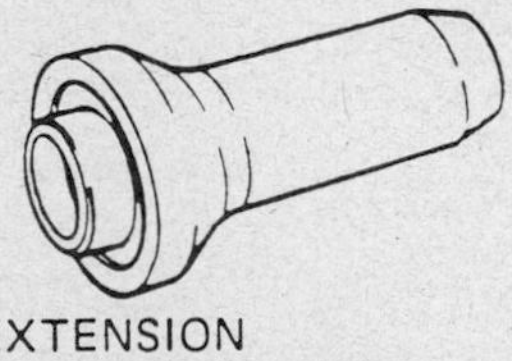
EXTENSION HOUSING OIL SEAL INSTALLER

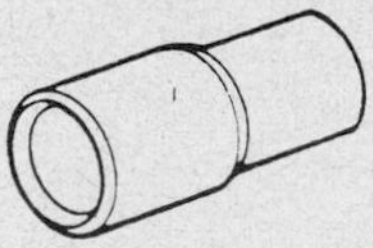
EXTENSION HOUSING BUSHING REMOVER/ INSTALLER

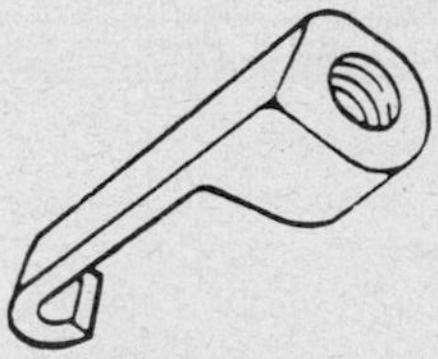
CONVERTER HOUSING SEAL REMOVER

Section 4

THM 200C Transmission
General Motors

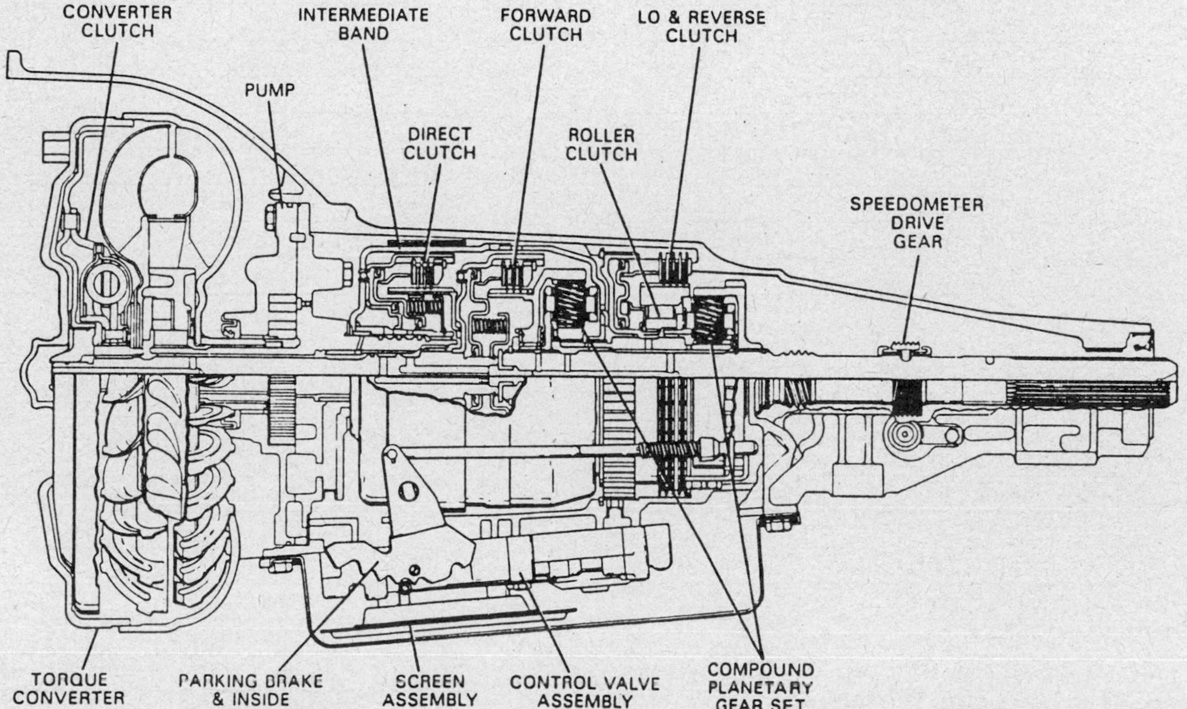

CONVERTER CLUTCH
INTERMEDIATE BAND
FORWARD CLUTCH
LO & REVERSE CLUTCH
PUMP
DIRECT CLUTCH
ROLLER CLUTCH
SPEEDOMETER DRIVE GEAR
TORQUE CONVERTER
PARKING BRAKE & INSIDE DETENT LEVER
SCREEN ASSEMBLY
CONTROL VALVE ASSEMBLY
COMPOUND PLANETARY GEAR SET

Cross section of General Motors THM 200C automatic transmission

REFERENCES

APPLICATION

1974–80 Chilton's Professional Automatic Transmission Manual #6927
1984–89 Chilton's Professional Transmission Manual—Domestic Vehicles #7959

GENERAL DESCRIPTION

1974–80 Chilton's Professional Automatic Transmission Manual #6927

MODIFICATIONS

1974–80 Chilton's Professional Automatic Transmission Manual #6927
1984–89 Chilton's Professional Transmission Manual—Domestic Vehicles #7959

TROUBLE DIAGNOSIS

1974–80 Chilton's Professional Automatic Transmission Manual #6927

ON CAR SERVICES

1974–80 Chilton's Professional Automatic Transmission Manual #6927

REMOVAL AND INSTALLATION

1974–80 Chilton's Professional Automatic Transmission Manual #6927
1984–89 Chilton's Professional Transmission Manual—Domestic Vehicles #7959

BENCH OVERHAUL

1974–80 Chilton's Professional Automatic Transmission Manual #6927

SPECIFICATIONS

1974–80 Chilton's Professional Automatic Transmission Manual #6927
1984–89 Chilton's Professional Transmission Manual—Domestic Vehicles #7959

SPECIAL TOOLS

1974–80 Chilton's Professional Automatic Transmission Manual #6927

APPLICATION

THM 200C

Year	Vehicle	Engine
1984	Electra	5.0L 4bbl
	Electra	5.7L Diesel
	LeSabre	3.8L 2bbl
	LeSabre	5.0L 4bbl
	LeSabre	5.7L Diesel
	Regal	3.8L 2bbl
	Regal	4.3L Diesel
	Camaro	2.5L TBI
	Camaro	2.8L 2bbl
	Camaro	5.0L 4bbl
	Caprice	3.8L 2bbl
	Caprice	5.7L Diesel
	Chevette	1.8L Diesel
	Monte Carlo	2.8L 2bbl
	Monte Carlo	5.7L Diesel
	Cutlass Supreme	3.8L 2bb.
	Cutlass Supreme	4.3L Diesel
	Cutlass Supreme	5.0L 4bbl
	Cutlass Supreme	5.7L Diesel
	Delta Eighty-eight	3.8L 2bbl
	Delta Eighty-eight	5.0L 4bbl
	Delta Eighty-eight	5.7L Diesel
	Ninety-eight	5.0L 4bbl
	Ninety-eight	5.7L Diesel
	1000	1.8L Diesel
	Firebird	2.5L TBI
	Firebird	2.8L 2bbl
	Firebird	5.0L 4bbl
	Grand Prix	3.8L 2bbl
	Grand Prix	5.7L Diesel
	Parisienne	3.8L 2bbl
	Parisienne	5.7L Diesel
1985	LeSabre	3.8L 2bbl
	Regal	2.8L 2bbl
	Regal	4.3L Diesel
	Regal	4.3L TBI
	Regal	5.0L 4bbl
	Caprice	3.8L 2bbl
	Caprice	4.3L TBI
	Caprice	5.0L 4bbl
	Caprice	5.7L Diesel
	Monte Carlo	3.8L 2bbl
	Monte Carlo	4.3L TBI
	Monte Carlo	5.0L 4bbl
	Cutlass Supreme	3.8L 2bbl
	Cutlass Supreme	4.3L Diesel

THM 200C

Year	Vehicle	Engine
	Cutlass Supreme	5.0L 4bbl
	Cutlass Supreme	5.7L Diesel
	Delta Eighty-eight	3.8L 2bbl
	Delta Eighty-eight	5.0L 4bbl
	Delta Eighty-eight	5.7L Diesel
	Grand Prix	3.8L 2bbl
	Grand Prix	5.0L 4bbl
	Grand Prix	5.7L Diesel
	Parisienne	3.8L 2bbl
	Parisienne	4.3L TBI
	Parisienne	5.0L 4bbl
	Parisienne	5.7L Diesel
1986	Regal	3.8L 2bbl
	Caprice	4.3L TBI
	Caprice	5.0L 4bbl
	Monte Carlo	3.8L 2bbl
	Monte Carlo	4.3L TBI
	Monte Carlo	5.0L 4bbl
	Cutlass Supreme	3.8L 2bbl
	Grand Prix	3.8L 2bbl
	Grand Prix	4.3L TBI
	Parisienne	4.3L TBI
1987	Regal	3.8L 2bbl
	Regal	5.0L 4bbl
	Estate Wagon	5.0L 4bbl
	Caprice	4.3L TBI
	Caprice	5.0L 4bbl
	Monte Carlo	3.8L 2bbl
	Monte Carlo	4.3L TBI
	Cutlass Supreme	3.8L 2bbl
	Cutlass Supreme	5.0L 4bbl
	Custom Cruiser	5.0L 4bbl
	Grand Prix	3.8L 2bbl
	Grand Prix	4.3L TBI

bbl—Barrel
TBI—Throttle Body Injection

TRANSMISSION MODIFICATIONS

Late Engagement or Slipping In Drive

Some early 1984 THM 200C transmissions may experience a late engagement or slipping in drive range. This condition is usually detected while the transmission is cold, and could be caused by a restricted forward clutch feed orifice hole.

Other causes for late engagement or slipping in drive are, low transmission oil level cut or nicked forward clutch piston seals and/or cut nicked Teflon® turbine shaft seals.

Serial and model numbers of the defective units are as follows, model number BH serial number 54628, model number OI serial number 21738, model number CZ serial number 1041, model number JY serial number 1075, model number OR serial number 1119 and model number OU serial number 1727.

Procedure

1. Check transmission oil level correct if required.
2. Visually inspect the forward clutch piston seals and turbine shaft Teflon® seals for cuts and nicks.
3. Using a 1/16 in. Allen wrench mark short end. Using the allen wrench as a gauge insert the small end into the forward clutch feed orifice hole and make sure the Allen wrench goes into the orifice hole (without any drag) until the mark on the wrench is flush with the ground diameter. If the mark on the Allen wrench does not go flush with the ground diameter, the forward clutch housing assembly must be replaced.

New Third Oil Pressure Switch

Beginning November 1, 1983 a new third oil pressure switch and bracket assembly went into production for all 1984 THM 200C transmissions, model OI.

When servicing any transmission that necessitates that the third oil pressure switch be replaced compare the oil pressure switch being replaced to the past and new oil switches.

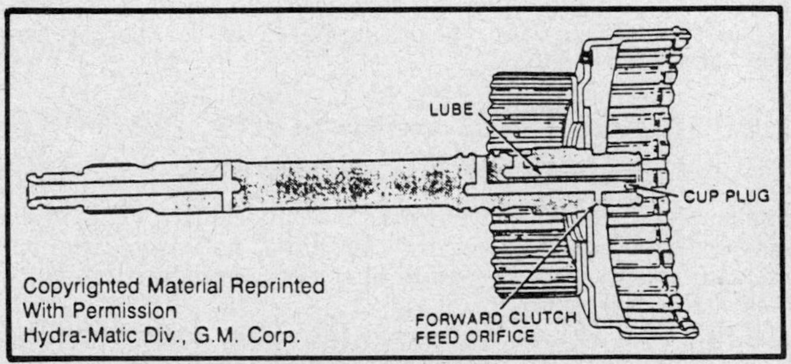
LUBE
CUP PLUG
Copyrighted Material Reprinted
With Permission
Hydra-Matic Div., G.M. Corp.
FORWARD CLUTCH
FEED ORIFICE

Forward clutch housing assembly

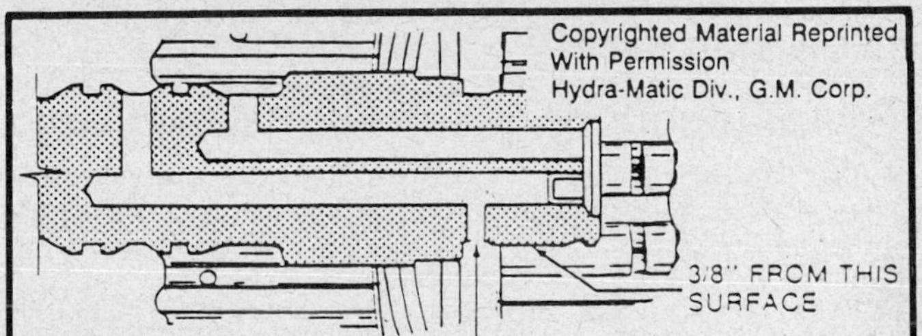
Copyrighted Material Reprinted
With Permission
Hydra-Matic Div., G.M. Corp.
3/8" FROM THIS
SURFACE
1/16" ALLEN WRENCH MUST GO 3/8" INTO
FORWARD CLUTCH FEED ORIFICE.

Forward clutch feed orifice

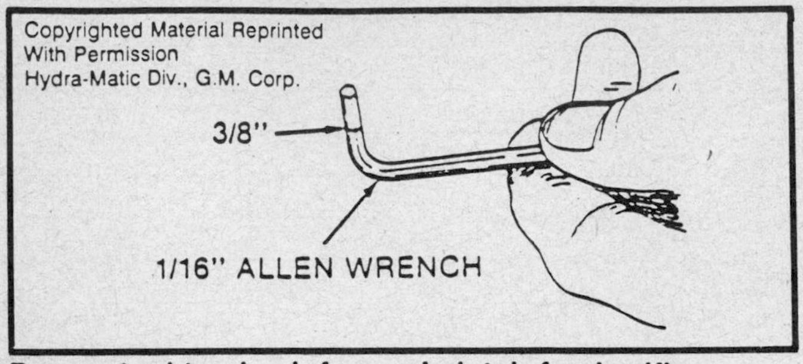
Copyrighted Material Reprinted
With Permission
Hydra-Matic Div., G.M. Corp.
3/8"
1/16" ALLEN WRENCH

Proper tool to check forward clutch feed orifice

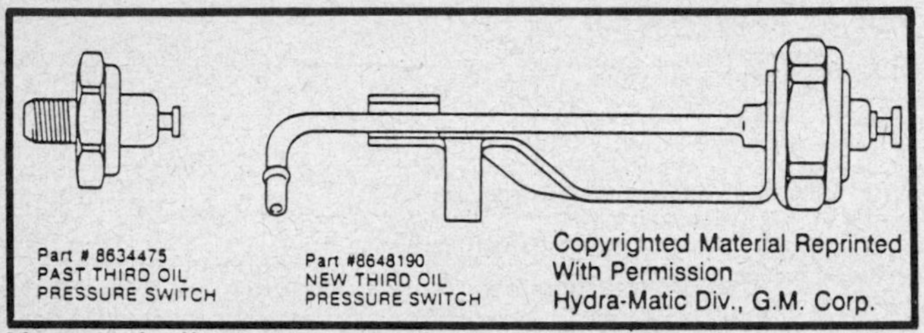
Part # 8634475
PAST THIRD OIL
PRESSURE SWITCH
Part #8648190
NEW THIRD OIL
PRESSURE SWITCH
Copyrighted Material Reprinted
With Permission
Hydra-Matic Div., G.M. Corp.

New 3rd oil pressure switch

If the oil switch required is the past switch, order part number 8634475. If the oil switch required is the new switch, order part number 8648190.

Transmission Oil Exhausting Out The Vent Pipe

If a 1984 THM 200C transmission exhibits oil exhausting out the vent pipe, possible causes may be due to out-of-flat conditions on the faces, nicked faces, porosity, or face damage on the pump body assembly or the oil pump cover assembly.

When servicing any 1984 THM 200C transmission for transmission oil exhausting out the vent pipe both the pump body assembly and the pump cover assembly should be replaced. At the same time, a new pump to case gasket should also be installed. Replacement of these parts will prevent transmission oil from exhausting out the vent pipe during harsh deceleration or during sharp turns.

Second or Third Gear Starts, No Third Gear or Erratic Shifts

Some 1984 THM 200C transmissions, models BH, CZ and OI, may experience second or third gear starts, no third gear or erratic shifts. Any of these conditions can be caused by a broken 2–3 throttle valve spring.

If this condition occurs, remove the control valve assembly. Thoroughly clean the control valve assembly before installing the new T.V. spring.

The new 2–3 T.V. spring is only available through the dealer parts system.

Occasional Long 1–2 Shift When Cold

Some 1984 through early 1985 THM 200C transmissions may experience an occasional long 1–2 shift when cold. To correct this condition, a new spacer plate when into production on September 24, 1984 (Julian date 268).

When servicing any 1984 through early 1985 THM 200C transmission built before the above Julian date, remove the valve body spacer plate. Drill the 2nd and 3rd clutch exhaust orifice out to $^{9}/_{64}$ in. (3.57mm). Using a 6.3mm drill bit, hand chamfer both sides of the hole, slightly, to remove burrs and rough edges.

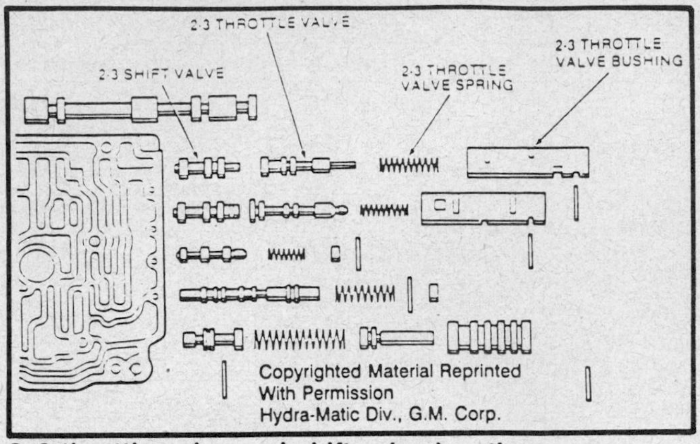
2-3 THROTTLE VALVE
2-3 SHIFT VALVE
2-3 THROTTLE VALVE SPRING
2-3 THROTTLE VALVE BUSHING
Copyrighted Material Reprinted
With Permission
Hydra-Matic Div., G.M. Corp.

2–3 throttle valve and shift valve locations

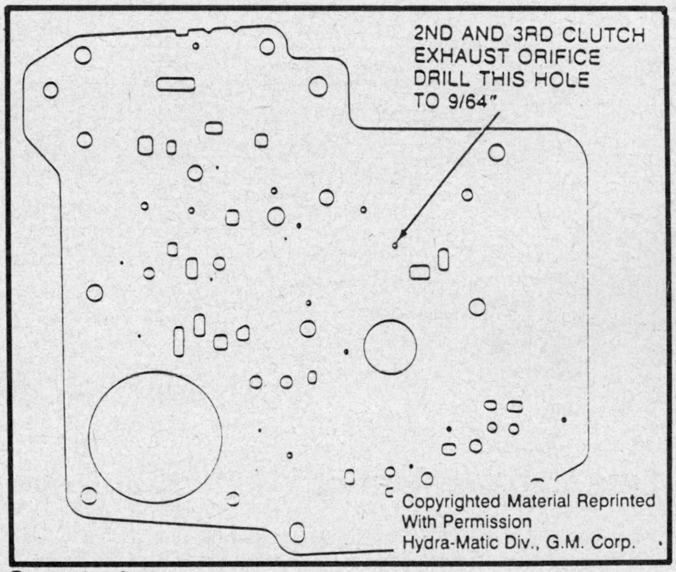
2ND AND 3RD CLUTCH
EXHAUST ORIFICE
DRILL THIS HOLE
TO 9/64"
Copyrighted Material Reprinted
With Permission
Hydra-Matic Div., G.M. Corp.

Spacer plate

Servo Cover Retainer Ring Information

During the 1985 model year, a new style servo cover retaining ring (with tangs) went into production on THM 200C transmission. In addition to the new style retaining ring, the notches on the case intermediate servo bore, used for access of retaining ring removal were omitted.

These changes will provide easier removal and installation of the servo cover retaining ring, as well as prevent damage to the servo cover and piston seals during assembly.

The past style servo cover retaining ring part number 8628137 (without tangs), can only be used with the past style (with notches) case intermediate servo bore. The new servo cover retaining ring part number 8657213 (with tangs) can be used with either style case intermediate servo bore.

Between Julian date 074 and Julian date 078 some new style (notches omitted) THM 200C case intermediate servo bores were assembled with past style servo cover retaining rings (without tangs). If service is required on the new style cases (notches omitted) that have been incorrectly assembled with the past style servo cover retaining ring use the following procedures:

1. Depress intermediate servo cover utilizing the servo cover depressor tool J-29714 or equivalent.

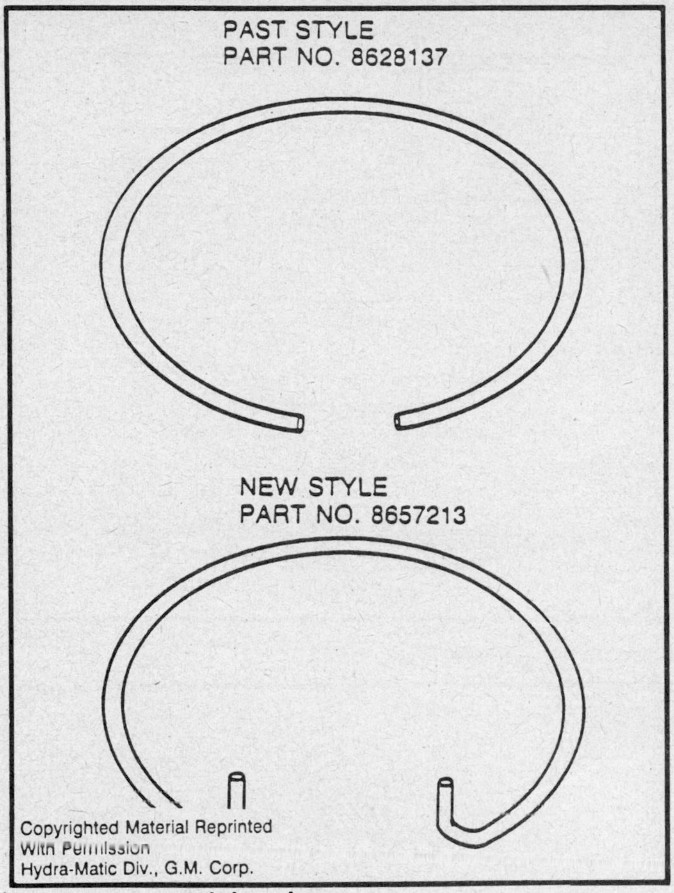
PAST STYLE
PART NO. 8628137
NEW STYLE
PART NO. 8657213
Copyrighted Material Reprinted
With Permission
Hydra-Matic Div., G.M. Corp.

Servo cover retaining ring

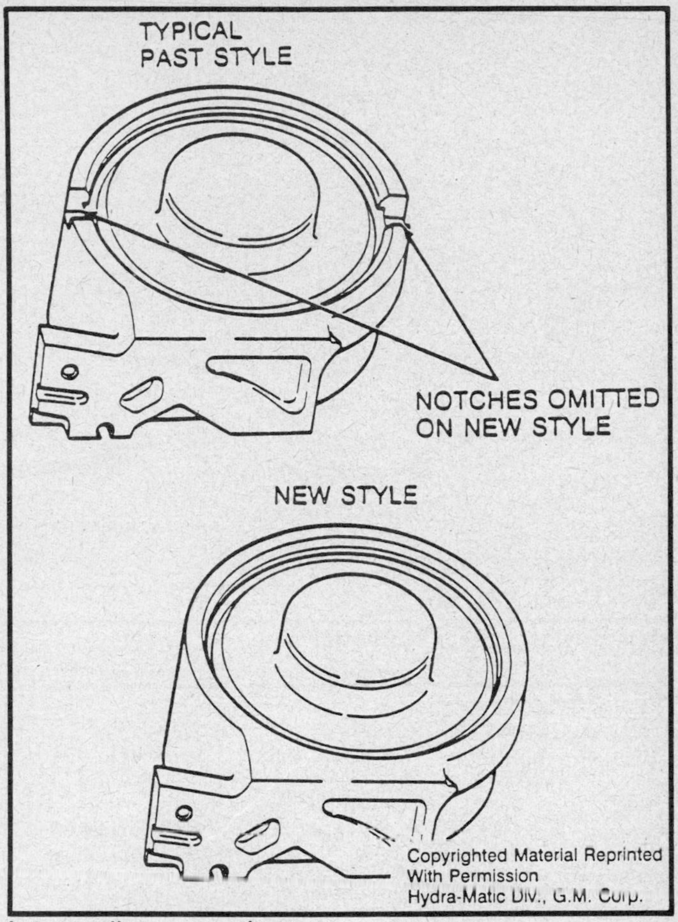
TYPICAL
PAST STYLE
NOTCHES OMITTED
ON NEW STYLE
NEW STYLE
Copyrighted Material Reprinted
With Permission
Hydra-Matic Div., G.M. Corp.

Intermediate servo bore

2.Use a small, flat bladed tool to remove the servo cover retaining ring.

3. Be certain to reassemble new style intermediate servo bore cases (notches removed) with new style servo retaining ring (with tangs).

New Design Torque Converter Lugs

Starting in early March, 1985, all 245mm and 298mm torque converter assemblies will be built with a new design converter lug.

Additionally, all 298mm torque converter assemblies used with diesel engine applications will have only 3 converter lugs instead of the 6 lugs that are currently being used.

Torque converter assemblies having the new design converter lugs will be optional for field service use.

Stripped Input Drum Splines Resulting in a Loss of Reverse, Second and/or Third Gear Ranges

Some THM 200C transmissions may have striped input drum splines resulting in a loss of reverse, second and third gear ranges. In mid 1984 production, the drum was hardened in the spline area to prevent the possibility of this occurrence. The hardened drum may be recognized by a black appearance from the heat treatment process.

If the above conditions are exhibited (or whenever a THM 200C is disassembled), inspect the drum and, if necessary, use the new hardened input spline drum.

Engine Stops Running When in Reverse Range

Some vehicles equipped with 1984, 1985, and early 1986 THM 200C transmissions may exhibit a condition which may cause the engine to stop running when the transmission is placed in reverse range. This condition may be caused by transmission fluid leaking across the pump gear causing the torque converter clutch (TCC) to engage. This condition can be aggravated by hot transmission oil, high line pressure demand, low idle speed, and large pump drive gear side clearance.

A pressure regulator valve with 2 flats ground between line and converter feed lands will prevent transmission fluid leakage across the pump. It is scheduled for mid 1986 production.

If this condition exists, order and install service package part number 8638951 which contains a pressure regulator valve part number 8638950 and instruction sheet part number 8638952.

No Upshift Condition After Downshift

Some 1984–86 THM 200C transmissions may exhibit a no upshift condition after a part or full throttle downshift.

This condition may be caused by internal leakage hydraulically locking the throttle valve plunger. If this occurs, the line pressure will remain at boosted level, resulting in late upshifts.

To repair this condition, a new throttle valve plunger and throttle valve bushing, and on some transmissions, a throttle valve lever and bracket assembly should be installed.

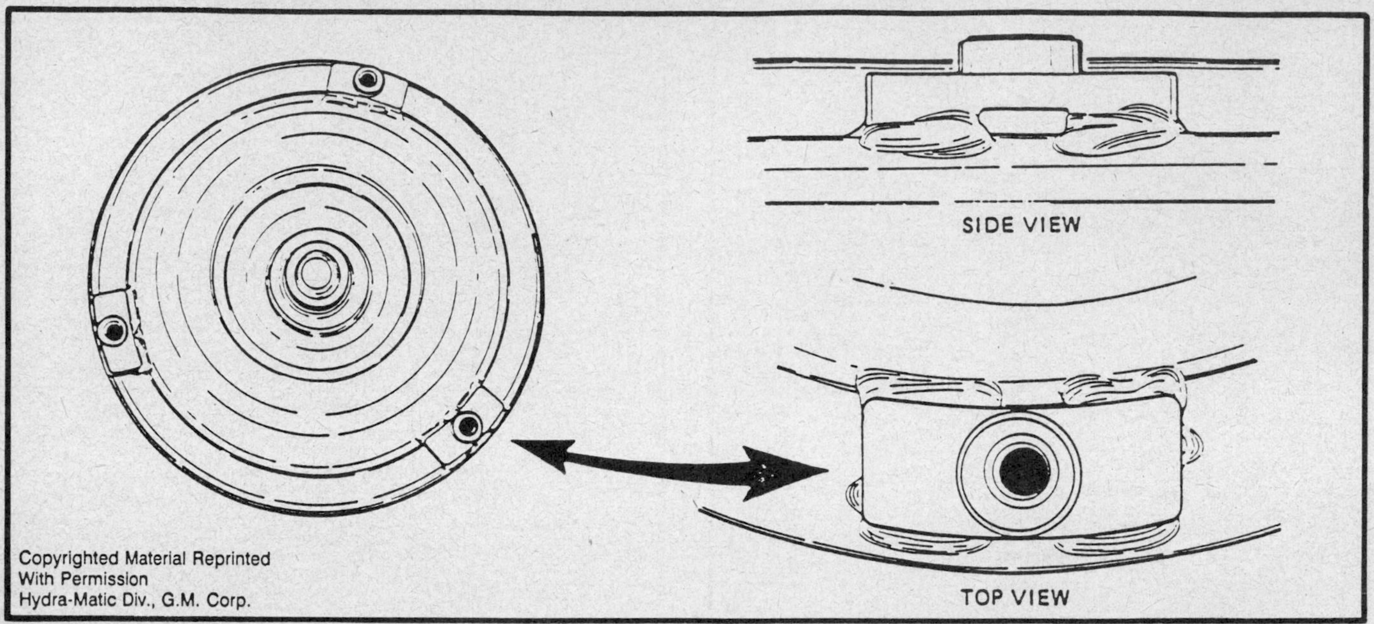
SIDE VIEW
TOP VIEW

New design torque converter lugs

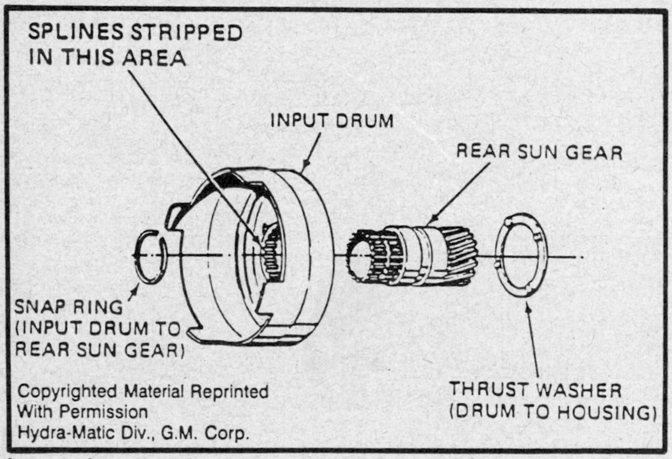
SPLINES STRIPPED IN THIS AREA
INPUT DRUM
REAR SUN GEAR
SNAP RING (INPUT DRUM TO REAR SUN GEAR)
THRUST WASHER (DRUM TO HOUSING)

Input drum and rear sun gear assembly

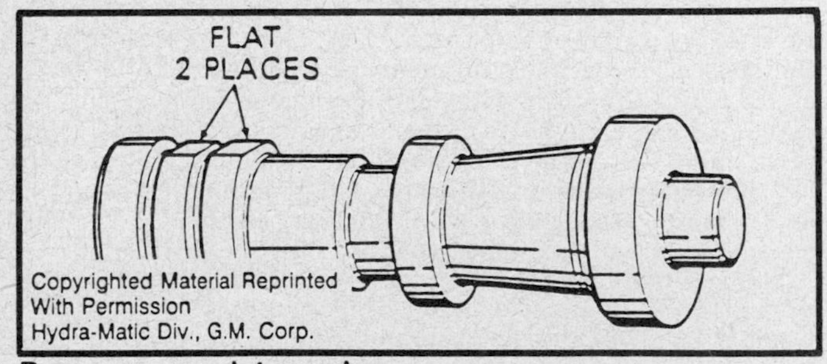
FLAT 2 PLACES

Pressure regulator valve

2-1 Coastdown Clunk

To service a 1984 through mid 1986 (prior to Julian date 016) THM 200C transmission equipped vehicle:

1. Check the engine idle speed and set to specifications.
2. Check the vehicle's differential back lash and repair if necessary.
3. On the 1984 or 1985 transmissions, BH model only, remove the primary spring from the governor assembly.

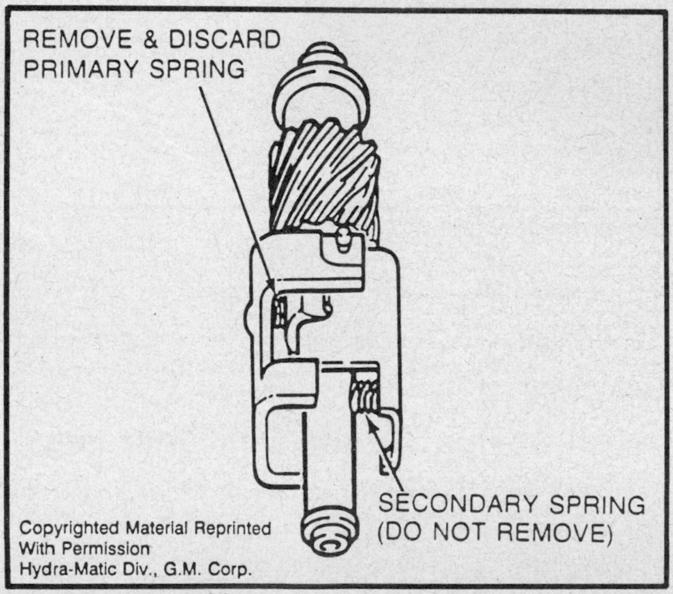
REMOVE & DISCARD PRIMARY SPRING
SECONDARY SPRING (DO NOT REMOVE)

Governor assembly

4. On a 1984 or 1985 THM 200C transmission built prior to Septemper 25, 1985 (Julian date 268), remove the valve body spacer plate. Drill the 2nd and 3rd clutch exhaust orifice out to $\frac{9}{64}$ in. (3.57mm). Using a 6.3mm drill bit, hand chamfer both sides of the hole, slightly, to remove burrs and rough edges.
5. On a THM 200C transmission built prior to January 10, 1986, (Julian date 016), replace the rear internal gear and output shaft with the new pressed on assembly.
6. Upon reassembly of the transmission, reduce the transmission output shaft rear endplay to 0.004–0.015 in. (0.010–0.38mm) by using a thicker selective washer, if required. Do not let output shaft endplay get lower than 0.004 in. as transmission damage may result.

NOTE: All 1984–85 THM 200C transmissions, models OI, OR and OU, already have the pressed on rear internal gear and do not require replacement.

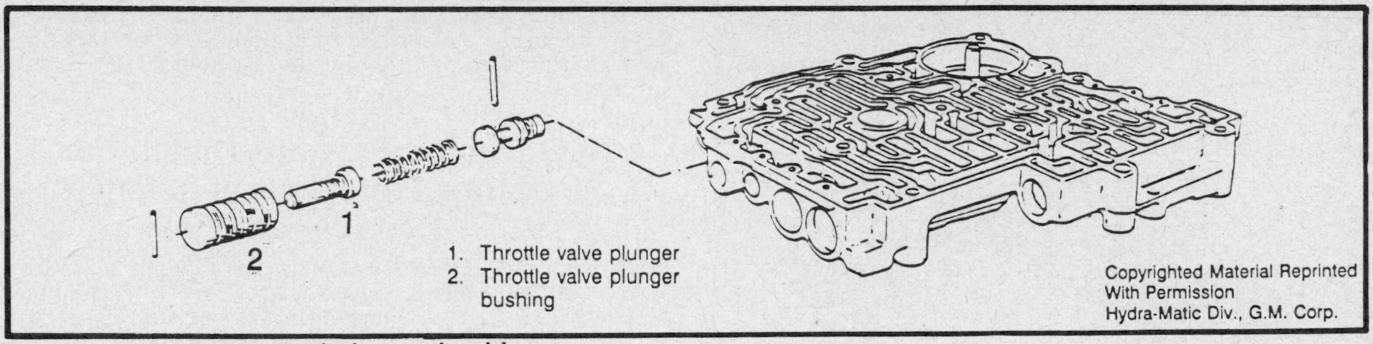
1
2
1. Throttle valve plunger
2. Throttle valve plunger bushing
Copyrighted Material Reprinted With Permission Hydra-Matic Div., G.M. Corp.

Throttle valve plunger and plunger bushing

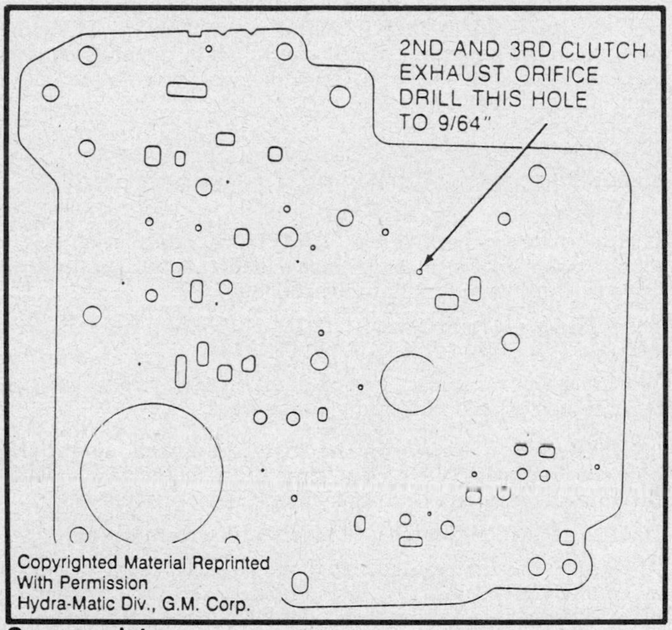
2ND AND 3RD CLUTCH EXHAUST ORIFICE DRILL THIS HOLE TO 9/64"
Copyrighted Material Reprinted With Permission Hydra-Matic Div., G.M. Corp.

Spacer plate

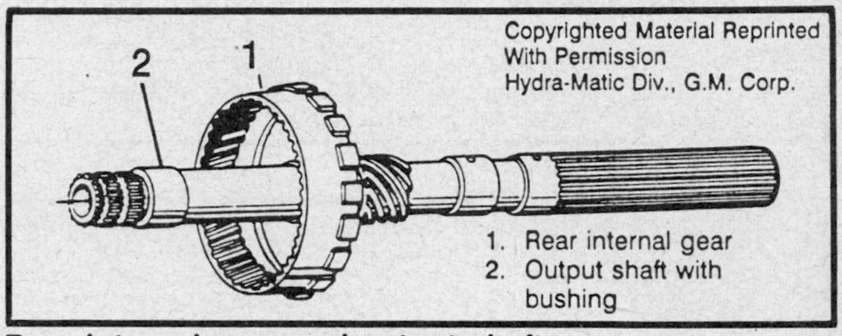
Copyrighted Material Reprinted With Permission Hydra-Matic Div., G.M. Corp.
2
1
1. Rear internal gear
2. Output shaft with bushing

Rear internal gear and output shaft

In some cases the above on vehicle procedure may not repair this condition; it will then be necessary to perform the following procedure:

1. Replace the rear internal gear and output shaft with part number 8633910. This service package will have the rear internal gear pressed onto the output shaft.

NOTE: The THM 200C transmissions, models OI, OR and OU, already have the press fit rear internal gear and output shaft and does not require replacement.

2. Upon reassembly of the transmission, reduce the transmission output shaft rear endplay to 0.004–0.015 in. (0.10–0.38mm) by using a thicker selective washer, if required.

NOTE: Do not let output shaft endplay get lower than 0.004 in. as transmission damage may result.

New Cellulose Coated Oil Pan Gasket

On May 29, 1986, all 1986 THM 200C transmissions were built using a new latex-coated cellulose (hard, fiber material) pan gasket. The new gaskets improve sealing capability by better withstanding fastener torques.

Any THM 200C transmission with a continuous raised rib pan can be serviced with the new cellulose gaskets.

Some service overhaul packages may contain the latex-coated cork (soft material) gasket, which should not be used. The new gasket can be ordered separately by ordering the appropriate part number.

NOTE: When replacing the oil pan, it is critical that a torque wrench is used to tighten the hex head pan screws. Hand torque all THM 200C hex head screws 7–12 ft. lbs. (10–16 Nm).

Erratic Shift Patterns—Damaged/Rolled Governor Seals

Some 1984–87 vehicles equipped with a THM 200C transmission may have erratic shift patterns (such as late or no upshifts) due to an O-ring seal on the governor cover assembly causing incorrect governor pressure.

NOTE: The erratic shift pattern may be more noticeable after the transmission reaches normal operating temperature.

Procedure

Remove the cover and inspect the internal governor cover seal. If damaged or rolled, replace the entire governor cover assembly. The seal is not available separately.

NOTE: Internal governor cover seals must be well lubricated with petroleum jelly to prevent damage or rolling during installation.

Slow TCC Apply—Plug In Pump Cover

When a new turbine shaft with a ball capsule was released for production, a plug should have been removed from the pump cover. This plug may not have been removed on all transmissions. If the plug is left in, a slightly slower torque converter clutch (TCC) apply may result. This slightly slower TCC apply is not a durability concern. This would include THM 200C transmissions built between October 8, 1986 (Julian date 281) and December 15, 1986 (Julian date 349).

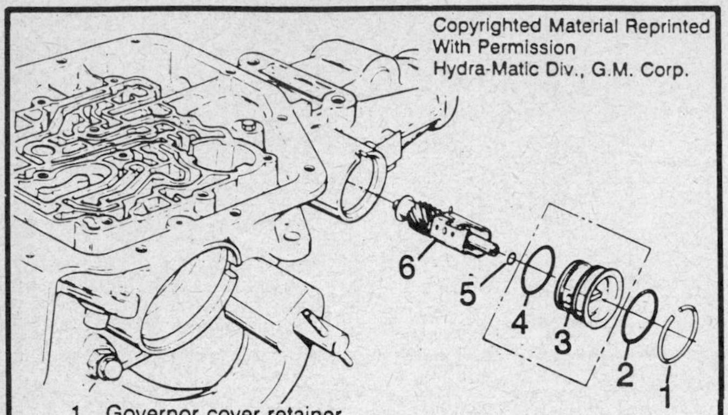
Copyrighted Material Reprinted With Permission Hydra-Matic Div., G.M. Corp.
6
5
4
3
2
1

1. Governor cover retainer ring
2. Governor cover seal
3. Governor cover with seal kit
4. Governor cover seal
5. Governor shaft seal ring
6. Governor

Governor cover assembly

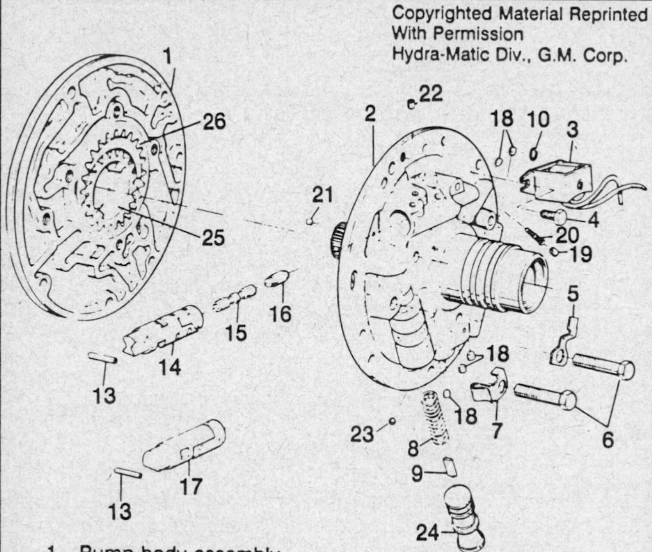
Copyrighted Material Reprinted With Permission Hydra-Matic Div., G.M. Corp.
1
26
21
25
16
15
14
13
17
13
22
2
18 10
3
4
20
19
5
18
18
7
6
23
8
9
24
11
12

1. Pump body assembly
2. Pump cover assembly
3. Solenoid and clip assembly
4. Hex bolt
5. Bushing pin retainer
6. Hex bolt
7. Solenoid wire clip
8. Pressure regulator spring
9. Pressure regulator guide pin
10. Solenoid O-ring seal
11. Valve bore pump plug
12. Retainer ring (type TA internal)
13. Coil spring pin
14. Converter clutch valve bushing
15. Converter clutch valve
16. Converter clutch actuator valve
17. Nonconverter clutch models bushing
18. Cup plug
19. Pump body plug
20. Screen
21. Pump cover plug
22. Converter clutch pump cover plug
23. Ball
24. Pressure regulator valve
25. Pump drive gear
26. Pump driven gear

Pump assembly

Procedure

When servicing the pump on a THM 200C, make sure the plug is removed from the pump cover.

NOTE: When installing or servicing neutral/drive shift package on a THM 200C built before December 15, 1986 (Julian date 349), make sure the plug is removed from the pump cover. The new turbine shaft assembly with ball capsule is included in the neutral/drive shift package.

New Design Forward Clutch For Smoother Park/Neutral to Drive Engagement

Beginning 1987 THM 200C transmissions are being built with a new design "feed/bleed" system for the forward clutch. This system gives more cushioning during forward clutch apply and can be used on certain past models to reduce a harsh park/neutral to drive engagement feel.

As of August 4, 1986 (Julian date 216), these changes were introduced for all THM 200C models. They include the following, forward clutch assembly, minus check ball in the housing, forward clutch piston assembly, an additional waved plate and a front internal gear.

Procedure

1984–87 TRANSMISSIONS BUILT BEFORE AUGUST 4, 1986 (JULIAN DATE 216)

If a vehicle equipped with a THM 200C transmission has a harsh park/neutral to drive engagement, install the neutral drive shift package (part number 8638957).

1987 TRANSMISSIONS BUILT ON OR AFTER AUGUST 4, 1986 (JULIAN DATE 216)

Transmissions built after implementation of the feed/bleed system can be serviced with individual parts.

NOTE: When servicing the forward clutch, select the correct backing plate. Backing plate selection affects piston travel, shift feel and clutch durability.

Backing Plate Selection (Clutch Fully Assembled)

Apply an evenly distributed load to remove slack from the clutch assembly. Use 20 lbs. (9 Kg) of pressure against the backing plate. Excessive pressure will cause an inaccurate measurement. With the clutch assembly compressed, use a feeler gauge to measure the distance between the backing plate and the snapring. Correct backing plate travel is 0.028–0.059 in. (0.7–1.5mm).

Following parts list is for 1987 transmissions built on or after August 4, 1986 (Julian date 216):

Forward clutch assembly, complete—part number 8648720
Forward clutch housing package—part number 8638956
Forward clutch piston—part number 8648721
Front internal gear and thrust washer package—part number 8639947
Selective backing plate-identification 7—part number 8657267
Selective backing plate-identification 6—part number 8657266
Selective backing plate-identification X—part number 8657618
Selective backing plate-identification 5—part number 8657265
Selective backing plate-identification 4—part number 8657264

Earlier design front internal gears (service package part number 8630909) can not be used with THM 200C transmissions built after August 4, 1986 (Julian date 216).

Fluid Leaks From the Speedometer Driven Gear/Adapter Area

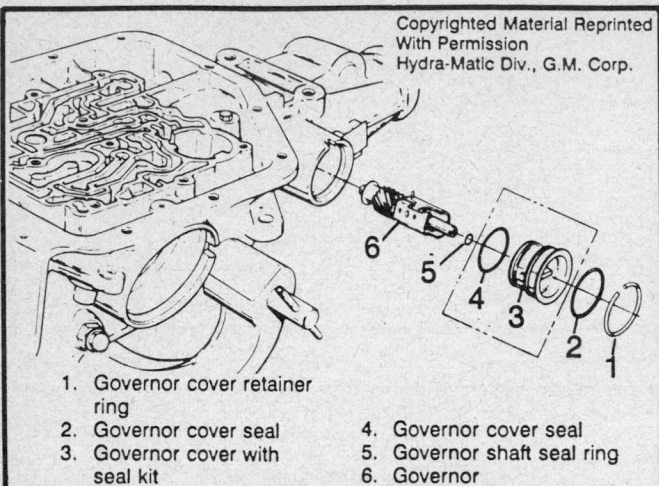

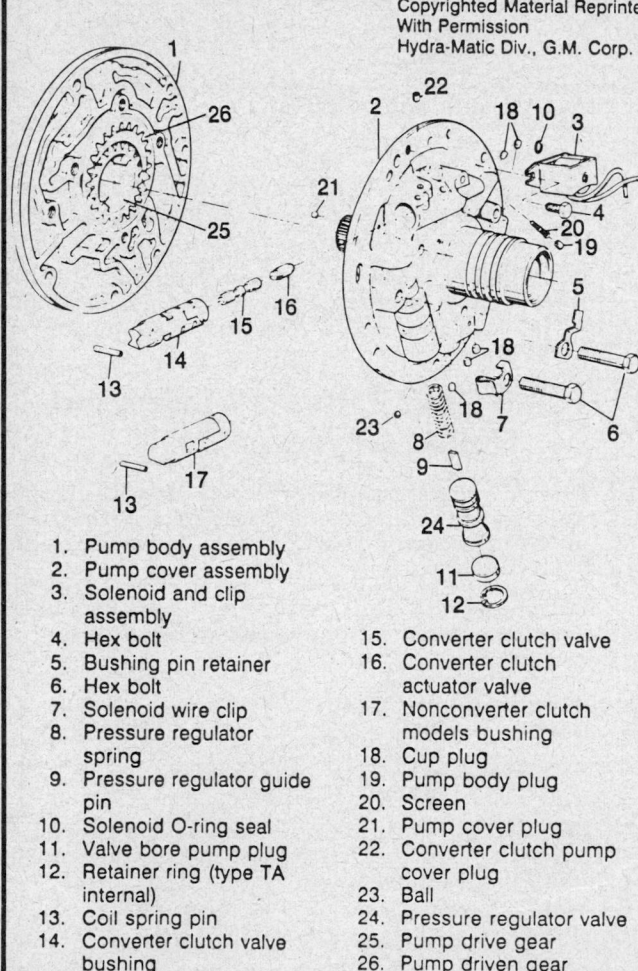

Transmission fluid leak at the plastic speedometer driven gear

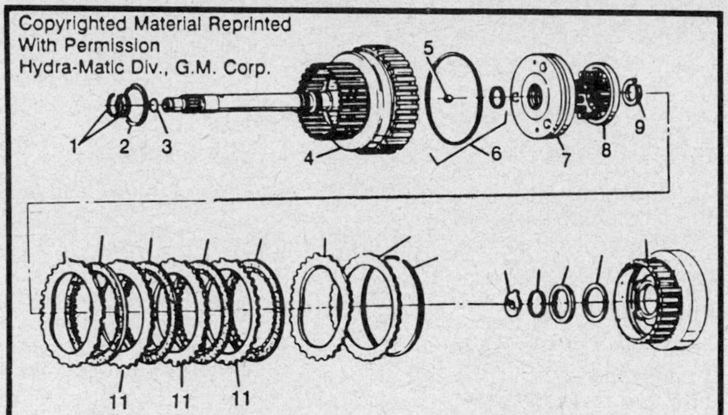
Copyrighted Material Reprinted
With Permission
Hydra-Matic Div., G.M. Corp.

1. Turbine shaft ring seals
2. Direct and forward clutch thrust washer
3. Turbine shaft to selective washer seal
4. Forward clutch housing assembly
5. Turbine shaft cup plug
6. Forward clutch piston seal kit
7. Forward clutch piston assembly
8. Forward clutch retainer with spring
9. Forward spring retainer snapring
10. Forward clutch waved plate
11. Forward clutch flat plate
12. Forward clutch backing plate (selective)
13. Direct clutch plate housing snapring
14. Front selective thrust washer
15. Output shaft snapring
16. Selective washer
17. Front internal gear washer
18. Front internal gear with bushing

Forward clutch assembly

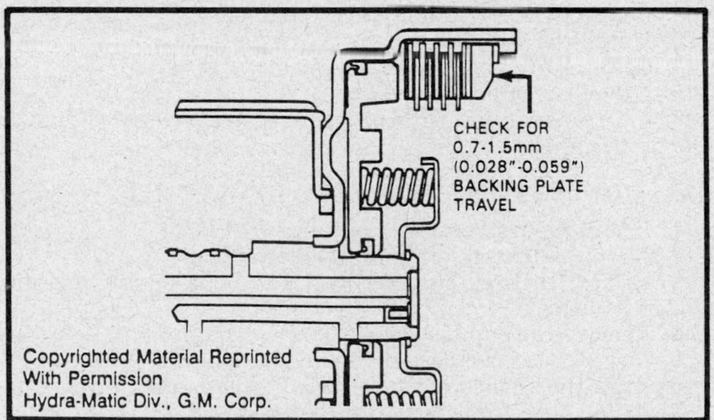
CHECK FOR
0.7-1.5mm
(0.028"-0.059")
BACKING PLATE
TRAVEL
Copyrighted Material Reprinted
With Permission
Hydra-Matic Div., G.M. Corp.

Checking backing plate travel

can be caused by a hole in the driven gear shaft. This condition could be misdiagnosed as a speedometer driven gear fitting leak.

Procedure

When servicing 1986–87 THM 200C transmissions for a fluid leak around the speedometer driven gear fitting or cable, check the speedometer driven gear shaft for a hole through the center of the shaft as follows:

1. Install a hand vacuum pump over the end of the gear (opposite gear teeth).
2. Pump up 10 in. Hg of vacuum.
3. Gauge should stay at 10 in. Hg for at least 30 seconds.
   a. If gear holds vacuum, inspect the seals in the adapter.
   b. If the gear does not hold vacuum, replace the gear.

NOTE: Check any new gear in the same manner, from parts stock before installing in the transmission.

BACKING PLATE TRAVEL = 0.7mm - 1.5mm (0.028" - 0.059")

PLATE THICKNESS	IDENTIFICATION
3.70mm - 4.15mm (.146" - .163")	7
4.25mm - 4.70mm (.167" - .185")	6
4.80mm - 5.25mm (.189" - .207")	X
5.35mm - 5.80mm (.211" - .228")	5
5.90mm - 6.35mm (.232" - .250")	4

Forward clutch backing plate selection

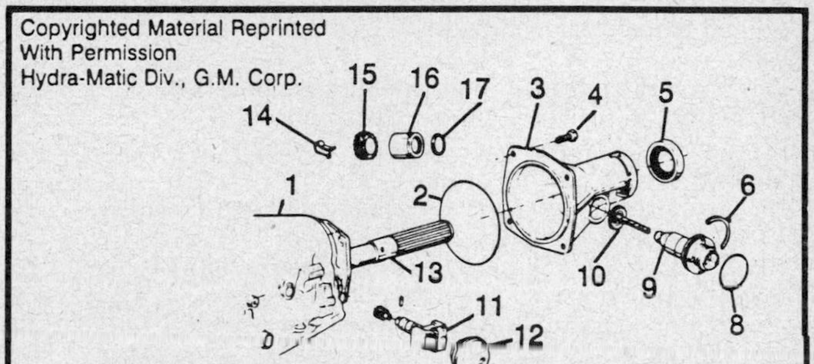
Copyrighted Material Reprinted
With Permission
Hydra-Matic Div., G.M. Corp.

1. Transmission case
2. Case extension to case seal
3. Case extension
4. Case extension to case bolt
5. Case extension oil seal assembly
6. Speedometer driven gear fitting retainer
7. Bolt and washer assembly
8. Speedometer fitting to case extension O-ring seal
9. Speedometer driven gear fitting assembly
10. Speedometer driven gear
11. Governor assembly
12. Governor cover
13. Output shaft
14. Speedometer drive gear clip
15. Speedometer drive gear
16. Output shaft sleeve
17. Output shaft seal

Extension and parts

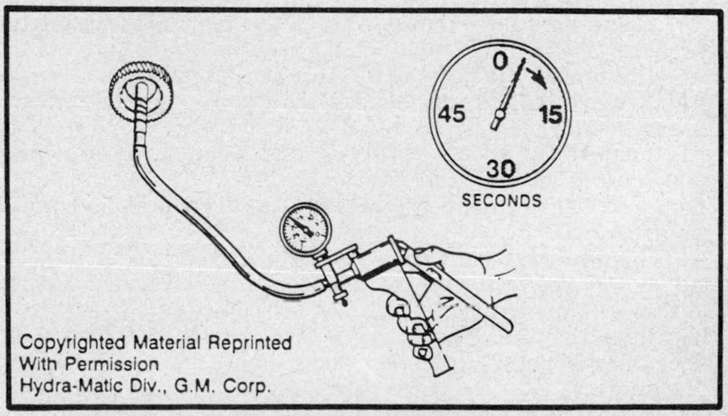
0
15
30
45
SECONDS
Copyrighted Material Reprinted
With Permission
Hydra-Matic Div., G.M. Corp.

Checking driven gear shaft for void or hole

REMOVAL AND INSTALLATION

TRANSMISSION REMOVAL

EXCEPT 1984 CHEVETTE WITH DIESEL ENGINE

1. Disconnect the negative battery cable.
2. Remove the air cleaner.
3. Disconnect the T.V. cable at its upper end.
4. Remove the transmission dipstick and (if accessible) the bolt holding the dipstick tube.
5. Raise the vehicle and support safely.
6. Remove the driveshaft. Remove the floor pan reinforcement (if used) if it interferes with the driveshaft removal.
7. Disconnect the speedometer cable at the transmission.
8. Disconnect the shift linkage at the transmission.
9. Disconnect all electrical connectors at the transmission and any clips that retain the leads to the transmission case.
10. Remove the flexplate cover bolts and cover.
11. Matchmark the flexplate and torque converter to aid in reassembly.
12. Remove the torque converter bolts and/or nuts.
13. Remove the catalytic converter support bracket.
14. Remove the transmission support to transmission mount bolt and transmission support to frame bolts (and insulators, if used).
15. Support and raise the transmission slightly.
16. Slide the transmission support rearward.
17. Lower the transmission to gain access to the oil cooler lines and T.V. cable attachments.
18. Remove the oil cooler lines and cap open ends.
19. Disconnect the T.V. cable.
20. Support the engine with a suitable support tool and remove the transmission to engine bolts.
21. Remove the transmission from the vehicle being careful not to damage any cables, lines or linkage.

1984 CHEVETTE WITH DIESEL ENGINE

1. Disconnect the negative battery cable.
2. Disconnect the T.V. cable/detent cable from the transmission.
3. Remove the transmission dipstick.
4. On vehicles with A/C, remove the heater core cover screws from the heater assembly and place it out of the way.
5. Raise the vehicle and support safely.
6. Remove the driveshaft.
7. Disconnect the speedometer cable.
8. Disconnect all electrical connectors.
9. Disconnect the oil cooler lines from the transmissions.
10. Disconnect the shift linkage and support the transmission.
11. Remove the rear transmission support bolts.
12. Remove the nuts holding the converter bracket to the support.
13. Disconnect the exhaust pipe at the rear of the catalytic converter.
14. Disconnect the exhaust pipe at the manifold.
15. Remove the exhaust pipe, catalytic converter and bracket as an assembly.
16. Remove the torque converter inspection cover attaching bolts and inspection cover.
17. Scribe the relationship between the flexplate and torque converter.
18. Remove the torque converter bolts.
19. Lower the transmission to expose the bell housing bolts and remove the bolts.
20. Raise the transmission to the normal position. Place a 2 in. block of wood between the rack and pinion housing and the engine oil pan.
21. Support the engine with a jack.
22. Remove the transmission from the vehicle.

TRANSMISSION INSTALLATION

EXCEPT 1984 CHEVETTE WITH DIESEL ENGINE

1. Raise the transmission into the vehicle and install the engine to transmission bolts.
2. Connect the oil cooler lines to the transmission.
3. Connect the T.V. cable to the transmission.
4. Install the transmission support and mount.
5. Remove the transmission jack.
6. Align the torque converter to flexplate match marks made during disassembly. Install the torque converter to flexplate bolts and tighten to 46 ft. lbs. (62 Nm).

NOTE: Make sure the weld nuts on the torque converter are flush with the flexplate. Test the torque converter for freedom of rotation. Tighten the 3 bolts finger tight, then torque to specification. Retorque the first bolt tightened.

7. Install the floor pan reinforcement, if removed.
8. Install the catalytic converter support bracket.
9. Install the flexplate cover and attaching bolts.
10. Connect the shift linkage.
11. Connect the speedometer cable.
12. Connect all electrical connectors and brackets.
13. Install the driveshaft.
14. Lower the vehicle.
15. Install the dipstick tube (with a new oil seal) and attaching bolt. Install the transmission dipstick.
16. Connect the T.V. cable at the engine.
17. Install the air cleaner.
18. Connect the negative battery cable.
19. Adjust the shift linkage and T.V. cable.
20. Fill transmission with proper type transmission fluid to the proper level.

1984 CHEVETTE WITH DIESEL ENGINE

1. Install the transmission into the vehicle.
2. Align the transmission on the dowel pins.
3. Remove the 2 in. block of wood from between the rack and pinion housing.
4. Remove the engine jack stand.
5. Lower the transmission and install the bell housing bolts.
6. Align the scribed marks on the flexplate and converter.
7. Install the torque converter bolts.
8. Install the torque converter inspection cover and attaching bolts.
9. Install the exhaust pipe, catalytic converter and bracket assembly.
10. Connect the exhaust pipe to the exhaust manifold.
11. Connect the exhaust pipe at the rear of the catalytic converter.
12. Install the nuts holding the converter bracket to the support.
13. Install the support bolts to the rear transmission.
14. Install the shift linkage to the transmission.
15. Connect the oil cooler lines to the transmission.
16. Connect all electrical connectors to the transmission.
17. Connect the speedometer cable to the transmission.
18. Lower the vehicle.
19. Install the transmission dipstick.
20. Connect the T.V. cable/detent cable to the transmission.
21. Adjust the T.V. cable/detent cable.
22. Connect the negative battery cable.
23. Fill transmission with proper type transmission fluid to the proper level.

SPECIFICATIONS

TORQUE SPECIFICATIONS

Item	ft. lbs.	Nm
Pump body to pump cover	15–20	20–27
Pump assembly to case	15–20	20–27
Parking lock bracket to case	15–20	20–27
Transmission oil pan to case	6–10	8–14
Manual shaft to inside detent lever	20–25	27–34
Cooler connector brass	20–25	27–34
Line pressure take off	5–10	7–14
Throttle lever, link and bracket to case	9–12	13–17
Control valve assembly to case	9–12	13–17
Oil screen to case	9–12	13–17
Speedometer driven gear retainer to case	6–10	8–14
Governor pressure switch	5–10	7–14
Solenoid assembly to pump	2–4	3–5

TORQUE SPECIFICATIONS

Item	ft. lbs.	Nm
Starter mounting bolts	30	41
Transmission to engine bolts	35	47
Flexplate to converter bolts	46	62
Converter housing cover screws	7	10
Dipstick tube brace to transmission	7	10
Oil cooler lines to transmission	16	22
Transmission support to frame	40	55
Mount to transmission support	25	34
Catalytic converter bracket	15	21
Floor pan reinforcement	19	25
Driveshaft (straps) bolts	16	22
T.V. cable to case	7	10

FRONT UNIT ENDPLAY THRUST WASHER THICKNESS

THICKNESS		IDENTIFICATION NUMBER AND/OR COLOR
1.66-1.77mm	(0.065"-0.070")	1 - —
1.79-1.90mm	(0.070"-0.075")	2 - —
1.92-2.03mm	(0.076"-0.080")	3 - BLACK
2.05-2.16mm	(0.081"-0.085")	4 - LIGHT GREEN
2.18-2.29mm	(0.086"-0.090")	5 - SCARLET
2.31-2.42mm	(0.091"-0.095")	6 - PURPLE
2.44-2.55mm	(0.096"-0.100")	7 - COCOA BROWN
2.57-2.68mm	(0.101"-0.106")	8 - ORANGE
2.70-2.81mm	(0.106"-0.111")	9 - YELLOW
2.83-2.94mm	(0.111"-0.116")	10 - LIGHT BLUE
2.96-3.07mm	(0.117"-0.121")	11 - BLUE
3.09-3.20mm	(0.122"-0.126")	12 - —
3.22-3.33mm	(0.127"-0.131")	13 - PINK
3.35-3.46mm	(0.132"-0.136")	14 - GREEN
3.48-3.59mm	(0.137"-0.141")	15 - GRAY

REAR UNIT ENDPLAY WASHER THICKNESS

THICKNESS		IDENTIFICATION NUMBER AND/OR COLOR
2.90-3.01mm	(0.114"-0.119")	1 - ORANGE
3.08-3.19mm	(0.121"-0.126")	2 - WHITE
3.26-3.37mm	(0.128"-0.133")	3 - YELLOW
3.44-3.55mm	(0.135"-0.140")	4 - BLUE
3.62-3.73mm	(0.143"-0.147")	5 - RED
3.80-3.91mm	(0.150"-0.154")	6 - BROWN
3.98-4.09mm	(0.157"-0.161")	7 - GREEN
4.16-4.27mm	(0.164"-0.168")	8 - BLACK
4.34-4.45mm	(0.171"-0.175")	9 - PURPLE

INTERMEDIATE BAND SERVO APPLY PIN SELECTION

DIAL INDICATOR TRAVEL		APPLY PIN IDENTIFICATION
.0 - .72mm	(.0" - .029")	1 RING
.72 - 1.44mm	(.029" - .057")	2 RINGS
1.44 - 2.16mm	(.057" - .086")	3 RINGS
2.16 - 2.88mm	(.086" - .114")	WIDE BAND

CLUTCH PLATE AND APPLY RING USAGE

	DIRECT CLUTCH					FORWARD CLUTCH								LO & REVERSE CLUTCH							
	FLAT STEEL PLATES		COMP. FACED PLATES	APPLY RING		WAVED STEEL PLATE		FLAT STEEL PLATES		COMP. FACED PLATES	APPLY RING		WAVED STEEL PLATE		FLAT STEEL PLATES		COMP. FACED PLATES	APPLY RING	
	NO.	THICK-NESS	NO.	I.D.	WIDTH*	NO.	THICK-NESS	NO.	THICK-NESS	NO.	I.D.	WIDTH*	NO.	THICK-NESS	NO.	THICK-NESS	NO.	I.D.	WIDTH*
ALL MODELS	5	2.32 mm (.091")	5	19	12.5 mm (.490")	1	1.59 mm (.062")	3	1.97 mm (.077")	4	18	13.5 mm (.530")	1	1.97 mm (.077")	7	1.97 mm (.077")	6	0	13.13 mm (.520")

The direct, forward, and lo and reverse clutch flat steel clutch plates and the forward clutch waved steel plate should be identified by their thickness.

The direct and forward clutch production installed composition-faced clutch plates must not be interchanged. For service, direct and forward clutch use the same composition-faced plates.

*Measure the width of the clutch apply ring for positive identification.

Section 4

THM 200-4R Transmission
General Motors

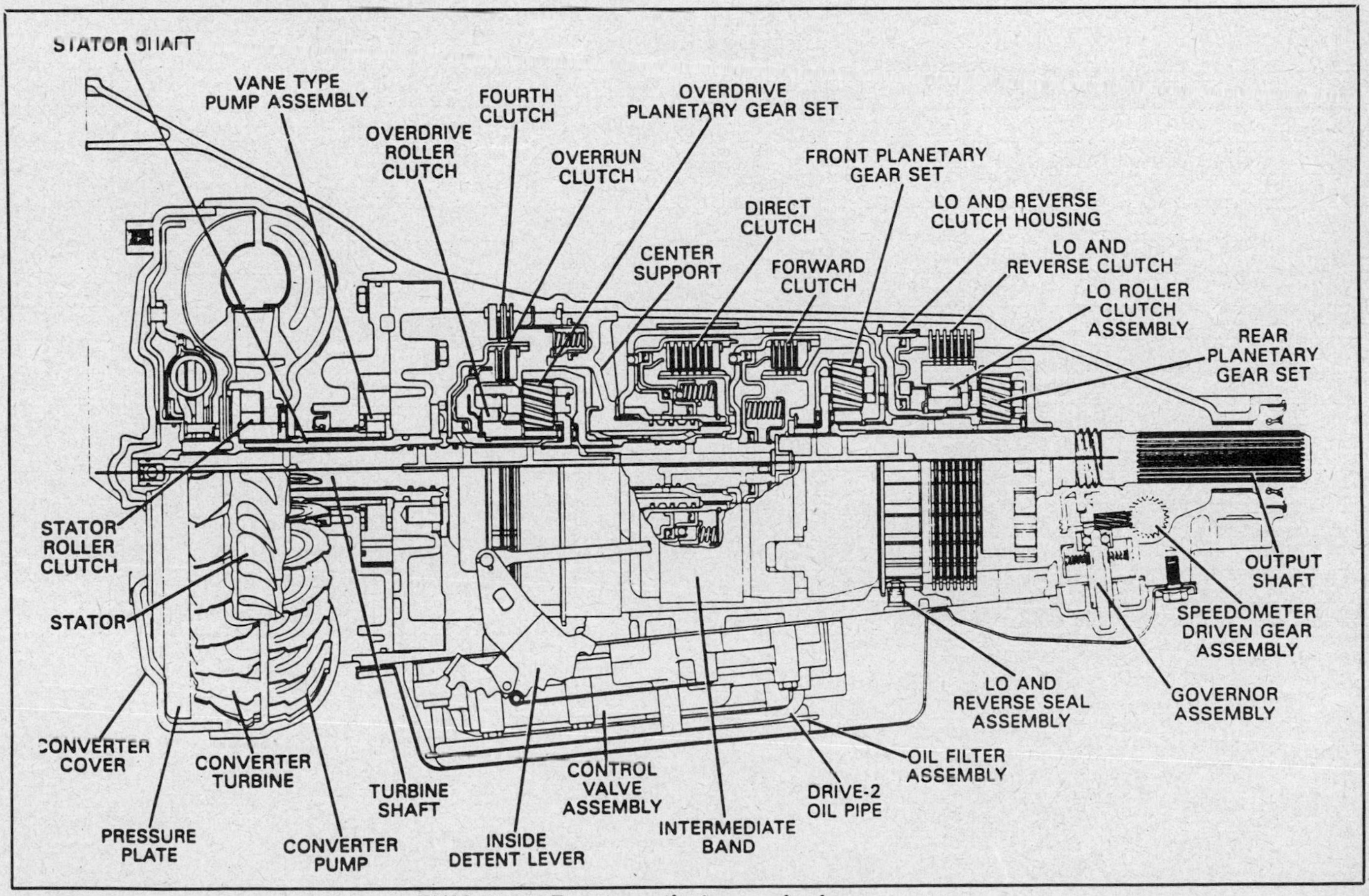

STATOR SHAFT
VANE TYPE PUMP ASSEMBLY
FOURTH CLUTCH
OVERDRIVE PLANETARY GEAR SET
OVERDRIVE ROLLER CLUTCH
OVERRUN CLUTCH
FRONT PLANETARY GEAR SET
LO AND REVERSE CLUTCH HOUSING
DIRECT CLUTCH
CENTER SUPPORT
FORWARD CLUTCH
LO AND REVERSE CLUTCH
LO ROLLER CLUTCH ASSEMBLY
REAR PLANETARY GEAR SET
STATOR ROLLER CLUTCH
STATOR
OUTPUT SHAFT
SPEEDOMETER DRIVEN GEAR ASSEMBLY
LO AND REVERSE SEAL ASSEMBLY
GOVERNOR ASSEMBLY
CONVERTER COVER
CONVERTER TURBINE
TURBINE SHAFT
CONTROL VALVE ASSEMBLY
OIL FILTER ASSEMBLY
DRIVE-2 OIL PIPE
INTERMEDIATE BAND
PRESSURE PLATE
CONVERTER PUMP
INSIDE DETENT LEVER

Cross section of General Motors THM 200-4R automatic transmission

REFERENCES

APPLICATION

1980–84 Chilton's Professional Automatic Transmission Manual #7390
1984–89 Chilton's Professional Transmission Manual—Domestic Vehicles #7959

GENERAL DESCRIPTION

1980–84 Chilton's Professional Automatic Transmission Manual #7390

MODIFICATIONS

1980–84 Chilton's Professional Automatic Transmission Manual #7390
1984–89 Chilton's Professional Transmission Manual—Domestic Vehicles #7959

TROUBLE DIAGNOSIS

1980–84 Chilton's Professional Automatic Transmission Manual #7390

ON CAR SERVICES

1980–84 Chilton's Professional Automatic Transmission Manual #7390

REMOVAL AND INSTALLATION

1980–84 Chilton's Professional Automatic Transmission Manual #7390
1984–89 Chilton's Professional Transmission Manual—Domestic Vehicles #7959

BENCH OVERHAUL

1980–84 Chilton's Professional Automatic Transmission Manual #7390

SPECIFICATIONS

1980–84 Chilton's Professional Automatic Transmission Manual #7390
1984–89 Chilton's Professional Transmission Manual—Domestic Vehicles #7959

SPECIAL TOOLS

1980–84 Chilton's Professional Automatic Transmission Manual #7390

APPLICATION

THM 200-4R

Year	Vehicle	Engine	Code
1986–87	Regal	3.8L	7
1984	Caprice, Impala, Monte Carlo	3.8L	9
1984–86	Regal	3.8L	9
1985–87	Cutlass	3.8L	A
1984	Regal, Estate Wagon	4.1L	4
1985	Fleetwood Brougham	4.1L	8
1984	Regal, Cutlass	4.3L	V
1985–88	Caprice, Impala, Monte Carlo	4.3L	Z
1985	Regal	4.3L	Z
1985–86	Parisienne	4.3L	Z
1985–87	Bonneville, Grand Prix	4.3L	Z
1984–86	Caprice, Impala, Parisienne	5.0L	H
1984–87	Regal, Cutlass, Safari, Grand Prix	5.0L	H
1984–88	Monte Carlo	5.0L	H
1985–88	Monte Carlo	5.0L	G
1986–88	Impala, Monte Carlo, Parisienne	5.0L	Y
1986–89	Caprice, Estate Wagon, Safari	5.0L	Y
1984	Park Avenue, Ninety Eight	5.0L	Y
1986–87	Regal	5.0L	Y
1984–85	Delta 88	5.0L	Y
1984–88	Cutlass	5.0L	Y
1984–89	Custom Cruiser	5.0L	Y
1984–89	Fleetwood Brougham	5.0L	Y
1984–89	Cutlass	5.0L	9
1989	Caprice	5.0L	E
1984	Ninety Eight, Park Avenue	5.7L	N
1984–85	Cutlass, Delta 88, Custom Cruiser	5.7L	N
1984–85	Estate Wagon, Monte Carlo	5.7L	N
1984–85	Caprice, Impala	5.7L	N
1984–85	Fleetwood Brougham	5.7L	N
1984–86	Parisienne, Grand Prix, Safari	5.7L	N
1984	Fleetwood Brougham	6.0L	9

TRANSMISSION MODIFICATIONS

Transmission Oil Pump Vane Rings

Beginning in 1987, the oil pump assemblies use modular iron vane rings. The past design or new design pump rings can be used on any year THM 200–4R transmission, but it is recommended that the new iron pump vane rings (part number 8657692) be used when replacement of the vane rings becomes necessary.

Endplay Adjustment for 1–2 Shift Gear Noise

Some vehicles may exhibit a gear noise prior to the 1–2 shift. This condition may be caused by a wear pattern on the front planetary gears. This noise may be reduced by replacing worn gearset, parts then checking and adjusting the forward clutch shaft endplay to 0.035–0.051 in. (0.90–1.30mm) and output shaft endplay to 0.013–0.025 in. (0.32–0.64mm).

New Filter and Seal

In August of 1987, a new filter design and a single piece seal was released. The new design assembly (part number 8639958) or the old design assembly with 2 O-rings can be used any transmission, but it is recommended to use the new lip seal (part number 8657767) with either design filter.

Oil Pan and Gasket

Beginning in January 1984 all transmissions are being produced with a ribbed flange transmission pan (part number 8639556) and special bolts with conical washers (part number 8643693). In June of 1986, a new cellulose coated transmission pan gasket was implemented for use on transmissions with raised rib pans. Some overhaul packages contain the latex-coated cork gasket, it should not be used. The new cellulose gasket (part number 8657387) should be install and pan bolts torqued to 6–10 ft. lbs. (8–14 Nm).

Engine Stops or Lugs

A condition where the engine stops running or lugs when shifted from **P** or **N** to **R** or **D**, may be a result of converter clutch engagement when either gear range is selected and may be caused by improper operation of the converter clutch valve, located in the pump cover. Check for valve not properly seated or missing retaining ring. When installing this ring make certain the rounded edge of ring faces the valve stop. Check the valve for sticking or burrs. Check for missing, broken or damaged converter clutch inner and outer springs.

Loss of Reverse, 2nd or 3rd Gear

Some transmissions may have stripped input drum splines. In 1984 a new heat hardened drum (part number 8648360) was introduced and should be used if this condition exists.

Harsh 1–2 Shift

Some 1984 vehicles may experience a harsh 1–2 upshift, which may be caused by the 2nd clutch check ball peening the 2nd clutch apply oil hole on the spacer plate. To correct this condition, verify the spacer plate as the cause and replace the spacer plate and gasket plate.

Some BT and BY model transmissions may experience a firm or harsh 1–2 shift. Beginning June 27, 1984 (Julian date 179) the parts in the service package number 8639912 went into production to correct this problem. This package contains: The control valve assembly, intermediate servo spring, 1–2 accumulator piston spring, pressure regulator valve spring, control valve body spacer plate and instructions. The basic procedures include replacing the above components, removing and discarding the 3–4 accumulator spring and replacing the 3–4 accumulator (without a spring).

Harsh 1–2 Shift Below Half Throttle

Some 1986, THM 200–4R CRF model transmissions may exhibit a harsh 1–2 shift at minimum to ½ throttle. This condition is may be caused by an incorrect (may be pink) intermediate inner servo spring. Replace with new spring (part number 8632147).

Harsh 1–2 or 3–4 Shift

Some 1984 vehicles may experience a harsh 1–2 or 3–4 shift. This condition may be caused by a damaged or mispositioned T.V. boost valve spring. Starting April 1984 a new control valve assembly, which has a new T.V. boost valve spring went into production on all transmission to correct this harsh shift. On transmissions prior April, if this problem exists, replace the control valve assembly.

No Upshift After Downshift

Some 1984–86 vehicles may exhibit a no upshift condition after a part or full throttle downshift. This condition may be caused by internal leakage hydraulically locking the throttle valve plunger. If this occurs, the line pressure remains at a boosted level, resulting in no upshift. To repair this condition, a new throttle valve plunger, throttle valve bushing and on some transmissions the lever and bracket assembly should be installed.

Servo Cover Retainer

In 1985, a new style servo cover retaining ring (with tangs) went into production. In addition to the new style ring, the notches on the case intermediate servo bore, used for access of the retaining ring, were omitted. The older style retaining ring (part number 8628137) can only be used with the past style case with notches. The new servo cover retaining ring (part number 8657213) can be used with either style case intermediate servo bore.

In the event an old style retaining ring was incorrectly install on a case without access notches, use the following removal procedure:

1. Depress the intermediate servo cover using servo cover depressor J-29714 or equivalent.
2. Use a small, flat blade tool to remove the servo cover retaining ring.
3. Be certain to reassemble new style servo bores (with notches) with the new style servo retaining ring (with tangs).

Torque Converter Lug Design

Starting March 1985, all 245mm and 298mm torque converters will use a new torque converter lug design. The 298mm torque converter used on diesel engines, will also use a new 3 lug torque converter in place of the older 6 lug design.

SERVICE PACKAGE PART NUMBER	TRANSMISSION	**1986 MODEL JULIAN DATE	TRANSMISSION MODEL AND YEAR		
			1986	1985	1984
8639933	THM 200-4R	057	BRF		BQ
8639934	THM 200-4R	336	HCF, KZF	HG, OZ	HG, OZ
8639935	THM 200-4R	057	CAF, CHF, CRF, CZF	CH, CQ, CR, CY, OM	CQ, OM
8639936	THM 200-4R	336	HFF, KBF, KCF, KJF	HE, OG, OJ	HE, OG, OJ
8639937	THM 200-4R	336	AAF, ABF, APF	AA, AP, AO	AA, AP, BT, BY, CH
***8639939	THM 200-4R	N/A		BQ	CR

T.V. leakage correction kit

Torque Converter Blade Design

Starting 1984, torque converters will use blades retained into the converter pump shell by a braze process. Blades which have moved in a converter may reduce torque multiplication causing a loss of power under 35 mph.

Design Forward Clutch

As of April 1986 (Julian date 115), some THM 200–4R model transmissions received a modified forward clutch assembly, minus the check ball in the housing, clutch piston, front internal gear and additional waved plates. On vehicles built prior to April 1986 (Julian date 115) exhibits a harsh **P/N** to **D** engagement, install service package number 8639943. This package cannot be installed on 1986–87 BRF or 1984–85 BQ model transmissions. Earlier design front internal gears (part number 8630909) cannot be used with models built before April 1986 (Julian date 115). Transmission built after April 1986 (Julian date 115) use complete clutch assembly number 8657257 and use front internal gear and thrust washer number 8639947 (can be used on all models). The correct backing plate travel is 0.028–0.059 in. (0.7–1.5mm).

Pump Bushing

As of January 1987, (Julian date 009) a machined step was added to the pump bushing bore to prevent walk-out. Because of the stepped bore in the pump body, removal/installation procedures and tool for this pump have been changed. Tool number J-25019-20 or equivalent must be used to service bushing.

Check Ball Location Data

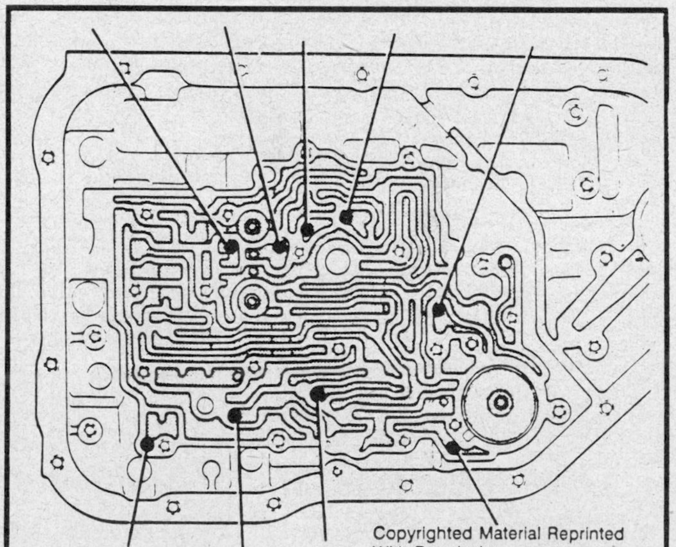
Copyrighted Material Reprinted
With Permission
Hydra-Matic Div., G.M. Corp.

Check ball locations

REMOVAL AND INSTALLATION

TRANSMISSION REMOVAL

1. Disconnect the negative battery cable.
2. Remove the air cleaner assembly. Disconnect the T.V./detent cable at its upper end.
3. Remove the transmission dipstick. Remove the bolt holding the dipstick tube, if accessible.
4. Raise the vehicle and support it safely.
5. Drain the transmission. Matchmark and remove the driveshaft.
6. Disconnect the speedometer cable at the transmission. Disconnect the shift linkage at the transmission.
7. Disconnect all electrical leads at the transmission. Disconnect any clips that retain these leads.
8. Remove the inspection cover. Mark the flexplate and torque converter, to aid in installation.
9. Remove the torque converter to flex plate retaining bolts. Remove the catalytic converter support bracket, if necessary.
10. Position the transmission jack under the transmission and remove the rear transmission mount.
11. Remove the floor pan reinforcement, if equipped. Remove the crossmember retaining bolts. Move the crossmember out of the way.
12. Remove the transmission to engine bolt on the left side. This is the bolt that retains the ground strap.
13. Disconnect and plug the transmission cooler lines.
14. With the transmission still lowered, disconnect the T.V. cable.
15. Support the engine, using the proper tools and remove the remaining engine to transmission bolts.
16. Carefully disengage the transmission assembly from the engine. Lower the transmission from the transmission jack.

TRANSMISSION INSTALLATION

1. Raise transmission using appropriate jack and install assembly to engine.
2. Connect the T.V. cable.
3. Connect the transmission cooler lines.
4. Replace transmission to engine bolts and torque to 35 ft. lbs. (47 Nm). Make certain to replace the grounding strap.
5. Install the crossmember and floor pan reinforcement. Tighten the support-to-frame bolts to 25 ft. lbs. (34 Nm).
6. Install the transmission mount bolts and torque to 35 ft. lbs. (47 Nm).
7. Before installing the flexplate to torque converter bolts, make sure that the weld nuts on the converter are flush with the flexplate and that the torque converter rotates freely by hand.
8. Be sure that the flexplate and torque converter matchmarks made during the removal procedure line up with each other.
9. Install the torque converter to flexplate bolts and torque to 46 ft. lbs. (62 Nm).
10. Install the catalytic converter support, if removed.
11. Install the inspection cover and torque the bolts to 53 inch lbs. (6 Nm).
12. Connect all electrical connections.
13. Connect the speedometer cable at the transmission.
14. Connect the shift linkage at the transmission.
15. Install the driveshaft, making certain to align marks.
16. Adjust the shift linkage and the T.V. cable as required.
17. Lower the vehicle and fill the transmission with the proper grade and type automatic transmission fluid.
18. Start the engine and check for leaks. Once the vehicle has reached operating temperature, recheck the fluid level.

SPECIFICATIONS

TORQUE SPECIFICATIONS

Item	ft. lbs.	Nm
Line pressure take-off	8	10
Direct clutch pressure take-off	8	10
Stator shaft to pump cover	9	12
Case to center support	18	24
Solenoid to case	9	12
Pressure switch	8	11
Accumulator housing to case	9	12
Governor cover to case	18	24
Transmission support to frame	41	55
Mount to transmission	25	34
Transmission to engine mounting bolts	35	47
Converter dust shield screws	8	11
Manual shaft nut	23	31
Speedometer driven gear attaching bolts	8	11
Detent cable attaching screw	6	9
Oil cooler line to transmission connector	15	20
Oil cooler line to radiator connector	20	27
Linkage swivel clamp nut	30	41
Shifter assembly to sheet metal screws	8	10
Converter bracket to adapter nuts	13	17
Catalytic converter to rear exhaust pipe nuts	17	23
Exhaust pipe to manifold nuts	12	16
Rear transmission support bolts	40	54
Mounting assembly to support nuts	21	29
Mounting assembly to support center nut	33	44
Adapter to transmission bolts	33	44
Pump cover bolts	18	24
Pump to case attaching bolts	18	24
Parking pawl bracket bolts	18	24
Control valve body bolts	11	15
Bottom pan attaching bolts	12	16
Converter to flywheel bolts	35	48

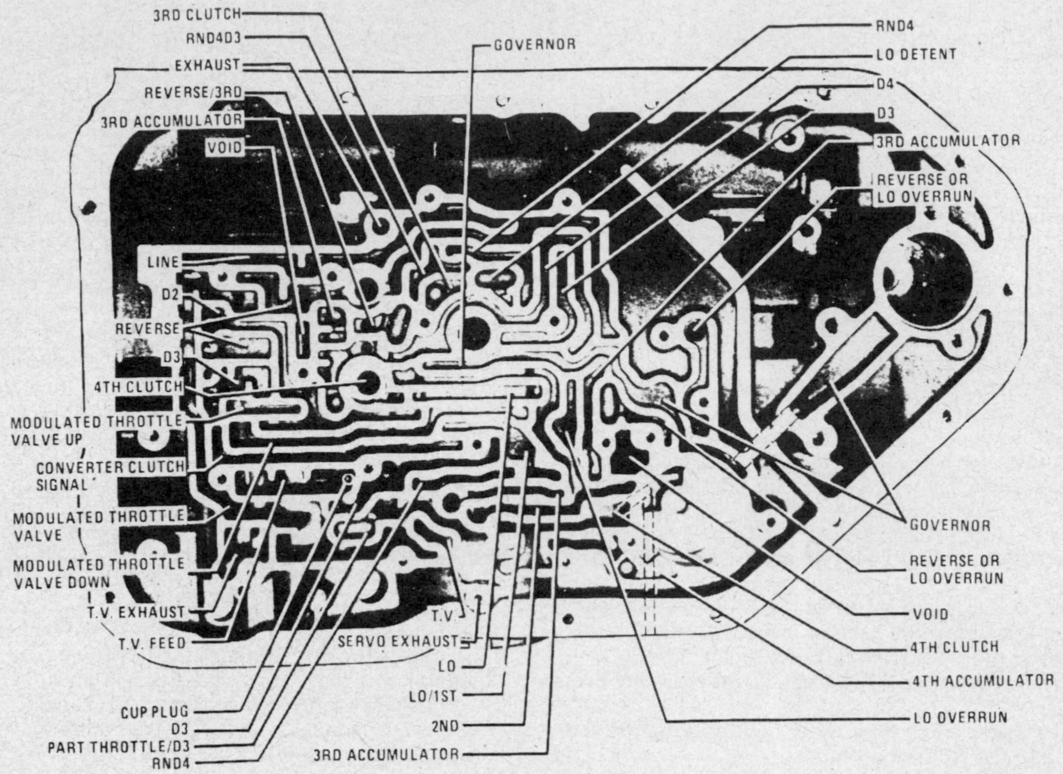
3RD CLUTCH
RND4D3
EXHAUST
REVERSE/3RD
3RD ACCUMULATOR
VOID
GOVERNOR
RND4
LO DETENT
D4
D3
3RD ACCUMULATOR
REVERSE OR
LO OVERRUN
LINE
D2
REVERSE
D3
4TH CLUTCH
MODULATED THROTTLE
VALVE UP
CONVERTER CLUTCH
SIGNAL
MODULATED THROTTLE
VALVE
MODULATED THROTTLE
VALVE DOWN
T.V. EXHAUST
T.V. FEED
T.V.
SERVO EXHAUST
LO
LO/1ST
2ND
3RD ACCUMULATOR
CUP PLUG
D3
PART THROTTLE/D3
RND4
GOVERNOR
REVERSE OR
LO OVERRUN
VOID
4TH CLUTCH
4TH ACCUMULATOR
LO OVERRUN

Front of case oil passages—200–4R automatic transmission

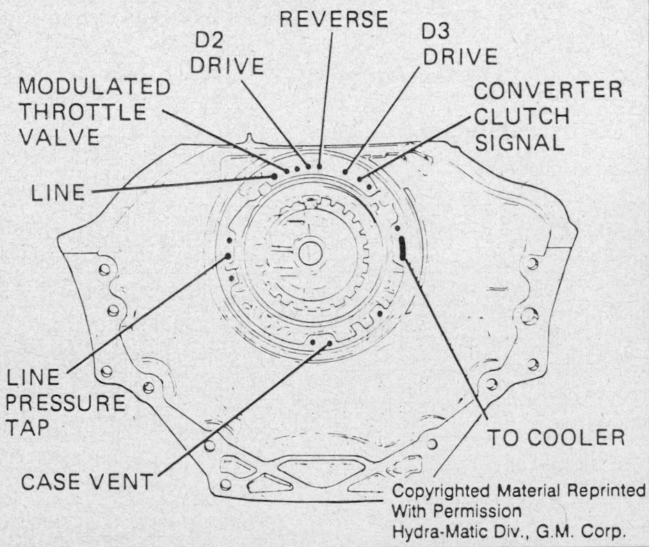
REVERSE
D2
DRIVE
D3
DRIVE
MODULATED
THROTTLE
VALVE
CONVERTER
CLUTCH
SIGNAL
LINE
LINE
PRESSURE
TAP
CASE VENT
TO COOLER

Bottom of case oil passages—200–4R automatic transmission

Section 4

THM 250C Transmission
General Motors

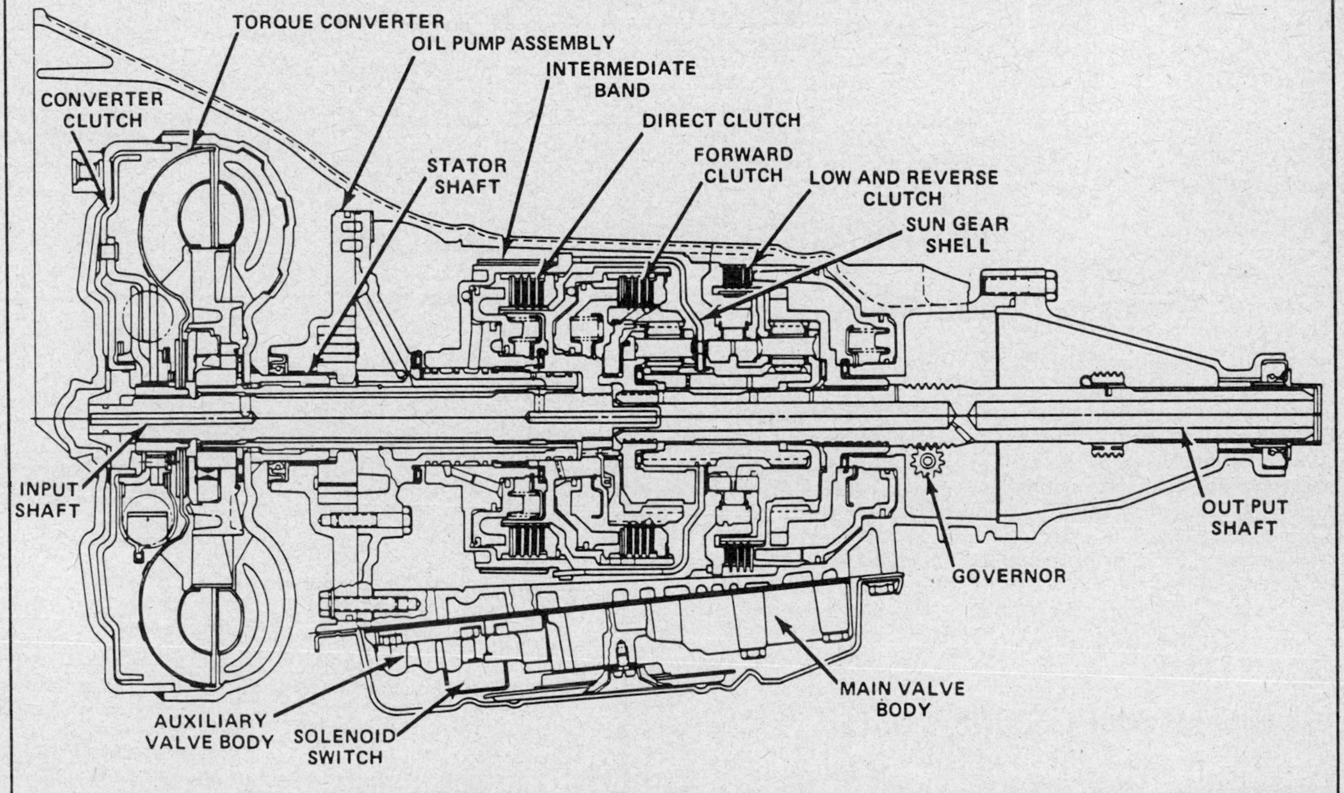

TORQUE CONVERTER
OIL PUMP ASSEMBLY
INTERMEDIATE BAND
CONVERTER CLUTCH
DIRECT CLUTCH
STATOR SHAFT
FORWARD CLUTCH
LOW AND REVERSE CLUTCH
SUN GEAR SHELL
INPUT SHAFT
OUT PUT SHAFT
GOVERNOR
MAIN VALVE BODY
AUXILIARY VALVE BODY
SOLENOID SWITCH

Cross section of General Motors THM 250C automatic transmission

REFERENCES

APPLICATION

1974–80 Chilton's Professional Automatic Transmission Manual #6927
1984–89 Chilton's Professional Transmission Manual—Domestic Vehicles #7959

GENERAL DESCRIPTION

1974–80 Chilton's Professional Automatic Transmission Manual #6927

MODIFICATIONS

1974–80 Chilton's Professional Automatic Transmission Manual #6927
1984–89 Chilton's Professional Transmission Manual—Domestic Vehicles #7959

TROUBLE DIAGNOSIS

1974–80 Chilton's Professional Automatic Transmission Manual #6927

ON CAR SERVICES

1974–80 Chilton's Professional Automatic Transmission Manual #6927

REMOVAL AND INSTALLATION

1974–80 Chilton's Professional Automatic Transmission Manual #6927
1984–89 Chilton's Professional Transmission Manual—Domestic Vehicles #7959

BENCH OVERHAUL

1974–80 Chilton's Professional Automatic Transmission Manual #6927

SPECIFICATIONS

1974–80 Chilton's Professional Automatic Transmission Manual #6927
1984–89 Chilton's Professional Transmission Manual—Domestic Vehicles #7959

SPECIAL TOOLS

1974–80 Chilton's Professional Automatic Transmission Manual #6927

APPLICATION

THM 250C

Year	Vehicle	Engine
1984	Monte Carlo	3.8L and 5.0L
	El Camino	3.8L and 5.0L
	Impala	3.8L
	Caprice	3.8L
	Cutlass Supreme	3.8L, 5.0L 4.3L and 5.7L Diesel
	Bonneville	3.8L and 5.0L
	Regal	3.8L

TRANSMISSION MODIFICATIONS

Reverse Clutch Rattle

Some vehicles equipped with floor console transmission selector levers, models: XD, WK, WN, XE, XH or XK, may experience a rattle, similar to a tin can, coming through the transmission shifter cable, which is noticable at speeds above 50 mph (80 km).

The noise is caused by the low/reverse piston and clutch plate assembly. To repair this condition, the low/reverse piston and clutch plate assembly must be replaced. Order service package number 8641962. The package contains a low/reverse clutch piston (green), 5 low/reverse clutch drive plates (yellow/green), 5 low/reverse clutch reaction plates and an instruction sheet.

Check Ball Location Data

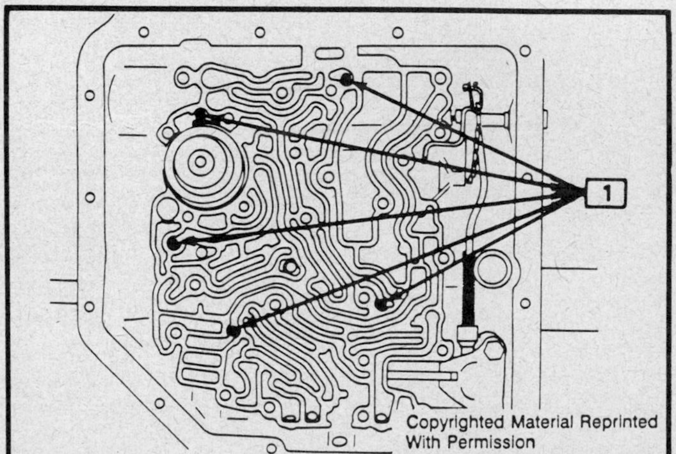
1
1. Check balls

View of the check ball locations

REMOVAL AND INSTALLATION

TRANSMISSION REMOVAL

1. Remove the air cleaner. Disconnect the negative battery cable.
2. Remove the T.V./detent cable retaining bolt, disconnect the cable from link and plug the hole. If the vehicle is equipped with a console, remove the shift control cable.
3. Disconnect the speedometer cable, manual shift linkage and driveshaft. Mark the driveshaft yoke so it may be reassembled the same way to avoid driveline problems later.
4. Remove the inspection cover.
5. Remove the flywheel cover pan and mark the converter-to-flexplate relationship so they may be reassembled in the same manner. Remove the converter-to-flexplate bolts.
6. Disconnect the electrical connectors and vacuum lines from the transmission.
7. Disconnect the cooler lines and any exhaust/converter brackets as necessary.
8. Using a floor jack, support engine and take the weight off the transmission. Remove the transmission mount and/or crossmember as necessary, depending on the application.
9. Remove the transmission-to-engine bolts; it may be necessary to lower the transmission slightly to gain access to the bellhousing bolts.
10. Lower the jack and remove the transmission from the vehicle.

NOTE: Be sure to support the converter as the transmission is being removed, as the converter may fall out.

TRANSMISSION INSTALLATION

1. Being careful of the converter, ease the transmission forward and upward, placing the unit onto the studs of the engine flange. Install the transmission-to-engine bolts and torque them to 40 ft. lbs. (54 Nm).
2. Turn the converter (through the cover pan opening) until the marks made at disassembly on the flexplate and converter align but do not install the torque converter-to-flexplate bolts.
3. Install the crossmember and/or transmission mounts that were loosened; torque the crossmember bolts to 25 ft. lbs. (34 Nm) and transmission mount bolts to 45 ft. lbs. (62 Nm). Remove the engine support jack.

NOTE: Make sure the parking brake cable is in its proper place, above the crossmember and routed properly.

4. Install the manual linkage.
5. With all the transmission supports in place and the engine jack removed, the transmission alignment should be correct. Install the torque converter-to-flexplate bolts and torque them to 30 ft. lbs. (40 Nm). Install flywheel cover.
6. Reconnect any exhaust parts that may have been loosened or removed. Torque the catalytic converter support brackets-to-transmission nuts to 35 ft. lbs. (48 Nm).
7. Connect the cooler lines, using care not to cross thread the fittings. Install the speedometer cable.
8. Install the driveshaft, being careful to keep the slip yoke clean and not damage the driveshaft seal; align the reference marks made at disassembly.
9. Connect the electrical connectors, the vacuum lines and the T.V./detent cable.
10. Lower the vehicle and refill transmission to the proper level.
11. Connect the negative battery cable and install the air cleaner.
12. Road test the vehicle for proper operation.

SPECIFICATIONS

TORQUE SPECIFICATIONS

Location	ft. lbs.	Nm
Oil pan to transmission case	13	17
Pump assembly to transmission case	20	27
Vacuum modulator retainer to case	12	16
Valve body assembly to case	13	17
Oil channel support plate to case	13	17
Pump body to pump cover	15	21
Parking lock bracket to case	29	39
Extension housing to case	35	45
Inside shift nut	30	40
External test plugs to case	8	11
Transmission mount to transmission	35	48
Speedo sleeve retainer on extension housing	150 ①	17
Detent cable to case	75 ①	8.5
Nut on end of selector lever sahft	20	27
Converter to flywheel	35	48

① inch lbs.

PRELIMINARY CHECKING PROCEDURE

CHECK TRANSMISSION OIL LEVEL

CHECK OUTSIDE MANUAL LINKAGE AND CORRECT

CHECK ENGINE TUNE

INSTALL OIL PRESSURE GAGE

CONNECT VACUUM GAGE TO MODULATOR LINE & TACHOMETER TO ENGINE

TOTAL RUNNING TIME FOR THIS COMBINATION NOT TO EXCEED 2 MINUTES. (applies to Drive, L2 or L1, and Reverse — brakes applied)

CHECK OIL PRESSURES IN FOLLOWING MANNER

RANGE	MODELS	— PSI — Modulator① Line Connected	Modulator② Line Disconnected
DRIVE — BRAKES APPLIED	XP WK XD XH, XK	51 - 58 55 - 64 51 - 58 55 - 66	137 - 158 143 - 164 137 - 158 119 - 139
L2 or L1 — BRAKES APPLIED	XP WK XD XH, XK	73 - 85 80 - 93 73 - 85 82 - 96	137 - 158 144 - 166 137 - 158 119 - 139
REVERSE — BRAKES APPLIED	XP WK XD XH, XK	77 - 89 83 - 97 77 - 89 88 - 106	219 - 250 216 - 249 219 - 250 236 - 269
NEUTRAL — BRAKES APPLIED	XP WK XD XH, XK	51 - 58 55 - 64 51 - 58 55 - 66	137 - 158 143 - 164 137 - 158 119 - 139
DRIVE IDLE — SET ENGINE IDLE TO SPECIFICATIONS BRAKES APPLIED	XP WK XD XH, XK	51 - 57 55 - 63 51 - 57 55 - 64	
DRIVE — 30 MPH CLOSED THROTTLE OR ON HOIST①	XP WK XD XH, XK	51 - 58 55 - 64 51 - 58 55 - 66	

① Modulator line connected—run engineto 1000 rpm, close throttle and check psi

② Modulator line disconnected—check psi at 1000 rpm, throttle open

Section 4

THM 350C Transmission
General Motors

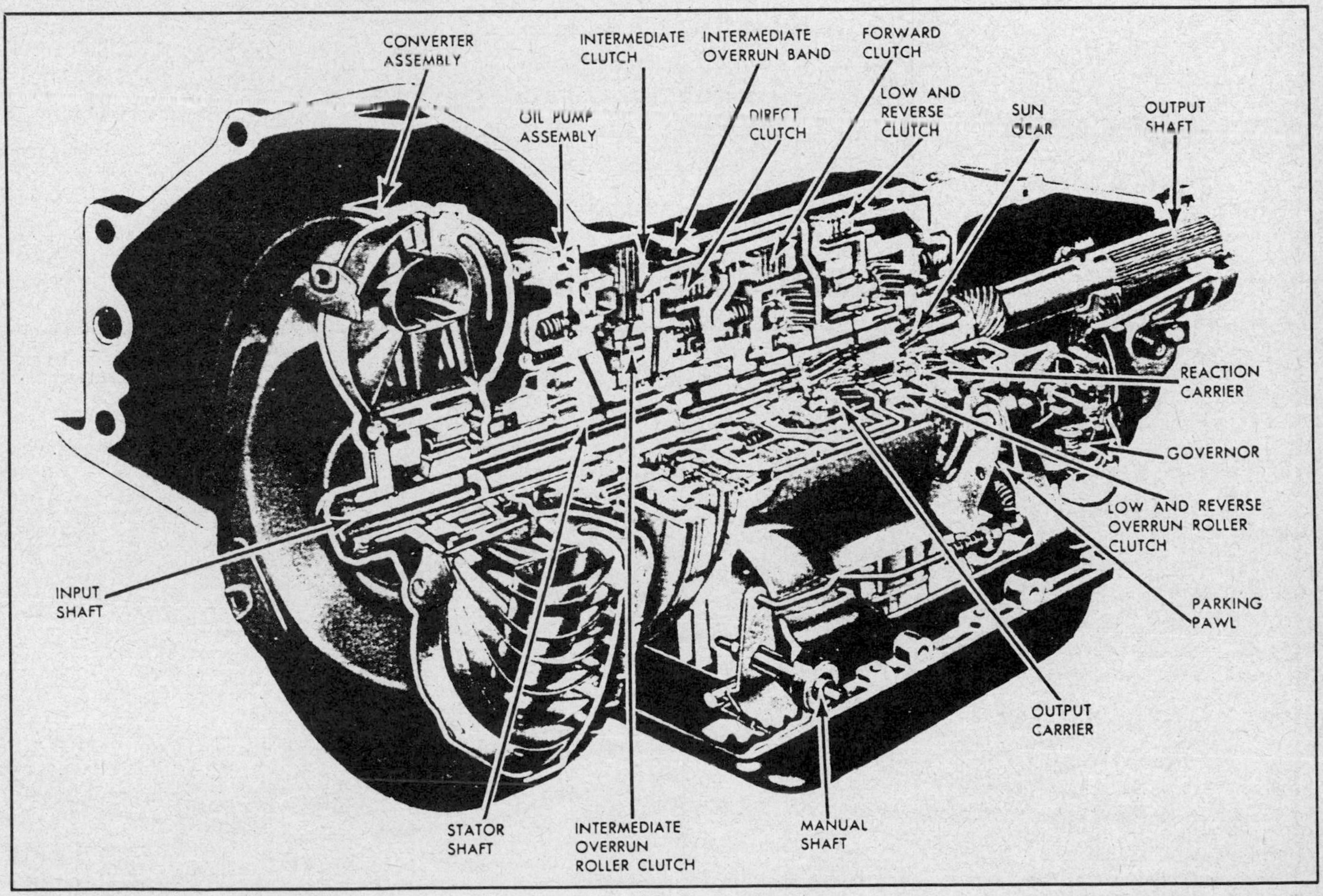

CONVERTER ASSEMBLY
INTERMEDIATE CLUTCH
INTERMEDIATE OVERRUN BAND
FORWARD CLUTCH
LOW AND REVERSE CLUTCH
OIL PUMP ASSEMBLY
DIRECT CLUTCH
SUN GEAR
OUTPUT SHAFT
REACTION CARRIER
GOVERNOR
LOW AND REVERSE OVERRUN ROLLER CLUTCH
PARKING PAWL
INPUT SHAFT
OUTPUT CARRIER
STATOR SHAFT
INTERMEDIATE OVERRUN ROLLER CLUTCH
MANUAL SHAFT

Turbo Hydra-Matic transmission

REFERENCES

APPLICATION

1974–80 Chilton's Professional Automatic Transmission Manual #6927
1984–89 Chilton's Professional Transmission Manual—Domestic Vehicles #7959

GENERAL DESCRIPTION

1974–80 Chilton's Professional Automatic Transmission Manual #6927

MODIFICATIONS

1974–80 Chilton's Professional Automatic Transmission Manual #6927
1984–89 Chilton's Professional Transmission Manual—Domestic Vehicles #7959

TROUBLE DIAGNOSIS

1974–80 Chilton's Professional Automatic Transmission Manual #6927

ON CAR SERVICES

1974–80 Chilton's Professional Automatic Transmission Manual #6927

REMOVAL AND INSTALLATION

1974–80 Chilton's Professional Automatic Transmission Manual #6927
1984–89 Chilton's Professional Transmission Manual—Domestic Vehicles #7959

BENCH OVERHAUL

1974–80 Chilton's Professional Automatic Transmission Manual #6927

SPECIFICATIONS

1974–80 Chilton's Professional Automatic Transmission Manual #6927
1984–89 Chilton's Professional Transmission Manual—Domestic Vehicles #7959

SPECIAL TOOLS

1974–80 Chilton's Professional Automatic Transmission Manual #6927

APPLICATION

THM 350C

Year	Vehicle	Engine
1984	Regal	5.0L 4bbl
	Caprice	5.0L 4bbl
	Monte Carlo	5.0L 4bbl
	Cutlass Supreme	5.0L 4bbl
	Grand Prix	5.0L 4bbl

TRANSMISSION MODIFICATIONS

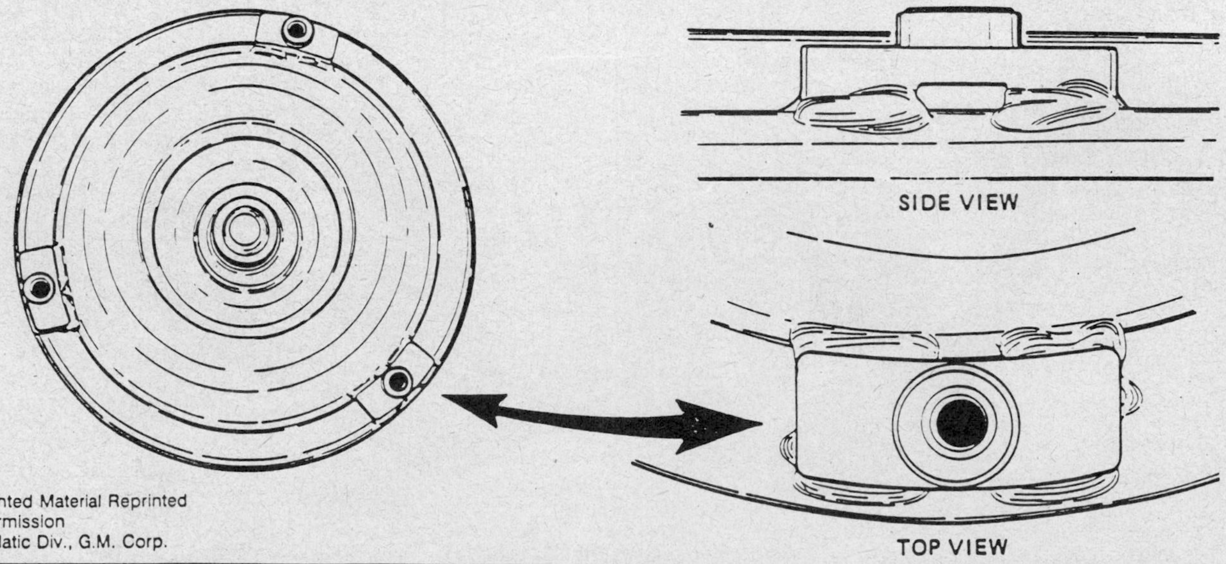
SIDE VIEW
TOP VIEW

New design torque converter lugs

New Design Torque Converter Lugs

Starting in early March, 1985, all 245mm and 298mm torque converter assemblies used on the transmission models listed above will be built with a new design converter lug.

Additionally, all 298mm torque converter assemblies used with diesel engine applications will have only 3 converter lugs instead of the 6 lugs that are currently being used.

Torque converter assemblies having the new design converter lugs will be optional for field service use.

Reverse Clutch Rattle Floor Console Only

Some 1984 THM 350C WS model transmissions may experience a rattle which is similar to a tin can rattle coming through the transmission shifter cable (floor console model only) which is most noticeable at speeds above 50 mph (80 km).

This noise may be caused by the low and reverse piston and clutch plate assembly. To repair this condition, replace the low and reverse piston and clutch plate assembly. Use service package part number 8641962. The package includes a low and reverse clutch piston, low and reverse plate assembly, low and reverse clutch plate and instruction sheet.

Check Ball Locations

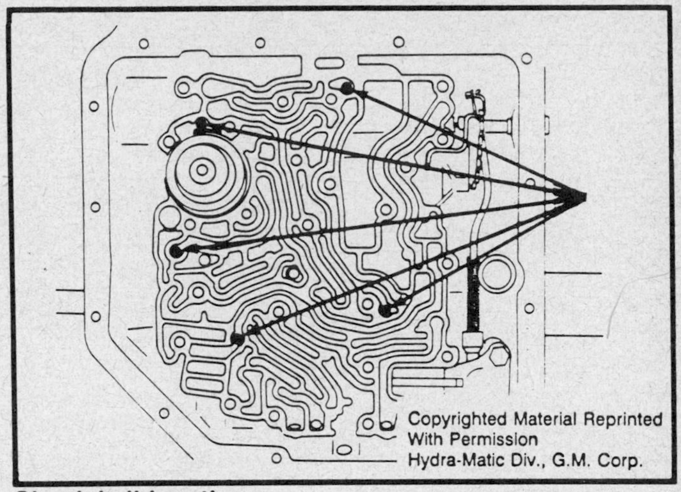
Copyrighted Material Reprinted
With Permission
Hydra-Matic Div., G.M. Corp.

Check ball locations

REMOVAL AND INSTALLATION

TRANSMISSION REMOVAL

1. Disconnect the negative battery cable.
2. Remove the air cleaner.
3. Remove the transmission dipstick and (if accessible) the bolt holding the dipstick tube.
4. Raise the vehicle and support safely.
5. Remove the driveshaft. Remove the floor pan reinforcement, if used and if it interferes with the driveshaft removal.
6. Disconnect the speedometer cable at the transmission.
7. Disconnect the shift linkage at the transmission.
8. Disconnect all electrical connectors and vacuum line at the transmission and any clips that retain the leads to the transmission case.
9. Remove the flexplate cover bolts and cover.
10. Matchmark the flexplate and torque converter to aid in reassembly.
11. Remove the torque converter bolts and/or nuts.
12. Remove the catalytic converter support bracket.
13. Remove the transmission support to transmission mount bolt and transmission support to frame bolts (and insulators, if used).
14. Support and raise the transmission slightly.
15. Slide the transmission support rearward.
16. Lower the transmission to gain access to the oil cooler lines.
17. Remove the oil cooler lines and cap open ends.
18. Support the engine with a suitable support tool and remove the transmission to engine bolts.
19. Remove the transmission from the vehicle being careful not to damage any cables, lines or linkage.

TRANSMISSION INSTALLATION

1. Raise the transmission into the vehicle and install the engine to transmission bolts.
2. Connect the oil cooler lines to the transmission.
3. Install the transmission support and mount.
4. Remove the transmission jack.
5. Align the torque converter to flexplate match marks made during disassembly. Install the torque converter to flexplate bolts and tighten to 46 ft. lbs. (62 Nm).

NOTE: Make sure the weld nuts on the torque converter are flush with the flexplate. Test the torque converter for freedom of rotation. Tighten the 3 bolts finger tight, then torque to specification. Retorque the first bolt tightened.

6. Install the floor pan reinforcement, if removed.
7. Install the catalytic converter support bracket.
8. Install the flexplate cover and attaching bolts.
9. Connect the shift linkage.
10. Connect the speedometer cable.
11. Connect all electrical connectors, vacuum line and brackets.
12. Install the driveshaft.
13. Lower the vehicle.
14. Install the dipstick tube (with a new oil seal) and attaching bolt. Install the transmission dipstick.
15. Install the air cleaner.
16. Connect the negative battery cable.
17. Adjust the shift linkage.
18. Fill transmission with proper type transmission fluid to the proper level.

SPECIFICATIONS

CLUTCH PLATE USAGE

TRANS-MISSION I.D. CODE	SPEEDOMETER DRIVE GEAR		INTERMEDIATE CLUTCH			DIRECT CLUTCH			FORWARD CLUTCH			LO & REVERSE CLUTCH		
	Number Of Teeth	Gear Color	No. Of Faced Plates	No. Of Steel Plates	Clutch Piston Thick-ness	No. Of Faced Plates	No. Of Steel Plates	Clutch Piston Thick-ness	No. Of Faced Plates	No. Of Steel Plates	Clutch Piston Thick-ness	No. Of Faced Plates	No. Of Steel Plates	Clutch Plate Thick-ness
XA	10	Purple	3	3	.992	4	4	.833	5	5	1.405	5	5	3.106
WS, WA XS, XX	9	Green	3	3	.992	4	4	.833	5	5	1.223	5	5	2.921
WC XC, XN	8	Orange	3	3	1.184	3	3	1.014	4	4	1.391	4	4	3.106

TORQUE SPECIFICATIONS

Item	Ft. lbs.	Nm
Oil pan to case	13	18
Pump assembly to case	20	27
Vacuum modulator retainer to case	12	17
Valve body assembly to case	13	18
Oil channel support plate to case	13	18
Pump body to pump cover	15	20
Parking lock bracket to case	29	40
Extension housing to case	35	48
Inside shift nut	30	41
External test plugs to case	8	11
Transmission mount to transmission	35	48
Speedometer sleeve retainer on extension housing	13	17
Detent cable to case	6	8.5
Selector lever nut	20	27
Converter to flywheel	35	47
Bellhousing bolts	35	47
Converter housing cover screws	5	6
Lever assembly	20	27
Oil cooler lines to transmission	10	13
Oil cooler lines to radiator	20	27

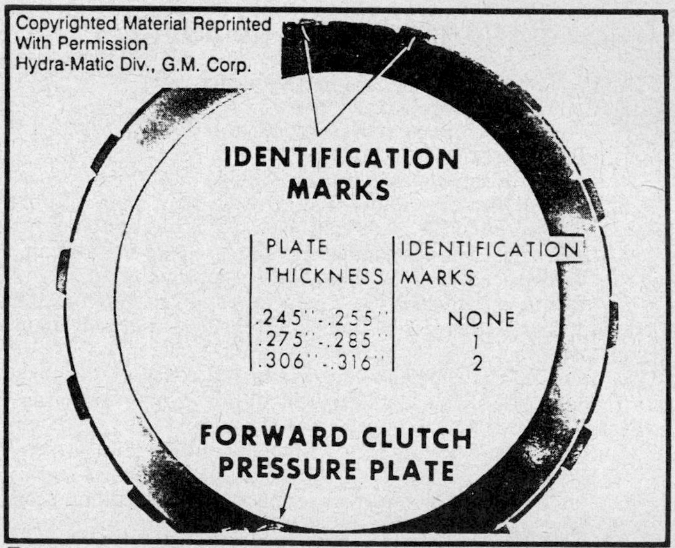
Copyrighted Material Reprinted
With Permission
Hydra-Matic Div., G.M. Corp.
IDENTIFICATION MARKS
PLATE THICKNESS
IDENTIFICATION MARKS
.245"-.255" NONE
.275"-.285" 1
.306"-.316" 2
FORWARD CLUTCH PRESSURE PLATE

Forward clutch pressure plate selection identification

Section 4

THM 400 and 475 Transmissions
General Motors

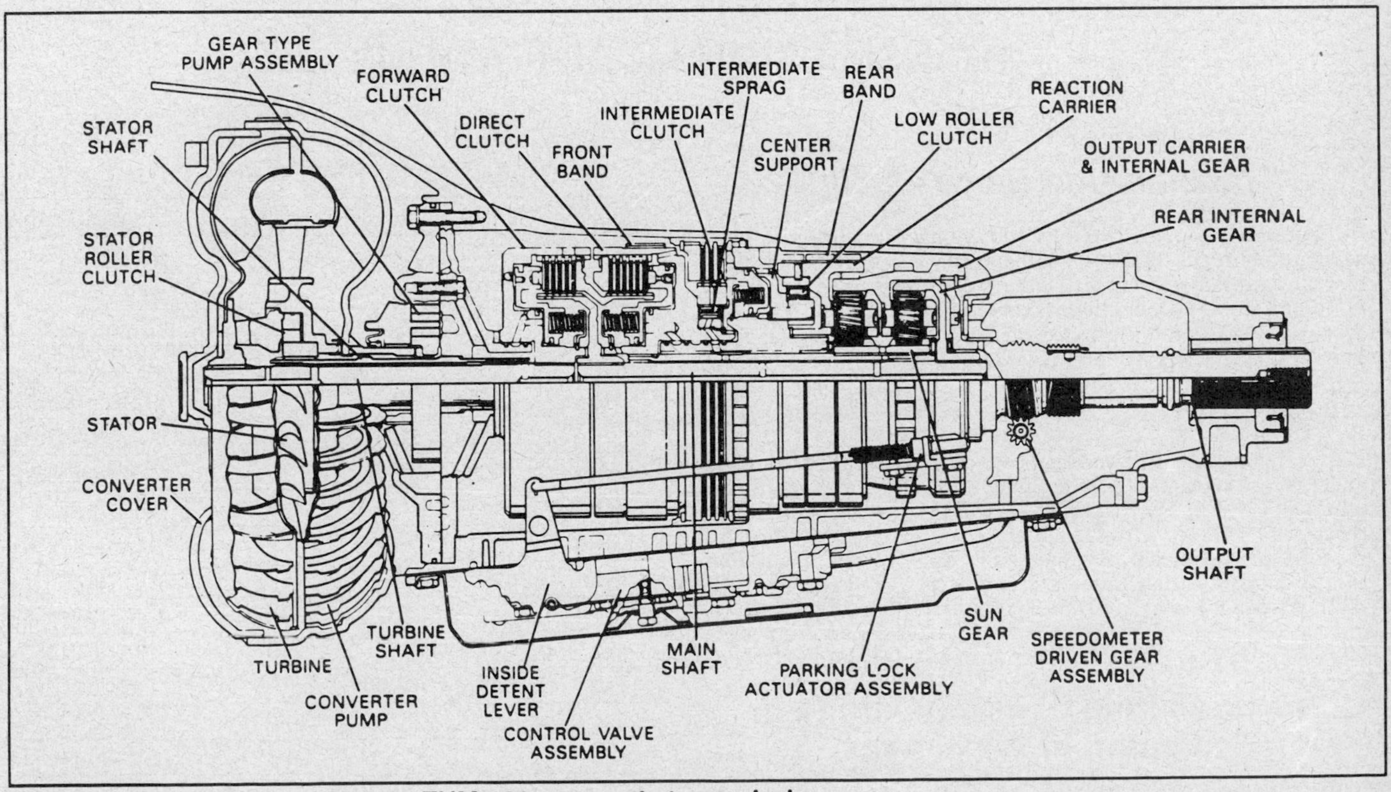

GEAR TYPE PUMP ASSEMBLY
FORWARD CLUTCH
INTERMEDIATE SPRAG
REAR BAND
REACTION CARRIER
STATOR SHAFT
DIRECT CLUTCH
INTERMEDIATE CLUTCH
LOW ROLLER CLUTCH
FRONT BAND
CENTER SUPPORT
OUTPUT CARRIER & INTERNAL GEAR
REAR INTERNAL GEAR
STATOR ROLLER CLUTCH
STATOR
CONVERTER COVER
OUTPUT SHAFT
SUN GEAR
SPEEDOMETER DRIVEN GEAR ASSEMBLY
TURBINE SHAFT
MAIN SHAFT
TURBINE
INSIDE DETENT LEVER
PARKING LOCK ACTUATOR ASSEMBLY
CONVERTER PUMP
CONTROL VALVE ASSEMBLY

Cross section of General Motors THM 400 automatic transmission

REFERENCES

APPLICATION

1980–84 Chilton's Professional Automatic Transmission Manual #7390
1984–89 Chilton's Professional Transmission Manual–Domestic Vehicles #7959

GENERAL DESCRIPTION

1980–84 Chilton's Professional Automatic Transmission Manual #7390

MODIFICATIONS

1980–84 Chilton's Professional Automatic Transmission Manual #7390
1984–89 Chilton's Professional Transmission Manual–Domestic Vehicles #7959

TROUBLE DIAGNOSIS

1980–84 Chilton's Professional Automatic Transmission Manual #7390

ON CAR SERVICES

1980–84 Chilton's Professional Automatic Transmission Manual #7390

REMOVAL AND INSTALLATION

1980–84 Chilton's Professional Automatic Transmission Manual #7390
1984–89 Chilton's Professional Transmission Manual–Domestic Vehicles #7959

BENCH OVERHAUL

1980–84 Chilton's Professional Automatic Transmission Manual #7390

SPECIFICATIONS

1980–84 Chilton's Professional Automatic Transmission Manual #7390
1984–89 Chilton's Professional Transmission Manual–Domestic Vehicles #7959

SPECIAL TOOLS

1980–84 Chilton's Professional Automatic Transmission Manual #7390

APPLICATION

1984–89 Chevrolet/GMC light duty trucks

TRANSMISSION MODIFICATIONS

Transmission Name Change

By September 1, 1991, Hydra-matic will have changed the name designation of the THM 400/475 automatic transmissions. The new designation for these transmissions will be Hydra-matic 3L80 and 3L80-HD. Transmissions built in 1989 and 1990 will serve as transitional years in which a dual system, made up of the old designation and the new designation will be in effect.

Loss of Power In Drive And Reverse

Beginning in 1984, a new torque converter went into production on THM400 transmissions equipped with diesel engines, model code FD, FF, FR, FS, FU, FV, FX and MA. This new torque has the blades retained into the converter pump shell by a brazing process. Blades which have moved in a converter may reduce torque multiplication causing a loss of power effect under 35 mph. When servicing any 1984 THM400 transmission which requires torque converter replacement the new part number is 8929917 and new identification number is 4M.

No Upshifting or Downshifting or Slipping In All Ranges

Some THM400 transmissions built between February 1, 1984

TRANSMISSION IDENTIFICATION DATA

MODEL	SERIAL NUMBERS	MODEL	SERIAL NUMBERS
AD	1230 to 1434	FP	9960 to 12837
AN	1790 to 2139	FQ	1714 to 1860
FA	9745 to 12760	FR	1069 to 1119
FB	8498 to 11466	FT	1255 to 2050
FC	1027 to 1051	FV	8735 to 12185
FD	7065 to 9425	FW	11270 to 14565
FE	1003 to 1004	FX	1949 to 2146
FG	1176 to 1177	FY	1608 to 1833
FF	3610 to 4254	FZ	1178 to 1225
FH	1701 to 2057	HZ	3915 to 4891
FJ	6335 to 7739	MF	43420 to 49554
FK	1136 to 1401	RC	1939 to 2118
FM	34782 to 47400	RR	1699 to 1861
FN	2372 to 2873	RT	1387 to 1458
		ZD	1164 to 1165
		ZV	3608 to 6011

and March 31, 1984, may not upshift or downshift or slip in all gear ranges. The models effected are listed by model identification and serial number.

When servicing the transmission, a mislocated forward clutch snapring or seals blown out of their seats may be found. These conditions may be caused by a sticking reverse boost valve in the reverse boost bushing. Inspect the reverse boost bushing which is located in the oil pump assembly for possible mismachining of the inner diameter. Service package number 8625902 contains 1 regulator boost valve bushing and 1 regulator boost valve.

Identification Information

Beginning with the start of 1986 production, all THM400 transmissions will have a 3 letter model designator as opposed to the 2 letter model designator used in previous years. The 3rd letter identifies the type of transmission. The code letter for the THM400/475 transmission is A.

Harsh 1-2 and 2-3 Upshifts

1985 THM400 transmissions, models VR, VS or VV and 1986 THM400 transmissions, models LRA and LSA transmissions may exhibit a harsh 1–2 and 2–3 upshift condition. This condition may be caused by improper engine tune, engine vacuum and/or vacuum supply to the modulator not within specification of minimum 18 in. Hg. vacuum at hot engine idle speed or transmission pump operation and line pressures not within specification.

If the cause for the firm shift condition cannot be found, replacement of the control valve and/or modulator assemblies may be necessary. At the start of production for the 1986 model year, all LRA and LSA models were built with the new control valve assembly part number 8655160. The vacuum modulator assembly, part number 3030843 started in production on November 19, 1985 (Julian date 323) for all 1986 LRA and LSA models.

Extension Housing Leak

Some 1984–86 THM400 transmissions, models RCA, RRA and RTA may exhibit a leak at the extension housing rear seal after extended high speed operation. This condition may be caused by extension housing bushing and seal wear due to a decrease of oil within the extension housing during high speed operation.

Beginning May 1, 1986 or (Julian date 121), all THM400 models RBA, RSA and RWA transmissions are built using a new lube and rear seal retainer package. THM400 model identifiers RCA, RRA and RTA were changed to RBA, RSA and RWA models with the introduction of this package. When servicing a transmission for extension housing bushing or seal wear, install service package number 8629927.

New Oil Pan Gasket

Starting July 22, 1986 (Julian date 203), all THM400/475 transmissions have a new latex coated cellulose pan gasket part number 8655625. The new gasket improves sealing because it will not crush like a cork gasket.

New Forward and Direct Clutch Assemblies

Certain 1987 THM400 transmission models have new forward and direct clutch assemblies with a dished steel clutch plate. The new design clutch plate improves forward and reverse engagement shift feel by cushioning the piston apply.

When servicing the forward and/or direct clutch, inspect the dished clutch plate for scoring, cracks or flatness. Note the position of the two 0.0915 in. (2.32mm) and three 0.0775 in.

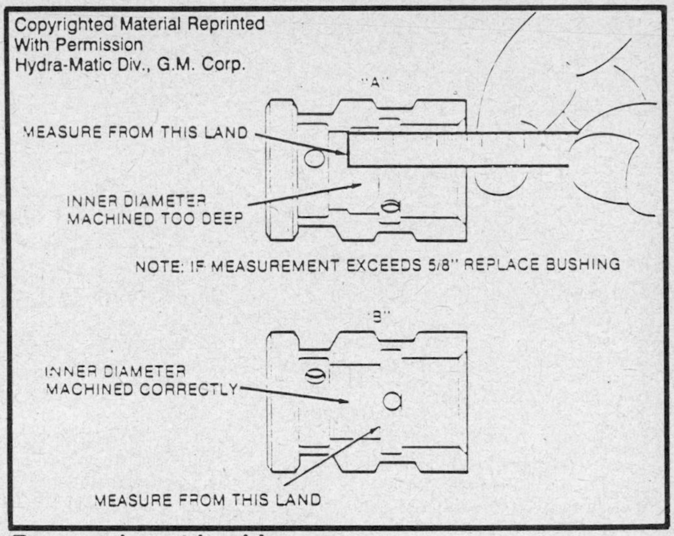
Copyrighted Material Reprinted
With Permission
Hydra-Matic Div., G.M. Corp.
"A"
MEASURE FROM THIS LAND
INNER DIAMETER
MACHINED TOO DEEP
NOTE: IF MEASUREMENT EXCEEDS 5/8" REPLACE BUSHING
"B"
INNER DIAMETER
MACHINED CORRECTLY
MEASURE FROM THIS LAND

Reverse boost bushing

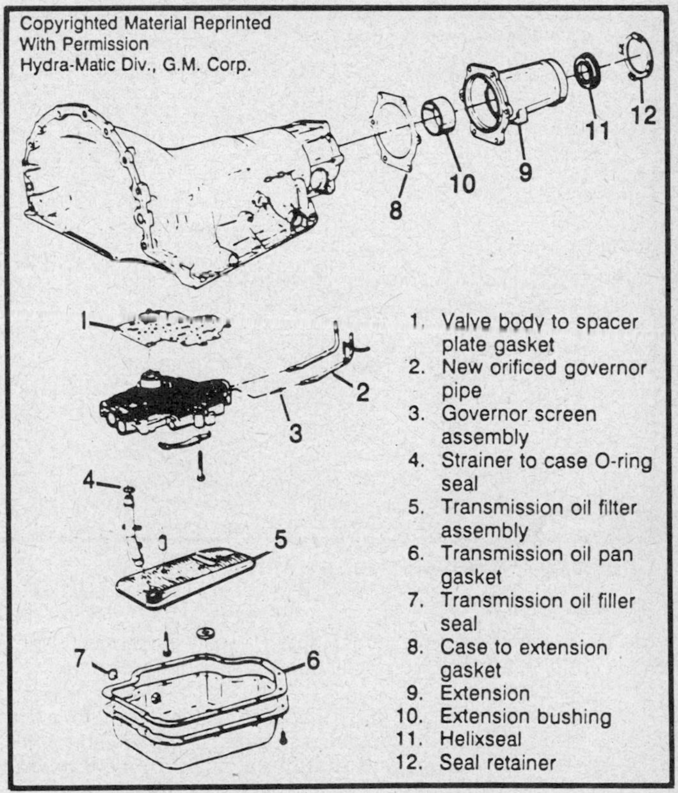
Copyrighted Material Reprinted
With Permission
Hydra-Matic Div., G.M. Corp.
1
2
3
4
5
6
7
8
9
10
11
12

1. Valve body to spacer plate gasket
2. New orificed governor pipe
3. Governor screen assembly
4. Strainer to case O-ring seal
5. Transmission oil filter assembly
6. Transmission oil pan gasket
7. Transmission oil filler seal
8. Case to extension gasket
9. Extension
10. Extension bushing
11. Helixseal
12. Seal retainer

Service package contents

(1.97mm) flat steel plates. To properly assemble the direct clutch pack, follow the dished steel plate with a 0.0915 in. steel plate, fiber plate, the 2nd 0.0915 in. steel plate and complete by alternating fiber and 0.0775 in. steel plates. The new forward and direct clutch assemblies cannot be used for past model service as other 1987 calibration changes contribute to the improved engagement shift feel.

Speedometer Driven Gear Fluid Leak

A transmission fluid leak at the plastic speedometer driven gear

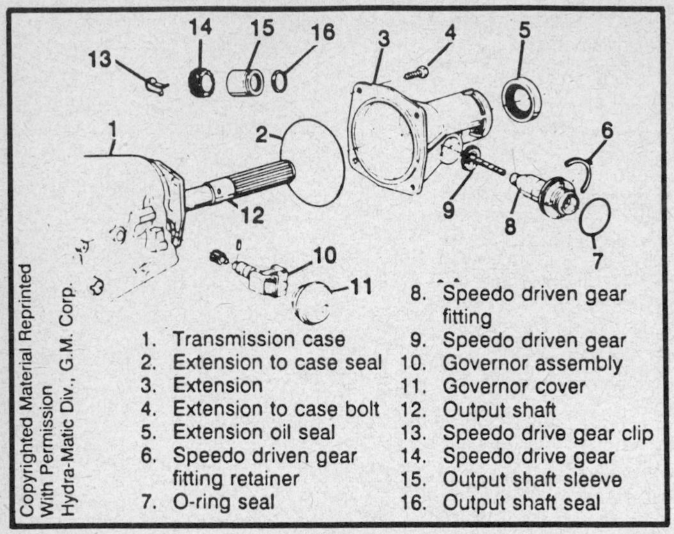
Copyrighted Material Reprinted
With Permission
Hydra-Matic Div., G.M. Corp.
1. Transmission case
2. Extension to case seal
3. Extension
4. Extension to case bolt
5. Extension oil seal
6. Speedo driven gear fitting retainer
7. O-ring seal
8. Speedo driven gear fitting
9. Speedo driven gear
10. Governor assembly
11. Governor cover
12. Output shaft
13. Speedo drive gear clip
14. Speedo drive gear
15. Output shaft sleeve
16. Output shaft seal

Extension and associated parts

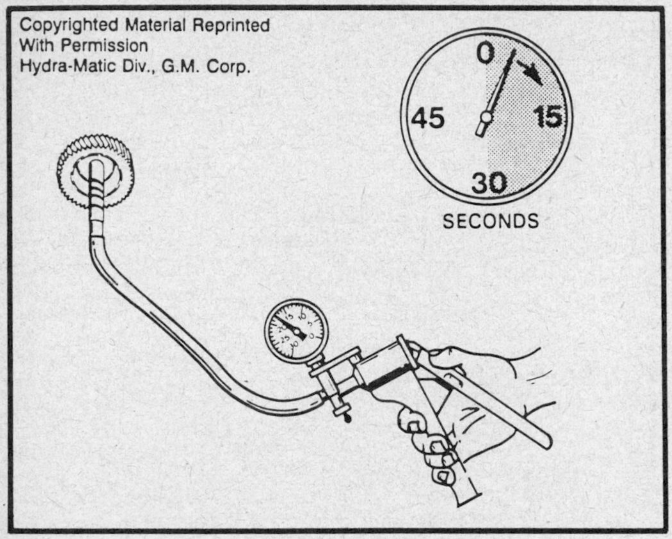
Copyrighted Material Reprinted
With Permission
Hydra-Matic Div., G.M. Corp.
0
15
30
45
SECONDS

Checking the driven gear shaft for hole

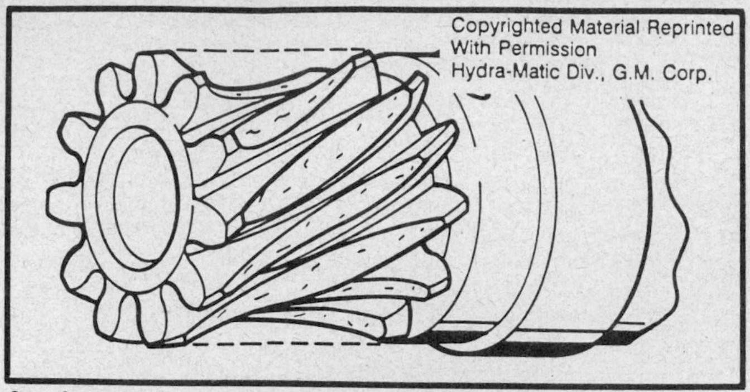
Copyrighted Material Reprinted
With Permission
Hydra-Matic Div., G.M. Corp.

Applecored governor driven gear

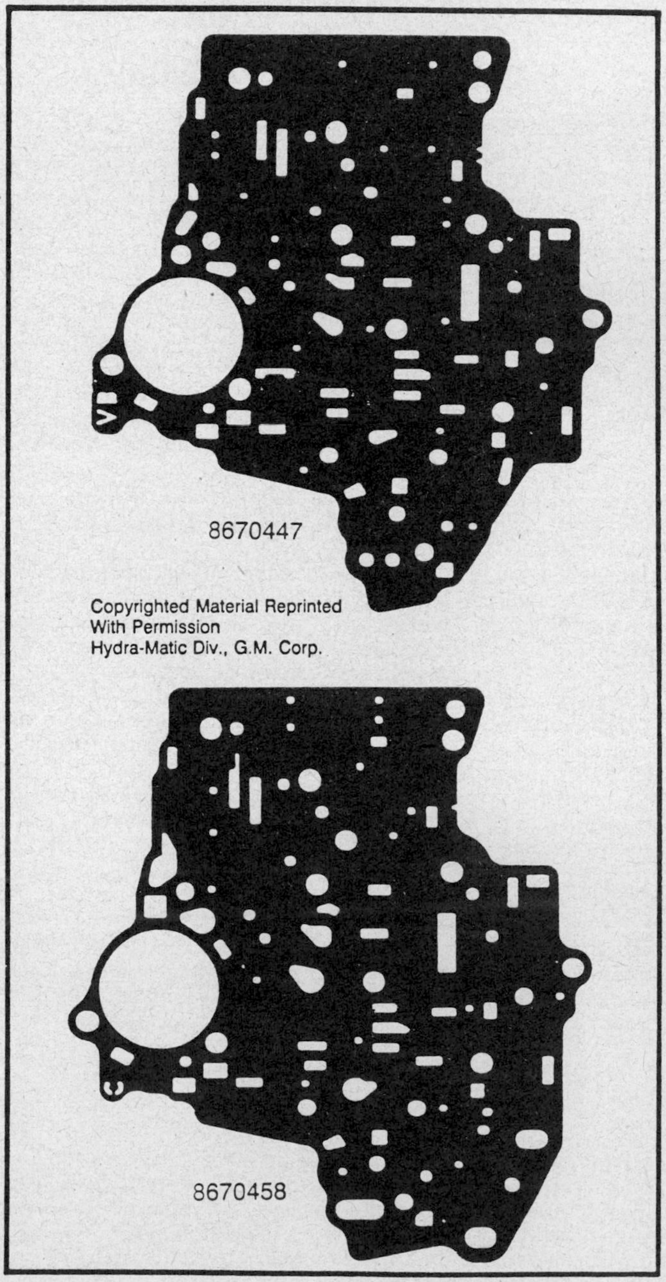
8670447
Copyrighted Material Reprinted
With Permission
Hydra-Matic Div., G.M. Corp.
8670458

Valve body gasket change

can be caused by a hole in the speedometer driven gear shaft. This condition could be misdiagnosed as a speedo driven gear fitting leak.

When servicing a 1986 or 1987 THM400 transmission for a fluid leak around the speedometer driven gear fitting or cable, check the speedometer driven gear shaft for a hole through the center of the shaft as follows:

1. Install a hand vacuum pump hose over the end of the gear.
2. Pump up 10 in. hg. of vacuum.
3. The gauge should hold for at least 30 seconds.
4. If the gear holds vacuum, inspect the seals in the adaptor.
5. If the gear does not hold vacuum, replace the gear.

Piston Seals Leaking

As of July 21, 1987 (Julian date 202), all THM400/475 transmissions have new aluminum die cast pistons used in the forward, direct and intermediate clutches. The new pistons reduce the occurrence of a cut and/or leaking seal. All other clutch assembly components, including the piston seals, remain the same.

Erratic or No Upshift

A THM400/475 transmission with an erratic or no upshift condition could be due to applecoring of the governor driven gear. Applecoring of the governor driven gear may be caused by nicks and/or burrs on the governor drive gear. The new governor driven gear has a higher fiberglass content. Inspect the output shaft for nicks or burrs on the governor drive gear. Replace the output shaft if nicks or burrs are found.
Beginning December 11, 1987 (Julian date 345), all THM400/475 transmissions were produced with a new governor driven gear. When servicing a transmission governor driven gear, use the new governor driven gear and pin package part number 8629977

Valve Body Gasket

Beginning November 23, 1987 all THM400/475 transmissions were built with new valve body gaskets. This change and additional changes done to the case, valve body and separator plates were done to relieve a harsh park/neutral to reverse engagement shift feel.

REMOVAL AND INSTALLATION

TRANSMISSION REMOVAL

1. Disconnect the negative battery cable.
2. Remove the air cleaner.
3. Disconnect the T.V. cable from the throttle linkage.
4. Raise the vehicle and support it safely.
5. Drain the transmission fluid.
6. Disconnect the shift linkage.
7. Remove the driveshaft. Remove the driveshaft from the transfer case, if equipped.
8. Remove the support bracket at the catalytic converter.
9. Remove any other components as needed for clearance.
10. Support the transmission and the transfer case, if used, with a suitable transmission jack.
11. Remove the transmission crossmember.
12. If necessary, lower the transmission far enough for clearance to reach other components.
13. Remove the dipstick tube and the seal. Cover the opening in the transmission.
14. Disconnect the speedometer cable.
15. Disconnect the vacuum modulator line.
16. Disconnect all electrical connectors from the transmission.
17. Disconnect the oil cooler lines. Cap the openings in the transmission and the lines.
18. Disconnect the transfer case shifter, if equipped and move it aside.
19. Remove the dampener and the support, if used.
20. Remove the transmission support braces.
21. Remove the converter housing cover. Mark the flywheel and the torque converter alignment.
22. Remove the torque converter bolts.
23. Support the engine with a suitable jack before disconnecting the transmission.
24. Remove the transmission case to engine bolts.
25. Slide the transmission straight back and off the locating pins and install converter holding strap tool J-21366.
26. Remove the transmission from the vehicle.

TRANSMISSION INSTALLATION

1. Install the torque converter onto the mainshaft and install the torque converter holding strap tool.

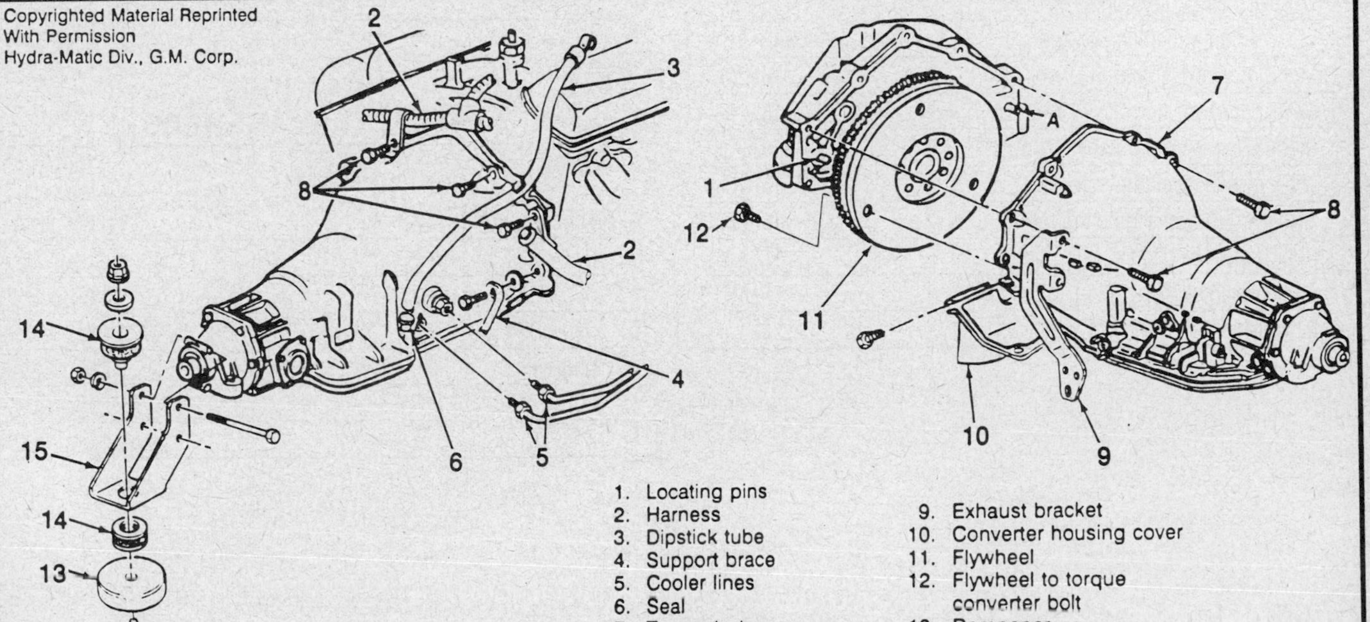
1
2
3
4
5
6
7
8
9
10
11
12
13
14
15
A

1. Locating pins
2. Harness
3. Dipstick tube
4. Support brace
5. Cooler lines
6. Seal
7. Transmission
8. Transmission to engine bolts
9. Exhaust bracket
10. Converter housing cover
11. Flywheel
12. Flywheel to torque converter bolt
13. Dampener
14. Insulator
15. Support

Transmission and components

2. Support the transmission and the transfer case, if used with a suitable transmission jack.
3. Raise the transmission into place and remove the torque converter holding tool.
4. Slide the transmission straight onto the locating pins.
5. Line up the marks on the flywheel and the torque converter. The torque converter must be flush onto the flywheel and rotate freely by hand.
6. Install the transmission case to engine bolts with all brackets, clips and harnesses positioned as they were removed. Do not install the dipstick tube or the transmission support brace bolts.
7. Install the torque converter bolts and tighten them finger tight to insure proper converter seating.
8. Tighten the torque converter bolts to 50 ft. lbs. (68 Nm).
9. Remove the engine support jack.
10. Install the converter housing cover. Hook the cover under the lip of the engine oil pan.
11. Install the support and dampener, if used.
12. Install the transmission support braces.
13. Connect the transfer case shifter, if equipped.
14. Uncover the openings and connect the cooler lines.
15. Connect the vacuum modulator line.
16. Connect the speedometer cable.
17. Connect the electrical connectors to the transmission.
18. Uncover the opening and install a new seal and the dipstick tube.
19. Install the crossmember and mount.
20. Remove the transmission jack.
21. Install the support bracket at the catalytic converter.
22. Install the driveshaft. Install the front driveshaft to the transfer case, if used.
23. Connect the shift linkage.
24. Lower the vehicle.
25. Fill the transmission with new Dexron® II automatic transmission fluid.
26. Connect the T.V. cable. Install the air cleaner.
27. Connect the negative battery cable. Start the engine, check for leaks, check fluid level and correct as required.

SPECIFICATIONS

TORQUE SPECIFICATIONS

	ft. lbs.
Pump cover bolts	18–20
Parking pawl bracket bolts	18–20
Center support bolt	20–25
Pump-to-case attaching bolts	22–25
Extension-to-case attaching bolts	22–25
Rear servo cover bolts	18–20
Detent solenoid bolts	8–10
Control valve body bolts	8–10
Bottom pan attaching screws	10–12
Modulator retaining bolt	18–20
Governor cover bolts	18–20
Manual lever-to-manual shaft nut	18–20
Linkage swivel clamp screw	18–20
Transmission-to-engine mounting bolts	35–40
Rear mount-to-transmission bolts	35–40
Oil cooler line	16–18
Filter retainer bolt	8–10

FRONT UNIT SELECTIVE WASHERS

Thickness (in.)	Color
.060–.064	Yellow
.071–.074	Blue
.082–.086	Red
.093–.097	Brown
.104–.108	Green
.115–.119	Black
.126–.130	Purple

REAR UNIT SELECTIVE WASHERS

Thickness (in.)	Notches and/or Numeral
.074–.078	None, 1
.082–.086	1 tab, side 2
.090–.094	2 tabs, side 3
.098–.102	1 tab, OD 4
.106–.110	2 tabs, OD 5
.114–.118	3 tabs, OD 6

Section 4

THM 700-R4 Transmission
General Motors

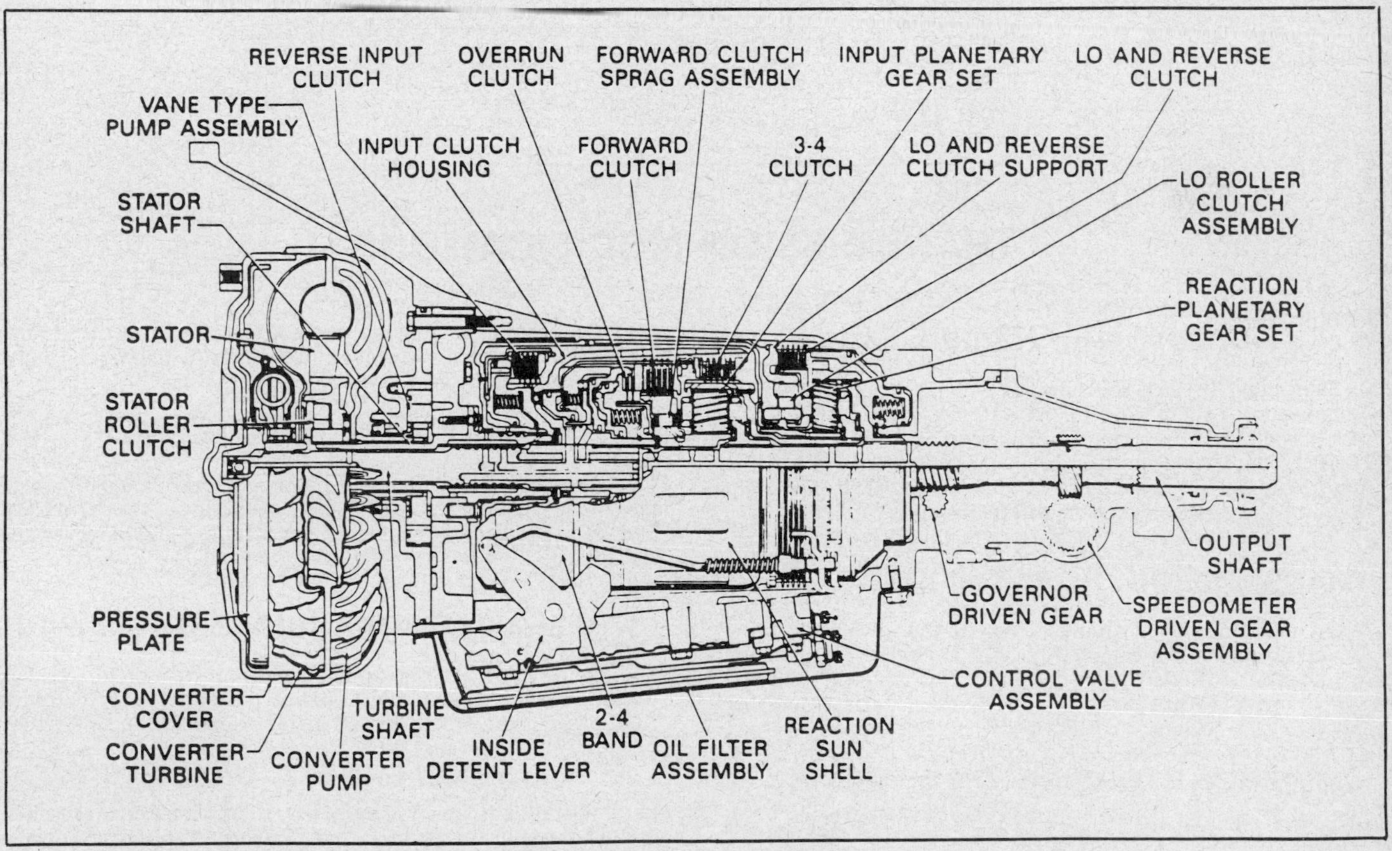

REVERSE INPUT CLUTCH
OVERRUN CLUTCH
FORWARD CLUTCH SPRAG ASSEMBLY
INPUT PLANETARY GEAR SET
LO AND REVERSE CLUTCH
VANE TYPE PUMP ASSEMBLY
INPUT CLUTCH HOUSING
FORWARD CLUTCH
3-4 CLUTCH
LO AND REVERSE CLUTCH SUPPORT
LO ROLLER CLUTCH ASSEMBLY
STATOR SHAFT
REACTION PLANETARY GEAR SET
STATOR
STATOR ROLLER CLUTCH
OUTPUT SHAFT
GOVERNOR DRIVEN GEAR
SPEEDOMETER DRIVEN GEAR ASSEMBLY
PRESSURE PLATE
CONTROL VALVE ASSEMBLY
CONVERTER COVER
TURBINE SHAFT
2-4 BAND
REACTION SUN SHELL
CONVERTER TURBINE
CONVERTER PUMP
INSIDE DETENT LEVER
OIL FILTER ASSEMBLY

Cross section of General Motors THM 700–R4 automatic transmission

REFERENCES

APPLICATION

1980–84 Chilton's Professional Automatic Transmission Manual #7390
1984–89 Chilton's Professional Transmission Manual—Domestic Vehicles #7959

GENERAL DESCRIPTION

1980–84 Chilton's Professional Automatic Transmission Manual #7390

MODIFICATIONS

1980–84 Chilton's Professional Automatic Transmission Manual #7390
1984–89 Chilton's Professional Transmission Manual—Domestic Vehicles #7959

TROUBLE DIAGNOSIS

1980–84 Chilton's Professional Automatic Transmission Manual #7390

ON CAR SERVICES

1980–84 Chilton's Professional Automatic Transmission Manual #7390

REMOVAL AND INSTALLATION

1980–84 Chilton's Professional Automatic Transmission Manual #7390
1984–89 Chilton's Professional Transmission Manual—Domestic Vehicles #7959

BENCH OVERHAUL

1980–84 Chilton's Professional Automatic Transmission Manual #7390

SPECIFICATIONS

1980–84 Chilton's Professional Automatic Transmission Manual #7390
1984–89 Chilton's Professional Transmission Manual—Domestic Vehicles #7959

SPECIAL TOOLS

1980–84 Chilton's Professional Automatic Transmission Manual #7390

APPLICATION

THM 700—R4

Year	Vehicle	Engine
1984–85	Impala	4.3L, 5.0L
1984–86	Parisenne	4.3L, 5.0L
1984–89	Corvette	5.7L
1984–89	Camaro, Firebird	2.5L, 2.8L, 5.0L
1984–89	Chevrolet and GMC Trucks and Vans	4.3L, 5.0L, 5.7L

TRANSMISSION MODIFICATIONS

Transmission Name Change

By September 1, 1991, Hydra-matic will have changed the name designation of the THM 700-R4 automatic transmission. The new designation for this transmission will be Hydra-matic 4L60. Transmissions built in 1989 and 1990 will serve as transitional years in which a dual system, made up of the old designation and the new designation will be in effect.

TRANSMISSION OIL FILTER INSTALLATION

When installing the oil filter assembly to the valve body, be very careful not to cut the O-ring. Press the O-ring into its bore with firmness but do not strike the filter. A cut O-ring will show up as a valve buzz, low line pressure, soft shifts or a converter clutch chatter.

FRONT TORRINGTON BEARING INSTALLATION

When installing the front Torrington bearing between the pump and the input housing, be sure the black bearing surface is against the pump. If the black surface is not against the pump damage to the unit case can occur.

VALVE BODY BORE PLUGS

Valve body bore plugs with a recess on one side should always be assembled with the recess towards the outside of the valve body. Failure to do so can result in mispositioning of the valves and springs.

RTV SEALANT—TRANSMISSION OIL PAN

Do not use RTV sealant on the fluid pan; use of this sealant could cause a possible block to the servo exhaust port. Use only a fluid pan gasket, when servicing this transmission.

LATE OR ERRATIC UPSHIFTS

Late or erratic upshifts may occur, if this happens, especially with changes in the temperature. To correct, remove the valve body and inspect the T.V. sleeve. The sleeve can rotate in the valve body and cause the passages to become blocked. To correct the problem, remove the sleeve and the valve and clean thoroughly. When reinstalling the sleeve, align it so all of the ports are open and not restricted. Insert the pin and make sure it is seated to help prevent rotation of the sleeve.

TORQUE CONVERTER RATTLE DIESEL ENGINE

On some vehicles equipped with a diesel engine, a rattle may occur coming from the torque converter area when the torque converter clutch is applied. If this problem exists, check the lockup clutch throttle alignment cable. If it is found to be defective, replace it with part number 8642970. The new part will raise the lockup torque converter function to 40 miles per hour, in 4th gear only. If the rattle still persists, the torque converter should be checked and replaced with part number 8647323 or 8642964.

BURNT 3–4 CLUTCH AND BANK ASSEMBLY

When servicing transmission and a burnt 3–4 clutch/band is encountered, inspect the 3rd accumulator check valve. The accumulator valve is located behind servo assembly in the transmission case. To inspect the 3rd accumulator, pour some solvent into the capsule; the solvent should only go into the servo area. If there is any trace of solvent inside the case barrel, the capsule assembly should be replaced. Follow the procedure outlined below when making this repair.

1. Remove the old capsule using a No. 4 Easy-Out®.
2. Install new capsule, small end first, so one of the 4 holes will align with the passage into the servo when the capsule is fully installed.
3. Use a ⅜ in. rod to drive the capsule into the case, far enough, so the capsule feed hole is completely open into the servo bore. This can be approximated by marking the rod at 1⅝ in. from the end that goes into the case. Again, recheck with solvent.

PARK TO REVERSE—HARSH SHIFT

Changes have been made to improve the park to reverse shift feel. If a park to reverse harsh shift feel exists, it can be corrected by installing a new style valve body spacer gasket, case gasket and spacer plate. Also, a new style oil pump cover and waved reverse clutch plate must be installed.

The new design parts are being used beginning with transmission serial number 9M61894D. All earlier units can be updated by using these new parts. Refer to the park/reverse change chart for the proper information when ordering the required parts.

PARK/REVERSE CHANGE

Component	Old Part Number	New Part Number
Valve body to spacer gasket	8642920	8642952
Spacer-to-case gasket	8642920	8642952
Spacer plate	8642655-A	8647066-B
Oil pump cover (TN,TP,M6,MH units)	8647043	8647190
Oil pump cover (all other units)	8647042	8647189
Waved reverse clutch plate	8642060	8647067

BURNED 3–4 CLUTCH DIAGNOSIS MODEL 3YH

When diagnosing for a burned 3–4 clutch pack, check the spacer

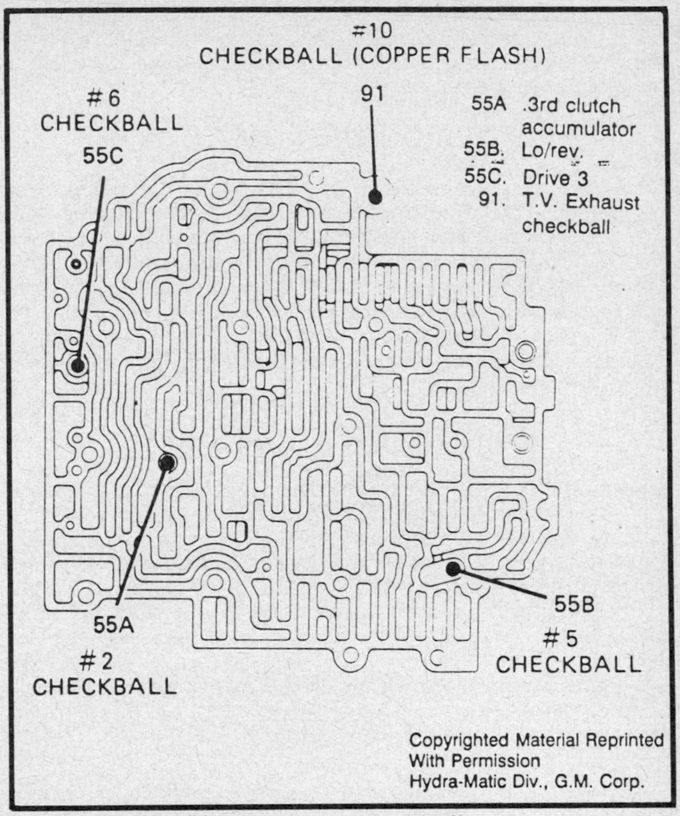
#10
CHECKBALL (COPPER FLASH)
#6
CHECKBALL
55C
91
55A .3rd clutch accumulator
55B. Lo/rev.
55C. Drive 3
91. T.V. Exhaust checkball
55B
#5
CHECKBALL
55A
#2
CHECKBALL
Copyrighted Material Reprinted With Permission Hydra-Matic Div., G.M. Corp.

Location of the valve body checkballs

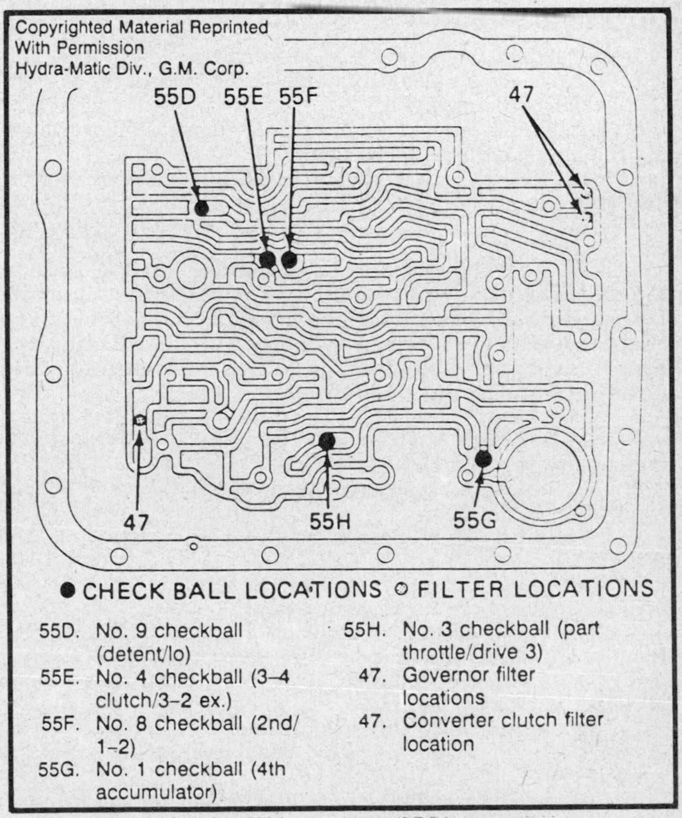
Copyrighted Material Reprinted With Permission Hydra-Matic Div., G.M. Corp.
55D 55E 55F
47
47
55H
55G
● CHECK BALL LOCATIONS ○ FILTER LOCATIONS
55D. No. 9 checkball (detent/lo)
55E. No. 4 checkball (3–4 clutch/3–2 ex.)
55F. No. 8 checkball (2nd/ 1–2)
55G. No. 1 checkball (4th accumulator)
55H. No. 3 checkball (part throttle/drive 3)
47. Governor filter locations
47. Converter clutch filter location

Location of the case checkballs and filters

plate for correct identification. If the plate is identified by the letter **C** change the plate. Use part number 8654083 which is identified with the letter **E**. The new spacer plate has a larger 3rd clutch feed hole.

NEW DESIGN OIL PUMP COVER

Some transmissions are being fitted with a new oil pump cover and a new oil pump to transmission case gasket. This new design prevents the transmission oil from exhausting out of the breather during harsh deceleration or on sharp turns.

When servicing the transmission, be sure to use the new design components. The new pump to case gasket cannot be used with the past oil pump cover. If the past pump to case gasket is used with the new oil pump cover oil will continue to exhaust out of the breather.

ERRATIC OR NO UPSHIFT

Beginning July 1987, all transmissions were produced with a new governor driven gear. The erratic or no upshift condition may be due to "applecoring" of the governor driven gear. Applecoring may be caused by nicks and/or burrs on the grovernor drive gear. The new governor driven gear has a higher fiberglass content. Also, inspect the output shaft for nicks and/or burrs on the governor drive gear; if necessary, replace the output shaft.

When servicing the governor driven gear, note that the new gear is green in color; the old one was blue. The new governor drive gear/pin package service part No. is 8663995.

PROPER FLUID LEVEL CHECKING PROCEDURE

The 1987–89 transmissions show the fluid level to be below the dipstick indicators.

To obtain the proper **COLD** fluid level check, perform the following procedures:

1. Start the engine and allow it to idle for 3 minutes.
2. With the vehicle on a level surface, accessories turned **OFF** and the engine idling in **P**, check the fluid level.
3. The fluid level should be below the **ADD** mark on the dipstick.

NOTE: The cold level check cannot take the place of the hot level check; it merely lets the person know there is enough fluid in the transmission to perform any other test.

To obtain the proper HOT fluid level check, perform the following procedures:

1. Drive the vehicle in all ranges until the fluid temperature reaches 180–200°F (80–90°C).
2. Stop the vehicle, place the transmission in **P** and allow it to idle for 3 minutes.
3. With the vehicle on a level surface, accessories turned **OFF** and the engine idling in **P**, check the fluid level.
4. The fluid level should be within the cross-hatched area.

REMOVAL AND INSTALLATION

TRANSMISSION REMOVAL

EXCEPT TRUCKS

1. Disconnect the negative battery cable.
2. Remove the air cleaner assembly. Disconnect the T.V./detent cable at it's upper end.
3. Remove the transmission dipstick. Remove the bolt holding the dipstick tube, if accessible.
4. Raise the vehicle and support it safely. Remove the torque arm to transmission retaining bolts, on Camaro and Firebird vehicles.

NOTE: Rear spring force will cause torque arm to move toward floor pan when arm is disconnected from transmission. Carefully place a piece of wood between the floor pan and torque arm when disconnecting the arm to avoid possible damage to floor pan and to avoid possible injury to hand or fingers.

5. Drain the transmission. Remove the driveshaft; matchmark the driveshaft to aid in installation.
6. Disconnect the speedometer cable and the shift linkage from the transmission.
7. Disconnect all electrical leads and any clips from the transmission.
8. Remove the inspection cover. Mark the flexplate and torque converter, to aid in installation.
9. Remove the torque converter to flexplate retaining bolts. Remove the catalytic converter support bracket, if necessary.
10. Position the transmission jack under the transmission and remove the rear transmission mount.
11. Remove the floor pan reinforcement. Remove the crossmember retaining bolts and move the crossmember out of the way.
12. Remove the transmission to engine bolt on the left side; this bolt retains the ground strap.
13. Disconnect and plug the transmission oil cooler lines; it may be necessary to lower the transmission in order to gain access to the cooler lines.
14. With the transmission still lowered, disconnect the T.V. cable.
15. Support the engine, using the proper tools and remove the remaining engine to transmission bolts.
16. Carefully disengage the transmission assembly from the engine. Using the proper equipment, lower and remove the transmission from the vehicle.

TRUCKS

NOTE: If the vehicle is equipped with four wheel drive, the transfer case must first be removed.

1. Disconnect the negative battery cable. Remove the air cleaner assembly.
2. Disconnect the T.V. cable at the upper end. Raise the vehicle and support it safely.
3. Remove the driveshaft. Matchmark the assembly to aid in installation.
4. Disconnect the speedometer cable at the transmission. Disconnect the shift linkage at the transmission.
5. Disconnect all electrical leads at the transmission and remove any clips that retain the leads to the transmission.
6. Remove the brake line to crossmember clips and remove the crossmember, four wheel drive vehicles only.
7. Remove the transmission support brace attaching bolts at the torque converter, if equipped.
8. Remove the exhaust crossover pipe and catalytic converter attaching bolts. Remove the components as an assembly.
9. Remove the inspection cover and mark the torque converter in relation to the flywheel. This will assure proper alignment upon installation.
10. Remove the torque converter-to-flywheel retaining bolts. Disconnect the catalytic converter support bracket.
11. Position the transmission jack under the transmission and raise the unit slightly.

12. Remove the transmission support retaining bolts.
13. Remove the left body mounting bolts and loosen the radiator support mounting bolt.
14. Raise the vehicle on the left side to gain the necessary clearance to remove the upper attaching bolt. Support the vehicle by placing a block of wood between the frame and the first body mount.
15. Slide the transmission support rearward.
16. Lower the transmission jack in order to gain access to the oil cooler lines. Disconnect and plug the oil cooler lines.
17. Using a floor jack and a block of wood, support the engine and remove the transmission-to-engine retaining bolts.
18. Slide the transmission back and lower it from the vehicle. Install the torque converter retaining tool, as required.

TRANSMISSION INSTALLATION

EXCEPT TRUCKS

1. Using a transmission jack, raise the transmission into position and install the engine-to-transmission bolts.
2. Connect the T.V. cable to the transmission.
3. Unplug the oil cooler lines and install them to the transmission.
4. Install the left side transmission-to-engine bolt; this bolt retains the ground strap.
5. Install the crossmember, the retaining bolts, the rear transmission mount and floor pan reinforcement. Remove the transmission jack.
6. Install the catalytic converter support bracket, if necessary. Install the torque converter to flexplate retaining bolts.

NOTE: Before installing the flexplate to torque converter bolts, make sure the weld nuts on the converter are flush with the flexplate and the torque converter rotates freely by and in this position. Be sure the flexplate and torque converter marks made during the removal procedure align with each other.

7. Install the inspection cover. Connect all electrical leads and any clips to the transmission.
8. Connect the speedometer cable and the shift linkage to the transmission. Adjust the shift linkage and the T.V. cable, as required.
9. Install the driveshaft; align the driveshaft marks.
10. Install the torque arm to transmission retaining bolts, on Camaro and Firebird vehicles.
11. Install the dipstick. Lower the vehicle and fill the transmission with the proper grade and type automatic transmission fluid.
12. Connect the T.V./detent cable at it's upper end. Install the air cleaner assembly.
13. Connect the negative battery cable.
14. Start the engine and check for leaks. Once the vehicle has reached operating temperature recheck the fluid level.

TRUCKS

1. Using a transmission jack, raise the transmission into position and install the engine-to-transmission bolts.
2. Install the engine-to-transmission bolts.
3. Unplug the oil cooler lines and install them to the transmission.
4. Install the left body mounting bolts and tighten the radiator support mounting bolt.
5. Connect the catalytic converter support bracket. Install the torque converter-to-flywheel retaining bolts.

NOTE: Before installing the flexplate to torque converter bolts, make sure the weld nuts on the converter are flush with the flexplate and the torque converter rotates freely by hand in this position. Be sure the flexplate and torque converter marks made during the removal procedure align with each other.

6. Install the inspection cover, the exhaust crossover pipe and catalytic converter attaching bolts.
7. Install the transmission support brace attaching bolts at the torque converter, if equipped.
8. Install the crossmember and the brake line to crossmember clips, four wheel drive vehicles only.
9. Install any clips that retain the leads to the transmission and connect all electrical leads at the transmission.
10. Connect the speedometer cable and the shift linkage at the transmission.
11. Align the matchmarks and install the driveshaft.
12. Connect the T.V. cable at the upper end. Adjust the shift linkage and the T.V. cable as required.
13. Install the air cleaner and connect the negative battery cable.
14. Lower the vehicle and fill the transmission with the proper grade and type automatic transmission fluid.
15. Start the engine and check for leaks. Once the vehicle has reached operating temperature, recheck the fluid level.

SPECIFICATIONS

SERVO PIN SELECTION

Pin Identification	Pin Length (in.)
Two rings	2.61–2.62
Three rings	2.67–2.68
Wide band	2.72–2.73

TRANSMISSION ENDPLAY SELECTIVE WASHER

Identification	Thickness (in.)
67	0.074–0.078
68	0.080–0.084
69	0.087–0.091
70	0.094–0.098
71	0.100–0.104
72	0.107–0.111
73	0.113–0.118
74	0.120–0.124

AUTOMATIC TRANSMISSION OIL PRESSURES

Range @ 1000 rpm	Model	Normal Oil Pressure at Minimum T.V. (psi)	Normal Oil Pressure at Full T.V. (psi)
Park & Neutral	TC, MB, MC, MJ, VN	55–65	130–170
	TE, TH, TK, MD, ME, MK, MW, VH	55–65	130–170
	VA, ML, T7, MP, MS, PQ, YN, YK, YP, YG, YF, T2	55–65	130–170
	T8, TZ, TP, TS, MH, VJ, TL	65–75	130–170
	YH	65–75	140–180
	Y9	55–65	140–180
Reverse	TC, MB, MC, MJ, VN	90–105	210–285
	TE, TH, TK, MD, ME, MK, MW, VH	90–105	210–285
	VA, ML, T7, MP, MS, PQ, YN, YK, YP, YG, YF, T2	90–105	210–285
	T8, TZ, TP, TS, MH, VJ, TL	110–120	210–285
	YH	110–120	225–300
	Y9	90–105	225–300
Drive and Manual Third	TC, MB, MC, MJ, VN	55–65	130–170
	TE, TH, TK, MD, ME, MK, MW, VH	55–65	130–170
	VA, ML, T7, MP, MS, PQ, YN, YK, YP, YG, YF, T2	55–65	130–170
	T8, TZ, TP, TS, MH, VJ, TL	65–75	130–170
	YH	65–75	140–180
	Y9	55–65	140–180
Manual Second and Lo	TC, MB, MC, MJ, VN	100–120	100–120
	TE, TH, TK, MD, ME, MK, MW, VH	100–120	100–120
	VA, ML, T7, MP, MS, PQ, YN, YK, YP, YG, YF, T2	100–120	100–120
	T8, TZ, TP, TS, MH, VJ, TL	100–120	100–120
	YH	100–120	100–120
	Y9	100–120	100–120

TORQUE SPECIFICATIONS

Location	Torque ft. lbs.	Nm
Accumulator cover to case	8	11
Accumulator cover to case	8	11
Detent spring to valve body	18	22
Valve body to case	8	11
Oil passage cover to case	8	11
Solenoid assembly to pump	8	11
Transmission oil pan to case	18	24
Pressure switches	8	11

Location	Torque ft. lbs.	Nm
Park brake bracket to case	31	41
Pump cover to body	18	22
Pump assembly to case	18	22
Case extension to case	23	31
Manual shaft to inside det. lever	23	31
Pressure plugs	8	11
Connector cooler pipe	28	38

Section 5

Manual Transmissions and Transaxles
General Information

GENERAL INFORMATION

Introduction

With this edition of Chilton's Professional Domestic Transmission Manual, we continue to assist the professional transmission repair trade to perform quality repairs and adjustments for that "like new" dependability of the transmission/transaxle assemblies.

These concise, but comprehensive service sections place emphasis on diagnosing, troubleshooting, adjustments, disassembly and assembly of the manual transmission/transaxle.

Manual Transmission

The purpose of the manual transmission is to provide the driver with a selection of gear ratios between the vehicle engine and the vehicle drive wheels, so that the vehicle can operate at its best efficiency under a variety of driving conditions and loads.

The manual transmissions of today are fully synchronized in all forward gears, but usually not in the reverse gear. The synchronization process allows gears to be selected without gear clashing, by equalizing the speeds of the mating parts before they engage.

The gearshifting process is accomplished by either the steering column method or the floor shifter method. Regardless of where the gearshift lever is located, the function is the same. The gearshift lever selects the engaging gear either remotely or positively and moves it either forward or backward to engage the desired gear position. When the gearshift lever is moved, the movement is carried by linkage to the transmission.

Todays transmissions are either 3 speed, 4 speed, 5 speed or 6 speed (1989 Corvette) assemblies. The 5 speed transmission is usually a 4 speed transmission with an additional overdrive gear assembly. The 5 speed transmission that is used in the Corvette is a 4 speed unit with an automatic overdrive unit attached to it.

Manual Transaxle

The purpose of the manual transaxle is to provide the driver with a selection of gear ratios between the vehicle engine and the vehicle drive wheels, so that the vehicle can operate at its best efficiency under a variety of driving conditions and loads.

Manual transaxles are representative of the constant mesh design transmission, combined with a differential unit and assembled in a single case. All forward gears are in constant mesh. For ease of shifting and selection of the desired gear range, synchronizers with blocker rings, controlled by shift forks are used. Reverse gear uses a sliding idler gear arrangement.

The differential is a convential arrangement of gears that divide the torque between the drive axle shafts and allows them to rotate at different speeds. A basic differential consists of a set of 4 gears. Two of these gears are called differential side gears and the other 2 are called differential pinion gears. Each side gear is splined to a drive axle shaft, consequently, each drive axle shaft must turn when its side gear rotates, thus transfering power to the wheels.

Clutch

A clutch is a mechanism designed to connect or disconnect, the delivery of power from one working part to another, namely the engine and the transmission.

When the clutch is in the connecting (normal running) position, power flows through it from the engine to the transmission. If the transmission is in gear, then power flows on through

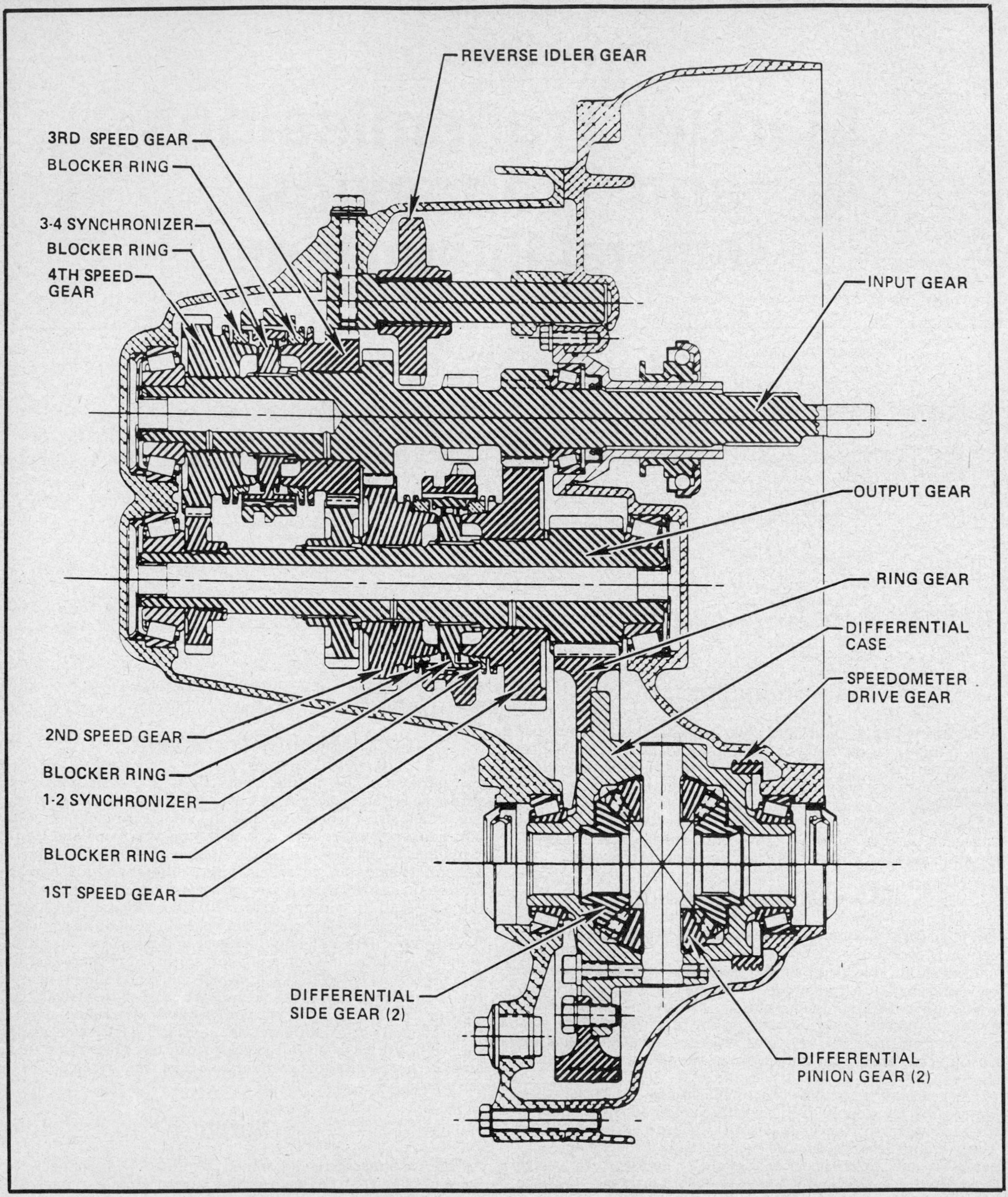
REVERSE IDLER GEAR
3RD SPEED GEAR
BLOCKER RING
3-4 SYNCHRONIZER
BLOCKER RING
4TH SPEED GEAR
INPUT GEAR
OUTPUT GEAR
RING GEAR
DIFFERENTIAL CASE
SPEEDOMETER DRIVE GEAR
2ND SPEED GEAR
BLOCKER RING
1-2 SYNCHRONIZER
BLOCKER RING
1ST SPEED GEAR
DIFFERENTIAL SIDE GEAR (2)
DIFFERENTIAL PINION GEAR (2)

Cross section of a typical manual transaxle assembly

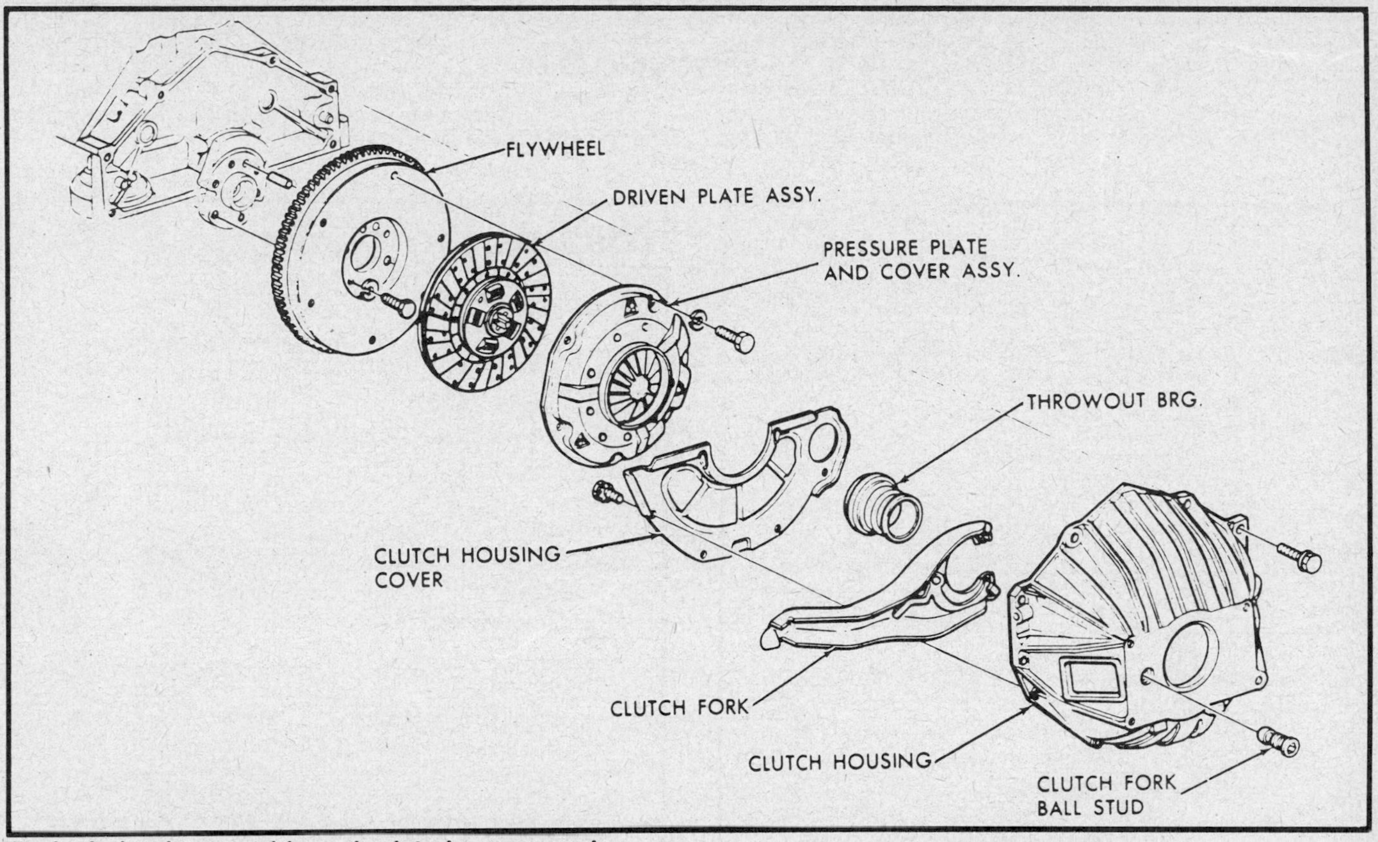
FLYWHEEL
DRIVEN PLATE ASSY.
PRESSURE PLATE AND COVER ASSY.
THROWOUT BRG.
CLUTCH HOUSING COVER
CLUTCH FORK
CLUTCH HOUSING
CLUTCH FORK BALL STUD

Typical clutch assembly and related components

to the wheels. The clutch permits the driver to uncouple the engine temporarily so that the gears can be shifted from one position to another. It is necessary to interrupt the flow of power (by coupling) before gears are shifted. Otherwise gear shifting would be extremely difficult and parts could be damaged.

The delivery of engine power must be gradual, to provide smooth engagement and to lessen the shock on the driving parts. After the engine and clutch are connected, the clutch must transmit all the engine power to the transmission without slipping. Again, it is desirable to disconnect the engine from the power train during the time the gears in the transmission are being shifted from one gear to another.

The most popular type clutch used in todays vehicles is of the single plate, dry disc type. It consists basically of 6 major parts, a flywheel, a clutch disc (driven plate), a pressure plate, springs, a pressure plate housing or cover and the linkage needed to operate the clutch assembly.

To describe basic construction and operation, the clutch will be divided into 3 main sections. These sections are, the driving members which are attached to the engine and turn with it, the driven members which are attached to the transmission and turn with it and the operating members which include the spring or springs and the linkage required to apply and release the pressure which holds the driving and driven members in contact with each other.

Clutches may be classified according to the number of plates or discs used. The single plate clutch contains 1 driven disc operating between the flywheel and the pressure plate. The flywheel is not considered to be a plate, even though it acts as one of the driving surfaces. A double plate clutch is almost the same except that another driven disc and an intermediate driving plate are added. A clutch having more than 3 discs is referred to as a multiple disc clutch. A further classification based on whether or not oil is supplied to the friction surfaces, provides a positive method of identifying the many types of clutches in use. If oil is supplied, the clutch is known as the wet type, if oil is not supplied, the clutch is the dry type.

Metric Fasteners and Inch System Fasteners

Metric bolt sizes and thread pitches are more commonly used for all fasteners on the manual transmissions/transaxles now being manufactured. The metric bolt sizes and thread pitches are very close to the dimensions of the similar inch system fasteners and for this reason, replacement fasteners must have the same measurement and strength as those removed.

Do not attempt to interchange metric fasteners for inch system fasteners. Mismatched and incorrect fasteners can result in damage to the transmission/transaxle unit through malfunction, breakage or possible personal injury. Care should be exercised to reuse the fasteners in their same locations as removed when every possible. If any doubt exists in the reuse of fasteners, install new ones.

To avoid stripped threads and to prevent metal warpage, the use of the torque wrench becomes more important, as the gear box assembly and internal components are being manufactured from light weight material. The torque conversion charts should be understood by the repairman, to properly service the requirements of the torquing procedures. When in doubt, refer to the specifications for the transmission/transaxle being serviced or overhauled.

Critical Measurements

With the increase use of transaxles and the close tolerances needed throughout the drive train, more emphasis is placed upon making the critical bearing and gear measurements correctly and being assured that correct preload and turning torque exists before the unit is reinstalled in the vehicle. Should a comeback occur because of the lack of proper clearances or torque, a costly rebuild can result. Rather than rebuilding a unit by "feel", the repairman must rely upon precise measuring tools, such as the dial indicator gauge, micrometers, torque wrenches and feeler gauges to insure that correct specifications are adhered to. At the end of each transmission/transaxle section, specification data is provided so that the repairman can measure important clearances that will effect the outcome of the transmission/transaxle rebuild.

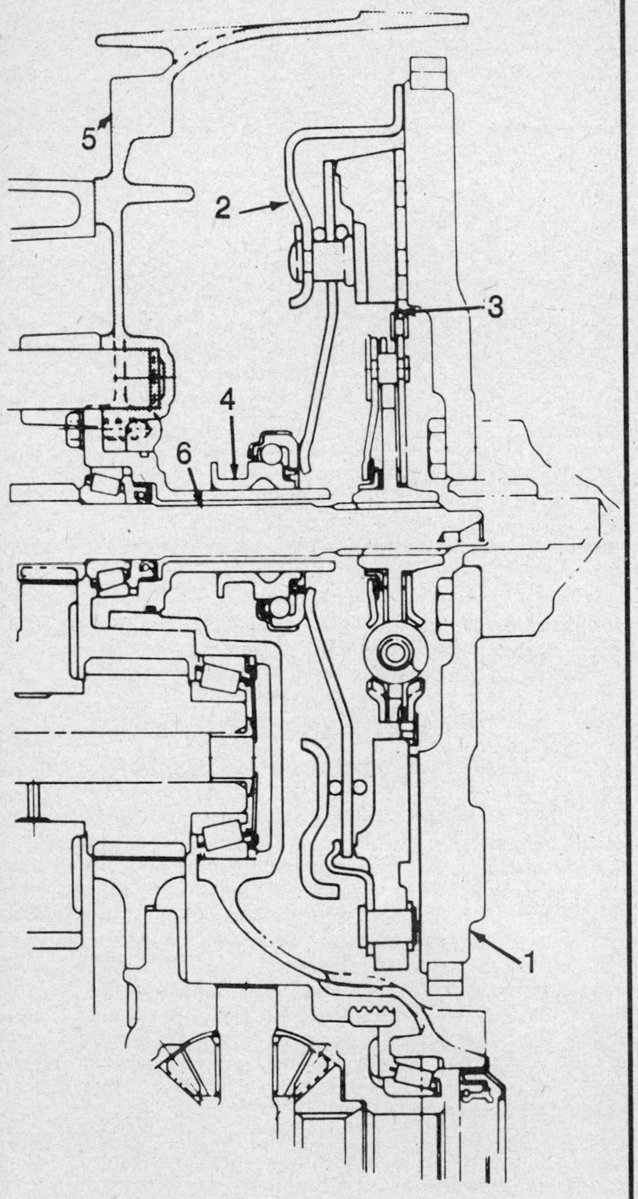
5
2
3
4
6
1

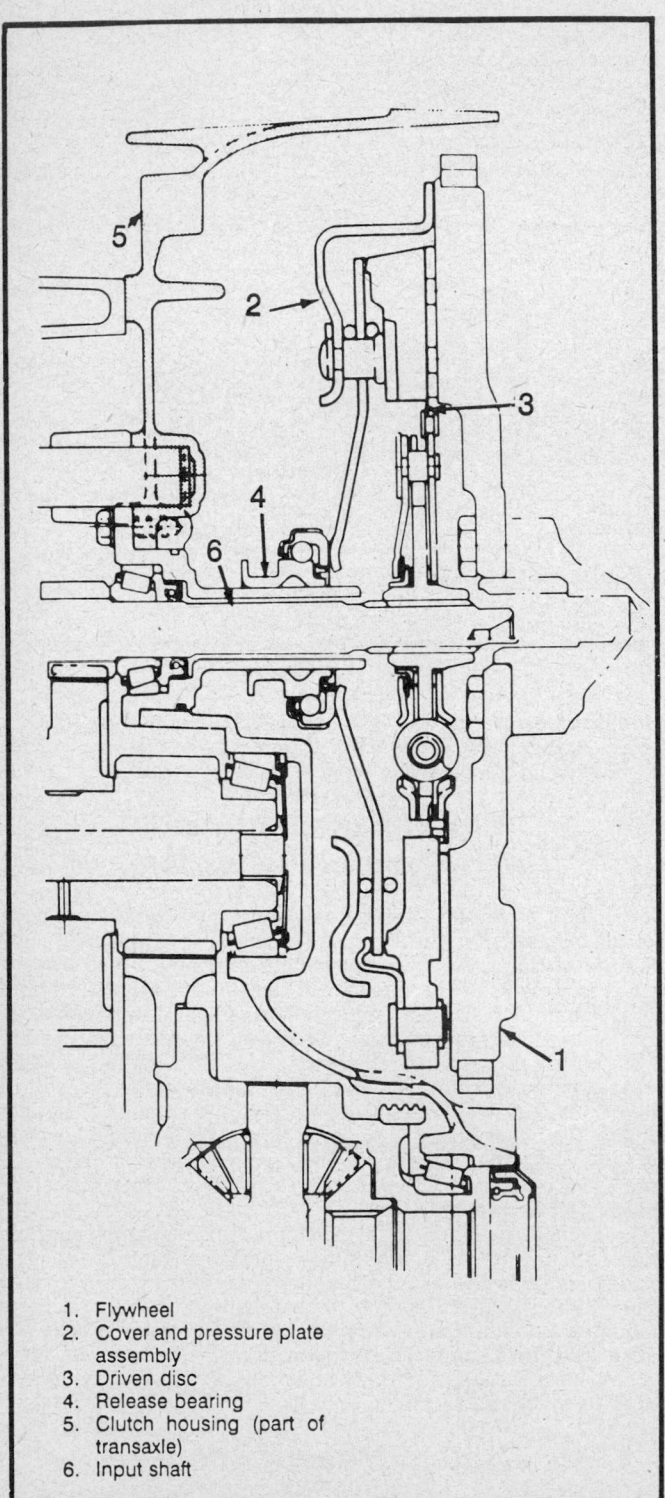

1. Flywheel
2. Cover and pressure plate assembly
3. Driven disc
4. Release bearing
5. Clutch housing (part of transaxle)
6. Input shaft

Single plate type clutch assembly

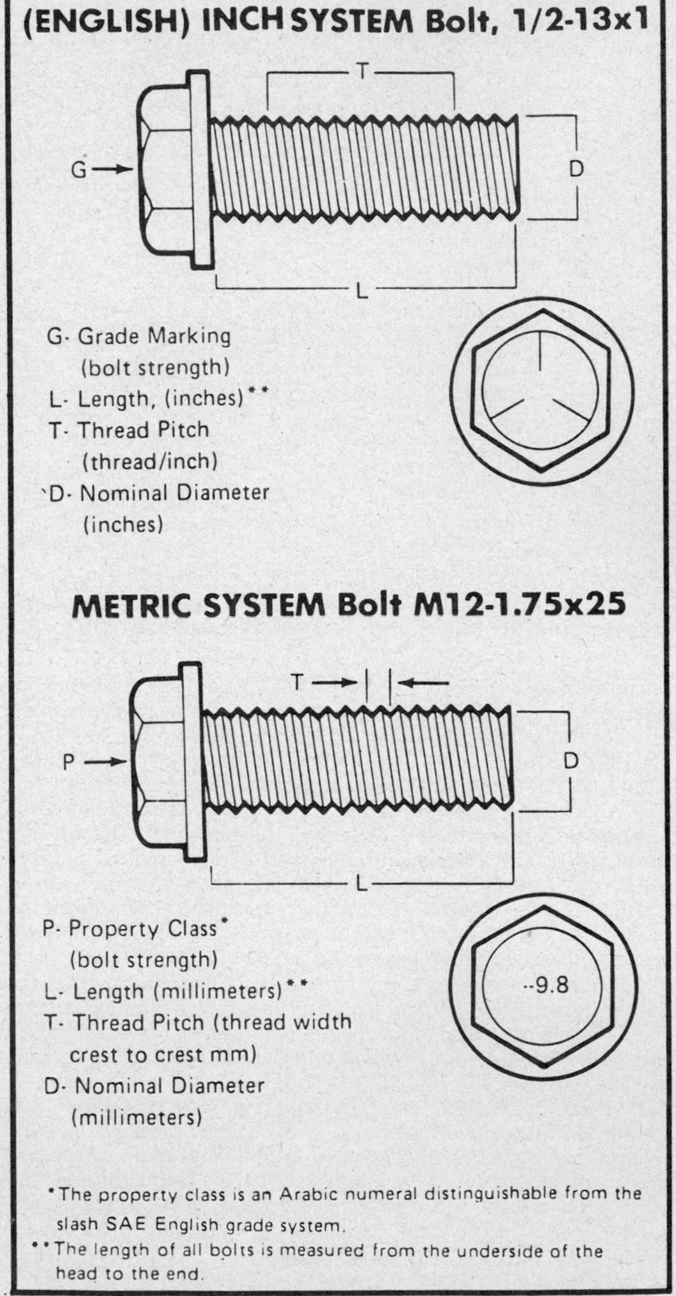
(ENGLISH) INCH SYSTEM Bolt, 1/2-13x1
T
G
D
L
G- Grade Marking (bolt strength)
L- Length, (inches)**
T- Thread Pitch (thread/inch)
D- Nominal Diameter (inches)
METRIC SYSTEM Bolt M12-1.75x25
T
P
D
L
P- Property Class* (bolt strength)
L- Length (millimeters)**
T- Thread Pitch (thread width crest to crest mm)
D- Nominal Diameter (millimeters)
9.8
*The property class is an Arabic numeral distinguishable from the slash SAE English grade system.
**The length of all bolts is measured from the underside of the head to the end.

Comparison of the English Inch and Metric system bolt and thread nomenclature

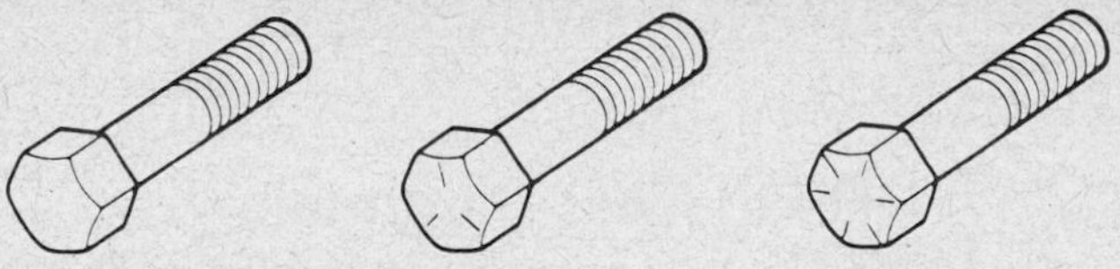
GRADE 1 or 2
GRADE 5
GRADE 8

English (Inch) Bolts—Identification Marks Correspond To Bolt Strength—Increasing Number Of Slashes Represent Increasing Strength.

Typical English Inch bolt head identification marks

4.6
4.8
5.8
8.8
9.8
10.9

Metric Bolts—Identification Class Numbers Correspond To Bolt Strength—Increasing Numbers Represent Increasing Strength. Common Metric Fastener Bolt Strength Property Are 9.8 And 10.9 With The Class Identification Embossed On The Bolt Head.

Typical Metric bolt head identification marks

(ENGLISH) INCH SYSTEM

Grade	Identification
Hex Nut Grade 5	 3 Dots
Hex Nut Grade 8	 6 Dots

Increasing dots represent increasing strength.

METRIC SYSTEM

Class	Identification
Hex Nut Property Class 9	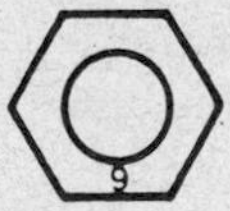 Arabic 9
Hex Nut Property Class 10	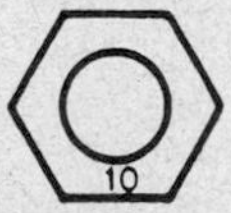 Arabic 10

May also have blue finish or paint daub on hex flat. Increasing numbers represent increasing strength.

Comparison of the English Inch and Metric hex nut strength identification marks

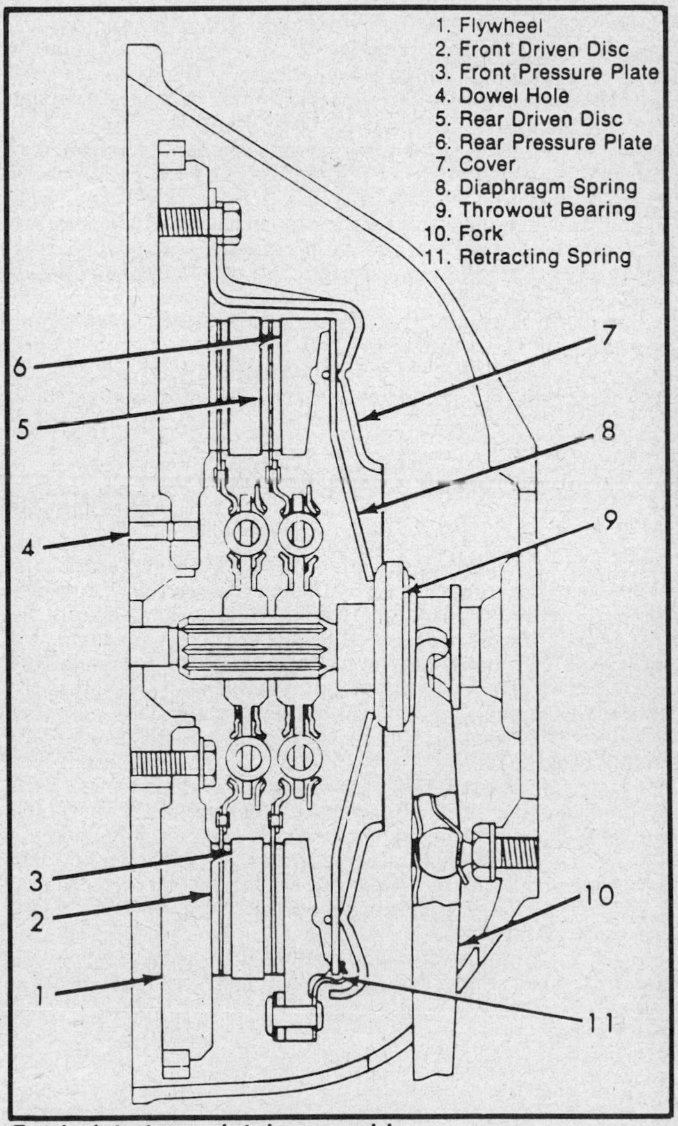
1. Flywheel
2. Front Driven Disc
3. Front Pressure Plate
4. Dowel Hole
5. Rear Driven Disc
6. Rear Pressure Plate
7. Cover
8. Diaphragm Spring
9. Throwout Bearing
10. Fork
11. Retracting Spring

Dual plate type clutch assembly

DIAGNOSING MANUAL TRANSMISSION/TRANSAXLE MALFUNCTIONS

Introduction

Diagnosing transmission/transaxle problems is simplified following a definite procedure and understanding the basic operation of the individual transmission/transaxle that is being inspected or serviced. Do no attempt to short-cut the procedure or take for granted that another technician has performed the adjustments or the critical checks. It may be an easy task to locate a defective or damaged unit, but the technician must be skilled in locating the primary reason for the unit failure and must repair the malfunction to avoid having the same failure occur again.

Each manual transmission/transaxle manufacturer has developed a diagnostic procedure for their individual transmissions/transaxles. Although the operation of the units are basically the same, many differences will appear in the construction and method of unit application.

The same model transmission/transaxle can be installed in different makes of vehicles and are designed to operate under different load stresses, engine applications and road conditions. Each make of vehicle will have specific adjustments or use certain outside manual controls to operate the individual gearing, but may not interchange with another transmission/transaxle vehicle application from the same manufacturer.

The identification of the transmission/transaxle is most important so that the proper preliminary inspections and adjustments may be done and if in need of a major overhaul, the correct parts may be obtained and installed to avoid costly delays.

Road Test Diagnosis

Prior to driving the vehicle on a road test, have the vehicle operator explain the malfunction of the transmission/transaxle as fully and as accurate as possible. Because the operator may not have the same technical knowledge as the diagnostician, ask questions concerning the malfunction in a manner that the operator can understand. It may be necessary to have the operator drive the vehicle, on a road test and to identify the problem to the diagnostician. The diagnostician can observe the manner in which the transmission/transaxle is being operated and can point out constructive driving habits to the operator to improve operation reliability.

Be aware of the engine's performance. If the engine is in need of a tune up, it should be taken care of before suspecting that the transmission/transaxle is at fault.

Noise Diagnosis

In diagnosisng transmission/transaxle noises, the diagnostician must be alert to any abnormal noises from the transmission/transaxle area or any excessive movement of the engine or the transmission/transaxle assembly during torque application or transmission/transaxle shifting.

Before attempting to diagnose transmission/transaxle noises, be sure the noises do not originate from the engine components, such as the water pump, alternator, air conditioner compressor, power steering or the air injection pump. Isolate these components by removing the proper drive belt and operate the engine.

SPECIAL TOOLS

There are an unlimited amount of special tools and accessories available to the transmission rebuilder to lessen the time and effort required in performing the diagnosing and overhaul of the transmission/transaxles. Specific tools are necessary during the disassembly and assembly of each unit and its subassemblies. Certain tools can be fabricated, but it becomes the responsibility of the repair shop operator to obtain commercially manufactured tools to insure quality rebuilding and to avoid costly "come backs".

For specific information concerning the various tools, a parts and tool supplier should be consulted. The use of the basic measuring tools has become more critical in the rebuilding process. The increased use of front drive transaxles, in which both the transaxle and the final drive gears are located, has required the rebuilder to adhere to specifications and tolerances more closely than ever before.

Bearings must be torqued or adjusted to specific preloads in order to meet the rotating torque drag specifications. The endplay and backlash of the varied shafts and gears must be measured to avoid excessive tightness or looseness. Critical tensioning bolts must be torqued to specification.

Dial indicators must be protected and used as a delicate measuring instrument. A mutilated or un-calibrated dial indicator invites premature unit failure and destruction. Torque wrenches are available in many forms, some cheaply made and others, accurate and durable under constant use. To obtain accurate readings and properly applied torque, recalibration should be applied to the torque wrenches periodically, regardless of the type used. Micrometers are used as precise measuring tools and should be properly stored when not in use. Instructions on the recalibration of the micrometers and a test bar usually accompany the tool when it is purchased.

Other measuring tools are available to the rebuilder and each in their own way, must be protected when not in use to avoid causing mis-measuring in the fitting of a component to the unit.

Section 6

Manual Transaxle Applications

Vehicle Manufacturer	Year	Vehicle Model	Transaxle Manufacturer	Transaxle Identification	Speeds
Chrysler Corp.	1984–89	All	Chrysler Corp.	A460	4
	1984–89	All	Chrysler Corp.	A465, A520, A525, A555	5
Jeep-Eagle	1988–89	Medallion	—	NG9	5
Ford Motor Co.	1984–89	Tempo, Topaz, EXP, Escort, Lynx, Taurus, Sable	Mazda	RGT	4
	1984–89	Tempo, Topaz, EXP, Escort, Lynx, Taurus, Sable	Mazda	RWB	5
	1988–89	Festiva	Mazda	MTX4	4
	1988–89	Festiva	Mazda	MTX5	5
	1989	Probe	Mazda	MTX5	5
	1987–89	Tracer	Mazda	MTX4	4
	1987–89	Tracer	Mazda	MTX5	5
General Motors	1984–87	Cavalier, Celebrity, Citation, Fiera, Firenza, Phoenix, Sunbird, Skylark, Omega, Skyhawk	Muncie	—	4
	1988–89	LeMans	—	M21	4
	1988–89	LeMans	—	M79	5
	1985–89	Sprint	—	MV2	5
	1985–89	Spectrum, Skyhawk, Firenza, Fiero, Cimarron, Sunbird, Grand Am, Calais, Somerset, Corsica, Beretta	Isuzu	MR8, MK7 MY7, MT2, MR3	5
	1985–89	Nova, Geo Prizm	—	—	5

Section 6

A460, A465, A520, A525 and A555 Transaxles

Chrysler Corp.

APPLICATION

A460, A465, A520, A525, A555

Year	Vehicle	Engine	Vin Code
1984	Aries	2.2L/2.6L	D
	Caravan	2.2L	K
	Caravelle	2.2L/2.6L	E
	Charger	1.6L/2.2L	Z
	Daytona	2.2L/2.6L	V
	E-Class	2.2L/2.6L	T
	Horizon	1.6L/2.2L	M
	Laser	2.2L/2.6L	C
	LeBaron	2.2L/2.6L	C
	Mini Ram Van	2.2L/2.6L	K
	New Yorker	2.2L/2.6L	T
	Omni	1.6L/2.2L	Z
	Rampage	1.6L/2.2L	Z
	Reliant	2.2L/2.6L	P
	Turismo	1.6L/2.2L	M
	Voyager	2.2L	H
	600	2.2L/2.6L	E
1985	Aries	2.2L/2.6L	D
	Caravan	2.2L	K
	Caravelle	2.2L/2.6L	J
	Caravelle①	2.2L/2.6L	L
	Charger	1.6L/2.2L	Z
	Daytona	2.2L/2.6L	V
	Horizon	1.6L/2.2L	M
	Lancer	2.2L/2.6L	D
	Laser	2.2L/2.6L	C
	LeBaron	2.2L/2.6L	C
	Mini Ram Van	2.2L/2.6L	K
	New Yorker	2.2L/2.6L	T
	Omni	1.6L/2.2L	Z
	Reliant	2.2L/2.6L	P
	Turismo	1.6L/2.2L	M
	Voyager	2.2L	H
	600	2.2L/2.6L	E
	600 premium	2.2L/2.6L	V
1986	Aries	2.2L/2.6L	D

A460, A465, A520, A525, A555

Year	Vehicle	Engine	Vin Code
1986	Caravan	2.2L	K
	Caravelle	2.2L/2.6L	J
	Charger	1.6L/2.2L	Z
	Daytona	2.2L/2.6L	V
	Horizon	1.6L/2.2L	M
	Lancer	2.2L/2.6L	D
	Laser	2.2L/2.6L	C
	LeBaron	2.2L/2.6L	C
	Mini Ram Van	2.2L/2.6L	K
	New Yorker	2.2L/2.6L	T
	Omni	1.6L/2.2L	Z
	Reliant	2.2L/2.6L	P
	Turismo	1.6L/2.2L	M
	Voyager	2.2L	H
	600	2.2L/2.6L	E
	600 premium	2.2L/2.6L	V
1987	Aries	2.2L/2.5L	D
	Caravan	2.2L	K
	Caravelle	2.2L/2.5L	J
	Charger	2.2L	Z
	Daytona	2.2L/2.5L	V
	Horizon	2.2L	M
	Lancer	2.2L/2.5L	D
	LeBaron	2.2L/2.5L	C
	Mini Ram Van	2.2L/2.6L	K
	New Yorker	2.2L/2.5L	T
	Omni	2.2L	Z
	Reliant	2.2L/2.5L	P
	Shadow	2.2L	D
	Sundance	2.2L	P
	Turismo	2.2L	M
	Voyager	2.2L	H
	600	2.2L/2.5L	E

Year	Vehicle	Engine	Vin Code
1988	Aries	2.2L/2.5L	D
	Caravan	2.5L/3.0L	K
	Caravelle	2.2L/2.5L	J
	Daytona	2.2L/2.5L	V
	Dynasty	2.5L/3.0L	D
	Horizon	2.2L	M
	Lancer	2.2L/2.5L	D
	LeBaron	2.2L/2.5L	C
	Mini Ram Van	2.5L/3.0L	K
	New Yorker	2.5L/3.0L	C
	Omni	2.2L	Z
	Reliant	2.2L/2.5L	P
	Shadow	2.2L	D
	Sundance	2.2L	P
	Voyager	2.5L/3.0L	H
	600	2.2L/2.5L	E
1989	Acclaim	2.5L/3.0L	P
	Aries	2.2L/2.5L	D
	Caravan	2.5L/3.0L	K
	Daytona	2.2L/2.5L	V
	Dynasty	2.5L/3.0L	D
	Horizon	2.2L	M
	Lancer	2.2L/2.5L	D
	LeBaron	2.2L/2.5L	C
	Mini Ram Van	2.5L/3.0L	K
	New Yorker	2.5L/3.0L	C
	Omni	2.2L	Z
	Reliant	2.2L/2.5L	P
	Shadow	2.2L/2.5L	D
	Spirit	2.5L/3.0L	D
	Sundance	2.2L/2.5L	P
	Voyager	2.5L/3.0L	H

① Canada

GENERAL DESCRIPTION

The A–460, A–465, and A–525 are fully-synchronized manual transaxles that combine gear reduction, ratio selection, and differential functions in a single unit housed in a die-cast aluminum case. The A–460 is a 4 speed while both the A–465 and A–525 are 5 speeds. The A–525 has a close-ratio gearset with different 2nd, 3rd/4th-gear ratios than the A–465, to provide better performance through the gears, while 1st and 5th gear ratios are the same as the A–465 to maintain the same launch and top-gear characteristics.

To accept the greater power and torque output of the Turbo II engine, the A–555 heavy-duty 5 speed manual transaxle was introduced to the Chrysler front wheel drive line. The A–555 features a new die cast case and a strengthened coarse-pitch gear set. The transaxle has 5 forward speed ratios and reverse. The forward gear ratios are as follows: 1st – 3.00, 2nd – 1.89, 3rd – 1.28, 4th – 0.94, 5th – 0.72. The reverse gear ratio is 3.14:1. The final drive ratio is 3.85. All the forward gears are synchronized. A new stop-ring actuation system has been incorporated to provide positive and precise shifting. The conventional 2-pinion differential is replaced by an enclosed 4-pinion design.

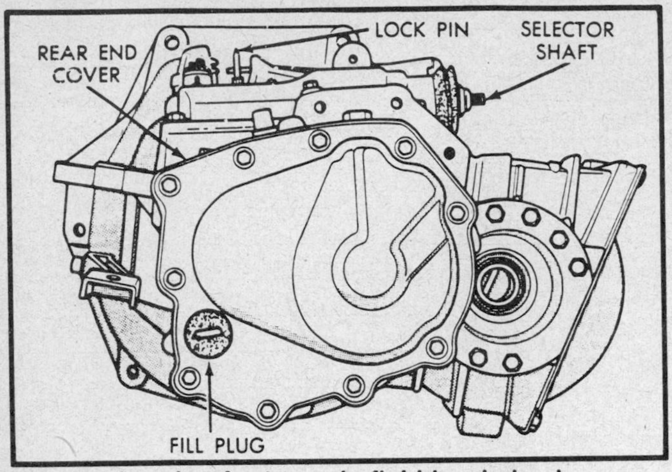
LOCK PIN
SELECTOR SHAFT
REAR END COVER
FILL PLUG

Fill plug location for tranaxle fluid level check

The A-520 5 speed transaxle is used on all applications except Turbo II equipped and L-Body vehicles. This transaxle has been designed for improved shifting quality and sealing. The final drive ratio is 3.50:1. The differential has been improved to accept the increased torque output of the 2.5L engine.

Both the A-520 and A-555 transaxles feature new dual-cone synchronizers for the 1st/2nd gears. The new synchronizer configuration provides greater shifting capacity and also reduces the 1st/2nd gear shifting effort.

Metric Fasteners

The metric fastener dimensions are very close to the dimensions of the familiar inch system fasteners. For this reason, replacement fasteners must have the same measurement and strength as those removed.

CAUTION

Do not attempt to interchange metric fasteners for inch system fasteners. Mismatched or incorrect fasteners can result in damage to the transmission unit through malfunctions, breakage or possible personal injury.

Care should be taken to re-use the fasteners in the same locations as removed.

Common metric fastener strength property classes are 9.8 and 12.9 with class identification embossed on the head of each bolt. The inch strength classes range from grade 2 to 8 with the line identification embossed on each bolt head. Markings correspond to 2 lines less than actual grade (for example grade 8 bolt will exhibit 6 embossed lines on the bolt head). Some metric nuts will be marked with a single digit strength identification numbers on the nut face.

CAPACITIES CHART

Year	Transaxle	Fluid Type	Pan Capacity	Overhaul Capacity ①
1984	A-460 ②	ATF	NA	1.9
	A-460 ③	ATF	NA	2.0
	A-465	ATF	NA	2.3
	A-525	ATF	NA	2.3
1985	A-460	ATF	NA	2.0
	A-525	ATF	NA	2.3
1986	A-460 ②	ATF	NA	2.0
	A-525	ATF	NA	2.3
1987	A-525	SAE 5W-30	NA	2.3
	A-520	SAE 5W-30	NA	2.5
	A-555	SAE 5W-30	NA	2.5
1988	A-525	SAE 5W-30	NA	2.3
	A-520	SAE 5W-30	NA	2.5
	A-555	SAE 5W-30	NA	2.5
1989	A-525	SAE 5W-30	NA	2.3
	A-520	SAE 5W-30	NA	2.5
	A-555	SAE 5W-30	NA	2.5

① Fill all transaxles to the bottom of the hole in the fill cover.
② With 1.6L engine
③ With 2.2L engine

Capacities
Checking Fluid Level

Check the fluid level in the transaxle by removing the fill plug located on the left side of the tranaxle. The fluid level should be between the bottom of the fill opening and at a point not more than $^{3}/_{16}$ in. below the bottom of the fill opening.

TRANSAXLE MODIFICATIONS

Grinding Noise in Clutch

1985 Omni/Horizons equipped with 1.6L engines have experienced a grinding noise in the clutch when the clutch pedal is fully depressed. This problem is attributed to excessive clutch pedal travel caused by the existing clutch pedal bumper stop. A newly designed clutch pedal bumper stop, part number 433658, has replaced the existing stop.

The problem is diagnosed by running the engine at curb idle and depressing the clutch pedal all the way to the floor while listening for a grinding noise. Partially release the clutch pedal approximately 1½–2 in. and again listen for the grinding noise. If the noise disappears when the clutch pedal is released, the problem may be a result of excessive clutch pedal play. If a problem does exist, remove the existing stop and install the modified stop. Reset the clutch adjustment by lifting the clutch pedal enough to allow the cable positioner to separate from the adjuster. Stroke the pedal 3–4 times to allow the self-adjuster mechanism to reach the proper adjustment.

Drive Gear Noise

1984–85 front wheel drive vehicles (excluding imports) have experienced drive gear noise (whine) originating from the transfer

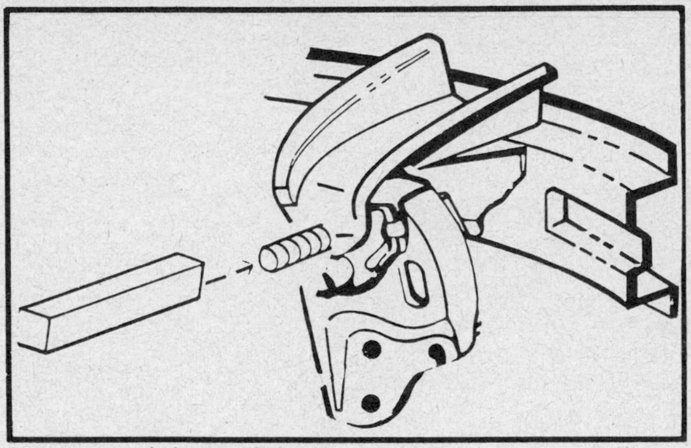

Damper weight modification

shaft and the ring gear in the 40–50 mph range. To correct the problem, a damper weight, part number 4431107, has been added onto the end of left hand mount through bolt.

To determine if the vehicle requires the modification, evalute the performance of the vehicle on a smooth road surface with the transaxle at normal operating temperature in the 40–50 mph range. If a drive gear noise is verified, perform the following procedure:

1. Remove the air inlet hose from the air cleaner.
2. Thread the damper weight onto the end of the left hand motor mount through bolt. The through bolt threads protrude into the engine compartment.
3. Torque the damper weight to 170–230 inch lbs.

Press Fit Fifth Speed Synchronizer Hub

A press fit 5th speed synchronizer hub was introduced to the A-525 transaxle in May of 1984. The press fit type hub reduces gear rattle and overall transaxle noise. The hub modification affects 1984 Ram Van/Caravan, Voyager, Horizon/Omni, Aries/Relient and Daytona/Laser.

The service procedure for the press fit hub does not change except for the actual removal and installation from and back on to the shaft. To remove the press fit type hub, use tool C–4621–1 or equivalent. To install the press fit hub, use tool C–4888 or equivalent.

The press fitted hub is interchangable with the existing A-465 and A-525 hubs.

Transaxle Rattling Noise

1984–86 passenger vehicles with 2.2L Turbo or 2.5L engines and Voyager, Caravan and Ram Van with 2.2L engines have experienced a rattling noise from the area of the transaxle with the clutch pedal out, transaxle in **N** and the engine at curb idle. The problem is attributed to existing clutch disc design. To correct the problem, the existing clutch discs are replaced with modified clutch discs, part number 4377521 for passenger vehicles, 4377472 for Voyager, Caravan and Ram Van with 2.2L engines.

To diagnose the problem, the engine and the transaxle should both be at normal operating temperature. With the engine idling, the transaxle in **N** and the clutch pedal out, listen for the rattling noise. Raise the idle speed slightly to see if the noise disappears. If the noise disappears or is lessened when the idle speed is raised, the clutch disc is the source of the problem and must be replaced with the modified disc.

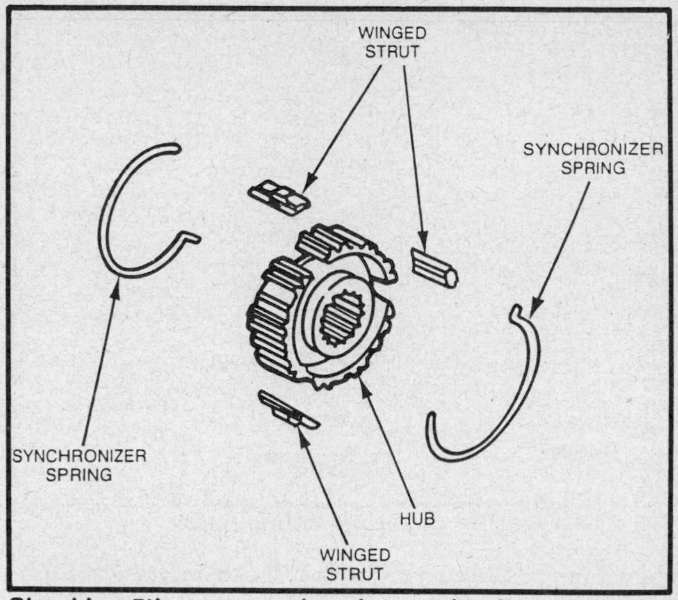
WINGED STRUT
SYNCHRONIZER SPRING
SYNCHRONIZER SPRING
HUB
WINGED STRUT

Checking 5th gear synchronizer spring location for scratchy and hard fifth gear shift complaints on A–525 transaxles

Shifter Noise

1985 Omni/Horizon, Relient/Aries with 4 and 5 speed transaxles and rod shift linkage have experienced a buzzing and/or squaking noise in the shifter. The noise can occur in any gear but is most pronounced in the 2000–3000 rpm range. The problem is corrected by injecting Mopar multi-purpose grease, part number 4318063, through a drilled hole in the shifter ball with the shifter in the third gear position.

To diagnose the problem, place the transaxle in **N** and raise the engine idle speeed to 3500 rpm while moving the shifter side-to-side in ¼ in. increments. Listen for a harsh roar. Shut off the engine and move the shifter through the shift pattern. Listen for a squawk particularly in the fore and aft travel. If either noise is indicated, perform the following procedure:

1. Pry the shift tube end fitting from the clevis ball stud.
2. Apply a heavy coat of Mopar multi-purpose grease or its equivalent to the ball stud and the end fitting cavity.
3. Reconnect the ball fitting to the ball stud.
4. Make sure that the connetor bushings are fully lubricated with multi-purpose grease.
5. De-trim the interior until the upper plate of the shift mechanism is accessible.
6. Move the shifter to the third gear position and drill a $^3/_{16}$ in. hole in the hollow plastic ball at its rearmost visible position. All drilling chips and shavings must be completely removed.
7. Slowly inject a small amount of multi-purpose grease into the hole.
8. Cycle the shifter rapidly up and down about 10 times.
9. Install the trim.
10. Perform a road test to ensure proper shift lever operation and verify that noise has beenm eliminated.

Scratchy and Hard Fifth Gear Shifts

1984–86 domestic front wheel drive vehicles equipped with A-525 5 speed manual transaxles have experienced hard shifting into and out of 5th gear and scratchy 5th gear shifts. The problem is corrected by replacing the existing 5th gear synchronizer strut retainer plate with a new plate, part number 4431894.

To diagnose the problem, perform a road test and check the 5th gear shift effort for abnormal difficulty or scratchy feeling

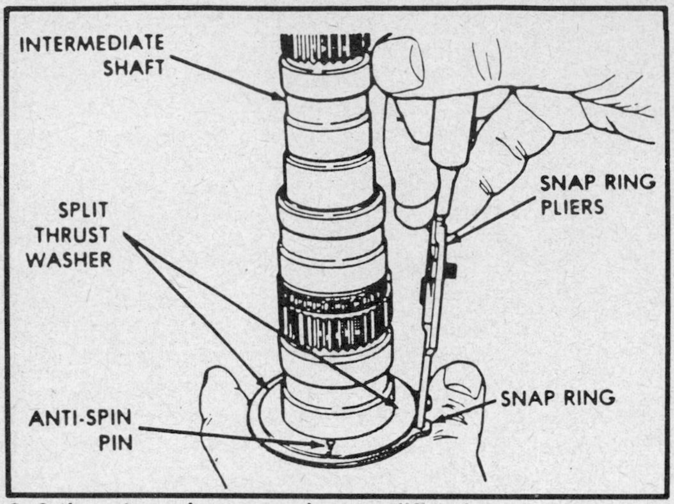
INTERMEDIATE SHAFT
SNAP RING PLIERS
SPLIT THRUST WASHER
SNAP RING
ANTI-SPIN PIN

2–3 thrust washer snapring modification

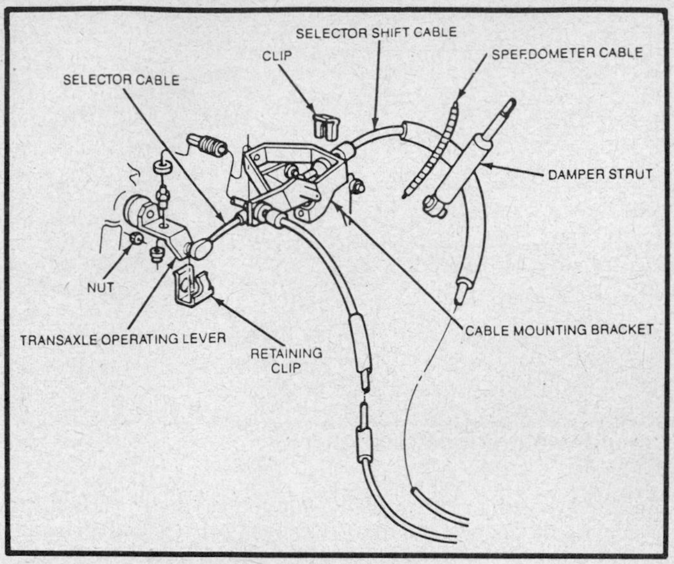
SELECTOR SHIFT CABLE
CLIP
SPEEDOMETER CABLE
SELECTOR CABLE
DAMPER STRUT
NUT
TRANSAXLE OPERATING LEVER
RETAINING CLIP
CABLE MOUNTING BRACKET

Proper selector shift cable routing

when shifting slowly out of 5th gear. If either condition is indicated, perform the following procedure:

1. Remove the left front wheel.
2. Remove the transaxle oil fill plug.
3. Remove the end cover.
4. Remove the 5th speed synchronizer ring.
5. Check the location of the 5th gear synchronizer spring. The spring should be properly positioned in the synchronizer hub. Reposition the spring as required.
6. Install the new strut retainer plate so that the 3 raised portions are equally spaced between the struts.
7. Install a new snapring.
8. Install the end cover and torque the cover bolts to 17–21 ft. lbs.
9. Fill the transxle to the proper level.
10. Perform a road test and check for proper operation.

Vehicle Jumps Out of Third Gear or Second or Third Gear Does Not Engage

The 1987 G, H, J, K, L and S Body vehicles that were equipped with manual transaxles have experienced 2nd or 3rd jump out or no 2nd or 3rd gear engagement. This problem may be the result of a failed 2/3 thrust washer snapring. The existing snapring should be removed and replaced with a modified snapring, part number 6033358. Regardless of the reason for disassembly of the transmission, the existing snapring should never be reused. To install the modified snapring, perform the following procedure:

1. Remove the the intermediate shaft gearset from the vehicle.
2. Disassemble the intermediate shaft and make sure that the snapring is not in the proper postion, deformed, over expanded or missing.
3. If the snapring ring is missing, the transaxle should be removed, disassembled and each component inspected for damage.
4. Replace the existing snapring with the new snapring. The new snapring must be completely seated in the groove and must not be allowed to become over expanded during installation.

Spongy First to Second or Second to First Shifts

The 1987 S, K, G, H, P and J Body vehicles equipped with A–520 or A–555 manual transaxles have experienced spongy 1st/2nd or 2nd/1st shifts characterized by the lever tending to spring back. This problem is the result of the 1–2 shift rail rail being assembled without oil relief grooves or flats in the case bore end of the rail. The oil relief groove or flat should be in the opposite end of the rail from the screwdriver slot. The problem is corrected by either repositioning or replacing the existing shift rail with a new rail transaxle gearshift fork, part number 4471308. To perform the modifiaction, perform the following procedure:

1. Remove the gear selector housing and rear cover.
2. Remove the shift rail and verify that it is not installed backwards. The shift rail is not installed backward if the longitudinal oil relief groove or flat is in the bearing plate side of the rail. If this is the case, re-position the rail so that the relief groove or flat is in the transaxle case bore.

NOTE: Be advised that the oil relief groove or flat is not the same as the screwdriver slot found at the end of the rail.

3. If there is no relief groove or flat in the rail, or it is on the same side as the screwdriver slot, then it must be replaced with the new shift fork.
4. Replace the gear selector housing and rear cover.

High Shift Effort or Gear Blockage

1984–86 Daytona/Lazer, Lancer and GTS with A–525 manual tranaxles have experieced a high shift effort or gear blockage due to improper crossover adjustment.

To diagnose the problem, with the engine off, engage the parking brake and depress the clutch pedal. Place the shifter in the 3rd gear. Make rapid shifts betwen 3rd and 2nd gears while applying a moderate pre-load toward the reverse-neutral position. Place the shifter in the 2nd gear position. Shift slowly between 2nd/1st gear while applying a minimum crossover load toward reverse-neutral (just against the 1–2 stop). Any obstruction or total blockage of a gear in the above test, after initial synchronizer line-up, indicates crossover misadjustment. If crossover misadjustment is indicated, perform the following procedure:

1. Place the transaxle in 1st gear and keep the gear engaged through Steps 2–6 of this proecedure.

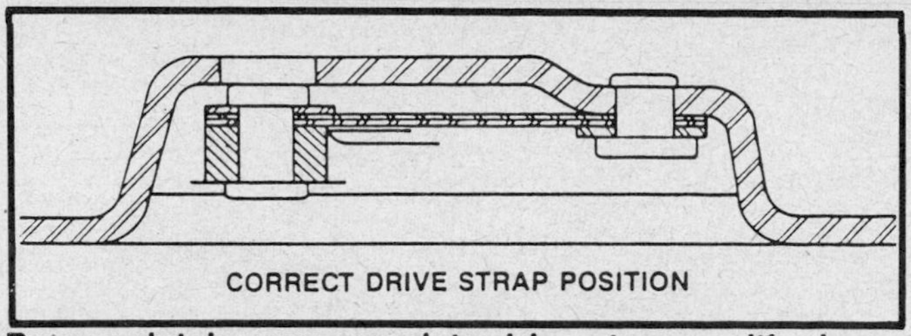
CORRECT DRIVE STRAP POSITION

Proper clutch pressure plate drive strap positioning

2. Slide the drivers seat back as far as it will go and remove the carpet pad from the driver side of the floor console.
3. Locate the crossover cable adjustment screw through the left side opening of the console.
4. Loosen the adjustment screw until the shift lever is free in the crossover direction.
5. Push the shift lever toward reverse until it contacts the reverse lockout device. Move the knob away from this position, toward 5th gear, approximately ⅛ in. Hold the knob in this position.
6. Tighten the adjusting screw to 55 inch lbs.
7. Replace the floor console carpet pad and move the seat to the desired position.
8. Repeat the diagnostic procedure to verify that the condition is corrected.

Increased Shifting Effort and Premature Core Wire Failure of Selector Shift Cable

1984–86 domestic passenger front wheel drive vehicles, Ram Van, Caravan and Voyager with cable operated manual transaxles have experienced increased shifting efforts and premature core wire failure of the shift selector cable. This condition is due to mis-routing of the cable.

To determine if mis-routing is present, inspect the shift cable in the area of the speedometer cable and/or damper strut, if equipped. The cable should be located forward of both components, if equipped. If mis-routing is indicated, re-route the selector cable by performing the following procedure:

1. Disconnect the selector shift cable from the transaxle operating lever and mounting bracket.
2. Re-route the selector shift cable to pass forward of the speedometer cable and the damper strut, if equipped.
3. Connect the cable to the mounting bracket and the operating lever.
4. Adjust the shift linkage.

Clash In Reverse, Excessive Spin Time, Clutch Will Not Release

1984–86 domestic front wheel drive vehicles have experienced gear clash during reverse engagement, excessive spin time and no-clutch release problems. These conditions are due to damaged drive strap in the clutch pressure plate.

To diagnose the problem, warm up the engine at curb idle. Set the parking brake and depress the clutch fully to the floor. Wait 5 seconds, shift into reverse and check for gear clash. If gear clash is indicated, inspect the clutch pressure plate by performing the following procedure:

1. Remove the transaxle from the vehicle.
2. Remove the clutch disc and pressure plate.

NOTE: The drive straps are the straps connecting the pressure plate to the cover. The straps are contructed of 2 long strips of steel with 2 small square steel washers attached at the rivets on the cover and the pressure plate.

3. Inspect the straps for uniformity. All straps should look the same. No strap should be bent or straighter than the other.
4. Inspect the straps for separation. The 2 straps should have no gaps betwen them. Any visible gaps will cause clutch performance problems.
5. Inspect the placement of the steel washers. The washer must be flush with the long strips. If a piece of paper can be slipped under the washers, the pressure plate is defective.
6. If any of the above conditions are noted, replace the clutch pressure plate.

POWER FLOW

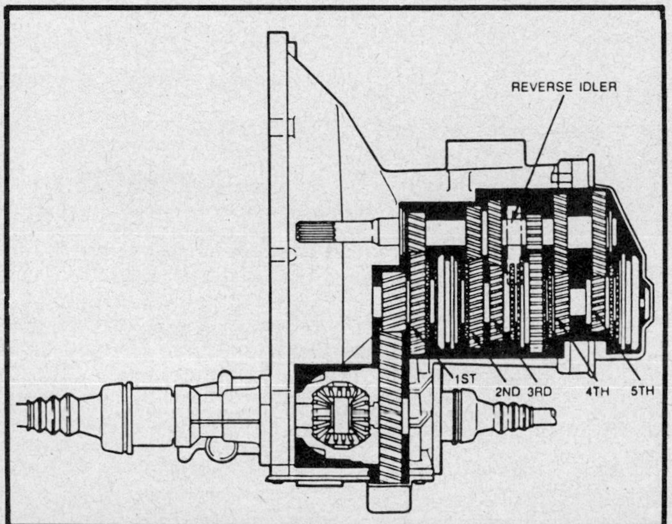
REVERSE IDLER
1ST
2ND
3RD
4TH
5TH

Constant mesh power flow

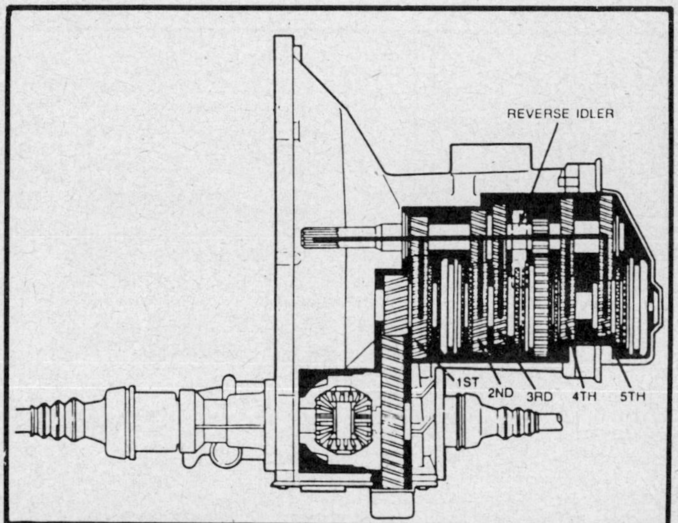
REVERSE IDLER
1ST
2ND
3RD
4TH
5TH

Neutral power flow

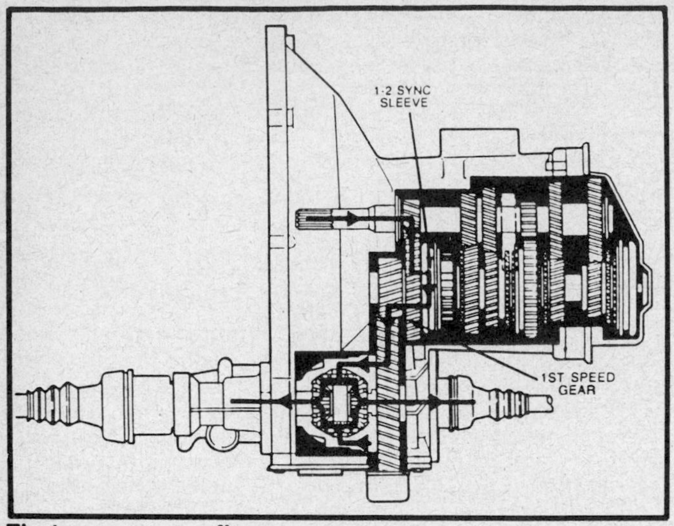
1-2 SYNC
SLEEVE
1ST SPEED
GEAR

First gear power flow

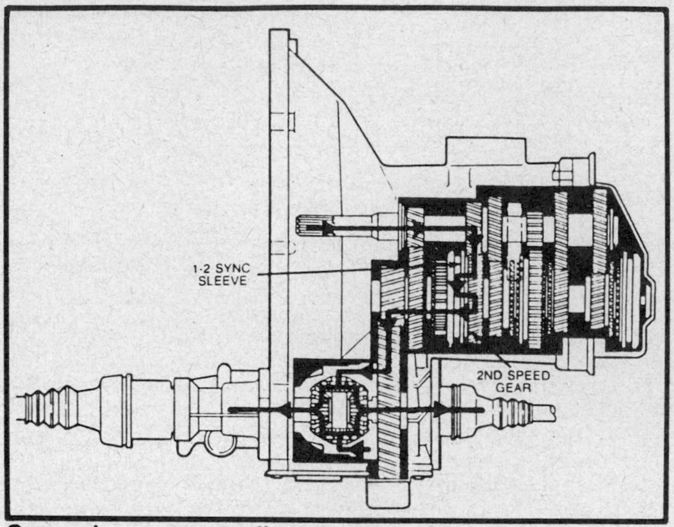
1-2 SYNC
SLEEVE
2ND SPEED
GEAR

Second gear power flow

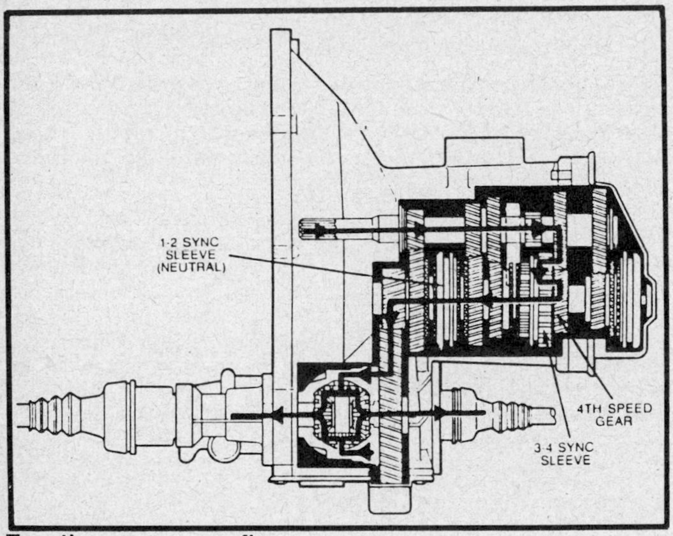
1-2 SYNC
SLEEVE
(NEUTRAL)
4TH SPEED
GEAR
3-4 SYNC
SLEEVE

Fourth gear power flow

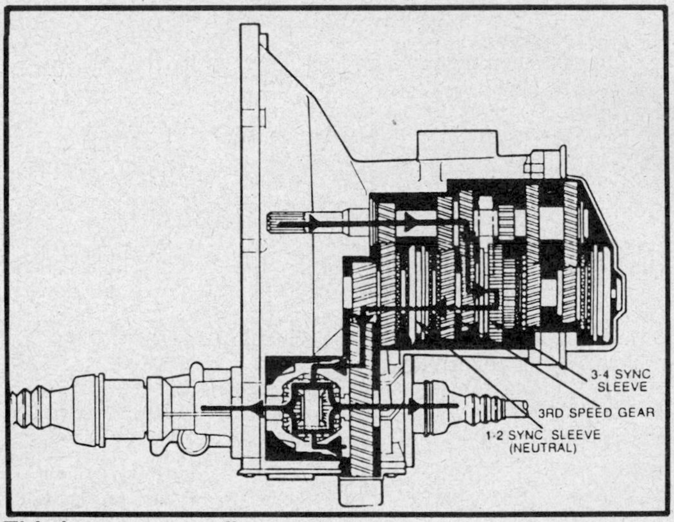
3-4 SYNC
SLEEVE
3RD SPEED GEAR
1-2 SYNC SLEEVE
(NEUTRAL)

Third gear power flow

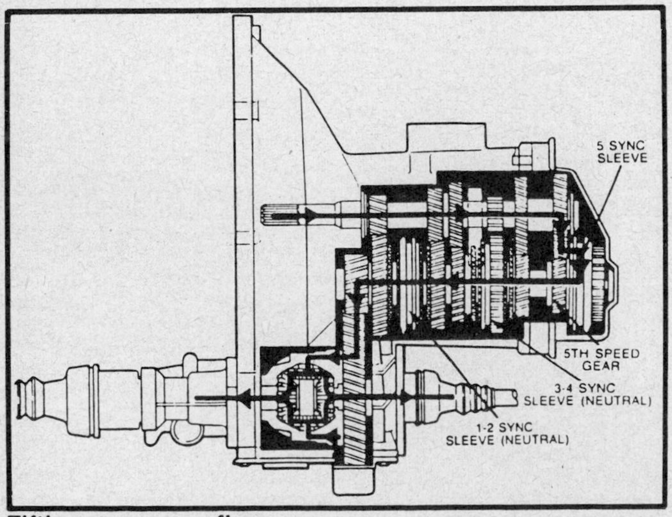
5 SYNC
SLEEVE
5TH SPEED
GEAR
3-4 SYNC
SLEEVE (NEUTRAL)
1-2 SYNC
SLEEVE (NEUTRAL)

Fifth gear power flow

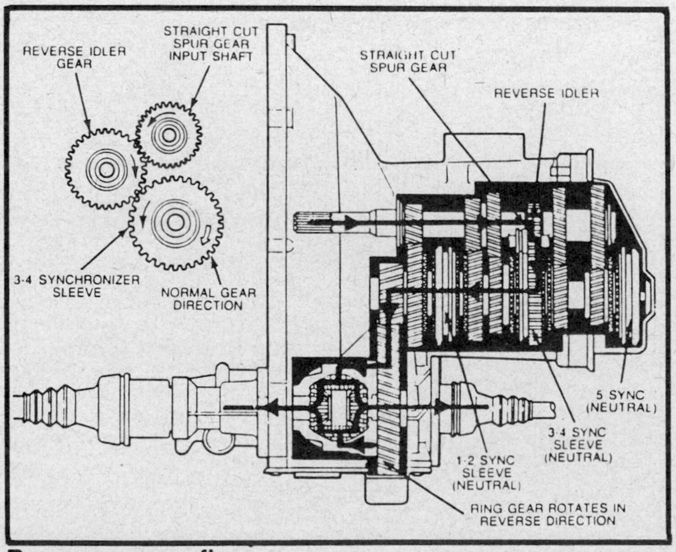
REVERSE IDLER
GEAR
STRAIGHT CUT
SPUR GEAR
INPUT SHAFT
3-4 SYNCHRONIZER
SLEEVE
NORMAL GEAR
DIRECTION
STRAIGHT CUT
SPUR GEAR
REVERSE IDLER
5 SYNC
(NEUTRAL)
3-4 SYNC
SLEEVE
(NEUTRAL)
1-2 SYNC
SLEEVE
(NEUTRAL)
RING GEAR ROTATES IN
REVERSE DIRECTION

Reverse power flow

ON CAR SERVICE

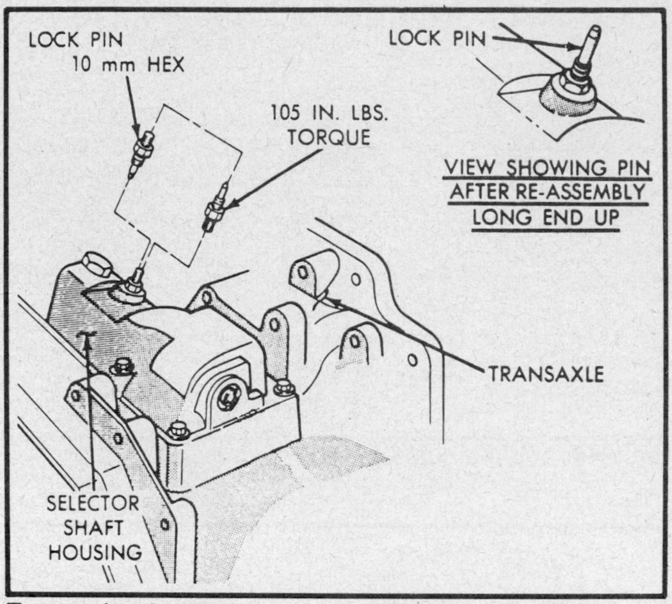
LOCK PIN
10 mm HEX
LOCK PIN
105 IN. LBS.
TORQUE
VIEW SHOWING PIN
AFTER RE-ASSEMBLY
LONG END UP
TRANSAXLE
SELECTOR
SHAFT
HOUSING

Transaxle pinned in the Neutral postion to adjust gearshift linkage

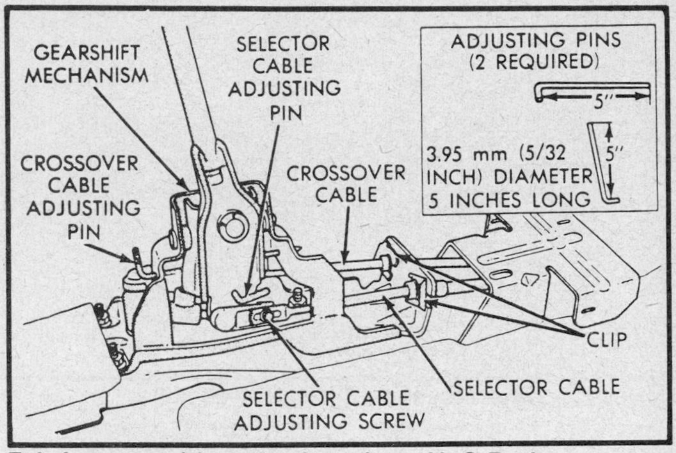
GEARSHIFT
MECHANISM
SELECTOR
CABLE
ADJUSTING
PIN
ADJUSTING PINS
(2 REQUIRED)
5″
5″
3.95 mm (5/32
INCH) DIAMETER
5 INCHES LONG
CROSSOVER
CABLE
ADJUSTING
PIN
CROSSOVER
CABLE
CLIP
SELECTOR CABLE
ADJUSTING SCREW
SELECTOR CABLE

Fabricate 2 cable adjusting pins—H, S-Body

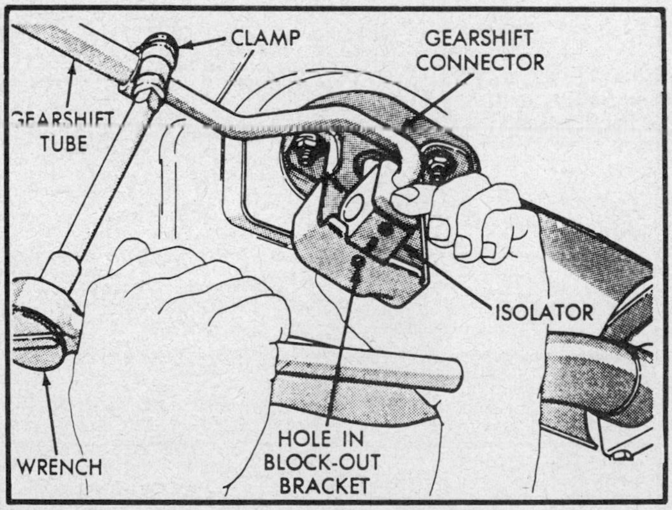
CLAMP
GEARSHIFT
CONNECTOR
GEARSHIFT
TUBE
ISOLATOR
WRENCH
HOLE IN
BLOCK-OUT
BRACKET

Gearshift linkage adjustment—L-Body

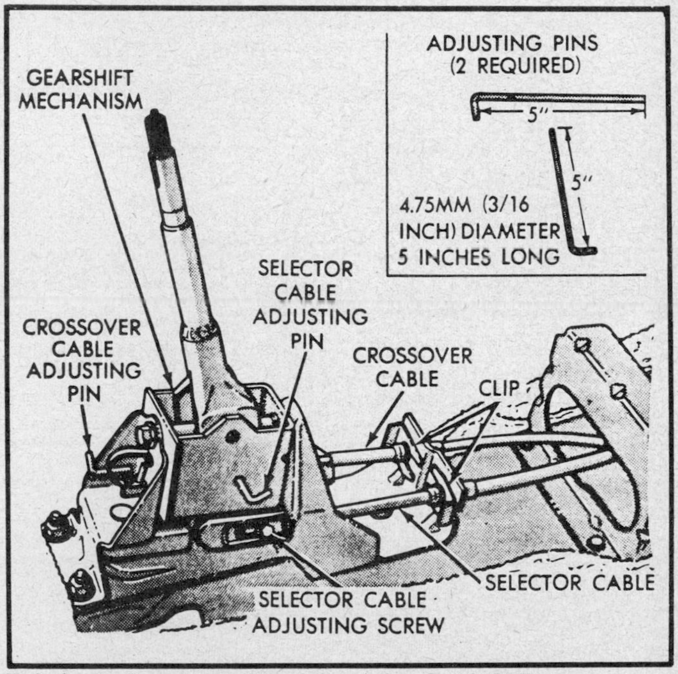
GEARSHIFT
MECHANISM
ADJUSTING PINS
(2 REQUIRED)
5″
5″
4.75MM (3/16
INCH) DIAMETER
5 INCHES LONG
SELECTOR
CABLE
ADJUSTING
PIN
CROSSOVER
CABLE
ADJUSTING
PIN
CROSSOVER
CABLE
CLIP
SELECTOR CABLE
SELECTOR CABLE
ADJUSTING SCREW

Fabricate 2 cable adjusting pins—E, K, P-Body

Adjustments

GEARSHIFT LINKAGE

ROD OPERATED

1. Working over the left front fender, remove the lock pin from the transaxle selector shaft housing.
2. Reverse the lock pin (so long end is down) and insert the lock pin into the same threaded hole while pushing the selector shaft into the selector housing. A hole in the selector shaft will align with the lock pin to be screwed into the housing. This locks the selector shaft in the 1–2 neutral position.
3. Raise the vehicle and support safely.
4. Loosen the clamp bolt that attaches the gearshift tube to the gearshift connector
5. Check to see that the gearshift connector slides and turns freely in the gear shift tube.
6. Position the shifter mechanism connector assembly so that the isolator is contacting the upstanding flange. Make sure the rib on the isolator is aligned fore and aft with the hole in the block out bracket. Hold the connector isolator in this position and tighten the clamp bolt on the gear shift tube to 170 inch lbs.

NOTE: Do not exert any force on the linkage while tightening the clamp bolt.

7. Lower the vehicle.
8. Remove the lock pin from the selector shaft housing and install the lock pin (long end up) in the selector shaft housing. Torque the lock pin to 105 inch lbs.
9. Check for proper shifting into 1st and reverse. Check for blockout into reverse.

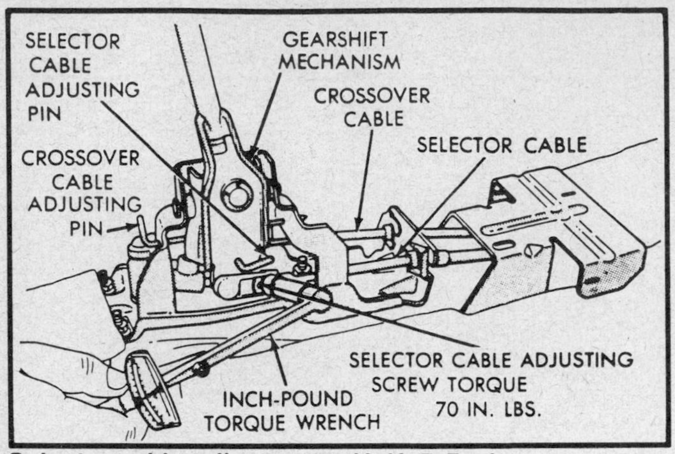
SELECTOR CABLE ADJUSTING PIN
GEARSHIFT MECHANISM
CROSSOVER CABLE
SELECTOR CABLE
CROSSOVER CABLE ADJUSTING PIN
SELECTOR CABLE ADJUSTING SCREW TORQUE 70 IN. LBS.
INCH-POUND TORQUE WRENCH
Selector cable adjustment—H, K, P-Body

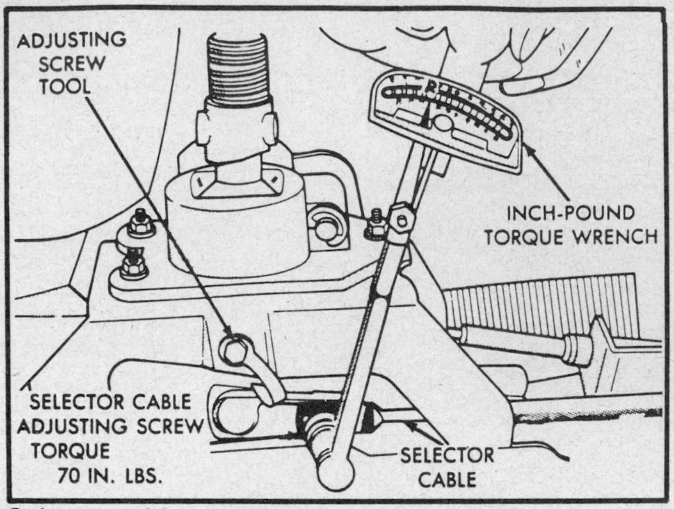
ADJUSTING SCREW TOOL
INCH-POUND TORQUE WRENCH
SELECTOR CABLE ADJUSTING SCREW TORQUE 70 IN. LBS.
SELECTOR CABLE
Selector cable adjustment—G, J, A-Body

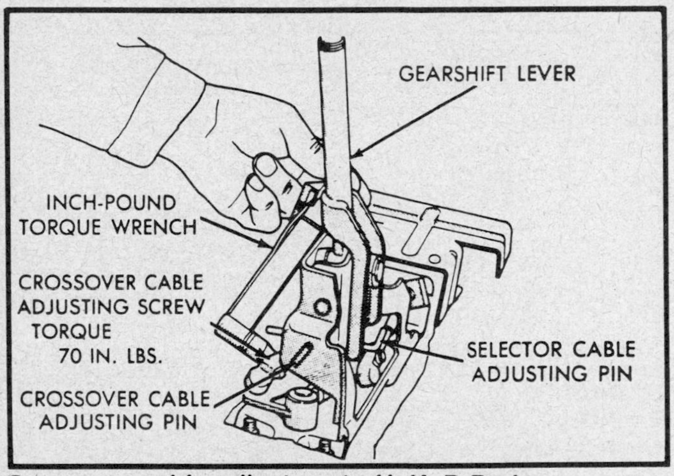
GEARSHIFT LEVER
INCH-POUND TORQUE WRENCH
CROSSOVER CABLE ADJUSTING SCREW TORQUE 70 IN. LBS.
SELECTOR CABLE ADJUSTING PIN
CROSSOVER CABLE ADJUSTING PIN
Crossover cable adjustment—H, K, P-Body

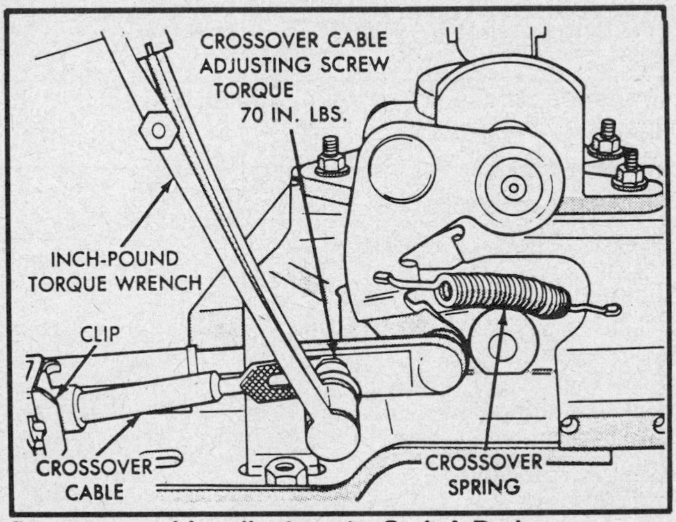
CROSSOVER CABLE ADJUSTING SCREW TORQUE 70 IN. LBS.
INCH-POUND TORQUE WRENCH
CLIP
CROSSOVER CABLE
CROSSOVER SPRING
Crossover cable adjustment—G, J, A-Body

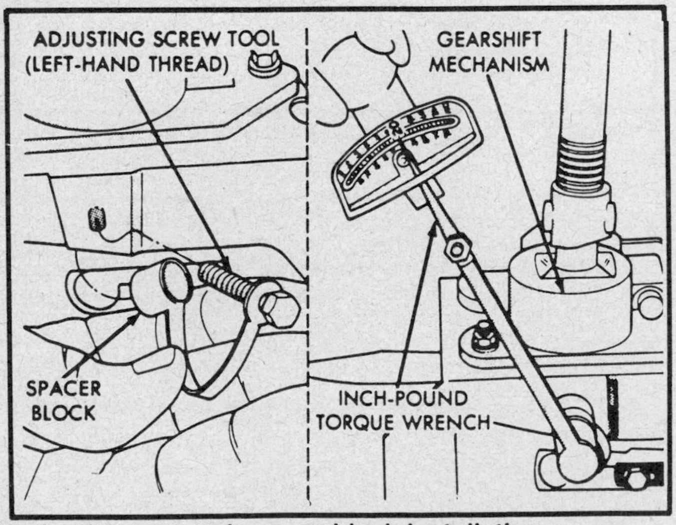
ADJUSTING SCREW TOOL (LEFT-HAND THREAD)
GEARSHIFT MECHANISM
SPACER BLOCK
INCH-POUND TORQUE WRENCH
Adjusting tool and spacer block installation

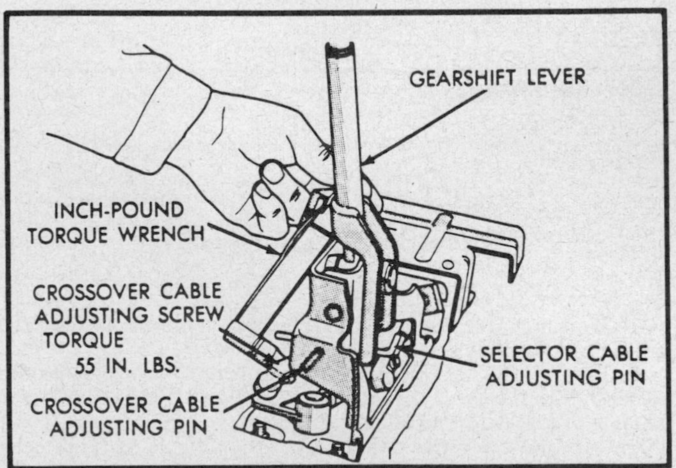
GEARSHIFT LEVER
INCH-POUND TORQUE WRENCH
CROSSOVER CABLE ADJUSTING SCREW TORQUE 55 IN. LBS.
SELECTOR CABLE ADJUSTING PIN
CROSSOVER CABLE ADJUSTING PIN
Crossover cable adjustment—D, S-Body

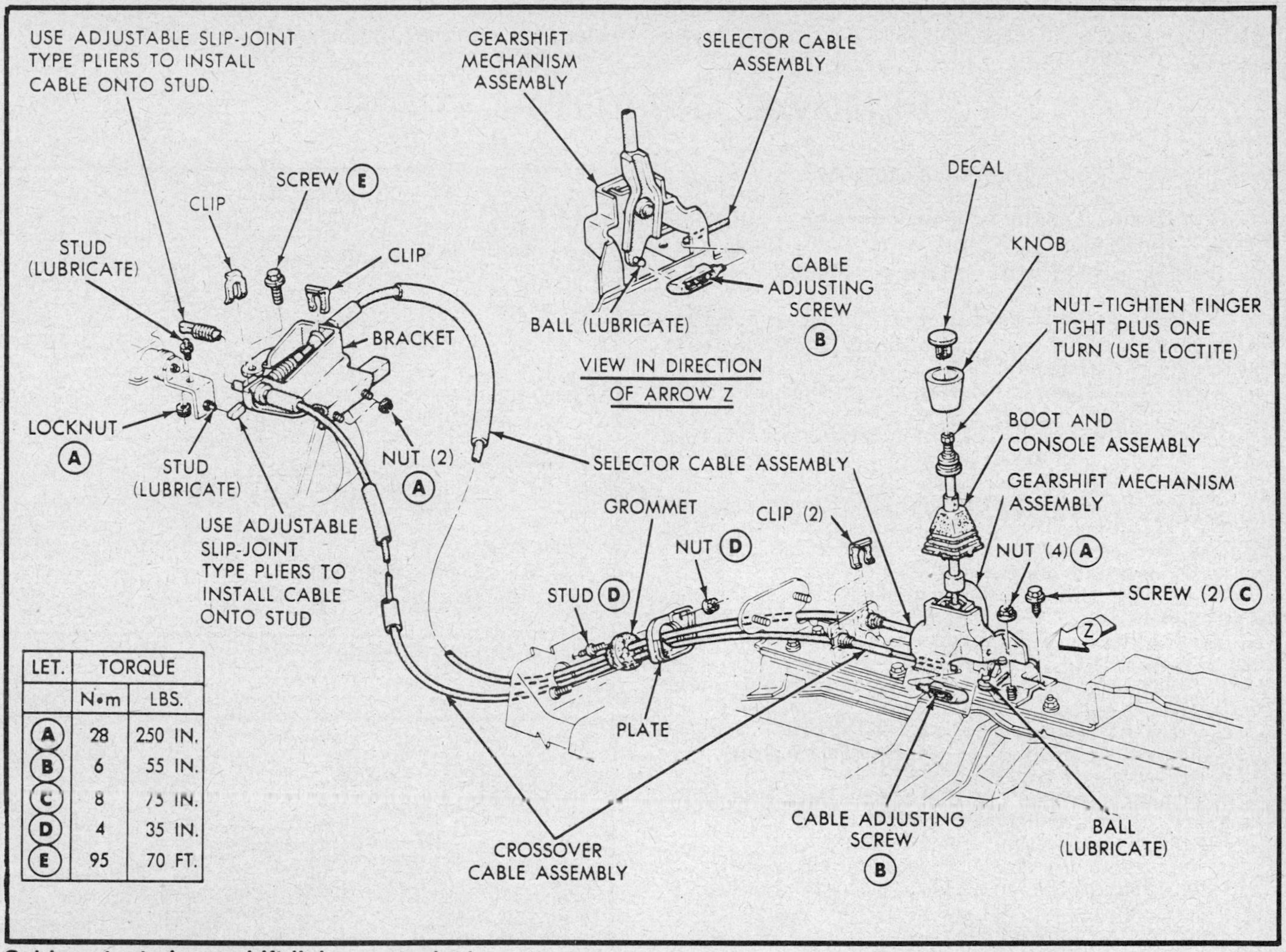
USE ADJUSTABLE SLIP-JOINT TYPE PLIERS TO INSTALL CABLE ONTO STUD.
GEARSHIFT MECHANISM ASSEMBLY
SELECTOR CABLE ASSEMBLY
DECAL
SCREW (E)
CLIP
STUD (LUBRICATE)
CLIP
KNOB
CABLE ADJUSTING SCREW (B)
BALL (LUBRICATE)
BRACKET
NUT-TIGHTEN FINGER TIGHT PLUS ONE TURN (USE LOCTITE)
VIEW IN DIRECTION OF ARROW Z
LOCKNUT (A)
BOOT AND CONSOLE ASSEMBLY
STUD (LUBRICATE)
NUT (2) (A)
SELECTOR CABLE ASSEMBLY
GEARSHIFT MECHANISM ASSEMBLY
GROMMET
CLIP (2)
USE ADJUSTABLE SLIP-JOINT TYPE PLIERS TO INSTALL CABLE ONTO STUD
NUT (D)
NUT (4) (A)
STUD (D)
SCREW (2) (C)
Z
PLATE
CABLE ADJUSTING SCREW (B)
BALL (LUBRICATE)
CROSSOVER CABLE ASSEMBLY

LET.	TORQUE	
	N•m	LBS.
(A)	28	250 IN.
(B)	6	55 IN.
(C)	8	75 IN.
(D)	4	35 IN.
(E)	95	70 FT.

Cable actuated gearshift linkage—typical

CABLE OPERATED

NOTE: Before replacing the gearshift selector or crossover cable for "hard shifting" complaint, disconnect both cables at the transaxle. From the driver's seat, manually operate the shifter lever though all ranges. If the gearshift lever moves freely, the cable(s) should not be replaced.

1. Working over the left front fender, remove the lock pin from the transaxle selector shaft housing.
2. Reverse the lock pin (so long end is down) and insert it into the same threaded hole while pushing the selector shaft into the selector housing. A hole in the selector shaft will align with the lock pin to be screwed into the housing. This locks the selector shaft in the 1–2 neutral position.
3. Unscrew the gearshift knob and remove the retaining nut.
4. Remove the console boot and console.
5. Loosen the cable adjusting screws before installing the adjusting pins or adjusting screw tool. The cable attachment clips must be installed from the inside. Install the cable fittings to the ball studs using pliers.

CAUTION

Torqueing the selector and crossover cable adjusting screw to the proper value is very important. Torque the adjusting screw tool to 20 inch lbs.

6. Remove the adjusting screw tool and fasten the shifter support bracket.
7. Install the console and boot.

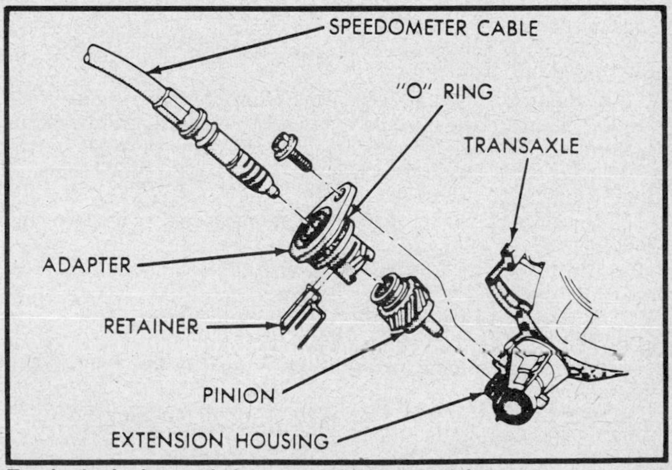
SPEEDOMETER CABLE
"O" RING
TRANSAXLE
ADAPTER
RETAINER
PINION
EXTENSION HOUSING

Exploded view of the speedometer drive

8. Remove the lock pin from the selector shaft housing and install it (long end up) in the selector shaft housing. Torque the lock pin to 105 inch lbs.

9. Install the retaining nut and gearshift knob.

10. Check for proper shifting into 1st and reverse. Check for blockout into reverse.

REMOVAL AND INSTALLATION

TRANSAXLE REMOVAL

NOTE: Removal of the transaxle does not require removal of the engine.

1. Disconnect the negative battery cable and drain the transaxle.
2. Position a shop crane and connect a lifting eye under the No. 4 cylinder exhaust manifold bolt or battery ground strap bolt to support the engine safely.
3. Disconnect the gearshift operating lever from the selector shaft.
4. Unbolt and disconnect the gearshift cable brackets from the transaxle.
5. Remove both front wheels.
6. Remove the left front splash shield and left engine mount.
7. On 1987–89 vehicles, disconnect the anti-rotation link (or anti-hop) damper from the crossmember bracket. Leave the bracket connected to the transaxle.
8. Support the bottom of the engine safely between the front fenders.
9. Remove the upper bolts that are accessible from above from the bell housing.
10. On 1984–85 vehicles, remove the driveshafts by performing the following procedure:
    a. Drain the differential and remove the cover.
    b. Remove the speedometer adapter, cable and gear.
    c. Remove the sway bar.
    d. Remove both lower ball joint-to-steering knuckle bolts.
    e. Pry the lower ball joint from the steering knuckle.
    f. Remove the driveshaft from the hub.
    g. Rotate both driveshafts to expose the circlip ends. Note the flat surface on the inner ends of both axle tripod shafts. Pry the circlip out.
    h. Remove both driveshafts.
9. On 1986–89 vehicles, remove the driveshafts by performing the following:
    a. Loosen the hub nut.
    b. Drain the transaxle differential and remove the cover.
    c. To remove the right hand driveshaft, disconnect the speedometer cable and remove the cable and gear before removing the driveshaft.
    d. Rotate the driveshaft to expose the circlip tangs.
    e. Compress the circlip and push the shaft into the side gear cavity.
    f. Remove the clamp bolt from the ball stud and steering knuckle.
    g. Separate the ball joint stud from the steering knuckle.
    h. Separate the outer CV-joint splined shaft from the hub. Do not pry on the slinger or outer CV-joint.
    i. Support the shaft at the CV-joints and remove them. Do not pull on the shaft.
11. Disconnect the plug for the neutral safety/backup light switch.
12. Remove the engine mount bracket from the front crossmember.
13. Support the transmission from underneath.
14. Remove the front mount insulator through-bolts.
15. Remove the long through-bolt from the left hand engine mount.
16. Remove the starter. Remove any bell housing bolts that are still in position.
17. Slide the transaxle directly away from the engine so the transmission input shaft will slide smoothly out of the bearing in the flywheel and the clutch disc. Lower the transaxle and remove it from the engine compartment.
18. Position the transaxle on a suitable holding fixture.

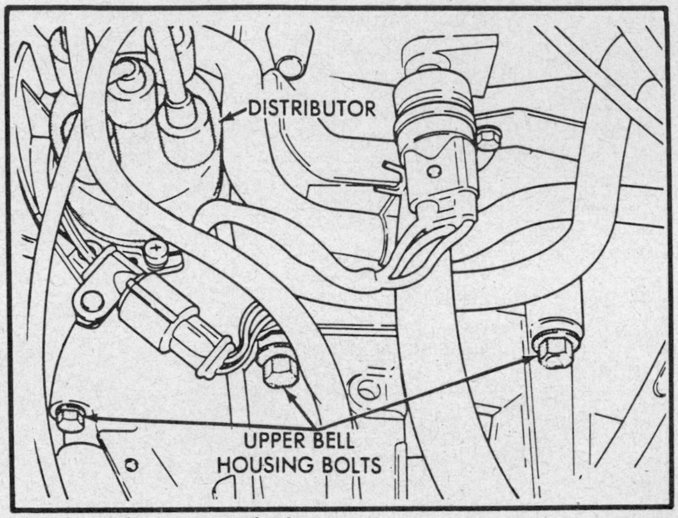
DISTRIBUTOR
UPPER BELL
HOUSING BOLTS
Bell housing upper bolts

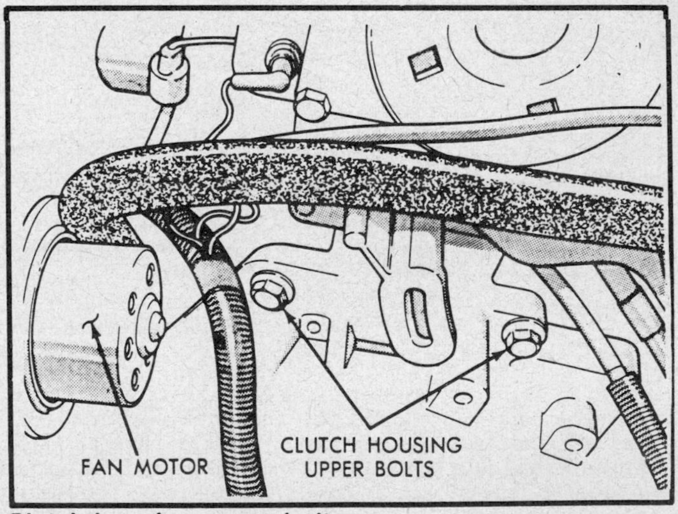
FAN MOTOR
CLUTCH HOUSING
UPPER BOLTS
Clutch housing upper bolts

TRANSAXLE INSTALLATION

1. Remove the transaxle from the holding fixture.
2. Before installing the transmission onto the engine block, install 2 locating pins into the top 2 engine block holes. Fabricate the pins from old or new bell housing bolts if needed.
3. Support the transaxle safely and raise it into precise alignment with the engine block. Then, move it toward the block, guiding the transmission input shaft into the clutch disc. Turn the input shaft slightly, if necessary, to allow the splines to align and engage.

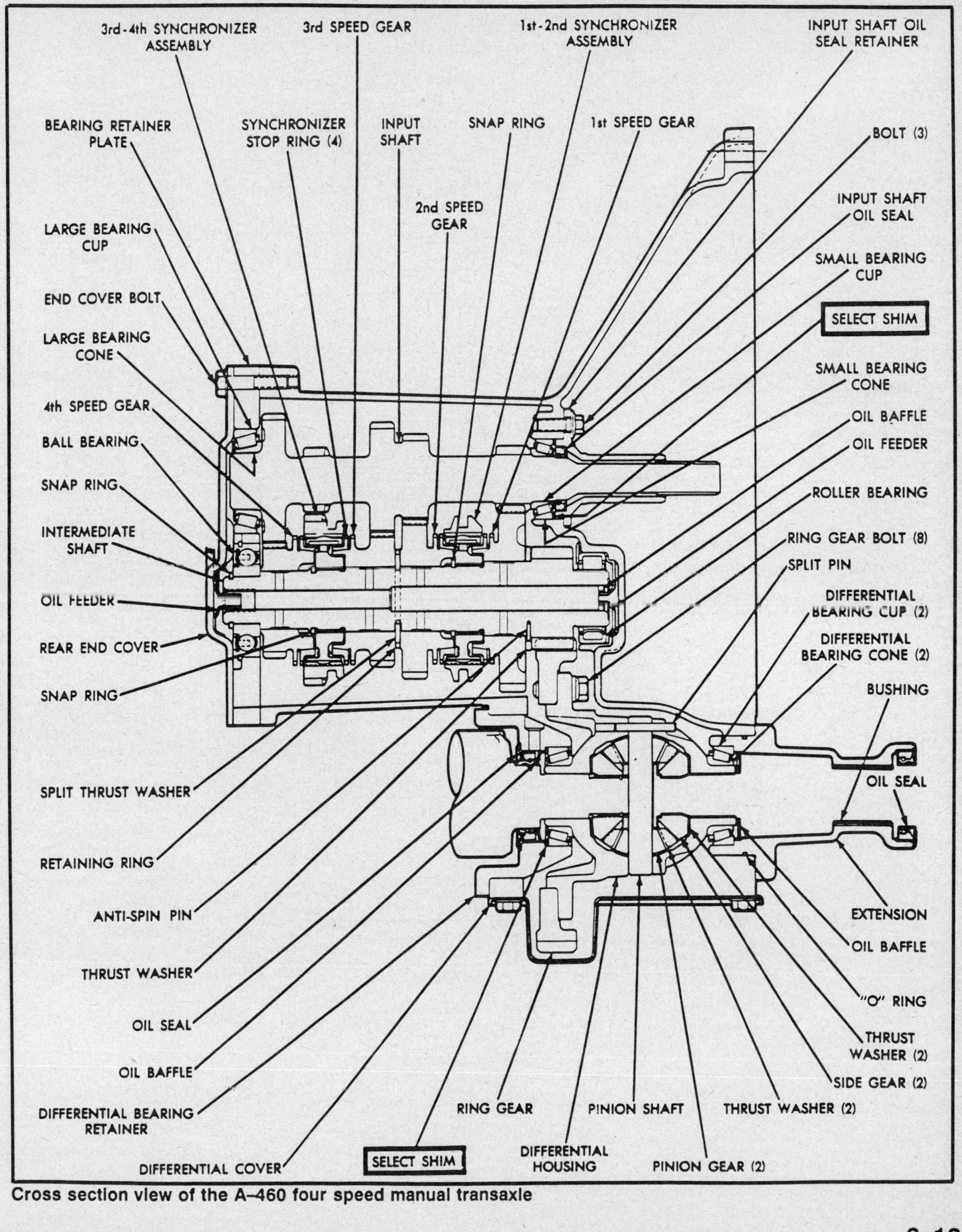
3rd-4th SYNCHRONIZER ASSEMBLY
3rd SPEED GEAR
1st-2nd SYNCHRONIZER ASSEMBLY
INPUT SHAFT OIL SEAL RETAINER
BEARING RETAINER PLATE
SYNCHRONIZER STOP RING (4)
INPUT SHAFT
SNAP RING
1st SPEED GEAR
BOLT (3)
2nd SPEED GEAR
INPUT SHAFT OIL SEAL
LARGE BEARING CUP
SMALL BEARING CUP
END COVER BOLT
SELECT SHIM
LARGE BEARING CONE
SMALL BEARING CONE
4th SPEED GEAR
OIL BAFFLE
BALL BEARING
OIL FEEDER
SNAP RING
ROLLER BEARING
INTERMEDIATE SHAFT
RING GEAR BOLT (8)
SPLIT PIN
OIL FEEDER
DIFFERENTIAL BEARING CUP (2)
REAR END COVER
DIFFERENTIAL BEARING CONE (2)
SNAP RING
BUSHING
SPLIT THRUST WASHER
OIL SEAL
RETAINING RING
ANTI-SPIN PIN
EXTENSION
OIL BAFFLE
THRUST WASHER
"O" RING
OIL SEAL
THRUST WASHER (2)
OIL BAFFLE
SIDE GEAR (2)
DIFFERENTIAL BEARING RETAINER
RING GEAR
PINION SHAFT
THRUST WASHER (2)
DIFFERENTIAL HOUSING
DIFFERENTIAL COVER
SELECT SHIM
PINION GEAR (2)

Cross section view of the A-460 four speed manual transaxle

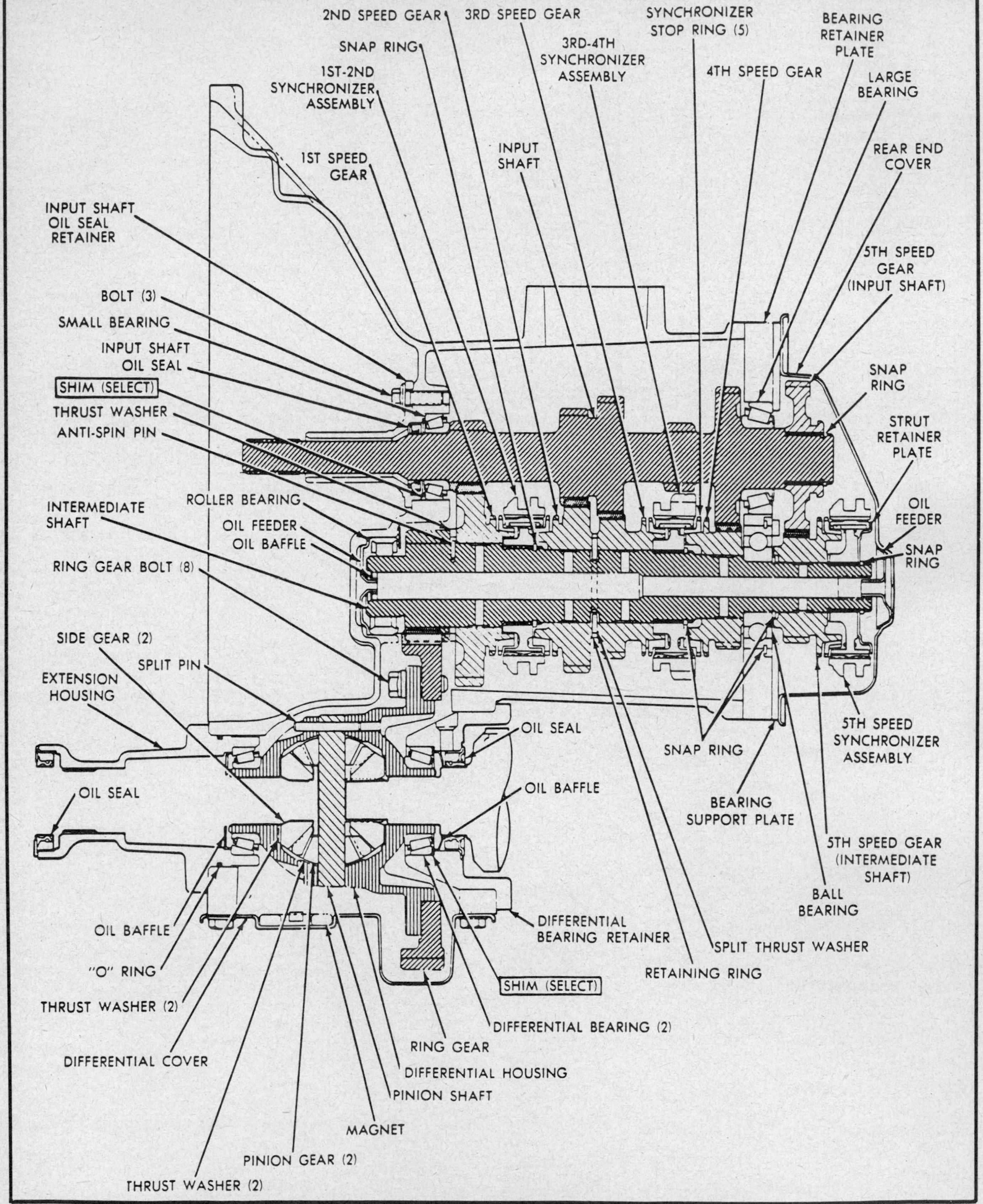
2ND SPEED GEAR
3RD SPEED GEAR
SYNCHRONIZER STOP RING (5)
BEARING RETAINER PLATE
SNAP RING
3RD-4TH SYNCHRONIZER ASSEMBLY
1ST-2ND SYNCHRONIZER ASSEMBLY
4TH SPEED GEAR
LARGE BEARING
1ST SPEED GEAR
INPUT SHAFT
REAR END COVER
INPUT SHAFT OIL SEAL RETAINER
5TH SPEED GEAR (INPUT SHAFT)
BOLT (3)
SMALL BEARING
INPUT SHAFT OIL SEAL
SNAP RING
SHIM (SELECT)
THRUST WASHER
STRUT RETAINER PLATE
ANTI-SPIN PIN
ROLLER BEARING
INTERMEDIATE SHAFT
OIL FEEDER
OIL FEEDER
OIL BAFFLE
SNAP RING
RING GEAR BOLT (8)
SIDE GEAR (2)
SPLIT PIN
EXTENSION HOUSING
OIL SEAL
5TH SPEED SYNCHRONIZER ASSEMBLY
SNAP RING
OIL SEAL
OIL BAFFLE
BEARING SUPPORT PLATE
5TH SPEED GEAR (INTERMEDIATE SHAFT)
BALL BEARING
OIL BAFFLE
DIFFERENTIAL BEARING RETAINER
SPLIT THRUST WASHER
"O" RING
RETAINING RING
SHIM (SELECT)
THRUST WASHER (2)
DIFFERENTIAL BEARING (2)
RING GEAR
DIFFERENTIAL COVER
DIFFERENTIAL HOUSING
PINION SHAFT
MAGNET
PINION GEAR (2)
THRUST WASHER (2)

Cross section view of the A–465, A–525, A-520, A–555 five speed manual transaxle

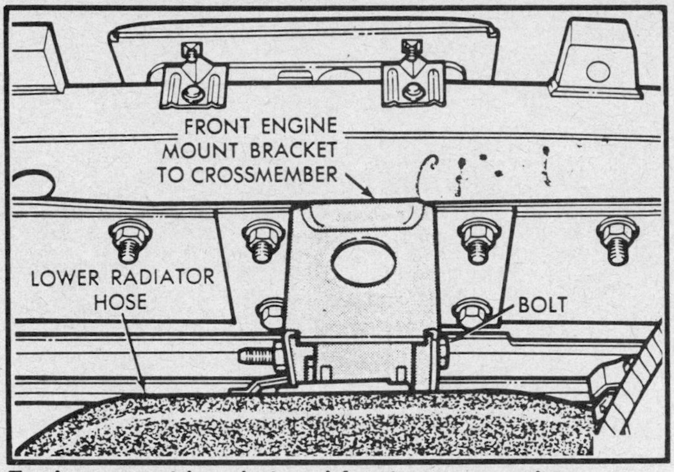
FRONT ENGINE
MOUNT BRACKET
TO CROSSMEMBER
LOWER RADIATOR
HOSE
BOLT
Engine mount bracket and front crossmember

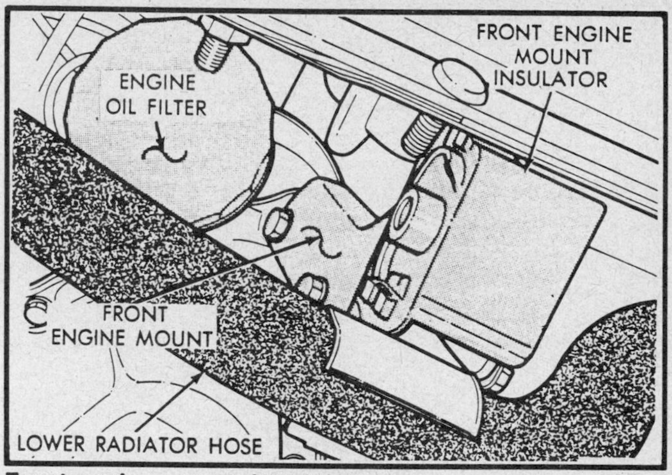
ENGINE
OIL FILTER
FRONT ENGINE
MOUNT
INSULATOR
FRONT
ENGINE MOUNT
LOWER RADIATOR HOSE
Front engine mount insulator

4. Install the lower bell housing bolts and the starter. Bell housing bolts are torqued to 70 ft. lbs.
5. Install the long through-bolt into the left hand engine mount.
6. Install the front mount insulator through-bolts.
7. Remove transaxle support and install the engine mount bracket onto the front crossmember.
8. Reconnect the electrical connector for the backup light/neutral safety switch.
9. Install the driveshafts by performing the following procedure:
   a. Thoroughly clean and lubricate the seal located between the outer end of the driveshaft and the steering knuckle.
   b. Inspect units on turbocharged cars to make sure the rubber washer seal is in place on the right inner joint. Relocate the seal, if necessary.
   c. Support the driveshaft by both CV-joints. Hold the inner joint assembly at the housing. Align the splined joint with the splines in the differential side gear, guide it into the housing, and insert it until it locks.
   d. Push the hub and knuckle assembly outward and insert the outer splined CV-joint shaft into the hub.
   e. Insert the ball joint stud into the knuckle assembly. Install the clamp bolt and nut and torque to 70 ft. lbs. If replacing the the clamp bolt, it is a prevailing torque type and must be replaced with an equivalent part.
   f. If removed, insert the speedometer pinion back into the transaxle. The bolt hole in the retaining collar should be aligned with the threaded hole in the transaxle. Install the retaining bolt.
10. Fill the transaxle.
11. Install the front wheels and lower the vehicle.
12. Install the washer and hub nut. Torque the hub nut to 180 ft. lbs. Install the lock fingertight and install a new cotter pin.
13. Install the left side splash shield.
14. Install the remaining bell housing bolts and torque them to 70 ft. lbs.
15. Remove the engine support fixture.
16. Connect the anti-rotational link (or anti-hop damper) to the crossmember bracket.
17. Connect the gearshift bracket to the transaxle.
18. Connect the shift linkage using a new self-locking nut. On 1984–85 vehicles torque the nut to 21 ft. lbs and on 1986–89 vehicles 27 ft. lbs.
19. Remove the lifting eye.
20. Connect the negative battery cable.

BENCH OVERHAUL

Transaxle Disassembly

A–460 TRANSAXLE

1. Remove the transaxle from the vehicle and position on suitable holding fixture.
2. Remove the differential cover bolts and the stud nuts, then remove the cover.
3. Remove the differential bearing retainer bolts.
4. Using spanner tool L–4435 or equivalent, rotate the differential bearing retainer to remove it.
5. Remove the extension housing bolts. Remove the differential assembly and extension housing.
6. Unbolt and remove the selector shaft housing.
7. Remove the stud nuts and the bolts from the rear end cover then pry off the rear end cover.
8. Remove the large snapring from the intermediate shaft rear ball bearing.
9. Remove the bearing retainer plate by tapping it with a plastic hammer.
10. Remove the 3rd/4th shift fork rail.
11. Remove the reverse idler gear shaft and gear.
12. Remove the input shaft gear assembly and the intermediate shaft gear assembly.
13. To remove the clutch release bearing, remove the E-clips from the clutch release shaft, then disassemble the clutch shaft components.
14. Remove the input shaft seal retainer bolts, the seal, the retainer assembly and the select shim.
15. Remove the reverse shift lever E-clip and flat washer and disassemble the reverse shift lever components.

A–465 AND A–525 TRANSAXLES

1. Remove the transaxle from the vehicle and position on suitable holding fixture.
2. Remove the extension housing bolts then remove the differential assembly and extension housing.
3. Remove the differential cover bolts, stud nuts and remove the cover.

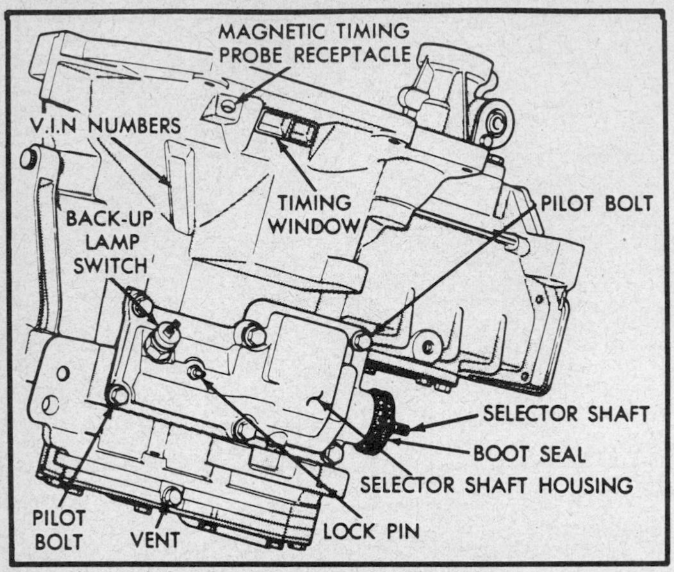
MAGNETIC TIMING
PROBE RECEPTACLE
V.I.N NUMBERS
TIMING
WINDOW
PILOT BOLT
BACK-UP
LAMP
SWITCH
SELECTOR SHAFT
BOOT SEAL
SELECTOR SHAFT HOUSING
PILOT
BOLT
VENT
LOCK PIN

Top view of the A–460 manual transaxle

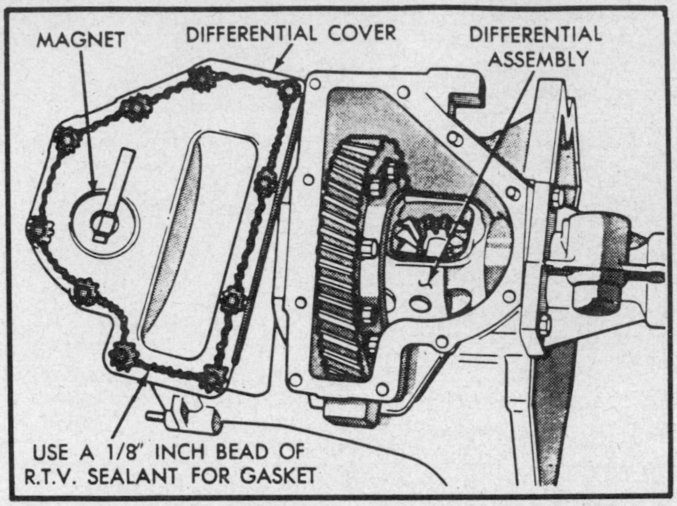
MAGNET
DIFFERENTIAL COVER
DIFFERENTIAL
ASSEMBLY
USE A 1/8" INCH BEAD OF
R.T.V. SEALANT FOR GASKET

Differential cover removal and installation—A-460, A-465, A–525 transaxles

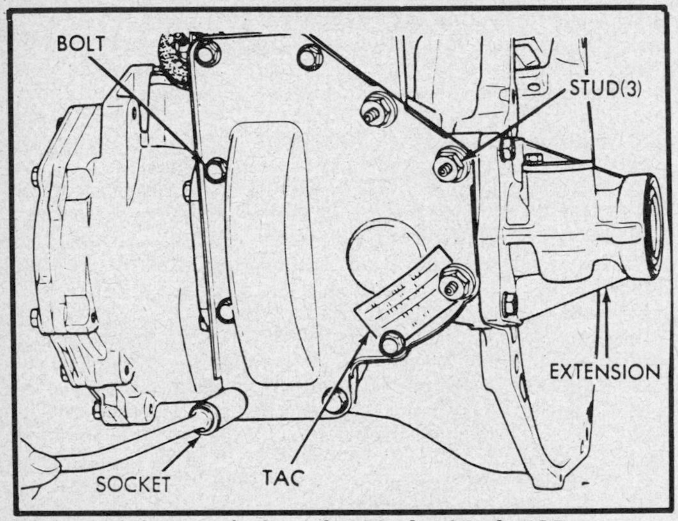
BOLT
STUD(3)
EXTENSION
SOCKET
TAC

Differential cover bolts—A-460, A-465, A–525 transaxles

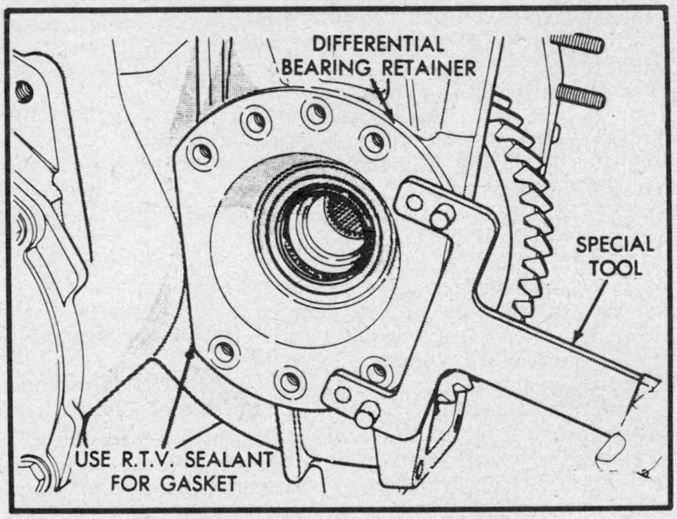
DIFFERENTIAL
BEARING RETAINER
SPECIAL
TOOL
USE R.T.V. SEALANT
FOR GASKET

Differential bearing retainer removal and installation—A-460, A-465, A–525 transaxles

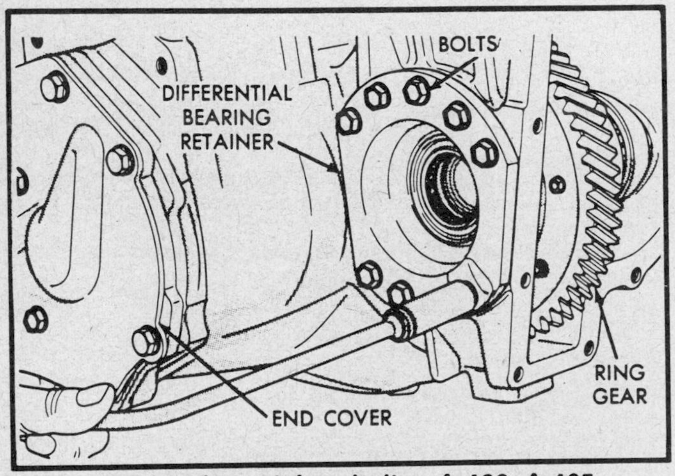
BOLTS
DIFFERENTIAL
BEARING
RETAINER
RING
GEAR
END COVER

Differential bearing retainer bolts—A-460, A-465, A–525 transaxles

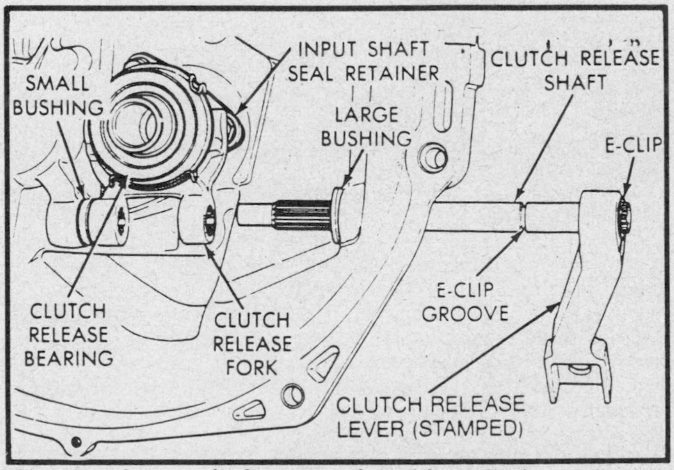
INPUT SHAFT
SEAL RETAINER
CLUTCH RELEASE
SHAFT
SMALL
BUSHING
LARGE
BUSHING
E-CLIP
E-CLIP
GROOVE
CLUTCH
RELEASE
BEARING
CLUTCH
RELEASE
FORK
CLUTCH RELEASE
LEVER (STAMPED)

Clutch release shaft removal and installation—all transaxles

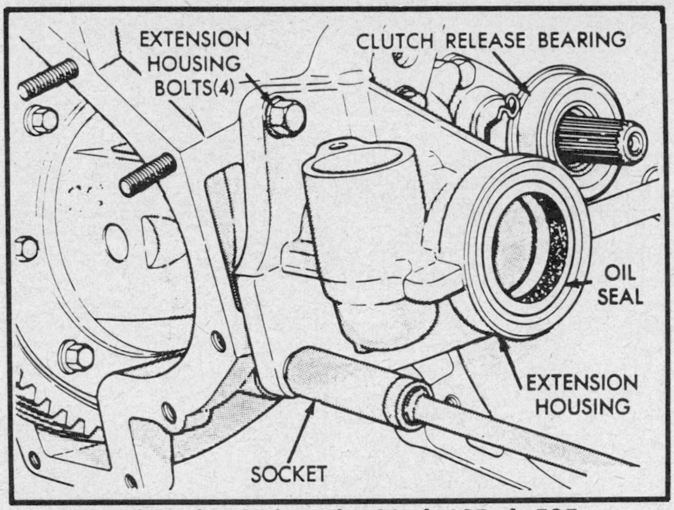
EXTENSION HOUSING BOLTS(4)
CLUTCH RELEASE BEARING
OIL SEAL
EXTENSION HOUSING
SOCKET

Extension housing bolts—A-460, A-465, A–525 transaxles

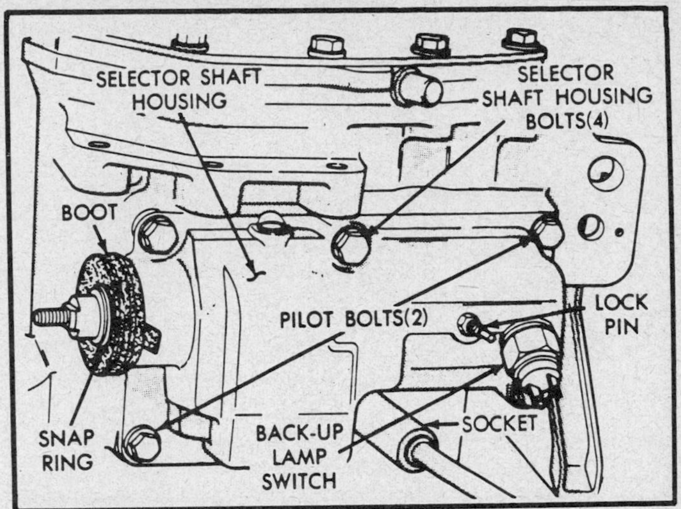
SELECTOR SHAFT HOUSING
SELECTOR SHAFT HOUSING BOLTS(4)
BOOT
PILOT BOLTS(2)
LOCK PIN
SNAP RING
BACK-UP LAMP SWITCH
SOCKET

Selector shaft housing removal and installation—A-460, A-465, A–525 transaxles

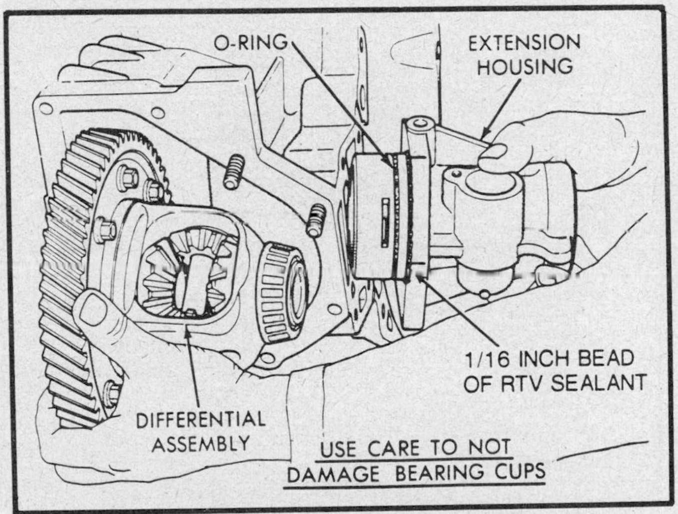
O-RING
EXTENSION HOUSING
1/16 INCH BEAD OF RTV SEALANT
DIFFERENTIAL ASSEMBLY
USE CARE TO NOT DAMAGE BEARING CUPS

Differential/extension housing removal and installation—A-460, A-465, A–525 transaxles

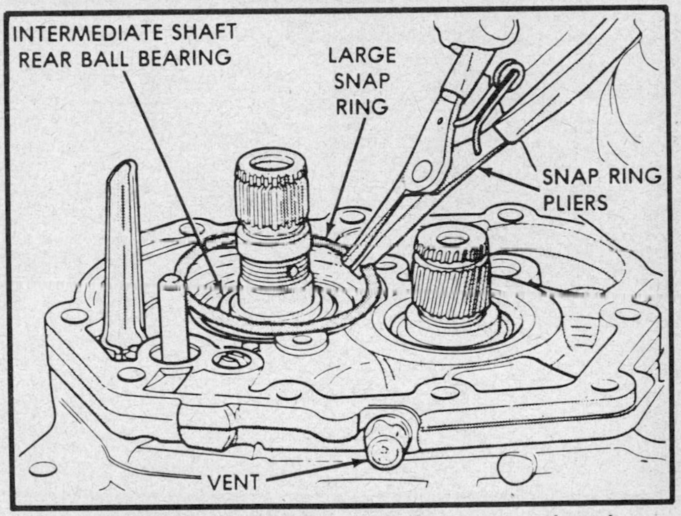
INTERMEDIATE SHAFT REAR BALL BEARING
LARGE SNAP RING
SNAP RING PLIERS
VENT

Intermediate shaft bearing snapring removal and installation—A–460 transaxles

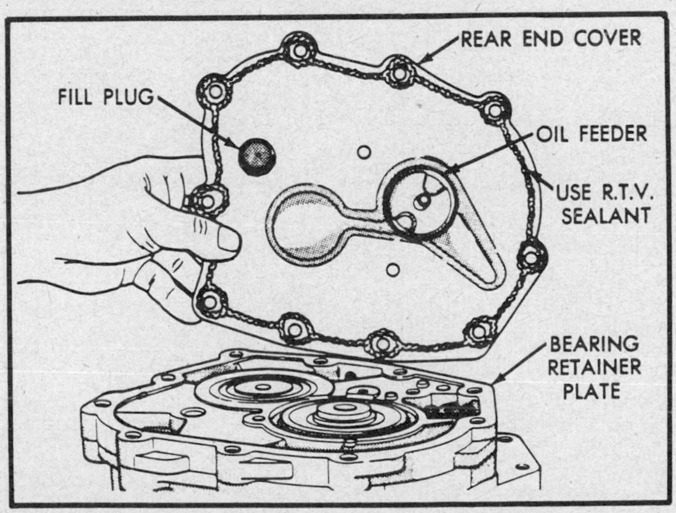
REAR END COVER
FILL PLUG
OIL FEEDER
USE R.T.V. SEALANT
BEARING RETAINER PLATE

End cover removal and installation—A–460 transaxles

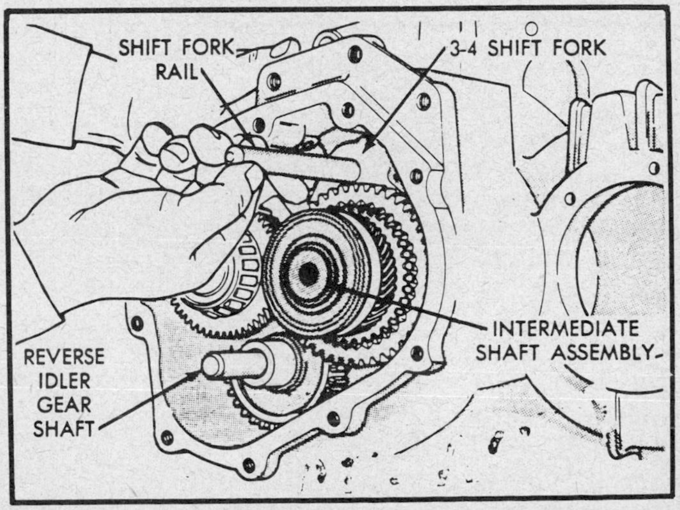
SHIFT FORK RAIL
3-4 SHIFT FORK
INTERMEDIATE SHAFT ASSEMBLY
REVERSE IDLER GEAR SHAFT

Shift fork rail removal and installation—A–460 transaxles

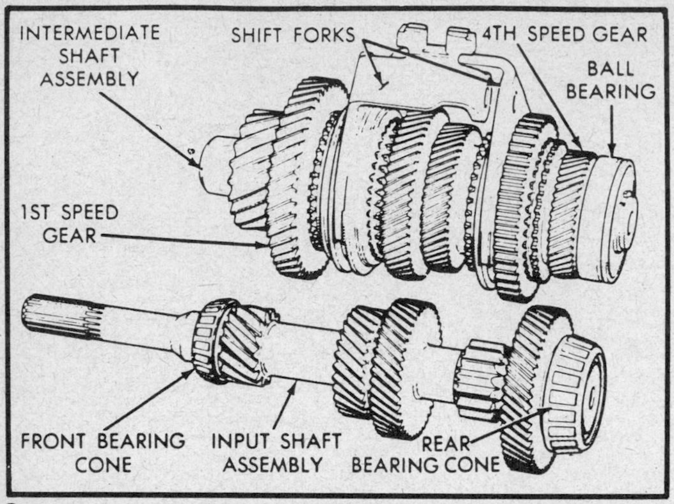
INTERMEDIATE SHAFT ASSEMBLY
SHIFT FORKS
4TH SPEED GEAR
BALL BEARING
1ST SPEED GEAR
FRONT BEARING CONE
INPUT SHAFT ASSEMBLY
REAR BEARING CONE
Gear set removal and installation—A-460 transaxles

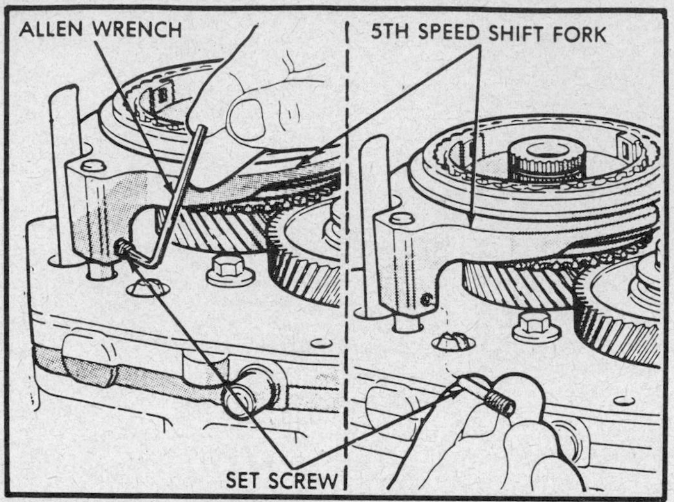
ALLEN WRENCH
5TH SPEED SHIFT FORK
SET SCREW
5th fork set screw removal and installation—A-465, A-525, A-520, A-555 transaxles

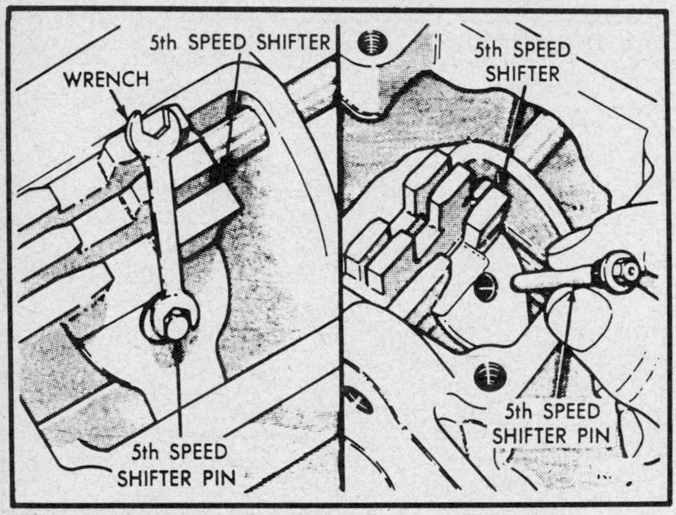
5th SPEED SHIFTER
WRENCH
5th SPEED SHIFTER
5th SPEED SHIFTER PIN
5th SPEED SHIFTER PIN
5th speed shifter pin removal and installation—A-465, A-525 transaxles

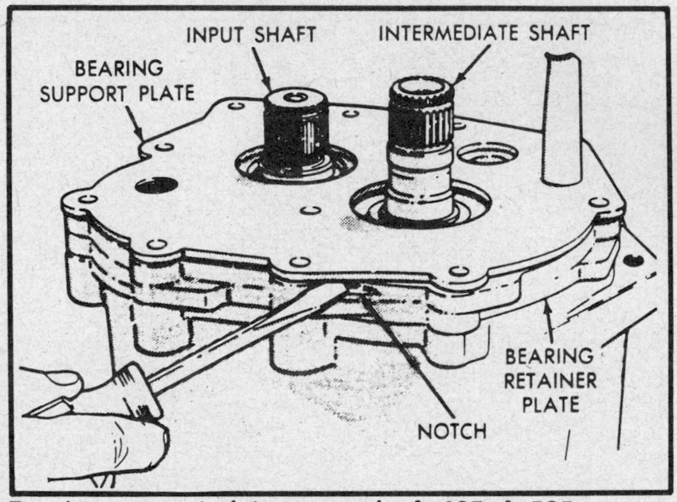
INPUT SHAFT
INTERMEDIATE SHAFT
BEARING SUPPORT PLATE
BEARING RETAINER PLATE
NOTCH
Bearing support plate removal—A-465, A-525, A-520, A-555 transaxles

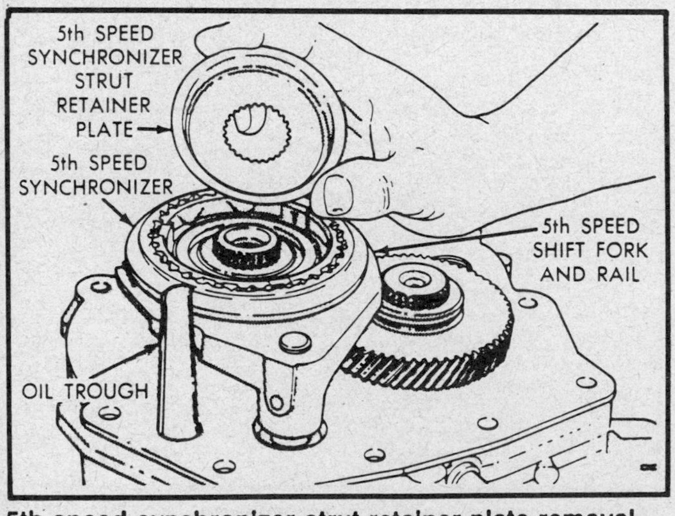
5th SPEED SYNCHRONIZER STRUT RETAINER PLATE
5th SPEED SYNCHRONIZER
5th SPEED SHIFT FORK AND RAIL
OIL TROUGH
5th speed synchronizer strut retainer plate removal and installation—A-465, A-525 transaxles

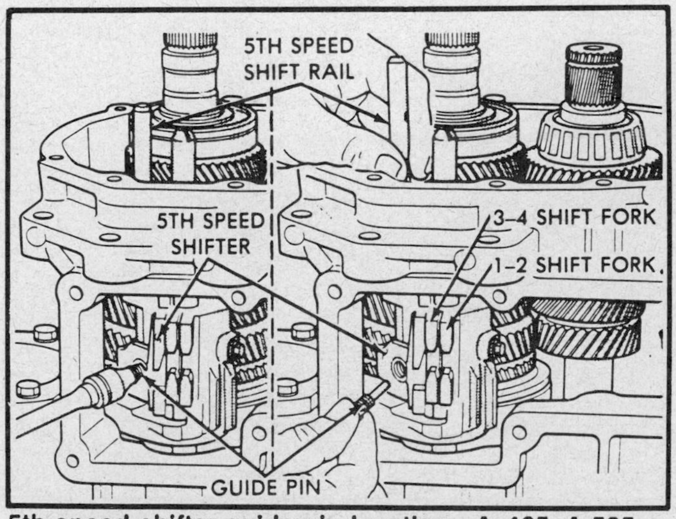
5TH SPEED SHIFT RAIL
5TH SPEED SHIFTER
3-4 SHIFT FORK
1-2 SHIFT FORK
GUIDE PIN
5th speed shifter guide pin location—A-465, A-525, A-520, A-555 transaxles

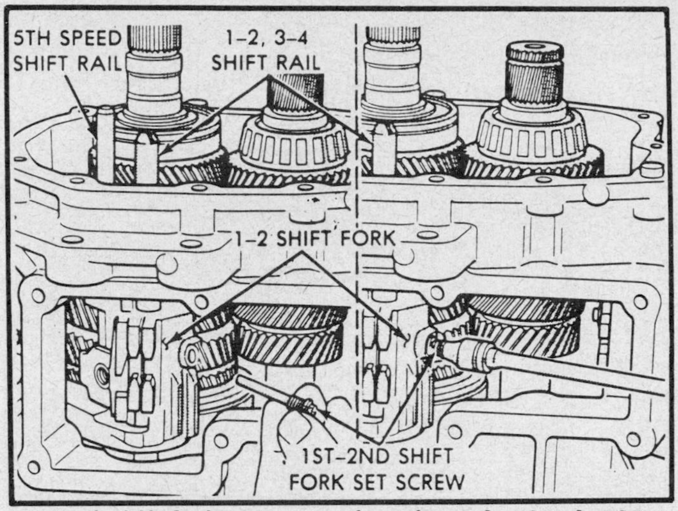
5TH SPEED
SHIFT RAIL
1–2, 3–4
SHIFT RAIL
1–2 SHIFT FORK
1ST–2ND SHIFT
FORK SET SCREW

1st-2nd shift fork set screw location—A-465, A-525, A-520, A-555 transaxles

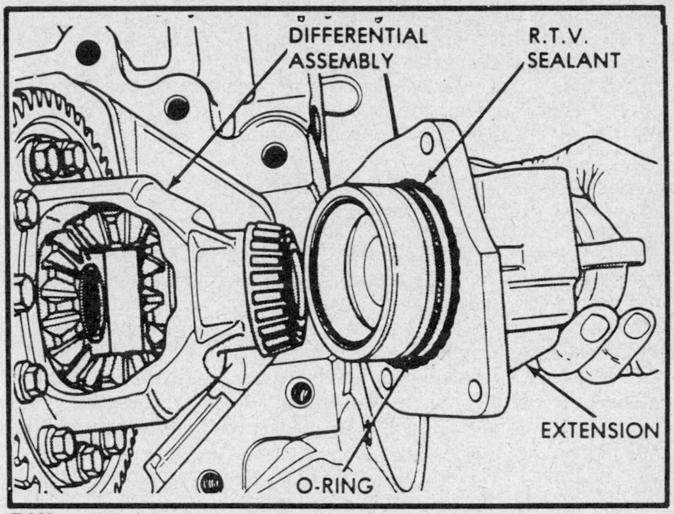
DIFFERENTIAL
ASSEMBLY
R.T.V.
SEALANT
EXTENSION
O-RING

Differential extension removal and installation—A-525, A-555 transaxles

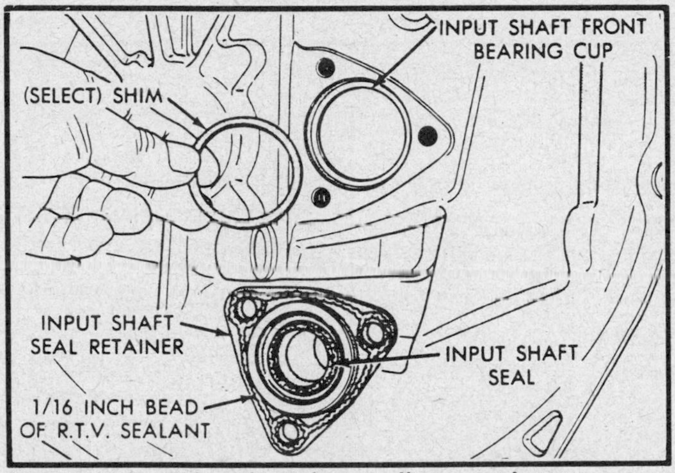
INPUT SHAFT FRONT
BEARING CUP
(SELECT) SHIM
INPUT SHAFT
SEAL RETAINER
INPUT SHAFT
SEAL
1/16 INCH BEAD
OF R.T.V. SEALANT

Input shaft seal and retainer—all transaxles

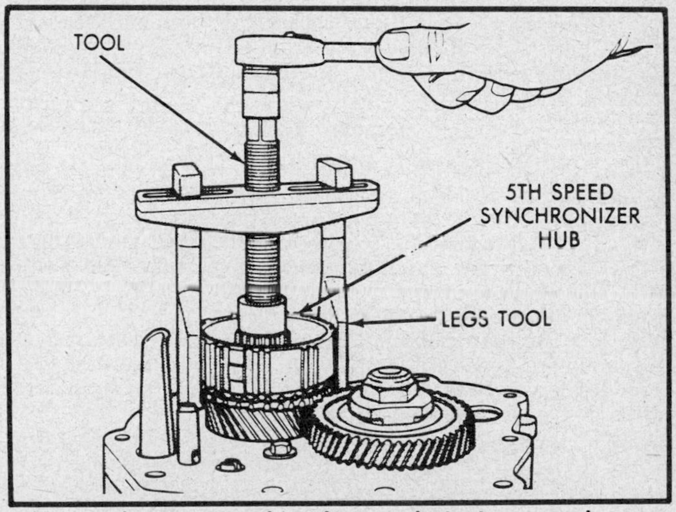
TOOL
5TH SPEED
SYNCHRONIZER
HUB
LEGS TOOL

5th speed gear, synchronizer and strut removal—A-465, A-525, A-520, A-555 transaxles

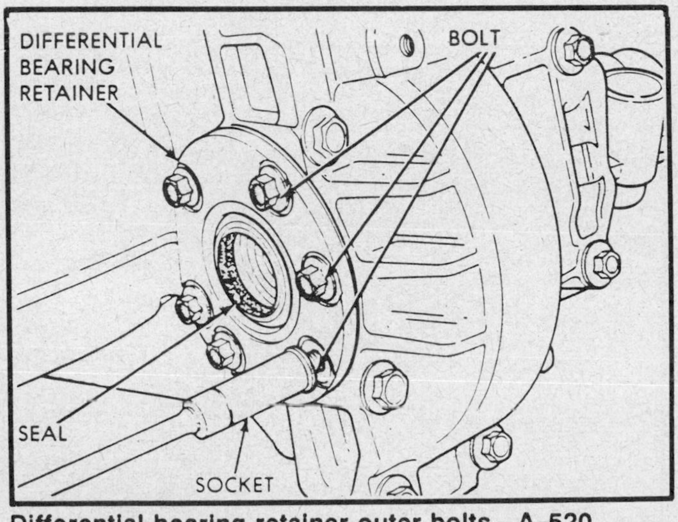
DIFFERENTIAL
BEARING
RETAINER
BOLT
SEAL
SOCKET

Differential bearing retainer outer bolts—A-520, A-555 transaxles

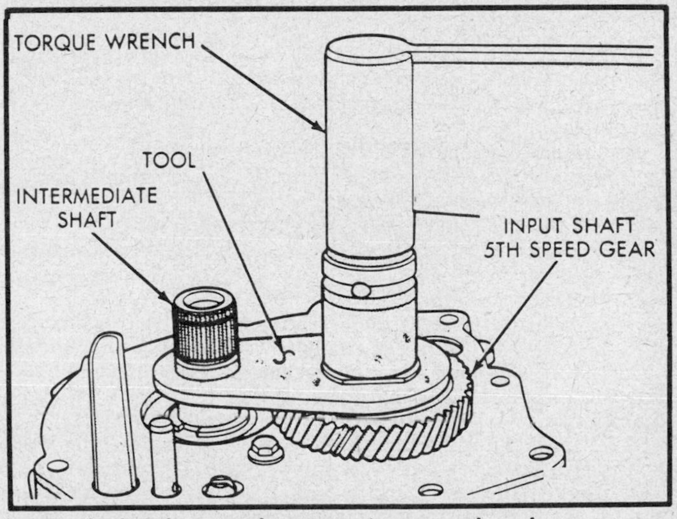
TORQUE WRENCH
TOOL
INTERMEDIATE
SHAFT
INPUT SHAFT
5TH SPEED GEAR

Input shaft 5th speed gear nut removal and installation

4. Remove the differential bearing retainer bolts.
5. Using spanner tool L–4435 or equivalent, rotate the differential bearing retainer to remove it.
6. Remove the selector shaft housing assembly bolts and remove the selector shaft housing assembly. On 1984–85 transaxles, remove the 5th speed shifter pin bolt and rear end cover fill plug.
7. Unbolt and remove the rear end cover.
8. Using snapring pliers, remove the snapring from the 5th speed synchronizer strut retainer plate to remove it, if installed.
9. Unscrew the 5th speed shift fork set screw. Lift the 5th speed synchronizer sleeve and shift fork off the synchronizer hub. Retrieve the 3 winged struts and top synchronizer spring.
10. Use a puller to remove the 5th speed synchronizer hub. Retrieve the remaining synchronizer spring.
11. Slide the 5th speed gear off the intermediate shaft. Remove the copper colored snapring retaining the 5th speed gear to the input shaft. Using a pulley puller, pull the 5th speed gear off the input shaft.
12. Remove the remaining bearing support plate bolts. Gently pry off the bearing support plate.
13. Remove the large snapring from the intermediate shaft rear ball bearing. Gently tap the lower surface of the bearing retainer plate with a plastic hammer to free it and lift it off the transaxle case. Clean the RTV sealer from both surfaces.
14. Unscrew the 5th speed shifter guide pin. Then, remove it. Do the same with the 1st/2nd shift fork setscrew. Withdraw the 1st/2nd, 3rd/4th shift fork rail.
15. Withdraw the reverse idler gear shaft, gear, and plastic stop.
16. Rotate the 3rd/4th shift fork to the left, and the 5th gear shifter to the right. Pull out the 5th speed shift rail. Withdraw the input shaft and intermediate shaft assemblies.
17. Remove the 1st/2nd, 3rd/4th and 5th speed shift forks.
18. To remove the clutch release bearing, remove the E-clips from the clutch release shaft, then disassemble the clutch shaft components.
19. Remove the input shaft seal retainer bolts, seal, retainer assembly and the select shim.
20. Remove the reverse shift lever E-clip and flat washer and disassemble the reverse shift lever components.

A-520 AND A-555 TRANSAXLES

1. Remove the transaxle from the vehicle and position on suitable holding fixture.
2. Remove the extension outer bolts.
3. Remove the differential retainer outer bolts.
4. Remove the differential cover bolts and gently pry the cover from the extension.
5. Remove the extension housing bolts then separate the differential assembly and extension housing. Remove the O-ring seal and clean the RTV from extension housing. Discard the O-ring seal.
6. Unbolt and remove the differential retainer.
7. Remove the selector shaft housing assembly bolts and remove the selector shaft housing.
8. Unbolt and remove the rear end cover.
9. Remove the snapring from the 5th speed synchronizer strut retainer plate.
10. Unscrew the 5th speed shift fork set screw. Lift the 5th speed synchronizer sleeve and shift fork off the synchronizer hub. Retrieve the (3) winged struts and top synchronizer spring.
11. Use a puller to remove the 5th speed synchronizer hub. Retrieve the remaining synchronizer spring.
12. Slide the 5th speed gear off the intermediate shaft.
13. Using holding tool 6252 or equivalent, remove the input shaft 5th speed gear nut. This nut is not to be reused.
14. Remove the remaining bearing support plate bolts. Gently pry off the bearing support plate.
15. Remove the large snapring from the intermediate shaft rear ball bearing.

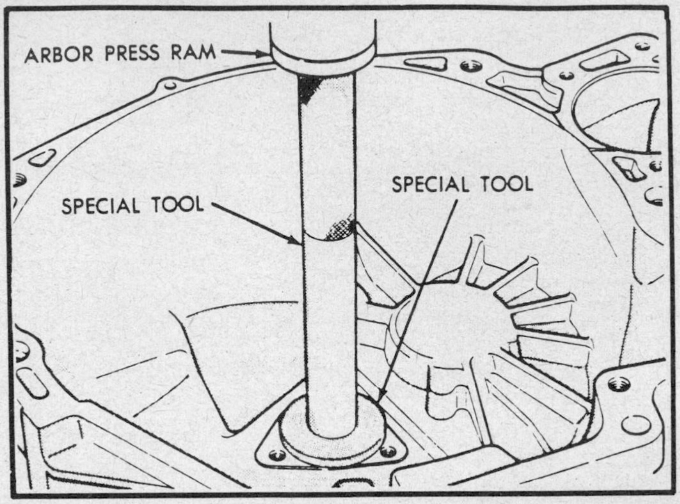
ARBOR PRESS RAM
SPECIAL TOOL
SPECIAL TOOL
Input shaft front bearing cup removal and installation

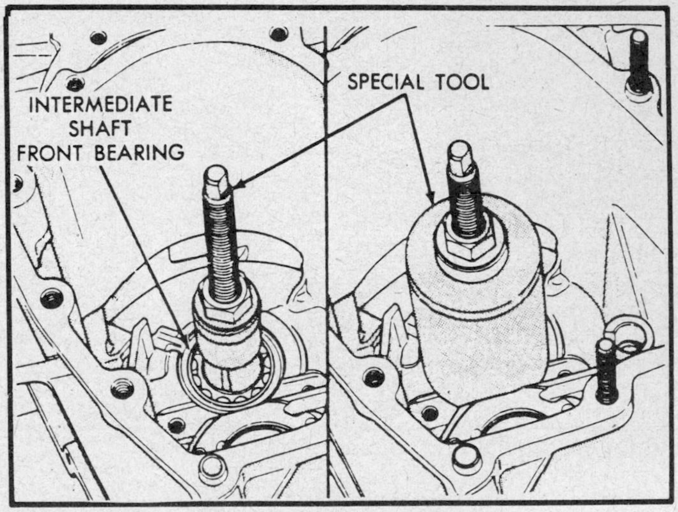
INTERMEDIATE SHAFT FRONT BEARING
SPECIAL TOOL
Intermediate shaft front bearing removal

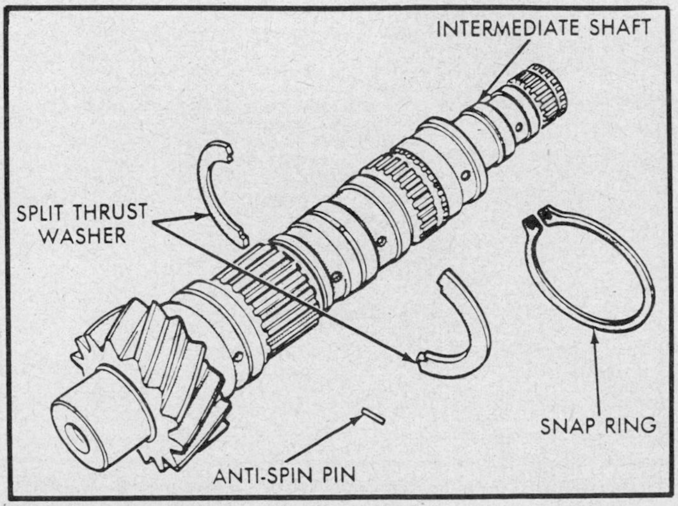
INTERMEDIATE SHAFT
SPLIT THRUST WASHER
SNAP RING
ANTI-SPIN PIN
Split thrust washer with snapring

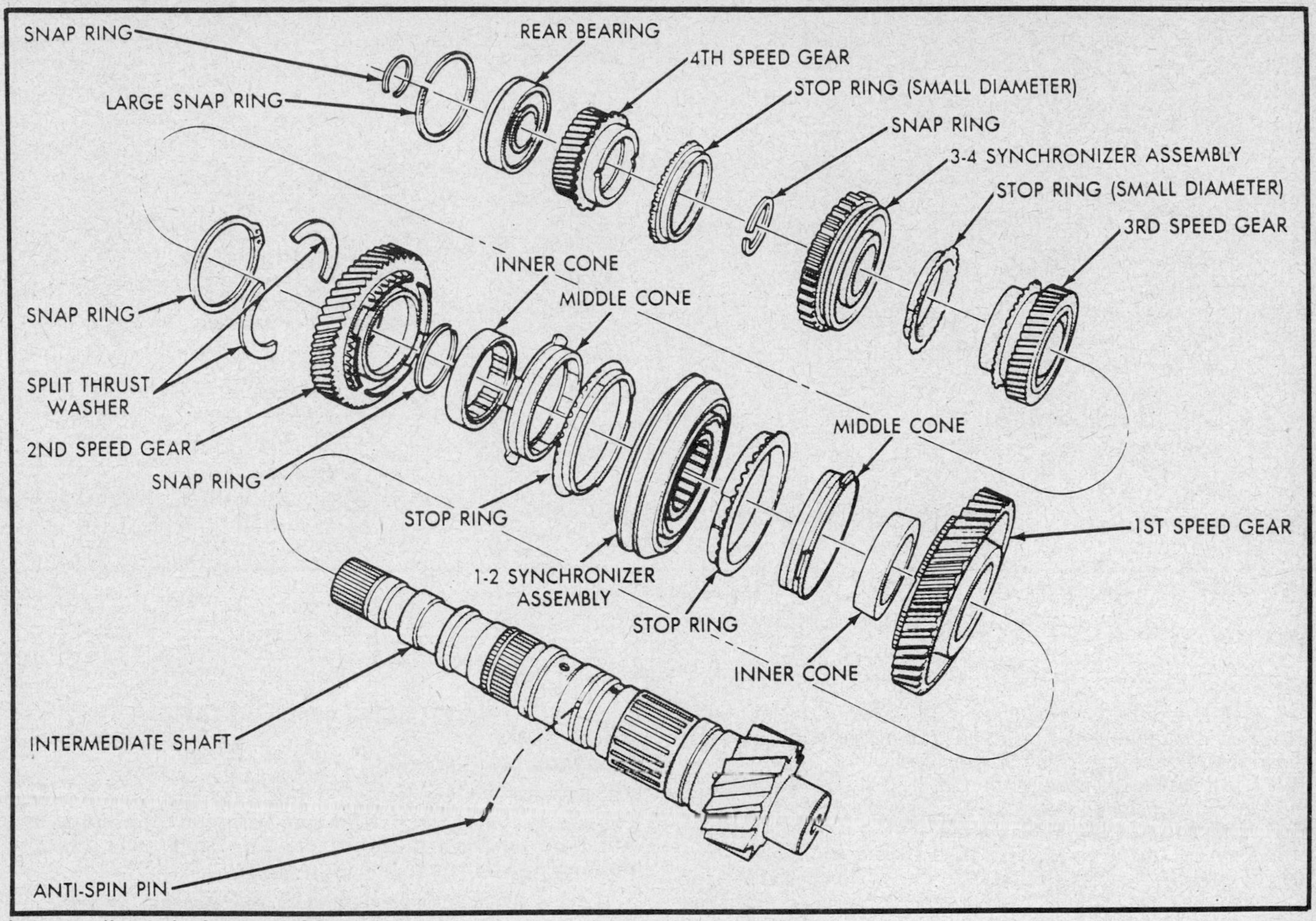
SNAP RING
REAR BEARING
4TH SPEED GEAR
LARGE SNAP RING
STOP RING (SMALL DIAMETER)
SNAP RING
3-4 SYNCHRONIZER ASSEMBLY
STOP RING (SMALL DIAMETER)
3RD SPEED GEAR
INNER CONE
MIDDLE CONE
SNAP RING
SPLIT THRUST WASHER
2ND SPEED GEAR
SNAP RING
MIDDLE CONE
STOP RING
1ST SPEED GEAR
1-2 SYNCHRONIZER ASSEMBLY
STOP RING
INNER CONE
INTERMEDIATE SHAFT
ANTI-SPIN PIN

Intermediate shaft assembly

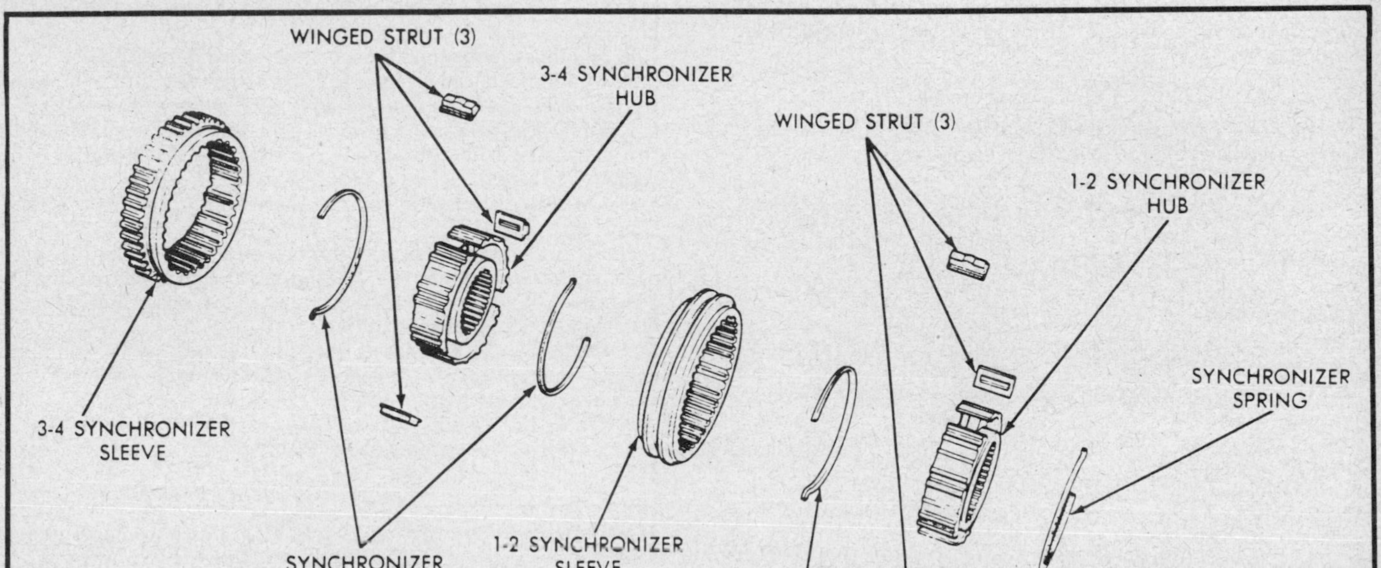
WINGED STRUT (3)
3-4 SYNCHRONIZER HUB
WINGED STRUT (3)
1-2 SYNCHRONIZER HUB
SYNCHRONIZER SPRING
3-4 SYNCHRONIZER SLEEVE
SYNCHRONIZER SPRINGS (2)
1-2 SYNCHRONIZER SLEEVE
SYNCHRONIZER SPRING

1–2, 3–4 synchronizer sleeves and hubs – A-525, A-520, A-555 tranasxles

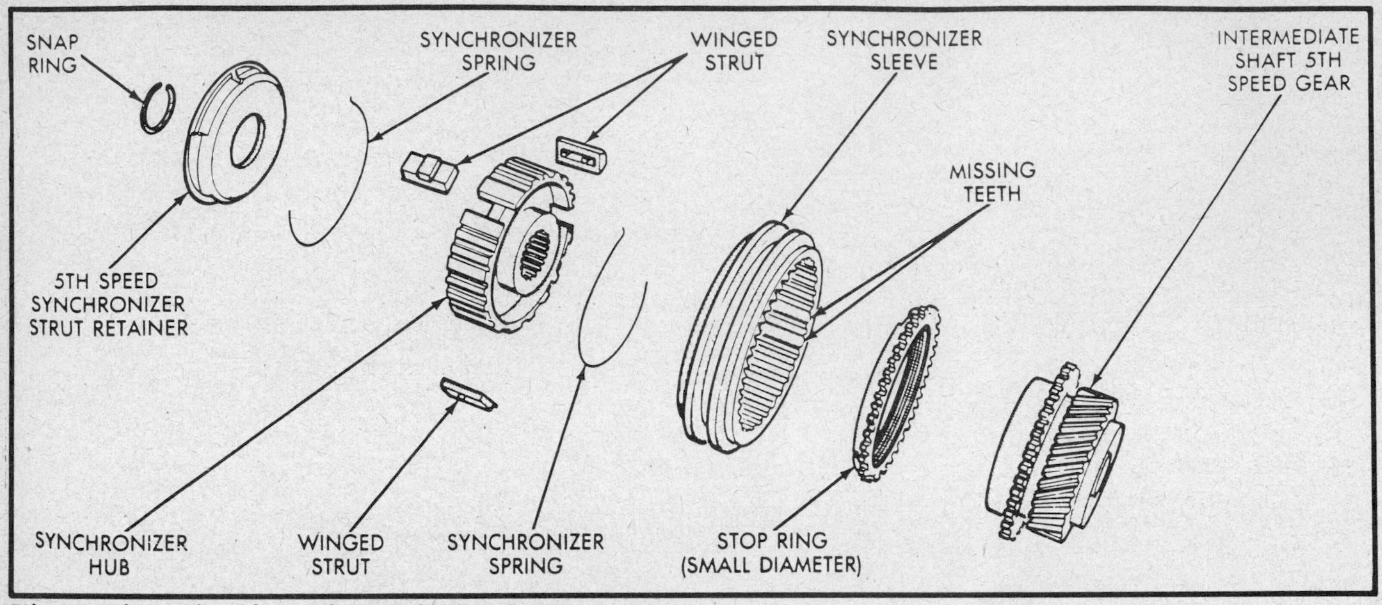
SNAP RING
SYNCHRONIZER SPRING
WINGED STRUT
SYNCHRONIZER SLEEVE
INTERMEDIATE SHAFT 5TH SPEED GEAR
MISSING TEETH
5TH SPEED SYNCHRONIZER STRUT RETAINER
SYNCHRONIZER HUB
WINGED STRUT
SYNCHRONIZER SPRING
STOP RING (SMALL DIAMETER)

5th speed synchronizer—A-525, A-520, A-555 tranasxles

16. Gently tap the lower surface of the bearing retainer plate with a plastic hammer to free it and lift it off the transaxle case. Clean the RTV sealer from both surfaces.
17. Unscrew the 5th speed shifter guide pin. Do the same with the 1st/2nd shift fork setscrew. Withdraw the 1st/2nd, 3rd/4th shift fork rail.
18. Slide out the reverse idler gear shaft, gear, and plastic stop.
19. Rotate the 3rd/4th shift fork to the left, and the 5th gear shifter to the right. Pull out the 5th speed shift rail. Pull out the input shaft and intermediate shaft assemblies.
20. Remove the 1st/2nd, 3rd/4th, and 5th speed shift forks.
21. To remove the clutch release bearing, remove the E-clips from the clutch release shaft, then disassemble the clutch shaft components.
22. Remove the input shaft seal retainer bolts, the seal, the retainer assembly and the select shim.
23. Remove the reverse shift lever E-clip and flat washer and disassemble the reverse shift lever components.

Unit Disassembly and Assembly

TRANSAXLE CASE

Disassembly

1. Press the input shaft front bearing cup from the transaxle case.
2. Unbolt and remove the intermediate shaft front bearing retaining strap.
3. Remove the intermediate shaft front bearing and oil feeder using a bearing puller.

Assembly

1. Press the intermediate shaft front bearing with the oil feeder into the transaxle case. The bearing identification letters must be facing upward during installation.
2. Install the intermediate shaft front bearing retaining strap.
3. Press the input shaft front bearing cup into the transaxle case.

INTERMEDIATE SHAFT

Disassembly

NOTE: The 1st/2nd, the 3rd/4th shift forks are interchangeable, however, the synchronizer stop rings are not. The 1st/2nd synchronizer stop rings have a larger diameter than the other stop rings.

1. Remove the intermediate shaft rear bearing snapring.
2. Remove the intermediate shaft rear bearing with a bearing puller. Note the color of the shaft ball bearing seal for replacement purposes: A-525 is black, A-520 is black and A-555 is blue.
3. Remove the 3rd/4th synchronizer hub snapring.
4. Matchmark and remove the 3rd/4th synchronizer hub and the 3rd speed gear using a puller.
5. On L-Body vehicles, remove the retaining ring, the split thrust washer, the 2nd speed gear and the synchronizer stop ring.
6. On L-Body vehicles, remove the 1st/2nd synchronizer hub snapring.
7. On L-Body vehicles, matchmark and remove the 1st speed gear, the stop ring and the 1st/2nd synchronizer assembly.
8. On L-Body vehicles, remove the 1st speed gear thrust washer and the anti-spin pin.
9. On A-520, A-555 transaxles (dual cone), remove the 2nd speed gear from the intermediate shaft and remove the 1-2 synchronizer hub snapring.
10. On A-520, A-555 transaxles (dual cone), pull the 1st speed gear and 1-2 synchronizer assembly from the intermediate shaft.

CAUTION

The 1-2 synchronizer assembly components are not interchangable with other synnchronizers or with previous year transxles.

Assembly

The assembly of the intermediate shaft is the reverse of the disassembly; however, please note the following:

1. When assembling the intermediate shaft, make sure the

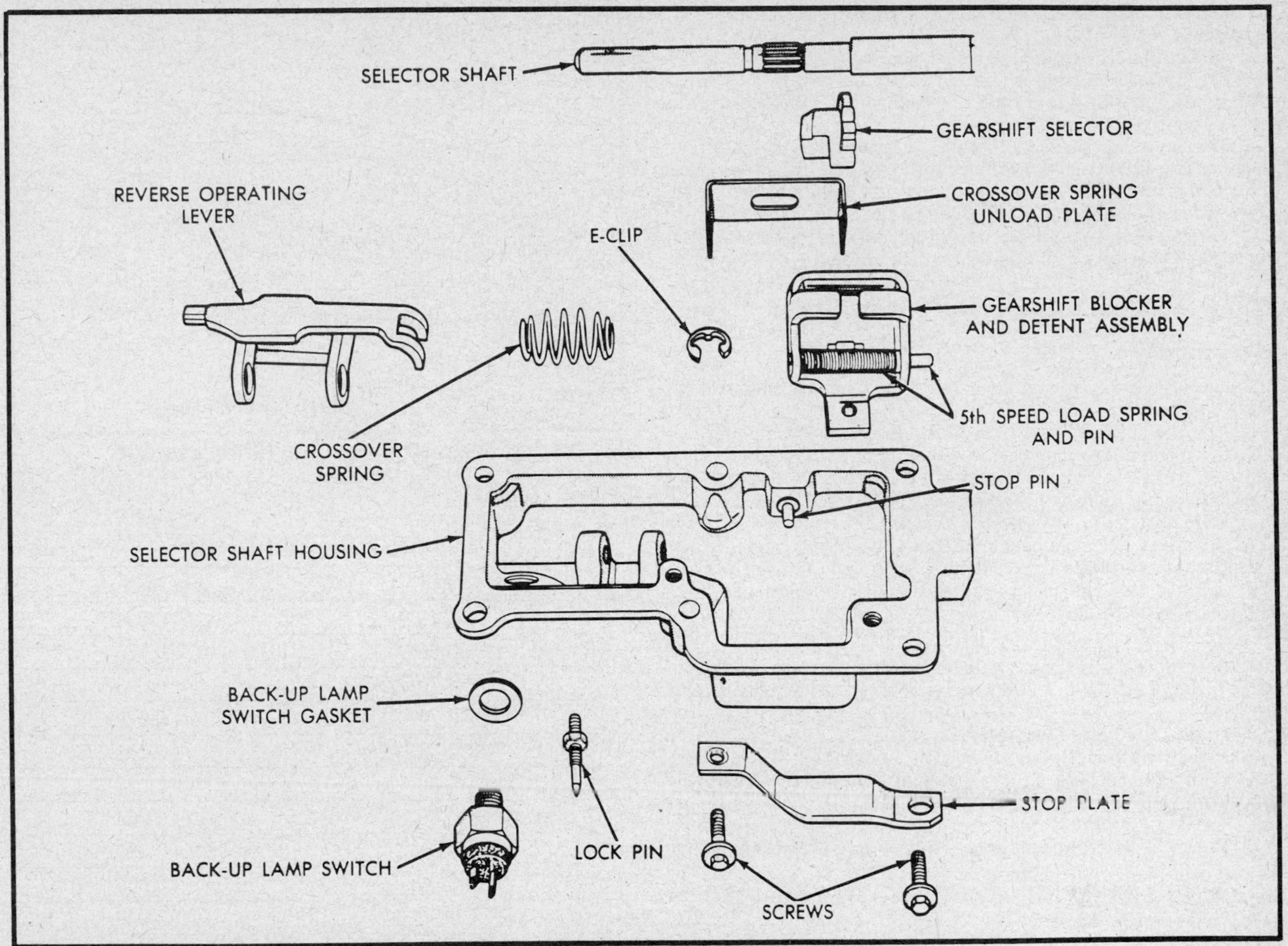
SELECTOR SHAFT
GEARSHIFT SELECTOR
REVERSE OPERATING LEVER
CROSSOVER SPRING UNLOAD PLATE
E-CLIP
GEARSHIFT BLOCKER AND DETENT ASSEMBLY
5th SPEED LOAD SPRING AND PIN
CROSSOVER SPRING
STOP PIN
SELECTOR SHAFT HOUSING
BACK-UP LAMP SWITCH GASKET
STOP PLATE
LOCK PIN
BACK-UP LAMP SWITCH
SCREWS
Selector shaft components—A-465, A-525, A-520, A-555 transaxles

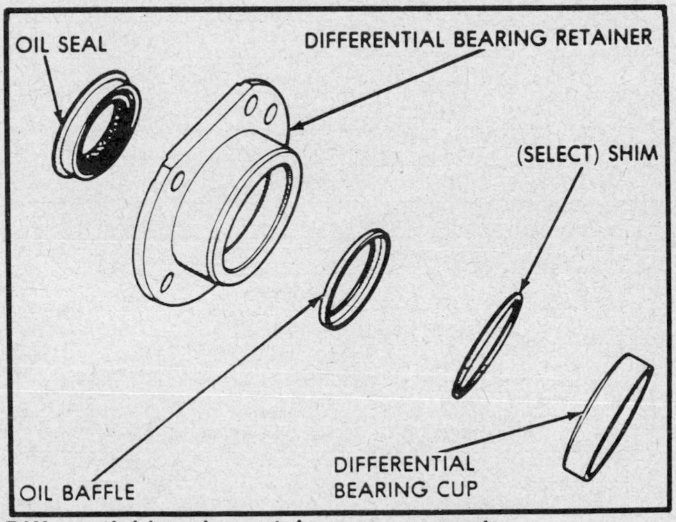
OIL SEAL
DIFFERENTIAL BEARING RETAINER
(SELECT) SHIM
DIFFERENTIAL BEARING CUP
OIL BAFFLE
Differential bearing retainer components

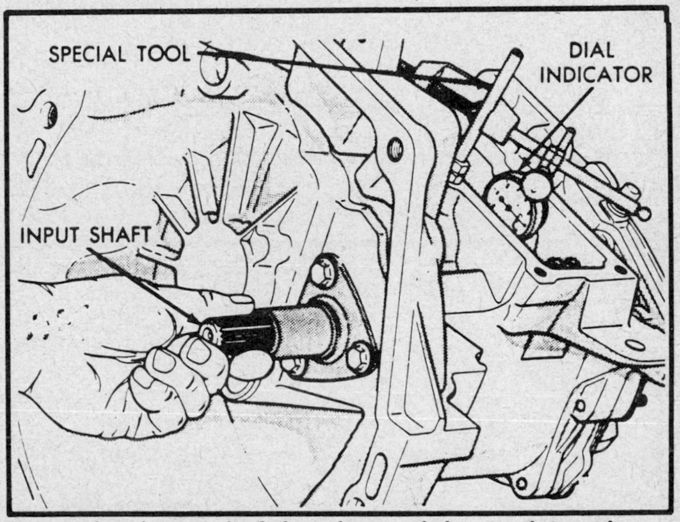
SPECIAL TOOL
DIAL INDICATOR
INPUT SHAFT
Measuring input shaft bearing endplay to determine shim thickness

speed gears turn freely and have a minimum of 0.003 in. endplay.

2. When installing the 1st speed gear thrust washer make sure the chamfered edge is facing the pinion gear.

3. When installing the 1st/2nd synchronizer make sure the relief faces the 2nd speed gear.

4. Use an arbor press to install the intermediate shaft rear bearing, the 3rd/4th synchronizer hub and the 3rd speed gear.

5. When installing the 3–4 synchronizer hub and 3rd speed gear, index the snapring 90 degrees to the split washer.

6. During the installation of synchronizer ring assemblies, make sure that all the matchmarks are aligned.

SELECTOR SHAFT HOUSING

Disassembly

1. Remove the snapring from the selector shaft boot and remove the boot.
2. Pry the shaft oil seal from the selector shaft housing.
3. With a small prybar positioned against the gearshift selector, compress the crossover and 5th speed load (all except A-460) spring and push the E-clip from the selector shaft. Remove the E-clip to release the selector shaft.
4. Withdraw the selctor shaft from the selector housing.
5. Remove the plate stop retaining bolts and remove the stop.
6. Disassemble the selector shaft housing components.

Assembly

1. Assemble the selector shaft housing components in the reverse order of removal. Use a new back-up lamp switch gasket if needed.
2. Install the plate stop with the retaining bolts.
3. Insert the selector shaft into the housing.
4. Compress the gearshift selector and install the E-clip.
5. Clean the bore and drive a new oil seal into the housing using the proper tool.
6. Place the boot onto the shaft and retain with the snapring.

DIFFERENTIAL BEARING RETAINER

Disassembly

1. Pry the oil seal from the retainer.
2. Remove the retainer cup with a puller. Be careful not to damage the oil baffle and the select shim .
3. Remove the oil baffle and the select shim from the retainer cup.

Asssembly

1. Drive the oil baffle and select shim into the retainer cup using the proper tool.
2. Drive the retainer cup into the retainer using the proper tool.
3. Drive in a new retainer oil seal.

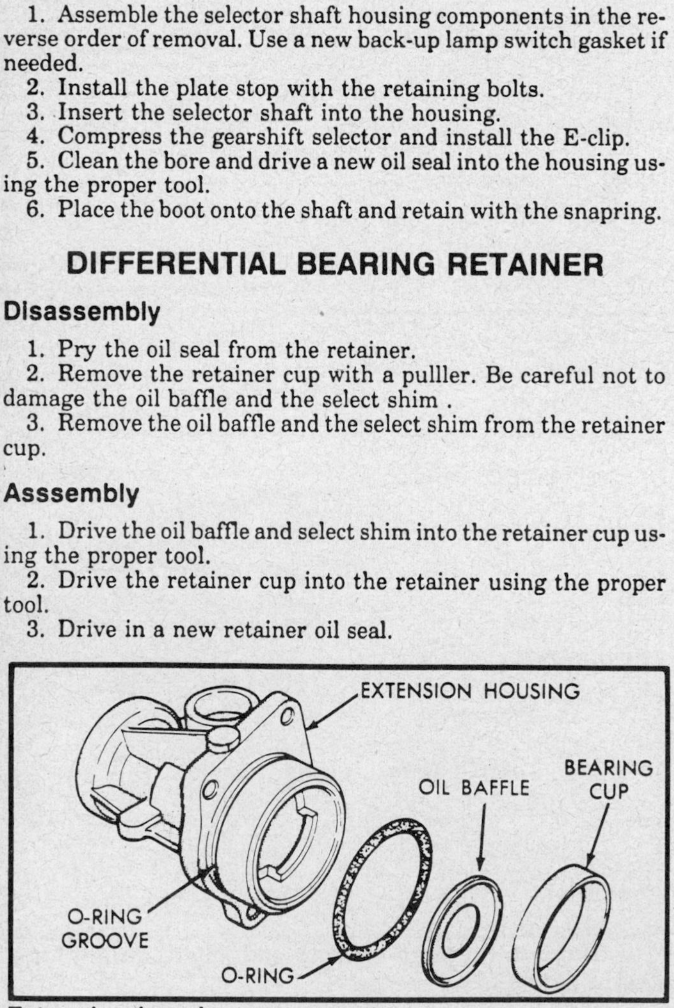
EXTENSION HOUSING
OIL BAFFLE
BEARING CUP
O-RING GROOVE
O-RING

Extension housing components

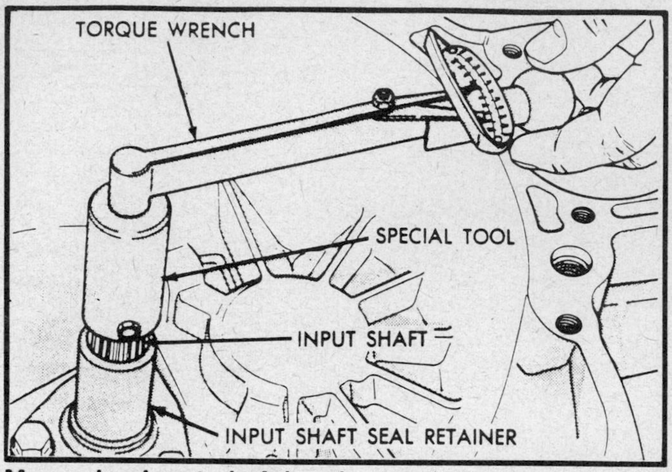
TORQUE WRENCH
SPECIAL TOOL
INPUT SHAFT
INPUT SHAFT SEAL RETAINER

Measuring input shaft bearing turning torque

INPUT SHAFT SHIM

inch	mm
0.62	0.024
0.66	0.026
0.70	0.028
0.74	0.029
0.78	0.031
0.82	0.032
0.86	0.034
0.90	0.035
0.94	0.037
0.98	0.039
1.02	0.040
1.06	0.042
1.10	0.043
1.14	0.045
1.18	0.046
1.22	0.048
1.26	0.050
1.30	0.051
1.34	0.053
1.36 (.66 + .70)	0.054
1.40 (.66 + .74)	0.055
1.44 (.70 + .74)	0.057
1.48 (.70 + .78)	0.059
1.52 (.74 + .78)	0.060
1.56 (.74 + .82)	0.061
1.60 (.78 + .82)	0.063
1.64 (.78 + .86)	0.065
1.68 (.82 + .86)	0.066
1.72 (.82 + .90)	0.068
1.76 (.86 + .90)	0.069

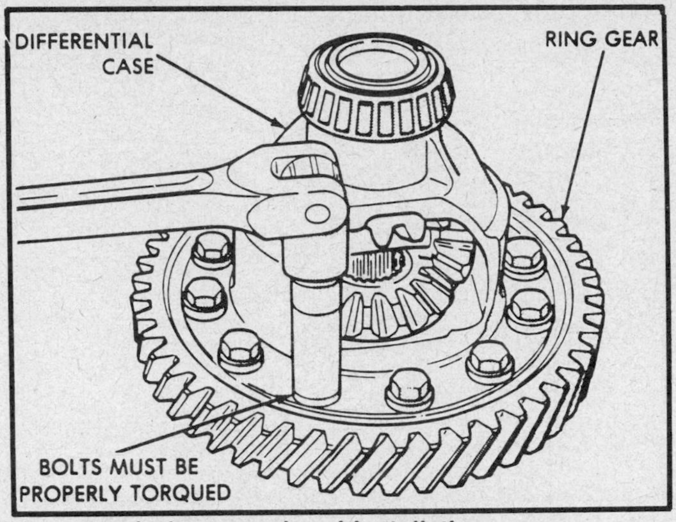
DIFFERENTIAL
CASE
RING GEAR
BOLTS MUST BE
PROPERLY TORQUED

Ring gear bolt removal and installation

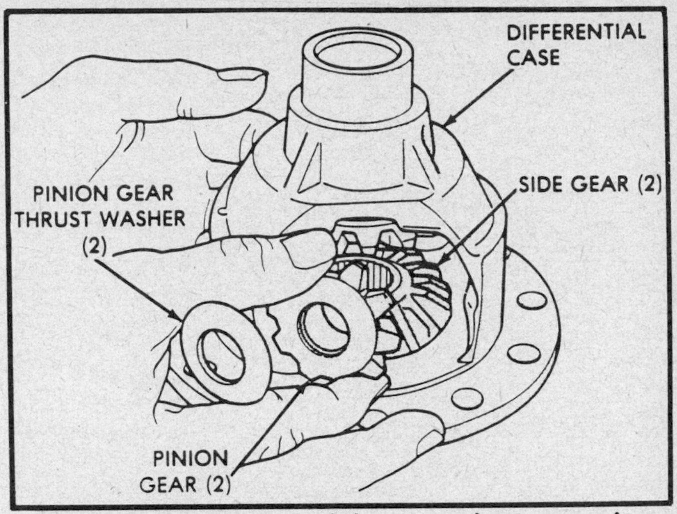
DIFFERENTIAL
CASE
PINION GEAR
THRUST WASHER
(2)
SIDE GEAR (2)
PINION
GEAR (2)

Pinion gear, side gear and thrust washer removal and installation

EXTENSION HOUSING

Disassembly

1. Pry the oil seal from the extension housing.
2. Pull the extension cup from the extension housing using the proper tool.
3. Remove the O-ring and oil baffle from the housing.

Assembly

1. Install a new O-ring into the groove on the outside of the housing.
2. Press the oil baffle in the housing using the proper tool.
3. Press the bearing cup into the extension housing using the proper tool.
4. Install a new housing oil seal.

INPUT SHAFT

Disasssembly

1. Remove the input shaft rear and front bearing cones using a suitable puller.
2. Mount the bearing retainer plate on wood blocks and press the input shaft rear bearing cup from the plate.

Assembly

1. Before pressing in the bearing cup, bolt the support plate onto the retainer plate.
2. With the support plate in place, press the bearing cup into the retainer plate.
3. Press the front and rear bearing cones onto the input shaft using the proper tool.

Bearing Endplay Adjustment

Shim thickness calculation and endplay adjustment need only be done if any of the following parts are replaced: transaxle case, input shaft seal retainer, bearing retainer plate, rear end cover, input shaft or input shaft bearings.

If any of the above components were replaced, use the follwing procedure to adjust the bearing preload and proper bearing turning torque.

1. Using special tool L–4656 or equivalent, with handle C–4171 or equivalent, press the input shaft front bearing cup slightly forward in the case. Then, using tool L–4655 or equivalent, with handle C–4171 or equivalent, press the bearing cup back into the case, from the front, to properly position the bearing cup before checking the input shaft endplay.

NOTE: This step is not necessary if special tool L–4655 was previously used to install the input shaft front bearing cup in the case and no input shaft select shim has been installed since pressing the cup into the case.

2. Select a gauging shim which will give 0.001–0.010 in. of endplay.

NOTE: Measure the original shim from the input shaft seal retainer and select a shim 0.010 in. thinner than the original for the gauging shim.

3. Install the gauging shim on the bearing cup and the input shaft seal retainer.
4. Alternately tighten the input shaft seal retainer bolts until the retainer is bottomed against the case. Torque the bolts to 21 ft. lbs.

CAUTION

The input shaft seal retainer is used to draw the input shaft front bearing cup the proper distance into the case bore.

5. Oil the input shaft bearings with automatic transmission fluid (1984–87) or SAE 5W-30 engine oil (1988–89) and install the input shaft in the case. Install the bearing retainer plate with the input shaft rear bearing cup pressed in and the end cover (A–460) or support plate (A–465, A–525, A–520, A–555) installed. Torque all bolts and nuts to 21 ft. lbs.
6. Position the dial indicator to check the input shaft endplay. Apply moderate load, by hand, to the input shaft splines. Push toward the rear while rotating the input shaft back and forth a number of times and to settle out the bearings. Zero the dial indicator. Pull the input shaft toward the front while rotating the input shaft back and forth a number of times to settle out the bearings. Record the endplay.
7. The shim required for proper bearing preload is the total of the gauging shim thickness, plus endplay, plus (constant) preload of 0.002–0.003 in. Combine shims, if necessary, to obtain a shim within 0.0016 in. of the required shim.
8. Remove the input shaft seal retainer and gauging shim. Install the shim(s) selected in Step 6 and reinstall the input shaft seal retainer with a $^1/_{16}$ in. bead of RTV sealant.

CAUTION

Keep RTV sealant out of the oil slot.

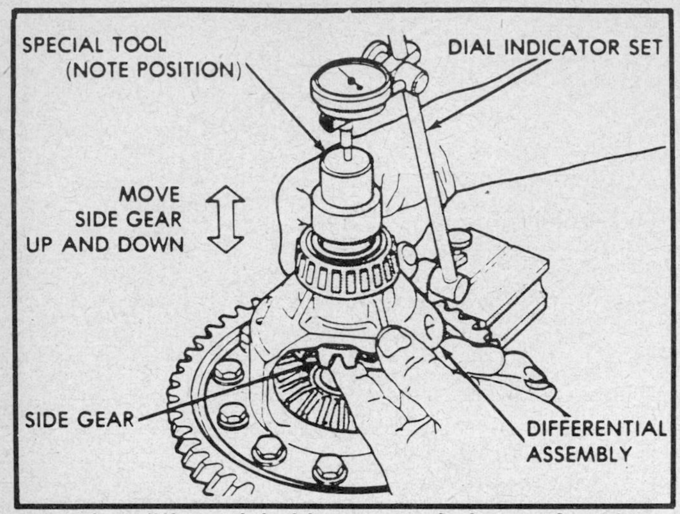
SPECIAL TOOL
(NOTE POSITION)
DIAL INDICATOR SET
MOVE
SIDE GEAR
UP AND DOWN
SIDE GEAR
DIFFERENTIAL
ASSEMBLY

Checking differential side gear end play on A-520 and A-555 transaxles

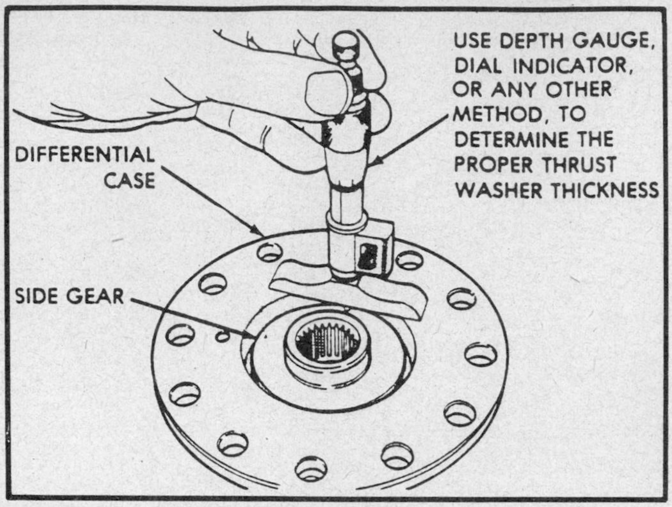
USE DEPTH GAUGE,
DIAL INDICATOR,
OR ANY OTHER
METHOD, TO
DETERMINE THE
PROPER THRUST
WASHER THICKNESS
DIFFERENTIAL
CASE
SIDE GEAR

Determining side gear thrust washer thickness on A-555 transaxles

9. Tighten the input shaft seal retainer bolts to 21 ft. lbs.

CAUTION

The input shaft seal retainer is used to draw the input shaft front bearing cup the proper distance into the case bore.

10. Using special tool L–4508 or equivalent and an inch lb. torque wrench, check the input shaft turning torque. The turning torque should be 1-5 inch lbs. for new bearings or a minimum of 1 inch lb. for used bearings. If the turning torque is too high, install a 0.0016 in. thinner shim. If the turning torque is too low, install a 0.0016 in. thicker shim.

CAUTION

Step 1 must be repeated every time a thinner shim in installed. This will ensure that the input shaft front bearing cup is pressed the proper distance into the case.

11. Recheck the input shaft turning torque. Repeat Step 10 until the proper bearing turning torque is obtained.

DIFFERENTIAL

Disassembly

1. Using special tool L–4406-1 or equivalent with adapters L–4406-3 or equivalent remove the bearing cone from the differential case.
2. Remove the ring gear bolts and separate the gear from the differential case. The ring gear bolts are epoxy patch type bolts and not to be reused.
3. Using a steel punch and hammer, knock the pinion shaft split pin from the ring gear and differential case. On A-555 tranaxles, drive out the 3 roll pins.
4. Withdraw the pinion shaft(s) from the differential case.
5. Rotate the side gears to align them with the case opening and remove the thrust washers, side gears and pinion gears.

Assembly

NOTE: Shim thickness calculation and bearing preload adjustment need only be done if any of the following parts are replaced: transaxle case, input shaft seal retainer, bearing retainer plate, rear end cover, input shaft or input shaft bearings. If any of the those components were replaced, refer to the appropriate section to adjust the bearing preload, proper bearing turning torque or side gear endplay.

1. In their original order, install the side gears, pinion gears and pinion gear washers. On A–520 and A–555 transaxles, leave the thrust washer out until after the side gear endplay is adjusted.
2. Insert the pinion shaft(s) into the differential case making sure that the hole in the shaft is aligned with the roll pin opening in the case.
3. Insert the pinion shaft roll pin(s) into the notched opening(s) on the side of the differential case and drive them into place.
4. Connect the ring gear to the differential case using new bolts. Torque the bolts in a criss-cross pattern to the proper specification.
5. Attach special tool L–4410 or equivalent to a suitable extension handle and press the bearing cone onto the diffential case.
6. Refer to the appropriate section to check and adjust the bearing preload, if necessary. On A–520 and A–555 check the side gear endplay and select the proper dimension thrust washer.

Side Gear Endplay Adjustment

A–520 TRANSAXLE

1. Once assembled, rotate the gears 2 complete revolutions in both a clockwise and counterclockwise direction.
2. Install special tool C–4996 on the bearing cone and mount a dial indicator so that the stylus of the dial rests on the surface of the tool.
3. Move 1 of the side gears up and down by hand and record the endplay.
4. Zero the dial and rotate the side gear in 90 degree increments an repeat Step 3.
5. Use the smallest endplay reading recorded and shim the side gear to within 0.001–0.013 in. For shimming, 4 select thrust washers are available, in the following sizes: 0.032 in. 0.037 in., 0.042 in. and 0.047 in.
6. Repeat Steps 1–4 for the other side gear.

A–555 TRANSAXLE

1. With a suitable depth gauge, measure and record the depth from the differential case to the machined surface of 1 side gear. Do this in 3 places.
2. Measure and record the height of the raised "step" on the ring gear.
3. Take the difference of the readings and subtract the ac-

ceptable side gear endplay range (0.001–0.013) to determine the correct thrust washer thickness.

4. For the other side gear, use the procedure described in Steps 1–4 for A-520 transaxles to determine the proper shim size.

5. After the differential is assembled, insert a suitable "dummy" spline shaft into the side gear. Turn the side gear by hand and check for freedom of movement. If the side will not turn or if it feels tight or binds, it will be necessary to remove the ring gear and install a thinner side gear thrust washer.

Bearing Preload Adjustment Procedure

1. Remove the bearing cup and existing shim from the differential bearing retainer.

2. Select a gauging shim which will give 0.001–0.010 in. endplay.

NOTE: Measure the original shim from the differentail bearing retainer and select a shim 0.015 in. thinner than the original for the gauging shim.

3. Install the gauging shim in the differential bearing retainer and press in the bearing cup. Installation of the oil baffle is not necessary when checking differential assembly endplay.

4. Lubricate the differential bearings with A.T.F. (1984–87) or SAE 5W-30 engine oil (1988–89) and install the differential assemby in the transaxle case.

5. Inspect the extension housing for damage and replace it as necessary. Apply a $^{1}/_{16}$ in. bead of RTV sealant to the extension flange. Install the extension housing and differential bearing retainer. Tighten the bolts to 21 ft. lbs.

6. Position the transaxle with the bell housing facing down on the workbench and secure with C-clamps. Position the dial indicator.

7. Apply a medium load to the ring gear, by hand, in the downward direction while rolling the differential assembly back and forth a number of times to settle the bearings. Zero the dial indicator. To obtain endplay readings, apply a medium load upward by hand while rolling the differential assembly back and forth a number of times to settle out the bearings. Record the endplay.

8. The shim required for proper bearing preload is the total of the gauging shim thickness, plus endplay, plus (constant) preload of 0.010 in. Combine shims if necessary, to obtain a shim within 0.002 in. of the shim(s).

9. Remove the differential bearing retainer. Remove the bearing cup and gauging shim. Properly install the oil baffle. Be sure the oil baffle is not damaged. Install the shim(s) selected in Step 8 and press the bearing cup into the differential bearing retainer.

10. Using a $^{1}/_{16}$ in. bead of RTV sealant for gaskets, install the differential bearing retainer and extension housing. Tighten the bolts to 21 ft. lbs.

11. Using special tool L–4436 or equivalent and an inch lb. torque wrench, check the turning torque of the differential assembly. The turning torque should be 9–14 inch lbs. for new bearings or a minimum of 6 inch lbs. for used bearings. If the turning torque is too low, install a 0.002 in. thicker shim.

12. Recheck the turning torque. Repeat Step 11 until the proper turning torque is obtained.

Transaxle Assembly

A–460 TRANSAXLE

1. Assemble the reverse shift lever components to the transaxle case and lock it in place with the E-clip.

2. Clean the input shaft bore and press in a new oil seal with special tool C–4674 or equivalent. Place the select shim into the retainer race and bolt the input shaft seal retainer onto the transaxle case. The drain hole on the retainer sleeve must be facing downward.

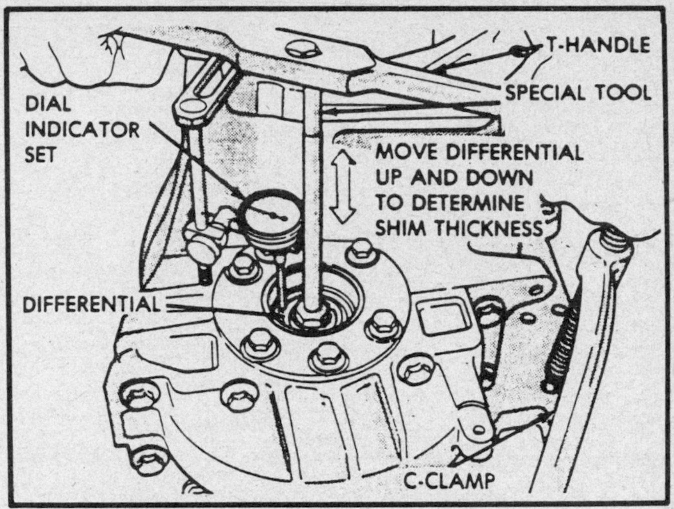
T-HANDLE
SPECIAL TOOL
DIAL INDICATOR SET
MOVE DIFFERENTIAL UP AND DOWN TO DETERMINE SHIM THICKNESS
DIFFERENTIAL
C-CLAMP

Checking differential end play to determine shim thickness

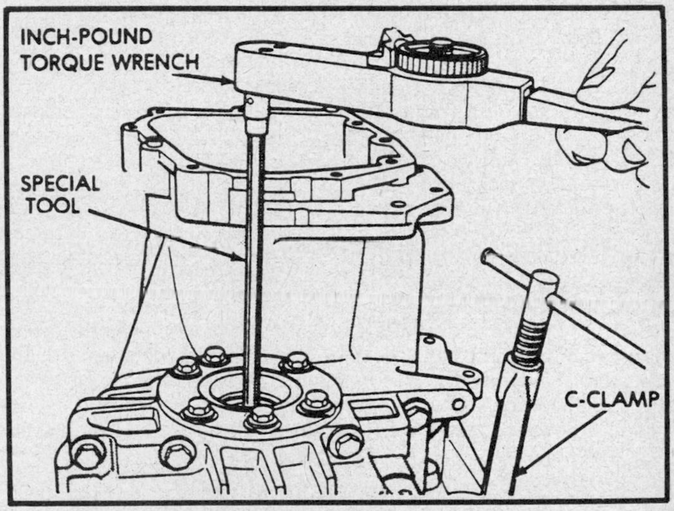
INCH-POUND TORQUE WRENCH
SPECIAL TOOL
C-CLAMP

Checking differential bearing turning torque

3. Assemble the clutch release shaft components in the reverse order of disassembly and secure the release lever with the E-clip. Insert the release shaft spline end through the bushing and engage it with the release shaft fork. Install the E-clip on the shaft groove to secure the shaft.

4. Install the shift fork and shift fork pads onto the intermediate shaft gear set. Install the intermediate and input shaft gear sets.

5. Install the reverse idler gear (with plastic stop) so that the roll pin on the end of the gear shaft aligns with the roll pin notch in the transaxle case. Lock the gear and engage the reverse shift lever. Make sure the plastic stop is firmly seated on the gear.

6. Install the 3–4 shift fork rail into the locating hole above the intermediate shaft assembly.

7. Remove all the excess sealant from the bearing retainer plate and run an ⅛ in. bead of RTV around the plate's seating surface. Keep the RTV away from the bolt holes. Align the locating dowel on the plate with the dowel on the transaxle case and install by tapping the plate with a rubber mallet. Install the intermediate shaft rear bearing snapring once the plate is in place.

8. Clean the excess sealant from the end cover and make sure the oil feeder hole is clear. Run a bead of RTV around the cover's seating surface and place the cover on the bearing retainer

DIFFERENTIAL BEARING SHIM CHART

Required Shim Combination	Total Thickness		Required Shim Combination	Total Thickness		Required Shim Combination	Total Thickness	
mm	mm	Inch	mm	mm	Inch	mm	mm	Inch
.50	.50	.020	.50 + .70	1.20	.047	1.00 + .70	1.70	.067
.75	.75	.030	.50 + .75	1.25	.049	1.00 + .75	1.75	.069
.80	.80	.032	.50 + .80	1.30	.051	1.00 + .80	1.80	.071
.85	.85	.034	.50 + .85	1.35	.053	1.00 + .85	1.85	.073
.90	.90	.035	.50 + .90	1.40	.055	1.00 + .90	1.90	.075
.95	.95	.037	.50 + .95	1.45	.057	1.00 + .95	1.95	.077
1.00	1.00	.039	.50 + 1.00	1.50	.059	1.00 + 1.00	2.00	.079
1.05	1.05	.041	.50 + 1.05	1.55	.061	1.00 + 1.05	2.05	.081
.50 + .60	1.10	.043	1.00 + .60	1.60	.063	1.05 + 1.05	2.10	.083
.50 + .65	1.15	.045	1.00 + .65	1.65	.065			

plate. Install the end cover bolts and torque to specification.

9. Clean the excess sealant from the selector shaft housing. If the back-up light switch was removed, install the switch with a new gasket. Run a $^{1}/_{16}$ in. bead of RTV sealant around the cover's seating surface and install the housing with the housing bolts. Torque the bolts to specification.

10. Connect the differential to the extension housing. Use a new extension housing O-ring seal. Attach the housing with the housing bolts and torque the bolts to specification.

11. Seal the differential bearing retainer with RTV and tighten the retainer with special tool L–4435 or equivalent spanner wrench.

12. Install the differential bearing retainer bolts and torque them to specifcation.

13. If the magnet was removed from the differential cover, install it at this time. Clean the excess sealant from the differential cover and run a ⅛ in. bead of RTV around the cover's seating surface. Install the differential cover with the cover bolts. Torque the bolts to specification.

14. Remove the transaxle from the holding fixture and install it in the vehicle.

A-465 AND A-525 TRANSAXLES

1. Assemble the reverse shift lever components to the transaxle case and lock it in place with the E-clip.

2. Clean the input shaft bore and press in a new oil seal with special tool C–4674 or equivalent. Place the select shim into the retainer race and bolt the input shaft seal retainer onto the transaxle case. The drain hole on the retainer sleeve must be facing downward.

3. Assemble the clutch release shaft components in the reverse order of disassembly and secure the release lever with the E-clip. Insert the release shaft spline end through the bushing and engage it with the release shaft fork. Install the E-clip on the shaft groove to secure the shaft.

4. Install the shift forks and shift rails onto the intermediate gear shaft assembly.

5. Install the intermediate and input shaft gear sets into the transaxle case and make sure that the gears are in proper mesh. Once the gear sets are in place, rotate the 5th speed shifter to the right and the 1–2 and 3–4 shift forks to the left.

6. Install the reverse idler gear (with plastic stop) so that the roll pin on the end of the gear shaft aligns with the roll pin notch in the transaxle case. Lock the gear and engage the reverse shift lever. Make sure the plastic stop is firmly seated on the gear.

7. Install the 1–2, 3–4 and 5th speed shift fork rails into their respective locating holes.

8. Install and tighten the 1–2 shift fork set screw and 5th speed selector guide pin.

9. Remove all the excess sealant from the bearing retainer plate and run an ⅛ in. bead of RTV around the plate's seating surface. Keep the RTV away from the bolt holes. Align the plate with the transaxle case and install it. The plate will align with the oil trough and the 5th speed shift rail. Install the intermediate shaft rear bearing snapring once the plate is in place and seated.

10. Install the bearing support plate with the retaining bolts. Torque the bolts to specification.

11. Install the input shaft 5th speed gear using special tool C–4810 or suitable gear installer. Retain the gear with the snapring.

12. Install the intermediate shaft 5th speed gear, synchronizer hub and struts using special tool C–4888 or equivalent gear installer.

13. Position the 5th speed synchronizer sleeve and shift fork over the 5th speed shift rail and install it. Lock the fork to the rail with the set screw.

14. Install the 5th speed synchronizer strut retainer plate with the snapring.

15. Clean the excess sealant from the end cover and make sure the oil fill plug hole is clear. Run a bead of RTV around the cover's seating surface and place the cover on the bearing retainer plate. Install the end cover bolts and torque to specification.

16. Clean the excess sealant from the selector shaft housing. If the back-up light switch was removed, install the switch with a new gasket. Run a $^{1}/_{16}$ in. bead of RTV sealant around the cover's seating surface and install the housing with the housing bolts. Torque the bolts to specification.

17. Connect the differential to the extension housing. Use a new extension housing O-ring seal. Attach the housing with the housing bolts and torque the bolts to specification.

18. Seal the differential bearing retainer with RTV and tighten the retainer with special tool L–4435 or equivalent spanner wrench.

19. Install the differential bearing retainer bolts and torque them to specifcation.

20. If the magnet was removed from the differential cover, install it at this time. Clean the excess sealant from the differential cover and run a ⅛ in. bead of RTV around the cover's seating surface. Install the differential cover with the cover bolts. Torque the bolts to specification.

21. Remove the transaxle from the holding fixture and install it in the vehicle.

A-520 AND A-555 TRANSAXLES

1. Assemble the reverse shift lever components to the transaxle case and lock it in place with the E-clip.

2. Clean the input shaft bore and press in a new oil seal with

special tool C–4674 or equivalent. Place the select shim into the retainer race and bolt the input shaft seal retainer onto the transaxle case. The drain hole on the retainer sleeve must be facing downward.

3. Assemble the clutch release shaft components in the reverse order of disassembly and secure the release lever with the E-clip. Insert the release shaft spline end through the bushing and engage it with the release shaft fork. Install the E-clip on the shaft groove to secure the shaft.

4. Install the shift forks and shift rails onto the intermediate gear shaft assembly.

5. Install the intermediate and input shaft gear sets into the transaxle case and make sure that the gears are in proper mesh. Once the gear sets are in place, rotate the 5th speed shifter to the right and the 1–2 and 3–4 shift forks to the left.

6. Install the reverse idler gear (with plastic stop) so that the roll pin on the end of the gear shaft aligns with the roll pin notch in the transaxle case. Lock the gear and engage the reverse shift lever. Make sure the plastic stop is firmly seated on the gear.

7. Install the 1–2, 3–4 and 5th speed shift fork rails into their respective locating holes.

8. Install and tighten the 1–2 shift fork set screw and 5th speed selector guide pin.

9. Remove all the excess sealant from the bearing retainer plate and run an ⅛ in. bead of RTV around the plate's seating surface. Keep the RTV away from the bolt holes. Align the plate with the transaxle case and install it. The plate will align with the oil trough and the 5th speed shift rail. Install the intermediate shaft rear bearing snapring once the plate is in place and seated.

10. Install the bearing support plate with the retaining bolts. Torque the bolts to specification.

11. Install the 5th speed gear onto the input shaft. Install a new gear nut with special holding tool 6252. The holding tool must be used to install the gear nut and the old nut must not be reused. Torque the nut to 190 ft. lbs. and remove the holding tool.

12. Install the intermediate shaft 5th speed gear, synchronizer hub and struts using special tool C–4888 or equivalent gear installer.

13. Position the 5th speed synchronizer sleeve and shift fork over the 5th speed shift rail and install it using the alignment marks for reference. Lock the fork to the rail with the set screw.

14. Install the 5th speed synchronizer strut retainer plate with the snapring.

15. Clean the excess sealant from the end cover and make sure the oil fill plug hole is clear. Run a bead of RTV around the cover's seating surface and place the cover on the bearing retainer plate. Install the end cover bolts and torque to specification.

16. Clean the excess sealant from the selector shaft housing. If the back-up light switch was removed, install the switch with a new gasket. Run a $^{1}/_{16}$ in. bead of RTV sealant around the cover's seating surface and install the housing with the housing bolts. Torque the bolts to specification.

17. Apply RTV sealant to the portion of the differential bearing retainer that bolts to the ring gear to form a gasket.

18. Position the differential in the support saddles.

19. Connect the bearing retainer to the differential with the inner bolts and torque the bolts to specification.

20. Remove the O-ring seal from the extension housing and replace it with a new one. Remove the old sealant from the base of the extension and run a bead of new sealant.

21. Connect the extension to the other side of the differential with the outer bolts. Torque the bolts to specification.

22. Clean the excess sealant from the differential cover and run a ⅛ in. bead of RTV around the cover's seating surface. Install the differential cover with the cover bolts. Torque the bolts to specification.

23. Install the remaining extension and differential bearing retainer (outer) bolts and torque them to specifcation.

24. Remove the transaxle from the holding fixture and install it in the vehicle.

SPECIFICATIONS

TORQUE SPECIFICATIONS
A-460 and A-525 Transaxle

	Torque	
	ft. lbs.	Nm
1st-2nd shift fork set screw	4	5
5th fork to rail set screw	4	5
5th speed shifter guide pin	4	5
Gearshift housing to case bolt	21	28
Gearshift operating lever attaching nut	21 ①	28 ①
Fork stop plate to gearshift housing bolt	9	12
Shift linkage adjusting pin	9	12
Dust covers to case screw	9	12
Strut to block bolt	70	95
Strut to case bolt	70	95
Flywheel to crankshaft bolt	70	95
Clutch pressure plate to flywheel bolt	21	28
Case to engine block bolt	70	95
Mount to block and case bolt	70	95
Anti-rotational strut bracket to stud nut	17	23
Differential ring gear bolt	70 ②	95 ②
Differential bearing retainer bolt	40	54
Differential extension bolt	21	28
Differential oil pan screw and washer	14	19
Differential oil pan stud nut	14	19
Fill plug		
Intermediate shaft bearing strap screw	9	12
Input shaft seal retainer bolt	21	28
Steel end cover to case bolt	21	28
Steel end cover to bearing retainer bolt	21	28

① This is a prevailing-torque nut. If removed, install a new prevailing-torque nut.
② This is an epoxy-patch prevailing torque bolt. If removed, install new bolts of the same part number.

TORQUE SPECIFICATIONS
A-525/A-520/A-555 Transaxle

	Torque	
	ft. lbs.	Nm
1st-2nd shift fork set screw	4	5
5th fork to rail set screw	4	5
5th speed gear (input shaft) nut	190 ③	258 ③
5th speed shifter guide pin	4	5
Gearshift housing to case bolt	21	28
Gearshift operating lever attaching bolt (A-520/A-555)	27 ①	37 ①
Gearshift operating lever attaching nut (A-525)	25 ②	34 ②
Fork stop plate to gearshift housing bolt	9	12
Shift linkage adjusting pin	9	12
Dust covers to case screw	9	12
Strut to block bolt	70	95
Strut to case bolt	70	95
Flywheel to crankshaft bolt	70	95
Clutch pressure plate to flywheel bolt	21	28
Case to engine block bolt	70	95
Mount to block and case bolt	70	95

	Torque	
	ft. lbs.	Nm
Anti-rotational strut bracket to stud nut	17	23
Differential cover to case bolt (A-520/A-555)	40	54
Differential ring gear bolt (A-525)	70 ②	95 ②
Differential ring gear bolt (A-520)	65 ②	88 ②
Differential ring gear bolt (A-555)	80 ②	108 ②
Differential bearing retainer bolt	40	54
Differential extension bolt	21	28
Differential oil pan screw and washer	14	19
Differential oil pan stud nut	14	19
Fill plug		
Intermediate shaft bearing strap screw	9	12
Input shaft seal retainer bolt	21	28
Steel end cover to case bolt	21	28
Steel end cover to bearing retainer bolt	21	28

① This is a lock-patch bolt. If removed, install a new lock-patch bolt.
② This is an epoxy-patch prevailing torque bolt (or nut). If removed, install new bolts (or nut) of the same part number.
③ This is a prevailing-torque nut. If removed, install a new nut of the same part number.

SPECIAL TOOLS

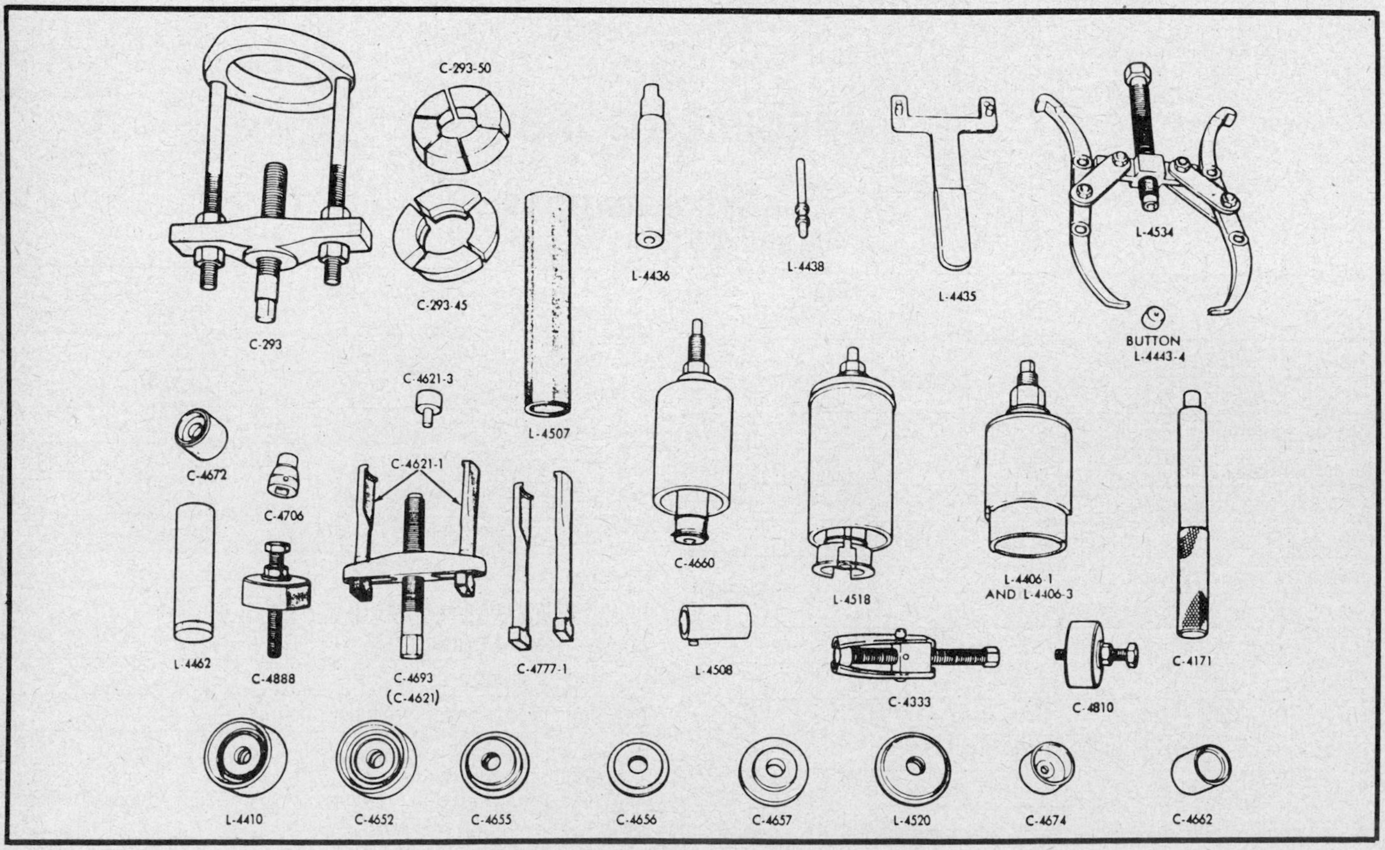
C-293
C-293-50
C-293-45
L-4436
L-4438
L-4435
L-4534
BUTTON
L-4443-4
C-4621-3
L-4507
C-4672
C-4706
C-4621-1
L-4462
C-4888
C-4693
(C-4621)
C-4777-1
C-4660
L-4508
L-4518
L-4406-1
AND L-4406-3
C-4333
C-4810
C-4171
L-4410
C-4652
C-4655
C-4656
C-4657
L-4520
C-4674
C-4662

Section 6

NG9 Transaxle
Jeep-Eagle

APPLICATION

1988–89 Medallion

GENERAL DECRIPTION

The NG9 manual transaxle is a 5 speed, fully synchronized transaxle providing overdrive operation in 4th and 5th gear ranges. The transaxle and differential gears are contained in a 2 piece gear case. The NG9 transaxle is equipped with 4 differential pinion gears and is used with turbo and larger displacement engine applications. On the NG9 transaxles, the 5th gear synchronizer and shift fork are serviced only when the transaxle is disassembled or removed from the vehicle.

The gear shift lever and linkage used with the NG9 transaxle has a reverse gear lock out mechanism. Lift the sleeve under the gear shift knob to override the lock out mechanism when shifting into reverse.

Transaxle Identification

The transaxle identification tag is attached to the rear housing of the transaxle. The tag provides the transaxle type, suffix and fabrication number. The information on the tag is needed to obtain the correct service parts. If the tag is removed during service operations, be sure the tag is reinstalled before repairs are completed.

Metric Fasteners

The metric fastener dimensions are very close to the dimensions of the familiar inch system fasteners. For this reason, replacement fasteners must have the same measurement and strength as those removed.

Do not attempt to interchange metric fasteners for inch system fasteners. Mismatched or incorrect fasteners can result in damage to the transmission unit through malfunctions, breakage or possible personal injury.

Care should be taken to re-use the fasteners in the same locations as removed.

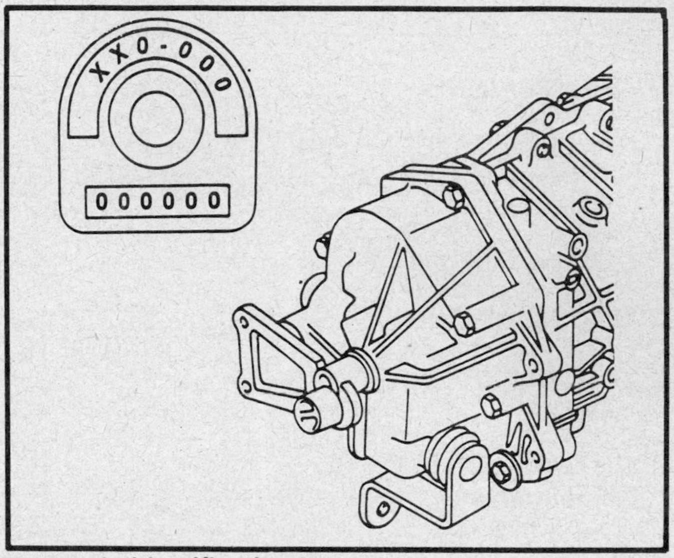
XX0-000
000000
Transaxle identification tag

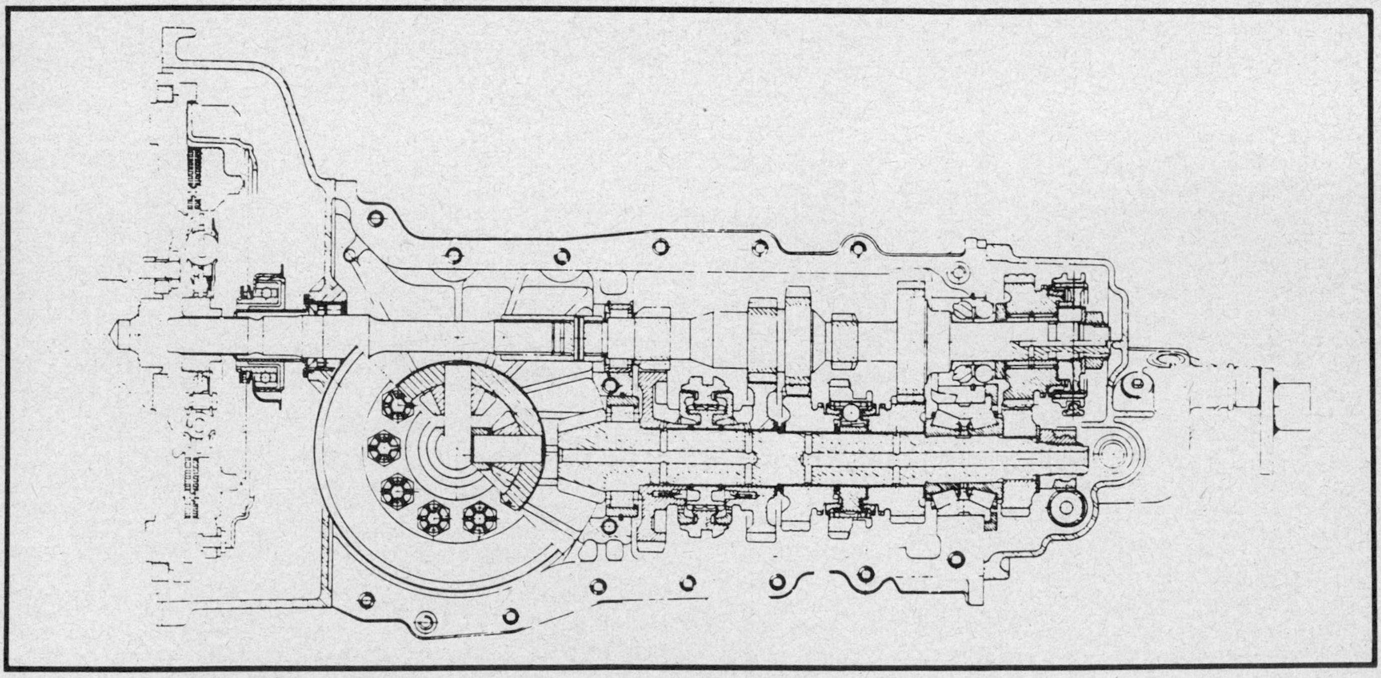
Internal view of the NG9 transaxle

Common metric fastener strength property classes are 9.8 and 12.9 with class identification embossed on the head of each bolt. The inch strength classes range from grade 2 to 8 with the line identification embossed on each bolt head. The markings correspond to 2 lines less than actual grade (for example grade 8 bolt will exhibit 6 embossed lines on the bolt head). Some metric nuts will be marked with a single digit strength identification numbers on the nut face.

Capacities

The transaxle lubricant capacity is 2.5 quarts (2.25L). The correct recheck or top-off level is to the bottom edge of the fill plug hole. The recommended lubricant for the NG9 transaxle is AMC/Jeep/Renualt gear lubricant or an equivalent SAE 75W–90, API grade GL5 gear lubricant.

Checking Fluid Level

1. Raise and support the vehicle safely.
2. Remove the transaxle oil fill plug.
3. The correct level should be to the bottom edge of the fill plug hole.
4. Fill the proper level with the recommended gear oil and reinstall the oil fill plug.
5. Lower the vehicle.

Transaxle Modifications

There are no transaxle modifications for the NG9 transaxle at the time of this publication.

TROUBLE DIAGNOSIS

CHILTON THREE C'S DIAGNOSIS
NG9 Manual Transaxle

Condition	Cause	Correction
Gear clash (One Gear Only)	a) Check the synchronizer and synchro spring of the gear in question	a) Repair or replace synchro and or synchro spring
Gear clash (All Gears)	a) Check the clutch and clutch release mechanism for proper operation b) Remove the gearbox; check the synchronizers and synchro hubs	a) Repair or replace clutch or clutch release mechanism b) Repair or replace synchronizers and or synchro hubs
Will not shift into any gear	a) Check the clutch and clutch release mechanism for proper operation b) Check condition of the shift linkage components	a) Repair or replace clutch or clutch release mechanism b) Adjust or repair the shift linkage and or components
Transaxle slips out of gear	a) Check condition of engine and transaxle mounts b) Check condition of the shift linkage components c) Check the hub, synchronizers, shift forks, gear locking mechanism	a) Repair or replace the mounts b) Adjust or repair the shift linkage and or components c) Repair or replace as necessary
Transaxle difficult to shift out of gear or locks into gear	a) Check condition of the shift linkage components b) Check detent balls, shift forks and detent plungers	a) Adjust or repair the shift linkage and or components b) Repair or replace as necessary

ON CAR SERVICES

Adjustments

SHIFT LINKAGE

The shift linkage is not adjustable. However, if diagnosis indicates that a shift problem is caused by worn or damaged components, the linkage bushings, isolators, linkage rods and other components can be serviced individually.

The lockout cable, linkage rods, shift lever, bushings and isolators are all accessible from the underside of the vehicle. The inner boot is serviced from above. The reverse lockout components are serviced after removing the shift lever from the vehicle.

The shift lever locators and are colored coded to indicate location. The green isoloator is used at the transaxle end of the linkage rod. Service set to tightening torques for the linkage fasteners are as follows:

a. Torque the shift lever nut to 15 ft. lbs.
b. Torque the base plate bolts to 11 ft. lbs.

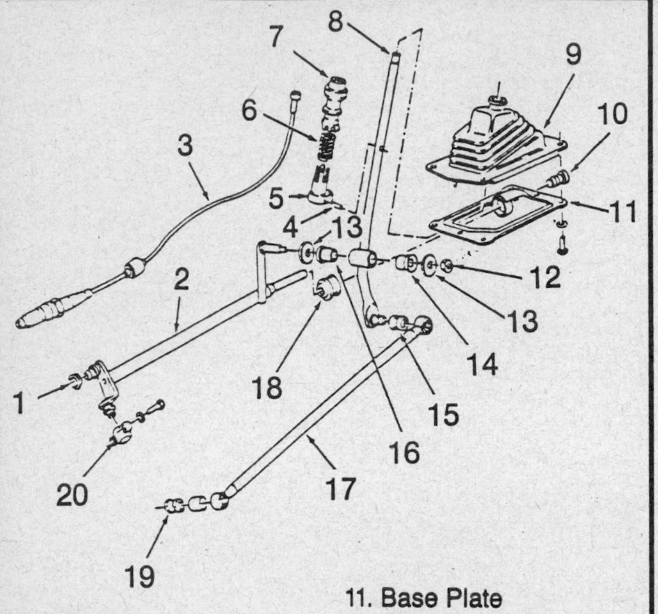
1
2
3
4
5
6
7
8
9
10
11
12
13
13
14
15
16
17
18
19
20

1. Bushing
2. Shift Rod
3. Reverse Lockout Cable
4. Roll Pin
5. Reverse Lockout Base
6. Lockout Spring
7. Reverse Lockout Handle
8. Shift Lever
9. Inner Shift Boot
10. End Piece
11. Base Plate
12. Shift Lever Nut
13. Washer
14. Outer Bushing
15. Isolator, Shift Rod-to-Shift Lever
16. Inner Bushing
17. Shift Rod
18. Isolator, End Piece
19. Isolator, Shift Rod End
20. Retainer Cup

Exploded view of the shifting lever assembly

c. Torque the retaining cup bolt to 11 ft. lbs.
d. Torque the lockout cable nut to 15 ft. lbs.

Services

REVERSE LOCKOUT CABLE

Removal and Installation

It is not necessary to drain the transaxle lubricant in order to replace the reverse lockout cable.

1. Unsnap the outer shift boot from the console opening and move the boot aside.
2. Pull down on the cable sheath and unsnap it from the base of the reverse lockout cable.
3. Unsnap the cable end from the bracket on the rear of the reverse lockout handle.
4. Raise and support the vehicle safely.
5. Unscrew the lockout cable retaining nut and pull cable out of the rear housing.
6. Remove the cable from the vehicle.
7. Feed the cable upward through the small hole in the inner shift boot.
8. Apply a small amount of thread sealer to the cable retaining nut threads and connect the cable to the rear housing. Torque the cable nut to 15 ft. lbs.
9. Lower the vehicle.
10. Snap the cable into position on the rear side of the lockout handle.
11. Pull downward on the cable sheath and snap the cable into place on the lockout base.
12. Install the outer shifter boot. in the console opening.

REAR HOUSING/SPEEDOMETER GEAR

Removal and Installation

On the NG9 transaxle, only the rear housing and speedometer gear are serviced with the transaxle in the vehicle. Do not attempt to service the 5th gear, synchronizer or shift fork with the transaxle in the vehicle. The 5th gear shift rail lock ball will disengage and fall into the case if the rail is moved even a small amount.

1. Raise and support the vehicle safely. Drain the gear oil from the transaxle.
2. Disconnect the speedometer cable.
3. Remove the shift rod bolts. Pry the shift rod off the shift lever ball stud.
4. Unscrew the reverse lockout cable nut and pull the cable out of the rear housing.
5. Shift the transaxle into 4th gear. This is necessary to prevent the lock ball on the 5th gear shift rail from falling into the case.
6. Remove the crossover shaft bracket from the housing shaft. Remove the roll pin from the shaft with a suitable punch set.
7. Remove the bolts attaching the rear housing to the case.
8. Disconnect the 5th speed switch wire harness at the top of the transaxle. Carefully loosen and remove the rear housing from the gear case.
9. Loosen and remove the secondary shaft nut. The secondary shaft nut and the speedometer drive gear are serviced as an assembly only. The 2 components cannot be separated.
10. Remove the speedometer driven gear from the shaft. Grasp the splined end of the shaft with a suitable pair of pliers and pull the gear off of the shaft.
11. Replace the speedometer driven gear shaft seal. Using a puller remove the seal. Use a suitable seal installer and install a new seal into the housing.
12. Clean the gasket surfaces of the rear housing and transaxle case. Lubricate the speedometer driven gear shaft with gear lubricant and install the shaft into the rear housing. Align and snap the speedometer driven gear on the shaft. Be sure the gear tangs are seated on the shaft.
13. Check the alignment dowels in the transaxle case. Be sure the dowels are in place before installing the housing.
14. Check that the transaxle is still in 4th gear. This keeps the 5th gear shift rail lock ball in place and makes alignment of the shift arm easier by avoiding the load on the arm from the spring.
15. Apply some suitable thread sealer to the threads of the speedometer drive gear/secondary shaft. Torque the nut to 110 ft. lbs. Then secure the nut by peening it over into the flat on the shaft.
16. Position a replacement gasket on the housing and install the housing onto the gear case. Torque the case bolts to 11 ft. lbs.
17. Install the washer on the housing shaft and install the crossover bracket on the shaft.
18. Connect the speedometer cable and shift linkage. Apply some suitable thread sealer to the threads of the nut on the lockout release cable and install the cable. Torque the cable locknut to 15 ft. lbs.
19. Refill the transaxle with the required gear lubricant and lower the vehicle.

FLUID CHANGES

1. Raise and support the vehicle safely.

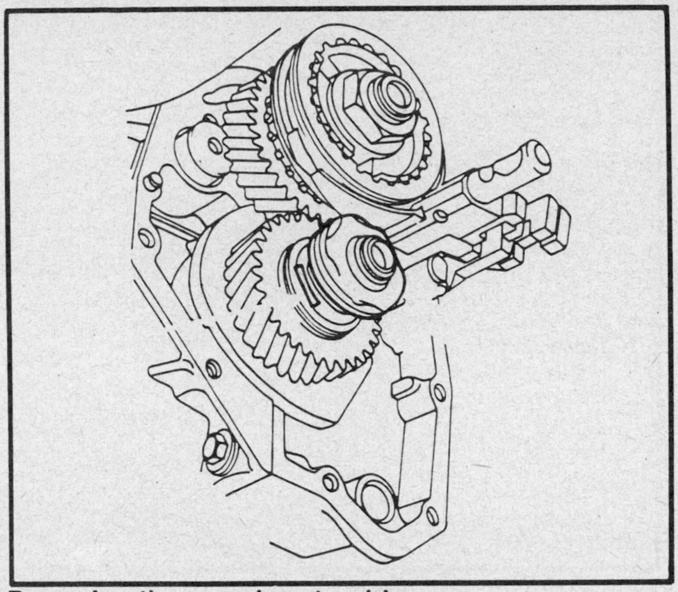

Removing the speedometer drive gear

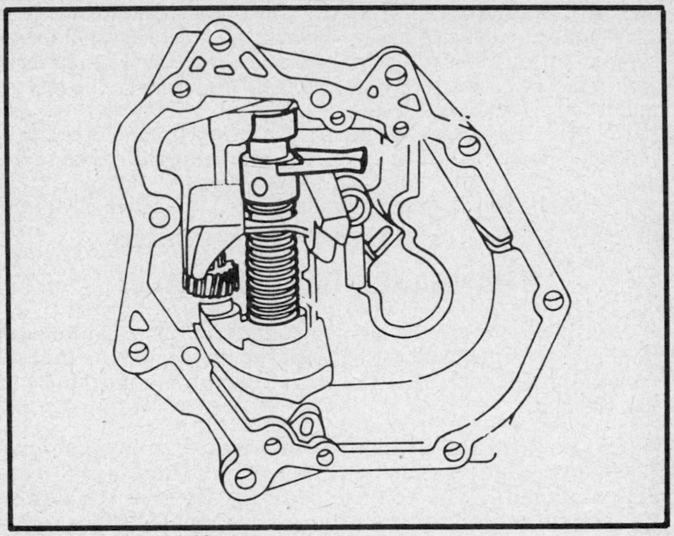

Aligning the shift arm

2. Remove the transaxle oil fill plug and allow the gear oil to drain into a suitable drain pan.
3. The correct level should be to the bottom edge of the fill plug hole.
4. Fill the proper level with the recommended gear oil and reinstall the oil fill plug.
5. Lower the vehicle.

REMOVAL AND INSTALLATION

TRANSAXLE REMOVAL

1. Disconnect the negative battery cable.
2. Disconnect the flexible hose from the heat tube and remove the bolt.
3. Remove the engine bracket bolt located at the rear of the engine.
4. Remove the remaining heat tube and tube bracket bolts/nuts and remove the heat tube and bracket. The remaining bolts and nuts are located under the intake manifold.
5. Remove the bolts attaching the top dead center sensor to the clutch housing and remove the sensor.
6. Remove the steering bracket. The bracket is attached to the steering rack and the steering tie rod with nuts and bolts.
7. On vehilces equipped with air conditioning, discharge the refrigerant from the A/C system. Disconnect the A/C lines at the expansion valve and retainer.
8. Remove the bolts attaching the crossmember to the side still and body.
9. Raise and support the vehicle safely. Remove the front wheels. Disconnect the passenger side tie rod ball stud from the knuckle with tool T.Av. 476 or equivalent.

NOTE: Run the tie rod ball joint nut to the end of the ball stud before installing the removal tool. This protects the stud threads when loosening the stud.

10. Remove the passenger side steering tie rod.
11. Loosen the coolant expansion tank retaining strap. Pull the tank out of the strap and move the tank aside. Move the tank far enough away to permit the A/C lines to be moved foward and away from the crossmember. Mount engine support tool M.S.1900 onto the fender flanges. Attach a lifting chain to the engine and support tool. Take up the slack in the chain, but do not left the engine at this time.
12. Remove the nuts attaching the exhaust pipe to the exhaust manifold. Disconnect the oxygen sensor.
13. Remove the 2 bolts connecting the front and rear exhaust pipe sections. Remove the hanger nut. Remove the front exhaust pipe section from the vehicle.
14. Turn the crossmember and remove it through the wheel well opening on the passenger side of the vehicle. Disengage the clutch cable adjusting mechanism by propping the clutch pedal up and completely against the upper stop.
15. Remove the steering knuckle upper mounting bolt. Then loosen (but do not remove) the lower bolt.

NOTE: The bolts are splined just below the head so it is necessary to remove the nut and tap the bolt out with a brass or lead mallet.

16. Remove the driveshaft roll pins with a suitable pin drift. Swing each rotor and steering knuckle outward. Then slide the driveshafts off of the transaxle output shafts.
17. Remove the plastic cover attached to the bottom of the transaxle. The cover is attached by a bolt and clip built into the cover.
18. Remove the transaxle drain plug and drain the transaxle lubricant. Install the drain plug.
19. Disconnect the reverse lockout cable. Unscrew the cable nut and pull the cable out of the transaxle.
20. Disconnect the shift rod from the lever. Remove the crossover bracket bolts and remove the shift rod from the bracket.
21. Disconnect the speedometer cable. Disconnect the ground strap at the transaxle. Support the transaxle with a suitable transaxle jack.

22. Remove the wiring from the starter solenoid. Remove the rear mounting bolts. Remove the bolts that attach the 2 transaxle mounting brackets to the transaxle and remove the brackets. Remove the starter mounting bolts and remove the starter and locating bushing.
23. Lower the the engine enough for access to the clutch housing bolts. Remove the nuts and bolts attaching the clutch to the engine.
24. Pull the transaxle straight back until the clutch clears the pressure plate and lower the transaxle carefully.

TRANSAXLE INSTALLATION

1. Using the transaxle jack, raise the transaxle up into position onto the engine dowels. Install the bolts and nuts that attach the clutch housing to the engine and torque them to 37 ft. lbs. (50 Nm).

NOTE: Hold the release lever inward to keep the release fork aligned. Also be sure the fork fingers are properly engaged in the release bearing. Be sure the clutch housing is seated on the alignment dowels before tightening the clutch housing bolts.

2. Install the starter mounting bolts on the clutch housing and tighten them. Reconnect the wire connections to the starter solenoid.
3. Connect the transaxle wire harness connectors. Install the grounding strap to the transaxle case. Raise the transaxle and install the 2 mounting brackets on the transaxle and torque the bolts to 30 ft. lbs. (40 Nm).
4. Align the transaxle support cushion studs and lower the transaxle. Install and tighten the support cushion nuts to 30 ft. lbs. (40 Nm).
5. Connect the speedometer cable, shift rods and reverse lockout cable. Be sure to apply some suitable thread sealer to the threads of the lockout cable retaining nut before tightening it.
6. Install the plastic cover on the bottom of the transaxle. Connect the clutch cable to the release lever and bracket on the clutch housing.
7. Remove the device used to prop up the clutch pedal. Then press and release the pedal several times to check cable operation.

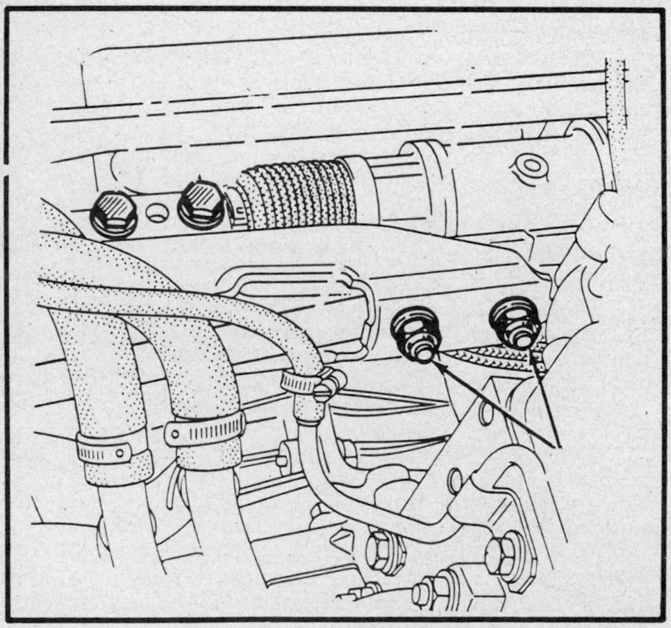

Removing the steering bracket assembly

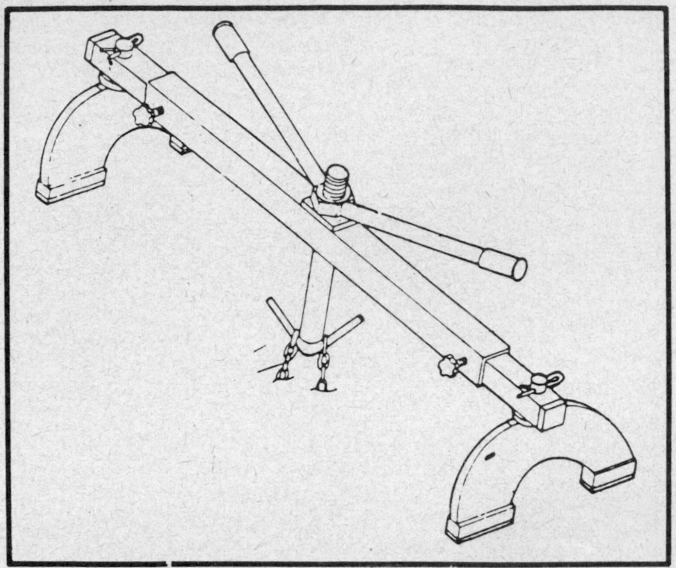

Typical engine support tool

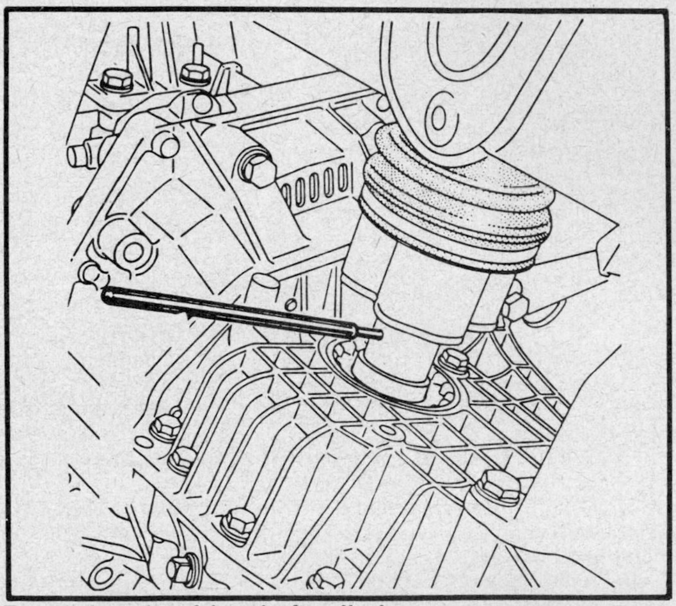

Removing the driveshaft roll pins

8. Before installing the driveshafts, identify the type of differential side gear shaft used in the transaxle. There are 3 different designs being used:
   a. Type 1 shafts have a shoulder that is 0.039 in. (1mm) long. A 0.118 in. (3mm) thick rubber washer is used between the ends of the side gear shaft and driveshaft.
   b. Type 2 shafts have a shoulder that is 0.118 in. (3mm) long. A 0.197 in. (5mm) thick rubber washer is used between the ends of the side gear shaft and driveshaft.
   c. Type 3 shafts do not have a machined shoulder nor do they require a rubber washer.
9. Verify the type side gear shaft and install the proper thickness rubber washer before proceeding.

NOTE: All side gear shaft types have a chamfer machined into one side of the shaft roll pin hole. The chamfer is included to make installation of the double pins easier.

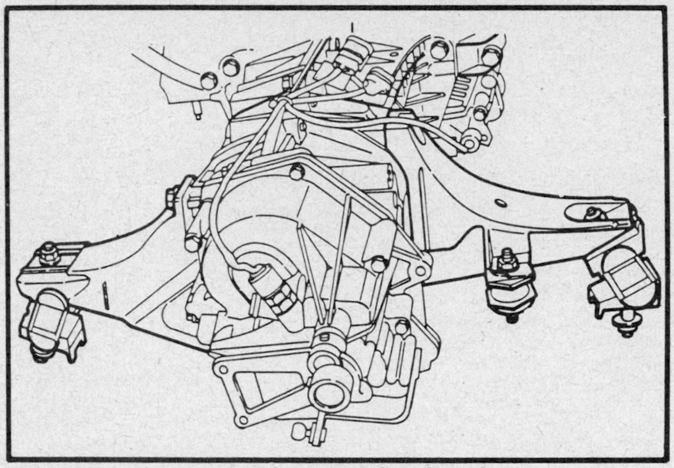

Removing the transaxle mounting brackets

10. Install the driveshafts onto the transaxle output shafts. Align the roll pin holes in the shafts.
11. Tilt the steering knuckle and rotor assemblies inward and align them in the shock bracket. Install the upper bolt. Torque the upper and lower bolt to 148 ft. lbs. (200 Nm).
12. Line up the driveshaft using tool B.Vi. 31–01 or equivalent and install the roll pins with a suitable drift pin. Place a small amount of silicone sealant at each end of the roll pins.
13. Install the crossmember through the wheel well opening on the passenger side of the vehicle. Position the crossmember on the side sills and body and install the crossmember attaching bolts and nuts. Tighten the crossmember to body bolts first and tighten the remaining bolts.
14. On the vehicles with A/C, connect the A/C lines to the connector block on the dash panel. Secure the lines to the side sill with the retainer.
15. Connect the steering tie rods to the knuckles. Torque the tie rod nuts to 25 ft. lbs. (35 Nm).
16. Connect the steering tie rods to the steering gear bracket. Tighten the attaching bolts and nuts to 25 ft. lbs. (35 Nm). Connect the steering gear bracket to the steering gear rack. Torque the bracket bolts to 30 ft. lbs. (40 Nm). Install the locknuts and washers on the bolts and torque the nuts to 25 ft. lbs. (35 Nm).

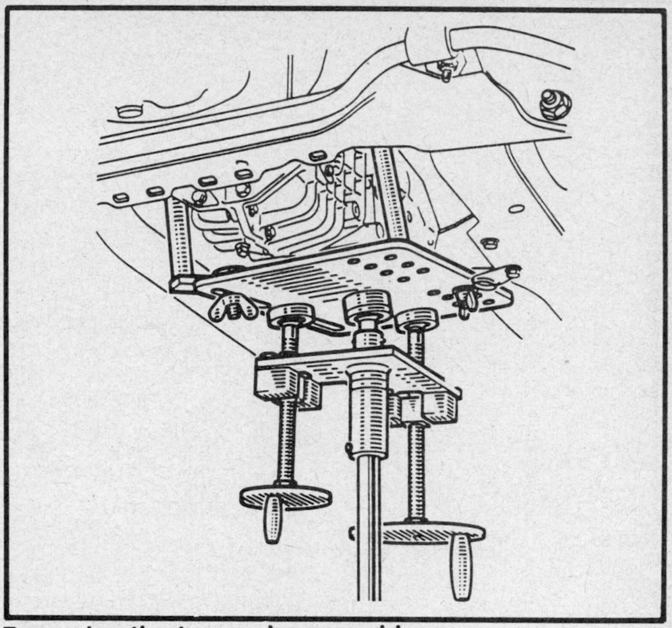

Removing the transaxle assembly

17. Install the front wheels and torque the bolts to 66 ft. lbs. (90 Nm).
18. Install the top dead center sensor. Install the warm up tube, brackets and hoses. Connect the exhaust head pipe to the exhaust manifold.
19. Install a replacement seal on the exhaust head pipe and connect the pipe to the converter. Tighten the bolts that attach the head pipe and converter until the spring coils are touching then back off each nut 1½ turns.
20. Connect the oxygen sensor wire. Install the vacuum capsule tube. Fill the transaxle with the recommended gear oil to the proper level and install the fill plug. Remove the engine support tool.
21. Connect the negative battery cable. Lower the vehicle.
22. On vehicles equipped with air conditioning recharge the system with fresh freon. Road test the vehicle and make all and any necessary adjustments.

BENCH OVERHAUL

Before Disassembly

Cleanliness during disassembly and assembly is necessary to avoid further transaxle trouble after overhaul. Before removing any of the transaxle subassemblies, plug all the openings and clean the outside of the of the transaxle thoroughly. Steam cleaning or car wash type high pressure equipment is preferable. During disassembly, clean all parts in suitable solvent and dry each part. Do not use cloth or paper towels to dry parts. Use compressed air only.

Transaxle Disassembly

1. With the transaxle removed from the vehicle and mounted on a suitable transmission jack, begin the disassembly as follows.
2. Remove the bolts attaching the clutch housing to the gear case and remove the clutch housing.
3. Remove the 5th gear shift rail detent plug, spring and interlock ball.
4. Remove the rear housing attaching bolts and remove the housing from the gear case.
5. Shift the transaxle into 1st and 5th to lock the transaxle. Remove the primary shift nut and washer.
6. Remove the 5th gear and synchronizer components and the shift fork with rail as an assembly.
7. Loosen the speedometer gear and secondary shaft nut. Remove the bolts attaching the 2 halves of the transaxle case and separate the case halves.
8. Remove the 5th gear shift rail interlock ball from the gear case.

NOTE: The 5th gear shift rail interlock will drop into the gear case when the 5th gear fork and rail are removed. Be sure to retrieve and retain the interlock ball after separating the gear case halves.

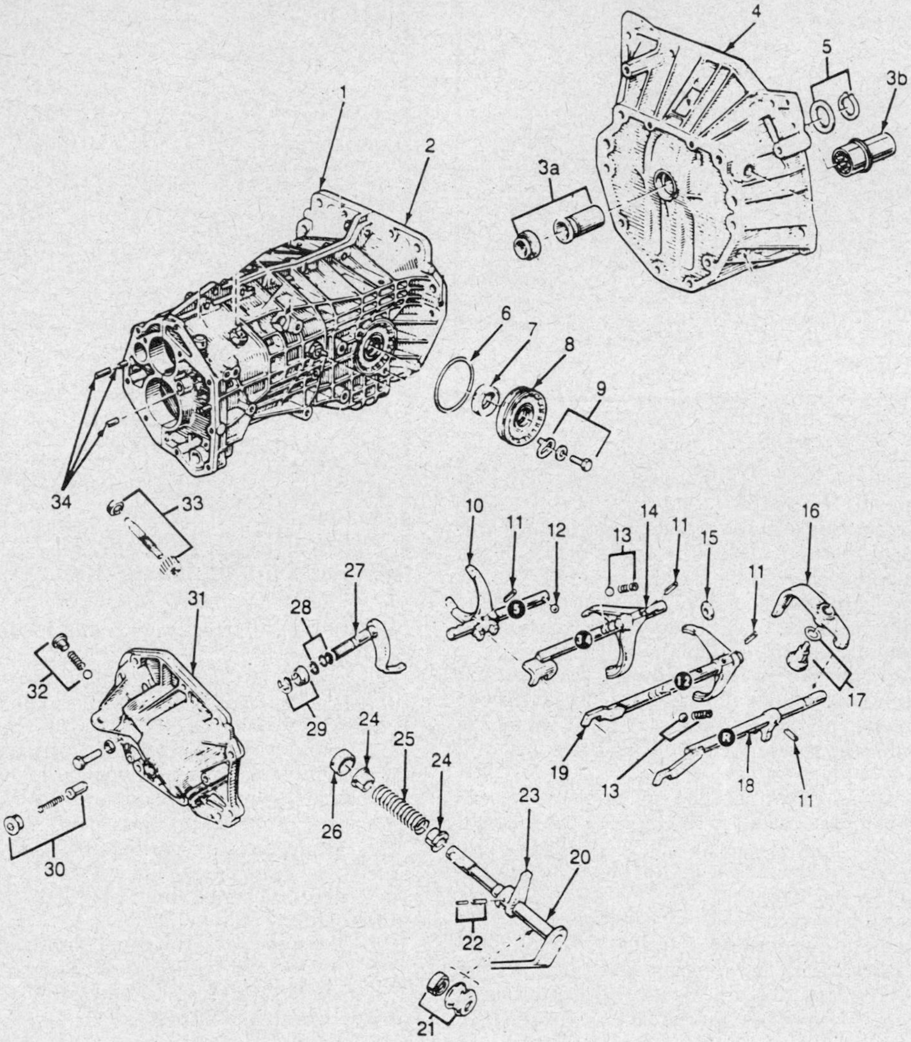
1
2
3a
3b
4
5
6
7
8
9
10
11
12
13
14
15
16
17
18
19
20
21
22
23
24
25
26
27
28
29
30
31
32
33
34

1. Gear Case - Left Half
2. Gear Case - Right Half
3a. Oil Seal and Guide Tube - Type 1
3b. Oil Seal, Guide Tube and Bearing - Type 2
4. Clutch Housing
5. Guide Tube Spacer and Retainer - Type 1
6. Adjusting Nut O-Ring
7. Adjusting Nut Oil Seal
8. Differential Adjusting Nut
9. Adjusting Nut Retaining Clip, Washer and Bolt
10. Fifth Gear Shift Fork and Rail
11. Roll Pins
12. Fifth Gear Shift Rail Interlock Ball
13. Detent Balls and Springs
14. 3-4 Shift Fork and Rail
15. Interlock Disc
16. Reverse Lever
17. Reverse Lever Pivot Bolt and Washer
18. Reverse Fork and Rail
19. 1-2 Shift Fork and Rail
20. Selector Shaft
21. Selector Shaft Boot and Oil Seal
22. Roll Pins
23. Selector Lever
24. Bushings
25. Spring
26. Cup Plug
27. Control Arm
28. Control Arm Shim and O-Rings (2)
29. Control Arm Retaining Clip and Washer
30. Stop Plunger, Spring and Plug
31. Rear Housing
32. Fifth Rail Detent Plug, Spring and Ball
33. Speedometer Driven Gear, Shaft and Seal
34. Dowel Pins

Exploded view of the NG9 transaxle case and its components

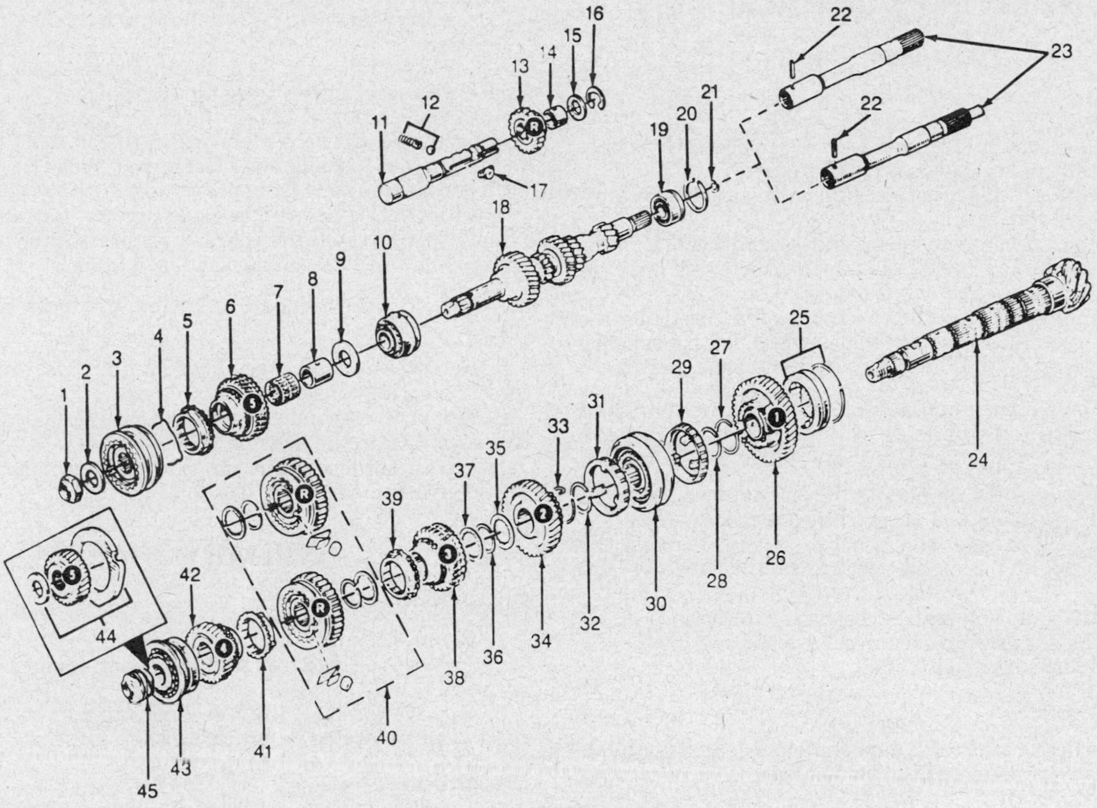

1. Primary Shaft Nut
2. Washer
3. Synchronizer Sleeve - Fifth Gear
4. Synchronizer Spring
5. Synchro Ring
6. Fifth Gear
7. Bearing
8. Bearing Race
9. Bearing Washer
10. Primary Shaft Rear Bearing
11. Reverse Idler Shaft
12. Detent Ball and Spring
13. Reverse Gear
14. Bushing
15. Thrust Washer
16. E-Clip
17. Reverse Idler Shaft Interlock
18. Primary Shaft
19. Primary Shaft Front Bearing
20. Snap Ring
21. Washer
22. Roll Pin
23. Clutch Shaft
24. Secondary Shaft
25. Bearing and Locating Ring
26. First Gear
27. Synchro Spring
28. Snap Ring
29. Synchro Ring
30. 1-2 Synchronizer Hub and Sleeve
31. Synchro Ring
32. Snap Ring
33. Synchro Spring
34. Second Gear
35. Splined Washer (Large)
36. Snap Ring
37. Splined Washer (Small)
38. Third Gear
39. Synchro Ring
40. 3-4 Synchronizer/Reverse Gear Assembly (See Note)
41. Synchro Ring
42. Fourth Gear
43. Double Row Bearing
44. Fifth Fixed Gear, Spacer Plate and Retaining Nut
45. Speedometer Gear/Secondary Shaft Nut

NOTE: Position of the 3-4 Synchro/reverse gear snap ring and splined washer is reversed in some NG transaxles.

Exploded view of the primary and secondary shafts

9. Remove the secondary shaft and gear assembly. Then remove the primary shaft and gear assembly.

Unit Disassembly and Assembly

SECONDARY SHAFT

Disassembly

1. Mount the secondary shaft in a vise equipped with protective jaws. Clamp the vise jaws on the 1st gear.
2. Move the 3rd/4th synchronizer hub downward into gear to the lock the gears in position.
3. Remove the speedometer gear and secondary shaft nut. Remove the washer, fixed 5th gear and spacer. Use a puller if the gear is a tight fit on the shaft.
4. Remove the rear ring, 4th gear and the synchro ring.
5. Remove the splined washer, snapring and 3rd/4th synchro reverse gear assembly. Retain the springs and roller cages.

NOTE: On some transaxles, the snapring and splined washer are installed opposite side of the synchro/gear assembly.

6. Remove the synchro ring and 3rd gear. Remove the small splined washer, snapring and large splined washer.
7. Remove the 2nd gear and synchro spring. Remove the snapring, synchro ring, 1st/2nd synchro hub with sleeve and synchro ring.
8. Remove the snapring and 1st gear synchro spring.
9. Remove the 1st gear and remove the front bearing and locating ring from the secondary shaft.

Inspection

1. Clean and inspect the secondary shaft and gear assemblies.
2. Replace any and all worn or damaged components as necessary.

Assembly

Do not reuse any of the shaft snaprings. Install replacement rings only. To avoid distorting snaprings, use snapring pliers to spread them. Whenever space permits, also use a pair of standard pliers to further support the snaprings during installation.

1. Lubricate the shaft and gears with gear lubricant prior to assembly. Petroleum jelly can also be used to hold parts in place during assembly operations.
2. Install the locating ring on the front bearing and install the assembled bearing and ring onto the secondary shaft.
3. Install the 1st gear, synchro spring and snapring on the shaft. Assemble the 1st/2nd synchro hub and sleeve and the synchros. Install the assembly onto the shaft. The stepped side of the synchro sleeve faces away from the shaft pinion gear.
4. Install the snapring and synchro spring. Install the 2nd gear, large splined washer and snapring. Install the small splined washer and 3rd gear with synchro ring.
5. Install the 3rd/4th synchro/reverse gear assembly, snapring and spacer.

NOTE: There are 2 different 3rd/4th synchro/reverse gear assemblies being used. If the transaxle has a type 1 assembly, install the snapring and spacer before installing the synchro/gear assembly. If the transaxle has a type 2 assembly, install the snapring and washer after installing the sleeve/gear assembly.

6. Install the 4th gear, synchro ring and rear bearing on the shaft. Install the spacer.
7. Apply some suitable thread sealer to the splines of the fixed 5th gear. Then install the gear and washer on the shaft.
8. Adjust the 5th speed fixed 5th gear and shaft bearing preload with a suitable press as follows:

   a. Mount the secondary shaft assembly in a press with the fixed gear facing up. Insert a tube or suitable tool between the press arbor and the gear to prevent the arbor from contacting the shaft. Press force must only be applied to the gear.
   b. Apply just enough press force to hold the shaft assembly in place.
   c. Wrap a length of cord around the rear bearing. Then attach a spring tension scale tool to the cord.
   d. Apply 200 lbs. press force to the gear and measure prelaod with the cord and spring tension tool.
   e. Preload should be 3–9 lbs. If the preload is less than specified, continue increasing press force until the desired preload is achieved. However, if preload exceeds specified limits, remove and reinstall the gear. Apply press force in smaller increments until desired preload is obtained.

NOTE: Do not exceed 3300 lbs. of press force at any time.

9. Remove the shaft assembly from the press and remove the cord and spring tension cord.
10. Mount the secondary shaft in a vise equipped with protective jaws. Clamp the jaws on the 1st gear.
11. Install the speedometer gear and secondary shaft nut. Torque the nut to 110 ft. lbs.

PRIMARY SHAFT

Disassembly

1. Remove the clutch shaft roll pin.
2. Remove the clutch shaft and washer. Remove the snapring or spacer.
3. Remove the shaft front bearing and rear bearing from the shaft with a suitable puller and a shop press.

Inspection

1. Clean and inspect the clutch shaft, primary shaft and bearing.
2. Inspect the 5th fixed gear and synchro assembly.
3. Replace any and all worn or damaged components as necessary.

Assembly

1. Lubricate the shaft and gears with gear lubricant prior to assembly. Petroleum jelly can also be used to hold parts in place during assembly operations.
2. Install a replacement rear bearings on the shaft with a shop press.
3. Install a replacement front bearing, spacer or snapring and washer. Do not install the 5th fixed gear and synchro assembly at this time.

REVERSE IDLER GEAR AND SHAFT

Disassembly

The differential must be removed in order to service the reverse idler gear and shaft.

1. Remove the shaft E-clip. Remove the shaft, gear and thrust washer.
2. Remove the interlock and detent ball with spring from the case bore.

Inspection

1. Clean and inspect the shaft and gear components.
2. Replace the thrust washer and E-clip, they are not reusable.
3. Replace any and all worn or damaged components as necessary.

Assembly

1. Lubricate the shaft and gears with gear lubricant prior to assembly. Petroleum jelly can also be used to hold parts in place during assembly operations.
2. Install the detent ball and spring into the case.
3. Slide the shaft part way into the case. Install the gear and thrust washer onto the shaft. The bronze side of the thrust washer faces up.
4. Insert the shaft guide into the case. Install the shaft and push the shaft guide into place. Install the replacement E-clip.

SHIFT FORK

Disassembly

1. Remove the shift rail detent springs and balls.
2. Remove the shift fork roll pins. Remove the interlock disc.
3. Remove the interlock ball. Slide the shift fork off the shift rails and remove the rails from the case.
4. Remove the reverse lever pivot bolt and washer and remove the reverse lever.

Inspection

1. Clean and inspect the shift rail and its components.
2. Replace any and all worn or damaged components as necessary.

Assembly

1. Lubricate the shift rail with gear lubricant prior to assembly. Petroleum jelly can also be used to hold parts in place during assembly operations.
2. Install the reverse shift rail and roll pin. Install all of the roll pins with the open side of the pin facing the rear case.
3. Position the reverse lever in the case. Be sure the lever is seated in the reverse shift rail notch. Install the pivot bolt and washer.
4. Install the remaining shift rails and forks in the case.

NOTE: A unique 1st/2nd shift rail is used on the NG9 transaxles. The rail is equipped with a spring loaded ball and the roll pin hole is oval shaped. When installing the shift fork, use a length of steel rod to align the holes in the fork and rail to ease roll pin installation.

5. Install the 5th gear interlock ball in the case. Insert the 5th gear shift rail in the case to hold the ball in position for final assembly.

REAR CASE

Disassembly

1. Remove the stop plunger, spring and plug.
2. Remove the control arm clip and washer.
3. Remove the roll pins that secures the selector lever to the shaft.
4. Slide the selector shaft out of the case. Remove the oil seal and boot from the selector shaft. Discard the seal.
5. Remove the selector lever, shaft bushings and spring. Inspect the cup plug. Replace the plug if lose or leaking. Remove the speedometer gear assembly.

NOTE: If the speedometer gear must be replaced, spread the gear retaining tabs to remove the gear from the shaft. Then discard the gear, because it is not reuseable once it has been removed.

Inspection

1. Clean and inspect the rear case and its components.
2. Replace any and all worn or damaged components as necessary.

Assembly

1. Install a replacement selector shaft cup plug if required. Apply sealer to the plug before tapping it into the case.
2. Install replacement O-rings on the control arm. The install the arm and spacer in the case.
3. Install a replacement selector shaft oil seal in the case. Coat the seal lip with pertroleum jelly or equivalent.
4. Install the boot on the selector shaft. Coat the end of the selector shaft with petroleum jelly or equivalent. Insert the shaft through the oil seal and part way into the case.
5. Slide the selector lever, bushings and spring onto the selector shaft. Seat the selector shaft into the case.
6. Align the selector lever and shaft and install the 2 roll pins. Position the openings in the pins opposite one another (180 degrees apart).
7. Install the stop plunger, spring and plug in the case. Install a replacement seal on the plug and coat the threads with a suitable thread sealer before installation.
8. Install the speedometer gear assembly in the case. Install a replacement speedometer gear oil seal.

DIFFERENTIAL

Disassembly

1. Remove and discard the ring gear bolts.
2. Remove the differential bearing with a suitable puller. Remove the puller tools, remove the ring gear and remove the roll pin with a suitable punch.
3. Remove and discard the pinion retaining collar. Remove the pinion shaft.
4. Remove the pinion gears, washers, pinion pins, thrust block, pinion gear washers, side gears and thrust washers.
5. Remove the differential adjusting nuts from the case halves. Remove the oil seals from the adjusting nuts. Discard the seals.
6. Remove the differential bearing races with a length of steel tube that has an outside diameter of 2.79 in. (71mm).

Inspection

1. Clean and inspect the differential components.
2. Replace any and all worn or damaged components as necessary.

Assembly

1. Lubricate the shift rail with gear lubricant prior to assembly. Petroleum jelly can also be used to hold parts in place during assembly operations.
2. Install the thrust washer in the differential case. Be sure the washer oil groove faces the side gears.
3. Install the first side gear in the differential case.

NOTE: The pinion gear washers have locating tabs. When installing the washers be sure the washer tabs are seated in the differential case tab slots.

4. On transaxles equipped with 2 differential pinions, proceed as follows:
   a. Install the pinion gears and washers.
   b. Start the pinion shaft into the case.
   c. Align the roll pin holes in the pinion shaft and case. Then tap the shaft into place.
   d. Start the roll pin in the case. Tap the pin into the case through the shaft.
5. On transaxles equipped with 4 differential pinions, proceed as follows:
   a. Install the pinion gears and washers.
   b. Install the pinion pins and thrust block.
   c. Align and install the pinion shaft.
6. Install the remaining side gear in the case.

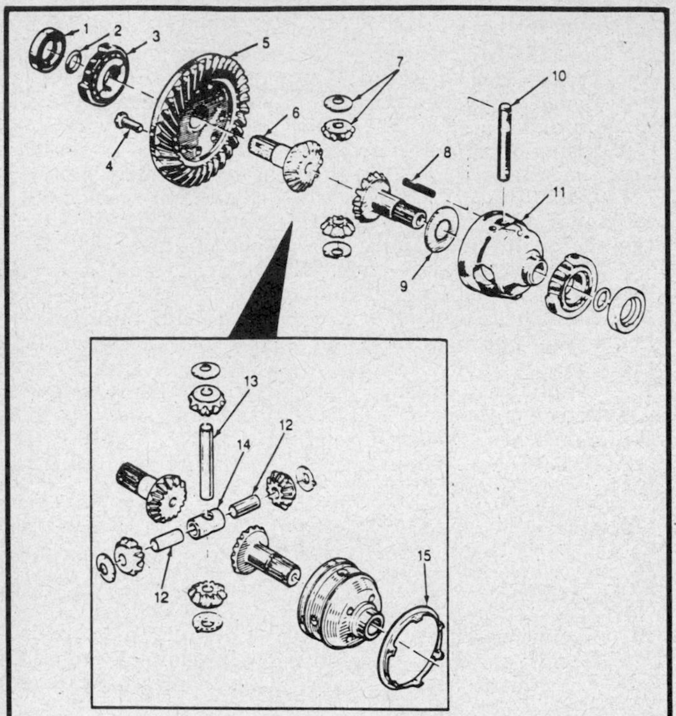

1. Oil Seal
2. Seal Washer
3. Differential Bearing and Race
4. Ring Gear Bolts
5. Ring Gear
6. Side Gear
7. Differential Pinion and Washer
8. Roll Pin
9. Thrust Washer
10. Pinion Shaft
11. Differential Case
12. Pinion Pins
13. Pinion Shaft
14. Thrust Block
15. Pinion Retaining Collar

Exploded view of the NG9 differential assembly

7. Install the ring gear. Secure the ring gear with replacement bolts only. Install a replacement collar on the case with the aid of a shop press.
8. Install the differential bearing. Install the differential bearing races in the gear case halves. Use a 2.79 in. (71mm) diameter tube to install the races.
9. Install the adjusting nuts in the gear case halves. Thread the nut facing the differential case 2–3 threads deeper than the nut facing the ring gear. Do not install replacement differential case oil seals at this time.

Adjust the differential preload.

CHECKING DIFFERENTIAL BEARING PRELOAD

1. Position the differential assembly into the gear case and assemble the gear case halves. Tighten the case attaching bolts to the specified torque.

NOTE: Do not install secondary shaft (and pinion gear) in the case at this time. Preload adjustment is performed with the shaft out of the case.

2. If the original differential bearing are being used, adjust the preload as follows:
   a. Tighten both adjusting nuts until most of the endplay is eliminated.
   b. Tighten the adjusting nut facing the differential case until all the endplay is removed.
   c. Rotate the differential. It should rotate freely after eliminating the endplay.
3. If replacement differential bearings were installed, adjust the preload as follows:
   a. Rotate the differential and tighten the adjusting nuts equally until a very slight darg is felt.
   b. Tighten the turn the left hand nut until the differential becomes slightly hard to turn.
   c. Rotate the differential several times to seat the bearings.
   d. Wrap a length of cord around the differential case. Then attach a spring tension gauge to the end of the cord.
   e. Pull on the gauge and note the amount of force needed to rotate the differential. Preload is correct when the rotating force is 7–22 ft. lbs.

NOTE: If the preload is too low, tighten the adjusting nut. If the preload is too high, loosen the adjusting nut.

Mark the position of the adjusting nuts and separate the gear case halves for final assembly and adjustment.

CLUTCH HOUSING SHAFT BEARING

There are 2 different clutch housing, clutch shaft and support bearing assemblies being used. Type 1 assemblies are used with engines having either a cast iron or aluminum cylinder block. Type 2 assemblies are used only on aluminum block engines.

On transaxles equipped with the type 1 assembly, the clutch shaft is supported by a pilot bearing in the crankcase. The release bearing rides on a guide tube mounted in the housing. A separate oil seal is used at the rear of the housing clutch shaft bore.

On transaxles equipped with type 2 assembly, the shorter clutch shaft is supported by a bearing in the combination guide tube, bearing and seal assembly. The release bearing rides on the tube portion of the assembly.

OIL SEAL AND GUIDE TUBE (TYPE 1)

Replacement

1. Tap the tube and seal out of the housing with a hammer and punch.
2. Install the replacement oil seal with tool seal installer and install the guide tube.

GUIDE TUBE ASSEMBLY (TYPE 2)

Replacement

The combination guide tube, oil seal and bearing assembly used for type 2 assemblies, is not a reuseable part. The assembly must be replaced once it has been removed.

1. Remove the clutch fork retaining pins by removing the fork roll pins.

NOTE: If the clutch fork retaining pins are the solid type, it may be necessary to use a slide hammer to remove them.

2. Remove the clutch fork and shaft from the housing. Remove the guide tube assembly with a shop press and suitable diameter length of steel tube.

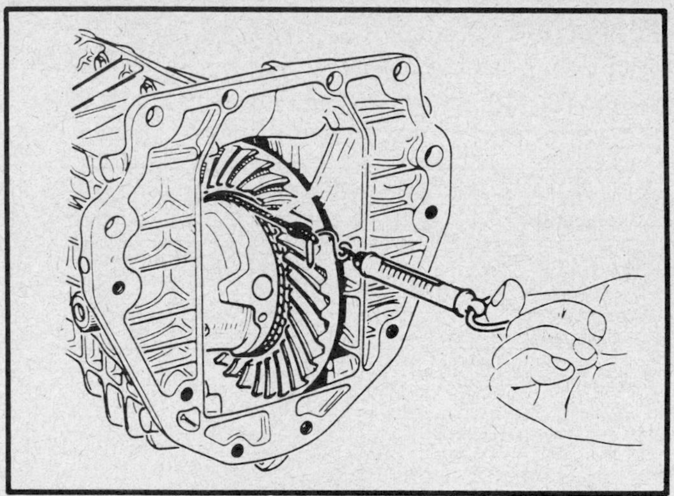

Measuring the differential preload

3. Install a new O-ring on the replacement guide tube assembly.
4. Align the oil feed hole in the tube with the feed in the housing. Press the guide tube into the housing until firmly seated.
5. Lubricate the clutch fork and shaft with a suitable lubricant and install the clutch fork and shaft.
6. If the fork is secured with roll pins, tap the pins into place with a hammer and punch.
7. If the fork is secured with solid type pins, start the pins in the fork. Tap the pins into the fork until the first shoulder of the pin reaches 0.040 in. (1mm) from the fork surface.

Transaxle Assembly

1. Install the primary and secondary shaft assemblies in the appropriate gear case half.
2. Install the differential assembly in the gear case.
3. Coat the seal surfaces of the gear case halves with RTV type sealant and assemble the case halves. Torque the case attaching bolts to 19 ft. lbs.
4. Make the ring and pinion backlash adjustment as follows:
   a. Mount a dial indicator on the gear case. Position the indicator stylus on the flank of one gear tooth.
   b. Move the ring gear back and fourth and measure the backlash which should be 0.004–0.009 in. (0.12–0.25mm).
   c. If the backlash is excessive, loosen the left hand adjusting nut and tighten the right hand adjusting nut.
   d. If the backlash is too little, loosen the right hand adjusting nut and tighten the left hand adjusting nut.
   e. Be sure to loosen or tighten the adjusting nuts in equal amounts when setting backlash.
   f. Remove the dial indicator and mark position of the adjusting nuts.
5. Install the differential adjusting nut seal as follows:
   a. Remove only one adjusting nut at a time when replacing the seals. If both nuts are removed at the same time, ring gear backlash and differential bearing preload will have to be reset.
   b. Mark position of both adjusting nuts for installation reference.
   c. Remove the first adjusting nut, count the number of turns required to unthread the nut from the case.
   d. Lubricate the replacement adjusting nut O-ring and oil seal with the proper gear lube.
   e. Carefully roll the O-ring onto the nut. Then install the oil seal. The seal should be flush with the inner face of the nut.
   f. Install the seal protector over the side gear.
   g. Install the adjusting nut. Thread the nut into the case the same number of turns counted during removal.
   h. Align the reference marks on the adjusting nut and case and install the nut retaining clip.
   i. Remove the seal protector and repeat the seal replacement procedure on the opposite adjusting nut.
   j. Recheck and adjust the ring gear backlash if necessary.
6. Install the 5th fixed gear spacer on the secondary shaft. Apply some suitable thread sealer to the bore of the 5th fixed gear and install the gear and washer on the secondary shaft. The raise shoulder of the 5th fixed gear faces upward.
7. Coat the threads of the combination speedometer/secondary shaft nut with thread sealer and install the nut on the secondary shaft. Do not fully tighten the nut at this time.
8. Install the washer, race, bearing and 5th gear onto the primary shaft.
9. Assemble the the 5th gear synchronizer ring, spring and sleeve. Insert the 5th gear shift fork in the synchronizer sleeve. Then align and install the fork sleeve, on the shift rail primary shaft.
10. Install the roll pin into the 5th gear shift fork.
11. Shift the transaxle into 1st and reverse to lock the primary and secondary shafts.
11. Torque the primary shaft nut to 96 ft. lbs. (130 Nm). Torque the secondary shaft gear/nut to 110 ft. lbs. (150 Nm).
12. Coat the rear housing gasket with sealer and position the gasket on the gear case.
13. Install the rear housing on the gear case. Torque the housing bolts to 7–11 ft. lbs. (10–15mm).
14. Install the 5th gear shift rail detent plug, spring and interlock ball.
15. Coat the clutch housing with sealer and position the gasket on the gear case.
16. Insert the seal protector in the clutch seal to protect the seal lip.
17. Install the clutch housing on the gear case. Keep the clutch shaft and seal protector aligned during installation.
18. Remove the seal protector and tighten the clutch housing bolts. Remove the transaxle from the transaxle support stand.
19. After installing the transaxle in the vehicle, fill the transaxle with the proper gear lubricant.

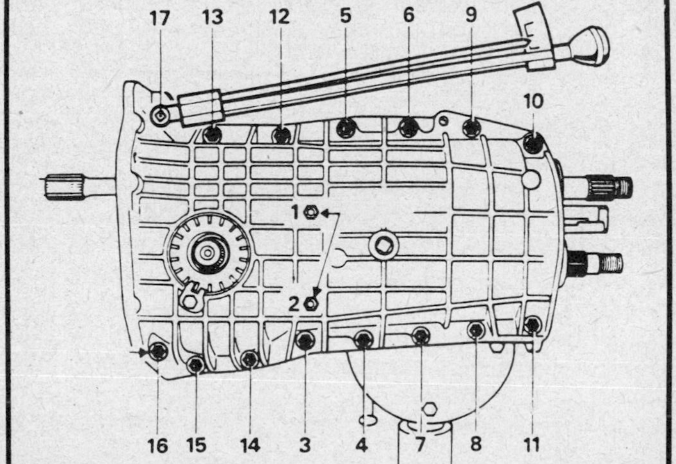
17
13
12
5
6
9
10
1
2
16
15
14
3
4
7
8
11

Bolt tightening sequence for the transaxle case assembly

SPECIFICATIONS

TORQUE SPECIFICATIONS

Component	ft. lbs.	Nm
Transaxle support cushion nuts	30	40
Transaxle mounting bracket bolt	30	40
Shock absorber lower nuts	148	200
Steering tie rod ball joint nut	30	40
Wheel retaining bolts	66	90
Steering bracket bolts	30	40
Tie rod bracket nuts	26	35
Steering bracket lock nuts	26	35
Transaxle-to-engine bolts	37	50
Primary shaft nut	96	130
Secondary shaft nut	110	150
Rear housing bolts	11	15
Reverse lockout cable nut	15	20

SPECIAL TOOLS

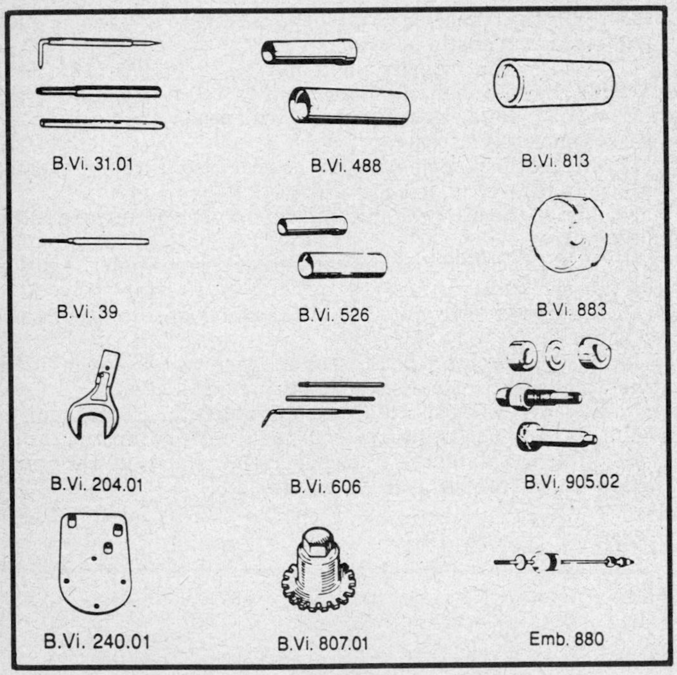
B.Vi. 31.01
B.Vi. 488
B.Vi. 813
B.Vi. 39
B.Vi. 526
B.Vi. 883
B.Vi. 204.01
B.Vi. 606
B.Vi. 905.02
B.Vi. 240.01
B.Vi. 807.01
Emb. 880

Tool Number	Tool Description
B.Vi. 31.01	Roll pin punch set
B.Vi. 39	Roll pin punch
B.Vi. 204.01	Secondary shaft nut wrench
B.Vi. 240	Gearbox support
B.Vi. 488	Clutch shaft oil seal installer
B.Vi. 526	Clutch shaft oil seal installer/protector
B.Vi. 606	Roll pin punch set
B.Vi. 807.01	Differential nut wrench
B.Vi. 813	Oil seal protector
B.Vi. 883	Differential collar
B.Vi. 905.02	Speedometer shaft oil seal tool set
Emb. 880	Slide hammer

Section 6

RGT and RWB Transaxle
Ford Motor Co.

APPLICATION

RGT and RWB

Year	Vehicle	Engine
1984	Topaz	2.3L, 2.0L Diesel
	Tempo	2.3L, 2.0L Diesel
	EXP	1.6 L
	Escort	1.6L, 2.0L Diesel
	Lynx	1.6L, 2.0L Diesel
1985	EXP	1.6L, 1.9L, 2.0L Diesel
	Escort	1.6L, 1.9L, 2.0L Diesel
	Lynx	1.6L, 1.9L, 2.0L Diesel
	Tempo	1.6L, 2.3L, 2.0L Diesel
	Topaz	1.6L, 2.3L, 2.0L Diesel
1986	EXP	1.9L, 2.0L Diesel
	Escort	1.9L, 2.0L Diesel
	Lynx	1.9L, 2.0L Diesel
	Tempo	2.3L, 2.0L Diesel
	Topaz	2.3L, 2.0L Diesel

Year	Vehicle	Engine
1986	Taurus	2.5L
	Sable	2.5L
1987	EXP	1.9L
	Escort	1.9L, 2.0L Diesel
	Lynx	1.9L, 2.0L Diesel
	Tempo	2.3L
	Topaz	2.3L
	Taurus	2.5L
	Sable	2.5L
1988–89	Escort	1.9L
	Tempo	2.3L
	Topaz	2.3L
	Taurus	2.5L
	Sable	2.5L

GENERAL DESCRIPTION

Because of the similarities in function, design and service, the 4 and 5 speed manual transaxle have been combined into this single section.

The 4 speed manual transaxle is similar in construction to the 5 speed manual transaxle except for the deletion of a 5th gear driveshaft assembly and a 5th gear shift fork assembly. Although similar in appearance, the gear set of the 4 speed transaxle cannot be interchanged with those of a 5 speed manual transaxle.

The transaxle houses the transaxle and differential assemblies in a 2 piece aluminum alloy case. The transaxle is fully synchronized in all forward gears with reverse provided by a sliding gear. All gears, except reverse, are helical cut for quiet operation. Both shafts, the input cluster and mainshaft, are supported in the case on tapered roller bearings. Preload is maintained on the bearings by a shim located behind each bearing cup in the transaxle housing. Lubrication is provided by oil transfer from the ring gear teeth to a funnel located behind the mainshaft bearing cup. The funnel directs oil into a hole bored through the center of the mainshaft, providing lubrication to rotating gears on the shaft. Further lubrication is provided by the splash created by the moving gears.

To prevent pressure buildup in the case, which could result in leaking seals, a vent is installed in the top of the transaxle housing. Automatic transmission fluid is used as a lubricant to ensure shifting ease under all driving conditions.

The differential and transaxle assemblies are connected as a unit. The lubrication fluid is the same for both the transaxle and the differential.

Metric Fasteners

The metric fastener dimensions are very close to the dimensions of the familiar inch system fasteners, and for this reason, replacement fasteners must have the same measurement and strength as those removed.

CAUTION

Do not attempt to interchange metric fasteners for inch system fasteners. Mismatched or incorrect fasteners can result in damage to the transaxle unit through malfunctions or breakage and possible personal injury.

Capacities

The 4 and 5 speed transaxles use Dexron®II automatic transmission fluid. The capacity of the 4 and 5 speed transaxles is 6.1 pints or 2.9L. The correct fluid level is to the bottom of the filler hole.

Checking Fluid Level

The transaxle fluid level check must be made with the vehicle level and the engine must not be running. The fluid level can be checked by removing the fill plug.The correct fluid fill will be even with the bottom edge of the filler plug opening or within ¼ inch of this level. If fluid is low, added the specified fluid to correct level.

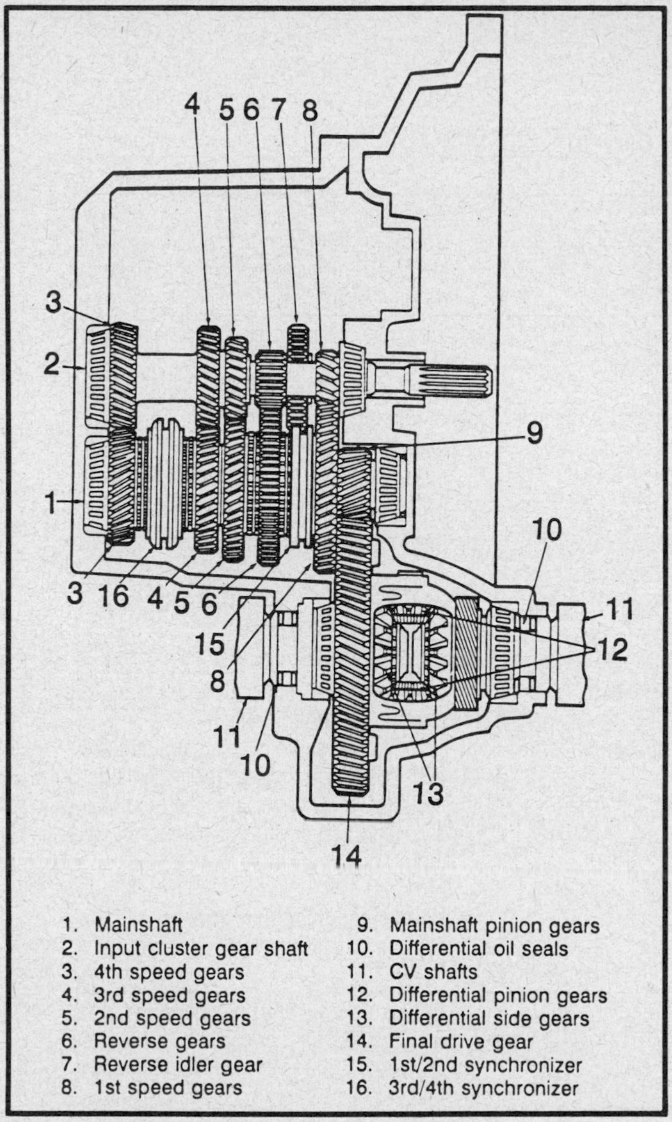
1. Mainshaft
2. Input cluster gear shaft
3. 4th speed gears
4. 3rd speed gears
5. 2nd speed gears
6. Reverse gears
7. Reverse idler gear
8. 1st speed gears
9. Mainshaft pinion gears
10. Differential oil seals
11. CV shafts
12. Differential pinion gears
13. Differential side gears
14. Final drive gear
15. 1st/2nd synchronizer
16. 3rd/4th synchronizer

View of 4 speed transaxle components

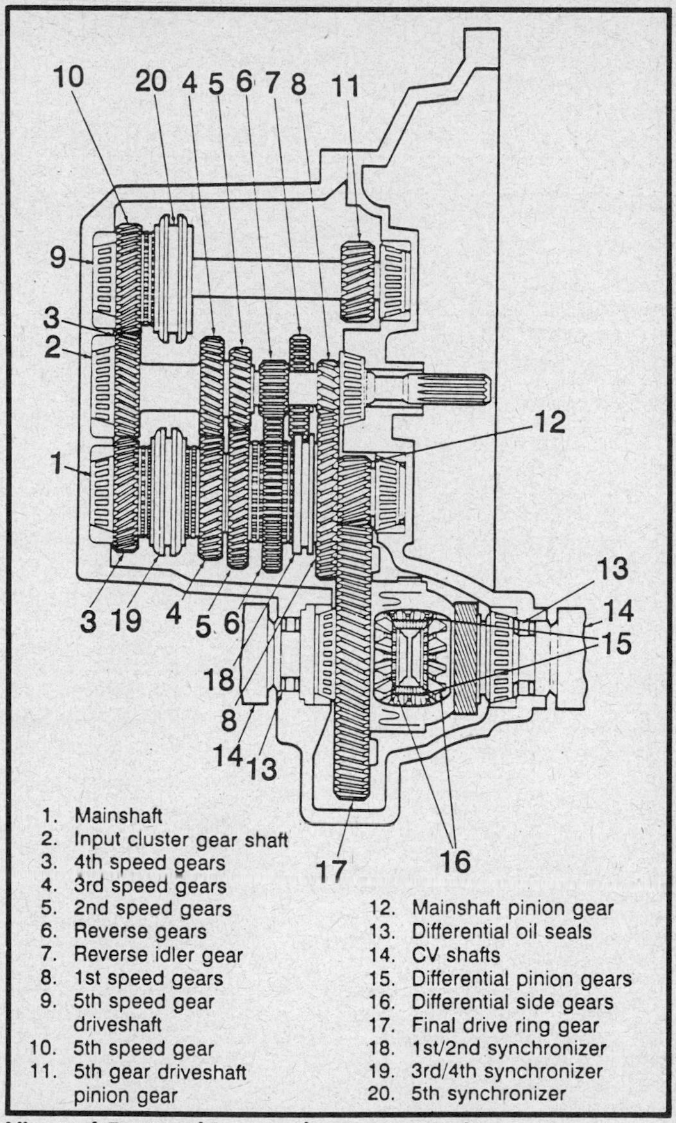
1. Mainshaft
2. Input cluster gear shaft
3. 4th speed gears
4. 3rd speed gears
5. 2nd speed gears
6. Reverse gears
7. Reverse idler gear
8. 1st speed gears
9. 5th speed gear driveshaft
10. 5th speed gear
11. 5th gear driveshaft pinion gear
12. Mainshaft pinion gear
13. Differential oil seals
14. CV shafts
15. Differential pinion gears
16. Differential side gears
17. Final drive ring gear
18. 1st/2nd synchronizer
19. 3rd/4th synchronizer
20. 5th synchronizer

View of 5 speed transaxle components

TRANSAXLE MODIFICATIONS

MANUAL TRANSAXLE JUMPS OUT OF 2ND OR 4TH GEAR

Determine whether vehicle was manufactured between March 1, 1985 and February 3, 1986. If it was, make sure that the shift lever boot is seated in the top spring of the console.

1. If the transaxle is a 4 speed unit, inspect the paddle that operated the upshift light top gear switch. Bend the paddle so the switch can be depressed 0.16 inch.
2. Lift and safely support the engine just enough to relieve tension on the engine mounts. Then loosen the front and rear mounts and the rear shifter to body mounts.
3. Position the engine as far forward as the slots in the mounts allow. The mounts should then be tightened.
4. Remove the transaxle detent plunger retainer screw and install a differently designed transaxle detent plunger and spring assembly (part number E6FZ–7233–A). Coat the threads of the assembly with a sealant that contains Teflon. The assembly should be tightened to 66–96 inch lbs.
5. Check the transaxle fluid level and replenish the fluid, if necessary. If vehicle is equipped with the 2L diesel engine, the shift boot assembly should also be replaced with part number E6FZ–7277–B.

TROUBLE DIAGNOSIS

CHILTON THREE C'S TRANSAXLE DIAGNOSIS

Condition	Cause	Correction
Clicking noise in reverse gear	a) Damaged or rough gears b) Damaged linkage preventing complete gear travel	a) Replace damaged gears b) Check for damaged or misaligned shift linkage or other causes of shift linkage travel restrictions
Gear clash into reverse	a) Owner not familiar with manual transmission shift techniques b) Damaged linkage preventing complete gear travel	a) Instruct customer on non-synchronous reverse and clutch spin-time-lapse required before a shift into reverse b) Check for damaged or misaligned shift linkage or other causes of shift linkage bind
Gears clash when shifting from one forward gear to another	a) Improper clutch disengagement b) Clutch disc installed improperly with damper springs toward flywheel c) Worn or damaged shift forks, synchro-teeth (usually high mileage phenomenon). Forward gears only	a) Check clutch system and adjustment b) Check clutch system c) Check for damage, and service or replace as required
Leaks	a) Excessive amount of lubricant in transaxle—wrong type b) Worn or damaged internal components c) Slight mist from vent	a) Check lube level and type. Fill to bottom of filler plug opening b) Remove transaxle clutch housing lower dust cover and inspect for lube inside housing. Inspect for leaks at the shift lever shaft seal, differential seals and input shift shaft seal. Service as required c) Normal condition that does not require service. If dripping, check lubricant level
Locked in one gear—it cannot be shifted out of that gear	a) Damaged external shift mechanism b) Internal shift components worn or damaged c) Synchronizer damaged by burrs which prevent sliding action	a) Check external shift mechanism for damage. Service or replace as required b) Disconnect external shift mechanism and verify problem by trying to shift input shift rail. Remove transaxle. Inspect the problem gear, shift rails, and fork and synchronizer assemblies for wear or damage. Service or replace as required c) Replace synchronizer assembly
Noise in Neutral	a) Neutral rollover rattle	a) Normal condition exists
Noisy in forward gears	a) Low lubricant level b) Contact between engine/transaxle and chassis c) Transaxle to engine block bolts loose d) Worn or damaged input/output bearings. Worn or damaged gear teeth (usually high mileage phenomenon) e) Gear rattle	a) Fill to bottom of filler plug opening with proper lubricant b) Check for contact or for broken engine motor mounts c) Tighten to specification d) Remove transaxle. Inspect bearings and gear teeth for wear or damage. Replace parts as required e) Normal condition exists
Shifts hard	a) Improper clutch disengagement b) External shift mechanism binding c) Clutch disc installed improperly with damper springs toward flywheel	a) Check clutch system and adjustment b) Check shift mechanism c) Check clutch system

CHILTON THREE C'S TRANSAXLE DIAGNOSIS

Condition	Cause	Correction
Shifts hard	d) Internal damage to synchronizers or shift mechanism	d) Check for damage to internal components
	e) Incorrect lubricant or sticking blocker ring	e) Verify that ATF type lube is present. Do not use gear lube or hypoid type lubricants
Walks out of gear	a) Damaged linkage preventing complete travel into gear	a) Check for damaged shift mechanism
	b) Floor shift stiff or improperly installed boot	b) Verify jumpout with boot removed, replace boot if necessary
	c) Floor shift interference between shift handle and console	c) Adjust console to eliminate interference
	d) Broken or loose engine mounts	d) Check for broken or loose engine mounts and service as required
	e) Loose shift mechanism stabilizer bar	e) Check stabilizer bar attaching bolt and torque to specification
	f) Worn or damaged internal components	f) Check shift forks, shift rails and shift rail detent system for wear or damage, synchronizer sliding sleeve and gear clutching teeth for wear or damage. Repair or replace as required
	g) Bent top gear locknut switch actuator	g) With shift lever in fourth gear, check actuator position with shift rod. Actuator should be positioned at a 90 degree angle to shift rod. Bend actuator to proper position, if required
Will not shift into one gear—all other gears OK	a) Damaged external shift mechanism	a) Check for damaged shift mechanism. Service or replace as necessary
	b) Floor shift. Interference between shift handle and console or floor cut out	b) Adjust console or cut out floor pan to eliminate interference
	c) Restricted travel of internal shift components	c) Disconnect external shift mechanism and shift the input shift rail through the gears to verify problem. Remove transaxle. Inspect fork system, synchronizer system and gear clutch teeth for restricted travel. Service or replace as required
Will not shift into reverse	a) Damaged external shift mechanism	a) Check for damaged external shift mechanism. Remove shift mechanism at input shift rail and try shifting into reverse at the rail
	b) Worn or damaged internal components	b) Remove transaxle. Check for damaged reverse gear train, misaligned reverse relay lever, shift rail and fork system. Check the gear clutching teeth and synchronizer system for restricted travel or damage. Service or replace as required

POWER FLOW

4 Speed Transaxle

From the clutch, engine torque is transferred to the mainshaft through the input cluster gear. Each gear on the input cluster is in constant mesh with a matching gear on the mainshaft. It is these matching gear sets which will provide the 4 forward gear ratios. The transaxle gear ratio is determined by the number of teeth on the input cluster gear and the number of teeth on the mainshaft gear.

Reverse is accomplished by sliding a spur gear into mesh with the input cluster shaft gear and the reverse idler gear. The reverse idler gear acts as an idler and reverses the direction of mainshaft rotation. In neutral, none of the gears on the mainshaft are locked to their shafts. Then, no torque from the engine to the input cluster gear shaft is transferred to the differential assembly and to the wheels through the halfshafts.

5 Speed Transaxle

Engine torque is transferred from the clutch to the input cluster gear shaft. The 4 forward gears on the input cluster gear shaft are in constant mesh with a matching gear on the mainshaft. The 4th gear on the input cluster gear shaft is simultaneously meshed with the 5th speed gear on the 5th gear shaft. These meshed gearsets provide the 5 available forward gear ratios.

Both the mainshaft and the 5th gear shaft have a pinion gear, which is constantly engaged with the final drive ring gear of the differential assembly. If a single gear (1st through 4th) on the mainshaft is selected and that gear is locked to the shaft by its shift synchronizer, then the input cluster shaft gear will drive the mainshaft pinion gear; driving the differential final drive ring gear. If the 5th gear is selected, the input cluster shaft 4th gear will drive the 5th gear shaft pinion gear, driving the differential final drive ring gear. At this time, the mainshaft gears will rotate freely.

Reverse is accomplished by sliding a spur gear into mesh with the input cluster shaft gear and the reverse idler gear. The reverse idler gear acts as an idler and reverses the direction of mainshaft rotation.

ON CAR SERVICE

Adjustment

SHIFT LINKAGE

The external gear shift mechanism consists of a gear shift lever, transaxle shift rod, stabilizer rod and shift housing. Adjustment of the external linkage is not necessary.

Services

BEARING CUPS AND PRELOAD SHIMS

Removal and Installation

The input cluster shaft, the mainshaft and the 5th gear driveshaft are supported at each end by tapered roller bearings. The cups, which support the bearings in the case, are located in the transaxle case and in the clutch housing. Shims, to preload the tapered roller bearings are located behind the bearing cups in the transaxle case only. It is important to keep the shim with its matching cup during the disassembly. It is equally important to label the bearing cups if they are removed from the case. After removal of the mainshaft bearing cups from the clutch housing, the funnels can be removed from the bearing cup bores. The funnels direct lubricant to a drilled hole in the center of the mainshaft. The lubricant flows through these shafts, where it lubricates the rotating gears.

A replacement bearing preload shim should be installed in place of the original shim when servicing any components listed

TRANSAXLE SERVICE SHIM—4 SPEED

	Shims Replaced With Service Shim		
Parts Replaced	Input Cluster Shaft	Main Shaft	5th Gear Shaft
1 Input Cluster Bearing	Yes	No	No
2 Input Cluster Bearings	Yes	No	No
1 Input Cluster Bearing 1 Mainshaft Bearing	Yes	Yes	Yes
2 Input Cluster Bearings 2 Mainshaft Bearings	Yes	Yes	Yes
1 Mainshaft Bearing	No	Yes	No
2 Mainshaft Bearings	No	Yes	No
Clutch Housing Assembly	Yes	Yes	Yes
Transaxle Case Assembly	Yes	Yes	Yes

NOTE: The shims must be installed only under the bearing cups at the transaxle case end of the three shafts.

NOTE: The use of a nominal thickness service shim eliminates the need for gauging bearing clearances prior to reassembly. While this method produces wider variations of bearing settings than are present in factory assembled units, the extreme possible settings have been tested and found to be acceptable.

TRANSAXLE SERVICE SHIM—5 SPEED

	Shims Replaced With Service Shim		
Parts Replaced	Input Cluster Shaft	Main Shaft	5th Gear Shaft
1 Input Cluster Bearing	Yes	No	No
2 Input Cluster Bearings	Yes	No	No
1 Input Cluster Bearing 1 Mainshaft Bearing 1 5th Gear Shaft Bearing	Yes	Yes	Yes
2 Input Cluster Bearings 2 Mainshaft Bearings 2 5th Gear Shaft Bearings	Yes	Yes	Yes
1 Mainshaft Bearing	No	Yes	No
2 Mainshaft Bearings	No	Yes	No
1 5th Gear Shaft Bearing	No	No	Yes
2 5th Gear Shaft Bearings	No	No	Yes
Clutch Housing Assembly	Yes	Yes	Yes
Transaxle Case Assembly	Yes	Yes	Yes

NOTE: The shims must be installed only under the bearing cups at the transaxle case end of the three shafts.

NOTE: The use of a nominal thickness service shim eliminates the need for gauging bearing clearances prior to reassembly. While this method produces wider variations of bearing settings than are present in factory assembled units, the extreme possible settings have been tested and found to be acceptable.

in the Service Shim chart. Do not use more than 1 shim per shaft. If any parts are replaced othan than the parts listed in the Service Shim chart, use the original shims.

1. Gently tap the bearing cup out of the transaxle case or the clutch housing, using a suitable tool.
2. Keep the bearing cup and shim in correct order.
3. Thoroughly clean the bearing cups, bores, shims and funnels.
4. Lightly grease the bearing cup lip.
5. Install bearing cup in the transaxle case or clutch housing. Tap bearing cup in with suitable tool until lip is flush with case or housing.

SPEEDOMETER DRIVEN GEAR

Removal and Installation

1. Using a 7mm socket or equivalent, remove the retaining screw from the speedometer driven gear retainer assembly.
2. Using a tool, carefully pry on the speedometer retainer to remove both the speedometer gear and retainer assembly from the clutch housing case bore. Be careful not to make contact with teeth on the speedometer gear.
3. Lightly grease the O-ring seal on the speedometer driven gear retainer.
4. Align the relief in the retainer with the attaching screw bore and using a tool, tap the assembly into its bore.
5. Tighten the retaining screw to 12–24 inch lbs.

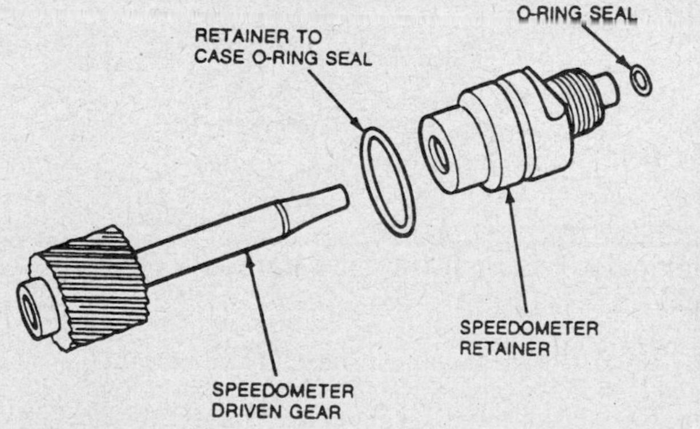
O-RING SEAL
RETAINER TO
CASE O-RING SEAL
SPEEDOMETER
RETAINER
SPEEDOMETER
DRIVEN GEAR

Speedometer gear assembly

REMOVAL AND INSTALLATION

TRANSAXLE REMOVAL

NOTE: The transaxle is removed and installed separately from the the engine.

1. Wedge a wood block approximately 7 in. in length under the clutch pedal to hold the clutch pedal up slightly beyond its normal position.
2. Grasp the clutch cable and pull forward, disconnecting it from the clutch release shaft assembly. Remove the clutch cable casing from the rib on the top surface of the transaxle case.
3. Remove the 2 top transaxle to engine mounting bolts.
4. Remove the the air management valve bracket to the transaxle.
5. Raise the vehicle and safely support. Remove the nut and bolt that secures the lower control arm ball joint to the steering knuckle assembly. Discard the and bolt. Repeat procedure on the opposite side.
6. Using a suitable tool, pry the lower control arm away from the knuckle. Exercise care not to damage or cut the ball joint boot. Pry bar must not contact the lower arm. Repeat the procedure on the opposite side.

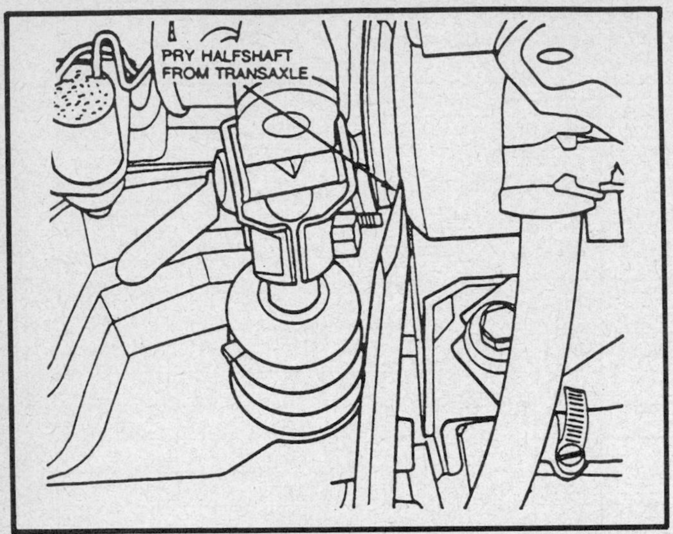
PRY HALFSHAFT
FROM TRANSAXLE

Removing halfshaft from the transaxle using pry bar

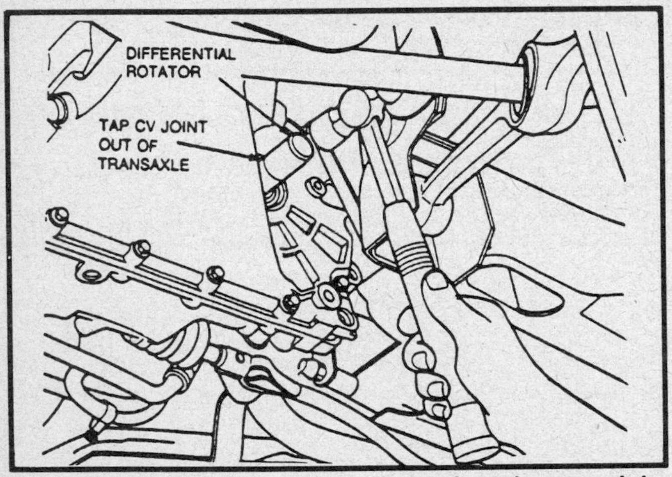
DIFFERENTIAL
ROTATOR
TAP CV JOINT
OUT OF
TRANSAXLE

Removing halfshaft from the transaxle using special tool

7. Place a drain pan under the left hand inboard CV-joint assembly at the transaxle. Using a large pry bar or equivalent, pry the left hand inboard CV-joint assembly from the transaxle. Install transaxle plug T81P–1177–B or equivalent. Remove the inboard CV-joint from the transaxle by grasping the left hand steering knuckle and swinging the knuckle and halfshaft outward from the transaxle. Repeat the procedure on the right hand side.

NOTE: Use care during the use of the pry bar and removal of the CV-joint assembly, to prevent damage to the differential oil seal.

8. If the CV-joint assembly cannot be removed from the transaxle, insert differential rotator T81P–4026–A or a brass punch through the left hand side and tap the joint out. The tool can be used from either side of transaxle.
9. Wire or tie up both halfshaft assemblies in a near level position to prevent damage to the assemblies.
10. Using a small tool, remove the backup lamp switch connector from the transaxle backup lamp switch, located on top of the transaxle.
11. Remove the 3 starter bolts.
12. Remove the shift mechanism to shift shaft attaching nut and bolt and control selector indicator switch arm. Remove from the shift shaft.
13. Remove the shift mechanism stabilizer bar to transaxle attaching bolt and the switch/bracket assembly control selector indicator.
14. Remove the speedometer cable from the transaxle.
15. Remove the 2 stiffener brace attaching bolts from the lower position of the clutch housing.
16. Position a safe, suitable jack under the transaxle.
17. Remove the 2 bolts securing the rear mount to the floorpan brace.
18. Loosen the nut in the bottom of the front mount. Then, remove the 3 bolts securing the front mount to the transaxle.
19. Lower the transmission jack until the transaxle clears the rear mount. Support the engine with a screw jackstand or equivalent under the oil pan. Use a piece of wood on top of the screw jack to prevent damage to the engine.
20. Remove the 4 engine to transaxle attaching bolts.
21. Remove the transaxle from the rear face of the engine and lower it safely from the vehicle.

TRANSAXLE INSTALLATION

NOTE: Do not attempt to start the engine prior to complete installation of the transaxle.

1. Using a safe suitable jack, raise the transaxle into position. Engage the input shaft spline into the clutch disc and work the transaxle onto the dowel sleeves. Ensure that the transaxle assembly is flush with the rear face of the engine.

NOTE: If the clutch assembly mounting bolts were loosen, then the clutch assembly will have to be align up with the use of a dummy input shaft or equivalent.

2. Install the 4 engine to transaxle attaching bolts. Tighen to 28–31 ft. lbs.
3. Install the 3 bolts that secure the front mount to the transaxle. Tighten to 25–35 ft. lbs. and the nut on the bottom of the front transaxle mount.
4. Install the top bolt that secures the air management valve to the transaxle, finger tight only. Install the bottom bracket bolt and tighten to 28–31 ft. lbs.
5. Install the 2 bolts that secure the rear mount to the floorpan brace. Tighten to 40–51 ft. lbs. Remove the jacks supporting the engine and the transaxle.
6. Install the speedometer cable. Use care when threading the cable nut onto the retainer to prevent crossthreading.
7. Install the bolt that attached the stabilizer bar and control selector indicator switch to the transaxle. Tighten to 23–35 ft. lbs.
8. Install the shift mechanism to the shift shaft. Install the switch actuator bracket, clamp, bolt and tighten the nut to 7–10 ft. lbs.
9. Install the 2 bolts that attach the stiffener brace to the lower portion of the clutch housing. Tighten bolts to 15–21 ft. lbs.
10. Install the 3 starter bolts. Tighten to 30–40 ft. lbs.
11. Install the backup lamp switch connector to the transaxle switch.
12. Remove transaxle plugs or equivalent and install the inner CV-joints into the transaxle.

NOTE: New circlips are required on both inner joints prior to installation. Use care while inserting shaft into the transaxle to avoid damage to the oil seals. Ensure that both joints are fully seated in the transaxle. Lightly pry outward to confirm that the retaining rings are seated. If rings are not seated, the joint will move out of the transaxle.

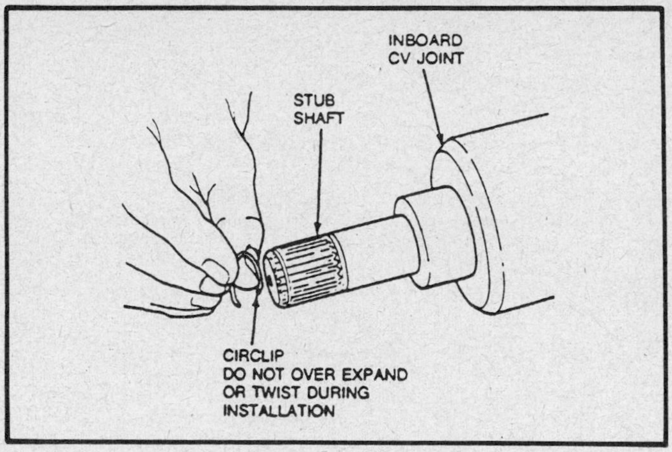
INBOARD CV JOINT
STUB SHAFT
CIRCLIP
DO NOT OVER EXPAND
OR TWIST DURING
INSTALLATION

Installing new circlip on the halfshaft

13. Attach the lower ball joint to the steering knuckle, taking care not to damage or cut the ball joint boot. Insert a new pinch bolt and nut. Tighten the nut to 37–44 ft. lbs. Do not tighten the bolt.
14. Fill the transaxle with specified fluid. Lower the vehicle.
15. Tighten the top bolt securing the air management valve bracket to the transaxle to 28–31 ft. lbs.
16. Install the 2 top transaxle to engine mounting bolts. Tighten to 28–31 ft. lbs.
17. Connect the clutch cable to the clutch release shaft assembly. Remove the wood block from under the clutch pedal. Prior to starting the engine, set the hand brake and pump the clutch pedal a minimum of 2 times to ensure proper clutch adjustment.
18. Road test for proper operation.

BENCH OVERHAUL

Before Disassembly

When servicing the unit, it is recommended that as each part is disassembled, it is cleaned in solvent and dried with compressed air. Disassembly and reassembly of this unit and its parts must be done on a clean work bench. Also, before installing bolts into aluminum parts, always dip the threads into clean transmission oil. Anti-seize compound can also be used to prevent bolts from galling the aluminum and seizing. Always use a torque wrench to keep from stripping the threads. Take care with the seals when installing them, especially the smaller O-rings. The slightest damage can cause leaks. Aluminum parts are very susceptible to damage so great care should be exercised when handling them. The internal snaprings should be expanded and the external snaprings compressed if they are to be reused. This will help insure proper seating when installed. Be sure to replace any O-ring, gasket, or seal that is removed.

Transaxle Disassembly

1. Shift the transaxle into neutral using a drift in the input shaft hole. Pull or push the shaft into the center detent position.
2. Remove the 2 transaxle plugs T81P–1177–B or equivalent from the transaxle and drain the fluid.

NOTE: Place the transaxle on a bench with the clutch housing facing down to facilitate draining and service.

3. Remove the reverse idler shaft retaining bolt.
4. Remove the detent plunger retaining screw. Then using a magnet, remove the detent spring and the detent plunger.

NOTE: Label these parts, as they appear similar to the input shift shaft plunger and spring contained in the clutch case.

5. Remove the shift fork interlock sleeve retaining pin and fill plug.
6. Remove the clutch housing to transaxle case attaching bolts.
7. Tap the transaxle case with a plastic tipped hammer to break the seal between the case halves. Separate the halves.

NOTE: Do not insert pry bars between case halves. Be careful not to drop out the tapered roller bearing cups or shims from the transaxle case housing.

8. Remove the detent plunger retaining screw. Then, using a pencil magnet or equivalent, remove the detent spring and the detent plunger.
9. Remove the case magnet.
10. Using a small tool, remove the C-clip retaining ring from the 5th relay lever pivot pin. Remove the 5th gear shift relay lever.
11. Lift the reverse idler shaft and reverse idler gear from the case.
12. Using a punch, drive the spring pin from the shift lever shaft.
13. Using a suitable tool, gently pry on the shift lever shaft so that the hole in the shaft is exposed. Be careful not to damage mainshaft gear teeth or pedestal when prying with tool.

NOTE: On vehicles equipped with the 1.9L engine, remove 2 screws holding the shift lever cover to the shift lever and remove the inhibitor ball and spring

14. Hold a shop towel over the hole in the lever to prevent the ball and the 5th/reverse inhibitor spring from shooting out and remove the shift lever shaft.
15. Remove the inhibitor ball and spring from the hole in the shift lever using a pencil magnet or equivalent. Remove the shift lever, 5th/reverse kickdown spring, and 3–4 bias spring.
16. Remove the mainshaft assembly, input cluster shaft assembly and the main shift control shaft assembly as a complete unit. Be careful not to drop bearings or gears.
17. On 4 speed transaxles, rotate shaft and remove the reverse actuator arm and shaft assembly from its bore in the case.
18. On 5 speed transaxles, remove the 5th gear shaft assembly and the 5th gear fork assembly from their bores in the case.
19. Lift the differential and final drive gear assembly from the clutch housing case.
20. Remove the 2 bolts retaining shift relay lever support bracket assembly.

1. 2nd speed gear
2. Synchronizer blocking ring
3. Synchronizer spring
4. 1st/2nd synchronizer aasy.
5. Synchronizer hub 1st/2nd insert
6. Input shaft seal
7. Input shaft bearing
8. Input cluster shaft
9. Input shaft seal
10. Mainshaft funnel
11. Mainshaft
12. 1st speed gear
13. 2nd/3rd gear thrust washer retaining ring
14. 2nd/3rd gear thrust washer
15. 3rd speed gear
16. 4th speed gear
17. 3rd/4th fork
18. Fork selector arm
19. Fork interlock sleeve
20. 1st/2nd fork
21. Main shift shaft
22. Reverse idler shaft
23. Reverse idler gear
24. Reverse relay lever
25. Reverse relay lever pivot pin
26. Back-up lamp switch
27. Dowel
28. Shift lever shaft
29. Pinion shaft
30. Pinion thrust washer
31. Side gear
32. Side gear thrust washer
33. Shim
34. Differential bearing assembly
35. Finial drive output gear
36. Differential aasy.
37. Speedometer drive gear
38. Differential pinion gear
39. Input shift shaft selector plate arm
40. Case magnet
41. Input detent shift shaft spring
42. Input shift shaft detent plunger
43. Input shift shaft
44. Transaxle case
45. Main shift shaft detent plunger
46. Main shift shaft detent spring

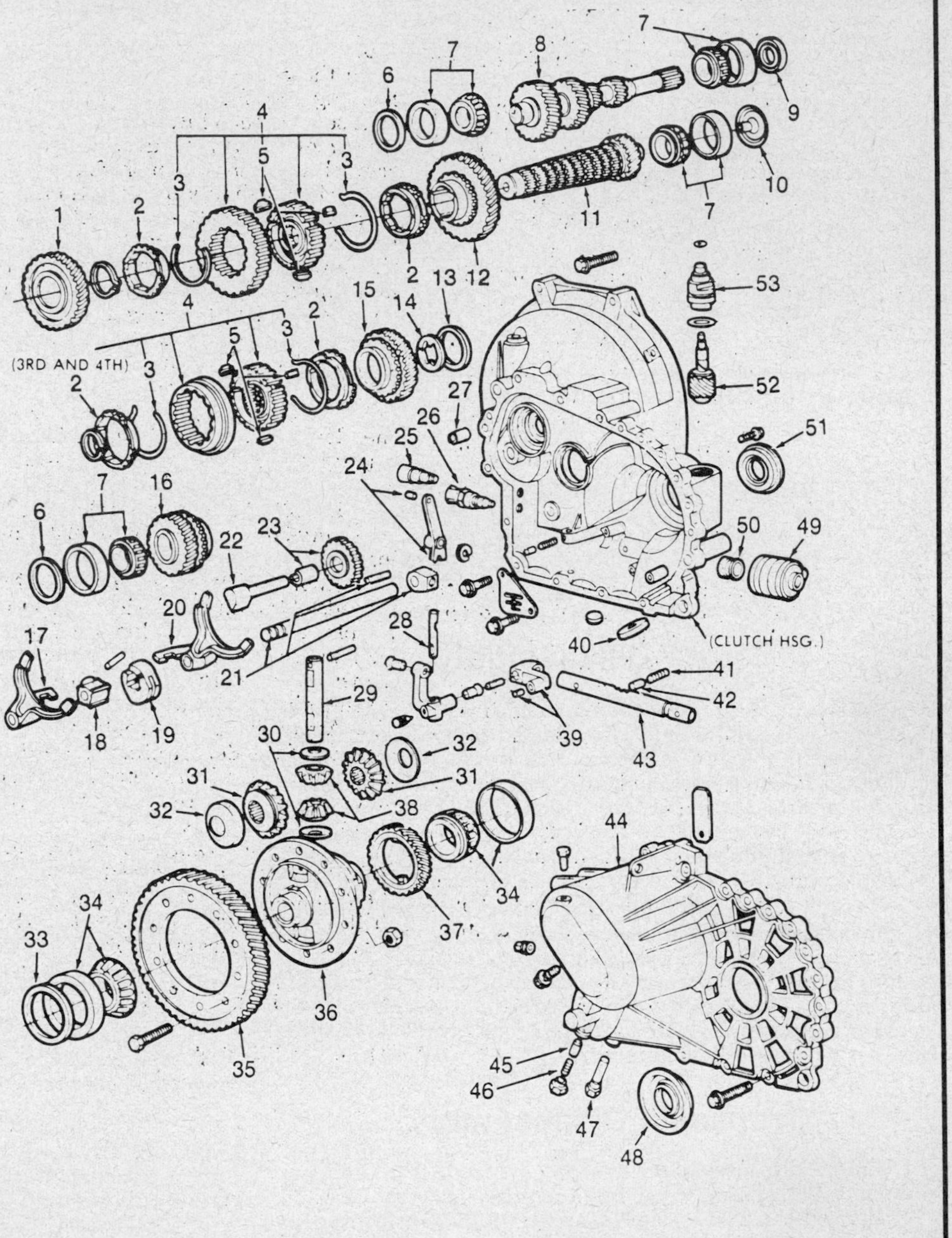
(3RD AND 4TH)
(CLUTCH HSG.)

47. Fork interlock sleeve retaining spring
48. Differential seal assembly
49. Shift shaft boot
50. Shift shaft oil seal
51. Differential oil seal
52. Speedometer driven gear
53. Gear retainer

Exploded view 4 speed—Ford transaxle (RGT)

Unit Disassembly And Assembly

MAINSHAFT ASSEMBLY

Disassembly

1. Remove the slip fit roller bearing on the 4th speed gear end of the shaft. Mark or tag the bearing for proper installation.
2. Remove the 4th speed gear and synchronizer blocker ring.
3. Remove the 3rd/4th synchronizer retaining ring. Slide the 3rd/4th gear synchronizer assembly, blocker ring and 3rd speed gear from the shaft.
4. Remove the 2nd/3rd thrust washer retaining ring and the 2 piece thrust washer.
5. Remove the 2nd speed gear and its blocker ring.

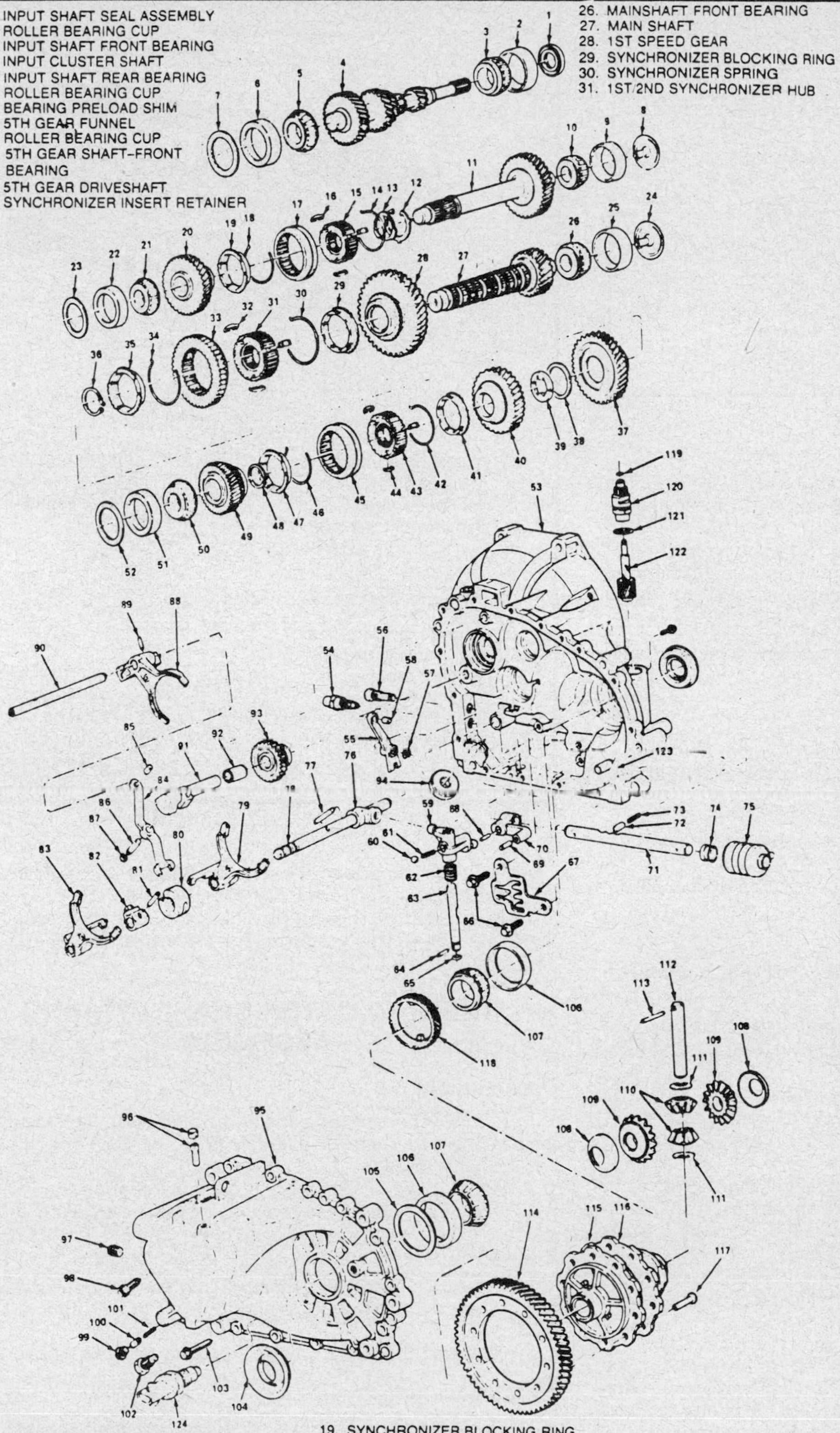

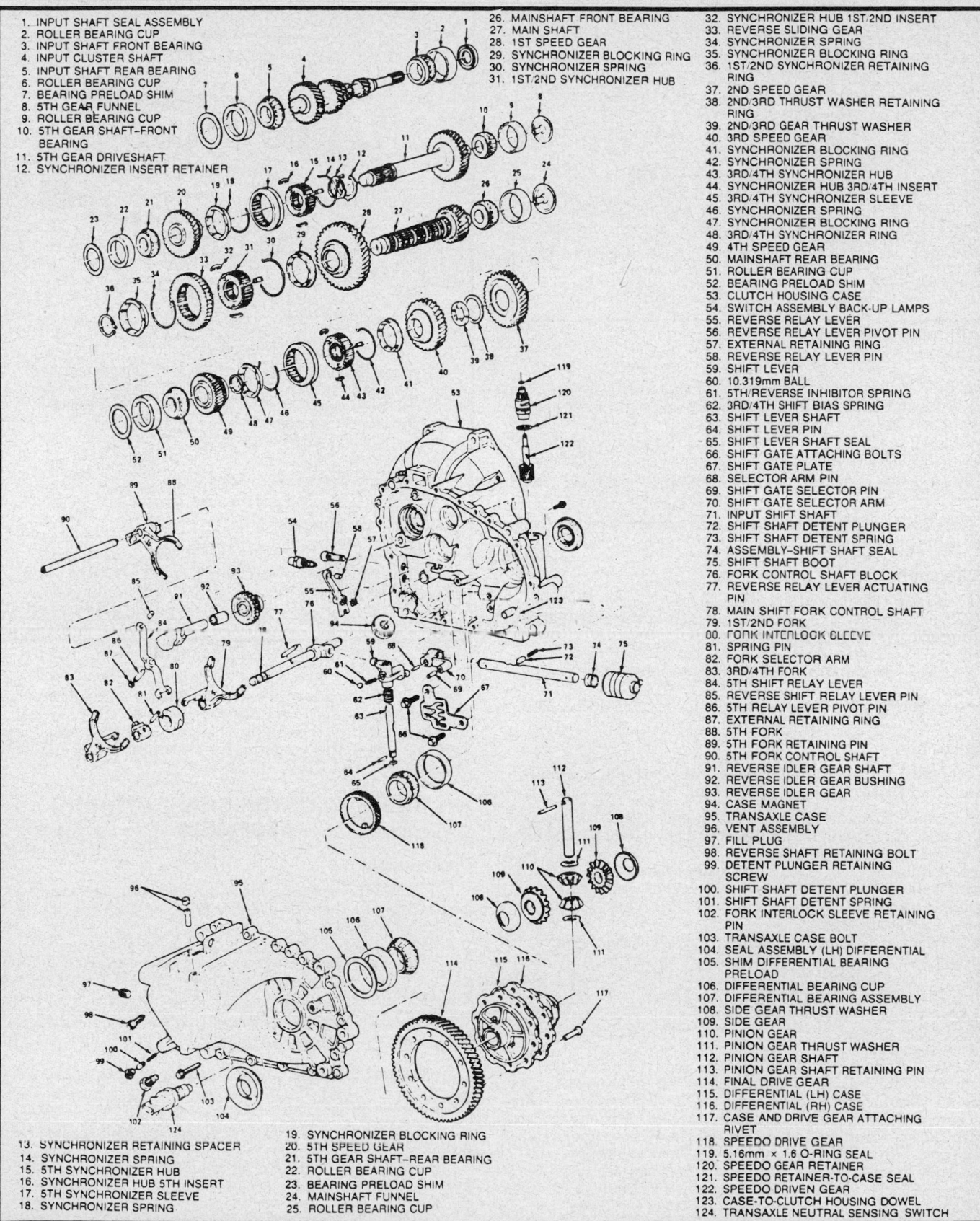

1. INPUT SHAFT SEAL ASSEMBLY
2. ROLLER BEARING CUP
3. INPUT SHAFT FRONT BEARING
4. INPUT CLUSTER SHAFT
5. INPUT SHAFT REAR BEARING
6. ROLLER BEARING CUP
7. BEARING PRELOAD SHIM
8. 5TH GEAR FUNNEL
9. ROLLER BEARING CUP
10. 5TH GEAR SHAFT–FRONT BEARING
11. 5TH GEAR DRIVESHAFT
12. SYNCHRONIZER INSERT RETAINER
13. SYNCHRONIZER RETAINING SPACER
14. SYNCHRONIZER SPRING
15. 5TH SYNCHRONIZER HUB
16. SYNCHRONIZER HUB 5TH INSERT
17. 5TH SYNCHRONIZER SLEEVE
18. SYNCHRONIZER SPRING
19. SYNCHRONIZER BLOCKING RING
20. 5TH SPEED GEAR
21. 5TH GEAR SHAFT–REAR BEARING
22. ROLLER BEARING CUP
23. BEARING PRELOAD SHIM
24. MAINSHAFT FUNNEL
25. ROLLER BEARING CUP
26. MAINSHAFT FRONT BEARING
27. MAIN SHAFT
28. 1ST SPEED GEAR
29. SYNCHRONIZER BLOCKING RING
30. SYNCHRONIZER SPRING
31. 1ST/2ND SYNCHRONIZER HUB
32. SYNCHRONIZER HUB 1ST/2ND INSERT
33. REVERSE SLIDING GEAR
34. SYNCHRONIZER SPRING
35. SYNCHRONIZER BLOCKING RING
36. 1ST/2ND SYNCHRONIZER RETAINING RING
37. 2ND SPEED GEAR
38. 2ND/3RD THRUST WASHER RETAINING RING
39. 2ND/3RD GEAR THRUST WASHER
40. 3RD SPEED GEAR
41. SYNCHRONIZER BLOCKING RING
42. SYNCHRONIZER SPRING
43. 3RD/4TH SYNCHRONIZER HUB
44. SYNCHRONIZER HUB 3RD/4TH INSERT
45. 3RD/4TH SYNCHRONIZER SLEEVE
46. SYNCHRONIZER SPRING
47. SYNCHRONIZER BLOCKING RING
48. 3RD/4TH SYNCHRONIZER RING
49. 4TH SPEED GEAR
50. MAINSHAFT REAR BEARING
51. ROLLER BEARING CUP
52. BEARING PRELOAD SHIM
53. CLUTCH HOUSING CASE
54. SWITCH ASSEMBLY BACK-UP LAMPS
55. REVERSE RELAY LEVER
56. REVERSE RELAY LEVER PIVOT PIN
57. EXTERNAL RETAINING RING
58. REVERSE RELAY LEVER PIN
59. SHIFT LEVER
60. 10.319mm BALL
61. 5TH/REVERSE INHIBITOR SPRING
62. 3RD/4TH SHIFT BIAS SPRING
63. SHIFT LEVER SHAFT
64. SHIFT LEVER PIN
65. SHIFT LEVER SHAFT SEAL
66. SHIFT GATE ATTACHING BOLTS
67. SHIFT GATE PLATE
68. SELECTOR ARM PIN
69. SHIFT GATE SELECTOR PIN
70. SHIFT GATE SELECTOR ARM
71. INPUT SHIFT SHAFT
72. SHIFT SHAFT DETENT PLUNGER
73. SHIFT SHAFT DETENT SPRING
74. ASSEMBLY–SHIFT SHAFT SEAL
75. SHIFT SHAFT BOOT
76. FORK CONTROL SHAFT BLOCK
77. REVERSE RELAY LEVER ACTUATING PIN
78. MAIN SHIFT FORK CONTROL SHAFT
79. 1ST/2ND FORK
80. FORK INTERLOCK SLEEVE
81. SPRING PIN
82. FORK SELECTOR ARM
83. 3RD/4TH FORK
84. 5TH SHIFT RELAY LEVER
85. REVERSE SHIFT RELAY LEVER PIN
86. 5TH RELAY LEVER PIVOT PIN
87. EXTERNAL RETAINING RING
88. 5TH FORK
89. 5TH FORK RETAINING PIN
90. 5TH FORK CONTROL SHAFT
91. REVERSE IDLER GEAR SHAFT
92. REVERSE IDLER GEAR BUSHING
93. REVERSE IDLER GEAR
94. CASE MAGNET
95. TRANSAXLE CASE
96. VENT ASSEMBLY
97. FILL PLUG
98. REVERSE SHAFT RETAINING BOLT
99. DETENT PLUNGER RETAINING SCREW
100. SHIFT SHAFT DETENT PLUNGER
101. SHIFT SHAFT DETENT SPRING
102. FORK INTERLOCK SLEEVE RETAINING PIN
103. TRANSAXLE CASE BOLT
104. SEAL ASSEMBLY (LH) DIFFERENTIAL
105. SHIM DIFFERENTIAL BEARING PRELOAD
106. DIFFERENTIAL BEARING CUP
107. DIFFERENTIAL BEARING ASSEMBLY
108. SIDE GEAR THRUST WASHER
109. SIDE GEAR
110. PINION GEAR
111. PINION GEAR THRUST WASHER
112. PINION GEAR SHAFT
113. PINION GEAR SHAFT RETAINING PIN
114. FINAL DRIVE GEAR
115. DIFFERENTIAL (LH) CASE
116. DIFFERENTIAL (RH) CASE
117. CASE AND DRIVE GEAR ATTACHING RIVET
118. SPEEDO DRIVE GEAR
119. 5.16mm × 1.6 O-RING SEAL
120. SPEEDO GEAR RETAINER
121. SPEEDO RETAINER-TO-CASE SEAL
122. SPEEDO DRIVEN GEAR
123. CASE-TO-CLUTCH HOUSING DOWEL
124. TRANSAXLE NEUTRAL SENSING SWITCH

Exploded view 5 speed—Ford transaxle (RWB)

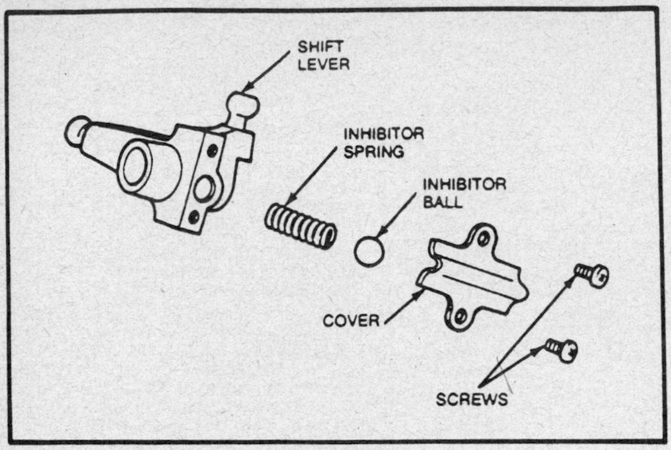
SHIFT LEVER
INHIBITOR SPRING
INHIBITOR BALL
COVER
SCREWS

Shift lever assembly—1.9L engine

6. Remove the 1st/2nd synchronizer retaining ring. Slide the 1st/2nd synchronizer assembly, blocker ring and 1st speed gear off the shaft.

7. Remove the tapered roller bearing from the pinion end of the mainshaft using a socket or extension and pinion bearing cone remover tool D79L–4621–A or equivalent and an arbor press. Mark or tag bearing.

NOTE: Bearing does not have to be removed to disassemble the mainshaft.

Inspection

1. Inspect the tapered roller bearing for wear or damage.
2. Check the teeth, splines and journals of the mainshaft for damage.
3. Check all gears for chipped, broken or worn teeth.
4. Check synchronizer sleeves for free movement on their hubs.
5. Inspect the synchronizer blocking rings for wear marks.

Assembly

NOTE: Lightly oil gear bores and other parts with the specified fluid before installation.

1. Install the bearing on the pinion end of the shaft using a $1^{1}/_{16}$ inch socket, pinion bearing cone remover tool D79L–4621–A or equivalent and an arbor press.
2. Slide the 1st speed gear and blocker ring onto the mainshaft. Slide the 1st/2nd synchronizer assembly into place, making sure the shift fork groove on the reverse sliding gear faces the 1st speed gear.
3. When installing the synchronizer, align the 3 grooves in the 1st gear blocker ring with the synchronizer inserts. This allows the synchronizer assembly to seat properly in the blocker ring. Install the synchronizer retaining ring.
4. Install the 2nd speed blocker ring and the 2nd speed gear. Align the 3 grooves in the 2nd gear blocker ring with the synchronizer inserts.
5. Install the thrust washer halves and retaining ring.
6. Slide the 3rd speed gear onto the shaft followed by the 3rd gear synchronizer blocker ring and the 3rd/4th gear synchronizer assembly. Align the 3 grooves in the 3rd gear blocker ring with the synchronizer inserts. Install the synchronizer retaining ring.
7. Install the 4th gear blocker ring and the 4th speed gear. Align the 3 grooves in the 4th gear blocker ring with the synchronizer inserts.
8. Install the slip fit roller bearing on the 4th gear end of the shaft.

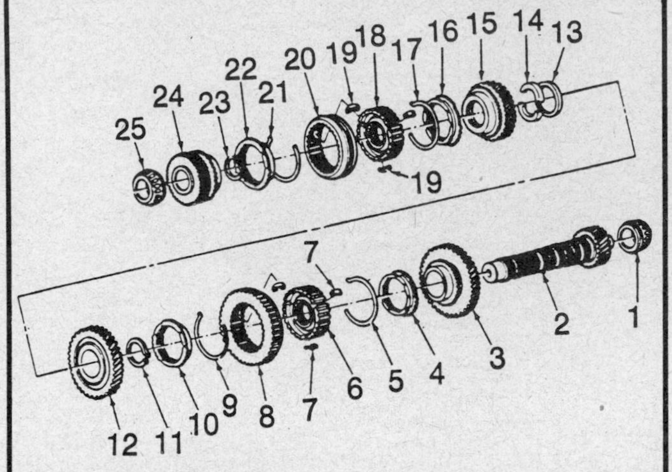
1
2
3
4
5
6
7
8
9
10
11
12
13
14
15
16
17
18
19
20
21
22
23
24
25

1. Mainshaft front bearing
2. Mainshaft
3. 1st speed gear
4. Synchro blocker ring
5. Synchronizer spring
6. 1st/2nd synchro hub
7. Synchro hub 1st/2nd insert
8. Reverse slinding gear
9. Synchronizer spring
10. Synchro blocker ring
11. 1st/2nd synchro retaining ring
12. 2nd speed gear
13. 2nd/3rd thrust washer retaining ring
14. 2nd/3rd gear thrust washer
15. 3rd speed gear
16. Synchro blocker ring
17. Synchronizer spring
18. 3rd/4th synchro hub
19. Synchro hub 3rd/4th insert
20. 3rd/4th synchro sleeve
21. Synchronizer spring
22. Synchro retaining ring
23. Retaining ring
24. 4th speed gear
25. Mainshaft rear bearing

Exploded view of mainshaft assembly

9. Make sure bearings are seated against the shoulder of the mainshaft. Make sure bearings are placed on the proper end. Rotate each gear on the shaft to check for binding or roughness. Make sure that the synchronizer sleeves are in the neutral position.

INPUT CLUSTER SHAFT BEARING ASSEMBLY

Disassembly

1. Remove the bearing cone and roller assemblies using pinion bearing cone remover/installer D79L–4621–A or equivalent, and an arbor press.
2. Mark or tag bearings for proper installation.
3. Thoroughly clean the bearings and inspect their condition.

Inspection

1. Inspect the tapered roller bearing for wear or damage.
2. Check the teeth, splines and journals of the input shaft for damage.
3. Check all gears for chipped, broken or worn teeth.

Assembly

1. Lightly oil the bearings with the specified transmission fluid.
2. Using pinion bearing cone remover/installer D79L–4621–A or equivalent and an arbor press, install the bearing on the shaft.
3. Make sure the bearings are pressed on the proper end as marked during the disassembly.

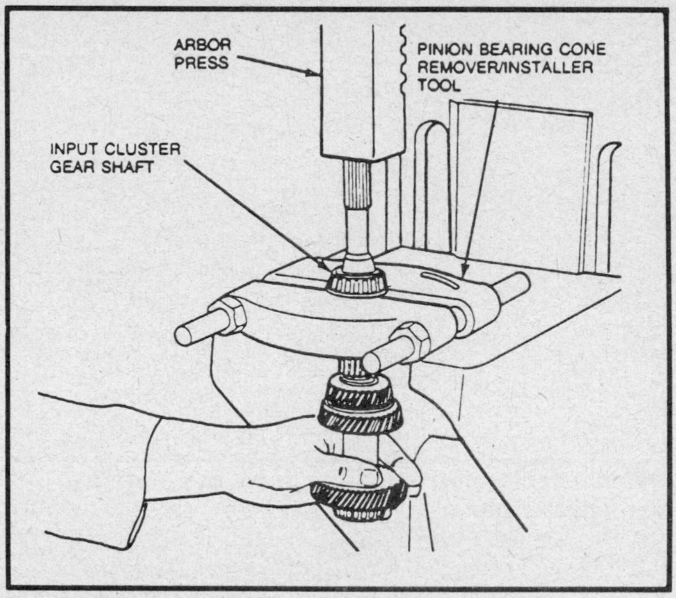
ARBOR
PRESS
PINION BEARING CONE
REMOVER/INSTALLER
TOOL
INPUT CLUSTER
GEAR SHAFT

Removing input cluster shaft bearing

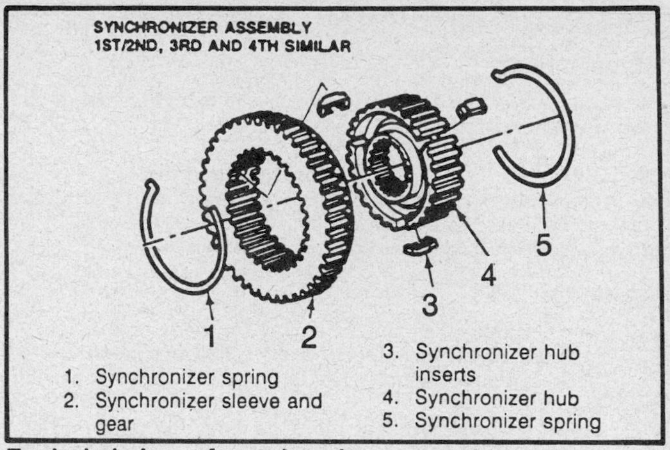
SYNCHRONIZER ASSEMBLY
1ST/2ND, 3RD AND 4TH SIMILAR
1
2
3
4
5
1. Synchronizer spring
2. Synchronizer sleeve and gear
3. Synchronizer hub inserts
4. Synchronizer hub
5. Synchronizer spring

Exploded view of synchronizer assembly

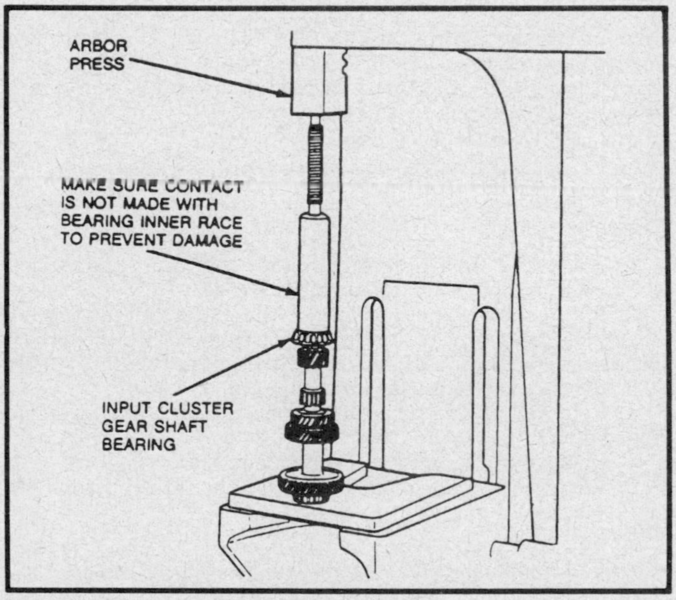
ARBOR
PRESS
MAKE SURE CONTACT
IS NOT MADE WITH
BEARING INNER RACE
TO PREVENT DAMAGE
INPUT CLUSTER
GEAR SHAFT
BEARING

Installing input cluster shaft bearing

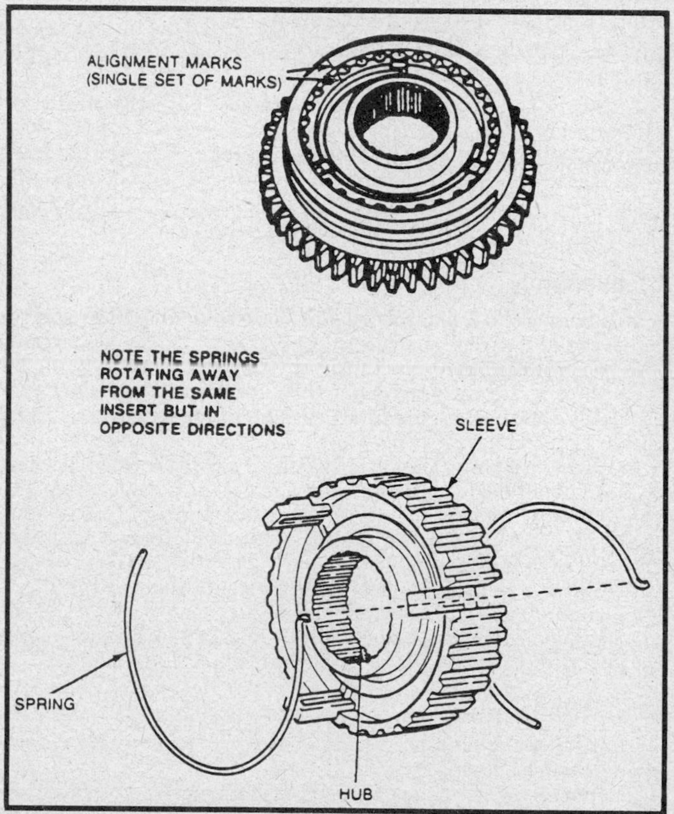
ALIGNMENT MARKS
(SINGLE SET OF MARKS)
NOTE THE SPRINGS
ROTATING AWAY
FROM THE SAME
INSERT BUT IN
OPPOSITE DIRECTIONS
SLEEVE
SPRING
HUB

Synchronizer installation

SYNCHRONIZER

Disassembly

1. Note position of the index marks.
2. Remove the synchronizer springs with a small tool. Do not compress the springs more than is necessary.
3. Remove the 3 hub inserts.
4. Slide the hub and sleeve apart.

Inspection

1. Check synchronizer sleeves for free movement on their hubs.
2. Check insert springs.
3. Inspect the synchronizer blocking rings for wear marks.

Assembly

1. Slide the sleeve over the hub. The shorter end of hub shoulder must face alignment mark on sleeve.
2. Place the 3 inserts into their slots. Place the tab on the synchronizer spring into the groove of one of the inserts and snap the spring into place.
3. Place the tab of the other spring into the same insert (on the other side of the synchronizer assembly) and rotate the spring in the opposite direction and snap into place.
4. When assembling synchronizers, notice that the sleeve and the hub have an extremely close fit and must be held square to prevent jamming. Do not force the sleeve onto the hub.

5TH GEAR SHAFT ASSEMBLY

Disassembly

1. Remove the slip fit bearing from the 5th gear end of the shaft and label it for correct installation.
2. Remove the 5th gear and blocking ring.
3. Remove the 5th gear synchronizer assembly.
4. Remove the press fit bearing from the pinion end of the shaft, using bearing remover/installer tool D79L–4621–A or equivalent bearing removal adapter.

Inspection

1. Inspect the tapered roller bearing for wear or damage.
2. Check all gears for chipped, broken or worn teeth.
3. Check synchronizer sleeves for free movement on their hubs.
4. Inspect the synchronizer blocking rings for wear marks.

Assembly

NOTE: Lightly oil gear bores and other parts with the specified fluid before installation.

1. Press the bearing onto the pinion gear end of the 5th gear shaft.
2. Install the 5th synchronizer assembly with the plastic insert retainer facing the pinion gear.
3. Install the 5th gear and blocking ring.
4. Install the slip fit bearing on the 5th gear end of the shaft.

CLUTCH HOUSING

Disassembly

1. Remove the 2 control selector plate attaching bolts and remove the plate from the case.
2. With the input shift shaft in the center detent position, using a drift, drive the spring pin through the selector plate arm assembly and through the input shift shaft into the recess in the clutch housing case.
3. Remove the shift shaft boot. Using a drift, rotate the input shift shaft 90 degrees, depressing the detent plunger from the shaft detent notches inside the housing and pull input shift shaft out. Remove the input shift selector plate arm assembly and the spring pin.
4. Using a pencil magnet or equivalent, remove the input shift shaft detent plunger and spring.
5. Using sector shaft seal tool T77F–7288–A or equivalent, remove the transaxle input shift shaft oil seal assembly.

Inspection

1. Inspect the clutch housing case for cracks, wear or damaged bearing bores.
2. Inspect for damaged threads in housing.
3. Inspect clutch housing case mating surfaces for small nicks or burrs that could cause misalignment of the 2 halves.

Assembly

NOTE: Lightly oil all parts and bores with the specified fluid.

1. Lubricate the seal lip of a new shift shaft oil seal. Using sector seal tool T77F–7288–A or equivalent, install a new input shift shaft oil seal assembly.
2. Install the input shift shaft detent spring and plunger in the clutch housing case.
3. Using a small drift, force the spring and plunger down into its bore while sliding the input shift shaft into its bore and over the plunger. Be careful not to cut the shift shaft oil seal when inserting the shaft.
4. Install the selector plate arm in its working position and slide the shaft through the selector plate arm. Align the hole in the selector plate arm with the hole in the shaft and install the spring pin. Install the input shift shaft boot.

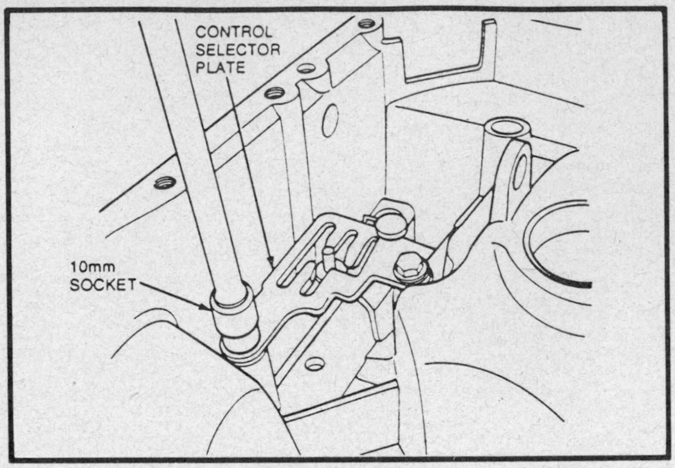
CONTROL
SELECTOR
PLATE
10mm
SOCKET

Pin in selector arm must ride in cut out of gate in the selector plate

NOTE: Make sure the notches in the shift shaft face the detent plunger.

5. Install the control selector plate. The pin in selector arm must ride in cutout of gate in the selector plate. Move input shift shaft through the selector plate positions to make sure everything works properly.

MAIN SHIFT CONTROL SHAFT

Disassembly

1. Rotate the 3rd/4th shift fork on the shaft until the notch in the fork is located over the interlock sleeve.
2. Rotate the 1st/2nd shift fork on the shaft until the notch in the fork is located over the selector arm finger. With the forks in position, slide the 3rd/4th fork and interlock sleeve off the shaft.
3. Remove the selector arm spring pin.
4. Remove the selector arm and the 1st/2nd shift fork from the shaft.
5. Remove the fork control spring pin.
6. Remove the fork control block from the shift control shaft.

Inspection

Check all components for wear or damage. Check the shift forks for proper alignment on the selector arm.

Assembly

NOTE: Lightly oil all parts with the specified fluid.

1. Slide the fork control block onto the shift control shaft. Align the hole in the block with the hole in the shaft and install the fork control block spring pin.

NOTE: With pin installed in control block, offset must point toward end of shaft. Also, check position of flat on shaft when installing control block.

2. Install the 1st/2nd shift fork and the selector arm on the shaft. The 1st/2nd shift fork is thinner than the 3rd/4th shift fork.
3. Align the hole in the selector arm with the hole in the shaft and install the spring pin.
4. Position the slot in the 1st/2nd fork over the fork selector arm finger.

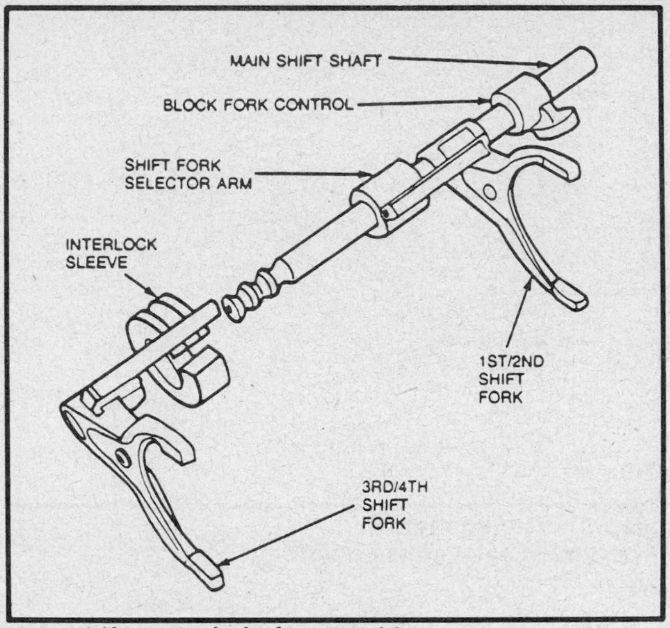
MAIN SHIFT SHAFT
BLOCK FORK CONTROL
SHIFT FORK
SELECTOR ARM
INTERLOCK
SLEEVE
1ST/2ND
SHIFT
FORK
3RD/4TH
SHIFT
FORK

Main shift control shaft assembly

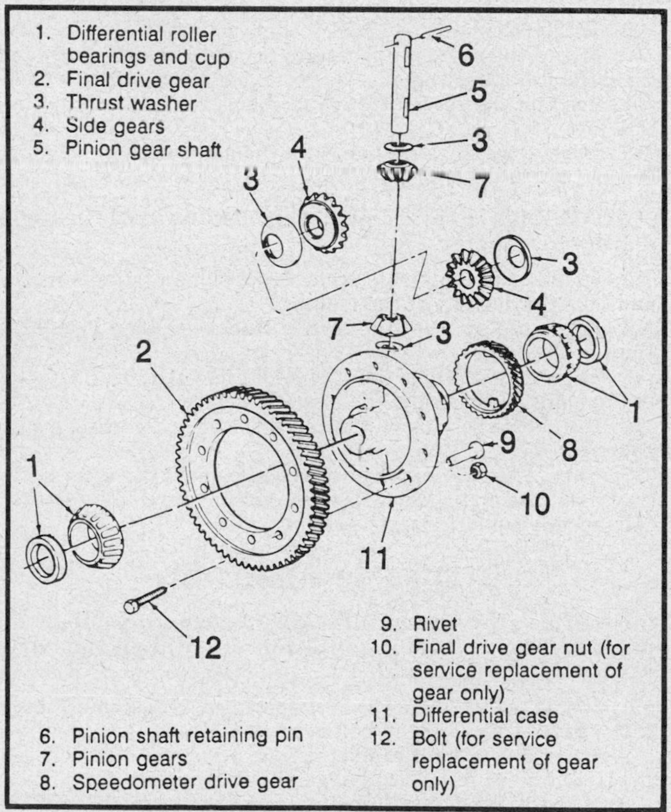
1. Differential roller bearings and cup
2. Final drive gear
3. Thrust washer
4. Side gears
5. Pinion gear shaft
6. Pinion shaft retaining pin
7. Pinion gears
8. Speedometer drive gear
9. Rivet
10. Final drive gear nut (for service replacement of gear only)
11. Differential case
12. Bolt (for service replacement of gear only)

Exploded view of differential assembly

5. Position the slot in the 3rd/4th fork over the interlock sleeve.
6. Slide the 3rd/4th fork and interlock sleeve onto the main shift control shaft.
7. Align the slot in the interlock sleeve with the splines on the fork selector arm and slide the sleeve and 3rd/4th fork into position. When assembled, the forks should be aligned.

5TH GEAR SHIFT CONTROL

Disassembly

1. Remove the spring pin.
2. Slide the fork from the shaft.

Assembly

1. Position the shaft with the hole on the left. Install the 5th gear shift fork so that the protruding arm is positioned toward the long end of the shaft.
2. Install the spring pin.

DIFFERENTIAL

Disassembly

1. Remove the left hand differential roller bearing using a suitable tool.
2. Remove the right hand differential bearing cup from the case and install over the right hand differential bearing.
3. With bearing cup in position, remove bearing from the speedometer side of the differental using suitable tool. Failure to use the bearing cup will result in damage to the bearing.
4. Remove the speedometer drive gear from the case.
5. Remove the differential side gears by rotating the gears toward the case window.
6. Remove the pinion shaft retaining pin.
7. Remove the pinion shaft, gears and thrust washer.
8. If final drive gear is to be replaced, drill out the rivets. To prevent distortion of the case, drill the preformed side of rivet only.

Inspection

Examine the pinion and side gears for scoring, excessive wear, nicks and chips. Worn, scored and damage gears cannot be serviced and must be replaced.

Assembly

1. Lubricate all components with the specified fluid before installation.
2. Install the pinion shaft, gears and thrust washer.
3. Install the pinion shaft retaining pin.
4. Install the differential side gears.
5. Install the speedometer drive gear. Install the drive gear with the bevel on the inside diameter facing the differential case.
6. Install the left and right differential roller bearings using a suitable tool.

DIFFERENTIAL BEARING PRELOAD

The differential preload is set at the factory and need not be checked or adjusted unless one of the following components are replaced.

Transaxle case
Differential case
Differential bearings
Clutch housing

1. Remove the differential seal from the transaxle case.
2. Remove the differential bearing cup from the transaxle case using a suitable tool.
3. Remove the preload shim which is located under the bearing cup.
4. If removed install the differential in the clutch housing.
5. Install special tool height gauge spacers on the clutch housing dowels.
6. Position the bearing cup removed from the transaxle case on the differential bearing.

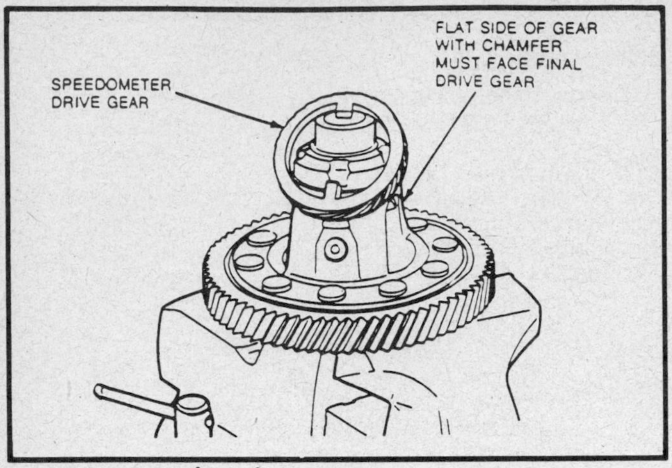
FLAT SIDE OF GEAR
WITH CHAMFER
MUST FACE FINAL
DRIVE GEAR
SPEEDOMETER
DRIVE GEAR

Installing speedometer gear

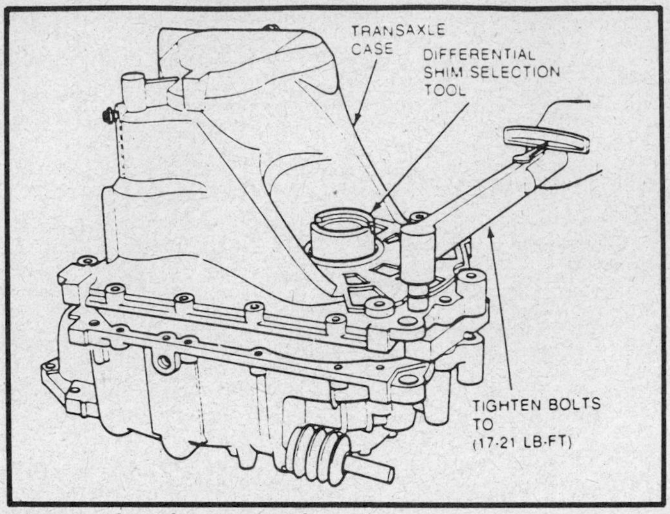
TRANSAXLE
CASE
DIFFERENTIAL
SHIM SELECTION
TOOL
TIGHTEN BOLTS
TO
(17-21 LB-FT)

Step 3—bearing preload
Positioning case on the height gauge

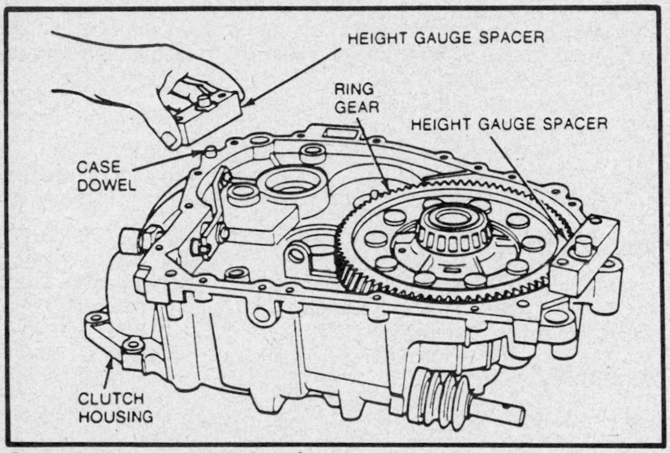
HEIGHT GAUGE SPACER
RING
GEAR
HEIGHT GAUGE SPACER
CASE
DOWEL
CLUTCH
HOUSING

Step 1—bearing preload
Installing height gauge spacers

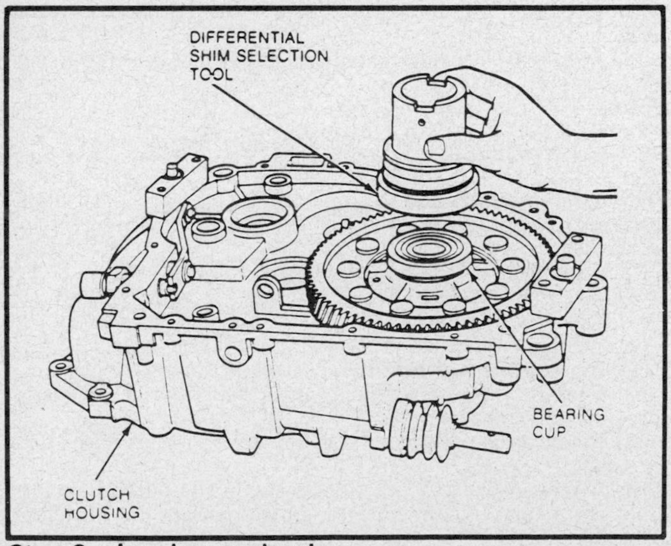
DIFFERENTIAL
SHIM SELECTION
TOOL
BEARING
CUP
CLUTCH
HOUSING

Step 2—bearing preload
Installing differential shim selection tool

7. Install the differential shim selection special tool over the bearing cup.
8. Position the transaxle case on the height spacer tool and install the 4 bolts supplied with the tool.
9. Torque the bolts to 17–21 ft. lbs.
10. Rotate the differential several times to ensure seating of the differential bearing.
11. Position the special tool gauge bar across the shim selection tool.
12. Using a feeler gauge, measure the gap between the gauge bar and the selector tool gauge surface.

NOTE: This measurement can also be made using a depth micrometer.

13. Obtain measurements from 3 positions around the tool and take the average of the readings.
14. Check the shim for the correct thickness, then install the shim in the transaxle case.
15. Apply a light film of the specified fluid to the bearing bores in the transaxle case and the clutch housing.
16. Install the bearing cup in the transaxle case using a suitable tool.
17. Check that the cup is fully seated against the shim in the transaxle case and against the shoulder in the clutch housing.
18. Install the differential seal.

Transaxle Assembly

NOTE: Prior to installation, thoroughly clean all parts and inspect their condition. Lightly oil the bores with the specified fluid.

1. Install the shift relay lever support bracket assembly to the case with 2 bolts. Tighten bolts to 6–9 ft. lbs.
2. Place the differential and the final drive gear assembly into the clutch housing case and align the differential gears.
3. If so equipped, install the 5th gear shaft assembly and the fork shaft assembly in the case. Be careful not to damage the 5th gear shaft oil funnel.
4. Position the main shift control shaft assembly so that the shift forks engage their respective slots in the synchronizer sleeves on the mainshaft assembly.
5. Bring the mainshaft assembly into mesh with the input cluster shaft assembly. Holding the 3 shafts (input cluster shaft, mainshaft and the main shift fork control shaft) in their respective working positions, lower them into their bores in the clutch

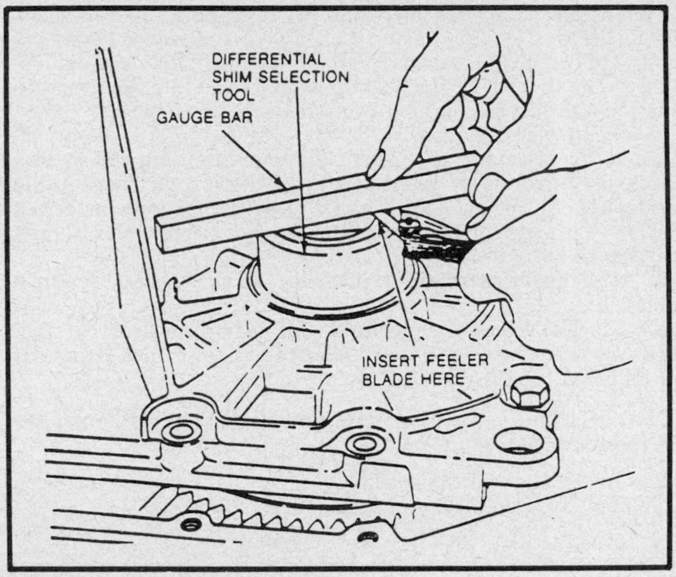
DIFFERENTIAL SHIM SELECTION TOOL
GAUGE BAR
INSERT FEELER BLADE HERE

Step 4 —bearing preload. Measure gap

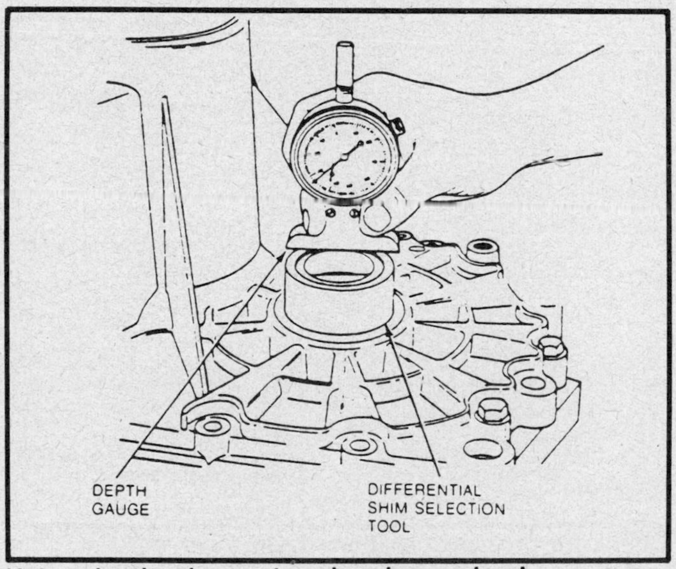
DEPTH GAUGE
DIFFERENTIAL SHIM SELECTION TOOL

Using depth micrometer—bearing preload

DIFFERENTIAL SHIM SIZE

in. (mm)	in. (mm)	in. (mm)
0.012 (0.30)	0.026 (0.65)	0.039 (1.00)
0.014 (0.35)	0.028 (0.70)	0.041 (1.05)
0.016 (0.40)	0.030 (0.75)	0.043 (1.10)
0.018 (0.45)	0.032 (0.80)	0.045 (1.15)
0.020 (0.50)	0.033 (0.85)	0.047 (1.20)
0.022 (0.55)	0.035 (0.90)	0.049 (1.25)
0.024 (0.60)	0.037 (0.95)	

The shim is located behind the differential bearing cup in the transmission case.

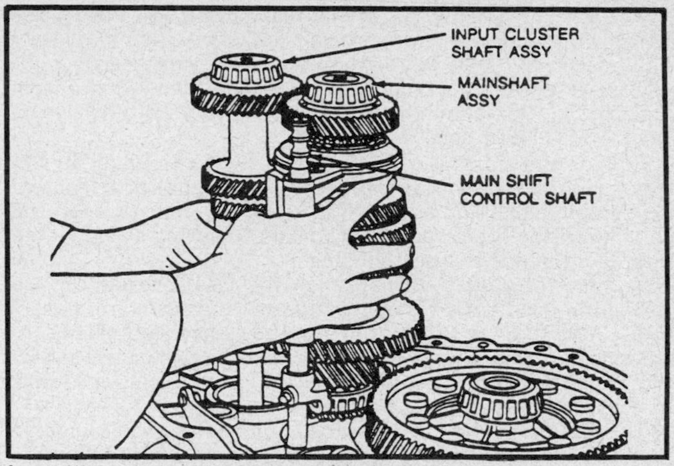
INPUT CLUSTER SHAFT ASSY
MAINSHAFT ASSY
MAIN SHIFT CONTROL SHAFT

Installing the mainshaft assembly

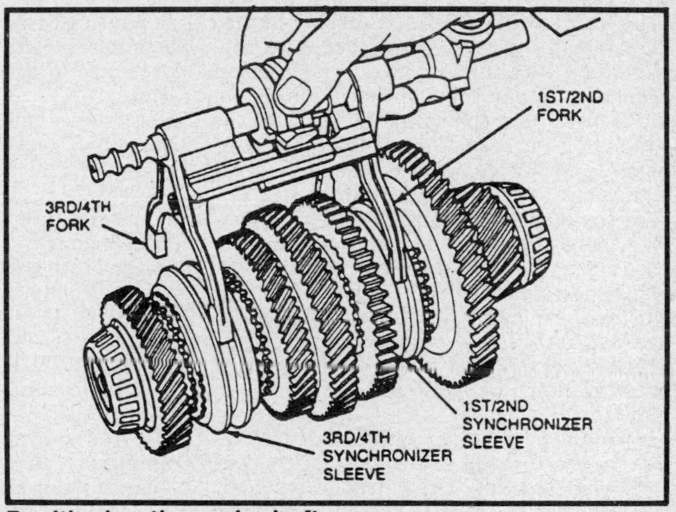
1ST/2ND FORK
3RD/4TH FORK
1ST/2ND SYNCHRONIZER SLEEVE
3RD/4TH SYNCHRONIZER SLEEVE

Positioning the mainshaft

CASE MAGNET

Installing the case magnet

housing case as a unit. Be careful not to damage the input shaft oil seal or mainshaft oil funnel.

NOTE: While performing this operation, care should be taken to avoid any movement of the 3rd/4th synchronizer sleeve, which may result in an overtravel of the synchronizer sleeve to hub allowing inserts to pop out of position.

6. Position the shift lever, 3–4 bias spring and 5th/reverse kickdown spring in their working positions (with 1 shift lever ball located in the socket of the input shift gate selector plate arm assembly and the other in the socket of the main shift control shaft block). Install the spring and ball in the 5th/reverse inhibitor shift lever hole.

7. Slide the shift lever shaft (notch down) through the 3/4 bias spring and the shift lever. Then, using a small drift, depress the inhibitor ball and spring. Tap the shift shaft through the shift lever, the 5th/reverse gear kickdown spring and then tap into its bore in the clutch housing.

8. Align the shift shaft bore with the case bore and tap the spring pin in, slightly below the case mating surface.

9. Check that the selector pin is in the neutral gate of the control selector plate and the finger of the fork selector arm is partially engaged with the 1st/2nd fork and partially engaged with the 3rd/4th fork.

10. Position reverse idler gear to clutch housing while aligning reverse shift relay lever to the slot in the gear. Slide the reverse idler shaft through the gear and into its bore. Place the reverse idler gear groove in engagement with the reverse relay lever.

11. Install the magnet in its pocket in the clutch housing case.

12. Install 5th gear relay lever onto the reverse idler shaft, aligning it with the fork interlock sleeve and reverse gear actuating arm slot and install the retaining ring C-clip.

13. Check that the gasket surfaces of the transaxle case and clutch housing are perfectly clean and free of burrs or nicks. Apply a $\frac{1}{16}$ inch wide bead of gasket eliminator E1FZ–19562–A or equivalent to the clutch housing.

14. Install the detent spring and plunger in their bore in the case. Carefully lower the transaxle case over the clutch housing, then using a punch, depress the spring and plunger. Move the transaxle case until the shift control shaft, mainshaft, input cluster shaft and reverse or 5th gear shaft align with their respective bores in the transaxle case.

15. Gently slide the transaxle case over the dowels and flush onto the clutch housing case. Make sure that the case does not bind on the magnet.

16. Apply pipe sealant with Teflon® D8AZ–19554–A or equivalent to the threads of the interlock sleeve retaining pin, in a clockwise direction. Use a drift or eqivalent to align the slot in the interlock sleeve with the hole in the transaxle case and install the retaining pin. Tighten to 12–15 ft. lbs.

NOTE: If the hole in the case does not align with the slot in the interlock sleeve, remove the case half and check for proper installation of the interlock sleeve.

17. Install the transaxle case to clutch housing bolts. Tighten to 13–17 ft. lbs.

18. Use a drift to align the bore in the reverse idler shaft with the retaining screw hole in the transaxle case.

19. Install the reverse idler shaft retaining bolt. Tighten to 16–20 ft. lbs.

20. Apply pipe sealant with Teflon® D8AZ–19554–A or equivalent to the threads of the backup lamp switch in a clockwise direction and install. Tighten the switch to 12–15 ft. lbs.

21. Apply pipe sealant with Teflon® D8AZ–19554–A or equivalent to the treads of the detent plunger retaining screw, in a clockwise direction. If applicable, install detent cartridge spring and plunger. Coat threads of cartridge with pipe sealant D8AZ–19554–A or equivalent. Install the retaining screw and tighten to 6–8 ft. lbs.

22. Tap the differential seal into the transaxle case with a suitable tool.

23. Place the transaxle upright and position a drift through the hole in the input shift shaft. Shift the transaxle into and out of all gears to verify proper installation.

NOTE: The transaxle will not shift directly into reverse from 5th gear.

24. Install the transaxle fill plugs after the transaxle has been installed in the vehicle and fluid has been added.

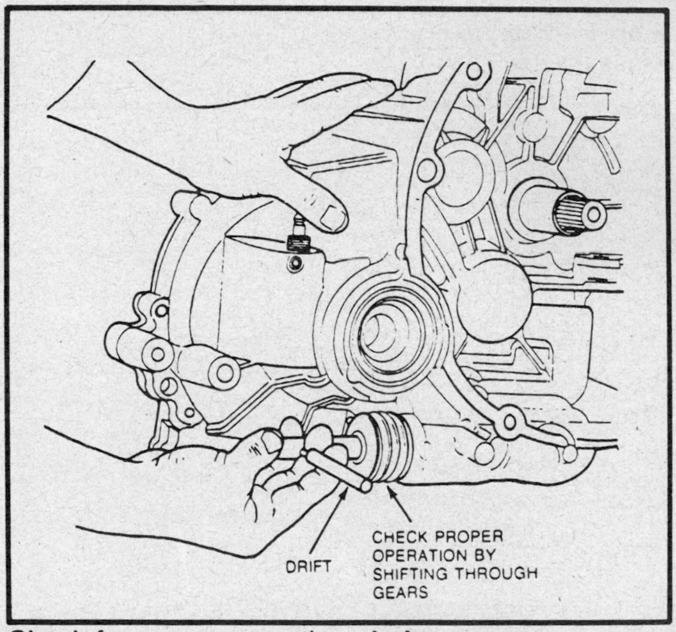
DRIFT
CHECK PROPER
OPERATION BY
SHIFTING THROUGH
GEARS

Check for proper operation—before vehicle installation

SPECIFICATIONS

TORQUE SPECIFICATIONS

Component	ft. lbs.	Nm
ESCORT/LYNX/EXP		
Transaxle to engine bolts	25–35	34–47
Air manage valve bracket bolt to transaxle	28–31	38–42
Switch actuator bracket bolt	7–10	9–13
Control arm to steering knuckle	37–44	50–60
Rear mounting bolts	35–50	47–68
Transaxle mounting stud	38–41	52–56
Front mount bracket bolts	25–35	34–47
Stiffener brace bolts	28–38	38–51
Starter stud bolts	30–40	41–54
Roll restrictor nuts	25–30	34–40
Shift stabilizer bar to transaxle case	23–35	3–47
Speedometer	2.5–3.5	3.4–4.5
Transaxle case-to-clutch housing	13–18	18–24
Reverse idler shaft-to-case	15–20	21–27
Fork interlock sleeve pin	12–15	16–20
Detent plunger retainer screw	6–8	7.5–11
Backup lamp switch	12–15	16–20
Control selector plate	6–8	8–11
Speedo retaining	1.5–2	2–3
Reverse shift relay lever bracket	6–8	8–11
Filler plug	9–15	12–20
Clutch release fork to shaft	30–41	40–41
Shift lever cover screws	1.5–2.0	2–3
Final drive gear to differential case attaching bolts and nuts (service only)	55–70	75–95
Transmission oil fill plug	9–14	12–20
TEMPO/TOPAZ		
Front stabilizer bar to control arm	107–125	145–169
Transaxle to engine bolts	25–35	34–47
Front stabilizer bar bracket bolts	47–55	64–74
Lower control arm ball joint to steering knuckle nut	37–44	50–60
Engine roll restrictor attaching nuts	14–20	19–27
Starter cable	70–130 ①	7.9–14.7
Starter stud bolts	30–40	41–54
Shift mechanism to shift shaft	7–10	9–13
Shift mechanism stabilizer to transaxle bolt	23–35	31–47
Speedometer cable	30–40 ①	3.4–4.5
Left hand rear no. 4 insulator to body bracket	35–50	47–68
Left hand front no. 1 insulator to body bracket	25–35	34–47
Oil pan to transaxle	28–38	38–51
Transaxle case-to-clutch housing	13–18	18–24
Reverse idler shaft-to-case	15–20	21–27
Fork interlock sleeve pin	12–15	16–20
Detent plunger retainer screw	6–8	7.5–11
Backup lamp switch	12–15	16–20
Control selector plate	6–8	8–11
Speedo retaining	1.5–2	2–3
Reverse shift relay lever bracket	6–8	8–11
Filler plug	9–15	12–20
Clutch release fork to shaft	30–41	40–41
Shift lever cover screws	1.5–2.0	2–3
Final drive gear to differential case attaching bolts and nuts (service only)	55–70	75–95
Transmission oil fill plug	9–14	12–20
TAURUS/SABLE		
Transaxle to engine bolts	①	①
Air manage valve bracket bolt to transaxle	28–31	38–42
Switch actuator bracket bolt	7–10	9–13
Control arm to steering knuckle	37–44	50–60
Rear mounting bolts	35–50	47–68
Transaxle mounting stud	38–41	52–56
Front mount bracket bolts	25–35	34–47
Stiffener brace bolts	38–38	38–51
Starter stud bolts	30–40	41–54
Roll restrictor nuts	25–30	34–40
Shift stabilizer bar to transaxle case	23–35	31–47
Speedometer	2.5–3.5	3.4–4.5
Transaxle case-to-clutch housing	13–18	18–24
Reverse idler shaft-to-case	15–20	21–27
Fork interlock sleeve pin	12–15	16–20
Detent plunger retainer screw	6–8	7.5–11
Backup lamp switch	12–15	16–20
Control selector plate	6–8	8–11
Speedo retaining	1.5–2	2–3
Reverse shift relay lever bracket	6–8	8–11
Filler plug	9–15	12–20
Clutch release fork to shaft	30–41	40–41
Shift lever cover screws	1.5–2.0	2–3
Final drive gear to differential case attaching bolts and nuts (service only)	55–70	75–95
Transmission oil fill plug	9–14	12–20

① RWB-BF, RWB-BJ models: All bolts 34–47 Nm (25–35 ft. lbs.).

SPECIAL TOOLS

Tool Number	Description
T50T-100-A	Impact slide hammer
T81P-1177-A	Differential seal replacer
T81P-1177-B	Transaxle plugs
D83P-4026-A	Halfshaft remover
T81P-4026-A	Differential rotator
D79L-4621-A	Pinion bearing cone remover/installer
T77F-7050-B	Input shaft seal remover
T77F-7288-A	Sector shaft seal tool
TOOL-4201-C	Dial indicator
T77F-1176-A	Draw bolt
T81P-1177-A	Differential seal replacer
T75T-1225-A	Stop differential bearing cup replacer
T57L-4220-A	Differential bearing cone remover
T77F-4220-B1	Differential bearing cone remover/installer
T81P-4220-A	Step plate differential bearing removal
T77F-4222-A	Differential bearing cup replacer
T77F-4222-B	Differential bearing cup remover
T80L-77003-A	Gauge bar
T83P-4220-CH	Bearing installer
T83P-4451-AH2	Height gauge spacer
T83P-4451-AH1	Shim selector tool
T81P-4451-B2 014-00210	Height gauge spacer Hi-lift jack
014-00225	Manual transaxle adapter

Section 6

Festiva MTX4 and MTX5 Transaxles
Ford Motor Co.

APPLICATION

1988–89 Festiva 1.3L

GENERAL DESCRIPTION

The Ford Festiva 4 and 5 speed transaxles are manufactured for Ford by Mazda. Because of the similarities in service and construction, the 4 and 5 speed transaxles have been combined into this single section. These transaxles provide fully synchronized forward speeds and a single non-synchronized reverse gear. All forward gears are helical cut and are in constant mesh. These transaxles are manufactured with metric measurements and all replacement parts must be the correct metric dimension. The transaxle case, cover and adapter housing case are aluminum.

Metric Fasteners

Metric bolt sizes and thread pitches are used for all fasteners on the Festiva MTX 4/5 Speed transaxles. The use of metric tools is mandatory in the service of these transaxles.

Do not attempt to interchange metric fasteners for inch system fasteners. Mismatched or incorrect fasteners can result in damage to the transaxle unit through malfunctions, breakage or possible personal injury. Care should be taken to reuse the fasteners in the same location as removed, whenever possible. Due to the large number of alloy parts used and the aluminum casing, torque specifications should be strictly observed. Before installing cap screws into aluminum parts, always dip screws into oil to prevent the screws from galling the aluminum threads and to prevent seizing.

Fluid Capacities

All Ford Festiva MTX 4/5 speed transaxles require 2.6 qts. (2.5L) of Dexron®II automatic transmission fluid.

Checking Fluid Level

The vehicle must be level before an accurate fluid level reading can be obtained. The speedometer driven gear is used as a dipstick to check the transaxle level, use the following procedure:

1. Remove the boot from the gear sleeve and slide it up the speedometer cable.
2. Disconnect the speedometer cable from the sleeve.
3. Remove the speedometer sleeve and gear assembly.
4. Remove the O-ring from the speedometer gear sleeve. Clean all traces of oil from the sleeve.
5. Insert the sleeve into its bore in the transaxle.
6. Remove the sleeve and check the oil level using the indicator mark on the sleeve. If the oil level is below the full mark, add the required amount of oil through the speedometer gear bore.
7. Inspect the O-ring and if it is in good condition, install it on the speedometer sleeve. Connect the speedometer cable and position the dust boot over the sleeve.

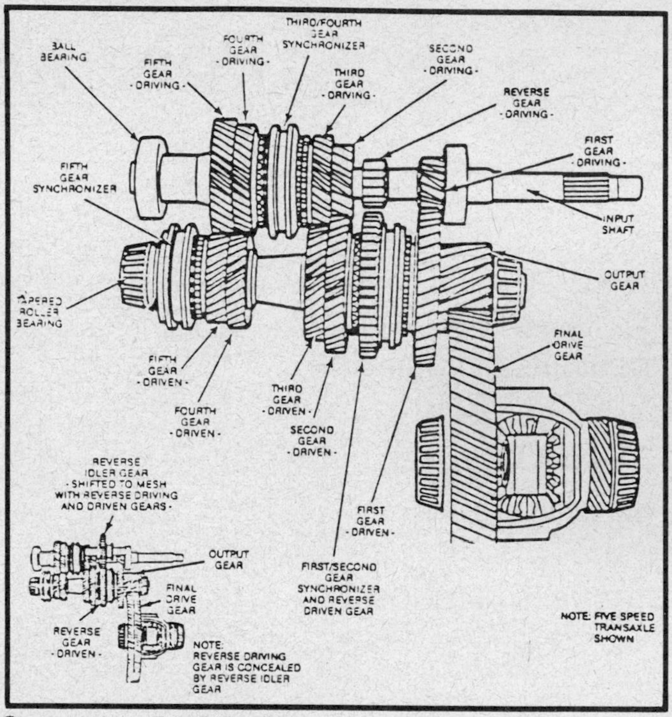
BALL BEARING
FIFTH GEAR DRIVING
FOURTH GEAR DRIVING
THIRD/FOURTH GEAR SYNCHRONIZER
THIRD GEAR DRIVING
SECOND GEAR DRIVING
REVERSE GEAR DRIVING
FIRST GEAR DRIVING
INPUT SHAFT
FIFTH GEAR SYNCHRONIZER
OUTPUT GEAR
TAPERED ROLLER BEARING
FINAL DRIVE GEAR
FIFTH GEAR DRIVEN
THIRD GEAR DRIVEN
FOURTH GEAR DRIVEN
SECOND GEAR DRIVEN
REVERSE IDLER GEAR SHIFTED TO MESH WITH REVERSE DRIVING AND DRIVEN GEARS
FIRST GEAR DRIVEN
OUTPUT GEAR
FINAL DRIVE GEAR
FIRST/SECOND GEAR SYNCHRONIZER AND REVERSE DRIVEN GEAR
REVERSE GEAR DRIVEN
NOTE: REVERSE DRIVING GEAR IS CONCEALED BY REVERSE IDLER GEAR
NOTE: FIVE SPEED TRANSAXLE SHOWN

Gear train—Ford Festiva

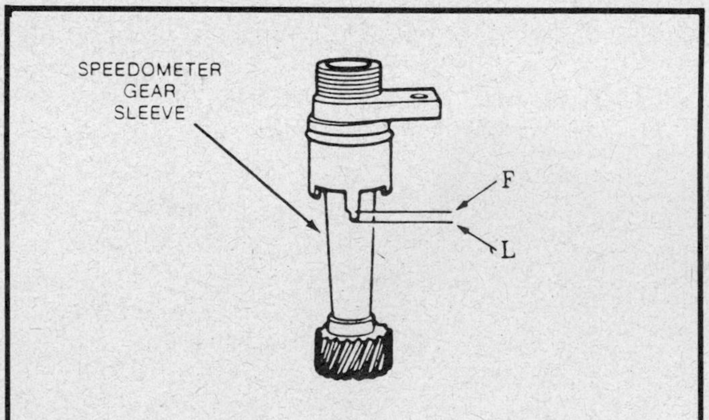
SPEEDOMETER GEAR SLEEVE
F
L

Checking fluid level—Ford Festiva MTX 4/5 speed

TRANSAXLE MODIFICATIONS

There were no modifications for the Ford Festiva MTX at the time of this publication.

TROUBLE DIAGNOSIS

CHILTON'S THREE C'S DIAGNOSIS
Festiva 4/5 MTX

Condition	Cause	Correction
Shift lever does not operate smoothly or binds or cannot be operated at all	a) Selector rod joint stiff	a) Repair or replace
	b) Selector rod bent	b) Replace bent rod
	c) Lack of lubrication on shift linkage pivots	c) Clean and lubricate with Molybdenum Disulphide grease
	d) Shift lever ball unit stiff	d) Repair or replace
	e) Gearshift gate incorrectly adjusted	e) Adjust gate
Excessive shift lever play	a) Selector rod bushing worn	a) Replace
	b) Loose or worn selector rod clamping bolts	b) Tighten or replace as necessary
	c) The spring in the shift lever ball unit is fatigued	c) Replace
	d) The bushing in shift lever ball unit worn	d) Replace
Hard shifting	a) Insufficient oil in transaxle	a) Add oil
	b) Incorrect oil quality	b) Drain and refill with specified oil
	c) Selector rod bent	c) Replace
	d) Transmission shifting mechanism insufficiently lubricated	d) Lubricate
	e) Excessive clutch pedal free play	e) Adjust
	f) Shift fork and shift rod worn	f) Replace
	g) Synchronizer ring worn	g) Replace
	h) Worn cone surface of gear	h) Replace
	i) Improper contact between synchronizer ring and cone surface	i) Replace
	j) Excessive play in the axial direction of each gear	j) Replace worn component
	k) Bearings worn	k) Adjust or replace
	l) Synchronizer key is fatigued	l) Replace
Locked in gear	a) Shift gate out of adjustment or worn	a) Repair or replace as necessary and adjust
	b) Worn interlock sleeve or bent or damaged shift fork	b) Check interlock sleeve for wear and repair or replace as necessary
	c) Gear seizure	c) Replace defective parts
	d) Synchronizer keys out of position	d) Repair or replace as necessary
Jumping out of gear	a) Worn or improperly installed engine mount	a) Repair or replace
	b) Loose or worn control rod clamping bolts or linkage	b) Repair or replace and tighten as necessary
	c) Bent shift control rod	c) Replace
	d) Worn shift control rod bushing	d) Replace
	e) Fatigued lever ball spring	e) Replace
	f) Improper installation of stabilizer bar	f) Fit correctly and tighten
	g) Worn synchronizer clutch hub	g) Replace
	h) Worn synchronizer clutch hub sleeve	h) Replace
	i) Worn steel ball sliding groove on control rod end	i) Replace
	j) Fatigued steel ball spring	j) Replace
	k) Excessive backlash	k) Replace
	l) Worn bearings	l) Adjust or replace

CHILTON'S THREE C'S DIAGNOSIS
Festiva 4/5 MTX

Condition	Cause	Correction
Noise	a) Insufficient oil in transaxle	a) Add oil
	b) Poor oil quality	b) Drain and refill with specified oil
	c) Worn sliding surfaces at synchronizer	c) Repair or replace
	d) Excessive backlash	d) Replace
	e) Surface of a gear is damaged	e) Replace
	f) Foreign matter in transmission	f) Repair or replace as necessary
	g) Differential gear is damaged. Backlash is excessive	g) Repair or replace as necessary
	h) Ring gear bolts are loose	h) Repair or replace as necessary
	i) Bearings worn or out of adjustment	i) Repair or adjust as necessary
Gear clash	a) Excessive engine idle speed	a) Adjust engine idle rpm
	b) Inadequate clutch pedal reserve resulting excessive spin time. Inadequate clutch disengagement	b) Check clutch adjustment, operating mechanism or for excessive clutch disc runout—replace parts as required
	c) Disc binding on transmission input shaft	c) Check for burrs on splines, replace if necessary
	d) Excessive disc runout	d) Replace
	e) Flywheel housing misalignment	e) Realign
	f) Oil or grease on clutch facings	f) Replace disc and correct cause of contamination
	g) Damaged or contaminated clutch lining	g) Replace disc
	h) Weak or broken insert keys in the synchronizer assembly	h) Replace parts as required
	i) Worn synchronizer rings and/or cone surfaces	i) Replace parts as required
	j) Broken synchronizer rings	j) Replace

POWER FLOW

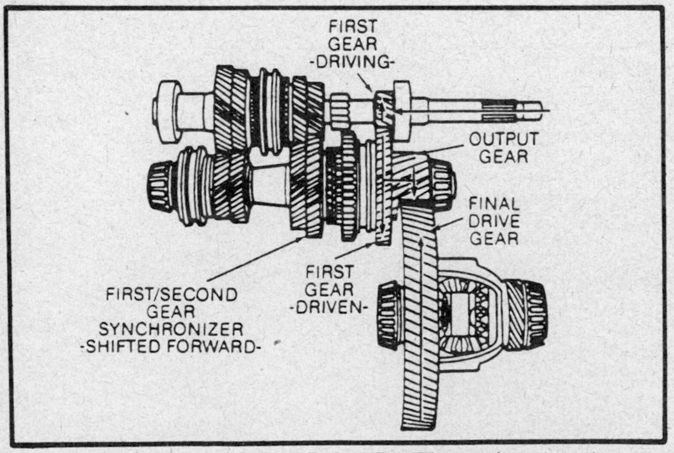
FIRST
GEAR
-DRIVING-
OUTPUT
GEAR
FINAL
DRIVE
GEAR
FIRST
GEAR
-DRIVEN-
FIRST/SECOND
GEAR
SYNCHRONIZER
-SHIFTED FORWARD-

Power flow 1st gear—Ford Festiva MTX 4/5 speed

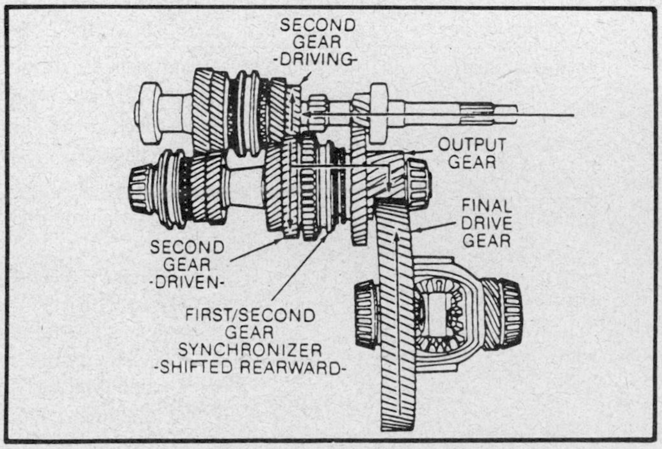
SECOND
GEAR
-DRIVING-
OUTPUT
GEAR
FINAL
DRIVE
GEAR
SECOND
GEAR
-DRIVEN-
FIRST/SECOND
GEAR
SYNCHRONIZER
-SHIFTED REARWARD-

Power flow 2nd gear—Ford Festiva MTX 4/5 speed

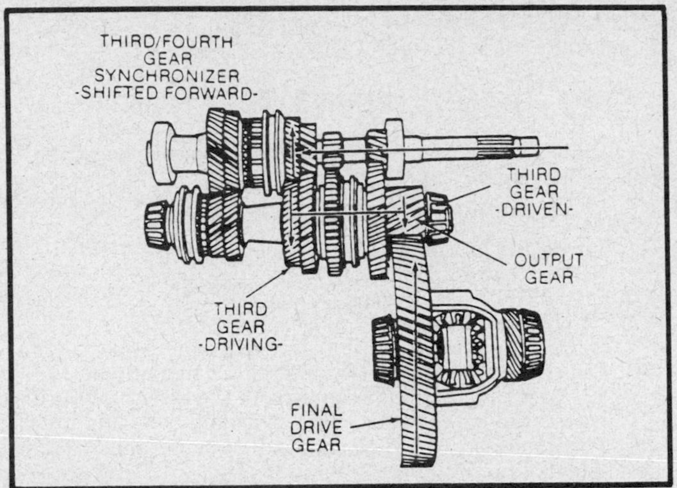
THIRD/FOURTH
GEAR
SYNCHRONIZER
-SHIFTED FORWARD-
THIRD
GEAR
-DRIVEN-
OUTPUT
GEAR
THIRD
GEAR
-DRIVING-
FINAL
DRIVE
GEAR
Power flow 3rd gear—Ford Festiva MTX 4/5 speed

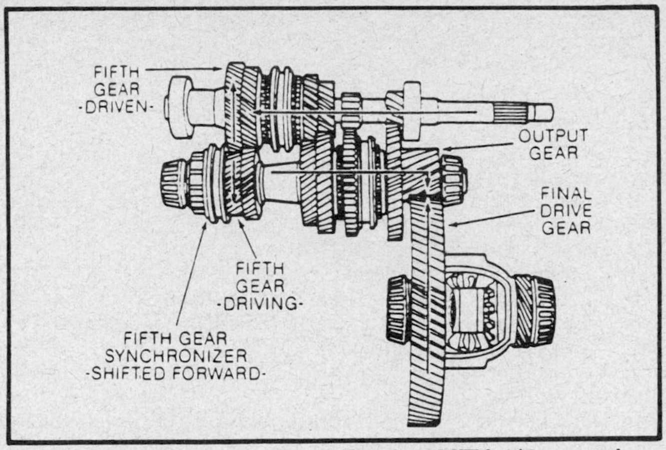
FIFTH
GEAR
-DRIVEN-
OUTPUT
GEAR
FINAL
DRIVE
GEAR
FIFTH
GEAR
-DRIVING-
FIFTH GEAR
SYNCHRONIZER
-SHIFTED FORWARD-
Power flow 5th gear—Ford Festiva MTX 4/5 speed

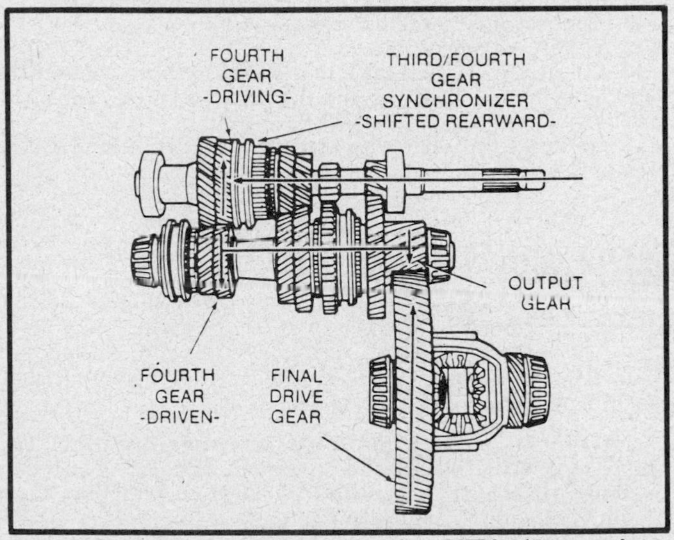
FOURTH
GEAR
-DRIVING-
THIRD/FOURTH
GEAR
SYNCHRONIZER
-SHIFTED REARWARD-
OUTPUT
GEAR
FOURTH
GEAR
-DRIVEN-
FINAL
DRIVE
GEAR
Power flow 4th gear—Ford Festiva MTX 4/5 speed

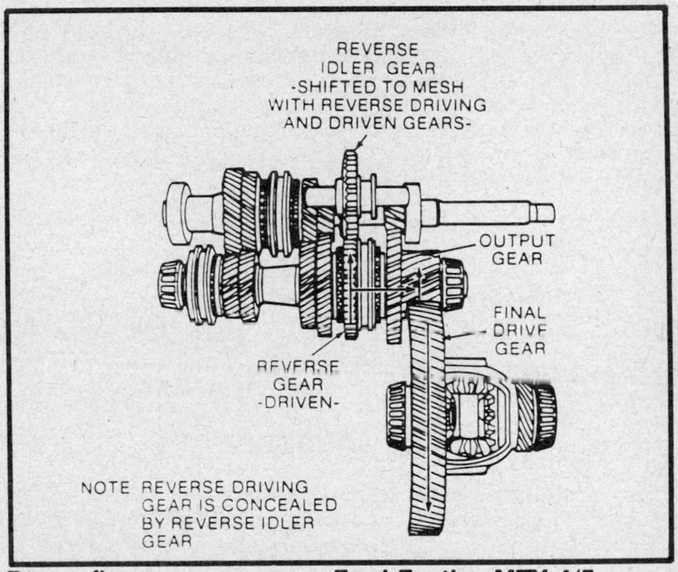
REVERSE
IDLER GEAR
-SHIFTED TO MESH
WITH REVERSE DRIVING
AND DRIVEN GEARS-
OUTPUT
GEAR
FINAL
DRIVE
GEAR
REVERSE
GEAR
-DRIVEN-
NOTE REVERSE DRIVING
GEAR IS CONCEALED
BY REVERSE IDLER
GEAR
Power flow reverse gear—Ford Festiva MTX 4/5 speed

ON CAR SERVICE

Services

FLUID CHANGE

1. Remove the boot from the gear sleeve and slide it up the speedometer cable.
2. Disconnect the speedometer cable from the sleeve.
3. Remove the speedometer sleeve and gear assembly.
4. Siphon the transmission fluid into a pan.
5. Remove the O-ring from the speedometer gear sleeve. Clean all traces of oil from the sleeve.
6. Fill to the transaxle with the proper type and amount of transmission fluid.
7. Insert the speedometer sleeve into its bore in the transaxle.
8. Remove the sleeve and check the oil level using the indicator mark on the sleeve. If the oil level is below the full mark, add the required amount of transmission fluid through the speedometer gear bore.
9. Inspect the O-ring and if it is in good condition, install it on the speedometer sleeve. Connect the speedometer cable and position the dust boot over the sleeve.

REMOVAL AND INSTALLATION

TRANSAXLE REMOVAL

1. Disconnect the negative battery cable. Disconnect the backup lamp switch at the wiring connector.
2. Loosen the clutch adjuster nut and disengage the cable from the release lever.
3. Remove the starter motor.
4. Disconnect the speedometer cable.
5. Remove the 2 bolts located at the top of the clutch housing.
6. Install an engine support bar D79P–6000–B or equivalent.
7. Raise and support the vehicle safely.
8. Remove the nut and bolt attaching the shift rod to the input shift rail.
9. Remove the bolts and nuts attaching the lower control arms to the steering knuckles.
10. Disengage the halfshafts from the transaxle. Install differential plugs T87C–7025–C or equivalent.
11. Remove the bracket attaching bolts.
12. Remove the crossmember. Position a suitable transmission jack under the transaxle and secure with a safety chain.
13. Remove the remaining transaxle attaching bolts.
14. Pull the transaxle away from the engine and lower it out of the vehicle.

TRANSAXLE INSTALLATION

1. Raise the transaxle into position and seat it against the back of the engine.
2. Install the transaxle lower attaching bolts. Tighten the bolts to 47–66 ft. lbs. (64–89 Nm).
3. Install the brackets.
4. Remove the transmission jack.
5. Install the crossmember.
6. Remove the differential plugs.
7. Install new circlips on the halfshafts.
8. Seat the halfshafts in the differential side gears.
9. Connect the lower control arms to the steering knuckles.
10. Install the lower control arm attaching bolts and nuts.
11. Position the shift rod on the input shift rail and install the attaching bolt and nut.
12. Lower the vehicle. Remove the engine support bar.
13. Install the 2 bolts located at the top of the clutch housing. Tighten the attaching bolts to 47–66 ft. lbs. (64–89Nm).
14. Connect the speedometer cable.
15. Install the starter motor.
16. Connect the clutch cable to the release lever. Connect the backup lamp switch wiring and the negative battery cable.
17. Adjust the clutch free play.
18. Refill the transaxle with the proper transaxle fluid. Check the fluid level.

BENCH OVERHAUL

Before Disassembly

Clean the exterior of the transaxle assembly before any attempt is made to disassemble it as to prevent dirt or other foreign materials from entering the transaxle assembly or its internal parts.

NOTE: If steam cleaning is done to the exterior of the transaxle, immediate disassembly should be done to avoid rusting, caused by condensation forming on the internal parts.

Transaxle Disassembly

NOTE: The 5 speed transaxle disassembly is included in the following procedures. With the exception of the 5th gear components, the procedures are identical for the 4 speed transaxle disassembly.

1. Mount the transaxle to the bench mounting fixture T57L–500–B or other suitable transaxle stand.
2. Remove the transaxle mount.
3. Remove the 5th/reverse detent plug, spring and ball.
4. Remove the 1st/2nd detent plug, spring and ball.
5. Remove the backup lamp switch and the 5th gear switch.
6. Remove the transaxle case-to-clutch housing attaching bolts.
7. Separate the transaxle case from the clutch housing and remove the transaxle case.

NOTE: It may be necessary to tap the transaxle case with a plastic-tipped hammer to break the sealant bond.

8. Remove the bearing race from the output shaft bearing.
9. Remove the bearing preload shims from the output shaft and the endplay shims from the input shaft.

NOTE: Tag the shims for assembly reference. The bearing preload shims and output shaft bearing cup may remain in the transmission case. If so, remove them from the bearing bores in the case.

10. Remove the case magnet.
11. Remove the reverse shift lever detent plate.
12. Remove the reverse idler gear and shaft.
13. Remove the reverse shift lever pivot pin retainer. Remove the pivot pin and the reverse shift lever.
14. Remove the 3rd/4th detent plug, spring and ball.
15. Make sure all the shift rails are in the neutral detent and remove the following roll pins:
    3rd/4th shift fork
    5th shift fork
    3rd/4th relay lever
    1st/2nd shift fork
16. Remove the 3rd/4th shift rail. While raising and lowering the 5th/reverse shift rail in slight amounts, pull upward on the 3rd/4th shift rail. When the interlocks are properly positioned , the 3rd/4th shift rail will slide upward out of the relay lever and shift fork.
17. Remove the 3rd/4th shift fork and the 3rd/4th shift lever.
18. Remove the interlock plug.
19. Remove the 1st/2nd shift rail.

While raising and lowering the 5th/reverse shift rail in slight amounts, pull upward on the 1st/2nd shift rail. When the interlocks are properly positioned, the 1st/2nd shift rail will slide upward out of the shift fork.

20. Remove the input gear train, output gear train and fifth/reverse shift rail, as an assembly.

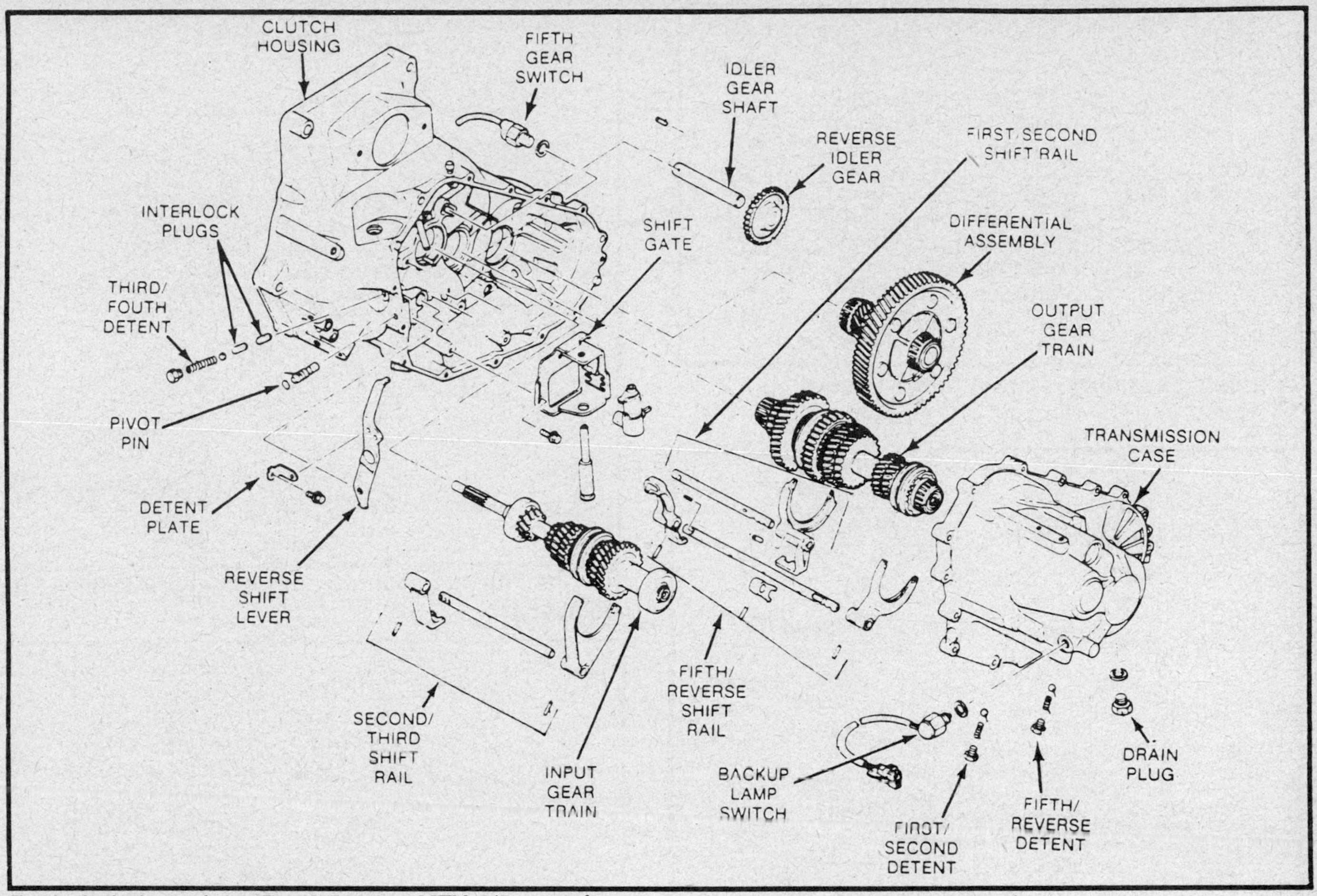
CLUTCH HOUSING
FIFTH GEAR SWITCH
IDLER GEAR SHAFT
REVERSE IDLER GEAR
FIRST/SECOND SHIFT RAIL
INTERLOCK PLUGS
SHIFT GATE
DIFFERENTIAL ASSEMBLY
THIRD/ FOUTH DETENT
OUTPUT GEAR TRAIN
PIVOT PIN
TRANSMISSION CASE
DETENT PLATE
REVERSE SHIFT LEVER
FIFTH/ REVERSE SHIFT RAIL
SECOND/ THIRD SHIFT RAIL
INPUT GEAR TRAIN
BACKUP LAMP SWITCH
DRAIN PLUG
FIFTH/ REVERSE DETENT
FIRST/ SECOND DETENT

Exploded view of the Ford Festiva MTX 4/5 speed

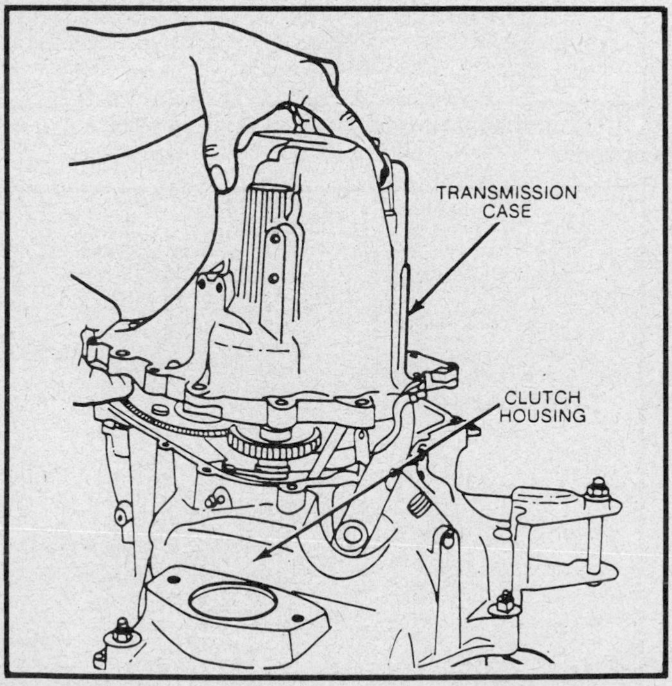
TRANSMISSION CASE
CLUTCH HOUSING

Separating the transaxle case from the clutch housing

NOTE: The interlock pin is located in the end of the 5th/reverse shift rail. Use care to prevent its loss.

21. Remove the interlock plugs using a magnet.
22. Remove the attaching bolts and the shift gate assembly. Note the location of the long attaching bolt.
23. Remove the speedometer gear attaching bolt. Remove the speedometer driven gear.
24. Remove the differential assembly.
25. Remove the attaching bolt and the oil baffle plate.
26. Remove the attaching bolt and the oil channel.
27. Remove the oil funnel from the transmission case. A length of wire with a hook at the end can be used to pull the funnel out of the bearing bore.

NOTE: Before removing the selector arm roll pin, move the input shift rail until the pin is located over the removal pocket molded into the case.

28. Remove the selector arm roll pin using a pin punch.
29. Remove the selector arm from the input shift rail.
30. Remove the input shift rail from the case.

Unit Disassembly and Assembly

OUTPUT SHAFT

Disassembly

1. Remove the 5th/reverse shift rail and 1st/2nd shift fork from the gear train.
2. Remove the interlock pin from the 5th/reverse shift rail.

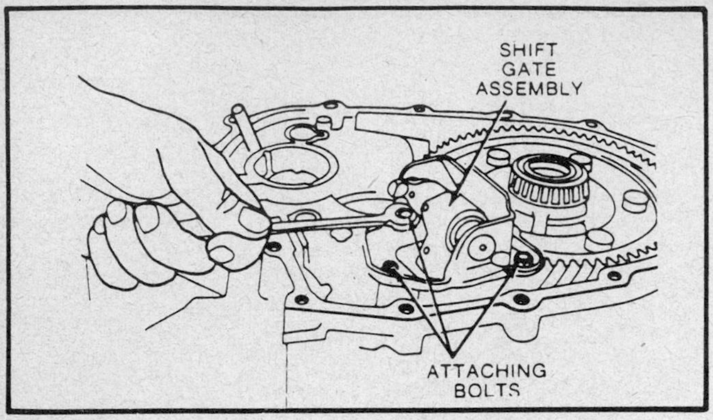
SHIFT
GATE
ASSEMBLY
ATTACHING
BOLTS

Shift gate assembly removal – Ford Festiva MTX 4/5 speed

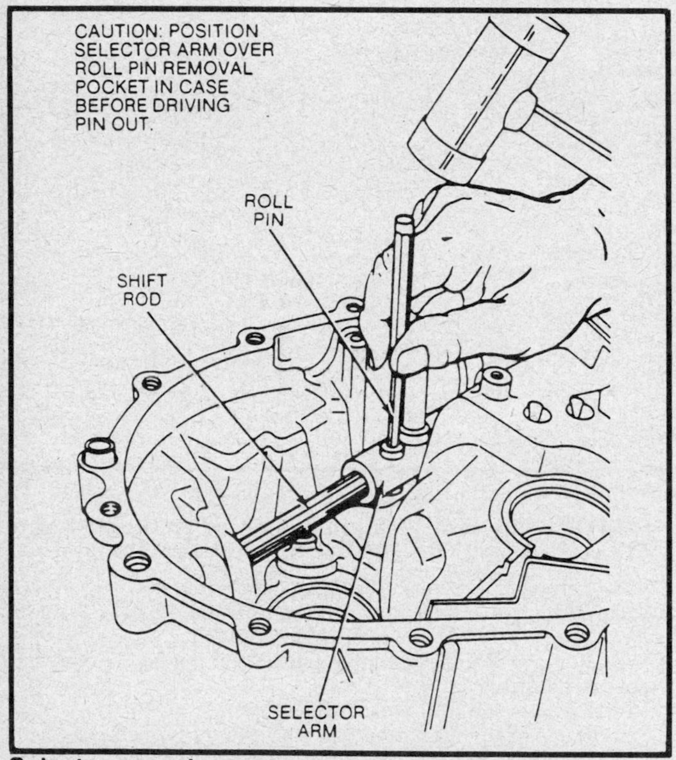
CAUTION: POSITION
SELECTOR ARM OVER
ROLL PIN REMOVAL
POCKET IN CASE
BEFORE DRIVING
PIN OUT.
ROLL
PIN
SHIFT
ROD
SELECTOR
ARM

Selector arm pin removal – Ford Festiva MTX 4/5 speed

3. Remove the bearing from the 5th gear synchronizer end of the output shaft. The bearing can be removed using a press and bearing pulling attachment tool D84L–1123–A, or equivalent.
4. Remove the 5th gear synchronizer snapring and stop washer.

NOTE: If the synchronizer is tight on the shaft, it can be removed using a press and bearing pulling attachment tool D84L–1123–A, or equivalent. Position the puller behind the gear.

5. Remove the 5th gear synchronizer and 5th gear, as an assembly.

NOTE: If the synchronizer is tight on the shaft, it can be removed using a press and bearing pulling attachment tool D81L–1123–A or equivalent. Position the puller behind the gear.

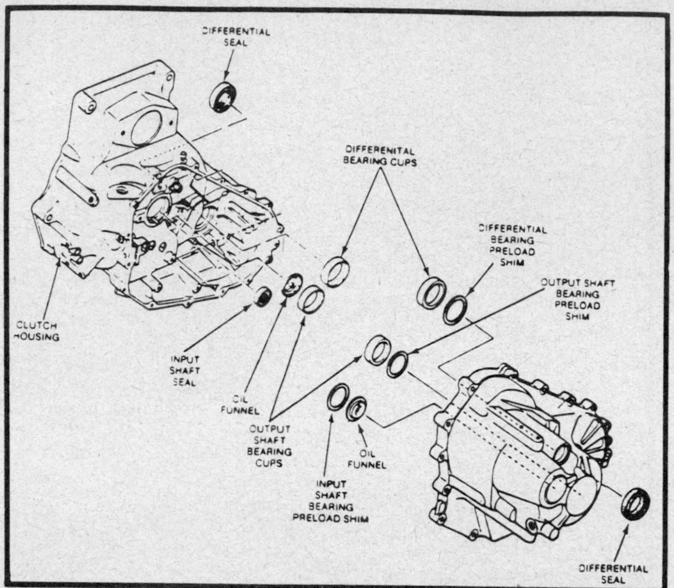
DIFFERENTIAL
SEAL
DIFFERENITAL
BEARING CUPS
DIFFERENTIAL
BEARING
PRELOAD
SHIM
OUTPUT SHAFT
BEARING
PRELOAD
SHIM
CLUTCH
HOUSING
INPUT
SHAFT
SEAL
OIL
FUNNEL
OUTPUT
SHAFT
BEARING
CUPS
OIL
FUNNEL
INPUT
SHAFT
BEARING
PRELOAD SHIM
DIFFERENTIAL
SEAL

Bearing cup and seal replacement – Ford Festiva MTX 4/5 speed

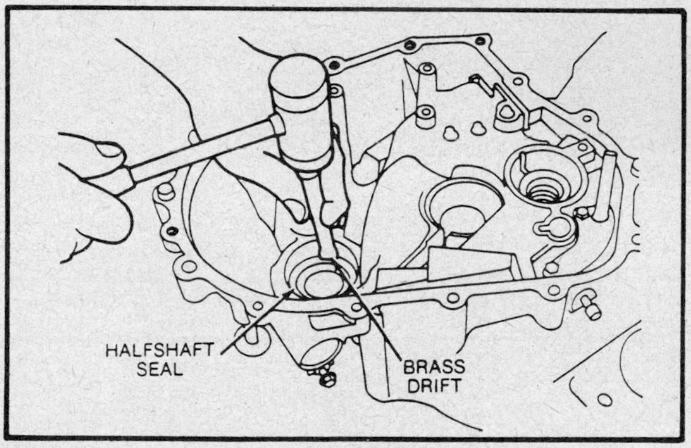
HALFSHAFT
SEAL
BRASS
DRIFT

Differential seal removal – Ford Festiva MTX 4/5 speed

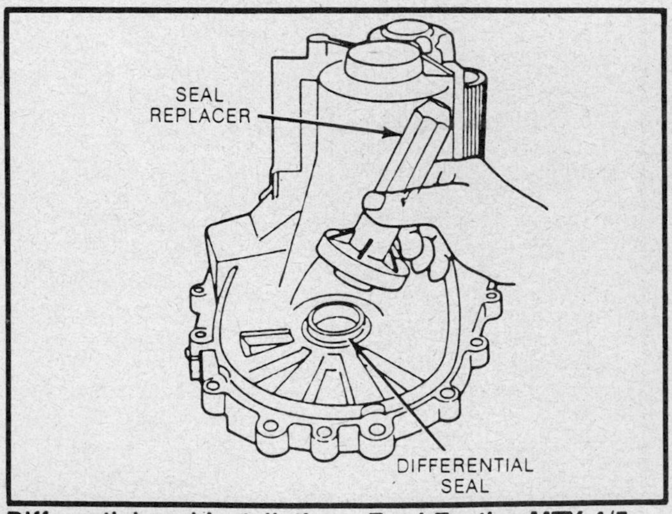
SEAL
REPLACER
DIFFERENTIAL
SEAL

Differential seal installation – Ford Festiva MTX 4/5 speed

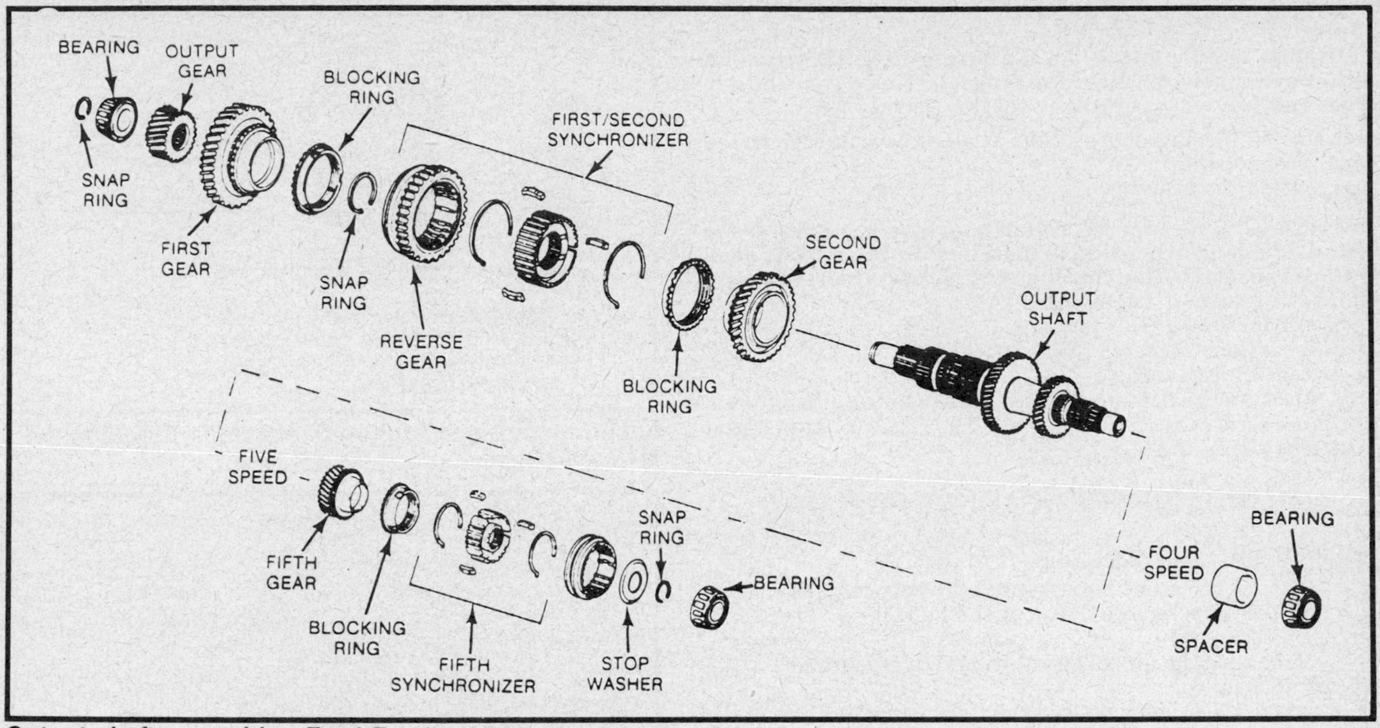
BEARING
OUTPUT GEAR
BLOCKING RING
FIRST/SECOND SYNCHRONIZER
SNAP RING
FIRST GEAR
SNAP RING
REVERSE GEAR
SECOND GEAR
OUTPUT SHAFT
BLOCKING RING
FIVE SPEED
FIFTH GEAR
BLOCKING RING
FIFTH SYNCHRONIZER
SNAP RING
STOP WASHER
BEARING
BEARING
FOUR SPEED
SPACER

Output shaft assembly – Ford Festiva MTX 4/5 speed

6. Remove the bearing snapring from the output gear end of the shaft.
7. Remove the bearing from the output gear end of the shaft. The bearing can be removed using a press and bearing pulling attachment tool D84L–1123–A or equivalent. Position the puller behind the gear.
8. Remove the blocking ring from the 1st/2nd gear synchronizer.
9. Remove the 1st/2nd gear synchronizer snapring.
10. Remove the 1st/2nd gear synchronizer.

NOTE: If the synchronizer is tight on the shaft, it can be removed using a press and bearing pulling attachment tool D81L–1123–A or equivalent. Position the puller behind 2nd gear.

11. Remove the blocking ring and 2nd gear.

Inspection

1. Clean all components, except plastic or nylon, in solvent.
2. Inspect for broken, chipped or worn gear teeth.
3. Inspect for bent or broken inserts.
4. Check for damaged roller thrust or needle bearings and inspect the bearing bores for cracks or damage.
5. Check for worn or loss of surface metal from the counter shaft and hub, clutch shaft or reverse idler gear shaft.
6. Check the thrust washers for damage and wear.
7. Check for nicked, broken or worn output or clutch shaft splines.
8. Check for bent, weak or distorted snaprings.
9. Check the reverse idler gear busing for wear.
10. Inspect front and rear bearings for roughness, wear and damage.
11. Inspect the shift rails, selector arms, plates, interlock for worn, bent or damaged parts.
12. Inspect synchronizers damage, wear and rough surfaces.
13. If any of the above are found, replace the appropriate part(s).

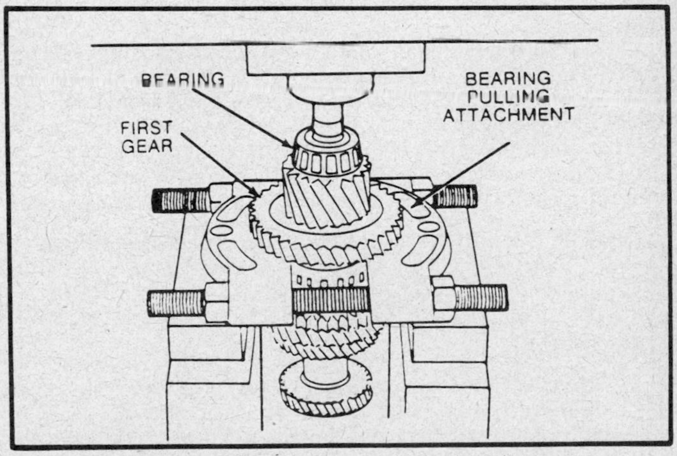
BEARING
FIRST GEAR
BEARING PULLING ATTACHMENT

Removing the output gear bearing – Ford Festiva MTX 4/5 speed

Assembly

1. Install 5th gear.
2. Install the blocking ring and the 5th gear synchronizer. Make sure the oil grooves on the synchronizer face 5th gear and that the beveled end of the synchronizer sleeve faces the bearing.

NOTE: If the synchronizer is tight on the shaft, it can be installed using a press and a suitable installer.

3. Install the synchronizer stop washer and snapring. Note that the dished side of the washer faces the synchronizer.
4. Install 2nd gear.
5. Install the blocking ring and the 1st/2nd gear synchronizer. Note that the reverse gear faces 2nd gear.

NOTE: If the synchronizer is tight on the shaft, it can be installed using a press and a suitable installer.

6. Install the blocking ring and 1st gear.

NOTE: The 1st gear blocking ring is different from the other synchronizer blocking rings. In three positions, it does not have synchronizer teeth.

7. Install the output gear. Note that the beveled edge on the gear faces 1st gear.
8. Install the bearings.

WARNING: The tapered roller bearings are 2 different sizes. The larger bearing is installed on the output gear end of the shaft. The bearings can be installed using the following, or equivalent, tools:
Step Plate D80L–630–3
Bearing Cone Replacer T87C–7025–B
Axle Bearing/Seal Plate T75L–1165–B

9. Measure the clearance between the snapring and the bearing inner race. If the clearance is not 0.0039 in. (0.1mm), install a selective snapring.

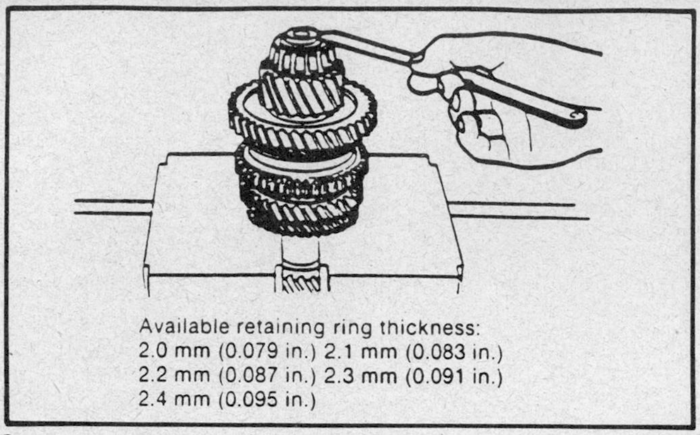
Available retaining ring thickness:
2.0 mm (0.079 in.) 2.1 mm (0.083 in.)
2.2 mm (0.087 in.) 2.3 mm (0.091 in.)
2.4 mm (0.095 in.)

Snapring measurement and selection – Ford Festiva MTX 4/5 speed

OUTPUT SHAFT BEARING PRELOAD

Adjustment

1. Remove the attaching bolts and the transmission case.
2. If necessary, remove the input shaft and install the correct endplay shim.
3. Make sure the funnel and bearing race are installed in the clutch housing.
4. Install the original preload shim(s) and bearing race in the transmission case.
5. Install the output shaft.
6. Shift the 3rd/4th gear synchronizer into the 4th gear position.
7. Install the transmission case to clutch housing attaching bolts. Tighten the attaching bolts to 14–19 ft lbs. (19–25 Nm).
8. Using torque adapter tool T88C–7025–E and an inch-pound torque wrench, measure the amount of torque required to turn the input shaft. If the torque wrench reads 4.34 inch lbs. (0.49 Nm), preload is within specifications. If preload is not within specifications, install a thicker or thinner preload shim.

NOTE: Do not install more than 2 preload shims.

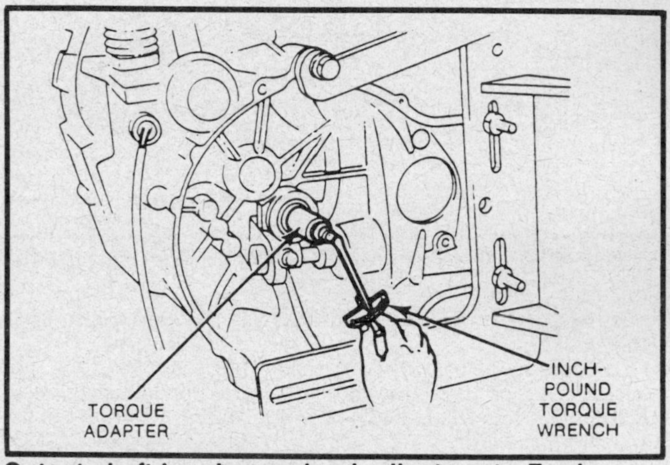
TORQUE ADAPTER
INCH-POUND TORQUE WRENCH

Output shaft bearing preload adjustment – Ford Festiva MTX 4/5 speed

INPUT SHAFT

Disassembly

1. Remove the snapring from the 5th gear end of the shaft.
2. Remove the bearing, spacer, 5th gear and 4th gear. The bearing can be removed using a press and bearing puller attachment tool D84L–1123–A, or equivalent. Position the puller behind 4th gear.
3. Remove the spacer locator ball from the shaft.
4. Remove the blocking ring from the 3rd/4th gear synchronizer.
5. Remove the 3rd/4th gear synchronizer snapring.
6. Remove the 3rd/4th gear synchronizer and blocking ring.

NOTE: If the synchronizer is tight on the shaft, it can be removed using a press and bearing pulling attachment tool D84L–1123–A or equivalent.

7. Position the puller behind 3rd gear. Remove 3rd gear.
8. Remove the bearing from the input end of the shaft. The bearing can be removed using a press and bearing pulling attachment tool D84L–1123–A, or equivalent.

Inspection

1. Clean all components, except plastic or nylon, in solvent.
2. Inspect for broken, chipped or worn gear teeth.
3. Inspect for bent or broken inserts.
4. Check for damaged roller thrust or needle bearings and inspect the bearing bores for cracks or damage.
5. Check for worn or loss of surface metal from the counter shaft and hub, clutch shaft or reverse idler gear shaft.
6. Check the thrust washers for damage and wear.
7. Check for nicked, broken or worn output or clutch shaft splines.
8. Check for bent, weak or distorted snaprings.
9. Check the reverse idler gear busing for wear.
10. Inspect front and rear bearings for roughness, wear and damage.
11. Inspect the shift rails, selector arms, plates, interlock for worn, bent or damaged parts.
12. Inspect synchronizers damage, wear and rough surfaces.
13. If any of the above are found, replace the appropriate part(s).

Assembly

1. Install 3rd gear.
2. Install the blocking ring and the 3rd/4th gear synchronizer. Make sure the oil grooves on the synchronizer hub face 3rd gear.

NOTE: If the synchronizer is tight on the shaft, it can be installed using a press and a suitable installer.

3. Install the blocking ring and 4th gear.
4. Install 5th gear.
5. Install the locator ball and spacer.
6. Install the bearings using a suitable driver.

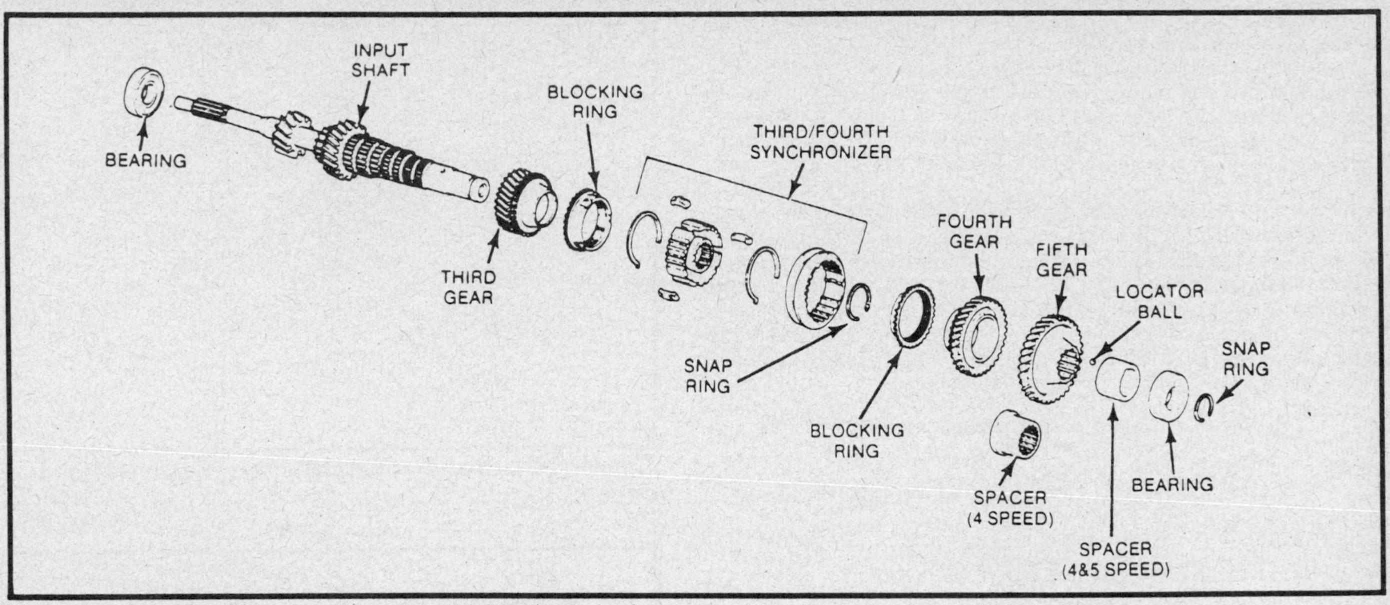
INPUT SHAFT
BEARING
BLOCKING RING
THIRD/FOURTH SYNCHRONIZER
THIRD GEAR
FOURTH GEAR
FIFTH GEAR
LOCATOR BALL
SNAP RING
SNAP RING
BLOCKING RING
SPACER (4 SPEED)
BEARING
SPACER (4&5 SPEED)

Input shaft assembly—Ford Festiva MTX 4/5 speed

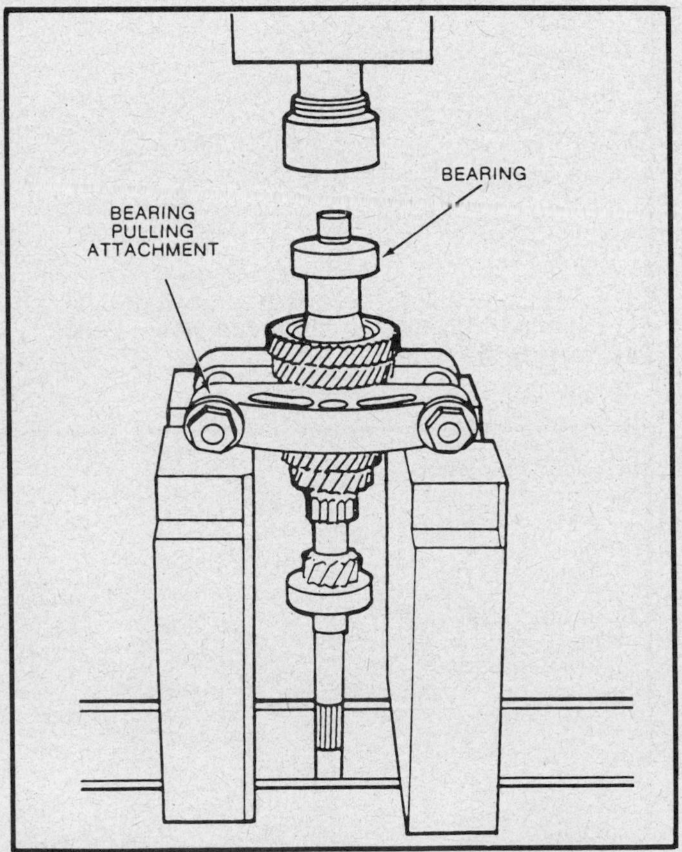
BEARING
BEARING PULLING ATTACHMENT

Bearing, spacer, 5th and 4th gear removal—Ford Festiva MTX 4/5 speed

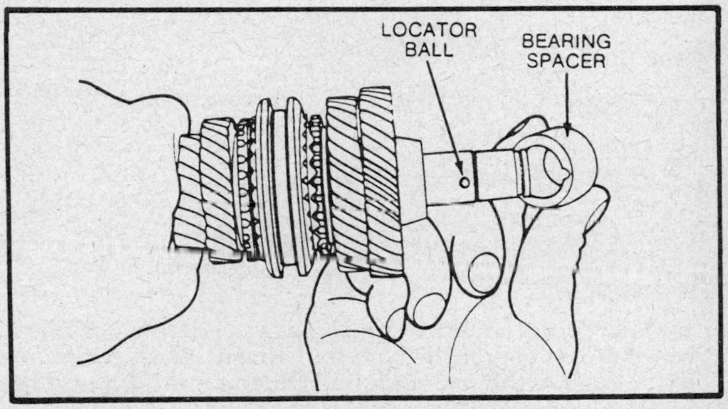
LOCATOR BALL
BEARING SPACER

Locator ball and spacer location—Ford Festiva MTX 4/5 speed

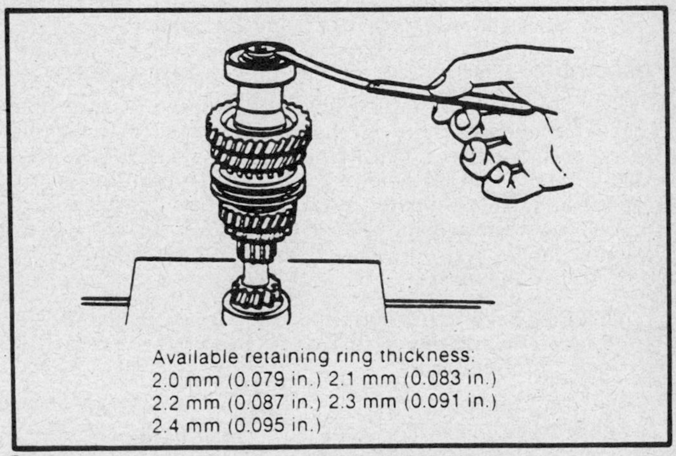
Available retaining ring thickness:
2.0 mm (0.079 in.) 2.1 mm (0.083 in.)
2.2 mm (0.087 in.) 2.3 mm (0.091 in.)
2.4 mm (0.095 in.)

Snapring measurement and selection—Ford Festiva MTX 4/5 speed

7. Install the bearing snapring on the 5th gear end of the shaft.
8. Measure the clearance between the snapring and the bearing inner race. If the clearance is not 0.1 mm (0.0039 in.), install a selective snapring.

INPUT SHAFT ENDPLAY

Inspection

1. Make sure the oil funnel is in position in the transmission case bore.

2. Install the original endplay shim(s) and the bearing race.
3. Install the input shaft.
4. Install the transmission housing.
5. Install the transmission case to clutch housing attaching bolts. Tighten the attaching bolts to 14–19 ft.lbs. (19–26 Nm).
6. Tap the end of the input shaft with a plastic-tipped hammer to make sure the shaft, bearing race and funnel are properly seated.
7. Mount a dial indicator 4201–C or equivalent on the clutch housing with the stylus contacting the end of the input shaft.
8. Raise the input shaft and note the dial indicator reading. If the reading is 0.00–0.0039 in. (0.00–0.1mm), the input shaft endplay is within specifications. If endplay is not within specifications, install a thicker or thinner shim. The available shim thicknesses are 0.0012 in. (0.3mm), 0.016 in. (0.4mm) and 0.020 in. (0.5mm).

NOTE: Do not install more than 2 endplay shims.

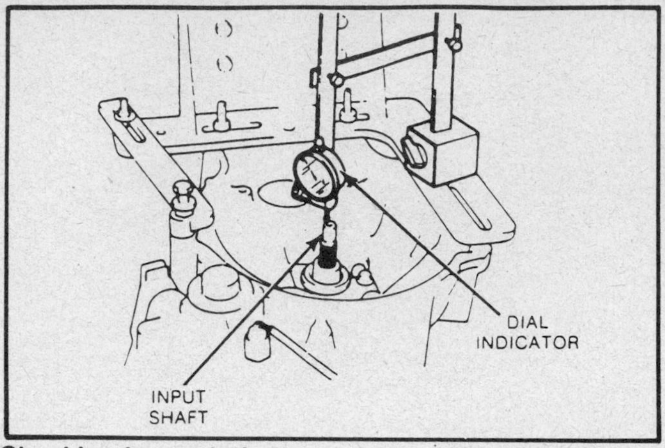
DIAL
INDICATOR
INPUT
SHAFT
Checking input shaft endplay – Ford Festiva MTX 4/5 speed

SPEEDOMETER DRIVEN GEAR

Disassembly

1. Remove the O-ring from the speedometer gear sleeve.
2. Remove the roll pin using a pin punch.
3. Remove the speedometer shaft and gear assembly from the sleeve.

Assembly

1. Install the speedometer driven gear and shaft assembly in the sleeve.
2. Install the roll pin.
3. Install a new O-ring on the speedometer sleeve.

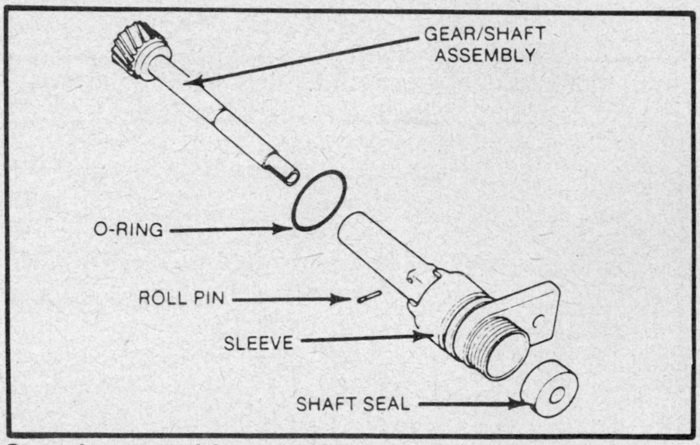
GEAR/SHAFT
ASSEMBLY
O-RING
ROLL PIN
SLEEVE
SHAFT SEAL
Speedometer driven gear, exploded view – Ford Festiva MTX 4/5 speed

SYNCHRONIZER

Disassembly

NOTE: Before disassembling a synchronizer, make an alignment mark on the hub and sleeve. The synchronizer sleeve and hub are matched during manufacture and should be reassembled in the exact same position. The marks will provide the necessary alignment reference.

1. Slide the hub and inserts out of the hub.
2. Remove the insert springs from the hub.

Assembly

1. Install the hub in the synchronizer sleeve. Use the marks made during disassembly for alignment reference. If alignment marks are not present, make sure the hub and sleeve are assembled with the hub oil grooves facing the proper direction. Because the outside diameter of the 3rd/4th synchronizer is concentric, the relationship of the oil grooves to the sleeve is unimportant.
2. Install the inserts.

NOTE: There are 2 synchronizer insert sizes. Be sure to select the proper size insert that belongs with the proper synchronizer.

3. Install the insert springs. Make sure each spring engages the hole in the hub as shown and that they rotate away from the same holes in opposite directions.

DIFFERENTIAL

Disassembly

1. Rotate the side gears until they can be removed through the opening in the differential case.
2. Using a pin punch, remove the pinion shaft roll pin.

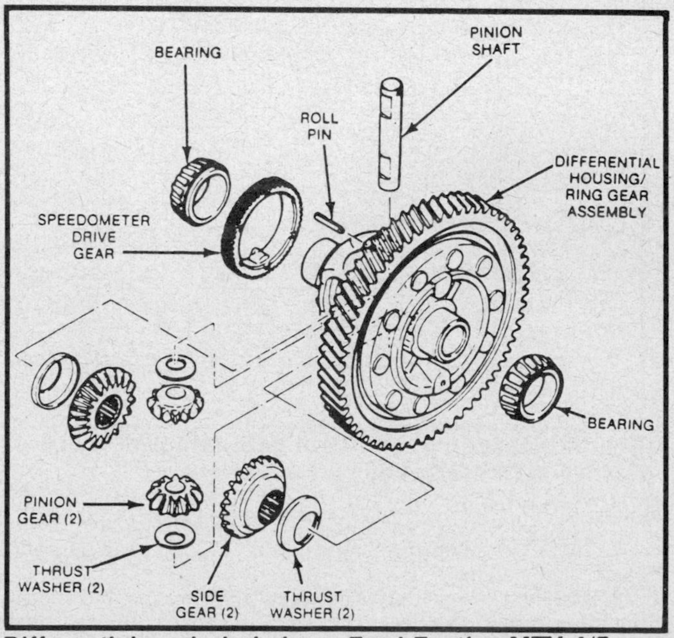
BEARING
PINION
SHAFT
ROLL
PIN
DIFFERENTIAL
HOUSING/
RING GEAR
ASSEMBLY
SPEEDOMETER
DRIVE
GEAR
BEARING
PINION
GEAR (2)
THRUST
WASHER (2)
SIDE
GEAR (2)
THRUST
WASHER (2)
Differential exploded view – Ford Festiva MTX 4/5 speed

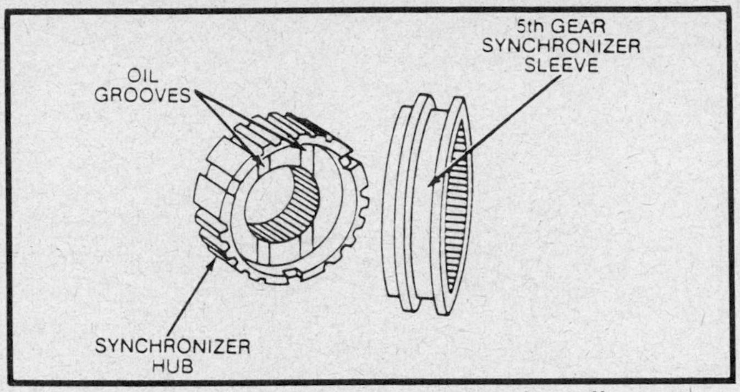
5th GEAR
SYNCHRONIZER
SLEEVE
OIL
GROOVES
SYNCHRONIZER
HUB

5th gear synchronizer assembly showing oil groove location—Ford Festiva MTX 4/5 speed

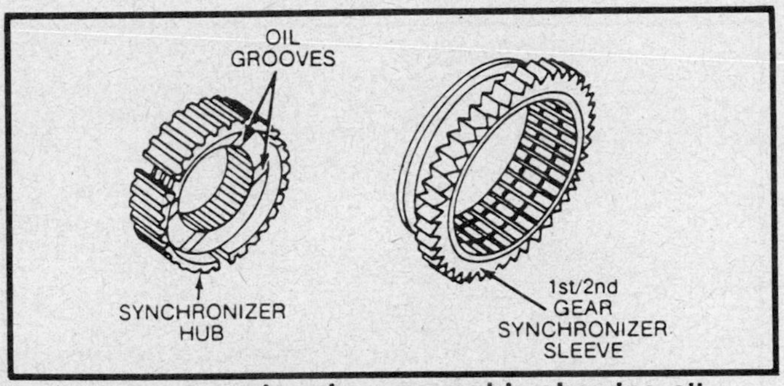
OIL
GROOVES
SYNCHRONIZER
HUB
1st/2nd
GEAR
SYNCHRONIZER
SLEEVE

1st/2ng gear synchronizer assembly showing oil groove location—Ford Festiva MTX 4/5 speed

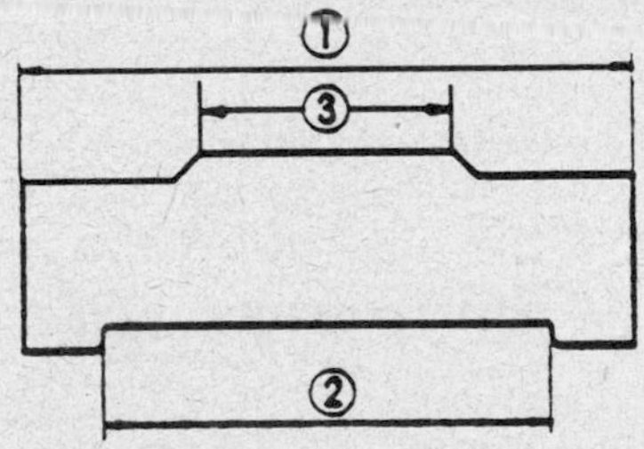
①
③
②

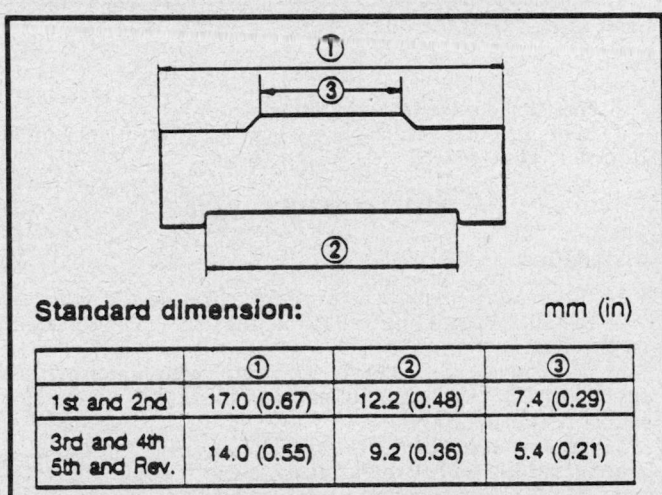

Standard dimension: mm (in)

	①	②	③
1st and 2nd	17.0 (0.67)	12.2 (0.48)	7.4 (0.29)
3rd and 4th 5th and Rev.	14.0 (0.55)	9.2 (0.36)	5.4 (0.21)

Synchronizer location—Ford Festiva MTX 4/5 speed

3. Remove the pinion shaft, pinion gears and thrust washers.
4. Remove the bearing from the side opposite the speedometer gear. The bearing can be removed using differential side bearing puller tool T77F–4220–B1 or equivalent, and a step plate.
5. Remove the bearing from the speedometer side of the differential. The bearing can be removed using a press and bearing pulling attachment tool D84L–1123–A, or equivalent. To avoid damage to the speedometer gear and bearing, alternately tighten the nuts until the puller seats under the bearing inner race.
6. Remove the speedometer gear.

Assembly

1. Install the speedometer drive gear.
2. Install the differential bearings, using driver tool T80T–

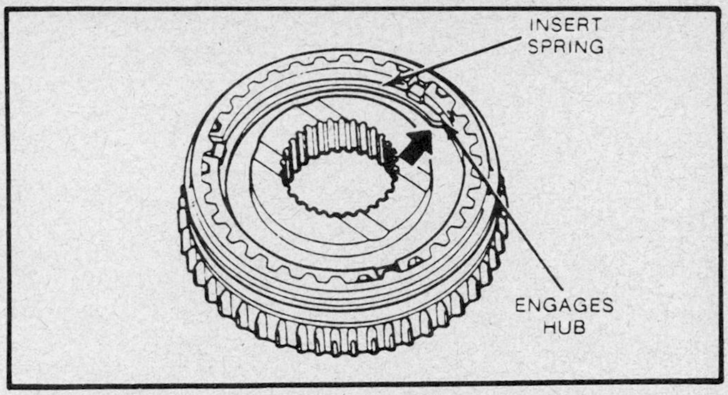
INSERT
SPRING
ENGAGES
HUB

Synchronizer insert spring location—Ford Festiva MTX 4/5 speed

4000–W and bearing cone replacer tool T87C–7025–B or equivalent.
3. Install the thrust washers, pinion gears, pinion shaft and roll pin.
4. Install the thrust washers and side gears.

DIFFERENTIAL BEARING PRELOAD

Adjustment

NOTE: Shim selection tool T88C–77000–J or equivalent and differential rotator tool T88C–77000–L or equivalent are required for this procedure.

1. Remove the transmission case, input shaft and output shaft.
2. Shift the 3rd/4th gear synchronizer to the neutral position.
3. If necessary, install the correct output shaft bearing preload shim
4. Install the differential assembly in the clutch housing.
5. Install the bearing race on the differential bearing.
6. Install the collars supplied with shim selection tool T88C–77000–J on the differential bearing race.
7. Install the transmission case using the spacers and bolts supplied with shim selection tool T88C–77000–J. Install the bolts and spacers in the locations shown. Tighten the tool bolts to 14–19 ft. lbs. (19–26 Nm).
8. Using the adjusting rods supplied with shim selection tool T88C–77000–J, spread the shim selection tool collars. When the collars begin to tighten, measure the amount of torque required to turn the differential using differential rotator T88C–77000–L and an inch-pound torque wrench. Continue tightening the collars until the torque wrench reads 4.3–6.6 inch lbs. (5–7.5 Nm).
9. Using a feeler gauge, measure the gap between the shim selector collars in three places. The average of the three readings is the shim thickness required to properly preload the differential bearings.

NOTE: Do not install more than 2 shims.

10. Remove the transmission case and the shim selection tools.
11. Install the preload shims and bearing race in the transmission case.
12. Install the transmission case and recheck the bearing preload.

5TH/REVERSE SHIFT RAIL

Disassembly

1. Remove the 5th gear shift fork from the rail.
2. Remove the reverse intermediate shift lever roll pin.
3. Remove the reverse intermediate shaft lever from the rail.

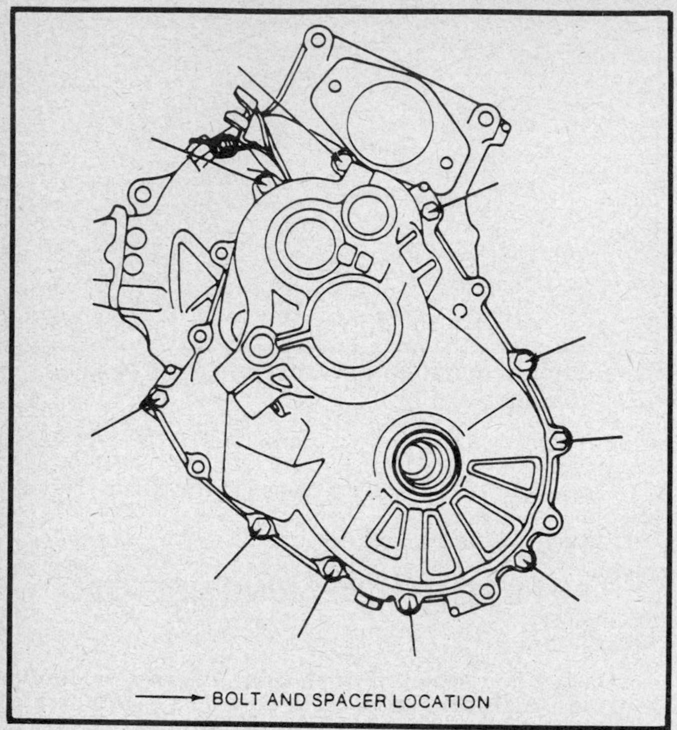
BOLT AND SPACER LOCATION

Transmission case bolt and spacer locations—Ford Festiva MTX 4/5 speed

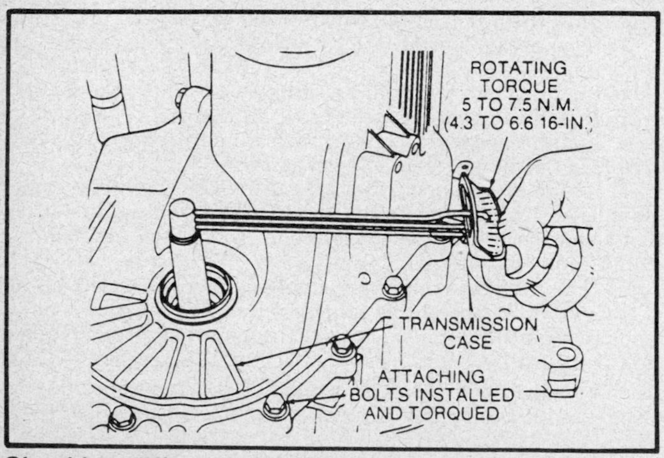
ROTATING
TORQUE
5 TO 7.5 N.M.
(4.3 TO 6.6 16-IN.)
TRANSMISSION
CASE
ATTACHING
BOLTS INSTALLED
AND TORQUED

Checking differential bearing preload—Ford Festiva MTX 4/5 speed

Assembly

1. Install the reverse intermediate shift lever on the rail. Note the direction of installation.
2. Install the reverse intermediate shift lever roll pin.

Install the 5th gear shift fork on the rail. Note the direction of installation.

SHIFT GATE

Disassembly

1. Remove the C-clip from the selector lever pin.
2. Remove the selector lever pin and the shift selector lever.

Assembly

1. Position the shift selector in the shift gate frame and install the selector lever pin.

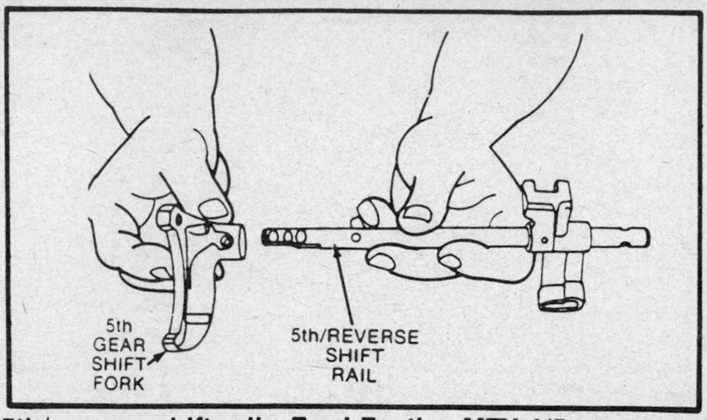
5th
GEAR
SHIFT
FORK
5th/REVERSE
SHIFT
RAIL

5th/reverse shift rail—Ford Festiva MTX 4/5 speed

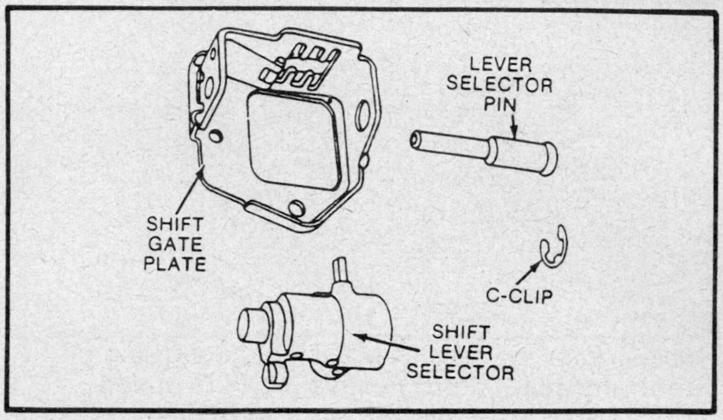
LEVER
SELECTOR
PIN
SHIFT
GATE
PLATE
C-CLIP
SHIFT
LEVER
SELECTOR

Shift gate exploded view—Ford Festiva MTX 4/5 speed

2. Install the selector lever pin clip.
3. Make sure the return spring is properly positioned on the reverse lockout pawl.

INPUT SHIFT RAIL

Installation

1. If necessary, install the protective boot on the shift rail.
2. Install the input shift rail. Note the location of the detents on the shaft.
3. Position the shift selector arm in the clutch housing, slide the rail through the arm and seat it in the case. Note the direction of the selector arm.
4. Install the selector arm roll pin.
5. Install the end of the protective boot over the input shift rail seal. Make sure the drain tube on the boot is facing the bottom of the transaxle.

DIFFERENTIAL

Installation

1. Install the differential assembly in the clutch housing with the speedometer gear facing the clutch housing.
2. Position the shift gate and install the attaching bolts. The longer bolt is installed in the position shown.
3. Install the speedometer gear and the attaching bolt.

Transmission Assembly

1. Install the lubrication funnel in the transmission case bearing bore.
2. Position the oil channel and install the attaching bolt.

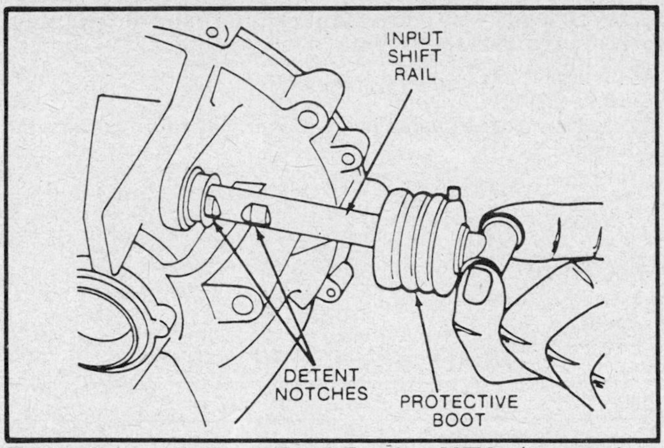
INPUT
SHIFT
RAIL
DETENT
NOTCHES
PROTECTIVE
BOOT

Shift rail installation—Ford Festiva MTX 4/5 speed

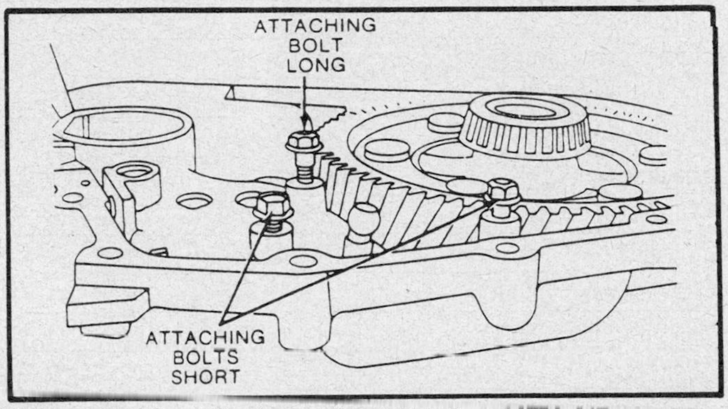
ATTACHING
BOLT
LONG
ATTACHING
BOLTS
SHORT

Differential installation—Ford Festiva MTX 4/5 speed

3. Position the baffle plate and install the attaching bolt.
4. Install an interlock plug. Position the plug in the bore so that it is located between the 1st/2nd and 5th/reverse shift rail bores.
5. Using a suitable tool, position the interlock plug so that its end does not extend into the 5th/reverse shift rail bore.
6. Make sure all synchronizers are in the neutral position.
7. Install the interlock pin in the 5th/reverse shift rail. A small dab of grease should be used to hold the pin in position.
8. Install the 1st/2nd shift fork on the output gear train. Note the installation direction.
9. If necessary, install the 5th gear shift fork on the shift rail. Note the installation direction.
10. Install the 5th/reverse shift rail on the output gear train.
11. Mesh the output gear train with the input gear train, grasp as an assembly and install in the clutch housing. Use care to prevent damage to the input shaft seal.
12. Carefully raise and lower the 5th/reverse shift rail in slight amounts while pushing the interlock plug back into its bore. When the rail is properly positioned, the plug will no longer extend into the 1st/2nd shift rail bore.
13. Install the 1st/2nd shift rail through it shift fork and into its bore. Note the installation direction of the detent notches. If the rail will not enter the bore, raise and lower the 5th/reverse shift rail in slight amounts, while pushing downward on the rail. When the 5th/reverse shift rail is properly positioned, the interlock will move in its bore, allowing installation of the 1st/2nd shift rail.
14. Install the remaining interlock plug.
15. Install the 3rd/4th shift fork and rail on the synchronizer. Note the installation direction of the fork and the detent on the rail.
16. While holding the 3rd/4th relay arm in position, install the shift rail through the relay arm. Rest the end of the shift rail on the edge of the clutch housing.
17. Carefully raise and lower the 5th/reverse shift rail in slight amounts while pushing the interlock plug back into its bore. When the rail is properly positioned, the plug will not longer extend into the 3rd and 4th shift rail bore.
18. Pivot the 3rd/4th shift rail into position and push it into the bore. If the rail will not enter the bore, raise and lower the 5th/reverse shift rail in slight amounts, while pushing downward on the rail. When the 5th/reverse shift rail is properly positioned, the interlock will move in its bore, allowing installation of the 1st/2nd shift rail.
19. Install the relay arm and shift fork roll pins.

NOTE: There are 3 roll pin sizes. Be sure to select the proper roll pin from the selection chart.

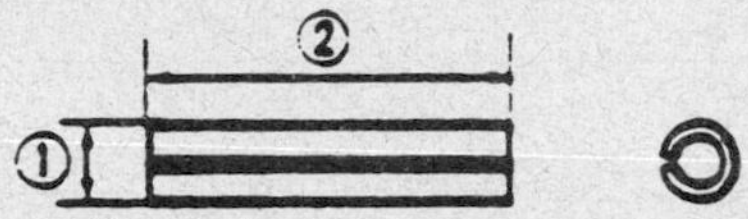
②
①

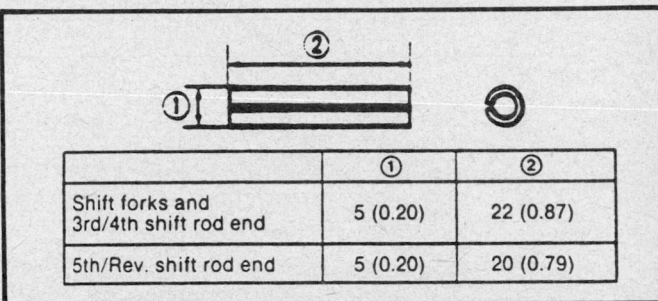

	①	②
Shift forks and 3rd/4th shift rod end	5 (0.20)	22 (0.87)
5th/Rev. shift rod end	5 (0.20)	20 (0.79)

Shift fork roll pin—Ford Festiva MTX 4/5 speed

20. Install a new O-ring on the reverse intermediate lever pivot pin.
21. Position the reverse intermediate lever and install the pivot pin.
22. Install the intermediate lever pivot pin retainer.
23. Install reverse gear on the idler shaft. Note the direction of installation.
24. Install the reverse idler gear and shaft. To gain the required clearance, raise the intermediate lever.
25. Install the reverse intermediate lever detent plate and attaching bolt. Tighten the attaching bolt to 6–7 ft. lbs. (8–9 Nm).
26. Install the case magnet.
27. Apply a 1/16 in. bead of RTV, or equivalent on transaxle housing. Make sure the sealant encircles the attaching bolt holes.
28. Install the transmission case on the clutch housing.
29. Install the transmission to clutch housing attaching bolts. Tighten the attaching bolts to 14–19 ft. lbs. (19–26 Nm).
30. Install the 5th gear switch. Before install the switch, coat the threads with pipe sealant D8AZ–195540–A, or equivalent. Tighten the switch to 15–22 ft. lbs. (20–29 Nm).

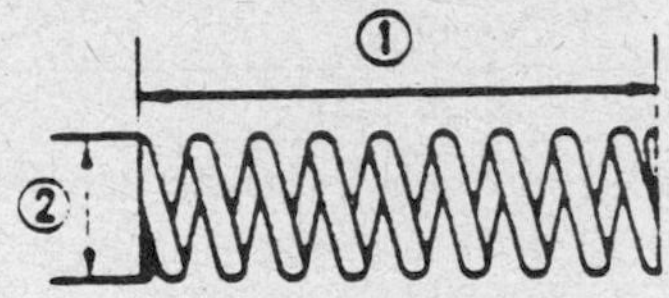
①
②

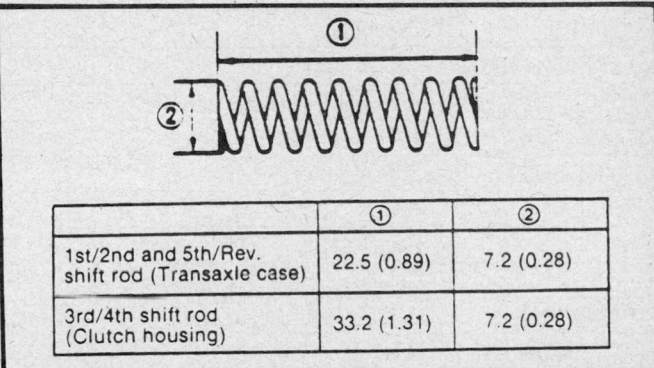

	①	②
1st/2nd and 5th/Rev. shift rod (Transaxle case)	22.5 (0.89)	7.2 (0.28)
3rd/4th shift rod (Clutch housing)	33.2 (1.31)	7.2 (0.28)

Detent spring length chart—Ford Festiva MTX 4/5 speed

31. Install the backup lamp switch. Before installing the switch, coat the threads with pipe sealant D8AZ–195540–A, or equivalent. Tighten the switch to 15–22 ft. lbs. (20–29 Nm).
32. Install the detent balls, springs and plugs.

NOTE: There are 2 lengths of detent springs. Before installing the plugs, coat the threads with pipe sealant D8AZ–195540–A, or equivalent. Tighten the detent plugs to 11–15 ft. lbs. (15–21 Nm).

33. Install the transaxle mount. Tighten the attaching bolts to 14–18 ft. lbs. (19–26 Nm)
34. Make sure the bushings are properly installed in the input shift rail.

SPECIFICATIONS

TRANSAXLE SPECIFICATIONS

Item		Specification
Transaxle control		Floor shift
Synchromesh system		Forward . . . Synchromesh, Reverse . . . Selective sliding
Gear ratio	First	3.454
	Second	1.944
	Third	1.275
	Fourth	0.861
	Fifth	0.692 (5 speed only)
	Reverse	3.583
Final gear ratio		3.777
Speedometer gear ratio		0.91
Oil	Type	API: GL-4, GL-5 (75W-80) ATF: M2C33F or Dexron® II
	Capacity	2.5 liters (2.6 US qt.)

TORQUE SPECIFICATION

Item	ft. lbs.	Nm
Backup lamp switch	15–22	20–29
Baffle plate	6–8	10–13
Clutch cover	13–20	18–26
Clutch pedal pivot bolt	14–25	20–34
Detent plug (1-2)	11–15	15–21
Detent plug (3-4)	15–22	20–29
Detent plug (4-5)	11–15	15–21
Fifth gear switch	15–22	20–29
Oil guide	6–8	8–11
Release bearing fork	26–30	35–41
Reverse detent plate	6–8	8–11
Shift gate	6–8	8–11
Shift lever housing	5–7	7–10
Shift rod	12–17	16–22
Stabilizer rod	23–34	31–46
Transmission case	14–19	19–26

SPECIAL TOOLS

Number	Description
D79P-6000-B	Engine support bar
D80L-100-M	Blind hole puller
D80L-100-S	Blind hole puller
D80L-630-3	Step plate
D84L-1123-A	Bearing splitter
T50T-100-A	Slide hammer
T57L-500-B	Bench mounted holding fixture
T75L-1165-B	Bearing/seal press plate
T77F-4220-B1	Bearing cone puller
T77F-1102-A	Bearing cup puller
T77F-1217-B	Bearing cup installer

Number	Description
T80T-4000-W	Universal drive handler
T87C-7025-B	Bearing cone replacer
T87C-7025-C	Differential plugs
T87C-770000-H	Differential seal replacer
T88C-7025-E	Rotating torque adaptor
T88C-7025-F	Countershaft rear bearing cone replacer
T88C-77000-L	Differential rotator
T88C-77000-J	Shim selection tool
T88C-77000-N	Differential bearing cone replacer
T00L-4201-C	Dial indicator

Section 6

Probe MTX5 Transaxle
Ford Motor Co.

APPLICATION

1989 Ford Probe with 2.2L engine and 2.2L engine with turbocharger

GENERAL DESCRIPTION

There are 2 variations (G-type non-turbocharged and H-type turbocharged) of this transaxle. The H-type transaxle has special design features which enable it to handle the higher torque output of the turbocharged engine. Each transaxle has it own gear ratio which are engineered to match the varying performance characteristics of the 2 engine designs.

Both types have reverse gear synchromesh shifting and the 5th/reverse clutch hub assembly on the input shaft. The helical cut forward gears are in constant mesh with the corresponding gears on the opposing shaft. The forward gears are selected by means of a synchronizer mechanism. 3rd, 4th and 5th gears are mounted on the input shaft. First, 2nd and reverse gears have straight cut teeth and are engaged through the reverse idler gear by means of a synchronizer mechanism.

The turbocharged transaxle has 3 separate shift rods for 1st/2nd, 3rd/4th and 5th/reverse. Needle bearings are used on the turbocharged transaxle to reduce the sliding resistance of the forward gears.

Metric Fasteners

The metric fastener dimensions are very close to the dimensions of the familiar inch system fasteners and for this reason, replacement fasteners must have the same measurement and strength as those removed.

CAUTION

Do not attempt to interchange metric fasteners for inch system fasteners. Mismatched or incorrect fasteners can result in damage to the transaxle unit through malfunctions or breakage and possible personal injury.

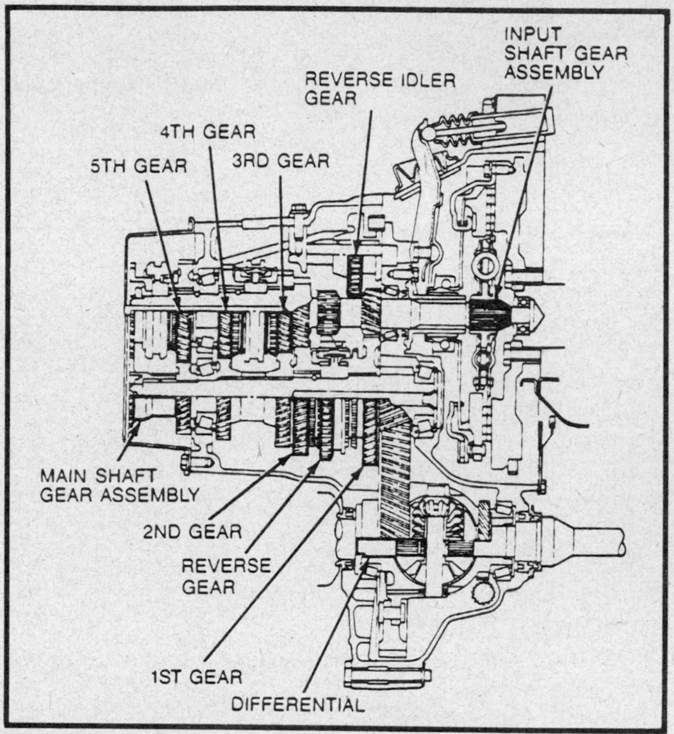
INPUT SHAFT GEAR ASSEMBLY
REVERSE IDLER GEAR
4TH GEAR
5TH GEAR
3RD GEAR
MAIN SHAFT GEAR ASSEMBLY
2ND GEAR
REVERSE GEAR
1ST GEAR
DIFFERENTIAL

Turbocharged transaxle componets (H-type)

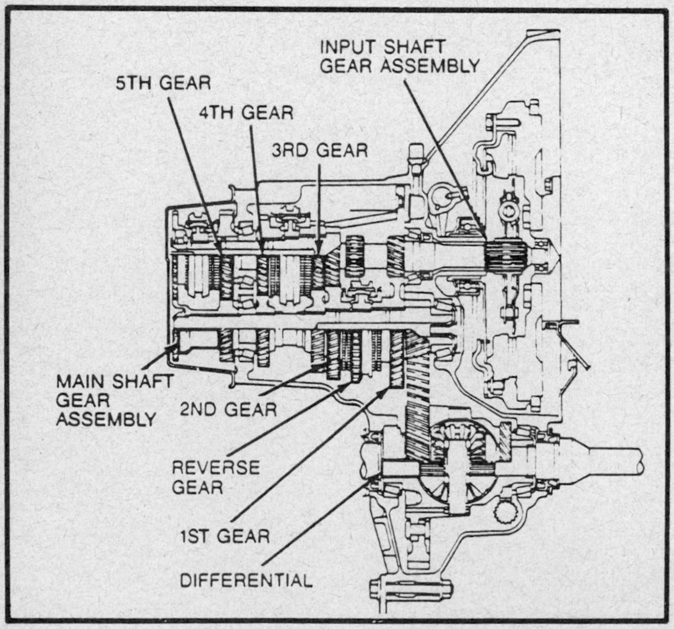
INPUT SHAFT GEAR ASSEMBLY
5TH GEAR
4TH GEAR
3RD GEAR
MAIN SHAFT GEAR ASSEMBLY
2ND GEAR
REVERSE GEAR
1ST GEAR
DIFFERENTIAL

Non-turbocharged transaxle componets (G-type)

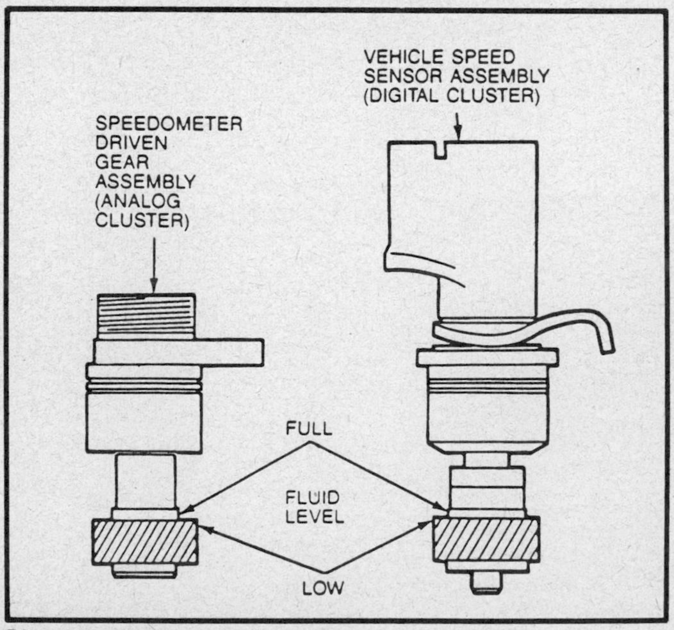
VEHICLE SPEED SENSOR ASSEMBLY (DIGITAL CLUSTER)
SPEEDOMETER DRIVEN GEAR ASSEMBLY (ANALOG CLUSTER)
FULL
FLUID LEVEL
LOW

Checking transaxle fluid level

Capacities

The 5 speed transaxle uses Dexron®II automatic transmission fluid. The capacity of the transaxle is 3.6 quarts or 3.35L for the G-type (non-turbocharged) and 3.9 quarts or 3.65L for the H-type (turbocharged).

Checking Fluid Level

1. Transaxle fluid level should only be checked after the vehicle has been standing on level ground for some time.
2. On vehicles equipped with a digital cluster, disconnect the harness from the vehicle speed sensor assembly located on the transaxle.
3. On vehicles equipped with an analog cluster, remove the speedometer cable dust cover and disconnect the cable from the speedometer driven gear.
4. After removing the retaining bolt, pull the gear case to remove it from the housing. Insert a suitable tool between the speedometer gear case and the clutch housing and use it to pry the gear case loose if necessary. On digital cluster vehicles, remove the vehicle speed sensor assembly in the same manner.
5. Check that the oil level is between the **F** and **L** on the dipstick.
6. Add the necessary amount of the specified oil through the gear case hole to correct the fluid level.

TRANSAXLE MODIFICATIONS

There have been no modifications to the 5 speed transaxle at the time of this printing.

TROUBLE DIAGNOSIS

CHILTON'S THREE "C'S" DIAGNOSIS

Condition	Cause	Correction
Change lever won't shift smoothly, or is hard to shift	a) Seized change lever ball b) Seized change control rod joint c) Bent change control rod	a) Replace change lever ball b) Replace change control rod joint c) Replace control rod
Too much play in change lever	a) Worn change control rod bushing b) Weak spring of ball or change lever c) Worn bushing of ball or change lever	a) Replace change control rod bushing b) Replace spring or change lever c) Replace bushing of ball or change lever
Difficult to shift	a) Bent change rod b) No grease in transmission control c) Insufficient oil d) Deterioration of oil quality e) Wear or play or shift fork or shift rod f) Wear of synchronizer ring g) Wear of synchronizer cone of gear h) Bad contact of synchronizer ring and cone of gear i) Excessive longitudinal play of gears j) Wear of bearing k) Wear of synchronizer key spring l) Excessive primary shaft gear bearing preload m) Improperly adjusted change guide plate	a) Replace change rod b) Lubricate with grease c) Add correct amount of oil d) Change transaxle oil e) Replace shift fork or shift rod f) Replace synchronizer ring g) Replace synchronizer or gear h) Replace synchronizer or gear i) Replace gears j) Replace bearing k) Replace synchronizer key spring l) Adjust bearing preload m) Adjust change guide plate
Won't stay in gear	a) Bent change control rod b) Worn change control rod bushing c) Weak change lever ball spring d) Improperly installed extension bar e) Wear of shift fork f) Wear of clutch hub g) Worn clutch hub sleeve h) Worn gear sliding part of both shaft gears i) Worn gear sliding part of each gear j) Worn steel sliding groove of control end k) Weak spring pressing against steel ball l) Excessive thrust clearance m) Worn bearing	a) Replace change control rod b) Replace control rod bushing c) Replace lever ball spring d) Replace extension bar e) Replace shift fork f) Replace clutch hub g) Replace hub sleeve h) Replace shaft gears i) Replace gears j) Replace control end k) Replace spring l) Replace necessary components m) Replace bearing

CHILTON'S THREE "C'S" DIAGNOSIS

Condition	Cause	Correction
Won't stay in gear	n) Improperly installed engine mount	n) Install properly and tighten to correct specifications
Abnormal noise	a) Insufficient oil b) Deterioration of oil quality c) Worn bearing d) Worn gear sliding surface of both shaft gears e) Wear of sliding surfaces of gears f) Excessive gear backlash g) Damaged gear teeth h) Foreign material in gears i) Damaged differential gear, or excessive backlash	a) Add correct amount of oil b) Change oil in transaxle c) Adjust or replace bearing d) Replace shaft gears e) Replace gears f) Repalce necessary component g) Replace gear h) Change oil and transaxle i) Adjust or replace differential gear

ON CAR SERVICE

Services

FLUID CHANGE

Drain and Refill

1. Park the vehicle on level ground and apply the parking brake.
2. Remove the speedometer driven gear assembly on vehicles equiped with an analog cluster or the vehicle speed sensor assembly on vehicles equipped with a digital cluster.

NOTE: It may be necessary to raise and support the vehicle safely to gain access to the drain plug.

3. Remove the drain plug and drain the oil into a suitable pan.
4. Replace the drain plug and torque to 29–43 ft. lbs. Then add the necessary amount of the specified oil through the speedometer gear case hole.
5. Start engine, road test and check for leaks.

DIFFERENTIAL OIL SEALS

Removal and Installation

1. Raise vehicle and safely support the vehicle. Drain oil from the transaxle assembly.
2. Remove the front wheels and remove the splash shields as necessary.
3. Separate the front stabilizer from the lower arm.
4. Remove the clinch bolt and pull the lower arm downward. Separate the knuckle from the lower arm ball joint. Be careful not to damage the ball joint dust boot.
5. Remove the cotter pin then disconnect the tie rod end with special tool.
6. Separate the halfshaft by pulling the front hub outward. Do not use too much force at once, increase the force gradually.

NOTE: When removing the right side halfshaft remove the right joint shaft bracket.

7. Do not allow the halfshaft joint to bent to its maximum extent as damage to the joint could result. Support the halfshaft using string or equivalent.
8. Remove the oil seal from the transaxle using a suitable tool.
9. Coat the new oil seal lip with transaxle oil. Tap the new seal until the oil seal installer or equivalent contacts the case.

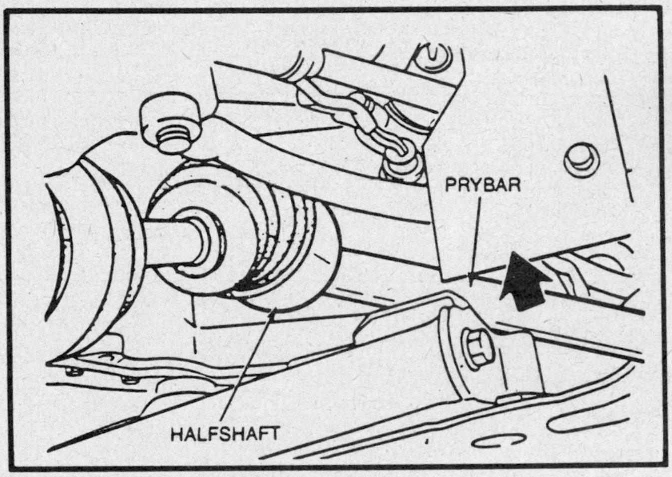
PRYBAR
HALFSHAFT
Removing halfshaft from vehicle

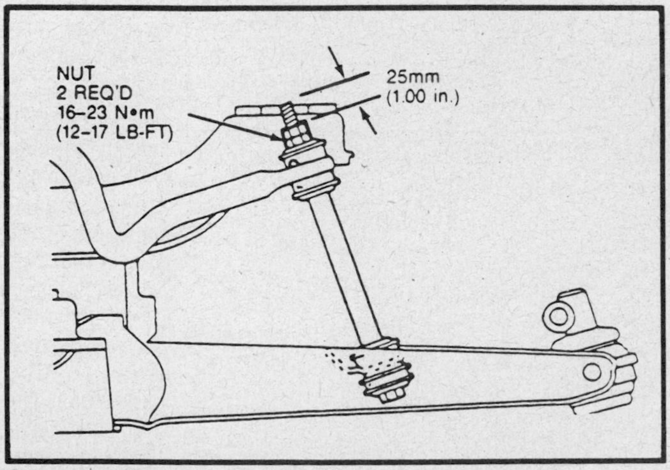
NUT
2 REQ'D
16–23 N•m
(12–17 LB-FT)
25mm
(1.00 in.)
Correct adjustment of stabilizer bolt assembly

10. Replace the halfshaft end clip. Insert the clip with the gap at the top of the groove.
11. Pull the front hub outward, then fit the halfshaft into the transaxle.
12. Insert the halfshaft into the transaxle by pushing on the wheel hub assembly. Be careful not to damage the oil seal when pushing on the wheel hub.
13. After installation is finished, pull the front hub slowly outward to check that the halfshaft is held securely by the clip.
14. Install the lower arm ball joint to the knuckle and tighten the clinch bolt to 32–40 ft. lbs.
15. Install the tie rod end and insert a new cotter pin. Tighten the tie rod end to 22–33 ft. lbs.
16. Adjust and tighten the stabilizer. The correct torque for the stabilizer is 12–17 ft. lbs. The correct adjustment of the stabilizer bolt should be tighten so that 1.00 in. of bolt is exposed.
17. Install the drain plug, splash shields and install wheel/tire assembly.
18. Add the correct quantity of the specified oil to the transaxle.
19. Start engine, road test and check for leaks.

REMOVAL AND INSTALLATION

TRANSAXLE REMOVAL

NOTE: The transaxle is removed separately from the engine.

1. Disconnect the battery cables. Remove the battery and battery carrier.
2. Disconnect the main fuse block.
3. Disconnect the center distributor lead.
4. Disconnect the air flow meter connector and remove the air cleaner assembly.
5. On the turbocharged vehicles remove the inter cooler hose to throttle body and air cleaner to turbocharger. On non-turbocharged vehicles just remove the resonance chamber and bracket.
6. Disconnect the speedometer cable on analog cluster or harness on digital cluster.
7. Disconnect the grounds from the transaxle case.
8. Raise and safely support the vehicle. Remove the front wheels.
9. Remove the splash shields.
10. Drain the transaxle oil into a suitable pan.
11. Remove the slave cylinder from the transaxle.
12. Disconnect the tie rod ends using a suitable tool.
13. Remove the stabilizer bar control links.
14. Remove the bolts and nuts at the left and right lower arm ball joints.
15. Pull the lower arms downward to separate them from the knuckles. Do not damage the ball joint dust boots.
16. Separate the left halfshaft from the transaxle by prying with a bar or equivalent inserted between the shaft and the transaxle case. Be careful not to damage the oil seal.
17. Remove the joint shaft bracket.
18. Separate the right halfshaft together with the joint shaft by prying with a bar or equivalent inserted between the shaft and the transaxle case. Be careful not to damage the oil seal.
19. Install the special tools to hold the differential side gear in the proper position.

NOTE: Failure to install the differential side gears holding tool may cause the differential side gears to become mispositioned.

20. Remove the gusset plate to transaxle bolts.
21. Remove the extension bar and the control rod.
22. Remove the flywheel inspection cover.
23. Mark or tag the electrical connections if necessary and remove the starter and access brackets.
24. Support the engine with the engine suppport fixture or equivalent.
25. Remove the center transaxle mount and bracket.
26. Remove the left transaxle mount and bracket.
27. Remove the nut and bolt which attaches the right transaxle mount to the frame.
28. Remove the crossmember and the left side lower arm as an assembly.

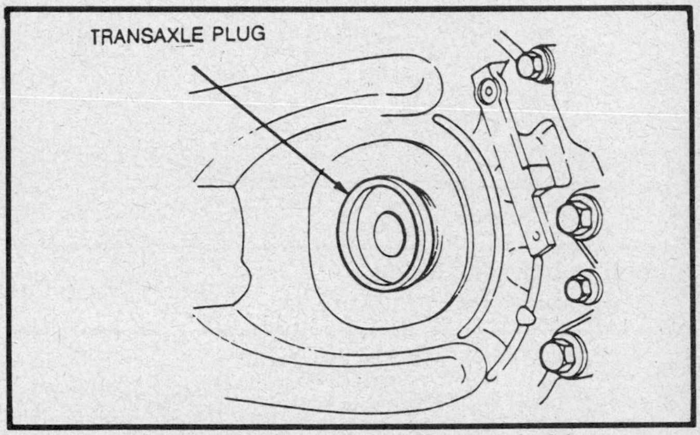
TRANSAXLE PLUG
Installing special tool to hold differential side gears

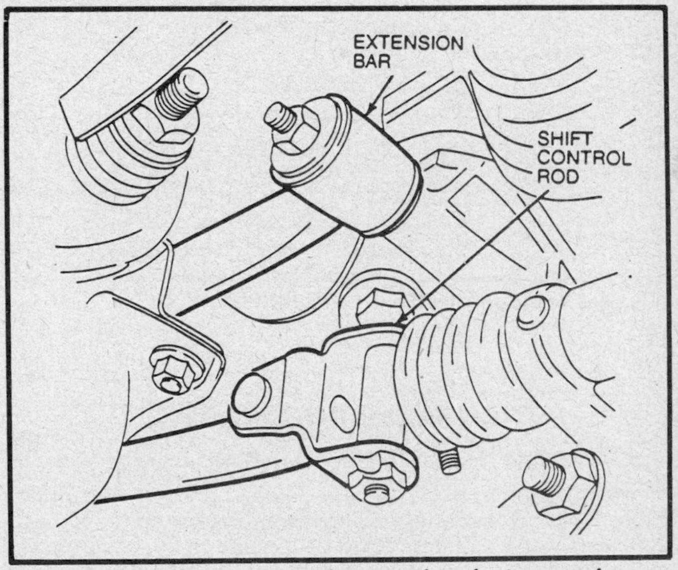
EXTENSION
BAR
SHIFT
CONTROL
ROD
Remove extension bar and control rod—transaxle removal

29. Position a transmission jack under the transaxle and secure the transaxle to the jack.
30. Remove the transaxle mounting bolts.
31. Remove the transaxle from the vehicle. The engine must always be properly supported while the transaxle is out of the vehicle.

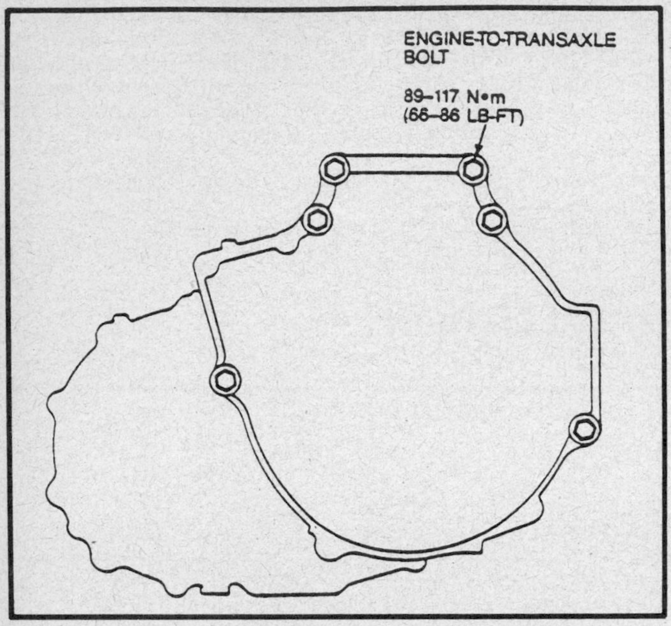
ENGINE-TO-TRANSAXLE BOLT
89–117 N•m
(66–86 LB-FT)
Mounting bolts locations

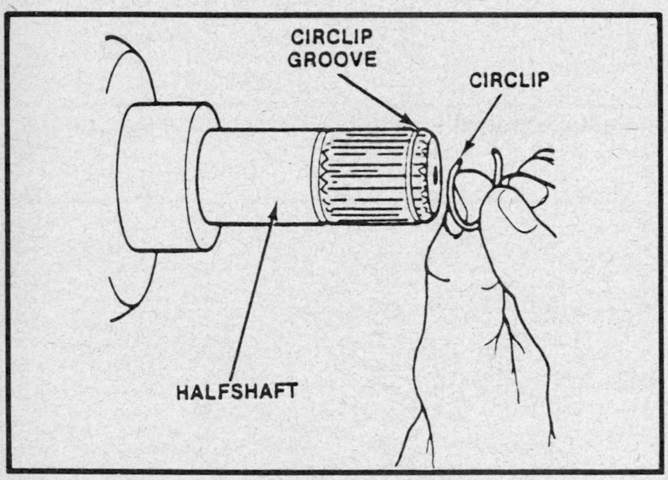
CIRCLIP GROOVE
CIRCLIP
HALFSHAFT
Always replace circlip on the halfshaft

TRANSAXLE INSTALLATION

NOTE: The transaxle is installed separately from the engine.

1. Apply a thin coating of clutch grease to the spline of the input shaft. Attach a safety device at 2 places on the transaxle and place a board on the jack and position the transaxle on it. The transaxle is not well balanced and be careful when positioning it on the jack.
2. Install the transaxle onto the engine. Tighten the transaxle mounting bolts to 66–86 ft. lbs.
3. Install the center transaxle mount and bracket. Tighten the bolts 27–40 ft. lbs. and the nuts to 47–66 ft. lbs. Do not install the nut which braces the throttle air hose bracket.
4. Install the left transaxle mount and tighten bolts on the mount to 27–38 ft. lbs. for non-turbocharged vehicles and 49–69 ft. lbs. for turbocharged vehicles. The retaining nuts on this mount are torque to 49–69 ft. lbs. for both engines.

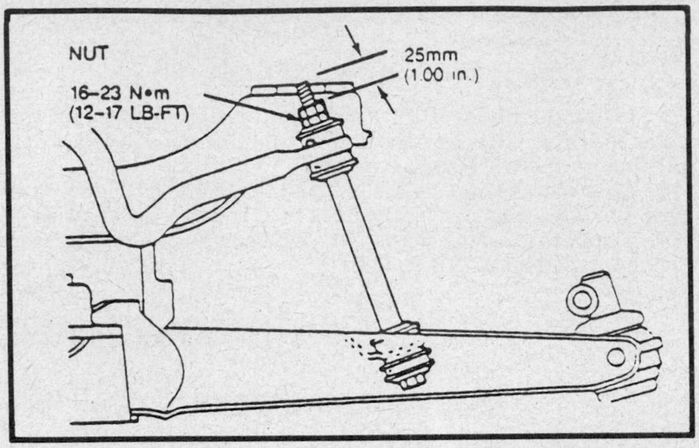
NUT
16–23 N•m
(12–17 LB-FT)
25mm
(1.00 in.)
Correct adjustment of stabilizer bolt assembly

5. Install the crossmember and the left side lower arm as an assembly. The tightening torque is 27–40 ft. lbs. on the bolts and 55–69 ft. lbs. on the nut.
6. Install the right transaxle mount bolt and nut and tigten to 63–86 ft. lbs.
7. Install the starter, electrical connections and access brackets.
8. Install the flywheel inspection cover.
9. Install the slave cylinder.
10. Install the gusset plate to transaxle bolts and torque the bolts to 27–38 ft. lbs.
11. Replace the circlips at the end of each halfshafts.
12. Remove the special tools holding the differential side gears in the proper position and install the halfshafts.

NOTE: After complete installation, pull the front hub outward to confirm that the halfshaft will not come out. Be careful not damage the oil seal.

13. Attach the lower arm ball joints to the knuckle.
14. Install the tie rod ends and new cotter pins. Tighten the tie rod end bolts to 22–33 ft. lbs.
15. Install the bolts and nuts to the lower arm ball joints. Tighten to 32–40 ft. lbs.
16. Install the stabilizer bar control link. Tighten nut/bolt to 12–17 ft. lbs. Make sure, that at least 1 in. of the bolt remains exposed after torque is reached.
17. Install the splash shields.
18. Install the front wheels.
19. Install the grounds to the transaxle case.
20. On non-turbocharged vehicles, install the resonance chamber and bracket and tighten to 69–95 inch lbs. On turbocharged vehicles, install the throttle body to intercooler air hose and the air cleaner to turbocharger air hose. Tighten the bracket to mount nut to 47–66 ft. lbs.
21. Install the air cleaner assembly and reconnect the air flow meter connector. The correct torque for the air cleaner assembly is 69–95 inch lbs.
22. Reconnect the center distributor lead.
23. Connect the main fuse block and tighten retaining bolts to 69–95 inch lbs.
24. Install the battery carrier and battery. Reconnect the battery cables.
25. Remove the engine support fixture or equivalent.
26. Add the correct quantity of the specified fluid.
27. Connect the speedometer cable on analog cluster equipped vehicles or harness for digital cluster equipped vehicles.
28. Start the engine, road test for proper operation and check for leaks.

BENCH OVERHAUL

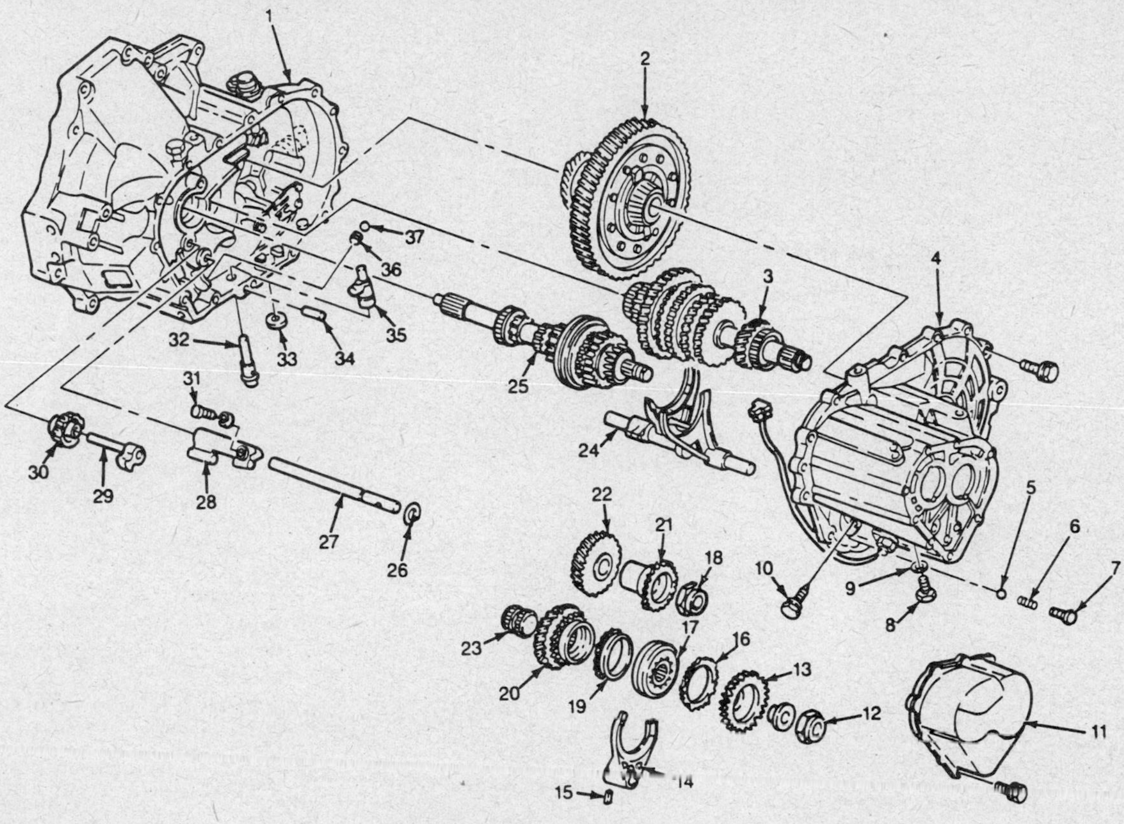

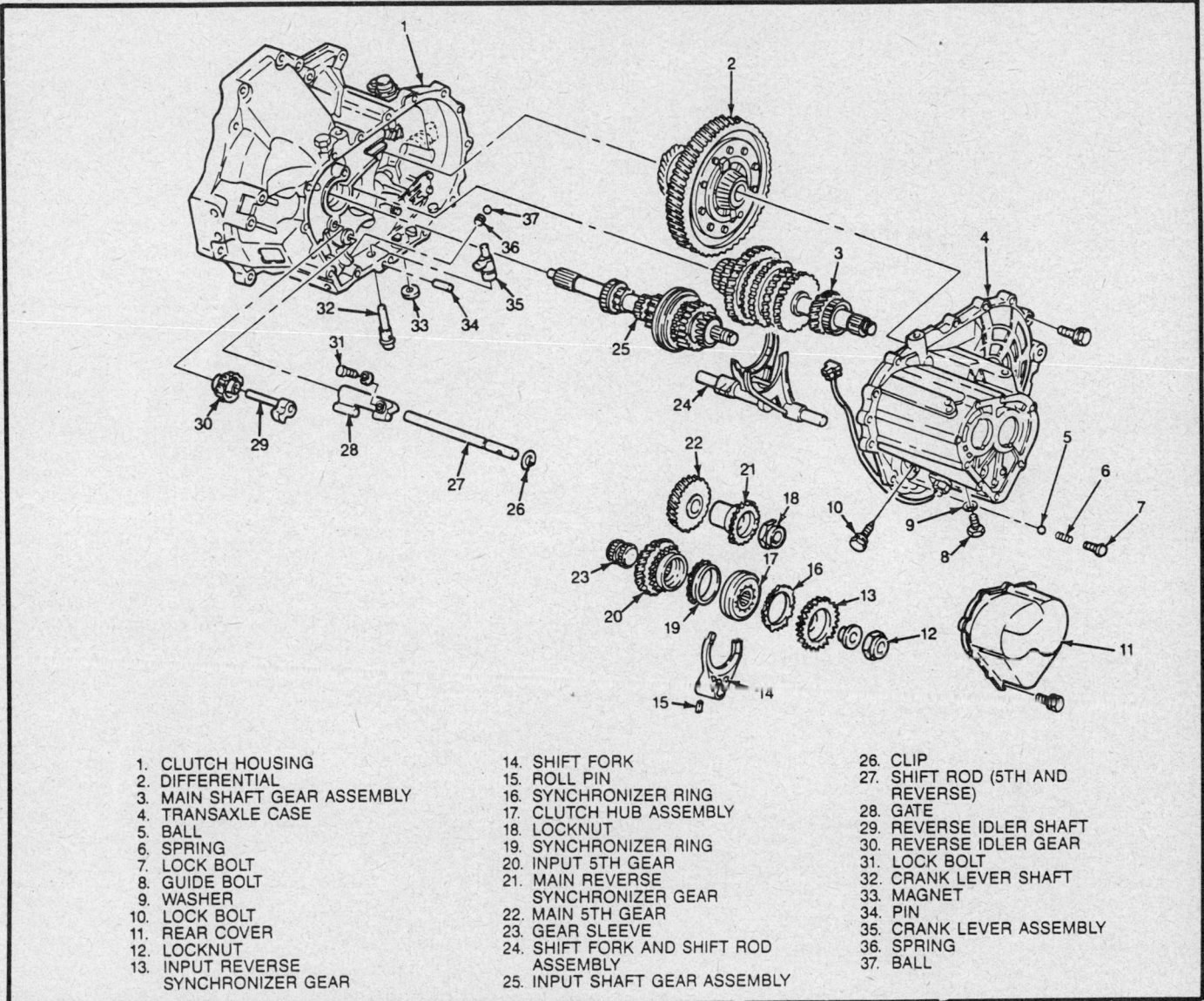

1. CLUTCH HOUSING
2. DIFFERENTIAL
3. MAIN SHAFT GEAR ASSEMBLY
4. TRANSAXLE CASE
5. BALL
6. SPRING
7. LOCK BOLT
8. GUIDE BOLT
9. WASHER
10. LOCK BOLT
11. REAR COVER
12. LOCKNUT
13. INPUT REVERSE SYNCHRONIZER GEAR
14. SHIFT FORK
15. ROLL PIN
16. SYNCHRONIZER RING
17. CLUTCH HUB ASSEMBLY
18. LOCKNUT
19. SYNCHRONIZER RING
20. INPUT 5TH GEAR
21. MAIN REVERSE SYNCHRONIZER GEAR
22. MAIN 5TH GEAR
23. GEAR SLEEVE
24. SHIFT FORK AND SHIFT ROD ASSEMBLY
25. INPUT SHAFT GEAR ASSEMBLY
26. CLIP
27. SHIFT ROD (5TH AND REVERSE)
28. GATE
29. REVERSE IDLER SHAFT
30. REVERSE IDLER GEAR
31. LOCK BOLT
32. CRANK LEVER SHAFT
33. MAGNET
34. PIN
35. CRANK LEVER ASSEMBLY
36. SPRING
37. BALL

Exploded view of non-turbocharged engine transaxle

Before Disassembly

When servicing the unit, it is recommended that as each part is disassembled, it is cleaned in solvent and dried with compressed air. Disassembly and reassembly of this unit and its parts must be done on a clean work bench. Also, before installing bolts into aluminum parts, always dip the threads into clean transmission oil. Anti-seize compound can also be used to prevent bolts from galling the aluminum and seizing. Always use a torque wrench to keep from stripping the threads. Take care with the seals when installing them, especially the smaller O-rings. The slightest damage can cause leaks. Aluminum parts are very susceptible to damage so great care should be exercised when handling them. The internal snaprings should be expanded and the external snaprings compressed if they are to be reused. This will help insure proper seating when installed. Be sure to replace any O-ring, gasket, or seal that is removed.

Transaxle Disassembly

TRANSAXLE WITH NON-TURBOCHARGED ENGINE (TYPE G)

1. Mount the transaxle on a bench mounting fixture or equivalent.
2. Remove the drain plug and drain any remaining fluid from the transaxle.
3. Remove the bolts that secure the rear cover to the transaxle case. Tap the cover with a rubber or plastic mallet to loosen the gasket seal. Remove the rear cover. Shift the transaxle to 1st gear.
4. Use special tool to lock up the input shaft. Uncrimp the tabs and remove the locknuts.
5. Remove the input reverse synchronizer gear.

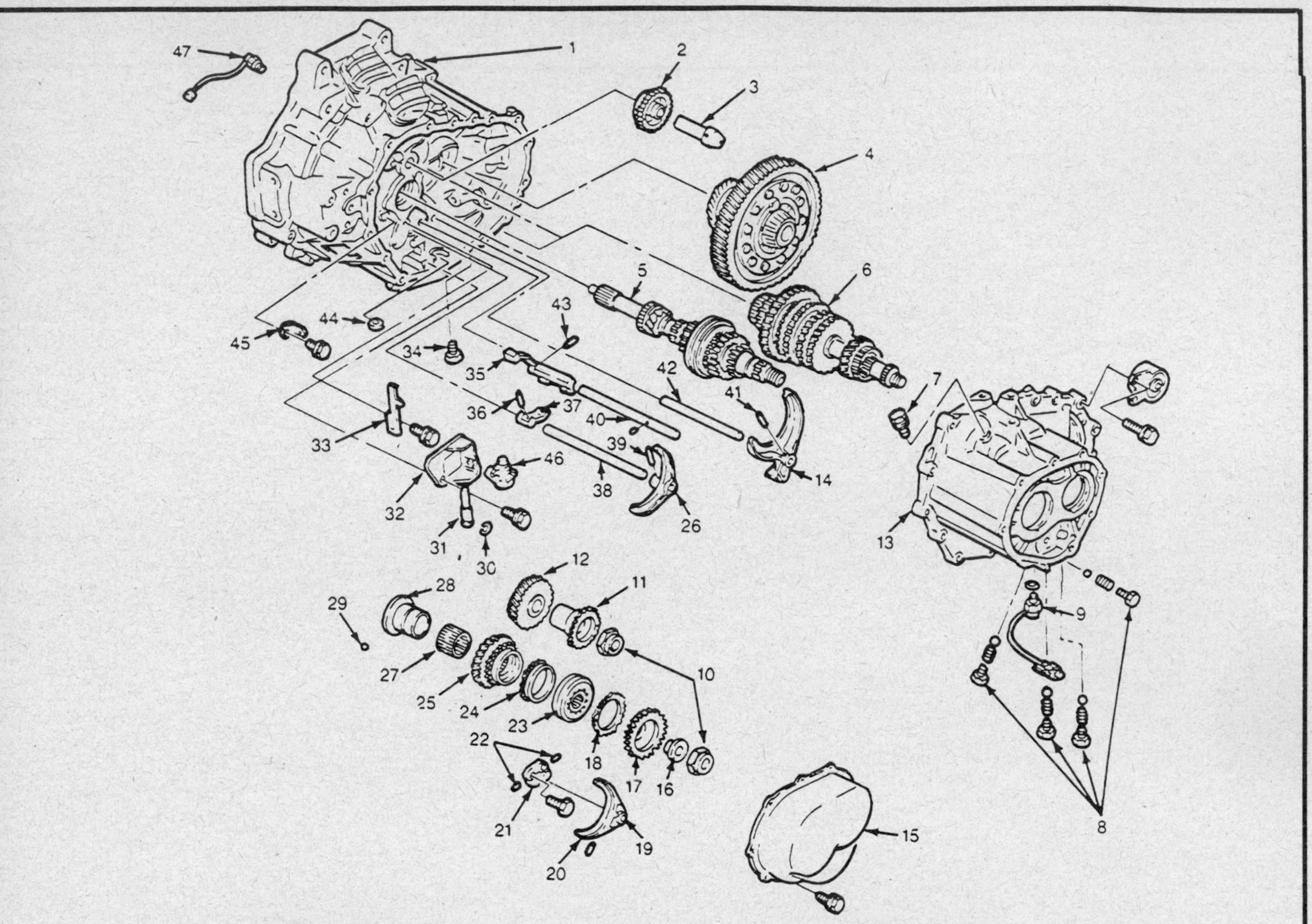

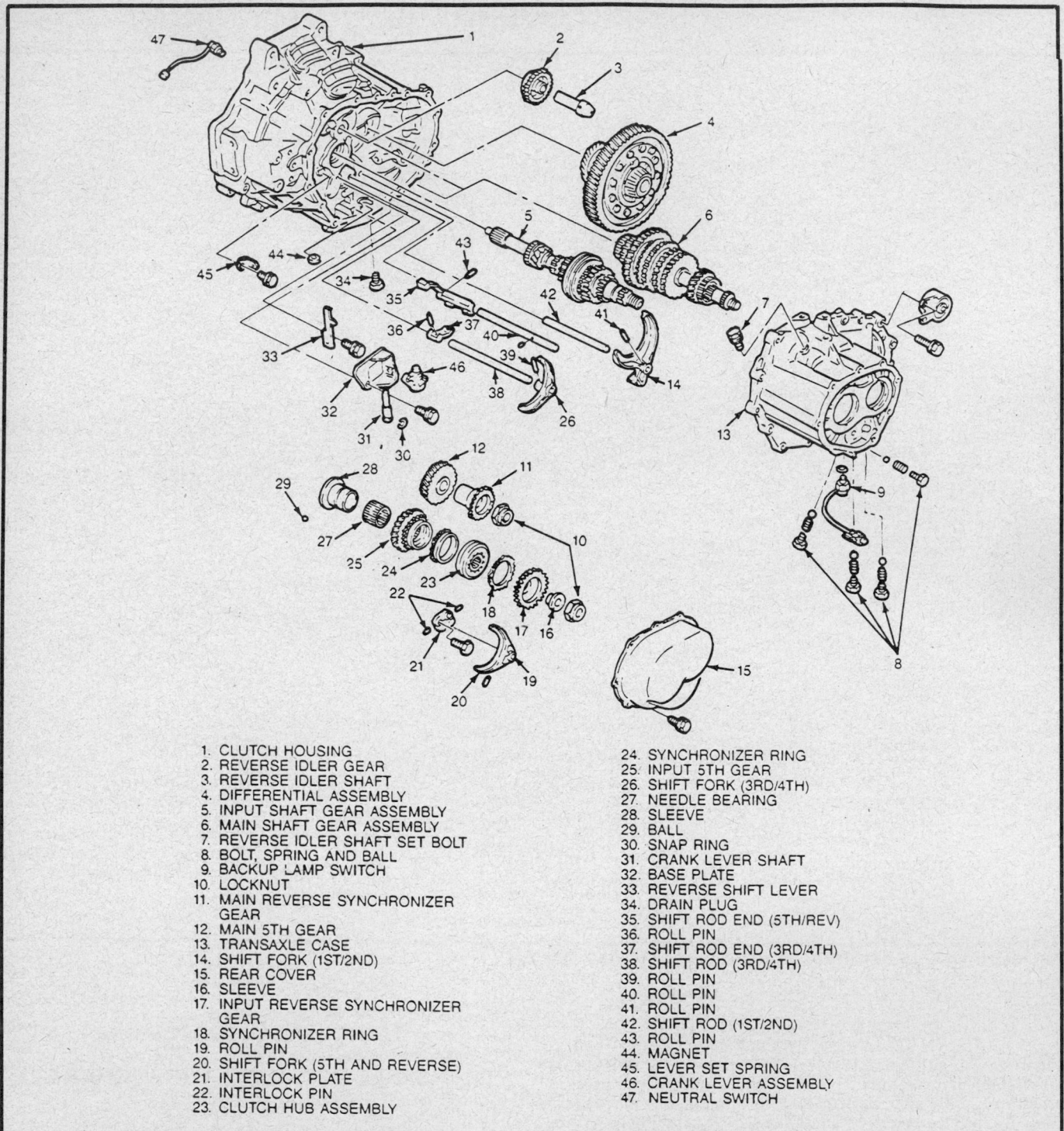

1. CLUTCH HOUSING
2. REVERSE IDLER GEAR
3. REVERSE IDLER SHAFT
4. DIFFERENTIAL ASSEMBLY
5. INPUT SHAFT GEAR ASSEMBLY
6. MAIN SHAFT GEAR ASSEMBLY
7. REVERSE IDLER SHAFT SET BOLT
8. BOLT, SPRING AND BALL
9. BACKUP LAMP SWITCH
10. LOCKNUT
11. MAIN REVERSE SYNCHRONIZER GEAR
12. MAIN 5TH GEAR
13. TRANSAXLE CASE
14. SHIFT FORK (1ST/2ND)
15. REAR COVER
16. SLEEVE
17. INPUT REVERSE SYNCHRONIZER GEAR
18. SYNCHRONIZER RING
19. ROLL PIN
20. SHIFT FORK (5TH AND REVERSE)
21. INTERLOCK PLATE
22. INTERLOCK PIN
23. CLUTCH HUB ASSEMBLY
24. SYNCHRONIZER RING
25. INPUT 5TH GEAR
26. SHIFT FORK (3RD/4TH)
27. NEEDLE BEARING
28. SLEEVE
29. BALL
30. SNAP RING
31. CRANK LEVER SHAFT
32. BASE PLATE
33. REVERSE SHIFT LEVER
34. DRAIN PLUG
35. SHIFT ROD END (5TH/REV)
36. ROLL PIN
37. SHIFT ROD END (3RD/4TH)
38. SHIFT ROD (3RD/4TH)
39. ROLL PIN
40. ROLL PIN
41. ROLL PIN
42. SHIFT ROD (1ST/2ND)
43. ROLL PIN
44. MAGNET
45. LEVER SET SPRING
46. CRANK LEVER ASSEMBLY
47. NEUTRAL SWITCH

Exploded view of turbocharged engine transaxle

6. Remove the main reverse synchronizer gear.
7. Use a suitable tool to drive out the roll pin.
8. Remove the shift fork, synchronizer ring and clutch hub asssembly.
9. Remove the synchronizer ring.
10. Remove the input 5th gear.
11. Remove the gear sleeve.
12. Remove the main 5th gear.
13. Remove the lock bolts, washer, guide bolt, springs and ball.
14. Tap the transaxle case lightly with a rubber or plastic hammer to loosen the gasket seal. Remove the transaxle case.
15. Remove the magnet.
16. Remove the reverse idler shaft and gear.
17. Remove the lock bolt from the gate.

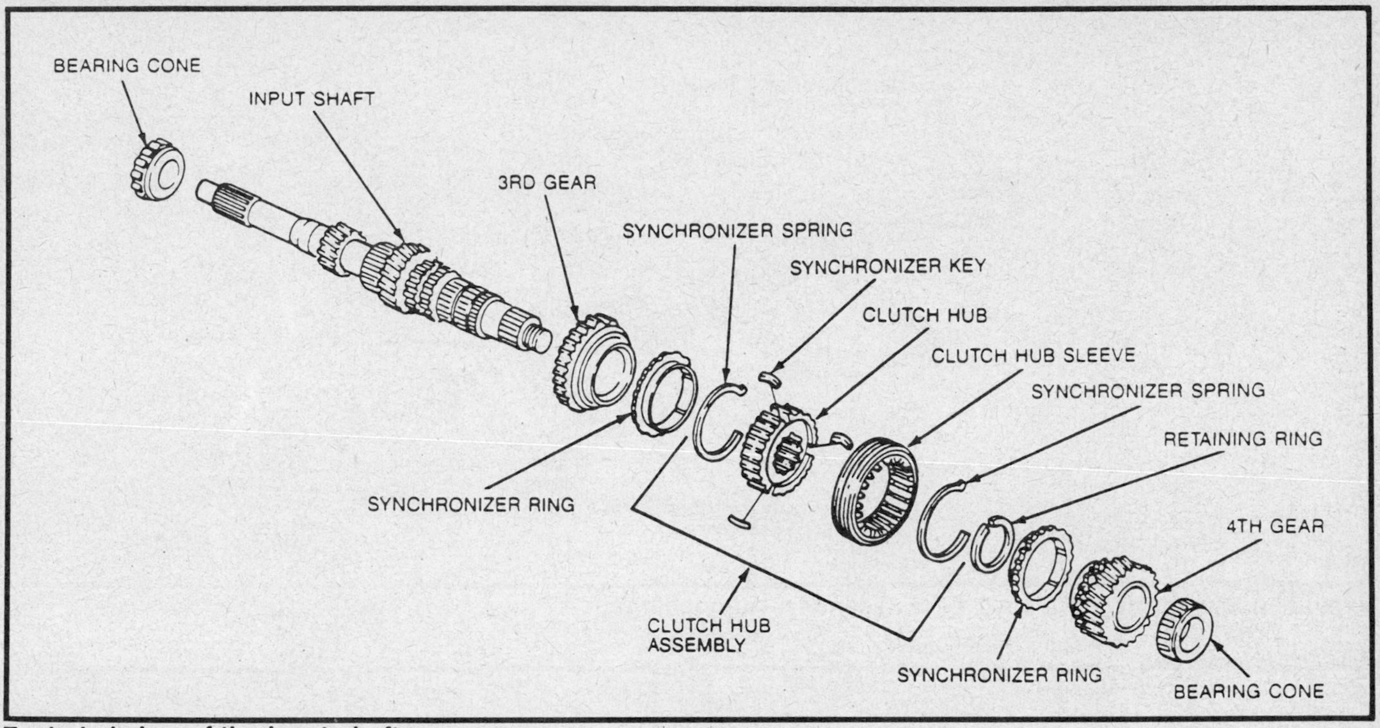
BEARING CONE
INPUT SHAFT
3RD GEAR
SYNCHRONIZER SPRING
SYNCHRONIZER KEY
CLUTCH HUB
CLUTCH HUB SLEEVE
SYNCHRONIZER SPRING
RETAINING RING
4TH GEAR
SYNCHRONIZER RING
CLUTCH HUB ASSEMBLY
SYNCHRONIZER RING
BEARING CONE

Exploded view of the input shaft gear assembly—non-turbocharged

18. Remove the clip and 5th and reverse shift rod.
19. Remove the gate.
20. Remove the pin.
21. Remove the crank lever shaft and assembly.
22. Align the ends of the interlock sleeve and control lever, then turn the shift rod counterclockwise.
23. Raise both shift forks and shift the clutch hub sleeves.
24. Lift the control end and remove the steel ball.
25. Remove the shift fork and shift rod assembly.
26. Remove the input shaft gear assembly.
27. Remove the mainshaft gear assembly.
28. Remove the differential assembly.
29. Remove the input shaft bearing cups, diaphragm spring and adjustment shims.
30. Remove the input shaft seal using special tool or equivalent.
31. Remove the mainshaft bearing cups, adjustment shims, diaphragm spring and funnel.
32. Remove the guide plate and spacer.
33. Remove the change arm.
34. Remove the selector pin using a suitable tool.
35. Remove the change rod and boot.
36. Remove the spring.
37. Remove the reverse gate.
38. Remove the selector.
39. Remove the change arm oil seal using special tools or equivalent.
40. Remove the breather cover and breather.
41. Remove the speedometer driven gear assembly on vehicles equipped with analog cluster or vehicle speed sensor assembly from vehicles equipped with digital cluster from the case.
42. Remove the differential bearing cups using special tools or equivalent. Remove the adjustment shim(s).
43. Remove the reverse lever shaft and reverse lever.
44. Remove the neutral switch and gasket.
45. Remove the differential seals using special tools or equivalent.

TRANSAXLE WITH TURBOCHARGED ENGINE (TYPE H)

1. Mount the transaxle on a bench mounting fixture or equivalent.
2. Remove the drain plug and drain any remaining fluid from the transaxle.
3. Remove the bolts that secure the rear cover to the transaxle case. Tap the cover with a rubber or plastic mallet to loosen the gasket seal. Remove the rear cover.
4. Remove the roll pin with a suitable tool.
5. Shift the transaxle into 5th gear by pressing down the shift fork and rod.
6. Shift the transaxle into 1st gear.
7. Uncrimp the tabs and remove the locknuts.
8. Remove the sleeve.
9. Remove the input reverse synchronizer gear.
10. Remove the main reverse synchronizer gear.
11. Remove the shift fork, synchronizer ring and clutch hub assembly.
12. Remove the synchronizer ring and input 5th gear.
13. Remove the needle bearing.
14. Remove the sleeve.
15. Remove the ball.
16. Remove the main 5th gear.
17. Remove the interlock plate and bolts.
18. Remove the interlock pins.
19. Remove the backup light switch.
20. Remove the bolts, springs and balls.
21. Remove the transaxle case attaching bolts and the reverse idler shaft set bolt.
22. Tap the transaxle case lightly with a rubber or plastic hammer to loosen the gasket seal. Remove the transaxle case.
23. Remove the magnet.
24. Remove the reverse idler shaft.
25. Remove the reverse idler gear.
26. Remove the base plate assembly.

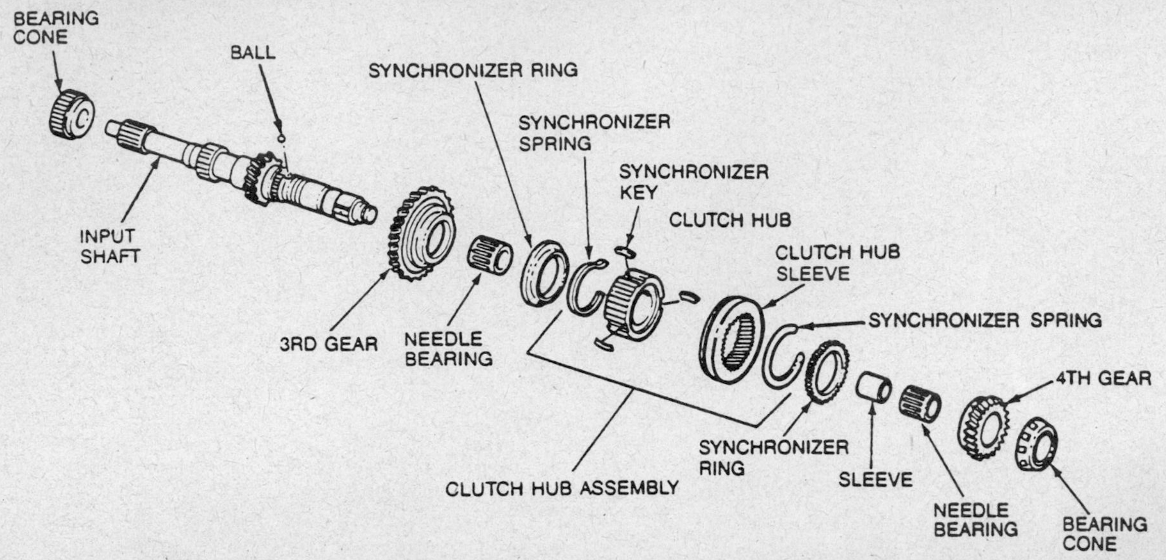
BEARING CONE
BALL
SYNCHRONIZER RING
SYNCHRONIZER SPRING
SYNCHRONIZER KEY
CLUTCH HUB
INPUT SHAFT
CLUTCH HUB SLEEVE
SYNCHRONIZER SPRING
3RD GEAR
NEEDLE BEARING
4TH GEAR
SYNCHRONIZER RING
SLEEVE
CLUTCH HUB ASSEMBLY
NEEDLE BEARING
BEARING CONE

Exploded view of the input shaft gear assembly—turbocharged

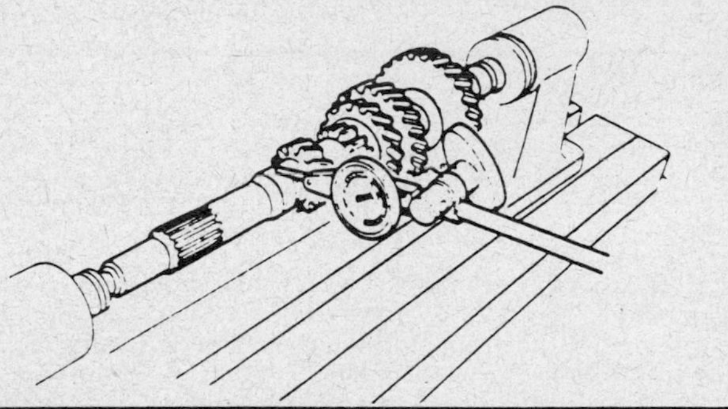

Checking the input shaft for runout

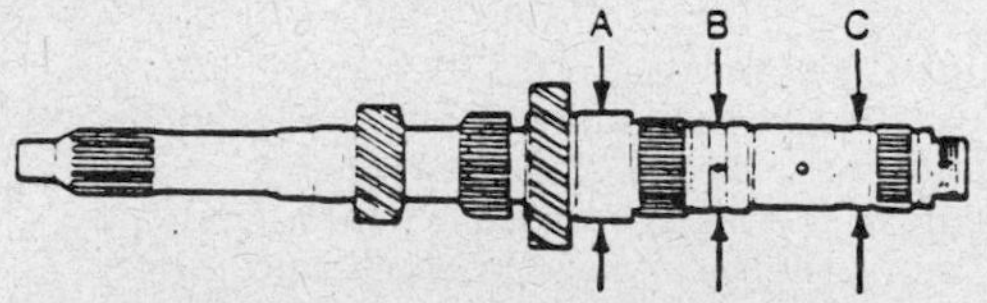
A
B
C

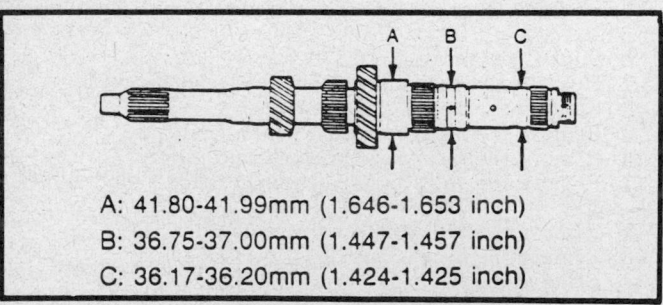

A: 41.80-41.99mm (1.646-1.653 inch)
B: 36.75-37.00mm (1.447-1.457 inch)
C: 36.17-36.20mm (1.424-1.425 inch)

Checking turbocharged input shaft diameter

27. Remove the snap ring, crank lever shaft and crank lever assembly from the base plate.
28. Remove the reverse shift lever.
29. Remove the lever set spring.
30. Remove the roll pin from the 3rd/4th shift fork and shift rod.
31. Remove the roll pin from the 3rd/4th shift rod end.
32. Remove the 3rd/4th shift fork, rod and rod end.
33. Remove the roll pin from the 1st/2nd shift fork and shift rod.

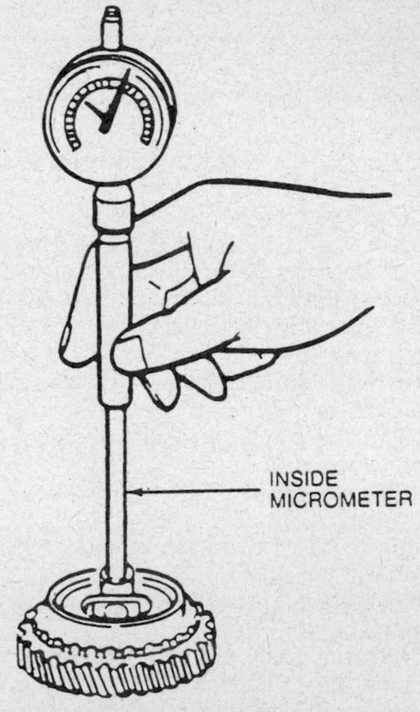
INSIDE MICROMETER

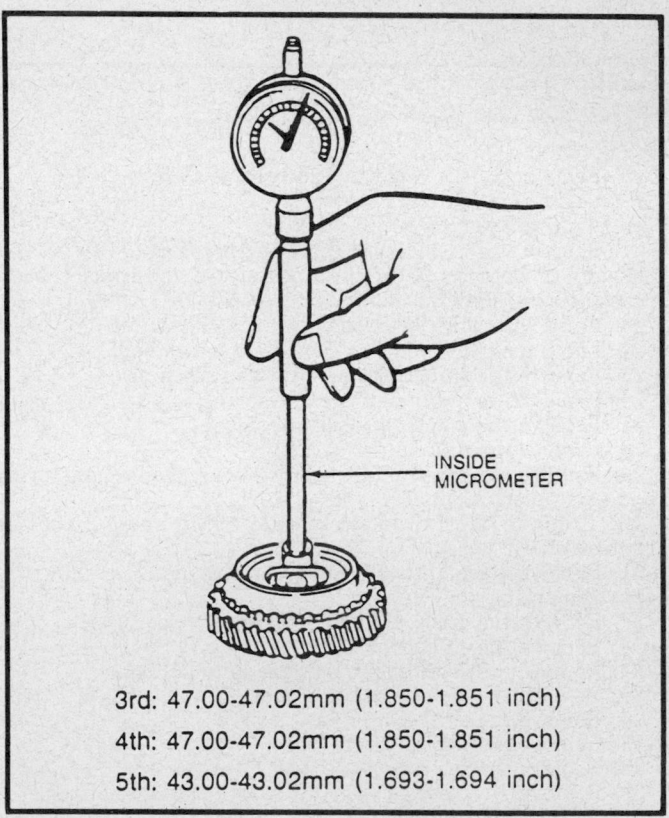

3rd: 47.00-47.02mm (1.850-1.851 inch)
4th: 47.00-47.02mm (1.850-1.851 inch)
5th: 43.00-43.02mm (1.693-1.694 inch)

Checking turbocharged input shaft inner gear diameter

34. Remove the 1st/2nd shift rod and shift fork.
35. Remove the 5th/reverse shift rod and shift rod end. Remove the interlock pin from the 5th/reverse shift rod.
36. Remove the main and input shaft gear assemblies as complete units.
37. Remove the differential assembly.
38. Remove the neutral safety switch.

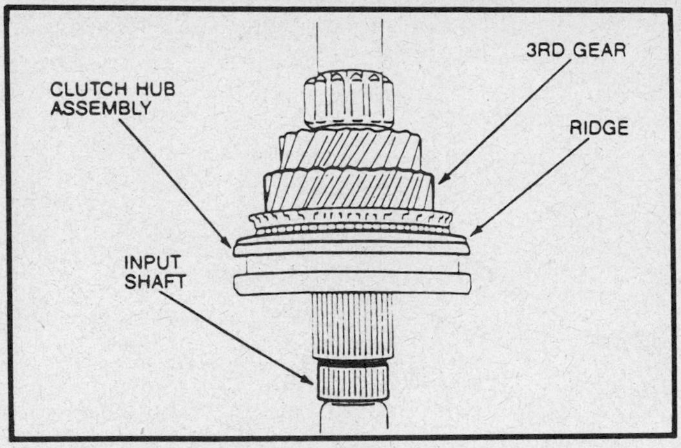
3RD GEAR
CLUTCH HUB
ASSEMBLY
RIDGE
INPUT
SHAFT

Installing clutch hub assembly

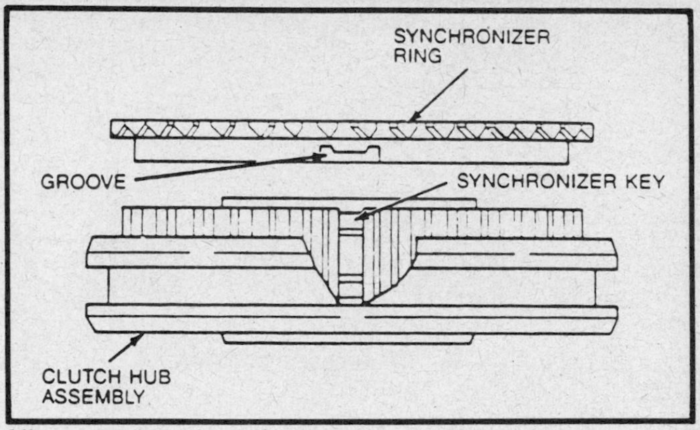
SYNCHRONIZER
RING
GROOVE
SYNCHRONIZER KEY
CLUTCH HUB
ASSEMBLY

Align synchronizer ring groove with synchronizer key

39. Remove the mainshaft bearing cups, adjustment shims, diaphragm spring and funnel.
40. Remove the input shaft bearing cups, diaphragm spring and adjustment shim(s).
41. Remove the differential bearing cups using special tools. Remove the adjustment shim(s).
42. Remove the input shaft seal using special tools.

NOTE: When removing this seal be careful not to split or damage the transaxle case. If necessary use special tools.

43. Remove the differential oil seals using special tools or equivalent.
44. Remove the change arm roll pin with suitable tool.
45. Remove the baffle plate, change arm and change rod.
46. Remove the change arm oil seal using special tools.
47. Remove the speedometer driven gear assembly on vehicles equipped with analog cluster or vehicle speed sensor assembly from vehicles equipped with digital cluster from the case.
48. Remove the oil passage, breather cover and breather.

Unit Disassembly and Assembly

INPUT SHAFT GEAR ASSEMBLY

Disassembly

TRANSAXLE WITH NON-TURBOCHARGED ENGINE

NOTE: The synchronizer rings are not interchangeable. Be sure to keep them in the order that they were removed.

1. Press off the bearing cone 4th gear end side using shaft protector and suitable tools. Hold the input shaft.
2. Remove the 4th gear and synchronizer ring.
3. Remove the retaining ring.
4. Press off the clutch hub assembly, synchronizer ring and 3rd gear using shaft protector and suitable tools. Hold the input shaft.
5. Press off the bearing cone 3rd gear end side using shaft protector and suitable tools.
6. Remove the synchronizer springs and keys from the clutch hub.
7. Remove the clutch hub from the clutch hub sleeve.

TRANSAXLE WITH TURBOCHARGED ENGINE

NOTE: The synchronizer rings are not interchangeable. Be sure to keep them in the order that they were removed.

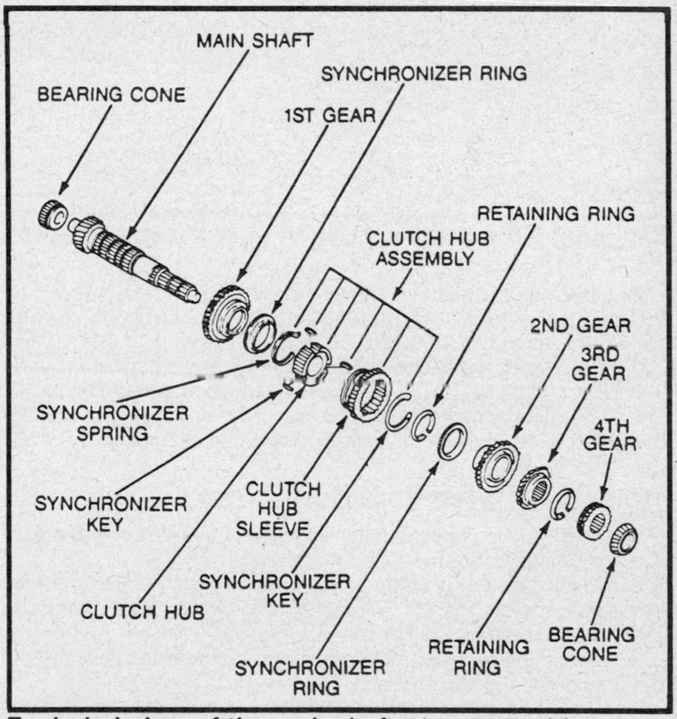
MAIN SHAFT
BEARING CONE
SYNCHRONIZER RING
1ST GEAR
RETAINING RING
CLUTCH HUB
ASSEMBLY
2ND GEAR
3RD
GEAR
4TH
GEAR
SYNCHRONIZER
SPRING
SYNCHRONIZER
KEY
CLUTCH
HUB
SLEEVE
SYNCHRONIZER
KEY
CLUTCH HUB
BEARING
CONE
RETAINING
RING
SYNCHRONIZER
RING

Exploded view of the mainshaft gear assembly—non-turbocharged

1. Press off the bearing cone 4th gear end side using shaft protector and suitable tools. Hold the input shaft.
2. Remove the 4th gear, needle bearing and sleeve.
3. Remove the ball and synchronizer ring.
4. Press off the clutch hub assembly, synchronizer ring and 3rd gear using shaft protector and suitable tools. Hold the input shaft.
5. Remove the needle bearing.
6. Press off the bearing cone 3rd gear end side using shaft protector and suitable tools. Hold the input shaft.
7. Remove the synchronizer springs and keys from the clutch hub.
8. Remove the clutch hub from the clutch hub sleeve.

Inspection

TRANSAXLE WITH NON-TURBOCHARGED ENGINE

1. Check for worn or damaged synchronizer cone, hub sleeve coupling or gear teeth.

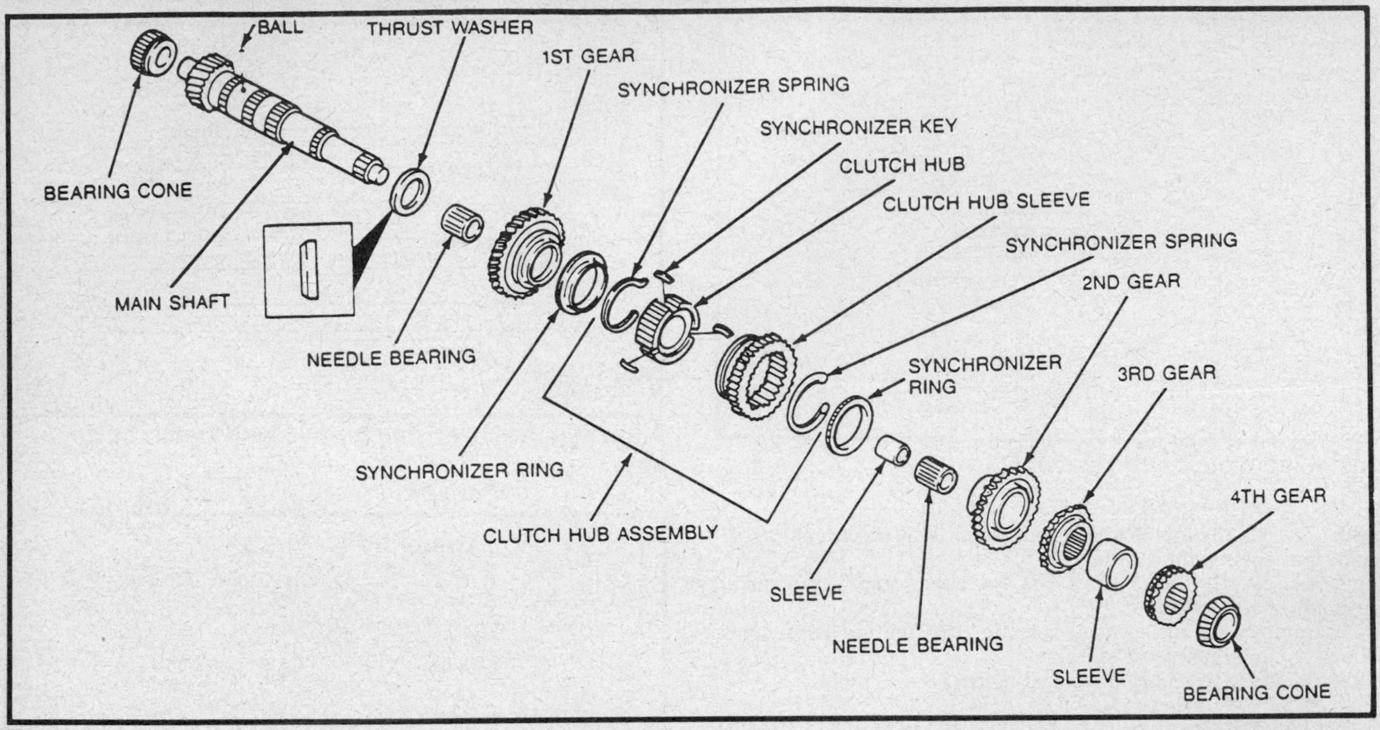
BALL
THRUST WASHER
1ST GEAR
SYNCHRONIZER SPRING
SYNCHRONIZER KEY
CLUTCH HUB
CLUTCH HUB SLEEVE
SYNCHRONIZER SPRING
2ND GEAR
BEARING CONE
MAIN SHAFT
NEEDLE BEARING
SYNCHRONIZER RING
3RD GEAR
SYNCHRONIZER RING
CLUTCH HUB ASSEMBLY
4TH GEAR
SLEEVE
NEEDLE BEARING
SLEEVE
BEARING CONE

Exploded view of the mainshaft gear assembly—turbocharged

2. Check the input shaft gear runout. Check the runout by mounting the gear shaft in a lathe or V-blocks. The maximum allowable runout is 0.002 in.
3. The clearance between the 3rd and 2nd gears is 0.0020–0.0079 in. The maximum allowable clearance is 0.0098 in.
4. The clearance between the 4th gear and bearing cone is 0.0064–0.0144 in. The maximum allowable clearance is 0.0163 in.

TRANSAXLE WITH TURBOCHARGED ENGINE

1. Check for worn or damaged synchronizer cone, hub sleeve coupling or gear teeth.
2. Check the input shaft gear runout. Check the runout by mounting the gear shaft in a lathe or V-blocks. The maximum allowable runout is 0.0014 in.
3. Check the input shaft diameter and the inner gear diameter for correct specifications.
4. Check the clearance between the 3rd and input shaft gears it should be 0.0059–0.0103 in.
5. Check the clearance between the 4th gear and bearing cone it should be 0.0059–0.0103 in.

Assembly

TRANSAXLE WITH NON-TURBOCHARGED ENGINE

1. Slide the clutch hub into the clutch hub sleeve.
2. Install the synchronizer keys and springs.

NOTE: Whenever a bearing cone is removed, it must be replaced.

3. Press on a new bearing cone 3rd gear end side using shaft protector and suitable tools.
4. When installing the synchronizer ring and clutch hub assembly, align the synchronizer ring groove and synchronizer key. Be sure to have the ridge facing the 3rd gear when installing the clutch hub assembly.
5. Press on the 3rd gear, synchronizer ring and clutch hub assembly using shaft protector and suitable tools.
6. Install the retaining ring.
7. Press on the synchronizer ring, 4th gear and new bearing cone using shaft protector and suitable tools.

TRANSAXLE WITH TURBOCHARGED ENGINE

1. Slide the clutch hub into the clutch hub sleeve.
2. Install the synchronizer keys and springs.
3. Lubricate needle bearing assembly with the specified transaxle fluid and install the needle bearing assembly.
4. When installing the synchronizer ring and clutch hub assembly, align the synchronizer ring groove and synchronizer key.
5. Press on the 3rd gear, synchronizer ring and clutch hub assembly using shaft protector and suitable tools.
6. Install the ball, sleeve and needle bearing. Lubricate the needle bearing assembly with the specified transaxle fluid before installing.

NOTE: Whenever a bearing cone is removed, it must be replaced.

7. Press on the synchronizer ring, 4th gear and new bearing cone using shaft protector and suitable tools.
8. Press on new bearing cone 3rd gear end side using shaft protector and suitable tools.

MAINSHAFT GEAR ASSEMBLY

Disassembly

TRANSAXLE WITH NON-TURBOCHARGED ENGINE

NOTE: The synchronizer rings are not interchangeable. Be sure to keep them in the order that they were removed.

1. Press off the bearing cone 4th gear end side using shaft protector and suitable tools. Hold the mainshaft.
2. Press off the 4th gear using shaft protector and suitable tools. Hold the mainshaft.
3. Remove the retaining ring.

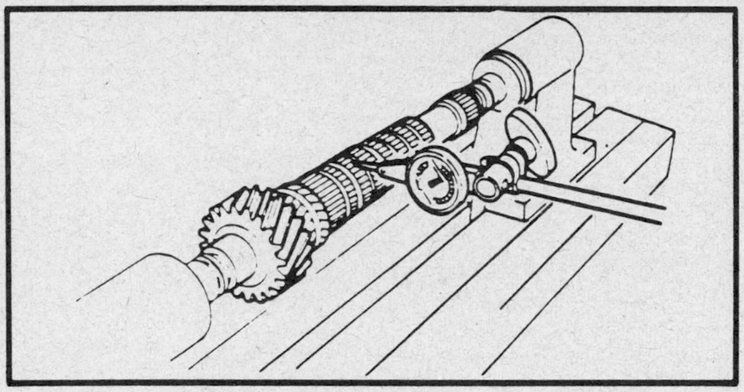
Checking mainshaft for runout

Checking mainshaft and gears for oil clearance – non-turbocharged

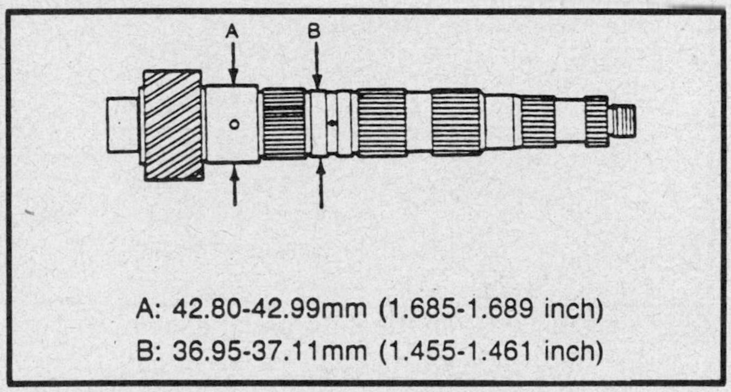
A
B
A: 42.80-42.99mm (1.685-1.689 inch)
B: 36.95-37.11mm (1.455-1.461 inch)
Checking turbocharged mainshaft diameter

4. Press off the 3rd and 2nd gears using shaft protector and suitable tools. Hold the mainshaft.
5. Remove the synchronizer ring.
6. Remove the retaining ring.
7. Press off the clutch hub assembly, synchronizer ring and 1st gear using shaft protector and suitable tools. Hold the mainshaft.
8. Press off the bearing cone 1st gear end side using shaft protector and suitable tools. Hold the mainshaft.
9. Remove the synchronizer springs and keys from the clutch hub.
10. Remove the clutch hub from the clutch hub sleeve.

TRANSAXLE WITH TURBOCHARGED ENGINE

NOTE: The synchronizer rings are not interchangeable. Be sure to keep them in the order that they were removed.

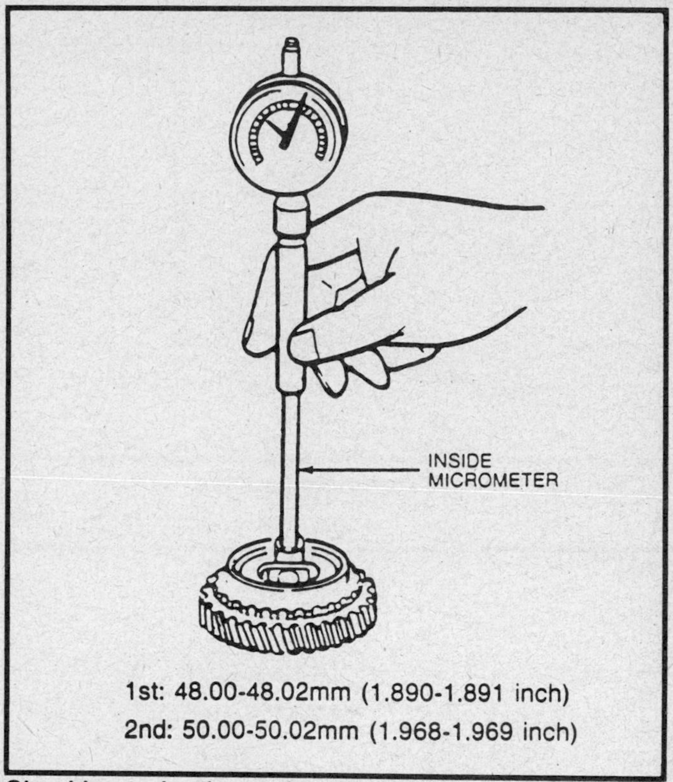
INSIDE MICROMETER
1st: 48.00-48.02mm (1.890-1.891 inch)
2nd: 50.00-50.02mm (1.968-1.969 inch)
Checking turbocharged mainshaft inner gear diameter

1. Press off the bearing cone 4th gear end side using shaft protector and suitable tools. Hold the mainshaft.
2. Remove the 4th gear and sleeve.
3. Remove the 3rd and 2nd gears.
4. Remove the needle bearing, sleeve and ball.
5. Remove the synchronizer gear.
6. Press off the clutch hub assembly, synchronizer ring and 1st gear using shaft protector and suitable tools. Hold the mainshaft.
7. Remove the needle bearing and thrust washer.
8. Press off the bearing cone 1st gear end side using shaft protector and suitable tools. Hold the mainshaft.
9. Remove the synchronizer springs and keys from the clutch hub.
10. Remove the clutch hub from the clutch hub sleeve.

Inspection

TRANSAXLE WITH NON-TURBOCHARGED ENGINE

1. Check for worn or damaged gear contact surfaces, splines or gear teeth.
2. Check for clogged oil passage.
3. Check the mainshaft gear runout. Mount gear shaft in a lathe or V-blocks and measure the runout. The maximum allowable runout is 0.0006 in.
4. Check the clearance between the 1st gear and differential drive gear it should be 0.002–0.011 in. The maximum allowable clearance is 0.019 in.
5. Check the clearance between the 2nd and 3rd gears it should be 0.0069–0.0179 in. The maximum allowable clearance is 0.0199 in.
6. Check the oil clearance between mainshaft and gears.
7. Measure the diameter of the gear shaft where the gear is installed. Measure the inside diameter of the gear. The difference between the two measurements is the clearance. If the

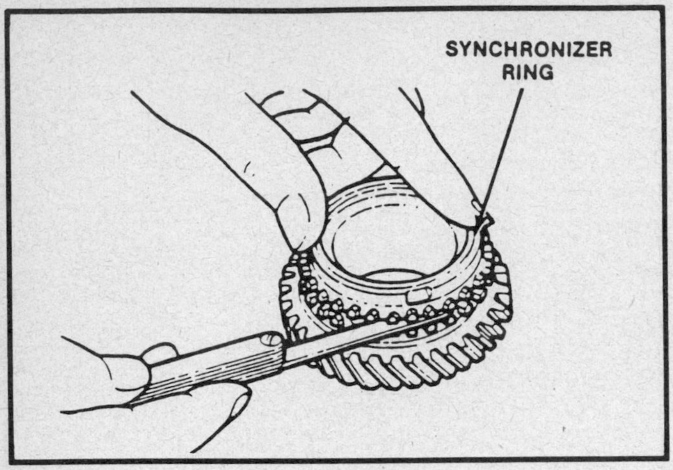
SYNCHRONIZER RING

Checking synchronizer ring clearance

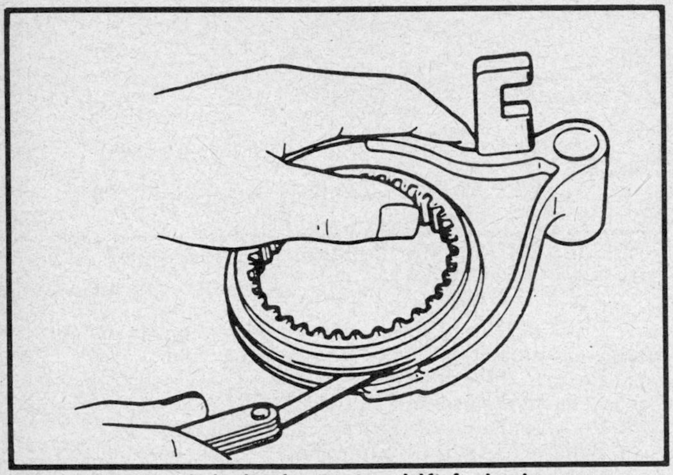

Checking clutch hub sleeve to shift fork clearance

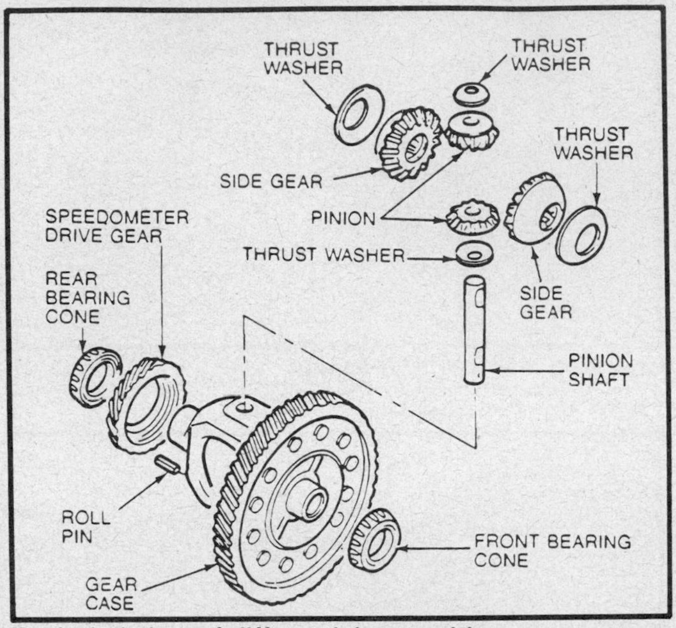
THRUST WASHER
THRUST WASHER
THRUST WASHER
SIDE GEAR
SPEEDOMETER DRIVE GEAR
PINION
THRUST WASHER
REAR BEARING CONE
SIDE GEAR
PINION SHAFT
ROLL PIN
FRONT BEARING CONE
GEAR CASE

Exploded view of differential assembly

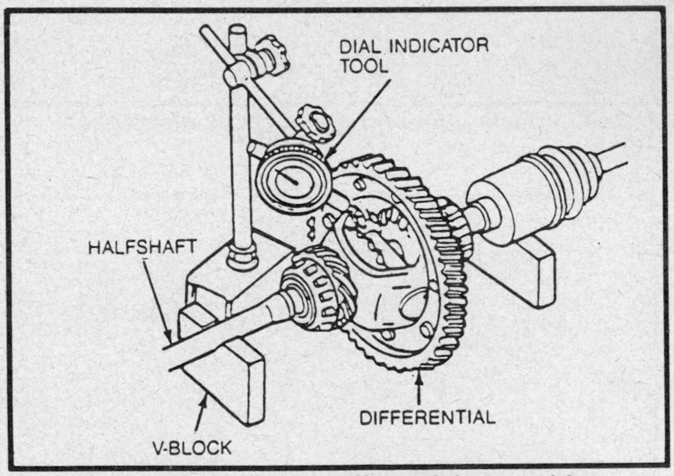
DIAL INDICATOR TOOL
HALFSHAFT
DIFFERENTIAL
V-BLOCK

Checking backlash in the differential assembly

clearance is more than allowable, replace the gear and/or shaft as necessary. The standard clearance should be 0.001–0.003 in.

8. Check the synchronizer ring engagement with gear. The ring must engage smoothly with gear.

9. Check the synchronizer ring clearance from the side of gear. Press the synchronizer ring uniformly against the gear and measure around the circumference. The standard clearance is 0.059 in. The minimum allowable clearance is 0.031 in.

10. Check clutch hub sleeve for excessive clearance between sleeve and shift fork. The maximum allowable clearance is 0.020 in.

TRANSAXLE WITH TURBOCHARGED ENGINE

1. Check for worn or damaged gear contact surfaces, splines or gear teeth.
2. Check for clogged oil passage.
3. Check the mainshaft gear runout. Mount gear shaft in a lathe or V-blocks and measure the runout. The maximum allowable runout is 0.0012 in.
4. Check the mainshaft diameter and the inner gear diameter for correct specifications.
5. The clearance between the 1st gear and thrust washer is 0.0051–0.0139 in.
6. The clearance between the 2nd and 3rd gears is 0.0059–0.0103 in.
7. Check the synchronizer ring engagement with gear. The ring must engage smoothly with gear.
8. Check the synchronizer ring clearance from the side of gear. Press the synchronizer ring uniformly against the gear and measure around the circumference. The standard clearance is 0.059 in. The minimum allowable clearance is 0.031 in.
9. Check clutch hub sleeve for excessive clearance between sleeve and shift fork. The maximum allowable clearance is 0.018 in.

Assembly

TRANSAXLE WITH NON-TURBOCHARGED ENGINE

1. Slide the clutch hub into the clutch hub sleeve.
2. Install the synchronizer keys and springs.

NOTE: Whenever a bearing cone is removed, it must be replaced.

3. Press on the new bearing cone 1st gear end side using shaft protector and suitable tools.
4. When installing the synchronizer ring and clutch hub assembly, align the synchronizer ring groove and synchronizer key.

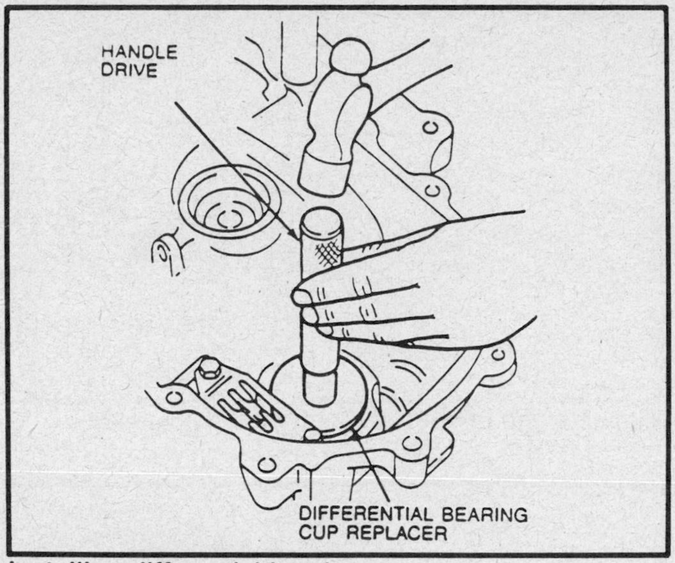
HANDLE DRIVE
DIFFERENTIAL BEARING CUP REPLACER
Installing differential bearing cup

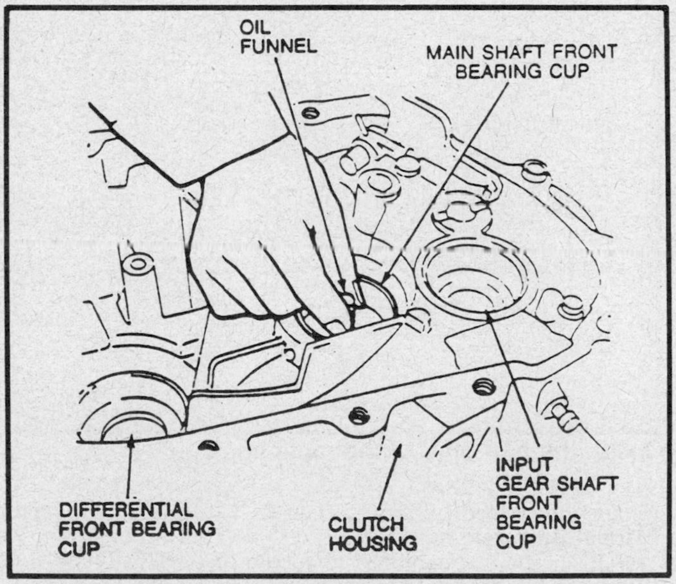
OIL FUNNEL
MAIN SHAFT FRONT BEARING CUP
DIFFERENTIAL FRONT BEARING CUP
CLUTCH HOUSING
INPUT GEAR SHAFT FRONT BEARING CUP
Installing the oil funnel and mainshaft front bearing cup into the clutch housing

5. Press on the 1st gear, synchronizer ring and clutch hub assembly using shaft protector and suitable tools.
6. Install the retaining ring.
7. Press on the synchronizer ring, 2nd gear and 3rd gear using shaft protector and suitable tools.
8. Install the retaining ring.
9. Press on the 4th gear using shaft protector and suitable tools.
10. Press on the new bearing cone 4th gear end side using shaft protector and suitable tools.

TRANSAXLE WITH TURBOCHARGED ENGINE

1. Slide the clutch hub into the clutch hub sleeve.
2. Install the synchronizer keys and springs.
3. Install the thrust washer and needle bearing. Before installing the needle bearing lubricate the bearing with the specified transaxle fluid.
4. When installing the synchronizer ring and clutch hub assembly, align the synchronizer ring groove and synchronizer key.
5. Press on the 1st gear, synchronizer ring and clutch hub assembly using shaft protector and suitable tools.
6. Install the synchronizer ring.
7. Install the ball, sleeve and needle bearing. Before installing the needle bearing lubricate the bearing with the specified transaxle fluid.
8. Install the 2nd and 3rd gears.
9. Install the sleeve and 4th gear.

NOTE: Whenever a bearing cone is removed, it must be replaced.

10. Press on the new bearing cone, 4th gear end side using shaft protector and suitable tools. Make sure that the sleeve does not rotate.
11. Press on the new bearing cone, 1st gear end side using shaft protector and suitable tools.

DIFFERENTIAL

NOTE: The differential assemblies are the same for both engines.

Disassembly

1. Remove the knock or roll pin.
2. Remove the pinion shaft.
3. Remove the pinions and thrust washers by rotating them out of the gear case.
4. Remove the side gears and thrust washers.
5. Remove the bearing cone, speedometer drive gear end side using suitable tools.
6. Remove the speedometer drive gear.
7. Remove the bearing cone, ring gear end side using suitable tools.

Inspection

1. Check for damaged or worn gears.
2. Check for cracked or damaged gear case.
3. Check the side gear and pinion backlash.
   a. Install the left and right halfshafts into the differential.
   b. Support the halfshafts on V-blocks.
   c. Use dial indicator or equivalent with magnetic base/flex arm to measure the backlash of both pinion gears. If the backlash is more than allowable, select a thrust washer with a different thickness.
   d. The maximum allowable amount of backlash is 0.000–0.004 in.

Assembly

NOTE: Whenever a bearing cone is removed, it must be replaced.

1. Install the speedometer drive gear and new bearing cone using suitable tools.
2. Install the new bearing cone, ring gear end side using suitable tools.
3. Install the thrust washers and pinions.
4. Install the pinion shaft.
5. Install the knock or roll pin, them crimp it so that it cannot come out of the gear case.
6. Install the thrust washers and side gears.

SHIM SELECTION

1. Install the differential bearing cup into the clutch housing using suitable tools.
2. Install the oil funnel and mainshaft front bearing cup into the clutch housing.

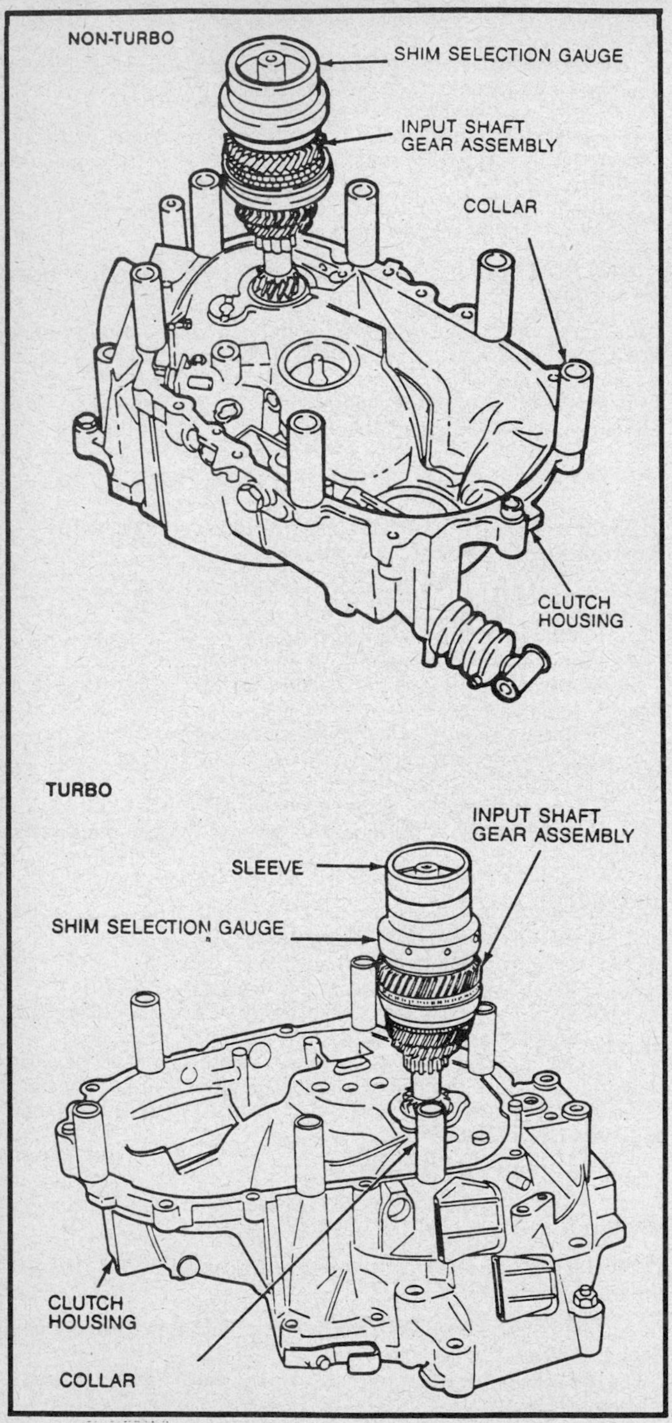
NON-TURBO
SHIM SELECTION GAUGE
INPUT SHAFT
GEAR ASSEMBLY
COLLAR
CLUTCH
HOUSING
TURBO
INPUT SHAFT
GEAR ASSEMBLY
SLEEVE
SHIM SELECTION GAUGE
CLUTCH
HOUSING
COLLAR

Installing special tool collars in the correct position

3. Install the input gear shaft front bearing cup into the clutch housing.

4. Inspect the bearing cups after installation to make sure that they are fully seated.

INPUT SHAFT

1. Install the input shaft gear assembly into the clutch housing.

2. Place the rear bearing cup on the input shaft bearing.

3. Position shim selection gauge tool T88C–77000–CH3 (non-turbocharged) or T88C–77000–CH2 (turbocharged) on top of input shaft gear assembly.

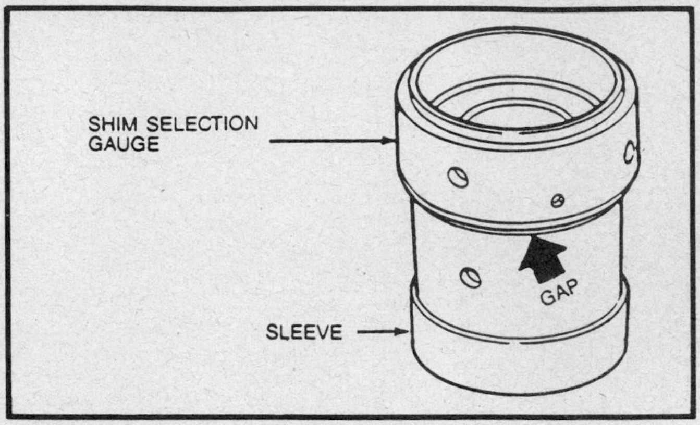
SHIM SELECTION
GAUGE
GAP
SLEEVE

Eliminate gap in special tool before adjustment

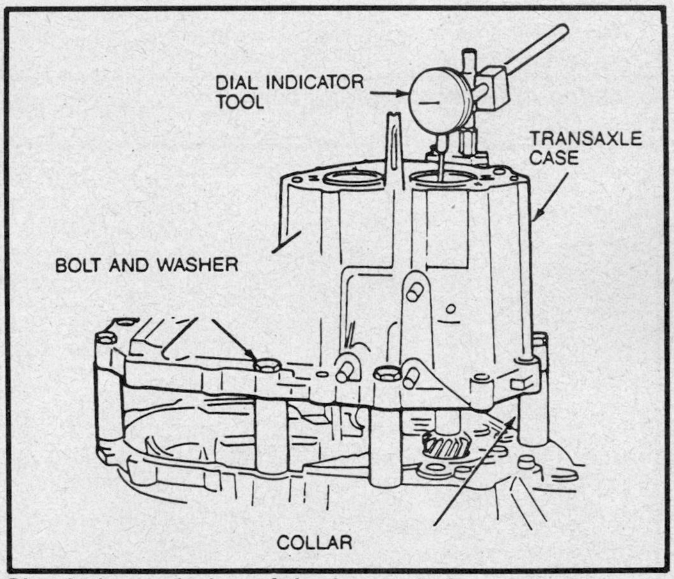
DIAL INDICATOR
TOOL
TRANSAXLE
CASE
BOLT AND WASHER
COLLAR

Check the end play of the input shaft

4. Place the 6 collars part of special tool T87C–77000–J on the clutch housing at the correct positions.

NOTE: The 2 halves of the shim selection gauge tool T88C–77000–CH2 (turbocharged) must be turned to eliminate any gap between them before adjustment.

5. Place the transaxle case on the collars, then install the washers and bolts part of special tool T87C–77000–J. Tighten the non-turbocharged version to 13–14 ft. lbs. Tighten the turbocharged version to 27–38 ft. lbs.

6. Mount a dial indicator or equivalent, to check the endplay of the input shaft.

7. Rotate the input shaft gear assembly several times to help seat the bearings.

8. Adjust the dial indicator to **0** at the lowest point on the end of the input shaft. Do not disturb dial indicator tool until at least 3 endplay readings have been taken.

9. Raise the input shaft gear assembly by hand and read endplay. Lower the input shaft gear assembly.

10. The input shaft gear assembly must be lifted equally on both sides or it will tend to cock to one side, resulting in an incorrect reading.

11. Turn the input shaft gear assembly several times until dial

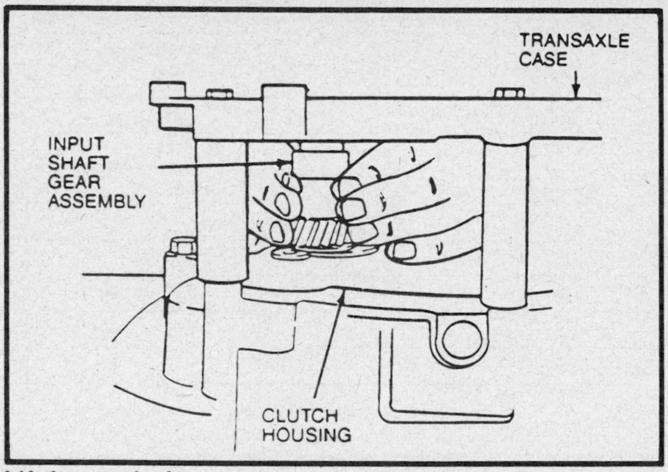
TRANSAXLE CASE
INPUT SHAFT GEAR ASSEMBLY
CLUTCH HOUSING
Lift input shaft gear assembly equally on both sides

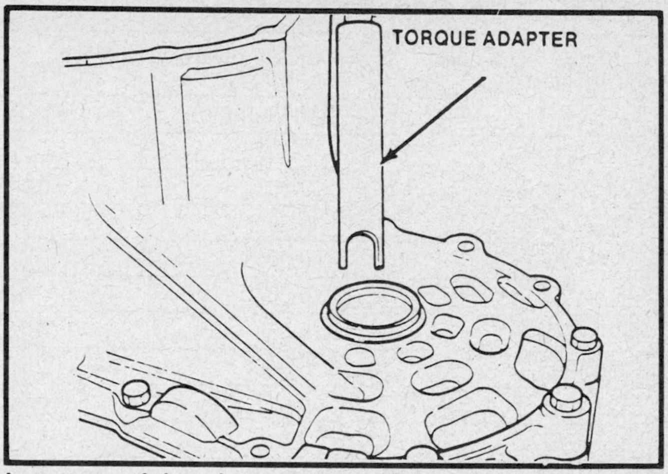
TORQUE ADAPTER
Insert special tool to engage differential pinion shaft

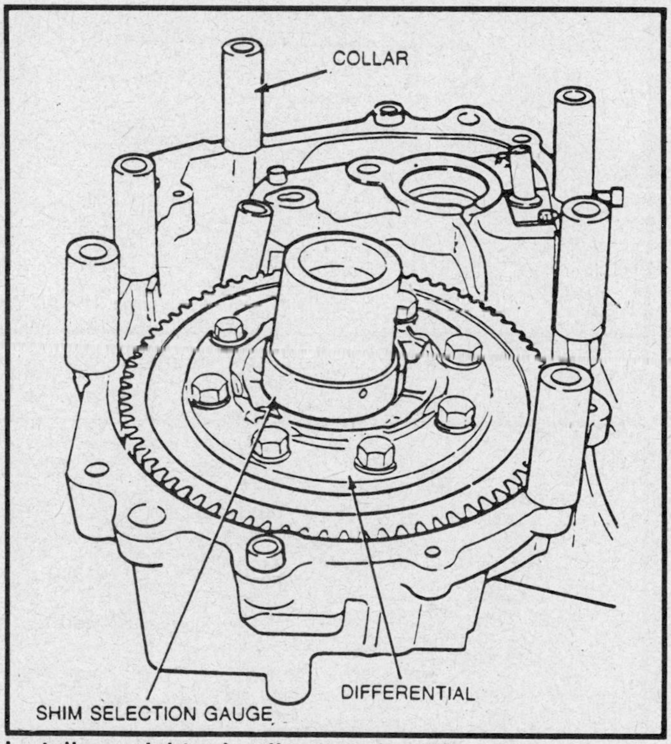
COLLAR
DIFFERENTIAL
SHIM SELECTION GAUGE
Install special tool collars on the differential assembly

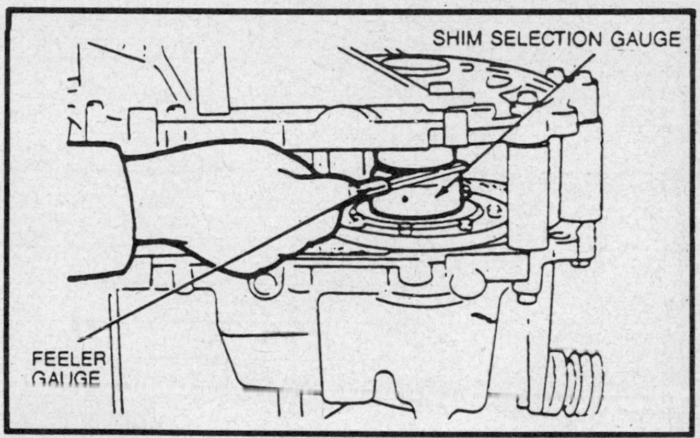
SHIM SELECTION GAUGE
FEELER GAUGE
Use feeler gauge to measure gap in the shim selection gauge

indicator returns to **0**. Raise input shaft gear assembly by hand to take a second endplay measurement.

12. Repeat Step 11 to obtain at least 3 readings within 0.004 in. of each other. Average at least 3 of the measurements to obtain an endplay reading.
13. Subtract 0.028 in. from the endplay reading to account for the thickness of the diaphragm spring which goes between the shim and the cup. This result is the final shim size.
14. From shim selection kit E92Z–7L172–A (non-turbocharged) or E92Z–7L172–C (turbocharged), select the shim(s) that is closest or slightly larger to the final shim size. Do not use more than 2 shims.
15. Remove the bolts and washers securing the transaxle case to the clutch housing. Remove the transaxle case, gauge, rear bearing cup and input shaft gear assembly. Do not remove the bearing cups in the clutch housing.

MAINSHAFT

Repeat the procedures given above for the mainshaft. Be sure to remove the sleeve from shim selection gauge tool T88C–77000–CH2 (turbocharged). There should be no gap between the gauge halves. Select the appropriate shim(s) from shim kit E92Z–7L172–A (non-turbocharged) or E92Z–7L172–B (turbocharged).

DIFFERENTIAL

1. Install the differential into the clutch housing.
2. Place the rear bearing cup on the differential bearing.
3. Position shim selection gauge tool T88C–77000–CH1 (non-turbocharged) or T88C–77000–CH2 (turbocharged) on top of the differential.

NOTE: The 6 collars part of special tool T87C–77000–J should be placed on the clutch housing at the same locations as described in the input shaft shim selection procedure.

4. Turn The 2 halves of the gauge tool to eliminate any gap between them.
5. Place the transaxle case on the collars, then install the washers and bolts part of special tool T87C–77000–J. Tighten the non-turbocharged to 13–14 ft. lbs. Tighten the turbocharged to 27–38 ft. lbs.
6. Adjust the gauge tool using the pins provided in special tool T87C–77000–J until all of the free play is removed and the bearing cup is seated. Then thread the gauge tool halves back together.

SHIM SELECTION

Part Number	Shim Thickness in.	mm
NON-TURBOCHARGED ENGINE		
E92Z-4067-A	0.004	0.10
E92Z-4067-B	0.005	0.12
E92Z-4067-C	0.006	0.14
E92Z-4067-D	0.0063	0.16
E92Z-4067-E	0.007	0.18
E92Z-4067-F	0.008	0.20
E92Z-4067-G	0.010	0.25
E92Z-4067-H	0.012	0.30
E92Z-4067-J	0.014	0.35
E92Z-4067-K	0.016	0.40
E92Z-4067-L	0.018	0.45
E92Z-4067-N	0.020	0.50
E92Z-4067-P	0.022	0.55
E92Z-4067-Q	0.024	0.60
E92Z-4067-R	0.026	0.65
E92Z-4067-S	0.028	0.70
E92Z-4067-T	0.030	0.75
E92Z-4067-U	0.032	0.80
E92Z-4067-V	0.034	0.85
E92Z-4067-W	0.036	0.90
E92Z-4067-X	0.038	0.95
E92Z-4067-Y	0.040	1.00
E92Z-4067-Z	0.042	1.05
E92Z-4067-AA	0.044	1.10
E92Z-4067-AB	0.046	1.15
E92Z-4067-AC	0.048	1.20
TURBOCHARGED ENGINE		
E92Z-4067-AD	0.004	0.10
E92Z-4067-AE	0.005	0.12
E92Z-4067-AF	0.006	0.14
E92Z-4067-AG	0.0063	0.16
E92Z-4067-AH	0.008	0.20
E92Z-4067-AJ	0.010	0.25
E92Z-4067-AK	0.012	0.30
E92Z-4067-AL	0.014	0.35
E92Z-4067-AM	0.016	0.40
E92Z-4067-AN	0.018	0.45
E92Z-4067-AP	0.020	0.50
E92Z-4067-AR	0.022	0.55
E92Z-4067-AS	0.024	0.60
E92Z-4067-AT	0.026	0.65
E92Z-4067-AU	0.028	0.70
E92Z-4067-AV	0.030	0.75
E92Z-4067-AW	0.032	0.80

SHIM SELECTION

Part Number	Shim Thickness in.	mm
E92Z-4067-AX	0.034	0.85
E92Z-4067-AY	0.036	0.90
E92Z-4067-AZ	0.038	0.95
E92Z-4067-BA	0.040	1.00
E92Z-4067-BB	0.042	1.05
E92Z-4067-BC	0.044	1.10
E92Z-4067-BD	0.046	1.15
E92Z-4067-BE	0.048	1.20

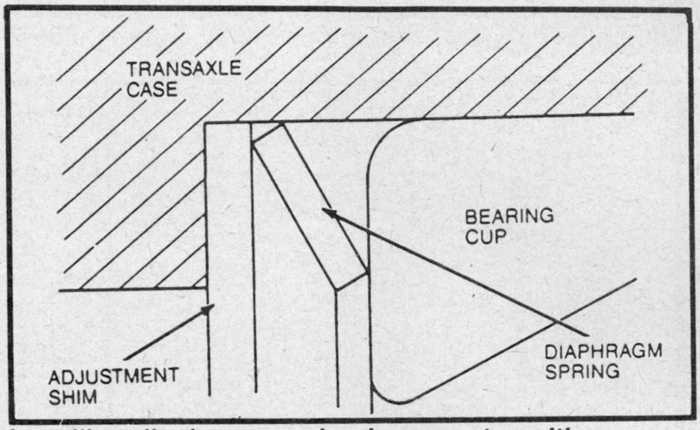
TRANSAXLE CASE
BEARING CUP
DIAPHRAGM SPRING
ADJUSTMENT SHIM
Installing diaphragm spring in correct position

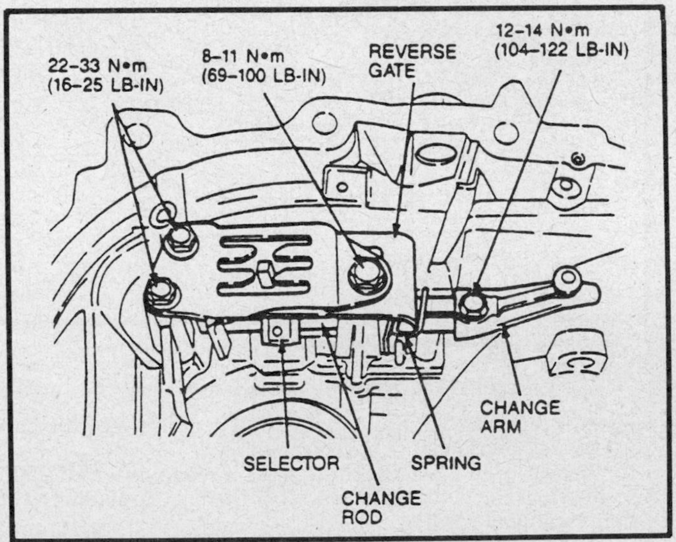
22–33 N•m (16–25 LB-IN)
8–11 N•m (69–100 LB-IN)
REVERSE GATE
12–14 N•m (104–122 LB-IN)
CHANGE ARM
SELECTOR
SPRING
CHANGE ROD
Installing the guide plate – non-turbocharged

7. Insert the torque adapter tool T88C–77000–L through the transaxle case and engage the differential pinion shaft.
8. Attach a torque wrench to the tool.
9. Turn the gauge using the adjusting rods part of special tool T87C–77000–J until a reading of 4.3 inch lbs. is obtained on the torque wrench.
10. Use a feeler gauge to measure the gap in the shim selection

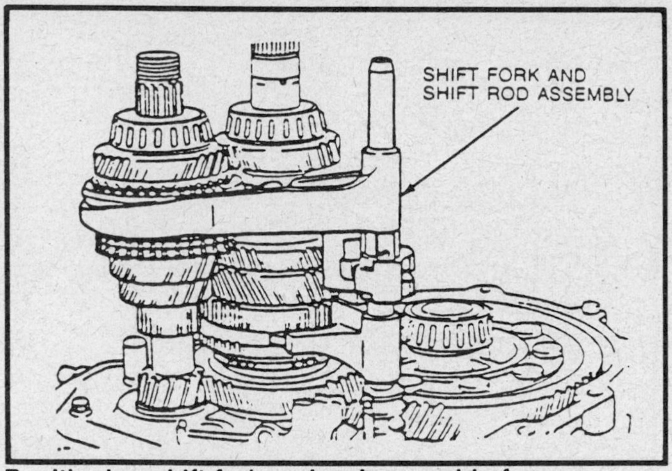
SHIFT FORK AND
SHIFT ROD ASSEMBLY
Positioning shift fork and rod assembly for installation—non-turbocharged

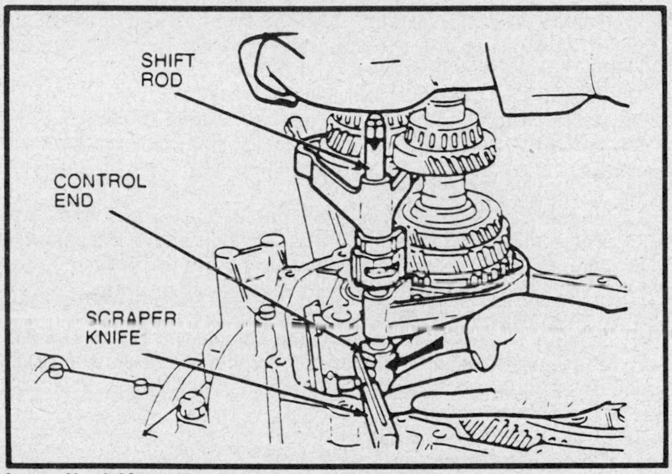
SHIFT
ROD
CONTROL
END
SCRAPER
KNIFE
Install shift rod assembly—non-turbocharged

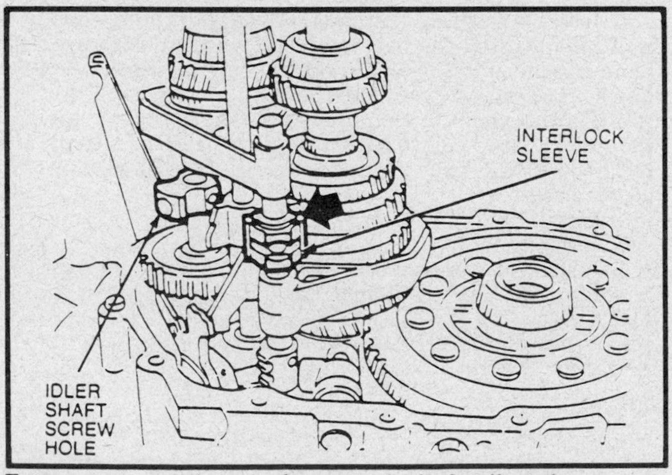
INTERLOCK
SLEEVE
IDLER
SHAFT
SCREW
HOLE
Face reverse idler shaft screw hole in direction shown—non-turbocharged

gauge tool. Measure the gap at 4 spots, at 90 degree intervals. Use the largest measurement.

11. Use the shim selection chart to select the shim(s) that is closest (or slightly larger than) the measured value of the gauge gap. Do not use more than 2 shims.

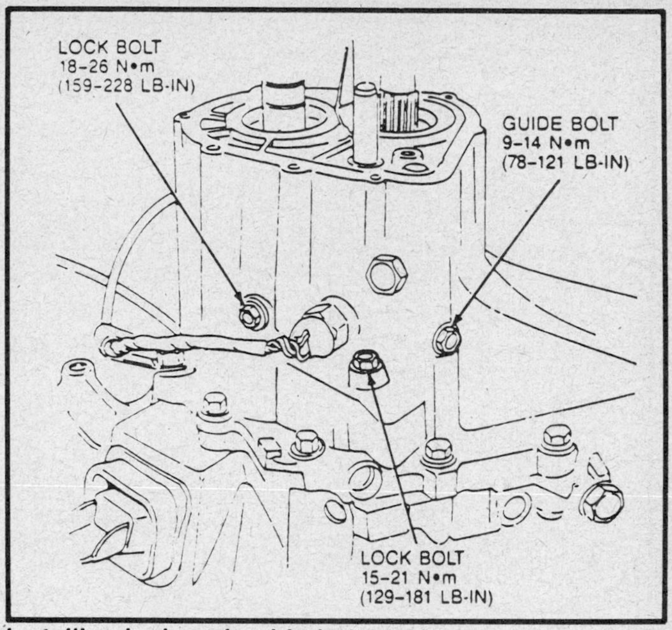
LOCK BOLT
18–26 N•m
(159–228 LB-IN)
GUIDE BOLT
9–14 N•m
(78–121 LB-IN)
LOCK BOLT
15–21 N•m
(129–181 LB-IN)
Installing lock and guide bolt—non-turbocharged

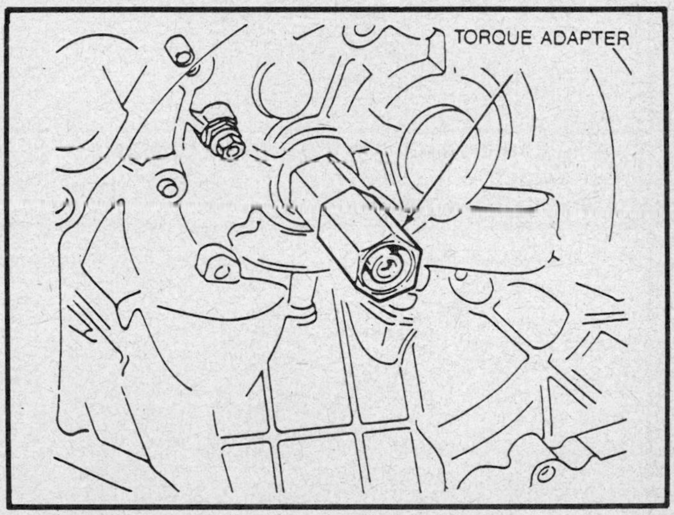
TORQUE ADAPTER
Lock up input shaft with special tool—non-turbocharged

12. Remove the bolts and washers securing the transaxle case to the clutch housing. Remove the transaxle case, collars, gauge, rear bearing cup and differential. Do not remove the bearing cups in the clutch housing.

Transaxle Assembly

TRANSAXLE WITH NON-TURBOCHARGED ENGINE (TYPE G)

1. Install the neutral switch, gasket and backup light switch. Tighten to 14–22 ft. lbs.
2. Install the washer and drain plug and tighten to 29–43 ft. lbs.
3. Install the differential seals using suitable tools.
4. Install the input shaft seal using suitable tools.

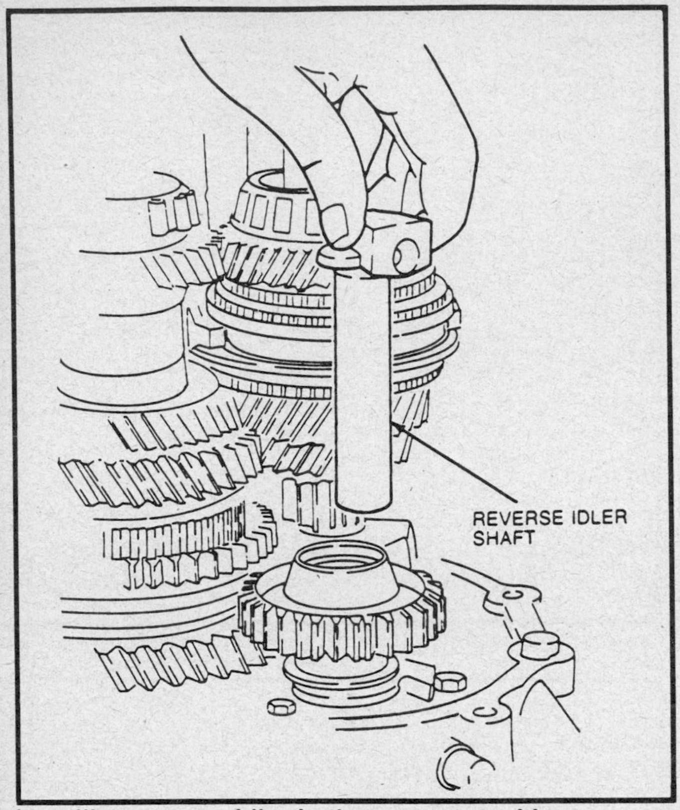
REVERSE IDLER SHAFT

Installing reverse idler in the correct position—turbocharged

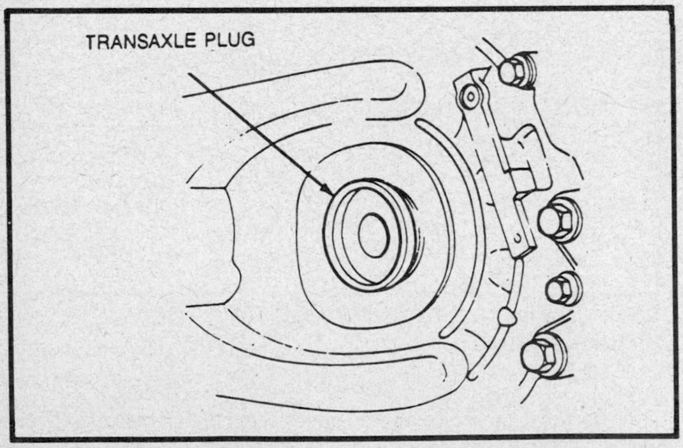
TRANSAXLE PLUG

Installing transaxle plug

NOTE: Whenever the transaxle is assembled, the bearing preload must be adjusted. The input shaft, mainshaft and differential bearing preload can be adjusted by selecting shims to insert between the rear bearing cups and transaxle case. To determine the correct thickness shim, use shim selection information (at the end of this section) and shim selection part sets T87C-77000-J and T88C-77000-CH or equivalent.

5. Install the adjustment shim(s) and rear differential bearing cup to the transaxle case using suitable tools.
6. Install the adjustment shim(s), diaphragm spring and input shaft bearing cup into the transaxle case. Install the diaphragm spring in the correct position.

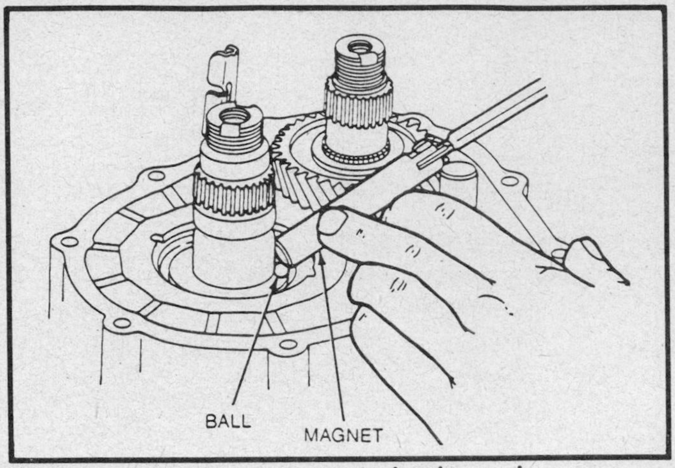
BALL
MAGNET

Installing ball with magnet—turbocharged

7. Install the adjustment shim(s), diaphragm spring and mainshaft bearing cup into the transaxle case. Install the diaphragm spring in the correct position.
8. Install the breather and breather cover.
9. Install the change rod seal using suitable tools.
10. Install the change rod and boot, spring, reverse gate and selector.
11. Install a new roll pin.
12. Install the change arm and tighten to 104–122 inch lbs.
13. Install the guide plate. Tighten the bolt above the spring to 69–100 inch lbs. Tighten the remaining bolts to 16–25 inch lbs.
14. Install the reverse lever and reverse lever shaft.
15. Install a new roll pin.
16. Install the speedometer driven gear assembly to the case on vehicles with analog cluster or vehicle speed sensor to the case on vehicles with digital cluster.
17. Install the magnet.
18. Install the differential.
19. Install the input and mainshaft.
20. Shift to 2nd gear and 4th gear and correctly position the shift fork and shift rod assembly.
21. Insert the spring seat and springs into the reverse lever shaft, then install the ball and hold it in place with a scraper knife or equivalent. Push the control end in a clockwise direction so that the ball goes into the shaft.
22. Position each clutch hub sleeve to the **N** positon, then tap the shift rod from above so that the steel ball goes into the center groove. Swivel the control end until the ball goes into the groove detent.
23. Install the crank lever shaft and assembly.
24. Insert the pin.
25. Install the gate and 5th/reverse shift rod. Tighten the gate mounting bolt. Be sure to align the shift rod and gate mounting hole.
26. Install the reverse idler gear and shft.
27. Align the end of the interlock sleeve with the control lever. Position the reverse idler shaft screw hole.
28. Clean the contact surfaces on the clutch housing and transaxle case, then apply a thin coat of sealant.
29. Place the transaxle case on the clutch housing and tighten the attaching bolts to 27–38 ft. lbs.
30. Install 2 transaxle plugs or equivalent, between the differential side gears. Failure to install the transaxle plugs may allow the differential side gears to become mispositioned.
31. Install the lock bolt and tighten to 159–228 inch lbs.
32. Install the guide bolt and tighten to 78–121 inch lbs.
33. Install the ball, springs and lock bolt and tighten to 129–181 inch lbs.

34. Install the main 5th gear.
35. Install the gear sleeve.
36. Install the input 5th gear.
37. Install the synchronizer ring.
38. Install the shift fork, clutch hub assembly and synchronizer ring.
39. Install the roll pin.
40. Install the main reverse synchronizer gear.
41. Install the input reverse synchronizer gear.
42. Shift to 1st gear. Use torque adapter tool T88C–7025–GH or equivalent to lock up the input shaft.
43. Install the locknuts and tighten to 94–145 ft. lbs. Stake the locknuts.
44. Clean the contact surfaces on the transaxle case and rear cover, than apply a thin coat of sealant.
45. Install the rear cover and tighten to 68–95 inch lbs.

TRANSAXLE WITH TURBOCHARGED ENGINE (TYPE H)

1. Install the breather and breather cover.
2. Install the oil passage and tighten to 69–100 inch lbs.
3. Install the differential oil seals using suitable tools.
4. Install the input shaft seal using suitable tools.

NOTE: Whenever the transaxle is assembled, the bearing preload must be adjusted. The input shaft, mainshaft and differential bearing preload can be adjusted by selecting shims to insert between the rear bearing cups and transaxle case. To determine the correct thickness shim, use shim selection information (at the end of this section) and selection part sets T87C–77000–J and T88C–77000–CH or equivalent.

5. Install the adjustment shim(s) and rear differential bearing cup into the transaxle case using suitable tools.
6. Install the adjustment shim(s), diaphragm spring and input shaft bearing cup into the transaxle case. Install the diaphragm spring in the correct position.
7. Install the adjustment shim(s), diaphragm spring and mainshaft bearing cup into the transaxle case. Install the diaphragm spring in the correct position.
8. Install the change arm oil seal using suitable tools.
9. Install the change rod and change arm.
10. Install the baffle plate and tighten to 87–113 inch lbs.
11. Install a new roll pin into the change arm. The roll pin length is 1.10 inch or 28mm.
12. Install the differential assembly.
13. Install the main and input shaft gear assemblies.
14. Install the interlock pin to the 5th/reverse shift rod.
15. Install the shift rod end and the 5th/reverse shift rod.
16. Install the 1st/2nd shift rod and shift fork.
17. Shift to 2nd gear and install the roll pin.
18. Install the lever set spring and the reverse shift lever. Tighten to 69–95 inch lbs.
19. Install the crank lever assembly and crank lever shaft into the base plate and install the snap ring.
20. Install the base plate assembly.
21. Install the reverse idler gear.
22. Install the reverse idler shaft in the correct position.
23. Install the magnet.
24. Clean the contact surfaces on the clutch housing and transaxle case, then apply a thin coat of sealent.
25. Place the transaxle case on the clutch housing and tighten the attaching bolts to 27–38 ft. lbs.
26. Install 2 transaxle plugs or equivalent, between the differential side gears. Failure to install the transaxle plugs may allow the differential side gears to become mispositioned.
27. Install the right transaxle mount and tighten bolts to 58–86 ft. lbs.
28. Install the reverse idler shaft set bolt and tighten to 13–19 ft. lbs.
29. Install the balls, springs and bolts. Tighten to 14–22 ft. lbs.
30. Install the backup light switch and tighten to 14–22 ft. lbs.
31. Install the interlock pins.
32. Install the interlock plate and tighten to 13–19 ft. lbs.
33. Install the neutral switch and tighten to 14–22 ft. lbs.
34. Install the washer and drain plug and tighten to 29–43 ft. lbs.
35. Install the speedometer driven gear assembly to the case on vehicles with analog cluster or vehicle speed sensor to the case on vehicles with digital cluster.
36. Install the main 5th gear.
37. Install the ball.
38. Install the sleeve.
39. Lubricate the needle beaering assembly with the specified fluid and install the needle bearing.
40. Install the input 5th gear and synchronizer ring.
41. Install the shift fork, clutch hub assembly and synchronizer ring.
42. Install the main reverse synchronizer gear.
43. Install the input reverse synchronizer gear.
44. Install the sleeve.
45. Shift to 5th gear and then to 1st gear.
46. Install the locknuts and tighten to 94–145 ft. lbs. Stake the locknuts.
47. Install the roll pin.
48. Clean the contct surfaces on the transaxle case and rear cover, then apply a thin coat of sealant. Install the rear cover and tighten to 68–95 inch lbs.

SPECIFICATIONS

TORQUE SPECIFICATIONS

	ft. lbs.	Nm
Gearshift housing assembly	60–84 ①	7–10
Extension bar to transaxle	23–34	31–46
Shift control rod	60–84 ①	7–10
Transaxle case to clutch housing (non-turbo)	13–14	18–20
Transaxle case to clutch housing (turbo)	27–38	37–52
Transaxle to engine	66–86	89–17
Center transaxle mount bolts	27–40	36–54
Center transaxle mount nuts	47–66	64–69
Transaxle to left mount (non-turbo)	27–38	37–52
Transaxle to left mount (turbo)	49–69	67–93
Left mount to bracket	49–69	67–93
Crossmember bolts	27–40	36–54
Crossmember nuts	55–69	75–93

	ft. lbs.	Nm
Right transaxle mount	63–86	85–117
Flywheel inspection cover	69–95 ①	8–11
Slave cylinder	14–19	19–26
Gusset plate to transaxle	27–38	37–52
Neutral switch	14–22	20–29
Backup lamp switch	14–22	20–29
Change arm (non-turbo)	104–122 ①	12–14
Rear cover	68–95 ①	8–11
Reverse shift lever (turbo)	69–95 ①	8–11
Reverse idler shaft (turbo)	13–19	18–26
Interlock plate	13–19	18–26
Locknuts (input and main shaft)	94–145	128–196

① Inch lbs.

SPECIAL TOOLS

Tool Number	Description
T87C-77000-H	Differential seal replacer
T88C-7025-AH	Transaxle plug set
T88C-77000-CH	Shim selection set
T88C-77000-EH	Differential bearing cone replacer
T88C-77000-FH	Differential bearing cup replacer
T88C-7025-BH	Differential bearing cup installer
T88C-7025-CH	Differential bearing cone replacer
T88C-7025-DH	Bearing cone replacer
T88C-7025-EH	Pilot bearing installer
T88C-7025-FH	Input shaft seal replacer
T88C-77000-CH2	Shim selection gauge
T88C-7025-GH	Torque adapter
T88C-77000-CH3	Shim selection gauge
D80L-100-A	Blind hole puller set
D80L-522-A	Gear and pulley puller
D80L-625-3	Shaft protector
D80L-625-4	Shaft protector
D80L-625-6	Shaft protector
D80L-630-3	Step plate
D80L-630-4	Step plate
D80L-927-A	Push puller set
D84L-1122-A	Bearing pulling attachment
014-00210	Hi-Lift transmission jack

Tool Number	Description
D84L-1123-A	Bearing pulling attachment
D87L-6000-A	Engine support bar
T50T-100-A	Slide hammer
T53T-4621-B	Bearing cone replacer
T57L-500-B	Bench mounting fixture
T57L-4621-B	Bearing cone replacer
T58L-101-B	Puller
T71P-4621-B	Puller plate
T73L-2196-A	Disc brake piston remover
T74P-7137-K	Clutch aligner
T75L-1165-B	Bearing installation plate
T75L-1165-DA	Bearing cone replacer
T77F-1102-A	Puller
T77F-4220-B1	Puller
T77J-7025-G	Installer
T78P-3504-N	Locknut pin remover
T80T-4000-W	Handle driver
T86P-70043-A	Puller jaws
T87C-77000-D	Bearing cone replacer
T87C-77000-J	Shim selection set
T88T-7025-B	Bearing cone replacer
TOOL-4201-C	Dial indicator tool
T77J-7025-B	Locknut staking tool

Section 6

Tracer MTX4 and MTX5 Transaxles
Ford Motor Co.

APPLICATION

1987–89 Mercury Tracer 1.6L engine

GENERAL DESCRIPTION

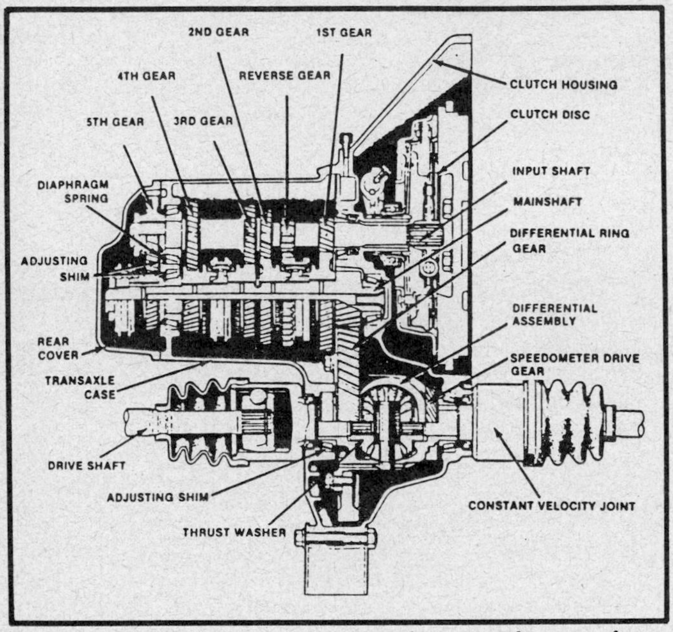
2ND GEAR
1ST GEAR
4TH GEAR
REVERSE GEAR
CLUTCH HOUSING
5TH GEAR
3RD GEAR
CLUTCH DISC
INPUT SHAFT
DIAPHRAGM SPRING
MAINSHAFT
DIFFERENTIAL RING GEAR
ADJUSTING SHIM
DIFFERENTIAL ASSEMBLY
REAR COVER
SPEEDOMETER DRIVE GEAR
TRANSAXLE CASE
DRIVE SHAFT
ADJUSTING SHIM
CONSTANT VELOCITY JOINT
THRUST WASHER

Cross sectional view of the 4 and 5 speed manual transaxle

The Mercury Tracer 4 and 5 speed manual transaxle's are similar in design, basically the major difference is the addition of a 5th gear, shift fork and rear cover, providing a 5th gear range feature for the 5 speed. Both incorporate helical cut gears in all forward gear ranges for quiet operation. All forward gears are synchronized for ease of shifting. For additional shift quality and precise operation a single rail gearshift mechanism has also been incorporated.

Metric Fasteners

Care must be exercised during the disassembly and assembly of the manual transaxle due to the usage of metric nuts and bolts. Proper wrenches and sockets should be used to avoid damage to the transmission and fasteners. Do not attempt to interchange metric fasteners for inch system fasteners. Mismatched or incorrect fasteners can result in damage to the transaxle unit through malfunctions, breakage or possible personal injury.

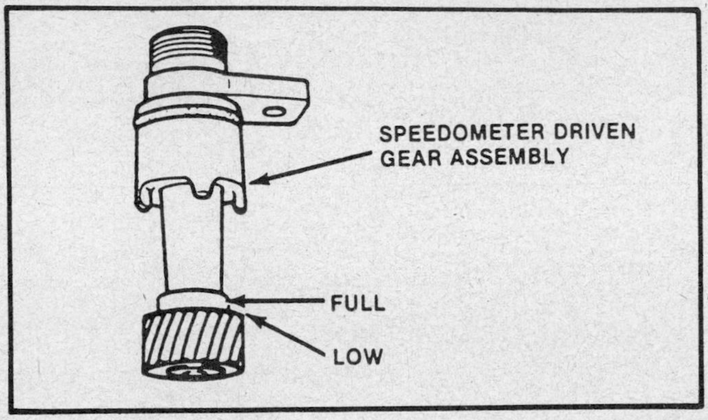
SPEEDOMETER DRIVEN GEAR ASSEMBLY
FULL
LOW

Transaxle fluid check

Capacities

The transaxle fluids used in both the 4 and 5 speed transaxles are Motorcraft, Dexron®II and Ford Motor Co. ESW-M2C-138-CJ or equivalent. These fluid types are recommended to ensure the ease of gear shifting as well as improved fuel economy. The same fluid types are used in the differential. Vehicles built before January 7, 1987 require the use of type F automatic transmission fluid instead of Dexron®II. The fluid capacity is 3.4 qts. (2.8 Liters).

Incorrect use or mixing of the 2 different transmission fluids may cause the synchronizer rings to wear and/or result in poor manual transaxle durability.

Checking Fluid Level

Procedure

1. Remove the speedometer cable from the transaxle, and check the level as indicated on the speedometer driven gear.
2. Place a funnel into the speedometer driven gear mounting hole.
3. Add fluid to the level indicated on the speedometer driven gear.
4. Install the speedometer cable into the speedometer mounting hole in the transaxle case.

TRANSAXLE MODIFICATIONS

There were no modifications for the Mercury Tracer manual transaxle at the time of this publication.

TROUBLE DIAGNOSIS

CHILTON'S THREE C'S DIAGNOSIS

Condition	Cause	Correction
Shift lever does not operate smoothly or binds or cannot be operated at all	a) Selector rod joint stiff b) Selector rod bent c) Lack of lubrication on shift linkage pivots d) Shift lever ball unit stiff e) Gearshift gate incorrectly adjusted	a) Repair or replace rod joint b) Replace bent rod c) Clean and lubricate with Molybdenum Disulphide grease d) Repair or replace shift lever ball e) Adjust gate
Excessive shift lever play	a) Selector rod bushing worn b) Loose or worn selector rod clamping bolts c) The spring in the shift lever ball unit is fatigued d) The bushing in shift lever ball unit worn	a) Replace selector rod bushing b) Tighten or replace as necessary c) Replace spring d) Replace bushing
Hard shifting	a) Insufficient oil in transaxle b) Incorrect oil quality c) Selector rod bent d) Transmission shifting mechanism insufficiently lubricated e) Excessive clutch pedal free play f) Shift fork and shift rod worn g) Synchronizer ring worn h) Worn cone surface of gear i) Improper contact between synchronizer ring and cone surface j) Excessive play in the axial direction of each gear k) Bearings worn l) Synchronizer key is fatigued	a) Add oil b) Drain and refill with specified oil c) Replace selector rod d) Lubricate shifting mechanism e) Adjust clutch freeplay f) Replace shift fork and rod g) Replace synchronizer ring h) Replace gear i) Replace synchronizer & cone j) Replace worn component k) Adjust or replace l) Replace synchronizer key
Noise	a) Insufficient oil in transaxle b) Poor oil quality c) Worn sliding surfaces at synchronizer d) Excessive backlash e) Surface of a gear is damaged f) Foreign matter in transmission g) Differential gear is damaged. Backlash is excessive h) Ring gear bolts are loose i) Bearings worn or out of adjustment	a) Add oil b) Drain and refill with specified oil c) Repair or replace synchronizer d) Repair or adjust as necessary e) Replace gear f) Repair or replace as necessary g) Repair or replace as necessary h) Repair or replace as necessary i) Repair or adjust as necessary
Gear clash	a) Excessive engine idle speed b) Inadequate clutch pedal reserve resulting excessive spin time. Inadequate clutch disengagement c) Disc binding on transmission input shaft d) Excessive disc runout e) Flywheel housing misalignment	a) Adjust engine idle rpm b) Check clutch adjustment, operating mechanism or for excessive clutch disc runout—replace parts as required c) Check for burrs on splines, replace if necessary d) Replace disc e) Realign flywheel housing

CHILTON'S THREE C'S DIAGNOSIS

Condition	Cause	Correction
Gear clash	f) Oil or grease on clutch facings	f) Replace disc and correct cause of contamination
	g) Damaged or contaminated clutch lining	g) Replace disc
	h) Weak or broken insert keys in the synchronizer assembly	h) Replace parts as required
	i) Worn synchronizer rings and/or cone surfaces	i) Replace parts as required
	j) Broken synchronizer rings	j) Replace synchronizer rings
Locked in gear	a) Shift gate out of adjustment or worn	a) Repair or replace as necessary and adjust
	b) Worn interlock sleeve or bent or damaged shift fork	b) Check interlock sleeve for wear and repair or replace as necessary
	c) Gear seizure	c) Replace defective parts
	d) Synchronizer keys out of position	d) Repair or replace as necessary
Jumping out of gear	a) Worn or improperly installed engine mount	a) Repair or replace engine mount
	b) Loose or worn control rod clamping bolts or linkage	b) Repair or replace and tighten as necessary
	c) Bent shift control rod	c) Replace shift control rod
	d) Worn shift control rod bushing	d) Replace bushing
	e) Fatigued lever ball spring	e) Replace spring
	f) Improper installation of stabilizer bar	f) Fit correctly and tighten
	g) Worn synchronizer clutch hub	g) Replace clutch hub
	h) Worn synchronizer clutch hub sleeve	h) Replace clutch hub sleeve
	i) Worn steel ball sliding groove on control rod end	i) Replace components as necessary
	j) Fatigued steel ball spring	j) Replace spring
	k) Excessive backlash	k) Check and repair as necessary
	l) Worn bearings	l) Adjust or replace bearings

POWER FLOW

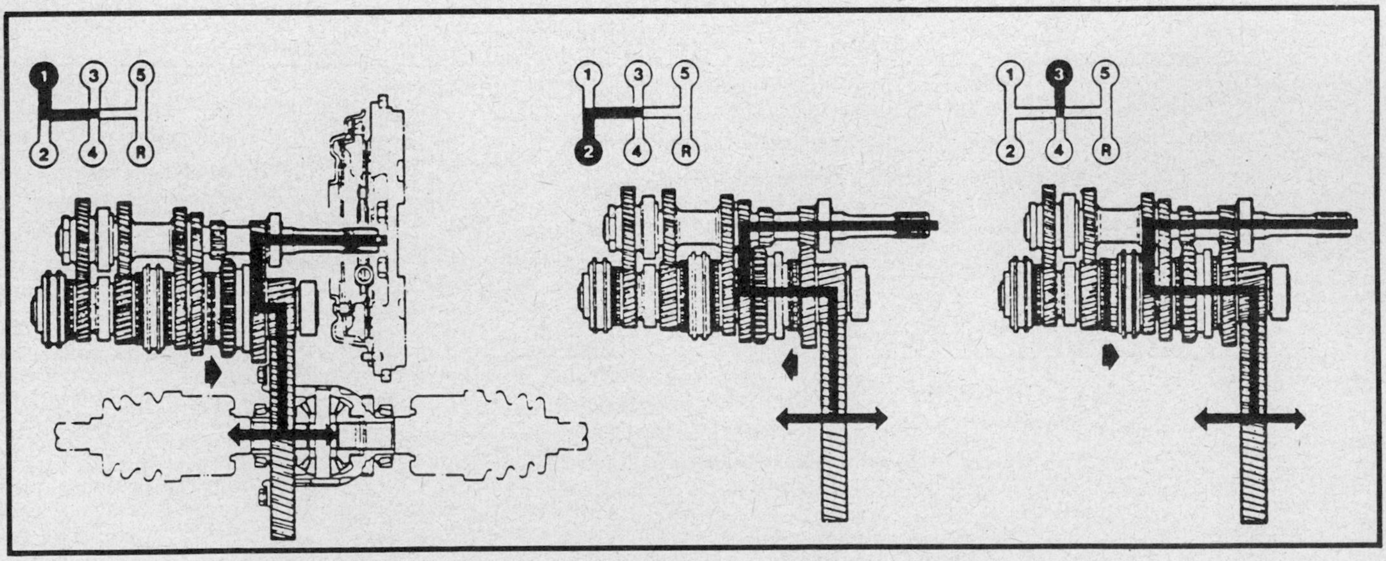
1 3 5
2 4 R
1 3 5
2 4 R
1 3 5
2 4 R

Power flow

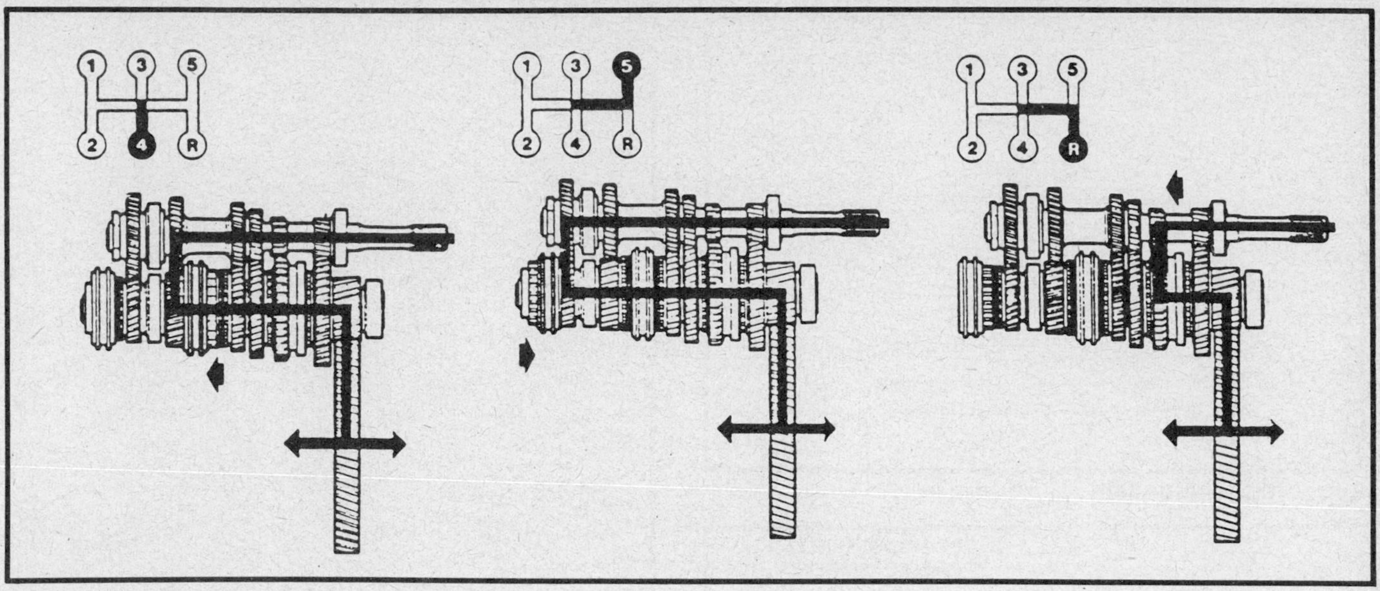
1
3
5
2
4
R
Power flow

REMOVAL AND INSTALLATION

TRANSAXLE REMOVAL

1. Disconnect the negative battery cable.
2. Remove the air cleaner. Loosen the front wheel lug nuts.
3. Disconnect the speedometer cable from the transaxle.
4. Remove the adjustment nut, pin and clutch cable from the clutch release lever.
5. Remove the clutch cable bracket-to-transaxle bolts and the bracket.
6. Remove the ground wire bolt and ground wire.
7. Remove the coolant pipe bracket bolt and the bracket.
8. Remove the secondary air pipe, the EGR pipe bracket and the electrical harness clip.
9. Disconnect the neutral safety switch/back-up light switch coupler and the body ground connector.
10. Remove the upper 2 transaxle-to-engine bolts.
11. Using the engine support bar tool D79P–6000–B or equivalent attach it to the rear engine lifting hook and support the engine's weight.
12. Raise and support the front of the vehicle safely.
13. Place a drain pan under the transaxle, remove the drain plug and drain the transaxle.
14. Remove the tire and wheel assembly. Remove the engine undercover and side covers.
15. Remove the front stabilizer bar from both sides, remove the lower control arm ball joint-to-steering knuckle nut/bolt, pull the control arm downward and separate the lower control arm from the steering knuckle.

NOTE: When separating the ball joint, be careful not to damage the ball joint seal cup.

16. Using both hands, grasp the steering knuckle/hub assembly and apply even pressure (gradually increasing), pull both halfshafts from the transaxle.

NOTE: When removing the halfshafts, withdraw them completely from the transaxle (to prevent damage to the oil seal lips). Do not move the CV-joints in excess of a 20 degree angle (damage to the booths and/joint may occur) and use a wire to support the halfshaft in the horizontal position.

17. From under the vehicle, remove the crossmember-to-chassis bolts and the crossmember.
18. Remove the shift control rod-to-transaxle nut/bolt and slide the control rod aside. Remove the shift extension bar-to-bracket bolt and slide the extension bar off the bracket.
19. Remove the starter's positive cable-to-solenoid nut and the solenoid wire.
20. Remove the starter-to-engine bolts and the starter bolts. Remove the dust cover-to-clutch housing bolts and the cover.
21. Loosen the bracket bar on the engine support tool to lower the transaxle. Using a floor jack, support the transaxle.
22. Remove the No. 2 engine mounting-to-transaxle nut/bolt, the transaxle-to-engine bolts and lower the transaxle from the vehicle.

TRANSAXLE INSTALLATION

1. Apply a thin coating of clutch grease to the spline of the input shaft.
2. Install the transaxle assembly by carefully aligning the input shaft through the clutch housing onto the engine guide dowel pins.

NOTE: The transaxle aluminum alloy construction requires that the torque specifications must be strictly adhered too.

3. Torque the tranaxle-to-engine bolts to 47–66 ft. lbs., the No. 2 engine mount-to-transaxle nut/bolt to 27–34 ft. lbs., the extension bar-to-transaxle bracket bolt to 23–34 ft. lbs., the control rod-to-transaxle nut/bolt to 12 ft. lbs., the crossmember-to-chassis bolts to 47–66 ft. lbs., the rear engine mount-to-crossmember nut to 20–34 ft. lbs.
4. Refill the transaxle with the proper fluid.

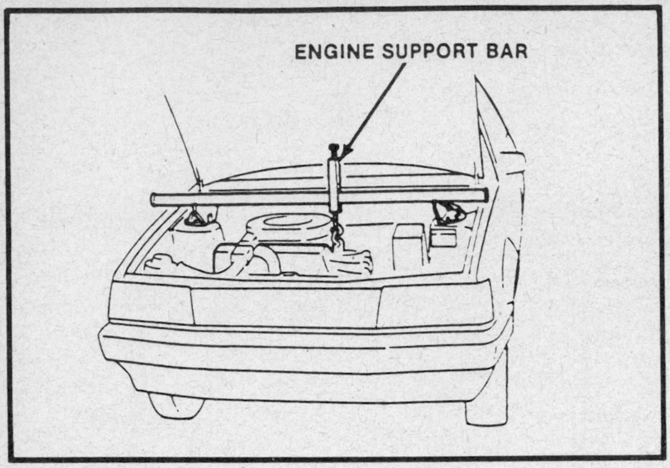
ENGINE SUPPORT BAR

Supporting the engine

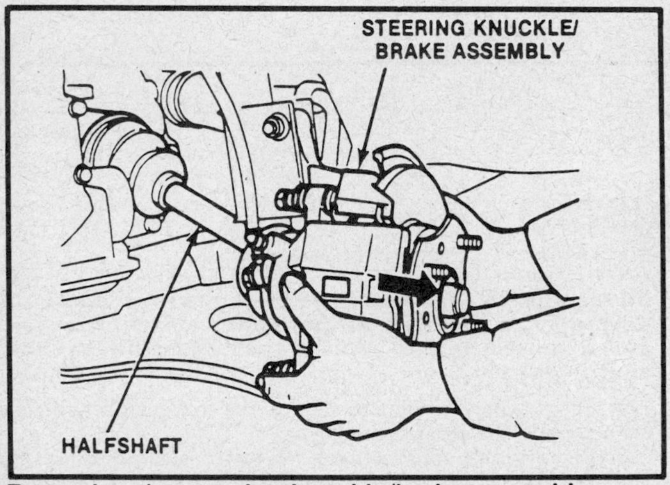
STEERING KNUCKLE/
BRAKE ASSEMBLY
HALFSHAFT

Removing the steering knuckle/brake assembly

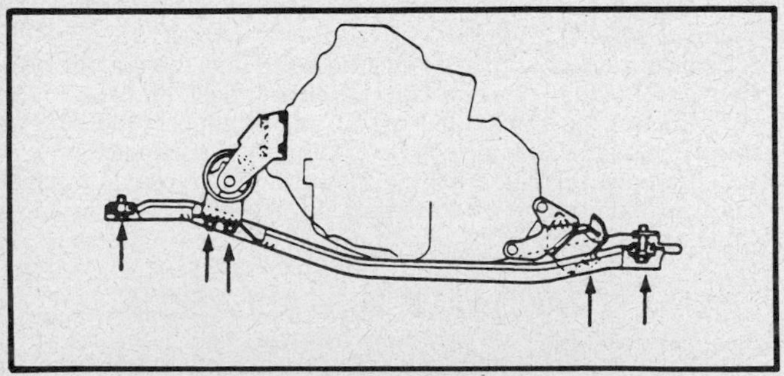

Removing the crossmember bolts

5. To install the halftshaft into the transaxle, perform the following procedures.

a. Install a new locking clip on the halfshaft spline. Be sure the gap in the clip is at the top of the clip groove.

b. Slide the halfshafts into the transaxle bore; be careful not to damage the oil seal lip.

c. Push firmly on the hub assembly, making sure the circlip snaps into place.

d. After installation, pull the front hub outward to confirm that the circlips are engaged.

6. Install the lower ball joints to the steering knuckle, install the attaching nut and tighten to 32–40 ft. lbs.

7. Install the stabilizer bar mounting brackets to the vehicle frame and tighten to 23–33 ft. lbs.

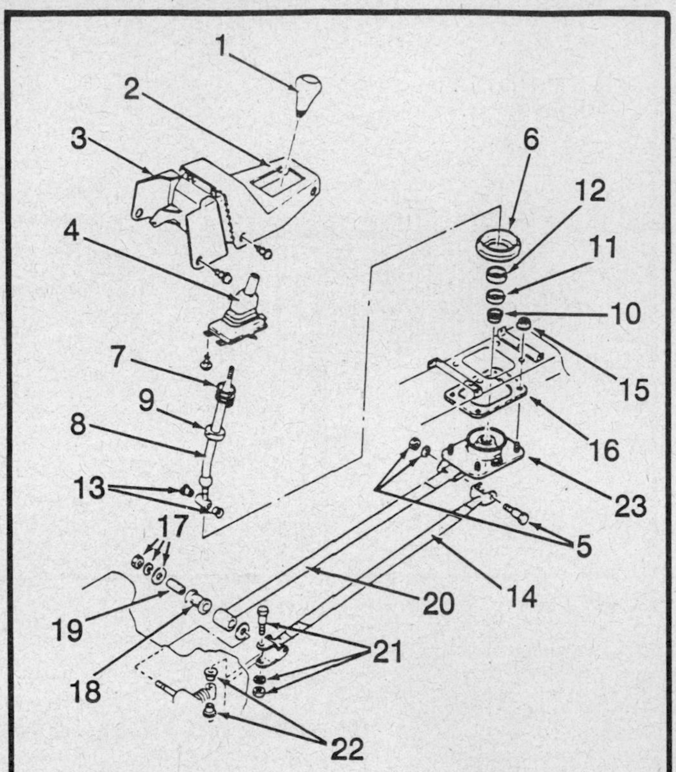
1
2
3
4
5
6
7
8
9
10
11
12
13
14
15
16
17
18
19
20
21
22
23

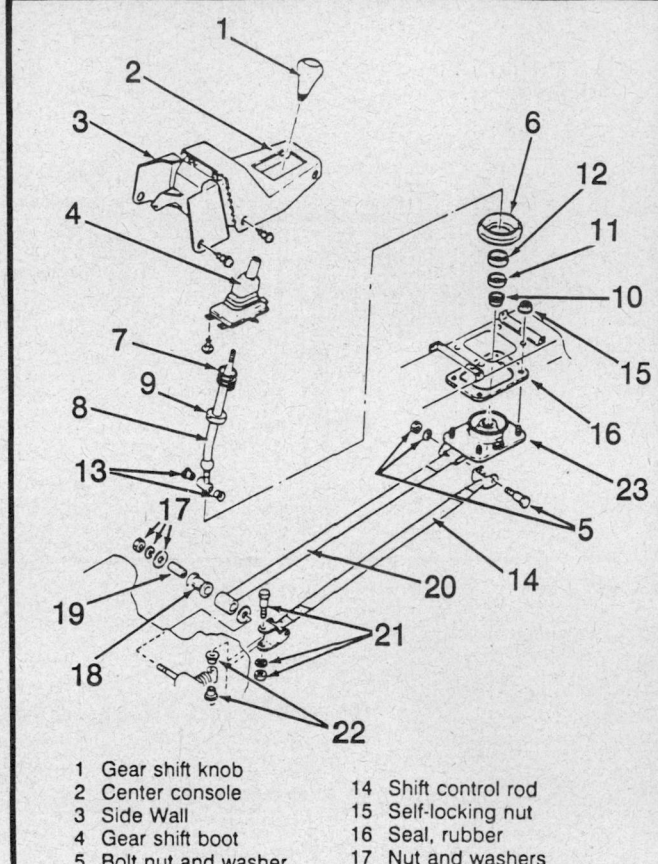

1 Gear shift knob
2 Center console
3 Side Wall
4 Gear shift boot
5 Bolt nut and washer
6 Mounting rubber
7 Shifter shaft spring
8 Gear shift lever
9 Ball seat (upper)
10 Boot, ball socket
11 Retainer
12 Ball seat (lower)
13 Bushing
14 Shift control rod
15 Self-locking nut
16 Seal, rubber
17 Nut and washers
18 Bushing, control rod-to-transaxle
19 Spacer, control rod
20 Extention Bar
21 Bolt and washer
22 Bushings shift control rod-to-transaxle
23 Housing assembly

Exploded view of the gear shift and rod assembly

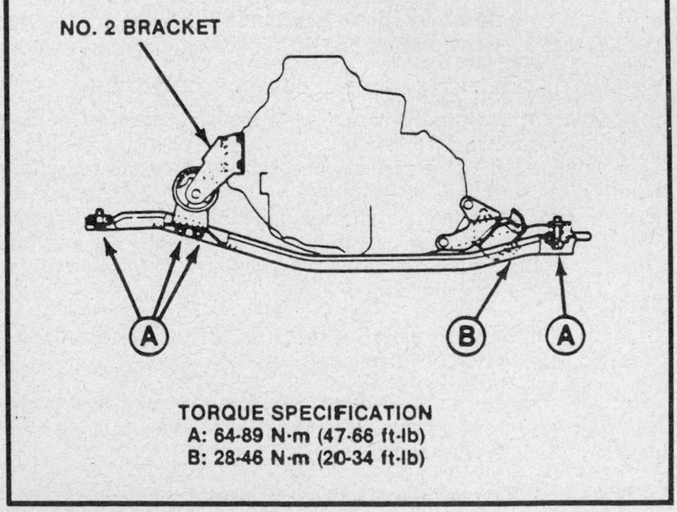
NO. 2 BRACKET
A
B
A
TORQUE SPECIFICATION
A: 64-89 N·m (47-66 ft-lb)
B: 28-46 N·m (20-34 ft-lb)

Crossmember bolt torque

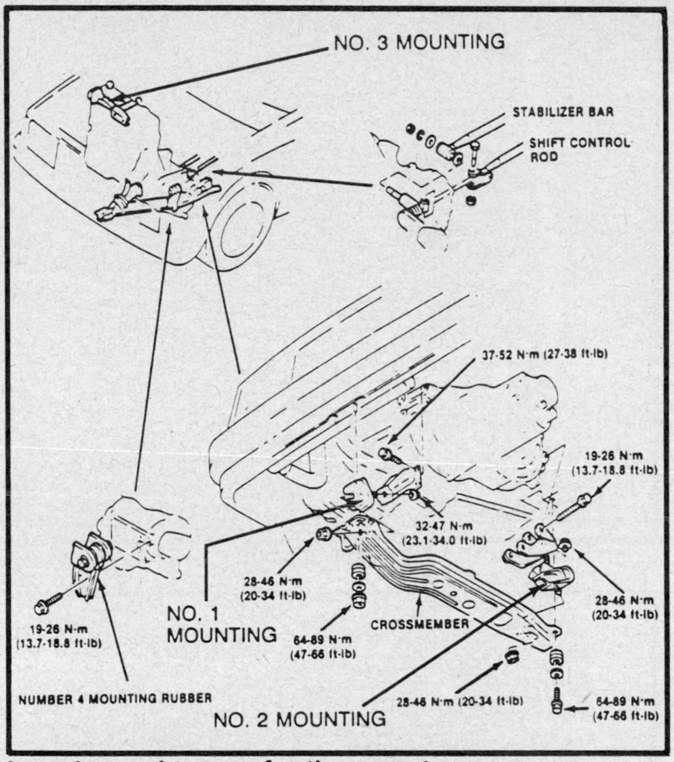
NO. 3 MOUNTING
STABILIZER BAR
SHIFT CONTROL ROD
37-52 N·m (27-38 ft-lb)
19-26 N·m (13.7-18.8 ft-lb)
32-47 N·m (23.1-34.0 ft-lb)
28-46 N·m (20-34 ft-lb)
28-46 N·m (20-34 ft-lb)
NO. 1 MOUNTING
19-26 N·m (13.7-18.8 ft-lb)
CROSSMEMBER
64-89 N·m (47-66 ft-lb)
NUMBER 4 MOUNTING RUBBER
NO. 2 MOUNTING
28-46 N·m (20-34 ft-lb)
64-89 N·m (47-66 ft-lb)

Location and torque for the mounts

8. Assemble the front stabilizer link by inserting the bolt through the bushings, washers and the spacer. Install the nuts and tighten to 9–13 ft. lbs. Tighten the nuts further, as necessary, until the threads exposed on the stabilizer link bolt past the nut are 0.43 in. (10.8mm) in length. Lock the nuts against each other.
9. Install the undercover and side covers.
10. Install the front wheels and hand tighten the lug nuts.

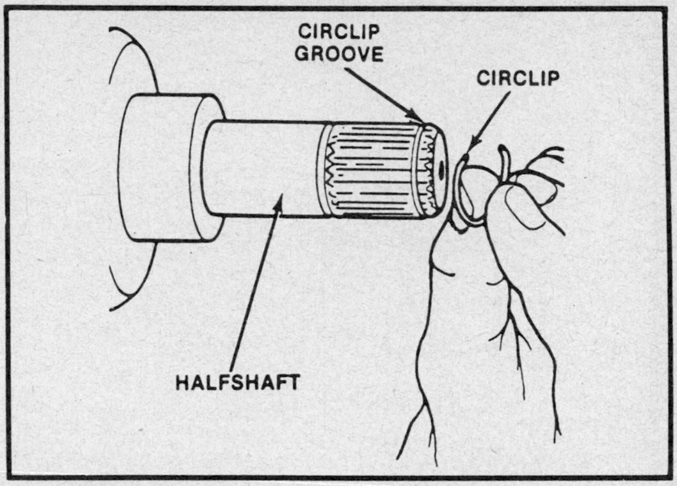
CIRCLIP GROOVE
CIRCLIP
HALFSHAFT

Installation of the circlip on the CV-joint

11. Lower the vehicle and tighten the front wheel lug nuts to 65–87 ft. lbs.
12. Install the 2 upper transxle-to-engine mounting bolts and tighten the bolts to 47–66 ft. lbs.
13. Remove the engine support bracket bar.
14. Connect the body ground connector.
15. Connect the coupler for the neutral switch and the back-up lamp.
16. Connect the wire harness clip.
17. Connect the secondary air pipe and the EGR pipe bracket.
18. Connect the coolant pipe bracket.
19. Install the ground wire and installation bolt and tighten the bolt.
20. Install the clutch cable mounting bracket to the transaxle.
21. Install the clutch cable pin and adjusting nut to the release lever and adjust the clutch pedal free play and pedal height.
22. Install the speedometer cable into the transaxle.
23. Connect the negative battery cable.

BENCH OVERHAUL

Before Disassembly

When servicing the unit, it is recommended that as each part is disassembled, it is cleaned in solvent and dried with compressed air. Disassembly and reassembly of this unit and its parts must be done on a clean work bench. Also, before installing bolts into aluminum parts, always dip the threads into clean transmission oil. Anti-seize compound can also be used to prevent bolts from galling the aluminum and seizing. Always use a torque wrench to keep from stripping the threads. Take care with the seals when installing them, especially the smaller O-rings. The slightest damage can cause leaks. Aluminum parts are very susceptible to damage so great care should be exercised when handling them. The internal snaprings should be expanded and the external snaprings compressed if they are to be reused. This will help insure proper seating when installed. Be sure to replace any O-ring, gasket, or seal that is removed.

Transaxle Disassembly

Disassembly

1. Mount the transaxle in a suitable holding fixture.
2. Remove the drain plug and drain any remaining fluid from the transaxle.
3. Shift to 1st or 2nd. Position the transaxle with the input shaft down, rear cover up.
4. Remove the bolts that secure the rear cover to the tranaxle case. Tap the cover with a fiber or plastic mallet to loosen the gasket seal. Remove the cover.
5. Bend down the tang on the lock washer under the fifth gear locknut.
6. Lock the input shaft with the input torque adapter tool T87C–7025–A or equivalent and remove the locknut. Apply even pressure and increase gradually. Do not strike or apply severe shocks to loosen the nut.
7. Drive out the roll pin and remove the shift fork (5th) together with the clutch hub assembly on 5 speed transaxles.
8. Remove the synchronizer ring, 5th gear and the gear sleeve as an assembly by sliding it off the shift rod.
9. Once again lock the input shaft with input torque adapter tool T87C–7025–A or equivalent. Remove the locknut on the input shaft gear and remove the input gear by sliding it off.
10. Remove the lock bolt, guide bolt and back-up light switch from the side of the transaxle case. Remove all of the transaxle

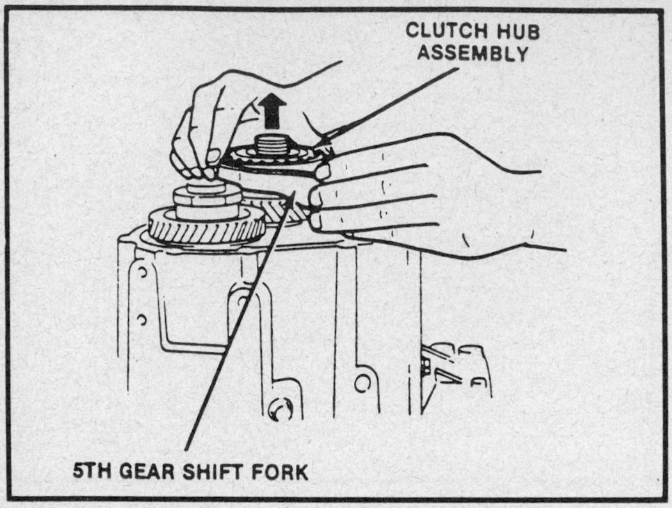
CLUTCH HUB
ASSEMBLY
5TH GEAR SHIFT FORK

Removing the shift fork and clutch assembly (5 speed only)

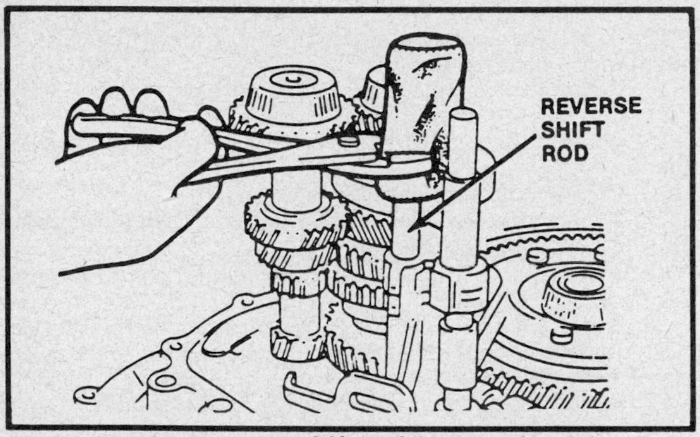
REVERSE
SHIFT
ROD

Removing the reverse shift rod (4 speed only)

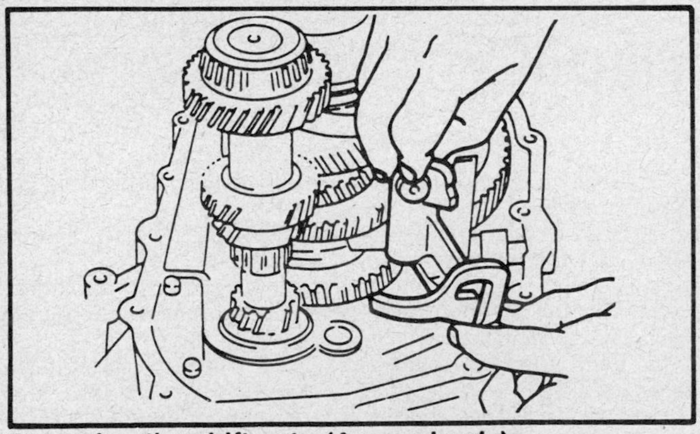

Removing the shift gate (4 speed only)

housing-to-clutch housing bolts. Number the bolts as they are removed.

11. Tap the transaxle case lightly with a plastic or fiber mallet to loosen the gasket seal. Remove the case by sliding it straight up from the clutch housing.

12. On 5 speed unit only; insert a pin punch or suitable rod into the roll pin hole of the shift rod. Pull out the shift rod while turning the pin punch or the rod.

13. On 4 speed unit only; to remove the reverse shift rod use a

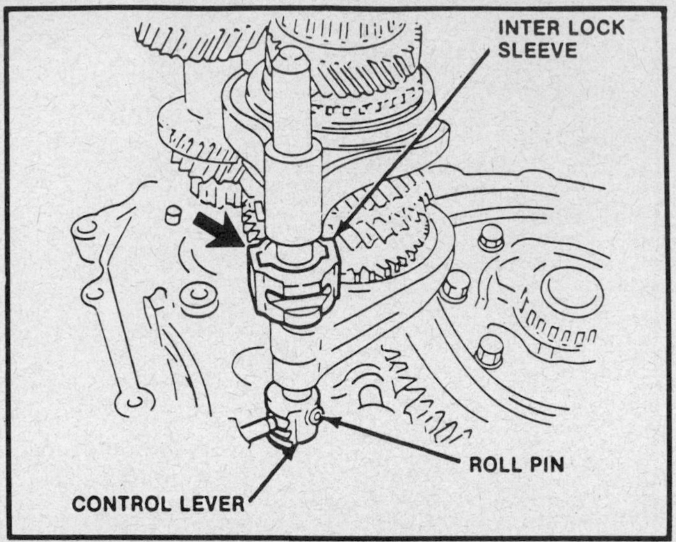
INTER LOCK
SLEEVE
ROLL PIN
CONTROL LEVER

Positioning of the interlock sleeve

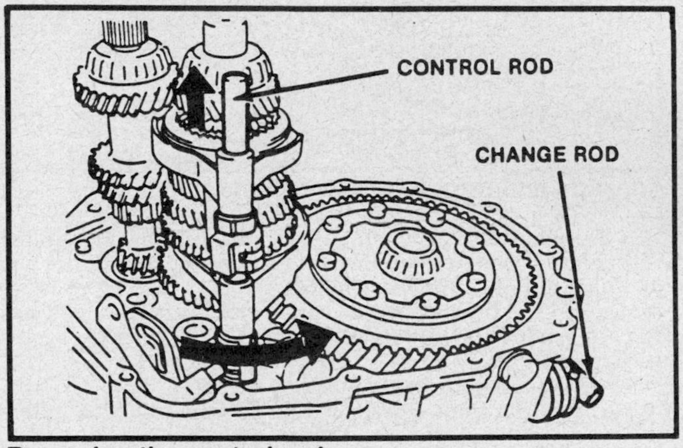
CONTROL ROD
CHANGE ROD

Removing the control rod

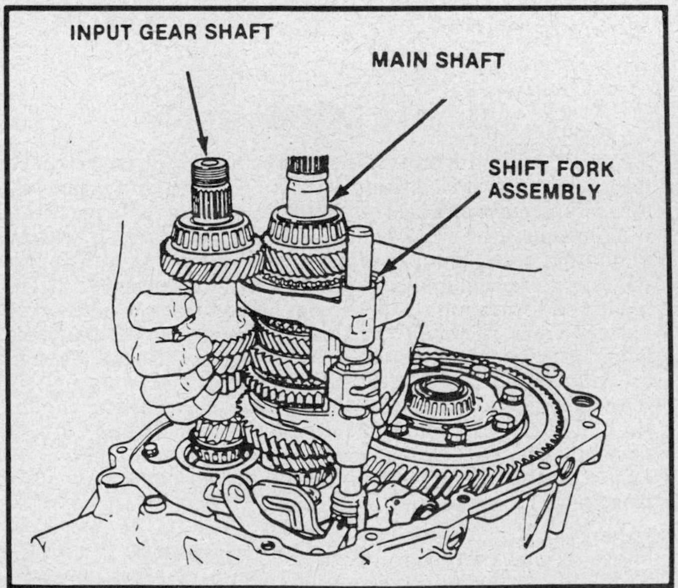
INPUT GEAR SHAFT
MAIN SHAFT
SHIFT FORK
ASSEMBLY

Removing the input shaft, mainshaft and shift fork assembly

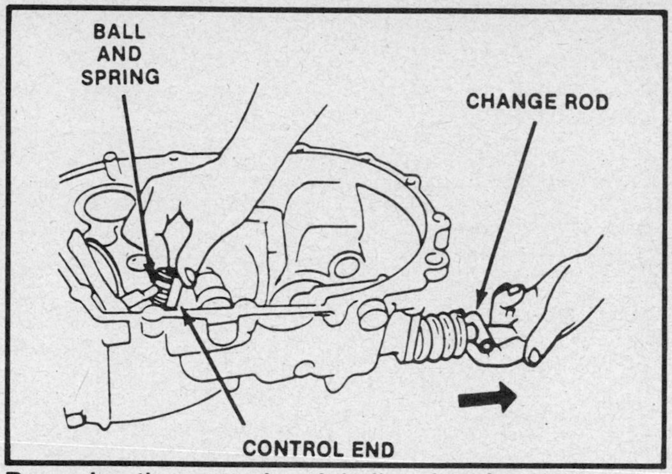
BALL AND SPRING
CHANGE ROD
CONTROL END
Removing the control end, ball and spring

cloth to prevent shaft damage and turn it with a suitable tool while pulling out.

14. On 4 speed unit only; remove the shift gate by lifting it out together with the lever.
15. Position the interlock sleeve and control lever in the position.
16. To gain access to and remove the roll pin attaching the control rod to the control end, use the following procedure.
    a. Move the change rod to turn the control rod counterclockwise.
    b. Hold the change rod in the turned position and push inward on it to raise the control rod upward.
    c. Remove the roll pin with a pin punch
17. Lift the input gear shaft, mainshaft and shift fork components as an assembly.
18. Pull the change rod rearward and remove the control end, ball and spring.

CAUTION

Be carefull not to loose the ball and spring.

19. Turn the lever with a suitable tool while pushing the lever out of the housing.
20. Remove the input gear shaft front bearing race using puller tool T77F–1102–A or equivalent and slide hammer tool T50T–1102–A or equivalent. Remove the mainshaft front bearing race by pulling up on the oil seal.
21. Remove bolts and washers securing the guide plate to the clutch housing and remove the guide plate.
22. Loosen and remove the change arm bolt and washer. Slide the change rod out of the clutch housing far enough to remove the change arm from the rod.
23. Remove the roll pin that secures the selector to the change rod. Match the pin's position with the removing groove and tap the pin out using a suitable drift and hammer.
24. Slide the change rod out of the clutch housing.
25. Remove the boot from the oil seal.
26. Retrieve the change rod oil seal from the clutch housing.
27. Remove the breather cover screws and remove the breather cover.
28. Remove the breather from the case by turning and pulling it out.
29. Remove the speedometer drive gear assembly from the case.
30. Remove the reverse lever shaft roll pin. Drive the reverse lever shaft out of the case using the proper size drift and plastic or fiber mallet.
31. Retrieve the reverse lever and lever set spring from the case.

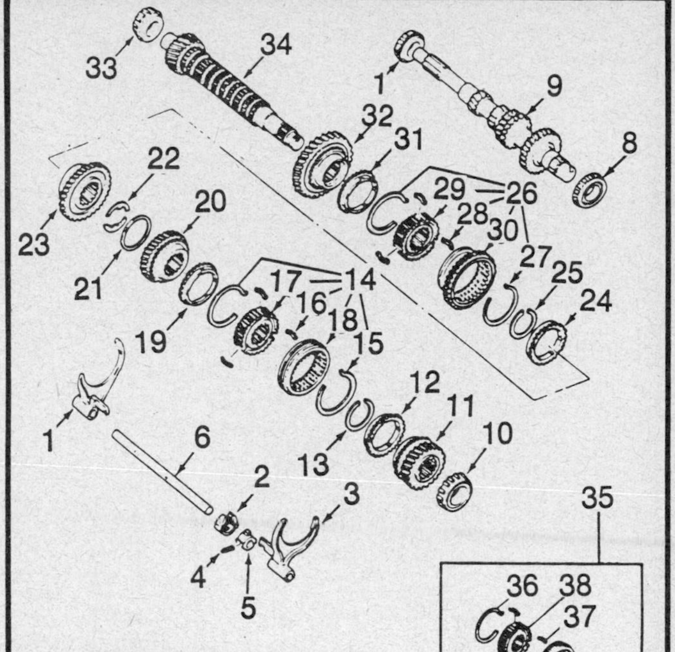
1
2
3
4
5
6
8
9
10
11
12
13
14
15
16
17
18
19
20
21
22
23
24
25
26
27
28
29
30
31
32
33
34
35
36
37
38
39

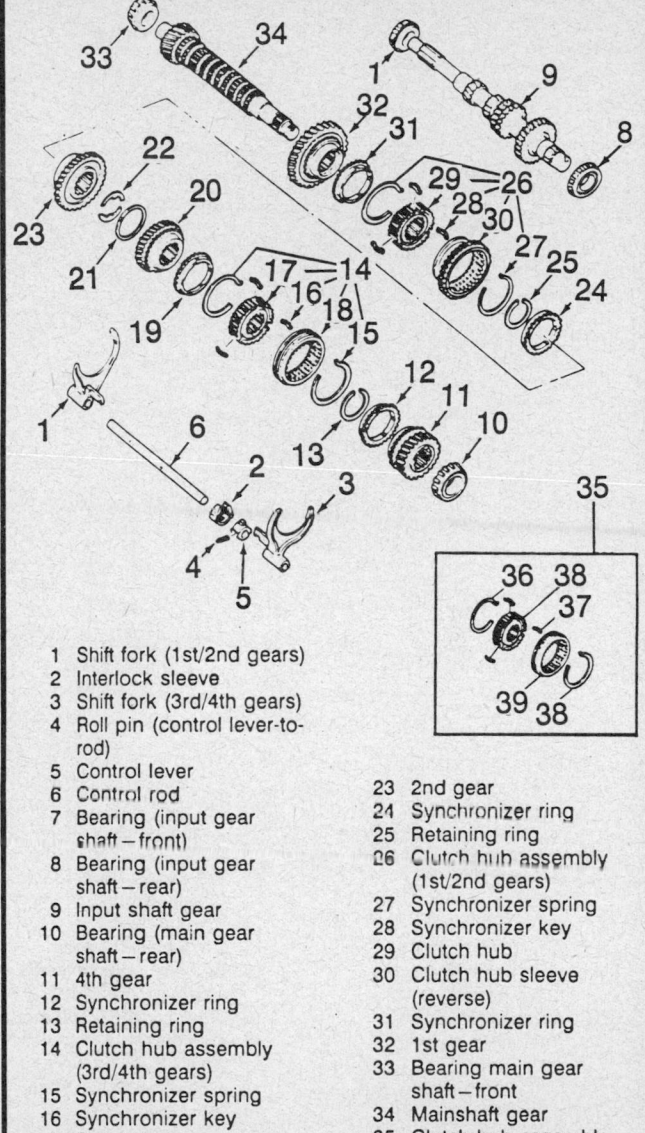

1 Shift fork (1st/2nd gears)
2 Interlock sleeve
3 Shift fork (3rd/4th gears)
4 Roll pin (control lever-to-rod)
5 Control lever
6 Control rod
7 Bearing (input gear shaft—front)
8 Bearing (input gear shaft—rear)
9 Input shaft gear
10 Bearing (main gear shaft—rear)
11 4th gear
12 Synchronizer ring
13 Retaining ring
14 Clutch hub assembly (3rd/4th gears)
15 Synchronizer spring
16 Synchronizer key
17 Clutch hub
18 Clutch hub sleeve
19 Synchronizer ring
20 3rd gear
21 Ring
22 Thrust washer
23 2nd gear
24 Synchronizer ring
25 Retaining ring
26 Clutch hub assembly (1st/2nd gears)
27 Synchronizer spring
28 Synchronizer key
29 Clutch hub
30 Clutch hub sleeve (reverse)
31 Synchronizer ring
32 1st gear
33 Bearing main gear shaft—front
34 Mainshaft gear
35 Clutch hub assembly (5th gear)
36 Synchronizer spring
37 Synchronizer key
38 Clutch hub
39 Clutch hub sleeve

Exploded view of the mainshaft and input shaft

NOTE: The following bearing races, diaphragm spring and adjustment shims should be identified upon removal so that they may be reinstalled exactly as removed if reused.

32. Remove the differential front bearing race from the transaxle case using puller tool T86P–70043–A or equivalent, extension tool T58L–101–A or equivalent and slide hammer tool T50T–100–A or equivalent.
33. Remove the input shaft rear bearing race from the transaxle case using puller tool T77F–1102–A or equivalent and slide hammer tool T50T–100–A or equivalent.
34. After removal of the input shaft rear bearing race, remove

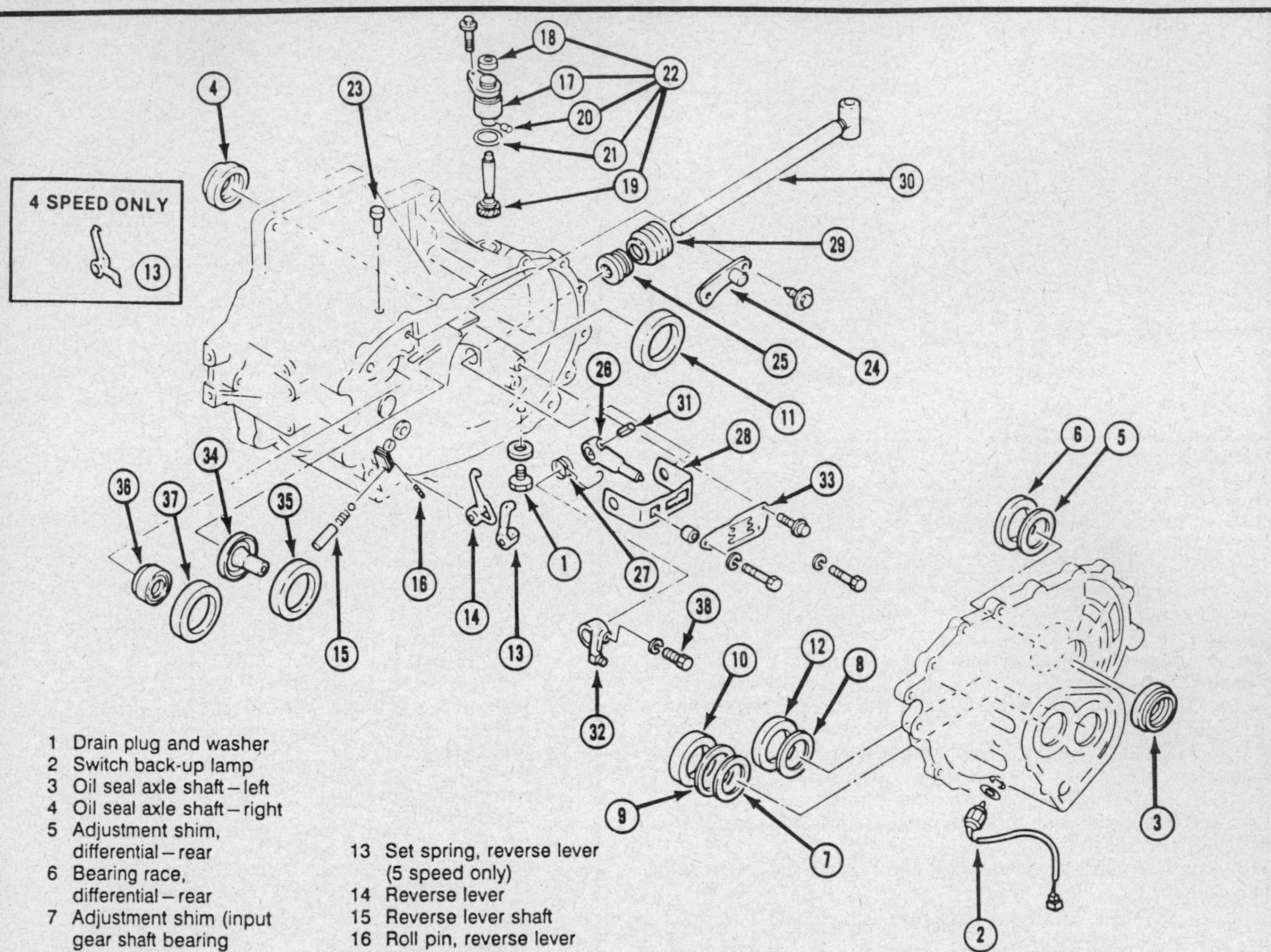
4 SPEED ONLY

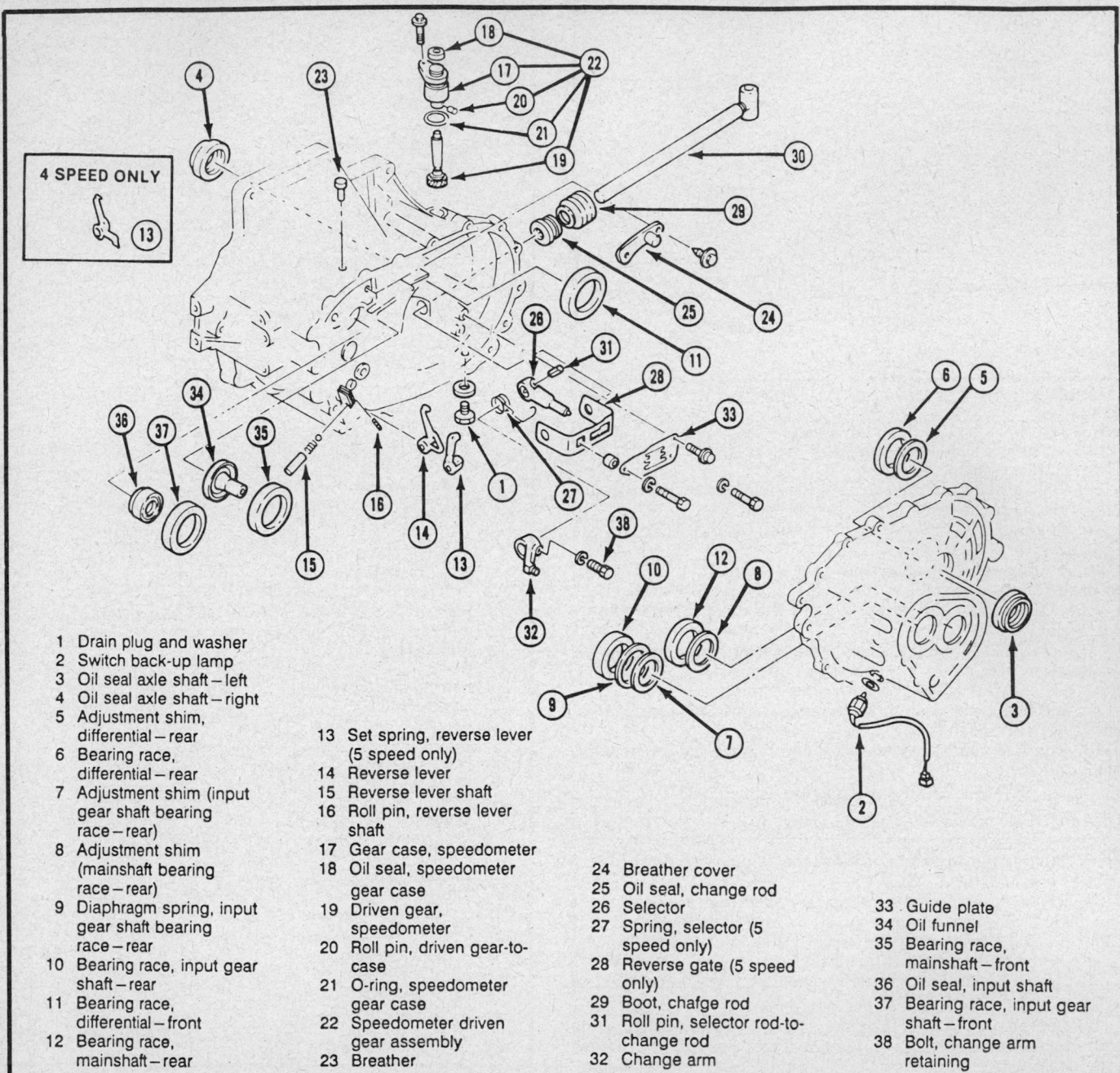

1 Drain plug and washer
2 Switch back-up lamp
3 Oil seal axle shaft—left
4 Oil seal axle shaft—right
5 Adjustment shim, differential—rear
6 Bearing race, differential—rear
7 Adjustment shim (input gear shaft bearing race—rear)
8 Adjustment shim (mainshaft bearing race—rear)
9 Diaphragm spring, input gear shaft bearing race—rear
10 Bearing race, input gear shaft—rear
11 Bearing race, differential—front
12 Bearing race, mainshaft—rear
13 Set spring, reverse lever (5 speed only)
14 Reverse lever
15 Reverse lever shaft
16 Roll pin, reverse lever shaft
17 Gear case, speedometer
18 Oil seal, speedometer gear case
19 Driven gear, speedometer
20 Roll pin, driven gear-to-case
21 O-ring, speedometer gear case
22 Speedometer driven gear assembly
23 Breather
24 Breather cover
25 Oil seal, change rod
26 Selector
27 Spring, selector (5 speed only)
28 Reverse gate (5 speed only)
29 Boot, chafge rod
31 Roll pin, selector rod-to-change rod
32 Change arm
33 Guide plate
34 Oil funnel
35 Bearing race, mainshaft—front
36 Oil seal, input shaft
37 Bearing race, input gear shaft—front
38 Bolt, change arm retaining

Exploded view—Ford transaxle

the diaphragm spring and adjustment shim, noting their original position.

35. Remove the differential rear bearing race from the transaxle housing using puller tool T86P–70043–A or equivalent, extension tool T58L–101–A or equivalent and slide hammer tool T50T–100–A or equivalent.

36. Remove the adjusting shim for the differential rear being race.

37. Remove the left differential oil seal from the transaxle case using puller tool T77F–1102–A or equivalent and slide hammer tool T50T–100–A or equivalent.

38. Remove the right differential oil seal from the clutch housing using puller tool T77F–1102–A or equivalent and slide hammer tool T50T–100–A or equivalent.

39 Remove the backup lamp switch and drain plug.

40. Remove the mainshaft rear bearing from the transaxle case using puller tool T77F–1102–A or equivalent and slide hammer tool T50T–100–A or equivalent.

Unit Disassembly And Assembly

SHIFT FORK

Disassembly

1. Remove the shift fork from the mainshaft assembly.
2. Disassemble the 1st and 2nd shift fork assembly interlock sleeve and the 3nd and 4th shift fork and interlock sleeve off the shaft.

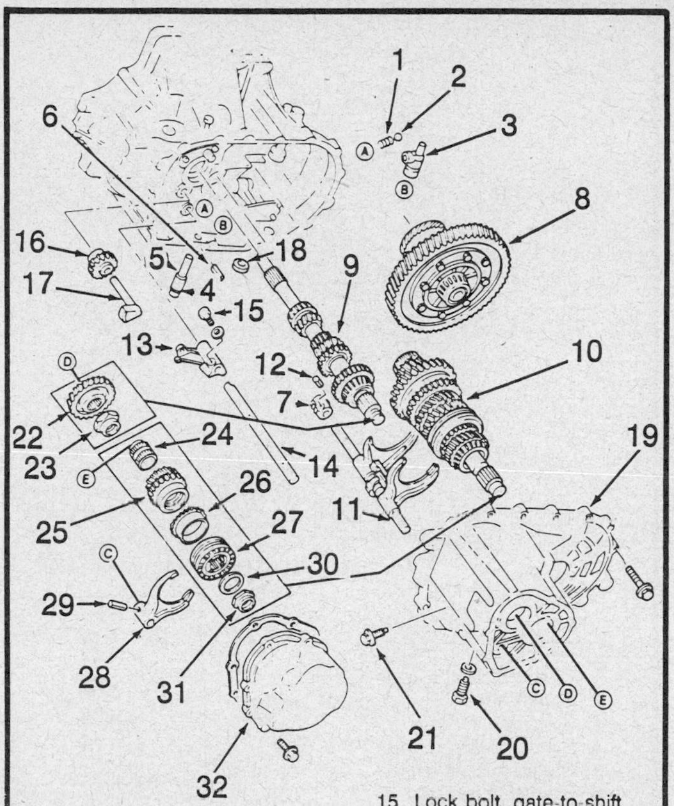
1
2
3
4
5
6
7
8
9
10
11
12
13
14
15
16
17
18
19
20
21
22
23
24
25
26
27
28
29
30
31
32

1 Spring
2 Steel ball
3 Crank lever assembly
4 O-ring
5 Crank lever shaft
6 Roll pin, crank lever shaft-to-clutch housing
7 Control end
8 Differential assembly
9 Input gear shft assembly
10 Mainshaft assembly
11 Shift fork assembly
12 Roll pin, control end-to-shift fork
13 Gate
14 Shift rod, 5th and reverse
15 Lock bolt, gate-to-shift rod
16 Reverse idler gear
17 Reverse idler shaft
18 Magnet
19 Transaxle case
20 Guide bolt and washer
21 Lock bolt
22 Input gear
23 Locknut
24 Gear sleeve (5th gear)
25 5th gear
26 Synchronizer ring
27 Clutch hub assembly
28 Shift fork (5th gear)
29 Roll pin, shift fork-to-shaft
30 Lock washer
31 Locknut
32 Rear cover

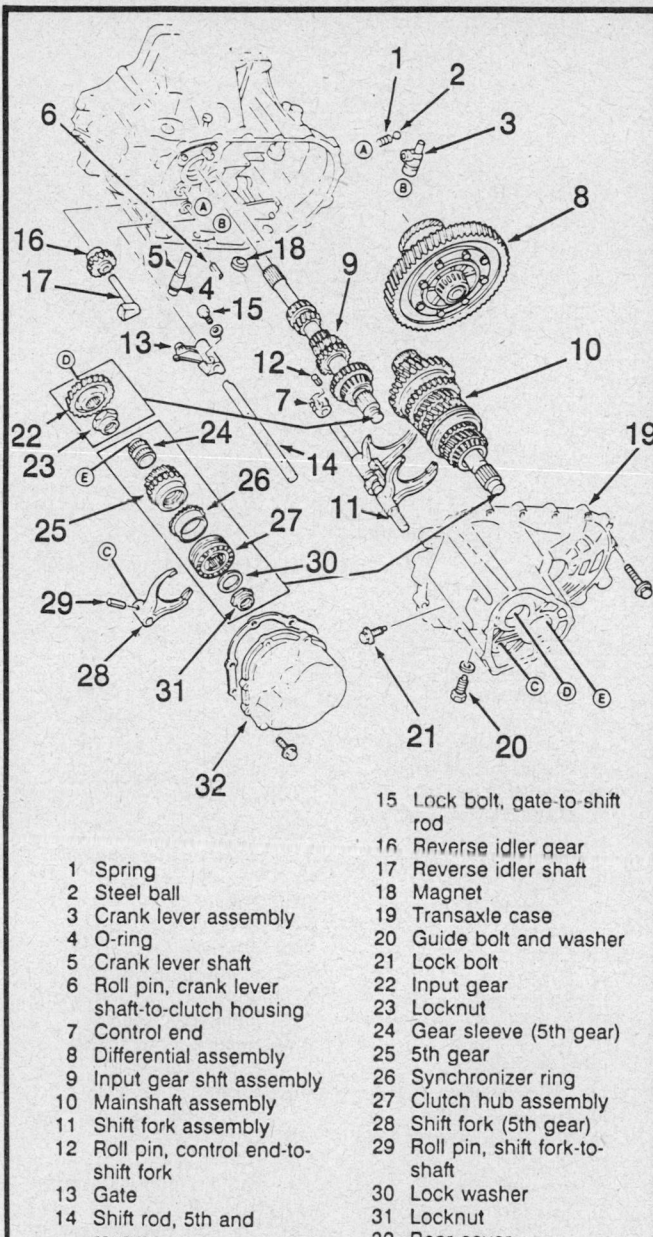

Exploded view – Ford transaxle

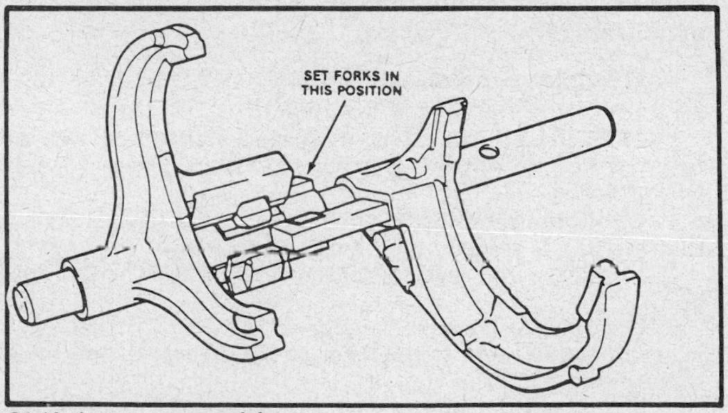
SET FORKS IN THIS POSITION

Shift lever assembly

Inspection

1. Inspect the shift fork interlock sleeve and shaft for wear.
2. Inspect the shift fork for excessive looseness on the interlock sleeve.
3. Replace defective parts as needed.

Assembly

Assemble the 1st and 2nd shift fork assembly interlock sleeve and the 3nd and 4th shift fork and interlock sleeve onto the shaft.

INPUT SHAFT

Disassembly

Press the front and rear bearings from the shaft using bearing puller attachment tool D84L–1123–A or equivalent and shaft protector tool D80L–625–3 or equivalent.

NOTE: Hold the gear shaft with one hand so that it does not fall.

Inspection

1. Inspect the all gears for worn teeth, cracks or chipping.
2. Position the input shaft on a lathe or V-blocks. Check the shaft run-out.
3. Standard run-out should be 0.000–0.002 in. (0–0.05mm).

Assembly

1. Press the input gear shaft into the front bearing using protector tool D80L–625–3 or equivalent between the press and the gear shaft, bearing cone replacer tool T87C–7025–B or equivalent and axle bearing/seal plate tool T75FL–1165–B or equivalent.
2. Press the input gear shaft into the rear bearing using protector tool D80L–625–3 or equivalent between the press and the gear shaft, bearing cone replacer tool T62F–4621–A or equivalent and axle bearing/seal plate tool T75L–1165–B or equivalent.

MAINSHAFT

Disassembly

1. Remove the mainshaft rear bearing by pressing it off the shaft using shaft protector tool D80L–625–4 or equivalent and holding fixture tool D84L–1123–A or equivalent.
2. Remove the 3rd and 4th gears and the clutch hub from the mainshaft by pressing the shaft out. Install bearing puller attachment D84L–1123–A onto the 3rd gear, positioning the lips of the fixture between the 2 sets of teeth on the 3rd gear. Install the shaft protector tool D80L–625–3 or equivalent.

NOTE: Hold the gear shaft with one hand so that it does not fall.

3. Support the 1st gear of the mainshaft on a press bed. Use shaft protector tool D80L–625–3 or equivalent and press the mainshaft through the hub and 1st gear.

Inspection

1. Check for worn or damaged contact surfaces, splines or gear teeth.
2. Check for clogged oil passage.
3. Check the oil clearance between the mainshaft gears.
   a. Measure the diameter of the gear shaft where the gear is installed. Measure the inside diameter of the gear. The difference between the 2 measurements is the clearance. If the clearance is more than allowable, replace the gear and/or shaft as necessary.

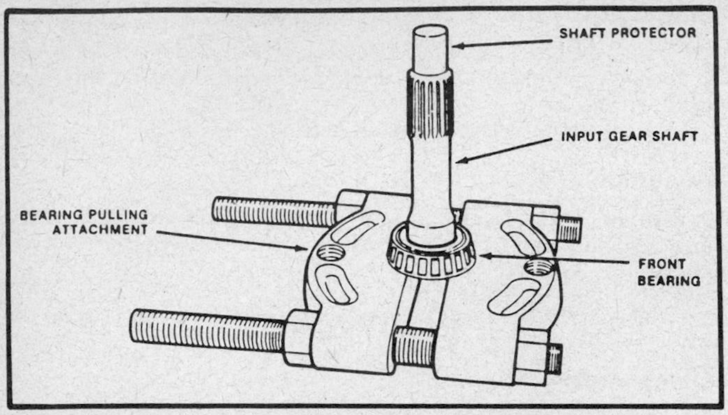
SHAFT PROTECTOR
INPUT GEAR SHAFT
BEARING PULLING ATTACHMENT
FRONT BEARING

Pressing the front bearing from the input shaft

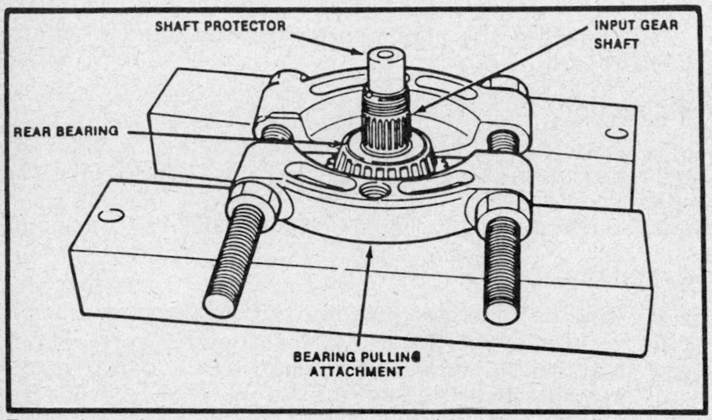
SHAFT PROTECTOR
INPUT GEAR SHAFT
REAR BEARING
BEARING PULLING ATTACHMENT

Pressing the rear bearing from the input shaft

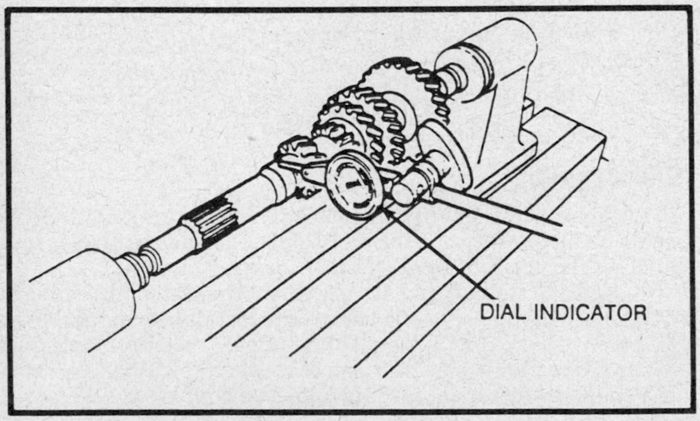
DIAL INDICATOR

Checking input shaft run-out

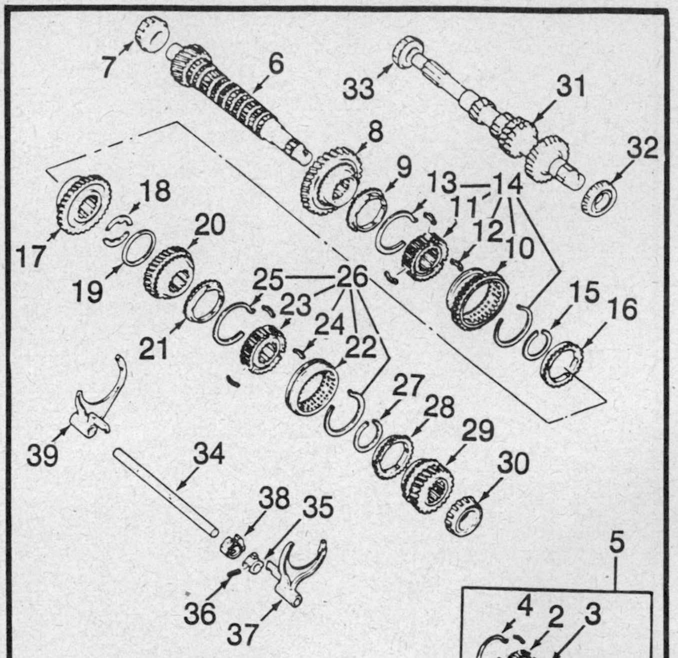

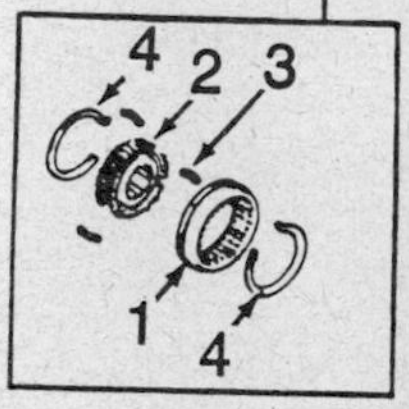

1 Clutch hub sleeve
2 Clutch hub
3 Synchronizer key
4 Synchronizer spring
5 Clutch hub assemby
6 Mainshaft gear
7 Bearing (mainshaft—front)
8 1st gear
9 Synchronizer ring
10 Clutch hub sleeve (reverse)
11 Clutch hub
12 Synchronizer key
13 Synchronizer spring
14 Clutch hub assemby (1st/2nd gears)
15 Retaining ring
16 Synchronizer ring
17 2nd gear
18 Thrust washer
19 Ring
20 3rd gear
21 Synchronizer ring
22 Clutch hub sleeve
23 Clutch hub
24 Synchronizer key
25 Synchronizer spring
26 Clutch hub assemby (3rd/4th gears)
27 Retaining ring
28 Synchronizer ring
29 4th gear
30 Bearing (mainshaft—rear)
31 Input shaft gear
32 Bearing input shaft gear—rear
33 Bearing input shaft gear—front
34 Control rod
35 Control lever
36 Roll pin (control lever-to-rod)
37 Shift fork (3rd/4th gears)
38 Interlock sleeve
39 Shift fork (1st/2nd gears)

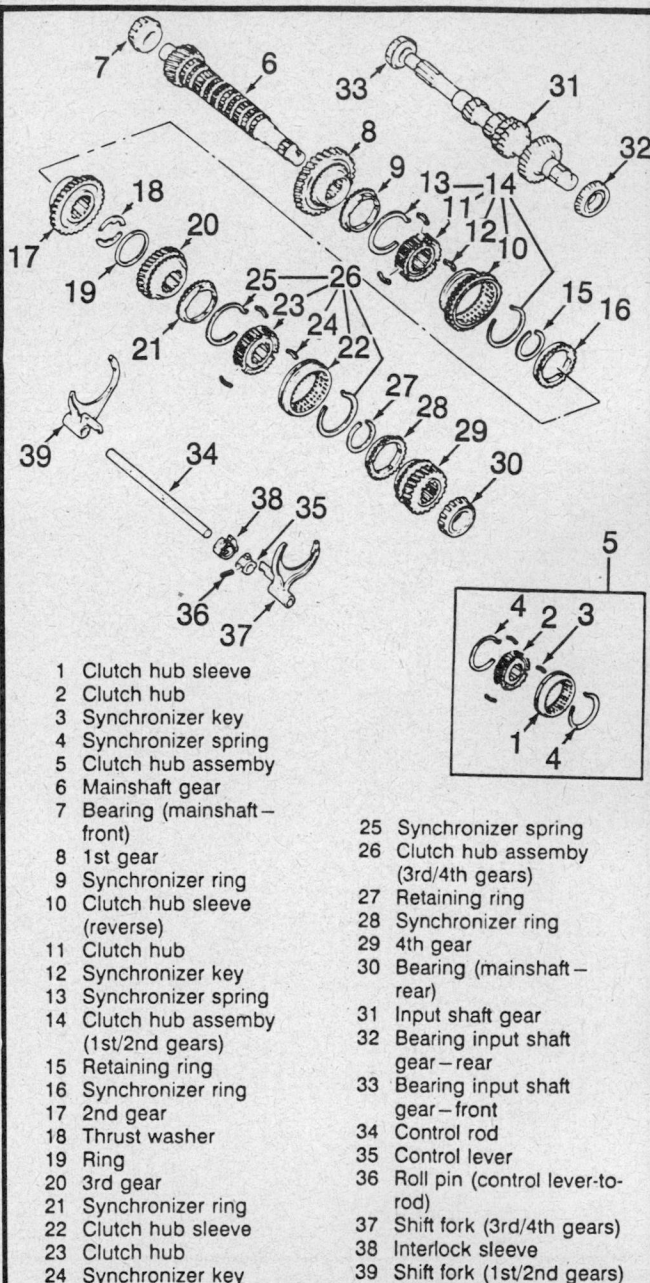

Exploded view of the mainshaft, input shaft and synchronizers

b. Standard Clearance is 0.001–0.003 in. (0.03–0.08mm).

4. Position the mainshaft on a lathe or V-blocks and measure the shaft run-out.
5. Standard run-out should be 0.006 in. (0.015mm).
6. Replace defective parts where necessary.

Assembly

1. Locate the 1st gear synchronizer ring. Take notice that the 1st gear synchronizer ring is different than the other synchronizer rings.
2. Slide the 1st gear and synchronizer ring onto the mainshaft. Slide the 1st/2nd clutch hub assembly into place, making sure the shift fork groove on the reverse sliding gear faces the 1st speed gear.
3. Press the mainshaft into the assembled components using protector tool D80L–625–3 or equivalent.

NOTE: Make sure the hub and shaft splines are aligned before applying press pressure. Press to 4,400 lbs. of force.

4. Install the retaining ring with a suitable tool. Make sure that the ring is seated properly in the groove.
5. Install the 2nd speed synchronizer ring and the 2nd speed gear.
6. Coat the thrust washer tangs (2 washers) with petroleum jelly and install them to the holes in the groove. Install the retaining ring.
7. Slide the 3rd gear onto the shaft followed by the 3rd gear synchronizer ring and the 3rd/4th gear clutch hub assembly.

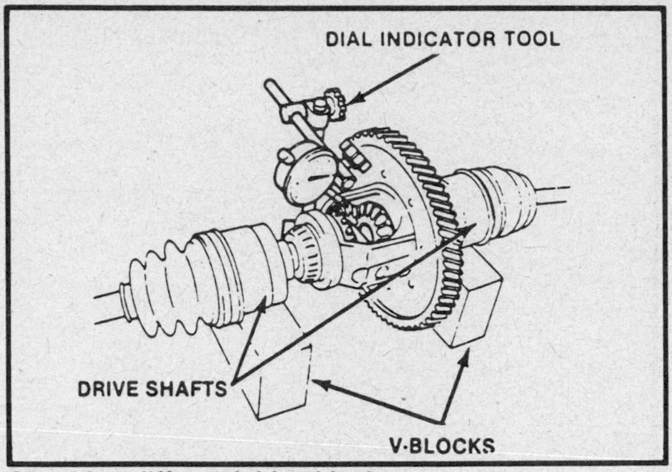
DIAL INDICATOR TOOL
DRIVE SHAFTS
V-BLOCKS
Checking differential backlash

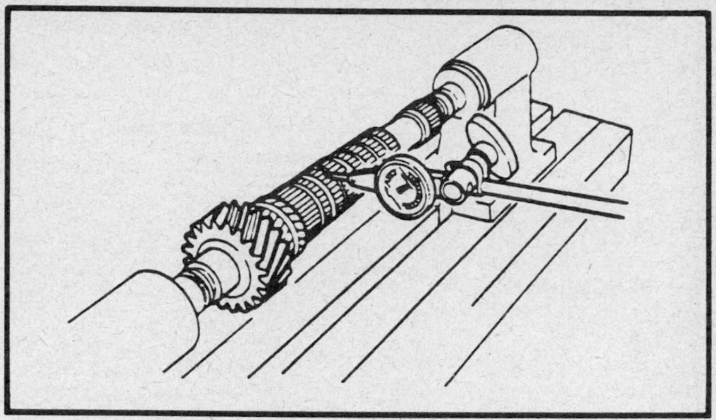
Checking mainshaft gear run-out

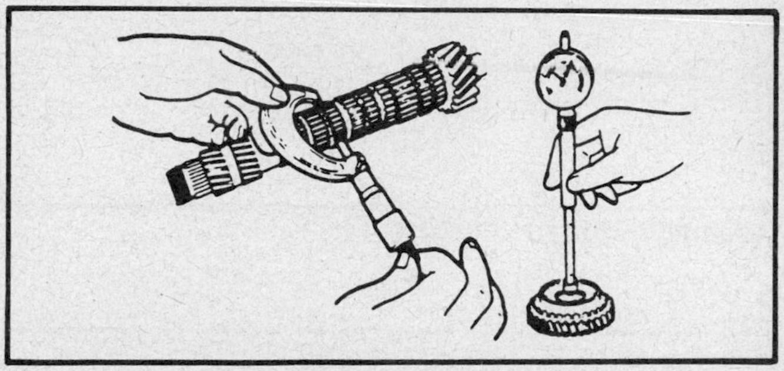
Measuring mainshaft oil clearance

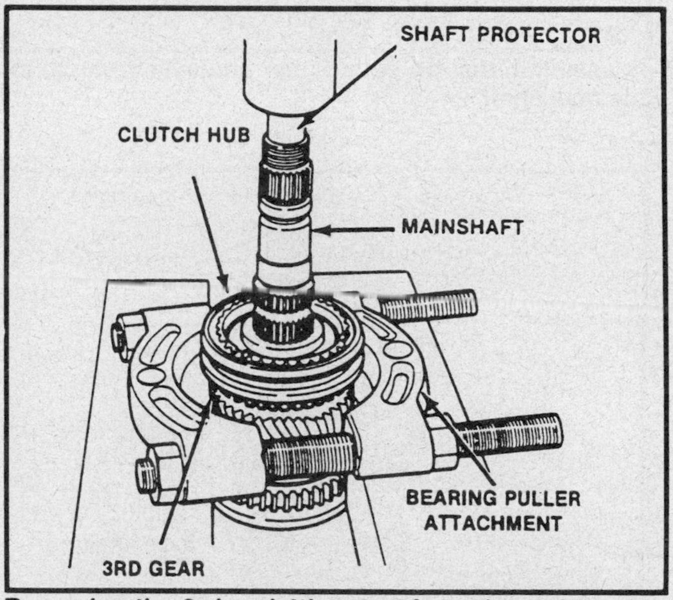
SHAFT PROTECTOR
CLUTCH HUB
MAINSHAFT
BEARING PULLER ATTACHMENT
3RD GEAR
Removing the 3rd and 4th gears from the mainshaft

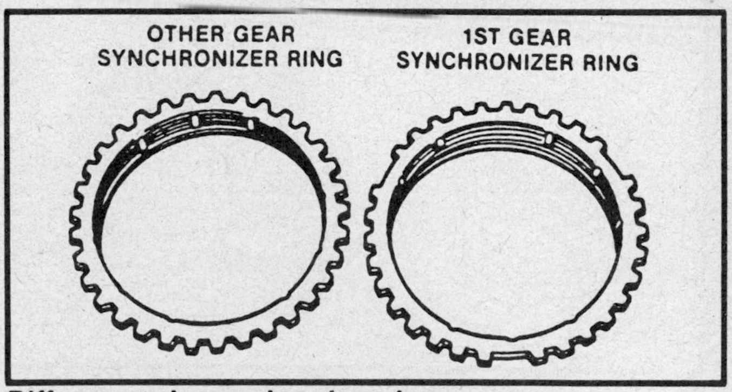
OTHER GEAR SYNCHRONIZER RING
1ST GEAR SYNCHRONIZER RING
Differences in synchronizer rings

8. Press the gear shaft into the components using protector tool D80AL–625–3 or equivalent.
9. Install the retaining ring using a suitable tool.
10. Install the 4th gear synchronizer ring and the 4th speed gear.
11. Press the 4th gear components and the mainshaft rear bearing onto the mainshaft using protector tool D80S–625–3 or equivalent, bearing cone replacer tool T62F–4621–A or equivalent and axle bearing/seal plate tool T75L–1165–B or equivalent.

Gear Thrust Clearance Measurement

1. Measure the clearance between the 1st gear and the differential drive gear.
   Standard: 0.006–0.015 in. (0.14–0.37mm).
   Limit: 0.017 in. (0.42mm).
2. Measure the clearance between the 2nd gear and the thrust washer.
   Standard: 0.010–0.023 in. (0.245–0.580mm).
   Limit: 0.025 in. (0.63mm).
3. Measure the clearance between the 3rd gear and the thrust washer.
   Standard: 0.004–0.015 in. (0.095–0.380mm).
   Limit: 0.17 in. (0.043mm).
4. Measure the clearance between the 4th gear and the bearing inner race.
   Standard: 0.004–0.016 in. (0.09–0.4mm).
   Limit: 0.18 in. (0.045mm).

SYNCHRONIZER RING

Disassembly

NOTE: Before disassembling a synchronizer, make an alignment mark on the hub and sleeve. The synchronizer sleeve and hub are matched during manufacture and should be reassembled in the exact same position. The marks will provide the necessary alignment reference.

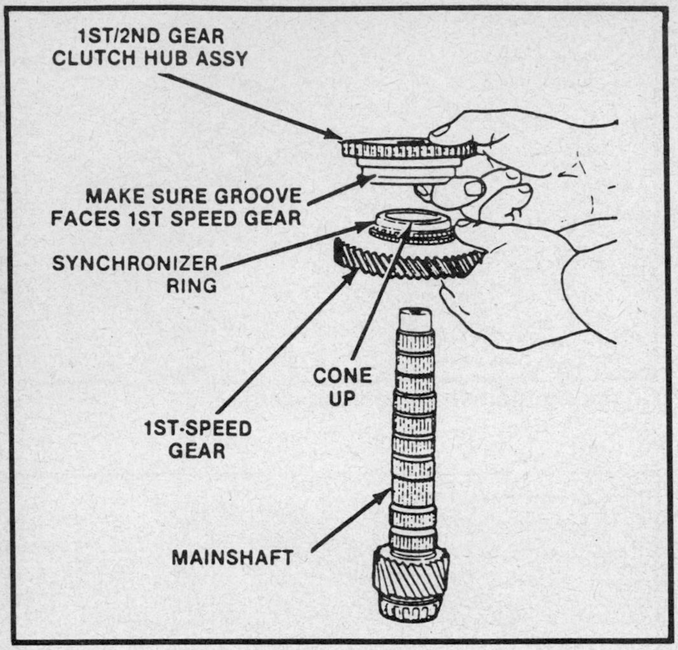
1ST/2ND GEAR
CLUTCH HUB ASSY
MAKE SURE GROOVE
FACES 1ST SPEED GEAR
SYNCHRONIZER
RING
CONE
UP
1ST-SPEED
GEAR
MAINSHAFT

Assembling the mainshaft

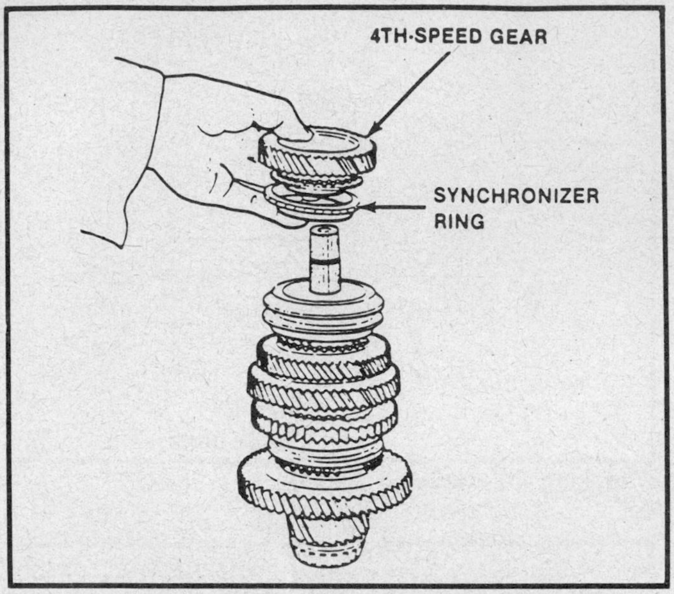
4TH-SPEED GEAR
SYNCHRONIZER
RING

Assembling the 4th speed gear and synchronizer to the mainshaft

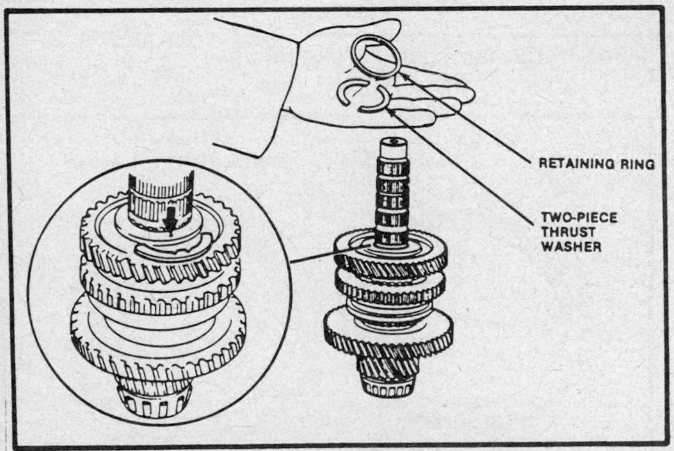
RETAINING RING
TWO-PIECE
THRUST
WASHER

Installing the 2 piece thrust washer on the mainshaft

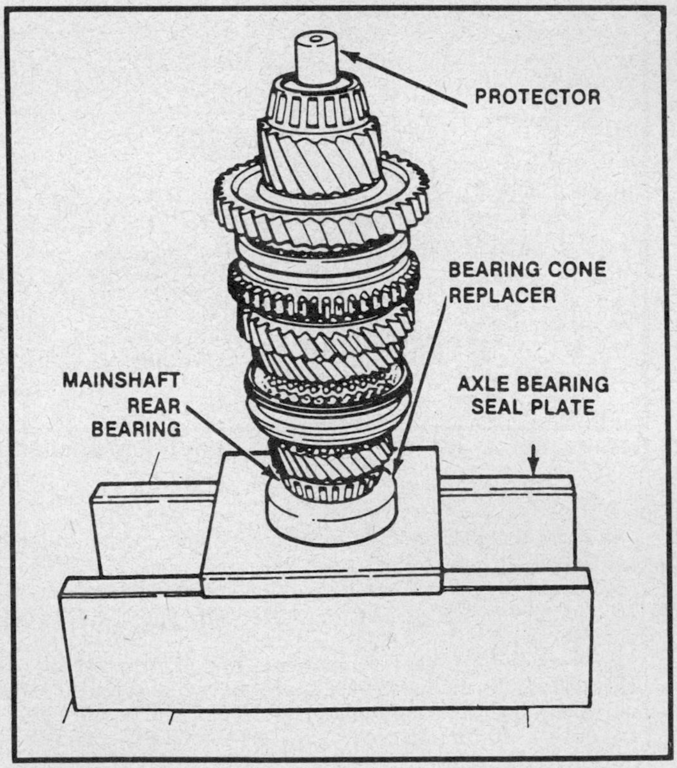
PROTECTOR
BEARING CONE
REPLACER
MAINSHAFT
REAR
BEARING
AXLE BEARING
SEAL PLATE

Preparing to press the mainshaft components in place

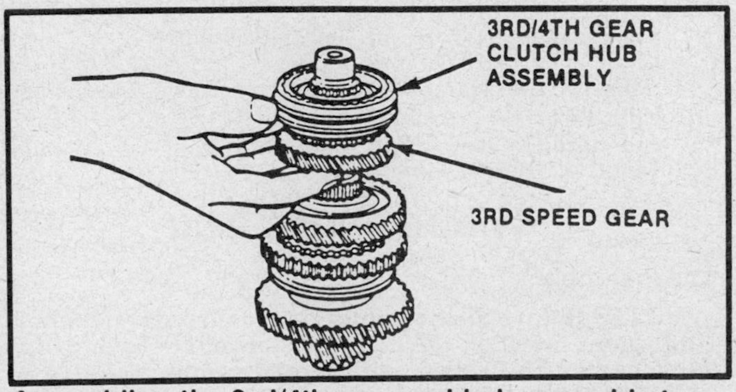
3RD/4TH GEAR
CLUTCH HUB
ASSEMBLY
3RD SPEED GEAR

Assembling the 3rd/4th gear and hub assembly to the mainshaft

1. Slide the hub and inserts out of the hub.
2. Remove the insert springs from the hub.

Inspection

1. Check that the ring and gear engage smoothly.
2. Check for worn teeth or tapered surface.

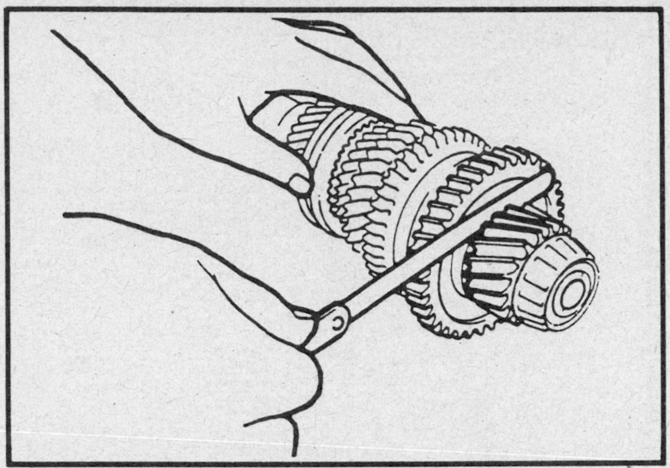

Checking clearance between the 1st gear and the differential drive gear

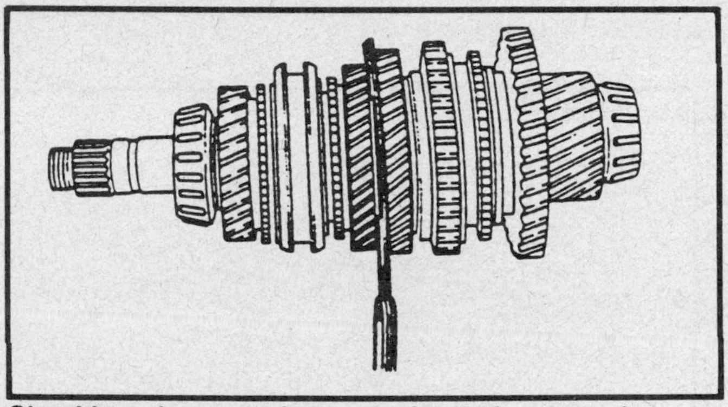

Checking clearance between the 2nd gear and thrust washer

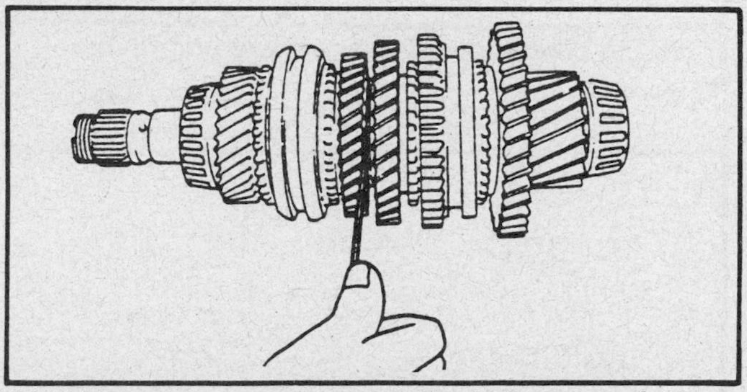

Checking clearance between the 3rd gear and thrust washer

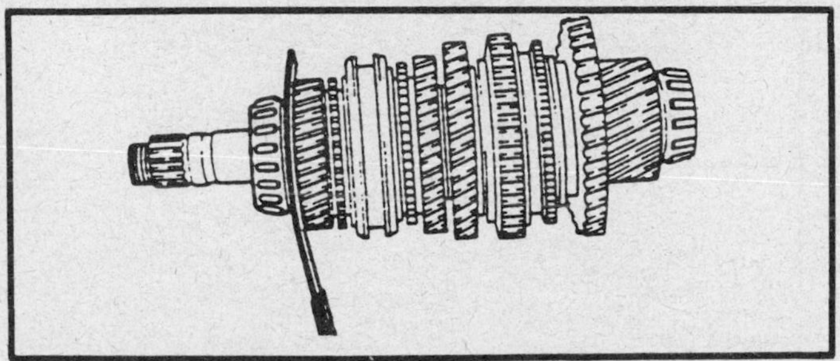

Checking clearance between the 4th gear and the bearing inner race

3. Check the clearance from the side of the gear. Press the synchronizer ring uniformly against the gear and measure around the circumference. If the measured value is less than the limit, replace the synchronizer.

Standard value: 0.059 in. (1.5mm).
Limit: 0.031 in. (0.8mm).

4. Replace defective parts where necessary.

Assembly

Install the hub in the synchronizer sleeve. Use the marks made during disassembly for alignment reference. If alignment marks are not present, make sure the hub and sleeve are assembled with the hub oil grooves facing the proper direction. Because the outside diameter of the 3rd/4th synchronizer is concentric, the relationship of the oil grooves to the sleeve is unimportant.

CLUTCH HUB AND SLEEVE

Disassembly

1. Slide the clutch hub sleeve off of the clutch hub.
2. Rotate the synchronizer key retainer spring so as to expose the tab on the end to the key groove.
3. Prior to removing the spring and synchronizer keys, take notice to the position and type of the synchronizer keys.
4. Remove the spring using a suitable tool and remove the synchronizer keys. Keep all parts with its respective assembly.

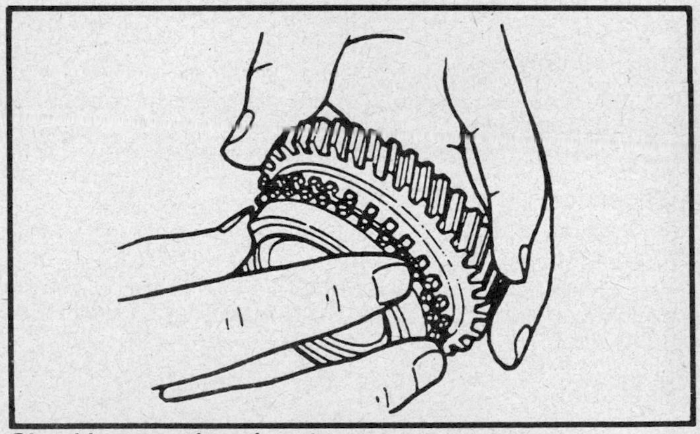

Checking synchronizer to gear engagement

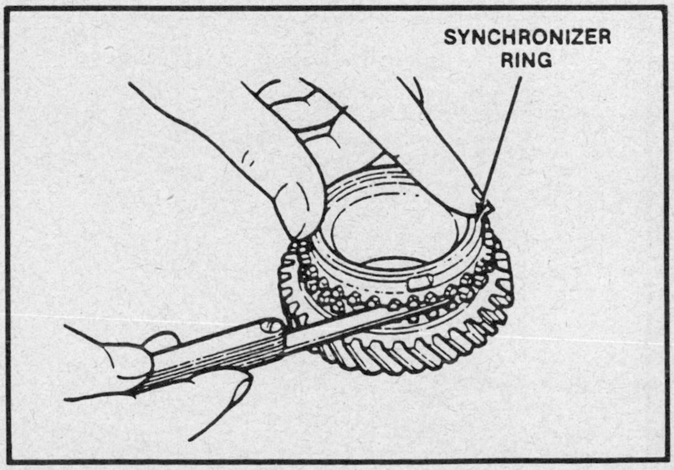
SYNCHRONIZER RING

Checking synchronizer side clearance

Inspection

1. Inspect the hub for worn or damaged splines, synchronizer key groove or end surface.
2. Check for smooth hub and sleeve engagement when it is installed.
3. Check the hub sleeve splines and sleeve fork groove for wear or damage.
4. Check clearance between the sleeve and shift fork:
Standard: 0.008–.018 in. (0.2–0.458mm).
Limit: 0.020 in. (0.5mm).
5. Check the synchronizer keys and springs for wear or distortion.
6. Replace defective parts where necessary.

Assembly

1. Position the 3 synchronizer keys into their slots in the clutch hub.

NOTE: The synchronizer keys for each clutch hub assembly are different. When installing the 5th gear clutch hub assembly, the larger end of the synchronizer must face the locknut.

2. Place the tab on the synchronizer spring into the groove of 1 of the keys and snap the spring into place.
3. Place the tab of the other spring into the same key on the other side of the synchronizer assembly and rotate the spring in the opposite direction and snap into place.

REVERSE IDLER GEAR

Disassembly

1. Remove the reverse idler and shaft as an assembly from the clutch housing.
2. Slide the reverse idler gear from the shaft and inspect.

Inspection

1. Inspect for worn or damaged bushing, gear teeth or release lever coupling groove.
2. Check clearance between the sleeve and reverse lever, the standard is 0.004–0.013 in. (0.095–0.318mm). The limit is 0.020 in. (0.5mm).

Assembly

Assemble the reverse idler gear onto the idler shaft and install the assembly into its bore in the clutch housing.

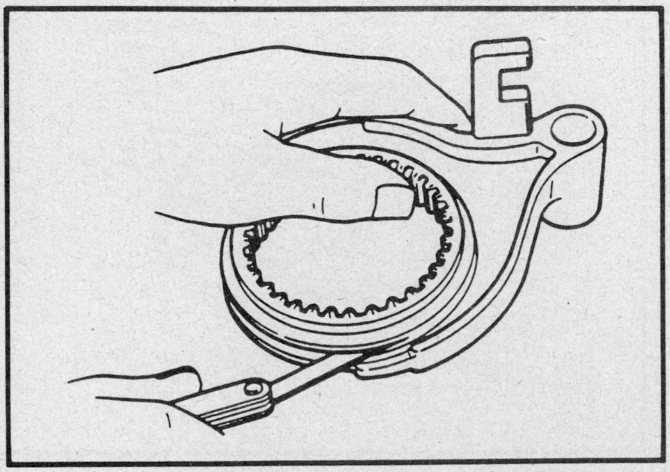

Checking clearance between sleeve and shift fork

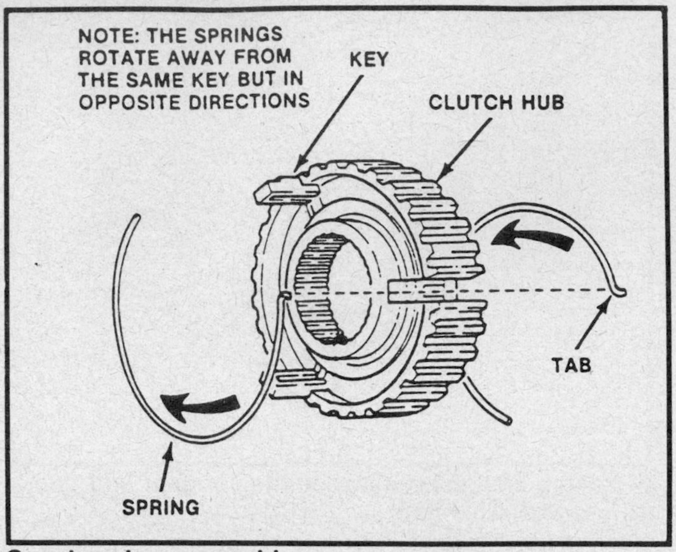
NOTE: THE SPRINGS ROTATE AWAY FROM THE SAME KEY BUT IN OPPOSITE DIRECTIONS
KEY
CLUTCH HUB
TAB
SPRING

Synchronizer assembly

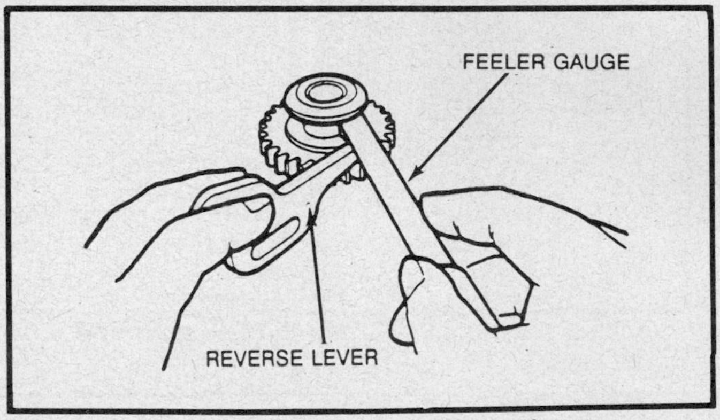
FEELER GAUGE
REVERSE LEVER

Checking clearance between sleeve and reverse lever

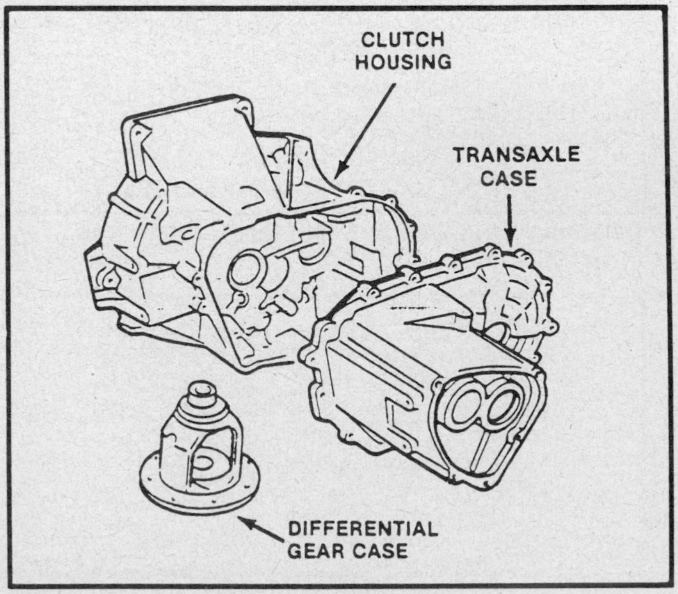
CLUTCH HOUSING
TRANSAXLE CASE
DIFFERENTIAL GEAR CASE

Inspect components for cracks

SPEEDOMETER DRIVE GEAR ASSEMBLY

1. Place the speedometer gear assembly securely in a soft jawed vise.
2. Remove the roll pin from the speedometer housing, using a small drift and hammer.
3. Slide the drive gear from the speedometer housing.
4. Remove the seal and inspect.

Inspection

1. Inspect the condition of the O-ring.
2. Inspect condition of the oil seal.
3. Check for damage or worn lip.
4. Inspect the ring gear speedometer drive gear.
5. Inspect the speedometer gear for worn or damaged teeth.
6. Replace defective parts where necessary.

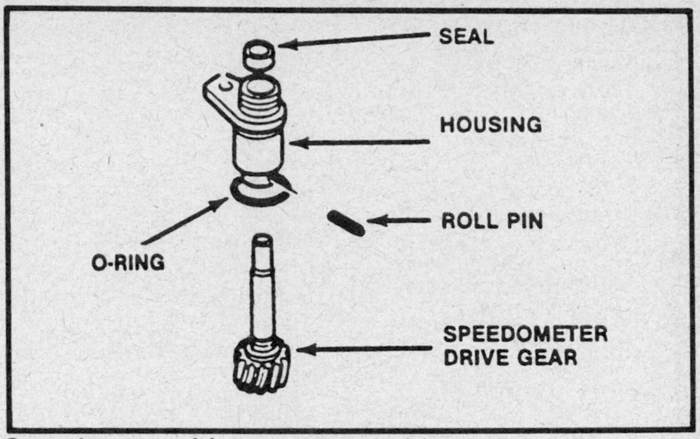
SEAL
HOUSING
ROLL PIN
O-RING
SPEEDOMETER DRIVE GEAR

Speedometer drive gear assembly

Assembly

1. Place the speedometer gear housing in a soft jawed vise.
2. Lubricate the speedometer shaft with transmission gear oil and assemble it into the speedometer housing.
3. Install the roll pin. Turn the gear and check for binding.

DIFFERENTIAL

Disassembly

Prior to disassembling the differential, measure and record the backlash of the pinion gears.

1. Install the left and right driveshafts on the differential assembly.
2. Support the driveshafts on V-blocks.
3. Measure and record the backlash of both pinion gears: Standard: 0.000–0.004 in. (0–0.1mm).
4. Remove the bolts securing the ring gear to the gear case. Tap the ring gear around its circumference with a plastic or fiber mallet to loosen it. Remove the ring gear from the gear case.
5. Mount the gear case in a soft jawed vise. Do not exert excess pressure on the vise.
6. Remove the pinion shaft roll pin using a 5⁄16 in. diameter rod at least 3 in. long and a hammer. Drive the roll pin free of the gear case.

NOTE: Whenever a differential bearing is removed from the gear case, it must be replaced with a new bearing and race.

7. Press the front differential bearing from the gear case using bearing puller attachment tool D84L–1123–A or equivalent and protector tool D80L–630–33 or equivalent.
8. Remove the rear differential bearing from the gear case using puller tool T77F–4220–B1 or equivalent and protector tool D80L–630–3 or equivalent.
9. Remove the pinion shaft by sliding it out of the gear case.
10. Remove the pinion gears and thrust washers by rotating them out of the gear case.

NOTE: The pinion and side gear thrust washers should be kept with their respective gears for possible reinstallation.

11. Remove the side gears and thrust washers from the gear case.
12. Remove the speedometer drive gear from the gear case.

Inspection

Before assembly wash all parts thoroughly and dry with compressed air. Examine the pinion and side gears for scoring, excessive wear, nicks and chips. Worn, scored and damage gears cannot be serviced and must be replaced.

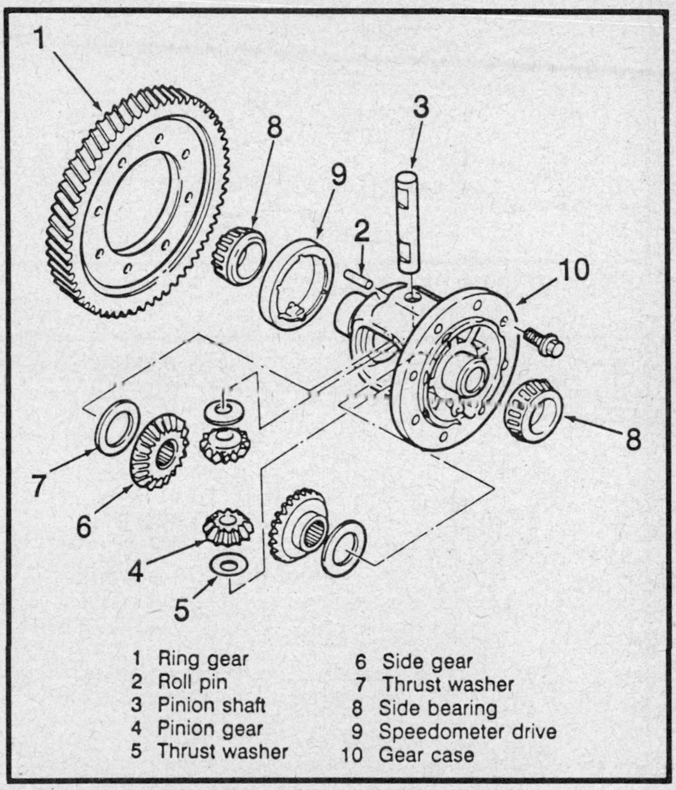
1
2
3
4
5
6
7
8
9
10
8
1 Ring gear
2 Roll pin
3 Pinion shaft
4 Pinion gear
5 Thrust washer
6 Side gear
7 Thrust washer
8 Side bearing
9 Speedometer drive
10 Gear case

Exploded view of the differential assembly

Assembly

1. Lubricate all components with Dexron®II or Ford Motor Co. ESW-M2C–138–CJ fluid or equivalent before installation.
2. Install the speedometer drive gear to the gear case, aligning the location tang on the gear with the groove in the gear case.
3. Install the front differential bearing to the gear case with a press, using handle tool T80T–4000–W or equivalent and protector tool T87C–77000–C or equivalent.
4. Install the rear differential bearing to the gear case with a press, using handle tool T80T–4000–W or equivalent and protector tool T87C–77000–C or equivalent.
5. Locate and record the identification number on each side gear thrust washer (curved washers). Select the proper thickness by the following:

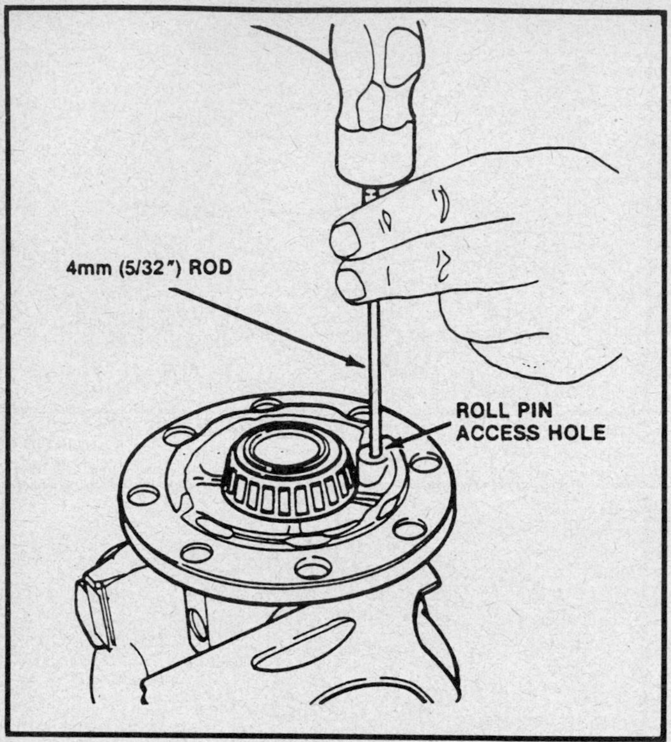
4mm (5/32") ROD
ROLL PIN
ACCESS HOLE

Removing the pinion roll pin

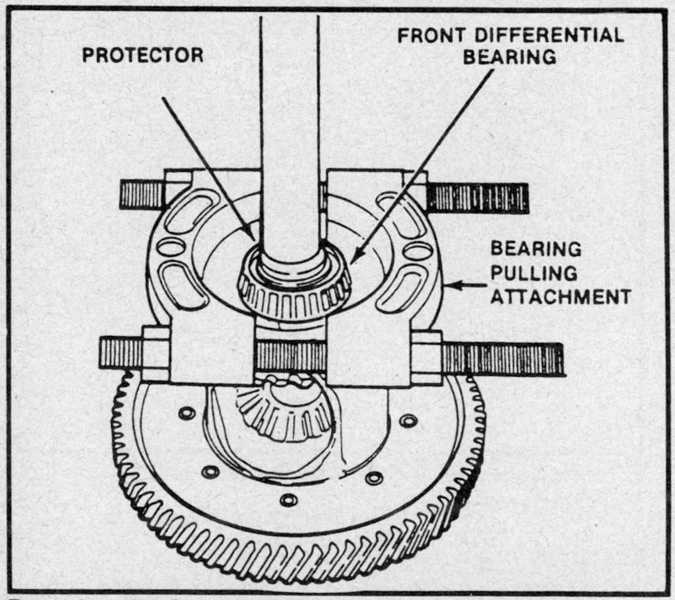
PROTECTOR
FRONT DIFFERENTIAL
BEARING
BEARING
PULLING
ATTACHMENT

Pressing the front bearing from the differential

Identification mark: – 0 Thickness: 0.079 in. (2.0mm).
Identification mark: – 1 Thickness: 0.083 in. (2.1mm).
Identification mark: – 2 Thickness: 0.087 in. (2.2mm).

6. Coat the pinion, flat thrust washers with transaxle fluid and install the thrust washers to the pinion gears. Install the pinion gear assemblies to the gear case.
7. Coat the pinion gear thrust washers with clean transaxle case so that they are parallel to each other. Install the gears and thrust washers in order.
8. After installing the thrust washers on the pinion gears,

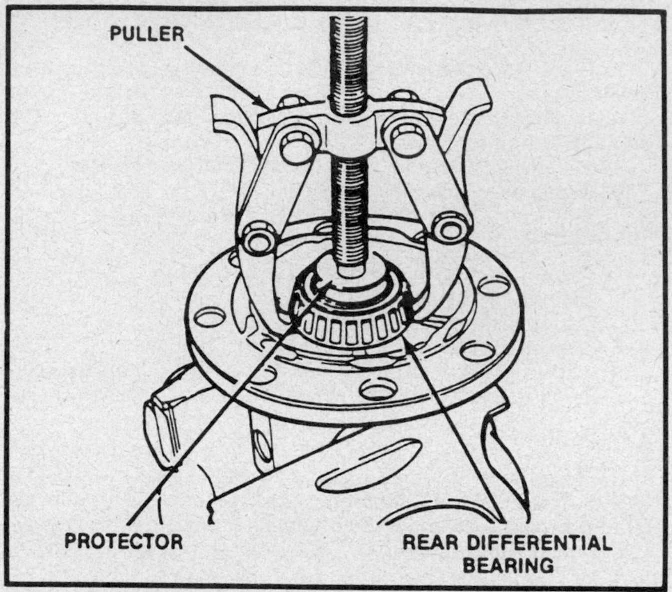
PULLER
PROTECTOR
REAR DIFFERENTIAL
BEARING

Removing the rear bearing from the differential

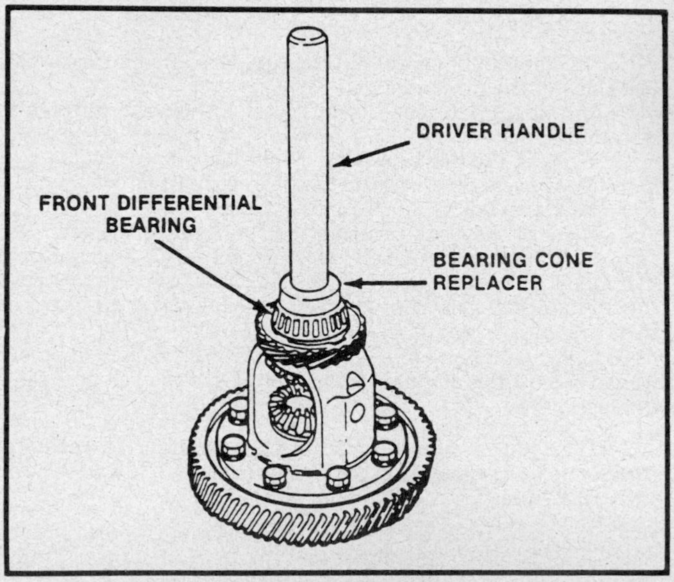
DRIVER HANDLE
FRONT DIFFERENTIAL
BEARING
BEARING CONE
REPLACER

Installing the speedometer drive gear

turn the gears back on the side gear and install them into the case. The pinion gears and pinion shaft hole must be aligned on both sides of the gear case. If the gears and gear case shaft hole do not line up, remove the pinion gears and install them into the case again.
9. After installing the pin, stake the gear case to prevent the pin from coming out.
10. Install the ring gear to the gear case, with the depression on the gear toward the gear case.
11. Align the threaded holes in the gear with the holes in the case. Install the ring gear bolts and hand tighten.
12. Tighten the bolts to 51–61 ft. lbs. (67–83 Nm). Tighten in 2 stages, marking the first bolt, tighten and working in a clockwise direction until all the bolts have been properly torqued.
13. Check and adjust (if necessary) the side gear and pinion gear back lash as follows:

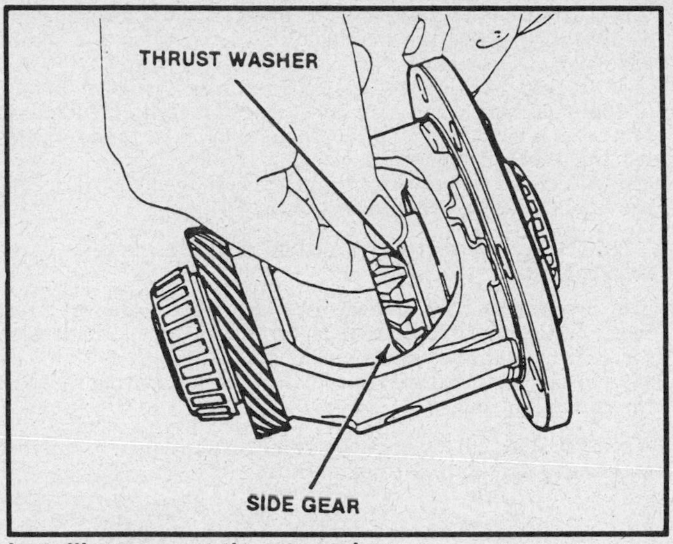
THRUST WASHER
SIDE GEAR
Installing correct thrust washer

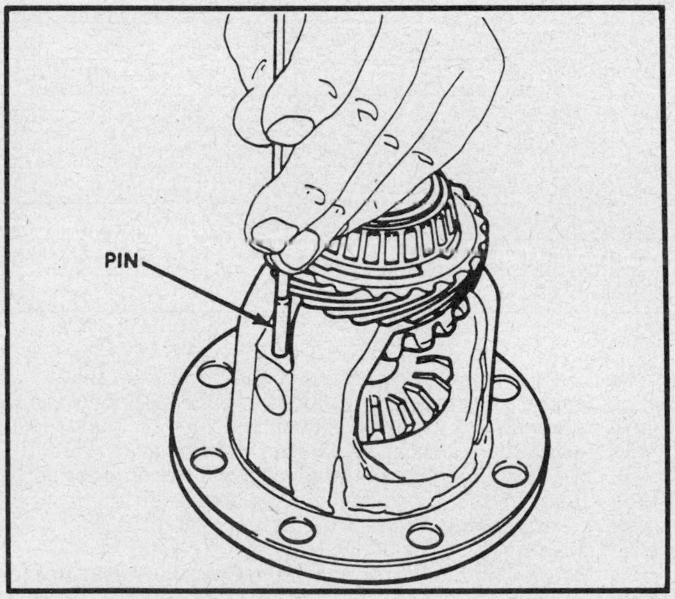
PIN
Installing the roll pin

a. Install the left and right driveshafts into the differential assembly.
b. Support the driveshafts on V-blocks.
c. Measure the backlash of both pinion gears. Standard backlash: 0.000–0.004 (0–0.1mm).
d. If the backlash is more than allowable, adjust it by selecting a thrust washer from the following:
Indentification – 0 Thickness: 0.79 in (2.0mm).
Indentification – 0 Thickness: 0.83 in (2.1mm).
Indentification – 0 Thickness: 0.87 in (2.2mm).

Transaxle Assembly

Assembly

NOTE: The mainshaft, input gear shaft and differential bearing preload must be adjusted by selecting shims to insert between the rear bearing races and transaxle case. To determine the correct thickness shim, use the shim selection tool T87C–77000–J or equivalent.

1. Install the differential bearing race into the clutch housing using driver tool T77F–1217–B or equivalent and handle tool T80T–4000–W or equivalent. Inspect the bearing race after it is fully seated.
2. Install the oil funnel and mainshaft front bearing race into the clutch housing.
3. Install the input gear shaft front bearing race into the clutch housing.
4. Install the mainshaft with its rear bearing race into the clutch housing.

NOTE: The 2 halves of the selector must be turned to eliminate any gap between them.

5. Position the selector tool T87C–77000–J or equivalent on top of the mainshaft.
6. Place 6 collars (part of tool T87C–77000–J) between the transaxle case and the clutch housing.
7. Install the transaxle case onto the mainshaft.
8. Install a flat washer of the appropriate size onto each of the bolts from the tool kit. Install the bolts through the transaxle case, collar and into the threaded holes in the clutch housing. Tighten the bolts to 13–14 ft. (18–20 Nm).
9. Mount a dial indicator tool 4201–C or equivalent to check shaft endplay.
10. Rotate the mainshaft several times to help seat the bearings.
11. Adjust the dial indicator to 0 at the lowest point on the end of the mainshaft.

CAUTION

Do not disturb the dial indicator until at least 3 endplay readings have been taken.

12. Raise the mainshaft by hand and read endplay. Lower the mainshaft.

CAUTION

The mainshaft must be lifted equally on both sides or it will tend to cock to one side which will result in an erroneous reading.

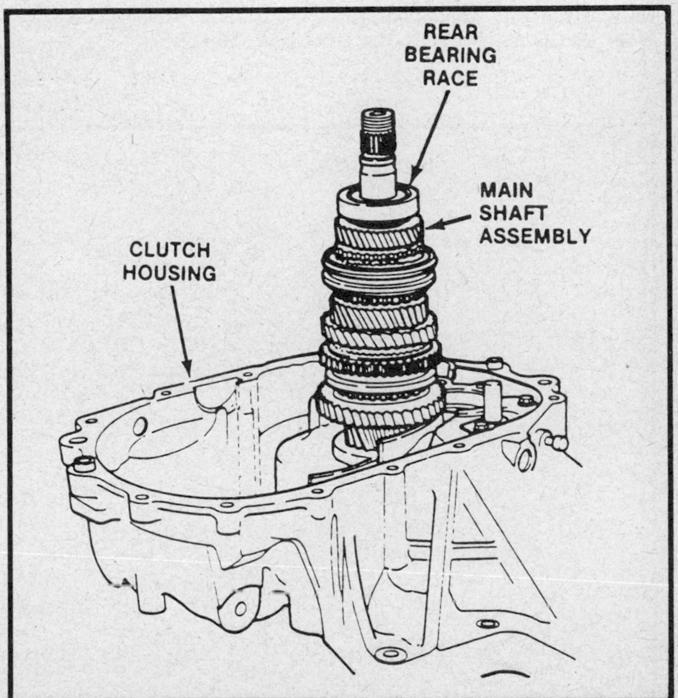
REAR
BEARING
RACE
MAIN
SHAFT
ASSEMBLY
CLUTCH
HOUSING
Installing the mainshaft

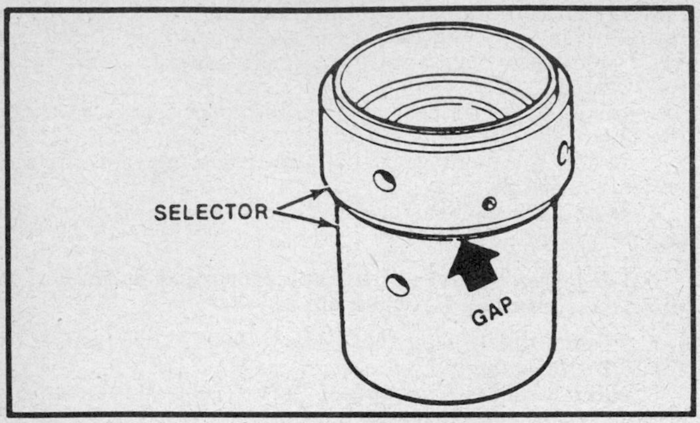
SELECTOR
GAP
Correct position of the selector halfs

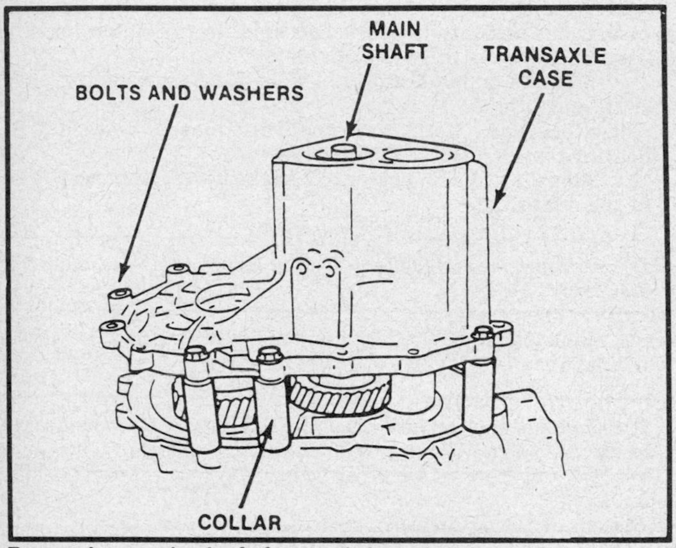
MAIN SHAFT
TRANSAXLE CASE
BOLTS AND WASHERS
COLLAR
Preparing mainshaft for endplay check

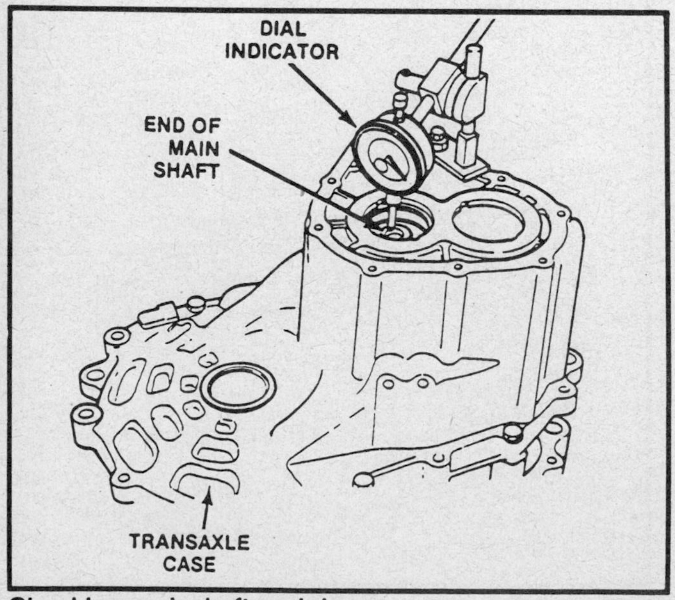
DIAL INDICATOR
END OF MAIN SHAFT
TRANSAXLE CASE
Checking mainshaft endplay

13. Turn the mainshaft several times until the dial indicator returns to 0. Raise the mainshaft by hand to take a second endplay measurement.
14. Repeat Step 13 at least 3 readings within 0.004 in. (0.10mm) of each other. Average at least 3 measurements to obtain an endplay reading. Add 0.003 in. (0.075mm) to the endplay reading to obtain final shim size.
15. Select the proper shim that is closest or slightly larger to the final shims size determined in Step 14.

NOTE: No more than 3 shims may be used under a bearing race.

16. Remove the bolts and washers securing the transaxle case to the clutch housing. Remove the transaxle case, collars selector, rear bearing race and mainshaft.
17. Install the input gear shaft with its rear bearing race into the clutch housing.

NOTE: The 2 halves of the selectors must be turned to eliminate any gap between them.

MAINSHAFT SHIM THICKNESS

	Thickness	
Part Number	in.	mm
E7GZ-4067-B	0.008	0.20
E7GZ-4067-L	0.010	0.25
E7GZ-4067-C	0.012	0.30
E7GZ-4067-M	0.014	0.35
E7GZ-4067-D	0.016	0.40
E7GZ-4067-N	0.018	0.45
E7GZ-4067-F	0.020	0.50
E7GZ-4067-P	0.022	0.55

18. Position the selector on top of the input gear shaft.
19. Place 6 collars tool T8–77000–J or equivalent between the transaxle case and the clutch housing.
20. Install the transaxle case onto the input gear shaft.
21. Install a flat washer of the appropriate size onto each of the bolts from the tool kit. Install the bolts through the transaxle case, collar housing. Tighten the bolts to 13–14 ft. lbs. (18–20 Nm).
22. Mount a dial indicator tool 4201–C or equivalent to check shaft endplay.
23. Rotate the input gear several times to help seat the bearings.
24. Adjust the dial indicator to 0 at the lowest point on the end of the input gear shaft.

CAUTION

Do not disturb the dial indicator until at least 3 endplay readings have been taken.

25. Raise the input gear shaft by hand and read endplay.

CAUTION

The input gear shaft must be lifted equally on both sides or it will tend to cock to one side which will result in an erroneous reading.

26. Turn the input gear shaft several times until the dial indicator returns to 0. Raise the input gear shaft by hand to take a second endplay measurement.
27. Repeat Step 26 at least 3 readings within 0.004 in. (0.10mm) of each other. Average at least 3 measurements to obtain an endplay reading. Add 0.003 in. (0.075mm) to the endplay reading to obtain final shim size.

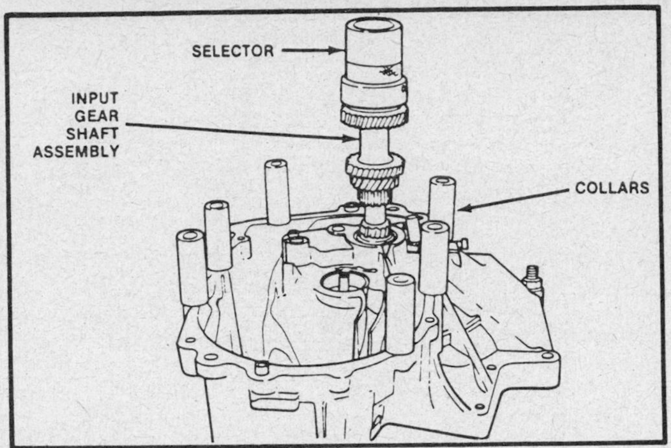
SELECTOR
INPUT GEAR SHAFT ASSEMBLY
COLLARS
Preparing the input shaft for endplay check

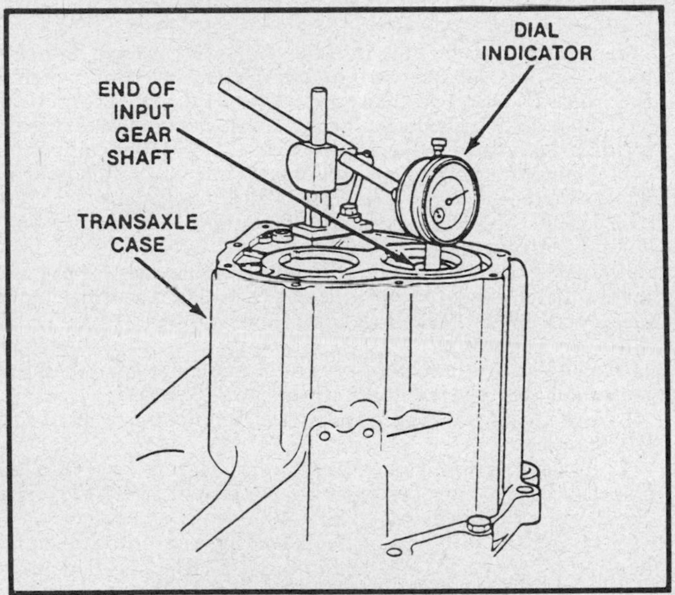
DIAL INDICATOR
END OF INPUT GEAR SHAFT
TRANSAXLE CASE
Checking the input shaft endplay

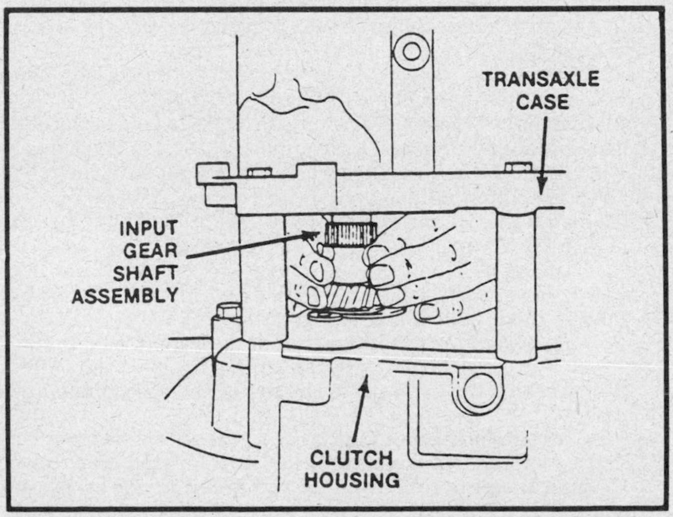
TRANSAXLE CASE
INPUT GEAR SHAFT ASSEMBLY
CLUTCH HOUSING
Seating the bearings by rotating the input shaft

28. Select the proper shim that is closest or slightly larger to the final shims size determined in Step 27.

NOTE: No more than 3 shims may be used under a bearing race.

29. Remove the bolts and washers securing the transaxle case to the clutch housing. Remove the transaxle case, collars selector, rear bearing race and input gear shaft.
30. Install the differential with its rear bearing race it to the clutch housing.

INPUT SHAFT SHIM THICKNESS

Part Number	Thickness in.	mm
E7GZ-4067-B	0.008	0.20
E7GZ-4067-L	0.010	0.25
E7GZ-4067-C	0.012	0.30
E7GZ-4067-M	0.014	0.35
E7GZ-4067-D	0.016	0.40
E7GZ-4067-N	0.018	0.45
E7GZ-4067-F	0.020	0.50
E7GZ-4067-P	0.022	0.55

NOTE: The 2 halves of the selector must be turned to eliminate any gap between them.

31. Position the selector on top of the differential, between the transaxle case and the clutch housing.
32. Install the transaxle case onto the differential.
33. Install a flat washer of the appropriate size onto each of the bolts from the tool kit. Install the bolts through the transaxle case, collar and into the threaded holes in the clutch housing. Tighten the bolts to 13–14 ft. lbs. (18–20 Nm).
34. Insert the torque adapter tool T88C–77000–L or equivalent through the transaxle case and engage the differential pinion shaft.
35. Attach a torque wrench to the tool.
36. Turn the selector for the differential using the adjusting rods, until a reading of 4.3–6.6 (0.5–0.75 Nm).
37. Use a feeler gauge to measure the gap in the differential selector. Measure the gap at 4 places, at 90 degree intervals. Average the 4 readings.
38. Select an appropriate adjustment shim to be used under the differential bearing race.

DIFFERENTIAL ADJUSTMENT SHIM

Part Number	Thickness in.	mm
E7GZ-4067-B	0.004	0.10
E7GZ-4067-K	0.006	0.15
E7GZ-4067-C	0.008	0.20
E7GZ-4067-L	0.010	0.25
E7GZ-4067-D	0.012	0.30
E7GZ-4067-M	0.014	0.35
E7GZ-4067-E	0.016	0.40
E7GZ-4067-N	0.018	0.45
E7GZ-4067-F	0.020	0.50
E7GZ-4067-P	0.022	0.55
E7GZ-4067-G	0.024	0.60
E7GZ-4067-R	0.026	0.65

DIFFERENTIAL ADJUSTMENT SHIM

Part Number	Thickness in.	Thickness mm
E7GZ-4067-H	0.028	0.70
E7GZ-4067-S	0.030	0.75
E7GZ-4067-I	0.032	0.80
E7GZ-4067-T	0.034	0.85
E7GZ-4067-J	0.036	0.90

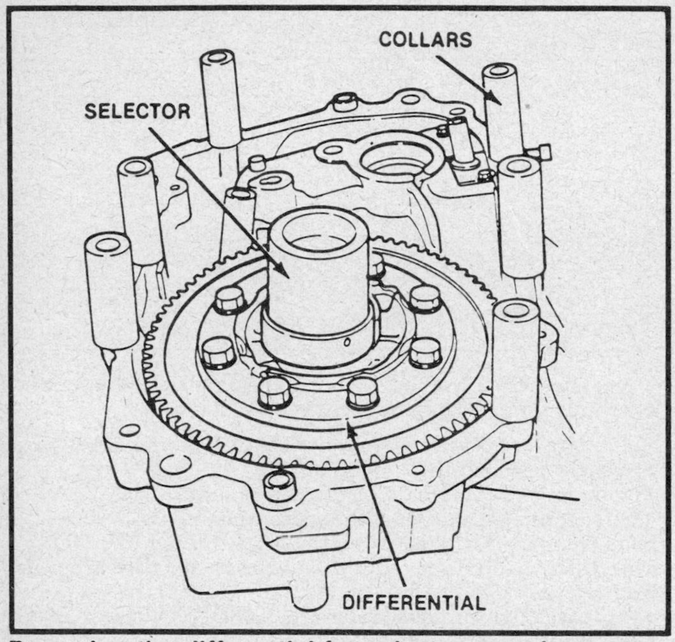
COLLARS
SELECTOR
DIFFERENTIAL

Preparing the differential for selector gap clearance check

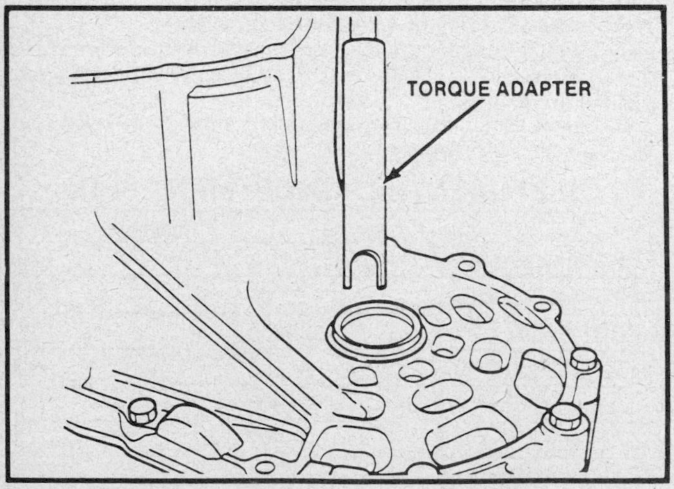
TORQUE ADAPTER

Positioning the torque adapter tool into the differential pinion shaft

NOTE: No more than 3 shims may be used under a bearing race.

39. Remove the bolts and washers securing the transaxle case to the clutch housing. Remove the transaxle case, collars, selector, rear bearing race and differential.

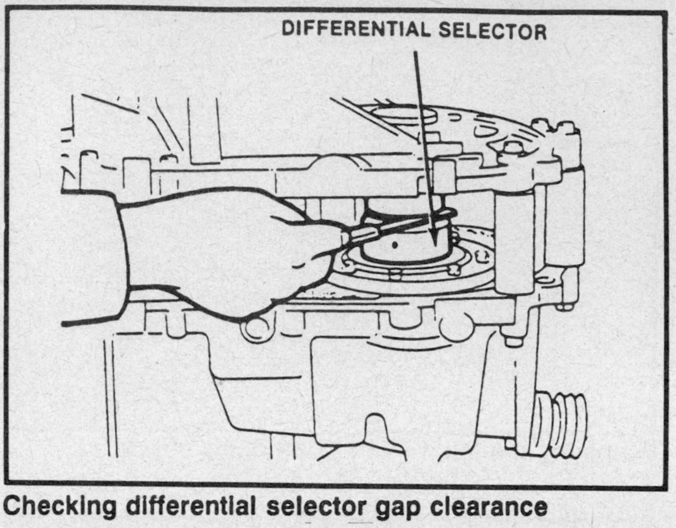
DIFFERENTIAL SELECTOR

Checking differential selector gap clearance

40. Install the drain plug into the transaxle case and tighten it to 29–40 ft. lbs. (39–54 Nm). Install the back-up lamp switch to the transaxle case and tighten it to 18–25 ft. lbs. (25–34 Nm).
41. Install the left axle shaft oil seal into the transaxle case using differential seal replacer tool T87C–77000–H or equivalent.
42. Install the right axle shaft oil seal into the clutch housing using differential tool T87C–77000–H or equivalent.
43. Install the input gear shaft seal to the clutch housing using a suitable driver tool.
44. Install the adjustment shim and bearing race for the rear differential bearing to the transaxle case, using bearing cup replacer tool T77F–1217–B or equivalent and handle tool T80T-4000–W or equivalent.
45. Install the adjustment shims for the input gear shaft and mainshaft rear bearing races to the transaxle case.
46. Install the diaphragm spring for the input gear shaft rear bearing race.
47. Install the input gear shaft rear bearing race into the transaxle case using, bearing cup replacer tool T77F–1217–B or equivalent and handle tool T80T–4000–W or equivalent.
48. Install the differential front bearing race into the clutch housing using bearing cup replacer tool T77F–1217–B or equivalent and handle tool T80T4000–W or equivalent.
49. Install the mainshaft rear bearing race into the transaxle case, using bearing cup replacer tool T77F–1217–B or equivalent and handle tool T80T–4000–W or equivalent.
50. Install the reverse lever set spring to the reverse lever (5-speed only).
51. Position the reverse lever and set spring in the clutch housing in their normal location.
52. Install the reverse lever shaft through its hole in the clutch housing (beveled end first), through the reverse lever and set spring. Align the hole in the reverse lever shaft with the roll pin hole in the clutch housing.
53. Install the roll pin through the set spring, clutch housing, and into the reverse lever shaft using a drift.
54. Coat the cup that seals the reverse lever shaft hole in the clutch housing with a suitable sealer and install it to the hole until it is flush with the housing.
55. Assemble the speedometer driven gear assembly as follows:
   a. Install a new oil seal to the top of the speedometer gear case.
   b. Install the speedometer driven gear shaft up through the bottom of the speedometer gear case. Install the roll pin through the gear case and into the speedometer driven gear shaft.
   c. Install a new O-ring to the speedometer gear case.

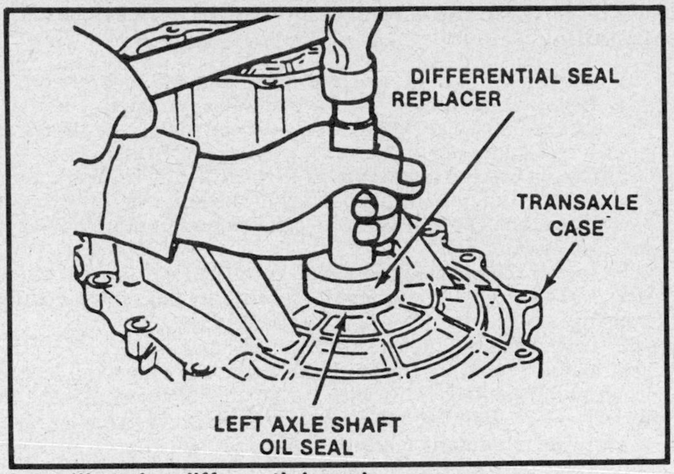
DIFFERENTIAL SEAL
REPLACER
TRANSAXLE
CASE
LEFT AXLE SHAFT
OIL SEAL

Installing the differential seal

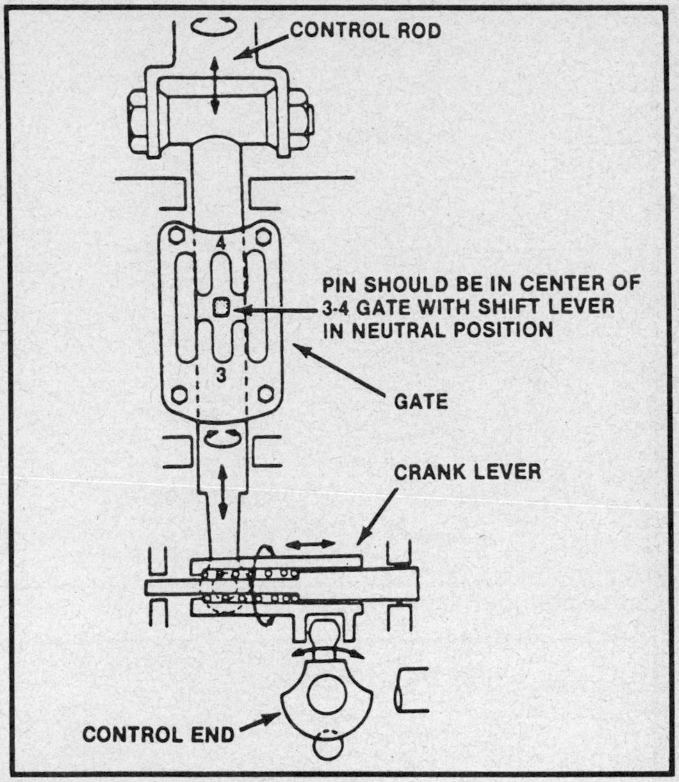
CONTROL ROD
4
PIN SHOULD BE IN CENTER OF
3-4 GATE WITH SHIFT LEVER
IN NEUTRAL POSITION
3
GATE
CRANK LEVER
CONTROL END

Shift gate adjustment position

56. Install the speedometer driven gear assembly to the clutch housing. Install the bolt and tighten to 69–95 inch lbs. (7.8–10.8 Nm).
57. Install the breather to the clutch housing by tapping it in with a fiber or plastic mallet.
58. Install the breather cover to the clutch housing and secure it with the retaining bolts. Tighten the bolts to 69–95 inch lbs. (7.8–10.8 Nm).
59. Install the change rod oil seal to the clutch housing using a driver or socket of the appropriate size.
60. Install the selector spring (5 speed only) to the selector. Install the selector and reverse gate (5 speed only) to the clutch housing and position them so they will accept the change rod when it is installed.
61. Install the change rod shift boot to the change rod. Insert the change rod through the seal and clutch housing. Feed the reverse gate and selector onto the rod and then the change arm. Align the roll pin hole in the change rod with the hole in the selector.
62. Install the roll pin through the selector and into the hole in the change rod. Drive the pin flush with the selector using a suitable drift and hammer.
63. Install the change arm bolt through the arm and into the threaded hole in the change rod. Tighten the bolt to 9–12 ft. lbs. (12–16 Nm).
64. Index the pin of the change arm into the center slot of the guide plate and install the guide plate, bolts and spacer to the clutch housing.
65. Adjust the guide plate so that the gate pin is in the center of the 3rd/4th gear slot when the control rod is in the neutral position.
66. While holding the gate in the proper position, tighten the guide plate bolts to 6–8 ft. lbs. (8–11 Nm).
67. Check the input gear shaft and differential bearing preload. It is necessary to confirm that the correct adjustment shims for both gear shafts and differential bearing cups are selected.
68. Install the input gear shaft and the differential assembly into the clutch housing.
69. Install the transaxle case to the clutch housing. Install all of the attaching bolts and tighten them to 13–19 ft. lbs. (19–26 Nm). Mark the first bolt to be tightened and work in a circle until all bolts are tight.

NOTE: The transaxle case and clutch housing are aluminum. To prevent component damage, do not overtighten the attaching bolts.

70. Install torque adapter tool T87C–77000–K or equivalent through the oil seal and onto the pinion shaft. Usa a suitable torque wrench to measure the preload by turning the tool and reading the torque wrench as the differential is rotating. Do not use the initial torque reading as it will be higher than the actual turning torque reading.

Preload: 0.26–6.6 inch lbs. (0.03–0.75 Nm).

71. Remove the turning tool and torque wrench.
72. With the input shaft facing up, install the input torque adapter tool T87C–7025–A or equivalent. Use an appropriate socket and install a torque wrench to the end of the shaft holder.
73. Measure the turning torque of the input shaft by rotating it with the torque wrench.

Preload: 0.87–3.0 inch lbs. (0.03–0.75 Nm).

NOTE: If the bearing preload measurements are not within limits, the adjustment shims are not within limits, the adjustment shims are not correct. Recheck shim adjustment again.

74. Install the spring and steel ball to the reverse lever shaft.
75. Install the crank lever assembly to the gear case.
76. Install a new O-ring to the crank lever shaft. Coat the shaft and O-ring with clean transaxle fluid.
77. Install the crank lever shaft through the clutch housing and into the crank lever with the roll pin hole positioned up.
78. Install the roll pin through the clutch housing and into the crank lever shaft, until it is just below the surface of the clutch housing.
79. Install the control end between the ball end of the crank lever and the ball end of the reverse lever shaft.
80. Install the differential assembly to the clutch housing.
81. Assemble the input gear shaft, mainshaft and shift fork assembly.
82. Install the gear shaft and shift fork assembly to the clutch housing, installing the shift fork control rod into the control end as the unit is lowered into place.

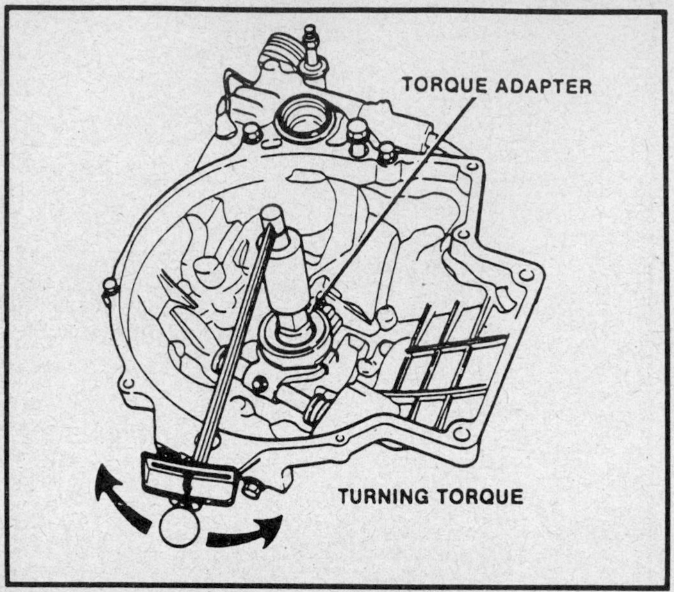
TORQUE ADAPTER
TURNING TORQUE

Checking input shaft preload torque

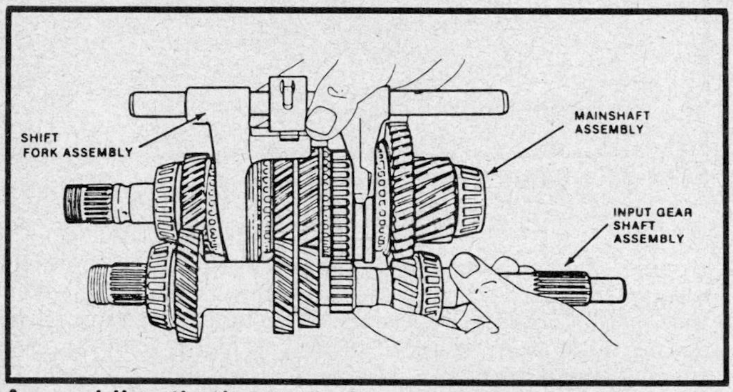
SHIFT FORK ASSEMBLY
MAINSHAFT ASSEMBLY
INPUT GEAR SHAFT ASSEMBLY

Assembling the input shaft, mainshaft and shift fork assembly

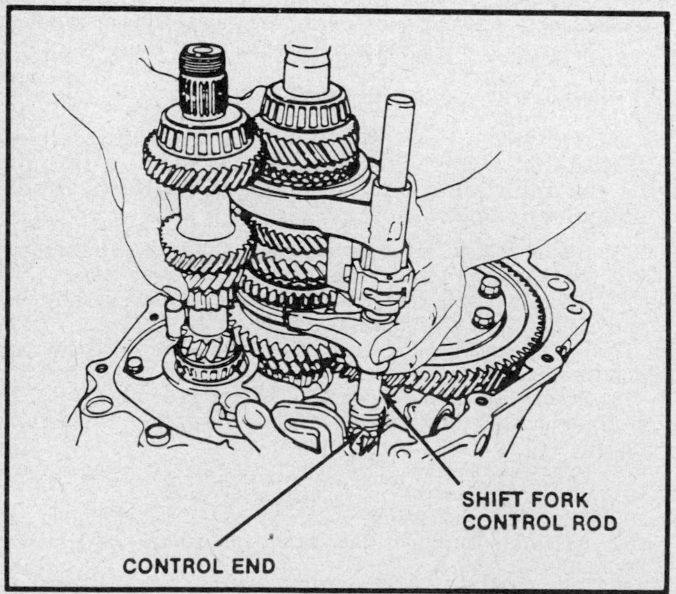
SHIFT FORK CONTROL ROD
CONTROL END

Installing all related components to the clutch housing

NOTE: Keep the assembly as vertical as possible while installing it.

83. Align the holes in the control rod and the control end.
84. Install the roll pin with a drift and hammer. On the 4 Speed transaxle, raise the reverse lever and install the gate in its groove and guide pin.
85. On the 5 speed install the shift rod (5th and reverse), make sure that the alignment mark on the rod is in the correct position. Install the gate-to-shift rod lock bolt and tighten to 9–12 ft. lbs. (12–16 Nm).
86. Install the reverse idler gear onto the reverse idler shaft.
87. Install the idler gear assembly into its bore in the clutch housing.
88. Install the magnet into the clutch housing.
89. Before installing the transaxle case, make sure the control lever (arrow) is kept flush with the surface of the end of the interlock sleeve. Point the threaded hole of the reverse idler shaft toward the alignment mark of the clutch housing.
90. Make sure mating surfaces are clean and free of grease or oil. Surfaces should also be free of nicks and burrs.
91. Apply continuous 1⁄16 in. beads of gasket eliminator E1FZ–19562–A or equivalent to the mating surfaces of the clutch housing and transaxle case. Run the bead between bolt holes and inside edge of gasket surface. Do not allow material to get inside of the transaxle.
92. Install the transaxle case to the clutch housing. Install the retaining bolts and tighten to 14–19 ft. lbs. (19–26 Nm). Mark the first bolt. Torque the bolts working in a circle until all the bolts are tightened.

NOTE: The transaxle case and clutch housing are aluminum. To prevent component damage, do not overtighten the retaining bolts.

93. Install the guide bolt to the transaxle case. Tighten the bolt to 7–9 ft. lbs. (9–12 Nm).
94. Install the lock bolt to the transaxle case and into the reverse idler shaft threaded hole. Tighten the bolt to 19–26 ft. lbs. (14–19 Nm).
95. Install the input gear to the end of the input gear shaft.
96. Install a new locknut. Put the transaxle in 1st or 2nd gear. Lock the input gear shaft with the torque adapter tool T87C–7025–A or equivalent.
97. Tighten the input gear locknut to 94–152 ft. lbs. (128–206 Nm). Stake the locknut to the groove in the input gear shaft after tightening.
98. Measure the 5th gear endplay by measuring the width of both the 5th gear and the 5th gear sleeve. The endplay equals the difference between these 2 measurements.
   Standard: 0.006 in. 0.010 in. (0.15–0.262mm).
   Limit: 0.012 in. (0.31mm).
99. Assemble the 5th gear sleeve to 5th gear. Install the synchronizer ring and clutch hub assembly to the 5th gear.
100. Install the 5th gear shift fork to the clutch hub.
101. Install the 5th gear assembly to the mainshaft while installing the shift fork to the 5th and reverse shift rod.
102. Install the roll pin through the 5th gear shift fork and into the 5th and reverse shift rod using a suitable drift and hammer. Sink the pin until it is just below the surface of the shift fork.
103. Install a new lock washer and locknut to the end of the mainshaft.
104. Place the transaxle in 1st or 2nd gear and lock the input shaft with the torque adapter tool T87C–7025–A or equivalent.
105. Tighten the mainshaft locknut to 94–152 ft. lbs. (128–206 Nm).
106. Using a new gasket, install the rear cover onto the transaxle case. Install the retaining bolts and tighten to 6–8 ft. lbs. (8–11 Nm). Do not overtighten.
107. Install the transaxle into the vehicle.

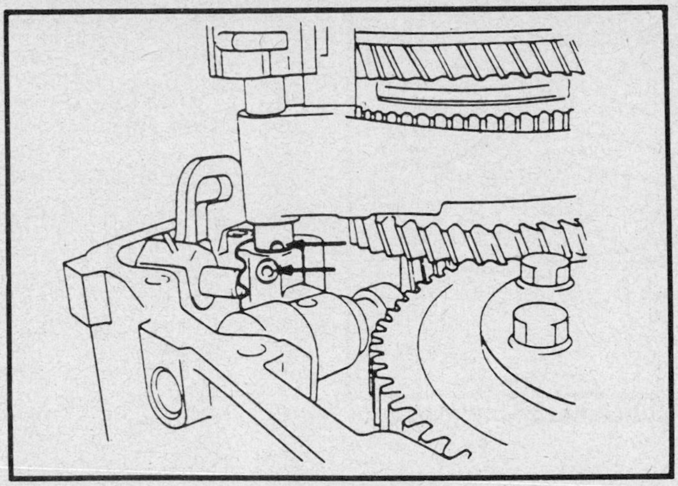

Aligning the holes in the control rod and control rod end

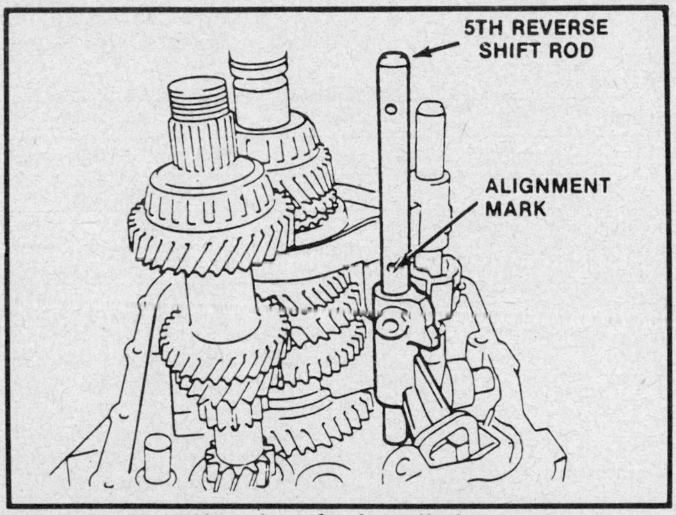
5TH REVERSE SHIFT ROD
ALIGNMENT MARK

Aligning the shift rod up for installation

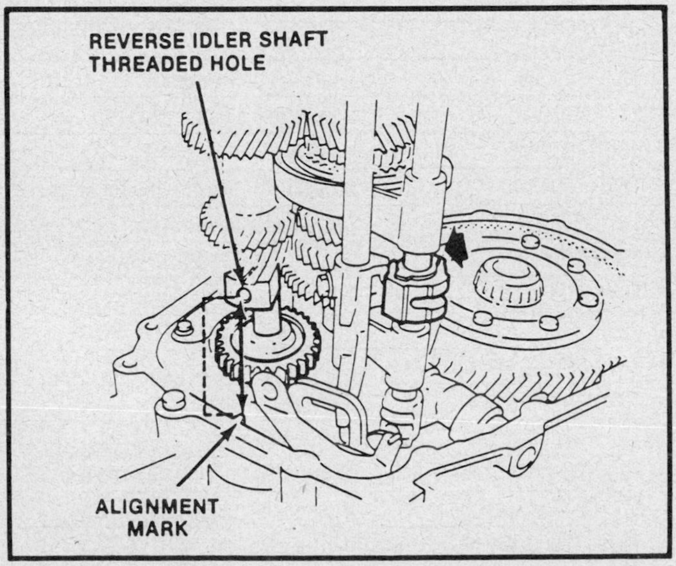
REVERSE IDLER SHAFT THREADED HOLE
ALIGNMENT MARK

Installation and alignment of the reverse idler shaft

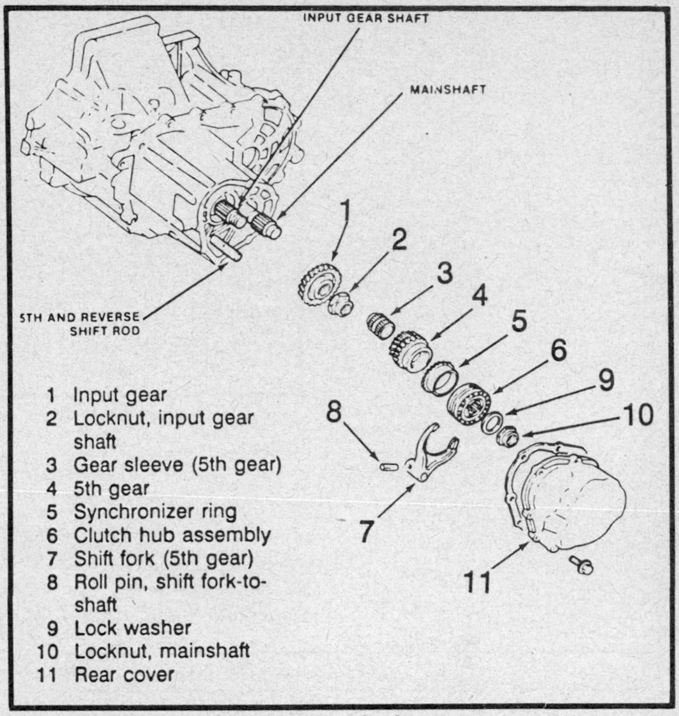
INPUT GEAR SHAFT
MAINSHAFT
5TH AND REVERSE SHIFT ROD
1 Input gear
2 Locknut, input gear shaft
3 Gear sleeve (5th gear)
4 5th gear
5 Synchronizer ring
6 Clutch hub assembly
7 Shift fork (5th gear)
8 Roll pin, shift fork-to-shaft
9 Lock washer
10 Locknut, mainshaft
11 Rear cover

Exploded view of the input shaft gear, clutch hub assembly, 5th gear and related components

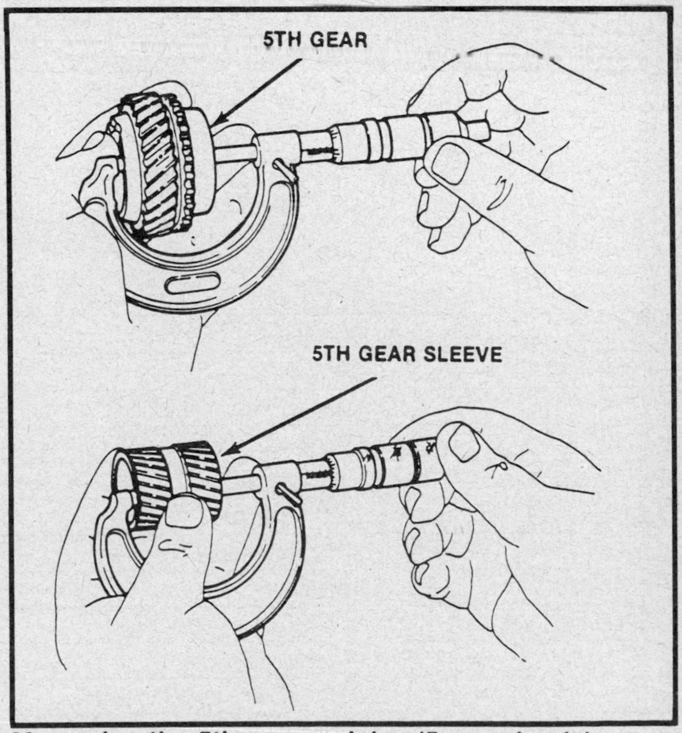
5TH GEAR
5TH GEAR SLEEVE

Measuring the 5th gear endplay (5 speed only)

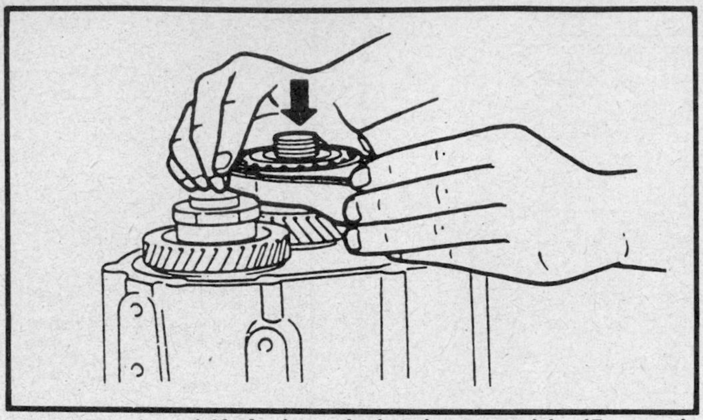

Installing the shift fork and clutch assembly (5 speed only)

SPECIFICATIONS

TORQUE SPECIFICATIONS

Item	ft. lbs.	Nm
Transaxle case to clutch housing bolts	14–20	20–27
Clutch pressure plate retaining bolts	13–20	18–27
Differential crown wheel retaining bolts	45–54	61–74
Gate lock bolt	9–12	12–16
Transcase	13–19	18–26
Rear cover	6–8	8–11

Item	ft. lbs.	Nm
Gear shaft locknut	96–155	130–210
Guide bolt	6.5–10	9–14
Reverse idle shaft lock bolt	15–22	21–31
Ring gear	51–62	69–84

SPECIAL TOOLS

Number	Description
TOOL-4201-C	Dial indicator
D78P-4201-C	Magnetic base for dial indicator
D79P-6000-B	Engine support bar
D80L-100-S	Input seal remover/collet
D80L-625-3	Shaft protector
D80L-625-4	Shaft protector
D80L-630-3	Adapter
D84L-1123-A	Bearing pulling attachment
T50T-100-A	Slide hammer
T57L-500-B	Bench mounted holding fixture
T58L-101-A	Extension
T62F-4621-A	Rear bearing cone replacer
T75L-1165-B	Axle bearing/seal plate
T77F-1102-A	Seal remover
T77F-1202-A	Bearing cup replacer

Number	Description
T77F-1217-B	Bearing cup installer
T77F-4220-B1	Differential side bearing remover
T80T-4000-W	Handle
T86P-70043-A	Differential bearing cup remover
T87C-7025-A	Torque adapter
T87C-7025-B	Bearing cone replacer
T87C-7025-C	Differential plugs
T87C-77000-C	Bearing cone replacer
T87C-77000-H	Differential seal replacer
T87C-77000-J	Shim selection tool
T87C-77000-K	Torque adapter
T86P-70043-A	Differential bearing cup remover
077-00033	Transmission jack
T87C-77000-A	Torque adapter

Section 6

Muncie 4 Speed Transaxle
General Motors

APPLICATION

MUNCIE 4 SPEED

Year	Vehicle	Engine	VIN Code
1984	Cavalier	2.0L	B,P
	Celebrity	2.5L	R,5
	Citation	2.5L, 2.8L	R,5,X,Z
	Firenza	2.0L	P
	Omega	2.5L, 2.8L	R,X,Z
	Skyhawk	2.0L, 1.8L	P,0,J
	Skylark	2.5L, 2.8L	R,X,Z
	Sunbird	1.8L, 2.0L	J,0,P
	Phoenix	2.5L, 2.8L	R,2,5
1985	Cavalier	2.0L, 2.8L	P,W
	Celebrity	2.5L	R
	Citation	2.5L, 2.8L	R,W,X
	Fiero	2.8L	X,W,S,9

Year	Vehicle	Engine	VIN Code
1985	Firenza	2.0L, 2.8L	P,W
	Skyhawk	2.0L, 1.8L	P,0,J
	Skylark	2.5L, 2.8L	R,X,Z
	Sunbird	1.8L	J,0
1986	Cavalier	2.0L, 2.8L	P,W
	Celebrity	2.5L	R
	Fiero	2.8L	W,X,S,9
	Firenza	2.0L, 2.8L	P,W
	Skyhawk	2.0L, 1.8L	P,J,0
	Sunbird	1.8L	J,0
1987	Cavalier	2.0L	P
	Skyhawk	2.0L	K,1,M

GENERAL DESCRIPTION

The 4 speed transaxle assembly is a constant mesh design, combined with a differential unit and assembled in a single case; the forward gears are in constant mesh. For ease of shifting and gear selecting, synchronizers with blocker rings are controlled by shifting forks. A sliding idler gear arrangement is used for reverse gear.

The components consists of: an aluminum transaxle case, an aluminum clutch cover, input gear (shaft), output gear (shaft) and the differential assembly. Preloaded tapered roller bearings support the input gear, output gear and differential. Selective shims, used to establish the correct preload, are located beneath the right hand bearing cups.

The halfshafts which are attached to the front wheels, are turned by the differential and ring gear which are controlled by the final output gear.

The differential, consisting of a set of 4 gears, is a conventional arrangement that divides the torque between the halfshafts, allowing them to rotate at different speeds. Of the 4 gear set, 2 are known as differential side gears and the others are differential pinion gears.

The differential pinion gears, mounted on a differential pinion shaft, are free to rotate on the shaft. The pinion shaft, placed in the differential case bore, is at a right angle to the drive axle shaft.

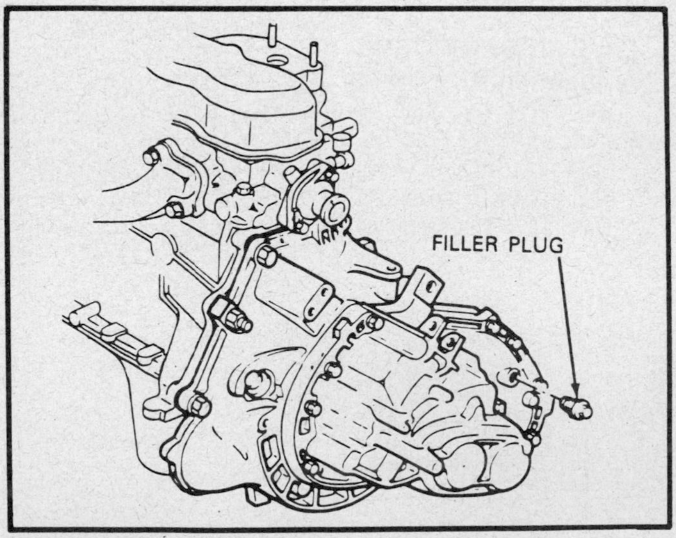
FILLER PLUG
Location of the filler plug

Metric Fasteners

The transaxle is of a metric design; all bolt sizes and thread pitches are metric. Metric fastener dimensions are very close to the customary inch system fastener dimensions; replacement of the fasteners must be of the same measurement and strength as those removed.

Do not attempt to interchange metric fasteners with the customary inch system fasteners. Mismatched or incorrect fasteners can result in damage to the transaxle. Care should be taken to reuse the same fasteners in the location from which they were removed.

NOTE: Be sure to check the case holes for the quality of their threads. It is rather difficult to rethread bolt holes after the transaxle is installed in the vehicle.

Capacities

The fluid quantities are approximate and the correct fluid level is determined to be correct by being level with the bottom of the oil filler hole. Dry fill is 3 qts. (2.8L). Use only GM manual transaxle fluid 1052931 or equivalent.

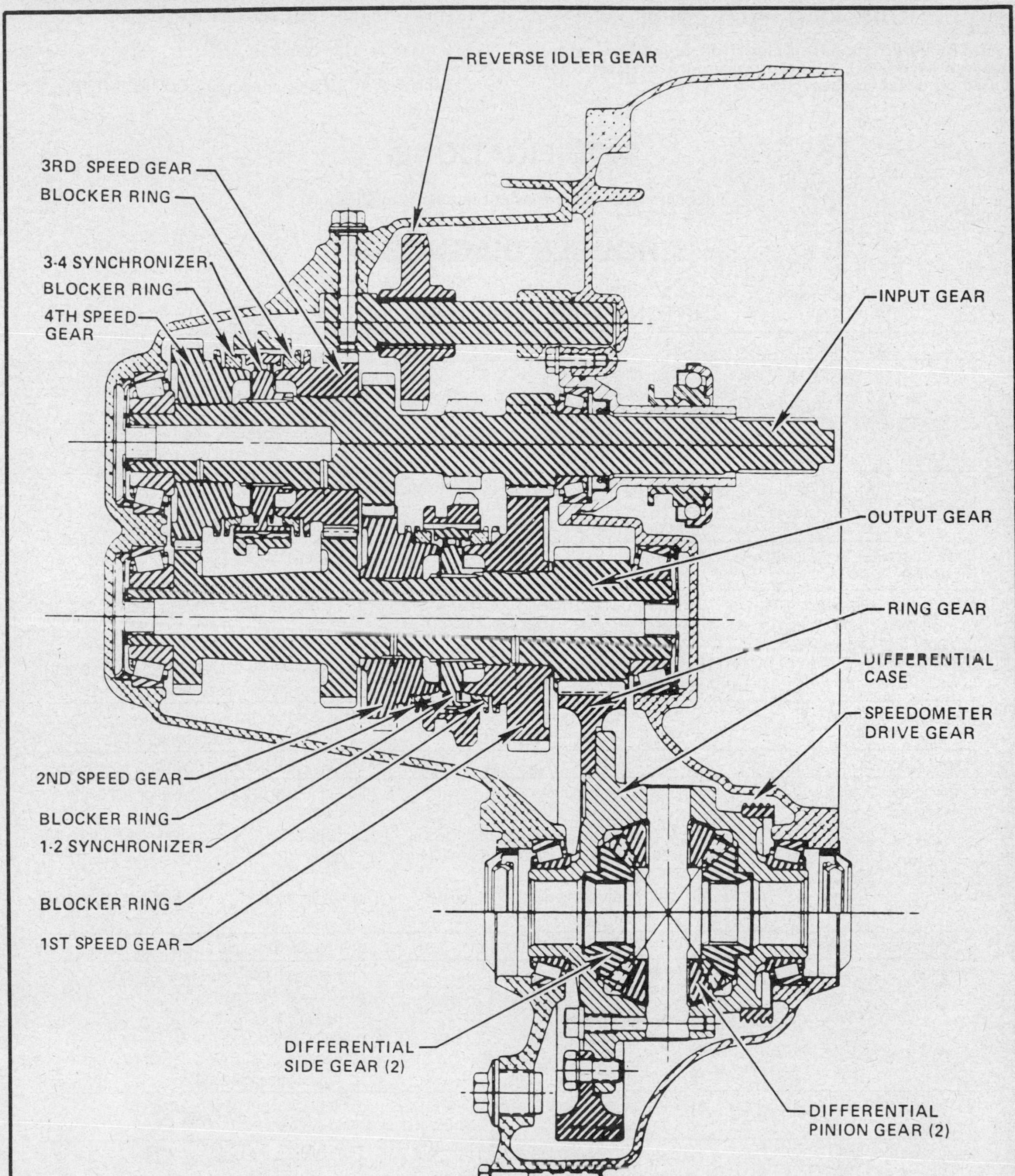
REVERSE IDLER GEAR
3RD SPEED GEAR
BLOCKER RING
3-4 SYNCHRONIZER
BLOCKER RING
4TH SPEED GEAR
INPUT GEAR
OUTPUT GEAR
RING GEAR
DIFFERENTIAL CASE
SPEEDOMETER DRIVE GEAR
2ND SPEED GEAR
BLOCKER RING
1-2 SYNCHRONIZER
BLOCKER RING
1ST SPEED GEAR
DIFFERENTIAL SIDE GEAR (2)
DIFFERENTIAL PINION GEAR (2)

Cross-sectional view of the transaxle

Checking Fluid Level

NOTE: When checking the fluid level, the vehicle must be at normal operating temperatures and positioned on a flat surface.

1. Remove the filler plug located in the middle of the rear case assembly.
2. If the fluid is low, add fluid until it is level with the bottom of the filler plug hole.
3. Replace the filler plug and torque it to 19–29 ft. lbs. (26–40 Nm).

MODIFICATIONS

There are no modifications at the time of publication.

TROUBLE DIAGNOSIS

CHILTON'S THREE C'S DIAGNOSIS

Condition	Cause	Correction
Noise is the same in drive or coast	a) Road noise b) Tire noise c) Front wheel bearing noise d) Incorrect drive axle angle (Standing Height)	a) Road test on a smooth road b) Select a different tire design c) Repair front wheel bearing d) Adjust vehicle height
Noise changes on a different type of road	a) Road noise b) Tire noise	a) Road test on a smooth road b) Select a different tire design
Noise tone lowers as vehicle speed is lowered	a) Tire noise	a) Select a different tire design
Noise is produced with engine running vehicle stopped and/or driving	a) Engine noise b) Transaxle noise c) Exhaust noise	a) Determine engine problem and repair it b) Determine transaxle problem and repair it c) Determine exhaust problem and repair it
A knock at low speed	a) Worn drive axle joints b) Worn side gear hub counterbore	a) Replace worn drive axle joints b) Replace differential assembly
Noise most pronounced on turns	a) Differential gear noise	a) Replace differential gear(s) or assembly
Clunk on acceleration or deceleration	a) Loose engine mounts b) Worn differential pinion shaft in case or side gear hub counterbore in case worn oversize c) Worn or damaged drive axle inboard joints	a) Tighten or replace loose engine mounts b) Replace differential pinion shaft or assembly c) Replace inboard drive axle joint
Clicking noise in turns	a) Worn or damaged outboard joint	a) Replace damaged outboard joint
Vibration	a) Rough wheel bearing b) Damaged drive axle shaft c) Out of round tires d) Tire unbalance e) Worn joint in drive axle shaft f) Incorrect drive axle angle	a) Replace damaged wheel bearing b) Replace axle shaft c) Replace the tires d) Balance the tires e) Replace worn drive axle joint f) Adjust vehicle height
Noisy in Neutral with engine running	a) Damaged input gear bearings	a) Replace input gear bearings
Noisy in First only	a) Damaged or worn first-speed constant mesh gears b) Damaged or worn 1-2 synchronizer	a) Replace damaged 1st gears b) Replace damaged 1-2 synchronizer
Noisy in Second only	a) Damaged or worn second-speed constant mesh gears b) Damaged or worn 1-2 synchronizer	a) Replace damaged 2nd gears b) Replace damaged 1-2 synchronizer

CHILTON'S THREE C'S DIAGNOSIS

Condition	Cause	Correction
Rattle noise in 2nd gear while making left turn	a) Bent reverse fork	a) Replace bent reverse fork
Noisy in Third only	a) Damaged or worn third-speed constant mesh gears b) Damaged or worn 3-4 synchronizer	a) Replace damaged 3rd gears b) Replace damaged 3-4 synchronizer
Noisy in High Gear	a) Damaged 3-4 synchronizer b) Damaged 4th speed gear or output gear	a) Replace damaged 3-4 synchronizer b) Replace damaged 4th gear
Noisy in Reverse only	a) Worn or damaged reverse idler gear or idler bushing b) Worn or damaged 1-2 synchronizer sleeve	a) Replace damaged reverse idler gear or bushing b) Replace damaged 1-2 synchronizer sleeve
Noisy in All Gears	a) Insufficient lubricant b) Damaged or worn bearings c) Worn or damaged input gear (shaft) and/or output gear (shaft)	a) Add lubricant b) Replace damaged bearings c) Replace damaged input and/or output gear/shaft
Slips out of Gear	a) Worn or improperly adjusted linkage b) Transmission loose on engine housing c) Shift linkage does not work freely; binds d) Bent or damaged cables e) Input gear bearing retainer broken or loose f) Dirt between clutch cover and engine housing g) Stiff shift lever seal h) Worn shift fork	a) Replace or adjust linkage b) Tighten transaxle housing bolts c) Replace, lubricate or adjust linkage d) Replace damaged cables e) Replace or tighten input gear bearing retainer f) Remove dirt between clutch cover and engine housing g) Replace or lubricate shift lever seal h) Replace worn shift fork
Leaks Lubricant	a) Axle shaft seals b) Excessive amount of lubricant in transmission c) Loose or broken input gear (shaft) bearing retainer d) Input gear bearing retainer "O" ring and/or lip seal damaged e) Lack of sealant between case and clutch cover or loose clutch cover f) Shift lever seal leaks	a) Replace axle shaft seals b) Remove excessive lubricant c) Replace or tighten input gear bearing retainer d) Replace O-ring or seal e) Add sealant between case and cover or tighten cover f) Replace shift lever seal.

ON CAR SERVICE

Adjustments

SHIFT CABLE

1. Disconnect the negative battery cable.
2. Place the transaxle's shift lever into **1st** gear.
3. At the transaxle's shifting levers **D** and **F**, loosen the shift cable nuts **E**.
4. At the console, remove the trim plate, slide the shifter boot up the shifter handle and remove the console.
5. Place the shift lever into the **1st** gear position (pulled to the left and held against the stop). Using a yoke clip, insert it into the shift lever to hold it hard against the reverse lockout stop. Using a $^5/_{32}$ in. or No. 22 drill bit, place it into the side alignment hole of the shifter lever.
6. Rotate shift lever **D** (in the direction of the arrow) to remove the lash (do not force lever) and tighten adjusting nut **E**.
7. At shift lever **F**, tighten adjusting nut **E**.
8. From the shifter assembly, remove the drill bit and yoke.
9. Install the console, the shifter boot and the trim plate.
10. Connect the negative battery cable.
11. Road test the vehicle and check for a good neutral gate feel during shifting; it may be necessary to fine turn the adjustment after road testing.

SHIFTER SHAFT WASHER

If a hang-up is experienced in the 1–2 gear range and the shift cables are properly adjusted, it may be necessary to change the shift selector washer.

1. From the end of the housing, remove the reverse inhibitor fitting spring and washer.
2. Position the shifter shaft into the **2nd** gear position.
3. Measure the end of housing-to-end of shifter shaft (dimension **A**).
4. On the opposite end of the shaft, apply a load of 8.8–13.3 lbs. (4–6 Kg) and measure the end of housing-to-end of shifter shaft (dimension **B**).

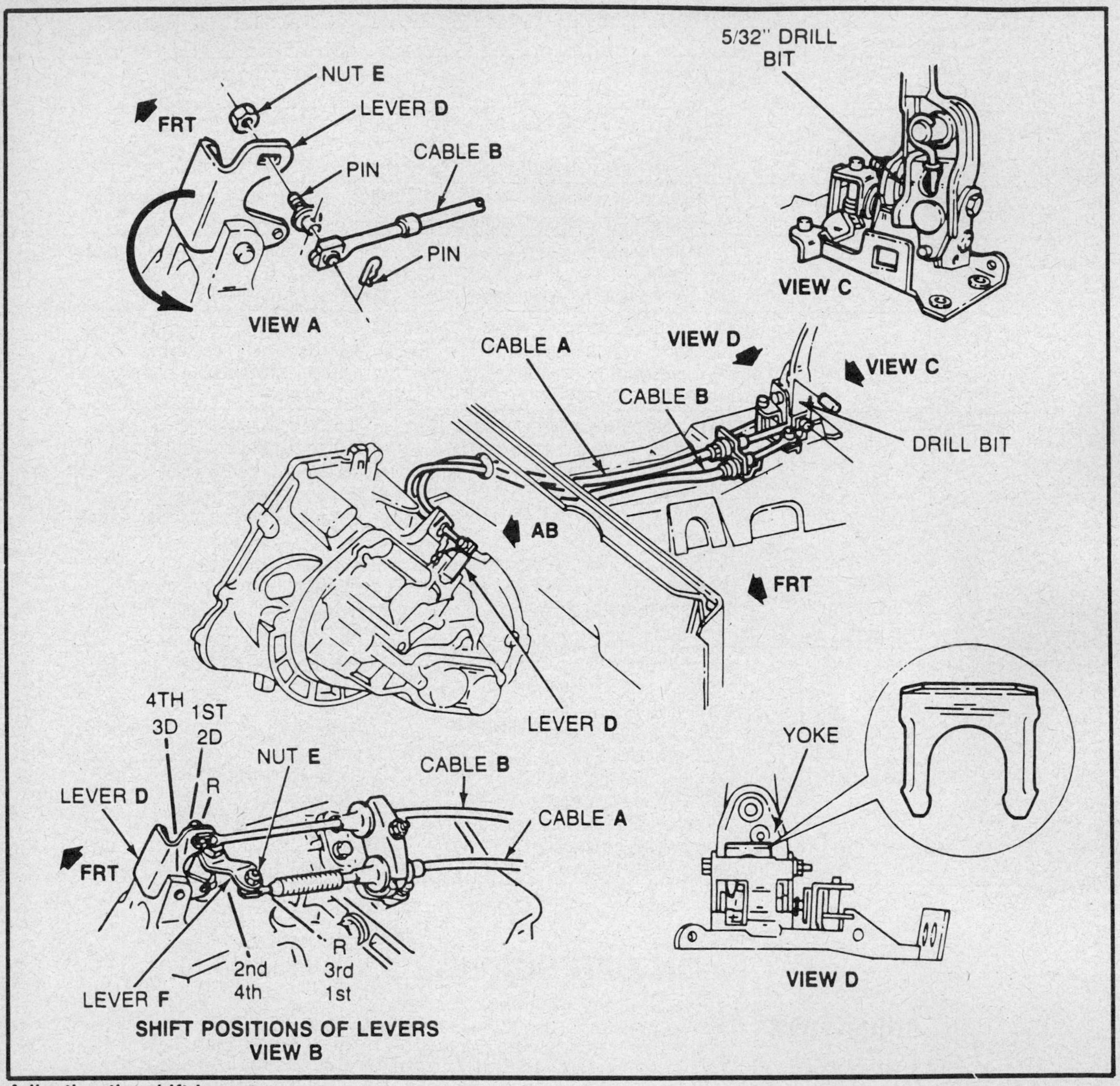
NUT E
LEVER D
FRT
PIN
CABLE B
PIN
VIEW A
5/32" DRILL BIT
VIEW C
CABLE A
VIEW D
VIEW C
CABLE B
DRILL BIT
AB
FRT
LEVER D
4TH 3D
1ST 2D
R
NUT E
CABLE B
LEVER D
CABLE A
FRT
R
2nd 4th
3rd 1st
LEVER F
SHIFT POSITIONS OF LEVERS
VIEW B
YOKE
VIEW D

Adjusting the shift lever

5. Subtract dimension **B** from dimension **A** and secure dimension **C**.

6. Using the dimension **C** and shifter shaft selective washer chart, select the correct shim washer to be used on the reinstallation.

Services

CONTROL ASSEMBLY

Removal and Installation

1. Disconnect the negative battery cable.
2. At the transaxle's shifting levers, loosen the shift cables.
3. At the shift control, remove the knob, the control cover and boot or console, if equipped.
4. Disconnect the shift cables from the shift control.
5. Remove the shift control assembly-to-chassis nuts and the assembly.
6. To install, position the shift control assembly into the vehicle and torque the mounting nuts to 15–20 ft. lbs. (20–27 Nm).
7. Connect and adjust the shift cables at the control assembly.
8. Install the control cover, boot or console, if equipped and the shift knob.
9. Connect the negative battery cable.

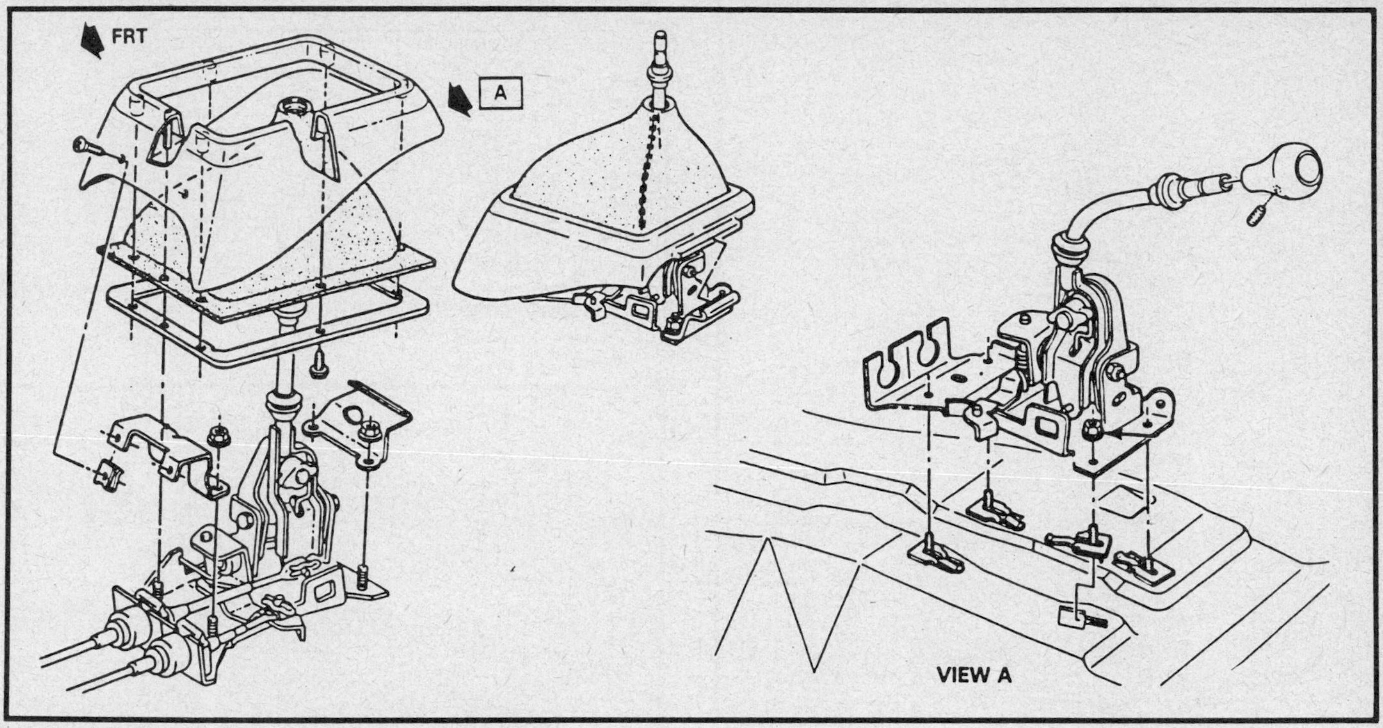
FRT
A
VIEW A

Exploded view of the shift control assembly

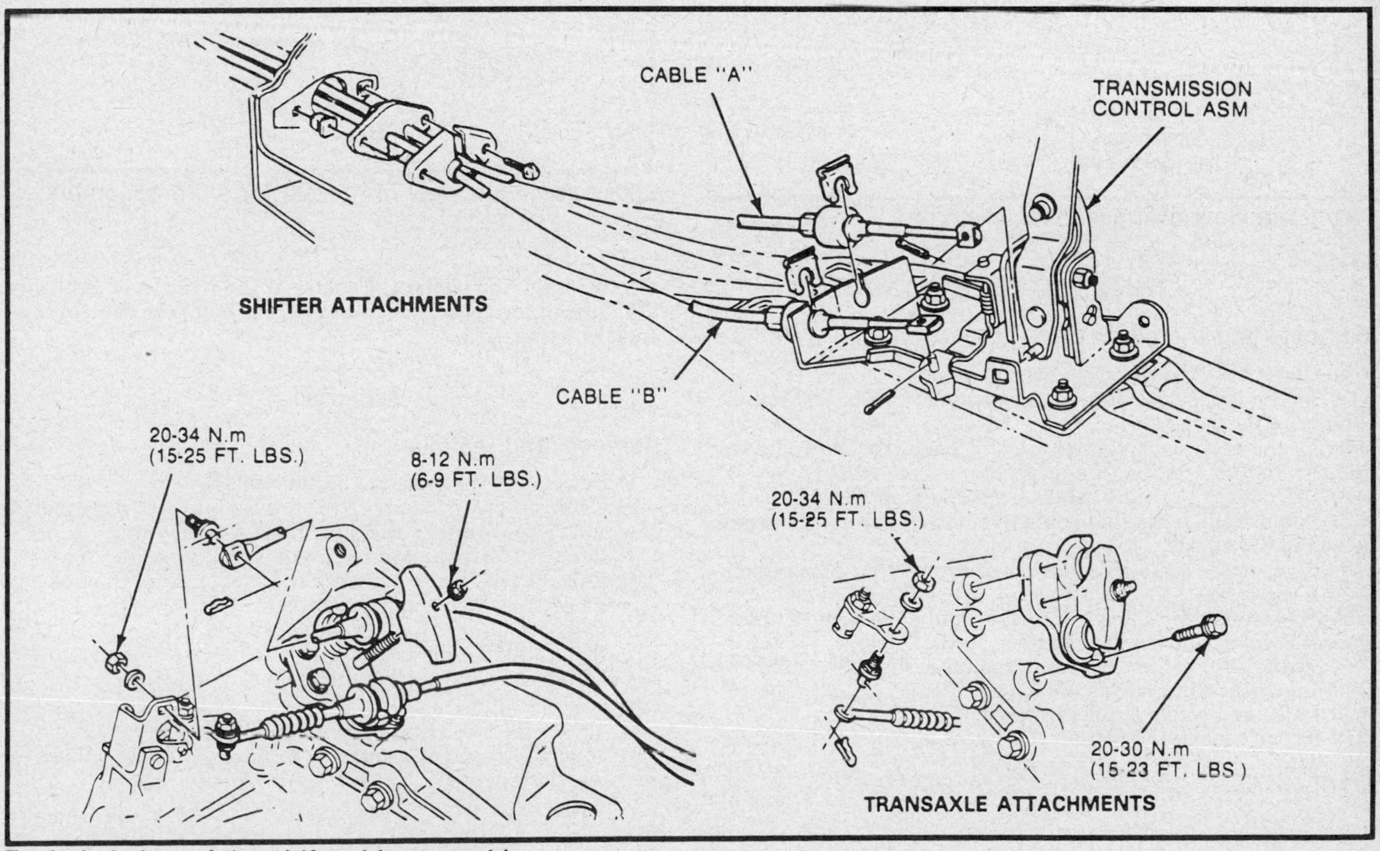
CABLE "A"
TRANSMISSION CONTROL ASM
SHIFTER ATTACHMENTS
CABLE "B"
20-34 N.m (15-25 FT. LBS.)
8-12 N.m (6-9 FT. LBS.)
20-34 N.m (15-25 FT. LBS.)
20-30 N.m (15-23 FT. LBS.)
TRANSAXLE ATTACHMENTS

Exploded view of the shift cable assembly

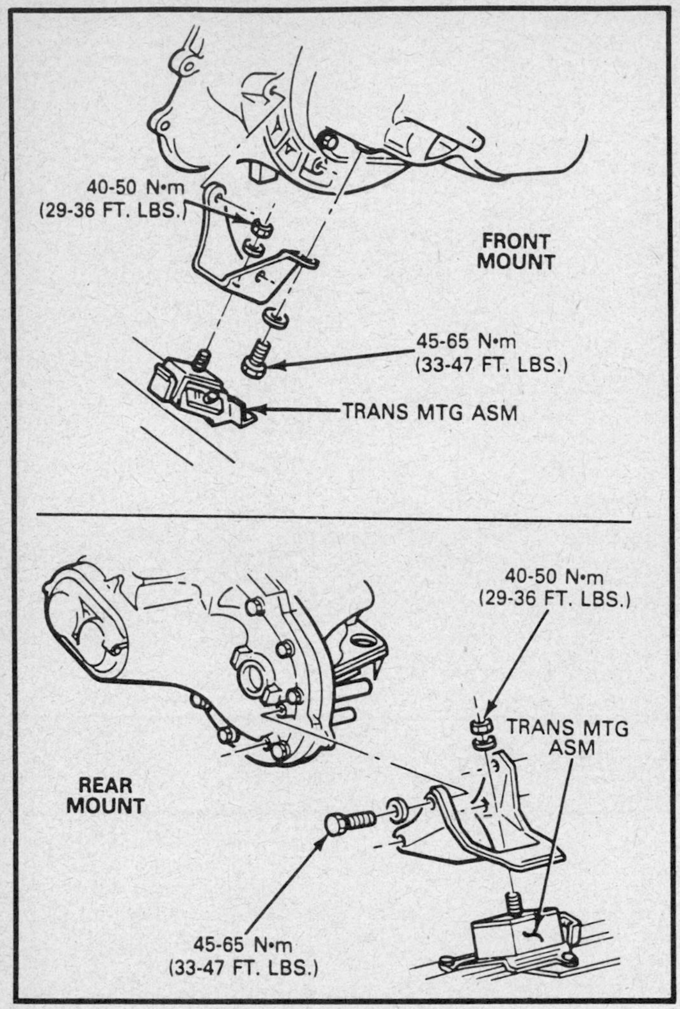
40-50 N•m
(29-36 FT. LBS.)
FRONT
MOUNT
45-65 N•m
(33-47 FT. LBS.)
TRANS MTG ASM
40-50 N•m
(29-36 FT. LBS.)
TRANS MTG
ASM
REAR
MOUNT
45-65 N•m
(33-47 FT. LBS.)

Exploded view of the transaxle mounts

SHIM PART NO.	DIM C (MM)	COLOR & NO OF STRIPES
14008235	1.8	3 WHITE
476709	2.1	1 ORANGE
476710	2.4	2 ORANGE
476711	2.7	3 ORANGE
476712	3.0	1 BLUE
476713	3.3	2 BLUE
476714	3.6	3 BLUE
476715	3.9	1 WHITE
476716	4.2	2 WHITE

Shifter shaft selective washer chart

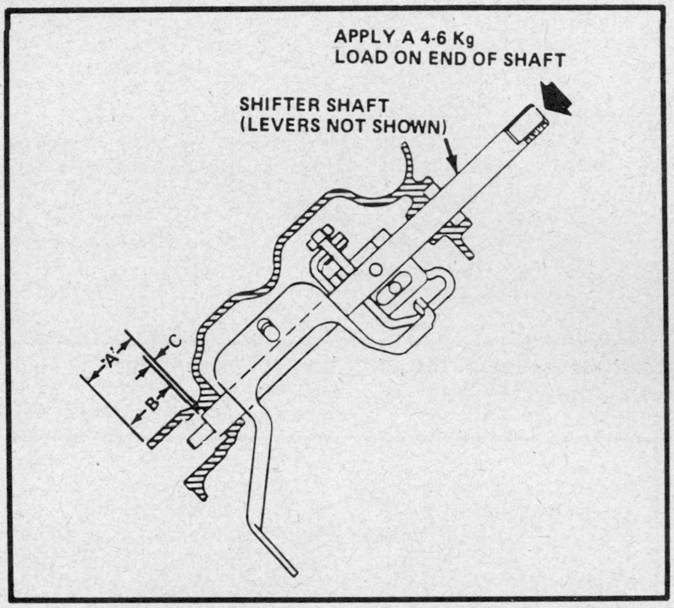
APPLY A 4-6 Kg
LOAD ON END OF SHAFT
SHIFTER SHAFT
(LEVERS NOT SHOWN)
A
B
C

Cross-sectional view of the shifter shaft assembly

SHIFT CABLES

Removal and Installation

1. Disconnect the negative battery cable.
2. At the transaxle, disconnect the shift cables and remove the retaining clamp.
3. At the shift control, remove the knob, the control cover and the boot or console (if equipped).
4. Disconnect the shift cables from the shift control.
5. Remove the front sill plate and pull the carpet back to gain access to the cables.
6. Remove the cable grommet cover screws, the floor pan cover and the cables.
7. To install the cables, route them into the vehicle, install the cable cover and screws (at floor pan).
8. At the shift control assembly, connect the shift cables. Reposition the carpet and install the sill plate.
9. Raise and safely support the vehicle.
10. At the transaxle, position the cables and install the retaining clamps.
11. Lower the vehicle.
12. Connect and adjust the cables at the shift lever assembly.
13. Install the console, if equipped, the boot, the control cover and the knob.

MOUNTS

Removal and Installation

1. Disconnect the negative battery cable.
2. Using an engine support fixture tool, secure it to the engine and raise it to take the weight off the mounts.
3. Remove the mount-to-transaxle bracket nut.
4. Raise the engine to separate the mount from the bracket.
5. Raise and safely support the vehicle.
6. Remove the mount-to-cradle nuts and the mount.
7. Install the mount and torque the mount-to-cradle nuts to 29–36 ft. lbs. (40–50 Nm) and the mount-to-transaxle bolts to 33–47 ft. lbs. (45–65 Nm).
8. Lower the vehicle and the engine.
9. Remove the engine support fixture tool and connect the negative battery cable.

REMOVAL AND INSTALLATION

TRANSAXLE REMOVAL

NOTE: The transaxle is removed without the engine.

1. Disconnect the negative battery cable from the transaxle and attach it the upper radiator hose with tape or a wire.
2. Disconnect the electrical connector from the horn and remove the horn from the vehicle.
3. Remove the air cleaner assembly.
4. To disconnect the clutch cable from the transaxle, perform the following procedures:
   a. While supporting the clutch pedal (upward) against the bumper stop, release the pawl from the quadrant.
   b. At the transaxle, disconnect the clutch cable from the clutch release lever.
3. If equipped with a V6 engine, disconnect the fuel lines, the fuel clamps from the clutch cable bracket and the exhaust crossover pipe.
4. Remove the clutch cable bracket from the transaxle.
5. At the transaxle, remove the shift linkage retaining clips, the shift cable clips from the mounting bosses and the speedometer cable or sensor.
6. Remove the upper engine-to-transaxle bolts.
7. Using an engine support fixture tool, secure it to the engine and support it.
8. Raise and safely support the vehicle. Drain the fluid from the transaxle.
9. Install halfshaft boot seal protectors on the left halfshaft to keep the boot from becoming damaged.
10. Remove the left front wheel/tire assembly.
11. To remove the left side cradle and crossmember, perform the following procedures:
   a. Rotate the steering wheel so the intermediate shaft-to-steering gear stub shaft attaching bolt is in the up position; remove the bolt.
   b. Raise the vehicle and position a jackstand under the engine to act as a support.
   c. Remove the power steering pressure and return line bracket.
   d. Remove the steering gear-to-chassis bolts and support the steering gear.
   e. If equipped, disconnect the driveline vibration absorber.
   f. From the left hand lower control arm, disconnect the front stabilizer bar.
   g. From the left steering knuckle, disconnect the lower ball joint.
   h. Remove both front stabilizer bar reinforcements and bushings.
   i. Using a ½ in. drill bit at the left hand front stabilizer bar mounting, drill through the spot weld located between the rear holes.
   j. Disconnect the engine and transaxle mounts from the cradle.
   k. Remove the side member-to-crossmember bolts, the left hand body mount bolts, the left side member/front crossmember assembly; it may be necessary to pull or gently pry the crossmember loose.
12. Using the halfshaft remover tool and a slide hammer puller, disengage the halfshafts from the transaxle; the tool will prevent damaging the CV joints. Using a wire, support the left halfshaft.

NOTE: The right halfshaft can be removed when the transaxle is being removed from the vehicle.

13. Remove the flywheel and starter shield bolts and the shield.
14. Using a transmission jack, secure the transaxle to it.
15. Remove the last transaxle-to-engine bolt (located on engine side), slide the transaxle from the engine and lower it from the vehicle.

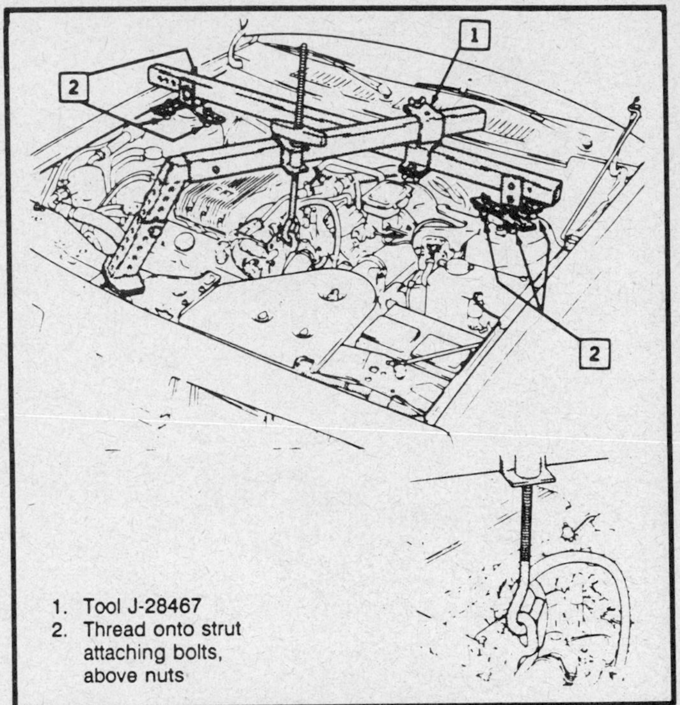
1
2
2
1. Tool J-28467
2. Thread onto strut attaching bolts, above nuts

Supporting the engine with the engine support fixture tool

TRANSAXLE INSTALLATION

1. When installing the transaxle, position the right halfshaft into the transaxle's bore.
2. Install the transaxle-to-engine bolts.
3. Install the flywheel and starter shield and secure with the bolts.
4. To install the side cradle and crossmember, perform the following procedures:
   a. Install the side member/crossmember assembly, the left hand body mount and secure with bolts.
   b. Using a ½ in. drill bit, insert it through the drilled hole in the rear cradle crossmember. Install the side member-to-crossmember bolts and torque to 22–29.6 ft. lbs. (30–40 Nm).
   c. Torque the left hand body mount bolts to 29.6–50.4 ft. lbs. (40–68 Nm).
   d. Connect the engine and transaxle mounts.
   e. Install the stabilizer bar bushings, reinforcements and remove the drill bit.
   f. Connect the front ball joint-to-steering knuckle, the front stabilizer gar-to-control arm and the drive vibration absorber (if equipped).
   g. Install the steering gear mount bolts, the power steering hose brackets and the steering shaft bolt.
5. Install the left wheel assembly.
6. Lower the vehicle and refill the transaxle with new fluid.
7. Remove an engine support fixture tool.
8. Install the upper engine-to-transaxle bolts and torque the bolts to 47–62 ft. lbs. (65–85 Nm).
9. At the transaxle, install the shift linkage retaining clips, the shift cable clips from the mounting bosses and the speedometer cable or sensor.

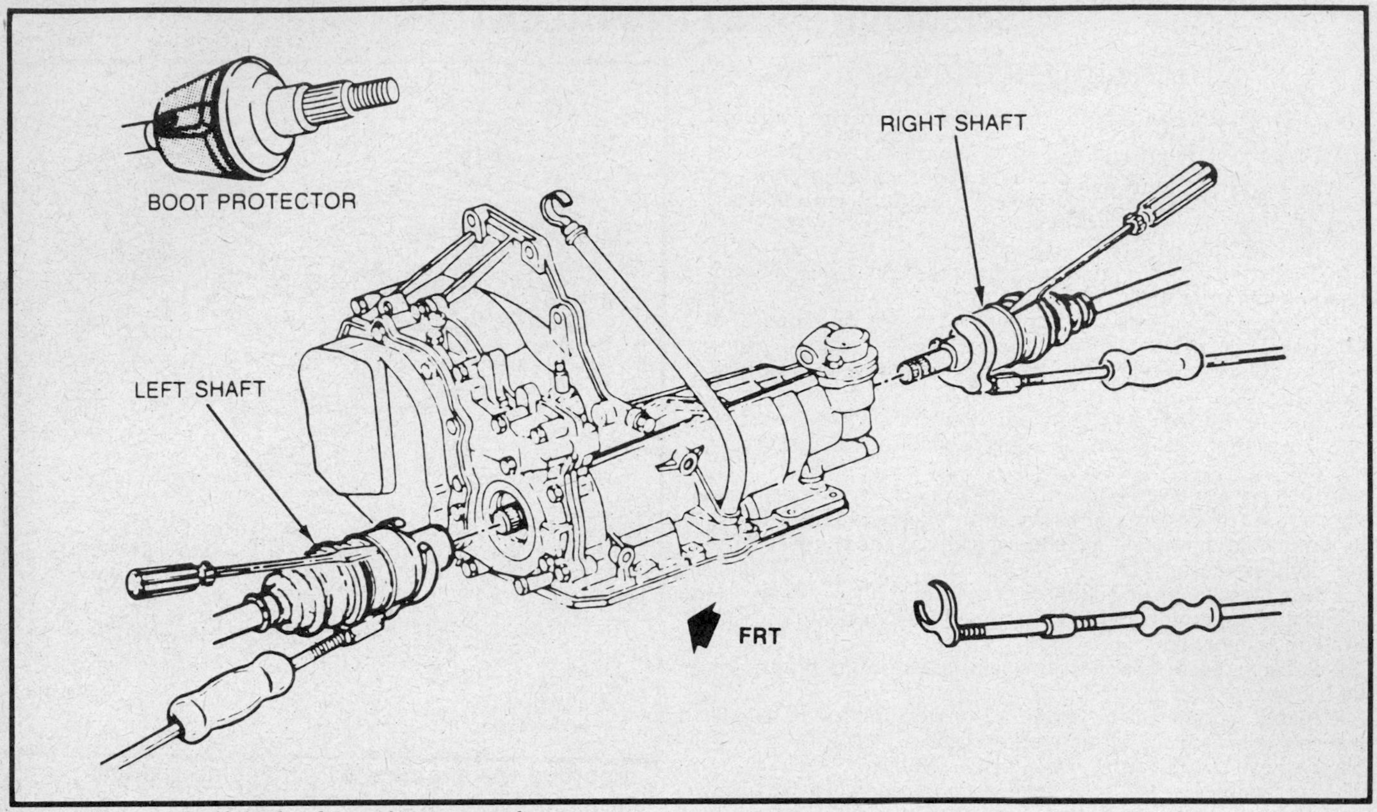
BOOT PROTECTOR
RIGHT SHAFT
LEFT SHAFT
FRT

Removing the halfshafts from the transaxle

10. Install the clutch cable bracket from the transaxle.
11. If equipped with a V6 engine, connect the fuel lines, the fuel clamps to the clutch cable bracket and the exhaust cross-over pipe.
12. To install the clutch cable, perform the following procedures:
    a. Attach the clutch cable to the clutch pedal quadrant; be sure to pass the cable under the pawl.
    b. While supporting the clutch pedal (upward) against the bumper stop (to release the pawl from the quadrant), attach the outer end to the cable release lever; be sure not to yank on the cable, for damage could occur to the quadrant.
    c. Check the clutch operation.
    d. To adjust, lift the clutch pedal (upward) to allow the mechanism to adjust the cable length. Depress the pedal slowly (several times) to set the pawl into mesh with the quadrant teeth.
13. Install the air cleaner assembly.
14. Install horn into the vehicle and connect the electrical connector to the horn.
15. Connect the negative battery cable to the transaxle.
16. Test drive the vehicle and check the operation of the transaxle.

BENCH OVERHAUL

Before Disassembly

When servicing the transaxle it is important to be aware of cleanliness. Before disassembling the transaxle, the outside should be throughly cleaned, preferably with a high-pressure spray cleaning equipment. Dirt entering the unit may negate all the effort and time spent on the overhaul.

During inspection and reassembly, all parts should be cleaned with solvent and dried with compressed air. Lubricate the seals with transaxle fluid and use petroleum jelly to hold the thrust washers; this will ease the assembly of the seals and not leave harmful residues in the system. Do not use solvent on neoprene seals, if they are to be reused.

Before installing bolts into aluminum parts, dip the threads into clean transmission fluid. Anti-seize compound may be used to prevent galling the aluminum or seizing. Be sure to use a torque wrench to prevent stripping the threads. Be especially careful when installing the seals, the smallest nick can cause a leak. Aluminum parts are very susceptible to damage; great care should be used when handling them. Reusing snaprings is not recommended but should they be: compress the internal ones and compress the external ones.

Transaxle Disassembly

1. Secure the transaxle to a work stand.
2. Remove the clutch cover-to-transaxle bolts. Using a plastic hammer, carefully tap the clutch cover from the transaxle case.
3. Remove the ring gear/differential assembly.
4. Move the shifter shaft into the **N** position; the shifter should move freely and not be engaged in any gear.
5. At the shifter shaft, bend back the lock tab and remove the bolt. Remove the shifter shaft and the shift fork shaft from the synchronizer forks.

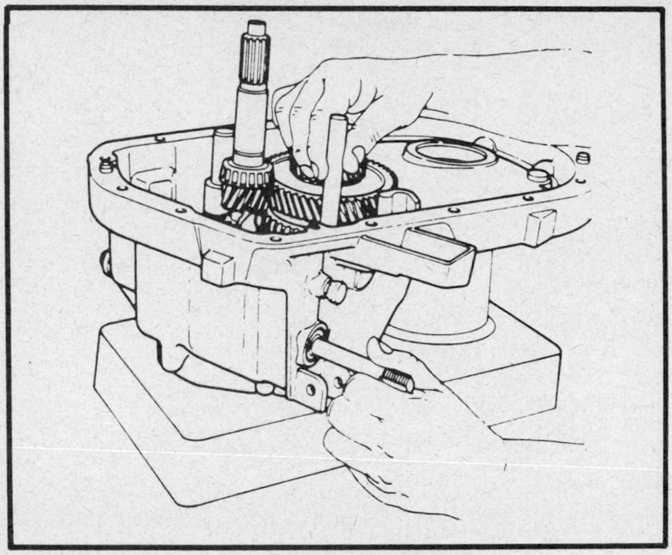

Removing the shifter shaft from the case

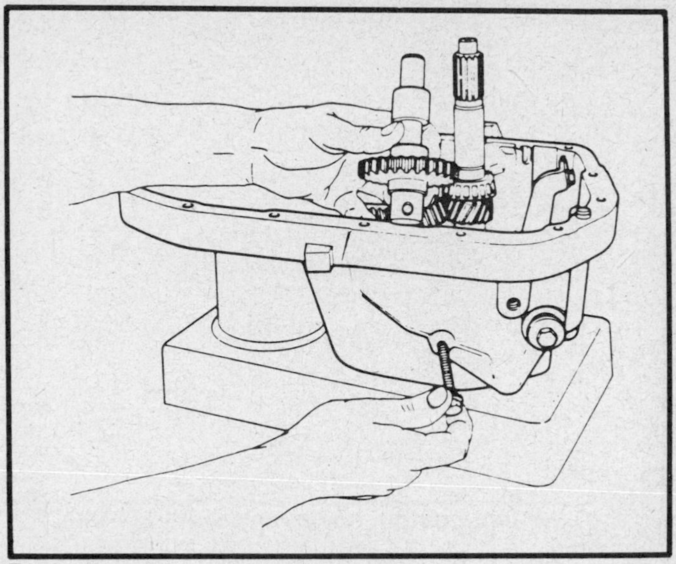

Removing the reverse idler gear

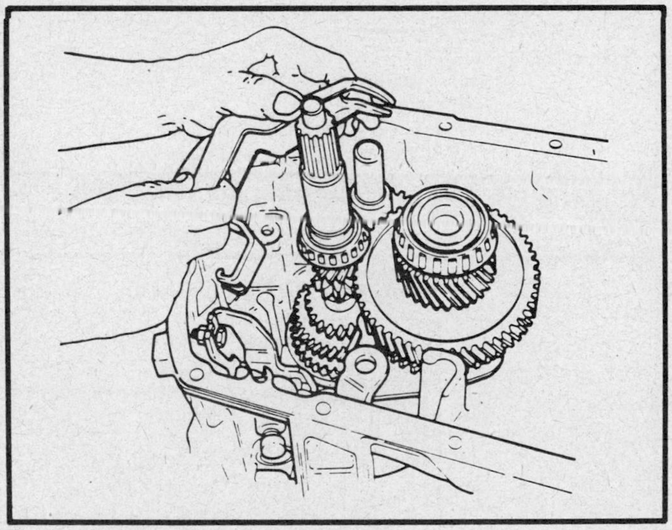

Removing the reverse shift fork

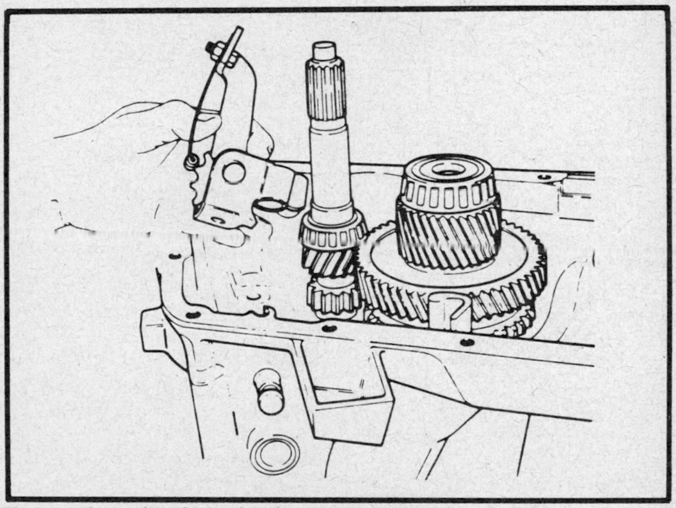

Removing the interlock assembly

6. Disengage the reverse shift fork from the guide pin and interlock bracket, then, remove the reverse shift fork.
7. Remove the reverse idler gear shaft lock bolt and the gear/shaft/spacer assembly.
8. Remove the detent shift lever and interlock assembly. Do not remove the shift forks from the synchronizer or the detent spring.
9. Grasp both the input and output shafts and lift them (as an assembly) from the case.

NOTE: When removing the input/output shaft assembly, note the position of the shift forks for reassembly purposes.

Unit Disassembly And Assembly

NOTE: The terms right hand and left hand refer to the installed positions on the vehicle. Right hand refers to the end nearest the clutch (passenger side); left hand refers to the end farthest from the clutch (drivers side).

INPUT SHAFT

Disassembly

1. From the input shaft, remove (slide) the left hand bearing and the 4th gear.
2. Remove the brass blocker ring and the snapring from the 3–4 synchronizer.
3. Using a shop press and support plates (behind the 3rd gear), press the 3rd gear and 3–4 synchronizer from the input shaft.
4. Using the input shaft right hand bearing remover tool or equivalent, press the right hand bearing from the shaft.

Inspection

1. Check the input shaft bearing surfaces for scoring and/or wear; if necessary, replace the shaft.
2. Check the input shaft splines for wear; if necessary, replace the shaft.
3. Inspect the bearings for scoring, wear and/or damage; if necessary, replace them.

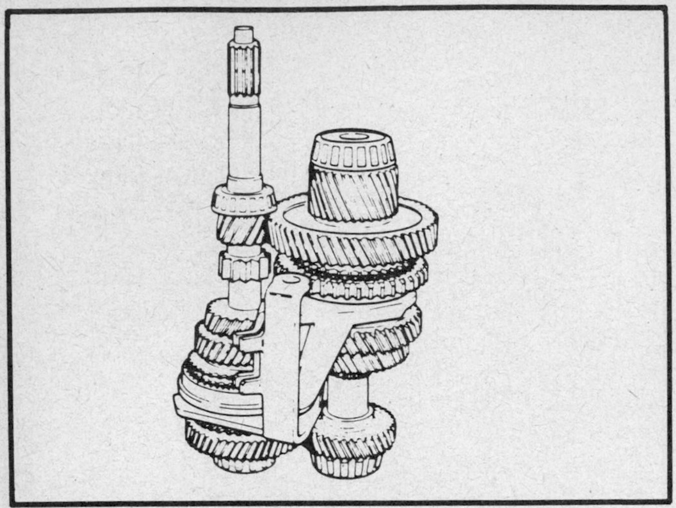

View of the input/output shafts and shifting forks assembly

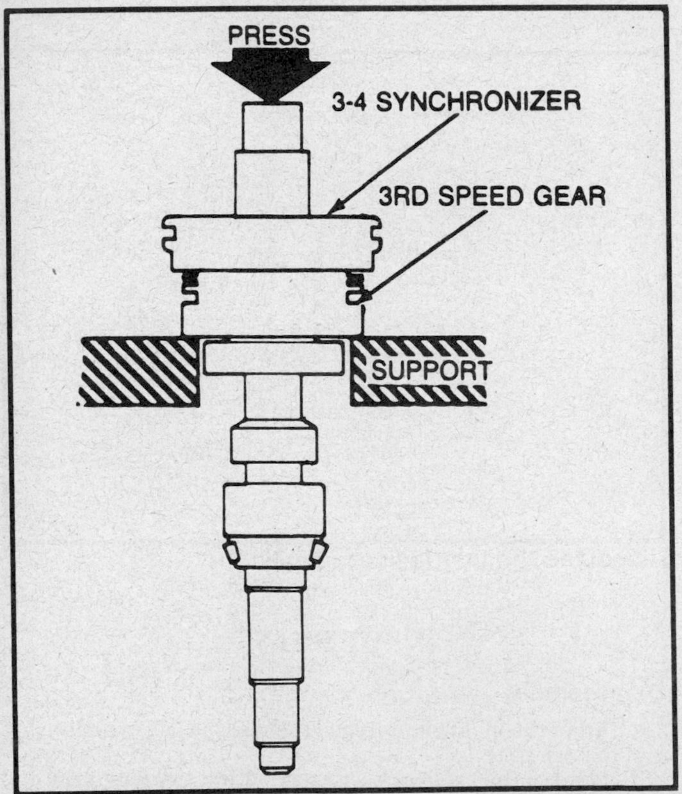
PRESS
3-4 SYNCHRONIZER
3RD SPEED GEAR
SUPPORT

Removing the 3–4 synchronizer and 3rd gear from the input shaft

4. Check the gears and synchronizer parts for damage and/or excessively worn teeth; if necessary, replace them.

Assembly

1. Using the input shaft right hand bearing installer tool or equivalent and a shop press, press the right hand bearing onto the input shaft.
2. Position the 3rd gear onto the shaft (aligned with the 3–4 synchronizer), the brass blocker ring onto the gear cone and the 3–4 synchronizer. Using a deep socket or a piece of pipe, position

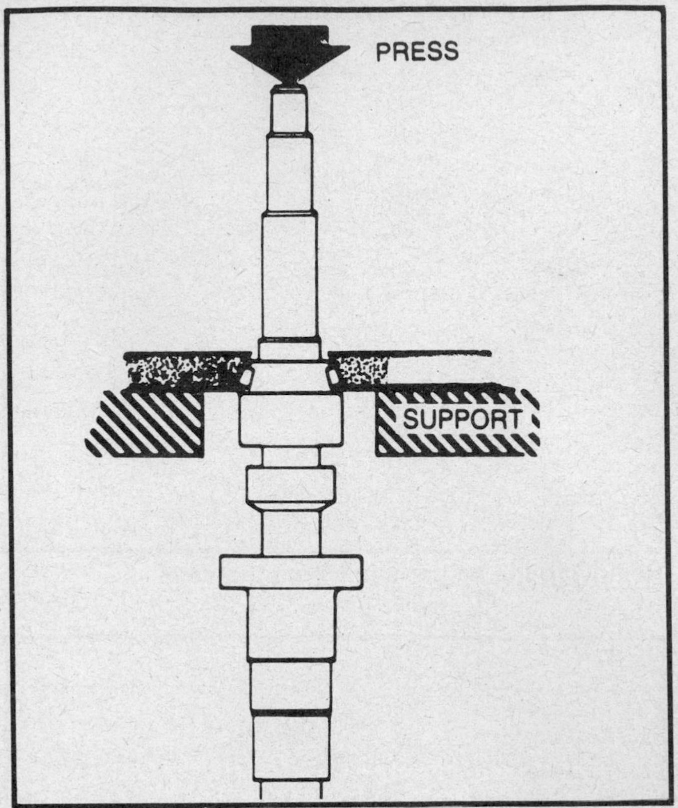
PRESS
SUPPORT

Removing the right hand bearing from the input shaft

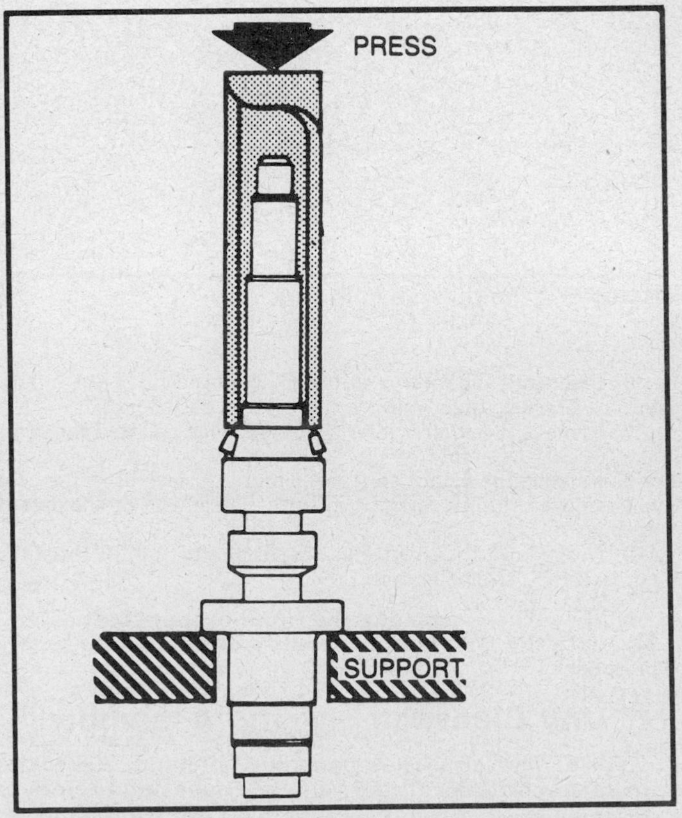
PRESS
SUPPORT

Installing the right hand bearing onto the input shaft

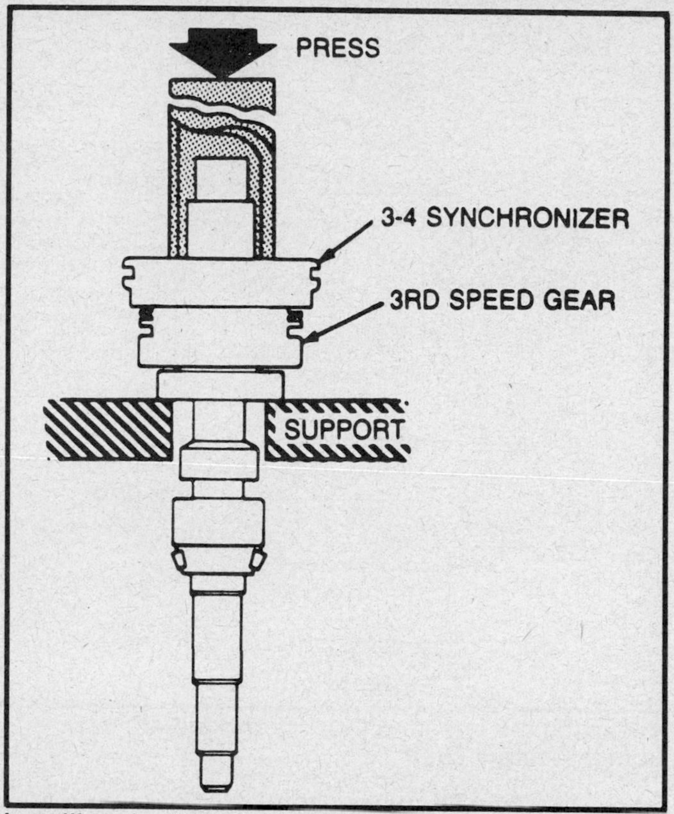
PRESS
3-4 SYNCHRONIZER
3RD SPEED GEAR
SUPPORT
Installing the 3rd and 3–4 synchronizer onto the input shaft

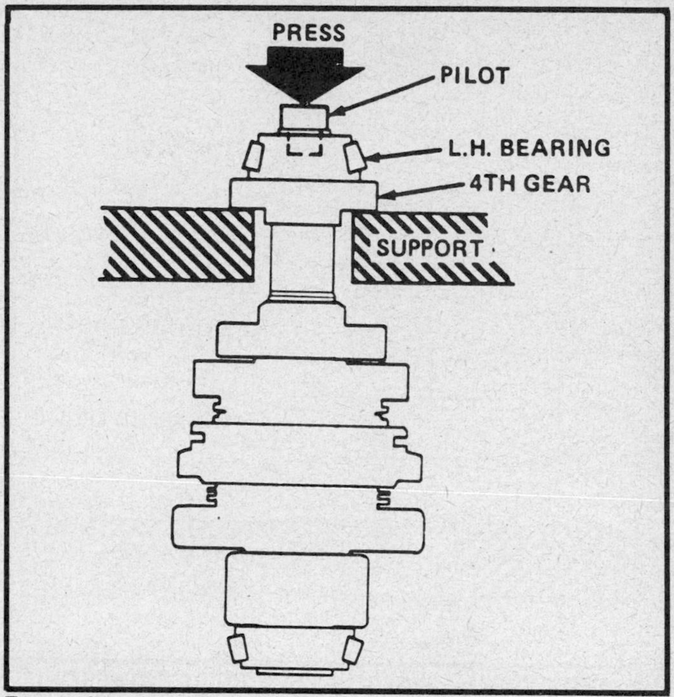
PRESS
PILOT
L.H. BEARING
4TH GEAR
SUPPORT
Removing the left hand bearing and 4th gear from the output shaft

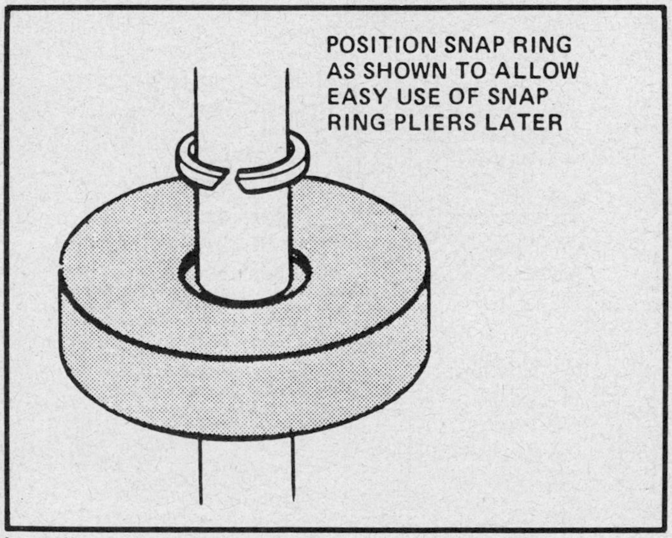
POSITION SNAP RING
AS SHOWN TO ALLOW
EASY USE OF SNAP
RING PLIERS LATER
Installing the snapring onto the input shaft

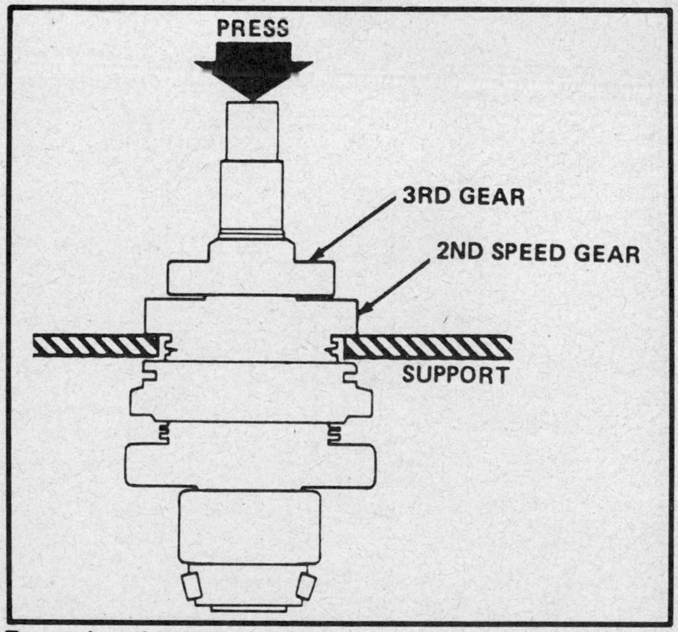
PRESS
3RD GEAR
2ND SPEED GEAR
SUPPORT
Removing the 3rd and 2nd gears from the output shaft

it on the synchronizer hub (not the sleeve) and press the synchronizer hub onto the shaft.

3. Install the snapring (beveled edges away from synchronizer) to secure the 3–4 synchronizer.

4. Install the brass blocker ring.

5. Position the 4th gear (aligned with the 3–4 synchronizer) and the left hand bearing onto the input shaft.

OUTPUT SHAFT

Disassembly

EXCEPT MX6

1. Using a shop press, the input/output shaft pilot tool (for bearing removal) or equivalent, and support plates (behind the

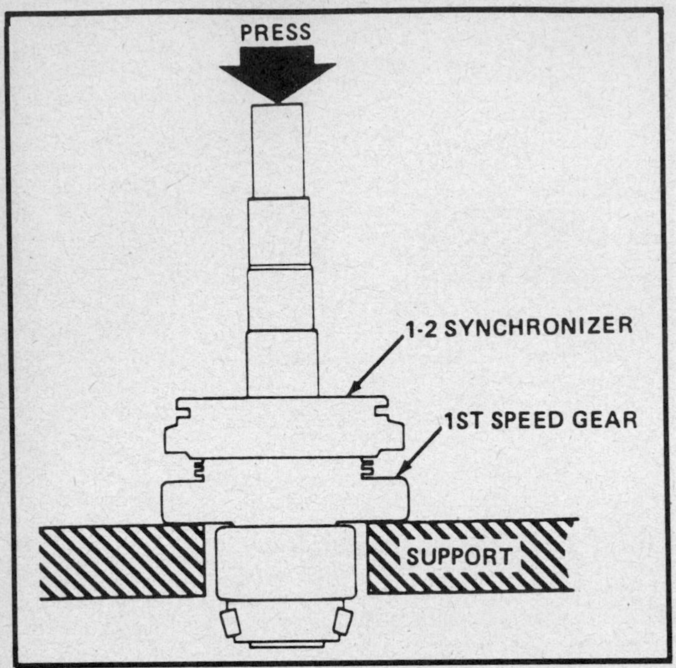
PRESS
1-2 SYNCHRONIZER
1ST SPEED GEAR
SUPPORT

Removing the 1–2 synchronizer and the 1st gear from the output shaft

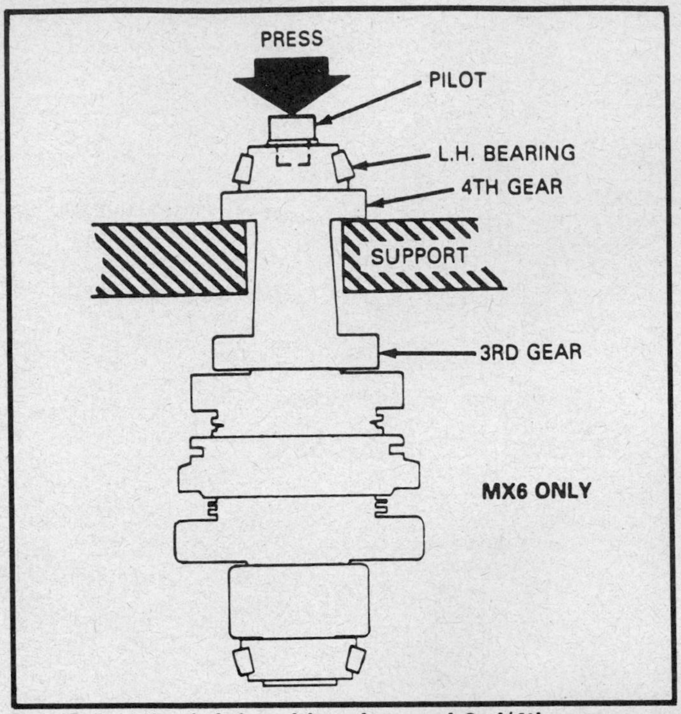
PRESS
PILOT
L.H. BEARING
4TH GEAR
SUPPORT
3RD GEAR
MX6 ONLY

Removing the left hand bearing and 3rd/4th gear from the output shaft

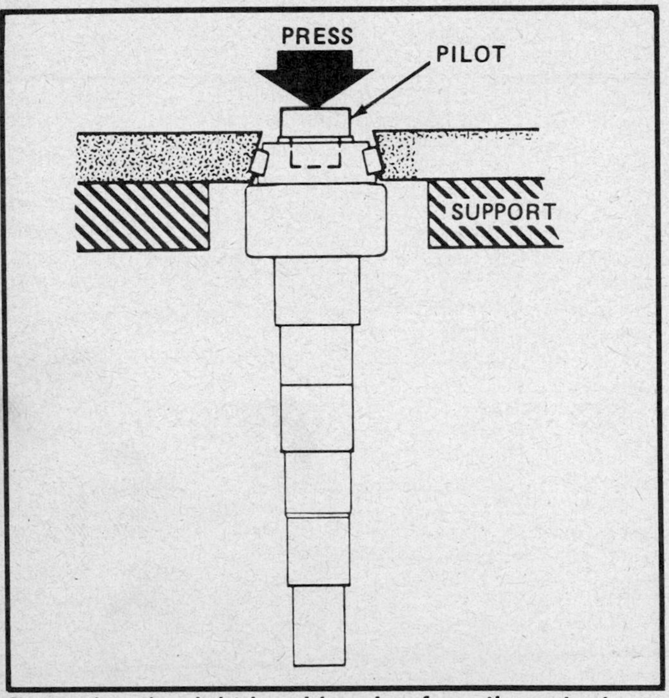
PRESS
PILOT
SUPPORT

Removing the right hand bearing from the output shaft

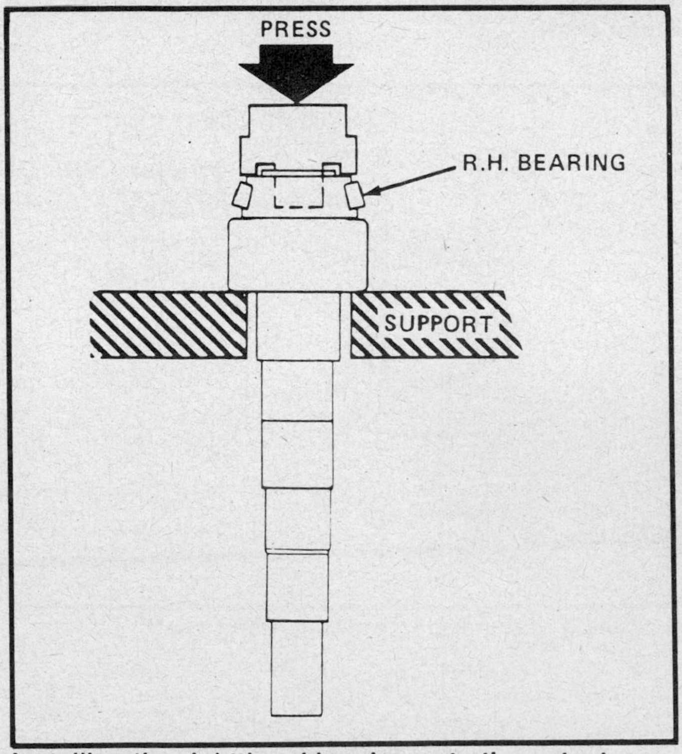
PRESS
R.H. BEARING
SUPPORT

Installing the right hand bearing onto the output shaft

4th gear), press the 4th gear and the left hand bearing from the output shaft.

2. Remove the 3rd gear snapring.
3. Move the 1–2 synchronizer into the 1st gear position and allow the press to support the 2nd gear. Press the 2nd and 3rd gears from the output shaft. Remove the brass blocker ring.
4. Remove the 1–2 synchronizer snapring.
5. Place press plates behind the 1st gear and press the 1st gear/1–2 synchronizer from the output shaft.
6. Using the output shaft right hand bearing remover tool or equivalent, and the input/output shaft pilot tool or equivalent, press the right hand bearing from the output shaft.

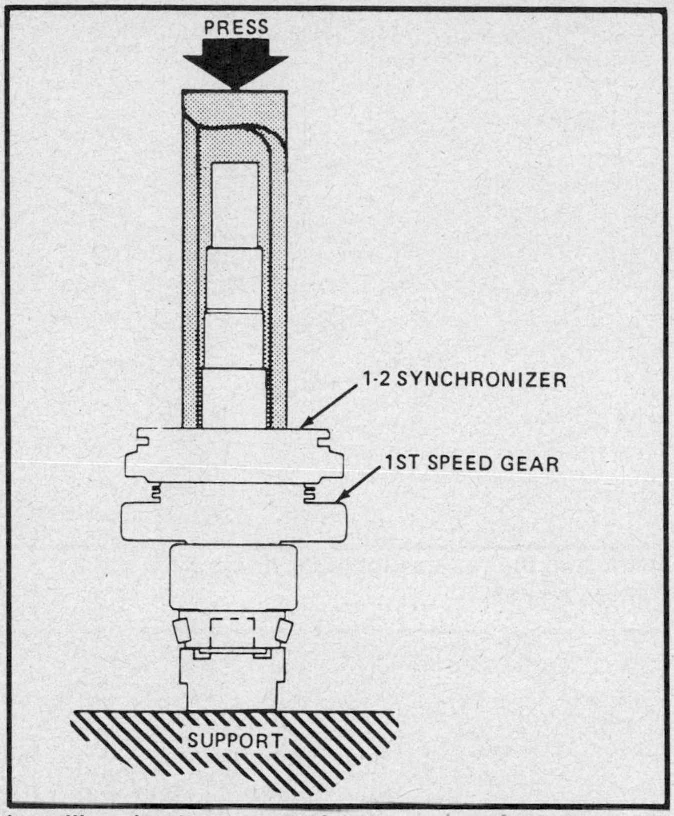
PRESS
1-2 SYNCHRONIZER
1ST SPEED GEAR
SUPPORT

Installing the 1st gear and 1–2 synchronizer onto the output shaft

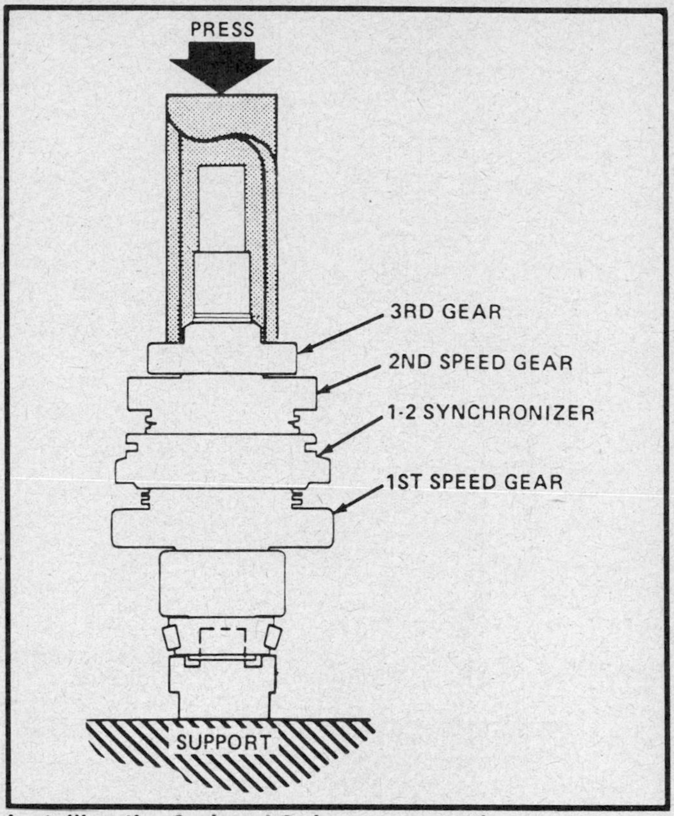
PRESS
3RD GEAR
2ND SPEED GEAR
1-2 SYNCHRONIZER
1ST SPEED GEAR
SUPPORT

Installing the 2nd and 3rd gears onto the output shaft

MX6

1. From the left end of the shaft, remove the retainer.

NOTE: The retainer and shaft are designed with left hand threads. The 3rd/4th gear is a one piece gear.

2. Using a shop press, the input/output shaft pilot tool (for bearing removal) or equivalent, and support plates (behind the 4th gear), press the 3rd/4th gear and the left hand bearing from the output shaft.
3. Remove the 2nd gear and the brass blocker ring from the output shaft.
4. Remove the 1–2 synchronizer snapring.
5. Using a shop press and press plates (behind the 1st gear), press the 1st gear and the 1–2 synchronizer from the output shaft.
6. Using the output shaft right hand bearing remover tool or equivalent, and the input/output shaft pilot tool or equivalent, press the right hand bearing from the output shaft.

Inspection

1. Check the output shaft bearing surfaces for scoring and/or wear; if necessary, replace the shaft.
2. Check the output shaft splines for wear; if necessary, replace the shaft.
3. Inspect the bearings for scoring, wear and/or damage; if necessary, replace them.
4. Check the gears and synchronizer parts for damage and/or excessively worn teeth; if necessary, replace them.

Assembly

EXCEPT MX6

1. Using a shop press and the input/output shaft pilot tool or equivalent, press the right hand bearing onto the output shaft.
2. Position the 1st gear onto the shaft (aligned with the 1–2 synchronizer), the brass blocker ring onto the gear cone and the 1–2 synchronizer. Using a deep socket or a piece of pipe, position it on the synchronizer hub (not the sleeve) and press the synchronizer hub onto the shaft.
3. Install the snapring to secure the 1–2 synchronizer.
4. Install the brass blocker ring.
5. Position the 2nd gear (aligned with the 1–2 synchronizer) and press the 3rd gear (aligned with the 4th gear hub) onto the output shaft; use a deep socket or a piece of pipe, positioned on the 3rd gear hub.
6. Install the 3rd gear snapring.
7. Using support plates, press the 4th gear (aligned with the 3rd gear) onto the output shaft. Using the input/output shaft inner race installer tool or equivalent, install the left hand bearing onto the output shaft.

MX6

1. Using a shop press and the input/output shaft pilot tool or equivalent, press the right hand bearing onto the output shaft.
2. Position the 1st gear onto the shaft (aligned with the 1–2 synchronizer), the brass blocker ring onto the gear cone and the 1–2 synchronizer. Using a deep socket or a piece of pipe, position it on the synchronizer hub (not the sleeve) and press the synchronizer hub onto the shaft.
3. Install the snapring to secure the 1–2 synchronizer.
4. Install the brass blocker ring.
5. Position the 2nd gear (aligned with the 1–2 synchronizer) onto the output shaft. Using support plates, press the 3rd/4th gear onto the output shaft.
6. Using the input/output shaft inner race installer tool or equivalent, press the left hand bearing onto the output shaft.
7. Install the left hand bearing retainer and torque it to 37–55 ft. lbs. (50–75 Nm).

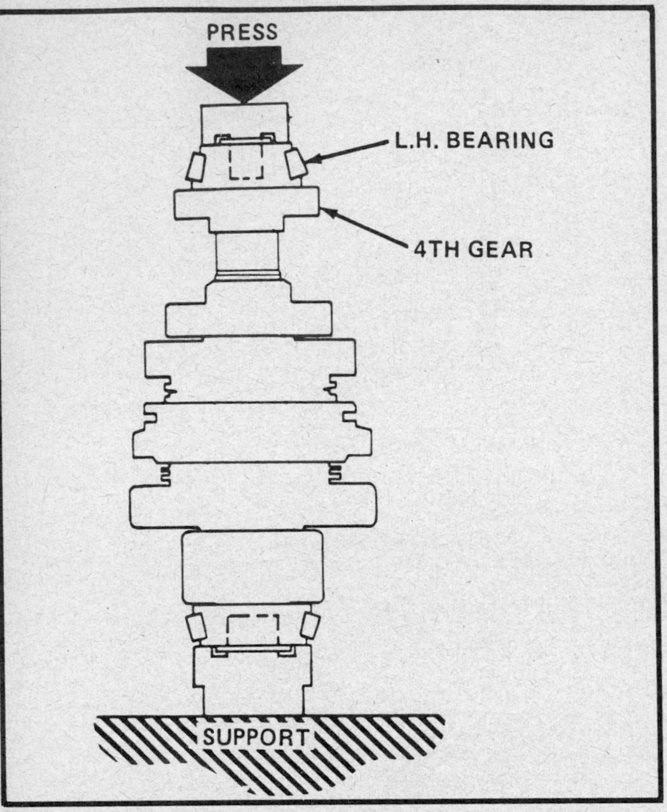
PRESS
L.H. BEARING
4TH GEAR
SUPPORT

Installing the 4th gear and left hand bearing onto the output shaft

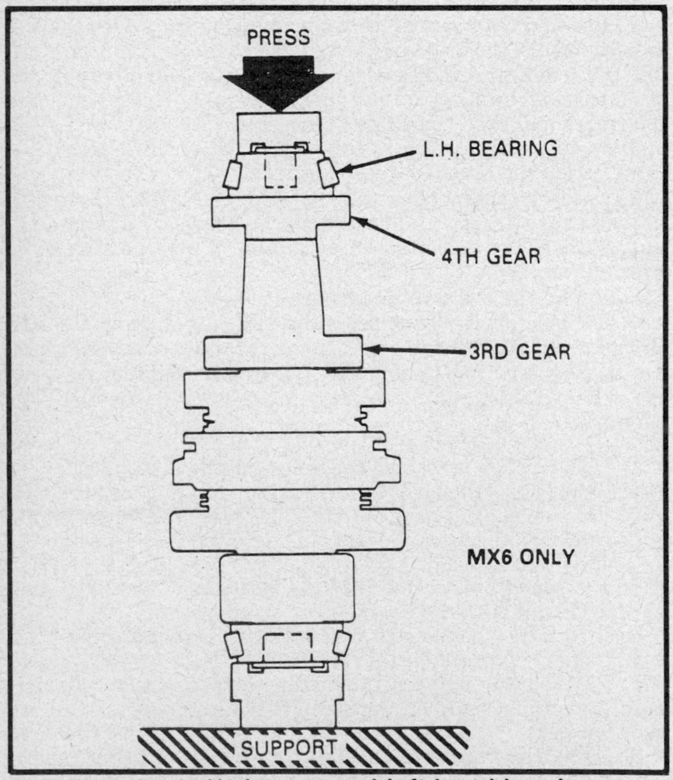
PRESS
L.H. BEARING
4TH GEAR
3RD GEAR
MX6 ONLY
SUPPORT

Installing the 3rd/4th gear and left hand bearing onto the output shaft

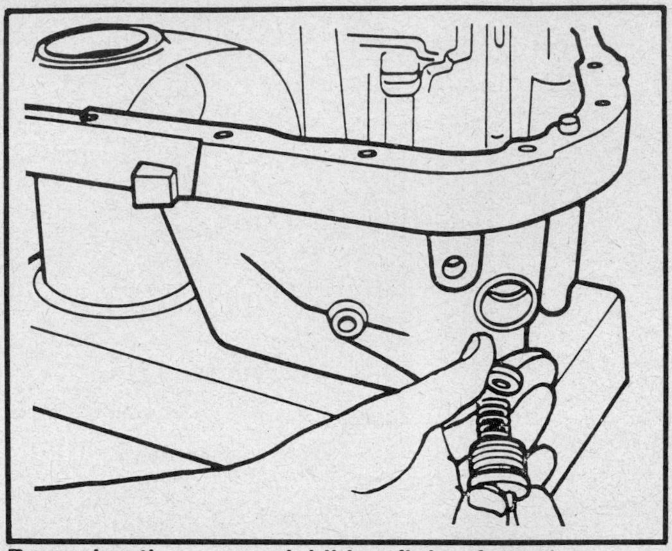

Removing the reverse inhibitor fitting from the transaxle case

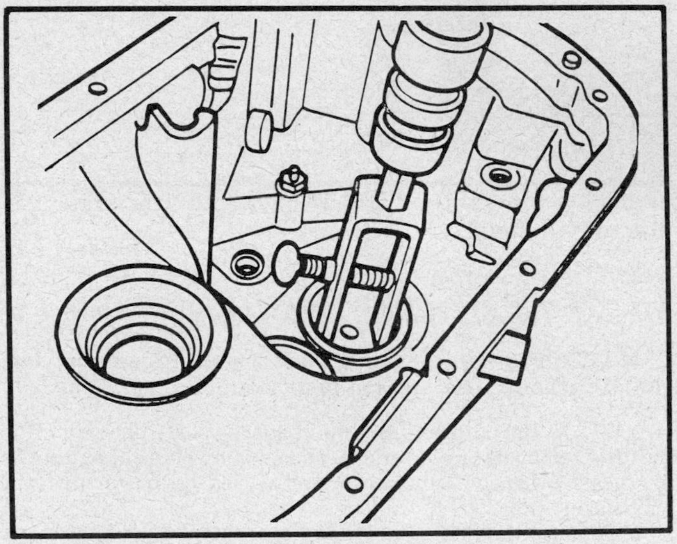

Removing the bearing cups from the transaxle case

SYNCHRONIZER

The transaxle is equipped with 2 synchronizers: a 1–2 and a 3–4.

Disassembly

1. From each synchronizer, remove both key springs.
2. Noting the relative positions, separate the hub, sleeve and keys; be sure to scribe the hub-to-sleeve location.

Inspection

Clean, inspect and/or replace any worn or damaged parts.

Assembly

1. Align the scribe marks and assemble the hub to the sleeve; the extruded hub lip should be directed away from the sleeve's shift fork groove.
2. Install a retaining ring. Carefully pry the ring back and insert the keys (one at a time); be sure to position the ring so it is captured by the keys.

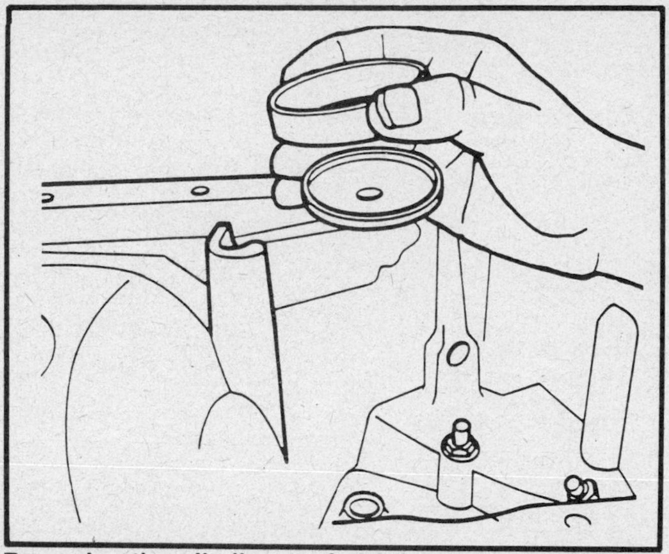
Removing the oil slingers from the transaxle case

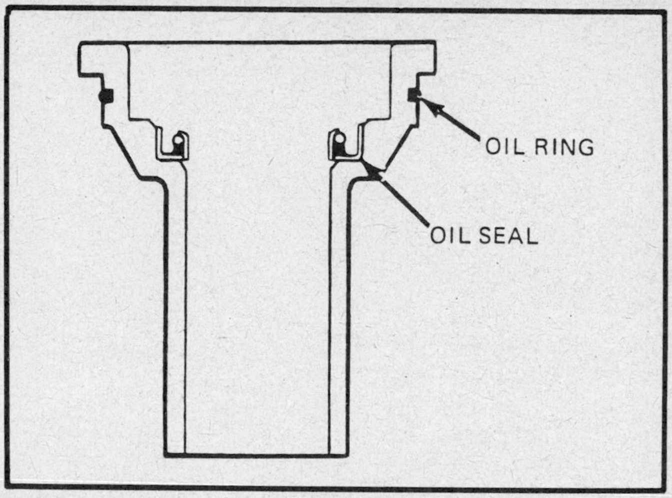
OIL RING
OIL SEAL
Cross-sectional view of the oil seal and oil ring

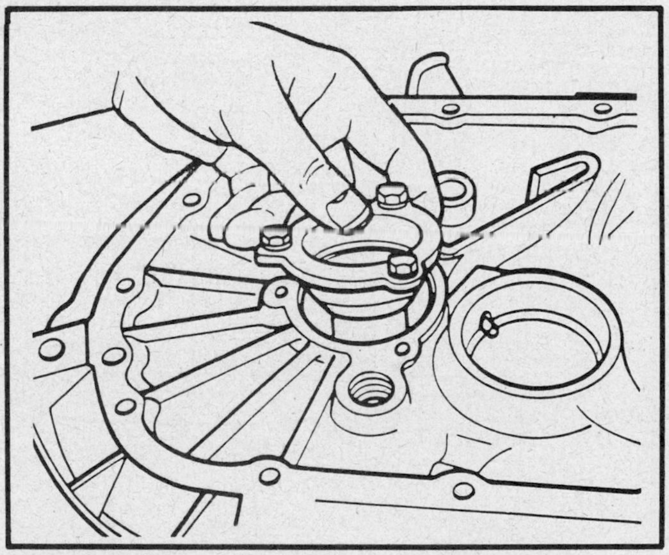
Removing the bearing release sleeve from the clutch cover

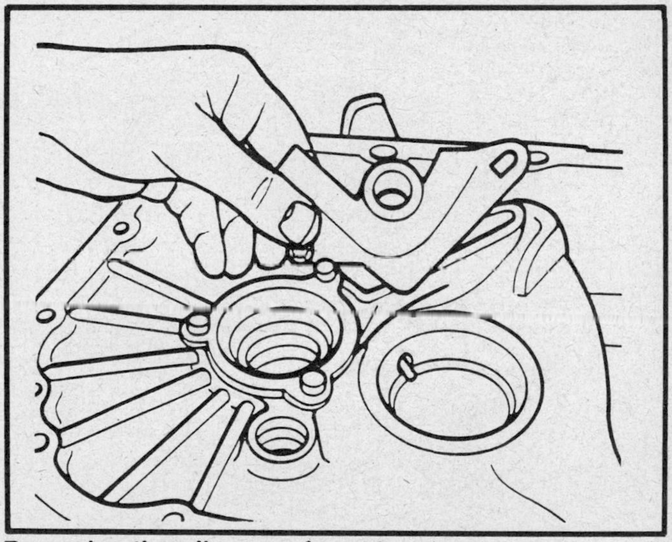
Removing the oil scoop from the clutch cover

3. Install the other retaining ring; be sure the ring's open segment is out of phase with the open segment of the other ring.

TRANSAXLE CASE

Disassembly

1. From the exterior of the case, remove the reverse inhibitor fitting.
2. From inside the case, remove the spring and pilot/spacer.
3. Using the transaxle case bearing cup remover tool or equivalent, and a slide hammer puller, press the input/output shaft left hand bearing cups and the differential side bearing cups from the transaxle case.
4. Remove the oil slingers.

Inspection

1. At the interlock bracket and the reverse shift fork, check both guide pins for wear; replace them, if necessary.
2. Check, clean and install the magnet.
3. Using a gasket remover tool or equivalent, remove the sealant from the mating surfaces; be careful not to gouge or damage the aluminum surfaces, for leaks can result.

Assembly

1. Using the transaxle case bearing cup installer tool or equivalent, install the differential side bearing cups into the transaxle case.
2. Install the oil slingers.
3. Using the axle shaft seal and bearing cup installer tool or equivalent, install the input/output shaft left hand bearing cups into the transaxle case.
4. Inside the case, install the spring and pilot/spacer.
5. Outside the case, install the reverse inhibitor fitting.

CLUTCH COVER

Disassembly

1. Using the transaxle case bearing cup remover tool or equivalent, remove the differential side bearing cup/shim and the input/output shafts right hand bearing cup. From the back

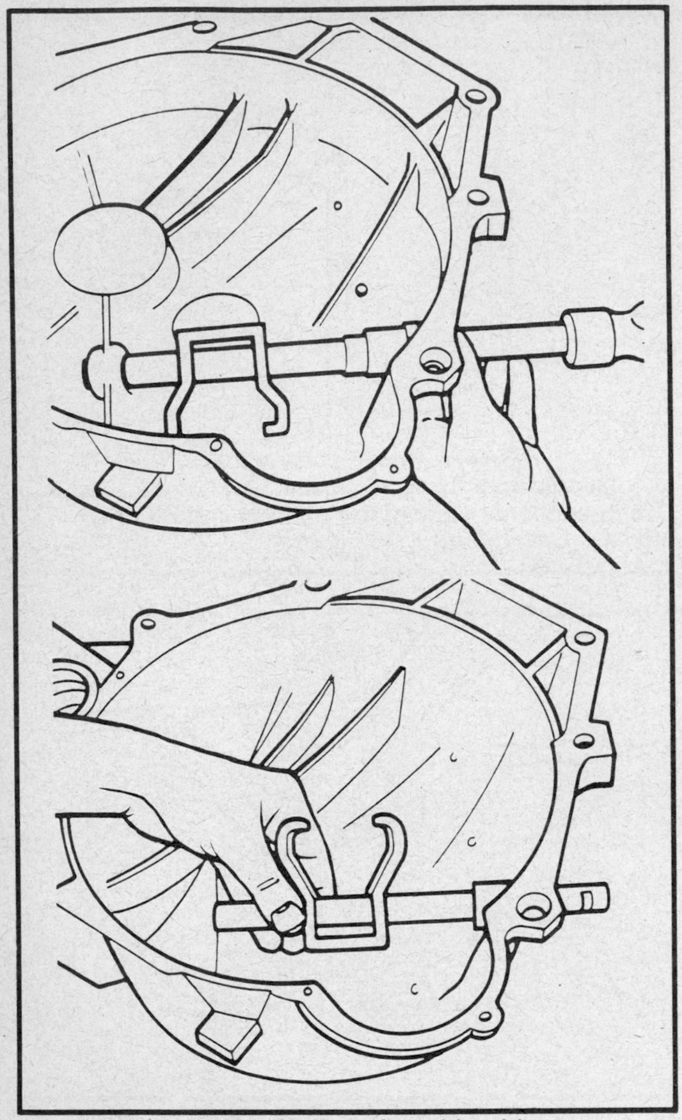

Replacing the clutch fork shaft and bushings

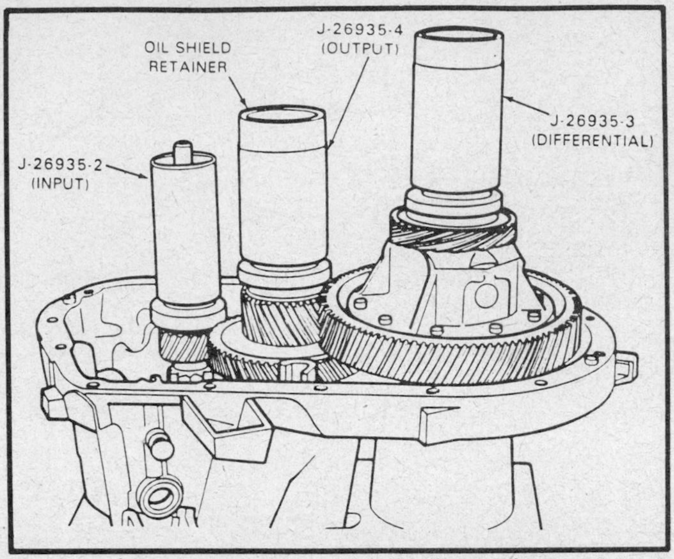
OIL SHIELD
RETAINER
J-26935-4
(OUTPUT)
J-26935-3
(DIFFERENTIAL)
J-26935-2
(INPUT)

Positioning the gauges on the shaft bearings

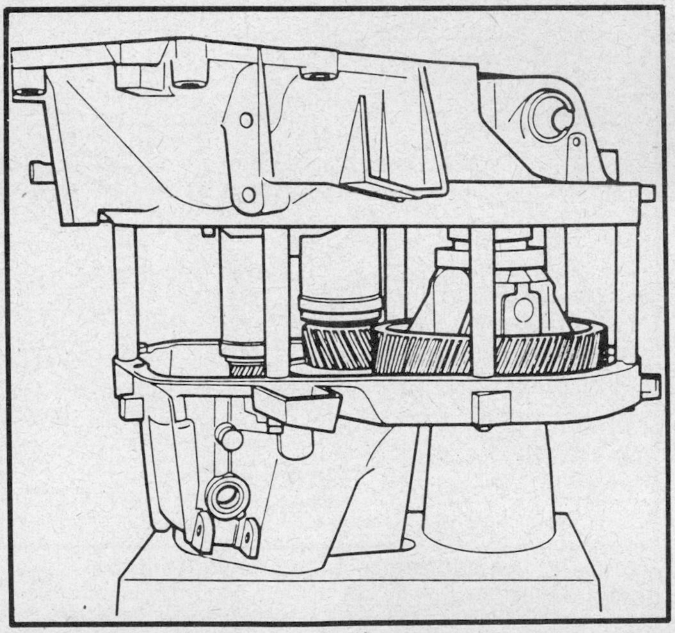

Compressing the gauges in the case

of the input bearing cup, remove the shim. From the back of the output shaft bearing cup, remove the oil shield, the shim and the retainer.

2. Remove the input gear bearing retainer bolts and tap the sleeve.
3. From the sleeve, remove the external and internal oil rings.
4. Remove the plastic oil scoop.

NOTE: If it is necessary to replace the clutch for shaft or bushing, do not break the weld on the clutch fork to remove it.

5. Using the clutch shaft bushing installer/remover tool or equivalent, remove the shaft bushings and slide the clutch shaft from the case at a slight angle.

Inspection

1. Check, clean and install the magnet.
2. Using a gasket remover tool or equivalent, remove the sealant from the mating surfaces; be careful not to gouge or damage the aluminum surfaces, for leaks can result.

Assembly

1. Install a new clutch fork shaft seal and the clutch fork shaft.
2. Install the plastic oil scoop.
3. On the bearing retainer, install a new oil ring.
4. Install the bearing retainer and torque the bolts to 7 ft. lbs. (9 Nm).
5. Using the right hand input shaft seal and bearing cup installer tool or equivalent, install the new internal oil seal and the bearing cups.

DIFFERENTIAL CASE/RING GEAR

Disassembly

1. Separate the ring gear from the differential case.

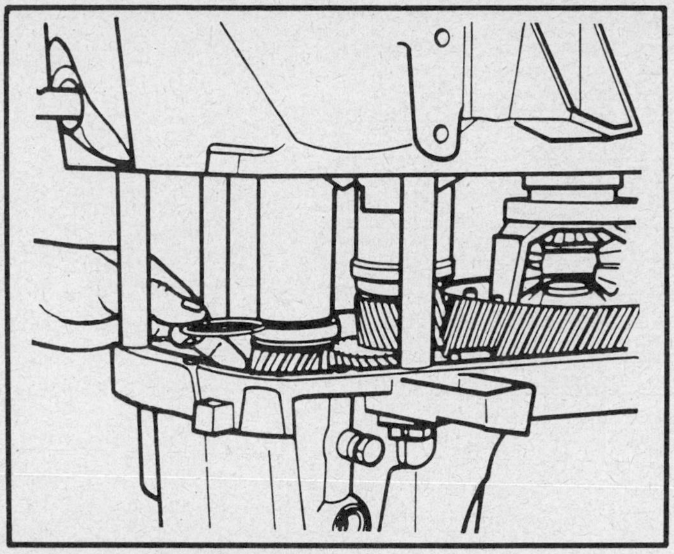

Measuring the shim gap

2. Remove the pinion shaft lock bolt, the pinion shaft, the gears and the thrust washers.
3. Using the differential side bearing puller leg tools or equivalent, and the puller, press the bearings from the differential case.

Inspection

Clean and inspect the parts for damage; replace the parts, if necessary.

Assembly

1. Using the cone installer tool or equivalent, install the side bearings into the differential case.
2. Install the gears and thrust washers into the case.
3. Install the pinion shaft and lock bolt; torque the lock bolt to 7 ft. lbs. (9 Nm).
4. Attach the ring gear to the differential case, apply sealant to the ring gear bolts and torque them to 54 ft. lbs. (73 Nm).

Transaxle Assembly

1. Position the input and output shafts on a bench, mesh the gears and install both shift forks.
2. Grasp both shafts (as an assembly) and carefully lower them into the case; be careful not to nick the gears.
3. Using a guide pin tool or equivalent, position the interlock bracket on it; be sure the bracket engages the shifting forks fingers.
4. Using a straight edge, position it on both sides of the interlock, to determine if the detent is out alignment with the interlock.

NOTE: The straight edge should rest on both sides of the interlock without interference from the detent paddle.

5. If interference is noted on either sides, perform the following procedures:
   a. With the alignment pin still in position, place the detent and interlock assembly in a vise. Using light pressure, press the detent paddle into alignment.
   b. Loosen the detent spring-to-interlock nut; the detent spring is slotted beneath the nut and will seek proper alignment.
   c. While exerting thumb pressure on the spring, tighten the detent spring-to-interlock nut.
   d. Using a straight edge, recheck the detent alignment.
6. Position the detent shift lever into the interlock.
7. Through the interlock bracket and the detent shift lever, install the shifter shaft; do not extend it any further at this time.
8. Onto the guide pin, install the reverse shift fork; be sure the fork engages with the interlock bracket.
9. Install the reverse idler gear and shaft into position; be sure the shaft's long end points upward and the gear teeth (large chamfered ends) are facing upward. Install the spacer onto the shaft.

NOTE: The reverse idler shaft's flat faces the input gear (shaft).

10. Through the reverse shift fork, install the shifter shaft until it pilots into the inhibitor spring spacer. Remove the dummy shaft. With the shaft in the **N** position, install the bolt and lock through the detent shift lever. Bend the lock tab over the bolt head.
11. Through the synchronizer forks and into the case bore, install the fork shaft.
12. Carefully install the ring gear/differential assembly into the case. Install the magnet.
13. To adjust the transaxle for the correct shim sizes, perform the following procedures:
   a. Place the transaxle case into holding fixture tool or equivalent.
   b. With the left hand bearing races installed in the case, place the input shaft, output shaft and differential assemblies into their installed positions. Position the right hand bearing races onto their respective bearings.
   c. Using the shim selector set or equivalent, position a gauge on the input bearing, output bearing and differential bearing; sure the bearing races fit smoothly into the bores of the gauge tools.
   d. On top of the output shaft gauge, install the oil shield retainer.
   e. Carefully assemble the clutch cover over the gauges and onto the case, using the spacers provided evenly around the perimeter. Retain the assembly with the bolts provided.
   f. Atlernately and gradually, tighten the cover-to-case bolts to 10 ft. lbs. (13 Nm); this will compress the gauge sleeves.
   g. Rotate each gauge to seat the bearings. Rotate the differential case through 3 revolutions in each direction.
   h. With the gauges compressed, the gap between the outer sleeve and the base pad is larger than the correct preload shim at each location. Carefully compare the gap to the available shims. Determine the largest shim that can be placed into the gap and drawn through without binding. Then, use the next size smaller on the output shaft and differential or reassembly. On the input shaft, use a shim 2 sizes smaller. If endplay occurs, use the next larger shim size.
   i. When each of the 3 shims have been selected, remove the clutch cover, spacers and gauges.
   j. Place the selected shims into their respective clutch cover bores and add the metal shield. Using the right hand input shaft seal and bearing cup installer tool or equivalent, install the bearing cups on the input shaft. Using the transaxle case bearing cup installer tool or equivalent, install the bearing cups onto the output shaft. Using the axle shaft seal and bearing cup installer tool or equivalent, install the bearing cup onto the differential side.
14. Using anaerobic sealant (not RTV), apply a thin bead onto the clutch cover. Using the dowel pins, install the clutch cover onto the transaxle case. Using a plastic hammer, gently tap the cover to insure the parts are seated.
15. Install the clutch cover-to-case bolts and torque to 16 ft. lbs. (21 Nm).
16. Torque the idler shaft lock bolt to 16 ft. lbs. (21 Nm).
17. Move the shift shaft through the gear ranges to test for freedom of movement of the internal parts.

SPECIFICATIONS

TORQUE SPECIFICATIONS

Description	ft. lbs.	Nm
Input shaft RH bearing retainer	7	9
Reverse idler shaft lock bolt	16	21
Reverse inhibitor fitting	26	35
Case to cover bolts	16	21
Ring gear bolts	54	73
Pinion shaft lock bolt	7	9
Output shaft LH bearing retainer	45	65

SPECIAL TOOLS

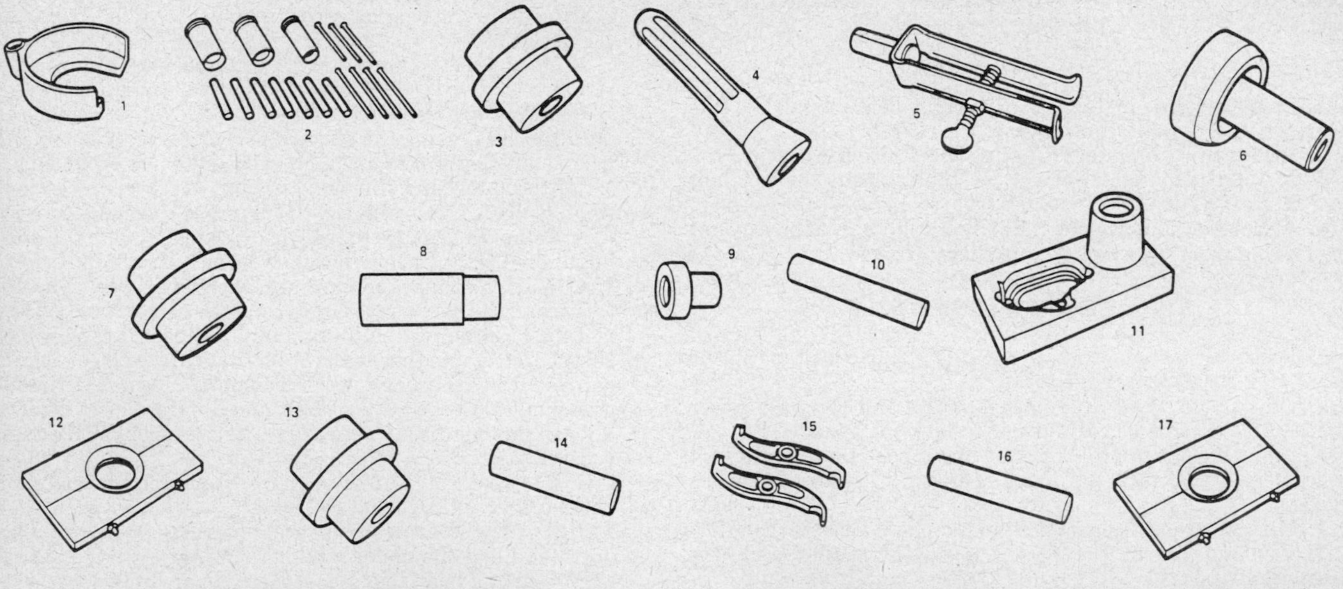
1
2
3
4
5
6
7
8
9
10
11
12
13
14
15
16
17

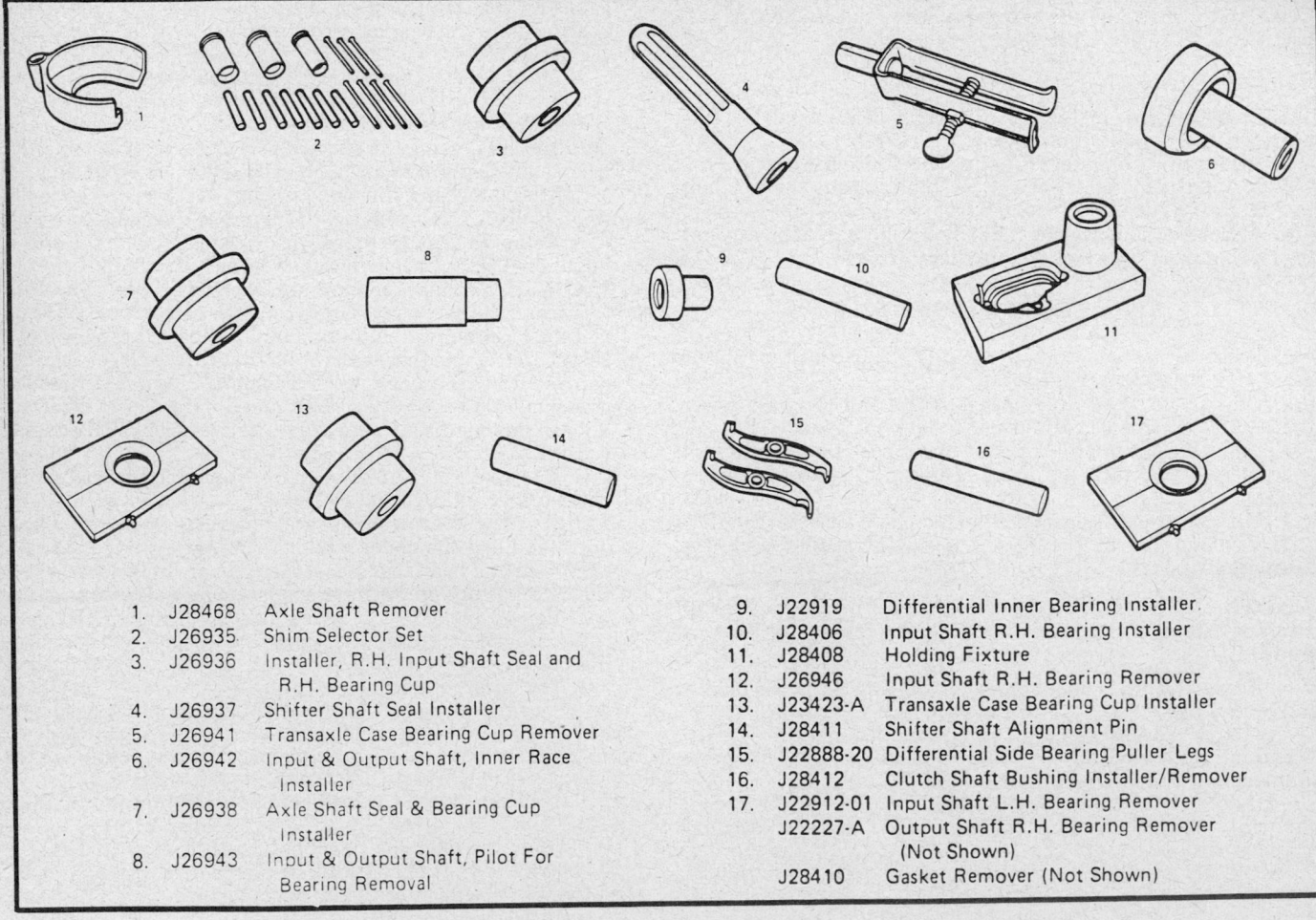

1. J28468 Axle Shaft Remover
2. J26935 Shim Selector Set
3. J26936 Installer, R.H. Input Shaft Seal and R.H. Bearing Cup
4. J26937 Shifter Shaft Seal Installer
5. J26941 Transaxle Case Bearing Cup Remover
6. J26942 Input & Output Shaft, Inner Race Installer
7. J26938 Axle Shaft Seal & Bearing Cup Installer
8. J26943 Input & Output Shaft, Pilot For Bearing Removal
9. J22919 Differential Inner Bearing Installer
10. J28406 Input Shaft R.H. Bearing Installer
11. J28408 Holding Fixture
12. J26946 Input Shaft R.H. Bearing Remover
13. J23423-A Transaxle Case Bearing Cup Installer
14. J28411 Shifter Shaft Alignment Pin
15. J22888-20 Differential Side Bearing Puller Legs
16. J28412 Clutch Shaft Bushing Installer/Remover
17. J22912-01 Input Shaft L.H. Bearing Remover
    J22227-A Output Shaft R.H. Bearing Remover (Not Shown)
    J28410 Gasket Remover (Not Shown)

Section 6

M21 and M79 Transaxles
General Motors

APPLICATION

1988–89 Pontiac LeMans

GENERAL DESCRIPTION

The 4 and 5 speed transaxle assemblies are of the constant mesh design. Combined in the assembly are all forward gears, reverse gear and the differential. Selection and shifting is accomplished by a combination of synchronizers and blocker rings controlled by sliding shift forks. Reverse gear is non-synchronized, using a sliding idler gear arrangement. The basic components of these units are the transaxle case, input shaft and gears, output shaft and gears and ring gear and differential assembly.

The differential is a conventional arrangement of gears that is supported by tapered roller bearings. The final output gear turns the ring gear and differential assembly which turn the drive axle shafts.

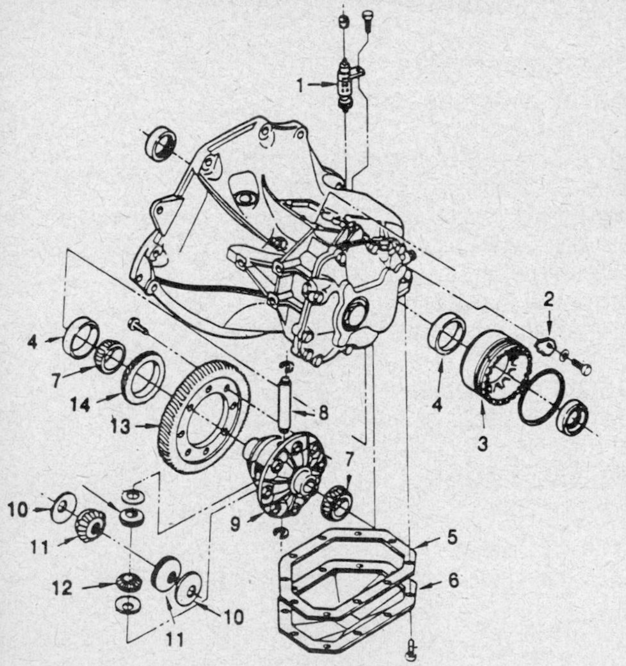
1
2
4
7
14
13
8
4
3
7
10
11
9
5
12
6
10
11

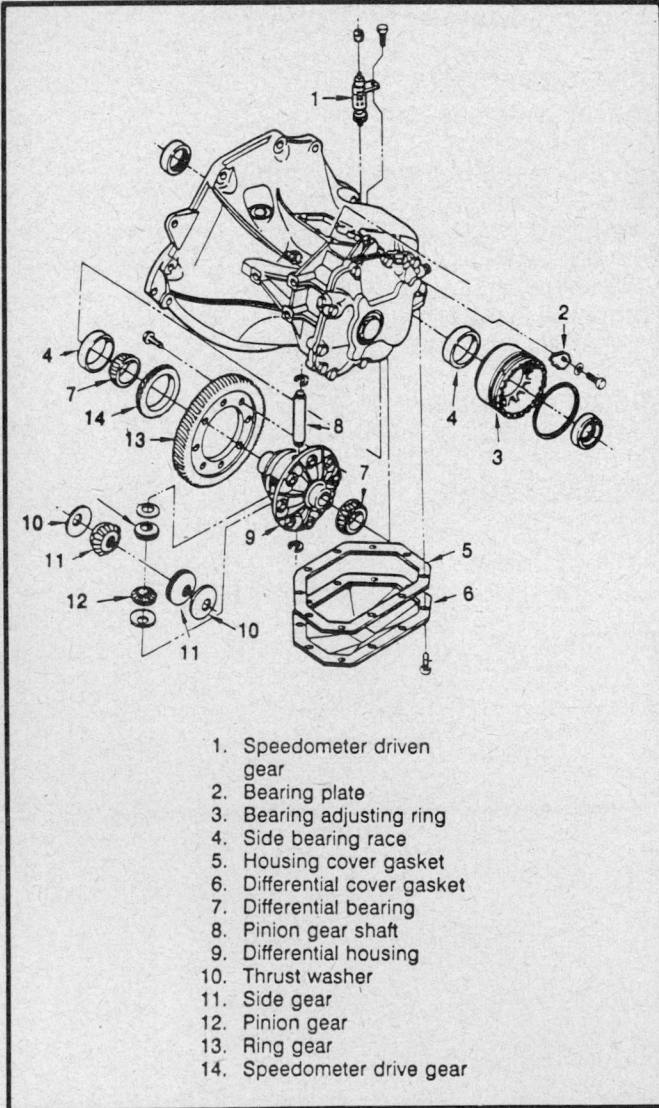

1. Speedometer driven gear
2. Bearing plate
3. Bearing adjusting ring
4. Side bearing race
5. Housing cover gasket
6. Differential cover gasket
7. Differential bearing
8. Pinion gear shaft
9. Differential housing
10. Thrust washer
11. Side gear
12. Pinion gear
13. Ring gear
14. Speedometer drive gear

Differential and case

Transaxle Identification

The transaxle identification tag is located on the rear transaxle pan. The tag, which is located in 2 places, contains the following information: the factory where the assembly was built, the day, year and shift that the assembly was built, the transaxle code number and drive ratio data.

Metric Fasteners

Metric fastener dimensions are very close to the dimensions of the inch system fasteners, and for that reason, replacement fasteners must have the same measurement and strength as those removed. Do not attempt to interchange metric fasteners for inch fasteners. Mismatched or incorrect fasteners can result in damage to the transaxle unit through malfunctions or breakage, or even personal injury. Care should be taken to reuse the fasteners in the same location as removed.

Capacities

The lubricant capacity for the M21 transaxle and the M79

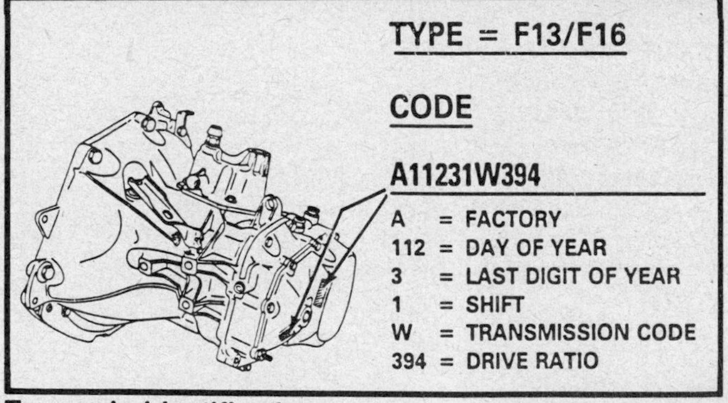
TYPE = F13/F16
CODE
A11231W394
A = FACTORY
112 = DAY OF YEAR
3 = LAST DIGIT OF YEAR
1 = SHIFT
W = TRANSMISSION CODE
394 = DRIVE RATIO

Transaxle Identification

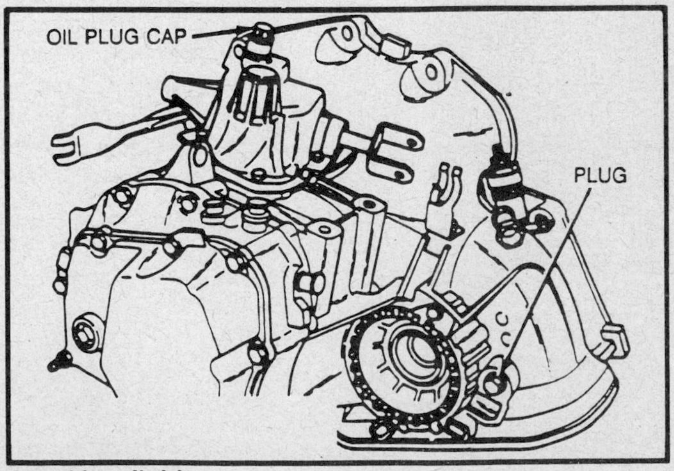
OIL PLUG CAP
PLUG

Checking fluid

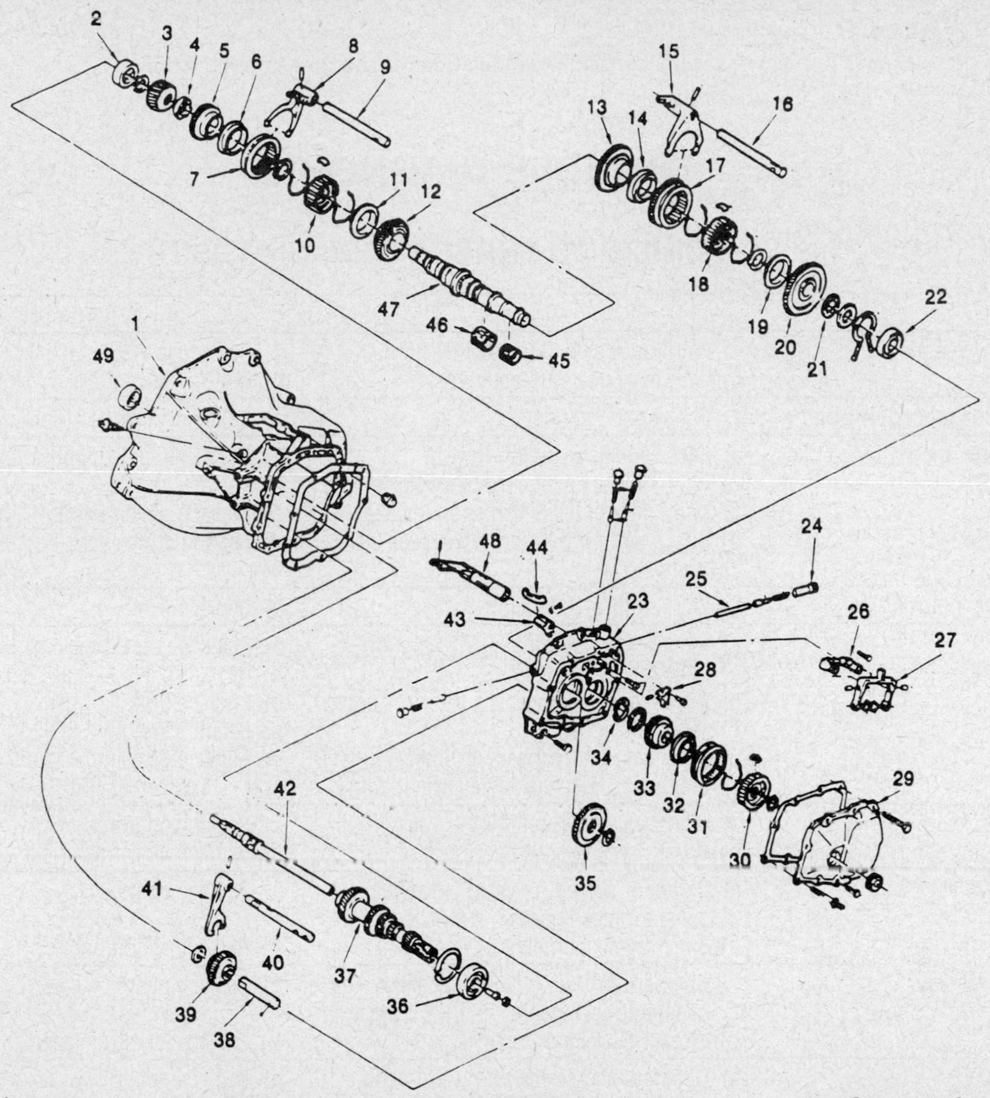

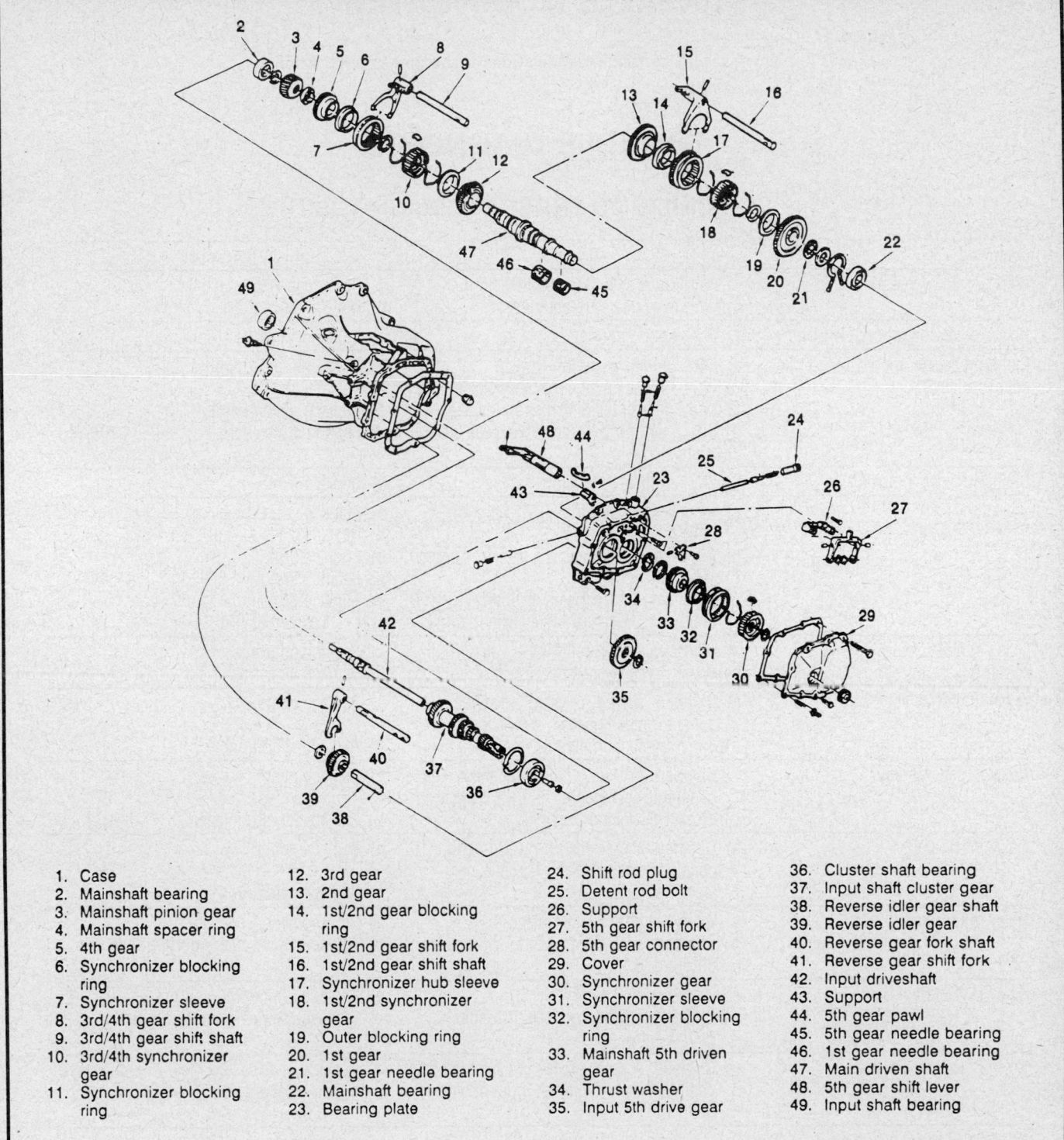

1. Case
2. Mainshaft bearing
3. Mainshaft pinion gear
4. Mainshaft spacer ring
5. 4th gear
6. Synchronizer blocking ring
7. Synchronizer sleeve
8. 3rd/4th gear shift fork
9. 3rd/4th gear shift shaft
10. 3rd/4th synchronizer gear
11. Synchronizer blocking ring
12. 3rd gear
13. 2nd gear
14. 1st/2nd gear blocking ring
15. 1st/2nd gear shift fork
16. 1st/2nd gear shift shaft
17. Synchronizer hub sleeve
18. 1st/2nd synchronizer gear
19. Outer blocking ring
20. 1st gear
21. 1st gear needle bearing
22. Mainshaft bearing
23. Bearing plate
24. Shift rod plug
25. Detent rod bolt
26. Support
27. 5th gear shift fork
28. 5th gear connector
29. Cover
30. Synchronizer gear
31. Synchronizer sleeve
32. Synchronizer blocking ring
33. Mainshaft 5th driven gear
34. Thrust washer
35. Input 5th drive gear
36. Cluster shaft bearing
37. Input shaft cluster gear
38. Reverse idler gear shaft
39. Reverse idler gear
40. Reverse gear fork shaft
41. Reverse gear shift fork
42. Input driveshaft
43. Support
44. 5th gear pawl
45. 5th gear needle bearing
46. 1st gear needle bearing
47. Main driven shaft
48. 5th gear shift lever
49. Input shaft bearing

Gears and case

transaxle is 3.5 pts. (1.6L) of 80W gear oil part number 12345371 or equivalent.

Checking Fluid Level

With the vehicle on a level surface and the fluid in the transaxle cold, remove the plug and check the fluid level. The fluid should come to the bottom edge of the plug hole. If the level is low, add manual transaxle fluid part number 12345371 or equivalent through the filler plug hole until it begins to run out of the plug hole. Reinstall and tighten the plug.

TRANSAXLE MODIFICATIONS

There are no transaxle modifications at the time of publication.

TROUBLE DIAGNOSIS

CHILTON'S THREE C'S DIAGNOSIS

Condition	Cause	Correction
A knock at low speeds	a) Worn drive axle CV or tri-pot joints b) Worn side gear hub counterbore	a) Replace joints b) Repair counterbore
Noise most pronounced on turns	a) Differential gear noise	a) Repair or replace differential
Clunk on acceleration or deceleration	a) Loose engine mounts b) Worn drive axle inboard tri-pot joints c) Worn differential pinion shaft in case d) Side gear hub counterbore in case worn oversize	a) Tighten engine mounts b) Replace joints c) Replace pinion shaft d) Repair counterbore
Clunking noise in turns	a) Worn outboard CV joint	a) Replace joint
Vibration	a) Rough wheel bearing b) Bent drive axle shaft c) Out of round tires d) Tire unbalance e) Worn CV joint in drive axle shaft f) Incorrect drive axle angle (Trim Height)	a) Replace wheel bearings b) Replace drive axle shaft c) Replace tires d) Balance tires e) Replace CV joint f) Correct trim height
Noisy in neutral with engine running	a) Worn input gear bearings b) Worn clutch release bearing	a) Replace bearings b) Replace bearings
Noisy in first only	a) Chipped, scored, or worn first-speed constant-mesh gears b) Worn 1-2 synchronizer	a) Replace gears b) Replace synchronizer
Noisy in second only	a) Chipped, scored, or worn second speed constant-mesh gears b) Worn 1-2 synchronizer	a) Replace gears b) Replace synchronizer
Noisy in third only	a) Chipped, scored, or worn third-gear constant-mesh gears b) worn 3-4 synchronizer	a) Replace gears b) Replace synchronizer
Noisy in fourth only	a) Worn 3-4 gear synchronizer b) Chipped, scored, or worn fourth-gear or output gear	a) Replace synchronizer b) Replace fourth gear or output gear
Noisy in fifth gear only	a) Worn 5th synchronizer b) Chipped, scored, or worn fifth-speed gear or output gear	a) Replace 5th gear synchronizer b) Replace 5th gear or output gear
Noisy in reverse only	a) Chipped, scored, or worn reverse idler gear, idler gear bushing, input or output gear(s)	a) Replace reverse idler gear bushing, input or output gears
Noisy in all gears	a) Insufficient lubricant b) Worn bearings c) Chipped, scored, or worn input gear (shaft) and/or output gear (shaft)	a) Add lubricant b) Replace bearings c) Replace input gear shaft or output gear shaft
Slips out of gear	a) Worn or improperly adjusted linkage b) Shift linkage does not work freely; binds c) Input gear bearing retainer broken or loose d) Worn or bent shift fork	a) Adjust linkage b) Repair linkage c) Repair or replace bearing retainer d) Repair or replace shift fork

CHILTON'S THREE C'S DIAGNOSIS

Condition	Cause	Correction
Leaks lubricant	a) Worn axle shaft seals b) Excessive amount of lubricant in transaxle c) Loose or broken input gear (shaft) bearing retainer d) Worn input gear bearing and/or lip seal damaged e) Worn shift lever seal leaks	a) Replace seals b) Correct lubricant level c) Replace bearing retainer d) Replace bearing and seal e) Replace seal

ON CAR SERVICE

Adjustments

SHIFT LINKAGE

1. Disconnect the negative battery cable.
2. Position the gear shift lever in **N**.
3. Loosen the rod clamp bolt.
4. Remove the adjustment hole plug from the shift lever cover.
5. Turn the shift rod to the left until a $^3/_{16}$ in. gauge pin can be inserted in the oil plug hole of the intermediate shift lever.
6. Remove the boot from the console and pull it upward to expose the shift control lever mechanism.
7. Position the gear shift lever to the 1st/2nd gear position while still in **N**.
8. With the lever against the stop and the arrow aligned with the notch, tighten the rod clamp bolt to 10 ft. lbs. (14 Nm).
9. After reaching 10 ft. lbs., turn the bolt an additional 90 degrees.
10. Check the clearance between the catch and the stop. The clearance should be 0.12 in. (3mm) after adjustment.
11. Remove the $^3/_{16}$ in. gauge pin and install the plug.
12. Bend back the 2 locking tabs at the nut and adjust the shifter to 0.457 in. (11.8mm). Secure the adjusting nut with the locking tabs.
13. Install the boot to the console and then the center console.
14. Connect the negative battery cable.

Services

GEAR SHIFT LEVER

Removal and Installation

1. Disconnect the negative battery cable.
2. Remove the center console.
3. Remove the gear shift lever knob.
4. Remove the boot from the console and lever.
5. Position the lever in **N**.
6. Rotate the clamp and pull the pin and clamp out of the gear shift lever.
7. Remove the gear shift lever.
8. For installation, position the shift lever in the shaft and install the pin and clamp. Rotate the clamp to retain the lever.
9. Install the boot on the lever and position it in the console.
10. Install the center console.
11. Install a new O-ring on the lever.
12. Apply a soap solution to the knob of the reverse gear block and install the rubber cap as far as the stop.
13. Preheat the shift lever knob in water to 176°F and install it on the lever.
14. Connect the negative battery cable.

GEAR SHIFT TUBE, BOOT, BUSHING AND BEARING RING

Removal and Installation

1. Disconnect the negative battery cable.
2. Loosen the rod clamp bolt and separate the linkage bolt from the gear shift tube.

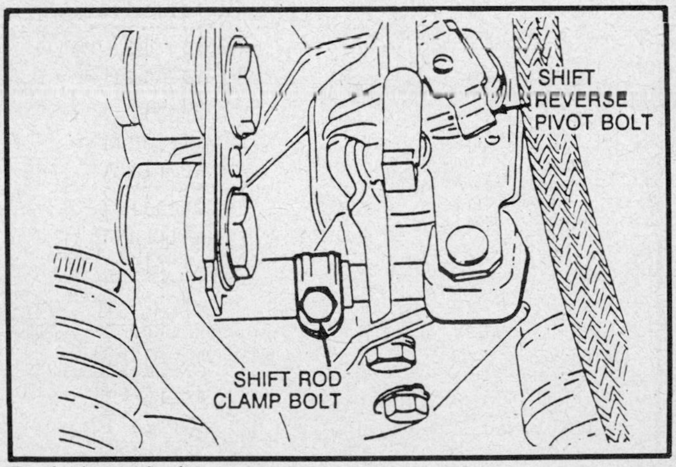
SHIFT REVERSE PIVOT BOLT
SHIFT ROD CLAMP BOLT

Rod clamp bolt

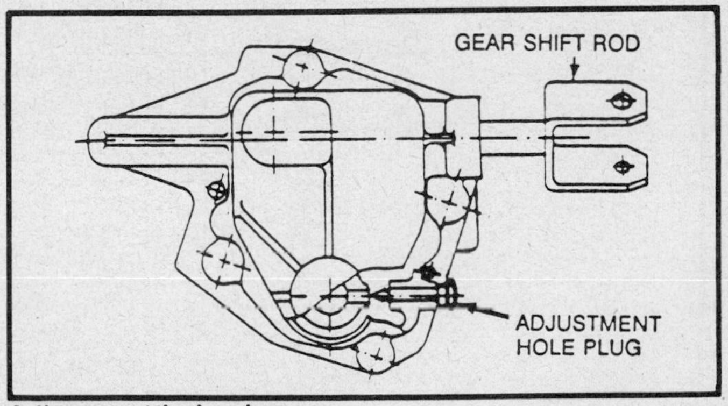
GEAR SHIFT ROD
ADJUSTMENT HOLE PLUG

Adjustment hole plug

1. Gear shift lever
2. Gear shift lever shaft
3. Gear shift lever stop bushing
4. Gear shift lever stop bushing
5. Gear shift housing
6. Linkage adjuster bolt
7. Gear shift control rod
8. Linkage ball socket
9. Linkage reverse lever
10. Rod U-joint bushing
11. Gear shift rod
12. Intermediate lever
13. Shift lever thrust spring
14. Intermediate gear shift lever
15. Oil filler plug
16. Oil plug cap
17. Gear shift lever cover
18. Gear shift adjuster linkage
19. Shift reverse pivot bolt
20. Gear shift tube
21. Gear shift tube
22. 5th gear plunger pin ball
23. Shift block
24. Gear shift tube bearing
25. Gear shift bearing stop

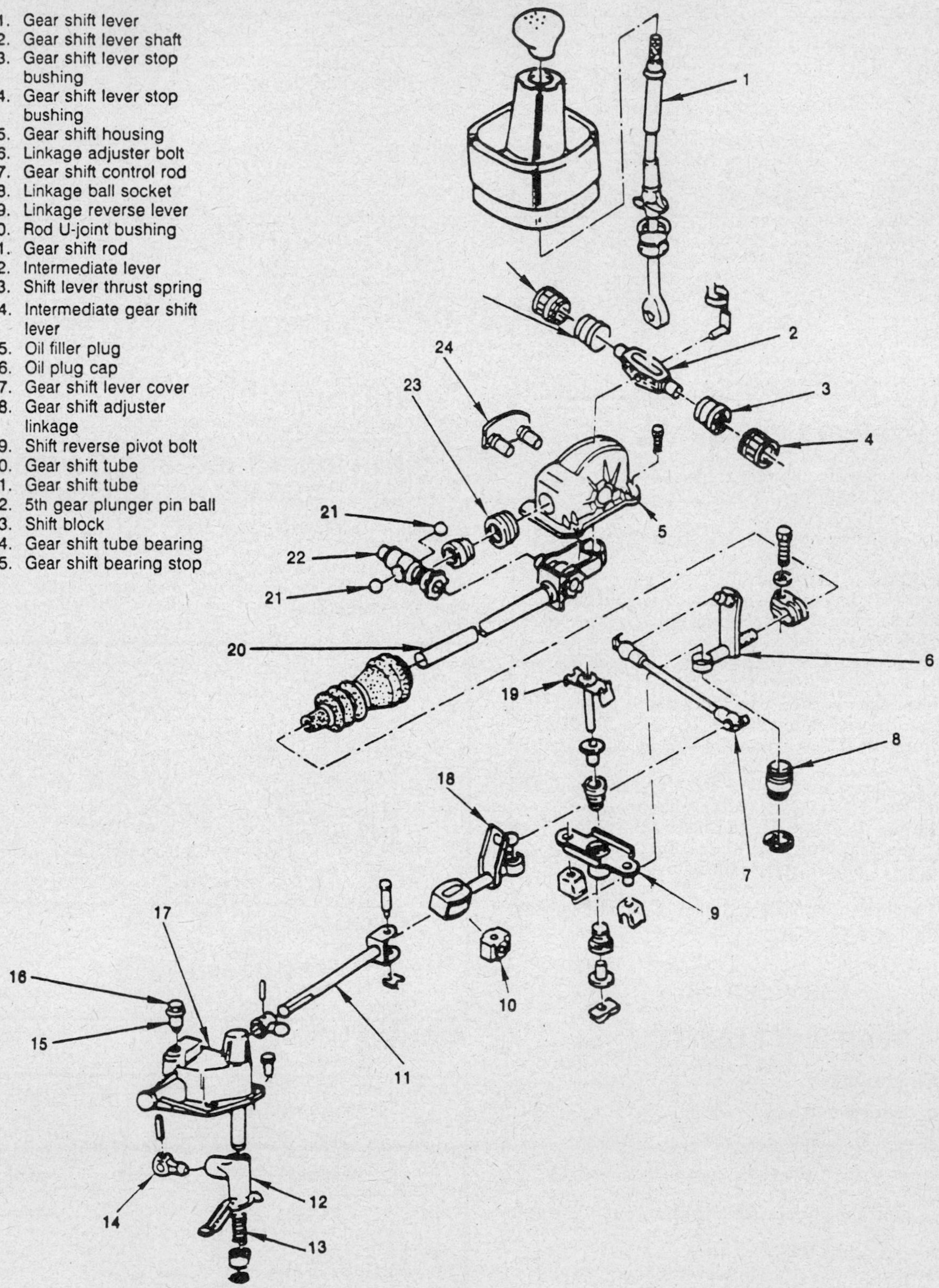

Shift linkage

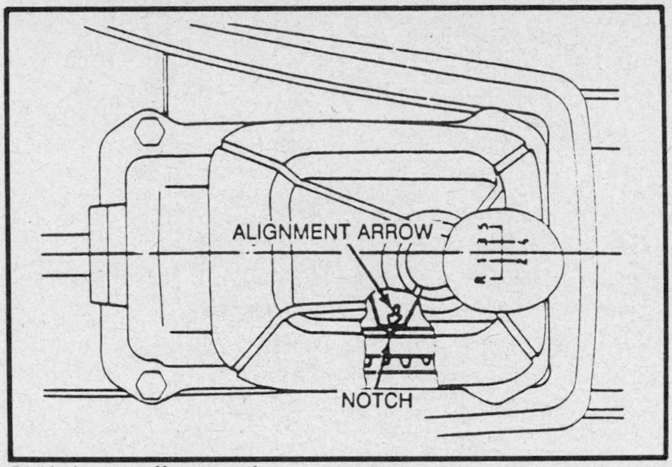
ALIGNMENT ARROW
NOTCH

Shift lever alignment

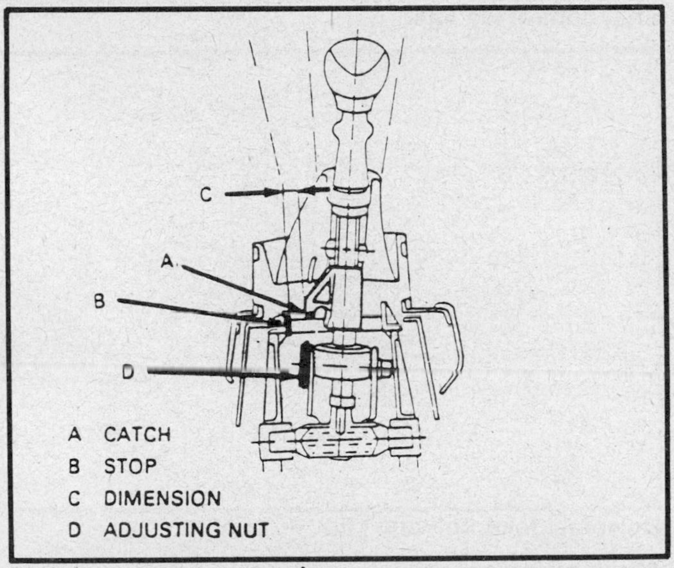
C
A
B
D
A CATCH
B STOP
C DIMENSION
D ADJUSTING NUT

Shift lever measurements

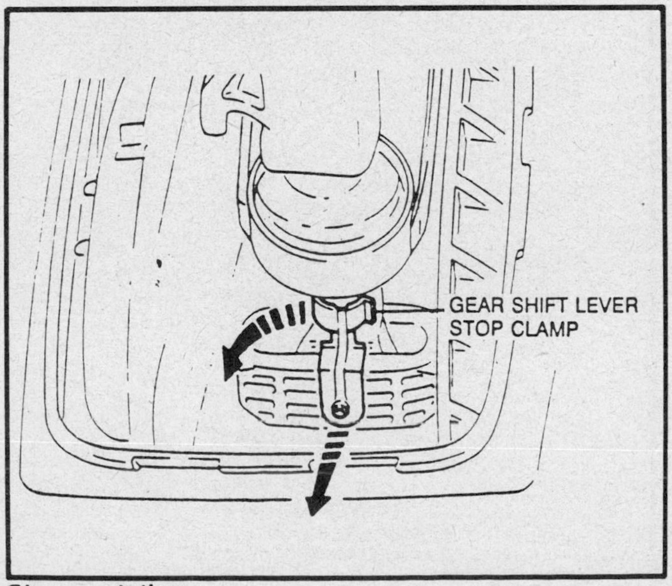
GEAR SHIFT LEVER
STOP CLAMP

Clamp rotation

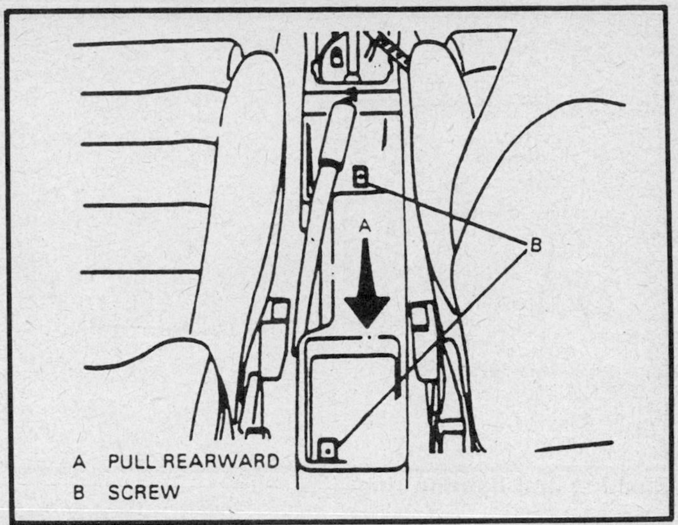
B
A
B
A PULL REARWARD
B SCREW

Full Console

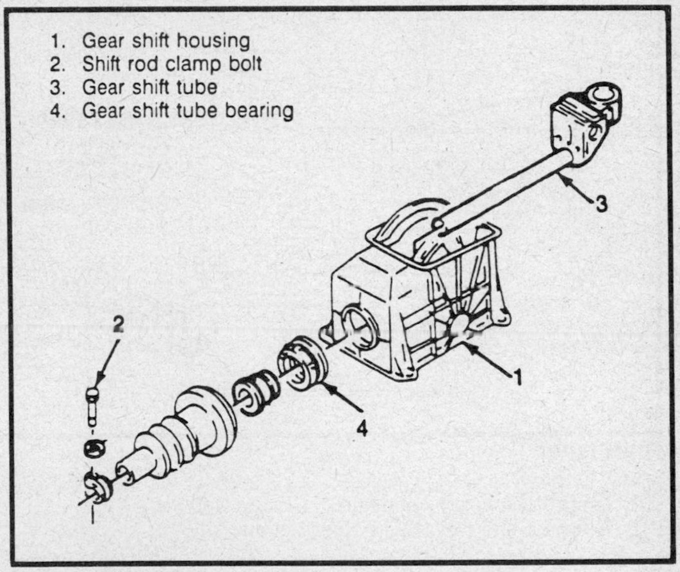
1. Gear shift housing
2. Shift rod clamp bolt
3. Gear shift tube
4. Gear shift tube bearing
3
2
1
4

Gear shift tube

3. Remove the center console.
4. Position the shift lever in **N**.
5. Rotate the clamp and pull the pin out of the lever. Remove the shift lever with the boot.
6. Remove the full console.
7. Remove the housing bolts.
8. Remove the gear shift tube with the housing.
9. Push out the tube bushing and bearing ring from the housing.
10. Detach the housing from the bearing ring.
11. For installation, install the bushing in the bearing ring.
12. Fill the grooves of the bushing inside with silicone grease and press the bearing ring in the housing from the inside.
13. Install the gear shift tube through the bushing in the housing.
14. Install the housing and tube.
15. Install the full console.
16. Install the lever in the shaft and install the pin and shaft. Rotate the clamp to retain the lever.
17. Install the gear shift tube boot.
18. Install the linkage bolt to the gear shift tube.
19. Adjust the shift linkage and tighten the rod clamp bolt.

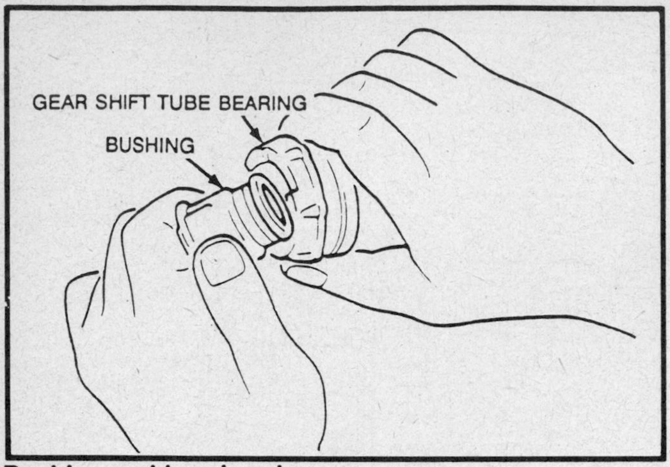
GEAR SHIFT TUBE BEARING
BUSHING

Bushing and bearing ring

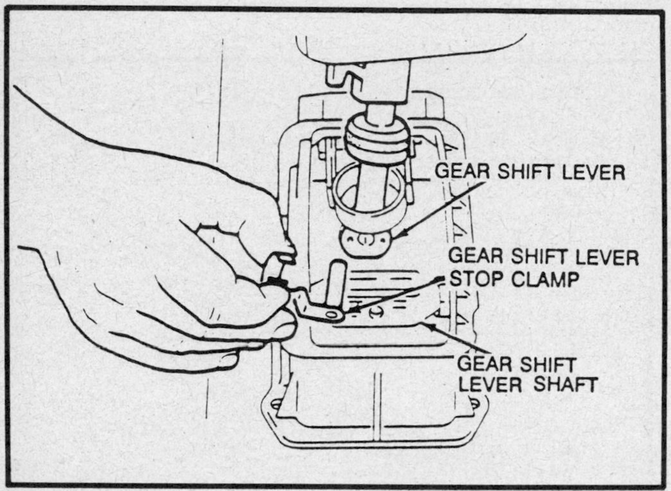
GEAR SHIFT LEVER
GEAR SHIFT LEVER STOP CLAMP
GEAR SHIFT LEVER SHAFT

Shift lever

20. Install the center console.
21. Connect the negative battery cable.

CONTROL SHIFT ROD

Removal and Installation

1. Disconnect the negative battery cable.
2. Press out the plastic clip at the end of each ball socket.
3. Separate the rod from the ball studs on the linkage bolts by pressing outward.
4. Remove the control shift rod.
5. Installation is the reverse of the removal procedure.

LINKAGE LEVER AND BUSHINGS

Removal and Installation

1. Disconnect the negative battery cable.
2. Remove the control shift rod.
3. Remove the clip and bolt from the universal joint.
4. Loosen the rod clamp bolt and separate the linkage bolt from the shift tube.
5. Remove the pivot bolt with the spring clips.
6. Remove the linkage lever and bolts.
7. Remove the bushings.
8. Remove the snaprings retaining the linkage bolts to the lever.

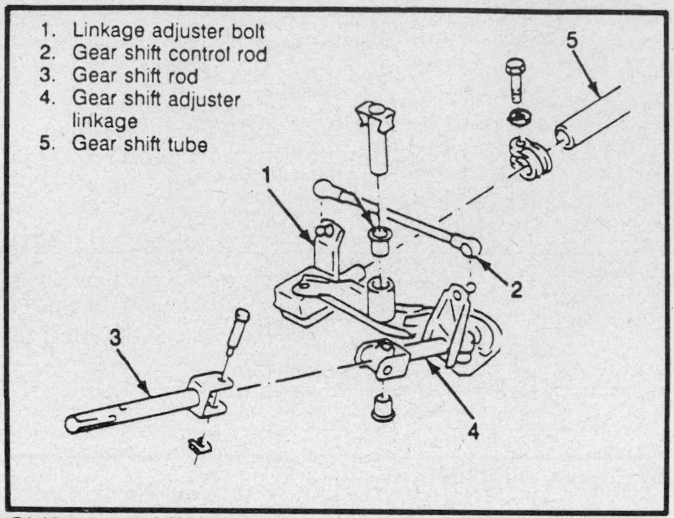
1. Linkage adjuster bolt
2. Gear shift control rod
3. Gear shift rod
4. Gear shift adjuster linkage
5. Gear shift tube
1
2
3
4
5

Shift control linkage

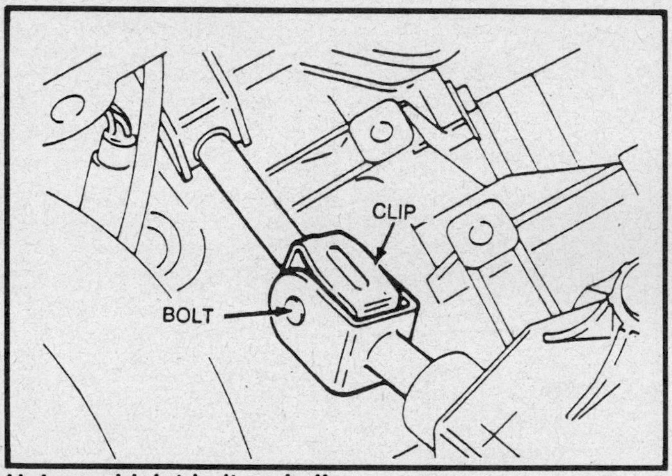
CLIP
BOLT

Universal joint bolt and clip

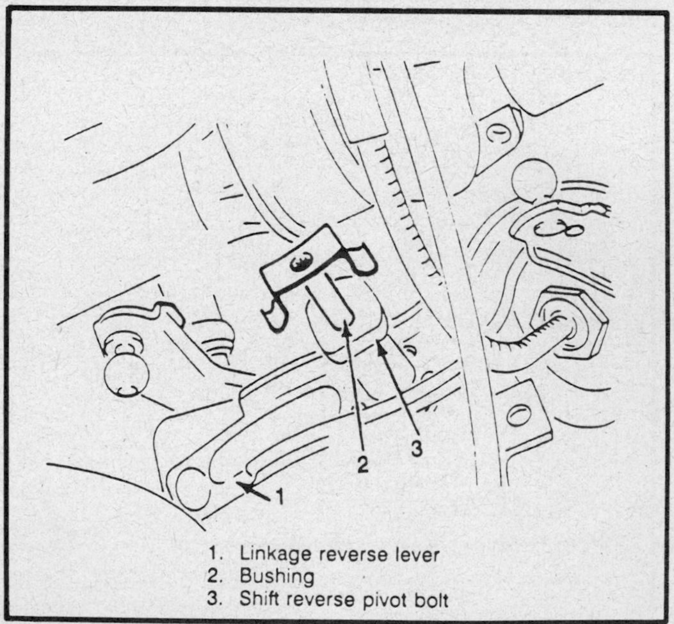
1
2
3
1. Linkage reverse lever
2. Bushing
3. Shift reverse pivot bolt

Linkage lever bolt

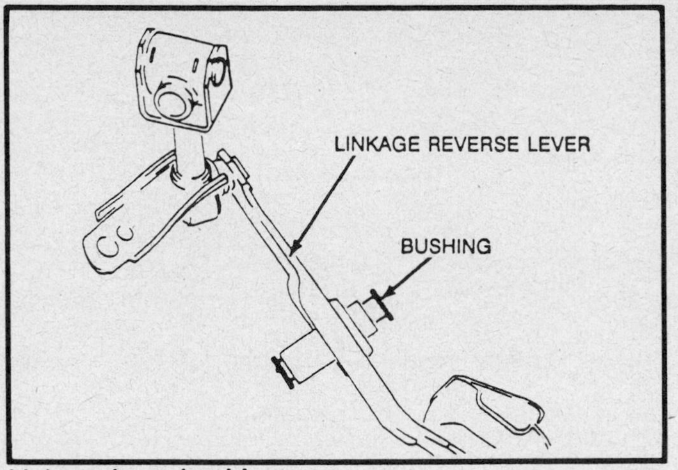
LINKAGE REVERSE LEVER
BUSHING
Linkage lever bushings

9. Installation is the reverse of the removal procedure. Coat the bushings and the universal joint bolt with silicone grease before installation.
10. Adjust the shift linkage and tighten the rod clamp bolt.

SHIFT LEVER COVER

Removal and Installation

1. Disconnect the negative battery cable.
2. Loosen the fill plug cap.
3. Remove the clip and bolt at the universal joint.
4. Remove the cover bolts and separate the cover from the housing.
5. Installation is the reverse of the removal procedure.
6. Tighten the cover bolts to 16 ft. lbs. (22 Nm). Check and adjust the fluid level.

SPEEDOMETER DRIVEN GEAR

Removal and Installation

1. Disconnect the negative battery cable.
2. Raise the vehicle and support it safely.
3. Disconnect the speedometer cable and speed sensor connection.
4. Remove the retaining bolt and plate.
5. Remove the driven gear and housing.
6. Remove the O-ring.
7. Installation is the reverse of the removal procedure.
8. Coat the O-ring with petroleum jelly before installation.
9. Tighten the retaining plate bolt to 45 inch lbs. (5 Nm).

DRIVE AXLE SEAL

Removal and Installation

1. Disconnect the negative battery cable.
2. Raise the vehicle and support it safely.
3. Remove the wheel and tire assembly.
4. Separate the lower ball joint and tie rod end.
5. Remove the drive axle from the transaxle.
6. Pry out the drive axle seal.
7. Coat the new axle seal lip with clean transaxle fluid before installation.
8. Install the seal into position, using the proper installation tool.
9. Continue the installation in the reverse order of the removal procedure.
10. Check and adjust the fluid level as necessary.

REMOVAL AND INSTALLATION

TRANSAXLE REMOVAL

1. Disconnect the negative battery cable.
2. Disconnect the clutch cable at the release lever.
3. Remove the clip and bolt at the universal joint.
4. Disconnect the speedometer cable, speed sensor and back-up lamp connections.
5. Remove the upper transaxle to engine bolts.
6. Install the engine support fixture no. J-28467-A or equivalent.
7. Raise the vehicle and support it safely.
8. Remove the left front wheel and tire assembly.
9. Remove the plug at the transaxle cover and ground wire.
10. Remove the internal snapring at the end of the input shaft.
11. Install tools J-36644 and J-6125-B or equivalent into the end of the input shaft and disengage the input shaft from the cluster gear.
12. Remove the flywheel cover.
13. Disconnect the left lower ball joint and tie rod end.
14. Disengage both drive axles at the transaxle. Pull out the left drive axle and support.

NOTE The right drive axle will come out, once the transaxle is removed from the vehicle, so properly support it.

15. Support the transaxle.
16. Remove the left front mount bracket.
17. Remove the left rear mount bracket bolts to transaxle.
18. Remove the lower transaxle to engine bolts.

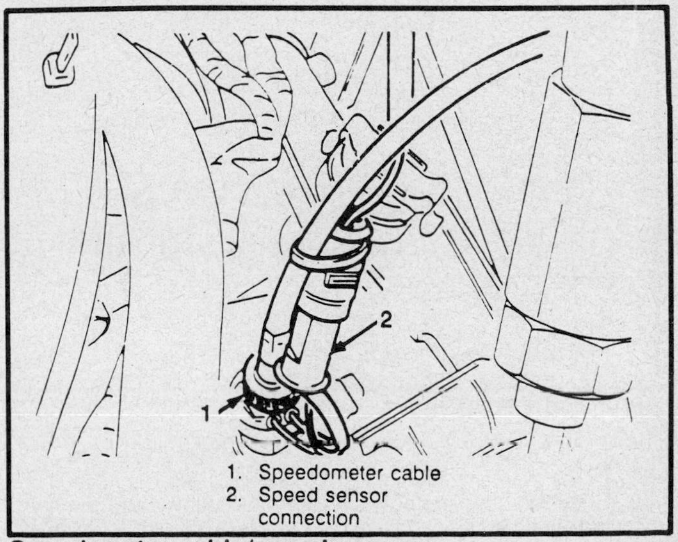
2
1
1. Speedometer cable
2. Speed sensor connection
Speedometer cable/speed sensor

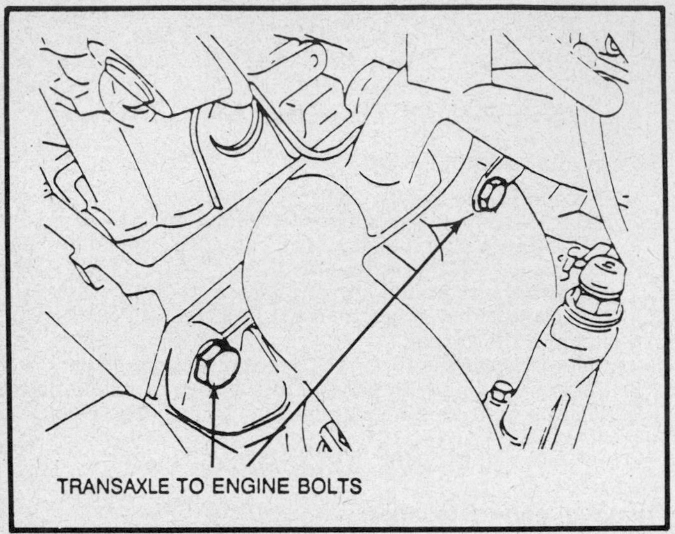
TRANSAXLE TO ENGINE BOLTS

Transaxle to engine bolts

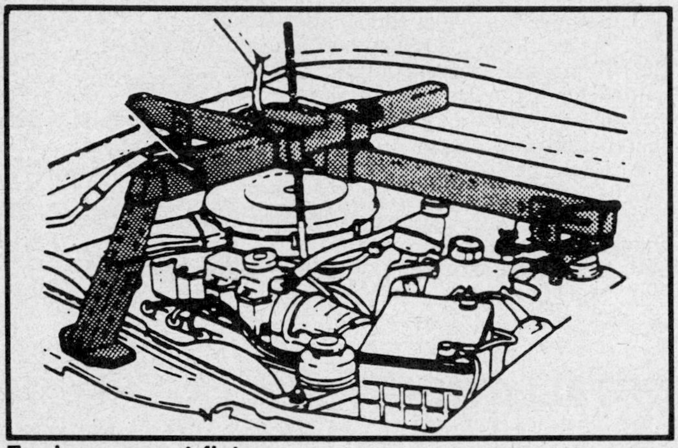

Engine support fixture

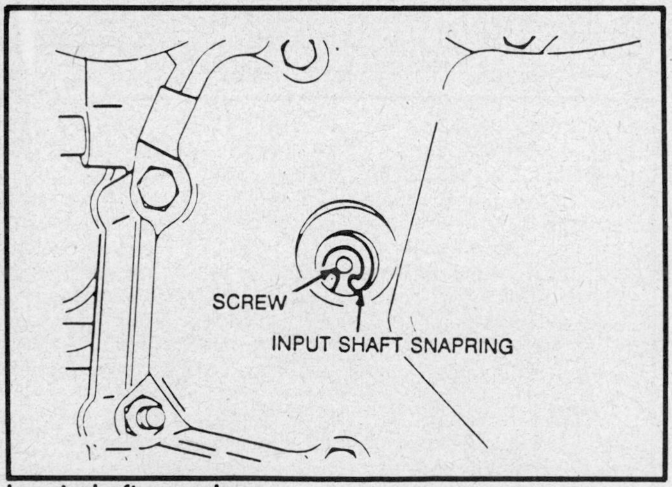
SCREW
INPUT SHAFT SNAPRING

Input shaft snapring

19. Slide the transaxle away from the engine and remove it from the vehicle.
20. Support the right drive axle.

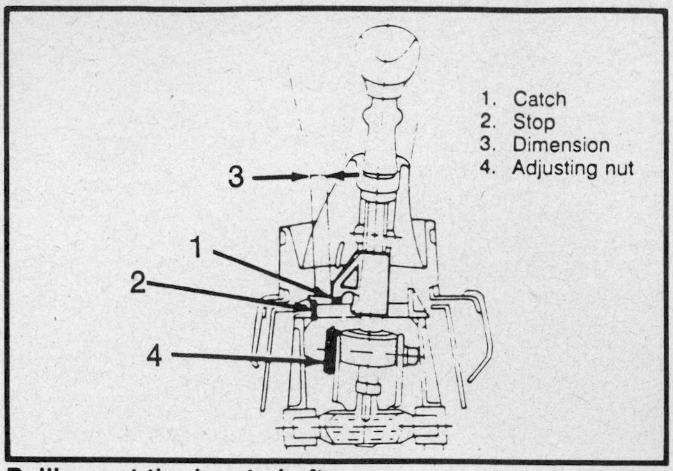
1. Catch
2. Stop
3. Dimension
4. Adjusting nut
3
1
2
4

Pulling out the input shaft

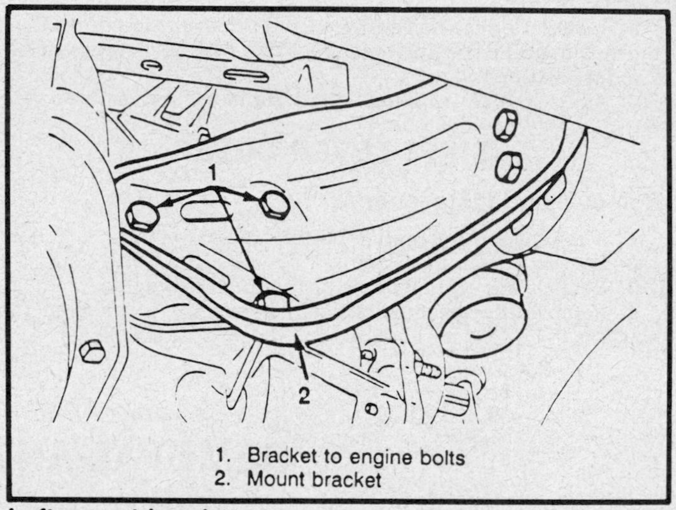
1
2
1. Bracket to engine bolts
2. Mount bracket

Left mount bracket

TRANSAXLE INSTALLATION

1. While installing the transaxle, guide the right drive axle into the transaxle. The right drive axle cannot be readily installed after the transaxle is connected to the engine.
2. Install the lower transaxle to engine bolts and tighten to 55 ft. lbs. (75 Nm).
3. Install the left rear mount bracket bolts and tighten to 55 ft. lbs. (75 Nm).
4. Install the left front mount bracket and bolts. Tighten the bolts at the transaxle to 47 ft. lbs. (65 Nm). Tighten the bolts at the mount to 55 ft. lbs. (75 Nm).
5. Install the left drive axle in the transaxle and seat both drive axles.
6. Connect the left lower ball joint and tie rod end.
7. Align the splines on the input shaft with the clutch disc and seat the input shaft. Install the screw and tighten it to 133 inch lbs. (15 Nm). Install the snapring.
8. Install the plug and ground wire at the cover.
9. Install the flywheel cover and tighten the bolts to 62 inch lbs. (7 Nm).
10. Install the left front wheel and tire assembly.
11. Lower the vehicle.
12. Remove the engine support fixture J-28467-A or equivalent.

13. Install the upper transaxle to engine bolt and tighten to 55 ft. lbs. (75 Nm).
14. Connect the speedometer cable, speed sensor and backup lamp connections.
15. Install the bolt and clip at the universal joint.
16. Connect the clutch cable to the release lever and adjust the cable.
17. Check the fluid level.
18. Connect the negative battery cable.

BENCH OVERHAUL

Before Disassembly

When servicing the unit, it is recommended that as each part is disassembled, it is cleaned in solvent and dried with compressed air. Disassembly and reassembly of this unit and its parts must be done on a clean work bench. Before installing bolts into aluminum parts, always dip the threads into clean transmission oil. Anti-seize compound can also be used to prevent bolts from galling the aluminum and seizing. Always use a torque wrench to keep from stripping the threads. Take care with the seals when installing them, especially the smaller O-rings. The slightest damage can cause leaks. Aluminum parts are very susceptible to damage so great care should be exercised when handling them. The internal snaprings should be expanded and the external snaprings compressed if they are to be reused. This will help insure proper seating when installed. Be sure to replace any O-ring, gasket, or seal that is removed.

Transaxle Disassembly

1. Remove the fill plug at the cover.
2. Remove the bolts from the cover and remove the shift lever from the case.
3. Bolt the cover to fixture J-36842 or equivalent, position the fixture in base plate J-328920 or equivalent.
4. Remove the shift lever cover components as follows: snapring, bushing, spring, intermediate lever and shift lever finger and pin.

NOTE: The remaining parts of the cover are not serviceable, do not remove.

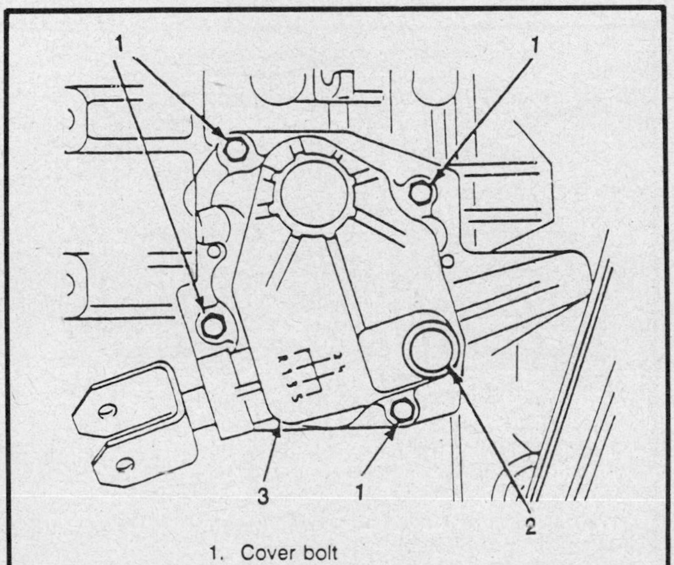
1
1
3
1
2
1. Cover bolt
2. Oil plug cap
3. Gear shift lever cover

Shift lever cover

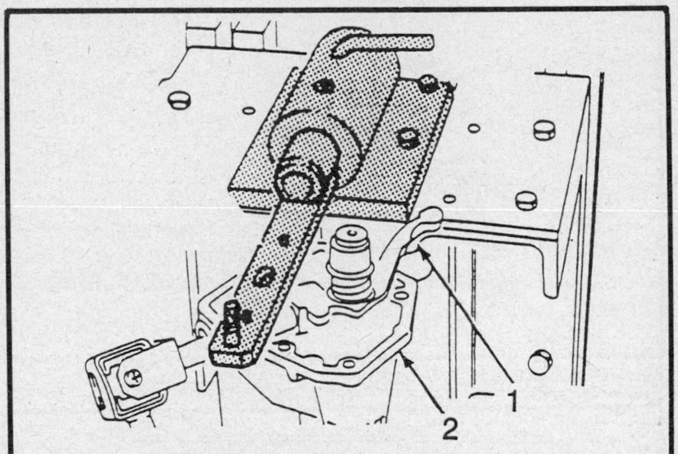
1
2
1. Intermediate lever
2. Gear shift lever cover

Shift lever cover fixture

5. Remove the bolt and the speedometer driven gear from the housing.
6. Remove the backup light switch.
7. Remove the cover and shift the transaxle into 2nd gear.
8. Remove the housing bearing plate bolts and the bearing plate with the shafts attached from the case.
9. Bolt the bearing plate to fixture J-36842 or equivalent and position the fixture in base plate J-328920 or equivalent.
10. Remove the 5th gear fork from the bearing plate. Remove the snapring.
11. Using puller J-22888-D or equivalent, remove the 5th speed driven gear, sleeve and gear. Remove the needle bearing, retaining ring and both halves of the thrust washer.
12. Remove the snapring and the input drive 5th gear using 5th gear remover pilot J-36555, puller bar J-35274 and puller legs J-36852 or equivalent.
13. Remove the bolts and the 5th gear shift connector with the pawl from the end shield. Press out the pin from the connector to replace the pawl.
14. Remove the 4 shift rod plugs.
15. Position the shift rails in 2nd, 5th and 3rd speed positions in that order to remove the bolts and the support. When 3rd gear is reached, the support will push out.
16. Remove the retaining pins for the 3rd/4th and reverse gear forks.
17. With the shifting sleeve in the neutral position, remove the shift rods and the shift forks for the 3rd/4th and reverse gear from the bearing plate.
18. Pull the shift driver for the 5th gear out of the bearing plate.
19. Remove the snapring retaining the cluster gear and the mainshaft at the bearing plate.
20. Remove the detent lock plate from the bearing plate.
21. Remove the mainshaft, cluster gear, reverse idler gear, shift fork and the 1st/2nd gear shift rod out of the bearing plate.

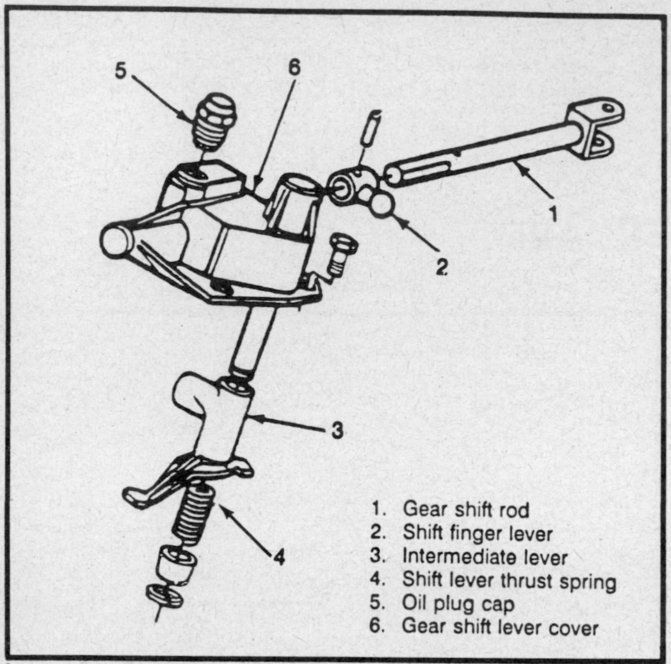
5
6
1
2
3
4
1. Gear shift rod
2. Shift finger lever
3. Intermediate lever
4. Shift lever thrust spring
5. Oil plug cap
6. Gear shift lever cover

Intermediate shift lever

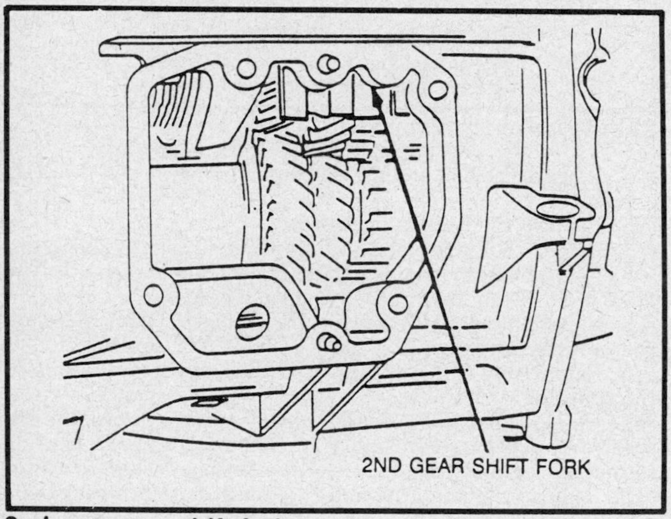
2ND GEAR SHIFT FORK

2nd gear rear shift fork

22. Position the reverse idler gear shaft between protective jaws in a vice. Tap the bearing plate to separate the shaft from the bearing plate. Disassemble using care not to lose the locking ball when the shaft is removed.

23. To remove the differential assembly, remove the cover and seals. Mark the bearing adjusting ring and the case. Remove the bearing plate. Remove the bearing adjusting ring using tool J-36635 or equivalent. Remove the differential assembly through the cover opening.

Unit Disassembly and Assembly

INPUT SHAFT/CLUSTER GEAR ASSEMBLY

Disassembly

1. Remove the input shaft from the cluster gear using a drift and a hydraulic press.

2. Remove the snapring and washer, then the bearing from the cluster gear.

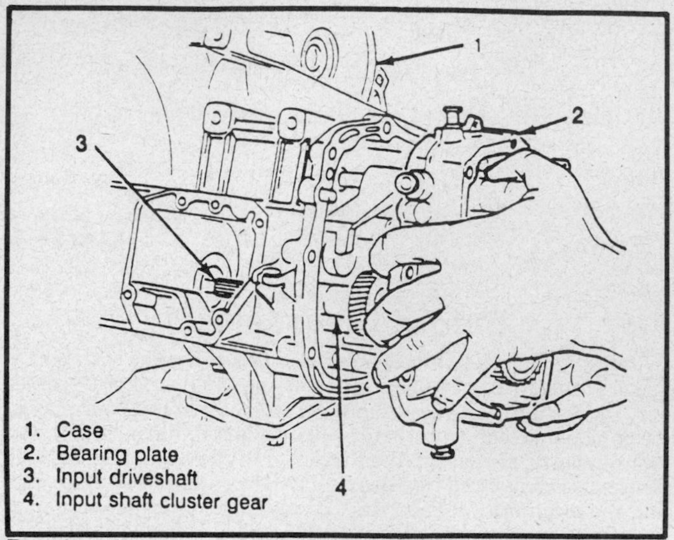
1
2
3
4
1. Case
2. Bearing plate
3. Input driveshaft
4. Input shaft cluster gear

Bearing plate

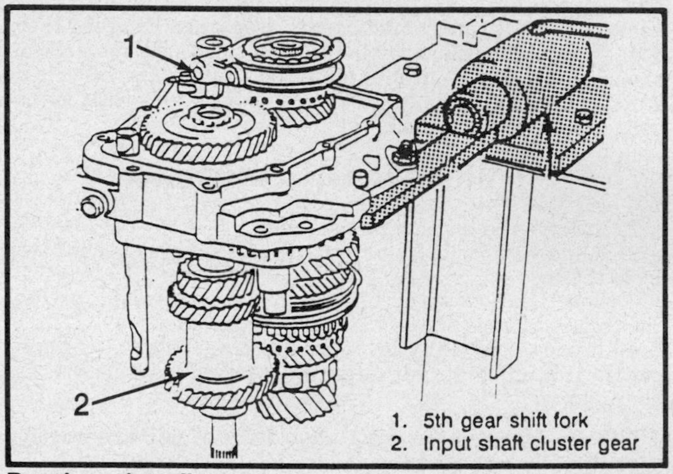
1
2
1. 5th gear shift fork
2. Input shaft cluster gear

Bearing plate fixture

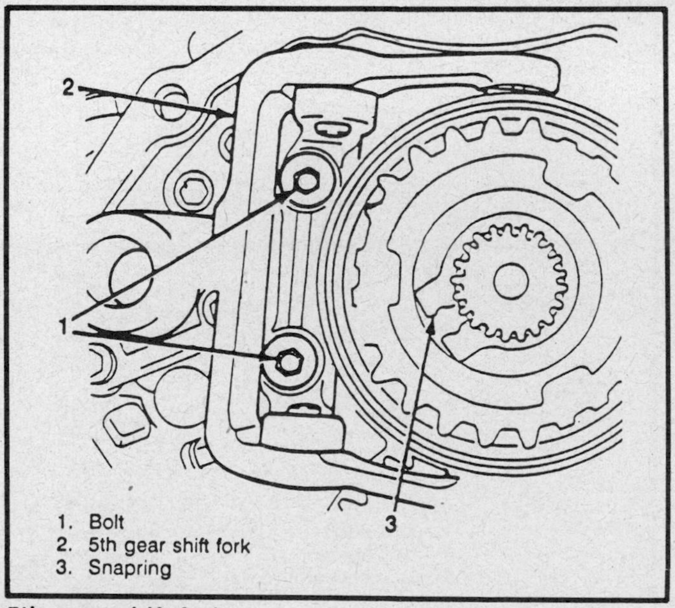
2
1
3
1. Bolt
2. 5th gear shift fork
3. Snapring

5th gear shift fork

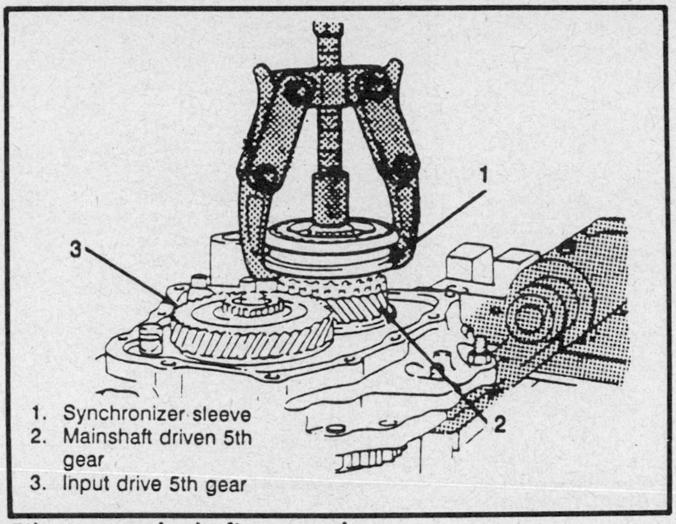
1
2
3
1. Synchronizer sleeve
2. Mainshaft driven 5th gear
3. Input drive 5th gear

5th gear mainshaft removal

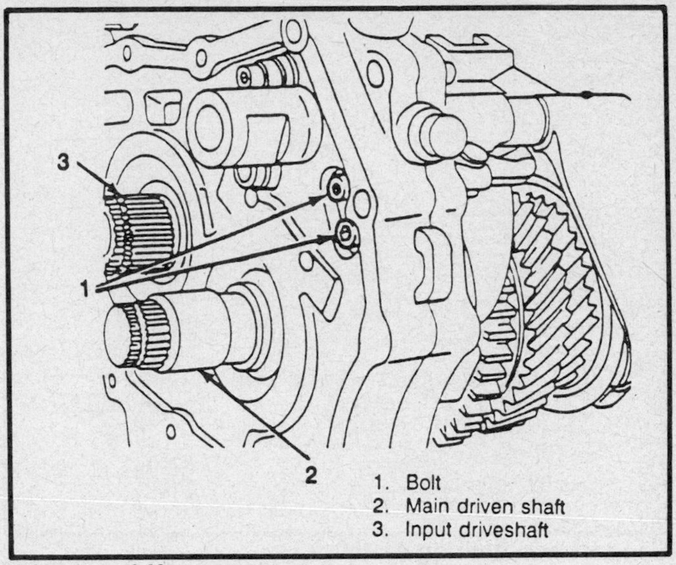
3
1
2
1. Bolt
2. Main driven shaft
3. Input driveshaft

5th gear shift connector

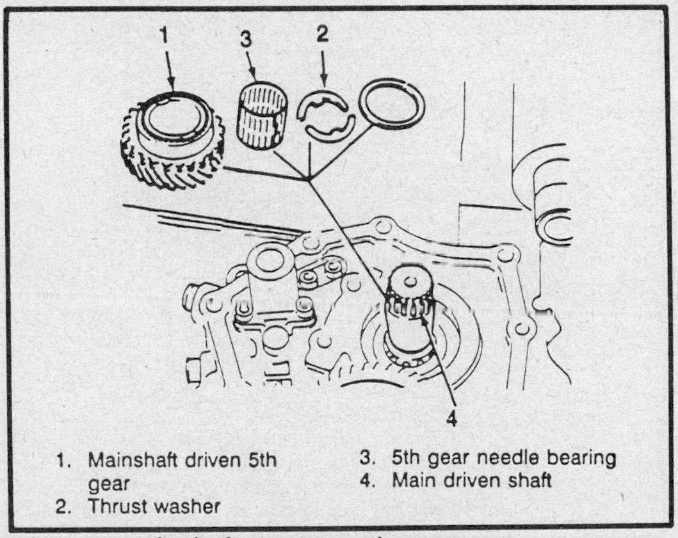
1
3
2
4
1. Mainshaft driven 5th gear
2. Thrust washer
3. 5th gear needle bearing
4. Main driven shaft

5th gear mainshaft components

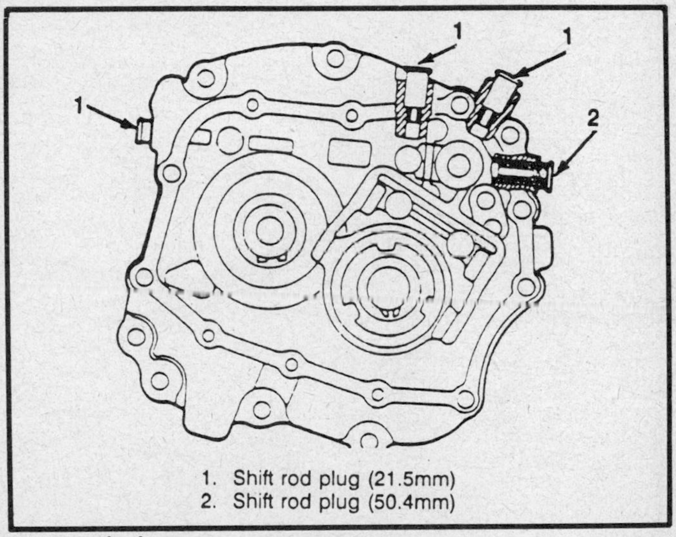
1
1
1
2
1. Shift rod plug (21.5mm)
2. Shift rod plug (50.4mm)

Shift rod plugs

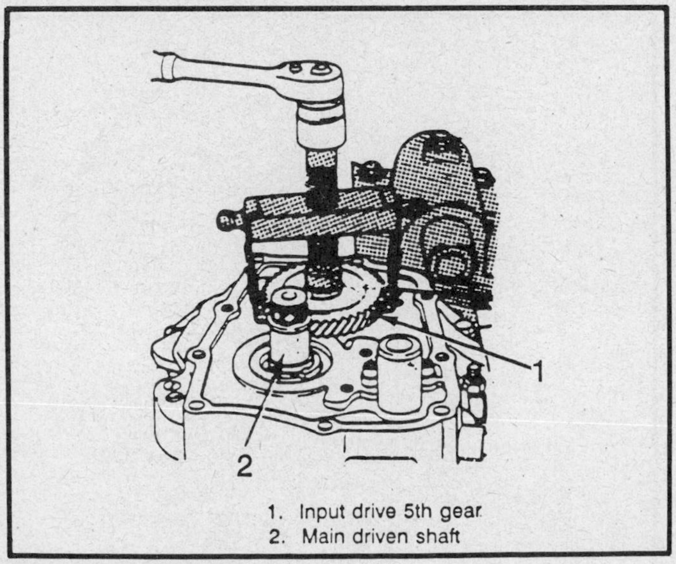
1
2
1. Input drive 5th gear
2. Main driven shaft

5th speed cluster gear removal

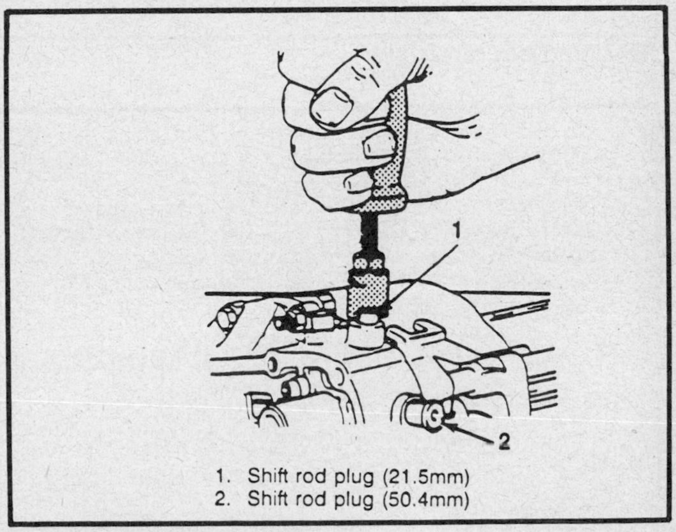
1
2
1. Shift rod plug (21.5mm)
2. Shift rod plug (50.4mm)

Removing shift rod plug

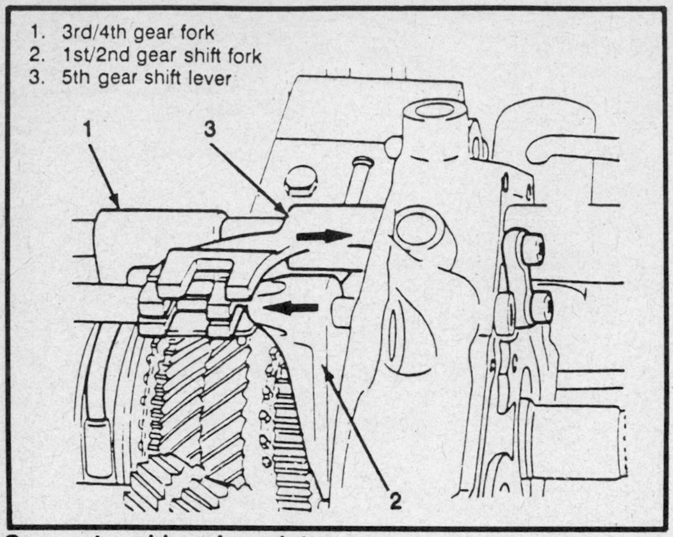
1. 3rd/4th gear fork
2. 1st/2nd gear shift fork
3. 5th gear shift lever
1
3
2

Support and bearing plate

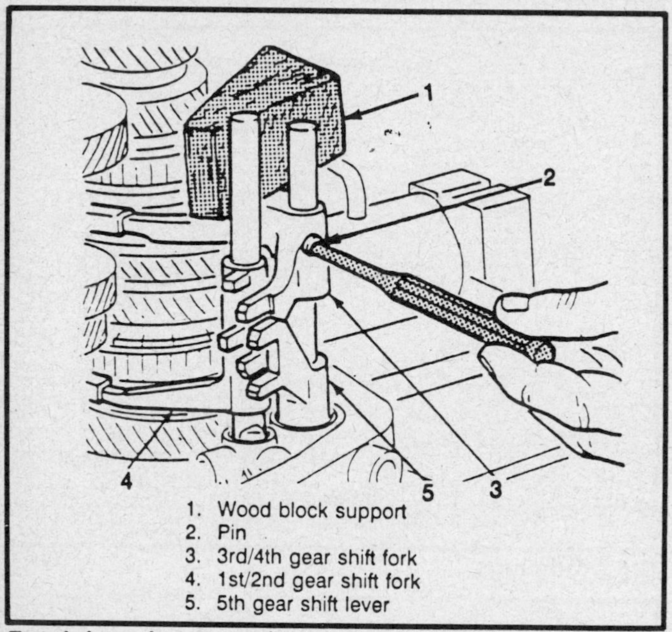
1
2
3
4
5
1. Wood block support
2. Pin
3. 3rd/4th gear shift fork
4. 1st/2nd gear shift fork
5. 5th gear shift lever

Retaining pin removal

5TH GEAR
SHIFT LEVER

5th gear lever

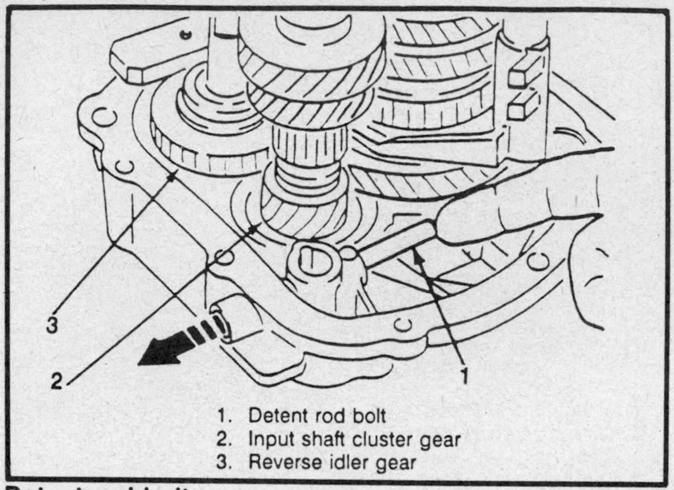
3
2
1
1. Detent rod bolt
2. Input shaft cluster gear
3. Reverse idler gear

Detent rod bolt

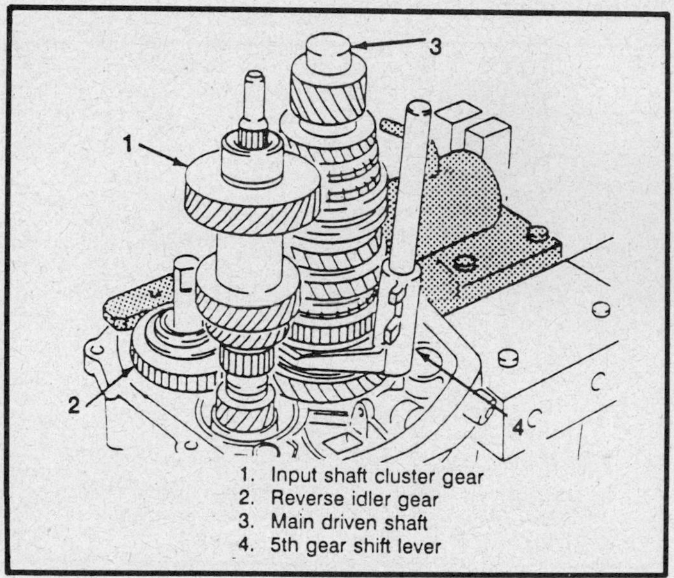
3
1
2
4
1. Input shaft cluster gear
2. Reverse idler gear
3. Main driven shaft
4. 5th gear shift lever

Gears and shafts

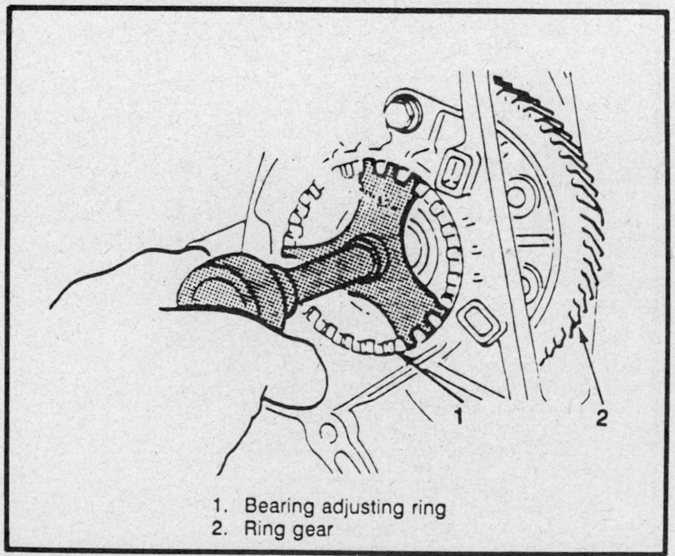
1
2
1. Bearing adjusting ring
2. Ring gear

Marking the bearing adjusting ring

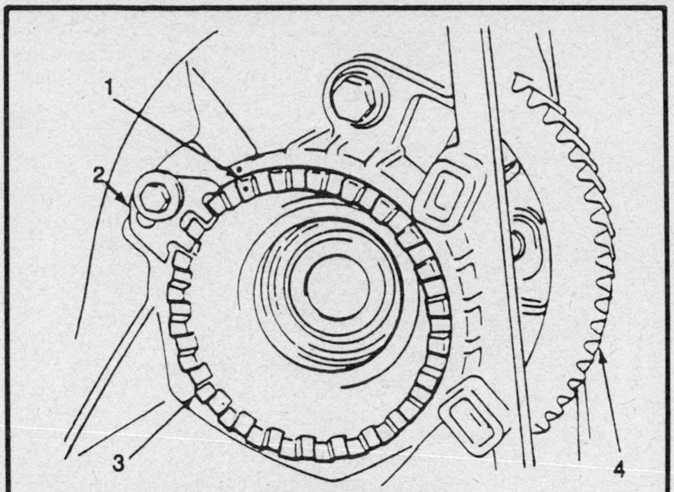
1
2
3
4

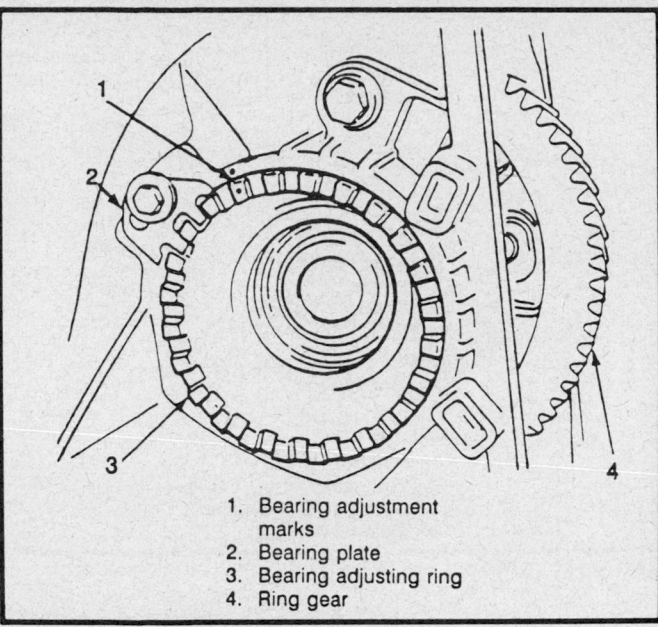

1. Bearing adjustment marks
2. Bearing plate
3. Bearing adjusting ring
4. Ring gear

Bearing adjusting ring

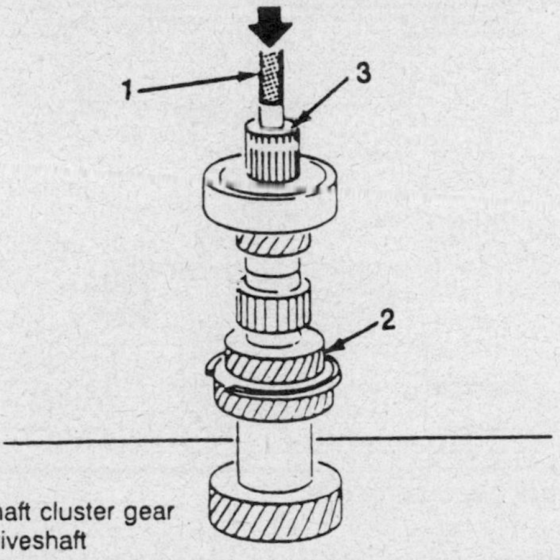
1
2
3

1. Drift
2. Input shaft cluster gear
3. Input driveshaft

Input shaft/cluster gear

Inspection

1. Check the shaft for spline wear or cracks. Replace if necessary.
2. Check the cluster gear for scuffed, nicked, burred or broken teeth.
3. Inspect the bearing for roughness of rotation. Replace if necessary.

Assembly

1. Install the snapring and the bearing on the cluster gear using tool J-36544 or equivalent and a hydraulic press.
2. Install the snapring and washer.
3. Install the input shaft with the small splines leading into the gear cluster. The input shaft should be in its backout position and should slide freely after being installed.

MAINSHAFT ASSEMBLY

Disassembly

1. Remove the 1st gear, thrust bearing, plate, snapring and

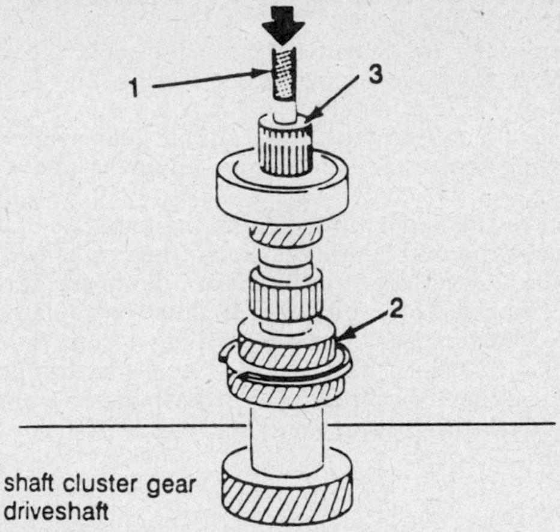
1
2
3

1. Drift
2. Input shaft cluster gear
3. Input driveshaft

Input shaft/cluster gear

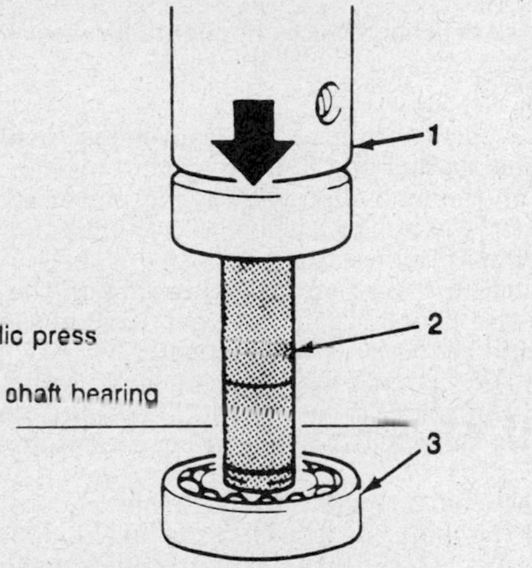
1
2
3

1. Hydraulic press
2. Drift
3. Cluster shaft bearing

Cluster gear bearing

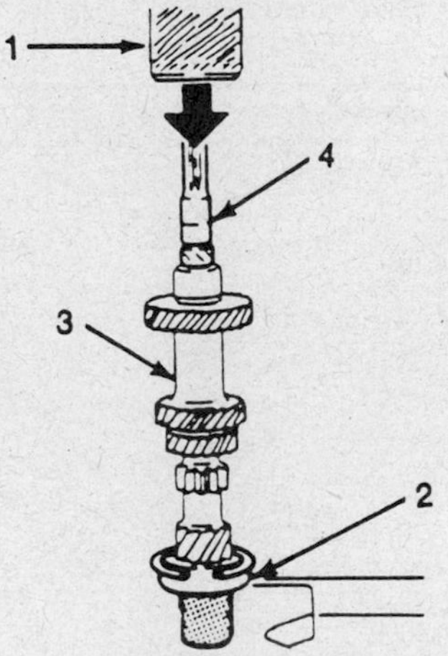
1
2
3
4

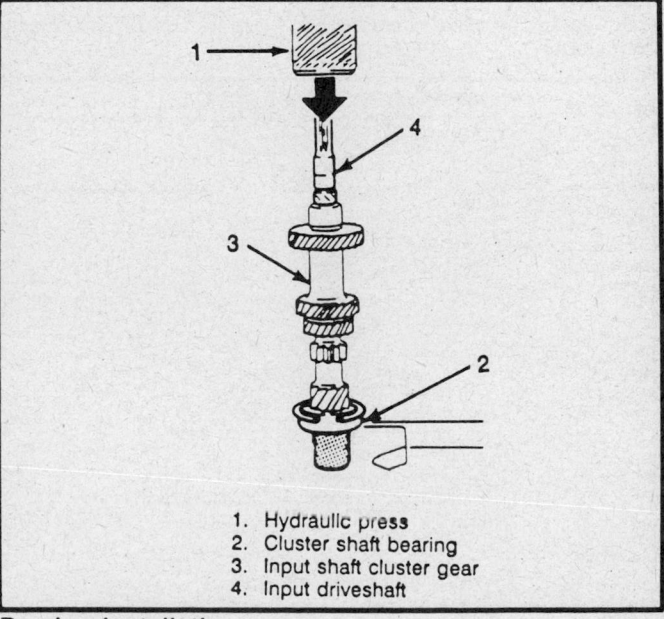

1. Hydraulic press
2. Cluster shaft bearing
3. Input shaft cluster gear
4. Input driveshaft

Bearing installation

bearing from the mainshaft using tool J-22912-01 or equivalent and a hydraulic press.

2. Remove the 1st gear needle bearing from the mainshaft.
3. Remove the snapring retaining the 1st/2nd gear synchronizer hub on the mainshaft.
4. Remove the 2nd gear and 1st/2nd gear synchronizer assembly using tool J-22912-01 or equivalent and a hydraulic press.
5. Remove the snapring from the mainshaft.
6. Remove the mainshaft gear from the mainshaft using tool J-22912-01 or equivalent and a hydraulic press. If the gear or shaft is damaged, both components must be replaced.
7. Remove the spacer and the 4th gear from the mainshaft using tool J-22912-01 or equivalent and a hydraulic press.
8. Remove the 3rd gear and the 3rd/4th gear synchronizer assembly from the mainshaft using tool J-22912-01 or equivalent and a hydraulic press.

Inspection

1. Inspect the gears for scuffed, nicked, burred or broken teeth.
2. Check the bearings for roughness of rotation. Replace if necessary.
3. Check all synchronizer components for wear or distortion.

Assembly

1. Coat all parts with gear lubricant before installing.
2. Heat the 1st/2nd and 3rd/4th synchronizer hubs to 212°F and install on the mainshaft. The synchronizer springs should be installed on the hubs with the hooked ends in the same key slot, but in opposite directions.
3. Install the 3rd gear and blocking ring on the mainshaft.
4. Press the 3rd/4th gear synchronizer assembly on the mainshaft and retain it with a snapring.
5. Install the 4th gear and blocking ring on the mainshaft.
6. Install the spacer ring and the mainshaft pinion gear on the mainshaft using tool J-36535 or equivalent and a hydraulic press.
7. Install the snapring on the mainshaft.
8. Install the 2nd gear and blocking on the mainshaft.
9. Press the 1st/2nd gear synchronizer assembly on the mainshaft and retain it with a snapring.
10. Install the 1st gear, roller bearing and blocking ring on the mainshaft.
11. Install the needle bearing and plate for the 1st gear on the mainshaft.

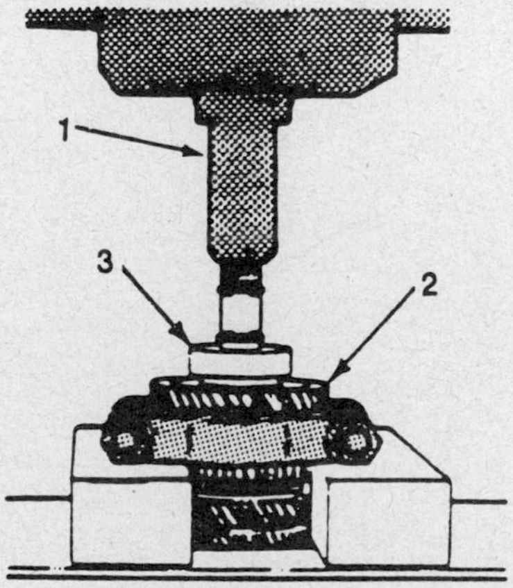
1
2
3

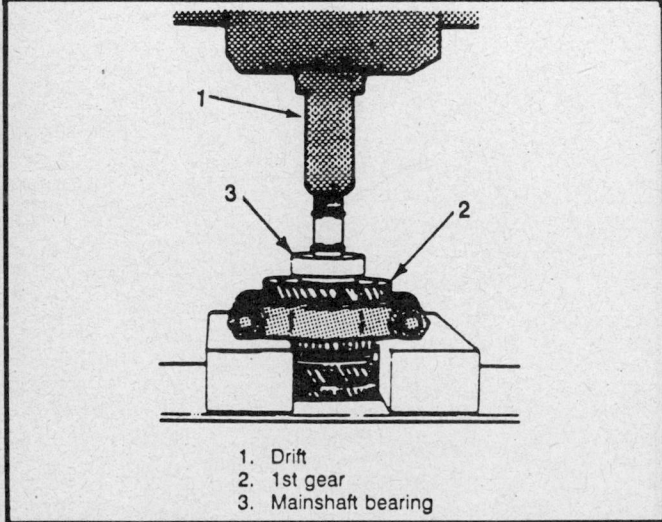

1. Drift
2. 1st gear
3. Mainshaft bearing

1st gear removal

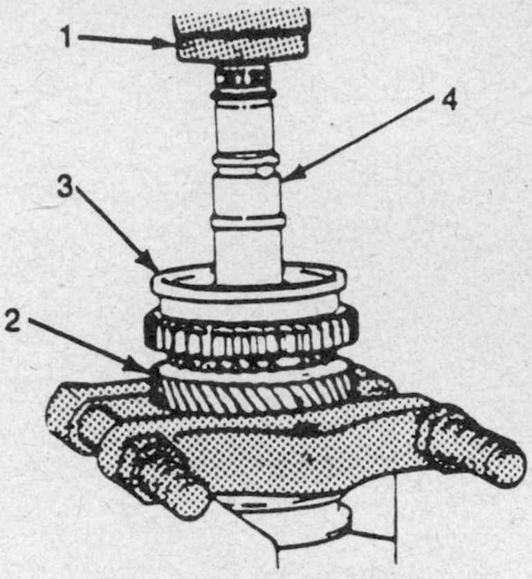
1
4
3
2

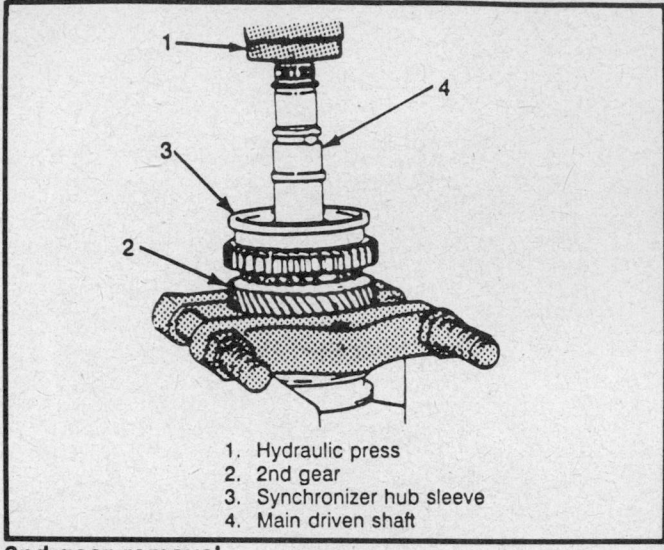

1. Hydraulic press
2. 2nd gear
3. Synchronizer hub sleeve
4. Main driven shaft

2nd gear removal

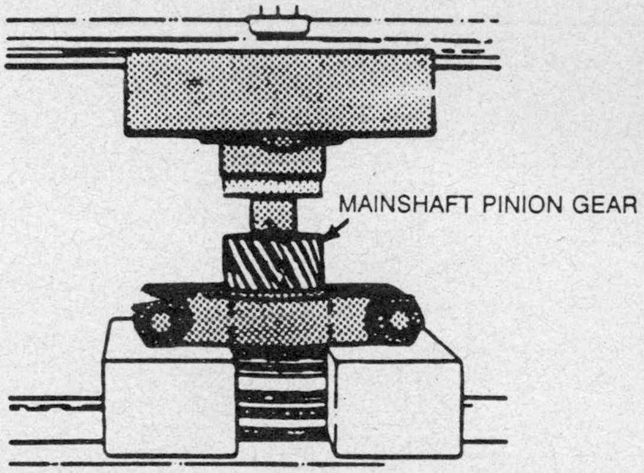
MAINSHAFT PINION GEAR

Mainshaft gear removal

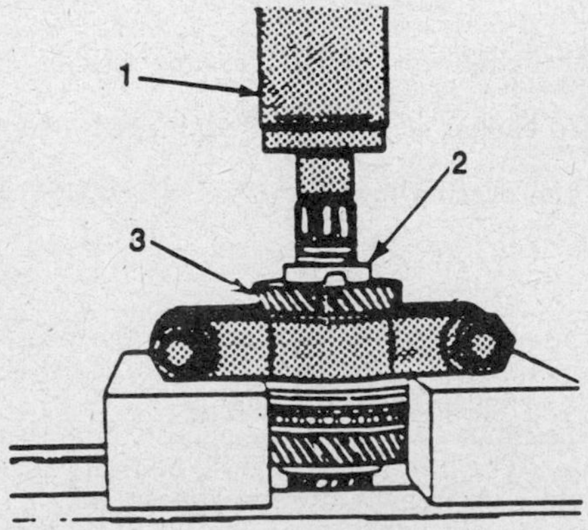
1
2
3

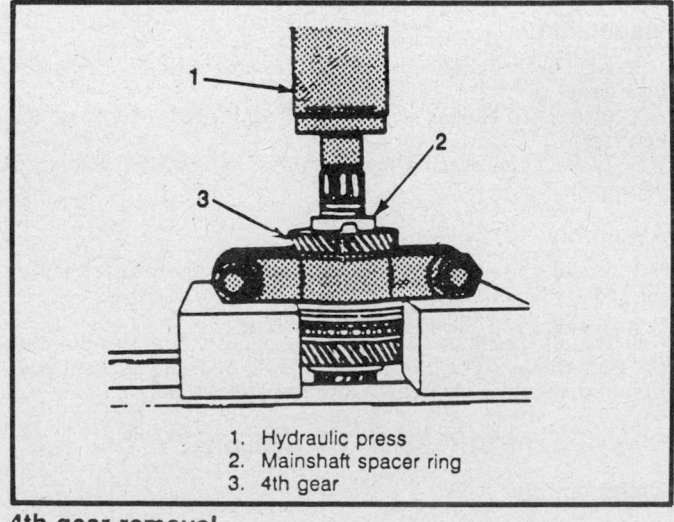

1. Hydraulic press
2. Mainshaft spacer ring
3. 4th gear

4th gear removal

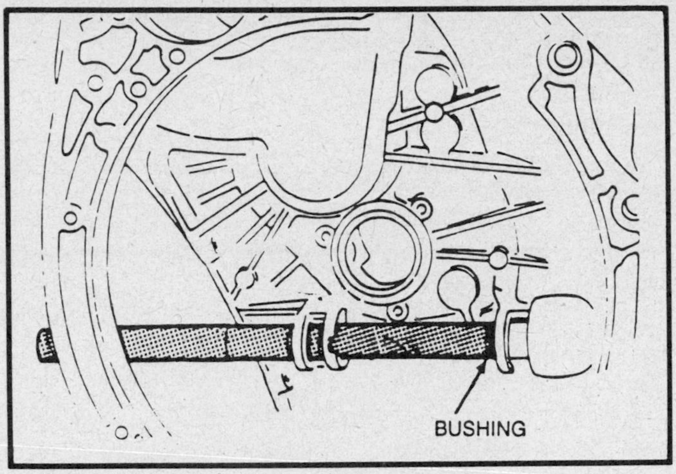
BUSHING

3rd gear removal

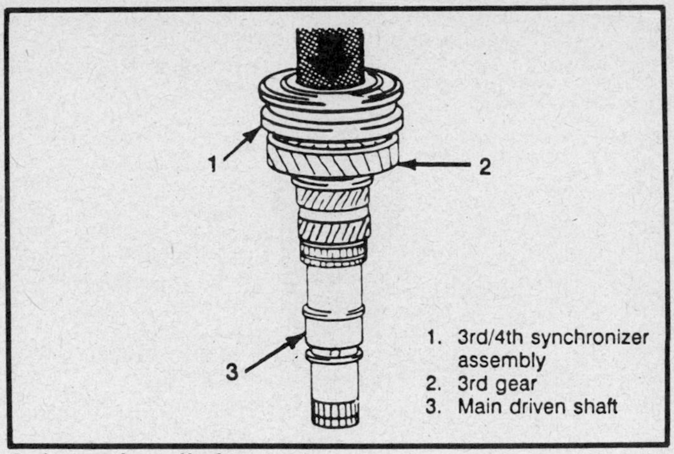
1
2
3
1. 3rd/4th synchronizer assembly
2. 3rd gear
3. Main driven shaft

3rd gear installation

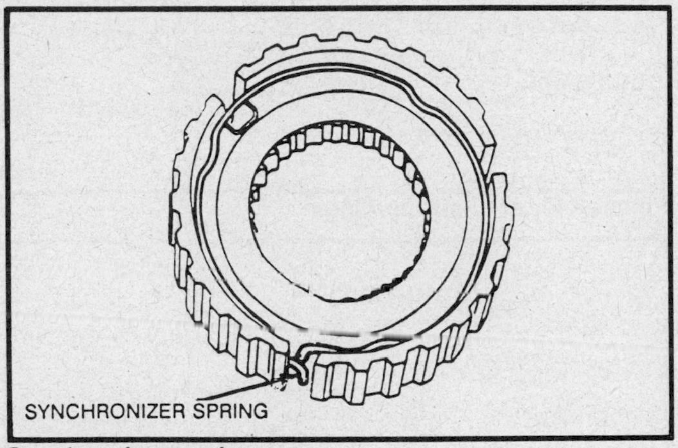
SYNCHRONIZER SPRING

Synchronizer springs

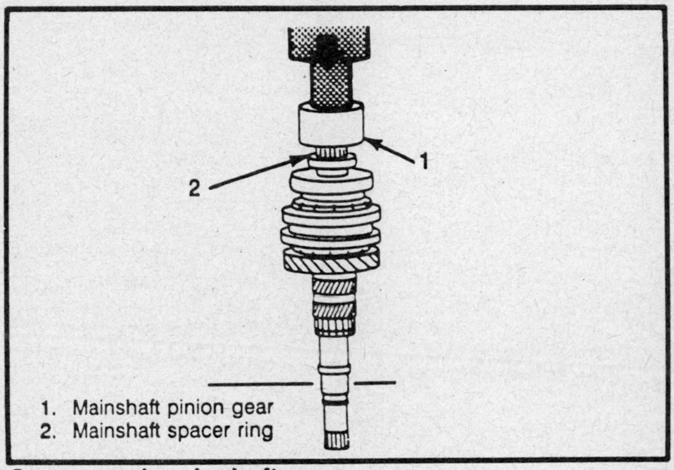
1
2
1. Mainshaft pinion gear
2. Mainshaft spacer ring

Spacer and mainshaft gear

12. Install the snapring and bearing on the mainshaft using tool J-36544 or equivalent and a hydraulic press.

HOUSING CASE

Disassembly

1. Remove the release fork bolt, release lever shaft, fork and the release bearing from the case.
2. Remove the bearing guide sleeve bolts, sleeve and the O-ring seal from the case.
3. Remove the release lever shaft bushings using tool J-36536 with J-36190 or equivalent to drive out the bushings.
4. Remove the the input shaft seal from the bearing guide sleeve.
5. Remove the input shaft bearing from the case using tool J-36550 with J-8092 or equivalent.
6. Remove the mainshaft bearing from the case using tool J-36669 and J-23907 or equivalent.
7. Slide the bearing race from the bearing adjusting ring. With tool J-36631 as a support, drive out the race using tools J-36548 and J-8092 or equivalent.
8. Slide the bearing race from the case using tool J-36548 and J-8092 or equivalent.

Inspection

1. Clean all parts with solvent and air dry.
2. Check the housing case for distortion, bores out of round or scoring.

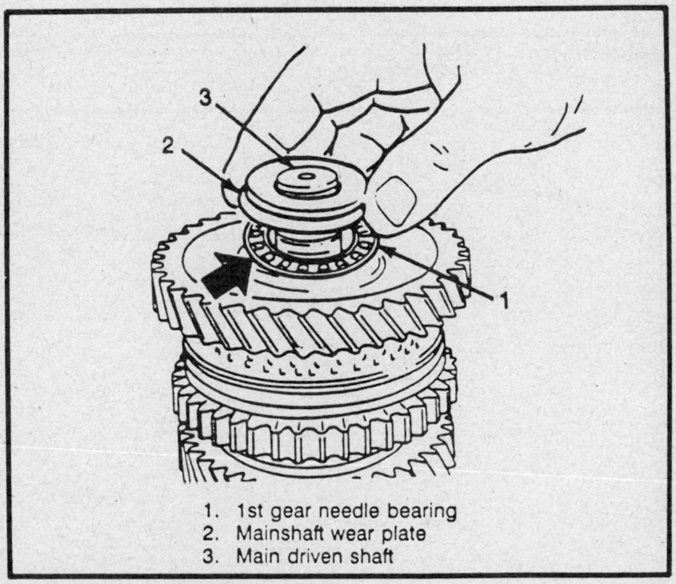
3
2
1
1. 1st gear needle bearing
2. Mainshaft wear plate
3. Main driven shaft

Needle bearing and plate

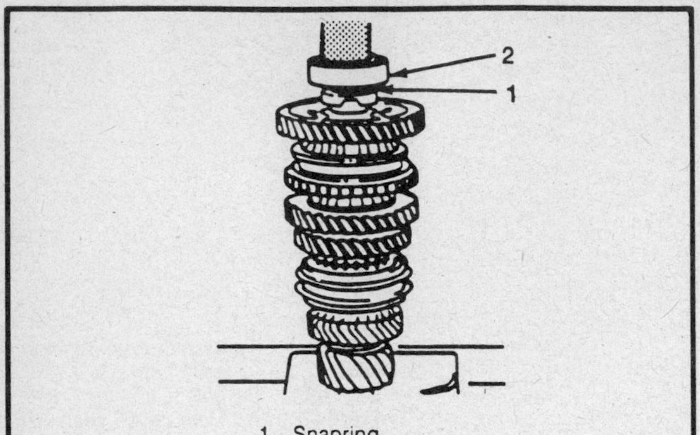
2
1

1. Snapring
2. Mainshaft bearing

Mainshaft bearing

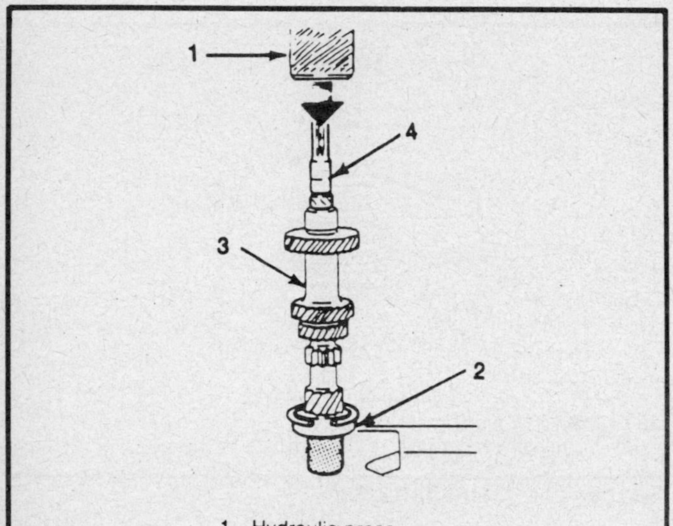
1
4
3
2

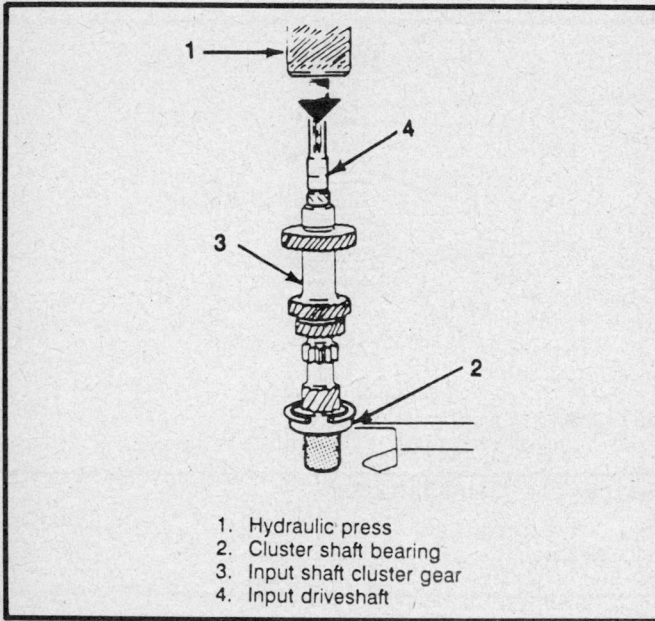

1. Hydraulic press
2. Cluster shaft bearing
3. Input shaft cluster gear
4. Input driveshaft

Release fork bolt

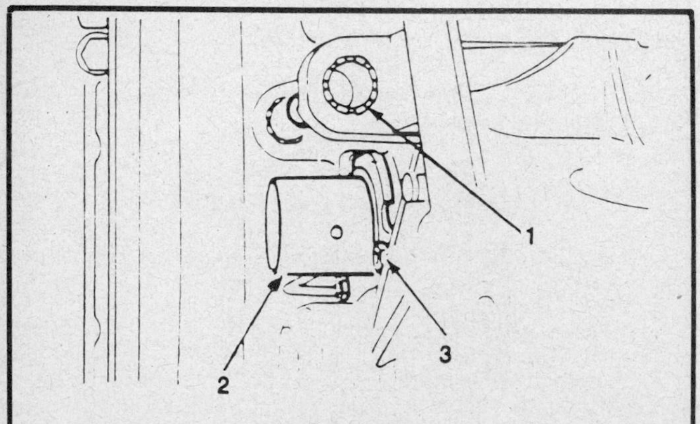
1
3
2

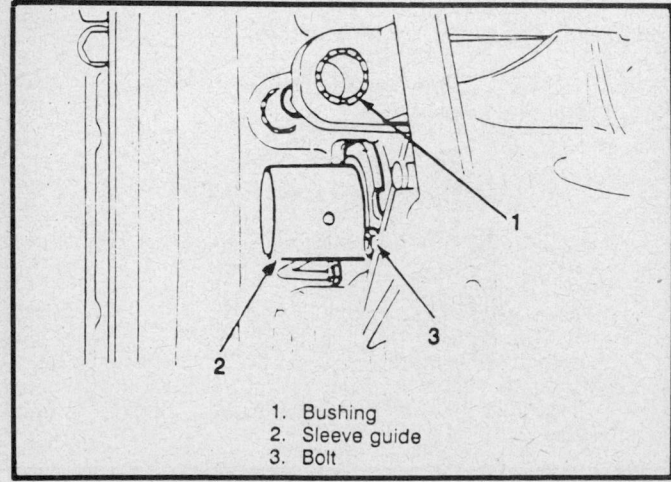

1. Bushing
2. Sleeve guide
3. Bolt

Bearing guide sleeve bolts

3. Check the bearings for roughness of rotation. Replace if necessary.
4. Inspect the thrust washers for wear, scuffed, nicked or burred condition.

Assembly

1. Install the side bearing race in the case using tools J–36548 and J–8092 or equivalent.

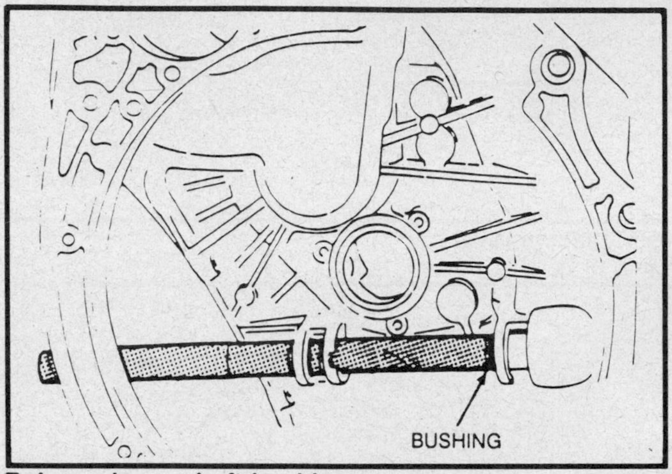
BUSHING

Release lever shaft bushings

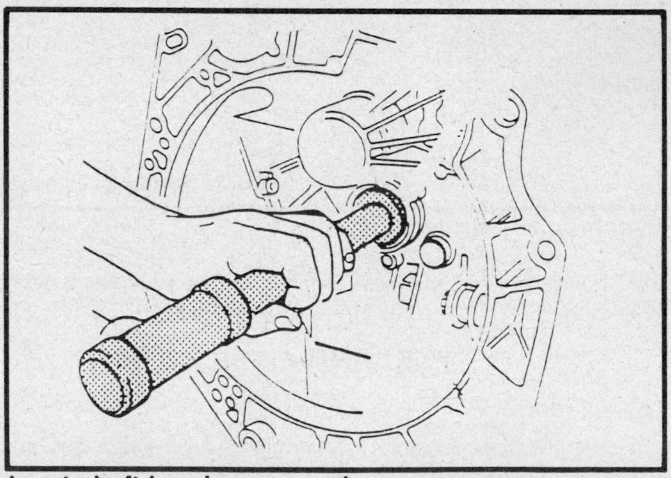

Input shaft bearing removal

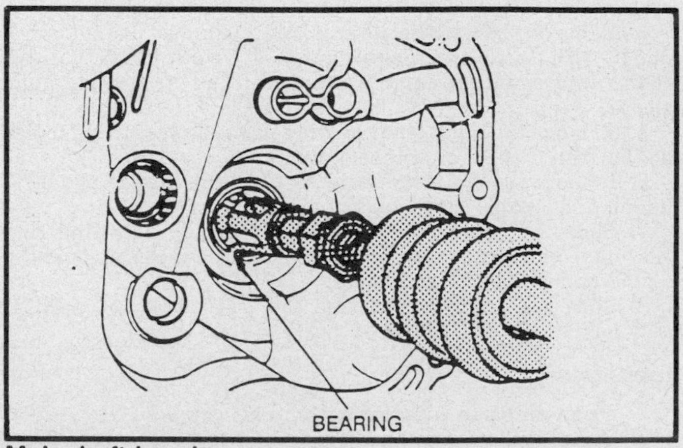
BEARING

Mainshaft bearing remover

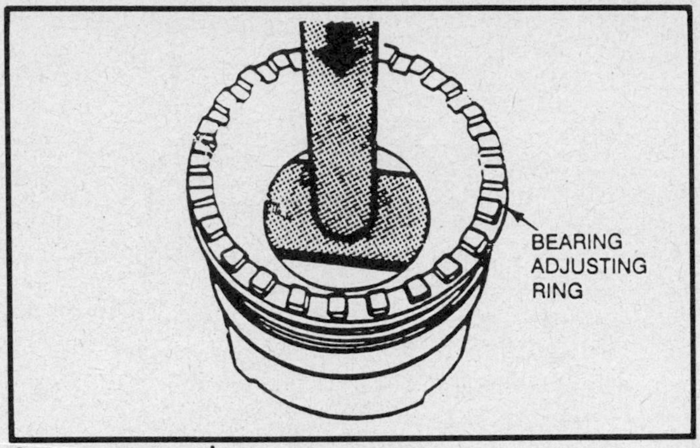
BEARING
ADJUSTING
RING

Outer race receiver

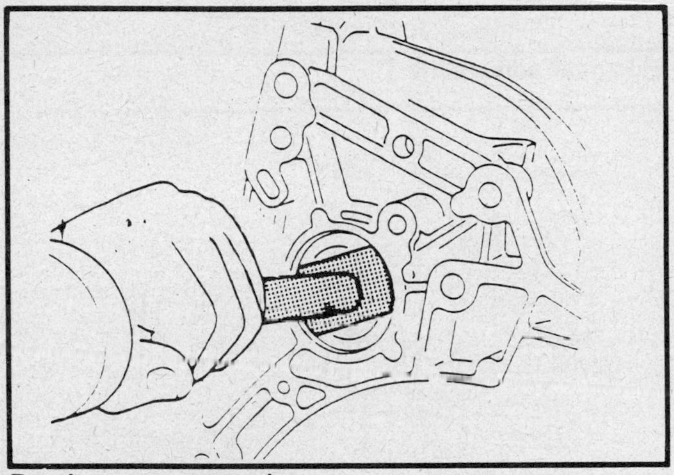

Bearing race removal

2. Install the side bearing race in the bearing adjusting ring using tools J-36548 and J-8092 or equivalent.
3. Install the drive axle seals in the case and the bearing adjusting ring using tool J-35647 or equivalent.
4. Install the mainshaft bearing flush into the case using tools J-36664 and J-8092 or equivalent.
5. Install the input bearing in the case using tools J-36550 and J-8092 or equivalent.
6. Install the input shaft seal in the bearing guide sleeve.
7. Install the O-ring seal to the case, then the bearing guide sleeve and the case bolts. Lightly coat the sleeve surface with grease before installation. Tighten the housing bolts to 45 inch lbs. (5 Nm).
8. Install the release lever shaft bushings using tools J-36536 and J-36190 or equivalent. Coat the bushing bores with grease after installation.
9. Coat the release bearing bore with grease. Install the release bearing and clutch fork on the guide sleeve.
10. Install the release lever shaft. Align the shaft to the fork and install the bolt. Tighten the bolt to 26 ft. lbs. (35 Nm).

DIFFERENTIAL ASSEMBLY

Disassembly

1. Remove the bearings from the differential housing using tool J-34851 and tool J-22888-D, or equivalents.
2. Remove the bolts and ring gear from the housing.
3. Remove the speedometer gear from the housing.
4. Remove the retaining rings from the pinion shaft. Drive the shaft out of the housing using a drift.

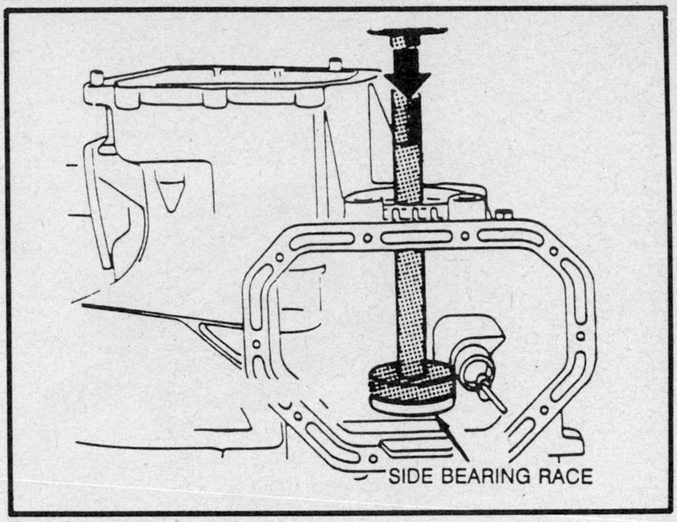
SIDE BEARING RACE

Bearing race installation

5. Remove the pinion gears and washers.
6. Remove the side gears and washers.

Inspection

1. Clean all parts with solvent and air dry.
2. Inspect the gears for scuffed, nicked, burred or broken teeth.
3. Check the differential case for distortion, bores out of round or scoring.
4. Check the bearings for roughness of rotation, burred or pitted condition.
5. Inspect the thrust washers for wear, scuffed, nicked or burred condition.

Assembly

1. Coat all parts with gear lubricant before installation.
2. Install the thrust washers and the side gears in the housing.
3. Install the thrust washers and pinion gears in the housing. Roll the gears into place by aligning the holes in the gears with the pinion shaft holes in the housing.
4. Install the pinion shaft in the housing.
5. Heat the speedometer gear to 176°F and install it on the housing using tool J-36552 or equivalent.
6. Heat the ring gear to 212°F and install it on the differential housing using a hydraulic press.
7. Install the ring gear and install the bolts to 63 ft. lbs. (85 Nm).
8. Install the bearings on the housing using tools J-36549 and J-8092 or equivalent.

Transaxle Assembly

1. Install the reverse idler shaft and locking ball in the bearing plate.
2. Install the pins into the shift fork and the shift rod of the 1st/2nd gear. Allow the new retaining pin to protrude approximately 0.08 in. (2mm).
3. Install the following components to the bearing plate: input shaft/cluster gear, mainshaft with the 1st/2nd shift rail and fork, reverse idler gear and retaining rings to the mainshaft and cluster gear.

NOTE: Use tool J-36633 to install the retaining rings. Make sure that the rings are seated in the grooves.

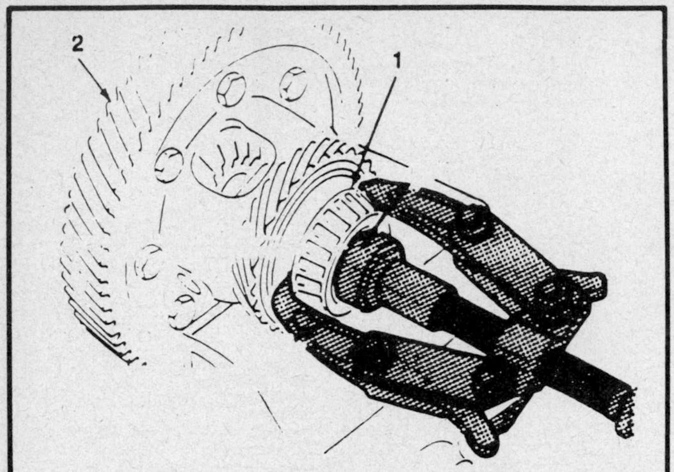
2
1

1. Differential bearing
2. Ring gear

Differential bearing removal

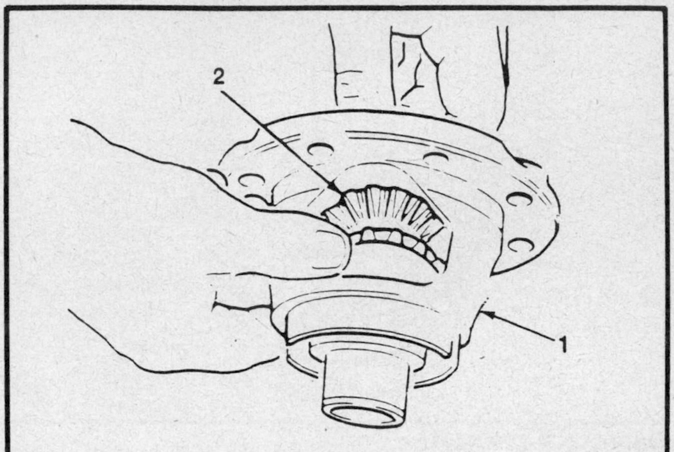
2
1

1. Differential housing
2. Side gear

Side gear and washer

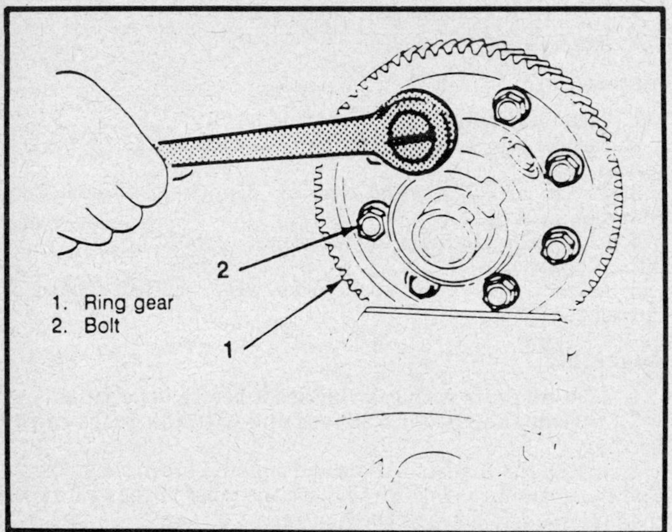
2
1

1. Ring gear
2. Bolt

Ring gear and case

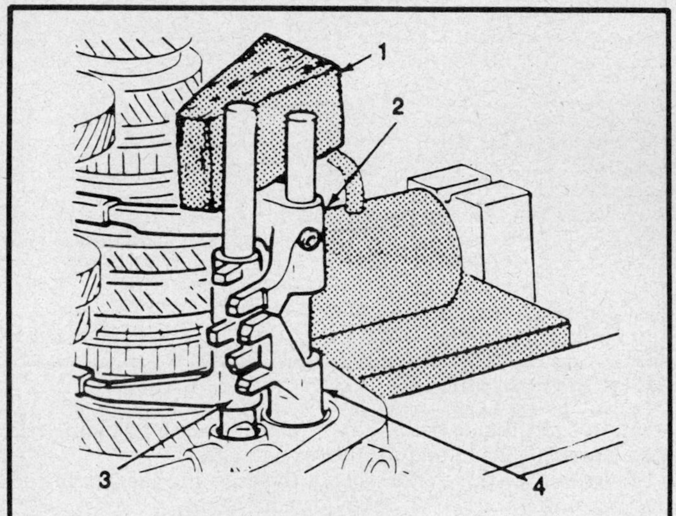
1
2
3
4

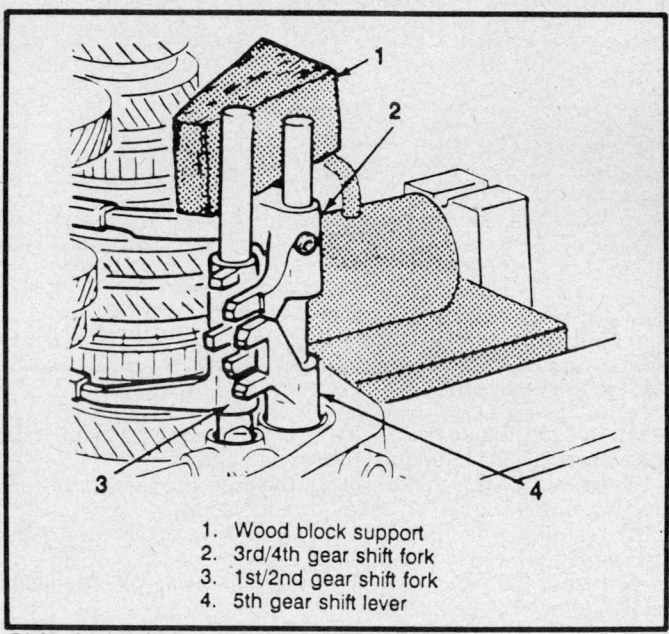

1. Wood block support
2. 3rd/4th gear shift fork
3. 1st/2nd gear shift fork
4. 5th gear shift lever

Shift fork pin

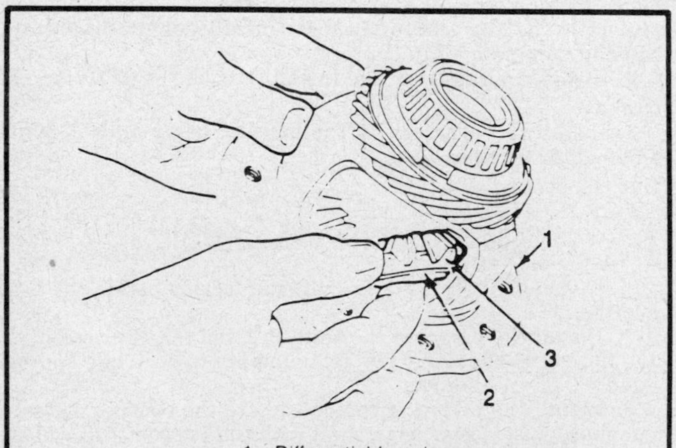
1
3
2

1. Differential housing
2. Washer
3. Pinion gear

Pinion gear and washer

4. Insert the reverse shift rod and reverse shift fork or reverse idler gear onto the 5th gear driveshaft.
5. Install the shift rod and shift fork for the 3rd/4th gear and new pins. Support the rails to prevent damage when installing the pins. Allow the new retaining pin to protrude approximately 0.08 in. (2mm).
6. Install the support for the locking pin to the bearing plate. Tighten the bolts to 63 ft. lbs. (7 Nm).
7. Install the bolt support with the pawl to the bearing plate. The slit in the shift rod for the 3rd/4th gear must align with the pawl. Tighten the bolts to 22ft. lbs. (30 Nm).
8. Install the pins, springs and plugs in the bearing plate.
9. Install the bearing plate with the shafts in tool J-36556.
10. Install the 5th gear on the cluster gear using tool J-36544 and a hydraulic press. The long end of the gear hub faces the bearing plate.
11. Install the snapring on the cluster gear.
12. Install the washer and retaining ring.
13. Install the roller bearing, 5th gear and blocking ring on the mainshaft.

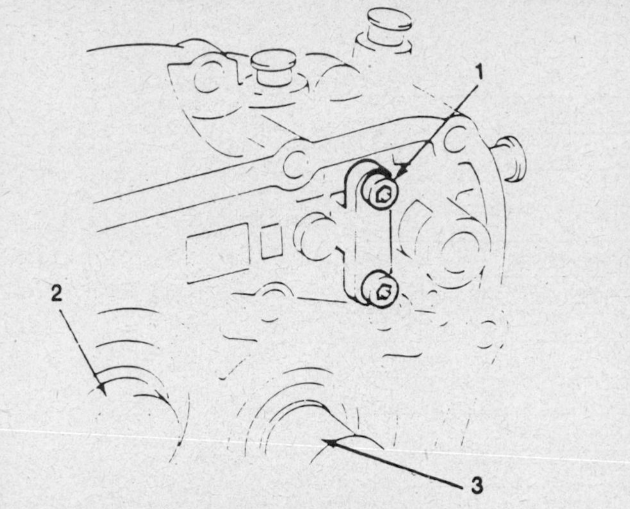
1
2
3

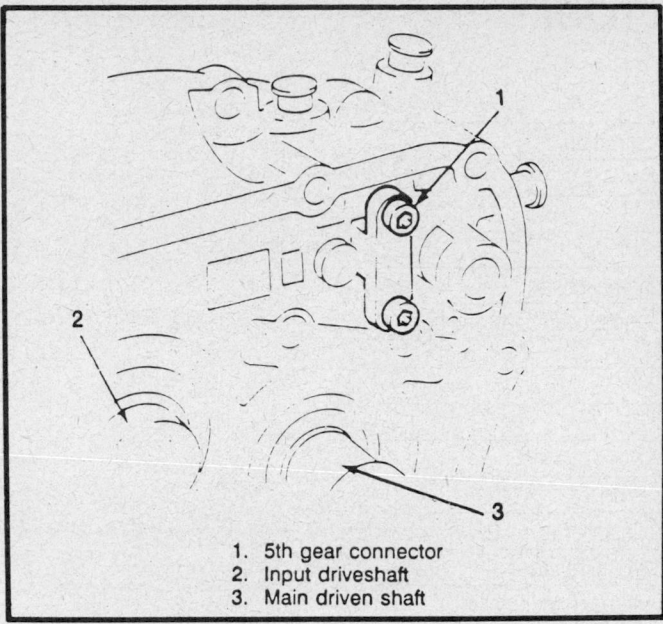

1. 5th gear connector
2. Input driveshaft
3. Main driven shaft

Locking pin support

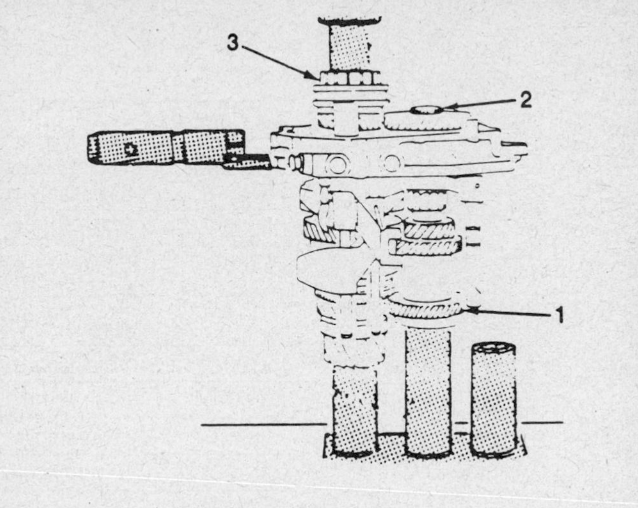
3
2
1

1. Input shaft cluster gear
2. Input driveshaft
3. Main driven shaft

5th gear installation

14. Heat the 5th gear sychronizer assembly to 212°F and install the it on the mainshaft using tool J–36545 and a hydraulic press.
15. Install the snapring on the mainshaft.
16. Install the 5th gear shift fork and bolts to the bearing plate. Tighten the bolts to 16 ft. lbs. (22 Nm).
17. Install the bearing plate with the shafts and gasket in the housing case.
18. Install the bolts through the bearing plate into the case and tighten to 11 ft. lbs. (15 Nm).
19. Install the differential assembly in the case.
20. Install the bearing adjusting ring and the O-ring using tool J–36635 until there is no endplay with the differential.
21. Adjust the preload on the differential bearings by tightening or loosening the bearing adjuster ring using tool J–36636 or equivalent and a torque wrench. For used bearings, 5–9 inch lbs. is required to rotate the differential 1 revolution per second. For new bearings, 15–17 inch lbs. is required to rotate the differential 1 revolution per second.
22. Install the lock plate and bolt to retain the bearing adjusting ring and tighten the bolt to 45 inch lbs. (22 Nm).
23. Install the cover, gasket and tighten the bolts to 22 ft. lbs. (30 Nm).
24. Install the shift lever cover using the following order: shift lever finger and pin, intermediate lever, spring, bushing and snapring.
25. Install the shift lever cover, gasket and bolts on the case with the transaxle in neutral. Tighten the bolts to 16 ft. lbs. (22 Nm).

SPECIFICATIONS

TORQUE SPECIFICATIONS

Component	ft. lbs.	Nm
Release bearing guide sleeve	26	35
Ring gear	63	85
Support pin lock	63①	7
Support pawl	22	30
Fifth gear fork	16	22
Bearing plate	11	15
Bearing ring retainer	45①	5
Differential cover	22	30
End cover (short)	11	15
End cover (long)	15	20
Shift lever cover	16	22

① inch lbs.

SPECIAL TOOLS

Tool Number	Description
J 3289-20	Base plate
J 6125-B	Slide hammer
J 8092	Driver handle
J 22888-D	Differential slide bearing puller set
J 22912-01	Split plate mainshaft syncromesh body and gear remover
J 23907	Slide hammer
J 35274	Fifth gear puller bar
J 35950	Gear remover legs
J 36190	Driver handle
J 36535	First gear bearing ring installer
J 36536	Clutch release lever bushing remover and installer
J 36544	Driveshaft and mainshaft bearing installer
J 36546	Driveshaft to thrust bearing seal installer
J 36548	Transmission case bearing cup remover and installer
J 36549	Differential side bearing installer
J 36550	Driveshaft needle bearing remover and installer
J 36552	Speedometer gear installer
J 36555	Fifth gear pilot
J 36631	Outer race receiver
J 36633	Snapring retainer
J 36635	Bearing ring remover and installer
J 36636	Spin torque measurer
J 36637	End shield plug remover
J 36638	Differential clamp
J 36664	Mainshaft bearing installer
J 36669	Mainshaft bearing remover
J 36842	Fixture
J 36852	Puller legs

Section 6

MV2 Transaxle
General Motors

TRANSAXLE APPLICATION

MV2 TRANSAXLE

Year	Vehicle	Engine
1985	Sprint	1.0L 2bbl
1986	Sprint	1.0L 2bbl
1987	Sprint	1.0L 2bbl
	Sprint	1.0L Turbocharged
1988	Sprint	1.0L 2bbl
	Sprint	1.0L Turbocharged
1989	Metro	1.0L 2bbl
	Metro	1.0L Turbocharged
	Metro	1.0L TBI

bbl—Barrell
TBI—Throttle Body Injection

GENERAL DESCRIPTION

The MV2 transaxle provides 5 forward speeds and 1 reverse speed by means of 3 synchronizers and 2 shafts—input shaft and countershaft.

All forward gears are in constant mesh and reverse uses a sliding idler gear arrangement.

The low-speed synchronizer and high-speed synchronizer are mounted on the countershaft. The low-speed synchronizer is engaged with the countershaft low or 2nd gear, while the high-speed synchronizer is engaged with 3rd or 4th gear.

The 5th gear synchronizer on the input shaft is engaged with the input shaft 5th gear.

The final output gear (an integral part of the countershaft) turns the final gear and differential assembly, thereby turning the driveshafts which are attached to the front wheels.

Transaxle Identification

The transaxle identification tag is located on the side of the transaxle assembly. This identification tag must be used when servicing this assembly.

Metric Fasteners

The metric fastener dimensions are very close to the dimensions of the familiar inch system fasteners and for this reason, replacement fasteners must have the same measurement and strength as those removed.

Do not attempt to interchange metric fasteners for inch system fasteners. Mismatched or incorrect fasteners can result in damage to the transaxle unit through malfunctions, breakage or possible personal injury.

Care should be taken to re-use the fasteners in the same locations as removed.

Capacities

The fluid capacity of this transaxle is 2.4 quarts (2.3L) of SAE-75W, SAE-80W or SAE 80W-90 GL-5 gear lube.

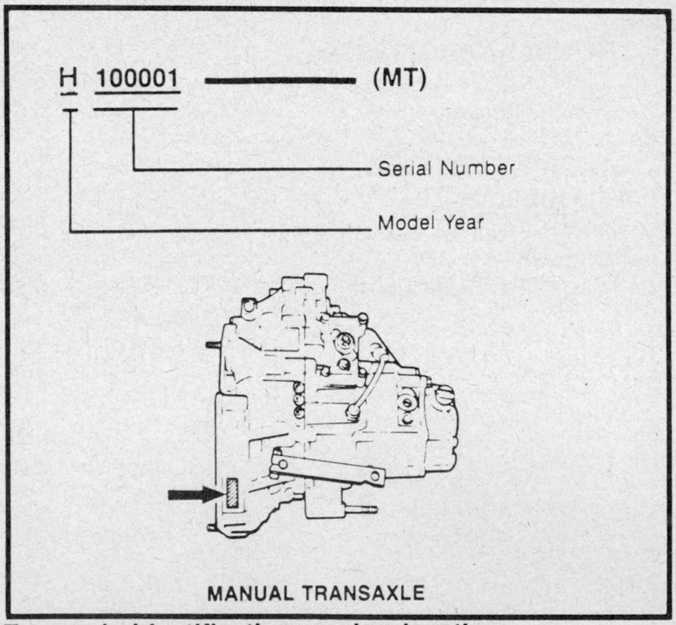
H 100001 (MT)
Serial Number
Model Year
MANUAL TRANSAXLE

Transaxle identification number location

Checking Fluid Level

1985–88 SPRINT

NOTE: Before checking the oil level, check the transaxle for oil leakage. If leakage is found, repair it.

1. With the engine stopped, remove the oil level gauge from the side case of the transaxle.
2. Wipe off the oil level gauge with clean cloth.
3. Fit the oil level gauge to the transaxle side case.
4. Remove the gauge and check the oil level. The level should be between **FULL** level line and **LOW** level line. If it is below the **LOW** level line, add oil until it reaches the **FULL** level line.

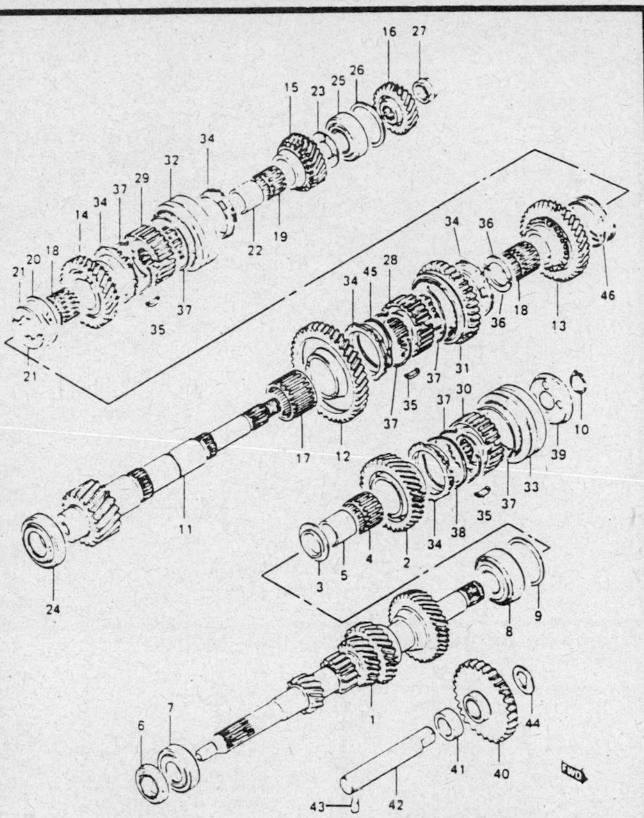
FWD

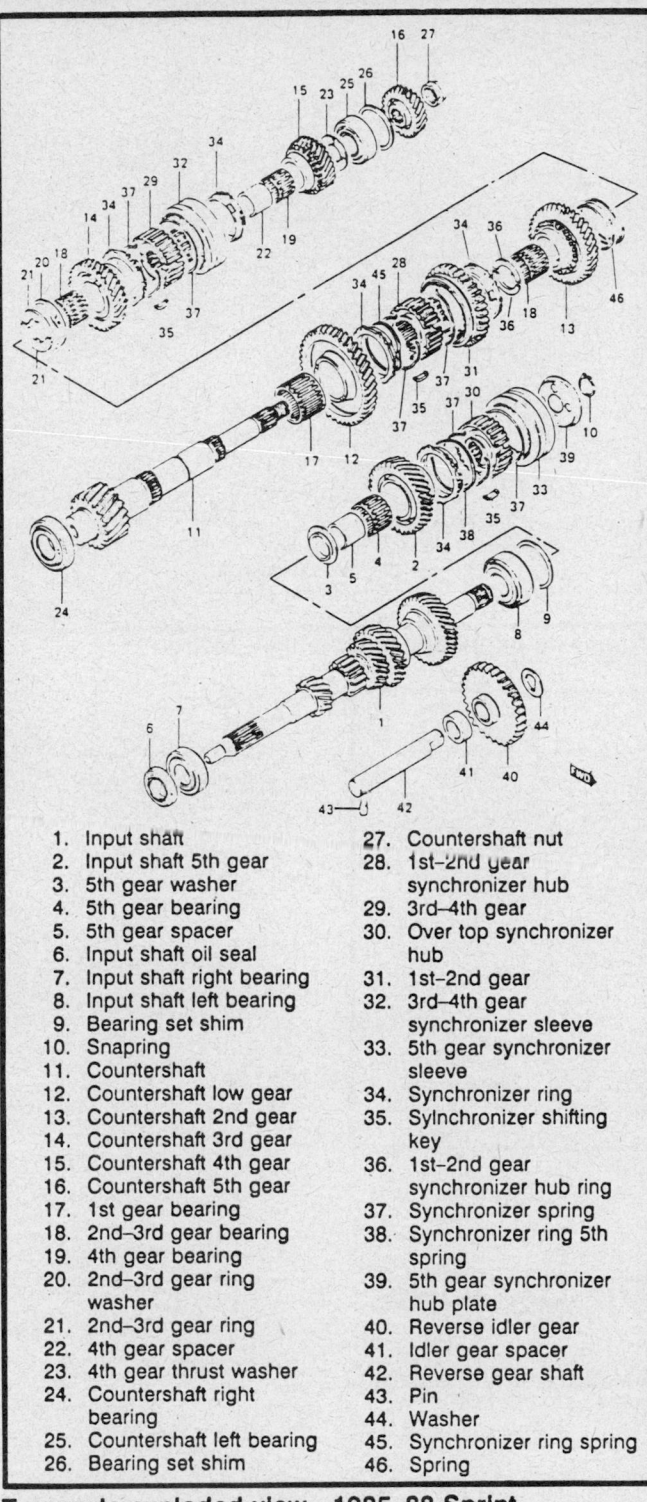

1. Input shaft
2. Input shaft 5th gear
3. 5th gear washer
4. 5th gear bearing
5. 5th gear spacer
6. Input shaft oil seal
7. Input shaft right bearing
8. Input shaft left bearing
9. Bearing set shim
10. Snapring
11. Countershaft
12. Countershaft low gear
13. Countershaft 2nd gear
14. Countershaft 3rd gear
15. Countershaft 4th gear
16. Countershaft 5th gear
17. 1st gear bearing
18. 2nd–3rd gear bearing
19. 4th gear bearing
20. 2nd–3rd gear ring washer
21. 2nd–3rd gear ring
22. 4th gear spacer
23. 4th gear thrust washer
24. Countershaft right bearing
25. Countershaft left bearing
26. Bearing set shim
27. Countershaft nut
28. 1st–2nd gear synchronizer hub
29. 3rd–4th gear
30. Over top synchronizer hub
31. 1st–2nd gear
32. 3rd–4th gear synchronizer sleeve
33. 5th gear synchronizer sleeve
34. Synchronizer ring
35. Sylnchronizer shifting key
36. 1st–2nd gear synchronizer hub ring
37. Synchronizer spring
38. Synchronizer ring 5th spring
39. 5th gear synchronizer hub plate
40. Reverse idler gear
41. Idler gear spacer
42. Reverse gear shaft
43. Pin
44. Washer
45. Synchronizer ring spring
46. Spring

Transaxle exploded view—1985–88 Sprint

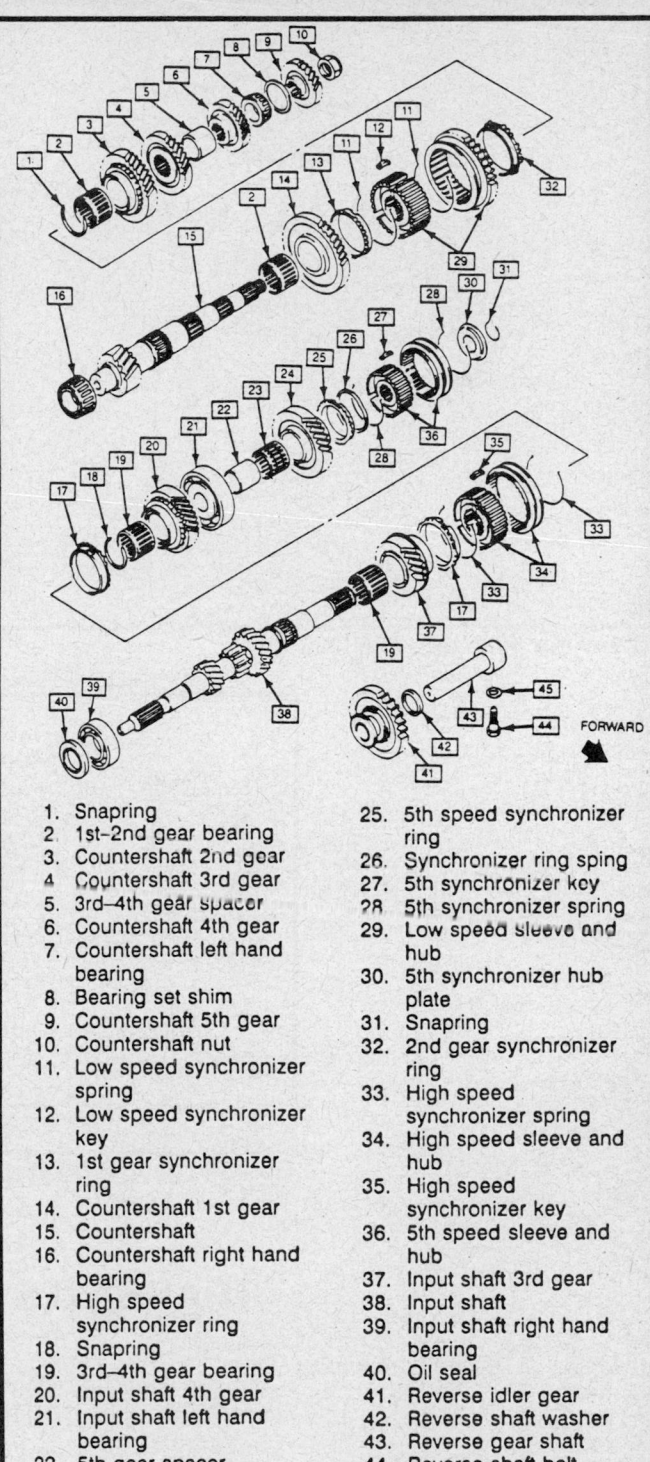

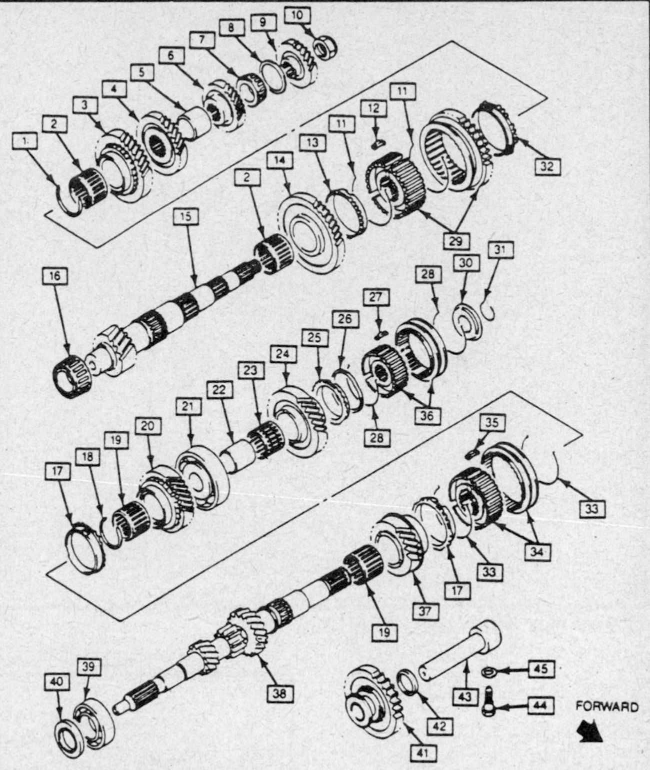
FORWARD

1. Snapring
2. 1st–2nd gear bearing
3. Countershaft 2nd gear
4. Countershaft 3rd gear
5. 3rd–4th gear spacer
6. Countershaft 4th gear
7. Countershaft left hand bearing
8. Bearing set shim
9. Countershaft 5th gear
10. Countershaft nut
11. Low speed synchronizer spring
12. Low speed synchronizer key
13. 1st gear synchronizer ring
14. Countershaft 1st gear
15. Countershaft
16. Countershaft right hand bearing
17. High speed synchronizer ring
18. Snapring
19. 3rd–4th gear bearing
20. Input shaft 4th gear
21. Input shaft left hand bearing
22. 5th gear spacer
23. 5th gear bearing
24. Input shaft 5th gear
25. 5th speed synchronizer ring
26. Synchronizer ring sping
27. 5th synchronizer key
28. 5th synchronizer spring
29. Low speed sleeve and hub
30. 5th synchronizer hub plate
31. Snapring
32. 2nd gear synchronizer ring
33. High speed synchronizer spring
34. High speed sleeve and hub
35. High speed synchronizer key
36. 5th speed sleeve and hub
37. Input shaft 3rd gear
38. Input shaft
39. Input shaft right hand bearing
40. Oil seal
41. Reverse idler gear
42. Reverse shaft washer
43. Reverse gear shaft
44. Reverse shaft bolt
45. Washer

Transaxle exploded view—1989 Metro

1989 METRO

NOTE: Before checking the oil level, check the transaxle for oil leakage. If leakage is found, repair it.

1. With the engine stopped, remove the oil level/filler plug from the transaxle.

2. Check to see that the oil level reaches the fill hole. If the oil level is low, fill the transaxle with oil.

3. Install the oil level/filler plug and torque to 33 ft. lbs. (45 Nm).

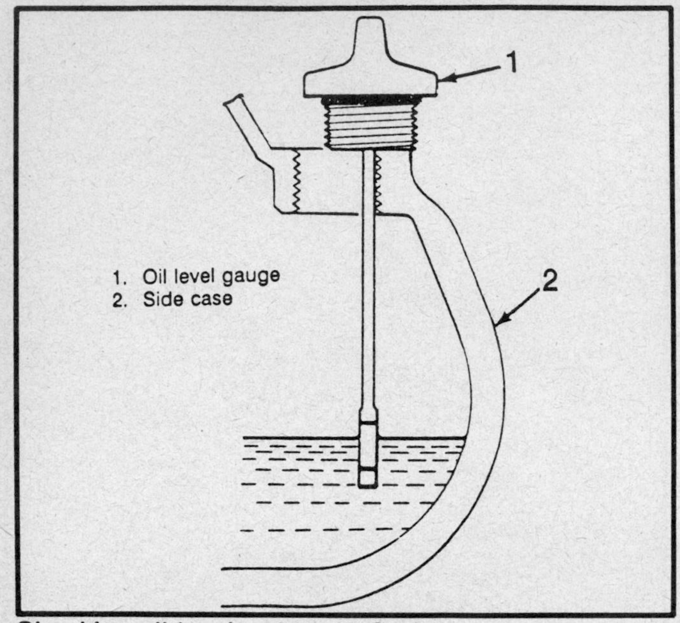
1
2
1. Oil level gauge
2. Side case

Checking oil level—1985–88 Sprint

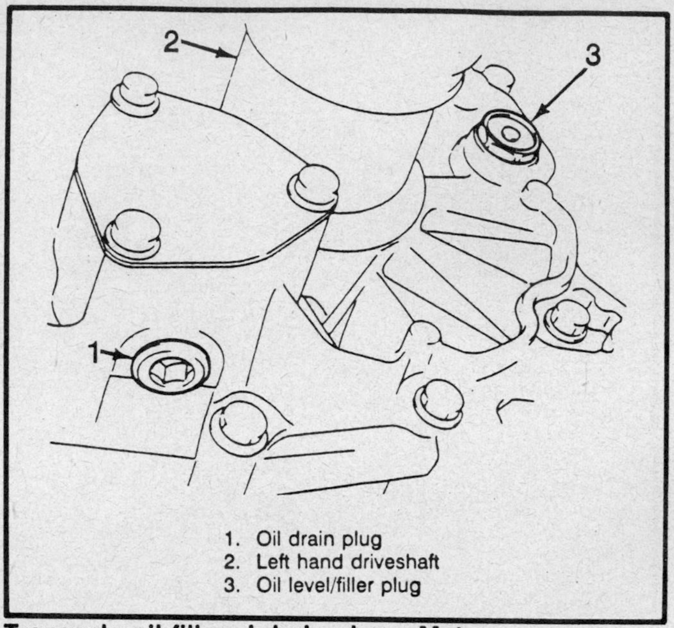
2
3
1
1. Oil drain plug
2. Left hand driveshaft
3. Oil level/filler plug

Transaxle oil fill and drain plug—Metro

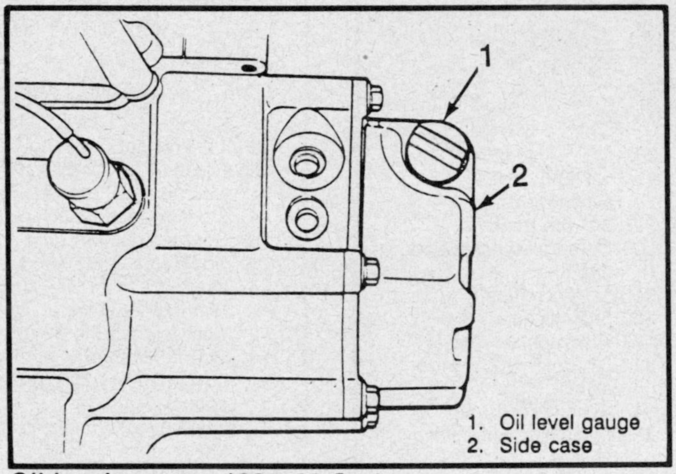
1
2
1. Oil level gauge
2. Side case

Oil level gauge—1985–88 Sprint

TRANSAXLE MODIFICATIONS

There are no modifications on the MV2 transaxle listed at the time of this publication printing.

TROUBLE DIAGNOSIS

CHILTON THREE C'S DIAGNOSIS
MV2 Transaxle

Condition	Cause	Correction
Shift lever slips out of gear	a) Worn shift fork shaft b) Worn shift fork or synchronizer sleeve c) Weak or damaged locating springs d) Worn bearings on input shaft or countershaft e) Worn chamfered tooth on sleeve or gear	a) Replace shaft b) Replace fork or synchronizer sleeve c) Replace locating springs d) Replace bearings e) Replace sleeve or gear
Hard Shifting	a) Inadequate lubricant b) Improper clutch pedal tree travel c) Distorted or broken clutch disc d) Damaged clutch pressure plate e) Worn synchronizer ring f) Worn chamfered tooth on sleeve or gear g) Worn gear shift control shaft joint bushing h) Distorted shift shaft	a) Replenish lubricant b) Adjust clutch pedal free travel c) Replace clutch disc d) Replace clutch cover e) Replace f) Replace sleeve or gear g) Replace bushing h) Replace shift shaft
Noise	a) Inadequate or insufficient lubricant b) Damaged or worn bearing(s) c) Damaged or worn gear(s) d) Damaged or worn synchronizer parts	a) Replenish lubricant b) Replace bearing(s) c) Replace gear(s) d) Replace synchronizer parts

POWER FLOW

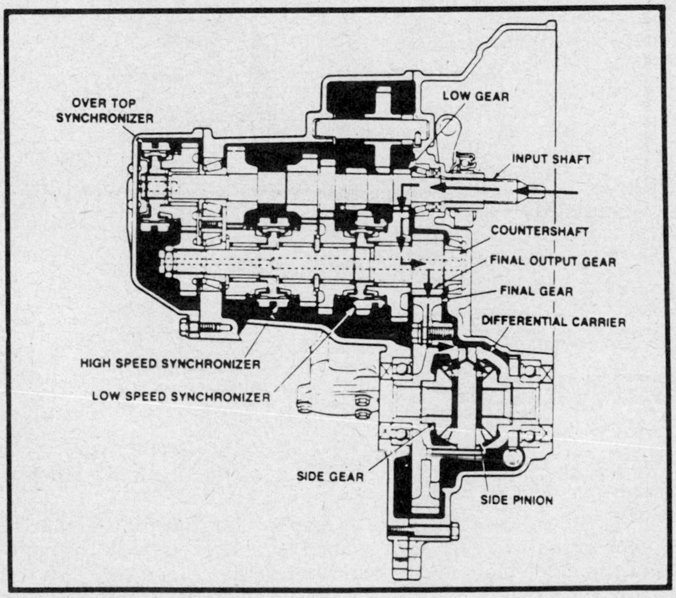
OVER TOP SYNCHRONIZER
LOW GEAR
INPUT SHAFT
COUNTERSHAFT
FINAL OUTPUT GEAR
FINAL GEAR
DIFFERENTIAL CARRIER
HIGH SPEED SYNCHRONIZER
LOW SPEED SYNCHRONIZER
SIDE GEAR
SIDE PINION

Power flow—1st gear

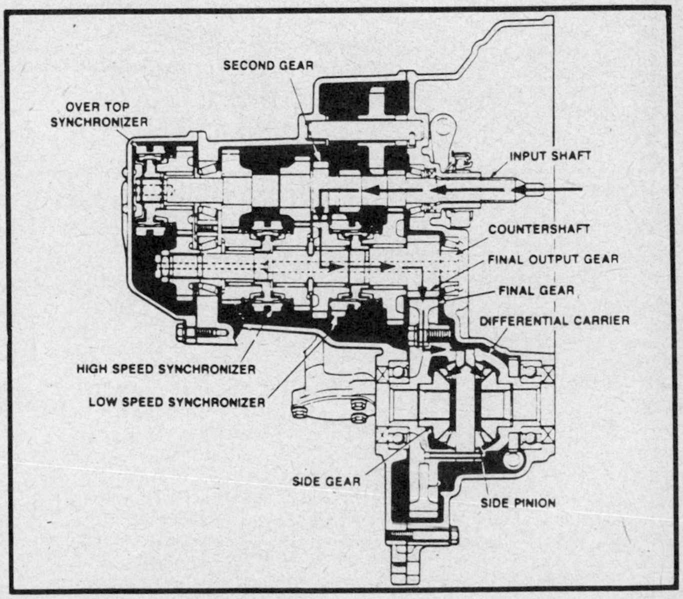
SECOND GEAR
OVER TOP SYNCHRONIZER
INPUT SHAFT
COUNTERSHAFT
FINAL OUTPUT GEAR
FINAL GEAR
DIFFERENTIAL CARRIER
HIGH SPEED SYNCHRONIZER
LOW SPEED SYNCHRONIZER
SIDE GEAR
SIDE PINION

Power flow—2nd gear

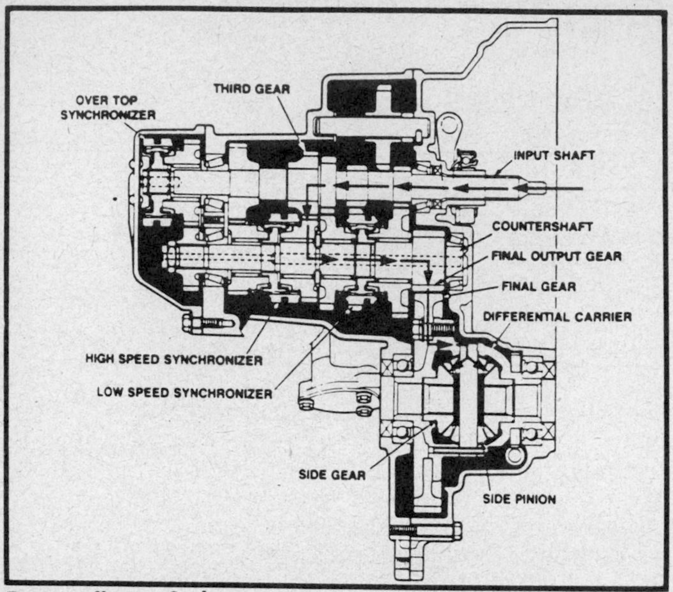
THIRD GEAR
OVER TOP SYNCHRONIZER
INPUT SHAFT
COUNTERSHAFT
FINAL OUTPUT GEAR
FINAL GEAR
DIFFERENTIAL CARRIER
HIGH SPEED SYNCHRONIZER
LOW SPEED SYNCHRONIZER
SIDE GEAR
SIDE PINION

Power flow–3rd gear

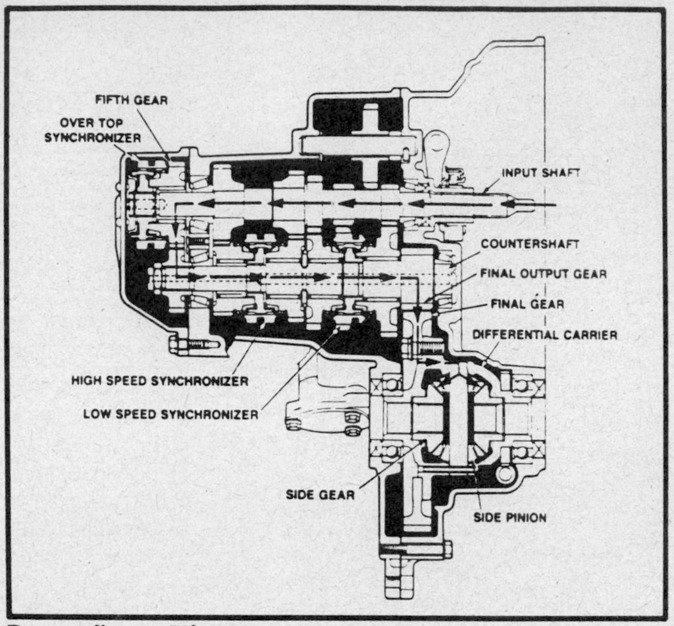
FIFTH GEAR
OVER TOP SYNCHRONIZER
INPUT SHAFT
COUNTERSHAFT
FINAL OUTPUT GEAR
FINAL GEAR
DIFFERENTIAL CARRIER
HIGH SPEED SYNCHRONIZER
LOW SPEED SYNCHRONIZER
SIDE GEAR
SIDE PINION

Power flow–5th gear

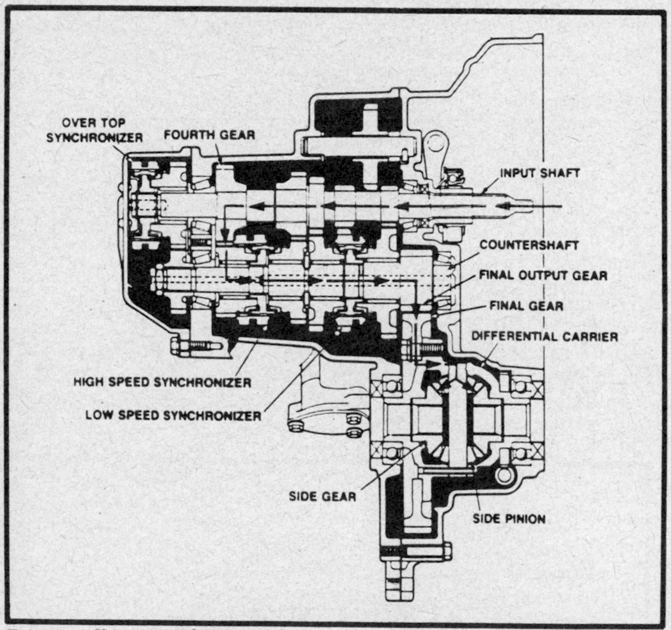
OVER TOP SYNCHRONIZER
FOURTH GEAR
INPUT SHAFT
COUNTERSHAFT
FINAL OUTPUT GEAR
FINAL GEAR
DIFFERENTIAL CARRIER
HIGH SPEED SYNCHRONIZER
LOW SPEED SYNCHRONIZER
SIDE GEAR
SIDE PINION

Power flow–4th gear

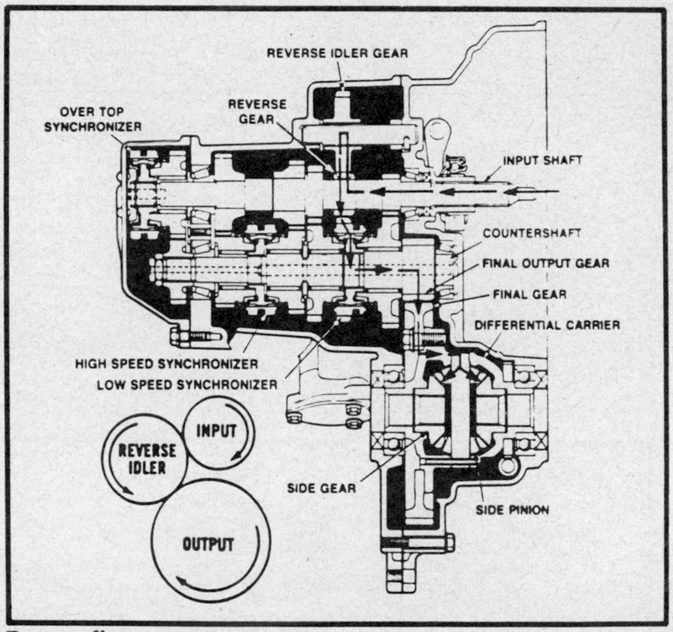
REVERSE IDLER GEAR
OVER TOP SYNCHRONIZER
REVERSE GEAR
INPUT SHAFT
COUNTERSHAFT
FINAL OUTPUT GEAR
FINAL GEAR
DIFFERENTIAL CARRIER
HIGH SPEED SYNCHRONIZER
LOW SPEED SYNCHRONIZER
INPUT
REVERSE IDLER
SIDE GEAR
SIDE PINION
OUTPUT

Power flow–reverse gear

ON CAR SERVICE

Adjustments

GEAR SHIFT CONTROL

When each shift stroke is short, or when the gears are not in complete mesh, adjust the gear shift lever position as follows:

1. Loosen the gear shift control housing nuts and guide plate bolts.
2. Adjust the guide plate by displacing it toward the front and rear so that the gear shift control lever is brought in the middle of the guide plate and at the right angle.
3. Once the guide plate is positioned properly, tighten the guide plate bolts and then housing nuts to specifications.

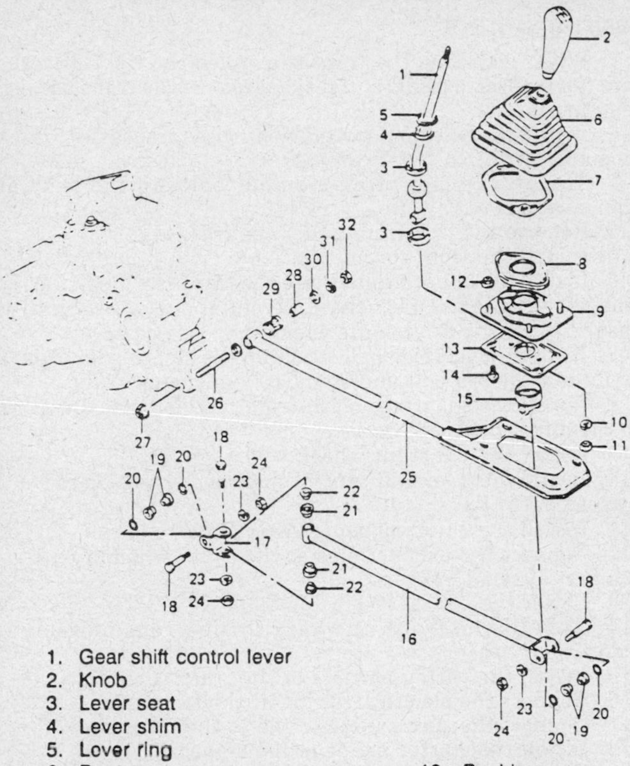

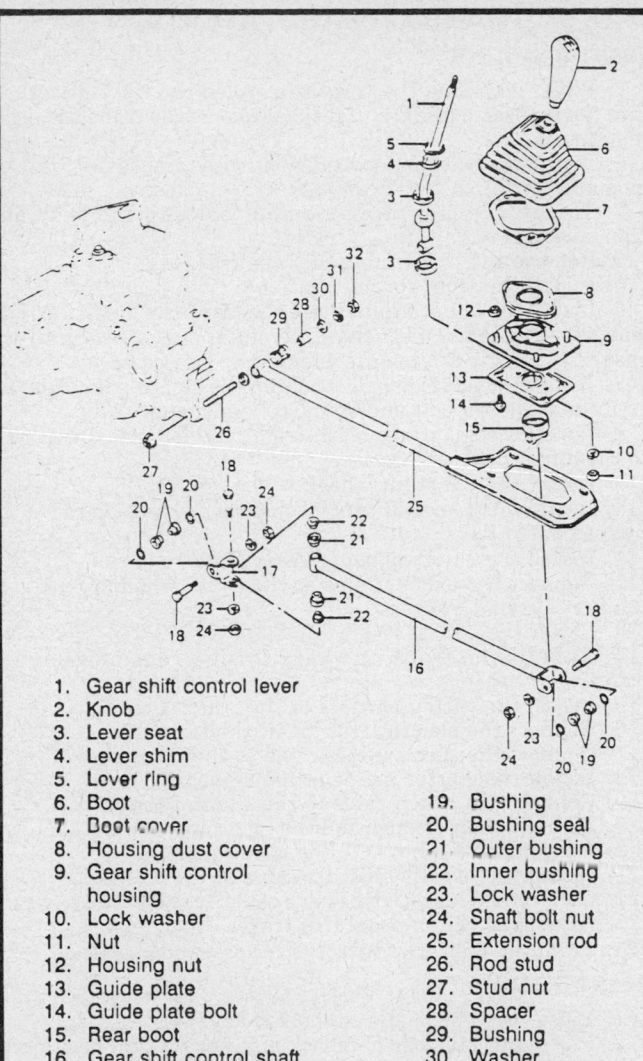

1. Gear shift control lever
2. Knob
3. Lever seat
4. Lever shim
5. Lever ring
6. Boot
7. Boot cover
8. Housing dust cover
9. Gear shift control housing
10. Lock washer
11. Nut
12. Housing nut
13. Guide plate
14. Guide plate bolt
15. Rear boot
16. Gear shift control shaft
17. Gear shift control joint
18. Control shaft bolt
19. Bushing
20. Bushing seal
21. Outer bushing
22. Inner bushing
23. Lock washer
24. Shaft bolt nut
25. Extension rod
26. Rod stud
27. Stud nut
28. Spacer
29. Bushing
30. Washer
31. Lock washer
32. Rod nut

Gear shift control components – 1985–88 Sprint shown, Metro similar

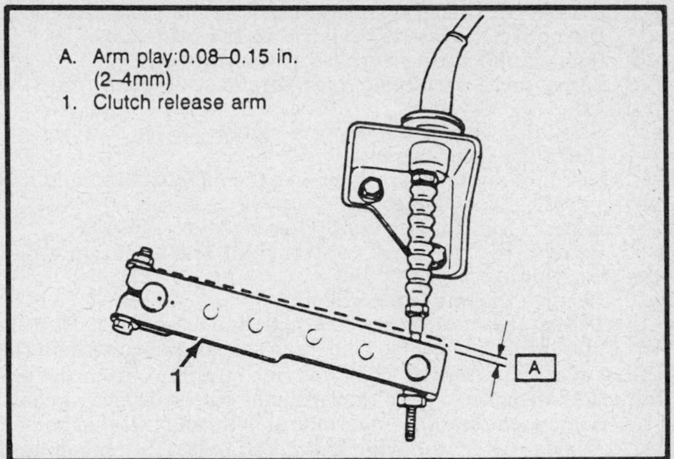
A. Arm play:0.08–0.15 in.
(2–4mm)
1. Clutch release arm
1
A

Clutch play adjustment – Metro

Services

GEAR SHIFT AND SELECT SHAFT ASSEMBLY

Removal and Installation

1. Disconnect the negative battery cable.
2. Place the gear shift lever in the **N** detent.
3. Drain the transaxle oil.
4. Raise the vehicle and support safely.
5. Remove the left case cap.
6. Remove the roll pin and gear shift yoke. Using a drift, drive out the roll pin and then remove the yoke.

NOTE: Do not forget to take out the driven-out roll pin from the transaxle case.

7. Remove the reverse check bolt, spring and ball. Remove the bolt and then take out the spring and ball from the screw hole. To take out the steel ball from the screw hole, insert a magnetic driver into the hole.
8. Remove the gear shift guide case by removing the 4 retaining bolts.
9. Remove the gear shift locating bolt.
10. Remove the gear shift and select shaft assembly. When the shaft is removed, the low speed select spring may drop into the transaxle. For this reason, take out this spring together with the shaft, supporting it by hand.
11. To install, place the gears in **N**.
12. Oil the shaft and then install the shaft assembly. Use new gaskets.
13. Install the gear shift locating bolt and washer. Tighten the bolt to specification.
14. Install the gear shift guide case. Tighten the 4 case bolts to specification
15. Install the reverse check ball, spring, washer and bolt.
16. Install the gear shift yoke and a new roll pin. Install the yoke to the select shaft and gear shift arm. Then, drive in the roll pin.
17. Install the O-ring and left case cap to the left case.
18. Refill the transaxle with recommended gear oil to the proper level.
19. Connect the negative battery cable.

REMOVAL AND INSTALLATION

TRANSAXLE REMOVAL

1985–88 SPRINT

1. Disconnect the negative battery cable at the battery.
2. Disconnect the negative battery cable at the transaxle.
3. Remove the air cleaner and heat pipe.
4. Disconnect the clutch cable from the clutch release lever.
5. Remove the starter motor from the transaxle.

6. Disconnect the speedometer cable at the transaxle.
7. Disconnect the electrical wires at the transaxle.
8. Remove the wiring harness at the transaxle.
9. Remove the front and rear torque rods bolts from the transaxle.
10. Raise the vehicle and support safely.
11. Drain the transaxle oil.
12. Disconnect the exhaust pipe at the exhaust manifold and at the first exhaust hanger.
13. Remove the clutch housing lower plate.
14. Remove the gear shift control shaft and extension rod at the transaxle.
15. Remove the left front wheel.
16. Detach the snaprings on the right and left driveshafts from the differential side gears. To detach the snapring fitted on the spline of the differential side joint (inboard joint) from the differential side gear, pry the inboard joint out by using a pry bar.
17. Remove the stabilizer bar mount bolts and ball stud bolt on the left side. After removing these bolts, detach the ball stud from the steering knuckle by pushing down on the stabilizer bar.
18. Draw out the inbound joint of the left driveshaft from the transaxle.
19. Remove the front torque rod.
20. Securely support the transaxle case with a jack for removal.
21. Remove the mounting member bolts from the body and transaxle.
22. Remove the bolts and nuts fastening transaxle to the engine.
23. Disconnect the transaxle from the engine by sliding towards the left side. Carefully lower the jack, along with the transaxle.

1989 METRO

1. Disconnect the negative battery cable.
2. Remove the clutch cable adjusting nuts, retaining clip from the cable and cable from the bracket.
3. Remove the wiring harness clamps and connectors.
4. Disconnect the speedometer cable boot, speedometer case clip, speedometer cable from case.
5. Remove the transaxle retaining bolts.
6. Remove the starter retaining bolts and starter. Also remove the starter motor plate.
7. Disconnect the vacuum hose from the pressure sensor.
8. Install an engine support to prevent engine from lowering excessively.
9. Raise the vehicle and support safely.
10. Drain the transaxle oil.
11. Remove the gear shift control shaft bolt and nut and detach the control shaft from the gear shift shaft.
12. Remove the extension rod nut and remove the rod with the washers.
13. Remove the exhaust pipe front flange bolts.
14. Remove the exhaust pipe rear flange nuts.
15. Remove the clutch housing lower plate.
16. Remove the left front wheel assembly.
17. Remove the left tie rod end using tool J-21687-02 or equivalent.
18. Remove the left ball joint by removing the joint stud bolt.
19. Remove both driveshafts at the transaxle.
20. Remove the transaxle to engine retaining bolts and nuts.
21. Support the transaxle with a jack.
22. Remove the 2 rear engine mounting nuts.
23. Remove the 3 bolts and 2 nuts from the transaxle mounting left hand bracket, remove the left hand bracket.
24. Lower the transaxle with the engine attached in order to detach it from the stud bolts at the engine rear mounting portion, pull the transaxle straight out toward the left side to disconnect the input shaft from the clutch cover, lower and remove the transaxle assembly.

TRANSAXLE INSTALLATION

1985–88 SPRINT

1. When installing the transaxle, guide the right driveshaft into the transaxle (differential side gear) as the transaxle is being raised.
2. Install the bolts and nuts fastening the transaxle to the engine and torque to specifications.
3. Install the mounting member bolts to the body and transaxle.
4. Remove the jack from under the transaxle.
5. Install the front torque rod.
6. Install the left driveshaft into the transaxle. Push the right and left driveshafts into the differential side gears until the snaprings on the driveshafts engage the side gears.
7. Install the left side ball stud into the steering knuckle. Install the ball stud bolt and stabilizer bar mount bolts.
8. Install the snaprings on the right and left side driveshafts to the differential side gears.
9. Install the left front wheel.
10. Install the gear shift control shaft and extension rod at the transaxle.
11. Install the clutch housing lower plate.
12. Connect the exhaust pipe at the exhaust manifold and at the first exhaust hanger.
13. Lower the vehicle.
14. Install the front and rear torque rods bolts at the transaxle.
15. Install the wiring harness at the transaxle.
16. Connect the electrical wiring at the transaxle.
17. Connect the speedometer cable to the transaxle.
18. Install the starter motor to the transaxle.
19. Connect the clutch cable to the clutch release lever.
20. Install the air cleaner and the heat pipe. Securely connect the emission control hoses to the air cleaner.
21. Install the negative battery cable at the transaxle.
22. Connect the negative battery cable at the battery.
23. Adjust the clutch pedal free travel.
24. Refill the transaxle with the recommended gear oil.

1989 METRO

1. Raise the vehicle and support safely.
2. Place the transaxle on a jack and raise it into position attaching the stud bolts at the engine rear mounting portion an input shaft clutch assembly.

NOTE: While the transaxle is being raised, connect the right hand driveshaft into the differential side.

3. Install the transaxle to engine bolts and nuts and torque to 37 ft. lbs. (50 Nm).
4. Install the transaxle mounting left hand bracket with 3 bolts and 2 nuts. Torque to 37 ft. lbs. (50 Nm).
5. Install the 2 rear engine mounting nuts and torque to 37 ft. lbs. (50 Nm).
6. Lower the transaxle supporting jack.
7. Install the left hand driveshaft to the transaxle.

NOTE: Be sure to push each driveshaft in fully to engage the snaprings with the differential gear.

8. Install the left ball joint and ball joint stud bolt. Torque ball joint bolt and nut to 44 ft. lbs. (60 Nm).
9. Install the left tie rod end, castle nut and split pin.
10. Install the left front wheel.
11. Install the clutch housing lower plate.
12. Install the exhaust pipe rear flange nuts.
13. Install the exhaust pipe front flange nuts.
14. Install the extension rod nut and washers. Torque to 24 ft. lbs. (33 Nm).
15. Install the control shaft to gear shift and install gear shift control shaft bolt and nut. Torque bolt and nut to 13 ft. lbs. (18 Nm).

16. Fill the transaxle with the recommended gear oil to the proper level.
17. Lower the vehicle.
18. Remove the engine support fixture.
19. Connect the vacuum hose to the pressure sensor.
20. Install the starter motor, starter motor plate and 2 bolts.
21. Install the transaxle retaining bolts. Torque to 37 ft. lbs. (50 Nm).
22. Connect the speedometer cable to case, speedometer case clip and speedometer cable boot.
23. Install the wiring harness clamps and connectors.
24. Connect the clutch cable to the bracket, retaining clip to cable and clutch cable adjusting nut.
25. Adjust the clutch pedal free travel.
26. Connect the negative battery cable.

BENCH OVERHAUL

Before Disassembly

Cleanliness is an important factor in the overhaul of the transaxle. Before attempting any disassembly operation, the exterior of the transaxle should be thoroughly cleaned to prevent the possibility of dirt entering the transaxle internal mechanism. During inspection and reassembly, all parts should be thoroughly cleaned with cleaning fluid and then air dried. Wiping cloths or rags should not be used to dry parts. All oil passages should be blown out and checked to make sure that they are not obstructed. Small passages should be checked with tag wire. All parts should be inspected to determine which parts are to be replaced.

Transaxle Disassembly

1985–88 SPRINT

1. Position the assembly in a suitable holding fixture.
2. Remove the backup light switch.
3. Remove the bolts from the side case and remove the side case.
4. Remove the snapring and hub plate from the input shaft.
5. Remove the shift fork screw and guide ball from the 5th gear shift fork.
6. With the transaxle in **N**, remove the roll pin from the 5th gear shift fork.
7. Push in on the gear shift shaft to engage the transaxle into gear. Slide the 5th gear synchronizer sleeve down to engage the 5th gear. This will lock the transaxle in 2 gears allowing for removal of the 5th gear (driven) retaining nut. The retaining nut is staked to the countershaft. Remove the stake mark before removing the retaining nut. Remove the nut.
8. Remove the 5th gear shift fork, sleeve, hub, synchronizer ring, spring and keys as an assembly from the input shaft.
9. Remove the 5th gear (drive), bearing, spacer and washer from the input shaft.
10. Remove the 5th gear (driven) from the countershaft.
11. Remove the screws retaining the bearing/shim retainer plate to the left case and remove the plate and shims. Mark or tag the shims for the input shaft and countershaft for reference during assembly.
12. Remove the bolts retaining the left case cap and remove the cap.
13. Remove the roll pin retaining the gear shift yoke to the gear shift/selector shaft.
14. Install a drift into the roll pin hole in the gear shift/selector shaft and raise the shaft for removal of the yoke and roll pin.
15. Remove the bolt retaining the reverse spring and detent ball. Remove the spring and ball from the case.
16. Remove the locating bolt for the gear shift/selector shaft.
17. Remove the bolts retaining the gear shift guide case.
18. Remove the gear shift/selector shaft assembly along with the low speed select spring from the case.
19. Remove the bolts retaining the detent balls and springs for the shift fork shafts.
20. Remove the bolts retaining the 2 case halves.
21. Insert a suitable tool into the slots between the 2 case halves, pry or lift to separate case halves.
22. Remove the left case. All inner parts such as input shaft, countershaft and differential, etc., should remain on the right case.
23. Raise the 5th and reverse shift shafts to gain clearance for removal of the reverse gear and shaft.
24. Remove the bolts retaining the reverse idler shift lever to the case and remove the lever.
25. Remove the 5th and reverse gear shift shafts together as an assembly.
26. Remove the input shaft, countershaft, 1st/2nd and 3rd/4th shift shafts together as an assembly. It may be necessary to raise the differential assembly to gain clearance for removal.
27. Remove the differential assembly from the right case.
28. Remove the input shaft and countershaft bearing cups from the right case using tool J–29369–2 or equivalent, with slide hammer J–23907 or equivalent.
29. Remove the input shaft seal from the right case using tool J–23907 or equivalent.
30. Remove the differential side oil seals from the right and left case.
31. Remove the roll pin from the gear shift arm. Position the gear shift arm just above the square recess of the right case by moving the shift shaft. Using a drift, drive the pin out.
32. Remove the retaining bolt for the shifter shaft detent ball and spring. Remove the ball and spring from the case.
33. Slide the shifter shaft out and remove the arm, roll pin and shifter shaft from the case.
34. Remove the shifter shaft boot from the flange of the oil seal.
35. Remove the shifter shaft seal from the case using pliers.
36. Remove the input shaft and countershaft bearing cups from the left case.

1989 METRO

1. Position the assembly in a suitable holding fixture.
2. Remove the transaxle side cover retaining bolts and cover.
3. Remove the snapring and hub plate.

NOTE: Care should be taken not to distort the side cover when it is removed from the left case.

4. Remove the shift fork plug and guide ball. Use of a magnet would make it easier in the removal of the guide ball.
5. Drive out the spring pin using a punch and hammer.
6. Remove the gear shift fork, sleeve and hub assembly, synchronizer ring spring, synchronizer ring and 5th gear all together.
7. Remove the staked portion of the countershaft nut, install the main shaft 5th gear and special tool J–35309 or equivalent, to lock the rotation of the shafts and then remove the countershaft nut.
8. Install special tool J–35309 or equivalent, then remove the main shaft 5th gear, needle bearing and countershaft 5th gear using tool J–22888–35 or equivalent.

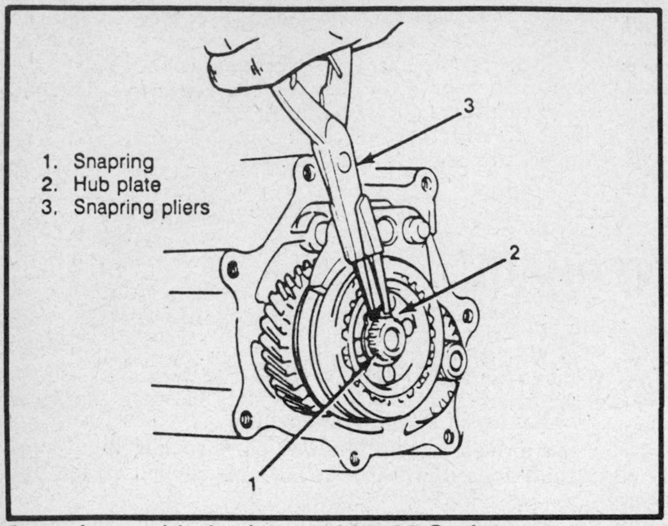
1. Snapring
2. Hub plate
3. Snapring pliers
1
2
3

Snapring and hub plate—1985–88 Sprint

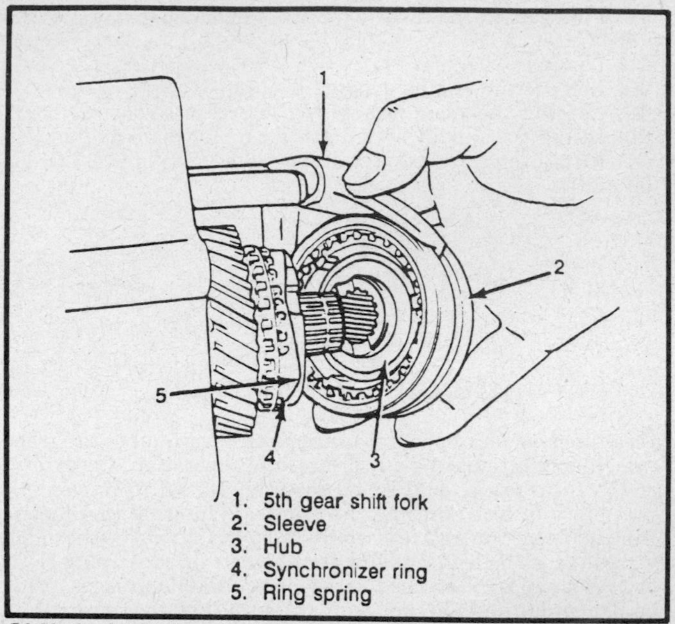
1
2
3
4
5
1. 5th gear shift fork
2. Sleeve
3. Hub
4. Synchronizer ring
5. Ring spring

Shift fork and associated parts servicing—1985–88 Sprint

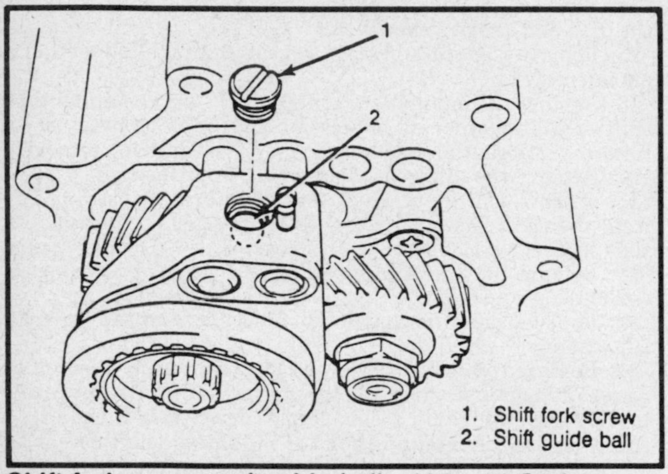
1
2
1. Shift fork screw
2. Shift guide ball

Shift fork screw and guide ball—1985–88 Sprint

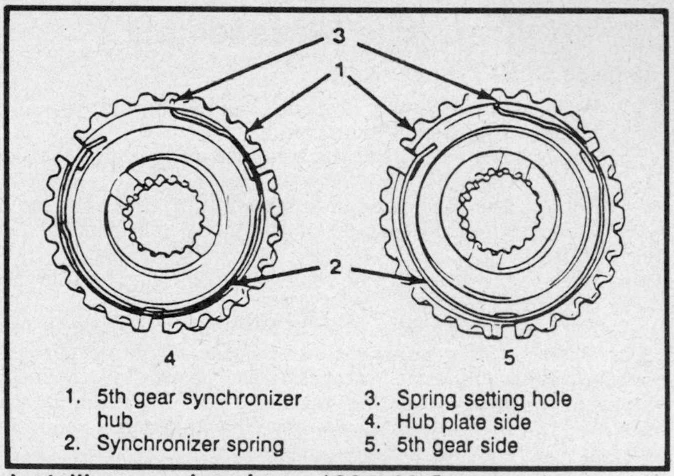
3
1
2
4
5
1. 5th gear synchronizer hub
2. Synchronizer spring
3. Spring setting hole
4. Hub plate side
5. 5th gear side

Installing synchronizer—1985–88 Sprint

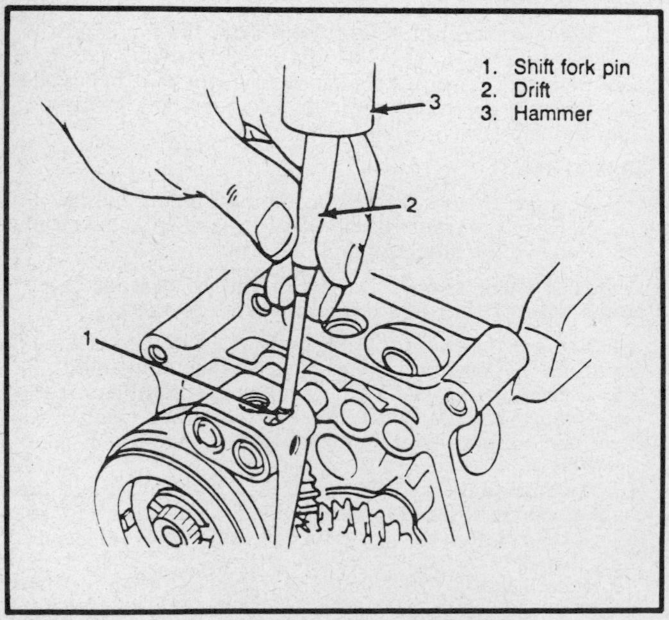
1. Shift fork pin
2. Drift
3. Hammer
1
2
3

Driving out shift fork pin—1985–88 Sprint

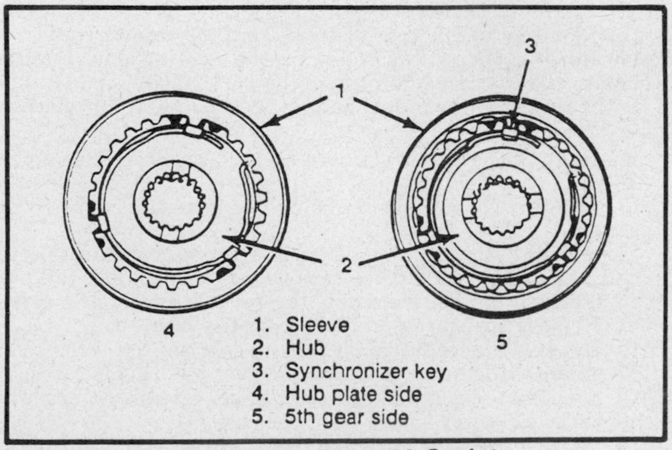
1
2
3
4
5
1. Sleeve
2. Hub
3. Synchronizer key
4. Hub plate side
5. 5th gear side

Install keys and sleeve—1985–88 Sprint

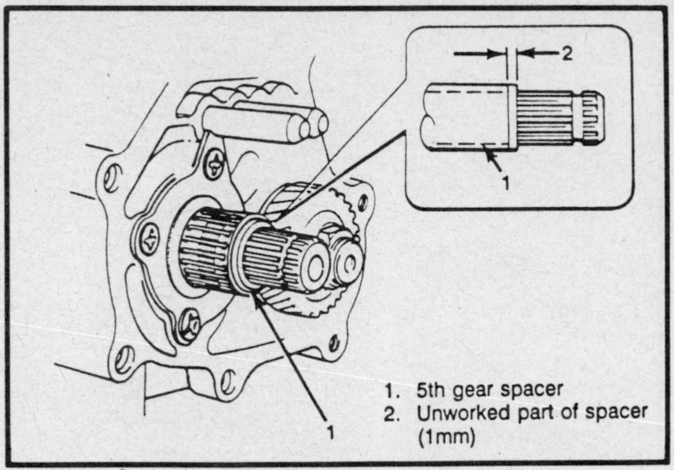
2
1
1
1. 5th gear spacer
2. Unworked part of spacer (1mm)

Installing spacer and bearing—1985–88 Sprint

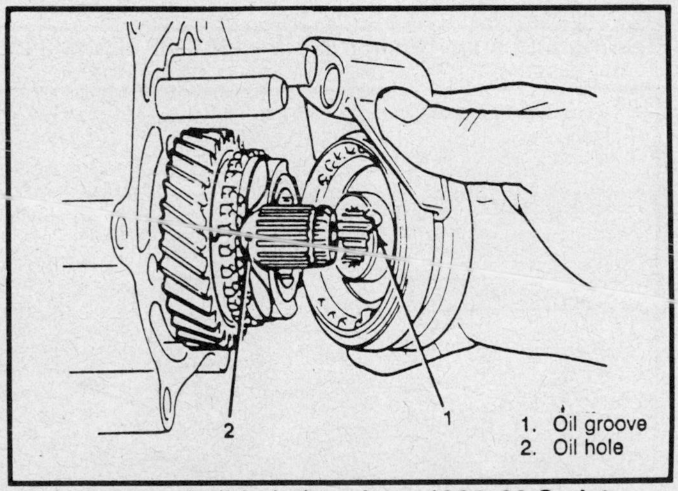
2
1
1. Oil groove
2. Oil hole

Oil groove and oil hole location—1985–88 Sprint

9. Remove the screws and take off the left case plate and remove the bearing shim.
10. Remove the bolts from the left case cap and remove the cap.
11. Remove the gear shift yoke bolt.
12. Remove the gear shift fork screws with washers, springs and steel balls.

NOTE: The springs are color coded. Make note of which hole each spring came from to aid in reassembly.

13. Remove the bolts from the gear shift guide case and remove the wiring harness clamp and guide case.
14. Remove the gear shift interlock bolt with washer.
15. Remove the backup light switch.

NOTE: Removal of the 5th to reverse interlock bolt is not necessary for removing the gear shift and select shaft assembly.

16. Remove the gear shift and select shaft assembly.
17. Remove the reverse shaft bolt and washer.
18. Remove the 11 case bolts from the outside of the transaxle and another 2 bolts from the clutch housing side.
19. Separate the case halves by tapping the left case flanges with a plastic hammer.
20. Remove the gear shift yoke.

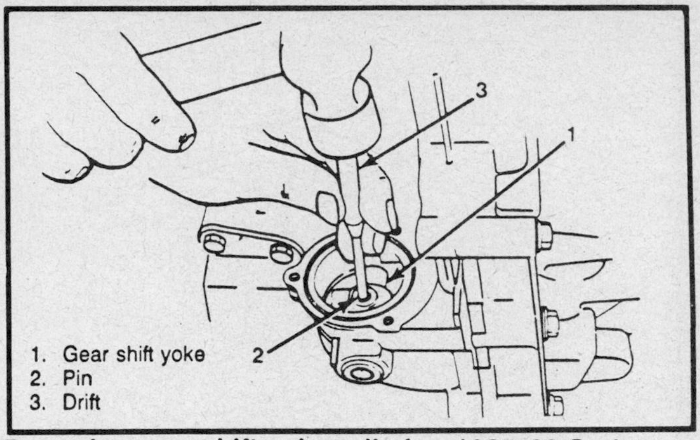
3
1
2
1. Gear shift yoke
2. Pin
3. Drift

Removing gear shift yoke roll pin—1985–88 Sprint

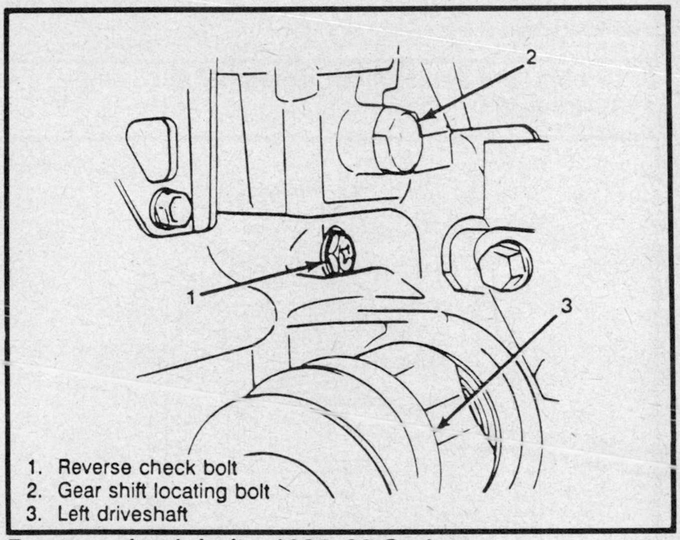
2
1
3
1. Reverse check bolt
2. Gear shift locating bolt
3. Left driveshaft

Reverse check bolt—1985–88 Sprint

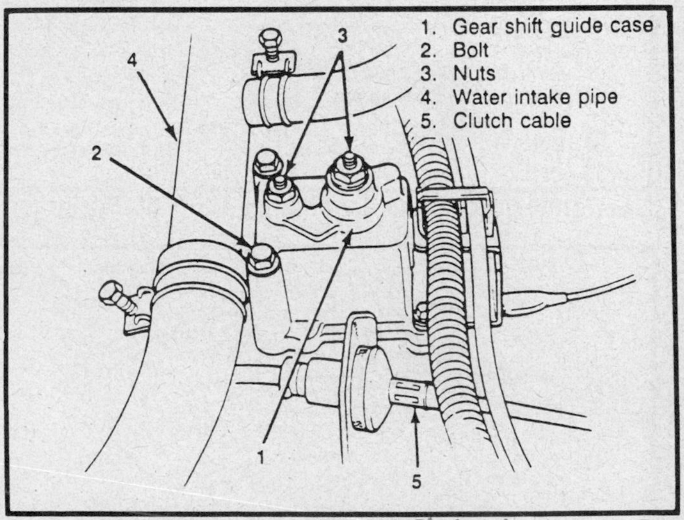
1. Gear shift guide case
2. Bolt
3. Nuts
4. Water intake pipe
5. Clutch cable
3
4
2
1
5

Gear shift guide case—1985–88 Sprint

21. Remove the reverse gear shaft with washer and remove the reverse idler gear.
22. Remove the 5th and reverse gear shift guide shaft together with 5th and reverse gear shift shaft.

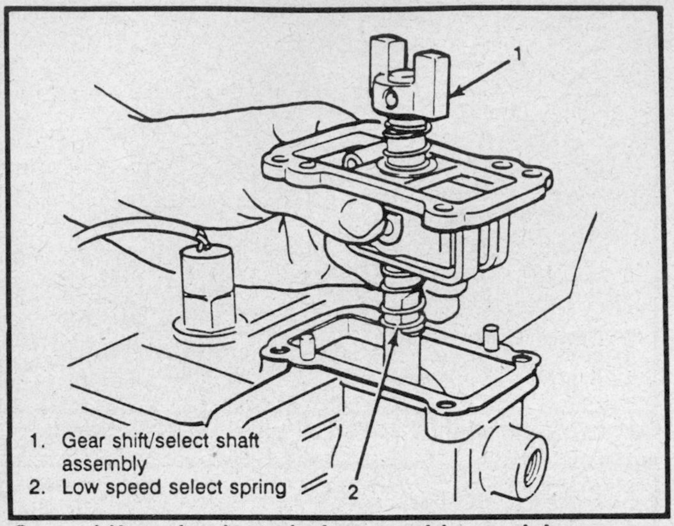
1. Gear shift/select shaft assembly
2. Low speed select spring
1
2
Gear shift and select shaft assembly servicing—1985–88 Sprint

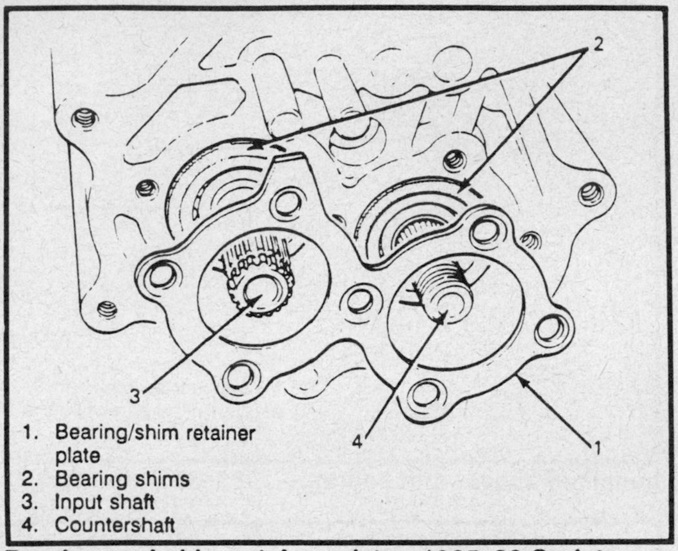
1. Bearing/shim retainer plate
2. Bearing shims
3. Input shaft
4. Countershaft
1
2
3
4
Bearing and shim retainer plate—1985–88 Sprint

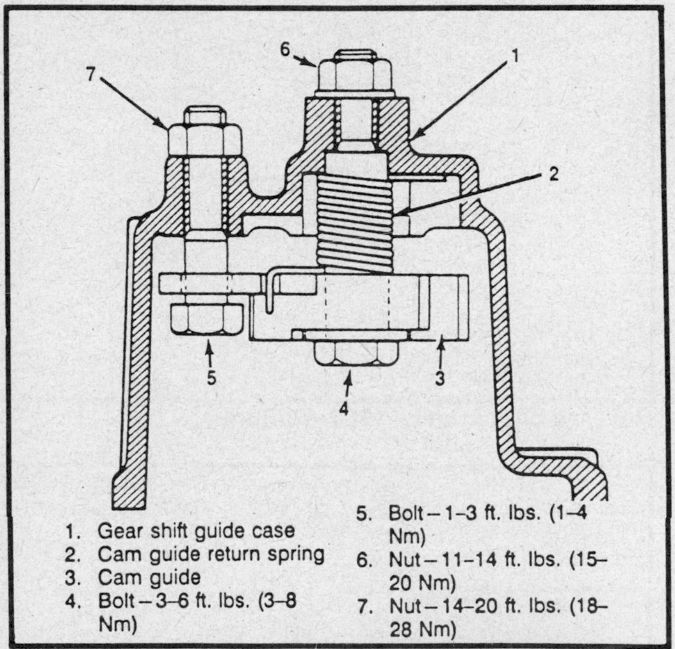
1. Gear shift guide case
2. Cam guide return spring
3. Cam guide
4. Bolt—3–6 ft. lbs. (3–8 Nm)
5. Bolt—1–3 ft. lbs. (1–4 Nm)
6. Nut—11–14 ft. lbs. (15–20 Nm)
7. Nut—14–20 ft. lbs. (18–28 Nm)
1
2
3
4
5
6
7
Gear shift guide case assembly—1985–88 Sprint

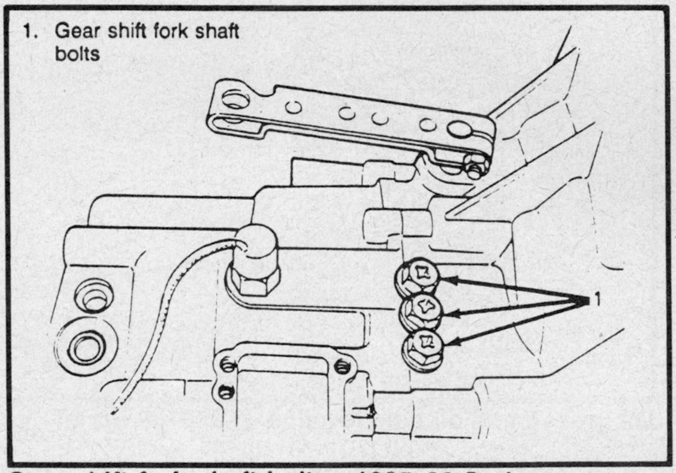
1. Gear shift fork shaft bolts
1
Gear shift fork shaft bolts—1985–88 Sprint

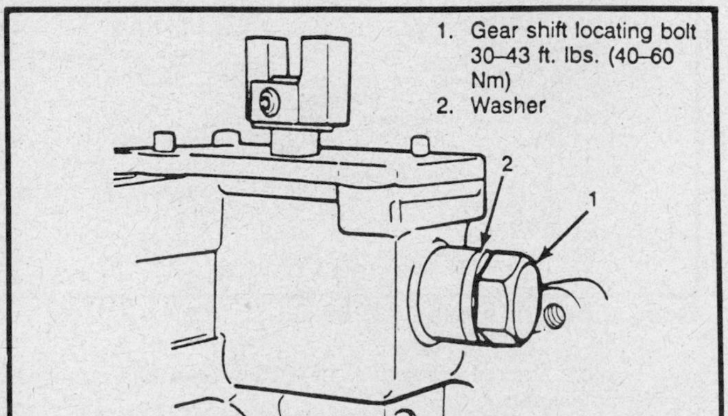
1. Gear shift locating bolt 30–43 ft. lbs. (40–60 Nm)
2. Washer
1
2
Gear shift locating bolt—1985–88 Sprint

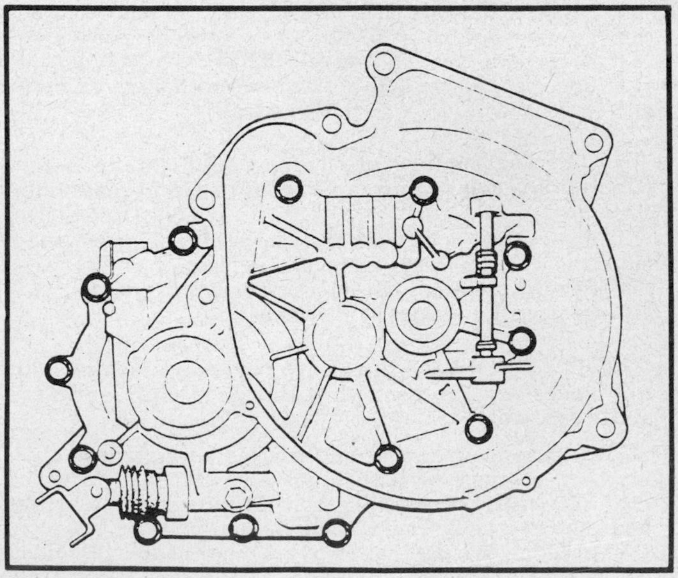
Case bolts locations—1985–88 Sprint

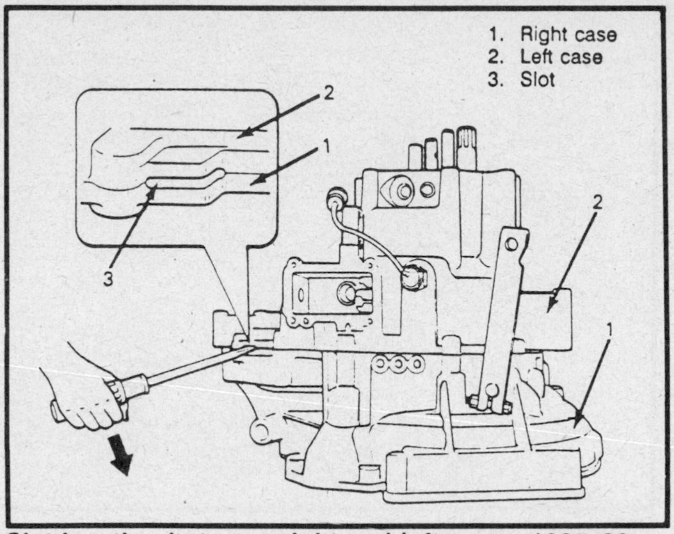
1. Right case
2. Left case
3. Slot

Slot location between right and left case—1985–88 Sprint

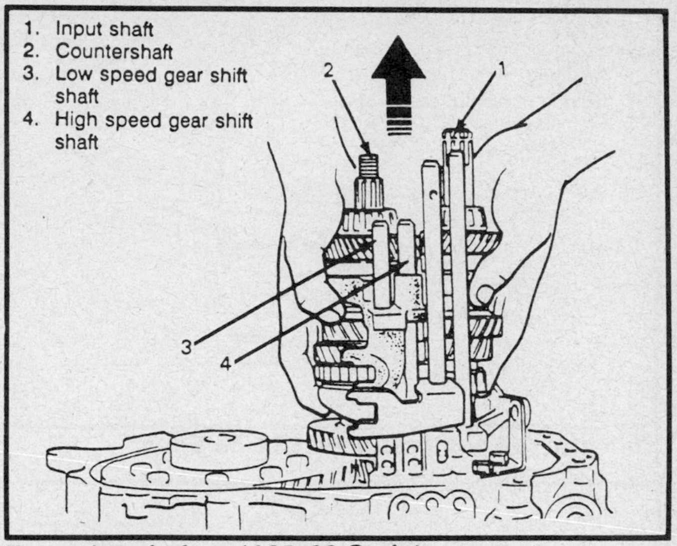
1. Input shaft
2. Countershaft
3. Low speed gear shift shaft
4. High speed gear shift shaft

Removing shafts—1985–88 Sprint

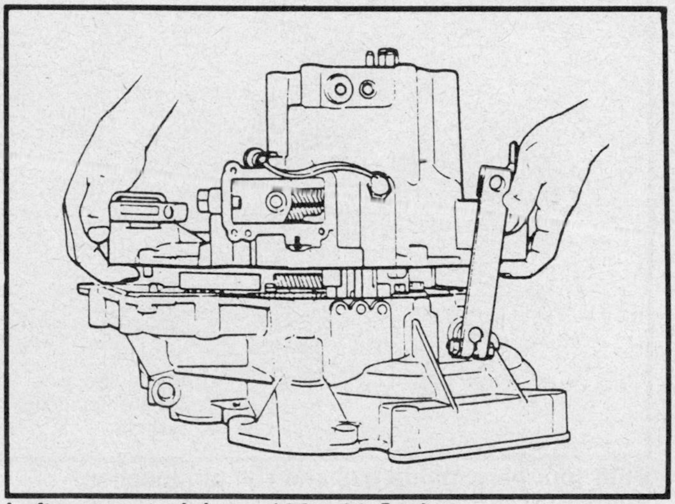

Left case servicing—1985–88 Sprint

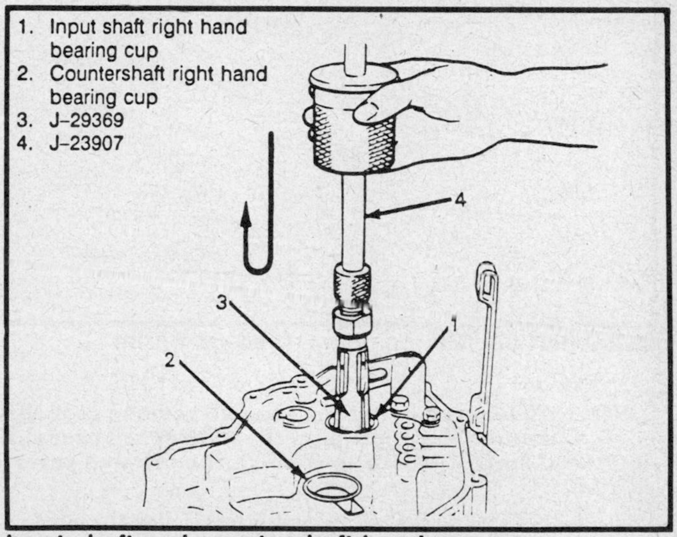
1. Input shaft right hand bearing cup
2. Countershaft right hand bearing cup
3. J-29369
4. J-23907

Input shaft and countershaft bearing cups removal—1985–88 Sprint

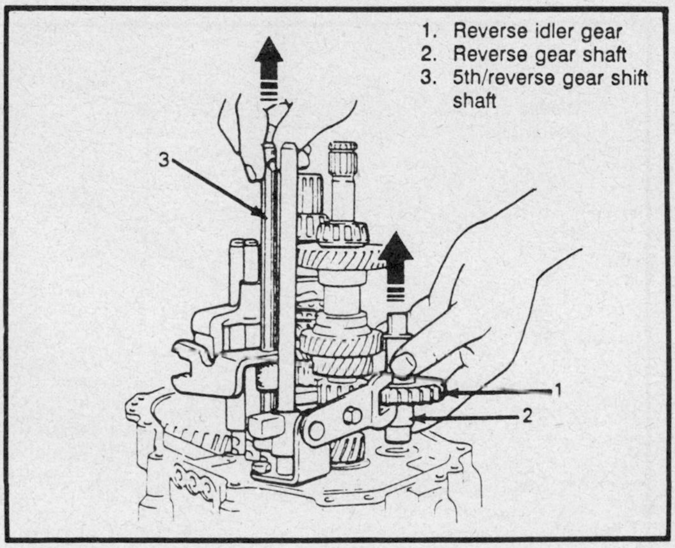
1. Reverse idler gear
2. Reverse gear shaft
3. 5th/reverse gear shift shaft

Removing idler gear and shaft—1985–88 Sprint

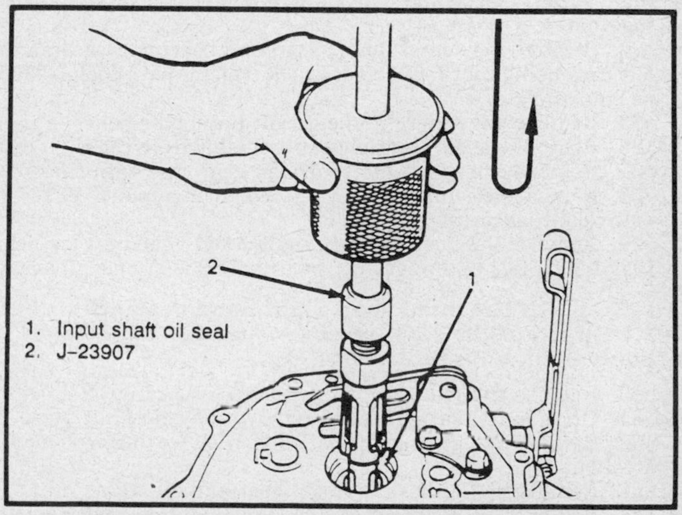
1. Input shaft oil seal
2. J-23907

Input shaft oil seal removal—1985–88 Sprint

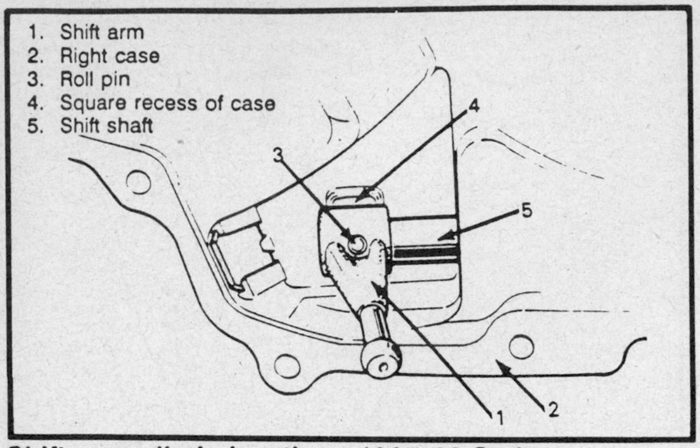
1. Shift arm
2. Right case
3. Roll pin
4. Square recess of case
5. Shift shaft
1
2
3
4
5

Shift arm roll pin location—1985–88 Sprint

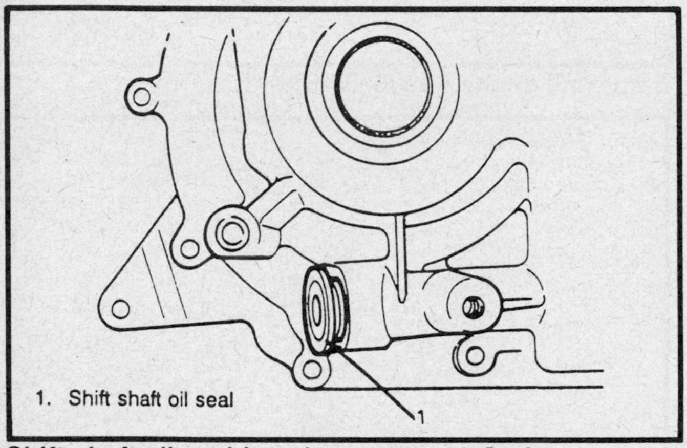
1. Shift shaft oil seal
1

Shift shaft oil seal location—1985–88 Sprint

NOTE: When removing the 5th and reverse gear shift shaft and guide shaft, push up the high speed gear shift shaft and shift it to 4th to remove the 5th and reverse shifter.

23. Remove the input shaft assembly by tapping the input end with a plastic hammer. Then push the assembly out from the case. Remove the input shaft assembly, countershaft assembly, high speed gear shift shaft and low speed gear shift shaft together.
24. Remove the countershaft left hand bearing race from the left case using tool J-29369-2 or equivalent, with tool J-23907 or equivalent.
25. Remove the differential side left hand oil seal.
26. Remove the differential gear assembly from the right case.
27. Remove the speedometer driven gear case with the gear.
28. Remove the input shaft oil seal using tool J-23907 or equivalent, with legs.
29. Remove the countershaft right hand bearing race using tool J-29369-2 or equivalent, with tool J23907 or equivalent.

NOTE: If the input shaft right hand bearing has been left in the right case, remove it using tool J-23907 or equivalent, with puller legs.

30. Remove the magnet from the case.
31. Drive out the roll pin and remove the gear shift arm.
32. Remove the gear shift shaft bolt with washer, spring and steel ball.
33. Remove the gear shift shaft, boot and oil seal.
34. Remove the differential side right hand oil seal from the right case.

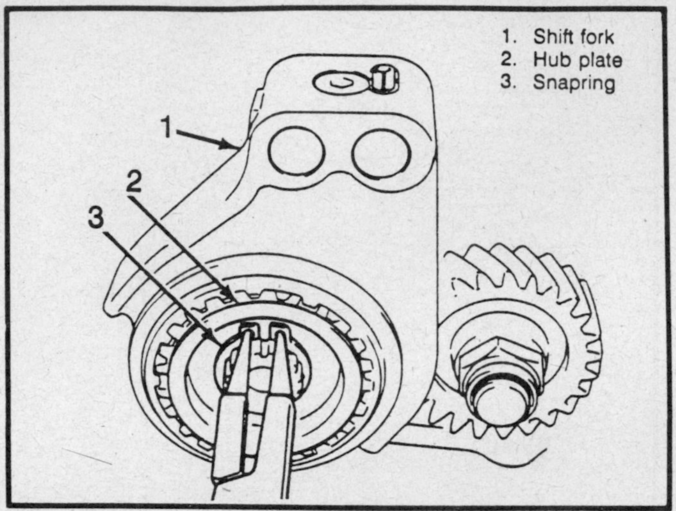
1. Shift fork
2. Hub plate
3. Snapring
1
2
3

Snapring and hub plate servicing—Metro

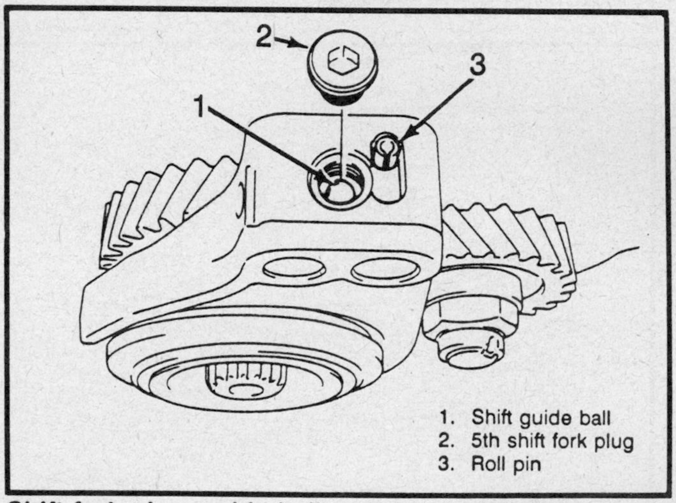
1. Shift guide ball
2. 5th shift fork plug
3. Roll pin
1
2
3

Shift fork plug, guide ball and roll pin location—Metro

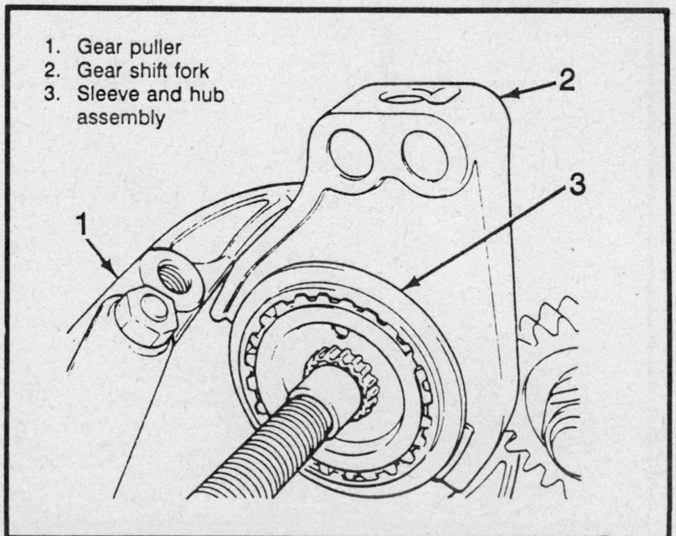
1. Gear puller
2. Gear shift fork
3. Sleeve and hub assembly
1
2
3

Sleeve and hub assembly servicing—Metro

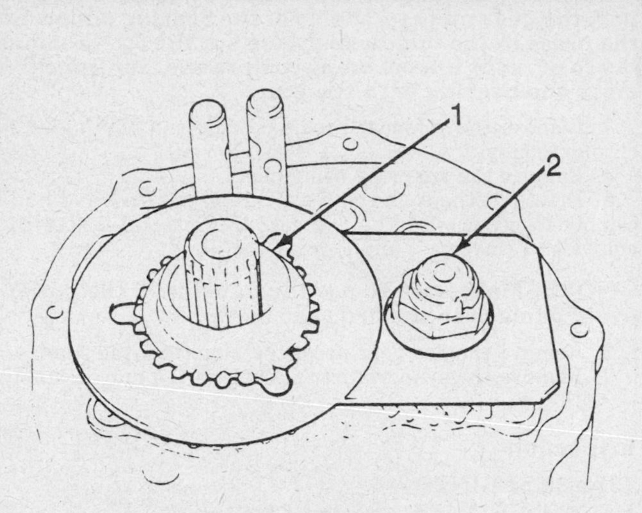
1
2

1. Input shaft 5th gear
2. Countershaft nut

Countershaft nut servicing—Metro

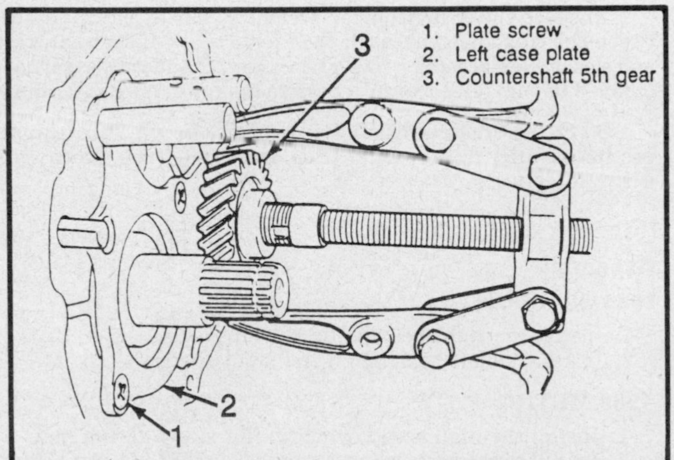
3
2
1

1. Plate screw
2. Left case plate
3. Countershaft 5th gear

5th gear and left case plate removal—Metro

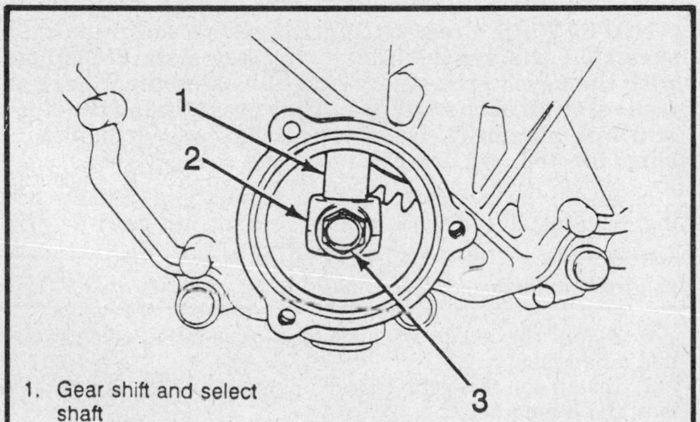
1
2
3

1. Gear shift and select shaft
2. Gear shift yoke
3. Yoke bolt

Gear shift yoke bolt location—Metro

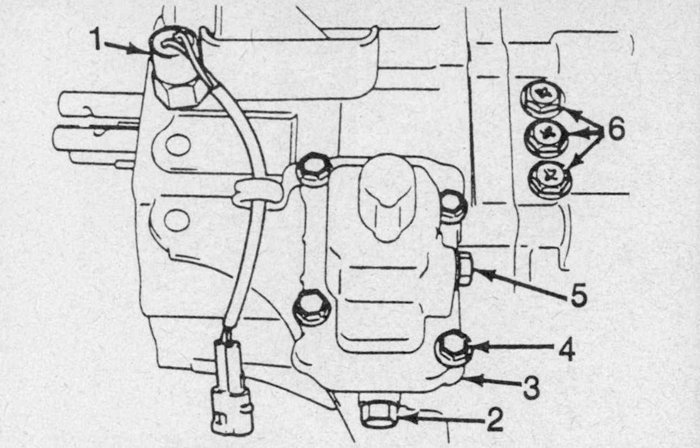
1
6
5
4
3
2

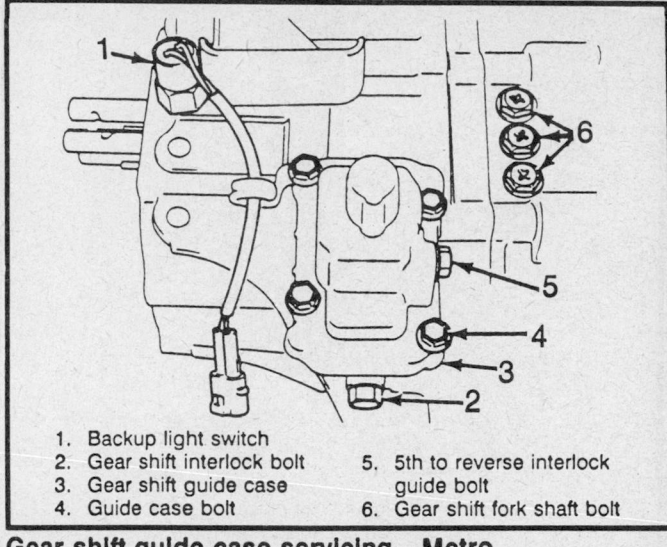

1. Backup light switch
2. Gear shift interlock bolt
3. Gear shift guide case
4. Guide case bolt
5. 5th to reverse interlock guide bolt
6. Gear shift fork shaft bolt

Gear shift guide case servicing—Metro

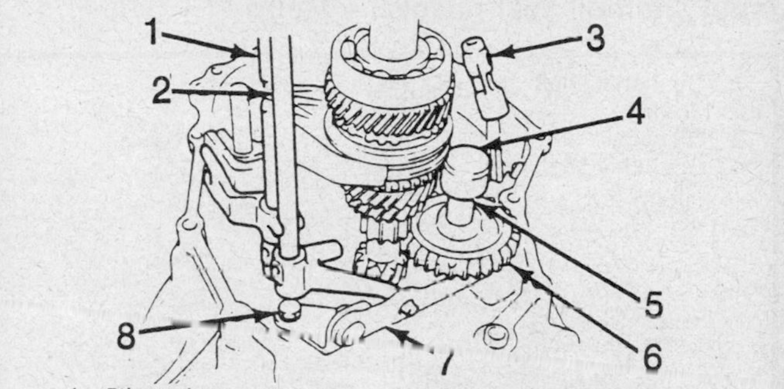
1
2
3
4
5
6
7
8

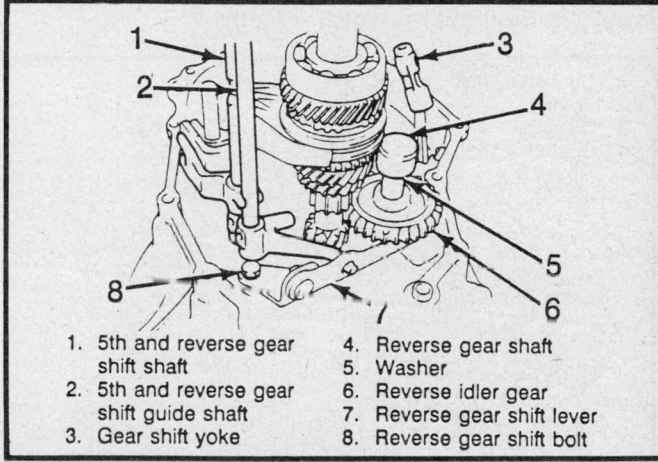

1. 5th and reverse gear shift shaft
2. 5th and reverse gear shift guide shaft
3. Gear shift yoke
4. Reverse gear shaft
5. Washer
6. Reverse idler gear
7. Reverse gear shift lever
8. Reverse gear shift bolt

Reverse idler gear servicing—Metro

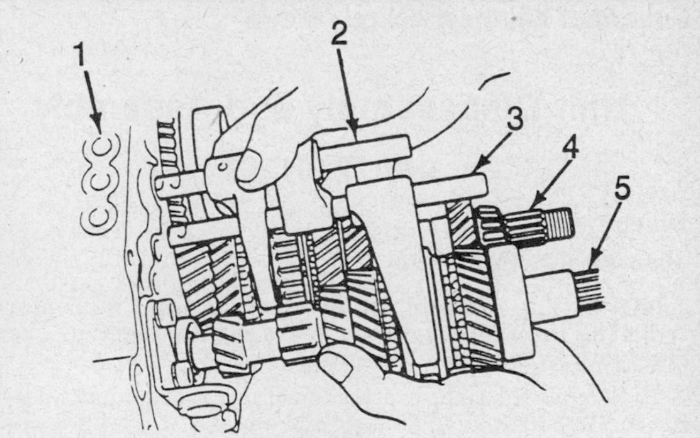
1
2
3
4
5

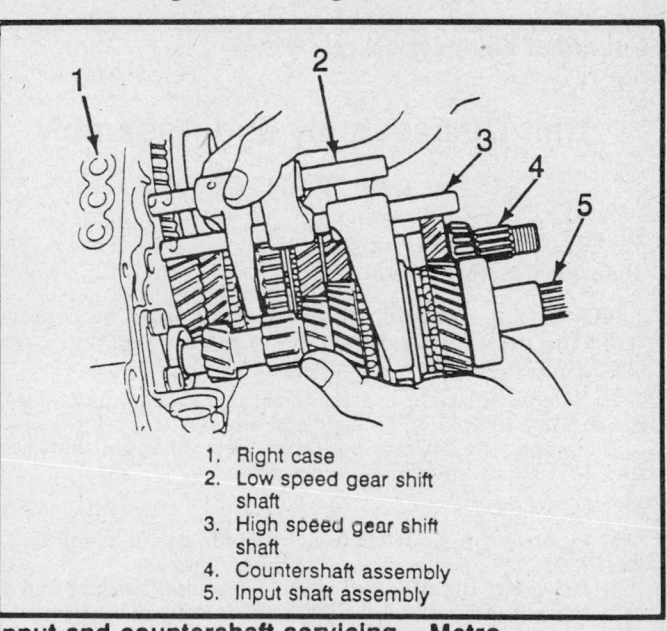

1. Right case
2. Low speed gear shift shaft
3. High speed gear shift shaft
4. Countershaft assembly
5. Input shaft assembly

Input and countershaft servicing—Metro

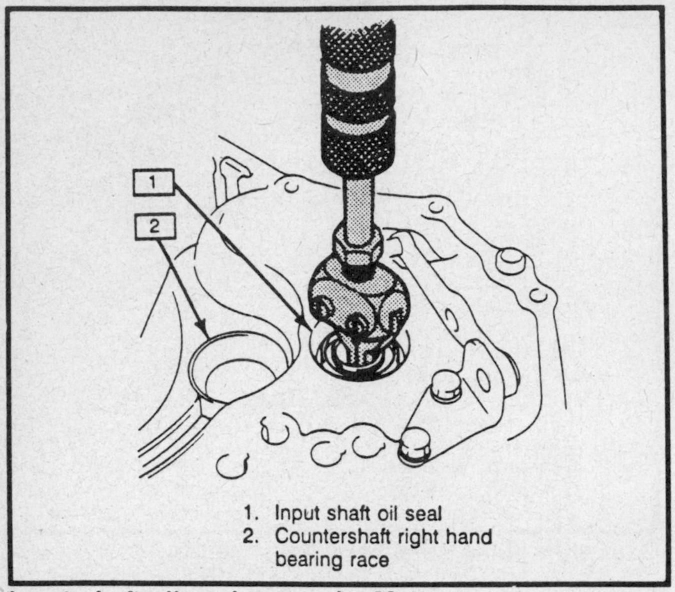
1
2
1. Input shaft oil seal
2. Countershaft right hand bearing race

Input shaft oil seal removal—Metro

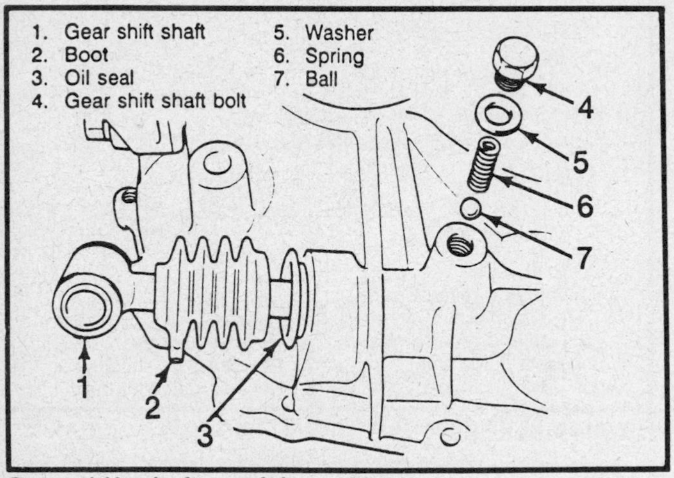
1. Gear shift shaft
2. Boot
3. Oil seal
4. Gear shift shaft bolt
5. Washer
6. Spring
7. Ball
1
2
3
4
5
6
7

Gear shift shaft servicing—Metro

Unit Disassembly and Assembly

INPUT SHAFT

Disassembly

1985–88 SPRINT

NOTE: The 3rd and 4th gears should not be removed from the input shaft. If the gears are damaged or excessively worn, replace the input shaft.

1. Remove the right bearing (small) from the input shaft using tool J-34843 or equivalent and a press.
2. Remove the left bearing (large) from the input shaft using tool J-34843 or equivalent and a press.

1989 METRO

1. Remove the input shaft right hand bearing using tool J-22912-01 or equivalent, (inverted) and a press.
2. Drive out the 5th gear spacer, left hand bearing and 4th gear together using tool J-29912-01 or equivalent, (inverted) and a press.

NOTE: To avoid gear teeth from being damaged, support the gear on the flat side of the bearing puller. Stop the press in the middle and take out the 5th gear bushing to prevent it from being compressed, continue to remove the bearing with the gear.

3. Remove the 4th gear needle bearing and high speed synchronizer ring.
4. Remove the snapring using pliers.
5. Drive out the high speed synchronizer sleeve and hub assembly together with the 3rd gear using tool J-22912-01 or equivalent, (inverted) and a press.

NOTE: Make sure to use the flat side of the puller to avoid damage to the 3rd gear teeth.

6. Remove the 3rd gear needle bearing from the shaft.
7. Remove the synchronizer sleeve and the hub assembly.

Inspection

1985–89 SPRINT

Inspect all components for any abnormality and replace, if necessary.

1989 METRO

1. Inspect all components for any abnormality and replace, if necessary.
2. Inspect the synchronizer assembly. Check clearance between the ring and gear, each chamfered tooth of the gear, ring and sleeve. Replace any parts if necessary. Standard clearance is 0.039–0.055 in. (1.0–1.4mm). Service limit is 0.019 in. (0.5mm).

NOTE: To ensure lubrication, blow the air through all oil holes and make sure that they are free from any obstruction.

Assembly

1985–88 SPRINT

1. Install the right bearing on the input shaft.
2. Install the left bearing on the input shaft.

1989 METRO

1. Install the high speed synchronizer sleeve to the hub.
2. Install the right bearing using tool J-34844 or equivalent.
3. Install the 3rd gear needle bearing, apply oil to it, then install the 3rd gear and synchronizer ring.
4. Drive in the high speed sleeve and hub assembly using tool J-35283 or equivalent and a hammer.

NOTE: While press fitting the sleeve and hub, make sure that the synchronizer ring key slots are aligned with the keys in the sleeve and hub assembly. Check the free rotation of the 3rd gear after press fitting the sleeve and hub assembly. Needle bearings and synchronizer rings for 3rd and 4th are identical respectively.

5. Install the snapring, needle bearing, apply oil to the bearing, then install the synchronizer ring and 4th gear.

CAUTION

Confirm that the snapring is installed in the groove securely.

6. Install the left bearing using tool J-34844 or equivalent and a hammer.
7. Install the 5th gear spacer using tool J-34844 or equivalent and a hammer.

NOTE: To prevent 5th gear spacer from being distorted because of excessive compression, do not press-fit it with the left hand bearing at once.

COUNTERSHAFT

Disassembly

1985–88 SPRINT

NOTE: The countershaft right bearing cage protrudes beyond the countershaft end face. Use care not to place the countershaft with the right bearing on a work bench or damage to bearing cage may result.

1. Remove the right bearing from the countershaft using tool J–34843 or equivalent and a press.
2. Remove the left bearing from the countershaft using tool J–34843 or equivalent and a press.
3. Remove the 4th gear thrust washer, 4th gear, bearing, spacer, 4th gear synchronizer ring, 3rd/4th synchronizer hub and sleeve assembly. Next, remove the 3rd gear synchronizer ring, 3rd gear and bearing.
4. Remove the ring washer from the 2nd gear retaining ring halves. Remove the 2 ring halves, 2nd gear, bearing and 2nd gear synchronizer ring.
5. Remove the 1st/2nd gear synchronizer assembly retaining ring halves. Remove the 1st/2nd synchronizer hub and sleeve assembly, 1st gear synchronizer ring, 1st gear and bearing.

1989 METRO

1. Drive out the left hand bearing with the 4th gear using tool J–22912–01 or equivalent,(inverted) and a press.

CAUTION

Use a puller and press that will bear at least 5 tons (11,000 lbs.) safely. To avoid tooth damage, support the 4th gear at the flat side of the puller.

2. Drive out the 3rd and 4th gear spacer together with the 2nd gear using tool J–22912–01 or equivalent, (inverted) and a press. The needle bearing comes out with the 2nd gear.

CAUTION

If compression exceeds 5 tons (11,000 lbs.) release compression once, reset the puller support and then continue press work again.

3. Remove the 2nd synchronizer ring.
4. Remove the snapring using snapring pliers.
5. Drive out the low speed synchronizer sleeve and hub assembly together with the 1st gear using tool J–22912–01 or equivalent, (inverted) and a press.
6. Disassembly the synchronizer sleeve and hub assembly.
7. Remove the needle bearing from the shaft.
8. Remove the right bearing using tool J–22912–01 or equivalent, (inverted), metal rod and a press.

Inspection

1985–88 SPRINT

Inspect all components for abnormalities and replace with new ones as necessary.

1989 METRO

1. Inspect all components for abnormalities and replace with new ones as necessary.
2. Inspect synchronizer parts, if repair is necessary, check clearance between the ring and gear, each chamfered tooth of gear, ring and sleeve, then determine parts replacement. Standard clearance is 0.039–0.055 in. (1.0–1.4mm). Service limit is 0.019 in. (0.5mm).

NOTE: To ensure lubrication, blow the air through all oil holes and make sure that they are free from any obstruction.

Assembly

1985–88 SPRINT

NOTE: Apply clean transaxle oil to the sliding surfaces of all parts before assembling.

1. Install the low gear bearing and then the gear on the shaft.
2. Install the low gear synchronizer ring and ring spring.
3. Install the 1st/2nd gear synchronizer assembly on the shaft.

NOTE: When installing the hub to the countershaft, align the oil groove on the hub with the oil hole on the shaft.

4. Install the low speed synchronizer hub rings. Install the 2 rings into the groove on the countershaft.
5. Install the 2nd gear bearing and then the gear on the shaft. Install the 2 rings with the protrusion of the ring into the hole on the countershaft.
6. Install the ring washer.
7. Install the spring to the 2nd gear.
8. Install the 3rd gear bearing and then the gear on the shaft.
9. Install the 3rd/4th gear synchronizer assembly on the shaft.

NOTE: Install the hub on the countershaft in such a way that the oil groove of the hub meets the oil hole of the shaft.

10. Install the 4th gear and then the thrust washer. The oil grooves on the washer goes against the 4th gear.
11. Install the left bearing on the shaft.
12. Install the right bearing on the shaft.

1989 METRO

1. Install the low speed synchronizer sleeve to the hub, insert the 3 keys, then the set of springs.

NOTE: No specific direction is assigned to the low speed synchronizer hub or each key, but it is assigned as an assembly. Size of the low speed synchronizer keys and springs are largest compared with those of the high speed and 5th speed ones.

2. Install the right bearing using tool J–34846 or equivalent and a hammer.

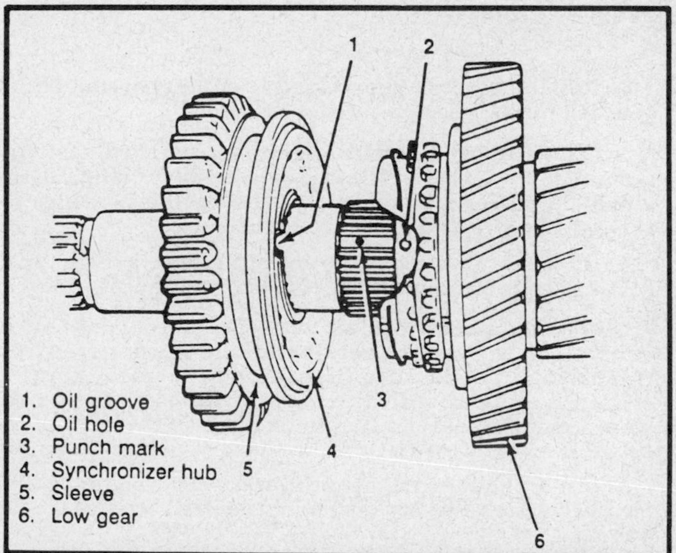
1
2
3
4
5
6
1. Oil groove
2. Oil hole
3. Punch mark
4. Synchronizer hub
5. Sleeve
6. Low gear

1st–2nd gear synchronizer assembly installation—1985–88 Sprint

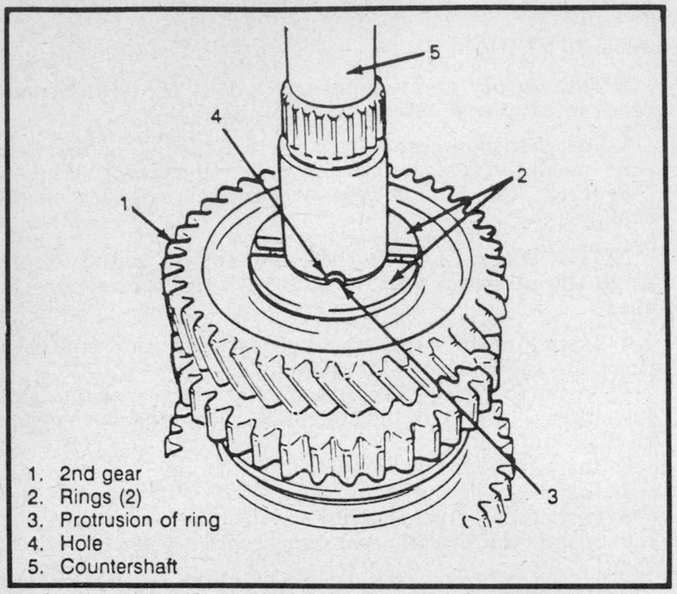
5
4
2
1
3
1. 2nd gear
2. Rings (2)
3. Protrusion of ring
4. Hole
5. Countershaft

2nd gear rings installation—1985–88 Sprint

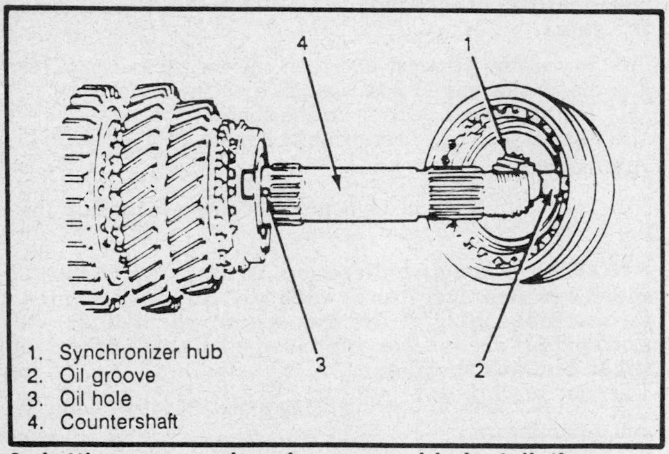
4
1
3
2
1. Synchronizer hub
2. Oil groove
3. Oil hole
4. Countershaft

3rd–4th gear synchronizer assembly installation—1985–88 Sprint

3. Install the needle bearing, apply oil, then install the 1st gear and the 1st gear synchronizer ring.

NOTE: The key slot width of the 1st synchronizer ring is smaller than that of the 2nd synchronizer ring. Distinguish the difference properly. The needle bearings for the 1st and the 2nd gear are identical.

4. Drive the low speed sleeve and hub assembly using tool J-35664 or equivalent and tool J-34846 or equivalent.

NOTE: Support the shaft with special tool J-34846 or equivalent, to keep the retainer of the bearing free from compression. Make sure that the synchronizer ring key slots are aligned with the keys while press-fitting the sleeve and hub assembly. Check for free rotation of the 1st gear after press-fitting the sleeve and hub assembly.

5. Install the snapring needle bearing, apply oil to the bearing, then install the 2nd gear synchronizer ring and the 2nd gear.

CAUTION

Confirm that the snapring is installed in the groove securely.

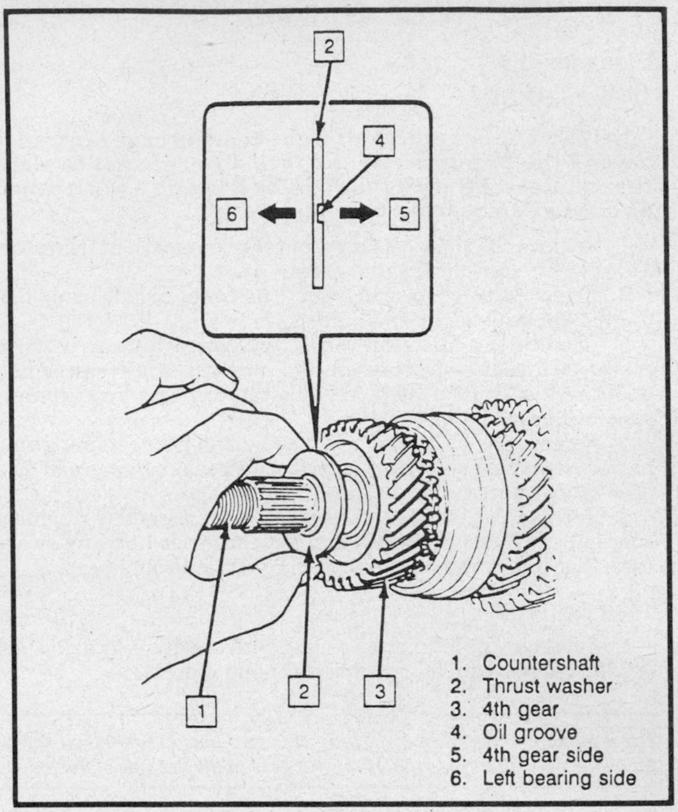
2
4
6
5
1
2
3
1. Countershaft
2. Thrust washer
3. 4th gear
4. Oil groove
5. 4th gear side
6. Left bearing side

4th gear thrust washer installation— Metro

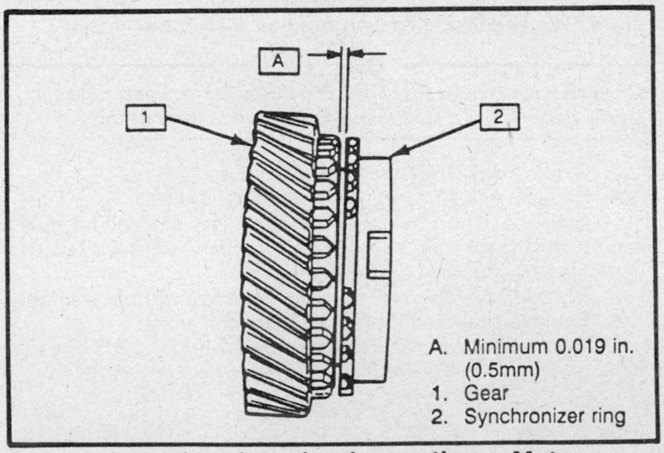
A
1
2
A. Minimum 0.019 in. (0.5mm)
1. Gear
2. Synchronizer ring

Gear and synchronizer ring inspection— Metro

6. Install the 3rd gear and spacer using tool J-35664 or equivalent, with tool J-37754 or equivalent, J-34846 or equivalent and a press.

NOTE: It is recommended to press-fit the spacer and 3rd gear first, then the 4th gear separately to prevent the countershaft from being compressed excessively.

7. Install the 4th gear using tool J-35664 or equivalent, with tool J-37754 or equivalent, J-34846 or equivalent and a press.
8. Install the left bearing using tool J-34844 or equivalent, with tool J-34846 or equivalent and a hammer.

NOTE: For protection of the bearing, always support the shaft with special tool J-34846 or equivalent.

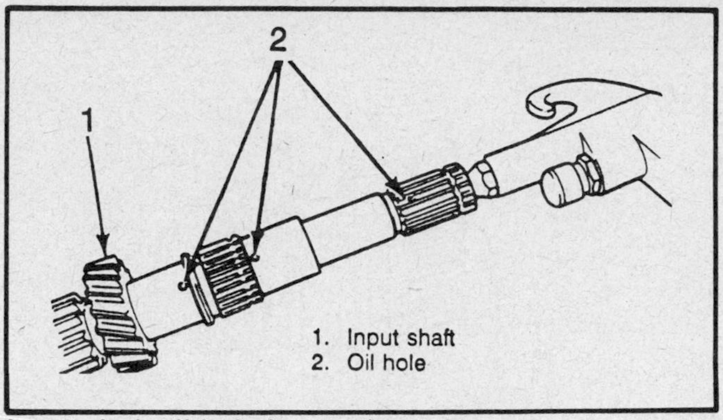
2
1
1. Input shaft
2. Oil hole

Input shaft oil hole inspection—Metro

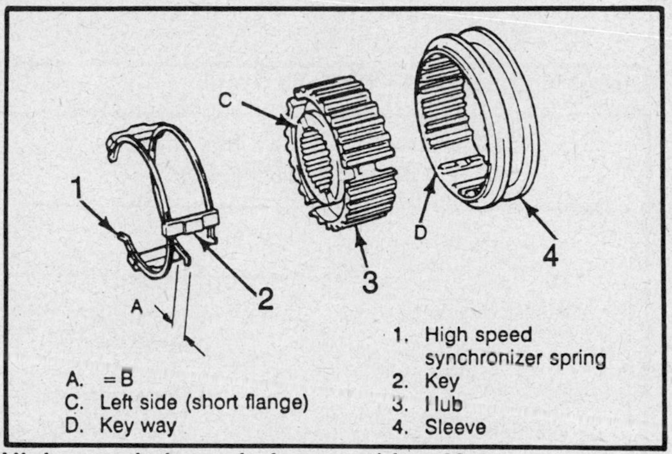
C
1
2
3
D
4
A
A. =B
C. Left side (short flange)
D. Key way
1. High speed synchronizer spring
2. Key
3. Hub
4. Sleeve

High speed sleeve hub assembly—Metro

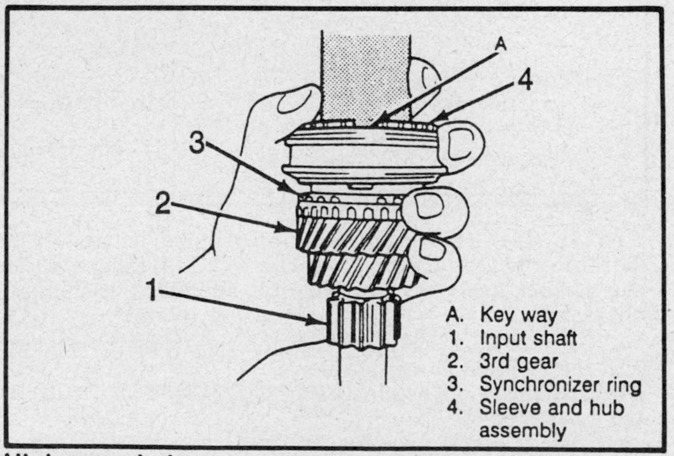
A
4
3
2
1
A. Key way
1. Input shaft
2. 3rd gear
3. Synchronizer ring
4. Sleeve and hub assembly

High speed sleeve assembly—Metro

DIFFERENTIAL CASE

Disassembly

1985–88 SPRINT

1. Remove the side bearings using puller tool J-28509-A or equivalent and plug tool J-34851 or equivalent.
2. Remove the speedometer drive gear using puller tool J-28509-A or equivalent and plug tool J-34851 or equivalent.
3. Remove the roll pin retaining the pinion shaft.
4. Remove the pinion shaft, pinion gears and side gears from the case.

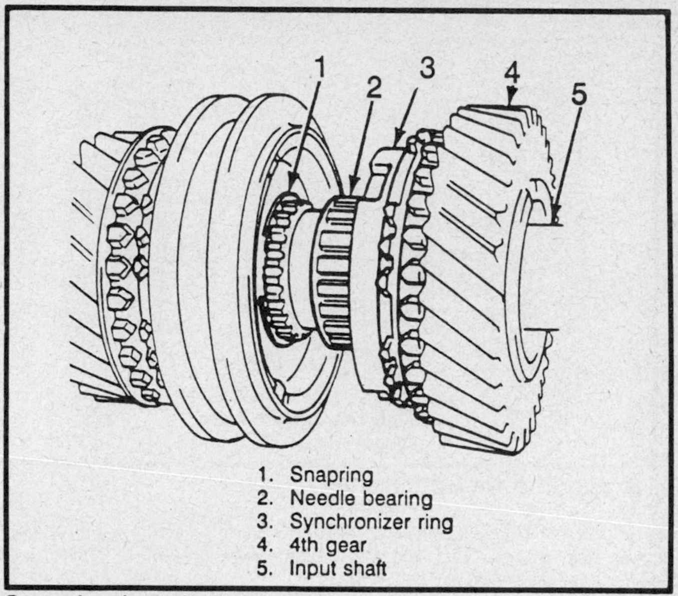
1
2
3
4
5
1. Snapring
2. Needle bearing
3. Synchronizer ring
4. 4th gear
5. Input shaft

Snapring installation—Metro

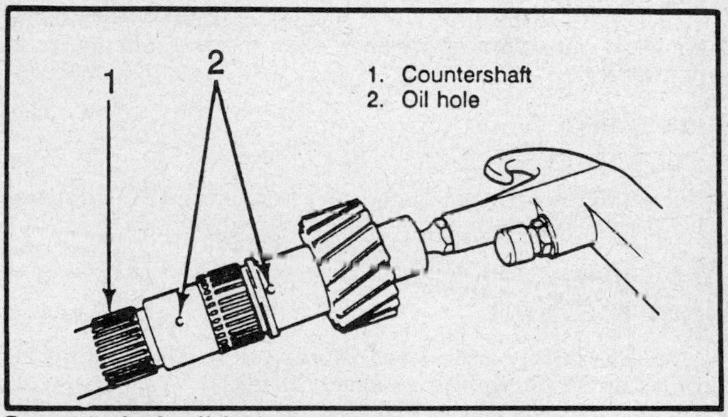
1
2
1. Countershaft
2. Oil hole

Countershaft oil hole inspection—Metro

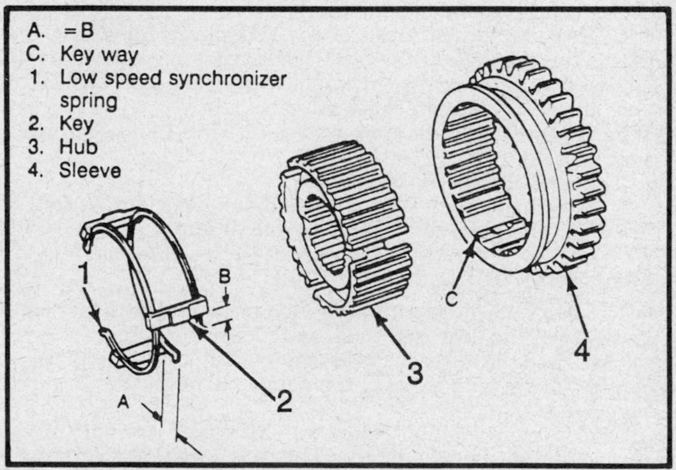
A. =B
C. Key way
1. Low speed synchronizer spring
2. Key
3. Hub
4. Sleeve
1
B
C
4
3
A
2

Low speed sleeve and hub assembly—Metro

5. Remove the ring gear retaining bolts and then remove the ring gear.

1989 METRO

1. Remove the right bearing using tool J-28509 or equivalent.

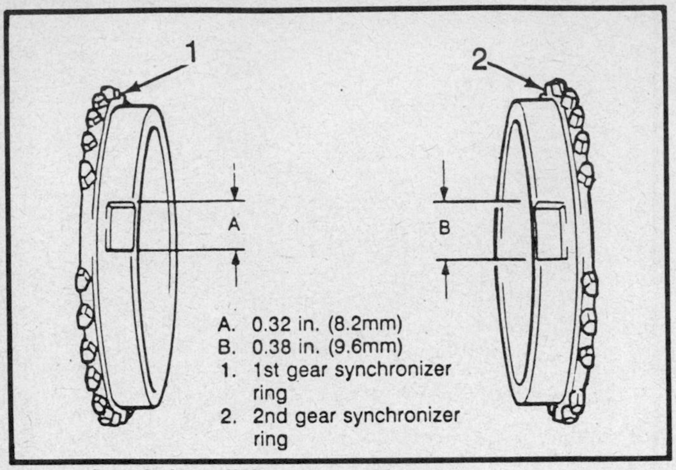
1
2
A
B
A. 0.32 in. (8.2mm)
B. 0.38 in. (9.6mm)
1. 1st gear synchronizer ring
2. 2nd gear synchronizer ring

1st and 2nd ring difference—Metro

2. Remove the speedometer drive gear.
3. Remove the left bearing using tool J-28509 or equivalent, with tool J-34851 or equivalent.
4. Remove the 8 final gear bolts, then the final gear while supporting the differential case with soft jawed vise.
5. Drive out the differential side pinion shaft roll pin using a standard punch and a hammer, then disassemble the component parts.

Inspection

ALL MODELS

Inspect for abnormality before disassembly and after disassembly. Replace components if necessary.

Assembly

1985–88 SPRINT

NOTE: Apply clean transaxle oil to the sliding surfaces of all parts before assembling.

1. Install the side washers and then the side gears.
2. Install the pinion washers and pinion gears.
3. Install the pinion gear shaft.
4. Measure the backlash of the side pinions and side gears with a feeler gauge. If the backlash exceeds the specified value, adjust it by varying the thickness of the side gear washers.
5. Install the pinion shaft roll pin.
6. Install the ring gear and retaining bolts. Torque to specification.
7. Install the speedometer gear.
8. Install the right and left hand side bearings. Install the bearings with the seal side facing the differential case.

1989 METRO

1. Install the differential gear and measure the thrust play of the differential gear (left side) as follows:
   a. Hold the differential assembly with a soft jawed vise and apply measuring tip of a dial gauge to the top surface of the gear.
   b. Using 2 small pry bars, move the gear up and down and read the movement of the dial gauge pointer. Thrust play should be between 0.002–0.013 in. (0.05–0.33mm).
2. Measure the thrust play of the differential gear (right side) as follows:
   a. Using a similar procedure to the above, set the dial gauge tip to the gear shoulder.
   b. Move the gear up and down by hand and read the dial gauge. Thrust play should be between 0.002–0.013 in. (0.05–0.33mm).

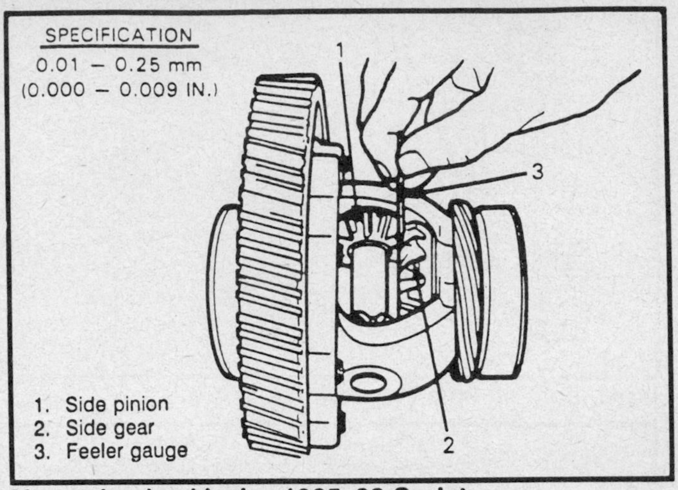
SPECIFICATION
0.01 – 0.25 mm
(0.000 – 0.009 IN.)
1
3
2
1. Side pinion
2. Side gear
3. Feeler gauge

Measuring backlash—1985–88 Sprint

SIDE GEAR WASHER SIZES 1985–88 SPRINT

	mm	IN.
Available side gear	0.90	0.035
	0.95	0.037
Washer sizes	1.00	0.039
	1.05	0.041
	1.10	0.043
	1.15	0.045
	1.20	0.047

NOTE: If the thrust play is out of specification, select suitable thrust washer from the available sizes, install the washer and check again that specified gear play is obtained.

3. Drive in the spring pin from the right side till it is flush with the differential case surface.
4. Install the left bearing using tool J-34851 or equivalent and a hammer.
5. Install the speedometer drive gear, support differential assembly to allow the left hand bearing to float and then press-fit the right hand bearing using tool J-29036 or equivalent, with tool J-8092 or equivalent and a hammer.
6. Install the ring gear while holding the differential assembly with soft jawed vise. Install the 8 final gear bolts and torque to 63 ft. lbs. (85 Nm).

CAUTION

Use of any other bolts than specified can cause internal transaxle damage.

Transaxle Assembly

1985–88 SPRINT

NOTE: Apply grease to the lips of the input shaft seal

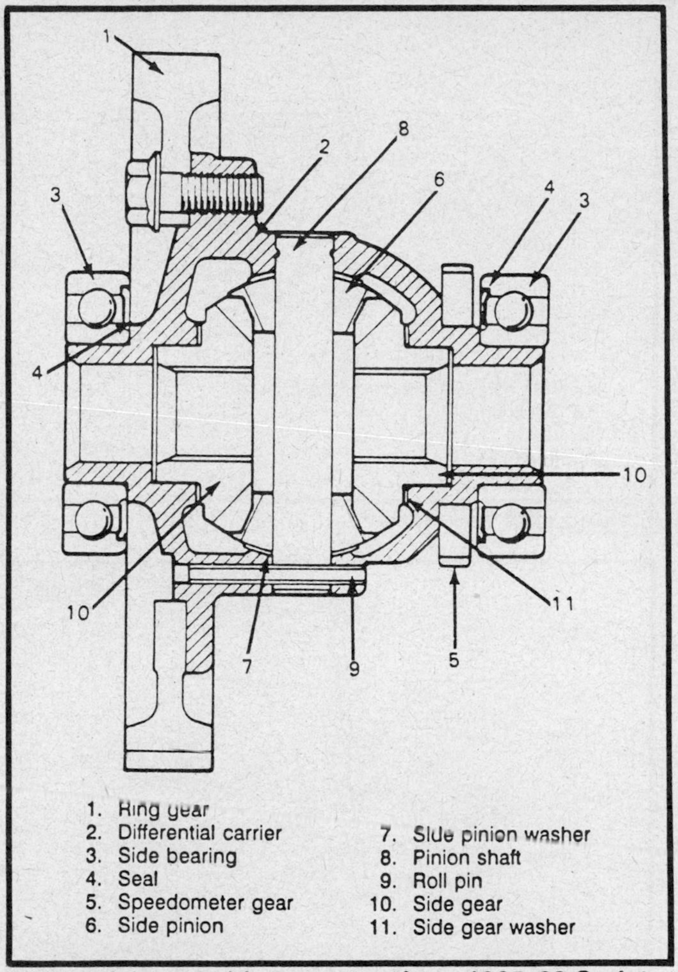
1
2
8
6
3
4
3
4
10
10
7
9
5
11
1. Ring gear
2. Differential carrier
3. Side bearing
4. Seal
5. Speedometer gear
6. Side pinion
7. Side pinion washer
8. Pinion shaft
9. Roll pin
10. Side gear
11. Side gear washer

Differential assembly cross-section—1985–88 Sprint

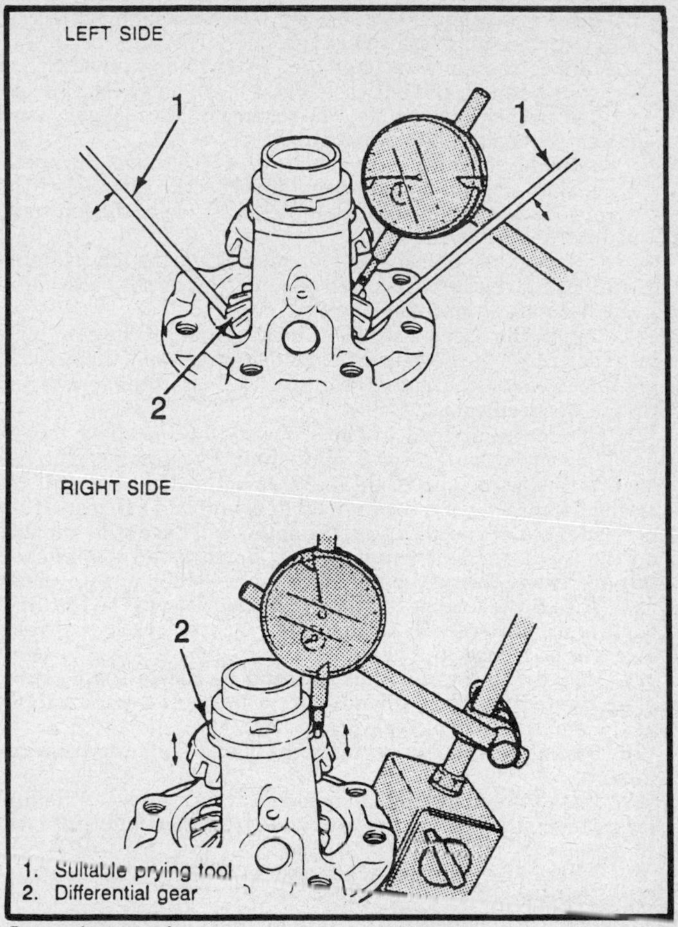
LEFT SIDE
1
1
2
RIGHT SIDE
2
1. Suitable prying tool
2. Differential gear

Gear thrust play measurement—Metro

and differential side oil seals after installation. Apply clean transaxle oil to the sliding surfaces of all parts before assembling.

1. Install the input shaft in the right case.
2. Install the input shaft bearing cup in the right case.
3. Install the countershaft bearing cup in the right case.
4. Install the right and left case differential oil seals.
5. Install the reverse gear shift lever. Apply thread locking cement to the bolts before installing. Torque bolts to specification.
6. Install the reverse gear shift lever. After installation, apply grease to the lip of the seal.
7. Install the gear shift shaft and boot. Apply grease to the shaft before installing.
8. Install the shift shaft detent ball, spring, gasket and bolt. Torque bolt to specification.
9. Install the gear shift arm and roll pin to the gear shift shaft.
10. Install the differential assembly into the right case.
11. Install the 5th and reverse gear shift and guide shaft to the right case.
12. Install the input shaft, countershaft, 1st/2nd and 3rd/4th gear shift shafts to the right case.
13. Install the idler gear, shaft, pin, spacer and washer to the right case.
14. Install the case magnet into the right case.
15. Apply Loctite® No. 518 sealant or equivalent, uniformly to the mating surface of the left case.
16. Place the left case on the right case and install the retaining bolts. Torque bolts to specification.

NOTE: After tightening the case bolts, check the input shaft and countershaft for smooth rotation by turning them by hand.

17. Install the 3 shift fork shaft detent balls, springs, gaskets and bolts. Torque bolts to specification.
18. Install the gear shift yoke on the gear shift arm. Install the gear shift/select shaft assembly, guiding the shaft into the hole in the gear shift yoke. Align the hole in the yoke with the hole in the shaft and install a new roll pin to retain.
19. Install the gear shift locating bolt and washer. Torque the bolt to specification.
20. Install the gear shift guide case. Torque the bolts to specification.
21. Install the reverse check ball, spring, washer and bolt. Torque the bolt to specification.
22. Install the case cap with a new O-ring to the left case. Install the 3 retaining bolts and torque to specification.
23. Measure and determine the bearing shim size for the input shaft and countershaft using tool J–34858 or equivalent, as follows:
    a. Install the left bearing cups for the input shaft and countershaft.
    b. Using finger pressure, press the countershaft left bearing cup against the bearing rollers.
    c. Rotate the countershaft 3–4 times by hand to seat the bearing. Install the nut from tool J–34858 or equivalent, on the shaft and then install the countershaft nut and torque to 44–58 ft. lbs. (60–80 Nm).
    d. Install tool J–29763 or equivalent, dial gauge on tool J–

34858 or equivalent, shim selector. Place the shim selector on a flat surface and zero the gauge.

e. Place the shim selector tool J-34858 or equivalent, on the countershaft and left case. Press down on the shim selector and read the dial gauge. The reading on the dial gauge will indicate the size shim required.

f. The shim stock is available in 12 selective thickness. Each shim has the thickness stamped on the side. Select the shim required and install it on the back side of the left hand countershaft bearing cup.

g. Repeat the above steps for measuring the input shaft.

h. Select the shim required and install it on the back side of the left hand input shaft bearing cup.

24. Install the case plate. Fit the protrusion of the case plate into the groove of the gear shift guide shaft. Apply thread locking cement to the threads of the screws. Install the screws and torque to specification.

25. Check the preload of the transaxle by installing tool J-34852 or equivalent, on the input shaft. Connect a spring balance to the wire on tool J-34852 or equivalent and pull the spring balance. Check the preload in **N** and in **4TH** gear. If the preload is out of specification, the shims will have to be changed on the input shaft or countershaft. Remove the countershaft nut and then remove the nut from tool J-34858 or equivalent.

26. With the transaxle in 4th gear, install 5th gear on the input shaft and countershaft. Engage the synchronizer so the transaxle will be locked in 2 gears.

27. Install the countershaft nut and torque to specification. After tightening, stake the nut with a chisel. If a crack is found at the staked portion, replace the nut.

28. Disengage 5th gear synchronizer and shift the transaxle into **N**.

29. Install the shift guide ball and shift fork screw to the shift fork. After tightening the screw to specification, stake the screw.

30. Install a new roll pin to 5th and reverse gear shift shaft, using a drift.

31. Install the synchronizer hub plate and a new snapring.

32. Apply Loctite® sealant No. 518 or equivalent to the mating surface of the side case. Install the side case. Install the oil receiver cup in the side case into the hole of the input shaft. Install the bolts and torque to specification.

33. Install the backup light switch.

1989 METRO

NOTE: The input shaft right hand bearing can be installed into the right case by tapping the shaft with tool J-34849 or equivalent and a plastic hammer.

1. Install the differential assembly into the right case using a rubber hammer to drive flush with the case.

2. Insert the speedometer driven gear case assembly applied with oil to its O-ring and gear, then tighten it with the bolt.

CAUTION

While inserting the case assembly into the transaxle, turn the final gear by hand slightly so that the gear can mesh easily. Never push or hit the slit portion of the case when inserting it. Such attempt may cause the case to break.

3. Install the input shaft, countershaft, low speed gear shift shaft assemblies all together; then install them into the right case.

NOTE: Check to make sure that the countershaft is engaged with the final gear while installing.

4. Install the 5th and reverse gear shift shaft with the 5th and reverse gear shift guide shaft into the right case. Reverse gear shift arm has to be joined with the reverse gear shift lever at the same time.

5. Install the reverse idler gear with the reverse gear shift le-

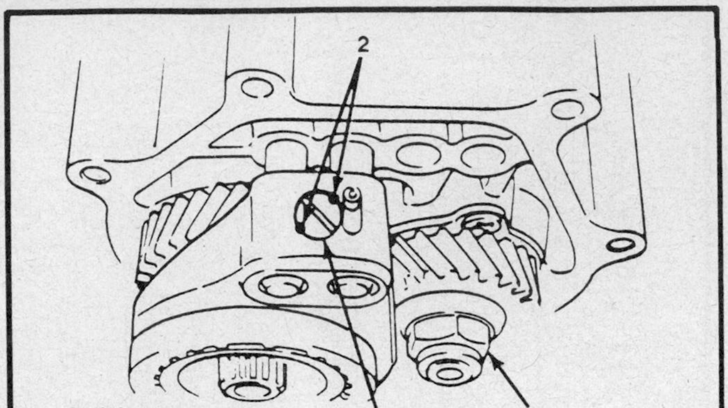
2
1
3

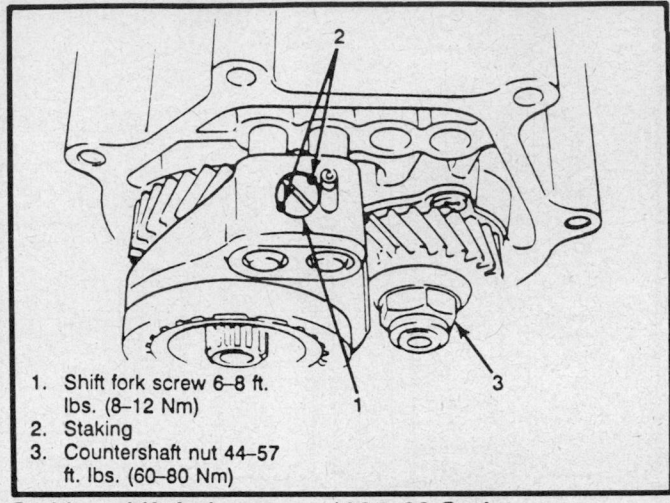

1. Shift fork screw 6–8 ft. lbs. (8–12 Nm)
2. Staking
3. Countershaft nut 44–57 ft. lbs. (60–80 Nm)

Staking shift fork screw—1985–88 Sprint

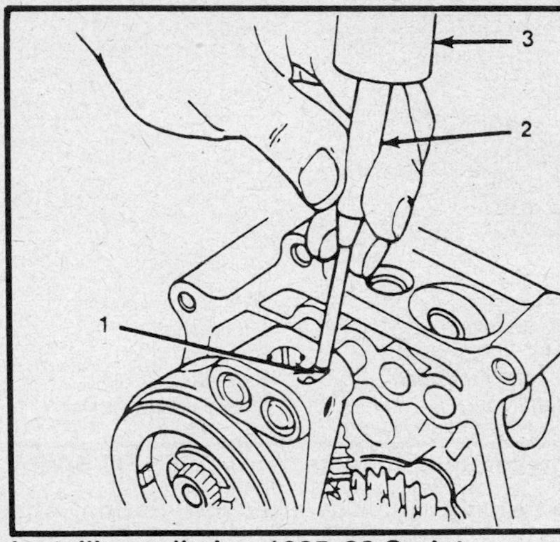
3
2
1

1. Roll pin
2. Drift
3. Hammer

Installing roll pin—1985–88 Sprint

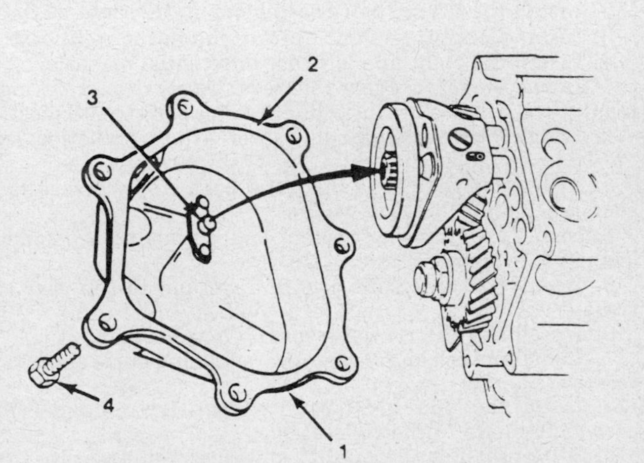
2
3
4
1

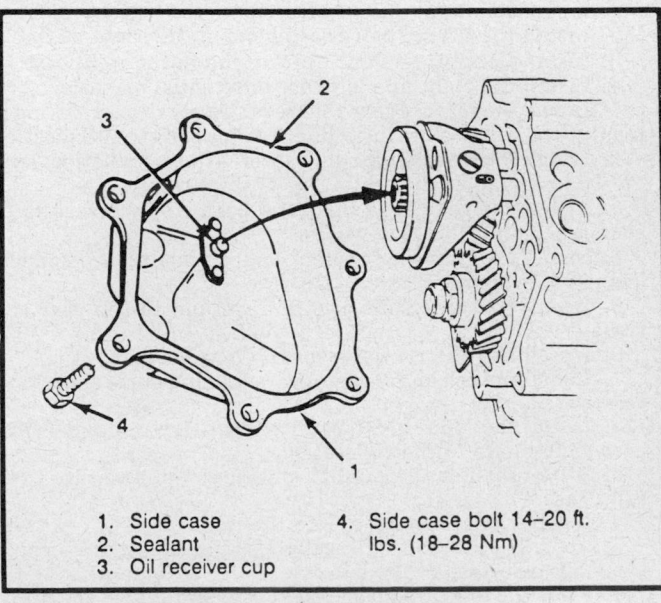

1. Side case
2. Sealant
3. Oil receiver cup
4. Side case bolt 14–20 ft. lbs. (18–28 Nm)

Installing side cover—1985–88 Sprint

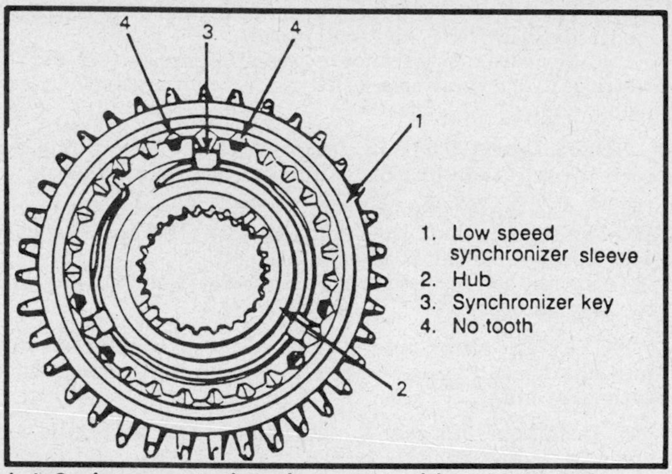
4
3
4
1
2
1. Low speed synchronizer sleeve
2. Hub
3. Synchronizer key
4. No tooth

1st–2nd gear synchronizer assembly—1985–88 Sprint

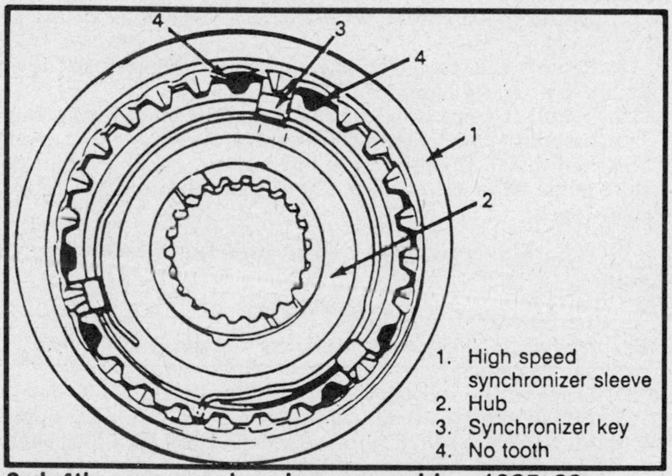
4
3
4
1
2
1. High speed synchronizer sleeve
2. Hub
3. Synchronizer key
4. No tooth

3rd–4th gear synchronizer assembly—1985–88 Sprint

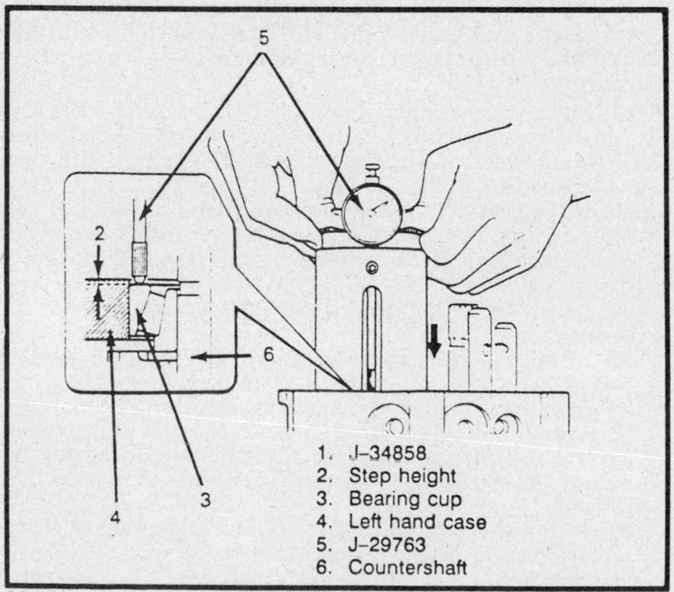
5
2
6
4
3
1. J-34858
2. Step height
3. Bearing cup
4. Left hand case
5. J-29763
6. Countershaft

Measuring step height (countershaft)—1985–88 Sprint

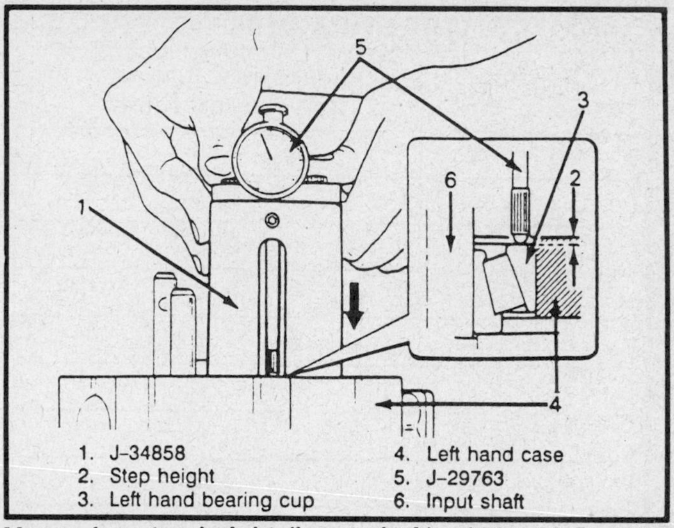
5
3
6
2
1
4
1. J-34858
2. Step height
3. Left hand bearing cup
4. Left hand case
5. J-29763
6. Input shaft

Measuring step height (input shaft)—1985–88 Sprint

COUNTERSHAFT SHIM—1985–88 SPRINT

Dial gauge reading (mm)	Thickness of bearing set shim to be installed (mm)
0.42 – 0.47	0.60
0.48 – 0.52	0.65
0.53 – 0.57	0.70
0.58 – 0.62	0.75
0.63 – 0.67	0.80
0.68 – 0.72	0.85
0.73 – 0.77	0.90
0.78 – 0.82	0.95
0.83 – 0.87	1.00
0.88 – 0.92	1.05
0.93 – 0.97	1.10
0.98 – 1.02	1.15

ver, insert the reverse gear shaft into the case through idler gear and then align the mark in the shift with the mark in the case.

NOTE: Make sure that the washer has been installed in the shaft above the gear. Check to confirm that the reverse gear shift lever end has clearance 0.04 in. (1mm) to the idler gear groove.

6. Clean the mating surfaces of both right and left cases. Coat mating surface of the left case with sealant No. 102751 or equivalent, evenly then mate it with the right case. Use a rubber hammer if necessary.
7. Install the 11 case bolts outside and torque to 14 ft. lbs. (19 Nm).
8. Install the reverse shaft bolt with an aluminum washer. Torque bolt to 10 ft. lbs. (14 Nm).

INPUT SHAFT SHIM—1985–88 SPRINT

Dial gauge reading (mm)	Thickness of bearing set shim to be installed (mm)
0.48 — 0.52	0.55
0.53 — 0.57	0.60
0.58 — 0.62	0.65
0.63 — 0.67	0.70
0.68 — 0.72	0.75
0.73 — 0.77	0.80
0.78 — 0.82	0.85
0.83 — 0.87	0.90
0.88 — 0.92	0.95
0.93 — 0.97	1.00
0.98 — 1.02	1.05
1.03 — 1.07	1.10

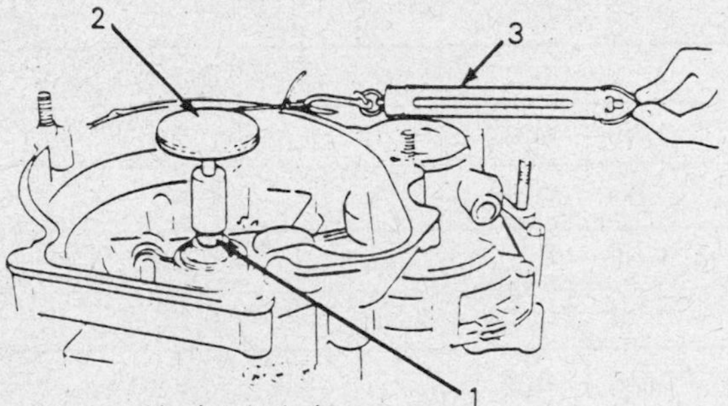
2
3
1

1. Input shaft
2. J-34852
3. Spring gauge (preload specification—3–15 Nm)

Checking input shaft preload—1985–88 Sprint

9. Install the 2 case bolts from the clutch housing side and tighten them to specification.

NOTE: Inspect the locating springs for deterioration and replace with new ones as necessary.

10. Install the steel ball and locating spring for respective gear shift shaft and tighten with the bolt.

NOTE: The springs are color coded. Be sure to place the springs in the correct position. The low speed spring standard length is 1.028 in. (26.1mm). The low speed spring service limit is 0.984 in. (25mm). The high speed spring standard length is 1.579 in. (40.1mm). The high speed service limit is 1.535 in. (39mm).

11. Install the countershaft left hand bearing race to the bearing using tool J-35425 or equivalent and a plastic hammer.
12. Select a suitable shim which adjusts the clearance between the bearing race and the case to specification; then install it onto the bearing race. Use the following procedure to determine the shim:

a. Place the shim on the bearing race, then place a straight edge flat bar over it and compress.
b. Measure the clearance between the case surface and the straight edge by using a feeler gauge. Shim protrusion clearance should be 0.0032–0.0047 in. (0.08–0.12mm).

NOTE: Insert 0.004 in. (0.1mm) feeler to quickly determine whether or not a shim fulfills specification.

13. Install the left case plate by inserting its end in the groove of the shift guide shaft; then tighten it with the 6 screws applied with Loctite® No. 1052624 or equivalent.
14. Install the 5th speed synchronizer sleeve and the hub with the keys and springs.

NOTE: The short side in the keys, long boss in the hub and chamfered spline in the sleeve should face inward (5th gear side).

15. Install the 5th gear to the countershaft facing the machined boss inward.
16. Install the needle bearing to the input shaft, apply oil, install the 5th gear and tool J-35309 or equivalent, to stop the shaft rotation.
17. Install the countershaft nut and torque to 52 ft. lbs. (70 Nm).
18. Remove the tool J-35309 or equivalent; then stake the nut with a drift and a hammer.
19. Install the synchronizer ring; then the snapring spring.
20. Install the 5th gear shift fork to the sleeve and hub assembly; then install them into the input shaft, 1st shift shaft and shift guide shaft all at once aligning the hub oil groove with shaft mark.

NOTE: The long boss of the hub faces inward (gear side).

CAUTION

Be careful not to punch the synchronizer ring spring by the hub.

21. Drive in the roll pin facing its slit outward.
22. Install the steel ball, tighten the shift fork plug applied with Loctite® No. 1052624 or equivalent. Shift fork plug should be torqued to 10 ft. lbs. (13 Nm).
23. Install the hub plate and fix it with the snapring.

NOTE: Coat shift fork plug with Loctite® No. 1052624 or equivalent lightly. Excess may interfere in the ball movement and cause hard shift to the 5th speed. Make sure the snapring is installed in the shaft groove securely.

24. Clean the mating surface of both left case and left cover. Coat the mating surface with sealant No. 1052751 or equivalent, evenly. Mate it with the left case; then tighten with 8 bolts and torque to 7 ft. lbs. (10 Nm).
25. If the gear shift guide case has been disassembled or replaced, tighten the gear shift case plate bolt to 4 ft. lbs. (6 Nm); torque the 5th to reverse interlock guide bolt to 17 ft. lbs. (23 Nm).
26. Clean the mating surface of the guide case and coat it with sealant evenly.
27. Install the gear shift yoke and join it with the gear shift arm.
28. Install the gear shift and select shaft assembly into the transaxle and join its bottom end with the gear shift yoke.
29. Tighten the yoke and shaft with the bolt coated with Loctite® No. 1052624 or equivalent. Torque the bolt to 17 ft. lbs. (23 Nm).
30. Install the gear shift interlock bolt with washer. Torque the bolt to 19 ft. lbs. (24 Nm).

NOTE: When installing the gear shift and select shaft assembly, position the gear in N so that the gear shift interlock plate will go in smoothly.

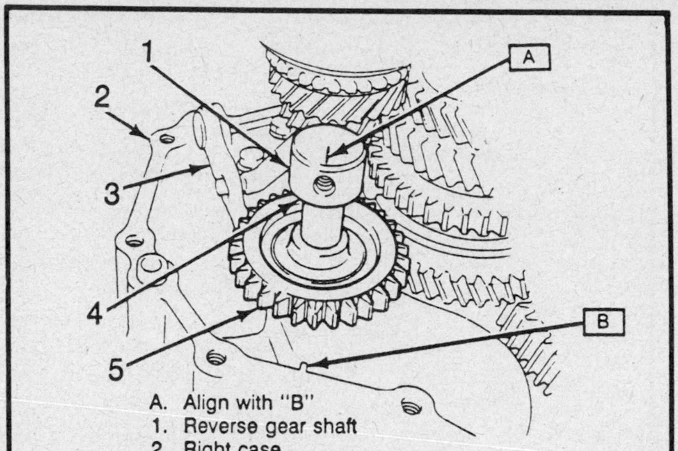
1
2
3
4
5
A
B

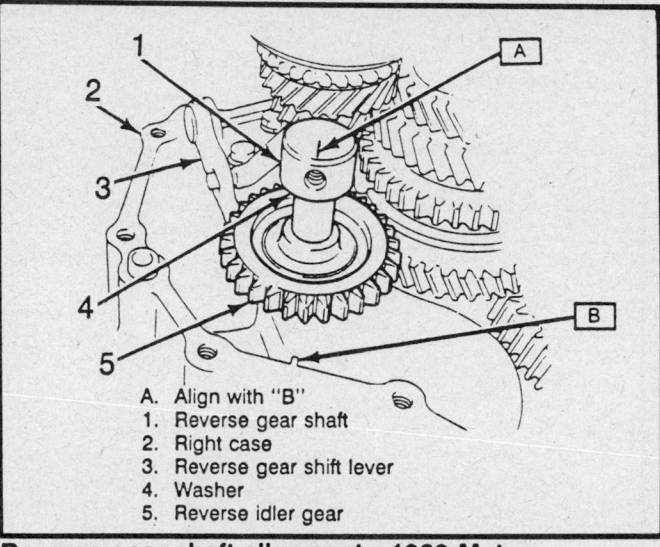

A. Align with "B"
1. Reverse gear shaft
2. Right case
3. Reverse gear shift lever
4. Washer
5. Reverse idler gear

Reverse gear shaft alignment—1989 Metro

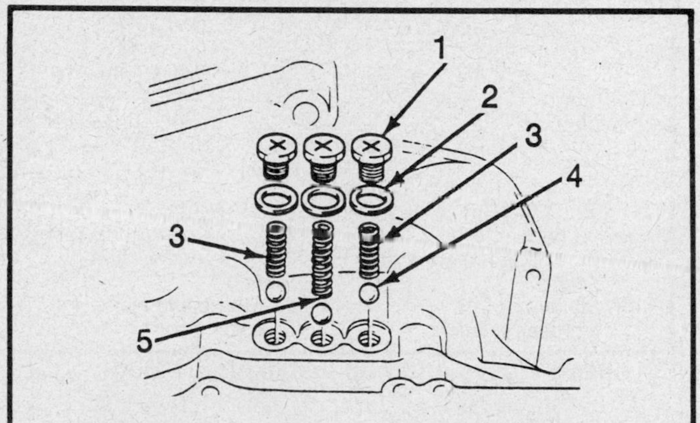
1
2
3
4
5

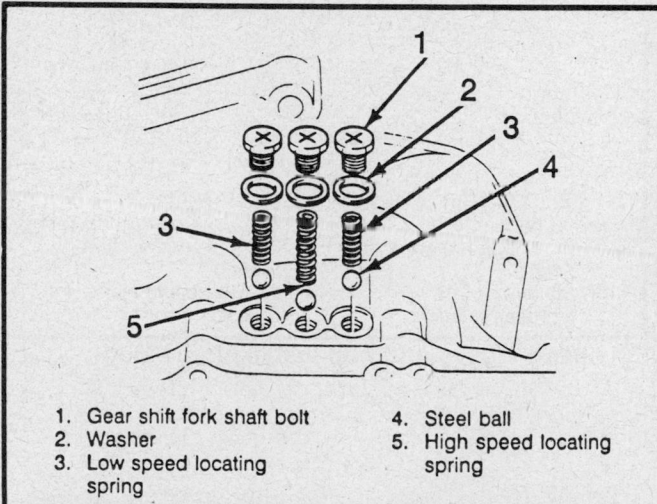

1. Gear shift fork shaft bolt
2. Washer
3. Low speed locating spring
4. Steel ball
5. High speed locating spring

Locating spring installation—1989 Metro

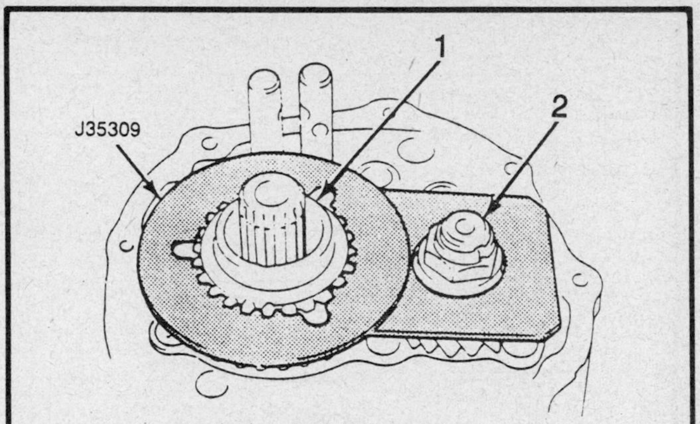
J35309
1
2

1. Input shaft 5th gear
2. Countershaft nut

Countershaft nut installation—1989 Metro

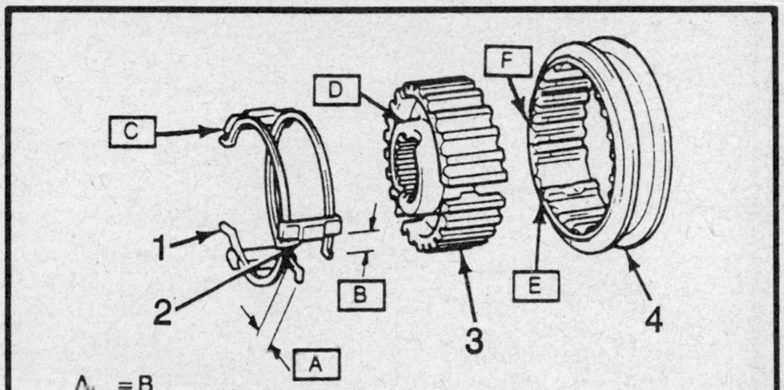
A
B
C
D
E
F
1
2
3
4

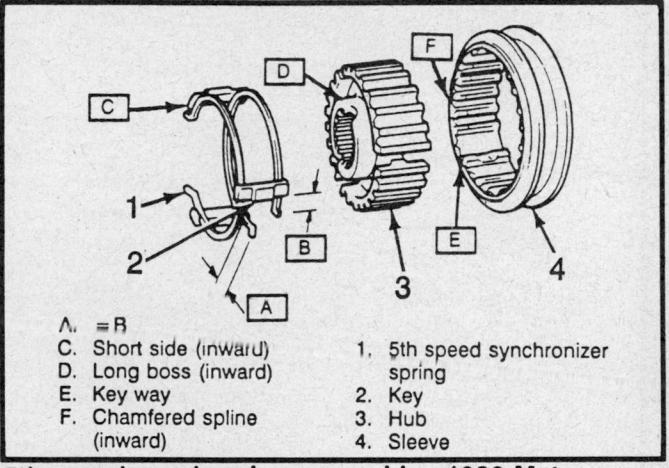

A. = B
C. Short side (inward)
D. Long boss (inward)
E. Key way
F. Chamfered spline (inward)

1. 5th speed synchronizer spring
2. Key
3. Hub
4. Sleeve

5th speed synchronizer assembly—1989 Metro

31. Clean the mating surface of the left case and put the gear shift guide case applied with sealant on it.

32. Install the wiring harness clamp and after applying sealant No. 1052751 or equivalent, to the bracket and fasten it together with the gear shift guide case. Torque the bolts to 7 ft. lbs. (10 Nm).

33. Install the backup light switch and clamp its lead. Torque it to 15 ft. lbs. (20 Nm).

34. Clean the left case cap and mating surface of the left case, check the condition of the ring and then fasten the cap with the 3 bolts. Torque bolts to 7 ft. lbs. (10 Nm).

35. Inspect the input shaft for rotation in each gear position. Check the function of the backup light switch in **R** position by using an ohmmeter.

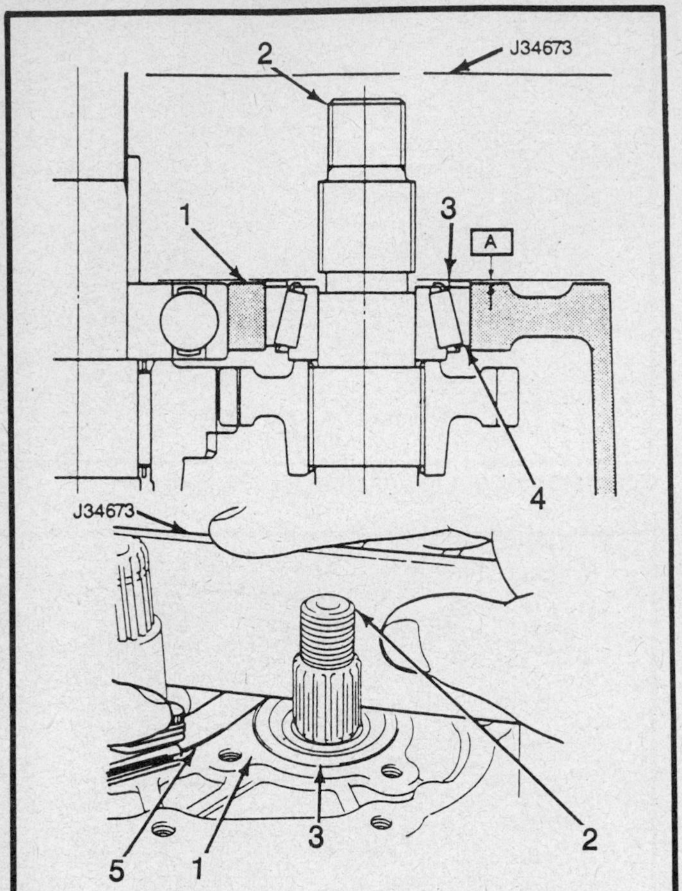
J34673
2
1
3
A
4
J34673
5
1
3
2

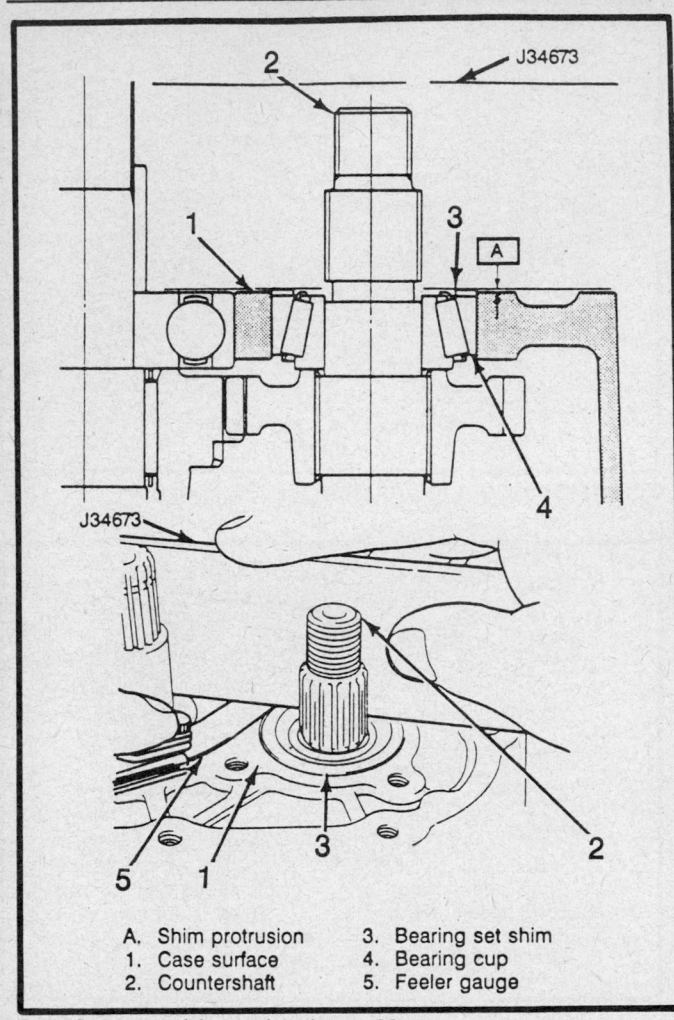

A. Shim protrusion
1. Case surface
2. Countershaft
3. Bearing set shim
4. Bearing cup
5. Feeler gauge

Bearing set shim selection—Metro

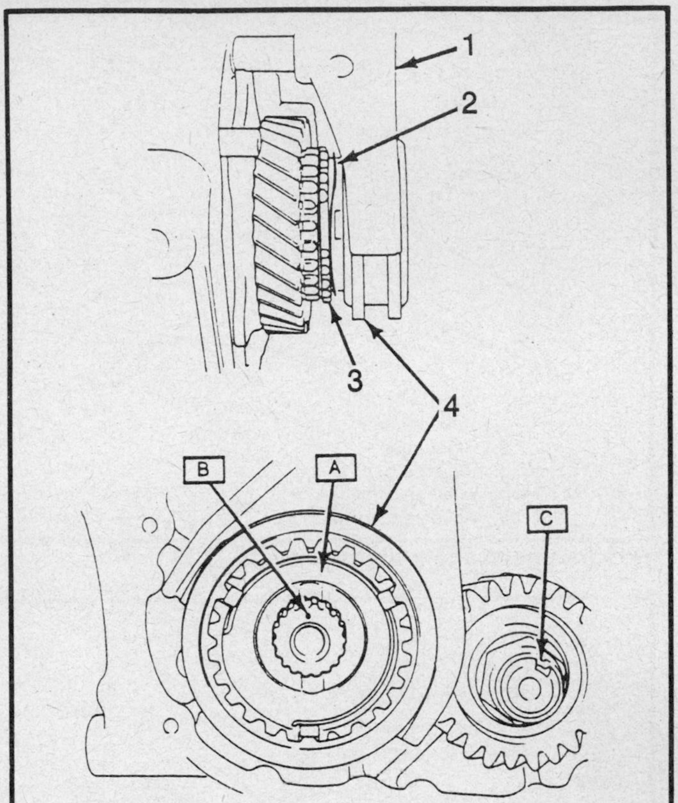
1
2
3
4
B
A
C

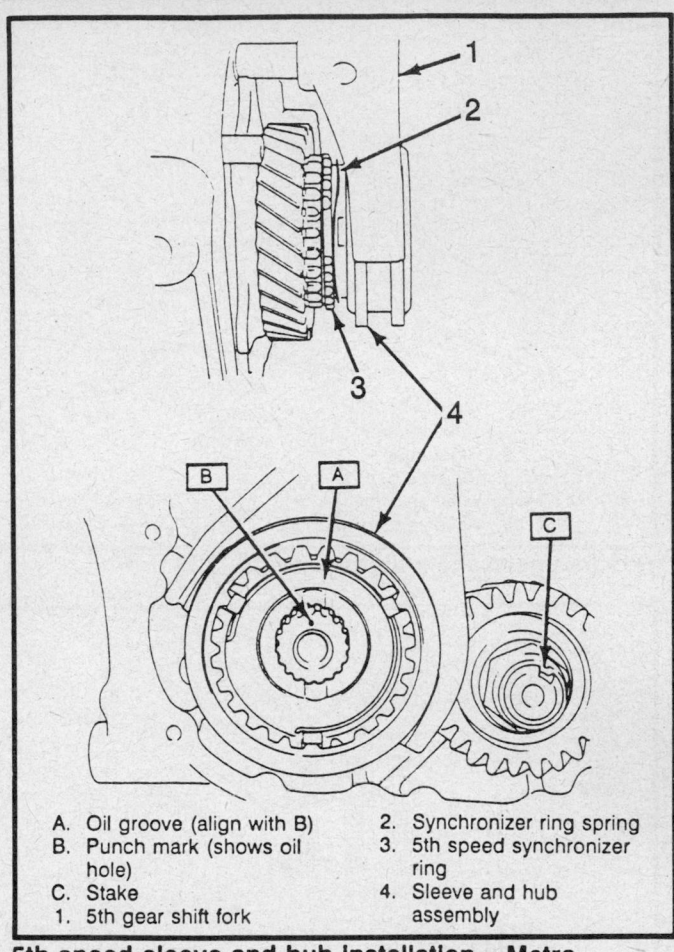

A. Oil groove (align with B)
B. Punch mark (shows oil hole)
C. Stake
1. 5th gear shift fork
2. Synchronizer ring spring
3. 5th speed synchronizer ring
4. Sleeve and hub assembly

5th speed sleeve and hub installation—Metro

SPECIFICATIONS

TORQUE SPECIFICATIONS

Item	ft. lbs.	Nm
Oil filler/level plug	3	45
Oil drain plug	21	28
Ball stud bolt and nut	44	60
Gear shift control housing nuts for boot cover	4	5
Housing nuts for control housing	13	18
Guide plate bolts	7	9
Control housing nuts	24	33
Extension rod stud bolt	13	18
Extension rod nut	24	33
Gear shift control shaft bolts and nuts	13	18
Speedometer driven gear case bolt	4	6
Transaxle to engine bolts and nuts	37	50

Item	ft. lbs.	Nm
Engine mounting nuts	37	50
Engine mounting left hand bracket bolts	37	50
Exhaust pipe to manifold bolts	37	50
Exhaust pipe to muffler nuts	37	50
Extension rod nut	24	33
Reverse gear shift lever bolts	17	23
Gear shift arm bolts	17	23
Oil gutter bolt	4	6
Final gear bolts	63	85
Transaxle case bolts	14	19
Reverse shaft bolt	14	19
Gear shift fork shaft bolts	10	13

TORQUE SPECIFICATIONS

Item	ft. lbs.	Nm
Left case plate screws	5	7
Counter shaft nut	52	70
5th shift fork plug	10	13
Left cover bolts	7	10
Gear shift case plate bolt	4	6
5th to reverse interlock guide case bolts	17	23
Gear shift yoke bolt	17	23
Gear shift interlock bolt	17	23
Gear shift guide case bolts	7	10
Left case cap bolts	7	10
Backup light switch	15	20

SPECIAL TOOLS

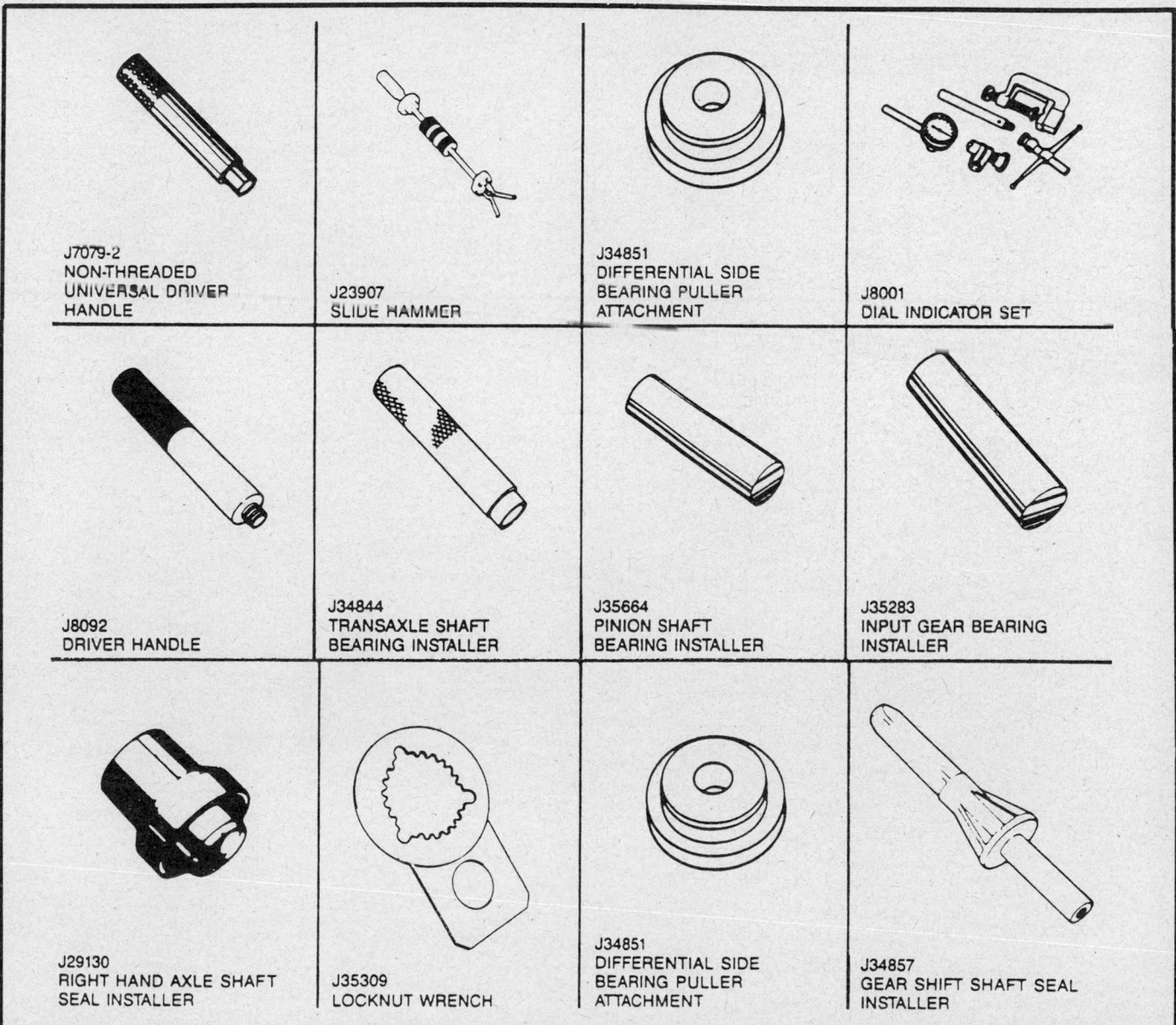

J7079-2
NON-THREADED
UNIVERSAL DRIVER
HANDLE

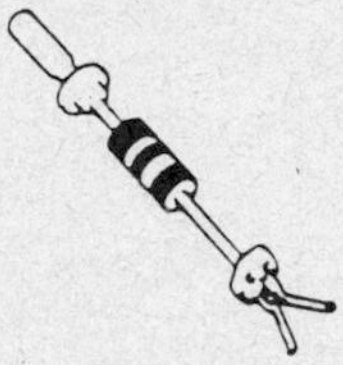
J23907
SLIDE HAMMER

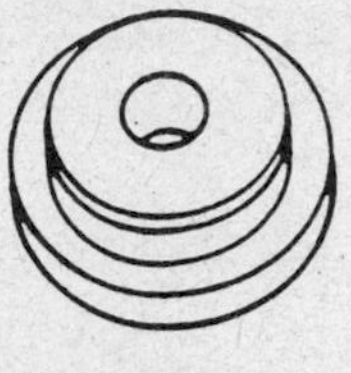
J34851
DIFFERENTIAL SIDE
BEARING PULLER
ATTACHMENT

J8001
DIAL INDICATOR SET

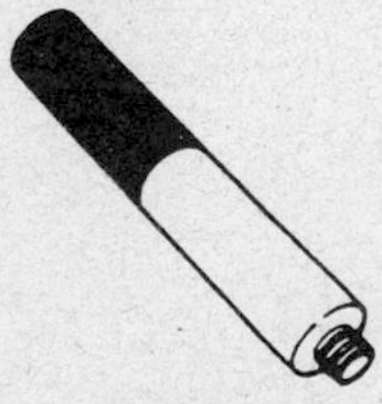
J8092
DRIVER HANDLE

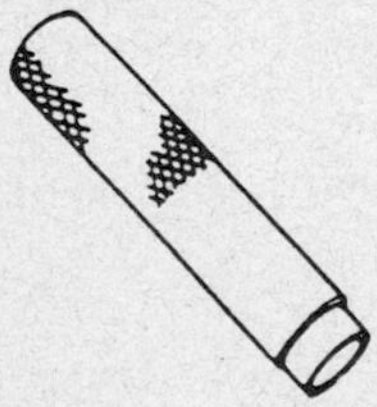
J34844
TRANSAXLE SHAFT
BEARING INSTALLER

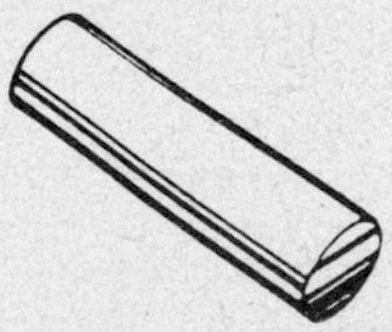
J35664
PINION SHAFT
BEARING INSTALLER

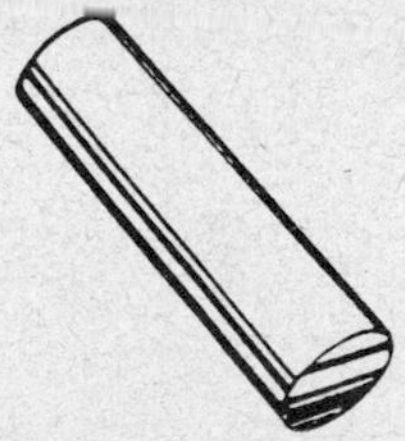
J35283
INPUT GEAR BEARING
INSTALLER

J29130
RIGHT HAND AXLE SHAFT
SEAL INSTALLER

J35309
LOCKNUT WRENCH

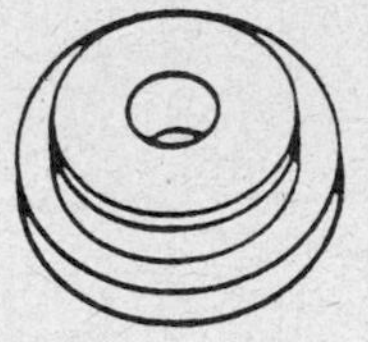
J34851
DIFFERENTIAL SIDE
BEARING PULLER
ATTACHMENT

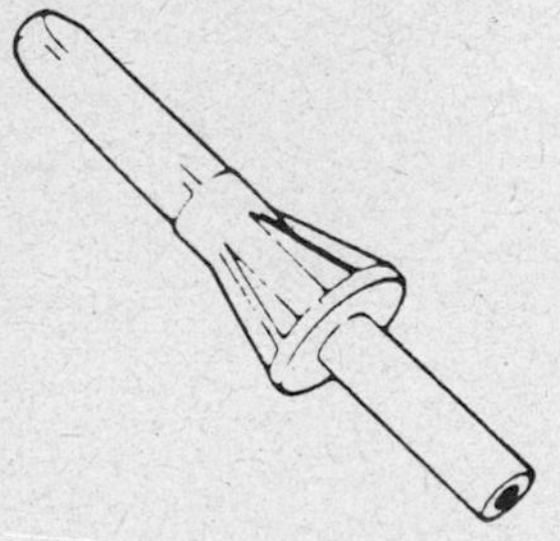
J34857
GEAR SHIFT SHAFT SEAL
INSTALLER

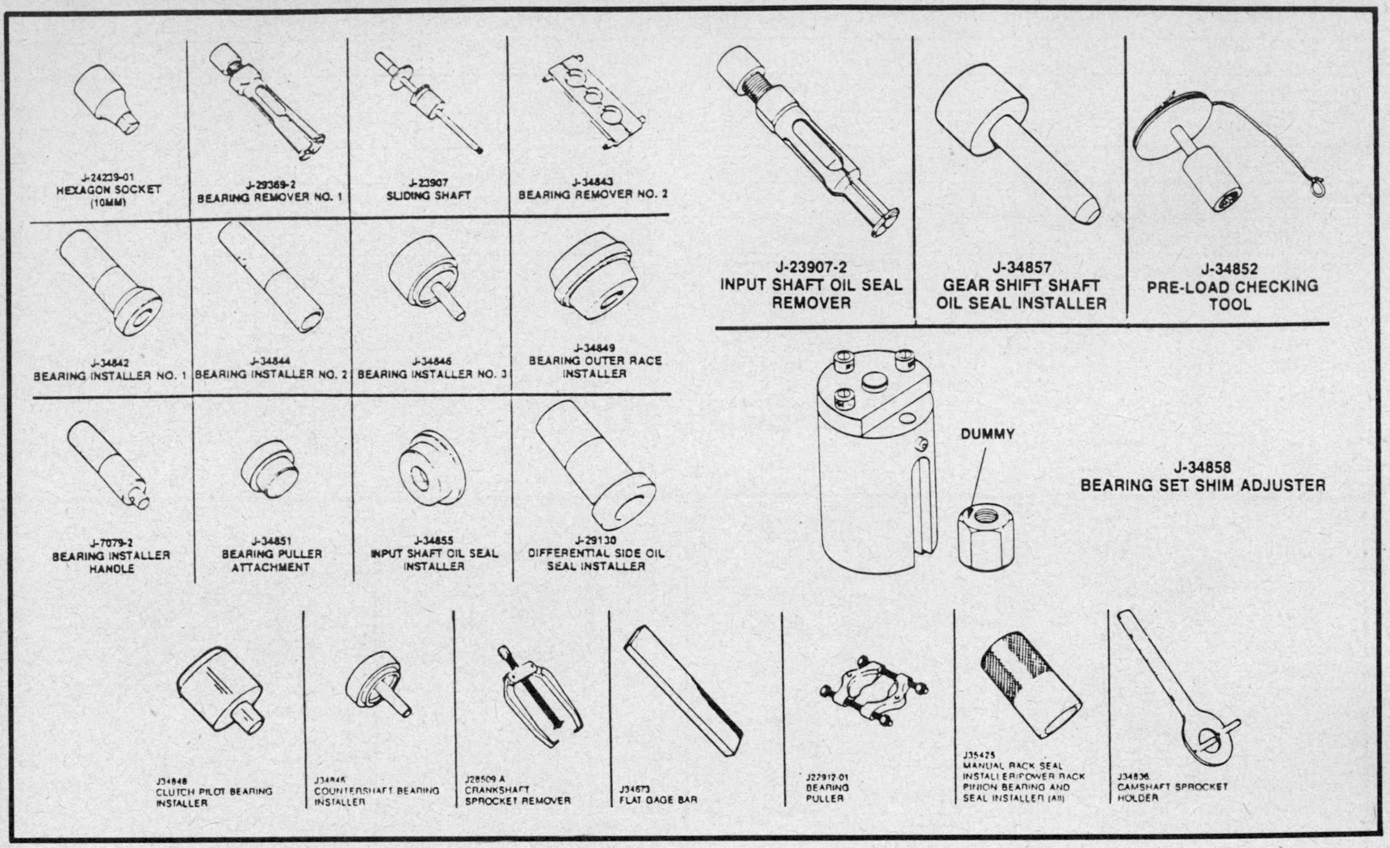
J-24239-01
HEXAGON SOCKET
(10MM)
J-29369-2
BEARING REMOVER NO. 1
J-23907
SLIDING SHAFT
J-34843
BEARING REMOVER NO. 2
J-23907-2
INPUT SHAFT OIL SEAL
REMOVER
J-34857
GEAR SHIFT SHAFT
OIL SEAL INSTALLER
J-34852
PRE-LOAD CHECKING
TOOL
J-34842
BEARING INSTALLER NO. 1
J-34844
BEARING INSTALLER NO. 2
J-34846
BEARING INSTALLER NO. 3
J-34849
BEARING OUTER RACE
INSTALLER
J-7079-2
BEARING INSTALLER
HANDLE
J-34851
BEARING PULLER
ATTACHMENT
J-34855
INPUT SHAFT OIL SEAL
INSTALLER
J-29130
DIFFERENTIAL SIDE OIL
SEAL INSTALLER
DUMMY
J-34858
BEARING SET SHIM ADJUSTER
J34848
CLUTCH PILOT BEARING
INSTALLER
J34846
COUNTERSHAFT BEARING
INSTALLER
J28509 A
CRANKSHAFT
SPROCKET REMOVER
J34673
FLAT GAGE BAR
J27912-01
BEARING
PULLER
J35425
MANUAL RACK SEAL
INSTALLER/POWER RACK
PINION BEARING AND
SEAL INSTALLER (All)
J34836
CAMSHAFT SPROCKET
HOLDER

Section 6

Isuzu 5 Speed Transaxle
General Motors

APPLICATION

Year	Vehicle	Engine
	MR8 TRANSAXLE	
1985	Spectrum	1.5L 2BBL
1986	Spectrum	1.5L
	Spectrum	1.5L Turbo
1987	Spectrum	1.5L
	Spectrum	1.5L Turbo
1988	Spectrum	1.5L
	Spectrum	1.5L Turbo
1989	Spectrum	1.5L
	Spectrum	1.5L Turbo
	MK7 TRANSAXLE	
1984	Skyhawk	1.8L TBI
	Sunbird	1.8L TBI
	Firenza	1.8L TBI
1985	Skyhawk	1.8L TBI
	Sunbird	1.8L TBI
	Firenza	1.8L TBI
1986	Skyhawk	1.8L TBI
	Sunbird	1.8L TBI
	Firenza	1.8L TBI
1987	Skyhawk	2.0L TBI
	Sunbird	2.0L TBI
	Firenza	2.0L TBI
1988	Skyhawk	2.0L TBI
	Sunbird	2.0L TBI
	Firenza	2.0L TBI
1989	Skyhawk	2.0L TBI
	Sunbird	2.0L TBI
	Cavalier	2.0L TBI
	MY7 TRANSAXLE	
1984	Skyhawk	1.8L TBI
	Skyhawk	2.0L TBI
	Skyhawk	2.0L 2 BBL
	Sunbird	1.8L TBI
	Sunbird	2.0L TBI
	Sunbird	2.0L 2BBL
	Cimarron	2.0L 2 BBL
	Cimarron	2.0L TBI

Year	Vehicle	Engine
	MR8 TRANSAXLE	
1985	Sunbird	1.8L TBI
1986	Sunbird	1.8L TBI
	MT2 TRANSAXLE	
1985	Grand Am	2.5L TBI
	Fiero	2.5L TBI
1986	Calais	2.5L TBI
	Grand Am	3.0L EFI
	Somerset	2.5L TBI
	Fiero	2.5L TBI
1987	Calais	2.5L TBI
	Grand Am	3.0L EFI
	Somerset	2.5L TBI
	Fiero	2.5L TBI
1988	Fiero	2.5L TBI
1989	Calais	2.5L TBI
	Grand Am	2.5L TBI
	MR3 TRANSAXLE	
1984	Cavalier	2.0L TBI
	Cavalier	2.0L 2BBL
	Firenza	1.8L TBI
	Firenza	2.0L TBI
	Firenza	2.0L 2BBL
1985	Cavalier	2.0L TBI
1986	Cavalier	2.0L TBI
	Cimarron	2.0L SFI
1987	Cavalier	2.0L TBI
	Beretta	2.0L EFI
	Beretta	2.8L MPI
	Corsica	2.0L EFI
	Corsica	2.8L MPI
1989	Beretta	2.0L EFI
	Corsica	2.0L EFI

BBL Barrel
EFI Electronic Fuel Injection
MPI Multi-Port Fuel Injection
TBI Throttle Body Injection

GENERAL DESCRIPTION

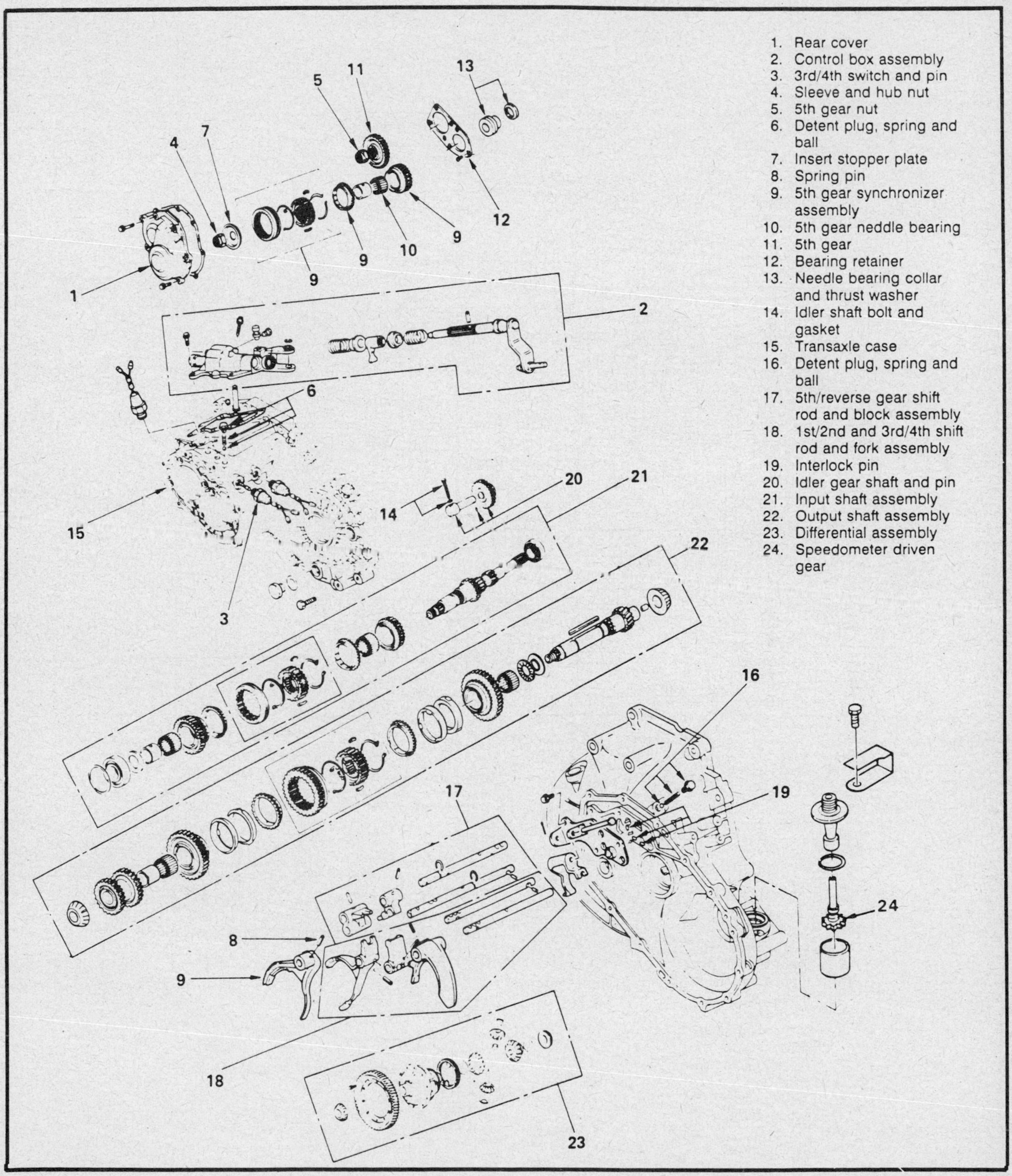
1. Rear cover
2. Control box assembly
3. 3rd/4th switch and pin
4. Sleeve and hub nut
5. 5th gear nut
6. Detent plug, spring and ball
7. Insert stopper plate
8. Spring pin
9. 5th gear synchronizer assembly
10. 5th gear neddle bearing
11. 5th gear
12. Bearing retainer
13. Needle bearing collar and thrust washer
14. Idler shaft bolt and gasket
15. Transaxle case
16. Detent plug, spring and ball
17. 5th/reverse gear shift rod and block assembly
18. 1st/2nd and 3rd/4th shift rod and fork assembly
19. Interlock pin
20. Idler gear shaft and pin
21. Input shaft assembly
22. Output shaft assembly
23. Differential assembly
24. Speedometer driven gear

Exploded view of the 5 speed transaxle

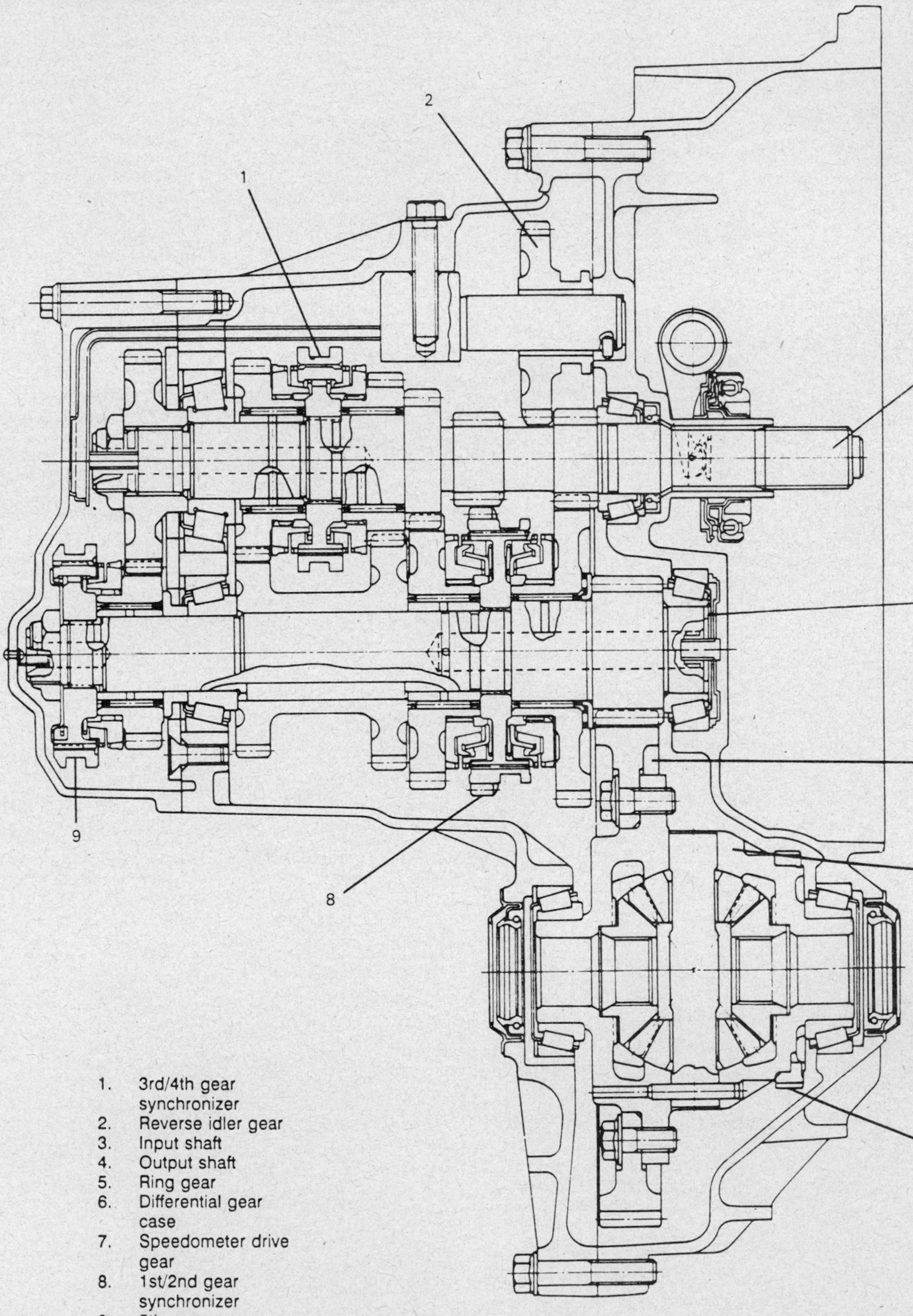
1
2
3
4
5
6
7
8
9

1. 3rd/4th gear synchronizer
2. Reverse idler gear
3. Input shaft
4. Output shaft
5. Ring gear
6. Differential gear case
7. Speedometer drive gear
8. 1st/2nd gear synchronizer
9. 5th gear synchronizer

5 speed transaxle cross section view

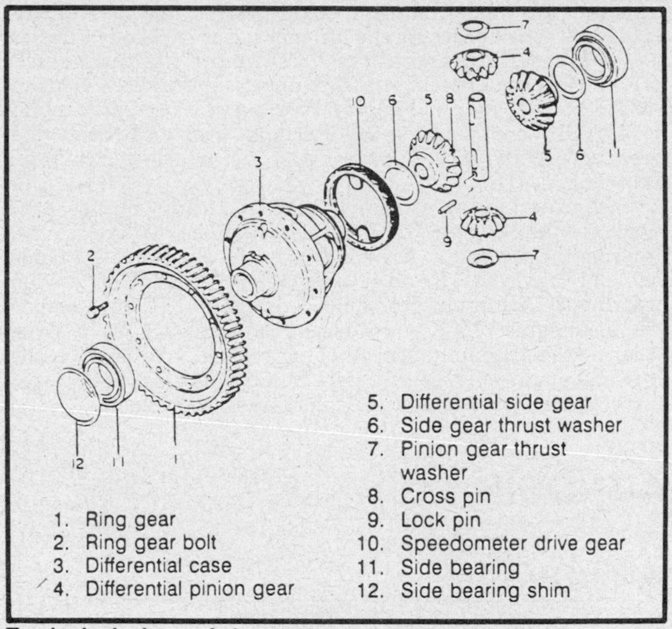
10
6
5
8
7
4
5
6
11
3
9
4
7
2
12
11
1
1. Ring gear
2. Ring gear bolt
3. Differential case
4. Differential pinion gear
5. Differential side gear
6. Side gear thrust washer
7. Pinion gear thrust washer
8. Cross pin
9. Lock pin
10. Speedometer drive gear
11. Side bearing
12. Side bearing shim

Exploded view of the differential assembly

The 5 speed transaxle assembly is a constant mesh design transaxle, combined with a different unit and assembled into a case with a rear cover. All forward gears are in constant mesh. For ease of shifting and selection of the desired gear range, synchronizers with blocker rings, controlled by shift forks are used. Reverse uses a sliding idler gear arrangement.

The components of the unit are the aluminum transaxle case, aluminum clutch housing, aluminum rear cover, input gear shaft, output gear shaft and the differential assembly. The input gear, output gear and differentail are all supported by tapered roller bearings. Selective shims are used behind the rear bearing outer races to establish the correct preload. The final output gear an integral part of the output shaft turns the ring gear and differential assembly, thereby turning the drive axle shaft which then transfers power to the front wheels.

The manual transaxle utilizes a cable to control the clutch. The clutch pedal and control system incorporate a constant no lash feature. There are 2 cable assemblies utilized to shift the gears. One is the transaxle selector cable and the other a transaxle shifter cable.

Transaxle Identification

The transaxle serial number is printed on the transmission bar code label, which is located on the transaxle clutch housing.

Metric Fasteners

The metric fastener dimensions are very close to the dimensions of the familiar inch system fasteners and for this reason, replacement fasteners must have the same measurement and strength as those removed.

Do not attempt to interchange metric fasteners for inch system fasteners. Mismatched or incorrect fasteners can result in damage to the transaxle unit through malfunctions, breakage or possible personal injury.

Care should be taken to reuse the fasteners in the same locations as removed.

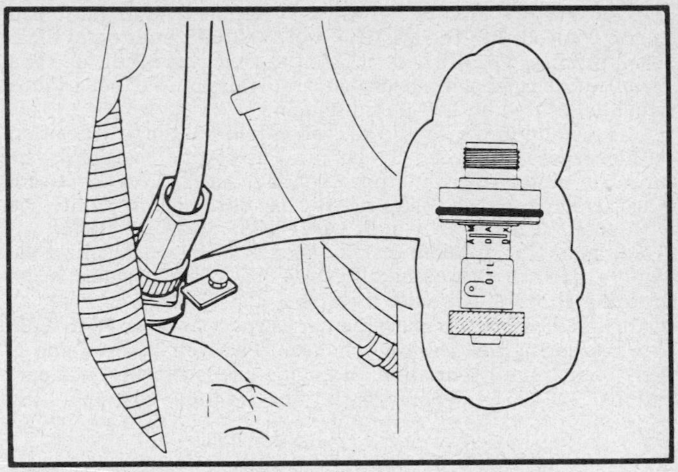

Fluid level gauge without dipstick assembly

Capacities

To fill a dry transaxle, fill the unit with approximately 3.0 quarts of 5W-30 manual transaxle oil, part number 1052931 or equivalent. On 1988–89 vehicles, use Syncromesh transaxle fluid, part number 12345349 or equivalent.

Checking Fluid Level

VEHICLES EQUIPPED WITH DIPSTICK ASSEMBLY

Check the fluid level only when the engine is **OFF**, the vehicle is level and the transaxle is cool. To check the fluid level, remove the dipstick and read the level indicated. Be sure the fluid level is between the **H** (hot) and **C** (cold) marks on the dipstick.

VEHICLES EQUIPPED WITHOUT DIPSTICK ASSEMBLY

Check the fluid level only when the engine is **OFF**, the vehicle is level and the transaxle is cool. To check, remove the speedometer driven gear assembly on the driver's side of the transaxle case, above the halfshaft. Make sure that the fluid level is between the **L** and **H** marks on the speedometer driven gear bushing. If necessary, add the specified fluid. Reinstall the speedometer driven gear assembly, making sure the gear is fully seated.

TRANSAXLE MODIFICATIONS

Shift Lever Rattle/Buzz Noise In 3rd Or 4th Gear

NOTE: Engine VIN 0 pertains to 1.8L engine. Engine VIN U and R pertain to 2.5L engine and other VIN numbers pertain to 2.0L engines.

On some 1985–88 Oldsmobile Firenza and 1985–89 Oldsmobile Calais equipped with Isuzu 76mm 5 speed, RPO codes MK7 and MT2 may exhibit a rattle or buzzing type noise through the shifter handle while driving in 3rd or 4th gears. This noise condition is normally caused by runout condition between the 3rd and 4th input gears.

This condition has been corrected during the 1989 model

year. Therefore any 1989 vehicles equipped with an Isuzu 76mm 5 speed, RPO codes MK7 and MT2 unit built on or after serial number 1X17J01, should not exhibit this condition. The serial number is printed on the transmission bar code label, which is located on the clutch housing.

The serial number should be read as follows, the first position is the corporation name (I = Isuzu). The second position is the month of built (1–9 = Jan. thru Sept. X = Oct. Y = Nov. Z = Dec.). The third and fourth position is the day of the month built. The fifth position is the year built (A–I = 1989 and J–Z = 1988).

To repair this problem install a new 3rd/4th synchronizer assembly, 3rd/4th fork assembly and a 3rd/4th input gear. When ordering the 3rd/4th shift fork on J and N body vehicles, all years and all engine displacements, use part number 94464752. When ordering the 3rd/4th synchronizer assembly on J and N body vehicles, all years and all engine displacements, use part number 94249573. When ordering the 3rd input gear on J body vehicles, 1988 model year and engine VIN 1, use part number 94149668. When ordering the 3rd input gear on J body vehicles, 1985–87 model year and engine VIN P and 1, use part number 94139574. When ordering the 3rd input gear on J body vehicles, 1985–88 model year and engine VIN K and 0 use part number 94149668. When ordering the 3rd input gear on N body vehicles, 1985–89 model year and engine VIN U, use part number 94149668.

When ordering the 4th input gear on J body vehicles, 1988 model year and engine VIN 1, use part number 94139575. When ordering the 4th input gear on J body vehicles, 1985–87 model year and engine VIN P and 1, use part number 94139576. When ordering the 4th input gear on J body vehicles, 1985–88 model year and engine VIN K and 0 use part number 94139575. When ordering the 4th input gear on N body vehicles, 1985–89 model year and engine VIN U, use part number 94139575. Parts are currently available from the dealership of the vehicle.

TROUBLE DIAGNOSIS

CHILTON THREE C's DIAGNOSIS

Condition	Cause	Correction
A knock at low speeds	a) Drive axle CV or TRI-POT joints worn b) Side gear hub counterbore worn	a) Replace CV or joints b) Replace hub counterbore
Noise most pronounced on turns	a) Differential gear noise	a) Check and repair or replace gear
Clunk on acceleration or deceleration	a) Engine mounts loose b) Drive axle inboard TRI-POT joints worn c) Differential pinion shaft in case worn d) Side gear hub counterbore in case worn oversize	a) Replace mounts b) Replace joints c) Replace shaft d) Repair or replace hub counterbore
Vibration	a) Wheel bearing rough b) Drive axle shaft bent c) Tires out-of-round d) Tire unbalance e) CV joint in drive axle shaft worn f) Drive axle angle (trim height) incorrect	a) Replace bearing b) Replace shaft c) Replace tire d) Balance tire e) Replace CV f) Adjust angle
Noisy in neutral with engine running	a) Input gear bearings worn b) Clutch release bearing worn	a) Replace bearings b) Replace bearing
Noisy in 1st only	a) 1st speed constant-mesh gears chipped, scored or worn b) 1–2 synchronizer worn	a) Replace gears b) Replace synchronizer
Noisy in 2nd only	a) 2nd speed constant-mesh gears chipped, scored or worn b) 1–2 synchronizer worn	a) Replace gears b) Replace synchronizer
Noisy in 3rd only	a) 3rd gear constant-mesh gears chipped, scored or worn b) 3–4 synchronizer worn	a) Replace gears b) Replace synchronizer
Noisy in 4th only	a) 4th gear or output gear chipped, scored or worn b) 3–4 synchronizer worn	a) Replace gears b) Replace synchronizer
Noisy in 5th only	a) 5th speed gear or output gears chipped, scored or worn b) 5th gear synchronizer worn	a) Replace gears b) Replace synchronizer
Noisy in reverse only	a) Reverse idler gear, idler gear bushing, input or output gear(s) chipped, scored or worn	a) Replace gear(s) and/or bushings

CHILTON THREE "C's" DIAGNOSIS

Condition	Cause	Correction
Noisy in all gears	a) Insufficient lubricant b) Bearings worn c) Input gear (shaft) and/or output gear (shaft) worn, chipped or scored	a) Add lubricant b) Replace bearings c) Replace gear(s)
Slips out of gear	a) Linkage worn or improperly adjusted b) Shift linkage does not work freely; binds c) Cables bent or worn d) Input gear bearing retainer broken or loose e) Shift fork worn or bent	a) Replace and/or adjust linkage b) Repair or replace linkage c) Replace cables d) Repair or replace retainer e) Replace shift fork
Leaks lubricant	a) Fluid level indicator not seated in fill port causing fluid leakage at vent plug b) Axle shaft seals worn c) Excessive amount of lubricant in transaxle d) Input gear (shaft) bearing retainer loose or broken e) Input gear bearing worn and/or lip seal damaged f) Worn shift lever seal leaks g) Lack of sealant between case and clutch cover or loose clutch cover	a) Seat fluid level indicator b) Replace seals c) Remove excessive fluid d) Repair or replace retainer e) Replace bearing and/or lip seal f) Replace seal g) Repair case and/or cover

ON CAR SERVICES

Adjustments

SHIFT AND SELECT CABLE

EXCEPT SPECTRUM AND FIERO

1. Disconnect negative battery cable.
2. Shift transaxle into **3rd** gear. Remove the lock pin (H) and reinstall with the tapered end down. This will lock the transaxle in 3rd gear.
3. Loosen the shift cable attaching nuts (E) at transaxle levers (G and F).

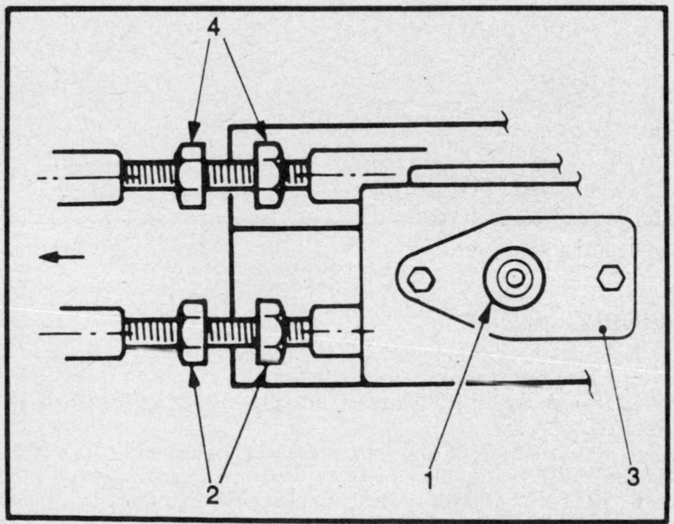
4
2
1
3

Shift linkage adjustment—Spectrum

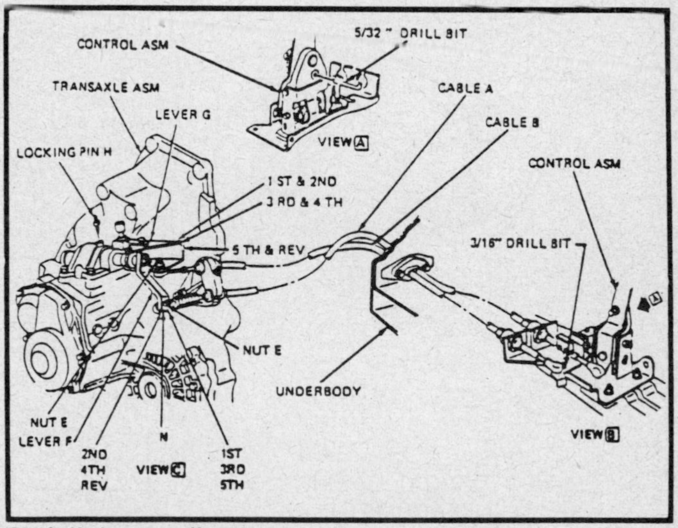
CONTROL ASM
5/32" DRILL BIT
TRANSAXLE ASM
CABLE A
LEVER G
CABLE B
VIEW A
LOCKING PIN H
CONTROL ASM
1 ST & 2ND
3 RD & 4 TH
5 TH & REV
3/16" DRILL BIT
NUT E
UNDERBODY
NUT E
LEVER F
N
2ND
4TH
REV
VIEW C
1ST
3RD
5TH
VIEW B

Shift linkage adjustment—J, L and N body

4. Remove the shift console trim plate and slide shifter boot up shifter handle. Remove the console.
5. Install a $^5/_{32}$ in. or No. 22 drill bit into the alignment hole at side of shifter assembly.
6. Align the hole in the select lever with the slot in the shifter plate and install a $^3/_{16}$ in. drill bit.
7. Tighten nuts (E) at levers (G and F). Remove drill bits from alignment holes at the shifter. Remove lockpin (H) and reinstall with tapered end up.
8. Install console, shifter boot and trim plate.
9. Connect the negative battery cable. Road test the vehicle to check for a good neutral gate feel during shifting.

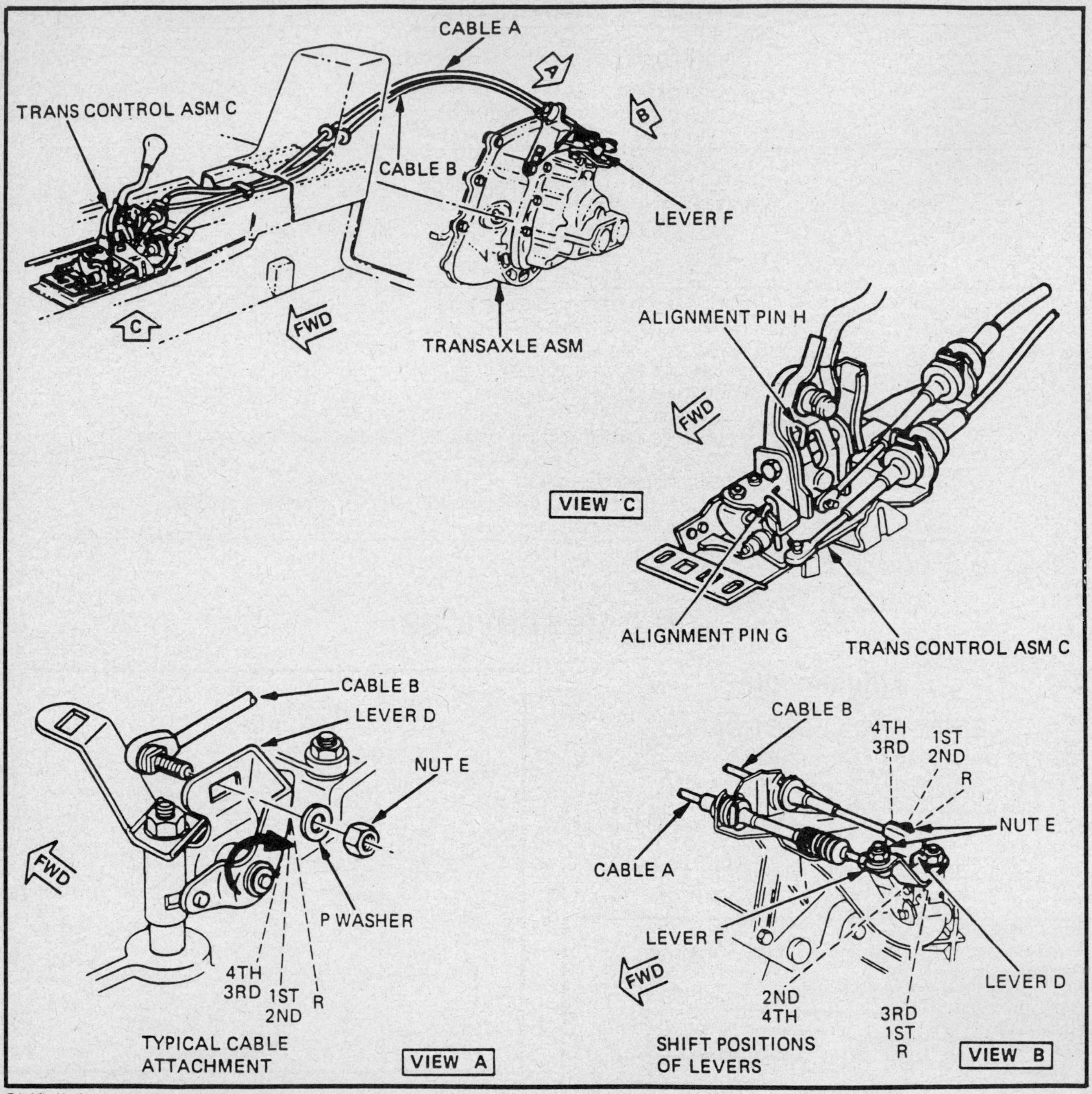
CABLE A
TRANS CONTROL ASM C
CABLE B
LEVER F
TRANSAXLE ASM
FWD
ALIGNMENT PIN H
VIEW C
ALIGNMENT PIN G
TRANS CONTROL ASM C
CABLE B
LEVER D
NUT E
P WASHER
4TH
3RD
1ST
2ND
R
TYPICAL CABLE ATTACHMENT
VIEW A
CABLE B
4TH
3RD
1ST
2ND
R
NUT E
CABLE A
LEVER F
LEVER D
2ND
4TH
3RD
1ST
R
SHIFT POSITIONS OF LEVERS
VIEW B
Shift linkage adjustment—Fiero

SPECTRUM

1. Place the transaxle in the **N** position.
2. Turn lower cable adjusting nuts (2) until the change lever (1) is at a right angle to the pivot case, as viewed from the side of the gear control.
3. Turn upper adjusting nuts (4) until the change lever (1) is at a right angle to the pivot case, as viewed from the rear of the gear control.
4. After correct adjustment, tighten upper (4) and lower (2) adjusting nuts securely.
5. Road test the vehicle for correct shifter operation.

FIERO

1. Disconnect the negative battery cable.
2. Place the transaxle in **1st** gear.
3. Loosen the shift cable attaching nuts (E) at transaxle levers (D) and (F).
4. Remove console and trim plates as required to gain access to the shifter assembly.
5. With shifter lever in the 1st gear position, pulled to the left and hold against stop, insert alignment pins in shifter at (H) and (G) positions.

6. Remove the play from the transaxle by rotating lever (D) in direction of arrow while tightening nut (E). Levers (D and F) should be kept from moving during this process. Similarly, tighten nut (E) on lever (F), levers (D and F) must remain stationary. The torque specification for the nut (E) on levers (D and F) is 20 ft. lbs.

7. Make sure that reverse inhibit cam is against roller and align if necessary.
8. Remove alignment pins at shifter assembly.
9. Replace console trim plate and console if removed.
10. Reconnect the negative battery cable. Road test the vehicle for correct shifter operation.

REMOVAL AND INSTALLATION

TRANSAXLE REMOVAL

J, L AND N BODY

NOTE: The transaxle is removed separately from the engine assembly.

1. Disconnect the negative battery cable.
2. Install the engine support fixture tool J-28467 or equivalent. Raise the engine enough to take pressure off the motor mounts.
3. Remove the left sound panel.
4. Remove the clutch master cylinder pushrod from the clutch pedal.
5. Remove clutch actuator cylinder from the transaxle support bracket annd position aside.
6. Remove the wiring harness (L Body vehicles) at the mount bracket.
7. Remove the transaxle mount attaching bolts and mount bracket attaching nuts/bolts.
8. Disconnect the shift cables and retaining clamp at the transaxle.
9. Remove the ground cables at the transaxle mounting studs. Disconnect the electrical connection for the shift light or back-up light as so equipped.
10. Raise the vehicle and support safely. Drain the transaxle fluid.
11. Remove the left front wheel assembly and the inner splash shield.
12. Remove the transaxle front strut and strut bracket.
13. Remove the transaxle clutch housing bolts.
14. Disconnect the speedometer cable or the sensor electrical connection (digital dash) at transaxle.
15. Disconnect stabilizer shaft at the left suspension support and control arm.
16. Disconnect the ball joint from the steering knuckle.
17. Remove the left suspension support attaching bolts and remove the support and control arm as an assembly.
18. Disconnect the driveshafts at the transaxle and remove the left driveshaft from the transaxle.
19. Attach a suitable transmission jack to the transaxle case.
20. Remove the remaining transaxle-to-engine bolts.
21. Remove the transaxle by sliding toward the drive side away from the engine. Carefully lower the jack, guiding the right driveshaft out of the transaxle.

FIERO

NOTE: The transaxle is removed separately from the engine. The deck lid assembly may have to be removed to gain additional access for the engine support fixture.

1. Disconnect the negative battery cable. Remove the drain plug and drain the transaxle.
2. Remove select and shift cable nuts securing the cables to the transaxle brackets.
3. Remove the backup light switch connector and switch.
4. Remove the shift cables and the nut on the stud securing the bracket to the transaxle.
5. Remove the 2 bolts securing the select cable mount.
6. Remove the 2 bolts attaching the clutch slave cylinder bracket.
7. Remove the exhaust crossover pipe.

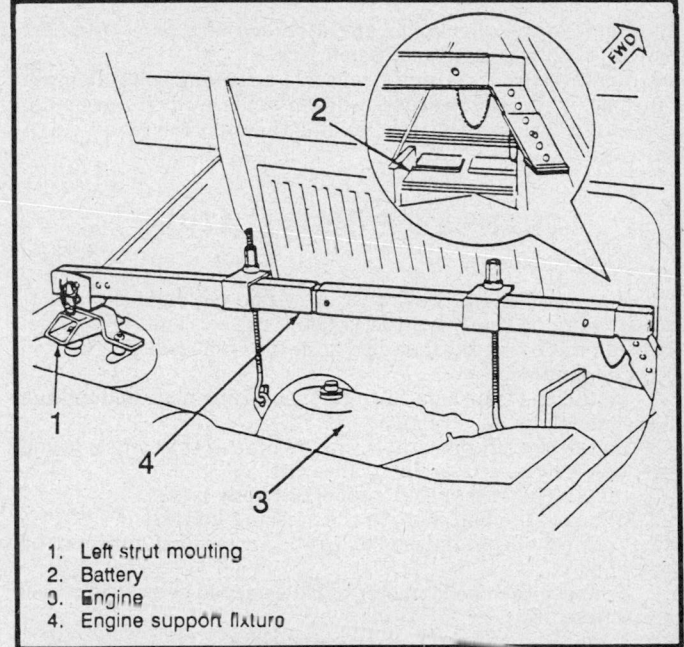

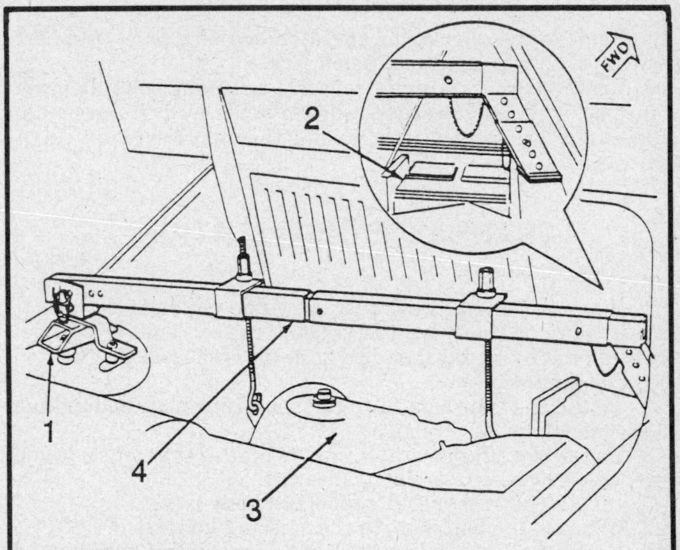
FWD
1
2
3
4
1. Left strut mouting
2. Battery
3. Engine
4. Engine support fixture

Typical engine support fixture installed—Fiero

8. Remove the nut, clip and wire from the center stud.
9. Remove the 3 upper bolts and 1 stud attaching the transaxle to the engine.
10. Install J-28467, J-35563, J-28467-60 (GT only) or equivalent, engine support fixtures. Attach the fixture hook to the engine lift ring and raise the engine enough to take the pressure off the mounts.
11. Remove the front and rear transaxle mounts.
12. Raise and support the vehicle safely.
13. Remove the 4 clutch inspection plate screws and the inspection plate.
14. Lower the frame and tilt.
15. Remove the driveshafts.
16. Remove the 2 nuts retaining the wire harness on the 2 lower studs.
17. Remove the 2 studs and the transaxle from the bottom of the vehicle, tilt the engine down for clearance removal.

SPECTRUM

NOTE: The transaxle is removed separately from the engine assembly.

1. Disconnect the negative cable from the battery.
2. Disconnect the negative cable retaining assembly from the transaxle case.
3. Disconnect the electrical connectors from the transaxle.
4. Disconnect the speedometer cable from the transaxle.
5. Dsconnect the clutch cable from the transaxle.
6. Disconnect the shift and select cables from the transaxle.
7. Remove air cleaner heat tube.
8. Remove the upper transaxle to engine attaching bolts.
9. Raise the vehicle and safely support.

10. Drain the transaxle oil.
11. Remove left front wheel assembly.
12. Remove splash shield.
13. Disconnect left side tie rod at the steering knuckle. Remove left side tension rod.
14. Disconnect the driveshafts from the transaxle. Remove the driveshafts from the transaxle case by pulling the shafts straight out.

NOTE: Use care when removing driveshafts from transaxle to prevent damage to oil seals.

15. Remove the dust cover at clutch housing. Safely support the transaxle with a transmission jack.
16. Remove the remaining transaxle attaching bolts. Remove transaxle by sliding towards the driver's side, away from engine.
17. Carefully lower the jack, guiding the right driveshaft out of the transaxle.

TRANSAXLE INSTALLATION

J, L AND N BODY

1. When installing the transaxle, guide the right driveshaft into its bore as the transaxle is being raised. The right driveshaft can not be readily installed after the transaxle is connected to the engine.
2. Install the transaxle-to-engine mounting bolts and tighten to specifications.
3. Install the left driveshaft into its bore at the transaxle and seat both driveshafts at the transaxle.
4. Install the suspension support-to-body bolts.
5. Connect the ball joint to the steering knuckle.
6. Connect the stabilizer bar to the suspension support and control arm.
7. Connect the speedometer cable or speed sensor electrical connection.
8. Install the clutch housing cover bolts.
9. Install the strut bracket to the transaxle.
10. Install the strut.
11. Install the inner splash shield.
12. Install the wheel assembly.
13. Lower the vehicle. Connect the ground cables at the transaxle mounting stud.
14. Install the transaxle mounting bracket. Connect the wiring harness (L Body vehicles) at the mount bracket.
15. Install the clutch slave cylinder to the transaxle bracket aligning the pushrod into the pocket of the clutch release lever. Install the retaining nuts and tighten evenly to prevent damage to the cylinder.
16. Install the transaxle mount and attaching bolt.
17. Connect the ball studs attaching cables to the transaxle levers. Connect the retaining clamp for shift cables to the transaxle.
18. Remove the engine support fixture. Reconnect the negative battery cable and install the left sound panel.
19. Fill transaxle with the specified fluid and to the correct level.
20. Road test the vehicle for proper operation.

FIERO

1. Install the transaxle through the bottom and attach the engine to transaxle with 2 studs. Do not torque the studs.
2. Raise the frame and install the driveshafts.
3. Install the clutch inspection plate and the screws and tighten to 116 inch lbs. (13 Nm).
4. Install the front and rear transaxle mounts.
5. Lower the vehicle.
6. Lower the engine and remove the engine support fixture.
7. Install the 3 upper bolts and 1 stud and tighten to 55 ft. lbs. (75 Nm). Also torque the 2 lower studs.
8. Install the wire harnesses on the 2 lower and center studs and 3 nuts. Torque to 152 inch lbs. (17 Nm).
9. Install the exhaust crossover pipe.
10. Install the 2 bolts attaching the clutch slave cylinder bracket and tighten to 27 ft. lbs. (50 Nm).
11. Install the 2 bolts retaining the select cable mount and tighten to 89 inch lbs. (10 Nm).
12. Install the select and shift cable to mount on the stud with the nut and tighten to 89 inch lbs. (10 Nm).
13. Install the select and shift cables nut and tighten to 223 inch lbs. (25 Nm).
14. Install the backup light switch and tighten to 25 ft. lbs. (34 Nm). Install the switch electrical connector.
15. Install the drain plug and tighten to 24 ft. lbs. (33 Nm).
16. Remove the screw and retainer from the speed sensor and fill the transaxle with the specified fluid and to the correct level. Install the retainer and screw and tighten to 45 inch lbs. (5 Nm).
17. Connect the negative battery cable and road test for proper operation.

SPECTRUM

1. When installing transaxle, guide the right driveshaft into the shaft bore as the transaxle is being raised. The right side driveshaft cannot be readily installed after the transaxle is connected to engine.
2. Install the transaxle to engine mounting bolts. Tighten bolts to 55 ft. lbs. (75 Nm).
3. Install left driveshaft into its bore at the transaxle. Seat the right and left driveshaft at transaxle.
4. Install left tension rod and torque bolts to 80 ft. lbs. (108 Nm).
5. Connect tie rod to the steering knuckle torque attaching nut to 42 ft. lbs. (57 Nm).
6. Install clutch housing dust cover bolts and install splash shield.
7. Install wheel assembly tighten wheel bolts.
8. Lower vehicle and install remaining transaxle to engine attaching bolts and tighten to 55 ft. lbs. (75 Nm).
9. Connect ground cable at transaxle. Reconnect the air cleaner heat tube.
10. Connect the clutch cable to the release lever and mounting bracket and adjust the clutch cable if necessary.
11. Connect negative cable at battery and fill the transaxle with the specified fluid and to the correct level.
12. Connect speedometer cable at the transaxle.
13. Adjust the shift linkage if necessary. Road test the vehicle for proper operation.

BENCH OVERHAUL

Before Disassembly

Cleanliness is an important factor in the overhaul of the transaxle. Before attempting any disassembly operation, the exterior of the transaxle should be thoroughly cleaned to prevent the possibility of dirt entering the transaxle internal mechanism. During inspection and reassembly, all parts should be thoroughly cleaned with cleaning fluid and then air dried. Wiping cloths or rags should not be used to dry parts. All oil passages should be blown out and checked to make sure that they are not obstructed. Small passages should be checked with tag wire. All parts should be inspected to determine which parts are to be re-

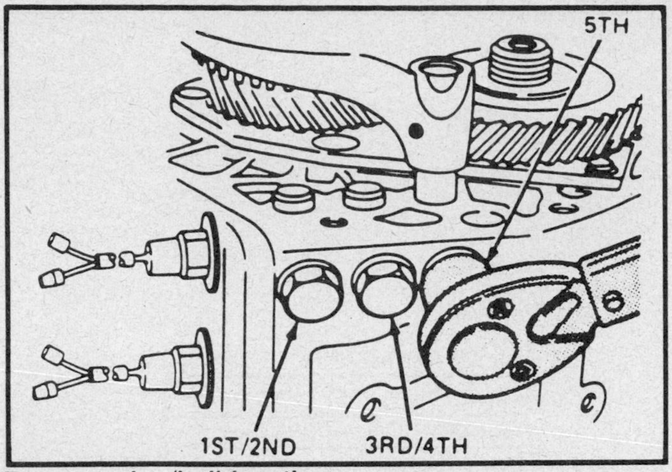
5TH
1ST/2ND
3RD/4TH
Detent spring/ball location

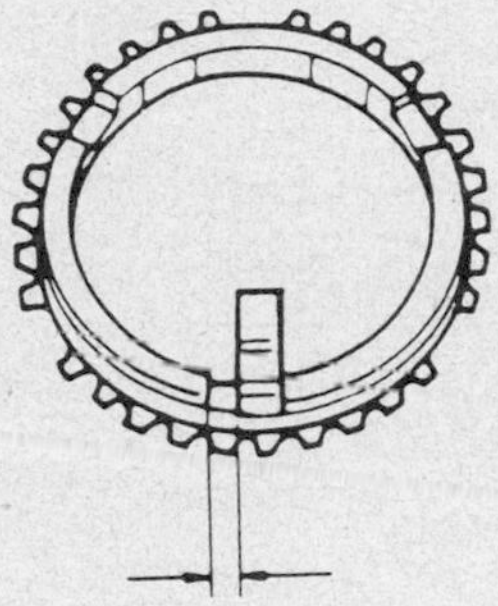

SPECIFICATIONS mm (in.)

1st 2nd GEAR	3.9 (0.154) OR LESS
3rd 4th AND 5th GEAR	3.7 (0.146) OR LESS

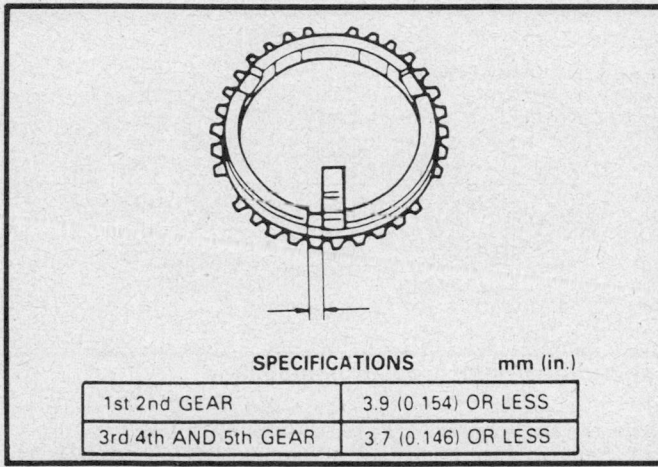

Measuring blocker ring to insert

placed. Try to kept all parts in order of there removal for correct installtion.

Transaxle Disassembly

1. Remove the clutch release bearing assembly from the transaxle. Attach the transaxle assembly to a suitable transaxle holding fixture.
2. Remove the rear cover assembly from the transaxle case.
3. Remove the control box assembly from the transaxle case.
4. Using a suitable tool, shift transaxle into gear. Remove the 5th speed drive and driven gear retaining nuts from the input and output shaft and discard the retaining nuts. Shift transaxle back into neutral, aligning the detents on the shift rails.
5. Remove the detent spring retaining bolts for 1st/2nd, 3rd/4th and 5th speeds and remove the detent springs and detent balls. Remove the reverse detent spring retaining bolts, spring and detent ball. Remove switches for 1st/2nd and 3rd/4th gears. Use a magnet and remove the 2 pins from the 3rd/4th gear switch hole, if so equipped with shift speed indicator light.
6. Place 5th speed synchronizer in neutral. Remove the roll pin at 5th gear shift fork and discard the roll pin. Remove 5th gear synchronizer hub, sleeve, roller bearing and gear with the shift fork as an assembly from the output shaft. Using a suitable gear puller, remove 5th speed gear from the input shaft.
7. Remove 7 bearing retainer screws with Torx® bit (No. 45) from the bearing retainer assembly.

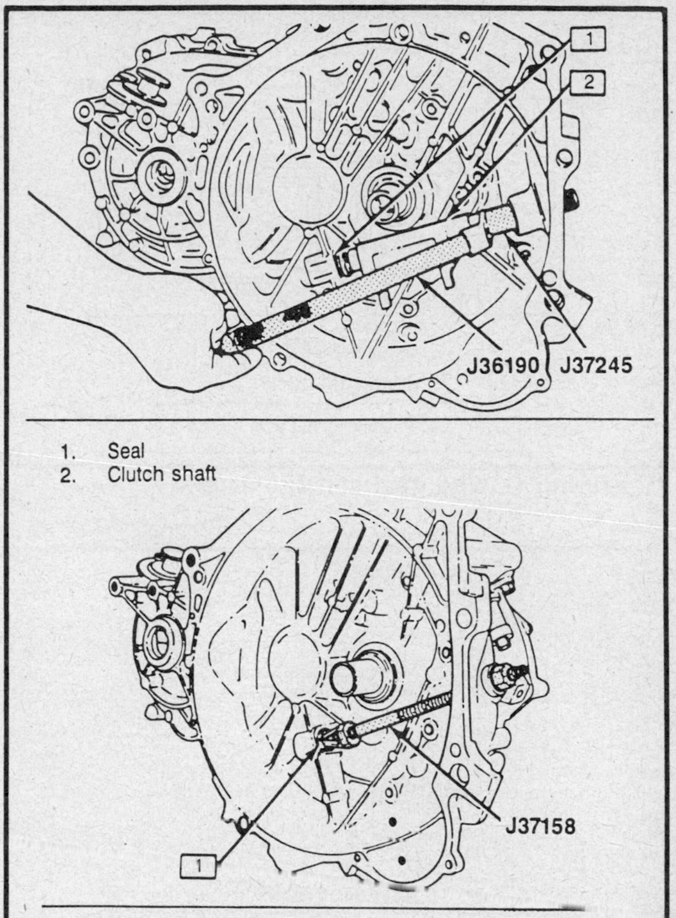
1
2
J36190
J37245
1. Seal
2. Clutch shaft
J37158
1
1. Needle bearing
Removing clutch shaft bushing and bearing

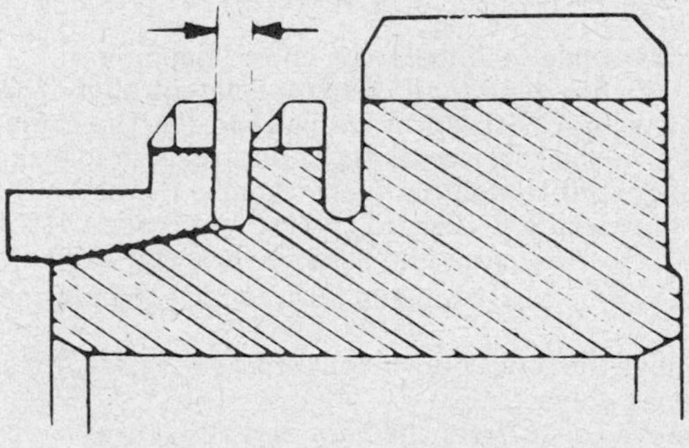

SPECIFICATIONS

MIN. LIMIT mm (in.)	0.8 (0.031)

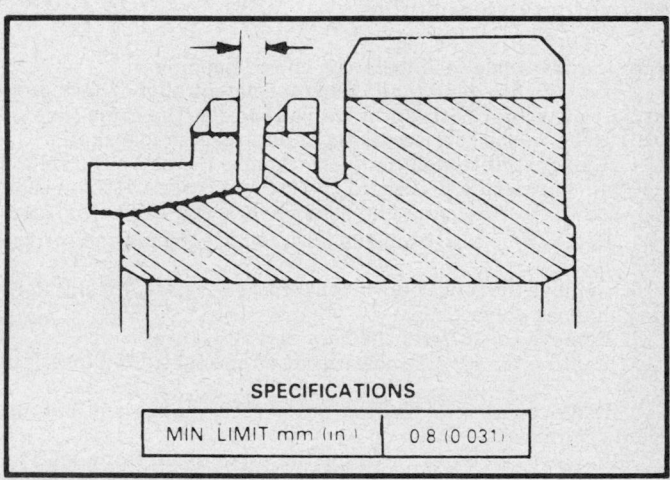

Measuring blocker ring to gear

8. Remove the bearing retainer and shims from the input and output shafts. Keep shims in order.
9. Remove the reverse idler retaining bolt used to retain the reverse idler shaft at the transaxle case.
10. Remove the collar and thrust washer from the output shaft using a suitable gear puller.
11. Remove the 14 bolts retaining the transaxle case and sepa-

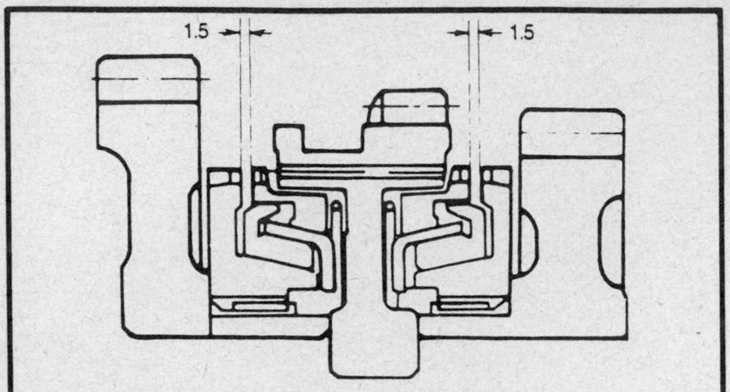
1.5
1.5

SPECIFICATIONS

1.5 mm (0.059 in.)

Measuring 1st/2nd gear synchronizer

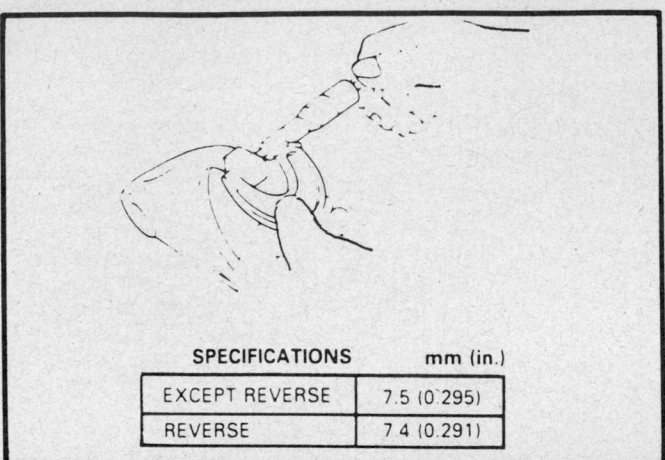

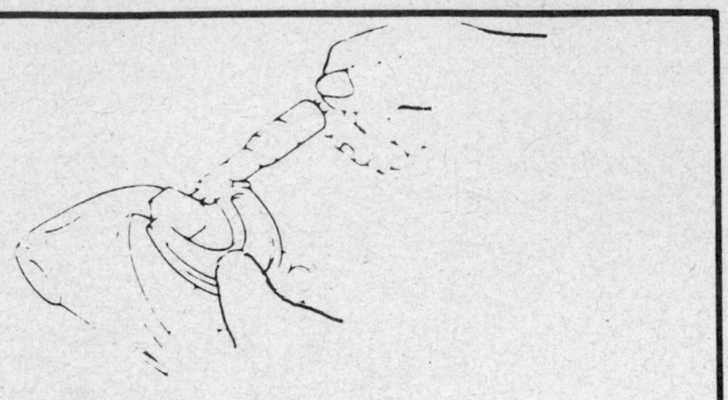

SPECIFICATIONS mm (in.)

EXCEPT REVERSE	7.5 (0.295)
REVERSE	7.4 (0.291)

Measuring shift fork pads

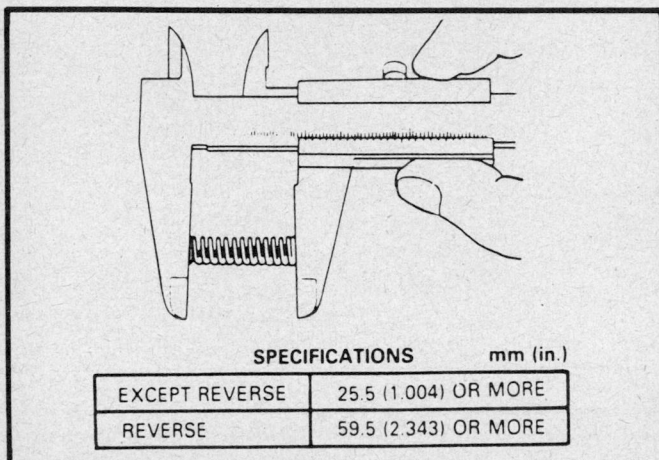

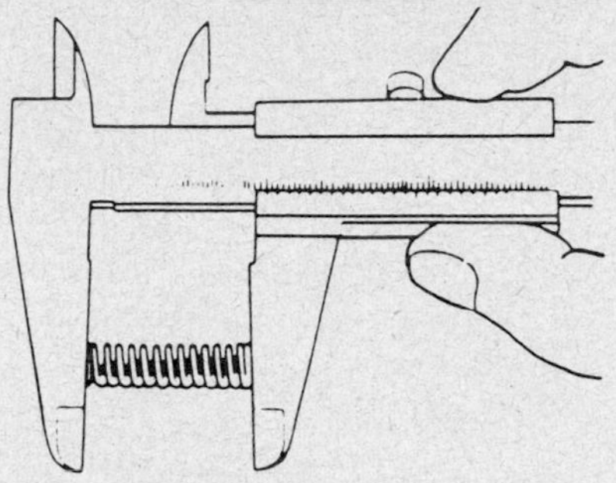

SPECIFICATIONS mm (in.)

EXCEPT REVERSE	25.5 (1.004) OR MORE
REVERSE	59.5 (2.343) OR MORE

Measuring detent spring

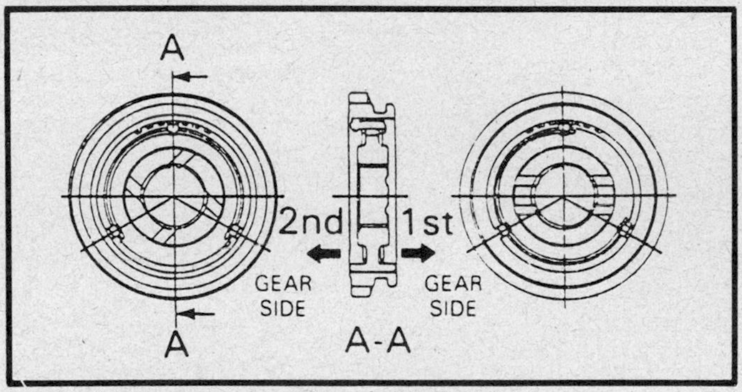
A
2nd
1st
GEAR SIDE
GEAR SIDE
A
A-A

1st/2nd gear synchronizer installation

rate the transaxle case from the clutch housing.

12. Lift the 5th gear shaft. With the detent aligned facing the same way, remove 5th and reverse shafts at the same time.
13. Remove the reverse idle gear and reverse idle shaft.
14. Using a punch and hammer, remove the roll pin from the 1st/2nd shift fork and discard the roll ping. Slide 1st/2nd shaft upward to clear housing and remove fork and shaft from case.
15. Remove the cotter pin and then remove the pin and reverse shift lever.
16. Remove the input and output shafts with 3rd/4th shift fork and shaft as an assembly.
17. Remove the differential case assembly.
18. Remove the reverse shaft bracket and take out 3 interlock pins.
19. Remove the rear bearing outer races (input and output) from the transaxle case.
20. Remove the outer races for the input shaft front bearing, output shaft front and differential side bearings. Use special tool J-26941 with J-33367 or equivalent for removing the input and output races in the housing and the differential race in the case. Use special tool J-26941 or equivalent with a slide hammer to remove the differential race in the housing.
21. Remove the input shaft seal from the housing. Remove the clutch shaft only if replacement of bushing, bearing or shaft is required.
22. Remove the outer clutch shaft seal and outer clutch shaft bushing using special tools J-37245 with J-36190 or equivalent. Drive the bushings towards the outside of clutch housing.
23. Remove the clutch shaft and inner clutch shaft seal.
24. Remove the inner clutch shaft needle bearing using special tool J-37158 or equivalent. Position the puller legs under bearing and thread rod down to expand legs. Hold rod while tightening nut to remove the bearing.

Unit Disassembly and Assembly

INPUT SHAFT

Disassembly

1. Remove the front bearing using special tool J-22912-01 or equivalent (inverted when installed on the shaft) with a suitable press.
2. Pull off the rear bearing, 4th gear, 3rd/4th synchronizer assembly and 3rd gear as a unit, using special tool J-22912-01 or equivalent (inverted when installed on the shaft) and a suitable press.
3. Remove the remaining parts from the input shaft.

Inspection

1. Clean parts with solvent and air dry.
2. Inspect shaft, spline wear for cracks.
3. Inspect gear teeth for scuffed, nicked, burred or broken teeth.
4. Inspect bearings for roughness of rotation, burred or pitted condition.
5. Inspect bearing races for scoring, wear or overheating.
6. Inspect synchronizers assembly for damage or wear.

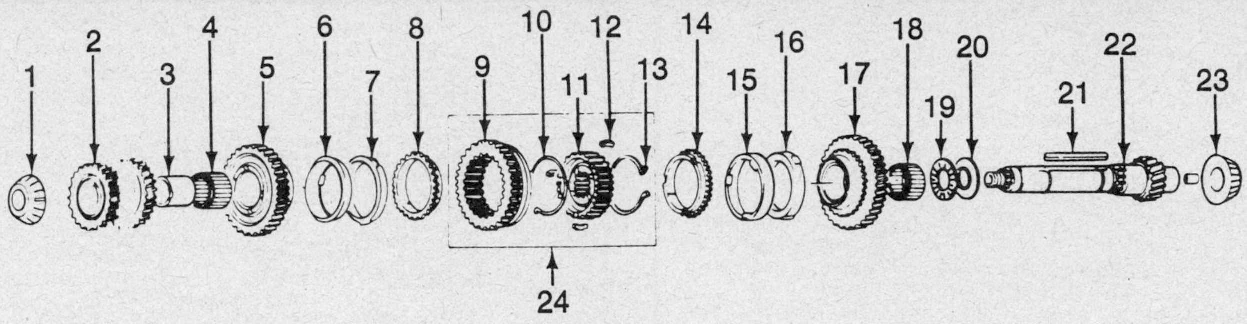
1
2
3
4
5
6
7
8
9
10
11
12
13
14
15
16
17
18
19
20
21
22
23
24

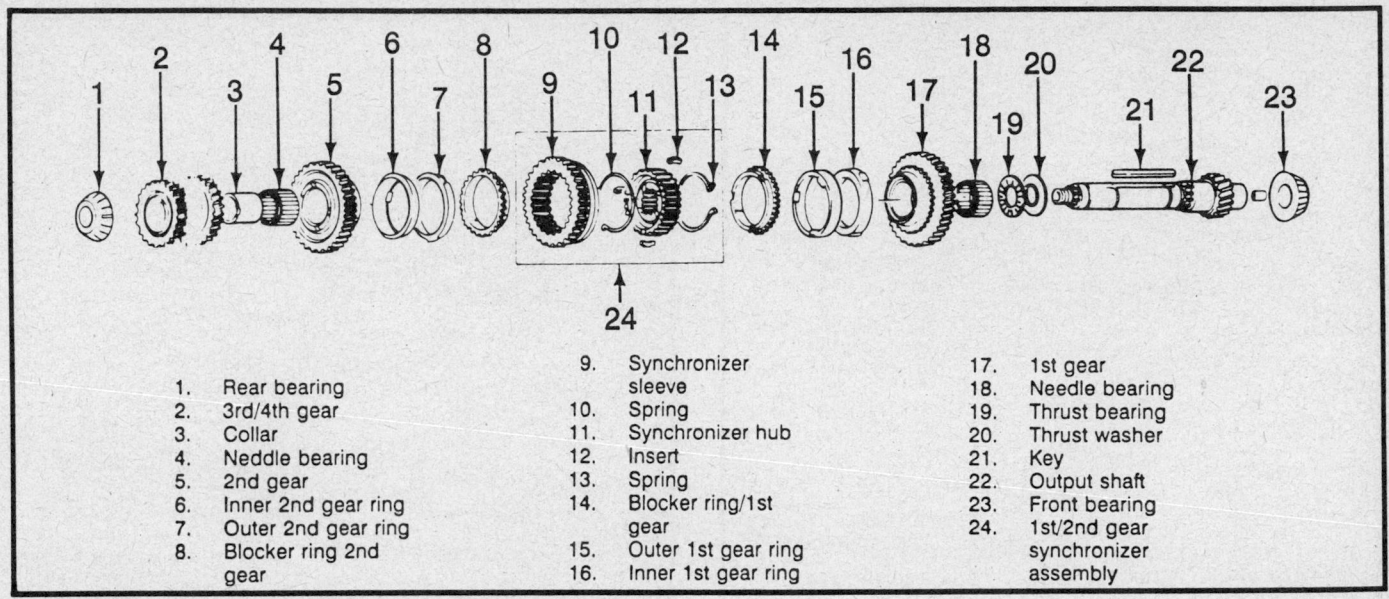

1. Rear bearing
2. 3rd/4th gear
3. Collar
4. Neddle bearing
5. 2nd gear
6. Inner 2nd gear ring
7. Outer 2nd gear ring
8. Blocker ring 2nd gear
9. Synchronizer sleeve
10. Spring
11. Synchronizer hub
12. Insert
13. Spring
14. Blocker ring/1st gear
15. Outer 1st gear ring
16. Inner 1st gear ring
17. 1st gear
18. Needle bearing
19. Thrust bearing
20. Thrust washer
21. Key
22. Output shaft
23. Front bearing
24. 1st/2nd gear synchronizer assembly

Exploded view of output shaft components

7. Measure the clearance between synchronizer blocker ring and inserts. Replace components if 0.146 in. (3.7mm) or less clearance exist.
8. Measure the clearance between blocker ring and gear. Replace components if not within 0.031 in. (0.8mm). This specification is the wear limit.
9. When reassembling synchronizer assemblies, each insert spring should support all 3 keys and each opening portion of the insert spring should face the opposite direction from the other.

Assembly

NOTE: Before assembling, apply oil to the thrust surfaces on all gears and washers.

1. Install needle bearing, 3rd gear and the blocker ring on the input shaft.
2. Match the inserts of the 3rd/4th sleeve and hub assembly with the grooves of the blocker ring and press the sleeve and hub assembly and collar using special tool J-33374 or equivalent and a suitable press. Before installing, apply oil to the collar and hub inside surface. After installation, apply oil to the circumference of the collar. Check to ensure the insert springs do not interfere with the hub after installation.
3. Install the blocker ring, needle bearing, 4th gear and thrust washer on the input shaft. When installing the thrust washer install with the recessed area facing 4th gear.
4. Install the front and rear bearings using special tool J–33374 or equivalent and a suitable press. Before installing, apply oil to the bearing inside and race surfaces.

OUTPUT SHAFT

Disassembly

1. Remove the front bearing using special tool J–22227–A and special tool J–33369 or equivalent and a suitable press.
2. Remove the rear bearing and 3rd/4th gear at the same time using special tool J–22912–01 (inverted when installed on the shaft) and a suitable press.
3. Remove the needle bearing, collar, 2nd gear, 1st/2d gear synchronizer assembly, 1st gear and key as a unit using a suitable press.
4. Remove the thrust bearing and washer from the output shaft.

Inspection

1. Clean parts with solvent and air dry.
2. Inspect shaft, spline wear for cracks.
3. Inspect gear teeth for scuffed, nicked, burred or broken teeth.
4. Inspect bearings for roughness of rotation, burred or pitted condition.
5. Inspect bearing races for scoring, wear or overheating.
6. Inspect synchronizers assembly for damage or wear.
7. Measure the clearance between synchronizer blocker ring and inserts. Replace components if 0.154 in. (3.9mm) or less clearance exist.
8. Measure the clearance between blocker ring and gear. Replace components if not within 0.031 in. (0.8mm). This specification is the wear limit.
9. Measure the clearance between blocker rings and synchronizing cones for 1st and 2nd gears by positioning 1st gear then 1st/2nd synchronizer assembly then 2nd gear in the correct position. Replace the 1st/2nd gear syncrhonizer components if the measurement exceeds 0.059 in. (1.5mm).
10. When reassembling synchronizer assemblies, each insert spring should support all 3 keys and each opening portion of the insert spring should face the opposite direction from the other.

Assembly

NOTE: Before assembling, apply oil to the thrust surfaces on all gears. Apply oil to all the bearing inside and race surfaces.

1. Install the thrust washer, thrust bearing, needle bearing, 1st gear, inner, outer ring and blocker ring for 1st gear on the output shaft.
2. Match the inserts of the sleeve and hub assembly with the grooves of the blocker ring and press the sleeve and hub assembly together with the collar using special tool J–8853–01, pilot tool J–33369 or equivalent and a suitable press. Before installing the sleeve and hub assembly, oil should be applied to the hub and collar inside surface. After installation, apply oil to the collar outside surface. Check to ensure that the insert springs do not interfere with the hub after installation.

NOTE: Install the 1st/2nd gear synchronizer in the correct position.

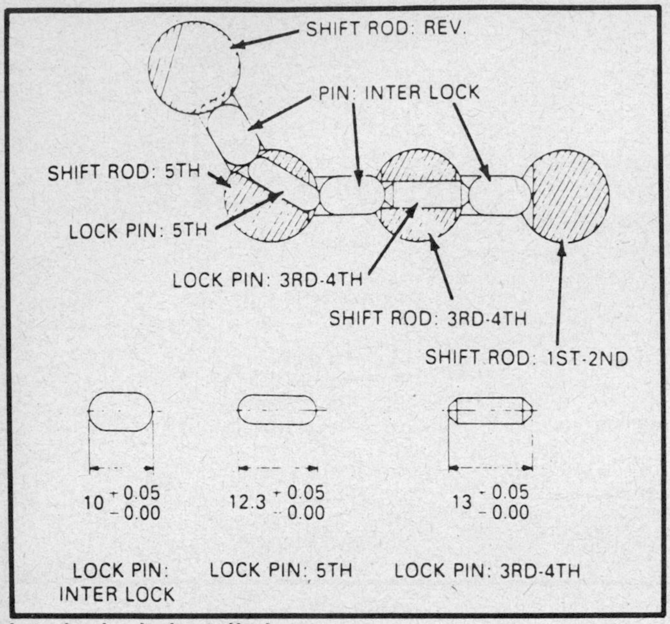
SHIFT ROD: REV.
PIN: INTER LOCK
SHIFT ROD: 5TH
LOCK PIN: 5TH
LOCK PIN: 3RD-4TH
SHIFT ROD: 3RD-4TH
SHIFT ROD: 1ST-2ND
10 + 0.05 − 0.00
12.3 + 0.05 − 0.00
13 − 0.05 − 0.00
LOCK PIN: INTER LOCK
LOCK PIN: 5TH
LOCK PIN: 3RD-4TH

Interlock pin installation

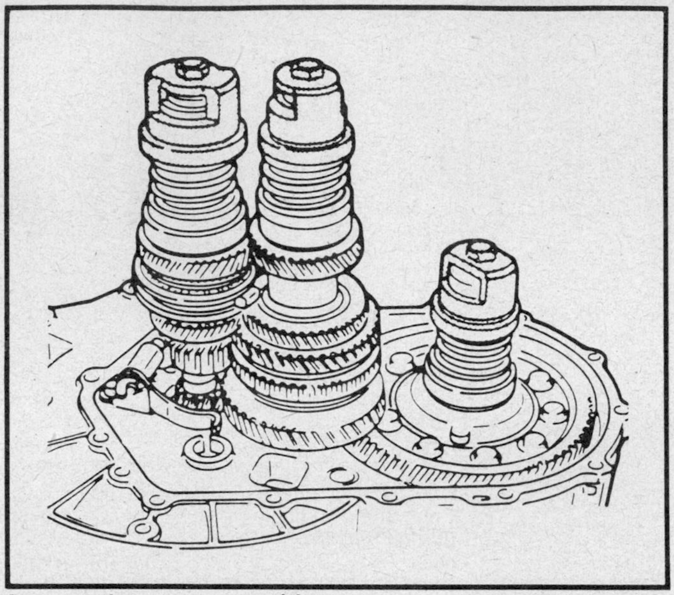

Gauges in correct position

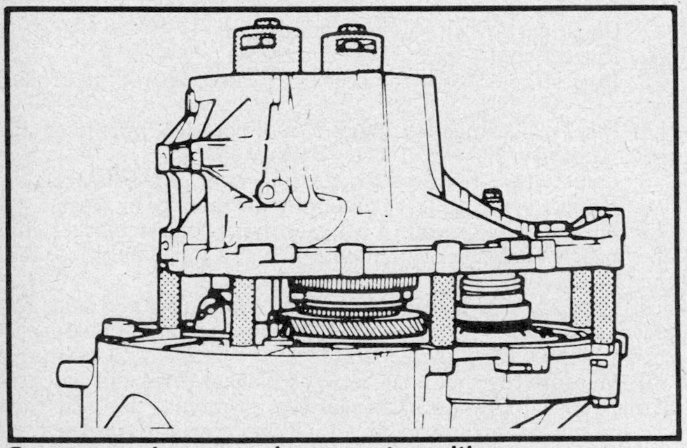

Gauges and spacers in correct position

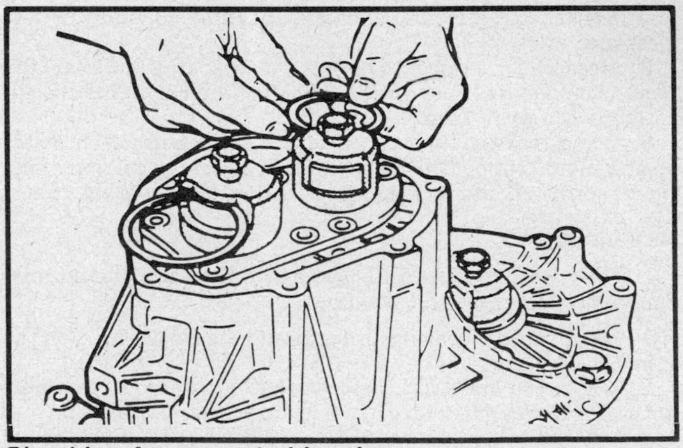

Checking for correct shim size

3. Install the blocker ring, outer and inner ring for 2nd gear, needle bearing, 2nd gear and then the key. Make sure the key is positioned properly in the key groove.

4. Apply oil to the 3rd/4th gear inner surface, match the key with the key groove and fit the key together with the rear bearing. Using special tool J–33374 or equivalent and a suitable press, press bearing on the shaft.

5. Press the front bearing on the shaft using special tool J–33368 or equivalent and a suitable press.

DIFFERENTIAL ASSEMBLY

Disassembly

1. Remove the side bearing using a suitable gear puller.
2. Remove the 10 retainig bolts and remove the ring gear. Discard the ring gear bolts.
3. Using a small tool, pry the speedometer drive gear and or rotor from the differential case. Do not reuse the speedometer drive gear and or rotor.
4. Using a punch, drive out the lock pin and pull out the cross pin.
5. Remove the pinion gears, thrust washers, side gears and thrust washers.

Inspection

1. Clean with solvent and air dry.
2. Inspect the housing bearing race bore for wear, scratches or grooves.
3. Inspect the housing bushings for scores, burrs, roundness or evidence of overheating.
4. Inspect housing for cracks, threaded openings for damaged threads, mounting faces for nicks, burrs or scratches.
5. Inspect gear teeth for scuffed, nicked, burred or broken teeth.

Assembly

NOTE: Before assembling, apply oil to the bearing inner and outer race surfaces.

1. Install the 2 side gears on the differential case together with the thrust washers. Next, position the 2 thrust washers and pinion gears opposite of each other and install them in their positions by turning the side gear.

THICKNESS mm(in.)	AVAILABLE INPUT	OUTPUT	DIFF	THICKNESS mm(in.)	AVAILABLE INPUT	OUTPUT	DIFF
1 00 0 0394	•			1 76 0 0693	•		•
1 04 0 0410	•			1 80 0 0709	•	•	•
1 08 0 0426	•		•	1 84 0 0725	•		•
1 12 0 0441	•		•	1 88 0 0741	•	•	•
1 16 0 0457	•	•	•	1 92 0 0756	•		•
1 20 0 0473	•		•	1 96 0 0772	•	•	•
1 24 0 0489	•	•	•	2 00 0 0788	•		•
1 28 0 0504	•		•	2 04 0 0804	•	•	
1 32 0 0520	•	•	•	2 08 0 0820	•		
1 36 0 0536	•		•	2 12 0 0835	•	•	
1 40 0 0552	•	•	•	2 16 0 0851	•		
1 44 0 0567	•		•	2 20 0 0867	•	•	
1 48 0 0583	•	•	•	2 24 0 0883	•		
1 52 0 0599	•		•	2 28 0 0899	•	•	
1 56 0 0615	•	•	•	2 32 0 0914	•		
1 60 0 0630	•		•	2 36 0 0930	•	•	
1 64 0 0646	•	•	•	2 40 0 0946	•		
1 68 0 0662	•		•	2 44 0 0951	•	•	
1 72 0 0678	•	•	•	2 48 0 0977	•		

Preload shim size

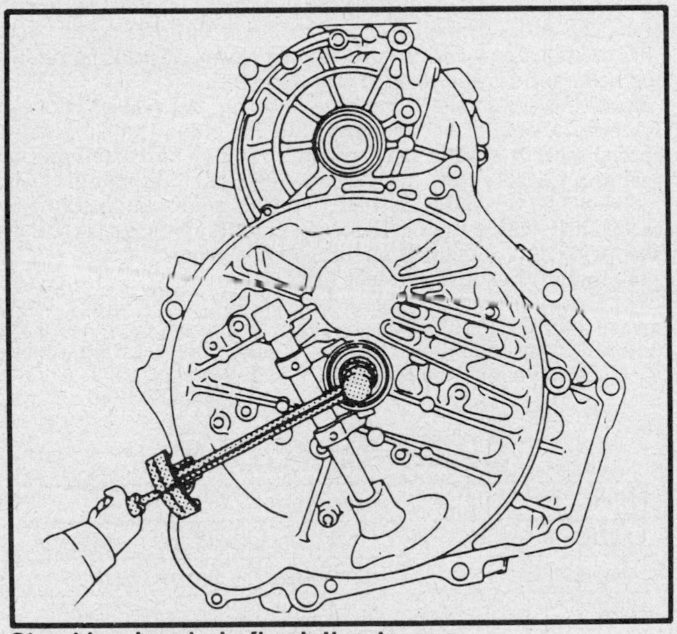

Checking input shaft rotating torque

2. Insert the cross pin and make sure the backlash is within the rated range 0.0012–0.0032 in. (0.03–0.08mm). If the backlash is outside the rated range, adjust it by installing different size thrust washers.

3. Install lock pin and cross pin.

4. Heat using a suitable tool (do not use hot water) a new speedometer drive gear and or rotor to about 200° F and then install it on the differential.

5. Heat using a suitable tool the ring gear to 122–212° F, then apply oil on the inside diameter of the ring gear then position the gear on the differential case. Apply a small amount of oil to the bottom side of the 10 new bolt heads only, then install bolts and tighten to 73–79 ft. lbs. in a diagonal sequence. Apply oil to the cross pin, differential gears, thrust portion, side gear shaft portion and side gear spline portion before installation.

6. Install the side bearing on the differential case using special tool J-22919 or equivalent and an arbor press.

Transaxle Assembly

1. Install the new input shaft seal in housing using special tool J-26540 or equivalent.

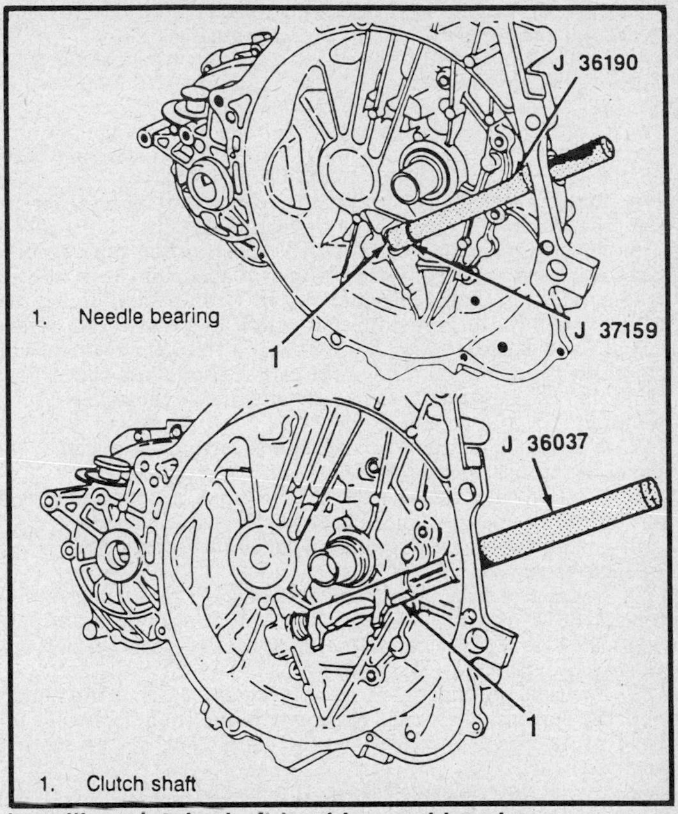
J 36190
1. Needle bearing
1
J 37159
J 36037
1
1. Clutch shaft

Installing clutch shaft bushing and bearing

2. Install the front outer bearing races for the input shaft, output shaft and differential into the clutch housing. Always apply oil to the bearing races before installation and use suitable tools.

3. Apply grease to the 3 interlock pins and install on the clutch housing.

4. Install the reverse shift bracket on the clutch housing. Use 3rd/4th shift shaft to align bracket to housing. Install retaining bolts and tighten to 13 ft. lbs. (17 Nm) make sure shaft operates smoothly after installation.

5. Install the differential assembly, then install the input, output shaft with the 3rd/4th shift fork and shaft together as an assembly into the clutch housing. Make sure the lock pin is in the 3rd/4th shifter shaft before installing. The 3rd/4th shift shaft is installed into the raised collar of reverse shift lever bracket.

6. Install the 1st/2nd gear shift fork onto the synchronizer sleeve and insert the shift shaft into the reverse shift lever bracket. Align hole in fork with the shaft and install a new double roll pin.

7. Install the reverse lever on shift bracket.

8. Install the reverse and 5th gear shifter shaft and at the same time, engage reverse shaft with reverse shift lever. Make sure lock pin is in the 5th gear shifter shaft before installing.

9. Install the reverse idle shaft with the gear into the clutch housing. Make sure reverse lever is engaged in collar of gear.

10. Measure and determine shim size using special tool J-33373. To determine corrrect shim size follow assembly procedure Steps 11–17.

11. Position the outer bearing races on the input, output and differential bearings. Position the shim selection gauges on the bearings races in the correct position.

NOTE: The 3 gauges are marked as input, output and differential.

12. Position 7 spacers provided with special tool J-33373 evenly around the perimeter of the clutch housing.
13. Install the bearing and shim retainer on transaxle case. Torque retaining screws to 13 ft. lbs. (17 Nm). After final torque on screws, stake screws to the retaining plate.
14. Carefully position the transaxle case over the gauges and on the spacer. Install the 7 bolts alternately until case is seated on spacers. Tighten bolts to 10 ft. lbs. (13 Nn).
15. Rotate each gauge to seat the bearings. Rotate the differential case through 3 revolutions in each direction.
16. With the 3 gauges compressed, measure the gap between the outer sleeve and base pad using available shim sizes in preload shim sizes chart. The input shaft shim should be 2 sizes smaller than the largest shim that will fit in the gap. The differential should use a shim 3 sizes larger than that which will smoothly fit in the gap. The output shaft should use the largest shim that can be placed into the gap and drawn through without binding.
17. When each of the 3 correct shims have been selected, remove the transaxle case, 7 spacers and 3 gauges.
18. Position the shim selected for the input, output and differential into the bearing race bores in the transaxle case.
19. Install the rear input and output shaft bearing races using suitable tools.
20. Install the rear differential case bearing race using special tools J-8611-01 with J-8092 or equivalent and a press. Apply oil to the bearing race before installation. Press the bearing until seated in its bore.
21. Position the magnet in clutch housing. Clean the clutch housing and transaxle case mating surface, then apply a ⅛ in. bead of Loctite No. 518 sealer or equivalent to the mating surfaces.
22. Install the transaxle case on the clutch housing. Install the reverse idle shaft bolt into the transaxle case. Tighten the retaining bolt to 28 ft. lbs. (38 Nm).
23. Install the 14 case bolts. Torque bolts to 28 ft. lbs. (38 Nm) in a diagonal sequence.
24. Install the drive axle seals using suitable tools.
25. Install the thrust washer and collar to the output shaft using special tool J-33374. Before installing, apply oil to the thrust surfaces and collar.
26. Install 5th gear to the input shaft. Install the needle bearing, 5th gear, blocker ring, hub/sleeve assembly with shift fork in its groove and back plate on the output shaft. Align shift fork on shift shaft and install a new double roll pin. Before installing apply oil to the output gear thrust surfaces.
27. Install the detent balls and detent springs for the reverse, 1st/2nd, 3rd/4th and 5th speeds. Install the retaining bolts and tighten to 18 ft. lbs. (25 Nm).
28. Install the 1st/2nd gear switch. Install the short pin and then the long pin into the 3rd/4th gear hole. Install 3rd/4th gear switch, if vehicle is equipped with a shift speed indicator light.
29. Apply Loctite No. 262 sealer or equivalent to the threads of the input and output shafts. Install new retaining nuts and torque to 94 ft. lbs. (128 Nm). Stake nuts after reaching final torque.
30. Install a new gasket with the control box assembly on the transaxle case and torque the 4 bolts to 13 ft. lbs. (17 Nm). Stake the retaining nuts after final torqe is applied and make sure transaxle shifts properly before installing the rear cover.
31. Install a new gasket with the rear cover. Install the retaining bolts and torque to 13 ft. lbs. (17 Nm).
32. If the clutch shaft, bushing, bearing and seals have been removed, installed new bearing into the clutch housing using special tools J-37159 with J-36190 or equivlent. Install new oil seal and clutch shaft. Install new outer bushing using special tool J-36037 or equivalent. Drive bushing inward until line on tool is flush with housing. Before installing, apply grease to both the inside and outside of bushing and bearing.
33. Install the clutch release bearing.
34. Measure the rotating torque on the input shaft. When measuring, the input shaft should be to the upper side and the differental assembly to the lower side. The rotating torque specification should be less than 7 inch lbs.

SPECIFICATIONS

TORQUE SPECIFICATIONS

Item	ft. lbs.	Nm
Reverse shift bracket	11–16	15–22
Ring gear bolts	73–79	98–107
Transaxle case to clutch housing bolts	22–33	30–45
Reverse idler shaft bolt	22–33	30–45
Detent spring retaining bolts	15–21	21–29
Input/output shaft retaining nuts	87–101	118–137
Control box to case bolts	11–16	15–22
Rear cover bolts	11–16	15–22
Clutch master cylinder retaining nuts	15–25	20–34
Slave cylinder retaining nuts	14–20	18–26
Clutch shaft release lever bolt	30–45	40–60

SPECIAL TOOLS

Tool No.	Description
J-33367	Front differential bearing race tool
J-26941	Rear differential bearing race tool
J-22912-01	Front input bearing tool
J-22227-A	Front output bearing tool
J-33366	Mounting fixture rim
J-33370	Bearing race tool
J-8611-01	Bearing race tool
J-7817	Bearing race tool
J-28412	Clutch shaft bushing tool
J-8092	Bearing installation handle tool
J-23598 OR J-2241-11	Bearing tool
J-22888 AND J-22888-30	3 leg puller
J-33371	Input shaft race tool
J-24256-A	Bearing race installation tool
J-22919	Press guide tool
J-8853-01	Press mounting tool
J-26540	Input shaft seal tool
J-33374	Press guide tool
J-33368	Press guide tool

Section 6

Chevrolet Nova and Geo Prizm Transaxle

General Motors

APPLICATION

1985–88 Nova
1989 Geo Prizm

GENERAL DESCRIPTION

The 5 speed transaxle assembly is a parallel/dual shaft, constant mesh unit, combined with a differential unit and assembled in a 3 piece case; the forward gears are in constant mesh. For ease of shifting and gear selecting, synchronizers with blocker rings are controlled by shifting forks. A sliding idler gear arrangement is used for reverse gear.

The components consists of: an aluminum transaxle case, an aluminum clutch housing/differential case, a case cover, input shaft, output shaft and the differential assemblies.

Needle bearings (to ensure durability) are used to support the forward gears. Single piece needle bearings support the 1st and 2nd gears, while, split needle bearings support the 3rd, 4th and 5th gears. The input and output shafts are supported by roller bearings at the engine side and tapered bearings at the opposite side. The differential is supported by tapered bearings. Selective shims are used to establish the correct preload.

The halfshafts which are attached to the front wheels, are turned by the differential and ring gear which are controlled by the final output gear.

The differential, consisting of a set of 4 gears, is a conventional arrangement that divides the torque between the halfshafts, allowing them to rotate at different speeds. Of the 4 gear set, 2 are known as differential side gears and the others are differential pinion gears.

The differential pinion gears, mounted on a differential pinion shaft, are free to rotate on the shaft. The pinion shaft, placed into the differential case bore, is at a right angle to the drive axle shaft.

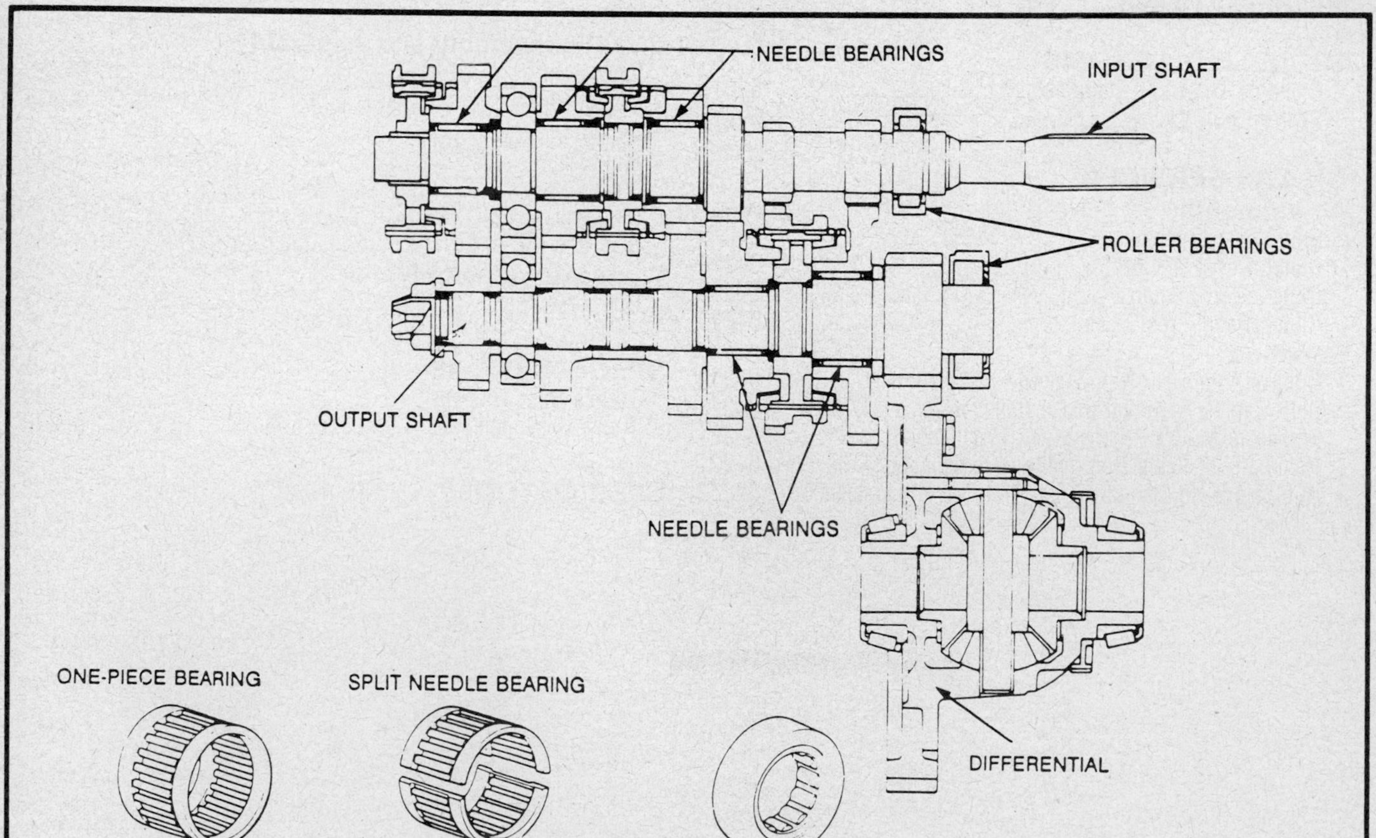
NEEDLE BEARINGS
INPUT SHAFT
ROLLER BEARINGS
OUTPUT SHAFT
NEEDLE BEARINGS
ONE-PIECE BEARING
SPLIT NEEDLE BEARING
DIFFERENTIAL
NEEDLE BEARINGS
ROLLER BEARINGS

View of the needle bearing locations

TRANSAXLE IDENTIFICATION

The transaxle identification number is stamped into the top of the front case. The 1st digit indicates the last digit of the manufacturing year, the 2nd and 3rd digits indicate manufacturing month and the last 5 digits indicate the production number.

Metric Fasteners

The transaxle is of a metric design; all bolt sizes and thread pitches are metric. Metric fastener dimensions are very close to the customary inch system fastener dimensions; replacement of the fasteners must be of the same measurement and strength as those removed.

Do not attempt to interchange metric fasteners with the customary inch system fasteners. Mismatched or incorrect fasteners can result in damage to the transaxle. Care should be taken to reuse the same fasteners in the location from which they were removed.

NOTE: Be sure to check the case holes for the quality of their threads. It is rather difficult to rethread bolt holes after the transaxle is installed in the vehicle.

Capacities

The fluid quantities are approximate and the correct fluid level is determined to be correct by being level with the bottom of the oil filler hole. Dry fill is 2.7 qts. (2.6L). Use only API GL-4 or SAE 75W-90 gear oil.

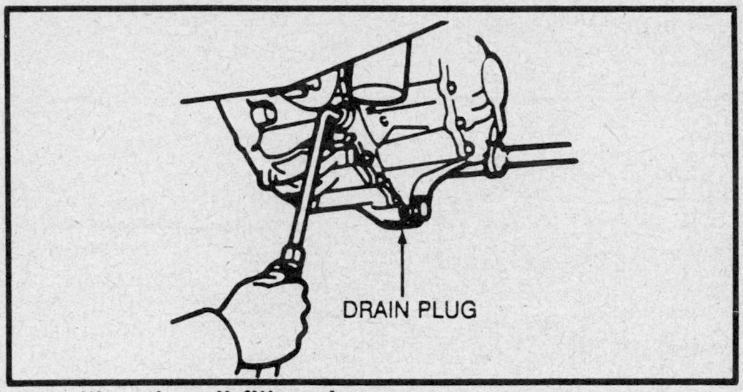
DRAIN PLUG
Installing the oil filler plug

Checking Fluid Level

NOTE: When checking the fluid level, the vehicle must be at normal operating temperatures and positioned on a flat surface.

1. Remove the filler plug located in the middle rear of the case assembly.
2. If the fluid is low, add fluid until it is level with the bottom of the filler plug hole.
3. Install the filler plug and torque it to 29 ft. lbs. (39 Nm).

TRANSAXLE MODIFICATIONS

There are no transmission modifications at the time of publication.

TROUBLE DIAGNOSIS

CHILTON'S THREE C'S TRANSAXLE DIAGNOSIS

Condition	Cause	Correction
Noise	a) Transmission or differential faulty	a) Disassemble and inspect transmission or differential
	b) Wrong oil grade	b) Replace oil
	c) Oil level low	c) Add oil
Oil leakage	a) Oil level too high	a) Drain oil
	b) Oil seal, O-ring or gasket worn or damaged	b) Replace oil seal, O-ring or gasket
Hard to shift or will not shift	a) Control cable faulty	a) Replace control cable
	b) Transmission faulty	b) Disassemble and inspect transmission
Jumps out of gear	a) Transmission faulty	a) Disassemble and inspect transmission

POWER FLOW

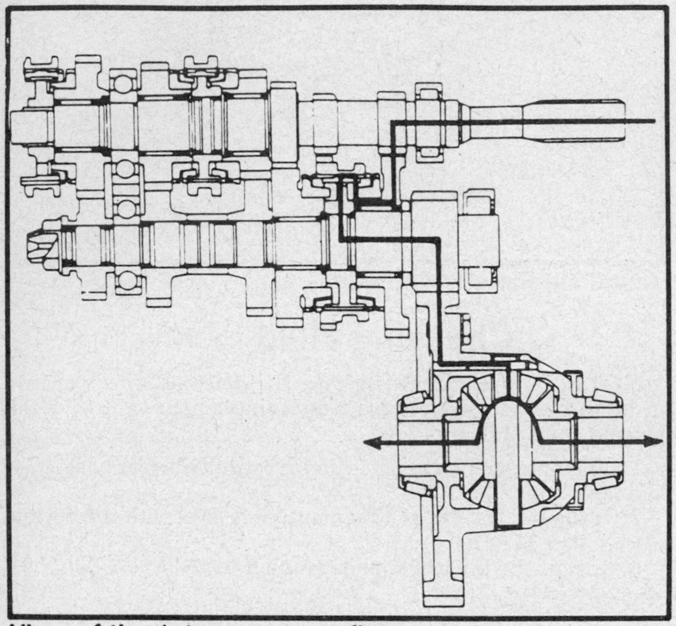

View of the 1st gear power flow

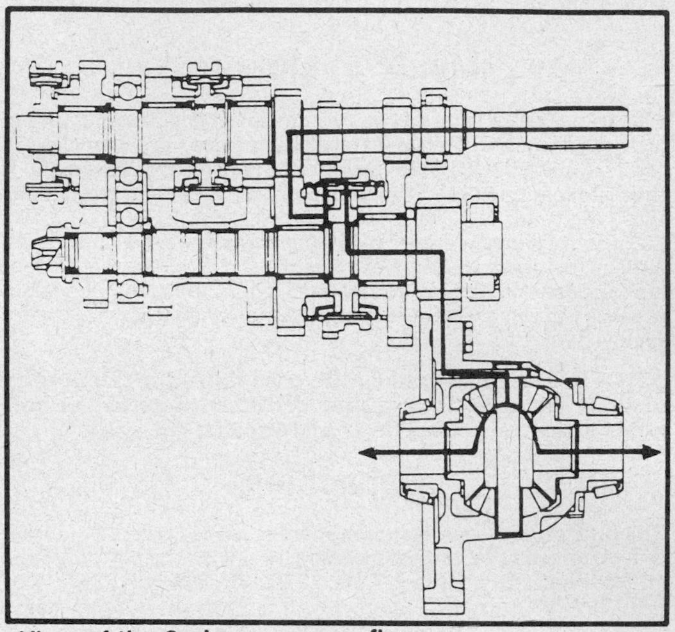

View of the 2nd gear power flow

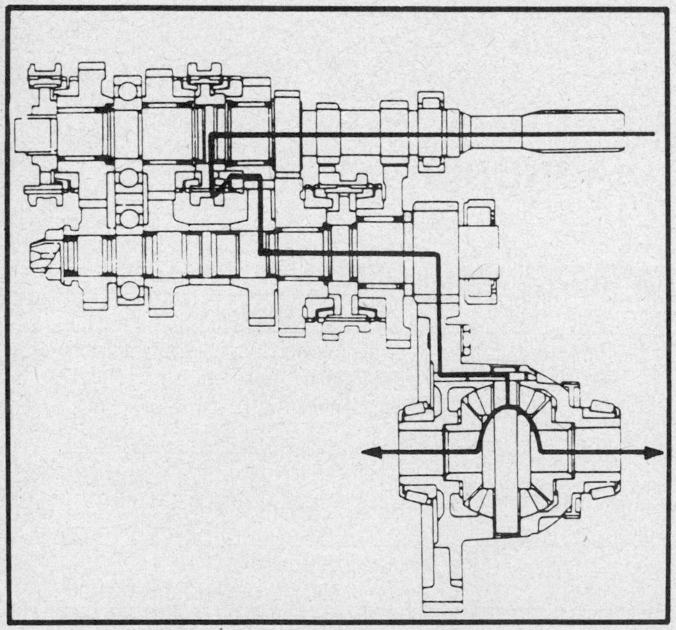

View of the 3rd gear power flow

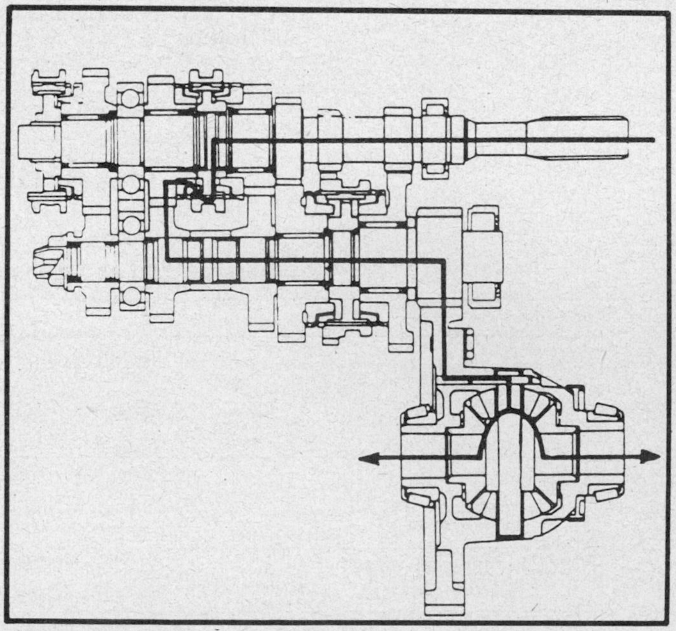

View of the 4th gear power flow

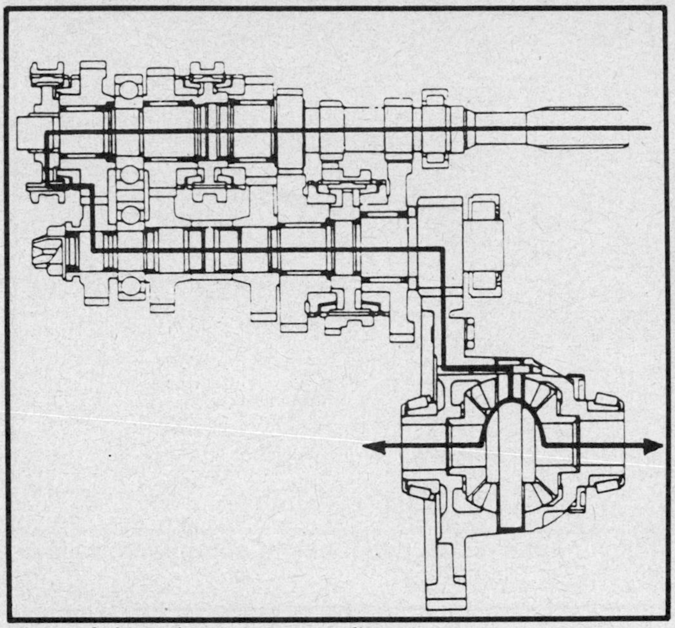

View of the 5th gear power flow

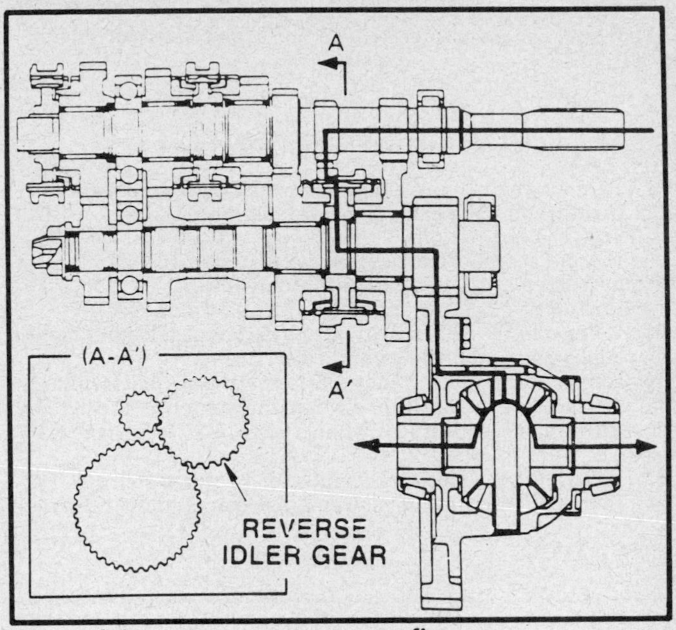
A
(A-A')
A'
REVERSE
IDLER GEAR

View of the reverse gear power flow

ON CAR SERVICE

Adjustments

SHIFT LEVER FREEPLAY

1. Remove the shift lever cover.
2. Using a feeler gauge, position it through the slot provided at the side of the shift lever cap and measure the shim-to-shift lever clearance; the allowable clearance is 0–0.004 in. (0–10mm).
3. If necessary, select a shim from the shim chart, which will provide a clearance of 0–0.004 in. (0–10mm) and install it at the bottom of the seat.
4. After adjustment, install the shift lever cover.

SHIFT LEVER SHIM THICKNESS

in.	mm	in.	mm
0.012	0.3	0.031	0.8
0.016	0.4	0.035	0.9
0.020	0.5	0.039	1.0
0.024	0.6	0.043	1.1
0.028	0.7	0.047	1.2

CLUTCH PEDAL HEIGHT

The clutch pedal height should be 5.650–6.043 in. (143.5–153.5mm) from the floor board; if not, perform the following.

1. Remove the instrument lower finish panel and air duct.
2. At the clutch pedal bracket, loosen the stopper bolt locknut and turn the stopper bolt until the pedal height is 5.650–6.043 in. (143.5–153.5mm).

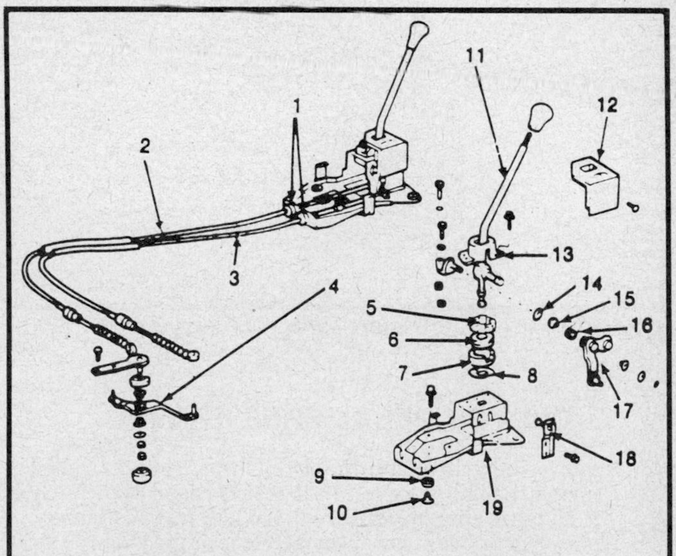
1
2
3
4
5
6
7
8
9
10
11
12
13
14
15
16
17
18
19

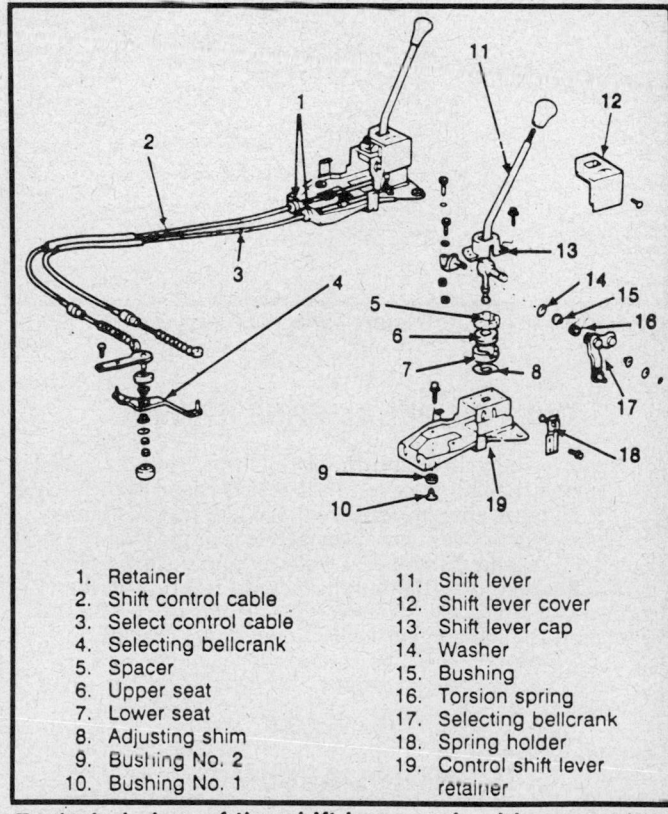

1. Retainer
2. Shift control cable
3. Select control cable
4. Selecting bellcrank
5. Spacer
6. Upper seat
7. Lower seat
8. Adjusting shim
9. Bushing No. 2
10. Bushing No. 1
11. Shift lever
12. Shift lever cover
13. Shift lever cap
14. Washer
15. Bushing
16. Torsion spring
17. Selecting bellcrank
18. Spring holder
19. Control shift lever retainer

Exploded view of the shift lever and cable assembly

3. Tighten the locknut.
4. Install the instrument lower finish panel and air duct.

CLUTCH START SWITCH

1. Remove the instrument lower finish panel and air duct.
2. Firmly apply the parking brake and place the shift control lever into the **N** position.
3. Disconnect the electrical connector from the clutch start switch, located at the top of the clutch pedal.
4. Loosen the locknut and back off the switch.
5. Using a feeler gauge, depress the clutch pedal and compare the pedal stroke with the switch clearance **A**.
6. Using a ohmmeter, connect it to the switch terminals. Make sure there is continuity when the switch is turned **ON** (pushed) and no continuity when the switch is turned **OFF** (free).
7. Reconnect the electrical connector to the switch.
8. Install the instrument lower finish panel and air duct.

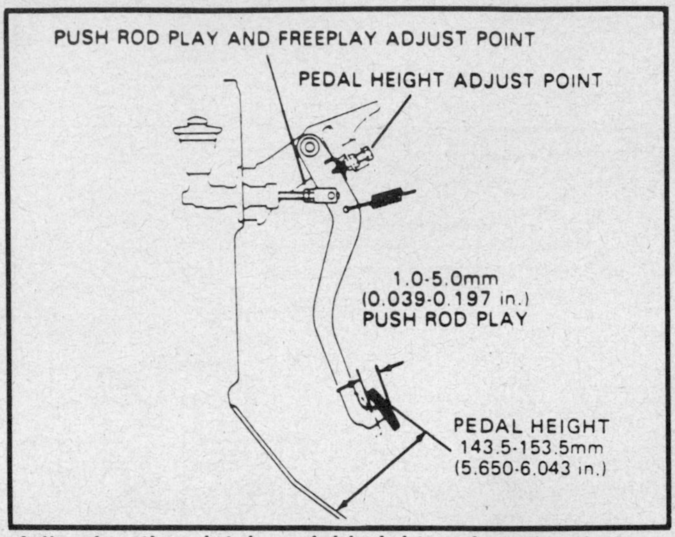
PUSH ROD PLAY AND FREEPLAY ADJUST POINT
PEDAL HEIGHT ADJUST POINT
1.0-5.0mm
(0.039-0.197 in.)
PUSH ROD PLAY
PEDAL HEIGHT
143.5-153.5mm
(5.650-6.043 in.)
Adjusting the clutch pedal height and pushrod play

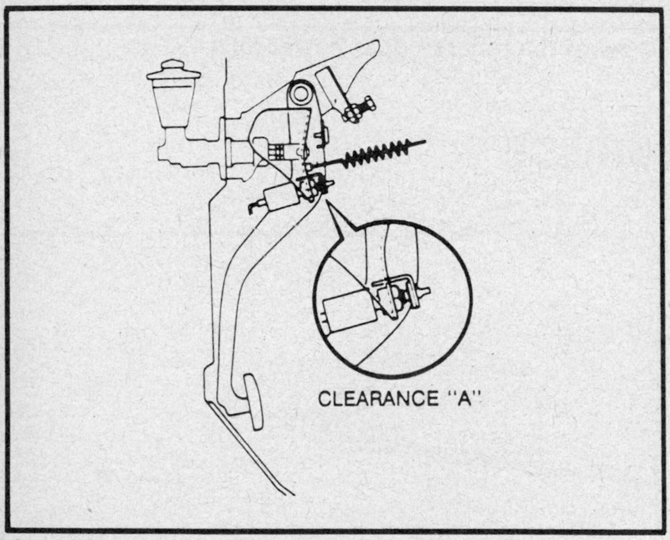
CLEARANCE "A"
Checking the clutch start switch clearance

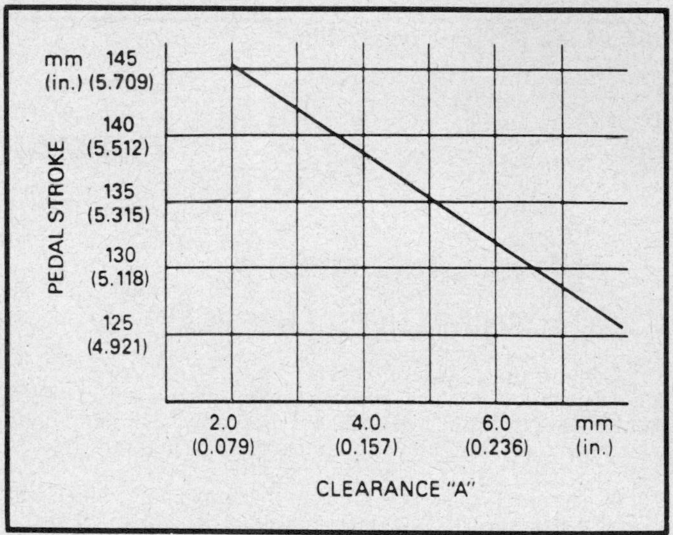
mm 145
(in.) (5.709)
140
(5.512)
135
(5.315)
130
(5.118)
125
(4.921)
PEDAL STROKE
2.0
(0.079)
4.0
(0.157)
6.0
(0.236)
mm
(in.)
CLEARANCE "A"
Clutch switch adjustment chart

CLUTCH PEDAL FREE TRAVEL

1. Remove the instrument lower finish panel and air duct.
2. Depress the clutch pedal until resistance of the clutch is felt; this distance should measure 0.20–0.59 in. (5–15mm).
3. If the measurement is not correct, loosen the pushrod locknut and adjust the clutch pedal freeplay and pushrod play.
4. After adjustment, tighten the locknut to 11–14 ft. lbs. (15–20 Nm).
5. Install the instrument lower finish panel and air duct.

Services

CONTROL ASSEMBLY

Removal and Installation

1. Disconnect the negative battery cable.
2. Remove the console and the shifter boot.

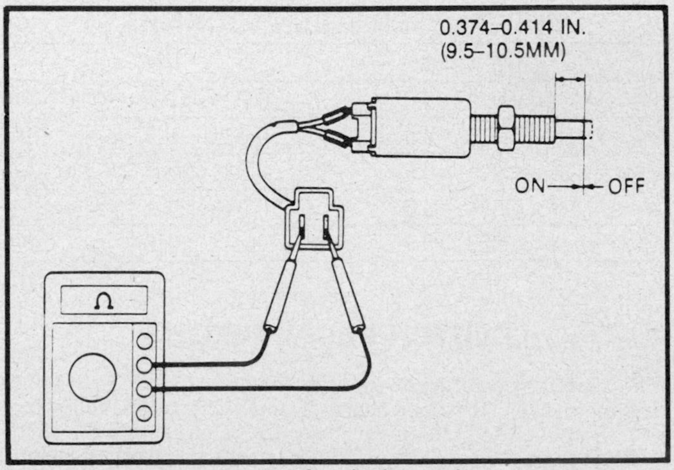
0.374–0.414 IN.
(9.5–10.5MM)
ON
OFF
Ω
Using an ohmmeter to check the switch continuity

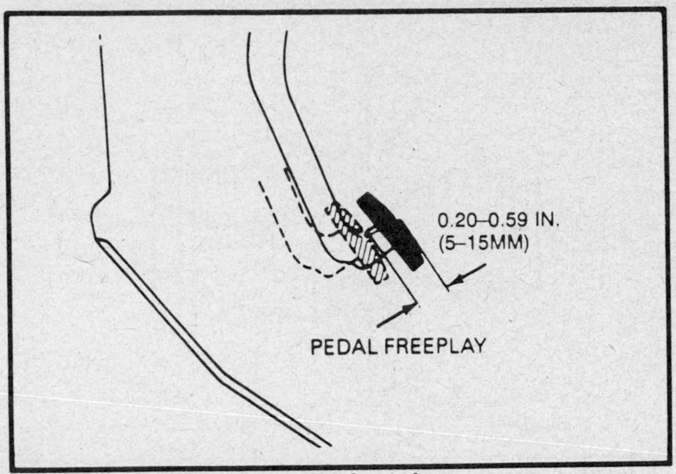
0.20–0.59 IN.
(5–15MM)
PEDAL FREEPLAY

Checking the clutch pedal freeplay

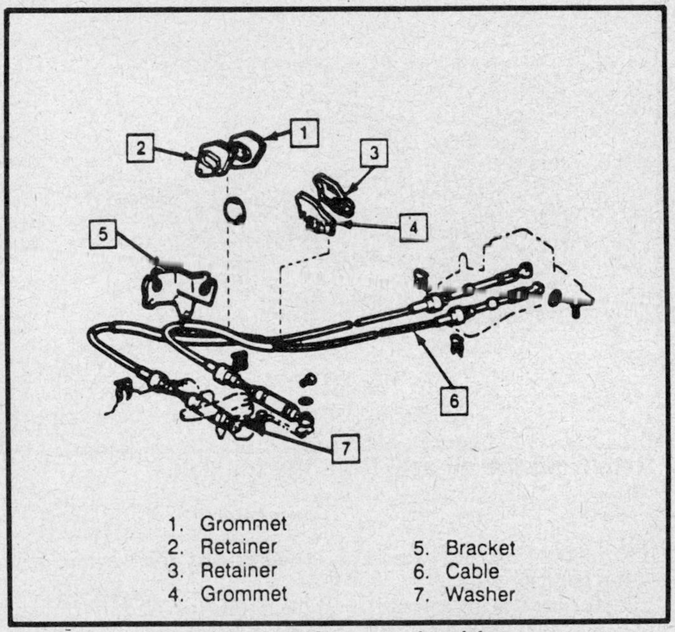
1
2
3
4
5
6
7
1. Grommet
2. Retainer
3. Retainer
4. Grommet
5. Bracket
6. Cable
7. Washer

Exploded view of the shift control cables

3. Disconnect the shift cables from the shift control assembly.
4. Remove the shift control assembly-to-chassis nuts and the assembly.
5. To install, position the shift control assembly into the vehicle and torque the mounting screws to 14 ft. lbs. (20 Nm).
6. Connect the shift cables to the shift control assembly.
7. Install the shifter boot and console.
8. Connect the negative battery cable.

SHIFT CABLES

Removal and Installation

1. Disconnect the negative battery cable.
2. At the transaxle, disconnect the shift cables and remove the retaining clips.
3. At the shift control, remove the console and the boot.
4. Disconnect the shift cables from the shift control assembly.
5. Remove the left, front sill plate and pull the carpet back to gain access to the cables.
6. At the floor pan, remove the cable retainer screws and the cables.
7. To install the cables, route them into the vehicle, install the cable retainer and screws, at floor pan.
8. At the shift control assembly, connect the shift cables. Reposition the carpet and install the left, front sill plate.
9. Raise and safely support the vehicle.
10. At the transaxle, position the cables and install the retaining clips.
11. Lower the vehicle.
12. Connect the cables to the shift control assembly.
13. Install the console and the boot.
14. Install the negative battery cable.

TRANSAXLE MOUNT

Removal and Installation

1. Disconnect the negative battery cable.
2. Using an engine support fixture tool, secure it to the engine and raise it to take the weight off the mount.
3. Raise and safely support the vehicle.
4. Remove the mount-to-transaxle bracket nut.
5. Raise the engine to separate the mount from the bracket and remove the mount.
6. Install the mount and tighten mount-to-transaxle bolts.
7. Remove the engine support fixture tool and connect the negative battery cable.

HALFSHAFT OIL SEALS

Removal and Installation

SINGLE OVERHEAD CAM ENGINE

1. Disconnect the negative battery cable.
2. Raise and safely support the vehicle.
3. Drain the gear oil from the transaxle.
4. To remove the halfshafts, perform the following procedures:
   a. Remove the wheel cover, the grease cup, the cotter pin, the hub nut cap, the hub nut and the washer.
   b. Loosen the wheel lug nuts.
   c. Remove the front wheels.
   d. Remove the lower control arm-to-ball joint nuts/bolts and separate the lower control arm from the steering knuckle.
   e. Remove the tie rod ball joint-to-steering knuckle cotter pin and nut. Using the ball joint separator tool or equivalent, separate the tie rod end from the steering knuckle.
   f. Remove the brake caliper-to-steering knuckle bolts, the brake caliper and suspend it on a wire. Remove the brake disc.
   g. Using a wheel puller tool or equivalent, press the half shaft from the wheel hub.
   h. Using the halfshaft removal tool or equivalent, and a slide hammer puller, pull the halfshaft from the transaxle.
   i. Remove the halfshaft from the vehicle.
5. Using an oil seal removal tool or equivalent, and a slide hammer puller, remove the oil seals from the transaxle.
6. Using an oil seal installer tool or equivalent, and a hammer, lubricate and drive the new oil seals into the transaxle until they seat.
7. To install the halfshafts, perform the following procedures:
   a. Check and/or replace the hub seal and grease the seal lips.

b. Install the halfshaft into the transaxle. Using a long brass drift (positioned on the tri-pot housing ribs) and a hammer, drive the halfshaft into the transaxle until it locks in place.

c. Install the halfshaft into the wheel hub.

d. Install the lower control arm-to-lower ball joint and torque the nuts/bolts to 59 ft. lbs. (80 Nm).

e. Install the tie rod end-to-steering knuckle and torque the nut to 36 ft. lbs. (49 Nm).

f. Install the brake disc onto the wheel hub and the brake caliper onto the steering knuckle. Torque the brake caliper-to-steering knuckle bolts to 65 ft. lbs. (103 Nm).

g. Install the wheels/tires and the hub nut and washer. Lower the vehicle.

h. Torque the lug nuts to 76 ft. lbs. (103 Nm) and the hub nut to 137 ft. lbs. (186 Nm). Install a new cotter pin.

8. Refill the transaxle with new gear oil.
9. Connect the negative battery cable.

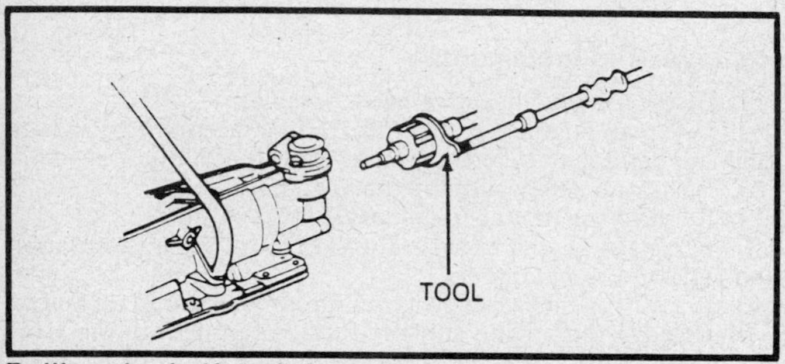
TOOL

Pulling the halfshaft from the transaxle—single overhead cam engine

DUAL OVERHEAD CAM ENGINE

1. Disconnect the negative battery cable.
2. Raise and safely support the vehicle.
3. Remove the wheel cover, the grease cup, the cotter pin, the hub nut cap, the hub nut and the washer.
4. Loosen the wheel lug nuts and remove the front wheels.
5. While an assistant depresses the brake pedal, remove the halfshaft-to-side gear shaft nuts. Pull the halfshaft from the wheel hub.
6. To remove the side gear shafts, perform the following procedures:
   a. Drain the gear oil from the transaxle.
   b. Push the side gear toward the differential and measure the distance between the case and the side gear shaft flange; this is to be used in installation.
   c. Using the side gear shaft removal tool or equivalent, and a slide hammer puller, pull the side gear shaft from the transaxle.
   d. Remove the side gear shaft from the vehicle.
7. Using an oil seal removal tool or equivalent, and a slide hammer puller, remove the oil seals from the transaxle.
8. Using an oil seal installer tool or equivalent, and a hammer, lubricate and drive the new oil seals into the transaxle until they seat.
9. Install the side gear shaft into the transaxle. Using a hammer, tap the side gear shaft into the transaxle until it locks in place.
10. Install the halfshaft into the wheel hub.
11. Connect the halfshaft to the side gear shaft. While an assistant is depressing the brake pedal, torque the nuts to 27 ft. lbs. (36 Nm).

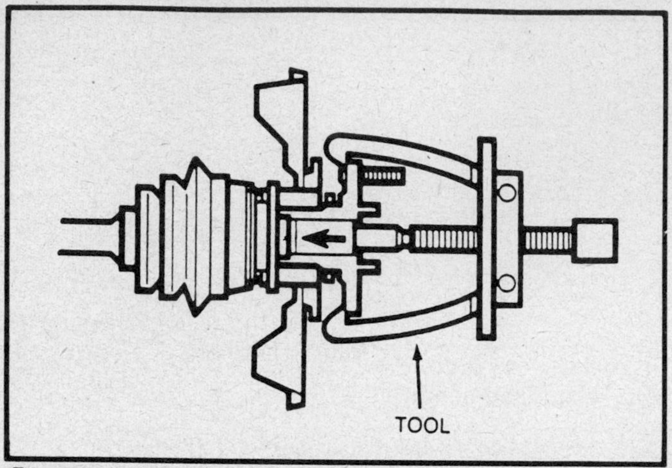
TOOL

Pressing the halfshaft from the wheel hub

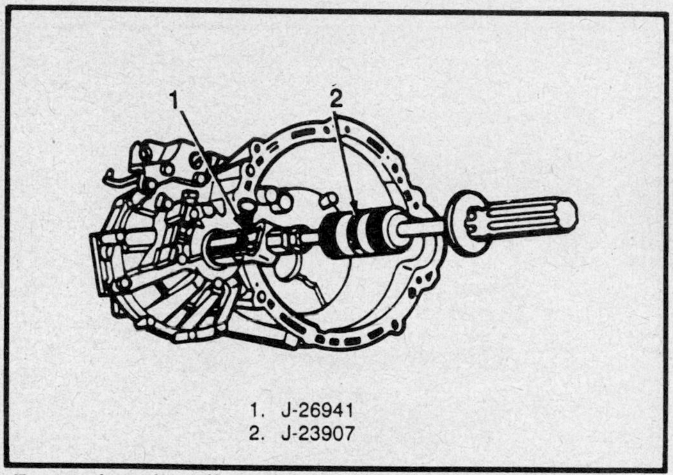
1
2
1. J-26941
2. J-23907

Removing the oil seal from the transaxle

12. Install the front wheels and the hub nut and washer. Lower the vehicle.
13. Torque the lug nuts to 76 ft. lbs. (103 Nm) and the hub nut to 137 ft. lbs. (186 Nm). Install a new cotter pin.
14. Refill the transaxle with new gear oil.
15. Connect the negative battery cable.

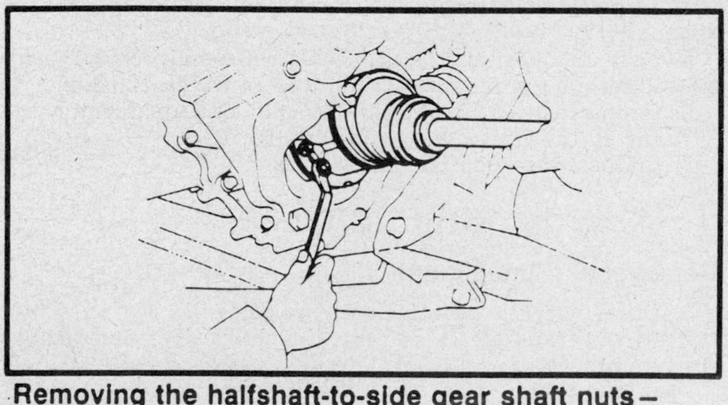

Removing the halfshaft-to-side gear shaft nuts—dual overhead cam engine

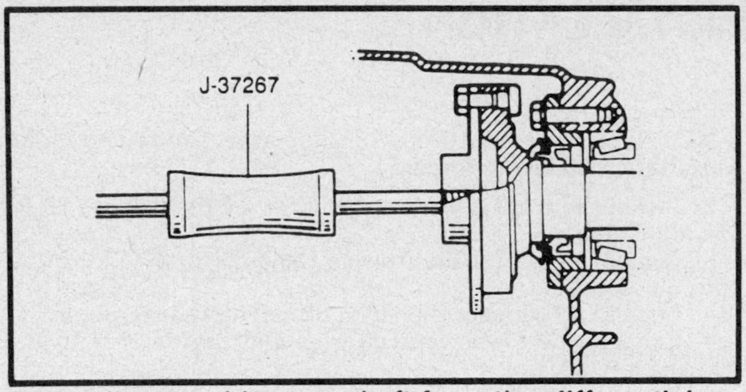
J-37267

Removing the side gear shaft from the differential—dual overhead cam engine

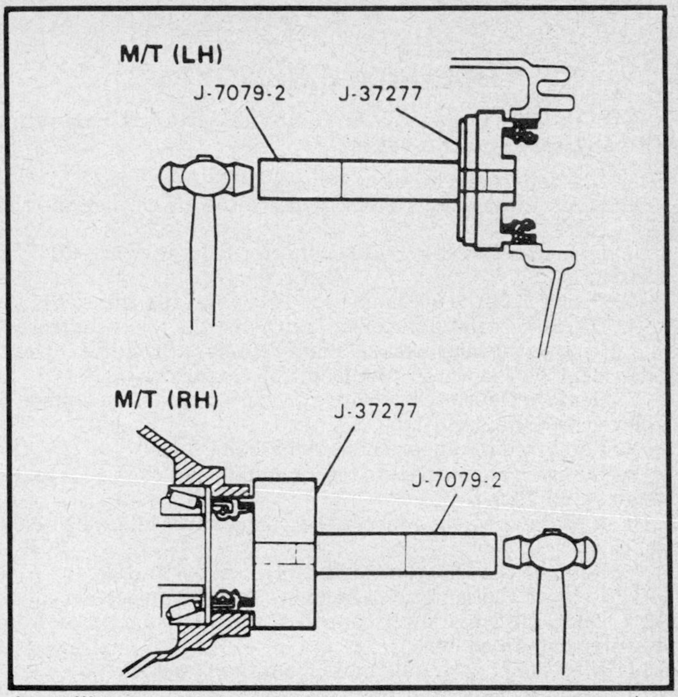
M/T (LH)
J-7079-2
J-37277
M/T (RH)
J-37277
J-7079-2

Installing new oil seals into the differential case—dual overhead cam engine

HUB SLEEVE NO. 3 and 5TH GEAR

Removal and Installation

1. Disconnect the negative battery cable.
2. Remove the transaxle-to-mount bolt.
3. Raise and support the vehicle.
4. Using a floor jack, support the transaxle.
5. Remove the transaxle bracket bolt and lower the transaxle to gain access to the cover bolts.
6. Remove the cover-to-transaxle bolts and the cover.
7. Remove the hub sleeve No. 3. Using a wheel puller tool or equivalent, press the 5th gear from the shaft.
8. Using the 5th gear installer tool or equivalent, drive the 5th gear onto the shaft. Install the hub sleeve No. 3.
9. Using sealant, apply it to the case cover and install the cover. Torque the cover-to-case screws to 13 ft. lbs. (18 Nm).
10. Raise the transaxle and install the lower bolt.
11. Lower the vehicle.

BACK-UP LIGHT SWITCH

Removal and Installation

1. From the back-up light switch, disconnect the electrical connector.

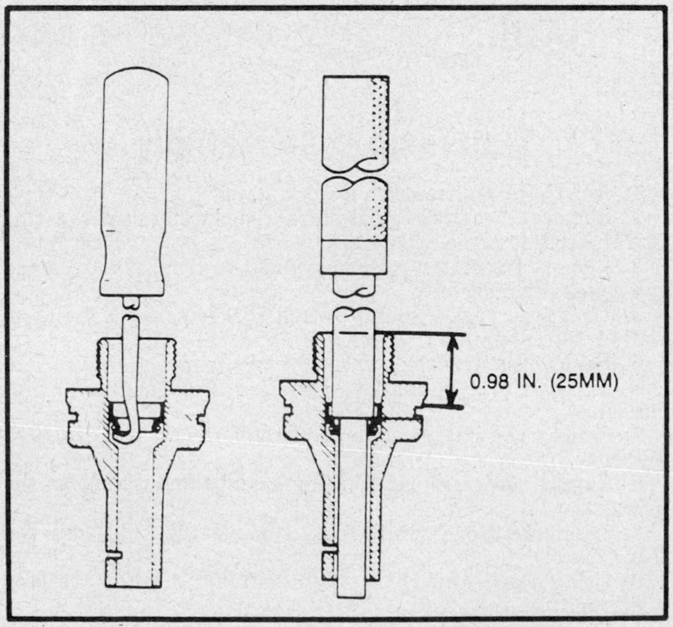
0.98 IN. (25MM)

Replacing the speedometer driven gear seal

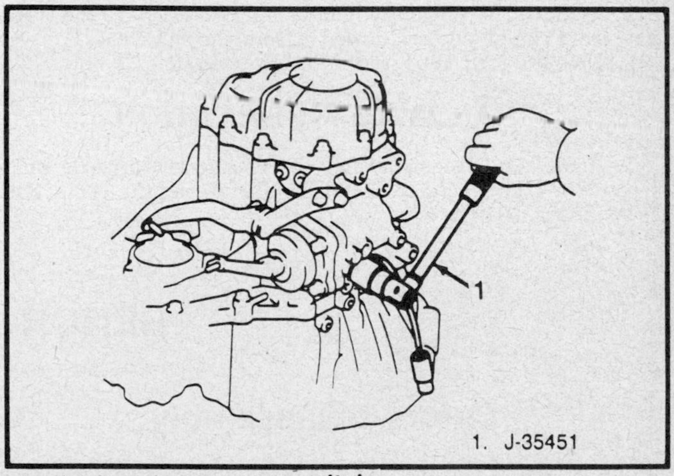
1
1. J-35451

Replacing the reverse switch

2. Using the back-up light switch socket or equivalent, and a breaker bar, remove the back-up light switch from the transaxle.
3. Using the back-up light switch socket or equivalent, and a breaker bar, install the back-up light switch onto the transaxle. Torque the back-up light switch to 30 ft. lbs. (40 ft. lbs.).

SPEEDOMETER DRIVEN GEAR SEAL

Removal and Installation

1. Remove the speedometer cable and the adapter housing from the transaxle.
2. Using a seal remover tool or equivalent, remove the seal from the from the housing.
3. Using a seal installation tool or equivalent, tap the new seal into the housing until it is seated 0.98 in. (25mm) from the top of the housing.

REMOVAL AND INSTALLATION

TRANSAXLE REMOVAL

NOTE: The transaxle is removed without removing the engine from the vehicle.

1. Disconnect the negative battery cable.
2. Drain the cooling system. Remove the air cleaner and the inlet tube.
3. Disconnect the electrical connector from the back-up light switch.
4. Disconnect the speedometer cable from the transaxle.
5. Disconnect the shift control cable and the select control cable. Remove the clips, washers and retainers from the cables.
6. Remove the water inlet from the transaxle.
7. Remove the clutch release cylinder-to-transaxle bolts and the release cylinder.
8. Remove both upper bellhousing bolts.
9. Using an engine fixture tool or equivalent, connect it to the engine and support it.
10. Raise and safely support the vehicle and remove the left wheel.
11. Remove the left, right and center splash shields.
12. Remove the center crossmember and the inspection cover.
13. Disconnect the lower control arms and the tie rods from the steering knuckles.
14. Disconnect both halfshafts from the transaxle.
15. Remove the starter bolts and the starter.
16. Using a floor jack, support the transaxle.
17. Remove the backside transaxle-to-engine bolts and the remaining mount bolt.
18. Remove the remaining bellhousing bolts, move the transaxle from the engine and lower it from the vehicle.
19. Using a wire, support the right halfshaft assembly.

TRANSAXLE INSTALLATION

1. Install the transaxle and the engine-to-transaxle bolts; torque the 10mm bolts to 34 ft. lbs. (46 Nm) and the 12mm bolts to 47 ft. lbs. (64 Nm).

NOTE: When installing the transaxle, guide the right halfshaft into the transaxle.

2. Install the transaxle mount and torque the bolts to 29 ft. lbs. (39 Nm).
3. Install the starter and torque the bolts to 29 ft. lbs. (39 Nm).
4. Install and connect the left halfshaft to the transaxle.
5. Connect the lower control arms and the tie rods to the steering knuckles.
6. Install the center crossmember and torque the bolts to 29 ft. lbs. (39 Nm). Install the inspection cover, the left, right and center splash shields.
7. Install the left wheel assembly.
8. Lower the vehicle and refill the transaxle with new fluid.
9. Remove the engine fixture tool or equivalent, from the engine.
10. Install the upper bellhousing bolts and torque the 10mm bolts to 34 ft. lbs. (46 Nm) and the 12mm bolts to 47 ft. lbs. (64 Nm).
11. Install the clutch release cylinder and the release cylinder-to-transaxle bolts.
12. Install the water inlet to the transaxle.
13. Install the clips, washers and retainers to the cables. Connect the shift control cable and the select control cable.
14. Connect the speedometer cable to the transaxle.
15. Connect the electrical connector to the back-up light switch.
16. Install the air cleaner and the inlet tube. Refill the cooling system.
17. Connect the negative battery cable.
18. Test drive the vehicle and check the operation of the transaxle.

BENCH OVERHAUL

Before Disassembly

When servicing the transaxle it is important to be aware of cleanliness. Before disassembling the transaxle, the outside should be throughly cleaned, preferably with a high-pressure spray cleaning equipment. Dirt entering the unit may negate all the effort and time spent on the overhaul.

During inspection and reassembly, all parts should be cleaned with solvent and dried with compressed air. Lubricate the seals with gear oil and use petroleum jelly to hold the thrust washers; this will ease the assembly of the seals and not leave harmful residues in the system. Do not use solvent on neoprene seals, if they are to be reused.

Before installing bolts into aluminum parts, dip the threads into clean gear oil. Anti-seize compound may be used to prevent galling the aluminum or seizing. Be sure to use a torque wrench to prevent stripping the threads. Be especially careful when installing the seals, the smallest nick can cause a leak. Aluminum parts are very susceptible to damage; great care should be used when handling them. Reusing snaprings is not recommended but should they be: compress the internal ones and compress the external ones.

Transaxle Disassembly

1. Secure the transaxle to a work stand.
2. Remove the clutch release fork, the clutch release bearing and the speedometer driven gear.
3. Remove the clutch release cylinder-to-transaxle bolts and the release cylinder.
4. Using the back-up light switch socket or equivalent, remove the back-up light switch.
5. Remove the transaxle case cover bolts and the cover.
6. Using a dial indicator, measure the 5th gear thrust clearance.
7. Remove the shift selection bellcrank, the set bolt, the shift and selecting lever.
8. Engage the gears, remove the locknut and disengage the gears.
9. From the No. 3 shift fork, remove the bolt with the lockwasher.
10. Using a pair of prybars and a hammer, tap the 5th gear snapring from the shaft.
11. Remove the 5th gear hub sleeve and driven gear. Using a

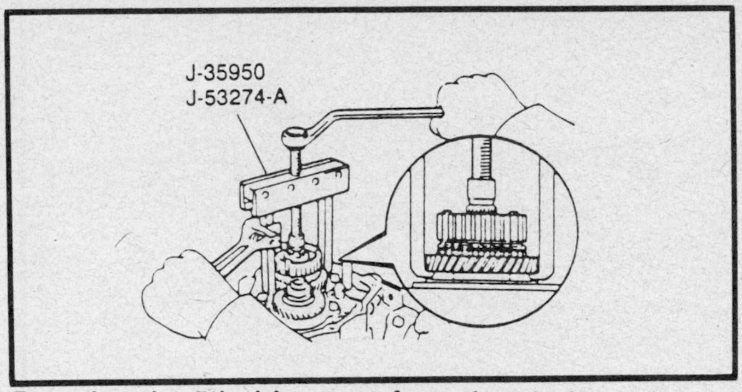
J-35950
J-53274-A
Pressing the 5th drive gear from the shaft

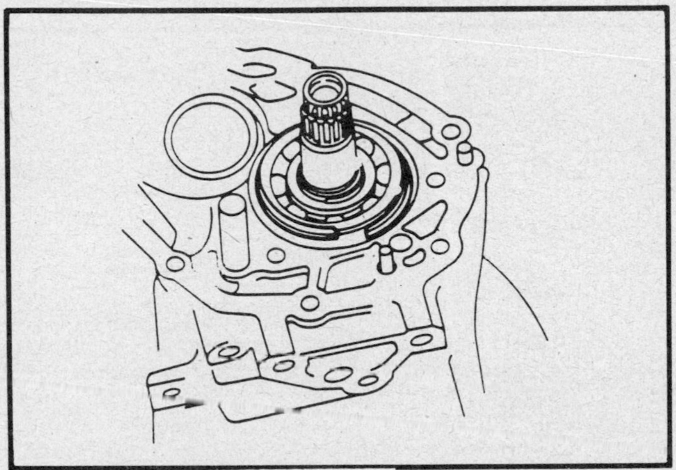
View of the bearing snaprings

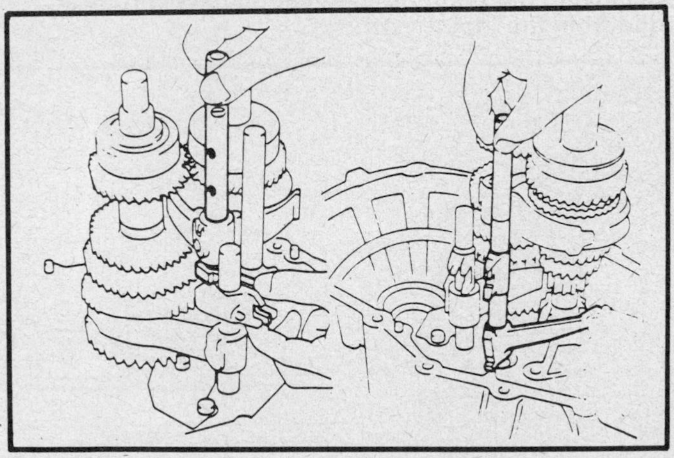
Removing the shift forks and the shafts

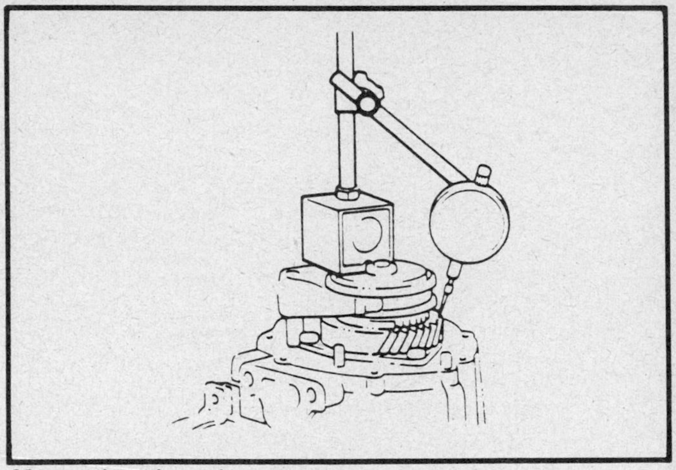
Measuring the 5th gear thrust clearance

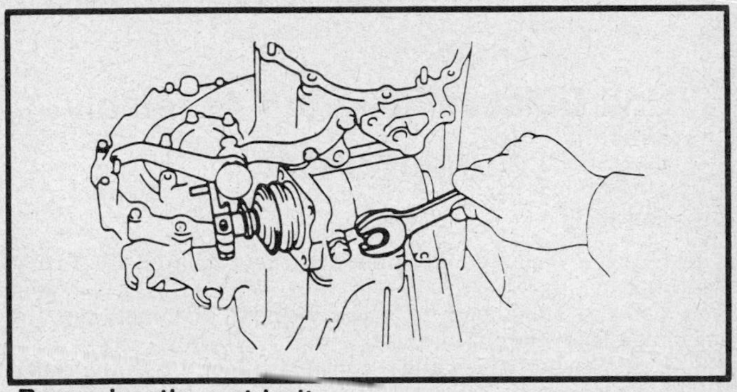
Removing the set bolt

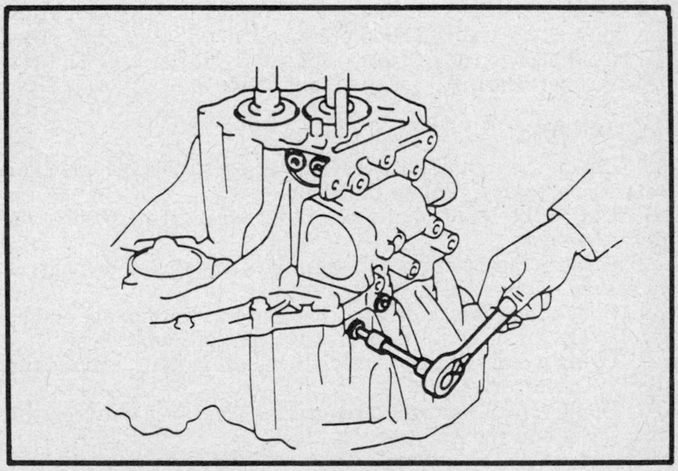
Removing the plugs and lock bolts

gear puller tool or equivalent, press the 5th drive gear from the shaft.

12. Remove the rear bearing retainer and both bearing snaprings. Remove the reverse idler gear shaft bolt and the No. 2 shift fork shaft snapring.

13. Using a No. 40 Torx® bit or equivalent, remove the 3 plug, seal, spring, ball and lockball assemblies; if necessary, use a magnet to remove the seats, springs and balls.

14. Remove the transaxle case-to-bellhousing case bolts and tap the cases apart with a plastic hammer; 3 bolts are located inside the bellhousing.

15. Remove the reverse idler gear and shaft. Using a pair of prybars and a hammer, tap out the shift shaft snaprings and remove the set bolts.

16. To remove the shift forks and shift fork shafts, perform the following procedures:

a. Remove the No. 2 fork shaft and shift head.
b. Using a magnet, remove both balls.
c. Remove the No. 3 fork shaft and the reverse shift fork.
d. Pull out the No. 1 fork shaft.
e. Remove the No. 1 and No. 2 shift forks.

17. Remove both the input and output shafts at the same time.

18. Remove the differential assembly, the magnet and the oil receiver.

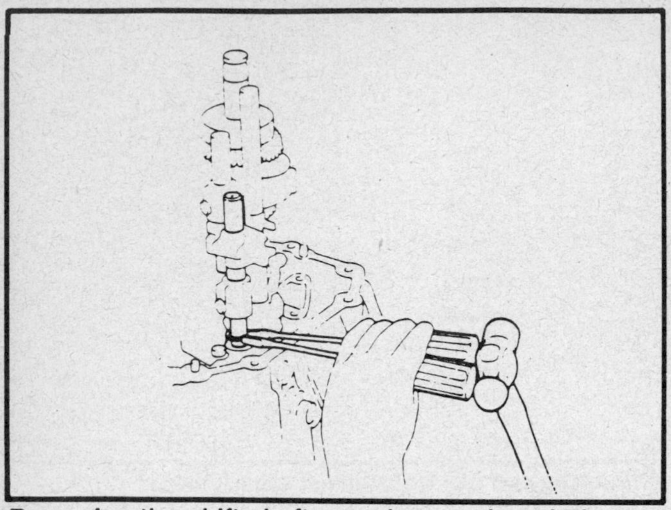

Removing the shift shaft snaprings and set bolts

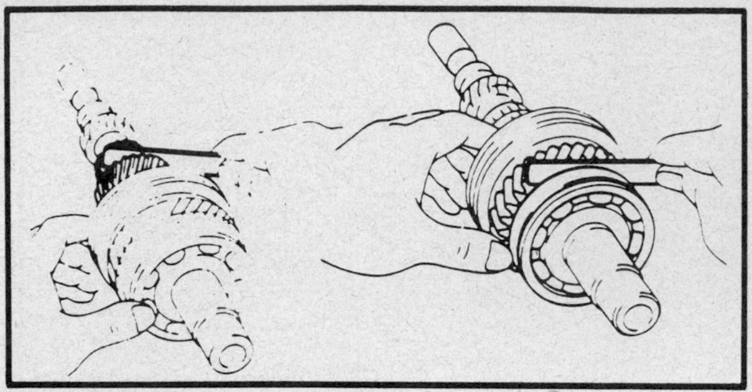

Measuring the 3rd and 4th gear thrust clearances

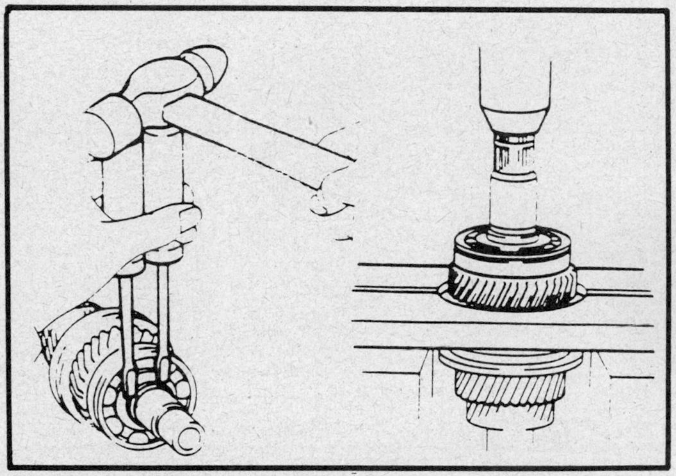

Removing the bearing, 4th gear and synchronizer ring from the input shaft

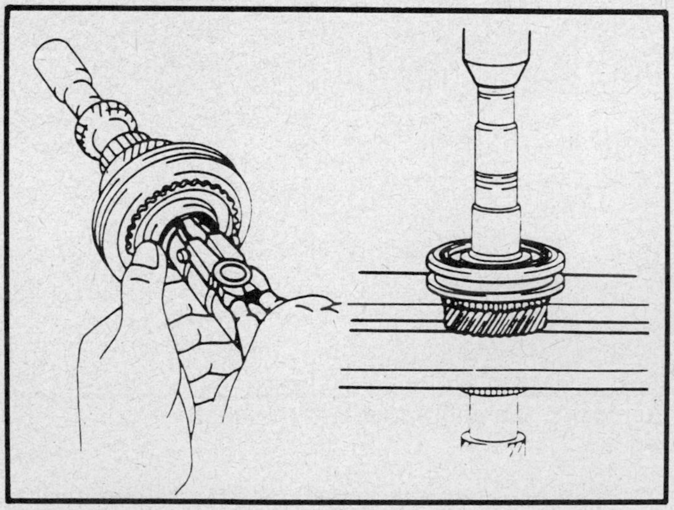

Removing the No. 2 hub sleeve and 3rd gear from the input shaft

Unit Disassembly And Assembly

INPUT SHAFT

Disassembly

1. Using a feeler gauge, measure the 3rd and 4th thrust clearance.
2. Using a small pair of prybars and a hammer, tap the snaprings from the input shaft.
3. Using a shop press and a split plate tool, press the radial ball bearing and 4th gear from the input shaft. Remove the needle roller bearings and the synchronizer ring.
4. Using a pair of snapring pliers, remove the snapring from the input shaft. Using a shop press and a split plate tool, press the No. 2 sleeve, the 3rd gear, the synchronizer ring and the needle roller bearing from the input shaft.

Inspection

1. Check the input shaft bearing surfaces for scoring and/or wear; if necessary, replace the shaft.
2. Check the input shaft splines for wear; if necessary, replace the shaft.
3. Inspect the bearings for scoring, wear and/or damage; if necessary, replace them.
4. Check the gears and synchronizer parts for damage and/or excessively worn teeth; if necessary, replace them.
5. Using a dial indicator, check the shaft runout; replace the shaft, if necessary.
6. Using a feeler gauge, measure the clearance between the shift forks and the hub sleeves.

Assembly

1. Using multi-purpose grease, lubricate the needle roller bearings.
2. Place the synchronizer ring onto the 3rd gear, align the ring slots with the shifting keys. Using a shop press and a split plate tool, press the 3rd gear/No. 2 hub sleeve onto the input shaft.
3. Select a snapring which will allow minimum axial play and install it onto the input shaft.
4. Using a feeler gauge, measure the 3rd gear thrust clearance; it should be 0.0039–0.0138 in. (0.10–0.35mm).
5. Using multi-purpose grease, lubricate the needle roller bearings.
6. Place the synchronizer ring onto the 4th gear and align the ring slots with the shifting keys.
7. Using a shop press, press the radial ball bearing onto the input shaft.
8. Select a snapring which will allow minimum axial play and install it onto the shaft.
9. Using a feeler gauge, measure the 4th gear thrust clearance; it should be 0.0039–0.0217 in. (0.10–0.55mm).

Mark	Thickness mm (in.)	Mark	Thickness mm (in.)
0	2.30 (0.0906)	3	2.48 (0.0976)
1	2.36 (0.0929)	4	2.54 (0.1000)
2	2.42 (0.0953)	5	2.60 (0.1024)

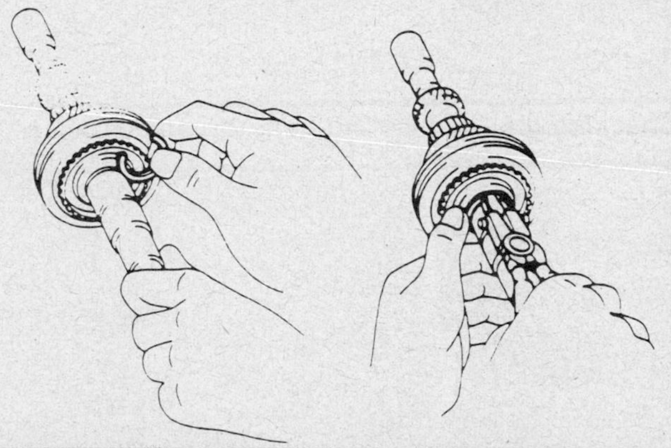

Selecting and installing the 3rd gear snapring onto the input shaft

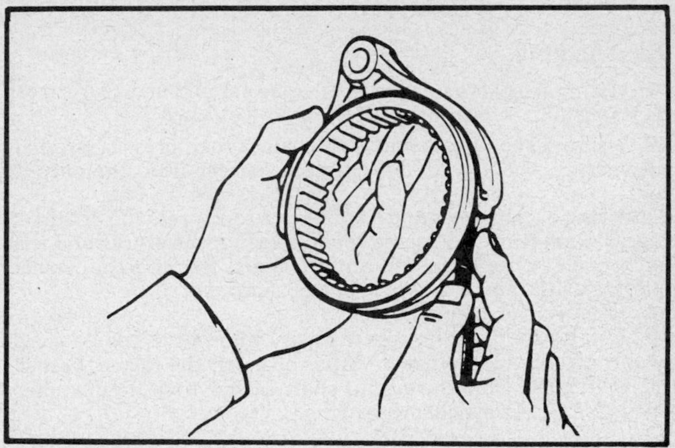

Measuring the shift fork-to-hub sleeve clearance

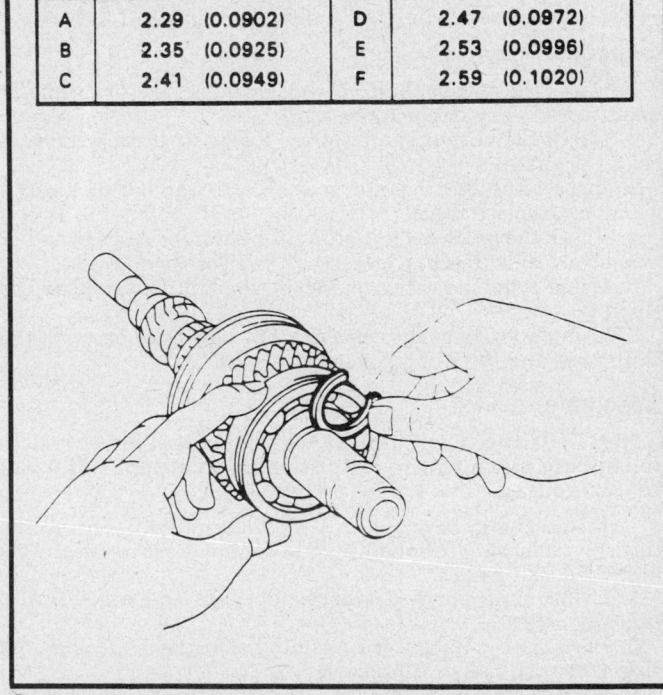

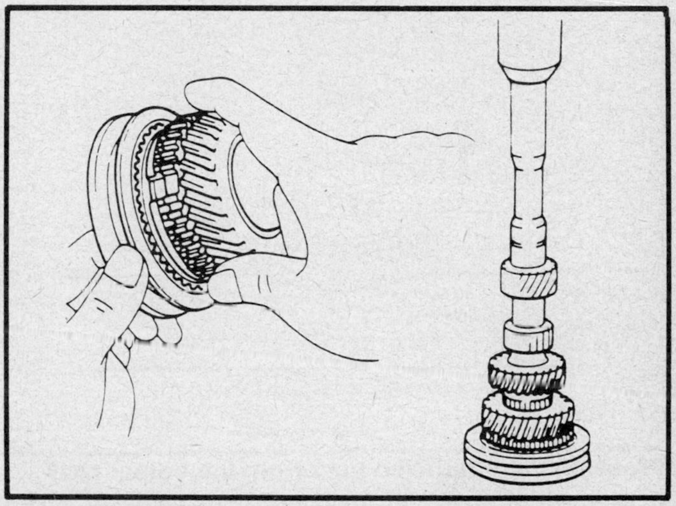

Installing the 3rd gear bearing synchronizer ring and No. 2 hub sleeve onto the input shaft

Mark	Thickness mm (in.)	Mark	Thickness mm (in.)
A	2.29 (0.0902)	D	2.47 (0.0972)
B	2.35 (0.0925)	E	2.53 (0.0996)
C	2.41 (0.0949)	F	2.59 (0.1020)

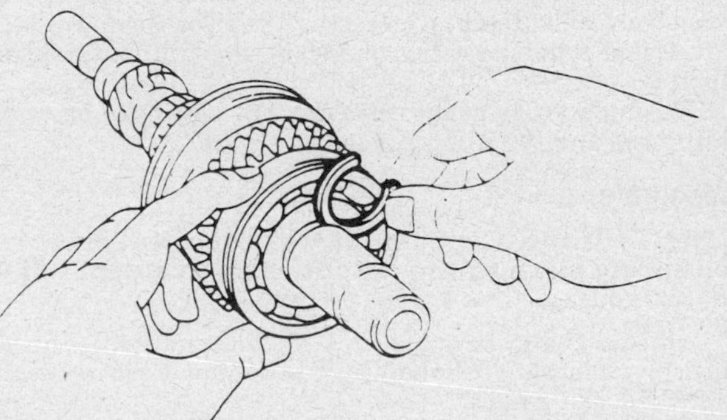

Selecting and installing the roller bearing snapring onto the input shaft

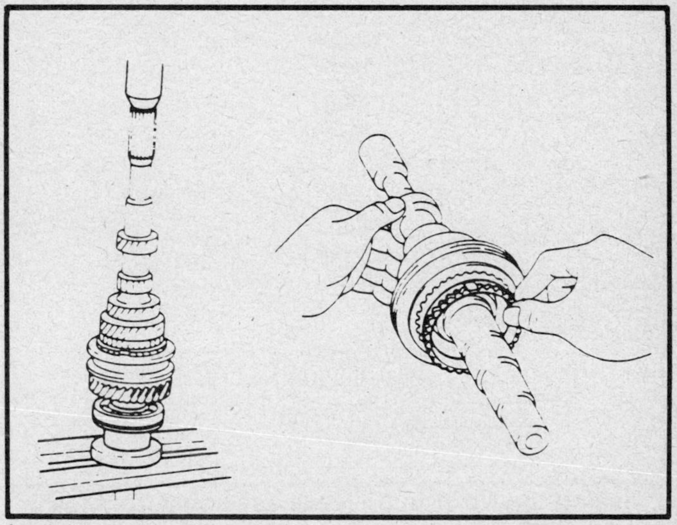

Installing the 4th gear bearing and synchronizer ring onto the input shaft

OUTPUT SHAFT

Disassembly

1. Using a feeler gauge, measure the 1st and 2nd gear thrust clearance on the output shaft.
2. Using a shop press and a split plate tool, press the radial ball bearing, the 4th driven gear and spacer from the output shaft.
3. Using a shop press and a split plate tool, press the 3rd driven gear, 2nd gear, the needle roller bearing, the spacer and the synchronizer ring from the output shaft. Remove the needle roller bearings and the synchronizer ring.
4. Remove the snapring.
5. Using a shop press, press the No. 1 hub sleeve, the 1st gear/synchronizer ring, the needle roller bearing, the thrust washer and locking ball from the output shaft. Remove the needle roller bearings and the synchronizer ring.

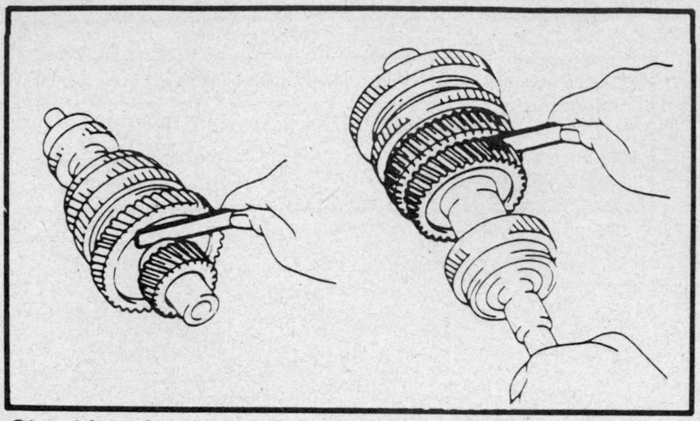
Checking the 1st and 2nd gear thrust clearance

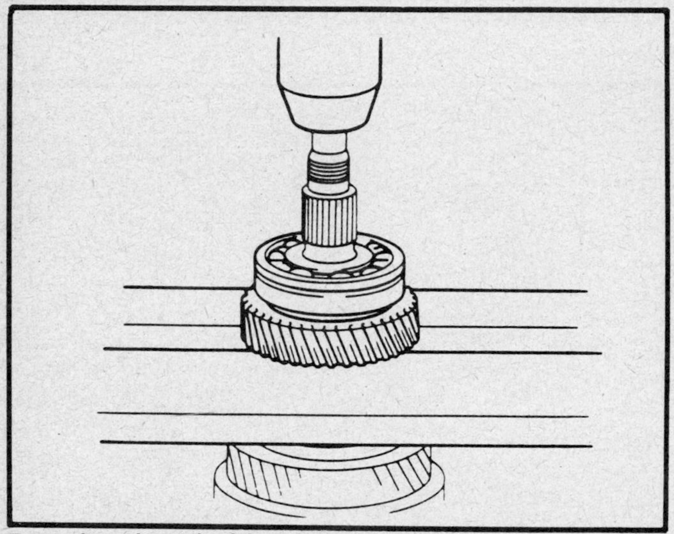
Pressing the 4th driven gear from the output shaft

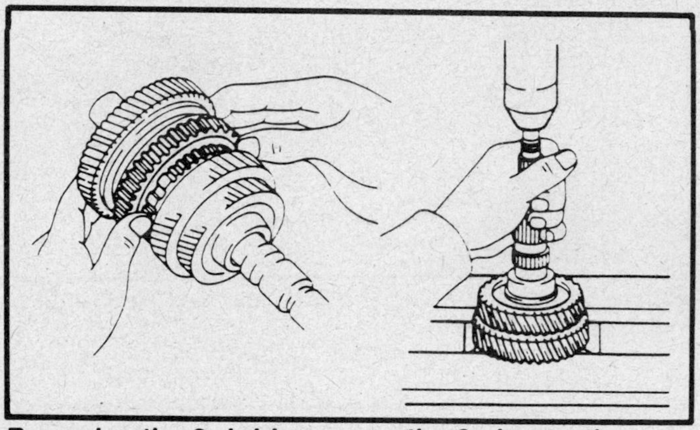
Removing the 3rd driven gear, the 2nd gear, the needle roller bearing, spacer and the synchronizer ring from the output shaft

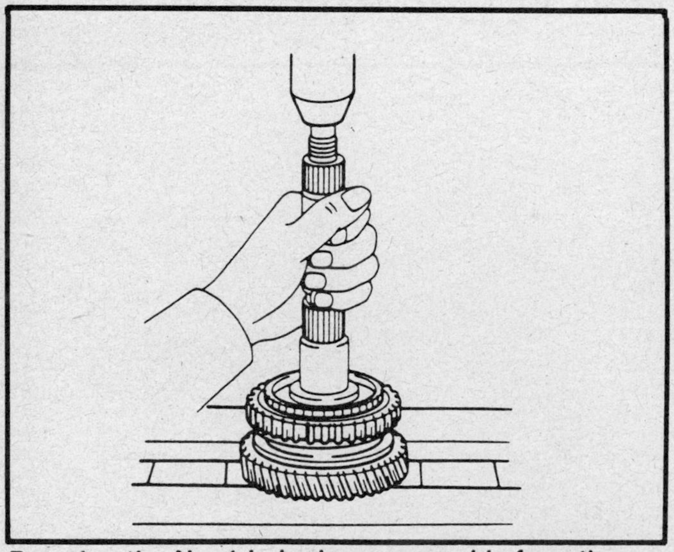
Pressing the No. 1 hub sleeve assembly from the output shaft

Inspection

1. Check the output shaft bearing surfaces for scoring and/or wear; if necessary, replace the shaft.
2. Check the output shaft splines for wear; if necessary, replace the shaft.
3. Inspect the bearings for scoring, wear and/or damage; if necessary, replace them.
4. Check the gears and synchronizer parts for damage and/or excessively worn teeth; if necessary, replace them.
5. Using a dial indicator, check the shaft runout; replace the shaft, if necessary.
6. Using a feeler gauge, measure the clearance between the shift forks and the hub sleeves.

Assembly

NOTE: If the output shaft was replaced, drive a slotted spring pin into the output shaft to a depth of 0.236 in. (6.0mm).

1. Install the locking ball into the output shaft. Install the thrust washer onto the shaft by positioning it's groove securely over the locking ball.
2. Using multi-purpose grease, lubricate the needle roller bearing.
3. Place the synchronizer ring onto the 1st gear and align the ring slots with the shifting keys.
4. Using a shop press, press the 1st gear/No. 1 hub sleeve assembly onto the output shaft.
5. Select a snapring which will allow minimum axial play and install it onto the shaft.

6. Using a feeler gauge, measure the 1st gear thrust clearance; is should be 0.0039–0.0157 in. (0.10–0.40mm).
7. Install the spacer. Position the synchronizer ring onto the gear and align the ring slots with the shifting keys.
8. Using multi-purpose grease, lubricate the needle roller bearings and install the 2nd gear.
9. Using a shop press and a split plate tool, press the 3rd driven gear onto the output shaft.
10. Using a feeler gauge, measure the 2nd gear thrust clearance; is should be 0.0039–0.0177 in. (0.10–0.45mm).
11. Install the spacer. Using a shop press and a split plate tool, press the 4th driven gear and bearing onto the output shaft.

Mark	Thickness mm (in.)	Mark	Thickness mm (in.)
A	2.50 (0.0984)	D	2.68 (0.1055)
B	2.56 (0.1008)	E	2.74 (0.1079)
C	2.62 (0.1031)	F	2.80 (0.1102)

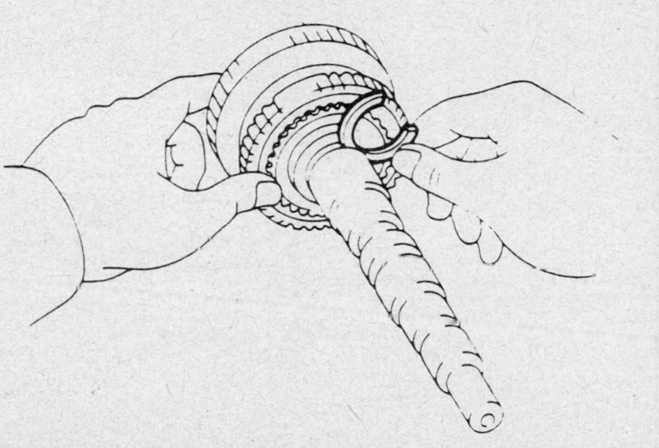

Selecting and installing the 1st gear snapring onto the output shaft

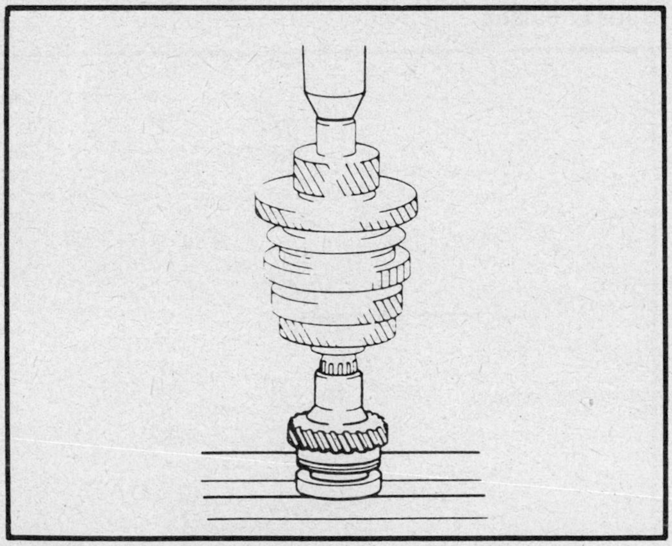

Installing the output gear spacer and 4th driven gear onto the output shaft

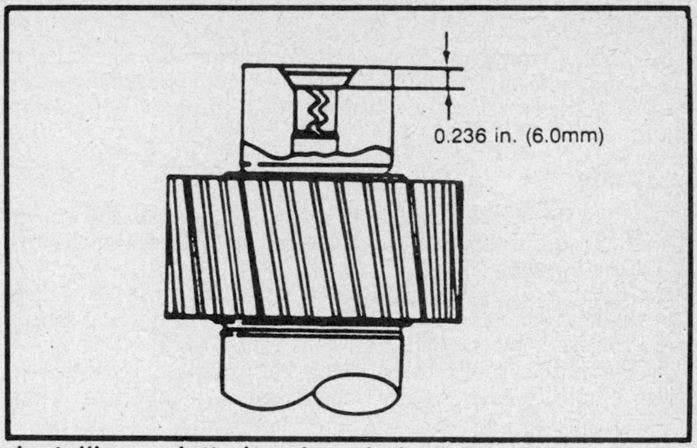
0.236 in. (6.0mm)

Installing a slotted spring pin into the output shaft

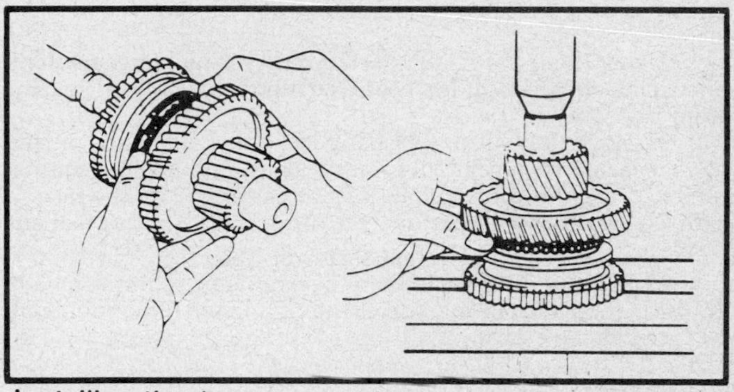

Installing the 1st gear and No. 1 hub sleeve onto the output shaft

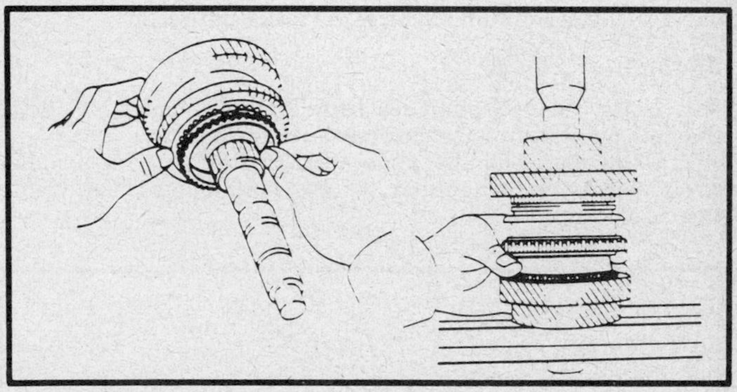

Installing the 2nd gear and 3rd driven gear onto the output shaft

SYNCHRONIZER

The transaxle is equipped with 3 synchronizers: 1–2, 3–4 and 5 speeds.

Disassembly

1. From each synchronizer, remove both key springs.
2. Noting the relative positions, separate the hub, sleeve and keys; be sure to scribe the hub-to-sleeve location.

Inspection

1. Clean, inspect and/or replace any worn or damaged parts.
2. Using a feeler gauge, measure the synchronizer ring back-to-gear spline end clearance; it should be a minimum of 0.024 in. (0.6mm).

Assembly

1. Align the scribe marks and assemble the hub to the sleeve; the extruded hub lip should be directed away from the sleeve's shift fork groove.
2. Install a retaining ring. Carefully pry the ring back and insert the keys (one at a time); be sure to position the ring so it is locked in place by the keys.
3. Install the other retaining ring; be sure the ring's open segment is out of phase with the open segment of the other ring.

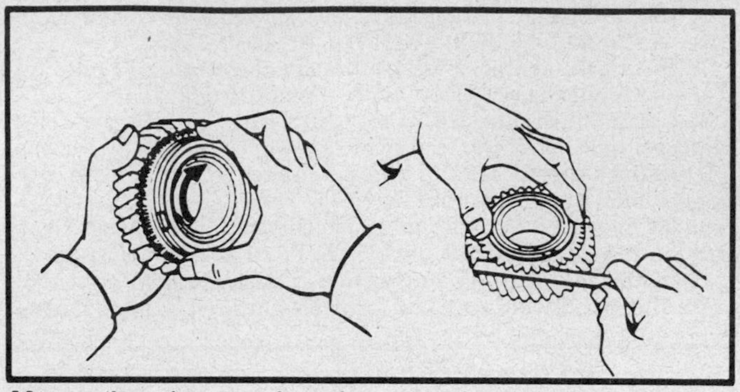

Measuring the synchronizer ring back-to-gear spline end

FRONT CASE

Disassembly

1. Using the input shaft bearing remover tool or equivalent, and a slide hammer puller, press the input shaft's front bearing from the case.
2. Remove the bolt and bearing lockplate. Using the output shaft bearing remover tool or equivalent, and a slide hammer puller, press the output shaft's front bearing from the case.
3. Using a small prybar, pry the input shaft front oil seal and the right hand differential oil seal from the case.
4. Using the differential side bearing cup removal tool or equivalent, and a hammer, drive the right hand differential outer race from the case.

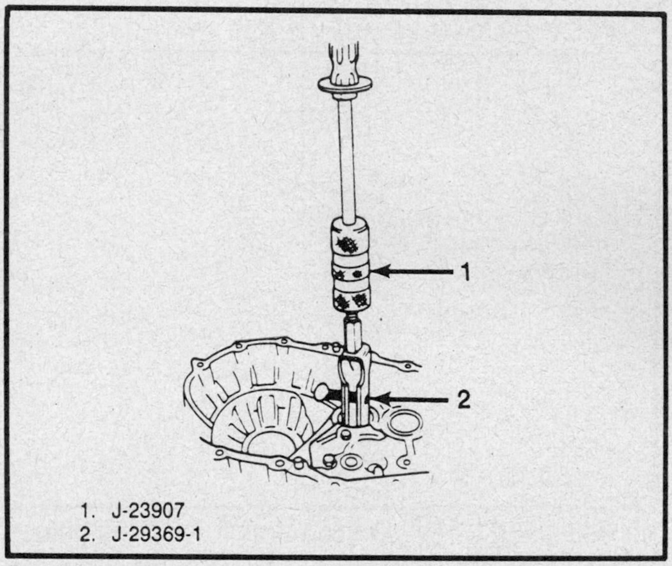
1
2
1. J-23907
2. J-29369-1

Removing the input shaft's front bearing from the transaxle case

Inspection

1. Check and clean the case.
2. Using a gasket remover tool or equivalent, remove the sealant from the mating surfaces; be careful not to gouge or damage the aluminum surfaces, as leaks can result.

Assembly

1. Using the input shaft seal installer tool or equivalent, drive the new oil seal into the case until it seats.
2. Using driver handle, an input shaft front bearing installer or equivalent, and a hammer, drive the input shaft bearing into the case.

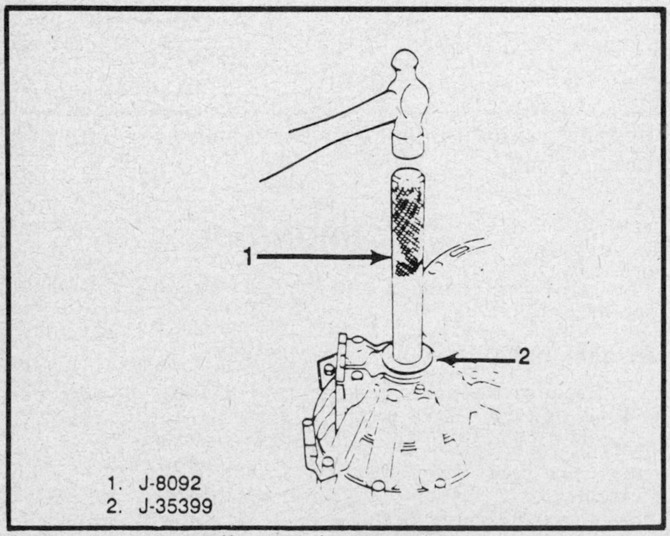
1
2
1. J-8092
2. J-35399

Removing the differential outer race

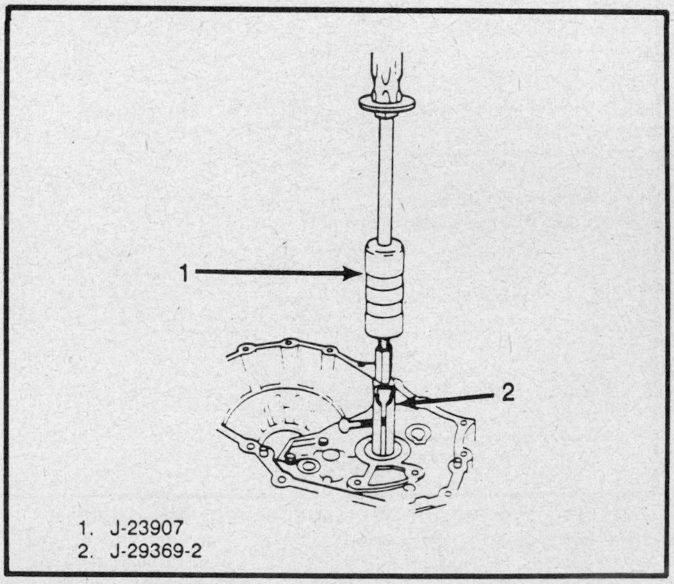
1
2
1. J-23907
2. J-29369-2

Removing the output shaft's front bearing from the transaxle case

3. Using driver handle, an output shaft front bearing installer or equivalent, and a hammer, drive the output shaft bearing into the case.
4. Using the differential seal installer tool or equivalent, and a hammer, drive the right hand differential outer race into the case.
5. Using multi-purpose grease, lubricate the new oil seal. Using the differential seal installer tool or equivalent, and a hammer, drive the new differential oil seal into the case.

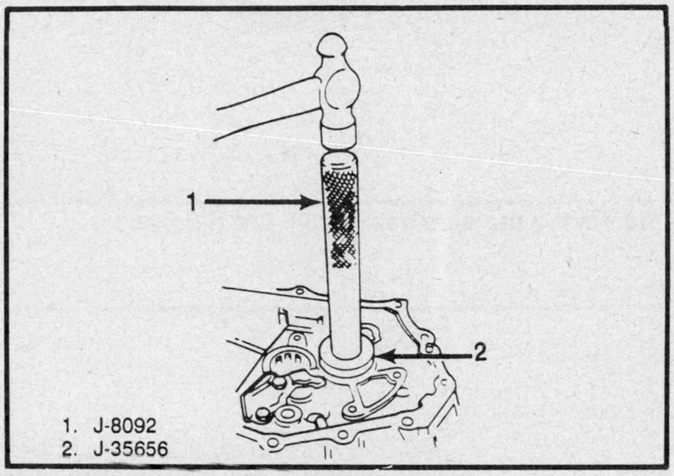
1
2
1. J-8092
2. J-35656
Installing the input shaft front bearing

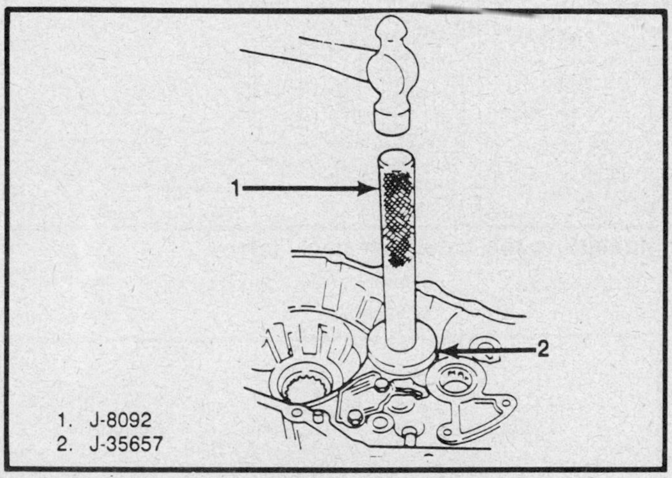
1
2
1. J-8092
2. J-35657
Installing the output shaft front bearing

MAIN CASE

Disassembly

1. Using a small prybar, pry the left hand differential oil seal from the case.
2. Using the differential side bearing cup removal tool or equivalent, and a hammer, drive the left hand differential outer race from the case.

Inspection

1. Check and clean the case and install the magnet.
2. Using a gasket remover tool or equivalent, remove the sealant from the mating surfaces; be careful not to gouge or damage the aluminum surfaces, as leaks can result.

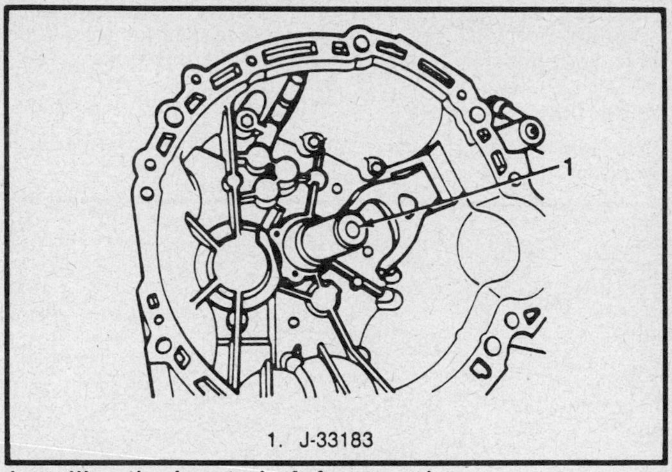
1
1. J-33183
Installing the input shaft front seal

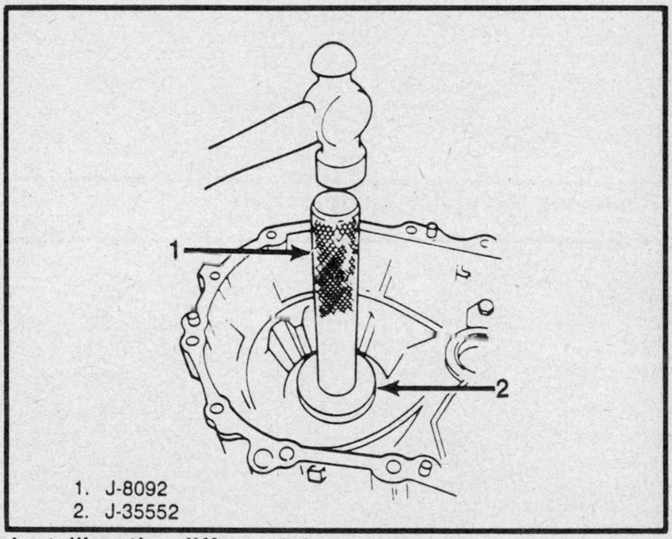
1
2
1. J-8092
2. J-35552
Installing the differential outer race

Assembly

1. Using the differential seal installer tool or equivalent, and a hammer, drive the left hand differential outer race into the case.
2. Using multi-purpose grease, lubricate the new oil seal. Using the differential seal installer tool or equivalent, and a hammer, drive the new differential oil seal into the case.

DIFFERENTIAL

Disassembly

1. Using the differential side bearing puller tool or equivalent, and a differential side bearing puller pilot or equivalent, press the side bearings from the differential case.
2. Loosen the staked part of the lockplate. Using a scribing tool, make alignment marks at the ring gear-to-case location.
3. Remove the ring gear-to-case bolts and tap the ring gear from the case.
4. Using a dial micrometer, hold a pinion and side gear toward the case and measure the backlash of the other side gear; it should be 0.0020–0.0079 in. (0.05–0.20mm). If the backlash is not within specifications, replace the thrust washer.

5. Using a punch and a hammer, drive the pinion shaft lockpin from the ring gear side of the case. Remove the pinion shaft, both pinions, both side gears and the thrust washers.

Inspection

Clean and inspect the parts for damage; replace the parts, if necessary.

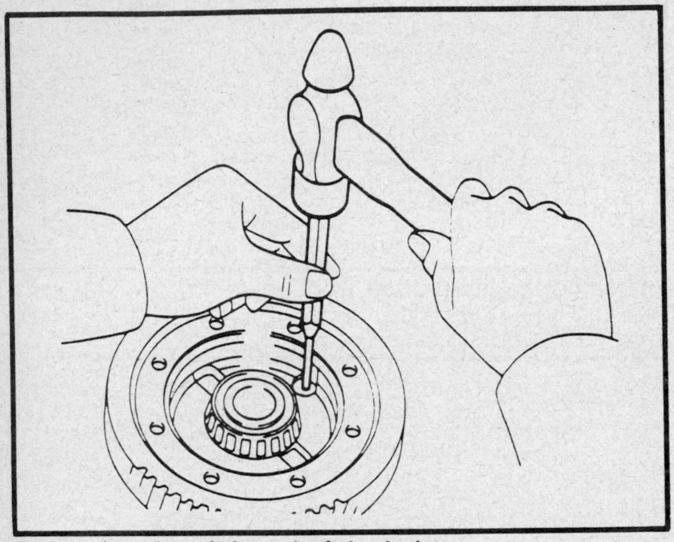

Removing the pinion shaft lockpin

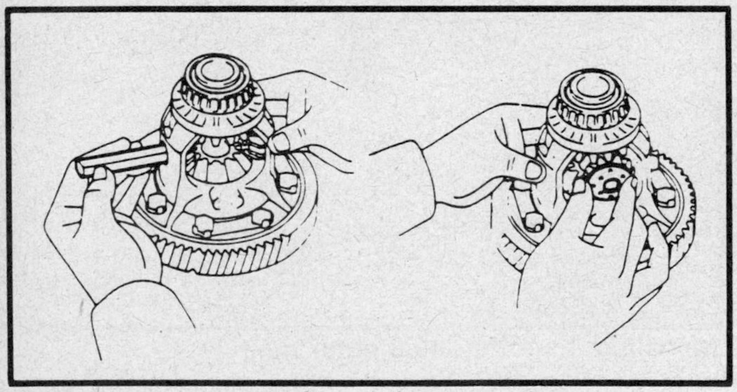

Removing pinion shaft and thrust washers

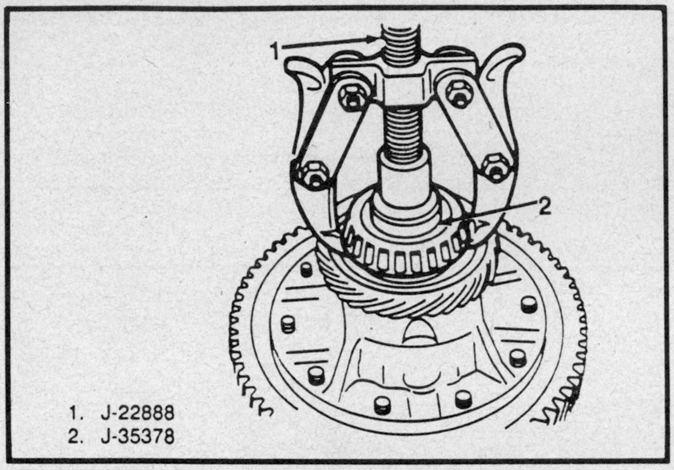
1
2
1. J-22888
2. J-35378

Removing the side gears from the differential

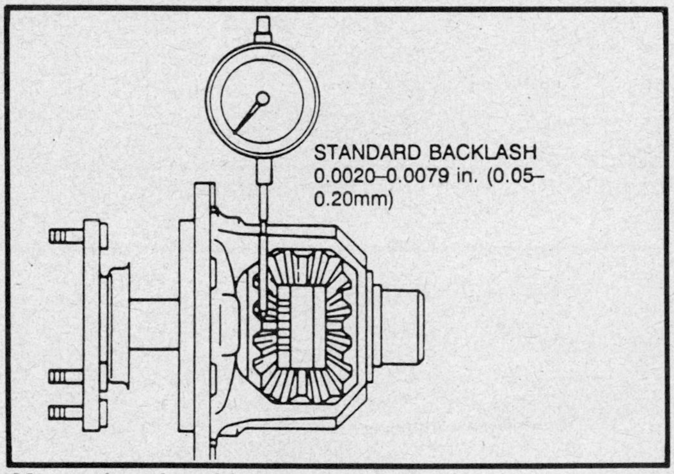
STANDARD BACKLASH
0.0020–0.0079 in. (0.05–0.20mm)

Measuring the side gear backlash

Assembly

1. Assemble the pinion shaft, both pinions, both side gears and the thrust washers into the case.
2. Using a dial micrometer, hold a pinion and side gear toward the case and measure the backlash of the other side gear; it should be 0.0020–0.0079 in. (0.05–0.20mm). If the backlash is not within specifications, replace the thrust washer.
3. Using a punch and a hammer, drive the pinion shaft lockpin into the ring gear side of the case.
4. Align the ring gear with the differential case, install a new locking plate and set bolts. Torque the bolts evenly, in several steps, to 71 ft. lbs. (97 Nm).
5. Using a hammer and a punch, stake the locking plates.

NOTE: Stake a claw flush with the flat surface of the nut. For the claw contacting the protruding portion of the nut, stake only ½ on the tightening side.

6. Using shop press and the differential side bearing installer tool or equivalent, press the differential side bearings onto the differential case.

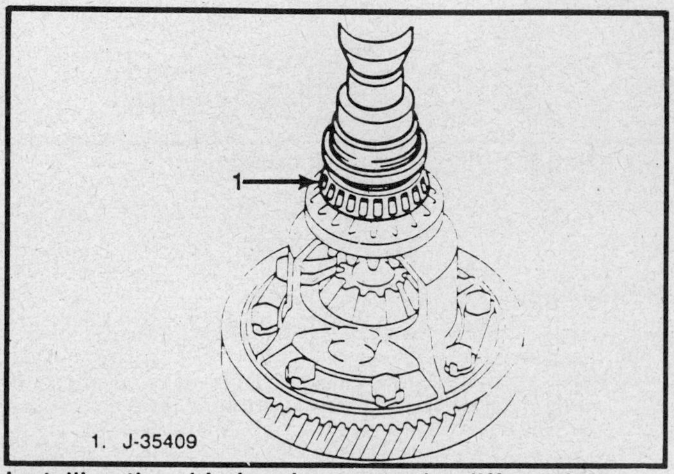
1
1. J-35409

Installing the side bearings onto the differential case

Thickness mm (in.)		Thickness mm (in.)	
0.95	(0.0374)	1.10	(0.0433)
1.00	(0.0394)	1.15	(0.0453)
1.05	(0.0413)	1.20	(0.0472)

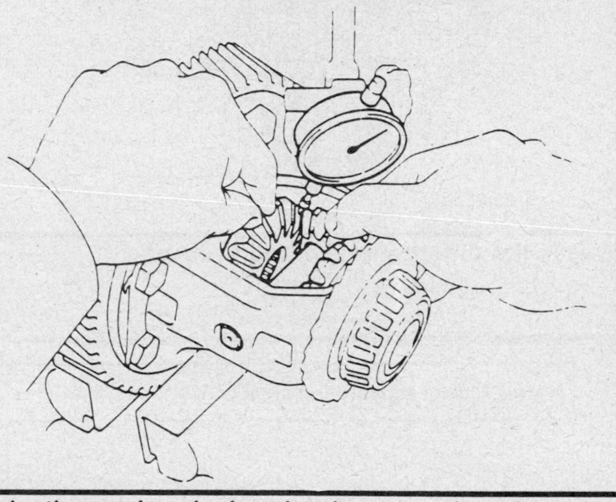

Selecting and replacing the thrust washers into the differential case

SHIFT AND SELECT LEVER

Disassembly

1. Remove the shift/select lever assembly from the case.
2. Remove the E-ring and compression ring.
3. Using a pin punch and a hammer, drive the slotted spring pins from the No. 1 and No. 2 inner shift levers.
4. Remove the No. 2 shift inner lever, the No. 1 shift inner lever and the shift interlock plate.
5. Using a pin punch and a hammer, drive the slotted spring pins from the select inner lever.
6. Remove the select inner lever, the compression spring and the spring seat.
7. Tap the snapring from the lever shaft and remove the lever shaft and boot.
8. Using a small prybar, pry the out seal and replace it, if necessary.

Inspection

Clean and inspect the parts for damage; replace the parts, if necessary.

Assembly

1. Grease the shaft, install the boot and shaft to the control shaft cover.

NOTE: When installing the boot, be sure to position the boot's air bleed downward.

2. Install the snapring, the spring seat, the compression spring and the select inner lever.
3. Using a pin punch and a hammer, drive in the slotted spring pin.
4. Align the interlock plate to the No. 1 inner shift lever and install it.
5. Install the No. 2 inner shift lever.
6. Using a pin punch and a hammer, drive in the slotted spring pins.
7. Install the compression spring, seat and E-ring.

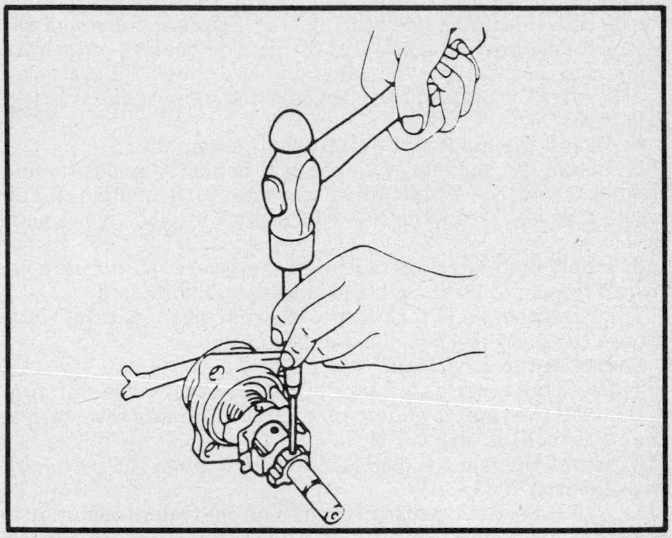

Removing the inner lever from the shift and select lever assembly

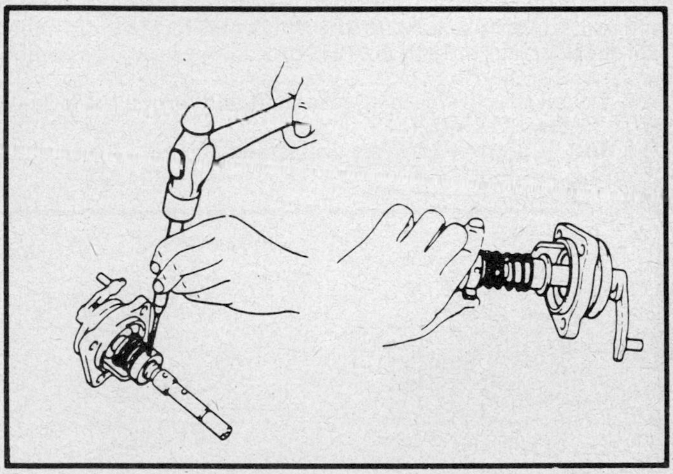

Removing the compression spring and seat from the shift and select lever assembly

Transaxle Assembly

1. Install the magnet into the bottom of the case.
2. Install the oil receiver and the differential into the case.
3. To adjust the differential side bearing preload, perform the following procedures:
   a. Install the thinnest side bearing shim into the transaxle case.
   b. Install the differential into the transaxle case.
   c. Install the transaxle case and torque the bolts to 22 ft. lbs. (29 Nm).
   d. Using the differential preload wrench or equivalent, and a torque wrench, measure the bearing preload; the new bearings should be 7–14 inch lbs. (0.8–1.6 Nm) or old bearings should be 4–9 inch lbs. (0.5–1.0 Nm).

NOTE: The preload will change about 2.6–3.5 inch lbs. (0.3–0.4 Nm) with each shim thickness.

e. If the preload is not correct, use the differential side bearing cup remover tool or equivalent, to remove the transaxle case side outer side bearing race. Select and install a different side bearing shim, install the side bearing outer race and the case. Torque the case bolts and check the preload.

f. After the preload has been adjusted, remove the transaxle case.

4. Install the input and output shaft assemblies.
5. Install the shift forks and shift fork shafts by positioning the No. 1 and No. 2 shift forks into the No. 1 and No. 2 hub sleeve grooves. Insert the No. 1 fork shaft into the No. 1 shift fork hole.
6. Insert both interlock balls into the reverse shift fork hole. Install the No. 3 fork shaft and the reverse shift fork.
7. Install the No. 2 fork shaft, the shift head and the bolts; torque the bolts to 12 ft. lbs. (16 Nm).
8. Install the snaprings.
9. Place the reverse shift fork pivot into the reverse shift arm and install the reverse shift arm into the transaxle case; torque the bolts to 13 ft. lbs. (17 Nm).
10. Install the reverse idler gear and shaft; align the gear mark with the case hole.
11. Using Loctite® sealant No. 518 or equivalent, apply it to the front and main transaxle case mounting surfaces and assemble the cases. Torque the main case-to-front case bolts to 22 ft. lbs. (29 Nm).
12. Into the lockball assembly holes, install the balls, springs and seats. Apply sealant to the plugs and lockball assembly. Torque the plugs to 18 ft. lbs. (25 Nm) and the lockball assembly to 29 ft. lbs. (39 Nm).
13. Install the reverse idler gear shaft and torque the lockbolt to 18 ft. lbs. (24 Nm).
14. Install both bearing snaprings and the No. 2 fork shaft snapring.

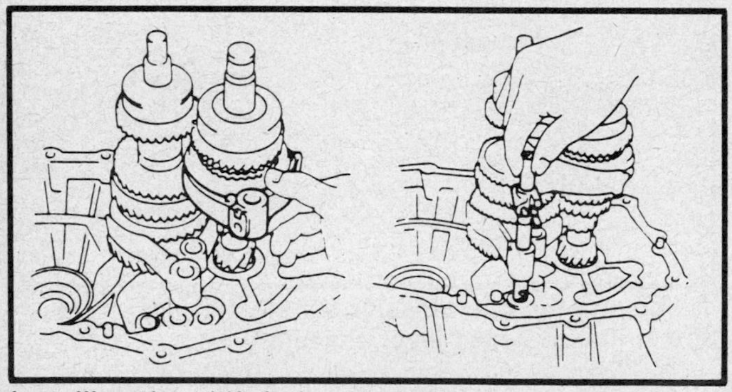

Installing the shift fork and shafts

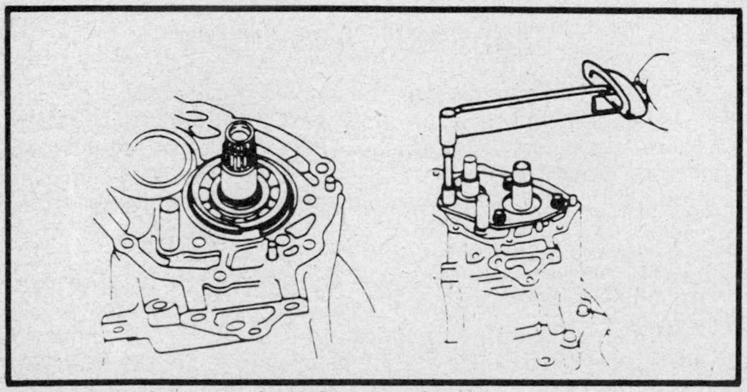

Installing the rear bearing retainer

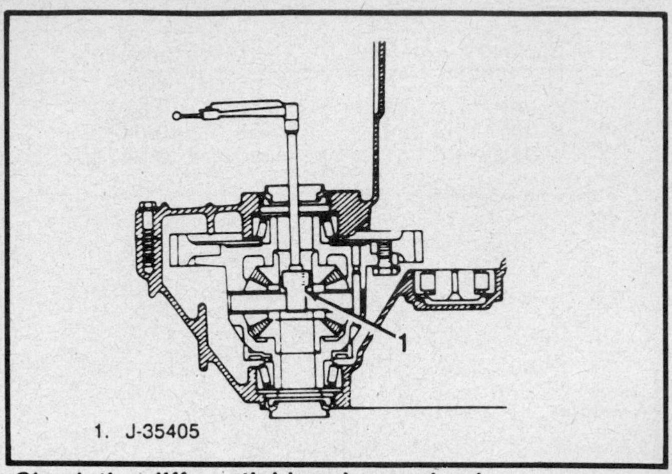
1
1. J-35405

Check the differential bearing preload

Mark	Thickness mm (in.)	Mark	Thickness mm (in.)
A	2.10 (0.0827)	L	2.60 (0.1024)
B	2.15 (0.0846)	M	2.65 (0.1043)
C	2.20 (0.0866)	N	2.70 (0.1063)
D	2.25 (0.0886)	P	2.75 (0.1083)
E	2.30 (0.0906)	Q	2.80 (0.1102)
F	2.35 (0.0925)	R	2.85 (0.1122)
G	2.40 (0.0945)	S	2.90 (0.1142)
H	2.45 (0.0965)	T	2.95 (0.1161)
J	2.50 (0.0984)	U	3.00 (0.1181)
K	2.55 (0.1004)		

Shim selection chart

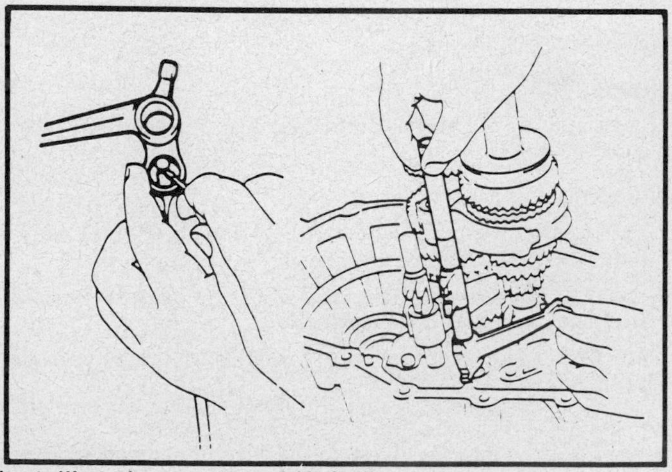

Installing the reverse shift fork

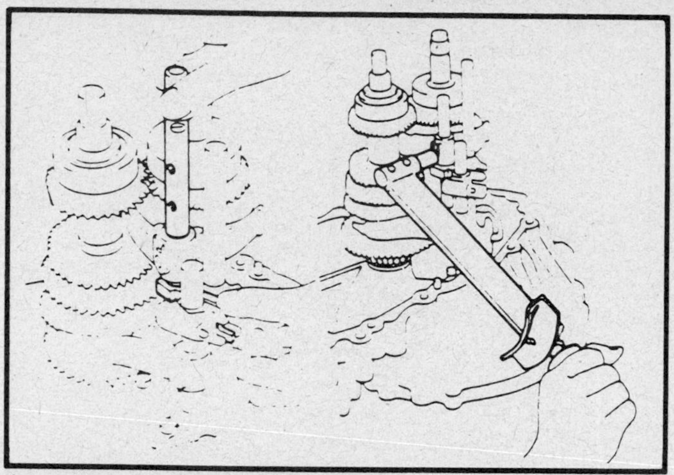

Installing the No. 2 fork shaft

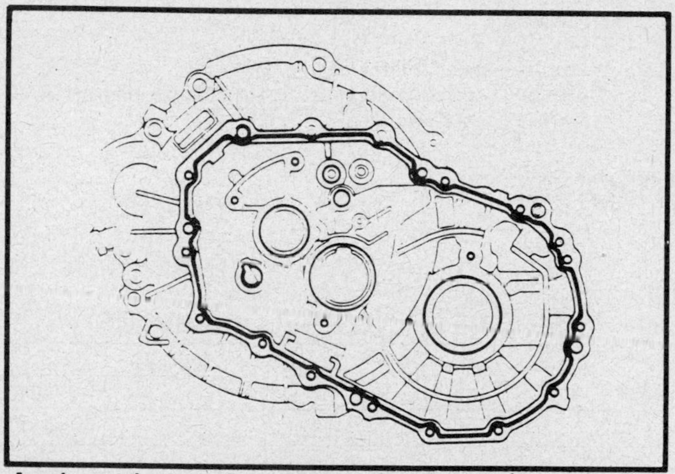

Apply sealant to the case mounting surfaces

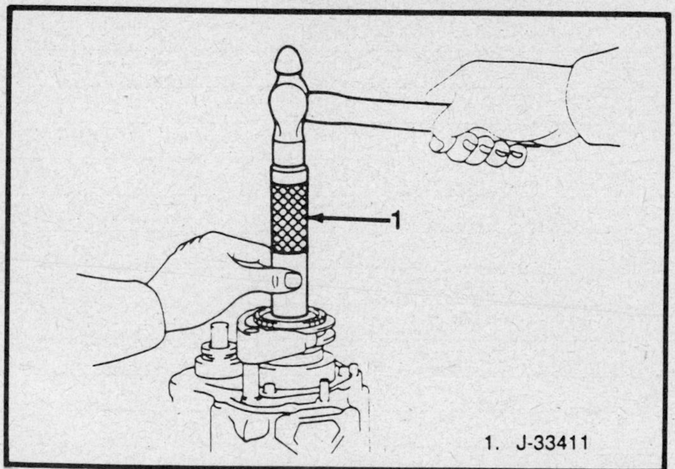
1
1. J-33411

Installing the No. 3 hub sleeve

15. Install the rear bearing retainer and torque the bolts to 14 ft. lbs. (19 Nm).
16. Using the 5th driven gear installer tool or equivalent, and a hammer, drive the 5th driven gear onto the shaft.
17. Install the spacer, the needle bearing, the 5th gear and the synchronizer ring.
18. Onto the hub sleeve, install the clutch hub and the shifting keys and the key springs.

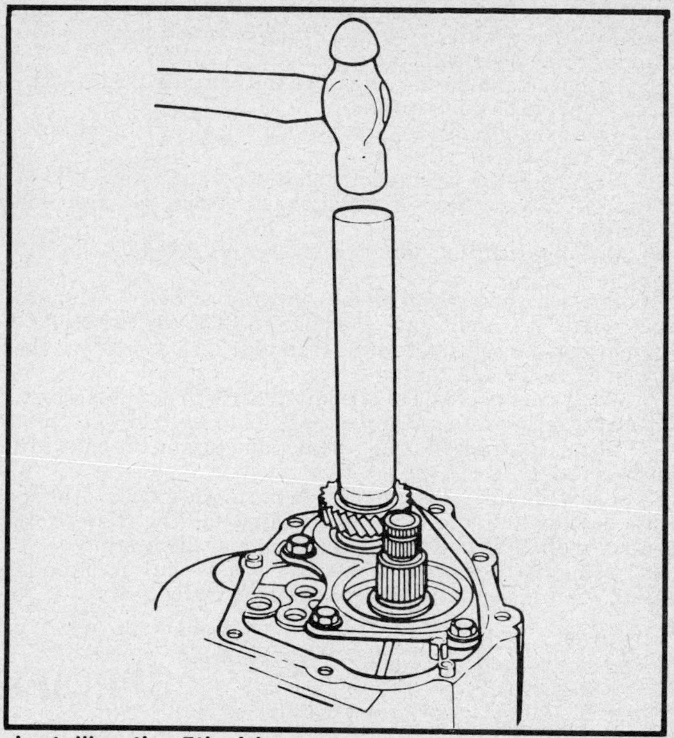

Installing the 5th driven gear

Mark	Thickness mm (in.)	Mark	Thickness mm (in.)
A	2.25 (0.0886)	E	2.49 (0.0980)
B	2.31 (0.0909)	F	2.55 (0.1004)
C	2.37 (0.0933)	G	2.61 (0.1028)
D	2.43 (0.0957)		

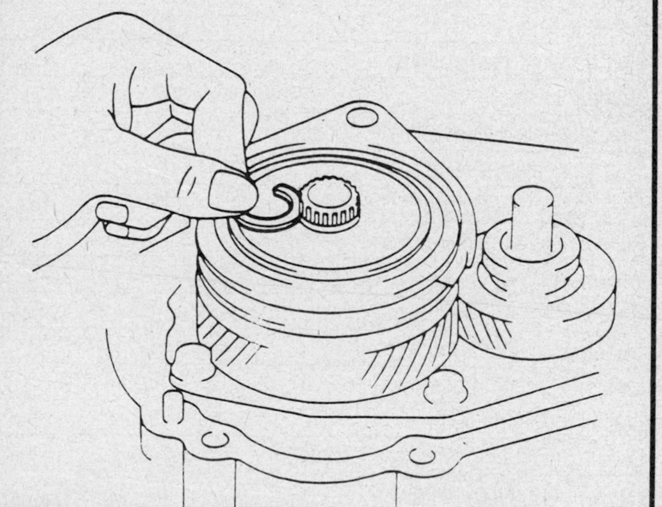

Selecting and installing the 5th gear snapring

NOTE: When installing the shifting keys, position them under the shifting keys and position them so their gaps do not align with each other.

19. Using the 5th gear installer tool, and a hammer, drive the No. 3 hub sleeve, with the No. 3 shifting fork, and align the synchronizer ring slots with the shifting keys.
20. Using a dial indicator, measure the 5th gear thrust clearance; it should be 0.004–0.022 in. (0.10–0.57mm).
21. Select and install a snapring which will allow minimum axial play, onto the output shaft.
22. Engage the 5th gear and install the locknut; torque it to 87 ft. lbs. (118 Nm). After installation, disengage the gear and stake the locknut.
23. Install the shifting fork bolt and torque it to 12 ft. lbs. (16 Nm).
24. Position a new gasket on the control shaft cover, install the shift/select lever and torque the bolts to 14 ft. lbs. (20 Nm).
25. Install the bellcrank to the transaxle case and torque the bolt to 22 ft. lbs. (29 Nm).
26. Apply Loctite® No. 518 or equivalent, to the transaxle case, install the case cover and torque the bolts to 13 ft. lbs. (18 Nm).
27. Install the front bearing retainer and torque the bolts to 8 ft. lbs. (11 Nm).
28. Using lithium grease, apply it to the inside groove of the release bearing hub, the input shaft spline and the release fork contact surface. Install the release fork and the bearing.
29. Using the back-up light switch socket or equivalent, and a socket wrench, install and torque the back-up light switch to 30 ft. lbs. (40 Nm).
30. Install the speedometer driven gear.
31. Shift the transaxle and check it for smooth operation.

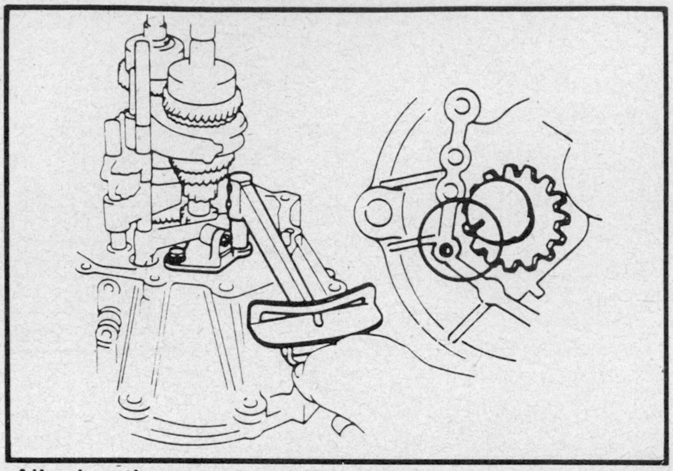
Aligning the reverse shift arm and idler gear

SPECIFICATIONS

TORQUE SPECIFICATIONS

	ft. lbs.	Nm
Output shaft front bearing lock plate	8	11
Transmission case bolts	22	29
Fork shaft set bolts	13	18
Reverse shift arm bracket	17	24
Lockball plugs	18	25
Reverse idler gear shaft lock bolt	29	39
Rear bearing retainer bolts	14	19
5th gear locknut	87	118
Shift and select lever assembly bolts	14	20
Shift and select lever assembly lock bolt	22	29
Transmission case cover bolts	13	18
Front bearing retainer bolts	8	11
Backup light switch	30	40
Transaxle to engine bolts (10 mm)	34	46
Transaxle to engine bolts (12 mm)	47	64
Left engine mount bolts	38	52
Starter mounting bolts	29	39
Engine mounting center member bolts	29	39
Front and rear mount bolts	29	39
Differential bearing preload—new ②	7–14 ①	0.8–1.6
Differential bearing preload—used ②	4–9 ①	0.5–1.0

① Inch lbs.
② For each change in shim thickness preload will change by 2.5–3.5 inch lbs. (0.3–0.4 Nm)

SIDE BEARING PRELOAD SHIMS

Mark	Thickness in.	Thickness mm
A	0.082	2.10
B	0.084	2.15
C	0.086	2.20
D	0.088	2.25
E	0.090	2.30
F	0.092	2.35
G	0.094	2.40
H	0.096	2.45
J	0.098	2.50
K	0.100	2.55
L	0.102	2.60
M	0.104	2.65
N	0.106	2.70
P	0.108	2.75
Q	0.110	2.80
R	0.112	2.85
S	0.114	2.90
T	0.116	2.95
U	0.118	3.00

RADIAL BALL BEARING TO INPUT SHAFT SNAPRING

Mark	Thickness in.	Thickness mm
A	0.090	2.29
B	0.092	2.35
C	0.094	2.41
D	0.097	2.47
E	0.099	2.53
F	0.102	2.59

1-2 HUB SLEEVE TO OUTPUT SHAFT SNAPRING

Mark	Thickness in.	Thickness mm
A	0.098	2.50
B	0.100	2.56
C	0.103	2.62
D	0.105	2.68
E	0.107	2.74
F	0.110	2.80

SHIFT LEVER FREEPLAY ADJUSTMENT SHIMS

Shim Thickness in.	mm	in.	mm
0.020	0.5	0.035	0.9
0.024	0.6	0.039	1.0
0.028	0.7	0.043	1.1
0.031	0.8	0.047	1.2

5TH GEAR HUB SLEEVE TO INPUT SHAFT SNAPRING

Mark	Thickness in.	Thickness mm
A	0.088	2.25
B	0.090	2.31
C	0.093	2.37
D	0.096	2.43
E	0.098	2.49
F	0.100	2.55
G	0.102	2.61

3-4 HUB SLEEVE TO INPUT SHAFT SNAPRING

Mark	Thickness in.	Thickness mm
0	0.090	2.30
1	0.092	2.36
2	0.095	2.42
3	0.097	2.48
4	0.100	2.54
5	0.102	2.60

DIMENSION AND CLEARANCE SPECIFICATIONS

	in.	mm
5th Gear Thrust Clearance		
Standard	.004–.022	.10–.57
Maximum	.025	.65
4th Gear Thrust Clearance		
Standard	.004–.022	.10–.55
Maximum	.023	.60
3rd Gear Thrust Clearance		
Standard	.004–.014	.10–.35
Maximum	.015	.40
2nd Gear Thrust Clearance		
Standard	.004–.018	.10–.45
Maximum	.020	.50
1st Gear Thrust Clearance		
Standard	.004–.016	.10–.40
Maximum	.018	.45
Input Shaft Dimensions		
Maximum Runout	.002	.05
Minimum Diameter		
Journal A	.979	24.87
Journal B	1.042	24.87
Journal C	1.2219	30.97
Journal D	.983	24.97
Output Shaft Dimensions		
Maximum Runout	.002	.05
Minimum Diameter		
Journal A	1.298	32.97
Journal B	1.495	37.97
Journal C	1.259	31.97

SPECIAL TOOLS

Number	Description
J8092	Universal driver handle
J2241-11	Side bearing puller adapter, differential
J22888	Bearing remover
J22888-35	Bearing remover leg (two)
J22888-50	Bearing remover leg set
J22912-01	Input/output shaft gears remover
J22919	Differential inner bearing installer
J23423-A	Differential and output shaft bearing cup installer
J23598	Side bearing puller adapter, differential
J23907	Slide hammer and adapter set
J26900-19	Metric dial depth gage
J26935	Shim selection set
J26935-3	Differential shimming gage
J26935-13	Shim selection set spacers (seven) Bolts, M8 x 1.25-6G/Length—160 mm (seven)
J26938	Seal and race, differential, installer
J35823	Shift shaft seal installer
J35824	Input bearing assembly remover/installer/shaft support bearing installer
J36027	Shift shaft bearing remover
J36029	Shift rail bushing remover/installer
J36029-1	Adapter to drive handle
J36029-2	Install/remover adapter car screw, $^{1}/_{4}$-20 x2$^{1}/_{2}$
J36030	Reverse shift rail bushing installer
J36031	Bearing retainer bolt hex socket
J36032	Clutch shaft inner reverse shift rail bushing remover
J36033	Clutch shaft inner bushing installer
J36034	Sliding sleeve bushing remover/installer
J36037	Clutch shaft upper bushing remover/installer
J36038	Output shaft bearing cup remover
J36039	Shift detent lever bushing remover/installer
J36181	Differential bearing cup remover
J36182	Input/output shaft remover/installer assembly set
J36182-1	Pallet
J36182-2	Disassembly adapter (two)
J36183	Input shaft remover/installer press tube
J36184	Output shaft remover/installer adapter press tube reducer
J36185	Input/output shaft bearing remover
J36189	Shift shaft bearing installer
J36190	Universal driver handle

Section 6

HM-282 Transaxle
General Motors

APPLICATION

MG1 TRANSAXLE

Year	Body	Vehicle	Engine
1987	N	Grand Am	2.0L TBI
	J	Sunbird	2.0L TBI
1988	N	Grand Am	2.0L TBI
	J	Sunbird	2.0L TBI
1989	N	Grand Am	2.0L TBI
	J	Sunbird	2.0L TBI

TBI—Throttle Body Injection

MG2 TRANSAXLE

Year	Body	Vehicle	Engine
1986	A	6000	2.8L MPI
	P	Fiero	2.8L MPI
	J	Cimarron	2.8L MPI
	J	Firenza	2.8L MPI
1987	A	6000	2.8L MPI
	P	Fiero	2.8L MPI
	J	Cimarron	2.8L MPI
	J	Firenza	2.8L MPI
	J	Cavalier	2.8L MPI
	L	Beretta	2.0L MPI
	L	Beretta	2.8L MPI
	L	Corsica	2.0L MPI
	L	Corsica	2.8L MPI
1988	A	6000	2.8L MPI
	A	Century	2.8L MPI

MG2 TRANSAXLE

Year	Body	Vehicle	Engine
1988	A	Celebrity	2.8L MPI
	J	Cavalier	2.8L MPI
	A	Ciera	2.8L MPI
	P	Fiero	2.8L MPI
	J	Cimarron	2.8L MPI
	W	Cutlass Supreme	2.8L MPI
	W	Grand Prix	2.8L MPI
1989	A	6000	2.8L MPI
	A	Celebrity	2.8L MPI
	J	Cavalier	2.8L MPI
	N	Calias	2.3L EFI
	W	Cutlass Supreme	2.8L MPI
	W	Grand Prix	2.8L MPI
	L	Beretta	2.8L MPI
	L	Corsica	2.8L MPI
	N	Grand Am	2.3L EFI

EFI—Electronic Fuel Injection
MPI—Multi-Port Fuel Injection

MG3 TRANSAXLE

Year	Body	Vehicle	Engine
1987	P	Fiero	2.8L MPI
	P	Fiero	2.5L TBI
	J	Sunbird	2.0L TBI

MPI—Multi-Port Fuel Injection
TBI—Throttle Body Injection

GENERAL DESCRIPTION

The Hydra-matic Muncie 282 (HM-282) transaxle is a 5 speed unit. The gearing provides for 5 synchronized forward speeds, a reverse speed, a final drive with differential output and speedometer drive.

The input and output gear clusters are nested very close together, requiring extremely tight tolerances of shafts, gears and synchronizers.

The input shaft is supported by a roller bearing in the clutch and differential housing and a ball bearing in the transaxle case.

The output shaft is supported by a roller bearing in the clutch and differential housing and a combination ball-and-roller bearing in the transaxle case.

The differential case is supported by opposed tapered roller bearings which are under preload.

The speed gears are supported by roller bearings. A bushing supports the reverse idler gear.

Transaxle Identification

The transaxle identification stamp is located at the center top of the case. The transaxle identification tag is located on the left side near the left side cover.

Metric Fasteners

The metric fastener dimensions are very close to the dimensions of the familiar inch system fasteners and, for this reason, replacement fasteners must have the same measurement and strength as those removed.

Do not attempt to interchange metric fasteners for inch system fasteners. Mismatched or incorrect fasteners can result in

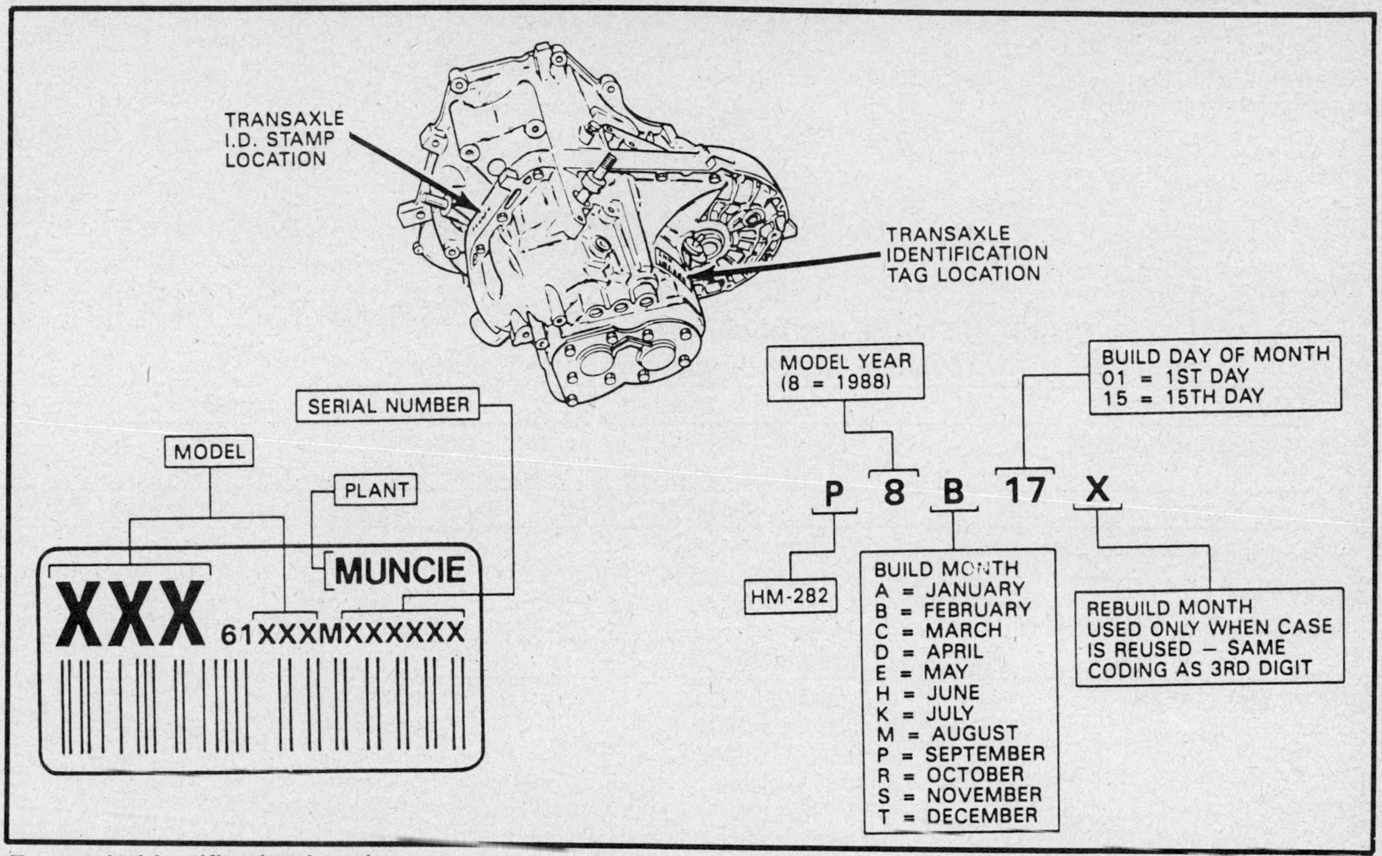
TRANSAXLE
I.D. STAMP
LOCATION
TRANSAXLE
IDENTIFICATION
TAG LOCATION
SERIAL NUMBER
MODEL
PLANT
MUNCIE
XXX
61XXXMXXXXXX
MODEL YEAR
(8 = 1988)
BUILD DAY OF MONTH
01 = 1ST DAY
15 = 15TH DAY
P 8 B 17 X
HM-282
BUILD MONTH
A = JANUARY
B = FEBRUARY
C = MARCH
D = APRIL
E = MAY
H = JUNE
K = JULY
M = AUGUST
P = SEPTEMBER
R = OCTOBER
S = NOVEMBER
T = DECEMBER
REBUILD MONTH
USED ONLY WHEN CASE
IS REUSED – SAME
CODING AS 3RD DIGIT

Transaxle identification location

damage to the transaxle unit through malfunctions, breakage or possible personal injury.

Care should be taken to re-use the fasteners in the same locations as removed.

Capacities

To fill a dry transaxle on A-body and P-body vehicles, fill the unit with 2 quarts (1.9L) of 5W–30 manual transaxle oil, part number 1052931 or equivalent. (On 1988–89 vehicles, use syncromesh transaxle fluid, part number 12345349 or equivalent.) To fill a dry transaxle on J-body, L-body, N-body and W-body vehicles, fill the unit with 5 pints (2.1L) of 5W–30 manual transaxle oil, part number 1052931 or equivalent. (On 1988–89 vehicles, use syncromesh transaxle fluid, part number 12345349 or equivalent.)

Checking Fluid Level

Check the fluid level only when the engine is **OFF**, the vehicle is level and the transaxle is cool. To check the fluid level, remove the dipstick and read the level indicated. Be sure the fluid level is between the **H** (hot) and **C** (cold) marks on the dipstick.

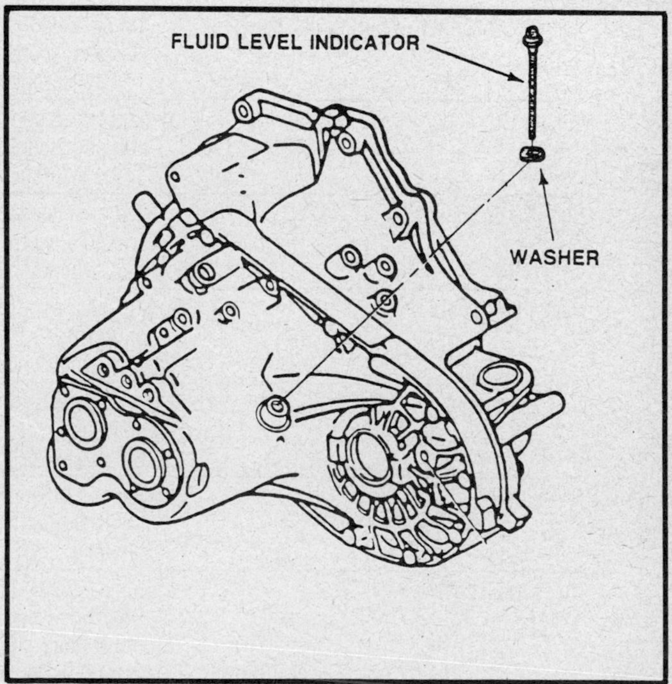
FLUID LEVEL INDICATOR
WASHER

Fluid level indicator/fill plug

MODIFICATIONS

There are no modifications on the HM-282 transaxle at the time of this publication printing.

TROUBLE DIAGNOSIS

CHILTON THREE C'S DIAGNOSIS CHART
Muncie HM 282 5-Speed Manual Transaxle

Condition	Cause	Correction
A knock at low speeds	a) Drive axle CV or TRI-POT joints worn b) Side gear hub counterbore worn	a) Replace CV or joints b) Replace hub counterbore
Noise most pronounced on turns	a) Differential gear noise	a) Check and repair or replace gear
Clunk on acceleration or deceleration	a) Engine mounts loose b) Drive axle inboard TRI-POT joints worn c) Differential pinion shaft in case worn d) Side gear hub counterbore in case worn oversize	a) Replace mounts b) Replace joints c) Replace shaft d) Repair or replace hub counterbore
Vibration	a) Wheel bearing rough b) Drive axle shaft bent c) Tires out-of-round d) Tire unbalance e) CV joint in drive axle shaft worn f) Drive axle angle (trim height) incorrect	a) Replace bearing b) Replace shaft c) Replace tire d) Balance tire e) Replace CV joint f) Adjust angle
Noisy in neutral with engine running	a) Input gear bearings worn b) Clutch release bearing worn	a) Replace bearings b) Replace bearings
Noisy in 1st only	a) 1st speed constant-mesh gears chipped, scored or worn b) 1/2 synchronizer worn	a) Replace gears b) Replace synchronizer
Noisy in 2nd only	a) 2nd speed constant-mesh gears chipped, scored or worn b) 1/2 synchronizer worn	a) Replace gears b) Replace synchronizer
Noisy in 3rd only	a) 3rd gear constant-mesh gears chipped, scored or worn b) 3/4 synchronizer worn	a) Replace gears b) Replace synchronizer
Noisy in 4th only	a) 4th gear or output gear chipped, scored or worn b) 3/4 synchronizer worn	a) Replace gears b) Replace synchronizer
Noisy in 5th only	a) 5th speed gear or output gears chipped, scored or worn b) 5th gear synchronizer worn	a) Replace gears b) Replace synchronizer
Noisy in reverse only	a) Reverse idler gear, idler gear bushing, input or output gear(s) chipped, scored or worn	a) Replace gear(s) and/or bushings
Noisy in all gears	a) Insufficient lubricant b) Bearings worn c) Input gear (shaft) and/or output gear (shaft) worn, chipped or scored	a) Add lubricant b) Replace bearings c) Replace gear(s)
Slips out of gear	a) Linkage worn or improperly adjusted b) Shift linkage does not work freely; binds c) Cables bent or worn	a) Replace and/or adjust linkage b) Repair or replace linkage c) Replace cables

CHILTON THREE C'S DIAGNOSIS CHART
Muncie HM 282 5-Speed Manual Transaxle

Condition	Cause	Correction
Slips out of gear	d) Input gear bearing retainer broken or loose	d) Repair or replace retainer
	e) Shift fork worn or bent	e) Replace shift fork
Leaks lubricant	a) Fluid level indicator not seated in fill port, causing fluid leakage at vent plug	a) Seat fluid level indicator
	b) Axle shaft seals worn	b) Replace seals
	c) Excessive amount of lubricant in transaxle	c) Remove excessive fluid
	d) Input gear (shaft) bearing retainer loose or broken	d) Repair or replace retainer
	e) Input gear bearing worn and/or lip seal damaged	e) Replace bearing and/or lip seal
	f) Worn shift lever seal leaks	f) Replace seal
	g) Lack of sealant between case and clutch cover or loose clutch cover	g) Repair case and/or cover

POWER FLOW

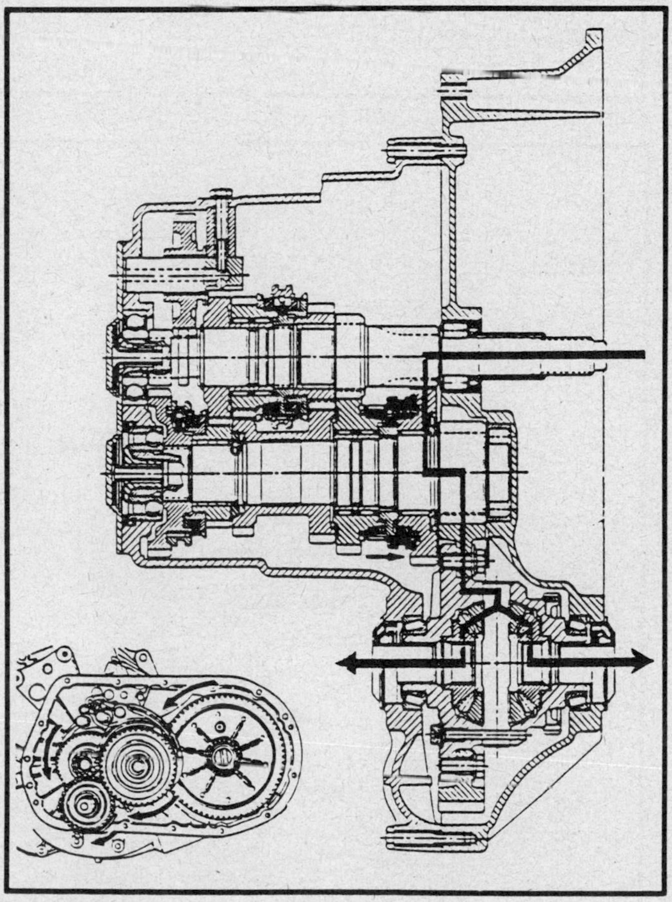
Manual transaxle power flow diagram—1st gear

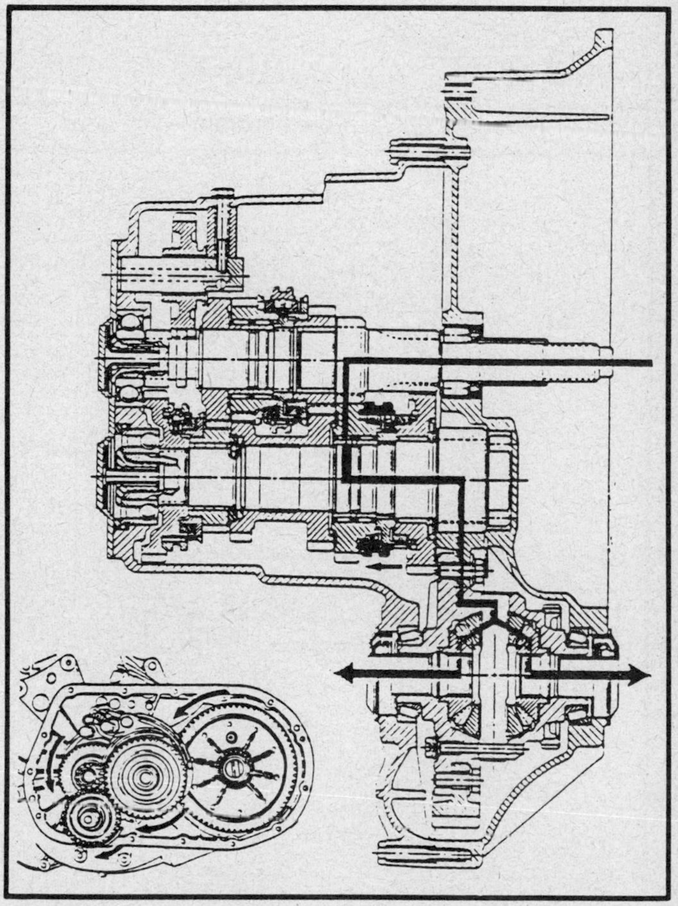
Manual transaxle power flow diagram—2nd gear

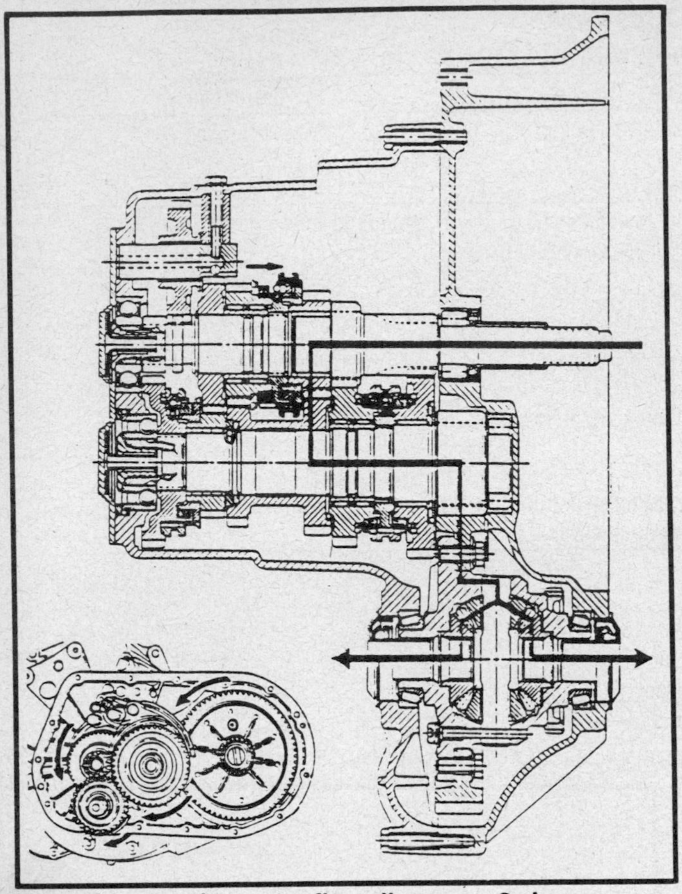
Manual transaxle power flow diagram—3rd gear

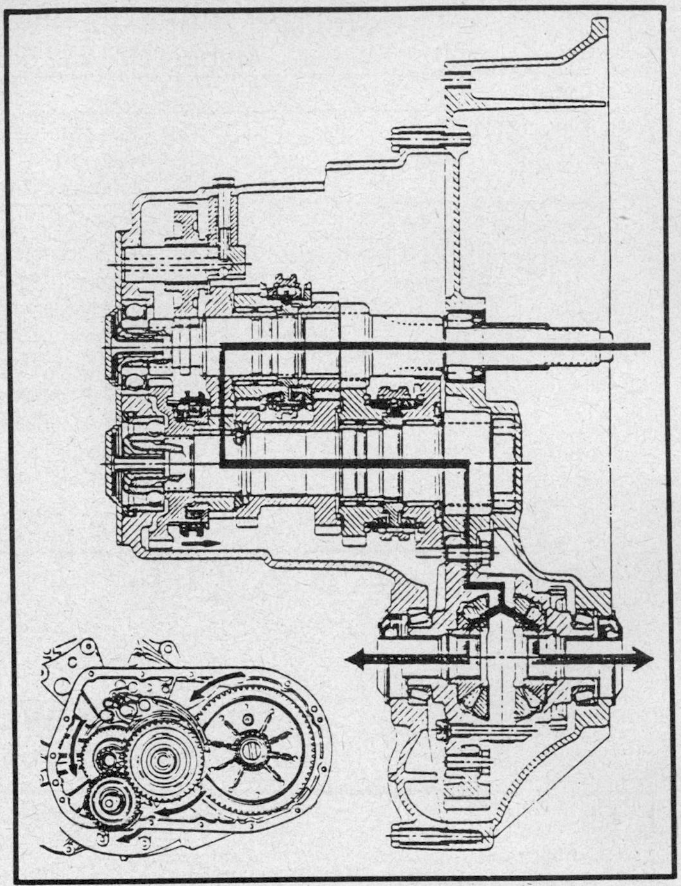
Manual transaxle power flow diagram—5th gear

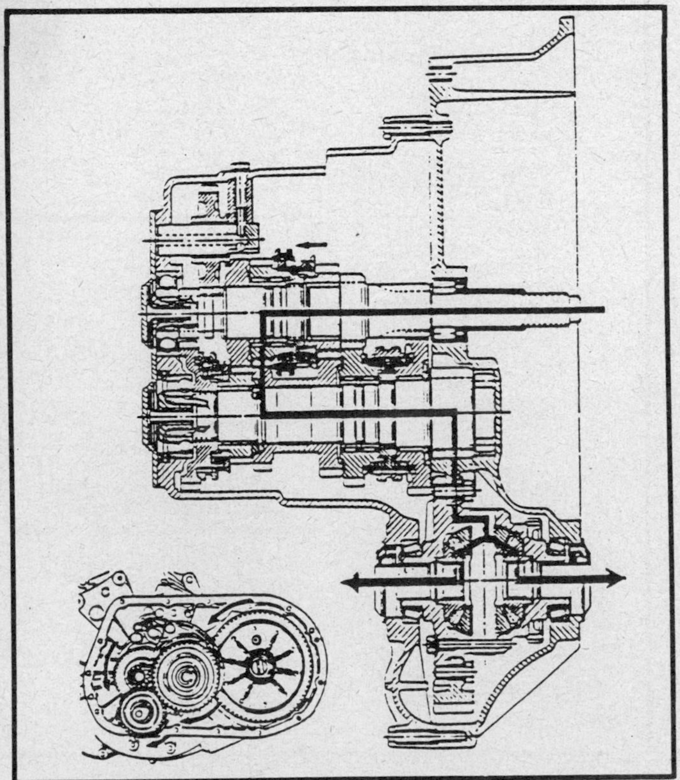
Manual transaxle power flow diagram—4th gear

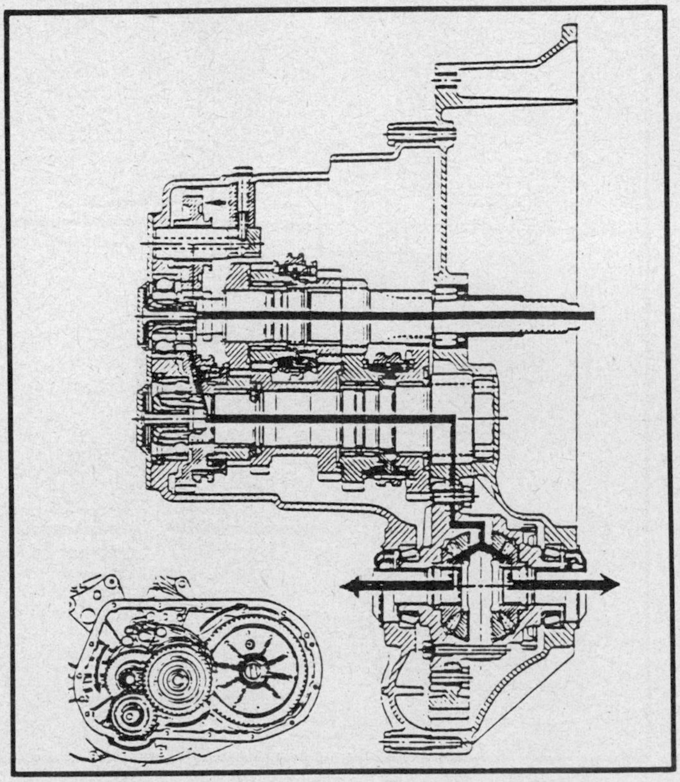
Manual transaxle power flow diagram—reverse

ON CAR SERVICE

Adjustments

SHIFT CABLE

FIERO

1. Loosen the nut on the transaxle shift lever ball stud on the shift cable only.
2. Place the transaxle in **3RD** gear.
3. Remove the screw and shift knob.
4. Remove the front trim plate.
5. Remove the shifter trim plate.
6. Remove the pin floor shift mechanism in **3RD** gear.
7. Tighten the nut to 19 ft. lbs. (25 Nm) on the shift cable ball stud.
8. Install the shifter trim plate.
9. Install the front trim plate.
10. Install the shift knob and screw.

Services

SHIFT CABLE

Removal and Installation

EXCEPT FIERO

1. Disconnect the negative battery cable.
2. Disconnect the shift cables from the transaxle by removing the cable clamp at the transaxle.
3. Remove the cable ends from the ball studs at the shift levers by twisting a large flat blade tool between the nylon socket and the lever.
4. Remove the knob, console and shift boot.
5. Remove the cable ends from the ball studs of the shifter by twisting a large flat blade tool between the nylon socket and the shifter lever. Do not pry the socket off the stud using the cable end for leverage.
6. Remove the spring clip holding the cables to the shifter base and remove the cables from the shifter.
7. Remove the right front sill plate and pull the carpet back to gain access to the cables.
8. Remove the cable grommet cover screws and cover at the floor pan and remove the cables.
9. To install, route the cables and install the cable grommet cover and attaching screws at the floor pan.
10. Install the shift cables at the control assembly. Snap the cable ends onto the ball studs using the channel locks.
11. Reposition the carpet and install the sill plate.
12. Raise and support the vehicle safely.
13. Route the cables to the transaxle.
14. Lower the vehicle.
15. Position the cables and install the retaining clamps at the transaxle. Connect the cables to the shift levers.
16. Install the shift boot, console and knob.
17. Connect the negative battery cable.

FIERO

1. Disconnect the negative battery cable.
2. Remove the screw and shift knob.
3. Remove the front trim plate and the shifter trim plates.
4. Remove the rear console pad assembly.
5. Remove the Electronic Control Module (ECM).
6. Disconnect the ECM electrical connector.
7. Remove the front carrier-to-instrument panel reinforcement.
8. Remove the carrier reinforcement.
9. Remove the carpet clips and rivets at the console.
10. Remove the heater control.
11. Remove the radio.
12. Remove the carrier assembly.
13. Remove the shift and select cable nuts from the cable ball studs and transaxle brackets.
14. Release the rubber grommet on the cable from the body.
15. Remove the bolt and retainer securing the shift cable-to-transaxle.
16. Remove the retainers from the select and shift cable.
17. Pull the cables through the body into the passenger's compartment.

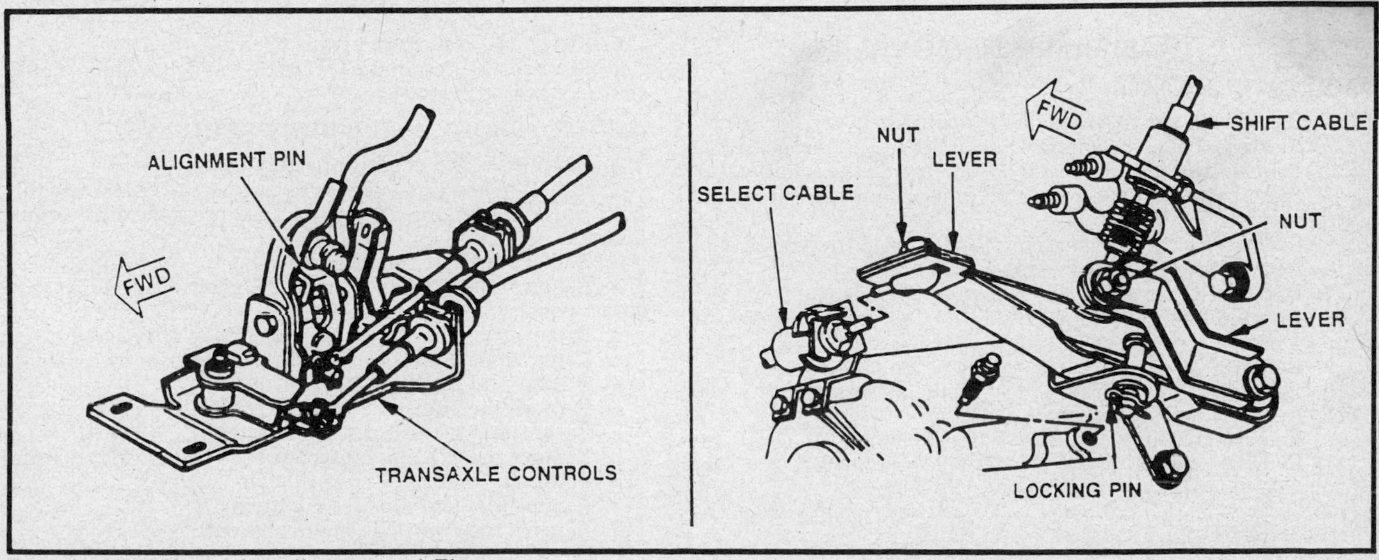
ALIGNMENT PIN
FWD
TRANSAXLE CONTROLS
FWD
NUT
LEVER
SELECT CABLE
SHIFT CABLE
NUT
LEVER
LOCKING PIN

Transaxle shift cable adjustment – Fiero

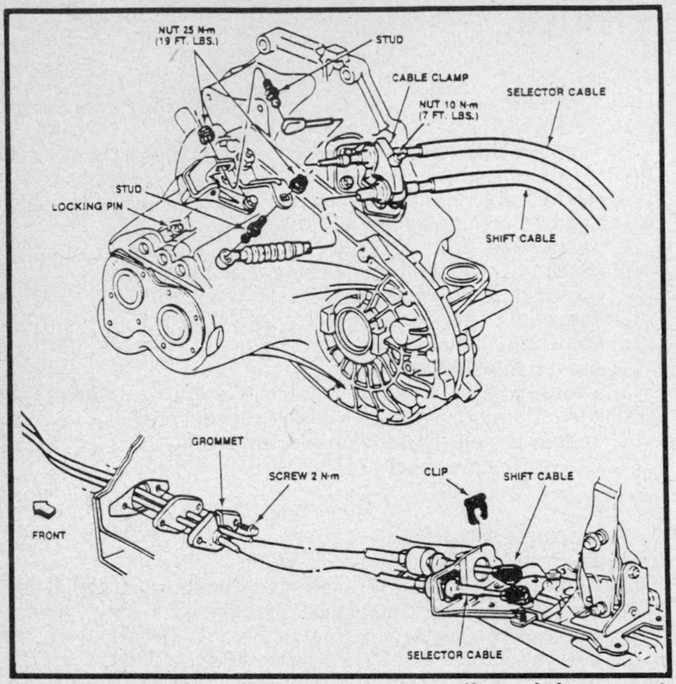
NUT 25 N·m
(19 FT. LBS.)
STUD
CABLE CLAMP
NUT 10 N·m
(7 FT. LBS.)
SELECTOR CABLE
STUD
LOCKING PIN
SHIFT CABLE
GROMMET
SCREW 2 N·m
CLIP
SHIFT CABLE
FRONT
SELECTOR CABLE

Transaxle shift cable replacement—all models except Fiero

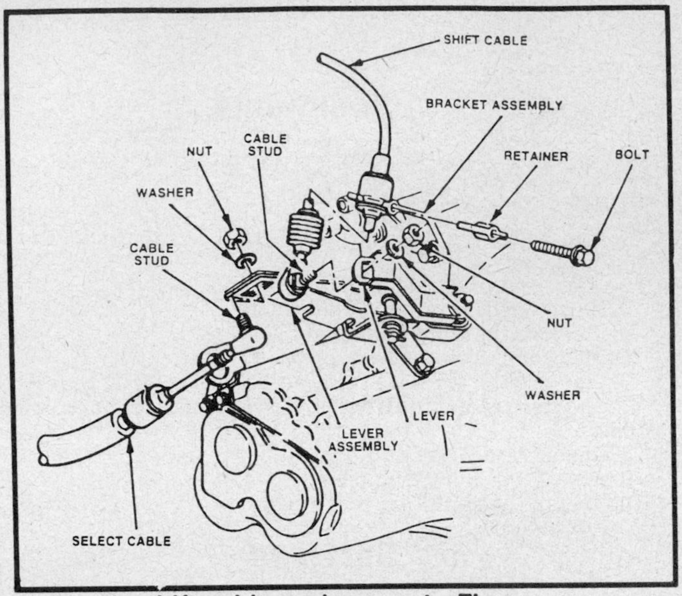
SHIFT CABLE
BRACKET ASSEMBLY
NUT
CABLE
STUD
RETAINER
BOLT
WASHER
CABLE
STUD
NUT
WASHER
LEVER
LEVER
ASSEMBLY
SELECT CABLE

Transaxle shift cable replacement—Fiero

18. To install, guide the cable from the passenger's side through the body into the engine compartment.
19. Install the select cable with the retainer.
20. Install the shift cable-to-retainer with the bolt tightened to 19 ft. lbs. (25 Nm).
21. Install the rubber grommet on the cable and the body.
22. Install the shift and select cables with the nuts and tighten to 19 ft. lbs. (25 Nm) on the cable studs to the transaxle brackets.
23. Install the carrier assembly.
24. Install the radio.
25. Install the heater control.
26. Install the carpet clips and rivets at the console.
27. Install the carrier reinforcements.
28. Install the front carrier-to-instrument panel reinforcement.
29. Connect the ECM electrical connector.
30. Install the ECM.
31. Install the rear console pad assembly.
32. Install the shifter and front trim plates.
33. Install the shift knob and screw.
34. Connect the negative battery cable.

REMOVAL AND INSTALLATION

TRANSAXLE REMOVAL

A-BODY VEHICLES

1. Disconnect the negative battery cable.
2. Remove the air cleaner and air intake duct assembly.
3. Remove the sound panel from the inside the vehicle.
4. Disconnect the clutch master cylinder pushrod from the clutch pedal.
5. Disconnect the clutch slave cylinder from the transaxle.
6. Remove the exhaust crossover pipe.
7. Disconnect the shift cables at the transaxle.
8. Install the engine support fixture tool J-28467 or equivalent.
9. Remove the top transaxle-to-engine bolts.
10. Raise and support the vehicle safely. Drain the transaxle.
11. Install the driveshaft boot seal protectors.
12. Remove the left hand front wheel assembly.
13. Remove the left hand side cradle. Disconnect the rear transaxle mount from the bracket.
14. Disengage the right and left hand driveshafts from the transaxle and support driveshafts.
15. Remove the flywheel cover and disconnect the speedometer cable.
16. Attach the jack to the transaxle case.
17. Remove the retaining transaxle-to-engine bolts. Remove the transaxle.

J-BODY, N-BODY AND L-BODY VEHICLES

1. Disconnect the negative battery cable.
2. Install the engine support fixture tool J-28467 or equivalent. Raise the engine enough to take pressure off the motor mounts.
3. Remove the left sound panel.
4. Remove the clutch master cylinder pushrod from the clutch pedal.
5. Remove the air cleaner and air intake duct assembly.
6. Remove the clutch slave cylinder from the transaxle support bracket and lay aside.
7. Remove the transaxle mount through bolt.
8. Raise the vehicle and support safely.
9. Remove the exhaust crossover bolts at the right manifold.
10. Lower the vehicle.
11. Remove the left exhaust manifold.
12. Remove the transaxle mount bracket.
13. Disconnect the shift cables.
14. Remove the upper transaxle-to-engine bolts.

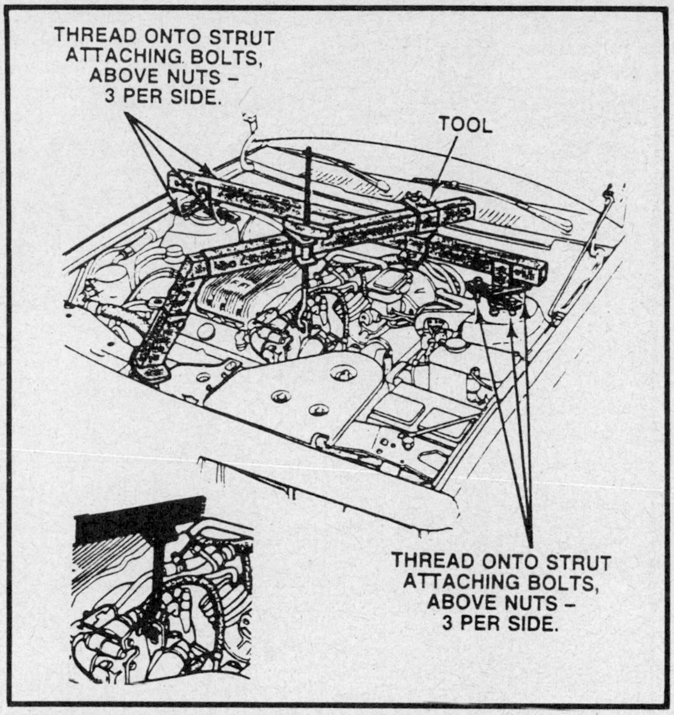
THREAD ONTO STRUT ATTACHING BOLTS, ABOVE NUTS – 3 PER SIDE.
TOOL
THREAD ONTO STRUT ATTACHING BOLTS, ABOVE NUTS – 3 PER SIDE.

Engine support fixture assembly—A-body

15. Raise the vehicle and support the safely.
16. Remove the left wheel assembly.
17. Remove the left front inner splash shield.
18. Remove the transaxle strut and bracket.
19. Drain the transaxle.
20. Remove the clutch housing cover bolts.
21. Disconnect the speedometer cable.
22. Disconnect the stabilizer bar at the left suspension support and control arm.
23. Disconnect the ball joint from the steering knuckle.
24. Remove the left suspension support attaching bolts and remove the support and control arm as an assembly.
25. Disconnect the driveshafts at the transaxle and remove the left driveshaft from the transaxle. Support the right driveshaft.
26. Attach the transaxle case to a jack.
27. Remove the remaining transaxle-to-engine bolts.
28. Remove the transaxle by sliding toward the drive side away from the engine. Carefully lower the jack, guiding the right driveshaft out of the transaxle.

P-BODY VEHICLES

1. Disconnect the negative battery cable. Remove the drain plug and drain the transaxle.
2. Remove select and shift cable nuts securing the cables to the transaxle brackets.
3. Remove the backup light switch connector and switch.
4. Remove the shift cables and the nut on the stud securing the bracket to the transaxle.
5. Remove the 2 bolts securing the select cable mount.
6. Remove the 2 bolts attaching the clutch slave cylinder bracket.
7. Remove the exhaust crossover pipe.
8. Remove the nut, clip and wire from the center stud.
9. Remove the 3 upper bolts and 1 stud attaching the transaxle to the engine.
10. Install J–28467, J–35563, J–28467–60 (GT only) or equivalent, engine support fixtures. Attach the fixture hook to the engine lift ring and raise the engine enough to take the pressure off the mounts.

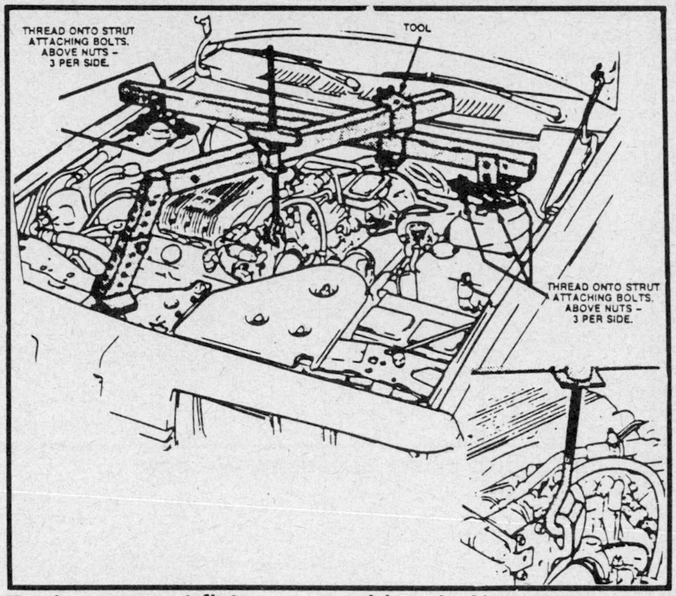
THREAD ONTO STRUT ATTACHING BOLTS, ABOVE NUTS – 3 PER SIDE.
TOOL
THREAD ONTO STRUT ATTACHING BOLTS, ABOVE NUTS – 3 PER SIDE.

Engine support fixture assembly—J-, N- and L-body

11. Remove the front and rear transaxle mounts.
12. Raise and support the vehicle safely.
13. Remove the 4 clutch inspection plate screws and the inspection plate.
14. Lower the frame and tilt.
15. Remove the driveshafts.
16. Remove the 2 nuts retaining the wire harness on the 2 lower studs.
17. Remove the 2 studs and the transaxle from the bottom of the vehicle, tilt the engine down for clearance removal.

W-BODY VEHICLES

NOTE: Prior to any vehicle service that requires the removal of the clutch actuator, the master cylinder pushrod must be disconnected from the clutch pedal and connection in the hydraulic line must be separated using tool J–36221 or equivalent. If not disconnected, permanent damage to the actuator will occur if the clutch pedal is depressed while the system is not resisted by clutch loads. Master cylinder pushrod bushing must also be replaced whenever it has once been removed from the clutch pedal.

1. Disconnect the negative battery cable.
2. Install the engine support fixture J–28467–A or equivalent, along with support leg J–36462 or equivalent.
3. Remove the air cleaner housing and intake tube.
4. Remove the clutch actuator from the transaxle.
5. Disconnect the electrical connection at the speedometer signal assembly.
6. Remove the nut and retaining clamp securing the shift and select cables to the transaxle.
7. Remove the 2 nuts from the cable ball studs at the transaxle levers.
8. Disconnect the exhaust crossover pipe at the left exhaust manifold.
9. Remove the EGR tube from the crossover pipe.
10. Remove the crossover to exhaust pipe bolts.
11. Loosen the crossover to the right exhaust manifold clamp.
12. Swing the crossover upward to gain clearance to the top transaxle bolts.
13. Remove the 2 upper transaxle mounting bolts and 2 upper transaxle mounting studs, leave 1 lower mounting stud and 1 lower mounting bolt attached.
14. Disconnect the electrical connection at the backup light switch.

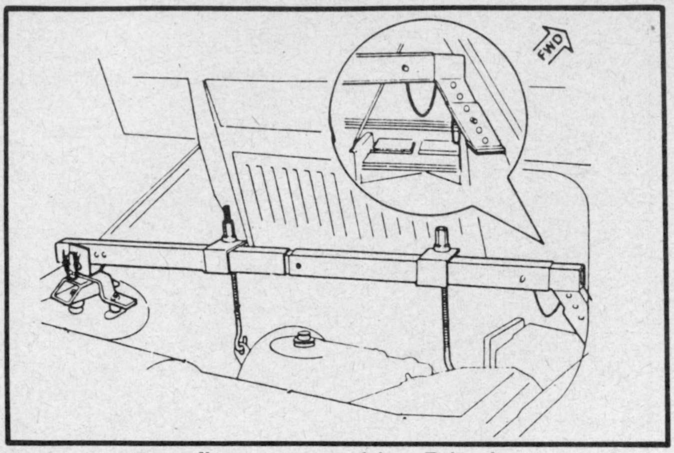
FWD

Engine support fixture assembly—P-body

15. Raise and support the vehicle safely.
16. Remove the drain plug and drain the transaxle.
17. Remove the 4 clutch housing cover retaining screws.
18. Remove both wheel assemblies.
19. Remove the right and left wheelhouse splash shields.
20. Remove the power steering cooler lines from the frame.
21. Remove the power steering rack and pinion heat shield.
22. Remove the power steering rack and pinion from the frame.
23. Disconnect the right and left ball joints at the steering knuckle.
24. Remove the transaxle mount upper retainer bolts.
25. Remove the engine mount lower retainer nuts.
26. Remove the frame retaining bolts from the body.
27. Remove the frame from the vehicle.
28. Remove the right and left driveshafts from the transaxle and support to the body.
29. Remove starter assembly and support to the body.
30. Securely attach the transaxle case to the jack for removal.
31. Remove the remaining transaxle mounting bolts and stud.
32. Remove the transaxle assembly.

TRANSAXLE INSTALLATION

A-BODY VEHICLES

1. When installing the transaxle, position the right side driveshaft into its bore as the transaxle is being installed. The right hand driveshaft cannot be readily installed after the transaxle is connected to the engine.
2. After the transaxle is fastened to the engine and the left hand driveshaft installed at the transaxle, position the cradle and install the cradle-to-body bolts.
3. Connect the transaxle to front and rear mounts.
4. Seat the driveshaft in the transaxle.
5. Install the flywheel cover.
6. Connect the speedometer cable.
7. Install the left hand wheel assembly.
8. Lower the vehicle.
9. Install the remaining transaxle-to-engine bolts.
10. Remove the engine support fixture.
11. Connect the shift cables at the transaxle.
12. Install the crossover pipe.
13. Install the clutch slave cylinder.
14. Connect the clutch master cylinder pushrod to the clutch pedal.
15. Install the sound panel.
16. Install the air cleaner and air intake duct assembly.
17. Fill the transaxle with 2 quarts (1.9L) of 5W-30 manual transaxle oil, part number 1052931 or equivalent. (On 1988–89

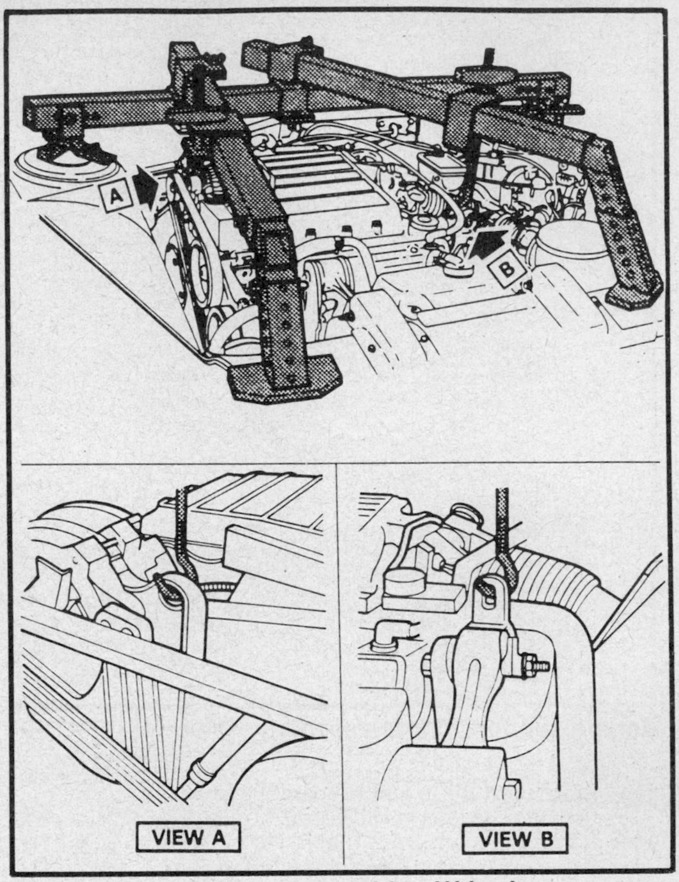
A
B
VIEW A
VIEW B

Engine support fixture assembly—W-body

vehicles, use syncromesh transaxle fluid, part number 12345349 or equivalent.)
18. Connect the negative battery cable.

J-BODY, N-BODY AND L-BODY VEHICLES

1. When installing the transaxle, guide the right driveshaft into its bore as the transaxle is being raised. The right driveshaft can not be readily installed after the transaxle is connected to the engine.
2. Install the transaxle-to-engine mounting bolts and tighten to specifications.
3. Install the left driveshaft into its bore at the transaxle and seat both driveshafts at the transaxle.
4. Install the suspension support-to-body bolts.
5. Connect the ball joint to the steering knuckle.
6. Connect the stabilizer bar to the suspension support and control arm.
7. Connect the speedometer cable.
8. Install the clutch housing cover bolts.
9. Install the strut bracket to the transaxle.
10. Install the strut.
11. Install the inner splash shield.
12. Install the wheel assembly and torque the lug nuts to specifications.
13. Lower the vehicle.
14. Install the upper transaxle-to-engine bolts.
15. Connect the shift cables.
16. Install the transaxle mount bracket.
17. Install the left exhaust manifold.
18. Raise and support the vehicle safely.
19. Install the exhaust crossover bolts at the right manifold.
20. Lower the vehicle.

21. Install the transaxle mount through bolt.
22. Install the clutch slave cylinder to the support bracket.
23. Install the air cleaner and air intake duct assembly.
24. Remove the engine support fixture.
25. Install the clutch master cylinder push rod to the clutch pedal.
26. Install the left sound panel.
27. Fill the transaxle with 5 pints (2.1L) of 5W–30 manual transaxle oil, part number 1052931 or equivalent. (On 1988–89 vehicles, use syncromesh transaxle fluid, part number 12345349 or equivalent.)
28. Connect the negative battery cable.

P-BODY VEHICLES

1. Install the transaxle through the bottom and attach the engine to transaxle with 2 studs. Do not torque the studs.
2. Raise the frame and install the driveshafts.
3. Install the clutch inspection plate and the 4 screws and tighten to 116 inch lbs. (13 Nm).
4. Install the front and rear transaxle mounts.
5. Lower the vehicle.
6. Lower the engine and remove the engine support fixtures.
7. Install the 3 upper bolts and 1 stud and tighten to 55 ft. lbs. (75 Nm). Also torque the 2 lower studs in Step 1.
8. Install the wire harnesses on the 2 lower and center studs and 3 nuts. Torque to 152 inch lbs. (17 Nm).
9. Install the exhaust crossover pipe.
10. Install the 2 bolts attaching the clutch slave cylinder bracket and tighten to 27 ft. lbs. (50 Nm).
11. Install the 2 bolts retaining the select cable mount and tighten to 89 inch lbs. (10 Nm).
12. Install the select and shift cable to mount on the stud with the nut and tighten to 89 inch lbs. (10 Nm).
13. Install the select and shift cables nut and tighten to 223 inch lbs. (25 Nm).
14. Install the backup light switch and tighten to 25 ft. lbs. (34 Nm). Install the switch electrical connector.
15. Install the drain plug and tighten to 24 ft. lbs. (33 Nm).
16. Remove the screw and retainer from the permanent magnet generator and fill the transaxle with 2 quarts (1.9L) of 5W–30 manual transaxle oil, part number 1052931 or equivalent. (On 1988 vehicles, use syncromesh transaxle fluid, part number 12345349 or equivalent.) Install the retainer and screw and tighten to 45 inch lbs. (5 Nm).
17. Connect the negative battery cable.

W-BODY VEHICLES

1. Install the transaxle assembly.
2. Install the transaxle to engine lower bolt and stud and tighten to 55 ft. lbs. (75 Nm).
3. Install the starter assembly and support to body.
4. Install the left and right driveshafts.
5. Install the frame assembly.
6. Install the frame retaining bolts.
7. Install the engine mount lower retaining nuts.
8. Install the transaxle upper retaining bolts.
9. Connect the right and left ball joints to the steering knuckle.
10. Install the power steering rack and pinion to the frame.
11. Install the power steering rack and pinion heat shield.
12. Install the power steering cooler lines to the frame.
13. Install the right and left wheelhouse splash shields.
14. Install both wheel assembly.
15. Install the transaxle drain plug.
16. Install the clutch housing cover and 4 retaining screws and tighten to 115 inch lbs. (13 Nm).
17. Lower the vehicle.
18. Connect the electrical connection at the backup light switch.
19. Install the 2 upper transaxle mounting bolts and 2 upper transaxle mounting studs. Torque to 55 ft. lbs. (75 Nm).
20. Move the crossover pipe to its proper position.
21. Install the crossover pipe to the right exhaust manifold clamp.
22. Install the crossover pipe to exhaust pipe bolts.
23. Install the EGR tube at the crossover pipe.
24. Install the exhaust crossover pipe at the left exhaust manifold.
25. Install the nut and retaining clamp securing the shift and select cables to the transaxle.
26. Install the 2 nuts to the cable ball studs at the transaxle levers.
27. Connect the electrical connection at the speedometer signal assembly.
28. Install the clutch actuator to the transaxle.
29. Install the air cleaner and intake tube.
30. Remove the engine support fixture.
31. Fill the transaxle with 5 pints (2.1L) of syncromesh transaxle fluid, part number 12345349 or equivalent.
32. Connect the negative battery cable.

BENCH OVERHAUL

Before Disassembly

Cleanliness is an important factor in the overhaul of the transaxle. Before attempting any disassembly operation, the exterior of the transaxle should be thoroughly cleaned to prevent the possibility of dirt entering the transaxle internal mechanism. During inspection and reassembly, all parts should be thoroughly cleaned with cleaning fluid and then air dried. Wiping cloths or rags should not be used to dry parts. All oil passages should be blown out and checked to make sure that they are not obstructed. Small passages should be checked with tag wire. All parts should be inspected to determine which parts are to be replaced.

Transaxle Disassembly

EXTERNAL TRANSAXLE MOUNTED LINKAGE

1. Remove the nut using a 21mm socket and driver.

NOTE: Do not allow the lever to move during removal of the nut. Use a ³⁄₈ in. drive ratchet to hold the external shift lever by the slot.

2. Remove the washer and lever.
3. Remove the pivot pin. Depending on the type of linkage,

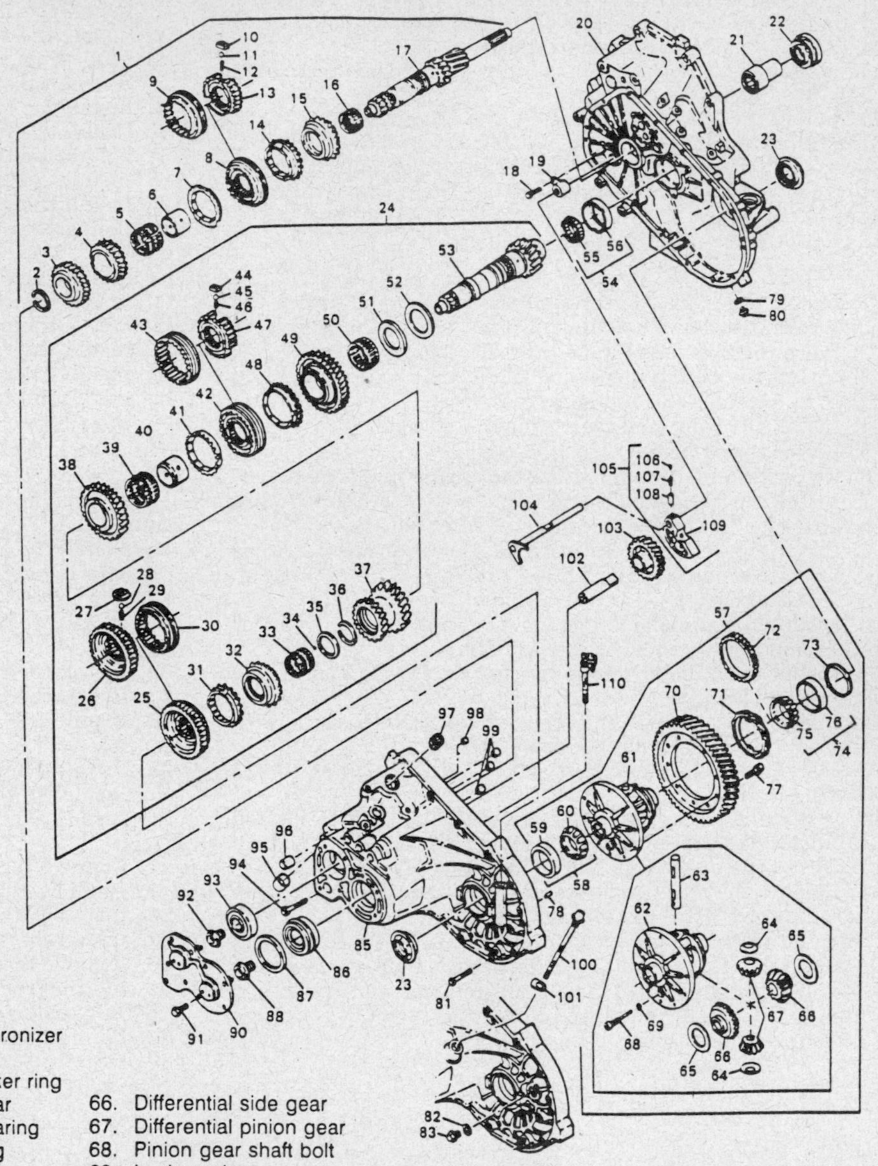

1. Input cluster shaft and gear assembly
2. Snapring
3. 5th input gear
4. 4th input gear
5. Cage bearing
6. Needle race
7. 4th blocker ring
8. 3rd/4th synchronizer assembly
9. 3rd/4th synchronizer sleeve
10. 3rd/4th synchronizer key (3)
11. 3rd/4th synchronizer ball (3)
12. 3rd/4th synchronizer spring (3)
13. 3rd/4th synchronizer hub clutch
14. 3rd blocker ring
15. 3rd input gear
16. Cage bearing (2)
17. Input shaft
18. Bolt
19. Reverse shift rail guide
20. Clutch and differential housing
21. Input shaft bearing/sleeve assembly
22. Clutch release bearing assembly
23. Driveshaft oil seal
24. Output cluster shaft and gear assembly
25. Reverse output/5th synchronizer assembly gear
26. Reverse gear
27. 5th synchronizer key (3)
28. 5th synchronizer ball (3)
29. 5th synchronizer spring (3)
30. 5th synchronizer sleeve
31. 5th gear blocker ring
32. 5th speed output bearing
33. 5th speed output bearing
34. Thrust washer positioner ball
35. Thrust washer
36. Snapring
37. 3rd/4th cluster gear
38. 2nd output gear
39. 2nd output bearing
40. 2nd output bearing race
41. 2nd gear blocker ring
42. 1st/2nd gear synchronizer assembly
43. 1st/2nd synchronizer sleeve
44. 1st/2nd synchronizer key (3)
45. 1st/2nd synchronizer ball (3)
46. 1st/2nd synchronizer spring (3)
47. 1st/2nd synchronizer hub
48. 1st gear blocker ring
49. 1st output gear
50. 1st output bearing
51. Thrust bearing
52. Thrust washer
53. Output shaft
54. Output shaft support bearing
55. Output bearing
56. Output bearing race
57. Gear and differential assembly
58. Differential assembly bearing
59. Differential bearing race
60. Differential bearing
61. Differential assembly case
62. Differential case
63. Differential cross pin
64. Pinion gear thrust washer
65. Side gear thrust washer
66. Differential side gear
67. Differential pinion gear
68. Pinion gear shaft bolt
69. Lock washer
70. Differential ring gear
71. Speedometer output gear (mechanical)
72. Speedometer output gear (electronic)
73. Differential selective shim
74. Differential assembly bearing
75. Differential bearing
76. Differential bearing race
77. Differential ring bolt (10)
78. Pin (2)
79. Oil drain plug
80. Washer
81. Transaxle case bolt (15)
82. Washer
83. Plug
84.
85. Transaxle case
86. Output gear bearing
87. Output gear selective shim
88. Output gear bearing retainer
89. Oil slinger washer
90. Transaxle case end plate
91. Bolt (9)
92. Input gear bearing retainer
93. Input gear bearing
94. Reverse idler bolt
95. Detent lever bushing
96. Sliding sleeve bushing
97. Shift shaft needle bearing
98. Reverse rail bushing
99. Shift rail bushing (3)
100. Fluid level indicator washer
101. Fluid level indicator
102. Reverse idler shaft
103. Reverse idler gear
104. Reverse shift idler gear rail
105. Reverse idler gear assembly bracket
106. Reverse idler gear bracket ball
107. Reverse idler gear bracket spring
108. Reverse idler gear detent bracket sleeve
109. Reverse idler gear bracket

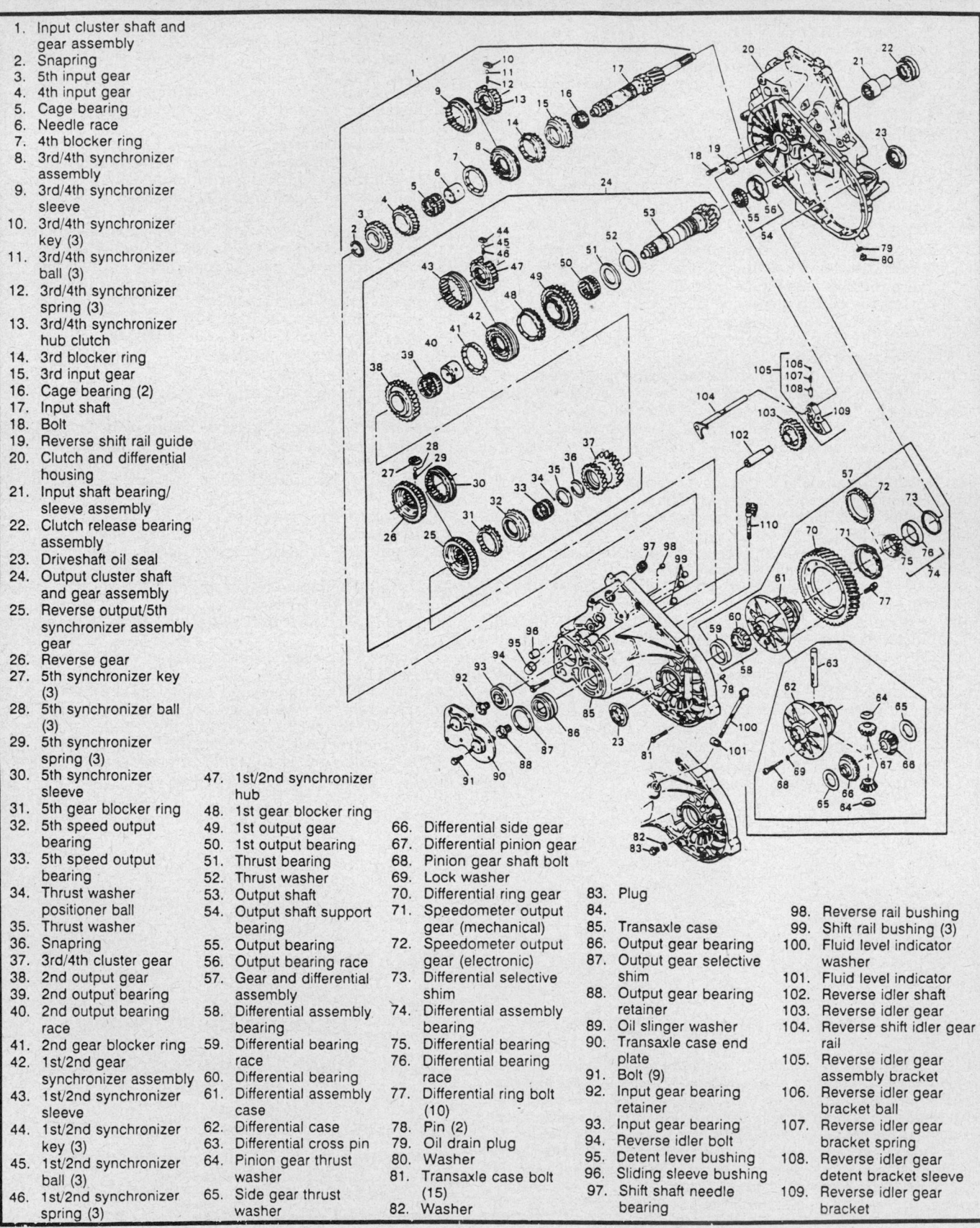

Transaxle gear and case exploded view

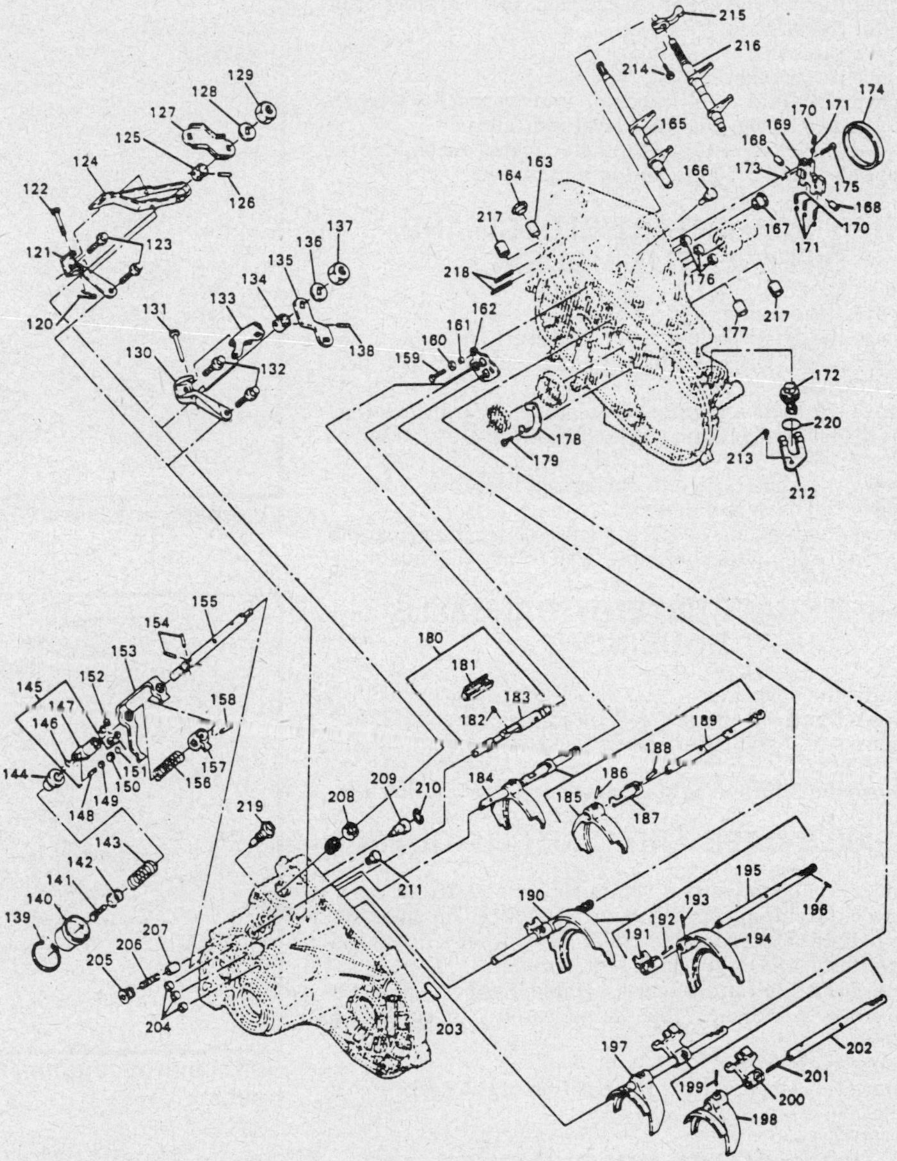
215
216
214
129
128
127
125
124
122
126
165
174
170
171
169
168
173
175
168
170
171
166
167
164
163
217
137
136
135
134
133
123
121
120
131
130
132
138
218
176
162
161
160
159
217
177
172
220
178
179
213
212
155
154
153
152
145
147
146
158
144
151
150
149
148
157
156
180
181
183
182
189
188
186
187
184
185
209
210
208
219
211
143
142
141
140
139
207
206
205
204
203
190
195
193
192
191
196
194
197
202
201
199
200
198

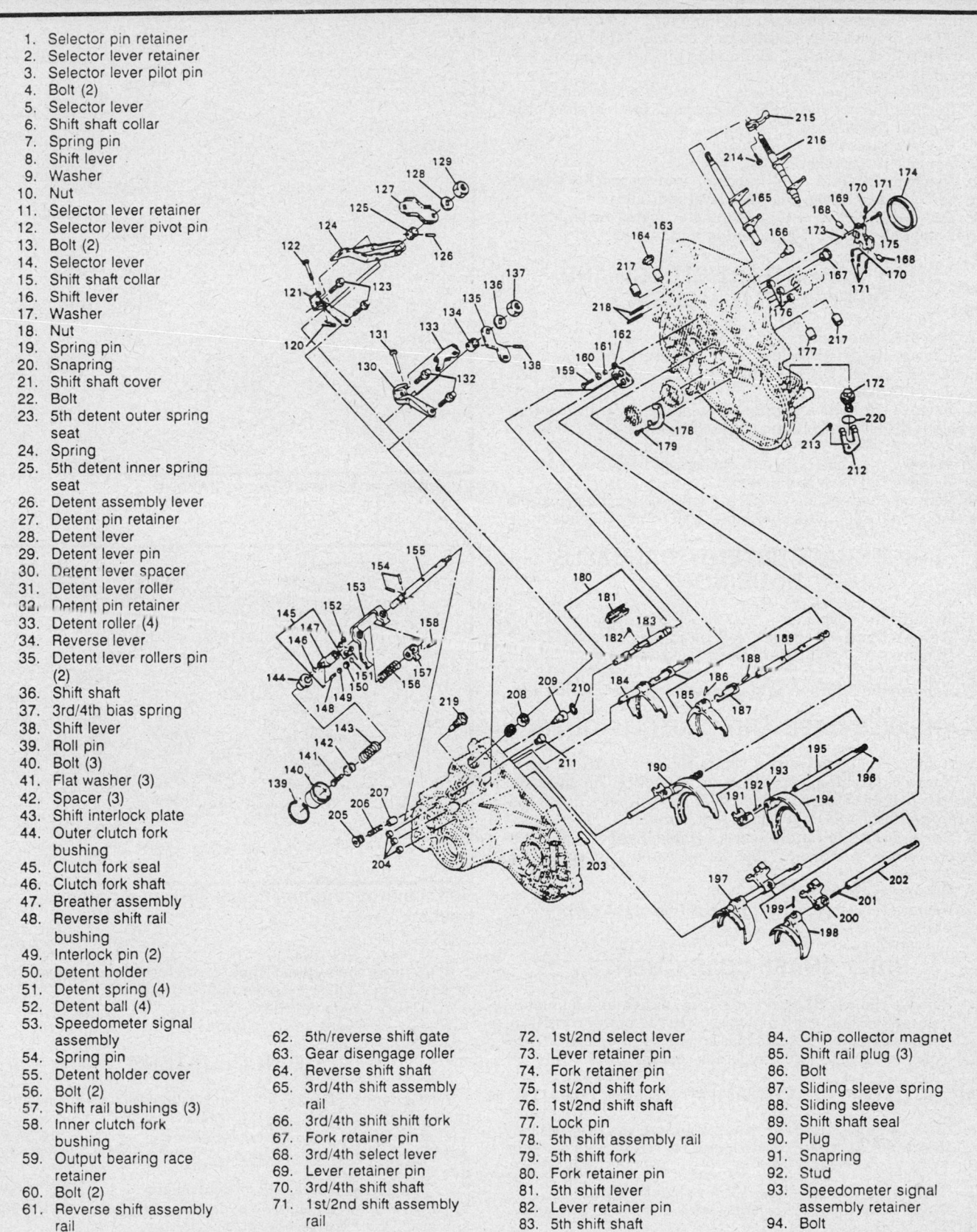

1. Selector pin retainer
2. Selector lever retainer
3. Selector lever pilot pin
4. Bolt (2)
5. Selector lever
6. Shift shaft collar
7. Spring pin
8. Shift lever
9. Washer
10. Nut
11. Selector lever retainer
12. Selector lever pivot pin
13. Bolt (2)
14. Selector lever
15. Shift shaft collar
16. Shift lever
17. Washer
18. Nut
19. Spring pin
20. Snapring
21. Shift shaft cover
22. Bolt
23. 5th detent outer spring seat
24. Spring
25. 5th detent inner spring seat
26. Detent assembly lever
27. Detent pin retainer
28. Detent lever
29. Detent lever pin
30. Detent lever spacer
31. Detent lever roller
32. Detent pin retainer
33. Detent roller (4)
34. Reverse lever
35. Detent lever rollers pin (2)
36. Shift shaft
37. 3rd/4th bias spring
38. Shift lever
39. Roll pin
40. Bolt (3)
41. Flat washer (3)
42. Spacer (3)
43. Shift interlock plate
44. Outer clutch fork bushing
45. Clutch fork seal
46. Clutch fork shaft
47. Breather assembly
48. Reverse shift rail bushing
49. Interlock pin (2)
50. Detent holder
51. Detent spring (4)
52. Detent ball (4)
53. Speedometer signal assembly
54. Spring pin
55. Detent holder cover
56. Bolt (2)
57. Shift rail bushings (3)
58. Inner clutch fork bushing
59. Output bearing race retainer
60. Bolt (2)
61. Reverse shift assembly rail
62. 5th/reverse shift gate
63. Gear disengage roller
64. Reverse shift shaft
65. 3rd/4th shift assembly rail
66. 3rd/4th shift shift fork
67. Fork retainer pin
68. 3rd/4th select lever
69. Lever retainer pin
70. 3rd/4th shift shaft
71. 1st/2nd shift assembly rail
72. 1st/2nd select lever
73. Lever retainer pin
74. Fork retainer pin
75. 1st/2nd shift fork
76. 1st/2nd shift shaft
77. Lock pin
78. 5th shift assembly rail
79. 5th shift fork
80. Fork retainer pin
81. 5th shift lever
82. Lever retainer pin
83. 5th shift shaft
84. Chip collector magnet
85. Shift rail plug (3)
86. Bolt
87. Sliding sleeve spring
88. Sliding sleeve
89. Shift shaft seal
90. Plug
91. Snapring
92. Stud
93. Speedometer signal assembly retainer
94. Bolt

Transaxle shift mechanism exploded view

this pin may be removed by using a hammer and punch to drive it out of the bracket (replace with part number 14091786) or removing a retaining clip and sliding the pin out of the bracket, which may be reused.

4. Remove the pivot using a $^3/_{16}$ in. punch and hammer.
5. Remove the pin and collar. Note the position of the slot in the collar for installation.
6. Remove the bolts.
7. Remove the bracket.
8. Remove the fluid level indicator and washer. A wrench may be needed to loosen the fluid level indicator.
9. Remove the electronic speedometer signal assembly, retainer and bolt using a 10mm socket and driver.

SHIFT RAIL DETENT/CLUTCH AND DIFFERENTIAL HOUSING

1. Remove the bearing.
2. Remove the detent holder cover. Puncture the cover in the middle and pry off. Discard this part and replace it with part number 14082039.
3. Remove the bolts and interlock plate (early transaxles). If the detent holder is not 18mm thick, interlock plate kit must be used.
4. Remove the holder, detent, springs and interlock pins.
5. Remove the balls and detent.
6. Remove the bushing by prying loose (2 small pry bars in the slots). Two allen wrenches work well to pry this bushing.

SHIFT SHAFT DETENT/TRANSAXLE HOUSING

1. Remove the snapring.
2. Remove the cover using a soft faced hammer.
3. Remove screw and outer spring seat using a 5mm bit and driver.
4. Remove the 5th/reverse bias spring and inner spring seat.

TRANSAXLE CASE AND CLUTCH HOUSING

1. Remove the bolts using a 13mm socket and driver.
2. Remove the clutch housing using a soft faced hammer. Remove the Loctite®518 anaerobic sealer with either a liquid gasket remover or J–28410 or equivalent, scraper.
3. Remove the differential gear assembly. Support the transaxle case on a workbench top, being careful to support it properly.
4. Remove the magnet.
5. Remove the bearing. Note the position of the bearing cage for installation.

SHIFT SHAFT COMPONENTS

1. Remove the pin using a size $^3/_{16}$ in. punch and hammer.
2. Remove the shift shaft assembly. This assembly consists of shaft, rollers and pins, 1st/2nd bias spring, shift and reverse levers. Take care not to lose the detent rollers.

GEAR CLUSTER SUPPORT COMPONENTS

NOTE: Engage the gear cluster in 4th and reverse by pushing down on the 3rd/4th gear rail and reverse gear rail.

1. Remove the bolts using a 13mm socket and driver.
2. Remove the cover. Tap gently with a soft faced hammer.
3. Remove the selective shim.
4. Remove the oil shield.
5. Remove the retainer, output gear cluster using tool J–36031 or equivalent.

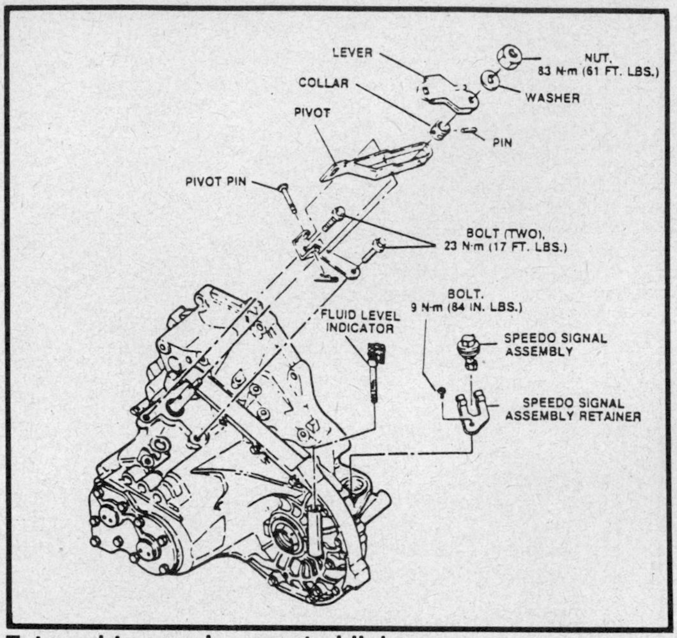
LEVER
NUT, 83 N·m (61 FT. LBS.)
COLLAR
WASHER
PIVOT
PIN
PIVOT PIN
BOLT (TWO), 23 N·m (17 FT. LBS.)
BOLT, 9 N·m (84 IN. LBS.)
FLUID LEVEL INDICATOR
SPEEDO SIGNAL ASSEMBLY
SPEEDO SIGNAL ASSEMBLY RETAINER

External transaxle mounted linkage

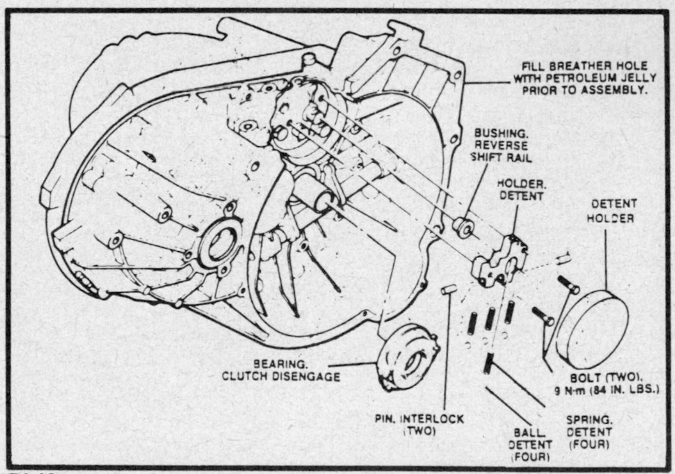
FILL BREATHER HOLE WITH PETROLEUM JELLY PRIOR TO ASSEMBLY.
BUSHING, REVERSE SHIFT RAIL
HOLDER, DETENT
DETENT HOLDER
BEARING, CLUTCH DISENGAGE
BOLT (TWO), 9 N·m (84 IN. LBS.)
PIN, INTERLOCK (TWO)
BALL, DETENT (FOUR)
SPRING, DETENT (FOUR)

Shift control components/clutch and differential housing

6. Remove the retainer, input gear cluster using tool J–36031 or equivalent. These retainers must not be reused.
7. Return the transaxle to **N**.

GEAR CLUSTERS

1. Position tool J–36182–1 and J–36182–2 or equivalent, in the hydraulic press.
2. Position the transaxle case/gear cluster assembly on tool J–36182–1 and –2 or equivalent. Align the shift rail and shaft pilots to the fixture.
3. Position tool J–36185 or equivalent, on the shaft support bearings and pilots. Using a hydraulic press, separate the shaft and gear clusters from the transaxle case.
4. After this operation, the input and output shaft bearings should be discarded. Remove the gear clusters from the pallets, as an entire assembly.

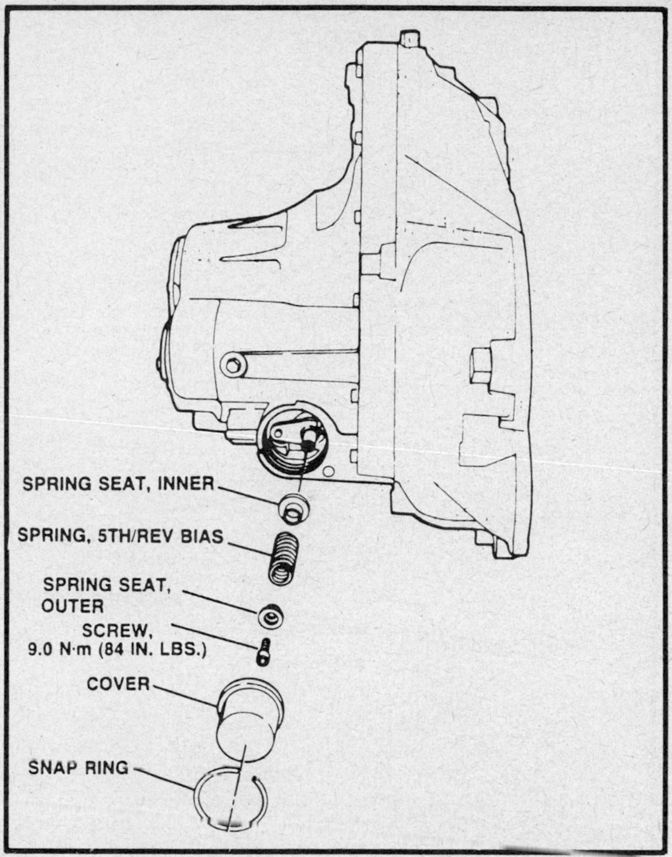
SPRING SEAT, INNER
SPRING, 5TH/REV BIAS
SPRING SEAT, OUTER
SCREW, 9.0 N·m (84 IN. LBS.)
COVER
SNAP RING
Shift shaft detent components/transaxle housing

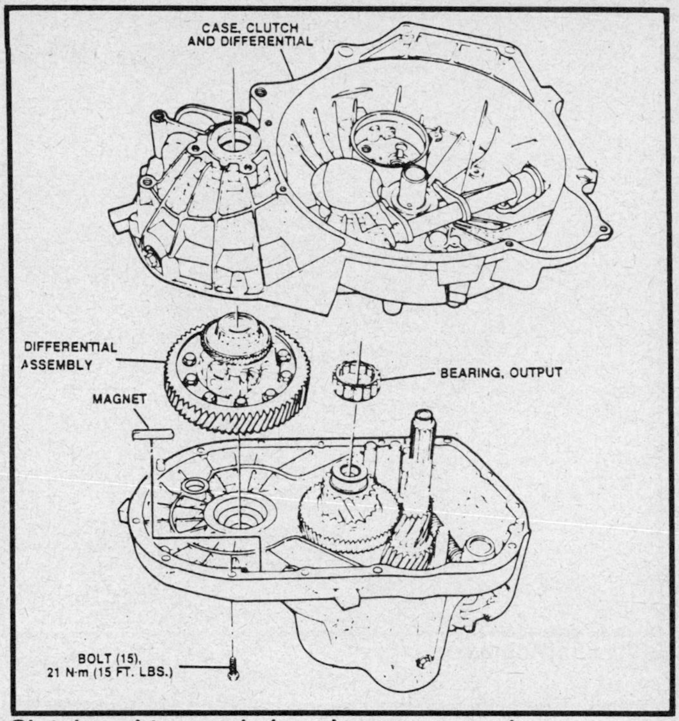
CASE, CLUTCH AND DIFFERENTIAL
DIFFERENTIAL ASSEMBLY
MAGNET
BEARING, OUTPUT
BOLT (15), 21 N·m (15 FT. LBS.)
Clutch and transaxle housing components

GEAR CLUSTERS AND SHIFT RAILS

NOTE: This should be done on a workbench after taking the gear clusters off the pallet.

1. Remove the 1/2 shift rail assembly and lock pin.
2. Remove the 3/4 rail assembly.
3. Remove the 5th rail assembly.
4. Remove the reverse rail assembly, consisting of the shift gate and disengage roller.

NOTE: Be careful not to lose both the lock pin and the gear disengage roller, as they are small parts.

Unit Disassembly and Assembly

NOTE: The following components will require heating prior to installation during assembly procedures. A suggested heating oven is a toaster oven (used as a kitchen appliance).

Heat the races, gear assembly and speedometer gear (electronic) for 7–10 minutes at 250°F (120°C).

Heat the speedometer gear (mechanical) in hot tap water for 5 minutes.

Heat the gear cluster for a minimum of 20 minutes at 250°F (120°C).

INPUT SHAFT

Disassembly

NOTE: Identify blocker ring for 3rd gear and blocker ring for 4th gear. Do not mix.

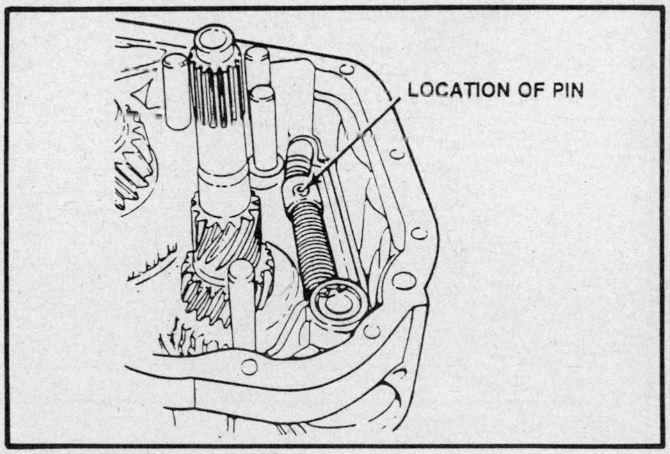
LOCATION OF PIN
Shift lever pin removal

1. Remove the snapring and discard, if stretched.
2. Remove the gear, bearing, race, blocker ring, synchronizer assembly and gear using tool J–36183, J–36184 or equivalent and a hydraulic press.
3. Remove the 3rd gear bearing.

Inspection

1. Clean parts with solvent and air dry.
2. Inspect shaft, spline wear for cracks.
3. Inspect gear teeth for scuffed, nicked, burred or broken teeth.
4. Inspect bearings for roughness of rotation, burred or pitted condition.
5. Inspect bearing races for scoring, wear or overheating.
6. Inspect snaprings for nicks, distortion or wear.
7. Inspect synchronizers assembly for wear.

Assembly

1. Install the 3rd gear bearing, 3rd gear (cone up) and blocker ring.

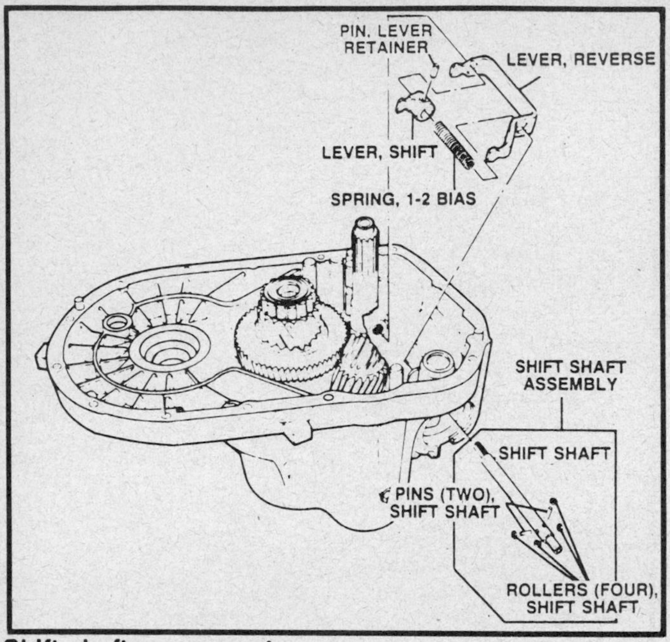
PIN, LEVER RETAINER
LEVER, REVERSE
LEVER, SHIFT
SPRING, 1-2 BIAS
SHIFT SHAFT ASSEMBLY
SHIFT SHAFT
PINS (TWO), SHIFT SHAFT
ROLLERS (FOUR), SHIFT SHAFT

Shift shaft components

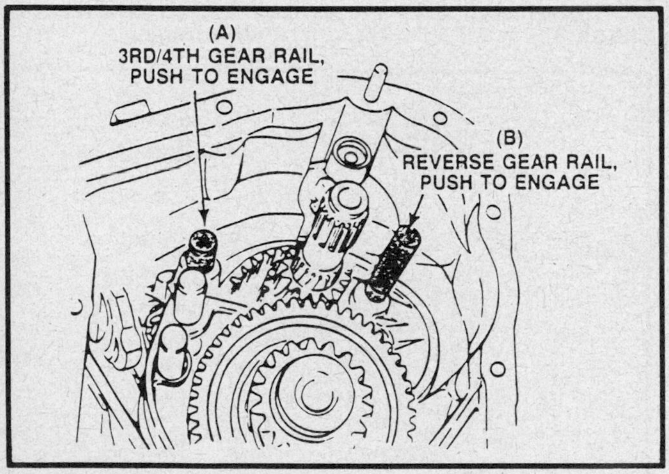
(A)
3RD/4TH GEAR RAIL, PUSH TO ENGAGE
(B)
REVERSE GEAR RAIL, PUSH TO ENGAGE

Engage 4th and reverse

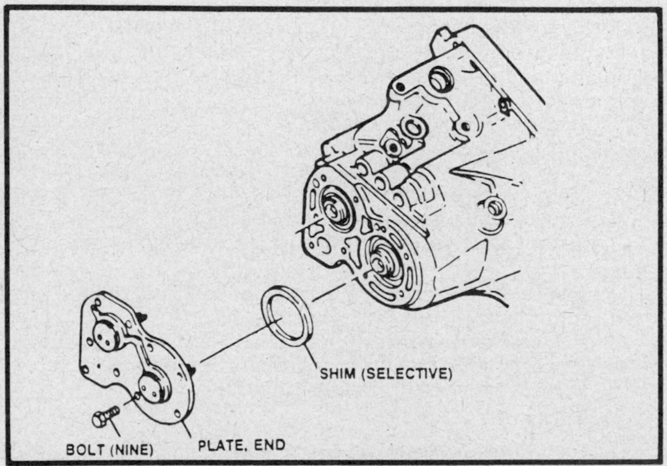
SHIM (SELECTIVE)
BOLT (NINE)
PLATE, END

Gear cluster support components

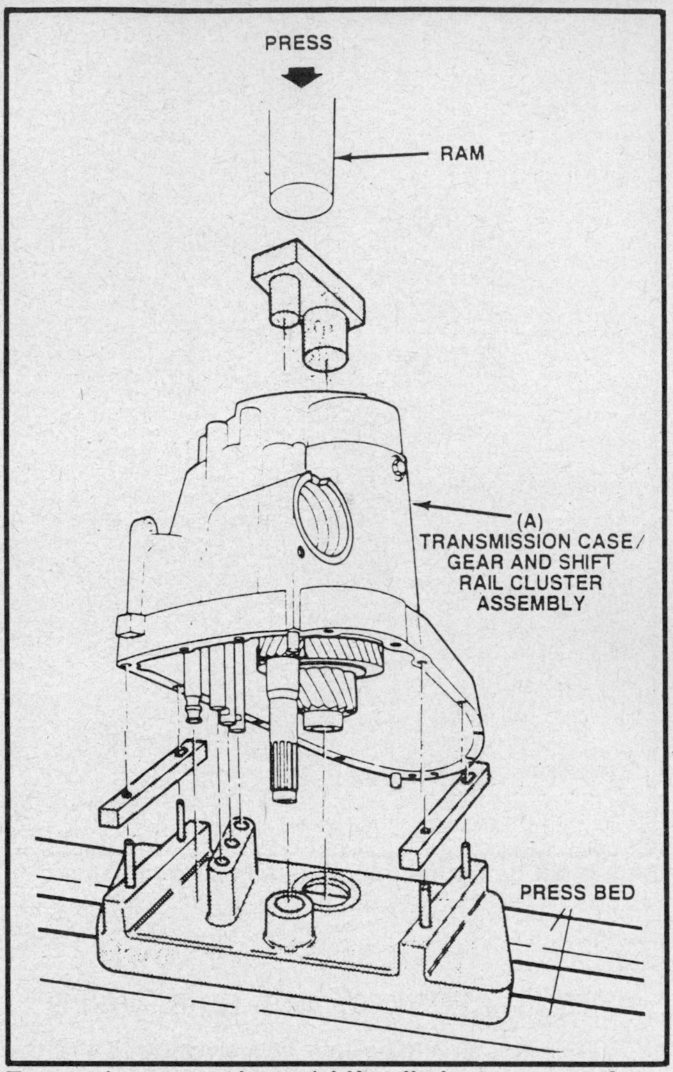
PRESS
RAM
(A)
TRANSMISSION CASE/
GEAR AND SHIFT
RAIL CLUSTER
ASSEMBLY
PRESS BED

Transaxle case and gear/shift rail cluster separation

NOTE: When pressing the 3rd/4th synchronizer assembly, start press operation, stop before tangs engage. Lift and rotate the 3rd gear into the synchronizer tangs. Continue to press until seated. Be sure all shavings are removed.

2. Install the 3/4 synchronizer, using tool J-22912-01, J-36183, J-36184 or equivalent and a hydraulic press. Tool J-22828 presses 4th gear bearing race on very well. The small outer diameter groove of the sleeve toward the 3rd gear and small end of the hub facing 4th.
3. Install the bearing race and bearing using gloves to handle the hot race. Check the temperature with Tempilstick® or thermometer.
4. Install the 4th gear blocker ring.
5. Install the 4th gear with the cone down.
6. Install the 5th gear (flat side down) using tool J-36183, J-36184 or equivalent and a hydraulic press.
7. Install the snapring.

OUTPUT SHAFT

Disassembly

NOTE: Identify the blocker ring for 5th gear, blocker ring for 2nd gear and blocker ring for 1st gear. Do not mix.

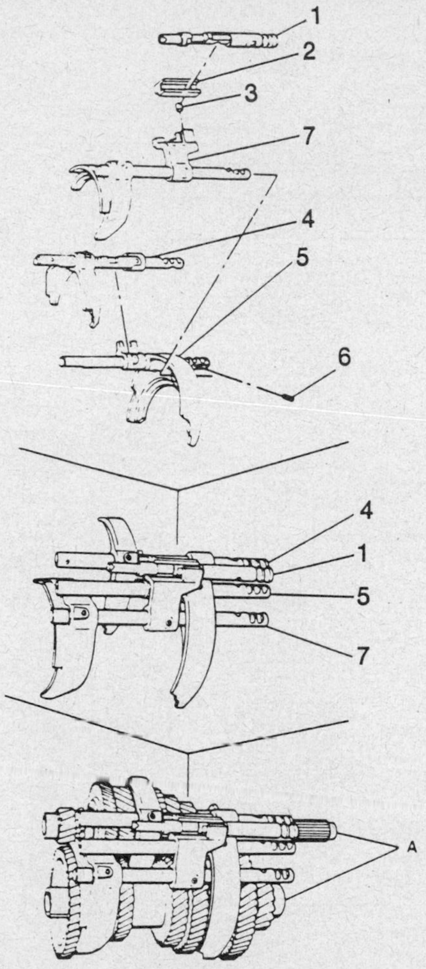
1
2
3
7
4
5
6
4
1
5
7
A

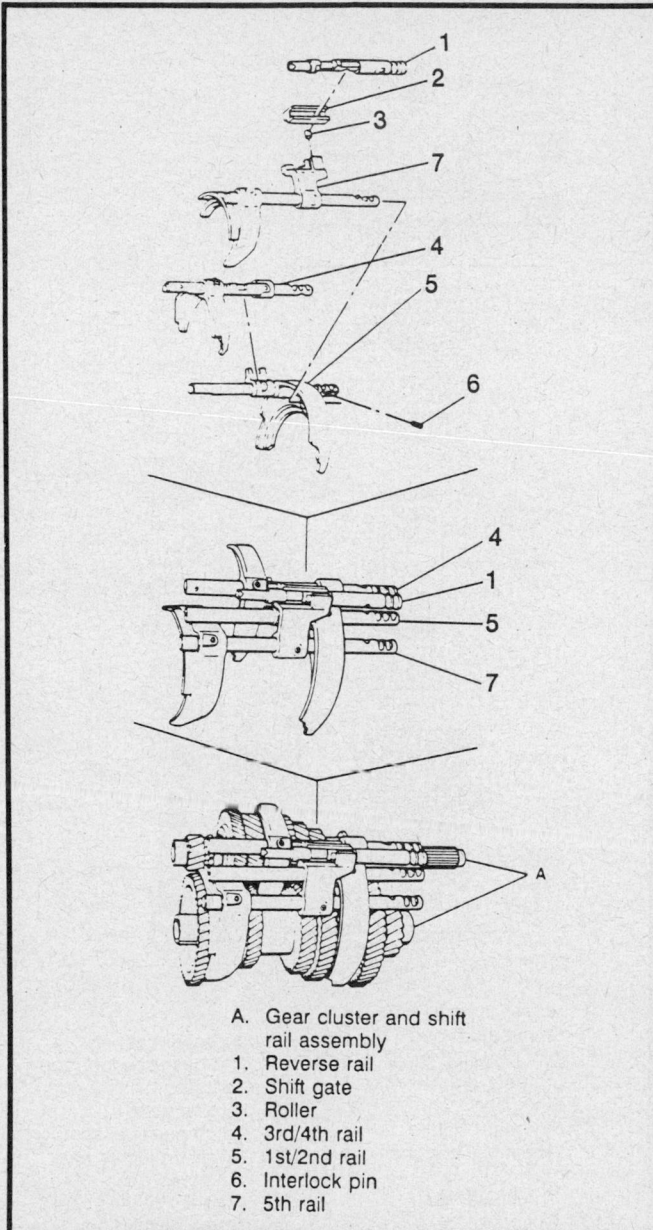

A. Gear cluster and shift rail assembly
1. Reverse rail
2. Shift gate
3. Roller
4. 3rd/4th rail
5. 1st/2nd rail
6. Interlock pin
7. 5th rail

Shift rail assemblies

1. Remove the reverse/5th gear synchronizer assembly using tool J-22912-01 or equivalent and a hydraulic press.
2. Remove the 5th gear blocker ring.
3. Remove the 5th speed gear.
4. Remove the 5th gear bearing.
5. Remove the thrust washer.
6. Remove the thrust washer positioner ball.
7. Remove the snapring. Discard the snapring if stretched.
8. Remove the 1st gear, bearing, caged thrust bearing and thrust washer using tool J-36183 or equivalent and a hydraulic press. The 2nd gear, bearing, race, 1/2 synchronizer, 1st and 2nd gear blocker rings and 3/4 gear cluster will press off with the 1st gear.

Inspection

1. Clean parts with solvent and air dry.
2. Inspect shaft, spline wear for cracks.

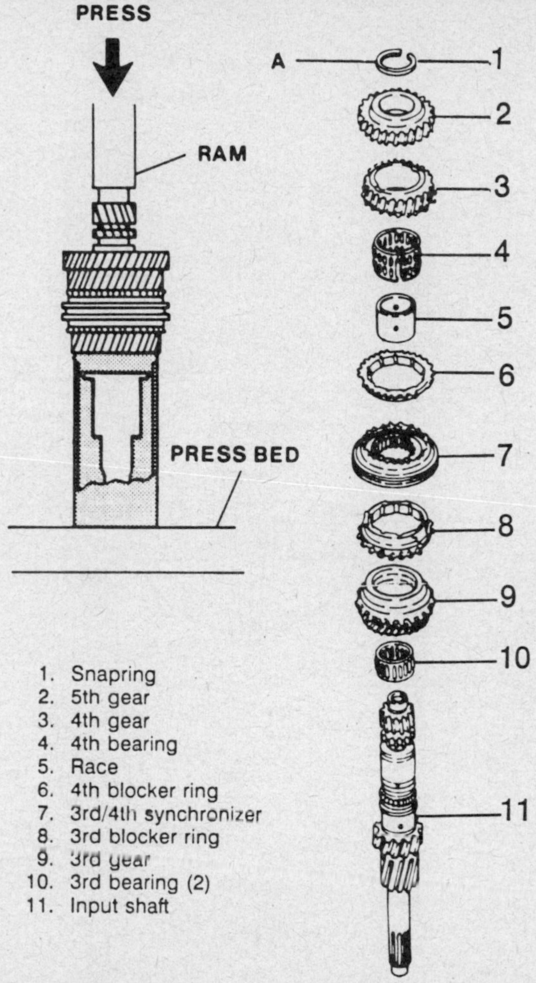
PRESS
RAM
PRESS BED
A
1
2
3
4
5
6
7
8
9
10
11

1. Snapring
2. 5th gear
3. 4th gear
4. 4th bearing
5. Race
6. 4th blocker ring
7. 3rd/4th synchronizer
8. 3rd blocker ring
9. 3rd gear
10. 3rd bearing (2)
11. Input shaft

Input shaft components removal

3. Inspect gear teeth for scuffed, nicked, burred or broken teeth.
4. Inspect bearings for roughness of rotation, burred or pitted condition.
5. Inspect bearing races for scoring, wear or overheating.
6. Inspect snaprings for nicks, distortion or wear.
7. Inspect synchronizers assembly for wear.

Assembly

NOTE: The 2nd gear bearing race requires heating 250°F (120°C) in oven, minimum of 7–10 minutes. The 3/4 gear cluster requires heating 250°F (120°C) in oven, minimum of 20 minutes. Lubricate all components as assembly progresses.

1. Install thrust washer with the chamfer down.
2. Install the thrust bearing with the needles down.
3. Install the 1st gear bearing.
4. Install the 1st gear with the cone up.
5. Install the 1st gear blocker ring.
6. Install the 1/2 synchronizer using tool J-36183, J-36184 or equivalent and a hydraulic press. Use tool J-22828 or equivalent, to do this. The small outer diameter groove on the sleeve (and small end of the hub) toward the 1st gear.

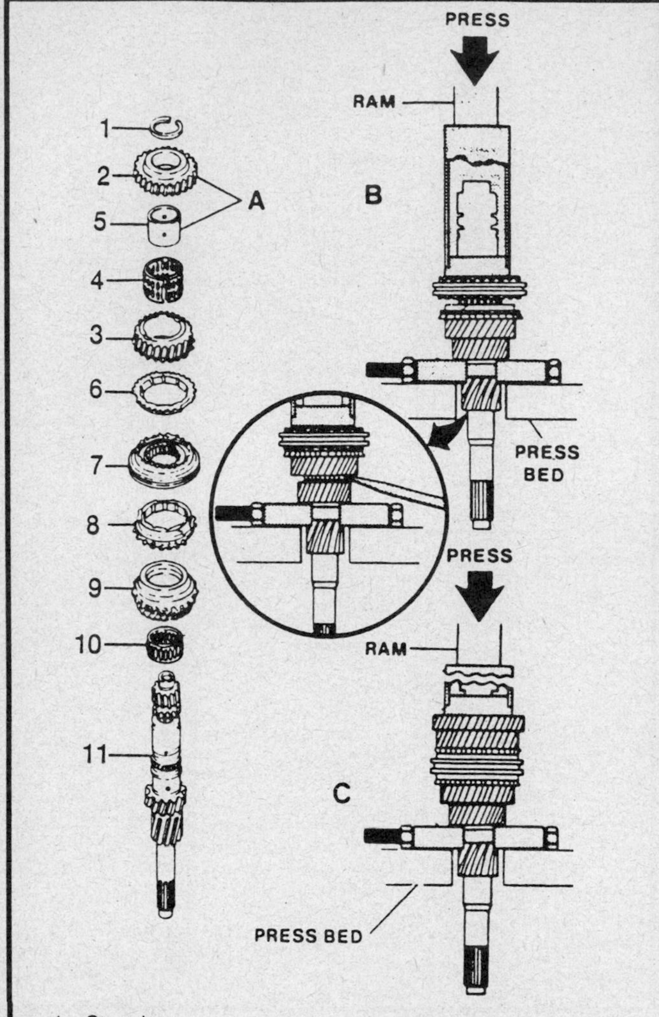
PRESS
RAM
PRESS BED
PRESS
RAM
PRESS BED
A
B
C
1
2
3
4
5
6
7
8
9
10
11

1. Snapring
2. 5th gear
3. 4th gear
4. 4th gear bearing
5. 4th gear bearing race
6. 4th gear blocker ring
7. 3rd/4th gear synchronizer assembly
8. 3rd gear blocker ring
9. 3rd gear
10. 3rd gear bearing
11. Input shaft

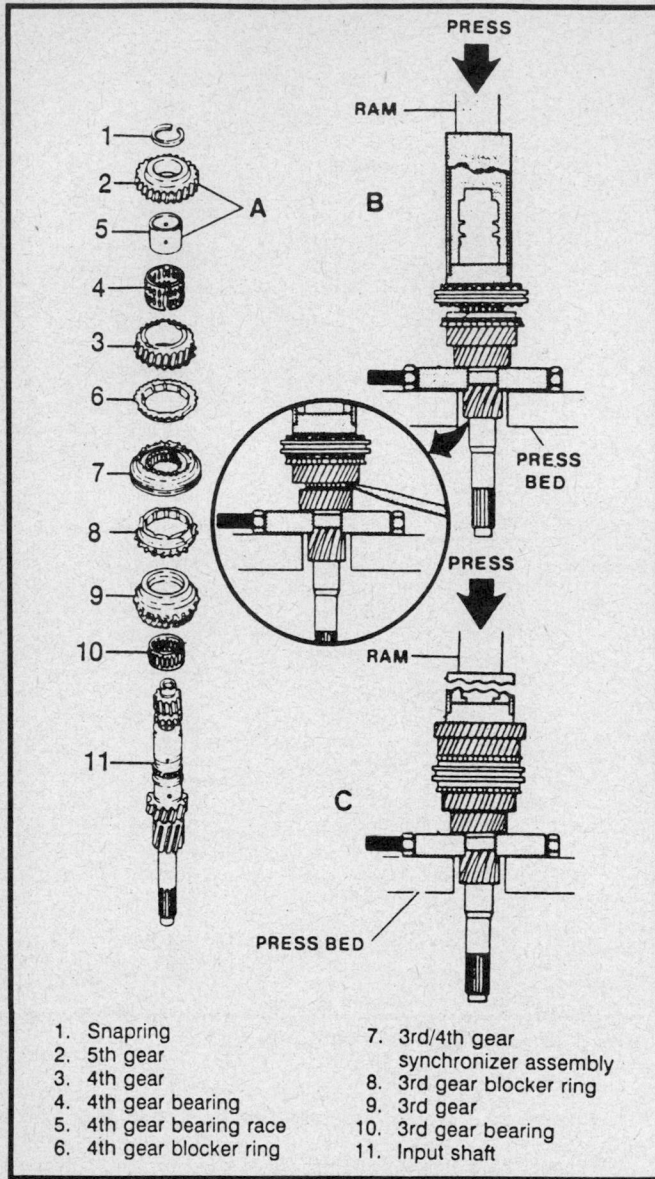

Input shaft components installation

NOTE: When pressing the 1/2 synchronizer assembly, start press operation, stop before tangs engage. Lift and rotate the 1st gear and 1st gear blocker ring, to engage the blocker ring tangs. Continue to press until seated. Be sure all the shavings are removed.

7. Install the 2nd gear bearing race (be careful handling the hot bearing race), 2nd gear bearing and 2nd gear (cone down).
8. Install the 3rd/4th gear cluster (be careful when handling hot cluster gear) using tool J-36183 or equivalent and a hydraulic press. The large outer diameter gear down.
9. Install the snapring, thrust washer positioning ball (retain with petroleum jelly) and slotted thrust washer. Align the inner diameter slot with the ball.
10. Install the 5th gear bearing and the 5th gear with the cone up.
11. Install the 5th gear blocker ring.
12. Install the reverse/5th gear synchronizer assembly using tool J-36183, J-36184 or equivalent and hydraulic press.

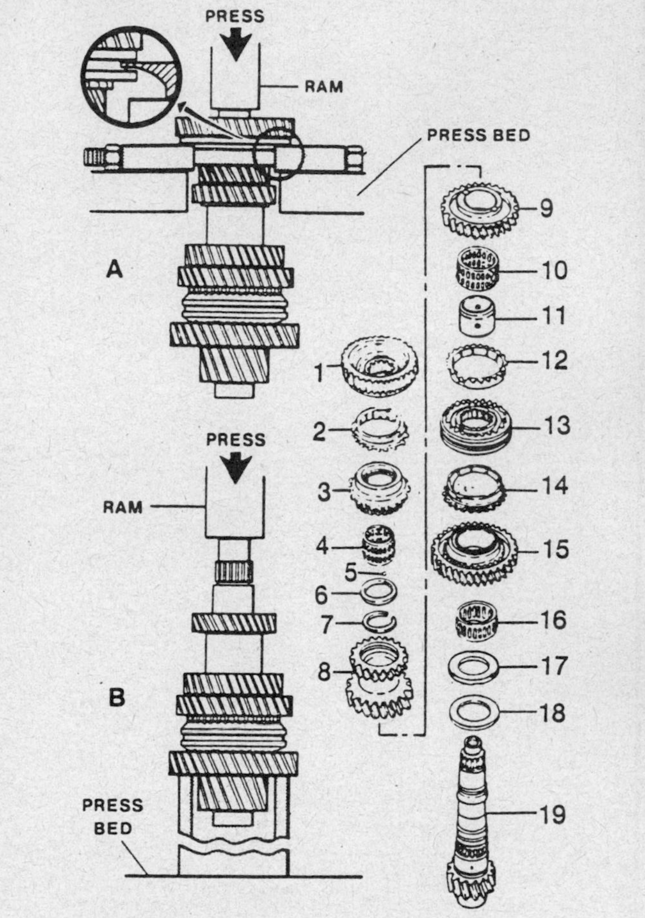
PRESS
RAM
PRESS BED
PRESS
RAM
PRESS BED
A
B
1
2
3
4
5
6
7
8
9
10
11
12
13
14
15
16
17
18
19

1. Reverse/5th gear synchronizer
2. 5th gear blocker ring
3. 5th speed gear
4. 5th gear bearing
5. Thrust washer positioner ball
6. Thrust washer
7. Snapring
8. 3rd/4th cluster gear
9. 2nd speed gear
10. 2nd gear bearing
11. 2nd gear bearing race
12. 2nd gear blocker ring
13. 1st/2nd gear synchronizer assembly
14. 1st gear blocker ring
15. 1st speed gear
16. 1st gear bearing
17. Thrust bearing
18. Thrust washer
19. Output shaft

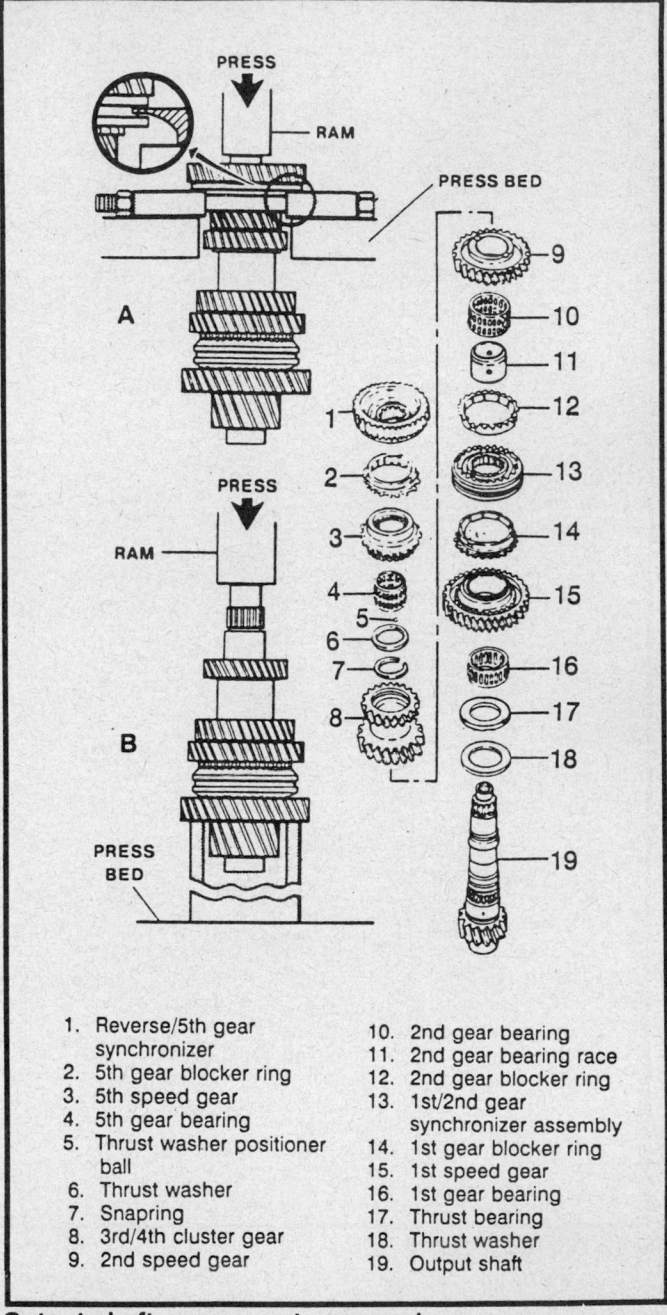

Output shaft components removal

NOTE: When pressing the reverse/5th synchronizer, start press operation, stop before tangs engage. Lift and rotate the 5th gear and 5th gear blocker ring (thrust washer must stay down), engaging tangs. Continue to press until seated. Be sure all shavings are removed.

REVERSE IDLER GEAR

Disassembly

1. Remove the bolt using a 13mm socket and driver.
2. Remove the shift rail, gear, shaft and bracket.
3. Remove the shift rail, detent ball and spring.

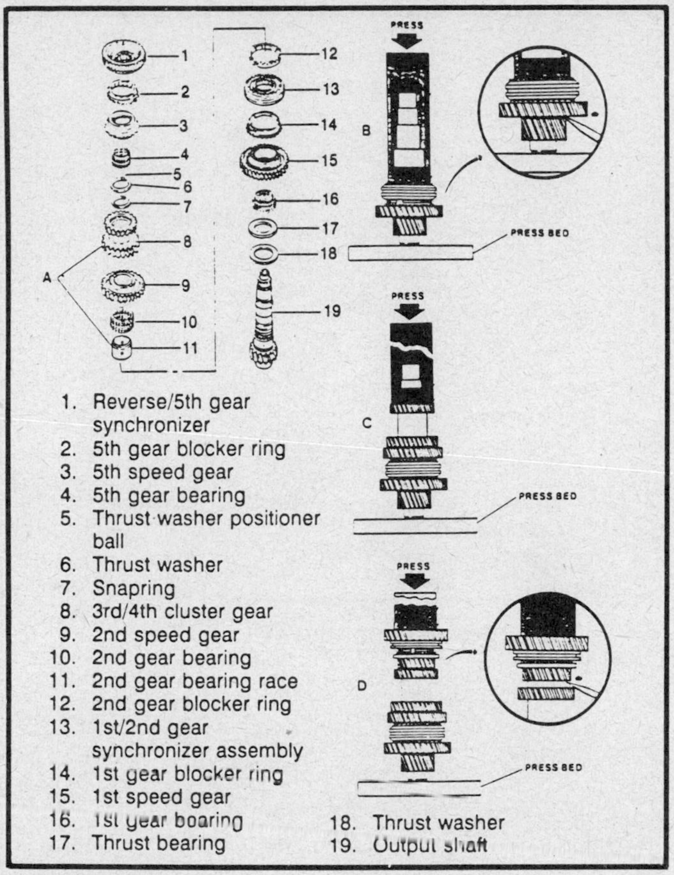
PRESS
PRESS BED
PRESS
PRESS BED
PRESS
PRESS BED
A
B
C
D
1. Reverse/5th gear synchronizer
2. 5th gear blocker ring
3. 5th speed gear
4. 5th gear bearing
5. Thrust washer positioner ball
6. Thrust washer
7. Snapring
8. 3rd/4th cluster gear
9. 2nd speed gear
10. 2nd gear bearing
11. 2nd gear bearing race
12. 2nd gear blocker ring
13. 1st/2nd gear synchronizer assembly
14. 1st gear blocker ring
15. 1st speed gear
16. 1st gear bearing
17. Thrust bearing
18. Thrust washer
19. Output shaft

Output shaft components installation

Inspection

1. Clean parts with solvent and air dry.
2. Inspect gear teeth for scuffed, nicked, burred or broken teeth.
3. Inspect bushings for roughness, burred, scores or overheating.
4. Inspect shaft for scoring, wear or overheating.

Assembly

1. Lubricate all components as assembly progresses.
2. Assemble spring and ball in bracket.
3. Install the shaft in the bracket assembly.
4. Install the gear on the shaft with the slot on the gear toward the threaded hole in the shaft.
5. Install the reverse idler gear assembly.
6. Install sealer to the bolt threads and install the bolt. Torque to 16 ft. lbs. (21 Nm).

TRANSAXLE CASE

Disassembly

NOTE: Remove the bearings and bushings only if there is evidence of damage or a mating part is being replaced.

1. Remove the snapring.
2. Remove the plug.
3. Remove the sliding sleeve screw, sliding sleeve spring and sliding sleeve.
4. Remove the sliding sleeve bushing using tool J-36034 and J-36190 or equivalent and a hammer.
5. Remove the detent lever.

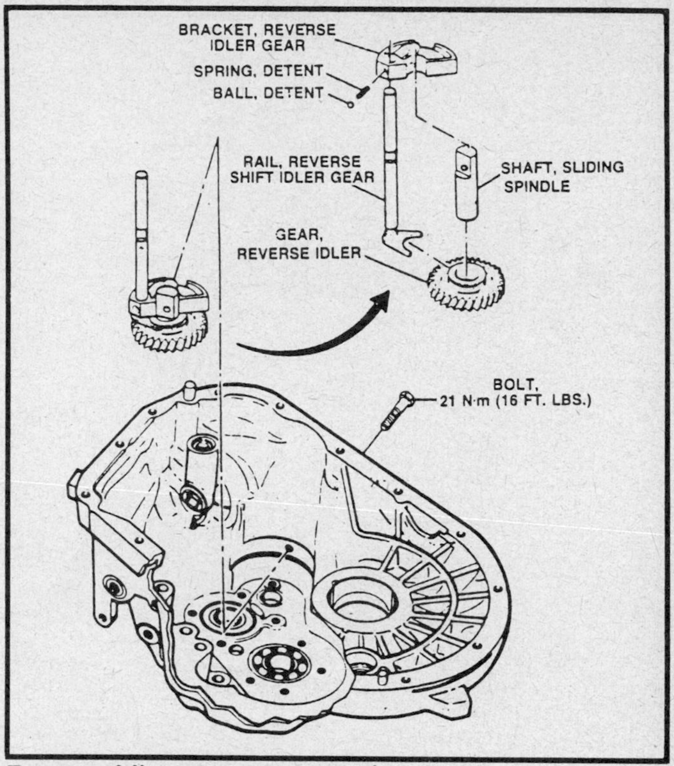
BRACKET, REVERSE IDLER GEAR
SPRING, DETENT
BALL, DETENT
RAIL, REVERSE SHIFT IDLER GEAR
SHAFT, SLIDING SPINDLE
GEAR, REVERSE IDLER
BOLT, 21 N·m (16 FT. LBS.)

Reverse idler gear components

6. Remove the bushing shift detent lever using tool J-36039 and J-36190 or equivalent and a hammer.
7. Remove the shift shaft seal using a small suitable tool.
8. Remove the bearing shift shaft using tool J-36027 and J-36190 or equivalent and a hammer.
9. Remove the axle seal using a punch and hammer.
10. Remove the differential case support outer race using tool J-36181 and J-8092 or equivalent and a hammer.
11. Remove the shift rails plugs using a punch and a hammer.
12. Remove the input shaft support bearing. Remember that these bearings are not to be reused.
13. Remove the output shaft support bearing using a hammer.
14. Remove the 3 shift rail bushings using tool J-36029 or equivalent, (small end ofJ-36029-2 or equivalent adapter in bushing) and J-36190 or equivalent and a hammer. Make sure the tool is positioned to clear the case.
15. Remove the reverse shift rail bushing using tool J-36032 and J-23907 or equivalent.
16. Remove the stud using a 13mm socket and driver.

Inspection

1. Clean with solvent and air dry.
2. Inspect the case bearing race bore for wear, scratches or grooves.
3. Inspect the case bushings for scores, burrs, roundness or evidence of overheating.
4. Inspect case for cracks, threaded openings for damaged threads, mounting faces for nicks, burrs or scratches.

Assembly

1. Install the shift shaft bearing using tool J-36189 and J-36190 or equivalent and a hammer.
2. Install the shift shaft seal using tool J-35823 or equivalent and a hammer.
3. Install the 3 shift rail bushings using tool J-36029 or equivalent, (place new bushing on -2 adapter and retain be-

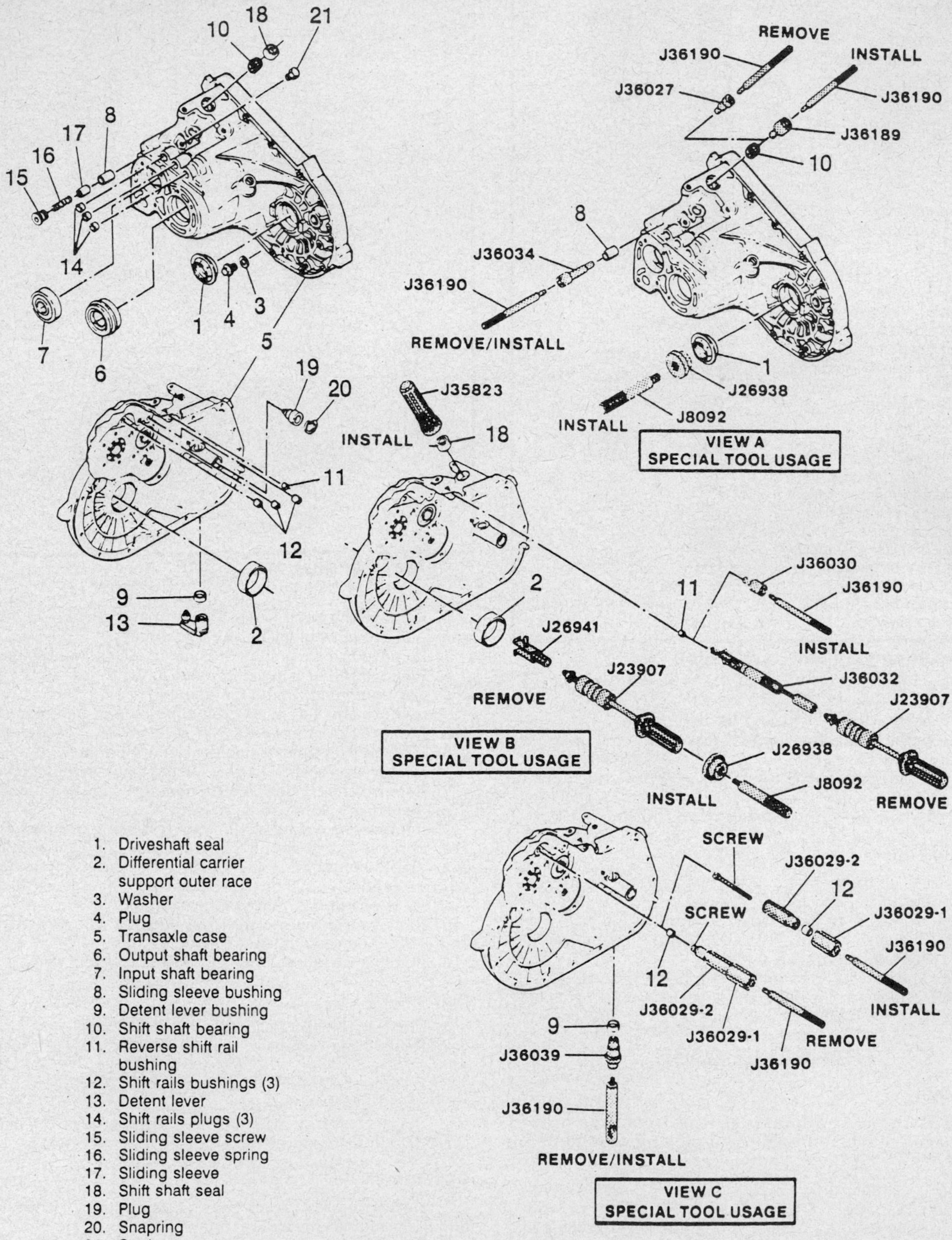
REMOVE
INSTALL
J36190
J36027
J36189
J36034
REMOVE/INSTALL
J35823
J26938
J8092
VIEW A
SPECIAL TOOL USAGE
J36030
J26941
J23907
J36032
VIEW B
SPECIAL TOOL USAGE
SCREW
J36029-2
J36029-1
J36039
VIEW C
SPECIAL TOOL USAGE

1. Driveshaft seal
2. Differential carrier support outer race
3. Washer
4. Plug
5. Transaxle case
6. Output shaft bearing
7. Input shaft bearing
8. Sliding sleeve bushing
9. Detent lever bushing
10. Shift shaft bearing
11. Reverse shift rail bushing
12. Shift rails bushings (3)
13. Detent lever
14. Shift rails plugs (3)
15. Sliding sleeve screw
16. Sliding sleeve spring
17. Sliding sleeve
18. Shift shaft seal
19. Plug
20. Snapring
21. Stud

Transaxle case components

tween the -1 and -2 tool parts) and J-36190 or equivalent and a hammer.

4. Install the reverse rail bushing using tool J-36030 and J-36190 or equivalent and a hammer.

5. Install the differential carrier support outer race using tool J-26938 or equivalent and a hammer.

6. Install the axle seal using tool J-26938 or equivalent and a hammer.

7. Install the plugs even with the bore surface using suitable socket and hammer.

8. Install the detent lever bushing using tool J-36039 or equivalent and a hammer.

9. Install the detent lever.

10. Install the sliding sleeve bushing using tool J-36034 and J-36190 or equivalent and a hammer.

11. Install the sliding sleeve, spring and screw. Use sealer on the screw and tighten to 32 ft. lbs. (44 Nm).

12. Install the plug and snapring with the flat side up.

13. Install the stud with the chamfer end out and tighten to 15 ft. lbs. (21 Nm).

CLUTCH AND DIFFERENTIAL HOUSING

Disassembly

1. Remove the bolts and retainer using a 10mm socket and driver.

2. Remove the output shaft race using tool J-36038 and J-23907 or equivalent and a hammer.

3. Remove the bolts, washers, spacer and plate using a 10mm socket and driver.

4. Remove the reverse rail guide bolt and guide. This may be difficult to remove. Use a 10mm socket and driver.

5. Remove the axle seal using a punch and hammer.

6. Remove the differential race and shim using tool J-36181 and J-8092 or equivalent and a hammer.

7. Remove the clutch shaft seal using small pry bar.

8. Remove the outer clutch shaft upper bushing using tool J-36037 or equivalent and a hammer.

9. Remove the clutch shaft.

10. Remove the inner clutch shaft bushing using tool J-36032 and J-23907 or equivalent.

11. Remove the input shaft bearing sleeve assembly using tool J-35824 or equivalent and a hydraulic press.

12. Remove the shift rail bushings using tool J-36029 or equivalent, (small end of -2 adapter in bushing) and hammer.

13. Remove the drain plug and washer using a 15mm socket and hammer.

14. Remove the breather assembly. Pry with a suitable tool.

Inspection

1. Clean with solvent and air dry.

2. Inspect the housing bearing race bore for wear, scratches or grooves.

3. Inspect the housing bushings for scores, burrs, roundness or evidence of overheating.

4. Inspect housing for cracks, threaded openings for damaged threads, mounting faces for nicks, burrs or scratches.

Assembly

NOTE: Do not install the differential bearing race and axle seal or shim. Installation will be after differential bearing selective shimming.

1. Install the drain plug and new washer and tighten to 18 ft. lbs. (24 Nm).

2. Install the shift rail bushings using tool J-36029 or equivalent, (place the new bushings on -2 adapter and retain between the -1 and -2 tool parts) and J-36190 or equivalent and a hammer. Bushings must not protrude into the case side of the clutch housing.

3. Install the input shaft bearing sleeve (oil seepage hole installed down in the clutch housing) using tool J-35824 or equivalent and a hydraulic press.

4. Install the inner clutch shaft bushing using tool J-36033 and J-36190 or equivalent and a hammer.

5. Install the clutch shaft.

6. Install the outer clutch shaft upper bushing using tool J-36037 or equivalent and a hammer.

7. Install the clutch shaft seal using a suitable socket and hammer.

8. Install the reverse rail guide using hammer. Short side in the bore and tighten bolt to 15 ft. lbs. (21 Nm).

9. Install the output shaft race using tool J-23423-A and J-8092 or equivalent and a hammer. Be sure to use side A of the driver. Align the race cutouts with the slots in the case.

10. Install the retainer and bolts and tighten to 15 ft. lbs. (21 Nm). Use Loctite® 242 on the bolt.

11. Install the interlock plate, spacers, washers and bolts. Use Loctite® 242 and torque bolts to 15 ft. lbs. (21 Nm).

12. Install the breather assembly using a hammer.

SYNCHRONIZERS

Disassembly

1. Place the 1-2, 3-4 and 5th speed synchronizers in separate shop towels, wrap the assemblies and press against the inner hub.

2. Mark the sleeve and hub for installation.

Inspection

1. Clean the assembly with solvent and air dry.

2. Inspect the synchronizer teeth for wear, scuffed, nicked, burred or broken teeth.

3. Inspect the synchronizer keys for wear or distortion.

4. Inspect the synchronizer balls and springs for distortion, cracks or wear.

Assembly

1. Install the 1st/2nd gear synchronizer assembly.

2. Install the 3rd/4th gear synchronizer assembly.

3. Install the 5th gear synchronizer assembly.

DIFFERENTIAL AND RING GEAR

1. Remove the differential carrier assembly bolts using a 15mm socket and driver.

2. Remove the differential ring gear.

3. Remove the differential bearings using tool J-22888 or equivalent, J-22888-35 or equivalent and J-2241-11 or equivalent or J-23598 or equivalent.

4. Remove the speedometer gear using a prybar. Do not reuse; the removal will destroy the gear.

5. Remove the differential cross pin locking bolt and washer.

6. Remove the differential cross pin.

7. Remove the differential pinion gear and washer, differential side gear and side gear thrust washer. Identify the parts for same installation.

Inspection

1. Clean the parts with a solvent and air dry.

2. Inspect the gears for scuffed, nicked, burred or broken teeth.

3. Inspect the carrier for distortion, bores out of round or scoring.

4. Inspect the bearings for roughness of rotation, burred or pitted condition.

5. Inspect the thrust washers for wear, scuffed, nicked or burred condition.

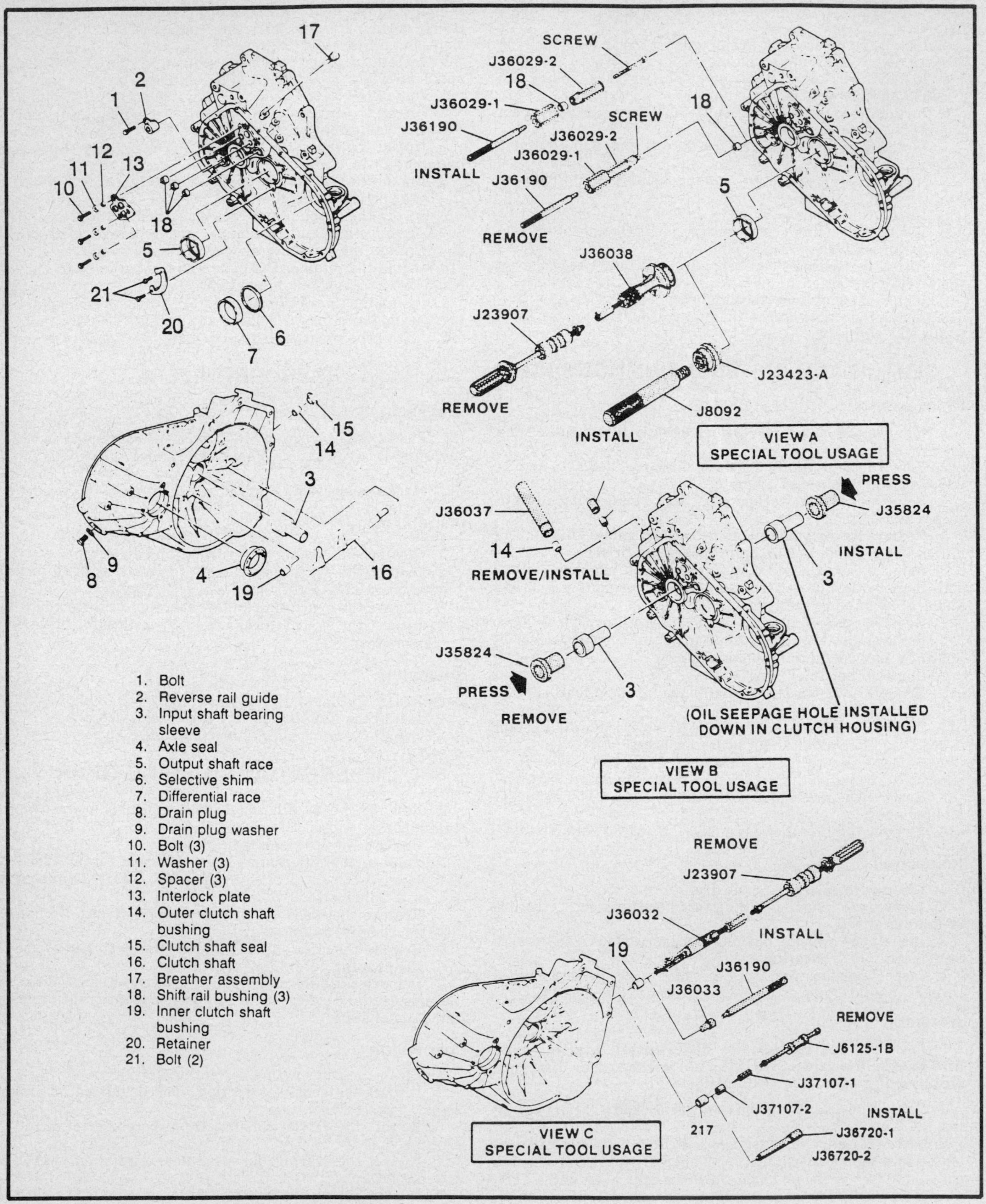
SCREW
J36029-2
J36029-1
J36190
INSTALL
REMOVE
J36038
J23907
J23423-A
J8092
VIEW A
SPECIAL TOOL USAGE
PRESS
J35824
J36037
REMOVE/INSTALL
(OIL SEEPAGE HOLE INSTALLED DOWN IN CLUTCH HOUSING)
VIEW B
SPECIAL TOOL USAGE
J36032
J36033
J6125-1B
J37107-1
J37107-2
217
J36720-1
J36720-2
VIEW C
SPECIAL TOOL USAGE
1. Bolt
2. Reverse rail guide
3. Input shaft bearing sleeve
4. Axle seal
5. Output shaft race
6. Selective shim
7. Differential race
8. Drain plug
9. Drain plug washer
10. Bolt (3)
11. Washer (3)
12. Spacer (3)
13. Interlock plate
14. Outer clutch shaft bushing
15. Clutch shaft seal
16. Clutch shaft
17. Breather assembly
18. Shift rail bushing (3)
19. Inner clutch shaft bushing
20. Retainer
21. Bolt (2)

Clutch and differential housing components

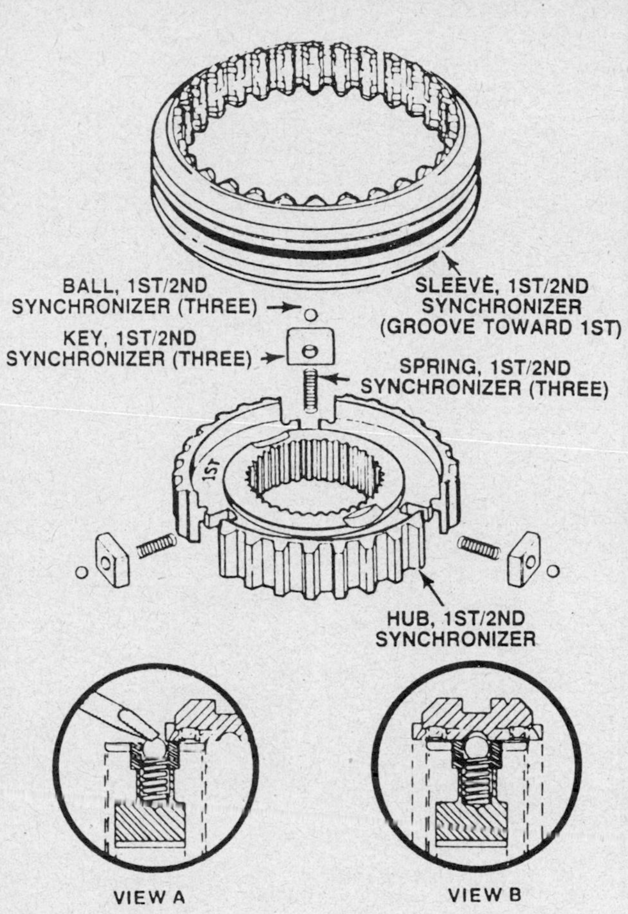
BALL, 1ST/2ND SYNCHRONIZER (THREE)
SLEEVE, 1ST/2ND SYNCHRONIZER (GROOVE TOWARD 1ST)
KEY, 1ST/2ND SYNCHRONIZER (THREE)
SPRING, 1ST/2ND SYNCHRONIZER (THREE)
1ST
HUB, 1ST/2ND SYNCHRONIZER
VIEW A
VIEW B

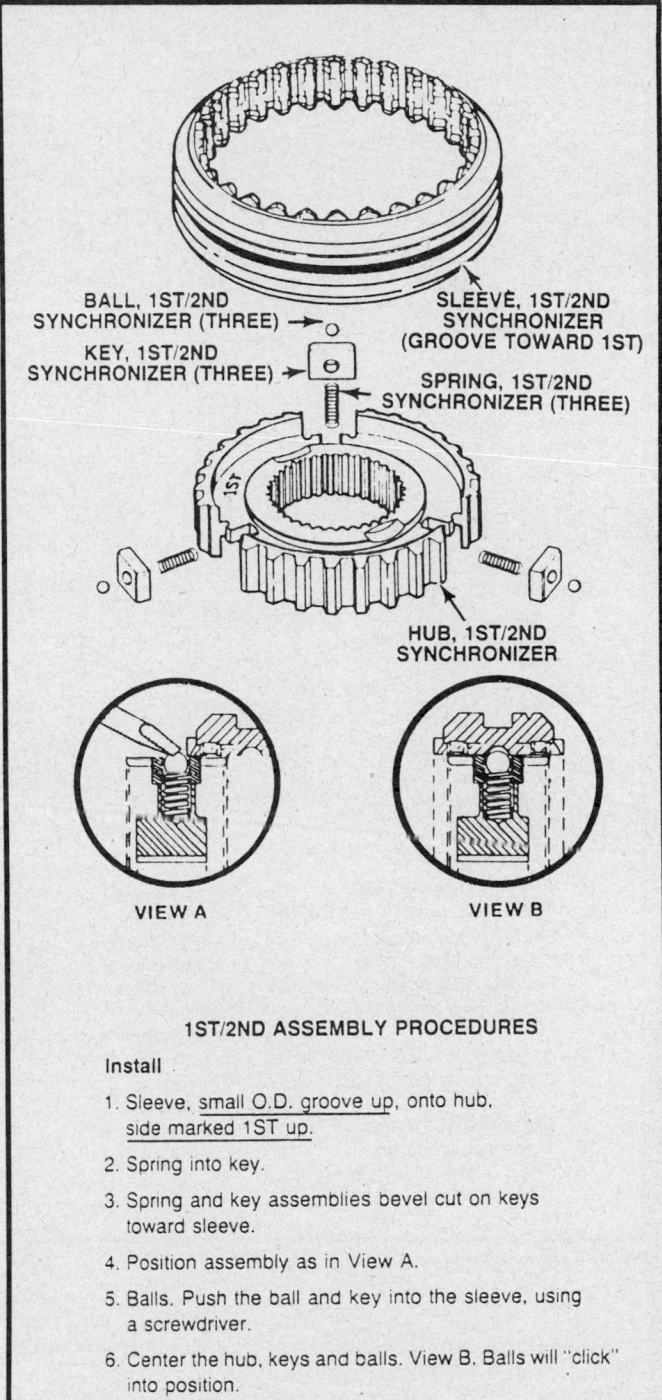

1ST/2ND ASSEMBLY PROCEDURES

Install

1. Sleeve, small O.D. groove up, onto hub, side marked 1ST up.
2. Spring into key.
3. Spring and key assemblies bevel cut on keys toward sleeve.
4. Position assembly as in View A.
5. Balls. Push the ball and key into the sleeve, using a screwdriver.
6. Center the hub, keys and balls. View B. Balls will "click" into position.

1st/2nd gear synchronizer components

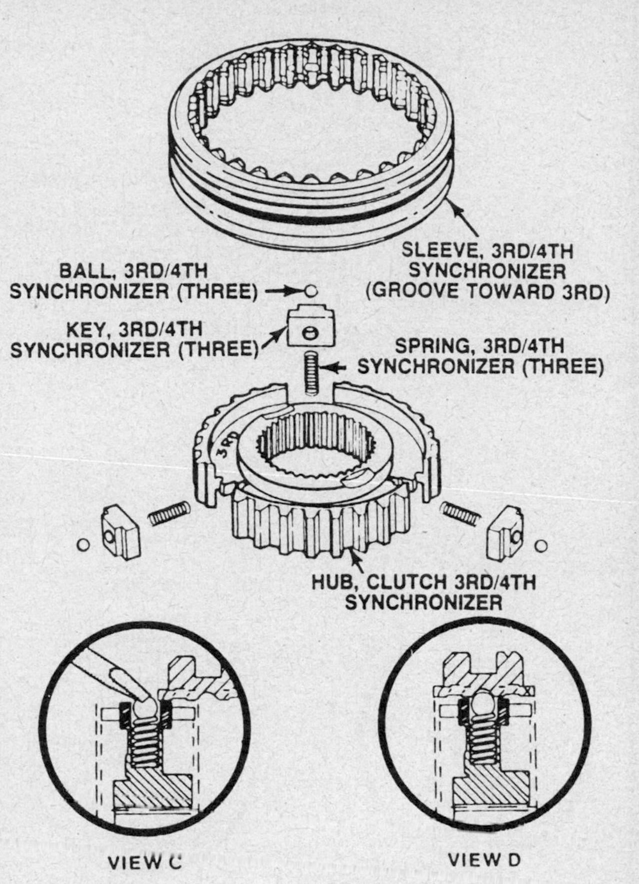
BALL, 3RD/4TH SYNCHRONIZER (THREE)
SLEEVE, 3RD/4TH SYNCHRONIZER (GROOVE TOWARD 3RD)
KEY, 3RD/4TH SYNCHRONIZER (THREE)
SPRING, 3RD/4TH SYNCHRONIZER (THREE)
3RD
HUB, CLUTCH 3RD/4TH SYNCHRONIZER
VIEW C
VIEW D

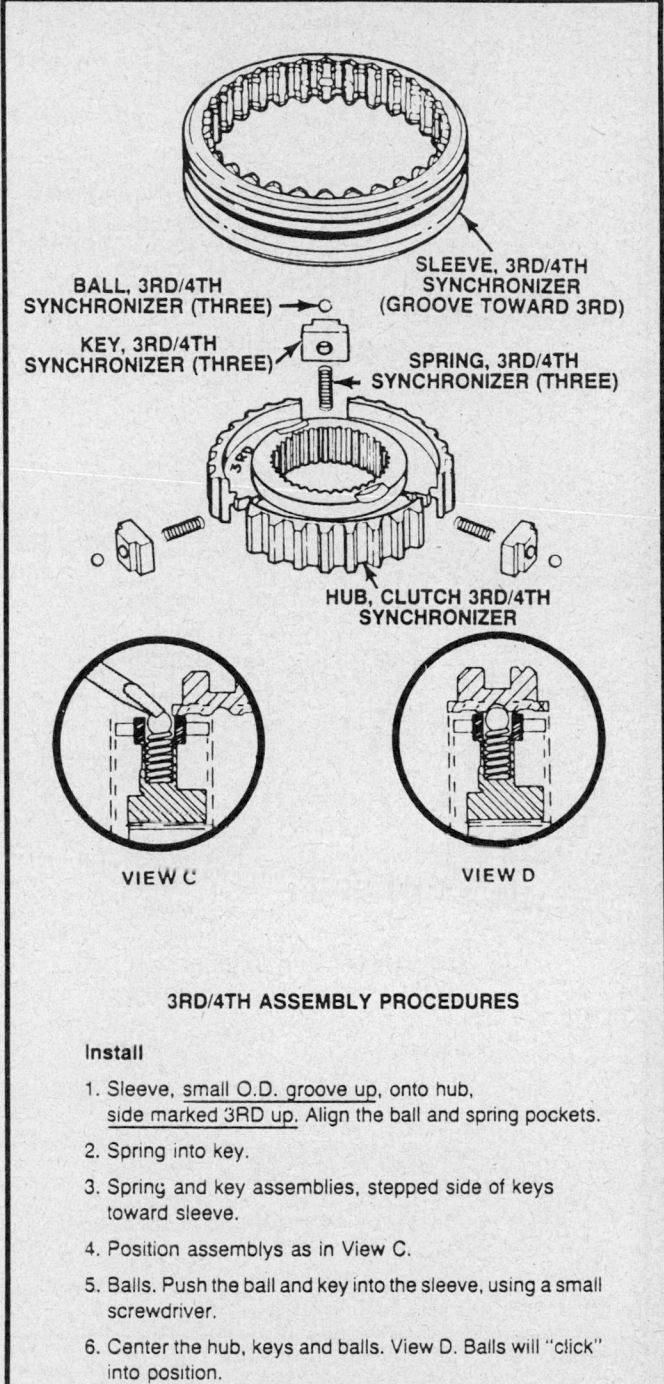

3RD/4TH ASSEMBLY PROCEDURES

Install

1. Sleeve, small O.D. groove up, onto hub, side marked 3RD up. Align the ball and spring pockets.
2. Spring into key.
3. Spring and key assemblies, stepped side of keys toward sleeve.
4. Position assemblys as in View C.
5. Balls. Push the ball and key into the sleeve, using a small screwdriver.
6. Center the hub, keys and balls. View D. Balls will "click" into position.

3rd/4th gear synchronizer components

Assembly

NOTE: Heat the mechanical configuration nylon speedometer drive gear in hot tap water for 5 minutes prior to installation. Heat the electronic configuration steel speedometer drive gear in an oven at 250°F (120°C) for 7–10 minutes prior to installation. Do not reuse bolts.

1. Install the speedometer gear and allow to cool.
2. Install the differential bearings using tool J–22919 or equivalent and a hydraulic press.
3. Install the differential side gear and side gear thrust washer and differential pinion gear and pinion gear side thrust washer.
4. Install the differential cross pin.
5. Install the differential cross locking bolt and washer. Torque to 84 inch lbs. (9 Nm).

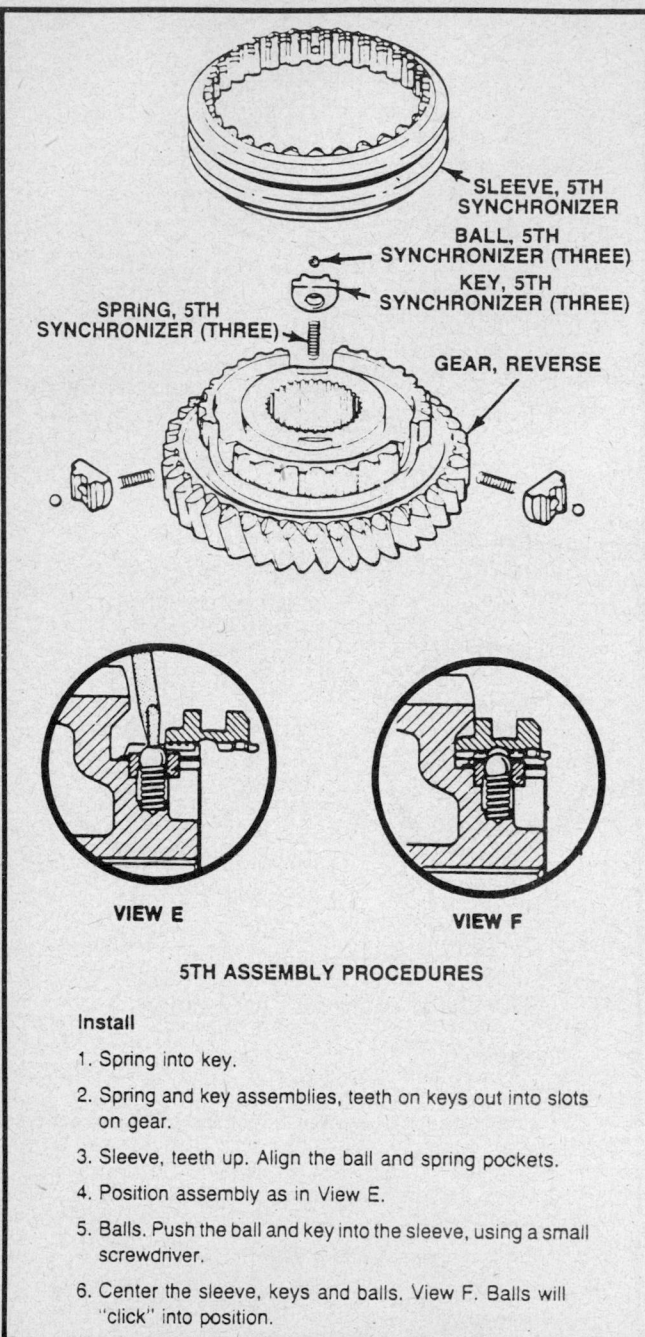

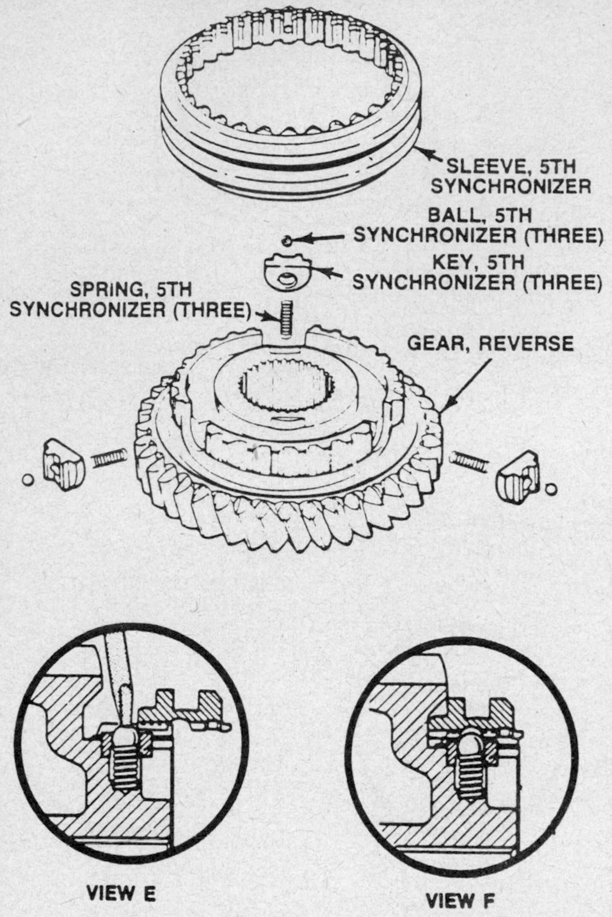
SLEEVE, 5TH SYNCHRONIZER
BALL, 5TH SYNCHRONIZER (THREE)
KEY, 5TH SYNCHRONIZER (THREE)
SPRING, 5TH SYNCHRONIZER (THREE)
GEAR, REVERSE
VIEW E
VIEW F

5TH ASSEMBLY PROCEDURES

Install

1. Spring into key.
2. Spring and key assemblies, teeth on keys out into slots on gear.
3. Sleeve, teeth up. Align the ball and spring pockets.
4. Position assembly as in View E.
5. Balls. Push the ball and key into the sleeve, using a small screwdriver.
6. Center the sleeve, keys and balls. View F. Balls will "click" into position.

5th gear synchronizer components

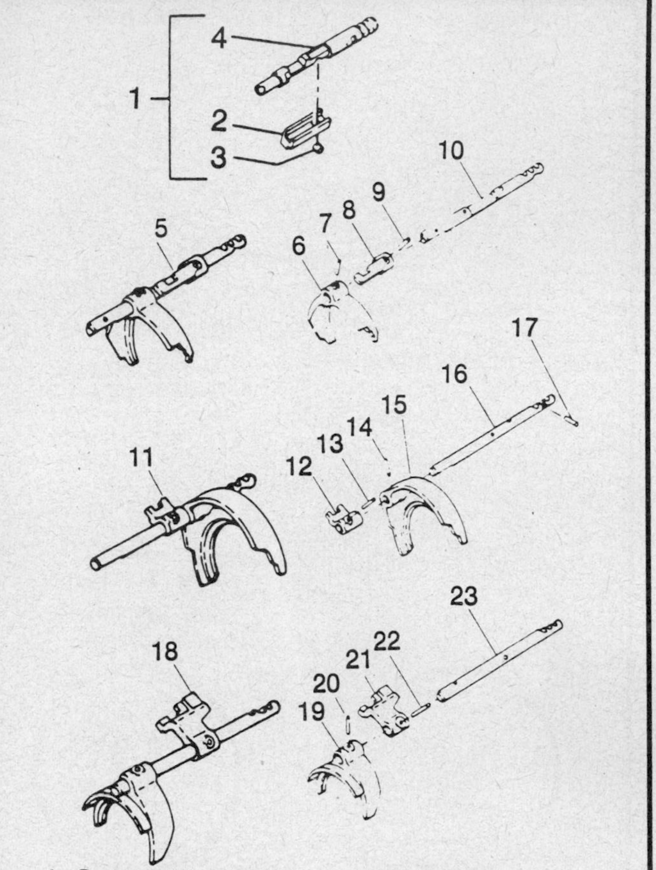
1
2
3
4
5
6
7
8
9
10
11
12
13
14
15
16
17
18
19
20
21
22
23

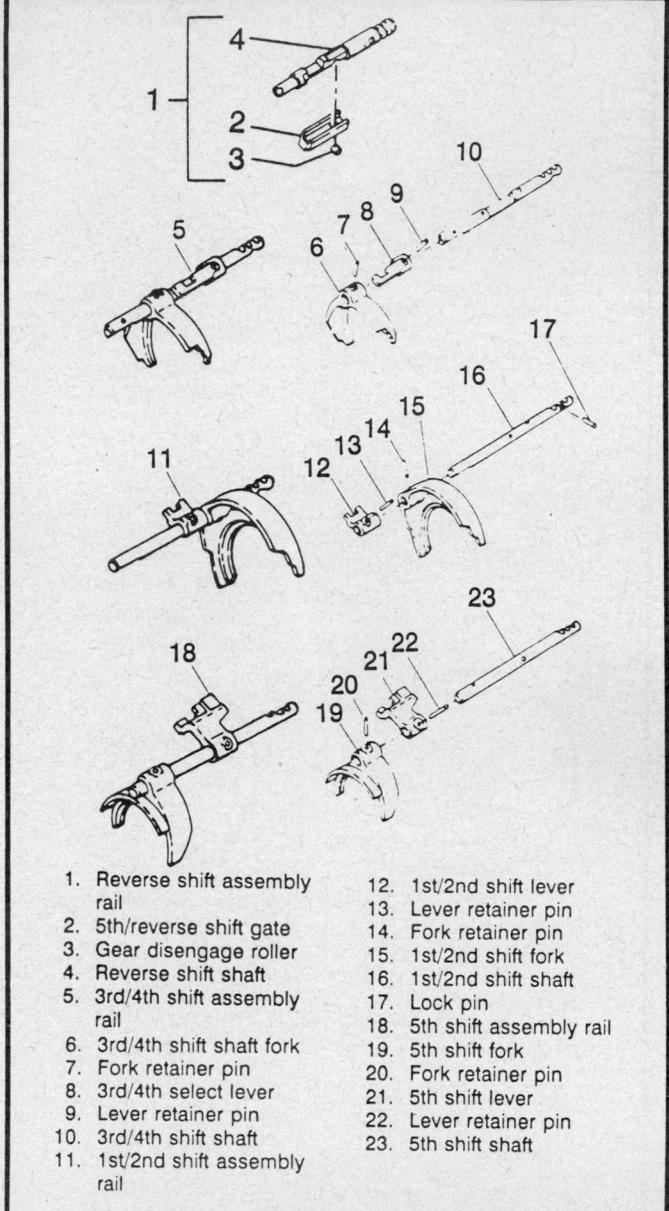

1. Reverse shift assembly rail
2. 5th/reverse shift gate
3. Gear disengage roller
4. Reverse shift shaft
5. 3rd/4th shift assembly rail
6. 3rd/4th shift shaft fork
7. Fork retainer pin
8. 3rd/4th select lever
9. Lever retainer pin
10. 3rd/4th shift shaft
11. 1st/2nd shift assembly rail
12. 1st/2nd shift lever
13. Lever retainer pin
14. Fork retainer pin
15. 1st/2nd shift fork
16. 1st/2nd shift shaft
17. Lock pin
18. 5th shift assembly rail
19. 5th shift fork
20. Fork retainer pin
21. 5th shift lever
22. Lever retainer pin
23. 5th shift shaft

Shift rail and fork assemblies

6. Install the differential ring gear with the identification chamfer to the carrier.
7. Install new bolts and torque to 61 ft. lbs. (83 Nm).

DIFFERENTIAL ASSEMBLY SELECTIVE SHIM PRELOAD PROCEDURE

1. Install tool J-26935 or equivalent, to the clutch housing and transaxle housing.
2. Measure the largest shim possible on tool J-26935-3; use the shim 2 sizes larger.
3. Install the selective shim.
4. Install the differential bearing race using tool J-26938 or equivalent and J-8092 or equivalent and a hammer.
5. Install the driveshaft seal using tool J-8092 or equivalent and a hammer.

Transaxle Assembly

GEAR/SHIFT RAIL ASSEMBLIES AND SUPPORT COMPONENT

1. Position the gear cluster/shift rail assembly on tool J-36182-1 or equivalent. Align the shift rail and shaft pilots to the fixture.
2. Install the transaxle case. Align the bearing bores in the case with the shaft pilots.

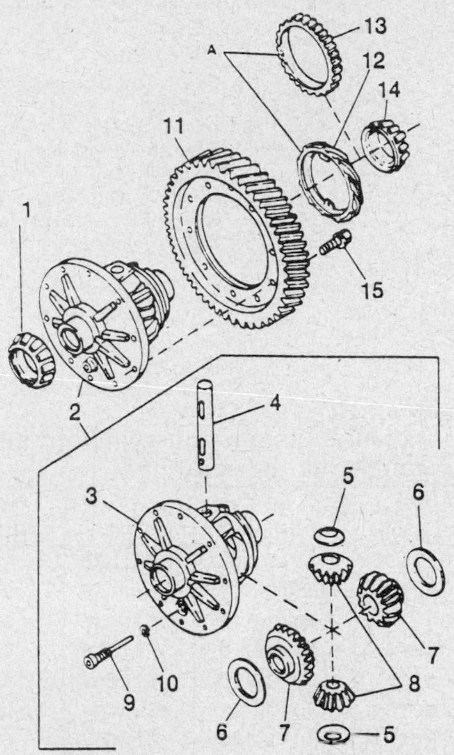
A
13
12
14
11
1
15
2
4
5
6
3
7
8
10
9
6
7
5

1. Differential bearing
2. Differential assembly carrier
3. Differential carrier
4. Differential cross pin
5. Pinion gear thrust washer
6. Side gear thrust washer
7. Differential side gear
8. Differential pinion gear
9. Screw
10. Lock washer
11. Differential ring gear
12. Speedometer gear (mechanical)
13. Speedometer gear (electronic)
14. Differential bearing
15. Bolt (10)

Differential and ring gear components

3. Install the new output shaft bearing using tool J–35824 or equivalent and a hydraulic press.
4. Install the new input shaft bearing using tool J–35824 or equivalent and a hydraulic press. Push the rails to engage and hold the transaxle in 4th and reverse gear. The bearings must be seated.
5. Install the new input and output retainers using tool J-36031 or equivalent and torque to 50 ft. lbs. (70 Nm). Return the transaxle to **N**.

SHIFT SHAFT

1. Assemble the pins and rollers on the shift shaft. Retain with petroleum jelly.
2. Install the shift shaft assembly. Tap in with a light hammer and align the hole in the shaft with the hole in the shift lever.
3. Install the lever retainer pin using a $^{3}/_{16}$ in. punch and a hammer. Install pin till it is even with the surface of the shift lever.

CLUTCH AND DIFFERENTIAL HOUSING

1. Apply sealant, part number 1052942 or equivalent, to the outside of the bolt hole pattern of the gear case flange.
2. Install the differential.
3. Install the output bearing noting the position of the cage. The small inner diameter of the bearing cage is toward the clutch housing.
4. Install the magnet.
5. Install the clutch housing.
6. Install the bolts and torque to 15 ft. lbs. (21 Nm).

OUTPUT SHAFT SUPPORT BEARING SELECTIVE SHIM PROCEDURE

NOTE: Be sure that the output bearing is seated in the bore by tapping the bearing into the case. Be sure that the bearing retainer is properly torqued. Selected shim can be 0.001 in. (0.03mm) above, or 0.004 in. (0.12mm) below the end plate mounting surface.

1. Using tool J–26900–19 metric dial depth gauge or equivalent, measure the distance between the end plate mounting surface and the outer race of the output shaft bearing.
2. Select the proper shim.

TRANSAXLE CASE END PLATE

1. Apply sealant, part number 1052942 or equivalent, to the outside of the end plate bolt hole pattern of the case.
2. Install the selective shim.
3. Install the oil shield.
4. Install the end cover plate.
5. Install the bolts and torque to 15 ft. lbs. (21 Nm).

SHIFT RAIL DETENT/CLUTCH AND DIFFERENTIAL HOUSING

1. Position the shift rails to expose the interlock notches in the **N** position.
2. Position the reverse shift rail to allow the detent ball to sit in the notch and on the reverse bushing.
3. Install the reverse bushing using a suitable socket.
4. Install the detent balls. Place them in the notched areas of the shift rails. Retain the ball positions with petroleum jelly.
5. Assemble the interlock pins and springs into the bores in the detent holder.
6. Install the detent holder and spring assembly. Position the detent balls over the springs using a small suitable tool. After all the detent balls are positioned over the springs, pry the reverse shift rail up to allow its detent ball to enter the spring pocket.
7. Position the detent holder using a pry to align the bolt holes with the threads.
8. Install the bolts and torque to 84 inch lbs. (9 Nm).
9. Install the protective cover by tapping with a hammer until seated in the bore.
10. Install the bearing. Apply high temperature grease to the inside of the bore.

SHIFT SHAFT DETENT/TRANSAXLE HOUSING

1. Install the inner spring seat.
2. Install the 5th/reverse bias spring.
3. Install the outer spring seat.
4. Install the spring screw and torque to 84 inch lbs. (9 Nm). Use a small amount of thread sealant, part number 11052624 or equivalent, to the screw.
5. Install the protective cover using a hammer. Position to below the snapring groove.
6. Install the snapring.

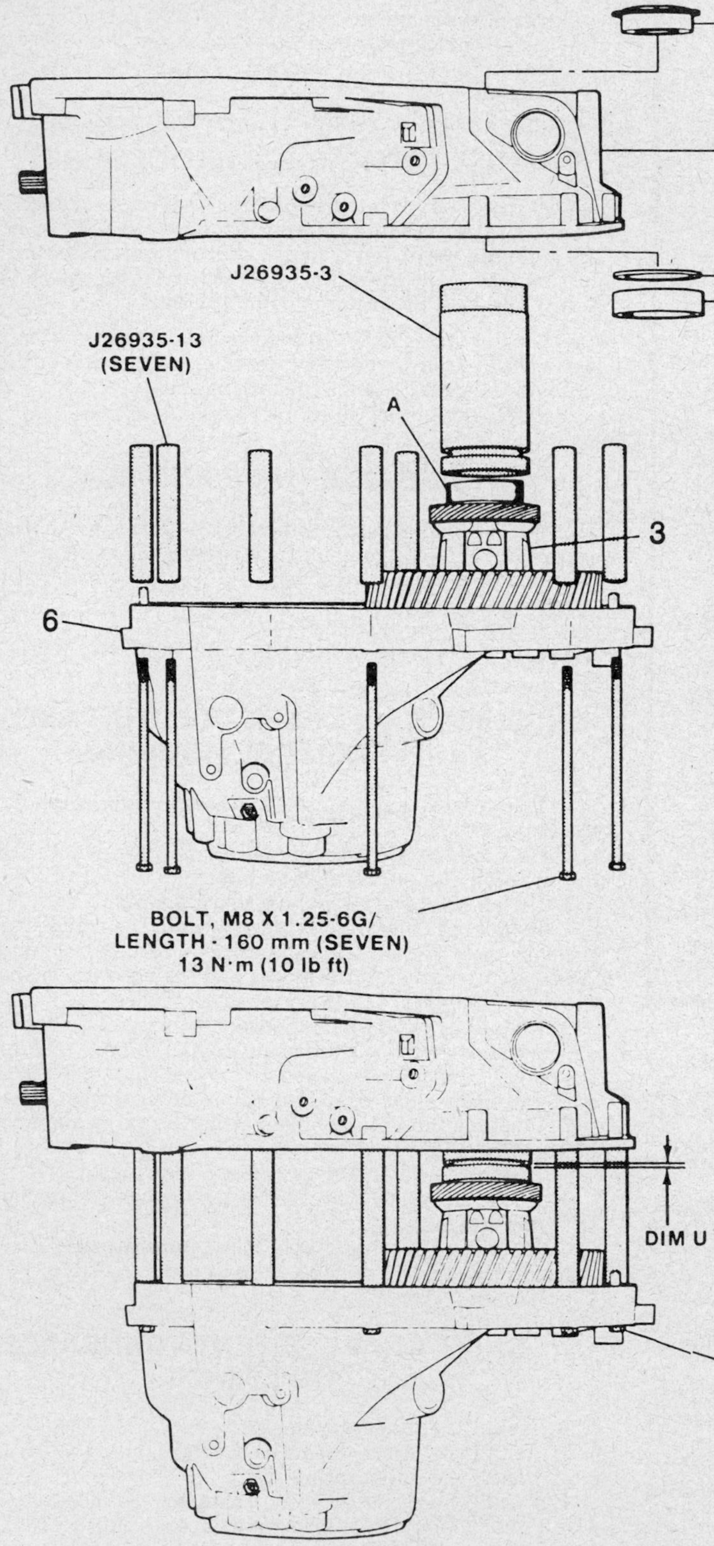
2
1
J26935-3
4
5
J26935-13
(SEVEN)
A
3
6
BOLT, M8 X 1.25-6G/
LENGTH - 160 mm (SEVEN)
13 N·m (10 lb ft)
DIM U
13 N·m (10 lb ft) (SEVEN)

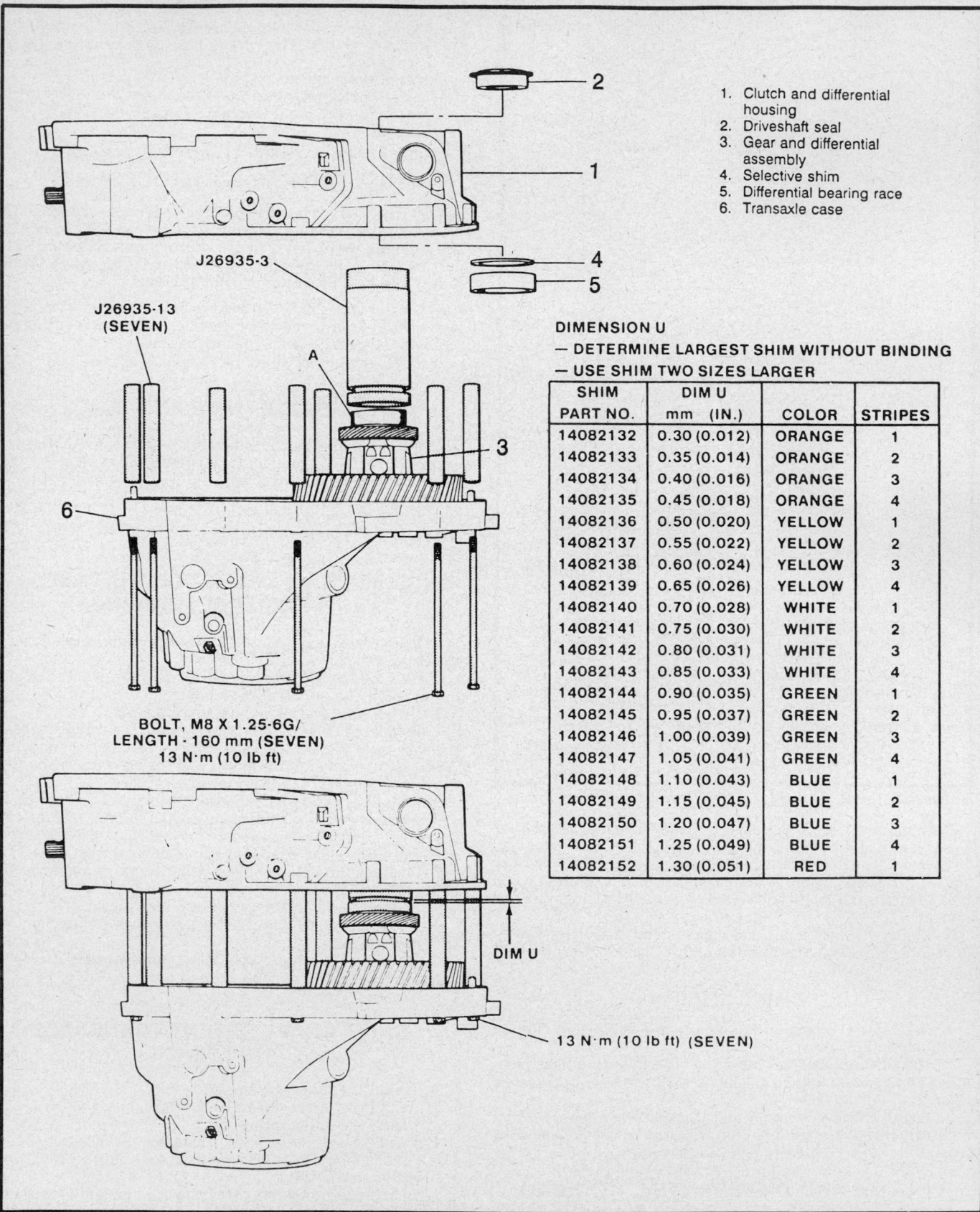

1. Clutch and differential housing
2. Driveshaft seal
3. Gear and differential assembly
4. Selective shim
5. Differential bearing race
6. Transaxle case

DIMENSION U
— DETERMINE LARGEST SHIM WITHOUT BINDING
— USE SHIM TWO SIZES LARGER

SHIM PART NO.	DIM U mm (IN.)	COLOR	STRIPES
14082132	0.30 (0.012)	ORANGE	1
14082133	0.35 (0.014)	ORANGE	2
14082134	0.40 (0.016)	ORANGE	3
14082135	0.45 (0.018)	ORANGE	4
14082136	0.50 (0.020)	YELLOW	1
14082137	0.55 (0.022)	YELLOW	2
14082138	0.60 (0.024)	YELLOW	3
14082139	0.65 (0.026)	YELLOW	4
14082140	0.70 (0.028)	WHITE	1
14082141	0.75 (0.030)	WHITE	2
14082142	0.80 (0.031)	WHITE	3
14082143	0.85 (0.033)	WHITE	4
14082144	0.90 (0.035)	GREEN	1
14082145	0.95 (0.037)	GREEN	2
14082146	1.00 (0.039)	GREEN	3
14082147	1.05 (0.041)	GREEN	4
14082148	1.10 (0.043)	BLUE	1
14082149	1.15 (0.045)	BLUE	2
14082150	1.20 (0.047)	BLUE	3
14082151	1.25 (0.049)	BLUE	4
14082152	1.30 (0.051)	RED	1

Differential assembly selective shim preload procedure

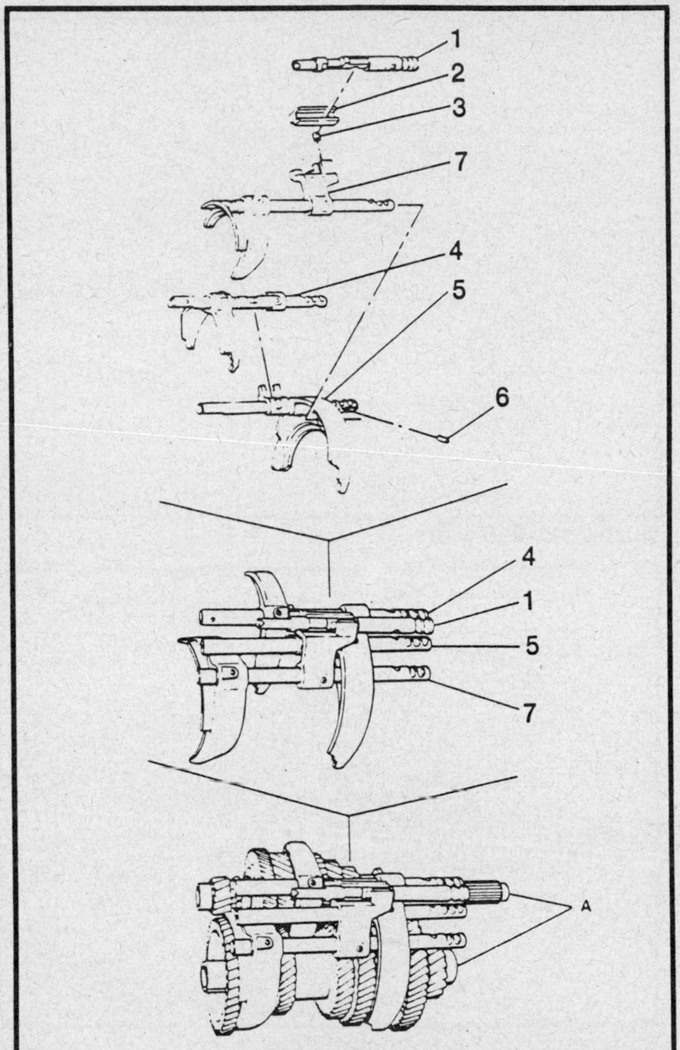
1
2
3
7
4
5
6
4
1
5
7
A

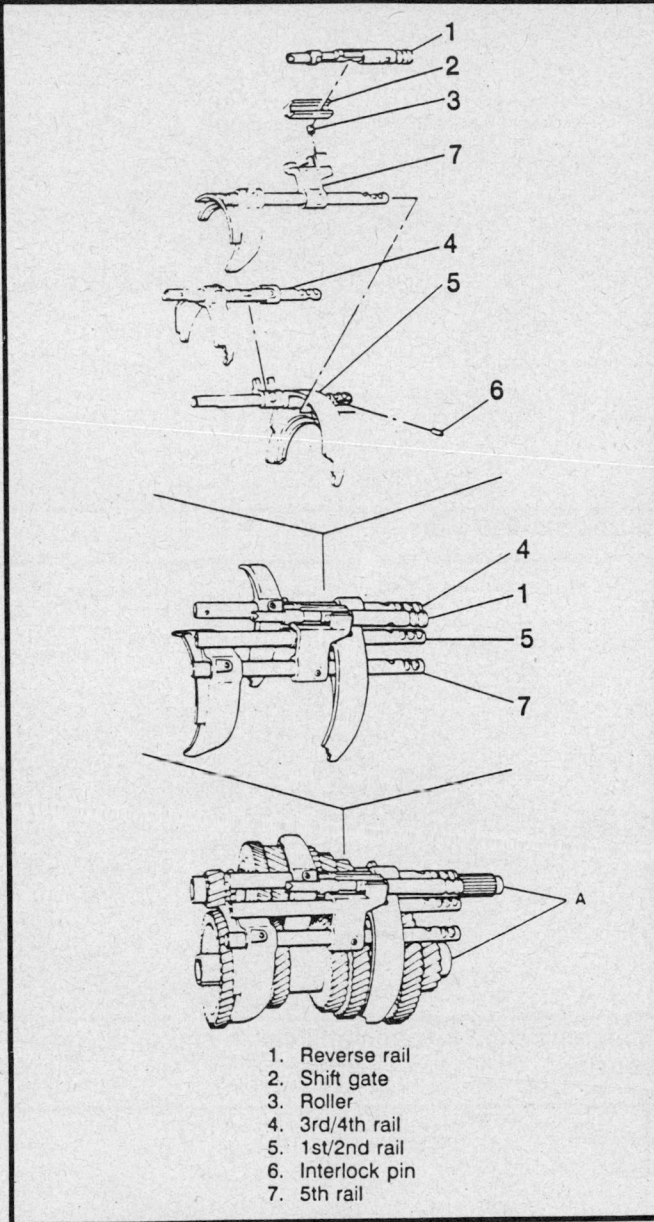

1. Reverse rail
2. Shift gate
3. Roller
4. 3rd/4th rail
5. 1st/2nd rail
6. Interlock pin
7. 5th rail

Gear and shift rail assemblies

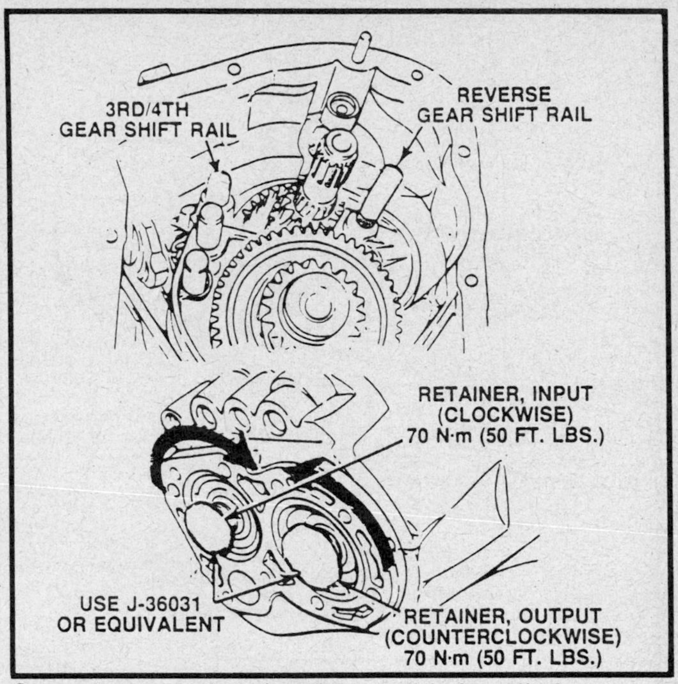
3RD/4TH
GEAR SHIFT RAIL
REVERSE
GEAR SHIFT RAIL
RETAINER, INPUT
(CLOCKWISE)
70 N·m (50 FT. LBS.)
USE J-36031
OR EQUIVALENT
RETAINER, OUTPUT
(COUNTERCLOCKWISE)
70 N·m (50 FT. LBS.)

Gear cluster and shift rail assembly installation

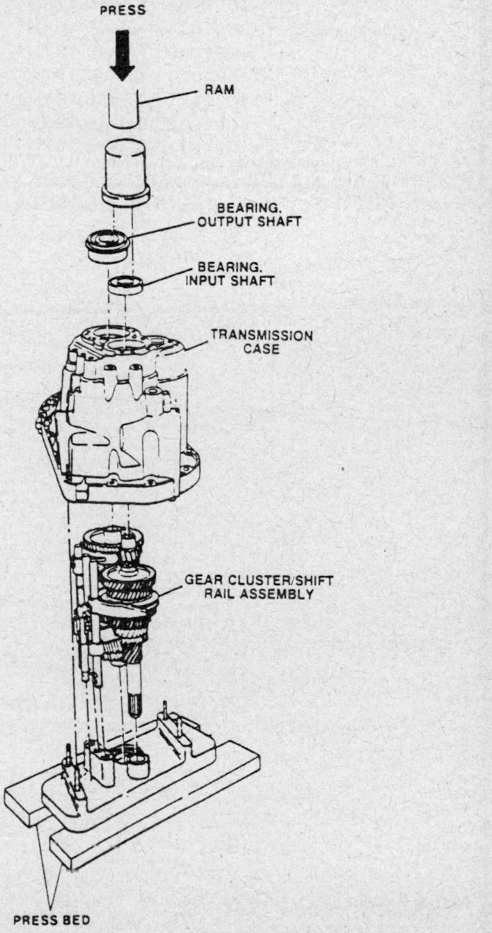
PRESS
RAM
BEARING,
OUTPUT SHAFT
BEARING,
INPUT SHAFT
TRANSMISSION
CASE
GEAR CLUSTER/SHIFT
RAIL ASSEMBLY
PRESS BED

Engaging transaxle in gear shaft support components installation

EXTERNAL TRANSAXLE MOUNTED LINKAGE

1. Install the bracket.
2. Install the bolts and torque to 17 ft. lbs. (23 Nm).
3. Install the collar and pin using a punch and a hammer.
4. Install the pivot.
5. Install a new pin.
6. Install the lever.
7. Install the washer and nut. Torque to 61 ft. lbs. (83 Nm). Do not allow the lever to move during installation of the nut.
8. Install the fluid level indicator and a new washer.
9. Install the electronic speedometer sensor assembly, retainer and bolt. Torque to 84 inch lbs. (9 Nm).

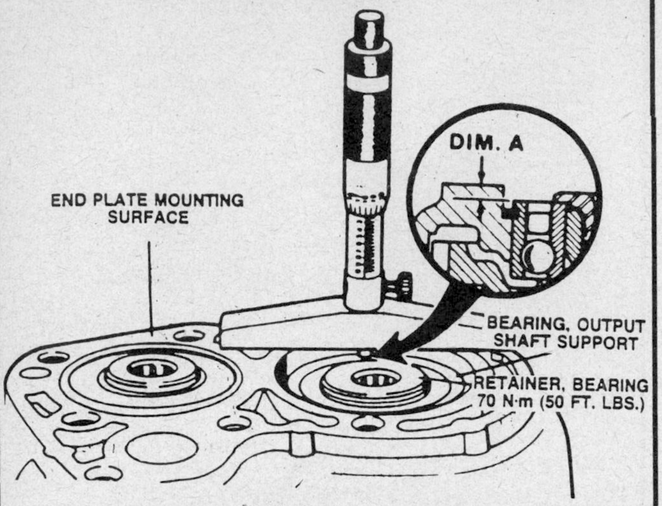
DIM. A
END PLATE MOUNTING SURFACE
BEARING, OUTPUT SHAFT SUPPORT
RETAINER, BEARING 70 N·m (50 FT. LBS.)

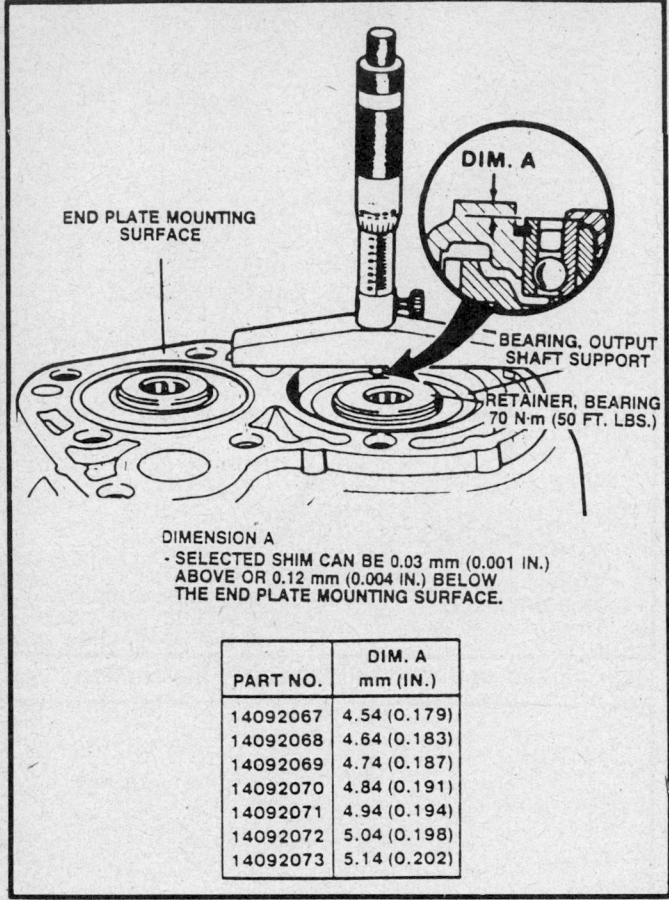

DIMENSION A
- SELECTED SHIM CAN BE 0.03 mm (0.001 IN.) ABOVE OR 0.12 mm (0.004 IN.) BELOW THE END PLATE MOUNTING SURFACE.

PART NO.	DIM. A mm (IN.)
14092067	4.54 (0.179)
14092068	4.64 (0.183)
14092069	4.74 (0.187)
14092070	4.84 (0.191)
14092071	4.94 (0.194)
14092072	5.04 (0.198)
14092073	5.14 (0.202)

Output shaft support bearing selective shim procedure

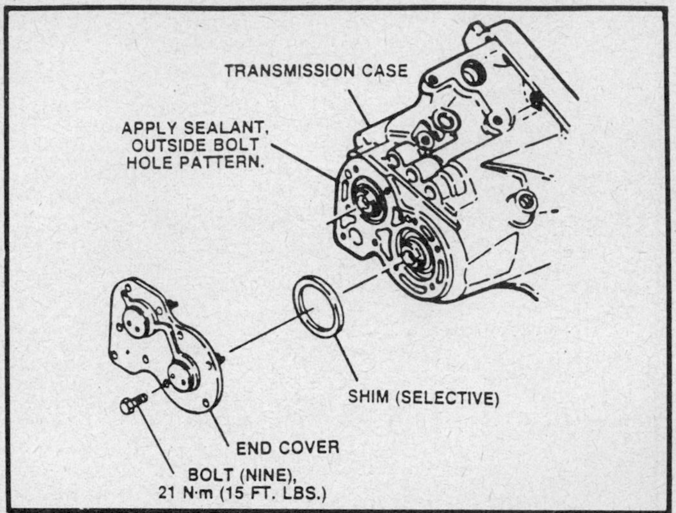
TRANSMISSION CASE
APPLY SEALANT, OUTSIDE BOLT HOLE PATTERN.
SHIM (SELECTIVE)
END COVER
BOLT (NINE), 21 N·m (15 FT. LBS.)

Gear case end plate

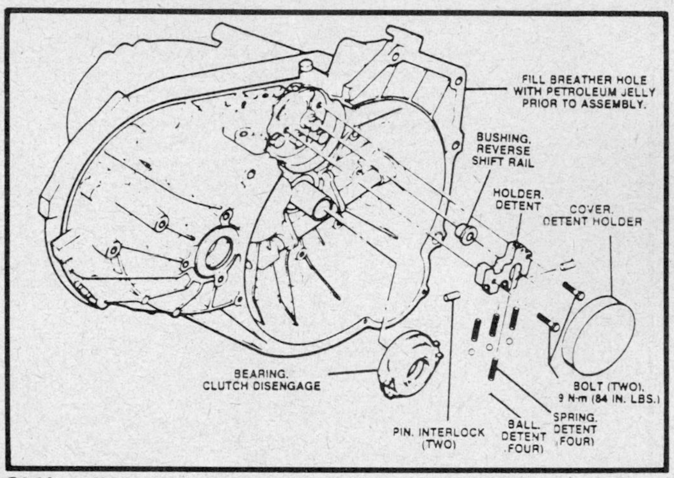
FILL BREATHER HOLE WITH PETROLEUM JELLY PRIOR TO ASSEMBLY.
BUSHING, REVERSE SHIFT RAIL
HOLDER, DETENT
COVER, DETENT HOLDER
BEARING, CLUTCH DISENGAGE
BOLT (TWO), 9 N·m (84 IN. LBS.)
SPRING, DETENT (FOUR)
BALL, DETENT (FOUR)
PIN, INTERLOCK (TWO)

Shift rail detent components/clutch and differential housing

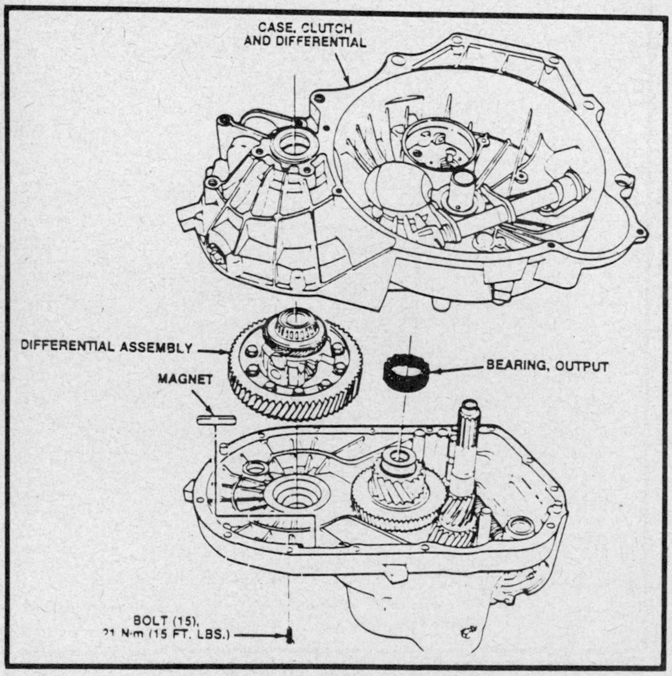
CASE, CLUTCH AND DIFFERENTIAL
DIFFERENTIAL ASSEMBLY
MAGNET
BEARING, OUTPUT
BOLT (15), 21 N·m (15 FT. LBS.)

Clutch and transaxle housing components

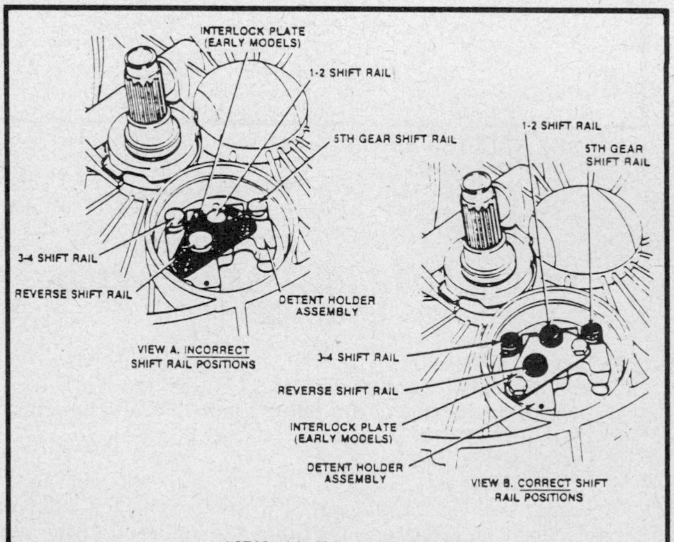
INTERLOCK PLATE (EARLY MODELS)
1-2 SHIFT RAIL
5TH GEAR SHIFT RAIL
3-4 SHIFT RAIL
REVERSE SHIFT RAIL
DETENT HOLDER ASSEMBLY
VIEW A. INCORRECT SHIFT RAIL POSITIONS
1-2 SHIFT RAIL
5TH GEAR SHIFT RAIL
3-4 SHIFT RAIL
REVERSE SHIFT RAIL
INTERLOCK PLATE (EARLY MODELS)
DETENT HOLDER ASSEMBLY
VIEW B. CORRECT SHIFT RAIL POSITIONS
NOTICE: 1-2 SHIFT RAIL MUST BE POSITIONED UP (VIEW B)
(VIEW A) POSITION WILL LOCK-UP SHIFT MECHANISM

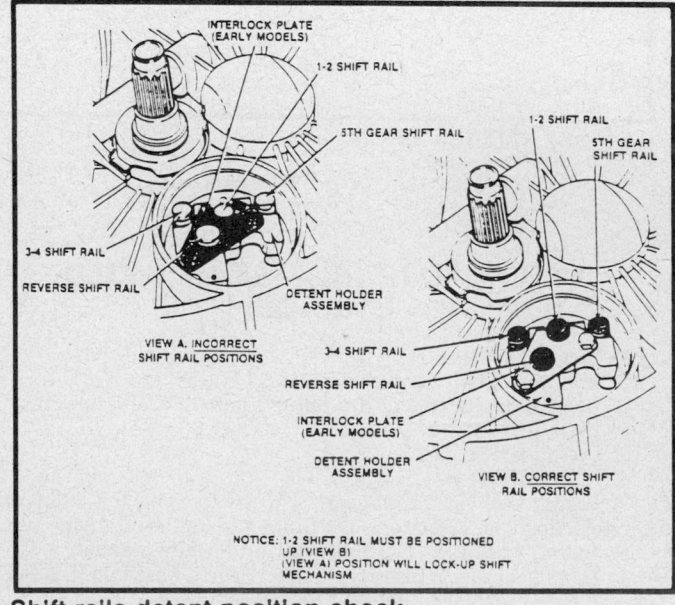

Shift rails detent position check

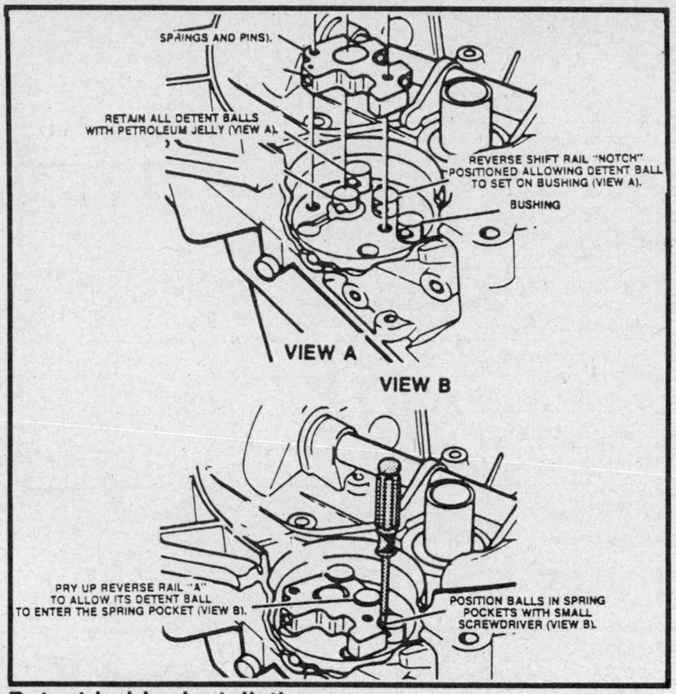
SPRINGS AND PINS).
RETAIN ALL DETENT BALLS WITH PETROLEUM JELLY (VIEW A).
REVERSE SHIFT RAIL "NOTCH" POSITIONED ALLOWING DETENT BALL TO SET ON BUSHING (VIEW A).
BUSHING
VIEW A
VIEW B
PRY UP REVERSE RAIL "A" TO ALLOW ITS DETENT BALL TO ENTER THE SPRING POCKET (VIEW B).
POSITION BALLS IN SPRING POCKETS WITH SMALL SCREWDRIVER (VIEW B).

Detent holder installation

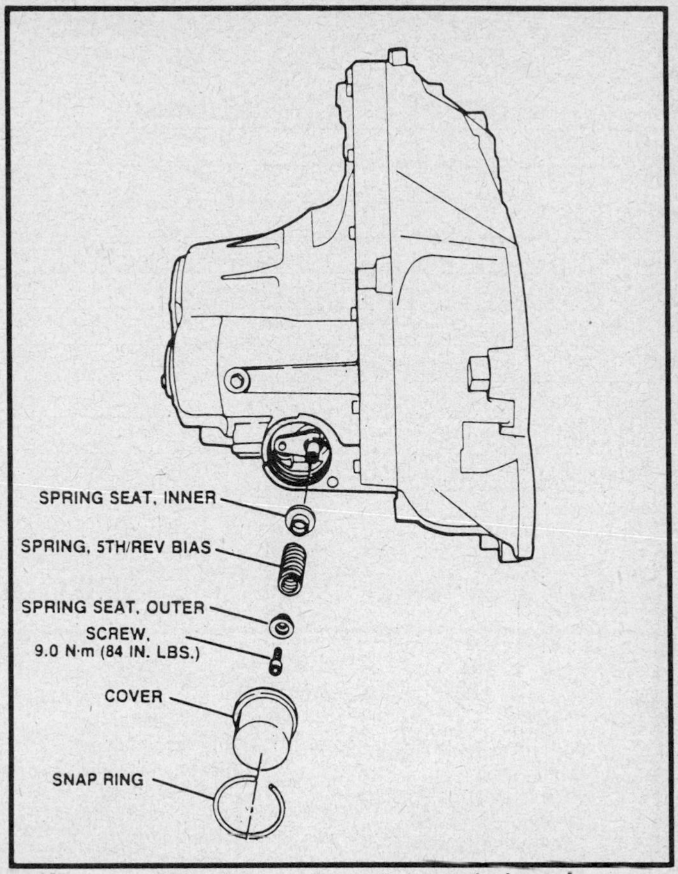
SPRING SEAT, INNER
SPRING, 5TH/REV BIAS
SPRING SEAT, OUTER
SCREW, 9.0 N·m (84 IN. LBS.)
COVER
SNAP RING

Shift shaft detent components/transaxle housing

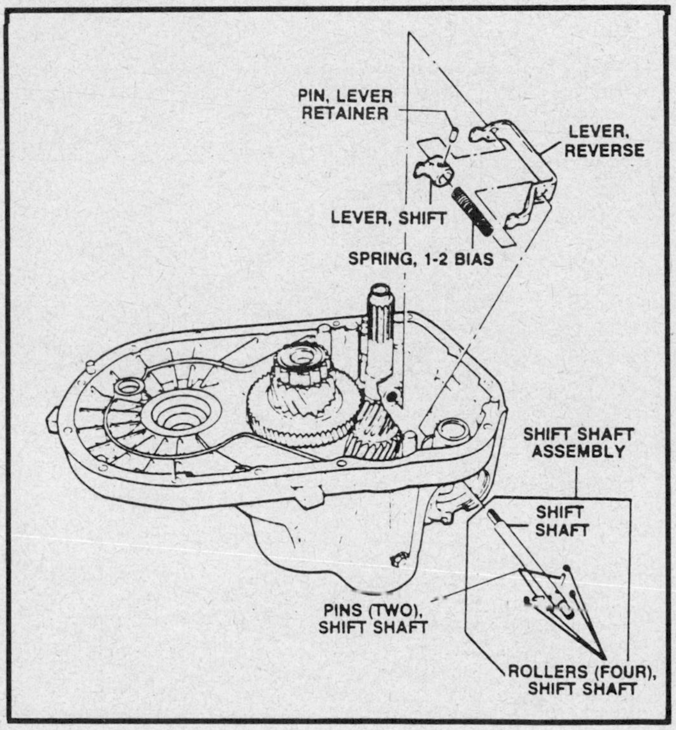
PIN, LEVER RETAINER
LEVER, REVERSE
LEVER, SHIFT
SPRING, 1-2 BIAS
SHIFT SHAFT ASSEMBLY
SHIFT SHAFT
PINS (TWO), SHIFT SHAFT
ROLLERS (FOUR), SHIFT SHAFT

Shift shaft components

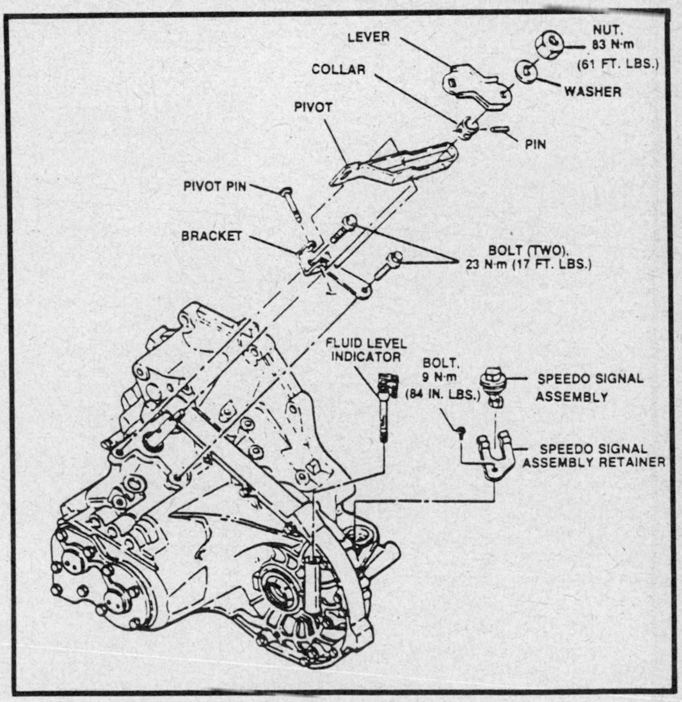
LEVER
NUT, 83 N·m (61 FT. LBS.)
COLLAR
WASHER
PIVOT
PIN
PIVOT PIN
BRACKET
BOLT (TWO), 23 N·m (17 FT. LBS.)
FLUID LEVEL INDICATOR
BOLT, 9 N·m (84 IN. LBS.)
SPEEDO SIGNAL ASSEMBLY
SPEEDO SIGNAL ASSEMBLY RETAINER

External transaxle mounted linkage

SPECIFICATIONS

TORQUE SPECIFICATIONS

Item	ft. lbs.	Nm
Reverse shift rail guide	15	21
Differential pin	84 ①	9
Differential gear	61	83
Fluid drain plug	18	24
Clutch housing to gear housing	15	21
Alternate oil level check/fill plug	18	24
Output shaft bearing support	50	70
End plate to gear housing	15	21
Input shaft bearing support	50	70

TORQUE SPECIFICATIONS

Item	ft. lbs.	Nm
Reverse idler gear bracket	15	21
Shift pivot bracket	17	23
Shift lever nut	61	83
Shift shaft detent	84 ①	9
Interlock plate	15	21
Shift rail detent holder	84 ①	9
Output bearing race	15	21
Sliding sleeve detent	32	44
Electronic speed sensor retainer	84 ①	9

① Inch lbs.

SPECIAL TOOLS

1. Output shaft bearing cup remover—J-36038
2. Clutch shaft inner reverse shift rail bushing remover—J-36032
3. Slide hammer and adapter set—J-23907
4. Input/output shaft gears remover—J-22912-01
5. Bearing remover—J-22888
6. Bearing remover leg set—J-22888-50
7. Bearing remover leg (2)—J-22888-35
8. Differential seal and race installer—J-26938
9. Differential bearing cup remover—J-36181
10. Differential inner bearing installer—J-22919
11. Input/output shaft remover/installer assembly set—J-36182
12. Pallet—J-36182-1
13. Disassembly adapter (2)—J-36182-2
14. Output shaft remover/installer adapter press tube reducer—J-36184
15. Differential and output shaft bearing cup installer—J-23423-A
16. Metric dial depth gauge—J-26900-19
17. Input/output shaft bearing remover—J-36185
18. Input shaft remover/installer press tube—J-36183
19. Clutch shaft upper bushing remover/installer—J-36037
20. Universal driver handle—J-36190
21. Universal driver handle—J-8092
22. Shift shaft seal installer—J-35823
23. Shim selection set—J-26935
24. Differential shimming gauge—J-26935-3
25. Bolts, M8 × 1.25-6G/length-160mm (7)
26. Shim selection set spacers (7)—J-26935-13
27. Shift shaft bearing remover—J-36027
28. Reverse shift rail bushing installer—J-36030
29. Sliding sleeve bushing remover/installer—J-36034
30. Shift detent lever bushing remover/installer—J-36039
31. Clutch shaft inner bushing installer—J-36033
32. Shift shaft bearing installer—J-36189
33. Bearing retainer bolt hex socket—J-36031
34. Differential side bearing puller adapter—J-2241-11
35. Differential side bearing puller adapter—J-23598

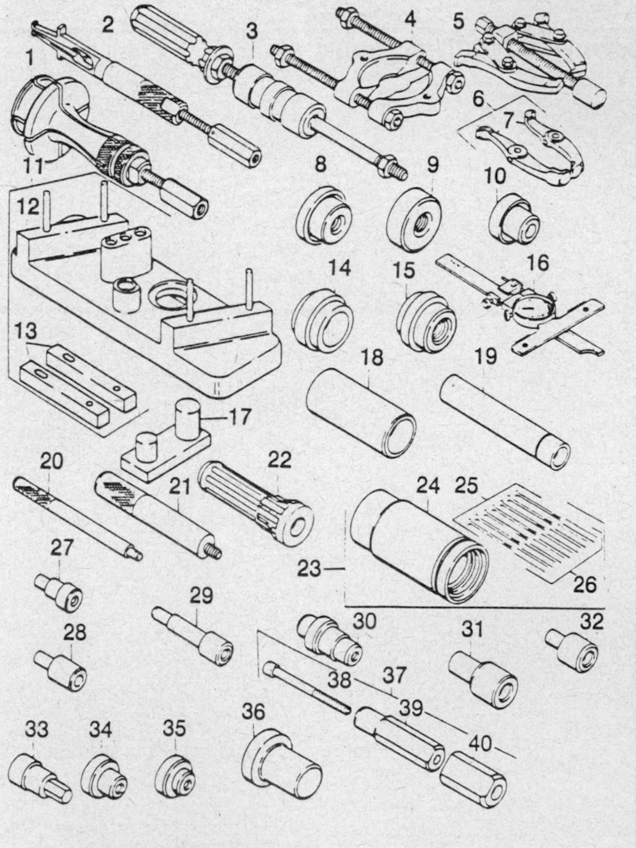
1
2
3
4
5
6
7
8
9
10
11
12
13
14
15
16
17
18
19
20
21
22
23
24
25
26
27
28
29
30
31
32
33
34
35
36
37
38
39
40

36. Input bearing assembly remover/installer-shaft support bearing installer—J-35824
37. Shift rail bushing remover/installer—J-36029
38. Cap screw ¼-20 × 2½
39. Installer/remover adapter—J-36029
40. Adapter to drive handle—J-36029-1

Section 7

Manual Transmission Applications

Vehicle Manufacturer	Year	Vehicle Model	Transmission Manufacturer	Transmission Identification	Speeds
AMC/Jeep-Eagle	1984–89	Cherokee, Wagoneer, Wrangler	—	AX4	4
	1984–89	Cherokee, Wagoneer, Wrangler	—	AX5	5
	1987–89	Cherokee, Comanche, Wagoneer, Wrangler	Peugeot	BA 10/5	5
AMC	1984–89	CJ7, Scrambler, J10, Wagoneer	Borg Warner	T176	4
Chrysler Corp.	1987–89	Dakota, D Series, W-350	New Process	NP2500	5
	1984–89	D Series, W Series	New Process	NP435	4
	1984–89	D-Series, W Series, B-Series, Ramcharger	Chrysler Corp.	OD4	4
Ford Motor Co.	1984–86	Mustang, Capri	Ford of Germany	83ET	4
	1984–89	Mustang, Capri, Cougar, Thunderbird	Borg Warner	T50D	5
	1984–87	E and F Series	Ford	3.03	3
	1984–87	E 150 Series	Ford	RUG	4
	1984–87	F-150-250 (2WD), F-150 (4WD), Bronco	Ford	TOD	4
	1987	Ranger, Bronco II	Mitsubishi	FM145	5
	1988–89	Ranger, Bronco II	Mitsubishi	FM146	5
	1988–89	Ranger, Bronco II, E and F Series, Bronco	Mazda	M50D	5
	1984–87	Ranger, Bronco II	Mazda	Toyo Kogyo	5
	1988–89	E-350, F105-350, F Series Super Duty, Bronco	ZF	S5-42 ZF	5
	1984–87	E & F Series, Bronco	Borg Warner	T18	4
	1984–87	F-250-350 Series	Borg Warner	T19	4
	1984–87	E-350, F-Series, Bronco	New Process	NP435	4
General Motors	1984–86	Chevette	Isuzu	M75	5
	1984–88	Corvette	—	4 Speed with Automatic OD	4
	1984–87	Chevette, T1000	Muncie	MY1	4
	1989	Geo Tracker	—	—	5
	1984–89	Light Duty Trucks	Muncie	HM282	5
	1984–89	Light Duty Trucks	Muncie	M62	3
	1984–89	Light Duty Trucks	Muncie	M64	3
	1984–89	Light Duty Trucks	Muncie	M20	4
	1984–89	Light Duty Trucks	New Process	M46	4
	1984	S10, S15, Blazer, Jimmy	Borg Warner	M77	4

MANUAL TRANSMISSION APPLICATION

Vehicle Manufacturer	Year	Vehicle Model	Transmission Manufacturer	Transmission Identification	Speeds
General Motors	1985–88	S10, S15, Blazer, Jimmy, Astro, Safari	Borg Warner	MC9	4
	1984	S10, S15, Blazer, Jimmy	Borg Warner	MB1	5
	1984	S10, S15, Blazer, Jimmy	Borg Warner	MB6	5
	1987–89	S10, S15, Jimmy, Blazer, Astro, Safari	Borg Warner	MH3	5
	1986–89	S10, S15, Jimmy, Blazer, Astro, Safari	Borg Warner	ML2	5
	1986–89	S10, S15, Jimmy, Blazer, Astro, Safari	Borg Warner	ML3	5
	1989	Camaro, Firebird, S10, S15, Astro, Safari	Muncie	HM260	4
	1988–89	Light Duty Trucks	Muncie	HM290	4, 5
AMC, Ford Motor Co., General Motors	1984–89	CJ-7, Eagle, Scrambler, Cougar, Mustang, Thunderbird, Capri, Astro, Safari, Chevette	Borg Warner	T4	4
		T1000, S10, S15, Camaro, Firebird	Borg Warner	T5	5

Section 7

AX4 and AX5 Transmissions
AMC/Jeep-Eagle

APPLICATION

1984–89 Cherokee
1984–89 Wagoneer
1987–89 Wrangler

GENERAL DESCRIPTION

Model AX–4 is a 4 speed manual transmission while AX–5 is a 5 speed manual transmission. Both transmissions have synchromesh engagement in all forward gears controlled by a floor shift mechanism integrated into the transmission top cover.

FLUID CAPACITY

All capacities given in quarts

Year	Vehicle	Transmission	Fluid Type	Overhaul Capacity
1984–89	Cherokee, Wagoneer	AX-4 2WD	75W-90	3.9
	Cherokee, Wagoneer	AX-5 2WD	75W-90	3.7
	Cherokee, Wagoneer	AX-5 4WD	75W-90	3.5
	Cherokee, Wagoneer	AX-4 4WD	75W-90	3.7
1987–89	Wrangler	AX-4 4WD	75W-90	3.7
	Wrangler	AX-5 4WD	75W-90	3.5

Metric Fasteners

Metric bolt sizes and thread pitches are used for all fasteners on the AX–4 and AX–5 transmission. The use of metric tools is mandatory in the service of this transmission.

Do not attempt to interchange metric fasteners for inch system fasteners. Mismatched or incorrect fasteners can result in damage to the transmission unit through malfunctions, breakage or possible personal injury. Care should be taken to reuse the fasteners in the same location as removed, whenever possible. Due to the large number of alloy parts used and the aluminum casing, torque specifications should be strictly observed. Before installing bolts into aluminum parts, always dip threads into oil, or sealant (if specifically required), to prevent the screws from galling the aluminum threads and to prevent seizing.

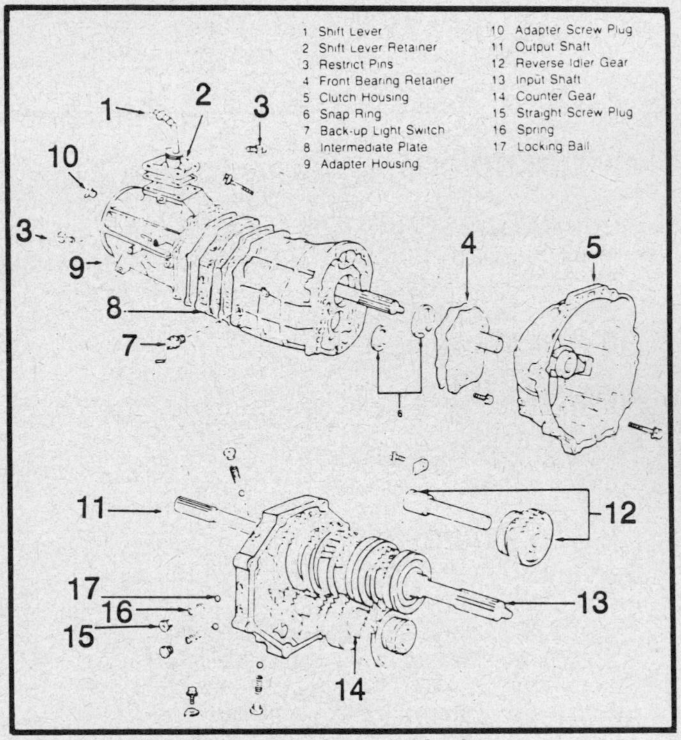
1 Shift Lever
2 Shift Lever Retainer
3 Restrict Pins
4 Front Bearing Retainer
5 Clutch Housing
6 Snap Ring
7 Back-up Light Switch
8 Intermediate Plate
9 Adapter Housing
10 Adapter Screw Plug
11 Output Shaft
12 Reverse Idler Gear
13 Input Shaft
14 Counter Gear
15 Straight Screw Plug
16 Spring
17 Locking Ball

AX–4 and AX–5 transmission—typical

Fluid Capacities

Checking Fluid Level

The vehicle must be level before an accurate reading can be obtained. The fluid level is considered full, when the lubricant is at the bottom of the fill hole.

TRANSMISSION MODIFICATIONS

There are no modifications at the time of this publication.

TROUBLE DIAGNOSIS

CHILTON'S THREE C'S TRANSMISSION DIAGNOSIS
AX-4 and AX-5

Condition	Cause	Correction
Transmission shifts hard	a) Clutch adjustment incorrect	a) Adjust clutch.
	b) Clutch linkage or cable binding	b) Lubricate or repair as necessary.
	c) Shift rail binding	c) Check for mispositioned selector arm roll pin, loose cover bolts, worn shift rail bores, worn shift rail, distorted oil seal, or extension housing not aligned with case. Repair as necessary.
	d) Internal bind in transmission caused by shift forks, selector plates, or synchronizer assemblies	d) Remove, disassemble and inspect transmission. Replace worn or damaged components as necessary.
	e) Clutch housing misalignment	e) Check runout at rear face of clutch housing.
	f) Incorrect lubricant	f) Drain and refill transmission.
	g) Block rings and/or cone seats worn	g) Blocking ring to gear clutch tooth face clearance must be 0.030 in. or greater. If clearance is correct, it may still be necessary to inspect blocking rings and cone seats for excessive wear. Repair as necessary.
	h) Air in hydraulic clutch system	h) Fill and bleed hydraulic clutch system.
	i) Leak in hydraulic clutch system	i) Repair or replace damaged components. Fill and bleed hydraulic system.
Gear clash when shifting from one gear to another	a) Clutch adjustment incorrect	a) Adjust clutch
	b) Clutch linkage or cable binding	b) Lubricate or repair as necessary.
	c) Clutch housing misalignment	c) Check runout at rear of clutch housing.
	d) Lubricant level low or incorrect lubricant	d) Drain and refill transmission and check for lubricant leaks if level was low. Repair as necessary.
	e) Gearshift components, or synchronizer assemblies worn or damaged	e) Remove, disassemble and inspect transmission. Replace worn or damaged components as necessary.
	f) Air in hydraulic clutch system	f) Fill and bleed hydraulic clutch system.
	g) Leak in hydraulic clutch system	g) Repair or replace damaged components. Fill and bleed hydraulic clutch system.
Transmission noisy	a) Lubricant level low or incorrect lubricant	a) Drain and refill transmission. If lubricant level was low, check for leaks and repair as necessary.
	b) Clutch housing-to-engine, or transmission-to-clutch housing bolts loose	b) Check and correct bolt torque as necessary.
	c) Dirt, chips, foreign material in transmission	c) Drain, flush, and refill transmission.
	d) Gearshift mechanism, transmission gears, or bearing components worn or damaged	d) Remove, disassemble and inspect transmission. Replace worn or damaged components as necessary.
	e) Clutch housing misalignment	e) Check runout at rear face of clutch housing.
Jumps out of gear	a) Clutch housing misalignment	a) Check runout at rear face of clutch housing.
	b) Gearshift lever loose	b) Check lever for worn fork. Tighten loose attaching bolts.

CHILTON'S THREE C'S TRANSMISSION DIAGNOSIS
AX-4 and AX-5

Condition	Cause	Correction
Jumps out of gear	c) Offset lever nylon insert worn or lever attaching nut loose	c) Remove gearshift lever and check for loose offset lever nut or worn insert. Repair or replace as necessary.
	d) Gearshift mechanism, shift forks, selector plates, interlock plate, selector arm, shift rail, detent plugs, springs or shift cover worn or damaged	d) Remove, disassemble and inspect transmission cover assembly. Replace worn or damaged components as necessary.
	e) Clutch shaft or roller bearings worn or damaged	e) Replace clutch shaft or roller bearings as necessary.
	f) Gear teeth worn or tapered, synchronizer assemblies worn or damaged, excessive end play caused by worn thrust washers or output shaft gears	f) Remove, disassemble, and inspect transmission. Replace worn or damaged components as necessary.
	g) Pilot bushing worn	g) Replace pilot bushing.
Will not shift into one gear	a) Gearshift selector plates, interlock plate, or selector arm, worn, damaged, or incorrectly assembled	a) Remove, disassemble, and inspect transmission cover assembly. Repair or replace components as necessary.
	b) Shift rail detent plunger worn, spring broken, or plug loose	b) Tighten plug or replace worn or damaged components as necessary.
	c) Gearshift lever worn or damaged	c) Replace gearshift lever.
	d) Synchronizer sleeves or hubs, damaged or worn	d) Remove, disassemble and inspect transmission. Replace worn or damaged components.
Locked in one gear—cannot be shifted out	a) Shift rail(s) worn or broken, shifter fork bent, setscrew loose, center detent plug missing or worn	a) Inspect and replace worn or damaged parts.
	b) Broken gear teeth on countershaft gear, clutch shaft, or reverse idler gear	b) Inspect and replace damaged part.
	c) Gearshift lever broken or worn, shift mechanism in cover incorrectly assembled or broken, worn damaged gear train components	c) Disassemble transmission. Replace damaged parts or assemble correctly.
Transfer case difficult to shift or will not shift into desired range	a) Vehicle speed too great to permit shifting.	a) Stop vehicle and shift into desired range. Or reduce speed to 2–3 mph (3–4 km/h) before attempting to shift.
	b) If vehicle was operated for extended period in 4H mode on dry paved surface, driveline torque load may cause difficult shifting.	b) Stop vehicle, shift transmission to neutral, shift transfer case to 2H mode and operate vehicle in 2H on dry paved surfaces.
	c) Transfer case external shift linkage binding.	c) Lubricate or repair or replace linkage, or tighten loose components as necessary.
	d) Insufficient or incorrect lubricant.	d) Drain and refill to edge of fill hole with SAE 85W–90 gear lubricant only.
	e) Internal components binding, worn, or damaged.	e) Disassemble unit and replace worn or damaged components as necessary.
Transfer case noisy in all drive modes	a) Insufficient or incorrect lubricant.	a) Drain and refill to edge of fill hole with SAE 85W–90 gear lubricant only. Check for leaks and repair if necessary. Note: If unit is still noisy after drain and refill, disassembly and inspection may be required to locate source of noise.
Noisy in—or jumps out of four wheel drive low range	a) Transfer case not completely engaged in 4L position.	a) Stop vehicle, shift transfer case in Neutral, then shift back into 4L position.

CHILTON'S THREE C'S TRANSMISSION DIAGNOSIS
AX-4 and AX-5

Condition	Cause	Correction
Noisy in—or jumps out of four wheel drive low range	b) Shift linkage loose or binding.	b) Tighten, lubricate, or repair linkage as necessary.
	c) Shift fork cracked, inserts worn, or fork is binding on shift rail.	c) Disassemble unit and repair as necessary.
Lubricant leaking from output shaft seals or from vent	a) Transfer case overfilled.	a) Drain to correct level.
	b) Vent closed or restricted.	b) Clear or replace vent if necessary.
	c) Output shaft seals damaged or installed incorrectly.	c) Replace seals. Be sure seal lip faces interior of case when installed. Also be sure yoke seal surfaces are not scored or nicked. Remove scores, nicks with fine sandpaper or replace yoke(s) if necessary.
Abnormal tire wear	a) Extended operation on dry hard surface (paved) roads in 4H range.	a) Operate in 2H on hard surface (paved) roads.

ON CAR SERVICES

FLUID CHANGE

1. Raise and support vehicle safely.
2. Remove drain plug, if equipped or siphon transmission fluid into a pan.
3. Remove filler plug.
4. Install drain plug.
5. Fill to the level of the filler plug hole with the proper type and gear lubricant.
6. Install fill plug and lower vehicle.

REAR SEAL

Removal and Installation

1. Raise and support vehicle safely.
2. Remove drain plug and drain gear lubricant into a pan.
3. Remove filler plug.
4. Install drain plug.
5. Mark driveshaft to yoke for reassembly and remove driveshaft.
6. Pry out old seal using an appropriate tool.
7. Install new seal, coated with locking compound, using tool J-35582 or equivalent.
8. Install the driveshaft, making certain to align marks and torque bolts to 15 ft. lbs. (20 Nm).
9. Fill to the level of the filler plug hole with 75W-90 gear lubricant.
10. Install fill plug and lower vehicle.

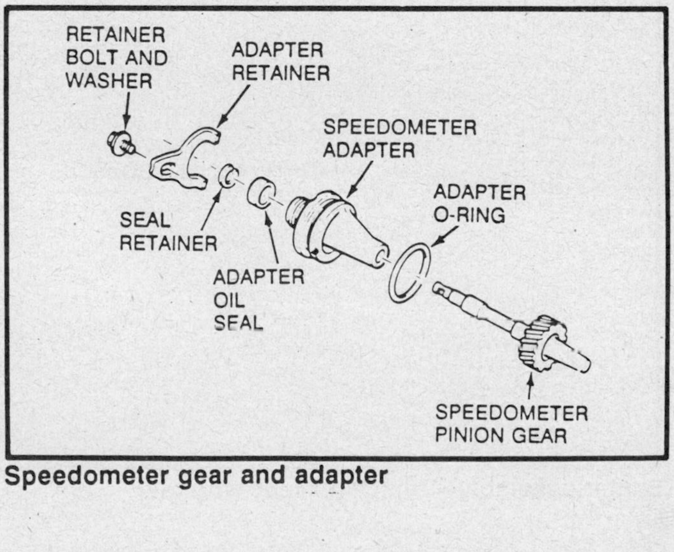
RETAINER BOLT AND WASHER
ADAPTER RETAINER
SPEEDOMETER ADAPTER
ADAPTER O-RING
SEAL RETAINER
ADAPTER OIL SEAL
SPEEDOMETER PINION GEAR

Speedometer gear and adapter

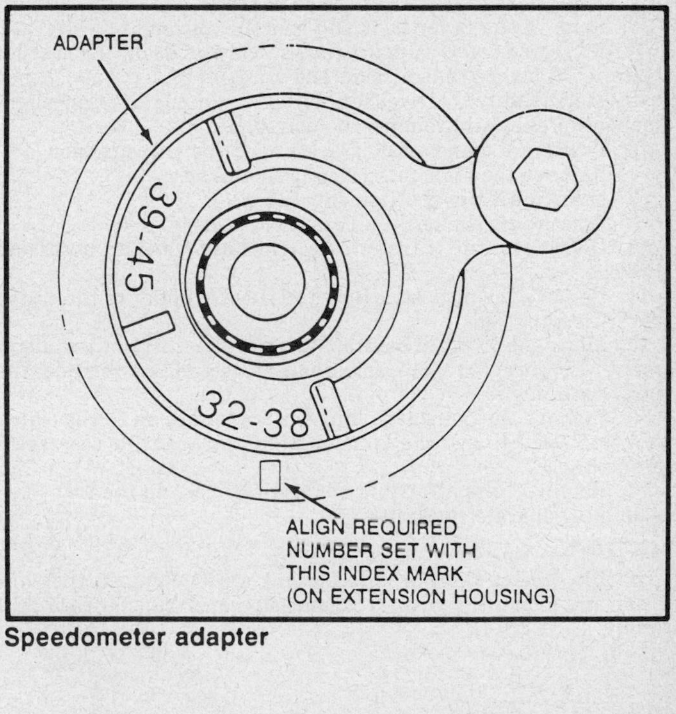
ADAPTER
39-45
32-38
ALIGN REQUIRED NUMBER SET WITH THIS INDEX MARK (ON EXTENSION HOUSING)

Speedometer adapter

SPEEDOMETER GEAR

Removal and Installation

1. Raise and support vehcile safely. Remove the speedometer adapter retainer.
2. Pull the speedometer adapter and pinion gear out of the case. Discard the O-ring, it is not reusable.
3. Inspect the interior of the speedometer cable cover. Replace the adapter oil seal if transmission fluid is found inside cable cover.
4. Install the speedometer pinion gear in the adapter. Make certain the gear shaft is properly engaged with the speedometer cable.
5. Lubricate the replacement adapter O-ring with transmission fluid and install it on the adapter.
6. Count the number of teeth on the pinion gear.
7. Locate the number of sets on the adapter face that corresponds to the number of teeth on the pinion gear.
8. Rotate the adapter until the desired number set on the adapter face is aligned with the index mark on the extension housing. The number set should be in a 6 o-clock position when aligned.
9. Insert the assembled adapter assembly and torque the retainer bolt to 100 inch lbs. (11 Nm).
10. Lower the vehicle.

REMOVAL AND INSTALLATION

TRANSMISSION REMOVAL

2WD VEHICLES

1. Disconnect the negative battery cable. Raise the outer gearshift lever boot and remove the upper part of the console.
2. Remove the lower part of the console.
3. Remove the inner boot and gearshift lever.
4. Raise and safely support the vehicle and drain the gear lubricant.
5. Mark the driveshaft to yoke, for installation alignment reference and remove the driveshaft.
6. Position a transmission jack under the transmission.
7. Remove the transmission crossmember.
8. Disconnect the speedometer cable.
9. Disconnect the backup light switch.
10. Remove the nuts attaching the slave cylinder to the clutch cover housing.
11. Remove the bolts attaching the transmission to the clutch housing and remove the transmission.

4WD VEHICLES EXCEPT WRANGLER

1. Disconnect the negative battery cable. Raise the outer gearshift lever boot and remove the upper part of the console.
2. Remove the lower part of the console.
3. Remove the inner boot and gearshift lever.
4. Raise and safely support the vehicle and drain the gear lubricant from the transmission and transfer case.
5. Mark the rear driveshaft to yoke, for installation alignment reference and remove the driveshaft.
6. Position a transmission jack under the transmission.
7. Remove the transmission crossmember.
8. Disconnect the speedometer cable.
9. Disconnect the backup light switch.
10. Disconnect the transfer case vent hose, vacuum hoses and linkage.
11. Remove the nuts attaching the slave cylinder to the clutch cover housing.
12. Mark the front driveshaft to yoke, for installation alignment reference. Move the driveshaft aside and secure under the body with a wire.
13. Remove the bolts attaching the transmission to the clutch housing and remove the transmission and transfer case as an assembly.
14. Remove bolts attaching the transfer case to the transmission and separate the units.

4WD WRANGLER

1. Disconnect the negative battery cable. Remove the shift knob and locknut from the transmission and transfer case shift levers.
2. Remove the screws attaching the transmission and transfer case shift lever boots and remove both boots.
3. Remove the transmission shift tower dust boot.
4. Press the stub shaft retainer downward. Rotate the retain-

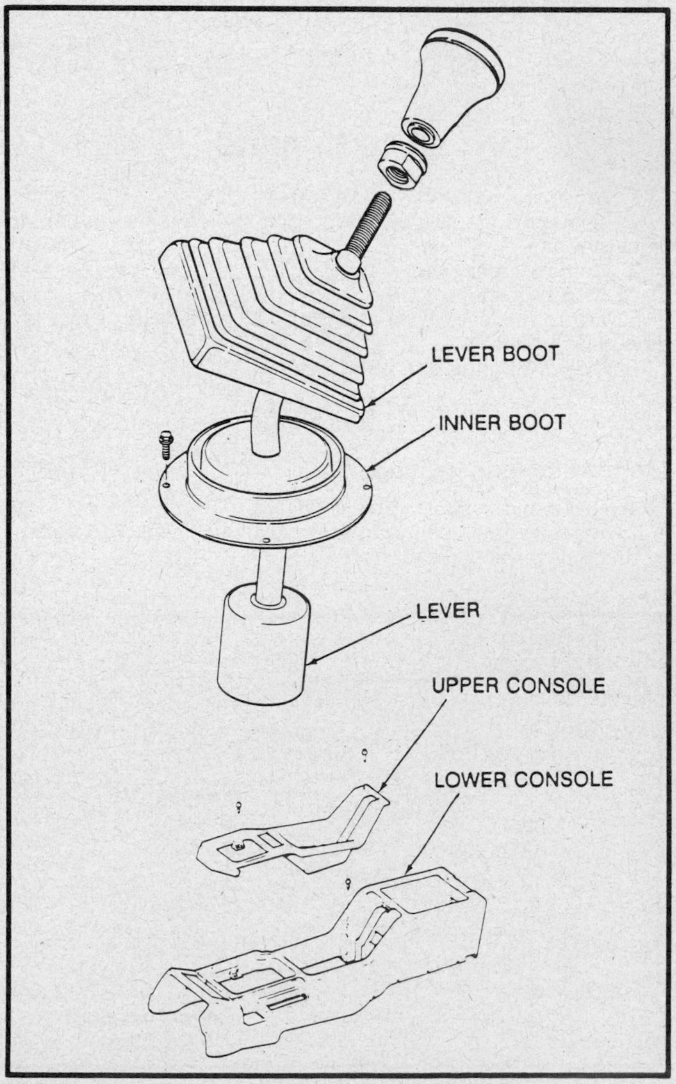
LEVER BOOT
INNER BOOT
LEVER
UPPER CONSOLE
LOWER CONSOLE

Shifter assembly—Typical except Wrangler

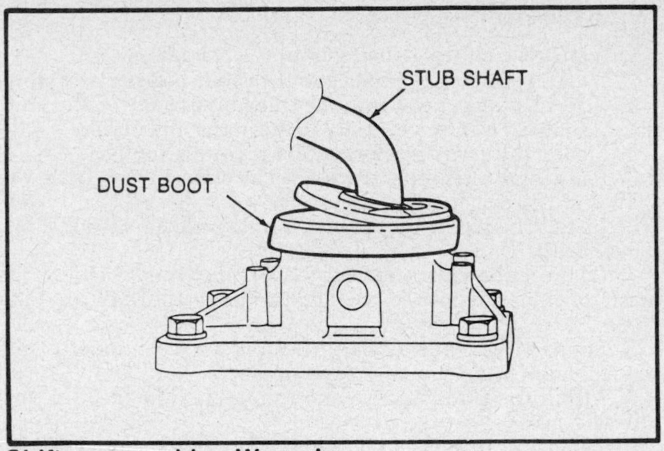
STUB SHAFT
DUST BOOT

Shifter assembly—Wrangler

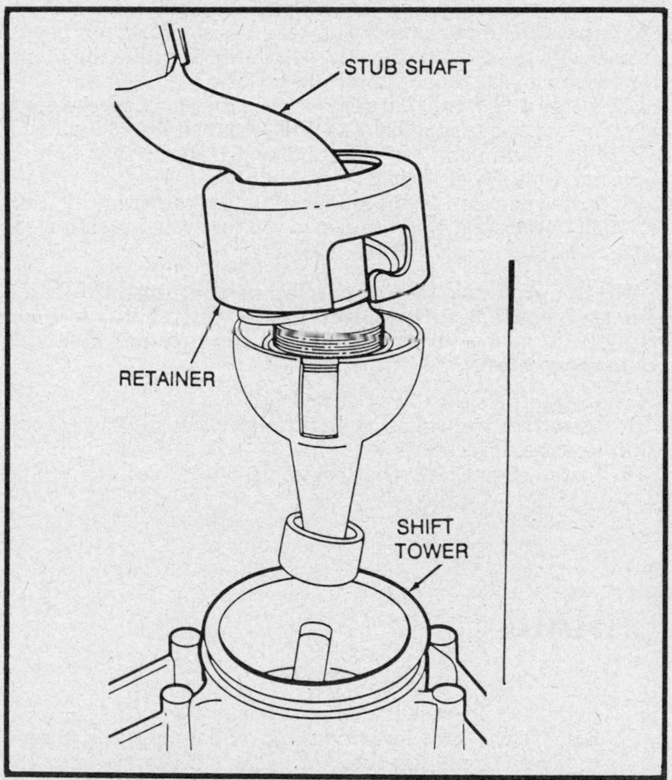
STUB SHAFT
RETAINER
SHIFT TOWER

Shifter assembly—Wrangler

er counterclockwise to release it from the lugs in the shift tower. Lift the retainer, stub shaft and shift lever up and out of the shift tower.

NOTE: Do not remove the shift lever from the stub shaft.

5. Raise and safely support the vehicle and drain the gear lubricant from the transmission and transfer case.
6. Mark the rear driveshaft to yoke, for installation alignment reference and remove the driveshaft.
7. Position a transmission jack under the transmission.
8. Remove the transmission crossmember.
9. Disconnect the speedometer cable.
10. Disconnect the backup light switch.

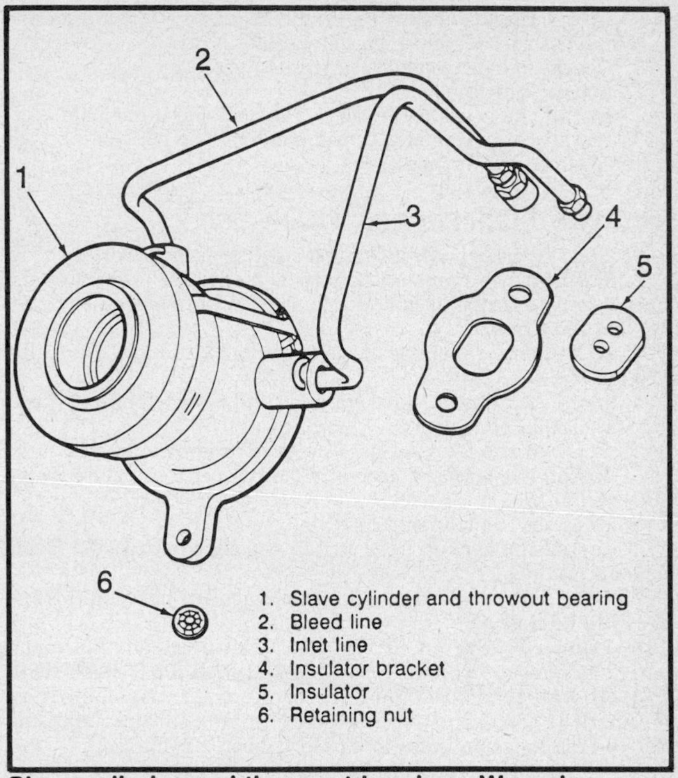
1
2
3
4
5
6
1. Slave cylinder and throwout bearing
2. Bleed line
3. Inlet line
4. Insulator bracket
5. Insulator
6. Retaining nut

Slave cylinder and throwout bearing—Wrangler

11. Disconnect the transfer case vent hose, vacuum hoses and linkage. Tag vacuum hoses for reassembly reference.
12. Mark the front driveshaft to yoke, for installation alignment reference and move the driveshaft aside and secure under the body with a wire.
13. Disconnect the clutch master cylinder hydraulic line from the slave cylinder inlet line.
14. Remove the bolts attaching the transmission to the clutch housing and remove the transmission and transfer case as an assembly.
15. Remove bolts attaching the transfer case to the transmission and separate the units.
16. Remove slave cylinder and throwout bearing assembly.
17. Remove the clutch housing from the transmission, if necessary.

TRANSMISSION INSTALLATION

2WD VEHICLES

1. Shift transmission into a gear using an appropriate tool.
2. Position the transmission on a transmission jack. Raise and align the transmission clutch shaft with the splines in the driven plate hub.
3. Install the transmission and torque bolts to 22–30 ft. lbs. (30–41 Nm).
4. Install the clutch slave cylinder and torque bolts to 13–19 ft. lbs. (18–26 Nm).
5. Install the backup light switch and torque to 19–35 ft. lbs. (26–48 Nm).
6. Connect the speedometer cable.
7. Install the crossmember, torque the crossmember-to-frame bolts to 20–35 ft. lbs. (27–41 Nm) and torque the insulator-to-transmission bolts to 27–38 ft. lbs. (37-52 Nm).
8. Remove the transmission jack.

9. Install the driveshaft making certain to align marks and torque bolts to 140–200 inch lbs. (16–23 Nm).
10. Check and add gear lubricant as necessary.
11. Lower vehicle.
12. Install the gearshift lever and install the inner boot.
13. Install the lower and upper parts of the console and the outer gearshift lever boot.
14. Install the negative battery cable.

4WD VEHICLES EXCEPT WRANGLER

1. Shift transmission into a gear using an appropriate tool.
2. Position the transmission on a transmission jack. Raise and align the transmission clutch shaft with the splines in the driven plate hub.
3. Install the transmission and torque bolts to 22–30 ft. lbs. (30–41 Nm).
4. Install the clutch slave cylinder and torque bolts to 13–19 ft. lbs. (18–26 Nm).
5. Position the transfer case on the transmission jack.
6. Install the transfer case and torque nuts to 22–30 ft. lbs. (30–41 Nm).
7. Connect the transfer case vent hose.
8. Install the backup light switch and torque to 19–35 ft. lbs. (26–48 Nm).
9. Connect the speedometer cable, vacuum hoses and transfer case linkage.
10. Connect the front driveshaft, making certain the marks are aligned and torque the bolts to 200 inch lbs. (16–23 Nm).
11. Install the crossmember, torque the crossmember-to-frame bolts to 20–35 ft. lbs. (27–41 Nm) and torque the insulator-to-transmission bolts to 27–38 ft. lbs. (37–52 Nm).
12. Remove the transmission jack.
13. Install the rear driveshaft, making certain to align marks and torque bolts to 140–200 inch lbs. (16–23 Nm).
14. Check and add gear lubricant and transfer case fluid, as necessary.
15. Lower vehicle.
16. Install the gearshift lever and install the inner boot.
17. Install the lower and upper parts of the console and the outer gearshift lever boot.
18. Install the negative battery cable.

4WD WRANGLER

1. Install the clutch housing onto the transmission.
2. Install the slave cylinder and throwout bearing assembly.
3. Insert the shift lever into the transmission tower and shift the transmission into any gear. Remove the shift lever.
4. Mount the transmission onto a transmission jack.
5. Raise and align the transmission clutch shaft with the splines in the driven plate hub.
6. Install the transmission and torque bolts to 22–30 ft. lbs. (30–41 Nm).
7. Connect the clutch slave cylinder inlet line to the clutch master cylinder hydraulic line and torque to 10–12 ft. lbs. (14–16 Nm).
8. Support the transmission, remove jack from under transmission and mount transfer case on jack.
9. Install the transfer case and torque nuts to 22–30 ft. lbs. (30–41 Nm).
10. Remove transmission jack and support transfer case with safety stand.
11. Connect the transfer case vacuum lines and vent hose.
12. Connect backup light switch and speedometer cable.
13. Install the crossmember, torque the crossmember-to-frame bolts to 20–35 ft. lbs. (27–41 Nm) and torque the insulator-to-transmission bolts to 27–38 ft. lbs. (37–52 Nm).
14. Install the front and rear driveshafts, making certain to align marks and torque bolts to 140–200 inch lbs. (16–23 Nm).
15. Check and add gear lubricant and transfer case fluid, as necessary and lower vehicle.
16. Install the shift lever, stub shaft and retainer in the transmission tower. Seat the retainer in the tower and turn it clockwise to lock it.

NOTE: Make certain the nylon insulator on the end of the stub shaft is fully seated in the shifter block inside the tower, and after installed that the retainer is securely locked into place.

17. Install the shift tower dust boot.
18. Install the transmission and transfer case shift lever boots, shift knobs and locknuts.
19. Install the negative battery cable.

BENCH OVERHAUL

Before Disassembly

Clean the exterior of the transmission assembly before any attempt is made to disassemble the unit, to prevent dirt or other foreign materials from entering the transmission assembly or its internal parts.

NOTE: If steam cleaning is done to the exterior of the transmission, immediate disassembly should be done to avoid rusting, caused by condensation forming on the internal parts.

Transmission Disassembly

NOTE: The following components and materials must be replaced whenever the transmission is overhauled. Lip-type oil seals, lock nuts, all roll pins and all snaprings.

1. Remove the clutch housing and slave cylinder if necessary.
2. Remove the straight screw plug, spring and ball using a Torx® bit to remove the screw plug and a magnet to remove spring and ball.
3. Remove 5 adapter housing bolts and the nut.
4. Remove the shift lever housing set bolt and lock plate.
5. Remove the plug at the rear of the shift fork shaft.
6. Remove the shaft, using a large magnet to pull it out.
7. Remove the select lever from the top while rotating it.
8. Remove the 5 adapter housing bolts, 2 studs and the nut.
9. Using a plastic hammer, tap and remove the extension housing. Leave the gasket attached to the intermediate plate.
10. Remove the front bearing retainer and outer snaprings from both front bearings.
11. Separate the intermediate plate from the transmission case using a small plastic hammer and remove the case.
12. Mount the intermediate plate in a vise. Be careful not to damage the plate.

NOTE: Before placing the intermediate plate in a vise, insert bolts, washers and nuts in the open holes at the bottom of plate. Tighten vise against these bolts to prevent damage to the plate.

13. Remove the straight screw plug, locking balls and springs using a Torx® bit and magnet.

14. Remove the 5 slotted spring pins using a hammer and punch and then remove the 2 E-rings from the shift rails The locking ball from the reverse shift head and locking ball and pin from the intermediate housing will fall from the holes so be sure to catch them with a magnet.
15. Pull out the shift fork shaft No. 4 from the intermediate plate and catch the locking ball.
16. Remove shift fork shaft No. 4 and the 5th gear fork.
17. Pull out shift fork shaft No. 5 from the intermediate plate, and remove it with the reverse shift head. The interlock pins will fall from their hole. If they do not come out, remove them with a magnet.
18. Remove the shift fork shaft No. 3 from the intermediate plate and catch the interlock pins. The interlock pin will fall from the hole so be sure to catch it. If it does not come out, remove it with a magnet.
19. Remove shift fork shaft No. 1 from the intermediate plate being careful not to drop the interlock pin.
20. Remove shift fork shaft No. 2, shift fork No. 2 and shift fork No. 1.
21. Remove the reverse idle gear shaft stopper, reverse idler gear and shaft.
22. Remove the reverse shift arm from the reverse shift arm bracket.
23. Using a feeler gauge, measure the counter 5th gear thrust clearance. Standard clearance should be 0.004–0.012 in.
24. Engage 2 gears to lock the output shaft. Using a hammer and chisel, loosen the staked part of the nut on the countershaft.
25. Remove the lock not disengage the gears.
26. Remove the gear spline piece No. 5, synchronizer ring, needle roller bearing and counter 5th gear using tool J-22888 or equivalent.
27. Remove the spacer and use a magnet to remove the ball.
28. Remove the reverse shift arm bracket.
29. Remove the rear bearing retainer bolts with a Torx® bit and the snapring using snapring pliers.
30. Remove the output shaft, counter gear and input shaft as a unit from the intermediate plate by pulling on the counter gear and tapping on the intermediate plate with a plastic hammer.
31. Remove the input shaft with 14 needle roller bearing from the output shaft.
32. Remove the counter rear bearing from the intermediate plate.
33. Measure the thrust clearance of each gear. Standard clearance should be 0.004–0.010 in.
34. Using a pair of awls and a hammer, tap out the snapring.
35. Using a press, remove the 5th gear, rear bearing, 1st gear and the inner race.
36. Remove the needle roller bearing.
37. Remove the synchronizer ring, 2nd gear.
38. Remove the needle roller bearing.
39. Remove the snapring from hub sleeve No. 2.
40. Using a press, remove the hub sleeve, synchronizer ring, and 3rd gear.
41. Remove the needle roller bearing.

Unit Disassembly and Assembly

OUTPUT SHAFT AND INNER RACE

Inspection

1. Check the output shaft and inner race for wear or damage.
2. Using calipers, measure the output shaft flange thickness. The minimum thickness is 0.189 in.
3. Using calipers, measure the inner face flange thickness. The minimum thickness is 0.157 in.
4. Using a micrometer, measure the outer diameter of the output shaft journal surface. The 2nd gear minimum allowed measurement is 1.495 in. and the 3rd gear minimum measurement is 1.377 in.

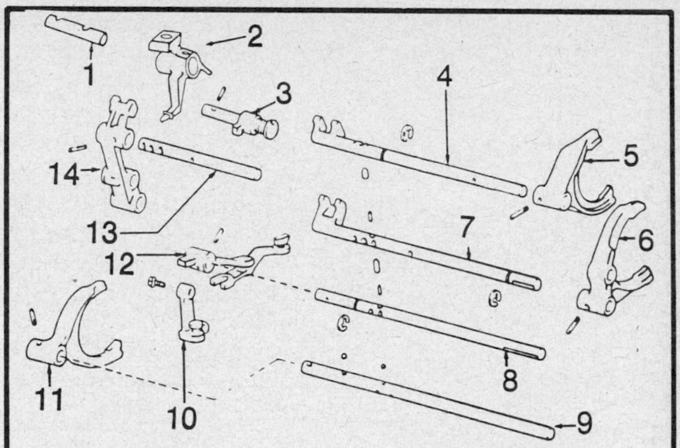
1
2
3
4
5
6
7
8
9
10
11
12
13
14

1. Shift lever shaft
2. Shift lever housing
3. Reverse restrict pin
4. No. 2 shift fork shaft
5. No. 2 shift fork
6. No. 1 shift fork
7. No. 1 shift fork shaft
8. No. 3 shift fork shaft
9. No. 4 shift fork shaft
10. Reverse shift arm bracket
11. No. 3 shift fork
12. Reverse shift arm and fork
13. No. 5 shift fork shaft
14. Reverse shift head

Shift shaft assembly

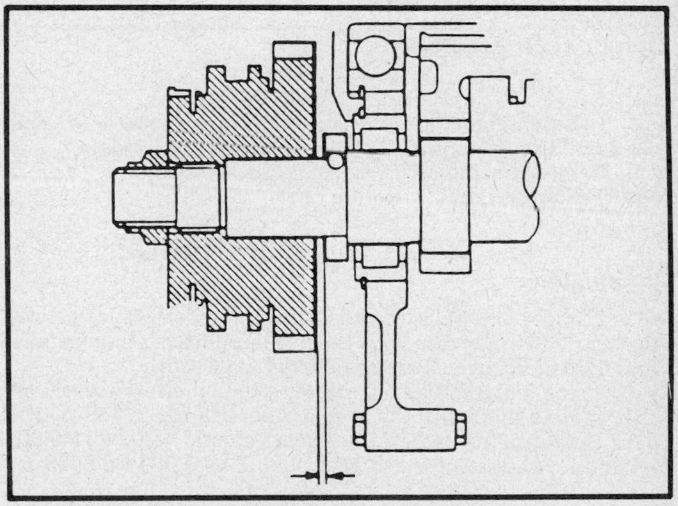

Checking 5th gear clearance

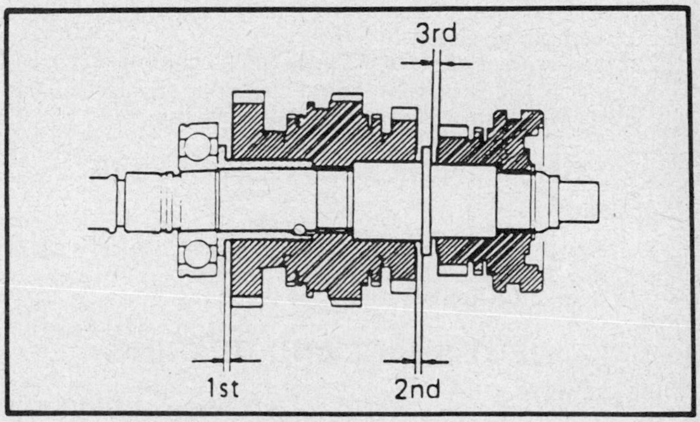
3rd
1st
2nd

Checking thrust clearance of gears

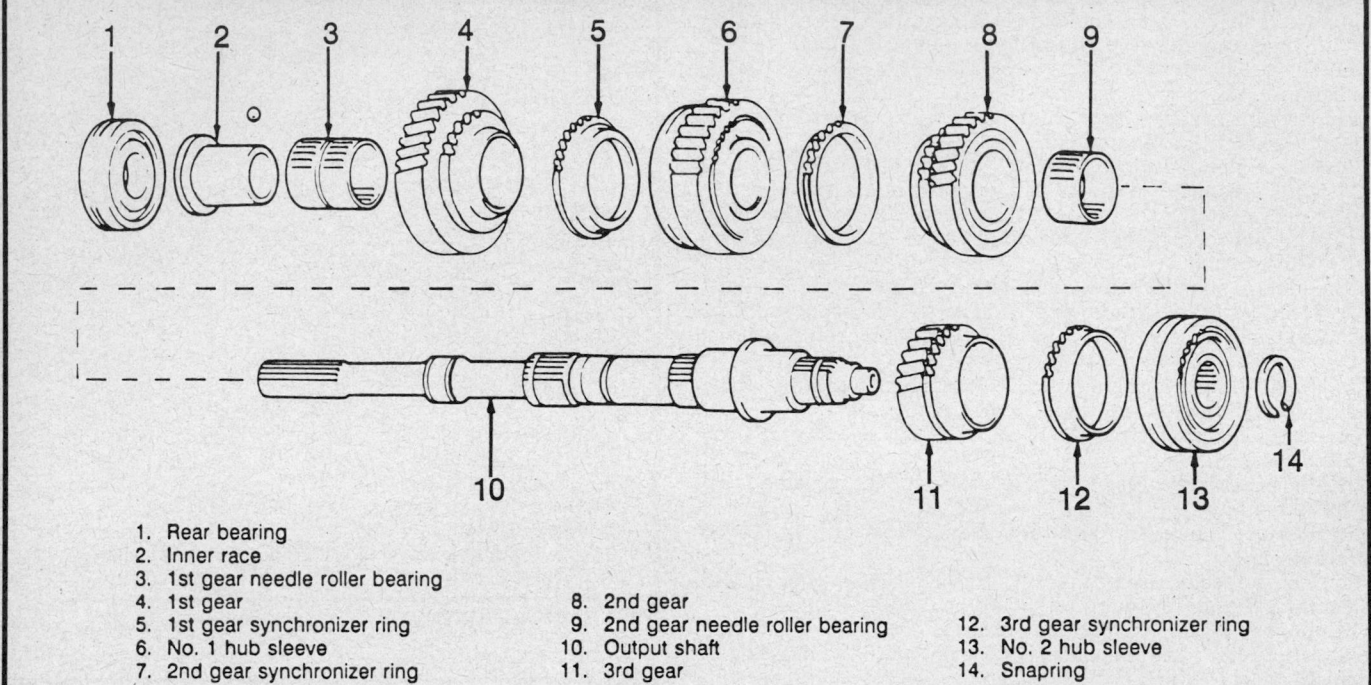

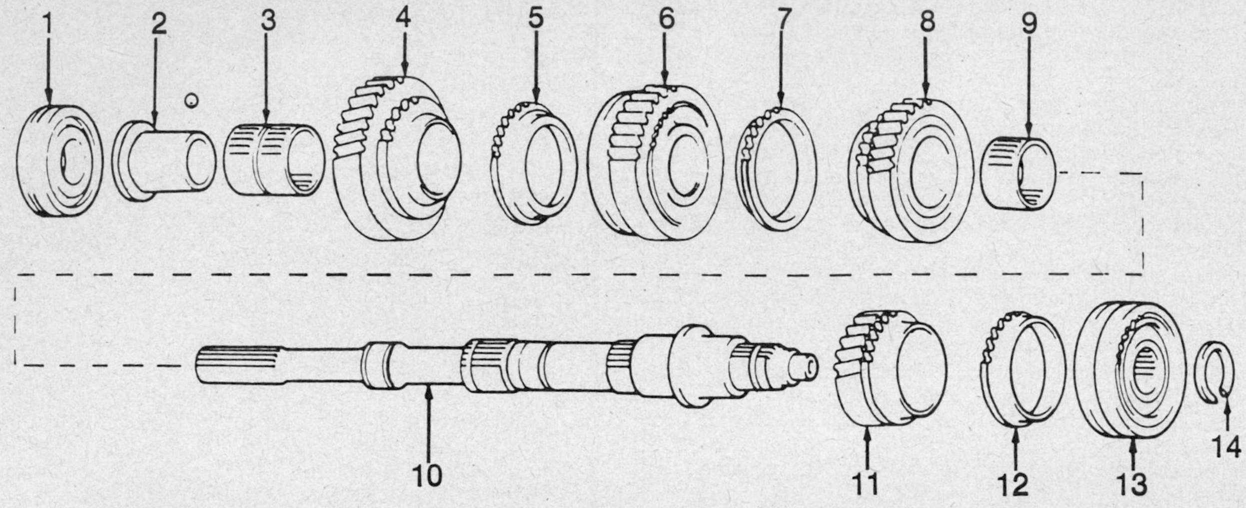
1
2
3
4
5
6
7
8
9
10
11
12
13
14
1. Rear bearing
2. Inner race
3. 1st gear needle roller bearing
4. 1st gear
5. 1st gear synchronizer ring
6. No. 1 hub sleeve
7. 2nd gear synchronizer ring
8. 2nd gear
9. 2nd gear needle roller bearing
10. Output shaft
11. 3rd gear
12. 3rd gear synchronizer ring
13. No. 2 hub sleeve
14. Snapring

Output shaft assembly

5. Using a micrometer, measure the outer diameter of the inner race. The minimum allowable diameter is 1.535 in.
6. Using a dial indicator, measure the shaft runout. The maximum allowable runout is 0.002 in.

1ST GEAR OIL CLEARANCE

Inspection

1. Using a dial indicator, measure the oil clearance between the gear and inner race with the needle roller bearing installed. Standard clearance should be 0.0004–0.0013 in.
2. Using a dial indicator, measure the oil clearance between the gear and shaft with the needle roller bearing installed. Standard clearance for 2nd and 3rd gears should be 0.0004–0.0013 in. and for the counter 5th gear should be 0.0004–0.0013 in.

SYNCHRONIZER RING

Inspection

1. Check for wear or damage. Turn the ring and push it in to check the braking action.
2. Measure the clearance between the synchronizer ring back and the gear spline end. Standard clearance should be 0.040–0.078 in., Minimum clearance must be 0.031 in.

SHIFT FORK AND HUB SLEEVE

Inspection

1. Using a feeler gauge, measure the clearance between the hub sleeve and shift fork.
2. Maximum clearance should be 0.039 in.

INPUT SHAFT AND BEARING

Disassembly

1. If bearing is damaged, remove the bearing snapring using snapring pliers.
2. Remove the bearing, using a press.

Inspection

1. Inspect bearing for roughness.
2. Inspect bearing for chips.
3. Inspect bearing for discoloration, caused by overheating or inadequate lubrication.
4. Replace bearing if any of the above are found.

Assembly

1. Lubricate shaft lightly with transmission lubricant.
2. Using a press, and tool J-34603 or equivalent, install the new bearing.
3. Select a snapring that will allow minimum axial play and install it on the shaft.

Mark	Thickness mm (in.)	
0	2.05-2.10	(0.0807-0.0827)
1	2.10-2.15	(0.0827-0.0846)
2	2.15-2.20	(0.0846-0.0866)
3	2.20-2.25	(0.0866-0.0886)
4	2.25-2.30	(0.0886-0.0906)
5	2.30-2.35	(0.0906-0.0925)

Snapring chart for input shaft bearing

COUNTER GEAR AND BEARING

Disassembly

1. If gear or bearing is damaged, remove the snapring using snapring pliers.
2. Press out the bearing using tool J-22912-01 or equivalent.

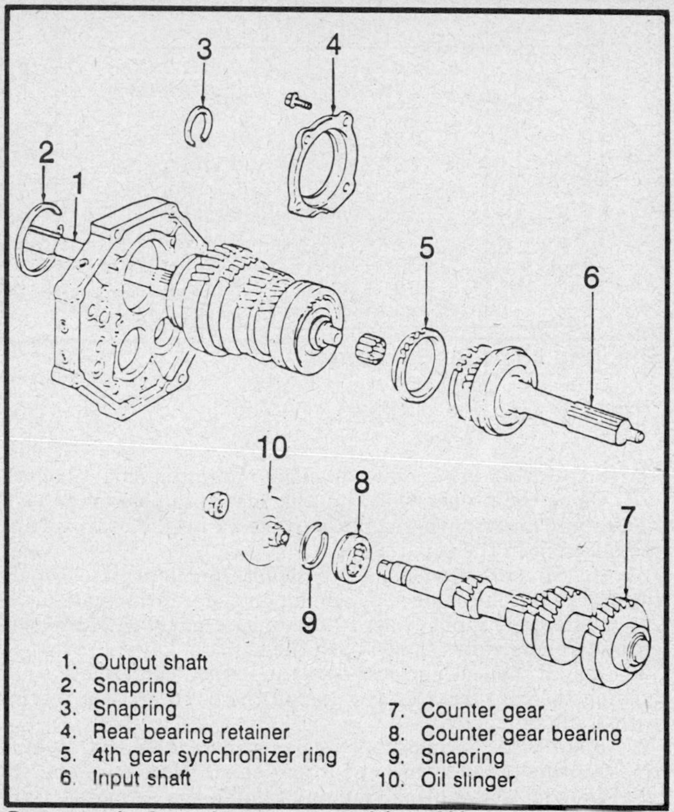
1. Output shaft
2. Snapring
3. Snapring
4. Rear bearing retainer
5. 4th gear synchronizer ring
6. Input shaft
7. Counter gear
8. Counter gear bearing
9. Snapring
10. Oil slinger

Countershaft and gear assembly—AX-4

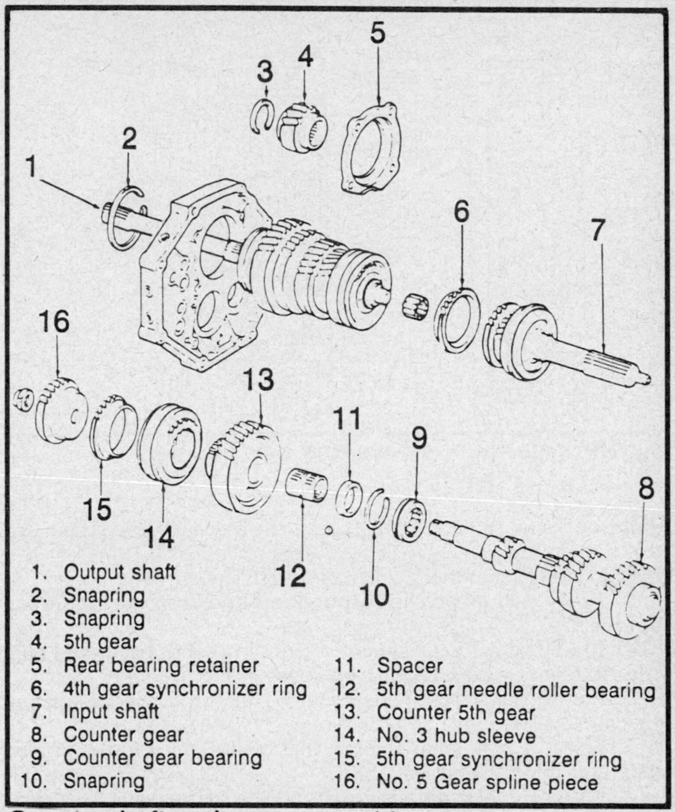
1. Output shaft
2. Snapring
3. Snapring
4. 5th gear
5. Rear bearing retainer
6. 4th gear synchronizer ring
7. Input shaft
8. Counter gear
9. Counter gear bearing
10. Snapring
11. Spacer
12. 5th gear needle roller bearing
13. Counter 5th gear
14. No. 3 hub sleeve
15. 5th gear synchronizer ring
16. No. 5 Gear spline piece

Countershaft and gear assembly—AX-5

Inspection

1. Check the gear teeth for wear or damage.
2. Inspect bearing for roughness and for chips.
3. Inspect bearing for discoloration, caused by overheating or inadequate lubrication.
4. Replace bearing if any of the above are found.

Assembly

1. Replace the side race.
2. Using tool J–28406 or equivalent press in the bearing and inner race.
3. Select a snapring that will allow minimum axial play and install it on the shaft.

FRONT BEARING RETAINER

Disassembly

1. Secure the bearing retainer in a soft jaw vise or equivalent.
2. Clean retainer and remove any old gasket material.
3. Using a awl, pry the old seal out of the housing.

Inspection

1. Check retainer for damage.
2. Check the oil seal lip for wear or damage.
3. If seal is not to be replaced, oil seal depth should be 0.441–0.480 in. from the housing surface to the top edge of the seal.

Assembly

1. Lubricate new seal with transmission lubricant.
2. Press in the new oil seal using tool J-34602 or equivalent.
3. Adjust oil seal depth to 0.441–0.480 in. from the housing-to-transmission surface to the top edge of the seal.

Mark	Thickness mm (in.)	
1	2.05-2.10	(0.0807-0.0827)
2	2.10-2.15	(0.0827-0.0846)
3	2.15-2.20	(0.0846-0.0866)
4	2.20-2.25	(0.0866-0.0886)
5	2.25-2.30	(0.0886-0.0906)
6	2.30-2.35	(0.0906-0.0925)

Snapring chart for countershaft

REVERSE RESTRICT PIN

Disassembly

1. Using a Torx® bit, remove the screw plug.
2. Using a hammer and pin punch, drive out the slotted spring pin.
3. Pull off the lever housing and slide out the shaft.

Inspection

Inspect the restrict pin for wear or damage.

Assembly

1. Install the lever housing.
2. Using a hammer and pin punch, drive out the slotted spring pin.
3. Using a Torx® bit, install and torque the screw plug to 27 ft. lbs. torque.

Transmission Assembly

1. Install the clutch hub No. 1 and No. 2 into hub sleeves

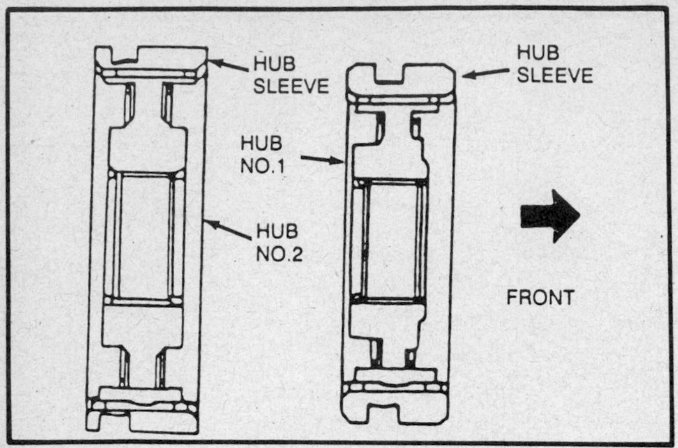
HUB SLEEVE
HUB SLEEVE
HUB NO.1
HUB NO.2
FRONT
Synchronizer hub, sleeve and key position

along with the shifting keys. Install the key springs so their gaps are not in line.

2. Install the shifting springs under the shifting keys.
3. Apply gear oil on the output shaft and 3rd gear needle roller bearing.
4. Place the 3rd gear synchronizer ring on the gear and align the ring slots with the shifting keys.
5. Install the needle roller bearing in the 3rd gear and hub sleeve No. 2.
6. Select a new snapring (2) that will allow minimum axial play and install it on the shaft.
7. Using a feeler gauge, measure the 3rd gear thrust clearance. Standard clearance should be 0.004–0.010 in.
8. Apply gear oil on the output shaft and 2nd gear needle bearing.

Mark	Thickness mm (in.)	
C-1	1.75-1.80	(0.0689-0.0709)
D	1.80-1.85	(0.0709-0.0728)
D-1	1.85-1.90	(0.0728-0.0748)
E	1.90-1.95	(0.0748-0.0768)
E-1	1.95-2.00	(0.0768-0.0787
F	2.00-2.05	(0.0788-0.0807)
F-1	2.05-2.10	(0.0807-0.0827)

Snapring chart for output shaft

9. Place the 2nd gear synchronizer ring on the 2nd gear and align the ring slots with the shifting keys.
10. Install the needle roller bearing in the 2nd gear.
11. Using a press install the 2nd gear and hub sleeve No.1.
12. Install the 1st gear locking ball in the output shaft.
13. Apply gear oil to the needle roller bearing.
14. Assemble the 1st gear, synchronizer ring, needle roller bearing and bearing inner race.
15. Install the assembly on the output shaft, with the synchronizer ring slots aligned with the shifting keys.
16. Turn the inner race to align it with the locking ball.
17. Install the output shaft rear bearing using tool J-34603 or equivalent and a press.
18. Install the bearing on the output shaft with the outer race snapring groove toward the rear. Hold the 1st gear inner race to prevent it from falling.
19. Measure the 1st and 2nd gear thrust clearance with a feeler gauge. Standard clearance should be 0.004–0.010 in.
20. Install 5th gear on the output shaft using tool J-35603 or equivalent and a press.

Mark	Thickness mm (in.)	
A	2.67-2.72	(0.1051-0.1071)
B	2.73-2.78	(0.1075-0.1094)
C	2.79-2.84	(0.1098-0.1118)
D	2.85-2.90	(0.1122-0.1142)
E	2.91-2.96	(0.1146-0.1165)
F	2.97-3.02	(0.1169-0.1189)
G	3.03-3.08	(0.1193-0.1213)
H	3.09-3.14	(0.1217-0.1236)
J	3.15-3.20	(0.1240-0.1260)
K	3.21-3.26	(0.1264-0.1283)
L	3.27-3.32	(0.1287-0.1307)

Snapring chart for 5th gear

21. Select a snapring that will allow minimum axial play.
22. Using the proper tools, tap the snap into position.
23. Apply multi-purpose grease to the 14 needle bearings and install them in the input shaft.
24. Install the output shaft into the intermediate plate by pulling on the output shaft and tapping on the intermediate plate.
25. Install the input shaft to the output shaft with the synchronizer ring slots aligned with the shifting keys.
26. Install the counter gear into the intermediate plate while holding the counter gear and install the counter rear bearing with a suitable driver.
27. Install the bearing snapring using snapring pliers. Be sure the snapring is flush with the intermediate plate surface.
28. Using a Torx® bit, install and tighten the screws to 13 ft. lbs. torque.
29. Install the reverse shift arm bracket and tighten the bolts to 13 ft. lbs. torque.
30. Install the ball and spacer.
31. Install the shifting keys and hub sleeve No. 3 onto the counter 5th gear. Install the key springs positioned so the end gaps are not in line.
32. Install shifting key springs under the shifting keys.
33. Apply gear oil to the needle roller bearing and install the counter 5th gear with hub sleeve No. 3 and needle roller bearings.
34. Install the synchronizer ring on gear spline piece.
35. Using tool J-28406 or equivalent drive in gear spline piece No. 5 with the synchronizer ring slots aligned with the shifting keys. When installing gear spline piece, support the counter gear in front with a 3–5 lb. hammer or equivalent.
36. Engage 2 gears to lock the output shaft.
37. Install and tighten the lock nut to 90 ft. lbs. torque on the countershaft.
38. Stake the lock nut.
39. Disengage the gears.
40. Measure the counter 5th gear thrust clearance using a feeler gauge. Standard clearance should be 0.004–0.012 in.
41. Install the reverse shift arm to the pivot of the reverse shift arm bracket.
42. Install the reverse idler gear on the shaft.
43. Align the reverse shift arm shoe to the reverse idler gear groove and insert the reverse idler gear shift to the intermediate plate.
44. Install the reverse idler gear shaft stopper and tighten the bolt to 13 ft. lbs. torque.
45. Place shift forks No. 1 and No. 2 into groove of hub sleeves No. 1 and No. 2 and install fork shaft No. 2 to the shift fork No. 1 and No. 2 through the intermediate plate.
46. Apply multi-purpose grease to the interlock pins.
47. Using a magnet and a suitable tool, install the interlock pin onto the intermediate plate.

48. Install the interlock pin into the shaft hole.
49. Install fork shaft No. 1 to shift fork No. 1 through the intermediate plate.
50. Using a magnet and a suitable tool, install the interlock pin into the intermediate plate.
51. Install the interlock pin into the shaft hole.
52. Install fork shaft No. 3 to the reverse shift arm through the intermediate plate.
53. Install the reverse shift head into fork shaft No. 5.
54. Insert fork shaft No. 5 to the intermediate plate and put in the reverse shift head to the shift fork No. 3.
55. Using a magnet and a suitable tool, install the locking ball into the reverse shift head hole.
56. Shift hub sleeve No. 3 to the 5th speed position.
57. Place shift fork No. 3 into the groove of hub sleeve No. 3 and install fork shaft No.4 to shift fork No. 3 and reverse shift arm.
58. Using a magnet and a suitable tool, install the locking ball into the intermediate plate and insert fork shaft No. 4 to the intermediate plate.
59. Check the interlock by positioning the shift fork shaft No.1 to the 1st speed position.
60. Fork shafts No. 2, No. 3, No. 4 and No. 5 should not move.
61. Using a pin punch and a hammer, drive in new slotted springs pins in each shift fork, reverse shift arm and reverse shift head.
62. Install 2 fork shaft E-rings.
63. Apply liquid sealer to the screw plugs.
64. Install the locking balls, springs and screw plugs with a Torx® bit and tighten to 14 ft. lbs. torque. Install the short spring into the tower of the intermediate plate.
65. Remove the intermediate plate from the vise.
66. Remove the bolts, nuts, washers and gasket.
67. Install the case onto assembly, align each bearing outer race, each fork shaft end and reverse idler gear with the holes in the case and install the case on the intermediate plate. If necessary, tap on the case with a plastic hammer.
68. Install 2 new bearing snaprings.
69. Install front bearing retainer with a new gasket.
70. Apply liquid sealer to the bolts.
71. Install and tighten the bolts to 12 ft. lbs. torque.
72. Install the new gasket to the intermediate plate.
73. Install the adapter housing.
74. Install and tighten the adapter bolts to 27 ft. lbs. torque.
75. Install the shift lever housing.
76. Insert the shift lever into the adapter and shift lever housing.
77. Install and tighten shift lever housing bolt with a lock plate to 28 ft. lbs. torque. Lock the lock plate.
78. Install and tighten the adapter screw plug to 13 ft. lbs. torque.
79. Apply liquid sealer to the plug.
80. Install the locking ball, spring and screw plug and tighten the plug to 14 ft. lbs. torque.
81. Check to see that the input shaft and output shafts rotate smoothly.
82. Check to see that shifting can be done smoothly to all positions.
83. Install the black restrict pin on the reverse gear/5th gear side.
84. Install the remaining pin and tighten the pins to 20 ft. lbs. (27 Nm) of torque.
85. Install the shift lever retainer with a new gasket and tighten the bolts to 13 ft. lbs. (18 Nm) torque.
86. Install the backup light switch and tighten to 19–35 ft. lbs (26–48 Nm). torque.
87. Install the oil level sensor, if equipped and tighten to 9–17 ft. lbs. (12–23 Nm) of torque.
88. Install the clutch housing and tighten the bolts to 27 ft. lbs. (36 Nm) of torque.

SPECIFICATIONS

TORQUE SPECIFICATIONS

Component	ft. lbs.	Nm
Clutch housing-to-transmission case	19–35	26–47
Front bearing retainer-to-transmission case	9–17	12–23
Adapter housing-to-transmission case	19–35	26–47
Rear bearing retainer-to-intermediate plate	9–17	12–23
Plug-shift lever shaft-to-adapter housing	9–17	12–23
Shift lever housing-to-shift lever shaft	19–35	26–47
Shift lever retainer-to-adapter housing	9–17	12–23
Reverse shift arm bracket-to-intermediate plate	7–17	10–23
Reverse idle gear shaft-to-intermediate plate stopper	7–17	10–23
Reverse restrict pin-to-adapter housing	19–35	26–47
Plug-lock ball spring	9–17	12–23
Counter gear lock nut	72–108	98–146
Filler/drain plug-transmission case	19–35	26 47
Back up lamp switch	19–35	26–47
Top gear switch	19–35	26–47
Plug-oil level sensor	9–17	12–23

SPECIAL TOOLS

Tool	Description
J-28406	Front counter bearing and 5th gear installer
J-29184	Adapter housing seal installer
J-34602	Front bearing seal installer
J-34603	Output shaft bearing and gear installer
J-22912-01	Bearing remover
J-22888-D	Gear and bearing puller
J-7818	Rear Retainer Bearing Installer
J-8614-01	Holder and Remover Companion Flange
J-26941	Needle Bearing Puller
J-29162	Rear Retainer Oil Seal Installer
J-29163	Front Output Shaft Rear Bearing Installer
J-29166	Rear Output Shaft Rear Bearing Installer
J-29167	Front Output Shaft Front Bearing Installer
J-29168	Front Output Shaft Front Bearing Remover
J-29169	Input Gear Bearing Installer

Tool	Description
J-29170	Input Gear Bearing Remover
J-29174	Mainshaft Bearing Installer
J-33826	Rear Output Bushing Installer and Mainshaft Sprocket Bearing Remover
J-33828	Mainshaft Sprocket Bearing Installer
J-33830	Front Input Bearing Installer
J-33831	Input Seal Installer
J-33832	Front Output Rear Bearing Installer
J-33833	Output Main Bearing Installer
J-33834	Front Output Seal Installer
J-33835	Pump Housing Seal Installer
J-33839	Rear Output Bushing Remover
J-33841	Front Input Bearing Remover
J-33843	Extension Housing Seal Installer
J-34635	Gearshift Lever Remover

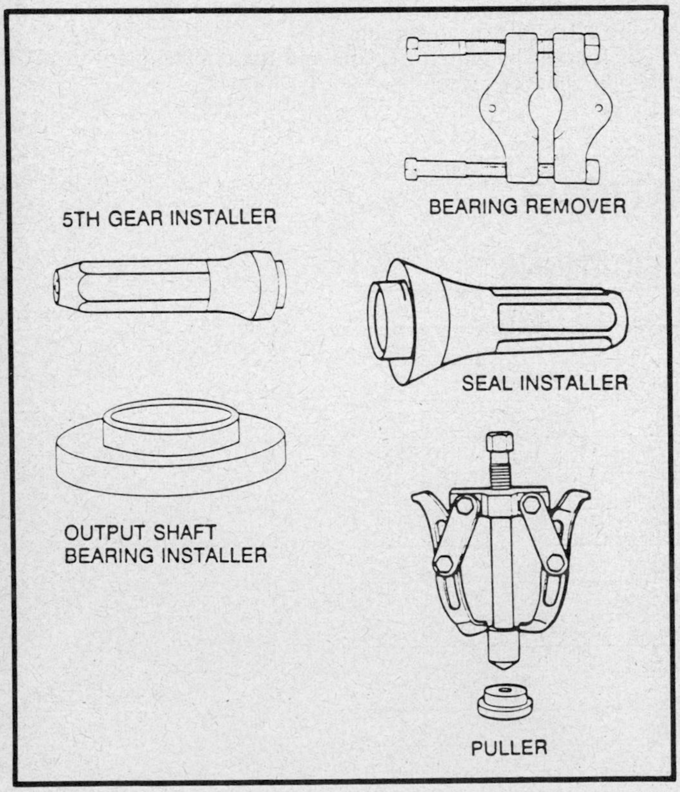
BEARING REMOVER
5TH GEAR INSTALLER
SEAL INSTALLER
OUTPUT SHAFT
BEARING INSTALLER
PULLER

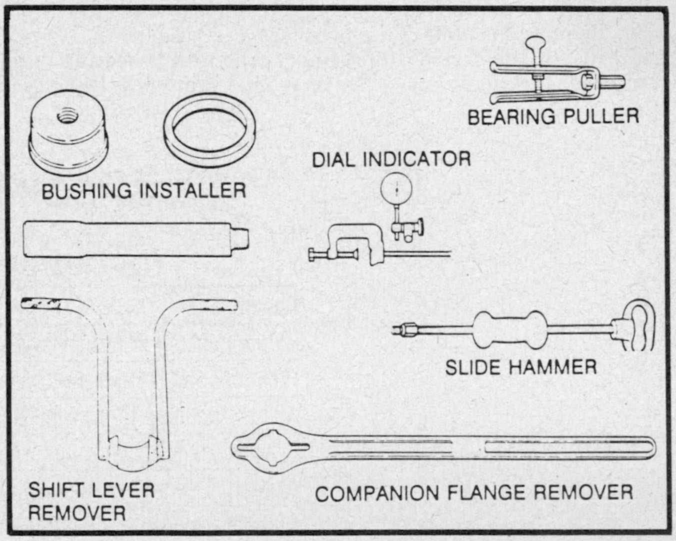
BEARING PULLER
DIAL INDICATOR
BUSHING INSTALLER
SLIDE HAMMER
SHIFT LEVER
REMOVER
COMPANION FLANGE REMOVER

Section 7

BA 10/5 Transmission
AMC/Jeep-Eagle

APPLICATION

BA10/5

Year	Vehicle	Engine	VIN Code
1987–88	Wrangler/CJ	4.2L	C
1989	Cherokee	4.0L, 4.2L	E, L
	Comanche	4.0L, 4.2L	E, L
	Wagoneer	4.0L, 4.2L	E, L
	Wrangler	4.0L, 4.2L	E, L

GENERAL DESCRIPTION

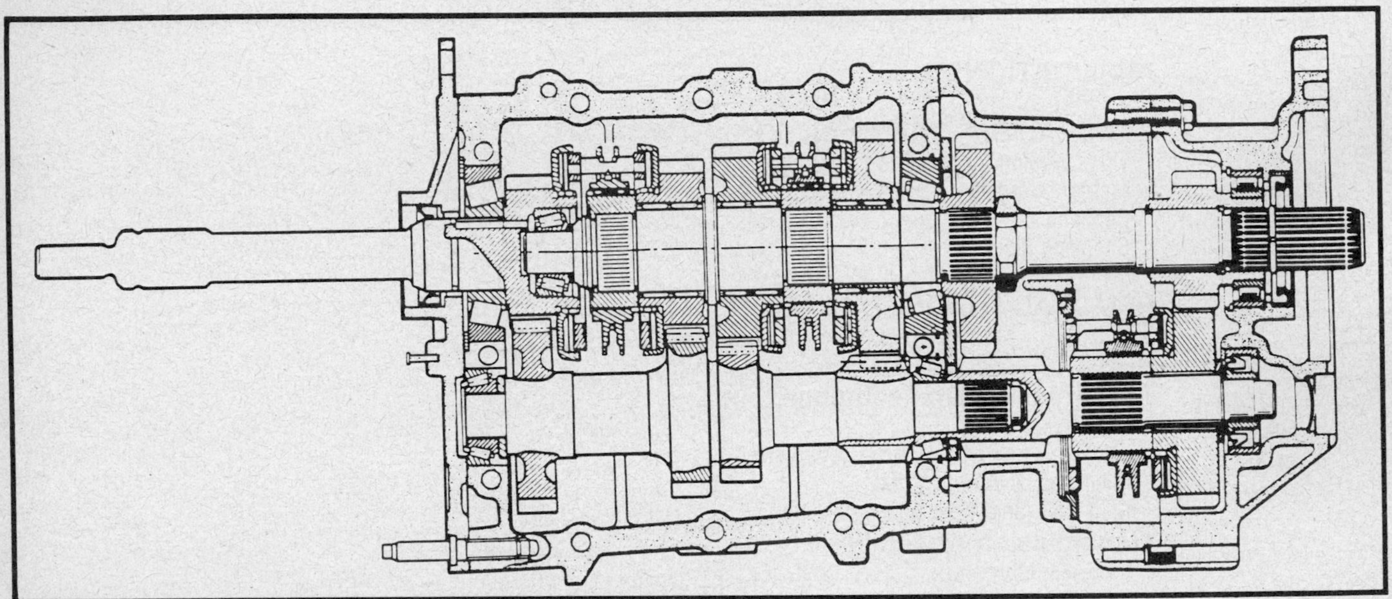

Cross section of BA10/5 speed manual transmission

The BA 10/5 transmission is a synchromesh 5 speed manual transmission. All forward gears are controlled through synchronizer engagement. Fifth gear is an overdrive gear. The transmission geartrain is divided into 3 sections. The input shaft, cluster gear and 1st through 4th gear assemblies are located in the front case. The 5th/reverse gears are located in both the intermediate and rear cases. The mainshaft is supported by a pilot bearing and by roller bearings installed in the front and the rear cases. All shift forks, rails and detent components are located in the front case. The gearshift lever is mounted to the top of the intermediate case.

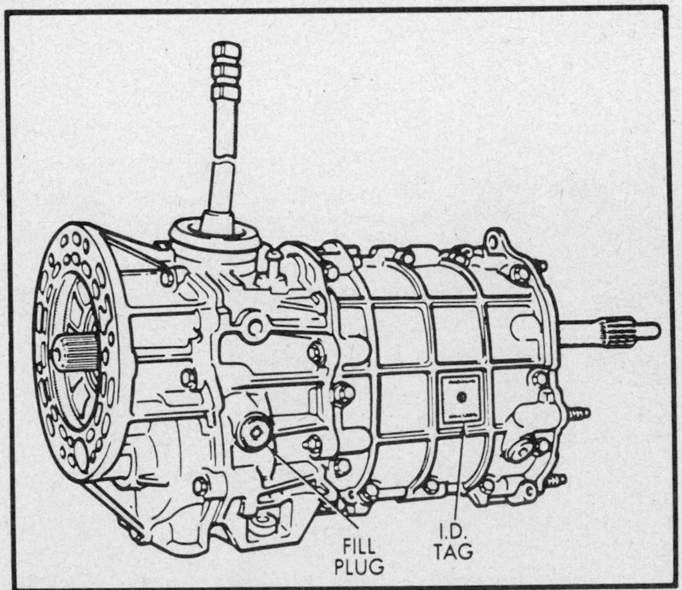
FILL
PLUG
I.D.
TAG

Location of the identification on late model transmissions

Transmission Identification

On earlier transmissions, the identification plate is bolted to the passenger side of the front case. The manufacturer's identification and part number is located on the top line of the plate. The manufacturer's reference number is located on the second line and transmission serial number on the third line.

On later transmissions, the identification plate is riveted to the front case on the passenger's side. The plate provides information on date of build, part number and serial number.

Metric Fasteners

The metric fastener dimensions are very close to the dimensions of the familiar inch system fasteners. For this reason, replacement fasteners must have the same measurement and strength as those removed.

CAUTION

Do not attempt to interchange metric fasteners for inch system fasteners. Mismatched or incorrect fasteners can result in damage to the transmission unit through malfunctions, breakage or possible personal injury.

Care should be taken to re-use the fasteners in the same locations as removed.

Common metric fastener strength property classes are 9.8 and 12.9 with class identification embossed on the head of each bolt. The inch strength classes range from grade 2–8 with the line identification embossed on each bolt head. Markings correspond to 2 lines less than actual grade (for example grade 8 bolt will exhibit 6 embossed lines on the bolt head). Some metric nuts will be marked with a single digit strength identification numbers on the nut face.

Capacities

Lubricant capacity varies according to year and whether the vehicle is equipped with 2WD or 4WD.

Early Wrangler transmissions have a fluid capacity of 3.5 pts. Later Wrangler transmissions have a 4.7 pint fluid capacity.

On Comanche, Cherokee and Wagoneer with 2WD, the fluid capacity is 5.2 pints and 4.9 pts if equipped with 4WD.

All transmissions use Mopar gear lubricant or equivalent SAE 75W–90 API Grade GL5 gear lubricant.

Checking Fluid Levels

The filler plug is located on the passenger side of the intermediate case and the drain plug is located at the back of the rear case.

Remove the filler plug and check that the fluid level is just below the bottom of the filler plug opening. Add fluid as required.

REMOVAL AND INSTALLATION

TRANSMISSION REMOVAL

1. Shift the transmission into **N**.
2. Unscrew the shift knob and remove the locknut.
3. Remove the boot and bezel from the shifter console.
4. Remove the shifter lever dust boot.
5. Unscrew the lever cover and remove the upper and lower bushings, washer, pin and lever.

NOTE: On some vehicles, the shifter lever is retained with a snapring and spring washer.

6. Raise the vehicle and support safely. Drain the transmission and transfer case, if equipped.

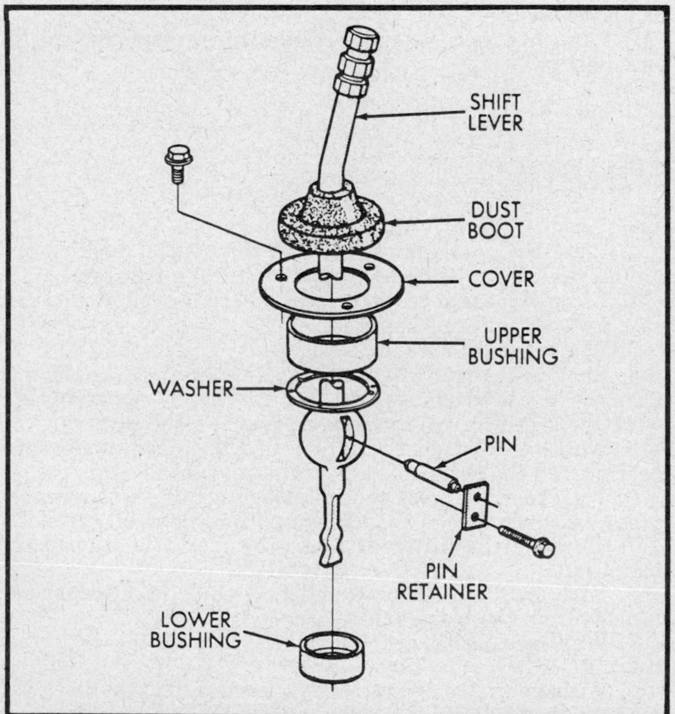
SHIFT LEVER
DUST BOOT
COVER
UPPER BUSHING
WASHER
PIN
PIN RETAINER
LOWER BUSHING

Shift lever components

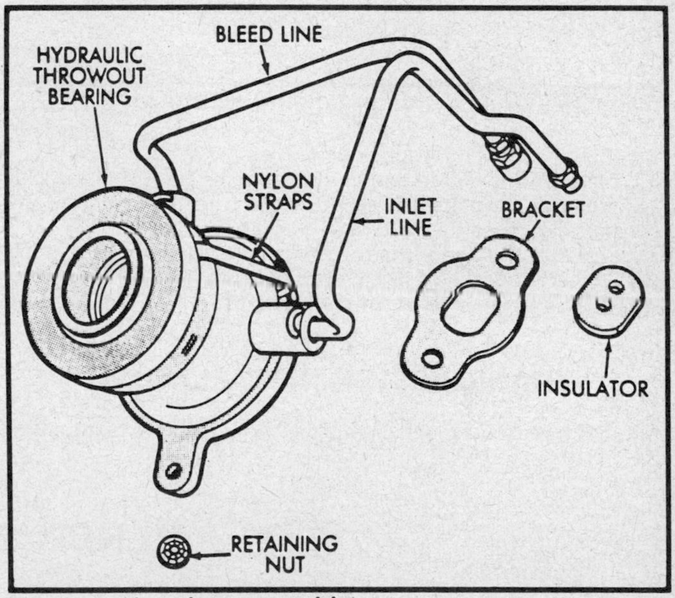
BLEED LINE
HYDRAULIC THROWOUT BEARING
NYLON STRAPS
INLET LINE
BRACKET
INSULATOR
RETAINING NUT

Throwout bearing assembly

7. Matchmark the rear driveshaft for assembly reference then remove it.
8. Position a safety stand or suitable support fixture under the transfer case, if equipped.
9. Unbolt and remove the rear crossmember.
10. Disconnect the speedometer cable and back-up switch light lead.
11. Disconnect the vent hoses from the transfer case and transmission, as required.
12. Disconnect the transfer case vacuum hoses, if equipped. Disconnect the transfer case range selector rod from the floor shift lever, if equipped.
13. Matchmark the front driveshaft for assembly reference and remove it.
14. Disconnect the clutch cylinder hydraulic line from the throwout bearing inlet line. Cap the line.
15. Position a suitable transmission jack under the transmission/transfer case. Secure assembly to the jack with safety chains or strap.

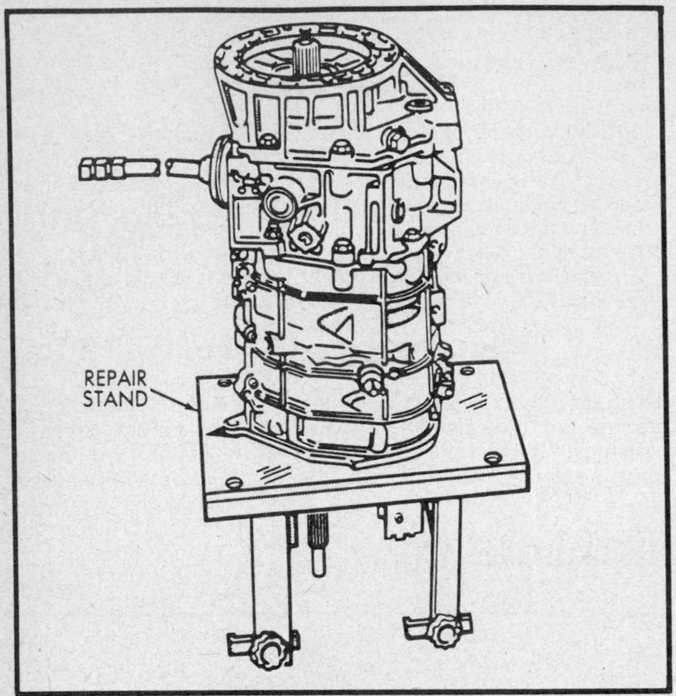
REPAIR STAND
Transmission installed in holding fixture

16. Remove the clutch housing bolts.
17. Remove the transmission and transfer case from vehicle.
18. Remove the bolts and separate the transfer case from the transmission, if equipped.
19. Remove the hydraulic throwout bearing assembly.
20. Unbolt and remove the clutch housing from the transmission case and position transmission on a suitable holding fixture.

TRANSMISSION INSTALLATION

1. Remove the transmission from the holding fixture and install the clutch housing. Torque the housing bolts to 28 ft. lbs. (38 Nm).
2. Install the hydraulic throwout bearing assembly.
3. Temporarily insert the shift lever into the transmission tower and shift the transmision into any forward gear. Remove the lever.
4. Raise the transmission with the transmission jack and align the input shaft with the clutch disc splines. Push the transmission forward until the transmission is seated against the engine.
5. Install the clutch housing-to-engine bolts and tighten as follows: M12x1.75 bolts to 55 ft. lbs. (75 Nm), 3/8–16 bolts to 27 ft. lbs. (37 Nm) and 7/16–14 bolts to 43 ft. lbs. (58 Nm).
6. Connect the throwout bearing hydraulic line.
7. Remove the transmission jack.
8. Mount the transfer case, if equipped, on the transmission jack and align and bolt it to the transmission. Torque the nuts and bolts to 26 ft. lbs. (35 Nm).
9. Remove the transmission jack and position a safety stand or equivalent under the transfer case for support, if equipped.
10. Connect the transfer case hose vacuum hose and linkage, if equipped.
11. Connect the transmission and transfer case vent hoses as required.
12. Connect the back-up switch lead and the speedometer cable.
13. Connect the transfer case range selector rod to the floor shift lever, if equipped.
14. Install the rear crossmember. Torque the crosmember-to-frame bolts to 30 ft. lbs. Torque the transmission-to-rear support bolts and nuts to 33 ft. lbs. (45 Nm).
15. Align and install the front and rear driveshafts. Torque the U-joint bolts to 170 inch lbs. (19 Nm).
16. Fill the transfer case and transmission to the proper level.
17. Lower the vehicle.
18. Install the shift lever and stub shaft into the shift tower. Secure the stub shaft with either the retainer plate or snapring (depending on what was originally installed).
19. Install the dust boot.
20. Install the shift lever extension, boot, bezel, locking nut and lever knob.
21. Perform a road test and check the transmission for smooth and quiet shifting.

BENCH OVERHAUL

Before Disassembly

Clean the transmission case with a suitable solvent and dry with compressed air. Inspect the case for cracks and stripped threads in the various bolt holes. Check the machined mating surfaces for burrs, nicks or any other surface imperfections that would make the transmission case unfit for service in the vehicle. The front mating surface should be smooth. Any minor burrs may be dressed with a fine mill file. Repair damaged threaded holes with an appropriate insert or by tapping and re-threading the hole.

If the front transmission case is being replaced, transfer the identification plate to the new case. On earlier models, the plate is bolted to the case. On later models, use a 5x10mm rivet to attach the plate to the new case.

Transmission Disassembly

NOTE: All of the service tools required to overhaul the BA10/5 transmission are contained in tool kit with part number B.Vi.FM.01.

1. Place the gear shift selector in the neutral position and mount the transmission on a suitable holding fixture.
2. Remove the rear case attaching bolts and separate the rear case from the intermediate case.
3. Remove and the rear case oil seal.
4. Remove the rear case bearing race snapring. Tap the race from the case with a brass drift. Tap alternately and evenly and work the race through the case grooves to remove it.
5. Working through the center of the intermediate shaft bearing, tap the access plug from the bore in the rear case.
6. With an arbor press and suitable driver and receiver tools, press the intermediate shaft bearing from the rear case.
7. Remove the 5th gear lock ring from the end of the mainshaft.
8. With a suitable 3-jawed puller, remove the 5th gear and bearing from the mainshaft.
9. With a suitable bearing puller, remove the fifth gear bearing.
10. With a suitable 3-jawed puller, remove the 5th intermediate gear, thrust washer, endplay shim and bearing race from the intermediate shaft.

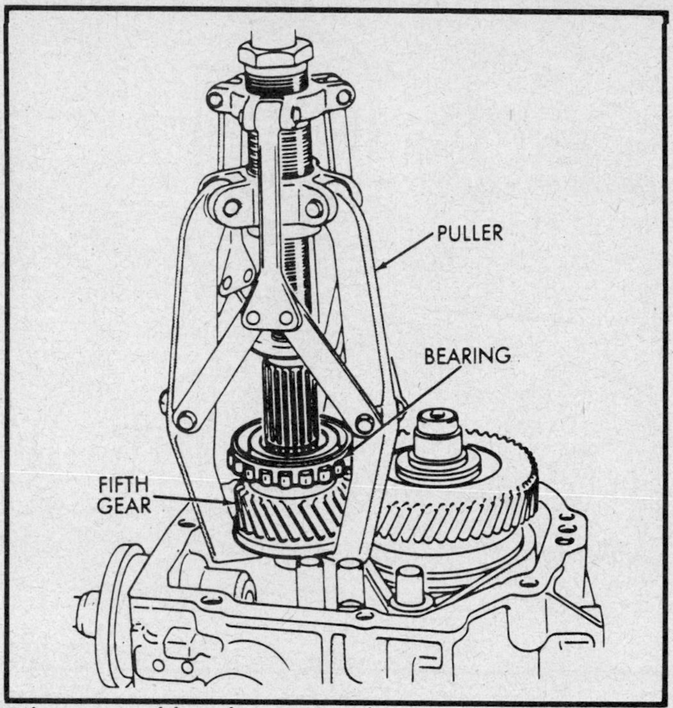
PULLER
BEARING
FIFTH GEAR
5th gear and bearing removal

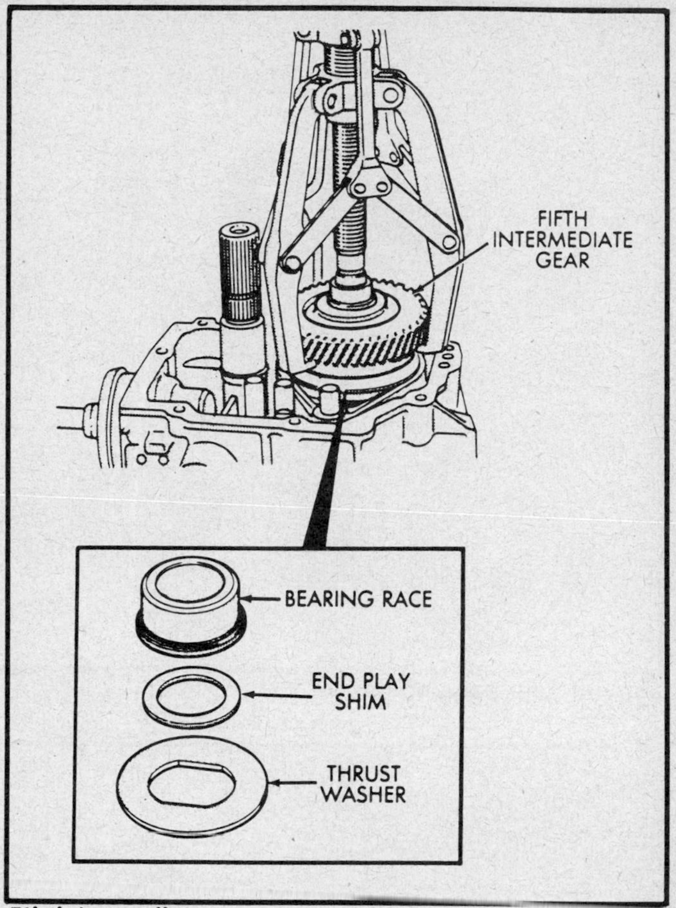
FIFTH INTERMEDIATE GEAR
BEARING RACE
END PLAY SHIM
THRUST WASHER
5th intermediate gear removal

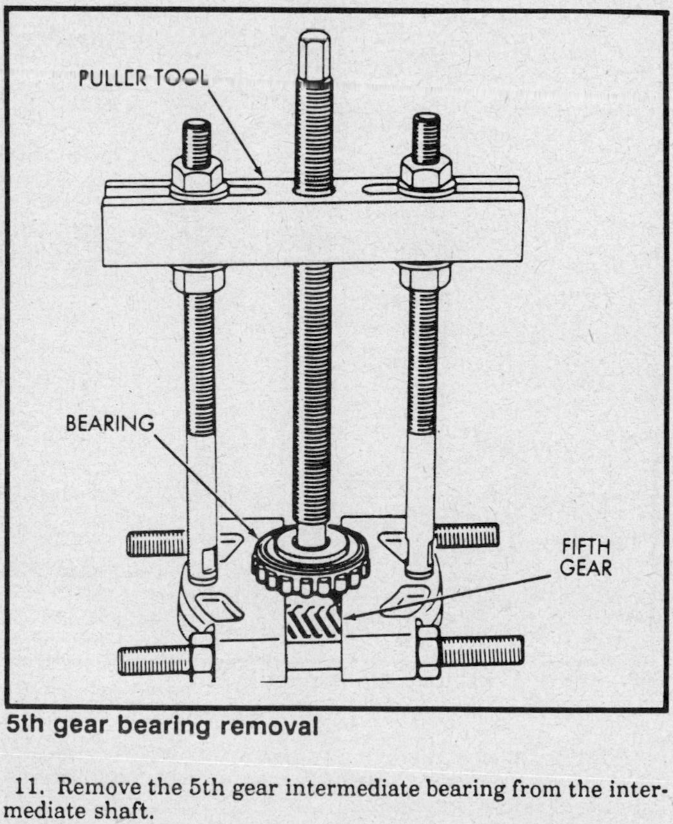
PULLER TOOL
BEARING
FIFTH GEAR
5th gear bearing removal

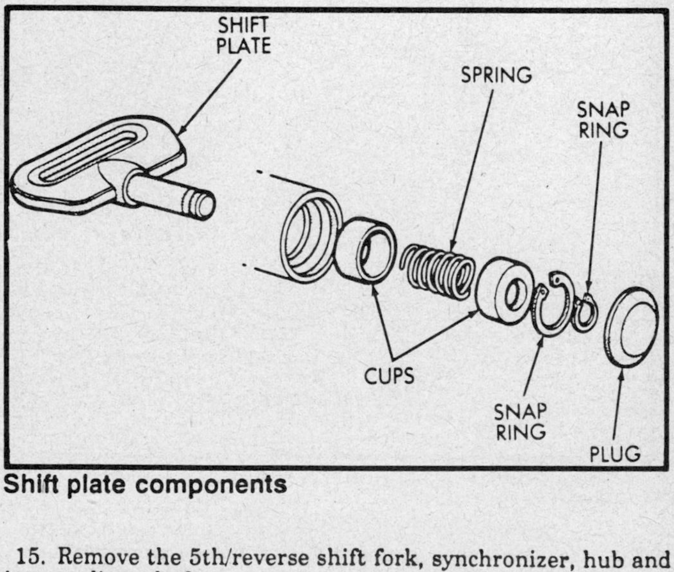
SHIFT PLATE
SPRING
SNAP RING
CUPS
SNAP RING
PLUG
Shift plate components

11. Remove the 5th gear intermediate bearing from the intermediate shaft.
12. Remove the 5th gear snapring from the mainshaft.
13. Matchmark the 5th/reverse synchronizer and hub for assembly reference.
14. Bolt a suitable holding tool onto the 5th/reverse shift rail and fork. Knock the roll pin from the shift rail with a pin punch.
15. Remove the 5th/reverse shift fork, synchronizer, hub and intermediate shaft.
16. Remove the shift plate access plug, cups, snaprings, spring and shift plate from the case.
17. Unbolt and separate the intermediate case from the front case.
18. Remove the transmission from the holding fixture. Reposition the transmission horizontally on the stand so that the left

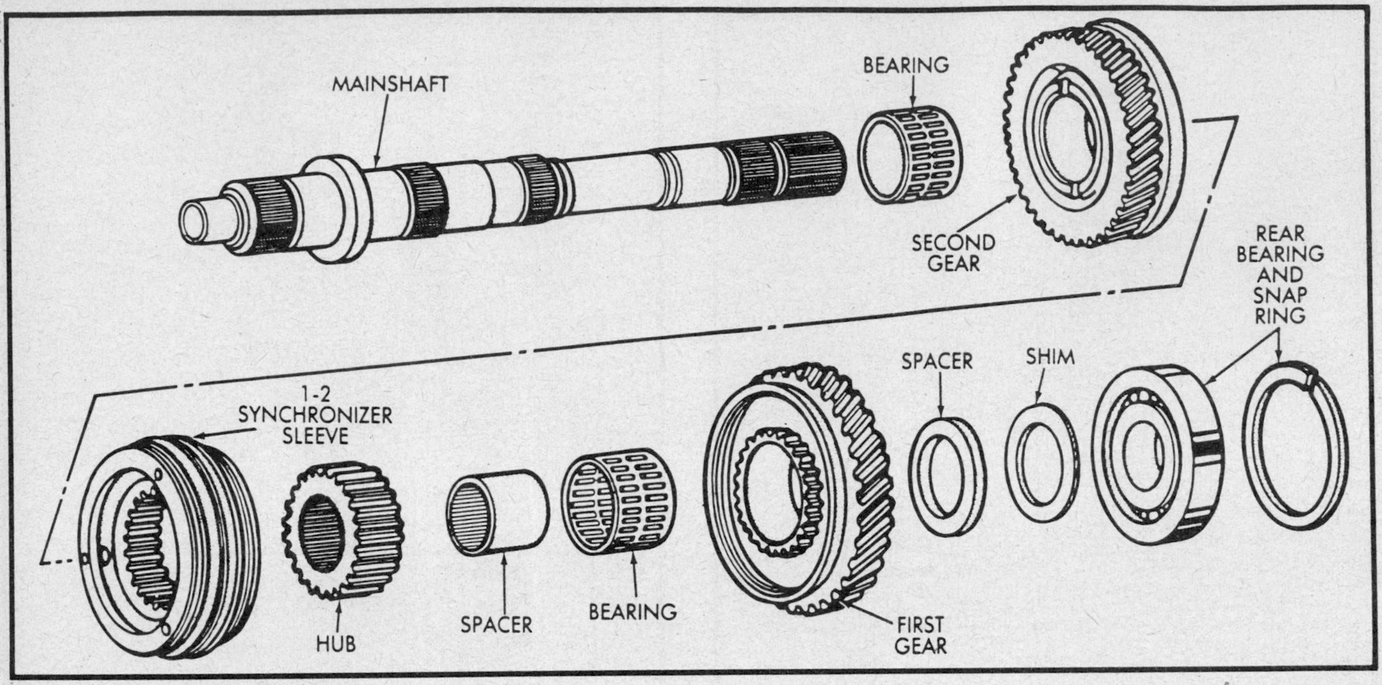
MAINSHAFT
BEARING
SECOND GEAR
REAR BEARING AND SNAP RING
SPACER
SHIM
1-2 SYNCHRONIZER SLEEVE
HUB
SPACER
BEARING
FIRST GEAR

1st/2nd gear components

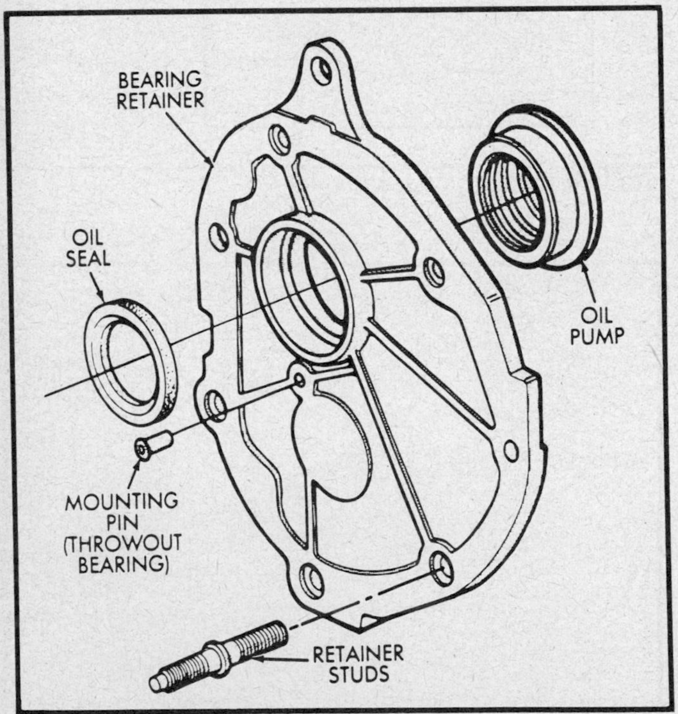
BEARING RETAINER
OIL SEAL
OIL PUMP
MOUNTING PIN (THROWOUT BEARING)
RETAINER STUDS

Bearing retainer assembly

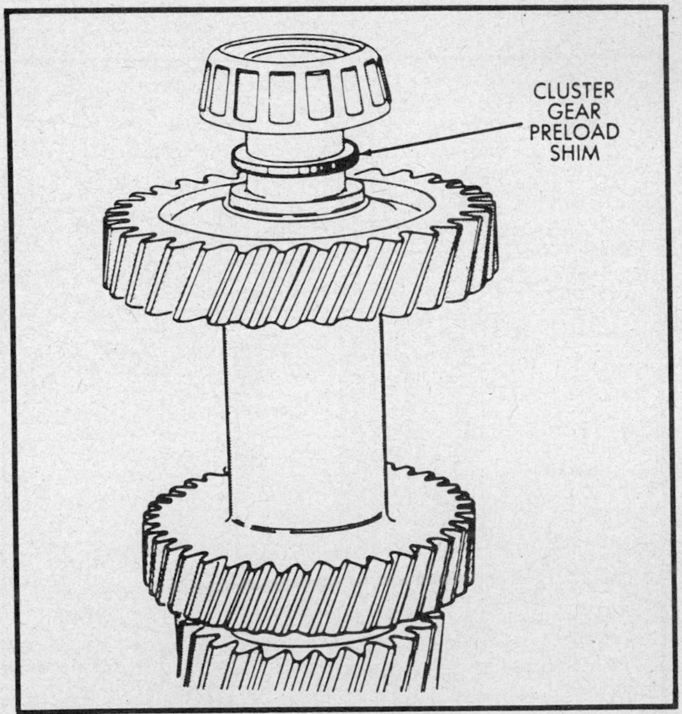
CLUSTER GEAR PRELOAD SHIM

Cluster gear preload shim installation

front case is properly supported (right case half facing up).

19. Hold the mainshaft spline stationary with a 12 point socket or dummy yoke and loosen the reverse gear nut.

20. Double nut the front case mounting studs and remove them.

21. Separate the front bearing retainer from the case and remove the preload shim. Tag the shim for assembly reference.

22. Remove the oil seal and oil pump from the retainer.

NOTE: Do not remove the throwout bearing mounting pin. If the pin or retainer are damaged, replace the entire retainer assembly.

23. Unbolt the front case halves. Lift the right case up and away from the gear sets.

24. Remove the cluster gear from the left case half.

25. With an arbor press and suitable receiver tools, press the front and rear bearings from the cluster gear shaft. Retrieve the

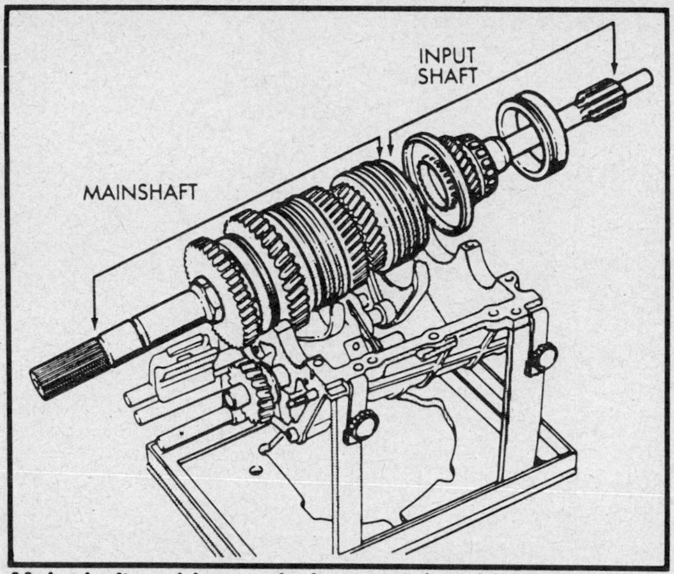
INPUT
SHAFT
MAINSHAFT

Mainshaft and input shaft removal and installation

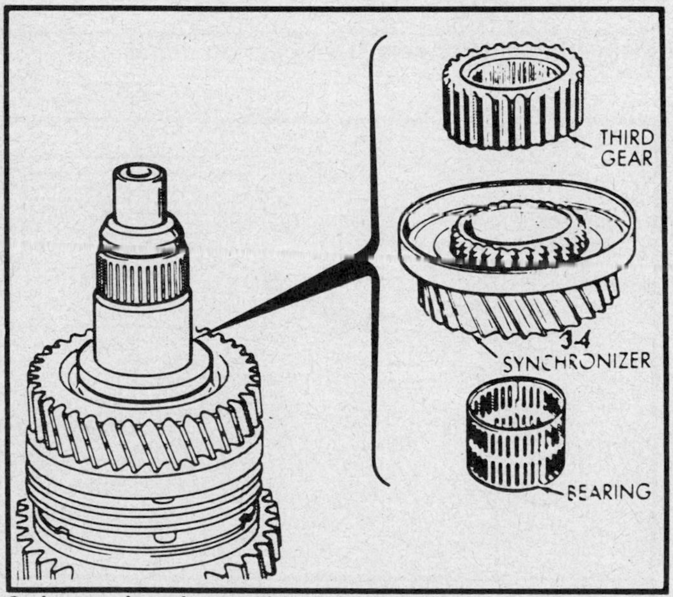
THIRD
GEAR
3-4
SYNCHRONIZER
BEARING

3rd gear, bearing and hub installation

cluster gear preload shim and tag it for assembly reference.

26. Remove the input and mainshaft assemblies from the lower case.

NOTE: If either of the front cases is being replaced, the front and rear bearings and stop ring must also be replaced.

Unit Disassembly and Assembly

MAINSHAFT AND INPUT SHAFT

Disassembly

1. Separate the mainshaft from the input shaft.
2. Using an arbor press with the proper receiving tools, press the bearing from the input shaft.
3. Remove the pilot bearing from the input shaft using the proper tool.

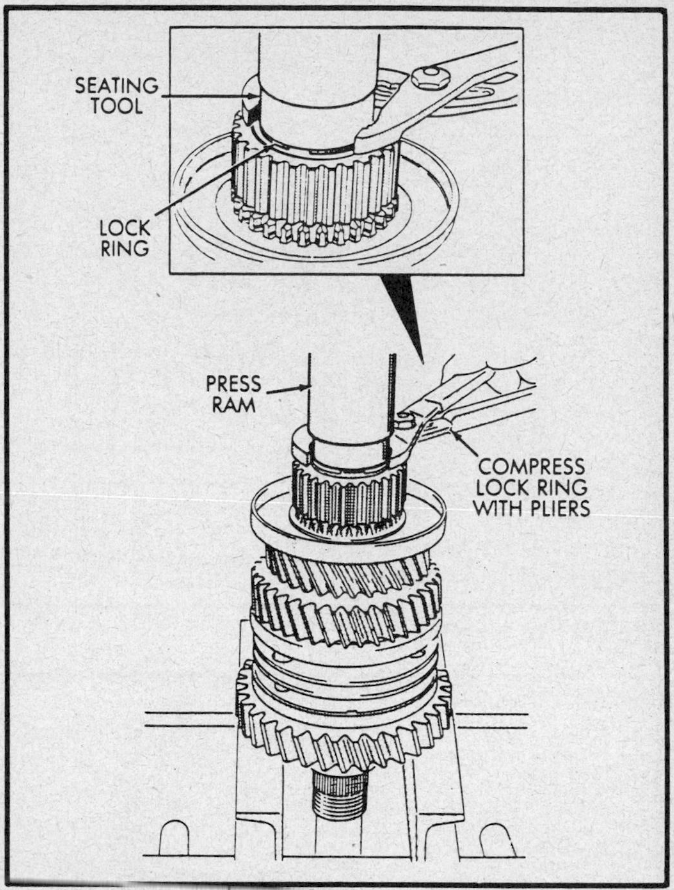
SEATING
TOOL
LOCK
RING
PRESS
RAM
COMPRESS
LOCK RING
WITH PLIERS

Seating the 3rd/4th synchronizer lock ring

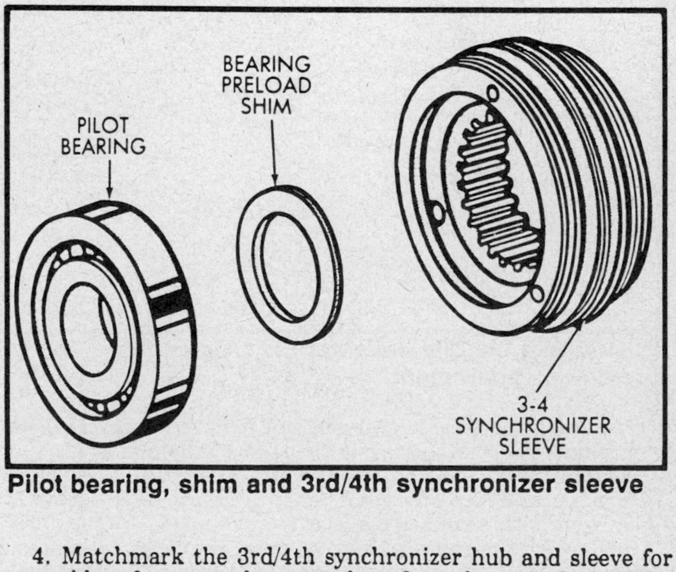
BEARING
PRELOAD
SHIM
PILOT
BEARING
3-4
SYNCHRONIZER
SLEEVE

Pilot bearing, shim and 3rd/4th synchronizer sleeve

4. Matchmark the 3rd/4th synchronizer hub and sleeve for assembly reference and remove them from the mainshaft. If the mainshaft pilot bearing and preload shim did not come off with the hub and sleeve, remove them. Tag the shim for assembly reference.
5. Remove the snapring and spring washer from the mainshaft.
6. Remove the 3rd/4th synchronizer hub, 3rd gear and needle bearing from the mainshaft.

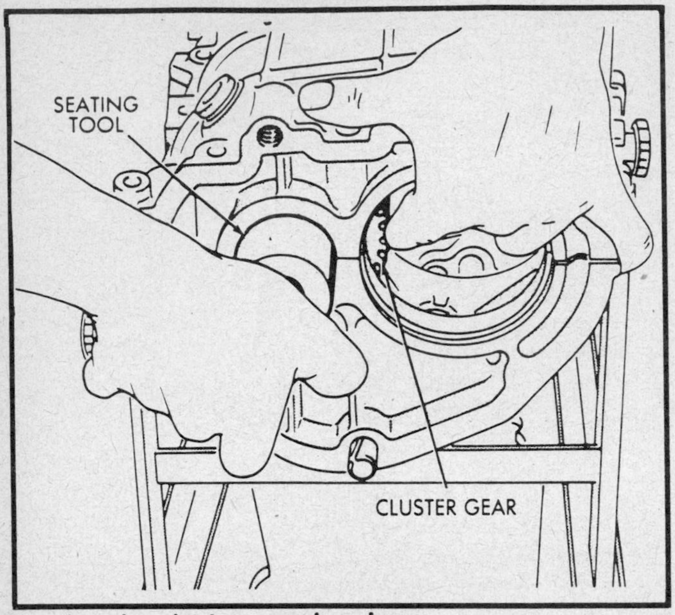
SEATING TOOL
CLUSTER GEAR

Seating the cluster gear bearings

MOVE STYLUS TO FRONT BEARING; THIS INDICATOR READING IS BEARING DEPTH

Measuring cluster gear front bearing depth

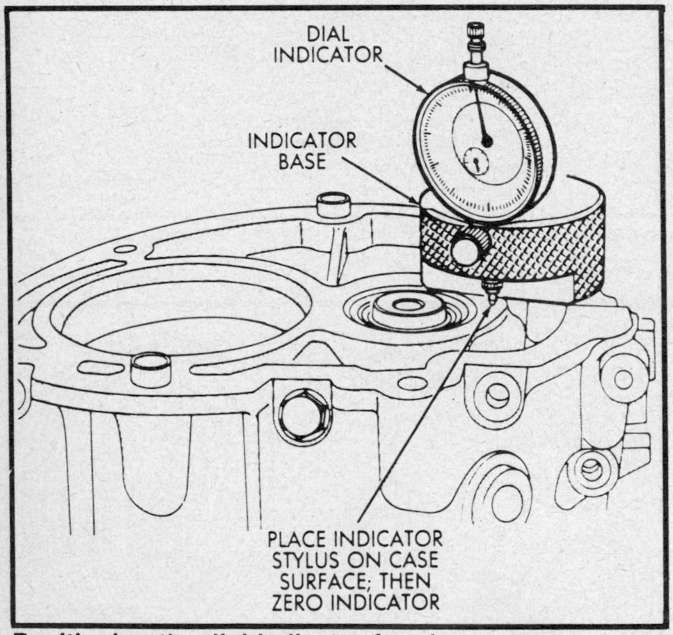
DIAL INDICATOR
INDICATOR BASE
PLACE INDICATOR STYLUS ON CASE SURFACE; THEN ZERO INDICATOR

Positioning the dial indicator for cluster gear preload measurement

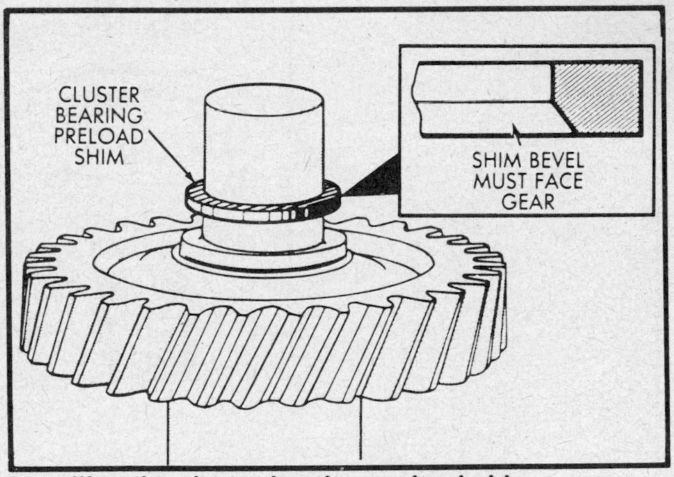
CLUSTER BEARING PRELOAD SHIM
SHIM BEVEL MUST FACE GEAR

Installing the cluster bearing preload shim

7. Loosen the reverse gear nut and remove the reverse gear.
8. Press the rear bearing ¾ the way off the mainshaft.
9. Matchmark the 1st/2nd symchronizer hub for assembly reference and tap the rear bearing completely from the mainshaft with a rubber mallet.
10. Remove the shim, spacer, 1st gear, first gear bearing and spacer, 1st/2nd synchronizer hub and sleeve, second gear and bearing.
11. Mark the relationship of the 1st/2nd synchronizer lugs for assembly reference.

Inspection

1. Clean all transmission components with solvent. Dry bearings with shop towels only.
2. Inspect the transmission gears and synchronizers for wear and damage. Replace any part showing signs of wear, chipping, cracking or any other damage. Synchronizers are serviced as an assembly. Replace any synchronizer that appears to be worn or damaged. Do not attempt to disassemble the synchronizers.
3. Check each transmission case for cracks and stripped threads, distortion or damaged bearing or sealing surfaces. Machine or replace as required.
4. If either the reverse idler gear or the shaft is worn or damaged, replace the whole assembly.

Assembly

1. Press the front bearing onto the input shaft.
2. Measure and record the thickness of the front bearing preload shim. Keep the shim with the shaft for the preload adjustment.
3. Install the second gear and bearing on the shaft with the spacer.

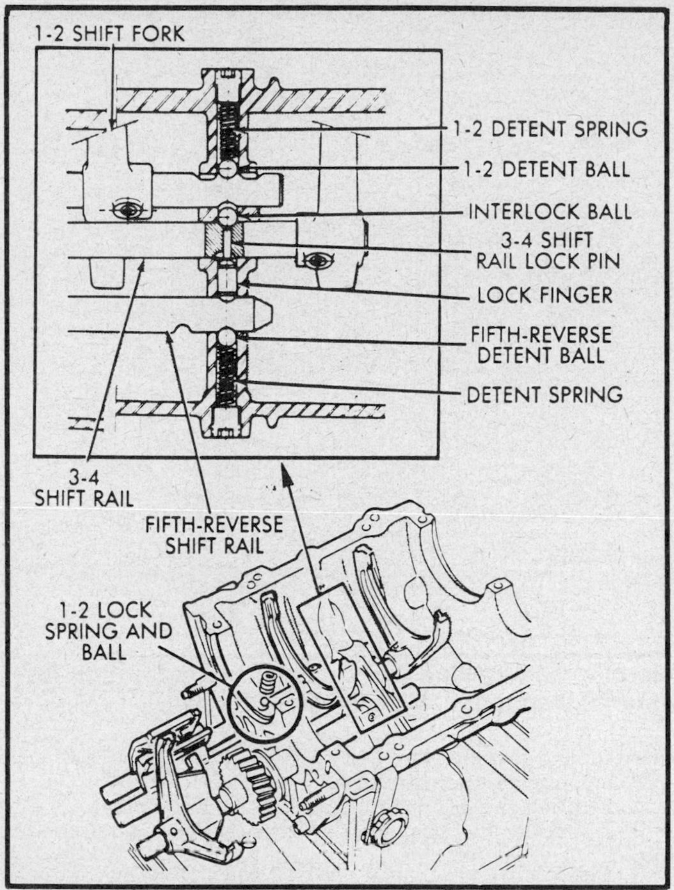
1-2 SHIFT FORK
1-2 DETENT SPRING
1-2 DETENT BALL
INTERLOCK BALL
3-4 SHIFT RAIL LOCK PIN
LOCK FINGER
FIFTH-REVERSE DETENT BALL
DETENT SPRING
3-4 SHIFT RAIL
FIFTH-REVERSE SHIFT RAIL
1-2 LOCK SPRING AND BALL
Shift rail components

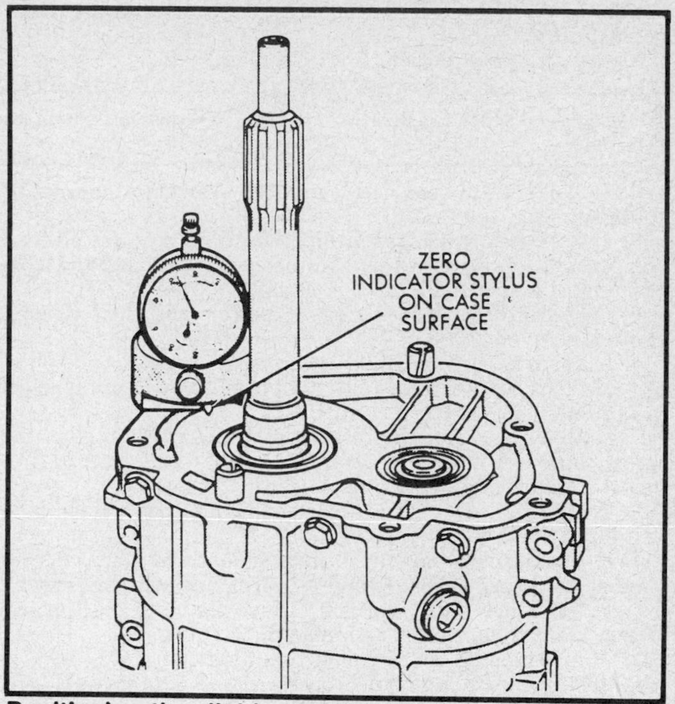
ZERO INDICATOR STYLUS ON CASE SURFACE
Positioning the dial indicator for input/mainshaft bearing preload measurement

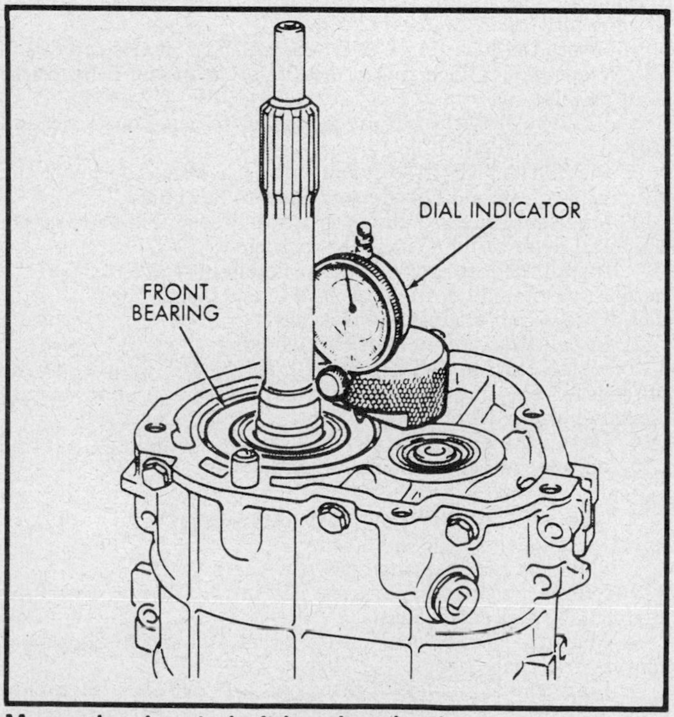
DIAL INDICATOR
FRONT BEARING
Measuring input shaft bearing depth

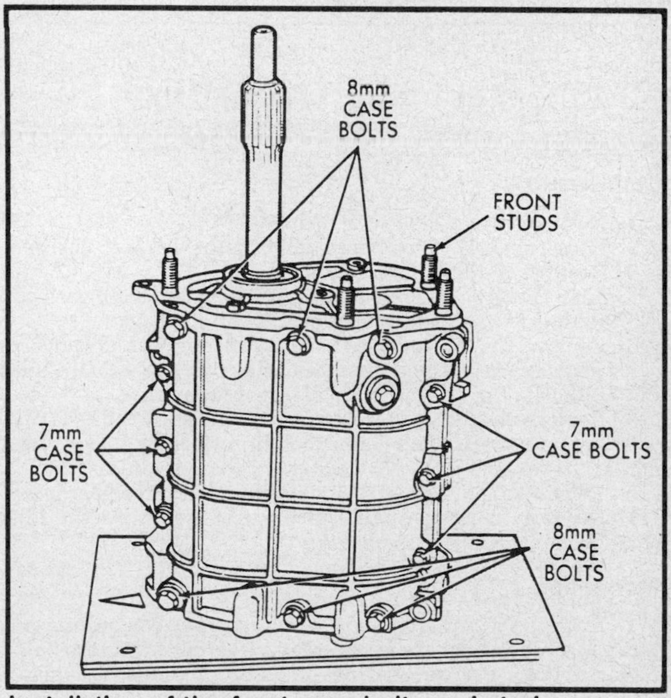
8mm CASE BOLTS
FRONT STUDS
7mm CASE BOLTS
7mm CASE BOLTS
8mm CASE BOLTS
Installation of the front case bolts and studs

4. Install the original or replacement rear bearing preload shim on the shaft by using the following guidelines:

a. Use original shim if when original transmission components (except races and bearings) are installed.

b. Use a new shim if any gear, shaft, synchronizer, shift fork/rail or transmission case was replaced. New shim(s) thickness must total 0.179 in. (4.55mm). New shims are available in 2 thicknesses: 0.063 in. (1.6mm) and 0.116 in. (2.95mm).

5. Press the rear bearing onto the mainshaft.

CAUTION

Do not exceed 6,600 lbs. of force when pressing the rear bearing onto the mainshaft.

6. Install the 3rd gear needle and 3rd gear onto the front of the mainshaft.
7. Install the 3rd/4th synchronizer hub without the sleeve.
8. Install a new 3rd/4th synchronizer spring washer and locknut onto the mainshaft.
9. Seat the 3rd/4th lock ring by performing the following procedure:
   a. Position the mainshaft on a press.
   b. Place a suitable installer tool over the end of the shaft and flush against the lock ring.
   c. Apply just enough pressure with the press to align the lock ring with groove in the mainshaft.
   d. Compress the lockring to seat it.
10. Install the original or replacement pilot bearing shim by using the following guidelines:
   a. Use the old shim only if the original third gear, synchronizer hub, spring washer and lock ring are being reused.
   b. Use a new shim that is 0.110 in. (2.80mm) thick if any gear train component was replaced.
11. Press the pilot bearing onto the mainshaft.
12. Install the 3rd/4th synchronizer sleeve onto its hub.

CAUTION

Do not install the rear bearing, reverse gear or reverse gear nut until after the cluster gear bearing preload is adjusted.

SHIFT RAIL, SHIFT FORK AND DETENT BALL

Disassembly

1. Remove the 1st/2nd lock ball and spring.
2. Move the 3rd/4th shift fork rail into the 4th gear position.
3. Remove the shift fork roll pins.
4. Slide the 3rd/4th shift rail back into the neutral position.
5. Withdraw the 1st/2nd detent plug and spring.
6. Remove the 5th/reverse detent plug, spring and ball.
7. Remove the 1st/2nd shift rail and fork and 3rd/4th shift rail and fork. Tag the parts for assembly reference.
8. Remove the lock pin from the 3rd/4th shift rail.
9. Remove the 1st/2nd detent ball, interlock ball, lock finger and 5th/reverse detent ball from the transmission case.
10. Remove the 5th/reverse shift rail and reverse idler gear.
11. Remove the idler shaft roll pin and withdraw the idler shaft from the transmission case.

Inspection

1. Inspect the shift rail bores in each case for scoring and wear. Machine or replace as necessary. If either front case half is replaced, the front/rear bearings and stop ring must be replaced also.
2. Replace the reverse gear idler shaft as an assembly if worn or damaged.
3. Inspect the shift rail forks for wear and damage. Replace any rail if worn, distorted or cracked.
4. Replace any detent balls that are chipped, cracked or worn.
5. Replace any detent springs that are distorted or broken.

Assembly

1. Lubricate all parts with SAE 75W-90 API Grade GL5 or equivalent gear lube before assembly. Petroleum jelly may be

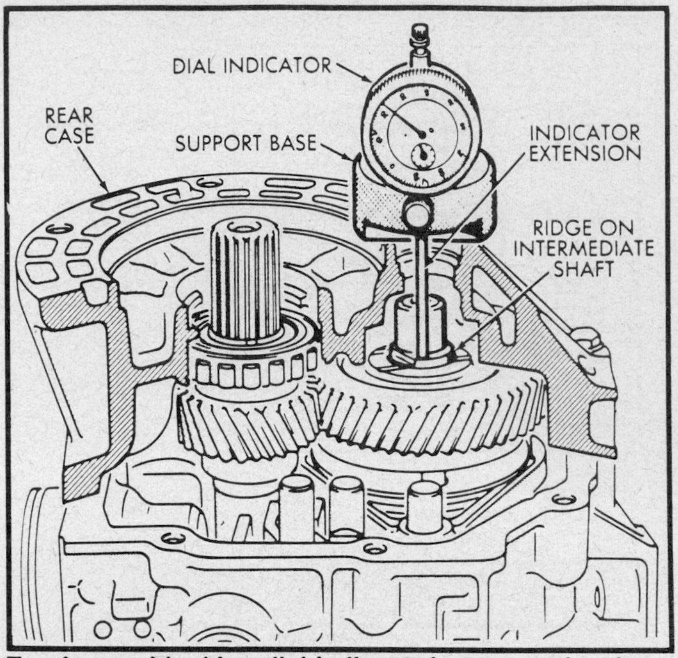
DIAL INDICATOR
REAR CASE
SUPPORT BASE
INDICATOR EXTENSION
RIDGE ON INTERMEDIATE SHAFT

Zeroing and locking dial indicator in preparation for intermediate shaft end play measurement

used to hold parts in place during assembly. Make sure all shift rail components are installed properly.

2. Position the left hand case in the holding fixture.
3. Insert the idler gear shaft into the case and install the roll pin.
4. Insert the reverse idler gear and and 5th/reverse shift rail into the transmission case.
5. Move the 5th/reverse sift rail to the neutral position.
6. Install the 5th/reverse detent ball and spring. Coat the threads of the detent plug with thread sealant and install. Torque plug to 10 ft. lbs. (13 Nm).
7. Insert the lock finger in the detent bore at the top of the case. Depress the lock finger until it seats in the 5th/reverse shift rail detent.
8. Coat the 3rd/4th shift rail lock pin with petroleum jelly and install it.
9. Install the 3rd/4th shift rail with fork. Move the rail to the neutral position. Do not allow the pin to fall out.
10. Make sure the lock finger is properly seated in the 3rd/4th shift rail detent and against the lock pin.
11. Install the interlock ball in the detent bore at the top of the case. Push the ball downward until it seats in the 3rd/4th shift rail dentent and against the lock pin.
12. Install the 1st/2nd shift fork and rail.
13. Install the 1st/2nd detent ball, spring and plug. Coat the threads of the detent plug with thread sealant and install. Torque plug to 10 ft. lbs. (13 Nm).
14. Install the shift fork roll pins.
15. Install the 1st/2nd lock ball and spring.

Transmission Assembly

1. Slide the rear bearing onto the cluster gear.
2. Install the cluster gear front bearing but do not install the preload shim at this time.
3. Place the cluster gear into the case and install alignment dowels into each corner.
4. Place the right half of the front case over the alignment dowels. Install the front case attaching bolts and make them finger tight.

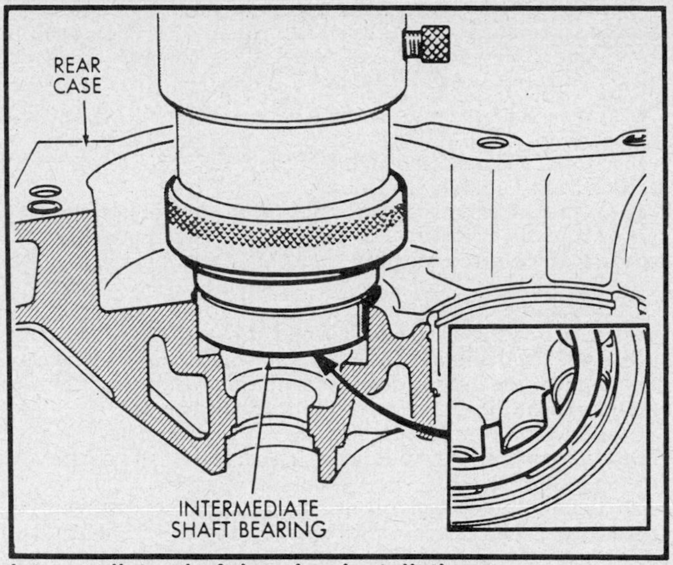
REAR CASE
INTERMEDIATE SHAFT BEARING

Intermediate shaft bearing installation

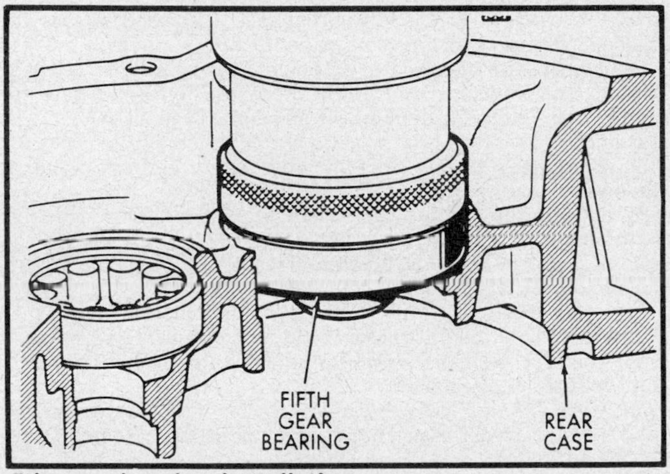
FIFTH GEAR BEARING
REAR CASE

5th gear bearing installation

5. Place the front case in a verticle position on the holding fixture so that the front bearing is facing up.
6. Insert a seating tool through the cluster gear opening and push the cluster gear set rearward. Maintain pressure and rotate the gear set to seat the bearing.
7. Torque the case center bolts to 4 ft. lbs. (5 Nm).
8. Turn the cluster by hand and make sure that the assembly rotates freely. If the gear does not rotate freely, loosen then tighten the center case bolts while rotating the gear.
9. Mount special setting gauge 80314G or equivalent with support base 80310Fz on the front case next to the cluster front bearing. Place the dial stylus on the case surface next to the bearing and zero the indicator.
10. Move the stylus to the surface of the cluster front bearing and record the indicator reading. The reading represents the front bearing depth.
11. To calculate the required preload shim thickness, add 0.004 in. (0.10 mm) to the dial indicator reading. For example, if the idicator reading was 0.104 in. (2.64 mm) the required shim thickness would be 0.108 in. (2.74 mm). Allowable shim tolerance is plus or minus 0.002 in. (0.05 mm).
12. Remove the right hand front case half and remove the cluster gear.

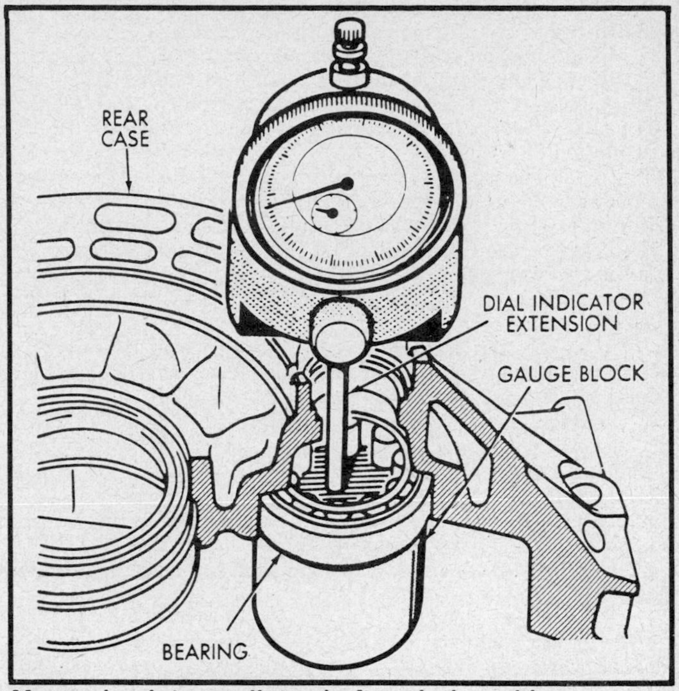
REAR CASE
DIAL INDICATOR EXTENSION
GAUGE BLOCK
BEARING

Measuring intermediate shaft end play with intermediate shaft bearing positioned on gauge block

13. Install the proper thickness preload shim on the end of the cluster gear. Install the shim so that the beveled side faces the gear.
14. Reinstall the cluster front bearing but do not install the cluster gear at this time.
15. Install the rear bearing, reverse gear and new nut onto the mainshaft. Hold the mainshaft stationary and torque the nut to 40 ft. lbs. (55 Nm). Stake the nut in place at the 2 small notches in the mainshaft.
16. Install the mainshaft and input shaft into the front case. Make sure that the shift forks are properly engaged with the synchronizer sleeves. Make sure the bearing snaprings are properly seated in the case.
17. Install the cluster gear into the case. If alignment dowels were not previously installed, install them at this time.
18. Apply sealing compound to front case mating surfaces and to the threads of the case bolts.
19. Install the right case half onto the left case half. Install the 0.313 in. (8mm) bolts only. Leave the 0.25 in. (7mm) bolts out at this time.
20. Position the front case in the verticle position on the stand with the input shaft facing up.
21. Check the runout of the input shaft bearing with a dial indicator. Position the stylus on the indicator on the bearing surface and zero it. Take 4 readings at 90 degree intervals. If the runout exceeds 0.001 in. (0.03 mm), reseat the bearing with the proper tool until the runout is within specification.
22. Position the dial indicator stylus on the case surface next to the input shaft bearing.
23. Move the stylus to the input shaft bearing surface and record the indicator reading. The reading represents the bearing depth.
24. To calculate the required preload shim thickness, add 0.004 in. (0.10 mm) to the dial indicator reading. For example, if the indicator reading was 0.085 in. (2.16 mm) the required shim thickness would be 0.088 in. (2.26 mm). Allowable shim tolerance is plus or minus 0.002 in. (0.05 mm).
25. Install the required preload shim on the input shaft. Use

petroleum jelly to hold the shim in place.
26. Install alignment dowels in the front case.
27. Install new oil seal and oil pump into the bearing retainer. Coat the bearing retainer mating surface with silicone sealant.
28. Align and install the bearing retainer on the case making sure that the preload shim is not disturbed during the installation.
29. Install the front case mounting studs. Torque the studs to 16 ft. lbs. (22 Nm) alternately and evenly. Rotate the mainshaft while tightening the studs to initially seat shaft bearings.
30. Perform the final seating of the case halves and gear assemblies as follows:
a. Loosen, but do not remove, the $^5/_{16}$ in. 8mm bolts installed previously.
b. Lightly tap the front case halves several times with a rubber mallet to help seat the case halves. Rotate the mainshaft several times to seat the bearings.
c. Retorque the 0.313 in. (8mm) bolts to 15 ft. lbs. (20 Nm).
d. Coat the threads of the 0.25 in. (7mm) bolts with silicone sealant. Install and torque the bolts to 11 ft. lbs. (15 Nm).
31. Install the intermediate case alignment dowels.
32. Reposition the transmission in the holding fixture so that the shift rail ends are facing up.
33. Carefully move the 5th/reverse shift rail into the 5th gear position.

CAUTION

Do not move the 5th/reverse shift rail past the 5th gear position. The 5th/reverse detent ball with disengage if the rail is moved too far. If the ball disengages, complete disassembly of the transmission will be necessary to reinstall the ball.

34. Coat the front and intermediate case mating surfaces with silicone sealant.
35. Align and install the intermediate case onto the front case. Install and tighten the intermediate case attaching bolts to 13 ft. lbs. (18 Nm).
36. Install the shift plate into the intermediate case.
37. Install the intermediate shaft and 5th/reverse synchronizer hub by aligning the reference marks. Make sure the shaft is fully seated on the cluster gear.
38. Assemble the 5th gear shift fork and synchronizer.
39. Start the 5th/reverse fork on the shift rail and slide the fork and synchronizer downward onto the intermediate shaft and synchronizer hub. Align the previously made reference marks.
40. Secure the shift fork to the shift rail with the roll pin using a suitable support tool to hold the fork stationary.
41. Install the 5th intermediate gear with needle bearing onto the intermediate shaft. Do not install the thrust washer, preload shim or bearing race at this time.
42. Install the 5th gear stop ring and press the bearing onto the 5th gear. Coat the bore of the gear and bearing rollers with transmission lubricant.
43. Install the 5th gear lock ring and make sure that it is fully seated in the groove of the shaft.

NOTE: Do not install the 5th gear bearing race or intermediate shaft or roller bearing in the rear case at this time. They will be installed after the intermediate shaft endplay is adjusted.

44. Clean the rear and intermediate case mounting surfaces thoroughly.
45. Mount the rear case on top of the intermediate case with 1 or 2 bolts just to hold it in place.
46. Install an extension onto the end of the dial indicator stylus. Mount the indiactor and extension on a support base and position the entire assembly over the access hole in the rear case.
47. Insert the dial indicator extension through the access hole until it contacts the ridge on the end of the intermediate shaft.
48. Zero the dial indicator. Lock the pointer at zero and carefully remove the indicator. Set the indicator aside until the next measurement. Do not disturb the setting.
49. Remove the rear case and install the intermediate shaft bearing in the case. Install the 5th gear bearing and secure with the snapring.
50. Install the rear case oil seal using the proper installer tool.
51. Invert the rear case and position the intermediate shaft bearing on the gauge block. The bearing must sit squarely on the block.
52. Position the previously locked and zeroed dial indicator over the access plug hole in the rear case. If the setting was disturbed, the indicator must be re-calibrated.
53. Insert the dial indicator extension through the access hole. Unlock the indicator and allow the extension to contact the gauge block. Note the dial indicator reading then lock the indicator and remove it.
54. Select the proper intermediate shaft endplay shim as follows:
a. Note the dial indicator reading that was obtained from the gauge block.
b. Subtract 0.002 in. (0.05 mm) from the gauge block reading to calculate the proper endplay.
c. For example, if the reading was 0.062 in. (1.57 mm), then the correct shim thickness would be 0.060 in. (1.52 mm).
55. Install the thrust washer, endplay shim and bearing race on the intermediate shaft.
56. Coat the rear case mating surfaces with silicone sealant and install the rear case. Install and torque the case bolts to 11 ft. lbs. (15 Nm).
57. Tap the rear-to-intermediate case alignment dowels into place.
58. Coat the edges of the access intermediate shaft access plug with silicone sealant and install the plug into the rear case.
59. Remove the transmission from the holding fixture and install in vehicle.
60. Coat the drain and fill plugs with silicone sealant. Install the drain plug and tighten to 20 ft. lbs. (28 Nm).
61. Install the gearshift lever components by reversing the removal procedure.
62. Fill the transmission to the proper lever.
63. Install and tighten the fill plug to 20 ft. lbs. (28 Nm).

SPECIFICATIONS

TORQUE SPECIFICATIONS

Item	ft. lbs.	Nm
Access cover bolts	21	28
Back-up lamp switch	15	20
Center support to case	40	54
Extension housing to case bolts	40	54
Gearshift lever	30	41
Input shaft bearing retaining bolts	21	28
Shifter assembly to extension bolts	21	28
Speedometer drive clamp screw	8	11
Transmission to clutch housing bolts	50	68
Transmission drain plug	30	41
Transmission fill plug	30	41

Section 7

Overdrive 4 Transmission
Chrysler Corp.

APPLICATION

OVERDRIVE 4

Year	Vehicle	Engine	VIN Code
1984	D-100	3.7L, 5.2L, 5.9L	H, T, W
	D-150	3.7L, 5.2L, 5.9L	H, T, W
	B-150	3.7L, 5.2L	H, T
	B-250	3.7L, 5.2L, 5.9L	H, T, V
1985	D-100	3.7L, 5.2L, 5.9L	H, T, W
	D-150	3.7L, 5.2L, 5.9L	H, T, W
	B-150	3.7L, 5.2L	H, T
	B-250	3.7L, 5.2L, 5.9L	H, T, V
1986	D-100	3.7L, 5.2L, 5.9L	H, T, W
	D-150	3.7L, 5.2L, 5.9L	H, T, W
	B-150	3.7L, 5.2L	H, T
	B-250	3.7L, 5.2L, 5.9L	H, T, V
1987	D-150	3.7L, 5.2L, 5.9L	H, T, W
	B-150	3.7L, 5.2L	H, T
	B-250	3.7L, 5.2L, 5.9L	H, T, V
1988	D-100	3.9L, 5.2L, 5.9L	X, Y, W
	D-150	3.9L, 5.2L, 5.9L	X, Y, W
	D-250	3.9L, 5.2L, 5.9L	X, Y, W
	D-350	5.9L	W
	W-100	5.2L, 5.9L	Y, W
	W-150	5.2L, 5.9L	Y, W
	W-250	5.2L, 5.9L	Y, W
	W-350	5.9L	W
	Ramcharger (4WD)	5.2L, 5.9L	Y, W
	B-150	3.9L, 5.2L, 5.9L	X, Y, W
	B-250	3.9L, 5.2L, 5.9L	X, Y, W
1989	D-150	5.2L, 5.9L	Y, W
	Ramcharger (D-100)	5.2L	Y
	Ramcharger (D-100)	5.2L, 5.9L	Y, W

GENERAL DESCRIPTION

The overdrive-4 transmission is a 4 speed unit with all forward gears synchronized. Third gear is direct, while the fourth gear is the overdrive ratio.

Transmission Identification

An identification tag is affixed to the right side of the transmission. This tag provides model and vehicle identification.

Metric Fasteners

The metric fastener dimensions are very close to the dimensions of the familiar inch system fasteners. For this reason, replacement fasteners must have the same measurement and strength as those removed.

CAUTION

Do not attempt to interchange metric fasteners for inch system fasteners. Mismatched or incorrect fasteners can result in damage to the transmission unit through malfunctions, breakage or possible personal injury.

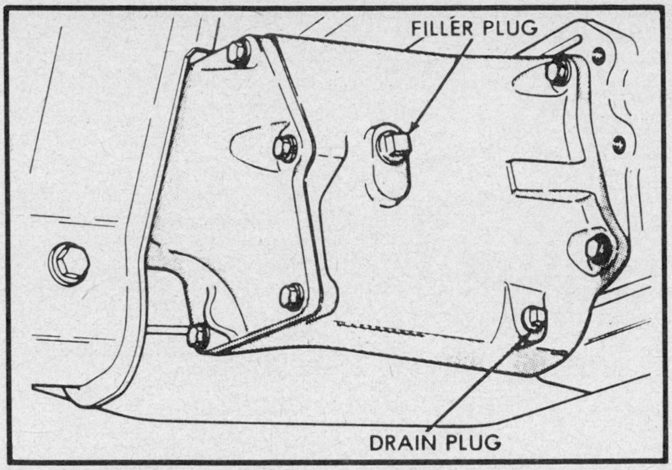
FILLER PLUG
DRAIN PLUG
Drain and filler locations

Care should be taken to re-use the fasteners in the same locations as removed.

Common metric fastener strength property classes are 9.8 and 12.9 with class identification embossed on the head of each bolt. The inch strength classes range from grade 2–8 with the line identification embossed on each bolt head. Markings correspond to 2 lines less than actual grade (for example grade 8 bolt will exhibit 6 embossed lines on the bolt head). Some metric nuts will be marked with a single digit strength identification numbers on the nut face.

Capacities

The fluid capacity of the unit is 3.5 qts. The transmission uses Dexron® II ATF or equivalent.

Checking Fluid Level

Check the fluid level in the transmission by removing the fill plug located under the identification tag. The fluid level should be at the bottom of the fill opening.

TROUBLE DIAGNOSIS

CHILTON'S THREE C'S TRANSMISSION DIAGNOSIS

Condition	Cause	Correction
Hard shifting	a) Synchronizer clutch sleeve damaged	a) Disassemble transmission and repair or replace defective parts, as required
	b) Synchronizer spring improperly installed	b) Disassemble transmission and repair or replace defective parts, as required
	c) Broken or worn synchronizer stop rings	c) Disassemble transmission and repair or replace defective parts, as required
Transmission slips out of gear	a) Linkage interference	a) Inspect and remove all linkage interferences
	b) Gearshift rods out of adjustment	b) Adjust gearshift rods
	c) Synchronizer clutch teeth worn	c) Disassemble transmission and replace parts as necessary
	d) Clutch housing bore or face out of alignment	d)
Transmission noises	a) Excessive end play in countershaft gear	a) Replace thrust washers
	b) Damaged, broken or excessively worn gear teeth	b) Replace worn gears
	c) Rough or pitted bearing races or balls	c) Replace worn bearing

ON CAR SERVICE

Adjustments

SHIFT LINKAGE

1. Raise the vehicle and support safely.
2. Fabricate a crossover alignment T-type tool from a ¼ in. diameter metal rod that is 2¼ in. in length. Insert the tool to hold the levers in the neutral-crossover position.
3. Disconnect all shift rods from the adjustable swivels on shift levers.
4. Place all shift levers in the **N** position.
5. Starting with the 1–2 shift rod, rotate the threaded shift rods to make the length align exactly with each lever. Remove the clips and washers (if installed) at the shifter ends as required to rotate the rods.
6. Install the rod end clips and washers if they were removed.
7. Remove the alignment tool and lower the vehicle.
8. Check the shifting for proper operation.

Services

SHIFT LINKAGE

Removal and Installation

1. Disconnect the negative battery cable.
2. Remove the floor pan boot retaining screws and slide the boot up and off the hand lever.
3. Remove the hand lever.
4. Remove the clips, washers and control rods from the shift levers.

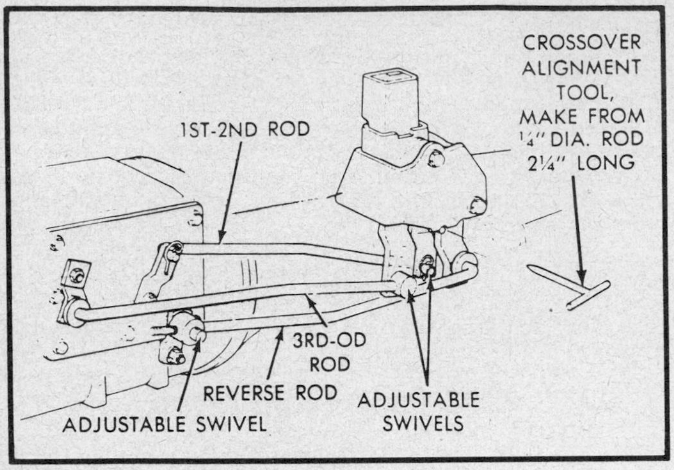
CROSSOVER
ALIGNMENT
TOOL,
MAKE FROM
¼" DIA. ROD
2¼" LONG
1ST-2ND ROD
3RD-OD
ROD
REVERSE ROD
ADJUSTABLE
SWIVELS
ADJUSTABLE SWIVEL
Adjusting gearshift linkage

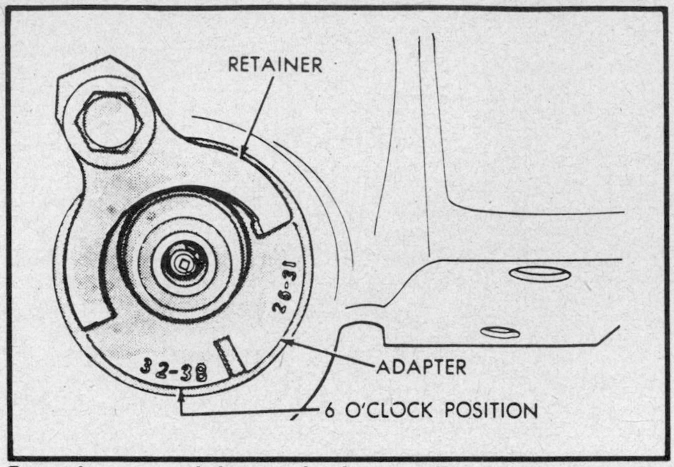
RETAINER
26-31
32-38
ADAPTER
6 O'CLOCK POSITION
Speedometer pinion and adapter alignment

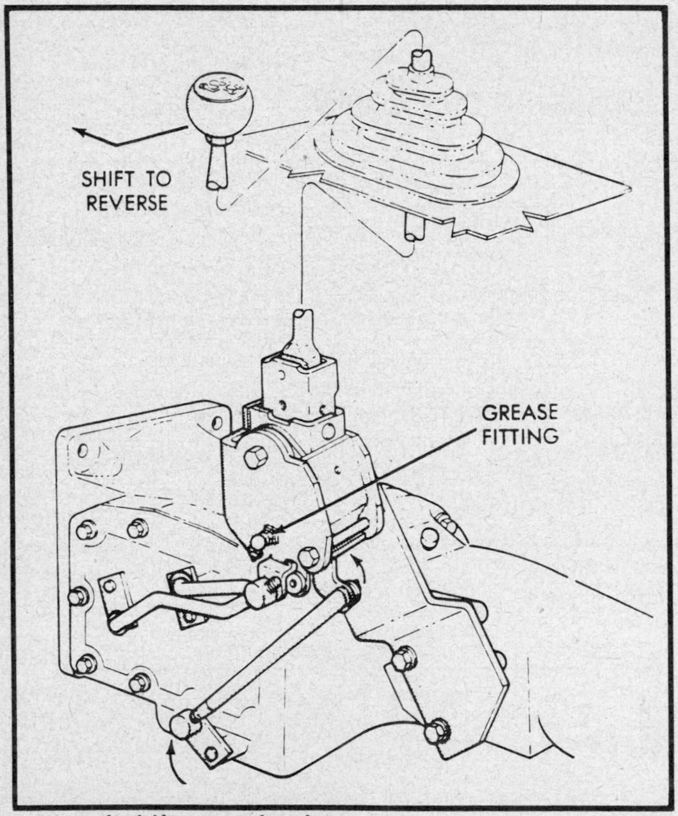
SHIFT TO
REVERSE
GREASE
FITTING
4 speed shifter mechanism

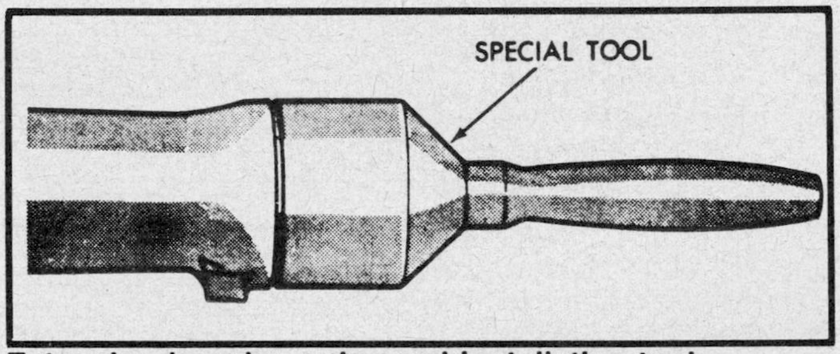
SPECIAL TOOL
Extension housing yoke seal installation tool

5. Remove the 2 bolts and washers that attach the shift unit to the mounting plate on the extension housing. Remove the unit.
6. To install, fasten the unit to the mounting plate and torque the bolts to 30 ft. lbs. (50 Nm).
7. Install the shift rods, washers and clips.
8. Install the hand lever. The hand lever clearance on the driver's side should be 0.010 in.
9. Slide the boot over the handle and attach it to the floor pan.
10. Connect the negative battery cable.
11. Adjust the linkage as required.

SPEEDOMETER PINION GEAR

Removal and Installation

NOTE: The rear axle gear ratio and the tire size determine the size of the pinion gear.

1. Remove the bolt and retainer attaching the speedometer pinion to the extension housing.
2. If transmission fluid is found in the cable housing, replace the oil seal as follows: Remove the lock ring and pry the old seal from the adapter bore. Start the seal and the lock ring into the adapter by hand then seat the seal and lock ring with special adpater tool C–4004 or equivalent.
3. Count the number of teeth on the pinion gear.

NOTE: Before the adapter and pinion are installed, the adapter flange and the extension housing mating surafaces must be cleaned thoroughly. Dirt or sand will cause mis-alignment of the gear and cause damage.

4. Rotate the speedometer pinion gear and adapter assembly so that the identification number on the adapter, corresponding to the number of teeth on the gear, is in the 6 o'clock position as the assembly is installed.
5. Install the retainer so that the tangs are aligned with the adapter positioning slots. Install the bolt and tap the adapter firmly into the extension housing. Torque the retainer bolt to 100 inch lbs.
6. Connect the speedometer cable to the adapter.
7. Fill the transmission to the proper level.

EXTENSION HOUSING YOKE SEAL

Removal and Installation

1. Raise and support the vehicle safely.

2. Position a drain pan under the yoke seal.
3. Matchmark the driveshaft with the rear U-joint. Disconnect the propeller shaft from the U-joint and support properly from a convenient location.
4. Withdraw the yoke shaft from the extension housing.
5. Using special seal removal tool C-3985 or equivalant, remove the yoke oil seal.
6. To install the new seal, position the seal in the extension housing opening and drive the seal into place using special tool C–3972 or equivalent.
7. Carefully guide the front U-joint yoke into the extension housing and engage with the mainshaft splines.
8. Align the matchmarks and connect the driveshaft to the rear axle pinion yoke shaft.
9. Lower the vehicle.
10. Fill the transmission to the proper level.

REMOVAL AND INSTALLATION

TRANSMISION REMOVAL

1. Remove the gearshift components. Raise the vehicle and support safely.
2. Drain the transmission fluid.
3. Matchmark the driveshaft with the rear U-joint. Disconnect the driveshaft from the U-joint and support properly from a convenient place. Carefully withdraw the yoke shaft from the extension housing.

CAUTION

Be careful not to scratch or nick the ground surface on the sliding spline yoke during the removal of the yoke shaft.

4. Disconnect the speedometer cable and back-up light switch leads.
5. Install engine support fixture tool C–3487–A or equivalent. Make sure the support tool ends are flush against the underside of the oil pan flange.
6. Raise the engine slightly with the support fixture. Disconnect the extension housing from the center crossmember.
7. Support the transmission safely with a transmission jack and unbolt the center crossmember from the frame.
8. Unbolt the transmission from the clutch housing.
9. Before lowering the transmission, slide the unit toward the rear until the drive pinion shaft clears the clutch disc.
10. Lower the transmission and remove it from under the vehicle.
11. Mount the transmission on a suitable holding fixture.

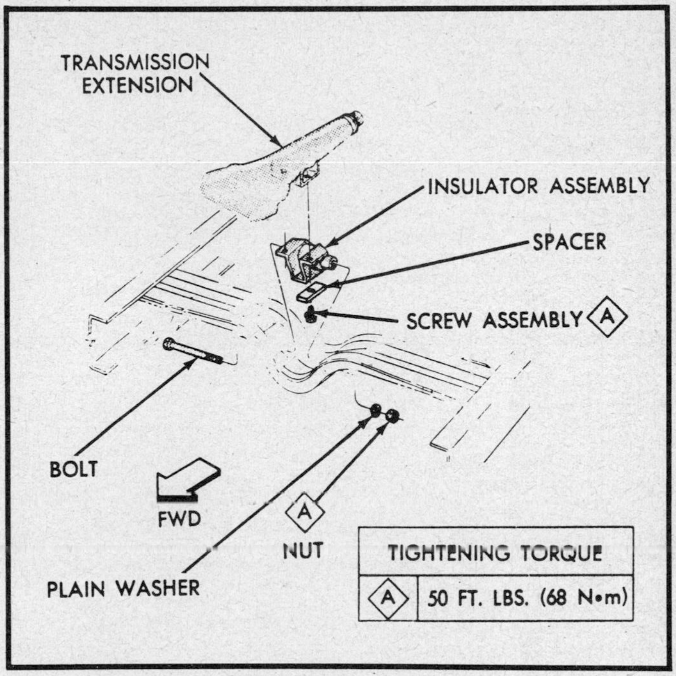
TRANSMISSION EXTENSION
INSULATOR ASSEMBLY
SPACER
SCREW ASSEMBLY (A)
BOLT
FWD
(A)
NUT
PLAIN WASHER
TIGHTENING TORQUE
(A) 50 FT. LBS. (68 N•m)

Extension and crossmember mounting

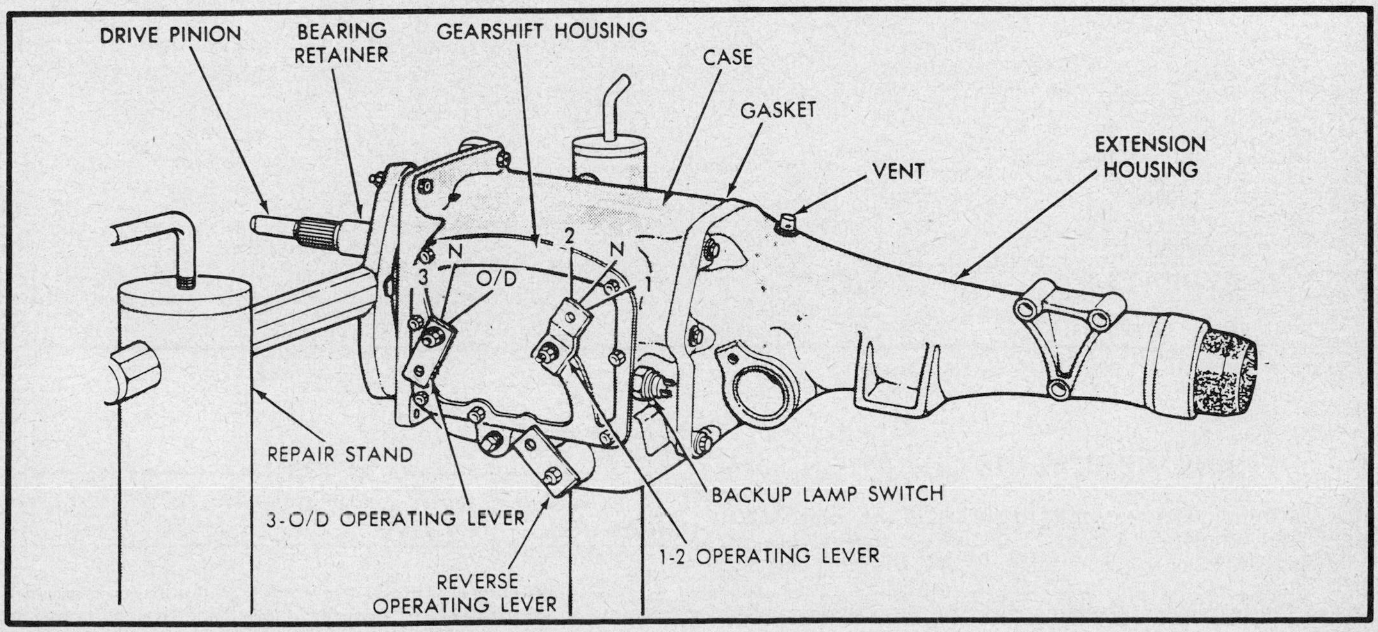
DRIVE PINION
BEARING RETAINER
GEARSHIFT HOUSING
CASE
GASKET
VENT
EXTENSION HOUSING
N
2
N
3
O/D
1
REPAIR STAND
BACKUP LAMP SWITCH
3-O/D OPERATING LEVER
1-2 OPERATING LEVER
REVERSE OPERATING LEVER

Transmission mounted in holding fixture

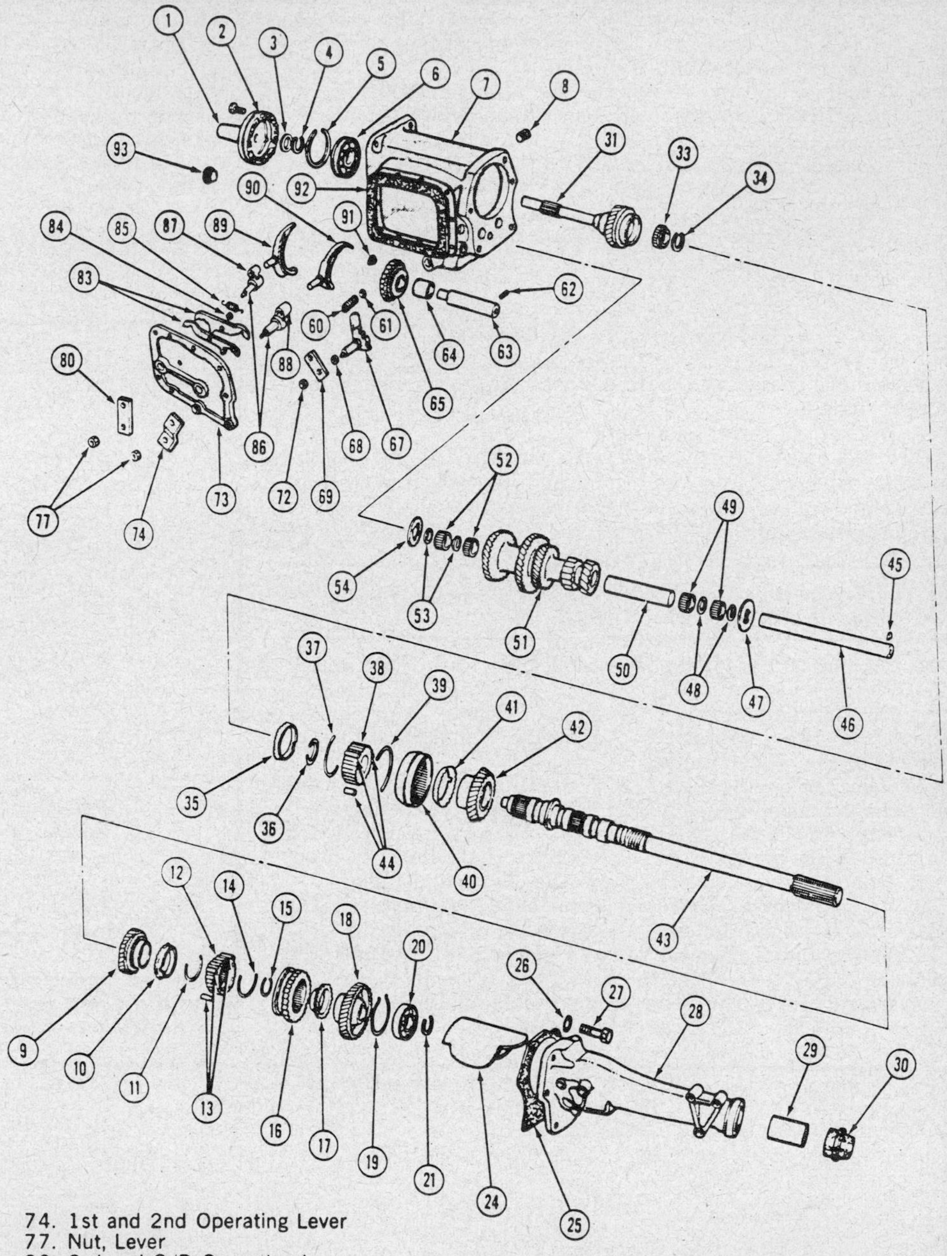

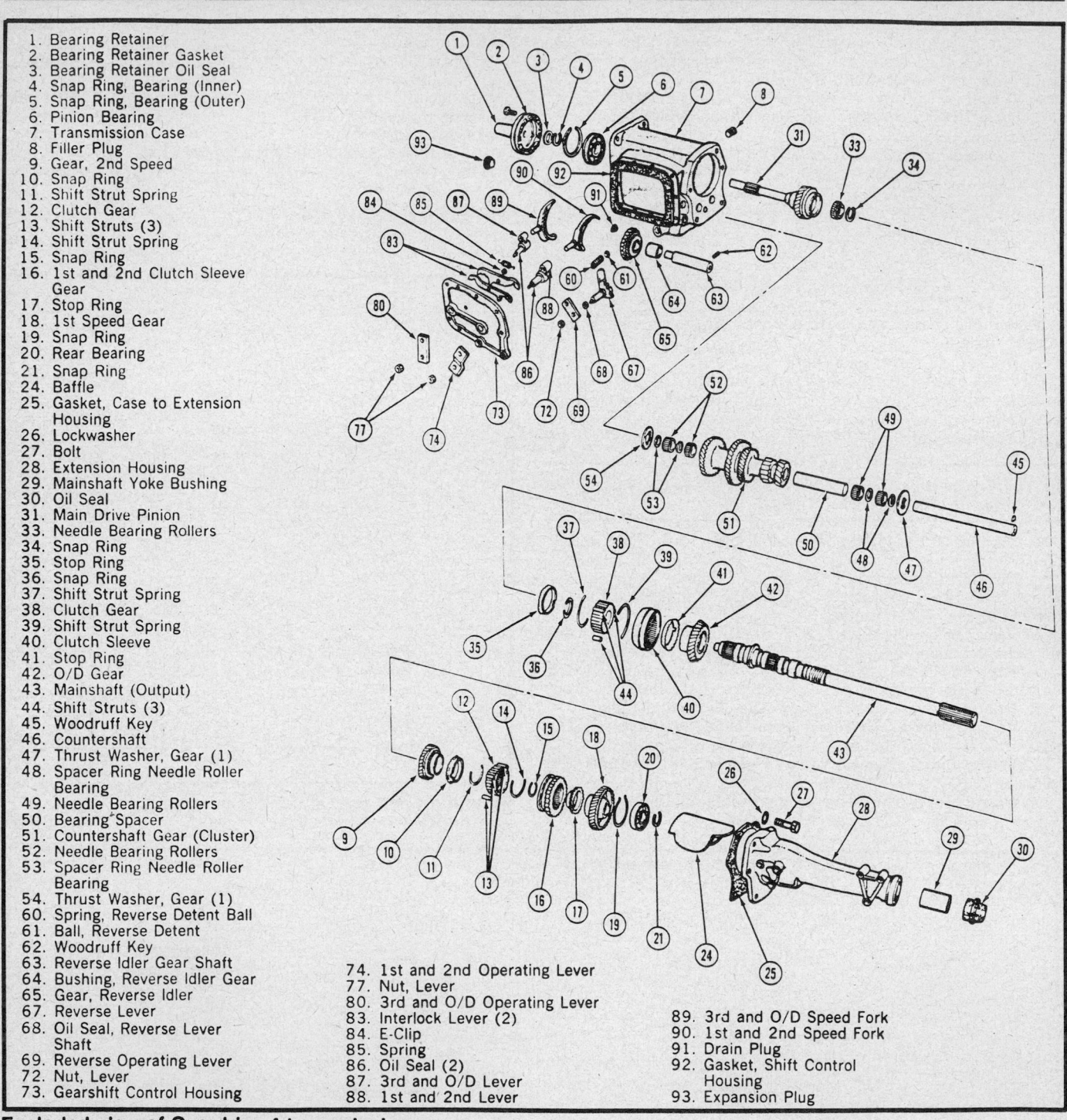

1. Bearing Retainer
2. Bearing Retainer Gasket
3. Bearing Retainer Oil Seal
4. Snap Ring, Bearing (Inner)
5. Snap Ring, Bearing (Outer)
6. Pinion Bearing
7. Transmission Case
8. Filler Plug
9. Gear, 2nd Speed
10. Snap Ring
11. Shift Strut Spring
12. Clutch Gear
13. Shift Struts (3)
14. Shift Strut Spring
15. Snap Ring
16. 1st and 2nd Clutch Sleeve Gear
17. Stop Ring
18. 1st Speed Gear
19. Snap Ring
20. Rear Bearing
21. Snap Ring
24. Baffle
25. Gasket, Case to Extension Housing
26. Lockwasher
27. Bolt
28. Extension Housing
29. Mainshaft Yoke Bushing
30. Oil Seal
31. Main Drive Pinion
33. Needle Bearing Rollers
34. Snap Ring
35. Stop Ring
36. Snap Ring
37. Shift Strut Spring
38. Clutch Gear
39. Shift Strut Spring
40. Clutch Sleeve
41. Stop Ring
42. O/D Gear
43. Mainshaft (Output)
44. Shift Struts (3)
45. Woodruff Key
46. Countershaft
47. Thrust Washer, Gear (1)
48. Spacer Ring Needle Roller Bearing
49. Needle Bearing Rollers
50. Bearing Spacer
51. Countershaft Gear (Cluster)
52. Needle Bearing Rollers
53. Spacer Ring Needle Roller Bearing
54. Thrust Washer, Gear (1)
60. Spring, Reverse Detent Ball
61. Ball, Reverse Detent
62. Woodruff Key
63. Reverse Idler Gear Shaft
64. Bushing, Reverse Idler Gear
65. Gear, Reverse Idler
67. Reverse Lever
68. Oil Seal, Reverse Lever Shaft
69. Reverse Operating Lever
72. Nut, Lever
73. Gearshift Control Housing
74. 1st and 2nd Operating Lever
77. Nut, Lever
80. 3rd and O/D Operating Lever
83. Interlock Lever (2)
84. E-Clip
85. Spring
86. Oil Seal (2)
87. 3rd and O/D Lever
88. 1st and 2nd Lever
89. 3rd and O/D Speed Fork
90. 1st and 2nd Speed Fork
91. Drain Plug
92. Gasket, Shift Control Housing
93. Expansion Plug

Exploded view of Overdrive-4 transmission

TRANSMISSION INSTALLATION

The transmission pilot bushing in the end of the crankshaft requires high temperature grease. Use multi-purpose grease, part number 4318063 or equivalent. Lightly coat the inner end of the pilot shaft bushing in the flywheel with the multi-purpose grease. Do the same for the pinion bearing retainer release bearing sleeve.

CAUTION

Do not lubricate the end of the pinion shaft, clutch disc splines or release levers.

1. Remove the transmission from the holding fixture and mount on a suitable transmission jack. Slide the transmission under the vehicle.

2. Raise the transmission until the drive pinion is centered in the clutch housing bore.
3. Roll the transmission forward slowly until the pinion shaft enters the clutch disc. Place transmission in gear and turn the output shaft until the splines are aligned then push the transmission forward until it seats against the clutch housing.

CAUTION

Do not allow the transmission to hang after the pinion has engaged the clutch disc.

4. Install the clutch housing bolts and torque them to 50 ft. lbs. (68 Nm). Remove the transmission jack.
5. Using a pointed drift pin or suitable alignment tool, raise the crossmember into position and align the bolt holes. Install the attaching bolts and torque them to 30 ft. lbs. (50 Nm).
6. Remove the engine support fixture. Bolt the extension housing to the rear mounting and center mounting. Leave the center mounting bolt and nut loose and torque the rear mounting bolts to to 50 ft. lbs. (68 Nm). Torque the center nut and bolt to 50 ft. lbs. (68 Nm).
7. Fasten the shift unit to the mounting plate and torque the 2 bolts and lockwashers to 24 ft. lbs. (40 Nm).
8. Connect the shift control rods to the shift levers.
9. Connect the back-up light switch lead and speedometer cable.
10. Carefully guide the front U-joint yoke into the extension housing and engage with the mainshaft splines. Align the matchmarks and connect the driveshaft to the rear axle pinion yoke shaft.

CAUTION

Be careful not to scratch or nick the ground surface on the sliding spline yoke during installation of the yoke shaft.

11. Fill the transmission to the proper level.
12. Lower the vehicle and install the gearshft lever and console. Make all the necessary adjustments.
13. Perform a road test to ensure that the transmission shifts smoothly and quietly.

BENCH OVERHAUL

Before Disassembly

Clean the transmission case with a suitable solvent and dry with compressed air. Inspect the case for cracks and stripped threads in the various bolt holes. Check the machined mating surfaces for burrs, nicks or any other surface imperfections that would make the transmission case unfit for service in the vehicle. The front mating surface should be smooth. Any minor burrs may be dressed with a fine mill file. Repair damaged threaded holes with an appropriate insert or by tapping and re-threading the hole.

Transmission Disassembly

1. Position the transmission on a suitable holding fixture.
2. Remove the reverse gear operating lever from the shaft.
3. Remove gearshift housing to transmission case attaching bolts.
4. With all levers in the neutral detent position, pull housing out and away from the case. If 1st/2nd or 3rd/4th shift forks remain in engagement with the synchronizer sleeves, move the sleeves and remove forks from the case.
5. If oil leakage is visible around the gearshift lever shafts or if the interlock levers are cracked, perform the following:
   a. Remove nuts, lock washers and flat washers that hold shift operating levers to the shafts. Disengage and remove the shift levers from the flats on the shafts. Make sure that the shafts are free of burrs or other imperfections before removal. If this is not done, the bores may become scored and cause leakage after reassembly.
   b. Pull the gearshift lever shafts out of the housing.
   c. Remove the O-rings and O-ring retainers from the housing. Discard the O-rings and set the retainers aside.
6. Remove the E-ring from the interlock lever pivot pin. Remove the interlock levers and spring from the housing.
7. Remove the reverse detent spring and ball from the bore in the side of the case.
8. Remove the bolt and retainer holding the speedometer pinion adapter in the extension housing, then remove the pinion adapter.
9. Remove the bolts attaching the extension housing to the transmission case.
10. Rotate the extension housing on the output shaft to expose the rear of the countershaft. This will allow sufficient clearance to install a bolt to hold the extension housing in place.

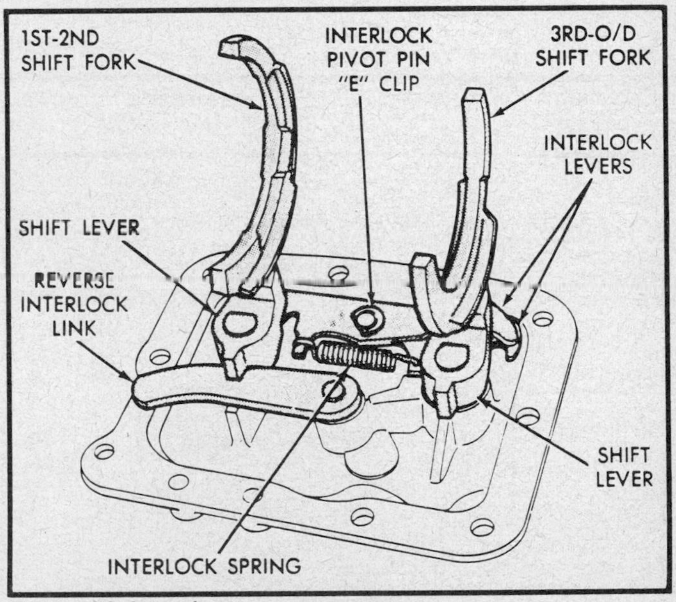
1ST-2ND SHIFT FORK
INTERLOCK PIVOT PIN "E" CLIP
3RD-O/D SHIFT FORK
INTERLOCK LEVERS
SHIFT LEVER
REVERSE INTERLOCK LINK
SHIFT LEVER
INTERLOCK SPRING

Gearshift housing components

11. Drill a hole or use a center punch to make a hole in the countershaft extension housing plug at the front of the case.
12. Reaching through this hole, push the countershaft to the rear to expose the Woodruff key, when exposed, remove it. Push the countershaft forward against the expansion plug and using a brass drift, tap the countershaft forward until the expansion plug is removed.
13. Using a countershaft arbor, push the countershaft out the rear of the case but don't let the countershaft washers fall out of position. Lower the cluster gear to the bottom of the transmission case.
14. Remove the bolt and rotate the extension back to the normal position.
15. Remove the drive pinion attaching bolts and slide the retainer and gasket from the pinion shaft, then pry the pinion or seal from the retainer. To avoid leakage during reassembly, don't nick or scratch the seal bore in the retainer or the seal seating surface.
16. Using a brass drift, tap the pinion and bearing assembly

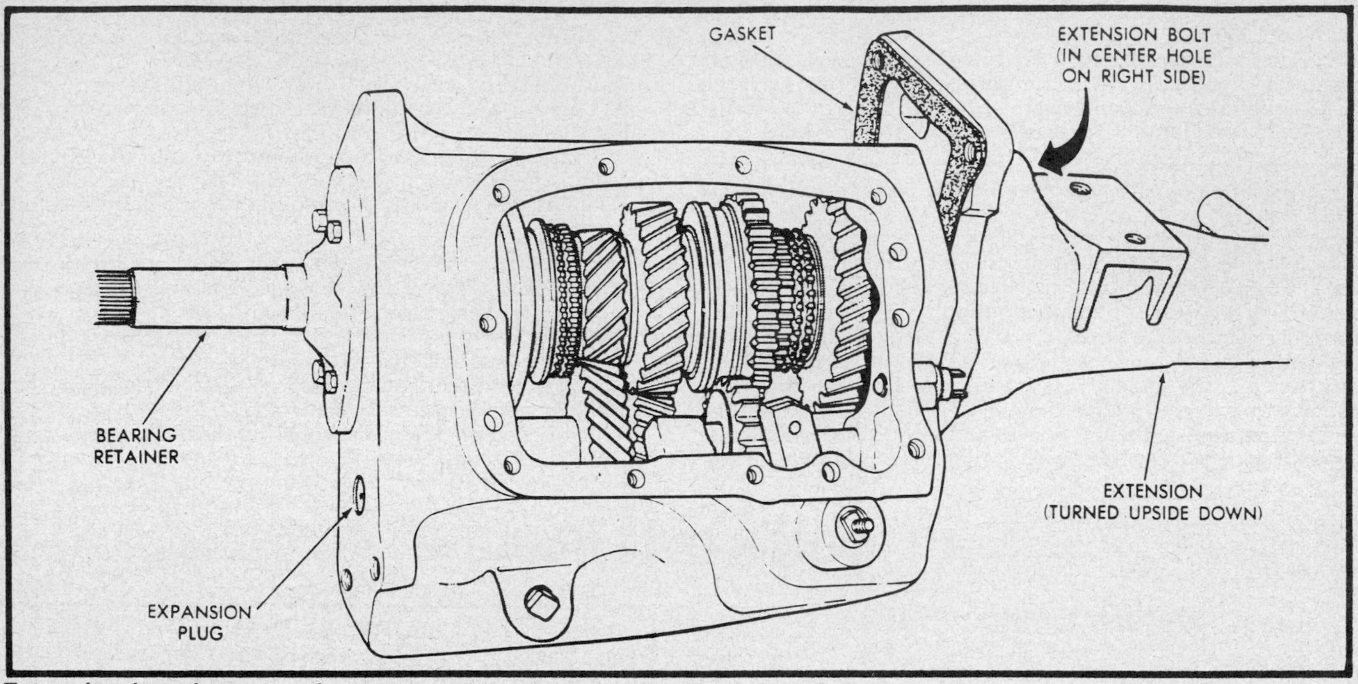
GASKET
EXTENSION BOLT
(IN CENTER HOLE
ON RIGHT SIDE)
BEARING
RETAINER
EXTENSION
(TURNED UPSIDE DOWN)
EXPANSION
PLUG

Extension housing rotated to expose rear of countershaft

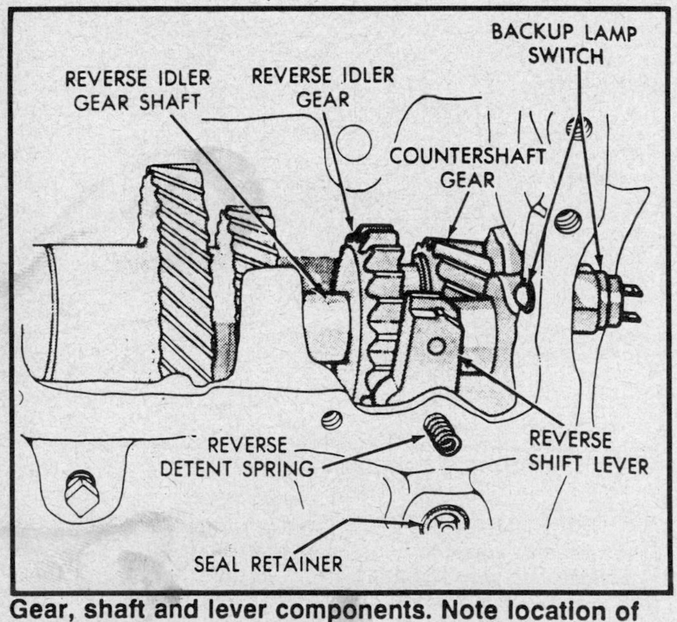
BACKUP LAMP
SWITCH
REVERSE IDLER
GEAR SHAFT
REVERSE IDLER
GEAR
COUNTERSHAFT
GEAR
REVERSE
DETENT SPRING
REVERSE
SHIFT LEVER
SEAL RETAINER

Gear, shaft and lever components. Note location of detent spring

forward and remove through the front of the case.

17. Slide the third and overdrive synchronizer sleeve slightly forward, slide the reverse idler gear to the center of its shaft, and tap the extension housing rearward. Slide the housing and mainshaft assembly out and away from the case.

Unit Disassembly and Assembly

EXTENSION HOUSING, MAINSHAFT AND MAIN DRIVE PINION

Disassembly

1. Remove the snapring holding the 3rd and overdrive synchronizer clutch gear and sleeve assembly to the mainshaft, then remove the synchronizer assembly.

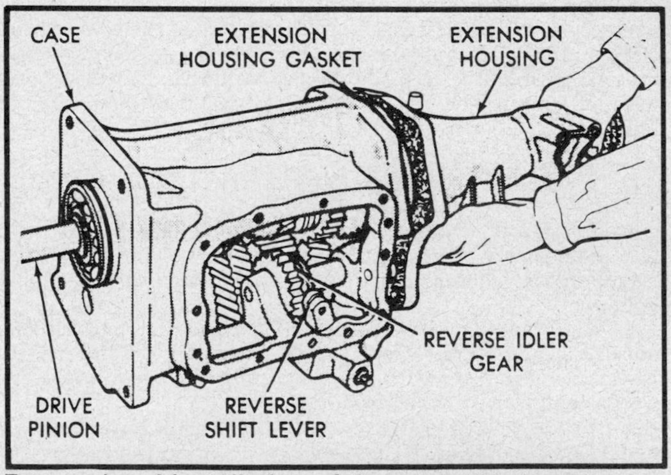
CASE
EXTENSION
HOUSING GASKET
EXTENSION
HOUSING
REVERSE IDLER
GEAR
DRIVE
PINION
REVERSE
SHIFT LEVER

Removal and installation of the extension housing and mainshaft assembly

2. Slide the overdrive gear and stop ring off the mainshaft. Compress the snapring holding the mainshaft bearing in the extension housing. Pull the mainshaft assembly and bearing out of the extension housing.
3. Remove the snapring holding the mainshaft bearing on the shaft. The bearing is removed by inserting steel plates on the front side of the 1st speed gear, then pressing the mainshaft through the bearing being careful not to damage the gear teeth.
4. Remove the bearing, retainer ring, first speed gear and stop ring from the shaft.
5. Remove the snapring and remove the 1st and 2nd clutch gear and sleeve assembly from the mainshaft.
6. Disassemble the synchronizers as required.

Inspection

1. Inspect the mainshaft bearing and gear surfaces for signs

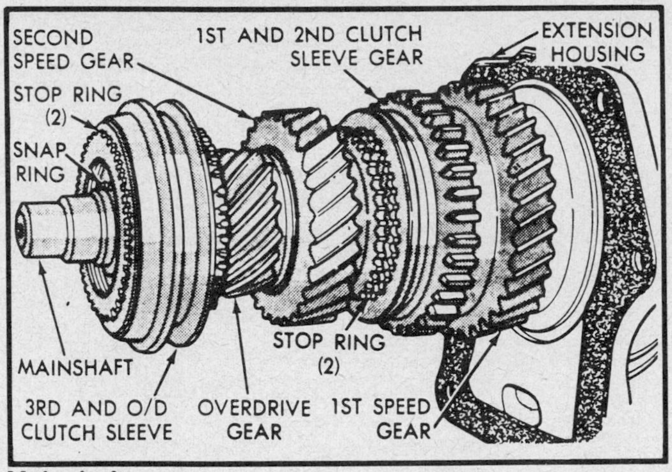
SECOND SPEED GEAR
1ST AND 2ND CLUTCH SLEEVE GEAR
EXTENSION HOUSING
STOP RING (2)
SNAP RING
MAINSHAFT
STOP RING (2)
3RD AND O/D CLUTCH SLEEVE
OVERDRIVE GEAR
1ST SPEED GEAR
Mainshaft gears

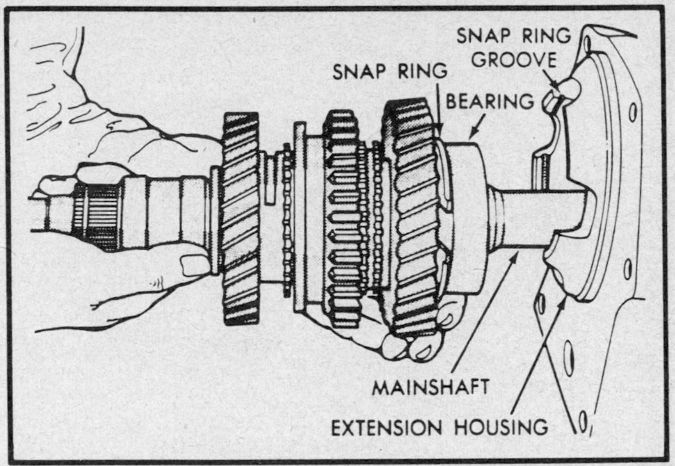
SNAP RING GROOVE
SNAP RING
BEARING
MAINSHAFT
EXTENSION HOUSING
Mainshaft removal and installation

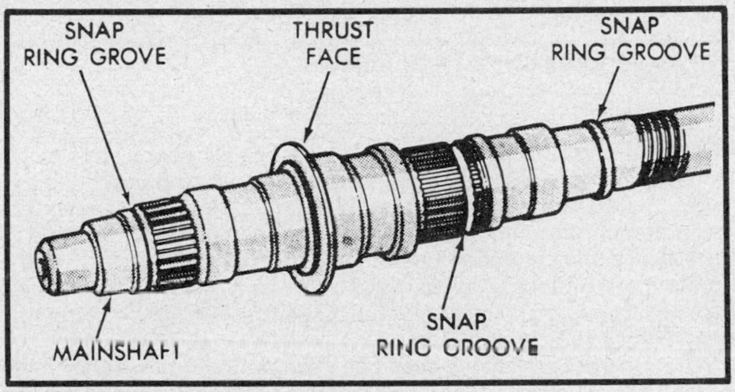
SNAP RING GROVE
THRUST FACE
SNAP RING GROOVE
MAINSHAFT
SNAP RING GROOVE
Mainshaft inspection points

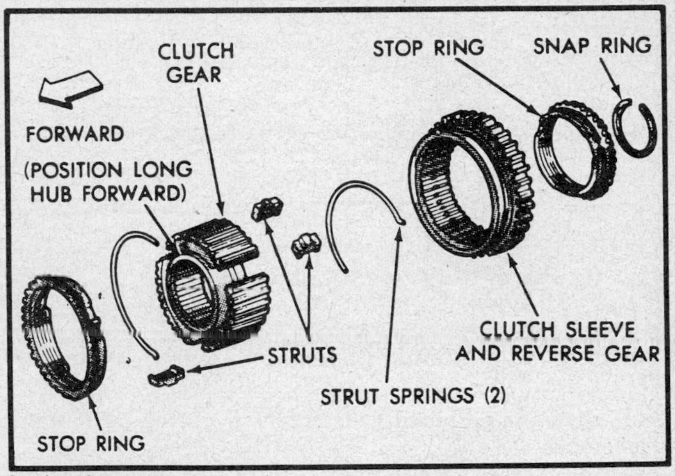
CLUTCH GEAR
STOP RING
SNAP RING
FORWARD
(POSITION LONG HUB FORWARD)
STRUTS
STRUT SPRINGS (2)
CLUTCH SLEEVE AND REVERSE GEAR
STOP RING
Ist/2nd synchronizer

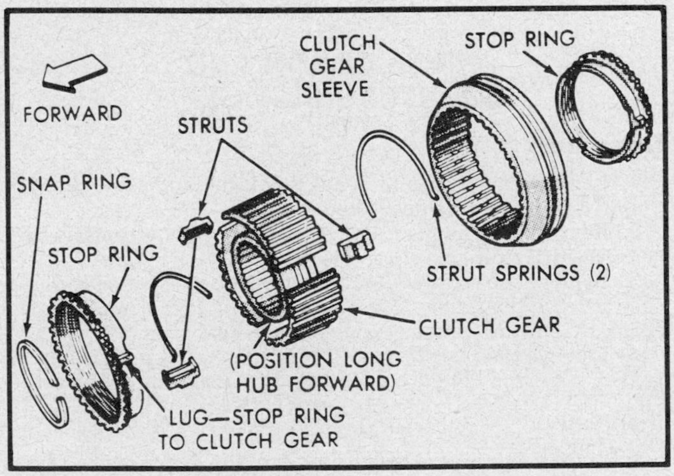
CLUTCH GEAR SLEEVE
STOP RING
FORWARD
STRUTS
SNAP RING
STOP RING
STRUT SPRINGS (2)
CLUTCH GEAR
(POSITION LONG HUB FORWARD)
LUG—STOP RING TO CLUTCH GEAR
3rd/overdrive synchronizer

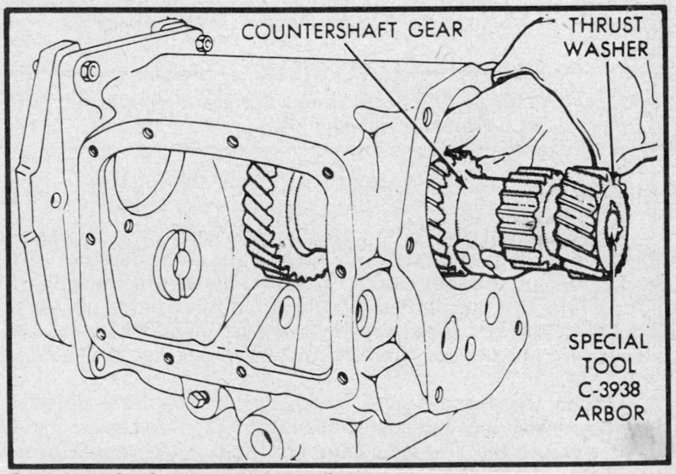
COUNTERSHAFT GEAR
THRUST WASHER
SPECIAL TOOL C-3938 ARBOR
Countershaft gear removal and installation

of wear, scoring or any other imperfection that would make the shaft unusable. They must be smooth and free of defects. Inspect the snapring grooves for burred edges. Polish the grooves with a fine file or crokus cloth as required. Inspect the splines on the synchronizer hub shaft for wear. Replace the mainshaft as necessary.

2. Inspect the synchronizer clutch gear spline teeth and stop rings. If there are sings of chipping or excessively worn teeth, make component replacements as required. Make sure that the clutch sleeve slides freely on the clutch gear.

3. Synchronizer springs should be in place with the tang inside the cavity of a strut. They should not show signs of interfer-

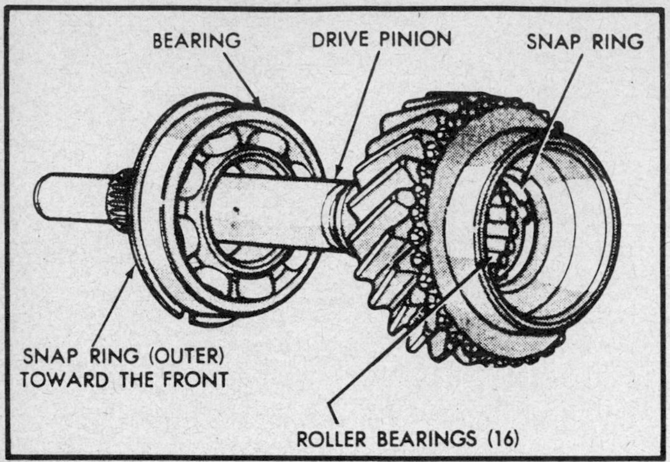
BEARING
DRIVE PINION
SNAP RING
SNAP RING (OUTER)
TOWARD THE FRONT
ROLLER BEARINGS (16)

Drive pinion assembly

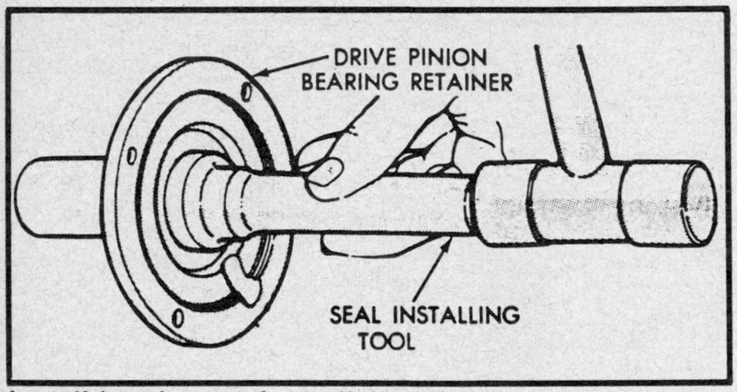
DRIVE PINION
BEARING RETAINER
SEAL INSTALLING
TOOL

Install bearing retainer oil seal using special tool

ence with the polished gear cones or clutch gear inside diameters.

4. Look for cracks and wear in the stop rings. Cracks and wear will usually appear in the threaded bore area. Replace the stop rings as required. During assembly, make sure that the new rings fit properly on the gear cones with minimum wobble. Inspect the synchronizer struts for wear or breakage.

CAUTION

Synchronizers should be serviced as an assembly. Stop rings may be interchanged but synchronizers should not.

Assembly

1. Place a stop ring flat on a bench followed by the clutch gear and sleeve, drop the struts in their slots and snap in a strut spring placing the tang inside the strut. Install the second strut spring tang in a different strut after turning the assembly over.
2. Slide the 2nd speed gear over the mainshaft with the synchronizer cone toward the rear and down against the shoulder on the shaft.
3. Slide the 1st/2nd gear synchronizer assembly including stop rings with lugs indexed in the hub slots, over the mainshaft down against the 2nd gear cone and hold it there with a new snapring. Slide the next snapring over the shaft and index the lugs into the clutch hub slots.
4. Slide the 1st speed gear with the synchronizer cone toward the clutch sleeve just installed over the mainshaft and into position against the clutch sleeve gear.
5. Install the mainshaft bearing retaining ring followed by the mainshaft rear bearing, press the bearing down into position and install a new snapring to secure it. There are several snapring thicknesses available for minimum endplay.
6. Install the partially assembled mainshaft into the extension housing far enough to engage the bearing retaining ring in the slot in the extension housing. Compress the ring so that the mainshaft ball bearing can move in and bottom against its thrust shoulder in the extension housing. Release the ring and make sure that it is seated.
7. Slide the overdrive gear over the mainshaft with the synchronizer cone toward the front followed by the gears snapring.
8. Install the 3rd/overdrive gear synchronizer clutch gear assembly on the mainshaft against the overdrive gear. Make sure to index the rear stop ring with the clutch gear struts.

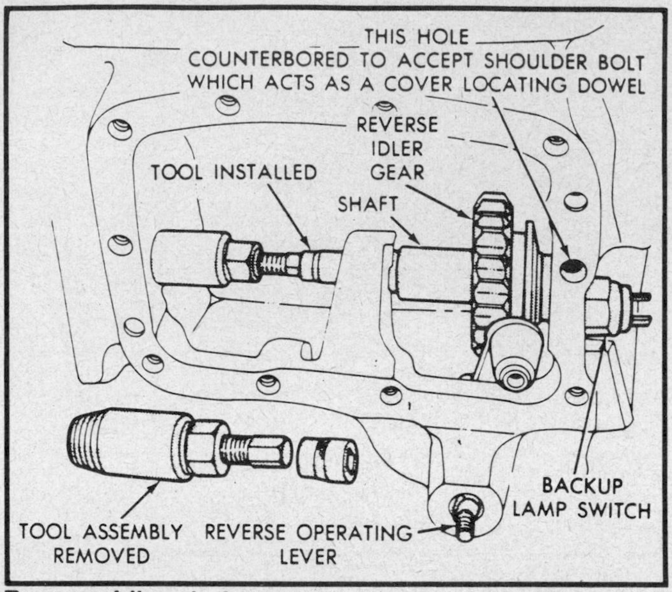
THIS HOLE
COUNTERBORED TO ACCEPT SHOULDER BOLT
WHICH ACTS AS A COVER LOCATING DOWEL
REVERSE
IDLER
GEAR
TOOL INSTALLED
SHAFT
BACKUP
LAMP SWITCH
TOOL ASSEMBLY
REMOVED
REVERSE OPERATING
LEVER

Reverse idler shaft removal

DRIVE PINION AND COUNTERSHAFT GEAR

Disassembly

1. Remove the drive pinion bearing inner snapring, then using an arbor press, remove the bearing.
2. Remove the snapring and 16 bearing rollers from the cavity in the drive pinion.
3. Remove the countershaft gear from the bottom of the case, then remove the arbor, 76 needle bearings, thrust washers and spacers from the center of the countershaft gear.
4. Remove the reverse gearshift lever detent spring retainer, gasket, plug and detent ball spring from the rear of the case.

Inspection

1. Inspect the countershaft gear for chipped or broken teeth and excessive wear. Small nicks or burrs may be removed with a polishing stone.
2. Inspect the main drive pinion gear teeth for broken teeth and wear. Replace the pinion gear as required. If the oil seal contact area is pitted, rusted or scratched a new drive pinion is recommended to ensure maximum seal life.

Assembly

1. Coat the inside bore of the countershaft gear with a thin film of grease and install the roller bearing spacer with an arbor into the gear, center the spacer and arbor.

2. Install the roller bearings and a spacer ring on each end.

3. Replace worn thrust washers, coat the new ones with grease and install them over the arbor with the tang side toward the case boss.

4. Install the countershaft assembly into the case and allow the gear assembly to sit on the bottom of the case so that the thrust washers won't come out of position.

5. Press the drive pinion bearing on the pinion shaft. Make sure the outer snapring groove is toward the front end and the bearing is seated against the shoulder on the gear.

6. Install a new snapring on the shaft to hold the bearing in place. Be sure the snapring is seated and that there is minimum endplay. There are several snapring thicknesses available for adjustment.

7. Place the pinion shaft in a soft jawed vise and install the roller bearings in the cavity of the shaft. Coat them with grease and install the bearing retaining snapring.

8. Install a new oil seal in the bore with special installer tool C–3789 or equivalent.

REVERSE GEAR, LEVER AND FORK

Disassembly

1. The reverse idler gear shaft is a tight fit in the case and will have to be pressed out using special tool C–3638 or equivalent.

2. If there is oil leakage visible around the reverse gearshift lever shaft, remove any burrs from the shaft then push the lever shaft in and remove it from the case.

3. Remove the O-ring and O-ring retainer from the case bore. Discard the O-ring and set the retainer aside.

Inspection

1. Inspect the shift fork for for pad and shaft wear.

2. Inspect the fork shaft bore in the shift lever for signs of galling.

Assembly

1. If the reverse shaft was removed due to an oil leak, install the reverse shift lever into the case bore. Install the O-ring into its retainer, coat light with grease and install.

2. Place the reverse idler gear shaft in position in the end of the case and drive it in far enough to position the reverse idler gear on the protruding end of the shaft with the fork slot toward the rear. While doing this, engage the slot with the reverse shift fork.

3. With the reverse idler gear correctly positioned, drive the reverse gear shaft into the case far enough to install the Woodruff key. Drive the shaft in flush with the end of the transmission case.

EXTENSION HOUSING BUSHING

Removal and Installation

1. Remove the yoke seal from the extension housing.

2. Drive out the old bushing with special tool C–3974 or equivalent.

3. Slide the new bushing on the end of special installer tool C–3974 or equivalent. Align the oil hole in the bushing with the slot in the housing and drive it into place.

5. To install a new seal, place the seal in the opening of the extension housing then drive it into place.

Transmission Assembly

1. Install the snapring and position the front stop ring over the clutch gear again lining up the ring lugs with the struts.

2. Coat a new extension gasket with grease and install it.

3. Slide the reverse idler gear to the center of its shaft and move the 3rd/overdrive synchronizer as far forward as possible without losing the struts.

4. Insert the mainshaft assembly in the case tilting it as necessary. Place the 3rd/overdrive sleeve in the neutral detent position.

5. Rotate the extension on the mainshaft to expose the rear of the countershaft and install a bolt to hold it in position.

6. Install the drive pinion and bearing assembly through the front of the case and position it in the front bore. Install the outer snapring in the bearing groove and tap lightly into place. If it doesn't bottom easily, check to see if a strut, pinion roller or stop ring is out of position.

7. Turn the transmission upside down while holding the countershaft gear to prevent damage. Then lower the countershaft gear assembly into position making sure that the teeth mesh with the drive pinion gear.

8. Start the countershaft into the bore at the rear of the case and push until it is in about halfway, then install the Woodruff key and push it in until it is flush with the rear of the case.

9. Rotate the extension back to normal position and install the retaining bolts. Turn the transmission upright and install the drive pinion bearing retainer and gasket. Coat the threads with sealing compound and tighten the attaching bolts to 30 ft. lbs. (41 Nm).

10. Install a new expansion plug in the countershaft bore in front of the case.

11. Install the interlock levers on the pivot pin and secure with the E-ring. Install the interlock lever hanger spring

12. Grease and install new O-ring seals (and retainers) on both shift shafts. Grease the housing bores and push the shafts through.

13. Install the operating levers and tighten the retaining nuts to 18 ft. lbs. (24 Nm). Make sure the 3rd/overdrive lever points down.

14. Rotate each shift shaft fork bore straight up and install the 3rd/overdrive shift fork in its bore and under both interlock levers.

15. Position both synchronizer sleeves in neutral and place the 1st/2nd gear shift fork in the groove of the 1st/2nd gear synchronizer sleeve. Slide the reverse idler gear to neutral. Turn the transmission on its right side and place the gearshift housing gasket in place holding it there with grease. Install the reverse detent ball and spring into the case bore.

16. As the shift housing is lowered in place, guide the 3rd/overdrive shift fork into its synchronizer groove then lead the shaft of the 1st/2nd shift lever.

17. Raise the interlock lever with a suitable tool to allow the 1st/2nd shift fork to slip under the levers. The shift housing will now seat against the case.

18. Install the bolts lightly and shift through all the gears to check for proper operation.

19. The reverse shift lever and the 1st/2nd gear shift lever have cam surfaces which mate in reverse position to lock the 1st/2nd lever, the fork and synchronizer in the neutral position.

20. To check for proper operation, put the transmission in reverse and, while turning the input shaft, move the 1st/2nd lever in each direction. If it locks up or becomes harder to turn, select a new shift lever size with more or less clearance. If there is too little cam clearance, it will be difficult or impossible to shift into reverse.

21. Grease the reverse shaft, install the operating lever and nut and install the speedometer drive pinion gear and adapter, making sure the range number is in the straight down (6 o'clock) position.

22. Install the transmission and make all necessary adjustments.

SPECIFICATIONS

TORQUE SPECIFICATIONS

Item	ft. lbs.	Nm
Back-up light switch	15	20
Drive pinion bearing retaining bolts	30	41
Extension housing to case bolts	50	68
Gearshift to mounting plate	24	33
Gearshift mounting plate to extension	12	16
Shift lever nuts	18	24
Transmission to clutch housing bolts	50	68
Transmission drain plug	25	34

Section 7

83ET-German Design Transmission
Ford Motor Co.

APPLICATION

1984–86 Ford Mustang/Capri with 2.3L engine

GENERAL DESCRIPTION

The 83ET 4 speed transmission is manufactured in Germany. Externally, the transmission can be identified by the indentification tag. The identification tag is located under the left extension housing-to-case bolt.

The 83ET 4 speed transmission is fully synchronized with all gears except the reverse gear being constant mesh. All forward speed changes are accomplished through blocker ring synchronized units.

The constant mesh gears, such as the output shaft gear, the 3 other forward gears on the output shaft and the corresponding gears on the countershaft gear, are helically cut. The reverse gear on the countershaft gear has straight cut spur teeth that mesh (through an idler gear) with spur teeth on the outside of the 1st/2nd gear synchronizer sleeve.

Gear selection is accomplished by means of a floor shift lever. The shift lever is attached directly to a single selector rail, which has a selector lever assembled to it. When the shift lever is move into a desired shift position, the selector rail causes the selector lever to locate in the approriate selector fork and thus moves it to the required position. Movement of the 1st/2nd or 3rd/4th selector fork causes the appropriate synchronizer sleeve to move as necessary to engage with the clutch teeth on the required gear. When reverse gear is required, movement of the reverse selector relay lever draws the reverse idler gear into mesh with the reverse gear on both the countershaft gear and the output shaft.

Engagement of the 2 gears at once is prevented by means of a selector interlock plate pivoted in the transmission. This plate engages with the selector forks that are not in use and holds them positively in the disengaged position. The 1st/2nd and 3rd/4th forks, while they are mounted on the rail, are free to slide, when not retained by the interlock plate.

Gear engagement only takes place when selector lever locates in appropriate fork and moves it into required position.

Selective snaprings compensate for tolerances which must be allowed in manufacture.

Metric Fasteners

This transmission is assembled with metric fasteners. Care must be exercised during the disassembly and assembly of the manual transmission due to the usage of metric nuts and bolts. Proper wrenches and sockets should be used to avoid damage to the transmission and fasteners. Do not attempt to interchange metric threaded fasteners with SAE fasteners as damage can result.

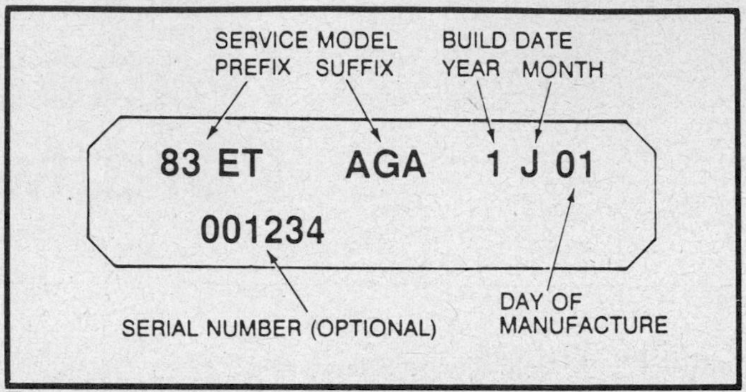
SERVICE PREFIX
MODEL SUFFIX
BUILD DATE
YEAR
MONTH
83 ET
AGA
1 J 01
001234
SERIAL NUMBER (OPTIONAL)
DAY OF MANUFACTURE

Transmission identification plate

Capacities

Type D8DZ–19C547–A gear oil, meeting Ford Motor Co. specifications, is used in the 83ET manual transmission. Fluid case capacity is 2.3 pints (1.3L). When checking or filling the transmission, the fluid level should be at the bottom of the filler plug hole.

Checking Fluid level

1. Raise the vehicle and support it safety.
2. Remove the fill plug from the side of the transmission.
3. Check the fluid level, it should be even to the bottom of the fill hole, if the fluid level is below the fill hole, top off the fluid level with gear oil meeting Ford Motor Co. specifications.
4. Install the fill plug and lower the vehicle.

TRANSMISSION MODIFICATIONS

There were no transmission modifications at time of printing.

TROUBLE DIAGNOSIS

CHILTON THREE "C'S" DIAGNOSIS

Condition	Cause	Correction
Transmission shifts hard	a) Improper clutch release	a) Check release linkage for damage; service as needed
	b) Internal shift mechanism binding	b) Remove transmission and free up shift mechanism
	c) Sleeve to hub fit	c) Remove and check for burrs or fit
	d) Binding condition between input shaft and crankshaft pilot bearing	d) Check alignment and service as required
	e) Improper fluid	e) Four-speed transmissions, drain and refill with ESP-M2C83-C
Gears clash when shifting from one forward gear to another	a) Engine idle speed too high	a) Adjust engine idle speed
	b) Improper clutch release	b) Assure complete clutch release
	c) Binding condition between transmission input shaft and crankshaft pilot bushing	c) Remove transmission and replace pilot bushing
	d) Worn or damaged shifter forks, synchronizer assembly, or gear clutch teeth	d) Service or replace as required
Transmission jumps out of gear	a) Floor shift—stiff or improperly installed boot	a) Verify jump-out with boot removed. Replace boot if necessary
	b) Floor shift—interference between shift handle and console	b) Adjust console to eliminate interference
	c) Transmission to engine mounting loose	c) Tighten bolts to specification
	d) Flywheel housing to engine crankshaft out of alignment	d) Replace parts as required
	e) Improper engagement between transmission pilot and flywheel housing	e) Replace parts as required
	f) Input shaft pilot bearing worn	f) Replace pilot bearing
	g) Worn or damaged internal components	g) Check input and output shafts for excessive end play, shift forks for loose mounting on shift rails, worn pads, shift rail detent system for wear or damage, synchronizer sliding sleeve and gear clutch teeth for wear or damage. Service or replace as required
Transmission will not shift into one gear—all other gears OK	a) Floor shift, interference between shift handle and console or floor cut out	a) Adjust console or cut out floor pan to eliminate interference
	b) Restricted travel of internal shifter components	b) Remove transmission. Inspect shift rail and fork system, synchronizer system and gear clutch-teeth for restricted travel. Service and replace as required
Transmission is locked in one gear. It cannot be shifted out of that gear	a) Internal shifter components worn or damaged	a) Remove transmission. Inspect the problem gear or gear shift rails and fork and synchronizer for wear or damage. Service or replace as required
	b) Selector arm finger broken	b) Remove transmission. Replace selector arm assembly
	c) Bent shifter forks at pads and selector slot	c) Service or replace as required
Transmission will not shift into reverse (All others OK)	a) Worn or damaged internal components	a) Remove transmission. Check for damaged reverse gear train, misaligned reverse relay lever, shift rail and fork system. Service or replace as required

CHILTON'S THREE C'S TRANSMISSION DIAGNOSIS

Condition	Cause	Correction
Transmission noisy in gear	a) Lube level low or wrong type	a) Fill to bottom of filler plug hole with proper lubricant
	b) Transmission to flywheel housing and flywheel housing to engine block bolts loose	b) Tighten bolts to specifications
	c) Pilot bushing worn or damaged	c) Remove transmission. If noise is howl during start-up, check pilot bushing. Check for loose flywheel and housing alignment. Service or replace as required
	d) Improper transmission pilot engagement into flywheel housing	d) Replace housing or input shaft bearing retainer as required
	e) Worn or damaged internal components	e) Disassemble transmission. Inspect input, output and countershaft bearings, gear and gear teeth for wear or damage. Service or replace as required
Transmission leaks	a) Excessive amount of lubricant in transmission—wrong type	a) Check level and type. Fill bottom of filler plug hole
	b) Other component(s) leaking	b) Identify leaking fluid at engine, power steering and/or transmission
	c) False report	c) Remove all traces of lube on exposed transmission surfaces. Check vent for free breathing. Operate transmission and inspect for new leakage. Service as required
	d) Worn or damaged internal components	d) Remove transmission. Inspect for leaks at the input shaft bearing retainer seal and gasket, shift rail expansion plug top cover gasket. Inspect case for sand holes. Service or replace as required
	e) Inadequate thread sealant on threads	e) Apply sealant to bolt threads

ON CAR SERVICES

Services

SHIFT LEVER

Removal and Installation

1. Place the gear shift lever in the **N** position.
2. Remove attaching screws at rear of the console coin tray. Lift the tray to release it from the front hold down notch on boot retainer. Lift it over the gear shift lever boot.
3. Remove the screws attaching the boot to the floor pan and move boot upward out of the way.
4. Remove the lever attaching bolts.
5. Remove the lever and boot assembly from the extension housing.
6. Remove the gear shift knob and slide the boot off the lever.
7. Make sure the shift lever insulator is in a straight downward position on the shift rail.
8. Position the shift lever in the extension housing so that the forked ends engage in the insulator properly.
9. Install the attaching bolts and tighten to 17–25 ft. lb. (23–34 Nm).
10. Slide the boot over the lever, install the attaching bolts and tighten to 3–7 ft. lb. (4–10 Nm).
11. Install gear shift knob and adjust.
12. Position coin tray over shift lever boot.
13. Secure the tray to front notch on boot retaining ring and attach at rear with screws.

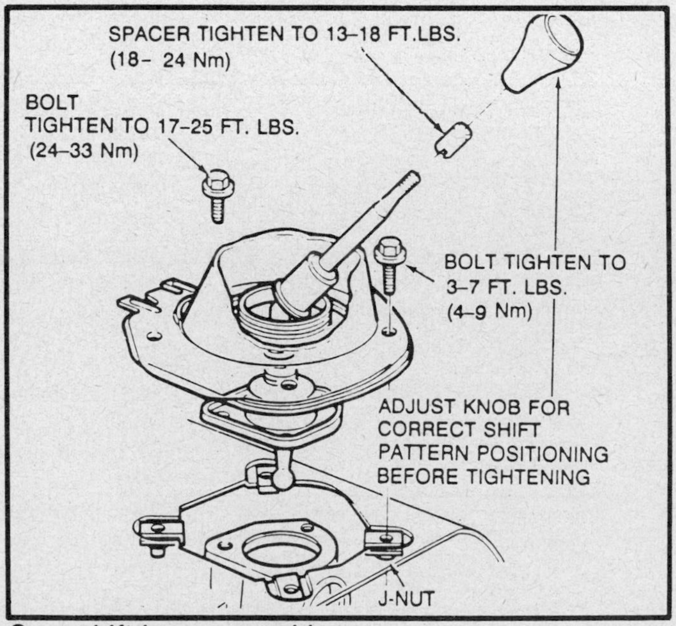
SPACER TIGHTEN TO 13–18 FT.LBS. (18– 24 Nm)
BOLT TIGHTEN TO 17–25 FT. LBS. (24–33 Nm)
BOLT TIGHTEN TO 3–7 FT. LBS. (4–9 Nm)
ADJUST KNOB FOR CORRECT SHIFT PATTERN POSITIONING BEFORE TIGHTENING
J-NUT
Gear shift lever assembly

REMOVAL AND INSTALLATION

TRANSMISSION REMOVAL

1. Disconnect the negative battery cable. Remove the shift lever.
2. Working from under the hood, remove upper bolts or stud nuts attaching the flywheel housing to engine.
3. Raise the vehicle and support it safely.
4. Mark the position of the driveshaft relative to the axle companion flange. Remove the driveshaft and install a suitable extension housing seal replacer tool in the extension housing to prevent lubricant leakage.
5. Remove the clutch release lever dust cover.
6. Disconnect the clutch release cable from the release lever.
7. Remove the starter motor attaching bolts and place the starter to the side.
8. Remove the speedometer cable attaching screw and lift the cable from the extension housing.
9. Disconnect the backup lamp switch.
10. Support the rear of the engine with a jack and remove the bolts that attach the crossmember to the body.
11. Remove the bolts attaching the crossmember to the extension housing, and remove crossmember.
12. Lower the engine as required to permit removal of bolts attaching the flywheel housing to the engine. Slide the clutch housing and transmission assembly away from the engine and from under the vehicle.

NOTE: It may be necessary to slide mounting bracket forward away from catalytic converter heat shield in order to move transmission rearward far enough to remove it.

13. Remove the cover attaching bolts and drain the lubricant into a container.
14. Remove the bolts that attach flywheel housing to the transmission and remove the housing.

TRANSMISSION INSTALLATION

1. Install a new shift rod seal in the flywheel housing.
2. Position the flywheel housing on transmission, and install and tighten the attaching bolts to 35–45 ft. lb. (47–61 Nm).
3. Install the clutch release lever and bearing.
4. Make certain that machined surfaces of flywheel housing and engine are free of dirt and foreign material.
5. Apply a film of transmission lubricant to the input shaft bearing retainer. Position the flywheel housing and transmission assembly on the engine block. It may be necessary to place the transmission in gear and rotate the output shaft to align the input shaft and clutch splines.
6. Slide the flywheel housing firmly and squarely onto the locating dowels, to be sure of a positive engagement. Then, holding the flywheel housing firmly in position on the dowels, thread the attaching bolts through the hollow portion of the dowels and into the housing. Tighten the bolts to 35–55 ft. lb. (52–75 Nm).
7. Install and tighten the center attaching bolts to 38–55 ft. lb. (52–75 Nm).
8. Lower the vehicle and install the upper attaching bolts or stud nuts. Tighten to 38–55 ft. lb. (52–75 Nm). Install the shift lever.
9. Raise the vehicle. Raise the engine until the transmission reaches its normal position. Secure the crossmember to the body with the attaching bolts, and tighten to 28–40 ft. lb. (38–54 Nm). Install the bolts that attach crossmember to extension housing, and tighten to 50–70 ft. lb. (68–95 Nm). Position the catalytic converter heat shield mounting bracket to the transmission mount.
10. Remove the extension housing seal replacer tool. Install the speedometer cable and tighten the attaching screw to 3–4.5 lb. ft. (4–6) Nm.
11. Position the starter, install the attaching bolts, and tighten to 15–20 lb. ft. (20–27 Nm).
12. Apply a multi-purpose lubricant to the ball end of the clutch release cable and connect it to the release lever. Install the clutch release fork dust cover.
13. Install the driveshaft, making sure that it is connected to the pinion flange in its original position.
14. Fill the transmission with transmission gear oil until it appears at the bottom of the filer plug hole. Install the filler plug and tighten to 24–27 ft. lb. (33–36 Nm).
15. Install the reverse lamp switch and connect the wire to switch.
16. Lower the vehicle.
17. Check the transmission for proper operation.

BENCH OVERHAUL

Before Disassembly

Clean the exterior of the transmission before any attempt is made to disassemble it, in order to prevent dirt or other foreign materials from entering the transmission assembly or its internal parts. If steam cleaning is done to the exterior of the transmission, immediate disassembly should be done to avoid rusting, caused by condensation forming on the internal parts.

Transmission Disassembly

1. Remove the clutch release bearing and the lever and detach the clutch housing.
2. Drain the lubricant and remove the cover and the gasket from the case.
3. Remove the threaded plug, the spring and the shift rail detent plunger from the front of the case.
4. Drive the access plug from the rear of the case. Drive the interlock retaining pin from the case and remove the interlock plate.
5. Remove the roll pin from the selector lever arm.
6. Tap the front end of the shift rail, to displace the plug at the rear of the extension housing. Remove the shift rail from the rear of the extension housing.
7. Remove the selector arm and shift forks from the case.
8. Remove the extension housing attaching bolts. Loosen the extension housing and rotate the housing to align the countershaft with the cutaway in the extension housing flange.
9. Drive the countershaft rearward until the shaft clears the front of the case. Install a dummy shaft through the case and into the gear until the countershaft gear can be lowered to the bottom of the case. Remove the countershaft.
10. Remove the extension housing and mainshaft assembly from the case.
11. Remove the input shaft bearing retainer bolts, the bearing retainer and the input shaft from the case.
12. Remove the reverse idler gear and shaft from the rear of the case.
13. Remove the bearing retainers, the bearings and the dummy shaft from the countershaft gear.

14. Remove the retainer and the pilot bearing from the input shaft gear.
15. Do not remove the ball bearing from the input shaft unless replacement is necessary. To remove it, remove the snapring and press the bearing off the shaft.
16. Pry the input shaft seal out of the bearing retainer.
17. Remove the 4th speed gear blocking ring from the front of the output shaft.
18. Remove the snapring from the forward end of the output shaft.
19. Support the 3rd speed gear (on press plates), the output shaft and the extension housing in a press. Push the output shaft out of the 3rd–4th synchronizer and the 3rd speed gear, while supporting the extension housing and the output shaft from beneath. Remove the snapring, washer, 2nd speed gear and the blocking ring from the output shaft.
20. Disassemble the synchronizer assembly by pulling the sleeve from the hub and remove the inserts and spring.
21. Remove the output shaft bearing-to-extension housing snapring.
22. Using a plastic hammer, tap the output shaft assembly from the extension housing.
23. Measure or scribe the speedometer gear location on the output shaft and press the gear off.

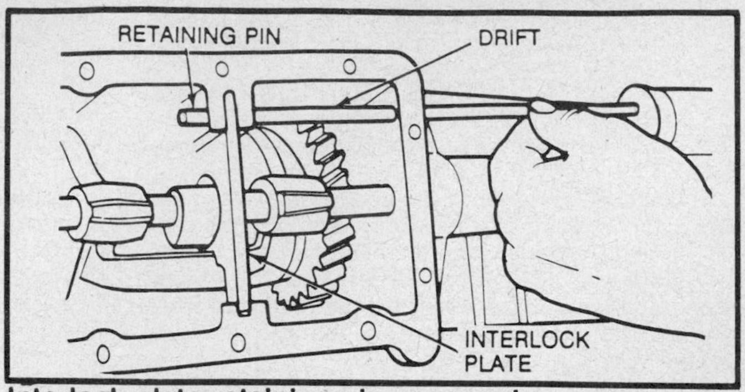
RETAINING PIN
DRIFT
INTERLOCK PLATE
Interlock plate retaining pin—removal

24. Position press plates behind 1st speed gear and place the assembly in a press. The 1st–2nd synchronizer are serviced as an assembly.

NOTE: No attempt should be made to separate the hub from the shaft. The only serviceable parts are the springs and inserts. If the hub or sleeve is worn, the shaft and synchronizer must be replaced as an assembly.

25. Using a $^{9}/_{16}$ in. socket, drive the shift rail bushing from the rear of the extension housing. Do not remove serviceable bushings.
26. Pry the shift rail seal from the rear of the case.
27. Remove the remaining shift linkage from the case.

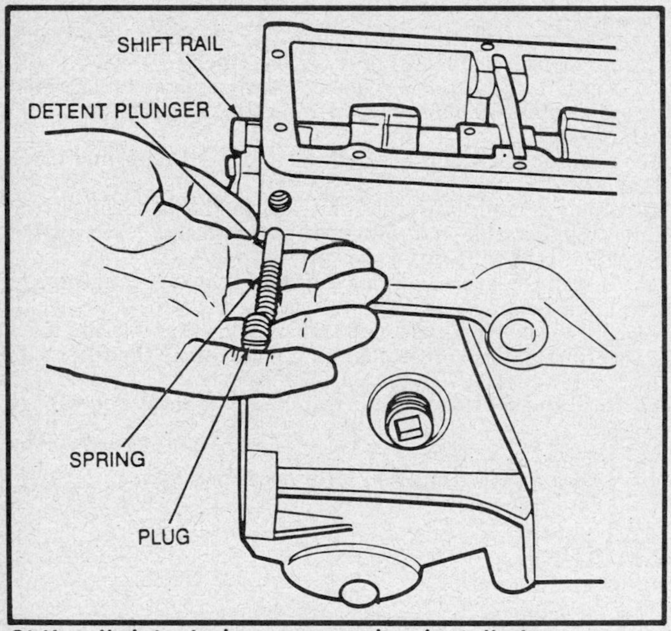
SHIFT RAIL
DETENT PLUNGER
SPRING
PLUG
Shift rail detent plug—removal or installation

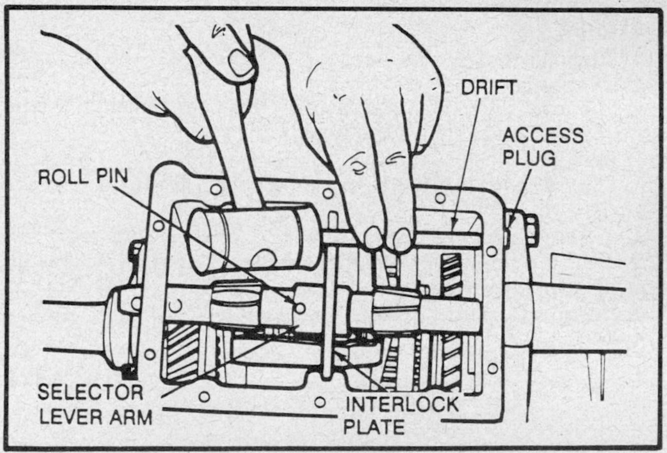
DRIFT
ACCESS PLUG
ROLL PIN
SELECTOR LEVER ARM
INTERLOCK PLATE
Interlock plate access plug—removal

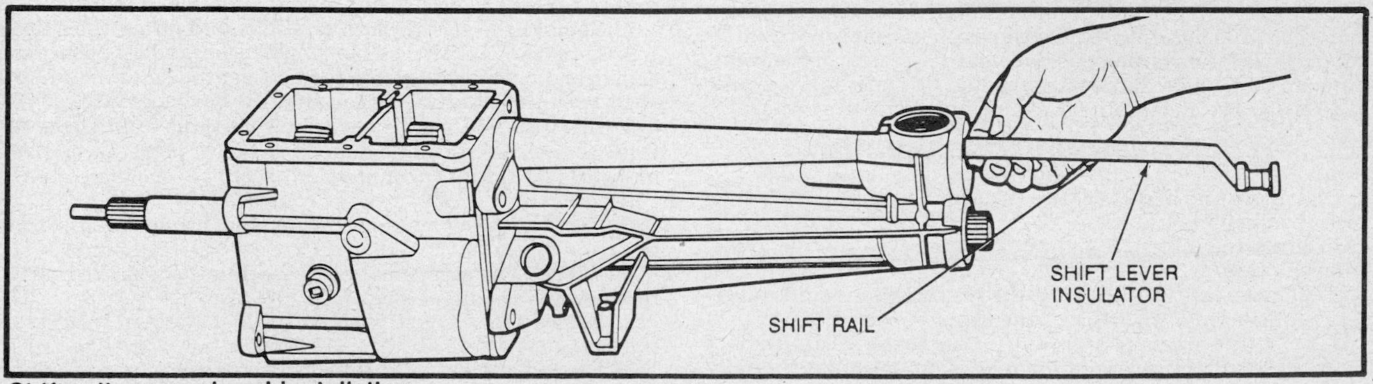
SHIFT LEVER INSULATOR
SHIFT RAIL
Shift rail removal and installation

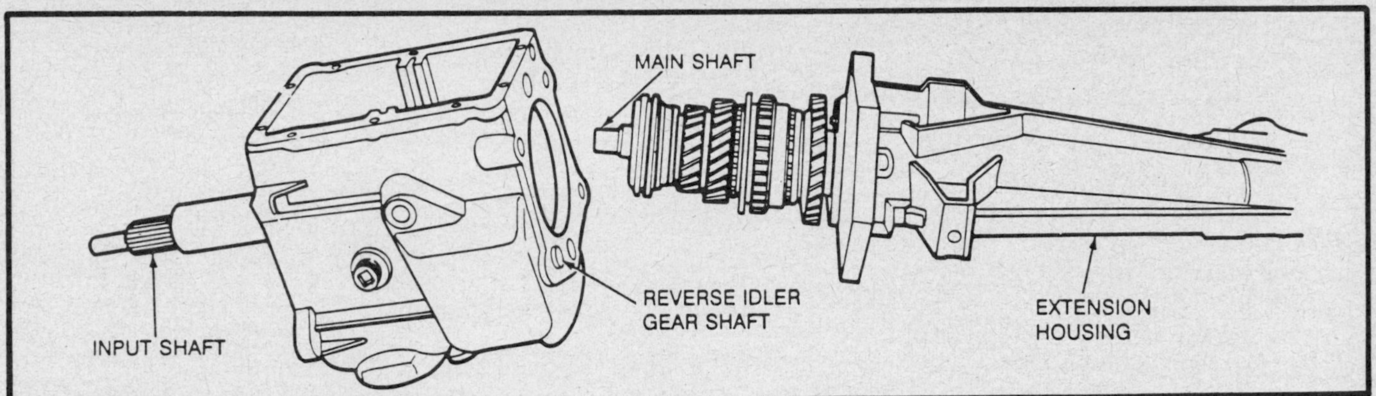
MAIN SHAFT
REVERSE IDLER GEAR SHAFT
EXTENSION HOUSING
INPUT SHAFT

Extension housing and mainshaft—removal or installation

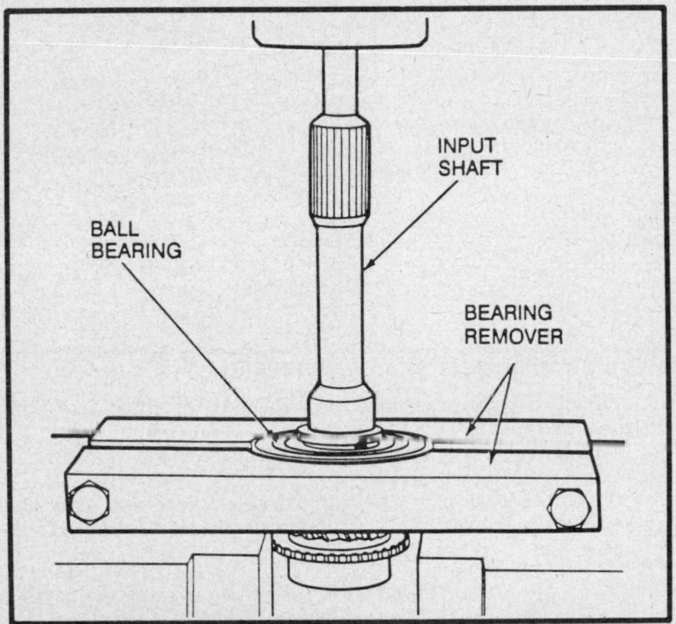
INPUT SHAFT
BALL BEARING
BEARING REMOVER

Input shaft bearing—removal

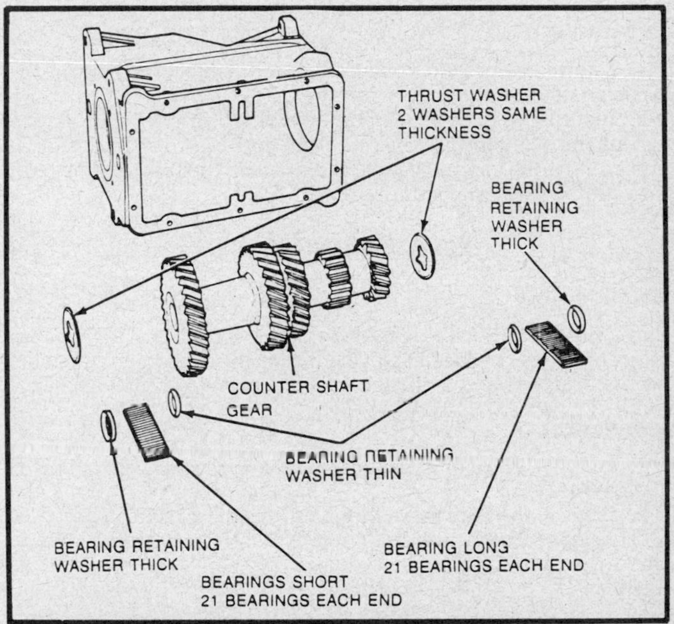
THRUST WASHER
2 WASHERS SAME
THICKNESS
BEARING
RETAINING
WASHER
THICK
COUNTER SHAFT
GEAR
BEARING RETAINING
WASHER THIN
BEARING RETAINING
WASHER THICK
BEARINGS SHORT
21 BEARINGS EACH END
BEARING LONG
21 BEARINGS EACH END

Countershaft gear disassembled

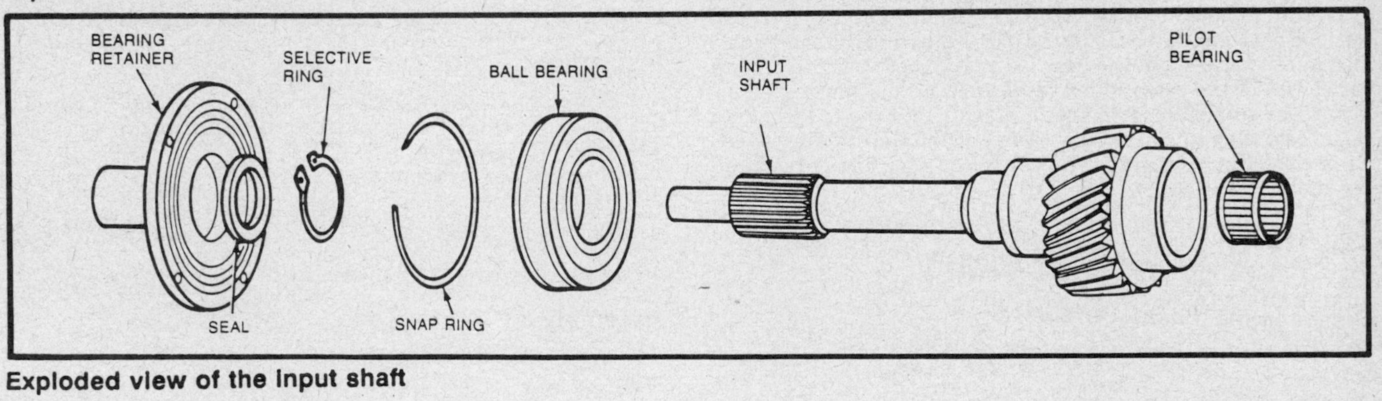
BEARING
RETAINER
SELECTIVE
RING
BALL BEARING
INPUT
SHAFT
PILOT
BEARING
SEAL
SNAP RING

Exploded view of the input shaft

Unit Disassembly and Assembly

INPUT SHAFT

Disassembly

1. Remove the retainer and the pilot bearing from the input shaft gear.
2. Do not remove the ball bearing from the input shaft unless replacement is necessary. To remove it, remove the snapring and press the bearing off the shaft.

Inspection

1. Carefully inspect the input shaft, ball bearing and pilot bearing for any signs of wear, stress, discoloration, cracks or warpage.
2. Inpect the input shaft splines for distortion.
3. Replace defective parts as necessary.

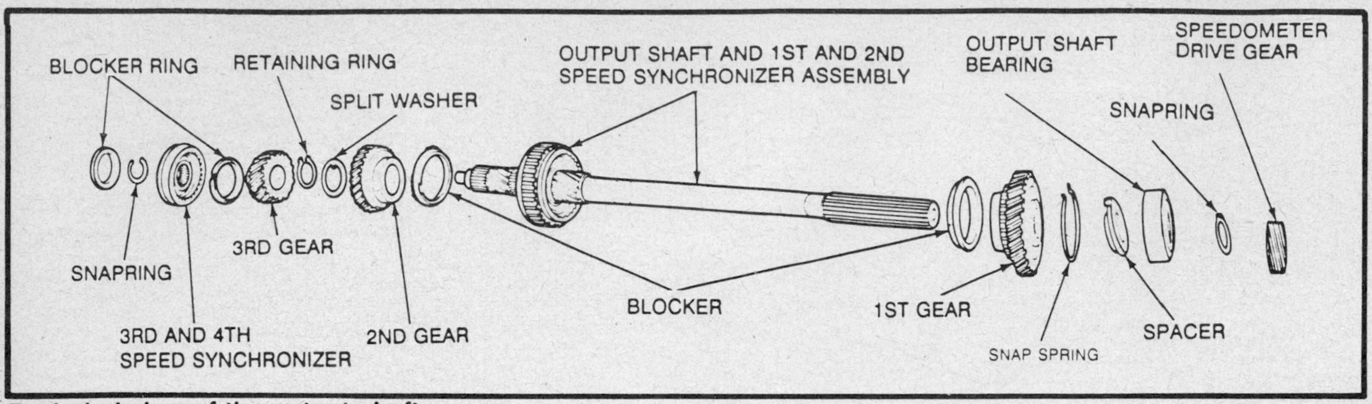
BLOCKER RING
RETAINING RING
SPLIT WASHER
OUTPUT SHAFT AND 1ST AND 2ND SPEED SYNCHRONIZER ASSEMBLY
OUTPUT SHAFT BEARING
SPEEDOMETER DRIVE GEAR
SNAPRING
3RD GEAR
SNAPRING
BLOCKER
1ST GEAR
SPACER
SNAP SPRING
3RD AND 4TH SPEED SYNCHRONIZER
2ND GEAR

Exploded view of the output shaft

Assembly

1. Lubricate the gear cone and place the blocking ring on the input shaft gear cone.
2. Install the pilot bearing and retainer on the input shaft.
3. If the ball bearing was removed, press the bearing onto the input shaft; the snapring groove must face the front of the shaft (use the thickest snapring that will fit).

COUNTERSHAFT

Disassembly

The countershaft is a single unit comprised of 5 gears which are not removable. It does however, incorporate 21 long and short needle bearings and 2 thrust washers that are removable.

Inspection

1. Inspect the countershaft gears for wear, chipping or distortion.
2. Inspect the thrust washers and needle bearings.
3. Replace the countershaft and/or related parts as necessary.

Assembly

1. Slide the spacer and the dummy shaft into the countershaft gear. Position a thin bearing retaining washer on each end of the dummy shaft. Lubricate the roller bearings and install the long bearings in the small end of the gear and the short bearings in the long end of the gear (21 needle bearings are used at either end of the gear, which total to 42 needle bearings).
2. Prior to installing the countershaft into the case, place a thick retaining washer over each end of the dummy shaft. Grease the thrust washers and place 1 on each end of the dummy shaft.

OUTPUT SHAFT

Disassembly

1. Remove the 4th speed gear blocking ring from the front of the output shaft.
2. Remove the snapring from the forward end of the output shaft.
3. Support the 3rd speed gear (on press plates), the output shaft and the extension housing in a press. Push the output shaft out of the 3rd–4th synchronizer and the 3rd speed gear, while supporting the extension housing and the output shaft from beneath. Remove the snapring, washer, 2nd speed gear and the blocking ring from the output shaft.
4. Disassemble the synchronizer assembly by pulling the sleeve from the hub and remove the inserts and spring.

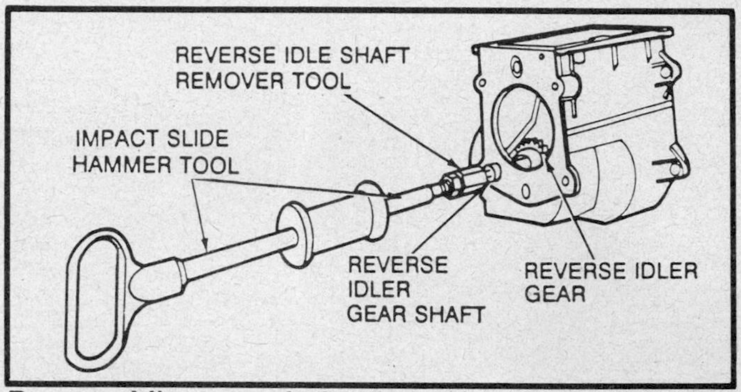
REVERSE IDLE SHAFT REMOVER TOOL
IMPACT SLIDE HAMMER TOOL
REVERSE IDLER GEAR SHAFT
REVERSE IDLER GEAR

Reverse idler gear shaft—removal

5. Using a plastic hammer, tap the output shaft assembly from the extension housing.
6. Measure or scribe the speedometer gear location on the output shaft and press the gear off.
7. Position press plates behind 1st speed gear and place the assembly in a press. The 1st–2nd synchronizer are serviced as an assembly.

NOTE: No attempt should be made to separate the hub from the shaft. The only serviceable parts are the springs and inserts. If the hub or sleeve is worn, the shaft and synchronizer must be replaced as an assembly.

Inspection

1. Carefully inspect all gears, thrust washers and synchronizers for any signs of wear, stress, discoloration, cracks or warpage.
2. Inspect the synchronizer sleeves for excessive wear or distortion.
3. Inspect the speedometer gear for excessive wear or distortion.
4. Replace defective parts as necessary.

Assembly

1. Assemble a blocking ring on the 1st gear side of the 1st–2nd synchronizer. Lubricate the cone surface of 1st gear and the output shaft gear journals and slide the cone onto the output shaft, so that the cone surface engages the blocking ring.
2. Position the spacer on the output shaft (the larger diameter rearward).
3. Install a snapring which will eliminate the endplay from the output shaft bearing. Position the output shaft bearing on the shaft and press it into place. Secure the bearing with the thickest snapring that will fit the groove.
4. Slide the synchronizer onto the hub and locate an insert in each of 3 slots in the sleeve.

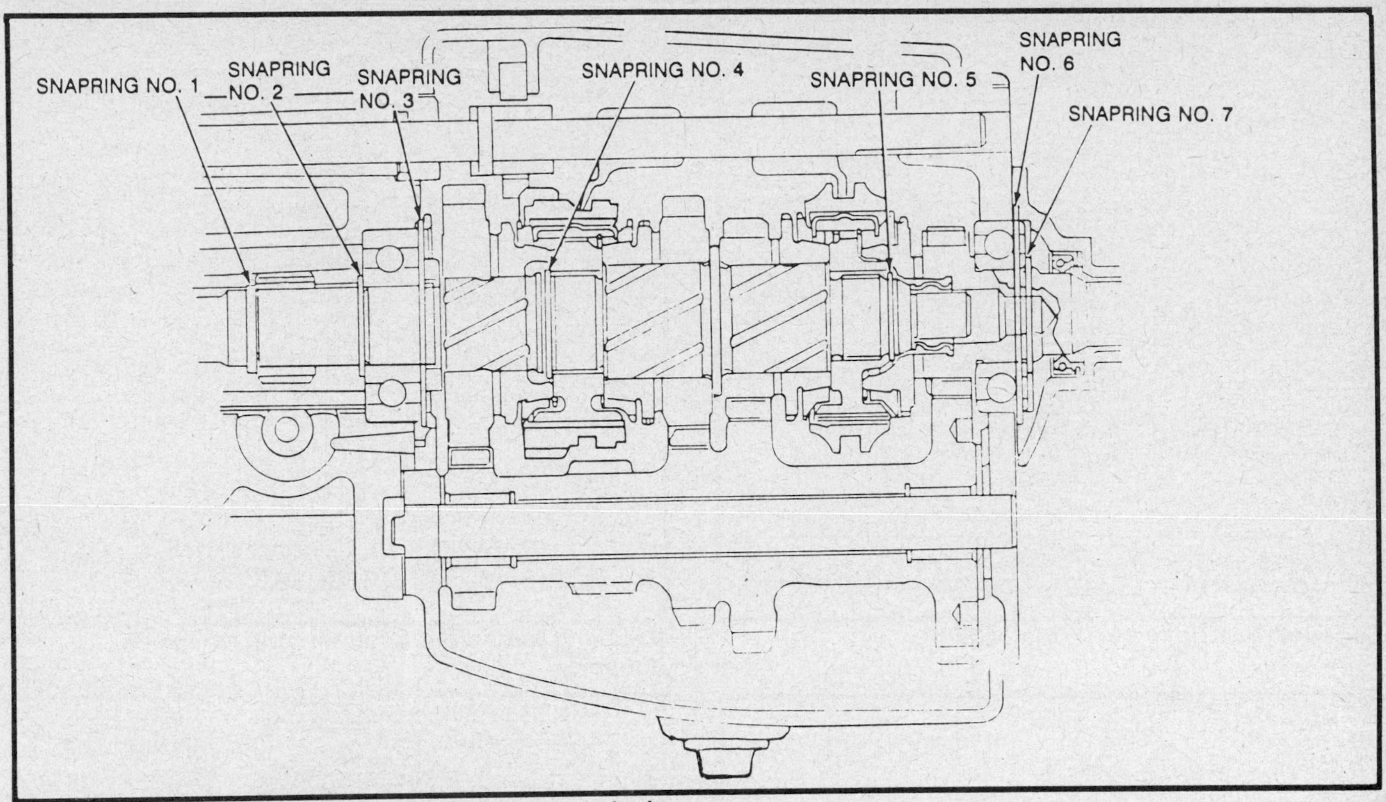
SNAPRING NO. 1
SNAPRING NO. 2
SNAPRING NO. 3
SNAPRING NO. 4
SNAPRING NO. 5
SNAPRING NO. 6
SNAPRING NO. 7

Snapring locations 83ET 4 speed manual transmission

OUTPUT SHAFT BEARING SNAPRING THICKNESS

Part Number	Thickness (in.)	Identification
D1FZ-7030-A	0.0679	Color Coded—Copper
D1FZ-7030-B	0.0689	Letter—W
D1FZ-7030-C	0.0699	Letter—V
D1FZ-7030-D	0.0709	Letter—U
D1FZ-7030-E	0.0719	None
D1FZ-7030-F	0.0728	Color Coded—Blue
D1FZ-7030-G	0.0738	Color Coded—Black
D1FZ-7030-H	0.0748	Color Coded—Brown

NOTE: When installing the sleeve to the hub, the etch marks must be in the same relative locations.

5. Lightly oil the parts and complete the assembly of the synchronizer by following directions in previous Steps 3 and 4.
6. Position the 2nd speed gear and the blocking ring on the output shaft (the dog teeth must face rearward. Install the washer and snapring, position the 3rd speed gear onto the output shaft (the dog teeth must face forward). Lubricate the gear cones and assemble a blocking ring onto the 3rd gear cone.
7. Position the 3rd–4th synchronizer assembly on the output shaft (the hub boss must face forward).
8. Install the press plates against the boss on the synchronizer hub, place the entire unit in a press (the extension end up) and press the synchronizer assembly onto the output shaft as far as possible.
9. Retain the 3rd–4th synchronizer assembly to the output shaft with a snapring. Pull up on the synchronizer so that the snapring is tight in the groove.

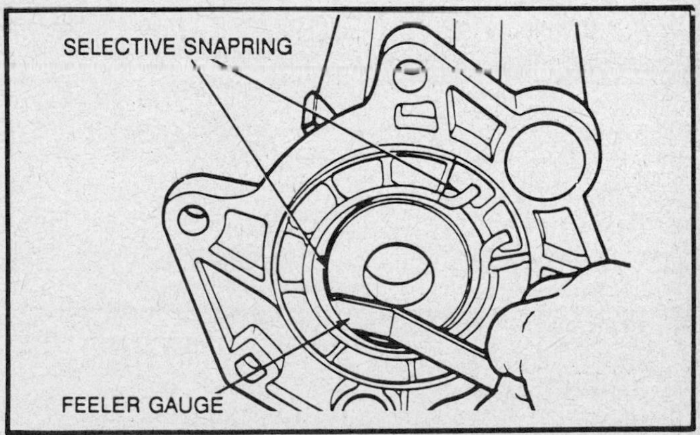
SELECTIVE SNAPRING
FEELER GAUGE

Determining snapring thickness

10. Lubricate the gear cone and place the blocking ring on the input shaft gear cone.
11. Press the speedometer drive gear onto the shaft to the marked location.

SHIFT RAIL, BUSHING AND SEAL

Disassembly

1. Drive the shift rail bushing from the rear of the extension housing, using a $^{9}/_{16}$ in. socket. Do not remove serviceable bushings.
2. Pry the shift rail seal from the rear of the case.
3. Remove the remaining shaft linkage from the case. Do not remove the seat belt sensing switch unless it is damaged.

Inspection

1. Inspect the shift rail and bushings for distortion.

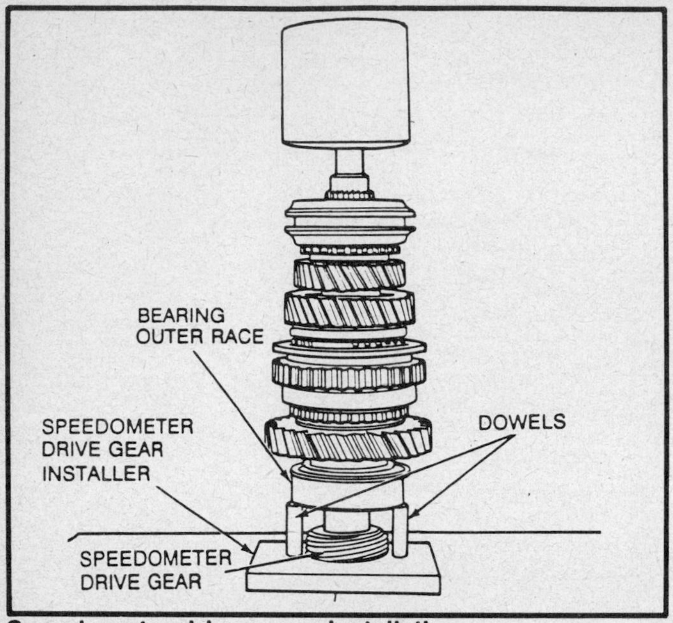
BEARING
OUTER RACE
SPEEDOMETER
DRIVE GEAR
INSTALLER
DOWELS
SPEEDOMETER
DRIVE GEAR

Speedometer drive gear—Installation

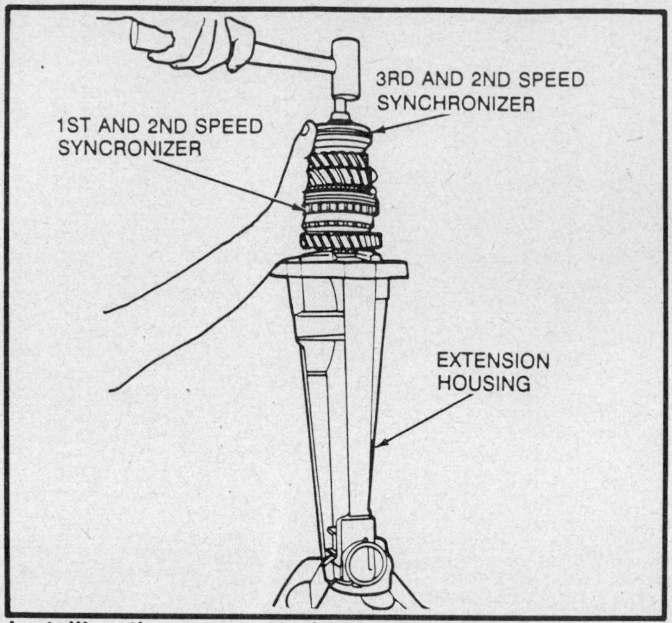
3RD AND 2ND SPEED
SYNCHRONIZER
1ST AND 2ND SPEED
SYNCRONIZER
EXTENSION
HOUSING

Installing the output shaft into the extension housing

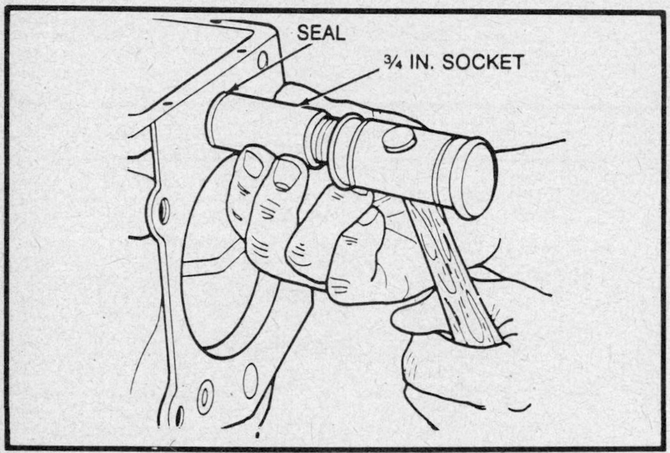
SEAL
¾ IN. SOCKET

Shift rail seal—Installation

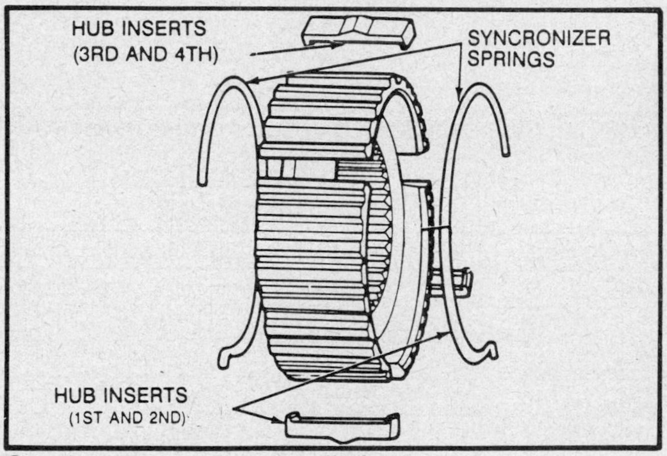
HUB INSERTS
(3RD AND 4TH)
SYNCRONIZER
SPRINGS
HUB INSERTS
(1ST AND 2ND)

Synchronizer spring installation

2. Replace the seal.

Assembly

1. Install a new shift rail seal in the rear of the case.
2. If the shift rail bushing was removed, drive a new bushing into position with a $^9/_{16}$ in. socket.

Transmission Assembly

1. Install a new shift rail seal in the rear of the case.
2. If the shift rail bushing was removed, drive a new one into position with a $^9/_{16}$ in. socket.
3. Slide the synchronizer hub onto the shaft, making sure that the shift fork groove is facing the front of the shaft.

NOTE: The sleeve and hub are select fit and must be assembled with the etch marks in the same relative locations.

4. Locate an insert in each of the 3 slots in the hub. Lubricate the parts and install an insert spring inside the sleeve; the spring tab must locate in a U-section of an insert. Fit the other spring to the opposite face, making sure that the tab locates in the same insert. Both springs should be in the same rotational direction. The tab end of 1 spring should be aligned with the tab of the spring on the opposite side.
5. Assemble a blocking ring on the 1st gear side of the 1st-2nd synchronizer. Lubricate the cone surface of 1st gear and the output shaft gear journals and slide the cone onto the output shaft, so that the cone surface engages the blocking ring.
6. Position the spacer on the output shaft (the larger diameter rearward).
7. Install a snapring which will eliminate the end play from the output shaft bearing. Position the output shaft bearing on the shaft and press it into place. Secure the bearing with the thickest snapring that will fit the groove.
8. Slide the synchronizer onto the hub and locate an insert in each of 3 slots in the sleeve.

NOTE: When installing the sleeve to the hub, the etch marks must be in the same relative locations.

9. Lightly oil the parts and complete the assembly of the synchronizer by following directions in previous Steps 3 and 4.
10. Position the 2nd speed gear and the blocking ring on the output shaft (the dog teeth must face rearward. Install the

washer and snapring, position the 3rd speed gear onto the output shaft (the dog teeth must face forward). Lubricate the gear cones and assemble a blocking ring onto the 3rd gear cone.

11. Position the 3rd–4th synchronizer assembly on the output shaft (the hub boss must face forward).

12. Install the press plates against the boss on the synchronizer hub, place the entire unit in a press (the extension end up) and press the synchronizer assembly onto the output shaft as far as possible.

13. Retain the 3rd–4th synchronizer assembly to the output shaft with a snapring. Pull up on the synchronizer so that the snapring is tight in the groove.

14. Lubricate the gear cone and place the blocking ring on the input shaft gear cone.

15. Press the speedometer drive gear onto the shaft to the marked location.

16. Lubricate the bearing bore of the extension housing. Install the output shaft in the housing; it may be necessary to tap the shaft while holding the synchronizer sleeves firmly. Secure the shaft to the housing with the snapring.

17. Press the bearing onto the input shaft; the snapring groove must face the front of the shaft (use the thickest snapring that will fit).

18. Slide the spacer and the dummy shaft into the countershaft gear. Position a thin bearing retaining washer on each end of the dummy shaft. Lubricate the roller bearings and install the long bearings in the small end of the gear and the short bearings in the long end of the gear (21 needle bearings are used at either end of the gear, for a total of 42 needle bearings).

19. Place a thick retaining washer over each end of the dummy shaft. Grease the thrust washers and place 1 on each end of the dummy shaft, then lower the gear into the case.

NOTE: When installing the thrust washers, the tabs must be in the same relative position to engage the slots in the case when the gear is lowered. Loop a piece of rope around each end of the gear and carefully install the gear and rope through the rear of the case.

20. Lubricate the reverse idler gear shaft. Position the selector lever relay on the pivot pin and secure with a spring clip. Hold the gear in the lever (with the long hub facing the rear of the case) and slide the reverse idler shaft into place. Seat the shaft in the case with a brass hammer.

21. Install a new seal in the input shaft bearing retainer and the input shaft in the case with a new bearing retainer O-ring. Tap on the outer race of the bearing to seat the outer snapring.

CAUTION

Use a soft hammer and do not tap on the input shaft itself.

22. Carefully slide the 3rd–4th synchronizer sleeve into the 4th speed position.

23. Place a new gasket on the extension housing.

24. Lubricate and install the input shaft pilot bearing on the shaft. Slide the extension housing and output shaft into place, being careful not to disturb the 3rd–4th synchronizer.

25. Align the cutaway in the extension housing flange with the countershaft bore in the rear of the case.

26. Move the countershaft gear into place and install the countershaft, making sure that the thrust washers remain in place. The flat on the countershaft should be parallel to the top of the case. Tap the shaft with a brass hammer until the front of the shaft is flush with the case.

27. Rotate the extension housing to align the bolt holes and loosely install the bolts. Apply sealer to the attaching bolts and torque to 33–36 ft. lbs.

NOTE: When installing the extension housing-to-case, make sure that the rail slides freely in its bore. Binding is remedied by slightly rotating the extension housing to free the rail, then push the housing into the case.

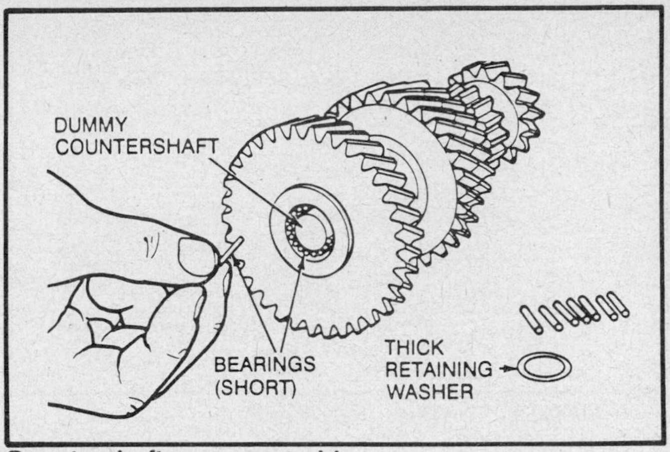
DUMMY
COUNTERSHAFT
BEARINGS
(SHORT)
THICK
RETAINING
WASHER

Countershaft gear assembly

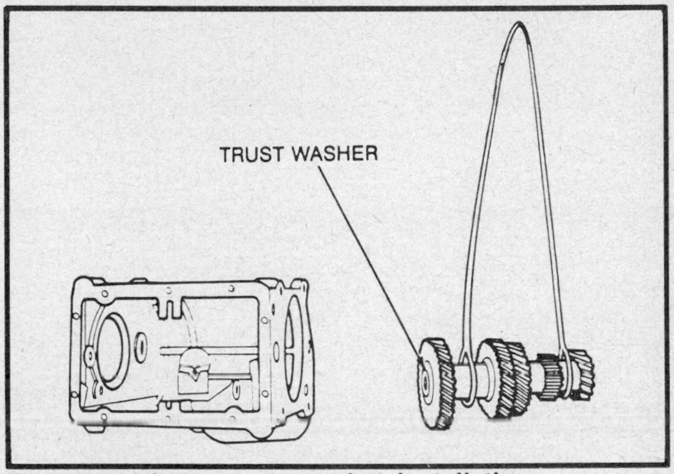
TRUST WASHER

Countershaft gear—removal or installation

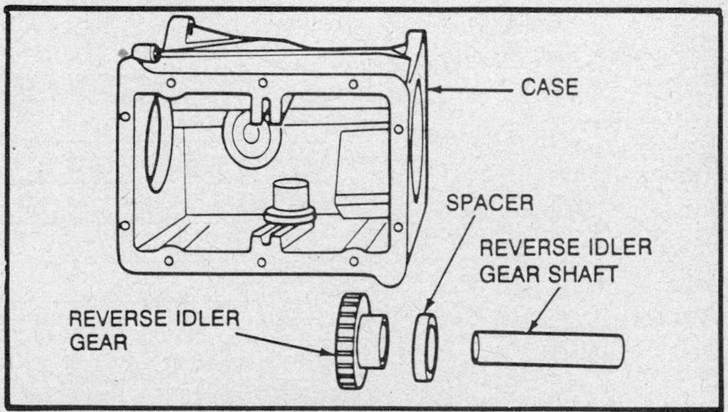
CASE
SPACER
REVERSE IDLER
GEAR SHAFT
REVERSE IDLER
GEAR

Reverse idler gear disassembled

28. Place the shift forks in the synchronizer sleeves and install the interlock lever and new retaining pin. Lubricate the shift rail oil seal and slide the shift rail through the extension housing, the case and the 1st–2nd speed shift fork. Position the selector arm on the rail and slide the rail through the 3rd–4th shift fork. Slide the shift rail through the front of the case until the center detent bore is aligned with the detent plunger bore, then install a new retaining pin in the selector arm.

29. Install the detent plunger, the spring and the plug with sealer. Install a new access plug in the rear of the case. Position a new oil seal with a tension spring and lip facing in the direction of the case and drive the seal in until it bottoms.

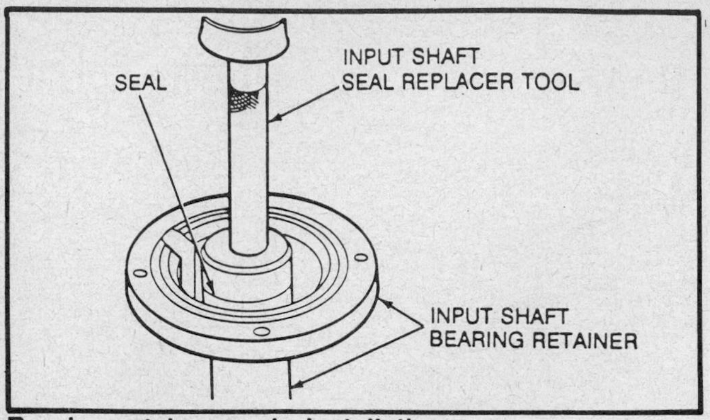
INPUT SHAFT
SEAL REPLACER TOOL
SEAL
INPUT SHAFT
BEARING RETAINER
Bearing retainer seal—installation

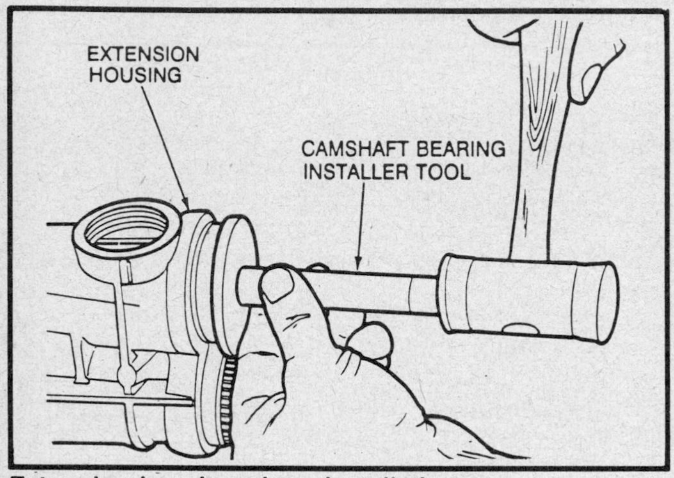
EXTENSION
HOUSING
CAMSHAFT BEARING
INSTALLER TOOL
Extension housing plug—installation

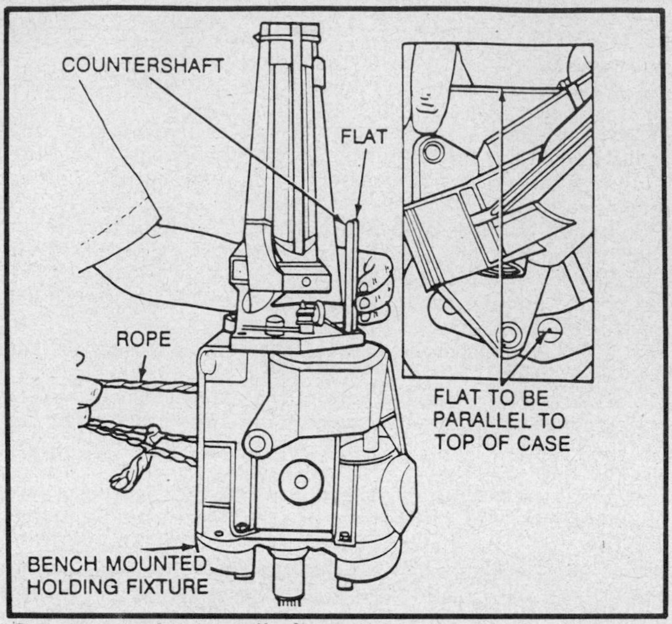
COUNTERSHAFT
FLAT
ROPE
FLAT TO BE
PARALLEL TO
TOP OF CASE
BENCH MOUNTED
HOLDING FIXTURE
Countershaft—Installation

30. Position a new O-ring in the groove in the case. Position the input shaft bearing retainer with the groove in the retainer aligned with the oil passage in the case and install the retaining bolts finger tight.
31. Install the flywheel housing, then torque the housing and the front bearing retainer bolts. Coat the retainer with grease.
32. Install the clutch release arm and the bearing.
33. Install a new extension housing plug, using sealer.
34. Install a new cover gasket and cover, with the vent facing the rear. Apply sealer to the left front cover attaching bolt and torque to 8–10 ft. lbs.

SPECIAL TOOLS

Tool Number	Description
T50T-100-A	Impact slide hammer
T57L-500-B	Bench mounted holding fixture
T71P-4621-B	Bearing remover
T70P-6011-A	Camshaft bearing plug installer
T71P-7050-A	Input shaft seal replacer
T71P-7095-A	Extension housing seal replacer
T71P-7111-B	Dummy countershaft
T71P-7140-A	Reverse idler shaft remover
T71P-17271-A	Speedometer drive gear installer

Section 7

T5OD Transmission
Ford Motor Co.

APPLICATION

FORD T5OD 5 SPEED WITH OVERDRIVE (REP)

Year	Vehicle	Engine
1984	Capri	2.3L Turbo
		5.0L 4V HO
	Cougar	2.3L Turbo
	Mustang	2.3L Turbo
		5.0L 4V HO
	Thunderbird	2.3L Turbo
1985	Capri	5.0L 4V HO
	Cougar	2.3L Turbo
	Mustang	2.3L Turbo
		5.0L 4V HO
	Thunderbird	2.3L Turbo

Year	Vehicle	Engine
1986	Capri	5.0L HO EFI
	Cougar	2.3L Turbo
	Mustang	2.3L Turbo
		5.0L HO EFI
	Thunderbird	2.3L Turbo
1987	Mustang	2.3L EFI
		5.0L HO SEFI
	Thunderbird	2.3L Turbo
1988–89	Mustang	2.3L EFI OHC
		5.0L HO EFI Plus
	Thunderbird	2.3L Turbo

GENERAL DESCRIPTION

The Ford 5 speed with overdrive (REP) transmission is a 5 speed transmission manufactured by Borg Warner. This transmissions provide fully synchronized forward speeds and a single non-synchronized reverse gear. All forward gears are helical cut and are in constant mesh. This transmission is manufactured with metric measurements and all replacement parts must be the correct metric dimension. The transmission case, cover and adapter housing case are aluminum.

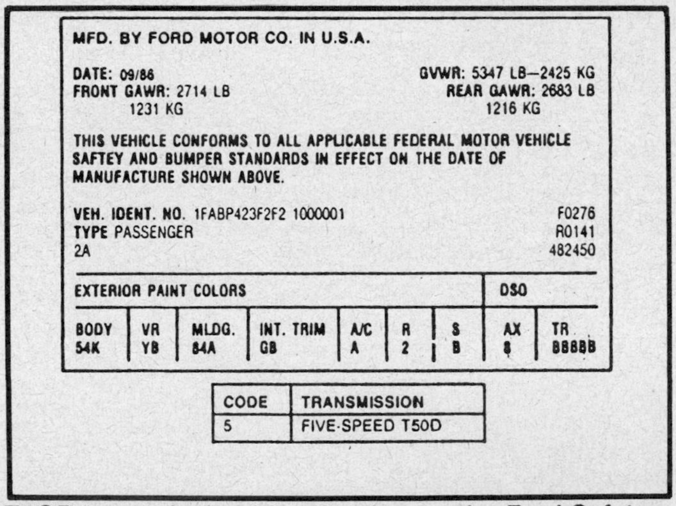
MFD. BY FORD MOTOR CO. IN U.S.A.
DATE: 09/86
GVWR: 5347 LB—2425 KG
FRONT GAWR: 2714 LB
1231 KG
REAR GAWR: 2683 LB
1216 KG
THIS VEHICLE CONFORMS TO ALL APPLICABLE FEDERAL MOTOR VEHICLE SAFTEY AND BUMPER STANDARDS IN EFFECT ON THE DATE OF MANUFACTURE SHOWN ABOVE.
VEH. IDENT. NO. 1FABP423F2F2 1000001
TYPE PASSENGER
2A
F0276
R0141
482450
EXTERIOR PAINT COLORS
DSO
BODY 54K
VR YB
MLDG. 84A
INT. TRIM GB
A/C A
R 2
S B
AX 8
TR BBBBB
CODE
TRANSMISSION
5
FIVE-SPEED T5OD

T5OD transmission identification on the Ford Safety Certification Label

Metric Fasteners

Metric bolt sizes and thread pitches are used for all fasteners on the T5OD transmission. The use of metric tools is mandatory in the service of this transmission.

Do not attempt to interchange metric fasteners for inch system fasteners. Mismatched or incorrect fasteners can result in damage to the transmission unit through malfunctions, breakage or possible personal injury. Care should be taken to reuse the fasteners in the same location as removed, whenever possible. Due to the large number of alloy parts used and the aluminum casing, torque specifications should be strictly observed. Before installing cap screws into aluminum parts, always dip screws into oil to prevent the screws from galling the aluminum threads and to prevent seizing.

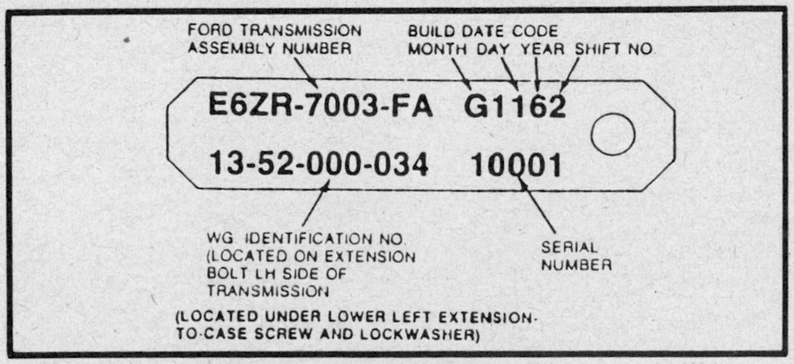
FORD TRANSMISSION ASSEMBLY NUMBER
BUILD DATE CODE
MONTH DAY YEAR SHIFT NO
E6ZR-7003-FA
G1162
13-52-000-034
10001
WG IDENTIFICATION NO (LOCATED ON EXTENSION BOLT LH SIDE OF TRANSMISSION
SERIAL NUMBER
(LOCATED UNDER LOWER LEFT EXTENSION-TO-CASE SCREW AND LOCKWASHER)

T5OD transmission identification tag

Capacities

All Ford models with the T5OD require 5.6 pts. (2.6L) of Dexron®II automatic transmission fluid.

Checking Fluid Level

The vehicle must be level before an accurate reading can be obtained. The fluid level considered full when the lubricant is to the bottom of the fill hole.

TRANSMISSION MODIFICATIONS

There were no modifications for the Ford models at the time of this publication.

TROUBLE DIAGNOSIS

CHILTON'S THREE C'S TRANSMISSION DIAGNOSIS
Ford T5OD 5 Speed with Overdrive (REP)

Condition	Cause	Correction
Transmission shifts hard	a) Clutch linkage binding	a) Lubricate or repair as necessary
	b) Shift rail binding	b) Check for mispositioned selector arm roll pin, loose cover bolts, worn shift rail bores, worn shift rail, distorted oil seal, or extension housing not aligned with case. Repair as necessary
	c) Internal bind in transmission caused by shift forks, selector plates, or synchronizer assemblies	c) Remove, disassemble and inspect transmission. Replace worn or damaged components as necessary
	d) Clutch housing misalignment	d) Check runout at rear face of clutch housing
	e) Incorrect lubricant	e) Drain and refill transmission
Gear clash when shifting from one gear to another	a) Air in hydraulic system	a) Bleed hydraulic control system
	b) Clutch housing misalignment	b) Lubricate or repair as necessary
	c) Clutch linkage or cable binding	c) Check runout at rear of clutch housing
	d) Lubricant level low or incorrect lubricant	d) Drain and refill transmission and check for lubricant leaks if level was low. Repair as necessary
	e) Gearshift components, or synchronizer assemblies worn or damaged	e) Remove, disassemble and inspect transmission. Replace worn or damaged components as necessary
Transmission noisy	a) Lubricant level low or incorrect lubricant	a) Drain and refill transmission. If lubricant level was low, check for leaks and repair as necessary
	b) Clutch housing-to-engine, or transmission-to-clutch housing bolts loose	b) Check and correct bolt torque as necessary
	c) Dirt, chips, foreign material in transmission	c) Drain, flush, and refill transmission
	d) Gearshift mechanism or transmission gear, or bearing components worn or damaged	d) Remove, disassemble and inspect transmission. Replace worn or damaged components as necessary
	e) Clutch housing misalignment	e) Check runout at rear face of clutch housing
Jumps out of gear	a) Clutch housing misalignment	a) Check runout at rear face of clutch housing
	b) Offset lever nylon insert worn or lever attaching nut loose	b) Remove gearshift lever and check for loose offset lever nut or worn insert. Repair or replace as necessary
	c) Gearshift mechanism, shift forks, selector plates, interlock plate, selector arm, shift rail, detent plugs, springs or shift cover worn or damaged	c) Remove, disassemble and inspect transmission cover assembly. Replace worn or damaged components as necessary

CHILTON'S THREE C'S TRANSMISSION DIAGNOSIS
Ford T5OD 5 Speed with Overdrive (REP)

Condition	Cause	Correction
Jumps out of gear	d) Clutch shaft or roller bearings worn or damaged	d) Replace clutch shaft or roller bearings as necessary
	e) Gear teeth worn or tapered, synchronizer assemblies worn or damaged, excessive end play caused by worn thrust washers or output shaft gears	e) Remove, disassemble, and inspect transmission. Replace worn or damaged components as necessary
	f) Pilot bushing worn	f) Replace pilot bushing
Will not shift into one gear	a) Gearshift selector plates, interlock plate, or selector arm, worn, damaged, or incorrectly assembled	a) Remove, disassemble, and inspect transmission cover assembly. Repair or replace components as necessary
	b) Shift rail detent plunger worn, spring broken, or plug loose	b) Tighten plug or replace worn or damaged components as necessary
	c) Gearshift lever worn or damaged	c) Replace gearshift lever
	d) Synchronizer sleeves or hubs, damaged or worn	d) Remove, disassemble and inspect transmission. Replace worn or damaged components
Locked in one gear—cannot be shifted out of that gear	a) Shift rail(s) worn or broken, shifter fork bent, setscrew loose, center detent plug missing or worn	a) Inspect and replace worn or damaged parts
	b) Broken gear teeth on countershaft gear, clutch shaft, or reverse idler gear	b) Inspect and replace damaged part
	c) Gearshift lever broken or worn, shift mechanism in cover incorrectly assembled or broken, worn damaged gear train components	c) Disassemble transmission. Replace damaged parts or assemble correctly

POWER FLOW

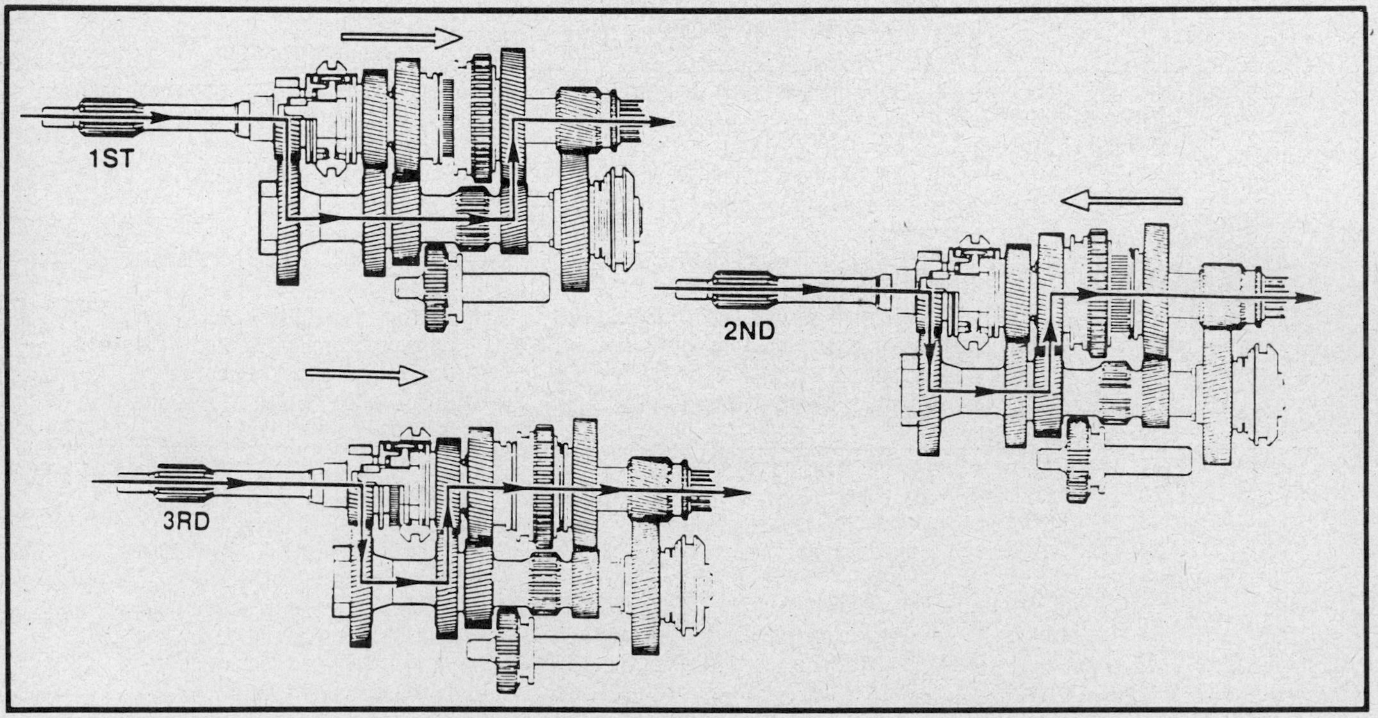
1ST
2ND
3RD

T5OD power flow

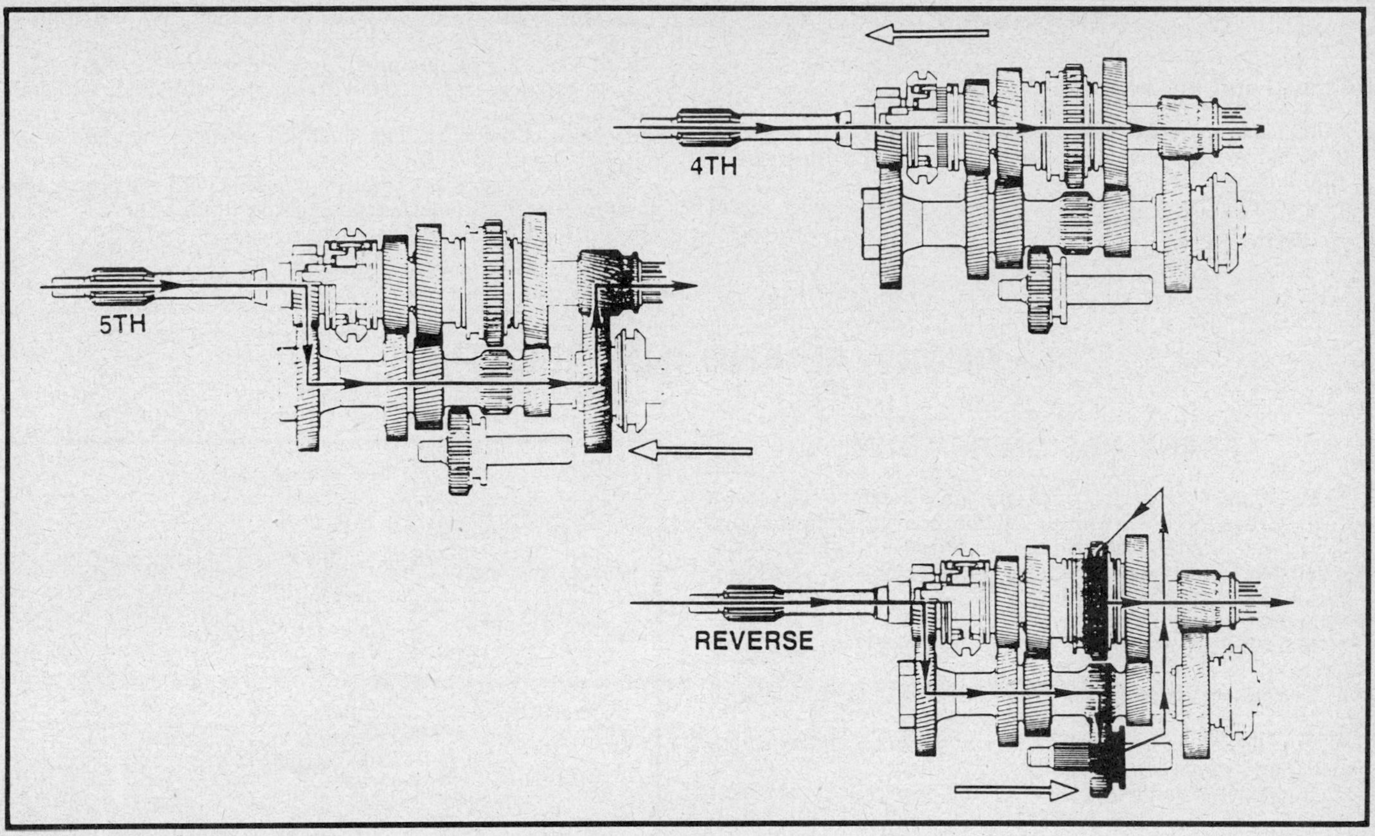
4TH
5TH
REVERSE
T5OD power flow

ON CAR SERVICE

Services

FLUID CHANGE

Procedure

1. Raise and support vehicle safely.
2. Remove the drain plug if equipped, or siphon transmission fluid into a pan.
3. Remove filler plug.
4. Install drain plug.
5. Fill to the level of the filler plug hole with the proper type and amount of automatic transmission fluid.
6. Install fill plug and lower vehicle.

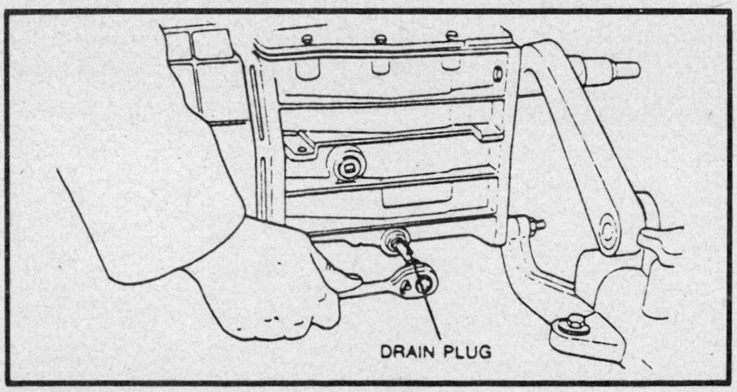
DRAIN PLUG
Transmission drain plug location

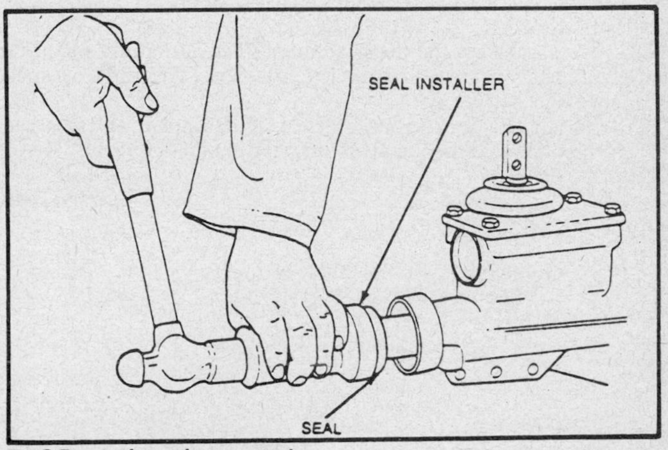
SEAL INSTALLER
SEAL
T5OD seal replacement

REAR SEAL

Removal and Installation

1. Raise and support vehicle safely.
2. Remove drain plug, if equipped or siphon transmission fluid into a pan.
3. Remove filler plug.
4. Install drain plug.
5. Mark driveshaft to yoke for reassembly and remove driveshaft.
6. Pry out old seal using an appropriate tool.
7. Install new seal, coated with locking compound, using tool J–21426 or equivalent.
8. Connect the driveshaft, making certain to align marks and torque bolts to 15 ft. lbs. (20 Nm).
9. Fill to the level of the filler plug hole with the proper type and amount of automatic transmission fluid.
10. Install fill plug and lower vehicle.

REMOVAL AND INSTALLATION

TRANSMISSION REMOVAL

1. Disconnect the negative battery cable. Remove the screws attaching the shift lever boot to the floorpan and slide boot up.
2. Remove the bolts attaching the shift lever housing to the transmission and remove the lever and housing.
3. Raise and support the vehicle safely and remove skid plate, if equipped.
4. Mark the driveshaft to output yoke for assembly alignment reference.
5. Disconnect the driveshaft and support it to vehicle with wire.
6. Support engine, using safety stand placed at the clutch housing and remove crossmember.
7. Remove the catalytic converter and, if necessary, the inlet pipe.
8. Mark and disconnect the speedometer adapter (if equipped) and cable.
9. Remove the back-up light wiring.
10. Support transmission with transmission jack and remove the crossmember if not already removed.
11. Remove the bolts attaching the transmission to the clutch housing and remove the transmission.

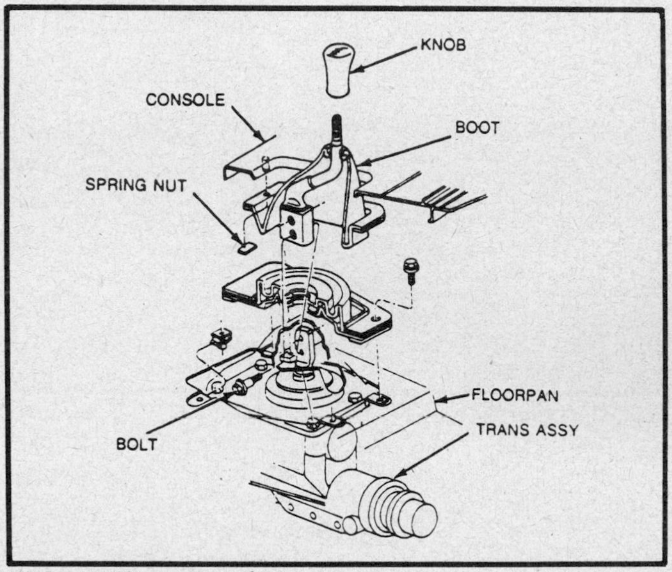
KNOB
CONSOLE
BOOT
SPRING NUT
FLOORPAN
TRANS ASSY
BOLT

T5OD shifter mounting

TRANSMISSION INSTALLATION

1. Align the throwout bearing with the splines in the driven plate hub.
2. Shift the transmission into gear, using a long appropriate tool.
3. Mount the transmission on the transmission jack, raise and position the transmission.
4. Make certain transmission is installed properly and torque bolts to 45–65 ft. lbs. (61–88 Nm).
5. Install the speedometer adapter if used, making certain to align marks on case. Install the speedometer cable or speed sensor.
6. Install the brace rods or rods, if equipped, and the rear crossmember and torque attaching bolts to 36–50 ft. lbs. (48–68 Nm).
7. Connect the backup lamp and neutral switch wiring.
8. Install the catalytic converter and torque the attaching bolts to 20–30 ft. lbs. (27–41 Nm).
9. Connect the driveshaft, making certain to align marks and torque the U-bolts to 42–57 ft. lbs. (56–77 Nm).
10. Check transmission lubrication level.
11. Lower the vehicle.
12. Apply RTV – type sealant to the gear shift lever mounting cover and install cover. Torque bolts to 13 ft. lbs. (18 Nm).
13. Install the gearshift lever on the mounting cover. Make certain the the lever is engaged with the shift rail before torquing bolts to 18 ft. lbs. (24 Nm).
14. Position the shift lever boot on the floorpan and install boot attaching screws.

BENCH OVERHAUL

Before Disassembly

Clean the exterior of the transmission assembly before any attempt is made to disassemble it in order to prevent dirt or other foreign materials from entering the transmission assembly or its internal parts.

NOTE: If steam cleaning is done to the exterior of the transmission, immediate disassembly should be done to avoid rusting, caused by condensation forming on the internal parts.

Transmission Disassembly

1. Remove drain bolt on transmission case and drain lubricant.
2. Thoroughly clean the exterior of the transmission assembly.
3. Position the shift lever in the **N** position and remove the turret cover attaching bolts.
4. Using a suitable prybar, break the turret cover-to-extension housing seal by prying carefully between the housing and the cover. Remove the turret cover.
5. Using a 5mm pin punch and hammer, remove roll pin attaching offset lever to shift rail.
6. Remove extension housing to transmission case bolts and remove housing and offset lever as an assembly. Do not attempt to remove the offset lever while the extension housing is still bolted in place. The lever has a positioning lug engaged in the housing detent plate which prevents moving the lever far enough for removal.
7. Remove detent ball and spring from offset lever and remove roll pin from extension housing or offset lever.
8. Remove bolts attaching transmission cover and shift fork assembly and remove cover. Take notice that 2 of the transmission cover attaching bolts are alignment-type dowel bolts. Note the location of these bolts for assembly reference.
9. Using long needlenose pliers, remove the C-clip attaching the 5th/reverse shift lever to the lever pivot pin.
10. Using a Torx® driver remove the 5th/reverse shift lever pivot pin. Do not remove the 5th/reverse shift lever at this point. Remove the backup light switch.
11. Remove 5th gear synchronizer snapring, shift fork, 5th gear synchronizer sleeve, blocking ring and 5th speed drive gear from rear of countershaft.
12. Remove snapring from 5th speed driven gear.
13. Using a hammer and punch, mark both bearing cap and case for assembly reference.
14. Remove front bearing cap bolts and remove front bearing cap. Remove front bearing race and endplay shims from front bearing cap.
15. Rotate drive gear until flat surface faces countershaft and remove drive gear from transmission case.

NOTE: Be careful not to drop the 15 roller bearings, thrust washer or race from the rear of the input shaft.

16. Remove reverse lever C-clip and pivot bolt.
17. Remove the output shaft rear bearing race and then tilt mainshaft assembly upward and remove assembly from transmission case.
18. Unhook the over center link spring from front of transmission case.
19. Rotate 5th/reverse gear shift rail to disengage rail from reverse lever assembly. Remove shift rail from rear of transmission case.
20. Remove reverse lever and fork assembly from transmission case.

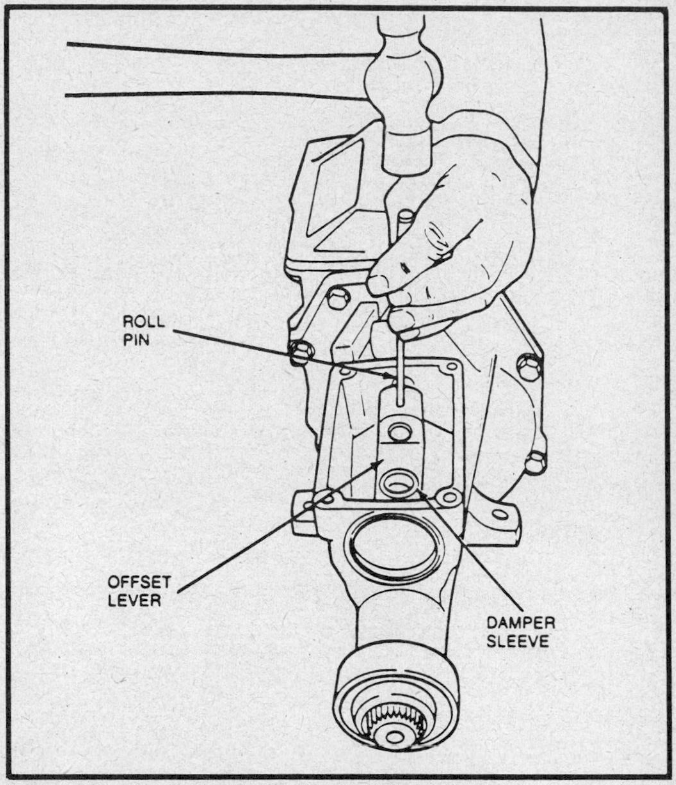
ROLL PIN
OFFSET LEVER
DAMPER SLEEVE
Removing roll pin from the shifter shaft

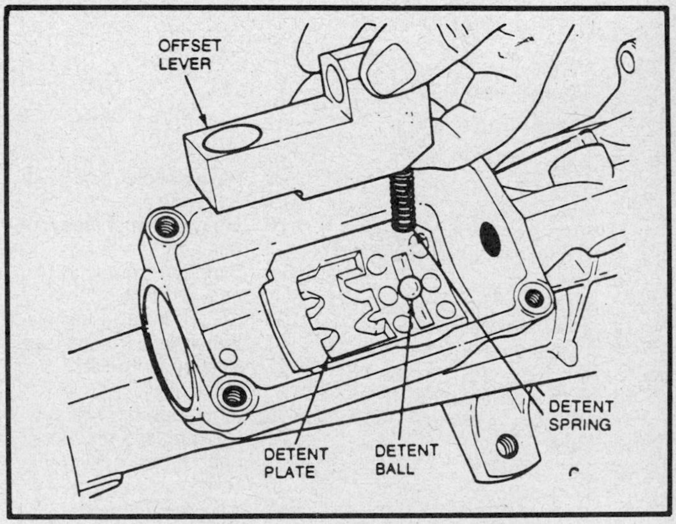
OFFSET LEVER
DETENT SPRING
DETENT PLATE
DETENT BALL
Removig the offset lever and related parts

21. Using hammer and punch, drive roll pin from forward end of reverse idler shaft and remove reverse idler shaft, rubber O-ring and gear from the transmission case.
22. Using a hammer and punch flatten out the tabs on the countershaft retainer at all four corners. Remove the bolts retaining the countershaft retainer and remove the retainer.

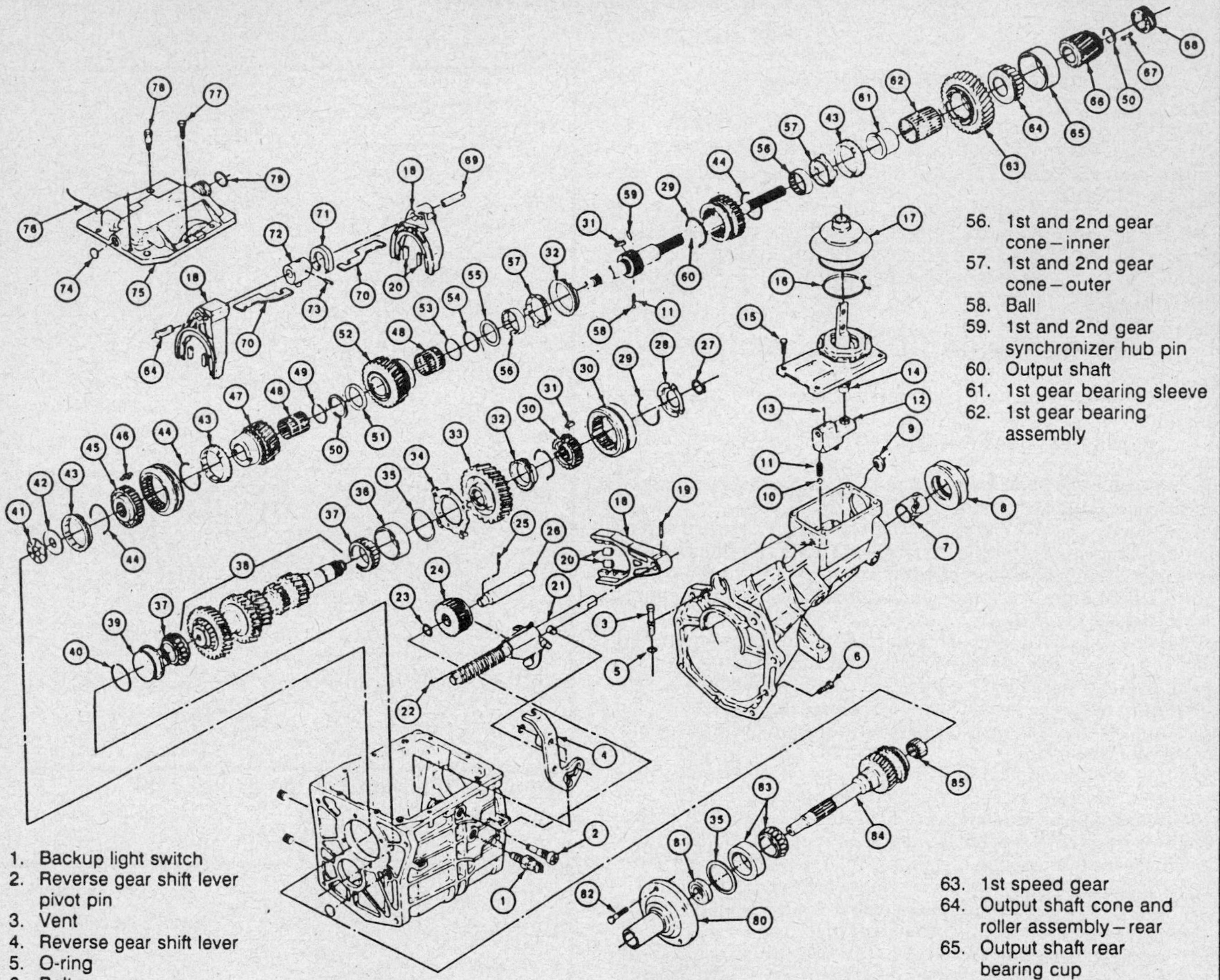

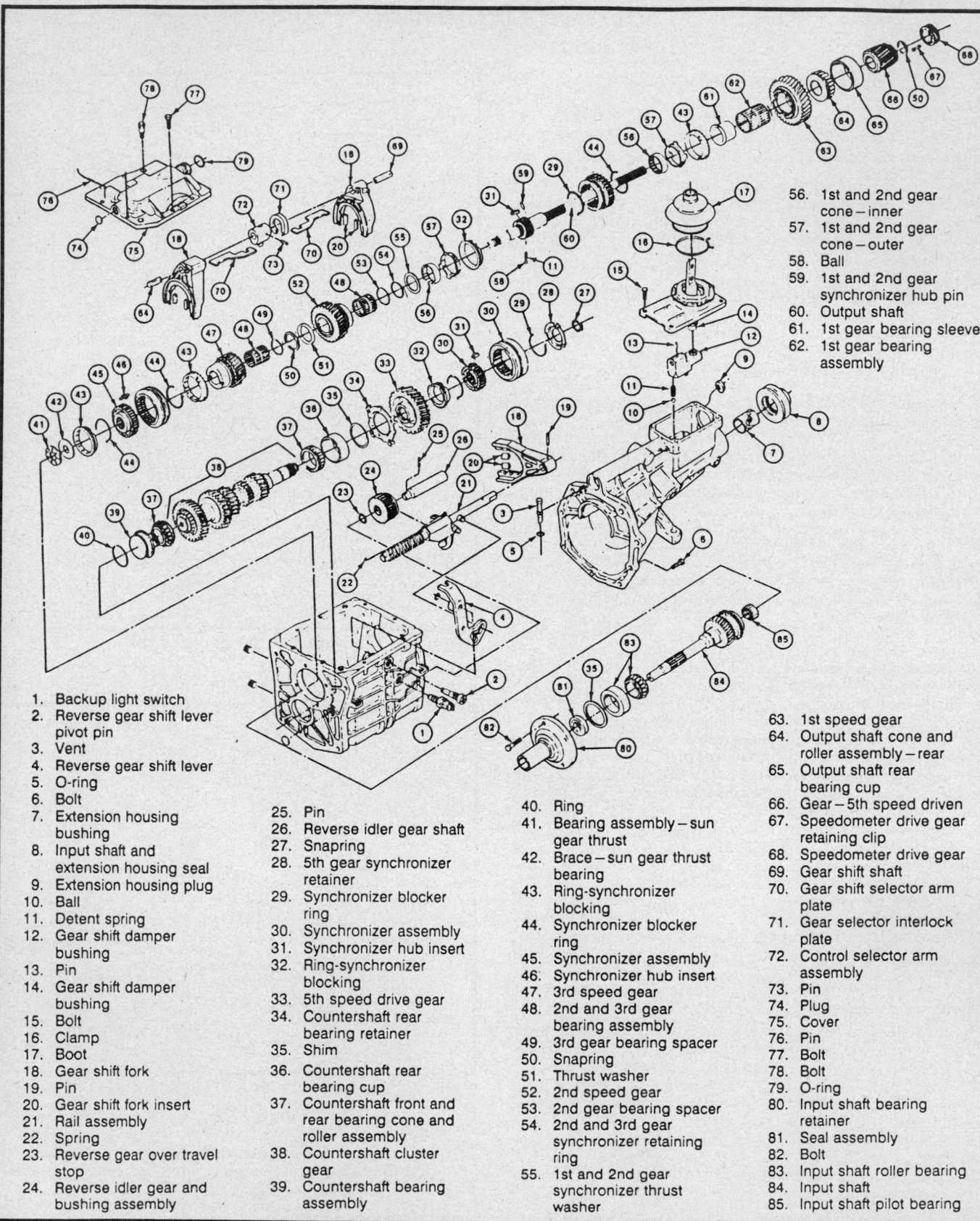

1. Backup light switch
2. Reverse gear shift lever pivot pin
3. Vent
4. Reverse gear shift lever
5. O-ring
6. Bolt
7. Extension housing bushing
8. Input shaft and extension housing seal
9. Extension housing plug
10. Ball
11. Detent spring
12. Gear shift damper bushing
13. Pin
14. Gear shift damper bushing
15. Bolt
16. Clamp
17. Boot
18. Gear shift fork
19. Pin
20. Gear shift fork insert
21. Rail assembly
22. Spring
23. Reverse gear over travel stop
24. Reverse idler gear and bushing assembly
25. Pin
26. Reverse idler gear shaft
27. Snapring
28. 5th gear synchronizer retainer
29. Synchronizer blocker ring
30. Synchronizer assembly
31. Synchronizer hub insert
32. Ring-synchronizer blocking
33. 5th speed drive gear
34. Countershaft rear bearing retainer
35. Shim
36. Countershaft rear bearing cup
37. Countershaft front and rear bearing cone and roller assembly
38. Countershaft cluster gear
39. Countershaft bearing assembly
40. Ring
41. Bearing assembly—sun gear thrust
42. Brace—sun gear thrust bearing
43. Ring-synchronizer blocking
44. Synchronizer blocker ring
45. Synchronizer assembly
46. Synchronizer hub insert
47. 3rd speed gear
48. 2nd and 3rd gear bearing assembly
49. 3rd gear bearing spacer
50. Snapring
51. Thrust washer
52. 2nd speed gear
53. 2nd gear bearing spacer
54. 2nd and 3rd gear synchronizer retaining ring
55. 1st and 2nd gear synchronizer thrust washer
56. 1st and 2nd gear cone—inner
57. 1st and 2nd gear cone—outer
58. Ball
59. 1st and 2nd gear synchronizer hub pin
60. Output shaft
61. 1st gear bearing sleeve
62. 1st gear bearing assembly
63. 1st speed gear
64. Output shaft cone and roller assembly—rear
65. Output shaft rear bearing cup
66. Gear—5th speed driven
67. Speedometer drive gear retaining clip
68. Speedometer drive gear
69. Gear shift shaft
70. Gear shift selector arm plate
71. Gear selector interlock plate
72. Control selector arm assembly
73. Pin
74. Plug
75. Cover
76. Pin
77. Bolt
78. Bolt
79. O-ring
80. Input shaft bearing retainer
81. Seal assembly
82. Bolt
83. Input shaft roller bearing
84. Input shaft
85. Input shaft pilot bearing

Ford T5OD transmission—exploded view

23. Remove the rear bearing outer race by pushing it rearward. Remove the bearing assembly from shaft using a puller and move the countershaft assembly rearward, tilt countershaft upward and remove from case.

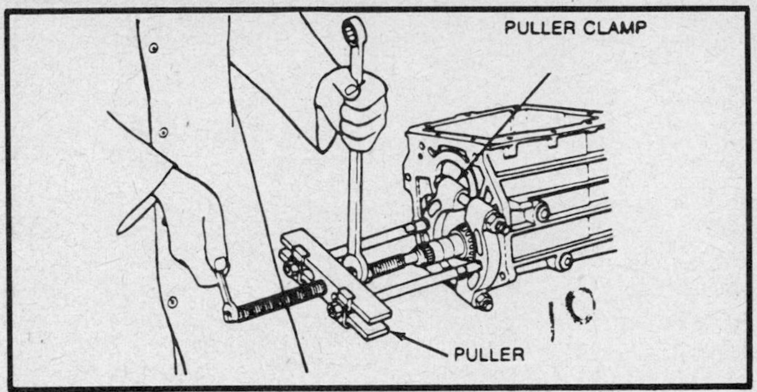
PULLER CLAMP
PULLER
Removing the rear countershaft bearing with gear puller

Unit Disassembly and Assembly

OUTPUT SHAFT

Disassembly

1. Using a press and puller plate, remove 3rd/4th synchronizer, blocking ring, sleeve, hub and 3rd gear as an assembly from the output shaft.
2. Remove the needle bearing assembly and spacer from the output shaft.
3. Remove the 5th gear snapring.
4. Using a press and puller plate, remove 5th gear. Remove the output shaft rear bearing.
5. Remove the 1st gear assembly. Remove the needle bearing assembly and sleeve.
6. Using a pencil magnet remove the roll pin from the 1st/2nd gear syncronizer.
7. Remove the 3-piece 1st gear blocker ring assembly from the output shaft.
8. Remove the 2nd gear snapring and thrust washer. Remove 2nd gear from the output shaft. Remove the 2nd gear needle bearing assembly and spacer.
9. Disengage the spiral snapring from the slot in the output shaft using a small suitable prybar. Remove the spiral snapring and thrust washer from the output shaft.
10. Remove the 3-piece 2nd gear blocker ring assembly. Remove the 1st/2nd synchronizer from the output shaft.
11. Remove the detent ball and spring from the hub on the output shaft.

NOTE: The 1st/2nd speed syncronizer and hub is serviced only as an assembly, with the output shaft. Therefore no attempt should be made to separate the hub from the shaft.

Inspection

1. Clean all components, except plastic or nylon, in solvent.

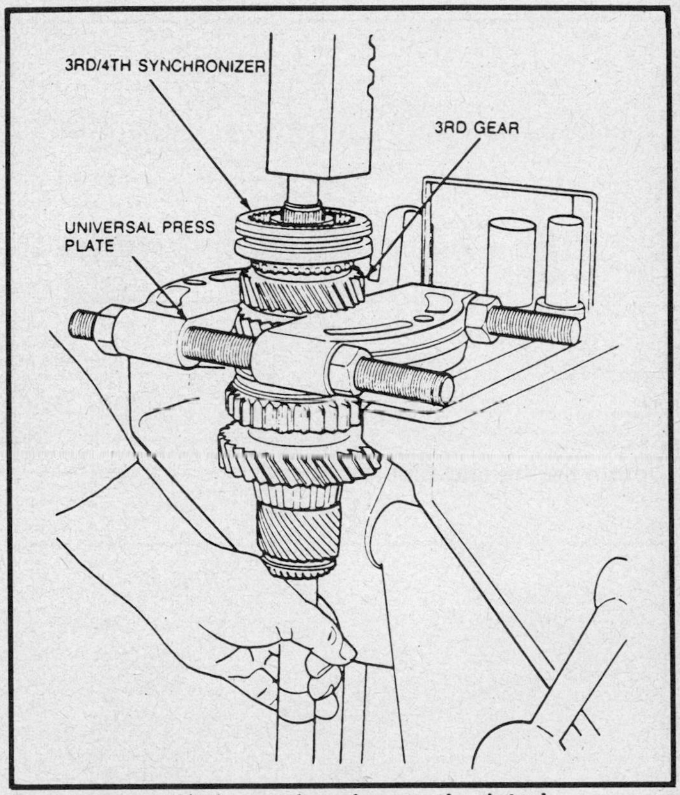
3RD/4TH SYNCHRONIZER
3RD GEAR
UNIVERSAL PRESS PLATE
Removing 3rd/4th synchronizer and related assemblies

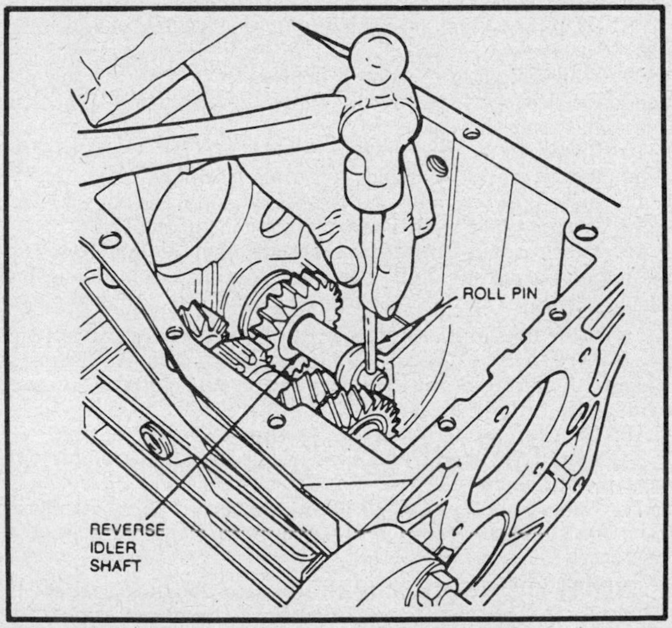
ROLL PIN
REVERSE IDLER SHAFT
Removing roll pin from reverse idler shaft

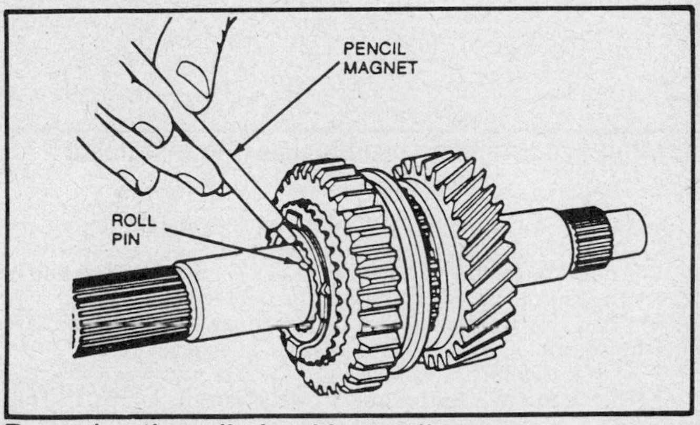
PENCIL MAGNET
ROLL PIN
Removing the roll pin with pencil magnet

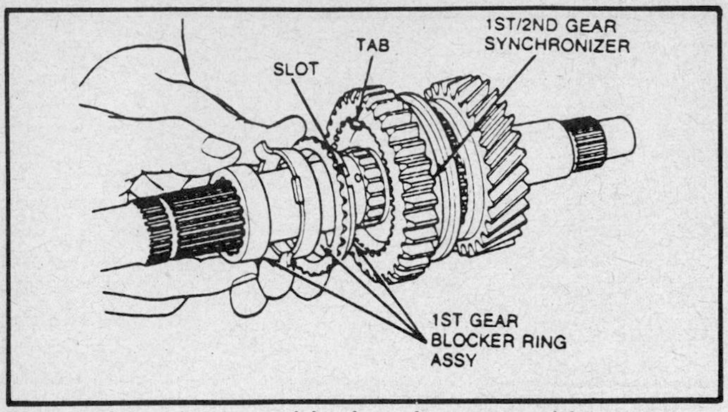
1ST/2ND GEAR SYNCHRONIZER
TAB
SLOT
1ST GEAR BLOCKER RING ASSY

The 3-piece 1st gear blocker ring assembly

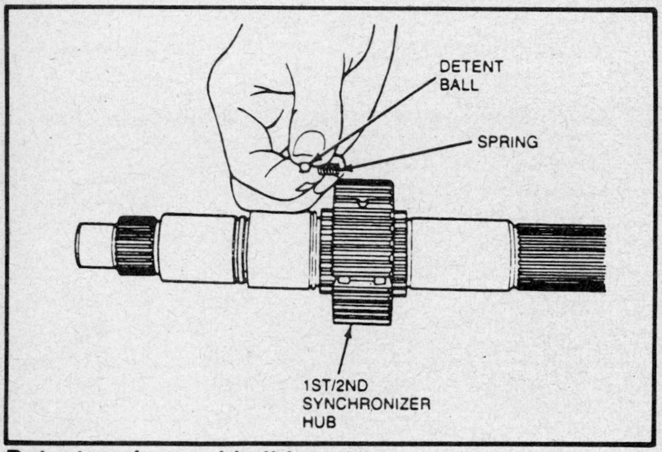
DETENT BALL
SPRING
1ST/2ND SYNCHRONIZER HUB

Detent spring and ball location

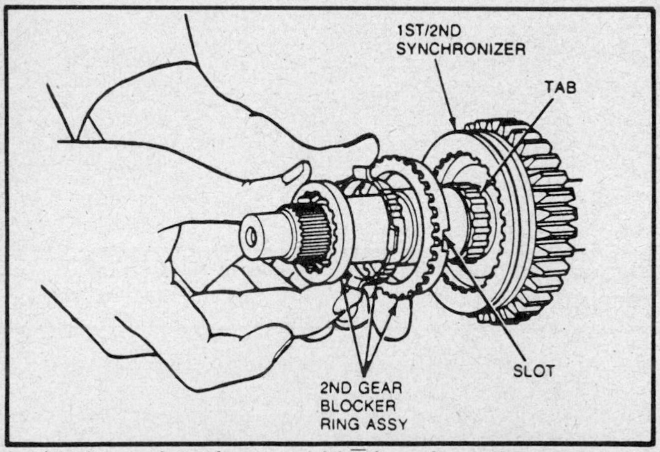
1ST/2ND SYNCHRONIZER
TAB
SLOT
2ND GEAR BLOCKER RING ASSY

1st/2nd synchronizer and blocker ring alignment

2. Inspect for broken, chipped or worn gear teeth.
3. Inspect for bent or broken inserts.
4. Check for damaged roller thrust or needle bearings and inspect the bearing bores for cracks or damage.
5. Check for worn or loss of surface metal from the counter shaft and hub, clutch shaft or reverse idler gear shaft.
6. Check the thrust washers for damage and wear.
7. Check for nicked, broken or worn output or clutch shaft splines.
8. Check for bent, weak or distorted snaprings.

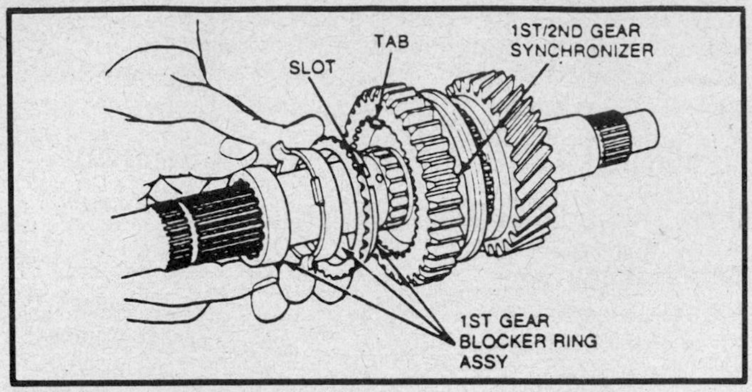
1ST/2ND GEAR SYNCHRONIZER
TAB
SLOT
1ST GEAR BLOCKER RING ASSY

1st/2nd synchronizer and blocker ring alignment

9. Check the reverse idler gear busing for wear.
10. Inspect front and rear bearings for roughness, wear and damage.
11. Inspect the shift rails, selector arms, plates, interlock for worn, bent or damaged parts.
12. Inspect synchronizers damage, wear and rough surfaces.
13. If any of the above are found, replace the appropriate part(s).

Assembly

1. Install the detent ball and spring in the 1st/2nd synchronizer hub on the output shaft.
2. Install the 1st/2nd synchronizer assembly, and 3-piece 2nd gear blocker ring assembly onto the output shaft. The slots in the blocker ring must be aligned with the tabs in the 1st/2nd synchronizer.
3. Install the thrust washer onto the output shaft. Install the spiral snapring using a suitable snapring tool.
4. Install the 2nd gear needle bearing and spacer.
5. Install the 2nd gear onto the output shaft, be certain to align the slots in the 2nd gear with the tabs on the blocker ring assembly.
6. Using snapring pliers, install the 2nd gear thrust washer and snapring.
7. Install the 3-piece 1st gear blocker ring assembly onto the output shaft, be certain to align the slots in the 1st gear blocking ring with the tab on the 1st/2nd synchronizer.
8. Drop the roll pin into the output shaft, using a pencil magnet if necessary.
9. While aligning the notch in the sleeve with the roll pin, slide the sleeve onto the output shaft. Install the needle bearing assembly over the sleeve
10. Install 1st gear onto the output shaft, be certain to align the tabs on the 1st gear blocking ring assembly.
11. Install the rear output shaft bearing onto the output shaft. The bearing taper must be facing the rear of the shaft.
12. Lubricate the 5th gear splines with petroleum jelly and position on the output shaft. Using a press and a capped length of pipe, 1¼ in. × 14 in. press 5th gear onto the output shaft.

NOTE: Be certain the output shaft and gear splines are carefully aligned. If the splines are improperly aligned, damage could occur when the gear is pressed onto the output shaft.

13. Using snapring pliers, install the 5th gear snapring.
14. Install the spacer and split needle bearing assembly onto the output shaft.
15. Using a press and a suitable 1½ in. deep well socket, install the 3rd/4th synchronizer, 3rd blocking ring and 3rd gear as an assembly.

NOTE: Before pressing the components together, be certain the synchronizer hub faces the short end of the output shaft. Hold the 3rd gear against the synchronizer to maintain the blocking ring alignment.

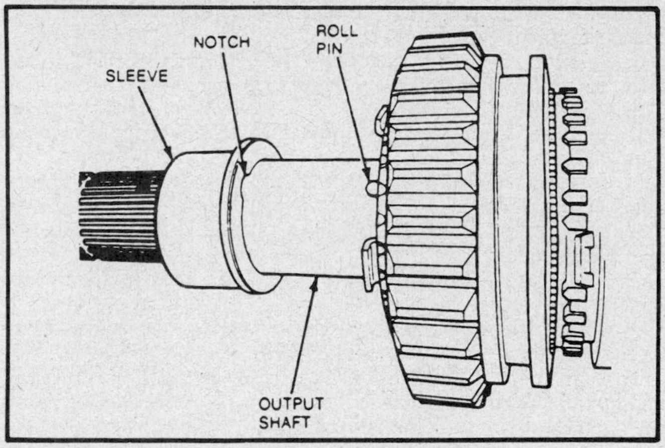
NOTCH
ROLL PIN
SLEEVE
OUTPUT SHAFT
Sleeve alignment on the output shaft

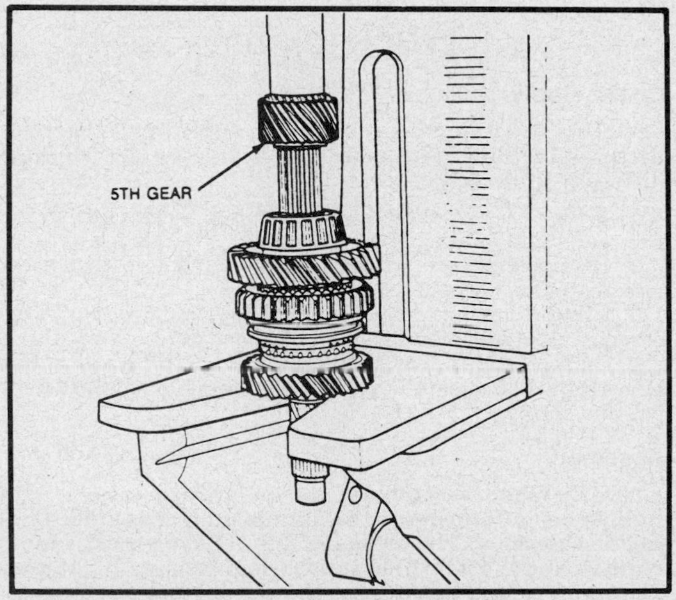
5TH GEAR
Pressing 5th gear onto the output shaft

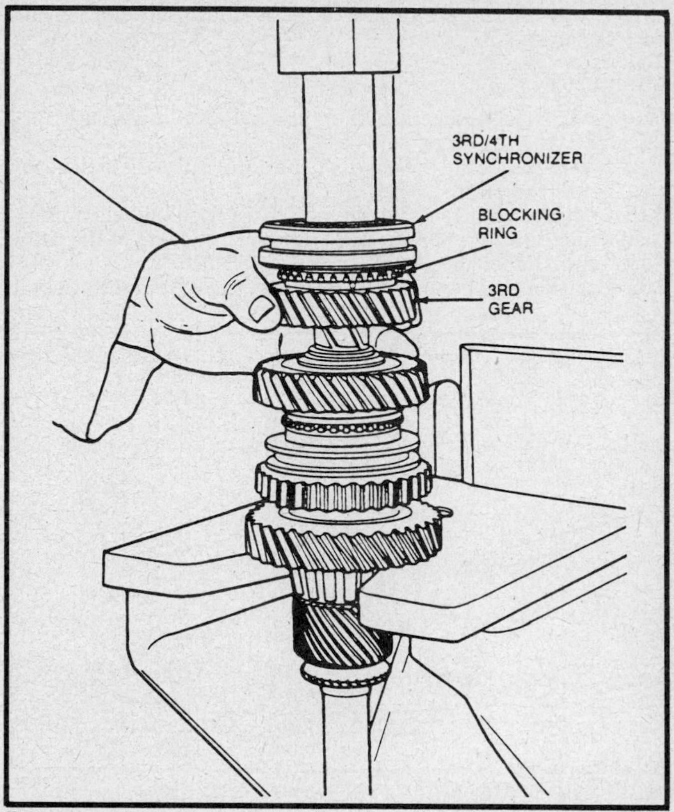
3RD/4TH SYNCHRONIZER
BLOCKING RING
3RD GEAR
Pressing 3rd gear and related components onto the output shaft

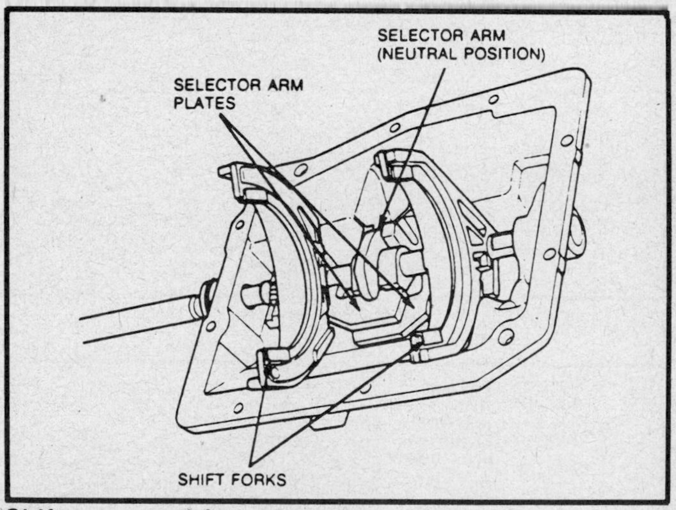
SELECTOR ARM (NEUTRAL POSITION)
SELECTOR ARM PLATES
SHIFT FORKS
Shift cover and fork assembly

SHIFT COVER AND FORKS

Disassembly

NOTE: Mark the position of the parts so that they may be properly installed. Note the order of components as disassembled for reference during assembly.

1. Place the selector arm plates and the shift rail centered in the neutral position.
2. Rotate the shift rail counterclockwise until the selector arm disengages from the selector arm plates; the selector arm roll pin should now be accessible.
3. Pull the shift rail rearward until the selector contacts the 1st/2nd shift fork.
4. Use a $^{3}/_{16}$ in. pin punch and remove the selector arm roll pin and the shift rail.
5. Remove the shift forks, the selector arm, the roll pin and the interlock plate.
6. Remove the shift rail oil seal and O-ring.
7. Remove the nylon inserts and the selector arm plates from the shift forks.

Inspection

1. Clean all components, except plastic or nylon, in solvent.
2. Inspect for cracks in the bores, sides, bosses or bolt holes.
3. Inspect for stripped threads in the bolt holes.
4. Check for nicks, burrs or rough surfaces in the shaft bores on the gasket surfaces.
5. Check for worn shift fork inserts and for broken, cracked or worn shift forks.
6. Inspect the shift rails, selector arms, plates, interlock for worn, bent or damaged parts, if not already done.

7. If any of the above are found, replace the appropriate part(s).

Assembly

1. Attach the nylon inserts to the selector arm plates and through the shift forks.
2. If removed, coat the edges of the shift rail plug with sealer and install the plug.
3. Coat the shift rail and the rail bores with petroleum jelly, then slide the shift rail into the cover until the end of the rail is flush with the inside edge of the cover.
4. Position the 1st/2nd shift fork into the cover; with the offset of the shift fork facing the rear of the cover. Push the shift rail through the fork. The 1st/2nd fork is the larger of the forks.
5. Position the selector arm and the C-shaped interlock plate into the cover, then push the shift rail through the arm. The widest part of the interlock plate must face away from the cover and the selector arm roll pin must face downward, toward the rear of the cover.
6. Position the 3rd/4th shift fork into the cover with the fork offset facing the rear of the cover. The 3rd/4th shift selector arm plate must be positioned under the 1st/2nd shift fork selector arm plate.
7. Push the shift rail through the 3rd/4th shift fork and into the front cover rail bore.
8. Rotate the shift rail until the forward selector arm plate faces away from parallel to the cover.
9. Align the roll pin holes of the selector arm and the shift rail and install the roll pin. The roll pin must be installed flush with the surface of the selector arm to prevent selector arm plate to pin interference.
10. Install the O-ring into the groove of the shift rail oil seal, then install the oil seal carefully after lubricating it.

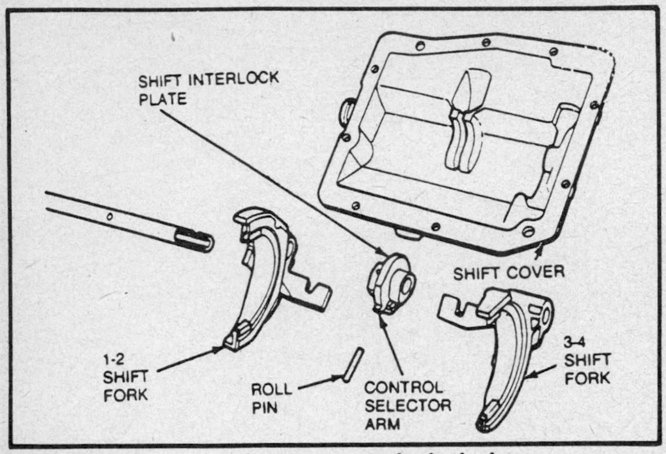
SHIFT INTERLOCK PLATE
SHIFT COVER
1-2 SHIFT FORK
ROLL PIN
CONTROL SELECTOR ARM
3-4 SHIFT FORK
Shift rail, forks, and cover—exploded view

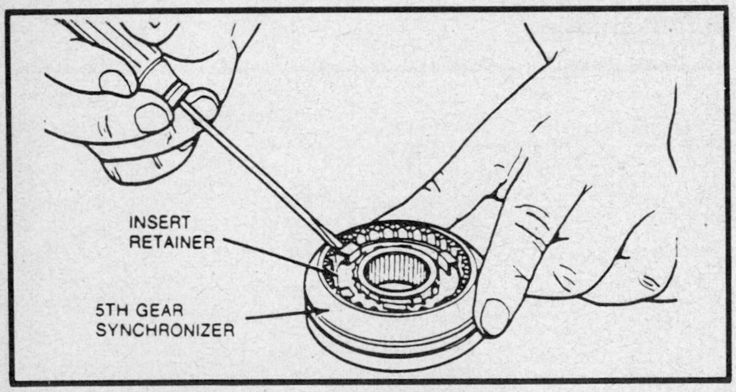
INSERT RETAINER
5TH GEAR SYNCHRONIZER
5th gear synchronizer insert retainer

SYNCHRONIZER

Disassembly

NOTE: The following procedures apply to all synchronizer assemblies. The gear synchronizers are slightly different in design.

1. Before disassembling the synchronizer, mark a line on the sleeve and hub for assembly reference.
2. On the 5th gear synchronizer only, remove the insert retainer.
3. Using a small prybar or suitable tool, remove the 2 synchronizer retaining springs. Remove the inserts.
4. On the 1st/2nd gear synchronizer only, the synchronizer sleeve should be removed carefully to prevent loss of the detent ball and spring.

Assembly

NOTE: When assembling a synchronizer sleeve and hub, be certain to match the alignment marks made during disassembly. The sleeve and hub have an extremely close fit and must be held square during assembly to prevent jamming. Do not force the sleeve onto the hub.

1. Assemble the inserts and synchronizer retaining springs onto the sleeve and hub. the retaining springs engage the same insert but rotate in opposite directions.
2. On the 5th gear synchronizer only, install the insert re-

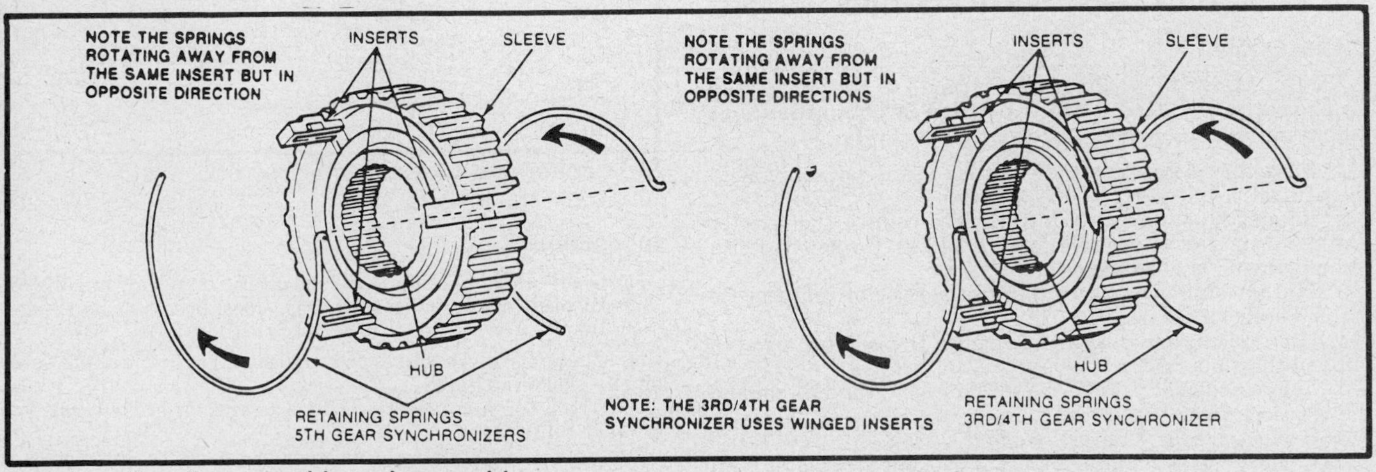
NOTE THE SPRINGS ROTATING AWAY FROM THE SAME INSERT BUT IN OPPOSITE DIRECTION
INSERTS
SLEEVE
HUB
RETAINING SPRINGS 5TH GEAR SYNCHRONIZERS
NOTE: THE 3RD/4TH GEAR SYNCHRONIZER USES WINGED INSERTS
NOTE THE SPRINGS ROTATING AWAY FROM THE SAME INSERT BUT IN OPPOSITE DIRECTIONS
INSERTS
SLEEVE
HUB
RETAINING SPRINGS 3RD/4TH GEAR SYNCHRONIZER
Synchronizer dissassembly and assembly

tainer. Be certain the retainer tabs are positioned over the synchronizer inserts.

3. On the 1st/2nd gear synchronizer only, install the detent ball and spring into the hub on the output shaft. Assemble the retaining springs, inserts and sleeve onto the hub.

Transmission Assembly

1. Using a press, install the front countershaft bearing onto the countershaft using a suitable bearing installation tool.
2. Install the countershaft into the case.
3. Using a press, install the countershaft rear bearing to the countershaft.

NOTE: Support the countershaft using 2 pieces of 1/4 in. barstock.

4. Install the countershaft rear bearing outer race and retainer (without shims). Tighten the retaining bolts to 11–15 ft. lbs. (15–20 Nm).
5. Mount a dial indicator on the case and set it up to measure countershaft endplay. Select and install shim(s) as necessary. Verify the endplay is within 0.001–0.005 inch. After installing the countershaft retainer, bend the tabs on the retainer over the 4 attaching bolts.
6. Install the reverse idler shaft, gear and rubber over travel stop. Install the shaft from the rear of the case and position the overtravel stop between the gear and the shaft support. Use a 3/16 in. diameter pin punch to install the reverse idler roll pin.
7. Install the reverse shift fork and the 5th/reverse lever in the case.
8. Install the output shaft into the case.
9. Coat the input shaft roller bearings, thrust bearing and race with Polyethylene grease. Install the components in the input shaft.
10. Align the notches in the blocking ring with the inserts in the 3rd–4th synchronizer and install the blocking ring.
11. Install the input shaft onto the front of the outputshaft. Align the flat synchronizer teeth with the countershaft before installing the input shaft. Be certain the 3rd–4th blocking ring is still in the proper position.
12. Temporarily, install the bearing retainer (without shims) onto the front of the case.
13. Torque the retaining bolts to 11–20 ft. lbs. (15–27 Nm).
14. Install the output shaft rear bearing race. If necessary, tap the race into position with a plastic tipped hammer. Install the 5th gear on the rear of the countershaft.
15. Slide the shift rail/5th gear fork assembly into the case. As the shift rail enters the case, align the reverse shift fork and slide the rail through the fork. Stop after the rod passes the fork.
16. Position the lever return spring in the case and slide the 5th/reverse shift rail through the spring. Be certain the long end of the spring faces the rear of the case.
17. Install the blocking ring and the 5th gear synchronizer in the 5th speed shift fork. Slide the fork rail assembly and synchronizer into position.
18. Using snapring pliers, install the 5th gear synchronizer retainer and snapring.
19. Using needlenose pliers, connect the lever return spring to the front of the case.
20. Apply Teflon® tape to the pivot pin threads and install the 5th/reverse shift lever pivot pin.

NOTE: Before installing the pivot pin, ensure the reverse shift fork pin and 5th gear shift rail pin are engaged with the shift lever.

21. Tighten the pivot pin to specifications using a T50 Torx® driver. Apply Teflon® tape to the backup lamp switch and install. Tighten the switch to 12–18 ft. lbs. (17-24)Nm.

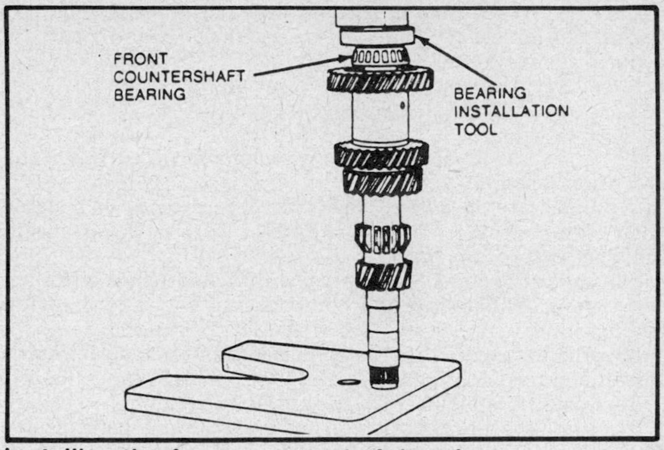
FRONT
COUNTERSHAFT
BEARING
BEARING
INSTALLATION
TOOL
Installing the front countershaft bearing

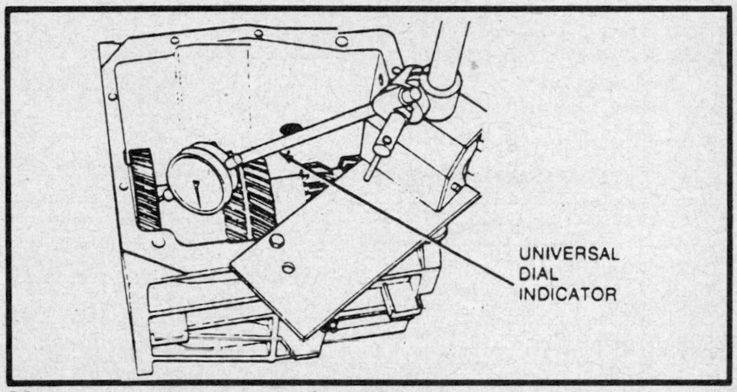
UNIVERSAL
DIAL
INDICATOR
Measuring countershaft end play

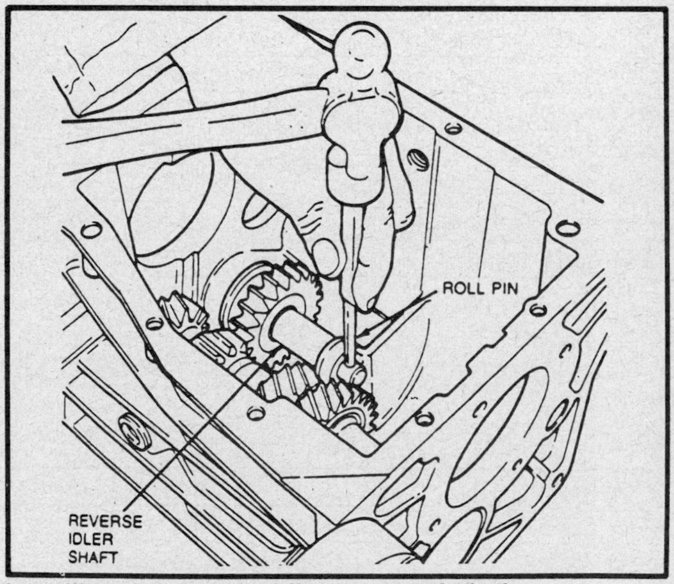
ROLL PIN
REVERSE
IDLER
SHAFT
Installing the reverse idler gear and shaft

22. Using needlenose pliers, install the C-clip on the shift lever pivot pin.
23. Install the speedometer drive gear onto the rear of the output shaft, making certain the speedometer gear retaining clip is fully seated.

24. Apply 1/8 in. bead of silicone sealer to the shift cover assembly. Lower the cover onto the case, allowing the 1-2 and 3-4 forks to slide onto their sleeves. Slide the cover toward the 5th/reverse lever side just before the cover contacts the case.
25. Install the cover bolts and torque to 6–11 ft. lbs. (8–15 Nm).
26. Apply a 1/8 in. bead of silicone sealant on the surface of the extension housing.
27. Lubricate the detent/guide plate in the extension housing with lithium grease. Position the offset lever in the extension housing with the spring over the detent ball.
28. Guide the offset lever onto the shift shaft while installing the extension housing and compressing the detent spring against its ball.
29. Apply sealer to the threads of the extension housing bolts, install and torque bolts to 20–45 ft. lbs. (27–61 Nm).
30. Install the offset lever-to-shift shaft roll pin.
31. Place the transmission on end and mount a dial indicator on the extension housing. Set the dial indicator to measure input/main shaft endplay. Move input/main shaft up and down.
32. Select a shim that is within 0.002 in. of the endplay measure. This will give the required 0 endplay.
33. Remove the input bearing retainer.
34. Install the shims behind the input bearing outer race in the retainer. If more than a single shim is used, install the smaller shims before the larger shims.
35. Install the retainer and torque bolts to 11–20 ft. lbs. (15–27 Nm).
36. Recheck endplay, if endplay is not less than 0.002 change shims as needed. Remove the bearing retainer.
37. Apply a 1/8 in. bead of RTV sealant to the retainer and apply sealer to the bolt threads.
38. Install the retainer and torque bolts to 11–20 ft. lbs. (15–27 Nm).
39. Apply a 1/8 in. bead of RTV sealant to the turret cover and apply sealer to the bolt threads.
40. Install the turret cover and torque the bolts to 11–15 ft. lbs. (15–20 Nm).
41. Install and tighten the drain plug.
42. Install the transmission and fill to proper level with Dexron®II automatic transmission fluid.

SPECIFICATIONS

TORQUE SPECIFICATIONS

Description	ft. lbs.	Nm
Bearing retainer	11–20	15–27
Shift cover	6–11	8–15
Extension housing	20–45	27–61
Turret cover	11–15	15–20
Drain plug	15–30	20–41
Transmission to flywheel housing	45–65	61–88
Speedometer cable retaining screw	36–54 ①	4–6
Transmission support	36–50	48–68
Transmission extension housing bolts	25–35	34–48
Driveshaft bolts	42–57	56–77
Catalytic converter attaching bolts	20–30	27–41
Shift lever to transmission	23–32	31–43
Shift boot to floor plan	3–7	4–9.5

TORQUE SPECIFICATIONS

Description	ft. lbs.	Nm
Cluster gear rear bearing retainer	11–15	15–20
Top gear sensing switch	12–18	17–24
Backup lamp switch	12–18	17–24
Neutral sensing switch	12–18	17–24

① inch lbs.

ENDPLAY SHIMS

(in.)

0.012	0.020	0.025	0.029	0.033	0.037	0.041
0.014	0.022	0.026	0.030	0.034	0.038	0.042
0.016	0.023	0.027	0.031	0.035	0.039	0.043
0.018	0.024	0.028	0.032	0.036	0.040	0.044

SPECIAL TOOLS

Tool Number	Description
T57L-500-B	Bench mounted holding fixture
D79L-4621-A	Universal press plate
T77F-4220-A	Bearing retainer seal installer
T57L-4621-B	Bearing installation tool
T81P-1104-C	Puller
D84L-1123-A	Puller clamp
T85P-7025-A	Snapring tool
T77J-7025-B	Bearing remover tube

Tool Number	Description
T82T-7003-DH	Bearing remover tube
T77J-7025-M	Case support
T77J-7025-L	Case support
T75L-7025-B	Bearing Remover/Installer
T83P-7025- AH	Bearing installer
D78P-4201-G	Universal dial indicator
D78P-4201-F	Bracketry

Section 7

3.03 Transmission
Ford Motor Co.

APPLICATION

1984–87 Ford E and F Series Vehicles

GENERAL DESCRIPTION

The Ford 3.03 is a fully synchronized 3 speed transmission. All gears except reverse are in constant mesh. Forward speed gear changes are accomplished with synchronizer sleeves.

Transmission Identification

Transmission identification codes are found on the Vehicle Safety Compliance Certification label located on the driver's door lock pillar.

Manual transmissions have service identification tags to identify assemblies for service purposes. Ford 3 speed transmissions have the identification tag located on the right side front of the transmission case.

Metric Fasteners

The Ford 3.03 transmission uses standard SAE fasteners. Do not attempt to interchange standard SAE fasteners for metric system fasteners. Mismatched or incorrect fasteners can result in damage to the transmission through malfunctions, breakage or possible personal injury. Care should be taken to reuse the fasteners in the same location as removed, whenever possible. Before installing cap screws into aluminum parts, always dip screws into oil to prevent the screws from galling the aluminum threads and to prevent seizing.

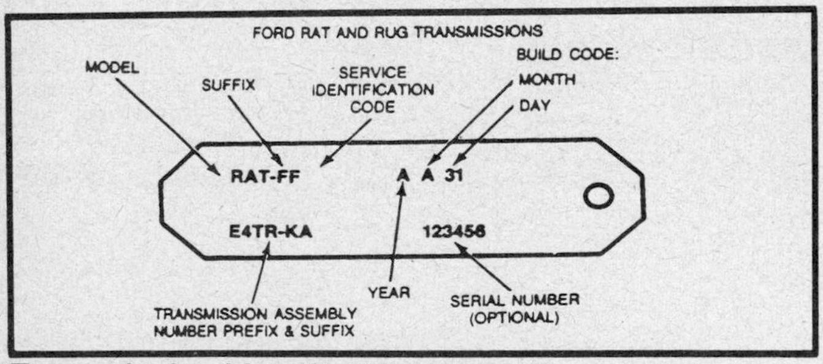
FORD RAT AND RUG TRANSMISSIONS
MODEL
SUFFIX
SERVICE IDENTIFICATION CODE
BUILD CODE:
MONTH
DAY
RAT-FF
A A 31
E4TR-KA
123456
YEAR
TRANSMISSION ASSEMBLY NUMBER PREFIX & SUFFIX
SERIAL NUMBER (OPTIONAL)

Transmission identification tag

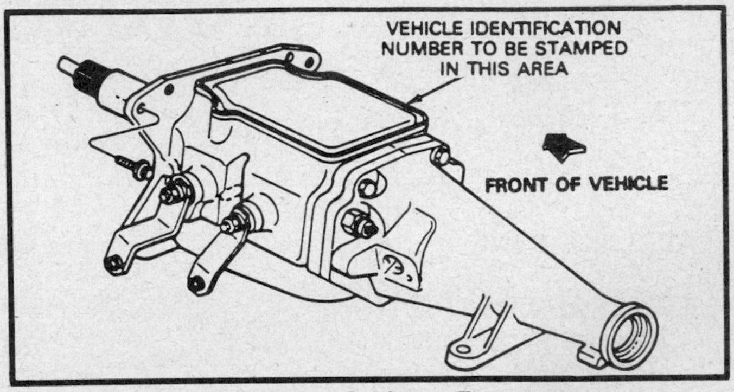
VEHICLE IDENTIFICATION NUMBER TO BE STAMPED IN THIS AREA
FRONT OF VEHICLE

Ford 3.03 manual transmission—F-150–250

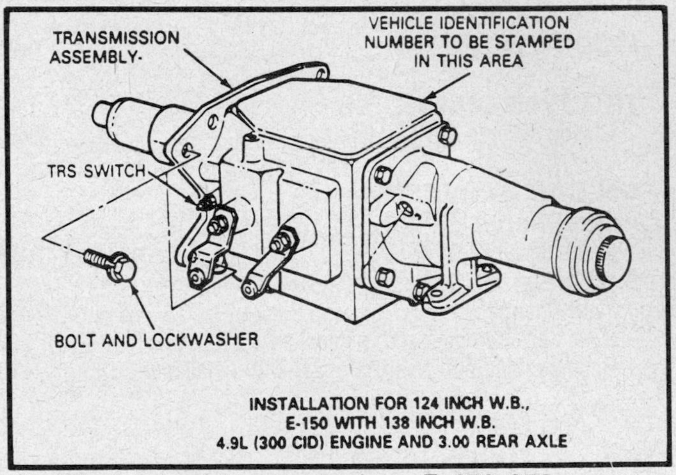
TRANSMISSION ASSEMBLY
VEHICLE IDENTIFICATION NUMBER TO BE STAMPED IN THIS AREA
TRS SWITCH
BOLT AND LOCKWASHER
INSTALLATION FOR 124 INCH W.B., E-150 WITH 138 INCH W.B. 4.9L (300 CID) ENGINE AND 3.00 REAR AXLE

Ford 3.03 manual transmission—E-150–350

Capacities

The Ford 3.03 transmission requires a total capacity of 3.5 qts. (1.6L) of DOAZ–19C547–A SAE 80 gear oil or equilvalent.

Checking Fluid Level

1. Raise the vehicle and support it safely.
2. Remove the fill plug located on the side of the case and check the level. The fluid level should be at the lower edge of the fill plug hole, if not, add fluid meeting specified recomendations.
3. Install the fill plug and lower the vehicle.

TRANSMISSION MODIFICATIONS

Rattle, Ringing Noise From Transmission Area

Some transmissions built before December 15, 1985 and equipped with 4.9L engine, a 10 in. clutch and have a gross vehicle weight under 8500 lbs. May have a rattle or ringing noise that can be heard during shifting coming from the transmission area. This condition is caused by the clutch disc vibration. The clutch disc must be removed and replaced with a different designed disc, part number E5TZ–7550–B.

TROUBLE DIAGNOSIS

CHILTON'S THREE C'S TRANSMISSION DIAGNOSIS
Ford 3.03 Manual Transmission

Condition	Cause	Correction
Shift lever—improper operation	a) Shift rods out of adjustment	a) Check crossover operation. Hand lever must move freely through neutral gate without catch or drag. Adjust shift rods
	b) Steering column shift tube out of alignment	b) Check lever position for proper transmission gear selection. Replace steering column shift tube if necessary
Transmission shifts hard	a) Clutch pedal free-travel out of adjustment	a) Adjust clutch pedal free-travel
	b) Clutch does not completely release	b) Inspect complete clutch system
	c) Transmission fluid low or improper type	c) Add lubricant or change lubricant as required
	d) Shift lever binding or worn	d) Remove cap from shift tower. Eliminate binding condition or replace components as required
	e) Worn or damaged internal shift mechanism	e) Remove transmission cover. Check internal shift mechanism by shifting into and out of all gears. Repair or replace as required
	f) Binding of sliding gears and/or synchronizers	f) Check for free movement of gears and synchronizers. Repair or replace as required
	g) Housings and/or shafts out of alignment	g) Remove transmission and check for binding condition between input shaft and engine crankshaft pilot bearing or bushing. Check flywheel housing alignment. Repair or replace as required
Noisy in forward gears.	a) Lubricant level low, or improper type	a) Add lubricant, or refill with specified lubricant
	b) Components grinding on transmission	b) Check for screws, bolts, etc., of cab or other components grounding out. Correct as required
	c) Component housing bolts loose	c) Check torque on transmission to flywheel housing bolts, output shaft flange nut and flywheel housing to engine block bolts. Tighten bolts to specification
	d) Flywheel housing to engine crankshaft alignment	d) Check and align flywheel housing to engine crankshaft
	e) Noisy bearings or gears	e) Remove and disassemble transmission. Inspect input, output and countershaft bearings. Inspect speedometer gear and gear teeth for wear or damage. Replace as required
Gears clash when shifting from one forward gear to another	a) Engine idle speed too high	a) Adjust engine idle speed
	b) Clutch pedal free-travel incorrect	b) Check clutch for pedal free-travel, complete release and spin time. Adjust as required
	c) Improper manual shift linkage	c) Adjust and repair manual shift linkage as required
	d) Pilot bearing binding	d) Remove transmission and check for a binding condition between input shaft and engine crankshaft pilot bearing. Replace as required
	e) Damaged gear teeth and/or synchronizer	e) Disassemble transmission, repair or replace as required

Condition	Cause	Correction
Transmission jumps out of gear	a) Manual-shift linkage binding, out of adjustment. Stiff shift boot	a) Adjust linkage. If necessary align steering column to vehicle body
	b) Improper fit of form isolation pad	b) Replace shift boot if exceptionally stiff. Replace or rework pad to provide clearance
	c) Loose transmission to engine mounting bolts, or loose levers	c) Tighten transmission to flywheel housing, and flywheel housing to engine block bolts to specifications. Loosen all bolts and reseat flywheel housing. Tighten all bolts. Tighten levers if necessary
	d) Flywheel housing to engine crankshaft out of line	d) Shim or replace housing as required
	e) Crankshaft pilot bearing worn	e) Replace bearing
	f) Interior components damaged	f) Disassemble transmission. Inspect the synchronizer sleeves for free movement on their hubs. Inspect the synchronizer blocking rings for widened index slots, rounded clutch teeth and smooth internal surface. Check countershaft cluster gear for excessive endplay. Check shift forks for worn or loose mounting on shift rails. Check fork pads for excessive wear. Inspect synchronizer sliding sleeve and gear clutch teeth for wear or damage. Repair or replace as required
	g) Worn gear teeth due to partial engagement	g) Replace worn or damaged gears
Transmission is locked in one gear. It cannot be shifted out of that gear	a) Manual-shift linkage out of adjustment, binding or damaged	a) Disconnect the problem shift rod from transmission shift lever. Try to shift the transmission lever into and out of gear position. If OK, repair or replace linkage parts
	b) Internal components	b) Remove transmission. Inspect problem gears, shift rails and forks and synchronizer for wear or damage. Repair as required. Check for broken fork slot tabs on TOD
	c) Loose fork on rail	c) Check shift rail interlock system
Transfer case makes noise	a) Incorrect tire inflation pressures and/or incorrect size tires and wheels	a) Assure that all tires and wheels are the same size, and that inflation pressures are correct
	b) Excessive tire tread wear	b) Check tire tread wear to see if there is more than .06 in. difference in tread wear between front and rear. Interchange one front and one rear wheel. Reinflate tires to specifications
	c) Internal components	c) Operate vehicle in all transmission gears with transfer case in 2HI, or Hi range. If there is noise in transmission in neutral gear, or in some gears and not in others, remove and repair transmission. If there is noise in all gears, operate vehicle in all transfer case ranges. If noisy in all ranges or HI range only, disassemble transfer case. Check input gear, intermediate and front output shaft gear for damage. Replace as required. If noisy in LO range only, inspect intermediate gear and sliding gears for damage. Replace as required

Condition	Cause	Correction
4WD transfer case jumps out of gear	a) Incomplete shift linkage travel	a) Adjust linkage to provide complete gear engagement
	b) Loose mounting bolts	b) Tighten mounting bolts
	c) Front and rear driveshaft slip yokes dry or loose	c) Lubricate and repair slip yokes as required. Tighten flange yoke attaching nut to specifications
	d) Internal components	d) Disassemble transfer case. Inspect sliding clutch hub and gear clutch teeth for damage. Replace as required
Transmission will not shift into one gear—all others OK	a) Manual linkage out of adjustment	a) Adjust manual linkage
	b) Manual shift linkage damaged or worn	b) Lubricate, repair or replace parts as required
	c) Back-up switch ball frozen	c) If reverse is problem, check back-up switch for ball frozen in extended position (if so equipped)
	d) Internal components	d) Remove transmission. Inspect shift rail and fork system synchronizer system and gear clutch teeth for restricted travel. Repair or replace as required

ON CAR SERVICE

Adjustments

GEAR SHIFT LINKAGE

1. Install a gauge pin of a $^{3}/_{16}$ in. (4.76mm) diameter through the locating hole in the steering column shift levers and the plastic spacer. Locate the levers in the center of the steermg column window.
2. Loosen the shift lever adjustment nuts and position the transmission shift levers in the **N** detents.
3. Tighten the shifter lever adjustment nuts to 12–18 ft. lbs. (17–24 Nm), using care to prevent motion between the stud and rod.
4. Remove the gauge pin.
5. Check the linkage operation. Make sure there is no interference between the steering column levers and window when shifting into gear.

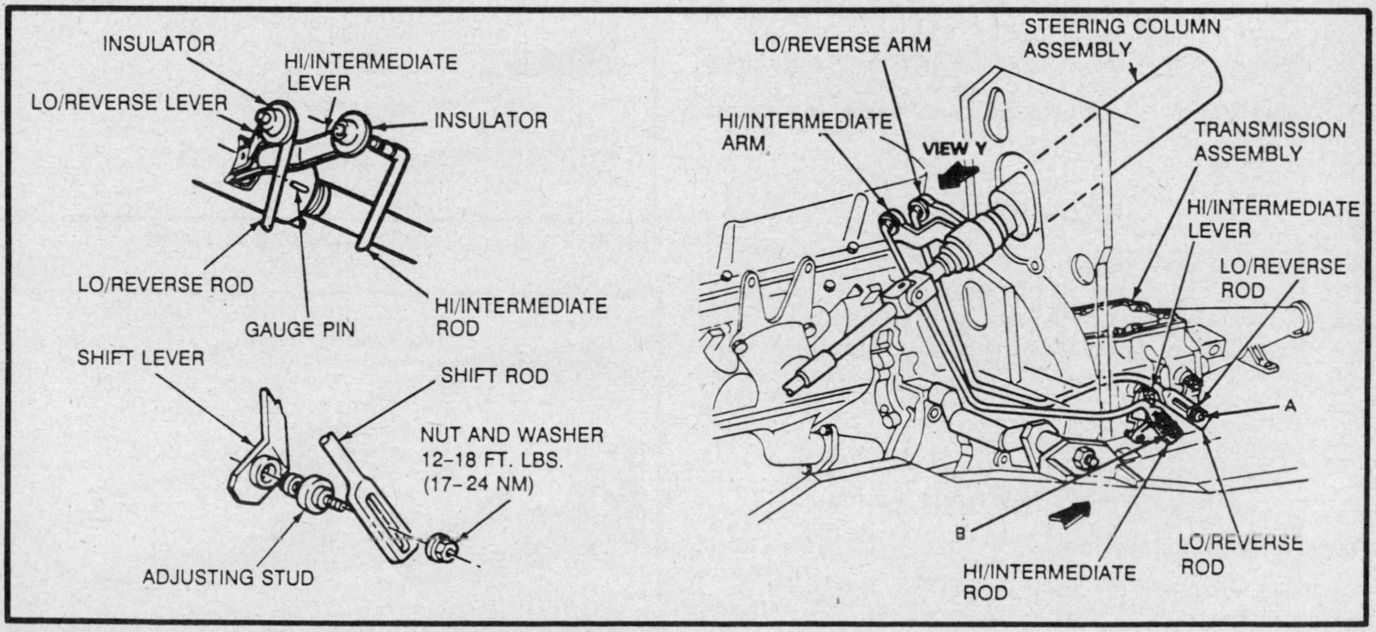
INSULATOR
HI/INTERMEDIATE LEVER
LO/REVERSE LEVER
INSULATOR
LO/REVERSE ROD
GAUGE PIN
HI/INTERMEDIATE ROD
SHIFT LEVER
SHIFT ROD
NUT AND WASHER
12–18 FT. LBS.
(17–24 NM)
ADJUSTING STUD
LO/REVERSE ARM
STEERING COLUMN ASSEMBLY
HI/INTERMEDIATE ARM
VIEW Y
TRANSMISSION ASSEMBLY
HI/INTERMEDIATE LEVER
LO/REVERSE ROD
A
B
LO/REVERSE ROD
HI/INTERMEDIATE ROD

Gear shift linkage adjustment—F-150 and F-250

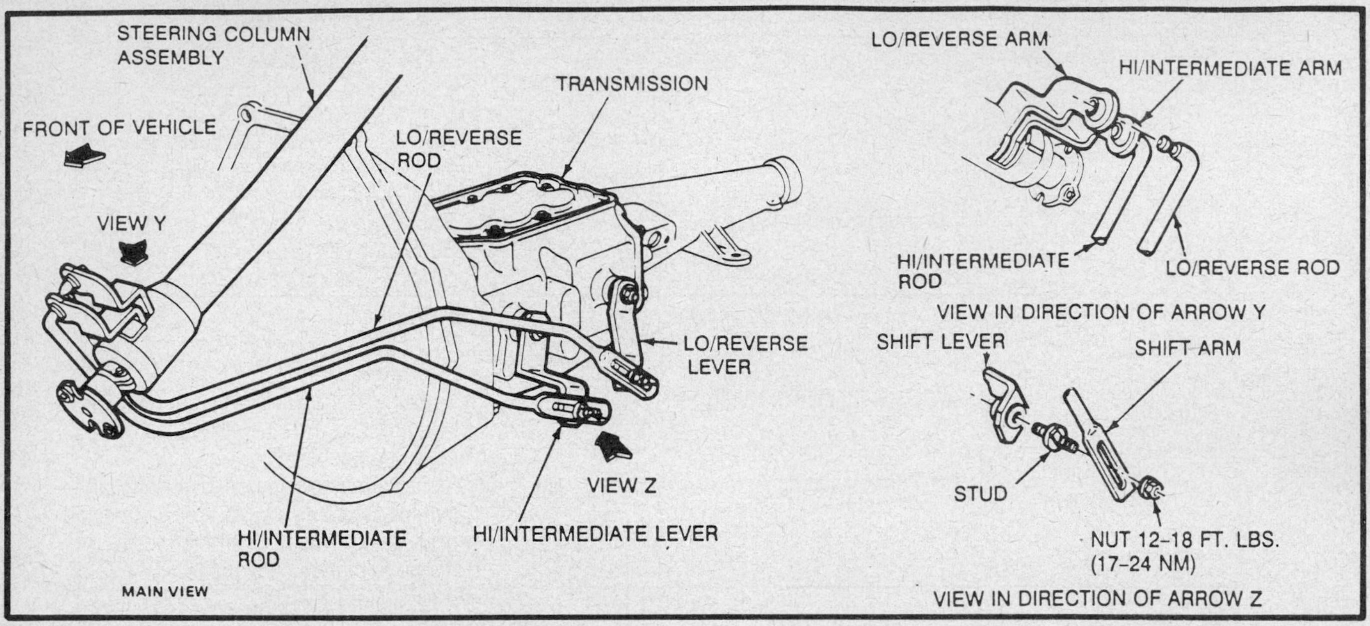
STEERING COLUMN
ASSEMBLY
FRONT OF VEHICLE
TRANSMISSION
LO/REVERSE
ROD
VIEW Y
LO/REVERSE
LEVER
VIEW Z
HI/INTERMEDIATE
ROD
HI/INTERMEDIATE LEVER
MAIN VIEW
LO/REVERSE ARM
HI/INTERMEDIATE ARM
HI/INTERMEDIATE
ROD
LO/REVERSE ROD
VIEW IN DIRECTION OF ARROW Y
SHIFT LEVER
SHIFT ARM
STUD
NUT 12–18 FT. LBS.
(17–24 NM)
VIEW IN DIRECTION OF ARROW Z

Gear shift linkage adjustment – E-150–E-350

NOTE: Always use new retaining rings and new insulators when making transmission control adjustments. New retaining rings should also be used whenever the existing retaining rings are removed. The rings as well as the plastic gromments where the shift rods are attached, should be replaced whenever excessive wear or looseness is noted during normal vehicle inspections.

SEAL REMOVER TOOL
IMPACT SLIDE HAMMER

Removing the extension housing seal

Services

FLUID CHANGE

1. Raise and support the vehicle safely.
2. Remove the transmission oil fill plug and allow the gear oil to drain into a suitable drain pan.
3. The correct level should be to the bottom edge of the fill plug hole.
4. Fill to the proper level with the recommended gear oil and reinstall the oil fill plug.
5. Lower the vehicle.

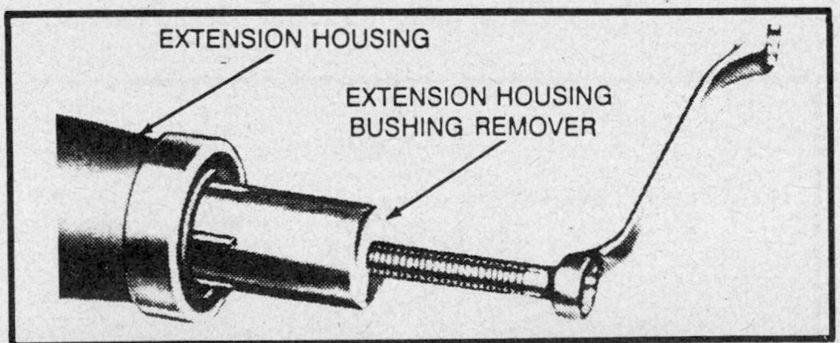
EXTENSION HOUSING
EXTENSION HOUSING
BUSHING REMOVER

Removing the extension housing bushing

REAR SEAL

Removal and Installation

1. Raise the vehicle and suppport it safely.
2. Remove the driveshaft or coupling shaft.
3. Remove the seal from the extension housing using a suitable seal puller tool.
4. Install a new seal in the extension housing using a suitable installer tool. Lubricate the seal with a muti-purpose lubricant.
5. Install the driveshaft and lower the vehicle.

REAR OIL SEAL
EXTENSION HOUSING
SEAL REPLACER
EXTENSION HOUSING

Installing the extension housing seal

REAR BUSHING AND SEAL

Removal and Installation

1. Raise the vehicle and support it safely.
2. Disconnect the driveshaft or coupling shaft from the transmission.
3. Remove the seal using a suitable seal remover tool.

4. Insert a suitable bushing remover tool into the extension housing until it grips on the front side of the bushing.
5. Turn the bushing remover screw clockwise until the bushing is free of the housing.
6. Lubricate the new bushing with a multi-purpose lubricant and drive it into the extension housing.
7. Install a new seal and lubricate it with a multi-purpose lubricant.
8. Connect the driveshaft or coupling shaft to the transmission. Lower the vehicle.

REMOVAL AND INSTALLATION

TRANSMISSION REMOVAL

F-150 AND F-250

1. Disconnect the negative battery cable. Raise the vehicle and support it safely.
2. Support the engine with a jack and wood block placed under the oil pan.
3. Drain the transmission lubricant by removing the lower extension housing-to-transmission bolt.
4. Position a transmission jack under the transmission and secure the transmission to the jack.
5. Disconnect the gear shift linkage at the transmission.
6. Disconnect the speedometer cable.
7. Disconnect the back-up lamp switch wiring.
8. Disconnect the driveshaft from the transmission.
9. Raise the transmission if necessary and remove the rear support, insulator and retainer assembly.
10. Remove the transmission-to-flywheel housing attaching bolts.
11. Move the transmission to the rear until the input shaft clears the clutch housing and and lower the transmission.

E-150 AND E-350

1. Disconnect the negative battery cable. Raise the vehicle and support it safely.
2. Drain the transmission lubricant by removing the lower extension housing-to-transmission bolt.
3. Disconnect the driveshaft from the slip yoke at the transmission. Secure the front end of the driveshaft out of the way with a lock wire.
4. Disconnect the speedometer cable from the extension housing and disconnect the gear shift rods from the transmission shift levers.
5. Position a transmission jack under the transmission and secure it in place.

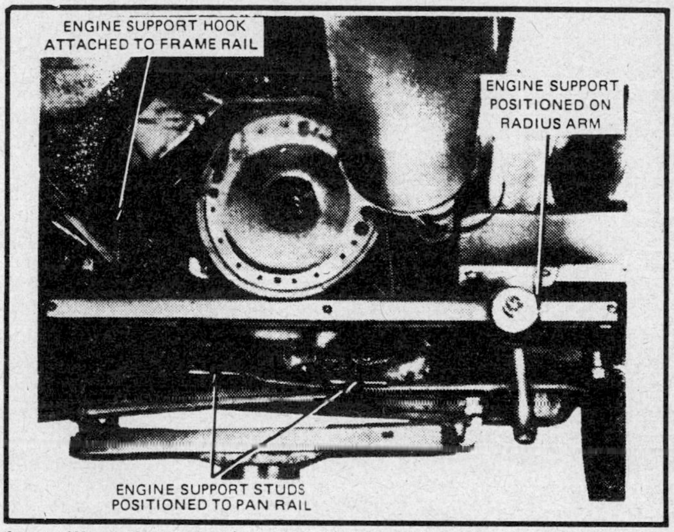
ENGINE SUPPORT HOOK
ATTACHED TO FRAME RAIL
ENGINE SUPPORT
POSITIONED ON
RADIUS ARM
ENGINE SUPPORT STUDS
POSITIONED TO PAN RAIL

Installing the engine support bar—E-150–E-350

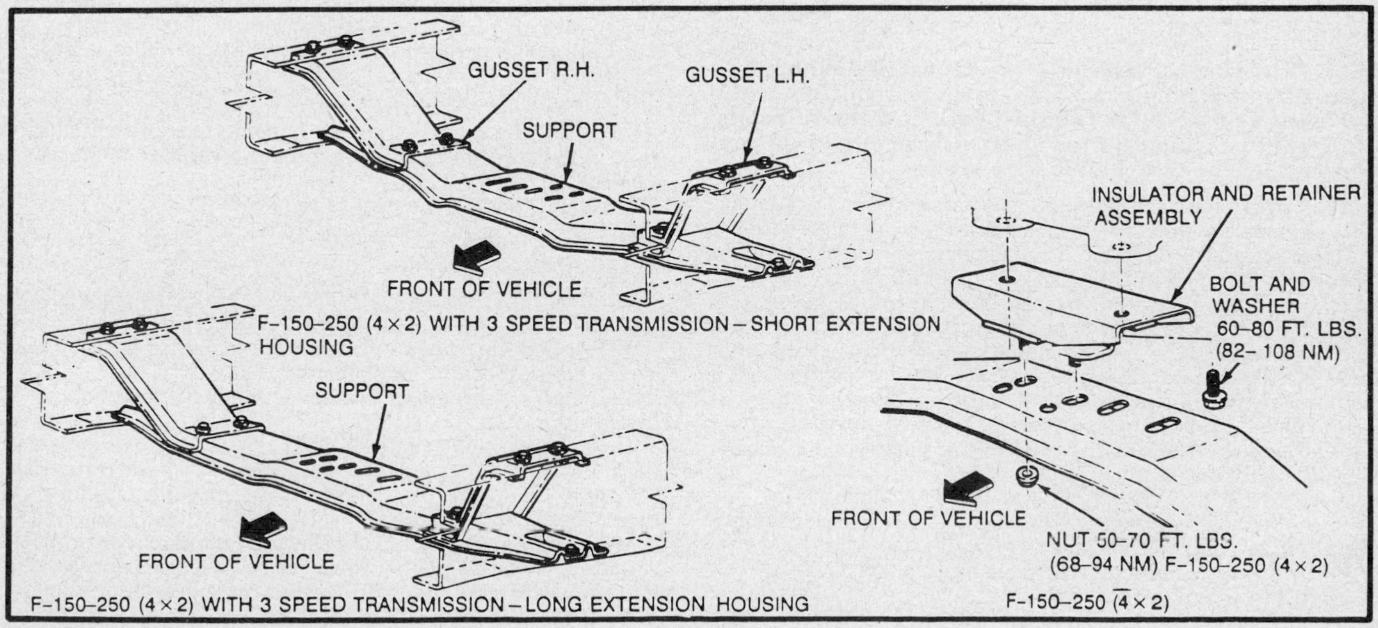
GUSSET R.H.
GUSSET L.H.
SUPPORT
FRONT OF VEHICLE
F-150–250 (4 × 2) WITH 3 SPEED TRANSMISSION—SHORT EXTENSION HOUSING
SUPPORT
FRONT OF VEHICLE
F-150–250 (4 × 2) WITH 3 SPEED TRANSMISSION—LONG EXTENSION HOUSING
INSULATOR AND RETAINER ASSEMBLY
BOLT AND WASHER
60–80 FT. LBS.
(82–108 NM)
FRONT OF VEHICLE
NUT 50–70 FT. LBS.
(68–94 NM) F-150–250 (4 × 2)
F-150–250 (4 × 2)

Transmission rear support—F-150 and F-250

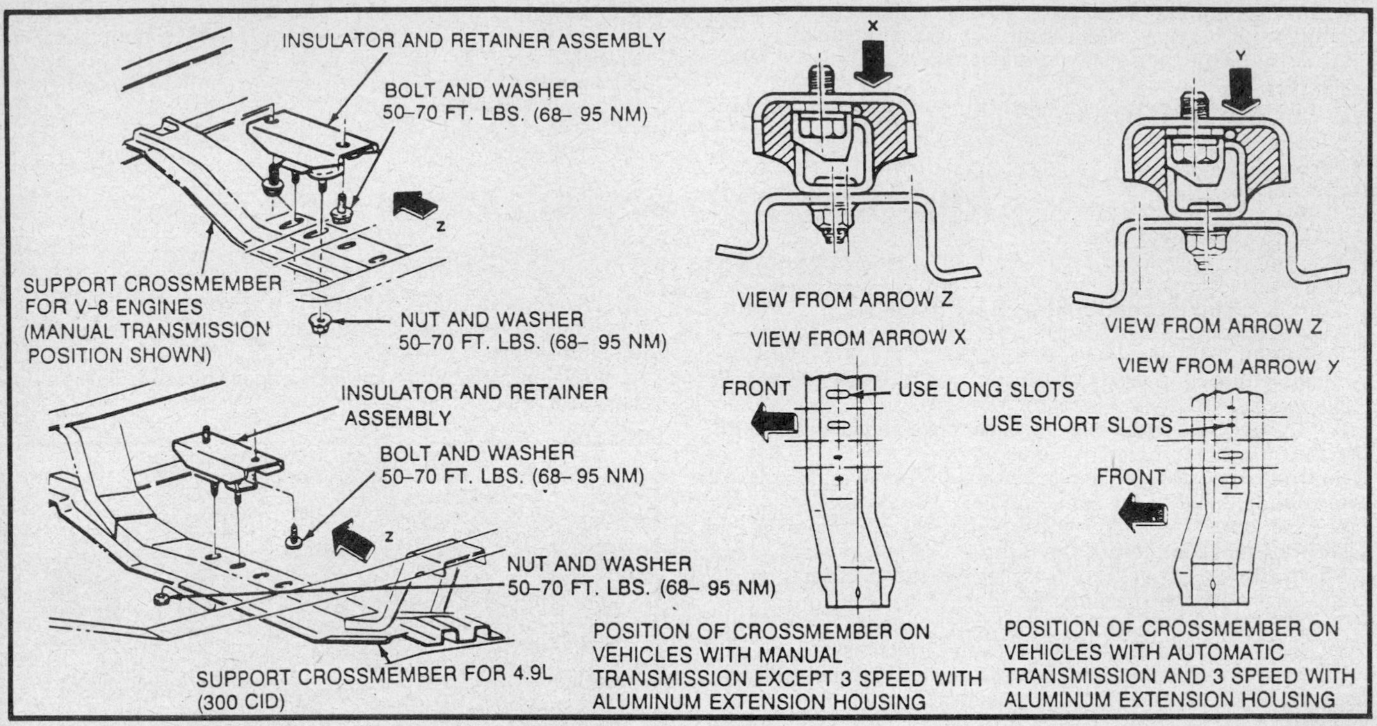
INSULATOR AND RETAINER ASSEMBLY
BOLT AND WASHER
50–70 FT. LBS. (68–95 NM)
Z
SUPPORT CROSSMEMBER FOR V-8 ENGINES (MANUAL TRANSMISSION POSITION SHOWN)
NUT AND WASHER
50–70 FT. LBS. (68–95 NM)
INSULATOR AND RETAINER ASSEMBLY
BOLT AND WASHER
50–70 FT. LBS. (68–95 NM)
Z
NUT AND WASHER
50–70 FT. LBS. (68–95 NM)
SUPPORT CROSSMEMBER FOR 4.9L (300 CID)
X
VIEW FROM ARROW Z
VIEW FROM ARROW X
FRONT
USE LONG SLOTS
POSITION OF CROSSMEMBER ON VEHICLES WITH MANUAL TRANSMISSION EXCEPT 3 SPEED WITH ALUMINUM EXTENSION HOUSING
Y
VIEW FROM ARROW Z
VIEW FROM ARROW Y
USE SHORT SLOTS
FRONT
POSITION OF CROSSMEMBER ON VEHICLES WITH AUTOMATIC TRANSMISSION AND 3 SPEED WITH ALUMINUM EXTENSION HOUSING

Rear crossmember and transmission supports—E-150–E-350

6. Raise the transmission slighty and remove the bolts retaining the transmission extension housing to the insulator and retainer assembly.
7. Remove the transmission-to-flywheel housing bolts.
8. Position the engine support bar to the frame and support the engine.
9. Lower the transmission from the vehicle.

TRANSMISSION INSTALLATION

F-150 AND F-250

NOTE: Prior to installing the transmission, apply a light film of lubricant to the release bearing inner hub surfaces, release lever fulcrum, fork and the transmission front bearing retainer. Exercise care to avoid contaminating the clutch disc with grease.

1. Position the transmission on a transmission jack. Raise the transmission until the input shaft splines are inline with the clutch disc splines. The clutch release bearing and hub must be properly positioned in the release lever fork.
2. Install a guide stud in each lower flywheel housing-to-transmission case bolt holes and align the splines on the input shaft with the splines on the clutch disc.
3. Slide the transmission forward on the guide studs until it contacts the clutch housing.
4. Install the transmission-to-flywheel housing upper mounting bolts. Remove the guide studs and install the lower mountiong bolts. Tighten all mounting bolts to 42–50 ft. lbs. (57–67 Nm).
5. Install the rear support, insulator and retainer assembly. Remove the transmission jack.
6. Connect the speedometer cable and the driven gear.
7. Install the driveshaft.
8. Connect each shift rod to its respective lever on the transmission.
9. Install the lower extension housing-to-insulator and retainer assembly.

6. Install the 2 upper mounting bolts and lockwashers that attach the flywheel housing to the transmission.
7. Remove the 2 guide pins and install the lower mounting bolts and lockwashers. Tighten the mounting bolts to 40–50 ft. lbs. (57–67 Nm).
8. Raise the jack slightly and remove the engine support bar.
9. Install the extension housing-to-insulator and retainer assembly retaining bolts.
10. Connect the gear shift rods and the speedometer cable.
11. Install the driveshaft.
12. Adjust the shift linkage as required.

10. Install the lower extension housing-to-transmission bolt and fill the transmission to the proper level with an approved lubricant.
11. Adjust the shift linkage as required.
12. Connect wiring to the back-up lamp switch.

E-150 AND E-350

1. Make sure that the machined surfaces of the transmission case and the flywheel housing are free of dirt, paint and burrs.
2. Position the transmission on a transmission jack and raise it into position behind the housing.
3. Install a guide pin in each lower mounting bolt hole of the flywheel housing.
4. Start the input shaft through the release bearing. Align the splines on the input shaft with the splines in the clutch disc.
5. Move the transmission forward on the guide pins until the input shaft pilot enters the bearing or bushing in the crankshaft. If the transmission front bearing retainer binds up on the clutch release bearing hub, work the release bearing lever until the hub slides onto the transmission front bearing retainer.

BENCH OVERHAUL

Before Disassembly

When servicing the unit, it is recommended that as each part is disassembled, it is cleaned in solvent and dried with compressed air. Disassembly and reassembly of this unit and its parts must be done on a clean work bench. Also, before installing bolts into aluminum parts, always dip the threads into clean transmission oil. Anti-seize compound can also be used to prevent bolts from galling the aluminum and seizing. Always use a torque wrench to keep from stripping the threads. Take care with the seals when installing them, especially the smaller O-rings. The slightest damage can cause leaks. Aluminum parts are very susceptible to damage so great care should be exercised when handling them. The internal snaprings should be expanded and the external snaprings compressed if they are to be reused. This will help insure proper seating when installed. Be sure to replace any O-ring, gasket, or seal that is removed.

Transmission Disassembly

1. Drain the lubricant by removing the lower extension housing bolt.

DUMMY SHAFT
COUNTERSHAFT

Removing the countershaft

2. Remove the case cover and gasket.
3. Remove the long spring that holds the detent plug in the case and remove the detent plug with a small magnet.
4. Remove the extension housing and gasket.
5. Remove the front bearing retainer and gasket.
6. Remove the filler plug on the right side of the transmission case. Working through the plug opening, drive the roll pin out of the case and countershaft with a ¼ in. punch.
7. Hold the countershaft gear with a hook. Install the dummy shaft and push the countershaft out of the rear of the case. As the countershaft comes out, lower the gear cluster to the bottom of the case. Remove the countershaft.

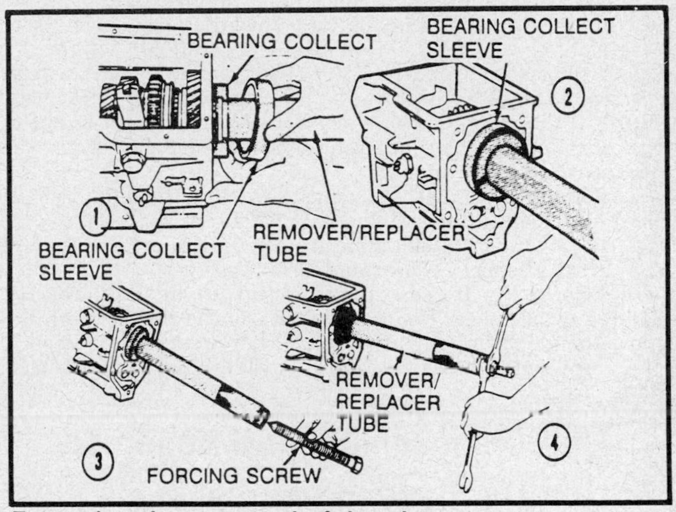
BEARING COLLECT
BEARING COLLECT SLEEVE
2
1
BEARING COLLECT SLEEVE
REMOVER/REPLACER TUBE
REMOVER/ REPLACER TUBE
3
FORCING SCREW
4

Removing the output shaft bearing

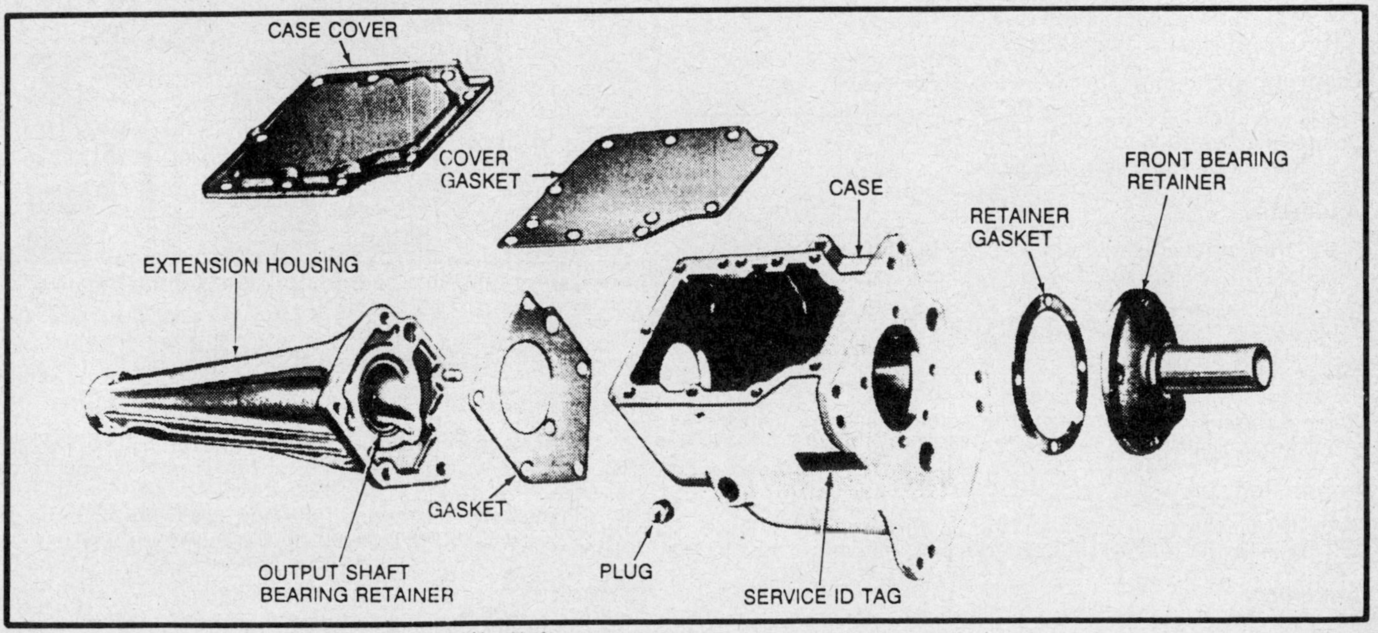
CASE COVER
COVER GASKET
FRONT BEARING RETAINER
CASE
RETAINER GASKET
EXTENSION HOUSING
GASKET
OUTPUT SHAFT BEARING RETAINER
PLUG
SERVICE ID TAG

Transmission case and related components

8. Remove the snapring that holds the speedometer drive gear on the output shaft. Slip the gear off the shaft and remove the gear lock ball.
9. Remove the snapring that holds the output shaft bearing. Using a special bearing puller, remove the output shaft bearing.
10. Place both shift levers in the neutral (center) position.
11. Remove the set screw that holds the 1st/reverse shift fork to the shift rail. Slip the 1st/reverse shift rail out through the rear of the case.
12. Move the 1st/reverse synchronizer forward as far as possible. Rotate the 1st/reverse shift fork upwards and lift it out of the case.
13. Place the 2nd/3rd shift fork in the 2nd position. Remove the set screw. Rotate the shift rail 90 degrees.
14. Lift the interlock plug out of the case with a magnet.
15. Remove the expansion plug from the 2nd/3rd shift rail by lightly tapping the end of the rail. Remove the 2nd/3rd shift rail.
16. Remove the 2nd/3rd shift rail detent plug and spring from detent bore.
17. Remove the input gear and shaft from the case.
18. Rotate the 2nd/3rd shift fork upwards and remove from case.
19. Using caution, lift the output shaft assembly out through the top of the case.
20. Lift the reverse idler gear and the thrust washers out of the case. Remove the countershaft gear, thrust washer and the dummy shaft the from case.
21. Remove the snapring from the front of the output shaft. Slip the synchronizer and 2nd gear off of the shaft.
22. Remove the 2nd snapring from output shaft and remove the thrust washer, 1st gear and blocking ring.
23. Remove the 3rd snapring from the output shaft. The 1st/reverse synchronizer hub is a press fit on the output shaft. Remove the synchronizer hub with an arbor press. Do not attempt to remove or install the synchronizer hub by prying or hammering.

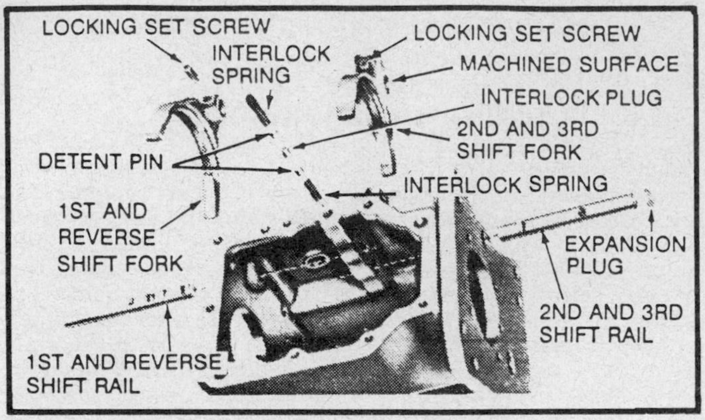
LOCKING SET SCREW
INTERLOCK SPRING
LOCKING SET SCREW
MACHINED SURFACE
INTERLOCK PLUG
2ND AND 3RD SHIFT FORK
DETENT PIN
INTERLOCK SPRING
1ST AND REVERSE SHIFT FORK
EXPANSION PLUG
2ND AND 3RD SHIFT RAIL
1ST AND REVERSE SHIFT RAIL
Exploded view of the shift rails and forks

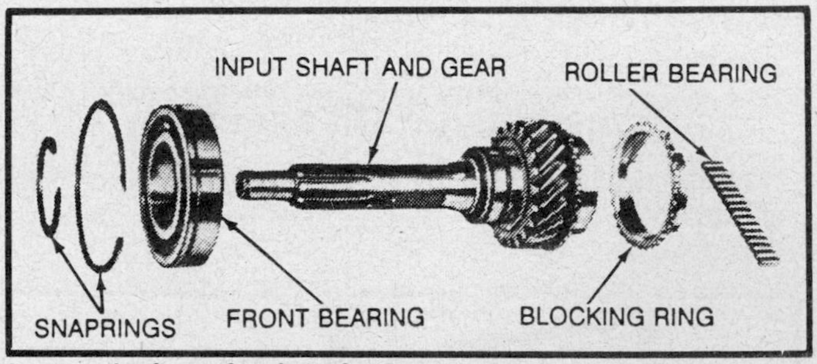
INPUT SHAFT AND GEAR
ROLLER BEARING
SNAPRINGS
FRONT BEARING
BLOCKING RING
Input shaft and related components

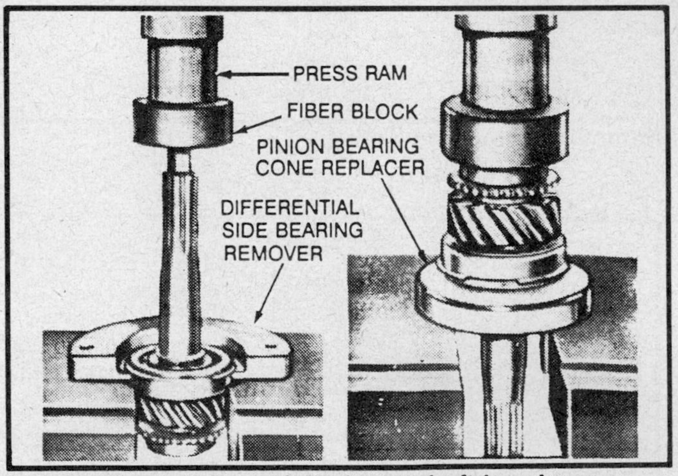
PRESS RAM
FIBER BLOCK
PINION BEARING CONE REPLACER
DIFFERENTIAL SIDE BEARING REMOVER
Removal/installation of the input shaft bearing

Unit Disassembly and Assembly

SHIFT LEVERS AND SEALS

Disassembly

1. Remove shift levers from the shafts.
2. Slip the levers out of the case.
3. Discard shaft sealing O-rings.

Inspection

1. Carefully check the shift levers for distortion or wear at linkage points.
2. Replace worn or defective parts.

Assembly

1. Lubricate and install new O-rings on shift shafts.
2. Install the shift shafts in the case and secure shift levers.

INPUT SHAFT

Disassembly

Remove the snapring securing the input shaft bearing, using an arbor press, remove the bearing.

Inspection

1. Inspect the input shaft bearing for wear or distortion.
2. Inspect the gear teeth for cracks or chips.

Assembly

Using an arbor press, press the input shaft bearing onto shaft.

SYNCHRONIZERS

Disassembly

1. Scribe alignment marks on synchronizer hubs before disassembly. Remove each synchronizer hub from the synchronizer sleeves.
2. Separate the inserts and insert springs from the hubs. Do not mix parts from the separate synchronizer assemblies.

Inspection

1. Check that the ring and gear engage smoothly.
2. Check for worn teeth or tapered surface.

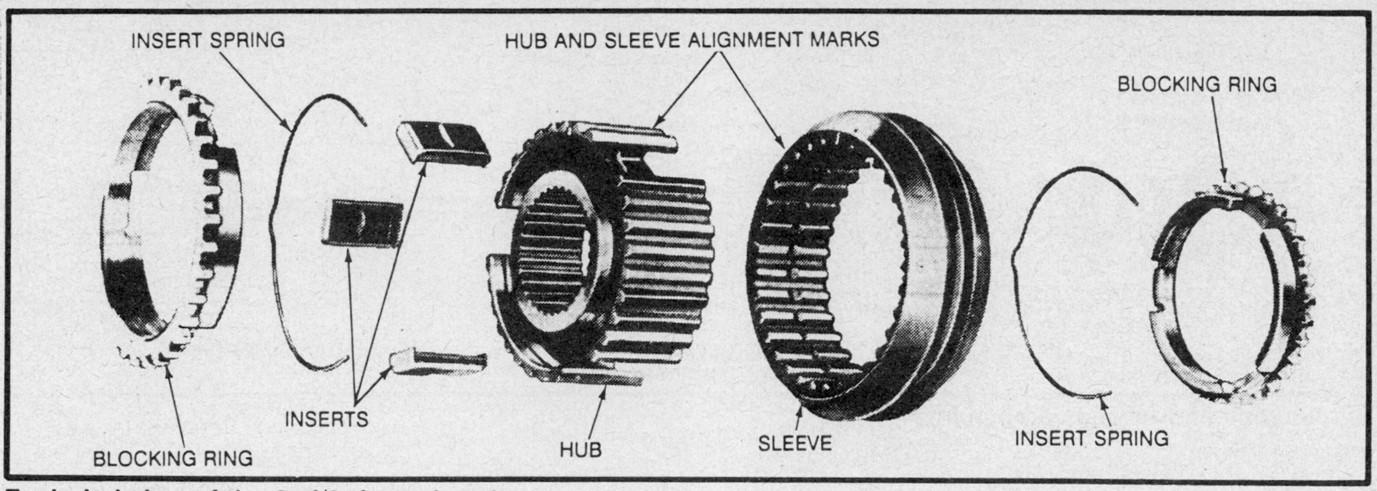
INSERT SPRING
HUB AND SLEEVE ALIGNMENT MARKS
BLOCKING RING
INSERTS
BLOCKING RING
HUB
SLEEVE
INSERT SPRING

Exploded view of the 2nd/3rd synchronizer

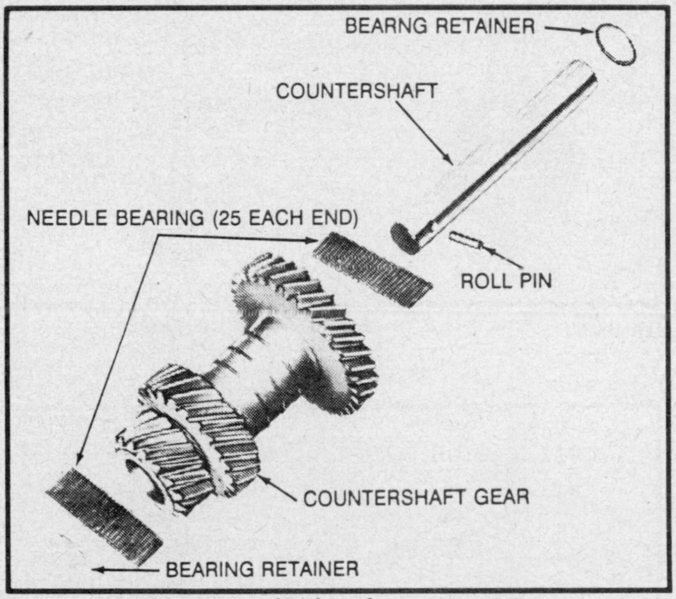
BEARNG RETAINER
COUNTERSHAFT
NEEDLE BEARING (25 EACH END)
ROLL PIN
COUNTERSHAFT GEAR
BEARING RETAINER

Countershaft gear and related components

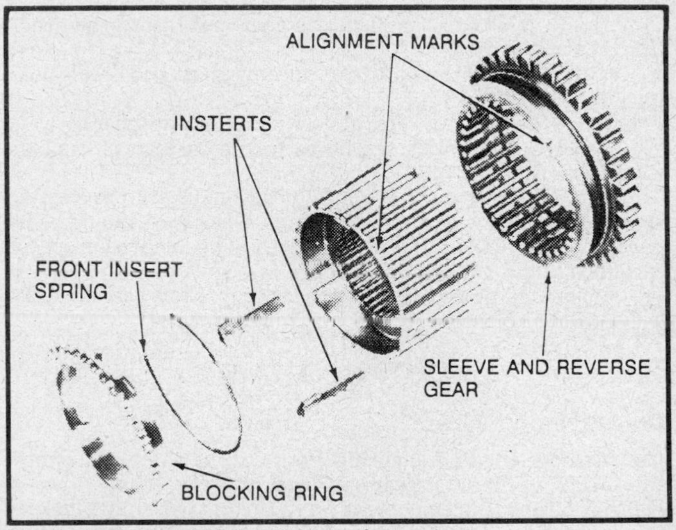
ALIGNMENT MARKS
INSTERTS
FRONT INSERT
SPRING
SLEEVE AND REVERSE
GEAR
BLOCKING RING

Exploded view of the 1st and reverse synchronizer

3. Inspect the synchronizer hub for worn or damaged splines, synchronizer key groove or end surface.

4. Check for smooth hub and sleeve engagement when it is installed.

5. Check the hub sleeve splines and sleeve fork groove for wear or damage.

Assembly

1. Install the insert spring in the hub of the 1st/reverse synchronizer.

2. Be sure that the spring covers all the insert grooves. Start the hub on the sleeve making certain that the scribed marks are properly aligned.

3. Place the 3 inserts in the hub, small ends on the inside. Slide the sleeve and reverse gear onto hub.

4. Install 1 insert spring into a groove on the 2nd/3rd synchronizer hub. Be sure that all 3 insert slots are covered.

5. Align the scribed marks on the hub and sleeve and start the hub into the sleeve. Position the 3 inserts on the top of the retaining spring and push the assembly together.

6. Install the remaining retainer spring so that the spring ends cover the same slots as the 1st spring. Do not stagger the springs.

7. Place a synchronizer blocking ring on the ends of the synchronizer sleeve.

COUNTERSHAFT GEAR

Disassembly

Remove the dummy shaft, needle bearings and bearing retainers from the countershaft gear.

Inspection

1. Inspect the roller bearing for wear or damage.
2. Check all gears for chipped, broken or worn teeth.
3. Inspect the countershaft for wear. Replace defective parts as needed.

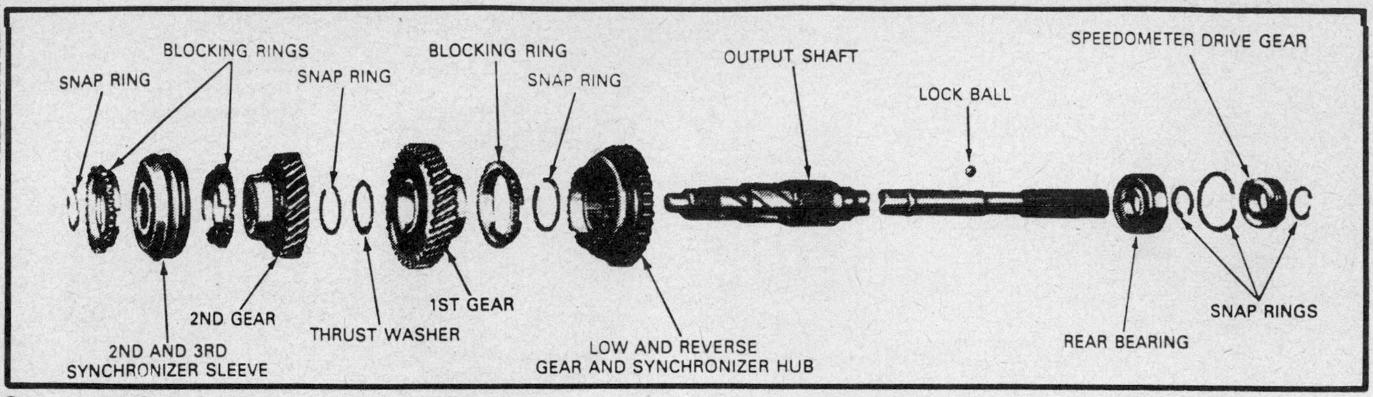
BLOCKING RINGS
SNAP RING
SNAP RING
BLOCKING RING
SNAP RING
OUTPUT SHAFT
LOCK BALL
SPEEDOMETER DRIVE GEAR
2ND GEAR
2ND AND 3RD
SYNCHRONIZER SLEEVE
THRUST WASHER
1ST GEAR
LOW AND REVERSE
GEAR AND SYNCHRONIZER HUB
REAR BEARING
SNAP RINGS

Output shaft and related components

Assembly

1. Coat the bore in each end of the countershaft gear with grease.
2. Hold the dummy shaft in the gear and install the needle bearings in the case.
3. Place the countershaft gear, dummy shaft and needle bearings in the case.
4. Place the case in a vertical position. Align the gear bore and the thrust washers with the bores in the case and install the countershaft.
5. Place the case in a horizontal position. Check the countershaft gear endplay with a feeler gauge. Clearance should be between 0.004–0.018 in. If clearance does not come within specifications, replace the thrust washers.
6. Install the dummy shaft in the countershaft gear and leave the gear at the bottom of the transmission case.

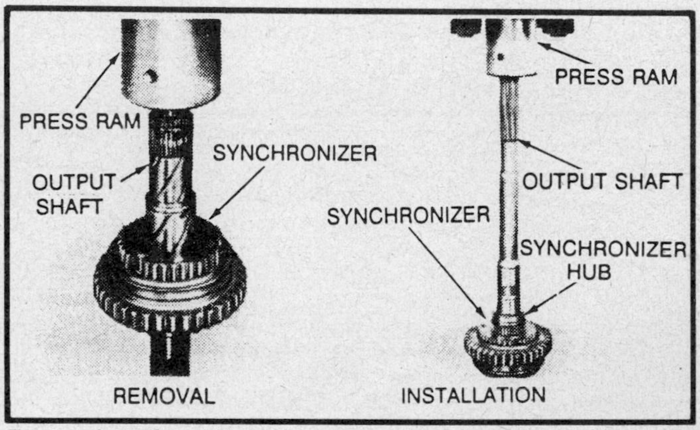
PRESS RAM
SYNCHRONIZER
OUTPUT SHAFT
REMOVAL
PRESS RAM
OUTPUT SHAFT
SYNCHRONIZER
SYNCHRONIZER HUB
INSTALLATION

Removing the 1st and reverse synchronizer

OUTPUT SHAFT

Disassembly

1. Remove the 2nd snapring from output shaft and remove the thrust washer, 1st gear and blocking ring.
2. Remove the 3rd snapring from the output shaft. The 1st/reverse synchronizer hub is a press fit on the output shaft.
3. Remove the synchronizer hub with an arbor press. Do not attempt to remove or install the synchronizer hub by prying or hammering.

Inspection

1. Inspect the bearings for wear or damage.
2. Check the teeth, splines and journals of the outputshaft for damage.
3. Check all gears for chipped, broken or worn teeth.
4. Check synchronizer sleeves for free movement on their hubs.
5. Inspect the synchronizer blocking rings for wear marks.

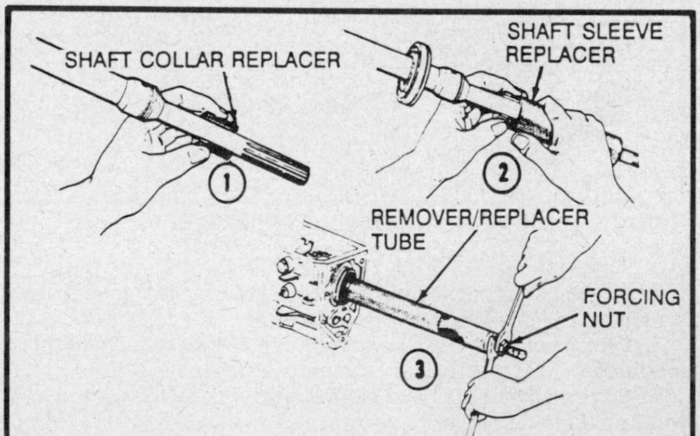
SHAFT COLLAR REPLACER
SHAFT SLEEVE REPLACER
1
2
REMOVER/REPLACER TUBE
FORCING NUT
3

Installing the output shaft rear bearing

Assembly

1. Lubricate the output shaft splines and machined surfaces with transmission oil.
2. The 1st/reverse synchronizer hub is a press fit on the output shaft. Hub must be installed in an arbor press. Install the synchronizer hub with the teeth-end of the gear facing towards the rear of the shaft. Do not attempt to install the 1st/reverse synchronizer with a hammer.
3. Place the blocking ring on the tapered surface of the 1st gear.
4. Slide the 1st gear on the output shaft with the blocking ring toward the rear of the shaft. Rotate the gear as necessary to engage the 3 notches in the blocking ring with the synchronizer inserts. Install the thrust washer and snapring.
5. Slide the blocking ring onto the tapered surface of the second gear. Slide the 2nd gear with blocking ring and the second/3rd synchronizer on the mainshaft. Be sure that the tapered surface of 2nd gear is facing the front of the shaft and that the notches in the blocking ring engage the synchronizer inserts. Install the snapring and secure assembly.

REVERSE IDLER GEAR SHAFT

Disassembly

Lift the reverse idler gear and thrust washers out of case.

Inspection

1. Inspect the roller bearing for wear or damage.
2. Check all gears for chipped, broken or worn teeth.
3. Inspect the idler shaft and trustwashers for wear. Replace defective parts as needed.

Assembly

1. Cover the reverse idler gear thrust surfaces in the case with a thin film of lubricant and install the 2 thrust washers in the case.
2. Install the reverse idler gear and shaft in the case. Align the case bore and thrust washers with gear bore and install the reverse idler shaft.
3. Measure the reverse idler gear endplay with a feeler gauge, clearance should be between 0.004–0.018 in. If endplay is not within specifications, replace the thrust washers. If clearance is correct, leave the reverse idler gear in case.

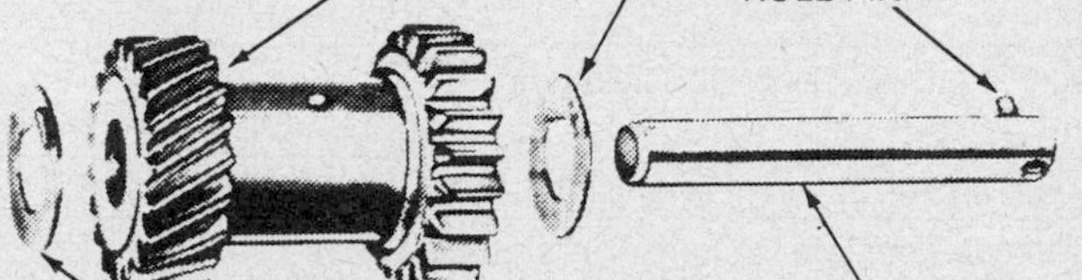
THRUST WASHER
REVERSE IDLER GEAR
ROLL PIN
REVERSE IDLER GEAR SHAFT
THRUST WASHER

Exploded view of the idler gear shaft

Transmission Assembly

1. Cover the reverse idler gear thrust surfaces in the case with a thin film of lubricant and install the 2 thrust washers in the case.
2. Install the reverse idler gear and shaft in the case. Align the case bore and thrust washers with gear bore and install the reverse idler shaft.
3. Measure the reverse idler gear endplay with a feeler gauge, clearance should be between 0.004–0.018 in. If endplay is not within specifications, replace the thrust washers. If clearance is correct, leave the reverse idler gear in case.
4. Lubricate the output shaft splines and machined surfaces with transmission oil.
5. The 1st/reverse synchronizer hub is a press fit on the output shaft. Hub must be installed in an arbor press. Install the synchronizer hub with the teeth-end of the gear facing towards the rear of the shaft. Do not attempt to install the 1st/reverse synchronizer with a hammer.
6. Place the blocking ring on the tapered surface of the 1st gear.
7. Slide the 1st gear on the output shaft with the blocking ring toward the rear of the shaft. Rotate the gear as necessary to engage the 3 notches in the blocking ring with the synchronizer inserts. Install the thrust washer and snapring.
8. Slide the blocking ring onto the tapered surface of the second gear. Slide the 2nd gear with blocking ring and the second/3rd synchronizer on the mainshaft. Be sure that the tapered surface of 2nd gear is facing the front of the shaft and that the notches in the blocking ring engage the synchronizer inserts. Install the snapring and secure assembly.
9. Cover the core of the input shaft with a thin coat of grease, be sure not to plug the lubricant holes, as damage to the bearings will resault.
10. Install the input shaft through the front of the case and insert snapring in the bearing groove.
11. Install the output shaft assembly in the case. Position the 2nd/3rd shift fork on the 2nd/3rd synchronizer.
12. Place a detent plug spring and a plug in the case. Place the 2nd/3rd synchronizer in the 2nd gear position (toward the rear of the case). Align the fork and install the 2nd/3rd shift rail. It will be necessary to depress the detent plug to install the shift rail in the bore. Move the rail forward until the detent plug enters the forward notch (2nd gear).
13. Secure the fork to the shift rail with a set screw and place the synchronizer in neutral.
14. Install the interlock plug in the case.
15. Place the 1st/reverse synchronizer in the 1st gear position (towards the front of the case). Place the shift fork in the groove of the synchronizer. Rotate the fork into position and install the shift rail. Move the shift rail inward until the center notch (neutral) is aligned with the detent bore. Secure shift fork with set screw.
16. Install a new shift rail expansion plug in the front of the case.
17. Hold the input shaft and blocking ring in position and move the output shaft forward to seat the pilot in the roller bearings on the input gear.
18. Tap the input gear bearing into place while holding the output shaft. Install the front bearing retainer and gasket. Torque attaching bolts to specifications.
19. Install the large snapring on the rear bearing. Place the bearing on the output shaft with the snapring end toward the rear of the shaft. Press the bearing into place using a special tool. Secure the bearing to the shaft with the snapring.
20. Hold the speedometer drive gear lock ball in the detent and slide the speedometer drive gear into position. Secure with snapring.
21. Place the transmission in the vertical position. Using a suitable tool insert it through the drain hole in the bottom of the case, align the bore of the countershaft gear and the thrust washer with the bore in the case.
22. Working from the rear of the case, push the dummy shaft out of the countershaft gear with the countershaft. Align the roll pin hole in the countershaft with the matching hole in the case. Drive the shaft into place and install the roll pin.
23. Position the new extension housing gasket on the case with sealer. Install the extension housing and torque to specification.
24. Place the transmission in gear and pour gear oil over entire gear train while rotating the input shaft.
25. Install the remaining detent plug and long spring in case.
26. Position cover gasket on case with sealer and install cover. Torque cover bolts to specifications.
27. Install the transmission into the vehicle.

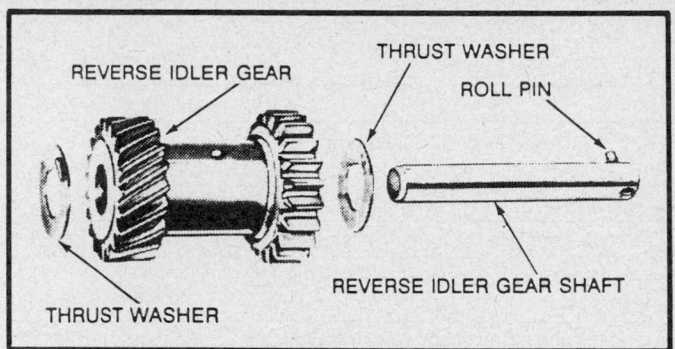

SPECIFICATIONS

TORQUE SPECIFICATIONS

Item	ft. lbs.	Nm
Input shaft gear bearing retainer to transmission case	30–36	41–48
Transmission to flywheel housing	42–50	57–67
Transmission cover to transmission case	20–25	28–33
Speedometer cable retainer to transmission extension	3–4.5	4.7–6.5
Transmission extension to transmission case	42–50	57–67
Flywheel housing to engine	40–50	55–67

Item	ft. lbs.	Nm
Back-up lamp switch	8–12	11–16
Gear shift lever to cam and shaft assembly locknuts	18–23	25–31
U-Joint flange to output shaft	60–80	82–108
Filler plug	10–20	14–27
Shifter fork set screw	10–18	14–24
Rear support to frame	48–65	65–88

SPECIAL TOOLS

Tool Number	Description
T57L-500-B	Bench mounted holding fixture
Tool-1175-AC	Seal remover
T00L-4201-C	Dial indicator with bracketry
D78P-4201-B	Dial indicator with base
T75L-4201-A	Clutch (flywheel) housing alignment adapter
T75L-4201-B	Clutch (flywheel) housing alignment adapter
T57L-4220-A	Input shaft bearing remover
T53T-4621-B	Input shaft bearing replacer
T75L-6392-A	Clutch (flywheel) housing alignment tool
T75L-7025-E	Bearing collet
T75L-7025-G	Bearing collet sleeve

Tool Number	Description
T75L-7025-B	Remover/replacer tube
T84T-7025-B	Forcing screw
T75L-7025-P	Shaft collar
T75L-7025-K	Shaft sleeve—replacing
T77L-7025-L	Bearing service set
T63P-7111-A	Dummy countershaft
T63P-7111-B	Dummy countershaft
T64P-7111-A	Dummy countershaft
T64P-7140-A	Dummy countershaft
T67P-7341-A	Shift linkage insulator tool

Section 7

RUG Transmission
Ford Motor Co.

APPLICATION

Ford E-150 Series Vehicles

GENERAL DESCRIPTION

This 4 speed overdrive transmission is fully synchronized with all gears except the reverse sliding gear which is in constant mesh. Forward gear changes are accomplished with synchronizer sleeves. All forward gears are helical type. The reverse sliding gear and the external teeth of the 1st and 2nd speeds synchronizer sleeve are spur type.

The 4 speed shift control unit of this transmission is serviced as a unit only. The shift control is not to be disassembled. The only parts to be removed from the shifter assembly are the 3 shift rods, the shift lever, back-up light switch bracket and the back-up light switch. The shift pattern for the control unit is imprinted on the shift knob.

Transmission Identification

The service identification tag is located on the right side of the case at the front. The first line shows the transmission model and service identification code. The second line shows the transmission serial number and this serial number is also stamped on the top side of the case flange.

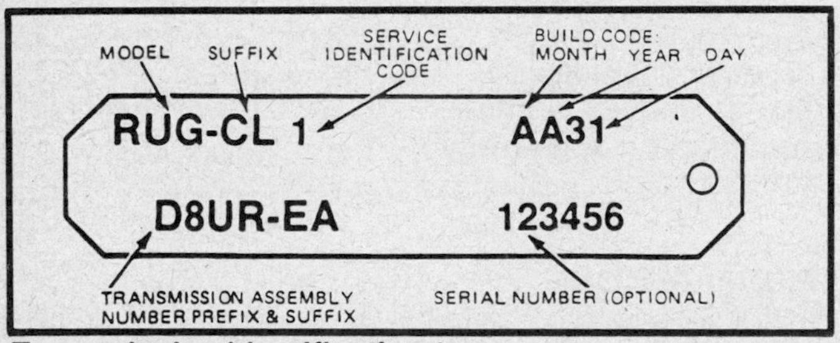
MODEL
SUFFIX
SERVICE IDENTIFICATION CODE
BUILD CODE: MONTH YEAR DAY
RUG-CL 1
AA31
D8UR-EA
123456
TRANSMISSION ASSEMBLY NUMBER PREFIX & SUFFIX
SERIAL NUMBER (OPTIONAL)

Transmission identification tag

Metric Fasteners

The metric fastener dimensions are very close to the dimensions of the familiar inch system fasteners and for this reason, replacement fasteners must have the same measurement and strength as those removed.

CAUTION

Do not attempt to interchange metric fasteners for inch system fasteners. Mismatched or incorrect fasteners can result in damage to the transaxle unit through malfunctions or breakage and possible personal injury.

Capacities

The 4 speed overdrive transmission uses standard transmission lubricant (SAE 80W). The capacity of the 4 speed overdrive transmission is 4.5 pints or 2.1L. The correct fluid level is to the bottom of the filler hole.

Checking Fluid Level

The transmission fluid level check must be made with the vehicle level. The fluid level can be checked by removing the fill plug. The correct fluid fill will be even with the bottom edge of the filler plug opening or within ¼ inch of this level. If fluid is low, add the specified fluid to correct level.

TRANSMISSION MODIFICATIONS

There are no modifications at the time of this printing.

TROUBLE DIAGNOSIS

CHILTON'S THREE C'S TRANSMISSION DIAGNOSIS
RUG Transmission

Condition	Cause	Correction
Shift lever—improper operation	a) Shift rods out of adjustment	a) Check crossover operation. Hand lever must move freely through neutral gate without catch or drag. Adjust shift rods
	b) Steering column shift tube out of alignment	b) Check lever position for proper transmission gear selection. Replace steering column shift tube if necessary

CHILTON'S THREE C'S TRANSMISSION DIAGNOSIS
RUG Transmission

Condition	Cause	Correction
Transmission shifts hard	a) Clutch pedal free-travel out of adjustment	a) Adjust clutch pedal free-travel
	b) Clutch does not completely release	b) Inspect complete clutch system
	c) Transmission fluid low or improper type	c) Add lubricant or change lubricant as required
	d) Shift lever binding or worn	d) Remove cap from shift tower. Eliminate binding condition or replace components as required
	e) Worn or damaged internal shift mechanism	e) Remove transmission cover. Check internal shift mechanism by shifting into and out of all gears. Repair or replace as required
	f) Binding of sliding gears and/or synchronizers	f) Check for free movement of gears and synchronizers. Repair or replace as required
	g) Housings and/or shafts out of alignment	g) Remove transmission and check for binding condition between input shaft and engine crankshaft pilot bearing or bushing. Check flywheel housing alignment. Repair or replace as required
Noisy in forward gears	a) Lubricant level low, or improper type	a) Add lubricant, or refill with specified lubricant
	b) Components grinding on transmission	b) Check for screws, bolts, etc., of cab or other components grounding out. Correct as required
	c) Component housing bolts loose	c) Check torque on transmission to flywheel housing bolts, output shaft flange nut and flywheel housing to engine block bolts. Tighten bolts to specification
	d) Flywheel housing to engine crankshaft alignment	d) Check and align flywheel housing to engine crankshaft
	e) Noisy bearings or gears	e) Remove and disassemble transmission. Inspect input, output and countershaft bearings. Inspect speedometer gear and gear teeth for wear or damage. Replace as required
Gears clash when shifting from one forward gear to another	a) Engine idle speed too high	a) Adjust engine idle speed
	b) Clutch pedal free-travel incorrect	b) Check clutch for pedal free-travel. Adjust as required
	c) Improper manual shift linkage	c) Adjust and repair manual shift linkage as required
	d) Pilot bearing binding	d) Remove transmission and check for a binding condition between input shaft and engine crankshaft pilot bearing. Replace as required
	e) Damaged gear teeth and/or synchronizer	e) Disassemble transmission, repair or replace as required
Transmission jumps out of gear	a) Manual shift linkage binding, out of adjustment. Stiff shift boot	a) Adjust linkage. If necessary align steering column to vehicle body
	b) Improper fit of form isolation pad	b) Replace shift boot if exceptionally stiff. Replace or rework pad to provide clearance
	c) Loose transmission to engine mounting bolts, or loose levers	c) Tighten transmission to flywheel housing, and flywheel housing to engine block bolts. Loosen all bolts and reseat flywheel housing. Tighten all bolts. Tighten levers if necessary

CHILTON'S THREE C'S TRANSMISSION DIAGNOSIS
RUG Transmission

Condition	Cause	Correction
Transmission jumps out of gear	d) Flywheel housing to engine crankshaft out of line	d) Shim or replace housing as required
	e) Crankshaft pilot bearing worn	e) Replace bearing
	f) Interior components damage	f) Disassemble transmission. Inspect the synchronizer sleeves for free movement on their hubs. Inspect the synchronizer blocking rings for widened index slots, rounded clutch teeth and smooth internal surface. Check countershaft cluster gear for excessive endplay. Check shift forks for worn or loose mounting on shift rails. Check fork pads for excessive wear. Inspect synchronizer sliding sleeve and gear clutch teeth for wear or damage. Repair or replace as required
	g) Worn gear teeth due to partial engagement	g) Replace worn or damaged gears
Transmission is locked in one gear. It cannot be shifted out of that gear	a) Manual shift linkage out of adjustment, binding or damaged	a) Disconnect the problem shift rod from transmission shift lever. Try to shift the transmission lever into and out of gear position. If OK, repair or replace linkage parts
	b) Internal components	b) Remove transmission. Inspect problem gears, shift rails and forks and synchronizer for wear or damage. Repair as required
	c) Loose fork on rail	c) Check shift rail interlock system
Transmission will not shift into one gear—all others OK	a) Manual linkage out of adjustment	a) Adjust manual linkage
	b) Manual shift linkage damaged or worn	b) Lubricate, repair or replace parts as required
	c) Back-up switch ball frozen	c) If reverse is problem, check back-up switch for ball frozen in extended position (if so equipped)
	d) Internal components	d) Remove transmission. Inspect shift rail and fork system synchronizer system and gear clutch teeth for restricted travel. Repair or replace as required

ON VEHICLE SERVICE

Adjustments

SHIFT LINKAGE

1. Raise and support the vehicle if necessary. Disconnect the 3 shift rods from the shifter assembly .
2. With the shift levers in the neutral position insert a ¼ inch (6.35mm) diameter pin through the alignment hole in the shift assembly.
3. Align the 3 transmission levers as follows:
   a. Foward lever (3rd/4th lever) in the mid-position
   b. Rearward lever (1st/2nd lever) in the mid-position
   c. Middle lever (reverse lever) rotate counterclockwise to the neutral position.
4. Rotate the output shaft to assure that the transmission is in neutral.
5. Shift the reverse lever (middle) clockwise to the reverse position. This causes the interlock system to align the 1/2 and 3/4 shift rails in their neutral position.
6. Position the 1/2 and 3/4 shift rods to their corresponding levers and tighten the locknuts to 15–20 ft. lbs.
7. Rotate the reverse lever counterclockwise back to the neutral position. Install the reverse shift rod and tigten the locknut to 15–20 ft. lbs.
8. Remove the alignment pin.
9. Check for proper operation.

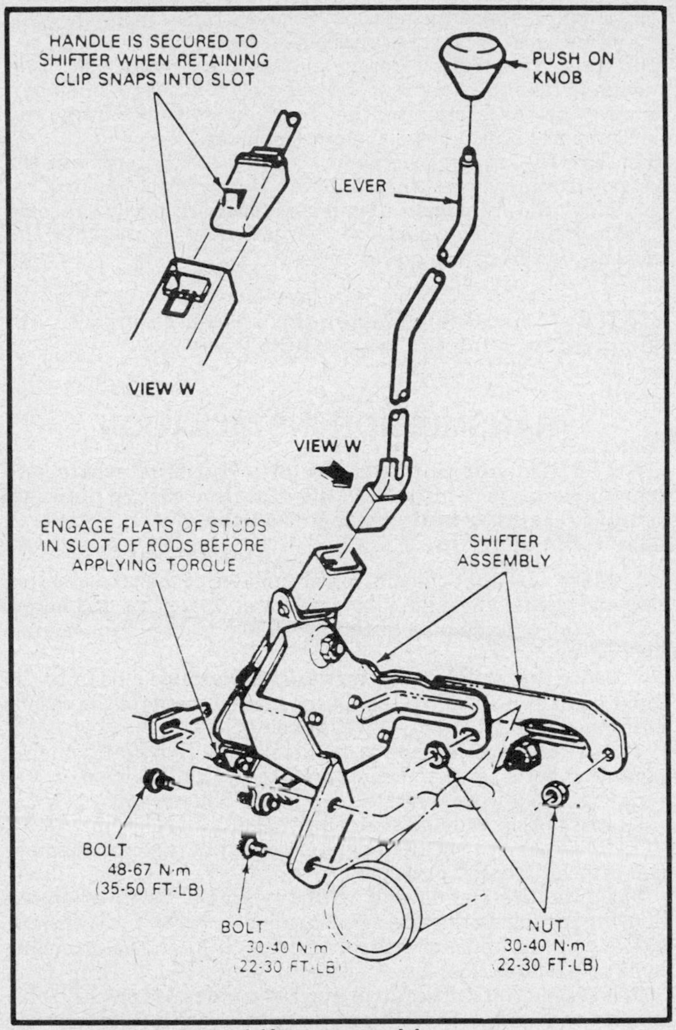
HANDLE IS SECURED TO SHIFTER WHEN RETAINING CLIP SNAPS INTO SLOT
PUSH ON KNOB
LEVER
VIEW W
VIEW W
ENGAGE FLATS OF STUDS IN SLOT OF RODS BEFORE APPLYING TORQUE
SHIFTER ASSEMBLY
BOLT 48-67 N·m (35-50 FT-LB)
BOLT 30-40 N·m (22-30 FT-LB)
NUT 30-40 N·m (22-30 FT-LB)

E–150 and E–250 shifter assembly

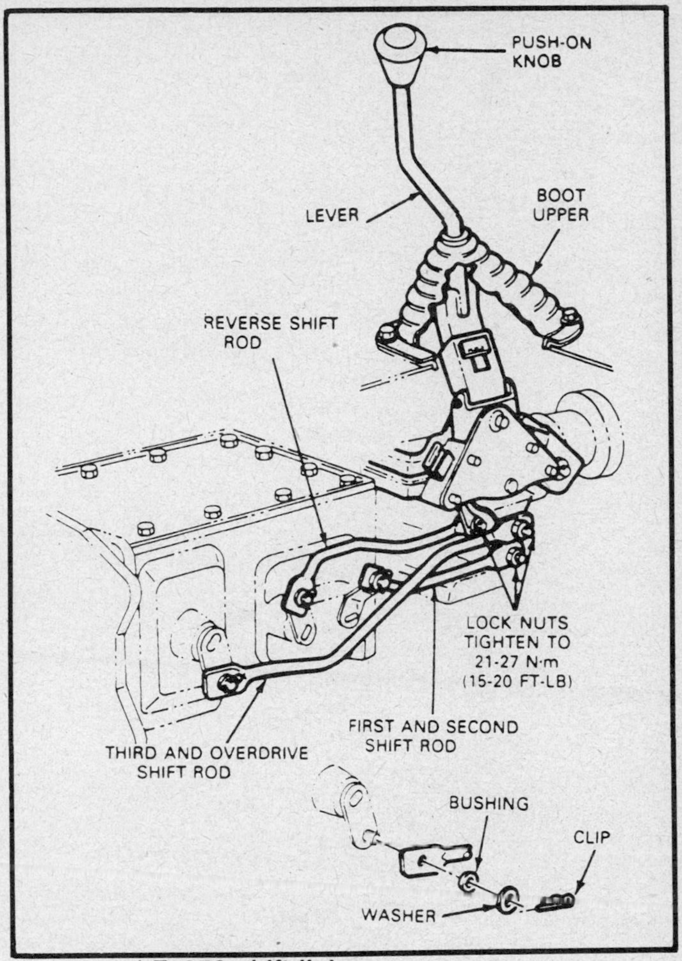
PUSH-ON KNOB
LEVER
BOOT UPPER
REVERSE SHIFT ROD
LOCK NUTS TIGHTEN TO 21-27 N·m (15-20 FT-LB)
FIRST AND SECOND SHIFT ROD
THIRD AND OVERDRIVE SHIFT ROD
BUSHING
CLIP
WASHER

E–150 and E–250 shift linkage

Services

REAR SEAL

Removal and Installation

1. Raise the vehicle and support it safety.
2. Mark the installed position then remove the drive shaft.
3. Remove the seal from the extension housing using a suitable seal puller tool.
4. Install a new seal in the extension housing using a suitable installer tool. Lubricate the seal with a muti-purpose lubricant.
5. Install the drive shaft and lower the vehicle.

REAR BUSHING AND SEAL

Removal and Installation

1. Raise the vehicle and support it safety.
2. Mark the installed position then remove the drive shaft.
3. Remove the seal using a suitable seal remover tool.
4. Insert a suitable bushing remover tool into the extension housing until it grips on the front side of the bushing.
5. Turn the bushing remover screw clockwise until the bushing is free of the housing.
6. Lubricate the new bushing with a multi-purpose lubricant and drive it into the extension housing.
7. Install a new seal and lubricate it with a multi-purpose lubricant.
8. Connect the driveshaft to the transmission. Lower the vehicle.

REMOVAL AND INSTALLATION

TRANSMISSION REMOVAL

NOTE: The transmission is removed separately from the engine. Do not depress the clutch pedal while the transmission is removed.

1. Place a wood block or equivalent under the clutch pedal to prevent it from being depressed.
2. Raise the vehicle and support it safely.
3. Mark the driveshaft so that it may be installed in the same relative position. Disconnect the driveshaft from the rear U-joint flange.
4. Slide the driveshaft off the trannsmission output shaft. Install an appropriate tool into the extension housing to prevent lubricant leakage.

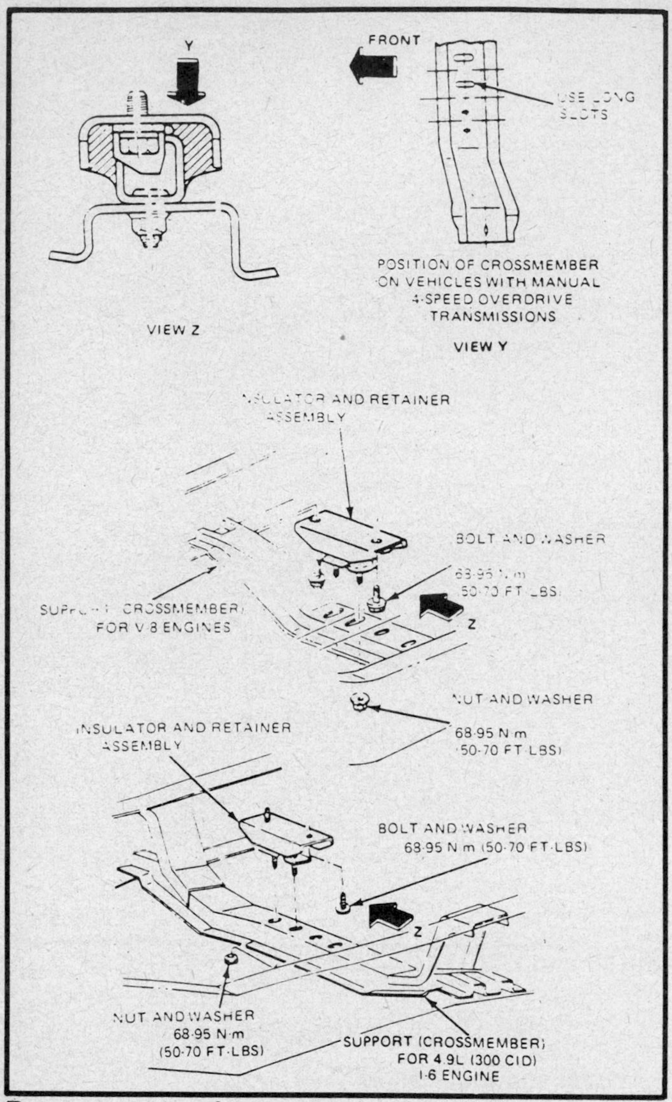
Y
FRONT
POSITION OF CROSSMEMBER
ON VEHICLES WITH MANUAL
4-SPEED OVERDRIVE
TRANSMISSIONS
VIEW Z
VIEW Y
Z
NUT AND WASHER
68-95 N·m
(50-70 FT-LBS)
INSULATOR AND RETAINER
ASSEMBLY
BOLT AND WASHER
68-95 N·m (50-70 FT-LBS)
Z
NUT AND WASHER
68-95 N·m
(50-70 FT-LBS)
SUPPORT (CROSSMEMBER)
FOR 4.9L (300 CID)
I-6 ENGINE

Rear support and crossmember installation – E-150 models

NOTE: A shop towel can be wrapped around the extension housing to prevent lubricant leakage while removing the transmission.

5. Disconnect the speedometer cable from the extension housing.
6. Remove the clips, flat washers and spring washers that secure the shift rods to the shift levers.
7. Remove the bolts connecting the shift control to the transmission extension housing. Remove the nuts connecting the shift control to the transmission case.
8. Remove the rear transmission support connecting bolts attaching the support on the crossmember to the transmission extension housing.
9. Support the transmission with a transmission jack or equivalent under the transmission case.
10. Raise the transmission, clutch and engine assembly enough to take the weight of the assembly off the crossmember. Remove the bolts retaining the crossmember to the frame rail side supports and remove the crossmember.
11. With the transmission supported on the jack, remove the bolts that attach the transmission to the flywheel housing.
12. Move the transmission and jack rearward until the transmission input shaft clears the flywheel housing. Remove the transmission from vehicle.

NOTE: If necessary, lower the engine enough to obtain clearance for the transmission removal.

TRANSMISSION INSTALLATION

NOTE: Do not depress the clutch pedal while the transmission is removed. If the clutch pressure plate assembly retaining bolts were loosen the clutch plate will have to be align up.

1. Make sure that the machined surfaces of the transmission case and the flywheel housing are free of dirt, paint and burrs.
2. Install 2 guide pins in the flywheel housing lower mounting bolt holes.
3. Move the transmission forward on the guide pins until the input shaft splines enter the clutch hub splines and the case is positioned against the flywheel housing.
4. Install the 2 upper transmission to flywheel housing mounting bolts snug and then remove the guide pins.
5. Tighten all transmission mounting bolts evenly.
6. Install the transmission support bolts and tighten evenly.
7. Raise the rear of the engine and install the crossmember and attaching bolts.
8. Lower the engine and with the transmission extension housing resting on the engine rear support, install the transmission extension housing attaching bolts. Tighten the attaching bolts to 42–50 ft. lbs.
9. Position the shift control bracket on the stud on the transmission case and on the bolt attaching holes on the transmission extension housing. The bracket must be placed in the proper position for correct shift control operation.
10. Tighten the nut connecting the bracket to the transmission case to 22–30 ft. lbs. Tighten the bolts to 22–30 ft. lbs.
11. Secure each shift rod to the correct lever with the spring washer, flat washer and retaining pin.
12. Connect the speedometer cable to the extension housing.
13. Remove the extension housing extension tool or shop towel and slide the forward end of the driveshaft over the transmission output shaft. Connect the driveshaft to the rear U-joint flange. Tighten nuts on U-bolts to 8–15 ft. lbs.

NOTE: Do not overtighten nuts on the U-bolts.

14. Adjust the shift linkage, if necessary.
15. Fill the transmission to the proper level with the specified fluid.
16. Lower the vehicle. Remove the wood block from under the clutch pedal.
17. Check the shift pattern for correct engagement.
18. Road test the vehicle for proper operation.

BENCH OVERHAUL

Before Disassembly

When servicing the unit, it is recommended that as each part is disassembled, it is cleaned in solvent and dried with compressed air. Disassembly and reassembly of this unit and its parts must be done on a clean work bench. Also, before installing bolts into aluminum parts, always dip the threads into clean transmission oil. Anti-seize compound can also be used to prevent bolts from galling the aluminum and seizing. Always use a torque wrench to keep from stripping the threads. Take care with the seals when installing them, especially the smaller O-rings. The slightest damage can cause leaks. Aluminum parts are very susceptible to damage so great care should be exercised when handling them. The internal snaprings should be expanded and the external snaprings compressed if they are to be reused. This will help insure proper seating when installed. Be sure to replace any O-ring, gasket, or seal that is removed.

Transmission Disassembly

1. Mount the transmission in a suitable holding fixture. Drain the lubricant by removing the lower extension to case screw. Remove the top cover attaching screws and lift the cover from the case.
2. Remove the screw, detent spring and detent plug from the case. A magnet or equivalent may be needed.
3. Remove the extension housing. Remove the input shaft bearing retainer attaching screws. Slide the retainer off the input shaft.
4. Support the countershaft gear with a wire hook. From the front of the case push the countershaft out the rear of the case with a dummy shaft and lower the countershaft to the bottom of the case then remove wire hook.
5. Remove the set screw from the 1st/2nd speed shift fork. Slide the shift rail out the rear of the case.
6. Using a magnet or equivalent remove the interlock detent from between the 1st/2nd and 3rd/overdrive shift rails.
7. Shift the transmission into the overdrive position. Remove the set screw from the 3rd/overdrive shift fork. Remove the side detent bolt, detent plug and spring.
8. Rotate the 3rd/overdrive shift rail 90 degrees clockwise and tap it out through the front of the case using suitable tools.
9. Remove the interlock pin from the top of the case with a magnet or equivalent.

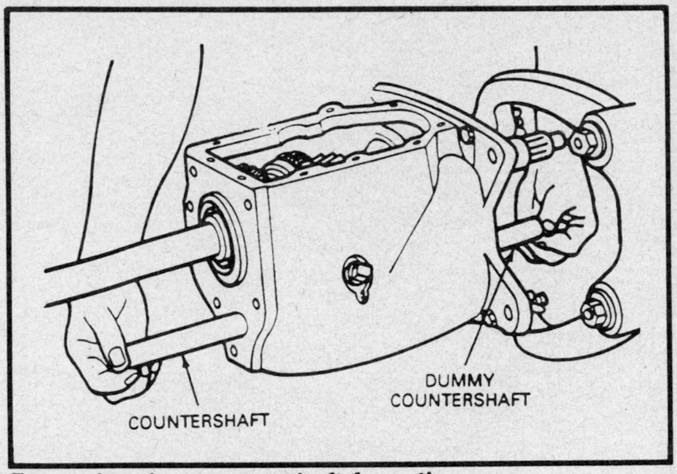
DUMMY
COUNTERSHAFT
COUNTERSHAFT
Removing the countershaft from the case

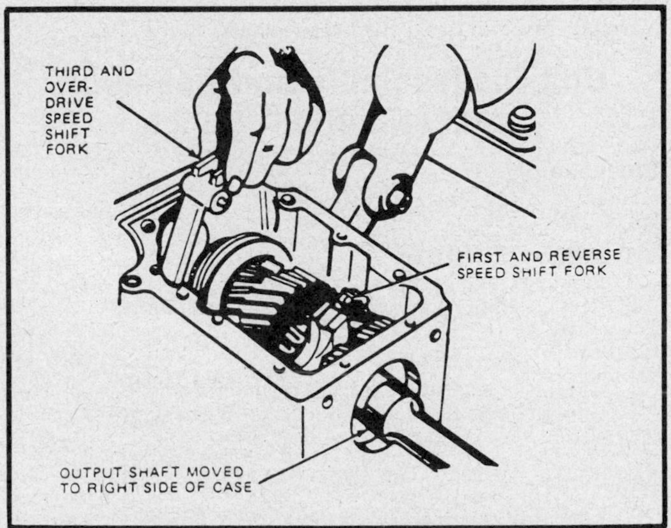
THIRD AND
OVER-
DRIVE
SPEED
SHIFT
FORK
FIRST AND REVERSE
SPEED SHIFT FORK
OUTPUT SHAFT MOVED
TO RIGHT SIDE OF CASE
Removing shift fork from transmission case

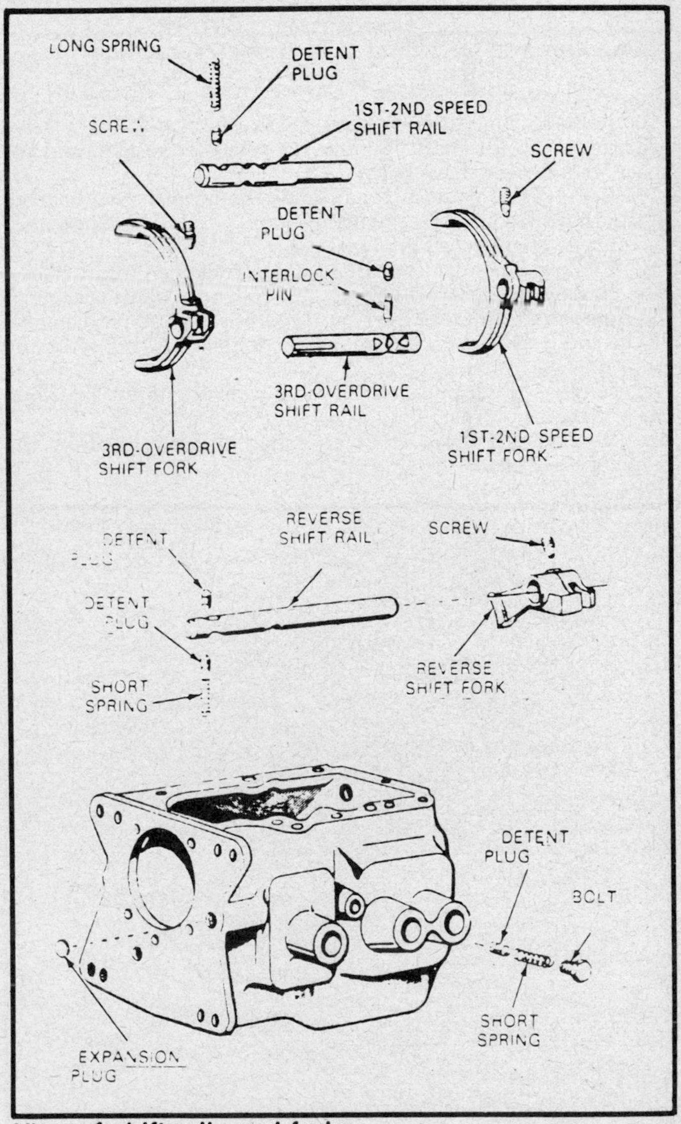
LONG SPRING
DETENT
PLUG
1ST-2ND SPEED
SHIFT RAIL
SCREW
DETENT
PLUG
3RD-OVERDRIVE
SHIFT RAIL
3RD-OVERDRIVE
SHIFT FORK
1ST-2ND SPEED
SHIFT FORK
REVERSE
SHIFT RAIL
SCREW
SHORT
SPRING
REVERSE
SHIFT FORK
DETENT
PLUG
BOLT
SHORT
SPRING
EXPANSION
PLUG
View of shift rails and forks

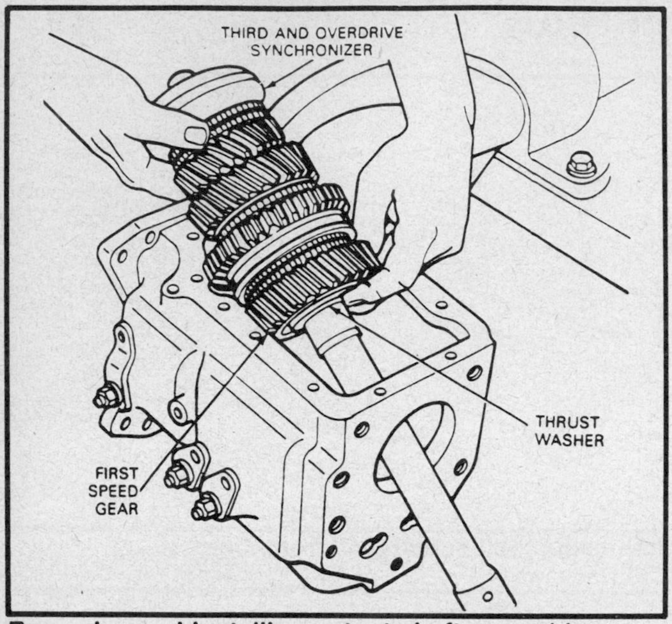
THIRD AND OVERDRIVE SYNCHRONIZER
THRUST WASHER
FIRST SPEED GEAR
Removing and installing output shaft assembly

10. Remove the snapring that holds the speedometer drive gear to the output shaft. Remove the gear off the shaft and the speedometer gear drive ball.
11. Remove the snapring that holds the output shaft bearing to the shaft. Remove the snapring from the outside diameter of the output shaft bearing.
12. Remove the output shaft bearing from the output shaft.
13. Remove the snapring that holds the input shaft bearing to the input shaft. Remove the snapring from the outside diameter of the input shaft bearing and then remove bearing retaining ring.
14. Remove the input shaft bearing from the input shaft and the transmission case.
15. Remove the input shaft and the blocking ring from the front of the case.

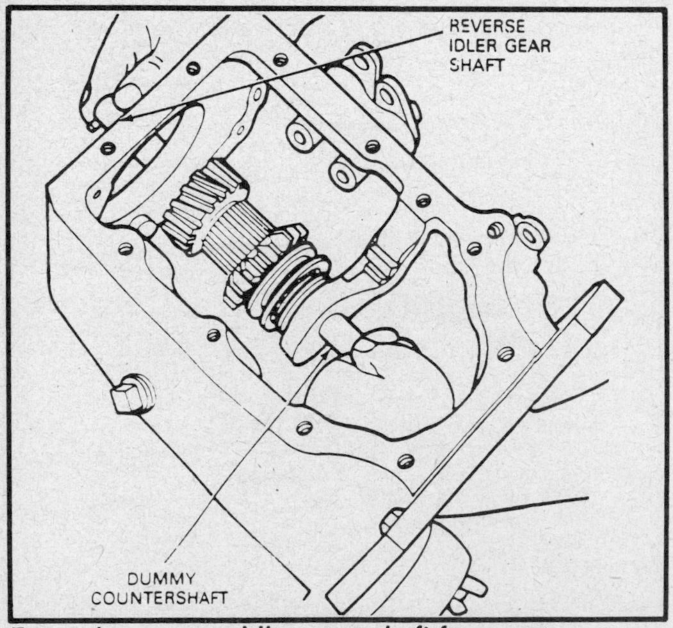
REVERSE IDLER GEAR SHAFT
DUMMY COUNTERSHAFT
Removing reverse idler gear shaft from case

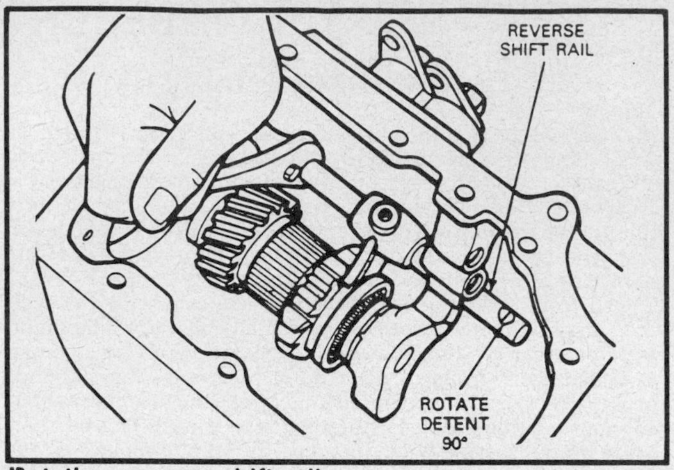
REVERSE SHIFT RAIL
ROTATE DETENT 90°
Rotating reverse shift rail

16. Move the output shaft to the right side of the case to provide clearance for the shift forks. Rotate the forks clockwise, then remove the forks from the case.
17. Support the thrust washer and the 1st speed gear to prevent them from sliding off the shaft, then lift the output shaft assembly from the case.
18. Remove the reverse gear shift fork set screw. Rotate the reverse shift rail 90 degrees and slide the shift rail out the rear of the case. Remove the reverse shift fork from the case.
19. Remove the reverse detent plug and spring from the case with a magnet or equivalent.
20. Remove the reverse idler gear shaft from the case, using a dummy countershaft or equivalent. Lift the countershaft gear and the thrust washers from the case.
21. Lift the reverse idler gear and the thrust washers from the case.

NOTE: Be careful not to drop the bearings and the dummy shaft from the gear.

22. Remove the snapring from the front of the output shaft. Slide the 3rd/overdrive synchronizer blocking ring and the gear off the shaft.
23. Remove the next snapring and the 2nd speed gear thrust washer from the shaft. Slide the 2nd speed gear and the blocking ring off the shaft.
24. Remove the snapring, thrust washer, 1st speed gear and blocking ring from the rear of the shaft. The 1st/2nd synchronizer hub is a slip fit on the output shaft.

Unit Disassembly and Assembly

CAM AND SHAFT SEALS

Disassembly

1. Remove the attaching nut, lockwasher and the flat washer from each shift lever.
2. Remove the 3 levers.
3. Remove the 3 cam and shafts from inside the case.
4. Remove the O-ring from each cam and shaft.

Inspection

1. Check the cam and shaft for wear or damage.
2. Check all the shift levers for wear at the transmission attaching slot.

Assembly

1. Lubricate the new O-rings in specified gear lubricant and install them on the cam and shafts.

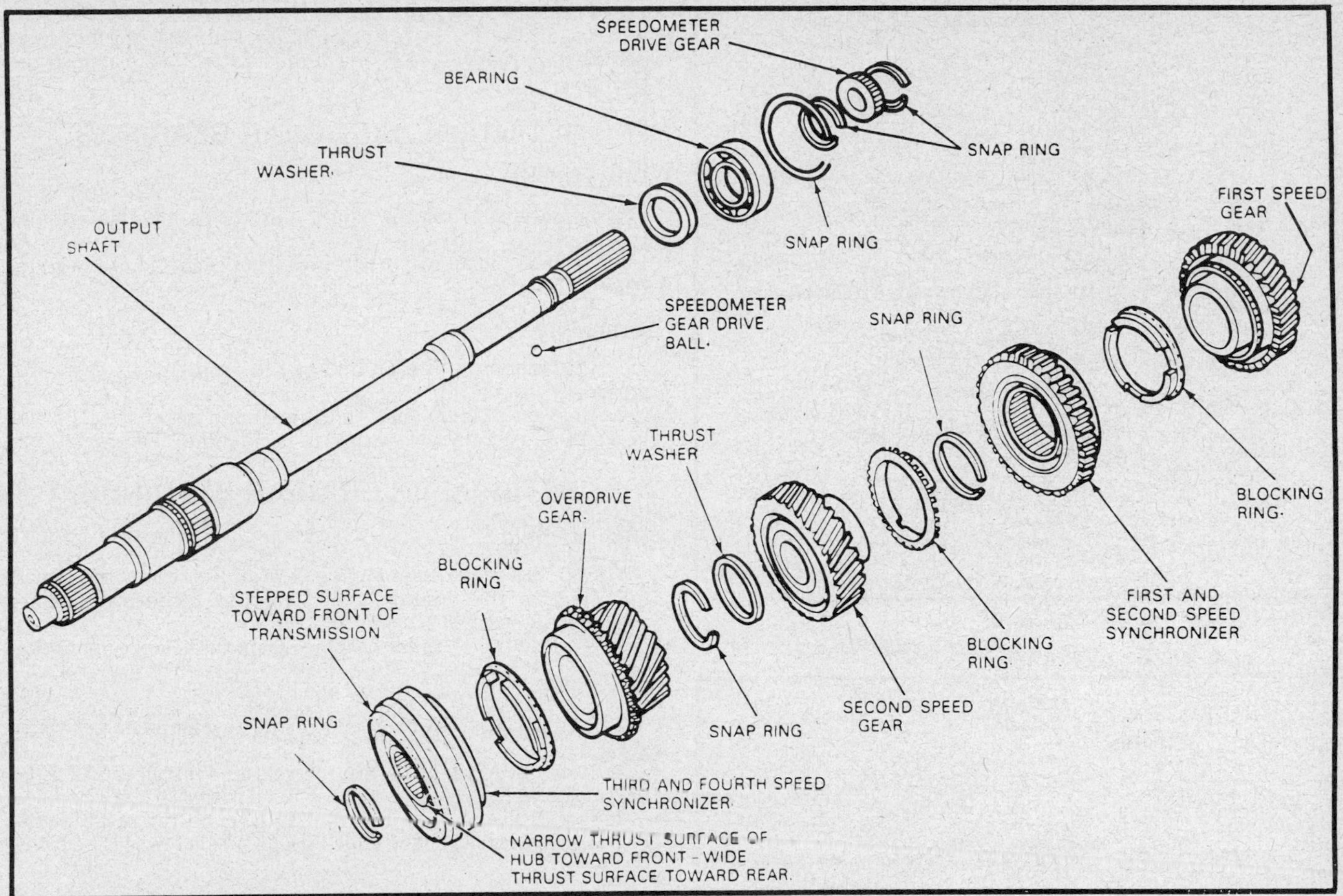
SPEEDOMETER
DRIVE GEAR
BEARING
SNAP RING
THRUST
WASHER
FIRST SPEED
GEAR
OUTPUT
SHAFT
SNAP RING
SPEEDOMETER
GEAR DRIVE
BALL
SNAP RING
THRUST
WASHER
BLOCKING
RING
OVERDRIVE
GEAR
BLOCKING
RING
FIRST AND
SECOND SPEED
SYNCHRONIZER
STEPPED SURFACE
TOWARD FRONT OF
TRANSMISSION
BLOCKING
RING
SECOND SPEED
GEAR
SNAP RING
SNAP RING
THIRD AND FOURTH SPEED
SYNCHRONIZER
NARROW THRUST SURFACE OF
HUB TOWARD FRONT - WIDE
THRUST SURFACE TOWARD REAR

Exploded view of output shaft

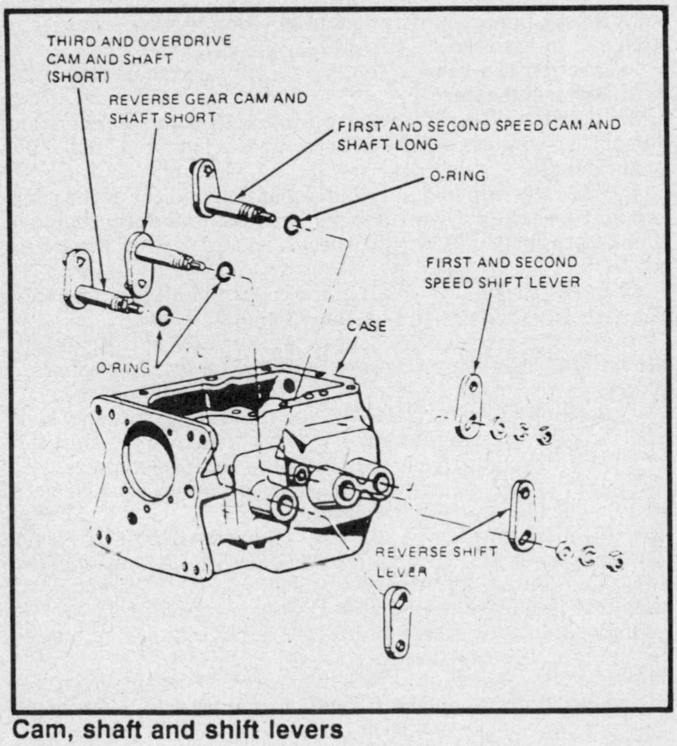
THIRD AND OVERDRIVE
CAM AND SHAFT
(SHORT)
REVERSE GEAR CAM AND
SHAFT SHORT
FIRST AND SECOND SPEED CAM AND
SHAFT LONG
O-RING
FIRST AND SECOND
SPEED SHIFT LEVER
CASE
O-RING
REVERSE SHIFT
LEVER

Cam, shaft and shift levers

2. Position each cam and shaft into its correct bore in the transmission case.
3. Secure each shaft lever with a flat washer, lockwasher and nut.

SYNCHRONIZER

Diassembly

1. Make alignment mark on the hub and sleeve of the synchronizer.
2. Remove the synchronizer hub from each synchronizer sleeve.
3. Remove the inserts and insert springs from the hubs.
4. Do not mix the parts of the 1st and 2nd speed synchronizer with the 3rd and 4th speed synchronizer.

Inspection

1. Check synchronizer discs for wear or damage.
2. Inspect the synchronizer blocking rings for widened index slots, rounded clutch teeth and smooth internal surfaces.
3. With the blocking ring on the cone, the distance between the face of the gear clutching teeth and the face of the blocking ring must not be less than 0.010 in.
4. Check the synchronizer sleeves for free movement on the hubs.

Assembly

1. Position the hub in the sleeve, making sure that the alignment marks are properly aligned.

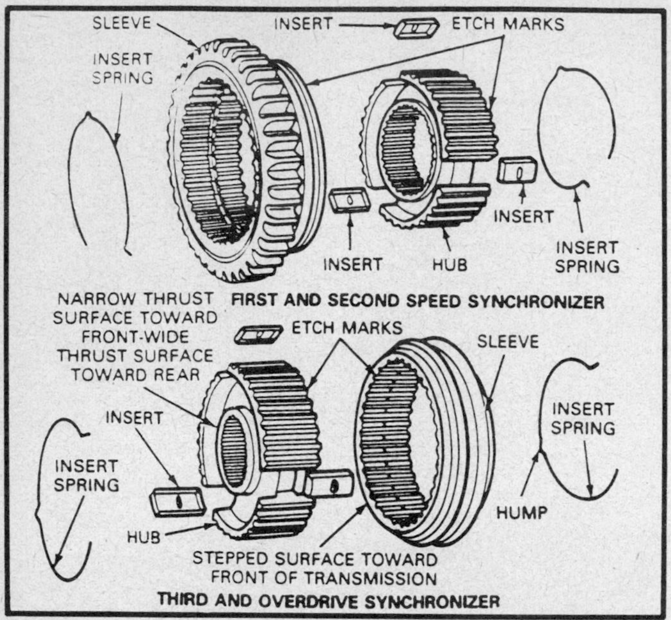
SLEEVE
INSERT
ETCH MARKS
INSERT SPRING
INSERT
INSERT
HUB
INSERT
INSERT SPRING
FIRST AND SECOND SPEED SYNCHRONIZER
NARROW THRUST SURFACE TOWARD FRONT-WIDE THRUST SURFACE TOWARD REAR
ETCH MARKS
SLEEVE
INSERT
INSERT SPRING
INSERT SPRING
HUMP
HUB
STEPPED SURFACE TOWARD FRONT OF TRANSMISSION
THIRD AND OVERDRIVE SYNCHRONIZER

Synchronizers disassembled

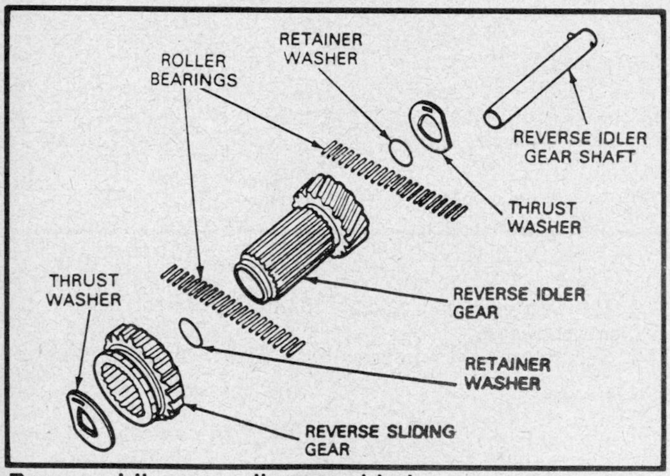
RETAINER WASHER
ROLLER BEARINGS
REVERSE IDLER GEAR SHAFT
THRUST WASHER
THRUST WASHER
REVERSE IDLER GEAR
RETAINER WASHER
REVERSE SLIDING GEAR

Reverse idler gear disassembled

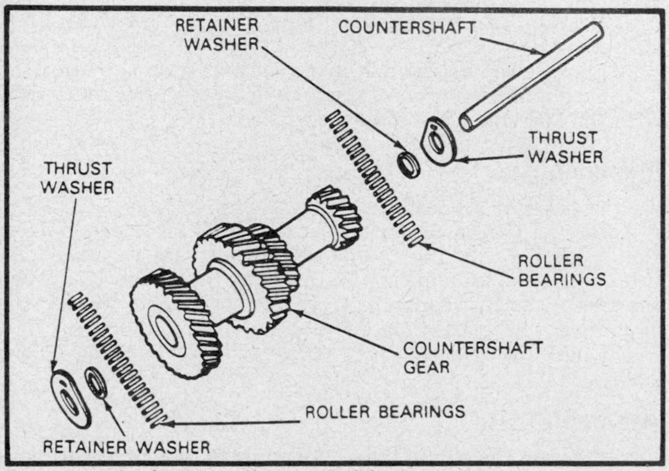
RETAINER WASHER
COUNTERSHAFT
THRUST WASHER
THRUST WASHER
ROLLER BEARINGS
COUNTERSHAFT GEAR
ROLLER BEARINGS
RETAINER WASHER

Countershaft gear disassembled

2. Place the 3 inserts into place on the hub.
3. Install the insert springs making sure that the irregular surface (hump) is seated in one of the inserts. Do not stagger the springs.

COUNTERSHAFT GEAR BEARINGS

Disassembly

1. Remove the dummy shaft and the 2 bearing retainer washers.
2. Remove the 21 roller bearings from each end of the countershaft gear.

Assembly

1. Coat the bore in each end of the countershaft gear with grease.
2. Hold the dummy shaft in the gear and install the 21 roller bearings and a retainer washer in each end of the gear.

REVERSE IDLER GEAR BEARINGS

Disassembly

1. Slip the reverse idler sliding gear off the reverse idler gear.
2. Remove the dummy shaft and the 2 bearing retainer washers.
3. Remove the 44 roller bearings from the reverse idler gear.

Assembly

1. Coat the bore in each end of the reverse idler gear with grease.
2. Hold the dummy shaft in the gear and install the 22 roller bearings and the retainer washer in each end of the gear.
3. Install the reverse idler sliding gear on the reverse idler gear making sure that the shift fork groove is toward the front.

Transmission Assembly

1. Lubricate the countershaft gear thrust surfaces in the case with a thin film of lubricant and position a thrust washer at each end of the case.
2. Position the countershaft gear, dummy shaft and roller bearings in the case.
3. Place the case in a vertical position. Align the gear bore and the thrust washers with the bores in the case and install the countershaft.
4. Place the case in a horizontal position and check the countershaft gear endplay with a feeler gauge. The endplay should be 0.004–.018 in. If not within specifications, replace the thrust washers.
5. Install the dummy shaft in the countershaft gear and allow the gear to remain at the bottom of the case.
6. Coat the reverse idler gear thrust surfaces in the case with a thin film of lubricant and position the 2 thrust washers in place.
7. Position the reverse idler gear, sliding gear, dummy shaft and the roller bearings in place making sure that the shift fork groove in the sliding gear is toward the front of the case.
8. Align the gear bore and thrust washers with the case bores and install the reverse idler shaft.
9. Meausre the reverse idler gear endplay with a feeler gauge. The endplay should be within 0.004–0.018 in. If the endplay is not within limits, replace the thrust washers. If the endplay is within limits, leave the reverse idler gear installed.
10. Position the reverse gear shift rail detent spring and detent plug in the case.
11. Hold the reverse shift fork in place on the reverse idler sliding gear and install the shift rail from the rear of the case. Secure the fork to the rail with the set screw.

12. Install the 1st and 2nd speed synchronizer into the front of the output shaft, making sure that the shift fork groove is toward the rear of the shaft.
13. The 1st and reverse synchronizer hub is a slip fit on the output shaft. Install the synchronizer hub with the teeth end of the gear facing toward the rear of the shaft.
14. Position the blocking ring on the 2nd speed gear. Lubricant the 2nd gear journal with the specified lubricant.
15. Slide the 2nd speed gear onto the front of the shaft, making sure that the inserts in the synchronizer engage the notches in the blocker ring.
16. Install the 2nd speed gear thrust washer and snapring. Lubricant the overdrive gear journal with the specified lubricant.
17. Slide the overdrive gear onto the shaft with the synchronizer coned surface toward the front.
18. Place a blocking ring on the overdrive gear.
19. Slide the 3rd and overdrive gear synchronizer onto the shaft making sure that the inserts in the synchronizer engage the notches in the blocking ring and the thrust surface is toward the overdrive gear.
20. Install the snapring on the front of the output shaft.
21. Position the blocking ring on the 1st speed gear. Lubricant the overdrive gear journal with the specified lubricant.
22. Slide the 1st speed gear onto the rear of the output shaft making sure that the notches in the blocking ring engage the synchronizer inserts.
23. Install the heavy thrust washer on the rear of the output shaft.
24. Support the thrust washer and 1st speed gear to prevent them from sliding off the shaft and carefully lower the output shaft assembly into the case.
25. Position the 1st and 2nd speed shift fork and the 3rd and overdrive shift fork in place on their respective gears and rotate them into place.
26. Place a spring and a detent plug in the detent bore. Place the reverse shift rail into neutral position.
27. Coat the 3rd and overdrive shift rail interlock pin tapered end with grease and position it in the shift rail.
28. Align the 3rd and overdrive shift fork with the shift rail bores and slide the shift rail into place making sure that the 3 detents are facing toward the outside of the case.
29. Place the front synchronizer into overdrive position and install the set screw in the 3rd and overdrive shift fork.
30. Position the synchronizer to the neutral position. Install the 3rd and overdrive shift rail detent plug, spring and bolt in the left side of the transmission case. Place the detent plug tapered end in the detent bore in the case.
31. Align the 1st and 2nd speed shift fork with the case bores and slide the shift rail into place. Secure the fork with the set screw.
32. Coat the input gear bore with a thin film of grease, then install the 15 roller bearings in the bore.

NOTE: A thick film of grease could plug the lubricant holes and restrict lubricant of the bearings.

33. Position the front blocking ring in the 3rd and overdrive synchronizer.
34. Position the dummy bearing tool T77L-7025-B or equivalent on the output shaft to support and align the shift assembly in the case.
35. Place the input shaft gear into the transmission case, making sure that the output shaft pilot enters the roller bearings in the input gear pockets.
36. Position the bearing into the shaft and into the transmission case.
37. Install the snaprings on the input shaft and the input shaft bearing.
38. Place a new gasket on the input shaft bearing retainer. Dip the attaching bolts in sealer and tighten to 19-25 ft. lbs.

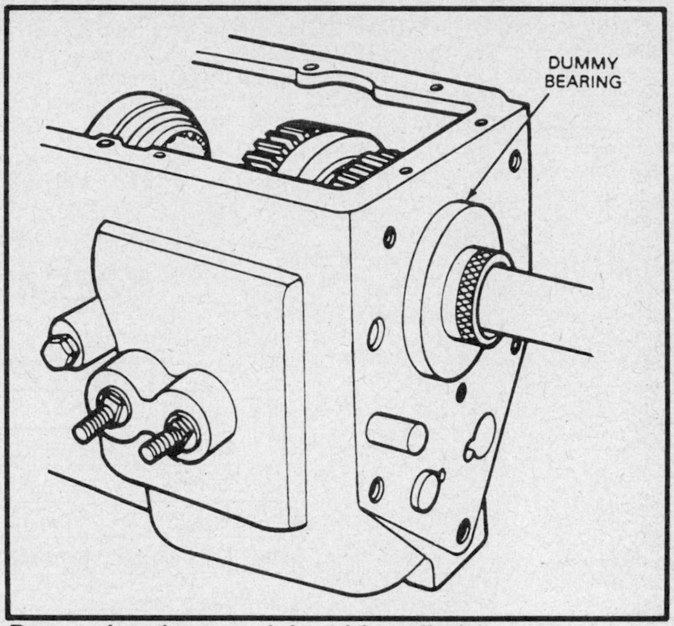
DUMMY
BEARING

Dummy bearing special tool installation

39. Remove the dummy bearing tool or equivalent from the output shaft.

NOTE: Be sure that the output shaft bearing is aligned with the transmission case bore and that the countershaft is not interfering with the output shaft assembly.

40. Install the output shaft bearing onto the output shaft and into the transmission case. Install the snaprings on the output shaft and the output shaft bearing.
41. Place the transmission in a vertical position. Align the countershaft gear bore and thrust washers with the bore and thrust washers with the bore in the case. Install the countershaft.

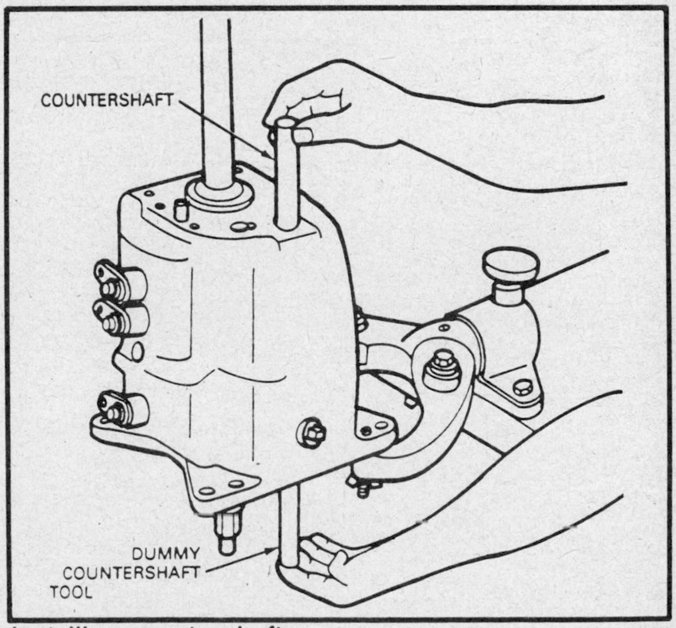
COUNTERSHAFT
DUMMY
COUNTERSHAFT
TOOL

Installing countershaft

42. Install the extension housing with new gasket to the case with the attaching bolts. Use a sealer on the extension housing attaching bolts. Tighten to 42–50 ft. lbs.
43. Install the filler plug in the case if it was removed.
44. Install specified lubricant (1 pint) over the entire gear train while rotating the input shaft. After installation of the transmission in the vehicle check for the correct oil level.
45. Place each shift fork in all positions to make sure that operation is correct.
46. Install the remaining detent plug in the case. Install the long spring, which is retained by the case, to secure the detent plug.
47. Install the cover with a new gasket. Coat the cover attaching screws with sealer and tighten them to 20–25 ft lbs.
48. Coat the 3rd and overdrive shift rail plug bore with sealer and install a new expansion plug.

SPECIFICATIONS

TORQUE SPECIFICATIONS

Item	ft. lbs.	Nm
Input shaft bearing retainer to case bolt	19–25	26–33
Extension housing to case bolt	42–50	57–67
Access cover to case screw	20–25	28–33
Outer gear shift levers to cam and shaft nut	18–23	25–31
Filler plug to case	10–20	14–27
Detent bolt to case	10–15	14–20

SPECIAL TOOLS

Number	Description
T59L-100-B	Impact slide hammer
T58L-101-A	Puller attachment
T64P-7111-A	Dummy countershaft
T64P-7140-A	Dummy countershaft
T57L-500-13	Bench mounted fixture holdings
022-00003	Transmission jack

Section 7

TOD Transmission
Ford Motor Co.

APPLICATION

F-150–250 Series (2WD)
F-150 Series (4WD)
Bronco

GENERAL DESCRIPTION

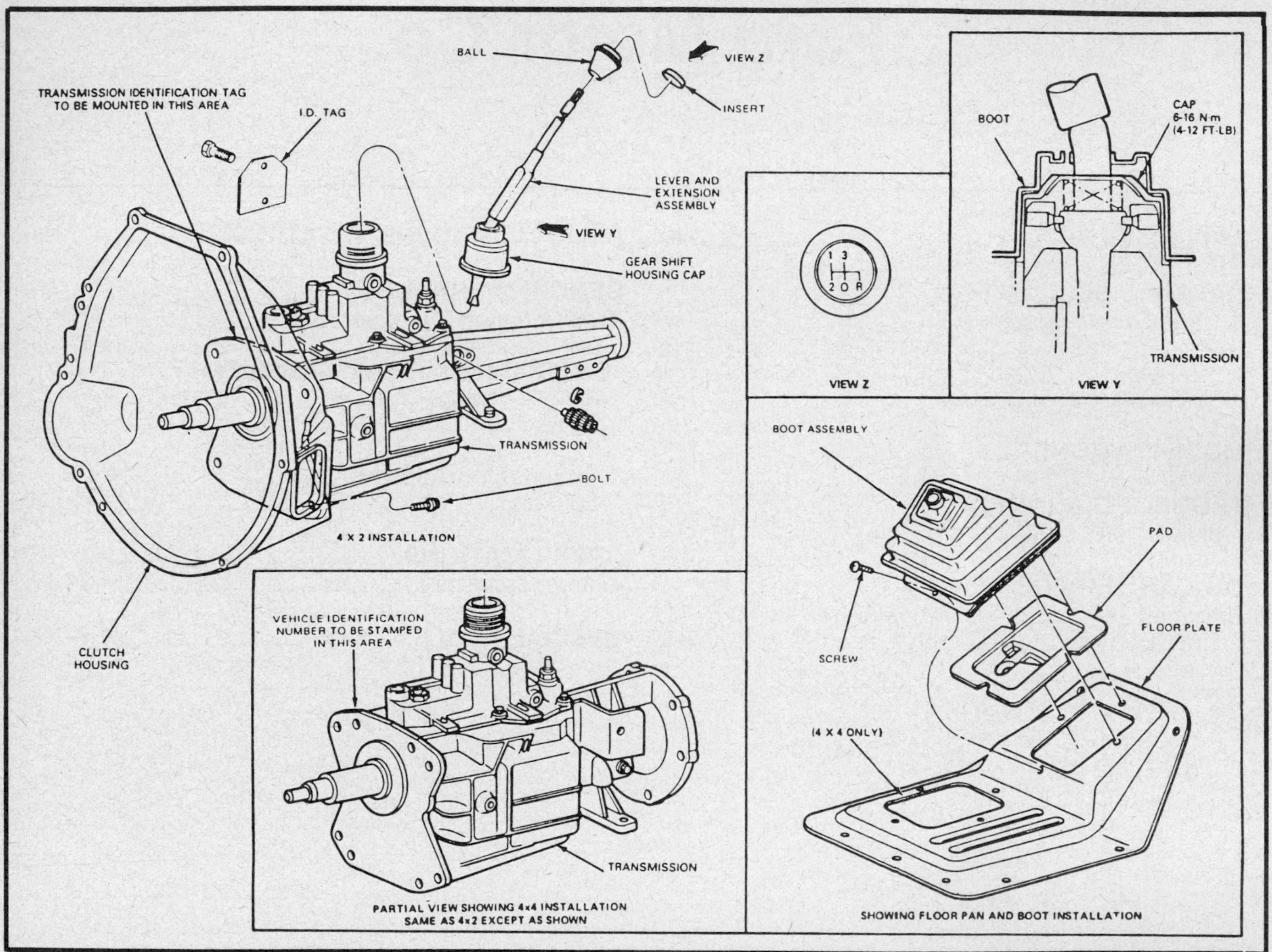
BALL
VIEW Z
INSERT
TRANSMISSION IDENTIFICATION TAG TO BE MOUNTED IN THIS AREA
I.D. TAG
LEVER AND EXTENSION ASSEMBLY
VIEW Y
GEAR SHIFT HOUSING CAP
TRANSMISSION
BOLT
4 X 2 INSTALLATION
CLUTCH HOUSING
VEHICLE IDENTIFICATION NUMBER TO BE STAMPED IN THIS AREA
TRANSMISSION
PARTIAL VIEW SHOWING 4x4 INSTALLATION SAME AS 4x2 EXCEPT AS SHOWN
BOOT
CAP 6-16 N·m (4-12 FT-LB)
TRANSMISSION
VIEW Z
VIEW Y
BOOT ASSEMBLY
PAD
FLOOR PLATE
SCREW
(4 X 4 ONLY)
SHOWING FLOOR PAN AND BOOT INSTALLATION

Top mounted shifter 4 speed overdrive transmission (TOD)

The top mounted shifter 4 speed overdrive transmission (TOD) is of the fully synchronized type with all gears except the reverse sliding gear being in constant mesh. All forward speed changes are accomplished with synchronizer sleeves. All forward speed gears in the transmission are helical type. However, the reverse sliding gear and the external teeth of the 1st and 2nd speed synchronizer sleeve are spur type.

The shift pattern for the control unit is imprinted on the shift lever knob. An automatically adjusted back-up lamp switch is installed in an indexing retainer at the front top portion of the gearshift housing assembly.

Transmission Identification

The transmission service identification tag is located on the right side of the case at the front. The first line on the tag will show the transmission model and service identification code when required. The second line will show the transmission serial number. Additionally, a serial identification number is stamped on the top side of the flange on the case for further identification.

Metric Fasteners

The metric fastener dimensions are very close to the dimensions of the familiar inch system fasteners and for this reason, replacement fasteners must have the same measurement and strength as those removed.

CAUTION

Do not attempt to interchange metric fasteners for inch system fasteners. Mismatched or incorrect fasteners can result in damage to the transaxle unit through malfunctions or breakage and possible personal injury.

Capacities

The 4 speed overdrive transmission uses standard transmission lubricant (SAE 80W). The capacity of the 4 speed overdrive transmission is 4.5 pints or 2.1L. The correct fluid level is to the bottom of the filler hole.

Checking Fluid Level

The transmission fluid level check must be made with the vehicle level. The fluid level can be checked by removing the fill plug.The correct fluid fill will be even with the bottom edge of the filler plug opening or within ¼ in. of this level. If fluid is low, added the specified fluid to correct level.

TRANSMISSION MODOFICATIONS

There are no modifications at the time of this printing.

TROUBLE DIAGNOSIS

CHILTON THREE C'S TRANSMISSION DIAGNOSIS

Condition	Cause	Correction
Shift lever—improper operation	a) Shift rods out of adjustment	a) Check crossover operation. Hand lever must move freely through neutral gate without catch or drag. Adjust shift rods.
	b) Steering column shift tube out of alignment	b) Check lever position for proper transmission gear selection. Replace steering column shift tube if necessary
Transmission shifts hard	a) Clutch pedal free-travel out of adjustment	a) Adjust clutch pedal free-travel
	b) Clutch does not completely release	b) Inspect complete clutch system
	c) Transmission fluid low or improper type	c) Add lubricant or change lubricant as required
	d) Shift lever binding or worn	d) Remove cap from shift tower. Eliminate binding condition or replace components as required
	e) Worn or damaged internal shift mechanism	e) Remove transmission cover. Check internal shift mechanism by shifting into and out of all gears. Repair or replace as required
	f) Binding of sliding gears and/or synchronizers	f) Check for free movement of gears and synchronizers. Repair or replace as required
	g) Housings and/or shafts out of alignment	g) Remove transmission and check for binding condition between input shaft and engine crankshaft pilot bearing or bushing. Check flywheel housing alignment. Repair or replace as required
Noisy in forward gears	a) Lubricant level low, or improper type	a) Add lubricant, or refill with specified lubricant
	b) Components grinding on transmission	b) Check for screws, bolts, etc., of cab or other components grounding out. Correct as required
	c) Component housing bolts loose	c) Check torque on transmission to flywheel housing bolts, output shaft flange nut and flywheel housing to engine block bolts. Tighten bolts to specification
	d) Flywheel housing to engine crankshaft alignment	d) Check and align flywheel housing to engine crankshaft

CHILTON THREE C'S TRANSMISSION DIAGNOSIS

Condition	Cause	Correction
Noisy in forward gears	e) Noisy bearings or gears	e) Remove and disassemble transmission. Inspect input, output and countershaft bearings. Inspect speedometer gear and gear teeth for wear or damage. Replace as required
Gears clash when shifting from one forward gear to another	a) Engine idle speed too high	a) Adjust engine idle speed
	b) Clutch pedal free-travel incorrect	b) Check clutch for pedal free-travel. Adjust as required
	c) Improper manual-shift linkage	c) Adjust and repair manual-shift linkage as required
	d) Pilot bearing binding	d) Remove transmission and check for a binding condition between input shaft and engine crankshaft pilot bearing. Replace as required
	e) Damaged gear teeth and/or synchronizer	e) Disassemble transmission, repair or replace as required
Transmission jumps out of gear	a) Manual-shift linkage binding, out of adjustment. Stiff shift boot	a) Adjust linkage. If necessary align steering column to truck body
	b) Improper fit of form isolation pad	b) Replace shift boot if exceptionally stiff. Replace or rework pad to provide clearance
	c) Loose transmission to engine mounting bolts, or loose levers	c) Tighten transmission to flywheel housing, and flywheel housing to engine block bolts. Loosen all bolts and reseat flywheel housing. Tighten all bolts. Tighten levers if necessary
	d) Flywheel housing to engine crankshaft out of line	d) Shim or replace housing as required
	e) Crankshaft pilot bearing worn	e) Replace bearing
	f) Interior components damaged	f) Disassemble transmission. Inspect the synchronizer sleeves for tree movement on their hubs. Inspect the synchronizer blocking rings for widened index slots, rounded clutch teeth and smooth internal surface. Check countershaft cluster gear for excessive end play. Check shift forks for worn or loose mounting on shift rails. Check fork pads for excessive wear. Inspect synchronizer sliding sleeve and gear clutch teeth for wear or damage. Repair or replace as required
	g) Worn gear teeth due to partial engagement	g) Replace worn or damaged gears
Transmission is locked in one gear. It cannot be shifted out of that gear	a) Manual-shift linkage out of adjustment, binding or damaged	a) Disconnect the problem shift rod from transmission shift lever. Try to shift the transmission lever into and out of gear position. If OK, repair or replace linkage parts
	b) Internal components	b) Remove transmission. Inspect problem gears, shift rails and forks and synchronizer for wear or damage. Repair as required. Check for broken fork slot tabs on TOD
	c) Loose fork on rail	c) Check shift rail interlock system

CHILTON THREE C'S TRANSMISSION DIAGNOSIS

Condition	Cause	Correction
Transfer case makes noise	a) Incorrect tire inflation pressures and/or incorrect size tires and wheels	a) Assure that all tires and wheels are the same size, and that inflation pressures are correct
	b) Excessive tire tread wear	b) Check tire tread wear to see if there is more than .06 inch difference in tread wear between front and rear. Interchange one front and one rear wheel. Reinflate tires to specifications
	c) Internal components	c) Operate vehicle in all transmission gears with transfer case in 2HL or Hi range. —If there is noise in transmission in neutral gear, or in some gears and not in others, remove and repair transmission —If there is noise in all gears, operate vehicle in all transfer case ranges. If noisy in all ranges or Hi range only, disassemble transfer case. Check input gear, intermediate and front output shaft gear for damage. Replace as required. If noisy in LO range only, inspect intermediate gear and sliding gears for damage. Replace as required
4-wheel drive transfer case jumps out of gear	a) Incomplete shift linkage travel	a) Adjust linkage to provide complete gear engagement
	b) Loose mounting bolts	b) Tighten mounting bolts
	c) Front and rear driveshaft slip yokes dry or loose	c) Lubricate and repair slip yokes as required. Tighten flange yoke attaching nut to specifications
	d) Internal components	d) Disassemble transfer case. Inspect sliding clutch hub and gear clutch teeth for damage. Replace as required
Transmission will not shift into one gear—all others OK	a) Manual linkage out of adjustment	a) Adjust manual linkage
	b) Manual-shift linkage damaged or worn	b) Lubricate, repair or replace parts as required
	c) Back-up switch ball frozen	c) If reverse is problem, check backup switch for ball frozen in extended position (if so equipped)
	d) Internal components	d) Remove transmission. Inspect shift rail and fork system synchronizer system and gear clutch teeth for restricted travel. Repair or replace as required

ON VEHICLE SERVICE

Service

SHIFT LEVER

Removal and Installation

NOTE: Remove the shift ball only if the shift ball, boot or lever is to be replaced. If any component is not being replaced, remove the ball, boot and lever as an assembly.

1. Remove the plastic insert from the shift ball. Warm the ball with a heat gun to 140–180 degrees F. Knock the ball off the lever with a block of wood and a hammer or equivalent, taking care not to damage the finish on the shift lever.
2. Remove the screws retaining the boot and pad to the floor plate.
3. Shift the transmission into **N**. Remove the boot from the cap. Place oil filter wrench or equivalent around the gearshift housing cap and twist off the cap.
4. Remove the shift lever.
5. Install the shift lever in the gearshift housing, making sure

that the slots in the lever align with the pins in the housing.

6. Twist the gearshift housing cap onto the housing using suitable tool. Tighten the housing cap to 12–18 ft. lbs. Install the boot over the cap.

7. Install the rubber boot and pad. Install the screws retaining the boot and pad to the floor plate.

8. Warm the ball with a heat gun to 140–180°F. Tap the ball on the lever with a suitable tool.

9. Install the plastic shift pattern insert.

REMOVAL AND INSTALLATION

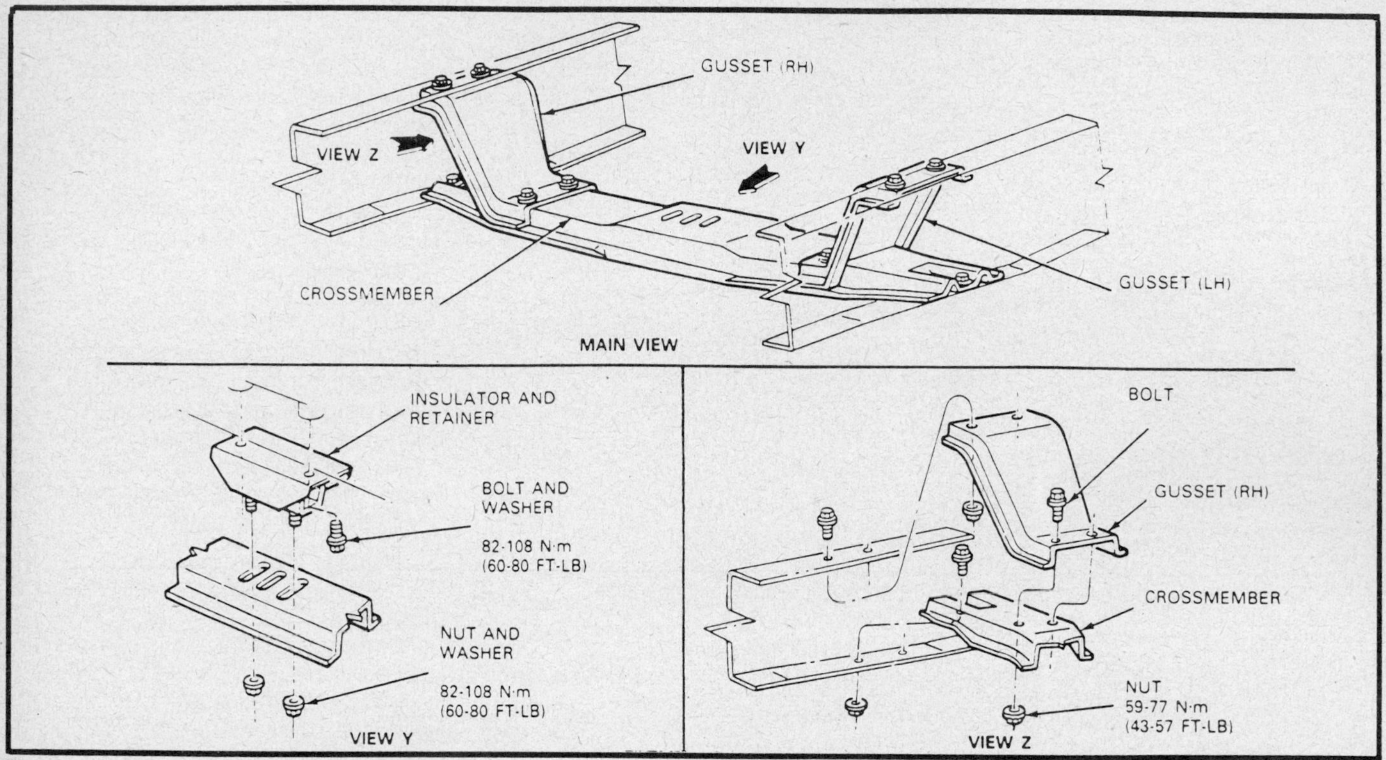
GUSSET (RH)
VIEW Z
VIEW Y
CROSSMEMBER
GUSSET (LH)
MAIN VIEW
INSULATOR AND RETAINER
BOLT AND WASHER
82-108 N·m (60-80 FT-LB)
NUT AND WASHER
82-108 N·m (60-80 FT-LB)
VIEW Y
BOLT
GUSSET (RH)
CROSSMEMBER
NUT
59-77 N·m (43-57 FT-LB)
VIEW Z

Transmission rear suport assembly

TRANSMISSION REMOVAL

F-150 AND F-250 (2WD)

NOTE: The transmission is removed separately from the engine. Do not depress the clutch pedal while the transmission is removed.

1. Raise the vehicle and support it safely. Drain the transmission.
2. Mark the driveshaft and disconnect the driveshaft from the rear U-joint flange.
3. Slide the driveshaft off the transmission output shaft. Install an appropriate tool into the extension housing to prevent lubricant leakage.
4. Disconnect the speedometer cable from the extension housing.
5. Disconnect back-up lamp switch and high gear switch (is so equipped) electrical connections.
6. Remove the shift lever from the transmission.
7. Support the engine with a transmission jack, or equivalent and remove the extension housing-to-engine rear support attaching bolts.
8. Raise the rear of the engine high enough to remove the weight from the crossmember. Remove the bolts retaining the crossmember to the frame side supports and remove the crossmember.
9. Support the transmission on a jack and remove the bolts that attach the transmission to the flywheel housing.
10. Move the transmission and jack rearward until the transmission input shaft clears the flywheel housing.
11. If necessary, lower the engine enough to obtain clearance for transmission removal.

F-150 (4WD) AND BRONCO

NOTE: The transmission is removed separately from the engine. On 4WD vehicles the transfer case is removed separately from the transmission. Do not depress the clutch pedal while the transmission is removed.

1. Raise the vehicle and support it safely.
2. Place a drain pan under transmission and or transfer case, remove drain plug from the transfer case and drain lubricant from the transmission by using a suction type pump or equivalent and or transfer case. Reinstall drain plug.
3. Disconnect 4WD drive indicator switch wire connector at transfer case. Disconnect back-up lamp switch wire connector at transmission.
4. Remove skid plate from frame, if so equipped.
5. Mark the front and rear driveshaft. Disconnect the rear driveshaft from the transfer case and position it out of the way.
6. Disconnect the front driveshaft from the transfer case and position it out of the way.
7. Remove the speedometer cable from the transfer case.
8. Remove retaining clips and shift rod from the transfer case control lever and transfer case shift lever.
9. Disconnect the transfer case vent hose.
10. Remove the transmission shift lever.

11. Support the transmission with a transmission jack or equivalent and remove the transmission housing rear support bracket.
12. Raise the rear of the transmission high enough to remove the weight from the crossmember.
13. Remove the 2 nuts connecting upper gusset to frame on both sides of the frame.
14. Remove gusset on left side.
15. Remove the bolts holding transmission to transmission support plate on crossmember.
16. Raise transmission with a transmission jack safely.
17. Remove nut and bolt assemblies connecting the support plate to crossmember. Remove support plate and the right gusset.
18. Remove nut and bolt assemblies connecting crossmember to frame. Remove crossmember.
19. Remove heat shield from transfer case. The catalytic converter is located beside heat shield.
20. Support the transfer case with a transmission jack or equivalent.
21. Remove the 6 bolts retaining transfer case to transmission adapter.
22. Slide the transfer case rearward off of the transmission output shaft and lower transfer case from vehicle.
23. Remove the gasket between transfer case and adapter.
24. Support the transmission on a jack or equivalent and remove the bolts that attach the transmission to the flywheel housing.
25. Move the transmission and jack rearward until the transmission input shaft clears the flywheel housing.
26. If necessary, lower the engine enough to obtain clearance for transmission removal.

TRANSMISSION INSTALLATION

F-150 AND F-250 (2WD)

NOTE: Do not depress the clutch pedal while the transmission is removed. If the clutch pressure plate assembly retaining bolts were loosen the clutch plate will have to be align up.

1. Position the transmission into the neutral position. Make sure that the machined surfaces of the transmission case and the flywheel housing are free of dirt, paint and burrs.
2. Install 2 guide pins in the flywheel housing lower mounting bolt holes.
3. Move the transmission forward on the guide pins until the input shaft splines enter the clutch hub splines and the case is positioned against the flywheel housing.
4. Install the 2 upper transmission to flywheel housing mounting bolts snug and then remove the guide pins.
5. Tighten all transmission mounting bolts evenly.
6. Install the transmission support bolts and tighten evenly.
7. Raise the rear of the engine and install the crossmember and attaching bolts.
8. Lower the engine and with the transmission extension housing resting on the engine rear support, install the transmission extension housing attaching bolts. Tighten the attaching bolts to 42–50 ft. lbs.
9. Install the transmission shift lever.
10. Connect the speedometer cable to the extension housing.
11. Connect the back-up lamp switch and high gear switch, if so equipped electrical connections.
12. Remove the extension housing extension tool or shop towel and slide the forward end of the driveshaft over the transmission output shaft. Connect the driveshaft to the rear U-joint flange. Tighten nuts on U-bolts to 8–15 ft. lbs.

NOTE: Do not overtighten nuts on the U-bolts.

13. Fill the transmission to the proper level with the specified fluid.

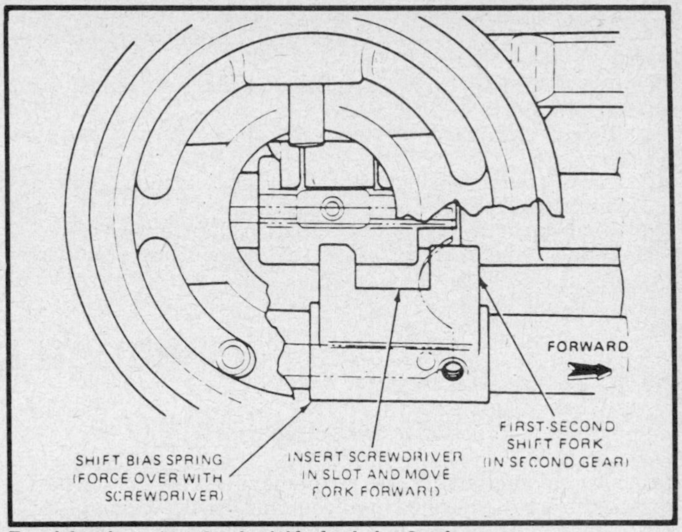
FORWARD
FIRST-SECOND SHIFT FORK (IN SECOND GEAR)
SHIFT BIAS SPRING (FORCE OVER WITH SCREWDRIVER)
INSERT SCREWDRIVER IN SLOT AND MOVE FORK FORWARD
Positioning 1st-2nd shift fork in 2nd gear

14. Lower the vehicle.
15. Check the shift pattern for correct engagement.
16. Road test the vehicle for proper operation.

F-150 (4WD) AND BRONCO

NOTE: Do not depress the clutch pedal while the transmission is removed. If the clutch pressure plate assembly retaining bolts were loosen the clutch plate will have to be align up.

1. Position the transmission into the neutral position. Place the transmission assembly on a transmission jack and raise the transmission until the input shaft splines are aligned with the clutch disc splines. The clutch release bearing and hub must be properly positioned in the release lever fork.
2. Make sure that the mounting surface of the transmission and the flywheel housing are free of dirt, paint and burrs.
3. Install 2 guide pins in the flywheel housing lower mounting bolt holes.
4. Move the transmission forward on the guide pins until the input shaft splines and the case is positioned against the flywheel housing.
5. Install the 2 upper transmission to flywheel housing mounting bolts snug and then remove the 2 guide pins.
6. Install the 2 lower mounting bolts. Tighten all mounting bolts evenly.
7. Install heat shield on the transfer case.
8. Place a new gasket between transfer case and transmission adapter.
9. Raise the transfer case with jack or equivalent so transmission output shaft aligns with splined transfer case input shaft.
10. Slide the transfer case forward on to transmission output shaft and onto the transmission adapter. Install the transfer case to adapter. Tighten all bolts evenly.
11. Position the shift rod on the transfer case shift lever and the transfer case control rod and attach with the retaining rings.
12. Install the transmission shift lever on the transmission.
13. Connect the speedometer cable to the transfer case.
14. Connect the 4WD drive indicator switch wire connector to the transfer case.
15. Connect the back-up switch wire connector to transmission.
16. Install crossmember and transmission support bracket and position right and left gussets to frame.
17. Install retaining nuts on the upper gusset to frame bolts and tighten evenly.

18. Install crossmember to frame.
19. Install nut and bolt assembly connecting gusset to crossmember.
20. Install nut and bolt assemblies connecting transmission support bracket to crossmember.
21. Lower the transmission assembly onto the support bracket.
22. Install transmission support bracket to transmission.
23. Install skid plate, if so equipped.
24. Remove the transmission jack from the transfer case.
25. Connect the rear driveshaft to rear output shaft yoke. Tighten retaining nuts to 20–28 ft. lbs.
26. Connect the front drivehsaft to front output yoke. Tighten retaining nuts to 8–15 ft. lbs.
27. Remove transfer case filler plug and install 6.5 pints (3.1L) of automatic transmission fluid, Dexron®II or equivalent. Install filler plug and tighten to 15–25 ft. lbs.
28. Fill the transmission to the proper lever with the specified lubricant.
29. Check the shift pattern for correct engagement.
30. Road test the vehicle for proper operation.

BENCH OVERHAUL

Before Disassembly

When servicing the unit, it is recommended that as each part is disassembled, it is cleaned in solvent and dried with compressed air. Disassembly and reassembly of this unit and its parts must be done on a clean work bench. Also, before installing bolts into aluminum parts, always dip the threads into clean transmission oil. Anti-seize compound can also be used to prevent bolts from galling the aluminum and seizing. Always use a torque wrench to keep from stripping the threads. Take care with the seals when installing them, especially the smaller O-rings. The slightest damage can cause leaks. Aluminum parts are very susceptible to damage so great care should be exercised when handling them. The internal snaprings should be expanded and the external snaprings compressed if they are to be reused. This will help insure proper seating when installed. Be sure to replace any O-ring, gasket, or seal that is removed.

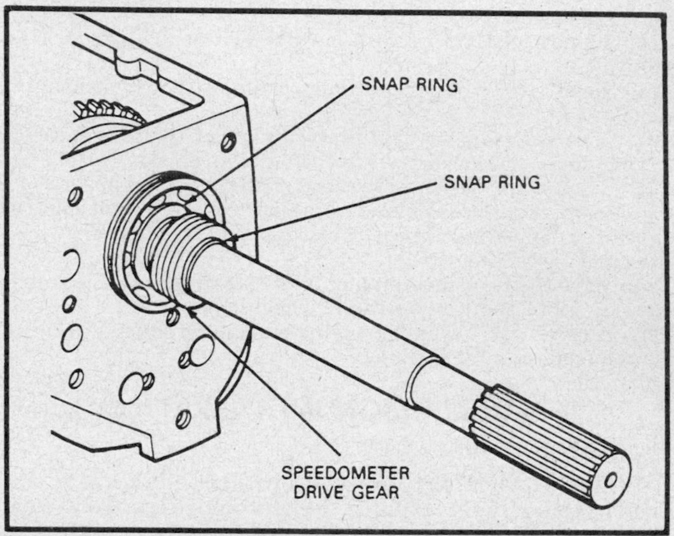
SNAP RING
SNAP RING
SPEEDOMETER DRIVE GEAR
Removing speedometer drive gear

Transmission Disassembly

1. Mount the transmission in a suitable holding fixture.
2. Unscrew the gearshift housing cap and remove the shift lever. Position a punch or equivalent into the shift lever bore. Force the shift bias spring over and engage the slot in the 1st/2nd fork and move the fork forward.

NOTE: Shifting the transmission into 2nd gear provides adequate clearance for the forks allowing the gearshift housing assembly to be removed.

3. Remove the 6 bolts and washers that retain the gearshift housing assembly to the transmission case. Remove the housing assembly from the case.
4. Shift the transmission synchronizer assemblies into the neutral position. Shifting the transmission into neutral position prevents the synchronizer inserts from accidentally slipping out of place.
5. Place a drain pan under the extension housing. Remove the 5 bolts and washers attaching the extension housing to the transmission case. Remove the extension housing from the case. If present remove the shipping seal from the shaft and discard the seal.
6. Remove the snapring that retains the speedometer drive gear to the output shaft and remove the drive gear. Remove the speedometer drive gear lock ball (0.25 in.) from the output shaft.
7. Drive out the roll pin that secures the 3rd/4th shift fork to the 3rd/4th shift rail.
8. Position a ⅜ in. outside diameter rod against the 3rd/4th shift rail in the case. Using a suitable punch, tap on the rod to remove the rail through the front of the case.
9. Remove the countershaft by positioning dummy countershaft special tool, T64P–7111–A or equivalent over the countershaft bore in the front of the case. Drive the dummy shaft into the case and countershaft gear bore until the countershaft is removed from the rear of the case.

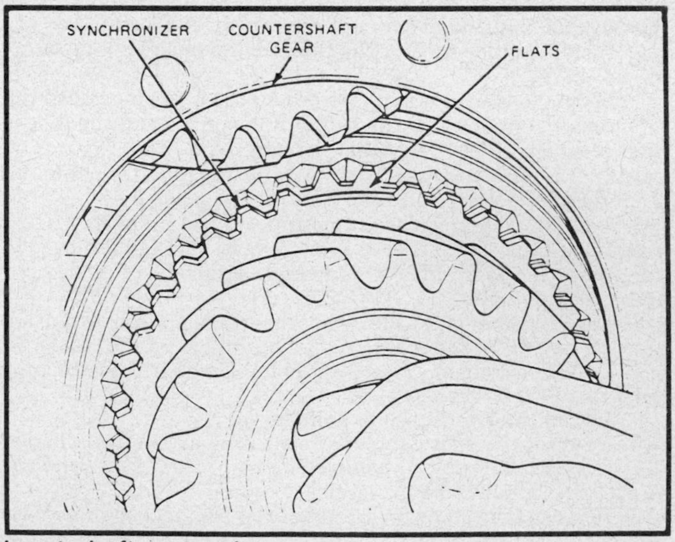
SYNCHRONIZER
COUNTERSHAFT GEAR
FLATS
Input shaft removal

10. When the countershaft is removed and the dummy shaft tool is in place, the countershaft gear can be lowered to the bot-

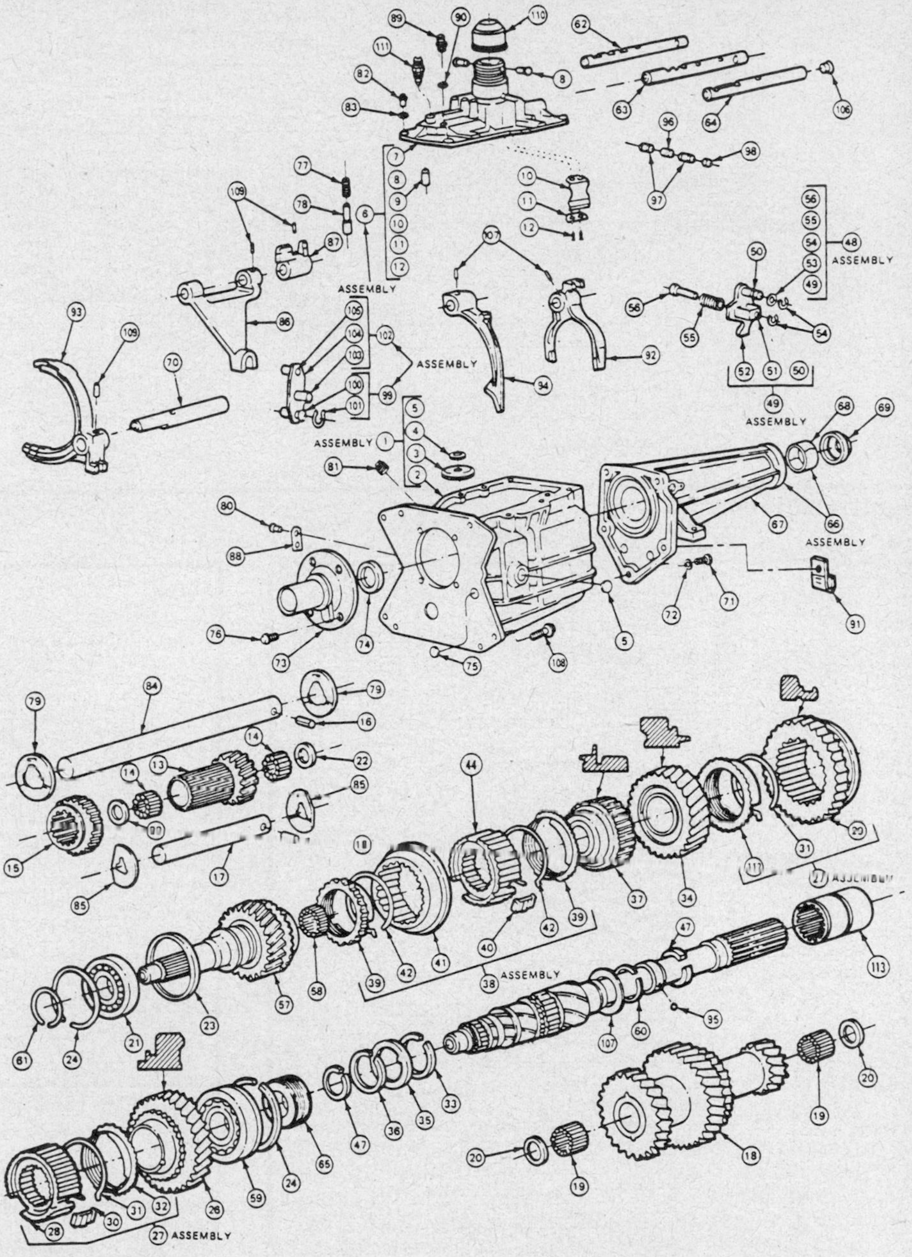

1. Transmission case assembly
2. Transmission case
3. Chip magnet
4. Spring push-on nut
5. Expansion cup plug
6. Gearshift housing assembly
7. Gearshift housing
8. Gearshift lever pin
9. Dowel
10. 3rd/overdrive shift bias spring
11. spring retainer plate
12. Rivet
13. Reverse idler gear
14. Idler shaft roller bearings (44)
15. Reverse idler sliding gear
16. Pin
17. Reverse idler gear shaft
18. Countershaft gear
19. Countershaft gear roller bearings (42)
20. 7/8 flatwasher
21. Front input shaft ball bearing
22. 3/4 flatwasher
23. Retainer ring
24. Snapring
25. Output shaft
26. 1st speed gear
27. 1st and 2nd gear synchronizer assy.
28. 1st and 2nd clutch hub
29. Reverse sliding gear
30. Synchronizer hub insert
31. Retaining spring
32. Blocking ring
33. Snapring
34. 2nd speed gear
35. Low gear thrust washer
36. Retaining ring
37. Overdrive gear
38. 3rd and 4th gear synchronizer assy.
39. Blocking ring
40. Hub insert
41. Clutch sleeve
42. Retaining spring
44. 3rd and 4th gear clutch hub
47. Snapring
48. Reverse rocker arm assy.
49. Pin and housing arm
50. Rocker arm pivot pin
51. Reverse plunger housing
52. Reverse rocker arm
53. O-Ring seal
54. 3/8 retaining ext. ring
55. Reverse plunger spring
56. Reverse rocker plunger
57. Input shaft
58. Input shaft roller bearings (15)
59. Output shaft rear ball bearing
60. Snapring
61. Snapring
62. 1st-2nd shifter rail
63. 3rd-overdrive shifter rail
64. Reverse gear shifter rail
65. Speedometer drive gear
66. Extension housing assy.
67. Extension housing
68. Extension housing bushing
69. Extension housing oil seal
70. 3rd-4th shift rail
71. Bolt
72. 7/16 ext. tooth washer
73. Input shaft bearing retainer
74. Input shaft oil seal
75. Expansion plug
76. Bolt
77. Shifter interlock spring
78. Meshlock plunger
79. Countershaft thrust washer
80. Drive screw
81. Filler plug
82. Bolt
83. Ext. tooth washer
84. Countershaft
85. Reverse idler gear thrust washer
86. 3rd-overdrive shift pawl
87. 3rd-overdrive shift gate
88. Service identification tag
89. Back-up lamp switch
90. Back up lamp switch gasket
91. Electrical wiring clip
92. 1st-2nd gearshift fork
93. 3rd-4th gearshift fork
94. Reverse gearshift fork
95. Ball 0.25 inch
96. Interlock pin
97. Interlock plunger
98. Plug
99. Overdrive shift control link and pin
100. Retaining ext. ring
101. Finger pin
102. Overdrive shift control link pin assy.
103. Overdrive shift control link shaft
104. Overdrive shift control link
105. Shift control finger pin
106. Cup plug
107. Output shaft thrust washer
108. Screw and washer
109. Pin
110. Housing cap
111. 4th gear sensing switch
112. Synchro blocking ring
113. Output shaft seal

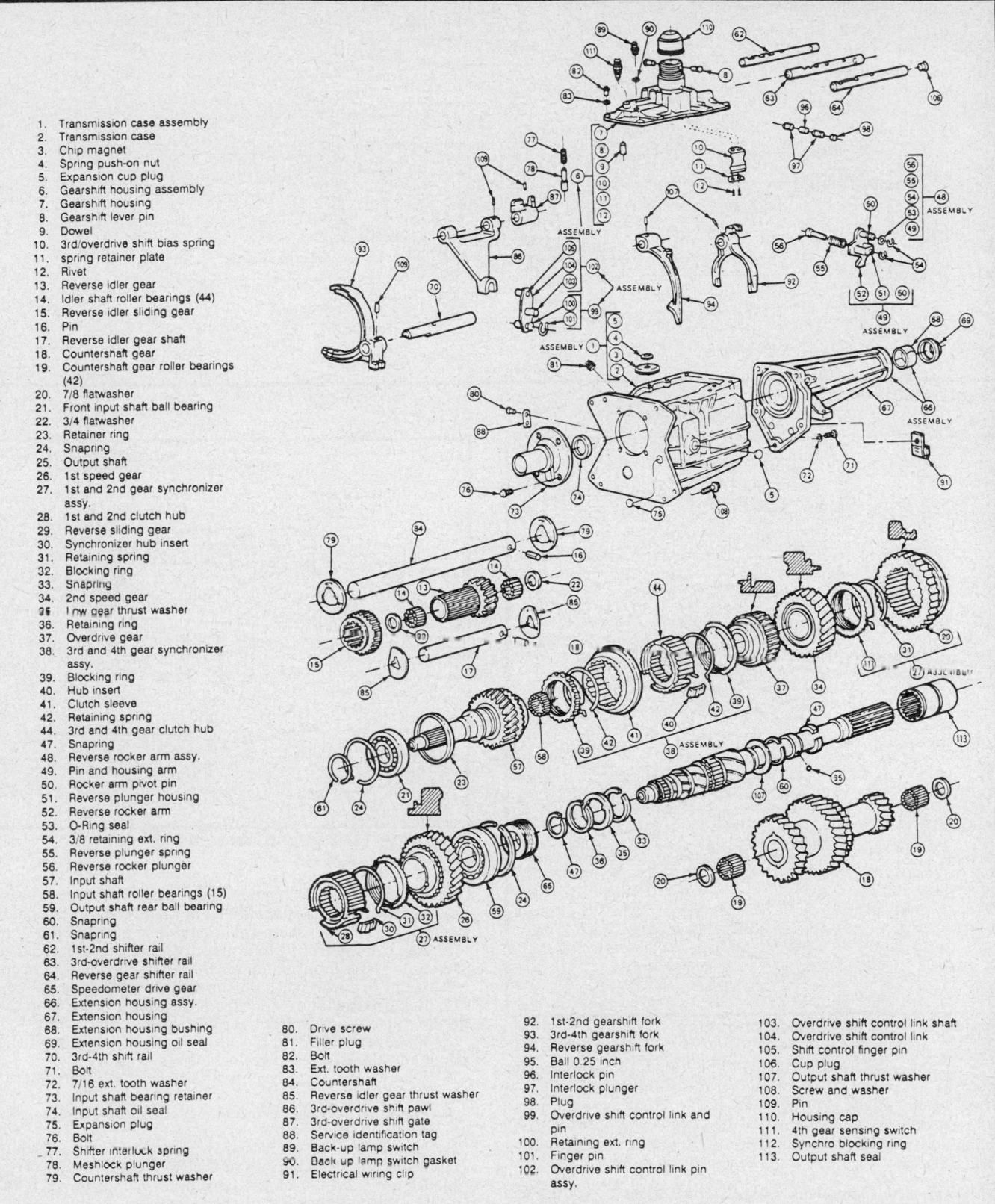

Exploded view of TOD transmission

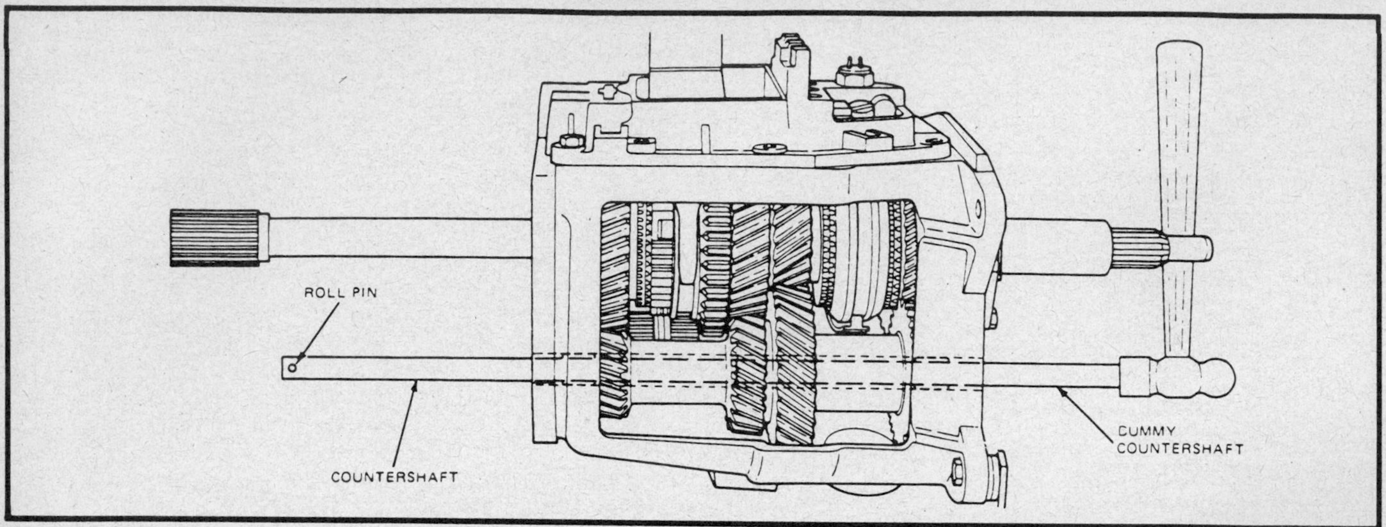
ROLL PIN
COUNTERSHAFT
DUMMY
COUNTERSHAFT

Countershaft removal

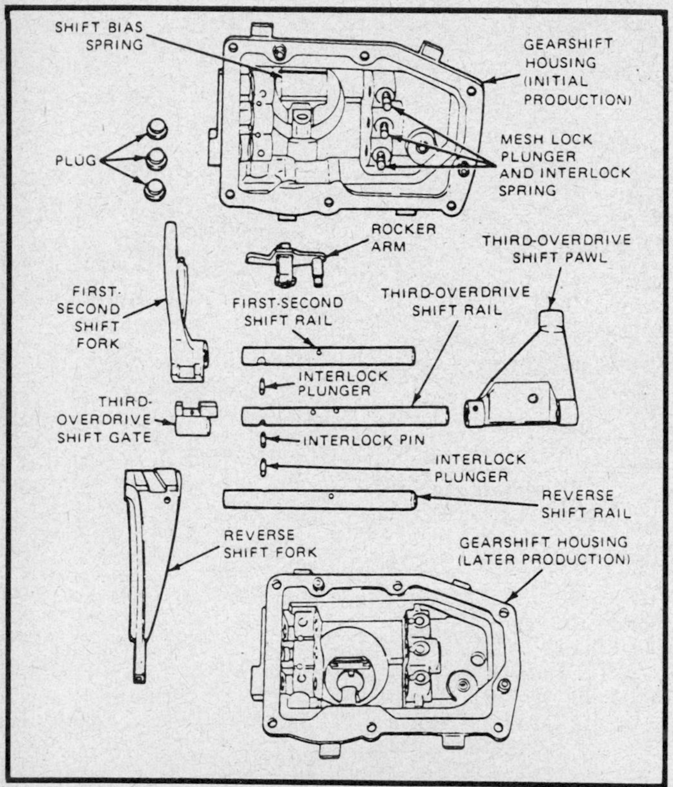
SHIFT BIAS
SPRING
GEARSHIFT
HOUSING
(INITIAL
PRODUCTION)
PLUG
MESH LOCK
PLUNGER
AND INTERLOCK
SPRING
ROCKER
ARM
THIRD-OVERDRIVE
SHIFT PAWL
FIRST-
SECOND
SHIFT
FORK
FIRST-SECOND
SHIFT RAIL
THIRD-OVERDRIVE
SHIFT RAIL
INTERLOCK
PLUNGER
THIRD-
OVERDRIVE
SHIFT GATE
INTERLOCK PIN
INTERLOCK
PLUNGER
REVERSE
SHIFT RAIL
REVERSE
SHIFT FORK
GEARSHIFT HOUSING
(LATER PRODUCTION)

Exploded view of gearshift housing

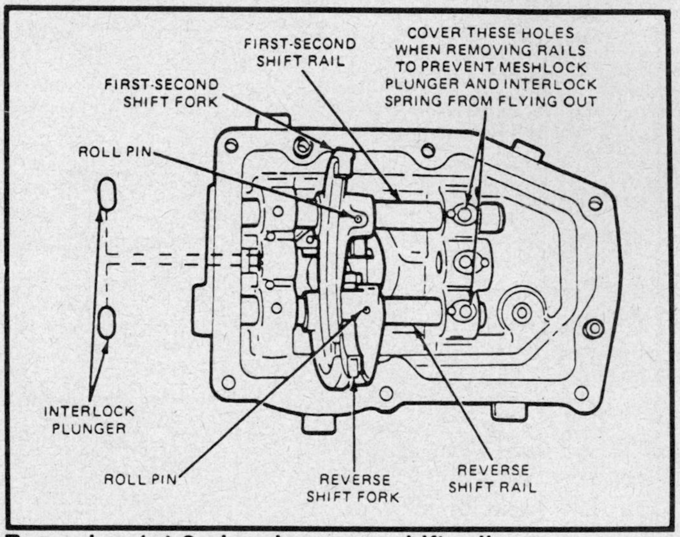
FIRST-SECOND
SHIFT RAIL
COVER THESE HOLES
WHEN REMOVING RAILS
TO PREVENT MESHLOCK
PLUNGER AND INTERLOCK
SPRING FROM FLYING OUT
FIRST-SECOND
SHIFT FORK
ROLL PIN
INTERLOCK
PLUNGER
ROLL PIN
REVERSE
SHIFT FORK
REVERSE
SHIFT RAIL

Removing 1st-2nd and reverse shift rail

tom of the case. This will give the required clearance for mainshaft removal. It is not necessary to remove the roll pin from the end of the countershaft.

11. Remove the snapring from the output shaft. Remove the snapring from the groove in the output shaft rear bearing.
12. Remove the output shaft rear bearing from the case.

NOTE: The input and output shaft rear bearings are of the slipfit design.

13. Remove the bolts that attach the input shaft bearing retainer to the case and remove the retainer. If necessary, pry the retainer free.
14. Remove the snapring that retains the front input shaft ball bearing. Remove the snapring from the groove in the bearing. Remove the retainer from the input shaft ball bearing.
15. Remove the front input shaft ball bearing from the case.
16. Rotate the input shaft until the flat on the synchronizer teeth align with the teeth on the countershaft gear.
17. Remove the input shaft from the case. Do not drop the 15 needle roller bearings from the rear of the input shaft.
18. Remove the 3rd/4th blocking ring from the rear of the input shaft. Mark or tag the blocking ring.
19. Remove the 3rd/4th shift fork from the intermediate and high clutch sleeve.
20. Position the output shaft and gear train assembly upwards and remove from the case.
21. Lift the countershaft gear, with dummy shaft special tool still inside, from the bottom of the transmission case.
22. Do not drop the 21 needle roller bearings on each end of the countershaft gear out of the assembly.
23. Remove the countershaft nylon thrust washers from each end of the case.
24. From inside the case, use a suitable punch to drive the reverse idler gear shaft out from the rear of the case.

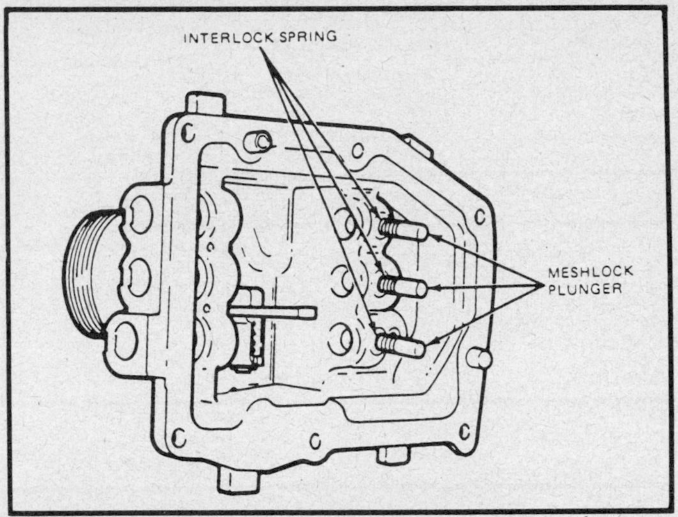
INTERLOCK SPRING
MESHLOCK PLUNGER

Meshlock plunger and interlock spring removal

25. Remove the reverse idler bronze thrust washers. Remove the reverse idler gear, reverse idler sliding gear, 2 flat washers and 22 idler shaft roller bearings on each end from the case.
26. Remove the 3rd/4th roll pin from the bottom of the case.
27. Remove the overdrive shift control link and pin assembly from the transmission case.

Unit Disassembly And Assembly

GEARSHIFT HOUSING ASSEMBLY

Disassembly

1. Remove the gearshift housing assembly from the transmission.
2. Position the gearshift housing is a vise.
3. Drive out the roll pins that retain the 3rd/overdrive shift pawl and 3rd/overdrive shift gate. When driving the roll pin avoid striking the bias spring.
4. Slide the 3rd/overdrive shift pawl to the front of the housing. Insert a suitable tool in the exposed roll pin hole in the rail and rotate the rail 90 degrees. Rotating the rail will prevent the meshlock plunger from locking the rail in place.
5. Position a suitable tool in the slots provided in the housing and remove pry the plugs from the housing.
6. Drive the 3rd/overdrive shaft rail from the housing. Remove the interlock pin from the rail. Remove the 3rd/overdrive shaft pawl and shift gate from the housing.
7. Remove each interlock plunger from the housing through the rear 3rd/ overdrive rail bore. By tilting the housing, the plungers should fall out.
8. Drive out the roll pin retaining the 1st/2nd shift fork to the rail. Avoid driving the pin into the inner wall of the housing.
9. Slide the 1st/2nd shift fork to the front of the case. Insert a suitable tool into the exposed roll pin hole and rotate the rail 90 degrees. Rotating the rail will prevent the meshlock plunger from locking the rail in place.
10. With a suitable punch, drive the 1st/2nd shift rail from the housing. Cover the plunger and spring bore to prevent the plunger and spring from flying out after the rail is removed.
11. Remove the 1st/2nd shift fork from the housing.
12. Slide the reverse fork and rail assembly forward to it is in the reverse position. Drive out the reverse fork roll pin.
13. Slide the reverse fork rearward to expose the roll pin hole. Insert a suitable tool in the roll pin hole and rotate the rail 90 degrees. Rotating the raill will prevent the meshlock plunger from locking the rail in place.

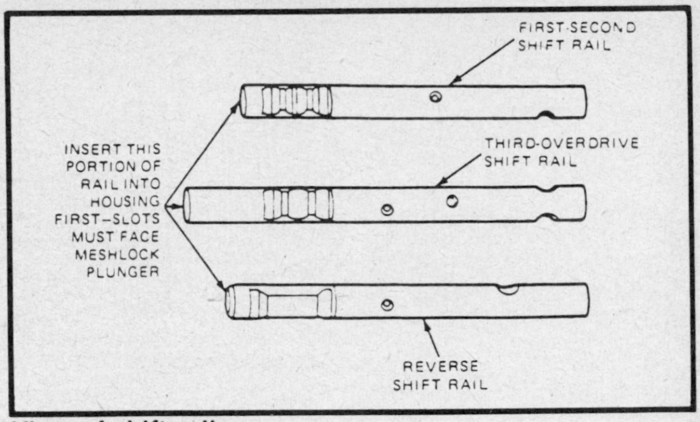
FIRST-SECOND SHIFT RAIL
INSERT THIS PORTION OF RAIL INTO HOUSING FIRST—SLOTS MUST FACE MESHLOCK PLUNGER
THIRD-OVERDRIVE SHIFT RAIL
REVERSE SHIFT RAIL

View of shift rails

14. With a suitable punch, drive the reverse rail from the housing. Cover the plunger and spring bore to prevent the plunger and spring from flying out after the rail is removed.
15. Remove the reverse fork from the housing.
16. Remove the 3 meshlock plungers and interlock springs from the bores in the housing.
17. Remove the C-clip that retains the reverse rocker arm assembly to the housing.
18. Position the shift bias spring outward only enough to remove the reverse rocker arm assembly. Remove the assembly from the housing.

NOTE: Do not disassemble the reverse rocker arm assembly.

19. Remove the back-up lamp switch and gasket from the housing.

Inspection

1. Check the operation and condition of the shift levers, forks and shift rails. Replace as necessary.
2. If binding occurs when the levers are operated, disassemble the housing assembly and replace the worn or damaged parts.
3. Replace the cover if it is bent or distorted.
4. Check the vent hole for restriction.

Assembly

1. Coat all shift rails with standard transmission lubricant prior to assembly.
2. Install the back-up lamp switch in the gearshift housing.
3. Lubricant the O-ring and shaft on the reverse rocker arm assembly with lubricant.
4. Position the shift bias spring outward only enough to install the reverse rocker arm assembly in the housing.
5. Install the C-clip that retains the reverse rocker arm assembly.
6. Position an interlock spring and meshlock plunger in each of the bores in the front of the housing.
7. Install the reverse shift rail in the housing. The detent slots in the rail must be inserted into the housing first with the slots facing the meshlock plunger.
8. Install the reverse shift fork so it engages the reverse rocker arm assembly and the fork pad faces the inside. Slide the rail through the fork bore and up to the meshlock plunger.
9. With a small tool, install the meshlock plunger down into the bore. Push the rail forward until it blocks the plunger. Withdraw the tool and push the rail through the bore.
10. Align the roll pin holes in the fork and rail and install the roll pin.
11. Position the 1st/2nd shift rail in the housing. The detent

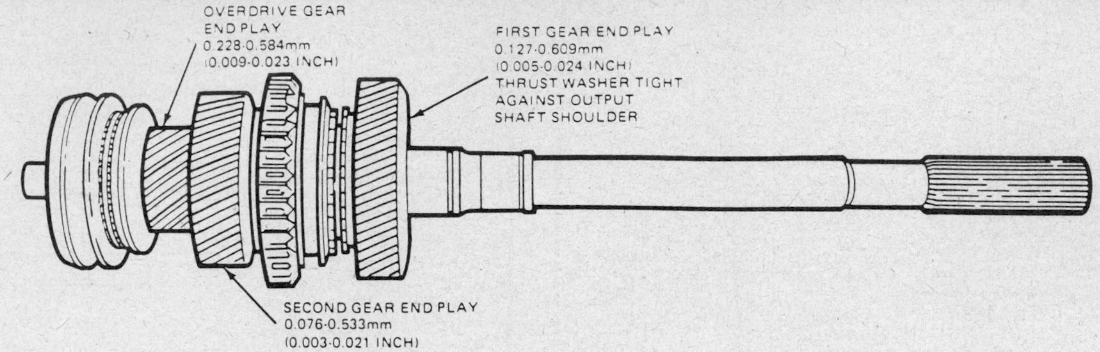
OVERDRIVE GEAR
END PLAY
0.228-0.584mm
(0.009-0.023 INCH)
FIRST GEAR END PLAY
0.127-0.609mm
(0.005-0.024 INCH)
THRUST WASHER TIGHT
AGAINST OUTPUT
SHAFT SHOULDER
SECOND GEAR END PLAY
0.076-0.533mm
(0.003-0.021 INCH)

Gear train endplay

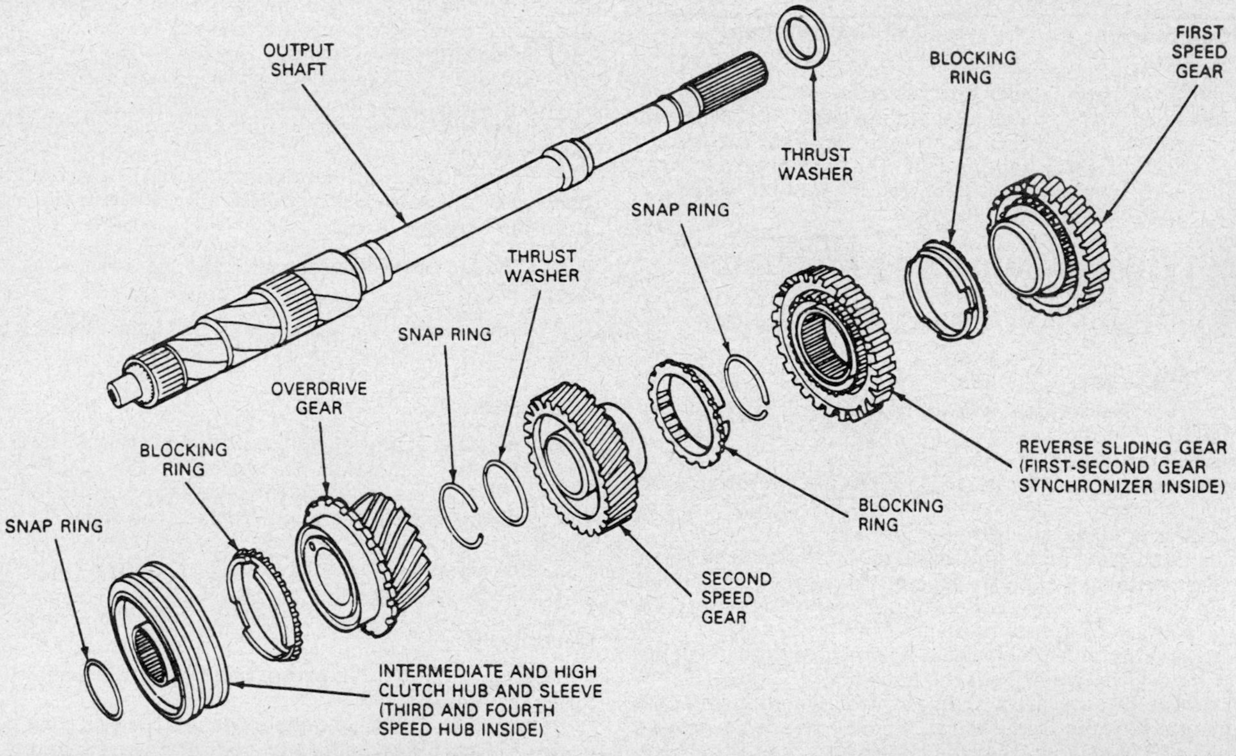
OUTPUT
SHAFT
THRUST
WASHER
BLOCKING
RING
FIRST
SPEED
GEAR
SNAP RING
THRUST
WASHER
SNAP RING
OVERDRIVE
GEAR
BLOCKING
RING
SNAP RING
REVERSE SLIDING GEAR
(FIRST-SECOND GEAR
SYNCHRONIZER INSIDE)
BLOCKING
RING
SECOND
SPEED
GEAR
INTERMEDIATE AND HIGH
CLUTCH HUB AND SLEEVE
(THIRD AND FOURTH
SPEED HUB INSIDE)

Exploded view of output shaft and gear train

slots in the rail must be inserted into the housing first with the slots facing the meshlock plunger.

12. Install the 1st/2nd shift fork so the gate on the fork faces the inside of the housing. Slide the rail through the fork and up to the meshlock plunger.

13. With a suitable tool, position the meshlock plunger down into the bore. Push the rail forward until it blocks the plunger. Withdraw the tool and push the rail through the bore.

14. Align the roll pin holes in the fork and rail and install the roll pin.

15. Check that the 1st/2nd and reverse shift forks are in the neutral position. The neutral position is when both forks are in alignment. The 1st/2nd fork should be in the center detent position. The reverse fork should be shifted fully rearward.

16. Install the interlock plungers through the 3rd/overdrive bore in the rear of the housing. One plunger will be positioned against the reverse shift rail. The other plunger will be positioned against the 1st/2nd shift rail.

17. Position the 3rd/overdrive shift rail in the housing. The detent slots in the rail must be inserted into the housing first with the slots facing the meshlock plunge.

18. Place the 3rd/overdrive shift gate in the housing so the slot in the gate faces down and the small tab is rearward in the housing. Slide the rail forward until it is just through the gate.

19. Install the interlock pin the rear of the 3rd/overdrive shift rail.

20. Place the 3rd/overdrive shift pawl in the housing. Position the pawl so that the slot is on the reverse rail side of the housing. Slide the rail through the pawl up to the plunger.

21. With a suitable tool, position the meshlock plunger down

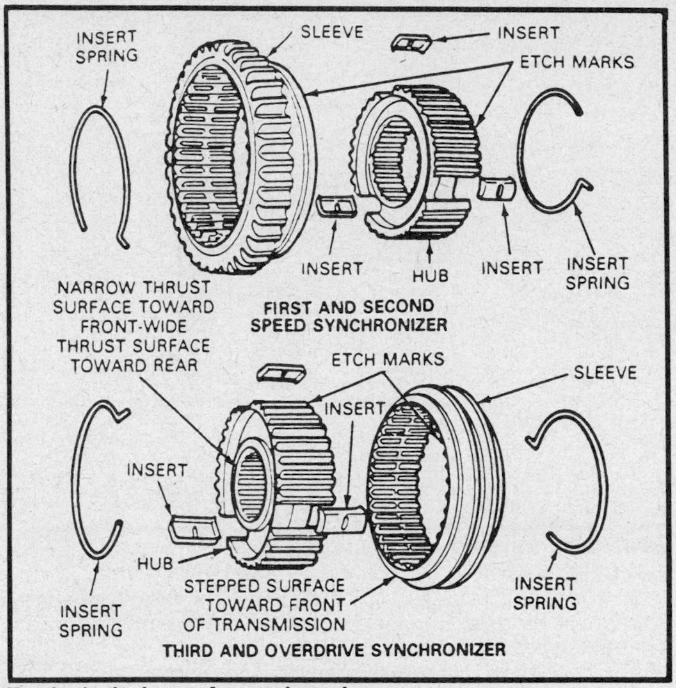
INSERT SPRING
SLEEVE
INSERT
ETCH MARKS
INSERT
HUB
INSERT
INSERT SPRING
NARROW THRUST SURFACE TOWARD FRONT-WIDE THRUST SURFACE TOWARD REAR
FIRST AND SECOND SPEED SYNCHRONIZER
ETCH MARKS
SLEEVE
INSERT
INSERT
HUB
STEPPED SURFACE TOWARD FRONT OF TRANSMISSION
INSERT SPRING
INSERT SPRING
THIRD AND OVERDRIVE SYNCHRONIZER

Exploded view of synchronizers

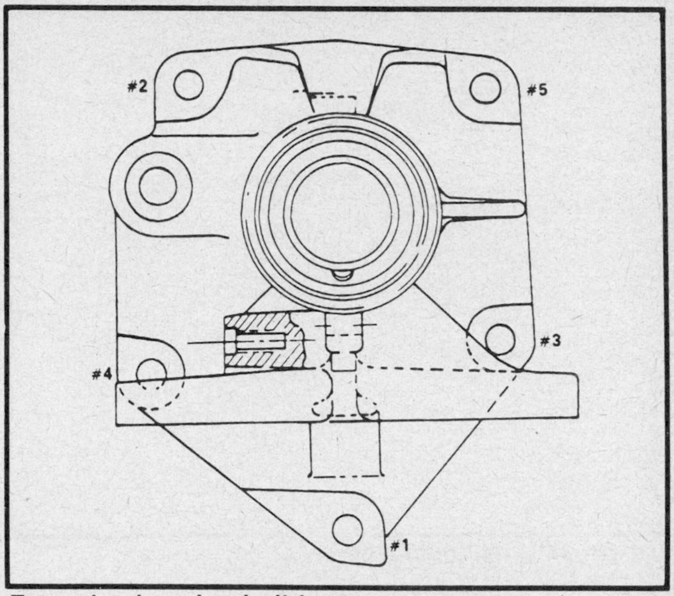
#2
#5
#3
#4
#1

Extension housing bolt torque sequence

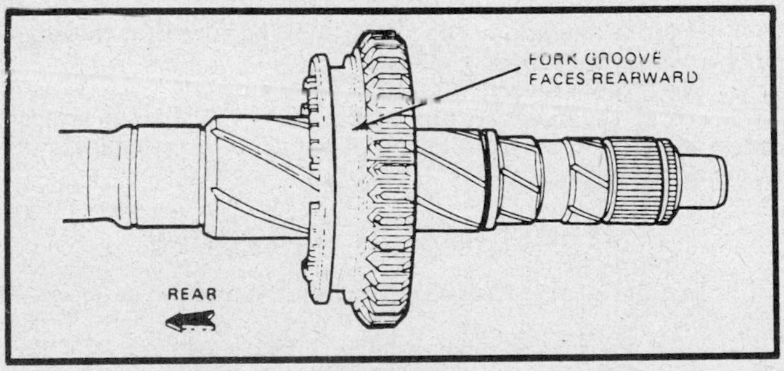
FORK GROOVE FACES REARWARD
REAR

1st-2nd synchronizer installation

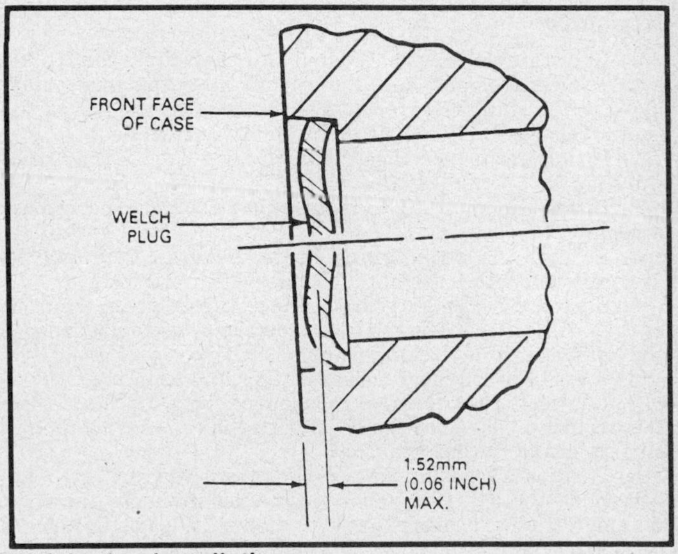
FRONT FACE OF CASE
WELCH PLUG
1.52mm (0.06 INCH) MAX.

Welch plug installation

into the bore. Push the rail forward until it blocks the plunger. Withdraw the tool and push the rail through the bore.

22. Align the roll pin holes in the pawl, gate, rail and install the roll pins.
23. Install the shift rail cup plugs into the bores located in the rear of the housing.
24. Place the gearshift housing assembly in 2nd gear for installation.
25. Install the gearshift housing assembly on the transmission.

OUTPUT SHAFT AND GEAR TRAIN

Before disassembling the output shaft and gear train assembly, the endplay of the 1st, 2nd and overdrive gears must be checked using a feeler gauge. If measurements exceed the specifications, rebuild the assembly with new components.

1. With the 1st gear thrust washer clamped tight against the shoulder of the output shaft, the 1st gear endplay should be 0.005–0.024 in.
2. The 2nd gear endplay should be 0.003–0.021 in.
3. The overdrive gear endplay should be 0.009–0.023 in.

Disassembly

1. Remove the snapring from the front of the output shaft. Remove the 3rd/4th synchronizer assembly which includes intermediate, high clutch hub, sleeve and blocking ring.
2. Remove the overdrive gear from the shaft.
3. Remove the next snapring and the 2nd gear thrust washer from the shaft. Slide the second gear and blocking ring from the shaft.
4. Remove the next snapring.
5. Using a suitable press or equivalent, remove the 1st/2nd synchronizer assembly which includes reverse sliding gear with 1st/2nd synchronizer inside and 2 blocking rings. The synchronizer assembly is a press fit on the output shaft.

Inspection

1. Inspect all output shaft gears for chips, nicks, or bad wear.

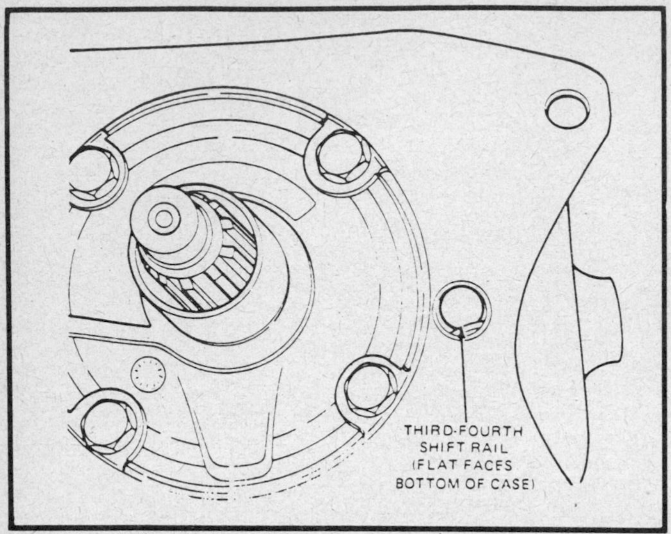
THIRD-FOURTH
SHIFT RAIL
(FLAT FACES
BOTTOM OF CASE)

3rd-4th shift rail installation

2. Inspect the output shaft for being warped.
3. Check the pilot bearing surface for being scored.
4. Inspect the speedometer gear teeth for wear or damage.

Assembly

1. Lubricate the 1st gear journal lubricant. Press the 1st/2nd synchronizer assembly onto the front of the output shaft, making sure the shift fork groove is facing towards the rear of the shaft. This assembly is a press fit on the output shaft.
2. Install the 1st speed gear and blocking ring on the rear of the shaft.
3. Install the snapring in front of the 1st/2nd synchronizer assembly.
4. Position the blocking ring on the 2nd gear. Lubricate the 2nd gear journal on the output shaft with lubricant.
5. Slide the 2nd speed gear with blocking ring onto the front of the shaft. Check that the inserts in the synchronizer engage the notches in the blocking ring.
6. Install the 2nd gear thrust washer and snapring.
7. Lubricate the overdrive gear journal with lubricant. Slide the overdrive gear onto the shaft with the coned synchronized surface facing toward the front.
8. Position a blocking ring on the overdrive gear.
9. Slide the 3rd/4th synchronizer assembly onto the shaft making sure the inserts in the syncrhonizer engage the notches in the blocking ring and the small thrust surface is facing forward.
10. Install the snapring on the front of the output shaft.

SYNCHRONIZER

Diassembly

1. Make alignment mark on the hub and sleeve of the synchronizer.
2. Remove the synchronizer hub from each synchronizer sleeve.
3. Remove the inserts and insert springs from the hubs.
4. Do not mix the parts of the 1st/2nd speed synchronizer with the 3rd/ overdrive speed synchronizer.

Inspection

1. Check synchronizer discs for wear or damage.
2. Inspect the synchronizer blocking rings for widened index slots, rounded clutch teeth and smooth internal surfaces.

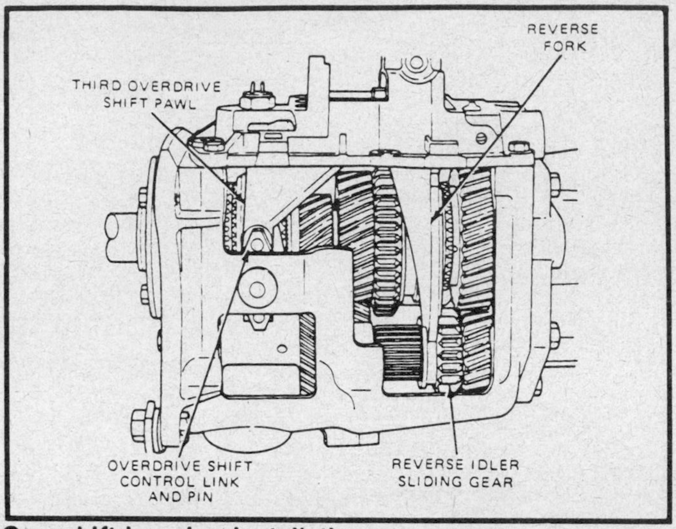
REVERSE
FORK
THIRD OVERDRIVE
SHIFT PAWL
OVERDRIVE SHIFT
CONTROL LINK
AND PIN
REVERSE IDLER
SLIDING GEAR

Gearshift housing installation

3. With the blocking ring on the cone, the distance between the face of the gear clutching teeth and the face of the blocking ring must not be less than 0.010 in.
4. Check the synchronizer sleeves for free movement on the hubs.

Assembly

1. Position the hub in the sleeve, making sure that the alignment marks are properly aligned.
2. Place the 3 inserts into place on the hub.
3. Install the insert springs making sure that the tab is located in a common insert rotating in opposite directions. Do not stagger the springs.

Transmission Assembly

1. Install the overdrive shift control link and pin assembly in the case.
2. Coat the reverse idler bronze thrust washers with lubricant. Install the thrust washers for the reverse idler gear assembly in position in the case. Align tabs on the washer with the slots in the case.
3. If removed, coat the reverse idler shaft needle roller bearings with lubricant. Install the 2 rows of 22 needle roller bearings in the reverse idler gear.
4. Slide the reverse idler sliding gear onto the reverse idler gear. The grooved portion of the sliding gear for the reverse fork must fact the front of the case when installed.
5. Position the reverse idler gear and sliding gear in the case. From the rear of the case, install the reverse idler gear shaft through the sliding gear, reverse idler gear assembly and thrust washers. The roll pin in the end of the shaft must face the rear of the case. Do not drop any of the neddle bearings.
6. Coat the countershaft nylon thrust washers with lubricant. Seat each thrust washer in position in the case notches.
7. If removed, coat the countershaft needle roller bearings with lubricant. Install the 2 sets of 21 needle roller bearings into the countershaft gear. Insert the dummy countershaft special tool into the countershaft gear.
8. Lower the countershaft to the bottom of the case. The countershaft gear must be at the bottom of the case with the countershaft removed to obtain the necessary clearance to install the input shaft, output shaft and gear train assembly.
9. Position the rear of the output shaft and gear train through the rear bearing bore in the case and tilt the assembly until it can be installed in the case.

10. Install the 3rd/4th shift fork with the slot facing down. Align the fork pads with the 3rd/4th synchronizer assembly and the slotted groove with the overdrive shift control link and pin assembly.

11. Install the 15 needle roller bearings in the bore of the input shaft. Coat the bearings with lubricant. Install the blocking ring in the intermediate and high clutch hub and sleeve on the output shaft and gear train.

12. Install the input shaft by aligning the flat of the synchronizer teeth with the countershaft teeth and move the input shaft into position on the output shaft and gear train assembly. Align the slots in the blocking ring with the inserts in the sleeve.

13. If removed, install the input shaft front bearing over the input shaft and into the correct position in the case.

14. Install the bearing retainer ring.

15. Install the snapring in the groove on the bearing and the snapring over the bearing.

16. Make sure the input bearing retainer and case mounting surfaces are free of all traces of gasket material. Coat the mating surface on the retainer with sealer. Form a mating surface gasket pattern on the retainer and form the gasket inside the bolt holes.

17. Position the retainer on the case so the slot inside the retainer is facing the bottom of the case. Coat the bolt threads with threadlock and sealer. Install and torque the bolts to 12–16 ft. lbs.

18. If removed, install the output shaft rear bearing over the output shaft and into position in the case.

19. Install the snapring in the groove in the output shaft bearing. Install the snapring on the output shaft.

20. Install the speedometer drive gear lock ball in the output shaft and install the speedometer drive gear over the lock ball. Retain the gear to the output shaft with the snapring.

21. Position the countershaft gear assembly with the dummy shaft special tool inside into the correct position in the case.

22. Make sure the assembly is in line with the thrust washers in the case. Install the countershaft in the rear of the case and drive the countershaft into the case as the dummy shaft special tool is removed.

23. Clean the extension housing and case mounting surfaces of all gasket material. Coat the mating surface of the extension housing with sealer. Form a mating surface gasket pattern on the housing.

24. Install the extension housing. Coat the retaining bolts with threadlock and sealer. Install all the bolts evenly and torque in sequence order to 42–50 ft. lbs.

25. Position the 3rd/4th shift rail in the bore in the front of the case so the flat on the rail faces the front and bottom of the case.

26. Install the shift rail into the case and through the 3rd/4th shift fork aligning the fork and roll pin holes. Drive the roll pin into the correct position.

27. Drive the welch plug into the 3rd fork shift rail bore in the case. The plug must not protrude above the front face of the case or more than 0.06 in. maximum below the front face of the case.

28. Place the transmission and the gearshift housing assembly in the second gear position. Make sure the reverse idler gear is positioned rearward.

29. Make sure the gearshift housing assembly and transmission case mounting surfaces are free of all gasket material. Coat the mating surface of the housing with sealer. Form a mating surface gasket pattern on the housing.

30. Install the gearshift housing assembly on the case. When properly installed the 3rd/overdrive shaft pawl engages the rounded portion of the overdrive shift control link and pin assembly, the reverse fork engages the reverse idler sliding gear and the 1st/2nd shift fork should engage the 1st/2nd synchronizer assembly.

31. Install the bolts and washers that retain the gearshift housing assembly to the case and torque to 18–22 ft. lbs.

32. Install the shift lever in the gearshift housing.

33. Remove the fill plug and fill the transmission with the correct amount of the specified lubricant. Install the fill plug.

34. Shift the transmission into all gear positions to verify correct operation.

SPECIFICATIONS

TORQUE SPECIFICATIONS

Description	ft. lbs.	Nm
Extension housing to transmission case	42–50	57–67
Gearshift housing to transmission case	18–22	25–29
Input shaft bearing retainer to transmission case	12–16	17–21

ENDPLAY SPECIFICATIONS

Gear	in.	mm
First gear	0.005–0.024	0.127–0.609
Second gear	0.003–0.021	0.076–0.533
Overdrive gear	0.009–0.023	0.228–0.584
Countershaft gear	0.004–0.018	0.101–0.457

SPECIAL TOOLS

Number	Description
T57L-500-B	Bench mounted holding fixture
D79L-6731-A or -B	Oil filter wrench
T75L-7025-B	Remover/replacer tube (short)
T75L-7025-C	Remover/replacer tube (long)
T75L-7025-D	Bearing collet
T75L-7025-F	Bearing collet
T75L-7025-H	Bearing collet sleeve
T84T-7025-A	Bearing remover/replacer
T84T-7025-B	Forcing screw
T75L-7025-K	Replacing shaft sleeve
T75L-7025-P	Shaft collar
T77L-7025-C	Bearing replacer tube
T77L-7025-D	Shaft clamp
T64P-7111-A	Dummy countershaft tool

Section 7

FM 145/FM 146 Transmissions
Ford Motor Co.

APPLICATION

1987–89 Ranger 4WD and Bronco II

GENERAL DESCRIPTION

The FM145 transmission was introduced in 1987 and replaced by the FM146 transmission in 1988–89. Internally both transmissions are identical, however the basic difference is in the shifter assemblies. Both transmissions incorporate 5 forward gears (the 5th gear being an overdrive gear) and 1 reverse gear. The FM145 and FM146 are fully synchronized in all forward gears and reverse. All gear changes are accomplished with the assistance of synchronizer sleeves. Reverse gear uses a reverse idler gear which is in constant mesh with the countershaft gear.

The top mounted shifter operates the shift rails through a set of shift forks. Shift forks mounted on the rails operate the synchronizer sleeves that allow shifting of the 1/2, 3/4 and overdrive/reverse.

A shift interlock system, located in the side of the transmission case, prevents the shift rails from engaging 2 gears at the same time.

The transmission assembly is composed of 5 major components: a front bearing retainer, a transmission case, a transfer case adapter and a control housing, all of which are cast aluminum. The 5th component is the under cover which is made from stamped aluminum.

Transmission Identification

The Mitsubishi FM145 and FM146 can be identified by a identification code, this code can be found on the transmission identification tag located on the side of the transmission case.

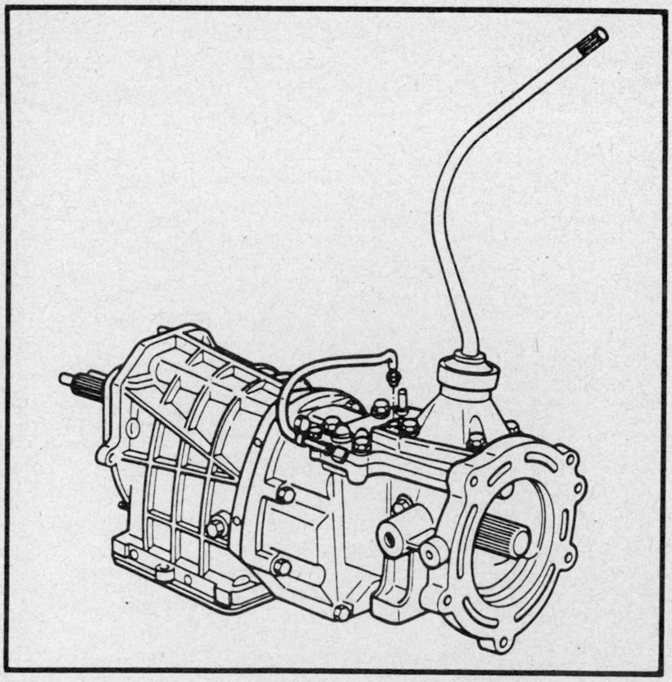

Side view of the FM145 manual transmission

Metric Fasteners

The metric fastener dimensions are very close to the dimensions of the familiar inch system fasteners and for this reason, replacement fasteners must have the same measurement and strength as those removed.

CAUTION

Do not attempt to interchange metric fasteners for inch system fasteners. Mismatched or incorrect fasteners can result in damage to the transaxle unit through malfunctions or breakage and possible personal injury.

Capacities

The transmission lubricant capacity is 2.4 quarts (1.3L). The correct recheck or top-off level is to the bottom edge of the fill plug hole. The recommended lubricant for the FM145 and FM146 transmission is standard transmission lubricant (SAE 80W) D8OZ–19C547–A, (ESP-M2C83–C) or equivalent.

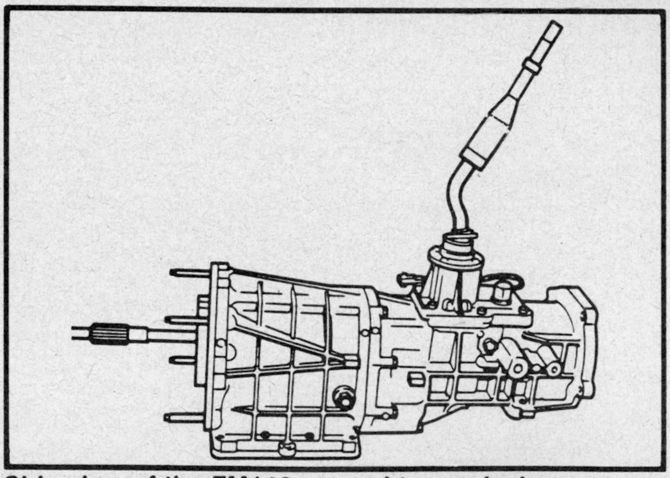

Side view of the FM146 manual transmission

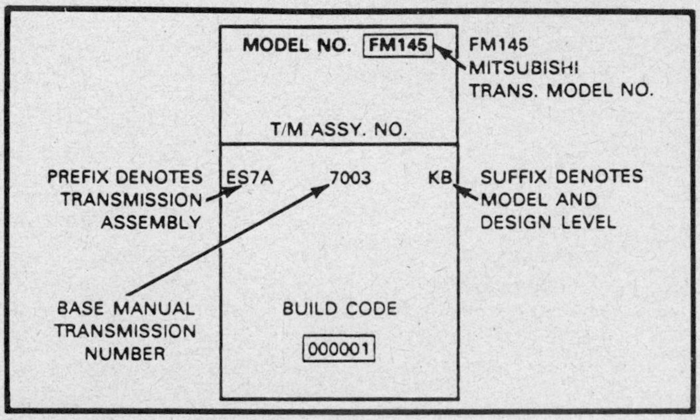
MODEL NO. FM145
FM145 MITSUBISHI TRANS. MODEL NO.
T/M ASSY. NO.
PREFIX DENOTES TRANSMISSION ASSEMBLY
ES7A
7003
KB
SUFFIX DENOTES MODEL AND DESIGN LEVEL
BASE MANUAL TRANSMISSION NUMBER
BUILD CODE
000001

Transmission identification tag

Checking Fluid Level

1. Raise and support the vehicle safely.
2. Remove the transmission oil fill plug.
3. The correct level should be to the bottom edge of the fill plug hole.
4. Fill the proper level with the recommended gear oil and re-install the oil fill plug.
5. Lower the vehicle.

TRANSMISSION MODIFICATIONS

There are no transmission modifications for the FM145 and FM146 transmissions at the time of this publication.

TROUBLE DIAGNOSIS

CHILTON'S THREE C'S TRANSMISSION DIAGNOSIS CHARTS

Condition	Cause	Correction
Transmission shifts hard	a) Transmission fluid low or improper type	a) Add lubricant or change lubricant as required
	b) Shift lever binding or worn	b) Remove cap from shift tower. Eliminate binding condition or replace components as required
	c) Worn or damaged internal shift mechanism	c) Remove transmission cover. Check internal shift mechanism by shifting into and out of all gears. Repair or replace as required
	d) Binding of sliding gears and/or synchronizers	d) Check for free movement of gears and synchronizers. Repair or replace as required
	e) Housings and/or shafts out of alignment	e) Remove transmission and check for binding condition between input shaft and engine crankshaft pilot bearing or bushing. Check flywheel housing alignment. Repair or replace as required
	f) Incomplete clutch disengagement	f) Inspect complete clutch system and repair
Noisy in forward gears	a) Lubricant level low, or improper type	a) Add lubricant, or refill with specified lubricant
	b) Components grinding on transmission	b) Check for screws, bolts, etc., of cab or other components grinding. Correct as required
	c) Component housing bolts loose	c) Check torque on transmission to flywheel housing bolts, output shaft flange nut and flywheel housing to engine block bolts. Tighten bolts to specification
	d) Flywheel housing to engine crankshaft alignment	d) Check and align flywheel housing to engine crankshaft
	e) Noisy bearings or gears	e) Remove and disassemble transmission. Inspect input, output and countershaft bearings. Inspect speedometer gear and gear teeth for wear or damage. Replace as required

CHILTON'S THREE C'S TRANSMISSION DIAGNOSIS CHARTS

Condition	Cause	Correction
Gears clash when shifting from one forward gear to another	a) Engine idle speed too high	a) Adjust engine idle speed
	b) Improper manual shift linkage	b) Adjust and repair manual shift linkage as required
	c) Pilot bearing binding	c) Remove transmission and check for binding condition between input shaft and engine crankshaft pilot bearing. Replace as required
	d) Damaged gear teeth and/or synchronizer	d) Disassemble transmission, repair or replace as required
	e) Incomplete clutch disengagement	e) Inspect complete clutch system and repair
Transmission jumps out of gear	a) Loose transmission to engine block bolts, or loose levers	a) Tighten transmission to engine block bolts to specifications. Loosen all bolts and reseat flywheel housing. Tighten all bolts. Tighten levers if necessary
	b) Crankshaft pilot bearing worn	b) Replace bearing
	c) Interior components damaged	c) Disassemble transmission. Inspect the synchronizer sleeves for free movement on their hubs. Inspect the synchronizer blocking rings for widened index slots, rounded clutch teeth and smooth internal surface. Check countershaft cluster gear for excessive endplay. Check shift forks for loose mounting on shift rails. Inspect synchronizer sliding sleeve and gear clutch teeth for wear or damage. Repair or replace as required
	d) Worn gear teeth due to partial engagement	d) Replace worn or damaged gears
Transmission will not shift into one gear—all others OK	a) Back-up switch ball frozen	a) If reverse is problem, check backup switch for ball frozen in extended position (if so equipped)
	b) Internal components	b) Remove transmission. If transmission will not shift into reverse, remove transmission, check for damaged reverse gear train, in single rail shift transmission. Also, check for misaligned reverse relay lever. Inspect shift rail and fork system synchronizer system and gear clutch teeth for restricted travel. Repair or replace as required
Transmission is locked in one gear. It cannot be shifted out of that gear	a) Internal components binding	a) Remove transmission. Inspect problem gears, shift rails and forks and synchronizer for wear or damage. Repair as required. Check for broken fork slot tabs on single rail transmissions
	b) Loose fork on rail	b) On single rail shift, check for broken selector arm pin or selector plate. Repair or replace as necessary
Transmission leaks	a) Improper amount of lubricant—wrong type	a) Check level and type. Fill to bottom of filler plug hole
	b) Other component leaking	b) Identify leaking fluid as engine, power steering or transmission. Repair as required

CHILTON'S THREE C'S TRANSMISSION DIAGNOSIS CHARTS

Condition	Cause	Correction
Transmission leaks	c) False report	c) Remove all traces of lube on exposed transmission surfaces. Check vent for free breathing. Operate transmission and inspect for new leakage. Repair as required
	d) Internal components	d) Remove transmission. Inspect for leaks at the input shaft bearing retainer seal and gasket, and shift rail expansion plug. Inspect for leaks at the top cover gasket. Inspect case for sand holes or cracks. Repair or replace as required
	e) Improper installation torque	e) Tighten to specified torque value
Shift lever loose	a) Retaining bolts loose	a) Tighten to specified torque
Shift lever tight	a) Worn or damaged shift lever lower bushing	a) Inspect shift lever assembly. Remove and replace parts as required
Transfer case makes noise	a) Incorrect tire inflation pressures and/or incorrect size tires and wheels	a) Assure that all tires and wheels are the same size, and that inflation pressures are correct
	b) Excessive tire tread wear	b) Check tire tread wear to see if there is more than .06 inch difference in tread wear between front and rear. Interchange one front and one rear wheel. Reinflate tires to specifications
	c) Internal components	c) Operate vehicle in all transmission gears with transfer case in 2HI, or HI range — If there is noise in transmission in neutral gear, or in some gears and not in others, remove and repair transmission — If there is noise in all gears, operate vehicle in all transfer case ranges. If noisy in all ranges, disassemble transfer case. Check input gear, planetary gear assembly, sprockets, chain and single cardan U-joint for damage. Replace as necessary
4-wheel drive transfer case jumps out of gear	a) Incomplete shift linkage travel	a) Check for interference to shift boot or body
	b) Loose mounting bolts	b) Tighten mounting bolts
	c) Front and rear driveshaft slip yokes dry or loose	c) Lubricate and repair slip yokes as required. Tighten flange yoke attaching nut to specifications
	d) Internal components	d) Disassemble transfer case. Inspect input gear, planetary gear assembly, sliding shift collar hubs and detent spring for damage. Replace as required

POWER FLOW

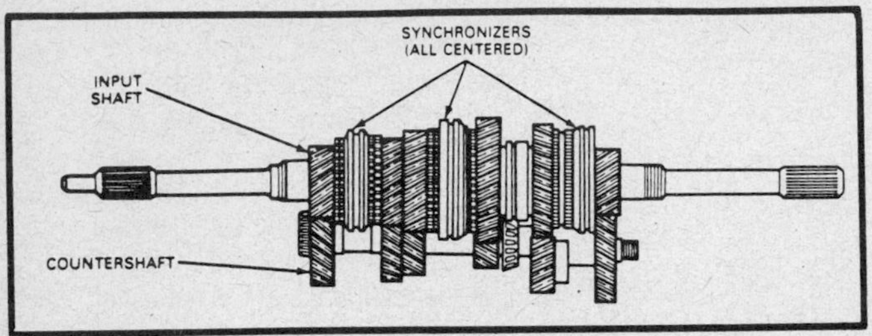
SYNCHRONIZERS
(ALL CENTERED)
INPUT
SHAFT
COUNTERSHAFT

Power flow—neutral position

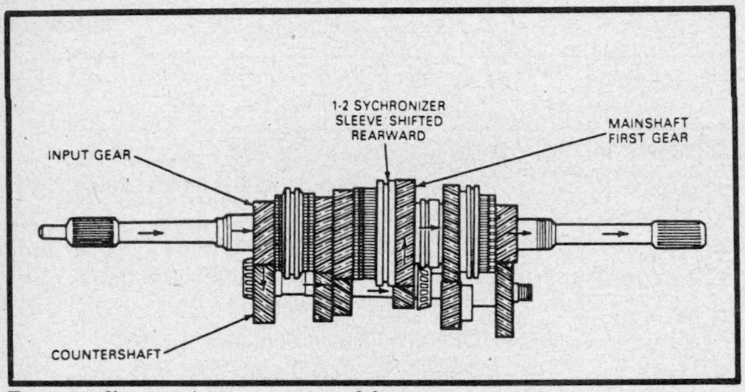
1-2 SYCHRONIZER
SLEEVE SHIFTED
REARWARD
MAINSHAFT
FIRST GEAR
INPUT GEAR
COUNTERSHAFT

Power flow—1st gear position

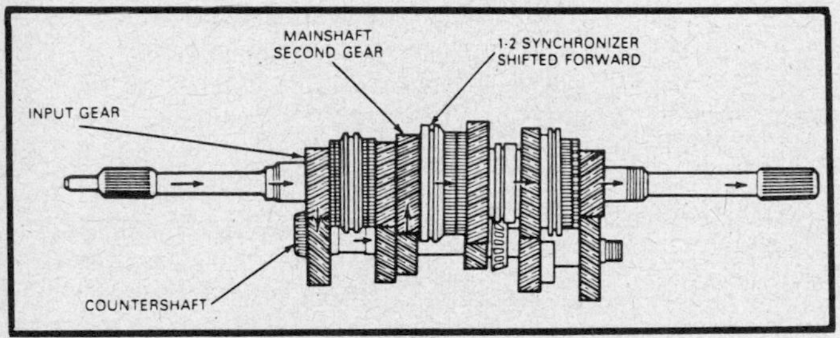
MAINSHAFT
SECOND GEAR
1-2 SYNCHRONIZER
SHIFTED FORWARD
INPUT GEAR
COUNTERSHAFT

Power flow—2nd gear position

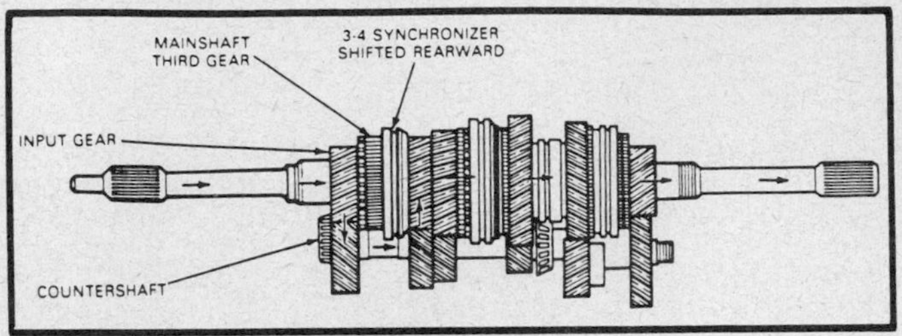
MAINSHAFT
THIRD GEAR
3-4 SYNCHRONIZER
SHIFTED REARWARD
INPUT GEAR
COUNTERSHAFT

Power flow—3rd gear position

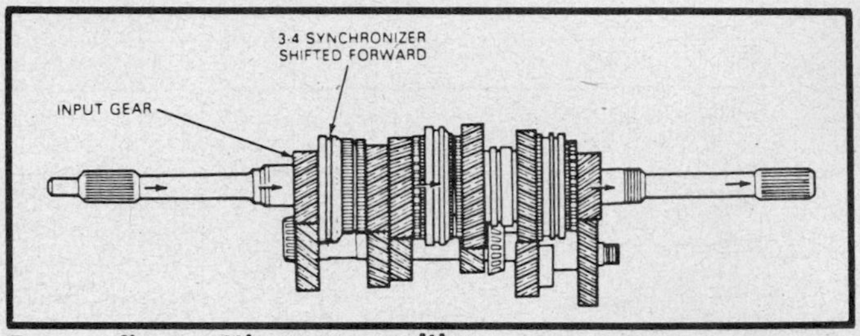
3-4 SYNCHRONIZER
SHIFTED FORWARD
INPUT GEAR

Power flow—4th gear position

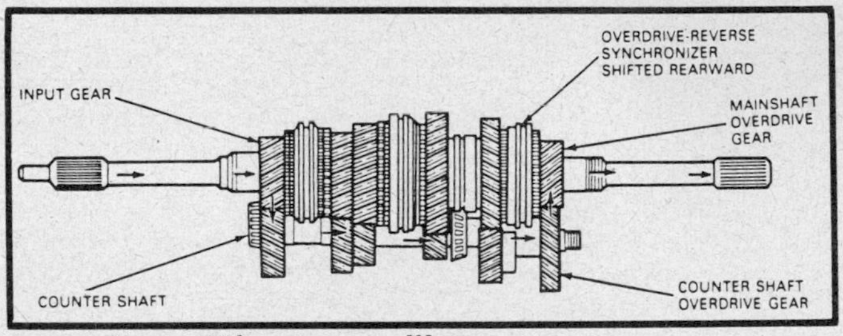
OVERDRIVE-REVERSE
SYNCHRONIZER
SHIFTED REARWARD
INPUT GEAR
MAINSHAFT
OVERDRIVE
GEAR
COUNTER SHAFT
COUNTER SHAFT
OVERDRIVE GEAR

Power flow—5th gear position

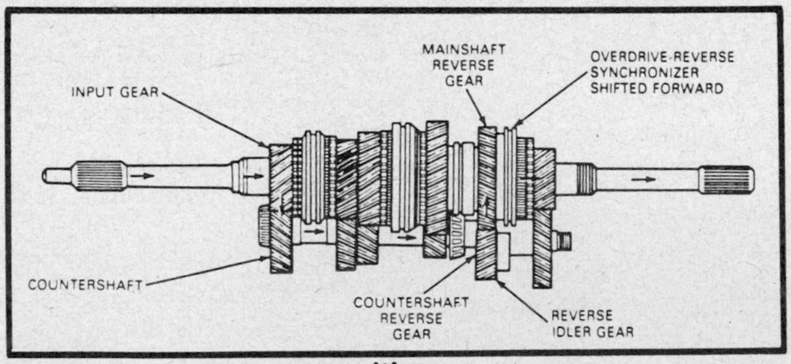
MAINSHAFT
REVERSE
GEAR
OVERDRIVE-REVERSE
SYNCHRONIZER
SHIFTED FORWARD
INPUT GEAR
COUNTERSHAFT
COUNTERSHAFT
REVERSE
GEAR
REVERSE
IDLER GEAR

Power flow—reverse position

Services

FLUID CHANGE

1. Raise and support the vehicle safely.
2. Remove the transmission oil fill plug and allow the gear oil to drain into a suitable drain pan.
3. The correct level should be to the bottom edge of the fill plug hole.
4. Fill to the proper level with the recommended gear oil and reinstall the oil fill plug.
5. Lower the vehicle.

REMOVAL AND INSTALLATION

TRANSMISSION REMOVAL

1. Place the gear shift selector in the **N** position.
2. Remove the boot retainer screws and the boot.
3. Pull the gearshift lever assembly out of the control housing.
4. Cover the opening in the control housing with a cloth to prevent dirt from falling into the transmission.
5. Disconnect the negative battery cable from the battery terminal.
6. Raise the vehicle and support it safely.
7. Disconnect the rear driveshaft at both the rear axle and transfer case flanges.
8. Disconnect the forward driveshaft from the front axle and remove by sliding forward. Install a suitable plug in the transfer case adapter to prevent fluid leakage.
9. Disconnect the hydraulic fluid line from the transfer case.
10. Disconnect the speedometer from the transfer case.
11. Disconnect the starter motor cable, the back-up lamp switch wire and the neutral safety switch wire.
12. Place a jack and a block of wood under the engine block, to protect the oil pan.
13. Support the transfer case using the proper equipment. Carefully remove the transfer case from the vehicle.
14. Remove the starter. Place a transmission jack under the transmission.
15. Remove the clutch housing bolts and remove the clutch housing.
16. Remove the nuts and bolts attaching the transmission mount and damper to the crossmember.
17. Remove the nuts and bolts attaching the crossmember to the frame side rails and remove the crossmember.
18. Lower the engine jack. Work the clutch housing off the locating dowels and slide the clutch housing and the transmission rearward until the input shaft clears the clutch disc.
19. Remove the transmission from the vehicle.
20. Remove the clutch housing from the transmission.

TRANSMISSION INSTALLATION

NOTE: Prior to installing the transmission, make sure that all machined mating surfaces and the locating dowels on the engine rear plate are free of burrs, dirt or paint.

1. Raise the vehicle and support it safely.
2. Mount the transmission on a transmission jack and position it under the vehicle, start the input shaft into the clutch disc.
3. Align the splines on the input shaft with the splines in the clutch disc.
4. Move the transmission forward and carefully seat the clutch housing on the locating dowels of the engine rear plate. The engine plate dowels must not shave or burr the clutch housing dowel holes.
5. Install the bolts and washers attaching the clutch housing to the rear plate and tighten to 28–38 ft. lbs. (38–51 Nm).
6. Install the starter motor. Tighten the attaching nuts to 15–20 ft. lbs. (21–27 Nm).
7. Raise the engine and install the rear crossmember, insulator and damper and attaching nuts and bolts.
8. Install the transfer case.
9. Install the driveshaft in the transfer case adapter and attach it to the rear axle flange. Make sure the marks made during removal are in alignment. Install the attaching nuts and bolts.
10. Connect the starter cable, back-up lamp switch wire, shift indicator wire and the neutral safety switch.
11. Connect the clutch hydraulic line to slave cylinder on the input shaft. Bleed the hydraulic clutch system.
12. Install the speedometer cable.
13. Remove the fill plug and check the fluid level. The fluid level should be level with the bottom of the fill hole. Standard transmission lubricant SAE 80W, D8DZ-19C547-A (ESP-M2C83-C) or equivalent is recomended.
14. Lower the vehicle.
15. Connect the negative battery cable.
16. Remove the cloth over the transfer case adapter opening. Avoid getting dirt in the adapter.
17. Install the gearshift lever assembly in the transfer case adapter. Make sure the ball on the lever is in the socket in the adapter. Install the attaching bolts and tighten to 6–10 ft. lbs. (8–14 Nm).
18. Install the boot cover and bolts.
19. Check for proper shifting and operation of the transmission.

BENCH OVERHAUL

Before Disassembly

Cleanliness during disassembly and assembly is necessary to avoid further transmission trouble after overhaul. Before removing any of the transmission subassemblies, plug all the openings and clean the outside of the transmission thoroughly. Steam cleaning or car wash type high pressure equipment is preferable. During disassembly, clean all parts in suitable solvent and dry each part. Do not use cloth or paper towels to dry parts. Use compressed air only.

Transmission Disassembly

1. Remove the transmission. Make sure the transmission is in the **N** position.
2. Remove the nuts retaining the clutch housing to the transmission and remove the housing. If not removed, remove the clutch slave cylinder from the input shaft.
3. Remove the back-up lamp switch and shift indicator switch from the transmission.
4. Drain the fluid from the transmission by removing the drain plug from the pan.
5. Remove the bolts retaining the pan to the case and remove the pan. Remove and discard the gasket. Remove all traces of the gasket from the mating surfaces of the pan and case.
6. On the FM145, remove the bolts retaining the cover to the transfer case adapter and remove the cover (with stopper bracket inside). Remove and discard the gasket. Clean all traces of gasket material from the mating surfaces of the adapter and cover.
7. On the FM146, remove the bolts retaining the control housing to the transfer case adapter and remove the control

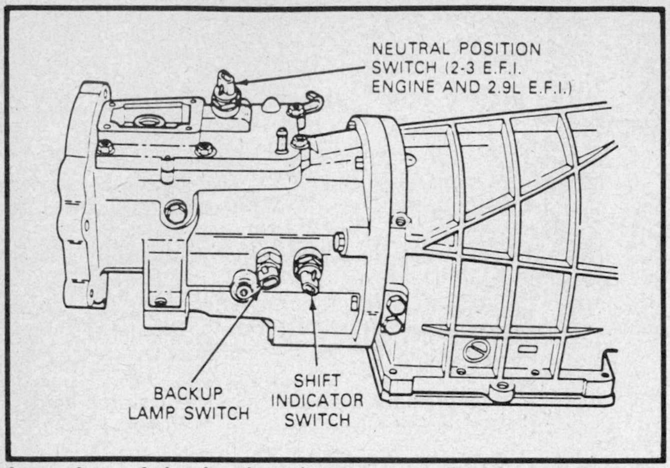
NEUTRAL POSITION
SWITCH (2-3 E.F.I.
ENGINE AND 2.9L E.F.I.)
BACKUP
LAMP SWITCH
SHIFT
INDICATOR
SWITCH

Location of the backup lamp, neutral safety and shift indicator switches

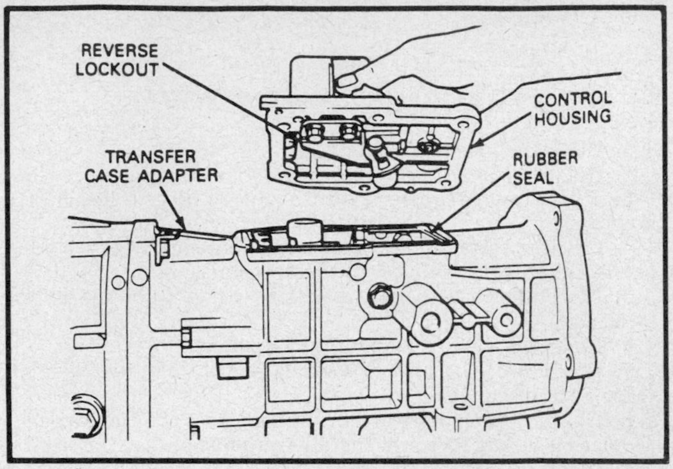
REVERSE
LOCKOUT
CONTROL
HOUSING
TRANSFER
CASE ADAPTER
RUBBER
SEAL

Removing the control housing and rubber seal—FM146

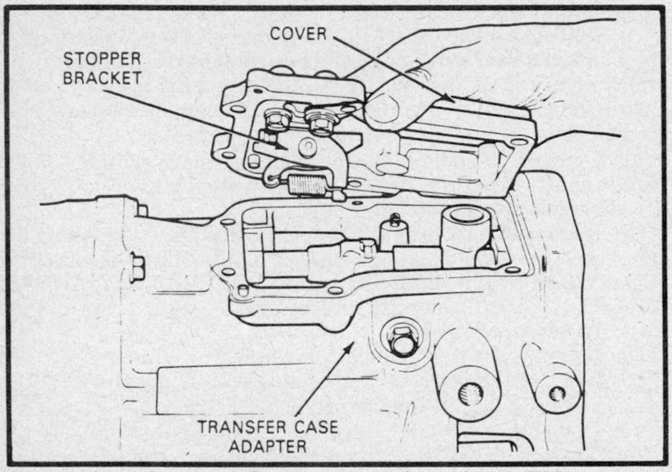
COVER
STOPPER
BRACKET
TRANSFER CASE
ADAPTER

Removing the transfer cover and gasket—FM145

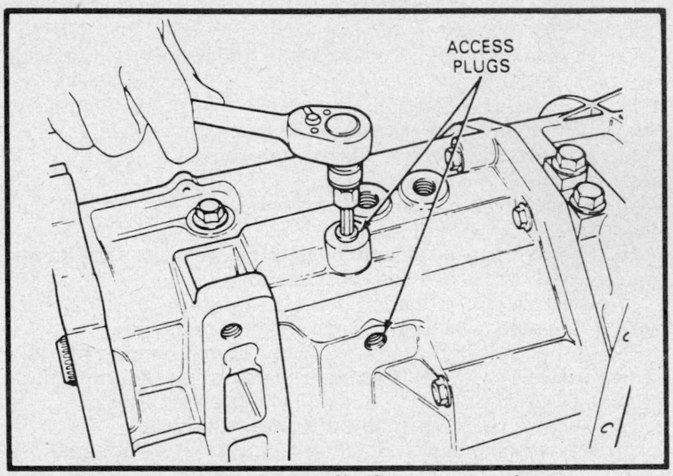
ACCESS
PLUGS

Removing the shift gate roll pin access plugs

housing (with reverse lockout assembly attached). Remove and discard the rubber seal. Clean the mating surfaces of the adapter and control housing.

8. Remove the detent spring and ball from the adapter.
9. Remove the 3 shift gate roll pin access plugs (2 on the side, 1 on the bottom) with a 6mm allen-head wrench.
10. Using a punch, drive the roll pins from the shift gates through the access holes.
11. From the right side of the adapter, remove the bolt, spring and neutral return plunger. Note that the plunger has a slot in the center for the detent ball.
12. From the left side of the adapter, remove the bolt, spring and neutral return plunger.
13. Lift the gate selector lever out of the shift gates. Move the lever to the rear of the adapter as far as it will go. This will allow clearance to remove the adapter from the case.
14. Remove the bolts retaining the transfer case adapter to the transmission case. Note that 3 different bolts lengths (35mm, 55mm and 110mm) are used to retain the case to the adapter. Mark the bolt holes accordingly.
15. Remove the adapter from the case. Make sure the shift gates do not bind in the adapter during removal. Rotate the gates on the rails as required. Remove and discard the gasket. Clean all traces of gasket material from the mating surfaces of the case and adapter.
16. Identify each shift rail and gate. Remove the gates from the rails.

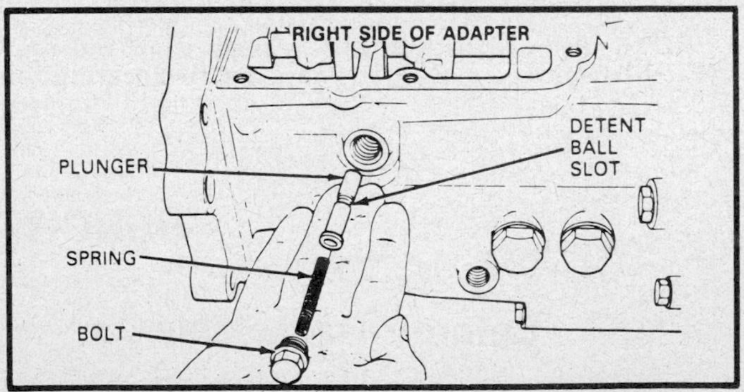
RIGHT SIDE OF ADAPTER
PLUNGER
DETENT
BALL
SLOT
SPRING
BOLT

Removing the neutral return plunger—right side

17. From inside the case, drive out the roll pins retaining the 1/2 and 3/4 shift forks to the rails. Remove the 1/2 shift fork.
18. Drive out the overdrive/reverse shift fork roll pin.

NOTE: The roll pin in the switch actuator does not need to be removed for transmission disassembly.

19. Remove the set screw on the top of the case and remove the poppet spring and steel ball. Remove the 2 bolts on the side of the case. Remove the poppet springs and steel balls.
20. Remove the overdrive/reverse shift rail and the 3/4 shift rail from the case. Remove the overdrive/reverse shift fork.

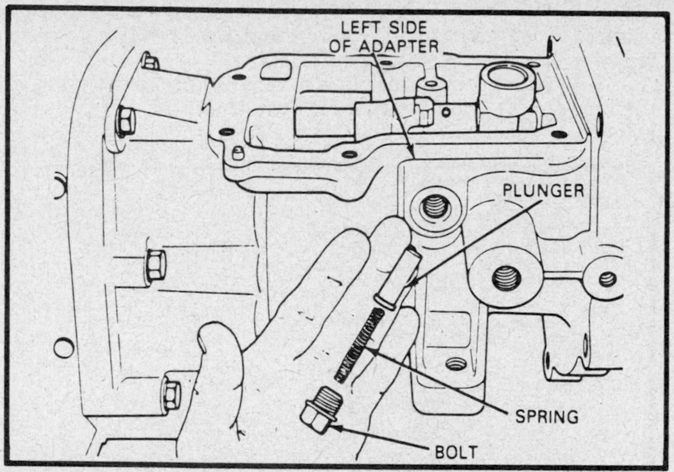
LEFT SIDE
OF ADAPTER
PLUNGER
SPRING
BOLT
Removing the neutral return plunger—left side

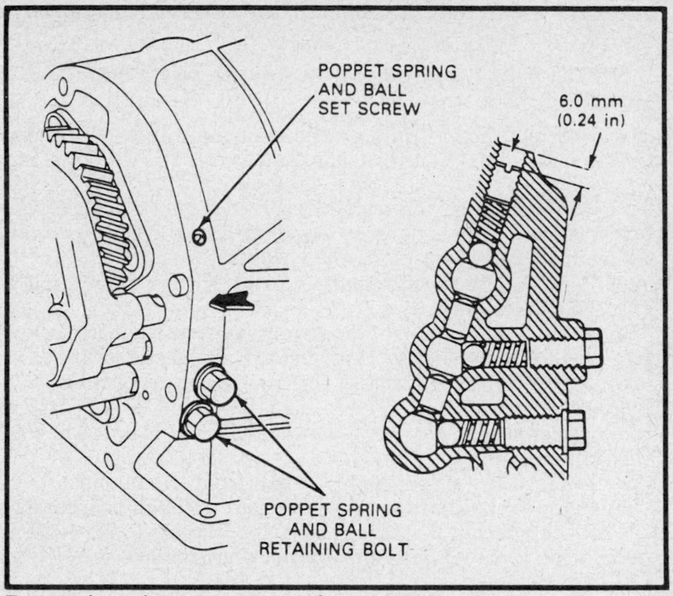
POPPET SPRING
AND BALL
SET SCREW
6.0 mm
(0.24 in)
POPPET SPRING
AND BALL
RETAINING BOLT
Removing the poppet spring and steel ball

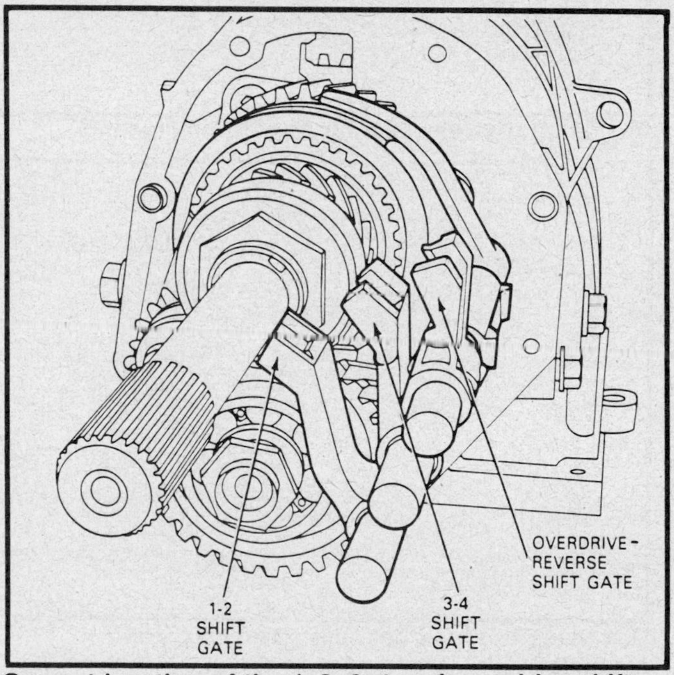
OVERDRIVE-
REVERSE
SHIFT GATE
1-2
SHIFT
GATE
3-4
SHIFT
GATE
Correct location of the 1–2, 3–4 and overdrive shift gates

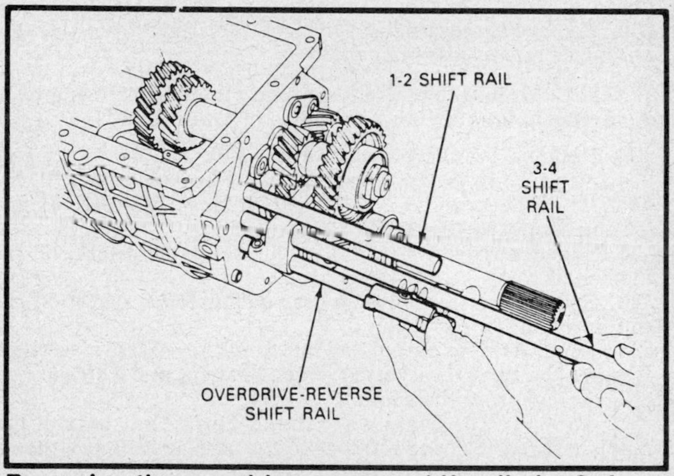
1-2 SHIFT RAIL
3-4
SHIFT
RAIL
OVERDRIVE-REVERSE
SHIFT RAIL
Removing the overdrive reverse shift rail, the 3–4 shift rail and the overdrive shift fork

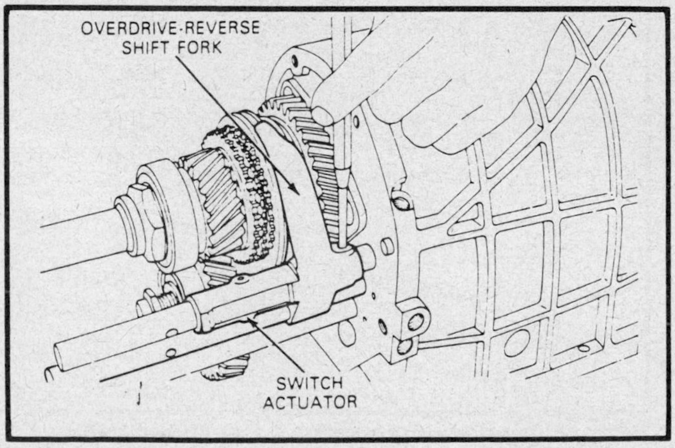
OVERDRIVE-REVERSE
SHIFT FORK
SWITCH
ACTUATOR
Removing the 1–2, 3–4 and overdrive shift forks

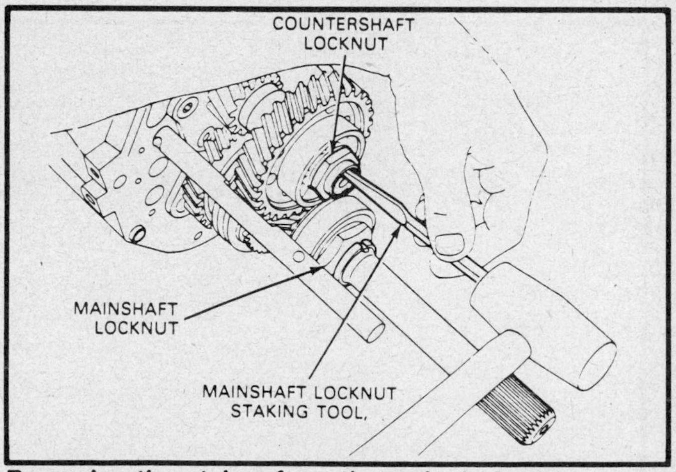
COUNTERSHAFT
LOCKNUT
MAINSHAFT
LOCKNUT
MAINSHAFT LOCKNUT
STAKING TOOL
Removing the stakes from the mainshaft and countershaft

When the 2 shift rails are removed, the interlock pins can be removed from the case.

NOTE: The 1/2 shift rail is unable to be removed at this time.

21. Unstake the locknuts on the mainshaft and countershaft using mainshaft locknut staking tool, T77J–7025–F or equivalent.
22. Double engage the transmission in 2 gears to lockup the transmission. This is done by engaging 2 of the synchronizers. This is necessary to remove the locknuts.
23. Remove the countershaft locknut with a 30mm socket. Discard the locknut.
24. Remove the mainshaft locknut with mainshaft locknut wrench, T77J–7025–C or equivalent. Discard the locknut.
25. Pull the rear bearing off the mainshaft using tube T77J–7025–B, forcing screw, T84T–7025–B, puller, T77J–7025–H, hand puller ring, T77J–7025–J or equivalents. Remove and discard the bearing.
26. Remove the spacer and lock ball from the mainshaft.
27. Remove the counter-overdrive gear and ball bearing from the countershaft by installing the jaws of puller, T77F–4220–B1 behind the gear. While removing the gear, remove the 1/2 shift rail from the case.
28. Remove the 1/2 and 3/4 shift forks from the case.
29. Remove the overdrive gear, needle bearing, spacer and synchronizer ring from the mainshaft.
30. Remove the overdrive synchronizer sleeve from the synchronizer hub on the mainshaft.

NOTE: Do not lose the 3 keys and 2 springs in the hub. A spring is located on each side of the hub.

31. Pull the overdrive synchronizer hub and overdrive gear bearing sleeve from the mainshaft using bearing puller, T77J–7025–H, puller ring, T77J–7025–J, tube, T75L–7025–B and forcing screw, T84T–7025–B or equivalents.
32. Slide the reverse gear and needle bearing assembly off the mainshaft.
33. Slide the counter/reverse gear and distance spacer off the countershaft.
34. Remove the cotter pin and nut from the reverse idler shaft. Remove the thrust washer, reverse idler gear and 2 sets of needle bearings from the shaft.
35. Remove the 6mm allen-head bolts that attach the mainshaft rear bearing retainer to the case and remove the retainer. Remove and discard the gasket. Clean any traces of gasket material from the mating surfaces of the case and retainer.
36. Remove the 6mm allen-head bolts that retain the reverse idler gearshaft assembly to the case.

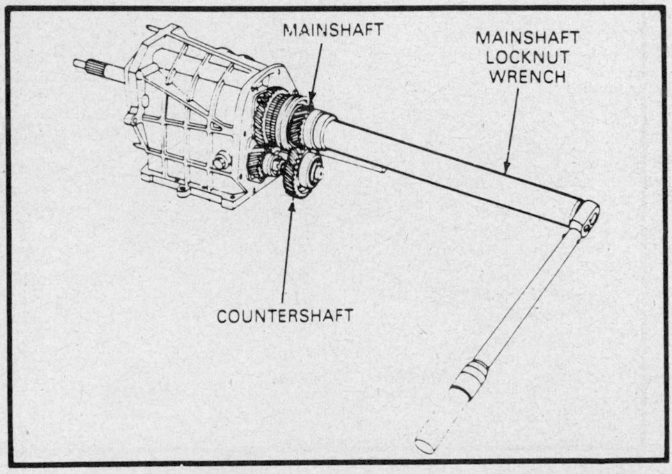
MAINSHAFT
MAINSHAFT LOCKNUT WRENCH
COUNTERSHAFT

Removing the locknut

37. Pull the reverse-idler gearshaft assembly out of the case using slide hammer, T50T–100–A and reverse idler gearshaft remover, T85T–7140–A or equivalents.
38. Use a double nut procedure to remove the studs that retain the input shaft front bearing retainer to the case. Remove the bolts that attach the retainer to the case.

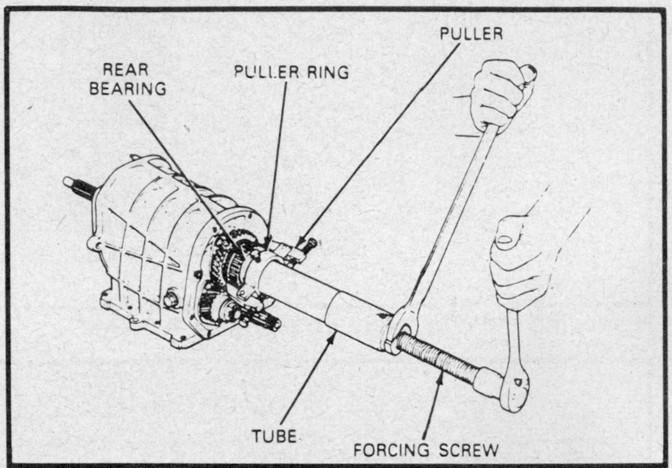
PULLER
REAR BEARING
PULLER RING
TUBE
FORCING SCREW

Removing the rear bearing from the mainshaft

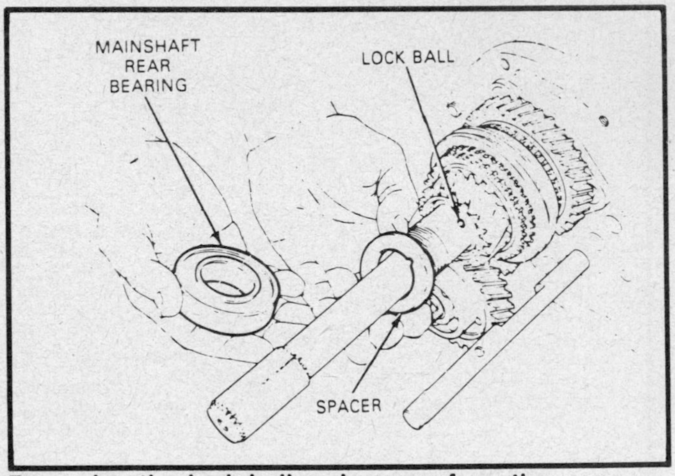
MAINSHAFT REAR BEARING
LOCK BALL
SPACER

Removing the lock ball and spacer from the mainshaft

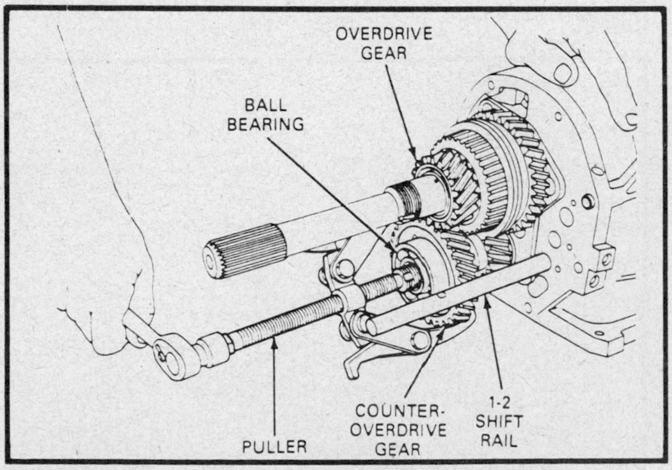
OVERDRIVE GEAR
BALL BEARING
PULLER
COUNTER-OVERDRIVE GEAR
1-2 SHIFT RAIL

Removing the counter-overdrive gear and ball bearing from the countershaft

39. Remove the input shaft front bearing retainer from the case. Remove and discard the gasket. Clean all traces of gasket material from the mating surfaces of the case and retainer.

40. Remove the selective shim from inside the input shaft front bearing retainer.

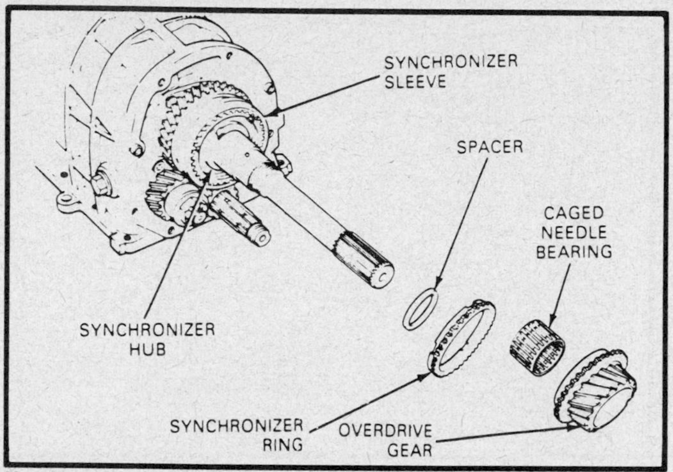
SYNCHRONIZER
SLEEVE
SPACER
CAGED
NEEDLE
BEARING
SYNCHRONIZER
HUB
SYNCHRONIZER
RING
OVERDRIVE
GEAR

Exploded view of the synchronizer sleeve and related components

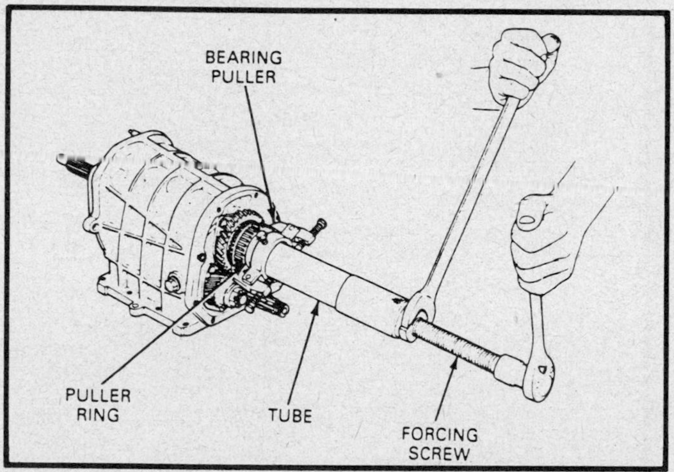
BEARING
PULLER
PULLER
RING
TUBE
FORCING
SCREW

Removing the synchronizer hub and over drive gear bearing sleeve

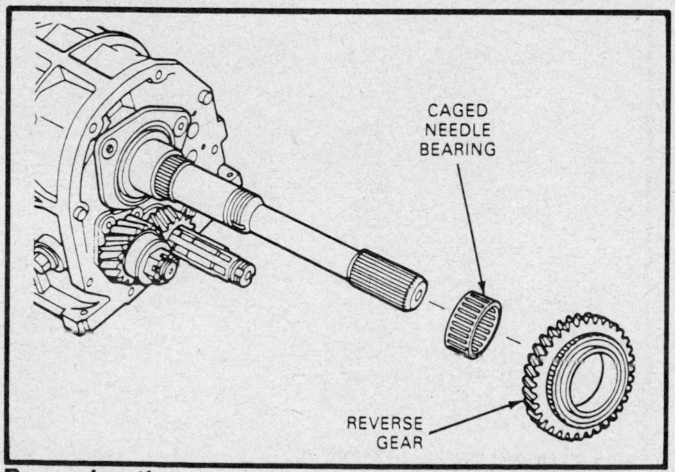
CAGED
NEEDLE
BEARING
REVERSE
GEAR

Removing the reverse gear and needle bearing assembly

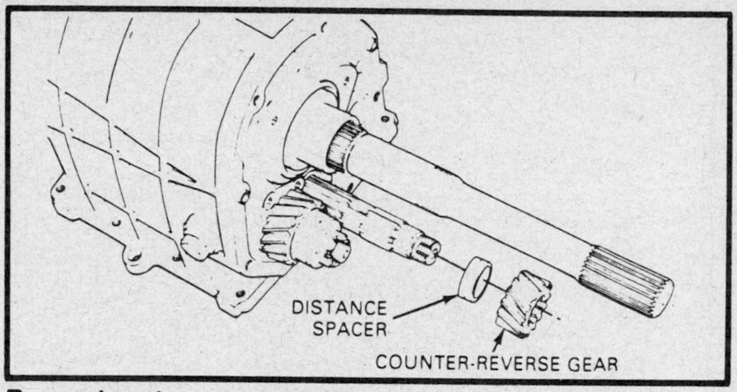
DISTANCE
SPACER
COUNTER-REVERSE GEAR

Removing the counter-reverse gear and distance spacer from the countershaft

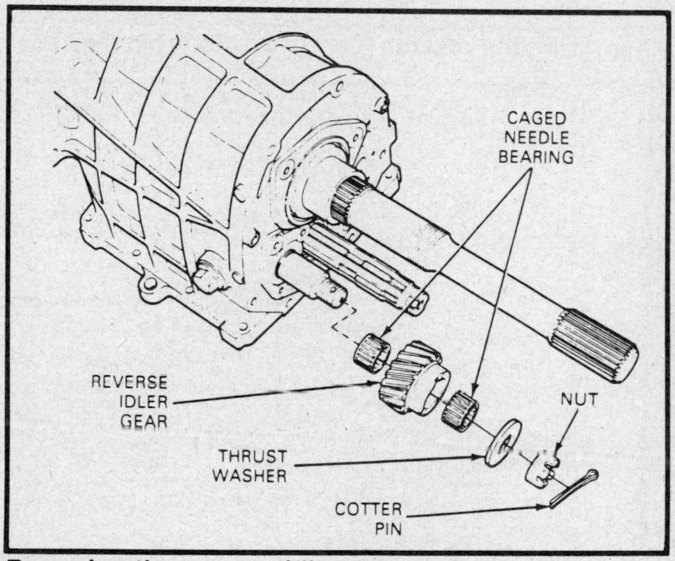
CAGED
NEEDLE
BEARING
REVERSE
IDLER
GEAR
NUT
THRUST
WASHER
COTTER
PIN

Removing the reverse idler gear and needle bearing from the shaft

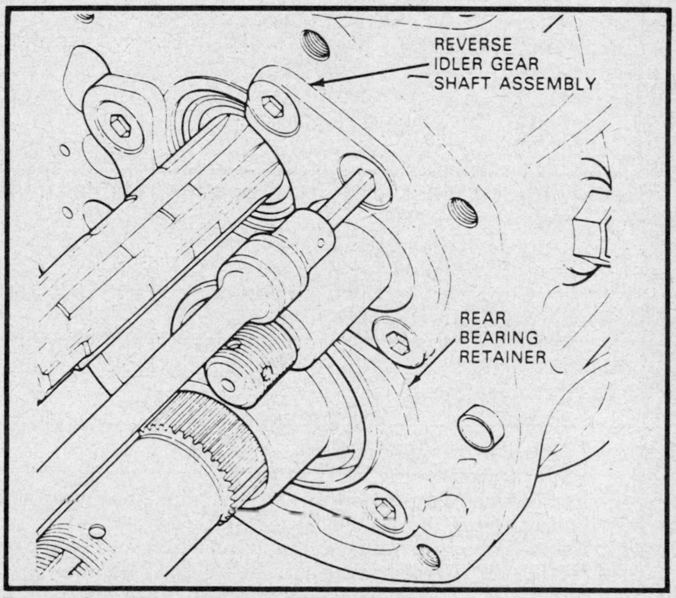
REVERSE
IDLER GEAR
SHAFT ASSEMBLY
REAR
BEARING
RETAINER

Removing the mainshaft bearing retainer and reverse idler gearshaft assembly

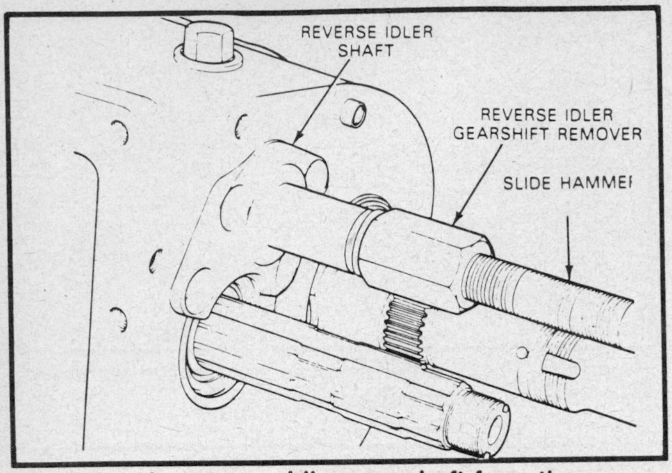
REVERSE IDLER SHAFT
REVERSE IDLER GEARSHIFT REMOVER
SLIDE HAMMER

Removing the reverse idler gearshaft from the case

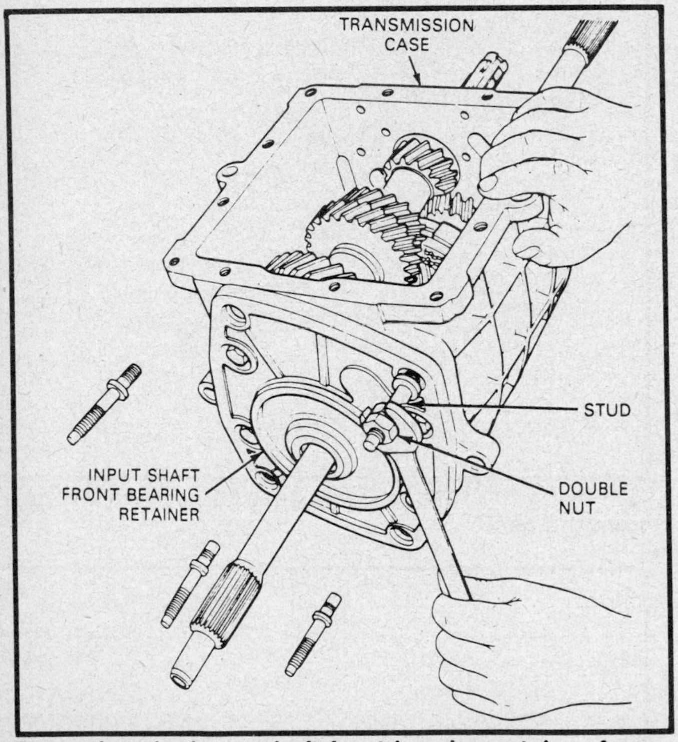
TRANSMISSION CASE
STUD
INPUT SHAFT FRONT BEARING RETAINER
DOUBLE NUT

Removing the input shaft front bearing retainer from the case

CAUTION

Do not discard the selective shim.

41. Remove the small selective snapring that retains the input shaft to the bearing.

CAUTION

Do not discard the selective snapring.

42. Remove the large snapring that retains the input shaft bearing to the case.
43. Remove the bearing from the input shaft and case using tube, T75L–7025–B, bearing collets, T75L–7025–D, bearing collet sleeve, T75L–7025–G and forcing screw, T84T–7025–B. Remove and discard the bearing.
44. Rotate the input shaft so the flats on the shaft face the

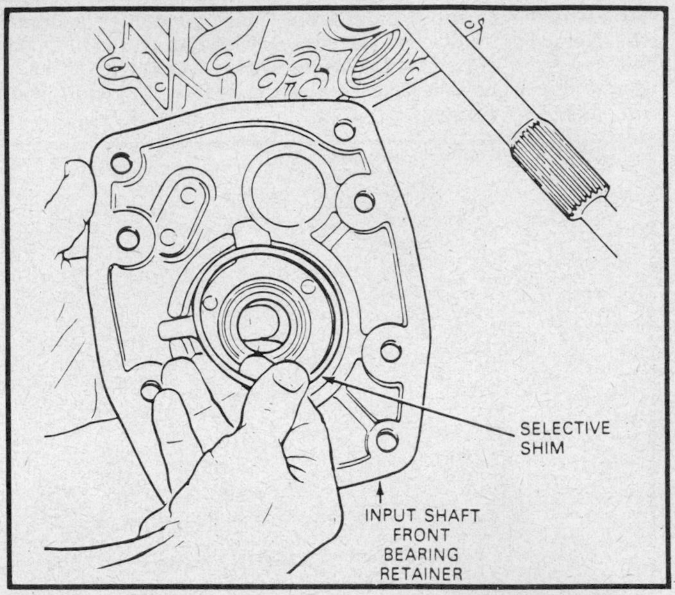
SELECTIVE SHIM
INPUT SHAFT FRONT BEARING RETAINER

Location of the input shaft selective shim—inside the front bearing retainer

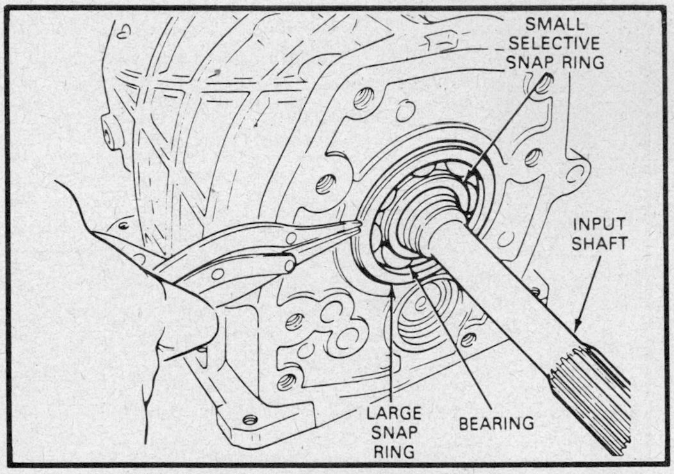
SMALL SELECTIVE SNAP RING
INPUT SHAFT
LARGE SNAP RING
BEARING

Removing the input shaft bearing selective snapring

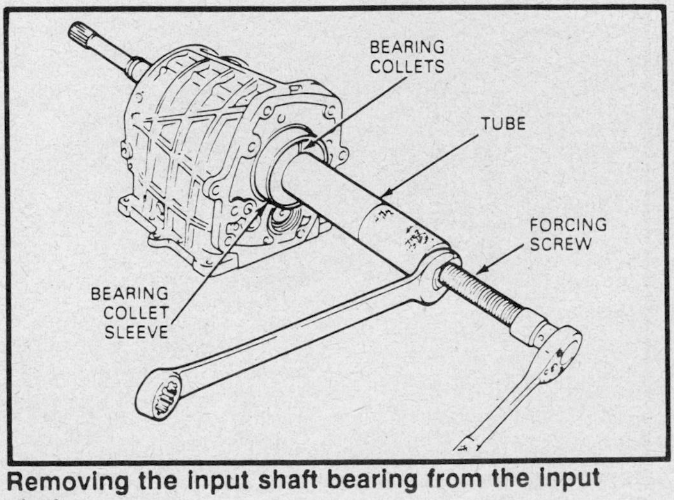
BEARING COLLETS
TUBE
FORCING SCREW
BEARING COLLET SLEEVE

Removing the input shaft bearing from the input shaft

countershaft, providing clearance to remove the input shaft. Remove the input shaft. The output shaft (mainshaft) may have to be pulled to the rear of the case. Remove the small caged needle bearing from the inside of the input gear.

45. Remove the snapring from the mainshaft (output shaft) outer bearing race.

46. Remove the outer mainshaft bearing race, ball bearing and bearing sleeve from the case using tube, T75L-7025-B, mainshaft bearing collet remover, T85T-7065-A, bearing collet sleeve, T77F-7025-C and forcing screw, T84T-7025-B. Discard the outer bearing race and ball bearing.

NOTE: The inner race of the front bearing will remain on the mainshaft.

47. Remove the countershaft front spacer and bearing race.

48. Remove the countershaft from the case. The mainshaft assembly may have to be moved slightly to the side to allow clearance for countershaft removal.

49. Remove the mainshaft assembly from the case.

Unit Disassembly And Assembly

MAINSHAFT

Disassembly

1. Remove the selective snapring that retains the 3/4 synchronizer assembly to the mainshaft. A new snapring will be used in the assembly.
2. Remove the 3/4 synchronizer assembly (hub, sleeve, spring and keys), synchronizer ring, 3rd speed gear and caged needle bearing from the front of the mainshaft. Note the position of the synchronizer hub and sleeve during disassembly.
3. Position the mainshaft assembly in a press so the 2nd speed gear is supported by the press bed. Press the mainshaft down and out from the 1/2 gear assembly.
4. Separate the inner ball bearing, bearing sleeve, 1st speed gear, caged needle bearing, 1/2 synchronizer assembly (hub, sleeve, rings and keys), 2nd speed gear and caged needle bearing. Note the direction of the 1/2 synchronizer hub and sleeve during disassembly. Discard the inner ball bearing.

Inspection

1. Carefully inspect all gears and synchronizers for any signs of wear, stress, discoloration, cracks or warpage.
2. Inspect the synchronizer sleeves for excessive wear or distortion.
3. Inspect all roller and needle bearings.
4. Replace defective parts as necessary.

Assembly

NOTE: Prior to assembly, lubricate all components with standard transmission lubricant, SAE 80W, D8DZ-19C547-A (ESP-M2C83-C) or equivalent.

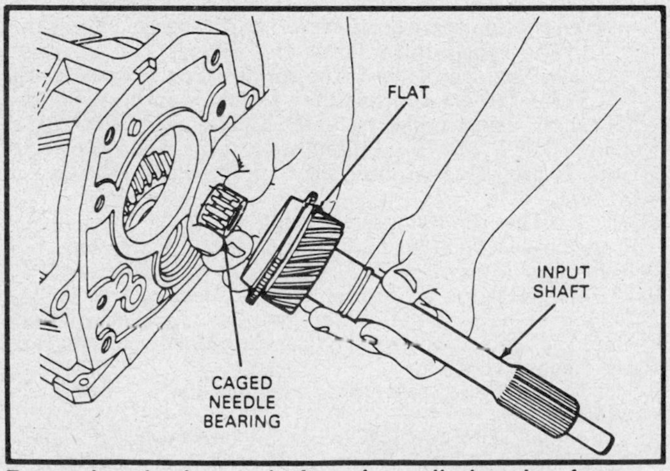
FLAT
INPUT SHAFT
CAGED NEEDLE BEARING

Removing the input shaft and needle bearing from the case

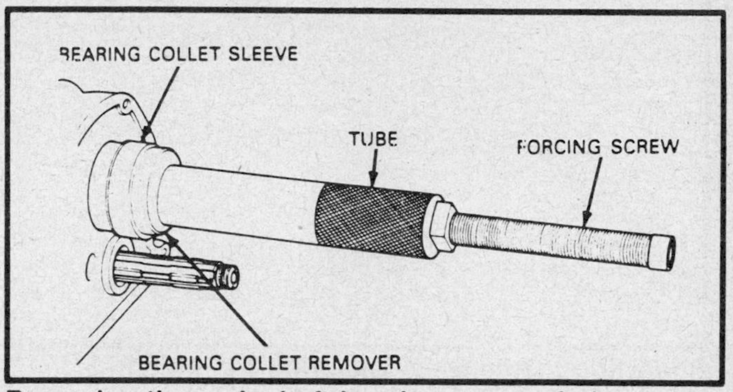
BEARING COLLET SLEEVE
TUBE
FORCING SCREW
BEARING COLLET REMOVER

Removing the mainshaft bearing race and sleeve from the case

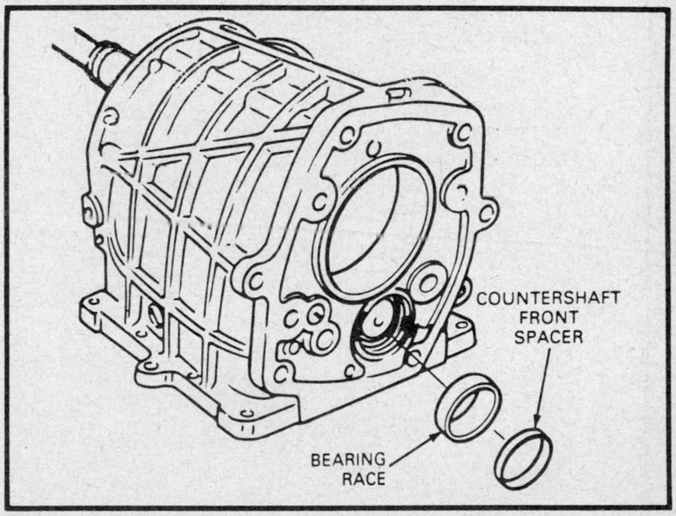
COUNTERSHAFT FRONT SPACER
BEARING RACE

Removing the countershaft front spacer and bearing race from the case

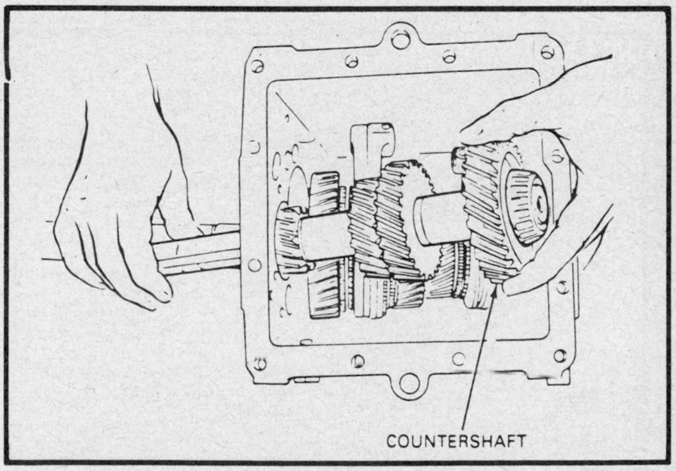
COUNTERSHAFT

Removing the countershaft from the case

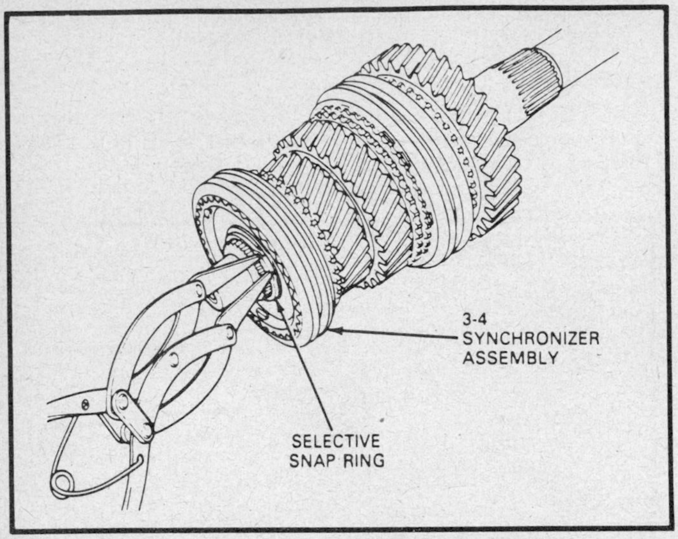
3-4
SYNCHRONIZER
ASSEMBLY
SELECTIVE
SNAP RING

Removing the selective snapring from the 3–4 synchronizer assembly

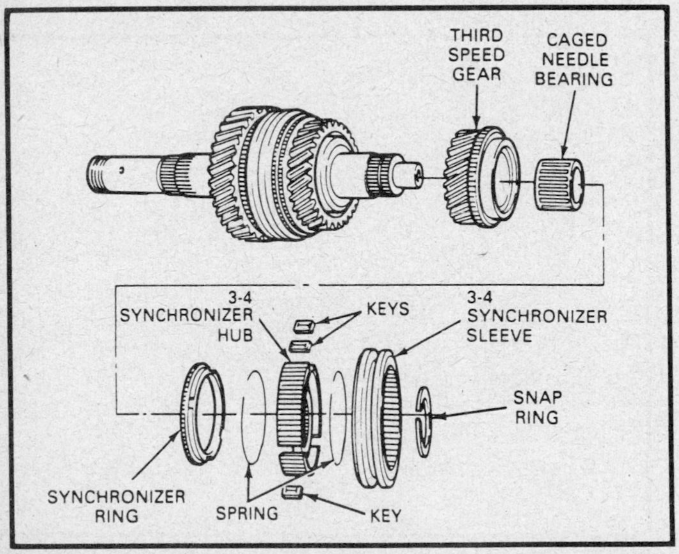
THIRD
SPEED
GEAR
CAGED
NEEDLE
BEARING
3-4
SYNCHRONIZER
HUB
KEYS
3-4
SYNCHRONIZER
SLEEVE
SNAP
RING
SYNCHRONIZER
RING
SPRING
KEY

Removing the 3–4 synchronizer assembly from the mainshaft

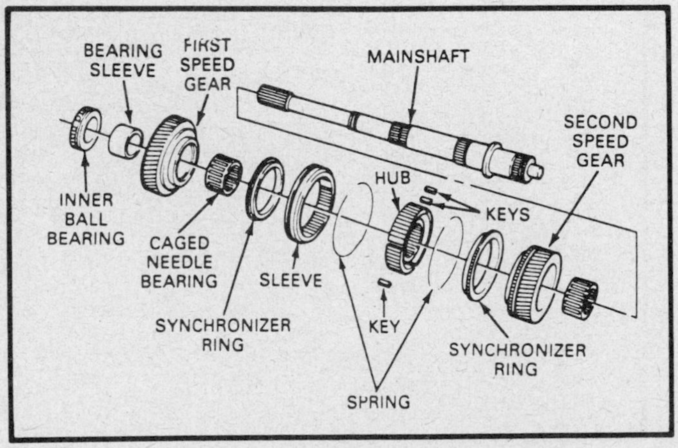
BEARING
SLEEVE
FIRST
SPEED
GEAR
MAINSHAFT
SECOND
SPEED
GEAR
HUB
KEYS
INNER
BALL
BEARING
CAGED
NEEDLE
BEARING
SLEEVE
SYNCHRONIZER
RING
KEY
SYNCHRONIZER
RING
SPRING

Exploded view of the mainshaft components

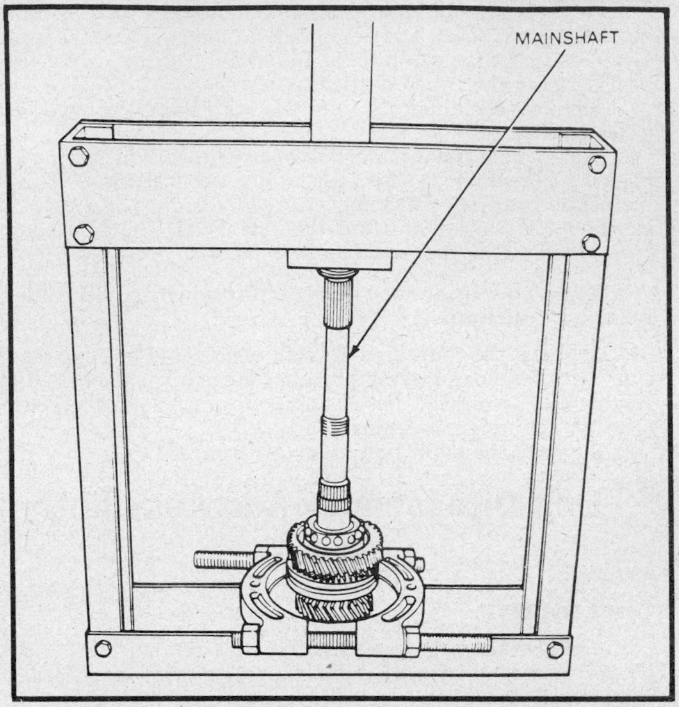
MAINSHAFT

Pressing the mainshaft from the 1-2 gear assembly

1. Check the clearance between the synchronizer rings and gears. Install the ring on the gear and insert a feeler gauge between the ring teeth and gear. If the clearance is less than 0.009 in. (0.23mm), replace the ring and/or gear.
2. From the rear of the mainshaft, install the caged needle bearing for the 2nd speed gear.
3. Position the 2nd speed gear on the mainshaft with the synchronizer ring surface facing the rear of the shaft.
4. Install the syncrhonizer ring on the 2nd speed gear.
5. Position the 1/2 synchronizer assembly on the rear of the mainshaft, making sure of the following:
   a. The splines of the mainshaft and synchronizer are properly aligned.
   b. The rear of the 1/2 hub is identified by a ridge machined on the rear surface. The ridge must face the front of the mainshaft.
   c. The synchronizer sleeve has a tooth missing at 6 positions. Assemble the hub to the sleeve so the single tooth between the 2 missing portions will touch the synchronizer key.
   d. The synchronizer keys and springs are properly installed. The open ends of the spring do not face each other.
6. Press the 1/2 synchronizer assembly in position on the mainshaft using replacing shaft sleeve tool, T75L–7025–K, shaft collar, T75L–7025–M and tube, T75L–7025–C or equivalents. If properly installed, the 2nd speed gear should rotate freely.
7. Position the 1st gear bearing sleeve on the mainshaft. Press the sleeve on the shaft using replacing shaft sleeve, T75L–7025–K, shaft collar, T75L–7025–M, rack bushing holder, T81P–3504–D (or an appropriate washer) and tube, T75L–7025–C or equivalents. When properly installed, the sleeve should be against the synchronizer hub. Make sure the gears rotate freely.
8. Install the synchronizer ring on the 1/2 synchronizer assembly.
9. Install the caged needle bearing and 1st speed gear.
10. Slide the inner ball in position on the mainshaft.
11. Press the inner ball bearing on the mainshaft using rack bushing holder, T81P–3504–D (or an appropriate size washer),

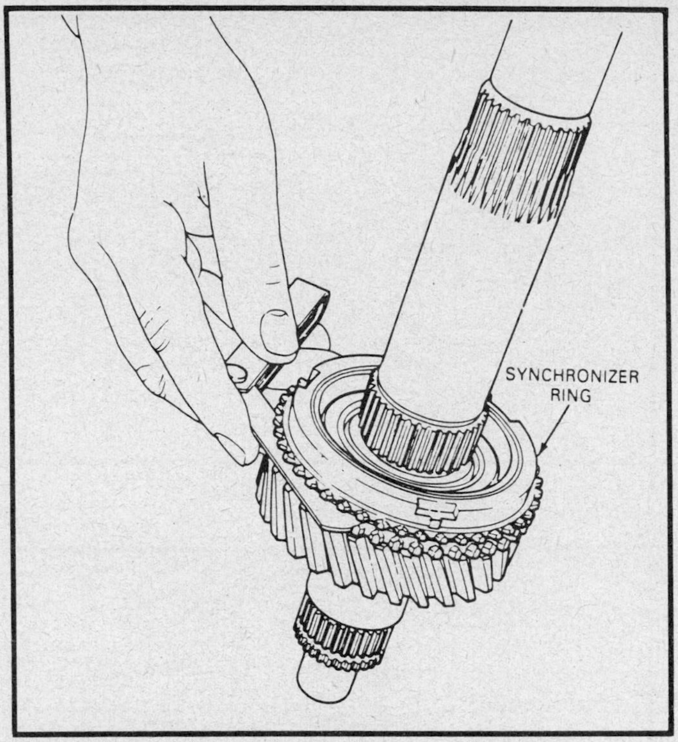
SYNCHRONIZER
RING
Checking synchronizer ring to gear clearance

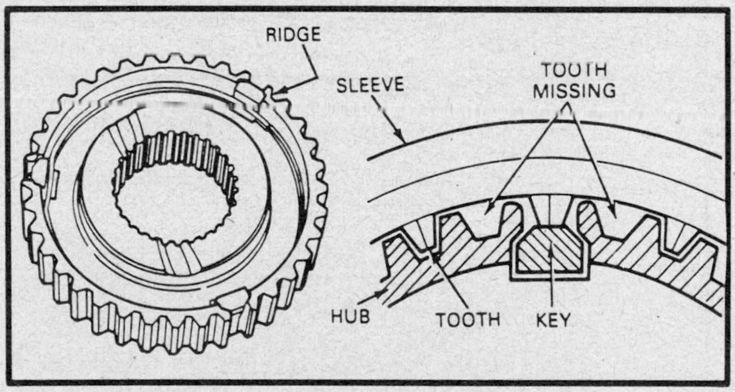
RIDGE
SLEEVE
TOOTH
MISSING
HUB
TOOTH
KEY
1–2 synchronizer hub and sleeve assembly positions

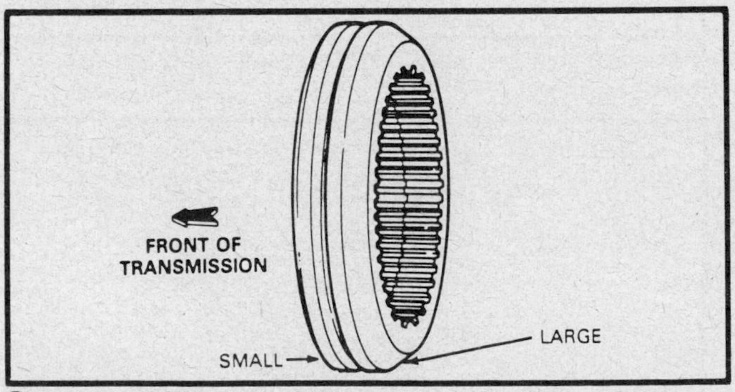
FRONT OF
TRANSMISSION
SMALL
LARGE
Correct positioning of the 1–2 synchronizer sleeve

tube, T85T–7025–A, replacing shaft sleeve, T75L–7025–K and shaft collar, T75L–7025–M or equivalents. When properly installed, the gears should rotate freely.

12. Install the 3rd speed gear and caged needle bearing over the front of the mainshaft.

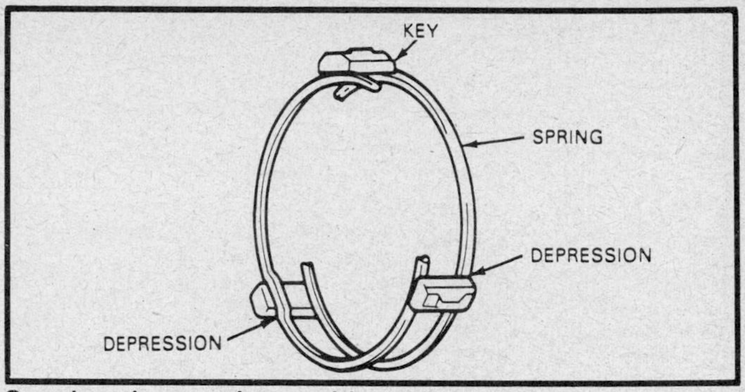
KEY
SPRING
DEPRESSION
DEPRESSION
Synchronizer spring and keys installation position

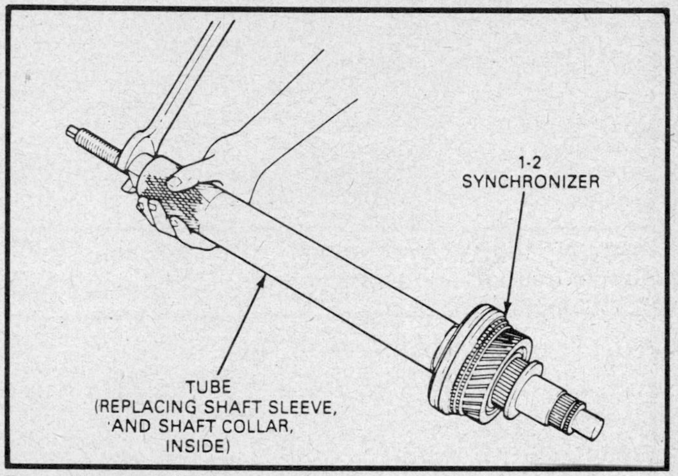
1-2
SYNCHRONIZER
TUBE
(REPLACING SHAFT SLEEVE,
AND SHAFT COLLAR,
INSIDE)
Pressing the 1–2 synchronizer assembly on the mainshaft

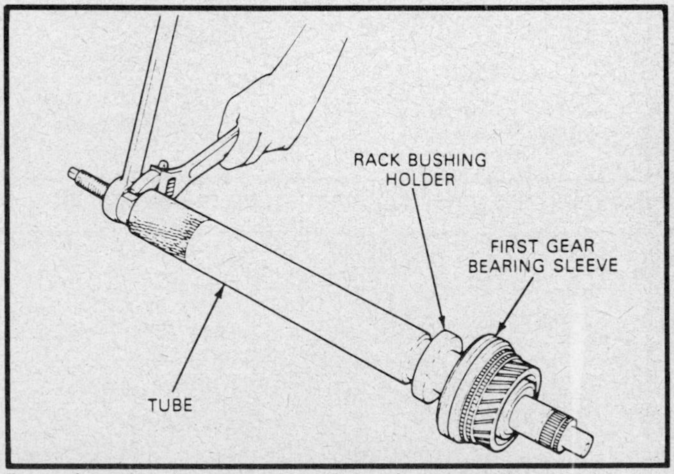
RACK BUSHING
HOLDER
FIRST GEAR
BEARING SLEEVE
TUBE
Pressing the 1st gear bearing sleeve on the mainshaft

13. Install the synchronizer ring against the 3rd speed gear.

14. Make sure the 3/4 synchronizer assembly is properly installed. Be sure of the following:

a. The splines of the mainshaft and synchronizer are properly aligned.

b. The small diameter boss of the hub faces the front of the mainshaft.

c. The small bevel angle of the sleeve faces the front of the mainshaft.

d. The synchronizer sleeve has a tooth missing at 6 posi-

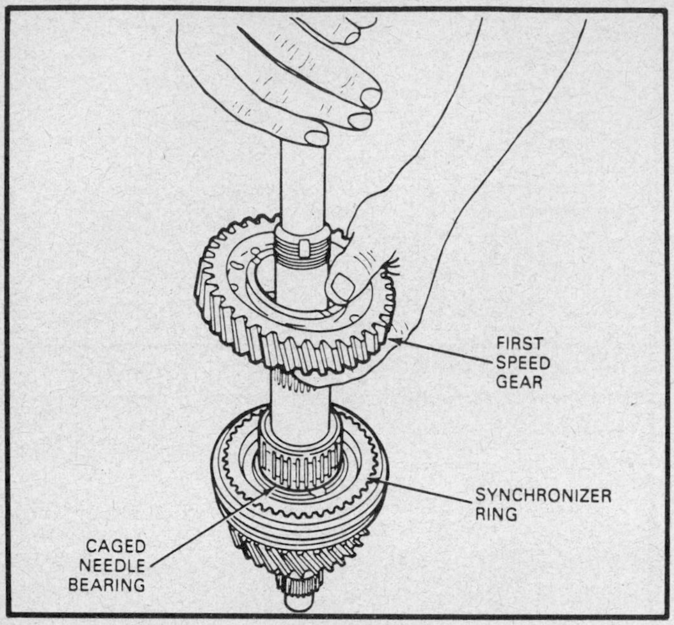
FIRST
SPEED
GEAR
SYNCHRONIZER
RING
CAGED
NEEDLE
BEARING

Installing the 1–2 synchronizer assembly and needle bearing on the mainshaft

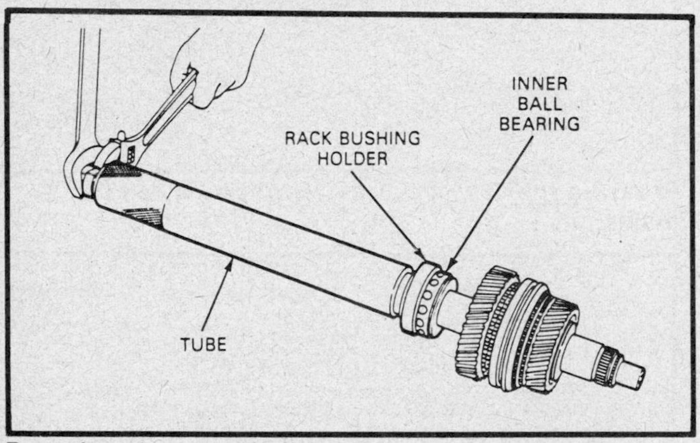
INNER
BALL
BEARING
RACK BUSHING
HOLDER
TUBE

Pressing the inner ball bearing on the mainshaft

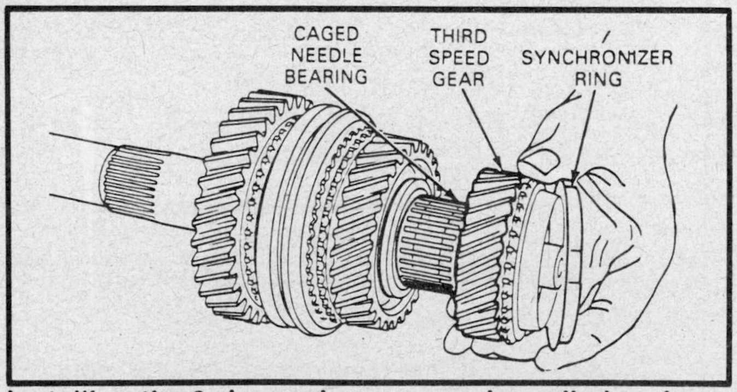
CAGED
NEEDLE
BEARING
THIRD
SPEED
GEAR
SYNCHRONIZER
RING

Installing the 3rd speed gear, caged needle bearing and ring on the mainshaft

tions. Assemble the hub to the sleeve so the single tooth between the 2 missing portions will touch the synchronizer key.

e. The synchronizer springs and keys are properly installed. The open ends of each spring do not face each other.

15. Install the 3/4 synchronizer assembly on the front of the mainshaft.

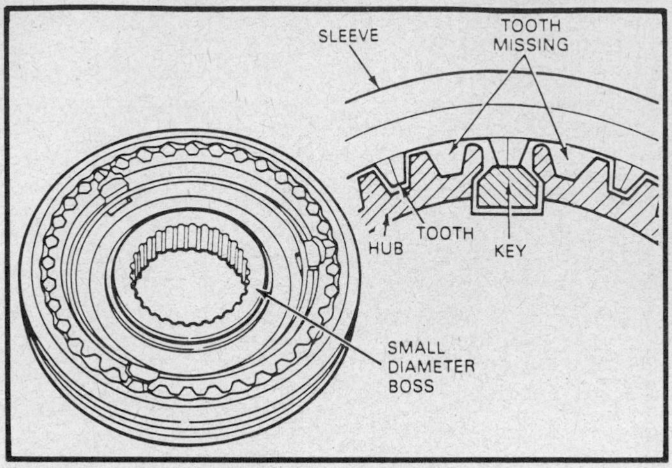
SLEEVE
TOOTH
MISSING
HUB
TOOTH
KEY
SMALL
DIAMETER
BOSS

Correct installation position for the 3–4 synchronizer hub and sleeve

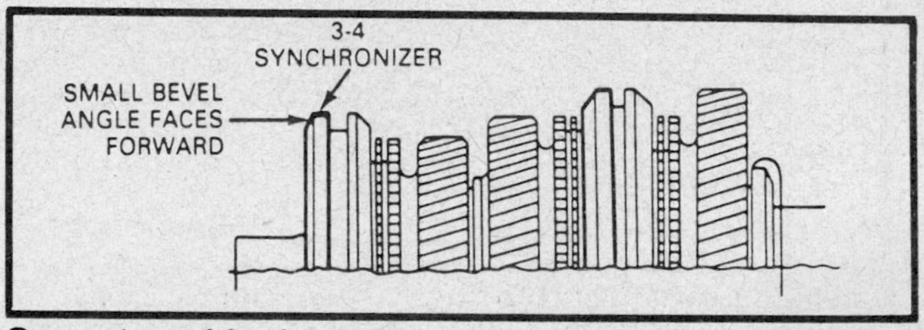
3-4
SYNCHRONIZER
SMALL BEVEL
ANGLE FACES
FORWARD

Correct positioning of the 3–4 synchronizer sleeve on the mainshaft

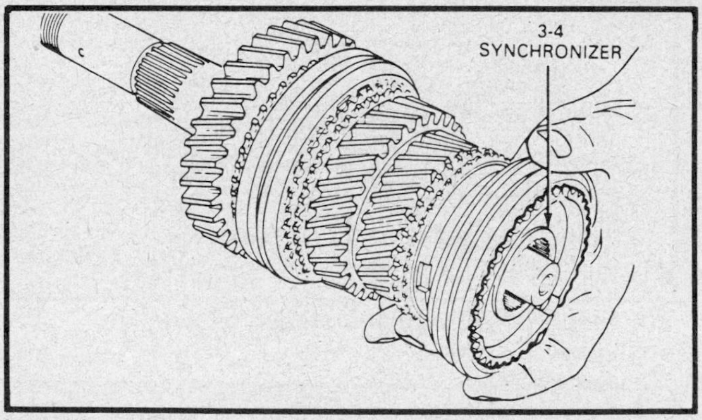
3-4
SYNCHRONIZER

Installing the 3–4 synchronizer assembly on the mainshaft

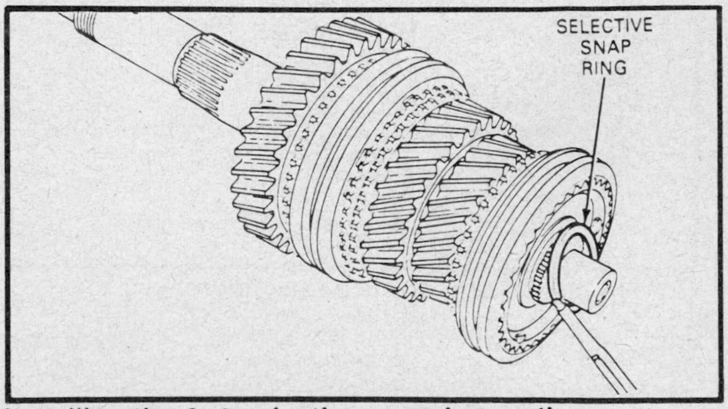
SELECTIVE
SNAP
RING

Installing the 3–4 selective snapring on the mainshaft

MAINSHAFT SELECTIVE SNAPRING

Snapring Thickness		Identification
in.	mm	Color
0.091	2.30	White
0.093	2.35	Brown
0.094	2.40	None
0.096	2.45	Blue
0.098	2.50	Yellow

16. Install a new selective snapring that retains the 3/4 synchronizer assembly to the mainshaft. Select the thickest snapring that fits in the groove.

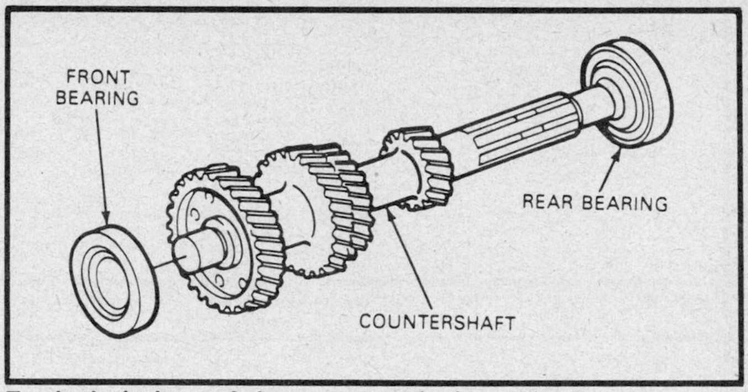
FRONT BEARING
REAR BEARING
COUNTERSHAFT
Exploded view of the countershaft assembly

SYNCHRONIZER

Disassembly

1. Make alignment mark on the hub and sleeve of the synchronizer.
2. Remove the synchronizer hub from each synchronizer sleeve.
3. Remove the inserts and insert springs from the hubs.
4. Do not mix the parts of the 1st and 2nd speed synchronizer with the 3rd and 4th speed synchronizer.

Inspection

1. Check synchronizer discs for wear or damage.
2. Inspect the synchronizer blocking rings for widened index slots, rounded clutch teeth and smooth internal surfaces.
3. With the blocking ring on the cone, the distance between the face of the gear clutching teeth and the face of the blocking ring must not be less than 0.010 in.
4. Check the synchronizer sleeves for free movement on the hubs.

Assembly

1. Position the hub in the sleeve, making sure that the alignment marks are properly aligned.
2. Place the 3 inserts into place on the hub.
3. Install the insert springs making sure that the irregular surface (hump) is seated in 1 of the inserts. Do not stagger the springs.

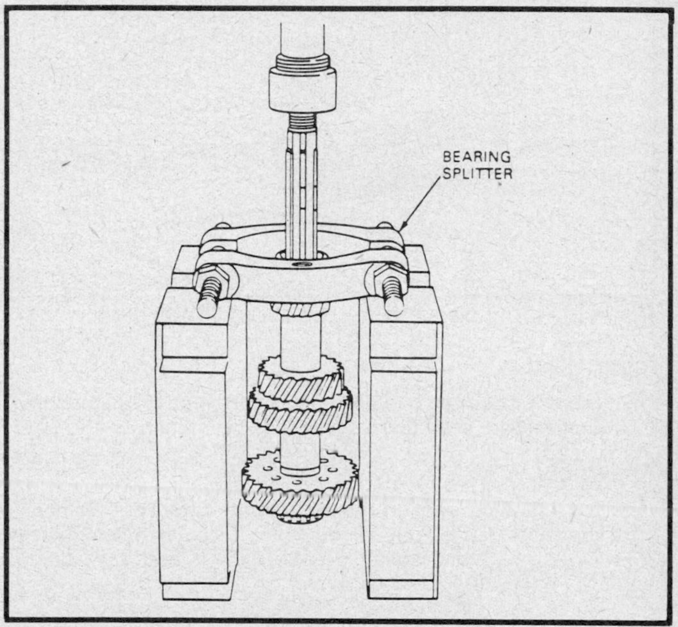
BEARING SPLITTER
Pressing the front and rear bearing off of the countershaft

COUNTERSHAFT

Disassembly

Press the front and rear bearing off the countershaft, using bearing splitter, D84L–1123–A or equivalent. Remove and discard the bearings.

Inspection

1. Inspect the roller bearing for wear or damage.
2. Check all gears for chipped, broken or worn teeth.
3. Inspect the countershaft for wear. Replace defective parts as needed.

Assembly

Press the new bearings on the countershaft using a press and countershaft bearing replacer tool, T85T–7121–A or equivalent.

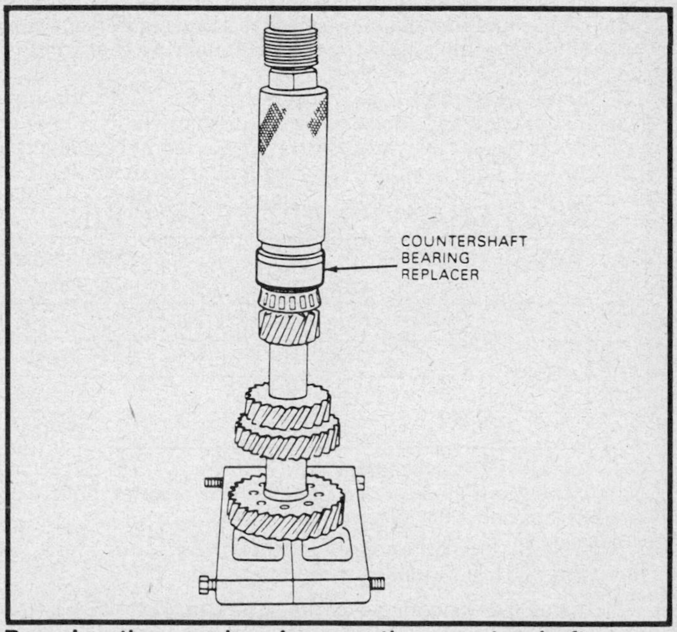
COUNTERSHAFT BEARING REPLACER
Pressing the new bearings on the countershaft

INPUT SHAFT

Disassembly

Position bearing splitter, D84L–1123–A or equivalent, behind the bearing and press the input shaft out of the bearing. Discard the bearing.

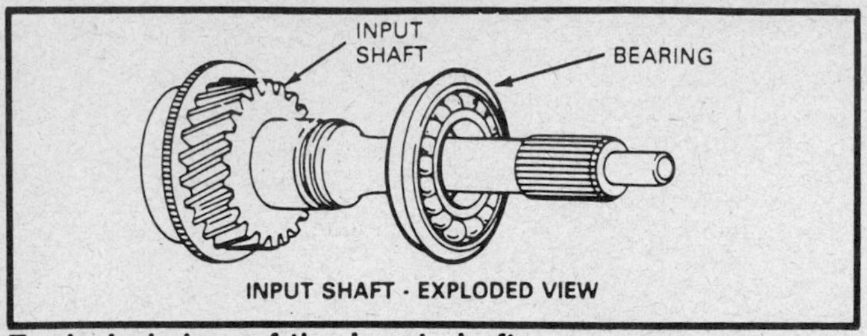
INPUT SHAFT
BEARING
INPUT SHAFT - EXPLODED VIEW
Exploded view of the input shaft

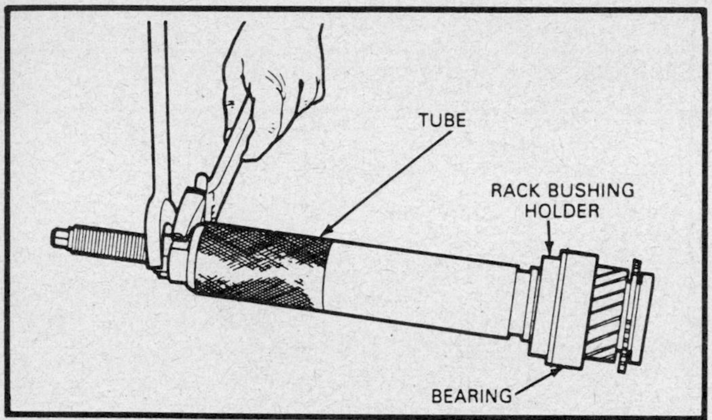
TUBE
RACK BUSHING HOLDER
BEARING
Pressing new bearing on the input shaft

Inspection

1. Inspect the input shaft bearing for wear or distortion.
2. Inspect the gear teeth for cracks or chips.

Assembly

Position a new bearing on the input shaft and press the bearing onto the shaft using tube, T75L-7025-B, replacing shaft sleeve, T75L-7025-K, shaft collar, T75L-7025-M and rack bushing holder, T81P-3504-D or equivalents.

Transmission Assembly

1. Install the mainshaft assembly in the case.
2. Choose and install a new selective snapring in front of the input shaft bearing. Select the thickest snapring that will fit in the groove.
3. Install the small caged needle bearing inside the input gear. Install the synchronizer ring on the input shaft. Check the clearance between the ring and gear. If the clearance is less than 0.009 in. (0.23mm), replace the ring and/or input shaft.

INPUT SHAFT SELECTIVE SNAPRING

Snapring Thickness		Identification
in.	mm	Color
0.085	2.15	Blue
0.087	2.22	None
0.090	2.29	Brown
0.093	2.36	White

4. Install the synchronizer ring and input shaft in the case. Rotate the input shaft so the flats face the countershaft to provide installation clearance.

NOTE: It may be necessary to tap the input shaft into position with a brass hammer.

5. Install a new snapring on the outer bearing race. The longer portion of the race must be installed in the case.

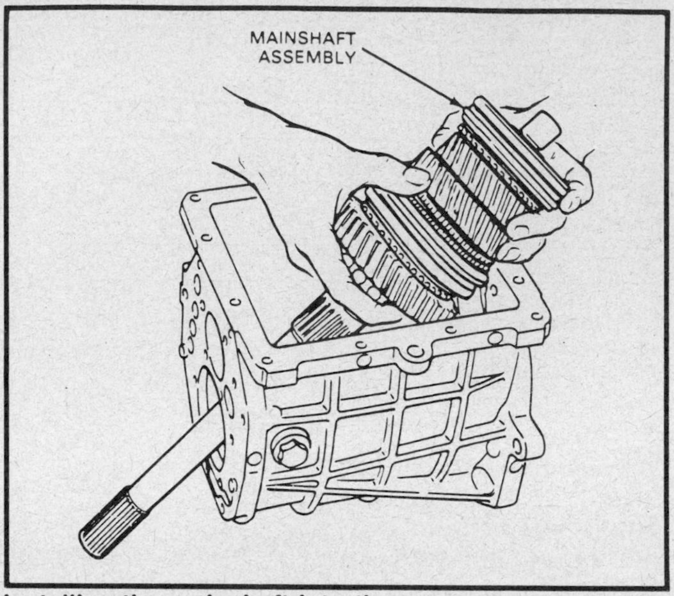
MAINSHAFT ASSEMBLY
Installing the mainshaft into the case

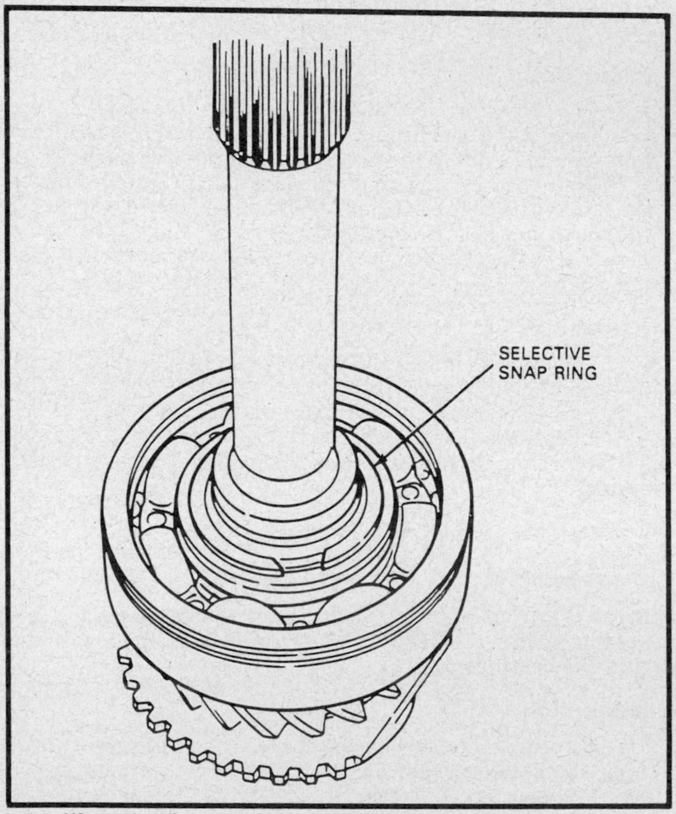
SELECTIVE SNAP RING
Installing the input shaft selective snapring

6. Slide the outer ball bearing on the mainshaft. Press the bearing on the mainshaft and in the race using tube, T85T-7025-A, replacing shaft sleeve, T75L-7-25-K and shaft collar, T75L-7025-M or equivalent. When pressed in position, all gears and shafts must rotate freely.
7. Install the 3/4 shift fork into its synchronizer sleeve. The roll pin boss on the fork must face to the rear.
8. Install the countershaft into the case. It may be necessary to move the mainshaft to one side in order for the countershaft to be easily inserted.

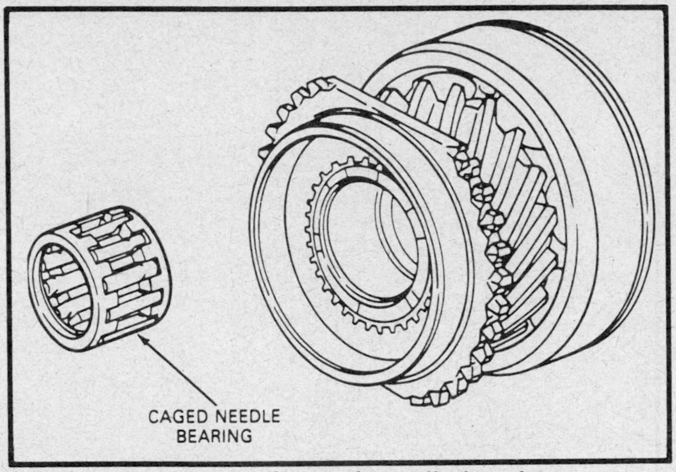
CAGED NEEDLE
BEARING

Synchronizer ring and caged needle bearing assembly

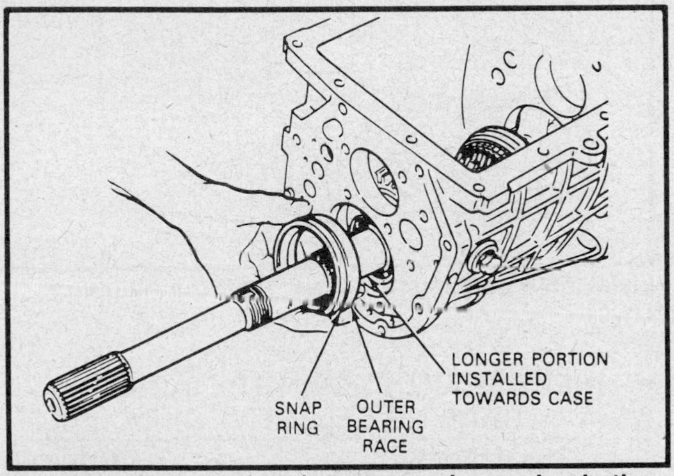
LONGER PORTION
INSTALLED
TOWARDS CASE
SNAP
RING
OUTER
BEARING
RACE

Installing the outer bearing race and snapring in the case

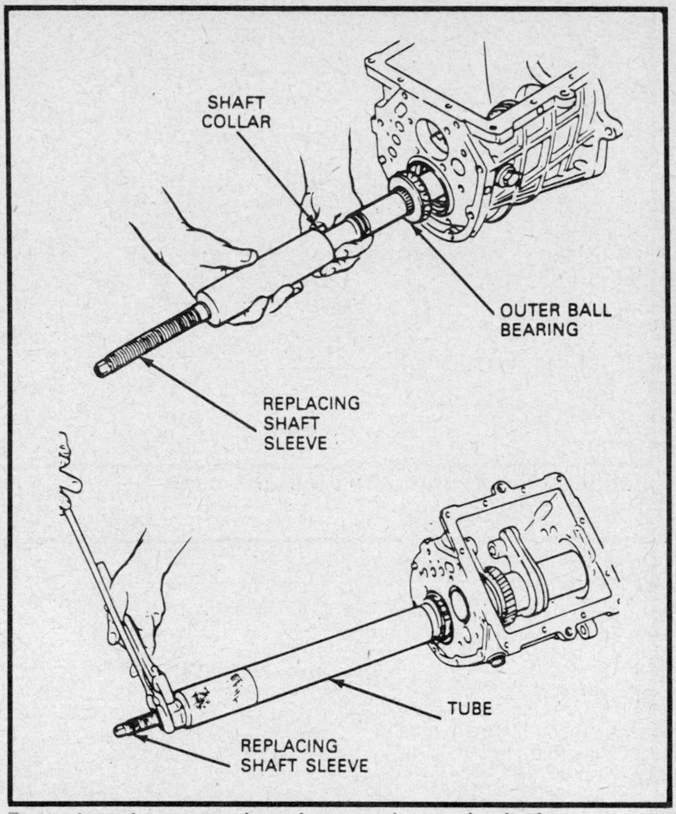
SHAFT
COLLAR
OUTER BALL
BEARING
REPLACING
SHAFT
SLEEVE
TUBE
REPLACING
SHAFT SLEEVE

Pressing the outer bearing on the mainshaft

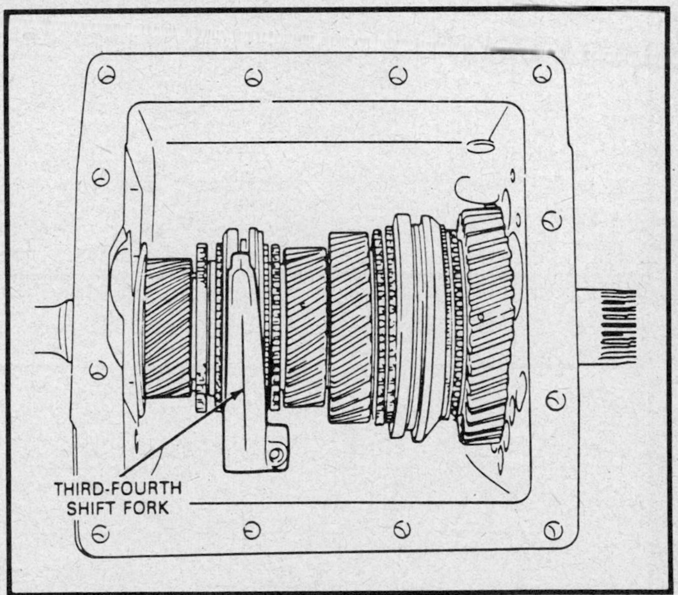
THIRD-FOURTH
SHIFT FORK

Correct installation position of the 3–4 shift fork

9. Install the 1/2 shift fork. The roll pin boss must face toward the 3/4 shift fork.
10. If removed, drive a new seal into the input shaft front bearing retainer using seal installer, T85T–7011–A and driver handle, T80T–4000–W or equivalents.
11. Install the large snapring that retains the input shaft bearing to the case.
12. Check the input shaft front bearing retainer-to-bearing clearance as follows:
   a. With the retainer selective shim removed, use a depth micrometer to measure the distance between the top machined surface to the spacer surface (second landing). Record the reading.
   b. Bottom the input shaft bearing so the snapring is flush against the case.
   c. Using a depth micrometer, measure the distance from the top of the outer front bearing race to the machined surface of the case.
   d. Subtract the distance of the bearing-to-case from the retainer dimensions. This will give the required maximum selective shim size to obtain a 0.000–0.004 in. (0.00–0.10mm) clearance.
   e. Measure and install the appropriate size selective shim in the front bearing retainer.
13. Install the countershaft front outer bearing race and non-selective spacer. Install the countershaft rear outer bearing race.
14. Install a new gasket between the front bearing retainer and case. Position the retainer on the case (with selective shim installed). Install the bolts and studs and tighten to 22–30 ft. lbs. (30–41 Nm).
15. Check and adjust the countershaft endplay as follows:
   a. Place the transmission so the rear of the mainshaft and

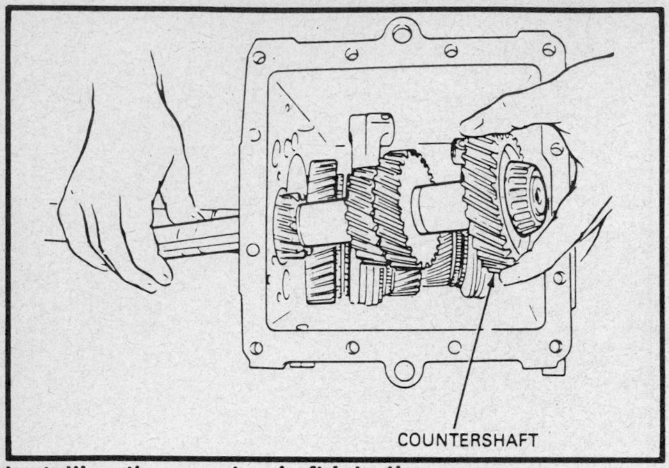
COUNTERSHAFT
Installing the countershaft into the case

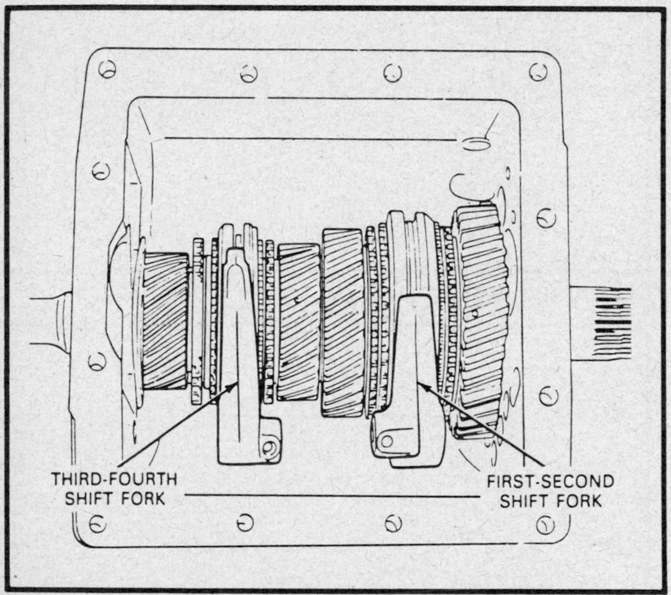
THIRD-FOURTH SHIFT FORK
FIRST-SECOND SHIFT FORK
Correct installation position of the 1–2 shift fork

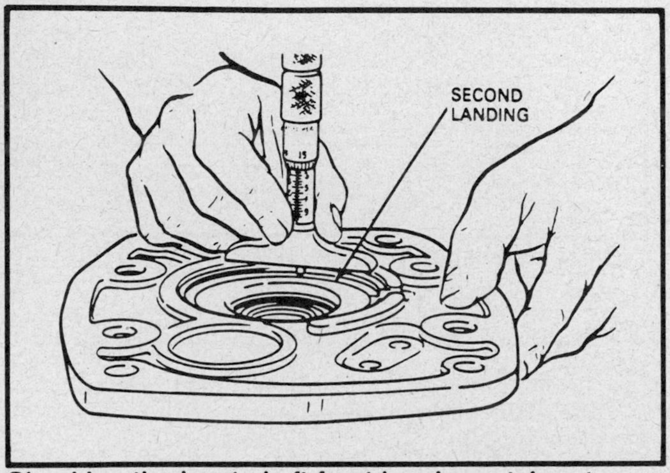
SECOND LANDING
Checking the input shaft front bearing retainer-to-bearing clearance

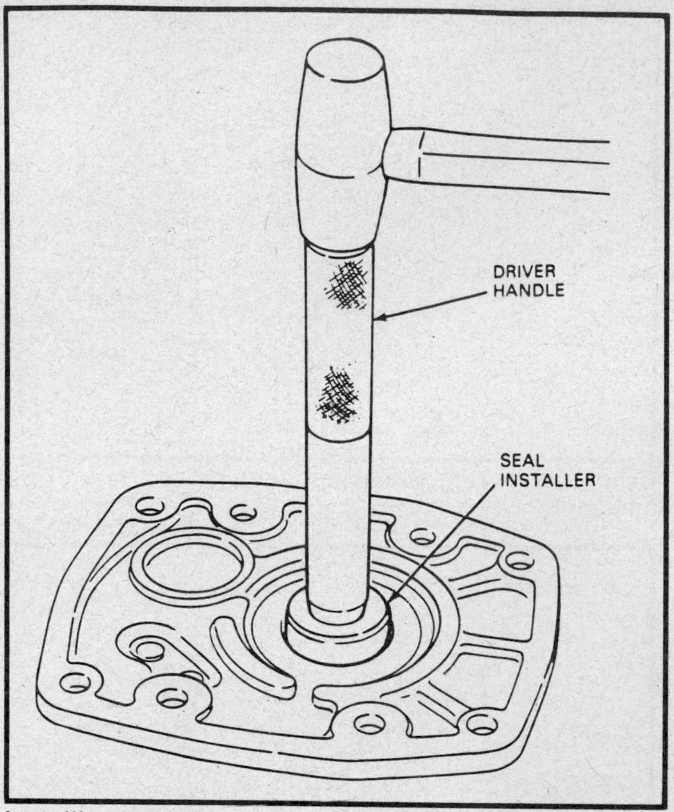
DRIVER HANDLE
SEAL INSTALLER
Installing a seal into the input shaft front bearing cover

INPUT SHAFT BEARING RETAINER-TO-BEARING SELECTIVE SHIM

Shim Thickness in.	mm	Identification Color
0.033	0.84	Black
0.037	0.93	None
0.040	1.02	Red
0.044	1.11	White
0.047	1.20	Yellow
0.051	1.29	Blue
0.054	1.38	Green

countershaft face upward. Install the countershaft rear selective spacer.

b. Force the countershaft downward so it bottoms against the front bearing retainer.

c. Place a straight edge across the rear countershaft selective spacer in the case.

d. Try to turn the spacer. If the spacer turns lightly, replace the spacer with the next larger size. Install a spacer so the clearance between the spacer and straight edge is 0.000–0.002 in. (0.00–0.05mm). Install the correct size spacer over the countershaft rear bearing cup.

16. Install the rear bearing retainer on the case with the 6mm bolts. Tighten to 11–16 ft. lbs. (15–21 Nm).

NOTE: Be sure the spacer installed in the previous Step does not fall out of place when installing the rear bearing retainer.

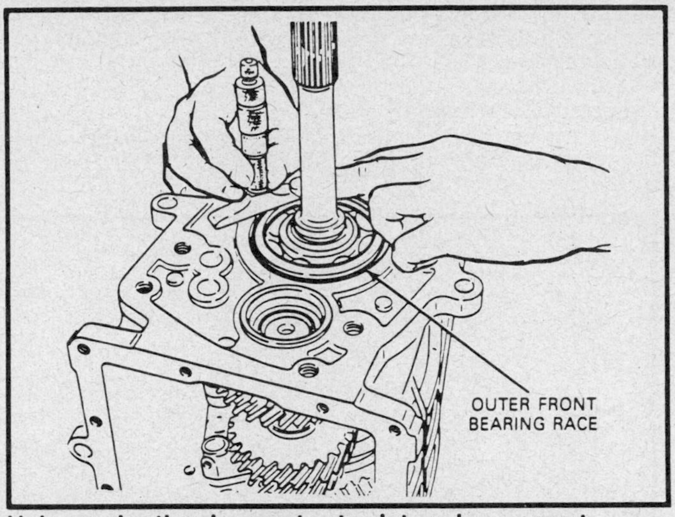
OUTER FRONT
BEARING RACE

Using a depth micrometer to determine correct selective shim, for bearing retainer

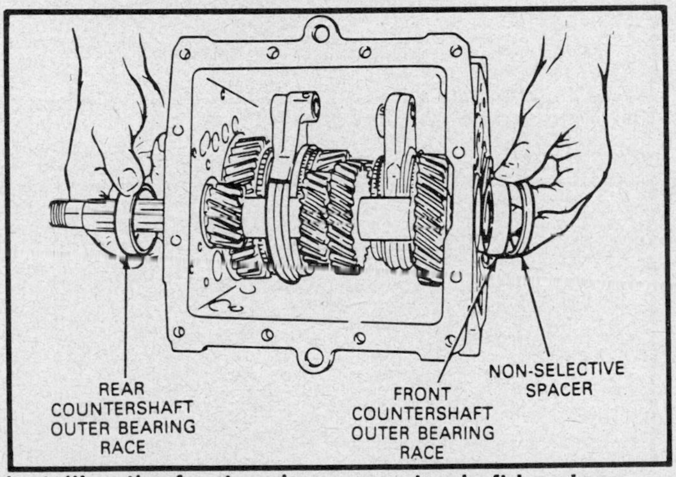
REAR
COUNTERSHAFT
OUTER BEARING
RACE
FRONT
COUNTERSHAFT
OUTER BEARING
RACE
NON-SELECTIVE
SPACER

Installing the front and rear countershaft bearings

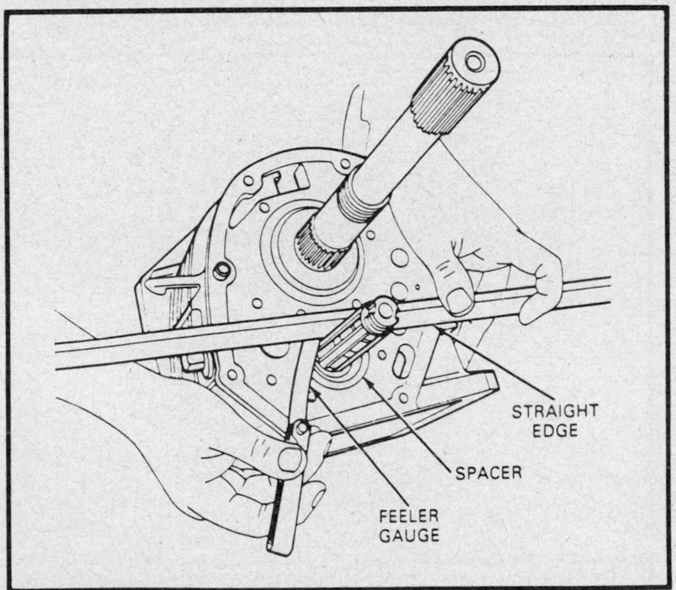
STRAIGHT
EDGE
SPACER
FEELER
GAUGE

Measuring countershaft endplay

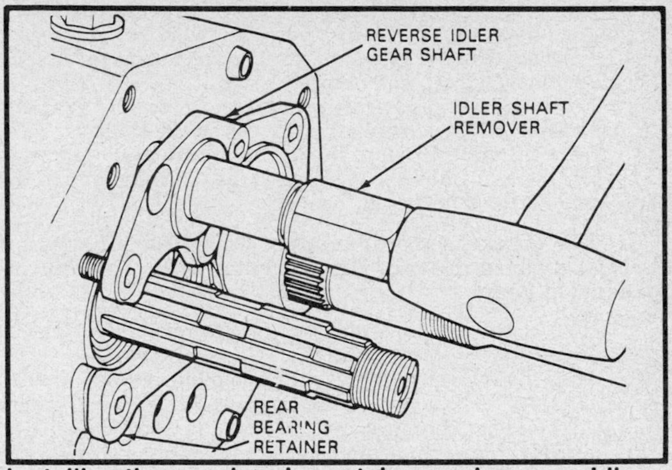
REVERSE IDLER
GEAR SHAFT
IDLER SHAFT
REMOVER
REAR
BEARING
RETAINER

Installing the rear bearing retainer and reverse idler shaft

COUNTERSHAFT ENDPLAY SELECTIVE SPACER

Spacer Thickness		Identification
in.	mm	Mark
0.0724	1.84	84
0.0736	1.87	87
0.0748	1.90	90
0.0760	1.93	93
0.0772	1.96	96
0.0783	1.00	99
0.0795	2.02	02
0.0807	2.05	06
0.0819	2.08	08
0.0831	2.11	11
0.0843	2.14	14
0.0854	2.17	17
0.0866	2.20	20
0.0878	2.23	23
0.0890	2.26	26
0.0902	2.29	29
0.0913	2.32	32
0.0925	2.35	35
0.0937	2.38	38
0.0949	2.41	41
0.0961	2.44	44
0.0972	2.47	47
0.0984	2.50	50
0.0996	2.53	53
0.1008	2.56	56
0.1020	2.59	59
0.1031	2.62	62
0.1043	2.65	65
0.1055	2.68	68

17. Position the reverse idler gear shaft on the case. Install the 6mm allen-head bolts to act as a pilot. Install reverse idler gear shaft remover, T85T–7140–A on the shaft and drive the assembly into place. Tighten the bolts to 11–16 ft. lbs. (15–21 Nm).

18. Install the 2 caged needle bearings, reverse idler gear and thrust washer on the idler shaft. The boss on the idler gear faces away from the transmission. Install the locknut and tighten to 15–42 ft. lbs. (20–58 Nm). If required, advance the nut to the next castellation and install the cotter pin.

NOTE: If required, cut 1 end of the cotter pin when it is bent over to prevent interference with the counter-overdrive gear.

19. Install the spacers and counter/reverse gear on the countershaft.

20. Press the reverse gear sleeve on the mainshaft using tube, T85T–7025–A, shaft sleeve replacer, T75L–7025–K, shaft collar T75L–7025–M and forcing screw, T84T–7025–B or equivalents.

21. Install the caged needle bearing and reverse gear on the mainshaft.

22. Assemble the overdrive synchronizer hub and sleeve as follows:

a. Install the hub in the sleeve. The recessed boss on the sleeve must face the front of the transmission. The large boss on the hub must face the front of the transmission.

b. When installing hub in the sleeve and the 3 keys, make sure that the single tooth between the 2 spaces will touch the key. Install the springs so the open ends do not face each other.

23. Install the overdrive synchronizer on the mainshaft. The recessed boss of the sleeve must face the front of the transmission.

24. Press the sleeve of the overdrive gear on the mainshaft using tube, T75L–7025–B, shaft sleeve replacer, T75L–7025–K, shaft collar, T75L–7025–M and overdrive gear bearing replacer, T85T–7061–A or equivalents.

25. Install the ring on the overdrive synchronizer.

26. Slide the small spacer, caged needle bearing and overdrive gear on the mainshaft. Check the clearance between the overdrive gear and synchronizer ring. If the clearance is less than 0.009 in. (0.23mm), replace the ring and/or overdrive gear.

27. Install the counter-overdrive gear and ball bearing onto the countershaft along with the 1/2 shift rail. Seat the bearing into position using countershaft bearing replacer collect, T85T–7121–A, rear countershaft bearing installer adapter, T85T–7111–A and remover and replacer tube, T77J–7025–B or equivalents. Make sure the rail engages the forks.

28. Install the lock ball and spacer on the mainshaft.

29. Place the rear bearing over the mainshaft and press the bearing in position using rack bushing holder, T81P–3504–D (or an appropriate size washer), tube T75L–7025–B, shaft sleeve replacer, T75L–7025–K and shaft collar, T75L–7025–M or equivalents.

30. Install new locknuts on the countershaft and mainshaft. Double engage the transmission in 2 gears to prevent the shafts from turning. Tighten the mainshaft locknut to 180–195 ft. lbs.

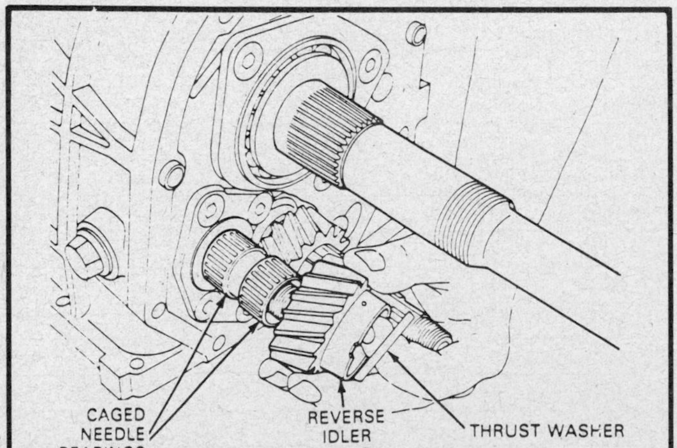
CAGED
NEEDLE
BEARINGS
REVERSE
IDLER
GEAR
THRUST WASHER
Installing the reverse idler gear, caged needle bearing and thrust washer

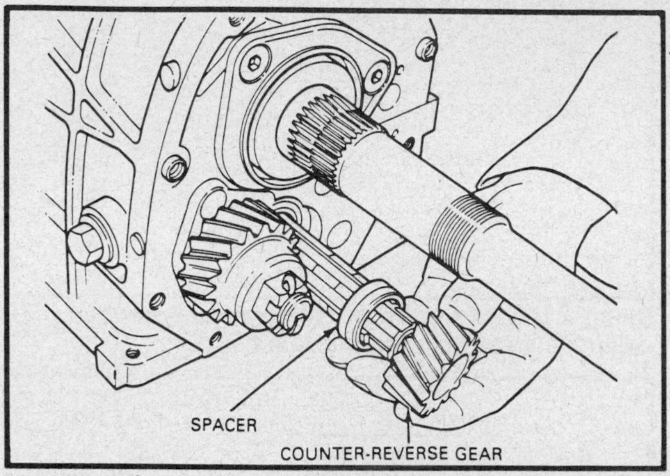
SPACER
COUNTER-REVERSE GEAR
Installing the counter-reverse gear on the countershaft

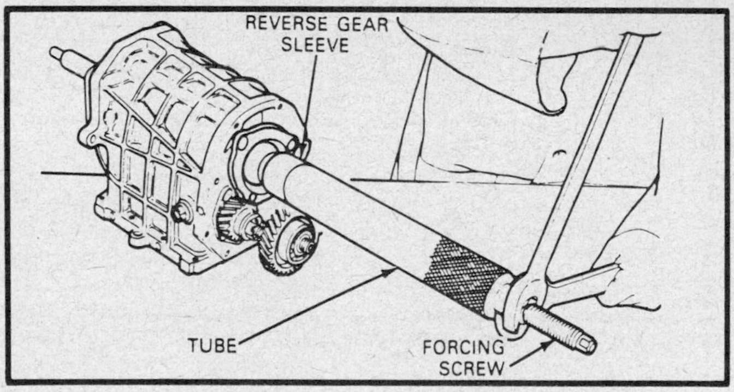
REVERSE GEAR
SLEEVE
TUBE
FORCING
SCREW
Pressing the reverse gear sleeve on the mainshaft

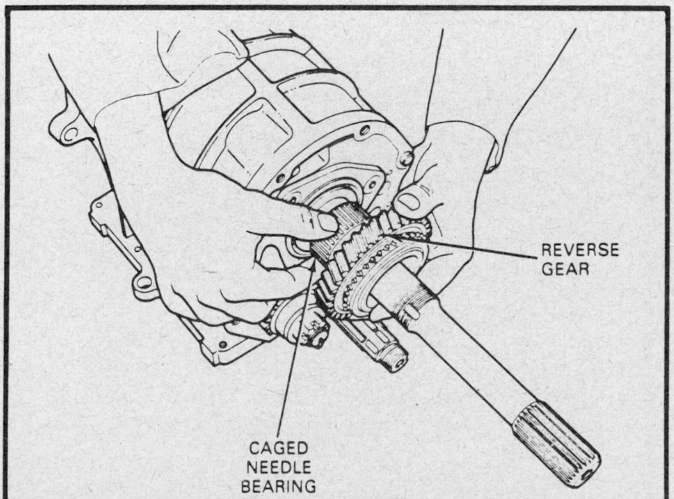
REVERSE
GEAR
CAGED
NEEDLE
BEARING
Installing the caged needle bearing and reverse gear on the shaft

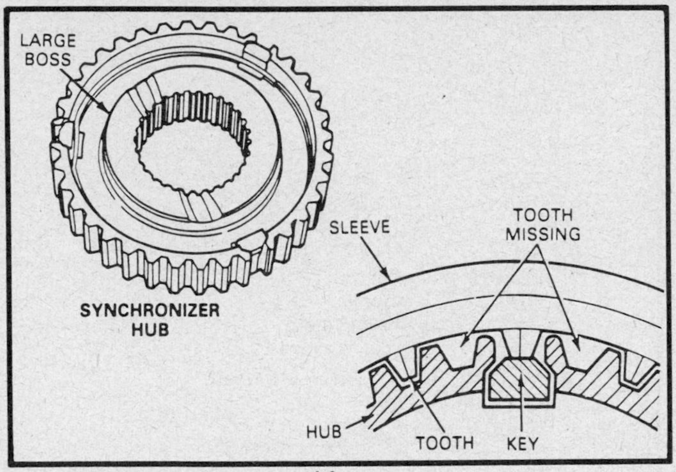
LARGE
BOSS
SYNCHRONIZER
HUB
SLEEVE
TOOTH
MISSING
HUB
TOOTH
KEY

Correct hub and key position

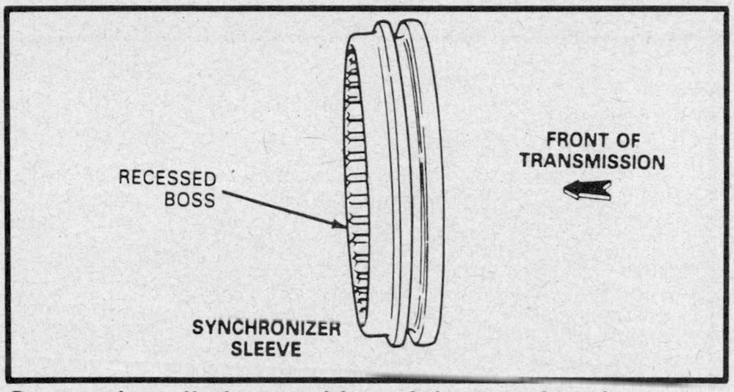
RECESSED
BOSS
FRONT OF
TRANSMISSION
SYNCHRONIZER
SLEEVE

Correct installation position of the synchronizer sleeve

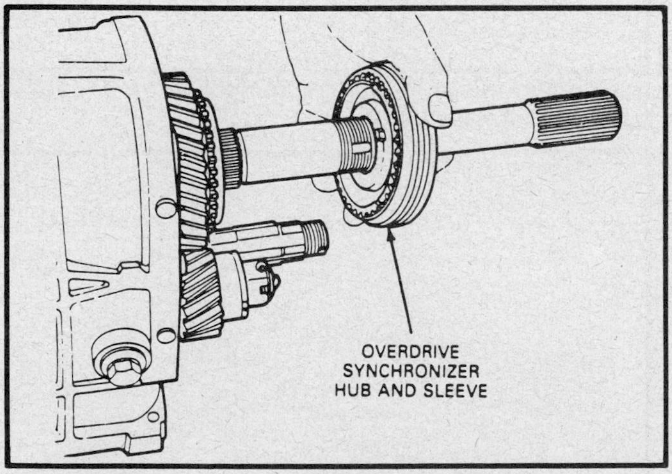
OVERDRIVE
SYNCHRONIZER
HUB AND SLEEVE

Installing the overdrive synchronizer on the mainshaft

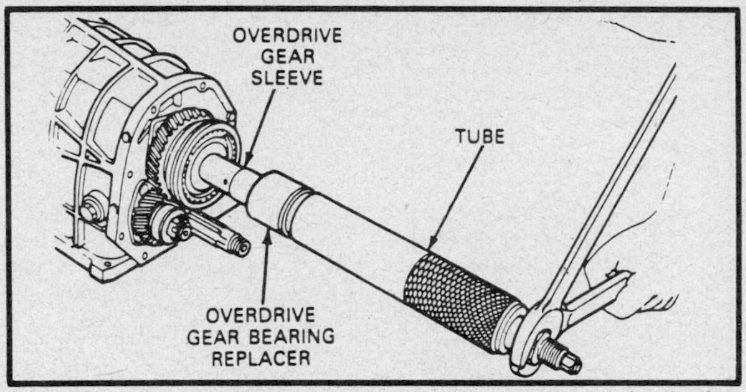
OVERDRIVE
GEAR
SLEEVE
TUBE
OVERDRIVE
GEAR BEARING
REPLACER

Pressing the overdrive gear sleeve on the mainshaft

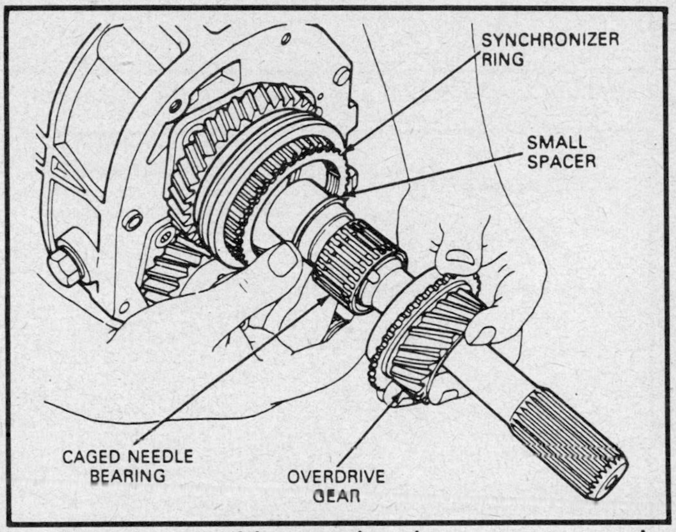
SYNCHRONIZER
RING
SMALL
SPACER
CAGED NEEDLE
BEARING
OVERDRIVE
GEAR

Installing the overdrive synchronizer, spacer, caged needle bearing and overdrive gear on the mainshaft

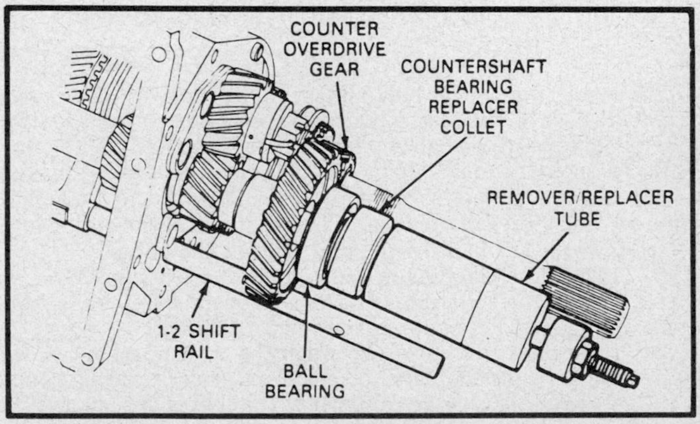
COUNTER
OVERDRIVE
GEAR
COUNTERSHAFT
BEARING
REPLACER
COLLET
REMOVER/REPLACER
TUBE
1-2 SHIFT
RAIL
BALL
BEARING

Installing the counter-overdrive gear, ball bearing and 1–2 shift rail on the countershaft

(245–265 Nm) using mainshaft locknut wrench, T77J–7025–C. Tighten the countershaft locknut to 115–135 ft. lbs. (157–186 Nm) using a 30mm socket. Disengage the transmission.

31. Stake the locknuts on the mainshaft and countershaft using locknut staking tool, T77J–7025–F or equivalent.

32. Install the interlock plunger in the bore between the 1/2 and 3/4 shift rails. Reposition the 1/2 shift rail so the flats for the poppet ball and spring and the interlock plunger are in the correct position. Make sure the roll pin holes for the shift forks are in alignment.

33. Install the overdrive/reverse shift fork on the synchronizer sleeve. Slide the 3/4 shift rail through the overdrive/reverse shift fork into the case and into the 3/4 shift fork inside the case. Position the shift rail flats to accept the poppet balls and interlock plunger. Insert the interlock plunger in the bore between the 3/4 shift rail and overdrive/reverse shift rail. Make sure the roll pin holes in the fork are in alignment.

34. Insert the overdrive/reverse shift rail so it engages the forks in the case. Make sure the roll pin holes in the fork and rail are in alignment.

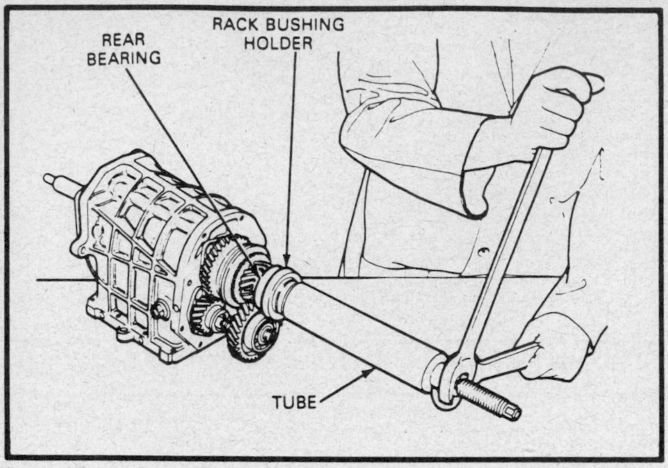
RACK BUSHING
HOLDER
REAR
BEARING
TUBE
Installing the lock ball and spacer on the mainshaft

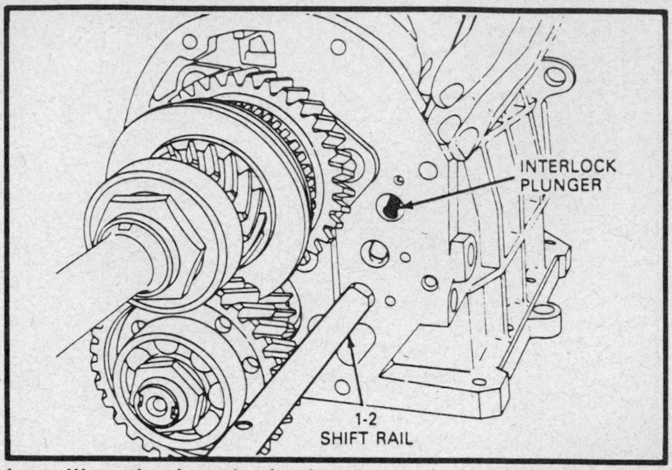
INTERLOCK
PLUNGER
1-2
SHIFT RAIL
Installing the interlock plunger and checking the alignment for the shift forks

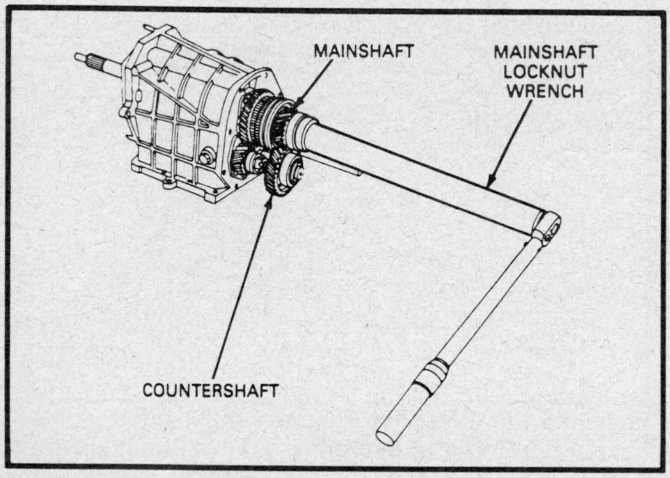
MAINSHAFT
MAINSHAFT
LOCKNUT
WRENCH
COUNTERSHAFT
Tighting the countershaft and mainshaft locknuts

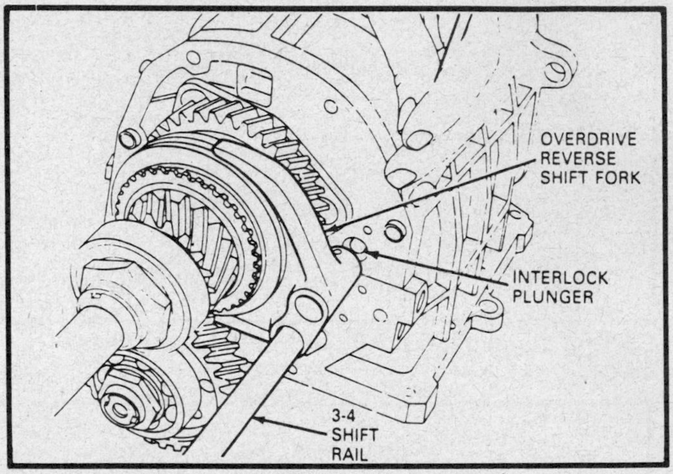
OVERDRIVE
REVERSE
SHIFT FORK
INTERLOCK
PLUNGER
3-4
SHIFT
RAIL
Positioning the shift rail flats to accept the poppet balls and interlock plunger

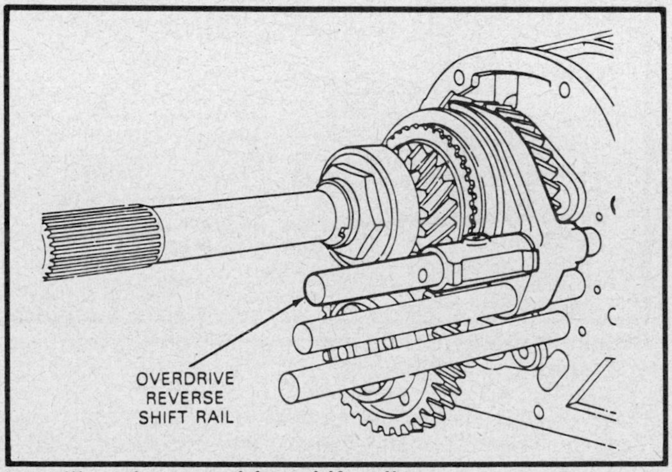
OVERDRIVE
REVERSE
SHIFT RAIL
Installing the overdrive shift rails

35. Insert the poppet ball and spring in the overdrive/reverse (upper) bore in the case. The small end of the spring should be installed toward the ball. Install the set screw and tighten until the set screw head is 0.24 in. (6mm) below the top of the bore.
36. Insert a poppet spring and poppet ball in the 3/4 and 1/2 bore (side 2 bores in the case). The small end of the spring must face towards the ball. Install and tighten the bolts.
37. Install the roll pins in the 1/2 and 3/4 shift forks.
38. Install the shift gates on the appropriate shift rails. Move the 1/2 gate and 3/4 gate to the rear of the rail.
39. Position a new gasket between the transmission case and the transfer case adapter. Make sure the selector arm is out of the gates and the change shifter is at the rear of the adapter. Position the adapter on the case making sure the shift gates clear the adapter. Make sure the shift rails and rear bearings line up with the bores in the adapter.
40. Install the 3 sizes of bolts (35mm, 55mm and 110mm) in the appropriate holes in the adapter. Tighten the bolts to 11–16 ft. lbs. (15–21 Nm).
41. Install the neutral return plungers, springs and bolts in the adapter. The longer plunger with the slot for the detent ball is installed on the right side of the adapter.
42. Position the shaft gates so the roll pin holes in the gates and rails are in alignment. Install the roll pins through the access holes. Install the access hole plugs.
43. Position the pan and new gasket on the case. Install the bolts and tighten to 11–16 ft. lbs. (15–21 Nm). Do not overtighten. Install the drain plug and tighten to 25–32 ft. lbs. (35–44 Nm).
44. Insert the plunger detent ball and spring in the hole above the neutral return plunger in the adapter case.
45. On the FM145, make sure the stopper bracket assembly on

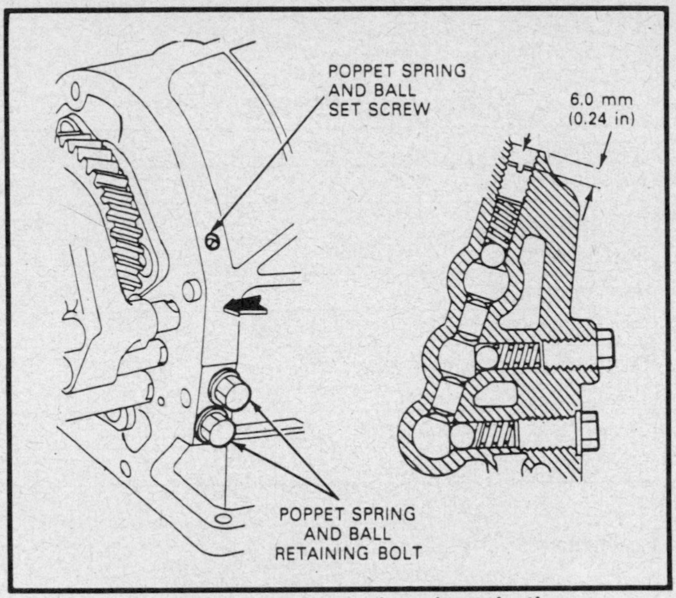
POPPET SPRING
AND BALL
SET SCREW
6.0 mm
(0.24 in)
POPPET SPRING
AND BALL
RETAINING BOLT

Installing the poppet balls and springs in the overdrive-reverse upper bore of the case

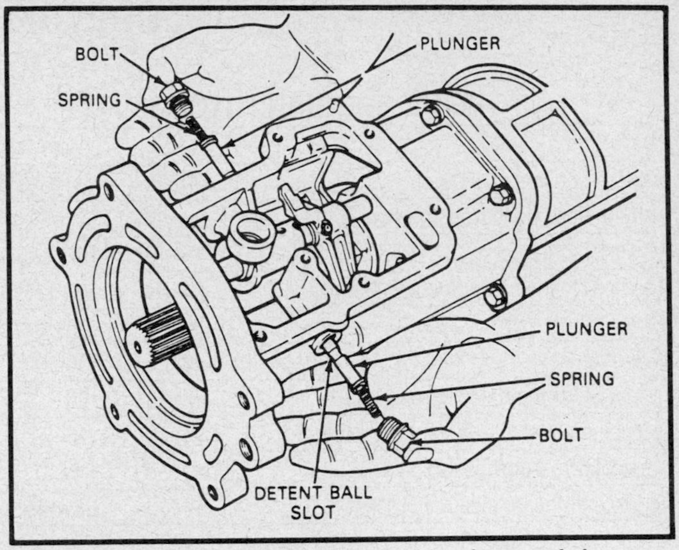
BOLT
SPRING
PLUNGER
PLUNGER
SPRING
BOLT
DETENT BALL
SLOT

Installing the return plungers and springs—right and left sides

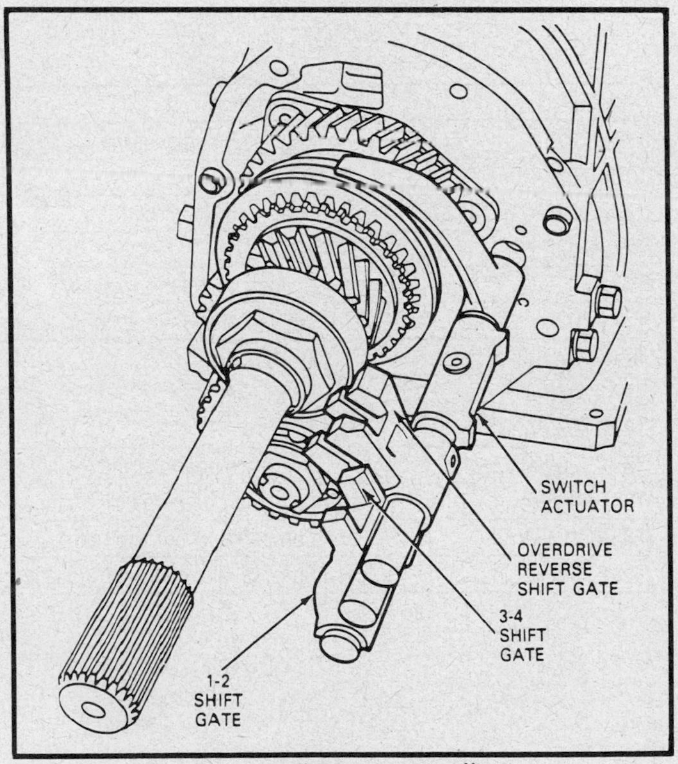
SWITCH
ACTUATOR
OVERDRIVE
REVERSE
SHIFT GATE
3-4
SHIFT
GATE
1-2
SHIFT
GATE

Installing the gates on the correct rails

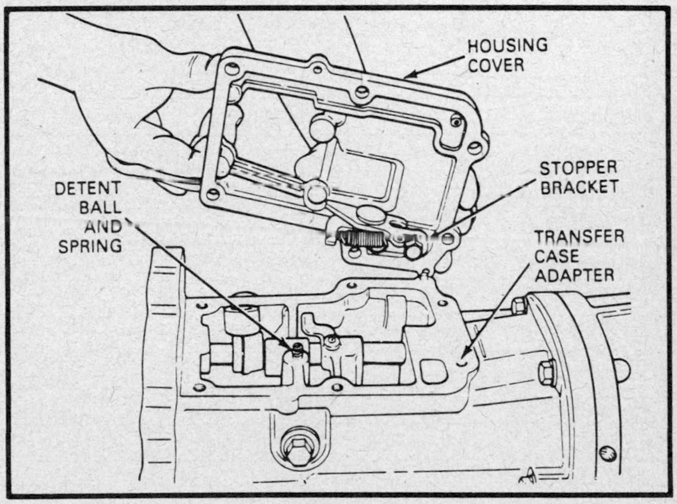
HOUSING
COVER
STOPPER
BRACKET
DETENT
BALL
AND
SPRING
TRANSFER
CASE
ADAPTER

Installing the stopper bracket and cover on the transfer case adapter—FM145 manual transmission

the cover for the transfer case adapter moves smoothly. Position a new gasket on the adapter and install the housing cover. Install and tighten to bolts to 11–16 ft. lbs. (15–21 Nm).

46. On the FM146, make sure the reverse lockout assembly on the control housing moves smoothly. Position a new seal on the adapter and install the control housing assembly. Install and tighten the retaining bolts to 11–16 ft. lbs. (15–21 Nm).

47. Install the back-up lamp switch and the shift indicator lamp switch in the adapter.

48. Remove the fill plug and fill the transmission to the bottom of the fill hole with standard transmission lube (SAE 80W) D8DZ–19C547–A (ESP-M2C83–C) or equivalent. The fluid capacity is 2.4 quarts (1.3L). Install the fill plug and tighten to 22–25 ft. lbs. (30–34 Nm).

49. Position the clutch slave cylinder on the input shaft. Position the clutch housing on the transmission case and install and tighten the nuts.

SPECIFICATIONS

TORQUE SPECIFICATIONS

Description	ft. lbs.	Nm
Clutch housing to engine	28–38	38–51
Clutch housing to transmission	30–40	41–54
Countershaft locknut	115–137	157–186
Damper to insulator on crossmember	71–94	97–127
Drain plug	25–32	35–44
Fill plug	22–25	30–34
Front bearing retainer to case	22–30	30–41
Housing cover to transfer case adapter	11–16	15–21
Insulator to transmission	60–80	81–108
Crossmember to side rail bracket	65–85	88–115

TORQUE SPECIFICATIONS

Description	ft. lbs.	Nm
Mainshaft locknut	180–195	245–265
Pan to case	11–16	15–21
Rear bearing to case	22–30	30–41
Reverse idler gear nut	15–42	20–58
Reverse idler gearshaft assembly to case	11–16	15–21
Shift lever assembly to transfer case adapter	6–10	8–14
Starter motor to clutch housing	15–20	21–27
Stud to front retainer and case	22–30	30–41

SPECIAL TOOLS

Number	Description
T50T-100-A	Impact slide hammer
D84L-1123-A	Bearing splitter
T81P-3504-D	Rack bushing holder
T80T-4000-W	Driver handle
D80P-4201-A	Depth micrometer
D82L-4201-C	Metric depth micrometer
D83L-4201-A	Straight edge
T77F-4220-B1	Puller
T85T-7011-A	Seal installer
T75L-7025-B	Tube
T75L-7025-C	Tube
T75L-7025-D	Bearing collet
T75L-7025-G	Bearing collet sleeve
T84T-7025-B	Forcing screw

Number	Description
T75L-7025-K	Replacing shaft sleeve
T75L-7025-M	Shaft collar
T77F-7025-C	Bearing collet sleeve
T77J-7025-B	Tube
T77J-7025-C	Mainshaft locknut wrench
T77J-7025-F	Mainshaft locknut staking tool
T77J-7025-H	Puller
T77J-7025-J	Puller ring
T85T-7025-A	Tube
T85T-7061-A	Overdrive gear bearing replacer
T85T-7065-A	Mainshaft bearing collet remover
T85T-7121-A	Countershaft bearing replacer
T85T-7140-A	Reverse idler gearshaft replacer

Section 7

M5OD Transmission

Ford Motor Co.

APPLICATION

1988–89 Ford F-150, F-250 2WD and 4WD, Bronco and Econoline
1988–89 Ranger and Bronco II

GENERAL DESCRIPTION

The M5OD transmission is a top shift fully synchronized 5 speed manual transmission, equipped with an overdrive 5th gear ratio. All gear changes including reverse are accomplished with the synchronizer sleeves.

The transmission main case, top cover and extension housing are constructed of aluminum alloy. Steel bearing race inserts provide durability in the appropriate areas.

The extension housing contains a bronze alloy bushing. The bushing cannot be serviced. The extension housing housing must be replaced as a unit if the bushing requires service.

Transmission Identification

The transmission is equipped with a service identification tag. The M5OD transmission service tag is located on the driver's side of the transmission.

Metric Fasteners

The metric fastener dimensions are very close to the dimensions of the familiar inch system fasteners. For this reason, replacement fasteners must have the same measurement and strength as those removed.

Do not attempt to interchange metric fasteners for inch system fasteners. Mismatched or incorrect fasteners can result in damage to the transmission unit through malfunctions, breakage or possible personal injury.

Care should be taken to re-use the fasteners in the same locations as removed.

Common metric fastener strength property classes are 9.8 and 12.9 with class identification embossed on the head of each bolt. The inch strength classes range from grade 2 to 8 with the line identification embossed on each bolt head. Markings correspond to 2 lines less than actual grade (for example grade 8 bolt will exhibit 6 embossed lines on the bolt head). Some metric nuts will be marked with a single digit strength identification numbers on the nut face.

Capacities

The M5OD transmission oil capacity is 7.6 pints. Mercon® multi-purpose automatic transmission fluid XT-2-QDX-E4AZ-19582-B or equivalent should be used.

Checking Fluid Level

1. Raise and support the vehicle safely.
2. Remove the transmission oil fill plug.
3. The correct level should be to the bottom edge of the fill plug hole.
4. Fill the proper level with the recommended gear oil and reinstall the oil fill plug.
5. Lower the vehicle.

TRANSMISSION MODIFICATIONS

There are no modifications listed for the M5OD transmission at the time of this publication.

TROUBLE DIAGNOSIS

CHILTONS THREE C'S TRANSMISSION DIAGNOSIS

Condition	Cause	Correction
Gears clash when shifting from one forward gear to another	a) Engine idle speed too high	a) Adjust engine idle speed
	b) Pilot bearing binding	b) Remove transmission and check for binding condition between input shaft and engine crankshaft pilot bearing. Replace as required
	c) Damaged gear teeth and/or synchronizer	c) Disassemble transmission, repair or replace as required

CHILTONS THREE C'S TRANSMISSION DIAGNOSIS

Condition	Cause	Correction
Transmission jumps out of gear	a) Stiff shift boot. Improper fit of form isolation pad	a) Replace shift boot if exceptionally stiff. Replace or rework pad to provide clearance
	b) Loose transmission to engine block bolts, or loose levers	b) Tighten transmission to engine block bolts to specifications. Loosen all bolts and reseat flywheel housing. Tighten all bolts. Tighten levers if necessary
	c) Flywheel housing to engine crankshaft out of line	c) Shim or replace housing as required
	d) Crankshaft pilot bearing worn	d) Replace bearing
	e) Interior components damaged	e) Disassemble transmission. Inspect the synchronizer sleeves for free movement on their hubs. Inspect the synchronizer blocking rings for widened index slots, rounded clutch teeth and smooth internal surface. Check countershaft cluster gear for excessive endplay. Check shift forks for loose mounting on shift rails. Inspect synchronizer sliding sleeve and gear clutch teeth for wear or damage. Repair or replace as required
	f) Worn gear teeth due to partial engagement	f) Replace worn or damaged gears
Transmission will not shift into one gear—all others OK	a) Manual-shift linkage damaged or worn	a) Lubricate, repair or replace parts as required
	b) Back-up switch ball frozen	b) If reverse is problem, check backup switch for ball frozen in extended position (if so equipped)
	c) Internal components	c) Remove transmission. Inspect shift rail and fork system synchronizer system and gear clutch teeth for restricted travel. Repair or replace as required
Transmission shifts hard	a) Clutch does not completely release	a) Check input shaft splines for lubrication
	b) Transmission fluid low or improper type	b) Add lubricant or change lubricant as required
	c) Worn or damaged internal shift mechanism	c) Remove transmission cover. Check internal shift mechanism by shifting into and out of all gears. Repair or replace as required
	d) Binding of sliding gears and/or synchronizers	d) Check for free movement of gears and synchronizers. Repair or replace as required
	e) Housings and/or shafts out of alignment	e) Remove transmission and check for binding condition between input shaft and engine crankshaft pilot bearing or bushing. Check flywheel housing alignment. Repair or replace as required
Noisy in forward gears	a) Lubricant level low, or improper type	a) Add lubricant, or refill with specified lubricant
	b) Components grinding on transmission	b) Check for screws, bolts, etc., of cab or other components grounding out. Correct as required

CHILTONS THREE C'S TRANSMISSION DIAGNOSIS

Condition	Cause	Correction
Noisy in forward gears	c) Component housing bolts loose	c) Check torque on transmission to flywheel housing bolts, output shaft flange nut and flywheel housing to engine block bolts. Tighten bolts to specification
	d) Flywheel housing to engine crankshaft alignment	d) Check and align flywheel housing to engine crankshaft
	e) Noisy bearings or gears	e) Remove and disassemble transmission. Inspect input, output and countershaft bearings. Inspect speedometer gear and gear teeth for wear or damage. Replace as required

While verifying the condition, determine whether the noise is gear roll-over noise, release bearing rub or some other transmission related noise.
Gear roll-over noise, inherent in manual transmissions, is caused by the constant mesh gears turning at engine idle speed, while the clutch is engaged and the transmission is in neutral; and release bearing rub are sometimes mistaken for mainshaft bearing noise. Gear roll-over noise will disappear when the clutch is disengaged or when the transmission is engaged in gear. Release bearing rub will disengage when the clutch is engaged in the event that a bearing is damaged, the noise is more pronounced while engaged in gear under load or coast than in neutral.

REMOVAL AND INSTALLATION

TRANSMISSION REMOVAL

The catalytic converter is located beside the transmission. Be careful when working around the catalytic converter because of the extremely high temperatures generated by the converter.

2WD

1. Shift the transmission into the **N** position. Disconnect the negative battery cable.
2. Remove the carpet or floor mat. Remove the shifter boot retainer screws and slide the boot up the shift lever shaft. Remove the shift lever retaining bolt and remove the shift lever.
3. Raise and support the vehicle safely. Disconnect the speedometer cable. Disconnect the backup lamp switch located at the top left hand side of the transmission.
4. Remove the drain plug from the transmission and drain the transmission fluid into a suitable drain pan. Position a suitable transmission jack under the transmission.
5. Disconnect the driveshaft from the transmission and wire it to one side. Disconnect the clutch hydraulic line.
6. Remove the transmission rear insulator and lower retainer. Remove the crossmember.
7. Remove the bolts that retain the transmission to the engine block. Move the transmission to the rear until the input shaft clears the engine flywheel. Lower the transmission from the vehicle.

4WD

1. Shift the transmission into the **N** position. Disconnect the negative battery cable.
2. Remove the carpet or floor mat. Remove the shifter boot retainer screws and slide the boot up the shift lever shaft. Remove the shift lever retaining bolt and remove the shift lever.
3. Raise and support the vehicle safely. Disconnect the driveshaft from the transmission and wire it to one side.
4. Disconnect the front driveshaft from the transfer case and wire it out of the way.
5. Disconnect the speedometer cable. Disconnect the backup lamp switch located at the top left hand side of the transmission. If equipped, remove the skid pad from underneath the transfer case.
6. Remove the drain plug from the transmission and drain the transmission fluid into a suitable drain pan. Position a suitable transmission jack under the transmission.
7. Support the transfer case, using a suitable transmission jack or equivalent. Remove the 6 bolts holding the transfer case to the transmission and carefully lower the transfer case from the vehicle, using care to ensure that the transfer case shift lever clears the opening in the floor pan.
8. Remove the transmission rear insulator and lower retainer. Remove the crossmember.
9. Remove the bolts that retain the transmission to the engine block. Move the transmission to the rear until the input shaft clears the engine flywheel. Lower the transmission from the vehicle.

TRANSMISSION INSTALLATION

2WD

1. With the transmission on a suitable transmission jack, install the guide studs into the engine block and raise the transmission up until the input shaft splines are aligned with the clutch disc splines.
2. Slide the transmission forward onto the guide studs until the transmission is in the correct position. Install the transmission retaining bolts and torque them to 40–50 ft. lbs. (54–67 Nm). Remove the guide studs and install the 2 remaining bolts. Torque the 2 bolts for the lower plate to 9–12 ft. lbs. (12–16 Nm).

3. Install the crossmember. Position the insulator between the transmission and crossmember and torque the bolts to 60–80 ft. lbs. (81–108 Nm). Remove the transmission jack.

4. Connect the speedometer cable and driven gear. Install the clutch hydraulic line. Connect the backup lamp switch.

5. Install the driveshaft. Fill the transmission with the recomended transmission fluid to the proper level. Install the oil fill plug and torque it to 29–43 ft. lbs. (40–58 Nm). Lower the vehicle.

6. Install the shift lever retaining bolt and tighten the retaining bolts. Slide the shifter boot into position on the shifter shaft and install the boot retaining screws.

7. Install the isolator pad assembly. Install the floor pan cover and floor mat.

4WD

1. With the transmission on a suitable transmission jack, install the guide studs into the engine block and raise the transmission up until the input shaft splines are aligned with the clutch disc splines.

2. Slide the transmission forward onto the guide studs until the transmission is in the correct position. Install the transmission retaining bolts and torque them to 40–50 ft. lbs. (54–67 Nm). Remove the guide studs and install the 2 remaining bolts. Torque the 2 bolts for the lower plate to 9–12 ft. lbs. (12–16 Nm).

3. Place the rear support bracket in position and install the retaining bolts. Install the crossmember. Position the insulator between the transmission and crossmember and torque the bolts to 60–80 ft. lbs. (81–108 Nm). Remove the transmission jack.

4. Position the transfer case on the transmission jack. Position the transfer case onto the transmission, using care to guide the transfer case shift lever through the opening in the floor pan. Install the gasket and 6 retaining bolts 24–34 ft. lbs. (32–46 Nm).

5. Connect the speedometer cable and driven gear. Connect the backup lamp switch.

6. Install the driveshaft to the transmission.

7. Connect the rear driveshaft to the transfer case. Fill the transmission with the recommended transmission fluid to the proper level. Install the oil fill plug and torque it to 29–43 ft. lbs. (40–58 Nm). Lower the vehicle.

8. Install the shift lever retaining bolt and tighten the retaining bolts. Slide the shifter boot into position on the shifter shaft and install the boot retaining screws.

9. Install the isolator pad assembly. Install the floor pan cover and floor mat. Install the shift ball on the transfer case shift lever.

BENCH OVERHAUL

Before Disassembly

Cleanliness during disassembly and assembly is necessary to avoid further transmission trouble after overhaul. Before removing any of the transmission subassemblies, plug all the openings and clean the outside of the of the transmission thoroughly. Steam cleaning or car wash type high pressure equipment is preferable. During disassembly, clean all parts in suitable solvent and dry each part. Do not use cloth or paper towels to dry parts. Use compressed air only.

Transmission Disassembly

1. Remove the transmission drain plug and drain the transmission fluid into a suitable drain pan.

2. Remove the shift lever and dust boot is necessary. Remove the 10 top cover assembly retaining bolts. Remove the top cover assembly.

3. Remove the 9 extension housing retaining bolts. Pry gently at the locations provided on the extension housing and transmission case. Remove the extension housing from the case.

NOTE: On the 2WD vehicles, if it is necessary to remove the rubber seal from the extension housing, the extension housing must be installed on the transmission case. Remove the extension housing rear seal using a suitable seal removal tool.

4. On 2WD vehicles, if necessary, remove rear oil passage from the extension housing. Remove and discard anti-spill oil seal from the output shaft. Remove the speedometer drive gear and seal.

NOTE: For reference during assembly, observe and record speedometer drive gear color. Depending upon application, 1 or 2 different speedometer drive gear may be installed. It will be color coded either green or white. Speedometer drive gear colors and corresponding part numbers are as follows, white E8TZ–17285–B and green E8TZ–17285–C.

5. Lock the transmission into 1st/3rd gears. Using countershaft locknut staking tool T–77J–7025–F or equivalent, release the staked areas securing the output shaft and countershaft locknut. The staked areas of locknuts must be fully released or damage to shaft threads will result.

6. Remove and discard the countershaft rear bearing locknut. Remove the countershaft bearing and thrust washer.

7. Using the mainshaft locknut wrench tool Y–88T–7025–A and remover tube tool T–75L–7025–B or equivalents, remove and discard output shaft locknut. Remove the reverse idler shaft fixing bolt. Remove the reverse idler gear assembly by grasping and pulling rearward.

8. Remove the output shaft rear bearing from the output shaft using remover/replacer tube tool T75L–7025–B, TOD forcing screw tool T84T–7025–B, bearing puller tool T77J–7025–H and puller ring tool T77J–7025–J or equivalents.

9. Using a suitable brass drift and hammer, drive the reverse gear from the output shaft. Remove the sleeve from the output shaft. Remove the counter reverse gear with the 2 needle bearings and reverse synchronizer ring.

10. Remove the thrust washer and split washer from the countershaft. Remove the 5th/reverse shift rod fixing bolt.

11. Remove the 5th/reverse synchronizer hub and sleeve as an assembly. Remove the 5th/reverse shift fork and rod. Do not separate the steel ball and spring (remove from the shift fork groove) unless necessary.

12. Remove the 5th gear synchronizer ring. Remove the 5th/reverse counter lever lockplate retaining bolt and inner circlip. Remove the counter lever assembly from the transmission case. Do not remove the Torx® nut retaining the counter lever pin at this time.

13. Remove the 5th gear (counter) with needle bearing. Remove the 5th gear from the output shaft using bearing collet sleeve for 3.5 inch bearing collet tool T75L–7025–G, remover/replacer tube tool T85T–7025–A, TOD forcing screw tool T84T–7025–B and gear removal collet tool T88T–7061–A or equivalents.

14. Remove the 5th gear sleeve and position ball, TOD forcing

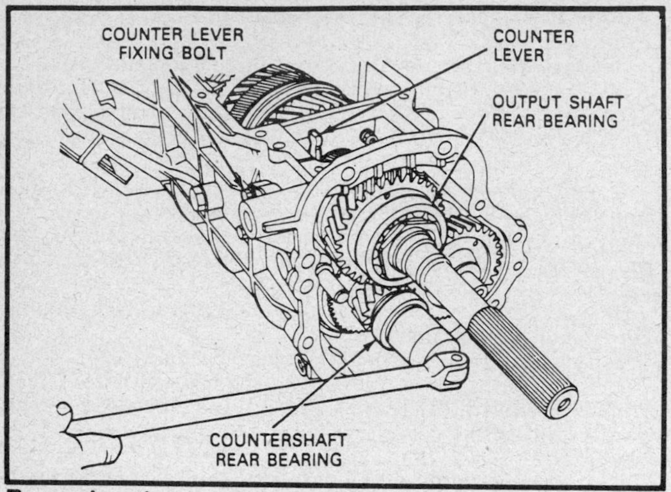
COUNTER LEVER
FIXING BOLT
COUNTER
LEVER
OUTPUT SHAFT
REAR BEARING
COUNTERSHAFT
REAR BEARING

Removing the countershaft bearing and thrust washer

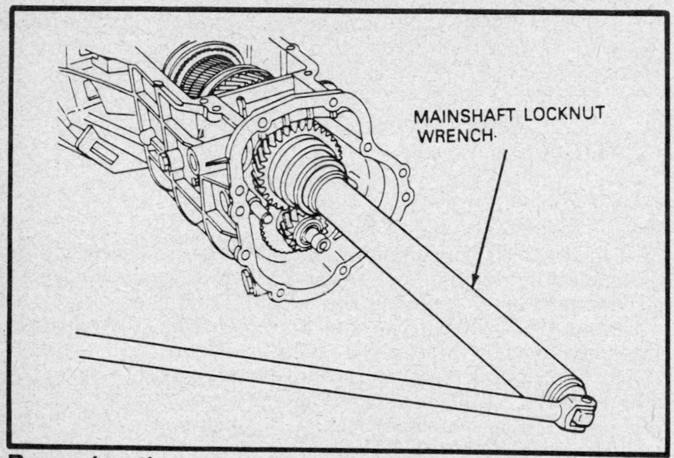
MAINSHAFT LOCKNUT
WRENCH

Removing the output shaft locknut

screw tool T84T–7025–B, countershaft 5th gear sleeve puller tool T84T–7025–J, gear removal collet tool T88T–7025–J1 and remover/replacer tube tool T77–7025–B or equivalents.

15. Remove the 6 center bearing cover retaining bolts. Remove the center bearing cover. For reference during assembly, observe that the reference arrow in the middle of the center bearing cover points upward. Observe the flanged side of the center bearing cover faces inward.

16. Remove the 6 front bearing cover attaching bolts. Remove the front bearing cover by threading 2 of the originally installed retaining bolts into the front bearing cover service bolt locations (9 o'clock and 3 o'clock). Alternately tighten bolts until the front bearing cover can be lifted away by hand. Remove and discard the front bearing cover oil baffle. The bolt threaded into the service bolt locations will bottom out and lift the front bearing cover away from the transmission case. Do not remove the plastic scoop ring from the input shaft at this time.

17. Remove the oil trough retaining bolt and oil trough from the upper transmission case. Pull the input shaft forward and remove the input bearing outer race. Pull the output shaft rearward.

18. Pull input shaft forward and separate it from the output shaft. Incline output shaft upward and lift it from the transmission case. Remove the input shaft from the transmission case.

19. Remove the countershaft bearing outer races (front and center) by moving the countershaft forward and rearward. Pull the countershaft rearward enough to permit the tool clearance

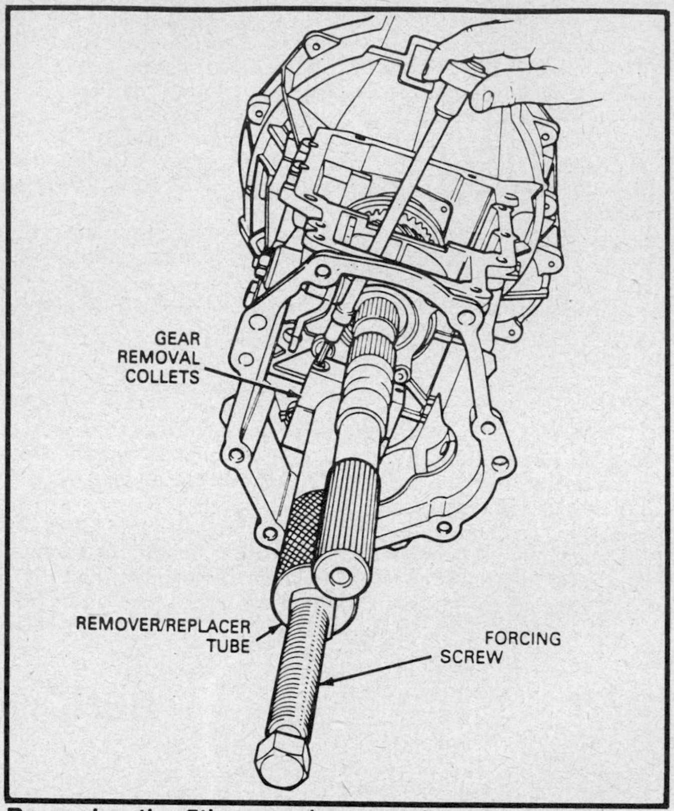
GEAR
REMOVAL
COLLETS
REMOVER/REPLACER
TUBE
FORCING
SCREW

Removing the 5th gear sleeve and positioning ball

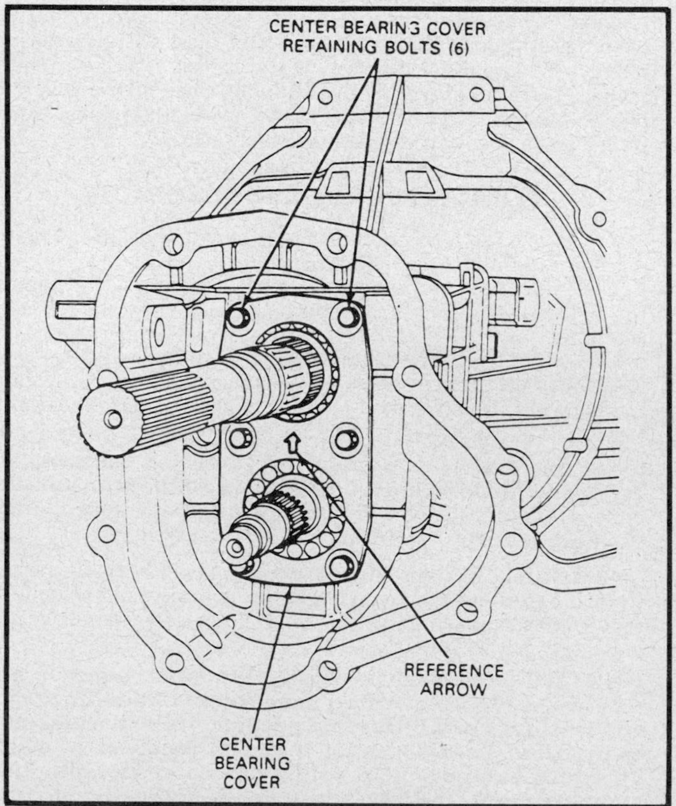
CENTER BEARING COVER
RETAINING BOLTS (6)
REFERENCE
ARROW
CENTER
BEARING
COVER

Removing the center bearing cover

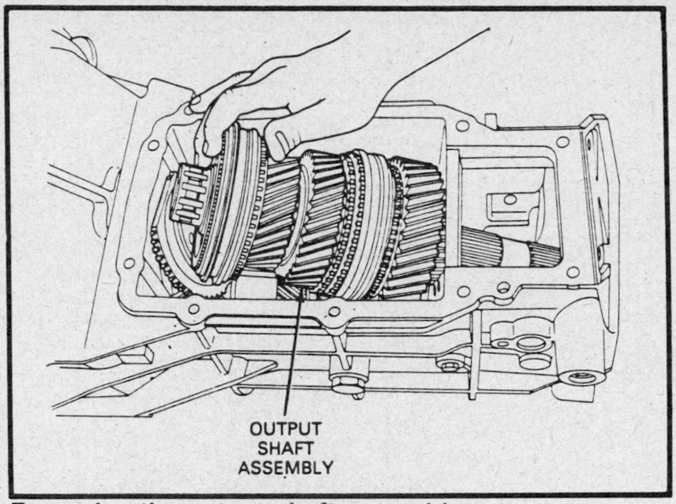
OUTPUT
SHAFT
ASSEMBLY

Removing the output shaft assembly

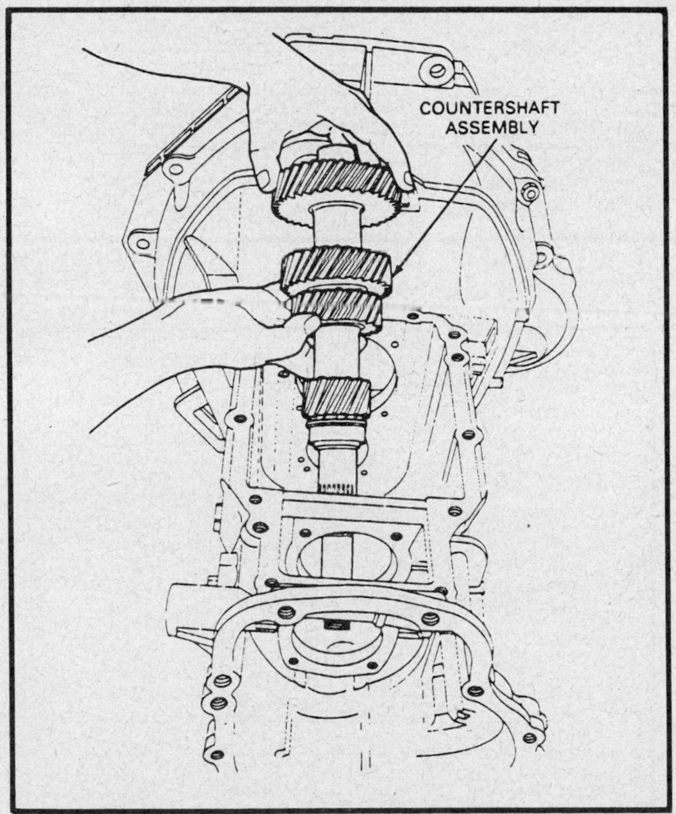
COUNTERSHAFT
ASSEMBLY

Removing the countershaft assembly

behind the front countershaft bearing. Using bearing race puller tool T88T-7120-A and slide hammer tool T50T-100-A or equivalents, remove the front countershaft bearing. Tap the bearing gently during removal. A forceful blow can cause damage to the bearing or transmission case.

CLEANING AND INSPECTION

Cleaning

During overhaul, all components of the transmission (except bearing assemblies) should be thoroughly cleaned with solvent and dried with air pressure prior to inspection and reassembly.

1. Clean the bearing assemblies as follows:

NOTE: Proper cleaning of bearings is of utmost importance. Bearings should always be cleaned separately from other parts.

a. Soak all bearing assemblies in clean solvent or fuel oil. Bearings should never be cleaned in a hot solution tank.

b. Slush bearings in solvent until all old lubricant is loosened. Hold races so that bearings will not rotate; then clean bearings with a soft bristled brush until all dirt has been removed. Remove loose particles of dirt by tapping bearing flat against a block of wood.

c. Rinse bearings in clean solvent; then blow bearings dry with air pressure.

CAUTION

Do not spin bearings while drying.

d. After drying, rotate each bearing slowly while examining balls or rollers for roughness, damage, or excessive wear. Replace all bearings that are not in first class condition.

NOTE: After cleaning and inspecting bearings, lubricate generously with recommended lubricant, then wrap each bearing in clean paper until ready for reassembly.

2. Remove all portions of old gaskets from parts, using a stiff brush or scraper.

Inspection

1. Inspect all parts for discoloration or warpage.
2. Examine all gears and splines for chipped, worn, broken or nicked teeth. Small nicks or burrs may be removed with a fine abrasive stone.
3. Inspect the breather assembly to make sure that it is open and not damaged.
4. Check all threaded parts for damaged, stripped, or crossed threads.
5. Replace all gaskets, oil seals and snaprings.
6. Inspect housings, retainers and covers for cracks or other damage. Replace the damaged parts.
7. Inspect keys and keyways for condition and fit.
8. Inspect shift forks for wear, distortion or any other damage.
9. Check detent ball springs for free length, compressed length, distortion or collapsed coils.
10. Check bearing fit on their respective shafts and in their bores or cups. Inspect bearings, shafts and cups for wear.

NOTE: If either bearings or cups are worn or damaged, it is advisable to replace both parts.

11. Inspect all bearing rollers or balls for pitting or galling.
12. Examine detent balls for corrosion or brinneling. If shift bar detents show wear, replace them.
13. Inspect the synchronizer ring for wear. To check the wear of the synchronizer ring, fit the synchronizer ring evenly to the gear cone. Measure the clearance between the side faces of the synchronizer ring and gear with a feeler gauge. If the clearance is less than 0.031 in. (0.8mm), replace the synchronizer ring or gear.
14. Check the contact surfaces of the shift fork and clutch hub sleeve for evidence of wear or damage. Measure from the shift fork to the clutch hub sleeve. If the clearance exceeds 0.031 in. (0.8mm), replace the shift fork/clutch hub sleeve.

a. The standard clearance for the 1st/2nd and 3rd/4th shift fork/clutch hub sleeves is 0.003–0.014 in. (0.1–0.358mm). The maximum is 0.314 in. (0.8 mm).

b. The standard clearance for the 5th/reverse shift fork/

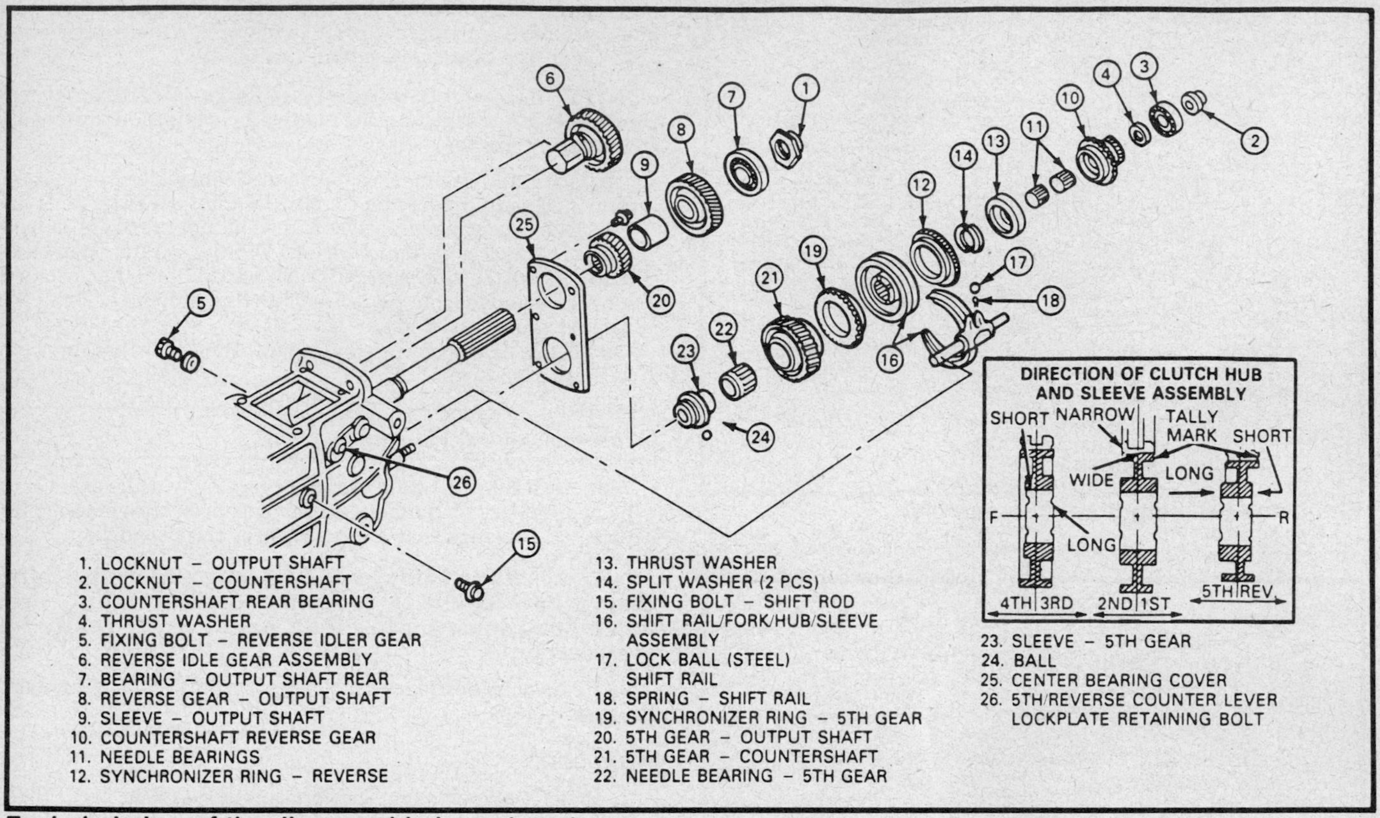
DIRECTION OF CLUTCH HUB AND SLEEVE ASSEMBLY
SHORT
NARROW
TALLY MARK
SHORT
WIDE
LONG
F
R
LONG
4TH 3RD
2ND 1ST
5TH REV
1. LOCKNUT – OUTPUT SHAFT
2. LOCKNUT – COUNTERSHAFT
3. COUNTERSHAFT REAR BEARING
4. THRUST WASHER
5. FIXING BOLT – REVERSE IDLER GEAR
6. REVERSE IDLE GEAR ASSEMBLY
7. BEARING – OUTPUT SHAFT REAR
8. REVERSE GEAR – OUTPUT SHAFT
9. SLEEVE – OUTPUT SHAFT
10. COUNTERSHAFT REVERSE GEAR
11. NEEDLE BEARINGS
12. SYNCHRONIZER RING – REVERSE
13. THRUST WASHER
14. SPLIT WASHER (2 PCS)
15. FIXING BOLT – SHIFT ROD
16. SHIFT RAIL/FORK/HUB/SLEEVE ASSEMBLY
17. LOCK BALL (STEEL) SHIFT RAIL
18. SPRING – SHIFT RAIL
19. SYNCHRONIZER RING – 5TH GEAR
20. 5TH GEAR – OUTPUT SHAFT
21. 5TH GEAR – COUNTERSHAFT
22. NEEDLE BEARING – 5TH GEAR
23. SLEEVE – 5TH GEAR
24. BALL
25. CENTER BEARING COVER
26. 5TH/REVERSE COUNTER LEVER LOCKPLATE RETAINING BOLT

Exploded view of the disassembled rear housing

clutch hub sleeves is 0.003–0.015 in. (0.1–0.37mm). The maximum is 0.314 in. (0.8 mm).

15. Replace all worn or damaged parts. When assembling the transmission, coat all moving parts with recommended lubricant.

Unit Disassembly and Disassembly

INPUT SHAFT

Disassembly

1. Remove and discard the plastic scoop ring.
2. Press the tappered roller bearing from the input shaft using bearing cone remover tool T71P–4621–B or equivalent.

Inspection

1. Clean and inspect the input shaft and gear assemblies.
2. Replace any and all worn or damaged components, as necessary.

Assembly

1. Install the input shaft tappered roller bearing onto the input shaft using a suitable press and bearing cone replacer tool T88T–7025–B or equivalent.
2. Install a plastic scoop ring onto the input shaft. Manually rotate the ring clockwise to ensure that the input oil holes properly engage the scoop ring. A click should be heard as the scoop ring notches align with the input shaft oil holes.

OUTPUT SHAFT

Disassembly

1. Pull back and separate the 3rd gear and the 2nd gear from the output shaft flange.

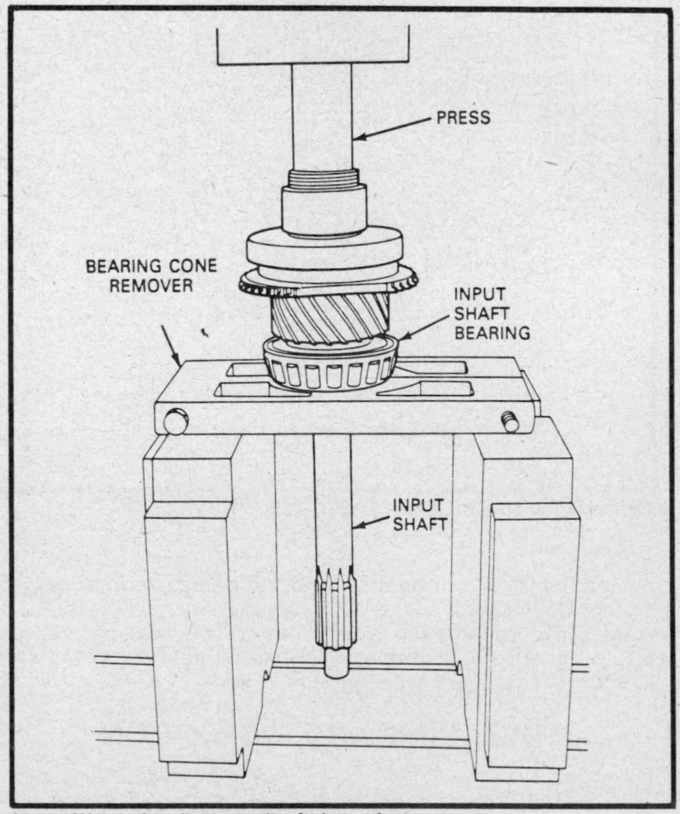
PRESS
BEARING CONE REMOVER
INPUT SHAFT BEARING
INPUT SHAFT

Installing the input shaft bearing

2. Remove the pilot bearing (needle roller), snapring, needle bearing and spacer from the front (short side of the flange) of the output shaft.

3. Position the front (short side of the flange) of the output shaft so that it faces upward. Lift off the following components as an assembly.
   a. Clutch hub and sleeve assembly (3rd/4th).
   b. Synchronizer ring (3rd).
   c. 3rd gear.
   d. Needle bearing.

4. Position the output shaft with the rear end (long side of the flange) facing upward. Position the output shaft into the press with the press cradle contacting the lower part of 2nd gear.

NOTE: Ensure that the output shaft flange does not contact or ride up onto the press cradle. Improper positioning of the output shaft can cause component damage.

5. Press off the following components as a unit, center bearing, 1st gear sleeve, 1st gear needle bearing, 1st/2nd clutch hub and sleeve assembly, 1st/2nd synchronizer rings, 2nd gear and the needle bearing using bearing replacer tool T53T–4621–B and bearing cone replacer tool T88T–7025–B or equivalents.

Inspection

1. Clean and inspect the output shaft and gear assemblies.
2. Replace any and all worn or damaged components as necessary.
3. Check the output shaft for run-out by mounting the shaft between V-blocks and applying a dial indicator tool to several places along the shaft.
4. The standard reading of the dial indicator for the run-out should be less than 0.002 in. (0.05mm). If the run-out exceeds the specifications, replace the mainshaft.
5. Replace the input shaft if the splines are damaged. If the needle bearing surface in the bore of the bearing is worn or rough, or if the cone surface is damaged, replace the shaft.

Assembly

During assembly apply the recommended transmission fluid to all rotating or sliding parts.

1. Position the output shaft so that the rear end (long side of the flange) faces upward. Install the following components in the order listed as follows:
   a. 2nd gear needle bearing.
   b. 2nd gear.
   c. 2nd gear synchronizer rings.
   d. 1st/2nd clutch hub and sleeve assembly.
   e. 1st gear synchronizer rings.
   f. 1st gear needle bearing.
   g. 1st gear.
   h. 1st gear sleeve.
   i. Center bearing (inner).

NOTE: To install the components onto the output shaft, press the components into position using bearing replacer tool T53T–4621–B and bearing plate tool T75L–1165–B or equivalents.

2. Ensure that the center bearing race is installed into the transmission case.
3. When installing the 1st/2nd clutch hub and sleeve, ensure that the smaller width of the sleeve faces 2nd gear (front) side. Ensure that the reference marks face the rear of the transmission, they reference the synchronizer key installation position.
4. Install the center bearing to the output shaft.
5. Position the output flange so that the front (short side) of the output shaft flange faces upward. Install the 3rd gear needle bearing, 3rd gear and 3rd gear synchronizer ring.
6. Install the 3rd/4th clutch hub and sleeve as follows:

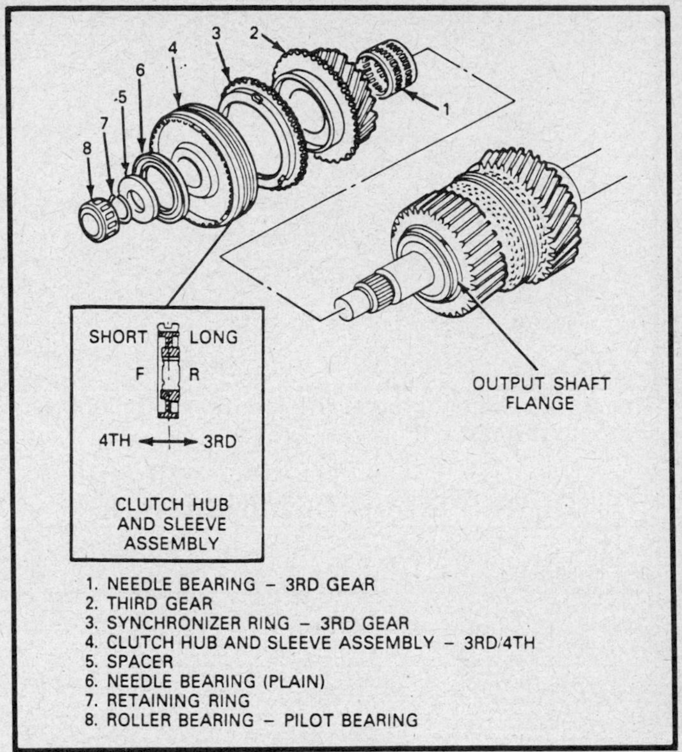
1
2
3
4
5
6
7
8
SHORT
LONG
F
R
4TH
3RD
CLUTCH HUB AND SLEEVE ASSEMBLY
OUTPUT SHAFT FLANGE
1. NEEDLE BEARING – 3RD GEAR
2. THIRD GEAR
3. SYNCHRONIZER RING – 3RD GEAR
4. CLUTCH HUB AND SLEEVE ASSEMBLY – 3RD/4TH
5. SPACER
6. NEEDLE BEARING (PLAIN)
7. RETAINING RING
8. ROLLER BEARING – PILOT BEARING

Disassembling the output shaft

   a. Mate the clutch hub synchronizer key groove with reference mark on the clutch hub sleeve. The mark should face rearward.
   b. Install the longer flange on the clutch hub sleeve toward 3rd gear (rear) side.

NOTE: The front and rear sides of the clutch hub are the same except for the reference mark on one side.

7. Install the spacer, needle bearing (install the rollers upward), retaining ring and pilot bearing (roller). Install the original retaining ring. Check the clutch hub endplay using a feeler gauge.
8. If necessary, adjust the 3rd/4th clutch hub endplay to 0.00–0.0019 in. (0.00–0.05mm) by selecting a required retaining ring.

COUNTERSHAFT

Disassembly

1. Using a suitable press and bearing cone remover tool T71P–4621–B or equivalent, remove the countershaft center bearing inner race.
2. Using a press and bearing spliter tool D84L–1123–A or equivalent, remove the countershaft front bearing inner race.

Inspection

1. Clean and inspect the output shaft and gear assemblies.
2. Replace any and all worn or damaged components, as necessary.

Assembly

1. Using a suitable press, a press plate and bearing replacer tool T53T–4621–B or equivalent, install the center bearing inner race.

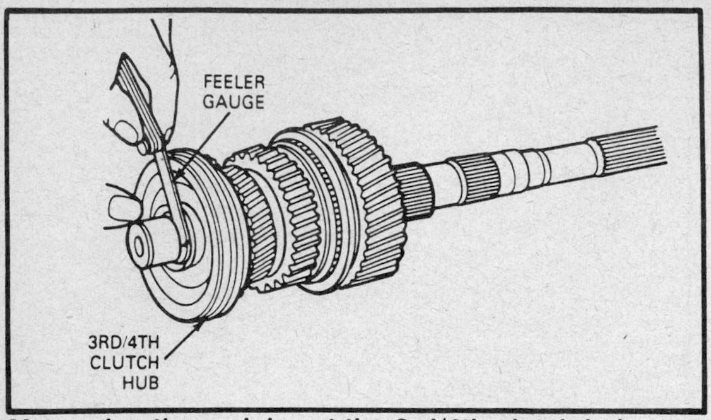
FEELER
GAUGE
3RD/4TH
CLUTCH
HUB

Measuring the endplay at the 3rd/4th clutch hub on the output shaft

OUTPUT SHAFT RETAINING RING

Part Number	Thickness in.	mm
E8TZ-7030-A	0.059	1.50
E8TZ-7030-B	0.061	1.55
E8TZ-7030-C	0.0629	1.60
E8TZ-7030-D	0.0649	1.65
E8TZ-7030-E	0.0669	1.70
E8TZ-7030-F	0.0688	1.75
E8TZ-7030-G	0.0708	1.80
E8TZ-7030-H	0.0728	1.85
E8TZ-7030-J	0.0748	1.90
E8TZ-7030-K	0.0767	1.95

REVERSE IDLER GEAR SHAFT RETAINING RING

Part Number	Thickness in.	mm
E8TZ-7156-F	0.059	1.5
E8TZ-7156-E	0.0629	1.6
E8TZ-7156-D	0.0669	1.7
E8TZ-7156-C	0.0708	1.8
E8TZ-7156-B	0.0748	1.9

2. Using a suitable press, a press plate and bearing replacer tool T53T–4621–B or equivalent, install the countershaft front bearing inner race.

REVERSE IDLER GEAR SHAFT

Disassembly

1. Remove the retaining ring, spacer, idler gear, needle bearing and thrust washer from the reverse idler gear shaft.

Inspection

1. Clean and inspect the reverse idler gear shaft and gear assemblies.
2. Replace any and all worn or damaged components, as necessary.

Assembly

1. Install the thrust washer onto the reverse idler gear shaft. Ensure that the tab on the thrust washer mates with the groove on the reverse idler shaft to prevent rotation of the thrust washer.
2. Install the needle bearing, idler gear and spacer.
3. Install the original retaining ring onto the reverse idler gear shaft. Insert a feeler gauge between the retaining ring and reverse idler gear to measure the reverse idler gear endplay.
4. Using the proper size retaining rings, adjust the endplay to 0.0039–0.0078 in. (0.1–0.2mm).

TOP COVER

Disassembly

1. If necessary, remove the dust boot and shift lever from the top cover. Remove the 3 dust cover screws and remove the dust cover.

NOTE: For reference during assembly, notice that the grooves in the bushing align with the slots in the lower shift lever pivot ball. Notice that the notch in the lower shift lever faces toward the front of the transmission.

2. Position the top cover assembly into a suitable holding fixture.
3. Remove the backup lamp switch from the top cover. Remove the backup lamp switch pin from the groove in the top cover. There is only 1 type of backup lamp pin used.
4. Using a $^{5}/_{32}$ in. drift punch, remove the spring pins retaining the shift forks to the shift rail. Discard the original spring pins.
5. Ensure that the 5th/reverse shift rail is in the fully forward position. Remove the spring pin from the end of the 5th/reverse rail.

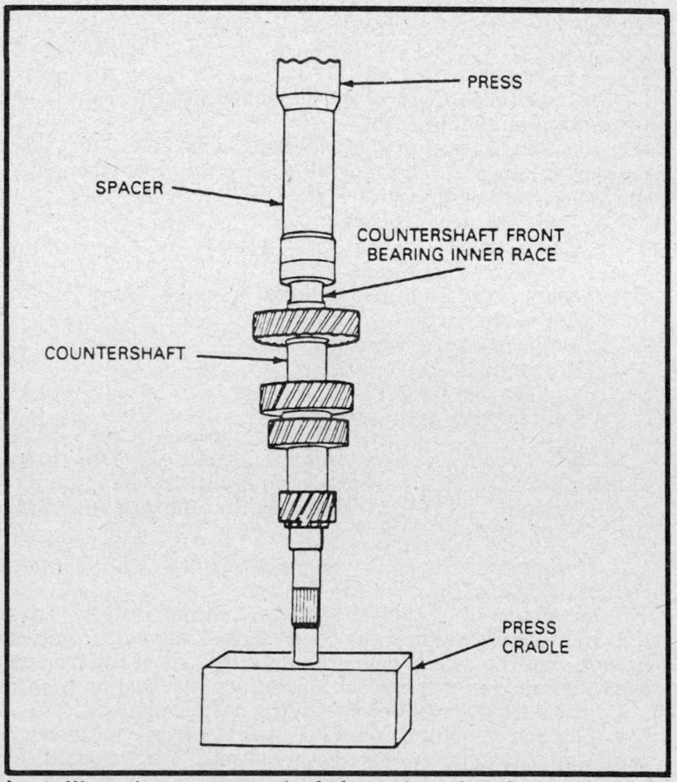
PRESS
SPACER
COUNTERSHAFT FRONT
BEARING INNER RACE
COUNTERSHAFT
PRESS
CRADLE

Installing the countershaft front bearing inner race

6. Remove the 3 rubber plugs blocking the shift rod service bores.

NOTE: Perform the following shift rail removal procedures with great care. Cover the lock ball bores and friction device and spring seats with a clean cloth held firmly in place during shift rail removal. Failure to firmly cover the lock ball bores and friction device can result in component loss when the ball/friction device and spring forcefully leave their installed positions. Be sure to wear safety goggles while performing the shift rail removal procedure.

7. Remove the 1st/2nd shift rail from the top cover through the service bore. If necessary, insert a $^5/_{16}$ in. drift punch through the spring pin bore and gently rock the shift rail from side to side while maintaining rearward pressure.

8. Remove the 3rd/4th shift rail from the top cover through the service bore. If necessary, insert a $^5/_{16}$ in. drift punch through the spring pin bore and gently rock the shift rail from side to side while maintaining rearward pressure.

9. Remove the 5th/reverse cam lockout plate retaining bolts and remove the 5th/reverse cam lockout plate.

Inspection

1. Clean and inspect the shift rail and shift fork assemblies.
2. Replace any and all worn or damaged components, as necessary.

Assembly

1. With the top cover in a suitable holding fixture. Position the 5th/reverse cam lockout plate to the top cover. Install the 5th/reverse cam lockout plate retaining bolts and torque them to 6–7 ft. lbs. (8–10mm).

2. Position the 3rd/4th shift rail into the top cover through the service bore. If necesary, insert a $^5/_{16}$ in. drift punch through the spring bore and gently rock the shift rail from side to side while maintaining forward pressure. Position the detent ball and spring into the top cover spring seats.

3. Compress the detent ball and spring assembly using a suitable tool and push the shift rail into position over the detent ball. Position the friction device and spring into the top cover seats. Compress the fiction device and spring assembly using a suitable tool and push the shift rail into position over the friction device. Install the spring pins, retaining the shift rail to top cover. Install the spring retaining 3rd/4th shift fork to the shift rail.

4. Position the 1st/2nd shift rail into the top cover through the service bore. If necesary, insert a $^5/_{16}$ in. drift punch through the spring bore and gently rock the shift rail from side to side while maintaining forward pressure. Position the detent ball and spring into the top cover spring seats.

5. Compress the detent ball and spring assembly using a suitable tool and push the shift rail into position over the detent ball. Position the friction device and spring into the top cover seats. Compress the fiction device and spring assembly using a suitable tool and push the shift rail into position over the friction device. Install the spring pins, retaining the shift rail to top cover. Install the spring retaining 1st/2nd shift fork to the shift rail.

6. Position the 5th/reverse shift rail into the top cover through the service bore. If necesary, insert a $^5/_{16}$ in. drift punch through the spring bore and gently rock the shift rail from side to side while maintaining forward pressure. Position the detent ball and spring into the top cover spring seats. Compress the detent ball and spring assembly using a suitable tool and push the shift rail into position over the detent ball. Position the friction device and spring into the top cover seats.

7. Compress the fiction device and spring assembly using a suitable tool and push the shift rail into position over the friction device. Install the spring pins, retaining the shift rail to top cover. Install the spring retaining 5th/reverse shift fork to the shift rail.

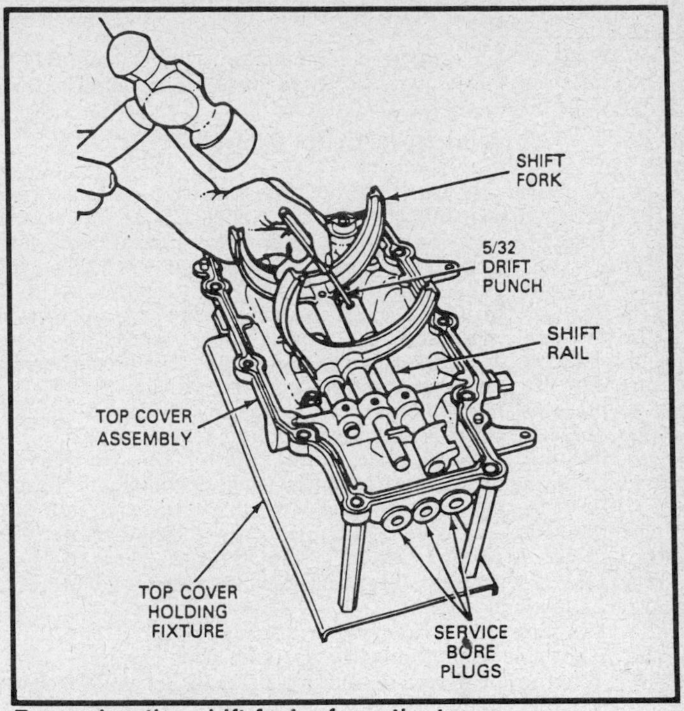
SHIFT FORK
5/32 DRIFT PUNCH
SHIFT RAIL
TOP COVER ASSEMBLY
TOP COVER HOLDING FIXTURE
SERVICE BORE PLUGS

Removing the shift forks from the top cover

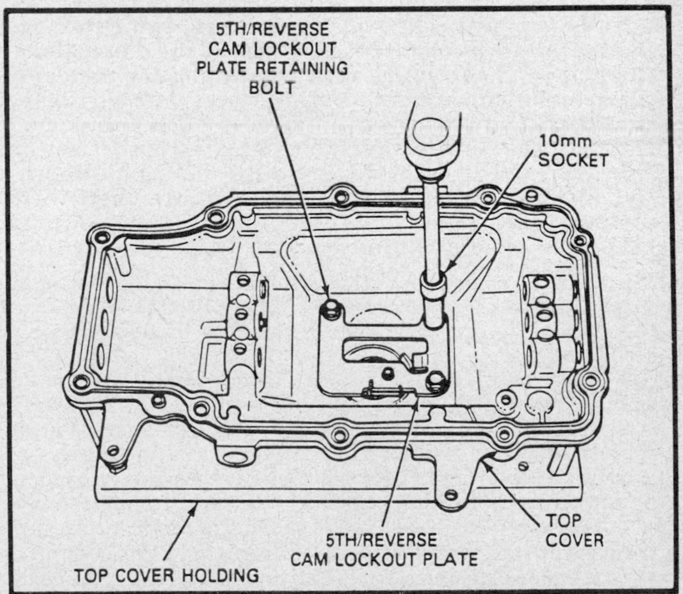
5TH/REVERSE CAM LOCKOUT PLATE RETAINING BOLT
10mm SOCKET
TOP COVER
5TH/REVERSE CAM LOCKOUT PLATE
TOP COVER HOLDING

Installing the the 5th/reverse cam lockout plate

8. Install the rubber plugs into the service bores. Install the interlock pins into the 1st/2nd and 3rd/4th shift rails. Ensure that the large and small interlock pins are installed into their original positions.

NOTE: Improper installation of the interlock pins will prevent activation of the backup lamp switch.

9. Apply a sealant to the backup lamp switch. Install the

switch to the top cover and torque the switch to 18–26 ft. lbs. (25–35 Nm).

10. Position the lower shift lever and dust cover assembly to the top cover. Install and tighten the 3 retaining screws.

Transmission Assembly

1. Position the countershaft into the transmission case through the top opening.
2. Position the input shaft into the transmission case through the top opening. Be sure that the needle roller bearing is installed into the input shaft.
3. Position the output shaft assembly into the transmission case. Mate the input shaft and output shaft assemblies by positioning them at an upward angle and setting them together. Be sure that 4th gear synchronizer ring is installed at this time.
4. Install the output shaft center bearing outer race using a brass drift. Seat the center bearing outer races.
5. Install the countershaft center bearing. Be sure that the center bearing outer races are squarley position in their bores.
6. Position the center bearing cover to the transmission case with reference arrow pointing upward. Install and tighten the center bearing cover retaining bolts and torque them to 14–19 ft. lbs. Be sure that all center bearing cover retaining bolt heads are marked with an 8.
7. Position the transmission vertically (input shaft and clutch housing facing upward). Be sure that the input shaft front bearing outer race is squarely position in the bore. If removed install the front cover oil seal using front cover seal installer tool T77J–7025–G or equivalent.
8. Install the countershaft front bearing by hand.

NOTE: If any related parts (such as outputshaft, bearing, etc.) have been replaced, measure the dimensions of the height of the input shaft bearing outer race above the transmission front bearing cover mating surface. Depth of the front cover outer race bore (input shaft). Depth of the countershaft front bearing race (transmission case to the front cover mating surface). Depth of the front cover outer race bore (countershaft). After measuring all dimensions, select bearing shim to maintain the endplay within specified limits.

9. Remove any sealant residue remaining on the mating surfaces of the transmission and front cover. To prevent damage to the oil seal lip during assembly, tape the input shaft splines along their entire length.
10. Apply a thin coat of oil to the front cover oil seal lip. Position the bearing shim and baffle plate into the front cover (install the shim with groove showing). Install the spacer to transmission case countershaft front bearing bore. If necessary apply a sufficient quantity of petroleum jelly to the shim, bearing cover and oil baffle to retain them in position during assembly.
11. Apply a ⅛ in. bead of RTV sealant to the front cover and front cover retaining bolt threads. Install the front bearing cover to the transmission case. Install and torque the front bearing cover retaining bolts to 9–17 ft. lbs. (12–22mm). Be sure the front bearing cover retaining bolt heads are marked with a 6.
12. Position the transmission horizontally in a holding fixture. Assemble the following parts in the following order, 5th gear sleeve positioning ball, 5th gear sleeve, output shaft locknut, countershaft locknut, countershaft rear bearing, thrust washer, reverse idler gear gear fixing bolt, reverse idle gear assembly, output shaft rear bearing, output shaft reverse gear, output shaft sleeve, countershaft reverse gear and needle bearing.
13. Install the reverse synchronizer ring, thrust washer, split washer, shift rod fixing bolt, shift rail/fork/hub/sleeve assembly, steel lock ball, shift rail, shift rail spring, 5th gear synchronizer ring, 5th gear output shaft, 5th gear countershaft, 5th gear needle bearing, 5th gear sleeve, ball, center bearing cover, 5th/reverse counter lever lockplate retaining bolt.

NOTE: Install the 5th gear sleeve using nut, shaft adapter replacing tool T75L–7025–L, adapter tool T88T–7025–J2, and remove and replacer tube tool T75L–7025–B or equivalents.

14. Install the 5th gear needle bearing onto the 5th gear (countershaft).
15. Install the 5th gear onto the output shaft using gear installing spacer tool T88T–7025–F, gear installation spacer tool T88T–7025–G, shaft adapter tool T75L–7025–P, shaft adapter screw tool T75L–7025–K, remover/replacer tube tool T75L–7025–B (2WD vehicles), or remover/replacer tube tool T85T–7025–A (4WD vehicles) nut and washer or equivalents. Be sure that the long flange on the 5th gear faces forward.

NOTE: On the 2WD vehicles, to install the 5th gear assembly, perform the process in 2 steps. First install the gear installing spacer tool T88T–7025–F, when the tool bottoms out add gear installing spacer tool T88T–7025–G and press the 5th gear assembly the rest of the way into position. On the 4WD vehicles, installation of the 5th gear assembly is similar to the 2WD installation except that remover/replacer tube tool T85T–7025–A and TOD bearing remover/replacer adapter tool T84T–7025–A are used.

16. Position the counter lever assembly to the transmission and install the thrust washer and retaining ring. Apply sealant to the counter lever fixing bolt threads. Install the counter lever fixing bolt and torque it to 6–7 ft. lbs. (8–10 Nm).
17. If removed, position the 5th/reverse shift fork and shift rail to the top cover. Insert the 5th/reverse shift rail through the top cover bore and 5th/reverse shift fork. Install the spring and detent ball to lower part of the rod.
18. Assemble the 5th/reverse synchronizer hub, sleeve and 5th gear synchronizer ring to the 5th/reverse shift fork and rod assembly. Be sure to install the longer flange (on the 5th/reverse hub, sleeve and synchonizer assembly) toward the front of the transmission. The reference mark on the synchronizer sleeve must be installed toward the reverse gear side.
19. Install the 5th/reverse shift fork and shift rail assembly (including 5th/reverse synchronizer hub, sleeve and 5th gear synchronizer ring) to countershaft. Mate the shift fork gate to the 5th/reverse counter lever end. Install the 5th/reverse fork and shift rail assembly with threaded fixing bolt bores (in rail and tranmission case) aligned with each other.

NOTE: For ease of assembly, position the 5th/reverse shift fork into the rearmost threaded bore of the 3 detent positions. Return the shift fork to the neutral gear position after installation.

20. Apply a suitable sealant to the 5th/reverse shift rail fixing bolt threads. Install the 5th/reverse shift rail fixing bolt to the transmission case and torque it to 16–22 ft. lbs. (20–30 Nm).
21. Apply a suitable sealant to the oil passage retaining bolt. Position the oil passage to transmission case and install the retaining bolt. Torque the oil passage retaining bolt to 6–7 ft. lbs. (8–9 Nm).
22. Install the split washer and thrust washer onto the countershaft.

NOTE: If the clutch hub and or counter reverse gear have been replaced, new split washers must be selected to maintain endplay within specified limits. Measure the endplay using a feeler gauge. Be sure the spilt washers are a matched set of identical thickness.

23. Install the reverse synchronizer ring and needle bearings onto the counter reverse gear. Install the counter reverse gear and needle bearing onto the countershaft as an assembly. Install the thrust washer to the countershaft.
24. Press the thrust washer forward against the shoulder on

the countershaft. Maintain forward pressure against the thrust washer and insert a feeler gauge between the thrust washer and counter reverse gear. Using the proper size thrust washer, bring the counter reverse gear endplay into specifications. The counter reverse gear endplay should be 0.009–0.013 in. (0.25–0.35mm).

25. Temporarily install a suitable spacer (inner bore larger than 21mm, outer bore smaller than 36mm, 15–20mm over all length) in place of the countershaft bearing. Loosely install the countershaft locknut to retain the components.

NOTE: The installation of a suitable spacer prevents the thrust washers from slipping off the shaft and avoids interference with the reverse idler gears.

25. Install the reverse idler gear assembly. Apply a suitable sealant to the threads of the reverse idler gear fixing bolt. Torque the bolt to 58–86 ft. lbs. (79–116 Nm).

27. Drive the sleeve and reverse gear assembly onto the output shaft using gear installing spacer tool T88T–7025–G, shaft adapter tool T75L–7025–P, shaft adapter screw tool T75L–7025–K, remover/replacer tube tool T75L–7025–B (2WD vehicles), or remover/replacer tube tool T85T–7025–A (4WD vehicles) nut and washer or equivalents. Install the reverse gear with the longer flange facing rearward.

28. Install the output shaft rear bearing using gear installing spacer T88T–7025–G, shaft adapter tool T75L–7025–P, shaft adapter screw tool T75L–7025–K, remover/replacer tube tool T75L–7025–B (2WD vehicles), or remover/replacer tube tool T85T–7025–A (4WD vehicles) nut and washer or equivalents. Remove the temporary spacer from the countershaft.

29. Install the countershaft rear bearing by hand. Tightening the shaft locknuts without fully seating the bearing can cause damage to the output shaft threads.

30. Lock the transmission into 1st/3rd gears. Install a new output and countershaft locknuts. Torque the output shaft locknut to 160–200 ft. lbs. (216–274 Nm) and the countershaft locknut to 94–144 ft. lbs. (128–196 Nm).

NOTE: Always install new output and countershaft locknuts when assembling the transmission. Locknuts unstaked during disassembly cannot be reused.

31. Stake the locknuts to the bottom shaft groove using countershaft locknut staking tool T77J–7025–F or equivalent.

32. Install the speedometer drive gear and steel ball to the output shaft. Install the snapring retaining the speedometer drive gear to the output shaft (2WD vehicles).

NOTE: The speedometer drive gear contains 3 detents into which the steel drive ball can be installed. The steel drive ball can be installed into any of the 3 detents.

NOTE: For reference during assembly, observe and record speedometer drive gear color. Depending upon application, 1 or 2 different speedometer drive gear may be installed. It will be color coded either green or white. Speedometer drive gear colors and corresponding part numbers are as follows, white E8TZ–17285–B and green E8TZ–17285–C.

33. Remove any sealant residue from the mating surfaces of the transmission case and extension housing. Apply a 1/8 in. bead of RTV sealant or equivalent to the transmission case.

NOTE: The extension housing bushing cannot be serviced. If the bushing requires service, the extension housing must be replaced as a unit.

34. Position the extension housing to the transmission case and install the extension housing retaining bolts. Torque the bolts to 24–34 ft. lbs. (32–46 Nm). Place the synchronizers into the neutral gear position. Be sure that the shift forks on the top cover assembly are in the neutral position.

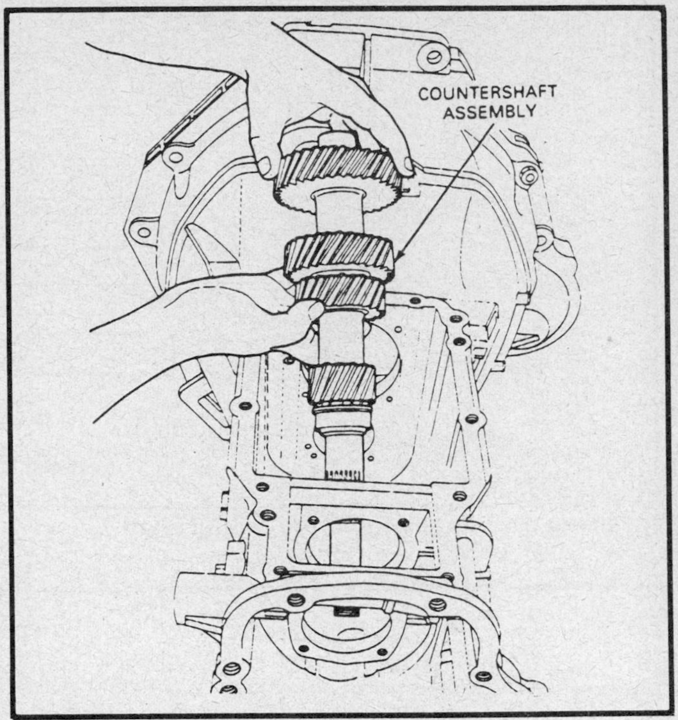
COUNTERSHAFT
ASSEMBLY

Installing the countershaft assembly into the transmission case

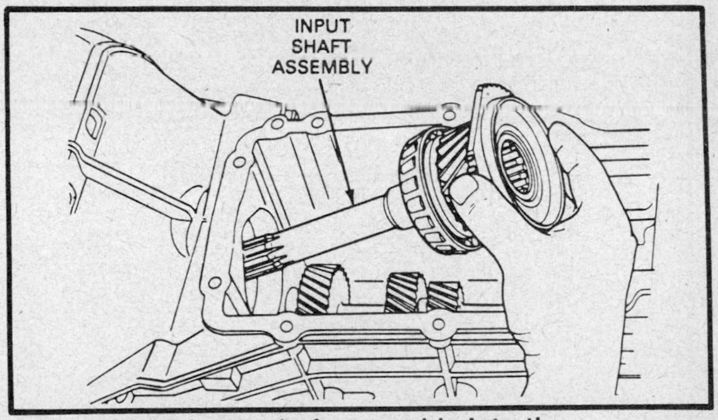
INPUT
SHAFT
ASSEMBLY

Installing the input shaft assembly into the transmission case

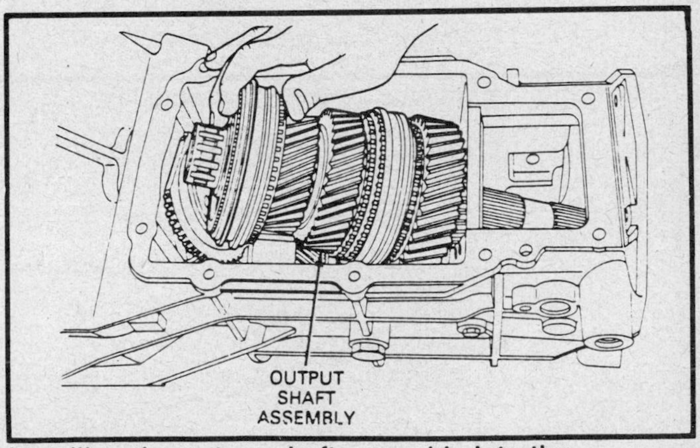
OUTPUT
SHAFT
ASSEMBLY

Installing the output shaft assembly into the transmission case

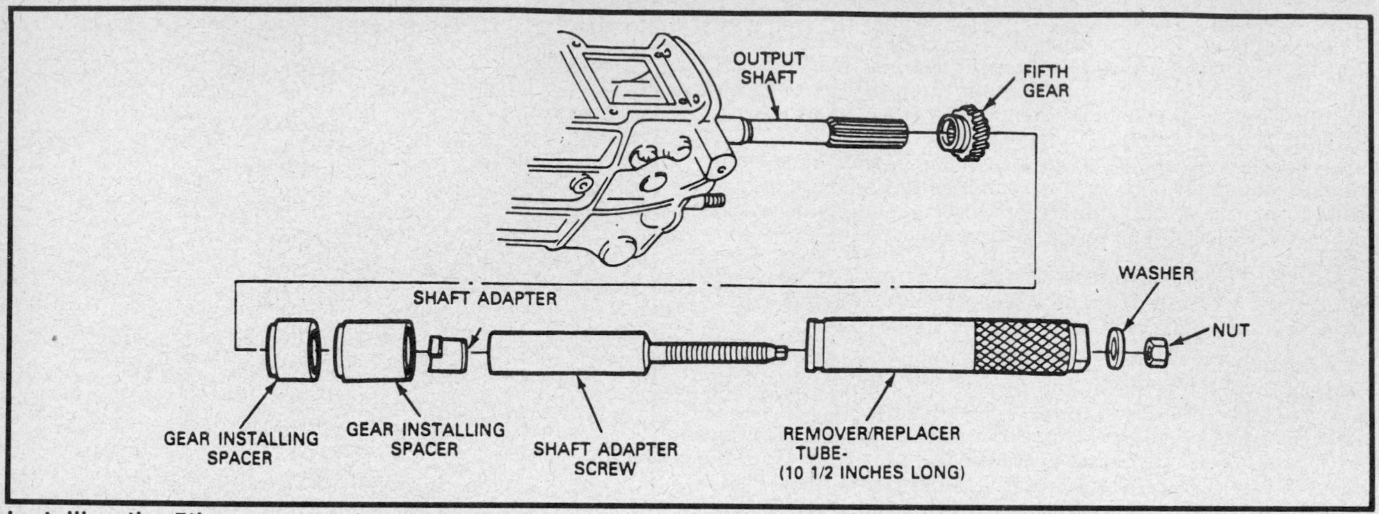
OUTPUT SHAFT
FIFTH GEAR
WASHER
NUT
SHAFT ADAPTER
GEAR INSTALLING SPACER
GEAR INSTALLING SPACER
SHAFT ADAPTER SCREW
REMOVER/REPLACER TUBE-
(10 1/2 INCHES LONG)

Installing the 5th gear onto the output shaft

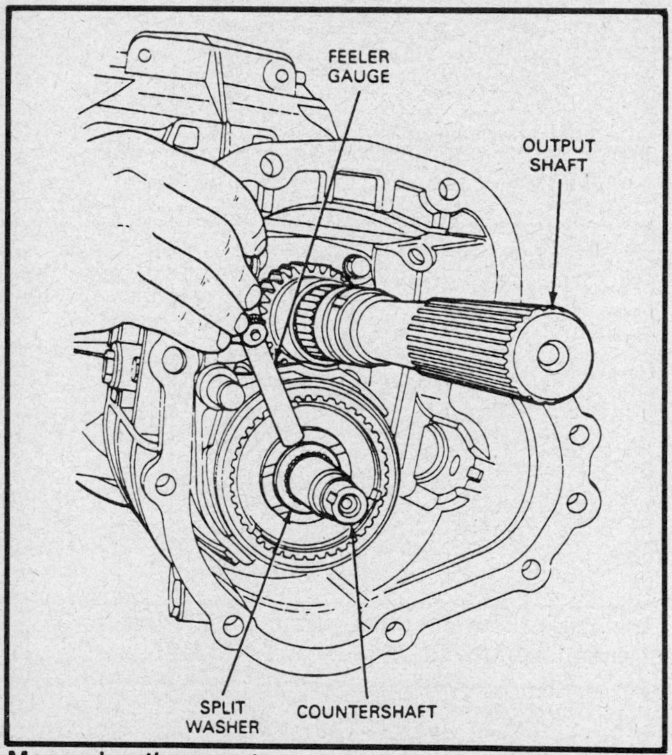
FEELER GAUGE
OUTPUT SHAFT
SPLIT WASHER
COUNTERSHAFT

Measuring the counter reverse gear endplay

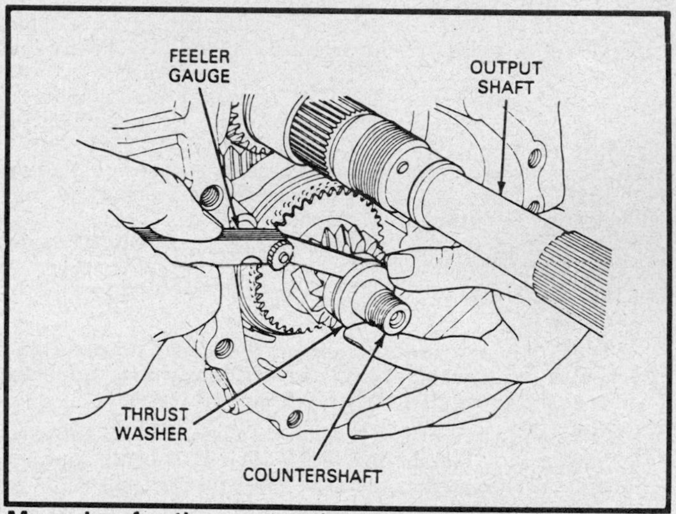
FEELER GAUGE
OUTPUT SHAFT
THRUST WASHER
COUNTERSHAFT

Meauring for the proper thrust washer thickness

35. Position the top cover on the transmission case. Carefully engage the shift forks with the synchronizers.

36. Apply a suitable sealant to the 2 rearmost top cover retaining bolts and install them into the top cover rear retaining bolt locations. Install the remaining top cover retaining bolts (no sealant) and torque them to 12–16 ft. lbs. (16–22 Nm).

NOTE: Do not apply sealant to the top cover or transmission mating surfaces. If necessary, apply a small quantity of grease to the sealing gasket to retain the gasket in position during assembly.

37. If removed, install the rear oil seal into the extension housing using extension housing seal replacer tool T61L–7657–A or equivalent. Be sure that the oil seal drain hole faces downward.

38. Refill the transmission with the recommended lubricant to the proper level. Install the transmission drain plug and torque it to 29–43 ft. lbs. (40–58 Nm).

SPECIFICATIONS

TRANSMISSION CASE SHIMS

Part Number	Thickness in.	Thickness mm
E8TZ-7029-FA	0.0551	1.4
E8TZ-7029-GA	0.0590	1.5
E8TZ-7029-Ha	0.0629	1.6
E8TZ-7029-Ja	0.0669	1.7
E8TZ-7029-S	0.0708	1.8
E8TZ-7029-T	0.0748	1.9
E8TZ-7029-U	0.0787	2.0
E8TZ-7029-V	0.0826	2.1
E8TZ-7029-W	0.0866	2.2
E8TZ-7029-X	0.0905	2.3
E8TZ-7029-Y	0.0944	2.4
E8TZ-7029-Z	0.0984	2.5
E8TZ-7029-AA	0.1023	2.6
E8TZ-7029-BA	0.1062	2.7
E8TZ-7029-CA	0.1102	2.8
E8TZ-7029-DA	0.1141	2.9
E8TZ-7029-EA	0.1181	3.0
E8TZ-7C434-K	0.122	3.1
E8TZ-7C434-L	0.125	3.2
E8TZ-7C404 M	0.129	3.3
E8TZ-7C434-N	0.133	3.4
E8TZ-7C434-P	0.137	3.5
E8TZ-7C434-R	0.141	3.6
E8TZ-7C434-S	0.145	3.7
E8TZ-7C434-T	0.1181	3.0

TORQUE SPECIFICATIONS

Description	ft. lbs.	Nm
Output shaft locknut	160–203	216–274
Countershaft locknut	94–144	128–196
Extension housing retaining bolts	24–34	32–46
Reverse idler shaft fixing bolt	58–86	79–116
Center bearing cover	14–19	18–26
Front bearing cover	12–17	16–22
Fifth/reverse cam lockout plate	6–7	8–10
Dust boot	6–8	8–11
Top cover	12–16	16–22
Filler plug	29–43	40–58
Front oil passage	6–8	8–10
Counter lever shaft fixing bolt	6–8	8–10
Rock plate	6–7	8–10
Drain plug	29–43	40–58
Backup lamp switch	18–26	25–35
Neutral switch (if equipped)	18–26	25–35
Rear oil passage (extension housing)	5–7	8–9
5th/reverse shift rail fixing bolt	16–22	20–30

COUNTER REVERSE GEAR SPLIT WASHER

Part Number	Thickness in.	Thickness mm
E8TZ-7R482-A	0.118	3.0
E8TZ-7R482-B	0.122	3.1
E8TZ-7R482-C	0.125	3.2
E8TZ-7R482-D	0.129	3.3
E8TZ-7R482-E	0.133	3.4
To be determined	0.120	3.05
To be determined	0.124	3.15
To be determined	0.127	3.25
To be determined	0.131	3.35
To be determined	0.135	3.45
To be determined	0.137	3.50

THRUST WASHER

Part Number	Thickness in.	Thickness mm
E8TZ-7C340-A	0.293	7.45
E8TZ-7C340-B	0.301	7.65
E8TZ-7C340-C	0.309	7.85
E8TZ-7C340-D	0.289	7.35
E8TZ-7C340-E	0.297	7.55
E8TZ-7C340-F	0.305	7.75

SPECIAL TOOLS

Number	Description
T74P-77248-A	Extension housing seal remover
T77J-7025-F	Countershaft locknut staking tool
T88T-7025-A	Mainshaft locknut wrench
T77J-7025-B	Remover/replacer tube
T75L-7025-B	Remover/replacer tube
T84T-7025-B	TOD forcing screw
T77L-7025-H	Bearing puller
T77J-7025-L	Puller ring
T75L-7025-G	Bearing collet sleeve for 3.5 in. bearing collets
T85T-7025-A	Remover/replacer tube
T84T-7025-A	TOD bearing remover/replacer adapter
T88T-7061-A	Gear removal collet
T88T-7120-A	Bearing race puller
T50T-100-A	Slide hammer
T53T-4621-B	Bearing replacer
T88T-7025-B	Bearing cone replacer
T75L-1165-B	Bearing plate
T71P-4621-B	Bearing cone remover
D84L-1123-A	Bearing splitter
T88T-7025-C	Top cover holding fixture
D84L-7000-B	Roll pin punch set
T88T-7025-F	Gear installing spacer
T88T-7025-G	Gear installing spacer
T75T-7025-P	Shaft adapter
T75L-7025-K	Shaft adapter screw
T61L-7657-A	Extension housing seal replacer
TOOL-4201-C	Dial indicator
T77J-7025-G	Front cover seal installer
T75L-7025-L	Shaft adapter—replacing
T88T-7025-J2	Adapter
T88T-7025-J	Countershaft fifth gear sleeve puller
T88T-7025-J1	Gear removal collets
D82L-4201-C	Depth micrometer

Section 7

Toyo Kogyo Transmission
Ford Motor Co.

APPLICATION

1984–87 Bronco II and Ranger

GENERAL DESCRIPTION

The 5 speed manual overdrive transmission is fully synchronized with all gears except reverse, which is in constant mesh. All forward speed gears are helical cut for quiet running. The reverse gear and reverse idler gear are spur cut. The gearshift mecahanism is a direct control with a floor shift. The floor shift mechanism is built into the extension housing. The transmission is available with both 2WD and 4WD vehicles.

Transmission Identification

The transmission can be identified by a tag attached to the upper right hand portion of the transmission case, just behind the bellhousing. The transmission case is of light metal construction with removable clutch and extension housings.

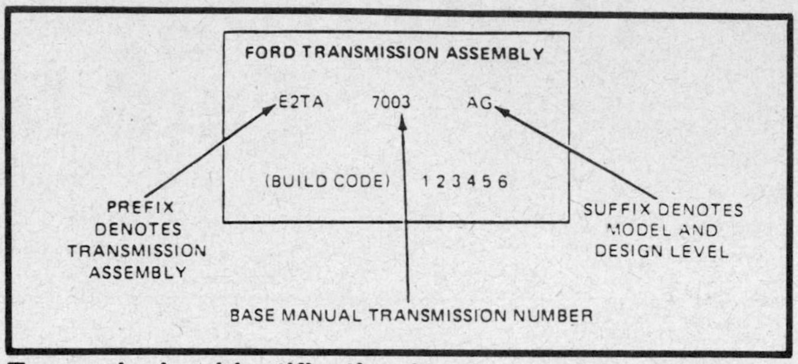
FORD TRANSMISSION ASSEMBLY
E2TA
7003
AG
(BUILD CODE)
1 2 3 4 5 6
PREFIX DENOTES TRANSMISSION ASSEMBLY
SUFFIX DENOTES MODEL AND DESIGN LEVEL
BASE MANUAL TRANSMISSION NUMBER
Transmission identification tag

Metric Fasteners

Metric fastener dimensions are very close to the dimensions of the inch system fasteners, and for that reason, replacement fasteners must have the same measurement and strength as those removed. Do not attempt to interchange metric fasteners for inch fasteners. Mismatched or incorrect fasteners can result in damage to the transaxle unit through malfunctions or breakage, or even personal injury. Care should be taken to reuse the fasteners in the same location as removed.

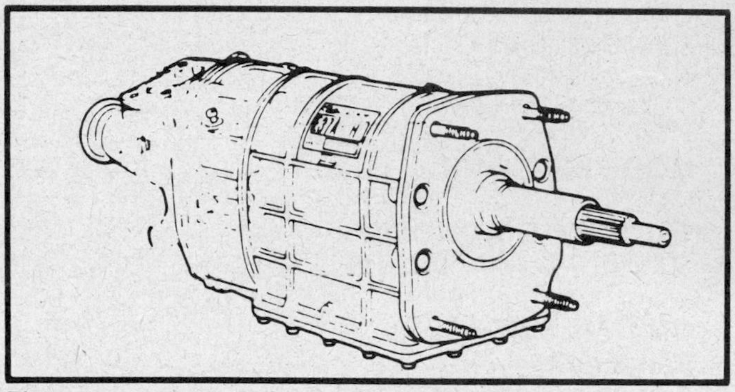
Five speed manual overdrive transmission

Capacities

The 5 speed manual overdrive transmission refill capacity is 3.6 pints (1.7L) of API GL4 or GL5 SAE 80W-90 gear lubricant.

Checking Fluid Level

With the vehicle on a level surface and the lubricant in the transmission cold, remove the plug and check the lubricant level. The lubricant should come to the bottom edge of the plug hole. If the level is low, add API GL4 or GL5 SAE 80W-90 gear lubricant through the filler plug hole until it begins to run out of the plug hole. Reinstall and tighten the plugs.

TRANSMISSION MODIFICATIONS

There are no transmission modifications at the time of publication.

TROUBLE DIAGNOSIS

CHILTON'S THREE C'S TRANSMISSION DIAGNOSIS
5 SPEED—TOYO KOGYO

Condition	Cause	Correction
Transmission shifts hard	a) Transmission fluid low or improper type	a) Add lubricant or change lubricant as required
	b) Shift lever binding or worn	b) Remove cap from shift tower. Eliminate binding condition or replace components as required

CHILTON'S THREE C'S TRANSMISSION DIAGNOSIS
5 SPEED—TOYO KOGYO

Condition	Cause	Correction
Transmission shifts hard	c) Worn or damaged internal shift mechanism	c) Remove transmission cover. Check internal shift mechanism by shifting into and out of all gears. Repair or replace as required
	d) Binding of sliding gears and/or synchronizers	d) Check for free movement of gears and synchronizers. Repair or replace as required
	e) Housings and/or shafts out of alignment	e) Remove transmission and check for binding condition between input shaft and engine crankshaft pilot bearing or bushing. Check flywheel housing alignment. Repair or replace as required
Noisy in forward gears ①	a) Lubricant level low, or improper type	a) Add lubricant, or refill with specified lubricant
	b) Components grinding on transmission	b) Check for screws, bolts, etc., of cab or other components grinding. Correct as required
	c) Component housing bolts loose	c) Check torque on transmission to flywheel housing bolts, output shaft flange nut and flywheel housing to engine block bolts. Tighten bolts to specification
	d) Flywheel housing to engine crankshaft alignment	d) Check and align flywheel housing to engine crankshaft
	e) Noisy bearings or gears	e) Remove and disassemble transmission. Inspect input, output and countershaft bearings. Inspect speedometer gear and gear teeth for wear or damage. Replace as required
Gears clash when shifting from one forward gear to another	a) Engine idle speed too high	a) Adjust engine idle speed
	b) Improper manual shift linkage	b) Adjust and repair manual shift linkage as required
	c) Pilot bearing binding	c) Remove transmission and check for a binding condition between input shaft and engine crankshaft pilot bearing. Replace as required
	d) Damaged gear teeth and/or synchronizer	d) Disassemble transmission, repair or replace as required
Transmission jumps out of gear	a) Loose transmission to clutch housing mounting bolts, or loose levers	a) Tighten transmission to flywheel housing, and flywheel housing to engine block bolts to specifications. Loosen all bolts and reseat flywheel housing. Tighten all bolts. Tighten levers if necessary
	b) Flywheel housing to engine crankshaft out of line	b) Shim or replace housing as required
	c) Crankshaft pilot bearing worn	c) Replace bearing
	d) Interior components damage	d) Disassemble transmission. Inspect the synchronizer sleeves for free movement on their hubs. Inspect the synchronizer blocking rings for widened index slots, rounded clutch teeth and smooth internal surface. Check countershaft cluster gear for excessive end play. Check shift forks for loose mounting on shift rails. Inspect synchronizer sliding sleeve and gear clutch teeth for wear or damage. Repair or replace as required

CHILTON'S THREE C'S TRANSMISSION DIAGNOSIS 5 SPEED—TOYO KOGYO

Condition	Cause	Correction
Transmission jumps out of gear	e) Worn gear teeth due to partial engagement	e) Replace worn or damaged gears
Transmission will not shift into one gear—all others OK	a) Back-up switch ball frozen	a) If reverse is problem, check back-up switch for ball frozen in extended position (if so equipped)
	b) Internal components	b) Remove transmission. If transmission will not shift into reverse, remove transmission, check for damaged reverse gear train, in single rail shift transmission. Also, check for misaligned reverse relay lever. Inspect shift rail and fork system synchronizer system and gear clutch teeth for restricted travel. Repair or replace as required
Transmission is locked in one gear. It cannot be shifted out of that gear	a) Manual-shift linkage out of adjustment, binding or damaged	a) Disconnect the problem shift rod from transmission shift lever. Try to shift the transmission lever into and out of gear position. If OK, repair or replace linkage parts
	b) Internal components	b) Remove transmission. Inspect problem gears, shift rails and forks and synchronizer for wear or damage. Repair as required. Check for broken fork slot tabs on single rail transmissions
	c) Loose fork on rail	c) On single rail shift, check for broken selector arm pin or selector plate. Repair or replace as necessary
Transmission leaks	a) Improper amount of lubricant—wrong type	a) Check level and type. Fill to bottom of filler plug hole
	b) Other component leaking	b) Identify leaking fluid as engine, power steering or transmission. Repair as required
	c) False report	c) Remove all traces of lube on exposed transmission surfaces. Check vent for free breathing. Operate transmission and inspect for new leakage. Repair as required
	d) Internal components	d) Remove transmission. Inspect for leaks at the input shaft bearing retainer seal and gasket, and shift rail expansion plug. Inspect for leaks at the top cover gasket. Inspect case for sand holes or cracks. Repair or replace as required
	e) Improper installation torque	e) Tighten to specified torque value
Shift lever loose	a) Retaining bolts loose	a) Tighten to specified torque
	b) Improper stack-up of attaching parts	b) Add shims to the top of the upper bushing. Refer to Adjustments in Section 16-22, Five-Speed Manual Transmission
Shift lever tight	a) Improper stack-up of attaching parts	a) Add shims below the lower boot. Refer to Adjustments in Section 16-22, Five-Speed Manual Transmission

CHILTON'S THREE C'S TRANSMISSION DIAGNOSIS
5 SPEED—TOYO KOGYO

Condition	Cause	Correction
Shift lever tight	b) Worn or damaged shift lever lower bushing	b) Inspect shift lever assembly. Remove and replace parts as required

① While verifying the condition, determine whether the noise is gear roll-over noise, release bearing rub or some other transmission related noise.

Gear roll-over noise, inherent in manual transmissions, is caused by the constant mesh gears turning at engine idle speed, while the clutch is engaged and the transmission is in neutral; and release bearing rub is sometimes mistaken for mainshaft bearing noise.

Gear roll-over noise will disappear when the clutch is disengaged or when the transmission is engaged in gear. Release bearing rub will disappear when the clutch is engaged. In the event that a bearing is damaged, the noise is more pronounced while engaged in gear under load or coast than in neutral.

ON CAR SERVICE

Services

GEARSHIFT LEVER

Removal and Installation

1. Disconnect the negative battery cable.
2. Place the gearshift lever in the **N** position.
3. Remove the boot retainer screws.
4. Remove the bolts attaching the gearshift lever assembly to the transmission.
5. Pull the gearshift lever assembly straight up and away from the gearshift lever retainer.
6. Cover the shift lever opening in the extension housing with a cloth to avoid dropping dirt into the transmission.
7. Installation is the reverse of the removal procedure.

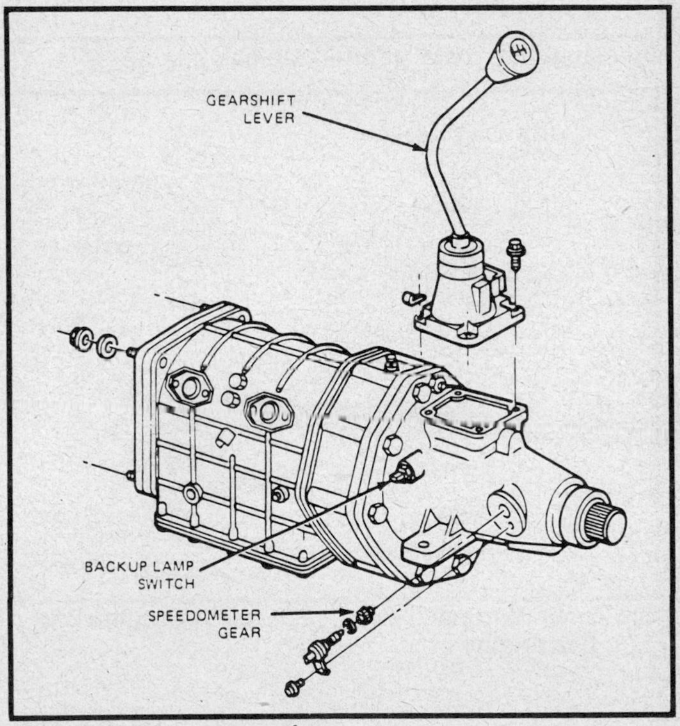
GEARSHIFT LEVER
BACKUP LAMP SWITCH
SPEEDOMETER GEAR
Gearshift lever removal

REMOVAL AND INSTALLATION

TRANSMISSION REMOVAL

1. Disconnect the negative battery cable.
2. Remove the gearshift lever assembly.
3. Raise the vehicle and support it safely.
4. Disconnect the driveshaft from the rear axle drive flange. Pull the driveshaft rearward and disconnect it from the transmission.
5. Install a suitable plug in the extension housing to prevent lubricant leakage.
6. Disconnect the clutch hydraulic line at the clutch housing. Plug both ends of the line to prevent fluid leakage and the entry of dirt.
7. Disconnect the speedometer cable from the extension housing.
8. Disconnect the starter motor and backup lamp switch wires.
9. Position a suitable jack under the engine, protecting the oil pan.
10. On 4WD vehicles, remove the transfer case.
11. Remove the starter motor.
12. Position a suitable jack under the transmission.
13. Remove the bolts attaching the transmission to the engine rear plate.
14. Remove the nuts and bolts attaching the transmission mount and damper to the crossmember.

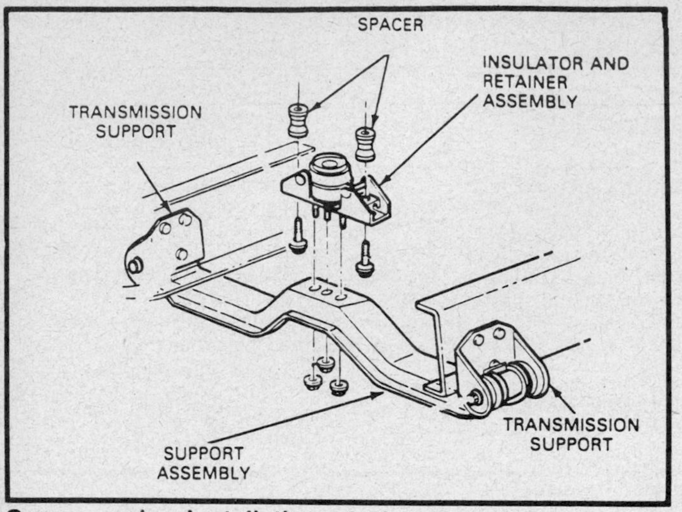
SPACER
INSULATOR AND RETAINER ASSEMBLY
TRANSMISSION SUPPORT
TRANSMISSION SUPPORT
SUPPORT ASSEMBLY

Crossmember installation—2.3L gas engine

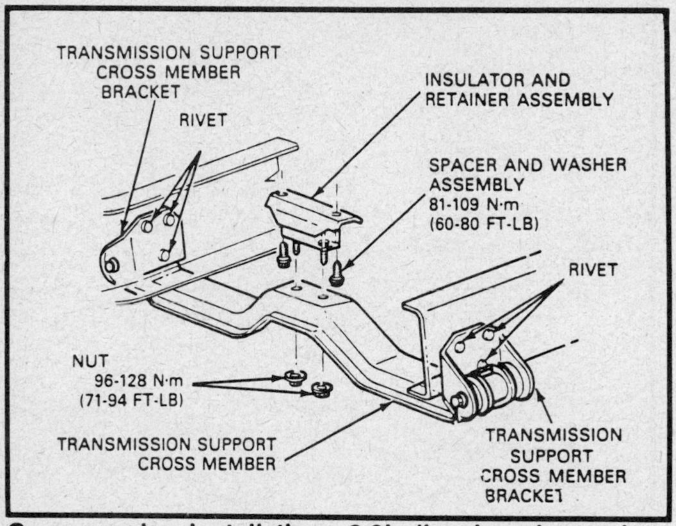
TRANSMISSION SUPPORT CROSS MEMBER BRACKET
RIVET
INSULATOR AND RETAINER ASSEMBLY
SPACER AND WASHER ASSEMBLY
81-109 N·m
(60-80 FT-LB)
RIVET
NUT
96-128 N·m
(71-94 FT-LB)
TRANSMISSION SUPPORT CROSS MEMBER
TRANSMISSION SUPPORT CROSS MEMBER BRACKET

Crossmember installation—2.3L diesel engine and 2.9L gas engine

15. Remove the nuts attaching the crossmember to the frame side rails and remove the crossmember.
16. Lower the engine jack. Work the clutch housing off of the locating dowels and slide the transmission rearward until the input shaft spline clears the clutch disc.
17. Remove the transmission from the vehicle.

TRANSMISSION INSTALLATION

1. Check that the machined mating surfaces and the locating dowels on the engine rear plate are free of burrs, dirt or paint.
2. Check the mating face of the clutch housing and the locating dowel holes for burrs, dirt or paint.
3. Mount the transmission on a suitable transmission jack.
4. Position the transmission under the vehicle and start the input shaft into the clutch disc. Align the splines on the input shaft with the splines on the clutch disc.
5. Move the transmission forward and seat the clutch housing on the locating dowels of the engine rear plate.
6. Install the bolts that attach the clutch housing to the engine rear plate and tighten to 28–38 ft. lbs. (38–51 Nm).
7. Remove the transmission jack.
8. Install the starter motor.
9. Raise the engine and install the rear crossmember, insulator and damper. Tighten the nuts to 65–85 ft. lbs. (88–115 Nm).
10. Install the bolts, nuts and washers attaching the transmission mount to the crossmember. Tighten the bolts to 60–80 ft. lbs. (81–109 Nm).
11. Remove the engine jack.
12. On 4WD vehicles, install the transfer case.
13. Install the driveshaft into the extension housing and install the center bearing attaching nuts, bolts and lockwashers.
14. Connect the driveshaft to the rear axle drive flange.
15. Connect the starter and backup lamp switch wires.
16. Connect the hydraulic clutch line and bleed the system.
17. Connect the speedometer cable.
18. Check the transmission fluid level.
19. Lower the vehicle.
20. Connect the clutch master cylinder pushrod to the clutch pedal.
21. Remove the cap from the shift lever opening in the extension housing.
22. Position the gear shift lever assembly straight up above the transmission, making sure that the ball end of the shift lever rod is inserted into the gear shift control lever socket.
23. Install the cover boot and retainer screws.

BENCH OVERHAUL

Before Disassembly

Clean the exterior of the transmission assembly before any attempt is made to disassemble it, in order to prevent dirt or other foreign material from entering the transmission assembly and damaging its internal parts. If steam cleaning is done to the exterior of the transmission, immediate disassembly should be done to avoid rusting, caused by condensation forming on the internal parts.

Transmission Disassembly

1. Remove the nuts attaching the bellhousing to the transmission case. Remove the bellhousing and gasket.
2. Remove the drain plug and drain the lubricant from the transmission into a suitable container. Clean the metal filings from the magnet of the drain plug, if necessary.
3. Secure the transmission in a suitable holding fixture.

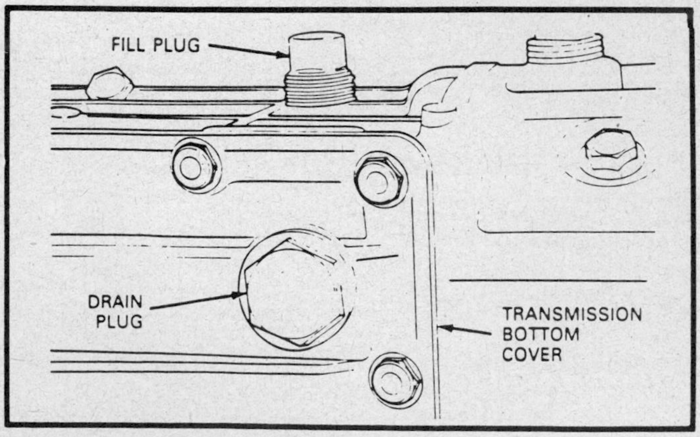
FILL PLUG
DRAIN PLUG
TRANSMISSION BOTTOM COVER

Transmission drain plug

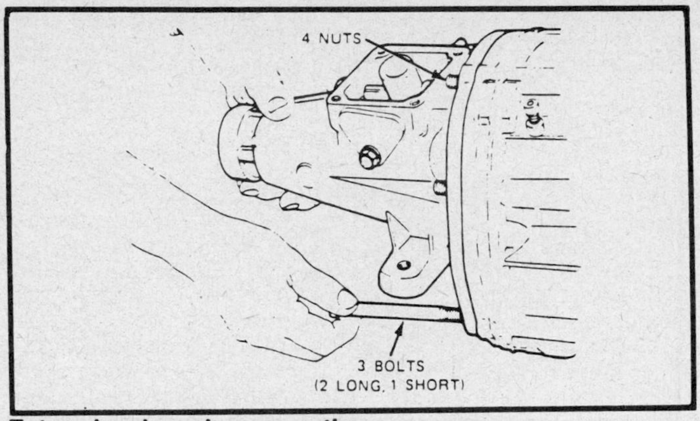
4 NUTS
3 BOLTS
(2 LONG, 1 SHORT)

Extension housing mounting

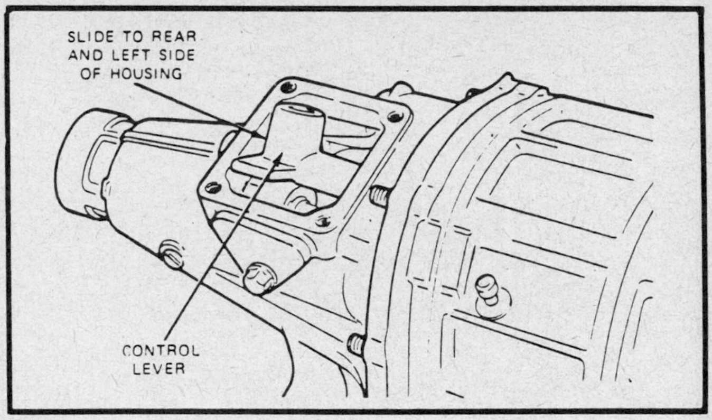
SLIDE TO REAR
AND LEFT SIDE
OF HOUSING
CONTROL
LEVER

Shift control lever

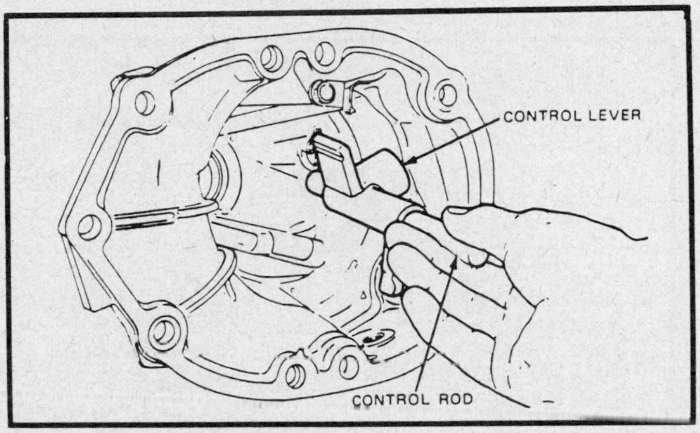
CONTROL LEVER
CONTROL ROD

Removing control lever and rod

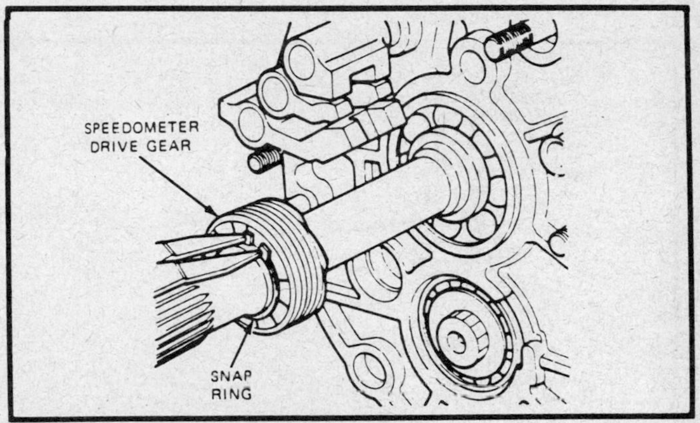
SPEEDOMETER
DRIVE GEAR
SNAP
RING

Removing speedometer drive gear

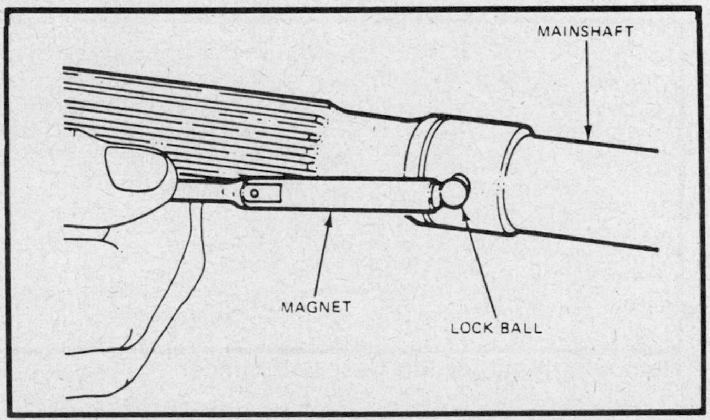
MAINSHAFT
MAGNET
LOCK BALL

Removing speedometer drive gear lock ball

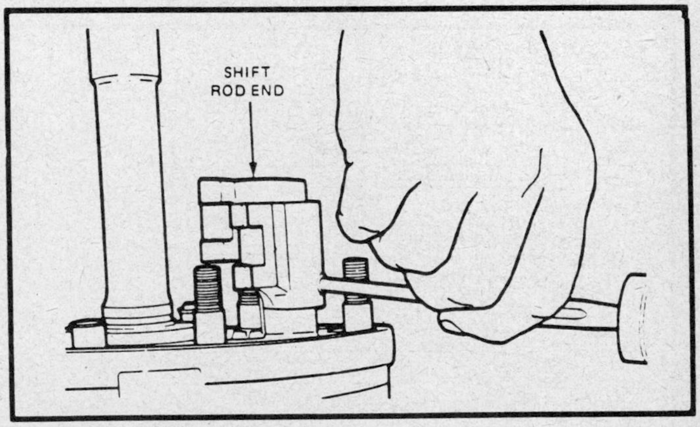
SHIFT
ROD END

Removing shift rod end

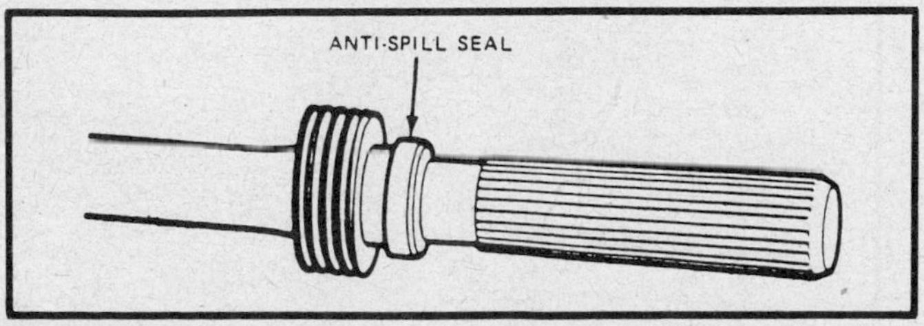
ANTI-SPILL SEAL

Anti-spill seal

4. Place the transmission in neutral.
5. Remove the speedometer sleeve and driven gear assembly from the extension housing.
6. Remove the bolts and nuts attaching the extension housing to the transmission case.

NOTE: There are 2 longer outer bolts and 1 shorter center (bottom) bolt used.

7. Raise the control lever to the left and slide it toward the rear of the transmission.
8. Slide the extension housing off of the mainshaft.

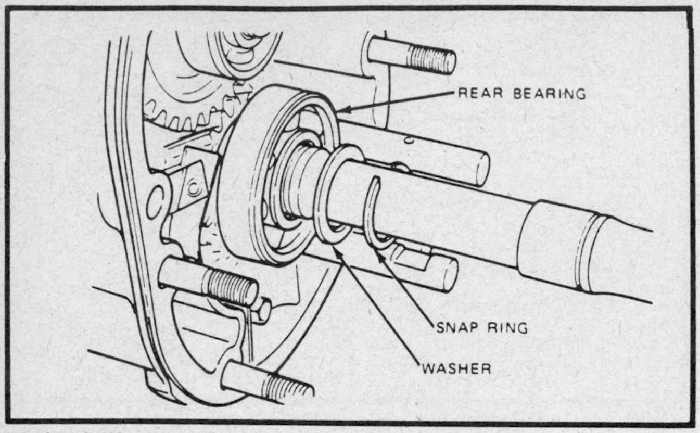
REAR BEARING
SNAP RING
WASHER
Mainshaft snapring and washer removal

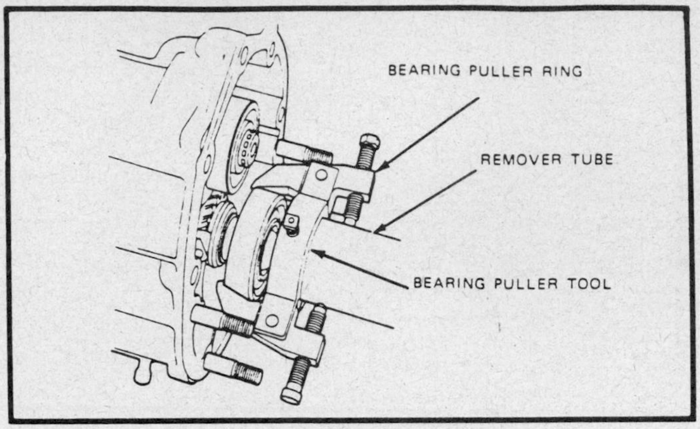
BEARING PULLER RING
REMOVER TUBE
BEARING PULLER TOOL
Installing mainshaft rear bearing puller tools

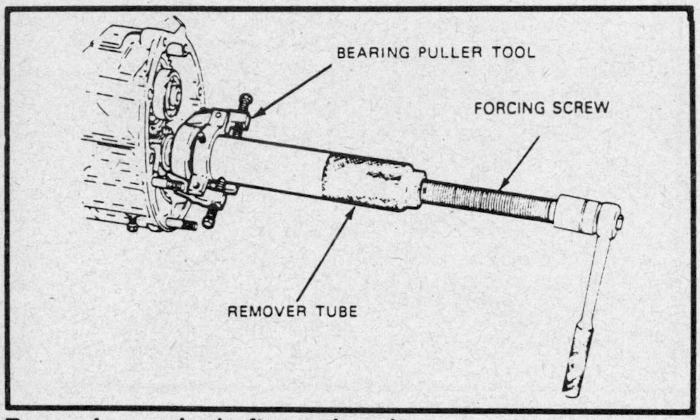
BEARING PULLER TOOL
FORCING SCREW
REMOVER TUBE
Removing mainshaft rear bearing

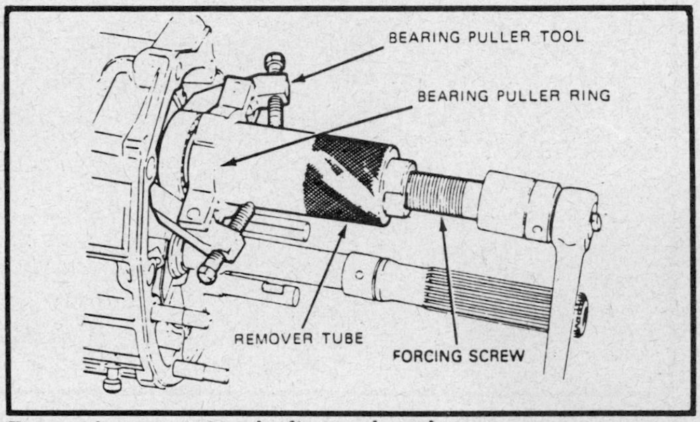
BEARING PULLER TOOL
BEARING PULLER RING
REMOVER TUBE
FORCING SCREW
Removing countershaft rear bearing

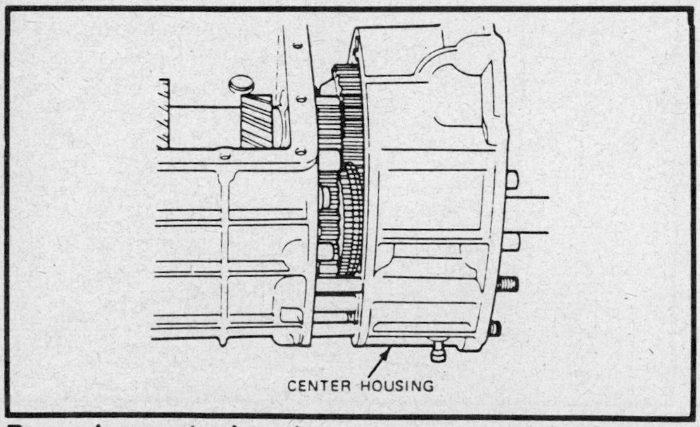
CENTER HOUSING
Removing center housing

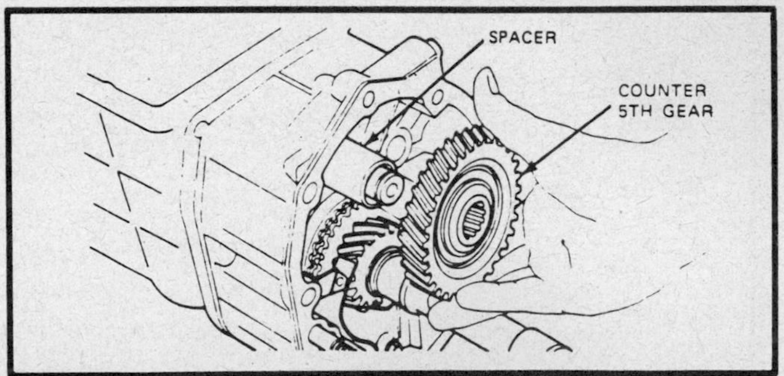
SPACER
COUNTER
5TH GEAR
Removing counter 5th gear and spacer

9. Pull the control lever and rod out of the front end of the extension housing.
10. If required, remove the backup lamp switch from the extension housing.
11. Remove the anti-spill seal from the mainshaft and discard it. A seal is not necessary for assembly.
12. Remove the snapring that secures the speedometer drive gear to the mainshaft. Slide the drive gear off the mainshaft and remove the lockball.
13. Evenly loosen the bolts securing the transmission case cover to the transmision case. Remove the cover and gasket.

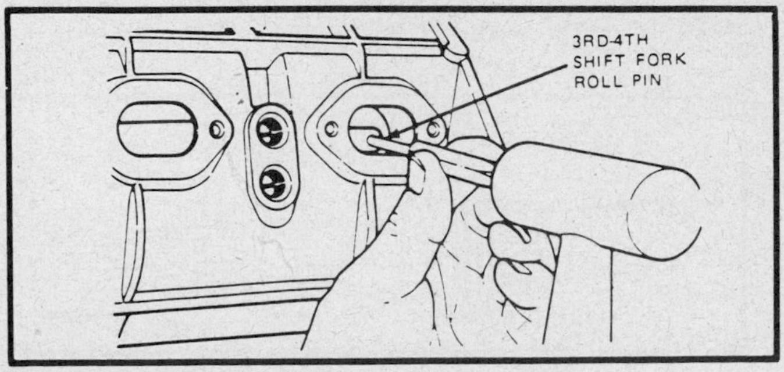
3RD-4TH
SHIFT FORK
ROLL PIN
Removing 3rd and 4th shift fork roll pin

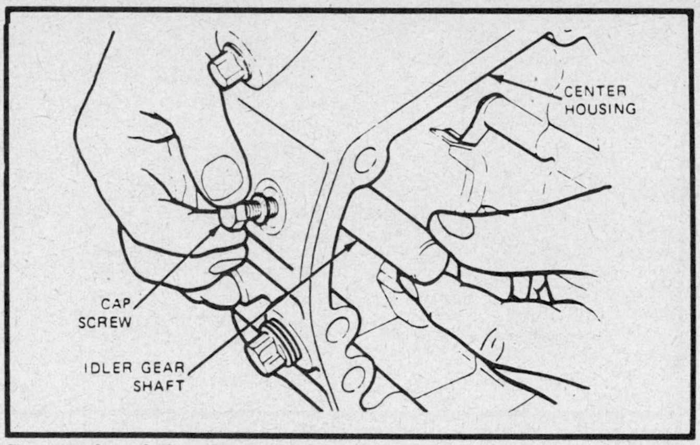
CENTER HOUSING
CAP SCREW
IDLER GEAR SHAFT
Removing idler gear shaft

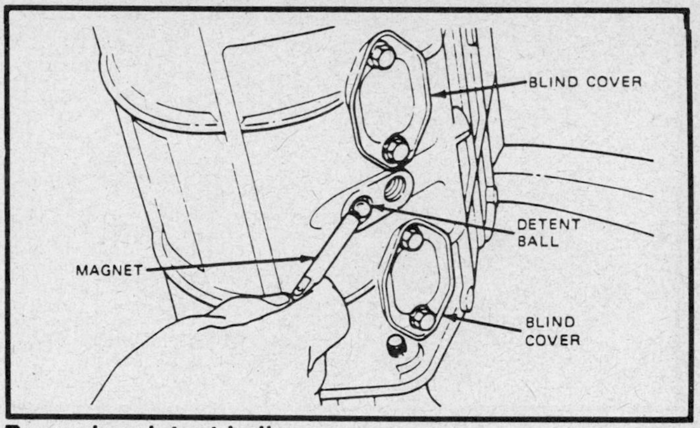
BLIND COVER
DETENT BALL
MAGNET
BLIND COVER
Removing detent balls

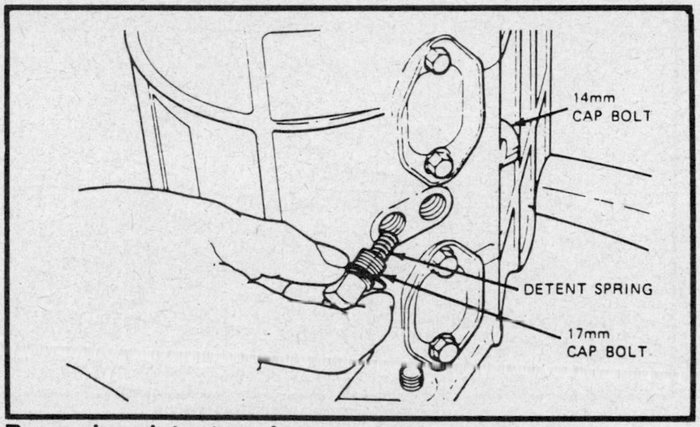
14mm CAP BOLT
DETENT SPRING
17mm CAP BOLT
Removing detent springs and cap bolts

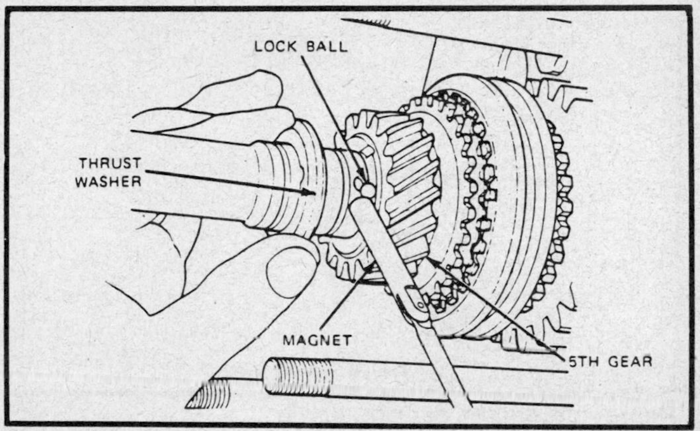
LOCK BALL
THRUST WASHER
MAGNET
5TH GEAR
Removing trust washer and lock ball

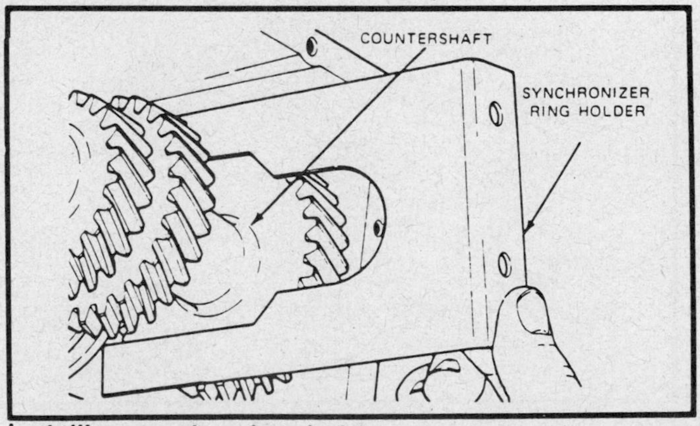
COUNTERSHAFT
SYNCHRONIZER RING HOLDER
Installing synchronizer holder

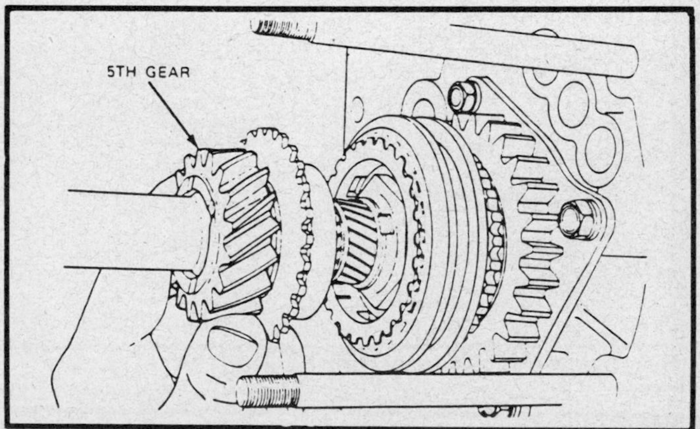
5TH GEAR
Removing 5th gear

14. Mark the shift rails and forks to aid during transmission assembly. Remove the roll pins attaching the shift rod ends to the shift rod and remove the shift rod ends.

15. Pry the bearing housing away from the transmission case without damaging the housing or case. Slide the bearing housing off of the mainshaft.

16. Remove the snapring and washer retaining the mainshaft rear bearing to the mainshaft.

17. Using the proper tools, remove the mainshaft rear bearing.

18. Remove the snapring from the rear end of the countershaft and using the proper tools, remove the countershaft rear bearing.

19. Remove the counter 5th gear and spacer from the rear of the countershaft.

20. Remove the center housing. Remove the reverse idler gear and 2 spacers with the housing.

21. Remove the screw from the center housing and remove the idler gear shaft.

22. Remove the 3 spring bolts, 2 on the case upper portion and 1 on the case side. Remove the detent springs and balls from the

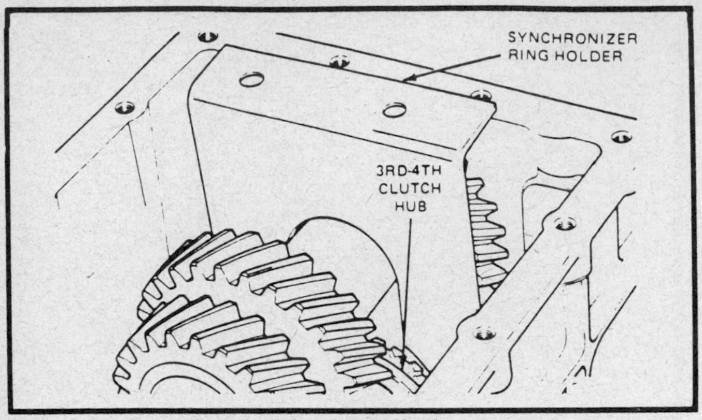
SYNCHRONIZER RING HOLDER
3RD-4TH CLUTCH HUB

Synchronizer ring holder installed

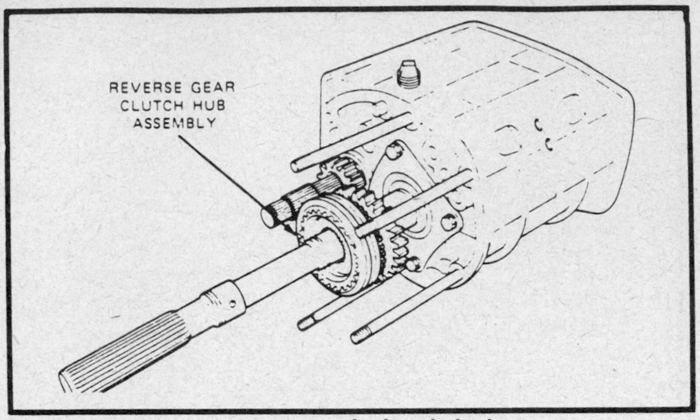
REVERSE GEAR CLUTCH HUB ASSEMBLY

Removing reverse gear and clutch hub

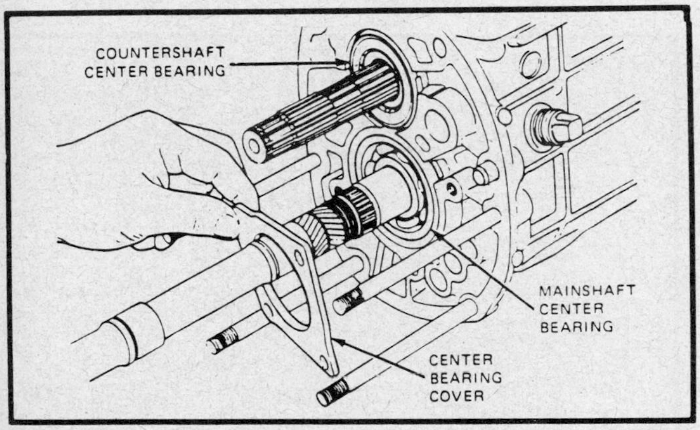
COUNTERSHAFT CENTER BEARING
MAINSHAFT CENTER BEARING
CENTER BEARING COVER

Removing center bearing cover

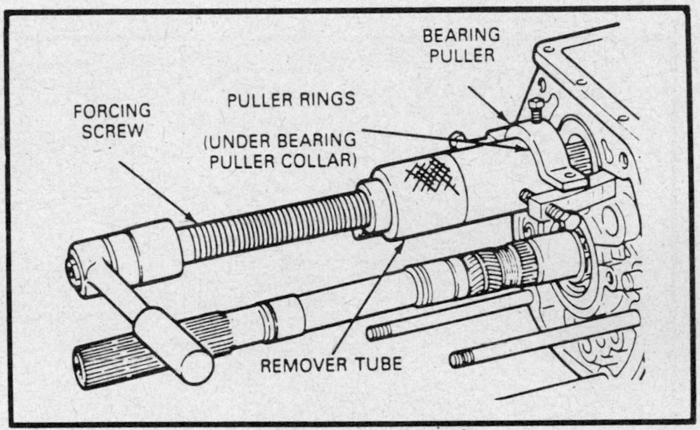
BEARING PULLER
PULLER RINGS
(UNDER BEARING PULLER COLLAR)
FORCING SCREW
REMOVER TUBE

Removing countershaft center bearing

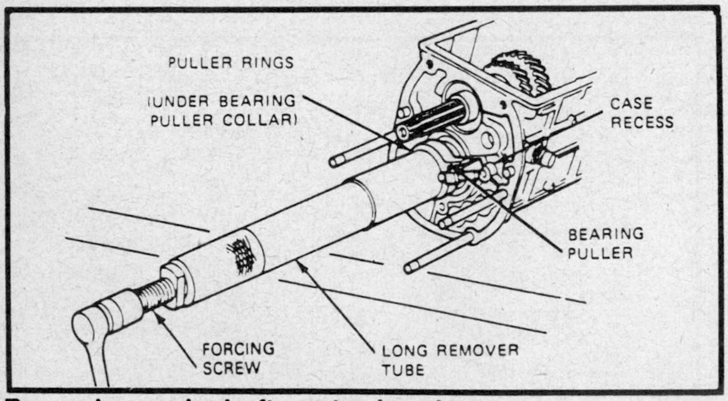
PULLER RINGS
(UNDER BEARING PULLER COLLAR)
CASE RECESS
BEARING PULLER
FORCING SCREW
LONG REMOVER TUBE

Removing mainshaft center bearing

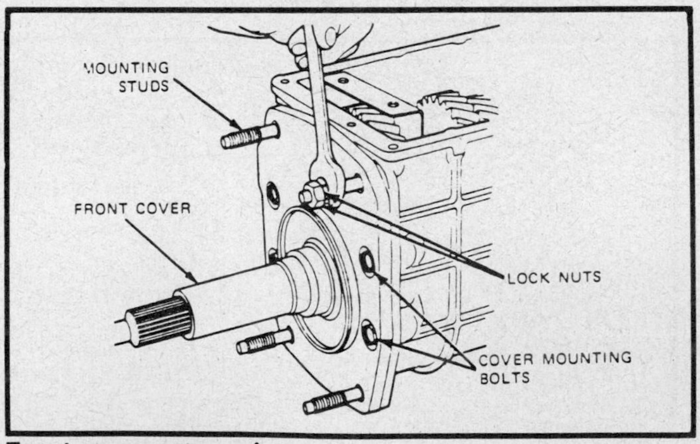
MOUNTING STUDS
FRONT COVER
LOCK NUTS
COVER MOUNTING BOLTS

Front cover removal

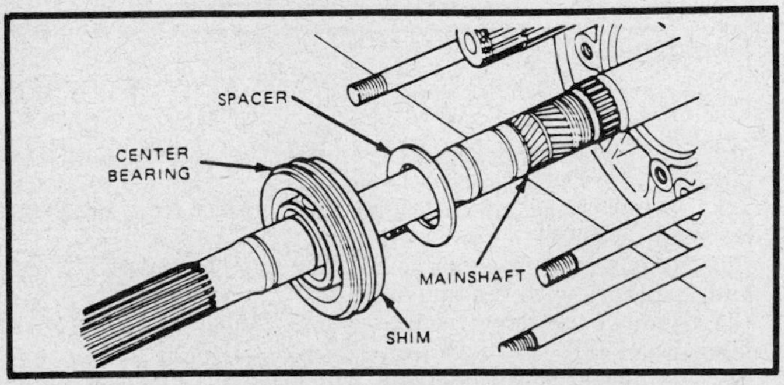
SPACER
CENTER BEARING
MAINSHAFT
SHIM

Mainshaft center bearing shim and spacer

transmission case.

23. Remove the 4 bolts attaching the blind covers to the transmission case and remove the blind covers and gaskets.

24. Remove the roll pin from the 5th/reverse shift fork and slide the shift fork shaft out of the transmission case.

25. Shift the transmission into 4th gear. This will provide adequate space to drive out the 3rd/4th shift fork roll pin. Slide the 3rd/4th shift fork shaft out of the rear of the transmssion case.

26. Remove the roll pin from the 1st/2nd shift fork and slide the shift fork shaft assembly out of the rear of the transmission case. Remove both interlock pins.

27. Remove the snapring that secures the 5th gear to the mainshaft.

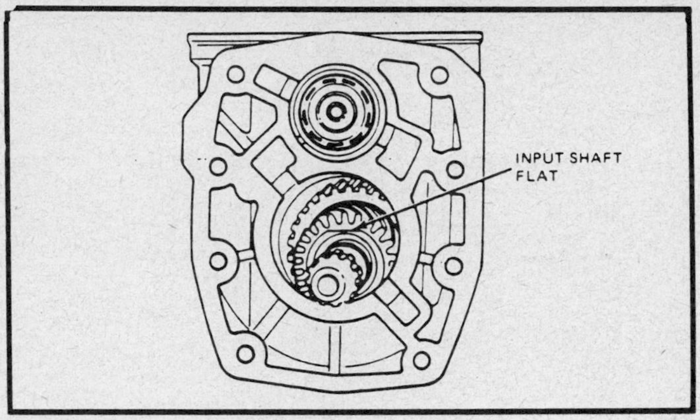
INPUT SHAFT FLAT

Positioning input shaft flats

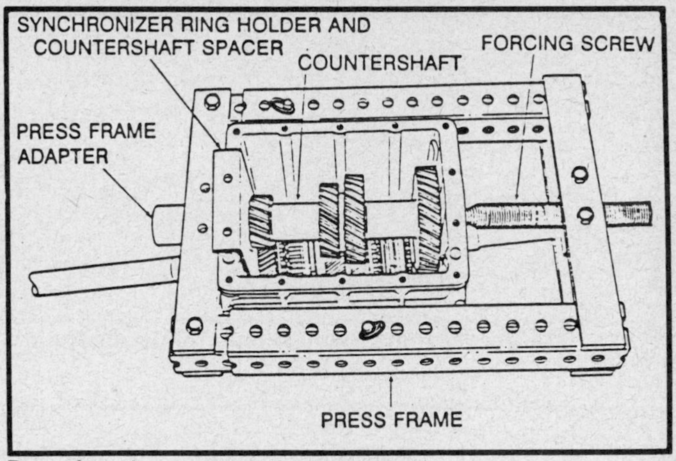
SYNCHRONIZER RING HOLDER AND COUNTERSHAFT SPACER
COUNTERSHAFT
FORCING SCREW
PRESS FRAME ADAPTER
PRESS FRAME

Pressing the countershaft reward

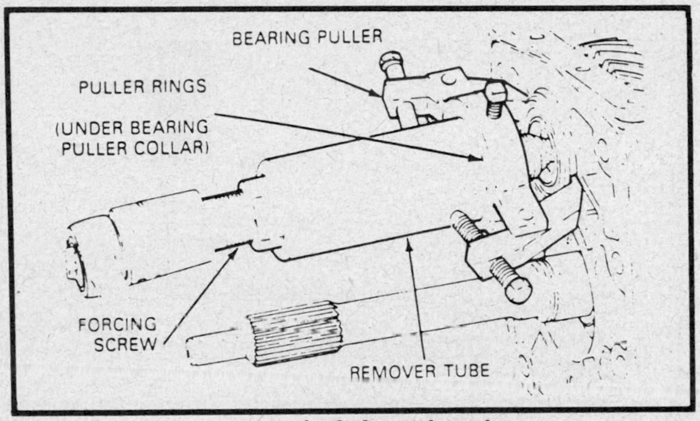
BEARING PULLER
PULLER RINGS
(UNDER BEARING PULLER COLLAR)
FORCING SCREW
REMOVER TUBE

Removing the countershaft front bearing

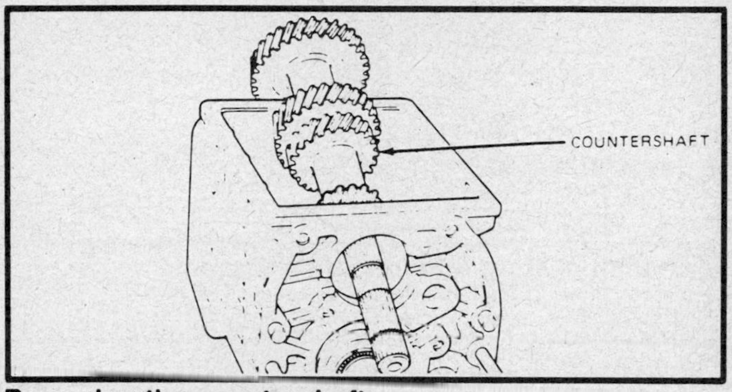
COUNTERSHAFT

Removing the countershaft

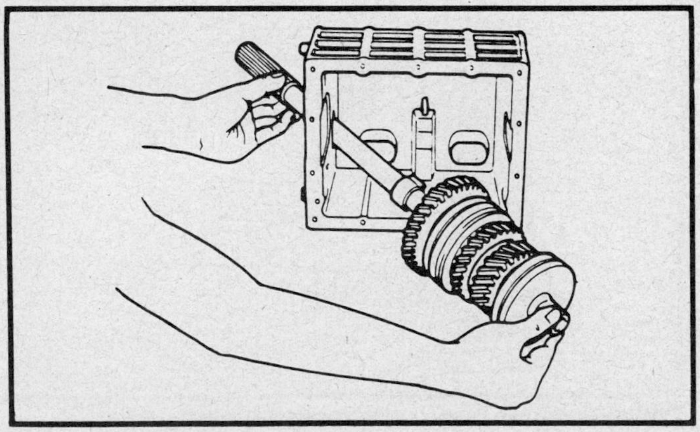

Mainshaft and gear removal

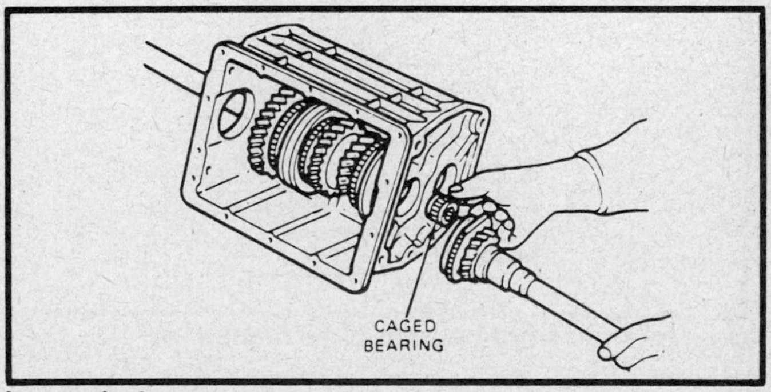
CAGED BEARING

Input shaft removal

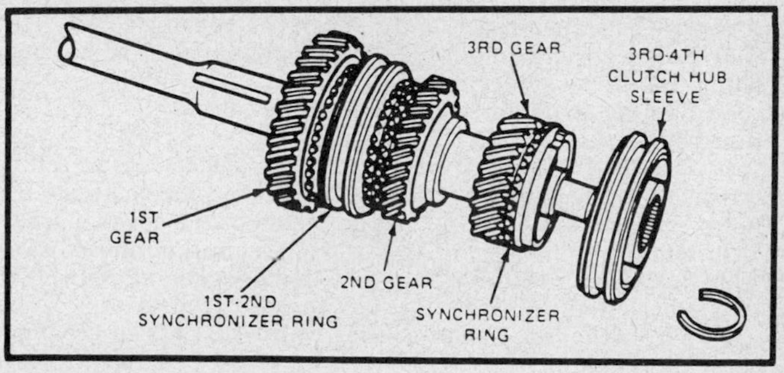
3RD GEAR
3RD-4TH CLUTCH HUB SLEEVE
1ST GEAR
2ND GEAR
1ST-2ND SYNCHRONIZER RING
SYNCHRONIZER RING

Mainshaft and gear assembly

28. Remove the thrust washer and lock ball, 5th gear and synchronizer ring from the rear of the mainshaft.
29. Install the synchronizer ring holder and countershaft spacer number T77J–7025–E between the 4th speed synchronizer ring and the synchromesh gear on the mainshaft. Shift the transmission into 2nd gear to lock the mainshaft and prevent the assembly from rotating.
30. Straighten the staked portion of the mainshaft bearing locknut and remove the locknut.
31. Slide the reverse gear clutch hub assembly off the mainshaft.

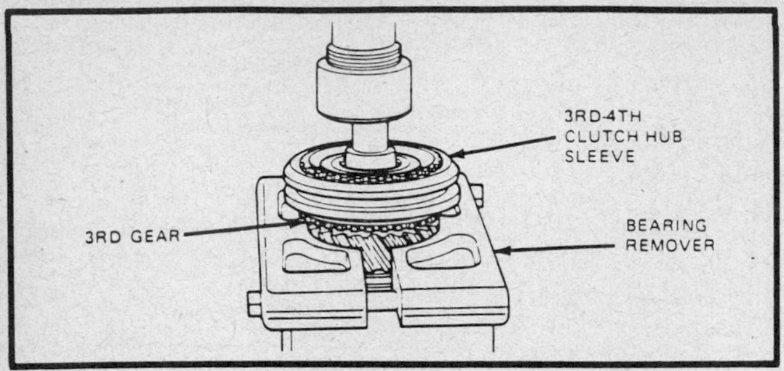
3RD-4TH
CLUTCH HUB
SLEEVE
3RD GEAR
BEARING
REMOVER

Removing 3rd gear and 3rd/4th clutch hub sleeve

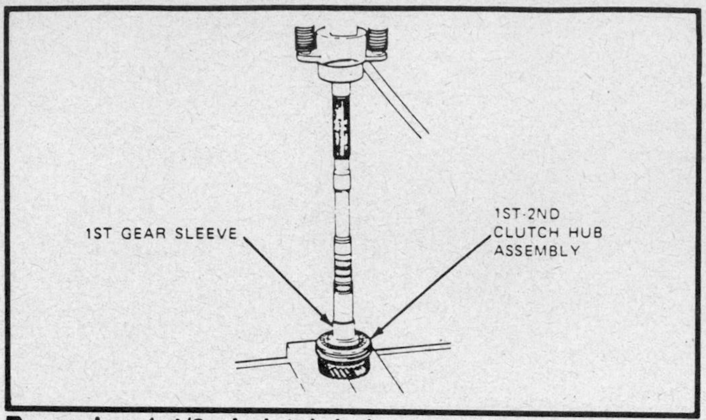
1ST GEAR SLEEVE
1ST-2ND
CLUTCH HUB
ASSEMBLY

Removing 1st/2nd clutch hub assembly from mainshaft

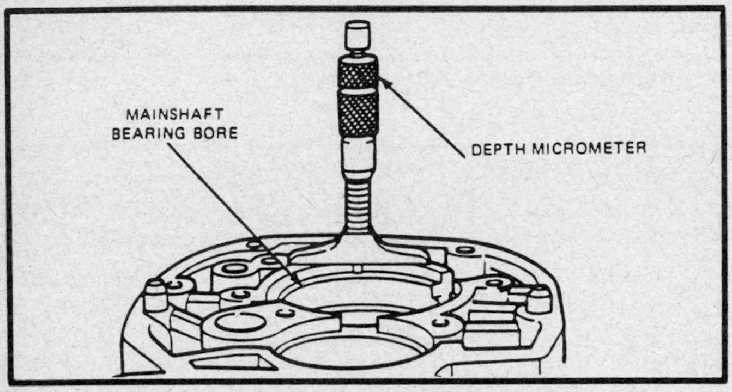
MAINSHAFT
BEARING BORE
DEPTH MICROMETER

Mainshaft bearing bore depth measurement

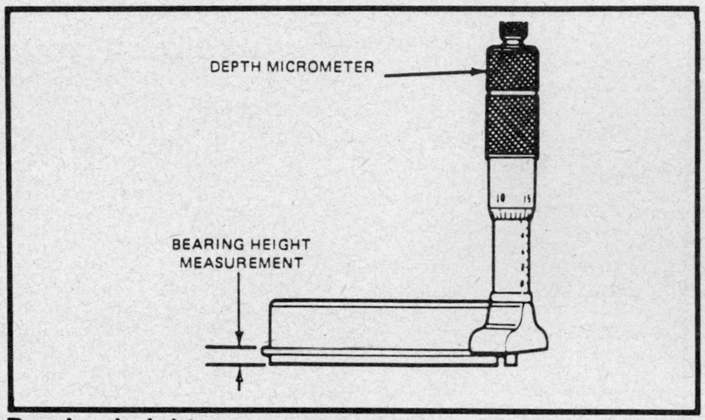
DEPTH MICROMETER
BEARING HEIGHT
MEASUREMENT

Bearing height measurement

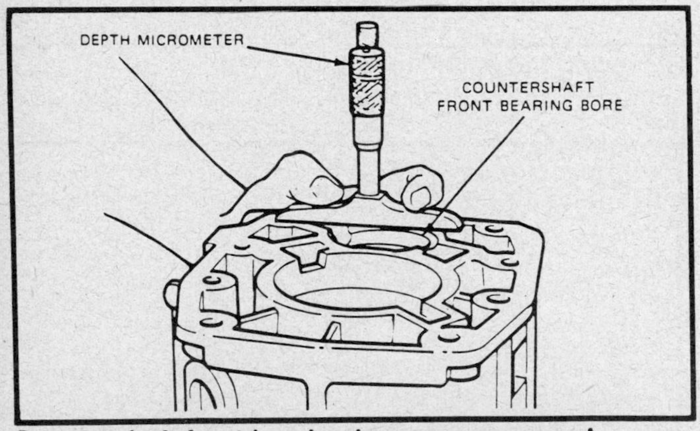
DEPTH MICROMETER
COUNTERSHAFT
FRONT BEARING BORE

Countershaft front bearing bore measurement

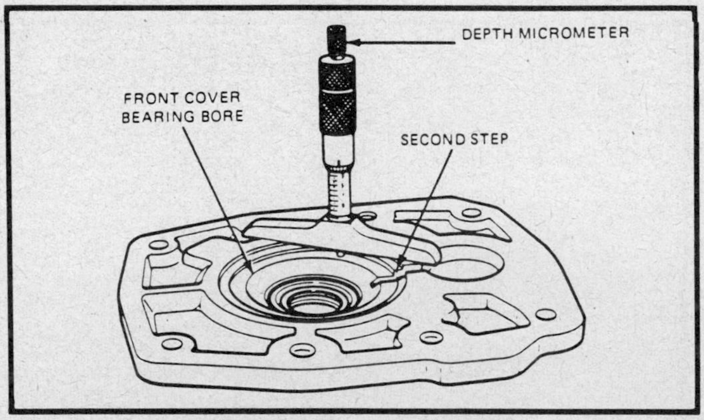
DEPTH MICROMETER
FRONT COVER
BEARING BORE
SECOND STEP

Clutch adapter plate bore measurement

32. Remove the counter reverse gear from the countershaft.
33. Remove the bolts attaching the mainshaft center bearing cover to the transmission and remove the bearing cover.
34. Remove the countershaft and mainshaft center bearings using the proper puller tools.
35. Remove the shim and spacer from behind the mainshaft rear bearing along with the bearing.
36. Remove the front cover by removing the 4 studs attaching the cover to the case then remove the 4 bolts and the cover. Do not discard the shim found on the inside of the cover.
37. Remove the snapring and bearing from the input shaft.
38. Rotate both shift forks so that the main gear train will fall to the bottom of the case. Remove the shift forks.
39. Rotate the input shaft so that 1 of the 2 flats on the input shaft faces upward.
40. Remove the snapring from the front of the countershaft.
41. Remove synchronizer ring holder number T77J–7025–E from the front of the case and install it between the 1st gear on the countershaft and the rear of the case.
42. Install forcing screw tool T84T–7025–B, press frame tool T77J–7025–N and press frame adapter tool T82T–7003–BH against the countershaft assembly.
43. Turn the forcing screw clockwise to press the countershaft rearward. Press the countershaft until it contacts the synchronizer ring holder and counterswhaft spacer.
44. Remove the countershaft front bearing and the shim from behind the bearing.
45. Remove the countershaft from behind the transmission case.
46. Remove the input shaft from the transmission case.

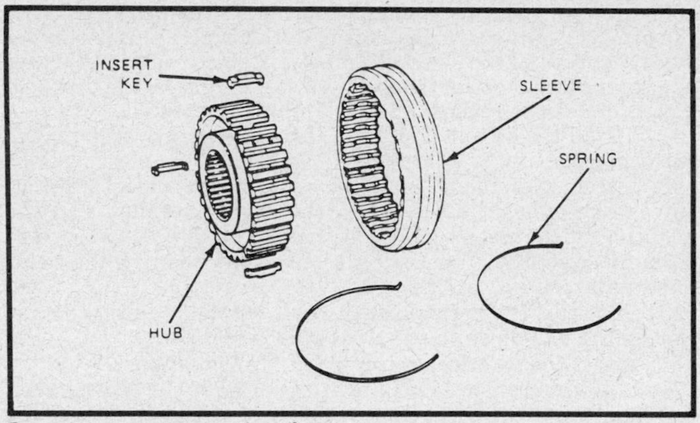
INSERT KEY
SLEEVE
SPRING
HUB
Synchromesh mechanism

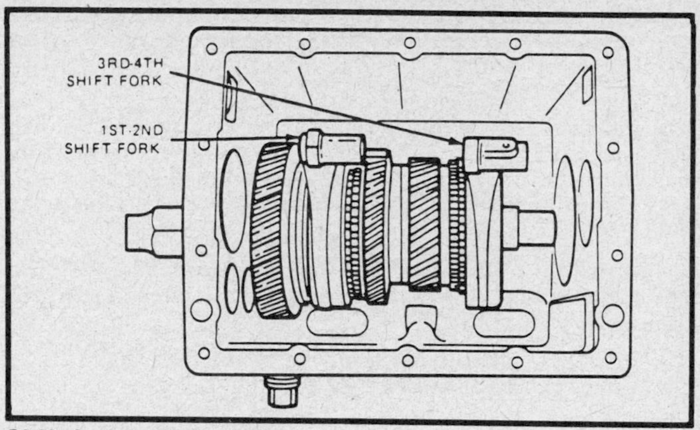
3RD-4TH SHIFT FORK
1ST-2ND SHIFT FORK
Shift fork installation

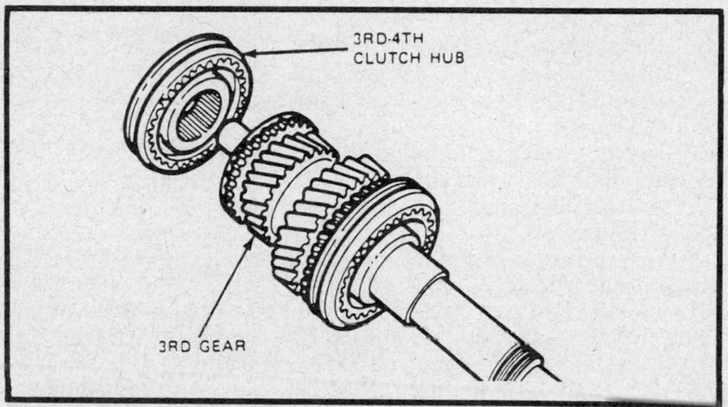
3RD-4TH CLUTCH HUB
3RD GEAR
Installing 3rd/4th clutch hub

47. Remove the synchronizer ring and the caged bearing from the mainshaft.
48. Remove the mainshaft and gear assembly from the transmission case.
49. Remove the countershaft center bearing inner race from the countershaft.
50. Remove the 1st gear and the 1st/2nd synchronizer ring. Remove the snapring retainer from the mainshaft.
51. Install bearing remover tool T71P–4621–B between the 2nd and 3rd gear.
52. Press the mainshaft out of the 3rd gear and the 3rd/4th clutch hub sleeve.
53. Press the 1st and 2nd clutch hub and sleeve assembly and the 1st gear sleeve from the mainshaft.

Cleaning and Inspection

1. Wash all parts except the ball bearings and seals, in a suitable cleaning solvent.
2. Dry all parts with compressed air.
3. Do not clean, wash or soak the transmission seals in cleaning solvent.
4. Lubricate the bearings and wrap them in a clean, lint free cloth until ready for use.
5. Inspect the transmission case and housing for cracks, worn or damaged bores, damaged threads or any other damage that could affect the operation of the transmission.
6. Inspect the machined mating surfaces for burrs, nicks or damage.
7. Inspect the front face of the case for small nicks or burrs that could cause misalignment of the transmission with the flywheel housing.
8. Inspect the bellhousing for cracks.
9. Check the condition of the shift levers, forks, shift rails and shafts.
10. Replace bearings that are broken, worn or rough.

NOTE: Before beginning the assembly procedure, the following 3 measurement checks must be performed:

MAINSHAFT THRUST PLAY

Check the mainshaft thrust play by measuring the depth of the mainshaft bearing bore in the transmission rear cage by using a depth micrometer. Measure the mainshaft rear bearing height. The difference between the 2 measurements indicates the required thickness of the adjusting shim. The standard thrust play is 0–0.0039 in. (0–0.1mm). Adjusting shims are available in 0.0039 in. (0.1mm) and 0.0118 in. (0.3mm) sizes.

COUNTERSHAFT THRUST PLAY

Check the countershaft thrust play by measuring the depth of the countershaft front bearing bore in the transmission case by using a depth micrometer. Measure the countershaft front bearing height. The difference between the 2 measurements indicates the required thickness of the adjusting shim. The standard thrust play is 0–0.0039 in. (0–0.1mm). Adjusting shims are available in 0.0039 in. (0.1mm) and 0.0118 in. (0.3mm) sizes.

MAINSHAFT BEARING CLEARANCE

Check the mainshaft bearing clearance by measuring the depth of the bearing bore in the clutch adapter plate with a depth micrometer. Make sure the micrometer is on the second step of the plate. Measure the bearing height. The difference between the 2 measurements indicates the required thickness of the adjusting shim. The standard clearance is 0–0.0039 in. (0–0.1mm). If an adjusting shim is required, select a shim to bring the clearance to within specifications.

NOTE: As each part is assembled, coat the part with gear lubricant.

Transmission Assembly

1. Assemble the 1st/2nd synchromesh mechanism and the 3rd/4th synchromesh mechanism by installing the clutch hub to the sleeve. Install the 3 synchronizer keys into the clutch hub key slots and install the key springs to the clutch hub.
2. Install the synchronizer ring on the 2nd gear and install the 2nd gear to the mainshaft with the synchronizer ring toward the rear of the shaft.
3. Slide the 1st/2nd clutch hub and sleeve assembly to the mainshaft with the oil grooves of the clutch hub toward the front of the mainshaft.

NOTE: Make sure that the 3 synchronizer keys in the synchromesh mechanism engage the notches in the 2nd synchronizer ring.

4. Insert the 1st gear sleeve on the mainshaft.
5. Install the synchronizer ring on the 3rd gear along with the caged roller bearing and slide the 3rd gear to the front of the mainshaft with the synchronizer ring toward the front.
6. Press the 3rd/4th clutch hub and sleeve assembly to the front of the mainshaft.

NOTE: Make sure that the 3 synchronizer keys in the synchromesh mechanism engage the notches in the synchronizer ring.

7. Install the snapring to the front of the mainshaft.
8. Slide the needle bearing for the 1st gear onto the mainshaft.
9. Install the synchronizer ring on the 1st gear. Slide the 1st gear onto the mainshaft with the synchronizer ring facing the front of the shaft. Rotate the 1st gear as necessary, to engage the 3 notches in the synchronizer ring with the synchronizer keys.
10. Install the original thrust washer to the mainshaft.
11. Install the mainshaft and gear assembly in the case.
12. Install the 1st/2nd and 3rd/4th shift forks in the groove of the clutch hub and sleeve assembly.
13. Install the caged bearing in the front end of the mainshaft.
14. Install the synchronizer ring on the input shaft and install the input shaft to the front end of the mainshaft.

NOTE: Make sure that the 3 synchronizer keys in the 3rd/4th synchromesh mechanism engage the notches in the synchronizer ring.

15. Press the inner race of the countershaft rear bearing onto the countershaft.
16. Install the countershaft gear in the case, making sure that the countershaft gear engages each gear of the mainshaft assembly.
17. Install the shim on the mainshaft center bearing.
18. Install the input shaft bearing and the mainshaft center bearing in their bearing bores. Make sure that the synchronizer and shifter forks have not been moved out of position.
19. Install the input shaft bearing snapring.
20. Install the shim in the countershaft front bearing bore.
21. Install the countershaft front and center bearings in the bores. Properly seat the bearing.
22. Install the snapring to secure tha countershaft front bearing.
23. Install the bearing cover to the transmission case and tighten the attaching bolts to 41–59 ft. lbs. (56–79 Nm).
24. Install the reverse idler gear and shaft with a spacer on each side of the shaft.
25. Slide the counter reverse gear (chamfer side forward) and spacer onto the countershaft.
26. Slide the thrust washer, reverse gear, caged roller bearings and clutch hub assembly onto the mainshaft. Install a new locknut hand tight.
27. Shift into 2nd gear and reverse gear to lock the rotation of the mainshaft. tighten the locknut to 115–172 ft. lbs. (156–233 Nm).
28. Install the 4th/3rd clutch sleeve in the 3rd gear.
29. If new synchronizers have been installed, check the clearance between the synchronizer key and the exposed edge of the synchronizer ring with a feeler gauge. If the measurement is greater than 0.079 in. (2.0mm), the synchronizer key can pop out of position. To correct this, change the thrust washer between the mainshaft center bearing and the 1st gear.
30. Position the 5th synchronizer ring on the 5th gear. Slide the 5th gear onto the mainshaft with the synchronizer ring toward the front of the shaft. Rotate the 5th gear as necessary, to engage the 3 notches in the synchronizer ring with the synchronizer keys in the reverse and clutch hub assembly.
31. Install the lock ball and the thrust washer on the rear of the 5th gear.
32. Install the snapring on the rear of the thrust washer. Check the clearance beetween the thrust washer and the snapring. If the clearance is not within 0.0039–0.0118 in. (0.1–0.3mm), select the proper size thrust washer to bring the clearance within specifications.
33. Slide the 1st/2nd shift fork shaft assembly into the case. Secure the shift fork to the shaft with a roll pin.
34. Insert the interlock pin into the transmission.
35. Shift the transmission into 4th gear. Slide the 3rd/4th shift fork shaft into the case, from the rear of the case. Secure the 3rd/4th shift fork to the shaft with a roll pin.
36. Shift the synchronizer hub into 5th gear. Install the reverse and 5th fork on the clutch hub and slide the reverse/5th fork shaft into the case.
37. Install the 2 blind covers and gaskets. Tighten the attaching bolts to 23–34 ft. lbs. (32–45 Nm).
38. Install the 3 detent balls and springs into the case and install the spring bolts.
39. Install the center housing on the case. Align the reverse idler gear shaft boss with the center housing attaching bolt boss. Tighten the idler shaft bolt to 41–59 ft. lbs. (56–79 Nm).
40. Slide the counter 5th gear to the countershaft.
41. Install the countershaft rear bearing on the countershaft and press it into position.
42. Install the thrust washer and snapring to the rear of the countershaft rear bearing. Check the clearance between the thrust washer and the snapring. If the clearance is not within 0–0.0059 in. (0–0.15mm), select the proper size thrust washer to bring the clearance within specifications.
43. Install the mainshaft rear bearing on the mainshaft.
44. Install the thrust washer and snapring to the rear of the mainshaft rear bearing. Check the clearance between the thrust washer and the snapring. If the clearance is not within 0–0.0039 in. (0–0.10mm), select the proper size thrust washer to bring the clearance within specifications.
45. Install the bearing housing on the center housing.
46. Install each shift fork shaft end onto the proper shift fork shaft and secure with roll pins.
47. Install the lock ball, speedometer drive gear and snapring onto the mainshaft.
48. If removed, install the control lever and rod in the extension housing.
49. Install the extension housing in the bearing housing with the gearshift control lever end loaid down to the left as far as it will go. Tighten the attaching bolts and nut to 60–80 ft. lbs. (82–108 Nm).
50. Install the speedometer driven gear assembly to the extension housing and secure it with a bolt.
51. Check to ensure that the gearshift control lever operates properly.
52. Install the transmission case cover gasket and the cover with the drain plug to the rear. Tighten the bolts to 23–34 ft. lbs. (32–45 Nm).
53. Install the shim on the 2nd step of the front cover.
54. Install the front cover to the transmission case and tighten the 4 bolts and 4 studs.
55. Fill with 3.6 pints (1.7L) of API GL4 or GL5 SAE 80W90 gear lubricant. Install the filler plugs and tighten to 18–29 ft. lbs. (25–39 Nm).

SPECIAL TOOLS

Tool Number	Description
T57L-500-B	Bench mounted holding fixture
T75L-1165-B	Axle bearing seal plate
D80P-4201-A	Depth micrometer
D79L-4621-A	Pinion bearing cone remover
T53T-4621-C	Pinion bearing cone replacer
T57L-4621-B	Pinion bearing cone replacer
T62F-4621-A	Pinion bearing cone replacer
T71P-4621-B	Pinion bearing cone replacer
T82T-7003-BH	Adjustable press frame adapter
T82T-7003-DH	Mainshaft front bearing replacer
T75L-7025-B	Remover/replacer tube (short)
T75L-7025-C	Remover and replacer tube (long)
T84T-7025-B	Forcing screw
T75L-7025-Q	Dummy bearing
T77J-7025-B	Remover tube (5½ in.)

Tool Number	Description
T77J-7025-C	Locknut wrench
T77J-7025-D	Bench holding fixture adapters
T77J-7025-E	Synchronizer ring holder and counter shaft spacer
T77J-7025-F	Staking tool
T77J-7025-H	Bearing puller
T77J-7025-J	Puller ring
T77J-7025-K	Center bearing replacer
T77J-7025-L	Countershaft front bearing replacer
T77J-7025-M	Mainshaft front bearing replacer
T77J-7025-N	Adjustable press frame
T71P-7095-A	Extension housing seal replacer
T72J-7280	Dummy shift rails (2) and interlock pins
T72J-7697	Extension housing bushing remover
T71P-7657-A	Extension housing seal remover
T72J-7697-A	Extension housing bushing replacer

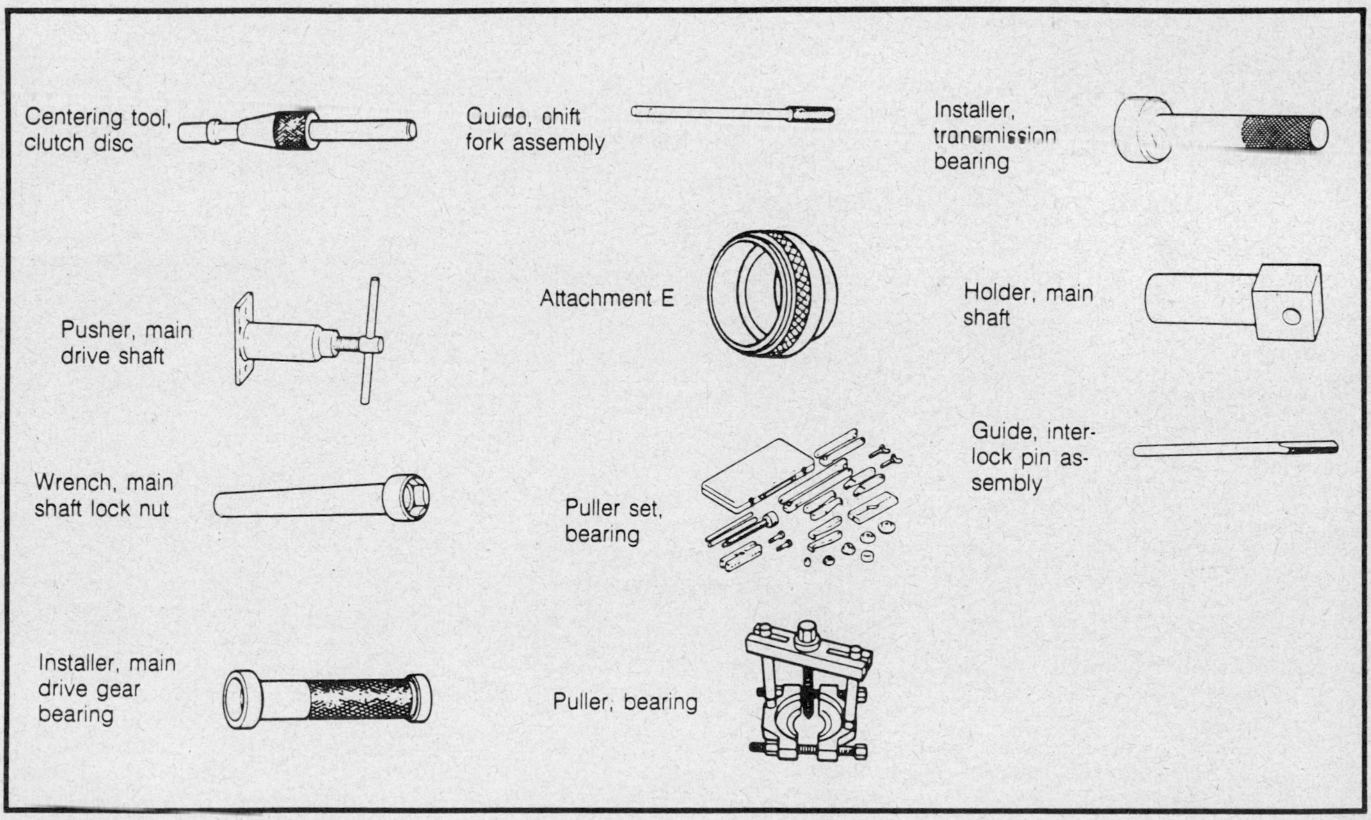
Centering tool, clutch disc
Guido, chift fork assembly
Installer, transmission bearing
Pusher, main drive shaft
Attachment E
Holder, main shaft
Wrench, main shaft lock nut
Puller set, bearing
Guide, interlock pin assembly
Installer, main drive gear bearing
Puller, bearing

SPECIFICATIONS

TORQUE SPECIFICATIONS

	ft. lbs.	Nm
Shift rail detent spring cap	29–43	40–58
Mainshaft gear retaining nut	145–203	197–275
Clutch release lever pivot	23–34	32–46
Interlock pin bore plug	7.5–11	11–14
Drain plug	29–43	40–58
Filler plug	18–29	25–39
Backup lamp switch	22–29	30–39
Shift indicator switch	22–29	30–39

Section 7

S5-42 ZF Transmission
Ford Motor Co.

APPLICATION

S5-42 ZF

Year	Vehicle	Engine
1988–89	E-350	4.9L, 5.8L, 7.3L, 7.5L
	F-150	4.9L, 5.8L, 7.3L, 7.5L
	F-250	4.9L, 5.8L, 7.3L, 7.5L
	F-350	4.9L, 5.8L, 7.3L, 7.5L
	F-Super Duty	4.9L, 5.8L, 7.3L, 7.5L
	Bronco	4.9L, 5.8L, 7.3L, 7.5L

GENERAL DESCRIPTION

The S5–42 ZF is a synchromesh 5 speed transmission. All forward gears are controlled through synchronizer engagement. The transmission uses an aluminum housing with an integral clutch housing. All gears are a shrink fit to the shaft and the countershaft is serviced as an assembly. There are 2 versions of this transmission; one used on gasoline engines that provides close ratio gearing and one used on diesel engines that provides wide ratio gearing.

Transmission Identification

An identification plate is affixed to the driver's side of the transmission just aft of the clutch housing. The plate provides information on serial number, manufacturer's part number, total gearing ratio, lubricant type and capacity, speedometer gear ratio and place of assembly.

Metric Fasteners

The metric fastener dimensions are very close to the dimensions of the familiar inch system fasteners. For this reason, replacement fasteners must have the same measurement and strength as those removed.

CAUTION

Do not attempt to interchange metric fasteners for inch system fasteners. Mismatched or incorrect fasteners can result in damage to the transmission unit through malfunctions, breakage or possible personal injury.

Care should be taken to re-use the fasteners in the same locations as removed.

Common metric fastener strength property classes are 9.8 and 12.9 with class identification embossed on the head of each bolt. The inch strength classes range from grade 2–8 with the line identification embossed on each bolt head. Markings correspond to 2 lines less than actual grade (for example grade 8 bolt will exhibit 6 embossed lines on the bolt head). Some metric nuts will be marked with a single digit strength identification numbers on the nut face.

Capcities

This transmission uses Motorcraft Mercon® multi-purpose automatic transmission fluid XT–2–QDX or DDX (ESP–M2C–166–H) or equivalent., with a fluid capacity of 7 pints.

Checking Fluid Level

The transmission fill plug is located on the driver's side of the transmission. To check the fluid level, remove the filler plug and check that the fluid level is just below the bottom of the filler plug opening. Add fluid as required.

TROUBLE DIAGNOSIS

CHILTON THREE C'S TRANSMISSION DIAGNOSIS

Condition	Cause	Correction
Gears clash when shifting from one forward gear to another	a) Engine idle speed too high	a) Adjust engine idle speed
	b) Pilot bearing binding	b) Remove transmission and check for a binding condition between input shaft and engine crankshaft pilot bearing. Replace as required
	c) Damaged gear teeth and/or synchronizer	c) Disassemble transmission, repair or replace as required
Transmission jumps out of gear	a) Stiff shift boot. Improper fit of form isolation pad	a) Replace shift boot if exceptionally stiff. Replace or rework pad to provide clearance

CHILTON THREE C'S TRANSMISSION DIAGNOSIS

Condition	Cause	Correction
Transmission jumps out of gear	b) Loose transmission to engine mounting bolts, or loose levers	b) Tighten transmission to engine block bolts to specifications. Loosen all bolts and reseat flywheel housing. Tighten all bolts. Tighten levers if necessary
	c) Flywheel housing to engine crankshaft out of line	c) Shim or replace housing as required
	d) Crankshaft pilot bearing worn	d) Replace bearing
	e) Interior components damage	e) Disassemble transmission. Inspect the synchronizer sleeves for free movement on their hubs. Inspect the synchronizer blocking rings for widened index slots, rounded clutch teeth and smooth internal surface. Check countershaft cluster gear for excessive end play. Check shift forks for worn or loose mounting on shift rails. Check fork pads for excessive wear. Inspect synchronizer sliding sleeve and gear clutch teeth for wear or damage. Repair or replace as required
	f) Worn gear teeth due to partial engagement	f) Replace worn or damaged gears
Transmission will not shift into one gear—all others OK	a) Manual-shift linkage damaged or worn	a) Lubricate, repair or replace parts as required
	b) Back-up switch ball frozen	b) If reverse is problem, check back-up switch for ball frozen in extended position (if so equipped)
	c) Internal components	c) Remove transmission. Inspect shift rail and fork system synchronizer system and gear clutch teeth for restricted travel. Repair or replace as required
Transmission shifts hard	a) Clutch does not completely release	a) Check input shaft splines for lubrication. Check clutch operation
	b) Transmission fluid low or improper type	b) Add lubricant or change lubricant as required
	c) Worn or damaged internal shift mechanism	c) Remove transmission cover. Check internal shift mechanism by shifting into and out of all gears. Repair or replace as required
	d) Binding of sliding gears and/or synchronizers	d) Check for free movement of gears and synchronizers. Repair or replace as required
	e) Housing and/or shafts out of alignment	e) Remove transmission and check for binding condition between input shaft and engine crankshaft pilot bearing or bushing. Check flywheel housing alignment. Repair or replace as required
Noisy in forward gears ①	a) Lubricant level low, or improper type	a) Add lubricant, or refill with specified lubricant
	b) Components grinding on transmission	b) Check for screws, bolts, etc., of cab or other components grounding out. Correct as required
	c) Component housing bolts loose	c) Check torque on transmission to flywheel housing bolts, output shaft flange nut and flywheel housing to engine block bolts. Tighten bolts to specification
	d) Flywheel housing to engine crankshaft alignment	d) Check and align flywheel housing to engine crankshaft

CHILTON THREE C'S TRANSMISSION DIAGNOSIS

Condition	Cause	Correction
Noisy in forward gears ①	e) Noisy bearings or gears	e) Remove and disassemble transmission. Inspect input, output and countershaft bearings. Inspect speedometer gear and gear teeth for wear or damage. Replace as required

① While verifying the condition, determine whether the noise is gear roll-over noise, release bearing rub or some other transmission related noise.

Gear roll-over noise, inherent in manual transmissions, is caused by the constant mesh gears turning at engine idle speed, while the clutch is engaged and the transmission is in neutral; and release bearing rub are sometimes mistaken for mainshaft bearing noise.

Gear roll-over noise will disappear when the clutch is disengaged or when the transmission is engaged in gear. Release bearing rub will disengage when the clutch is engaged in the event that a bearing is damaged, the noise is more pronounced while engaged in gear under load or coast than in neutral.

ON CAR SERVICES

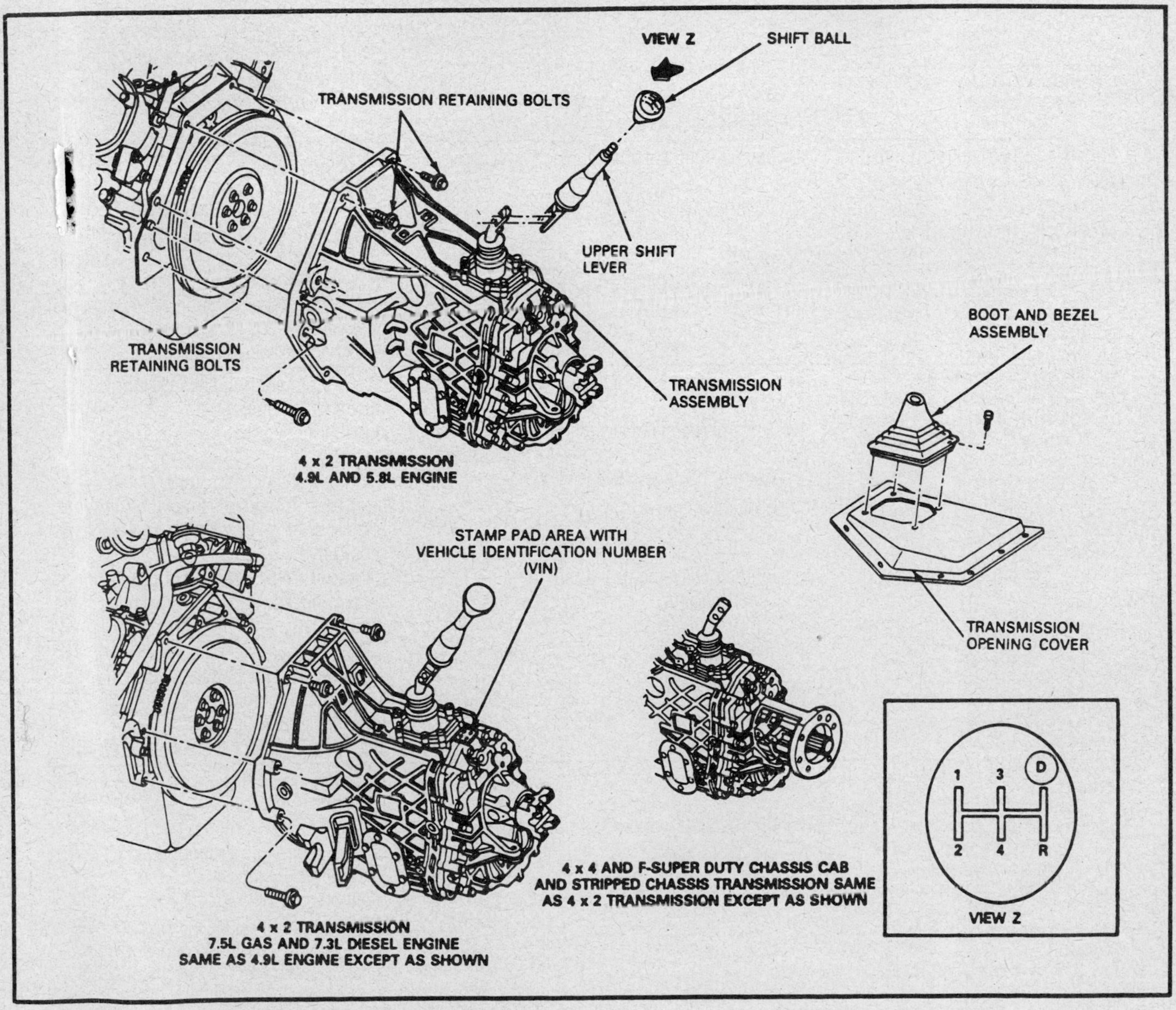
VIEW Z
SHIFT BALL
TRANSMISSION RETAINING BOLTS
UPPER SHIFT LEVER
TRANSMISSION RETAINING BOLTS
TRANSMISSION ASSEMBLY
BOOT AND BEZEL ASSEMBLY
4 x 2 TRANSMISSION
4.9L AND 5.8L ENGINE
STAMP PAD AREA WITH
VEHICLE IDENTIFICATION NUMBER
(VIN)
TRANSMISSION OPENING COVER
1 3 D
2 4 R
4 x 4 AND F-SUPER DUTY CHASSIS CAB
AND STRIPPED CHASSIS TRANSMISSION SAME
AS 4 x 2 TRANSMISSION EXCEPT AS SHOWN
VIEW Z
4 x 2 TRANSMISSION
7.5L GAS AND 7.3L DIESEL ENGINE
SAME AS 4.9L ENGINE EXCEPT AS SHOWN

Transmission mounting for 2WD and 4WD vehicles

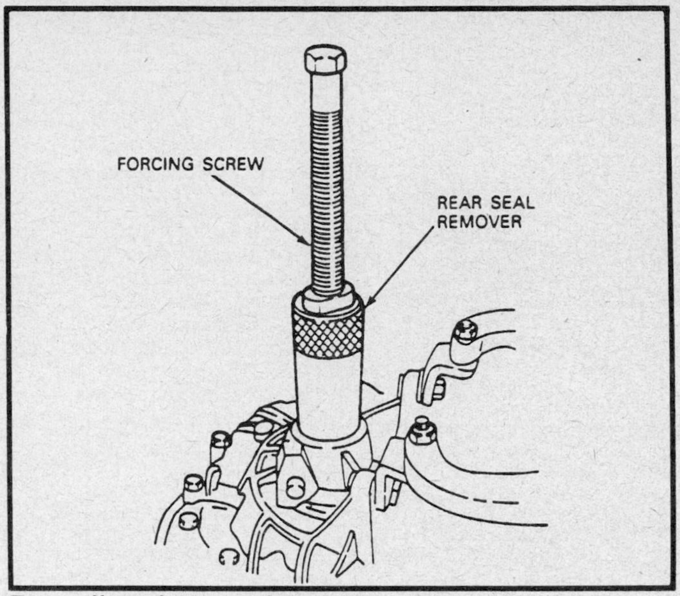
FORCING SCREW
REAR SEAL REMOVER
Rear oil seal removal on 2WD vehicles

Services

REAR OIL SEAL

Removal and Installation

2WD

1. Raise and support the vehicle safely.
2. Matchmark and disconnect the driveshaft from the output flange. Set the bolts aside.
3. Unstake the tab on the output flange retaining nut.
4. Attach special companion flange holding tool T78P–4851–A or equivalent to the output flange with 4 of the reserved bolts.
5. Loosen the nut that secures the output flange to the mainshaft using the proper tool. Remove the holding tool after loosening the nut.
6. Remove the output flange from the output end of the mainshaft.
7. Insert a suitable rear seal removal tool over the output end of the mainshaft and tighten the tool into the rear oil seal.
8. Install a suitable forcing screw into the removal tool. Turn the screw while holding the tool stationary to remove the seal.
9. To install, place the seal onto a suitable seal installer and position the tool with seal over the output end of the mainshaft.
10. With a rubber mallet, gently tap the seal until it seats in the opening. Use a small amount of liquid soap around the opening to minimize friction and facilitate installation.
11. Install the output shaft onto the end of the mainshaft.
12. Attach the companion flange holding tool to the output flange with the 4 bolts.
13. Install the output flange retaining nut onto the mainshaft. Torque the nut to 184 ft. lbs. (250 Nm). Remove the flange holding tool after tightening the nut.
14. Stake the tab on the retaining nut with the proper tool.
15. Connect the driveshaft to the output flange. On all vehicles except E–150, torque the U-bolt nuts to 8–15 ft. lbs. (11–20 Nm). On E–150, torque the axle flange yoke bolts to 61–87 ft. lbs. (83–117 Nm).
16. Lower the vehicle.

4WD AND F-SUPER DUTY

1. Raise and support the vehicle safely.
2. Remove the transfer case. On F-Super Duty vehicles, remove the transmission mounted parking brake.
3. Insert special rear seal remover tool T87T–7025–CH or equivalent over the output end of the mainshaft and tighten the tool into the rear seal.
4. Install a suitable forcing screw into the removal tool. Turn the screw while holding the tool stationary to remove the seal.
5. To install, place the seal onto a suitable seal installer and position the tool with seal over the output end of the mainshaft.
6. With a rubber mallet, gently tap the seal until it seats in the opening. Use a small amount of liquid soap around the opening to minimize friction and facilitate installation.
7. On F-Super Duty vehicles, install the transmission mounted parking brake.
8. Install the transfer case.
9. Lower the vehicle.

REMOVAL AND INSTALLATION

TRANSMISSION REMOVAL

2WD

1. Shift the transmission into **N** and remove the carpets, floor mats or isolator pad as required.
2. Unscrew the boot and bezel assembly from the transmission opening cover.
3. Unbolt and remove the upper shift lever from the lower shift lever.
4. Disconnect the negative battery cable. Raise the vehicle and support safely.
5. Disconnect the speedometer cable.
6. Disconnect the back-up lamp switch lead located at the top left hand side of the transmission.
7. Drain the transmission.
8. Support the transmission with a suitable transmission jack.
9. Disconnect the clutch linkage. Matchmark and disconnect the driveshaft.
10. On F-Super Duty vehicles, remove the transmission parking brake.
11. Remove the transmission rear insulator and lower retainer. Remove the crossmember.
12. Unbolt the transmission from the engine block. Chain the transmission to the proper removal equipment.
13. Move the transmission to the rear until the input shaft clears the flywheel. Lower the transmission from the vehicle and position on a suitable holding fixture.

4WD

1. Shift the transmission into **N** and remove the carpets, floor mats or isolator pad as required.
2. Unscrew the boot and bezel assembly from the transmission opening cover.
3. Unbolt and remove the upper shift lever from the lower shift lever.
4. Disconnect the negative battery cable. Raise the vehicle and support safely.
5. Drain the transmission and transfer case.
6. Disconnect the front driveshafts from the transfer case and wire them out of the way. Matchmark and disconnect the rear driveshaft.
7. Disconnect the back-up lamp switch from the transfer case.
8. Disconnect the speedometer cable from the transfer case.

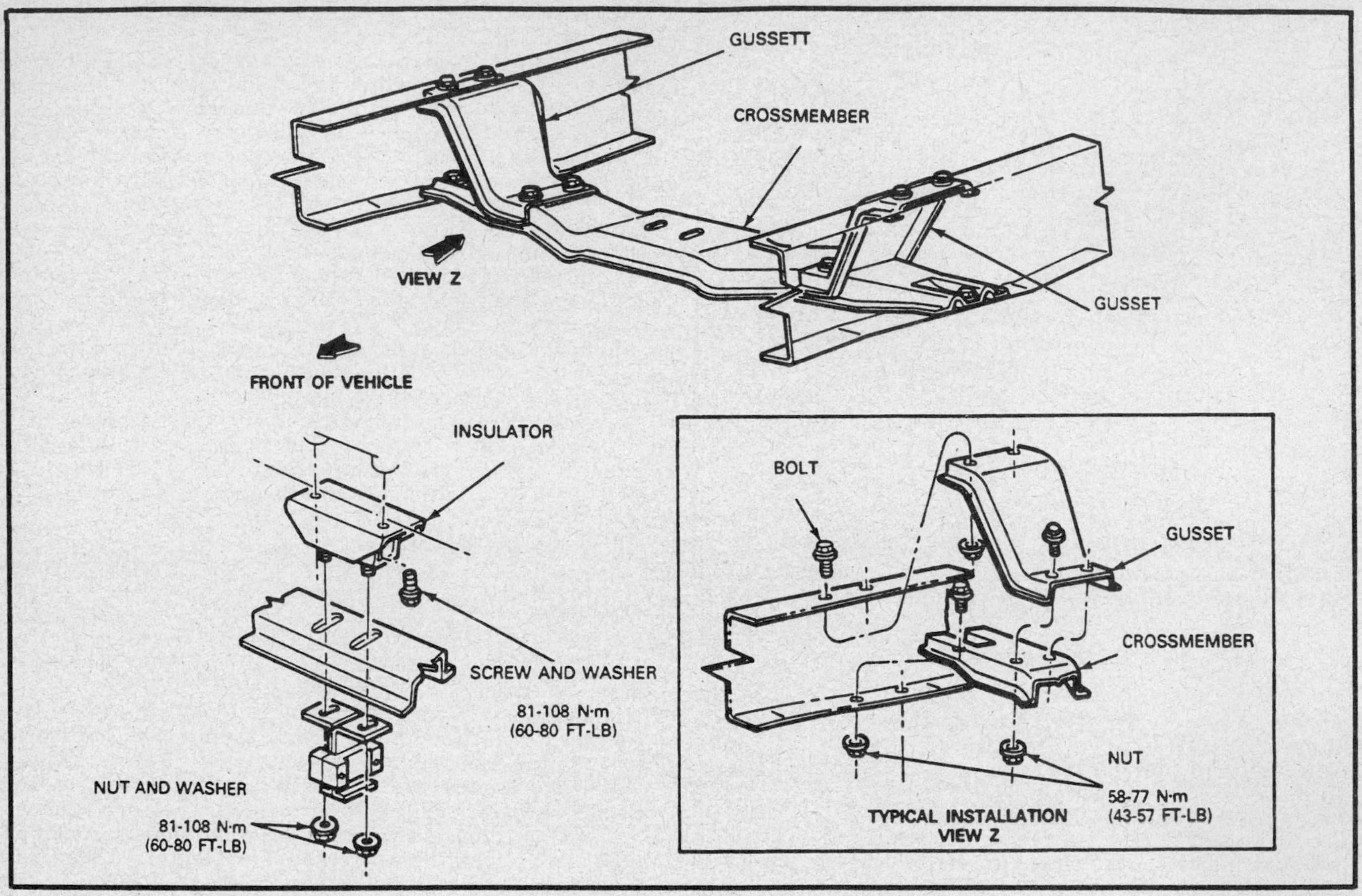
GUSSETT
CROSSMEMBER
VIEW Z
GUSSET
FRONT OF VEHICLE
INSULATOR
SCREW AND WASHER
81-108 N·m
(60-80 FT-LB)
NUT AND WASHER
81-108 N·m
(60-80 FT-LB)
BOLT
GUSSET
CROSSMEMBER
NUT
58-77 N·m
(43-57 FT-LB)
TYPICAL INSTALLATION
VIEW Z

Crossmember components

9. Remove the skid pad from the under the transfer case, if equipped.
10. Support the transfer case with a suitable transmission jack.
11. Unbolt the transfer case from the transmission and carefully lower the transfer case from the vehicle. Make sure that the transfer case shift lever clears the floor pan opening. Remove the gasket.
12. Remove the transmission rear insulator and lower retainer. Remove the crossmember.
13. Unbolt the transmission from the engine block.
14. Move the transmission to the rear until the input shaft clears the flywheel. Lower the transmission from the vehicle and position on a suitable holding fixture.

TRANSMISSION INSTALLATION

2WD

1. Remove the transmission from holding fixture and mount on a suitable transmission jack.
2. Install 2 guide studs onto the front case top holes and raise the transmission until the input shaft splines are aligned with the clutch disc splines. Make sure that the clutch release bearing and hub are properly positioned in the release fork when aligning the spline shafts.
3. Slide the transmission forward until the guide studs have located the transmission to the block. Install the bolts and torque to 40–50 ft. lbs. (54–67 Nm). Remove the guide studs and install the 2 remaining transmission-to-block bolts and torque them also to 40–50 ft. lbs. (54–67 Nm).
4. Install the crossmember.
5. Place the insulator and retainer between the transmission and crossmember. Install the bolts and torque to 45–60 ft. lbs. (54–67 Nm). Install the insulator retaining nut and torque to 50–70 ft. lbs. (68–94 Nm). Remove the transmission jack.
6. Connect the speedometer cable and driven gear and clutch linkage.
7. Connect the back-up light switch lead.
8. On F-Super Duty vehicles, install the transmission mounted parking brake.
9. Connect the driveshaft and fill the transmission.
10. Lower the vehicle.
11. Install the upper shift lever to the lower shift lever. Tighten the retaining screws to 16–24 ft. lbs. (22–33 Nm).
12. Install the boot and bezel assembly to the transmission opening cover.
13. Install the isolator pad, floor pan cover and floor mats.

4WD

1. Position the transmission on a suitable transmission jack.
2. Install 2 guide studs in the upper case holes. Raise the transmission until the transmission and engine block mating surfaces are joined. Install the 2 lower bolts and torque them to 40–50 ft. lbs. (54–67 Nm).
3. Remove the guide studs and install the upper bolts. Torque the bolts to 40–50 ft. lbs. (54–67 Nm).
4. Position the rear support bracket and install the retaining bolts. Torque the bolts to 45–60 ft. lbs. (60–80 Nm).
5. Install the rear support insulator bracket bolts and remove the transmission jack.
6. Position the transfer case onto the transmission jack and raise into position on the transmission. Take care to guide the

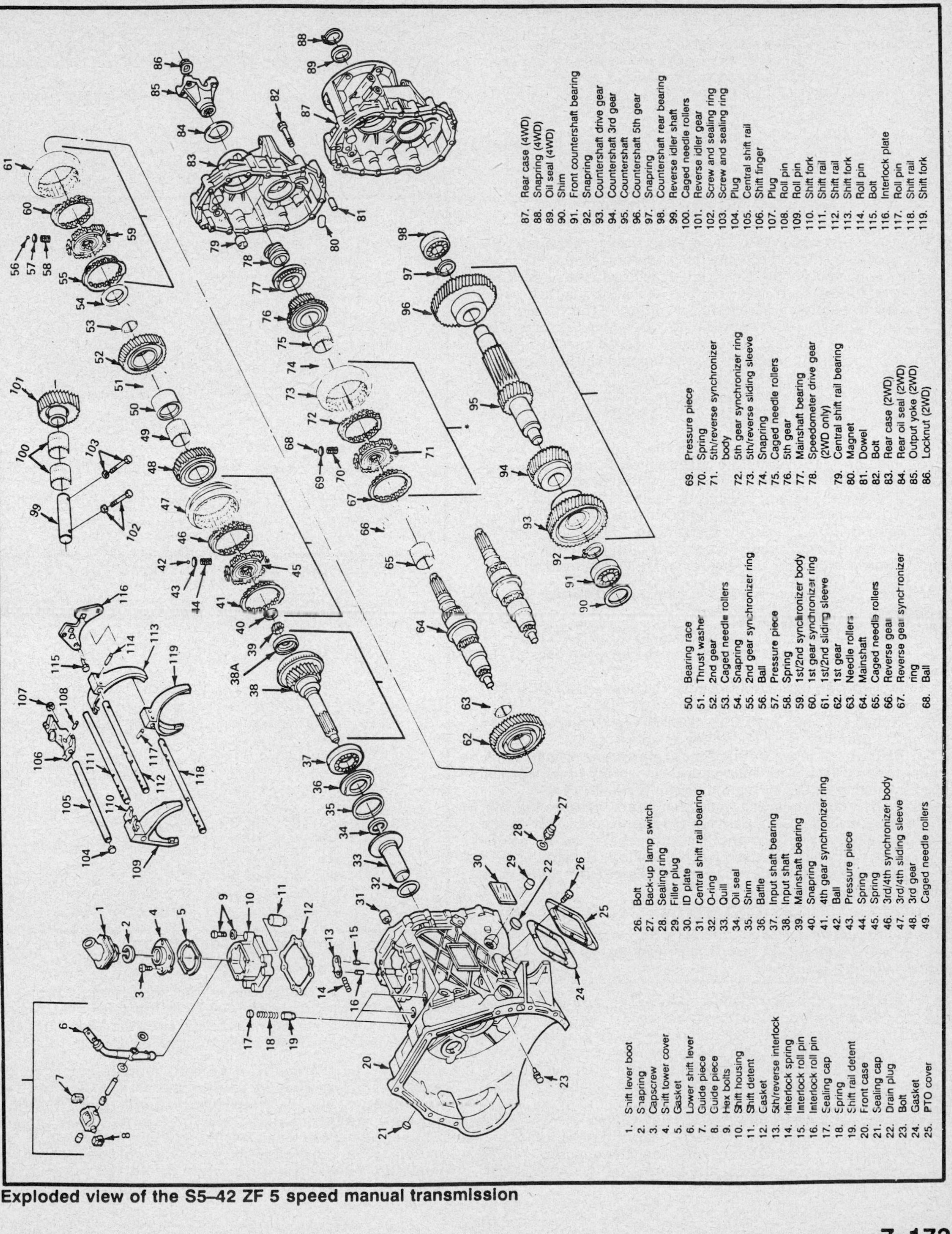

1. Shift lever boot
2. Snapring
3. Capscrew
4. Shift tower cover
5. Gasket
6. Lower shift lever
7. Guide piece
8. Guide piece
9. Hex bolts
10. Shift housing
11. Shift detent
12. Gasket
13. 5th/reverse interlock
14. Interlock spring
15. Interlock roll pin
16. Interlock roll pin
17. Sealing cap
18. Spring
19. Shift rail detent
20. Front case
21. Sealing cap
22. Drain plug
23. Bolt
24. Gasket
25. PTO cover
26. Bolt
27. Back-up lamp switch
28. Sealing ring
29. Filler plug
30. ID plate
31. Central shift rail bearing
32. O-ring
33. Quill
34. Oil seal
35. Shim
36. Baffle
37. Input shaft bearing
38. Input shaft
39. Mainshaft bearing
40. Snapring
41. 4th gear synchronizer ring
42. Ball
43. Pressure piece
44. Spring
45. Spring
46. 3rd/4th synchronizer body
47. 3rd/4th sliding sleeve
48. 3rd gear
49. Caged needle rollers
50. Bearing race
51. Thrust washer
52. 2nd gear
53. Caged needle rollers
54. Snapring
55. 2nd gear synchronizer ring
56. Ball
57. Pressure piece
58. Spring
59. 1st/2nd synchronizer body
60. 1st gear synchronizer ring
61. 1st/2nd sliding sleeve
62. 1st gear
63. Needle rollers
64. Mainshaft
65. Caged needle rollers
66. Reverse gear
67. Reverse gear synchronizer ring
68. Ball
69. Pressure piece
70. Spring
71. 5th/reverse synchronizer body
72. 5th gear synchronizer ring
73. 5th/reverse sliding sleeve
74. Snapring
75. Caged needle rollers
76. 5th gear
77. Mainshaft bearing
78. Speedometer drive gear (2WD only)
79. Central shift rail bearing
80. Magnet
81. Dowel
82. Bolt
83. Rear case (2WD)
84. Rear oil seal (2WD)
85. Output yoke (2WD)
86. Locknut (2WD)
87. Rear case (4WD)
88. Snapring (4WD)
89. Oil seal (4WD)
90. Shim
91. Front countershaft bearing
92. Snapring
93. Countershaft drive gear
94. Countershaft 3rd gear
95. Countershaft
96. Countershaft 5th gear
97. Snapring
98. Countershaft rear bearing
99. Reverse idler shaft
100. Caged needle rollers
101. Reverse idler gear
102. Screw and sealing ring
103. Screw and sealing ring
104. Plug
105. Central shift rail
106. Shift finger
107. Plug
108. Roll pin
109. Roll pin
110. Shift fork
111. Shift rail
112. Shift rail
113. Shift fork
114. Roll pin
115. Bolt
116. Interlock plate
117. Roll pin
118. Shift rail
119. Shift fork

Exploded view of the S5–42 ZF 5 speed manual transmission

shift lever through the floor pan opening. Install a new gasket and the retaining bolts.
7. Connect the speedometer cable to the transfer case. Connect the front driveshaft to the transfer case yoke or flange.
8. Fill the transfer case and transmission.
9. Connect the back-up light switch.
10. Lower the vehicle.
11. Connect the upper shift lever to the lower shift lever. Tighten the retaining screws to 16–24 ft. lbs. (22–33 Nm).
12. Install the boot and bezel assembly to the transmission opening cover.
13. Install the isolator pad, floor pan cover and floor mats.

BENCH OVERHAUL

Before Disassembly

Clean the transmission case with a suitable solvent and dry with compressed air. Inspect the case for cracks and stripped threads in the various bolt holes. Check the machined mating surfaces for burrs, nicks or any other surface imperfections that would make the transmission case unfit for service in the vehicle. The front mating surface should be smooth. Any minor burrs may be dressed with a fine mill file. Repair damaged threaded holes with an appropriate insert or by tapping and re-threading the hole.

Transmission Disassembly

1. Remove the transmission from the vehicle.
2. Position transmission on a suitable holding fixture in a verticle position with the front case pointing down.
3. Unstake the tab on the output shaft retaining nut.
4. Attach special companion flange holding tool T78P–4851–A or equivalent to the output flange with 4 bolts.
5. Loosen the nut that secures the output flange to the mainshaft. Remove the holding tool after loosening the nut.

NOTE: Do not remove the nut or output flange at this time.

6. Remove 15 of the 17 hex head screws that attach the front and rear cases. Leave the 2 screws at the opposite ends of the cases installed.
7. If equipped, unbolt and remove the power take off (PTO) covers from the front case. Discard the gaskets.
8. Unbolt the shift lever from the front case and remove the complete gearshift lever and tower as an assembly.
9. Remove the interlock plate and compression spring. Do not allow the interlock plate and compression spring to fall into the transmision. Tag the spring for assembly reference.
10. Place a drift punch against the detent bolt sealing cap. Hold the punch at an angle so that the point is off center on the cap. Drive the cap inward until the spring pressure on the underside of the cap forces the cap out of its hole. Repeat this procedure for the other 2 caps in the front case. Discard the caps and replace with new.

CAUTION

The sealing caps are under considerable spring pressure. Wear protective eyewear to prevent the caps from striking an eye when they are forced out of the case.

11. After the caps are removed, remove the springs from the cap openings and tag for assembly reference.
12. Drive out the 2 seal caps that cover the reverse idler shaft retaining screws then loosen and remove the screws.
13. If necessary, remove the back-up light switch that is located just forward of the transmission identification plate. Remove the switch sealing ring also.
14. There are 2 dowel pins located in the upper corners of the rear case. Drive them out toward the rear of the transmission.
15. Remove the 2 remaining bolts from the opposite ends of the front and rear cases. Make sure that the input shaft is facing up before removing the 2 bolts.

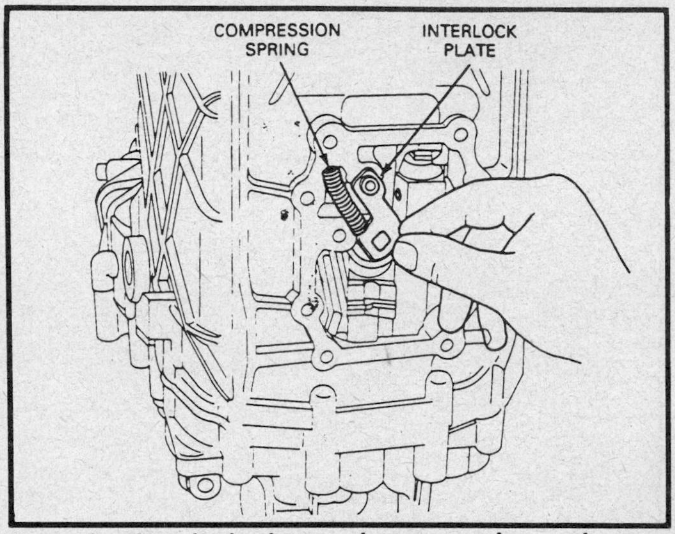
COMPRESSION SPRING
INTERLOCK PLATE

Removing interlock plate and compression spring

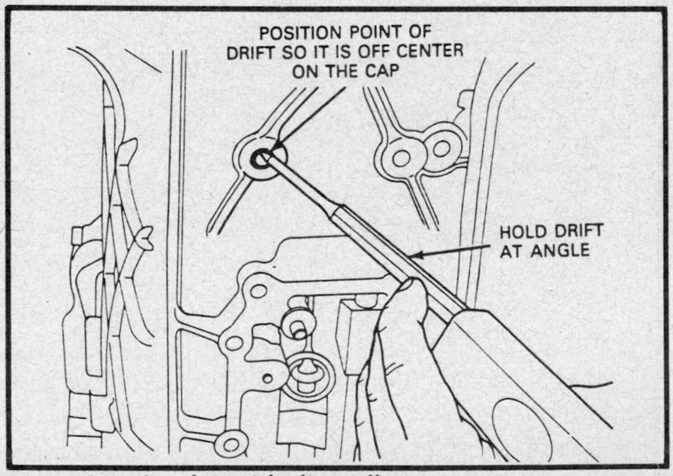
POSITION POINT OF DRIFT SO IT IS OFF CENTER ON THE CAP
HOLD DRIFT AT ANGLE

Removing the detent bolt sealing cap

16. Carefully lift the front case from the rear case. It may be necessary to push the central shift rail in to prevent it from hanging up on the front case during removal. Make sure that the central shift rail is not lifted off together with the front case.

NOTE: The front and rear case mating surfaces are coated with an adhesive sealing agent. This sealant can make separating the housings difficult. If difficulty is encountered because of the adhesive, carefully tap the front of the front case with a rubber mallet until the bond is loosened. Never use a prying tool to separate the housings. This could cause damage to the mating surfaces and cause the transmission to leak.

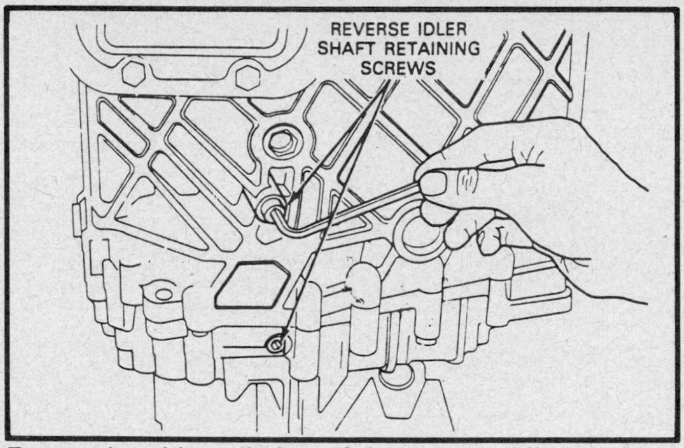
REVERSE IDLER
SHAFT RETAINING
SCREWS

Removal and installation of the reverse idler shaft retaining screws

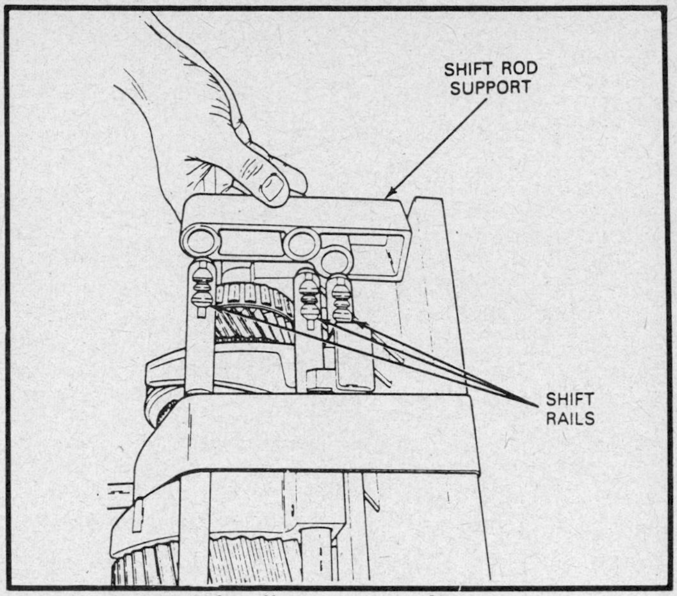
SHIFT ROD
SUPPORT
SHIFT
RAILS

Installing the shift rail support tool

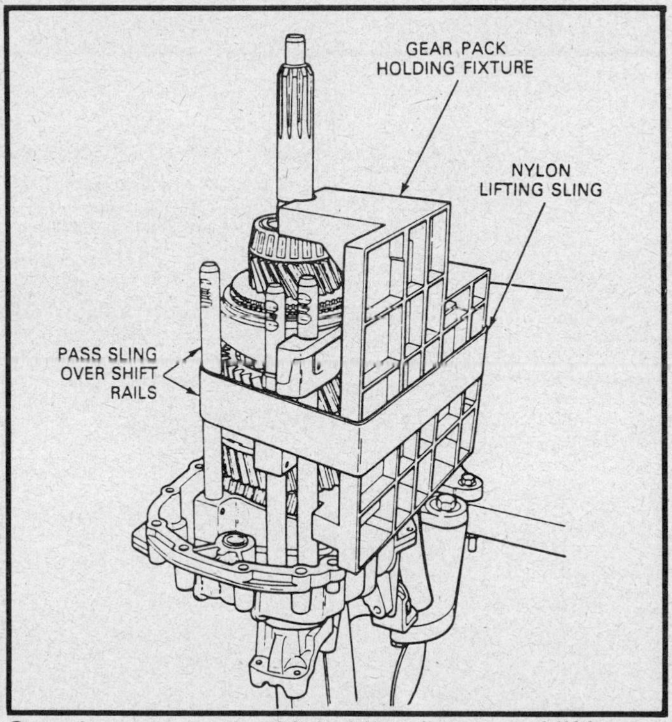
GEAR PACK
HOLDING FIXTURE
NYLON
LIFTING SLING
PASS SLING
OVER SHIFT
RAILS

Securing gear pack with holding fixture and lifting strap

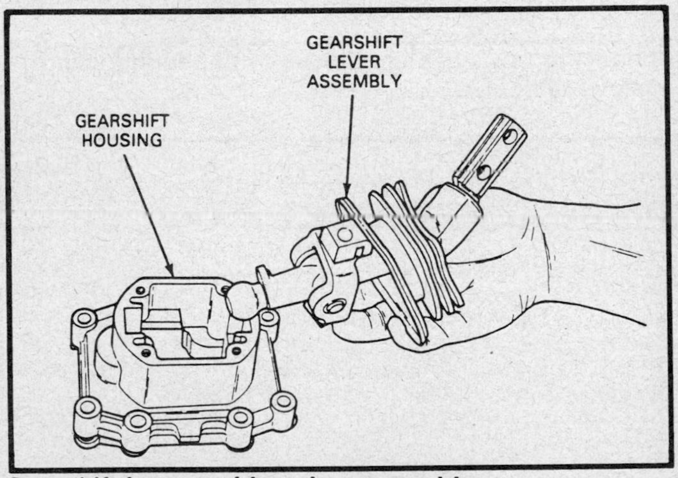
GEARSHIFT
LEVER
ASSEMBLY
GEARSHIFT
HOUSING

Gearshift lever and housing assembly

17. Remove the central shift rail and shift finger assembly from the rear case.
18. Lift the shaft from the reverse idler gear. Remove the gear and 2 caged roller bearings from the rear case.
19. Remove the 3 capscrews that attach the shift interlock to the rear case.
20. With the transmission in the verticle position (input shaft up), wrap a nylon lifting strap or equivalent with special gear pack holding fixture T87T–7025–HH or equivalent around the mainshaft and output shaft assemblies. The strap should be positioned so that it passes over the shift rails.
21. Place special shift rod support tool T87T–7025–JH or equivalent over the ends of the shift rails.
22. Carefully rotate the transmission with the gear pack holding device to the horizontal position so that the holding fixture is beneath the gear pack.
23. Remove the output shaft flange retaining nut and remove the flange from the output shaft. Tap the flange with a hammer if necessary to remove it.
24. Carefully pull the gear pack and shift rails (with their holding fixtures) to dislodge them from the rear case. Place the assembly on a work bench.
25. Slide the speedometer gear from the mainshaft, if equipped.
26. Remove the strap from the shift rails, gear packs and holding fixture.
27. Rotate the gearshift rails approximately 45 degrees to release them from the shift rails. Lift the rails, forks and interlock and shift rail support tool from the mainshaft.
28. Using the shift rail support tool as a base, set the shift rail assembly on the bench and remove the interlock from the shift rails.
29. While still in the holding tool, tag each shift rail and fork for assembly reference.
30. Lift the countershaft from the support fixture. Separate the mainshaft from the input shaft and remove the mainshaft from the holding fixture.

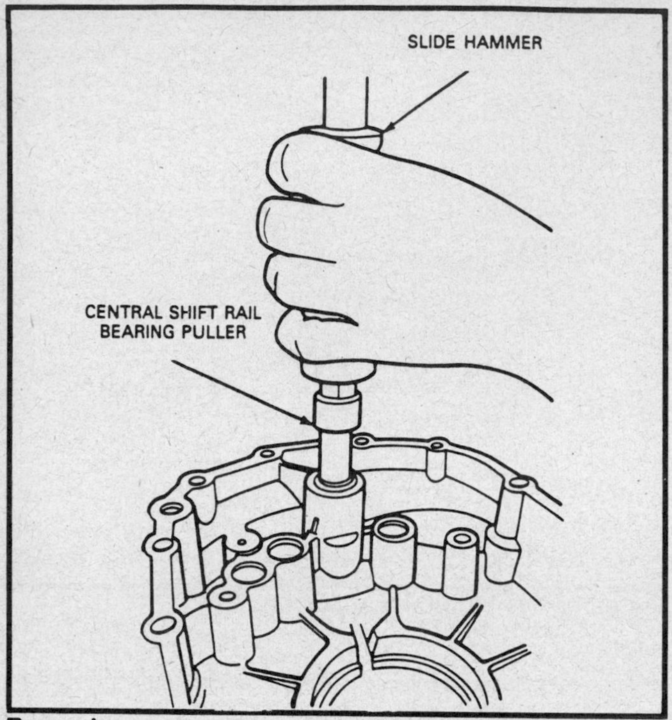
SLIDE HAMMER
CENTRAL SHIFT RAIL
BEARING PULLER

Removing rear case central shift rail bearing using special tool

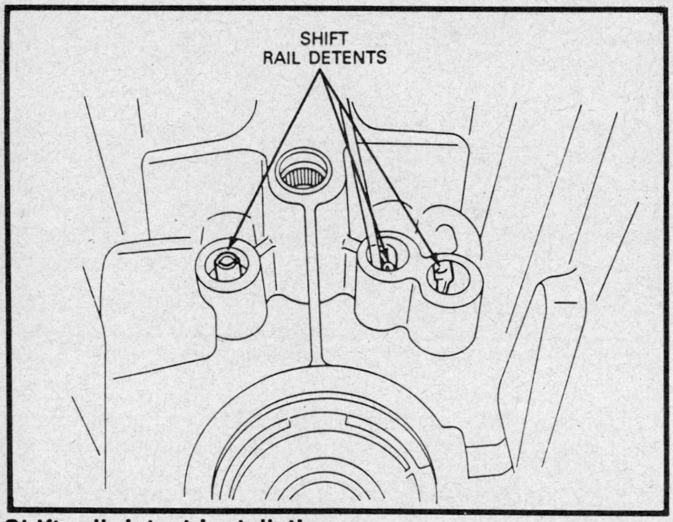
SHIFT
RAIL DETENTS

Shift rail detent installation

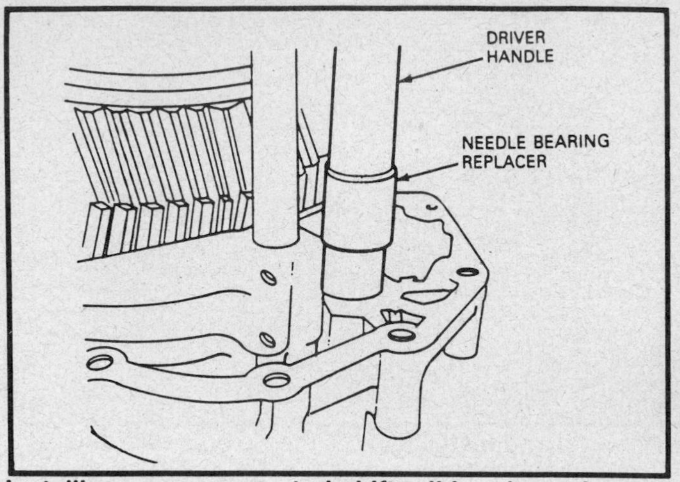
DRIVER
HANDLE
NEEDLE BEARING
REPLACER

Installing rear case central shift rail bearing using special tool

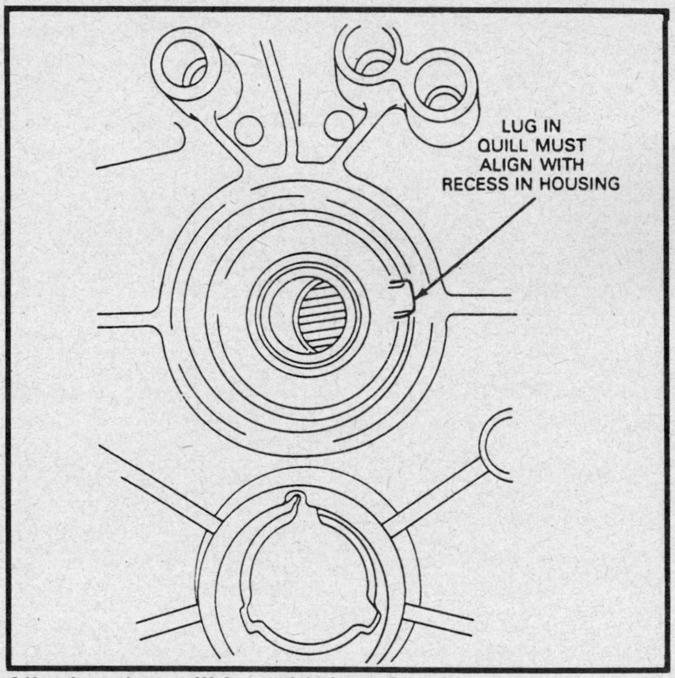
LUG IN
QUILL MUST
ALIGN WITH
RECESS IN HOUSING

Aligning the quill lug with housing recess

Unit Disassembly and Assembly

SHIFT TOWER

Disassembly

1. Unscrew the gearshift lever cover from the gearshift housing.
2. Lift the lever boot, cover and attached parts from the housing as an assembly.
3. Slide the 2 guide pieces from the Cardan joint and slide the boot and cover assembly past the top of the gearshift lever.
4. Invert the cover with the boot and remove the snapring that holds the parts together.

NOTE: A cross shaft passes through the Cardan joint and gearshift lever. This shaft rides inside the sleeved needle bearings that seat in the joint bores. Do not disassemble these components.

Inspection

1. Check the operation of the gear shift lever.
2. If binding occurs when the levers are operated, disassemble the shift housing assembly and replace all worn or damaged components as required.

Assembly

1. Insert the boot into the gearshift lever cover and install the snapring. The snapring must be installed so that the boot and ring are seated in the groove in the cover.
2. Slide the boot and cover over the gearshift lever.
3. Place the 2 guide pieces on the Cardan joint lugs. Install

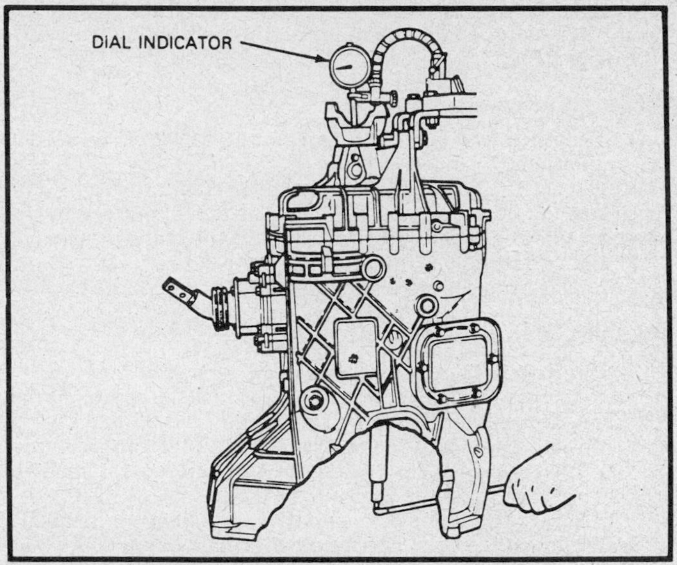
DIAL INDICATOR
Input shaft and mainshaft bearing preload measurement

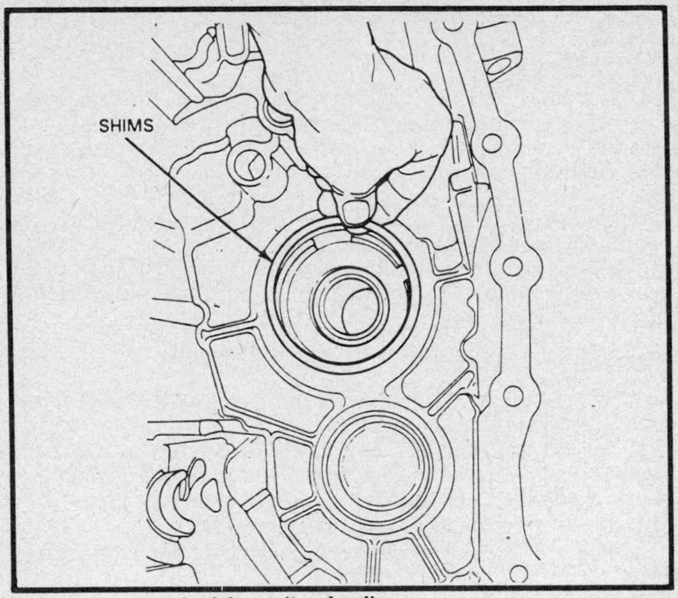
SHIMS
Fitting preload shims (typical)

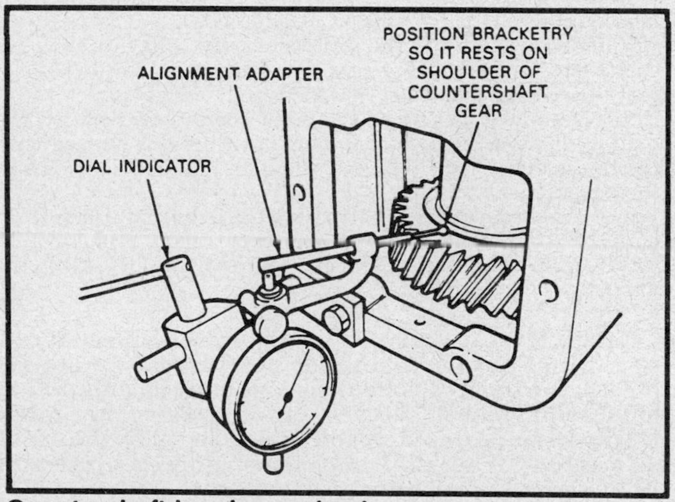
POSITION BRACKETRY SO IT RESTS ON SHOULDER OF COUNTERSHAFT GEAR
ALIGNMENT ADAPTER
DIAL INDICATOR
Countershaft bearing preload measurement

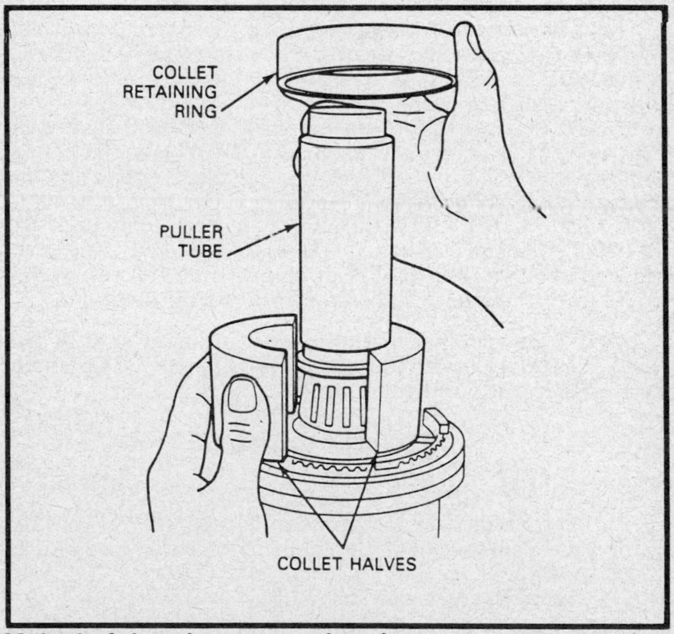
COLLET RETAINING RING
PULLER TUBE
COLLET HALVES
Mainshaft bearing removal tools

the guide pieces so that the slotted ends are pointed inward towards the cover.

4. If necessary, install new detent plug as follows:
   a. Press the old detent plug out of the mounting hole.
   b. Heat the detent mounting area to aproximately 248°F with a heat gun.
   c. Press the new detent plug into the mounting hole until it rests against the stop.
5. Place a new gasket on the gearshift housing. Do not use silicone sealant on the gasket.
6. Place the gearshift lever with guide pieces in the guide piece grooves. The slotted ends of the guide pieces must be facing upward.

NOTE: Place the nose of the lever toward the direction of the stop plate and compression spring.

7. Attach the shifter head assembly to the housing and torque the screws to 7 ft. lbs. (10 Nm).

SHIFT RAILS

Disassembly

1. Place each shift rail in a soft jawed vise.
2. Drive the 2 roll pins out of the 1st/2nd and 5th/reverse shift fork and gearshift lever on the central shift rail.
3. Slide each fork off the rail.
4. Slide the finger off the central shift rail.

Inspection

1. Inspect the shift rails for excessive wear or any other surface imperfection.
2. Inspect the fork pads for wear.
3. Replace the shift rails and/or forks as necessary.

Assembly

1. Slide the fork onto the proper shift rail.
2. Install the roll pins.

REAR CASE

Disassembly

1. If necessary, drive the 2 dowel pins out of the rear case.
2. Using a slide hammer and internal-jawed bearing puller, remove the mainshaft outer bearing from the rear case.
3. Drive the mainshaft rear seal from the case. Discard the seal.
4. Using a slide hammer and bearing cup puller, remove the countershaft rear bearing outer race.
5. If necessary, remove the central shift rail bearing from the rear case using a slide hammer and blind hole puller D80L–1000–Q or equivalent.

Inspection

1. Steam clean the cases to remove any grease or dirt accumulations.
2. Inspect the front and rear cases for cracks, worn or damaged bearing bores, damaged threads or any imperfection that would make the case unfit for service in the transmission.
3. If any defects are found, replace the case.
4. Replace the central shift rail bearing as necessary.

Assembly

1. If removed, install the central shift rail bearing as follows:
   a. Heat the rear case in the area of the central shift rail bearing bore to 320°F with a heat gun.
   b. Insert the ball sleeve and drive the bearing in until it seats against the stop. Use needle bearing replacer tool T87T–7025–DH or equivalent to install the bearing until it flush with the surface of the bore.
2. Heat the rear case in the area of the counter shaft outer race to 320°F. Using driver handle T80T–4000–W and bearing cup replacer tool T73T–4222–A or equivalents, install the race until fully seated against the stop.
3. Heat the rear case in the area of the mainshaft outer race to 320°F. Using driver handle T80T–4000–W and bearing cup replacer tool T87T–7025–PH or equivalents, tap the bearing cup into the bore until fully seated against the stop.

NOTE: Do not install the mainshaft rear seal at this time. The mainshaft rear seal will be installed just prior to installing the output flange.

FRONT CASE

Disassembly

1. Using a slide hammer and bearing cup puller T77F–1102–A or equivalents, remove the input shaft outer race fom the front case.
2. Remove the oil baffle and shims.

NOTE: When the race is removed, the oil baffle will be destroyed. Discard the oil baffle and install a new one during assembly.

3. Punch the input shaft from the base of the quill. If replacing the quill, the seal may be removed later when the quill is out of the housing.
4. If necessary, gently tap the quill with a rubber mallet to remove it.
5. Remove the O-ring from the base of the quill. If not already removed, remove the seal at this time.
6. Using a slide hammer and internal-jawed puller, remove the counter shaft front bearing outer race.
7. If necessary, remove the fill and drain plugs from the case.
8. If necessary, remove the sealing caps and 3 detents from the front case.
9. If necessary, remove the roll pins that hold the 5th/reverse interlock plate from their bores in the front case.
10. If necessary, remove the central shift rail bearing from the front case using a slide hammer and blind hole puller D80L–100–Q or equivalents.

Inspection

1. Steam clean the case to remove any grease or dirt accumulations.
2. Inspect the front and rear cases for cracks, worn or damaged bearing bores, damaged threads or any imperfection that would make the case unfit for service in the transmission.
3. If any defects are found, replace the case.
4. Replace the central shift rail bearing as necessary.

Assembly

1. If removed, tap the 5th/reverse roll pins into their bores in the front case until the larger pin protrudes approximately 0.315 in. and the smaller pin protrudes 0.16–0.20 in.
2. If removed, install the central shift rail bearing as follows:
   a. Heat the rear case in the area of the central shift rail bearing bore to 320°F with a heat gun.
   b. Insert the ball sleeve and drive the bearing in until it seats against the stop. Use needle bearing replacer tool T87T–7025–DH or equivalent to install the bearing until it is flush with the surface of the bore.
3. If removed, install the drain and fill plugs into the front bore. Torque the plugs to 29 ft. lbs. (40 Nm).
4. If removed, install the shift rail detent balls into their respective bores in the front case. The balls must seat in the detent of each rail and must move freely.
5. Place a new O-ring on the input shaft quill. Position the quill in the housing so that the lug in the quill is aligned with the recess in the housing. Push the quill in until it bottoms in the housing.
6. Place the new input shaft seal in the front case. Using driver handle tool T80T–4000–W and input shaft seal installer T87T–7025–EH or equivalent, drive the seal until it seats in the quill stop.

NOTE: If the countershaft, input shaft, mainshaft or 1 or more of the tapered roller bearings have been replaced, the tapered roller bearings must be adjusted to obtain a preload of 0.00079–0.00343 in. using the proper thickness shim. Countershaft preload is set by the use of shims alone. Input shaft and mainshaft preload is set by using shims and a baffle.

7. Heat the mounting bore for the countershaft bearing outer race to 320°F.
8. Place the proper thickness shim in the bore, using driver handle T80T–4000–W and bearing cup replacer T73T–4222–A or equivalents, seat the countershaft outer bearing race against the stop in the case.
9. Heat the front case in the area of the input shaft tapered roller bearing outer race to 320°F.
10. Place the proper thickness shim/sealing disc pack in the outer race bore. Using driver handle T80T–4000–W and mainshaft bearing cup replacer T87T–7025–PH or equivalents, seat the bearing cup against the stop in the bore.

Preload Measurements

Measurement and adjustment of the roller bearings is necessary if a housing, countershaft, mainshaft or input shaft has been replaced or if a tapered roller bearing per shaft has been replaced.

The mainshaft input shaft and countershaft should be adjusted to provide a preload of 0.00079–0.00434 in. After the shafts are adjusted, each shaft should be rotated several times to center the rollers in the bearings.

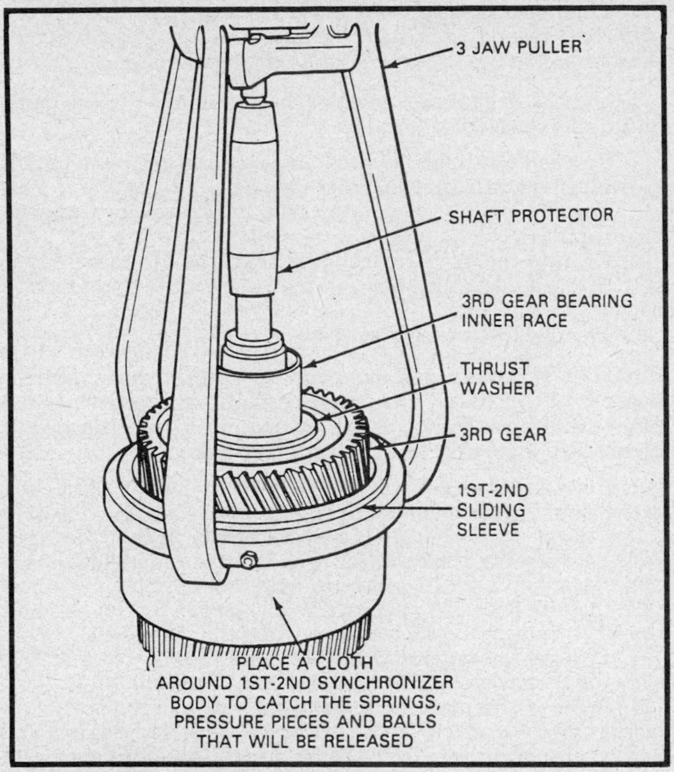
3 JAW PULLER
SHAFT PROTECTOR
3RD GEAR BEARING INNER RACE
THRUST WASHER
3RD GEAR
1ST-2ND SLIDING SLEEVE
PLACE A CLOTH AROUND 1ST-2ND SYNCHRONIZER BODY TO CATCH THE SPRINGS, PRESSURE PIECES AND BALLS THAT WILL BE RELEASED

1st/2nd sliding gear removal

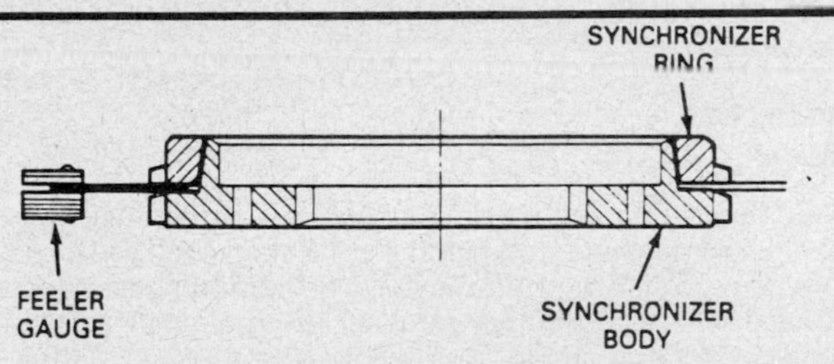
SYNCHRONIZER RING
FEELER GAUGE
SYNCHRONIZER BODY

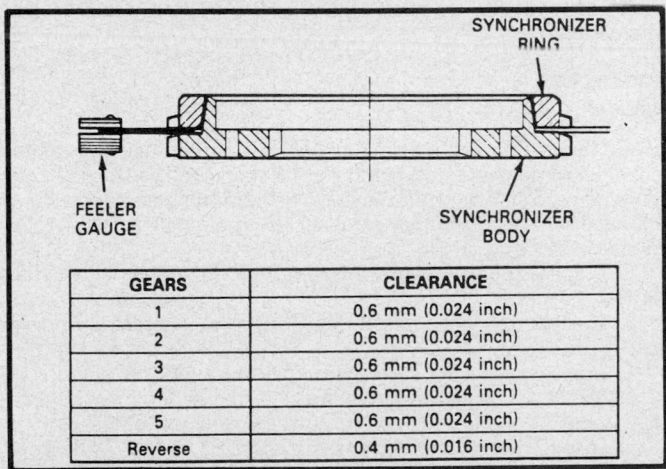

GEARS	CLEARANCE
1	0.6 mm (0.024 inch)
2	0.6 mm (0.024 inch)
3	0.6 mm (0.024 inch)
4	0.6 mm (0.024 inch)
5	0.6 mm (0.024 inch)
Reverse	0.4 mm (0.016 inch)

Measuring synchronizer ring-to-body clearance

INPUT SHAFT AND MAINSHAFT BEARING PRELOAD MEASUREMENT

NOTE: The shims and baffle must be removed prior to the preload measurement. The baffle is part of the shim package under the bearing race.

1. Place the transmission on the work bench so that the output flange is facing upward.
2. Attach dial indicator gauge and magnetic base set D78P–4201–B or equivalent onto the output flange. The dial indicator extension portion should be resting on the output end of the mainshaft.
3. Zero the indicator and using the proper tool, gently pry up on the input shaft and mainshaft.

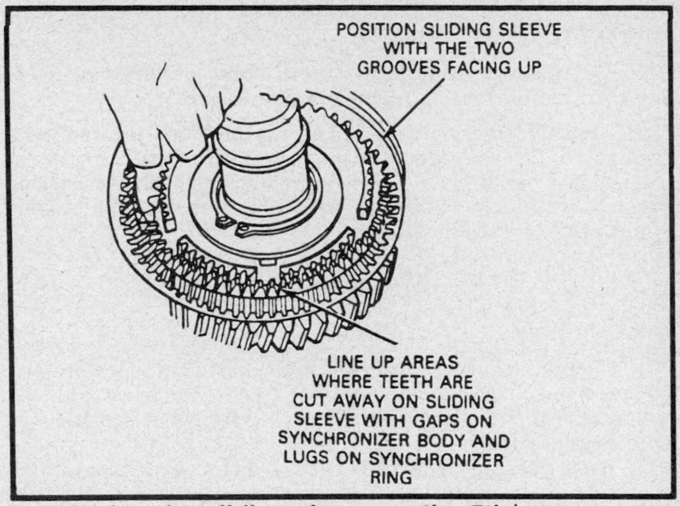
POSITION SLIDING SLEEVE WITH THE TWO GROOVES FACING UP
LINE UP AREAS WHERE TEETH ARE CUT AWAY ON SLIDING SLEEVE WITH GAPS ON SYNCHRONIZER BODY AND LUGS ON SYNCHRONIZER RING

Positioning the sliding sleeve on the 5th/reverse synchronizer ring

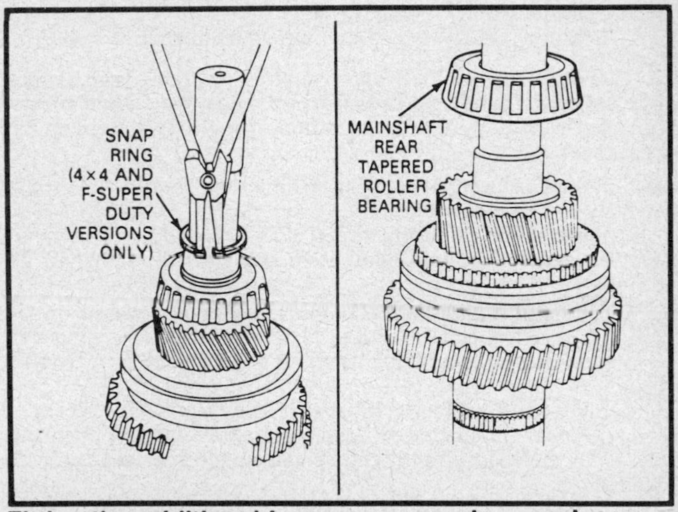
SNAP RING (4×4 AND F-SUPER DUTY VERSIONS ONLY)
MAINSHAFT REAR TAPERED ROLLER BEARING

Fitting the additional inner race snapring used on 4WD and F-Super Duty vehicles

4. Record the reading on the indicator.
5. Select a shim and baffle that is equal to the combined thickness of the indicator reading plus 0.00079–0.00434 in. This is the specified preload after the shims and shaft seal and mainshaft and countershaft are installed.

COUNTERSHAFT BEARING PRELOAD MEASUREMENT

1. With two 10mm hex screws, attach dial indicator set with bracket adapter tool 4201–C or equivalent to the power take off opening in the front case.
2. Position the dial indicator on the support bracket so that the measurement bar rests against the flat face of the 4th speed helical gear on the countershaft.
3. Zero the dial indicator needle.
4. Insert a suitable prying tool through each of the 2 power take off openings and position them below the 4th speed helical gear on the countershaft. Pry up on the countershaft.
5. Record the reading on the indicator.
6. Select a shim that is equal to the combined thickness of the indicator reading plus 0.00079–0.00434 in. This is the specified preload after the shim is installed on the countershaft.

MAINSHAFT AND INPUT SHAFT PRELOAD ADJUSTMENT

NOTE: The preload adjustments may be made only after the preload measurements are taken.

Using the information obtained in the previous preload measurements, fit each race with the proper shim washer or shim washer and baffle plate that will provide the specified bearing preload. Install the shims below the outer race of the bearing which seats in the front case.

MAINSHAFT

Disassembly

1. Clamp the output end of the mainshaft firmly in a protected jaw vise.
2. Remove the 4th synchronizer ring from the 3rd/4th synchronizer assembly.
3. Place bearing collets T87T–7025–FH or equivalent on either side of the mainshaft front bearing. Position puller tube T77J–7025–B or equivalent in the collets. Pass collet retaining ring T75L–7025–G or equivalent over the puller tool and on the collets so that they clamp firmly to the bearing. Pull the bearing from the mainshaft.
4. Remove the 3rd/4th sliding sleeve from the mainshaft.

NOTE: Wrap a cloth around the synchronizer assembly to catch the compression springs, pressure pieces and balls that will fall out when the sliding sleeves are removed.

5. Remove the snapring that retains the 3rd/4th synchronizer body to the mainshaft.
6. Place collet retaining ring T87T–7025–OH or equivalent over the mainshaft and allow it to rest on the mainshaft 1st gear.
7. Place collet halves T87T–7025–NH or equivalent on the 3rd/4th synchronizer body and slide the collet retaining ring over the collet halves to hold them in place on the synchronizer body.
8. Place shaft protector D80L–625–4 or equivalent spcaer on the end of the mainshaft. Install a suitable 3-jawed puller onto the collet halves and pull the synchronizer gear set from the mainshaft.
9. Remove the synchronizer ring, 3rd gear and needle roller bearings from the mainshaft.
10. Slide the 1st/2nd gear sliding sleeve as far up as will go.
11. Place collet retaining ring T87T–7025–OH or equivalent over the mainshaft and allow it to rest on the 1st gear.
12. Place collet halves T87T–7025–MH or equivalent on the 1st/2nd sliding sleeve so that they seat in the groove on the sleeve. Pass the retaining ring over the collets and secure to the sleeve.
13. Place shaft protector D80L–625–4 or equivalent on the end of the mainshaft. Install a suitable 3-jawed puller onto the collet retaining ring and pull the 1st/2nd sliding sleeve, 2nd gear, thrust washer and 3rd gear bearing inner race from the mainshaft.

NOTE: Wrap a cloth around the synchronizer assembly to catch the compression springs, pressure pieces and balls that will fall out when the sliding sleeve clears the 1st/2nd synchronizer body.

14. Remove the snapring that retains the 1st/2nd synchronizer to the mainshaft.
15. Reposition the mainshaft in the vise so that the output end of the shaft is facing up.

NOTE: On 4WD and F-Super Duty versions of this transmission, a snapring that retains the bearing inner race is also removed.

16. Place bearing gripper D81L–4220–A or equivalent on the mainshaft rear tapered roller bearing. Use the gripper to pull the bearing by the shoulder and no the bearing cage.

NOTE: If a bearing gripper is not used, the bearing will be destroyed.

17. Position a suitable 3-jawed puller on the gripper and pull the mainshaft rear bearing from the shaft.
18. Remove the 5th gear and caged needle rollers from the mainshaft.
19. Remove the synchronizer ring from the 5th/reverse synchronizer. Remove the snapring that retains the synchronizer body to the mainshaft.
20. Remove the 5th/reverse sliding sleeve.

NOTE: Wrap a cloth around the 5th/reverse synchronizer body to catch the compression springs, pressure pieces and balls that will fall out when the sliding sleeve clears the 5th/reverse synchronizer body.

21. Place collet retaining ring T87T–7025–OH or equivalent over the mainshaft and allow it to rest on the 1st gear. Position collet halves T87T–7025–NH or equivalents so that the ridge rests between the synchronizer body and the ring. Slide the retaining ring around the collets to hold them in place.
22. Install a 3-jawed puller on the collet retaining ring and pull the 5th/reverse synchronizer body from the mainshaft.
23. Remove the synchronizer ring from the reverse gear and slide the gear with caged needle bearings from the shaft.
24. Remove the mainshaft from the vise and support on 2 blocks of wood under an arbor press. Press the 1st gear and 1st/2nd synchronizer body from the mainshaft. An alternate method is to clamp the mainshaft at the output end and use a 3-jawed puller to pull the gear and synchronizer from the shaft.
25. Remove the 1st gear and caged needle rollers from the mainshaft.

Inspection

GEARS

Some forms of grind marks and tooth wear contact paterns are considered normal and should not be suspected as the source of gear noise. Factory grind marks are distinguished from wear or damage which appear as local swells (polished raised projections), nicks and chips.

1. Inspect the gear teeth for roughness or ridges on the face of the tooth.
2. Inspect the gear teeth for abnormal contact and wear patterns.
3. Inspect the gear teeth for chips and nicks.
4. Minor imperfections may be repaired as long as the damage is not in the location of the face-in gear pattern area.
5. Inspect the gear bushings for burrs or any other damage.
6. Inspect the gear shaft for wear and scoring.
7. Repair or replace the gears as necessary.

BEARINGS

1. Inspect the bearings for cracked races and cups.
2. Inspect the races for roughness.
3. Inspect the ball bearings for looseness, wear, chipping, flaking or other damage.
4. Inspect the bearings for binding on the shaft or looseness in the bore.
5. Replace the bearings as required.

SYNCHRONIZER BLOCKING RINGS AND COMPRESSION SPRINGS

1. Inspect the synchronizer blocking rings for widened index slots, rounded clutch teeth and smooth internal surfaces. The surfaces must have machined grooves.
2. Place the synchronizer ring on the body. With a feeler gauge, measure the distance between the face of the gear clutch-

ing teeth and the face of the blocking ring must not be less than 0.024 in. on 1st/5th synchronizers and 0.016 in. for the reverse synchronizer. Take the measurement at 2 opposite points. Replace the synchronizer ring and/or body to bring the clearance within specification.

3. Inspect the synchronizer sleeves for free movement on the hubs. Make sure the alignment marks are properly indexed.

4. Check the length and tension of all the synchronizer compression springs. Unloaded length should be 0.583 in., outside diameter should be 0.235 in. and the wire diameter should be 0.037 in. Replace any spring that does meet specification.

Assembly

1. Clamp the input end of the mainshaft in a protected jaw vise.

2. Slide the reverse gear caged needle rollers on the mainshaft and place the reverse gear over the rollers. Make sure that the clutch teeth on the gear are facing upwards.

NOTE: Before installing the original synchronizer and blocking ring, check the clearance between the ring and the body.

3. Position the reverse gear synchronizer ring on the taper of the reverse gear.

4. Heat the 5th/reverse gear body to 320°F. Do not heat the gear for more than 15 minutes.

5. Place the synchronizer body on the mainshaft splines so that the side with the deeper hub and the gaps on the synchronizer body align with the short lugs on the synchronizer ring.

6. Install the snapring on the mainshaft next to the 5th/reverse synchronizer body. The clearance between the snapring and the synchronizer should not be more than 0.004 in. Zero clearance is preferred. Make sure that the snapring is free of burrs before checking the clearance.

7. Check the reverse gear endplay. Endplay must be between 0.006–0.014 in.

8. There are 2 grooves on the 5th/reverse sliding sleeve. With the 2 grooves facing up, position the sliding sleeve over the synchronizer body. In 3 places on the sleeve, 3 teeth have been cut away. Align these areas with the 3 gaps in the synchronizer body and with the 3 lugs on the ring. Slide the sliding gear down until it rests against the reverse gear clutching teeth.

9. Insert the 3 compression springs with pressure pieces in the synchronizer body recesses.

NOTE: If the original springs are being used, make sure that they are inspected first.

10. Push back the pressure pieces using the proper tool, push in the balls and slide the pressure piece so that it rests against the ball.

11. Place the 5th gear synchronizer ring on the body. The short lugs on the ring should be located over the gaps in the body.

12. Push the 5th gear synchronizer ring downward while pulling the sliding sleeve into the center position.

13. Place the 5th gear caged needle rollers on the mainshaft and slide the 5th gear over the rollers.

14. Heat the inner race of the mainshaft rear taper roller bearing to 320°F. Slide the gear on the mainshaft until it seats against the stop on the shaft. If necessary, tap the gear to assist in seating.

NOTE: Do not heat the bearing for more than 15 minutes. If the bearing must be driven onto the shaft, drive against the inner race only and not against the bearing cone.

15. Check the 5th gear endplay. Endplay should be between 0.006–0.014 in.

NOTE: On 4WD and F-Super Duty vehicles, fit an additional snapring in the machined groove next to the taper roller bearing inner race. It must have an endplay between 0.0–0.004 in. Zero endplay is preferred.

16. Turn the mainshaft over and clamp it at the input end.

17. Place the 1st gear caged needle roller bearings on the mainshaft and install the 1st gear synchronizer on the taper of the 1st gear.

18. Heat the 1st/2nd synchronizer body in a suitable gear oven to 320°F. Do not heat the gear for more that 15 minutes.

19. Place the synchronizer body on the mainshaft splines so that the short lugs on the ring engage with the gaps in the body. Push or lightly tap the body until it seats against the ring.

20. Install the snapring on the mainshaft next to the 1st/2nd synchronizer body. The clearance between the snapring and the body should not be more than 0.004 in. Zero clearance is prefered. Make sure that the snapring is free of burrs before checking the clearance.

21. Check the 1st gear endplay. Endplay must be between 0.006–0.014 in.

22. Position the sliding sleeve over the synchronizer body so that the tapered collars are facing down (towards the output end of the mainshaft). In 3 places on the sleeve, 3 teeth have been cut away. Align these areas with the 3 gaps in the synchronizer body and with the 3 lugs on the ring. Slide the sliding gear down until it rests against the 1st gear clutching teeth.

23. Insert the 3 compression springs with pressure pieces in the synchronizer body recesses.

NOTE: If the original springs are being used, make sure that they are inspected first.

24. Push back the pressure pieces using the proper tool, push in the balls and slide the pressure piece so that it rests against the ball.

25. Place the 2nd gear synchronizer ring on the 1st/2nd body. The short lugs on the ring should be located over the gaps in the body.

26. Push the synchronizer ring downwards while pulling the sliding sleeve into the center position.

27. Place the 2nd gear gaged needle rollers on the mainshaft and slide the 2nd gear over the rollers.

28. Heat the 2nd gear thrust washer to 320°F. Do not heat the thrust washer for more that 15 minutes.

29. Place the thrust washer onto the mainshaft and push down until it seats on stop of the shaft. If necessary, gently tap the washer to seat it.

30. Heat the 3rd gear bearing inner race to 320°F. Do not heat the bearing race for more than 15 minutes.

31. Place the race on the mainshaft and push down until it seats on stop of the shaft. If necessary, gently tap the washer to seat it.

32. Heat the 3rd gear bearing inner race to 320°F.

33. Check the 2nd gear endplay. Endplay must be between 0.006–0.017 in.

34. Place the 3rd gear synchronizer ring on the taper of the 3rd gear.

35. Heat the 3rd/4th synchronizer body to 320°F. Do not heat the synchronizer body for more that 15 minutes.

36. Place the synchronizer body on the mainshaft splines so that the short lugs on the ring engage with the gaps in the body. Push or lightly tap the body until it seats against the ring. Make sure the recess in the body is facing up.

37. Install the snapring on the mainshaft next to the 3rd/4th synchronizer body. The clearance between the snapring and the body should not be more than 0.004 in. Zero clearance is preferred. Make sure that the snapring is free of burrs before checking the clearance and measuring the endplay.

38. Check the 3rd gear endplay. Endplay must be between 0.006–0.014 in.

39. Position the sliding sleeve over the synchronizer body so

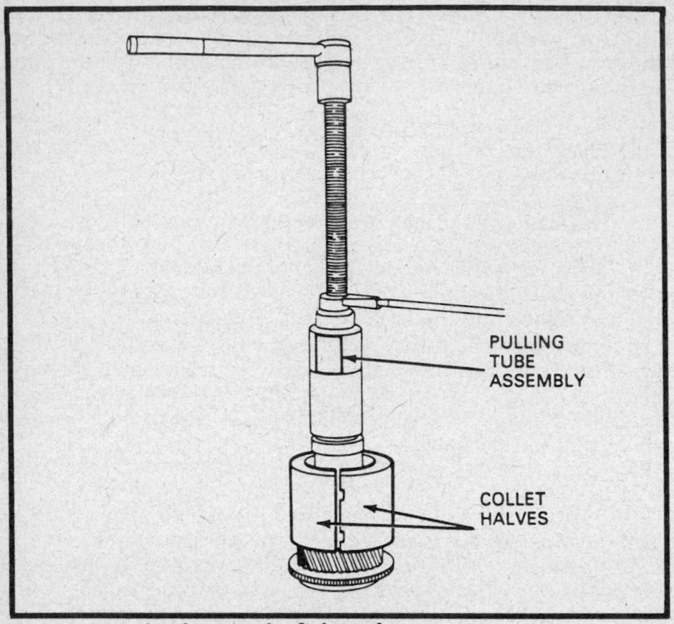
PULLING
TUBE
ASSEMBLY
COLLET
HALVES

Removing the input shaft bearing

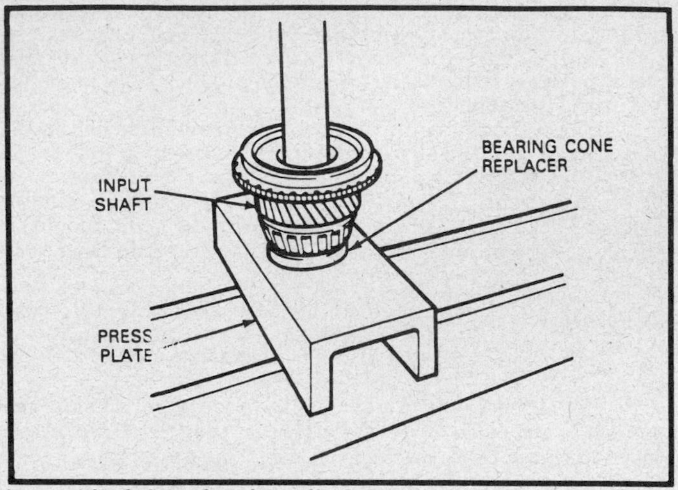
BEARING CONE
REPLACER
INPUT
SHAFT
PRESS
PLATE

Input shaft bearing installation

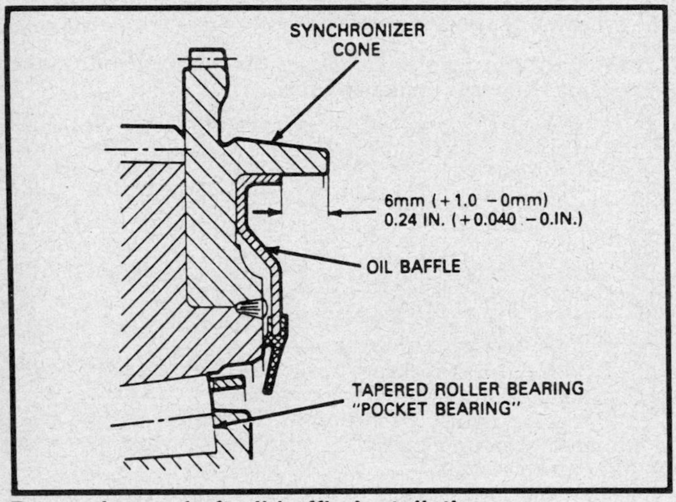
SYNCHRONIZER
CONE
6mm (+1.0 −0mm)
0.24 IN. (+0.040 −0.IN.)
OIL BAFFLE
TAPERED ROLLER BEARING
"POCKET BEARING"

Proper input shaft oil baffle installation

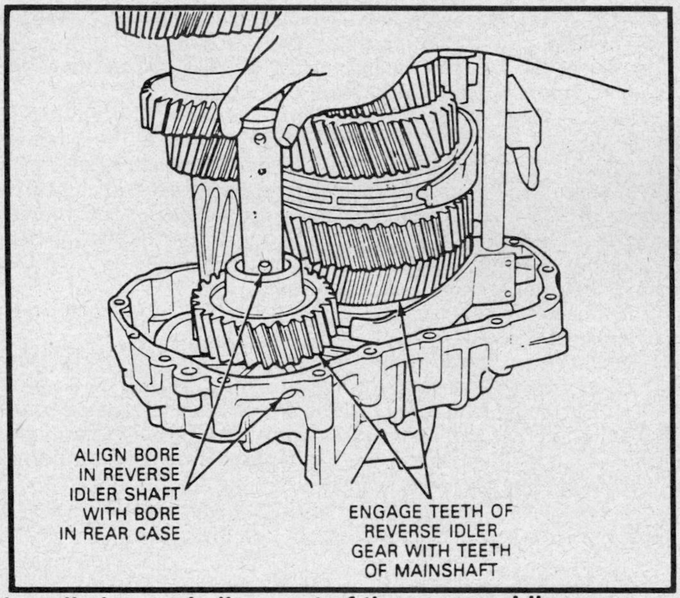
ALIGN BORE
IN REVERSE
IDLER SHAFT
WITH BORE
IN REAR CASE
ENGAGE TEETH OF
REVERSE IDLER
GEAR WITH TEETH
OF MAINSHAFT

Installation and alignment of the reverse idler gear shaft

that the smaller of the 2 grooves are facing down. In 3 places on the sleeve, 3 teeth have been cut away. Align these areas with the 3 gaps in the synchronizer body and with the 3 lugs on the ring. Slide the sliding gear down until it rests against the 3rd gear clutching teeth.

40. Insert the 3 compression springs with pressure pieces in the synchronizer body recesses.

NOTE: If the original springs are being used, make sure that they are inspected first.

41. Push back the pressure pieces using the proper tool, push in the balls and slide the pressure piece so that it rests against the ball.

42. Place the 4th gear synchronizer ring on the synchronizer body. The short lugs on the ring should be located over the gaps in the body.

43. Push the synchronizer ring downwards while pulling the sliding sleeve into the center position.

44. Heat the inner race of the mainshaft front taper roller bearing to 320°F. Do not heat the bearing for more than 15 minutes. If the bearing must be driven onto the shaft, drive against the inner race only and not against the bearing cone.

NOTE: An alternate method of installing the bearing is to press it on the shaft using bearing cone replacer T85T–4621–AH or equivalent.

45. Place the race on the shaft and push down until it seats on the mainshaft stop. If necessary, tap the gear to assist in seating.

INPUT SHAFT

Disassembly

1. Place collet halve tools 44803 and 44797 or equivalent around the input shaft bearing cone.

2. Install pulling tube D81L–4220–A or equivalent on the collet tools and pull the input shaft bearing from the input shaft.

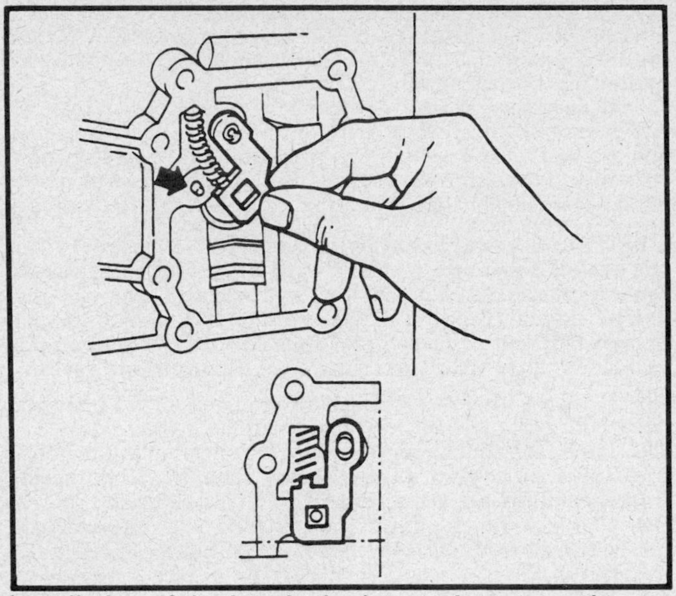

Installation of the interlock plate and compression spring

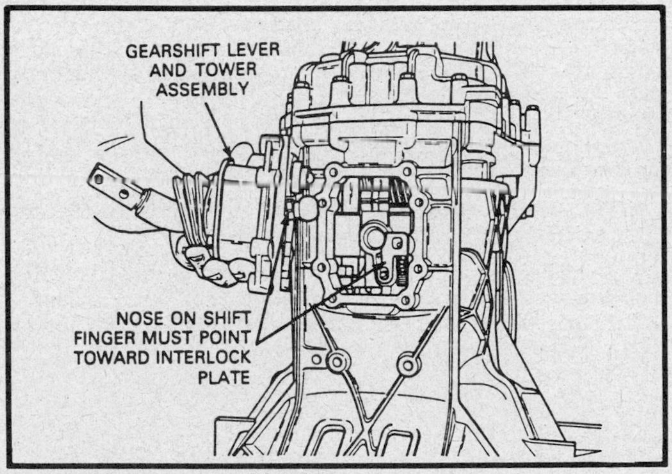
GEARSHIFT LEVER AND TOWER ASSEMBLY
NOSE ON SHIFT FINGER MUST POINT TOWARD INTERLOCK PLATE

Gearshift tower positioning

Inspection

1. Inspect the input shaft for wear, twisting, bending. The bearing race must be free of burrs and imperfections.
2. Inspect the gear for chipped, worn, nicked or missing teeth.
3. Inspect the guide splines for wear.
4. Verify that the input shaft oil baffle is properly installed in the synchronizer cone recess.
5. Visually inspect the oil baffle for tears or similar damage.
6. Make sure the lip of the oil baffle should be recessed 0.244 in. below the edge of the synchronizer cone.
7. Make sure the oil baffle fits tight and cannot be moved by hand.
8. Make sure the 3 oil holes in the shaft are clear of any obstruction. Also, the oil channel in the main case must be clear.
9. Inspect the oil ring (that is part of the shim pack) behind the outer race of the input bearing for damage.
10. If the pilot bearing bore is worn or scored, replace the gear or gear rollers.
11. Replace the input shaft bearing retainer seal.

Assembly

1. Place the new bearing on the shaft and position the bearing replacer tool T85T–4621–AH or equivalent over the bearing.
2. Place the shaft, bearing and cone a suitable press plate.

NOTE: Make sure that the oil baffle is properly installed and is not damaged.

3. Press the bearing on the shaft until it seats against the shaft stop.

COUNTERSHAFT

Do not attempt to service the countershaft. If repairs are required, the countershaft is replaced as an assembly.

Transmission Assembly

1. Using gear pack holding fixture T87T–7025–HH or equivalent, assemble the input shaft over the tapperd roller bearing on the input end of the mainshaft. Lay the assembly into the holding fixture.
2. Place the countershaft on the fixture and mesh the gears of the 2 shafts.
3. Place the 3 shift rail and fork assemblies into support tool T87T–7025–HJ or equivalent. Install the interlock.
4. Place the shift rails, forks and support tool on the gear sets so that the shift rails engage with their respective mainshaft sliding sleeves.
5. Place the shift interlock on the 3 rails and engage it in the grooves on the 5th/reverse upper rail.
6. If equipped, slide the speedometer worm gear onto the mainshaft until it seats onto the top of the mainshaft.
7. Position the rear case onto the bench mounted holding fixture.
8. Wrap a suitble lifting strap around the shift rails, holding fixture, mainshaft and countershaft.
9. Position the gear pack into the rear case and push the shafts and the rails forward until the bearings seat in the housing outer races and the shift rails slide into the housing retaining holes.
10. Rotate the gear set and the rear case so that the input shaft is pointing up.
11. If equipped, slide the output flange onto the output end of the mainshaft until it seats against the stop. Screw the retaining nut on finger tight.

NOTE: Make sure the mainshaft bearing is not forced off its race when the output flange is installed.

12. Remove the shift rod support tool from the ends of the shift rails.
13. Remove the lifting strap and the gear pack holding fixture.
14. Install the 3 screws that secure the shift interlock to the rear transmission housing. Torque the screws to 7 ft. lbs. (10 Nm). Make sure the interlock moves freely after the screws are tightened.
15. Engage the teeth of the reverse idler with the reverse gear on the mainshaft.
16. Slide the reverse idler gear down through the bearings and into the rear case. Align the lower of the 2 threaded holes in the shaft with the bore in the rear case. Install an O-ring onto the capscrew and insert the screw through the housing bore into the mainshaft. Make the screw finger tight.
17. Insert the central shift rail and finger assembly into its bore in the rear case.
18. If the magnet was removed, insert it into the housing recess.
19. If the mainshaft, input shaft or countershaft bearings did not require preload adjustment, apply a thin coat of gasket compound E2AZ–19562–B (ESE–M4G234–A2) or equivalent onto

the rear case sealing surface. Do not use silicon sealing compound or its equivalent.

NOTE: The sealant cures rapidly. Do not wait more than 15 minutes to install and tighten the case bolts. If the bearings require adjustment, do not apply sealant at this time.

20. If removed, push the 3 shift rail detents back into their respective holes in the front case. Make sure that the dentents do not interfere with the entry of the shift rails.
21. Carefully position the front case over the shafts and shift rails until it contacts the with the rear case mating surface. It may be necessary to push the central shift rail inward to clear the inner surfaces of the front case.

NOTE: Make sure the input shaft does not damage the front seal. Do not bend the shim pack oil baffle when positioning the case over the input shaft.

22. Drive in the 2 rear and front case alignment dowels that are located in the corners of the 2 cases. Install 2 case screws and make finger tight.
23. Install 2 more case screws into the rear case bore near the shafts and make finger tight.
24. If it is not required to adjust the mainshaft/input shaft preload, install all the remaining case attaching screws and torque them to 18 ft. lbs. (25 Nm).

NOTE: If either the front or rear cases, countershaft, mainshaft or input shaft were replaced, the bearing preload must be adjusted be proceeding with the remainder of the assembly procedure. Refer to the appropriate adjustment procedure to adjust bearing preload.

25. Install the O-rings onto the reverse idler shaft retaining screws and install the screws into the front and rear cases. Torque the screws to 16 ft. lbs. (22 Nm). After they are installed push the sealing caps onto the screws until they are flush with the surface of the screw head.
26. Rotate the transmission so that the input shaft is pointing down.
27. Install the speedometer drive gear on the mainshaft, if equipped.
28. Install a new rear oil seal as required using the proper procedure.
29. On 2WD vehicles, install the hex nut onto the mainshaft after the new oil seal is in place. Using the proper socket, torque the nut to 184 ft. lbs. (250 Nm). Secure the nut to the mainshaft by staking the locking tab on the nut.
30. If necessary, install new gaskets on the power take off (PTO) covers. Install the covers onto the front case openings and torque the cover retaining screws to 28 ft. lbs. (38 Nm).
31. Install the 5th/reverse gear interlock plate and place the gasket over the shift tower mating surface in the front case.

NOTE: Make sure that the stop plate moves freely and the gasket does not interfere with it. The interlock plate must be installed in a postion that is lower than the gasket to ensure freedom of movement and also to ensure proper 5th and reverse gear operation. Do not drop the interlock plate into the front case during installation.

32. Place the interlock spring above the nose of the plate and move both parts in their installed positions.
33. Install the shift tower assembly. The nose on the gearshift finger must point toward the interlock plate. Install the spring washers and tighten the screws to 18 ft. lbs. (25 Nm).
34. Check the functioning of the interlock. The shift lever cannot be moved from 5th to reverse, if the interlock is properly installed.

NOTE: When checking the operation of the interlock, do not force the lever to shift into reverse. Damage to the interlock and spring could result.

35. A compression spring is installed over each of the 3 detent bolts. Before installing the springs, make sure that the unloaded length is 1.736 in. If the unloaded length is not as specified, replace the spring(s).
36. Install a new sealing cap in each bore in the front case where the detents, bolts and springs were installed. Drive each cap into place so that it seats $^{3}/_{64}$ in. below the housing surface.

NOTE: If the caps are installed deeper than $^{3}/_{64}$ in., high shifting efforts will result.

36. If the back-up light switch was removed, install the switch with a new gasket. Torque the switch to 15 ft. lbs. (20 Nm).
37. Remove the transmission from the holding fixture and install it in the vehicle.

SPECIFICATIONS

TORQUE SPECIFICATIONS

Description	ft. lbs.	Nm
Drain plug	25–30	35–40
Oil filler plug	25–30	35–40
Extension adapter to main case	16	22
End yoke to mainshaft	184	250
Shift tower cover to main case	16	22
Power take off cover plate	28	38
Idler shaft retention	16	22
Shift rail plate	7	10
Shift cover to tower cover	7	10
Reverse switch	15	20

PRE-LOAD SPECIFICATIONS

Description	in.	mm
Countershaft	0.00079–0.00434	0.02–0.11
Input shaft and mainshaft	0.00079–0.00434	0.02–0.11
Mainshaft reverse gear	0.00591–0.01378	0.15–0.35
Mainshaft 1st gear	0.00591–0.01378	0.15–0.35
Mainshaft 2nd gear	0.00591–0.01717	0.15–0.45
Mainshaft 3rd gear	0.00591–0.01378	0.15–0.35
Mainshaft 4th gear	0.00591–0.01378	0.15–0.35
Mainshaft synchronizer body retention rings	0–0.00394	0–0.1

Section 7

M75 Transmission
General Motors

APPLICATION

M75 TRANSMISSION

Year	Vehicle	Engine
1984	Chevette	1.8L Diesel
	1000	1.8L Diesel
1985	Chevette	1.8L Diesel
	1000	1.8L Diesel
1986	Chevette	1.8L Diesel

GENERAL DESCRIPTION

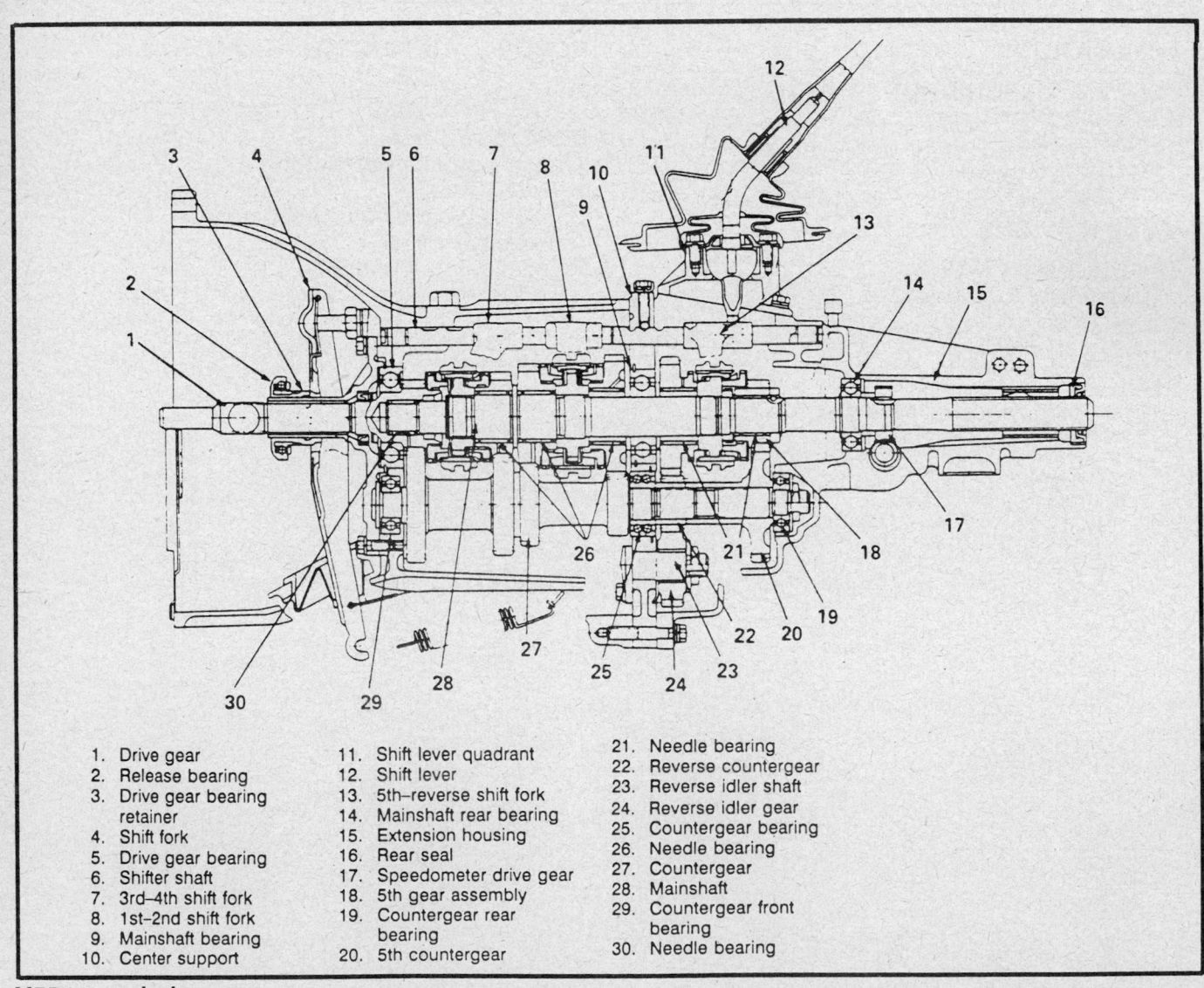

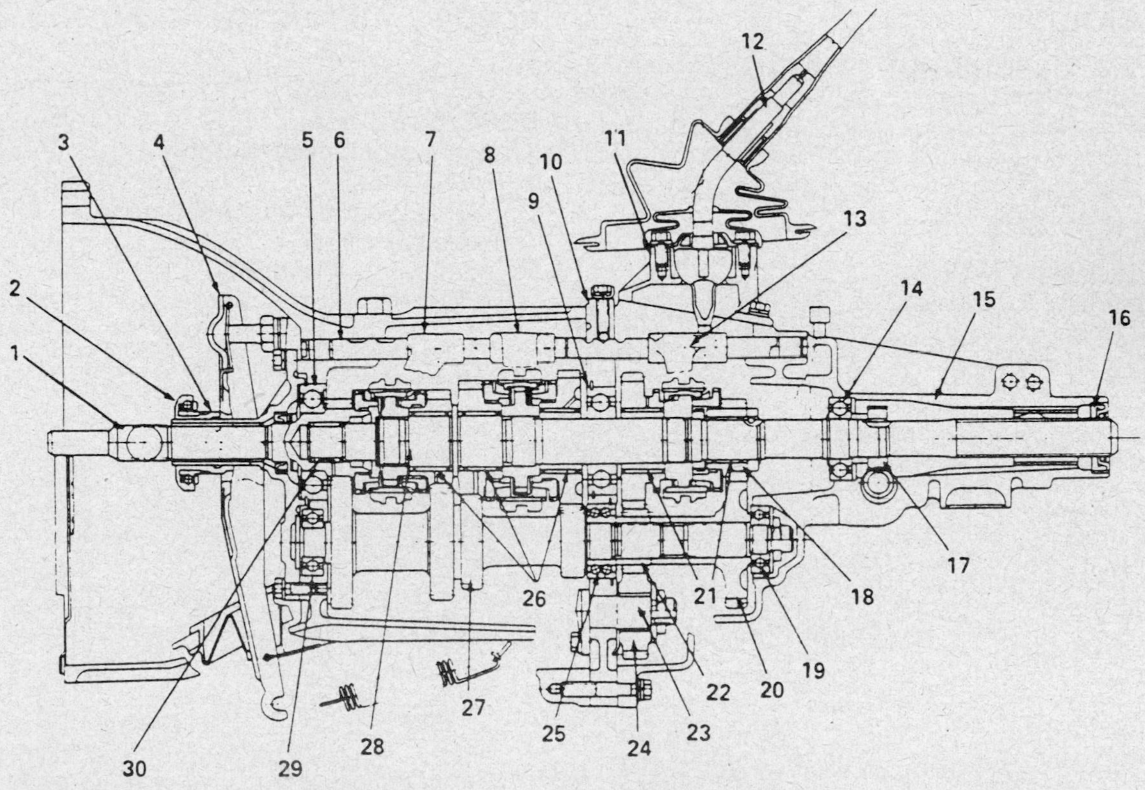
1
2
3
4
5
6
7
8
9
10
11
12
13
14
15
16
17
18
19
20
21
22
23
24
25
26
27
28
29
30

1. Drive gear
2. Release bearing
3. Drive gear bearing retainer
4. Shift fork
5. Drive gear bearing
6. Shifter shaft
7. 3rd–4th shift fork
8. 1st–2nd shift fork
9. Mainshaft bearing
10. Center support
11. Shift lever quadrant
12. Shift lever
13. 5th–reverse shift fork
14. Mainshaft rear bearing
15. Extension housing
16. Rear seal
17. Speedometer drive gear
18. 5th gear assembly
19. Countergear rear bearing
20. 5th countergear
21. Needle bearing
22. Reverse countergear
23. Reverse idler shaft
24. Reverse idler gear
25. Countergear bearing
26. Needle bearing
27. Countergear
28. Mainshaft
29. Countergear front bearing
30. Needle bearing

M75 transmission

FWD
V.I.N. LOCATION

Transmission identification location

The M75 5 speed manual transmission is a fully synchronized unit with block ring synchronizers and a constant mesh reverse gear.

It primarily consists of a case, center support and extension housing that house the various gears and bearings. The shift lever assembly is mounted on top of the extension housing.

In **N**, with engine clutch engaged, the main drive gear turns the countergear. The countergear then turns the 5th, 4th, 3rd, 2nd, 1st and reverse idler gears. But, because the clutch (sleeves) are neutrally positioned and the reverse clutch (sleeve) is neutrally positioned, power will not flow through the mainshaft.

In 1st speed, the 1st and 2nd speed clutch (sleeve) is moved rearwards to engage the 1st speed gear, which is being turned by the countergear. Because the 1st and 2nd speed clutch (hub) is splined to the mainshaft, torque is imparted to the mainshaft from the 1st speed gear through the clutch assembly.

The 2nd speed, the 1st and 2nd speed clutch (sleeve) is moved forward to engage the 2nd speed gear which is being turned by the countergear. This engagement of the clutch (sleeve) with the 2nd speed gear imparts torque to the mainshaft because the 1st and 2nd speed clutch (hub) is splined to the mainshaft.

The 3rd speed, the 3rd and 4th speed clutch (sleeve) moves rearward to engage the 3rd speed gear, which is being turned by the countergear. Because the 3rd and 4th speed clutch (hub) is splined to the mainshaft, torque is imparted to the mainshaft from the 3rd speed gears through the clutch assembly.

The 4th speed, or direct drive, the 3rd and 4th speed clutch (sleeve) is moved forward to engage the main drive gear. This engagement of the main shaft gear with the 3rd and 4th speed clutch assembly imparts torque directly to the mainshaft.

In 5th speed, or overdrive, the reverse and 5th speed clutch (sleeve) is moved rearward to engage the 5th speed gear. This engagement of the 5th speed gear with the 5th and reverse speed clutch assembly imparts torque directly to the mainshaft; in this case, an overdrive.

In reverse speed, the reverse and 5th clutch (sleeve) is moved forward to engage the reverse gear. This engagement of the reverse gear with the reverse and 5th speed clutch assembly imparts torque through the countershaft and reverse idler to the reverse gear and directly to the mainshaft.

Transmission Identification

The transmission identification tag is located on the right side of the transmission case. This tag must be used when ordering parts to service this transmission.

Metric Fasteners

The metric fastener dimensions are very close to the dimensions of the familiar inch system fasteners and for this reason, replacement fasteners must have the same measurement and strength as those removed.

Do not attempt to interchange metric fasteners for inch system fasteners. Mismatched or incorrect fasteners can result in damage to the transmission unit through malfunctions, breakage or possible personal injury.

Care should be taken to re-use the fasteners in the same locations as removed.

Capacities

The fluid capacity of this transmission is 3¼ pints (1.54L) of 5W 30SF engine oil.

Checking Fluid Level

The fluid level should be checked at regular interval (ie., every oil change). Fluid level should be level with the bottom of the fill plug on the side of the case.

TRANSMISSION MODIFICATIONS

There are no modifications on the M75 transmission listed at the time of this publication printing.

TROUBLE DIAGNOSIS

CHILTON THREE C'S TRANSMISSION DIAGNOSIS

Condition	Cause	Correction
Hard shifting	a) Clutch b) Worn or broken synchronizers c) Worn shift shafts or forks	a) Adjustment or malfunctioning b) Replace c) Replace
Slips out of gear	a) Worn shift shafts b) Worn bearings c) Broken or loose drive gear retainer d) Excessive play in synchronizers	a) Replace b) Replace as necessary c) Tighten or replace retainer d) Replace

CHILTON THREE C'S TRANSMISSION DIAGNOSIS CHART

Condition	Cause	Correction
Noisy in all gears	a) Insufficient lubricant b) Worn countershaft bearings c) Worn or damaged drive gear and countergear d) Damaged drive gear or main shaft e) Worn or damaged countergear	a) Add to correct level b) Replace countershaft bearings and shaft c) Replace worn or damaged gears d) Replace damaged drive gear or bearings e) Replace countergear
Noisy in neutral	a) Damaged drive gear bearing b) Damaged or loose pilot bearing c) Worn or damaged countergear d) Worn countergear bearings	a) Replace damaged bearing b) Replace pilot bearing c) Replace countergear d) Replace countergear bearings and shaft
Noisy in reverse	a) Worn or damaged reverse idler gear or idler bushing b) Worn or damaged reverse gear c) Damaged or worn countergear	a) Replace reverse idler gear assembly b) Replace reverse gear c) Replace countergear assembly
Leaks lubricant	a) Excessive amount of lubricant in transmission b) Loose or broken drive gear bearing retainer c) Damaged drive gear bearing retainer gasket d) Center support gaskets either side e) Rear extension seal f) Speedometer driven gear	a) Drain to correct level b) Tighten or replace retainer c) Replace gasket d) Replace gaskets e) Replace f) Replace O-ring seal

POWER FLOW

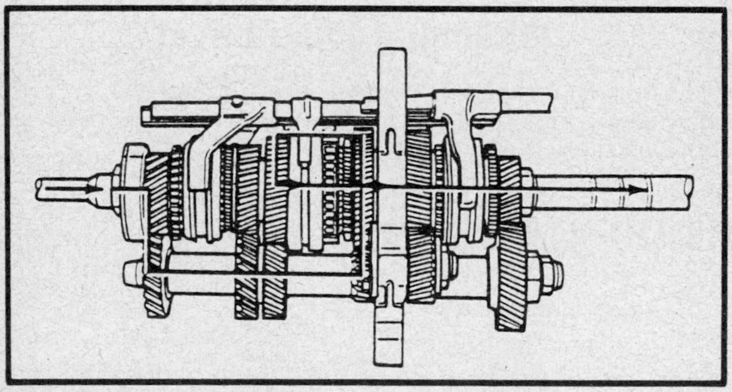
Power flow—1st gear

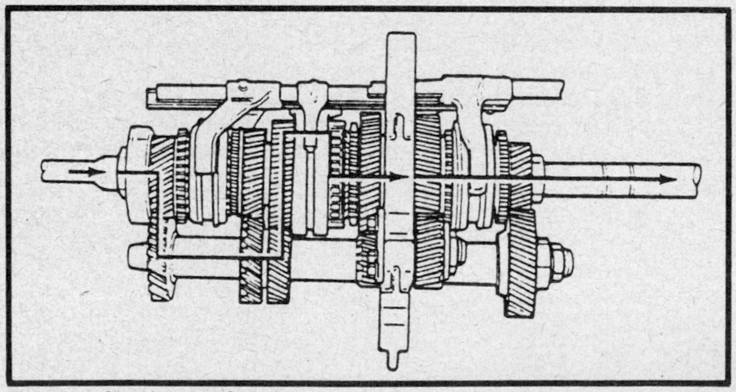
Power flow—2nd gear

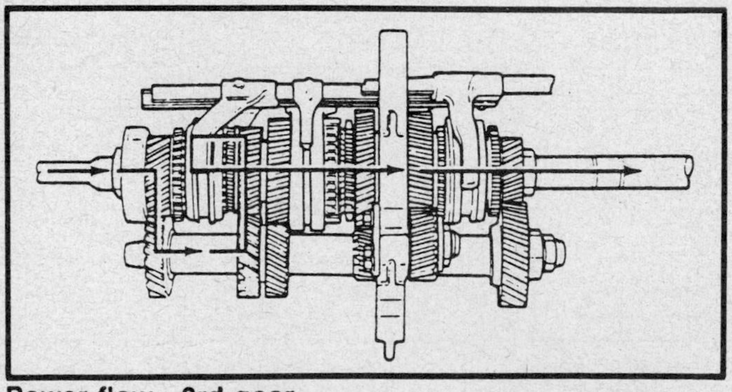
Power flow—3rd gear

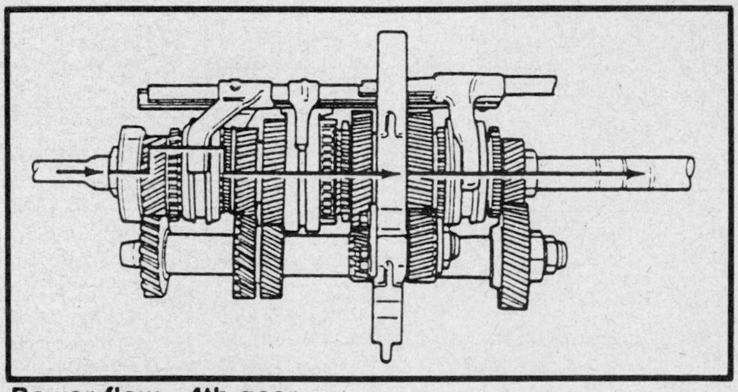
Power flow—4th gear

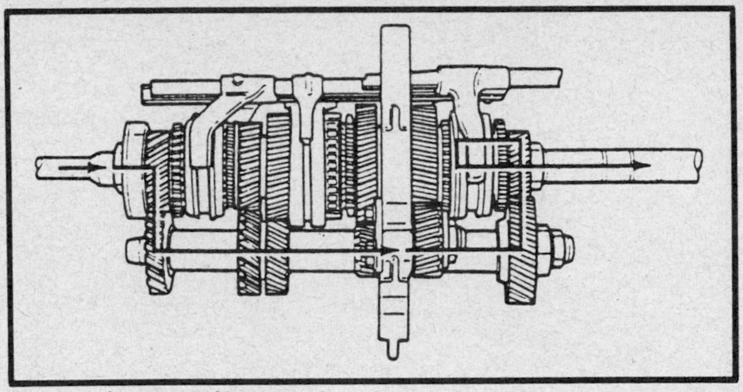

Power flow—5th gear

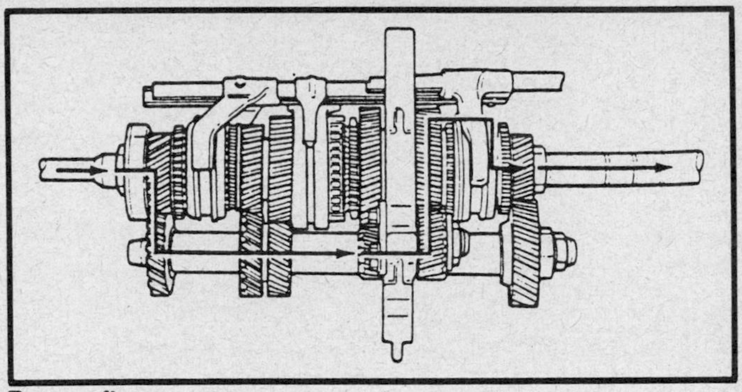

Power flow—reverse gear

ON CAR SERVICE

Services

SHIFTER CONTROL

Removal and Installation

1. Remove the screws on the left and right sides of the console.
2. Remove the console.
3. Remove the outer boot from the floor panel.
4. Remove the inner boot from the transmission.
5. Remove the bolts from the shifter and remove the shifter.
6. To install, position a new gasket between the shifter and the case.
7. Install the shifter and bolts to the shifter.
8. Install the inner boot to the transmission.
9. Install the outer boot to the floor panel.
10. Install the console.
11. Install the screws on the left and right sides of the console.

REMOVAL AND INSTALLATION

TRANSMISSION REMOVAL

1. Disconnect the negative battery cable.
2. From inside the vehicle, remove the shift lever assembly.
3. Remove the upper starter mounting nuts.
4. Raise and support the vehicle safely.
5. Remove the driveshaft.
6. Disconnect the speedometer cable.
7. Remove the clutch return spring and disconnect the cable from the transmission.
8. Remove the starter lower bolts and support starter.
9. Disconnect the exhaust pipe from the manifold.
10. Remove the flywheel inspection cover.
11. Remove the rear transmission support mounting bolts.
12. Support the transmission under the case and remove the rear transmission support from the frame.
13. Lower the transmission and position approximately 4 in. lower than when mounted.
14. Disconnect the backup light switch connector.
15. Remove the transmission housing to engine block bolts.
16. Move the transmission straight back and lower away from the vehicle.

TRANSMISSION INSTALLATION

1. Raise and install transmission onto the engine housing.
2. Install the transmission housing to engine block bolts and torque to specification.
3. Connect the backup light switch connector.
4. Install the rear transmission support to the frame.
5. Install the transmission support mounting bolts.
6. Install the flywheel inspection cover.
7. Connect the exhaust pipe to the manifold.
8. Install the support starter and the starter lower bolt.
9. Connect the clutch cable to the transmission and install the clutch return spring.
10. Connect the speedometer cable.
11. Install the driveshaft.
12. Fill the transmission to the level of the filler hole with specified lubricant.
13. Lower the vehicle.
14. Install the upper starter mounting nuts.
15. From inside the vehicle, install the shift lever assembly.
16. Readjust the clutch lash.
17. Connect the negative battery cable.

BENCH OVERHAUL

Before Disassembly

Cleanliness is an important factor in the overhaul of the transmission. Before attempting any disassembly operation, the exterior of the transmission should be thoroughly cleaned to prevent the possibility of dirt entering the transmission internal mechanism. During inspection and reassembly, all parts should be thoroughly cleaned with cleaning fluid and then air dried. Wiping cloths or rags should not be used to dry parts. All oil passages should be blown out and checked to make sure that they are not obstructed. Small passages should be checked with tag wire. All parts should be inspected to determine which parts are to be replaced.

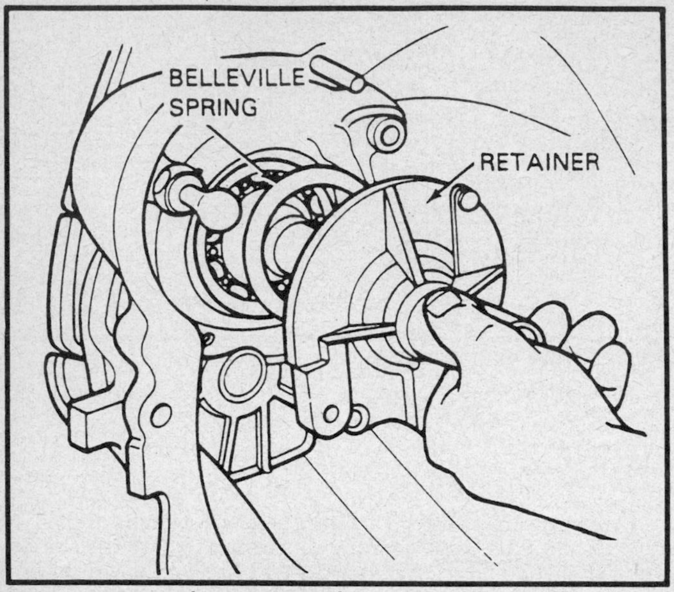
BELLEVILLE
SPRING
RETAINER
Drive gear retainer removal

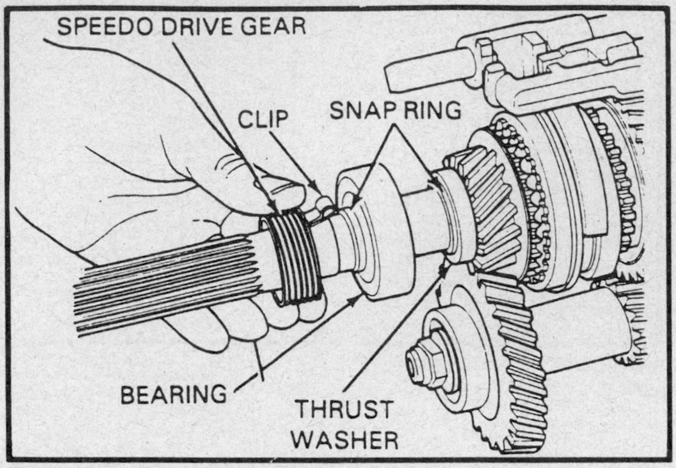
SPEEDO DRIVE GEAR
CLIP
SNAP RING
BEARING
THRUST
WASHER
Speedometer drive gear removal

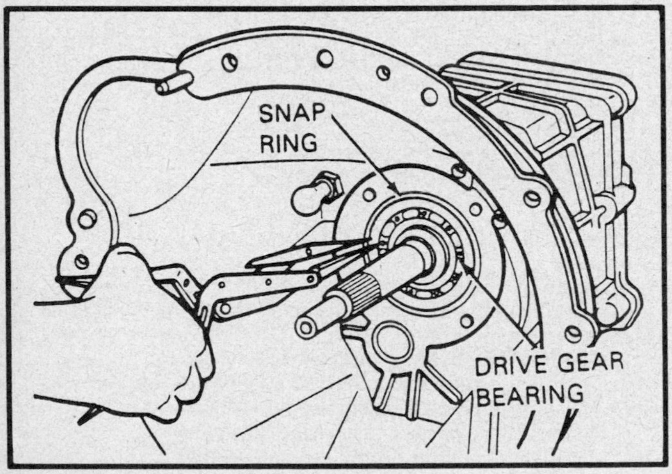
SNAP
RING
DRIVE GEAR
BEARING
Drive gear bearing snapring removal

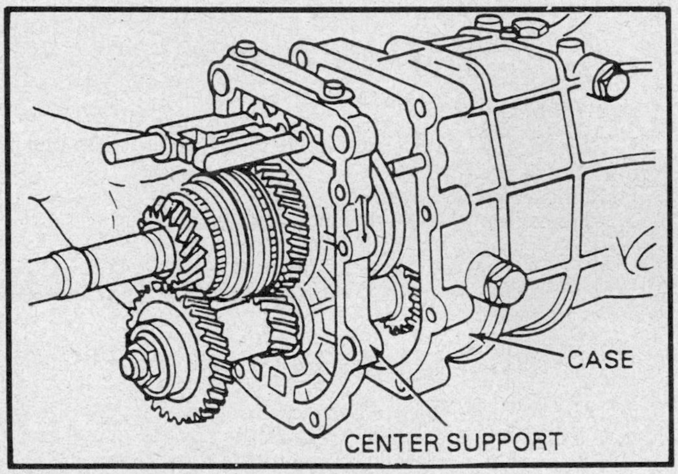
CASE
CENTER SUPPORT
Center support from case removal

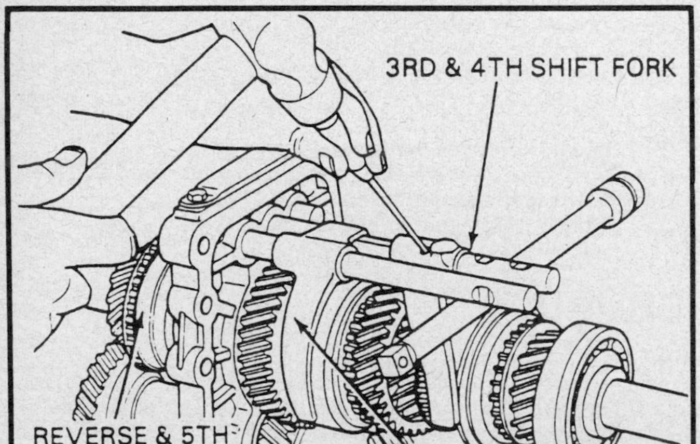
3RD & 4TH SHIFT FORK
REVERSE & 5TH
SHIFT FORK
1ST & 2ND SHIFT FORK
Shift fork pins removal

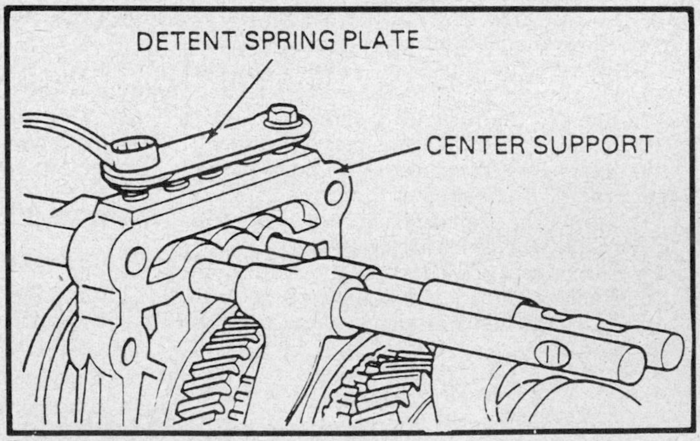
DETENT SPRING PLATE
CENTER SUPPORT
Detent spring plate removal

Transmission Disassembly

1. Remove the transmission drain plug and allow the lubricant to drain from the transmission.
2. Remove the throwout bearing and fork from the transmission.

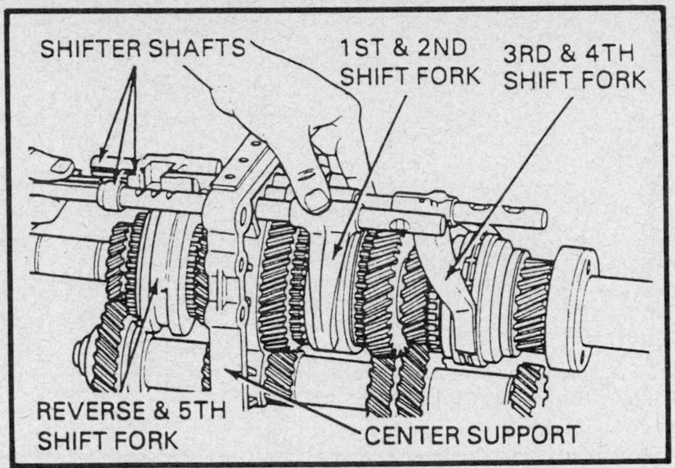
SHIFTER SHAFTS
1ST & 2ND SHIFT FORK
3RD & 4TH SHIFT FORK
REVERSE & 5TH SHIFT FORK
CENTER SUPPORT

Shifter shafts and forks removal

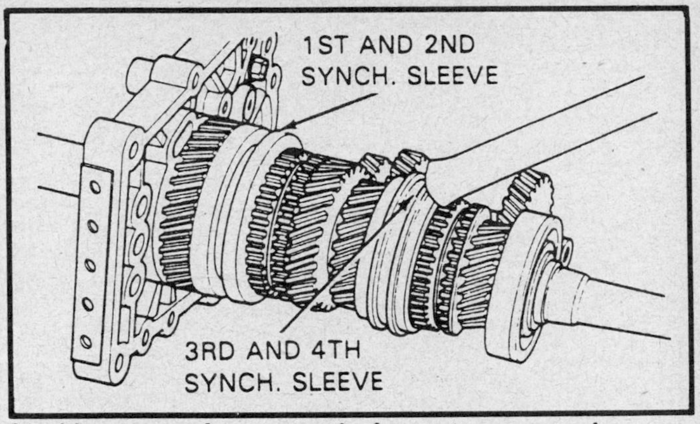
1ST AND 2ND SYNCH. SLEEVE
3RD AND 4TH SYNCH. SLEEVE

Locking gears for removal of countergear and mainshaft

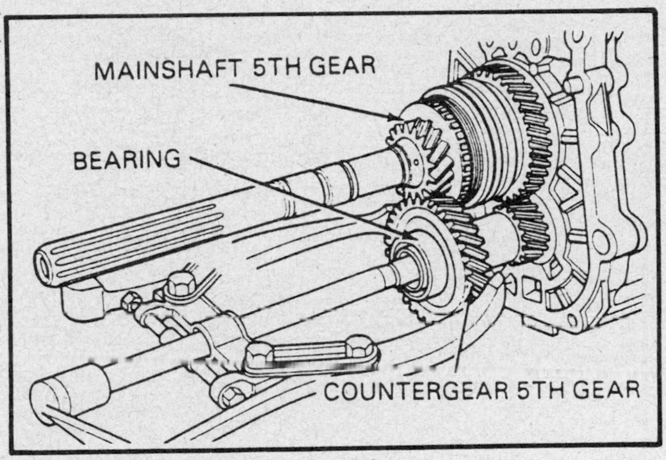
MAINSHAFT 5TH GEAR
BEARING
COUNTERGEAR 5TH GEAR

Countergear 5th gear removal

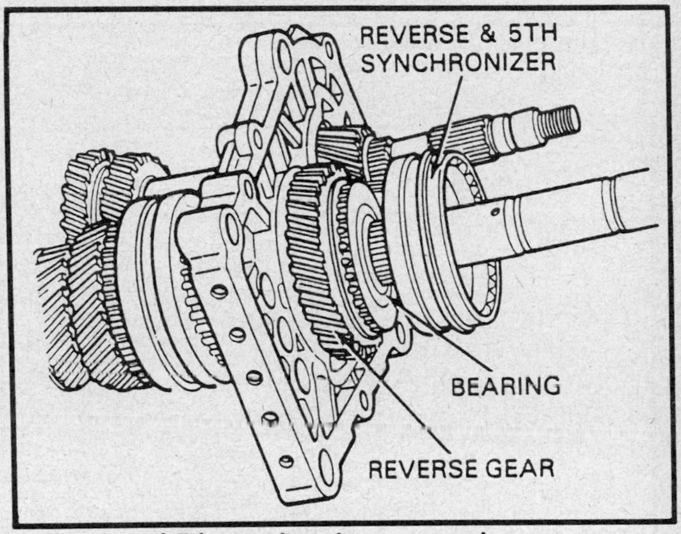
REVERSE & 5TH SYNCHRONIZER
BEARING
REVERSE GEAR

Reverse and 5th synchronizer removal

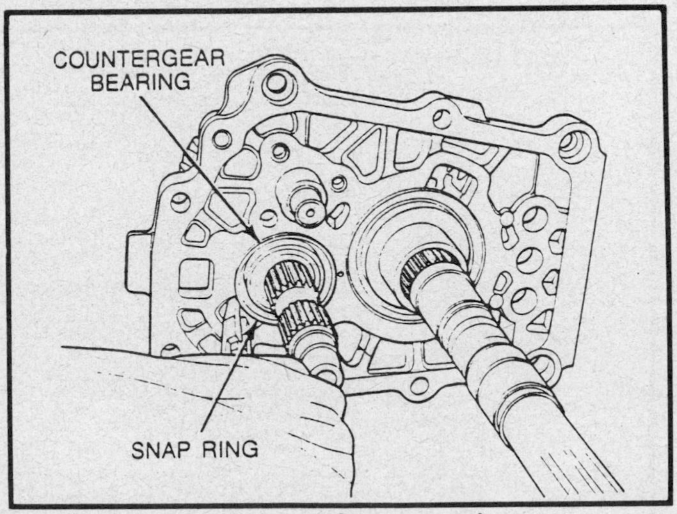
COUNTERGEAR BEARING
SNAP RING

Countergear bearing snapring removal

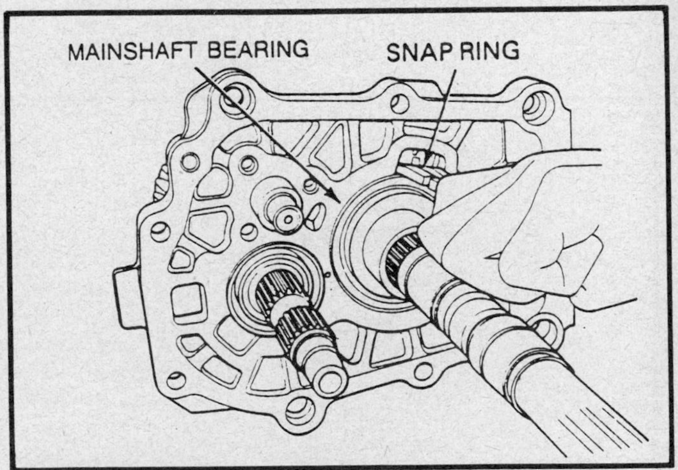
MAINSHAFT BEARING
SNAP RING

Mainshaft bearing snapring removal

3. Remove the drive gear bearing retainer. If damaged, remove the ball stud.
4. Remove the belleville spring from the front of the drive gear bearing.
5. Remove the bolt, retainer and speedometer driven gear from the side of the transmission.
6. Remove the shift lever quadrant from the extension housing.
7. Remove the backup light switch.

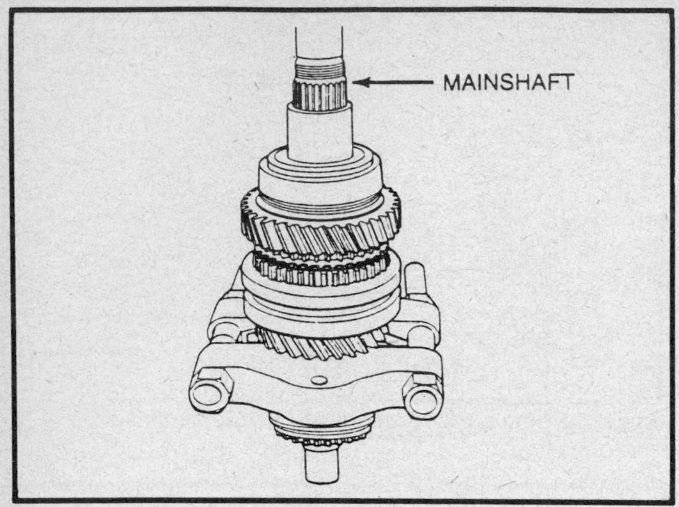
MAINSHAFT
Mainshaft rear bearing removal

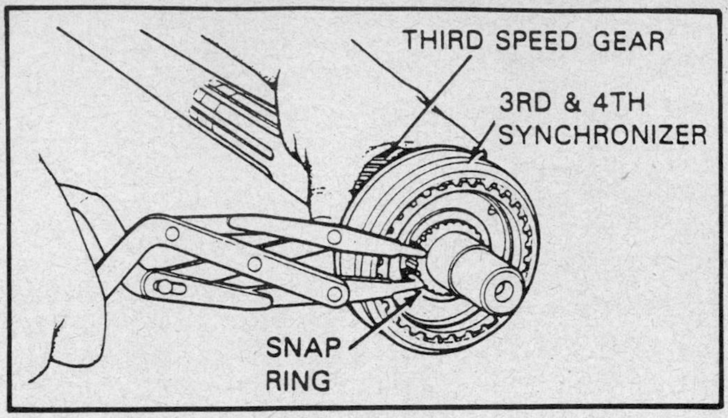
THIRD SPEED GEAR
3RD & 4TH SYNCHRONIZER
SNAP RING
3rd and 4th synchronizer removal

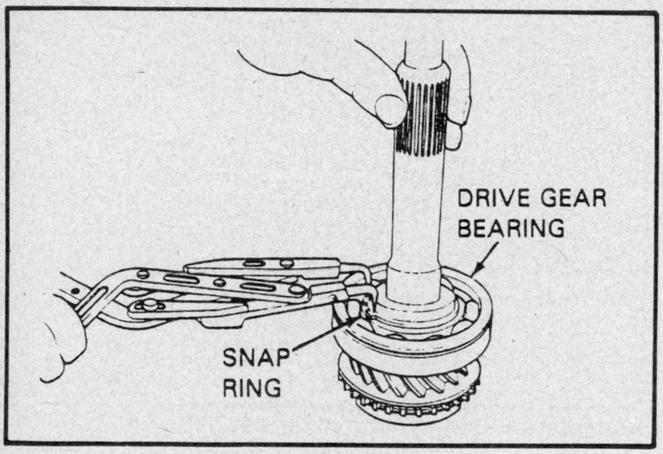
DRIVE GEAR BEARING
SNAP RING
Drive gear bearing snapring removal

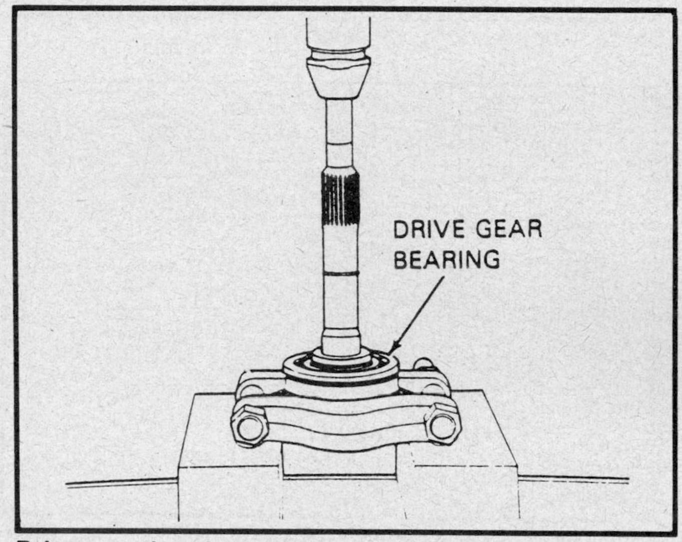
DRIVE GEAR BEARING
Drive gear bearing removal

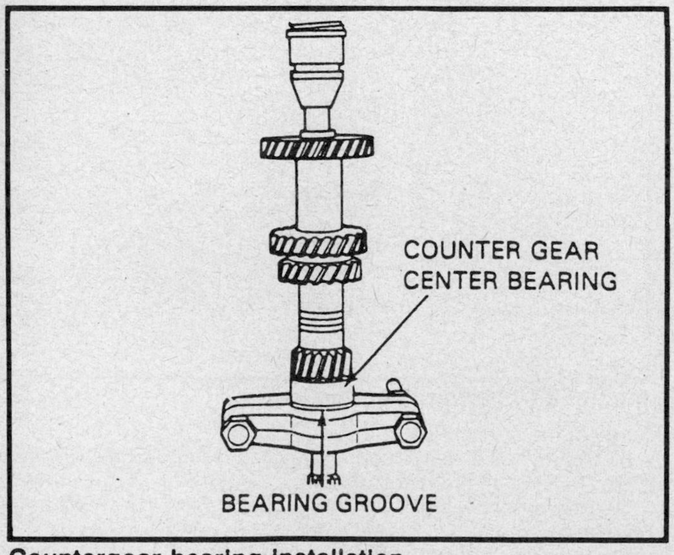
COUNTER GEAR CENTER BEARING
BEARING GROOVE
Countergear bearing installation

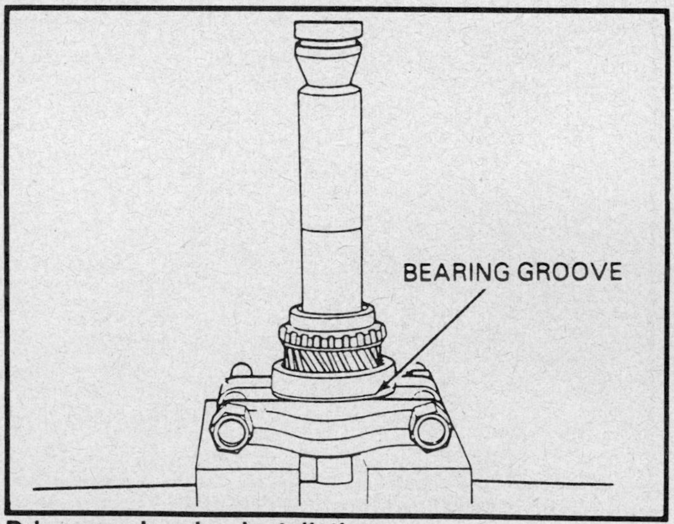
BEARING GROOVE
Drive gear bearing installation

8. Remove the extension housing bolts and the extension housing.
9. Remove the snaprings, speedometer drive gear, spacer and bearing from the mainshaft.

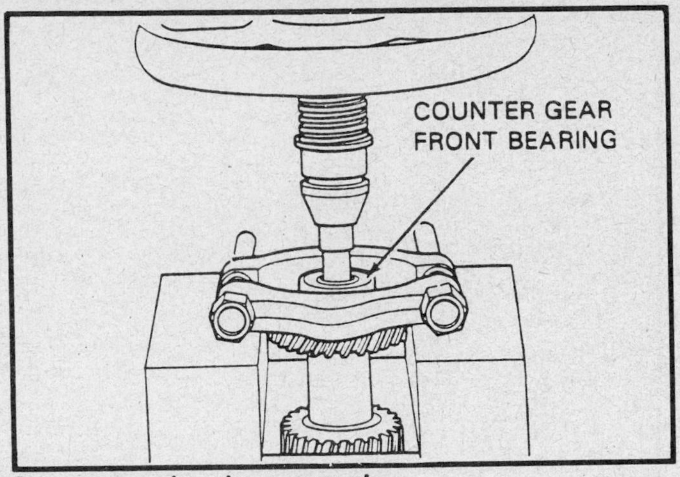
COUNTER GEAR
FRONT BEARING

Countergear bearing removal

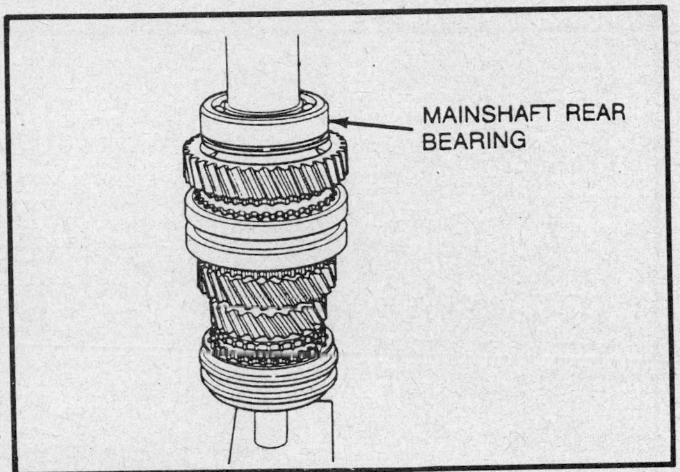
MAINSHAFT REAR
BEARING

Mainshaft component installation

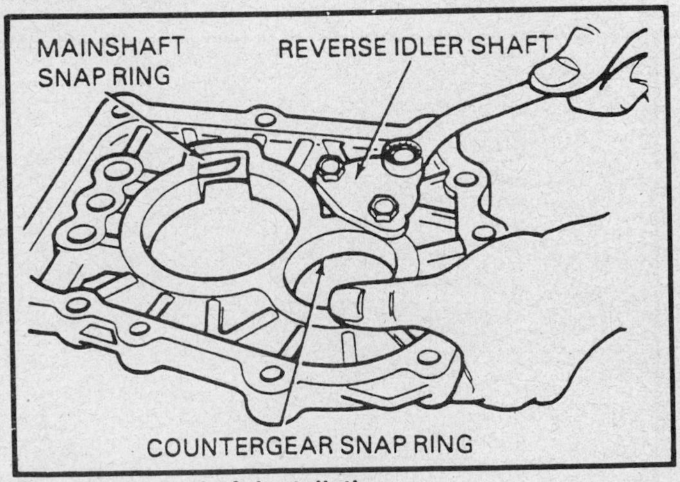
MAINSHAFT
SNAP RING
REVERSE IDLER SHAFT
COUNTERGEAR SNAP RING

Reverse idler shaft installation

10. Remove the snapring, then the thrust washer and lock ball from the mainshaft.

11. Remove the large snapring from the main drive gear bearing.

12. Remove the center support, mainshaft, countergear and drive gear from the case as an assembly.

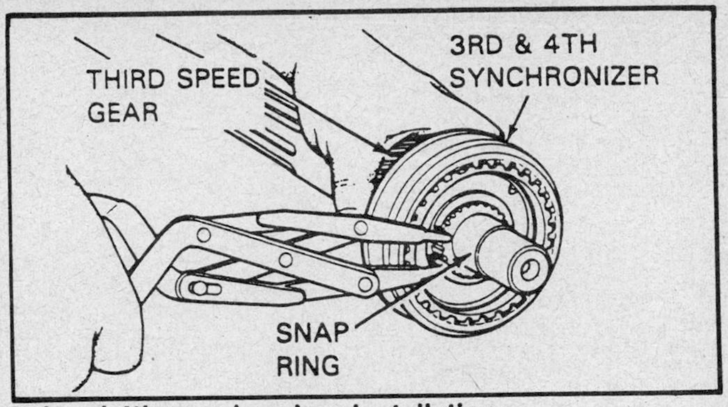
THIRD SPEED
GEAR
3RD & 4TH
SYNCHRONIZER
SNAP
RING

3rd and 4th synchronizer installation

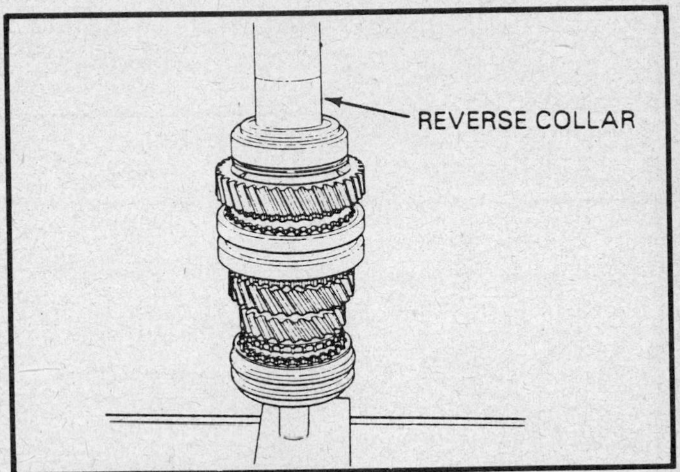
REVERSE COLLAR

Reverse collar installation

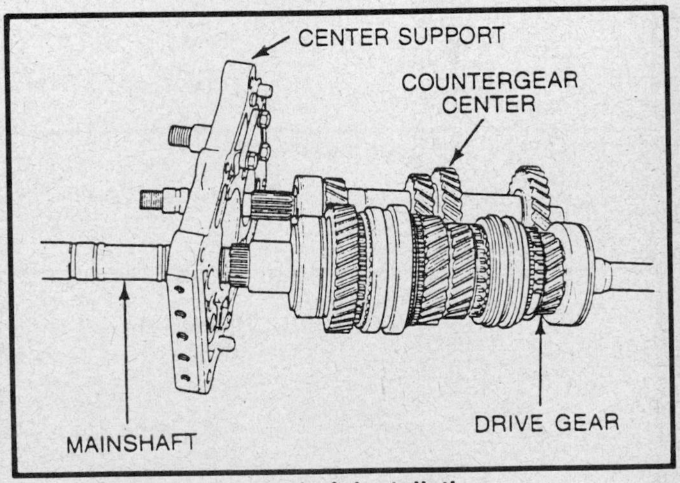
CENTER SUPPORT
COUNTERGEAR
CENTER
MAINSHAFT
DRIVE GEAR

Countergear and mainshaft installation

13. Using a drift punch and a hammer, carefully drive the roll pins from the 1st/2nd, 3rd/4th and 5th/reverse shift forks. Support the shaft ends with a bar or a block of wood to prevent damage to these components.

14. Remove the detent spring plate mounting bolts, detent spring plate and the 3 springs and balls from the center support.

15. Remove the shifter shafts from the center support, then remove the shift forks from the shafts. Remove the interlock pins from the center support.

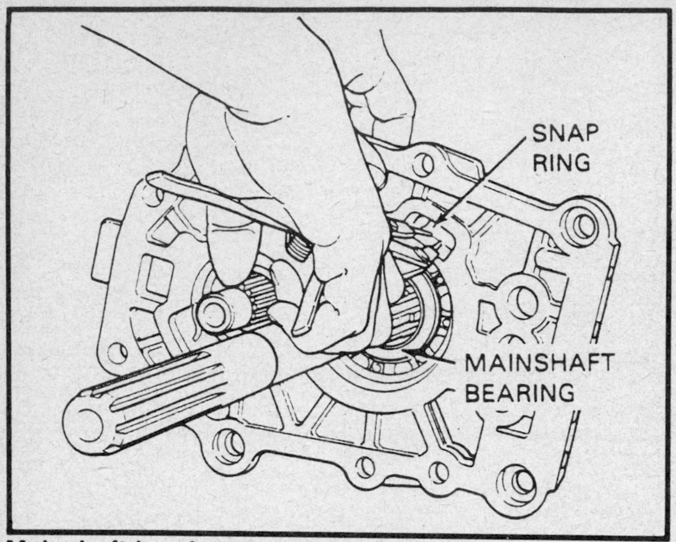
SNAP
RING
MAINSHAFT
BEARING

Mainshaft bearing snapring installation

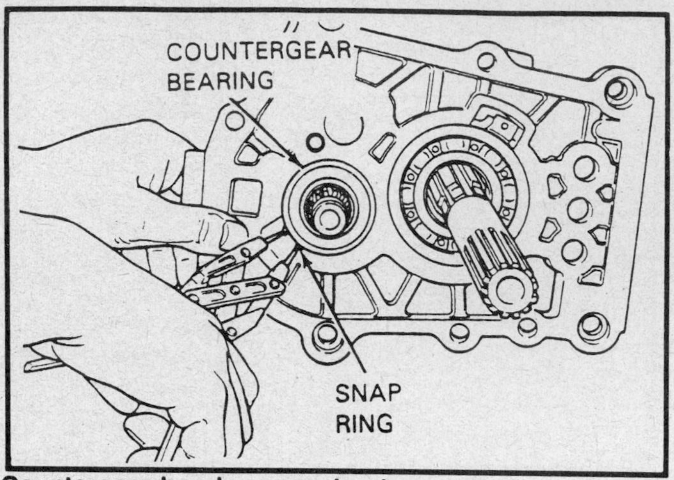
COUNTERGEAR
BEARING
SNAP
RING

Countergear bearing snapring installation

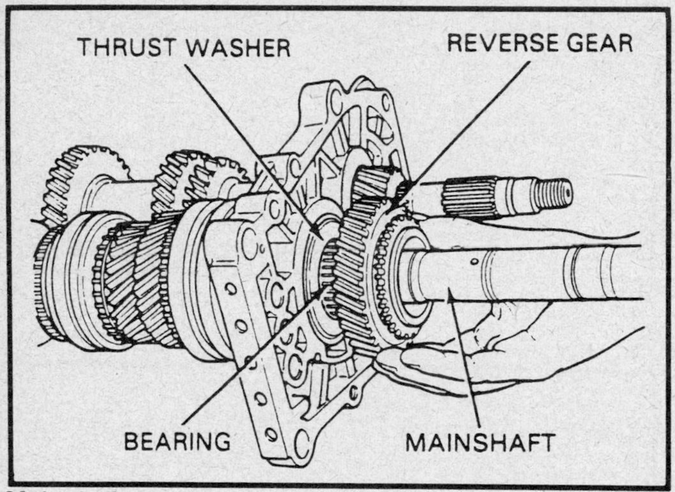
THRUST WASHER
REVERSE GEAR
BEARING
MAINSHAFT

Mainshaft–reverse gear installation

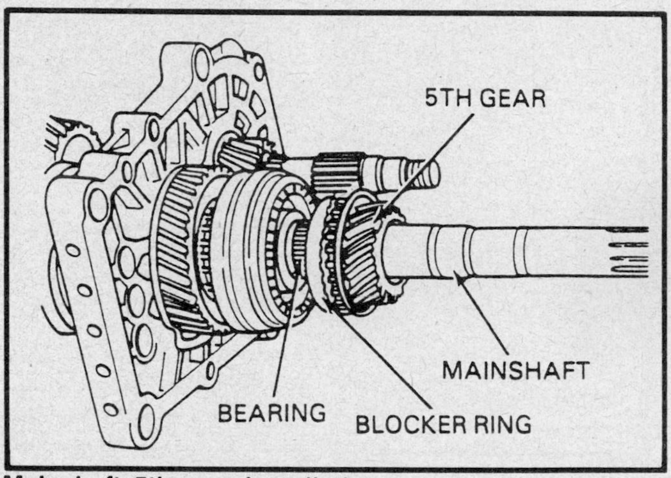
5TH GEAR
MAINSHAFT
BEARING
BLOCKER RING

Mainshaft–5th gear installation

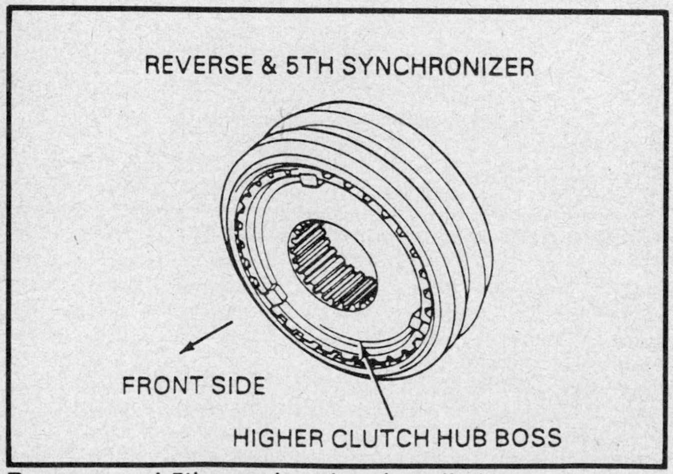
REVERSE & 5TH SYNCHRONIZER
FRONT SIDE
HIGHER CLUTCH HUB BOSS

Reverse and 5th synchronizer installation

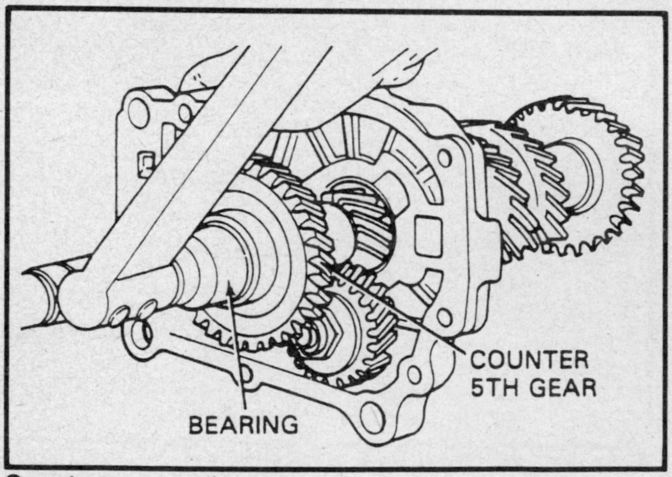
COUNTER
5TH GEAR
BEARING

Countergear nut installation

16. Move the 1st/2nd synchronizer sleeve to the 1st gear position and the 3rd/4th synchronizer sleeve to the 3rd gear position.

17. Install a holding fixture, tool J–29768 or equivalent, on the

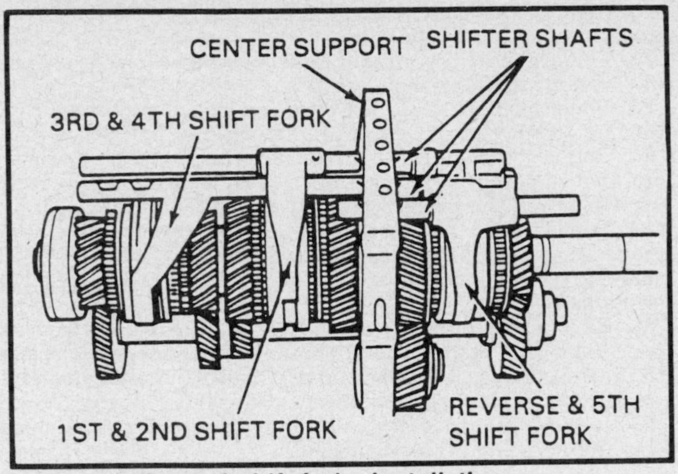
CENTER SUPPORT
SHIFTER SHAFTS
3RD & 4TH SHIFT FORK
1ST & 2ND SHIFT FORK
REVERSE & 5TH SHIFT FORK

Shifter shafts and shift forks installation

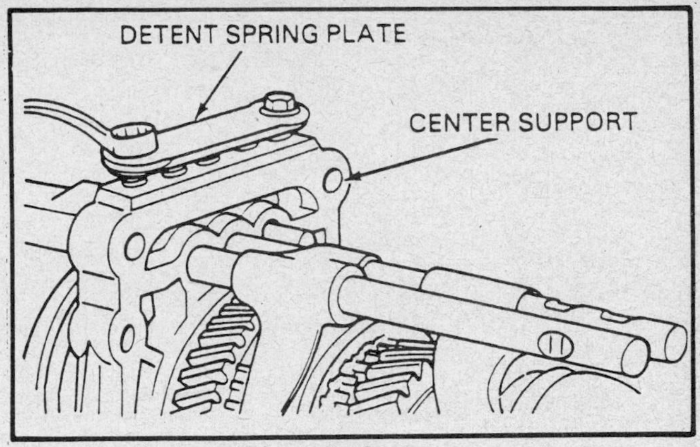
DETENT SPRING PLATE
CENTER SUPPORT

Detent spring plate installation

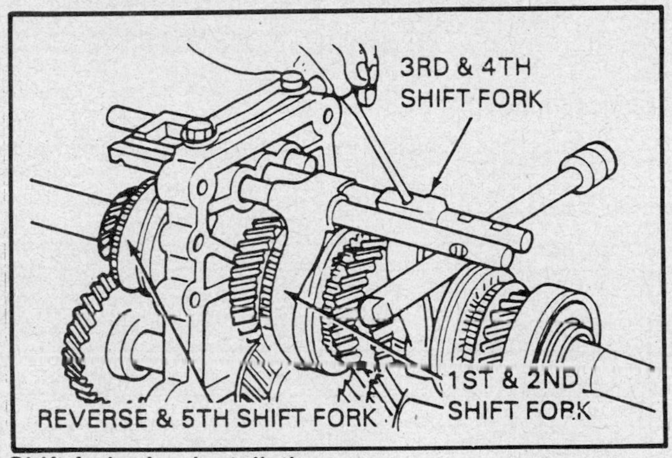
3RD & 4TH SHIFT FORK
1ST & 2ND SHIFT FORK
REVERSE & 5TH SHIFT FORK

Shift fork pins installation

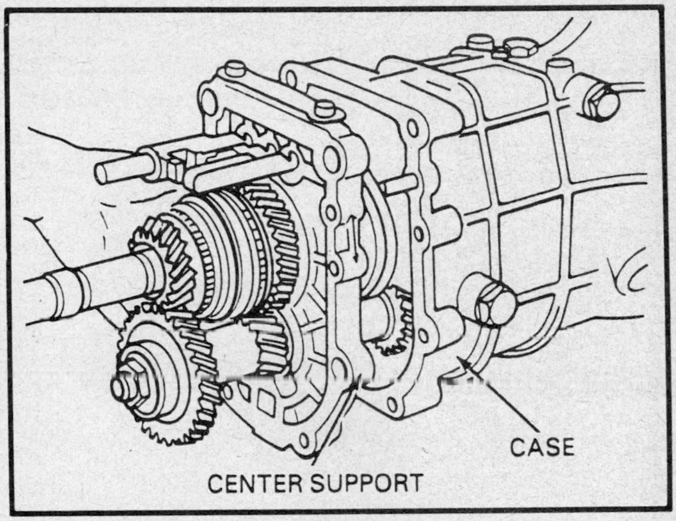
CASE
CENTER SUPPORT

Center support to case installation

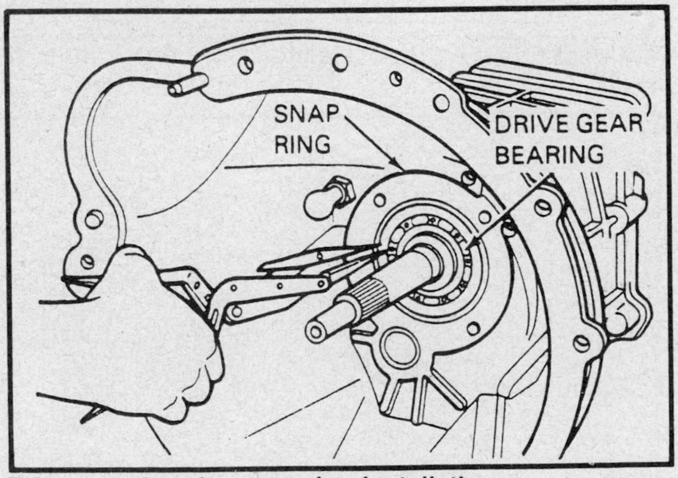
SNAP RING
DRIVE GEAR BEARING

Drive gear bearing snapring installation

end of the drive gear shaft and countergear. Remove the countergear retaining nut and the washer.

18. Using an appropriate puller, remove the ball bearing and 5th speed gear from the countershaft.

19. Remove the 5th gear, blocking ring and the needle bearing from the mainshaft.

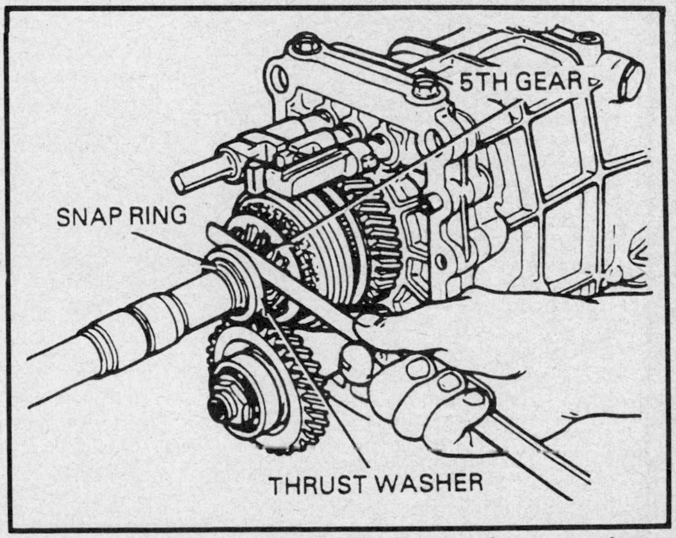
5TH GEAR
SNAP RING
THRUST WASHER

Checking clearance between 5th gear thrust washer

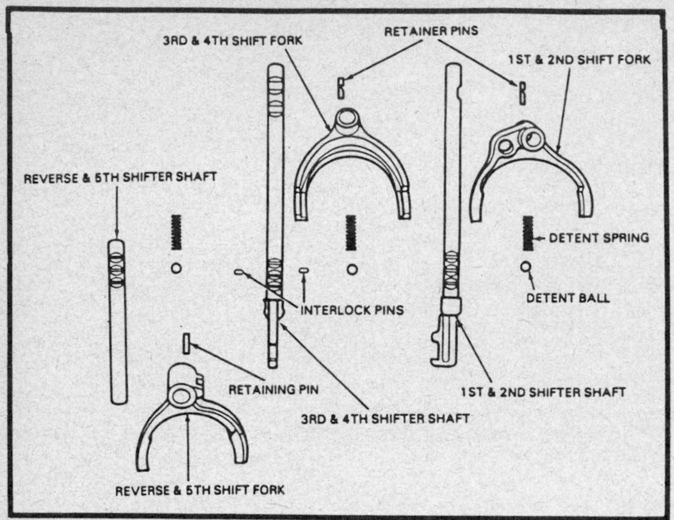
3RD & 4TH SHIFT FORK
RETAINER PINS
1ST & 2ND SHIFT FORK
REVERSE & 5TH SHIFTER SHAFT
DETENT SPRING
DETENT BALL
INTERLOCK PINS
RETAINING PIN
1ST & 2ND SHIFTER SHAFT
3RD & 4TH SHIFTER SHAFT
REVERSE & 5TH SHIFT FORK

Shifter shafts and shift forks

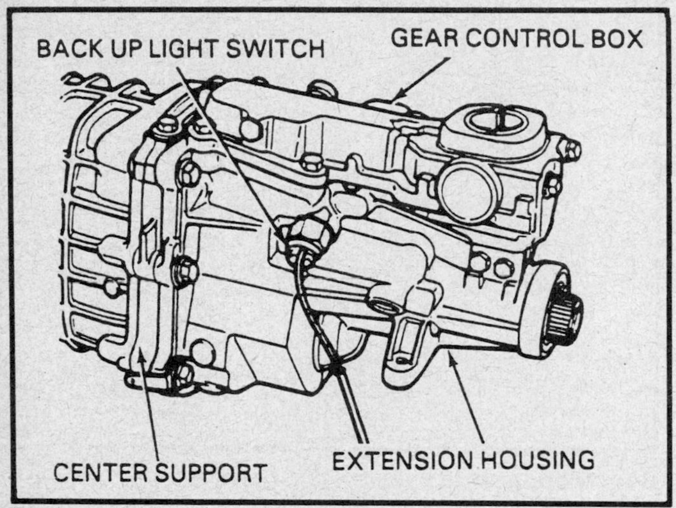
BACK UP LIGHT SWITCH
GEAR CONTROL BOX
CENTER SUPPORT
EXTENSION HOUSING

Backup light switch installtion

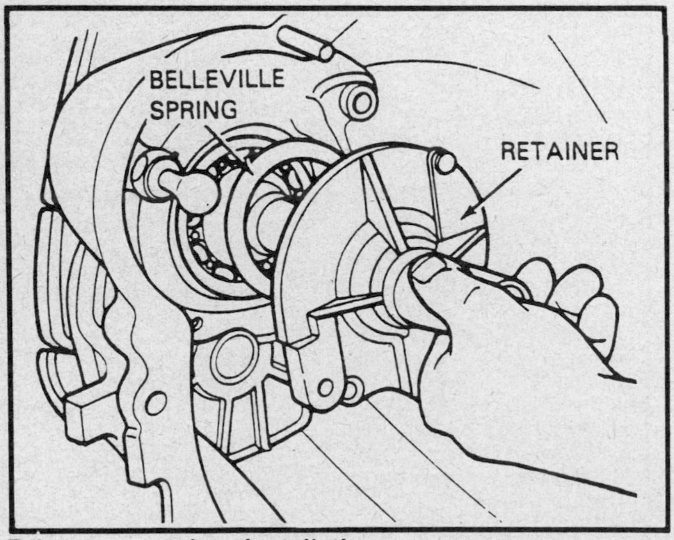
BELLEVILLE
SPRING
RETAINER

Drive gear retainer installation

20. Remove the self-locking nut from the reverse idler gear shaft.
21. Remove the thrust washers and the reverse idler gear from the reverse idler gear shaft.
22. Bend the locking retainer of the mainshaft nut away from the nut. Remove the mainshaft nut using the appropriate tool.
23. Remove the 5th/reverse synchronizer locking retainer, 5th/reverse synchronizer assembly, reverse gear, needle bearing, collar and the thrust washer from the mainshaft.
24. Remove the reverse gear from the countergear and remove the holding fixture installed during Step 17.
25. Move the synchronizer sleeves back to their neutral positions.
26. Expand the countergear bearing snapring (using snapring pliers) and gently tap on the front of the center support. Expand the mainshaft bearing snapring and move the mainshaft inward. Remove the countergear and mainshaft.
27. Remove the drive gear, needle bearing and the blocking ring from the end of the mainshaft.

Unit Disassembly and Assembly

MAINSHAFT

Disassembly

1. Remove the mainshaft rear bearing using an arbor press

Part Number	Thickness Inches	Thickness Millimeters
94027215	0.307	7.80
94027216	0.311	7.90
94027217	0.315	8.00
94027218	0.319	8.10
94027219	0.323	8.20
94027220	0.327	8.30

5th gear thrust washer selection

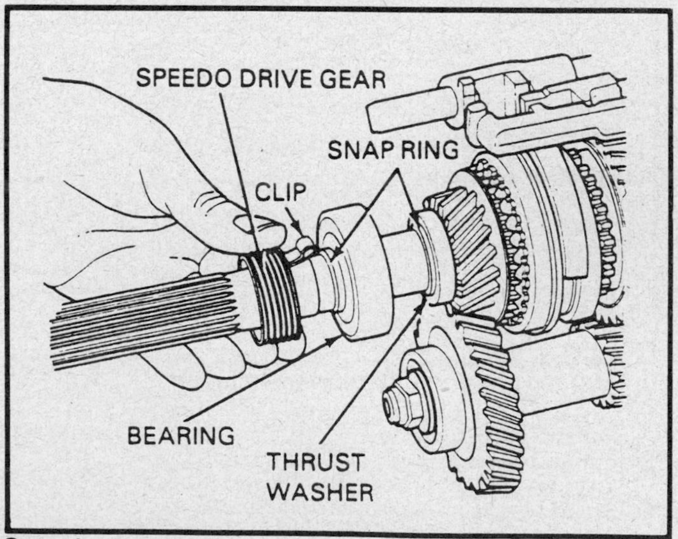
SPEEDO DRIVE GEAR
SNAP RING
CLIP
BEARING
THRUST
WASHER

Speedometer drive gear installation

Select the snap ring thickness so that the clearance between the snap ring and 3rd/4th synchronizer (synchronizer endwise movement) will be adjusted to 0.05mm (0.002 in.).

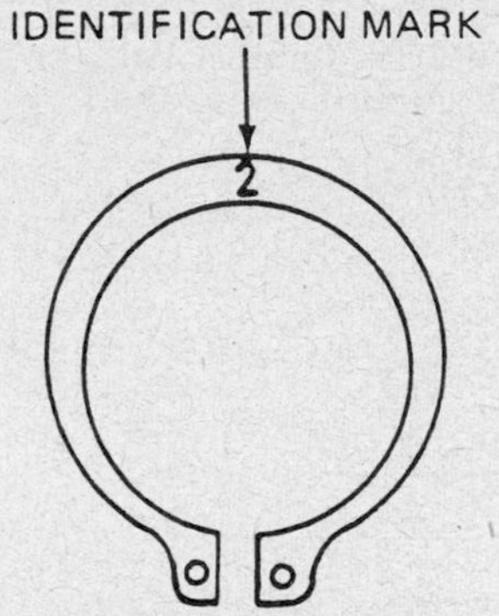
IDENTIFICATION MARK
2

Thickness		Identification Mark
Millimeters	Inches	
1.5	0.059	1
1.55	0.061	2
1.6	0.063	3
1.65	0.065	4

3rd–4th synchronizer snapring identification

and tool J-22912-01 or equivalent.

2. Remove the thrust washer, 1st speed gear, needle bearings and the spacer.
3. Remove the 1st/2nd synchronizer assembly, the 2nd speed gear and needle bearings.
4. Remove the snapring from in front of the 3rd/4th synchronizer and slide the synchronizer off of the mainshaft.

Inspection

1. Inspect all the gears and bearings for excessive wear and replace any that are scored, worn or damaged.
2. Check the clutch sleeves to see that they slide freely on their hubs.
3. Inspect the gears, blocker rings, sleeves and springs for wear or damaged.

Assembly

1. Install the 3rd speed gear (with needle bearings) onto the front of the mainshaft. Note that the coned side of the gear is installed toward the front of the mainshaft.
2. Install the 3rd/4th synchronizer assembly onto the mainshaft, with the large chamfered end toward the front of the transmission. Retain the synchronizer and the snapring.
3. Install the 2nd speed gear (with needle bearings) onto the rear of the mainshaft. Note that the coned side of the gear is installed toward the rear of the mainshaft.
4. Install the 1st/2nd synchronizer assembly onto the mainshaft, with the large chamfered end toward the rear of the transmission.
5. Install the 1st speed gear (with the spacer and needle bearings), with the coned end facing the front of the transmission.
6. Install the 1st gear thrust washer, with the slots of the washer facing the gear.
7. Press the rear bearing onto the mainshaft.

Transmission Assembly

1. If removed, install the countergear and mainshaft snaprings into the center support. Also install the reverse idler shaft into the center support.
2. Install the drive gear onto the front of the mainshaft and engage it with the countershaft.
3. Install a holding fixture, tool J-29768 or equivalent, on the end of the drive gear shaft and countergear. Remove the countergear retaining nut and the washer.
4. With the mainshaft and countergears meshed together, slide the center support onto the mainshaft. Expand the mainshaft snapring and continue to push the center support on until the mainshaft bearing groove aligns with the snapring. Release the mainshaft snapring to lock the mainshaft bearing in place. Repeat the same procedure to seat the countergear snapring.
5. Move the synchronizer sleeves to engage both 1st and 3rd gear ranges in order to lock the gears.
6. Install the reverse gear onto the countergear.
7. Install the thrust washer on the mainshaft (oil groove toward the rear), then install the collar, needle bearing and the reverse gear onto the mainshaft.
8. Install the 5th/reverse synchronizer, with the face of the higher clutch hub boss facing the reverse gears.
9. Install the locking retainer and the mainshaft nut onto the mainshaft. Torque the mainshaft nut to 94 ft. lbs. (127 Nm), then bend the locking retainer tabs in order to lock the nut.
10. Install the thrust washers and the reverse idler gear on the reverse idler gear shaft. Thread a new self-locking nut onto the reverse idler shaft and torque the nut to 80 ft. lbs. (108 Nm).

NOTE: The flange of the plate-side thrust washer must be fitted to the stopper on the center support.

11. Install the synchronizer blocking ring and the 5th speed gear (with needle bearings) onto the mainshaft.
12. Install the 5th speed gear (of the countergear), ball bearing, washer and a new self-locking nut onto the rear of the countergear. Torque the nut to 80 ft. lbs. (108 Nm).
13. Remove the holding fixture and move the synchronizer sleeves back to their eutral positions.
14. Grease the interlock pins and install them into the center support.
15. Place the shaft forks into position on the synchronizer sleeves. Install the shifter shafts through their respective forks from the rear of the center support, except the 5th/reverse shaft, which is installed from the front of the support.
16. Install the 3 detent balls, springs, detent plate gasket and the detent plate. Torque the detent plate bolts to 14 ft. lbs. (19 Nm).
17. Using a drift punch and a hammer, carefully install the retaining pins into the shift forks. Remember to support the shafts with a bar or a block of wood to prevent damage.
18. If removed, lubricate the countergear needle bearing and install it into the front of the case. The bearing should be driven into place while a socket is positioned on the outer bearing race.
19. Install a new center support-to-transmission case gasket on the transmission case and install the center support, mainshaft, countergear, drive gear assembly into the case.
20. Install the large snapring onto the shaft of the drive gear bearing.
21. Install the lock ball, thrust washer and the retaining snapring onto the mainshaft.
22. Using a feeler gauge, check the clearance between the 5th speed gear (of the mainshaft) and its thrust washer. The clearance should be 0.004–0.012 in. (0.10–0.30mm). If necessary, adjust the clearance by purchasing a thrust washer of the correct thickness which will replace the existing washer. Thrust washers are available in thickness ranging from 0.307–0.027 in. (7.80–8.30mm) in increments of 0.004 in. (0.10mm).

NOTE: Use care when removing/installing the snapring; it must be replaced if it becomes distorted.

23. Install ball bearings, snapring, speedometer gear clip and speedometer drive gear on the mainshaft behind the 5th gear snapring.
24. Attach the extension housing to the center support, using a new gasket. Torque the bolts to 27 ft. lbs. (37 Nm).
25. Install the shift lever quadrant onto the extension housing, using a new gasket. Torque the bolts to 14 ft. lbs. (19 Nm).
26. Install the speedometer driven gear and torque the bolt to 14 ft. lbs. (19 Nm).
27. Install the backup light switch into the extension housing.
28. Install the belleville washer in front of the drive gear bearing, noting that the dished side of the washer should face the bearing.
29. Install the drive gear bearing retainer, using a new gasket. Before installing the 3 lower bearing retainer bolts, coat the threads of the bolts with Permatex® No. 2 sealer or equivalent. Torque all of the bearing retainer bolts to 14 ft. lbs. (19 Nm).
30. Install the throwout bearing and fork.

SPECIFICATIONS

TORQUE SPECIFICATIONS

Component	ft. lbs.	Nm
Countershaft to center support	80	108
Reverse idler shaft to center support	14	19
Main shaft to center support	94	127
Extension housing to case	27	37
Shift quadrant to extension housing	14	19
Drive gear retainer to case	14	19
Reverse idler gear to shaft	80	108

Section 7

Corvette (84-88) Transmission

General Motors

APPLICATION

83MM WITH AUTOMATIC OVERDRIVE

Year	Vehicle	Engine
1984	Corvette	5.7L CFI
1985	Corvette	5.7L TPI
1986	Corvette	5.7L TPI
1987	Corvette	5.7L TPI
1988	Corvette	5.7L TPI

CFI—Cross Fire Injection
TPI—Tuned Port Injection

GENERAL DESCRIPTION

1984 VEHICLES

The manually-shifted transmission has 4 forward gears with an automatically engaged overdrive useable in 2nd, 3rd and 4th gears. The overdrive is engaged by a computer program which monitors road speed (mph), engine temperature (engine temperatures below 184°F (84°C) will prevent engagement), and throttle position. This program engages the overdrive under light acceleration and lower speed conditions in 2nd and 3rd gears and cruise conditions in 4th gear. Rapid acceleration and/or higher speeds in 2nd and 3rd gears eliminate overdrive engagement. When the overdrive is engaged, the lighted words **OVERDRIVE ENGAGED** will appear in the instrument panel.

The system also includes an overdrive switch. When the overdrive switch is in the **OFF** position ("o" symbol is depressed), the transmission will remain in direct drive. Under normal operating conditions the overdrive switch should be turned **ON** to maximize fuel economy and reduce engine speed. However, if additional engine braking is needed when going downhill or if the transmission repeatedly upshifts and downshifts between the direct and overdrive gears, the overdrive switch may be turned to the **OFF** position. Be sure to turn the switch back to the **ON** position immediately afterward. The overdrive engaged or **ON** position is indicated on the switch by a closed circle (●).

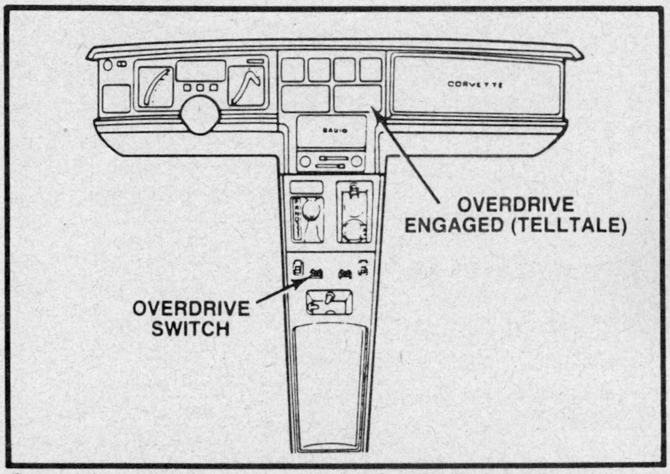
OVERDRIVE
ENGAGED (TELLTALE)
OVERDRIVE
SWITCH

Overdrive switch and engaged telltale locations—1984 models

1985 VEHICLES

The manually-shifted transmission has 4 forward gears with an automatically engaged overdrive useable in 2nd, 3rd and 4th gears. An overdrive light will appear in the information center when the overdrive is engaged. The overdrive is engaged by a computer program which monitors road speed, engine speed, coolant temperature, and throttle position. This programming engages the overdrive under light acceleration conditions. Rapid acceleration from a stop delays overdrive engagement.

Large throttle openings in 4th gear will cause the overdrive to downshift (kickdown) to direct drive if engine speed is below 3100 rpm. After accelerating, the overdrive will upshift after a 2 second delay. No kickdown will occur in 2nd or 3rd gears. The overdrive will automatically engage (upshift) at 110 mph.

The system also includes an overdrive switch (the overdrive switch may be located on the console or on top of the gearshift lever). When the overdrive switch is in the **OFF** position ("o" symbol is depressed, console switch or switch on top of gearshift lever), the transmission will remain in direct drive. Under normal operating conditions the overdrive switch should be turned **ON** to maximize fuel economy and reduce engine speed. However, if additional engine braking is needed when going downhill, the overdrive switch may be turned to the **OFF** position (console switch or top of gearshift depressed). Be sure to turn the switch back to the **ON** position immediately afterward (console switch or top of gearshift depressed). The overdrive engaged, or **ON** position, is indicated on the switch by a closed circle (●) (console only).

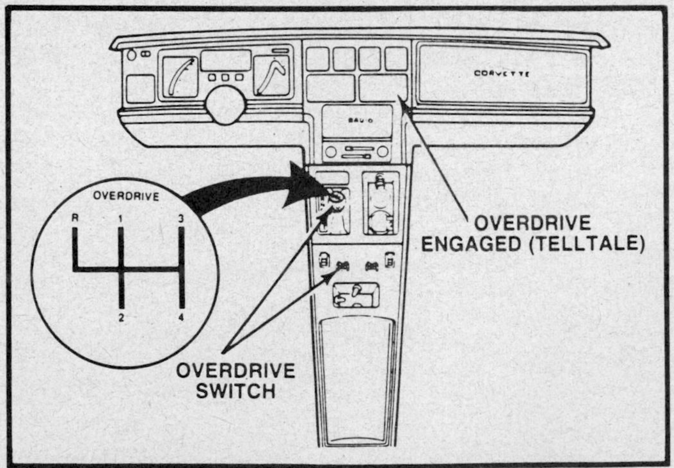
OVERDRIVE
R
1
3
2
4
OVERDRIVE
ENGAGED (TELLTALE)
OVERDRIVE
SWITCH

Overdrive switch and engaged telltale locations—1985 models

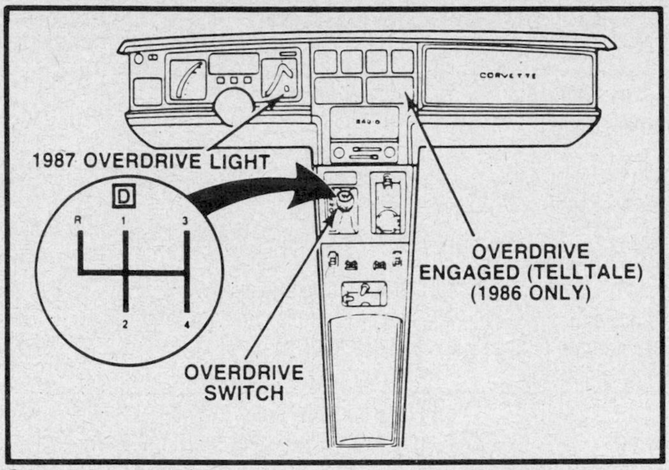
1987 OVERDRIVE LIGHT
OVERDRIVE
ENGAGED (TELLTALE)
(1986 ONLY)
OVERDRIVE
SWITCH
CORVETTE

Overdrive switch and engaged telltale locations–1986–88 models

1986–88 VEHICLES

The manually-shifted transmission has 4 forward gears with an automatically engaged overdrive useable in 2nd, 3rd and 4th gears. An overdrive light will appear in the information center when the overdrive is engaged. The overdrive is engaged by a computer program which monitors road speed, engine speed, coolant temperature, and throttle position. This programming engages the overdrive under light acceleration conditions. Rapid acceleration from a stop delays overdrive engagement.

Large throttle openings in 2nd, 3rd and 4th gears will cause the overdrive to downshift (kickdown) to direct drive if engine speed is below 2200 rpm. After accelerating, the overdrive will upshift in 4th only. No upshift will occur in 2nd or 3rd gears. After a downshift the overdrive will automatically engage (upshift) at 4500 rpm in 4th gear.

The system also includes an overdrive switch located in the shifter knob. When the overdrive is engaged, the symbol **D** will appear in the information center. The overdrive may be turned **OFF** by depressing the switch. Under normal operating conditions, the overdrive should be turned **ON** to maximize fuel economy and reduce engine speed. However, if additional engine braking is needed when going downhill, the overdrive may be turned **OFF**. Be sure to turn the overdrive back **ON** immediately afterward.

NOTE: On 1987 vehicles, the overdrive light is located on the instrument cluster, near the tachometer.

Metric Fasteners

The metric fastener dimensions are very close to the dimensions of the familiar inch system fasteners, and for this reason, replacement fasteners must have the same measurement and strength as those removed.

Do not attempt to interchange metric fasteners for inch system fasteners. Mismatched or incorrect fasteners can result in damage to the transmission unit through malfunctions, breakage or possible personal injury.

Care should be taken to re-use the fasteners in the same locations as removed.

Capacities

To fill a dry transmission (4 speed unit), fill the unit with 1.5 quarts (1.4L) of SAE–80W or SAE 80W–90 GL–5 gear lube. To fill an overdrive unit (including cooler and lines), fill the unit with 1.8 quarts (1.7L) of Dexron®II automatic transmission fluid.

Checking Fluid Level

When checking the fluid level, make sure the transmission is cold and the engine is turned **OFF**.

MANUAL TRANSMISSION (4 SPEED) UNIT

When checking the fluid level in the manual transmission (4 speed) unit, the fluid level should be filled to the level of the filler plug hole located on the passenger's side of the transmission case.

OVERDRIVE UNIT

When checking the fluid level in the overdrive unit, the fluid level should be filled to the level of the filler plug hole located on the driver's side of the overdrive case.

TRANSMISSION MODIFICATIONS

Loss of Overdrive

Some 1984–85 Corvettes equipped with 4 speed with automatic overdrive transmissions may experience a loss of overdrive due to the direct clutch thrust washer wearing grooves in the overdrive unit case. A new direct clutch thrust washer, part number 14091496 is to be used.

Noise in Direct Drive

Some 1984–85 Corvettes equipped with 4 speed with automatic overdrive transmissions may experience a noise in direct drive due to worn annular bearing on the planetary carrier. A new carrier and bearing, part number 14094571 is to be used.

Damage to Transmission Due to Improperly Installed Roll Pin

Some 1984–85 Corvettes equipped with 4 speed with automatic overdrive transmissions may experience damage to the transmission due to an improperly installed roll pin in the reverse idler shaft. Check the installation of ¼ in. diameter × 0.75 in. long roll pin in the reverse idler shaft.

CHILTON THREE C'S TRANSMISSION DIAGNOSIS CHART

Condition	Cause	Correction
Slips out of high gear	a) Transmission loose on clutch housing b) Shift linkage does not work freely; binds c) Damaged mainshaft pilot bearing d) Main drive gear retainer broken or loose e) Dirt between transmission case and clutch housing f) Misalignment of transmission g) Pilot bearing loose in crankshaft h) Worn or improperly adjusted linkage	a) Tighten mounting bolts b) Adjust and free up shift linkage c) Replace pilot bearing d) Tighten or replace main drive gear e) Clean mating surfaces f) Check transmission alignment g) Replace bearing h) Adjust or replace linkage
Noisy in all gears	a) Insufficient lubricant b) Worn countergear bearings c) Worn or damaged main drive gear and countergear d) Damaged main drive gear or main shaft bearings e) Worn or damaged countergear	a) Fill to correct level b) Replace countergear bearings and shaft c) Replace worn or damaged gears d) Replace bearings or main drive gear e) Replace countergear
Noisy in high gear	a) Damaged main drive gear bearing b) Damaged mainshaft bearing c) Damaged high speed gear synchronizer	a) Replace bearing b) Replace bearing c) Replace synchronizer
Noisy in neutral with engine running	a) Damaged main drive gear bearing b) Damaged or loose pilot bearing c) Worn or damaged countergear d) Worn countergear bearings	a) Replace bearing b) Replace pilot bearing c) Replace countergear d) Replace countergear bearing and shaft
Noisy in all reduction gears	a) Insufficient lubricant b) Worn or damaged main drive gear and countergear c) Damaged main drive gear bearing	a) Fill to correct level b) Replace worn or damaged gears c) Replace bearing
Noisy in 2nd only	a) Damaged or worn 2nd speed gears b) Worn or damaged countergear rear bearings c) Damaged or worn 2nd speed synchronizer	a) Replace gears b) Replace bearings and shaft c) Replace synchronizer
Noisy in 3rd only	a) Damaged or worn 3rd speed gears b) Worn or damaged countergear bearings	a) Replace gears b) Replace bearings and shaft
Noisy in reverse only NOTE: Reverse gear operation is noisier than forward gear operation due to spur gears	a) Worn or damaged reverse idler gear or idler bushing b) Worn or damaged reverse gear on mainshaft c) Worn or damaged reverse countergear d) Damaged shift mechanism	a) Replace reverse idler gear assembly b) Replace reverse gear c) Replace countergear assembly d) Inspect linkage and adjust or replace
Excessive backlash in all reduction gears	a) Worn countergear bearings b) Excessive endplay in countergear	a) Replace bearings b) Replace countergear thrust washer
Main drive gear bearing retainer burned or scored by input shaft	a) Loose or damaged pilot bearing b) Misalignment of transmission	a) Replace bearing b) Align transmission
Leaks lubricant	a) Excessive amount of lubricant in transmission b) Loose or broken drive gear bearing retainer c) Damaged drive gear bearing gasket d) Loose side cover or damaged gasket e) Loose countershaft in case	a) Drain to correct level b) Tighten or replace retainer c) Replace gasket d) Tighten cover or replace gasket e) Replace case

CHILTON THREE C'S TRANSMISSION DIAGNOSIS CHART

Condition	Cause	Correction
Overdrive inoperative	a) Insufficient lubricant	a) Fill to correct level
	b) Speedometer inoperative	b) Replace speedometer sensors. Replace speedometer drive or driven gears
	c) Blown fuse	c) Replace fuse
	d) Temperature gauge inoperative	d) Replace gauge or sending unit
	e) O/D switch at shifter knob	e) Replace switch or repair wires
	f) Solenoid inoperative (Check for 12 volts at O/D connector—Engine at operating temperature—wheels off the ground shift transmission from 1st to 2nd gear at 12–15 mph—solenoid should energize at this point)	f) Replace solenoid or repair wires
	g) Solenoid check ball missing	g) Install check ball
	h) Low pressure switch	h) Replace switch
	i) Shift valve inoperative	i) Free up valve or replace valve body
	j) Pressure regulator valve inoperative	j) Free up valve or replace valve body
	k) Drive pin broken or missing in pump	k) Replace pin or pump gears
Overdrive in all gear	a) Solenoid plunger stuck	a) Replace solenoid
	b) Solenoid exhaust hole plugged	b) Clear exhaust passage
	c) Shift valve stuck	c) Free up shift valve or replace valve body
Harsh upshifts	a) Stuck T.V. valve	a) Free up valve or replace valve body
	b) Stuck accumulator piston	b) Free up or repace
	c) Accumulator seal damaged	c) Replace seal
Harsh downshifts	a) Stuck T.V. valve	a) Free up valve or replace valve body
Soft upshift	a) Stuck T.V. valve	a) Free up valve or replace valve body
	b) O/D clutch plates burnt	b) Replace clutch plates and discs
No downshift	a) ECM, throttle position switch	a) Repair or replace ECM or throttle position switch
Slips on upshift	a) Direct clutch plates burnt	a) Replace direct clutch plates and discs
	b) Excessive clutch pack clearance	b) Adjust clutch pack clearance
Chatters on upshifts	a) Piston seals damaged	a) Replace seals
	b) Direct clutch plates burnt	b) Replace direct clutch plates and discs
	c) Low mainline pressure	c) Check lines for restrictions. Check pump
Overdrive overheats	a) Insufficient lubricant	a) Fill to correct level
	b) Stuck cooler valve	b) Free up cooler valve or replace valve
	c) Cooler line restriction	c) Flush lines or replace
	d) Restriction in radiator	d) Flush radiator or replace
No reverse	a) Direct clutch plates burnt	a) Replace direct clutch plates and discs
Noisy in direct drive	a) Front carrier bearing	a) Replace carrier cover
	b) Thrust bearing in direct clutch	b) Replace thrust bearing
Noisy in overdrive	a) Pinion roller bearings	a) Replace carrier cover or housing
	b) Pinion gears (scored, chipped or burnt)	b) Replace pinion gears
	c) Input sun gear (scored, chipped or burnt)	c) Replace input sun gear
	d) Output sun gear (scored, chipped or burnt)	d) Replace output sun gear
No direct drive	a) Sprag clutch damaged	a) Replace sprag clutch

TROUBLE DIAGNOSIS

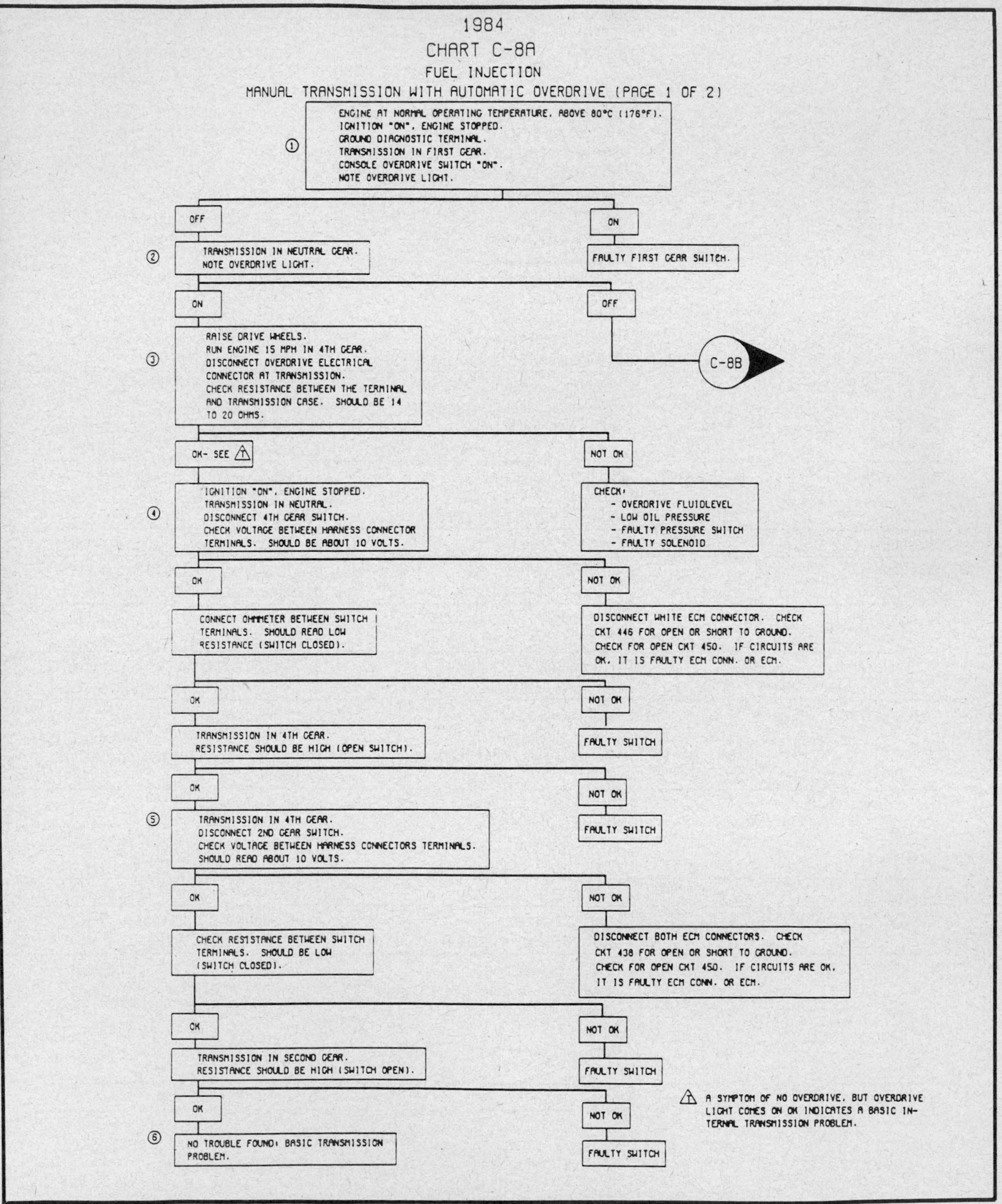
1984
CHART C-8A
FUEL INJECTION
MANUAL TRANSMISSION WITH AUTOMATIC OVERDRIVE (PAGE 1 OF 2)
① ENGINE AT NORMAL OPERATING TEMPERATURE, ABOVE 80°C (176°F). IGNITION "ON", ENGINE STOPPED. GROUND DIAGNOSTIC TERMINAL. TRANSMISSION IN FIRST GEAR. CONSOLE OVERDRIVE SWITCH "ON". NOTE OVERDRIVE LIGHT.
OFF
ON
FAULTY FIRST GEAR SWITCH.
② TRANSMISSION IN NEUTRAL GEAR. NOTE OVERDRIVE LIGHT.
ON
OFF
C-8B
③ RAISE DRIVE WHEELS. RUN ENGINE 15 MPH IN 4TH GEAR. DISCONNECT OVERDRIVE ELECTRICAL CONNECTOR AT TRANSMISSION. CHECK RESISTANCE BETWEEN THE TERMINAL AND TRANSMISSION CASE. SHOULD BE 14 TO 20 OHMS.
OK- SEE ⚠
NOT OK
CHECK: - OVERDRIVE FLUIDLEVEL - LOW OIL PRESSURE - FAULTY PRESSURE SWITCH - FAULTY SOLENOID
④ IGNITION "ON", ENGINE STOPPED. TRANSMISSION IN NEUTRAL. DISCONNECT 4TH GEAR SWITCH. CHECK VOLTAGE BETWEEN HARNESS CONNECTOR TERMINALS. SHOULD BE ABOUT 10 VOLTS.
OK
NOT OK
CONNECT OHMMETER BETWEEN SWITCH TERMINALS. SHOULD READ LOW RESISTANCE (SWITCH CLOSED).
DISCONNECT WHITE ECM CONNECTOR. CHECK CKT 446 FOR OPEN OR SHORT TO GROUND. CHECK FOR OPEN CKT 450. IF CIRCUITS ARE OK, IT IS FAULTY ECM CONN. OR ECM.
OK
NOT OK
TRANSMISSION IN 4TH GEAR. RESISTANCE SHOULD BE HIGH (OPEN SWITCH).
FAULTY SWITCH
OK
NOT OK
⑤ TRANSMISSION IN 4TH GEAR. DISCONNECT 2ND GEAR SWITCH. CHECK VOLTAGE BETWEEN HARNESS CONNECTORS TERMINALS. SHOULD READ ABOUT 10 VOLTS.
FAULTY SWITCH
OK
NOT OK
CHECK RESISTANCE BETWEEN SWITCH TERMINALS. SHOULD BE LOW (SWITCH CLOSED).
DISCONNECT BOTH ECM CONNECTORS. CHECK CKT 438 FOR OPEN OR SHORT TO GROUND. CHECK FOR OPEN CKT 450. IF CIRCUITS ARE OK, IT IS FAULTY ECM CONN. OR ECM.
OK
NOT OK
TRANSMISSION IN SECOND GEAR. RESISTANCE SHOULD BE HIGH (SWITCH OPEN).
FAULTY SWITCH
OK
NOT OK
⑥ NO TROUBLE FOUND: BASIC TRANSMISSION PROBLEM.
FAULTY SWITCH
⚠ A SYMPTOM OF NO OVERDRIVE, BUT OVERDRIVE LIGHT COMES ON OK INDICATES A BASIC INTERNAL TRANSMISSION PROBLEM.

Manual transmission with automatic overdrive diagnostic chart—1984 models

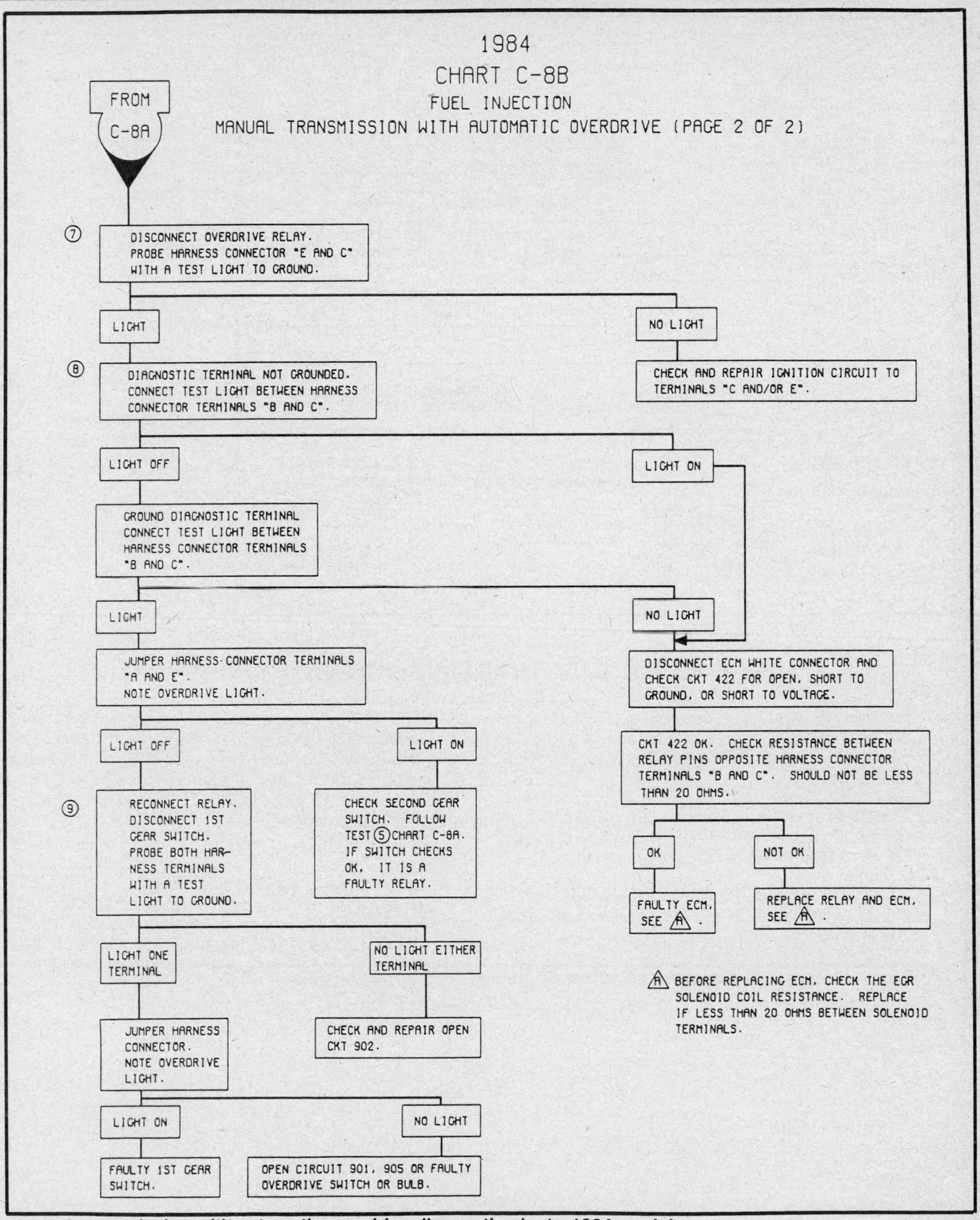
1984
CHART C-8B
FUEL INJECTION
MANUAL TRANSMISSION WITH AUTOMATIC OVERDRIVE (PAGE 2 OF 2)
FROM C-8A
7
DISCONNECT OVERDRIVE RELAY. PROBE HARNESS CONNECTOR "E AND C" WITH A TEST LIGHT TO GROUND.
LIGHT
NO LIGHT
CHECK AND REPAIR IGNITION CIRCUIT TO TERMINALS "C AND/OR E".
8
DIAGNOSTIC TERMINAL NOT GROUNDED. CONNECT TEST LIGHT BETWEEN HARNESS CONNECTOR TERMINALS "B AND C".
LIGHT OFF
LIGHT ON
GROUND DIAGNOSTIC TERMINAL CONNECT TEST LIGHT BETWEEN HARNESS CONNECTOR TERMINALS "B AND C".
LIGHT
NO LIGHT
JUMPER HARNESS CONNECTOR TERMINALS "A AND E". NOTE OVERDRIVE LIGHT.
DISCONNECT ECM WHITE CONNECTOR AND CHECK CKT 422 FOR OPEN, SHORT TO GROUND, OR SHORT TO VOLTAGE.
LIGHT OFF
LIGHT ON
CKT 422 OK. CHECK RESISTANCE BETWEEN RELAY PINS OPPOSITE HARNESS CONNECTOR TERMINALS "B AND C". SHOULD NOT BE LESS THAN 20 OHMS.
9
RECONNECT RELAY. DISCONNECT 1ST GEAR SWITCH. PROBE BOTH HARNESS TERMINALS WITH A TEST LIGHT TO GROUND.
CHECK SECOND GEAR SWITCH. FOLLOW TEST 5 CHART C-8A. IF SWITCH CHECKS OK, IT IS A FAULTY RELAY.
OK
NOT OK
FAULTY ECM, SEE A.
REPLACE RELAY AND ECM, SEE A.
LIGHT ONE TERMINAL
NO LIGHT EITHER TERMINAL
A BEFORE REPLACING ECM, CHECK THE EGR SOLENOID COIL RESISTANCE. REPLACE IF LESS THAN 20 OHMS BETWEEN SOLENOID TERMINALS.
JUMPER HARNESS CONNECTOR. NOTE OVERDRIVE LIGHT.
CHECK AND REPAIR OPEN CKT 902.
LIGHT ON
NO LIGHT
FAULTY 1ST GEAR SWITCH.
OPEN CIRCUIT 901, 905 OR FAULTY OVERDRIVE SWITCH OR BULB.

Manual transmission with automatic overdrive diagnostic chart—1984 models

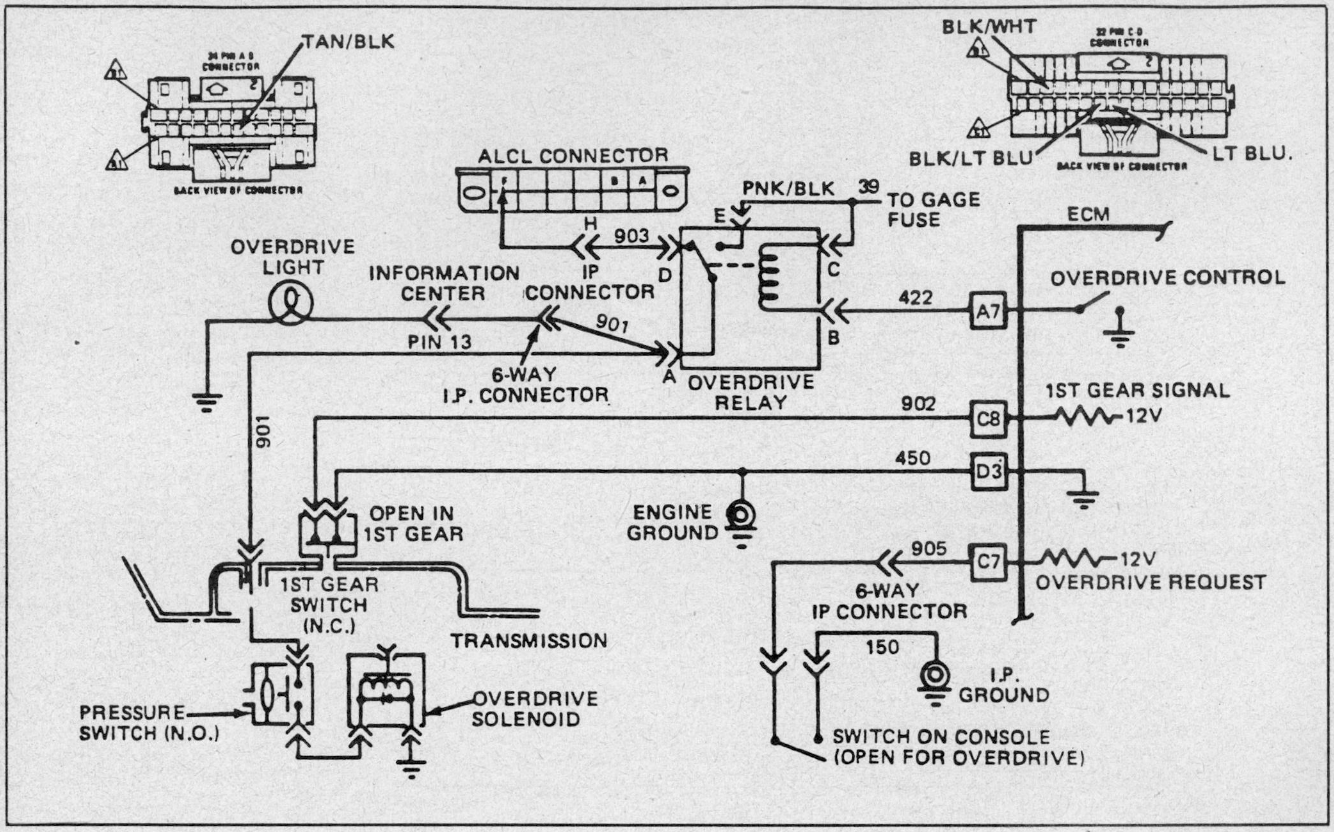
TAN/BLK
BACK VIEW OF CONNECTOR
BLK/WHT
BLK/LT BLU
LT BLU.
ALCL CONNECTOR
PNK/BLK
39
TO GAGE FUSE
ECM
OVERDRIVE LIGHT
INFORMATION CENTER
IP CONNECTOR
903
H
E
D
C
OVERDRIVE CONTROL
422
A7
PIN 13
901
B
6-WAY I.P. CONNECTOR
A
OVERDRIVE RELAY
1ST GEAR SIGNAL
902
C8
12V
450
D3
OPEN IN 1ST GEAR
ENGINE GROUND
905
C7
12V
OVERDRIVE REQUEST
6-WAY IP CONNECTOR
1ST GEAR SWITCH (N.C.)
TRANSMISSION
150
I.P. GROUND
PRESSURE SWITCH (N.O.)
OVERDRIVE SOLENOID
SWITCH ON CONSOLE (OPEN FOR OVERDRIVE)

CHART C-8B MANUAL TRANSMISSION WITH OVERDRIVE

6. An overdrive light that does come "on" while in 2nd gear above 15 mph indicates a fault in the 1st gear switch circuit, the overdrive request circuit or a faulty ECM.

7. The ECM supplies 12 volts on Circuit 902. When the ECM sees 12 volts at ECM Pin C8, it thinks the vehicle is in 1st gear.

8. The ECM supplies 12 volts to CKT 905 which causes overdrive to be requested when CKT 905 is not grounded.

9. The RPM may not drop at the instant the overdrive light comes "on" as the oil pressure switch may have not yet closed. The oil pressure switch should close at about 15 mph.

10. Checks continuity of CKT 901 between the relay and transmission.

11. If the voltage is present at the transmission connector and no RPM drop was detected, then the problem is an internal transmission problem. Because the overdrive solenoid is not directly tied into the ECM, its resistance value can be lower than ECM controlled solenoids or relays.

Manual transmission with automatic overdrive diagnostic chart – 1985 models

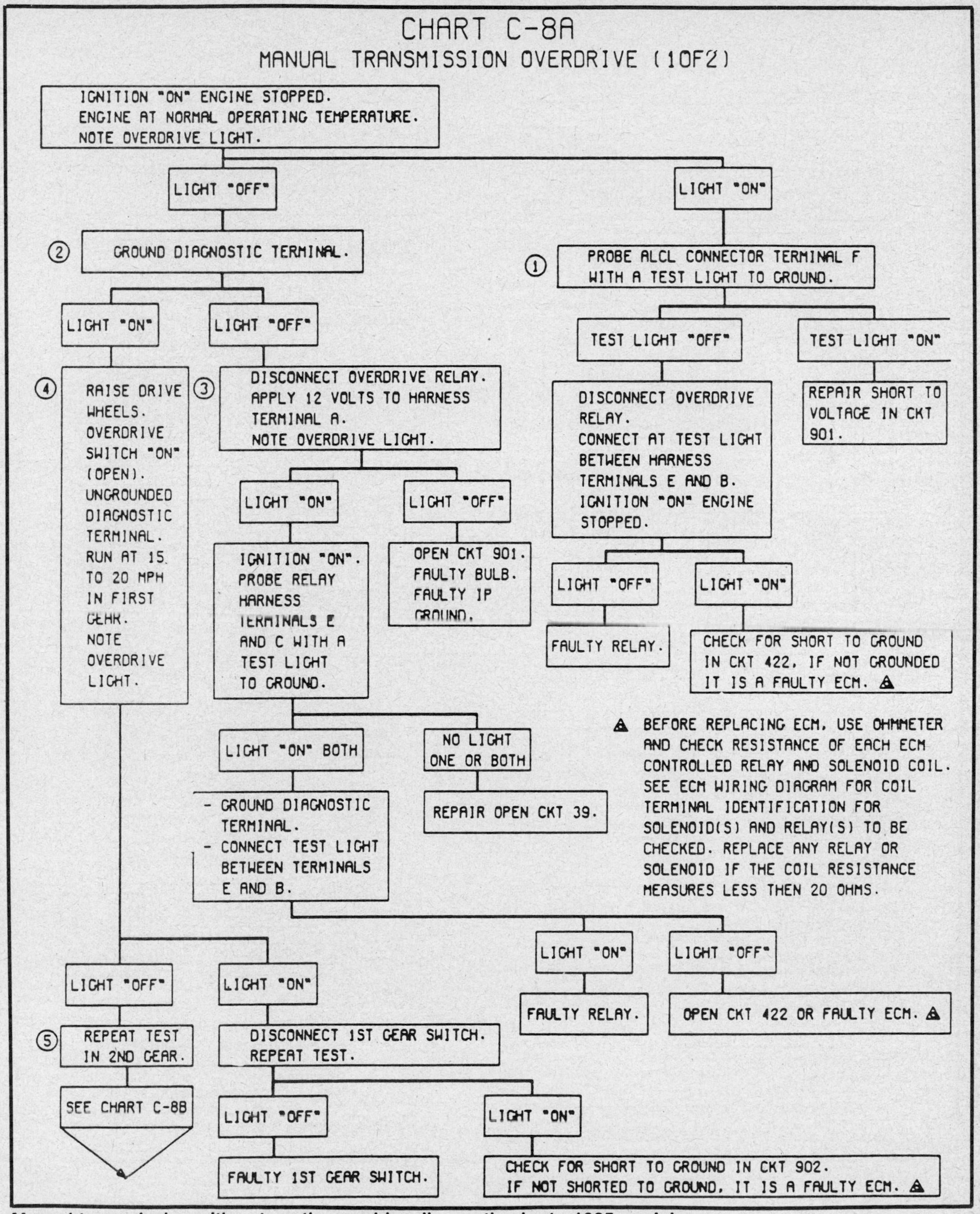
CHART C-8A
MANUAL TRANSMISSION OVERDRIVE (1OF2)
IGNITION "ON" ENGINE STOPPED. ENGINE AT NORMAL OPERATING TEMPERATURE. NOTE OVERDRIVE LIGHT.
LIGHT "OFF"
LIGHT "ON"
② GROUND DIAGNOSTIC TERMINAL.
① PROBE ALCL CONNECTOR TERMINAL F WITH A TEST LIGHT TO GROUND.
LIGHT "ON"
LIGHT "OFF"
TEST LIGHT "OFF"
TEST LIGHT "ON"
④ RAISE DRIVE WHEELS. OVERDRIVE SWITCH "ON" (OPEN). UNGROUNDED DIAGNOSTIC TERMINAL. RUN AT 15 TO 20 MPH IN FIRST GEAR. NOTE OVERDRIVE LIGHT.
③ DISCONNECT OVERDRIVE RELAY. APPLY 12 VOLTS TO HARNESS TERMINAL A. NOTE OVERDRIVE LIGHT.
DISCONNECT OVERDRIVE RELAY. CONNECT AT TEST LIGHT BETWEEN HARNESS TERMINALS E AND B. IGNITION "ON" ENGINE STOPPED.
REPAIR SHORT TO VOLTAGE IN CKT 901.
LIGHT "ON"
LIGHT "OFF"
IGNITION "ON". PROBE RELAY HARNESS TERMINALS E AND C WITH A TEST LIGHT TO GROUND.
OPEN CKT 901. FAULTY BULB. FAULTY IP GROUND.
LIGHT "OFF"
LIGHT "ON"
FAULTY RELAY.
CHECK FOR SHORT TO GROUND IN CKT 422, IF NOT GROUNDED IT IS A FAULTY ECM. ▲
▲ BEFORE REPLACING ECM, USE OHMMETER AND CHECK RESISTANCE OF EACH ECM CONTROLLED RELAY AND SOLENOID COIL. SEE ECM WIRING DIAGRAM FOR COIL TERMINAL IDENTIFICATION FOR SOLENOID(S) AND RELAY(S) TO BE CHECKED. REPLACE ANY RELAY OR SOLENOID IF THE COIL RESISTANCE MEASURES LESS THEN 20 OHMS.
LIGHT "ON" BOTH
NO LIGHT ONE OR BOTH
- GROUND DIAGNOSTIC TERMINAL. - CONNECT TEST LIGHT BETWEEN TERMINALS E AND B.
REPAIR OPEN CKT 39.
LIGHT "ON"
LIGHT "OFF"
LIGHT "OFF"
LIGHT "ON"
FAULTY RELAY.
OPEN CKT 422 OR FAULTY ECM. ▲
⑤ REPEAT TEST IN 2ND GEAR.
DISCONNECT 1ST GEAR SWITCH. REPEAT TEST.
SEE CHART C-8B
LIGHT "OFF"
LIGHT "ON"
FAULTY 1ST GEAR SWITCH.
CHECK FOR SHORT TO GROUND IN CKT 902. IF NOT SHORTED TO GROUND, IT IS A FAULTY ECM. ▲

Manual transmission with automatic overdrive diagnostic chart – 1985 models

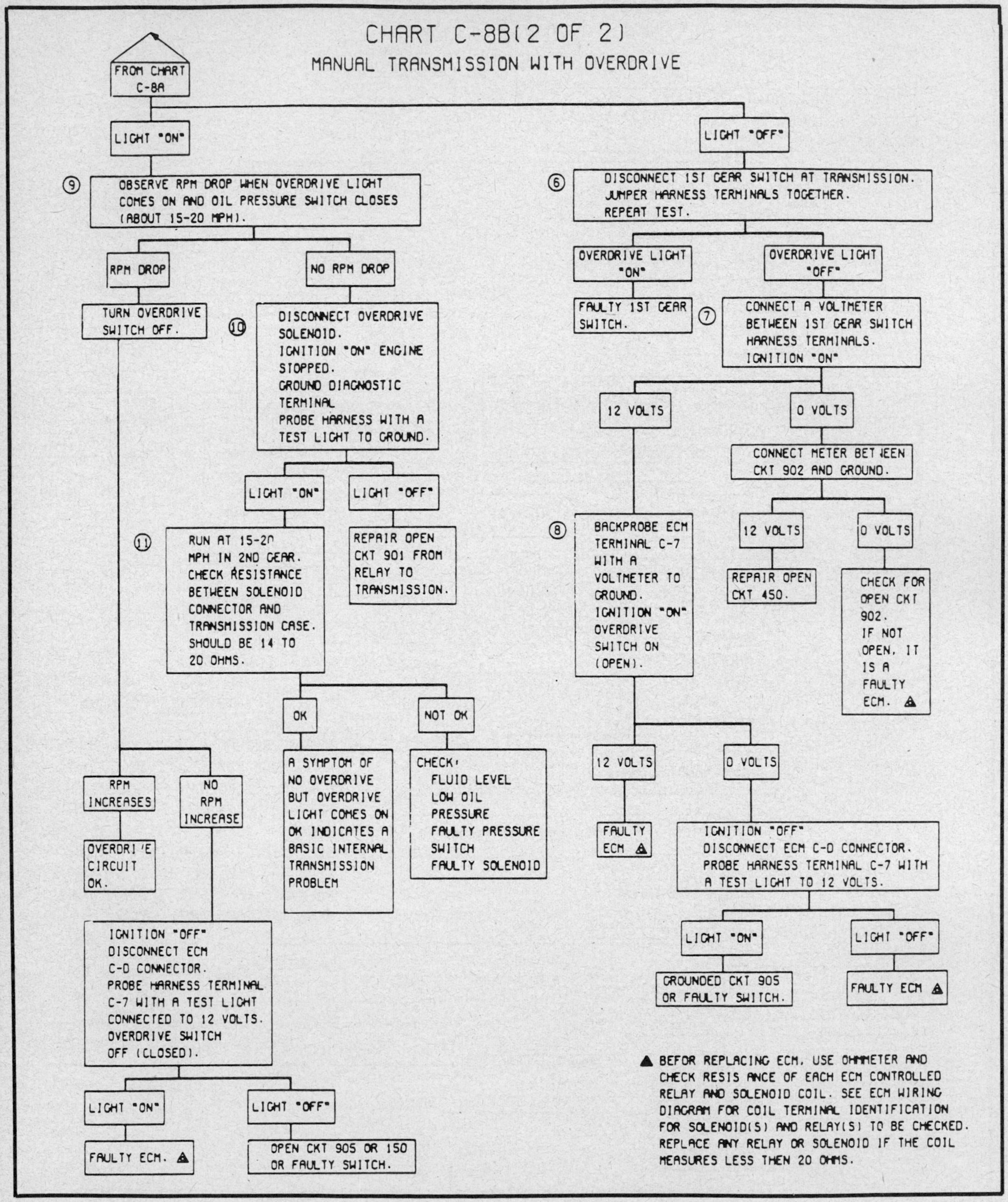
CHART C-8B(2 OF 2)
MANUAL TRANSMISSION WITH OVERDRIVE
FROM CHART C-8A
LIGHT "ON"
LIGHT "OFF"
9
OBSERVE RPM DROP WHEN OVERDRIVE LIGHT COMES ON AND OIL PRESSURE SWITCH CLOSES (ABOUT 15-20 MPH).
RPM DROP
NO RPM DROP
TURN OVERDRIVE SWITCH OFF.
10
DISCONNECT OVERDRIVE SOLENOID. IGNITION "ON" ENGINE STOPPED. GROUND DIAGNOSTIC TERMINAL PROBE HARNESS WITH A TEST LIGHT TO GROUND.
LIGHT "ON"
LIGHT "OFF"
11
RUN AT 15-20 MPH IN 2ND GEAR. CHECK RESISTANCE BETWEEN SOLENOID CONNECTOR AND TRANSMISSION CASE. SHOULD BE 14 TO 20 OHMS.
REPAIR OPEN CKT 901 FROM RELAY TO TRANSMISSION.
OK
NOT OK
A SYMPTOM OF NO OVERDRIVE BUT OVERDRIVE LIGHT COMES ON OK INDICATES A BASIC INTERNAL TRANSMISSION PROBLEM
CHECK: FLUID LEVEL LOW OIL PRESSURE FAULTY PRESSURE SWITCH FAULTY SOLENOID
RPM INCREASES
NO RPM INCREASE
OVERDRIVE CIRCUIT OK.
IGNITION "OFF" DISCONNECT ECM C-D CONNECTOR. PROBE HARNESS TERMINAL C-7 WITH A TEST LIGHT CONNECTED TO 12 VOLTS. OVERDRIVE SWITCH OFF (CLOSED).
LIGHT "ON"
LIGHT "OFF"
FAULTY ECM.
OPEN CKT 905 OR 150 OR FAULTY SWITCH.
6
DISCONNECT 1ST GEAR SWITCH AT TRANSMISSION. JUMPER HARNESS TERMINALS TOGETHER. REPEAT TEST.
OVERDRIVE LIGHT "ON"
OVERDRIVE LIGHT "OFF"
FAULTY 1ST GEAR SWITCH.
7
CONNECT A VOLTMETER BETWEEN 1ST GEAR SWITCH HARNESS TERMINALS. IGNITION "ON"
12 VOLTS
0 VOLTS
CONNECT METER BETWEEN CKT 902 AND GROUND.
8
BACKPROBE ECM TERMINAL C-7 WITH A VOLTMETER TO GROUND. IGNITION "ON" OVERDRIVE SWITCH ON (OPEN).
12 VOLTS
0 VOLTS
REPAIR OPEN CKT 450.
CHECK FOR OPEN CKT 902. IF NOT OPEN, IT IS A FAULTY ECM.
12 VOLTS
0 VOLTS
FAULTY ECM
IGNITION "OFF" DISCONNECT ECM C-D CONNECTOR. PROBE HARNESS TERMINAL C-7 WITH A TEST LIGHT TO 12 VOLTS.
LIGHT "ON"
LIGHT "OFF"
GROUNDED CKT 905 OR FAULTY SWITCH.
FAULTY ECM
BEFOR REPLACING ECM, USE OHMMETER AND CHECK RESIS ANCE OF EACH ECM CONTROLLED RELAY AND SOLENOID COIL. SEE ECM WIRING DIAGRAM FOR COIL TERMINAL IDENTIFICATION FOR SOLENOID(S) AND RELAY(S) TO BE CHECKED. REPLACE ANY RELAY OR SOLENOID IF THE COIL MEASURES LESS THEN 20 OHMS.

Manual transmission with automatic overdrive diagnostic chart—1985 models

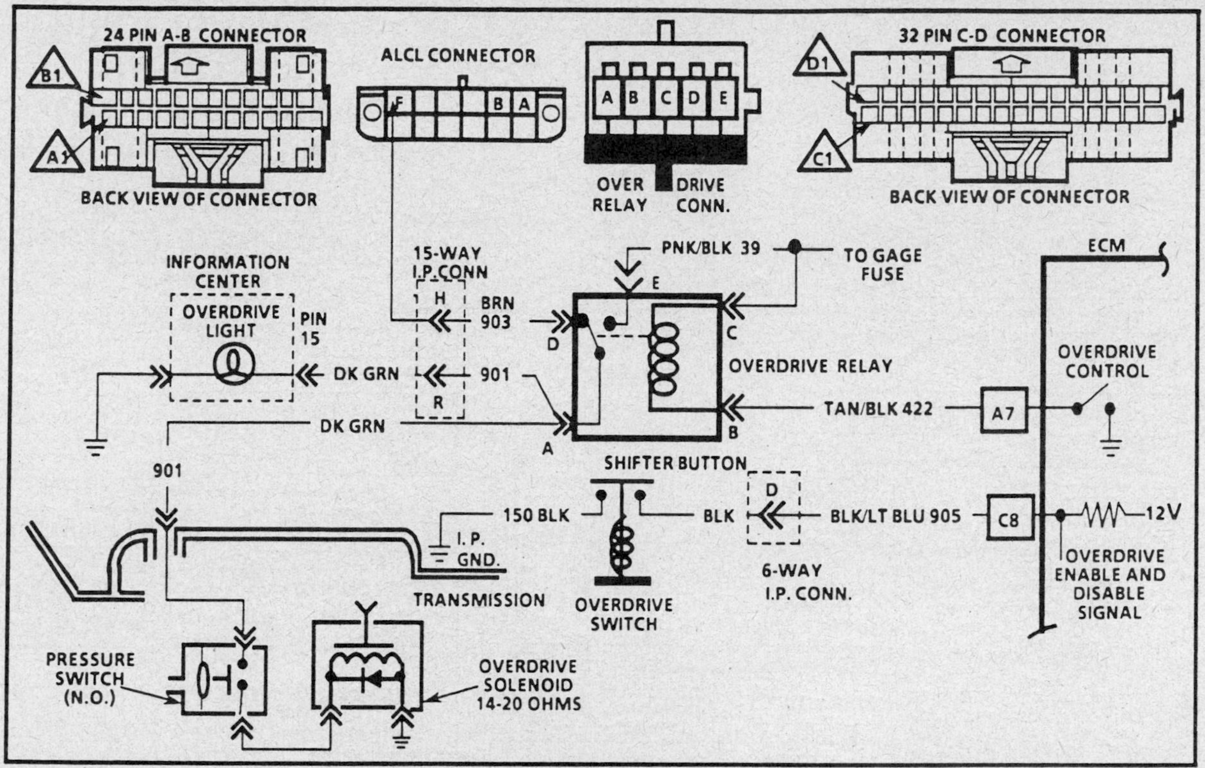
24 PIN A-B CONNECTOR
B1
A1
BACK VIEW OF CONNECTOR
ALCL CONNECTOR
F
B
A
A B C D E
OVER RELAY
DRIVE CONN.
32 PIN C-D CONNECTOR
D1
C1
BACK VIEW OF CONNECTOR
INFORMATION CENTER
OVERDRIVE LIGHT
PIN 15
DK GRN
15-WAY I.P.CONN
H
R
BRN 903
901
DK GRN
PNK/BLK 39
TO GAGE FUSE
E
C
D
A
B
OVERDRIVE RELAY
TAN/BLK 422
A7
ECM
OVERDRIVE CONTROL
SHIFTER BUTTON
150 BLK
I. P. GND.
BLK
BLK/LT BLU 905
C8
12V
OVERDRIVE ENABLE AND DISABLE SIGNAL
6-WAY I.P. CONN.
901
TRANSMISSION
OVERDRIVE SWITCH
PRESSURE SWITCH (N.O.)
OVERDRIVE SOLENOID 14-20 OHMS

CHART C-8B

MANUAL TRANSMISSION WITH OVERDRIVE

(1 OF 2)

5.7L "Y" SERIES

FUEL INJECTION (PORT)

The overdrive used in the manual transmission is controlled by the ECM when selected by the driver. The driver controls the overdrive with a button on the shifter knob which activates a switch at the base of the shifter.

The ECM supplies a current limiting 12 volts on CKT 905 and when the overdrive button is pushed the ECM will see 0 volts.

The switch is spring loaded with normally open contacts.

When the correct conditions are met to allow overdrive, the overdrive can be engaged or disengaged each time the button is pushed.

By monitoring the engine parameters, the ECM can determine what gear the vehicle is in. This will cause the ECM to keep the overdrive relay de-energized if the vehicle is in 1st gear or during WOT. If WOT is detected while in 2nd or 3rd gear, the ECM will keep the relay de-energized until 4th gear is selected or until the over-drive button is again pushed. If the ECM detects WOT while in 4th gear and the overdrive was selected, the ECM will again energize the relay when the WOT is no longer being detected. The ECM uses engine temperature, RPM, VSS, TPS, and the MAF sensor signal to determine when to allow overdrive.

1. If the overdrive light is "on" with the ignition turned "on", then the relay is energized or CKT 901 is shorted to voltage. If the relay is energized, the test light should not light at this point.
2. Grounding the diagnostic terminal with ignition "on" should energize the relay and turn "on" the "overdrive" light.
3. Connecting 12 volts to CKT 901 will check the continuity of the bulb circuit.
4. Grounding the diagnostic terminal with the ignition "on", engine "off", will cause the ECM to ground CKT 422. The test light should turn on, if the ECM has the ability to ground the circuit.

Manual transmission with automatic overdrive diagnostic chart – 1986 models

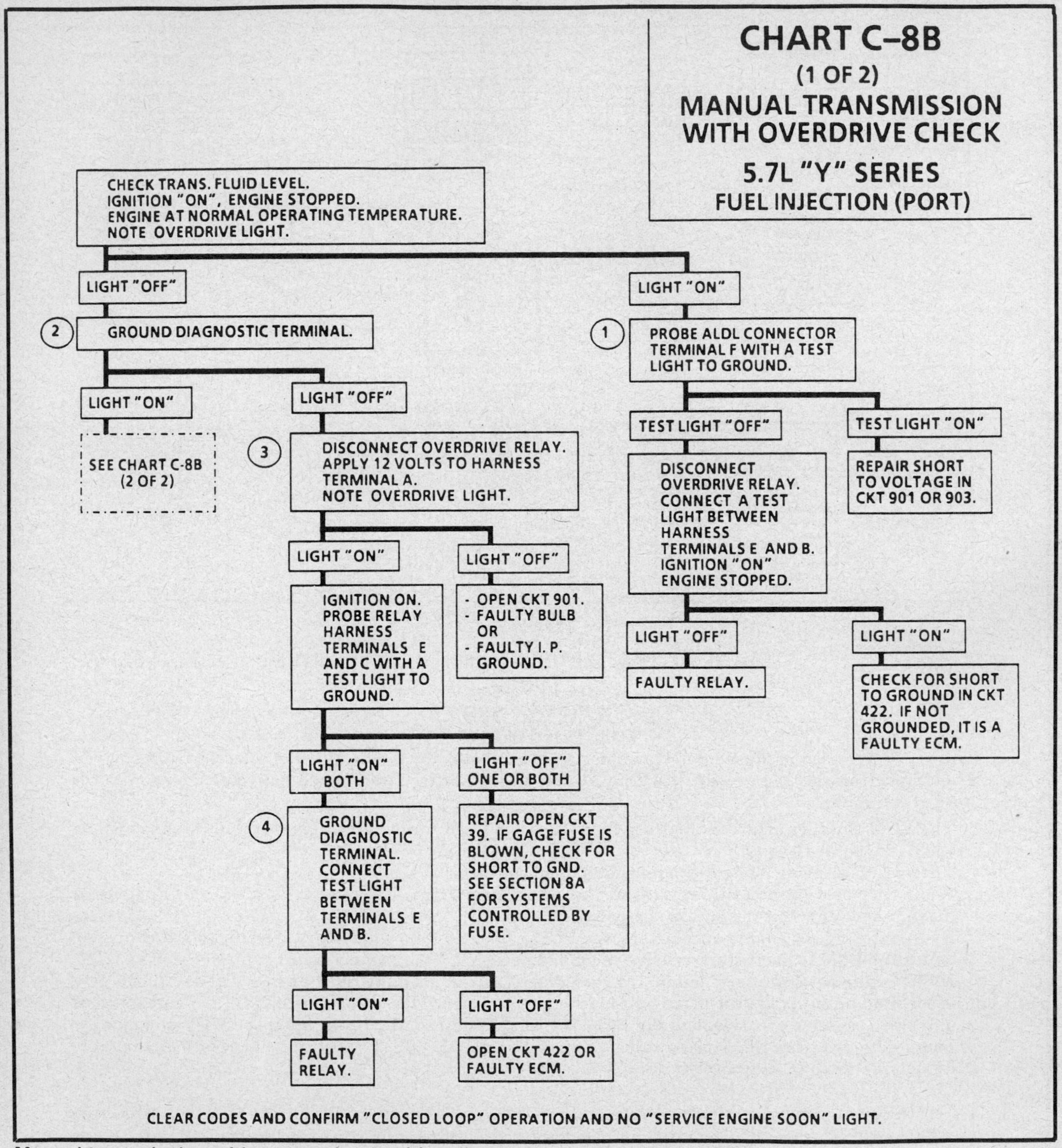
CHART C–8B
(1 OF 2)
MANUAL TRANSMISSION
WITH OVERDRIVE CHECK
5.7L "Y" SERIES
FUEL INJECTION (PORT)
CHECK TRANS. FLUID LEVEL.
IGNITION "ON", ENGINE STOPPED.
ENGINE AT NORMAL OPERATING TEMPERATURE.
NOTE OVERDRIVE LIGHT.
LIGHT "OFF"
LIGHT "ON"
2
GROUND DIAGNOSTIC TERMINAL.
1
PROBE ALDL CONNECTOR TERMINAL F WITH A TEST LIGHT TO GROUND.
LIGHT "ON"
LIGHT "OFF"
SEE CHART C-8B (2 OF 2)
3
DISCONNECT OVERDRIVE RELAY. APPLY 12 VOLTS TO HARNESS TERMINAL A. NOTE OVERDRIVE LIGHT.
TEST LIGHT "OFF"
TEST LIGHT "ON"
DISCONNECT OVERDRIVE RELAY. CONNECT A TEST LIGHT BETWEEN HARNESS TERMINALS E AND B. IGNITION "ON" ENGINE STOPPED.
REPAIR SHORT TO VOLTAGE IN CKT 901 OR 903.
LIGHT "ON"
LIGHT "OFF"
IGNITION ON. PROBE RELAY HARNESS TERMINALS E AND C WITH A TEST LIGHT TO GROUND.
- OPEN CKT 901.
- FAULTY BULB OR
- FAULTY I. P. GROUND.
LIGHT "OFF"
LIGHT "ON"
FAULTY RELAY.
CHECK FOR SHORT TO GROUND IN CKT 422. IF NOT GROUNDED, IT IS A FAULTY ECM.
LIGHT "ON" BOTH
LIGHT "OFF" ONE OR BOTH
4
GROUND DIAGNOSTIC TERMINAL. CONNECT TEST LIGHT BETWEEN TERMINALS E AND B.
REPAIR OPEN CKT 39. IF GAGE FUSE IS BLOWN, CHECK FOR SHORT TO GND. SEE SECTION 8A FOR SYSTEMS CONTROLLED BY FUSE.
LIGHT "ON"
LIGHT "OFF"
FAULTY RELAY.
OPEN CKT 422 OR FAULTY ECM.
CLEAR CODES AND CONFIRM "CLOSED LOOP" OPERATION AND NO "SERVICE ENGINE SOON" LIGHT.

Manual transmission with automatic overdrive diagnostic chart—1986 models

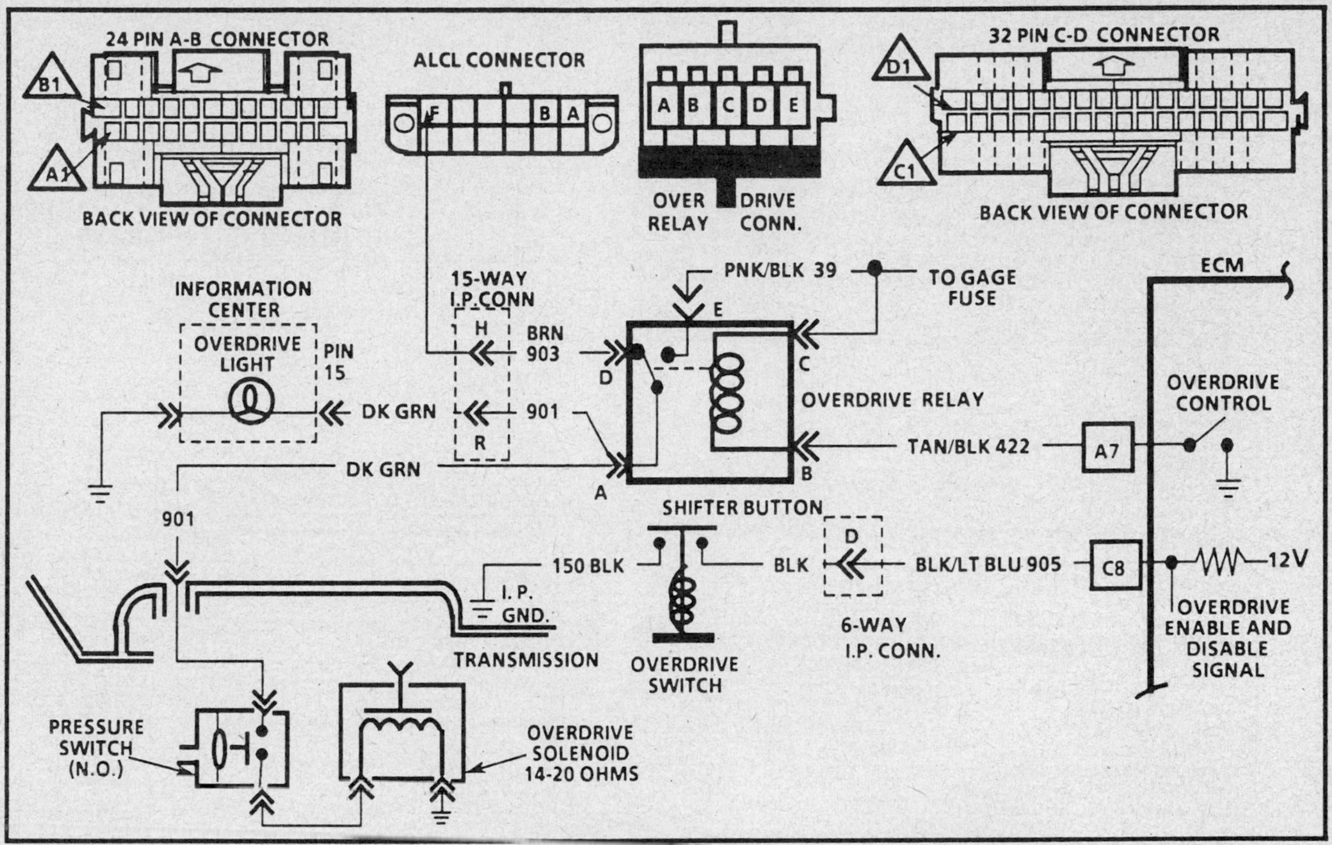
24 PIN A-B CONNECTOR
B1
A1
BACK VIEW OF CONNECTOR
ALCL CONNECTOR
F
B A
A B C D E
OVER DRIVE
RELAY CONN.
32 PIN C-D CONNECTOR
D1
C1
BACK VIEW OF CONNECTOR
PNK/BLK 39
TO GAGE FUSE
ECM
INFORMATION CENTER
15-WAY I.P. CONN
OVERDRIVE LIGHT
PIN 15
H
BRN 903
D
E
C
OVERDRIVE RELAY
OVERDRIVE CONTROL
DK GRN
901
R
TAN/BLK 422
A7
DK GRN
A
B
901
SHIFTER BUTTON
D
150 BLK
BLK
BLK/LT BLU 905
C8
12V
I. P. GND.
6-WAY I.P. CONN.
OVERDRIVE ENABLE AND DISABLE SIGNAL
TRANSMISSION
OVERDRIVE SWITCH
PRESSURE SWITCH (N.O.)
OVERDRIVE SOLENOID 14-20 OHMS

CHART C-8B

MANUAL TRANSMISSION WITH OVERDRIVE
(2 OF 2)

5.7L "Y" SERIES
FUEL INJECTION (PORT)

5. The ECM monitors the overdrive request on PIN C8. When the button is pushed, on the shifter, CKT 905 will be connected to ground, and when the button is released, the signal at the ECM goes back to 12 volts. If the conditions are met to allow overdrive, the overdrive relay will be energized, or de-energized, each time the button is pushed and released.
6. When the overdrive light comes "On", the overdrive solenoid should be energized as long as the oil pressure switch is closed. When the overdrive solenoid energizes a noticable RPM drop should be detected. A road test should be performed to be sure the trans. has the ability to shift into overdrive under road load conditions.
7. This test will determine if voltage is reaching the transmission connector. If the test light is on during this test, the cause of no RPM drop, when the overdrive light came on, is due to an internal transmission problem.
8. If pressure is over 70 psi (483 kPa), it indicates that the oil pressure switch and solenoid must be working correctly.

Manual transmission with automatic overdrive diagnostic chart—1986 models

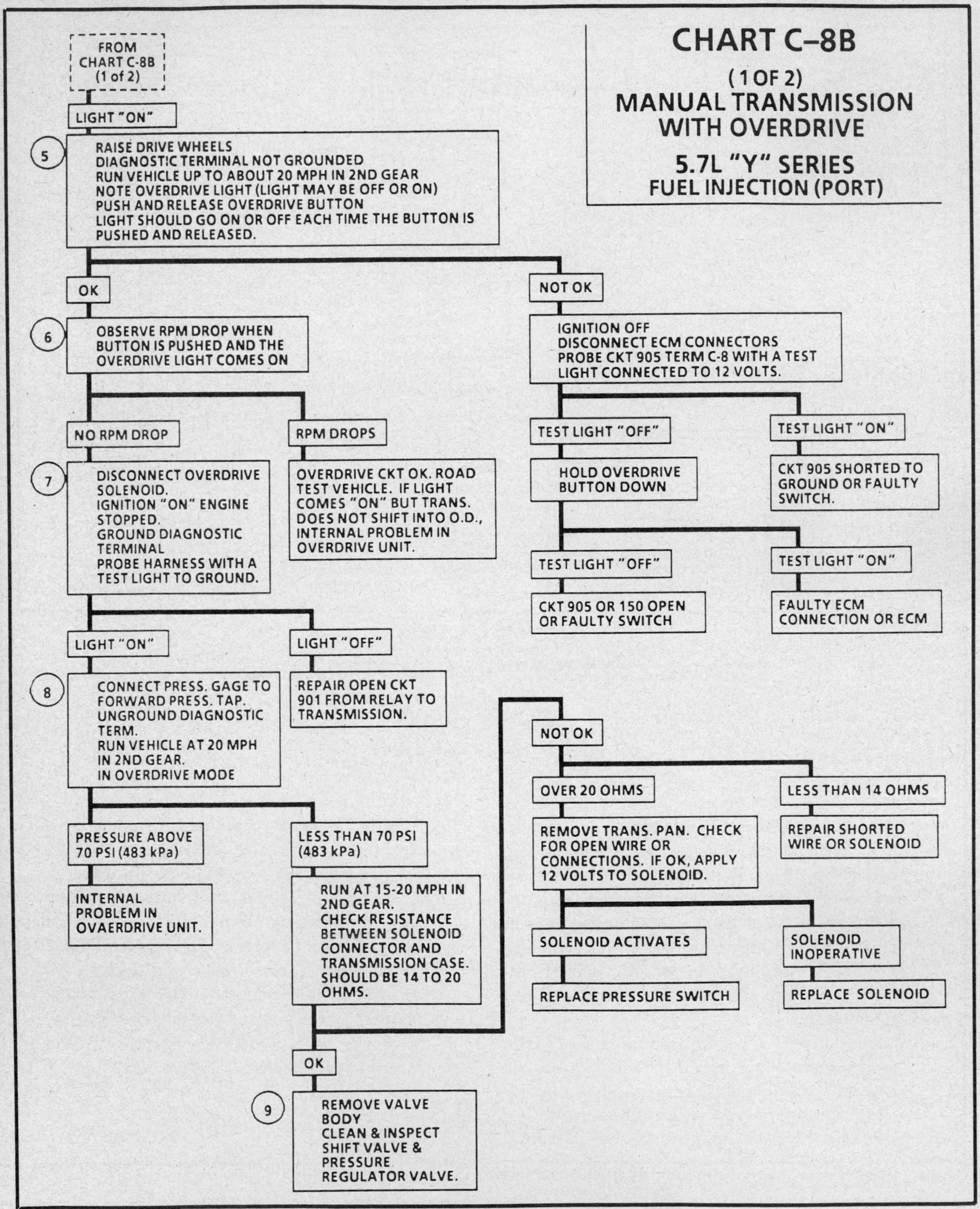
CHART C–8B
(1 OF 2)
MANUAL TRANSMISSION
WITH OVERDRIVE
5.7L "Y" SERIES
FUEL INJECTION (PORT)
FROM CHART C-8B (1 of 2)
LIGHT "ON"
5
RAISE DRIVE WHEELS
DIAGNOSTIC TERMINAL NOT GROUNDED
RUN VEHICLE UP TO ABOUT 20 MPH IN 2ND GEAR
NOTE OVERDRIVE LIGHT (LIGHT MAY BE OFF OR ON)
PUSH AND RELEASE OVERDRIVE BUTTON
LIGHT SHOULD GO ON OR OFF EACH TIME THE BUTTON IS PUSHED AND RELEASED.
OK
6
OBSERVE RPM DROP WHEN BUTTON IS PUSHED AND THE OVERDRIVE LIGHT COMES ON
NO RPM DROP
7
DISCONNECT OVERDRIVE SOLENOID.
IGNITION "ON" ENGINE STOPPED.
GROUND DIAGNOSTIC TERMINAL
PROBE HARNESS WITH A TEST LIGHT TO GROUND.
RPM DROPS
OVERDRIVE CKT OK. ROAD TEST VEHICLE. IF LIGHT COMES "ON" BUT TRANS. DOES NOT SHIFT INTO O.D., INTERNAL PROBLEM IN OVERDRIVE UNIT.
LIGHT "ON"
8
CONNECT PRESS. GAGE TO FORWARD PRESS. TAP.
UNGROUND DIAGNOSTIC TERM.
RUN VEHICLE AT 20 MPH IN 2ND GEAR.
IN OVERDRIVE MODE
LIGHT "OFF"
REPAIR OPEN CKT 901 FROM RELAY TO TRANSMISSION.
PRESSURE ABOVE 70 PSI (483 kPa)
INTERNAL PROBLEM IN OVAERDRIVE UNIT.
LESS THAN 70 PSI (483 kPa)
RUN AT 15-20 MPH IN 2ND GEAR.
CHECK RESISTANCE BETWEEN SOLENOID CONNECTOR AND TRANSMISSION CASE.
SHOULD BE 14 TO 20 OHMS.
OK
9
REMOVE VALVE BODY
CLEAN & INSPECT SHIFT VALVE & PRESSURE REGULATOR VALVE.
NOT OK
IGNITION OFF
DISCONNECT ECM CONNECTORS
PROBE CKT 905 TERM C-8 WITH A TEST LIGHT CONNECTED TO 12 VOLTS.
TEST LIGHT "OFF"
HOLD OVERDRIVE BUTTON DOWN
TEST LIGHT "OFF"
CKT 905 OR 150 OPEN OR FAULTY SWITCH
TEST LIGHT "ON"
FAULTY ECM CONNECTION OR ECM
TEST LIGHT "ON"
CKT 905 SHORTED TO GROUND OR FAULTY SWITCH.
NOT OK
OVER 20 OHMS
REMOVE TRANS. PAN. CHECK FOR OPEN WIRE OR CONNECTIONS. IF OK, APPLY 12 VOLTS TO SOLENOID.
SOLENOID ACTIVATES
REPLACE PRESSURE SWITCH
SOLENOID INOPERATIVE
REPLACE SOLENOID
LESS THAN 14 OHMS
REPAIR SHORTED WIRE OR SOLENOID

Manual transmission with automatic overdrive diagnostic chart—1986 models

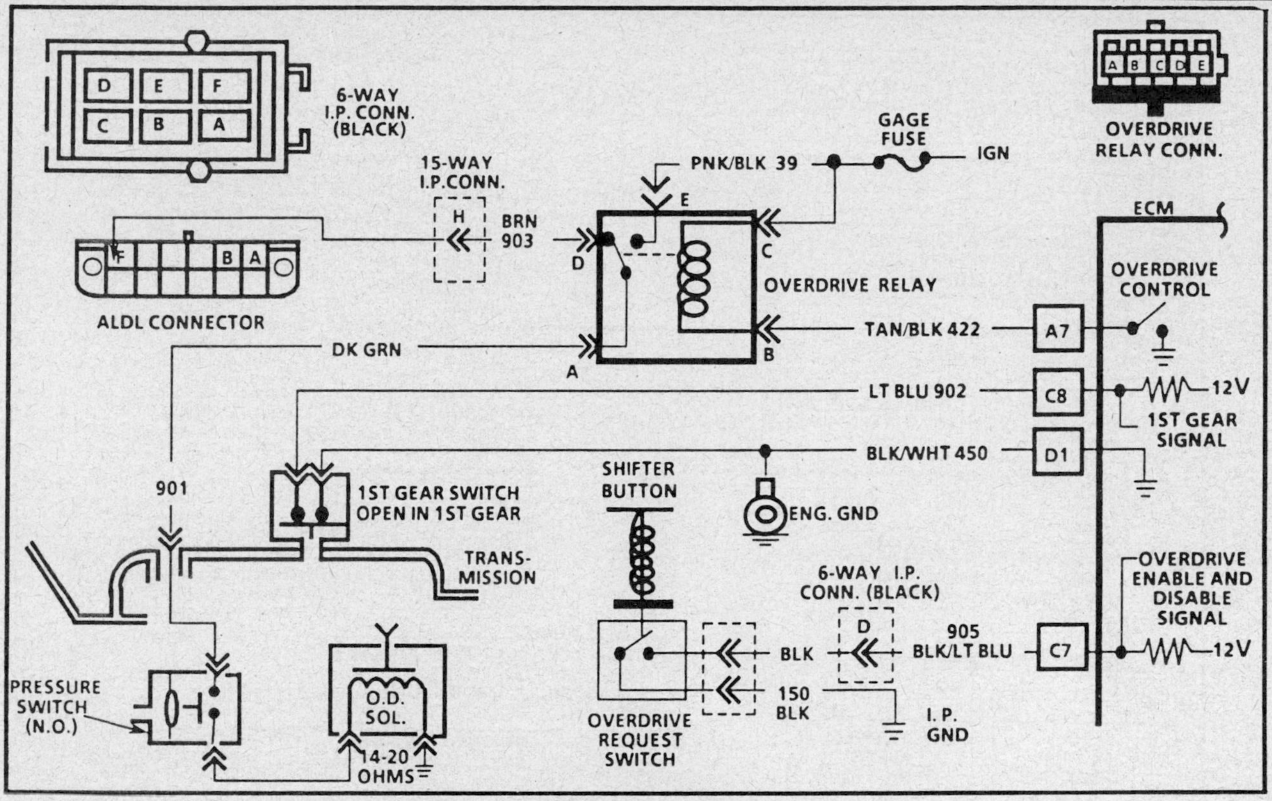
6-WAY
I.P. CONN.
(BLACK)
15-WAY
I.P.CONN.
BRN
903
ALDL CONNECTOR
GAGE
FUSE
PNK/BLK 39
IGN
OVERDRIVE
RELAY CONN.
ECM
OVERDRIVE RELAY
OVERDRIVE
CONTROL
DK GRN
TAN/BLK 422
A7
LT BLU 902
C8
12V
1ST GEAR
SIGNAL
BLK/WHT 450
D1
901
1ST GEAR SWITCH
OPEN IN 1ST GEAR
SHIFTER
BUTTON
ENG. GND
TRANS-
MISSION
OVERDRIVE
ENABLE AND
DISABLE
SIGNAL
6-WAY I.P.
CONN. (BLACK)
905
BLK/LT BLU
C7
12V
BLK
150
BLK
PRESSURE
SWITCH
(N.O.)
O.D.
SOL.
14-20
OHMS
OVERDRIVE
REQUEST
SWITCH
I.P.
GND

CHART C-8B
(Page 2 of 2)
MANUAL TRANSMISSION (WITH OVERDRIVE) ELECTRICAL DIAGNOSIS
5.7L "Y" SERIES (PORT)

Circuit Description:

The overdrive is activated when the ECM grounds CKT 422, which energizes the overdrive relay. The ECM supplies a current limiting 12 volts on CKT 905 and, when the overdrive button is pushed, the ECM will see 0 volts and each time the button is pushed and released the overdrive will be activated or deactivated, if all the correct parameters are met to allow overdrive. The first gear switch, which is open in first gear, is used to disable overdrive while in first gear, except during a first gear deceleration. The ECM also turns "ON" the overdrive light, located in the I.P., by sending a message via the serial data line.

Test Description: Step numbers refer to step numbers on diagnostic chart.

1. The ECM turns "ON" the overdrive light by sending a message to the I.P. via the serial data line. The message is sent to the I.P. whenever the ECM energizes the overdrive relay. The instant miles per gallon and shift light information is also transmitted to the I.P. from the ECM. With a "Scan" tool plugged into the ALDL connector, the serial data information does not get transmitted to the I.P.
2. The ECM will automatically enable overdrive when all the correct parameters are met, even if the overdrive button is not pushed. The "Scan" should display "closed", when the OD button is pushed and held down (switch contacts closed).
3. With the overdrive button pushed down and held, the switch contacts should be closed which should be displayed by the "Scan" as "closed".
4. With the shifter in 1st gear the switch contacts should be open and should be displayed as a "yes" on the "Scan".
5. With the shifter in 2nd, 3rd, or 4th gear, the first gear switch contacts should be closed and should be displayed as a "no" on the 1st gear position of a "Scan" tool.

Manual transmission with automatic overdrive diagnostic chart—1987 models

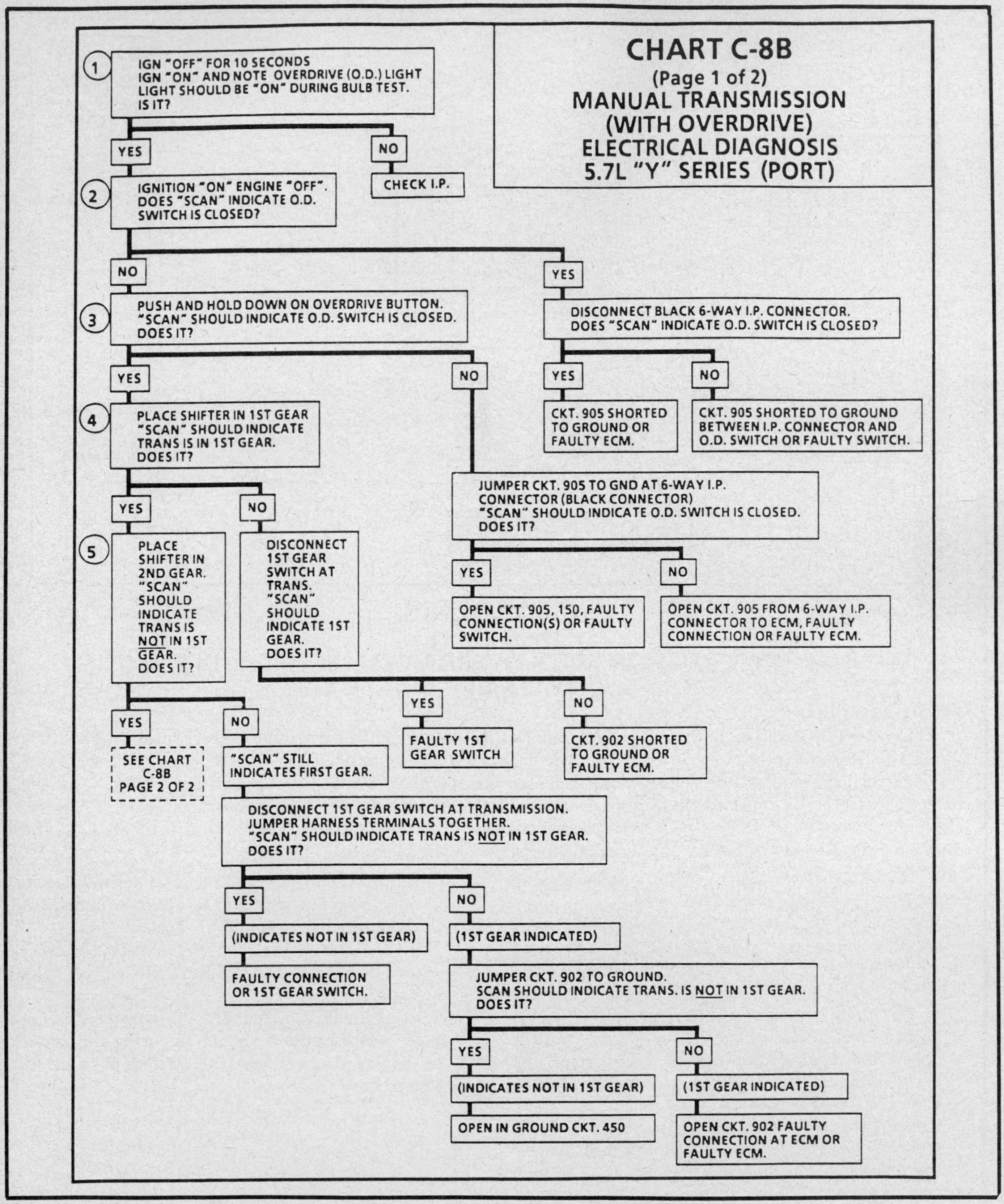
CHART C-8B
(Page 1 of 2)
MANUAL TRANSMISSION
(WITH OVERDRIVE)
ELECTRICAL DIAGNOSIS
5.7L "Y" SERIES (PORT)
1 IGN "OFF" FOR 10 SECONDS IGN "ON" AND NOTE OVERDRIVE (O.D.) LIGHT LIGHT SHOULD BE "ON" DURING BULB TEST. IS IT?
YES
NO
CHECK I.P.
2 IGNITION "ON" ENGINE "OFF". DOES "SCAN" INDICATE O.D. SWITCH IS CLOSED?
NO
YES
3 PUSH AND HOLD DOWN ON OVERDRIVE BUTTON. "SCAN" SHOULD INDICATE O.D. SWITCH IS CLOSED. DOES IT?
DISCONNECT BLACK 6-WAY I.P. CONNECTOR. DOES "SCAN" INDICATE O.D. SWITCH IS CLOSED?
YES
NO
YES
NO
CKT. 905 SHORTED TO GROUND OR FAULTY ECM.
CKT. 905 SHORTED TO GROUND BETWEEN I.P. CONNECTOR AND O.D. SWITCH OR FAULTY SWITCH.
4 PLACE SHIFTER IN 1ST GEAR "SCAN" SHOULD INDICATE TRANS IS IN 1ST GEAR. DOES IT?
JUMPER CKT. 905 TO GND AT 6-WAY I.P. CONNECTOR (BLACK CONNECTOR) "SCAN" SHOULD INDICATE O.D. SWITCH IS CLOSED. DOES IT?
YES
NO
YES
NO
5 PLACE SHIFTER IN 2ND GEAR. "SCAN" SHOULD INDICATE TRANS IS NOT IN 1ST GEAR. DOES IT?
DISCONNECT 1ST GEAR SWITCH AT TRANS. "SCAN" SHOULD INDICATE 1ST GEAR. DOES IT?
OPEN CKT. 905, 150, FAULTY CONNECTION(S) OR FAULTY SWITCH.
OPEN CKT. 905 FROM 6-WAY I.P. CONNECTOR TO ECM, FAULTY CONNECTION OR FAULTY ECM.
YES
NO
FAULTY 1ST GEAR SWITCH
CKT. 902 SHORTED TO GROUND OR FAULTY ECM.
YES
NO
SEE CHART C-8B PAGE 2 OF 2
"SCAN" STILL INDICATES FIRST GEAR.
DISCONNECT 1ST GEAR SWITCH AT TRANSMISSION. JUMPER HARNESS TERMINALS TOGETHER. "SCAN" SHOULD INDICATE TRANS IS NOT IN 1ST GEAR. DOES IT?
YES
NO
(INDICATES NOT IN 1ST GEAR)
(1ST GEAR INDICATED)
FAULTY CONNECTION OR 1ST GEAR SWITCH.
JUMPER CKT. 902 TO GROUND. SCAN SHOULD INDICATE TRANS. IS NOT IN 1ST GEAR. DOES IT?
YES
NO
(INDICATES NOT IN 1ST GEAR)
(1ST GEAR INDICATED)
OPEN IN GROUND CKT. 450
OPEN CKT. 902 FAULTY CONNECTION AT ECM OR FAULTY ECM.

Manual transmission with automatic overdrive diagnostic chart – 1987 models

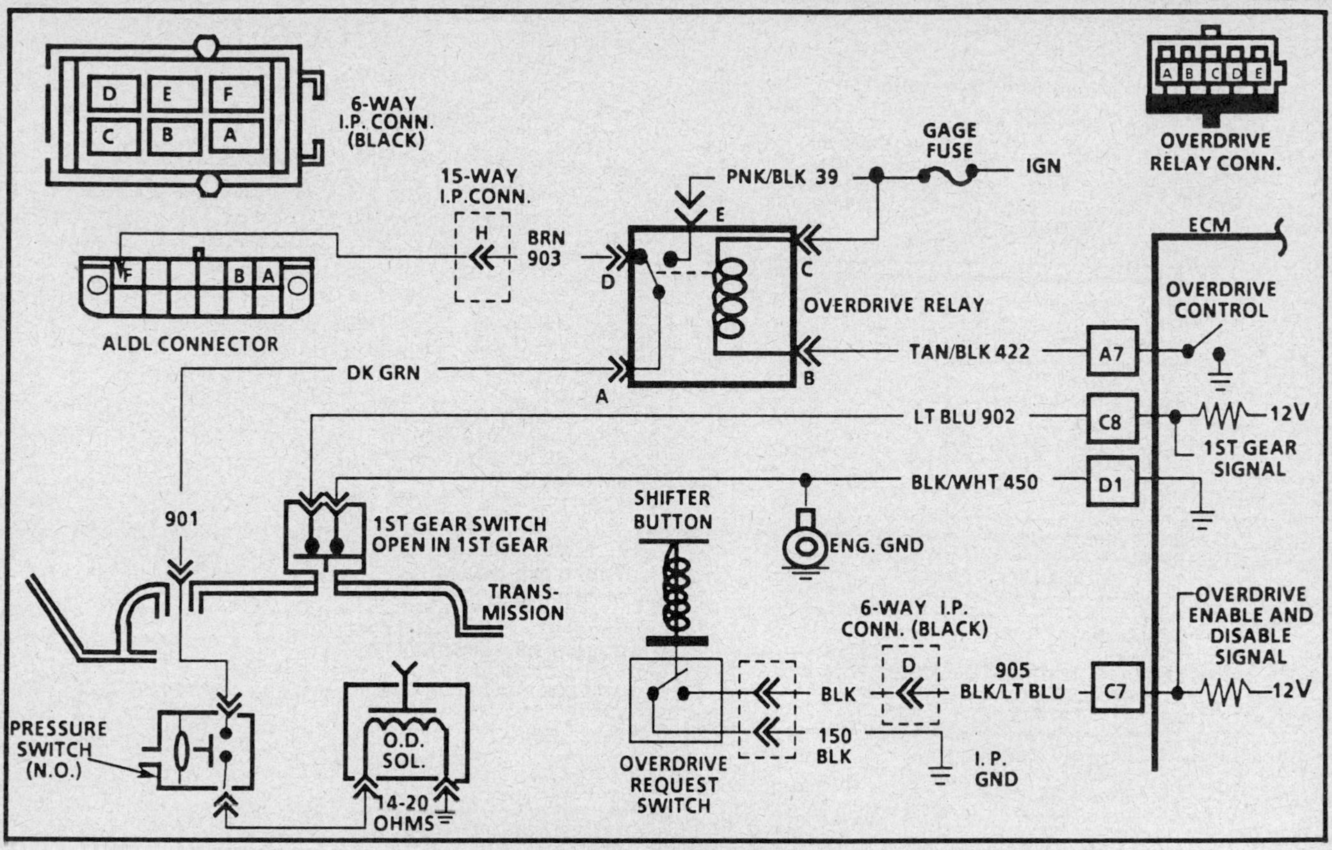
6-WAY I.P. CONN. (BLACK)
15-WAY I.P.CONN.
BRN 903
ALDL CONNECTOR
PNK/BLK 39
GAGE FUSE
IGN
OVERDRIVE RELAY CONN.
ECM
OVERDRIVE RELAY
OVERDRIVE CONTROL
DK GRN
TAN/BLK 422
A7
LT BLU 902
C8
12V
1ST GEAR SIGNAL
BLK/WHT 450
D1
901
1ST GEAR SWITCH OPEN IN 1ST GEAR
SHIFTER BUTTON
ENG. GND
TRANS-MISSION
6-WAY I.P. CONN. (BLACK)
OVERDRIVE ENABLE AND DISABLE SIGNAL
905
BLK
BLK/LT BLU
C7
12V
150 BLK
I. P. GND
PRESSURE SWITCH (N.O.)
O.D. SOL.
14-20 OHMS
OVERDRIVE REQUEST SWITCH

CHART C-8B

(Page 2 of 2)
MANUAL TRANSMISSION (WITH OVERDRIVE) ELECTRICAL DIAGNOSIS 5.7L "Y" SERIES (PORT)

Test Description: Step numbers refer to step numbers on diagnostic chart.

6. At 20 mph, the oil pressure switch should be closed, which should allow the overdrive solenoid to be energized. This step checks the entire electrical control portion of the overdrive system.
7. The overdrive electrical circuit is OK up to this point, but does not necessarily mean the overdrive is operational under a road load condition. The vehicle should be road tested, if possible, to verify that the internal portion of the overdrive assembly is OK.
8. This step checks for continuity of CKT 39 from the gage fuse.
9. With the ignition "ON" and the diagnostic terminal grounded, the overdrive control driver in the ECM should ground CKT 422. This step checks for the ability of the ECM to ground the relay control circuit.
10. This step checks the overdrive relay and continuity of CKT 901.
11. If the oil pressure switch and overdrive solenoid are both functioning properly, the oil pressure at this point should be above 70 psi (483 kPa). If the correct pressure is available at the pressure tap, but the overdrive is not engaging, there is an internal problem in the overdrive unit.
12. At 20 mph, the oil pressure siwtch should be closed and complete the circuit to the overdrive solenoid. A normal circuit should measure about 14-20 ohms, at this step.

Manual transmission with automatic overdrive diagnostic chart – 1987 models

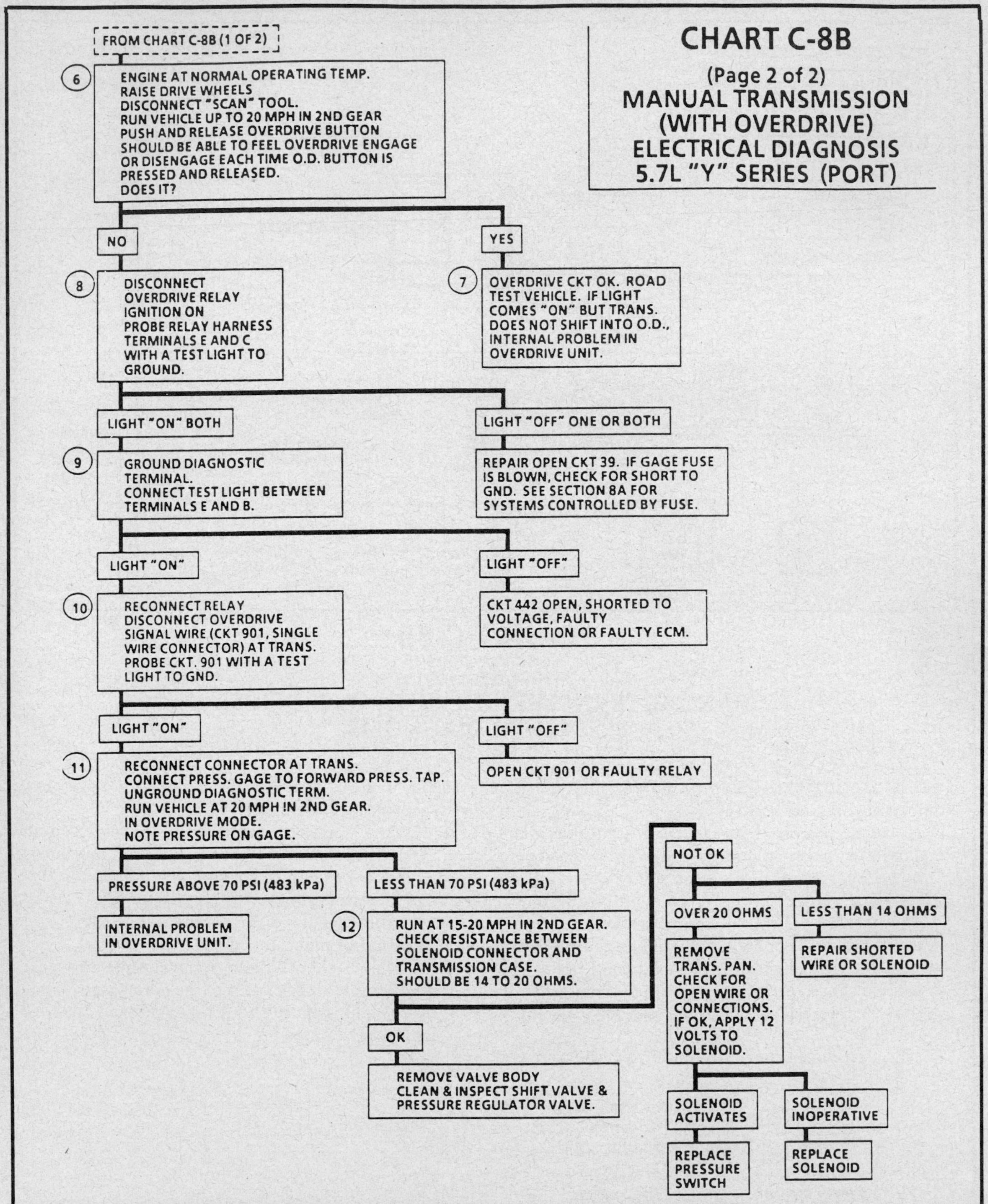
CHART C-8B
(Page 2 of 2)
MANUAL TRANSMISSION
(WITH OVERDRIVE)
ELECTRICAL DIAGNOSIS
5.7L "Y" SERIES (PORT)
FROM CHART C-8B (1 OF 2)
6
ENGINE AT NORMAL OPERATING TEMP. RAISE DRIVE WHEELS DISCONNECT "SCAN" TOOL. RUN VEHICLE UP TO 20 MPH IN 2ND GEAR PUSH AND RELEASE OVERDRIVE BUTTON SHOULD BE ABLE TO FEEL OVERDRIVE ENGAGE OR DISENGAGE EACH TIME O.D. BUTTON IS PRESSED AND RELEASED. DOES IT?
NO
YES
7
OVERDRIVE CKT OK. ROAD TEST VEHICLE. IF LIGHT COMES "ON" BUT TRANS. DOES NOT SHIFT INTO O.D., INTERNAL PROBLEM IN OVERDRIVE UNIT.
8
DISCONNECT OVERDRIVE RELAY IGNITION ON PROBE RELAY HARNESS TERMINALS E AND C WITH A TEST LIGHT TO GROUND.
LIGHT "ON" BOTH
LIGHT "OFF" ONE OR BOTH
REPAIR OPEN CKT 39. IF GAGE FUSE IS BLOWN, CHECK FOR SHORT TO GND. SEE SECTION 8A FOR SYSTEMS CONTROLLED BY FUSE.
9
GROUND DIAGNOSTIC TERMINAL. CONNECT TEST LIGHT BETWEEN TERMINALS E AND B.
LIGHT "ON"
LIGHT "OFF"
CKT 442 OPEN, SHORTED TO VOLTAGE, FAULTY CONNECTION OR FAULTY ECM.
10
RECONNECT RELAY DISCONNECT OVERDRIVE SIGNAL WIRE (CKT 901, SINGLE WIRE CONNECTOR) AT TRANS. PROBE CKT. 901 WITH A TEST LIGHT TO GND.
LIGHT "ON"
LIGHT "OFF"
OPEN CKT 901 OR FAULTY RELAY
11
RECONNECT CONNECTOR AT TRANS. CONNECT PRESS. GAGE TO FORWARD PRESS. TAP. UNGROUND DIAGNOSTIC TERM. RUN VEHICLE AT 20 MPH IN 2ND GEAR. IN OVERDRIVE MODE. NOTE PRESSURE ON GAGE.
PRESSURE ABOVE 70 PSI (483 kPa)
INTERNAL PROBLEM IN OVERDRIVE UNIT.
LESS THAN 70 PSI (483 kPa)
12
RUN AT 15-20 MPH IN 2ND GEAR. CHECK RESISTANCE BETWEEN SOLENOID CONNECTOR AND TRANSMISSION CASE. SHOULD BE 14 TO 20 OHMS.
OK
REMOVE VALVE BODY CLEAN & INSPECT SHIFT VALVE & PRESSURE REGULATOR VALVE.
NOT OK
OVER 20 OHMS
REMOVE TRANS. PAN. CHECK FOR OPEN WIRE OR CONNECTIONS. IF OK, APPLY 12 VOLTS TO SOLENOID.
LESS THAN 14 OHMS
REPAIR SHORTED WIRE OR SOLENOID
SOLENOID ACTIVATES
REPLACE PRESSURE SWITCH
SOLENOID INOPERATIVE
REPLACE SOLENOID

Manual transmission with automatic overdrive diagnostic chart—1987 models

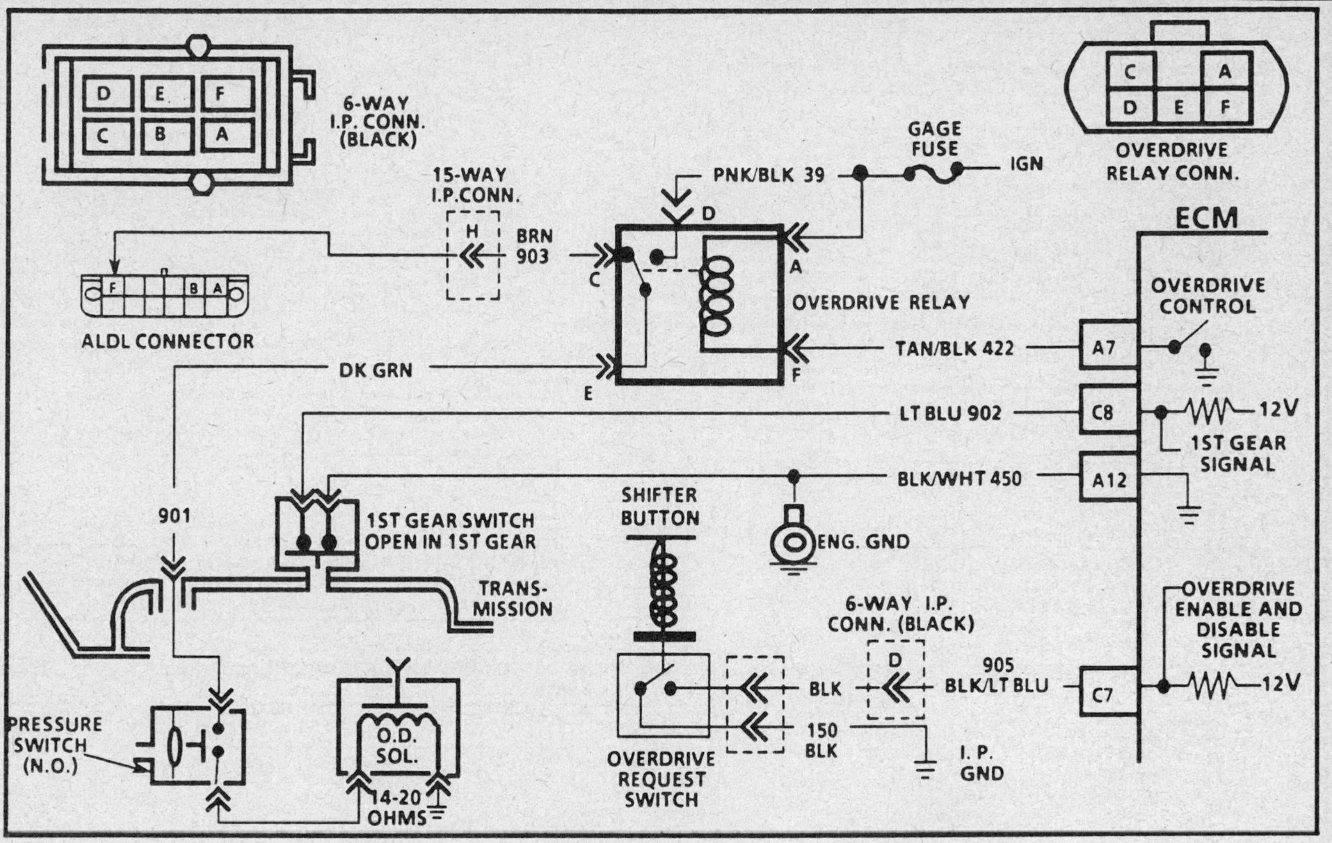
6-WAY
I.P. CONN.
(BLACK)
15-WAY
I.P.CONN.
GAGE
FUSE
IGN
PNK/BLK 39
OVERDRIVE
RELAY CONN.
BRN
903
ALDL CONNECTOR
OVERDRIVE RELAY
ECM
OVERDRIVE
CONTROL
TAN/BLK 422
A7
DK GRN
LT BLU 902
C8
12V
1ST GEAR
SIGNAL
BLK/WHT 450
A12
901
1ST GEAR SWITCH
OPEN IN 1ST GEAR
SHIFTER
BUTTON
ENG. GND
TRANS-
MISSION
6-WAY I.P.
CONN. (BLACK)
905
BLK/LT BLU
BLK
C7
OVERDRIVE
ENABLE AND
DISABLE
SIGNAL
PRESSURE
SWITCH
(N.O.)
O.D.
SOL.
14-20
OHMS
OVERDRIVE
REQUEST
SWITCH
150
BLK
I. P.
GND

CHART C-8B
(Page 1 of 2)
MANUAL TRANSMISSION (WITH OVERDRIVE)
(ELECTRICAL DIAGNOSIS)
5.7L "Y" SERIES (PORT)

Circuit Description:

The overdrive is activated when the ECM grounds CKT 422, which energizes the overdrive relay. The ECM supplies a current limiting 12 volts on CKT 905 and, when the overdrive button is pushed, the ECM will see 0 volts and each time the button is pushed and released the overdrive will be activated or deactivated, if all the correct parameters are met to allow overdrive. The first gear switch, which is open in first gear, is used to disable overdrive while in first gear, except during a first gear deceleration. The ECM also turns "ON" the overdrive light, located in the I.P., by sending a message via the serial data line.

Test Description: Numbers below refer to circled numbers on the diagnostic chart.

1. The ECM turns "ON" the overdrive light by sending a message to the I.P. via the serial data line. The message is sent to the I.P. whenever the ECM energizes the overdrive relay. The instant miles per gallon and shift light information is also transmitted to the I.P. from the ECM. With a "Scan" tool plugged into the ALDL connector, the serial data information does not get transmitted to the I.P.
2. The ECM will automatically enable overdrive when all the correct parameters are met, even if the overdrive button is not pushed. The "Scan" should display "closed", when the OD button is pushed and held down (switch contacts closed).
3. With the overdrive button pushed down and held, the switch contacts should be closed which should be displayed by the "Scan" as "closed".
4. With the shifter in 1st gear the switch contacts should be open and should be displayed as a "yes" on the "Scan".
5. With the shifter in 2nd, 3rd, or 4th gear, the first gear switch contacts should be closed and should be displayed as a "no" on the 1st gear position of a "Scan" tool.

Manual transmission with automatic overdrive diagnostic chart – 1988 models

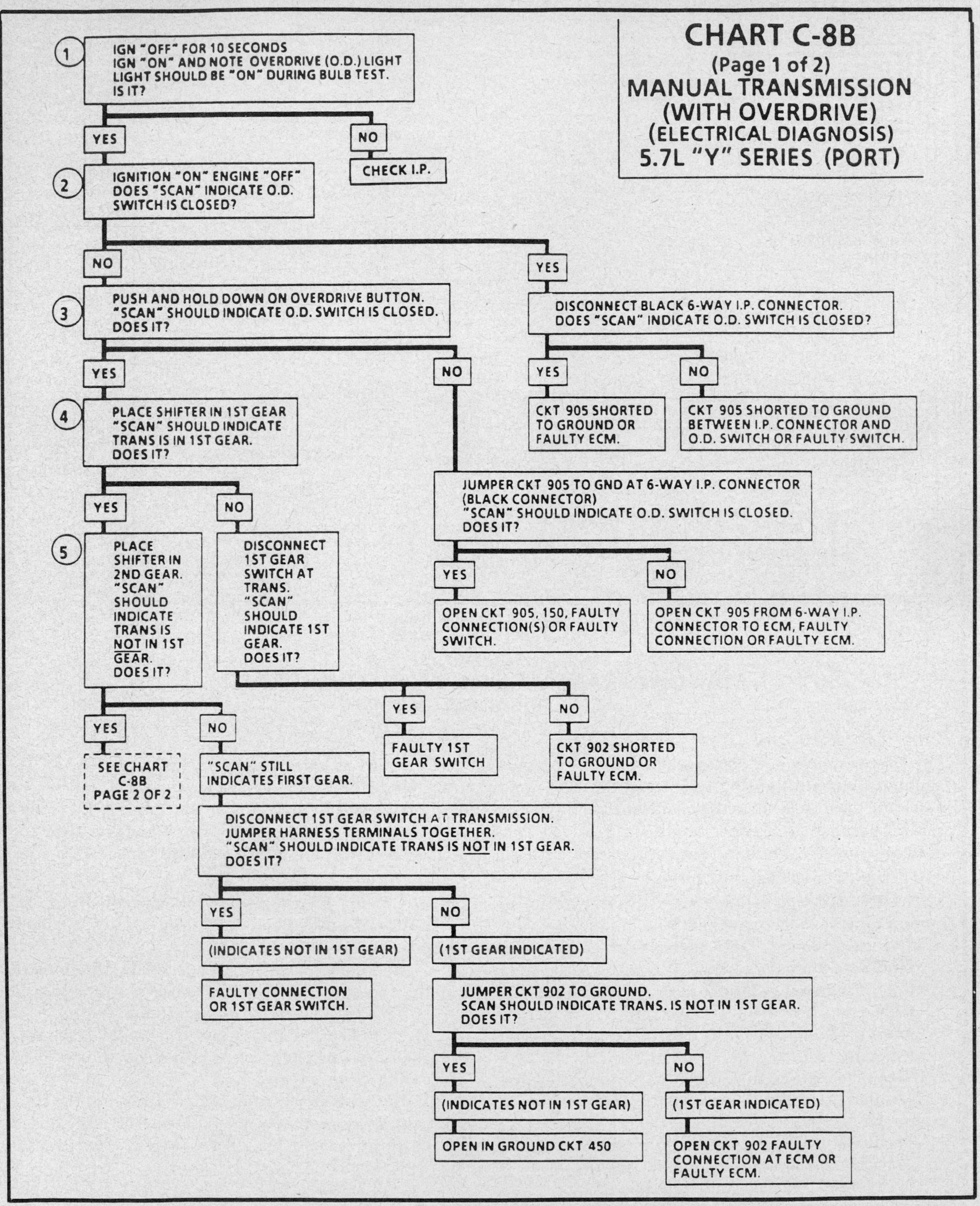
CHART C-8B
(Page 1 of 2)
MANUAL TRANSMISSION
(WITH OVERDRIVE)
(ELECTRICAL DIAGNOSIS)
5.7L "Y" SERIES (PORT)
1
IGN "OFF" FOR 10 SECONDS IGN "ON" AND NOTE OVERDRIVE (O.D.) LIGHT LIGHT SHOULD BE "ON" DURING BULB TEST. IS IT?
YES
NO
CHECK I.P.
2
IGNITION "ON" ENGINE "OFF" DOES "SCAN" INDICATE O.D. SWITCH IS CLOSED?
NO
YES
3
PUSH AND HOLD DOWN ON OVERDRIVE BUTTON. "SCAN" SHOULD INDICATE O.D. SWITCH IS CLOSED. DOES IT?
DISCONNECT BLACK 6-WAY I.P. CONNECTOR. DOES "SCAN" INDICATE O.D. SWITCH IS CLOSED?
YES
NO
YES
NO
CKT 905 SHORTED TO GROUND OR FAULTY ECM.
CKT 905 SHORTED TO GROUND BETWEEN I.P. CONNECTOR AND O.D. SWITCH OR FAULTY SWITCH.
4
PLACE SHIFTER IN 1ST GEAR "SCAN" SHOULD INDICATE TRANS IS IN 1ST GEAR. DOES IT?
JUMPER CKT 905 TO GND AT 6-WAY I.P. CONNECTOR (BLACK CONNECTOR) "SCAN" SHOULD INDICATE O.D. SWITCH IS CLOSED. DOES IT?
YES
NO
5
PLACE SHIFTER IN 2ND GEAR. "SCAN" SHOULD INDICATE TRANS IS NOT IN 1ST GEAR. DOES IT?
DISCONNECT 1ST GEAR SWITCH AT TRANS. "SCAN" SHOULD INDICATE 1ST GEAR. DOES IT?
YES
NO
OPEN CKT 905, 150, FAULTY CONNECTION(S) OR FAULTY SWITCH.
OPEN CKT 905 FROM 6-WAY I.P. CONNECTOR TO ECM, FAULTY CONNECTION OR FAULTY ECM.
YES
NO
YES
NO
SEE CHART C-8B PAGE 2 OF 2
"SCAN" STILL INDICATES FIRST GEAR.
FAULTY 1ST GEAR SWITCH
CKT 902 SHORTED TO GROUND OR FAULTY ECM.
DISCONNECT 1ST GEAR SWITCH AT TRANSMISSION. JUMPER HARNESS TERMINALS TOGETHER. "SCAN" SHOULD INDICATE TRANS IS NOT IN 1ST GEAR. DOES IT?
YES
NO
(INDICATES NOT IN 1ST GEAR)
(1ST GEAR INDICATED)
FAULTY CONNECTION OR 1ST GEAR SWITCH.
JUMPER CKT 902 TO GROUND. SCAN SHOULD INDICATE TRANS. IS NOT IN 1ST GEAR. DOES IT?
YES
NO
(INDICATES NOT IN 1ST GEAR)
(1ST GEAR INDICATED)
OPEN IN GROUND CKT 450
OPEN CKT 902 FAULTY CONNECTION AT ECM OR FAULTY ECM.

Manual transmission with automatic overdrive diagnostic chart – 1988 models

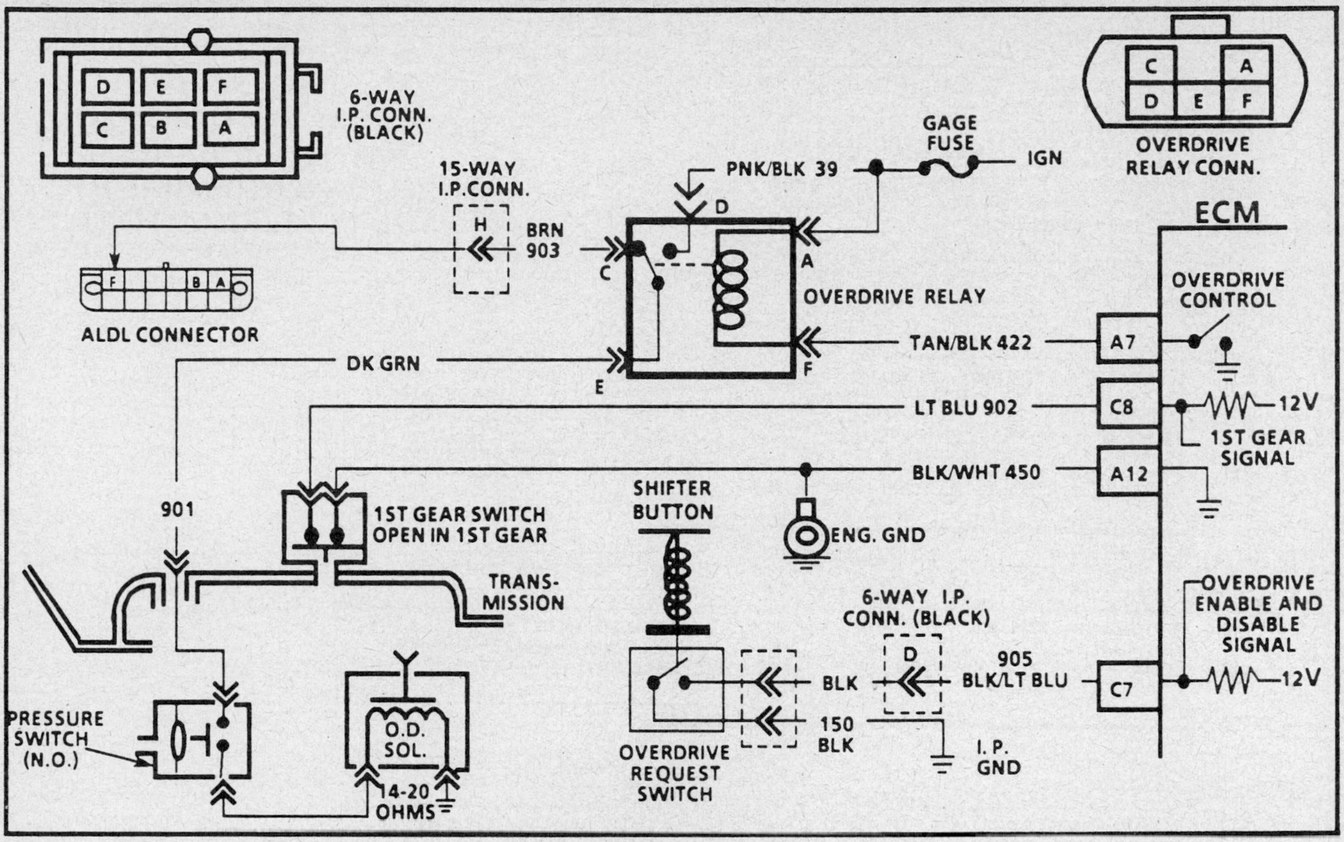
6-WAY
I.P. CONN.
(BLACK)
15-WAY
I.P. CONN.
ALDL CONNECTOR
BRN
903
GAGE
FUSE
PNK/BLK 39
IGN
OVERDRIVE
RELAY CONN.
ECM
OVERDRIVE RELAY
OVERDRIVE
CONTROL
TAN/BLK 422
DK GRN
LT BLU 902
1ST GEAR
SIGNAL
BLK/WHT 450
901
1ST GEAR SWITCH
OPEN IN 1ST GEAR
SHIFTER
BUTTON
ENG. GND
TRANS-
MISSION
6-WAY I.P.
CONN. (BLACK)
OVERDRIVE
ENABLE AND
DISABLE
SIGNAL
905
BLK
BLK/LT BLU
150
BLK
I.P.
GND
12V
PRESSURE
SWITCH
(N.O.)
O.D.
SOL.
14-20
OHMS
OVERDRIVE
REQUEST
SWITCH

CHART C-8B

(Page 2 of 2)
MANUAL TRANSMISSION (WITH OVERDRIVE)
(ELECTRICAL DIAGNOSIS)
5.7L "Y" SERIES (PORT)

Test Description: Numbers below refer to circled numbers on the diagnostic chart.

6. At 20 mph, the oil pressure switch should be closed, which should allow the overdrive solenoid to be energized. This step checks the entire electrical control portion of the overdrive system.
7. The overdrive electrical circuit is OK up to this point, but does not necessarily mean the overdrive is operational under a road load condition. The vehicle should be road tested, if possible, to verify that the internal portion of the overdrive assembly is OK.
8. This step checks for continuity of CKT 39 from the gage fuse.
9. With the ignition "ON" and the diagnostic terminal grounded, the overdrive control driver in the ECM should ground CKT 422. This step checks for the ability of the ECM to ground the relay control circuit.
10. This step checks the overdrive relay and continuity of CKT 901.
11. If the oil pressure switch and overdrive solenoid are both functioning properly, the oil pressure at this point should be above 70 psi (483 kPa). If the correct pressure is available at the pressure tap, but the overdrive is not engaging, there is an internal problem in the overdrive unit.
12. At 20 mph, the oil pressure switch should be closed and complete the circuit to the overdrive solenoid. A normal circuit should measure about 14-20 ohms, at this step.

Manual transmission with automatic overdrive diagnostic chart – 1988 models

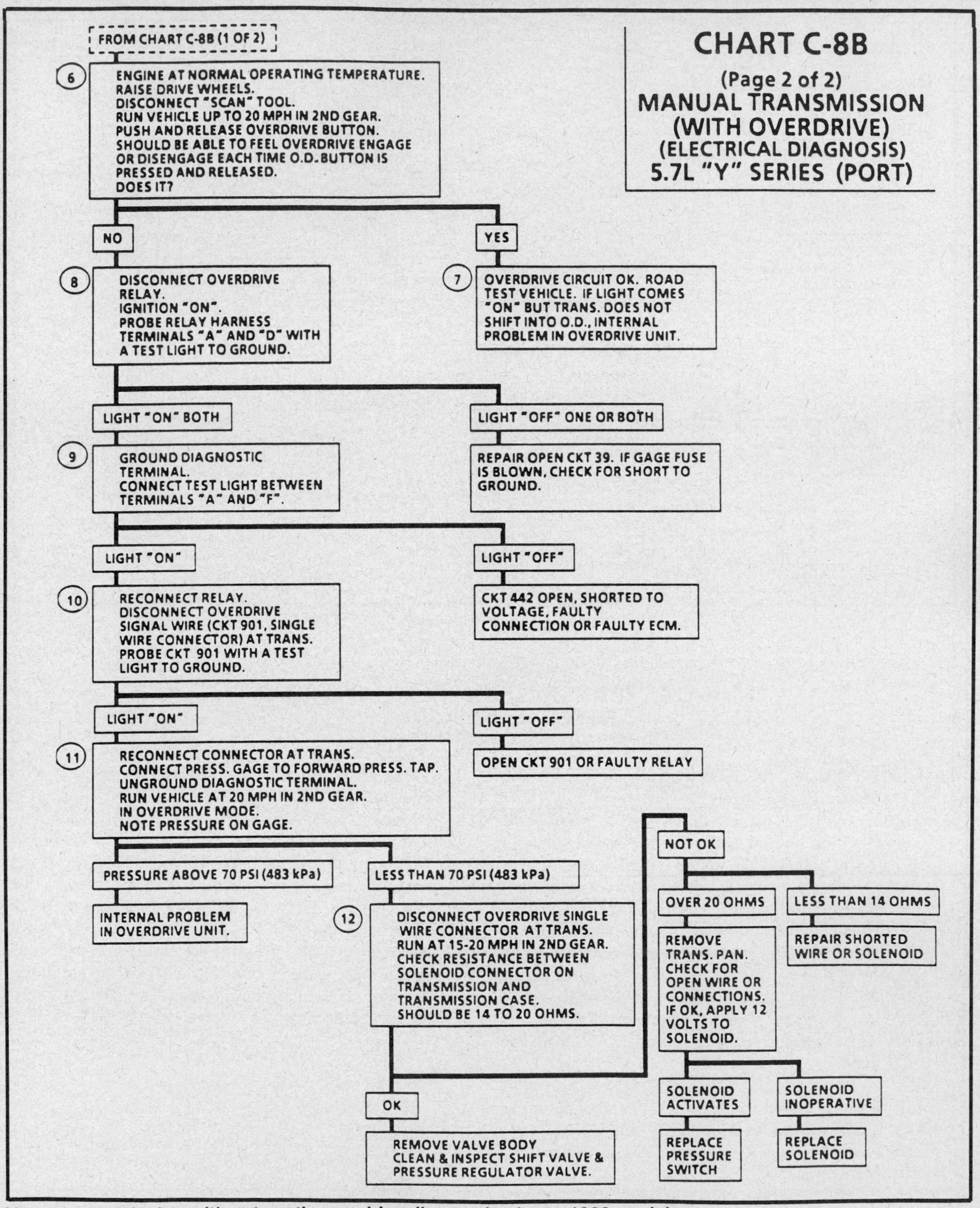
CHART C-8B
(Page 2 of 2)
MANUAL TRANSMISSION
(WITH OVERDRIVE)
(ELECTRICAL DIAGNOSIS)
5.7L "Y" SERIES (PORT)
FROM CHART C-8B (1 OF 2)
6
ENGINE AT NORMAL OPERATING TEMPERATURE. RAISE DRIVE WHEELS. DISCONNECT "SCAN" TOOL. RUN VEHICLE UP TO 20 MPH IN 2ND GEAR. PUSH AND RELEASE OVERDRIVE BUTTON. SHOULD BE ABLE TO FEEL OVERDRIVE ENGAGE OR DISENGAGE EACH TIME O.D. BUTTON IS PRESSED AND RELEASED. DOES IT?
NO
YES
8
DISCONNECT OVERDRIVE RELAY. IGNITION "ON". PROBE RELAY HARNESS TERMINALS "A" AND "D" WITH A TEST LIGHT TO GROUND.
7
OVERDRIVE CIRCUIT OK. ROAD TEST VEHICLE. IF LIGHT COMES "ON" BUT TRANS. DOES NOT SHIFT INTO O.D., INTERNAL PROBLEM IN OVERDRIVE UNIT.
LIGHT "ON" BOTH
LIGHT "OFF" ONE OR BOTH
9
GROUND DIAGNOSTIC TERMINAL. CONNECT TEST LIGHT BETWEEN TERMINALS "A" AND "F".
REPAIR OPEN CKT 39. IF GAGE FUSE IS BLOWN, CHECK FOR SHORT TO GROUND.
LIGHT "ON"
LIGHT "OFF"
10
RECONNECT RELAY. DISCONNECT OVERDRIVE SIGNAL WIRE (CKT 901, SINGLE WIRE CONNECTOR) AT TRANS. PROBE CKT 901 WITH A TEST LIGHT TO GROUND.
CKT 442 OPEN, SHORTED TO VOLTAGE, FAULTY CONNECTION OR FAULTY ECM.
LIGHT "ON"
LIGHT "OFF"
OPEN CKT 901 OR FAULTY RELAY
11
RECONNECT CONNECTOR AT TRANS. CONNECT PRESS. GAGE TO FORWARD PRESS. TAP. UNGROUND DIAGNOSTIC TERMINAL. RUN VEHICLE AT 20 MPH IN 2ND GEAR. IN OVERDRIVE MODE. NOTE PRESSURE ON GAGE.
PRESSURE ABOVE 70 PSI (483 kPa)
LESS THAN 70 PSI (483 kPa)
INTERNAL PROBLEM IN OVERDRIVE UNIT.
12
DISCONNECT OVERDRIVE SINGLE WIRE CONNECTOR AT TRANS. RUN AT 15-20 MPH IN 2ND GEAR. CHECK RESISTANCE BETWEEN SOLENOID CONNECTOR ON TRANSMISSION AND TRANSMISSION CASE. SHOULD BE 14 TO 20 OHMS.
NOT OK
OVER 20 OHMS
LESS THAN 14 OHMS
REMOVE TRANS. PAN. CHECK FOR OPEN WIRE OR CONNECTIONS. IF OK, APPLY 12 VOLTS TO SOLENOID.
REPAIR SHORTED WIRE OR SOLENOID
SOLENOID ACTIVATES
SOLENOID INOPERATIVE
OK
REMOVE VALVE BODY CLEAN & INSPECT SHIFT VALVE & PRESSURE REGULATOR VALVE.
REPLACE PRESSURE SWITCH
REPLACE SOLENOID

Manual transmission with automatic overdrive diagnostic chart – 1988 models

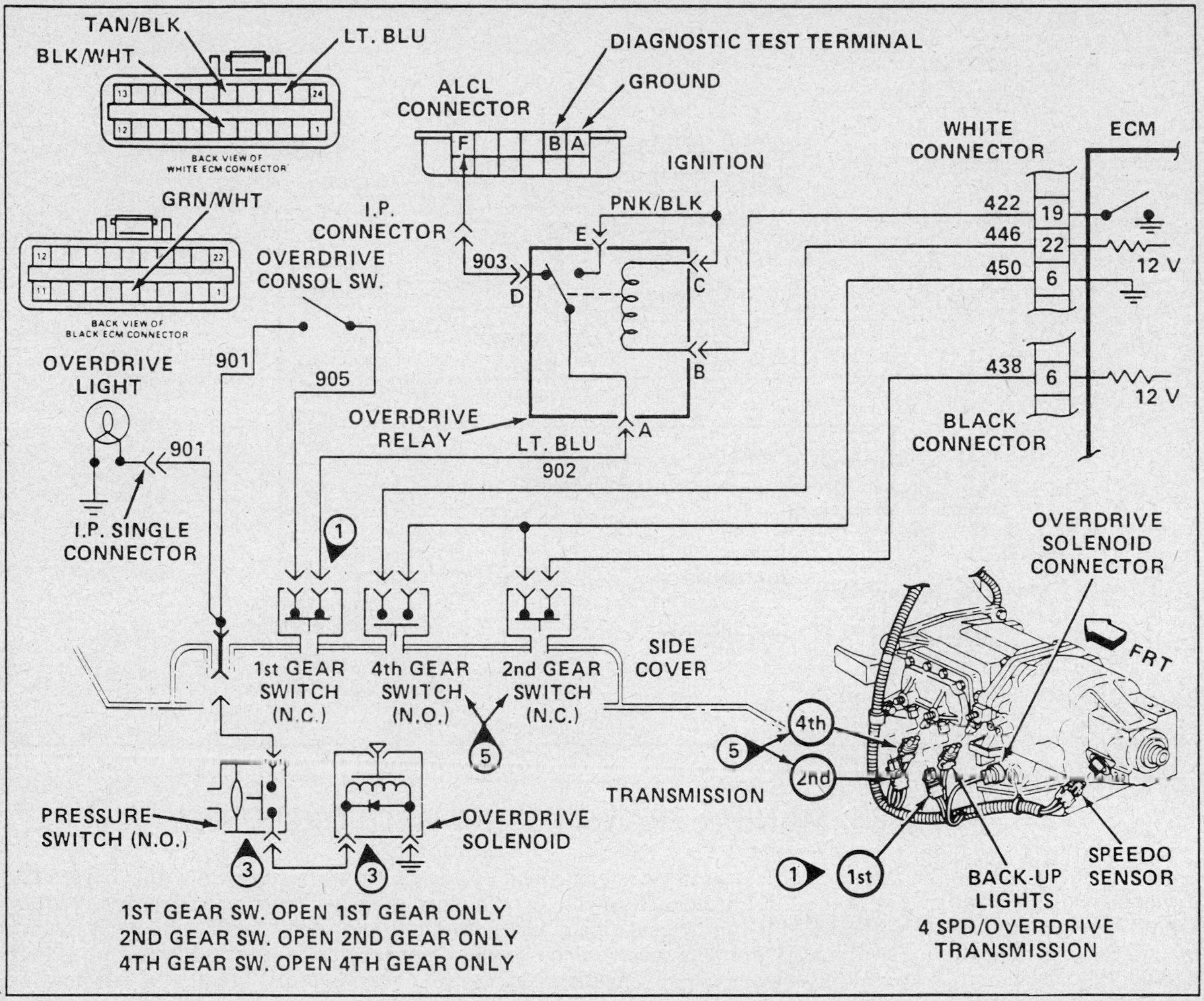
TAN/BLK
BLK/WHT
LT. BLU
BACK VIEW OF WHITE ECM CONNECTOR
GRN/WHT
BACK VIEW OF BLACK ECM CONNECTOR
DIAGNOSTIC TEST TERMINAL
GROUND
ALCL CONNECTOR
F B A
IGNITION
I.P. CONNECTOR
OVERDRIVE CONSOL SW.
PNK/BLK
E
903
D
C
B
A
WHITE CONNECTOR
ECM
422
446
450
19
22
6
12 V
438
6
12 V
BLACK CONNECTOR
OVERDRIVE LIGHT
901
905
OVERDRIVE RELAY
LT. BLU
902
901
I.P. SINGLE CONNECTOR
1
OVERDRIVE SOLENOID CONNECTOR
SIDE COVER
FRT
1st GEAR SWITCH (N.C.)
4th GEAR SWITCH (N.O.)
2nd GEAR SWITCH (N.C.)
5
4th
2nd
TRANSMISSION
PRESSURE SWITCH (N.O.)
OVERDRIVE SOLENOID
3
3
1
1st
BACK-UP LIGHTS
SPEEDO SENSOR
4 SPD/OVERDRIVE TRANSMISSION
1ST GEAR SW. OPEN 1ST GEAR ONLY
2ND GEAR SW. OPEN 2ND GEAR ONLY
4TH GEAR SW. OPEN 4TH GEAR ONLY

CHART C-8B, MANUAL TRANSMISSION WITH AUTOMATIC OVERDRIVE

7. Checks for ignition feed to relay.
8. Checks for ECM control of relay.
9. Checks for open first gear switch or faulty lamp circuit.

Manual transmission with automatic overdrive diagnostic chart—1984 models

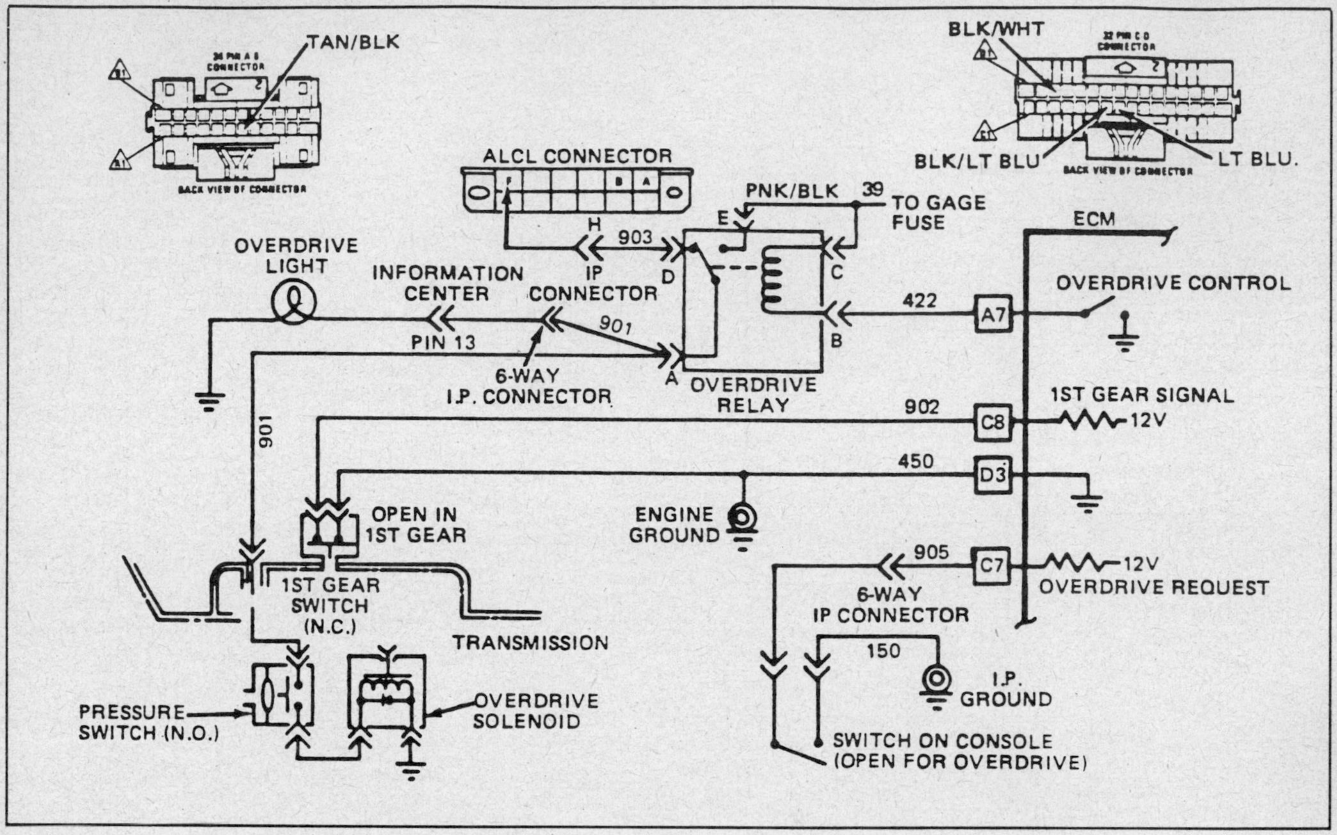
TAN/BLK
BACK VIEW OF CONNECTOR
BLK/WHT
BLK/LT BLU
LT BLU.
ALCL CONNECTOR
PNK/BLK
39
TO GAGE
FUSE
ECM
H
903
E
OVERDRIVE
LIGHT
INFORMATION
CENTER
IP
CONNECTOR
D
C
OVERDRIVE CONTROL
422
A7
901
PIN 13
B
6-WAY
I.P. CONNECTOR
A
OVERDRIVE
RELAY
902
1ST GEAR SIGNAL
C8
12V
901
450
D3
OPEN IN
1ST GEAR
ENGINE
GROUND
905
C7
12V
OVERDRIVE REQUEST
1ST GEAR
SWITCH
(N.C.)
6-WAY
IP CONNECTOR
TRANSMISSION
150
I.P.
GROUND
PRESSURE
SWITCH (N.O.)
OVERDRIVE
SOLENOID
SWITCH ON CONSOLE
(OPEN FOR OVERDRIVE)

CHART C-8A, MANUAL TRANSMISSION WITH OVERDRIVE

The overdrive used in the manual transmission is controlled by the ECM when selected by the driver. The driver controls the overdrive with the console mounted switch, which when open, will enable the overdrive request input. Another input used by the ECM is the 1st gear signal which does not allow the overdrive to engage while in 1st gear. The oil pressure switch also prohibits the overdrive solenoid engagement until the oil pressure is high enough (about 15 mph 24 km/h). As long as the transmission is not in first gear, overdrive request is selected and the oil pressure switch is closed. The overdrive light and solenoid should be energized.

The overdrive relay will be energized by the ECM while all conditions are met and will only be de-energized during a 4th gear WOT. The ECM uses RPM, road speed and the MAF sensor signal to determine that the vehicle is in 4th gear. Then as the TPS signal indicates WOT, the ECM will de-energize the overdrive relay. This should only occur while in 4th gear as the correct parameters to de-energize the relay are only met in 4th gear.

1. If the overdrive light is "on" with the ignition turned "on", then the relay is energized or CKT 901 is shorted to voltage. If the relay is energized, the test light should not light at this point.
2. Grounding the diagnostic terminal with ignition "on" should energize the relay and turn "on" the "overdrive" light.
3. Connecting 12 volts to CKT 901 will check the continuity of the bulb circuit.
4. The 1st gear switch input (open) should not allow the overdrive to be energized.
5. With the vehicle in 2nd gear and above 15 mph, the overdrive light should be "on". An RPM drop should be noticed when the oil pressure switch closes (about 15 mph) which completes the circuit to the overdrive solenoid.

Manual transmission with automatic overdrive diagnostic chart—1985 models

CHART C-8A, MANUAL TRANSMISSION WITH AUTOMATIC OVERDRIVE

The following symptoms will reduce diagnosis time by starting at a point within the chart close to the component which is the most likely to be at fault. Symptoms not listed will require use of the complete chart.

No overdrive but overdrive light comes "ON". Basic internal transmission problem. Begin at test 3 to check:

- Overdrive oil pressure switch or solenoid.

No overdrive, and overdrive light will not come on. Begin at test 7 to check:

- Relay control circuit
- First or second gear switch circuit

Third gear overdrive shift pattern same as fourth gear. Begin at test 4 to check:

- Fourth gear switch

Fourth gear overdrive shift pattern same as third gear. Begin at test 4 to check:

- Fourth gear switch

No overdrive in any range after shifting to direct. Begin at test 4 to check:

- Fourth gear switch

No direct drive any gear except first. Overdrive light "ON". Begin at test 7 to check:

- Relay control circuit

Overdrive and overdrive light "ON" in first gear above 12 MPH on light throttle. Begin at test 1 to check:

- First gear switch

1. Checks for first gear switch stuck closed in first gear.
3. Checks overdrive oil pressure switch. Resistance should be high (open switch) with wheels stopped. Resistance will be low (14 to 20 ohms) when switch closes above 15 MPH vehicle speed. If OK and the symptoms were "No overdrive but overdrive light operates OK", there is a basic internal transmission problem.
4. and 5. Checks second and fourth gear switches.

Manual transmission with automatic overdrive diagnostic chart – 1984 models

POWER FLOW

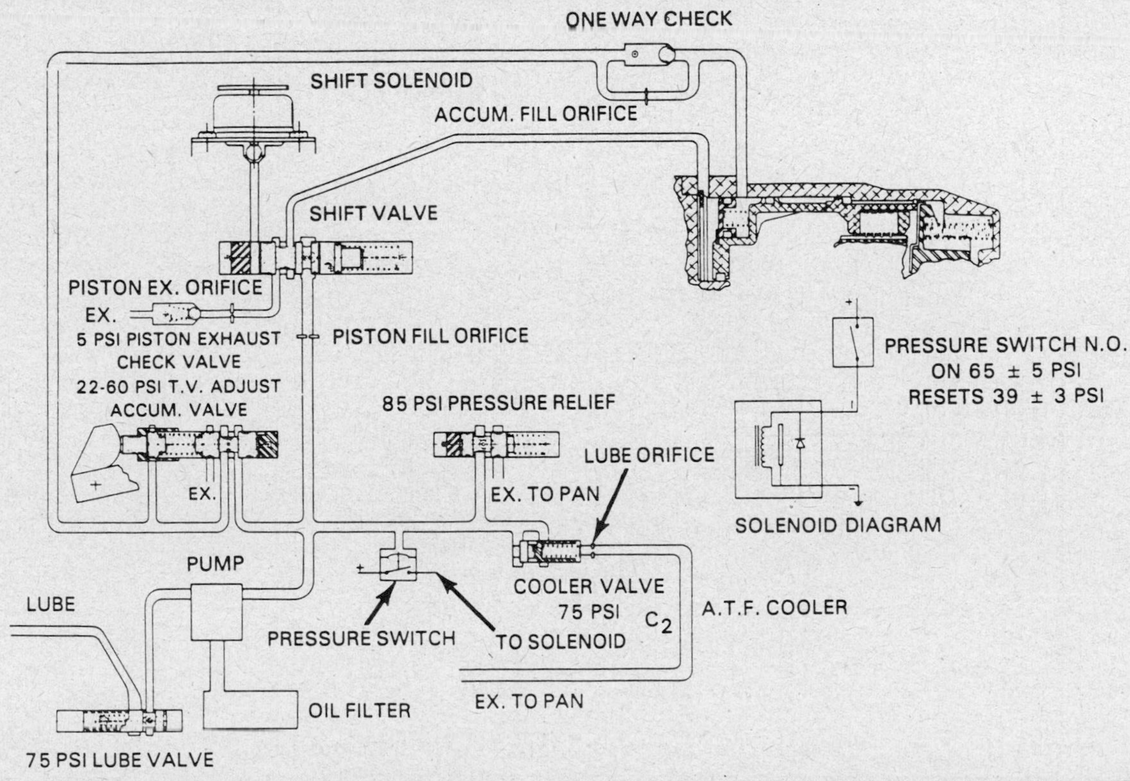
ONE WAY CHECK
SHIFT SOLENOID
ACCUM. FILL ORIFICE
SHIFT VALVE
PISTON EX. ORIFICE
EX.
5 PSI PISTON EXHAUST CHECK VALVE
PISTON FILL ORIFICE
22-60 PSI T.V. ADJUST ACCUM. VALVE
85 PSI PRESSURE RELIEF
PRESSURE SWITCH N.O.
ON 65 ± 5 PSI
RESETS 39 ± 3 PSI
LUBE ORIFICE
EX.
EX. TO PAN
SOLENOID DIAGRAM
PUMP
LUBE
COOLER VALVE
75 PSI
C2
A.T.F. COOLER
PRESSURE SWITCH
TO SOLENOID
OIL FILTER
EX. TO PAN
75 PSI LUBE VALVE

Overdrive unit hydraulic diagram – 1984–85 models

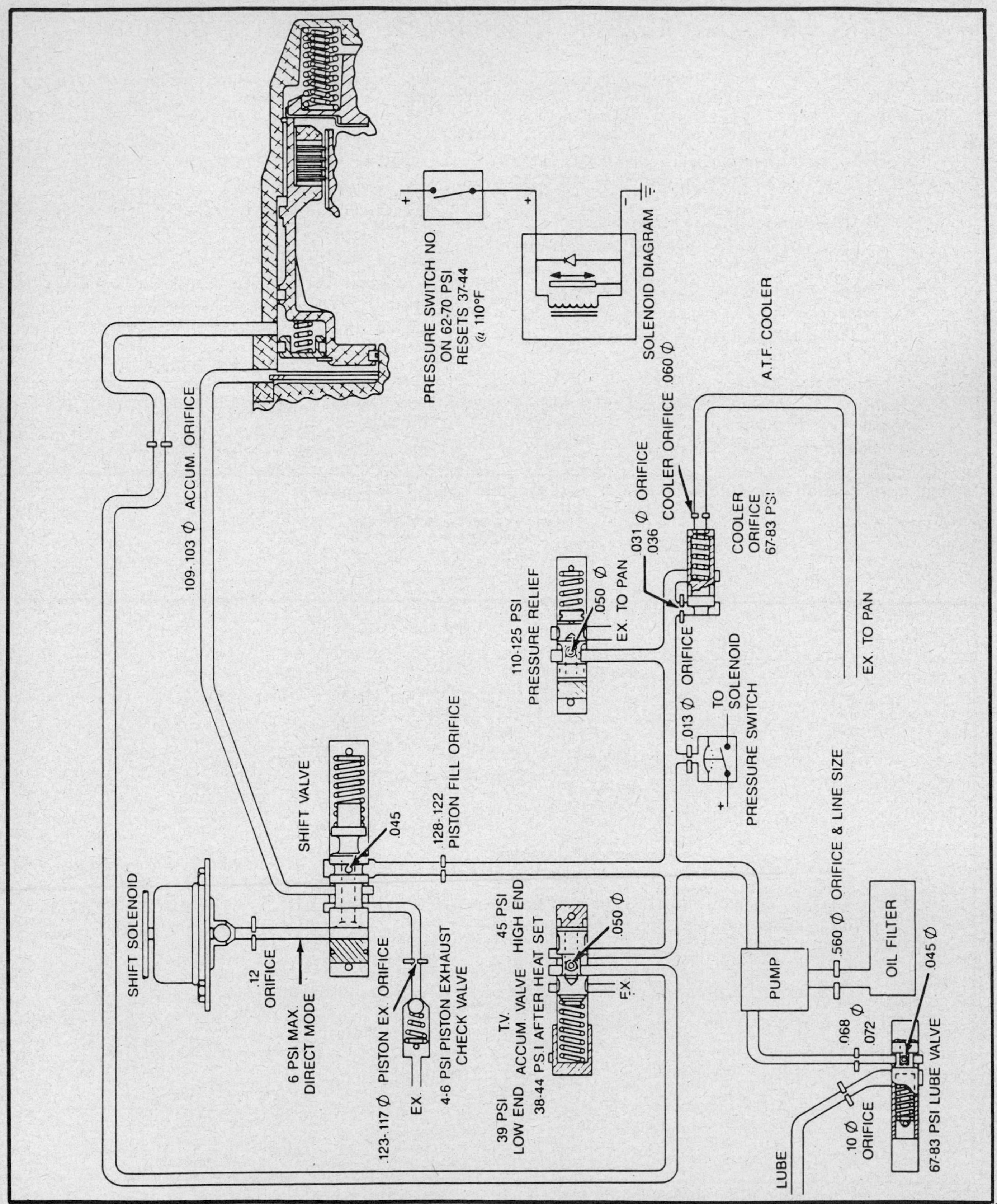
PRESSURE SWITCH NO.
ON 62-70 PSI
RESETS 37-44
@ 110°F
SOLENOID DIAGRAM
A.T.F. COOLER
109-.103 Ø ACCUM. ORIFICE
COOLER ORIFICE .060 Ø
.031 Ø ORIFICE
.036
COOLER
ORIFICE
67-83 PSI
110-125 PSI
PRESSURE RELIEF
.050 Ø
EX. TO PAN
.013 Ø ORIFICE
TO
SOLENOID
PRESSURE SWITCH
EX. TO PAN
SHIFT VALVE
.045
128-.122
PISTON FILL ORIFICE
SHIFT SOLENOID
.12
ORIFICE
6 PSI MAX.
DIRECT MODE
.123-.117 Ø PISTON EX. ORIFICE
EX.
4-6 PSI PISTON EXHAUST
CHECK VALVE
39 PSI
LOW END
T.V.
ACCUM. VALVE
38-44 PS.I AFTER HEAT SET
45 PSI
HIGH END
.050 Ø
EX.
.560 Ø ORIFICE & LINE SIZE
PUMP
OIL FILTER
.045 Ø
.068 Ø
.072
67-83 PSI LUBE VALVE
.10 Ø
ORIFICE
LUBE

Overdrive unit hydraulic diagram – 1986–88 models

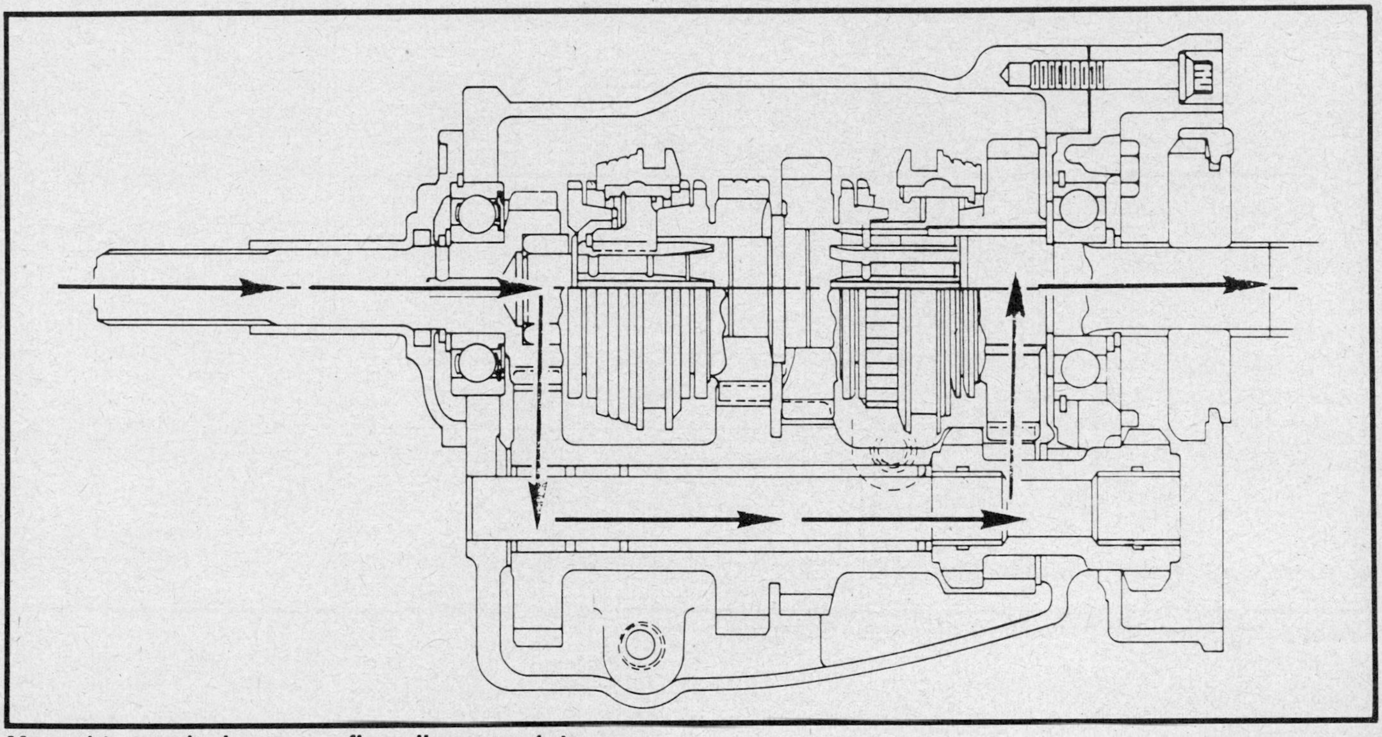

Manual transmission powerflow diagram – 1st gear

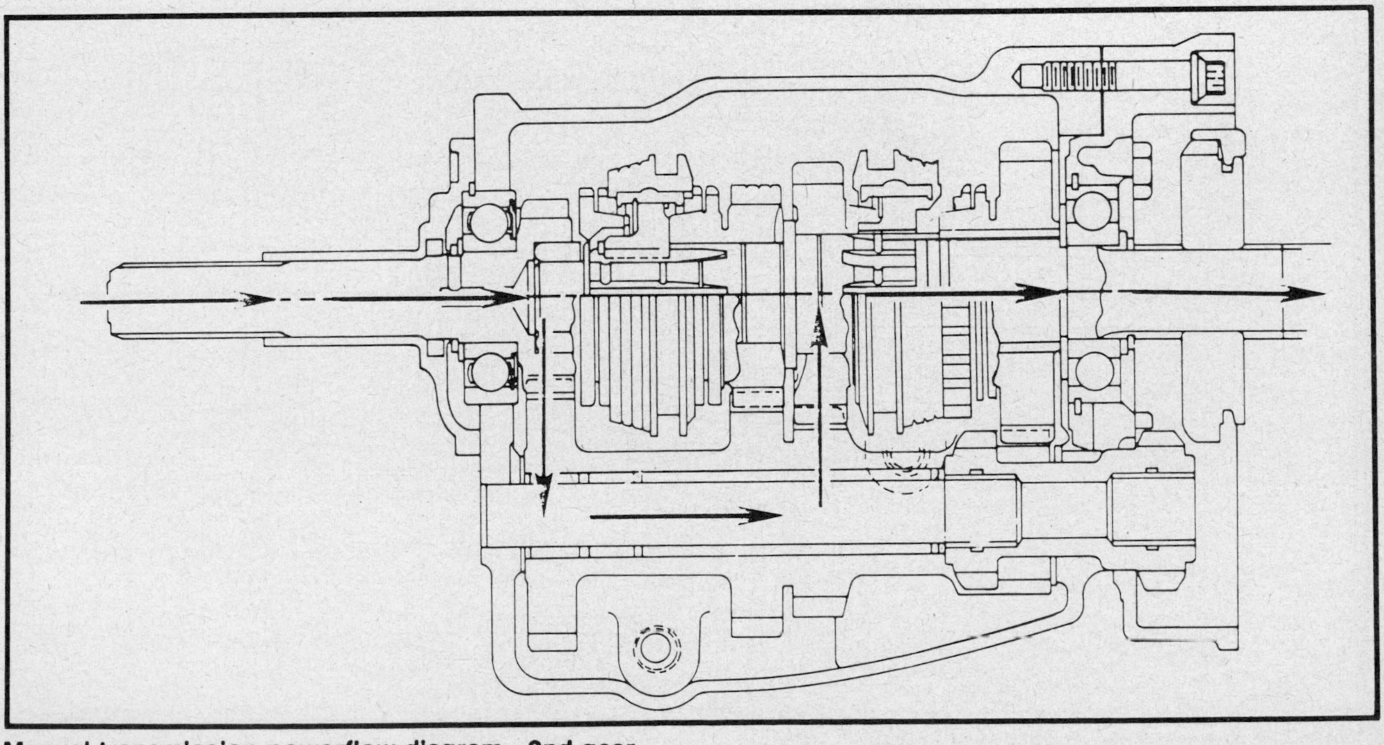

Manual transmission powerflow diagram – 2nd gear

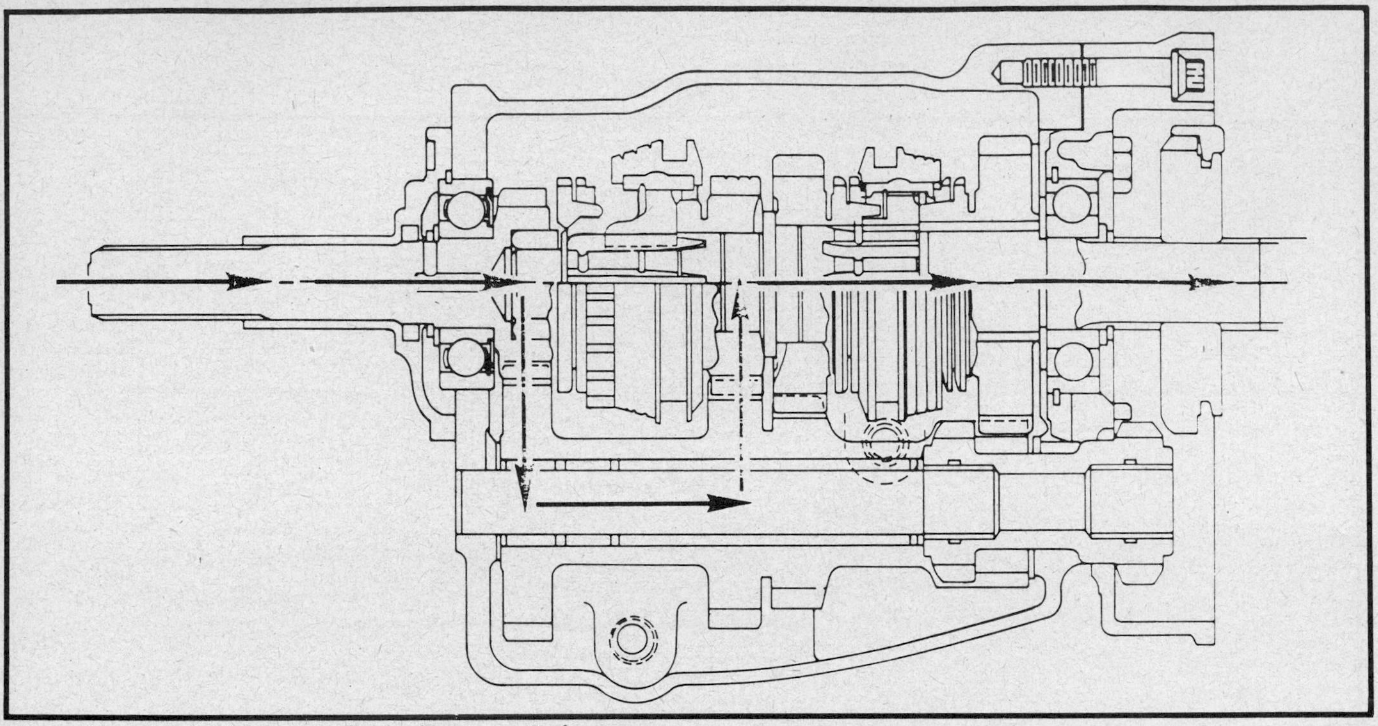

Manual transmission powerflow diagram – 3rd gear

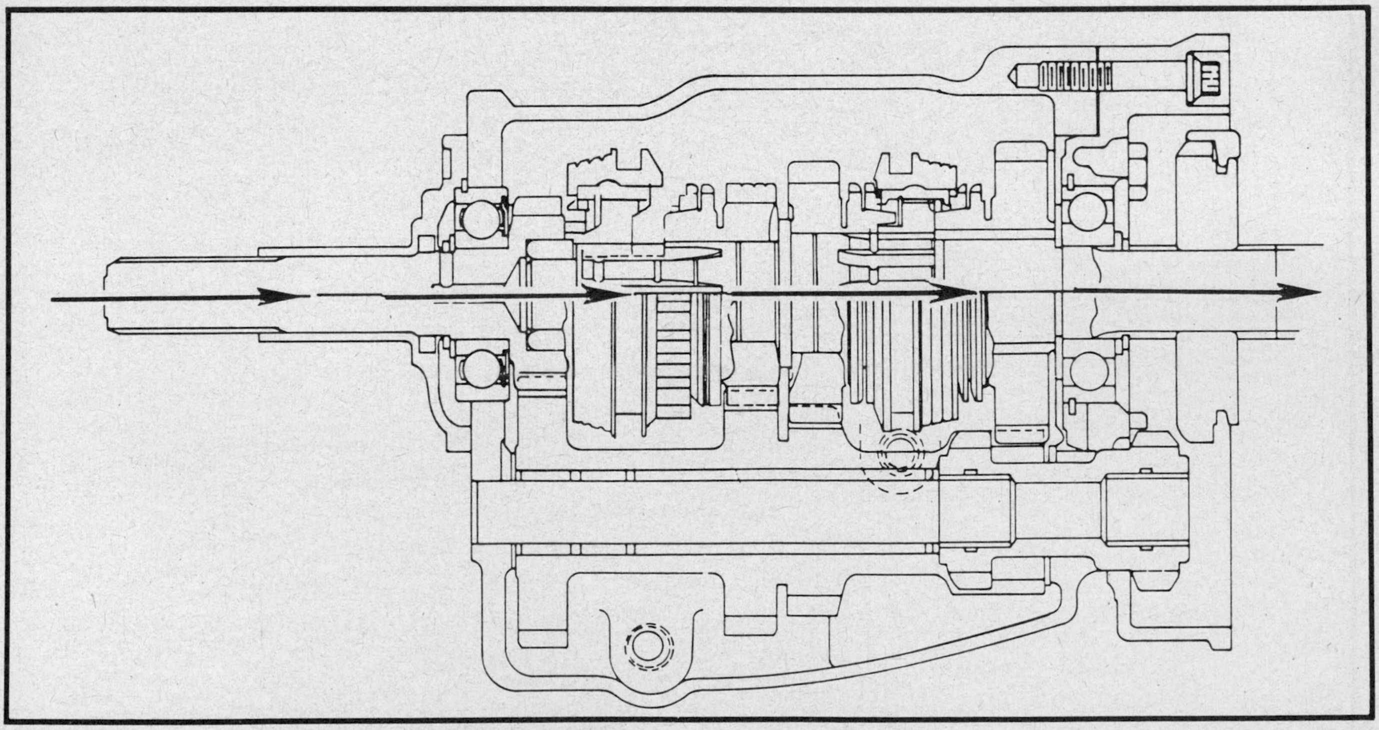

Manual transmission powerflow diagram – 4th gear

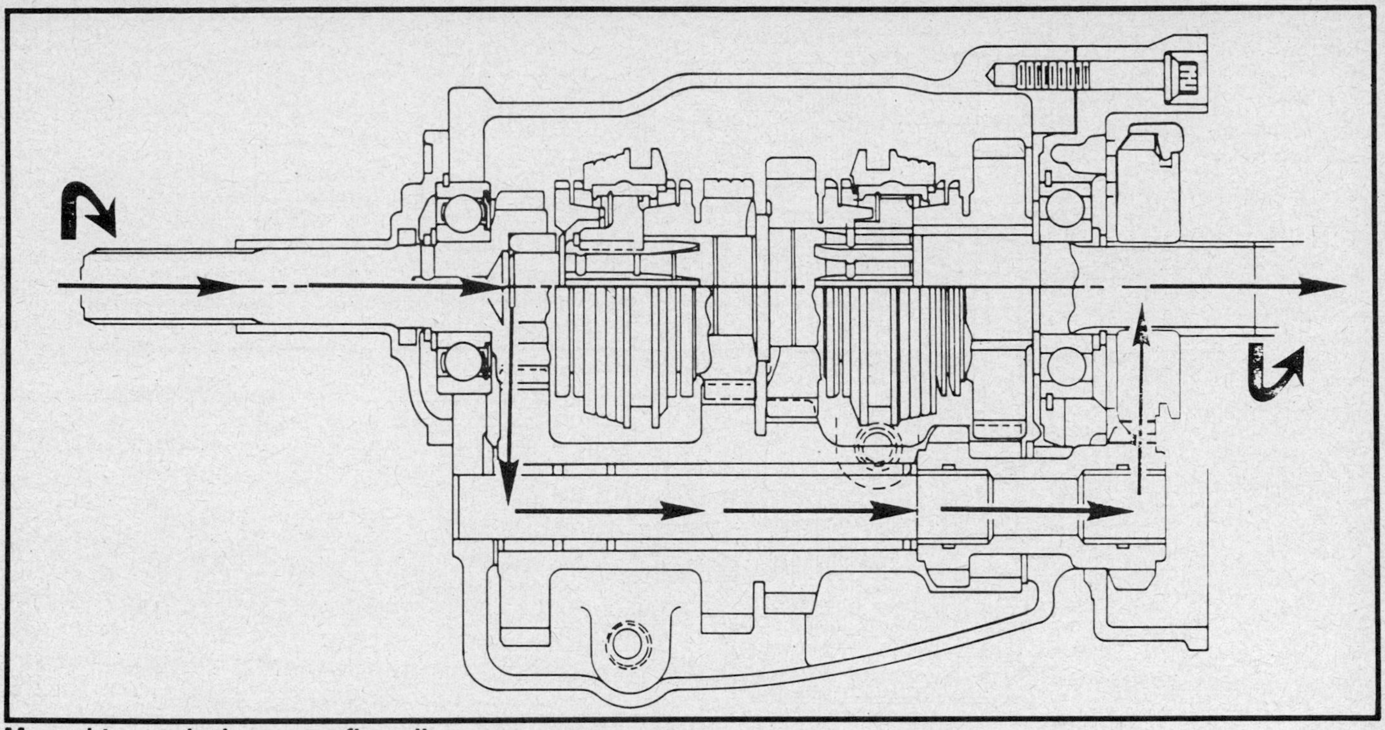

Manual transmission powerflow diagram—reverse gear

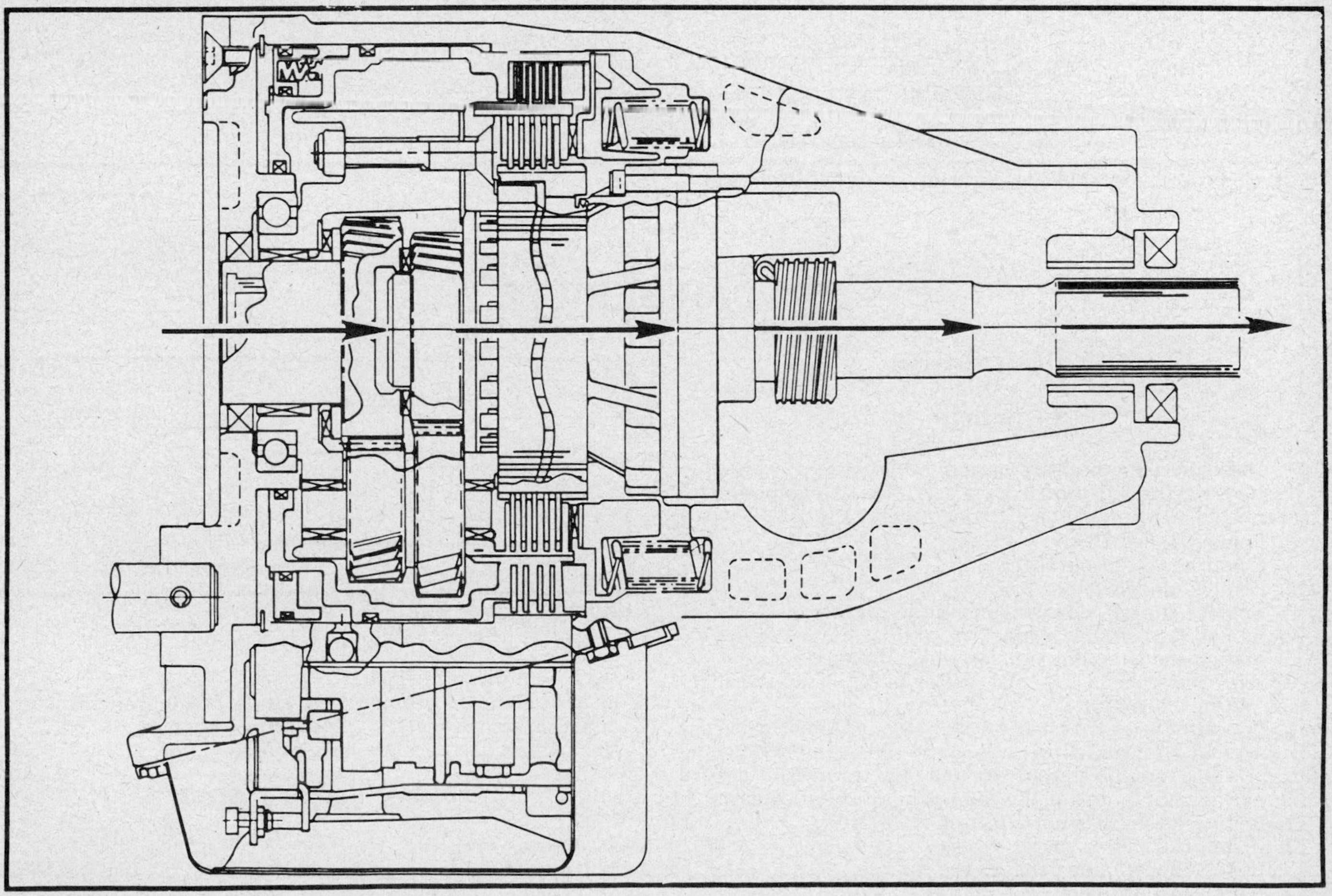

Overdrive unit powerflow diagram—direct mode

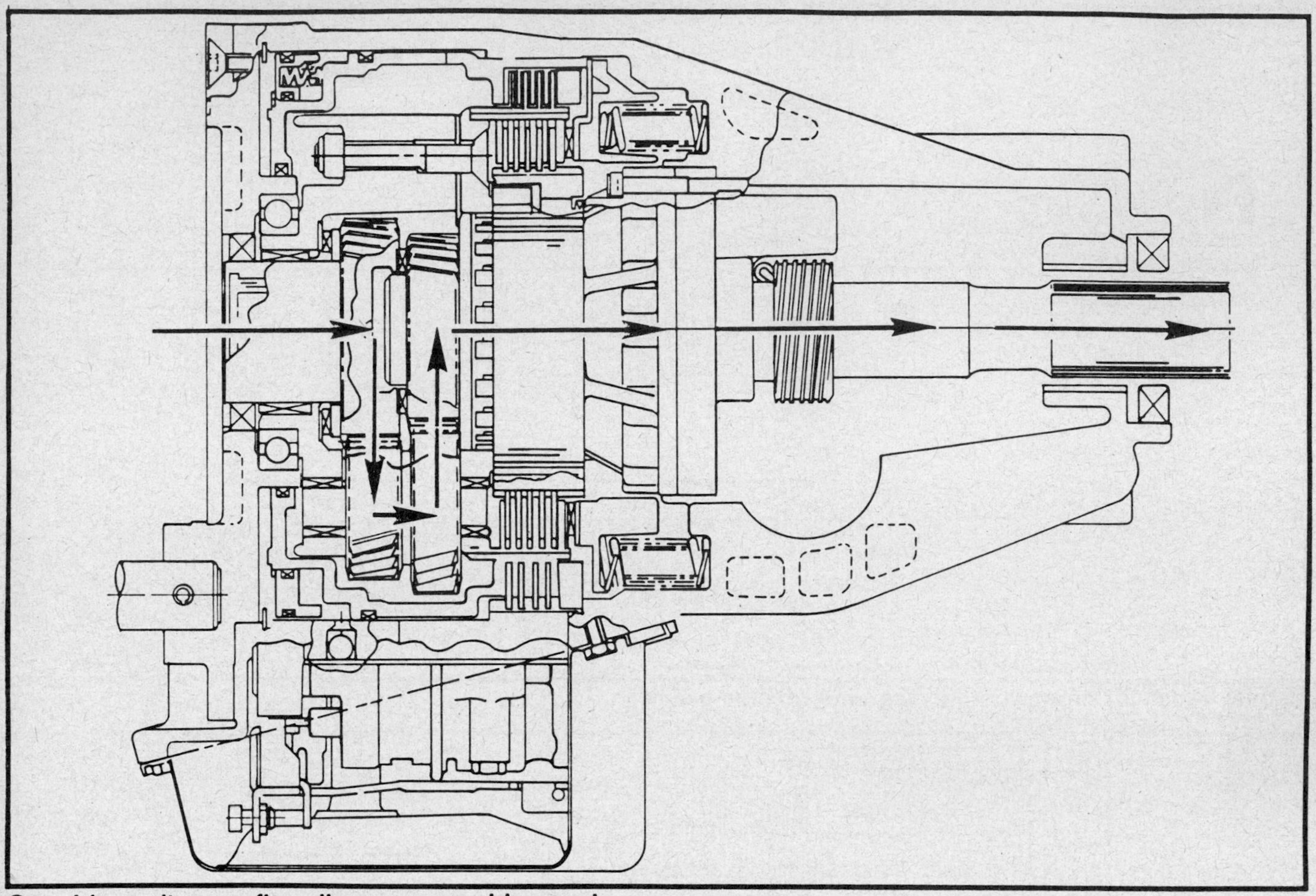

Overdrive unit powerflow diagram – overdrive mode

ON CAR SERVICE

Adjustments

LINKAGE

1. Disconnect the negative battery cable.
2. Remove the left seat from the vehicle. If equipped with power seats, disconnect the electrical leads.
3. Remove the shift knob.
4. Remove the console cover.
5. Remove the glove box lock.
6. Remove the left side panel from the console.
7. Remove the shifter cover.
8. Loosen the adjusting nuts on the shifter rods.
9. With the transmission and shifter in **N**, install the alignment pin in the shifter.
10. Equalize the swivels on all 3 shift levers. Hand tighten the forward and rear adjusting nuts at the same time with equal force. Do this for all 3 shifter rods and then torque the forward and rear adjusting nuts at the same time to specifications.
11. Remove the alignment pin from the shifter.
12. Install the shifter cover.
13. Install the left side panel to the console.
14. Install the glove box lock.
15. Install the console cover.

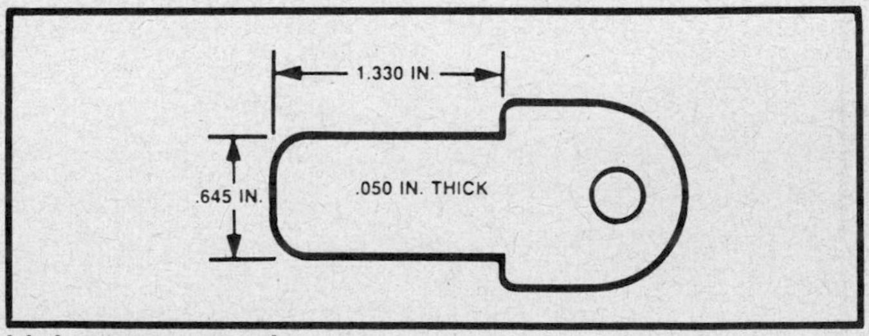
1.330 IN.
.645 IN.
.050 IN. THICK

Linkage gauge pin

16. Install the shift knob.
17. Install the left seat into the vehicle. If equipped with power seats, connect the electrical connector leads.
18. Connect the negative battery cable.

PARK LOCK CABLE

1. Lift up on the adjusting key to release the cable.
2. Place the steering column lock lever in the **LOCK PARK** position.
3. Shift the transmission into reverse gear.

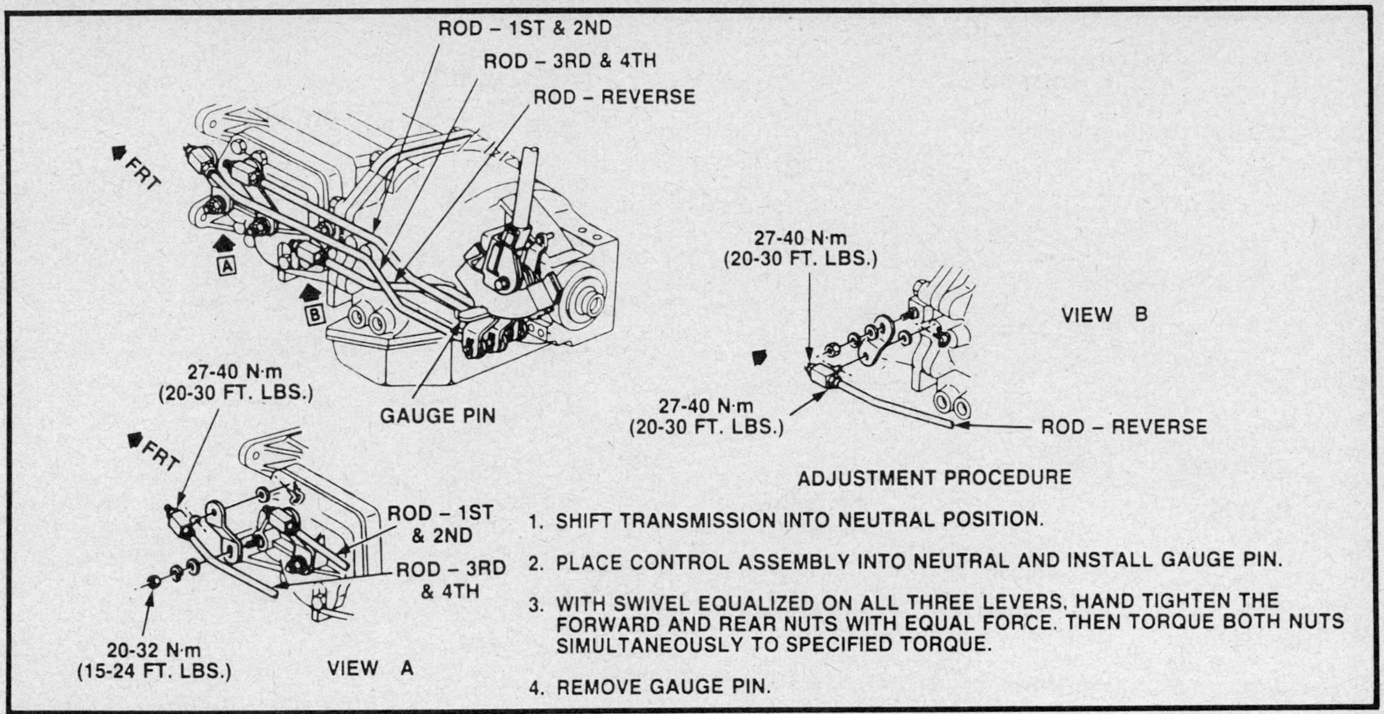
ROD – 1ST & 2ND
ROD – 3RD & 4TH
ROD – REVERSE
FRT
A
B
27-40 N·m
(20-30 FT. LBS.)
VIEW B
27-40 N·m
(20-30 FT. LBS.)
GAUGE PIN
27-40 N·m
(20-30 FT. LBS.)
ROD – REVERSE
FRT
ROD – 1ST & 2ND
ROD – 3RD & 4TH
20-32 N·m
(15-24 FT. LBS.)
VIEW A
ADJUSTMENT PROCEDURE
1. SHIFT TRANSMISSION INTO NEUTRAL POSITION.
2. PLACE CONTROL ASSEMBLY INTO NEUTRAL AND INSTALL GAUGE PIN.
3. WITH SWIVEL EQUALIZED ON ALL THREE LEVERS, HAND TIGHTEN THE FORWARD AND REAR NUTS WITH EQUAL FORCE. THEN TORQUE BOTH NUTS SIMULTANEOUSLY TO SPECIFIED TORQUE.
4. REMOVE GAUGE PIN.

Linkage adjustment

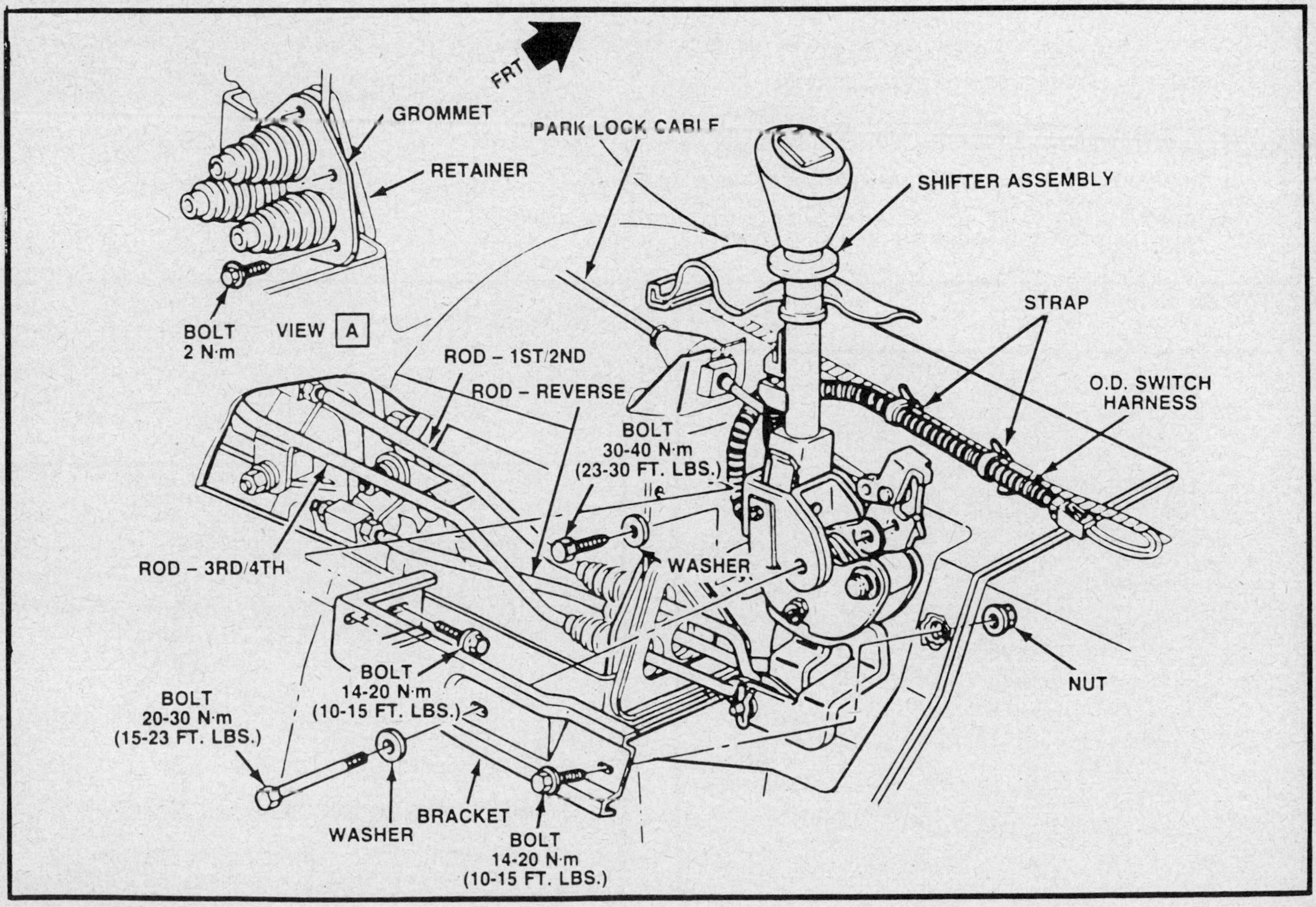
FRT
GROMMET
PARK LOCK CABLE
RETAINER
SHIFTER ASSEMBLY
BOLT
2 N·m
VIEW A
STRAP
ROD – 1ST/2ND
ROD – REVERSE
O.D. SWITCH HARNESS
BOLT
30-40 N·m
(23-30 FT. LBS.)
WASHER
ROD – 3RD/4TH
BOLT
14-20 N·m
(10-15 FT. LBS.)
NUT
BOLT
20-30 N·m
(15-23 FT. LBS.)
BRACKET
WASHER
BOLT
14-20 N·m
(10-15 FT. LBS.)

Shifter attachments

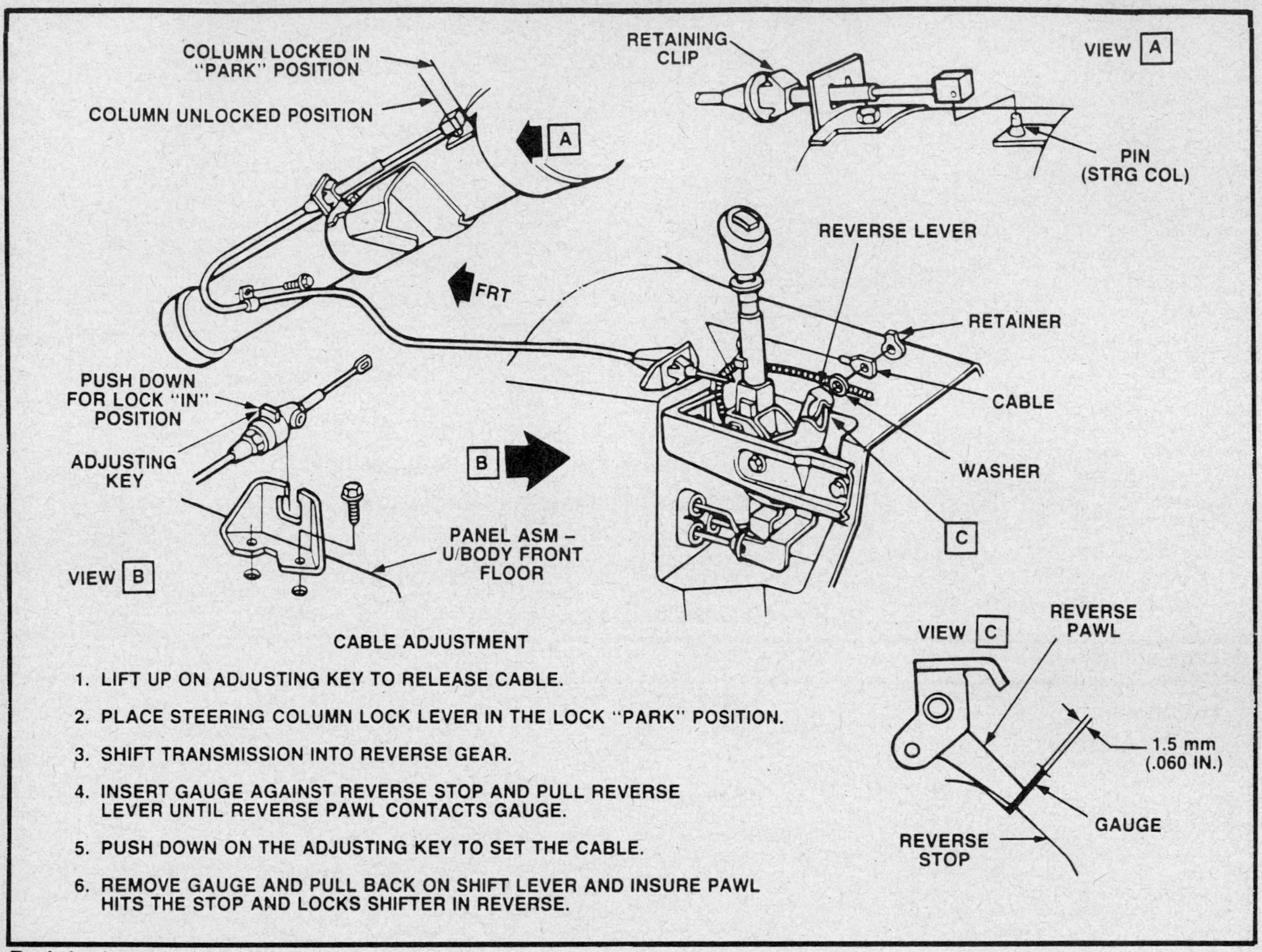
COLUMN LOCKED IN "PARK" POSITION
COLUMN UNLOCKED POSITION
RETAINING CLIP
VIEW A
PIN (STRG COL)
A
REVERSE LEVER
FRT
RETAINER
PUSH DOWN FOR LOCK "IN" POSITION
CABLE
ADJUSTING KEY
B
WASHER
C
PANEL ASM – U/BODY FRONT FLOOR
VIEW B
CABLE ADJUSTMENT
1. LIFT UP ON ADJUSTING KEY TO RELEASE CABLE.
2. PLACE STEERING COLUMN LOCK LEVER IN THE LOCK "PARK" POSITION.
3. SHIFT TRANSMISSION INTO REVERSE GEAR.
4. INSERT GAUGE AGAINST REVERSE STOP AND PULL REVERSE LEVER UNTIL REVERSE PAWL CONTACTS GAUGE.
5. PUSH DOWN ON THE ADJUSTING KEY TO SET THE CABLE.
6. REMOVE GAUGE AND PULL BACK ON SHIFT LEVER AND INSURE PAWL HITS THE STOP AND LOCKS SHIFTER IN REVERSE.
VIEW C
REVERSE PAWL
1.5 mm (.060 IN.)
GAUGE
REVERSE STOP

Park lock cable adjustment

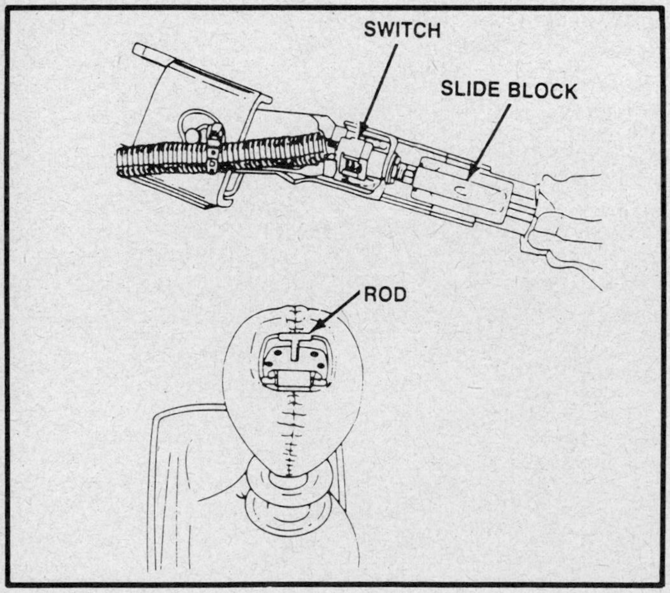
SWITCH
SLIDE BLOCK
ROD

Overdrive switch

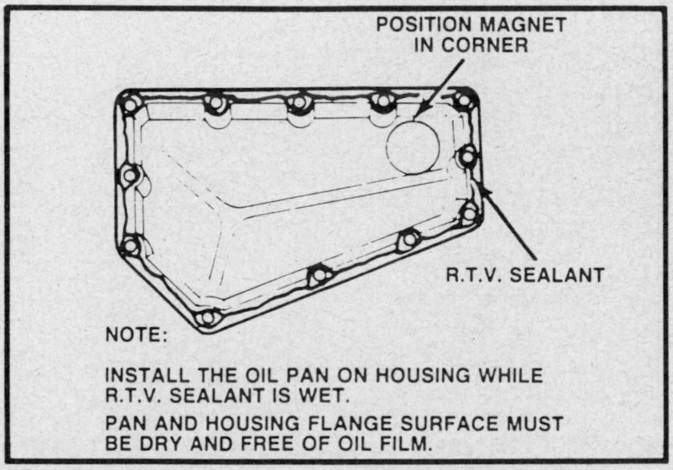
POSITION MAGNET IN CORNER
R.T.V. SEALANT
NOTE:
INSTALL THE OIL PAN ON HOUSING WHILE R.T.V. SEALANT IS WET.
PAN AND HOUSING FLANGE SURFACE MUST BE DRY AND FREE OF OIL FILM.

Oil pan magnet location and RTV application

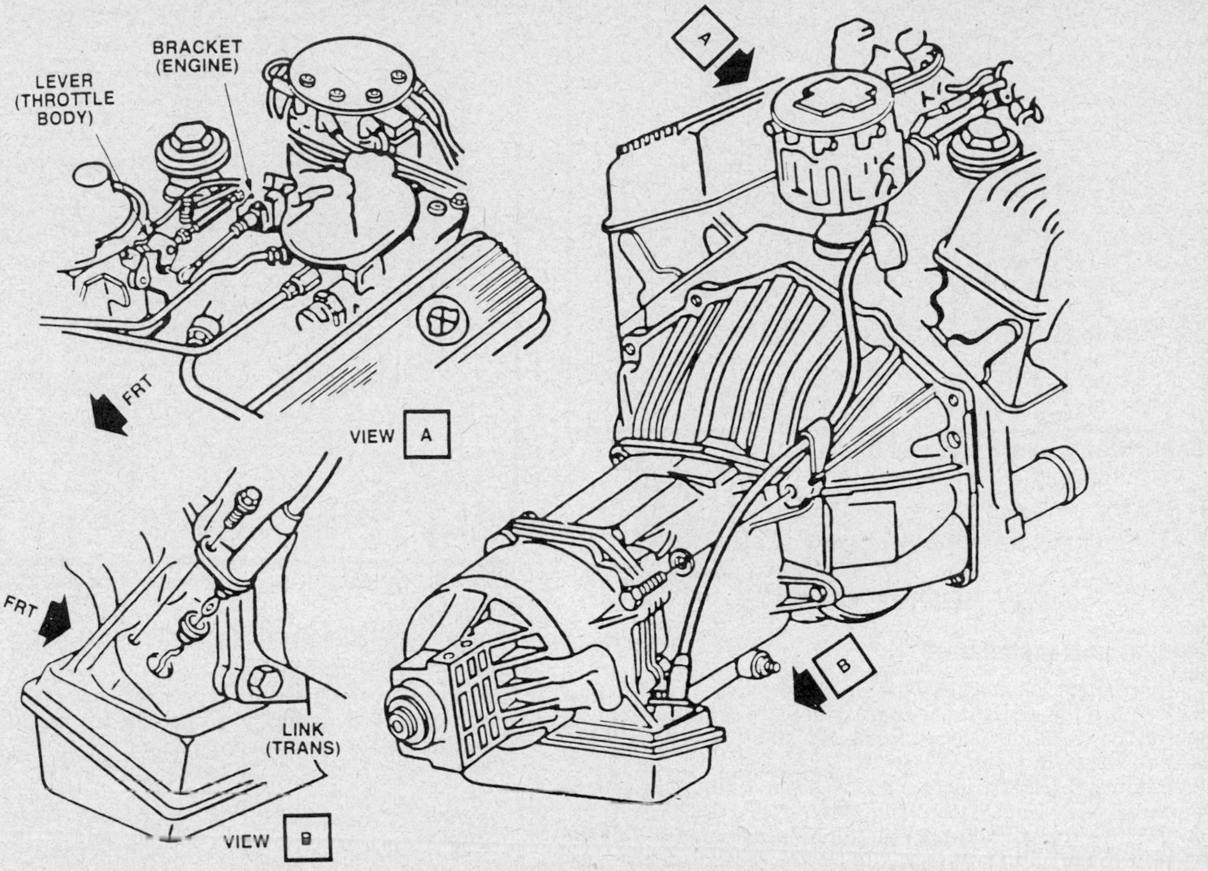

BRACKET (ENGINE)
LEVER (THROTTLE BODY)
FRT
VIEW A
A
FRT
LINK (TRANS)
VIEW B
B

THROTTLE VALVE CABLE ADJUSTMENT & READJUSTMENT PROCEDURE

THROTTLE VALVE CABLE ADJUSTMENT PROCEDURE

1. AFTER INSTALLATION OF CABLE TO THE TRANSMISSION, ENGINE BRACKET, AND THROTTLE LEVER, CHECK TO ASSURE THAT THE CABLE SLIDER IS IN THE ZERO OR FULLY READJUSTED POSITION. (IF NOT, REFER TO READJUSTMENT PROCEDURE.)
2. ROTATE THE THROTTLE LEVER TO THE "FULL TRAVEL STOP" POSITION TO OBTAIN A MINIMUM OF ONE CLICK ADJUSTMENT.

CAUTION LOCK TAB MUST NOT BE DEPRESSED DURING THIS OPERATION.

THROTTLE VALVE CABLE READJUSTMENT PROCEDURE

IN CASE READJUSTMENT IS NECESSARY BECAUSE OF INADVERTENT ADJUSTMENT BEFORE OR DURING ASSEMBLY, OR FOR REPAIR, PERFORM THE FOLLOWING:

1. DEPRESS AND HOLD METAL LOCK TAB.
2. MOVE SLIDER BACK THROUGH FITTING IN DIRECTION AWAY FROM THROTTLE BODY LEVER UNTIL SLIDER STOPS AGAINST FITTING.
3. RELEASE METAL LOCK TAB.
4. REPEAT STEP 2 OF ADJUSTMENT PROCEDURE.

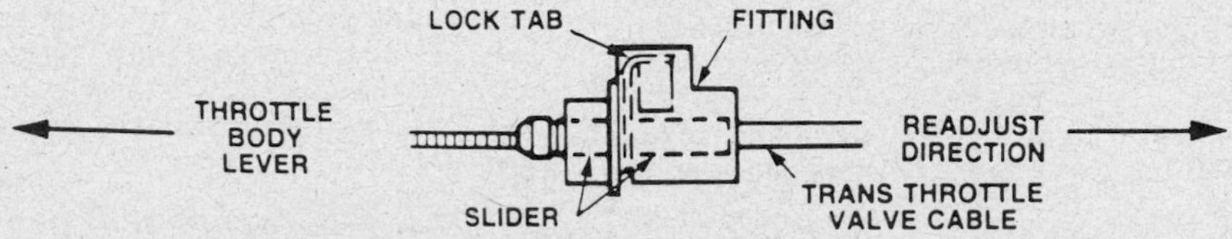

LOCK TAB
FITTING
THROTTLE BODY LEVER
READJUST DIRECTION
TRANS THROTTLE VALVE CABLE
SLIDER

Throttle valve (T.V.) cable adjustment – 1984–85 models

4. Insert the gauge against the reverse stop and pull the reverse lever until the reverse pawl contacts the gauge.
5. Push down on the adjusting key to set the cable.
6. Remove the gauge and pull back on the shift lever and insure the pawl hits the stop and locks the shifter in reverse.

Throttle Valve (T.V.) Cable

1. With the engine stopped, depress and hold down the readjust tab at the T.V. cable adjuster.
2. Move the cable conduit until it stops against the fitting.
3. Release the readjustment tab.
4. Rotate the throttle lever by hand to its full travel position.
5. The slider must move (ratchet) toward the lever when the lever is rotated to its full travel position.

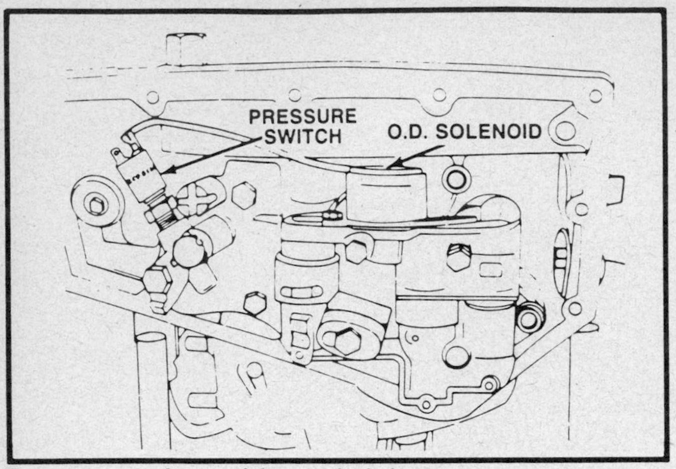
PRESSURE SWITCH
O.D. SOLENOID

Pressure and overdrive switch locations

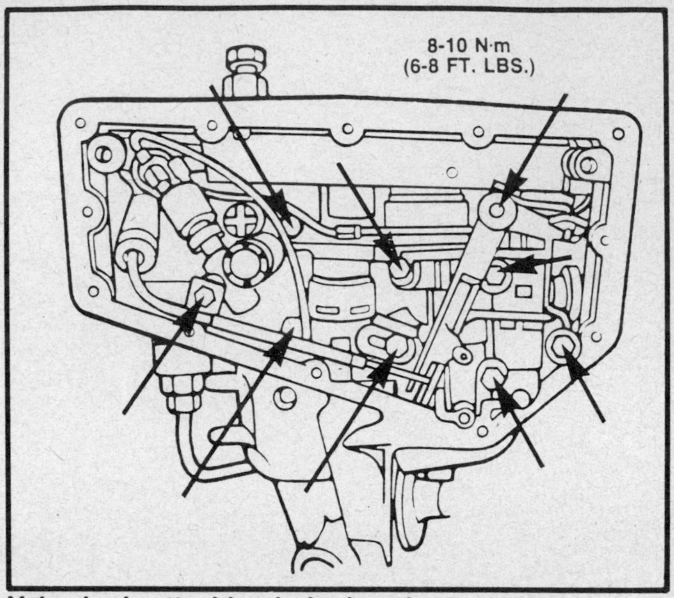
8-10 N·m
(6-8 FT. LBS.)

Valve body attaching bolts locations—1984–85 models

Services

OVERDRIVE SWITCH

Removal and Installation

1. Disconnect the negative battery cable.
2. Remove the instrument panel trim plate retaining screws. Remove the headlight switch knob, tilt steering column lever and then the trim plate.
3. Remove the instrument panel accessory trim plate retaining screws, then remove the trim plate.
4. Remove the console trim plate retaining screws. Pull the trim plate back and disconnect the electrical lead for the cigarette lighter. Remove the shift boot to trim plate retaining screws, then remove the trim plate.
5. Remove the overdrive switch button from the shifter knob. Use care when prying the button from the knob to prevent damage to the knob or button.
6. Remove the rod for the overdrive switch from the shifter. Count the number of turns required to remove the rod.
7. Loosen the switch to shifter retaining nut and remove the switch and actuator block from the shifter. Care must be taken to prevent the loss of the return spring located in the actuator block.
8. Disconnect the electrical lead for the switch.
9. Remove the switch to actuator block retaining pin and remove the switch.
10. To install, position the switch to the actuator block and install the retaining pin. Lubricate the sliding surface on the actuator block with grease.
11. Compress the return spring in the actuator block using a small suitable tool, then position the switch and block on the shifter.
12. Tighten the switch retaining nut. Apply Loctite®, part number 1052624, or equivalent, on the threads before tightening.
13. Install the overdrive switch rod in the shifter. Screw the rod into the actuator block the same amount of turns it took to remove. Install the switch button. The button should be flush with the top of the shift knob. If it is not, remove the button and turn the rod in or out to make it flush.
14. Connect the electrical lead for the switch.
15. Position the console trim plate to the shifter boot and install the retaining screws.
16. Connect the electrical lead for the cigarette lighter.
17. Position the console trim plate and install the retaining screws.
18. Position the instrument panel accessory trim plate and install the retaining screws.
19. Position the instrument cluster trim plate and install the retaining screws. Install the knob for the headlight switch and the lever for the tilt steering column.
20. Connect the negative battery cable.

OIL FILTER

Removal and Installation

1. Raise the vehicle and support safely.
2. Place a drain pan under the overdrive unit and remove the oil pan attaching bolts from the front and side of the pan.
3. Loosen the rear pan attaching bolts approximately 4 turns.
4. Carefully pry the oil pan loose, allowing the fluid to drain.
5. Remove the remaining pan bolts and remove the oil pan.
6. Drain the fluid from the oil pan. Clean the pan with solvent and dry thoroughly with clean compressed air.
7. Remove the filter from the transmission.
8. Install a new filter. Do not attempt cleaning an old filter. Always replace the filter. Coat the pickup tube and new filter grommet with automatic transmission fluid before installing to be sure the grommet is not pushed into the body of the filter.
9. Install the magnet in the oil pan. Apply a bead of RTV Sealant, part number 1052915 or equivalent to the oil pan flange and assemble wet. Install the oil pan bolts and torque to 6–8 ft. lbs. (8–10 Nm).

NOTE: The case and oil pan flange surfaces must be free of oil before applying RTV sealant.

10. Fill the overdrive unit with Dexron®II automatic transmission fluid.

OVERDRIVE SOLENOID

Removal and Installation

1. Raise the vehicle and support safely.
2. Remove the transmission oil pan and filter.
3. Remove the valve body.
4. Using tool J-34529 or equivalent, compress the shift valve spring and remove the pin.

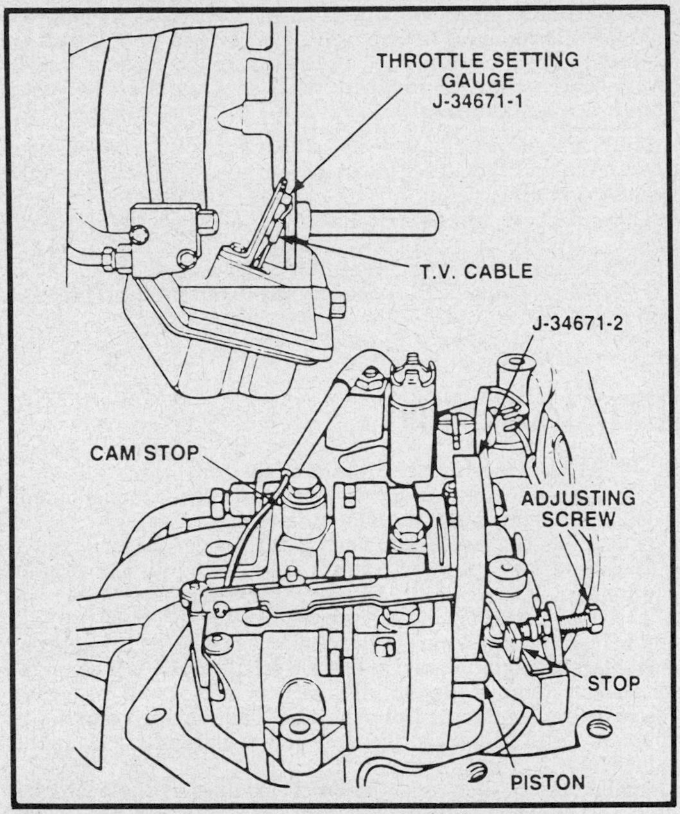
THROTTLE SETTING GAUGE J-34671-1
T.V. CABLE
J-34671-2
CAM STOP
ADJUSTING SCREW
STOP
PISTON

Cam stop and throttle valve lever adjustment – 1984–85 models

CAUTION

Removing the overdrive solenoid without first relieving the valve spring tension of the shift valve and pressure relief valve could damage the valve body and also cause personal injury.

5. Using tool J-34529 or equivalent, compress the relief valve springs and remove the pin.
6. Remove the bolts attaching the solenoid valve to the valve body. Remove the solenoid and check ball from the valve body.
7. To install, place the check ball and solenoid into the valve body. Install the attaching bolts to the valve body.
8. Using tool J-34529 or equivalent, compress the relief valve springs and install the pin.
9. Using tool J-34529 or equivalent, compress the shift valve spring and install the pin.
10. Install the valve body.
11. Install the transmission filter and oil pan.
12. Fill the overdrive unit with Dexron®II automatic transmission fluid.

PRESSURE SWITCH

Removal and Installation

1. Raise the vehicle and support safely.
2. Remove the transmission oil pan.
3. Remove the transmission oil filter.
4. Disconnect the electrical leads at the switch.
5. Unscrew the switch from the valve body and remove the switch.
6. To install, screw the switch into the valve body.
7. Connect the electrical leads to the switch.
8. Install the transmission oil filter and oil pan.
9. Fill the overdrive unit with Dexron®II automatic transmission fluid.

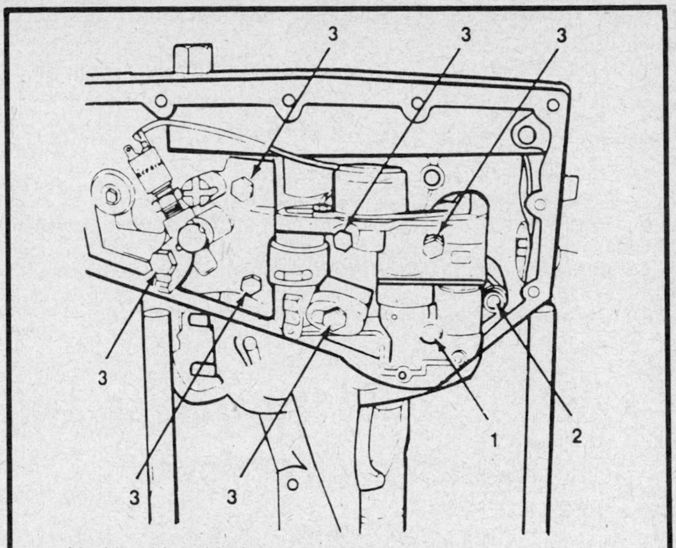
3
3
3
3
1
2
3
3

1. Machined head bolt
2. Allen head bolt
3. Bolts

Valve body attaching bolts locations – 1986–88 models

VALVE BODY

Removal and Installation

1984–85 VEHICLES

1. Raise the vehicle and support safely.
2. Disconnect the throttle valve (T.V.) cable at the throttle body lever.
3. Remove the transmission oil pan and filter.
4. Disconnect the tan electrical lead from the low pressure cut-off switch.
5. Disconnect the T.V. cable from the T.V. lever. Remove the bolt retaining the T.V. cable to the valve body.
6. Remove the remaining bolts attaching the valve body to the transmission case.
7. Remove the valve body with the spacer plate.

NOTE: Care must be taken to prevent loss of the (2) check balls located in the valve body. One check ball is spring loaded under the separator plate in the valve body. The other is located in the transmission case above the separator plate.

8. To install, position the valve body into the transmission case and install the attaching bolts. Torque bolts to 6–8 ft. lbs. (8–10 Nm).
9. Install throttle setting gauge J-34671-1 or equivalent, in the T.V. cable bore in the overdrive case. Hook the cable to the top step. Adjust cam stop.
10. Unhook the T.V. cable from the top step. Hook T.V. cable to the lower step. Position tool J-34671-2 or equivalent, between the piston and the solenoid bracket. Adjust the screw/bolt on the T.V. lever until the bolt makes contact with the stop on the cam.
11. Remove the throttle setting gauge tools from the overdrive unit.
12. Install the bolt retaining the T.V. cable to the valve body. Connect the T.V. cable to the T.V. lever.
13. Connect the tan electrical lead to the low pressure cut-off switch.
14. Install the transmission oil filter and pan.
15. Connect the T.V. cable at the throttle body lever.

16. Fill the overdrive unit with Dexron®II automatic transmission fluid.

1986–88 VEHICLES

1. Raise the vehicle and support safely.
2. Remove the transmission oil pan and filter.
3. Disconnect the tan electrical lead from the low pressure cut-off switch.
4. Remove the bolts attaching the valve body to the transmission case.
5. Remove the valve body with the spacer plate.

NOTE: Care must be taken to prevent loss of the (2) check balls located in the valve body. One check ball is spring loaded under the separator plate in the valve body. The other is located in the transmission case above the separator plate.

6. To install, position the valve body into the transmission case and install the attaching bolts. Torque the bolts to 6–8 ft. lbs. (8–10 Nm).
7. Connect the tan electrical lead to the low pressure cut-off switch.
8. Install the transmission oil filter and pan.
9. Fill the overdrive unit with Dexron®II automatic transmission fluid.

REMOVAL AND INSTALLATION

TRANSMISSION REMOVAL

1984–85 VEHICLES

1. Disconnect the negative battery cable.
2. Remove the air cleaner. Disconnect the throttle valve cable at the left of the throttle body unit. Remove the distributor cap.
3. Raise the vehicle and support it safely.
4. Remove the complete exhaust system as follows:
   a. Disconnect the A.I.R. pipe at the converter.
   b. Disconnect the A.I.R. pipe clamps at the exhaust manifold.
   c. Disconnect the oxygen sensor electrical lead.
   d. Remove the bolts attaching the mufflers to the hanger.
   e. Remove the hanger bracket at the converter.
   f. Disconnect the exhaust pipes from the exhaust manifolds and remove the exhaust system.
5. Remove the exhaust hanger at the transmission.
6. Support the transmission assembly using the proper equipment.
7. Remove the bolts attaching the driveline beam at the axle and transmission. Remove the driveline beam from the vehicle.
8. Mark the relationship of the driveshaft to the axle companion flange. Remove the trunnion bearing straps and disengage the rear universal joint from the axle. Slide the driveshaft slip yoke out from the overdrive unit and remove shaft from the vehicle.
9. Disconnect the transmission cooler lines at the overdrive unit. Disconnect the throttle valve at the overdrive unit. Disconnect the shift linkage at the side cover.
10. Disconnect the electrical connectors at the side cover switches, back up light switch, overdrive unit and speedometer sensor.
11. Lower the transmission and support the engine.
12. Remove the bolts attaching the transmission to the bellhousing. Slide the transmission rearward to disengage the input shaft from the clutch. Remove the transmission from the vehicle.

1986–88 VEHICLES

1. Disconnect the negative battery cable.
2. Remove the distributor cap and lay aside.
3. Raise the vehicle and support it safely.
4. Remove the upper and lower underbody braces on the vehicles equipped with a convertible top.
5. Remove the complete exhaust system as follows:
   a. Disconnect the A.I.R. pipe at the converter.
   b. Disconnect the A.I.R. pipe clamps at the exhaust manifold.
   c. Disconnect the oxygen sensor electrical lead.
   d. Remove the exhaust hanger bolts at the driveline support beam.
   e. Remove the bolts attaching the mufflers to the hanger.
   f. Remove the hanger bracket at the converter.
   g. Disconnect the exhaust pipes from the exhaust manifolds and remove the exhaust system.
6. Remove the exhaust hanger at the transmission.
7. Support the transmission assembly using the proper equipment.
8. Remove the bolts attaching the driveline beam at the axle and transmission. Remove the driveline beam from the vehicle.
9. Mark the relationship of the driveshaft to the axle companion flange. Remove the trunnion bearing straps and disengage the rear universal joint from the axle. Slide the driveshaft slip yoke out from the overdrive unit and remove shaft from the vehicle.
10. Disconnect the transmission cooler lines at the overdrive unit. Disconnect the shift linkage at the side cover.
11. Disconnect the electrical connectors at the transmission: Back up light switch, overdrive unit and speedometer sensor.
12. Lower the transmission and support the engine.
13. Remove the bolts attaching the transmission to the bellhousing. Slide the transmission rearward to disengage the input shaft from the clutch. Remove the transmission from the vehicle.

TRANSMISSION INSTALLATION

1984–85 VEHICLES

1. Install the transmission into the vehicle. Slide the transmission forward to engage the input shaft into the clutch. Install the bolts attaching the transmission to the bellhousing. Torque to 45–60 ft. lbs. (60–80 Nm).
2. Raise the transmission and remove the engine support.
3. Connect the electrical connectors at the side cover switches, back up light switch, overdrive unit and speedometer sensor.
4. Connect the shift linkage at the side cover. Connect the throttle valve at the overdrive unit. Connect the transmission cooler lines at the overdrive unit. Torque fittings to 8–12 ft. lbs. (11–16 Nm).
5. Align the matchmarks on the driveshaft and axle companion flange made during disassembly. Slide the driveshaft slip yoke into the overdrive unit. Engage the rear universal joint to the axle and install the trunnion bearing straps.
6. Install the driveline beam into the vehicle. Install the bolts attaching the driveline beam at the axle and torque to 60 ft. lbs. (80 Nm). Install the bolts attaching the driveline beam at the transmission and torque to 52 ft. lbs. (70 Nm).

NOTE: Do not over torque the bolts attaching the driveline beam to the transmission. Over torquing can damage the bushing and seal in the overdrive unit and result in fluid leakage.

7. Remove the transmission support.

8. Install the exhaust hanger at the transmission.
9. Install the complete exhaust system as follows:
   a. Install the exhaust system and connect the exhaust pipes to the exhaust manifolds.
   b. Install the hanger bracket at the converter.
   c. Install the bolts attaching the mufflers to the hanger.
   d. Connect the oxygen sensor electrical lead.
   e. Connect the A.I.R. pipe clamps at the exhaust manifold.
   f. Connect the A.I.R. pipe at the converter.
10. Lower the vehicle.
11. Install the distributor cap. Connect the throttle valve cable at the left of the throttle body unit. Install the air cleaner.
12. Connect the negative battery cable.
13. Refill the transmission (4 speed unit) with SAE-80W or SAE-80W-90 GL-5 gear lube. Refill the overdrive unit with Dexron®II automatic transmission fluid.
14. Adjust the T.V. cable.

1986–88 VEHICLES

1. Install the transmission into the vehicle. Slide the transmission forward to engage the input shaft into the clutch. Install the bolts attaching the transmission to the bellhousing. Torque to 45–60 ft. lbs. (60–80 Nm).
2. Raise the transmission and remove the engine support.
3. Connect the electrical connectors at the transmission: Back up light switch, overdrive unit and speedometer sensor.
4. Connect the shift linkage at the side cover. Connect the transmission cooler lines at the overdrive unit. Torque fittings to 8–12 ft. lbs. (11–16 Nm).
5. Align the matchmarks on the driveshaft and axle companion flange made during disassembly. Slide the driveshaft slip yoke into the overdrive unit. Engage the rear universal joint to the axle and install the trunnion bearing straps.
6. Install the driveline beam into the vehicle. Install the bolts attaching the driveline beam at the axle and torque to 60 ft. lbs. (80 Nm). Install the bolts attaching the driveline beam at the transmission and torque to 52 ft. lbs. (70 Nm).

NOTE: Do not over torque the bolts attaching the driveline beam to the transmission. Over torquing can damage the bushing and seal in the overdrive unit and result in fluid leakage.

7. Remove the transmission support.
8. Install the exhaust hanger at the transmission.
9. Install the complete exhaust system as follows:
   a. Install the exhaust system and connect the exhaust pipes to the exhaust manifolds.
   b. Install the hanger bracket at the converter.
   c. Install the bolts attaching the mufflers to the hanger.
   d. Install the exhaust hanger bolts at the driveline support beam.
   e. Connect the oxygen sensor electrical lead.
   f. Connect the A.I.R. pipe clamps at the exhaust manifold.
   g. Connect the A.I.R. pipe at the converter.
10. Install the upper and lower underbody braces on vehicles equipped with a converterable top.
11. Lower the vehicle.
12. Install the distributor cap.
13. Connect the negative battery cable.
14. Refill the transmission (4 speed unit) with SAE-80W or SAE-80W-90 GL-5 gear lube. Refill the overdrive unit with Dexron®II automatic transmission fluid.

BENCH OVERHAUL

Before Disassembly

Cleanliness is an important factor in the overhaul of the transmission. Before attempting any disassembly operation, the exterior of the transmission should be thoroughly cleaned to prevent the possibility of dirt entering the transmission internal mechanism. During inspection and reassembly, all parts should be thoroughly cleaned with cleaning fluid and then air dried. Wiping cloths or rags should not be used to dry parts. Do not use solvents on neoprene seals, composition-faced clutch plates or thrust washers. All oil passages should be blown out and checked to make sure that they are not obstructed. Small passages should be checked with tag wire. All parts should be inspected to determine which parts are to be replaced.

Transmission Disassembly

GM WARNER 83MM—4 SPEED

Disassembly

1. Thoroughly clean the exterior of the transmission assemblies.
2. Remove the 7 bolts attaching the overdrive unit to the reverse housing and then separate the 2 transmission units.
3. Remove the drain plug from the lower right side of the case and drain the lubricant from the transmission.
4. Shift the transmission into second gear. Remove the shift cover attaching bolts, cover, gasket and both shift forks from the transmission.
5. Remove the backup switch from the reverse housing.
6. Rotate the reverse shifter shaft and remove the shift fork and gear from the mainshaft.
7. Remove the lock pin from the reverse shift lever boss and pull the shaft from the housing.
8. Remove the drive gear bearing retainer bolts, retainer and gasket from the front of the transmission.
9. Remove front bearing snapring, selective fit snapring and spacer washer.
10. Using tool J-6654-01 and J-8433-1 or equivalent, pull the drive gear bearing from the transmission.
11. Remove the bolts attaching the reverse housing to the case. Using a small drift and a hammer, tap the locating pin for the reverse housing into the case.
12. Rotate the reverse housing on the mainshaft until the hole for the reverse idler gear shaft in the housing lines up with the countergear shaft.
13. Using tool J-24658, or a dummy shaft, drive the countergear shaft rearward out of the gear and through the reverse housing. The countergear will drop to the bottom of the case allowing clearance for the removal of the mainshaft.
14. Remove the mainshaft with the reverse housing and drive gear from the transmission case.
15. Remove the front reverse idler gear and thrust washer from the case.
16. Remove the countergear and two tanged thrust washers from the transmission case. Check the bottom of the transmission case for loose pilot bearings. Remove the locating pin for the reverse housing and any other loose components.

OVERDRIVE UNIT

Disassembly

1. Remove the fill plug and drain oil from the case.
2. Remove the retaining bolt and bracket for the speedometer sensor and drive gear. Remove the sensor and gear.

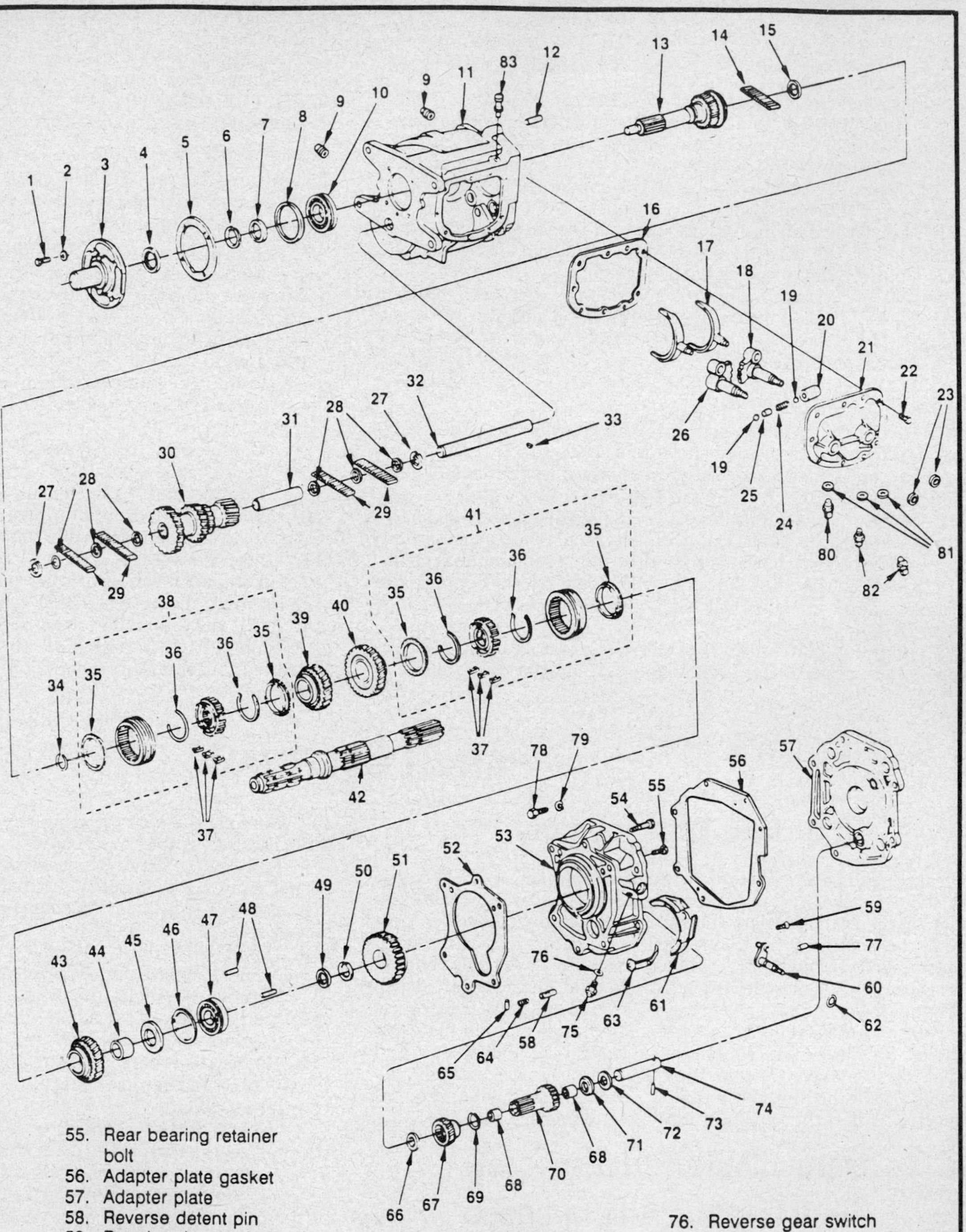

1. Hex bolt
2. External lock washer
3. Main drive gear front bearing retainer
4. Main drive gear front bearing oil seal
5. Main drive gear bearing retainer gasket
6. Main drive gear bearing retainer snapring
7. Main drive gear bearing spacer
8. Main drive gear bearing lockring
9. Magnetic drain plug
10. Main drive gear bearing
11. Transmission case
12. Dowel pin
13. Main drive gear
14. Main drive gear rear pilot roller
15. Main shaft pilot bearing spacer
16. Side cover gasket
17. Shift fork
18. 1–2 shift shaft
19. Steel ball
20. Interlock sleeve
21. Side cover
22. Transmission cover bolt
23. Shift lever shaft oil seal
24. Shift lever poppet spring
25. Interlock pin
26. 3–4 shift shaft
27. Counter gear rear washer
28. Counter shaft bearing roller washer
29. Counter gear bearing roller
30. Counter gear
31. Counter gear bearing spacer
32. Counter gear shaft
33. Woodruff key
34. Main shaft snapring
35. Synchronizer blocking ring
36. Synchronizer spring
37. Synchronizer key
38. 3–4 Synchronizer
39. 3rd speed gear
40. 2nd speed gear
41. 1–2 Synchronizer
42. Main shaft
43. 1st speed gear
44. 1st speed gear sleeve
45. 1st speed gear thrust washer
46. Rear bearing lock snapring
47. Main shaft rear bearing
48. Dowel pin
49. Main drive gear bearing spacer
50. Main shaft snapring
51. Reverse gear
52. Rear bearing retainer to transmission case gasket
53. Rear bearing retainer
54. Rear bearing retainer bolt
55. Rear bearing retainer bolt
56. Adapter plate gasket
57. Adapter plate
58. Reverse detent pin
59. Rear bearing retainer screw
60. Reverse shift shaft
61. Reverse shift fork
62. Reverse shaft oil seal
63. Electrical harness clip bracket
64. Reverse detent pin poppet spring
65. Reverse detent spring pin
66. Front reverse idler gear thrust washer
67. Reverse idler front gear
68. Reverse idler bushing
69. Reverse idler gear retainer ring
70. Reverse idler rear gear
71. Reverse idler thrust bearing
72. Rear reverse idler gear thrust washer
73. Spring pin
74. Reverse idler shaft
75. Overdrive override reverse gear switch
76. Reverse gear switch seal
77. Solid taper pin
78. Transmission to overdrive bolt
79. Transmission to overdrive bolt lockwasher
80. Overdrive override (3–4) switch
81. Overdrive override switch seal
82. Overdrive override (1–2) switch
83. Transmission ventilator

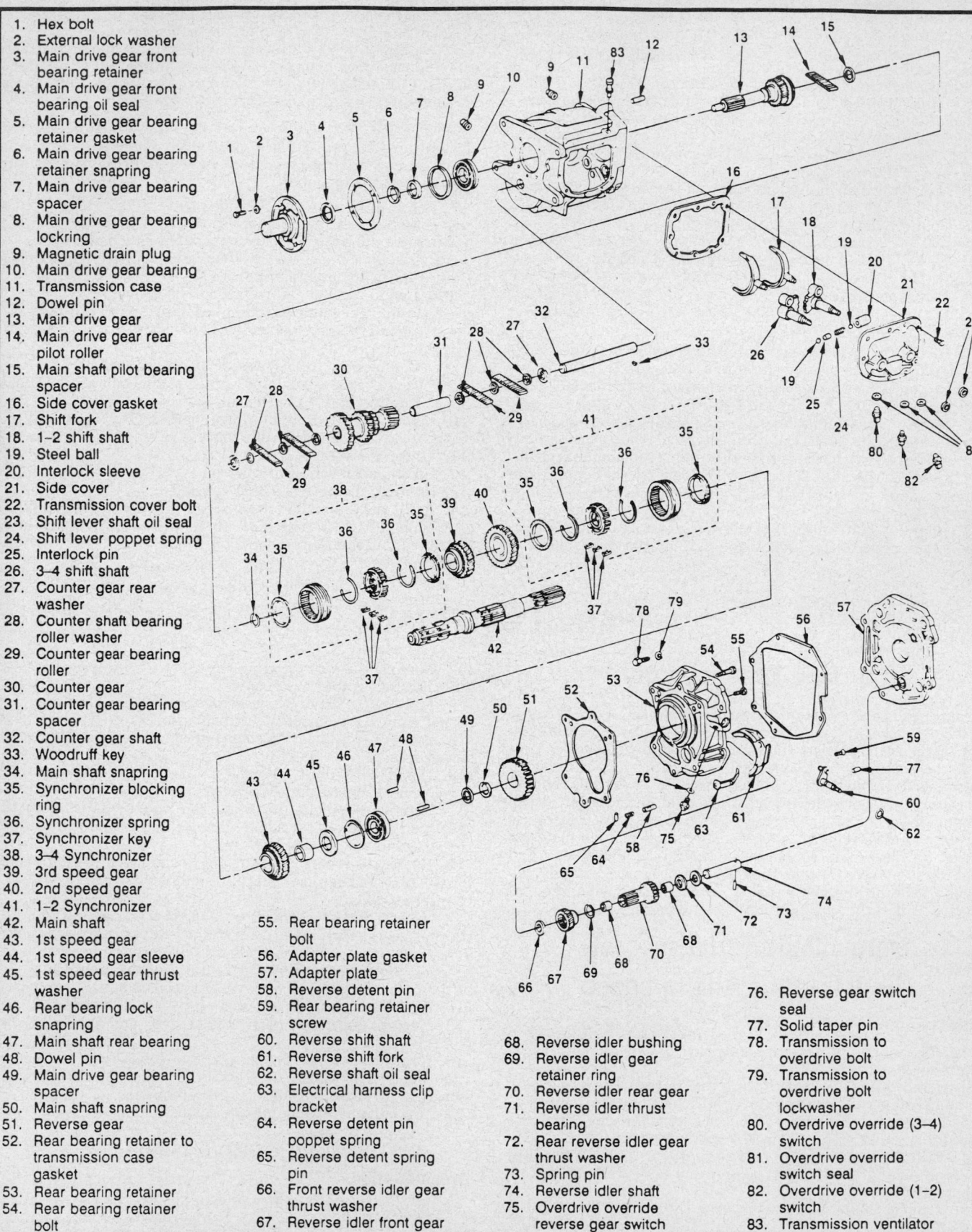

Manual 4-speed exploded view

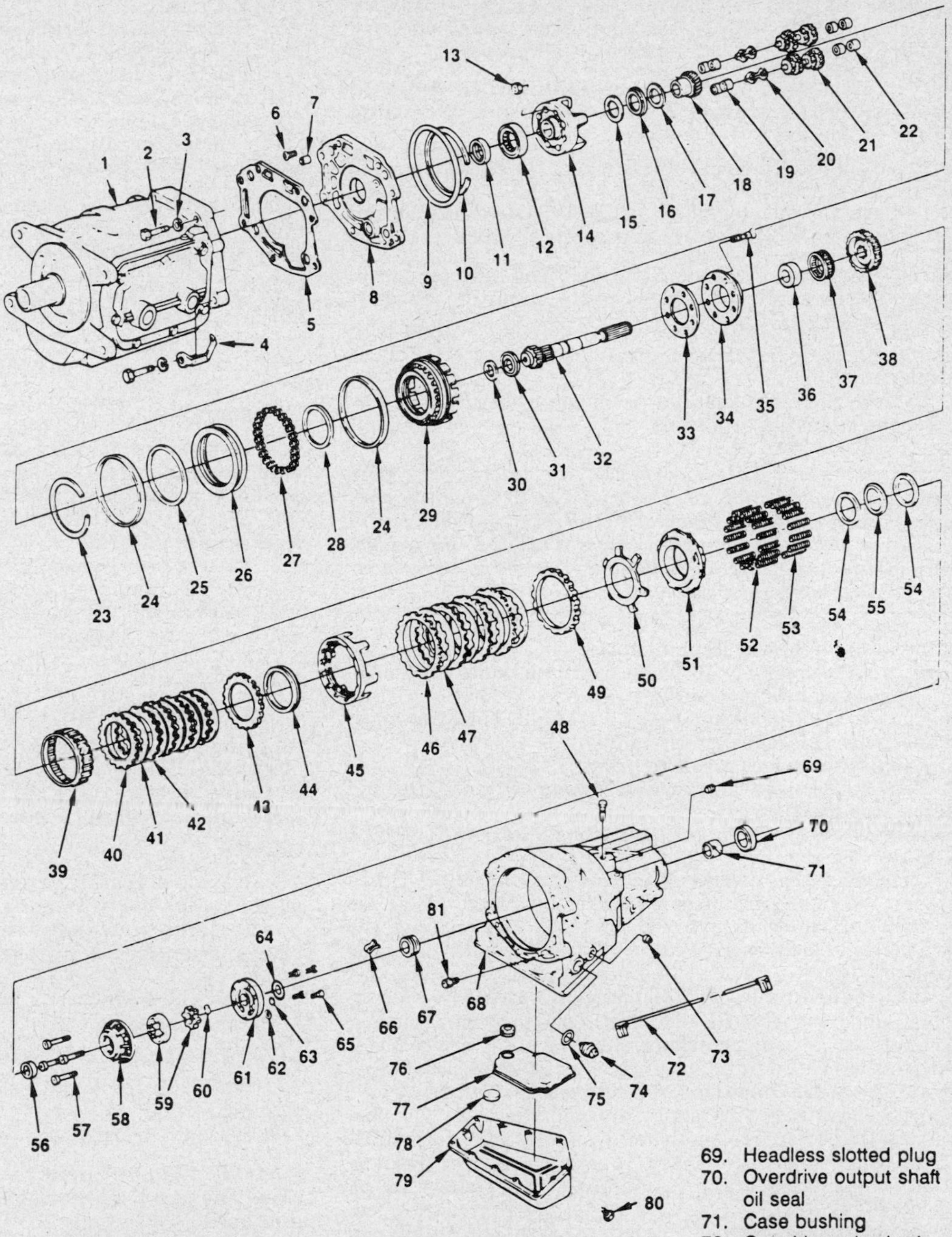
1
2
3
4
5
6
7
8
9
10
11
12
13
14
15
16
17
18
19
20
21
22
23
24
25
26
27
28
24
29
30
31
32
33
34
35
36
37
38
39
40
41
42
43
44
45
46
47
48
49
50
51
52
53
54
55
54
56
57
58
59
60
61
62
63
64
65
66
67
68
69
70
71
72
73
74
75
76
77
78
79
80
81

1. Transmission (less overdrive unit)
2. Transmission to overdrive bolt
3. Transmission to overdrive bolt lockwasher
4. Electrical harness clip bracket
5. Adapter plate gasket
6. Adapter plate screw
7. Dowel pin
8. Adapter plate
9. Adapter plate O-ring
10. Accumulator piston retainer ring
11. Input sun gear oil seal
12. Annular bearing
13. Carrier bearing locknut
14. Planetary gear carrier
15. Input sun gear thrust washer
16. Input sun gear thrust bearing
17. Input sun gear thrust (selective) washer
18. Input sun gear
19. Bearing cup
20. Planetary gear thrust washer
21. Planetary gear
22. Bearing cup
23. Accumulator piston retainer ring
24. Accumulator piston seal
25. Accumulator piston seal
26. Accumulator cushion piston
27. Accumulator piston spring
28. Accumulator piston seal
29. Accumulator piston
30. Output shaft thrust washer
31. Output shaft thrust bearing
32. Output shaft
33. Planetary gear thrust plate
34. Clutch drum plate
35. Clutch drum bolt
36. Inner race
37. Direct clutch sprag
38. Direct clutch hub
39. Direct clutch drum
40. Direct clutch inner driven (selective) plate
41. Direct clutch plate (6)
42. Direct clutch driven plate (6)
43. Direct clutch pressure plate
44. Direct clutch bearing
45. Overdrive clutch piston
46. Overdrive clutch driven plate (5)
47. Overdrive clutch plate (5)
48. Overdrive vent tube
49. Overdrive clutch pressure plate
50. Direct clutch thrust washer
51. Overdrive direct clutch piston
52. Overdrive direct clutch outer spring
53. Overdrive direct clutch inner spring
54. Overdrive direct clutch hub thrust washer
55. Overdrive direct clutch hub thrust bearing
56. Pump bearing cup
57. Overdrive pump and output shaft screw
58. Overdrive pump (gerotor) housing with bearing
59. Overdrive oil (gerotor) pump
60. Oil pump drive pin
61. Pump (gerotor) cover
62. Overdrive oil pump O-ring
63. Overdrive oil pump O-ring
64. Pump seal
65. Overdrive pump spool (gerotor) screw
66. Speedometer drive gear clip
67. Speedometer drive gear
68. Overdrive case
69. Headless slotted plug
70. Overdrive output shaft oil seal
71. Case bushing
72. Overdrive valve body pressure switch wire
73. Square head filler plug
74. Overdrive solenoid electrical connector
75. Overdrive solenoid electrical connector O-ring
76. Overdrive oil screen tube grommet
77. Overdrive oil screen
78. Overdrive oil pan magnet
79. Overdrive oil pan
80. Overdrive oil pan bolt
81. Oil cooler fitting

Overdrive unit exploded view

3. Remove 1/8 in. pipe plugs from the rear of the unit.
4. Install the pressure plate retaining bolts, tool J-34681 or equivalent, until flush with the case. Turn bolts additional turns, by rotating each bolt one turn at a time.

NOTE: This sequence must be followed in order to prevent the pressure plate from cocking and causing damage to the unit.

5. Remove the allen head bolts retaining the adapter plate to the case.
6. Remove the adapter plate, using a plastic hammer and suitable tool. Tap adapter plate to separate from case.

NOTE: Do not pry between the case and adapter plate, damage to the sealing surfaces could occur.

7. Bolt the overdrive unit to tool J-34162 or equivalent. Mount the holding fixture to the base plate tool J-3389-20 or equivalent.
8. Remove the large snapring from the overdrive unit forward of the accumulator piston.

CAUTION

If pressure is felt at snapring, do not remove. Check to insure the pressure plate retaining bolts are installed. If the bolts are installed, tighten each bolt one additional turn until pressure is relieved. The pressure plate is under a 1200 lbs. spring load. If the retaining bolts are not installed, personal injury could occur.

9. Remove the piston/accumulator assembly. Using an allen wrench, pry the assembly up evenly by lifting under the flange. Do not pry at or near seal surface.
10. Remove the carrier and bearing assembly (includes input sun and pinion gears) as an assembly.
11. Remove the finger pressure plate.
12. Remove the overdrive clutches, (4 composition, 4 steel and 1 clutch stop plate).
13. Remove the direct clutch plates, (5 composition, 5 steel and 1 steel bearing plate).
14. Measure each selective clutch plate in the direct clutch pack and record the readings. The selective clutch plates are used to control the clutch pack clearance. When replacing the clutch plates, replace each selective clutch plate with one of the same size.
15. Inspect the overdrive and direct clutch plates as follows:
   a. Compositioned plate – Dry plates and inspect for pitting, flaking, wear, glazing, cracking, charring and chips or metal particles imbedded in lining. If a compositioned plate shows any of the above conditions, replacement is required.
   b. Steel Plates – Wipe plates dry and check for discoloration. If the surface is smooth and even color smear is indicated, the plate should be reused. If severe heat spot discoloration or surface scuffing is indicated, the plate must be replaced.
16. Remove thrust washer and bearing from the output sun gear. Thrust washer may stick to the input sun gear hub.
17. Remove allen head pump housing retaining bolts by rotating the hub to gain access to the bolts.
18. Remove the output shaft assembly (including output sun gear, sprag clutch, clutch hub, gerotor pump and speedometer drive gear).
19. Remove the pressure plate and springs by positioning tool J-21420-2 or equivalent, on the pressure plate with the bolt from tool J-23327 or equivalent, through the center of the plate. Next position tool J-23327 or equivalent, on the rear of the case and install the retaining nut. Remove the retaining bolts, tool J-34681 or equivalent, from the rear of the case. Loosen the retaining nut on tool J-23327 or equivalent, bolt to relieve the spring pressure.
20. Remove the cooler valve assembly by loosening the nuts on the tube and then remove the bolts holding the valve to the case.
21. Remove the oil pan retaining bolts and then pry the pan from the case.
22. Remove the oil filter and tube from the valve body.
23. Disconnect the T.V. cable from the lever. Remove the cable retaining bolt and remove the cable assembly.
24. Remove the T.V. lever retaining bolt and then lever from the valve body.
25. Remove the remaining valve bolts and then remove the valve body with the spacer plate.

NOTE: There are 2 check balls, one on each side of spacer plate. One ball is located in the case and the other is spring loaded in the valve body.

Unit Disassembly

MAINSHAFT

Disassembly

1. Using snapring pliers, remove 3–4 synchronizer assembly retaining ring at front of mainshaft. Slide washer, synchronizer assembly, synchronizer ring 3rd speed gear from mainshaft.
2. Spread rear bearing retainer snapring and slide retainer from mainshaft.
3. Remove rear bearing-to-mainshaft snapring.
4. Support mainshaft under 2nd gear and press mainshaft from rear bearing, 1st gear and sleeve, 1–2 synchronizer assembly, and the second gear.

Inspection

ROLLER BEARINGS AND SPACERS

All main drive gear and countergear bearing rollers should be inspected closely and replaced if they show wear. Inspect countershaft and reverse idler shaft at the same time. Replace if necessary. Replace all worn spacers.

GEARS

1. Inspect all gears for excessive wear, chips or cracks and replace any that are worn or damaged.
2. Inspect reverse gear bushing and if worn or damaged, replace the entire gear (reverse gear bushing is not serviced separately).
3. Check both synchronizer sleeves to see that they slide freely on their hubs.

FRONT AND REAR BEARINGS

1. Wash the front and rear ball bearings thoroughly in a cleaning solvent.
2. Blow out bearing with compressed air.

NOTE: Do not allow the bearing to spin. Turn them slowly by hand. Spinning bearings may damage the race and balls.

Assembly

1. From rear of mainshaft, assemble the 2nd speed gear (with hub of gear toward rear of shaft).
2. Install 1–2 synchronizer assembly (sliding synchronizer sleeve taper toward rear, hub to front) on the mainshaft together with a synchronizer ring on both sides of the synchronizer assemblies.
3. Position the 1st gear sleeve on the shaft and press the sleeve onto the mainshaft until the 2nd gear, synchronizer assembly and sleeve bottom against the shoulder of the mainshaft.
4. Install 1st speed gear (with hub toward front) and supporting inner race, press the rear bearing onto the mainshaft with the snapring groove toward front of the transmission.
5. Install spacer and new correct selective fit (thickest that will assemble) snapring in mainshaft behind rear bearing.

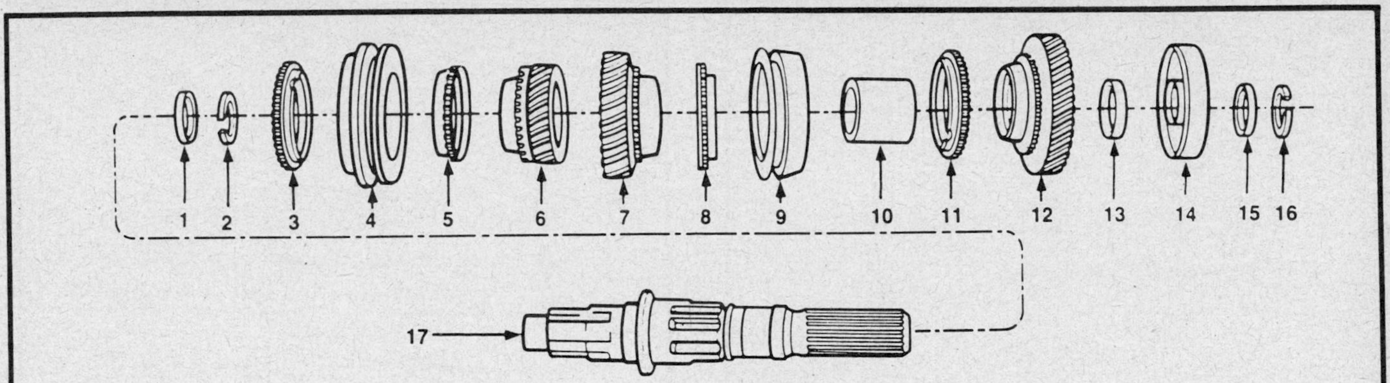
1
2
3
4
5
6
7
8
9
10
11
12
13
14
15
16
17

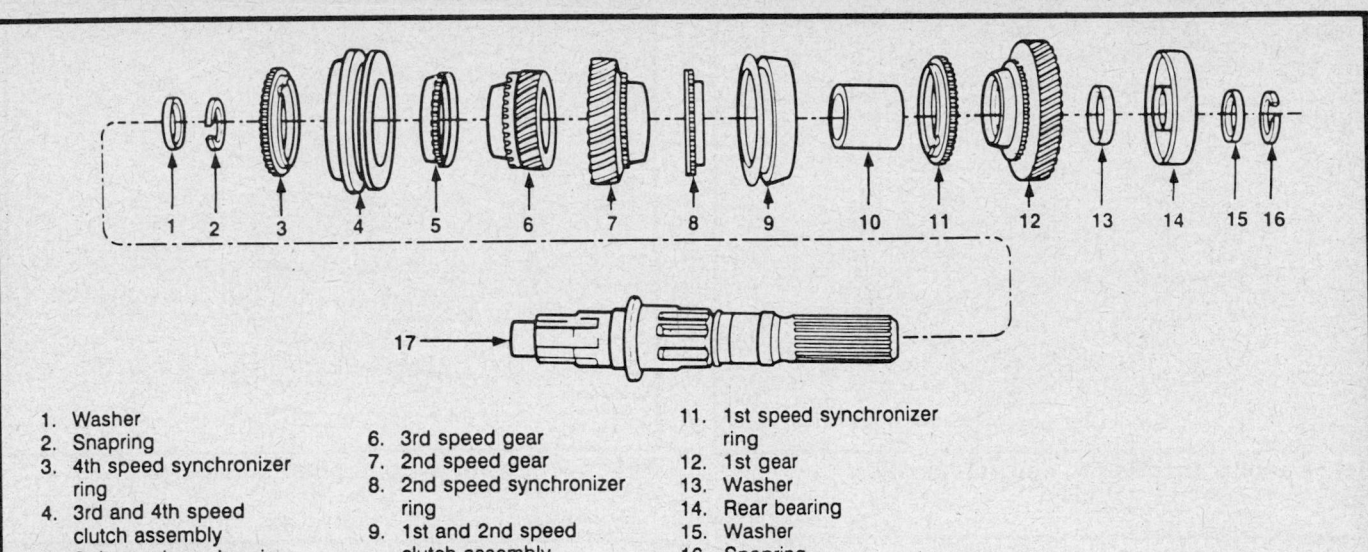

1. Washer
2. Snapring
3. 4th speed synchronizer ring
4. 3rd and 4th speed clutch assembly
5. 3rd speed synchronizer ring
6. 3rd speed gear
7. 2nd speed gear
8. 2nd speed synchronizer ring
9. 1st and 2nd speed clutch assembly
10. 1st gear bushing
11. 1st speed synchronizer ring
12. 1st gear
13. Washer
14. Rear bearing
15. Washer
16. Snapring
17. Mainshaft

Mainshaft exploded view

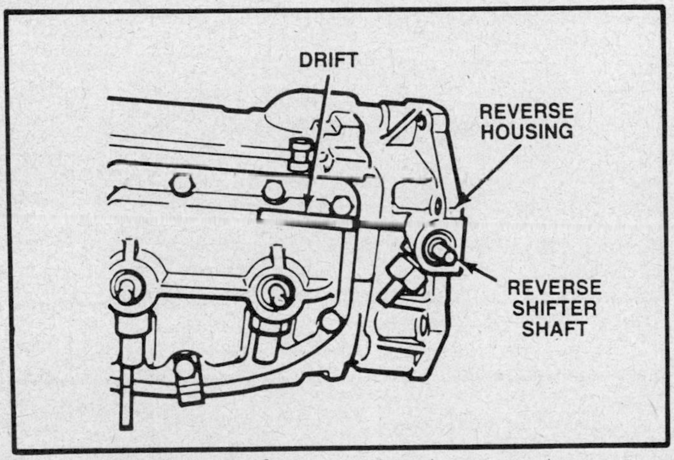
DRIFT
REVERSE HOUSING
REVERSE SHIFTER SHAFT

Reverse shaft lock pin

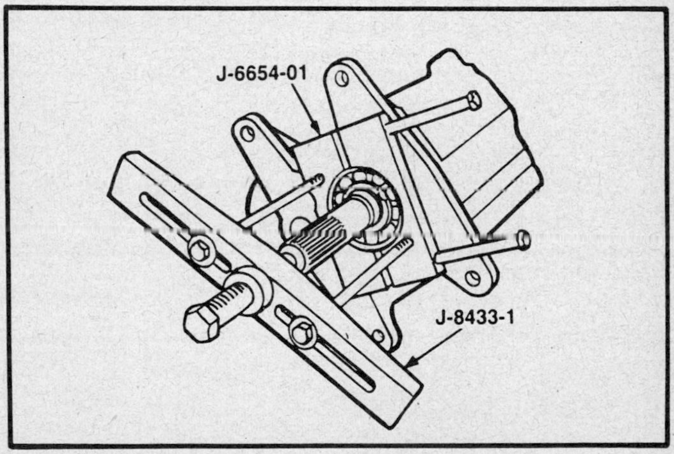
J-6654-01
J-8433-1

Front bearing removal

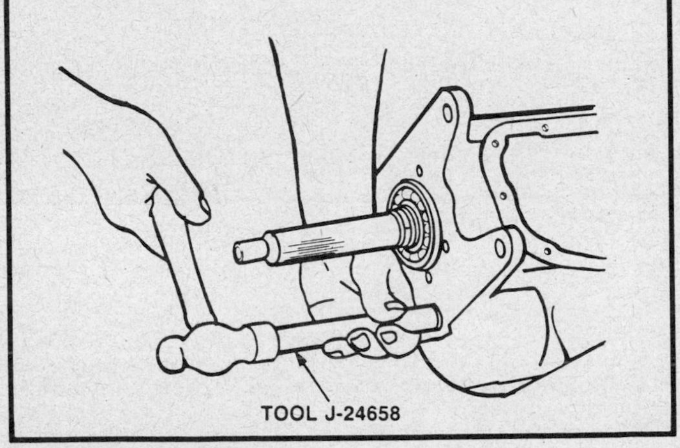
TOOL J-24658

Using tool J-24658

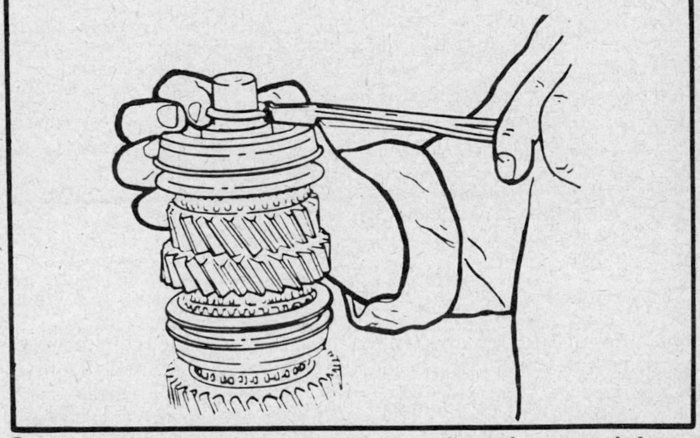

Synchronizer retainer snapring to housing servicing

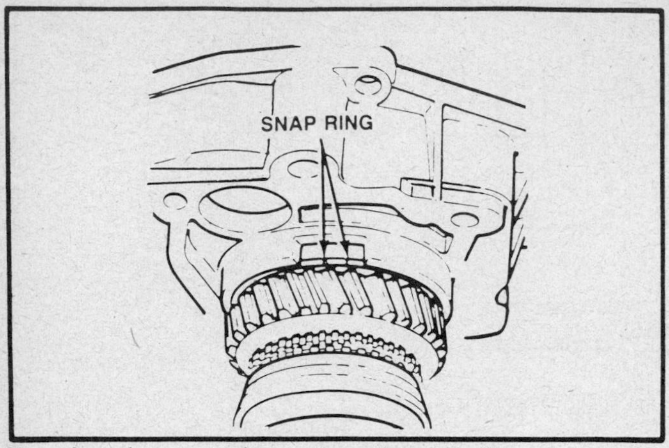
SNAP RING
Rear bearing snapring to housing location

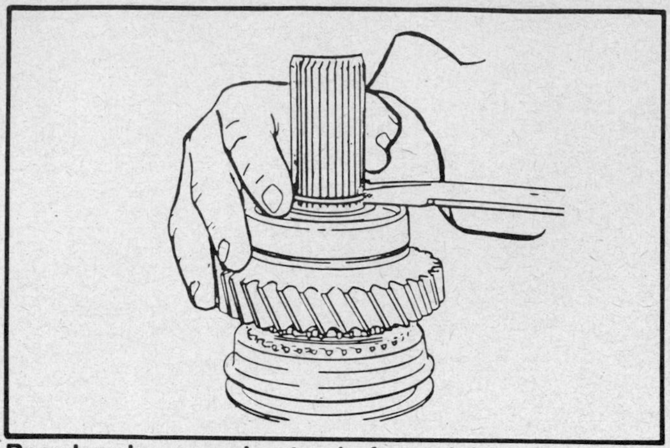
Rear bearing snapring to shaft servicing

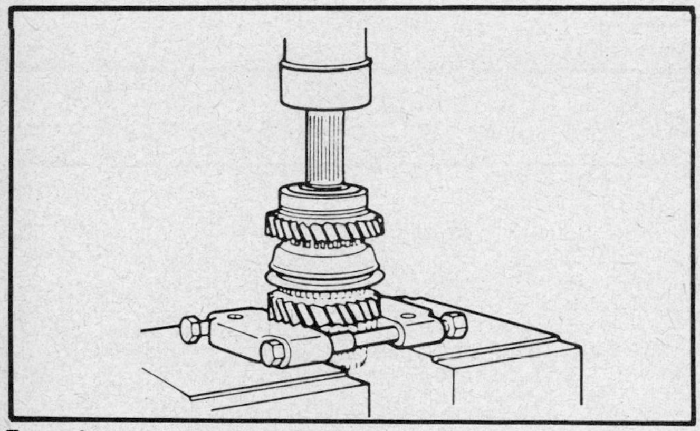
Pressing rear bearing from shaft

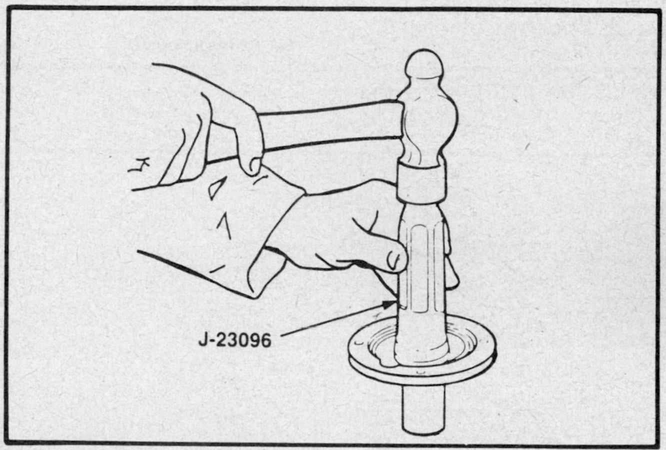
J-23096
Using tool J-23096

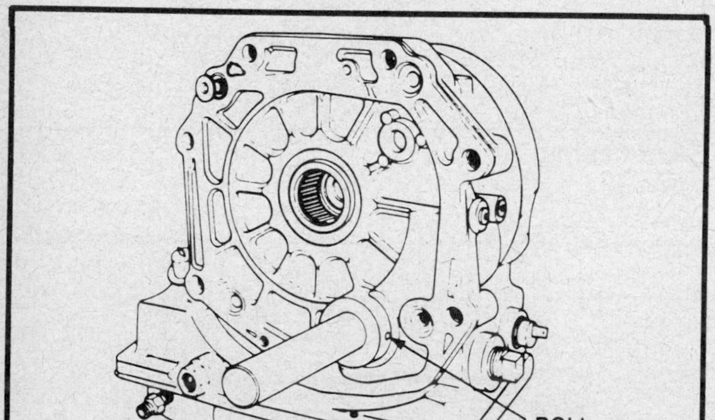
ROLL
PIN
Reverse idler shaft roll pin location

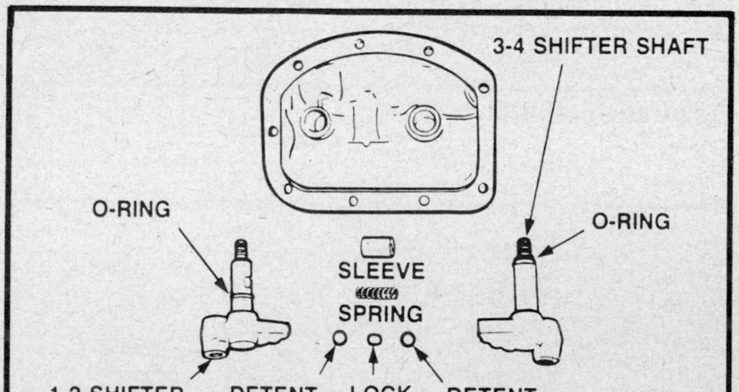
3-4 SHIFTER SHAFT
O-RING
O-RING
SLEEVE
SPRING
1-2 SHIFTER SHAFT
DETENT BALL
LOCK PIN
DETENT BALL
Side cover components

6. Install the 3rd speed gear (hub to front of transmission) and the 3rd speed gear synchronizing ring (notches to front of transmission).

7. Install the 3rd and 4th speed gear synchronizer assembly (hub and sliding sleeve) with taper toward the front making sure that the keys in hub correspond to the notches in the 3rd speed gear synchronizing ring.

8. Install new selective fit snapring (thickest that will install) in the groove in mainshaft in front of the 3rd and 4th speed synchronizer assembly.

9. Install the rear bearing retainer (reverse housing) over end of mainshaft. Spread the snapring to drop around the rear bearing. Release snapring when it aligns with groove in rear bearing.

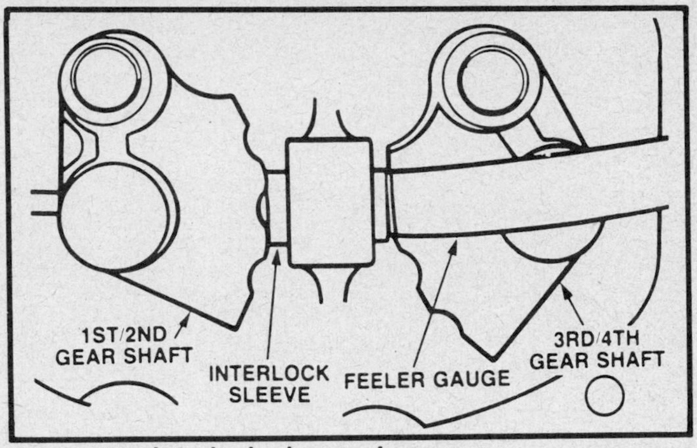
1ST/2ND
GEAR SHAFT
INTERLOCK
SLEEVE
FEELER GAUGE
3RD/4TH
GEAR SHAFT

Measuring interlock sleeve clearance

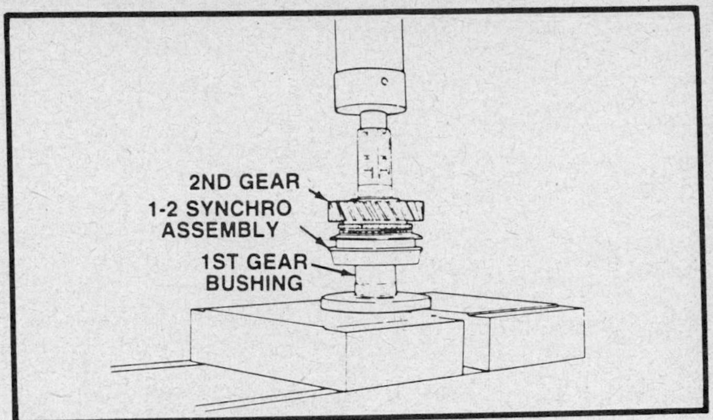
2ND GEAR
1-2 SYNCHRO
ASSEMBLY
1ST GEAR
BUSHING

First gear sleeve servicing

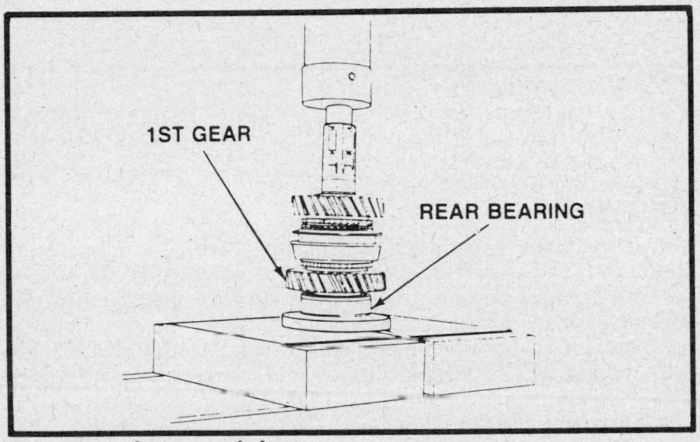
1ST GEAR
REAR BEARING

Rear bearing servicing

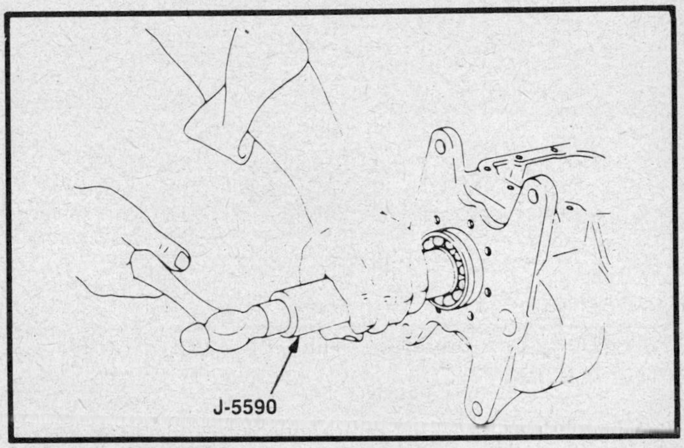
J-5590

Using tool J-5590

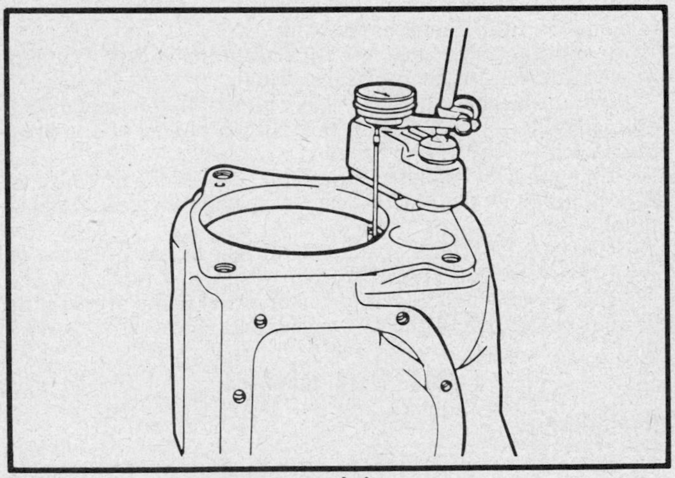

Measuring countergear endplay

COUNTERGEAR

Disassembly

1. Remove tool J-24658 or equivalent, from the countergear.
2. Tip the countergear on end and let the spacers, rollers and roller sleeve slide out from the gear.

Inspection

ROLLER BEARINGS AND SPACERS

All main drive gear and countergear bearing rollers should be inspected closely and replaced if they show wear. Inspect countershaft and reverse idler shaft at the same time. Replace if necessary. Replace all worn spacers.

GEARS

1. Inspect all gears for excessive wear, chips or cracks and replace any that are worn or damaged.
2. Inspect reverse gear bushing and if worn or damaged, replace the entire gear (reverse gear bushing is not serviced separately).
3. Check both synchronizer sleeves to see that they slide freely on their hubs.

FRONT AND REAR BEARINGS

1. Wash the front and rear ball bearings thoroughly in a cleaning solvent.
2. Blow out bearing with compressed air.

NOTE: Do not allow the bearing to spin. Turn them slowly by hand. Spinning bearings may damage the race and balls.

Assembly

1. Install roller spacer in countergear, if removed.
2. Insert a dummy shaft or loading tool J-24658 or equivalent, into countergear.

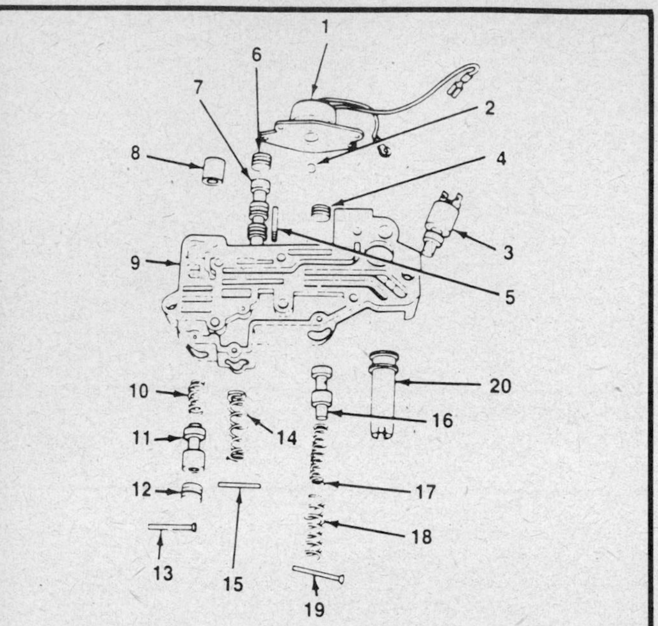
1
2
3
4
5
6
7
8
9
10
11
12
13
14
15
16
17
18
19
20

1. Solenoid
2. Solenoid check ball
3. Pressure switch
4. End plug
5. Pin
6. End plug
7. Shift valve
8. Spring seat
9. Valve body
10. Spring
11. T.V. valve
12. End plug
13. Pin
14. Spring
15. Pin
16. Relief valve
17. Inner spring
18. Outer spring
19. Pin
20. Pickup tube

Valve body components – 1986–88 shown, earlier models similar

3. Using heavy grease to retain rollers, install spacer, 28 rollers, spacer, 28 rollers, and spacer in either end of countergear. Repeat in other end of countergear.

Checking Endplay

1. Rest the transmission case on its side with the side cover opening toward the assembler. Put countergear tanged thrust washers in place, retaining them with heavy grease, making sure the tangs are resting in the notches of the case.
2. Set countergear in place in bottom of transmission case, making sure that tanged thrust washers are not knocked out of place.
3. Position the transmission case resting on its front face.
4. Lubricate and insert countergear (pushing loading tool J-24658 or equivalent, out front of case) until Woodruff key slot is in its relative installed position (do not install key).
5. Attach a dial indicator and check endplay of the countergear. If endplay is greater than 0.025 in., a new thrust washers must be installed.

VALVE BODY

Disassembly

1. Using tool J-34529 or equivalent, relieves the pressure on the shift valve. Remove the pin, spring and valve.
2. Using tool J-34529 or equivalent, relieves the pressure on pressure relief valve. Remove the pin, spring and valve.
3. Using tool J-34529 or equivalent, relieves the pressure on the accumulator valve. Remove the pin, spring, valve, plug, sleeve and plunger.
4. Disconnect the solenoid electrical lead at the pressure switch. Remove the solenoid attaching bolts. Remove the solenoid and check ball.

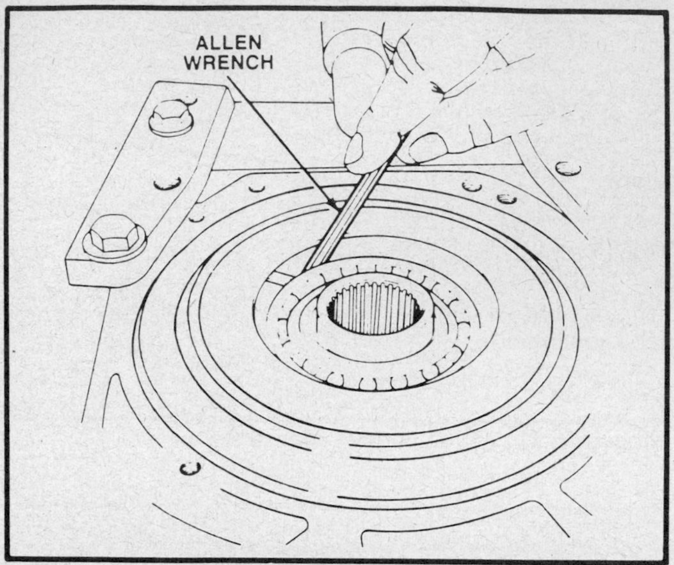
ALLEN
WRENCH

Piston/accumulator removal

5. Disconnect the other electrical lead at the pressure switch. Remove the switch from the valve body.

Inspection

1. Inspect the valve body casting for cracks, porosity, damaged machined surfaces, nicks or burrs in valve bores and/or flatness of valve body to case mating surface (using a straight edge or by inspecting the gaskets for uniform compression).
2. Inspect valves for burrs, nicks, scratches and/or scoring.
3. Inspect valve bushings for porosity, burrs, nicks, scratches and/or scoring.
4. Inspect springs for damaged and/or distorted coils.

Assembly

1. Coat all the components with clean Dexron®II automatic transmission fluid before assembling.
2. Install the valve body into the transmission case. Connect the electrical lead at the pressure switch.
3. Install the check ball and solenoid. Install the solenoid attaching bolts. Connect the solenoid electrical lead at the pressure switch.
4. Using tool J-34529 or equivalent, relieves the pressure on the accumulator valve. Install the pin, spring, valve, plug, sleeve and plunger.
5. Using tool J-34529 or equivalent, relieves the pressure on pressure relief valve. Install the pin, spring and valve.
6. Using tool J-34529 or equivalent, relieves the pressure on the shift valve. Install the pin, spring and valve.

OUTPUT SHAFT

Disassembly

1. Remove the speedometer gear retaining clip and gear.
2. Remove the allen head bolts retaining the pump cover to the pump housing. Remove the cover.
3. Mark pump gears with a grease pencil. Gears must be installed in same direction as removed.
4. Position the output shaft with the splines down. Rotate the pump housing until gears slide out.
5. Remove the drive pin from the output shaft.
6. Remove the pump housing from the output shaft.
7. Remove the thrust washer from the pump housing.
8. Remove the thrust bearing and washer from the clutch hub.

9. Remove the clutch hub from the output shaft. Note the direction of the hub on the shaft. The oil grooves face the sprag clutch or forward on shaft.
10. Remove the sprag clutch from the output shaft. Note direction of the sprag clutch. The lip on the sprag clutch cage goes toward oil grooves on the clutch hub.

Inspection

1. Clean all parts well and check for damage or excessive wear. Check the gears for burrs and cracks.
2. Inspect the washers and replace any that appear damaged or warped.
3. Check the bearing assemblies for any mutilation and replace as needed.

Assembly

Coat all parts before assembling with clean Dexron®II automatic transmission fluid.

1. Install the sprag clutch on the output shaft. The lip on the sprag clutch cage faces rearward or towards the oil grooves on the clutch hub.
2. Install the clutch hub on the output shaft. The oil grooves on the hub face the sprag clutch or forward on the shaft.
3. Install the thrust washer and then the thrust bearing on the clutch hub.
4. Install the thrust washer on the pump housing. Use petrolatum to retain the thrust washer to the housing.
5. Install the pump housing on the output shaft.
6. Install the pin in the output shaft.
7. Install the pump gears in the housing. Gears must be installed in same direction as removed.
8. Place pump cover on the housing. Align the bolt holes in cover to pump housing. Install the bolts and torque to specifications.
9. Install the speedometer gear on the output shaft. Install the retaining clip.
10. Install new O-rings on pump. Use petroleum jelly to retain the O-rings to the cover.

CARRIER

Disassembly

1. Remove the nuts retaining the carrier cover. Remove the cover.
2. Remove the thrust washer, thrust bearing, selective washer and input sun gear.
3. Remove the pinion gears.
4. Remove the steel thrust plate from the carrier.

Inspection

Clean and inspect parts. Replace any parts that are cracked, chipped or show excessive wear.

Assembly

NOTE: The carrier assembly must be reassembled in the transmission case.

1. Install the steel thrust plate into the carrier.
2. Install the pinion gears.
3. Install the input sun gear, selective washer, thrust bearing and thrust washer.
4. Install the cover. Install the carrier cover retaining nuts.

PISTON/ACCUMULATOR

Disassembly

1. Remove the snapring retaining the accumulator to the piston.
2. Remove the accumulator and springs from the piston.
3. Remove the O-rings from the accumulator.
4. Remove the O-rings from the piston.

Inspection

Clean and inspect parts. Replace any parts that are cracked, chipped or show excessive wear.

Assembly

1. Coat O-rings with clean Dexron®II automatic transmission fluid before installing.
2. Install the O-ring onto the piston.
3. Install the O-ring onto the accumulator.
4. Install the accumulator and springs onto the piston.
5. Install the snapring retaining the accumulator to the piston.

SYNCHRONIZER KEYS AND SPRINGS

Replacement

The synchronizer hubs and sliding sleeves are a selected assembly and should be kept together as originally assembled, but the keys and 2 springs may be replaced if worn or broken.

1. If relation of hub and sleeve are not already marked, mark for assembly purposes.
2. Push the hub from the sliding sleeve, the keys will fall free and the springs may be easily removed.
3. Place the 2 springs in position (one on each side of hub) so all three keys are engaged by both spring. Place the keys in position and while holding them in place, slide the sleeve onto the hub, aligning the marks made before disassembly.

DRIVE GEAR BEARING RETAINER OIL SEAL

Replacement

1. Pry out old seal.
2. Using a new seal, install new seal into retainer using tool J-23096 or equivalent, until it bottoms in bore. Lubricate inner diameter of seal with transmission lubricant.

REVERSE SHIFTER SHAFT AND/OR SEAL

Replacement

1. With the reverse housing removed from transmission, the reverse shifter shaft lock pin will already be removed.
2. Carefully drive shifter shaft into the reverse housing allowing ball detent to drop into case. Remove shaft and ball detent spring. Remove O-ring seal from shaft.
3. Place ball detent spring into detent spring hole and start reverse shifter shaft into hole in boss.
4. Place detent ball on spring and while holding ball down, push the shifter shaft into place and turn until the ball drops into place in detent on the shaft detent plate.
5. Install O-ring seal on shaft.
6. Install shift fork. Do not drive the shifter shaft lock pin into place until the reverse housing has been installed on the transmission case.

REVERSE IDLER SHAFT

Replacement

1. Place a small punch into hole in front cover of the overdrive unit and drive the pin into the shaft until the shaft can be pulled from front cover.
2. Insert new idler shaft into cover until hole in shaft lines up with hole in boss.

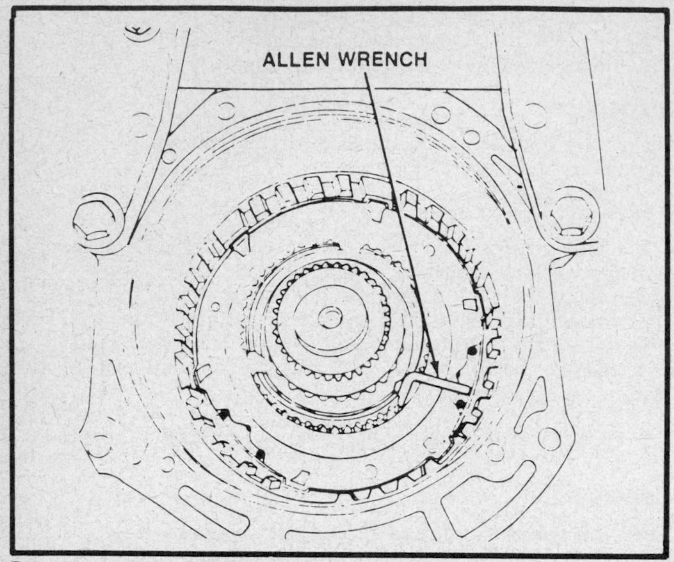
ALLEN WRENCH
Clutch plates removal

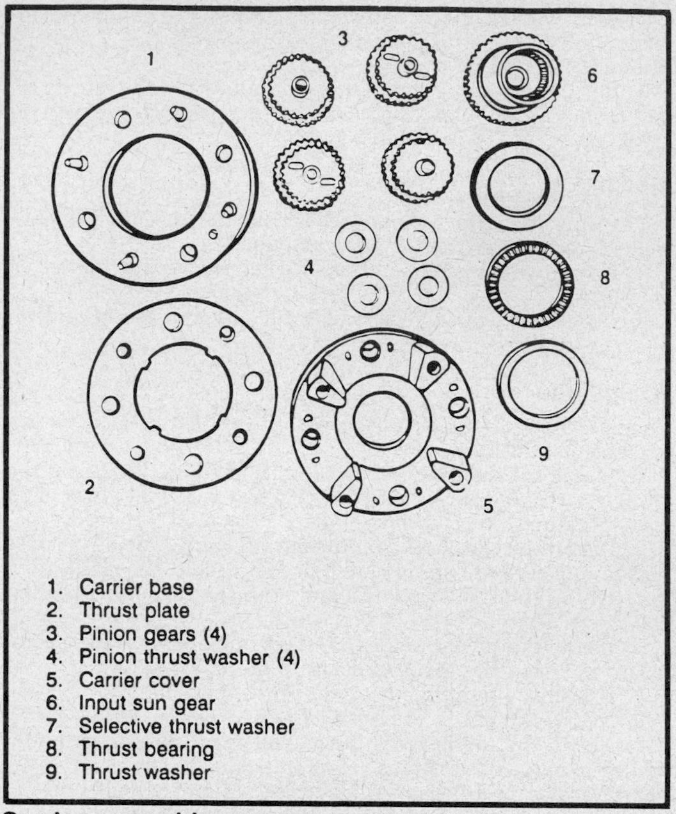
1
2
3
4
5
6
7
8
9
1. Carrier base
2. Thrust plate
3. Pinion gears (4)
4. Pinion thrust washer (4)
5. Carrier cover
6. Input sun gear
7. Selective thrust washer
8. Thrust bearing
9. Thrust washer
Carrier assembly components

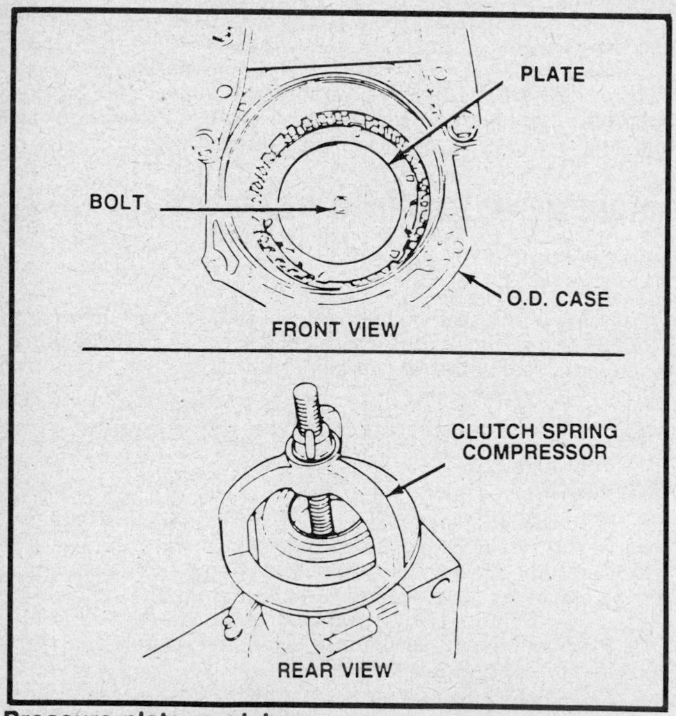
PLATE
BOLT
O.D. CASE
FRONT VIEW
CLUTCH SPRING
COMPRESSOR
REAR VIEW
Pressure plate servicing

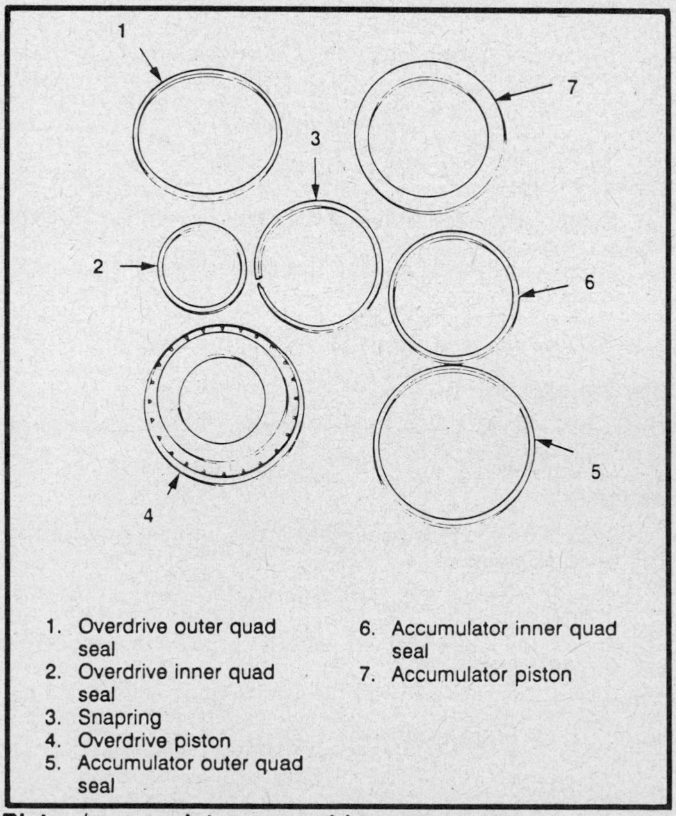
1
2
3
4
5
6
7
1. Overdrive outer quad seal
2. Overdrive inner quad seal
3. Snapring
4. Overdrive piston
5. Accumulator outer quad seal
6. Accumulator inner quad seal
7. Accumulator piston
Piston/accumulator assembly components

3. Insert roll pin into boss opening and drive the pin into the cover until the shaft is securely locked in place.

Transmission Assembly

GM WARNER 83MM – 4 SPEED

Assembly

1. Place the transmission case on its side with the shift cover opening toward the assembler. Position the countergear tanged washers in place, using a heavy grease to retain them.

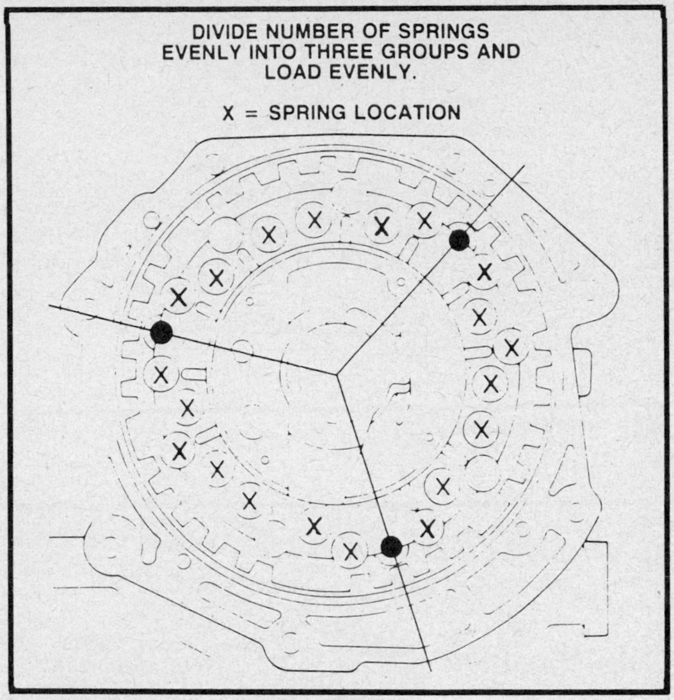
DIVIDE NUMBER OF SPRINGS
EVENLY INTO THREE GROUPS AND
LOAD EVENLY.
X = SPRING LOCATION

Pressure plate spring installation

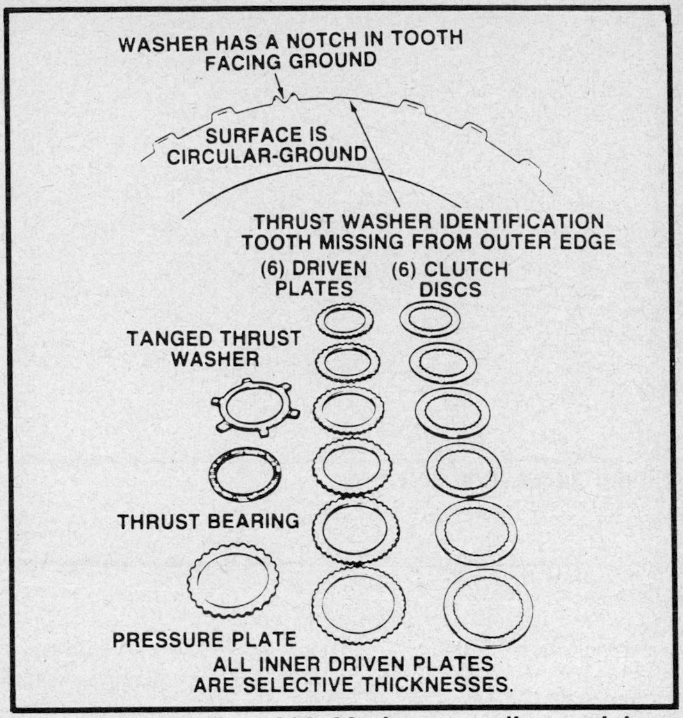
WASHER HAS A NOTCH IN TOOTH
FACING GROUND
SURFACE IS
CIRCULAR-GROUND
THRUST WASHER IDENTIFICATION
TOOTH MISSING FROM OUTER EDGE
(6) DRIVEN
PLATES
(6) CLUTCH
DISCS
TANGED THRUST
WASHER
THRUST BEARING
PRESSURE PLATE
ALL INNER DRIVEN PLATES
ARE SELECTIVE THICKNESSES.

Direct clutch pack – 1986–88 shown, earlier models similar

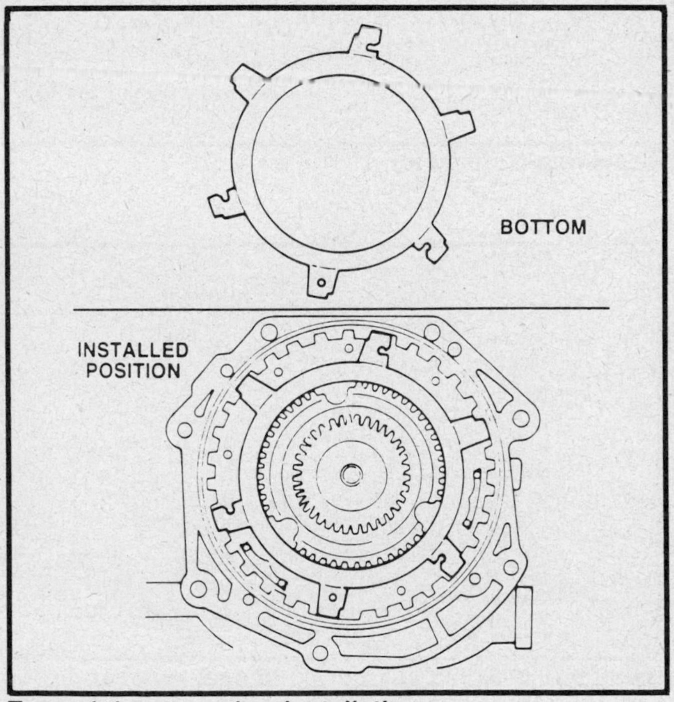
BOTTOM
INSTALLED
POSITION

Tanged thrust washer installation

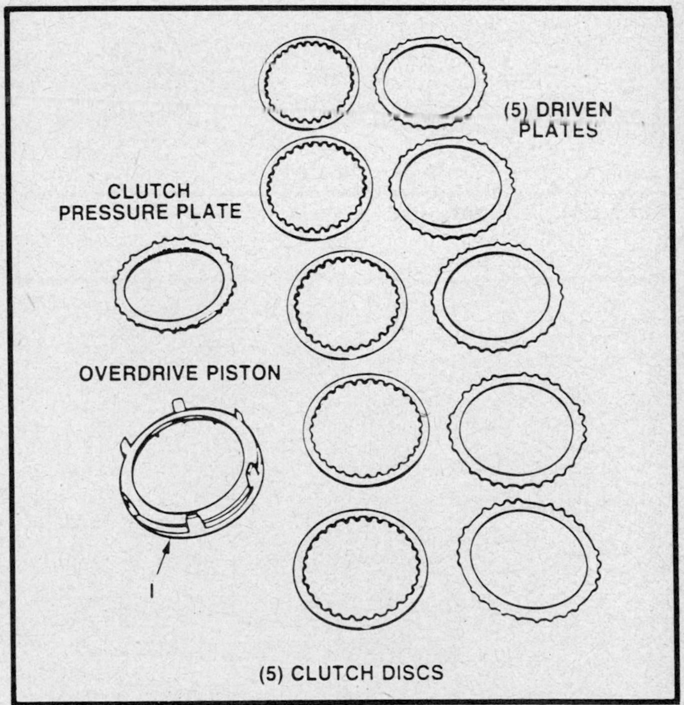
(5) DRIVEN
PLATES
CLUTCH
PRESSURE PLATE
OVERDRIVE PISTON
(5) CLUTCH DISCS

Overdrive clutch pack – 1986–88 shown, earlier models similar

NOTE: Be sure the tangs are in the notches of the thrust face.

2. Position the countergear in the bottom of the case.
3. Install front reverse idler gear (teeth forward) and thrust washer in case. Use a heavy grease to hold thrust washer in position.
4. Using a heavy grease, install 16 roller bearings and washer into main drive gear. Mate main drive gear with mainshaft assembly. Position 3–4 synchronizer sliding sleeve forward. This will provide clearance for installation as well as hold the assembly together.
5. Position a new reverse housing to case gasket on the rear of the case.

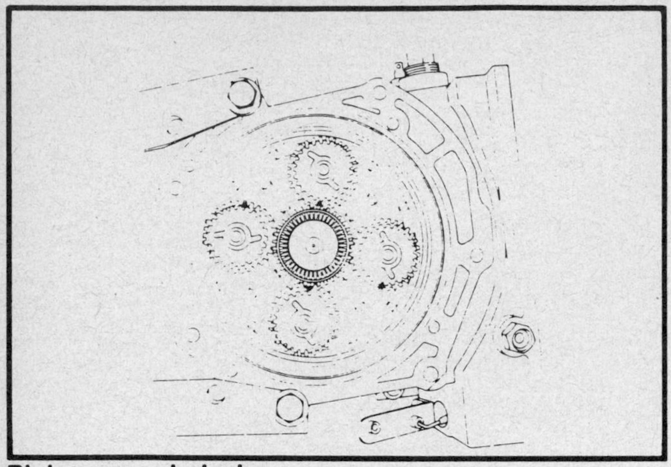
Pinion gears indexing

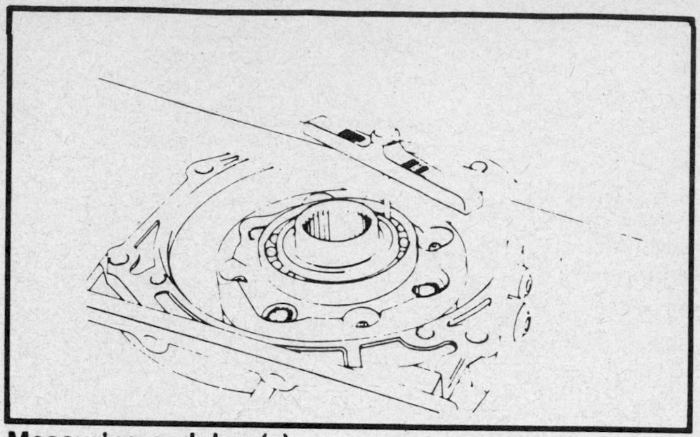
Measuring endplay (a)

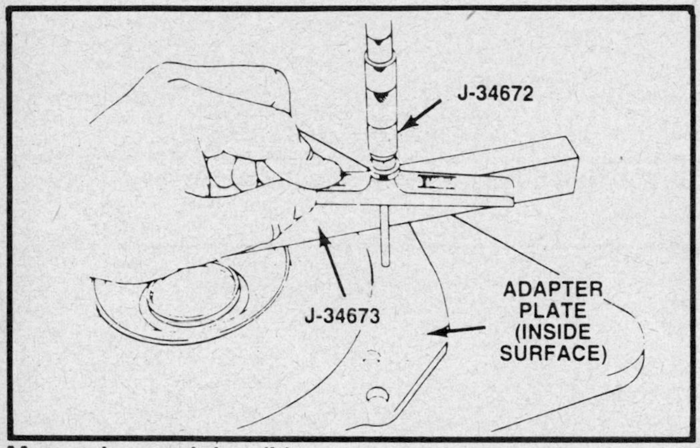
J-34672
J-34673
ADAPTER
PLATE
(INSIDE
SURFACE)
Measuring endplay (b)

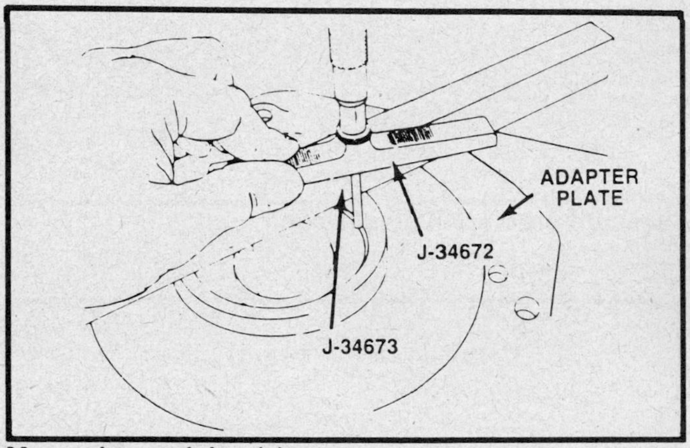
ADAPTER
PLATE
J-34672
J-34673
Measuring endplay (c)

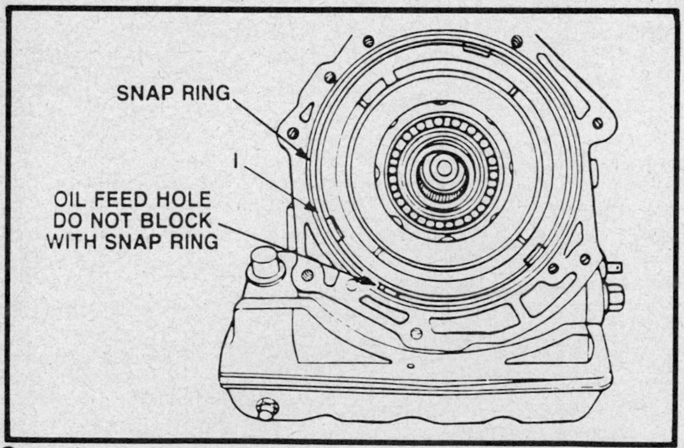
SNAP RING
OIL FEED HOLE
DO NOT BLOCK
WITH SNAP RING
Snapring installation

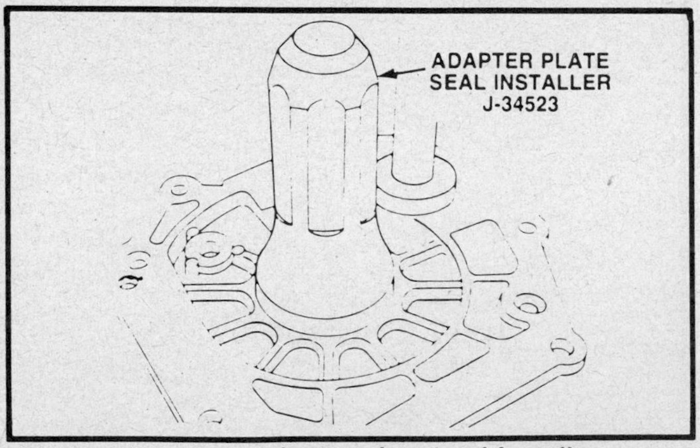
ADAPTER PLATE
SEAL INSTALLER
J-34523
Using tool J–34523 adapter plate seal installer

6. Install the mainshaft and drive gear assembly into the case.

7. Place bearing snapring on front main bearing. Position front main bearing to case opening and with a hollow shaft, or tool J–5590, tap bearing into case. Install spacer washer and selective fit snapring to secure main drive bearing.

8. Raise the countergear in the case, aligning the holes in the case with the center of the gear. With the thrust washers in place, slide the countershaft through the rear of the case. Install the woodruff key and tap the shaft into the case, until flush with the rear face of the transmission case.

9. Align the reverse housing and gasket to the transmission case. Install the locating pin for the reverse housing. Tap the pin in until flush with housing.

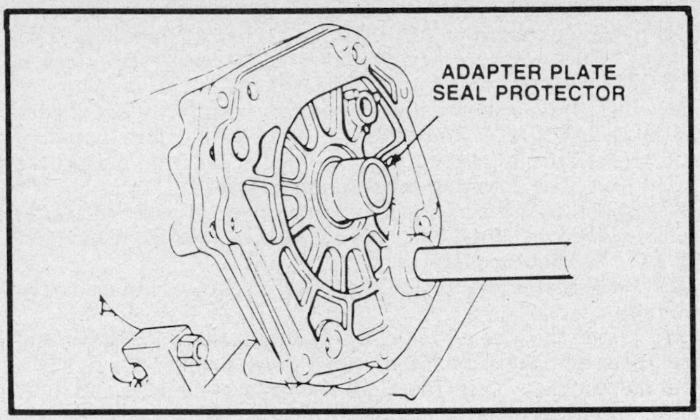
ADAPTER PLATE
SEAL PROTECTOR
Using tool J-34621 adapter plate seal protector

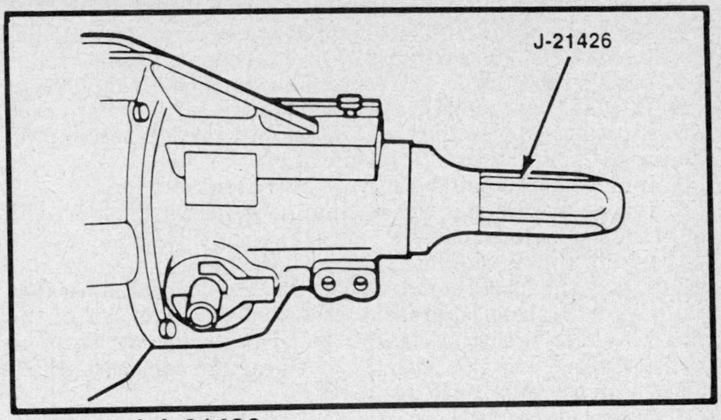
J-21426
Using tool J-21426

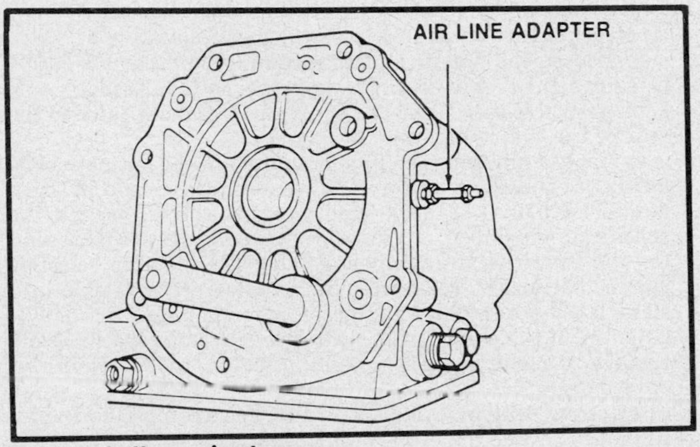
AIR LINE ADAPTER
Using air line adapter

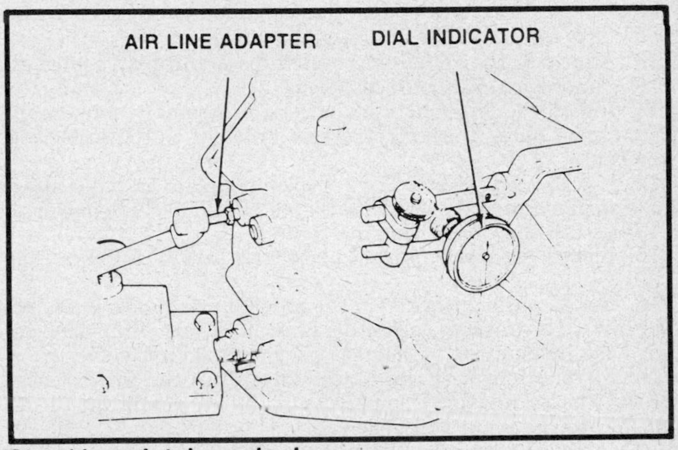
AIR LINE ADAPTER
DIAL INDICATOR
Checking clutch pack clearance

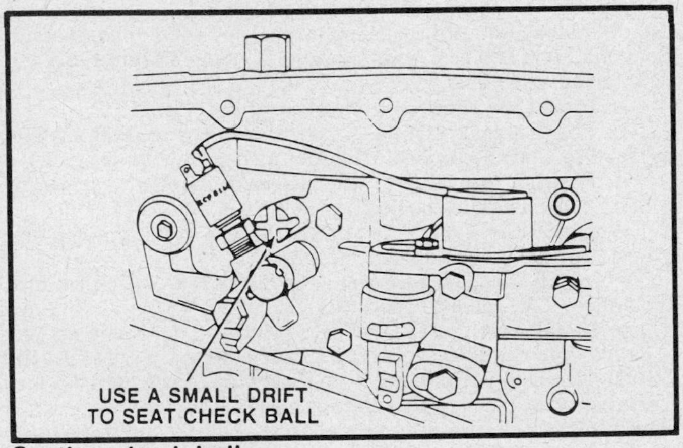
USE A SMALL DRIFT
TO SEAT CHECK BALL
Seating check ball

10. Install the bolts attaching the reverse housing to the case. Torque bolts to specifications.
11. Install the reverse shift shaft and O-ring into the housing. Install retaining pin.
12. Install the reverse gear and shift fork. Slide the gear and fork forward on the mainshaft until shift fork and shifter shaft can be indexed into position.
13. Position the drive gear bearing retainer and gasket to the front of the case. Apply sealer to the bolts. Install bolts and torque to specifications.
14. Install the rear reverse idler gear. Align the splines on the rear gear with the front gear and slide together.
15. Assemble the overdrive unit to the reverse housing. Guide the idler shaft on the overdrive unit into the idler gears and align the splines on the mainshaft with the splines in the input sun gear. Slide the units together and install the retaining bolts. Torque the bolts to specifications.
16. Slide the 1–2 synchronizer forward into 2nd gear. Install the shift forks into the grooves of the synchronizers. Place the side cover with a gasket on the transmission. Guide the shift forks into the cover and install the retaining bolts. Torque the bolts to specifications.
17. Check operation of transmission by manually shifting the transmission into all gears.

OVERDRIVE UNIT

Assembly

1. Install the pressure plate springs into the pockets of the transmission case.
2. Place the pressure plate on top of the springs. Seat the springs into the pockets of pressure plate.
3. Position plate tool J–21420–2 or equivalent, on top of the pressure plate with the bolt from tool J–23327 or equivalent, through the center of the plate. Next position tool J–23327 or equivalent, on the rear of the case and install the retaining nut. Tighten the nut until the pressure plate is drawn approximately

⅛ in. below the step for the overdrive clutch plates. Install the pressure plate retaining bolts, tool J-34681 or equivalent. Remove tools J-21420 and J-23327 or equivalent, for the case.

4. Install the output shaft assembly into the transmission case. Be sure the O-rings are positioned properly on pump cover before installing the output shaft assembly. Install the pump retaining bolts and torque to specifications.

5. Install the thrust bearing on the output sun gear.

6. Install the tanged direct clutch thrust washer with the tabs facing pressure plate.

7. Install the direct clutch thrust bearing.

8. Install the direct clutch thrust washer. The thrust washer will have a tooth missing from its outer edge. The side of the thrust washer with the circular grind pattern must face the thrust bearing. The side with the grind pattern can be identified by the notch ground into the tooth.

9. Install 1 composition clutch disc and then install a selective clutch plate. The selective clutch plates come in 5 sizes (0.080–0.120 in.) and is used to control clutch pack clearance. A 0.050–0.070 in. clearance must be maintained in the direct clutch pack. Excessive or insufficient amount of clutch travel will cause failure to the clutch plates and discs.

10. Alternate the remaining clutch discs and plates until all plates and discs are installed.

11. Install the lower half of the carrier assembly onto the direct clutch pack. Index the carrier until all clutch plates are engaged.

12. Install the steel overdrive stop clutch plate and then alternate with a disc and plate until all plates and disc are installed.

13. Install the finger pressure plate.

14. Install the carrier thrust plate with tabs facing the sprag clutch.

15. Install pinion gears with the index mark on the gears facing inward or towards each other. Install the other pinion gears with the index mark 90 degrees off from the first gears.

16. Install the thrust washer for the output sun into the rear of the input sun gear. Use petroleum jelly to retain the thrust washer to the input sun gear.

17. Install the input sun gear. If the input sun gear spreads the pinion gears when installing, the pinion gears are not indexed properly.

18. Install the selective thrust washer with the oil grooves on washer facing input sun gear.

19. Install the thrust bearing on the input sun gear.

20. Install the carrier thrust washer to the cover. Use petrolatum to retain the thrust washer to the cover.

21. Install the pinion gear thrust washers onto the carrier cover. Use petroleum jelly to retain washers to the cover.

22. Install the carrier cover. If the pinion gears are not indexed properly, the bolt holes in the cover will not align with the bolts in lower half of the carrier.

23. Install new retaining nuts and torque to specification.

24. Measure the endplay for the overdrive unit as follows:

a. Place the straight edge across the face of the overdrive unit. Use the depth micrometer J-34672 or equivalent, and measure the distance from the bearing to the top of the bar. Next, measure the thickness of the bar, tool J-34673 or equivalent, with a 0–1 micrometer and subtract this from reading of the depth micrometer, tool J-34672 or equivalent, and record this reading.

b. Place the straight edge across the rear of the adapter plate. Use the depth micrometer J-34672 or equivalent, and measure the distance from the top of the bar to the adapter plate mounting surface as shown and record the reading.

c. Next measure the distance from the top of the bar to the bearing seat in the adapter plate and record the reading.

d. Subtract the reading from Step c from Step b and record the difference.

e. Next, subtract the difference from Step d from Step a. The difference will be the endplay. Specification is 0 ± 0.003 in. If the results of your measurements are not within the specifications, it will be necessary to remove the carrier cover and change the input sun selective thrust washer. The selective thrust washers are available in (8) sizes. They are in 0.005 in. increments ranging from 0.123–0.158 in.

25. Install the accumulator and piston assembly. Coat the lips of the seals with automatic transmission fluid before installing.

26. Install the large snapring that goes in the front of the overdrive unit. The snapring must be installed.

27. Install a new seal in the adapter plate. Place the seal on tool J-34523 or equivalent, and install the seal from the front side of the adapter plate.

28. Place seal protector tool J-34621 or equivalent, on the input sun gear.

29. Install the adapter plate. Apply a light coating of RTV Sealant, part number 1052366 or equivalent, around the heads of the adapter plate bolts. Install the adapter plate bolts and tighten to specifications.

30. Remove the seal protector.

31. Remove the first ⅛ in. pipe plug from the left side of the overdrive unit. Install air line fitting tool J-34742 or equivalent, into plug hole and tighten.

32. Measure the clutch pack clearance as follows:

a. Loosen the pressure plate retaining bolts, tool J-34681 or equivalent,evenly until spring pressure is released.

b. Assemble tool J-8001 or equivalent, dial indicator to the rear of the overdrive unit.

c. Apply a minimum of 100 psi to the air line fitting tool J-34742 or equivalent, and read the dial indicator (specification is 0.050–0.070 in.). If the reading does not fall within the specification 0.050–0.070 in., it will be necessary to disassemble the overdrive unit to change the direct clutch selective clutch plates. The selective clutch plates are available in 5 sizes. They are in 0.010 in. increments ranging from 0.080–0.120 in. If the clutch pack clearance is within specification, remove the clutch pack retaining bolts, tool J-34681 or equivalent.

33. Coat the ⅛ in. pipe plugs with anti-seize compound and install plugs. Torque plugs to specification.

34. Remove the air line adapter tool J-34742 or equivalent. Coat the plug with anti-seize compound and install the plug. Torque plug to specifications.

35. Install the speedometer gear and sensor.

36. Install a new output seal using tool J-21426 or equivalent. Coat the lip of the seal with Dexron®II transmission fluid.

37. Install the valve body as follows:

a. Install the check ball into the case.

b. Position gaskets, one on each side of the separator plate.

c. Position the separator plate on the valve body.

d. Position the valve to the case and install the retaining bolts. Torque bolts to specifications.

e. Install the T.V. cable and install the retaining clip and bolt. Torque bolt to specifications.

f. Install the T.V. lever and torque bolt to specifications. Connect T.V. cable to the lever.

g. Install tool J-34671-1 or equivalent, (throttle setting gauge) into the T.V. cable bore on the side of the case. Set the hook on the T.V. cable onto the high step of the gauge. Place the cam stop on the valve body as close to the lever as possible and install the retaining bolt. Torque bolt to specifications.

h. Set the hook on the T.V. cable onto the lower step of the gauge. Place tool J-34671-2 or equivalent, between the piston and the solenoid bracket. Adjust the screw/bolt on the T.V. lever until the bolt makes contact with the stop on the cam.

38. Install the pickup tube and oil filter on the valve body.

39. Apply a bead of sealant, part number 1052366 or equivalent, to the oil pan flange and assemble wet. Install the magnet in the oil pan. The bead of RTV should be applied around the inside of the bolt holes. Install the pan bolts and torque to specifications.

40. Assemble the overdrive unit to the reverse housing. Guide

the idler shaft on the adapter plate into the idler gears and align the splines on the mainshaft with the splines in the input sun gear. Slide the units together and install the retaining bolts. Torque bolts to specifications.

SPECIFICATIONS

TORQUE SPECIFICATIONS

Item	ft. lbs.	Nm
Drive gear bearing retainer bolts	15–20	20–27
Side cover to case bolts	15–20	20–27
Reverse gear housing to case (1) bolt	30–40	40–54
Reverse gear housing to case (2) bolt	40–50	54–67
Reverse gear housing to case (3) bolt	35–45	47–61
Drain plug	15–25	20–33
Filler plug	25–35	33–47
Transmission to bell housing bolts	45-60	60–80
Cooler block to case bolts (2)	6–8	8–10
Pressure tap plugs (2)	55 ①	—

TORQUE SPECIFICATIONS

Item	ft. lbs.	Nm
Pressure plate access plugs (3)	55 ①	—
Valve body to case bolts	6–8	8–10
Adapter plate to case bolts	18–20	24–27
Oil pan to case bolts	6–8	8–10
Pump housing to case bolts	6–8	8–10
Pump cover to pump cavity	10–12	13–16
Overdrive case to reverse housing bolts	34–36	46–48
Carrier cover retaining nuts	16–18	21–24

① Inch lbs.

SPECIAL TOOLS

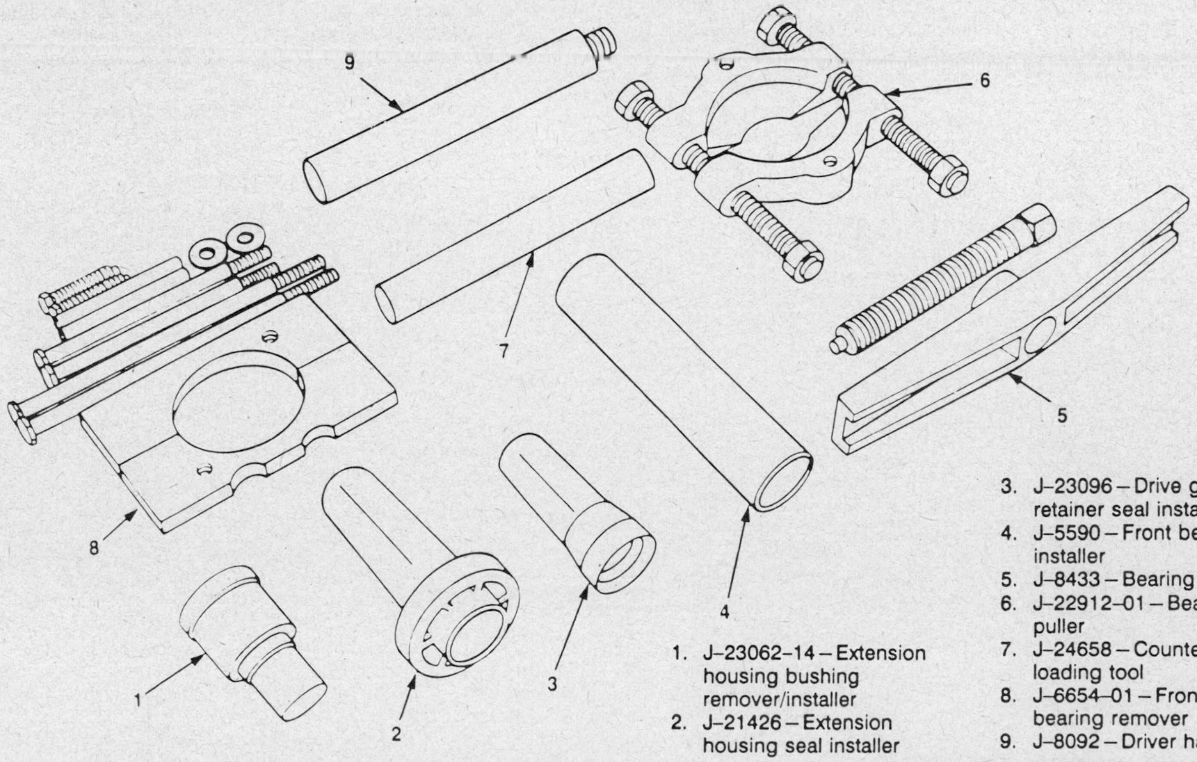
9
6
7
5
8
4
1
2
3

1. J-23062-14 – Extension housing bushing remover/installer
2. J-21426 – Extension housing seal installer
3. J-23096 – Drive gear retainer seal installer
4. J-5590 – Front bearing installer
5. J-8433 – Bearing puller
6. J-22912-01 – Bearing puller
7. J-24658 – Countergear loading tool
8. J-6654-01 – Front bearing remover
9. J-8092 – Driver handle

SPECIAL TOOLS

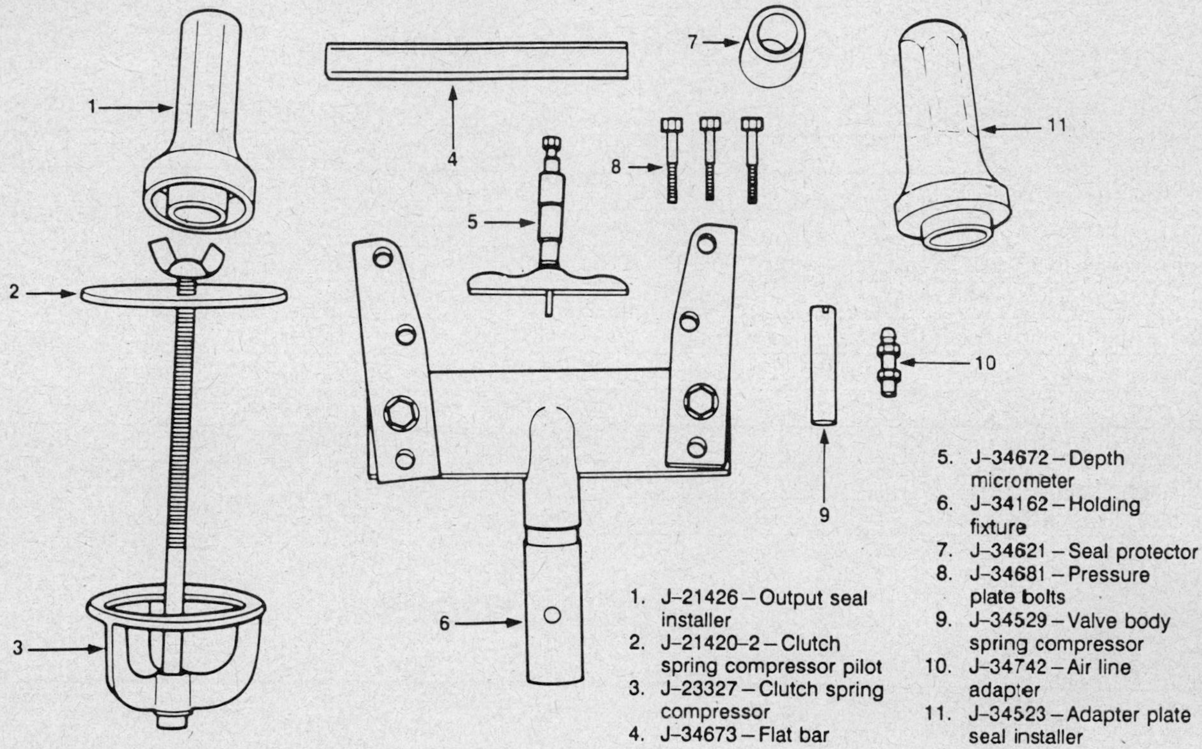
1
2
3
4
5
6
7
8
9
10
11

1. J-21426 – Output seal installer
2. J-21420-2 – Clutch spring compressor pilot
3. J-23327 – Clutch spring compressor
4. J-34673 – Flat bar
5. J-34672 – Depth micrometer
6. J-34162 – Holding fixture
7. J-34621 – Seal protector
8. J-34681 – Pressure plate bolts
9. J-34529 – Valve body spring compressor
10. J-34742 – Air line adapter
11. J-34523 – Adapter plate seal installer

Section 7

MY1 Transmission
General Motors

APPLICATION

1984–87 Chevrolet Chevette/Pontiac 1000

GENERAL DESCRIPTION

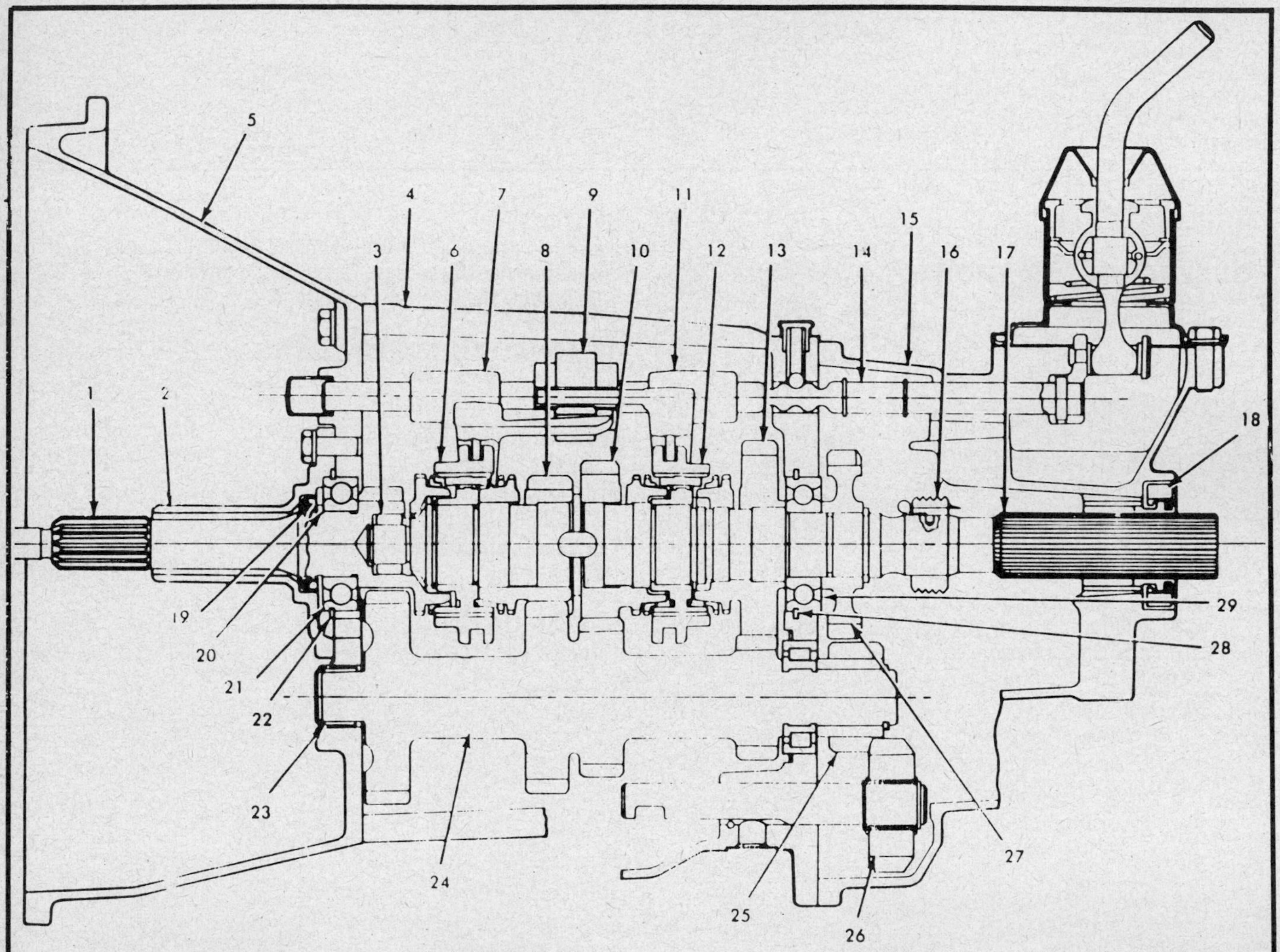
1
2
3
4
5
6
7
8
9
10
11
12
13
14
15
16
17
18
19
20
21
22
23
24
25
26
27
28
29

1. DRIVE GEAR
2. BEARING RETAINER
3. PILOT BEARINGS
4. CASE
5. BELLHOUSING
6. 3-4 SYNCHRONIZER ASSEMBLY
7. 3-4 SHIFTER FORK
8. THIRD SPEED GEAR
9. DETENT BUSHING
10. SECOND SPEED GEAR
11. 1-2 SHIFTER FORK
12. 1-2 SYNCHRONIZER ASSEMBLY
13. FIRST SPEED GEAR
14. SHIFTER SHAFT
15. EXTENSION
16. SPEEDOMETER DRIVE GEAR AND CLIP
17. MAINSHAFT
18. REAR OIL SEAL
19. RETAINER OIL SEAL
20. SNAP RING — BEARING TO GEAR
21. DRIVE GEAR BEARING
22. SNAP RING BEARING TO CASE
23. COUNTERGEAR ROLLER BEARINGS
24. COUNTERGEAR ASSEMBLY
25. COUNTER REVERSE GEAR
26. REVERSE IDLER GEAR
27. REVERSE GEAR
28. SNAP RING — BEARING TO EXTENSION
29. REAR BEARING

4 speed transmission

The 4 speed Muncie 70mm (MY1) transmission is a constant mesh design transmission. The input shaft has an inegral main drive gear and rotates with the clutch driven plate (disc), the shaft rotates all the time the clutch is engaged and the engine is running. The input shaft is supported in the case by a ball bearing and at the front end by an oil impregnated bushing mounted in the engine crankshaft.

The drive gear is in constant mesh with the crankshaft drive gear. Since all gears in the countershaft cluster are integral to the shaft, they also rotate at the time the clutch is engaged. The countergear is carried on a needle bearing at the front and a roller bearing at the rear and thrust is taken through a thrust washer at the front and a roller bearing at the rear. The transmission mainshaft is held in line with the input shaft by a pilot bearing at its front end, which allows it to rotate or come to rest independently of the input shaft. It is carried at the rear by a ball bearing mounted in the rear face of the case.

Helical gears are incorporated throughout the transmission. The mainshaft gears are free to rotate independently on the mainshaft and are in constant mesh with the countershaft gears. The reverse idler gear is carried on a bushing; finish bored in place.

The transmission is fully synchronized in all forward speeds, however, reverse is not. The synchronizer hubs are splined to the mainshaft and retained by snaprings. These assemblies permit gears to be selected without clashing, by synchronizing the speeds of mating parts before they engage.

Metric Fasteners

Do not attempt to interchange metric fasteners for inch system fasteners. Mismatched or incorrect fasteners can result in damage to the transmission unit through malfunctions, breakage or possible personal injury. Care should be taken to reuse the fasteners in the same location as removed, whenever possible. Due to the large number of alloy parts used and the aluminum casing, torque specifications should be strictly observed. Before installing bolts into aluminum parts, always dip threads into oil, or sealant (if specifically required), to prevent the bolts from galling the aluminum threads and to prevent seizing.

Capacities

The transmission holds 1.7 qts. (1.6L) of SAE 80W-90 GL5 gear lubricant.

Checking Fluid Level

With the vehicle on a level surface and the lubricant in the transaxle cold, remove the plug and check the lubricant level. The lubricant should come to the bottom edge of the plug hole. If the level is low, add SAE 80W-90 GL5 gear lubricant through the filler plug hole until it begins to run out of the plug hole. Reinstall and tighten the plugs.

TRANSMISSION MODIFICATIONS

There are no transmission modifications at the time of publication.

TROUBLE DIAGNOSIS

CHILTON'S THREE C'S TRANSMISSION DIAGNOSIS

Condition	Cause	Correction
Slips out of high gear	a) Transmission loose on engine b) Damaged mainshaft pilot bearing c) Main drive gear retainer broken or loose d) Misalignment of transmission e) Stiff shift lever boot f) Pilot bearing loose in crankshaft	a) Tighten mounting bolts b) Replace pilot bearing c) Tighten or replace main drive gear retainer d) Realign transmission e) Replace boot f) Repair or replace pilot bearing
Noisy in all gears	a) Insufficient lubricant b) Worn countergear bearings c) Worn or damaged main drive gear and countergear d) Damaged main drive gear or main shaft bearings	a) Fill to correct level b) Replace countergear bearings and shaft c) Replace worn or damaged gears d) Replace damaged bearings or main drive gear
Noisy in high gear	a) Damaged main drive gear bearing b) Damaged mainshaft bearing	a) Replace damaged bearing b) Replace damaged bearing
Noisy in neutral with engine running	a) Damaged main drive gear bearing b) Damaged or loose mainshaft pilot bearing c) Worn countergear bearings	a) Replace damaged bearing b) Replace pilot bearings c) Replace countergear bearings and shaft
Noisy in all reduction gears	a) Insufficient lubricant b) Worn or damaged main drive gear or countergear	a) Fill to correct level b) Replace faulty or damaged gears

CHILTON'S THREE C'S TRANSMISSION DIAGNOSIS

Condition	Cause	Correction
Noisy in all reduction gears	c) Damaged countergear or mainshaft bearings	c) Replace bearing
Noisy in second only	a) Damaged or worn second-speed constant mesh gears b) Worn or damaged countergear rear bearings c) Damaged or worn second-speed synchronizer	a) Replace damaged gears b) Replace countergear bearings c) Replace synchronizer
Noisy in third only (four speed)	a) Damaged or worn third-speed constant mesh gears b) Worn or damaged countergear bearings	a) Replace damaged gears b) Replace damaged countergear bearings
Noisy in reverse only	a) Worn or damaged reverse idler gear or idler bushing b) Worn or damaged reverse gear on mainshaft c) Damaged or worn reverse countergear d) Damaged shift mechanism	a) Replace reverse idler gear assembly b) Replace reverse gear c) Replace countergear reverse gear d) Inspect and replace damaged parts
Excessive backlash in all reduction gears	a) Worn countergear bearings b) Excessive end play in countergear	a) Replace bearings b) Replace countergear thrust washers
Main drive gear bearing retainer burned or scored by input shaft	a) Loose or damaged mainshaft pilot bearing b) Misalignment of transmission	a) Replace bearing b) Align transmission
Leaks lubricant	a) Excessive amount of lubricant in transmission b) Loose or broken main drive gear bearing retainer c) Main drive gear bearing retainer gasket damaged d) Case to clutch housing gasket leaking e) Extension to case gasket leaking f) Extension seal leaking	a) Drain to correct level b) Tighten or replace retainer c) Replace gasket d) Replace gaskets e) Replace gaskets f) Inspect bushing for wear. Replace seal and/or bushing as required

ON CAR SERVICE

Adjustments

BALL STUD

1. Disconnect the negative battery cable.
2. Raise the vehicle and support it safely.
3. Position gauge J–28449 or equivalent so that the flat end is against the front face of the clutch housing and the hooked end is aligned with the bottom depression of the clutch fork.
4. Turn the ball stud clockwise by hand until the clutch release bearing makes contact with the clutch spring and the fork is snug on the gauge.
5. Install the locknut and tighten to 25 ft. lbs. (33 Nm) being careful not to change the ball stud adjustment.
6. Remove the gauge by pulling outward at the housing end.

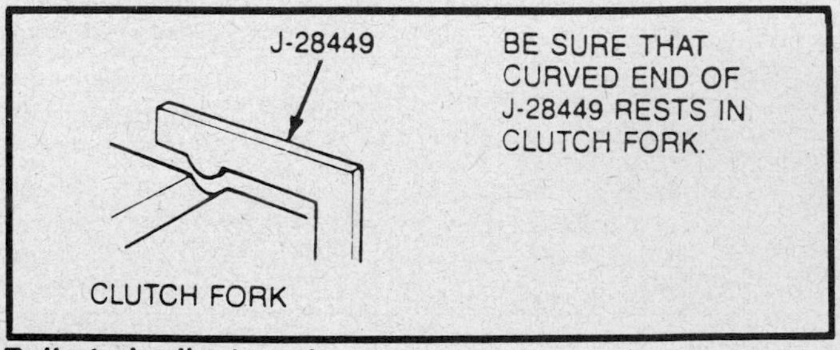
J-28449
BE SURE THAT CURVED END OF J-28449 RESTS IN CLUTCH FORK.
CLUTCH FORK
Ball stud adjustment

CLUTCH CABLE

The following adjustments are to made with the cable and loose parts assembled to the front of the dash and with the cable attached to the clutch pedal.

1. Disconnect the negative battery cable.
2. Raise the vehicle and support it safely.
3. Install the cable through the hole in the clutch fork and seat.
4. Install the return spring.
5. From the engine compartment, pull the cable away from the dash until the clutch pedal is firmly against the pedal bumper and hold it in position.
6. Install the ring in the first fully visible groove in the cable from the sleeve. Release the cable.

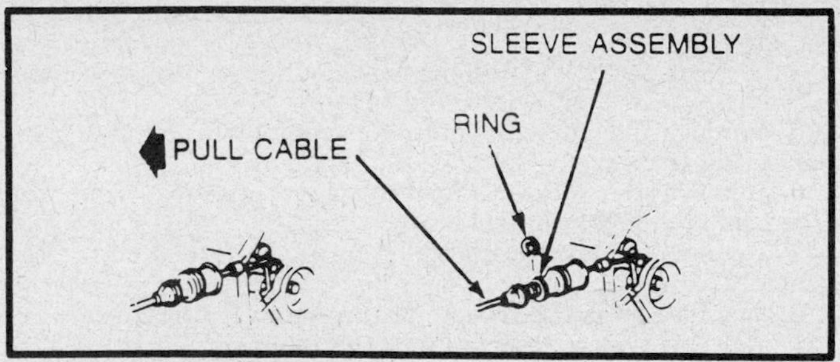
SLEEVE ASSEMBLY
RING
PULL CABLE
Clutch cable adjustment

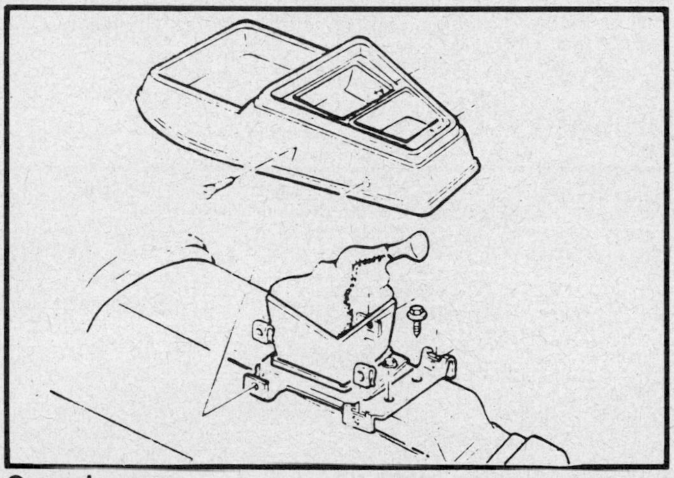
Console cover

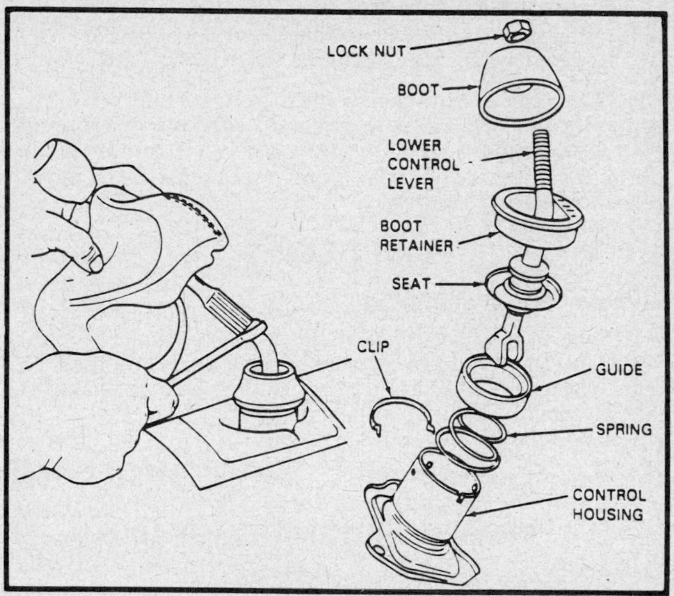
LOCK NUT
BOOT
LOWER CONTROL LEVER
BOOT RETAINER
SEAT
CLIP
GUIDE
SPRING
CONTROL HOUSING
Shift lever replacement

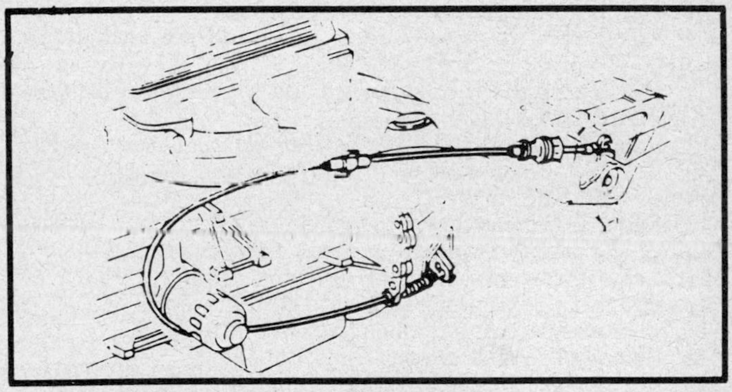
Clutch pedal free travel adjustment

7. Depress the clutch pedal to the floor 4 times minimum to insure all parts of the clutch control system are properly seated.
8. The cable adjustment procedure should produce 1 in. (25mm) of lash at the clutch pedal.

NOTE: If the clutch pedal lash does not fall within the 1 in. (25mm) range, the minor clutch pedal free travel adjustment will be necessary.

MINOR CLUTCH PEDAL FREE TRAVEL

1. If there is insufficient clutch pedal lash, remove the ring from the cable and allow the cable to move into the dash by 1 cable notch and reinstall the ring.
2. If there is excessive clutch pedal lash, remove the ring from the cable and pull the cable out of the dash by 1 cable notch and reinstall the ring.

Services

EXTENSION HOUSING OIL SEAL

Removal and Installation

1. Disconnect the negative battery cable.
2. Raise the vehicle and support it safely.
3. Remove the driveshaft. Position a pan under the extension housing to catch any lubricant that may leak out.
4. Remove or disconnect any necessary items to obtain clearance.
5. Pry the seal out of the extension housing.
6. Wash the counterbore with cleaning solvent and inspect for damage.
7. Using tool J–5158 or J–21426 or equivalent, tap the seal into the counterbore until the flange bottoms against the extension.
8. Reinstall the driveshaft and any items removed to obtain clearance.
9. Fill the transmission to the proper level with SAE 80W90 GL5 gear lubricant.
10. Lower the vehicle.
11. Connect the negative battery cable.

SPEEDOMETER DRIVEN GEAR

Removal and Installation

1. Disconnect the negative battery cable.
2. Raise the vehicle and support it safely.
3. Disconnect the speedometer cable.
4. Remove the lock plate to extension bolt and lockwasher. Remove the lock plate.
5. Pry the fitting, gear and shaft from the extension housing.
6. Pry the O-ring from the groove in the fitting.
7. Install a new O-ring in the groove in the fitting. Coat the O-ring and driven gear shaft with transmission lubricant and insert the shaft.
8. Hold the assembly so that the slot in the fitting is toward the lock plate boss on the extension housing and install the extension housing.

9. Push the fitting into the extension until the lock plate can be inserted in the groove and attach it to the extension.
10. Lower the vehicle and connect the negative battery cable.

NOTE: Speedometer driven gears used with this transmission are not common and will not interchange with speedometer driven gears used in other transmissions. Damage to the drive gear may result if the driven gears are interchanged.

SHIFT LEVER

Removal and Installation

1. Remove the floor console and the boot retainer.
2. Raise the boot to gain access to the locknut on the shift lever. Loosen the locknut and unscrew the upper portion of the shift lever with the knob attached.
3. Remove the foam insulator to gain access to the control assembly bolts.
4. Remove the bolts on the extension and remove the control assembly.
5. Remove the clip with caution as component parts are under spring pressure.
6. Remove the locknut, boot retainer and seat from the threaded end of the control lever.
7. Remove the spring and guide from the forked end of the control lever.
8. Installation is the reverse of the removal procedure.
9. Bolt the control assembly to the extension, making sure that the fork at the lower end of the control lever has properly engaged the shifter shaft lever arm pin.
10. Slide the boot below the threaded portion of the control lever, assemble the upper shift lever and tighten the locknut.

REMOVAL AND INSTALLATION

TRANSMISSION REMOVAL

1. Disconnect the negative battery cable.
2. Remove the shifter lever.
3. Raise the vehicle and support it safely.
4. Drain the lubricant from the transmission.
5. Matchmark and remove the driveshaft.
6. Disconnect the speedometer cable and the backup lamp switch.
7. Disconnect the return spring and the clutch cable at the clutch shift fork.
8. Remove the crossmember to transmission mount bolts.
9. Remove the exhaust manifold nuts and the converter to tailpipe bolts and nuts.
10. Remove the converter to transmission bracket bolts and remove the converter assembly from the vehicle.
11. Remove the crossmember to frame bolts.
12. Support the transmission on a suitable jack.
13. Remove the crossmember from the vehicle.
14. Remove the dust cover.
15. Remove the clutch housing to engine retaining bolts.
16. Slide the transmission and clutch housing rearward and remove it from the vehicle.

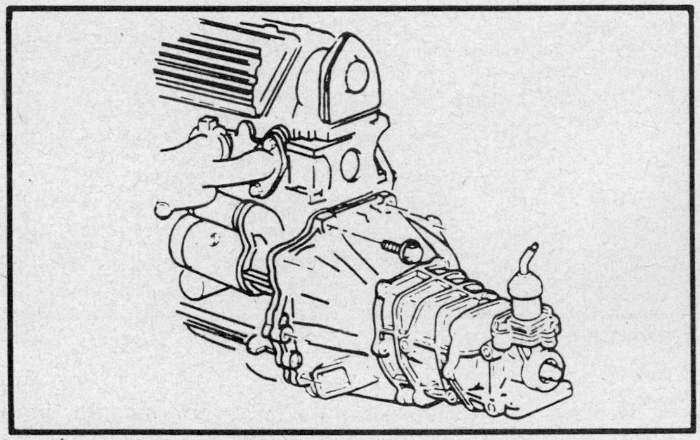
Transmission attachment

TRANSMISSION INSTALLATION

1. Support the transmission on a suitable transmission jack.
2. With the transmission in gear, raise the transmission and clutch housing into position and slide it forward.
3. Turn the output shaft to align the spline on the input shaft with the clutch disc.
4. Install the clutch housing retaining bolts and lockwashers. Torque the bolts to 25 ft. lbs. (34 Nm).
5. Install the dust cover.
6. Position the crossmember to the frame and loosely install the retaining bolts.
7. Install the crossmember to transmission mount bolts and tighten the center bolt to 33 ft. lbs. (45 Nm) and the end bolt to 21 ft. lbs. (28 Nm).
8. Tighten the crossmember to frame bolts to 40 ft. lbs. (55 Nm).
9. Install the exhaust pipe to the manifold and the converter bracket to the transmission.
10. Connect the clutch cable and adjust the clutch.
11. Connect the speedometer and backup lamp switch.
12. Install the driveshaft.
13. Fill the transmission to the proper level with SAE 80W90 GL5 gear lubricant.
14. Install the shift lever and lower the vehicle.

BENCH OVERHAUL

Before Disassembly

Clean the exterior of the transmission assembly before any attempt is made to disassemble it, in order to prevent dirt or other foreign materials from entering the transmission assembly or its internal parts. If steam cleaning is done to the exterior of the transmission, immediate disassembly should be done to avoid rusting, caused by condensation forming on the internal parts.

Transmission Disassembly

1. Position the transmission so that it is resting on the front of the bellhousing, on wooden blocks.
2. Drive the spring pin from the shifter shaft arm assembly and shifter shaft. Remove the shifter shaft arm assembly.
3. Remove the extension housing from the case.
4. Press down on the speedometer gear retainer and remove the gear and retainer from the mainshaft.

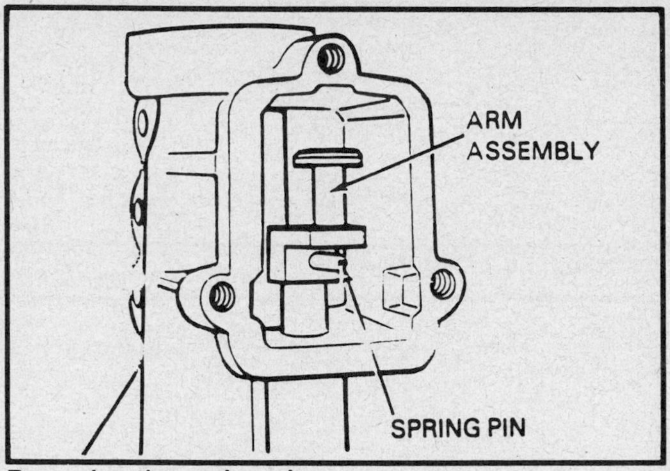
ARM
ASSEMBLY
SPRING PIN
Removing the spring pin

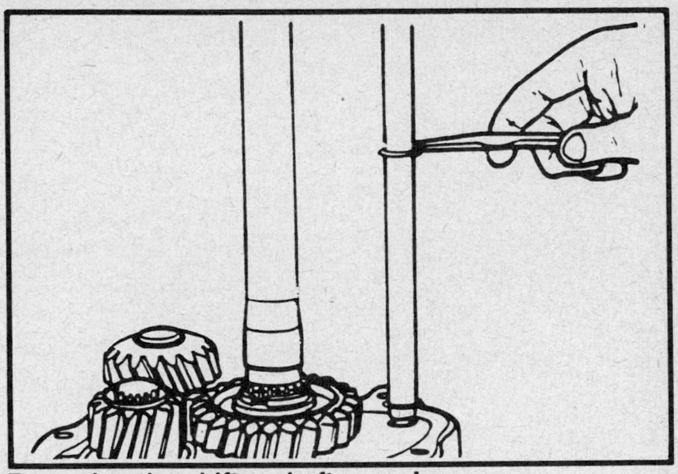
Removing the shifter shaft snaprings

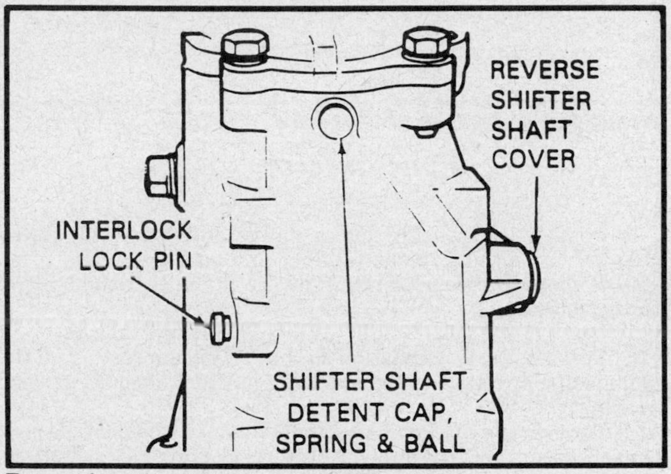
REVERSE
SHIFTER
SHAFT
COVER
INTERLOCK
LOCK PIN
SHIFTER SHAFT
DETENT CAP,
SPRING & BALL
Removing the cover, cap and lockpin

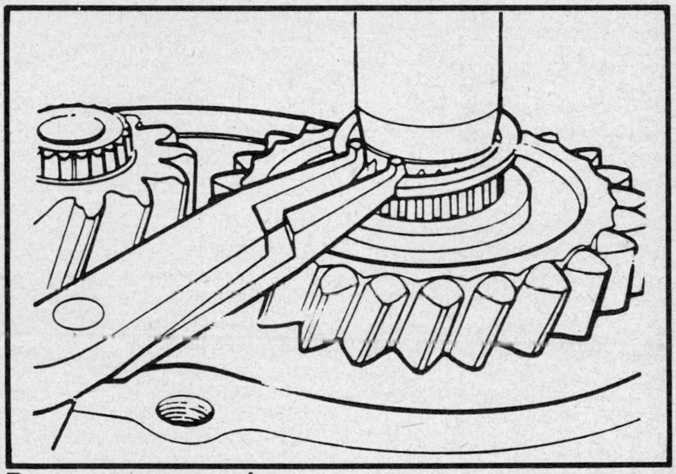
Reverse gear snapring

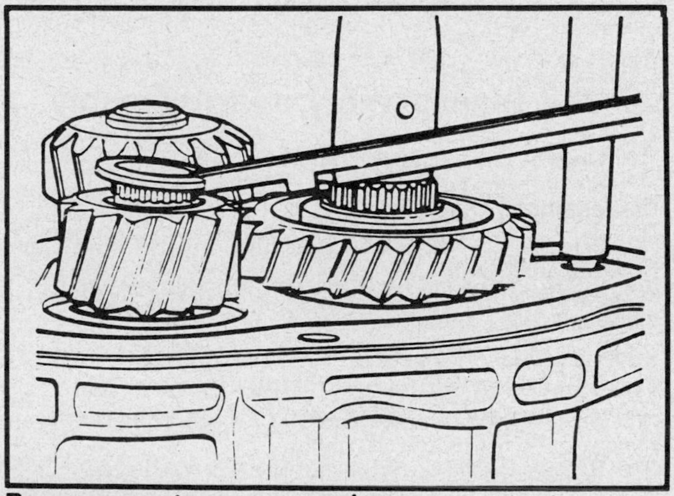
Reverse counter gear snapring

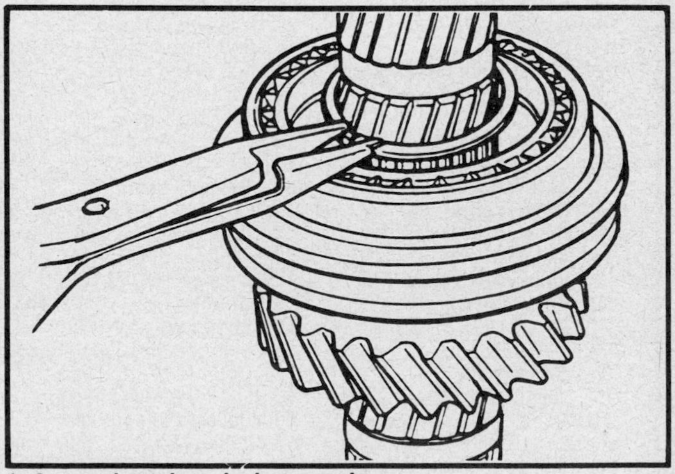
1–2 synchronizer hub snapring

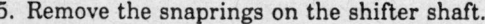

5. Remove the snaprings on the shifter shaft.
6. Remove the reverse shifter shaft cover, shifter shaft, detent cap, spring and ball, and the interlock pin.
7. Pull the reverse lever shaft outward to disengage the reverse idler. Remove the idler shaft with the gear attached.
8. Remove the snapring on the reverse gear and the reverse countershaft gear. Remove the gears.

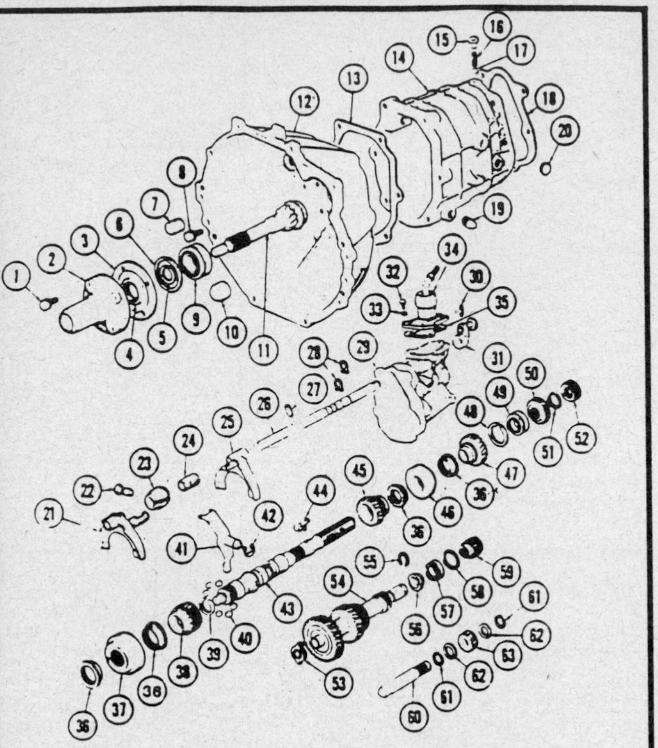

1. Bearing retainer bolts
2. Bearing retainer
3. Bearing retainer gasket
4. Bearing retrainer seal
5. Snapring
6. Bearing outer snapring
7. Shifter shaft stop plug
8. Bell housing to case bolts
9. Clutch gear bearing
10. Counter gear front needle bearings
11. Clutch gear
12. Bell housing
13. Bell housing to case gasket
14. Case
15. Shifter shaft detent cap
16. Shifter shaft detent spring
17. Shifter shaft detent ball
18. Case to extension gasket
19. Magnet plug
20. Reverse lever cap
21. 3–4 shift forks
22. Interlock lock pin
23. Detent bushing
24. Detent lever
25. 1–2 shift fork
26. Shifter shaft
27. Detent lever pin
28. Shifter shaft snaprings
29. Extension housing
30. Shifter shaft arm spring pin
31. Shifter shaft arm
32. Shift lever to extension bolts
33. Shift lever to extension washer
34. Shift lever assembly
35. Shift lever to extension gasket
36. Blocker rings
37. 3–4 synchronizer assembly
38. 3rd speed gear
39. Snapring hub to shaft
40. Clutch gear bearings
41. Reverse lever assembly
42. Reverse lever snapring
43. Mainshaft
44. Speedometer gear retainer
45. 2nd speed gear
46. 1–2 synchronizer assembly
47. 1st speed gear
48. Outer bearing snapring
49. Rear bearing
50. Reverse gear
51. Reverse gear snapring
52. Speedometer gear
53. Counter gear thrust washer
54. Counter gear
55. Snapring
56. Inner bearing race
57. Counter gear bearing
58. Outer bearing snapring
59. Reverse gear on the countergear
60. Idler gear shaft
61. Idler gear thrust washer
62. Idler gear snapring
63. Reverse idler gear

4 speed transmission exploded view

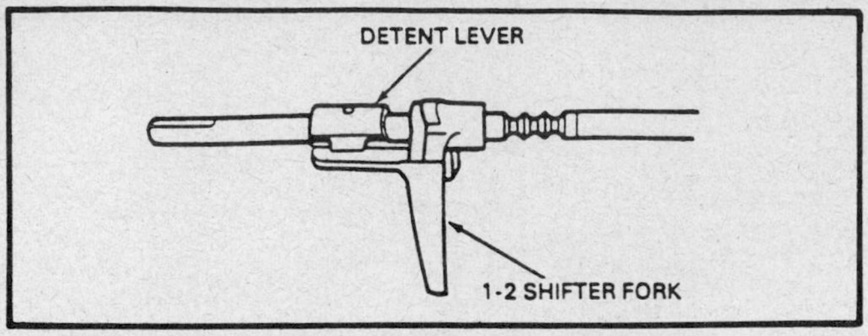
DETENT LEVER
1-2 SHIFTER FORK

Installing the 1–2 shifter fork

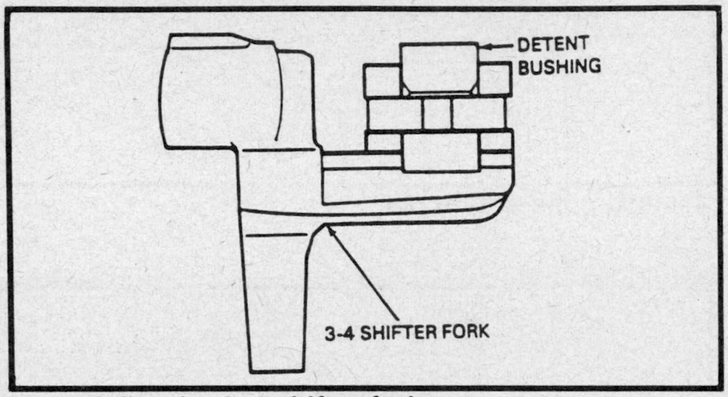
DETENT BUSHING
3-4 SHIFTER FORK

Assembling the 3–4 shifter fork

9. Position the transmission on its side and remove the clutch gear bearing retainer bolts, retainer and gasket.
10. Remove the snapring retaining the clutch gear ball bearing to the bellhousing.
11. Remove the bolts holding the bellhousing to the case.
12. Position the transmission so that it is again resting on the bellhousing and expand the snapring in the mainshaft bearing opening.
13. Remove the case by lifting it off the mainshaft. Make sure that the mainshaft assembly, countergear and shifter shaft assembly remain with the bellhousing.
14. Lift from the bellhousing as an assembly, the mainshaft with the shifter forks attached and the countergear meshed with the gear teeth in the mainshaft.

Unit Disassembly and Assembly

MAINSHAFT

Disassembly

1. Separate the shift shaft assembly and the countergear from the mainshaft.
2. Remove the clutch gear and blocker ring from the mainshaft.

NOTE: The clutch gear has 15 roller bearings. Be careful of them falling out during disassembly.

3. Remove the snapring before the 3–4 synchronizer hub and remove the synchronizer assembly.
4. Remove the blocker ring and the 3rd speed gear.
5. Using press plates, remove the ball bearing from the rear of the mainshaft.
6. Remove the 1st speed gear and the blocker ring.
7. Remove the snapring before the 1–2 synchronizer hub and remove the synchronizer assembly.
8. Remove the 2nd speed gear.

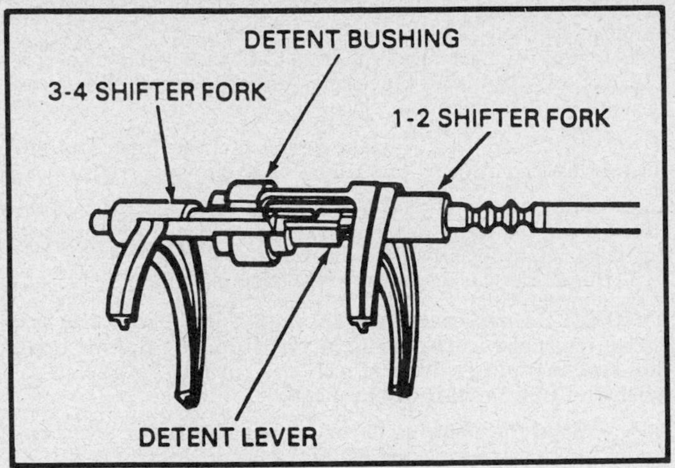
DETENT BUSHING
3-4 SHIFTER FORK
1-2 SHIFTER FORK
DETENT LEVER
1–2 and 3–4 shifter forks

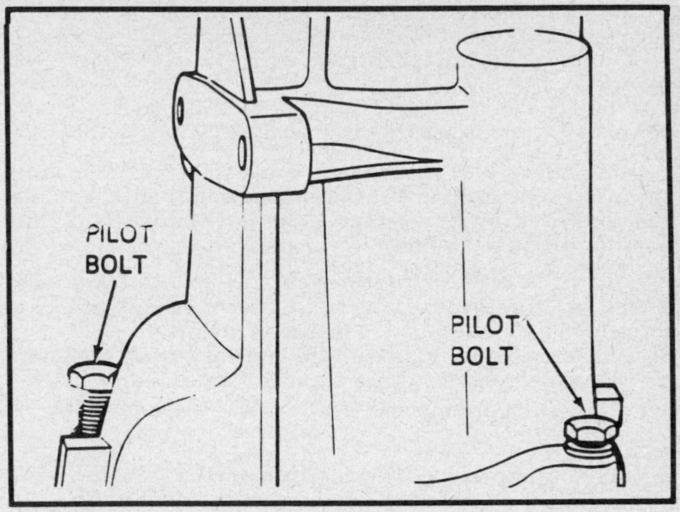
PILOT
BOLT
PILOT
BOLT
Pilot bolts

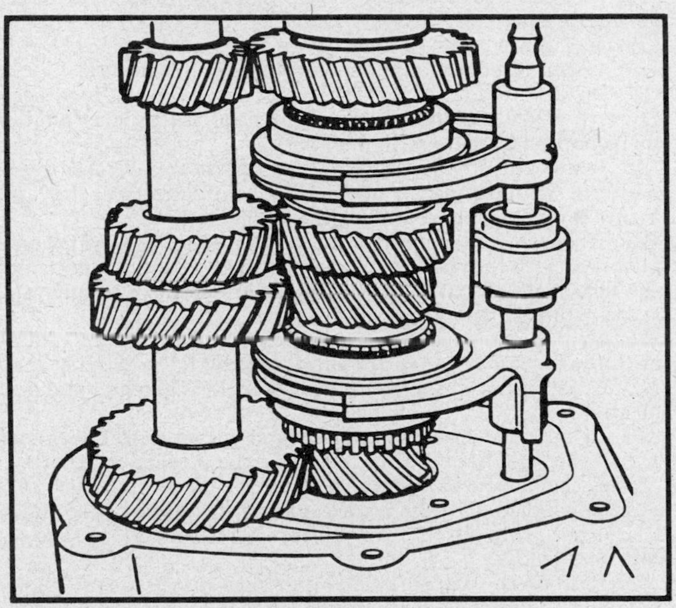
Counter gear and mainshaft to bellhousing

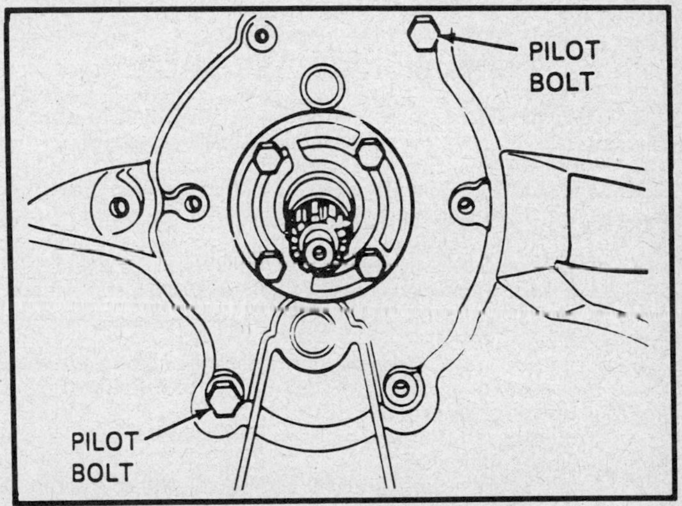
PILOT
BOLT
PILOT
BOLT
Installing the pilot bolts to the bellhousing

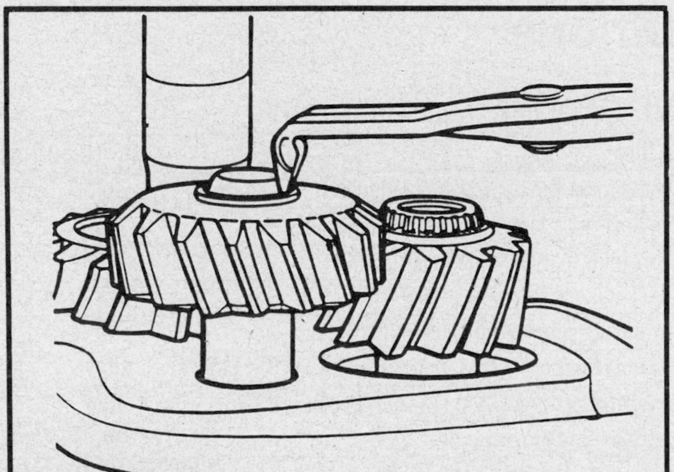
Reverse idler gear

Inspection

Make sure the bearings are clean, then lubricate them with engine oil and check them for roughness by turning the race by hand. Inspect all gears fo excessive wear, chips or cracks and replace any that are worn or damaged.

Assembly

1. Turn the rear of the mainshaft upward.
2. Install the 2nd speed gear with the clutching teeth upward. The rear face of the gear will butt against the flange on the mainshaft.
3. Install a blocker ring with the clutching teeth downward over the synchronizing surface of the 2nd speed gear.

NOTE: All 4 blocker rings used in this transmission are identical.

4. Install the 1st and 2nd synchronizer assembly with the fork slot downward. Press it on the splines on the mainshaft until it bottoms out.

NOTE: Make sure the notches of the blocker ring align with the keys of the synchronizer assembly.

5. Install the synchronizer hub to the mainshaft snapring.

6. Install a blocker ring with the notches downward so they align with the keys of the 1st and 2nd synchronizer assembly.
7. Install the 1st speed gear with the clutching teeth downward.
8. Install the rear ball bearing with the snapring groove downward. Press it onto the mainshaft.

NOTE: Two ball bearings are used in this transmission. The one used on the mainshaft is not shielded, but the one used on the clutch gear is shielded and they should not be switched.

9. Turn the front of the mainshaft upward. Install the 3rd speed gear with the clutching teeth upward. The front face of the gear will butt against the flange on the mainshaft.
10. Install a blocker ring with the clutching teeth downward over the synchronizer surface of the 3rd speed gear.
11. Install the 3rd and 4th synchronizer assembly with the fork slot downward.

NOTE: Make sure the notches of the blocker ring align with the keys of the synchronizer assembly.

12. Install a synchronizer hub to the mainshaft snapring.
13. Install a blocker ring with the notches downward so that they align with the keys of the 3–4 synchronizer assembly.

Transmission Assembly

1. Using an arbor press, install the shielded ball bearing to the clutch gear shaft with the snapring groove upward.
2. Install the snapring on the clutch gear shaft.
3. Load the 15 mainshaft pilot roller bearings into the clutch gear cavity. Use heavy grease or equivalent to hold them in place.
4. Install the clutch gear on the mainshaft.
5. Install the detent lever to the shift shaft with the roll pin. Slide the 1-2 shifter fork on the shaft so that it engages the detent lever.
6. Install the 3–4 shifter fork to the detent bushing and then slide the assembly on the shift shaft to locate it below the 1st and 2nd shifter fork arm.
7. Install the shifter assembly to the synchronizer sleeve grooves on the mainshaft.
8. Install a thrust washer over the hole for the countershaft gear in the bell housing.
9. Mesh the countershaft gears with the mainshaft gears and install the bell housing as an assembly.
10. Position the bell housing on its side and install the snapring to the ball bearing on the clutch gear.
11. Install the bearing retainer to the bell housing.
12. If removed, install the reverse lever to the case. Use grease or equivalent to hold it in place.

NOTE: When the reverse lever is installed, the slot should be parallel to the front of the case.

13. Install the reverse lever snapring.
14. Install the roller bearing to the countergear opening with the snapring groove inside the case.
15. Install the gasket to the bell housing.

NOTE: Before installing the case, make sure the synchronizers are in the neutral position, the detent bushing slot is facing outward and the reverse lever is flush with the inside wall of the case.

16. Expand the snapring in the mainshaft opening of the case and let it pilot over the mainshaft bearing.
17. Install the interlock lock pin to hold the shifter shaft in place.
18. Install the idler shaft so it will engage with the reverse lever inside the case.
19. Install the cover to hold the reverse lever in place.
20. Install the detent ball, spring and cap in the case.
21. Install the reverse gear with the chamfer on the gear teeth upward. Push the the reverse gear onto the splines on the mainshaft and secure with a snapring.
22. Install the smaller reverse gear on the countergear shaft with the shoulder resting against the countergear bearing and secure it with a snapring.
23. If removed, install the snapring, thrust washer and reverse idler gear with the chamfer of the gear teeth facing downward to the idler shaft. Secure with a thrust washer and a snapring.
24. Install the snaprings on the shifter shaft.
25. Heat the speedometer gear to 175°F and engage the retainer in the hole provided in the mainshaft with the retainer loop forward, slide the speedometer gear over the mainshaft and into position.
26. Install the extension housing and gasket on the case and install the pilot bolts before installing the 3 remaining bolts.
27. Assemble the shifter shaft arm over the shifter shaft to a position aligned with the drilled hole near the end of the shaft. Drive the spring pin into the shifter shaft arm and shaft to retain these parts.
28. Position the transmission on its side. Install the pilot bolts before installing the 4 remaining bolts to the bellhouse and case.

SPECIFICATIONS

TORQUE SPECIFICATIONS

Item	ft. lbs.	Nm
Bellhousing to engine bolts	25	34
Crossmember to transmission—center nut	33	45
Crossmember to transmission—end nut	21	28
Crossmember to frame bolts	40	55
Shift control lever retaining bolts	35	4
Transmission filler plug	25	34
Clutch cover to flywheel bolts	18	24
Extension housing to case bolts	26	35
Bellhousing to case bolts	26	35

Item	ft. lbs.	Nm
Bearing retainer to bellhousing bolts	105	12
Speedometer driven gear retaining bolts	44	5
Clutch fork ball stud locknut	24	33
Clutch cable locknut	53	6
Bellhousing lower cover bolts	90	10
Rear support to transmission bolts	32	43
Converter bracket to rear support nuts	150	17
Back-up lamp switch	25	34

Section 7

Geo Tracker Transmission
General Motors

APPLICATION

1989 Geo Tracker

GENERAL DESCRIPTION

The 5 speed manual transmission consists of an input shaft, mainshaft, countershaft and reverse gear, installed in an aluminum case. The 5 forward speed gears are synchro meshed and a reverse speed gear is constant meshed.

The mainshaft gears, held by the needle bearings, are assembled with synchronizer rings, sleeves and hubs.

The gear shift lever case, located behind the transmission cases, has a gear shift limit yoke to prevent direct gear shift from the 5th speed gear to reverse gear.

Since the aluminum case is sealed with a liquid type gasket, it is necessary to use GM silicone sealant part number 1052917 or equivalent, on its mating surfaces upon reassembly. Also, the case fastening bolts must be tightened to specified torque using a torque wrench; tightening over or below the specified torque should be avoided.

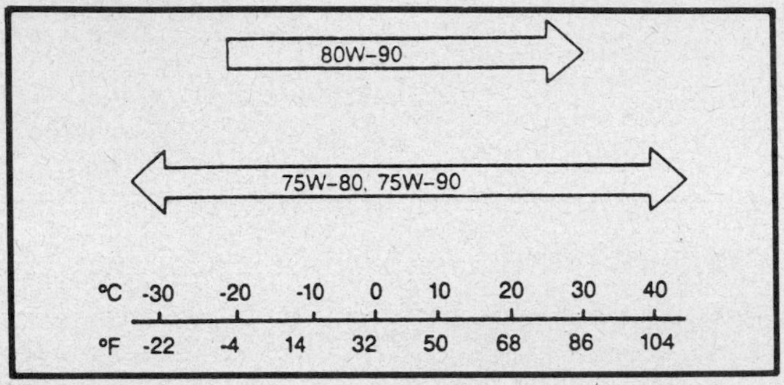
80W-90
75W-80, 75W-90
°C -30 -20 -10 0 10 20 30 40
°F -22 -4 14 32 50 68 86 104

Viscosity chart

Transmission identification

The transmission identification number is stamped into the top front main case.

The 1st digit (letter) indicates the year of manufacture and the next 6 digits indicate the serial number.

Metric Fasteners

The transaxle is of a metric design; all bolt sizes and thread pitches are metric. Metric fastener dimensions are very close to the customary inch system fastener dimensions; replacement of the fasteners must be of the same measurement and strength as those removed.

Do not attempt to interchange metric fasteners with the customary inch system fasteners. Mismatched or incorrect fasteners can result in damage to the transaxle. Care should be taken to reuse the same fasteners in the location from which they were removed.

NOTE: Be sure to check the case holes for the quality of their threads. It is rather difficult to rethread bolt holes after the transaxle is installed in the vehicle.

Capacities

The fluid quantities are approximate and the correct fluid level is determined to be correct by being level with the bottom of the oil filler hole. Dry fill is 3.2 pts. (1.5L). Use only SAE 75W-90 gear oil.

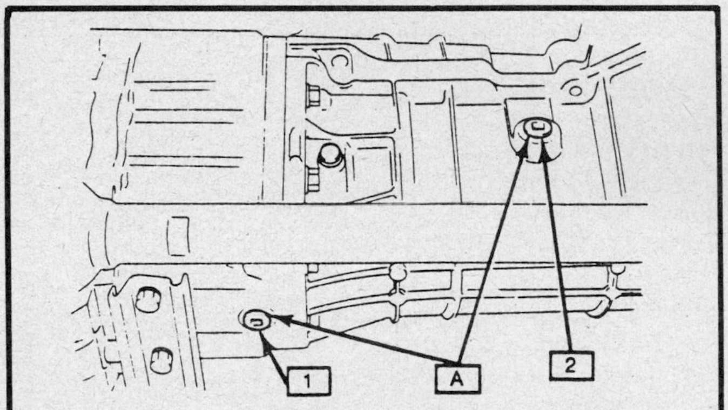
1
A
2

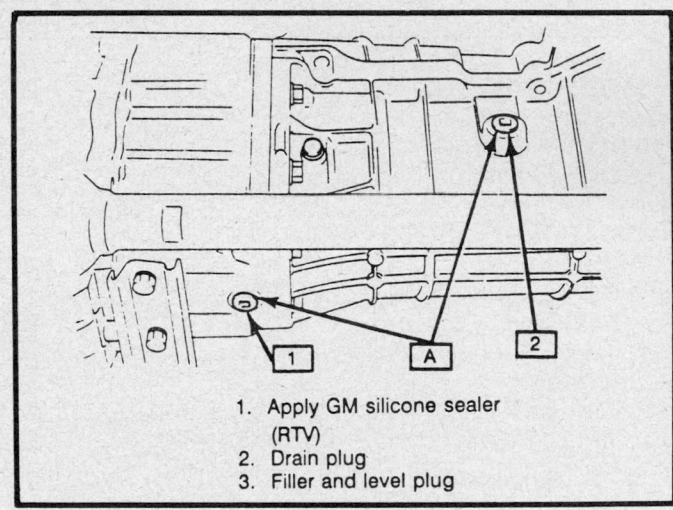

1. Apply GM silicone sealer (RTV)
2. Drain plug
3. Filler and level plug

View of the transmission's drain and filler plugs

Checking Fluid Level

NOTE: When checking the fluid level, the vehicle must be at normal operating temperatures and positioned on a flat surface.

1. Remove the filler plug located in the middle of the rear case assembly.
2. If the fluid is low, add fluid until it is level with the bottom of the filler plug hole.
3. Replace the filler plug and torque it to 33 ft. lbs. (45 Nm).

TRANSMISSION MODIFICATIONS

There are no modifications at the time of this printing.

TROUBLE DIAGNOSIS

CHILTON'S THREE C'S TRANSMISSION DIAGNOSIS

Condition	Cause	Correction
Slipping out of gear	a) Worn shift fork shaft b) Worn shift fork or synchronizer sleeve c) Weak or damaged location spring d) Worn bearings on input shaft or main shaft e) Worn chamfered tooth on sleeve or gear f) Missing disengaged circlips	a) Replace shift fork shaft b) Replace shift fork or synchronizer sleeve c) Replace location spring d) Replace input shaft or main shaft bearings e) Replace sleeve or gear f) Install circlips
Gears refusing to disengage	a) Weakened or broken synchronizer spring b) Distorted shift shaft or shift fork	a) Replace synchronizer spring b) Replace shift shaft or fork
Hard shifting	a) Improper clutch pedal free travel b) Distorted or broken clutch disc c) Damaged clutch pressure plate d) Worn synchronizer ring e) Worn chamfered tooth on sleeve or gear f) Distorted shift shaft	a) Adjust clutch pedal free travel b) Replace clutch disc c) Replace clutch pressure plate d) Replace synchronizer ring e) Replace sleeve and gear f) Replace shift shaft
Noise	a) Inadequate or insufficient lubricant b) Damaged or worn bearing(s) c) Damaged or worn gear(s) d) Damaged or worn synchronizer ring e) Damaged or worn chamfered tooth on sleeve or gear	b) Replenish lubricant b) Replace bearings c) Replace gear(s) d) Replace synchronizer ring e) Replace sleeve or gear

ON CAR SERVICE

Adjustments

CLUTCH PEDAL HEIGHT

The clutch pedal should be 0.2 in. (5mm) higher than the brake pedal; if not, perform the following procedures.

1. At the clutch pedal bracket, loosen the adjusting bolt locknut.
2. Adjust the bolt until the clutch pedal is 0.2 in. (5mm) higher than the brake pedal.
3. Tighten the locknut.

CLUTCH START SWITCH

1. Firmly apply the parking brake and place the shift control lever into the **N** position.
2. Disconnect the electrical connector from the clutch start switch, located at the top of the clutch pedal.
3. Loosen the locknut and back off the switch.
4. Press the clutch pedal to the floor and allow it to return 2.0–2.7 in. (50–70mm) from the floor.
5. Using a ohmmeter, connect it to the switch. Slowly screw in the switch until it turns **ON**. Secure the switch and tighten locknut to 7.5–10.5 ft. lbs. (10–15 Nm).
6. Reconnect the electrical connector to the switch.

NOTE: With the clutch pedal fully depressed, the remaining switch stroke (clearance B) should be 0.02–0.04 in. (0.5–1.0mm). If the stroke is greater than 2.7 in. (70mm), clearance B will be less than 0.02 in. (5mm) and may damage the switch bracket.

CLUTCH PEDAL FREE TRAVEL

NOTE: Be sure to check the clutch pedal free play and the clutch function with the engine running.

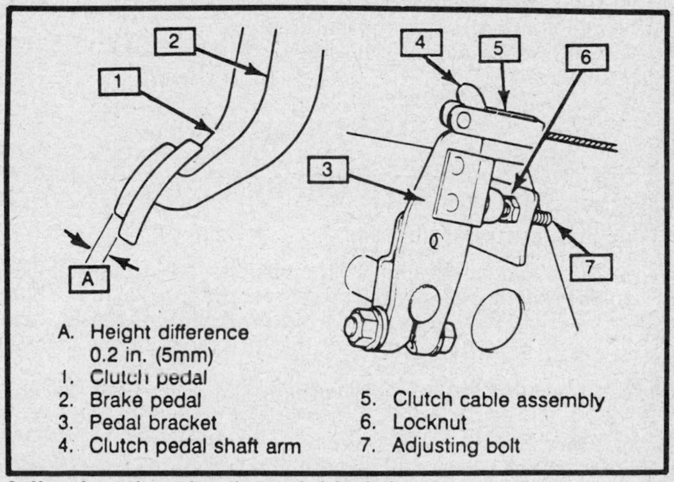
1
2
3
4
5
6
7
A
A. Height difference 0.2 in. (5mm)
1. Clutch pedal
2. Brake pedal
3. Pedal bracket
4. Clutch pedal shaft arm
5. Clutch cable assembly
6. Locknut
7. Adjusting bolt

Adjusting the clutch pedal height

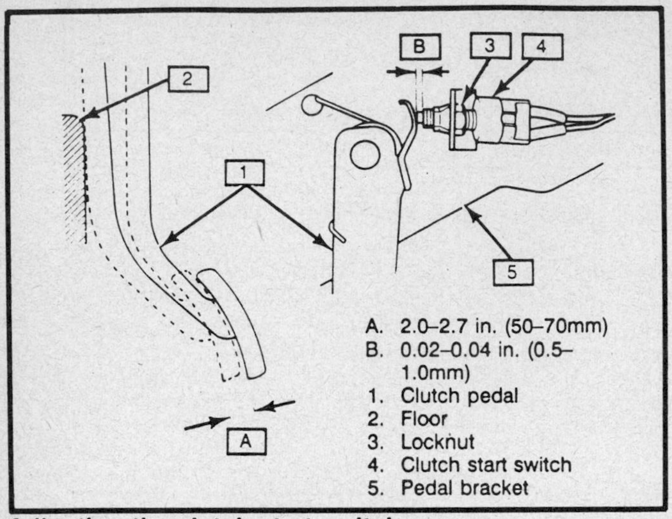
B
3
4
2
1
5
A
A. 2.0–2.7 in. (50–70mm)
B. 0.02–0.04 in. (0.5–1.0mm)
1. Clutch pedal
2. Floor
3. Locknut
4. Clutch start switch
5. Pedal bracket

Adjusting the clutch start switch

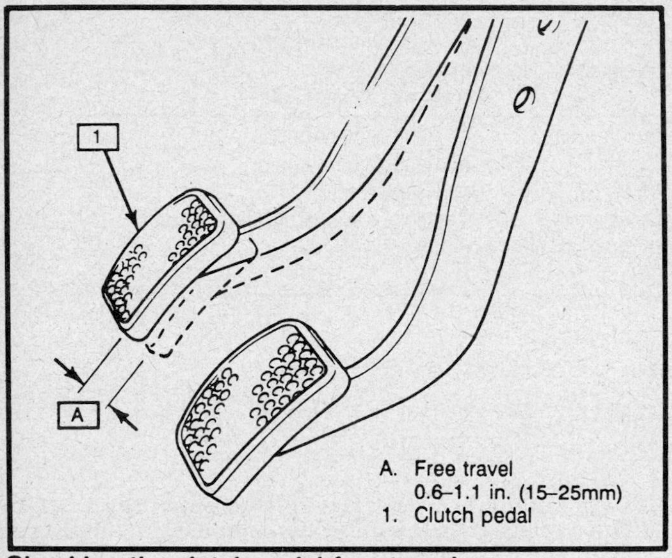
1
A
A. Free travel
0.6–1.1 in. (15–25mm)
1. Clutch pedal

Checking the clutch pedal free travel

1. Depress the clutch pedal until resistance of the clutch is felt; this distance should measure 0.6–1.1 in. (15–25mm).
2. If the measurement is not correct, loosen the cable's outer nuts and adjust the clutch pedal free play.
3. After adjustment, torque the outer nuts to 11–14 ft. lbs. (15–20 Nm).

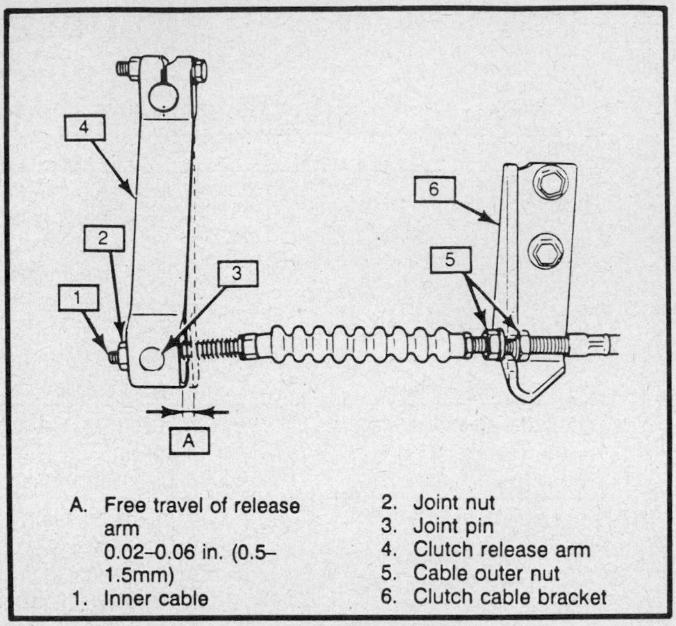
4
6
2
5
3
1
A
A. Free travel of release arm
0.02–0.06 in. (0.5–1.5mm)
1. Inner cable
2. Joint nut
3. Joint pin
4. Clutch release arm
5. Cable outer nut
6. Clutch cable bracket

Adjusting the clutch pedal free travel

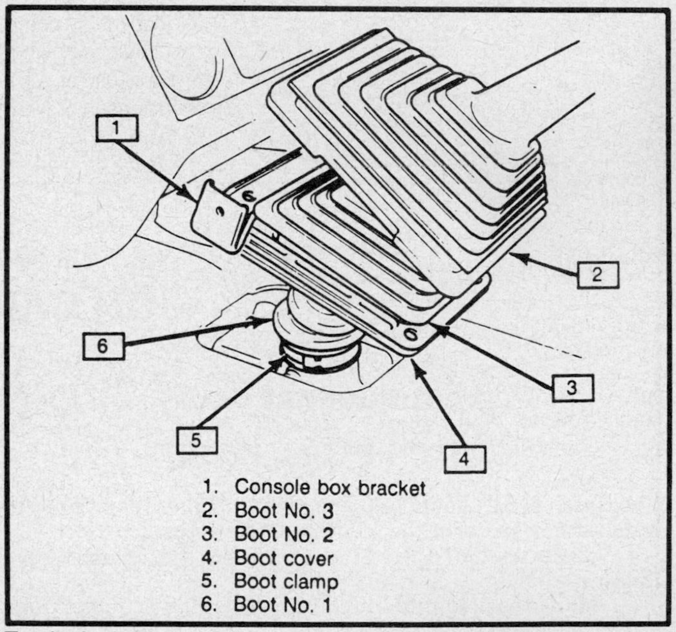
1
2
6
3
5
4
1. Console box bracket
2. Boot No. 3
3. Boot No. 2
4. Boot cover
5. Boot clamp
6. Boot No. 1

Exploded view of the control lever boots

Service

REAR OIL SEAL

Removal and Installation

1. Raise and safely support the vehicle in a level position.
2. Remove the rear driveshaft-to-rear differential flange bolts and the driveshaft. Slide the driveshaft from the transfer case.
3. Using a small prybar, pry the oil seal from the transfer case.
4. Clean the front of the driveshaft (sliding portion) that contacts the oil seal. Make sure it is free of dents or scratches.
5. Using the oil seal installer tool or equivalent, and a plastic hammer, drive the new oil seal into the rear of the transfer case. Lubricate the oil seal lip with grease.
6. Lubricate the driveshaft's inside splines with grease. Install the driveshaft's front portion into the transfer case and the rear to the differential flange. Torque the flange bolts to 37 ft. lbs. (50 Nm).

SHIFT CONTROL LEVERS

Removal and Installation

1. At the console, remove the front screws, the rear clips and the console box.
2. Remove the console box bracket bolts, lift the boot cover and No. 2 boot.

NOTE: To remove the clip, push in it's center pin.

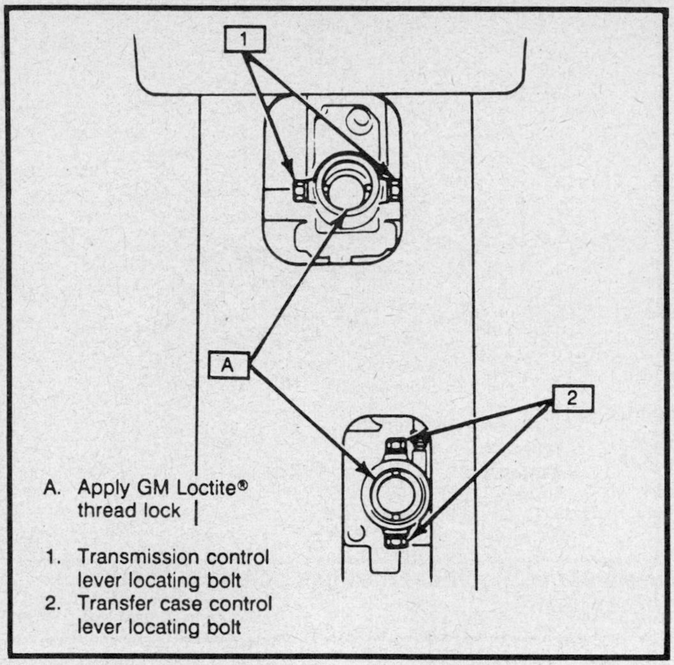
1
A
2
A. Apply GM Loctite® thread lock
1. Transmission control lever locating bolt
2. Transfer case control lever locating bolt

View of the control lever locating bolts

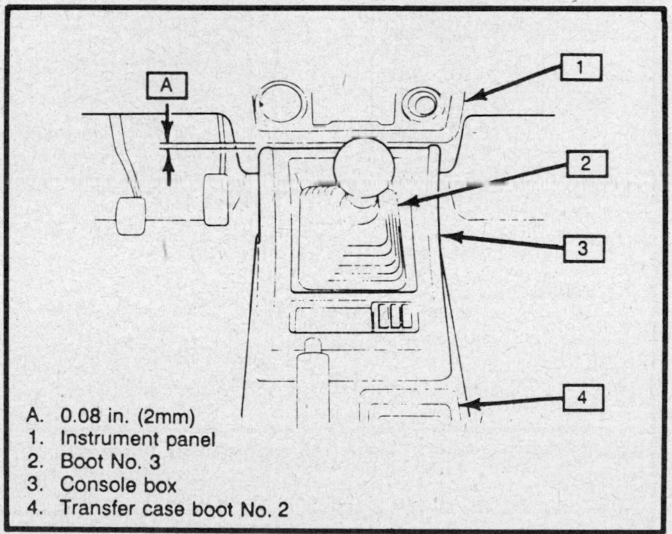
A
1
2
3
4
A. 0.08 in. (2mm)
1. Instrument panel
2. Boot No. 3
3. Console box
4. Transfer case boot No. 2

Installation of the console box

3. To remove the No. 1 boot, remove the boot clamp and the boot.
4. To remove the shift control lever, push the gear shift control case cover downward, turn it clockwise and remove the lever.
5. Remove the transfer case shift control lever in the same manner as described above.
6. Inspect the shift control lever lower portion and control lever locating sheet for excessive wear and the boot for damage; replace the parts, if necessary.
7. Lubricate the shift control levers pivot portions and seats, then, install them into the transmission.
8. If replacing or retightening the control lever locating bolts, coat them with Loctite® and torque them to 13 ft. lbs. (17 Nm).
9. Install both No. 1 boots and clamp them securely with new clamps. Tighten the No. 2 transmission boot with the boot cover and console box bracket.

NOTE: Be sure the No. 3 transmission boot's flared end is engaged with the console box.

10. When installing the console box, allow 0.08 in. (2mm) clearance between the console box and the instrument panel. Torque the boot cover bolts to 4.4 ft. lbs. (6 Nm).

TRANSFER CASE 4WD SWITCH

Removal and Installation

1. Remove the console box and lift the No. 2 and No. 3 transmission control lever boots.
2. Remove the No. 2 transfer case control lever boot. Unclamp the 4WD switch from the transfer case and disconnect the electrical connector.
3. Remove the 4WD switch and check it for damage or malfunction; replace it, if necessary. Do not allow dust or dirt to enter the transfer case.
4. Install the 4WD switch and torque it to 15 ft. lbs. (20 Nm). Secure the wire with the clamp and connect the electrical connector.
5. Install the No. 2 transfer case control lever boot.
6. Install the No. 2 and No. 3 transmission control lever boots and the console box.
7. Turn the ignition switch **ON**, shift the transfer case shift control lever to the **4WD** position and confirm that the indicator light turns **ON**. Turn the ignition switch **OFF**.
8. To check the 4WD switch, perform the following procedures:
   a. At the rear of the intake manifold, unclamp the wire and disconnect the electrical connector.
   b. Using an ohmmeter, check the electrical harness terminals No. 2 and No. 5. Move the transfer case's shift control lever into **4**, **N** and **4L** positions; there should be continuity.
9. After the checking procedure, connect the electrical harness connector and reclamp the wiring harness.

BACK-UP LAMP SWITCH AND 5TH GEAR SWITCH

Removal and Installation

1. Raise and safely support the vehicle.
2. From under the vehicle, unclamp the wiring and disconnect the electrical connectors from both switches; the front electrical connector is for the back-up lamp and the rear one is for the 5th gear switch.
3. Remove both switches from the transmission and check them for damage or malfunction; replace them, if necessary. Do not allow dirt or dust to enter the transmission.
4. Install both switches and torque them to 15 ft. lbs. (20 Nm). Secure the wires with the clamps and connect the electrical connectors.
5. To check the back-up lamp switch, turn the ignition switch **ON**, shift the transmission into the **R** position and check that the back-up lamps turn **ON**. After checking turn the ignition switch **OFF**.
6. To check the 5th gear switch, perform the following procedures:
   a. At the rear of the intake manifold, unclamp the wire and disconnect the electrical connector.
   b. Move the shift control lever into the **5TH** gear position.
   c. Using an ohmmeter, check the electrical harness terminals No. 1 and No. 2; there should be continuity. Move the shift control lever to any other gear position; there should be no continuity.
7. To check the back-up lamp switch, perform the following procedures:
   a. Move the shift control lever into the **R** gear position.
   b. Using an ohmmeter, check the electrical harness termi-

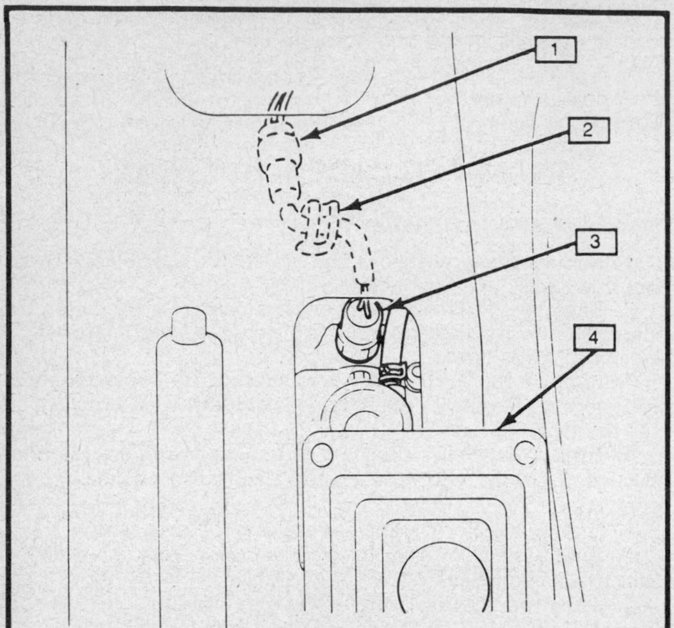
1
2
3
4
1. Connector
2. Clamp
3. 4WD switch
4. No. 2 boot

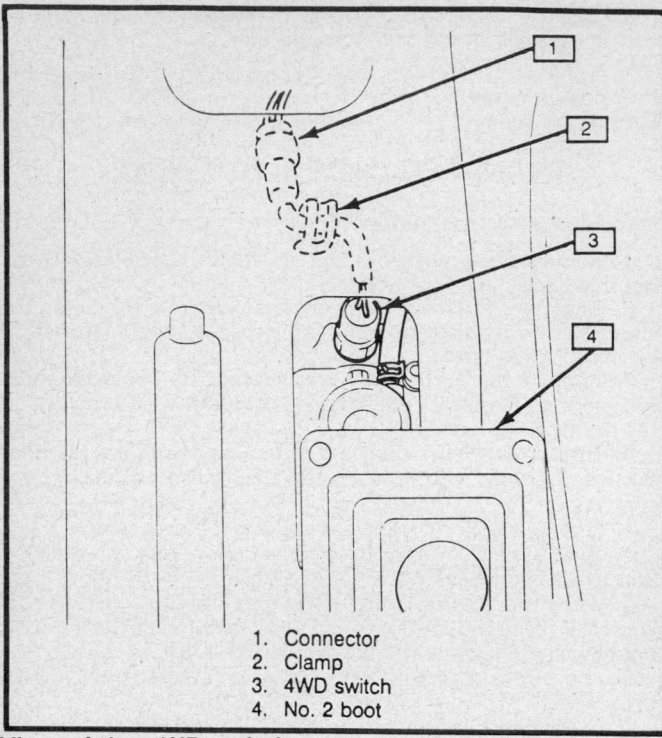

View of the 4WD switch

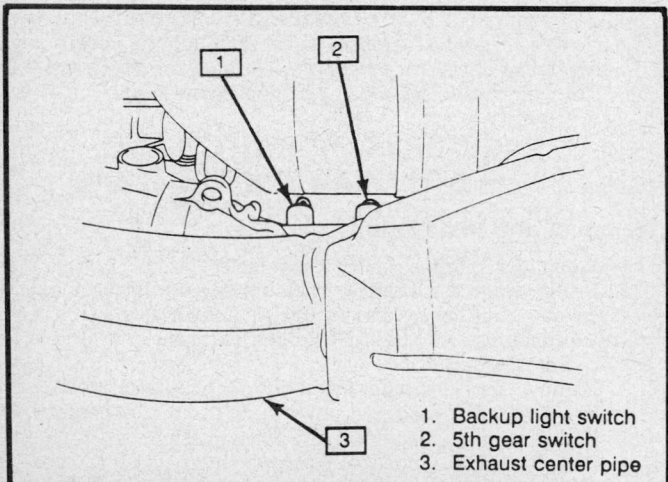

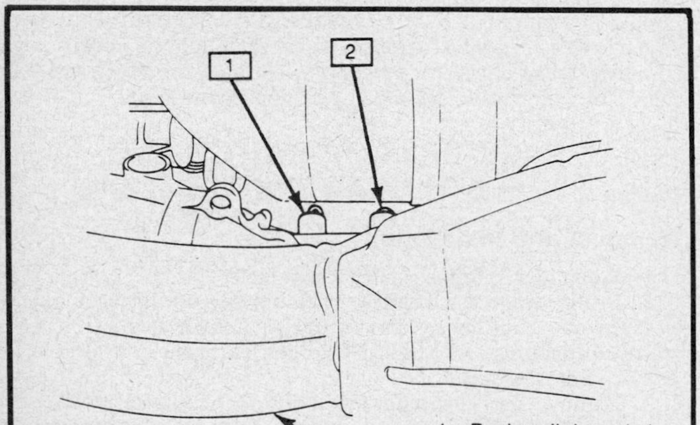
1
2
3
1. Backup light switch
2. 5th gear switch
3. Exhaust center pipe

View of the backup lamp and 5th gear switches

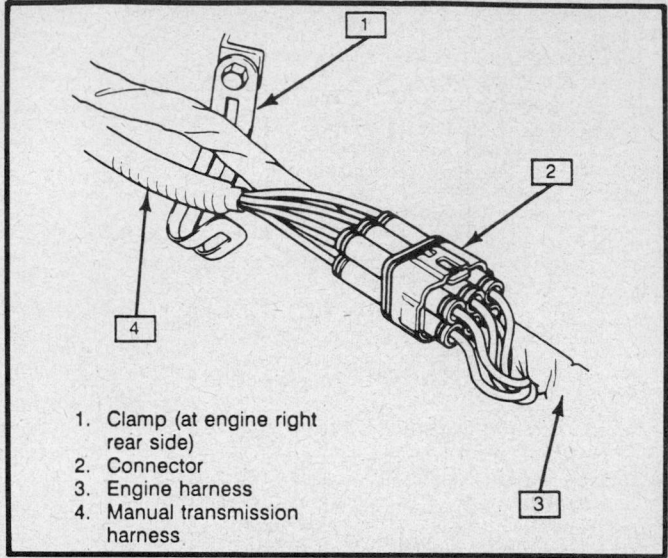

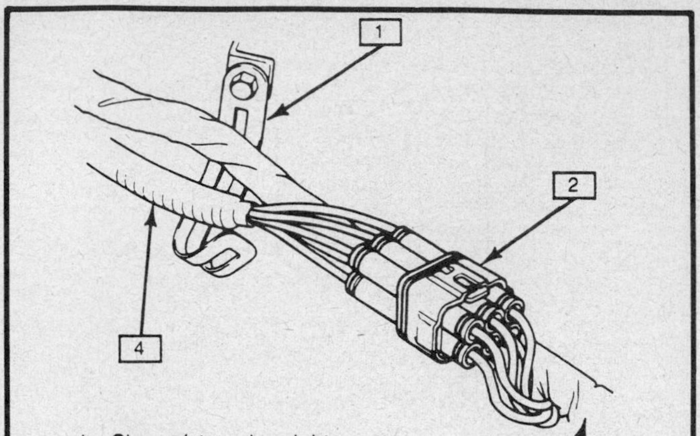
1
2
3
4
1. Clamp (at engine right rear side)
2. Connector
3. Engine harness
4. Manual transmission harness

View of the electrical harness connector at the intake manifold

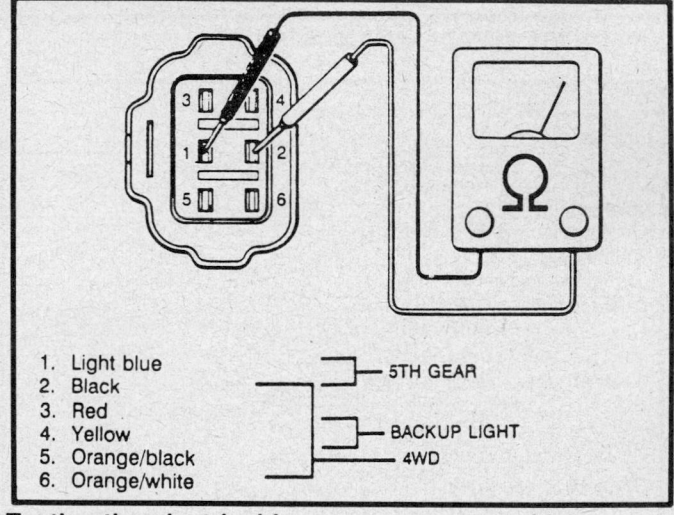

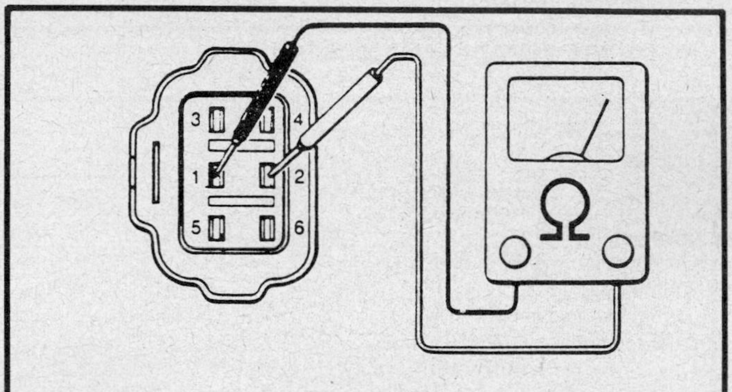
3
4
1
2
5
6
Ω
1. Light blue
2. Black
3. Red
4. Yellow
5. Orange/black
6. Orange/white
5TH GEAR
BACKUP LIGHT
4WD

Testing the electrical harness connector with an ohmmeter

nals No. 3 and No. 4; there should be continuity. Move the shift control lever to any other gear position; there should be no continuity.

8. After the checking procedures, connect the electrical harness connector and reclamp the wiring harness.

SPEEDOMETER DRIVEN GEAR

Removal and Installation

1. Raise and safely support the vehicle.
2. From the transfer case, remove the speedometer cable.
3. Remove the retaining bolt and the gear case.
4. Using a pin punch, drive the spring pin from the gear case and remove the speedometer driven gear.

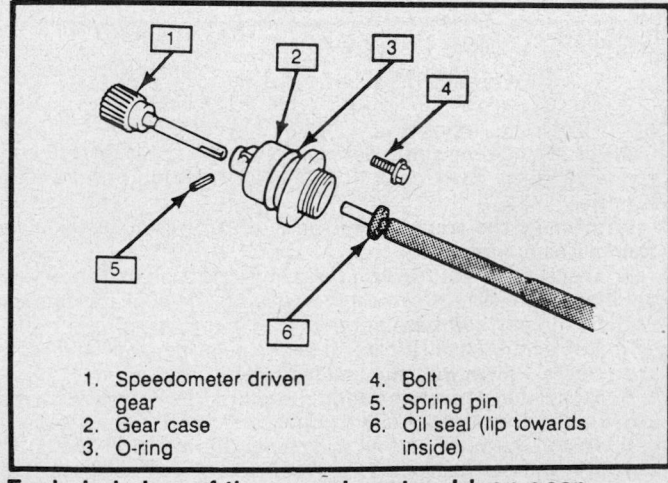

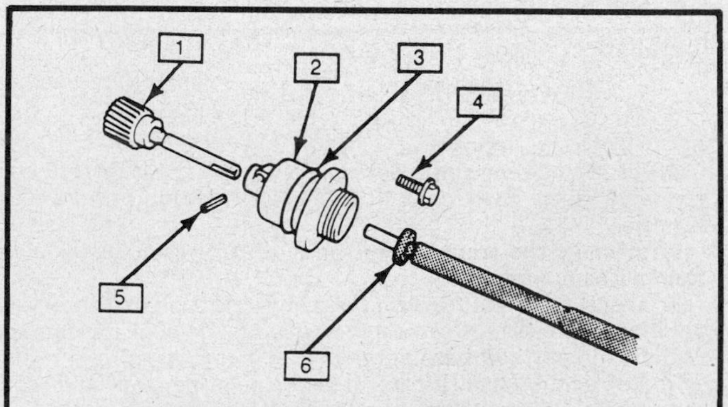
1
2
3
4
5
6
1. Speedometer driven gear
2. Gear case
3. O-ring
4. Bolt
5. Spring pin
6. Oil seal (lip towards inside)

Exploded view of the speedometer driven gear

5. Remove the oil seal and O-ring (if necessary) from the gear case.
6. Inspect the speedometer driven gear teeth for abnormal wear or the shaft for bending; replace the parts, if necessary.
7. Using the valve guide removal tool or equivalent, grease the new oil seal and install it (lip facing inside) into the gear case.
8. Assemble the driven gear into the gear case and secure it with the spring pin; make sure the driven gear rotates freely.
9. Install the gear case assembly into the transfer case, torque the retaining bolt to 7.5 ft. lbs. (10 Nm) and connect the speedometer cable.
10. Check the oil level and lower the vehicle.

REMOVAL AND INSTALLATION

TRANSMISSION REMOVAL

1. Remove the transmission shift control lever and the transfer case shift control knob.
2. Disconnect the negative battery cable.
3. From the rear of the cylinder head, remove the breather hose from the clamp.
4. From the rear of the intake manifold, separate the wiring harness from the clamp and disconnect the electrical harness connector.
5. Remove the starter nuts and the starter.
6. Remove the transmission-to-engine bolts.
7. Using a drain pan, drain the oil from the transmission and the transfer case.
8. From the rear differential, remove the driveshaft-to-pinion shaft flange bolts, separate the driveshaft and pull it from the transfer case.
9. From the front differential, remove the driveshaft-to-pinion shaft flange bolts and separate the driveshaft from the transfer case.
10. Disconnect the clutch cable from the clutch release arm and bracket.
11. Remove the lower plate from the clutch housing.
12. Remove the exhaust center pipe.
13. Using a transmission jack or equivalent, secure it to the transmission and support the weight.
14. Remove the rear engine mount crossmember-to-chassis bolts, the crossmember-to-transmission bolts and the crossmember.
15. Move the transmission assembly rearward and lower it from the vehicle.
16. Remove the wiring harness and the breather hose from the transmission assembly.
17. Separate the gear shift lever case and transfer case from the transmission.

TRANSMISSION INSTALLATION

1. Install the gear shift lever case and the transfer case to the transmission. Torque the gear shaft lever case-to-case bolt to 17 ft. lbs. (23 Nm) and the transfer case-to-case bolt to 17 ft. lbs. (23 Nm).
2. Connect the wiring harness and the breather hose to the transmission assembly.
3. Using the transmission jack, raise and position the assembly into the vehicle.
4. Install the rear engine mount and torque the mount-to-chassis bolts to 37 ft. lbs. (50 Nm).
5. Install the speedometer cable to the transmission assembly and torque the retaining bolt to 7.5 ft. lbs. (10 Nm).
6. Install the exhaust center pipe. Torque the pipe-to-manifold bolts to 37 ft. lbs. (50 Nm), the muffler-to center pipe bolts to 37 ft. lbs. (50 Nm) and the pipe bracket bolts to 37 ft. lbs. (50 Nm).

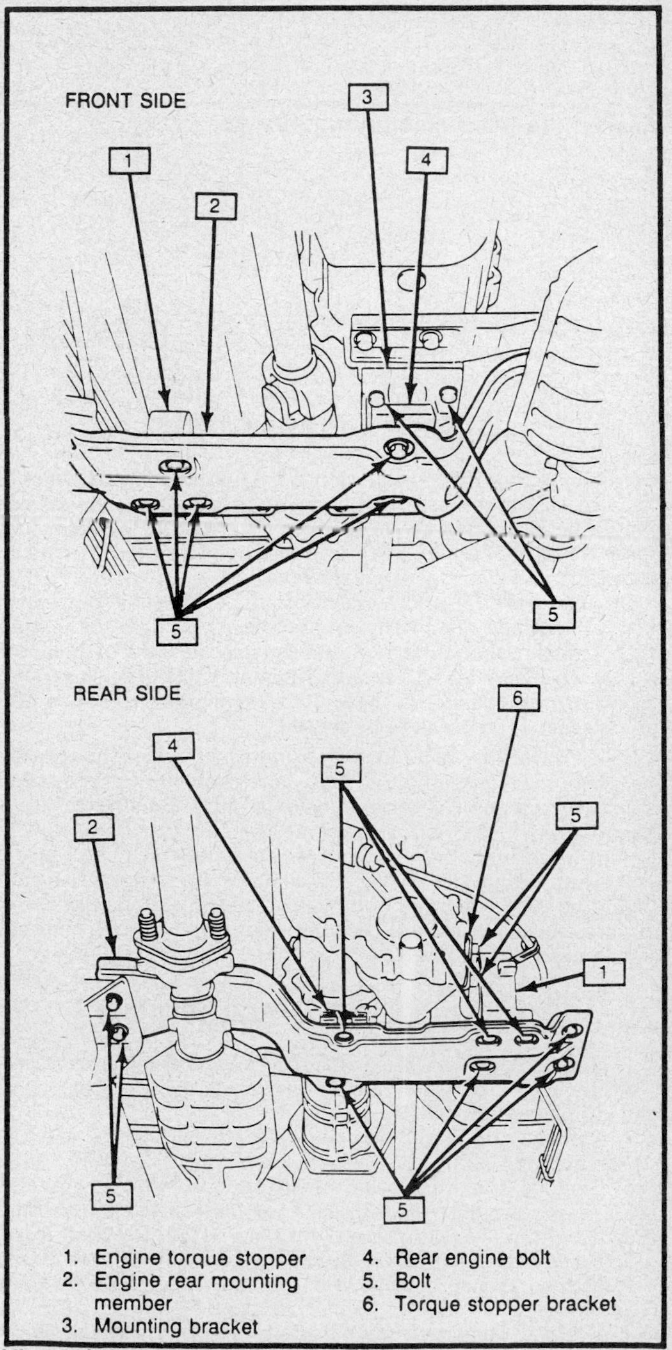
FRONT SIDE
1
2
3
4
5
5
REAR SIDE
1
2
4
5
5
5
5
6
1. Engine torque stopper
2. Engine rear mounting member
3. Mounting bracket
4. Rear engine bolt
5. Bolt
6. Torque stopper bracket

View of the rear engine mount

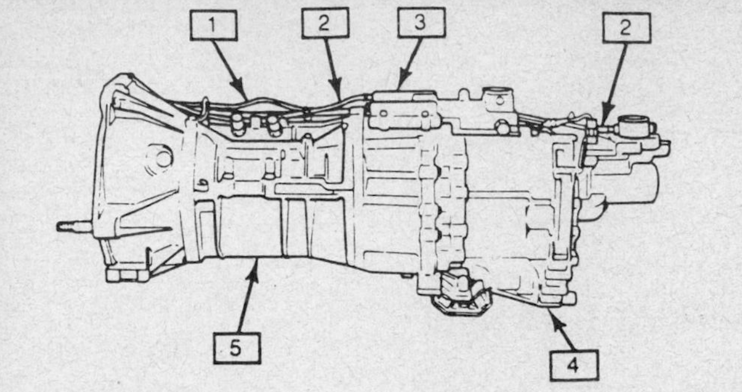
1
2
3
2
5
4

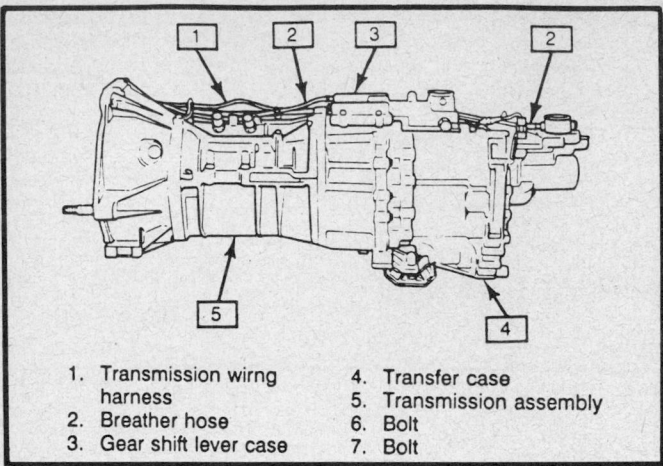

1. Transmission wirng harness
2. Breather hose
3. Gear shift lever case
4. Transfer case
5. Transmission assembly
6. Bolt
7. Bolt

View of the transmission/transfer assembly

7. Install the clutch housing lower plate and torque the bolts to 37 ft. lbs. (50 Nm).
8. Connect the clutch cable to the clutch release arm and bracket.
9. Install the front driveshaft to the transmission flange and torque the flange bolts to 37 ft. lbs. (50 Nm).
10. Slide the rear driveshaft into the transfer case, connect the driveshaft-to-rear differential flange and torque the flange bolts to 37 ft. lbs. (50 Nm).
11. Install the transmission-to-engine bolts and torque the bolts to 63 ft. lbs. (85 Nm).
12. Install the starter.
13. At the rear of the intake manifold, connect the electrical harness connector and clamp the wiring harness.
14. At the rear of the cylinder head, clamp the breather hose.
15. Connect the negative battery cable.
16. Install the transmission shift control lever and the transfer shift lever knob.
17. Refill the transmission and transfer case. Lower the vehicle.
18. Check and/or adjust the clutch arm play for 0.02–0.06 in. (0.5–1.5mm) with the clutch pedal play of 0.6–1.0 in. (15–25mm).
19. Check the engine, clutch and transmission operation.

BENCH OVERHAUL

Before Disassembly

When servicing the transmission it is important to be aware of cleanliness. Before disassembling the transmission, the outside should be throughly cleaned, preferably with a high-pressure spray cleaning equipment. Dirt entering the unit may negate all the effort and time spent on the overhaul.

During inspection and reassembly, all parts should be cleaned with solvent and dried with compressed air. Lubricate the seals with transmission fluid and use petroleum jelly to hold the thrust washers; this will ease the assembly of the seals and not leave harmful residues in the system. Do not use solvent on neoprene seals, if they are to be reused.

Before installing bolts into aluminum parts, dip the threads into clean transmission fluid. Anti-seize compound may be used to prevent galling the aluminum or seizing. Be sure to use a torque wrench to prevent stripping the threads. Be especially careful when installing the seals, the smallest nick can cause a leak. Aluminum parts are very susceptible to damage; great care should be used when handling them. Reusing snaprings is not recommended but should they be: compress the internal ones and compress the external ones.

Transmission Disassembly

1. Remove the rear case-to-main case bolts and the rear case.
2. Remove the clutch release bearing from the input shaft and clutch release fork.
3. Remove the clutch housing-to-main case bolts and the clutch housing.
4. Remove the input shaft bearing retainer-to-main case bolts. Using two 6mm bolts, place them into the bearing retainer and tighten them to press the retainer from the main case.
5. Remove the upper case-to-lower case bolts. Using the case separator tool or equivalent, separate the upper case from the lower case.
6. Remove the input/mainshaft assembly from the lower case.

Unit Disassembly And Assembly

LOWER CASE ASSEMBLY

Disassembly

1. Using the output shaft C-ring remover/installer tool or 2 prybars, remove the circlip from the countershaft.
2. Using a universal puller tool or equivalent, and 13mm socket, press the rear bearing from the countershaft.
3. Remove the 5th gear and the reverse gear.

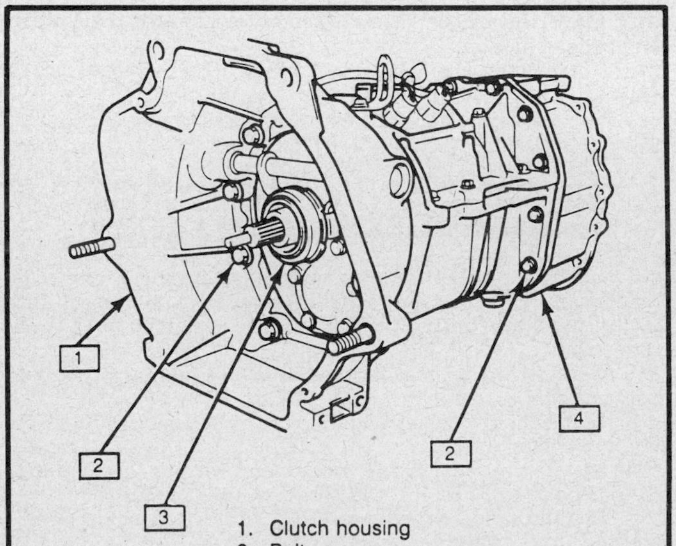
1
2
3
2
4

1. Clutch housing
2. Bolt
3. Clutch release bearing
4. Rear case

Removing the housings from the main case

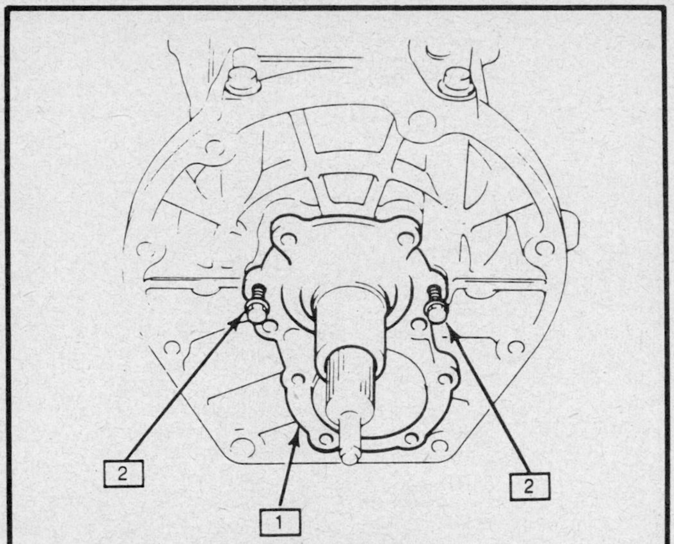
2
2
1

1. Input shaft bearing retainer
2. 6mm bolts as remover

Removing the input shaft bearing retainer from the main case

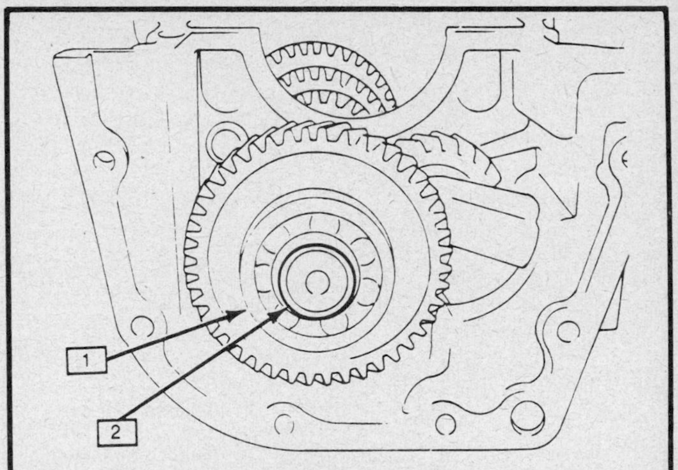
1
2

1. Rear bearing
2. Countershaft circlip

View of the circlip on the countershaft

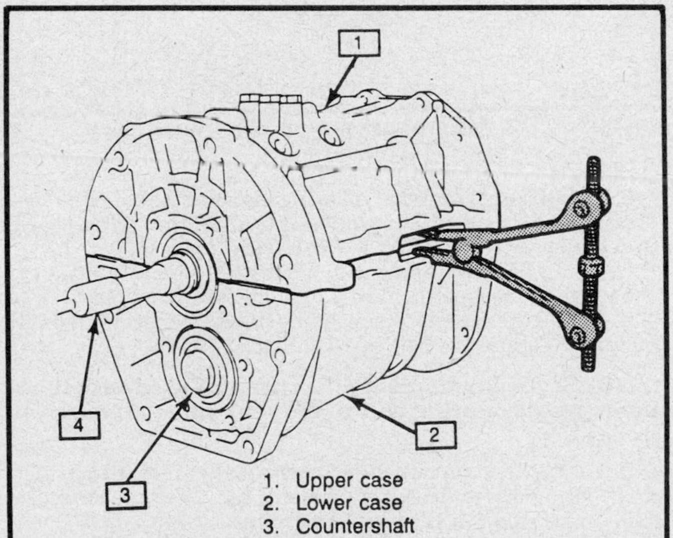
1
4
2
3

1. Upper case
2. Lower case
3. Countershaft
4. Input shaft

Separating the upper case from the lower case

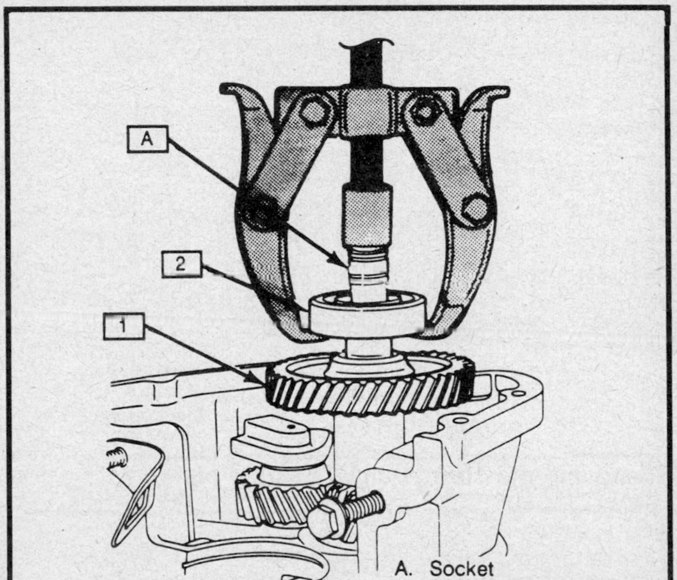
A
2
1

A. Socket
1. Countershaft 5th gear
2. countershaft rear bearing

Removing the rear bearing from the countershaft

4. Remove the reverse idler gear shaft bolt and the reverse idler gear assembly.
5. Remove the bearing plate bolts and the plate. Using modified vacuum grip tool or equivalent, remove the circlip from the front bearing.
6. Using a plastic hammer, tap the countershaft free of the lower case.
7. Using a universal puller tool or equivalent, press the countershaft center bearing from the countershaft.
8. Remove the countershaft from the lower case.
9. Using the output shaft C-ring remover/installer tool or equivalent, remove the circlip from the countershaft.

Inspection

1. Check the countershaft bearing surfaces for scoring and/or wear; if necessary, replace the shaft.
2. Inspect the bearings for scoring, wear and/or damage; if necessary, replace them.
3. Check the gears for damage and/or excessively worn teeth; if necessary, replace them.

Assembly

1. Position the countershaft into the lower case.
2. Using the modified vacuum grip tool or equivalent, install the outer clip onto the front bearing.
3. While supporting the countershaft in the lower case, install the front bearing onto it.
4. Using the output shaft C-ring remover/installer tool or equivalent, install the circlip onto the countershaft.
5. Support the lower case allowing the countershaft to move freely in a downward direction and install the front bearing. Using the input gear installer tool or equivalent, and a mallet, drive the front bearing into the lower case until the outer circlip is flush to the case.

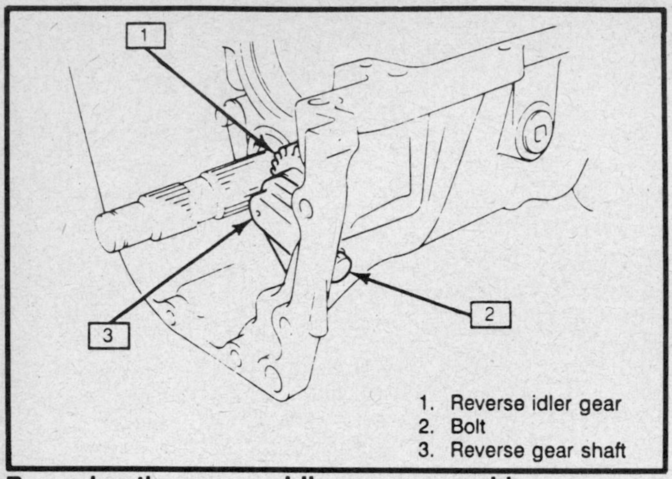
1
2
3
1. Reverse idler gear
2. Bolt
3. Reverse gear shaft
Removing the reverse idler gear assembly

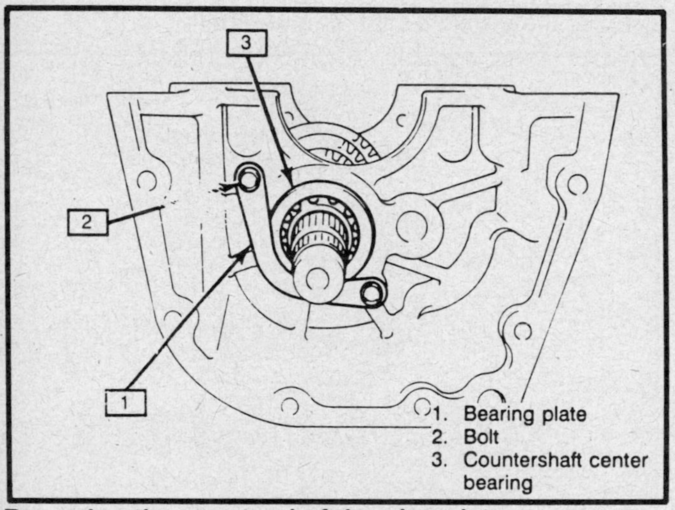
3
2
1
1. Bearing plate
2. Bolt
3. Countershaft center bearing
Removing the countershaft bearing plate

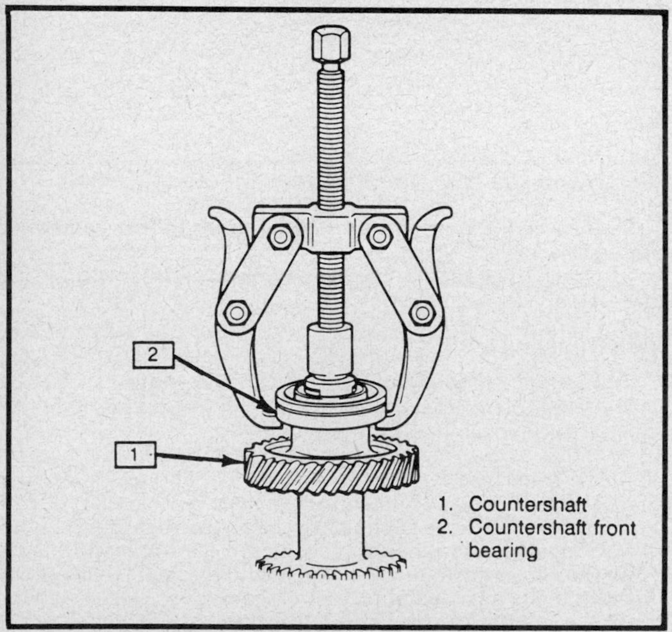
2
1
1. Countershaft
2. Countershaft front bearing
Removing the front countershaft bearing

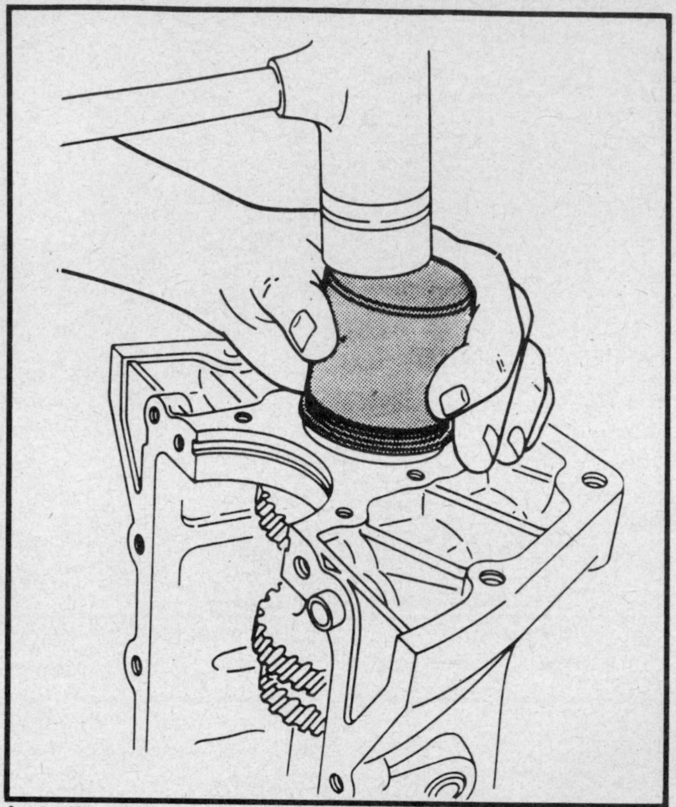
Installing the front bearing into the lower case

6. Using the countershaft bearing installer tool or equivalent, and bearing driver tool or equivalent, drive the center bearing into the lower case until it is flush with the case.

7. Using Loctite®, apply it to the bearing plate bolts, install the bolts and torque them to 7.5 ft. lbs. (10 Nm).

8. Install the reverse idler gear assembly, washer (aluminum) and bolt. Torque the bolt to 16 ft. lbs. (22 Nm).

NOTE: An aluminum washer must be used with the reverse gear mounting bolt to prevent damage to the lower case.

9. Using the countershaft bearing installer tool or equivalent, and bearing driver tool or equivalent, tap on the reverse gear.

10. Install the 5th gear onto the countershaft.

NOTE: When installing the rear bearing, support the countershaft to prevent the front bearing from shifting out of the lower case.

11. Using the input bearing installer tool or equivalent (to support the countershaft), the countershaft bearing installer tool or equivalent, the bearing driver tool or equivalent, and a hammer, tap on the rear bearing onto the countershaft.

12. Using the output shaft C-ring remover/installer tool or equivalent, install the circlip onto the countershaft.

INPUT SHAFT

Disassembly

NOTE: Since the input shaft rides on a needle bearing assembly on the mainshaft assembly, which may remain in the input shaft or on the mainshaft. The assembly should be removed and set aside to prevent damaging it.

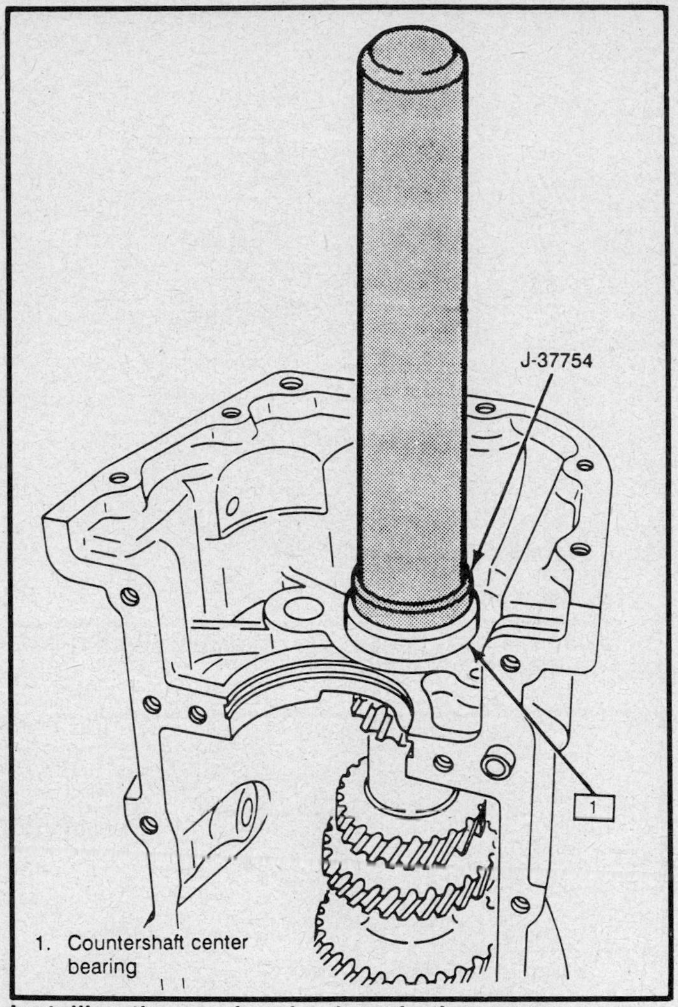
J-37754
1
1. Countershaft center bearing

Installing the rear bearing into the lower case

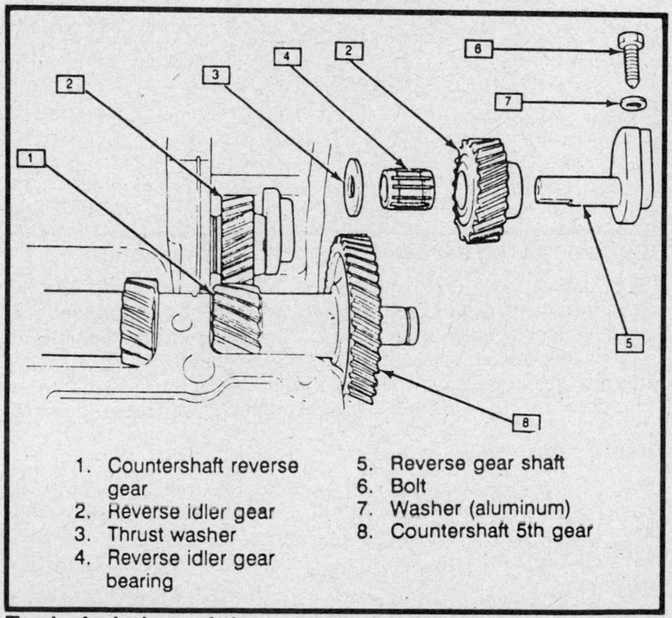
1
2
3
4
5
6
7
8
1. Countershaft reverse gear
2. Reverse idler gear
3. Thrust washer
4. Reverse idler gear bearing
5. Reverse gear shaft
6. Bolt
7. Washer (aluminum)
8. Countershaft 5th gear

Exploded view of the reverse idler gear and 5th gear assembly

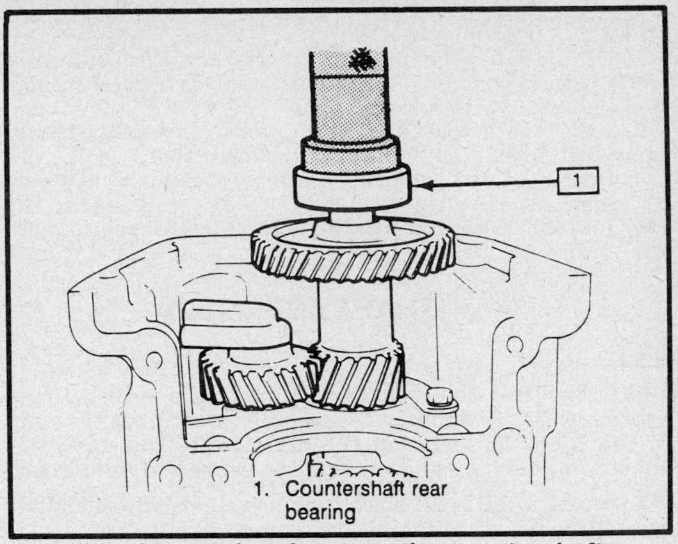
1
1. Countershaft rear bearing

Installing the rear bearing onto the countershaft

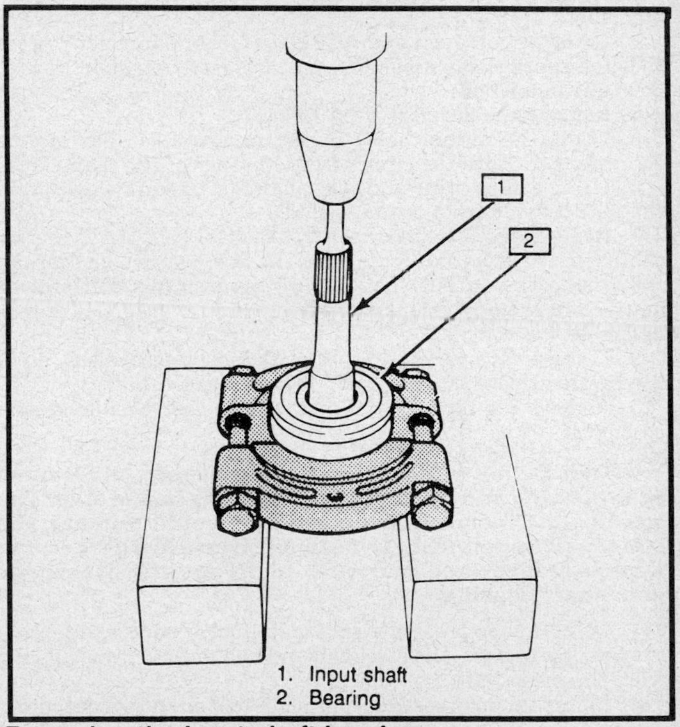
1
2
1. Input shaft
2. Bearing

Removing the input shaft bearing

1. Separate the input shaft from the mainshaft assemblies and set the input shaft needle bearings aside.
2. Using the output shaft C-ring remover/installer tool or equivalent, remove the circlip from the input shaft.
3. Using a shop press and split plate tool or equivalent, press the bearing from the input shaft.

Inspection

1. Check the input shaft bearing surfaces for scoring and/or wear; if necessary, replace the shaft.
2. Check the input shaft splines for wear; if necessary, replace the shaft.
3. Inspect the bearings for scoring, wear and/or damage; if necessary, replace them.
4. Check the gears for damage and/or excessively worn teeth; if necessary, replace them.

Assembly

1. Using a shop press, the bearing installer tool or equivalent, and the bearing driver tool or equivalent, press the bearing onto the input shaft until it seats.
2. Using the output shaft C-ring remover/installer tool or equivalent, install the circlip onto the input shaft.
3. Install the needle bearing assembly and input shaft onto the mainshaft assembly.
4. Install the sleeve onto the input shaft/mainshaft assembly.

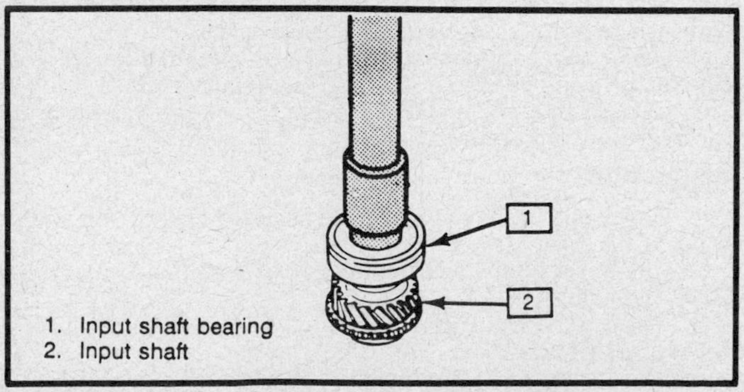
1
2
1. Input shaft bearing
2. Input shaft

Installing the bearing onto the input shaft

MAINSHAFT ASSEMBLY

Disassembly

NOTE: Since the input shaft rides on a needle bearing assembly on the mainshaft assembly, which may remain in the input shaft or on the mainshaft. The assembly should be removed and set aside to prevent damaging it.

1. Separate the input shaft from the mainshaft assemblies and set the input shaft needle bearings aside.
2. Using the output shaft C-ring remover/installer tool or equivalent, remove the circlip from the high speed hub assembly.
3. Using a shop press and split plate tool or equivalent, press the high speed sleeve assembly, synchronizer ring and 3rd gear from the mainshaft.
4. Remove the 3rd gear needle bearing.
5. Using the output shaft C-ring remover/installer tool or equivalent, remove the circlip from the rear of the mainshaft.
6. Using a shop press and split plate tool or equivalent, press the rear bearing from the mainshaft.
7. Remove the 5th gear washer, the steel ball, the 5th gear, the 5th gear synchronizer ring and the 5th gear needle bearing.
8. Using the output shaft C-ring remover/installer tool or equivalent, remove the circlip from the reverse sleeve/hub assembly.
9. Using a shop press and split plate tool or equivalent, press the reverse sleeve/hub assembly from the mainshaft.
10. Remove the reverse gear and reverse gear needle bearing assembly.

NOTE: Do not remove the bearing washer, the reverse gear bushing and the mainshaft bearing in one step. The steel ball, which locks the bearing, could damage the mainshaft or mainshaft bearing. Remove the bearing washer and reverse gear bushing in one step, then, the mainshaft bearing.

11. Using a shop press and split plate tool or equivalent, press the bearing washer, the steel ball and the reverse gear bushing from the mainshaft.
12. Using a shop press and split plate tool or equivalent, press the mainshaft bearing from the mainshaft.

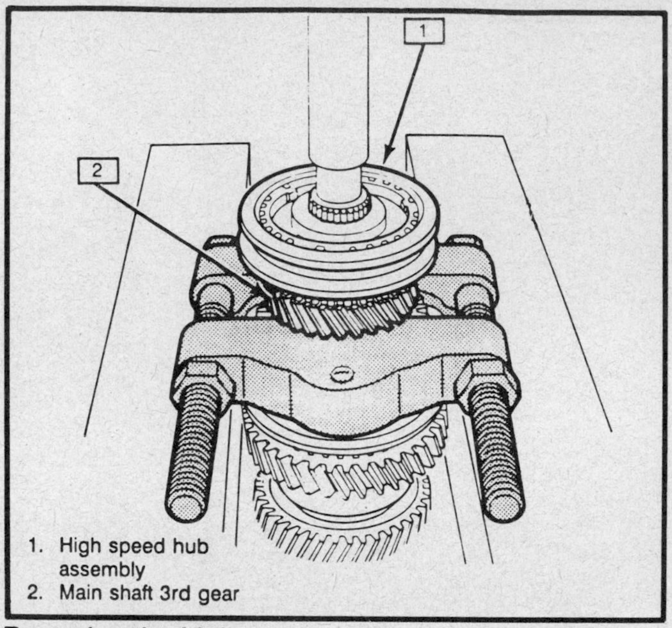
1
2
1. High speed hub assembly
2. Main shaft 3rd gear

Removing the high speed hub assembly and 3rd gear from the main shaft

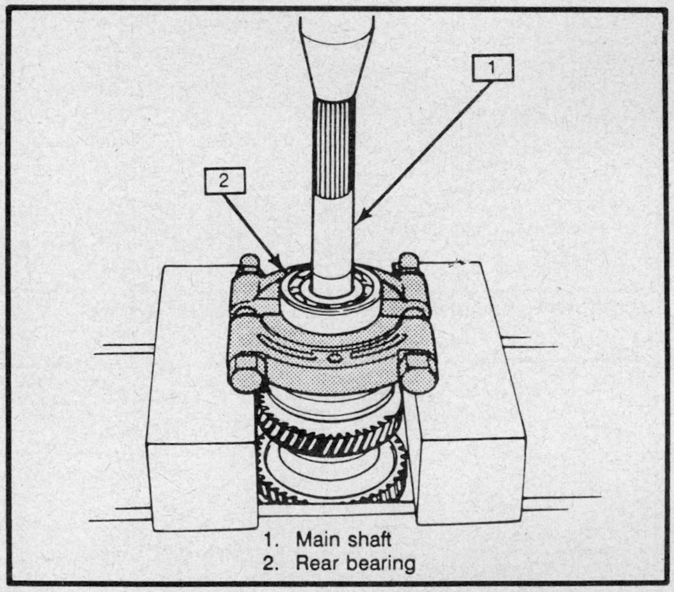
1
2
1. Main shaft
2. Rear bearing

Removing the rear bearing from the main shaft

13. Remove the bearing washer, the steel ball, the low gear, the low gear synchronizer and the low gear needle bearing.
14. Using a shop press and split plate tool or equivalent, press the low gear bushing, the low speed sleeve/hub assembly, the 2nd gear and the 2nd gear synchronizer from the mainshaft.

Inspection

1. Check each ball bearing for smooth rotation; replace the bearing, if necessary.
2. Check the needle bearings and the contacting (rotating) surfaces; replace the bearings, the gears and/or the shafts, if necessary.
3. If abnormal noise is heard before disassembly, carefully check the contacting (rotating) surfaces inside each gear by performing the following procedures:

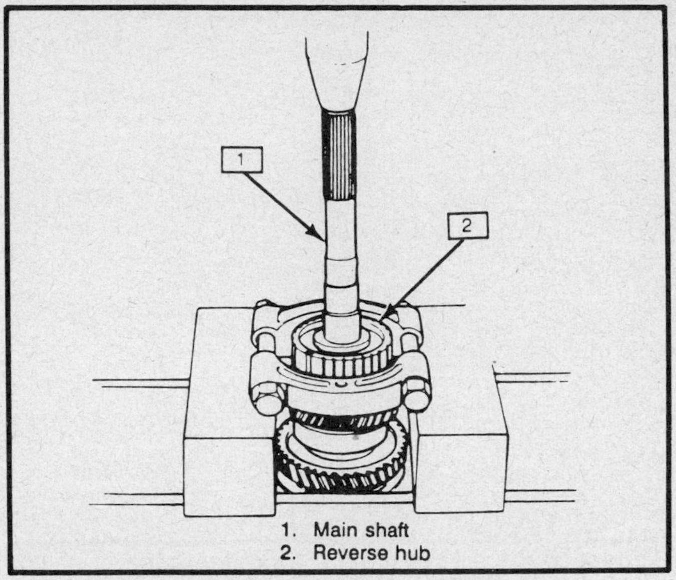
1
2
1. Main shaft
2. Reverse hub

Removing the reverse sleeve/hub assembly from the main shaft

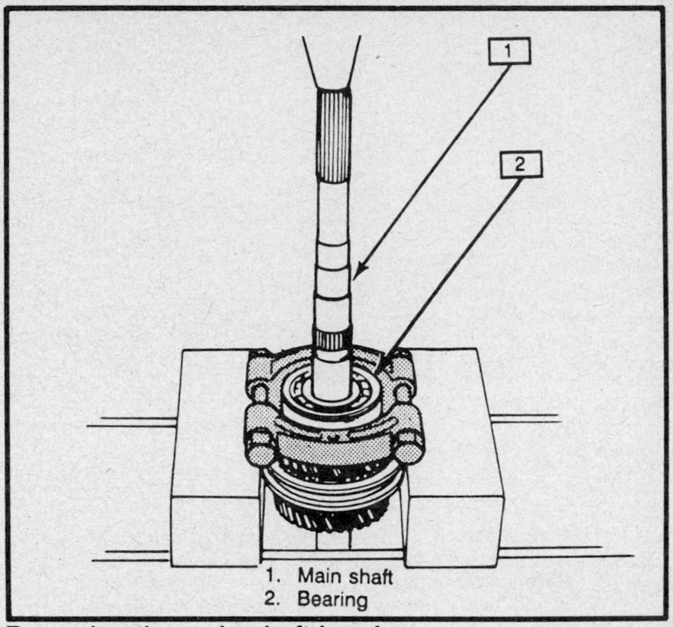
1
2
1. Main shaft
2. Bearing

Removing the main shaft bearing

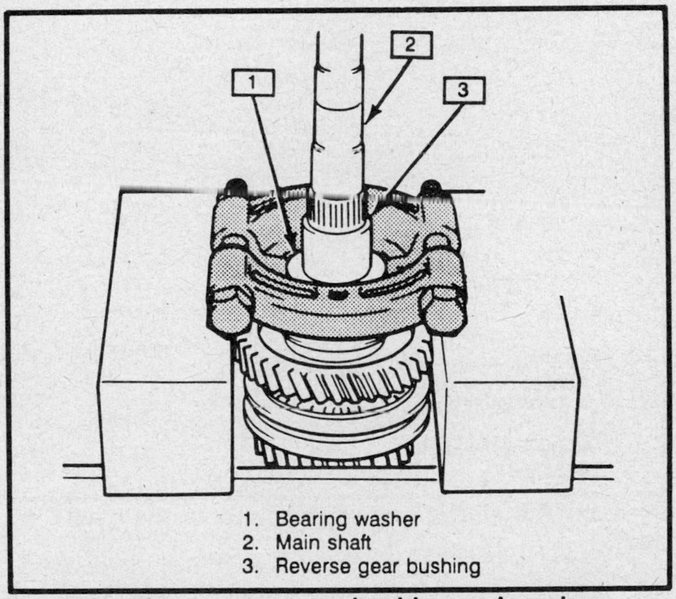
1
2
3
1. Bearing washer
2. Main shaft
3. Reverse gear bushing

Removing the reverse gear bushing and washer

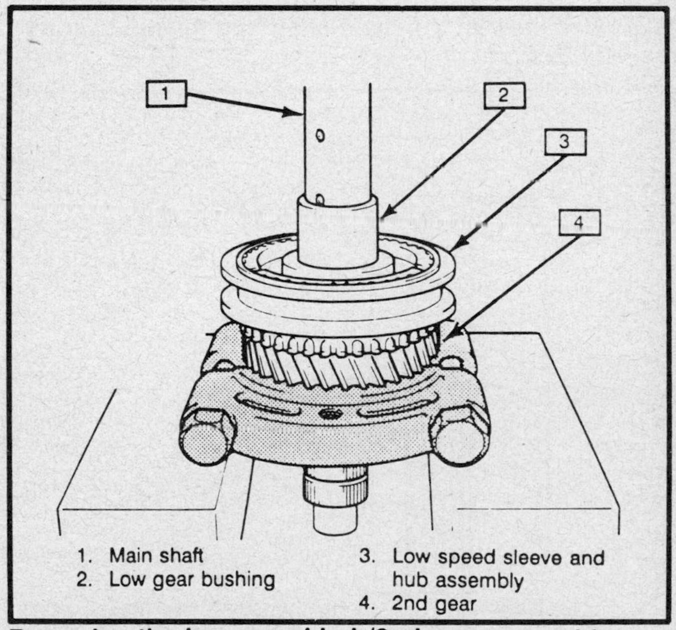
1
2
3
4
1. Main shaft
2. Low gear bushing
3. Low speed sleeve and hub assembly
4. 2nd gear

Removing the low speed hub/2nd gear assembly

a. Using a feeler gauge, check the clearance **A** between the synchronizer ring and gear and the key slot width **B** in the synchronizer ring. Check the chamfered teeth of the gear and synchronizer ring; replace the parts, if necessary.

b. Check the chamfered part of each sleeve for damage or excessive wear; replace the sleeve, if necessary.

c. Using a feeler gauge, check for the maximum sleeve-to-shifter fork clearance of 0.039 in. (1mm); replace the parts, if necessary.

4. Check the mainshaft bearing surfaces for scoring and/or wear; if necessary, replace the shaft.

5. Check the mainshaft splines for wear; if necessary, replace the shaft.

Assembly

1. Onto the mainshaft, install the 2nd gear needle bearing, the 2nd gear and the 2nd gear synchronizer ring.

2. Using a shop press and the bearing installer tool or equivalent, press the low speed sleeve/hub assembly onto the mainshaft, followed by the low gear bushing.

3. Install the low gear needle bearing, the low gear synchronizer ring, the low gear, the steel ball and the bearing washer.

4. Using a shop press, the bearing installer tool or equivalent, and the bearing driver tool or equivalent, press the mainshaft bearing into the shaft. Install the steel ball and bearing washer.

5. Using the bearing installer tool or equivalent, and the bearing driver tool or equivalent, tap on the reverse gear bushing. Install the reverse gear needle bearing and reverse gear.

6. Using the bearing installer tool or equivalent, and the bearing driver tool or equivalent, tap on the reverse sleeve/hub assembly. Using the output shaft C-ring remover/installer tool or equivalent, install the outer circlip.

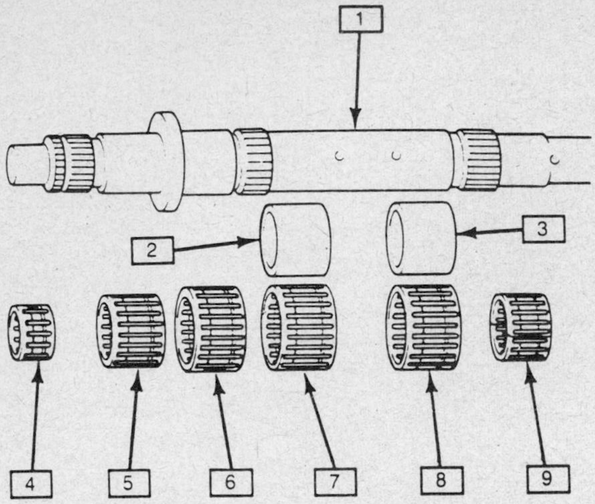
1
2
3
4
5
6
7
8
9

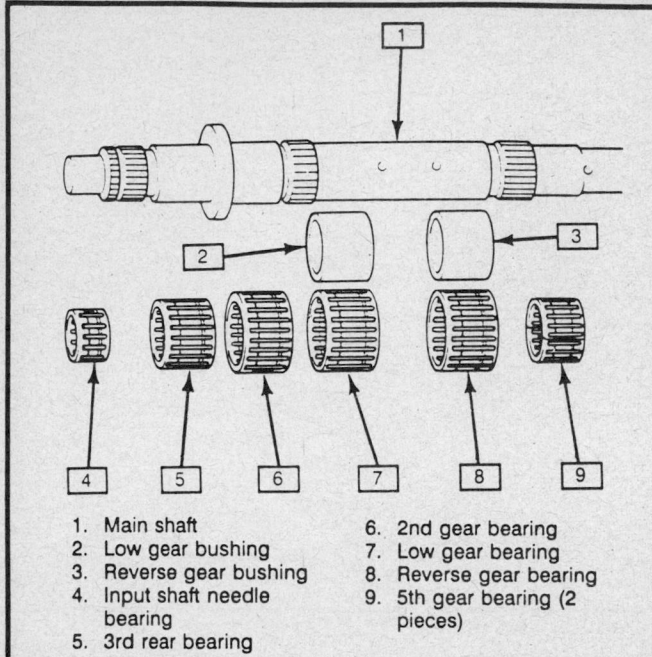

1. Main shaft
2. Low gear bushing
3. Reverse gear bushing
4. Input shaft needle bearing
5. 3rd rear bearing
6. 2nd gear bearing
7. Low gear bearing
8. Reverse gear bearing
9. 5th gear bearing (2 pieces)

Exploded view of the main shaft and needle bearing assemblies

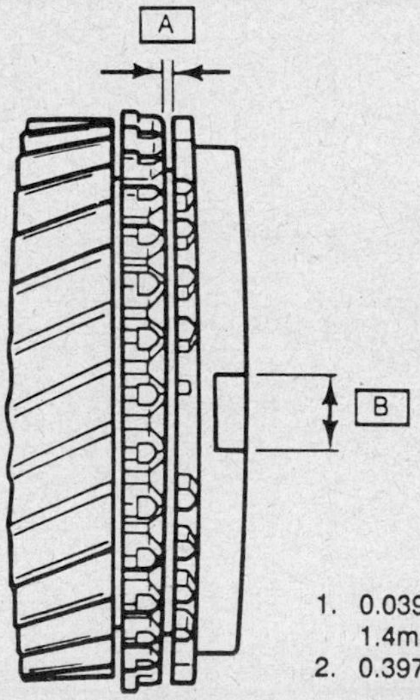
A
B

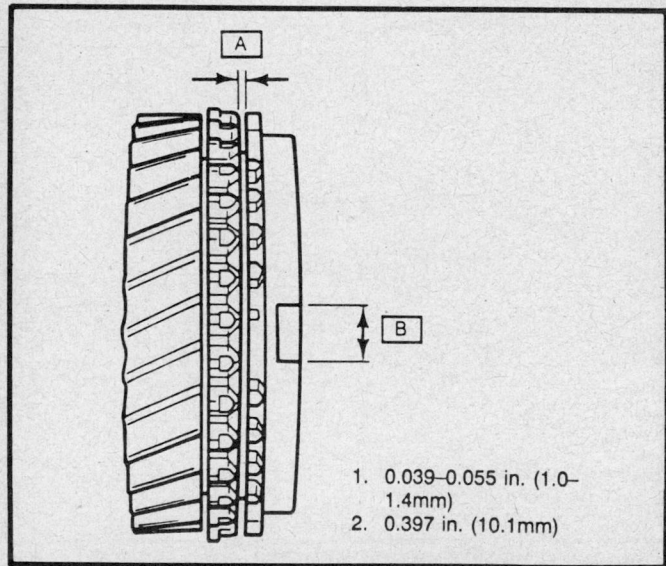

1. 0.039–0.055 in. (1.0–1.4mm)
2. 0.397 in. (10.1mm)

Inspecting the synchronizer ring and gear clearances

7. Install the 5th gear synchronizer ring, the 5th gear needle bearing, the 5th gear, the steel ball and the 5th gear washer (oil slot facing 5th gear).

8. Using a shop press and the bearing installer tool or equivalent, press the rear bearing onto the mainshaft. Using the output shaft C-ring remover/installer tool or equivalent, install the circlip.

9. Install the 3rd gear needle bearing, the 3rd gear and the high speed synchronizer ring. Using the high speed hub assembly tool or equivalent, tap on the high speed sleeve/hub assembly.

10. Using the output shaft C-ring remover/installer tool or equivalent, install the outer circlip.

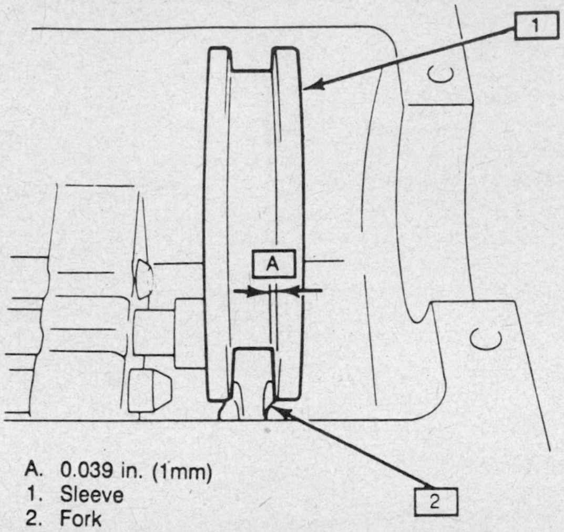
1
A
2

A. 0.039 in. (1mm)
1. Sleeve
2. Fork

Inspecting the sleeve-to-shift fork clearance

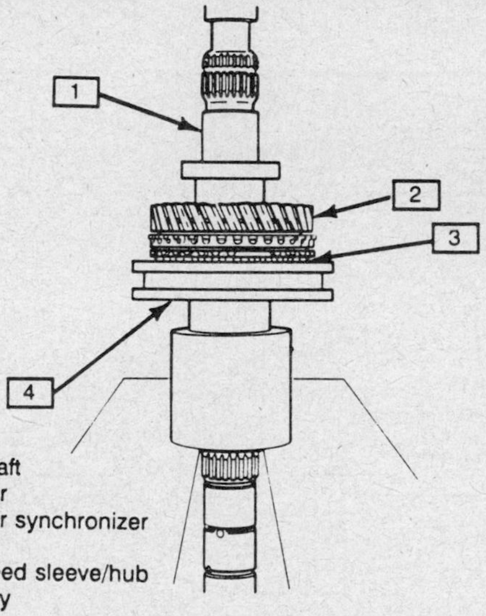
1
2
3
4

1. Main shaft
2. 2nd gear
3. 2nd gear synchronizer ring
4. Low speed sleeve/hub assembly

Pressing the sleeve/hub assembly onto the main shaft

11. Install the input shaft needle bearing, the high speed synchronizer ring and the input shaft.

UPPER CASE AND SHIFTER ASSEMBLY

Disassembly

1. Remove the back-up light and the 5th gear switches from the case.

2. Remove the locating spring bolts, the locating springs and the balls.

3. Remove the spring pin, the low speed gear shift shaft, the interlock ball and the low speed gear shift fork.

4. Remove the spring pin, the high speed gear shift shaft, the interlock pin, the interlock ball and the high speed shift fork.

NOTE: Due to the interlock system, the remaining gear shift shafts must be in the N position when removing a gear shift shaft.

5. Remove the reverse gear shift shaft, the spring pin and the reverse gear shift fork.

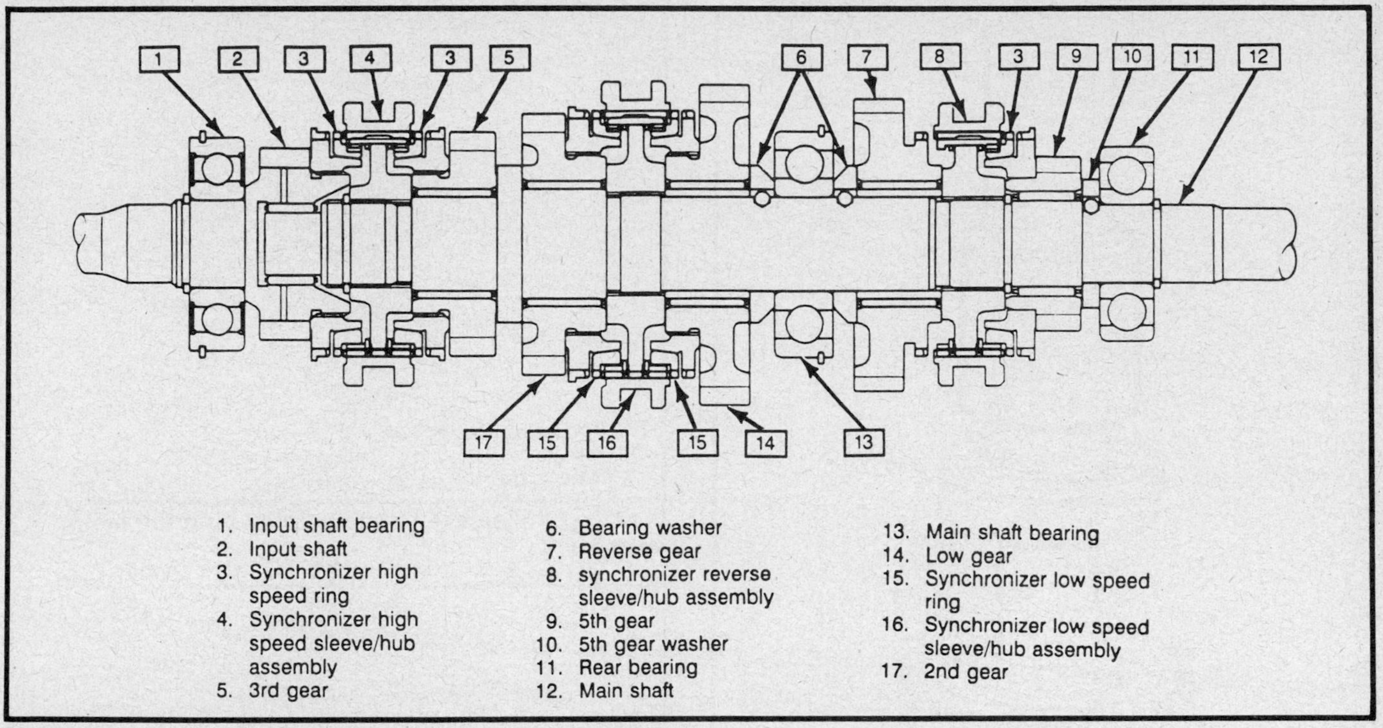
1
2
3
4
3
5
6
7
8
3
9
10
11
12
17
15
16
15
14
13
1. Input shaft bearing
2. Input shaft
3. Synchronizer high speed ring
4. Synchronizer high speed sleeve/hub assembly
5. 3rd gear
6. Bearing washer
7. Reverse gear
8. synchronizer reverse sleeve/hub assembly
9. 5th gear
10. 5th gear washer
11. Rear bearing
12. Main shaft
13. Main shaft bearing
14. Low gear
15. Synchronizer low speed ring
16. Synchronizer low speed sleeve/hub assembly
17. 2nd gear

Sectional view of the input shaft/main shaft assembly

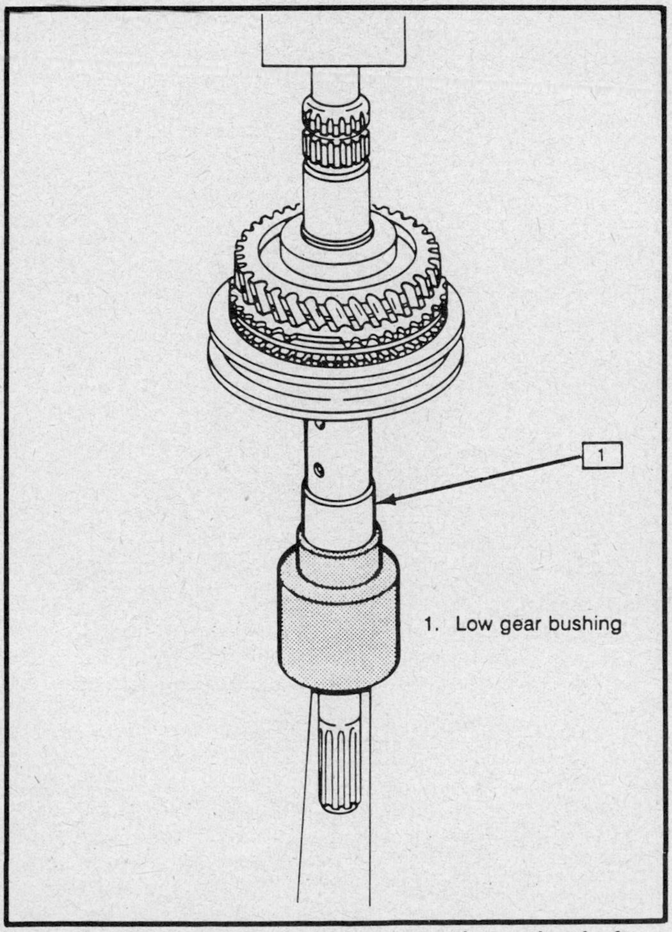
1
1. Low gear bushing

Pressing the low gear bushing onto the main shaft

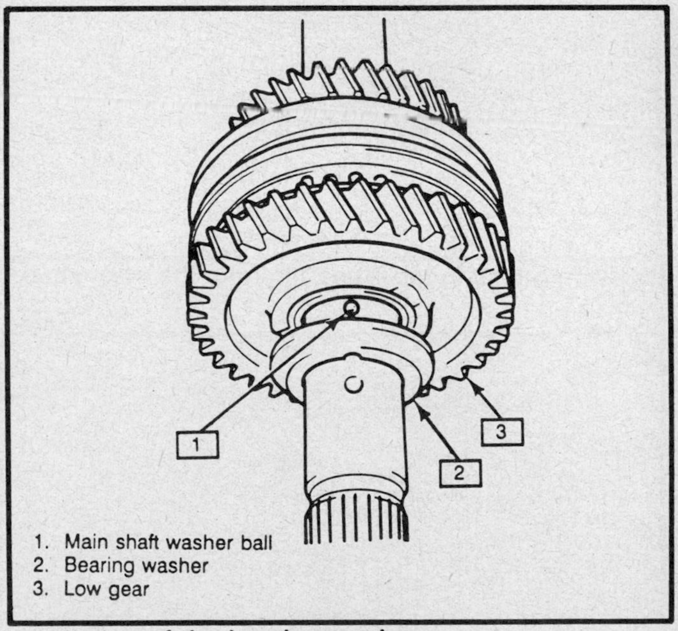
1
2
3
1. Main shaft washer ball
2. Bearing washer
3. Low gear

Installation of the bearing washer

Inspection

1. Using a feeler gauge, check the fork-to-sleeve clearance; if it exceeds 0.039 in. (1mm), replace the parts.
2. Make sure each gear shift shaft slides freely in the case; if not, use an oil stone to correct the problem.
3. Check the gear shift shaft for excessive wear; replace it, if necessary.
4. Check each locating spring's free length of 0.945–1.063 in. (24–27mm); if not within limits, replace it.

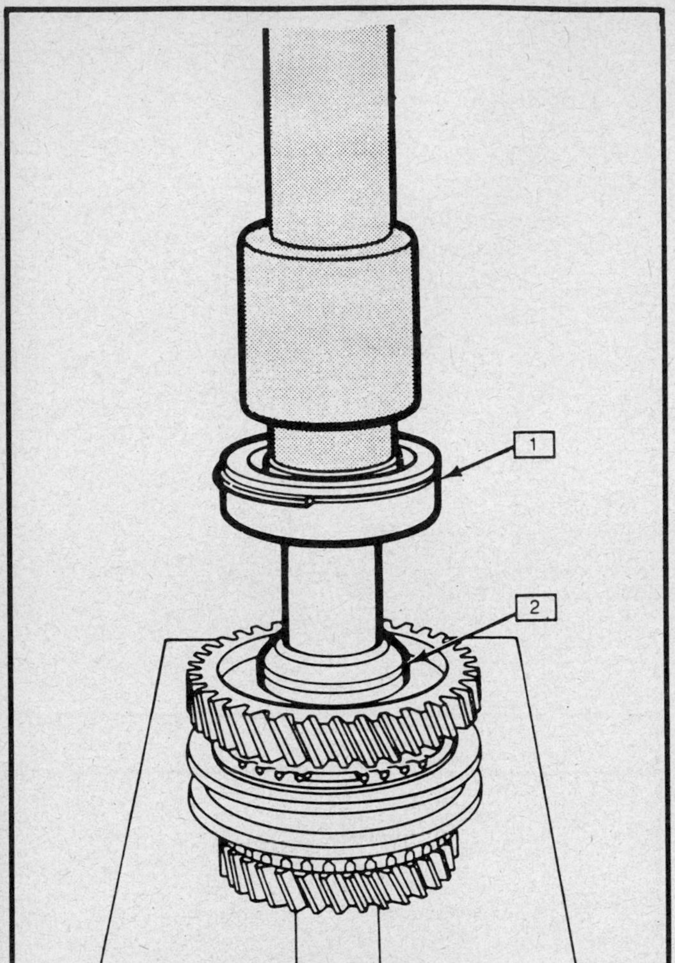
1
2

1. Main shaft bearing
2. Bearing washer

Pressing the main shaft bearing onto the main shaft

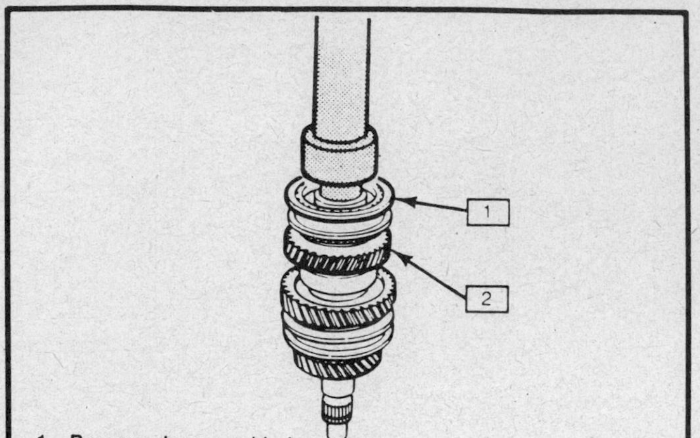
1
2

1. Reverse sleeve and hub assembly
2. Reverse gear

Pressing the sleeve/hub assembly onto the main shaft

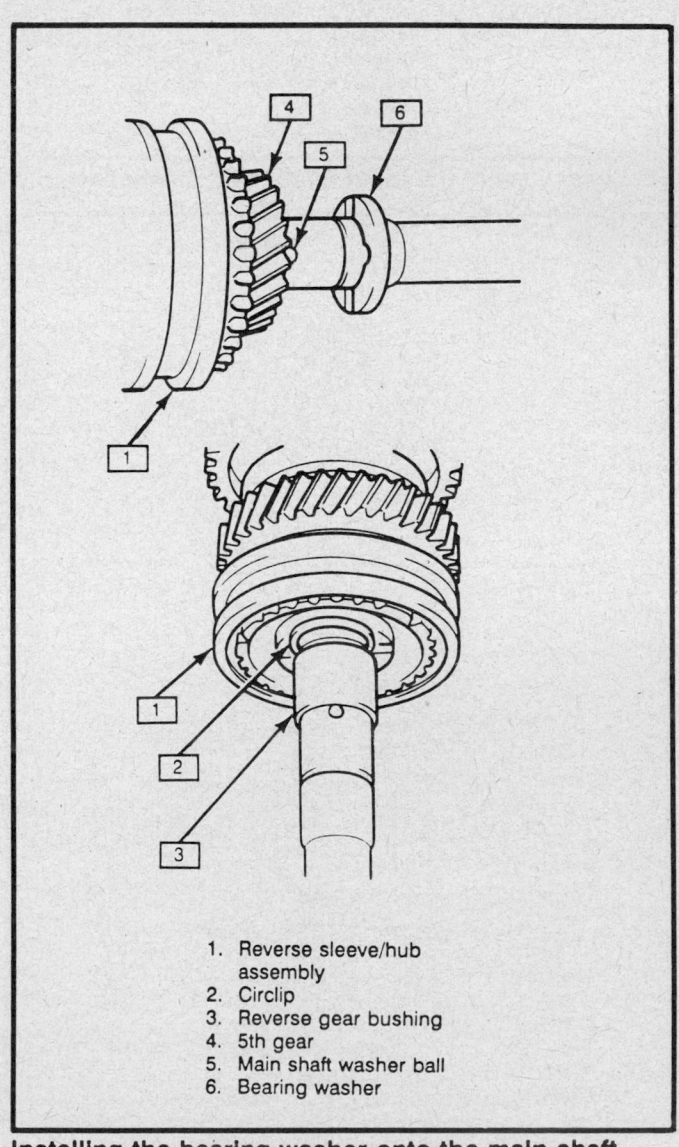

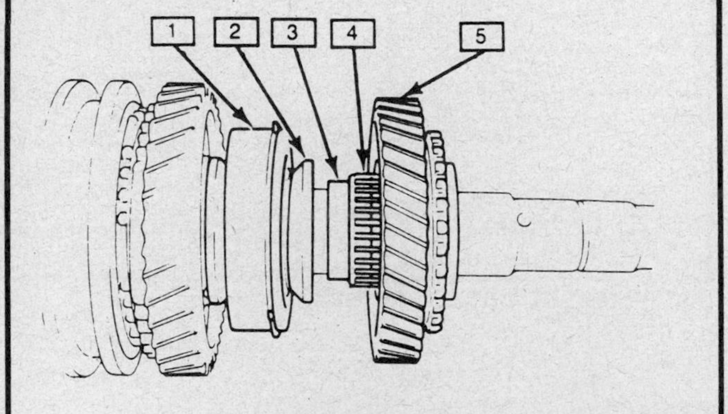
1
2
3
4
5

1. Main shaft bearing
2. Bearing washer
3. Main shaft reverse gear bushing
4. Main shaft reverse gear bearing
5. Main shaft reverse gear

Installing the reverse gear onto the main shaft

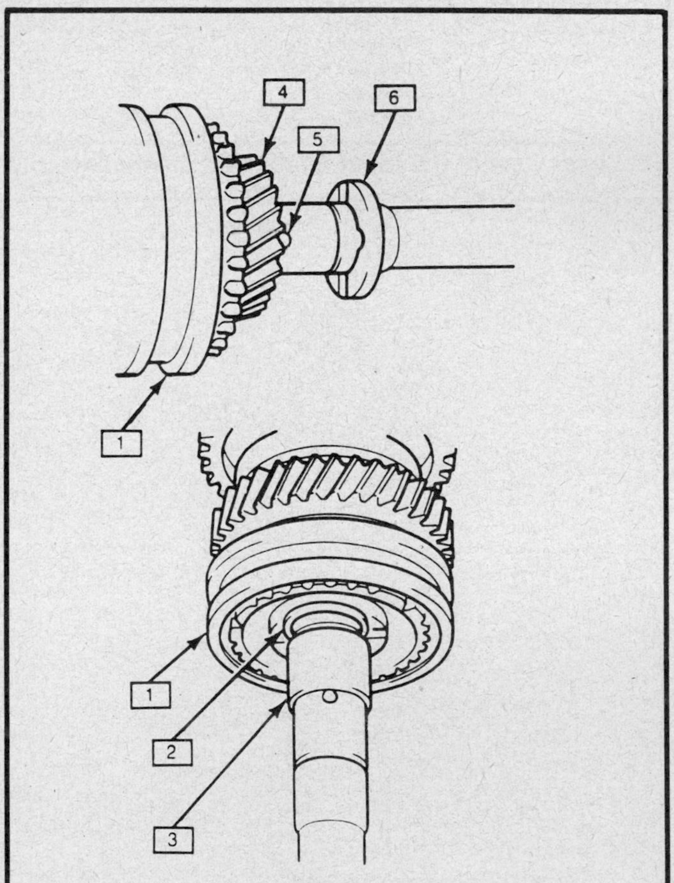
4
6
5
1
1
2
3

1. Reverse sleeve/hub assembly
2. Circlip
3. Reverse gear bushing
4. 5th gear
5. Main shaft washer ball
6. Bearing washer

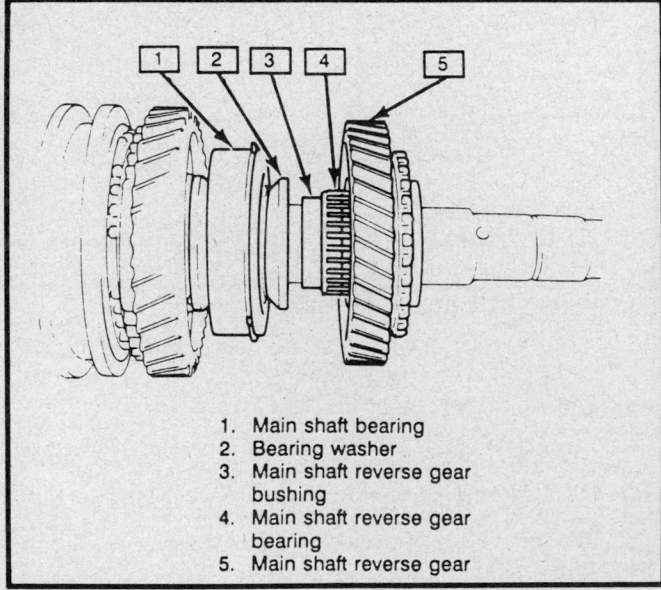

Installing the bearing washer onto the main shaft

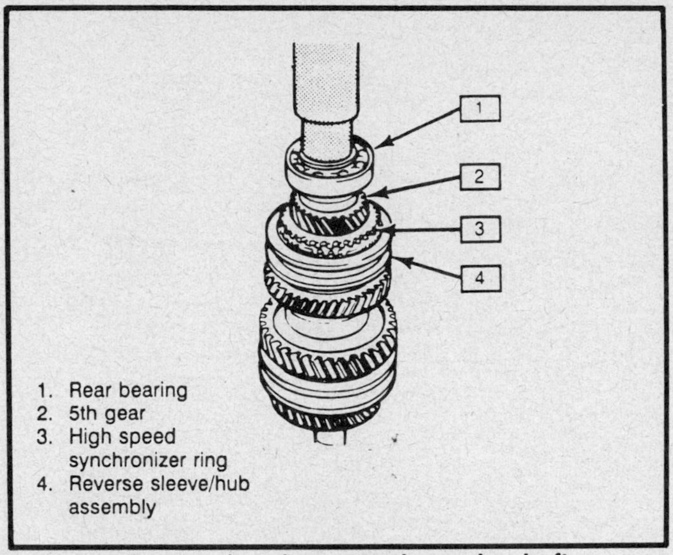
1
2
3
4
1. Rear bearing
2. 5th gear
3. High speed synchronizer ring
4. Reverse sleeve/hub assembly

Pressing the rear bearing onto the main shaft

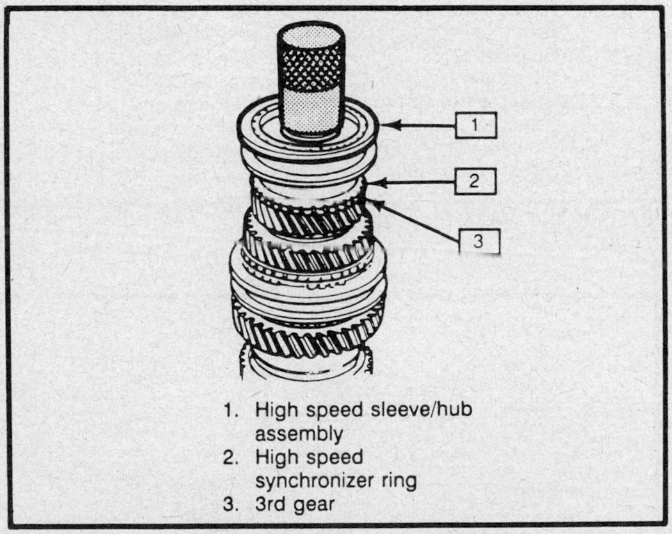
1
2
3
1. High speed sleeve/hub assembly
2. High speed synchronizer ring
3. 3rd gear

Pressing the sleeve/hub assembly onto the main shaft

Assembly

1. Install the reverse shift fork and spring pin to the reverse gear shift shaft.
2. Install the reverse gear shift shaft and interlock ball.

NOTE: Be sure to place the reverse gear shift shaft into the N position before installing the high speed gear shift shaft.

3. Install the interlock pin to the high speed shift shaft.
4. Install the high speed shift shaft, the high speed gear shift fork, the spring pin and the interlock ball.
5. Place the reverse gear and high speed gear shift shafts into the **N** positions. Install the low speed gear shift shaft, the low speed gear shift fork and the spring pin.
6. To confirm that the interlock system operates correctly, perform the following procedures:
   a. When shifted to the 1st or 2nd position, attempt to shift into the 3rd/4th speed or 5th/reverse speed; it must not be possible.

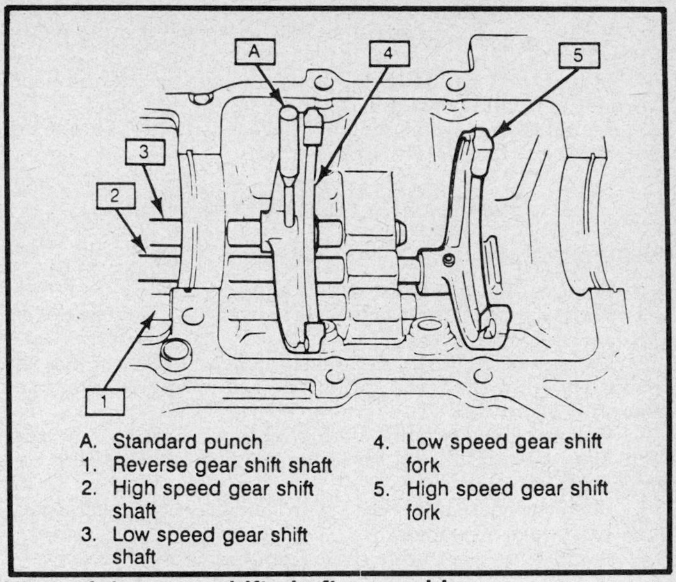
A
4
5
3
2
1
A. Standard punch
1. Reverse gear shift shaft
2. High speed gear shift shaft
3. Low speed gear shift shaft
4. Low speed gear shift fork
5. High speed gear shift fork

View of the gear shift shaft assembly

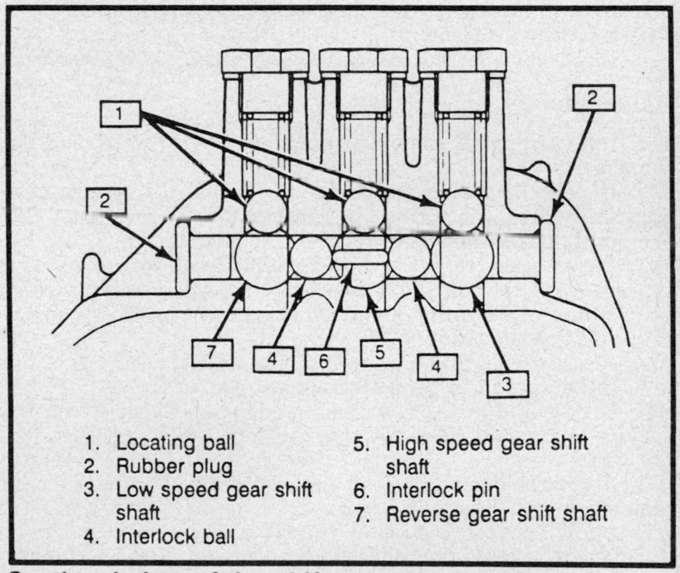
1
2
2
7
4
6
5
4
3
1. Locating ball
2. Rubber plug
3. Low speed gear shift shaft
4. Interlock ball
5. High speed gear shift shaft
6. Interlock pin
7. Reverse gear shift shaft

Sectional view of the shift assembly interlock system

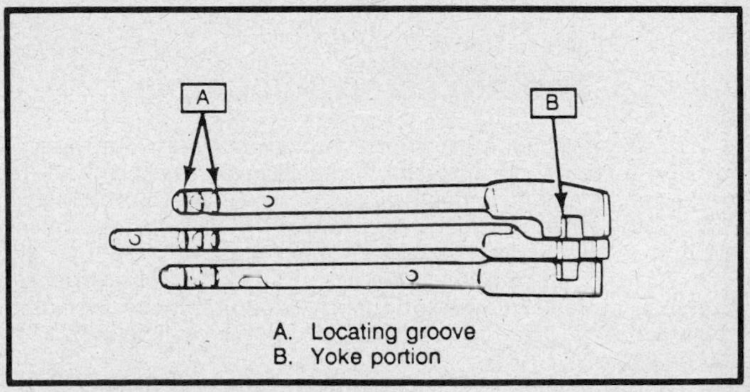
A
B
A. Locating groove
B. Yoke portion

Inspecting the gear shift shaft

b. When shifted to the 5th or reverse position, attempt to shift into the 1st/2nd speed or 3rd/4th speed; it must not be possible.

7. Install the locating spring bolts, the spring and the balls. Torque the locating spring bolts to 21 ft. lbs. (28 Nm).

8. Install the back-up light and 5th gear switches. Torque the switches to 15 ft. lbs. (20 Nm).

GEAR SHIFT LEVER CASE

Disassembly

1. Remove the case plate screws and the plate.
2. While support the case in a soft-jawed vise, remove the select return spring bolts.
3. Using a drift punch, drive out the gear shift arm pin, the spring pin from the lever spring and the spring pin from the gear shift lever.
4. Pull out the gear shift shaft slightly and remove the gear shift arm; be careful, the case plug may come off during this procedure.
5. Push the gear shift lever inward and drive the reverse gear shift limit yoke's spring pin.
6. Remove the gear shift shaft from the case.

Inspection

1. Make sure the shift shaft, the shift lever, the limit yoke and the shift arm slide freely in the case; if not, use an oil stone to correct the problem.
2. Check each part for excessive wear; replace it, if necessary.

Assembly

1. Insert the shift shaft (flat area facing downward) and the inner parts.
2. Install the spring pin (single short) for the reverse gear shift limit yoke, the spring pin (double) for the shift lever and the spring pin (single long) for the return spring.
3. Install the straight end of the shift limit spring into the line groove in the shift limit yoke and the hooked end turned in its winding direction by about 90 degrees from its free state. Hook it onto the spring pin.
4. Install the spring pin for the gear shift arm.
5. Caulk and install the case plug.
6. Grease and install the select guide pins, the select return springs and spring bolts. Torque the spring bolts to 17 ft. lbs. (23 Nm).
7. Apply sealant and install the case plate. Torque the case plate screws to 7 ft. lbs. (9 Nm).
8. Apply Loctite® and install the reverse gear shift limit bolt. Torque the reverse gear shift limit bolt to 24 ft. lbs. (32 Nm).
9. Install the gear shift lever case to the transmission, check its operation; make sure it shifts smoothly according the shift pattern and remove it.
10. Apply silicone to the mating surface of the gear shift lever, install it and torque the bolts to 12 ft. lbs.

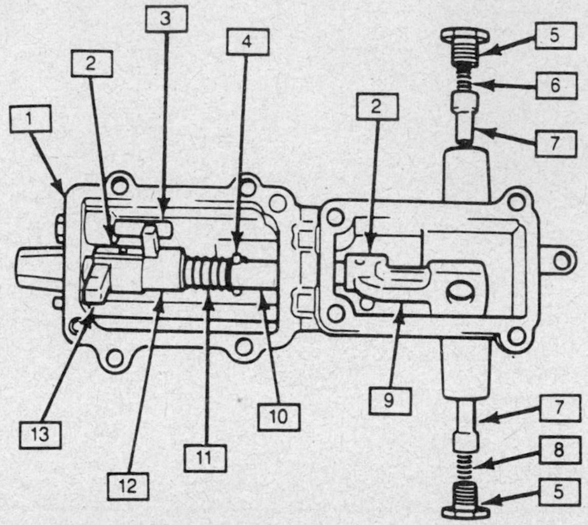
1
2
3
4
2
5
6
7
9
10
11
12
13
7
8
5

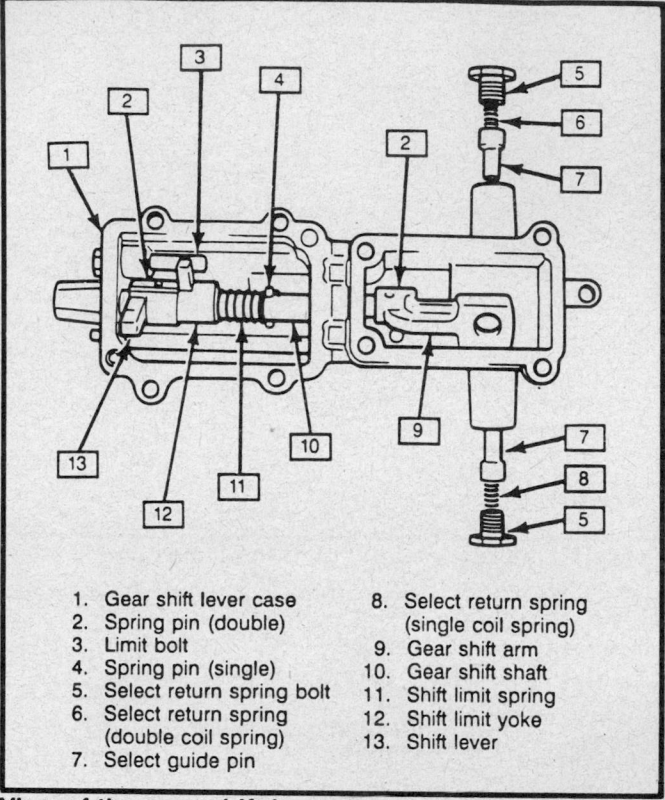

1. Gear shift lever case
2. Spring pin (double)
3. Limit bolt
4. Spring pin (single)
5. Select return spring bolt
6. Select return spring (double coil spring)
7. Select guide pin
8. Select return spring (single coil spring)
9. Gear shift arm
10. Gear shift shaft
11. Shift limit spring
12. Shift limit yoke
13. Shift lever

View of the gear shift lever case

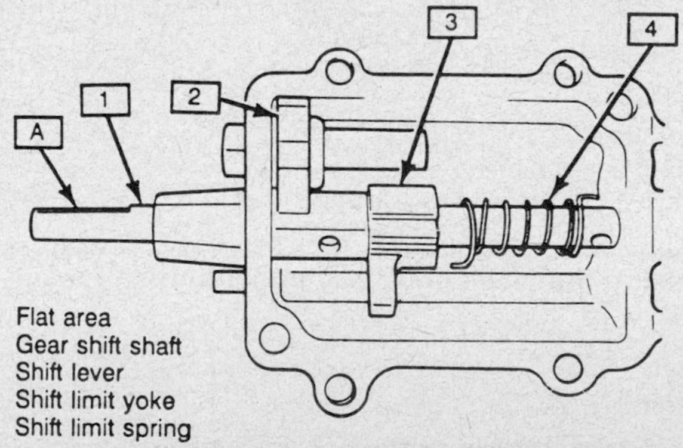
A
1
2
3
4

A. Flat area
1. Gear shift shaft
2. Shift lever
3. Shift limit yoke
4. Shift limit spring

View of the gear shift lever case assembly

Transmission Assembly

1. Using gear oil, lubricate the input shaft/mainshaft assembly and install them into the lower case; be sure to check for proper engagement (smooth rotation) with the countershaft.
2. Using silicone sealer part number 1052917 or equivalent, apply it to the mating surface of upper case.
3. Position all of the gears and forks into the **N** positions.
4. Assemble the cases and install the bolts; make sure all of the bolts are at the same height. Torque the case bolts to 17 ft. lbs. (23 Nm).
5. Using silicone sealer part number 1052917 or equivalent, apply it to the mating surface of input shaft bearing retainer. Install the bearing retainer, apply Loctite® to the bolts and torque them to 17 ft. lbs. (23 Nm).
6. Install the clutch housing and torque the bolts to 37 ft. lbs. (50 Nm).
7. Lightly grease the sliding part of the release bearing and install it.
8. Using silicone sealer part number 1052917 or equivalent, apply it to the mating surface of rear case. Install the rear case and torque them to 17 ft. lbs. (23 Nm).
9. Install the back-up light and the 5th gear switches; torque them to 15 ft. lbs. (20 Nm).
10. Install the gear shift lever case and torque the bolts to 17 ft. lbs. (23 Nm).
11. Check the transmission's operation.

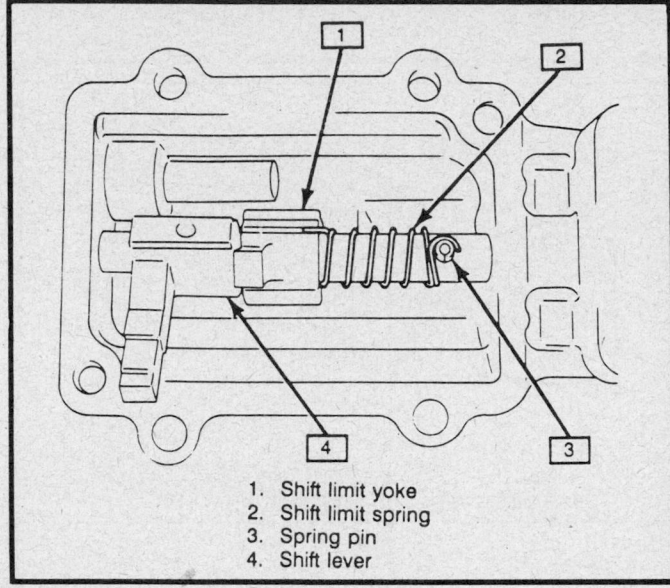

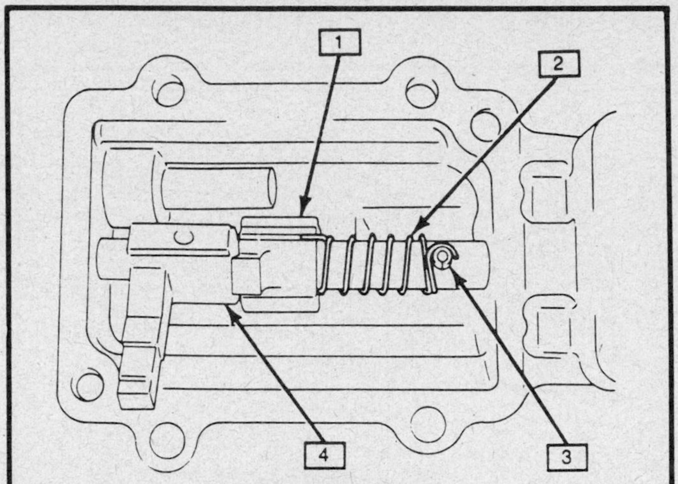
1
2
3
4
1. Shift limit yoke
2. Shift limit spring
3. Spring pin
4. Shift lever

Installing the return inner spring

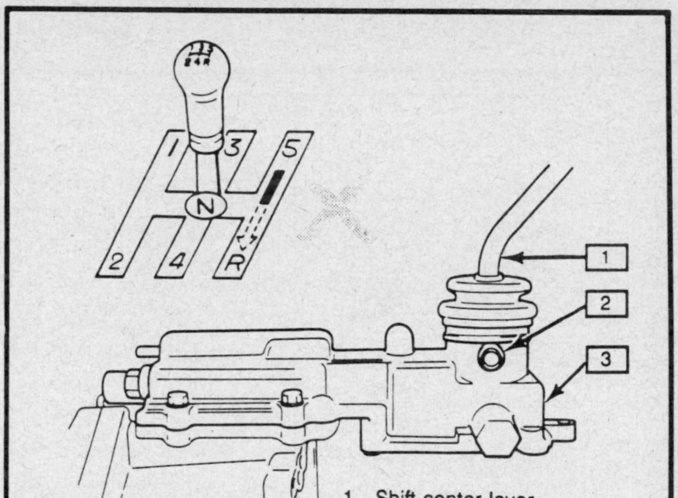
1
3
5
N
2
4
R
1
2
3
1. Shift contor lever
2. Control lever locating bolt
3. Gear shift lever case

Gear shift pattern

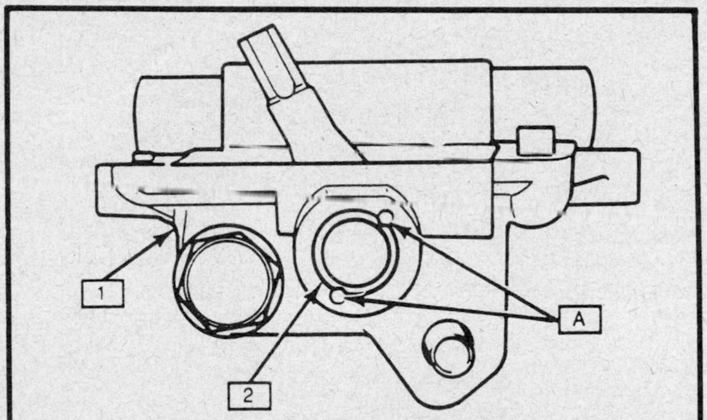
1
2
A
1. Caulking
2. Gear shift lever case
3. Case plug

Installing the gear shift lever case plug

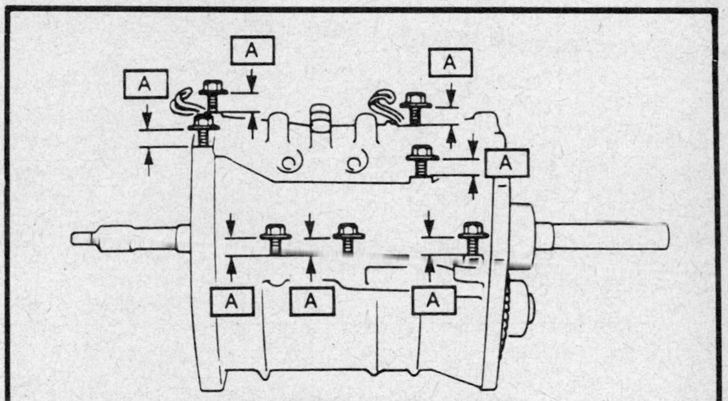
A
1. Bolt height before the screw is in

Check the case bolt lengths

SPECIFICATIONS

TORQUE SPECIFICATIONS

Item	ft. lbs.	Nm
Oil filler/level and drain plugs	33	45
Control lever boot cover bolts	4.4	6
Backup lamp, 5th and 4WD switches	15	20
Speedometer driven gear case bolt	7.5	10
Control lever locating bolts	13	17
Reverse gear shift limit bolt	24	32
Select return spring bolts	17	23
Shift lever case plate screws	7	9
Gear shift case bolts	17	23
Transmission to engine bolts and nuts	63	85
Enhaust center pipe to manifold nuts	37	50

TORQUE SPECIFICATIONS

Item	ft. lbs.	Nm
Engine rear mounting bolts and nut	37	50
Muffler to exhaust center pipe bolts	37	50
Exhaust center pipe bracket bolts	37	50
Universal joint flange bolts	37	50
Center bearing plate bolts	7.5	10
Reverse idle gear shaft bolts	16	22
Locating spring bolts	21	28
Upper case bolts	17	23
Input shaft retainer bolts	17	23
Clutch housing bolts	37	50
Rear case bolts	17	23

SPECIAL TOOLS

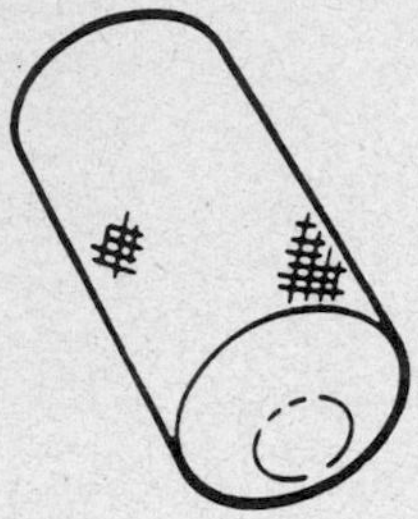

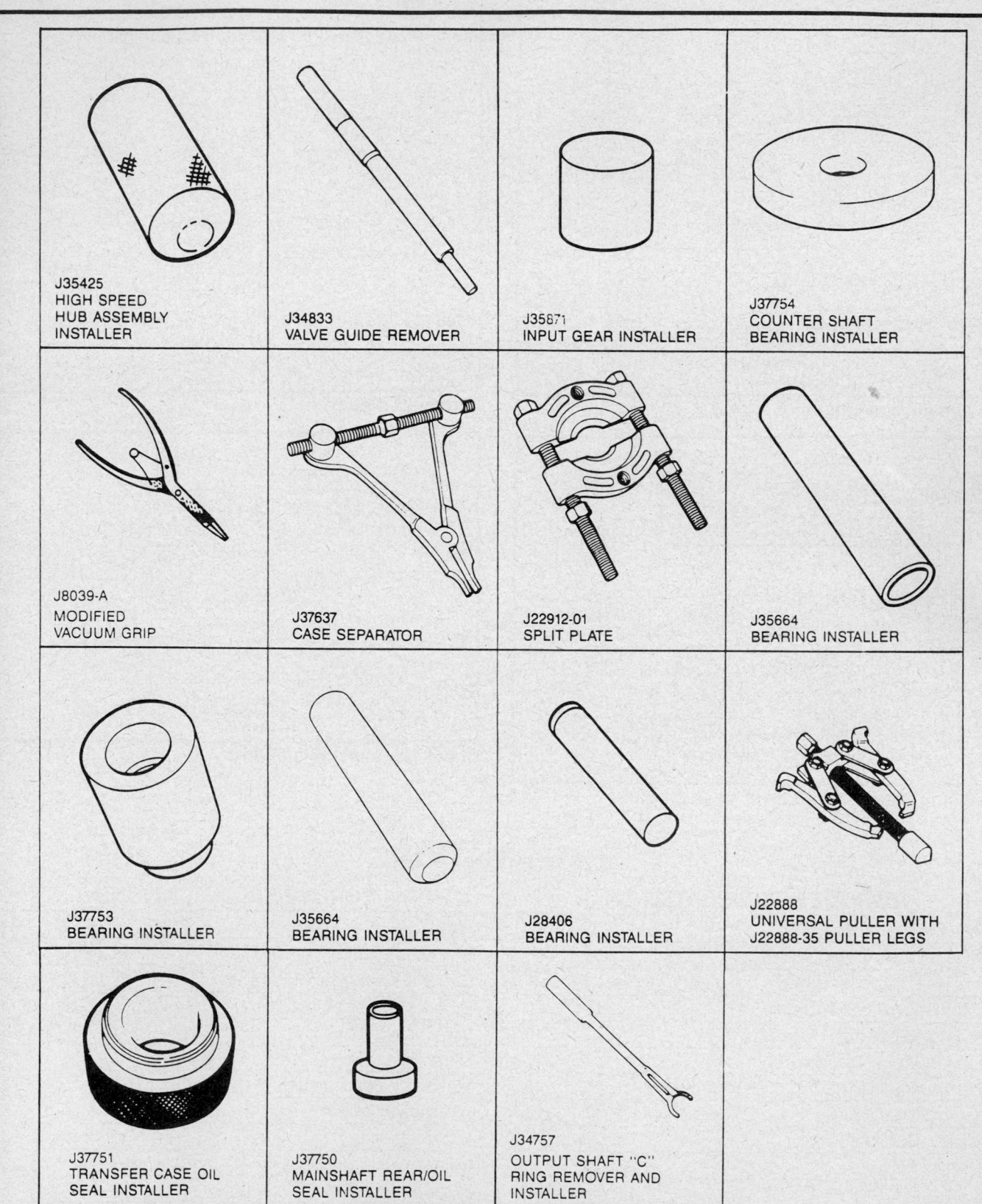

J35425
HIGH SPEED
HUB ASSEMBLY
INSTALLER

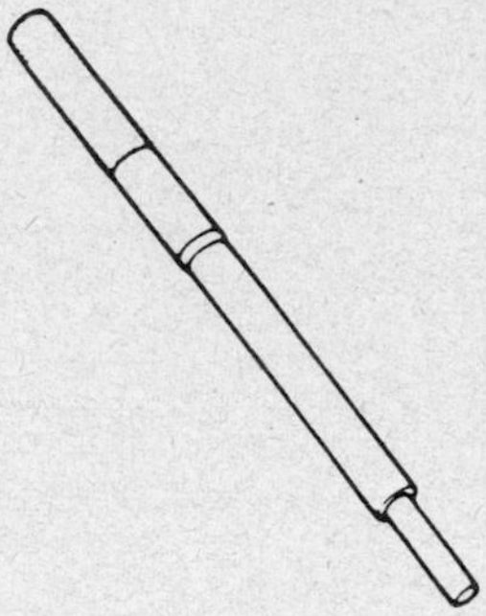

J34833
VALVE GUIDE REMOVER

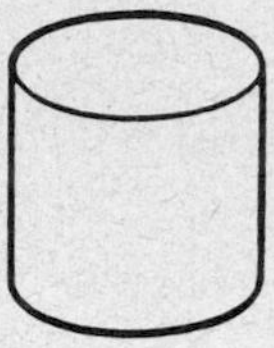

J35871
INPUT GEAR INSTALLER

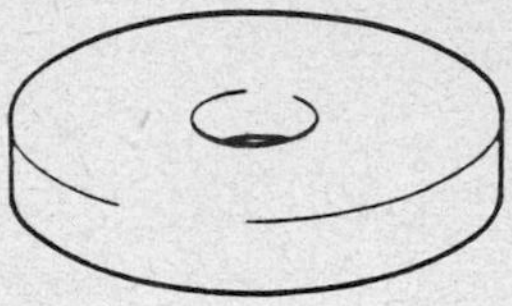

J37754
COUNTER SHAFT
BEARING INSTALLER

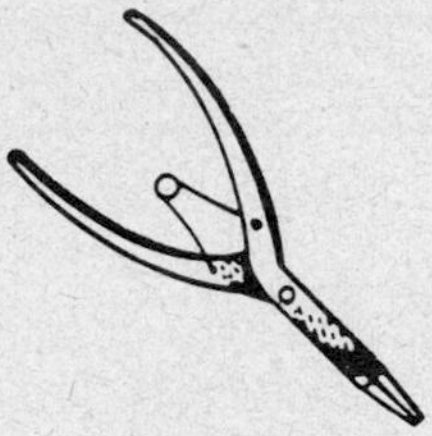

J8039-A
MODIFIED
VACUUM GRIP

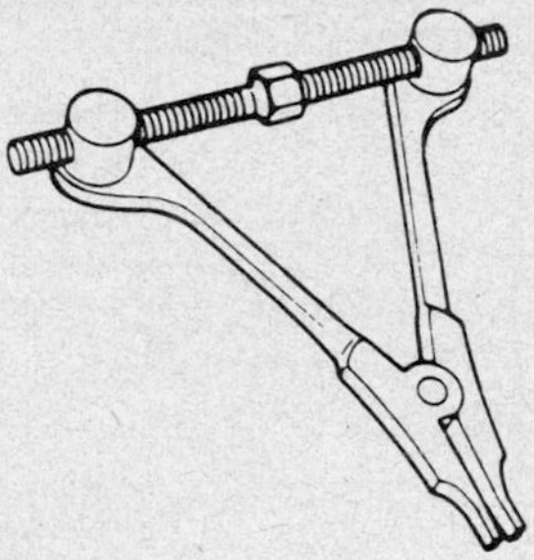

J37637
CASE SEPARATOR

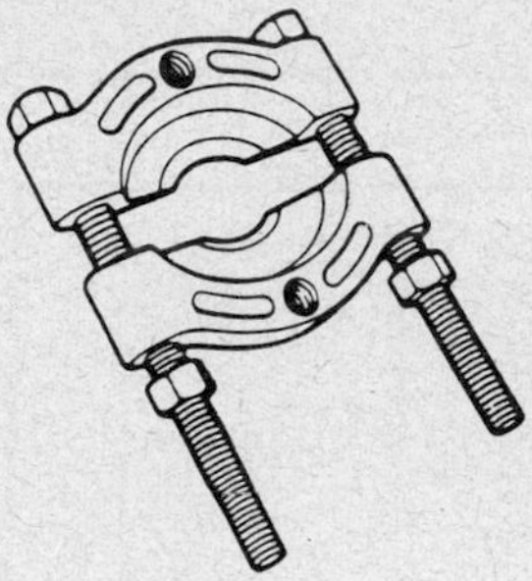

J22912-01
SPLIT PLATE

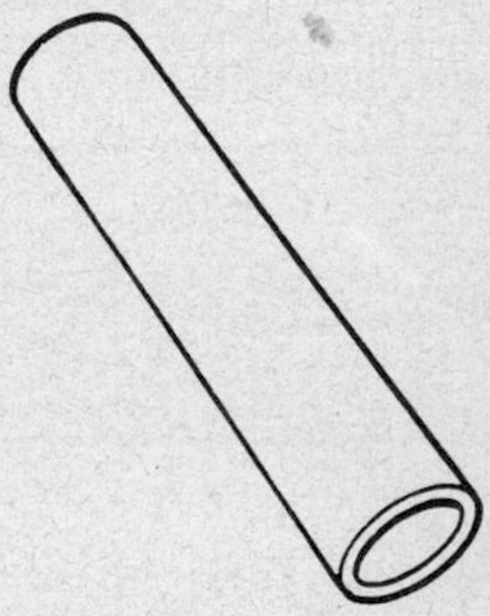

J35664
BEARING INSTALLER

J37753
BEARING INSTALLER

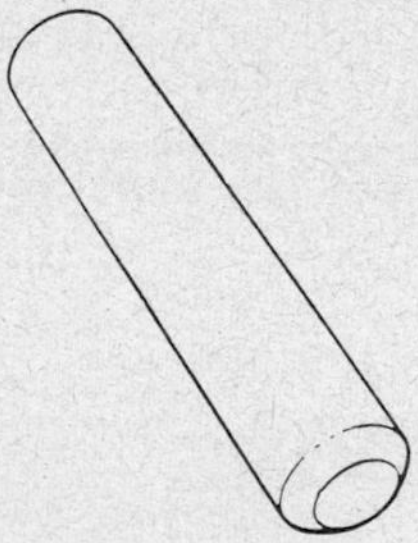

J35664
BEARING INSTALLER

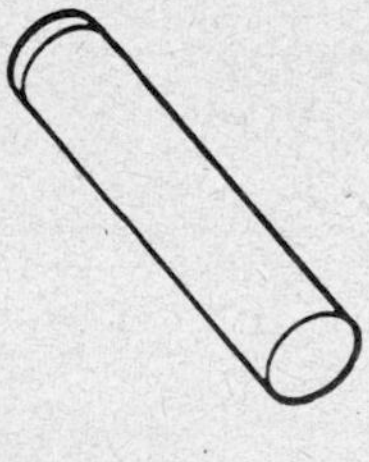

J28406
BEARING INSTALLER

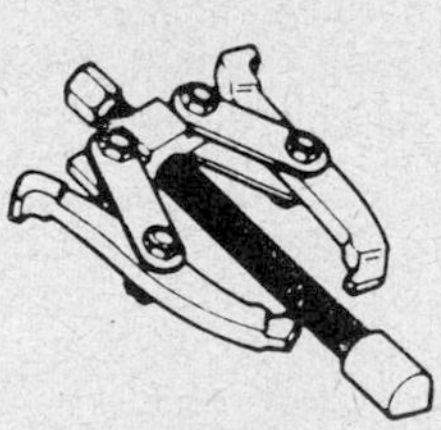

J22888
UNIVERSAL PULLER WITH
J22888-35 PULLER LEGS

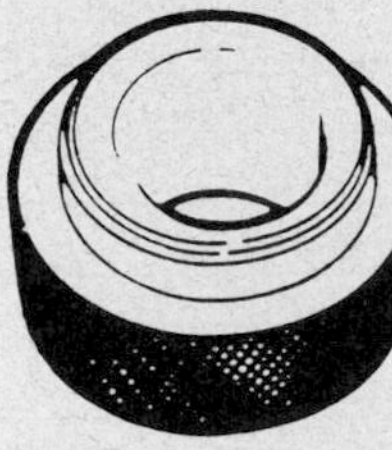

J37751
TRANSFER CASE OIL
SEAL INSTALLER

J37750
MAINSHAFT REAR/OIL
SEAL INSTALLER

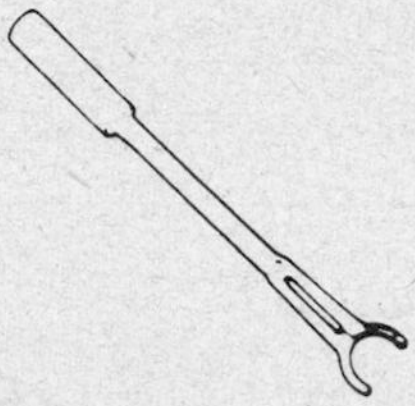

J34757
OUTPUT SHAFT "C"
RING REMOVER AND
INSTALLER

Section 7

Muncie Transmission
General Motors

APPLICATION

1984–89 GM light trucks

GENERAL DESCRIPTION

The 3 speed Muncie 76mm (M62 and M64) transmissions are in constant mesh and fully synchronized in all forward speeds with 2 sliding synchronizer sleeves. The gear shift levers are either a floor type or a steering column type shift lever.

The 4 speed Muncie 117mm (M20) transmission uses a constant mesh 1st gear and synchronized 2nd, 3rd and 4th gears. Gear shifting is controlled with a transmission cover mounted shift lever. The cover has a ball pin type interlock which stops the transmission from being shifted into 2 gears at the same time.

Transmission Identification

The 3 speed transmissions can be identified by a label attached to the top of the case. The 3 letter code is UUF, UUH, UUJ or UUK.

The 4 speed transmission can be identified by a label attached to the top of the cover near the tower. The 3 letter code is GGT, GGK, GBF, GBG, GAU or GAV.

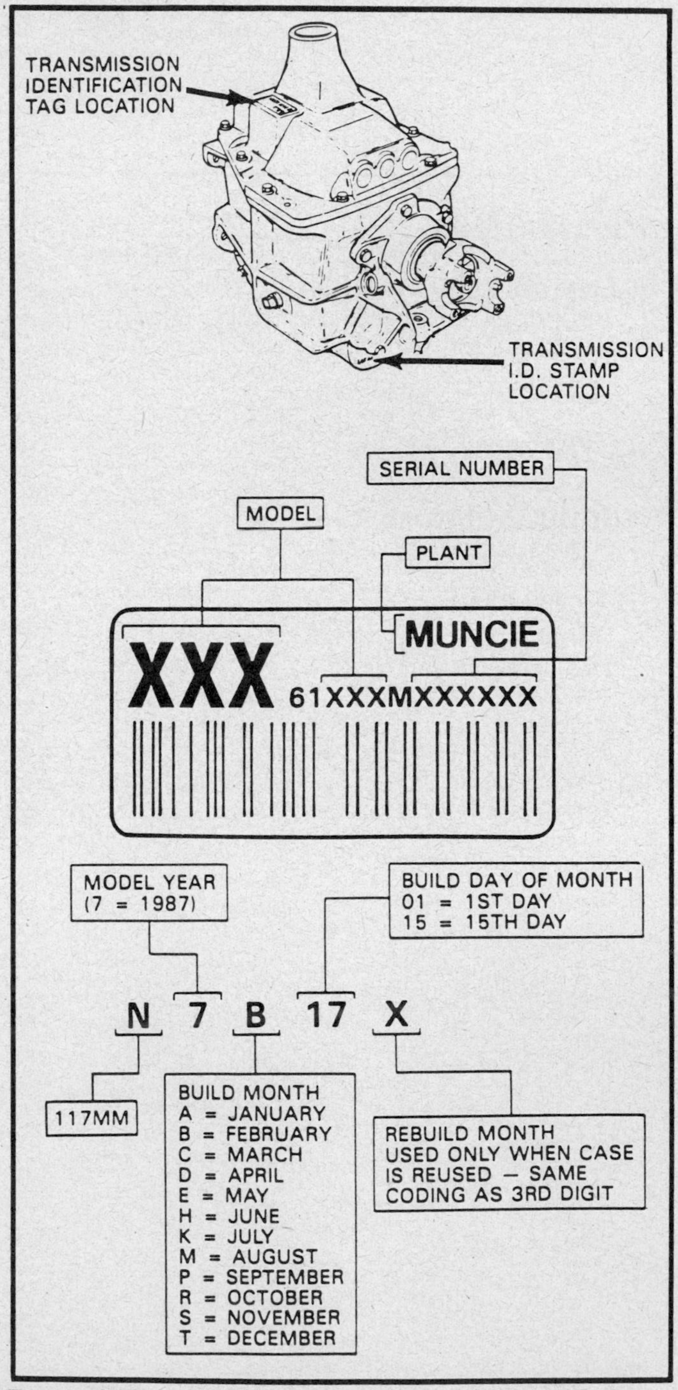
TRANSMISSION IDENTIFICATION TAG LOCATION
TRANSMISSION I.D. STAMP LOCATION
SERIAL NUMBER
MODEL
PLANT
MUNCIE
XXX
61XXXMXXXXXX
MODEL YEAR (7 = 1987)
BUILD DAY OF MONTH
01 = 1ST DAY
15 = 15TH DAY
N 7 B 17 X
117MM
BUILD MONTH
A = JANUARY
B = FEBRUARY
C = MARCH
D = APRIL
E = MAY
H = JUNE
K = JULY
M = AUGUST
P = SEPTEMBER
R = OCTOBER
S = NOVEMBER
T = DECEMBER
REBUILD MONTH USED ONLY WHEN CASE IS REUSED — SAME CODING AS 3RD DIGIT
Transmission identification—117mm (M20) 4 speed

Metric Fasteners

Do not attempt to interchange metric fasteners for inch system fasteners. Mismatched or incorrect fasteners can result in damage to the transmission unit through malfunctions, breakage or possible personal injury. Care should be taken to reuse the fasteners in the same location as removed, whenever possible. Due to the large number of alloy parts used and the aluminum casing, torque specifications should be strictly observed. Before installing bolts into aluminum parts, always dip threads into oil, or sealant (if specifically required), to prevent the bolts from galling the aluminum threads and to prevent seizing.

Capacities

The lubricant capacity for the 3 speed transmission is 1.6 qts. (1.5L) of API GL5 SAE 80W90 gear lubricant. The lubricant capacity for the 4 speed transmission is 4.2 qts. (4.0L) of API GL5 SAE 80W90 gear lubricant.

Checking Fluid Level

With the vehicle on a level surface and the lubricant in the transaxle cold, remove the plug and check the lubricant level. The lubricant should come to the bottom edge of the plug hole. If the level is low, add API GL5 SAE 80W90 gear lubricant through the filler plug hole until it begins to run out of the plug hole. Reinstall and tighten the plugs.

NOTE: Do not use automatic transmission fluid or engine oil, this will cause foaming out of the vent or durability problems.

TRANSMISSION MODIFICATIONS

Shift Lever Rattle or Buzz

Some 1986 C/K series trucks or 1987 R/V series trucks with the 117mm manual transmission, models GAU, GAV and GGJ, may have a shift lever rattle or buzz condition in 1st and 2nd gear. The rattle/buzz condition can be corrected by using a 1985 or prior shift lever assembly, part number 6274077.

NOTE: Only chrome plated shift lever assemblies are available for service. These are completely interchangeable with the original equipment black painted shift levers.

3rd Gear Hop-Out

Some 1985–89 R/V, C/K series trucks or P chassis vehicles with a 117mm manual transmission, may exhibit 3rd gear hop-out on acceleration or deceleration. If this condition occurs, it can be corrected by installing the service part package part number 8672944.

Fluid Leak Out The Vent

Some 1988–89 C/K series trucks with a 117mm manual transmission, may develop a fluid leak out of the vent. The addition of a remote vent will prevent fluid from leaking out of the cover vent. If the model code on the identification tag is GGD or GGG, perform the following procedure:

1. Remove the transmission and the transfer case, if applicable.
2. Remove the existing vent.
3. Install a vent sleeve and pipe assembly, part number 08652580 in the vent tube hole with the end of the fitting toward the clutch housing. Install it with a dull wide bladed tool.
4. Attach a 30 in. (762mm) piece of $^5/_{16}$ in. (8.00mm) inside diameter hose, part number 09439104 to the pipe assembly and secure it with a clamp, part number 25518880.
5. Install the transmission and the transfer case, if applicable.
6. Install a new vent assembly, part number 08640496 in the end of the hose.
7. Secure the hose to the large wire harness underhood on the passenger side of the engine cowl with a plastic tie wrap, part number 03816659.

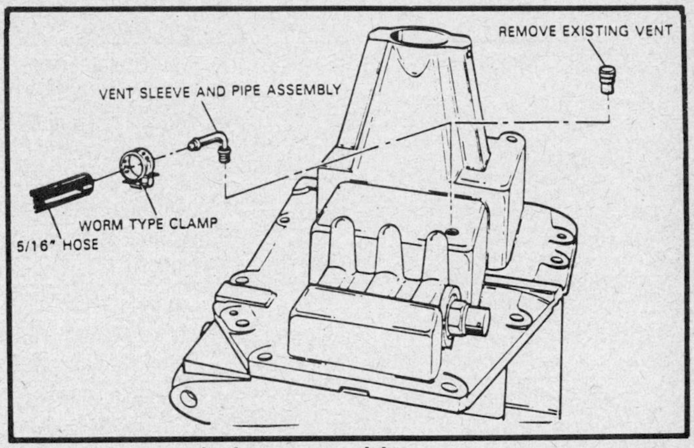
REMOVE EXISTING VENT
VENT SLEEVE AND PIPE ASSEMBLY
WORM TYPE CLAMP
5/16" HOSE

Vent sleeve and pipe assembly

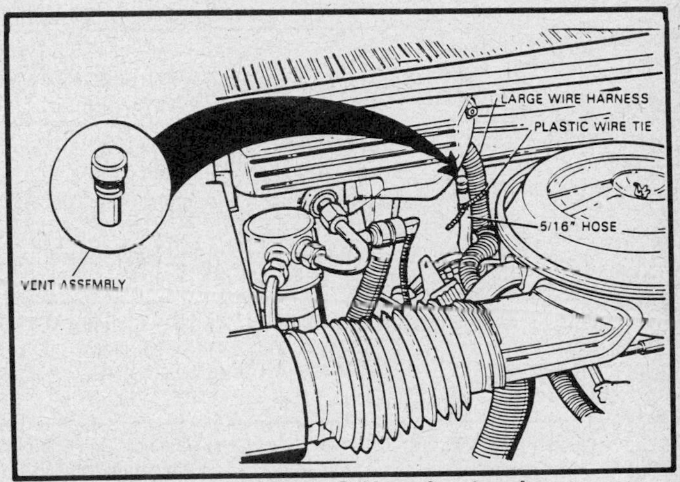
LARGE WIRE HARNESS
PLASTIC WIRE TIE
5/16" HOSE
VENT ASSEMBLY

Engine cowl of a 1988–89 C/K series truck

TROUBLE DIAGNOSIS

CHILTON'S THREE C'S TRANSMISSION DIAGNOSIS

Condition	Cause	Correction
Slips out of high gear	a) Transmission loose on clutch housing	a) Tighten mounting bolts
	b) Shift rods interfere with engine mounts or clutch throw-out lever	b) Replace or bend levers and rods to eliminate interference
	c) Shift linkage does not work freely; binds	c) Adjust and free up shift linkage
	d) Damaged mainshaft pilot bearing	d) Replace pilot bearing
	e) Main drive gear retainer broken or loose	e) Tighten or replace main drive gear
	f) Dirt between transmission case and clutch housing	f) Clean mating surfaces
	g) Misalignment of transmission	g) Align transmission
	h) Stiff shift lever seal	h) Replace seal

CHILTON'S THREE C'S TRANSMISSION DIAGNOSIS

Condition	Cause	Correction
Slips out of high gear	i) Pilot bearing loose in crankshaft j) Worn or improperly adjusted linkage	i) Replace bearing j) Adjust or replace linkage as required
Noisy in all gears	a) Insufficient lubricant b) Worn countergear bearings c) Worn or damaged main drive gear and countergear d) Damaged main drive gear or main shaft bearings e) Worn or damaged countergear anti-lash plate	a) Fill to correct level b) Replace countergear bearings and shaft c) Replace worn or damaged gears d) Replace damaged bearings or main drive gear e) Replace countergear
Noisy in high gear	a) Damaged main drive gear bearing b) Damaged mainshaft bearing c) Damaged high speed gear synchronizer	a) Replace damaged bearing b) Replace damaged bearing c) Replace synchronizer
Noisy in neutral with engine running	a) Damaged main drive gear bearing b) Damaged or loose mainshaft pilot bearing c) Worn or damaged countergear anti-lash plate d) Worn countergear bearings	a) Replace damaged bearing b) Replace pilot bearings c) Replace countergear d) Replace countergear bearings and shaft
Noisy in all reduction gears	a) Insufficient lubricant b) Worn or damaged main drive gear or countergear	a) Fill to correct level b) Replace faulty or damaged gears
Noisy in second only	a) Damaged or worn second-speed constant mesh gears b) Worn or damaged countergear rear bearings c) Damaged or worn second-speed synchronizer	a) Replace damaged gears b) Replace countergear bearings and shaft c) Replace synchronizer
Noisy in third only (four speed)	a) Damaged or worn third-speed constant mesh gears b) Worn or damaged countergear bearings	a) Replace damaged gears b) Replace damaged countergear bearings and shaft
Noisy in reverse only	a) Worn or damaged reverse idler gear or idler bushing b) Worn or damaged reverse gear on mainshaft c) Damaged or worn reverse countergear d) Damaged shift mechanism	a) Replace reverse idler gear assembly b) Replace reverse gear c) Replace countergrear assembly d) Inspect linkage and adjust or replace damaged parts
Excessive backlash in all reduction gears	a) Worn countergear bearings b) Excessive endplay in countergear	a) Replace bearings b) Replace countergear thrust washers
Main drive gear bearing retainer burned or scored by input shaft	a) Loose or damaged mainshaft pilot bearing b) Misalignment of transmission	a) Replace bearing b) Align transmission
Leaks lubricant	a) Excessive amount of lubricant in transmission b) Loose or broken main drive gear bearing retainer c) Main drive gear bearing retainer gasket damaged d) Side cover loose or gasket damaged e) Rear bearing retainer oil seal leaks f) Countershaft loose in case g) Shift lever seals leak	a) Drain to correct level b) Tighten or replace retainer c) Replace gasket d) Tighten cover or replace gasket e) Replace seal f) Replace case g) Replace seal

CHILTON'S THREE C'S TRANSMISSION DIAGNOSIS

Condition	Cause	Correction
High shift effort-column shift (effort exceeds 2 ft. lbs. at lever knob with transmission linkage disconnected)	a) Binding of column levers	a) Adjust column
		Clean and lubricate all rod and swivel connections
	a) Lever endplay exceeds .005 in. b) Misalignment of column control levers	a) Adjust levers b) Adjust levers
Gear clash and binding	a) Improper linkage adjustment	a) Adjust shift linkage
Lost motion	a) Loose or worn swivels and grommets. Deflection of Mounting Brackets. Loose shift levers. Damaged cordon shaft	a) Replace defective parts

ON CAR SERVICE

Adjustments

CLUTCH PEDAL FREE TRAVEL

Only one adjustment is necessary to compensate for all normal clutch wear. The clutch pedal should have free travel measured at the clutch pedal pad, before the release bearing engages the clutch diaphragm spring or levers. Lash is required to prevent clutch slippage which would occur if the bearing was held against the clutch fingers or to prevent the bearing from running continually. A clutch that has been slipping prior to free play adjustment may still slip right after the new adjustment due to previous heat damage.

NOTE: 1985–86 C/K and 1987–89 R/V series trucks use a hydraulic clutch. This does not require a clutch pedal free travel adjustment.

C/K AND P SERIES EXCEPT P30 WITH 4 WHEEL DISC HYDRAULIC POWER BRAKES

1. Raise the vehicle and support it safely.
2. Disconnect the return spring at the clutch fork.
3. Rotate the clutch lever and shaft assembly until the clutch pedal is against the rubber bumper on the brake pedal bracket.
4. Push the outer end of the clutch fork rearward until the release bearing lightly contacts the pressure plate fingers or levers.
5. Loosen the locknut and adjust the rod length so that the swivel slips freely into the gauge hole. Increase the pushrod length until all lash is removed from the system.
6. Remove the swivel from the gauge hole and insert it into the lower hole on the lever. Install 2 washers and the cotter pin. Tighten the locknut being careful not to change the rod length.
7. Reinstall the return spring and check the pedal free travel. The pedal travel should be 1⅜–1⅝ in. (35–41mm) on C and K series and 1¼–1½ in. (31–37mm) on P series.

P30 SERIES WITH 4 WHEEL DISC HYDRAULIC POWER BRAKES

1. Raise the vehicle and support it safely.
2. Disconnect the clutch fork return spring.
3. Loosen the locknut at the swivel.
4. Move the clutch fork rod against the fork to eliminate all clearance between the release bearing and the clutch fingers.
5. Rotate the shaft lever until the clutch pedal contacts the bumper mounted on the brake pedal bracket.
6. Rotate the fork rod until a clearance of about ¼–⁵⁄₁₆ in. (6.35–7.9mm) is obtained between the shoulder on the fork rod and the adjustment nut.

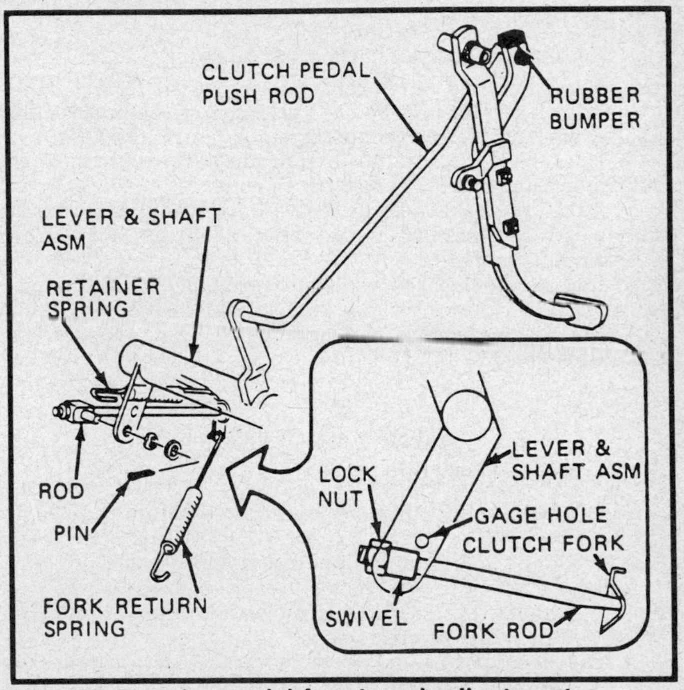
CLUTCH PEDAL
PUSH ROD
RUBBER
BUMPER
LEVER & SHAFT
ASM
RETAINER
SPRING
ROD
PIN
FORK RETURN
SPRING
LOCK
NUT
LEVER &
SHAFT ASM
GAGE HOLE
CLUTCH FORK
SWIVEL
FORK ROD

C/K and P series pedal free travel adjustment

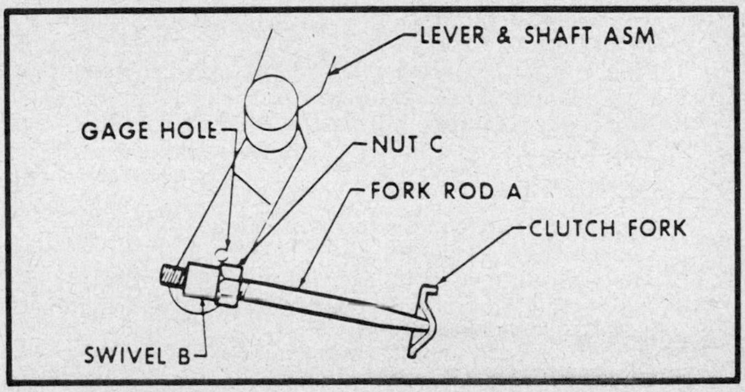
LEVER & SHAFT ASM
GAGE HOLE
NUT C
FORK ROD A
CLUTCH FORK
SWIVEL B

P30 (except motor home) pedal free travel adjustment

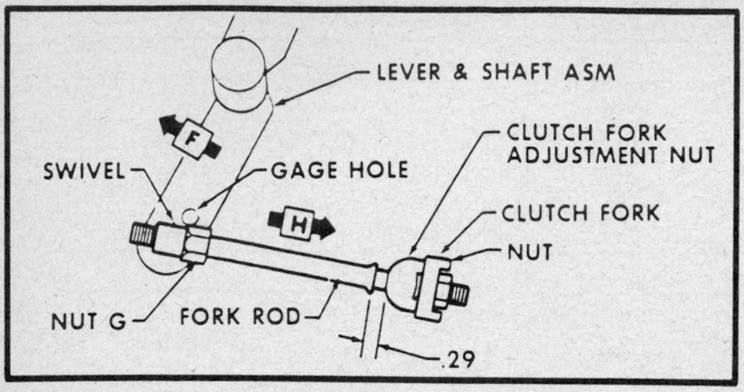
LEVER & SHAFT ASM
CLUTCH FORK
ADJUSTMENT NUT
SWIVEL
GAGE HOLE
CLUTCH FORK
NUT
NUT G
FORK ROD
.29

P30 with (JF9) pedal free travel adjustment

7. Tighten the locknut against the swivel and install the clutch return spring.
8. Check the free pedal clearance at the pedal. The pedal clearance should be 1⅜–1⅝ in. (35–41mm).

G SERIES

1. Raise the vehicle and support it safely.
2. Disconnect the clutch return spring at the fork.
3. Loosen the locknut on the outer end of the pushrod and back it off from the swivel approximately ½ in. (12.7mm).
4. Hold the clutch fork pushrod against the fork to move the release bearing against the clutch fingers.
5. Adjust the locknut at the inner (fork) end of the pushrod to obtain ¼ in. (6.35mm) clearance between the inner locknut and the swivel.
6. Release the pushrod, connect the return spring and tighten the outer locknut to lock the swivel against the inner locknut.
7. Check the free pedal clearance at the pedal. The clearance should be 1¼–1½ in. (31–37mm).

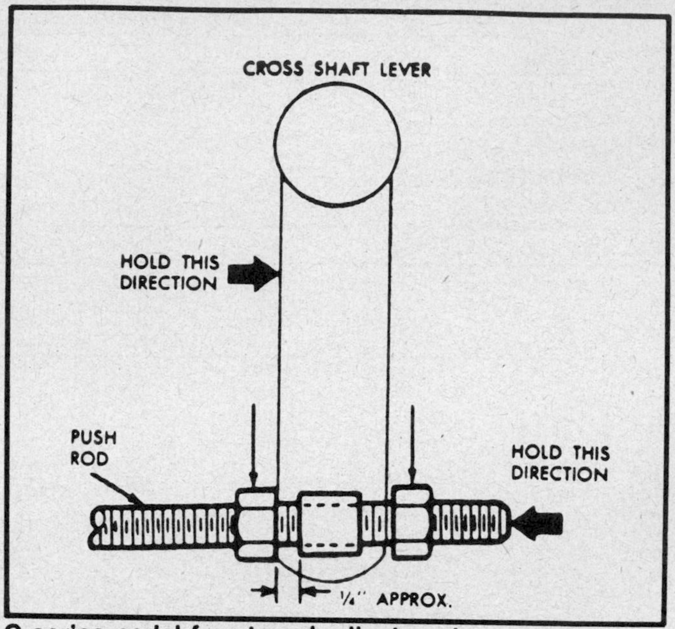
CROSS SHAFT LEVER
HOLD THIS DIRECTION
PUSH ROD
HOLD THIS DIRECTION
¼" APPROX.

G series pedal free travel adjustment

SHIFT LINKAGE

COLUMN SHIFT LINKAGE

The 1st/reverse shift rod must be adjusted before the 2nd/3rd shift rod.

1. Raise the vehicle and support it safely.
2. Loosen the screw at the shift lever.
3. On the 1st/reverse shift rod, move the shift rod to the front detent. This is the reverse detent.
4. On the 2nd/3rd shift rod, move the shift lever to the front detent then back a detent. This is **N** detent.
5. On the 1st/reverse shift rod, move the column lever to **R** and lock the steering column.
6. On the 2nd/3rd shift rod, move the column lever into **N** and at the column put a 0.249–0.250 in. gauge pin through the levers and the relay lever.
7. Hold the 1st/reverse or the 2nd/3rd shift rod down tightly in the swivel and tighten the screw at the shift lever.
8. Remove the gauge pin from the steering column.
9. Lubricate all rod and swivel connections.

FLOOR SHIFT LINKAGE

1. Raise the vehicle and support it safely.
2. Loosen the 2 swivel nuts.
3. Move the shift control lever into the **N** position.
4. Move the shift levers to the front detent and then back a detent. This is the **N** detent.
5. Install a 0.249–0.250 in. gauge pin through the shift levers.
6. Hold the shift rods forward tightly in the swivels and tighten the 2 locknuts.
7. Remove the gauge pin and lubricate the shift control.

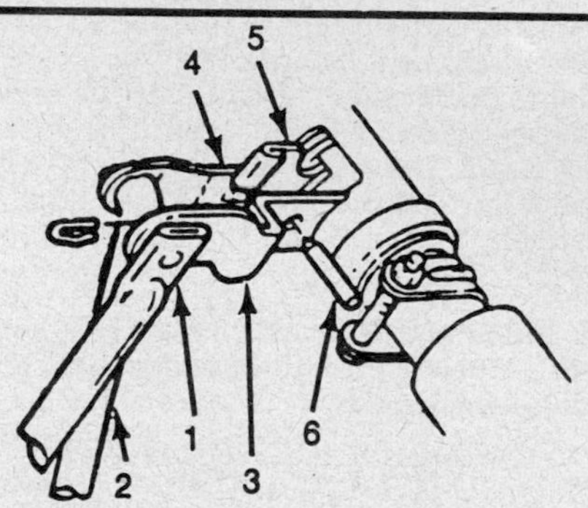
1
2
3
4
5
6

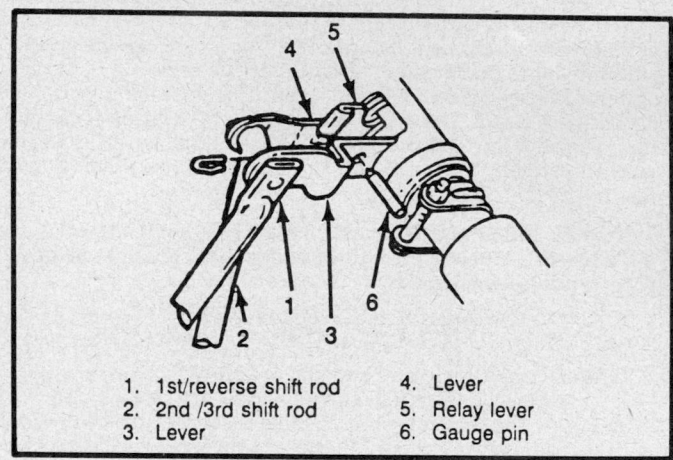

1. 1st/reverse shift rod
2. 2nd /3rd shift rod
3. Lever
4. Lever
5. Relay lever
6. Gauge pin

Steering column shift linkage adjustment

Services

REAR OIL SEAL

Removal and Installation

3 SPEED

1. Disconnect the negative battery cable.
2. Raise the vehicle and support it safely.
3. Drain the gear lubricant from the transmssion.
4. Matchmark and remove the driveshaft.
5. Remove the seal from the rear of the extension housing.
6. Apply locking compound on the outside of the new seal.
7. Install the new seal using tool J-21426 or equivalent and fill between the seal lips with petroleum jelly.
8. Install the driveshaft.
9. Fill the transmission with gear lubricant to the proper level. Lower the vehicle.

4 SPEED

1. Disconnect the negative battery cable.
2. Raise the vehicle and support it safely.

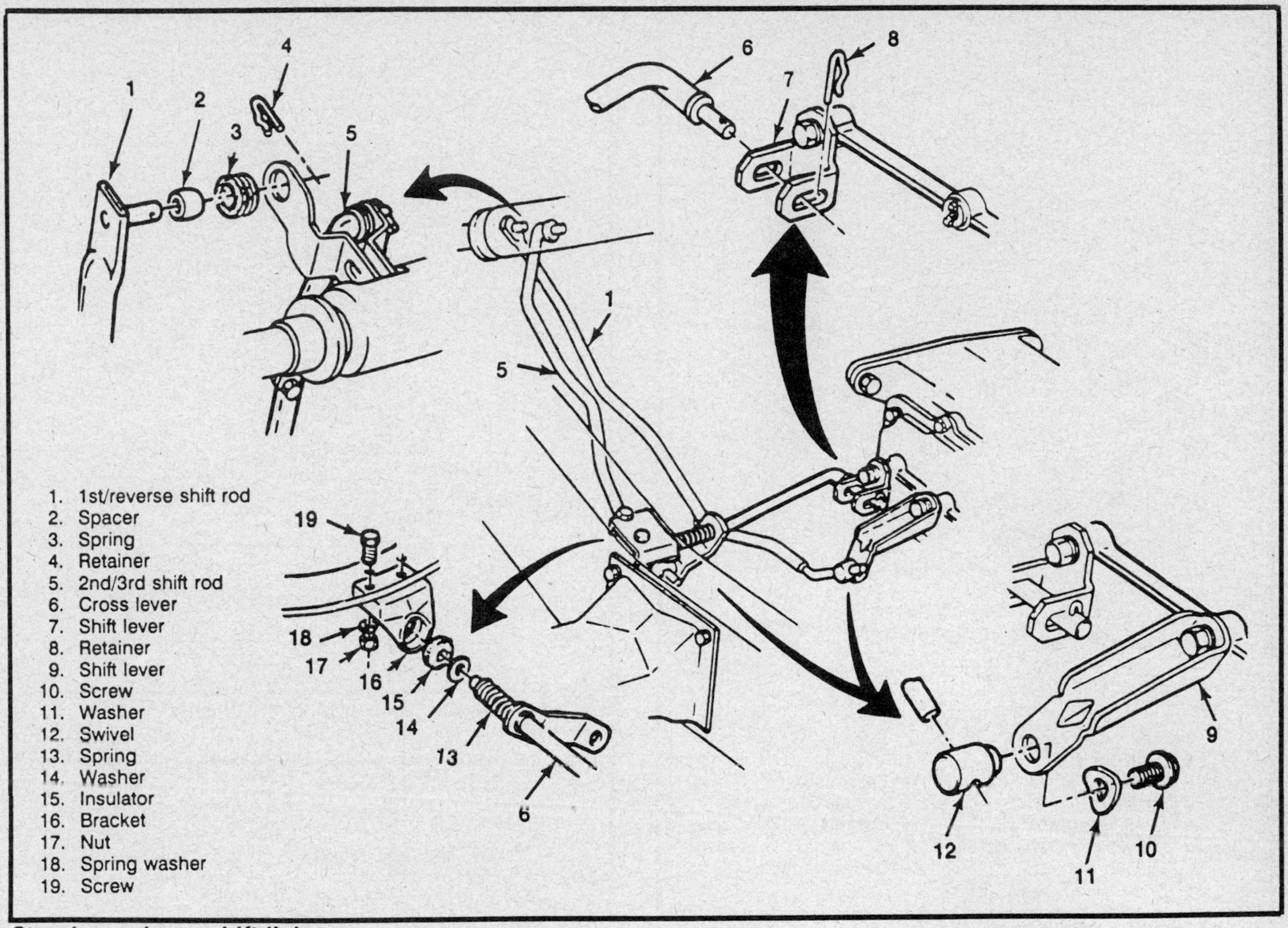
1
2
3
4
5
6
7
8
1
5
19
18
17
16
15
14
13
6
9
12
10
11
1. 1st/reverse shift rod
2. Spacer
3. Spring
4. Retainer
5. 2nd/3rd shift rod
6. Cross lever
7. Shift lever
8. Retainer
9. Shift lever
10. Screw
11. Washer
12. Swivel
13. Spring
14. Washer
15. Insulator
16. Bracket
17. Nut
18. Spring washer
19. Screw

Steering column shift linkage

3. Drain the gear lubricant from the transmssion.
4. Matchmark and remove the driveshaft.
5. Disconnect the parking brake, if used.
6. Disconnect the speedometer cable and seal.
7. Remove the nut and the flange from the rear bearing retainer.
8. Support the transmission with a suitable jack and remove the transmission mount.
9. Remove the bolts and the rear bearing retainer.
10. Remove the gasket and remove all gasket material from the retainer and the case.
11. Remove the seal.
12. Apply locking compound on the outside of the new seal.
13. Install the new seal using tool J-22834-2 or equivalent. If a parking brake is used, use tool J-22834-1 or equivalent.
14. Fill between the seal lips with petroleum jelly.
15. Install a new gasket between the rear bearing retainer and the case.
16. Install the rear bearing retainer.
17. Install the transmission mount and remove the jack.
18. Install the flange and the nut.
19. Install a new seal and connect the speedometer cable.
20. Connect the speedometer cable, if used.
21. Install the driveshaft.
22. Fill the transmission with gear lubricant to the proper level. Lower the vehicle.

SPEEDOMETER DRIVEN GEAR

Removal and Installation

1. Disconnect the negative battery cable.
2. Raise the vehicle and support it safely.
3. Disconnect the speedometer cable and remove the seal.
4. Remove the adapter, if used.
5. Remove the bolt and retainer.
6. Remove the sleeve, seal and gear.
7. Installation is the reverse of the removal procedure.
8. Apply a thin coat of transmission oil on the gear and the seal.

SHIFT LINKAGE

Removal and Installation

STEERING COLUMN LINKAGE

1. Disconnect the negative battery cable.
2. Raise the vehicle and support it safely.
3. Remove the retainer attaching the 1st/reverse shift rod to the column. Disconnect the shift rod from the column.
4. Remove the spring and spacer.
5. Remove the screw and spring washer at the shift lever.
6. Remove the shift rod and the swivel from the shift lever.

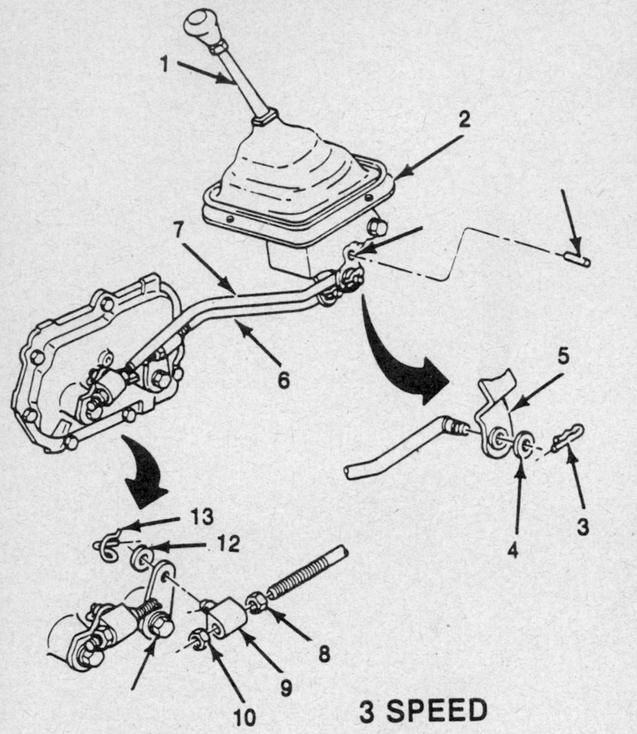
3 SPEED

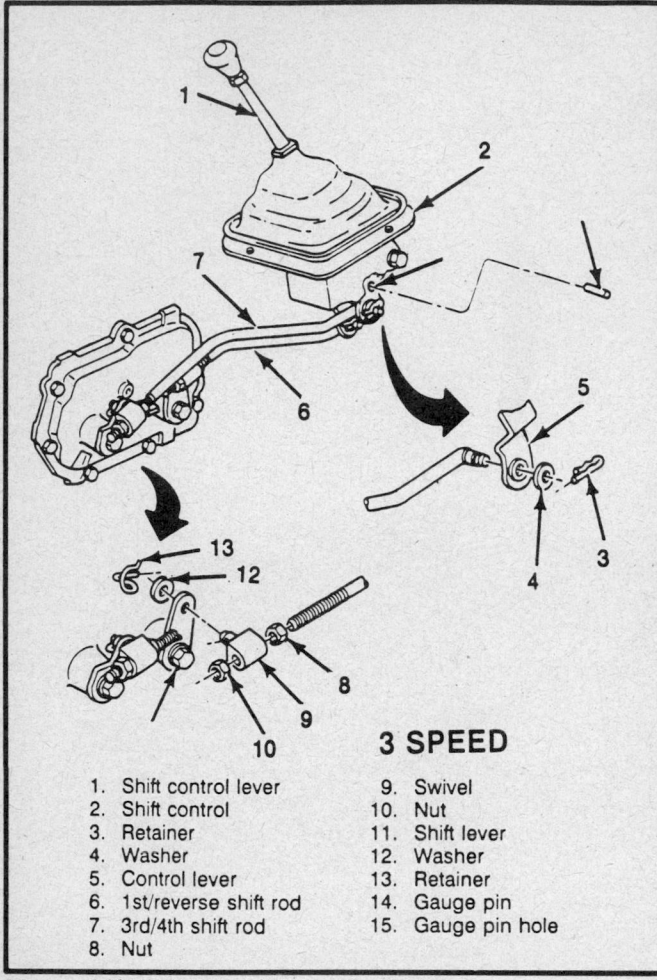

1. Shift control lever
2. Shift control
3. Retainer
4. Washer
5. Control lever
6. 1st/reverse shift rod
7. 3rd/4th shift rod
8. Nut
9. Swivel
10. Nut
11. Shift lever
12. Washer
13. Retainer
14. Gauge pin
15. Gauge pin hole

Floor shift linkage

7. Slide the swivel off of the shift rod.
8. Remove the retainer attaching the cross lever to the shift lever.
9. Remove the nuts, spring washers and bolts from the cross lever bracket.
10. Remove the bracket, insulator, washer and spring from the cross lever.
11. Remove the cross lever.
12. Installation is the reverse of the removal procedure. Apply a thin coat of grease on the insulator.
13. Adjust the shift linkage.

FLOOR SHIFT LINKAGE

1. Disconnect the negative battery cable.
2. Raise the vehicle and support it safely.
3. Remove the retainers attaching the shift rods to the control levers.
4. Remove the retainers attaching the shift rods to the shift levers.
5. Remove the nuts and the swivels from the shift rods.
6. Installation is the reverse of the removal procedure.
7. Adjust the shift linkage.

SHIFT CONTROL

Removal and Installation

FLOOR SHIFT CONTROL

1. Remove the knob and the nut from the shift control lever.

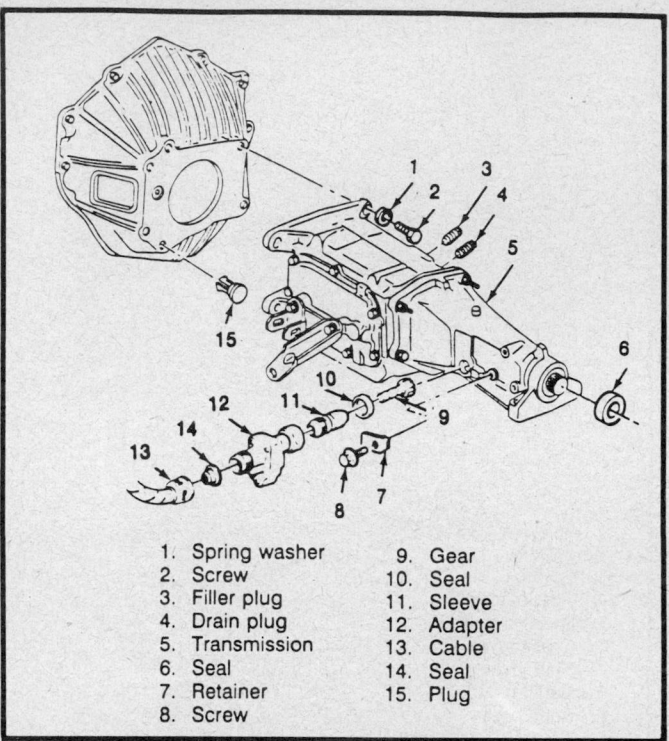

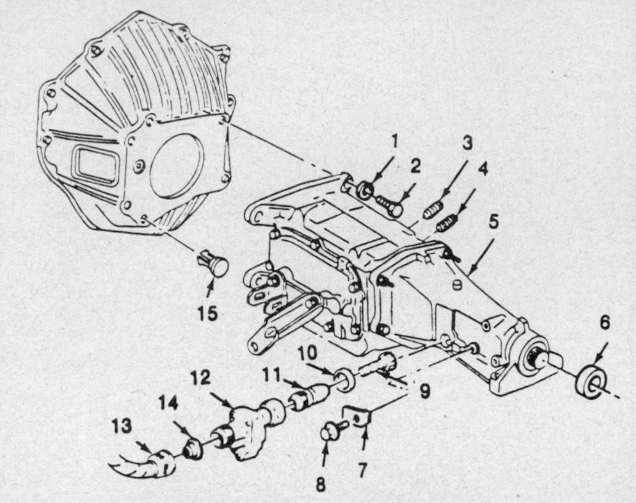

1. Spring washer
2. Screw
3. Filler plug
4. Drain plug
5. Transmission
6. Seal
7. Retainer
8. Screw
9. Gear
10. Seal
11. Sleeve
12. Adapter
13. Cable
14. Seal
15. Plug

3 speed transmission and components

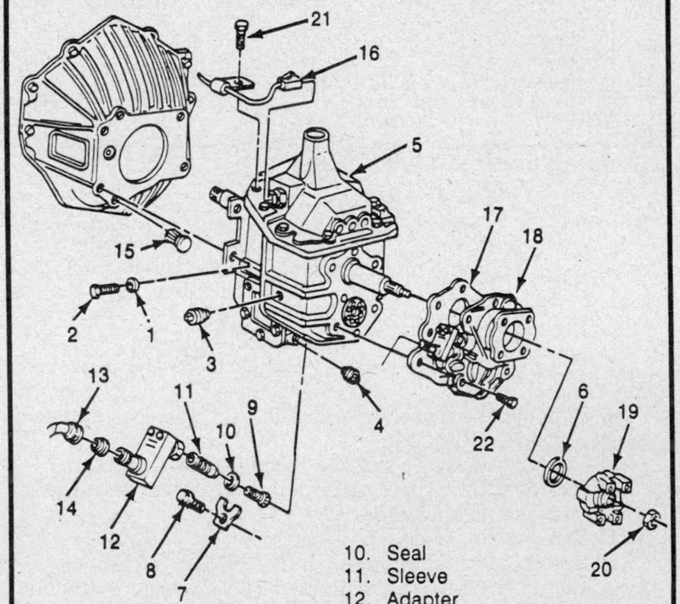

1. Spring washer
2. Screw
3. Filler plug
4. Drain plug
5. Transmission
6. Seal
7. Retainer
8. Screw
9. Speedometer driver gear
10. Seal
11. Sleeve
12. Adapter
13. Speedometer cable
14. Seal
15. Plug
16. Harness
17. Gasket
18. Rear bearing retainer
19. U-joint flange
20. Nut
21. Screw
22. Screw

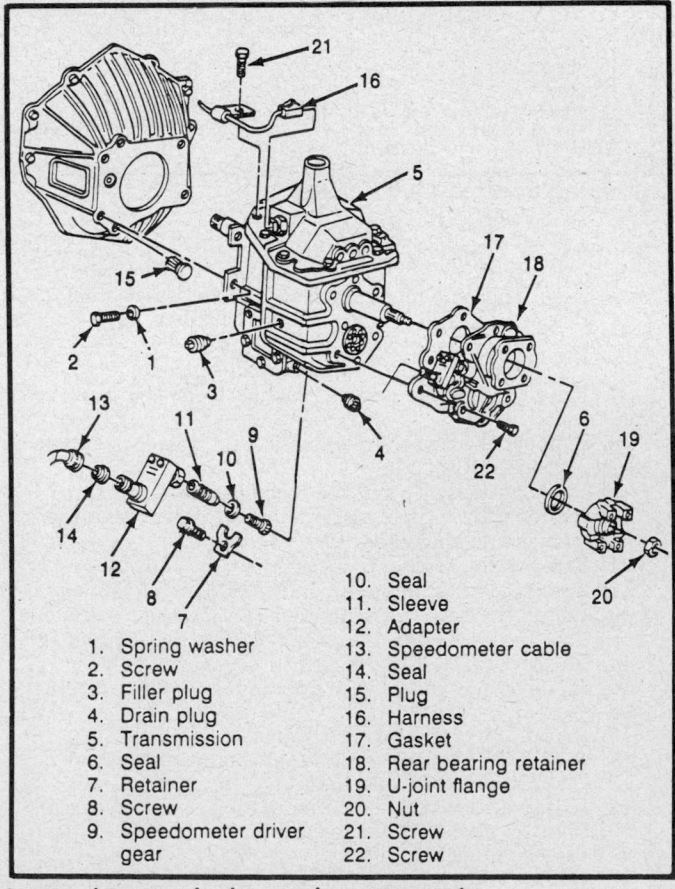

4 speed transmission and components

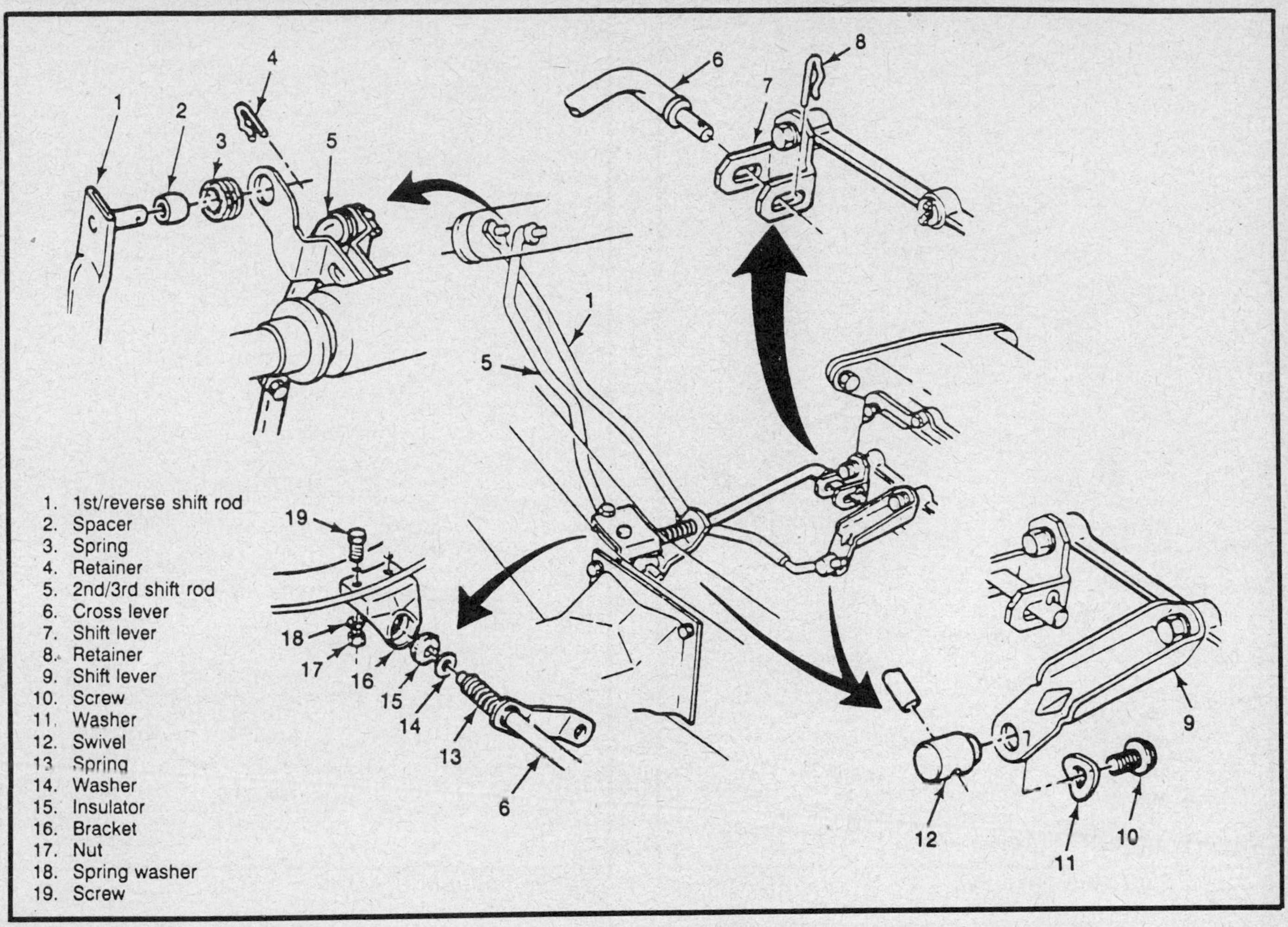
1. 1st/reverse shift rod
2. Spacer
3. Spring
4. Retainer
5. 2nd/3rd shift rod
6. Cross lever
7. Shift lever
8. Retainer
9. Shift lever
10. Screw
11. Washer
12. Swivel
13. Spring
14. Washer
15. Insulator
16. Bracket
17. Nut
18. Spring washer
19. Screw

Steering column shift linkage

2. Remove the screws, boot and the plate, if used.
3. Release the shift control lever from the control and remove the shift control lever.
4. Disconnect the shift rods at the control.
5. Remove the gear position switch, if used.
6. Remove the screws and the fitting from the shift control. Remove the shift control.
7. Remove the nuts, bolts and spring washers from the bracket and remove the bracket.
8. Installation is the reverse of the removal procedure.

NOTE: When the control is installed on the transmission, the screw with the lubrication fitting must go into the outer top hole in the control towards the rear of the transmission.

9. Lubricate the shift control and the shift linkage. Adjust the shift linkage.

SHIFT CONTROL LEVER (4 SPEED)

1. Remove the transfer case shift lever boot, if used.
2. Remove the screws and the retainer plate, if used.
3. Remove the screws and the boot.
4. Push the cap down and turn it counterclockwise to remove the lever from the transmission.
5. Installation is the reverse of the removal procedure.
6. To install the lever, push the cap down and turn it clockwise.

SIDE COVER

Removal and Installation

1. Disconnect the negative battery cable.
2. Raise the vehicle and support it safely.
3. Drain the gear lubricant from the transmssion.
4. Disconnect all wiring harnesses from the cover.
5. Move the shift levers into neutral and disconnect the shift rods.
6. Remove the bolts, spring washers and the brackets.
7. Note the positions of the shift forks while removing the side cover.
8. Remove the gasket and scrape all gasket material from the cover and the case.
9. Installation is the reverse of the removal procedure.

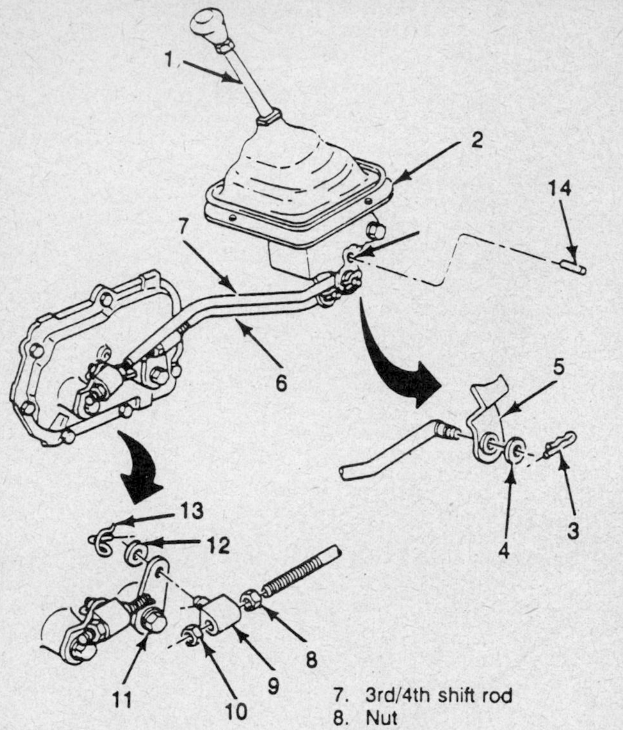

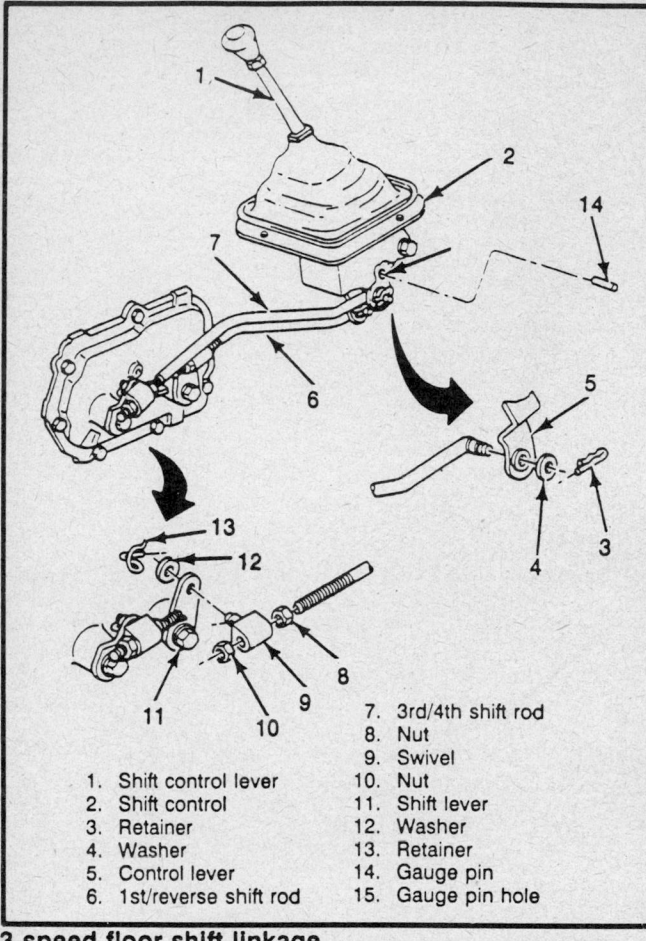

1. Shift control lever
2. Shift control
3. Retainer
4. Washer
5. Control lever
6. 1st/reverse shift rod
7. 3rd/4th shift rod
8. Nut
9. Swivel
10. Nut
11. Shift lever
12. Washer
13. Retainer
14. Gauge pin
15. Gauge pin hole

3 speed floor shift linkage

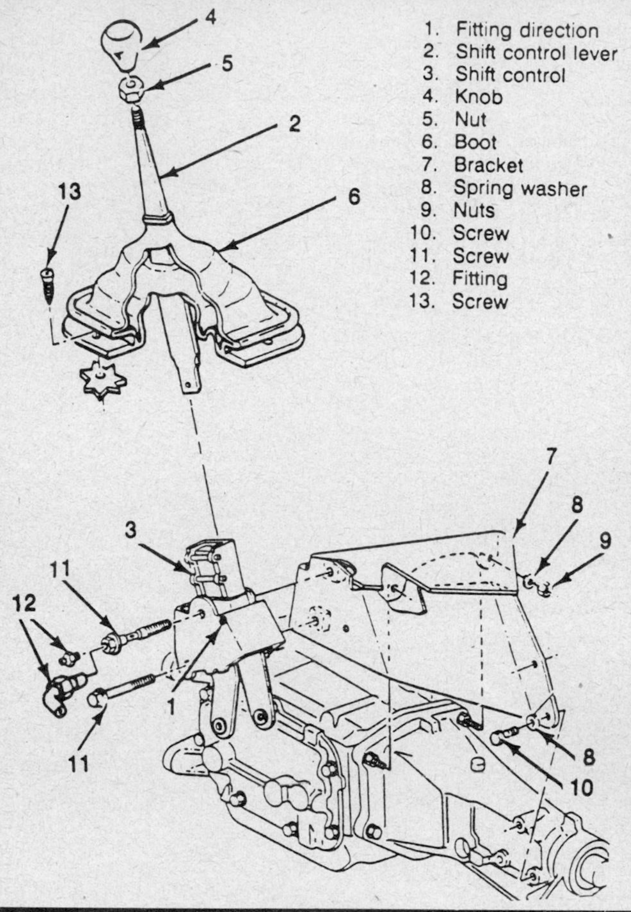

1. Fitting direction
2. Shift control lever
3. Shift control
4. Knob
5. Nut
6. Boot
7. Bracket
8. Spring washer
9. Nuts
10. Screw
11. Screw
12. Fitting
13. Screw

3 speed floor shift control

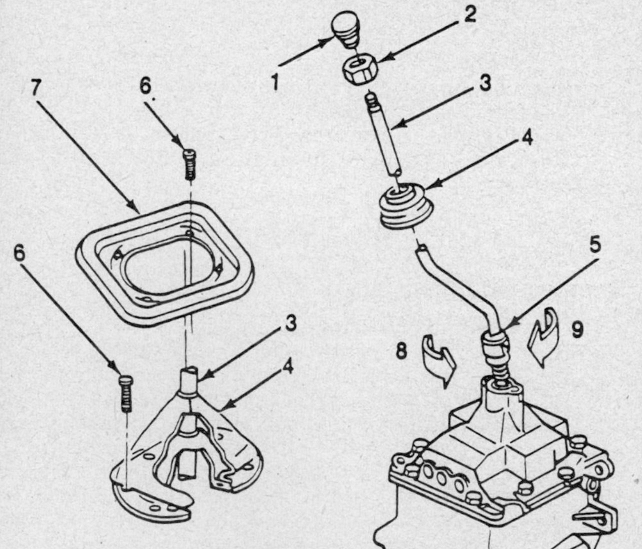

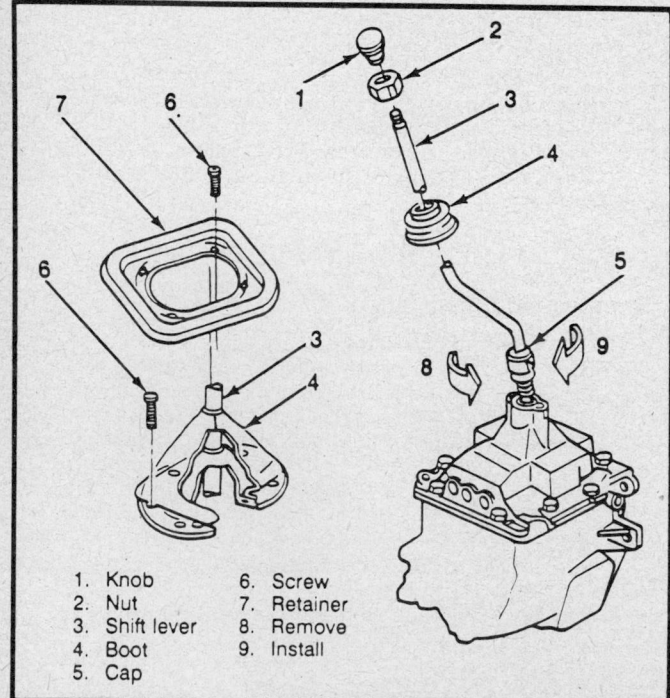

1. Knob
2. Nut
3. Shift lever
4. Boot
5. Cap
6. Screw
7. Retainer
8. Remove
9. Install

4 speed shift control lever

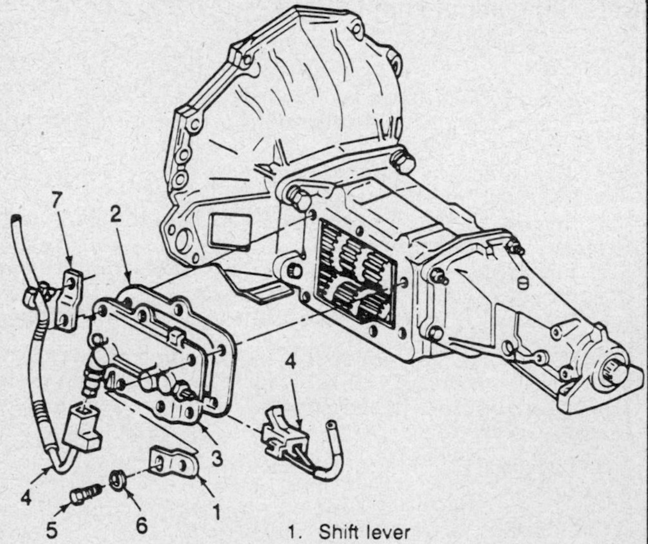

1. Shift lever
2. Gasket
3. Side cover
4. Harness
5. Screw
6. Spring washer
7. Bracket

3 speed side cover

REMOVAL AND INSTALLATION

TRANSMISSION REMOVAL

1. Disconnect the negative battery cable.
2. Raise the vehicle and support it safely.
3. Drain the gear lubricant from the transmssion.
4. On the 3 speed transmission, disconnect the shift control and rods. On the 4 speed transmission, remove the shift lever.
5. Matchmark and remove the driveshaft.
6. Remove the transfer case, if equipped.
7. Disconnect the parking brake and controls.
8. Disconnect the speedometer cable and remove the seal.
9. Disconnect all wiring harnesses from the transmission.
10. Disconnect the exhaust pipes.
11. Support the engine with a suitable jack.
12. Support the transmission with a suitable jack.
13. Disconnect or remove any parts if needed for clearance.
14. Remove the crossmember.
15. On the 4 speed transmission, remove the top 2 transmission to bellhousing bolts and spring washers and install guide pins.
16. Pull the transmission straight back from the clutch hub splines and support the clutch release bearing.

NOTE: Do not let the transmission hang from the clutch.

17. Note the location of the plugs in the bellhousing and remove them if they are loose or damaged.

TRANSMISSION INSTALLATION

1. Install new plugs in the bellhousing, if required.
2. Apply a thin coat of high temperature grease on the main drive gear splines.
3. Shift the transmission into high gear before installation.
4. On the 4 speed transmission, install guide pins in the top 2 holes.
5. Remove the clutch release bearing support.
6. Install the transmission.

NOTE: Do not force the transmission into the clutch.

7. Install new spring washers and the transmission to bellhousing bolts. On the 4 speed transmission, install the 2 bottom bolts before removing the guide pins.
8. Install the crossmember.
9. Remove the transmission jack.
10. Install any parts that were removed or disconnected for clearance.
11. Connect the exhaust pipes.
12. Connect the wiring harnesses to the transmission.
13. Install a new seal and connect the speedometer cable.
14. Connect the parking brake lever and controls.
15. Install the transfer case, if equipped.
16. Install the driveshaft.
17. On the 3 speed transmission, install the shift control and rods. On the 4 speed transmission, install the shift lever.
18. Adjust the shift linkage.
19. Fill the transmission with new gear lubricant to the proper level.
20. Lower the vehicle.

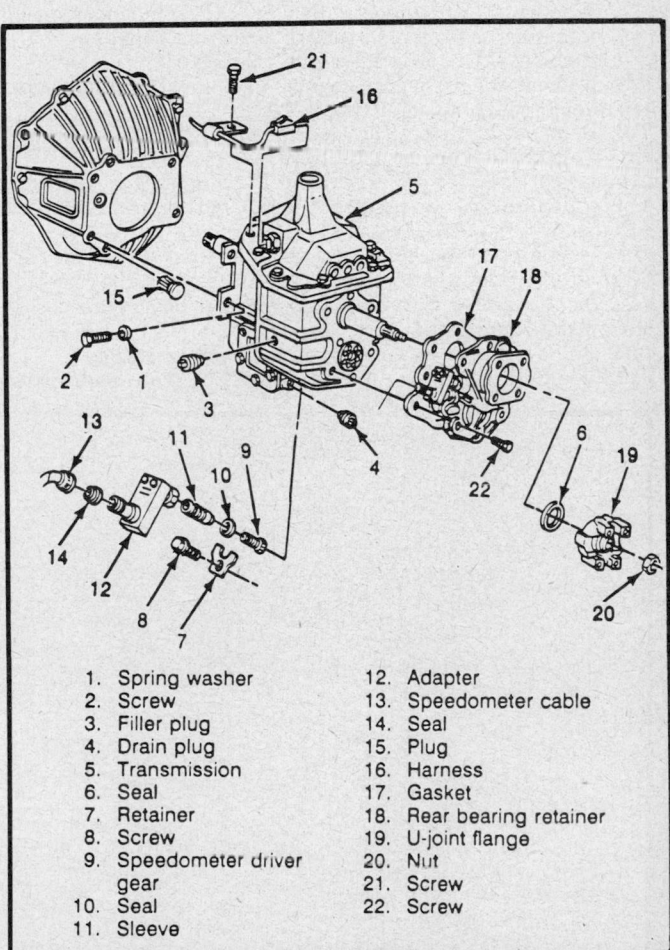

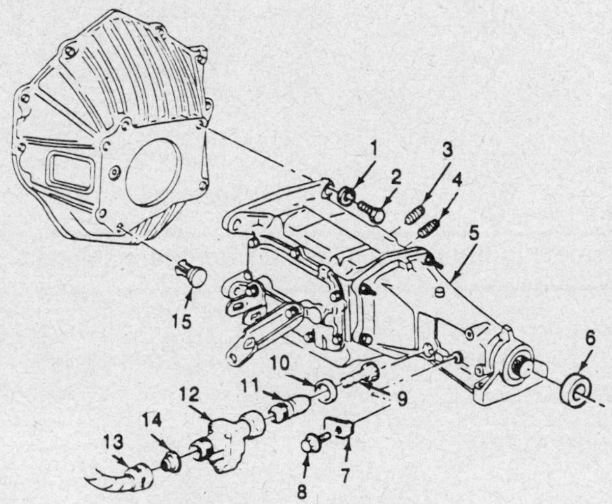
1
2
3
4
5
6
7
8
9
10
11
12
13
14
15

1. Spring washer
2. Screw
3. Filler plug
4. Drain plug
5. Transmission
6. Seal
7. Retainer
8. Screw
9. Gear
10. Seal
11. Sleeve
12. Adapter
13. Cable
14. Seal
15. Plug

3 speed transmission and components

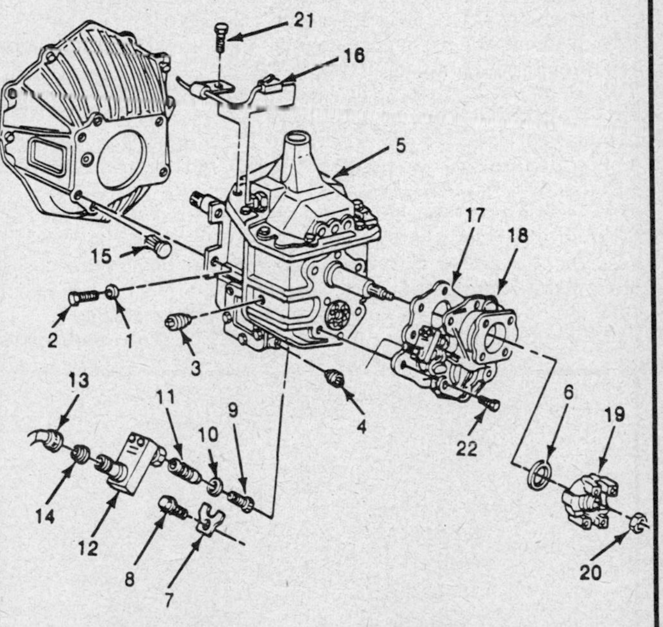
1
2
3
4
5
6
7
8
9
10
11
12
13
14
15
16
17
18
19
20
21
22

1. Spring washer
2. Screw
3. Filler plug
4. Drain plug
5. Transmission
6. Seal
7. Retainer
8. Screw
9. Speedometer driver gear
10. Seal
11. Sleeve
12. Adapter
13. Speedometer cable
14. Seal
15. Plug
16. Harness
17. Gasket
18. Rear bearing retainer
19. U-joint flange
20. Nut
21. Screw
22. Screw

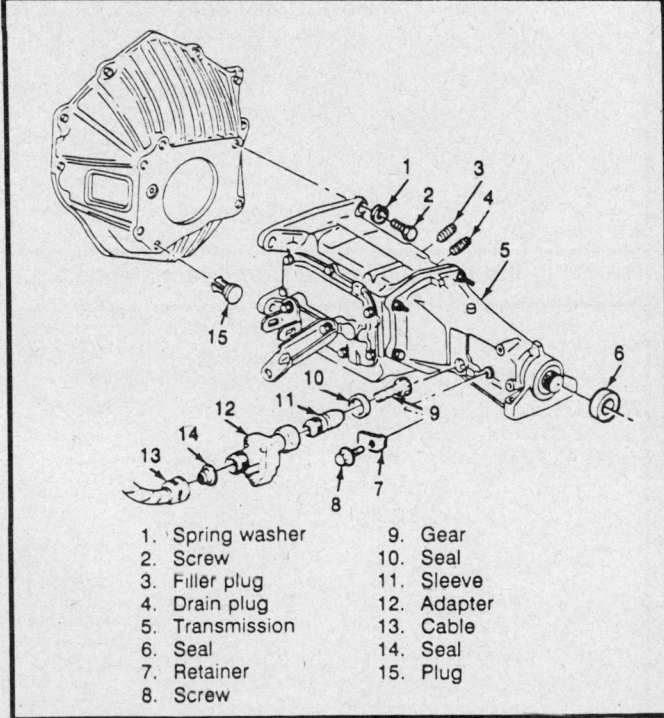

4 speed transmission and components

BENCH OVERHAUL

Before Disassembly

Clean the exterior of the transmission assembly before any attempt is made to disassemble it, in order to prevent dirt or other foreign material from entering the transmission assembly and damaging its internal parts. If steam cleaning is done to the exterior of the transmission, immediate disassembly should be done to avoid rusting, caused by condensation forming on the internal parts.

Transmission Disassembly

3 SPEED

1. Remove the side cover attaching bolts and the side cover assembly.
2. Remove the drive gear bearing retainer and gasket.
3. Remove the drive gear bearing-to-gear stem snapring.
4. Remove the clutch gear bearing by pulling outward on the clutch gear until a suitable tool can be inserted between the large bearing snapring and the case to complete the removal.

NOTE: The clutch gear bearing is a slip fit on the gear and into the case bore. This provides clearance for removal of the clutch gear and the mainshaft assembly.

5. Remove the speedometer driven gear from the extension.
6. Remove the extension to case attaching bolts.
7. Remove the reverse idler shaft E-ring.
8. Remove the drive gear, mainshaft and extension assembly together through the rear case opening.
9. Remove the drive gear, needle bearings and synchronizer ring from the mainshaft assembly.
10. Using snapring pliers, expand the snapring in the extension which retains the mainshaft rear bearing and remove the extension.
11. Using tool J-22246 or equivalent at the front of the countershaft, drive the shaft and its Woodruff key out the rear of the case. Tool J-22246 will now hold the roller bearing in position within the countergear bore.
12. Remove the gear, bearings and thrust washers.
13. Use a long drift or punch through the front bearing case

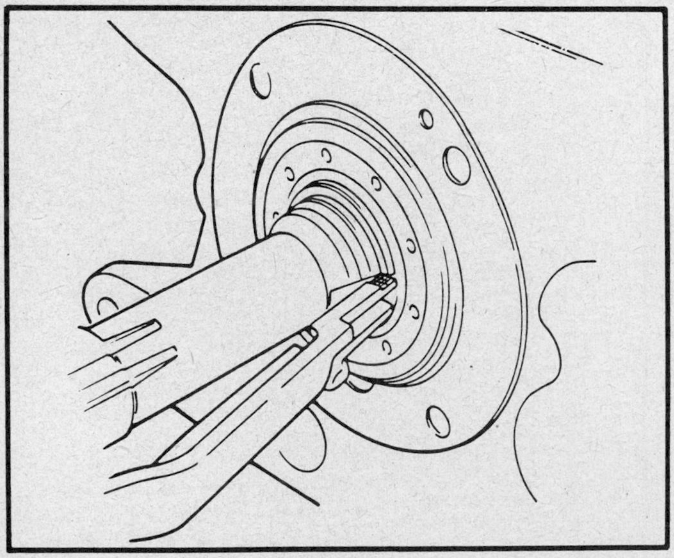

Clutch gear bearing retaining ring

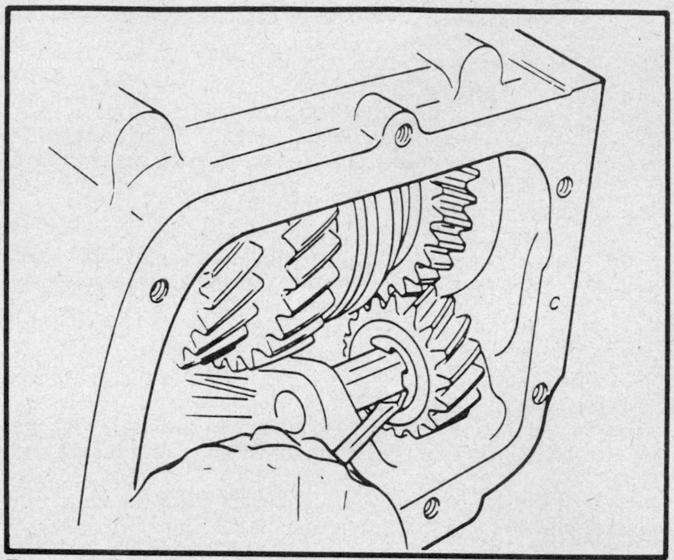

Reverse idler shaft E-ring

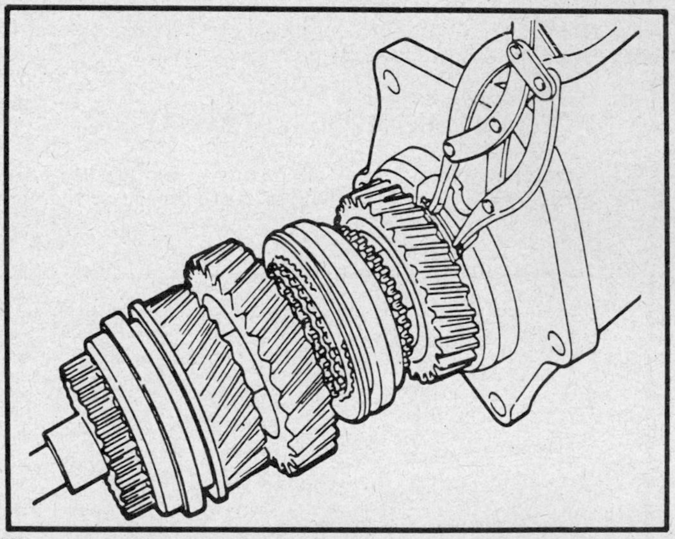

Removing the rear extension from the mainshaft

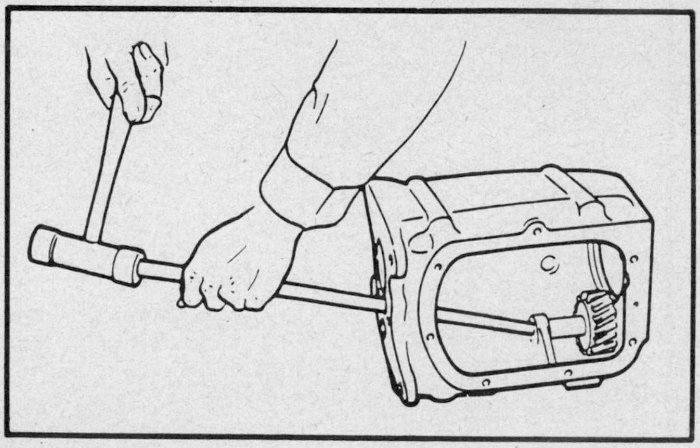

Reverse idler shaft

bore and drive the reverse idler shaft and Woodruff key through the rear of the case.

4 SPEED

1. Move the reverse shifter fork so that the reverse idler gear is partially engaged before attempting to remove the cover. The forks must be positioned so that the rear edge of the slot in the reverse fork is in line with the front edge of the slot in the forward forks as viewed through the tower opening.
2. Remove the bolts and the cover from the transmission case.
3. Place the transmission in 2 gears at the same time to lock the gears.
4. Remove the universal joint flange nut, universal joint front flange and brake drum assembly. On vehicles equipped with a transfer case, use tool J-23070 or equivalent to remove the mainshaft rear locknut.
5. Remove the parking brake and brake flange plate assembly on vehicles equipped with a driveshaft parking brake.
6. Remove the rear bearing retainer and gasket.
7. Slide the speedometer drive gear off the mainshaft.
8. Remove the drive gear bearing retainers and gasket.
9. Remove the countergear front bearing cap and gasket.
10. Pry the countergear front bearing out by inserting 2 pronged puller tool J-28509 or equivalent through the cast slots in the case.
11. Remove the countergear rear bearing retaining rings from

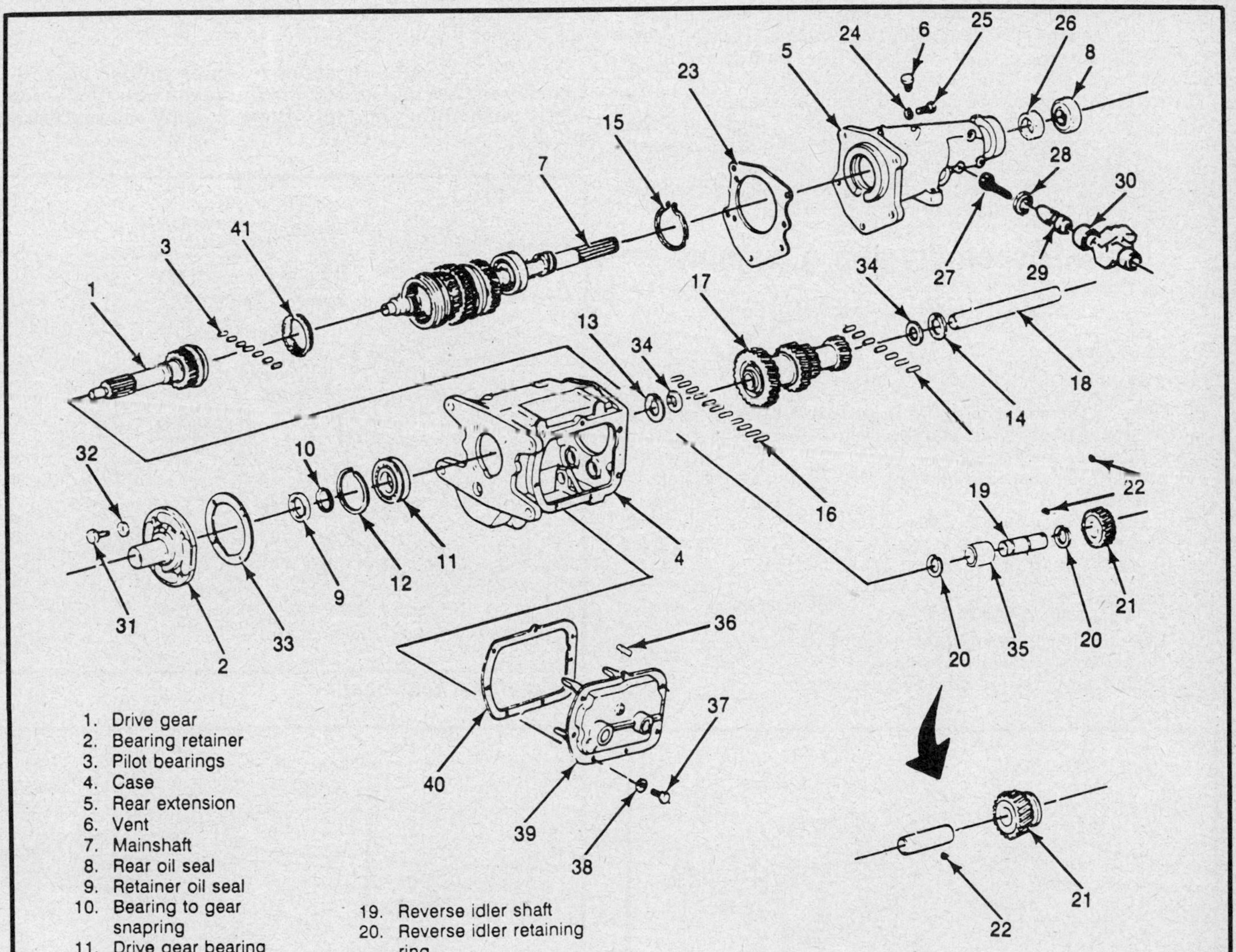

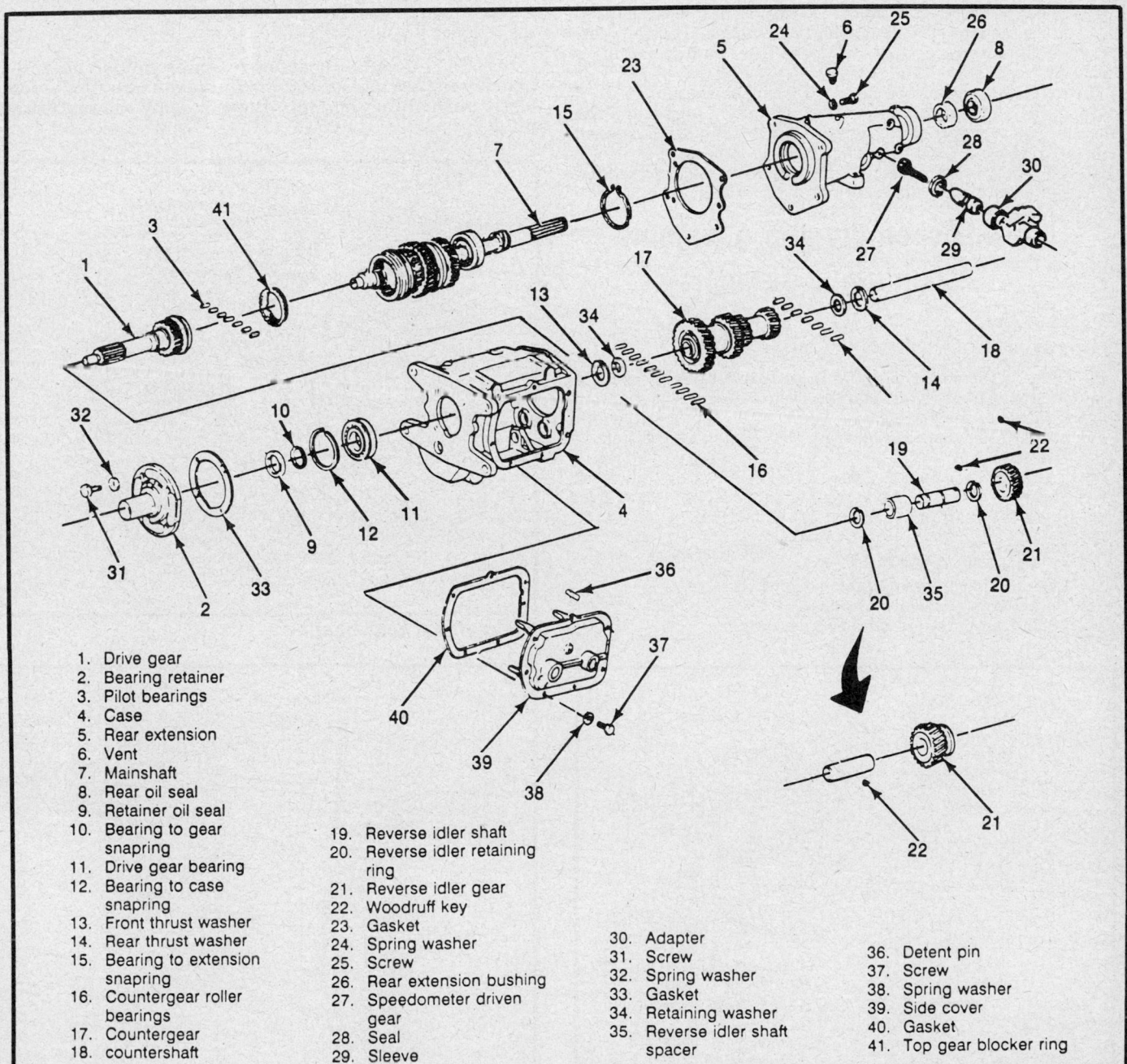

1. Drive gear
2. Bearing retainer
3. Pilot bearings
4. Case
5. Rear extension
6. Vent
7. Mainshaft
8. Rear oil seal
9. Retainer oil seal
10. Bearing to gear snapring
11. Drive gear bearing
12. Bearing to case snapring
13. Front thrust washer
14. Rear thrust washer
15. Bearing to extension snapring
16. Countergear roller bearings
17. Countergear
18. countershaft
19. Reverse idler shaft
20. Reverse idler retaining ring
21. Reverse idler gear
22. Woodruff key
23. Gasket
24. Spring washer
25. Screw
26. Rear extension bushing
27. Speedometer driven gear
28. Seal
29. Sleeve
30. Adapter
31. Screw
32. Spring washer
33. Gasket
34. Retaining washer
35. Reverse idler shaft spacer
36. Detent pin
37. Screw
38. Spring washer
39. Side cover
40. Gasket
41. Top gear blocker ring

Transmission and components—3 speed

the shaft and bearing. Using tool J-22832 and J-8433-1, remove the countergear rear bearings. This will allow the countergear assembly to rest on the bottom of the case.

12. Remove the drive gear bearing outer race to case retaining ring.

13. Remove the drive gear by tapping on the bottom side of the drive gear shaft and prying directly opposite against the case and bearing snapring groove at the same time.

14. Remove the 4th gear synchronizer ring.

15. Index the cutout section of the drive gear in the down position with the countergear to obtain clearance for removing the clutch gear.

16. Remove the rear mainshaft bearing retainer ring and using tool J-22832 and J-8433-1 or equivalent, remove the bearing from the case.

17. Slide the 1st speed gear thrust washer off the mainshaft.

18. Raise the rear of the mainshaft assembly and push it rearward in the case bore, then swing the front end up and lift it from the case.

19. Remove the synchronizer cone from the shaft.

20. Slide the reverse idler gear rearward and move the countergear rearward until the front end is free of the case, then lift and remove it from the case.

21. To remove the reverse idler gear, drive the reverse idler gear shaft out of the case from the front to rear.

Unit Disassembly and Assembly

MAIN DRIVE GEAR

Disassembly

4 SPEED

1. Remove the mainshaft pilot bearing rollers from the drive gear and remove the roller retainer. Do not remove the snapring on the inside of the drive gear.
2. Remove the snapring securing the bearing on the stem of the drive gear.
3. To remove the bearing, install tool J-22872 or equivalent to the bearing. Using an arbor press and tool J-358-1 or equivalent, press the gear and shaft out of the bearing.

Inspection

1. Clean all parts in a suitable solvent.
2. Do not spin the bearings dry.
3. Inspect all parts for damage and wear.
4. Oil the bearings and check for roughness.

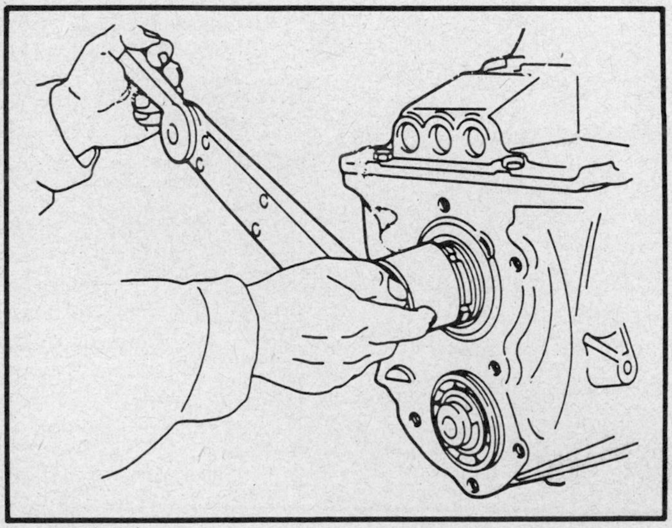

4 wheel drive

Assembly

4 SPEED

1. Press the bearing and the new oil slinger onto the drive gear shaft using tool J-22872 or equivalent.
2. The slinger should be located flush with the bearing shoulder on the drive gear. The concave side goes toward the gear.
3. Install the bearing retainer ring in the groove on the outside diameter of the bearing. The bearing must turn freely after it is installed on the shaft.
4. Install the snapring on the inside diameter of the mainshaft pilot bearing bore in the clutch gear.
5. Apply a small amount of grease to the bearing surface in the shaft recess. Install the mainshaft pilot roller bearings and the retainer.

NOTE: The roller bearing retainer holds the roller bearing in position and in the final transmission assembly is pushed forward into the recess by the mainshaft pilot.

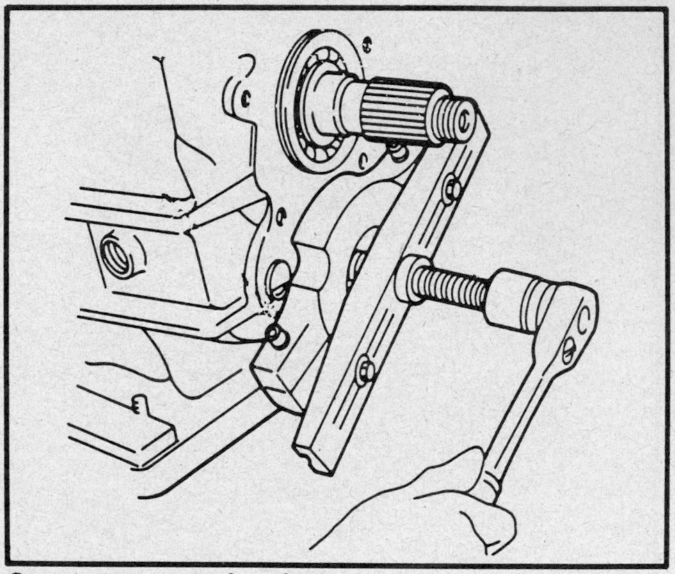

Countergear rear bearing

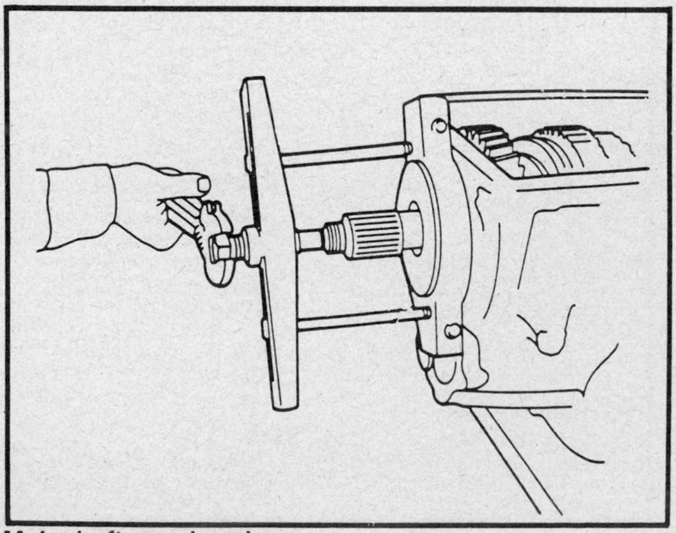

Mainshaft rear bearing

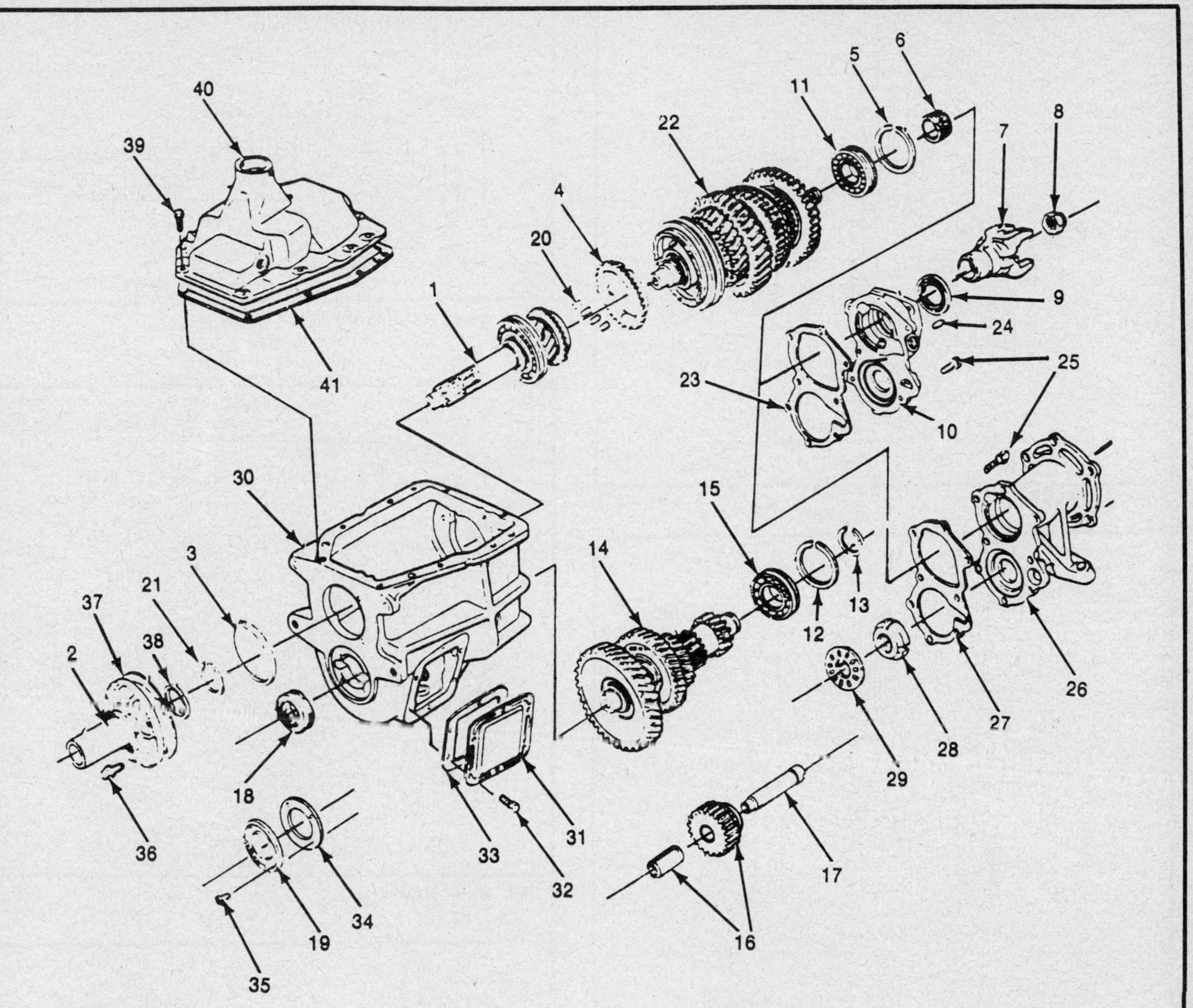

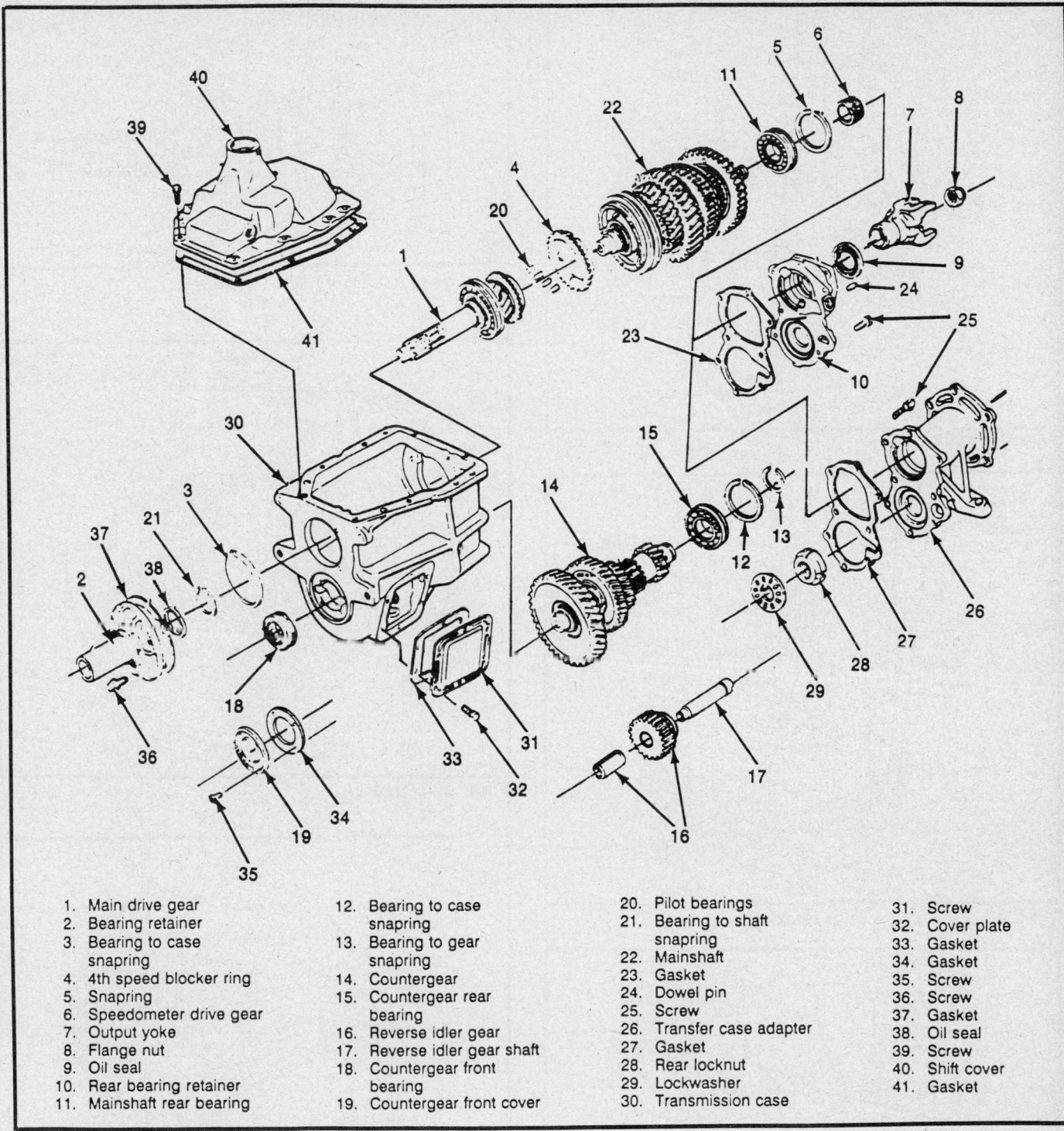

1. Main drive gear
2. Bearing retainer
3. Bearing to case snapring
4. 4th speed blocker ring
5. Snapring
6. Speedometer drive gear
7. Output yoke
8. Flange nut
9. Oil seal
10. Rear bearing retainer
11. Mainshaft rear bearing
12. Bearing to case snapring
13. Bearing to gear snapring
14. Countergear
15. Countergear rear bearing
16. Reverse idler gear
17. Reverse idler gear shaft
18. Countergear front bearing
19. Countergear front cover
20. Pilot bearings
21. Bearing to shaft snapring
22. Mainshaft
23. Gasket
24. Dowel pin
25. Screw
26. Transfer case adapter
27. Gasket
28. Rear locknut
29. Lockwasher
30. Transmission case
31. Screw
32. Cover plate
33. Gasket
34. Gasket
35. Screw
36. Screw
37. Gasket
38. Oil seal
39. Screw
40. Shift cover
41. Gasket

Transmission and components—4 speed

MAINSHAFT

Disassembly

3 SPEED

1. Using snapring pliers, remove the 2nd and 3rd speed sliding clutch hub snapring from the mainshaft. Remove the clutch assembly, 2nd speed blocker ring and the 2nd speed gear from the front of the mainshaft.

2. Depress the speedometer retaining clip and slide the gear from the mainshaft.

3. Remove the rear bearing snapring from the mainshaft groove.

4. Support the reverse gear with press plates and press on the rear of the mainshaft to remove the reverse gear, thrust washer, spring washer, rear bearing and snapring from the rear of the mainshaft.

5. Remove the 1st/reverse sliding clutch hub snapring from

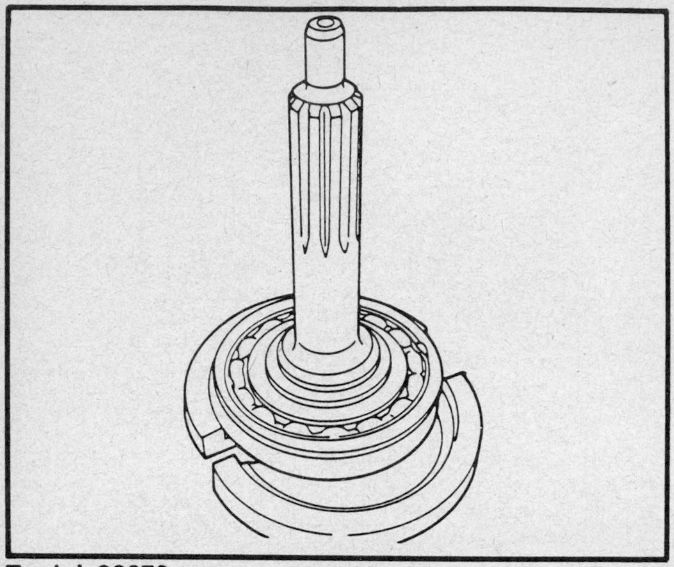

Tool J–22872

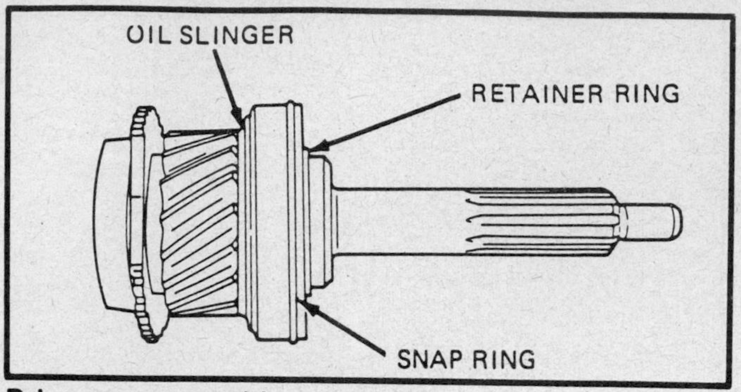
OIL SLINGER
RETAINER RING
SNAP RING

Drive gear assembly

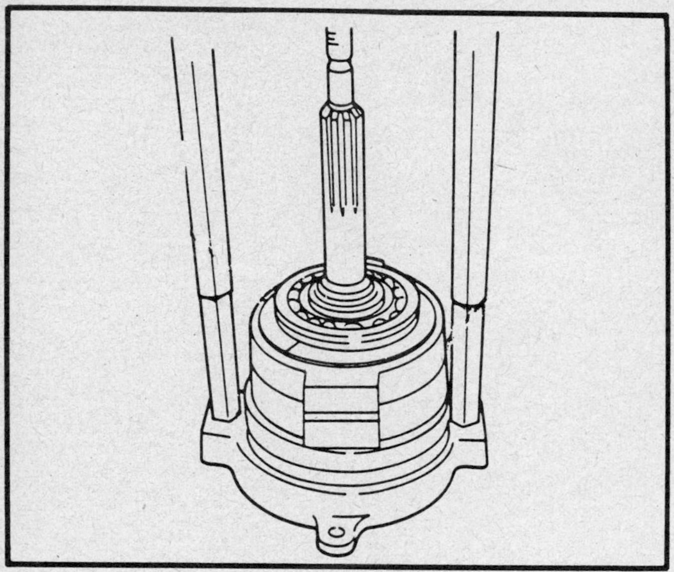

Removing the drive gear bearing

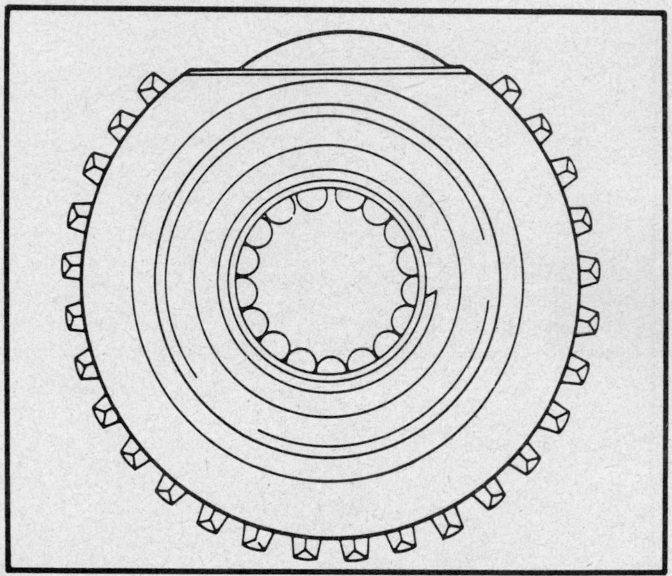

Pilot roller bearing

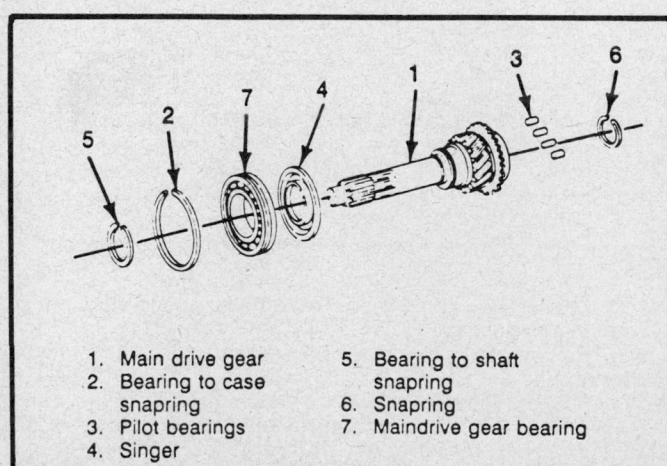

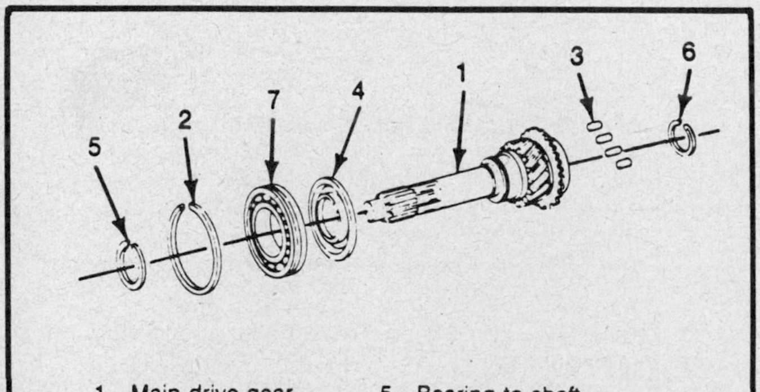
1
2
3
4
5
6
7
1. Main drive gear
2. Bearing to case snapring
3. Pilot bearings
4. Singer
5. Bearing to shaft snapring
6. Snapring
7. Maindrive gear bearing

Main drive gear components

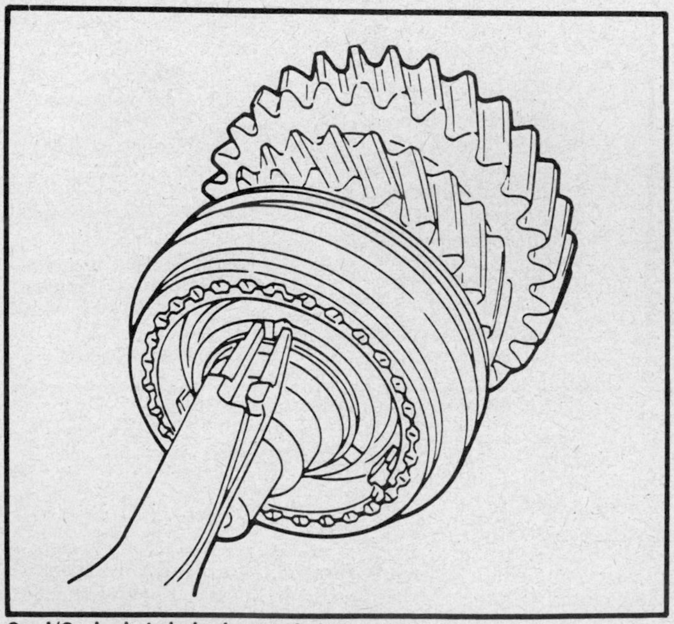

2nd/3rd clutch hub retaining ring

the mainshaft and remove the clutch assembly, 1st speed blocker ring and 1st speed gear from the rear of the mainshaft.

4 SPEED

1. Remove the 1st speed gear and thrust washer.
2. Remove the snapring in front of the 3rd/4th synchronizer assembly.
3. Remove the reverse driven gear.
4. Press behind the 2nd speed gear to remove the 3rd/4th synchronizer assembly, 3rd speed gear and 2nd speed gear along with the 3rd speed gear bushing and thrust washer.
5. Remove the 2nd speed synchronizer ring.

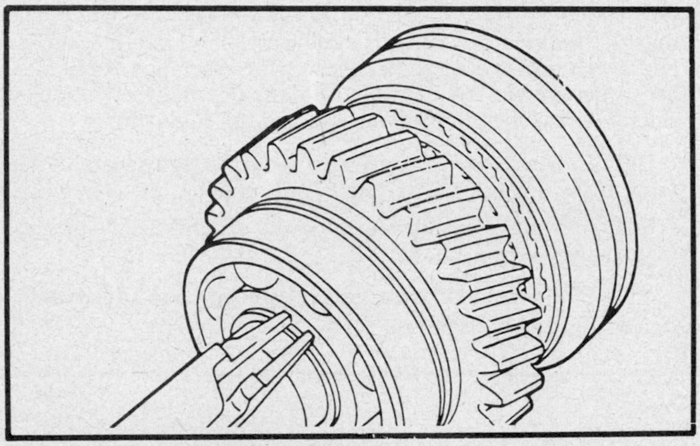

Rear bearing retaining ring

6. Support the 2nd speed synchronizer hub at the front face and press the mainshaft through removing the 1st speed gear bushing and the 2nd speed synchronizer hub.
7. Split the 2nd speed gear bushing and remove the bushing from the shaft.

Inspection

1. Inspect the gears for cracks, chipped gear teeth and other damage that could cause gear noise.
2. Inspect the thrust washers and bushings for damage and wear.
3. Inspect surfaces on the gears such as thrust faces and bearing surface diameters.
4. Inspect the reverse sliding gear for a sliding fit on the synchronizer hub without excess radial or circular play. If the sliding gear is not free on the hub, inspect it for burrs on the ends of the internal splines. Remove the burrs by honing.
5. Inspect the synchronizer sleeves for a sliding fit on the synchronizer hubs and the hubs for a snug fit on the mainshaft splines.
6. Inspect the synchronizer springs and keys for looseness or damage.
7. Inspect the brass synchronizer rings for wear or damage.
8. Inspect all gear teeth for wear.
9. Inspect the bearings and the bearing surfaces for nicks, burrs and wear.
10. Lubricate all roller bearings with light engine oil and check for rough rotation.

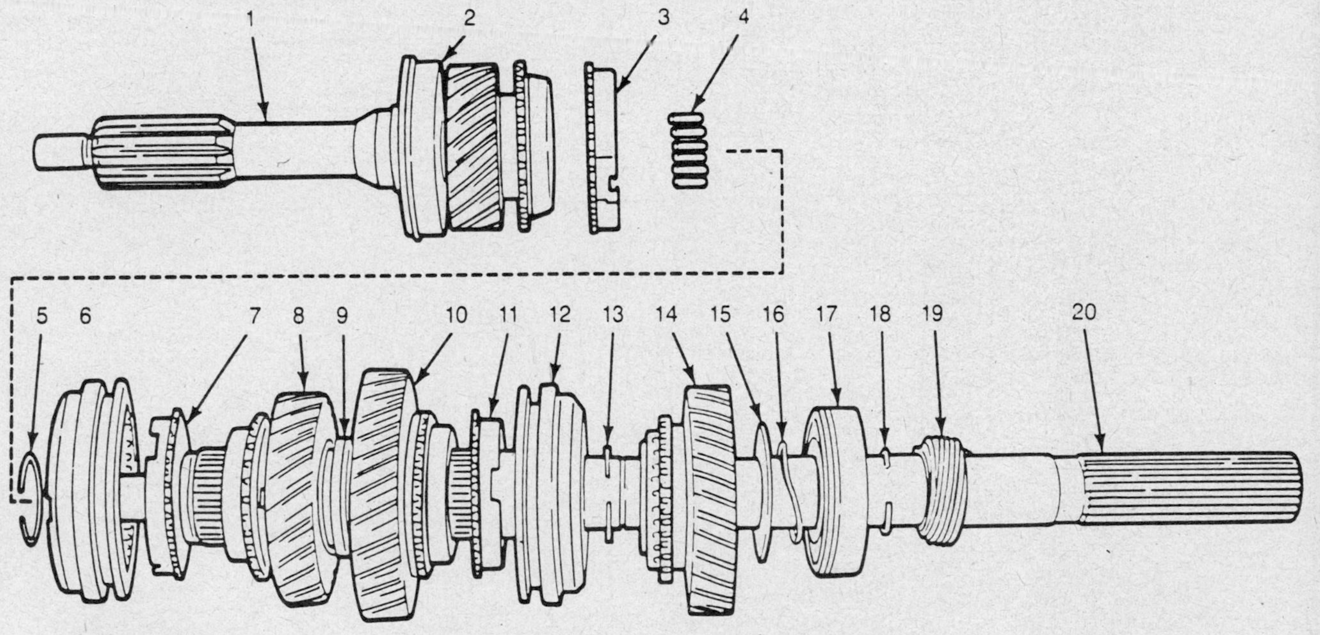
1
2
3
4
5
6
7
8
9
10
11
12
13
14
15
16
17
18
19
20
1. DRIVE GEAR
2. DRIVE GEAR BEARING
3. 3RD SPEED BLOCKER RING
4. MAINSHAFT PILOT BEARINGS (14)
5. SNAP RING
6. 2-3 SYNCHRONIZER ASSEMBLY
7. 2ND SPEED BLOCKER RING
8. 2ND SPEED GEAR
9. SHOULDER (PART OF MAIN SHAFT)
10. 1ST SPEED GEAR
11. 1ST SPEED BLOCKER RING
12. 1ST SPEED SYNCHRONIZER ASSEMBLY
13. SNAP RING
14. REVERSE GEAR
15. REVERSE GEAR THRUST WASHER
16. SPRING WASHER
17. REAR BEARING
18. SNAP RING
19. SPEEDO DRIVE GEAR AND CLIP
20. MAINSHAFT

Drive gear and mainshaft assembly

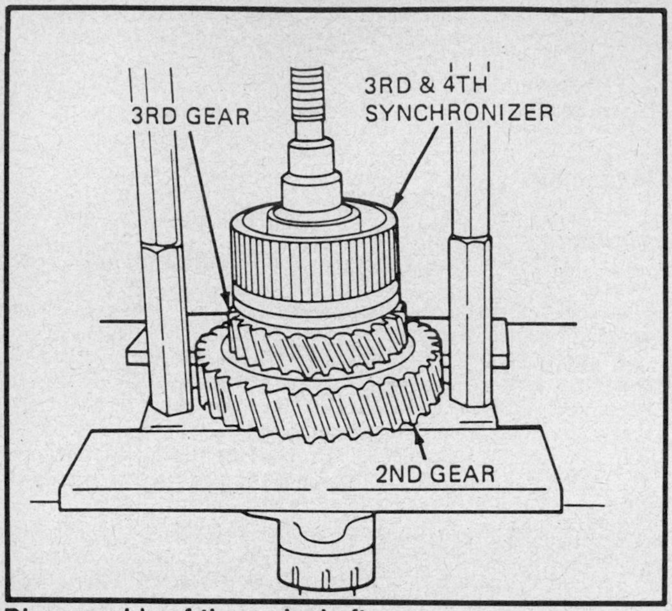
3RD GEAR
3RD & 4TH SYNCHRONIZER
2ND GEAR
Disassembly of the mainshaft

Assembly

3 SPEED

1. Turn the front of the mainshaft upward.
2. Install the 2nd speed gear with the clutching teeth upward. The rear face of the gear will butt against the flange on the mainshaft.
3. Install a blocking ring with the clutching teeth downward over the synchronizing surface of the 2nd speed gear.

NOTE: All 3 blocker rings used in this transmission are identical.

4. Install the 2nd and 3rd synchronizer assembly with the fork slot downward. Press it onto the splines on the mainshaft until it bottoms out. Be sure the notches of the blocker ring align with the keys of the synchronizer assembly. If the sleeve becomes removed from the 2/3 hub, the notches on the hub outside diameter face the forward end of the mainshaft.

NOTE: Both synchronizer assemblies used in this transmission are identical.

5. Install the snapring retaining the synchronizer hub to the mainshaft.

NOTE: Both synchronizer snaprings are identical.

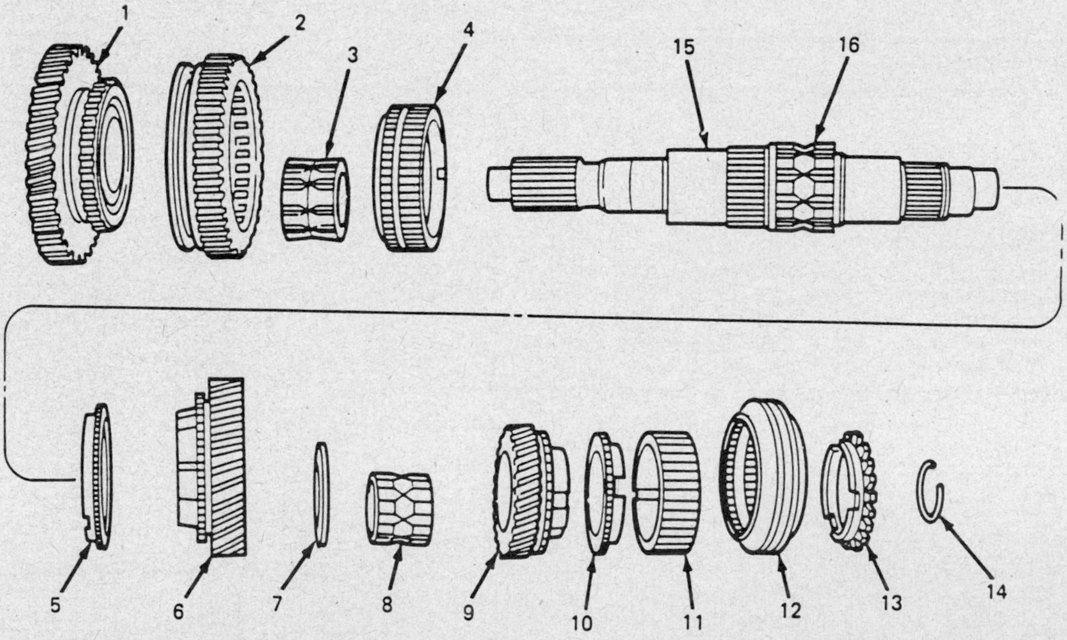
1
2
3
4
15
16
5
6
7
8
9
10
11
12
13
14

1. 1st Speed Gear
2. Reverse Driven Gear
3. 1st Gear Bushing
4. 1st-2nd Gear Synchronizer Hub Assembly
5. 2nd Speed Blocker Ring
6. 2nd Speed Gear
7. Thrust Washer
8. 3rd Speed Bushing
9. 3rd Speed Gear
10. 3rd Speed Blocker Ring
11. 3rd-4th Speed Synchronizer Hub Assembly
12. 3rd-4th Speed Synchronizer Sleeve
13. 4th Speed Blocker Ring
14. Snap Ring
15. Mainshaft
16. 2nd Speed Gear Bushing

Mainshaft assembly

6. Turn the rear of the mainshaft upward.
7. Install the 1st speed gear with the clutching teeth upward. The front face of the gear will butt against the flange on the mainshaft.
8. Install a blocker ring with the clutching teeth downward over the synchronizing surface of the 1st speed gear.
9. Install the 1st/reverse synchronizer assembly with the fork slot downward. Push it onto the splines on the mainshaft.
10. Install the synchronizer hub to the mainshaft snapring. Be sure the notches of the blocker ring align with the keys of the synchronizer assembly.
11. Install the reverse gear with the clutching teeth downward.
12. Install the reverse gear steel thrust washer.
13. Install the reverse gear spring washer.
14. Install the rear ball bearing with the snapring slot downward. Press it onto the mainshaft.
15. Install the rear bearing to the mainshaft snapring.
16. Install the speedometer drive gear and retaining clip.

4 SPEED

1. Using tool J-22873 or equivalent, press the 2nd speed bushing onto the mainshaft until it bottoms against the shoulder.

NOTE: 1st, 2nd and 3rd speed gear bushings are Sintered iron, use care when installing them.

2. Press the 1st/2nd speed synchronizer hub onto the mainshaft until it bottoms against the shoulder with the annulus toward the rear of the shaft.
3. Install the 1st/2nd synchronizer keys and springs.
4. Using tool J-22873 or equivalent, press the 1st speed gear bushing onto the mainshaft until it bottoms against the hub.
5. Install the synchronizer blocker ring and the 2nd speed gear onto the mainshaft and against the synchronizer hub. Index the synchronizer key slots with the keys in the synchronizer hub.
6. Install the 3rd speed gear thrust washer onto the mainshaft with the tang on the thrust washer in the slot on the shaft and against the 2nd speed gear bushing.
7. Press the 3rd speed gear bushing onto the mainshaft using tool J-22875 or equivalent, until it bottoms against the thrust washer.
8. Install the 3rd speed gear synchronizer blocker ring and 3rd speed gear onto the mainshaft against the 3rd speed gear thrust washer.
9. Index the synchronizer ring key slots with the synchronizer assembly keys and press the 3rd/4th synchronizer assembly onto the mainshaft, using tool J-22875 or equivalent and against the 3rd speed gear bushing thrust face toward the 3rd speed gear. Retain the synchronizer assembly with a snapring.
10. Install the reverse driven gear with the fork groove toward the rear.
11. Install the 1st speed gear onto the mainshaft and against the 1st/2nd synchronizer hub.
12. Install the 1st speed gear thrust washer.

COUNTERGEAR

Disassembly

4 SPEED

1. Remove the front countergear retaining ring and thrust washer. Discard the retaining ring.
2. Install tool J-22832 or equivalent, on the countershaft, open side to the spacer. Support the assembly in an arbor press and press the countershaft out of the clutch countergear assembly.
3. Remove the clutch countergear rear retaining ring.
4. Remove the 3rd speed countergear retaining ring.
5. Position the assembly on an arbor press and press the shaft from the 3rd speed countergear.

Inspection

1. Clean all parts in a suitable solvent.
2. Do not spin the bearings dry.
3. Inspect all parts for damage and wear.
4. Oil the bearings and check for roughness.

Assembly

4 SPEED

1. Position the 3rd speed countergear and shaft on an arbor press and press the gear onto the shaft. Install the gear with the machined surface to mate with the snapring toward the front. The rear side of the gear is undercut.

NOTE: The 3rd speed gear must be installed with a load of 1500 lb. pressure. If the gear requires less than 1500 lb. pressure to install, the gear must be replaced. The press fit is required for proper operation.

2. Install the spacer, then press the front gear on the countershaft and install the snapring.
3. Install a new clutch countergear rear retaining ring using tool J-22830-A, J-22873 or equivalent and a snapring pliers.
4. Install tool J-22830-A or equivalent on the shaft and position the snapring on the tool.
5. Using tool J-22873 or equivalent, push down on the snapring until it engages the groove on the shaft.
6. Using snapring pliers, expand the ring until it just slides

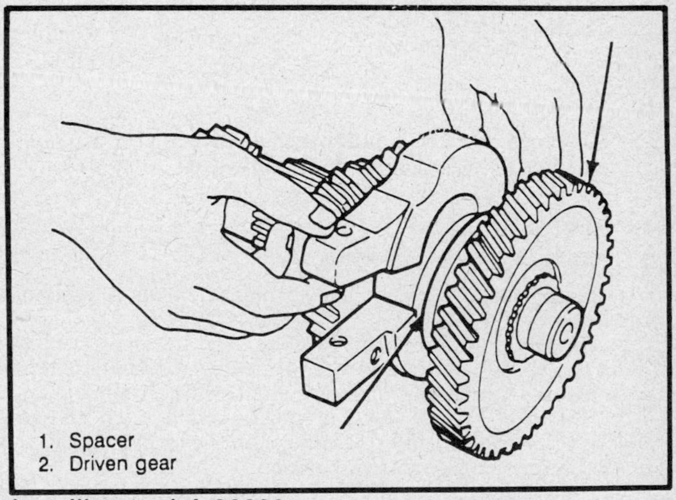
1. Spacer
2. Driven gear

Installing tool J-22832

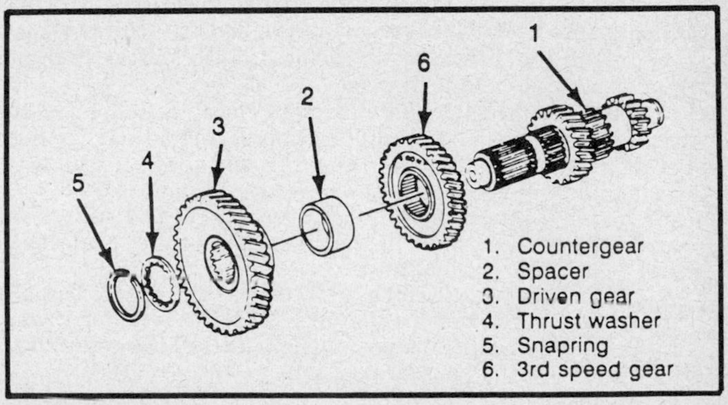
1
6
2
3
4
5
1. Countergear
2. Spacer
3. Driven gear
4. Thrust washer
5. Snapring
6. 3rd speed gear

Countergear

onto the splines, then push the ring down the shaft until it engages the groove on the shaft.

7. Install the clutch countergear and spacer on the shaft and press the countergear onto the shaft against the snapring using tool J-22873 or equivalent.

8. Install the clutch countergear thrust washer and front retaining ring using tool J-22830-A and J-22873 or equivalent.

TRANSMISSION COVER

Disassembly

4 SPEED

1. Using a small punch, drive out the pins retaining the 1st/2nd and 3rd/4th shifter forks to shifter shafts. Drive out the expansion plugs.
2. The pin retaining the 3rd/4th shifter fork to the shaft must be removed and the shifter fork removed from the cover before the reverse shifter head pin can be removed.
3. With the shifter shaft in the neutral position, drive the shafts out of the cover and the shifter forks.

NOTE: When removing the shifter shafts, avoid losing the shaft detent balls, springs and the interlock pin located in the cover.

4. Drive out the pin holding the reverse shifter head and drive out the shaft.

NOTE: During shaft removal, avoid losing the detent balls since they are under spring tension in the rear rail boss holes.

Inspection

1. Clean all parts in a suitable solvent.
2. Inspect all parts for damage and wear.

Assembly

4 SPEED

1. Install the fork detent ball springs and balls in position in the holes in the cover.
2. Install the shifter shafts into the cover in the proper order. Start the shifter shafts into the cover, depress the detent balls with a small punch and push the shafts on over the balls.
3. Hold the reverse fork in position and push the shaft through the yoke. Install the split pin in the fork and shaft then push the fork in the neutral position.
4. Hold the 3rd/4th shift fork in position and push the shaft through the yoke, but not through the front support bore.
5. Install the 2 interlock balls in the cross bore in the front support boss between reverse and the 3rd/4th shifter shaft.
6. Install the interlock pin in the 3rd/4th shifter shaft hole. Apply grease to hold it in place.
7. Push the 3rd/4th shaft through the fork and cover bore, keeping both balls and pin in position between the shafts until the retaining holes line up in the fork and shaft. Install the retaining pin and move it to the neutral position.
8. Install 2 interlock balls between the 1st/2nd shifter shaft and 3rd/4th shifter shaft in the cross bore of the front support boss.
9. Hold the 1st/2nd fork in position and push the shaft through the cover bore in the fork until the retainer hole and fork line up with the hole in the shaft. Install the retaining pin and move it to the neutral position.
10. Install new shifter shaft hole expansion plugs and expand in place.

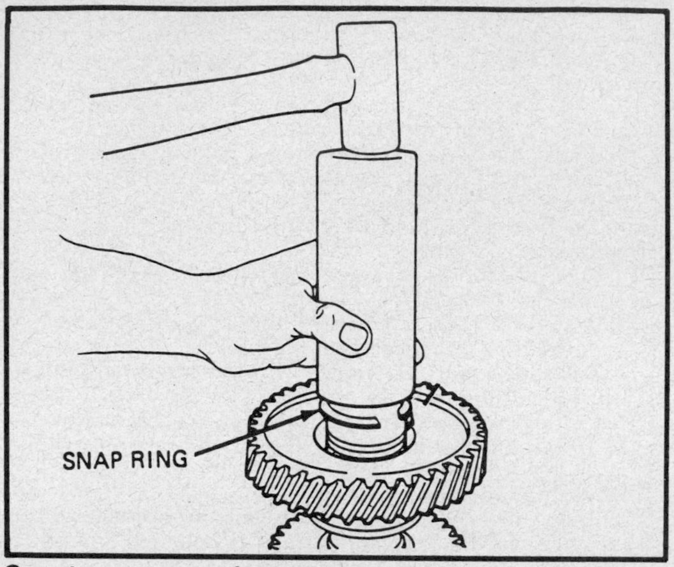
SNAP RING

Countergear snapring

Transmission Assembly

3 SPEED

1. Using tool J-22246 or equivalent, install a row of roller bearings and a thrust washer at each end of the countergear. Use petroleum jelly to hold them in place.
2. Install the countergear assembly through the case rear opening along with a tanged thrust washer, (tang away from the gear) at each end and install the countergear shaft and Woodruff key from the rear of the case.
3. Install the reverse idler gear and shaft with its Woodruff key from the rear of the case. Do not install the idler shaft E-ring yet.
4. Using snapring pliers, expand the snapring in the extension and assemble the extension over the rear of the mainshaft and onto the rear bearing. Seat the snapring in the rear bearing groove.
5. Load the mainshaft pilot roller bearings into the clutch gear cavity and assemble the 3rd speed blocker ring onto the clutch gear clutching surface with its teeth toward the gear.
6. Install the clutch gear, pilot bearings and 3rd speed blocker ring assembly over the over the front of the mainshaft assembly. Be sure the notches in the blocker ring align with the keys in the 2nd/3rd synchronizer assembly.
7. Install the extension to case gasket at the rear of the case holding it in place with grease. From the rear of the case, assemble the clutch gear, mainshaft and extension to case as an assembly.
8. Install the extension to case retaining bolts.
9. Install the front bearing outer snapring to the bearing and position the bearing over the stem of the clutch gear and into the front case bore.
10. Install the snapring to clutch gear stem and clutch gear bearing retainer and gasket to the case. The retainer oil return hole should be at the bottom.
11. Install the reverse idler gear retainer E-ring to the shaft.
12. Shift the synchronizer sleeves to the neutral positions and install the cover, gasket and fork assembly to the case. Be sure that the forks align with their synchronizer sleeve grooves.
13. Install the speedometer driven gear in the extension.
14. Rotate the clutch gear shaft and shift the transmission to free rotation in all gears.

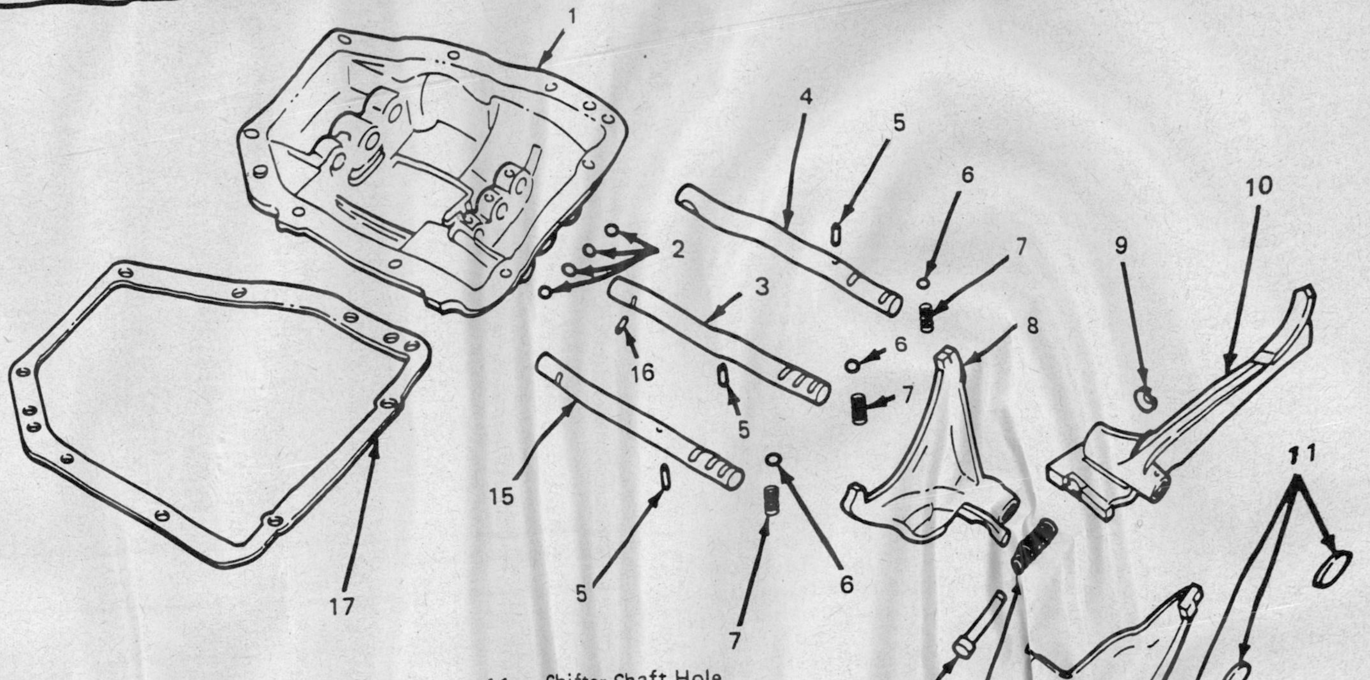
1
2
3
4
5
6
7
8
9
10
11
12
13
14
15
16
17

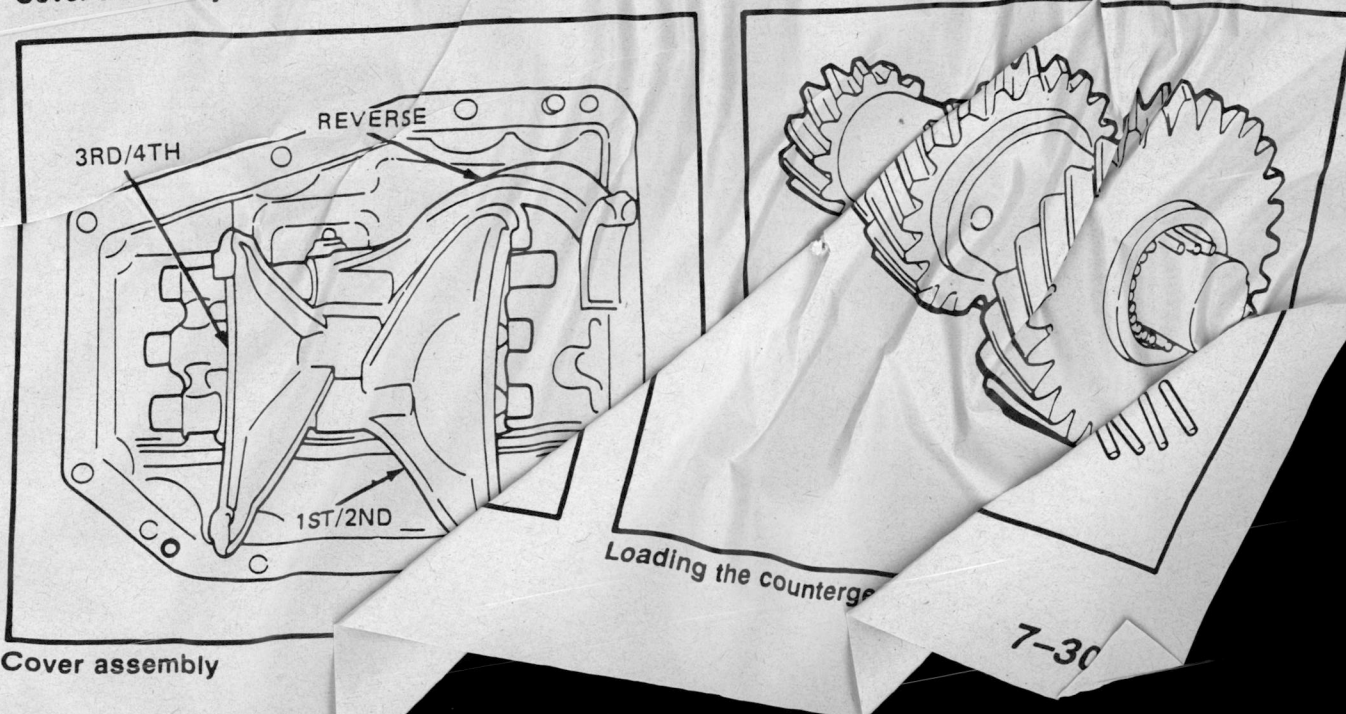

1. Transmission Cover
2. Interlock Balls
3. 3rd-4th Shifter Shaft
4. Reverse Shifter Shaft
5. Fork Retaining Pin
6. Detent Ball
7. Detent Spring
8. 3rd-4th Shifter Fork
9. "C" Ring Lock Clip
10. Reverse Shifter Fork
11. Shifter Shaft Hole Plugs
12. 1st-2nd Shifter Fork
13. Interlock Plunger Spring
14. Reverse Interlock Plunger
15. 1st-2nd Shifter Shaft
16. Interlock Pin
17. Cover Gasket

Cover assembly

REVERSE
3RD/4TH
1ST/2ND

Cover assembly

Loading the counterge

4 SPEED

1. Lower the countergear into the case until it rests on the bottom of the case.

2. Install the reverse idler gear into the case with the gear teeth toward the front. Install the idler gear shaft from rear to front and have the slot in the end of the shaft in facing down. The shaft slot face must be at least flush with the case.

3. Install the mainshaft assembly into the case with the rear of the shaft protruding out of the rear bearing hole in the case.

4. Install tool J-22874 or equivalent in the clutch gear case opening and engaging the front mainshaft, rotate the case onto the front end. Install the 1st speed gear thrust washer on the shaft.

5. Install the snapring on the bearing outside diameter and install the rear mainshaft bearing on the shaft. Using tool J-22874-1 or equivalent, drive the bearing onto the shaft and into the case. Rotate the case and remove tool J-22874-5 or equivalent.

6. Install the synchronizer cone on the pilot end of the mainshaft and slide it rearward toward the clutch hub. Make sure the 3 cut out sections of the 4th speed synchronizer cone align with the 3 clutch keys in the clutch assembly.

7. Install the snapring on the drive gear bearing outside diameter. Index the cut out portion of the drive gear teeth to obtain clearance over the countershaft drive gear teeth and install the clutch gear assembly onto the case. Rise the mainshaft to get the clutch gear started.

8. Install the drive gear bearing retainer using a new gasket and install the bolts.

9. Using tool J-22874 0 or equivalent, support the countergear and rotate the se onto the front end.

10. Install the snapring the countergear rear bearing outside diameter. Install the ing on the countergear and using tool J-22874-1 or equiva drive the bearing into place. Rotate the case, install th pring on the countershaft at the rear bearing and then re tool J-22874-1.

11. Tap the counterge nt bearing assembly into the case.

12. Install the counte front bearing cap and new gasket.

13. Slide the speedom drive gear over the mainshaft to the bearing.

14. Install the rear bg retainer with a new gasket. Be sure the snapring ends are e lube slot and cut out in the bearing retainer.

15. Install the bra backing plate assembly on vehicles equipped with a driveft br

16. On vehicles euied with WD, install the rear locknut and washer using t -23070 uivalent.

17. Install th pag brake and/or universal joint flange.

18. Lock the transmsion in 2 gea stall the universal joi flange locknu same time and in-

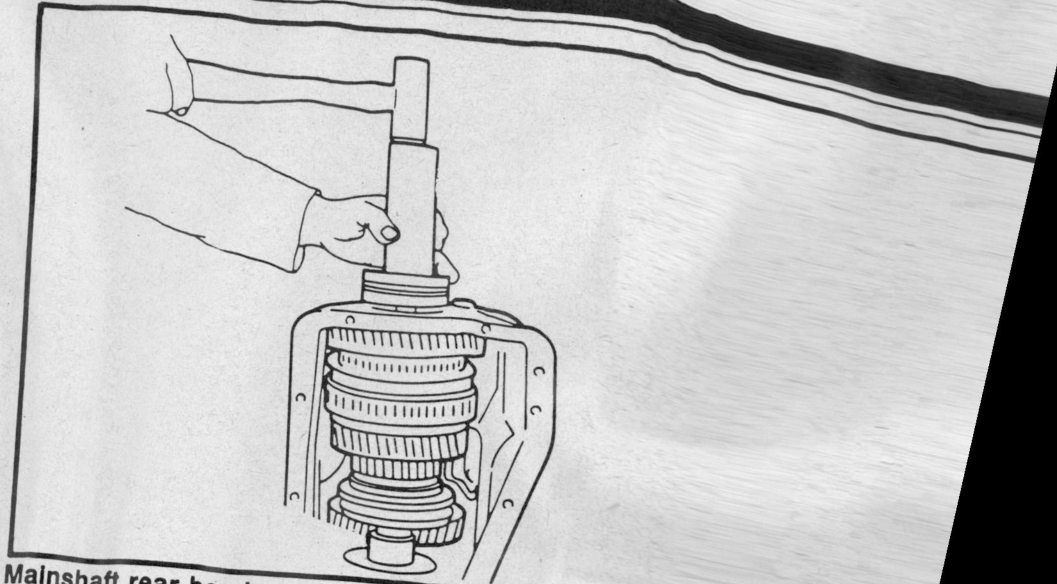

Mainshaft rear bearing

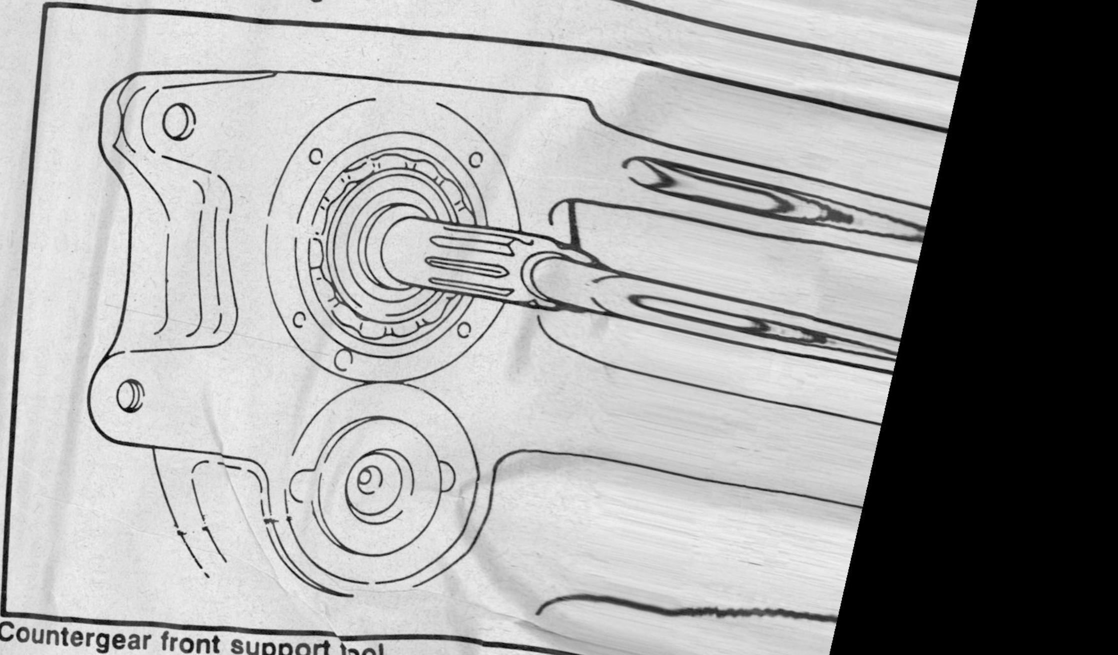

Countergear front support tool

SPECIFICATIONS

TORQUE SPECIFICATIONS

Item	ft. lbs.	Nm
3 SPEED		
Drive gear retainer to case bolts	15	20
Side cover to case bolts	15	20
Extension to case bolts	45	61
Shift lever to shifter shaft bolts	25	32
Lubrication filler plug	13	17
Transmission case to clutch housing bolts	75	101
Crossmember to frame bolts	55	75
Crossmember to mount bolts	40	54
2–3 cross over shaft bracket retaining nut	18	24
1—reverse swivel attaching bolts	20	27

Item	ft. lbs.	Nm
Mount to transmission bolt	40	55
4 SPEED		
Drive gear bearing retainer to case bolts	25	32
Cover to case bolts	20	27
Extension and retainer to case bolts—(upper)	20	27
Extension and retainer to case bolts—(lower)	30	40
Lubrication filler plug	30	40
Shift lever to shifter shaft nut	20	27
Crossmember to frame nuts	55	75
Crossmember to mount bolts	40	55

SPECIAL TOOLS

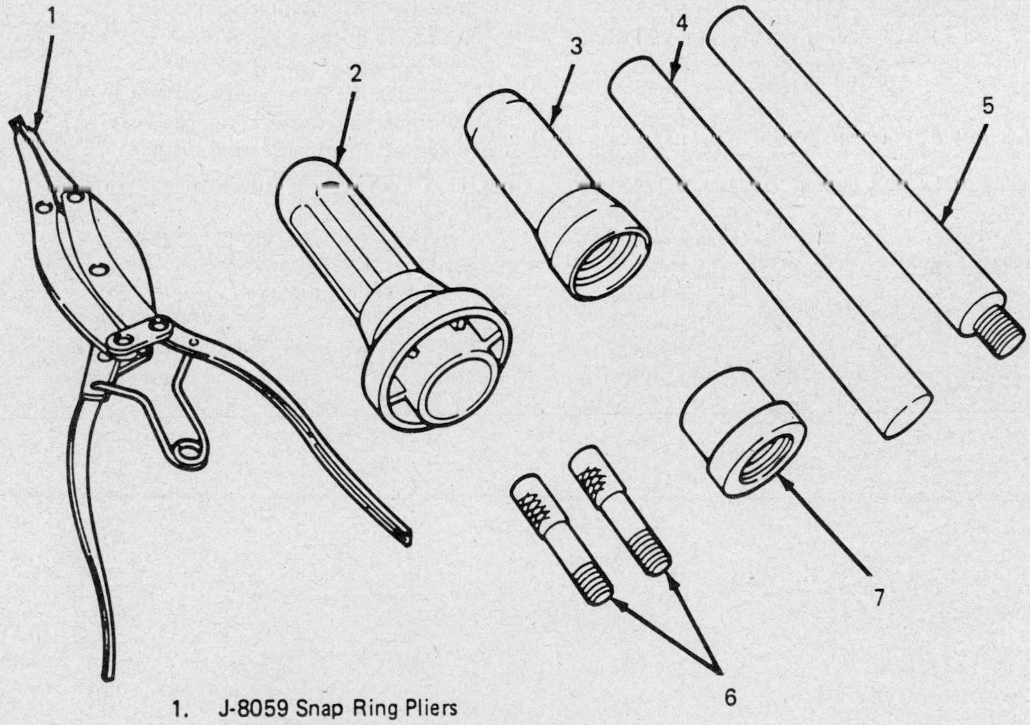
1
2
3
4
5
6
7

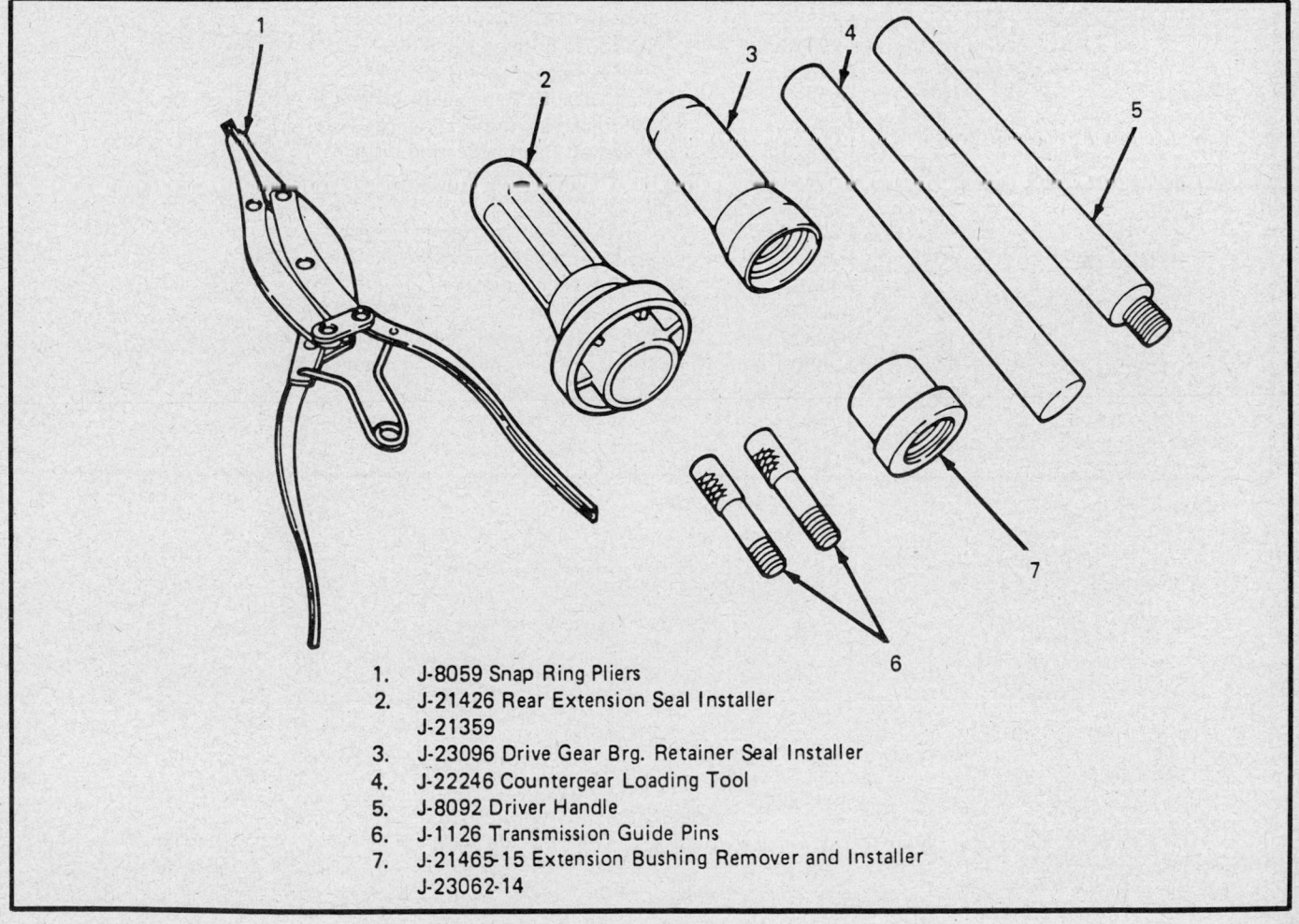

1. J-8059 Snap Ring Pliers
2. J-21426 Rear Extension Seal Installer
   J-21359
3. J-23096 Drive Gear Brg. Retainer Seal Installer
4. J-22246 Countergear Loading Tool
5. J-8092 Driver Handle
6. J-1126 Transmission Guide Pins
7. J-21465-15 Extension Bushing Remover and Installer
   J-23062-14

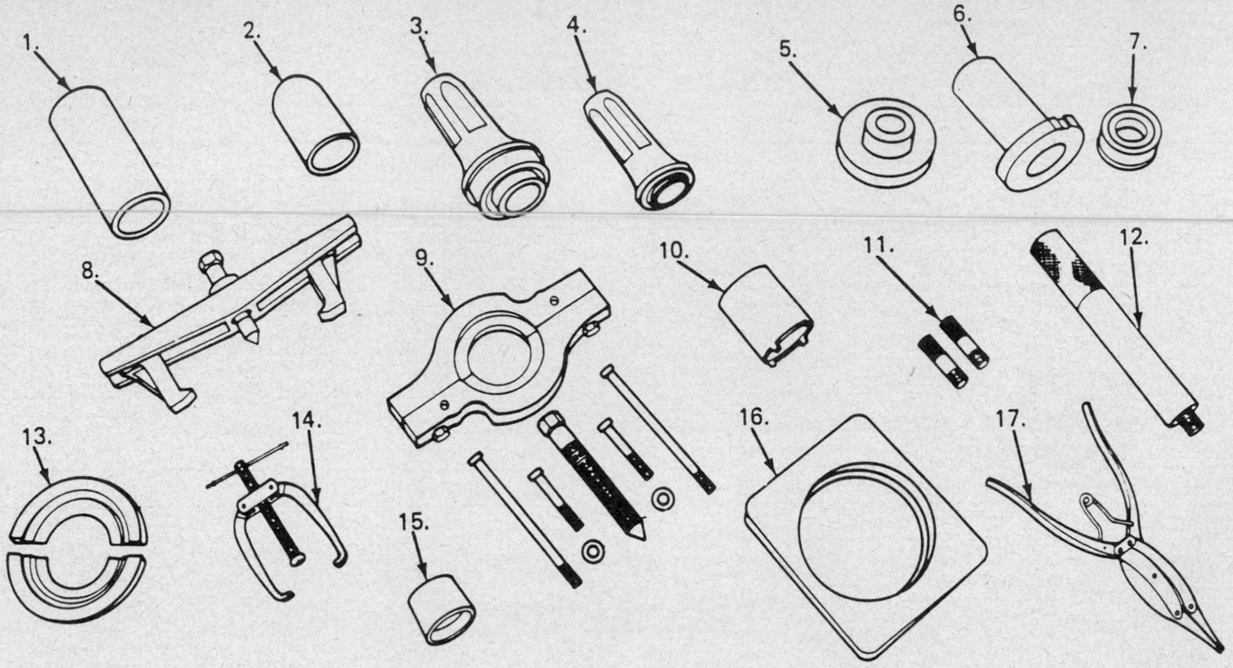
1.
2.
3.
4.
5.
6.
7.
8.
9.
10.
11.
12.
13.
14.
15.
16.
17.

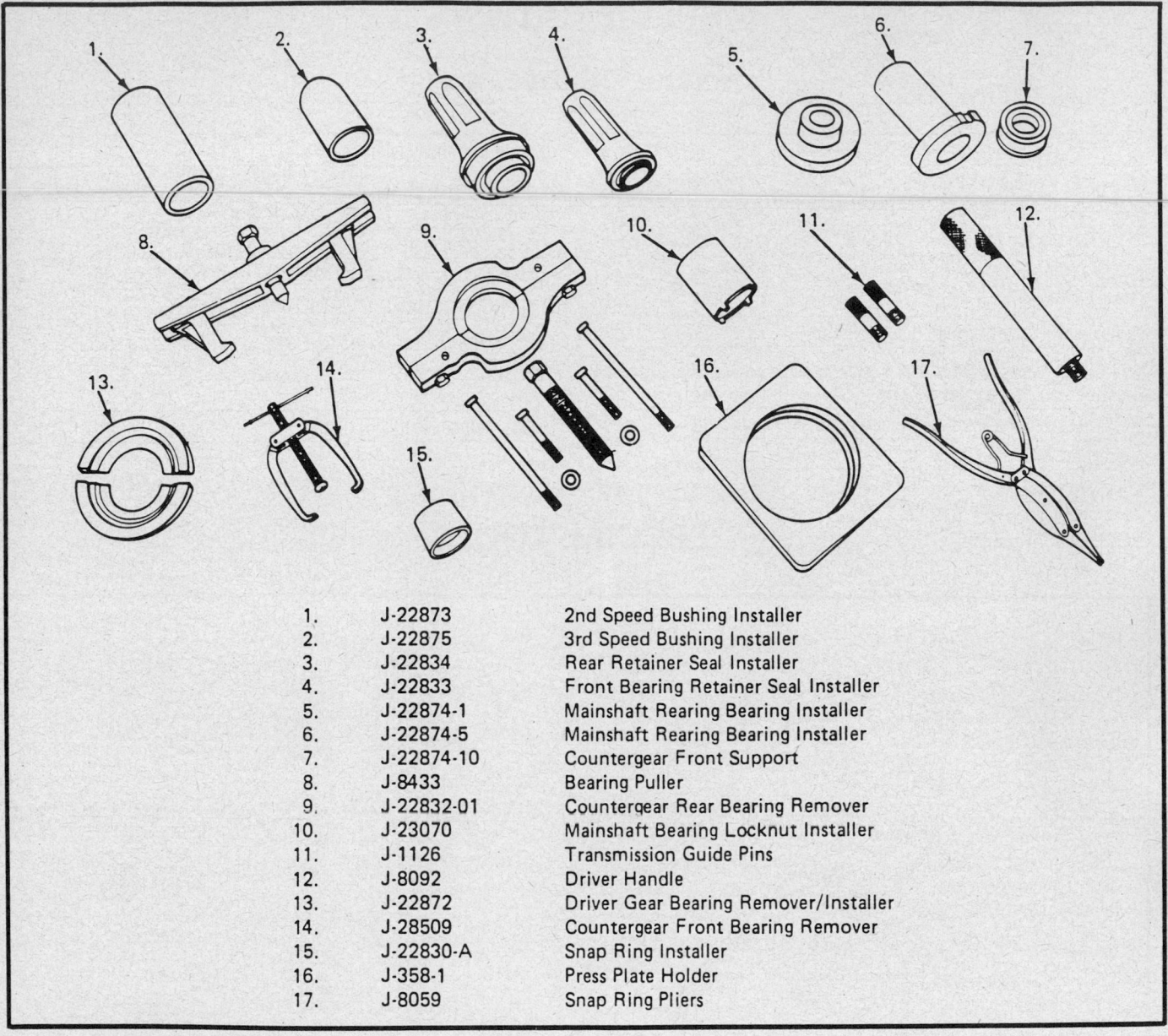

1.	J-22873	2nd Speed Bushing Installer
2.	J-22875	3rd Speed Bushing Installer
3.	J-22834	Rear Retainer Seal Installer
4.	J-22833	Front Bearing Retainer Seal Installer
5.	J-22874-1	Mainshaft Rearing Bearing Installer
6.	J-22874-5	Mainshaft Rearing Bearing Installer
7.	J-22874-10	Countergear Front Support
8.	J-8433	Bearing Puller
9.	J-22832-01	Countergear Rear Bearing Remover
10.	J-23070	Mainshaft Bearing Locknut Installer
11.	J-1126	Transmission Guide Pins
12.	J-8092	Driver Handle
13.	J-22872	Driver Gear Bearing Remover/Installer
14.	J-28509	Countergear Front Bearing Remover
15.	J-22830-A	Snap Ring Installer
16.	J-358-1	Press Plate Holder
17.	J-8059	Snap Ring Pliers

Section 7

New Process Transmission
General Motors

APPLICATION

1984–89 GM light trucks

GENERAL DESCRIPTION

The 4 speed New Process 89mm (MY-6) overdrive transmission is fully synchronized in all forward speeds with 2 sliding synchronizer sleeves. The countershaft is sealed in the front of the case with an expansion plug and is not a pressed fit. The gear shift lever is a floor type shift lever. The transmissions can be identified by a label attached to the top of the case.

Metric Fasteners

Do not attempt to interchange metric fasteners for inch system fasteners. Mismatched or incorrect fasteners can result in damage to the transmission unit through malfunctions, breakage or possible personal injury. Care should be taken to reuse the fasteners in the same location as removed, whenever possible. Due to the large number of alloy parts used and the aluminum casing, torque specifications should be strictly observed. Before installing bolts into aluminum parts, always dip threads into oil or sealant (if required), to prevent the bolts from galling the aluminum threads and to prevent seizing.

Capacity

The lubricant capacity is 4.2 qts. (4.0L) of Dexron®II automatic transmission fluid.

Checking Fluid Level

1. Raise and safely support the vehicle in a level position; the transmission should be cold.
2. From the right, middle side of transmission, remove the fluid plug and check the fluid level.
3. The fluid should be level with the bottom of the fluid plug hole; if not, add Dexron®II automatic transmission fluid until it reaches the bottom of the hole.
4. Install the filler plug and torque to 17 ft. lbs. (23 Nm).

TRANSMISSION MODIFICATIONS

There are no transmission modifications at the time of publication.

TROUBLE DIAGNOSIS

CHILTON'S THREE C'S TRANSMISSION DIAGNOSIS

Condition	Cause	Correction
Leaks lubricant	a) Lubricant level too high	a) Drain to correct level
	b) Main drive bearing retainer or gasket loose or damaged	b) Tighten or replace retainer or gasket
	c) Side cover or gasket loose or damaged	c) Tighten or replace cover or gasket
	d) Rear extension seal damaged	d) Replace rear extension seal
	e) Countershaft loose in case	e) Replace case
Hard shifting	a) Levers binding—dirty or damaged	a) Clean and lubricate or replace levers
	b) Lever endplay more than 0.13mm (0.005 in.)	b) Adjust lever endplay
Noisy shifting	a) Shift linkage out of adjustment or damaged	a) Adjust or replace shift linkage
	b) Clutch linkage out of adjustment or damaged	b) Adjust or replace clutch linkage
	c) Synchronizers or gears worn or damaged	c) Repair the transmission

CHILTON'S THREE C'S TRANSMISSION DIAGNOSIS

Condition	Cause	Correction
Noisy neutral	a) Shift linkage out of adjustment or damaged	a) Adjust or replace shift linkage
	b) Pilot bearing worn or damaged	b) Replace pilot bearing
	c) Main drive gear or countergear bearings worn or damaged	c) Repair the transmission
Noisy operation	a) Lubricant level low	a) Fill to correct level
	b) Shift linkage damaged	b) Replace shift linkage
	c) Synchronizers worn or damaged	c) Repair the transmission
	d) Bearings worn or damaged	d) Repair the transmission
	e) Gears worn or damaged	e) Repair the transmission
Slips out of gear	a) Shift lever seal stiff	a) Replace shift lever seal
	b) Shift linkage out of adjustment or binding	b) Adjust or replace shift linkage
	c) Pilot bearing loose or damaged	c) Replace pilot bearing
	d) Dirt between the clutch housing and transmission	d) Clean the mating surfaces
	e) Transmission loose	e) Tighten transmission bolts
	f) Main drive gear retainer loose or damaged	f) Tighten or replace retainer
	g) Transmission not aligned	g) Align transmission

ON CAR SERVICE

Adjustments

SHIFT LINKAGE

1. Raise and safely support the vehicle.
2. Loosen the shift linkage adjusting nuts.
3. Move the shift control lever into the **N** position.
4. Move the control levers 2 detents into **N**.
5. Using a 0.249–0.250 in. (6.314–6.350mm) gauge pin insert it through the control levers.
6. While holding the shift rods forward, tightly in the swivels, tighten the shift rod nuts.
7. Remove the gauge pin and lubricate the shift control assembly.

CLUTCH PEDAL FREE TRAVEL

1. Raise and safely support the vehicle.
2. Remove the pull back spring from the clutch fork.
3. Loosen the outer adjusting rod nut.
4. While holding the clutch pedal against the bumper, turn the inner adjusting rod nut until it is 0.28 in. (7.11mm) from the cross lever.
5. Tighten the outer adjusting nut against the cross lever.
6. Install the pull back spring to the clutch fork.
7. Check the clutch pedal free travel for 1.375 in. (34mm); if not correct, readjust it.
8. Lubricate the clutch linkage.

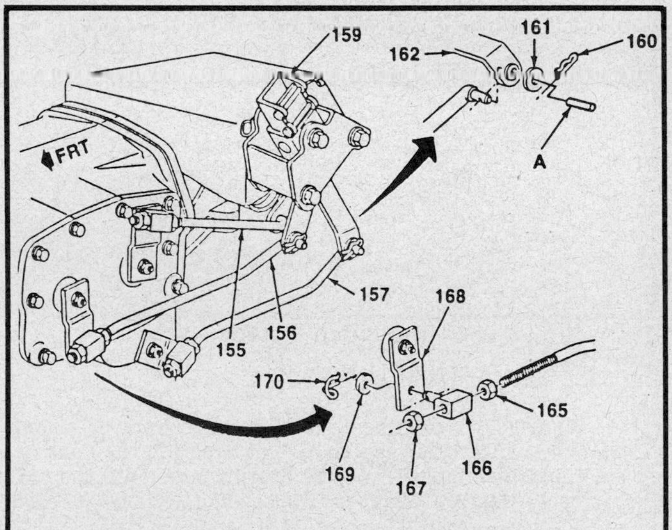
159
161
160
162
FRT
A
157
168
156
155
170
165
169
166
167

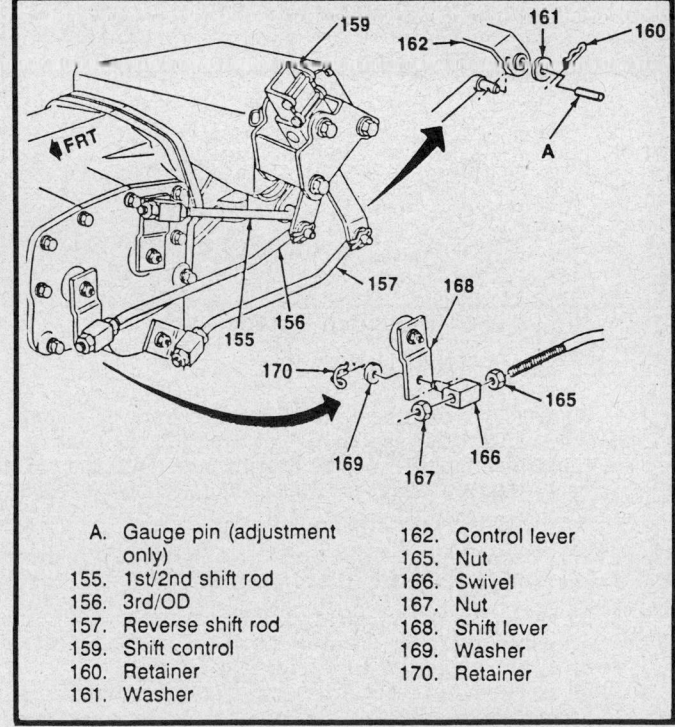

A. Gauge pin (adjustment only)
155. 1st/2nd shift rod
156. 3rd/OD
157. Reverse shift rod
159. Shift control
160. Retainer
161. Washer
162. Control lever
165. Nut
166. Swivel
167. Nut
168. Shift lever
169. Washer
170. Retainer

Exploded view of the floor shift linkage

Service

REAR OIL SEAL

Removal and Installation

1. Raise and safely support the vehicle.
2. Drain the fluid from the transmission.
3. Remove the one-piece driveshaft by performing the following procedure:
   a. If equipped with a skid plate, remove it.
   b. Using chalk or paint, make alignment marks at the driveshaft-to-pinion flange connection.

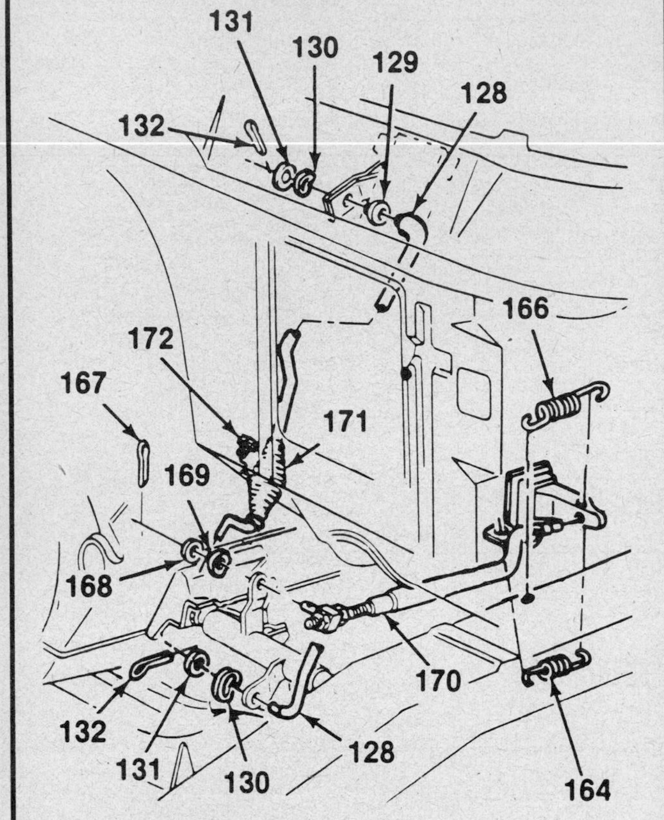
131
130
129
128
132
166
172
167
171
169
168
170
132
131
130
128
164

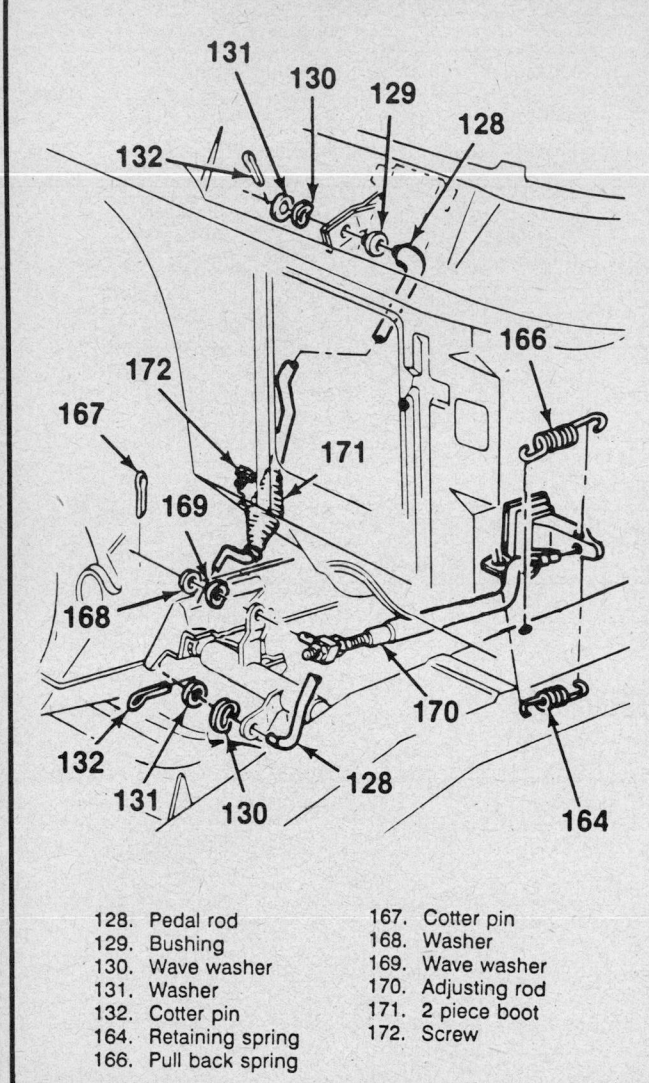

128. Pedal rod
129. Bushing
130. Wave washer
131. Washer
132. Cotter pin
164. Retaining spring
166. Pull back spring
167. Cotter pin
168. Washer
169. Wave washer
170. Adjusting rod
171. 2 piece boot
172. Screw

Exploded view of the clutch linkage

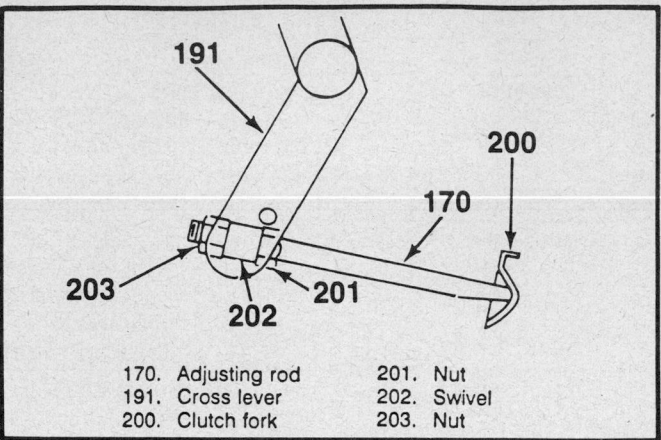

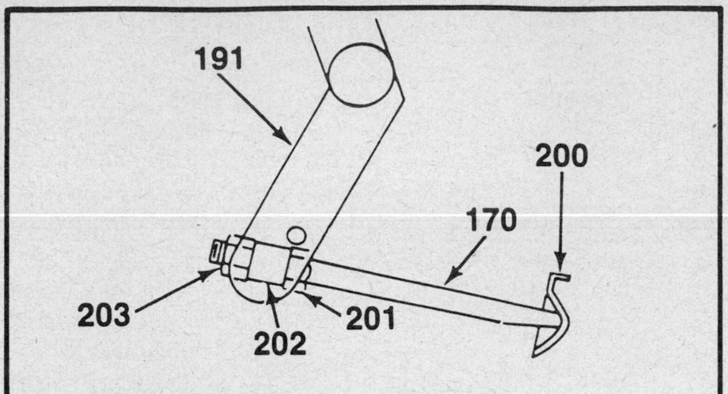
191
200
170
203
201
202

170. Adjusting rod
191. Cross lever
200. Clutch fork
201. Nut
202. Swivel
203. Nut

View of the clutch linkage adjustment

c. While supporting the driveshaft, remove the driveshaft-to-pinion flange bolts and retainers. Separate the yoke/universal joint assembly. Tape the bearing cups onto the yoke/universal assembly to prevent the needle rollers from becoming lost.

NOTE: Do not pound or pry on the original driveshaft yoke ears, for the plastic injection joints may fracture.

d. Move the driveshaft forward, lower the rear and withdraw it rearward (under the axle); do not allow the universal joint to incline greatly, for the joint may fracture.

4. Remove the multi-piece driveshaft by performing the following procedure:

a. If equipped with a skid plate, remove it.

b. Using chalk or paint, make alignment marks at the driveshaft-to-pinion flange connection.

c. While supporting the rear driveshaft, remove the driveshaft-to-pinion flange bolts and retainers. Separate the yoke/universal joint assembly. Tape the bearing cups onto the yoke/universal assembly to prevent the needle rollers from becoming lost.

NOTE: Do not pound or pry on the original driveshaft yoke ears, for the plastic injection joints may fracture.

d. Move the rear driveshaft forward, lower the rear and withdraw it rearward (under the axle); do not allow the universal joint to incline greatly, for the joint may fracture.

e. While supporting the front driveshaft, remove the center bearing support-to-hanger nuts, bolts and washers.

f. From the rear of the front driveshaft, remove the cap, washer and seal.

g. Remove the front driveshaft from the transmission; do not allow the universal joint to incline greatly, for the joint may fracture.

5. Using a small pry bar, pry the oil seal from the rear of the transmission.

6. Clean the seal seat.

7. Using GM sealant part number 1052942 or equivalent, apply it to the outside of the new seal.

8. Using the rear extension seal installer tool or equivalent, drive the new seal into the extension housing until it seats. Using chassis grease, lubricate the seal lips.

9. Install the one-piece driveshaft by performing the following procedure:

a. Using chassis grease, lubricate the driveshaft slip joint.

b. Install the driveshaft onto the transmission's output shaft.

c. Align and seat the yoke/universal joint assembly with the pinion flange.

d. Install the retainers and bolts. Torque the driveshaft-to-pinion flange bolts to 15 ft. lbs. (20 Nm).

10. Install the multi-piece driveshaft by performing the following procedure:

a. Using chassis grease, lubricate the front driveshaft's slip joint.

b. Install the front driveshaft onto the transmission's output shaft.

c. Install the center bearing support-to-hanger nuts, bolts and washers. Torque the center bearing support-to-hanger nuts/bolts to 25 ft. lbs. (34 Nm).

d. Position the front driveshaft so the front yoke bearing ears are in a vertical position.

e. Install the seal, washer and cap onto the rear of the front driveshaft.

f. Position the rear driveshaft so the front yoke bearing ears are in the horizontal position. Align the yoke's missing tooth (rear driveshaft) with the splined shaft's bridged tooth (front drivehsaft) and slide the rear driveshaft onto the front driveshaft.

g. While supporting the rear driveshaft, align the reference marks and fit the bearing caps into the pinion flange.

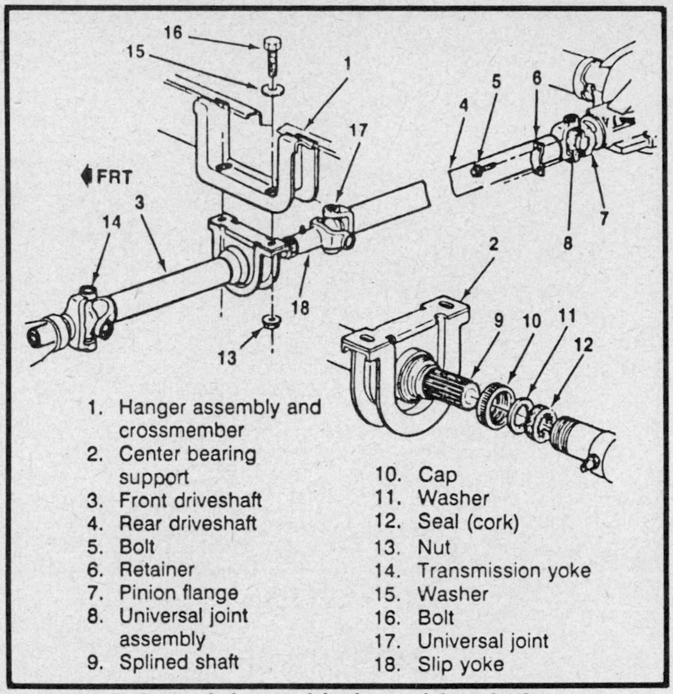
FRT
1. Hanger assembly and crossmember
2. Center bearing support
3. Front driveshaft
4. Rear driveshaft
5. Bolt
6. Retainer
7. Pinion flange
8. Universal joint assembly
9. Splined shaft
10. Cap
11. Washer
12. Seal (cork)
13. Nut
14. Transmission yoke
15. Washer
16. Bolt
17. Universal joint
18. Slip yoke

Exploded view of the multi-piece driveshafts

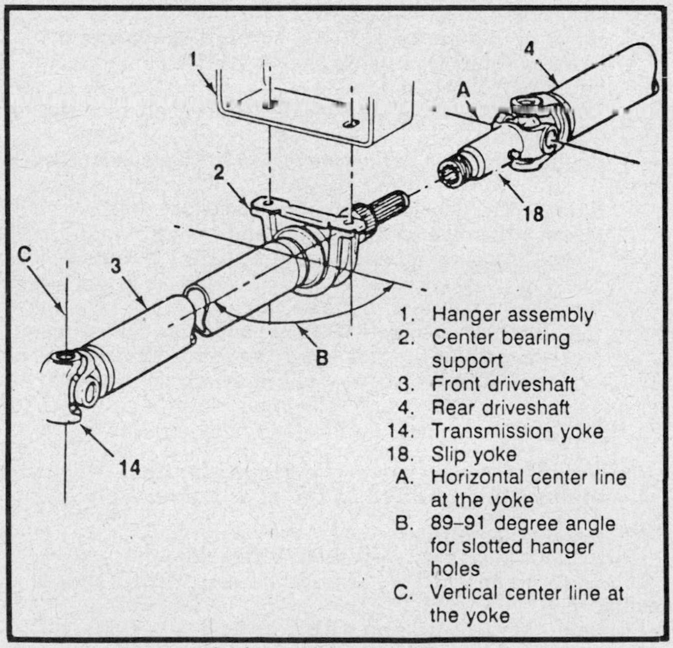
1. Hanger assembly
2. Center bearing support
3. Front driveshaft
4. Rear driveshaft
14. Transmission yoke
18. Slip yoke
A. Horizontal center line at the yoke
B. 89–91 degree angle for slotted hanger holes
C. Vertical center line at the yoke

Aligning the multi-piece driveshafts

h. Install the driveshaft-to-pinion flange retainers and bolts; torque the bolts to 15 ft. lbs. (20 Nm).

11. Refill the transmission with new fluid and lower the vehicle.

SPEEDOMETER DRIVEN GEAR

Removal and Installation

1. Raise and safely support the vehicle.

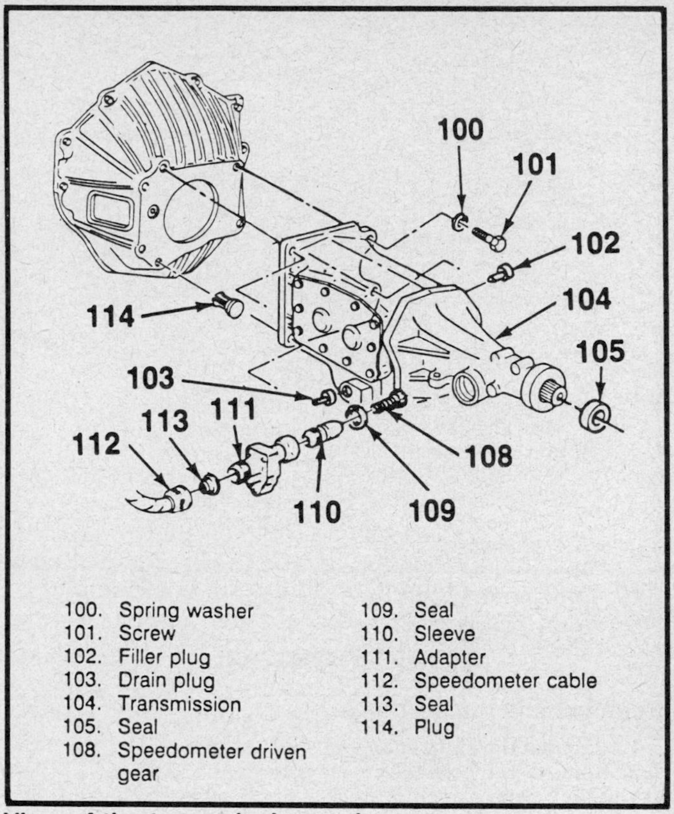
100. Spring washer
101. Screw
102. Filler plug
103. Drain plug
104. Transmission
105. Seal
108. Speedometer driven gear
109. Seal
110. Sleeve
111. Adapter
112. Speedometer cable
113. Seal
114. Plug

View of the transmission and components

2. Remove the speedometer cable and seal. If equipped, remove the adapter, the retainer and screw.
3. Remove the sleeve, the seal and the gear.
4. Using transmission fluid, lubricate the gear and new seal; install them into the transmission.
5. Install the sleeve. If equipped, install the retainer, the screw and the adapter.
6. Install the new speedometer seal and the speedometer cable.
7. Lower the vehicle.

SHIFT LINKAGE

Removal and Installation

1. Raise and safely support the vehicle.
2. Remove the shift rod(s)-to-control lever(s) retaining pin and washer.
3. Remove the shift rod(s)-to-shift lever(s) retaining pin and washer.
4. Remove the shift rod(s) from the vehicle.
5. If replacing the shift rod(s), remove the adjusting nuts and swivel from the rod.
6. If removed, install the adjusting nuts and swivel to the shift rod(s); do not tighten.
7. Install the shift rod(s) to the shift lever(s) and control lever(s).
8. Install the washers and retaining pins.

NOTE: If the adjusting nuts and swivel(s) were removed from the shift rod(s), perform the shift linkage adjustment.

9. Lower the vehicle.

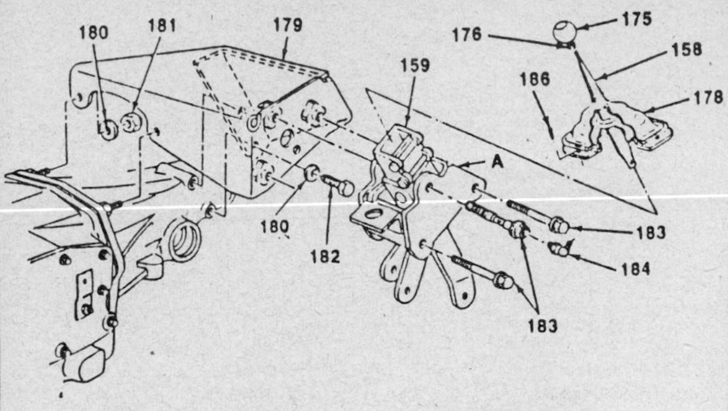

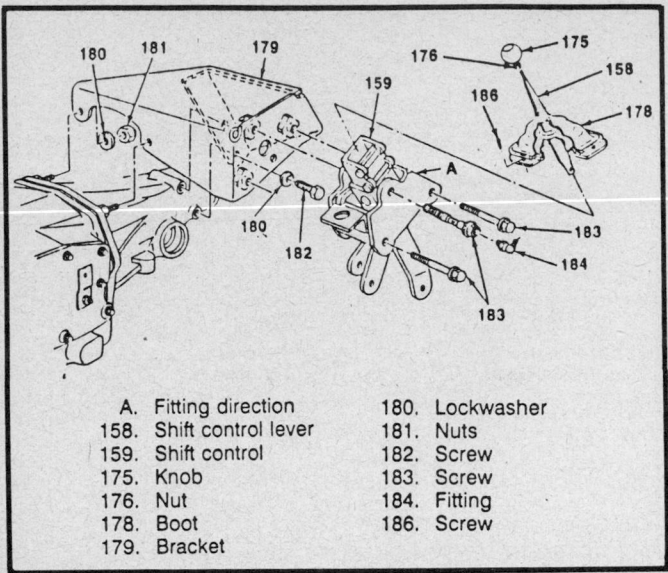

A. Fitting direction
158. Shift control lever
159. Shift control
175. Knob
176. Nut
178. Boot
179. Bracket
180. Lockwasher
181. Nuts
182. Screw
183. Screw
184. Fitting
186. Screw

Exploded view of the floor shift control assembly

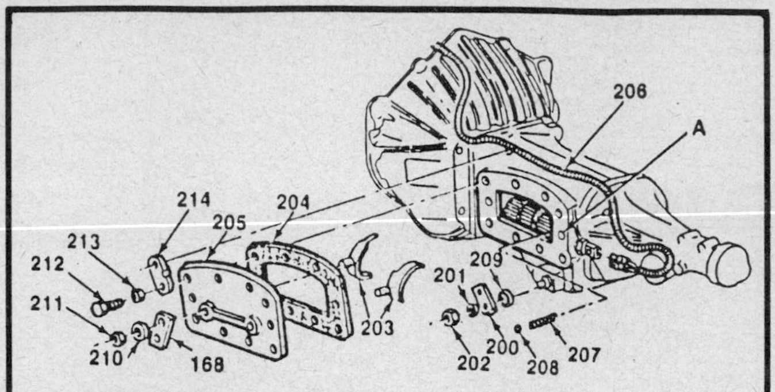

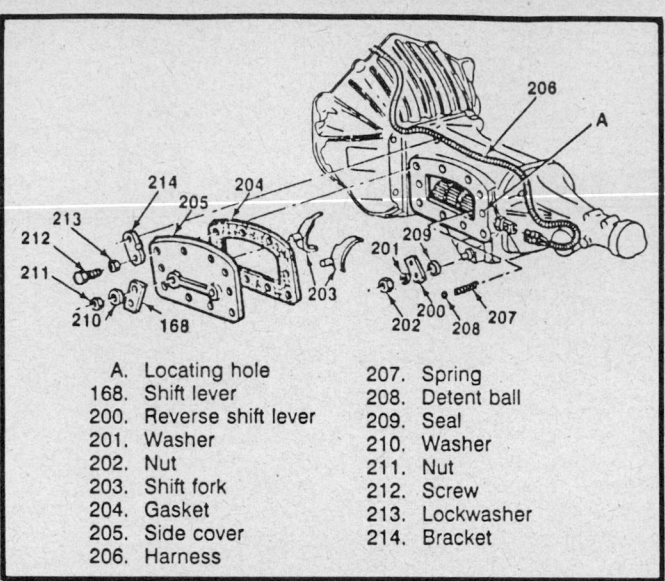

A. Locating hole
168. Shift lever
200. Reverse shift lever
201. Washer
202. Nut
203. Shift fork
204. Gasket
205. Side cover
206. Harness
207. Spring
208. Detent ball
209. Seal
210. Washer
211. Nut
212. Screw
213. Lockwasher
214. Bracket

Exploded view of the side cover

SHIFT CONTROL

Removal and Installation

1. Remove the shift knob and nut.
2. Remove the boot-to-console screws and the boot.
3. Using a piece of shim stock, slide it between the shift control lever and the shift control to release the lever and remove it.
4. Remove the shift rods-to-shift controls retaining pins and washers; disconnect the shift rods from the shift controls.
5. Remove the shift control assembly-to-bracket bolts, the fitting and the control assembly.
6. If necessary, remove the shift control bracket-to-transmission bolts, lockwashers, nuts and the bracket.
7. Install the bracket, the nuts, bolts and lockwashers; torque the bolts to 24 ft. lbs. (33 Nm).
8. Install the shift control assembly and torque the bolts to 33 ft. lbs. (45 Nm).

NOTE: When installing the lubricating fitting, be sure it points to the corner.

9. Install the shift rods, the washers and retaining pins.
10. Wipe the shift control lever with a rag and install it into the shift control assembly.
11. Install the boot, screws, nut and knob.
12. Lubricate the shift control and the shift linkage.

NOTE: It may be necessary to perform the shift linkage adjustment.

13. Lower the vehicle.

SIDE COVER

Removal and Installation

1. Raise and safely support the vehicle.
2. Drain the transmission fluid.
3. If equipped with an electrical harness, disconnect it.
4. Move the shift levers into the **N** positions and disconnect the shift rods from them.
5. Remove the reverse shift lever-to-shift shaft nut, washer, the lever and the seal.
6. Remove the bolts, lockwashers and brackets from the side cover.
7. Remove the side cover bolts and the side cover.
8. Remove the reverse detent ball and spring.
9. Clean the gasket material from the case and the side cover surfaces.
10. Install a new gasket onto the case.
11. Install the reverse detent spring and ball.
12. When installing the side cover, move the shift levers into the **N** position and lift the reverse interlock lever to seat the cover.
13. Install the brackets, new lockwashers and screws.

NOTE: The screw with the large shoulder (locating screw) must be installed in the rear center hole A.

14. Install a new seal on the reverse shift shaft.
15. Install the reverse shift lever, washer and nut.
16. Install the shift rods and the electrical harness, if equipped.
17. Refill the transmission with new fluid.
18. Adjust the shift linkage and lower the vehicle.

REMOVAL AND INSTALLATION

TRANSMISSION REMOVAL

1. Disconnect the negative battery cable. Raise and safely support the vehicle.
2. Drain the fluid from the transmission and transfer case, if equipped.
3. Remove the shift control assembly and the shift rods from the transmission.
4. Remove the one-piece driveshaft by performing the following procedure:
   a. If equipped with a skid plate, remove it.
   b. Using chalk or paint, make alignment marks at the driveshaft-to-pinion flange connection.
   c. While supporting the driveshaft, remove the driveshaft-to-pinion flange bolts and retainers. Separate the yoke/universal joint assembly. Tape the bearing cups onto the yoke/universal assembly to prevent the needle rollers from becoming lost.

NOTE: Do not pound or pry on the original driveshaft yoke ears, for the plastic injection joints may fracture.

   d. Move the driveshaft forward, lower the rear and withdraw it rearward (under the axle); do not allow the universal joint to incline greatly, for the joint may fracture.
5. Remove the multi-piece driveshaft by performing the following procedure:
   a. If equipped with a skid plate, remove it.
   b. Using chalk or paint, make alignment marks at the driveshaft-to-pinion flange connection.
   c. While supporting the rear driveshaft, remove the driveshaft to pinion flange bolts and retainers. Separate the yoke/universal joint assembly. Tape the bearing cups onto the yoke/universal assembly to prevent the needle rollers from becoming lost.

NOTE: Do not pound or pry on the original driveshaft yoke ears, for the plastic injection joints may fracture.

   d. Move the rear driveshaft forward, lower the rear and withdraw it rearward (under the axle); do not allow the universal joint to incline greatly, for the joint may fracture.
   e. While supporting the front driveshaft, remove the center bearing support-to-hanger nuts, bolts and washers.
   f. From the rear of the front driveshaft, remove the cap, washer and seal.
   g. Remove the front driveshaft from the transmission; do not allow the universal joint to incline greatly, for the joint may fracture.
6. Remove the transfer case, if equipped.
7. If necessary, disconnect the parking brake cable. Remove the lower clutch housing inspection plate, if equipped.
8. Disconnect the speedometer cable and seal from the transmission.
9. If equipped, disconnect the electrical harness and connector.
10. If the exhaust pipe(s) interfere with the transmission removal, remove it (them).
11. Using the proper equipment, support the transmission. Remove the crossmember-to-transmission bolts, the crossmember-to-chassis bolts and the crossmember.
12. Remove the transmission-to-clutch housing bolts and washers. Pull the transmission straight and lower it from the vehicle.

NOTE: Do not allow the transmission to hang from the clutch. Support the clutch release bearing.

TRANSMISSION INSTALLATION

1. Using high temperature grease, apply a light coat to the main drive gear splines.
2. Shift the transmission into 4th gear. If the clutch release bearing was supported, remove the support.
3. Using the proper equipment, raise the transmission into position, move it forward, align the input shaft splines with the clutch splines and seat the transmission against the clutch housing; be sure to support it with the jack.

NOTE: When installing the transmission, do not force it into the clutch and do not allow it to hang from the clutch.

4. Using new lockwashers, install the transmission-to-clutch housing bolts and torque them to 74 ft. lbs. (100 Nm).
5. Install the crossmember. Torque the crossmember-to-chassis bolts to 55 ft. lbs. (75 Nm) and the crossmember-to-transmission bolts to 40 ft. lbs. (54 Nm).
6. Remove the transmission jack.
7. Install the transfer case, if equipped.
8. If the exhaust pipe(s) was removed, install it (them).
9. If equipped with an electrical harness connector(s), connect it (them).
10. Lubricate the new speedometer seal with transmission fluid. Install the new seal and the speedometer cable.
11. Install the one-piece driveshaft by performing the following procedure:
   a. Using chassis grease, lubricate the driveshaft slip joint.
   b. Install the driveshaft onto the transmission's output shaft.
   c. Align and seat the yoke/universal joint assembly with the pinion flange.
   d. Install the retainers and bolts. Torque the driveshaft-to-pinion flange bolts to 15 ft. lbs. (20 Nm).
12. Install the multi-piece driveshaft by performing the following procedure:
   a. Using chassis grease, lubricate the front driveshaft's slip joint.
   b. Install the front driveshaft onto the transmission's output shaft.
   c. Install the center bearing support-to-hanger nuts, bolts and washers. Torque the center bearing support-to-hanger nuts/bolts to 25 ft. lbs. (34 Nm).
   d. Position the front driveshaft so the front yoke bearing ears are in a vertical position.
   e. Install the seal, washer and cap onto the rear of the front driveshaft.
   f. Position the rear driveshaft so the front yoke bearing ears are in the horizontal position. Align the yoke's missing tooth (rear driveshaft) with the splined shaft's bridged tooth (front drivehsaft) and slide the rear driveshaft onto the front driveshaft.
   g. While supporting the rear driveshaft, align the reference marks and fit the bearing caps into the pinion flange.
   h. Install the driveshaft-to-pinion flange retainers and bolts; torque the bolts to 15 ft. lbs. (20 Nm).
13. Install the shift control assembly and shift rods; adjust the shift linkage.
14. Refill the transmission and transfer case (if equipped) with new fluid and lower the vehicle.

BENCH OVERHAUL

Before Disassembly

When servicing the transmission it is important to be aware of cleanliness. Before disassembling the transmission, the outside should be throughly cleaned, preferably with a high-pressure spray cleaning equipment. Dirt entering the unit may negate all the effort and time spent on the overhaul.

During inspection and reassembly, all parts should be cleaned with solvent and dried with compressed air. Lubricate the seals with Dexron®II and use petroleum jelly to hold the thrust washers; this will ease the assembly of the seals and not leave harmful residues in the system. Do not use solvent on neoprene seals, if they are to be reused.

Before installing bolts into aluminum parts, dip the threads into clean gear oil. Anti-seize compound may be used to prevent galling the aluminum or seizing. Be sure to use a torque wrench to prevent stripping the threads. Be especially careful when installing the seals, the smallest nick can cause a leak. Aluminum parts are very susceptible to damage; great care should be used when handling them. Reusing snaprings is not recommended but should they be: compress the internal ones and compress the external ones.

Transmission Disassembly

1. Remove the speedometer driven gear, the switches, if equipped and any external components.
2. Move the shift levers to position the transmission into **N**.
3. Remove the reverse shift lever-to-shaft nut and the lever.
4. Remove the side cover bolts, the cover, the detent ball/spring, the shift forks (note positions they were removed) and the gasket.
5. Remove the extension housing-to-main case bolts, rotate the housing 180 degrees and install a bolt to hold the housing stationary.

NOTE: When removing the plug and countershaft, be careful not to damage the countershaft or bearings.

6. At the center of the countershaft plug, drill or punch a hole.
7. Insert a punch through the plug's hole, drive the countershaft rearward until the Woodruff key is exposed and remove the key.
8. Using a punch at the rear of the main case, drive the countershaft forward until the plug is removed from the case.

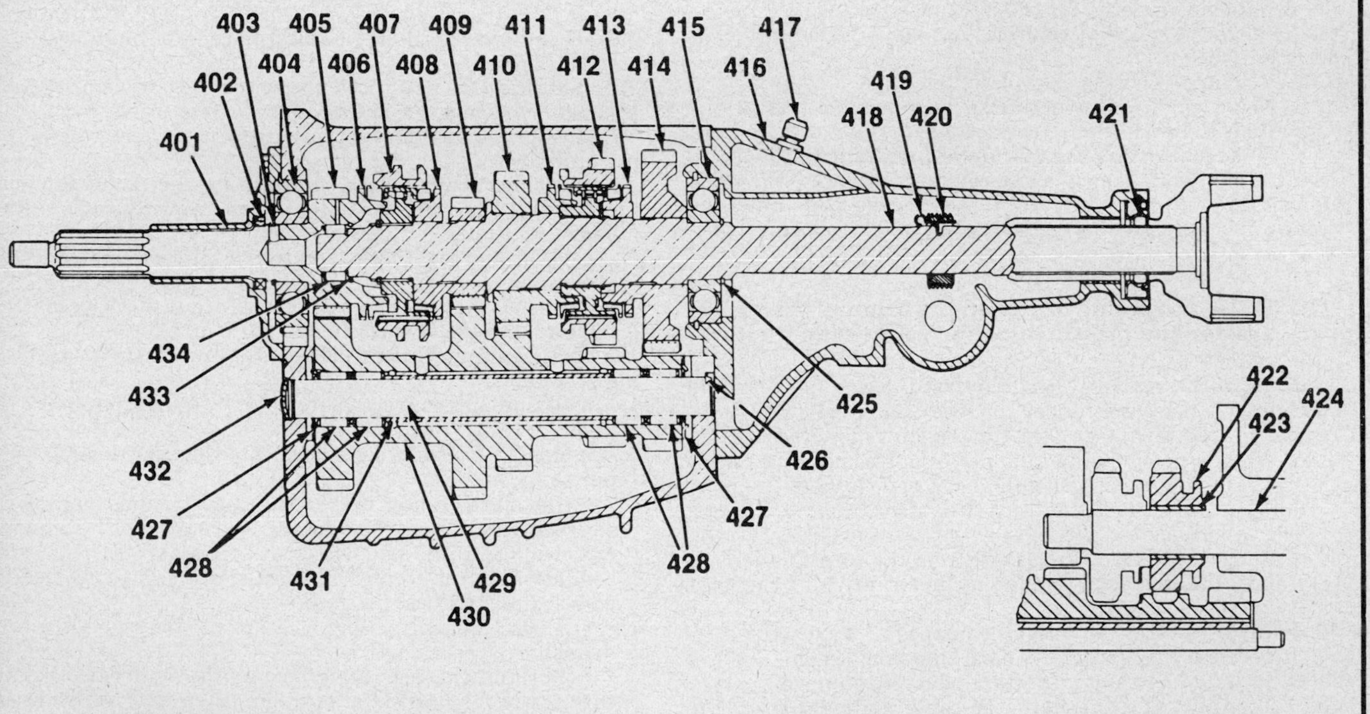
403
405
407
409
411
413
415
417
402
404
406
408
410
412
414
416
419
418
420
421
401
434
433
425
422
424
423
426
432
427
427
428
428
431
429
430

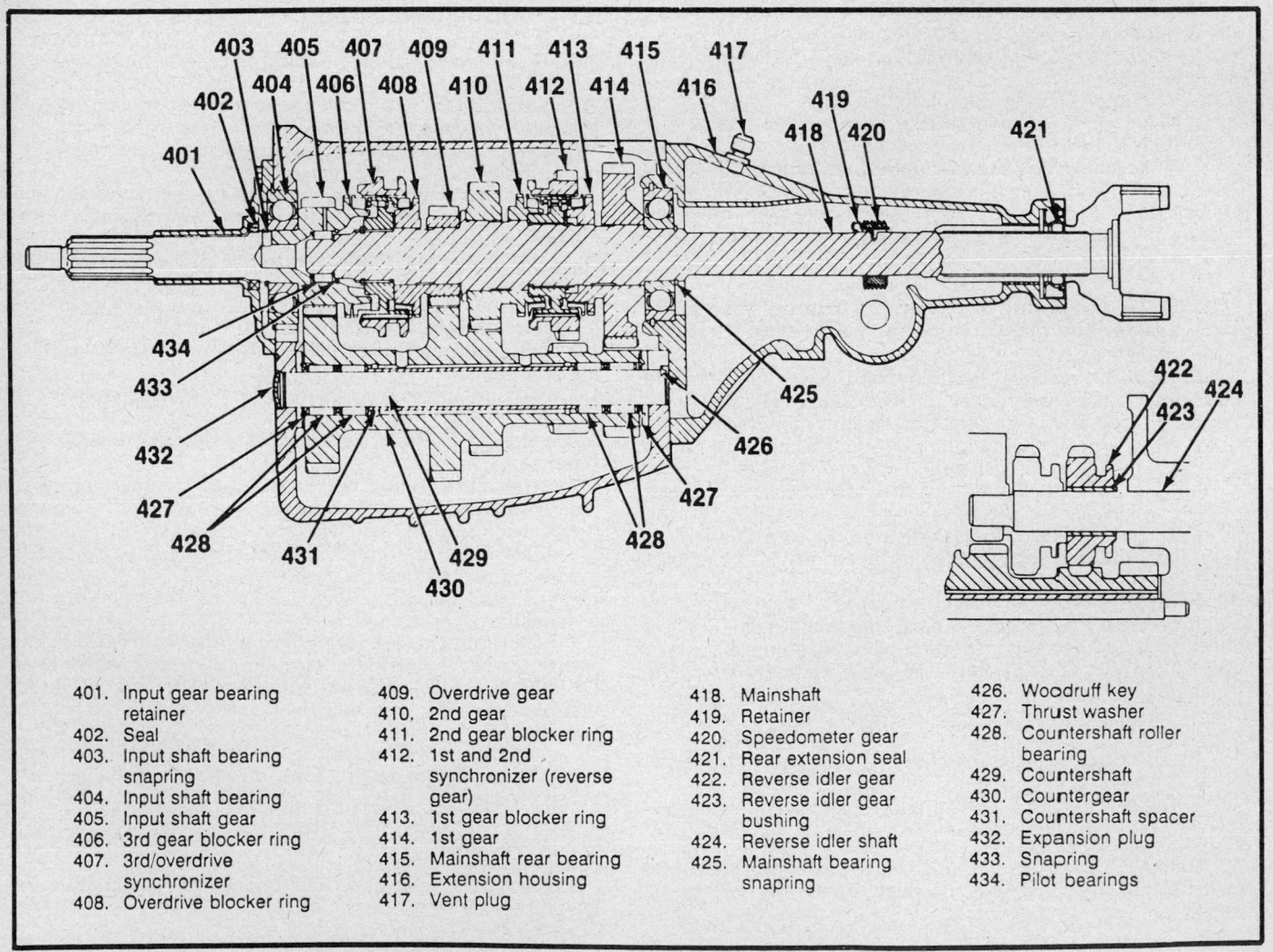

401. Input gear bearing retainer
402. Seal
403. Input shaft bearing snapring
404. Input shaft bearing
405. Input shaft gear
406. 3rd gear blocker ring
407. 3rd/overdrive synchronizer
408. Overdrive blocker ring
409. Overdrive gear
410. 2nd gear
411. 2nd gear blocker ring
412. 1st and 2nd synchronizer (reverse gear)
413. 1st gear blocker ring
414. 1st gear
415. Mainshaft rear bearing
416. Extension housing
417. Vent plug
418. Mainshaft
419. Retainer
420. Speedometer gear
421. Rear extension seal
422. Reverse idler gear
423. Reverse idler gear bushing
424. Reverse idler shaft
425. Mainshaft bearing snapring
426. Woodruff key
427. Thrust washer
428. Countershaft roller bearing
429. Countershaft
430. Countergear
431. Countershaft spacer
432. Expansion plug
433. Snapring
434. Pilot bearings

Cross-sectional view of the transmission assembly

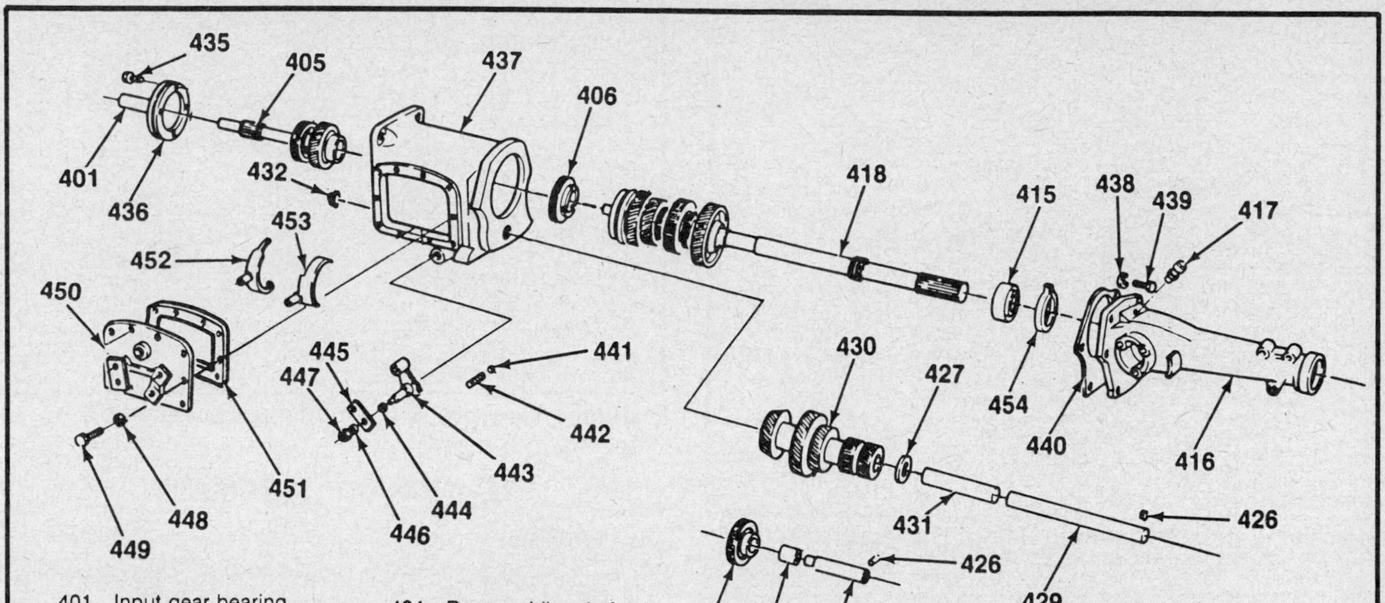
401. Input gear bearing retainer
405. Input shaft/drive gear assembly
406. Overdrive blocker ring
415. Mainshaft rear bearing
416. Extension housing
417. Vent plug
418. Mainshaft
422. Reverse idler gear
423. Reverse idler gear bushing
424. Reverse idler shaft
426. Woodruff key
427. Thrust washer
429. Countershaft
430. Countergear
431. Countershaft spacer
432. Expansion plug
435. Bearing retainer gasket
437. Transmission case
438. Lockwasher
439. Screw
440. Extension housing gasket
441. Detent ball
442. Detent spring
443. Reverse shift shaft
444. Seal
445. Reverse shift lever
446. Washer
447. Nut
448. Lockwasher
449. Screw
450. Side cover
451. Side cover gasket
452. 3rd/overdrive shift fork
453. 1st/2nd shift fork
454. Extension housing bearing snapring

Exploded view of the transmission assembly

9. Place the countershaft alignment tool or equivalent, on the front of the countershaft, drive the countershaft rearward and remove it from the case; the alignment tool will stay inside the countergear. Allow the countergear assembly to rest in the bottom of the case.

10. Remove the extension housing retaining bolt and rotate it to the normal position.

11. Remove the input shaft bearing retainer-to-main case bolts, the bearing retainer and the input shaft; it may be necessary to use a punch and a hammer (through the side cover opening) to drive the input shaft/gear from the main case.

12. Remove the extension housing by performing the following procedures:

a. Move the 3rd/overdrive synchronizer sleeve forward.

b. Move the reverse idler gear to the center of the idler shaft.

c. Move the extension housing/mainshaft assembly rearward and separate it from the main case.

13. Using a pair of snapring pliers, remove the bearing-to-housing snapring and pull the mainshaft assembly from the extension housing.

14. Remove the countergear/thrust washer assembly from the main case.

15. Remove the reverse idler shaft by performing the following procedures:

a. Using a ⅜ in. × 3½ in. bolt/nut assembly, position it into a $^{7}/_{16}$ in. deep socket.

b. Place the bolt head against the case and the socket against the reverse idler shaft.

c. Turn the nut against the socket to drive the idler shaft from it's bore.

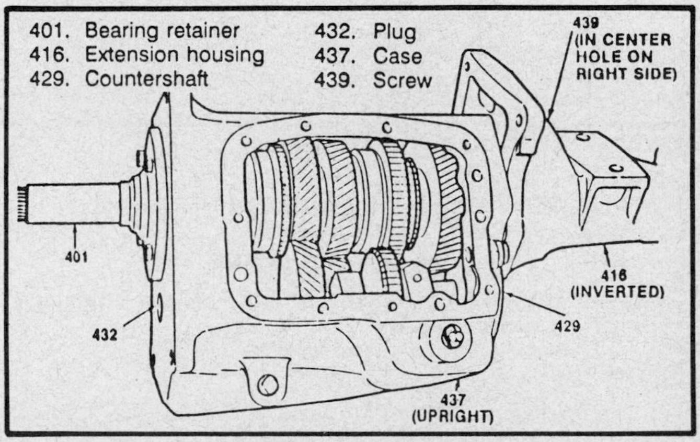
401. Bearing retainer
416. Extension housing
429. Countershaft
432. Plug
437. Case
439. Screw
439 (IN CENTER HOLE ON RIGHT SIDE)
401
416 (INVERTED)
429
432
437 (UPRIGHT)

Turning and securing the extension housing

d. Remove the reverse idler shaft, reverse idler gear and the key.

16. Remove the reverse shift shaft and seal.

Unit Disassembly And Assembly

INPUT SHAFT/GEAR ASSEMBLY

Disassembly

1. From the rear of the input shaft/gear, remove the roller bearings and snapring.

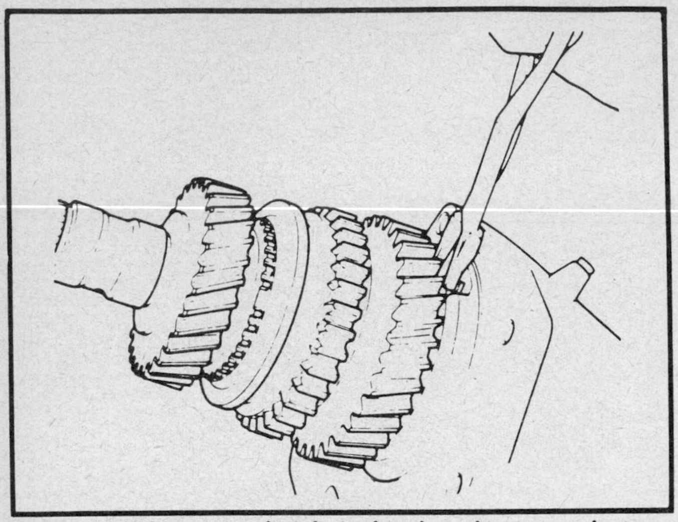
Removing the extension housing bearing snapring

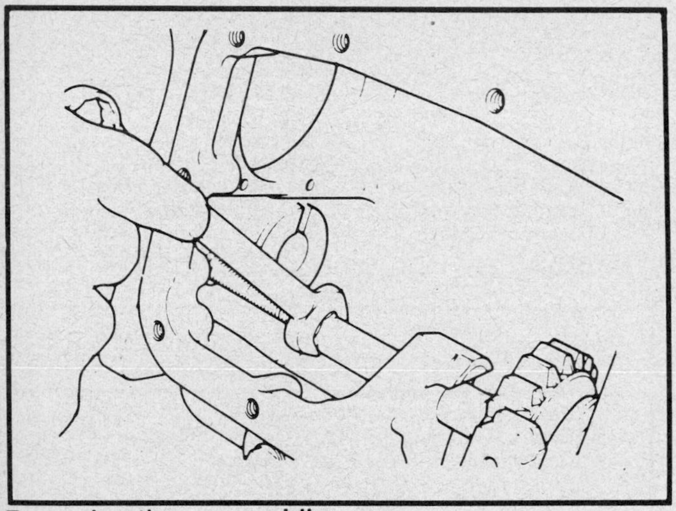
Removing the reverse idler gear

2. Remove the front bearing-to-shaft snapring and the bearing snapring.
3. Using a shop press, press the bearing from the shaft.

NOTE: If the bearing will not slide off the shaft, use a shop press to remove it.

Inspection

1. Using solvent, clean the parts and blow dry with compressed air; do not spin dry the bearings.
2. Check the parts for damage and/or wear; replace them, if necessary.

Assembly

1. Using new Dexron®II fluid, lubricate the input shaft bearing.
2. Using a shop press, press the bearing onto the input shaft until it seats against the shoulder.
3. Install a new front input shaft-to-bearing snapring and a new bearing-to-case snapring.
4. At the rear of the input shaft, lubricate the roller bearings with chassis grease and install them into the shaft and install the rear snapring.

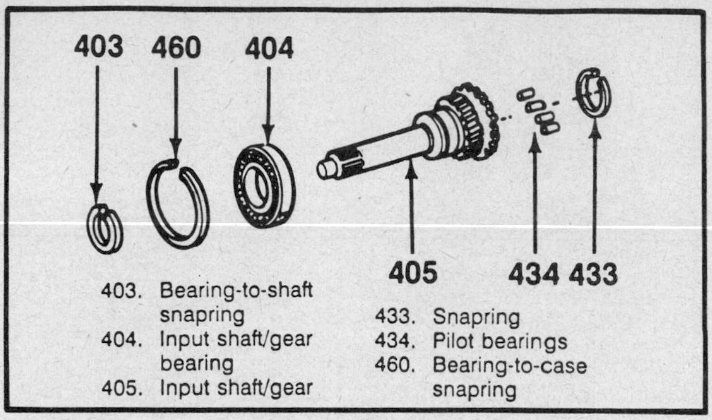
403
460
404
405
434
433
403. Bearing-to-shaft snapring
404. Input shaft/gear bearing
405. Input shaft/gear
433. Snapring
434. Pilot bearings
460. Bearing-to-case snapring
Exploded view of the input shaft/gear assembly

MAINSHAFT ASSEMBLY

Disassembly

1. Slide the 3rd gear blocker ring from the mainshaft.
2. Remove the snapring and slide the 3rd/overdrive synchronizer assembly from the shaft; do not allow the synchronizer to come apart.
3. Remove the overdrive blocker ring and the overdrive gear.
4. At the rear of the mainshaft, press the speedometer drive gear retaining clip down and tap the speedometer from the shaft with a soft hammer.

NOTE: When removing the speedometer drive gear, be careful not to allow the mainshaft to fall on the floor.

5. Remove the rear bearing and the 1st gear by performing the following procedures:
   a. Remove the bearing-to-shaft snapring.
   b. Using a shop press, support the 1st gear and press the mainshaft through the 1st gear/bearing.
6. Slide the 1st gear blocker ring from mainshaft assembly.
7. Remove the snapring and slide the synchronizer assembly from the shaft; do not allow the synchronizer assembly to come apart.
8. Remove the 2nd gear blocker ring and the 2nd gear.
9. Disassemble the synchronizers by performing the following procedures:
   a. Mark the hub and sleeve alignment for reassembly.
   b. While holding the spring and keys, push the hub out of the sleeve.

NOTE: Since the synchronizer hub and sleeve assemblies are a select fit, be careful not to mix the parts.

Inspection

1. Using solvent, clean the parts and blow dry with compressed air; do not spin dry the bearings.
2. Check the gears for cracks, chipped teeth and/or other damage; replace them, if necessary.
3. Check the thrust washers and bushing for damage and/or wear; replace them, if necessary.
4. Check the related gear surfaces and bearing surface diameters; replace them, if necessary.
5. Inspect the reverse sliding gear for a sliding fit on the synchronizer hub without excess circular or radial play. If the gear is not free on the hub, inspect it for burrs on the internal spline ends; use a honing stone to remove the burrs.
6. Inspect the synchronizer sleeves for a sliding fit on the hubs and the hubs for a snug fit on the mainshaft; replace them, if necessary.
7. Inspect the synchronizer spring and keys for looseness and/or damage; replace them, if necessary.

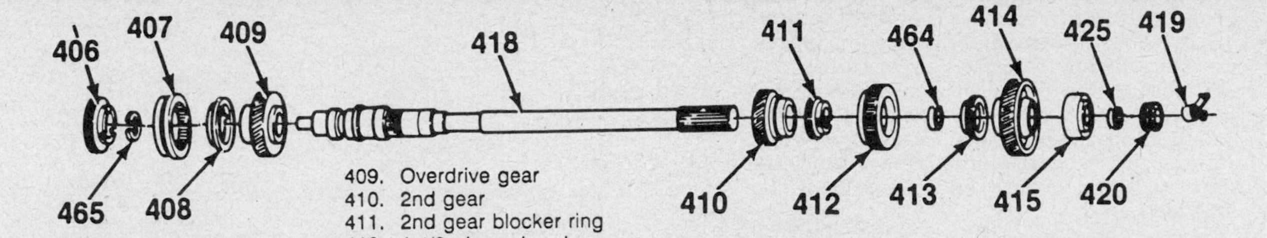
406
407
409
418
411
464
414
425
419
465
408
410
412
413
415
420

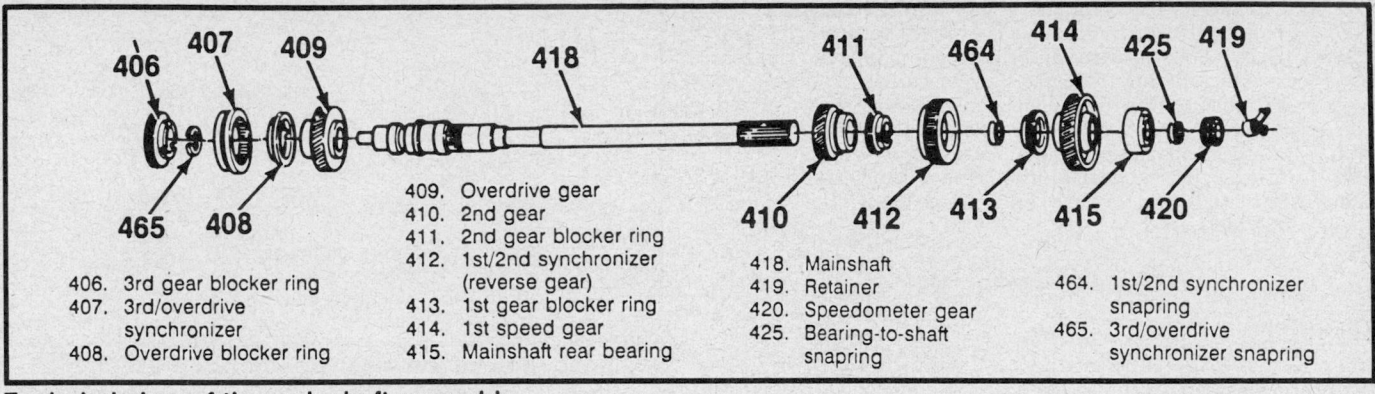

406. 3rd gear blocker ring
407. 3rd/overdrive synchronizer
408. Overdrive blocker ring
409. Overdrive gear
410. 2nd gear
411. 2nd gear blocker ring
412. 1st/2nd synchronizer (reverse gear)
413. 1st gear blocker ring
414. 1st speed gear
415. Mainshaft rear bearing
418. Mainshaft
419. Retainer
420. Speedometer gear
425. Bearing-to-shaft snapring
464. 1st/2nd synchronizer snapring
465. 3rd/overdrive synchronizer snapring

Exploded view of the mainshaft assembly

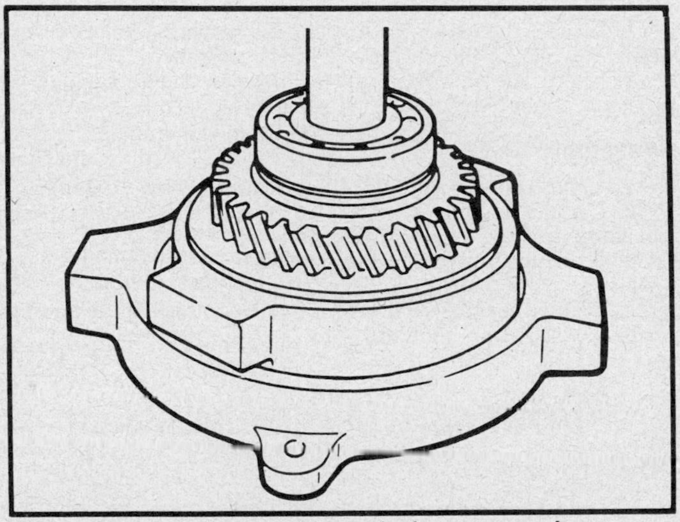

Pressing the mainshaft through the gear and bearing

8. Inspect the brass synchronizer rings for excessive wear and/or damage; replace them, if necessary.
9. Inspect the bearings and bearing surfaces for wear, nicks and/or burrs; replace them, if necessary.
10. Lubricate the roller bearings with new Dexron®II fluid and check for rough rotation; replace them, if necessary.

Assembly

1. Using new Dexron®II fluid, coat all of the parts before assembly.
2. Assemble the synchronizers by performing the following procedures:
   a. Position the keys onto the hub(s).
   b. On both sides of the hub(s), engage the springs so they support the keys.
   c. Align the sleeve(s) with the hub(s) and slide the sleeve(s) onto the hub(s).

NOTE: The long side, of the hub(s) center, must face forward.

3. Position the 2nd gear onto the mainshaft with it's cone surface facing rearward and the blocker ring onto the 2nd gear cone with the clutching teeth facing forward.
4. Install the 1st/2nd synchronizer and snapring onto the mainshaft; be sure the hub slots engage with the blocker ring.
5. Install the 1st gear blocker ring (clutching teeth facing rearward) and engage it with the synchronizer slots. Install the 1st gear onto the mainshaft with it's cone surface facing forward and into the blocker ring.
6. Using a shop press, press the rear bearing onto the mainshaft and install the snapring; be sure to use the correct snapring, it is a select fit used to limit the mainshaft endplay.

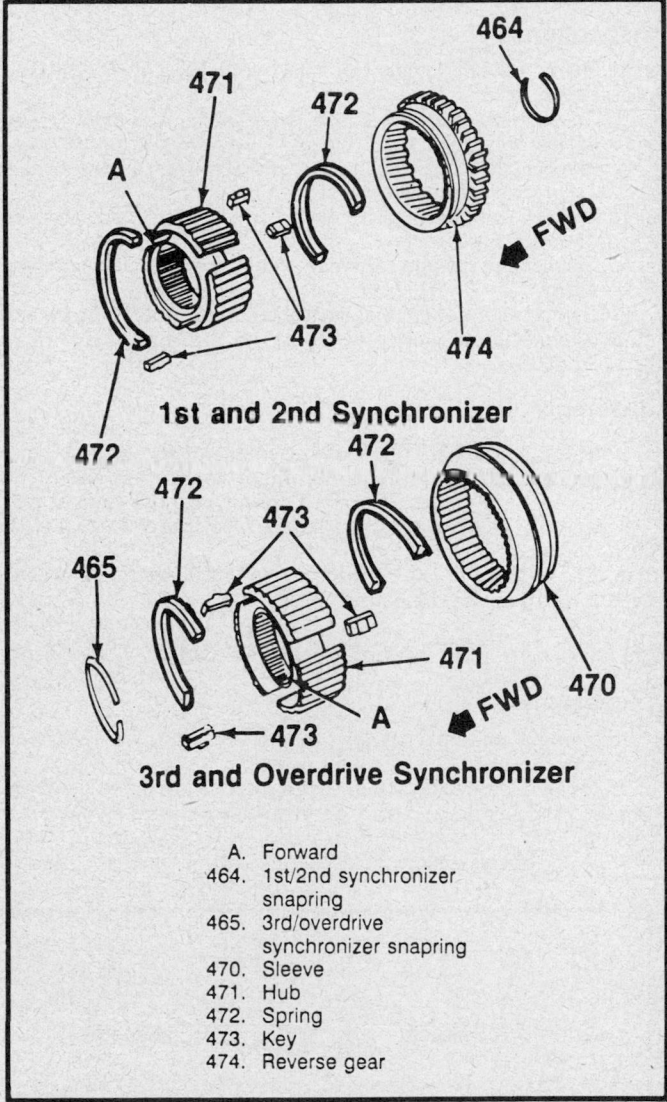

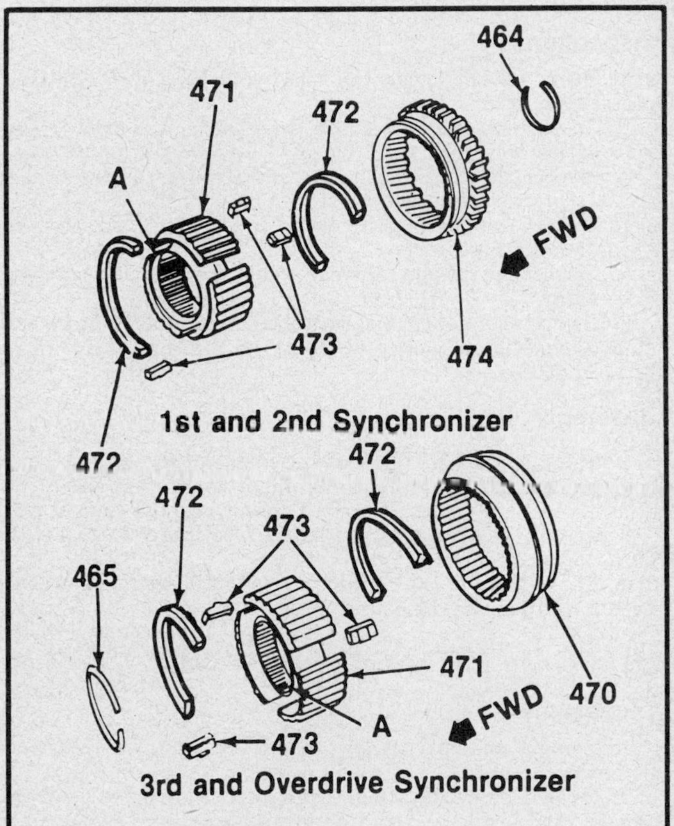
464
471
472
A
FWD
473
474
1st and 2nd Synchronizer
472
472
472
473
465
471
470
A
FWD
473
3rd and Overdrive Synchronizer

A. Forward
464. 1st/2nd synchronizer snapring
465. 3rd/overdrive synchronizer snapring
470. Sleeve
471. Hub
472. Spring
473. Key
474. Reverse gear

Exploded view of the synchronizer assemblies

7. Install the speedometer drive gear and the retainer.
8. At the front of the mainshaft, install the overdrive gear (cone surface facing forward) and the blocker ring onto the gear cone with the clutching teeth facing rearward.
9. Install the 3rd/overdrive synchronizer and a new snapring; the synchronizer's shift fork slot must face rearward and the hub slots must engage with the blocker ring.
10. Lubricate the 3rd gear blocker ring with grease, install it (clutching teeth facing forward) and engage it with the synchronizer's hub slots.

COUNTERGEAR

Disassembly

1. Remove the countershaft alignment tool from the countergear assembly.
2. Remove the bearings, the bearing spacers and the countershaft spacer.

Inspection

1. Using solvent, clean the parts and blow dry with compressed air.
2. Check the countershaft for wear, nicks and/or burrs; replace it, if necessary.
3. Inspect the countergear for nicks, burrs and/or broken teeth; replace it, if necessary.
4. Inspect the bearings for wear and/or rough surfaces; replace them, if necessary.
5. Inspect the spacers for wear and/or grooves; replace them, if necessary.
6. Using feeler gauge, measure the countershaft-to-case bore clearance; the clearance must be no more than 0.005 in. (0.127mm).

Assembly

1. Using chassis grease, coat the inside of the countergear.
2. Using the countershaft alignment tool or equivalent, install the tool and the countershaft spacer into the countergear.
3. Install the bearing and spacers into the countergear; be sure they are centered.
4. At the front of the countergear, install the thrust washer with the tab inside the countergear.

EXTENSION HOUSING

Disassembly

1. Using a small prybar, pry the oil seal from the rear of the extension housing.
2. If the bushing is worn or damaged, use the extension housing bushing remover tool or equivalent, and remove the bushing.
3. Remove the gasket from the mounting surface.

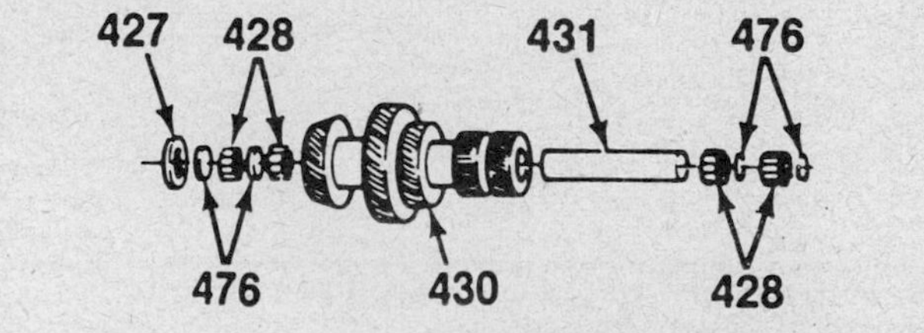
427
428
431
476
476
430
428

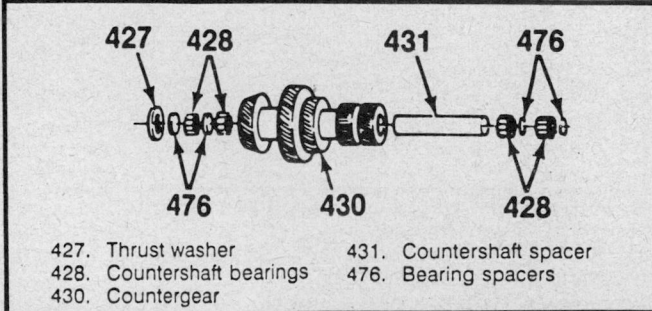

427. Thrust washer
428. Countershaft bearings
430. Countergear
431. Countershaft spacer
476. Bearing spacers

Exploded view of the countergear assembly

478
421
416

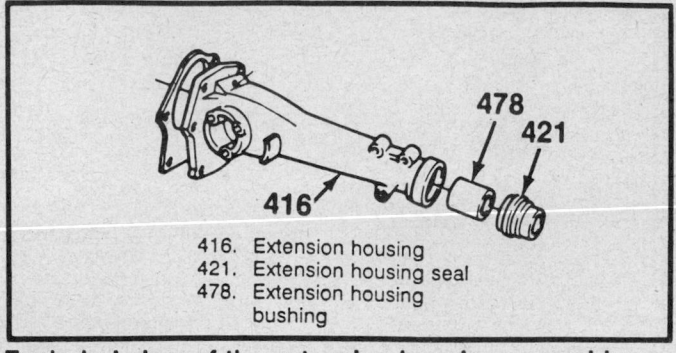

416. Extension housing
421. Extension housing seal
478. Extension housing bushing

Exploded view of the extension housing assembly

Inspection

1. Inspect the extension housing for scoring, wear or cracks, especially at the flange; replace it, if necessary.
2. Check the snapring groove for wear and/or damage; replace it, if necessary.

Assembly

1. Using transmission fluid, lubricate the new bushing.
2. Using a bushing installer tool or equivalent, drive the new bushing into the extension housing.
3. Using locking compound, apply it to the outside of the new seal.
4. Using a seal installer tool or equivalent, drive the new seal into the extension housing.

INPUT GEAR BEARING RETAINER

Disassembly

1. Using a small prybar, pry the oil seal from the input gear bearing retainer.
2. Remove the gasket from mounting surface.

Inspection

1. Inspect the retainer nose for scoring, wear or cracks, especially at the flange; replace it, if necessary.
2. Check the snapring groove for damage; replace it, if necessary.
3. If the retainer is worn or damaged, replace it.

Assembly

1. Using transmission fluid, coat the seal lips.
2. Using the input gear seal installer tool or equivalent, drive the new seal into the bearing retainer.

SIDE COVER

Disassembly

1. Remove the shift levers from the side cover.
2. Remove the shift shafts by performing the following procedures:
   a. If burrs are present, remove them to prevent scoring the cover bores.
   b. Pull the shafts from the cover.
   c. Label the shafts so they may be installed in the same location.
3. Remove the retaining clip, the detent cams and the spring.
4. Using a small prybar, pry the oil seals and the retainers from the side cover.

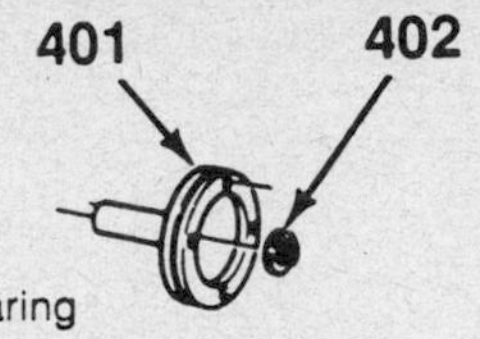
401
402

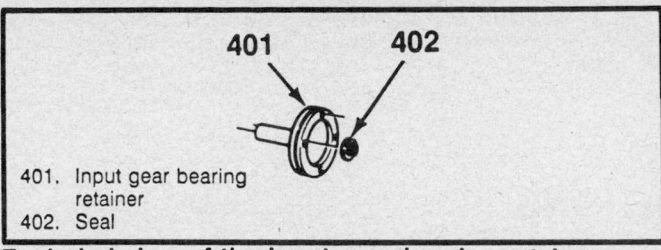

401. Input gear bearing retainer
402. Seal

Exploded view of the input gear bearing retainer

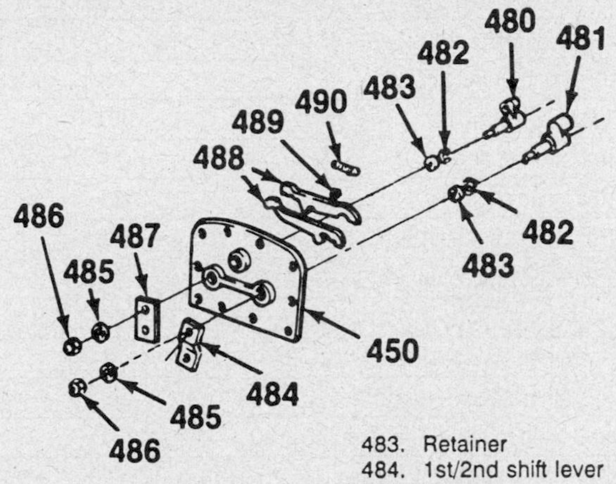
480
481
482
483
490
489
488
486
487
485
482
483
450
484
485
486

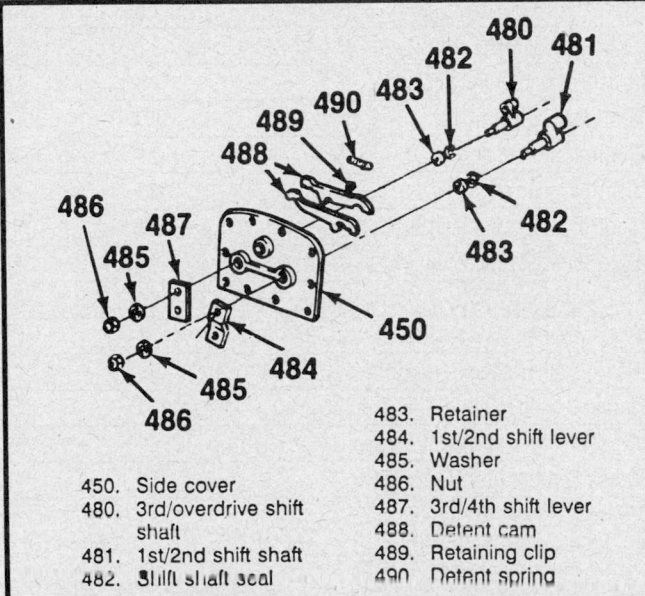

450. Side cover
480. 3rd/overdrive shift shaft
481. 1st/2nd shift shaft
482. Shift shaft seal
483. Retainer
484. 1st/2nd shift lever
485. Washer
486. Nut
487. 3rd/4th shift lever
488. Detent cam
489. Retaining clip
490. Detent spring

Exploded view of the side cover assembly

Inspection

1. Clean all of the parts in solvent and blow dry with compressed air.
2. Inspect the parts for wear and/or damage; replace them, if necessary.
3. Inspect the gasket surfaces for nicks and scratches.
4. Inspect the shift shafts and forks for burrs and/or wear; remove the burrs or replace the shafts, if necessary.
5. Inspect the seal bores for cracks and/or damage.

Assembly

1. Install the detent cams, a new retaining clip and the spring.
2. Using chassis grease, lubricate the cover bores and install the shift shafts into their original positions.
3. Using new seals, install them and the retainers into the side cover.
4. Install the shift levers and torque the nuts to 18 ft. lbs. (25 Nm); the 3rd/overdrive lever must point downward.

Transmission Assembly

1. Using transmission fluid, lubricate all of the parts that are being installed.
2. Lower the countergear assembly into the main case; make sure the front and rear thrust washer tabs are installed in the case slots.
3. Install the mainshaft assembly into the extension housing and secure the bearing with the snapring.
4. Using a new gasket, install the extension housing assembly into the main case, turn the housing 180 degrees and secure it with a bolt.
5. Tap the input gear assembly into the front of the main case until the bearings snapring bottoms against the case; make sure the roller bearings do not fall out during installation to the mainshaft.
6. Install the countershaft by performing the following procedures:
   a. Raise the countergear into mesh with the mainshaft gears; be sure the thrust washers are in place.
   b. From the rear, push the countershaft half way into the countergear and install the Woodruff key.
   c. Push the countershaft into position and remove the countershaft alignment tool or equivalent.
7. Using a new seal, install the reverse shift shaft.
8. Install the reverse idler gear and shaft by performing the following procedures:
   a. Insert the reverse idler shaft part way.
   b. Position the reverse idler gear onto the shaft with the fork slot facing rearward and engage the slot with the reverse shift fork.
   c. Install the Woodruff key and drive the shaft flush with the case.
9. Remove the extension housing bolt, rotate the housing to the normal position, install the extension housing-to-main case bolts and torque the bolts to 50 ft. lbs. (68 Nm); be careful not to damage the gasket.
10. Using a new gasket install the input gear bearing retainer and torque the bolts to 30 ft. lbs. (41 Nm).
11. Using a new countershaft plug, coat it with sealant and drive it into the front of the main case.
12. Install the 1st/2nd shift fork into the synchronizer slot and the 3rd/overdrive shift fork into the side cover.
13. Install the side cover by performing the following procedures:
   a. Position a new gasket onto the main case. Install the detent ball and spring.
   b. Position the synchronizer sleeves and the reverse idler gear in the **N** positions.
   c. Position the shift levers into the **N** (straight up) position.
   d. While holding the detent cam against the 1st/2nd shift lever, lift the cam over the fork to install the cover; be sure the detent ball and spring are in position.
   e. When installing the side cover bolts, install the side cover locating (long shoulder) bolt in the rear center hole; tighten the bolts finger tight. Torque the side cover-to-main case bolts to 15 ft. lbs. (20 Nm).

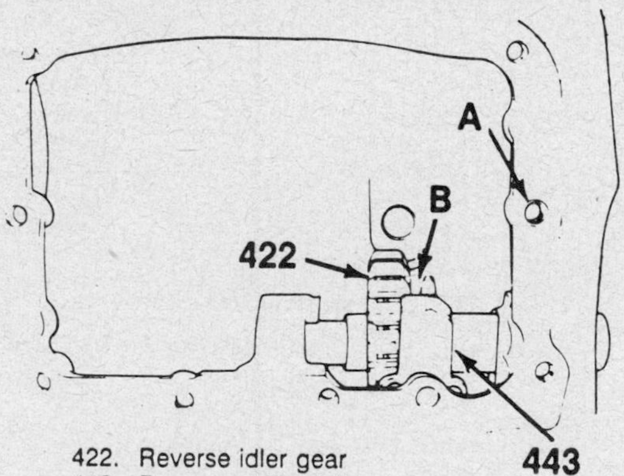
A
B
422
443

422. Reverse idler gear
443. Reverse shift shaft
A. Locating hole
B. Gear slot

View of the reverse idler gear in the main case

14. Install the reverse shift lever and torque the nut to 18 ft. lbs. (25 Nm).
15. Shift the transmission through all the gears and check the operation.
16. Install the speedometer driven gear, the switches (if equipped) and any other external components that were removed.

SPECIFICATIONS

TRANSMISSION SPECIFICATIONS

Transmission	
Make	New Process
Case material	Cast iron/alum.
Torque rating (ft. lb.)	260
Ratio (:1)	
1st gear	3.09
2nd gear	1.67
3rd gear	1.00
4th gear	0.73
5th gear	—
Reverse	3.00
Shafts center distance	3.5 in. (89 mm)
Clutch plate diameter	12.0 in. (302 mm)

TORQUE SPECIFICATIONS

	Ft. Lbs.	Nm
Drain and fill plug	17	23
Cross lever bracket screw	18	25
Shift rod swivel screw	18	25
Control bracket screw	24	33
Control mounting screw	33	45
Control lubrication screw	33	45
Side cover screw	15	20
U-Joint flange nut	100	135
Rear retainer screw—top	20	27
Rear retainer screw—bottom	30	40
Transmission to clutch housing screw	74	100
Crossmember to mount screw	40	54
Crossmember to frame screw	55	75

SPECIAL TOOLS

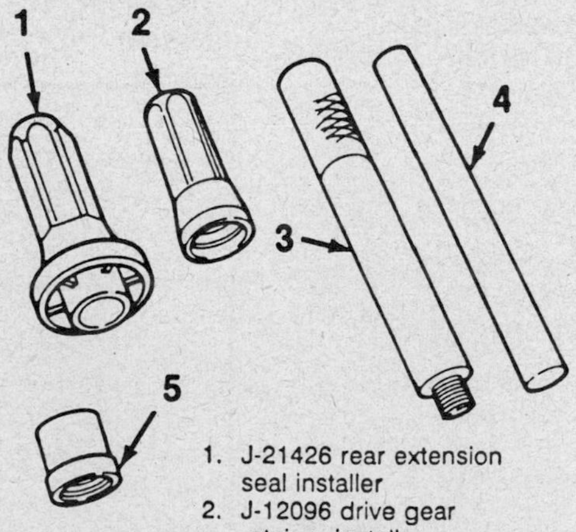
1
2
3
4
5

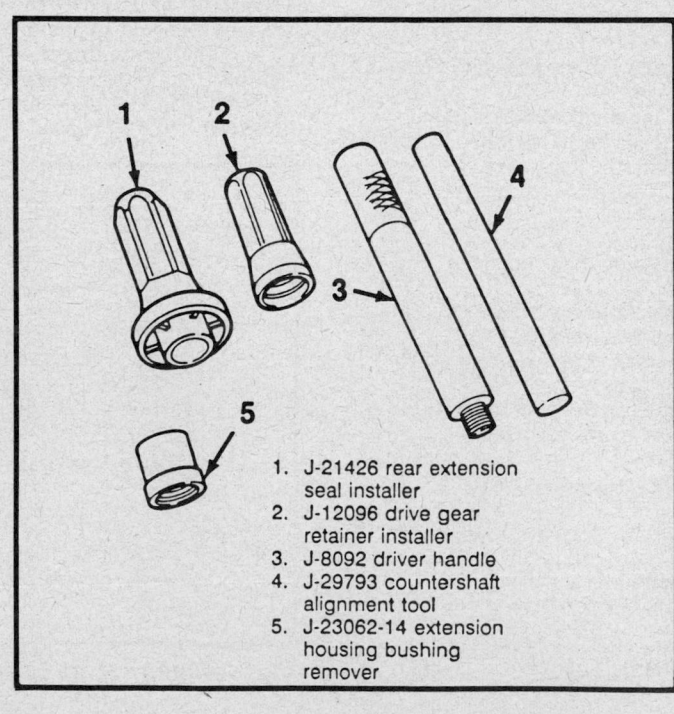

1. J-21426 rear extension seal installer
2. J-12096 drive gear retainer installer
3. J-8092 driver handle
4. J-29793 countershaft alignment tool
5. J-23062-14 extension housing bushing remover

Section 7

Borg Warner Transmission
General Motors

APPLICATION

TRANSMISSION APPLICATION

Year	Vehicle	Engine
	M77 TRANSMISSION	
1984	S-10, S-15,	All
	Blazer, Jimmy	All
	MC9 TRANSMISSION	
1985–88	S-10, S-15	All
	Blazer, Jimmy	All
	Astro, Safari	All
	MB1 TRANSMISSION	
1984	S-10, S-15	All
	Blazer, Jimmy	All
	MB6 TRANSMISSION	
1984	S-10, S-15	All
	Blazer, Jimmy	All

Year	Vehicle	Engine
	MH3 TRANSMISSION	
1987–89	S10, S-15	All
	Blazer, Jimmy	All
	Astro, Safari	All
	ML2 TRANSMISSION	
1986–89	S-10, S-15	All
	Blazer, Jimmy	All
	Astro, Safari	All
	ML3 TRANSMISSION	
1986–89	S-10, S-15	All
	Blazer, Jimmy	All
	Astro, Safari	All

GENERAL DESCRIPTION

The 4 and 5 speed 77mm transmissions are fully synchronized units with blocker ring synchronizers and a sliding mesh reverse gear. They have an aluminum transmission case that house the various gears and bearings and an extension housing. The gearshift lever assembly is floor mounted and is located on top of the extension housing. The shift mechanism does not require adjustment and can be serviced independently of the transmission.

Metric Fasteners

The metric fastener dimensions are very close to the dimensions of the familiar inch system fasteners and, for this reason, replacement fasteners must have the same measurement and strength as those removed.

Do not attempt to interchange metric fasteners for inch system fasteners. Mismatched or incorrect fasteners can result in damage to the transaxle unit through malfunctions, breakage or possible personal injury.

Care should be taken to reuse the fasteners in the same locations as removed.

Capacities

To fill a dry the 4 or 5 speed transmission, fill the unit with 2 quarts (2.0L) of Dexron® II automatic transmission fluid.

Checking Fluid Level

To check the fluid level in the transmission, remove the filler plug and check that the fluid is level with the bottom of the filler plug hole.

TRANSMISSION MODIFICATIONS

There are no modifications on the 4 or 5 speed 77mm transmission at the time of this publication printing.

TROUBLE DIAGNOSIS

CHILTON'S THREE C'S TRANSMISSION DIAGNOSIS

Condition	Cause	Correction
Hard shifting	a) Clutch b) Worn or broken synchronizers c) Worn shift shafts or forks	a) Adjustment or malfunctioning b) Replace c) Replace
Slips out of gear	a) Worn shift shafts b) Worn bearings c) Broken or loose drive gear retainer d) Excessive play in synchronizers	a) Replace b) Replace as necessary c) Tighten or replace retainer d) Replace

CHILTON'S THREE C'S TRANSMISSION DIAGNOSIS

Condition	Cause	Correction
Noisy in all gears	a) Insufficient lubricant b) Worn countershaft bearings c) Worn or damaged drive gear and countergear d) Damaged drive gear or mainshaft e) Worn or damaged countergear	a) Add to correct level b) Replace countershaft bearings and shaft c) Replace worn or damaged gears d) Replace damaged drive gear or bearings e) Replace countergear
Noisy in neutral	a) Damaged drive gear bearing b) Damaged or loose pilot bearing c) Worn or damaged countergear d) Worn countergear bearings	a) Replace damaged bearing b) Replace pilot bearing c) Replace countergear d) Replace countergear bearings and shaft
Noisy in reverse	a) Worn or damaged reverse idler gear or idler bushing b) Worn or damaged reverse gear c) Damaged or worn countergear	a) Replace reverse idle gear assembly b) Replace reverse gear c) Replace countergear assembly
Leaks lubricant	a) Excessive amount of lubricant in transmission b) Loose or broken drive gear bearing retainer c) Damaged drive gear bearing retainer gasket d) Center support gaskets either side e) Rear extension seal f) Speedometer driven gear	a) Drain to correct level b) Tighten or replace retainer c) Replace gasket d) Replace gaskets e) Replace f) Replace O-ring seal

ON CAR SERVICE

Services

SHIFT CONTROL LEVER

Removal and Installation

1. Remove the screws from the transmission shift lever boot retainer and slide the boot up the lever.
2. Remove the shift lever attaching bolts at the transmission and remove the lever.
3. To install, position the lever onto the transmission and install the attaching bolts. Torque the bolts to 10 ft. lbs. (13 Nm).
4. Slide the boot down the lever and install the shift lever boot retainer screws.

REMOVAL AND INSTALLATION

TRANSMISSION REMOVAL

S–10, S–15, BLAZER AND JIMMY WITHOUT 4WD

1. Disconnect the negative battery cable.
2. Remove the shift lever boot attaching screws and slide the boot up the shift lever.
3. Shift the transmission into **N** and remove the shift lever.
4. Raise and support the vehicle safely.
5. Remove the driveshaft.
6. Disconnect the speedometer cable and electrical connectors at the transmission.
7. Disconnect the slave cylinder from the transmission.
8. Support the transmission and remove the transmission mount attaching bolts.
9. Remove the catalytic converter hanger.
10. Remove the crossmember attaching bolts and crossmember.
11. Remove the dust cover bolts.
12. Remove the transmission-to-engine attaching bolts and remove the transmission.

S–10, S–15, BLAZER AND JIMMY WITH 4WD

1. Shift the transfer case into **4H**.
2. Disconnect the negative battery cable.
3. Raise and support the vehicle safely. Remove the skid plate.
4. Drain the lubricant from the transfer case.
5. Mark the transfer case front output shaft yoke and driveshaft for assembly reference. Disconnect the front driveshaft from the transfer case.
6. Mark the rear axle yoke and driveshaft for assembly reference. Remove the rear driveshaft.
7. Disconnect the speedometer cable and vacuum harness at the transfer case. Remove the shift lever from the transfer case.
8. Remove the catalytic converter hanger bolts at the converter.
9. Raise the transmission and transfer case and remove the transmission mount attaching bolts. Remove the mount and catalytic converter hanger and lower the transmission and transfer case.
10. Support the transmission and transfer case and remove from the vehicle.

11. Separate the transfer case from the extension housing of the transmission.

ASTRO AND SAFARI

1. Disconnect the negative battery cable.
2. Remove the shift lever boot attaching screws and slide the boot up the shift lever.
3. Shift the transmission into **N** and remove the shift lever.
4. Raise and support the vehicle safely.
5. Remove the driveshaft.
6. Disconnect the speedometer cable and electrical connectors at the transmission.
7. Support the transmission and remove the transmission mount attaching bolts.
8. Remove the crossmember attaching bolts and crossmember.
9. Remove the transmission support braces.
10. Remove the transmission-to-engine attaching bolts and remove the transmission.

TRANSMISSION INSTALLATION

S–10, S–15, BLAZER AND JIMMY WITHOUT 4WD

1. Install the transmission and transmission-to-engine attaching bolts. Torque bolts to specification.
2. Install the dust cover bolts.
3. Install the crossmember and crossmember attaching bolts. Torque the bolts to 25 ft. lbs. (30 Nm).
4. Install the catalytic converter hanger.
5. Install the transmission mount attaching bolts and torque to 35 ft. lbs. (50 Nm).
6. Connect the slave cylinder to the transmission.
7. Connect the speedometer cable and electrical connectors at the transmission.
8. Install the driveshaft.
9. Lower the vehicle safely.
10. Install the shift lever.
11. Slide the boot down the shift lever and install the shift lever boot attaching screws.
12. Connect the negative battery cable.

S–10, S–15, BLAZER AND JIMMY WITH 4WD

1. Install the transfer case to the extension housing of the transmission. Install attaching bolts and torque to 19–29 ft. lbs. (26–40 Nm).
2. Install the transmission and transfer case into the vehicle. Install the attaching bolts and torque to specification.
3. Raise the transmission and transfer case and install the transmission mount and mount attaching bolts. Torque bolts to 35 ft. lbs. (50 Nm). Install the catalytic converter hanger and lower the transmission and transfer case.
4. Install the catalytic converter hanger bolts at the converter.
5. Install the shift lever to the transfer case. Connect the speedometer cable and vacuum harness at the transfer case.
6. Install the rear driveshaft. Note the mark on the rear axle yoke and driveshaft.
7. Connect the front driveshaft to the transfer case. Note the mark on the transfer case front output shaft yoke and driveshaft.
8. Fill the transmission and transfer case with the specified lubricant to the proper level.
9. Install the skid plate. Lower the vehicle.
10. Connect the negative battery cable.

ASTRO AND SAFARI

1. Install the transmission and transmission-to-engine attaching bolts. Torque bolts to specification.
2. Install the transmission support braces.
3. Install the crossmember and crossmember attaching bolts. Torque the bolts to 25 ft. lbs. (30 Nm).
4. Install the transmission mount attaching bolts and torque to 35 ft. lbs. (50 Nm).
5. Connect the speedometer cable and electrical connectors at the transmission.
6. Install the driveshaft.
7. Lower the vehicle safely.
8. Install the shift lever to the transmission.
9. Slide the boot down the shift lever and install the shift lever boot attaching screws.
10. Connect the negative battery cable.

BENCH OVERHAUL

Before Disassembly

Cleanliness is an important factor in the overhaul of the transmission. Before attempting any disassembly operation, the exterior of the transmission should be thoroughly cleaned to prevent the possibility of dirt entering the transmission internal mechanism. During inspection and reassembly, all parts should be thoroughly cleaned with cleaning fluid and then air dried. Wiping cloths or rags should not be used to dry parts. All oil passages should be blown out and checked to make sure that they are not obstructed. Small passages should be checked with tag wire. All parts should be inspected to determine which parts are to be replaced.

Transmission Disassembly

FOUR SPEED

1. Remove drain plug and drain lubricant from transmission.
2. Thoroughly clean the exterior of the transmission assembly.
3. Using a hammer and punch, remove the roll pin that retains the offset lever to shift rail.
4. Remove extension housing attaching bolts. Separate the extension housing from the transmission case and remove housing and offset lever as an assembly.
5. Remove detent ball and spring from offset lever and remove roll pin from extension housing or offset lever.
6. Remove transmission shift cover attaching bolts. Pry the shift cover loose using the proper tool and remove cover from transmission case.
7. Remove clip that retains reverse lever to reverse lever pivot bolt.
8. Remove reverse lever pivot bolt and remove reverse lever and fork as an assembly.
9. Using a hammer and punch, mark position of front bearing cap to transmission case. Remove front bearing cap bolts and remove bearing cap.
10. Remove small retaining and large locating snaprings from front drive gear bearing.
11. Install bearing puller J–22912–01 or equivalent, on front bearing and puller J–8433–1 or equivalent, with 2 bolts on end of drive gear and remove and discard bearing. A new bearing must be used when assembling the transmission.
12. Remove retaining and locating snaprings from rear bearing and mainshaft. Install puller J–22912–01 or equivalent, on bearing and puller J–8433–1 or equivalent, with 2 bolts (J–

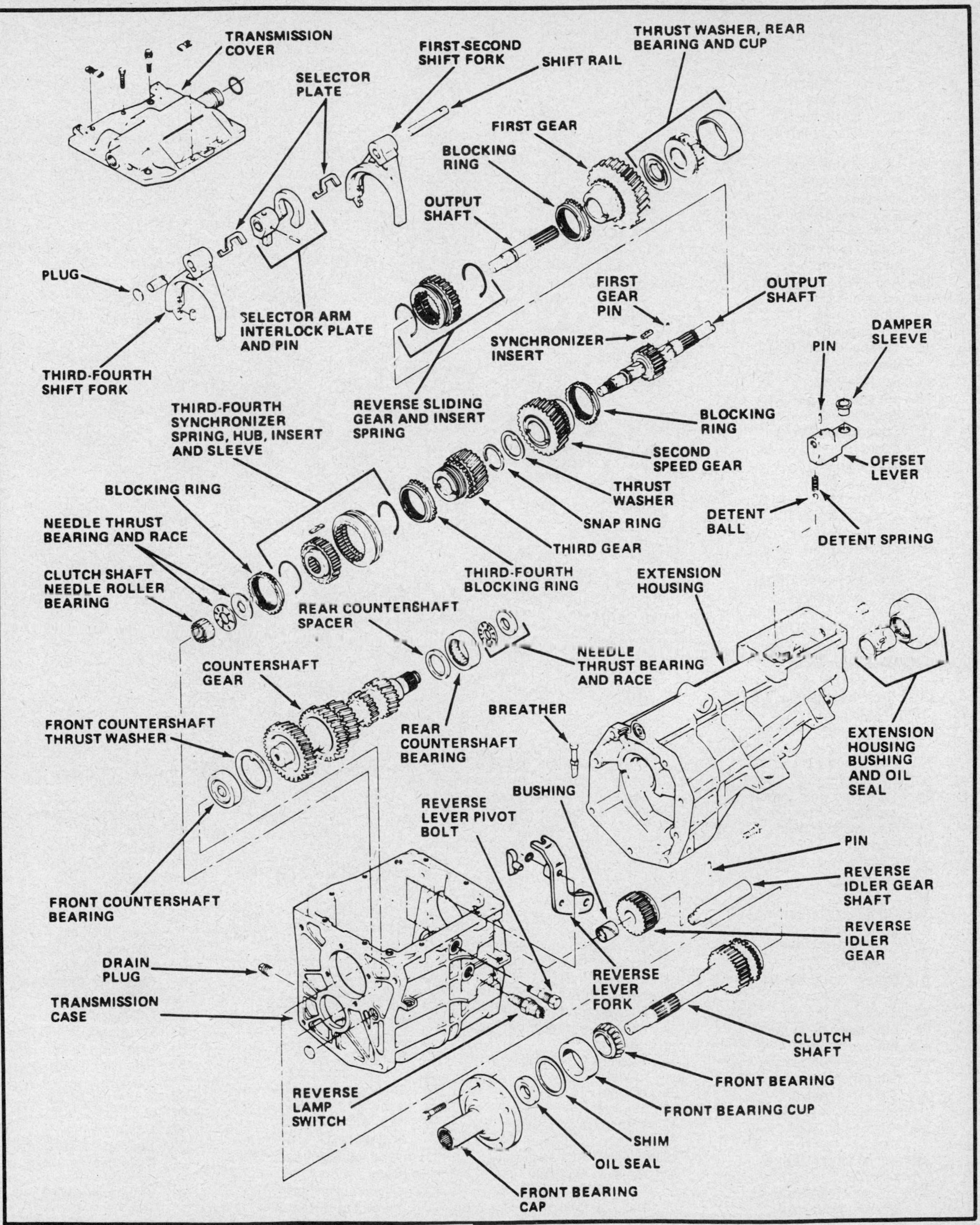
TRANSMISSION COVER
FIRST-SECOND SHIFT FORK
SHIFT RAIL
THRUST WASHER, REAR BEARING AND CUP
SELECTOR PLATE
FIRST GEAR
BLOCKING RING
OUTPUT SHAFT
PLUG
FIRST GEAR PIN
OUTPUT SHAFT
SELECTOR ARM INTERLOCK PLATE AND PIN
DAMPER SLEEVE
SYNCHRONIZER INSERT
PIN
THIRD-FOURTH SHIFT FORK
THIRD-FOURTH SYNCHRONIZER SPRING, HUB, INSERT AND SLEEVE
REVERSE SLIDING GEAR AND INSERT SPRING
BLOCKING RING
SECOND SPEED GEAR
OFFSET LEVER
THRUST WASHER
BLOCKING RING
DETENT BALL
SNAP RING
NEEDLE THRUST BEARING AND RACE
DETENT SPRING
THIRD GEAR
CLUTCH SHAFT NEEDLE ROLLER BEARING
THIRD-FOURTH BLOCKING RING
EXTENSION HOUSING
REAR COUNTERSHAFT SPACER
NEEDLE THRUST BEARING AND RACE
COUNTERSHAFT GEAR
BREATHER
FRONT COUNTERSHAFT THRUST WASHER
REAR COUNTERSHAFT BEARING
EXTENSION HOUSING BUSHING AND OIL SEAL
BUSHING
REVERSE LEVER PIVOT BOLT
PIN
REVERSE IDLER GEAR SHAFT
FRONT COUNTERSHAFT BEARING
REVERSE IDLER GEAR
DRAIN PLUG
REVERSE LEVER FORK
TRANSMISSION CASE
CLUTCH SHAFT
FRONT BEARING
REVERSE LAMP SWITCH
FRONT BEARING CUP
SHIM
OIL SEAL
FRONT BEARING CAP

Transmission exploded view—4 speed

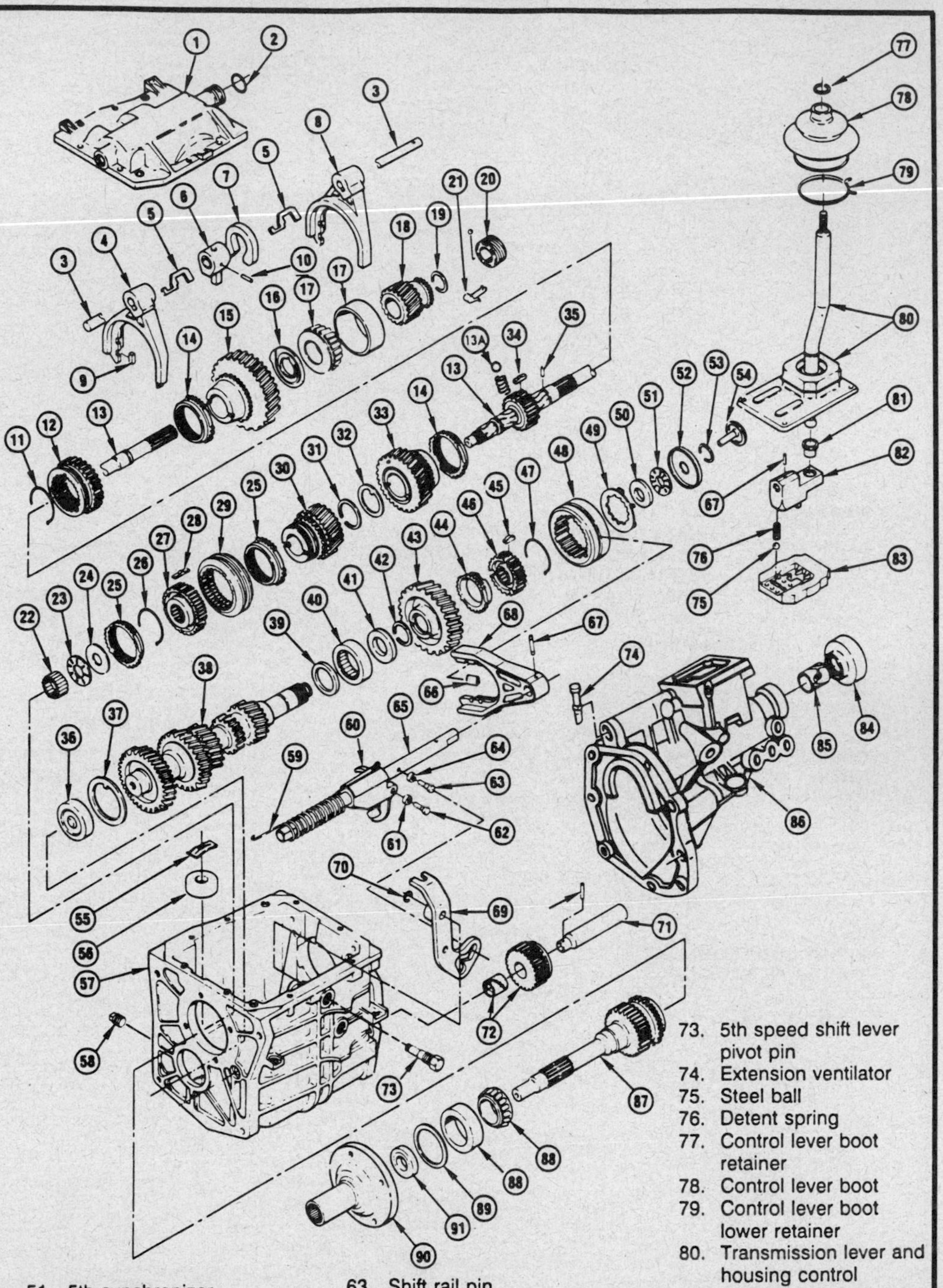

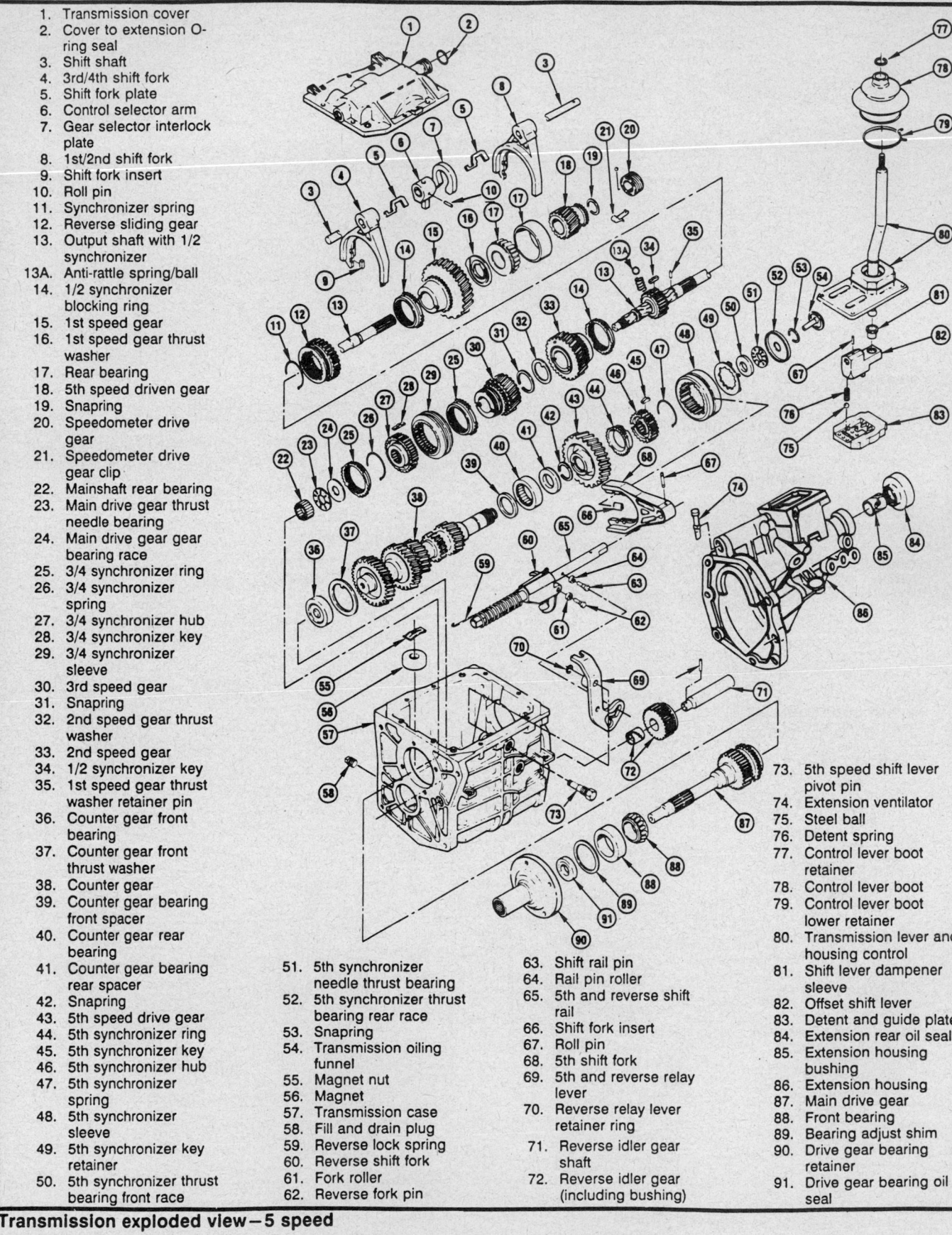

1. Transmission cover
2. Cover to extension O-ring seal
3. Shift shaft
4. 3rd/4th shift fork
5. Shift fork plate
6. Control selector arm
7. Gear selector interlock plate
8. 1st/2nd shift fork
9. Shift fork insert
10. Roll pin
11. Synchronizer spring
12. Reverse sliding gear
13. Output shaft with 1/2 synchronizer
13A. Anti-rattle spring/ball
14. 1/2 synchronizer blocking ring
15. 1st speed gear
16. 1st speed gear thrust washer
17. Rear bearing
18. 5th speed driven gear
19. Snapring
20. Speedometer drive gear
21. Speedometer drive gear clip
22. Mainshaft rear bearing
23. Main drive gear thrust needle bearing
24. Main drive gear gear bearing race
25. 3/4 synchronizer ring
26. 3/4 synchronizer spring
27. 3/4 synchronizer hub
28. 3/4 synchronizer key
29. 3/4 synchronizer sleeve
30. 3rd speed gear
31. Snapring
32. 2nd speed gear thrust washer
33. 2nd speed gear
34. 1/2 synchronizer key
35. 1st speed gear thrust washer retainer pin
36. Counter gear front bearing
37. Counter gear front thrust washer
38. Counter gear
39. Counter gear bearing front spacer
40. Counter gear rear bearing
41. Counter gear bearing rear spacer
42. Snapring
43. 5th speed drive gear
44. 5th synchronizer ring
45. 5th synchronizer key
46. 5th synchronizer hub
47. 5th synchronizer spring
48. 5th synchronizer sleeve
49. 5th synchronizer key retainer
50. 5th synchronizer thrust bearing front race
51. 5th synchronizer needle thrust bearing
52. 5th synchronizer thrust bearing rear race
53. Snapring
54. Transmission oiling funnel
55. Magnet nut
56. Magnet
57. Transmission case
58. Fill and drain plug
59. Reverse lock spring
60. Reverse shift fork
61. Fork roller
62. Reverse fork pin
63. Shift rail pin
64. Rail pin roller
65. 5th and reverse shift rail
66. Shift fork insert
67. Roll pin
68. 5th shift fork
69. 5th and reverse relay lever
70. Reverse relay lever retainer ring
71. Reverse idler gear shaft
72. Reverse idler gear (including bushing)
73. 5th speed shift lever pivot pin
74. Extension ventilator
75. Steel ball
76. Detent spring
77. Control lever boot retainer
78. Control lever boot
79. Control lever boot lower retainer
80. Transmission lever and housing control
81. Shift lever dampener sleeve
82. Offset shift lever
83. Detent and guide plate
84. Extension rear oil seal
85. Extension housing bushing
86. Extension housing
87. Main drive gear
88. Front bearing
89. Bearing adjust shim
90. Drive gear bearing retainer
91. Drive gear bearing oil seal

Transmission exploded view—5 speed

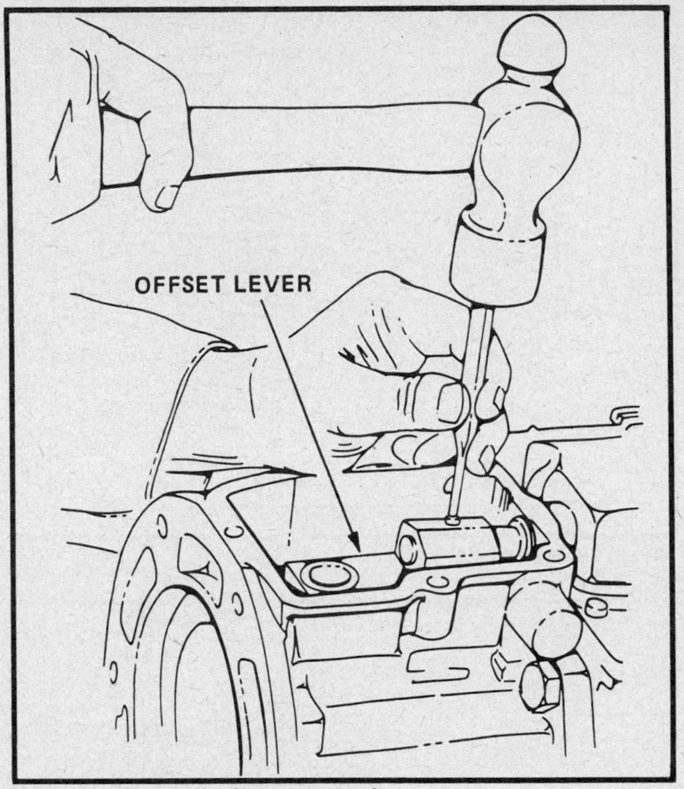
OFFSET LEVER
Offset lever roll pin—4 speed

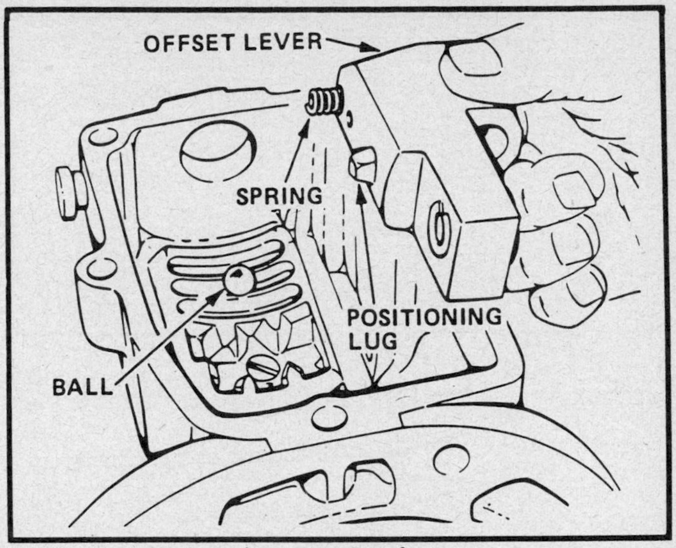
OFFSET LEVER
SPRING
POSITIONING LUG
BALL
Detent ball and spring—4 speed

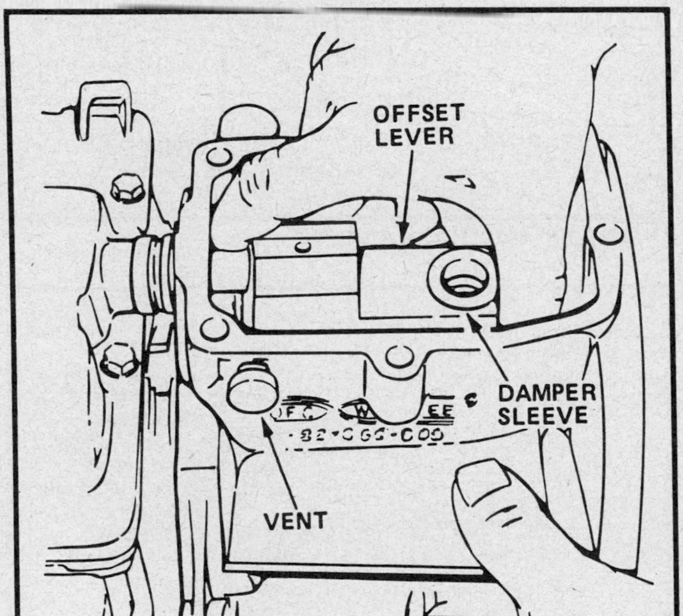
OFFSET LEVER
DAMPER SLEEVE
VENT
Extension housing to transmission case—4 speed

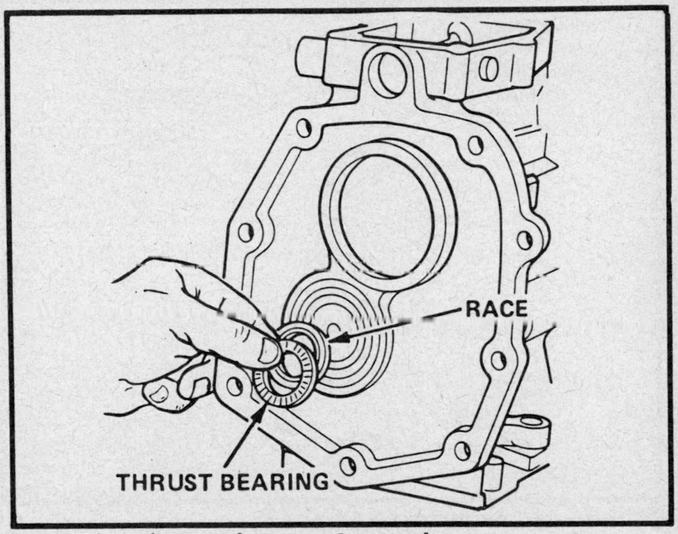
RACE
THRUST BEARING
Thrust bearing and race—4 speed

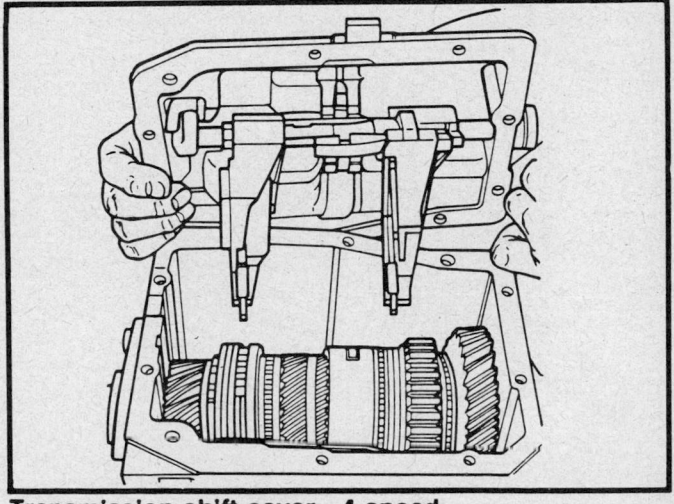

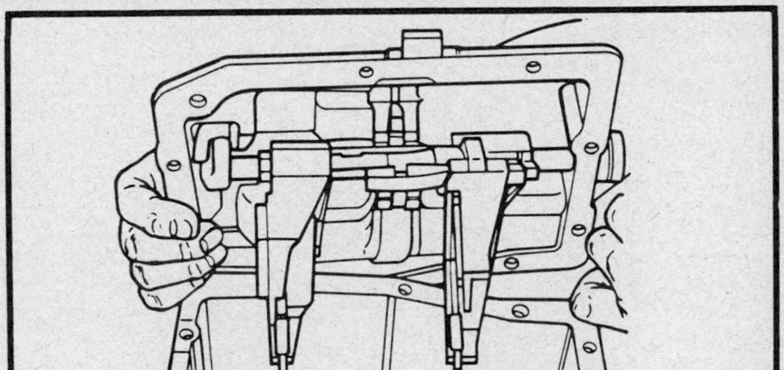
Transmission shift cover—4 speed

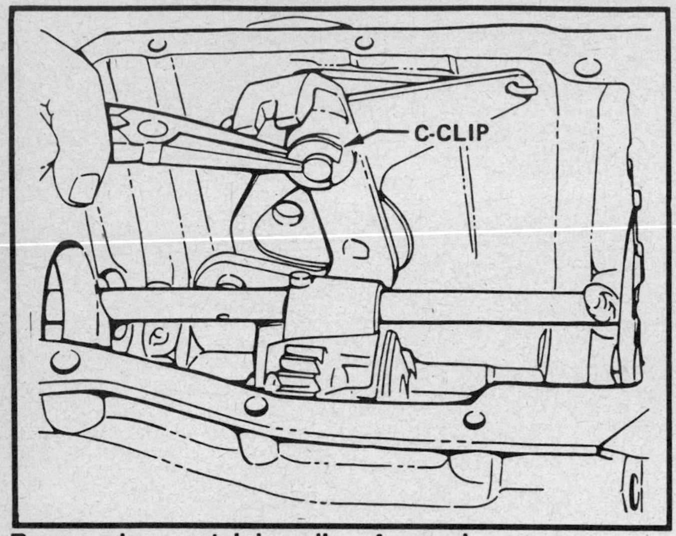
C-CLIP
Reverse lever retaining clip—4 speed

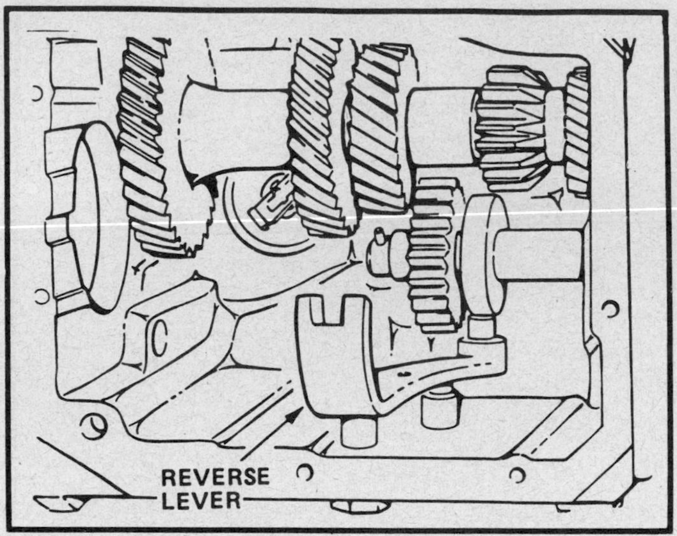
REVERSE
LEVER
Reverse lever and shift fork—4 speed

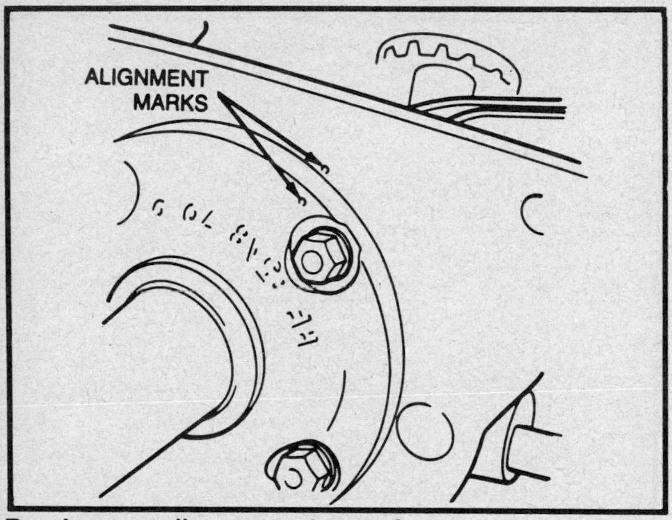
ALIGNMENT
MARKS
Bearing cap alignment—4 speed

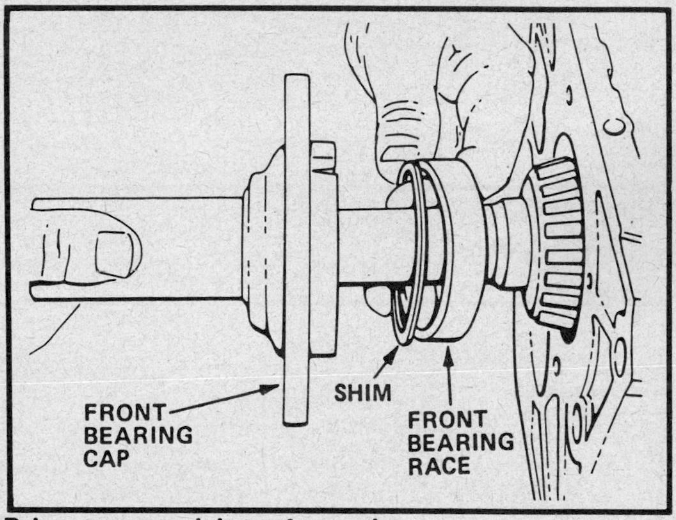
FRONT
BEARING
CAP
SHIM
FRONT
BEARING
RACE
Drive gear servicing—4 speed

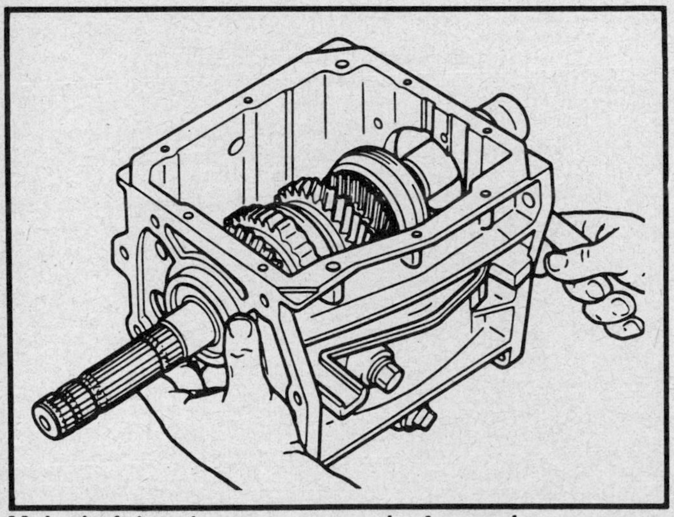
Mainshaft bearing race removal—4 speed

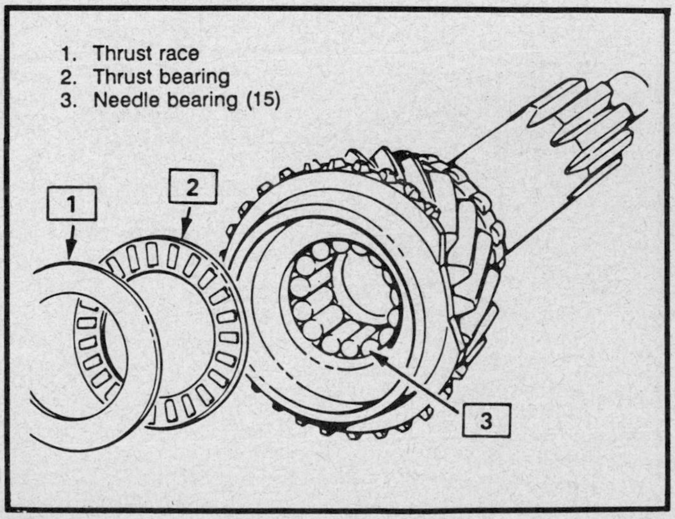
1. Thrust race
2. Thrust bearing
3. Needle bearing (15)
1
2
3
Drive gear roller bearing, thrust bearing and race—4 speed

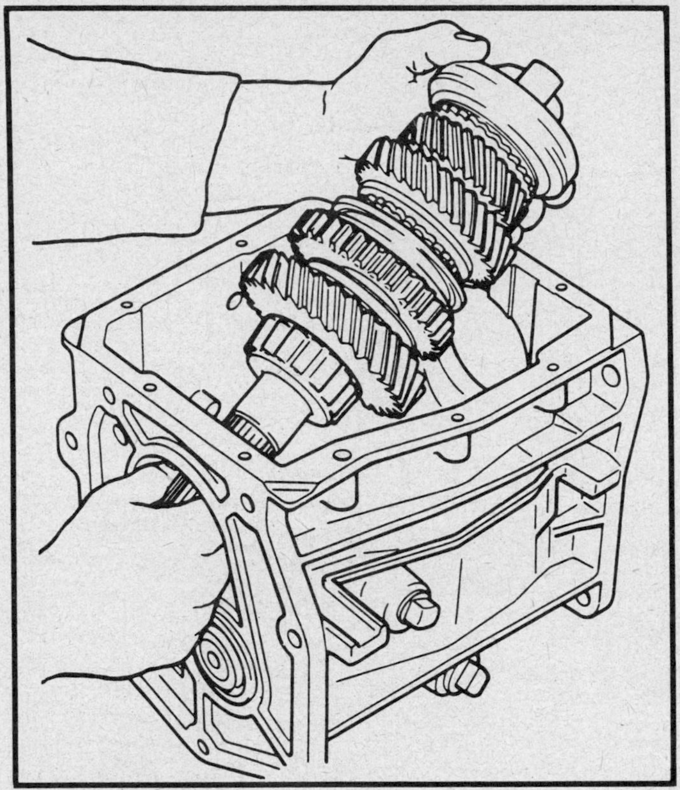

Mainshaft servicing—4 speed

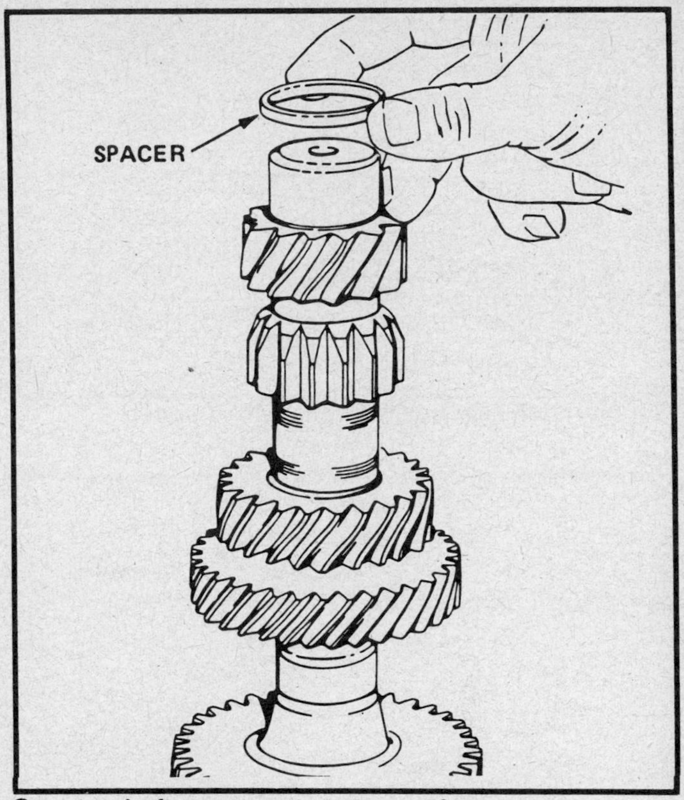
SPACER

Countershaft rear spacer—4 speed

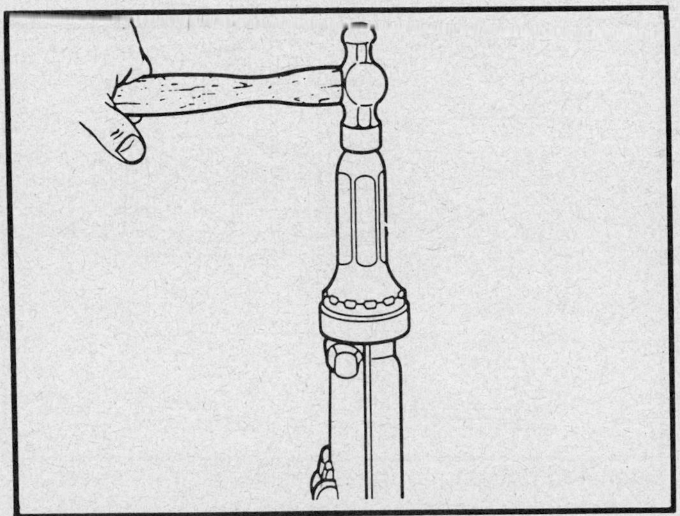

Extension housing seal servicing—4 speed

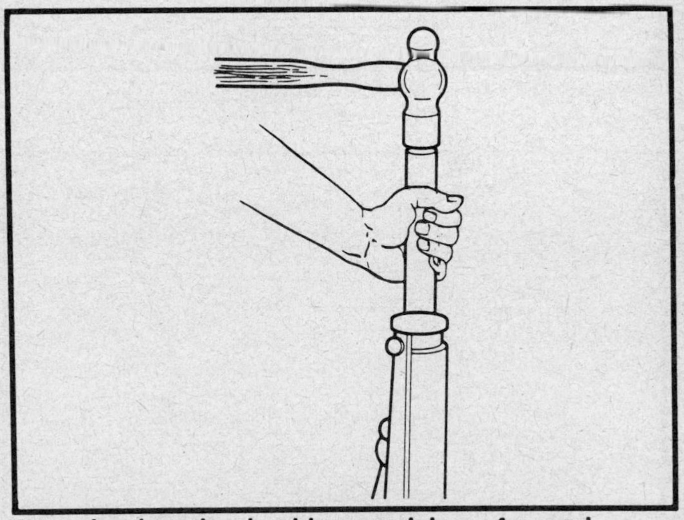

Extension housing bushing servicing—4 speed

33171 or equivalent) on end of mainshaft and remove and discard used bearing. A new bearing must be used when assembling transmission.

13. Remove drive gear from mainshaft and transmission case.
14. Remove mainshaft from transmission case by tipping mainshaft down at the rear and lifting shaft out through shift cover opening.
15. Using a hammer and punch, remove roll pin retaining reverse idler gear shift in transmission case. Remove idler gear and shaft from case.
16. Remove countershaft from rear of case using loading tool J-26624. Remove countershaft gear and loading tool as an assembly from case along with thrust washers.

FIVE SPEED

1. Remove drain plug on transmission case and drain lubricant.
2. Thoroughly clean the exterior of the transmission assembly.
3. Using pin punch and hammer, remove roll pin attaching offset lever to shift rail.

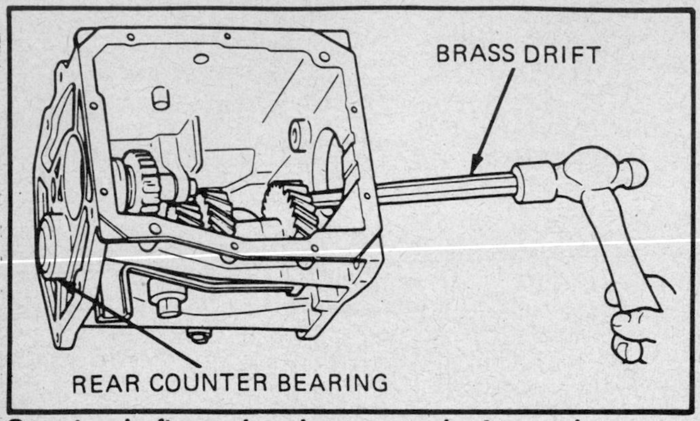
BRASS DRIFT
REAR COUNTER BEARING
Countershaft rear bearing removal—4 speed

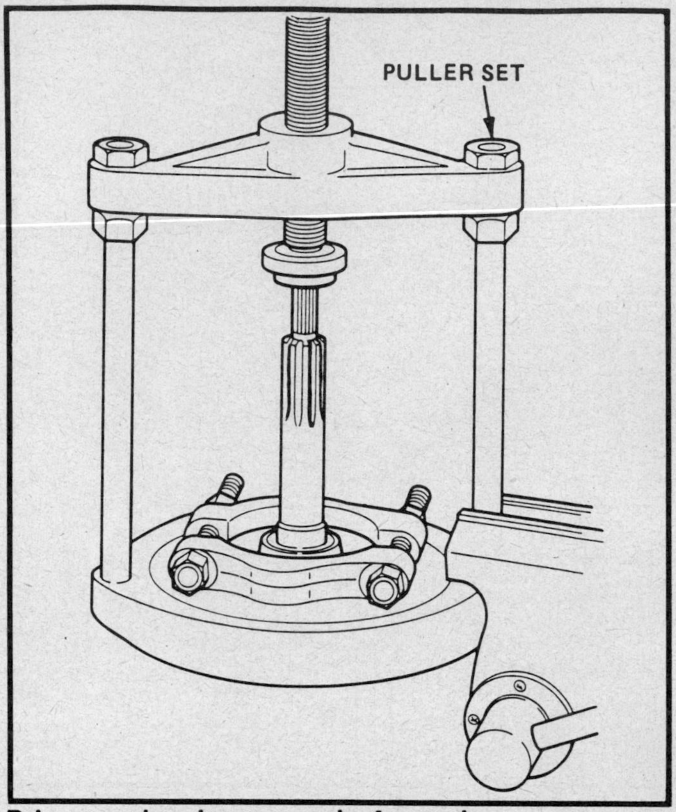
PULLER SET
Drive gear bearing removal—4 speed

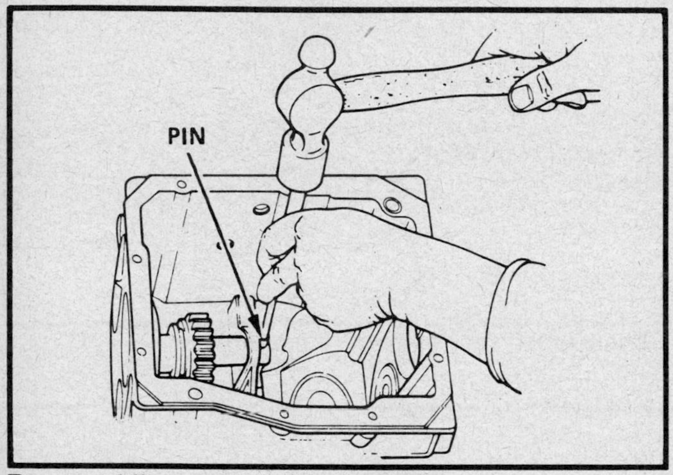
PIN
Reverse idler gear shaft—4 speed

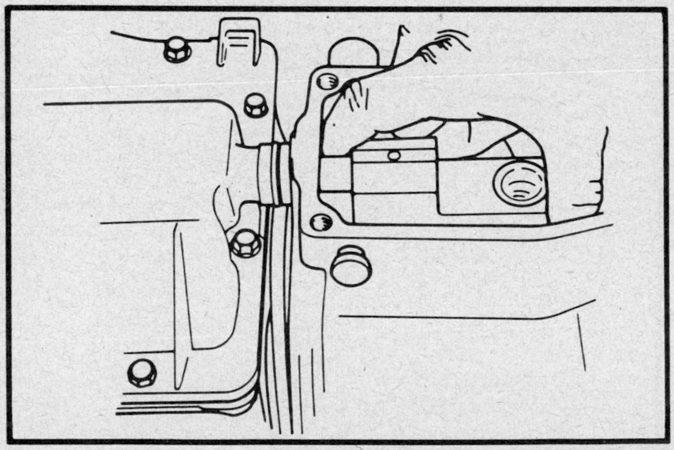
Extension housing to transmission case—5 speed

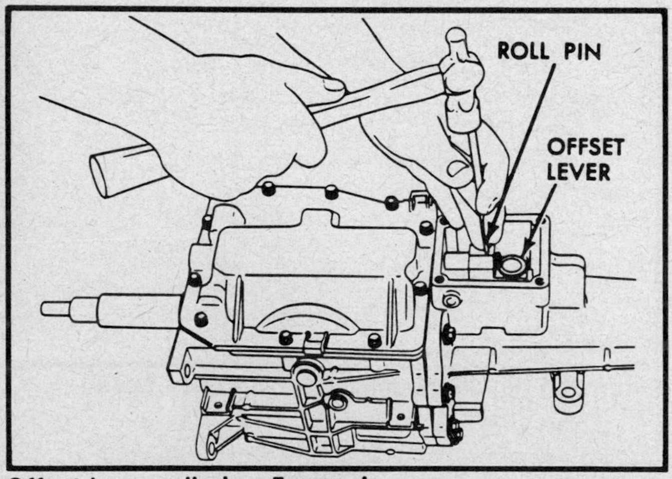
ROLL PIN
OFFSET LEVER
Offset lever roll pin—5 speed

4. Remove extension housing to transmission case bolts and remove housing and offset lever as an assembly. Do not attempt to remove the offset lever while the extension housing is still bolted in place. The lever has a positioning lug engaged in the housing detent plate which prevents moving the lever far enough for removal.

5. Remove detent ball and spring from offset lever and remove roll pin from extension housing or offset lever.

6. Remove plastic funnel, thrust bearing race and thrust bearing from rear of countershaft. The countershaft rear thrust bearing, bearing washer and plastic funnel may be found inside the extension housing.

7. Remove bolts attaching transmission cover and shift fork assembly and remove cover. Two of the transmission cover attaching bolts are alignment-type dowel bolts. Note the location of these bolts for assembly reference.

8. Using a punch and hammer, drive the roll pin from the 5th gearshift fork while supporting the end of the shaft with a block of wood.

9. Remove 5th synchronizer gear snapring, shift fork, 5th gear synchronizer sleeve, blocking ring and 5th speed drive gear from rear of countershaft.

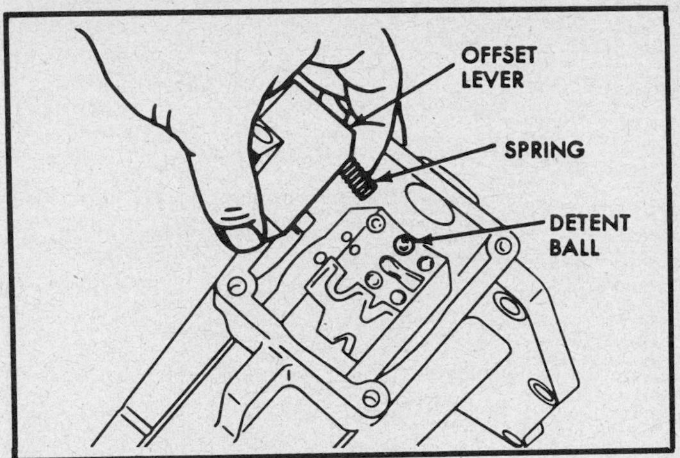
OFFSET LEVER
SPRING
DETENT BALL

Detent ball and spring—5 speed

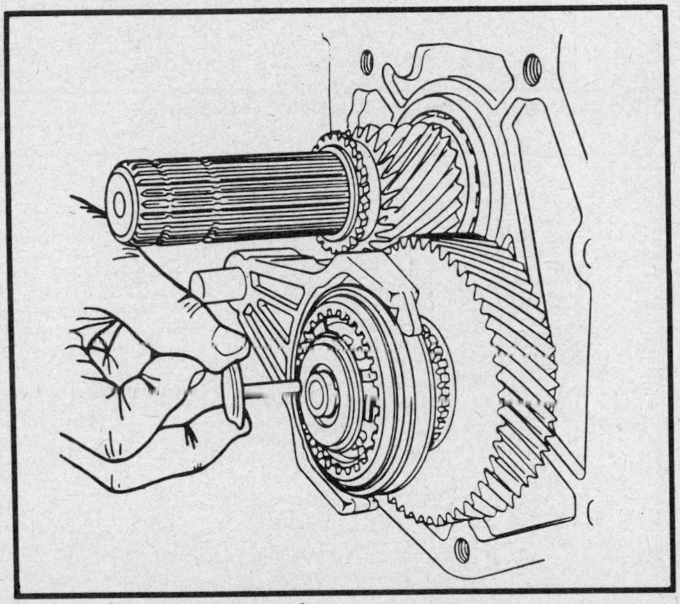

Plastic funnel—5 speed

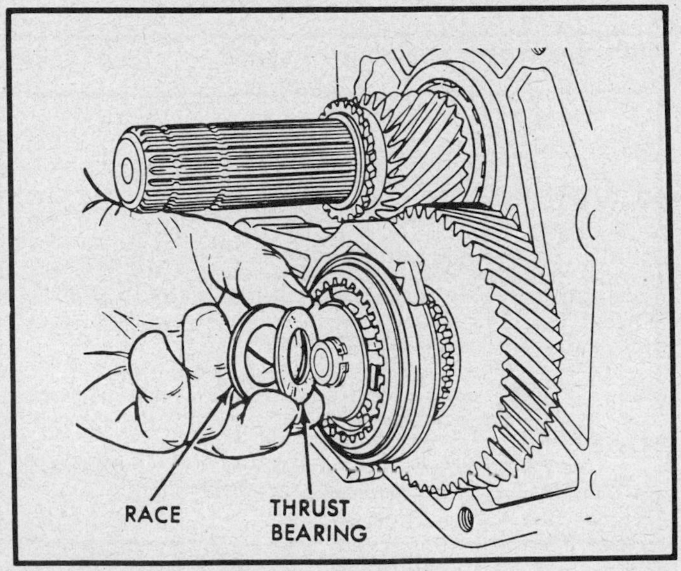
RACE
THRUST BEARING

Thrust bearing and race—5 speed

10. Remove snapring from 5th speed driven gear.
11. Using a hammer and punch, mark both bearing cap and case for assembly reference.
12. Remove front bearing cap bolts and remove front bearing cap. Remove front bearing race and endplay shims from front bearing cap.
13. Rotate drive gear until flat surface faces counter shaft and remove drive gear from transmission case.
14. Remove reverse lever C-clip and pivot bolt.
15. Remove mainshaft rear bearing race and then tilt mainshaft assembly upward and remove assembly from transmission case.
16. Unhook overcenter link spring from front of transmission case.
17. Rotate 5th gear/reverse shift rail to disengage rail from reverse lever assembly. Remove shift rail from rear of transmission case.
18. Remove reverse lever and fork assembly from transmission case.
19. Using hammer and punch, drive roll pin from forward end of reverse idler shaft and remove reverse idler shaft, rubber O-ring and gear from the transmission case.
20. Remove rear countershaft snapring and spacer.
21. Insert a brass drift through drive gear opening in front of transmission case and, using an arbor press, carefully press countershaft rearward to remove rear countershaft bearing.
22. Move countershaft assembly rearward, tilt countershaft upward and remove from case. Remove countershaft front thrust washer and rear bearing spacer.
23. Remove countershaft front bearing from transmission case using an arbor press.

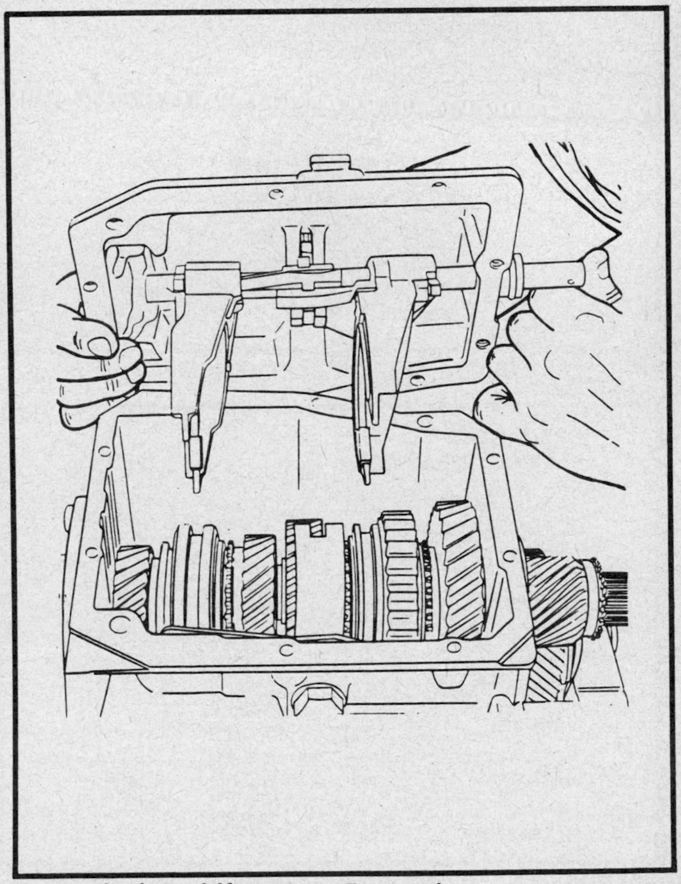

Transmission shift cover—5 speed

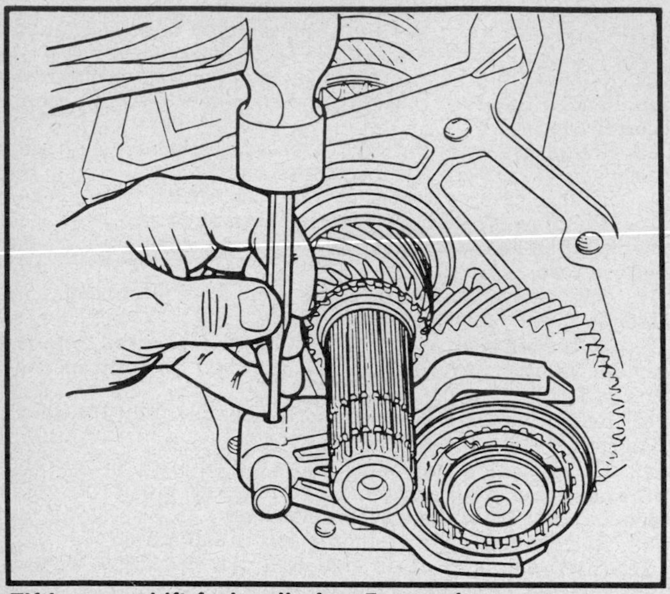
Fifth gear shift fork roll pin—5 speed

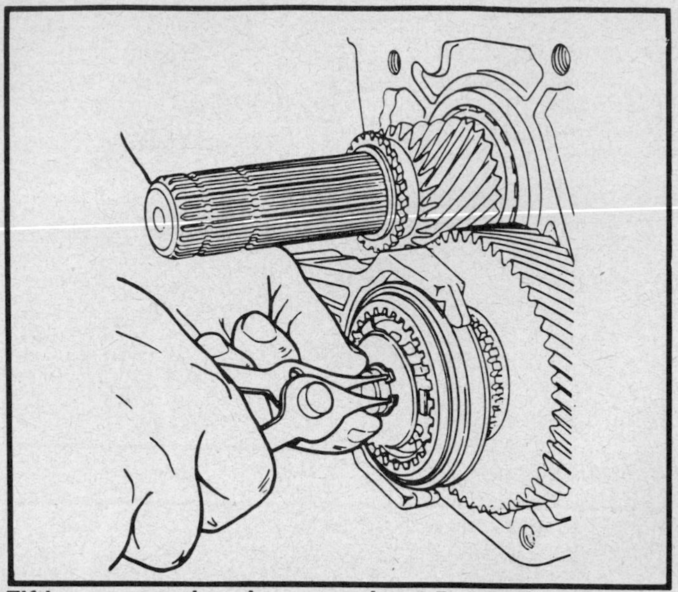
Fifth gear synchronizer snapring—5 speed

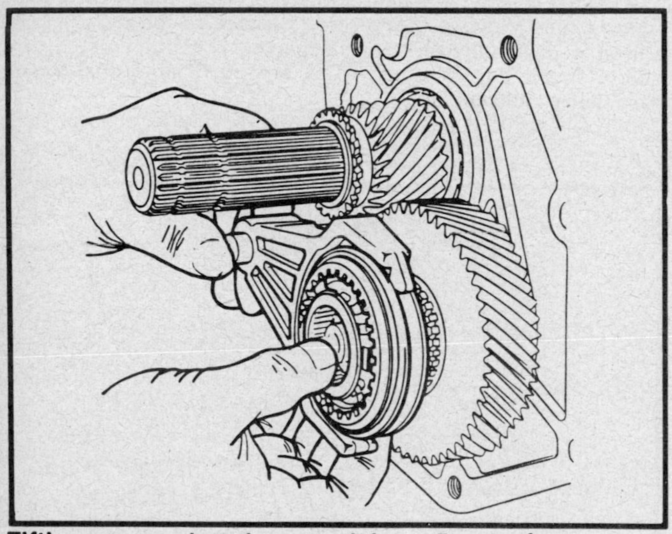
Fifth gear synchronizer servicing—5 speed

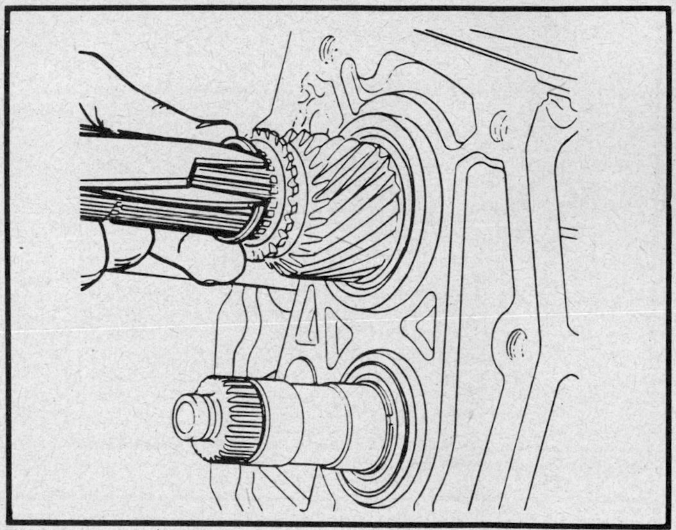
Fifth speed gear snapring—5 speed

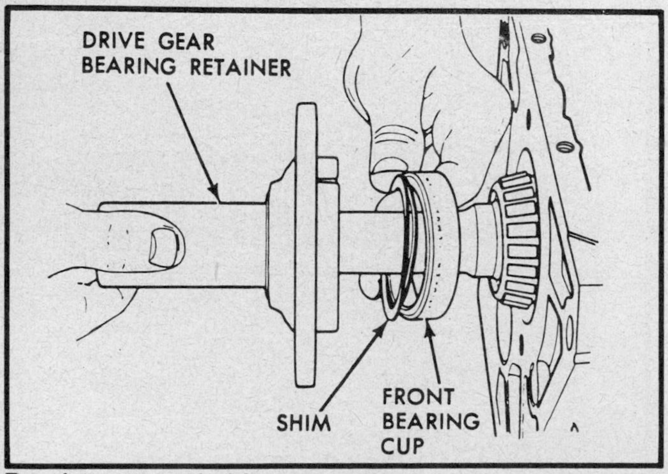
DRIVE GEAR
BEARING RETAINER
SHIM
FRONT
BEARING
CUP
Bearing cap and shims—5 speed

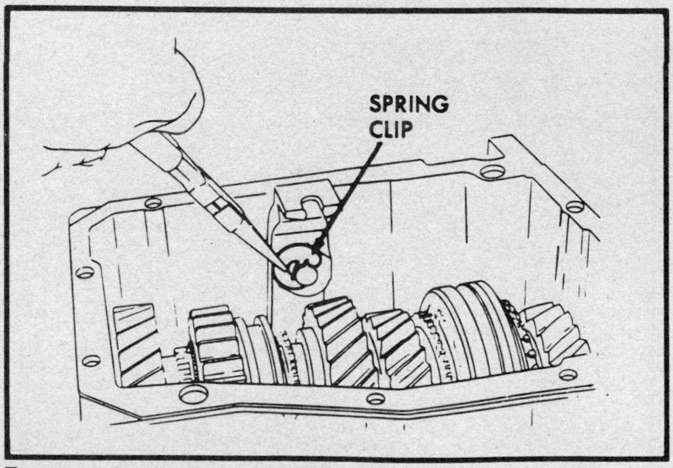
SPRING
CLIP
Reverse lever retaining clip—5 speed

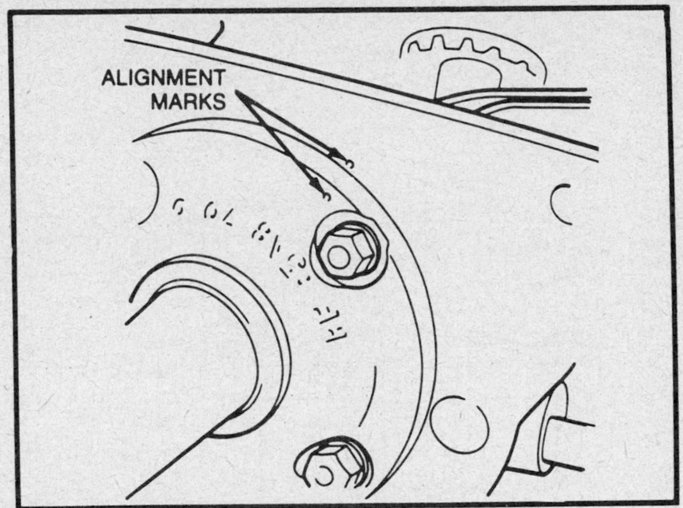
ALIGNMENT
MARKS

Bearing cap to case alignment—5 speed

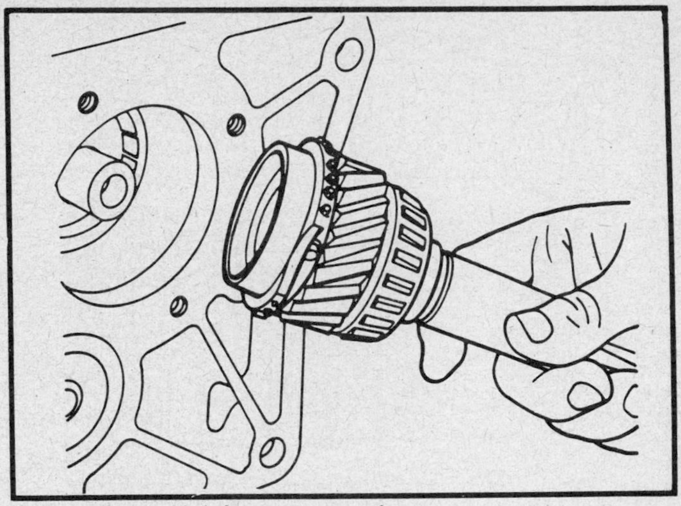

Drive gear servicing—5 speed

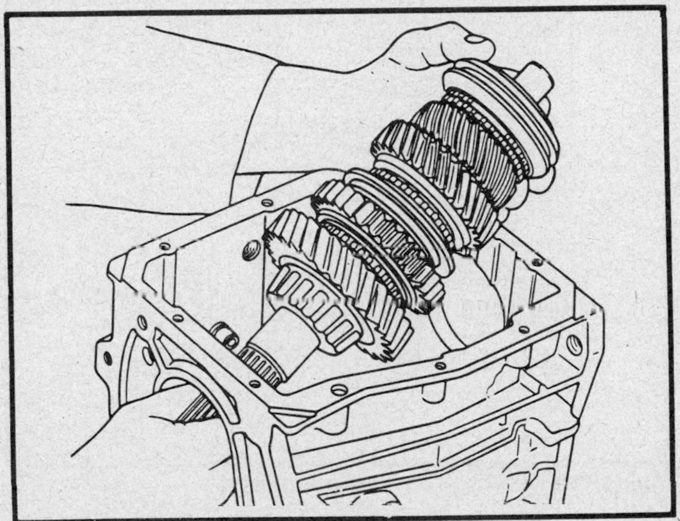

Mainshaft servicing—5 speed

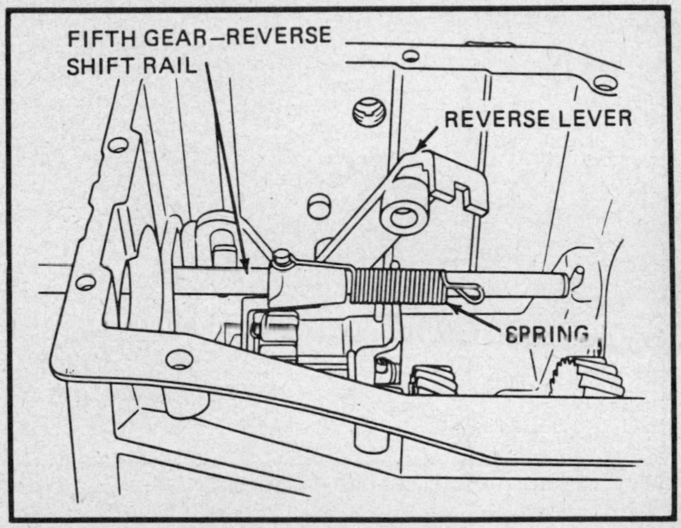
FIFTH GEAR—REVERSE
SHIFT RAIL
REVERSE LEVER
SPRING

Fifth gear/reverse shift rail—5 speed

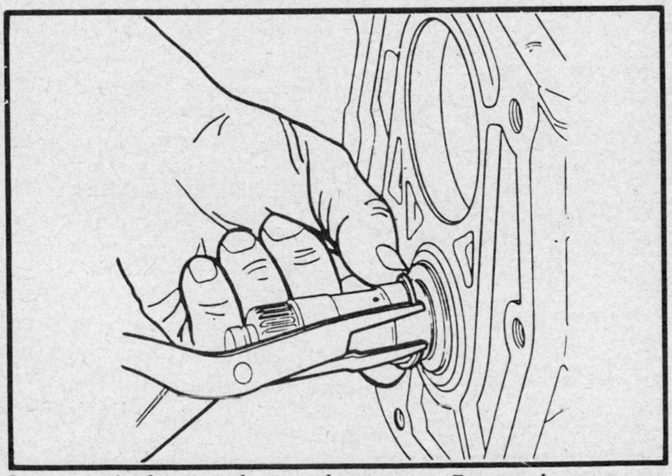

Countershaft snapring and spacer—5 speed

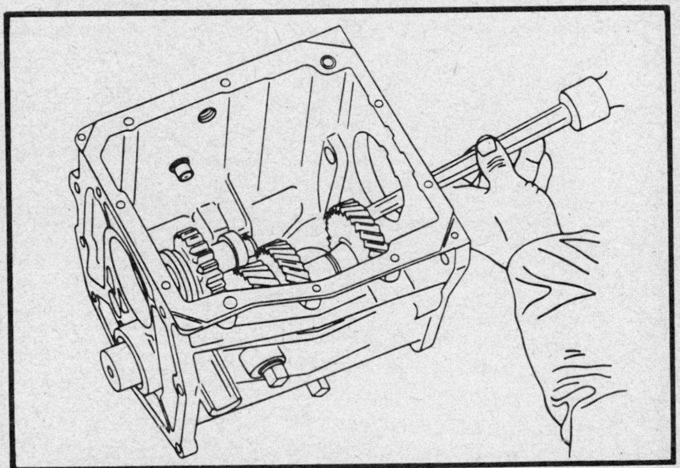

Rear countershaft bearing removal—5 speed

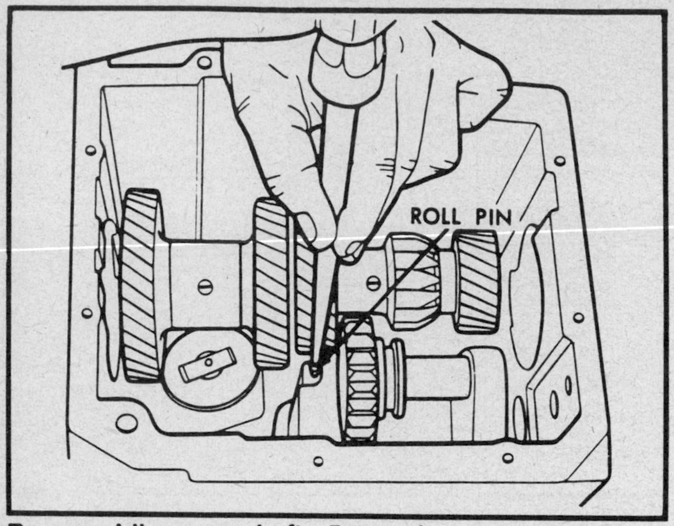
ROLL PIN
Reverse idler gear shaft—5 speed

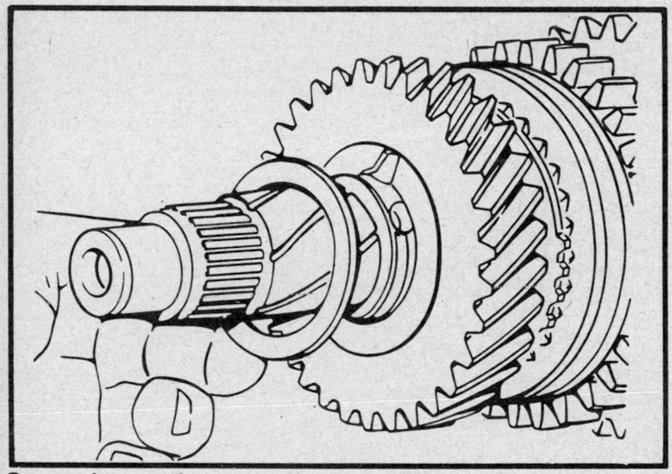
Second gear thrust washer—4 speed

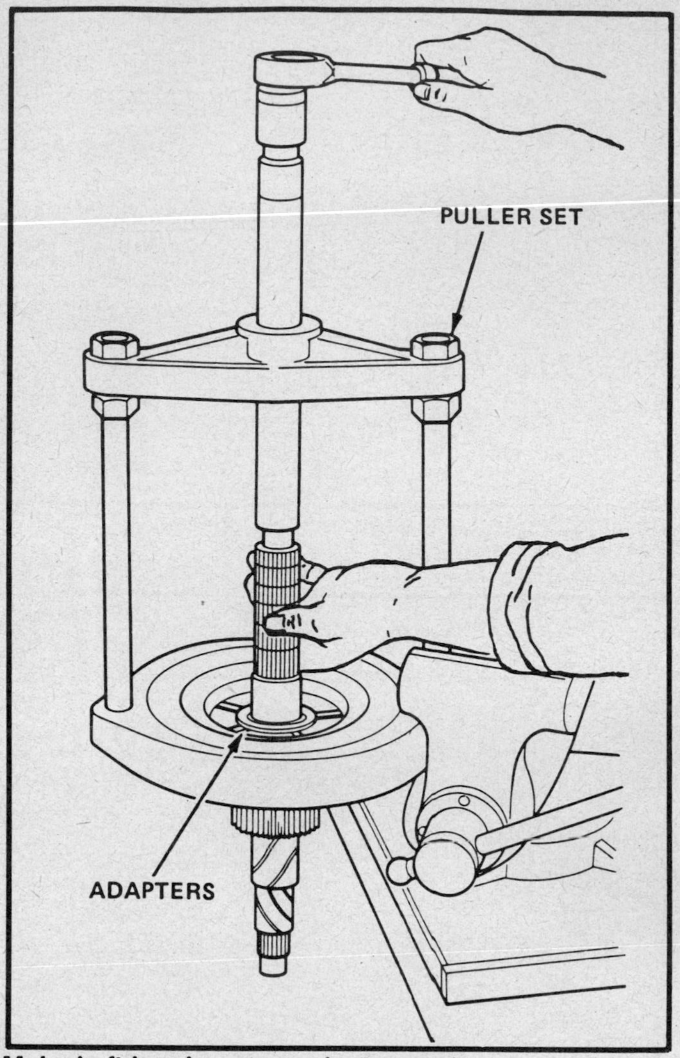
PULLER SET
ADAPTERS
Mainshaft bearing removal—4 speed

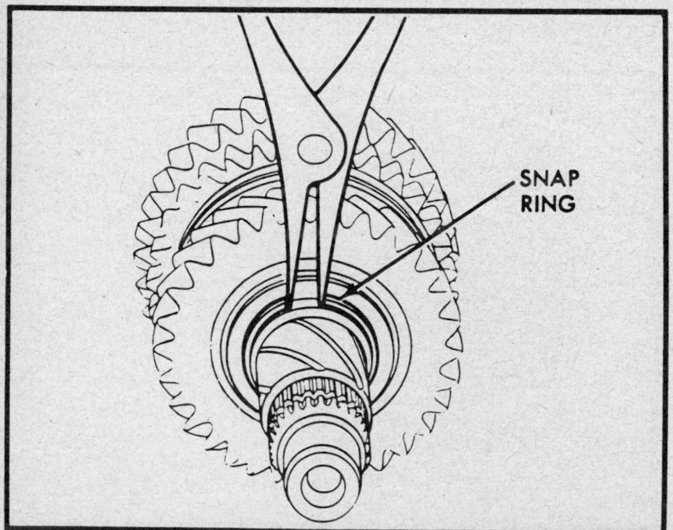
SNAP
RING
Second gear snapring—5 speed

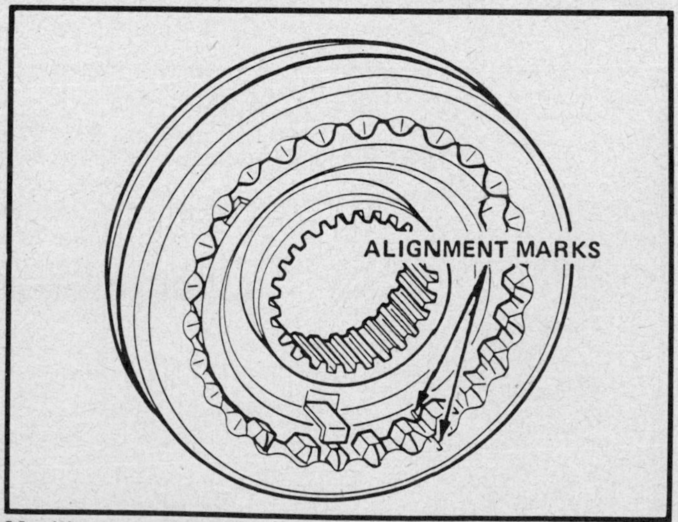
ALIGNMENT MARKS
Marking synchronizer assembly—4 speed

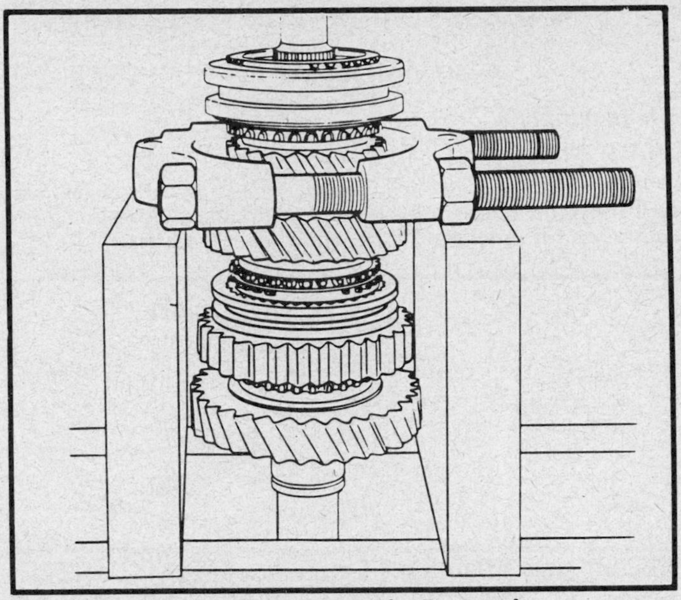

Third/4th synchronizer removal—5 speed

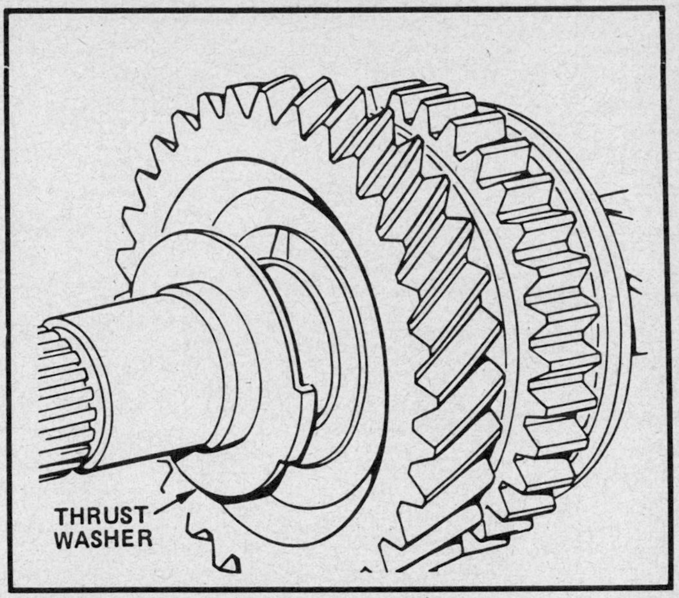
THRUST
WASHER

First gear thrust washer—4 speed

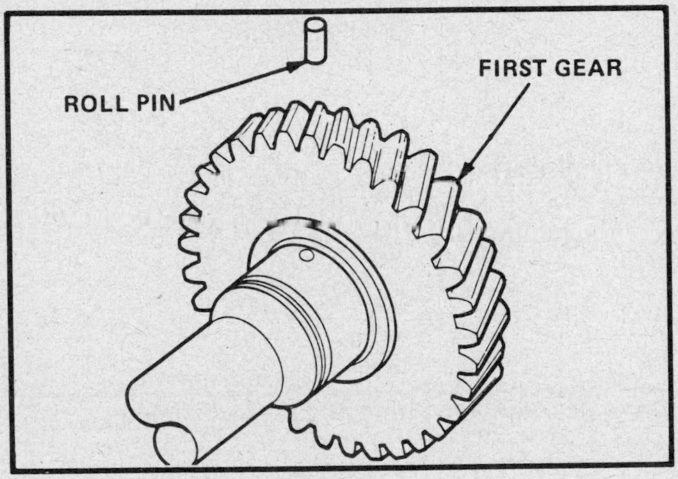
ROLL PIN
FIRST GEAR

First gear roll pin—4 speed

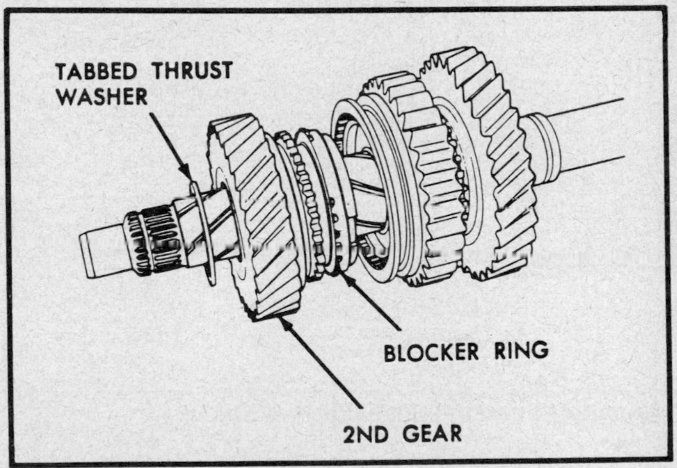
TABBED THRUST
WASHER
BLOCKER RING
2ND GEAR

Second gear and thrust washer—5 speed

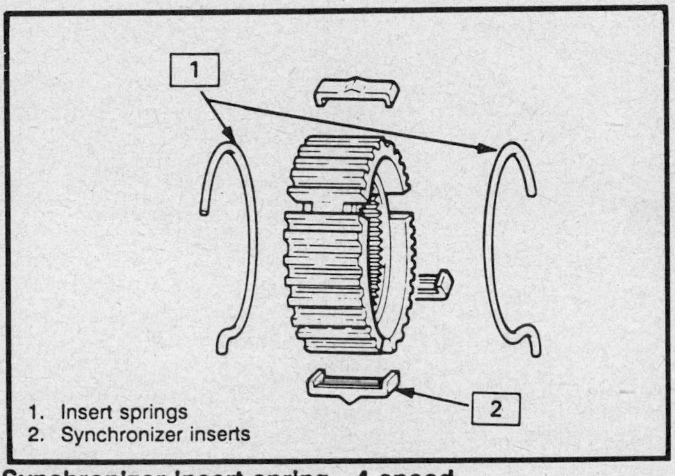
1
2
1. Insert springs
2. Synchronizer inserts

Synchronizer insert spring—4 speed

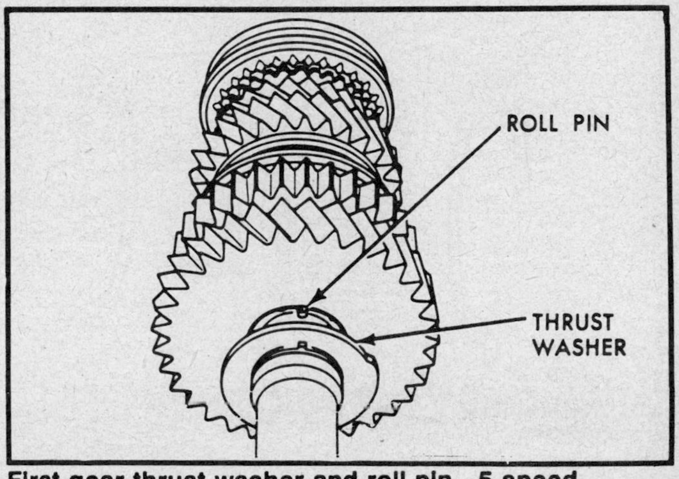
ROLL PIN
THRUST
WASHER

First gear thrust washer and roll pin—5 speed

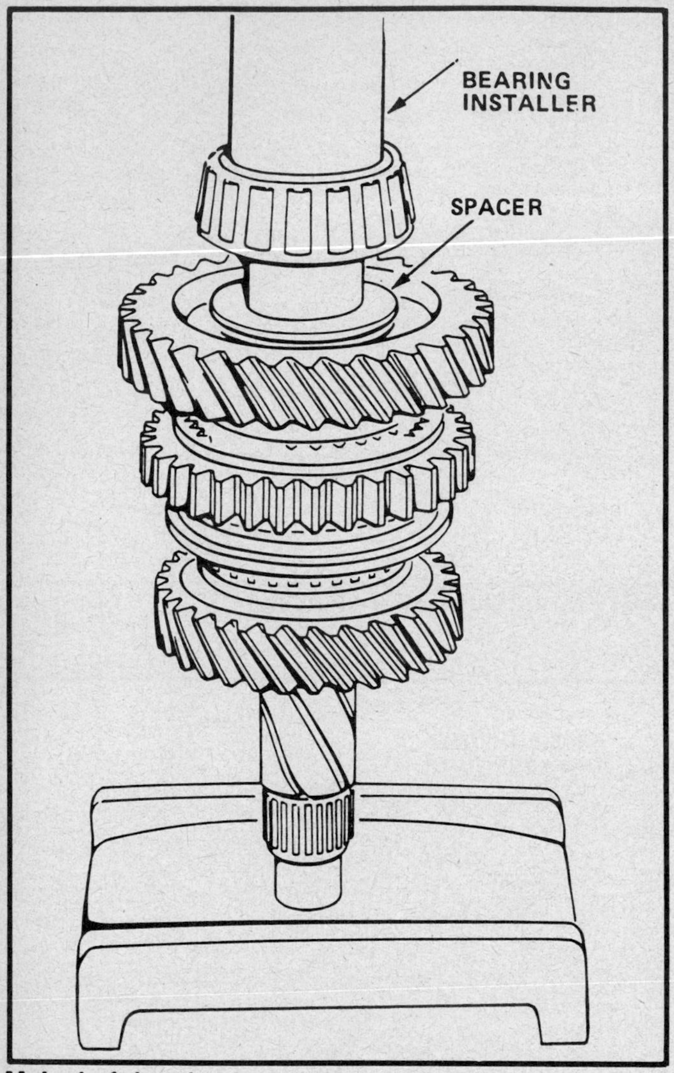
BEARING INSTALLER
SPACER
Mainshaft bearing installation—4 speed

Unit Disassembly and Assembly

MAINSHAFT

Disassembly

FOUR SPEED

1. Scribe an alignment mark on 3rd/4th synchronizer hub and sleeve for reassembly. Remove retaining snapring and remove 3rd/4th synchronizer assembly from mainshaft.

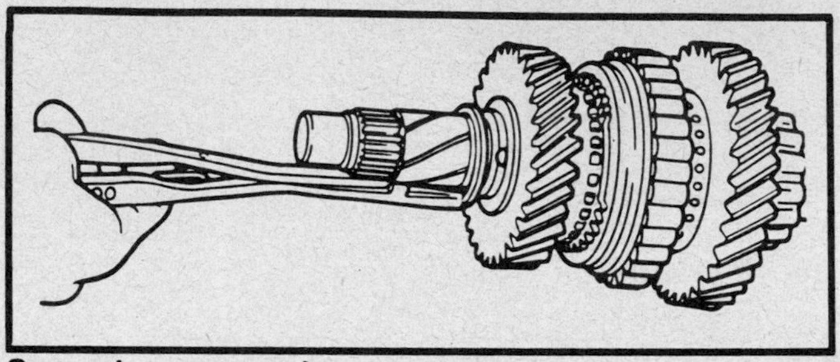
Second gear snapring—4 speed

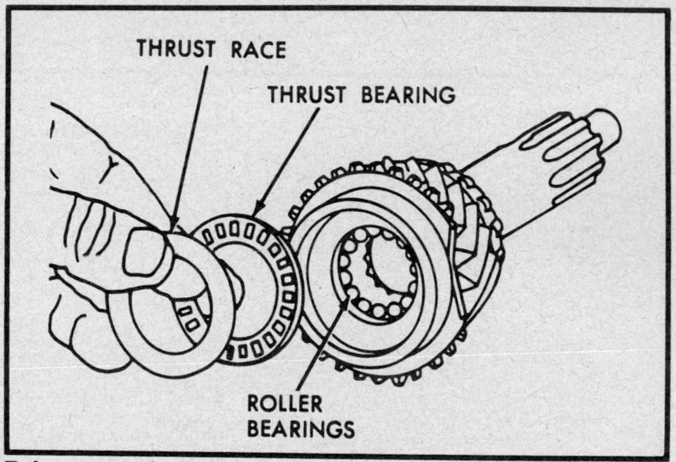
THRUST RACE
THRUST BEARING
ROLLER BEARINGS
Drive gear thrust bearing and race—5 speed

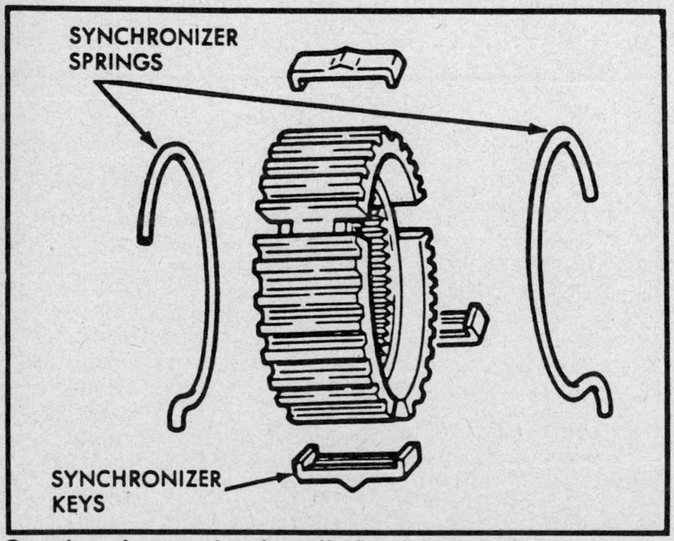
SYNCHRONIZER SPRINGS
SYNCHRONIZER KEYS
Synchronizer spring installation—5 speed

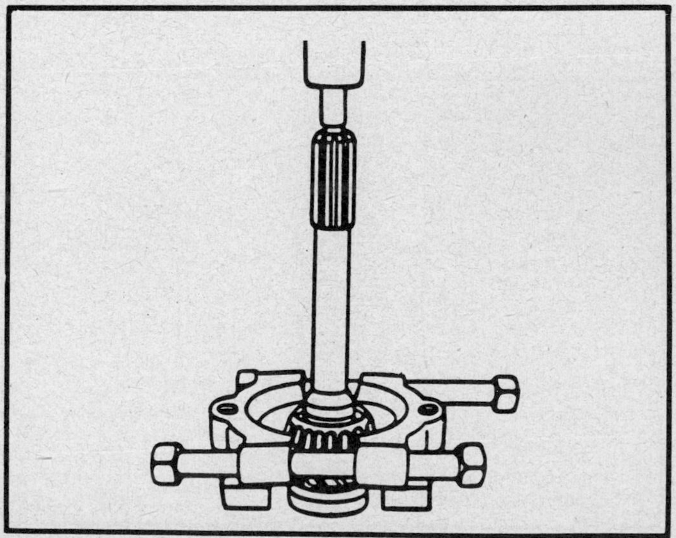
Drive gear bearing removal—5 speed

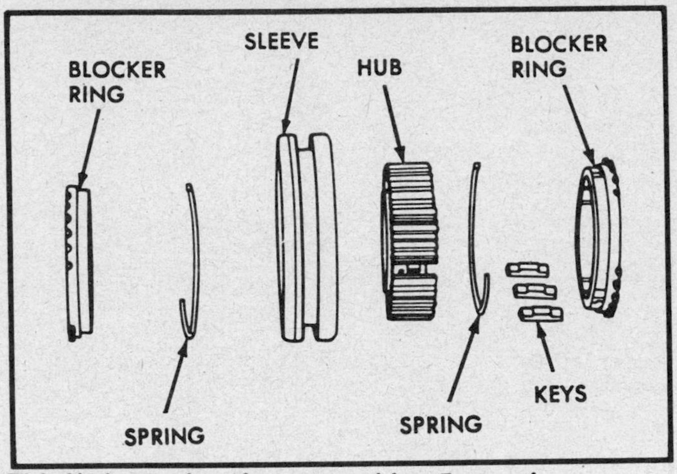
BLOCKER RING
SLEEVE
HUB
BLOCKER RING
SPRING
SPRING
KEYS

Third/4th synchronizer assembly—5 speed

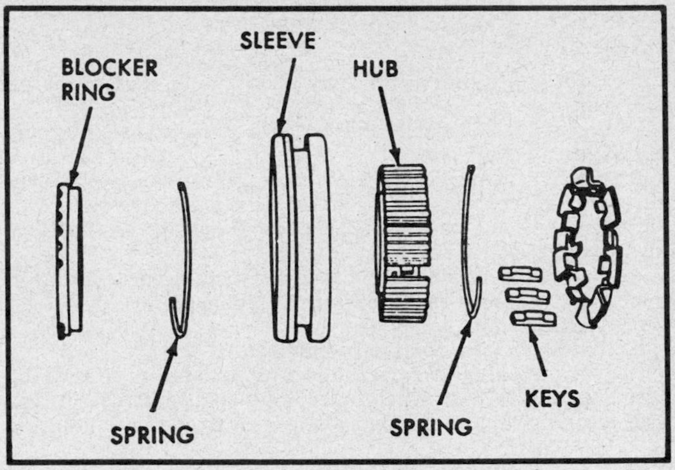
BLOCKER RING
SLEEVE
HUB
SPRING
SPRING
KEYS

Fifth gear synchronizer assembly—5 speed

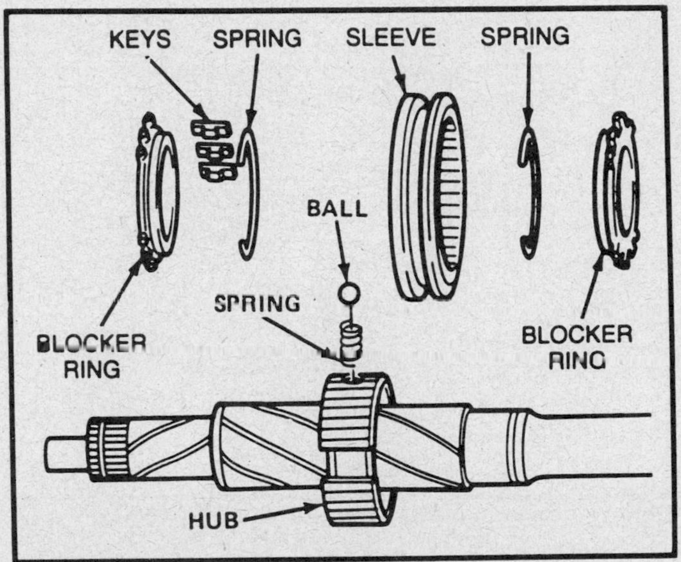
KEYS
SPRING
SLEEVE
SPRING
BALL
SPRING
BLOCKER RING
BLOCKER RING
HUB

First/2nd synchronizer assembly—5 speed

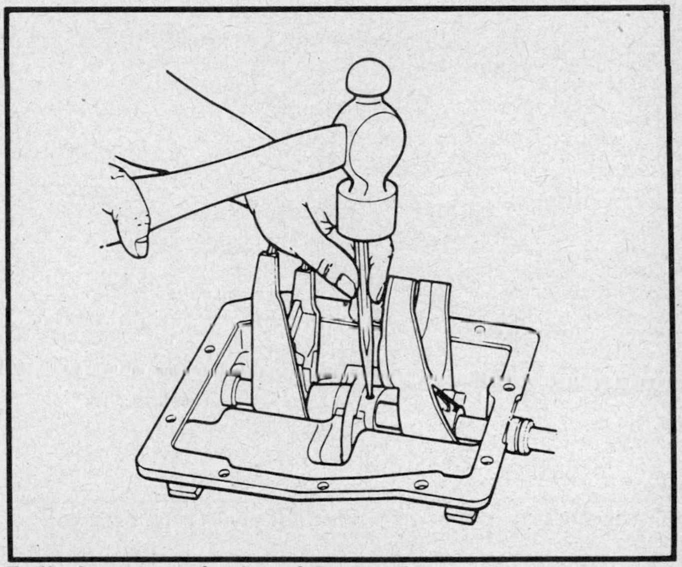

Roll pin removal—4 and 5 speed

2. Slide 3rd gear off mainshaft.
3. Remove 2nd gear retaining snapring. Remove tabbed thrust washer, 2nd gear and blocker ring from mainshaft.
4. Remove 1st gear thrust washer and roll pin from mainshaft. Use pliers to remove roll pin.
5. Remove 1st gear and blocker ring from mainshaft.
6. Scribe alignment mark on 1st/2nd synchronizer hub and sleeve for reassembly.
7. Remove synchronizer springs and keys from 1st/2nd sleeve and remove sleeve from shaft.

NOTE: Do not attempt to remove the 1st/2nd hub from the mainshaft. The hub and mainshaft are assembled and machined as a unit.

8. Remove loading tool J-26624 or equivalent, roller bearings, spacers and thrust washers from the countershaft gear.

FIVE SPEED

1. Remove thrust bearing washer from front end of mainshaft.
2. Scribe a reference mark on 3rd/4th synchronizer hub and sleeve for reassembly.
3. Remove 3rd/4th synchronizer blocking ring, sleeve, hub and 3rd gear as an assembly from mainshaft.
4. Remove snapring, tabbed thrust washer and 2nd gear from mainshaft.
5. Remove 5th gear with tool J-22912-01 or equivalent and arbor press. Slide rear bearing of mainshaft.
6. Remove 1st gear thrust washer, roll pin, 1st gear and synchronizer ring from mainshaft.
7. Scribe a reference mark on 1st/2nd synchronizer hub and sleeve for reassembly.
8. Remove synchronizer spring and keys from 1st/reverse sliding gear and remove gear from mainshaft hub. Do not attempt to remove the 1st/2nd reverse hub from mainshaft. The hub and shaft are assembled and machined as a matched set.

Inspection

1. Inspect all drive gear bearing rollers for wear.
2. Inspect all gears for excessive wear, chips or cracks and replace any that are worn or damaged.
3. Check the clutch sleeves to see that they slide freely on their hubs.
4. Inspect all synchronizers for wear.

Assembly

FOUR SPEED

1. Coat mainshaft and gear bores with transmission lubricant.
2. Install 1st/2nd synchronizer sleeve on mainshaft, aligning marks previously made.
3. Install synchronizer keys and springs into the 1st/2nd synchronizer sleeve. Engage tang end of springs into the same synchronizer key but position open ends of springs so they face away from one another.
4. Place blocking ring on 1st gear and install gear and ring on mainshaft. Be sure synchronizer keys engage notches in 1st gear blocking ring.
5. Install 1st gear roll pin in mainshaft.
6. Place blocking ring on 2nd gear and install gear and ring on mainshaft. Be sure synchronizer keys engage notches in 2nd gear blocking ring. Install 2nd gear thrust washer and snapring

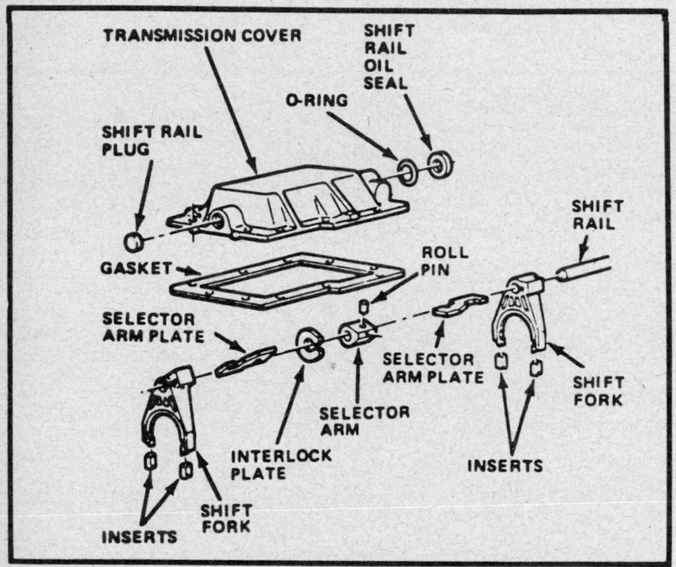
TRANSMISSION COVER
SHIFT RAIL OIL SEAL
O-RING
SHIFT RAIL PLUG
SHIFT RAIL
GASKET
ROLL PIN
SELECTOR ARM PLATE
SELECTOR ARM PLATE
SHIFT FORK
SELECTOR ARM
INTERLOCK PLATE
INSERTS
SHIFT FORK
INSERTS

Transmission cover exploded view—4 and 5 speed

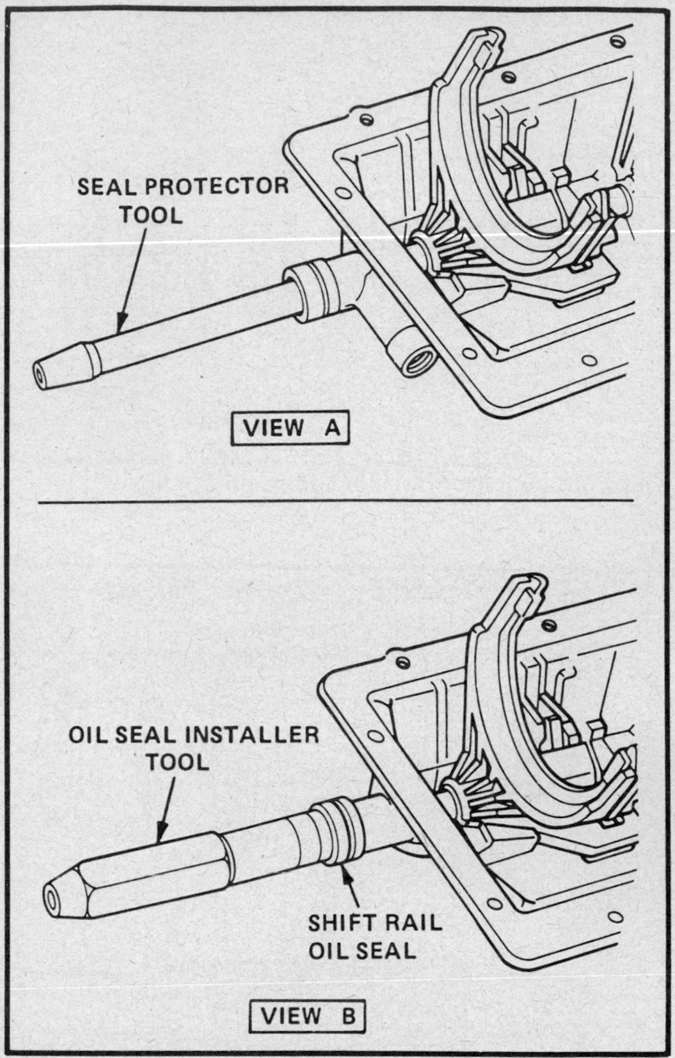
SEAL PROTECTOR TOOL
VIEW A
OIL SEAL INSTALLER TOOL
SHIFT RAIL OIL SEAL
VIEW B

Shift rail oil seal installation—4 and 5 speed

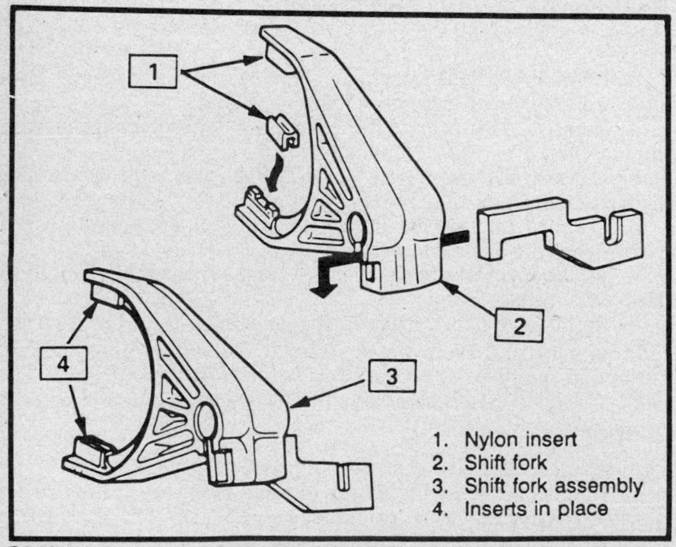
1
2
3
4
1. Nylon insert
2. Shift fork
3. Shift fork assembly
4. Inserts in place

Shift forks and selector arm plates—4 and 5 speed

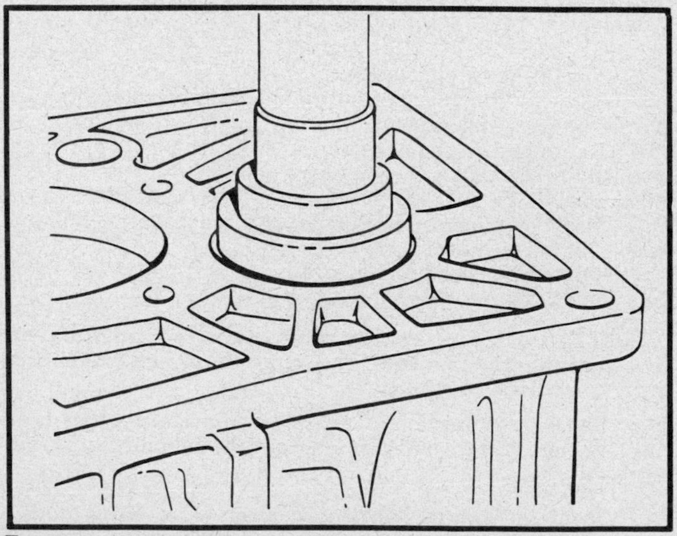

Front countershaft bearing installation—4 and 5 speed

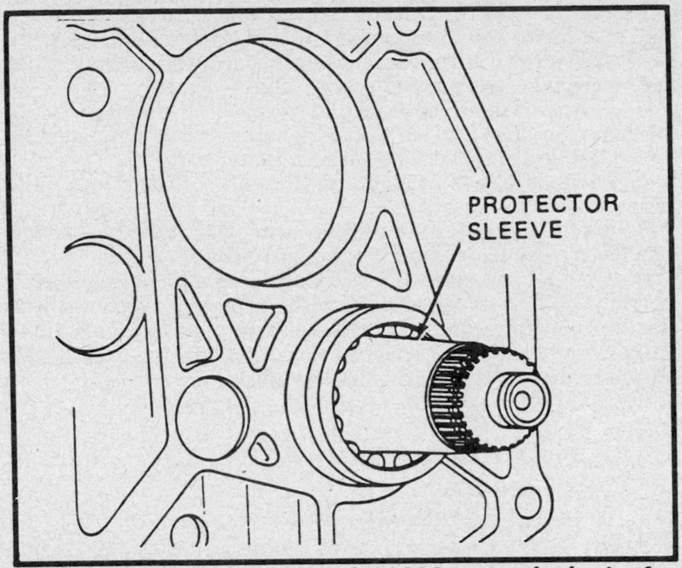
PROTECTOR
SLEEVE

Installing protector sleeve J-33032 or equivalent—4 and 5 speed

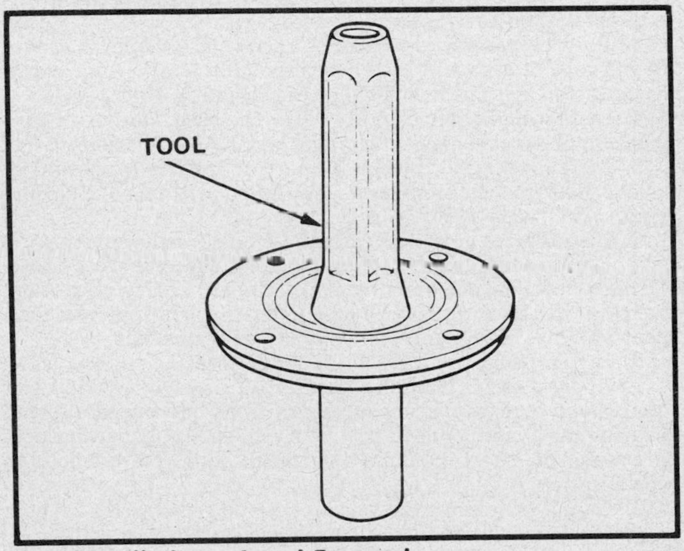
TOOL

Seal installation—4 and 5 speed

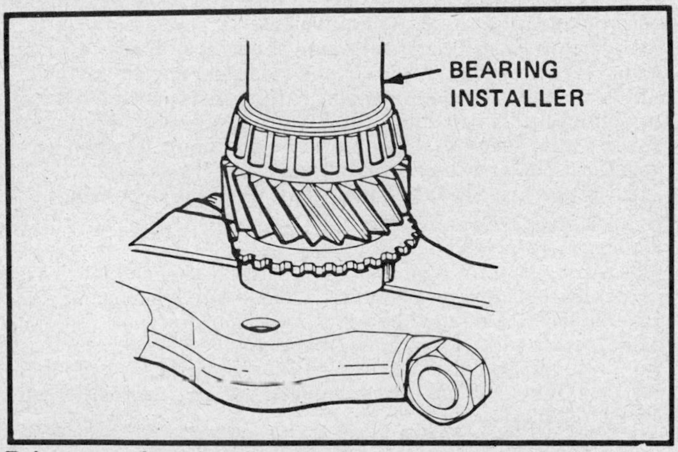
BEARING
INSTALLER

Drive gear bearing installation—4 and 5 speed

on mainshaft. Be sure thrust washer tab is engaged in mainshaft notch.

7. Measure 2nd gear endplay using feeler gauge. Insert gauge between gear and thrust washer. Endplay should be 0.004–0.014 in. If endplay is over 0.014 in., replace thrust washer and snapring and inspect synchronizer hub for excessive wear.

8. Place blocking ring on 3rd gear and install gear and ring on mainshft.

9. Install 3rd/4th synchronizer sleeve on hub, aligning marks previously made.

10. Install synchronizer keys and springs in 3rd/4th synchronizer sleeve. Engage tang end of each spring in same key but position open ends of springs so they face away from one another.

11. Install 3rd/4th synchronizer assembly on the mainshaft with machined groove in hub facing forward. Install snapring on mainshaft. Be sure synchronizer keys are engaged in notches in 3rd gear blocker ring.

12. Install tool J-26624 or equivalent, into countershaft gear. Using a light weight grease, lubricate roller bearings and install into bores at front and rear of countershaft gear. Install roller bearing retainers on tool J-26624 or equivalent.

FIVE SPEED

1. Coat mainshaft and gear bores with transmission lubricant.

2. Install 1st/2nd synchronizer sleeve on mainshaft hub aligning marks made at disasssembly.

3. Install 1st/2nd synchronizer keys and springs. Engage tang end of each spring in same synchronizer key but position open end of springs opposite of each other.

4. Install blocker ring and 2nd gear on mainshaft. Install tabbed thrust washer and 2nd gear retaining snapring on

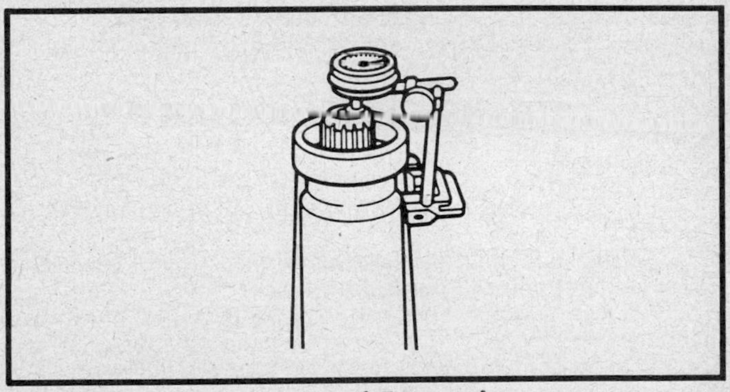

Measuring end play—4 and 5 speed

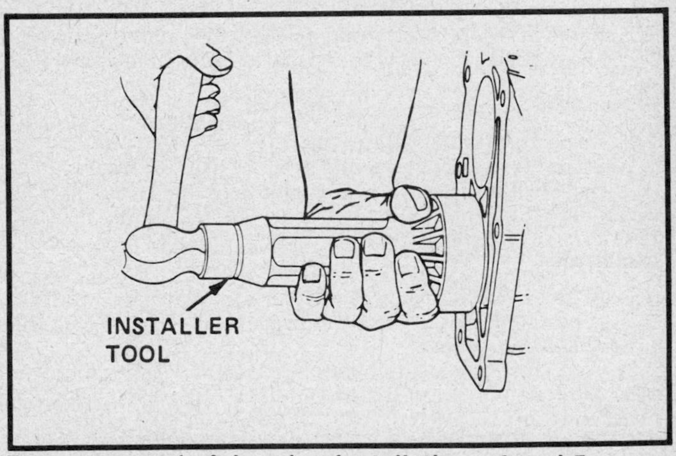
INSTALLER
TOOL

Rear countershaft bearing installation—4 and 5 speed

mainshaft. Be sure washer tab is properly seated in mainshaft notch.

5. Install blocker ring and 1st gear on mainshaft. Install 1st gear roll pin and then 1st gear thrust washer.
6. Slide rear bearing on mainshaft.
7. Install 5th speed gear on mainshaft using tool J-22912-01 or equivalent and arbor press. Install snapring on mainshaft.
8. Install 3rd gear, 3rd/4th synchronizer assembly and thrust bearing on mainshaft. Synchronizer hub offset must face forward.

DRIVE GEAR

Disassembly

FIVE SPEED

1. Remove bearing race, thrust bearing and roller bearings from cavity of drive gear.
2. Using tool J-22912-01 or equivalent and arbor press, remove bearing from drive gear.
3. Wash parts in a cleaning solvent.

Inspection

FIVE SPEED

1. Inspect gear teeth for wear.
2. Inspect drive shaft pilot for wear.

Assembly

FIVE SPEED

1. Using tool J-22912-01 or equivalent with an arbor press, install bearing on drive gear.
2. Coat roller bearings and drive gear bearing bore with grease. Install roller bearings into bore of drive gear.
3. Install thrust bearing and race in drive gear.

TRANSMISSION COVER

Disassembly

1. Place selector arm plates and shift rail in neutral position (centered).
2. Rotate shift rail until selector arm disengages from selector arm plates and roll pin is accessible.
3. Remove selector arm roll pin using a pin punch and hammer.
4. Remove shift rail, shift forks, selector arm plates, selector arm, interlock plate and roll pin.
5. Remove shift cover to extension housing O-ring seal using a suitable tool.
6. Remove nylon inserts and selector arm plates from shift forks. Note position of inserts and plates for assembly reference.

Inspection

1. Inspect the shift rail for wear.
2. Inspect the shift forks and selector arm for wear.
3. Inspect the selector arm plates and interlock plate for wear.

Assembly

1. Install nylon inserts and selector arm plates in shift forks.
2. If removed, install shift rail plug. Coat edges of plug with sealer before installing.
3. Coat shift rail and rail bores with light weight grease and insert shift rail in cover. Install rail until flush with inside edge of cover.
4. Place 1st/2nd shift fork in cover with fork offset facing rear of cover and push shift rail through fork. The 1st/2nd shift fork is the larger of the 2 forks.
5. Position selector arm and C-shaped interlock plate in cover and insert shift rail through arm. Widest part of interlock plate must face away from cover and selector arm roll pin hole must face downward and toward rear of cover.
6. Position 3rd/4th shift fork in cover with fork offset facing rear of cover. The 3rd/4th shift fork selector arm plate must be under 1st/2nd shift for selector arm plate.
7. Push shift rail through 3rd/4th shift fork and into front bore in cover.
8. Rotate shift rail until selector arm plate at forward end of rail faces away from, but is parallel to cover.
9. Align roll pin holes in selector arm and shift rail and install roll pin. Roll pin must be flush with surface of selector arm to prevent pin from contacting selector arm plates during shifts.
10. Install a new shift cover to extension housing O-ring seal. Coat O-ring seal with transmission lubricant.

Transmission Assembly

FOUR SPEED

1. Coat countershaft gear thrust washers with petroleum jelly and position washer in case.
2. Position countershaft gear in case and install countershaft from rear of case. Be sure that thrust washers stay in place during installation of countershaft and gear.
3. Position reverse idler gear in case with shift lever groove facing rear of case and install reverse idler shaft from rear of case. Install roll pin in shaft and center pin in shaft.
4. Install mainshaft assembly into the case. Do not disturb position of synchronizer assemblies during installation.
5. Install 4th gear blocking ring in 3rd/4th synchronizer sleeve. Be sure synchronizer keys engaged in notches in blocker ring.
6. Install drive gear into case and engage with mainshaft.
7. Position mainshaft 1st gear against the rear of the case. Using a new bearing, start front bearing onto drive gear. Align bearing with bearing bore in case and drive bearing onto drive gear and into case using tool J-25234 or equivalent.
8. Install front bearing retaining and locating snaprings.
9. Apply a ⅛ in. diameter bead of RTV sealant, No. 732 or equivalent, on case mating surface of front bearing cap. Install bearing cap aligning marks previously made. Apply non-hardening sealer on attaching bolts and install bolts. Torque bolts to specification.
10. Install 1st gear thrust washer with oil grove facing 1st gear on mainshaft, aligning slot in washer with 1st gear roll pin.
11. Using a new bearing, position rear bearing on mainshaft. Align bearing with bearing bore in case and drive bearing into case using tool J-25234 or equivalent.
12. Install locating and retaining snaprings on rear bearing.
13. Install speedometer gear and retaining clip on mainshaft.
14. Apply non-hardening sealer to threads of reverse lever pivot bolt and start bolt into case. Engage reverse lever fork in the reverse idler gear and reverse lever on pivot bolt. Tighten bolt to specifications and install retaining clip.
15. Rotate drive gear and mainshaft gear. If blocker rings tend to stick on gears, release the rings by gently prying them off the cones.
16. Apply a ⅛ in. diameter bead or RTV sealant, No. 732 or equivalent, on the cover mating surface of transmission. Place reverse lever in neutral and position cover on case.
17. Install 2 dowel type bolts first to align cover on case. Install remaining cover bolts and torque to specifications. The offset lever to shift rail roll pin hole must be in the vertical position after cover installation.
18. Apply a ⅛ in. diameter bead of RTV sealant, No. 732 or equivalent, on the extension housing to transmission case mating surface.

19. Place extension housing over mainshaft to a position where shift rail is in shift cover opening.
20. Install detent spring in offset lever. Place ball in neutral guide plate detent position. Apply pressure on the offset lever, slide offset lever onto shift rail and seat extension housing to transmission case.
21. Install extension housing retaining bolts. Torque bolts to specifications.
22. Align hole in offset lever and shift rail and install roll pin.
23. Fill transmission to its proper level with recommended lubricant.

FIVE SPEED

1. Coat countershaft front bearing bore with Loctite® 601, or equivalent and install front countershaft bearing flush with facing of case using an arbor press.
2. Coat countershaft tabbed thrust washer with grease and install washer so tab engages depression in case.
3. Tip transmission case on end and install countershaft in front bearing bore.
4. Install countershaft rear bearing spacer. Coat countershaft rear bearing with grease and install bearing using tool J–29895 or equivalent and sleeve J–33032, or its equivalent. The bearing when correctly installed will extend beyond the case surface 0.125 in.
5. Position reverse idler gear in case with shift lever groove facing rear of case and install reverse idler shaft from rear of case. Install roll pin in idler shaft.
6. Install assembled mainshaft in transmission case. Install rear mainshaft bearing race in case.
7. Install drive gear in case and engage in 3rd/4th synchronizer sleeve and blocker ring.
8. Install front bearing race in front bearing cap. Do not install shims in front bearing cap at this time.
9. Temporarily install front bearing cap.
10. Install 5th speed/reverse lever, pivot bolt and retaining clip. Coat pivot bolt threads with non-hardening sealer. Be sure to engage reverse lever fork in reverse idler gear.
11. Install countershaft rear bearing spacer and retaining snapring.
12. Install 5th speed gear on countershaft.
13. Insert 5th speed/reverse rail in rear of case and install into reverse 5th speed lever. Rotate rail during installation to simplify engagement with lever. Connect spring to front of case.
14. Position 5th gear shift fork on 5th gear synchronizer assembly and install synchronizer on countershaft and shift fork on shift rail. Make sure roll pin hole in shift fork and shift rail are aligned.
15. Support 5th gear shift rail and fork on a block of wood and install roll pin.
16. Install thrust race against 5th speed synchronizer hub and install snapring. Install thrust bearing against race on countershaft. Coat both bearing and race with petroleum jelly.
17. Install lipped thrust race over needle-type thrust bearing and install plastic funnel into hole in end of countershaft gear.
18. Temporarily install extension housing and attaching bolts. Turn transmission case on end and mount a dial indicator on extension housing with indicator on the end of mainshaft.
19. Rotate mainshaft and 0 dial indicator. Pull upward on mainshaft until endplay is removed and record reading. Mainshaft bearings require a preload of 0.001–0.005 in. to set preload, select a shim pack measuring 0.001–0.005 in. greater than the dial indicator reading recorded.
20. Remove front bearing cap and front bearing race Install necessary shims to obtain preload and reinstall bearing race.
21. Apply a ⅛ in. bead of RTV sealant, No. 732 or equivalent, on case mating surface of front bearing cap. Install bearing cap aligning marks made during disassembly and torque bolts to specification.
22. Remove extension housing.
23. Move shift forks on transmission cover and synchronizer sleeves inside transmission to the neutral position.
24. Apply a ⅛ in. bead of RTV sealant, No. 732 or equivalent, or cover mating surface of transmission.
25. Lower cover onto case while aligning shift forks and synchronizer sleeves. Center cover and install the 2 dowel bolts. Install remaining bolts and torque to specification. The offset lever to shift rail roll pin hole must be in the vertical position after cover installation.
26. Apply a ⅛ in. bead of RTV sealant, No. 732 or equivalent, on extension housing to transmission case mating surface.
27. Install extension housing over mainshaft and shift rail to a position where shift rail just enters shift cover opening.
28. Install detent spring into offset lever and place steel ball in neutral guide plate detent. Position offset lever on steel ball and apply pressure on offset lever and at the time seat extension housing against transmission case.
29. Install extension housing bolts and torque to specification.
30. Align and install roll pin in offset lever and shift rail.
31. Fill transmission to its proper level with lubricant.

SPECIFICATIONS

TORQUE SPECIFICATIONS

Item	ft. lbs.	Nm
Transmission to engine (1.9L)	25	35
Transmission to engine (2.5L)	37	50
Transmission to engine (all others)	55	75
Extension housing to case	25	30
Shift cover to case	10	13
Front bearing retainer to case	15	20
Reverse pivot bolt to case	20	27
Fill plug to case	20	27
Crossmember to frame	25	30
Transmission mount to transmission	35	50
Transmission mount to crossmember	25	30
Shiftrod swivel nut	18	25
Control bracket screw	23	32
Control mounting screw	23	32
Transmission brace	26	35
Shift lever nut	35	47

SPECIAL TOOLS

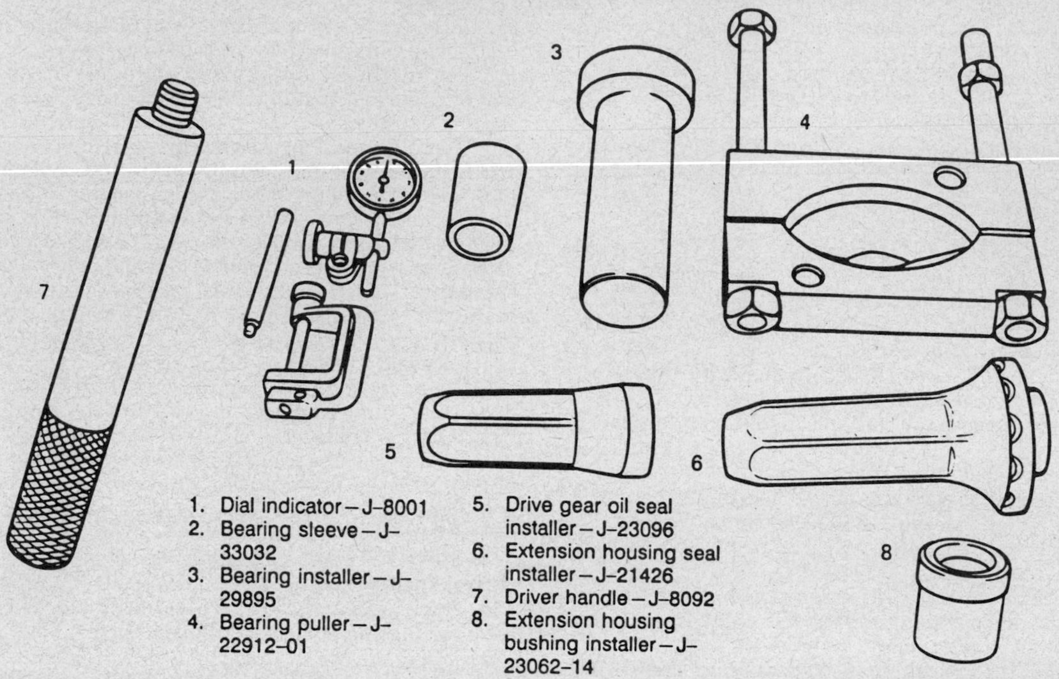
1
2
3
4
5
6
7
8

1. Dial indicator—J–8001
2. Bearing sleeve—J–33032
3. Bearing installer—J–29895
4. Bearing puller—J–22912–01
5. Drive gear oil seal installer—J–23096
6. Extension housing seal installer—J–21426
7. Driver handle—J–8092
8. Extension housing bushing installer—J–23062–14

Section 7

HM-260 and HM-290 Transmissions
General Motors

APPLICATION

1988–89 Chevrolet C/K Light Duty Trucks
1989 Chevrolet Camaro and Pontiac Firebird
1989 Chevrolet Astro and GMC Safari
1989 Chevrolet S–10 and GMC S–15

GENERAL DESCRIPTION

The HM–260 and HM–290 manual transmissions are identified by the number of forward gears and the measured distance between the centerlines of the mainshaft and countershaft.

The Muncie HM–290 manual transmission has an 85mm center distance between the mainshaft and countershaft. This transmission has 2 versions, the (MCO) 4 speed and the (MG5) 5 speed. Both the 4 speed and 5 speed units are used in 2WD and 4WD powertrains. The (MG5) 5 speed is an overdrive transmission.

A cover mounted shift lever provides the method of gear shifting. The cover has a ball pin type interlock which prevents the transmission from shifting into 2 gears at a time.

Unique features of the HM–290 include:

1. A 2-piece aluminum case, with an integral clutch housing.
2. Constant mesh helical gearing for all speed ranges, including reverse.
3. Construction to meet the highest standards of operating quietness, efficiency and ease of shifting.
4. Input torque capacity up to 300 ft. lbs.
5. Input speed limit of 6000 rpm.
6. A rear housing, which has an integral extension section fitting to the driveshaft (2WD) or the transfer case (4WD).

The HM–260 is a direct copy of the HM–290, with all parts made smaller.

The Muncie HM–260 has a 76mm centerline distance between the mainshaft and countershaft and an input torque capacity of 175 ft. lbs.

GEAR RANGE AND TRANSMISSION RANGE RATIO

Speed Range	4-Speed (MCO) ①	5-Speed (MG5) ①
1st	4.01:1	4.01:1
2nd	2.32:1	2.32:1
3rd	1.40:1	1.40:1
4th	1.00:1	1.00:1
5th	—	0.73:1
Reverse	3.74:1	3.74:1

① MCO and MG5 are regular production option (RPO) codes.

Metric Fasteners

Metric bolt sizes and thread pitches are used for all fasteners on this manual transmission. The use of metric tools is mandatory in the service of this transmission.

Do not attempt to interchange metric fasteners for inch system fasteners. Mismatched or incorrect fasteners can result in damage to the transmission unit through malfunctions, breakage or possible personal injury. Care should be taken to reuse the fasteners in the same location as removed, whenever possible. Due to the large number of alloy parts used, torque specifications should be strictly observed. Before installing capscrews into aluminum parts, always dip screws into oil to prevent the screws from galling the aluminum threads and to prevent seizing.

Capacities

This transmission uses 5W–30 oil (GM part number 1052931) for its lubrication. This oil has an additive that meets Hydra-Matic requirements for proper lubrication and designed operation of all internal parts. The HM–260 and HM–290 (MCO) 4 speed has a 2.13 qt. (2.03 liter) capacity. The HM–290 (MG5) 5 speed has a 2.08 qt. (1.98 liters) capacity.

Checking Fluid Level

This transmission, does not require regular fluid changes. A drain plug located on the bottom and a fill plug located on the right hand side of the transmission are used during servicing.

1. Raise the vehicle and support it safely.
2. Using hex bit J–36511 or equivalent, remove transmission fill plug.
3. Lubricant should be at level of the fill plug. If necessary, adjust oil level.
4. Install fill plug.
5. Lower vehicle.

1. Input shaft
2. Synchronizer ring
3. Pilot bearing
4. Mainshaft assembly (2WD—MCO & MG5)
5. Hex head bolt (M8 × 50)
6. Snapring (selective)
7. 3-4 synchronizer assembly
8. Synchronizer sleeve
9. Synchronizer detent ball
10. Synchronizer detent spring
11. Synchronizer key
12. 3-4 synchronizer hub
13. 3rd speed gear
14. 3rd speed gear needle bearing assembly
15. Snapring (selective)
16. 2nd speed gear inner needle bearing race
17. Speed gear needle bearing assembly
18. 2nd speed gear
19. 1/2 synchronizer assembly
20. 1/2 synchronizer sleeve
21. 1/2 synchronizer hub
22. 1st speed gear
23. Output shaft
24. Hex head bolt (M8 × 50)
25. Reverse speed gear
26. 5th/reverse synchronizer assembly (MG5)
27. Sprial lock ring (MG5)
28. 5th/reverse synchronizer hub
29. Snarping (selective)
30. 5th speed gear needle bearing assembly (MG5)
31. 5th speed gear assembly (MG5)
32. Reverse synchronizer assembly (MCO)
33. Output shaft bearing assembly
34. Roller bearing race
35. Roller bearing
36. Inner ball bearing race
37. Bearing assembly outer race
38. Ball bearing outer race
39. Outer threaded thrust ring
40. Ball (4mm)
41. Inner threaded thrust ring
42. Spril roll pin
43. Snapring
44. Speedometer gear thrust washer (2WD)
45. Speed sensor rotor
46. Reverse idler assembly
47. Snapring
48. Thrust washer
49. Reverse gear
50. Needle bearing assembly
51. Reverse idler shaft
52. O-ring
53. Countershaft assembly (MG5)
54. Countershaft (MCO)
55. 5th gear (MG5)
56. Front housing
57. Countershaft bearing assembly
58. Countershaft bearing race
59. Countershaft bearing
60. Shim (selective)
61. Rear housing (2WD)
62. Output shaft bearing retainer (2WD)
63. Hex head bolt (M8 × 50)
64. Hex head bolt (M8 × 50)
65. Idler shaft support

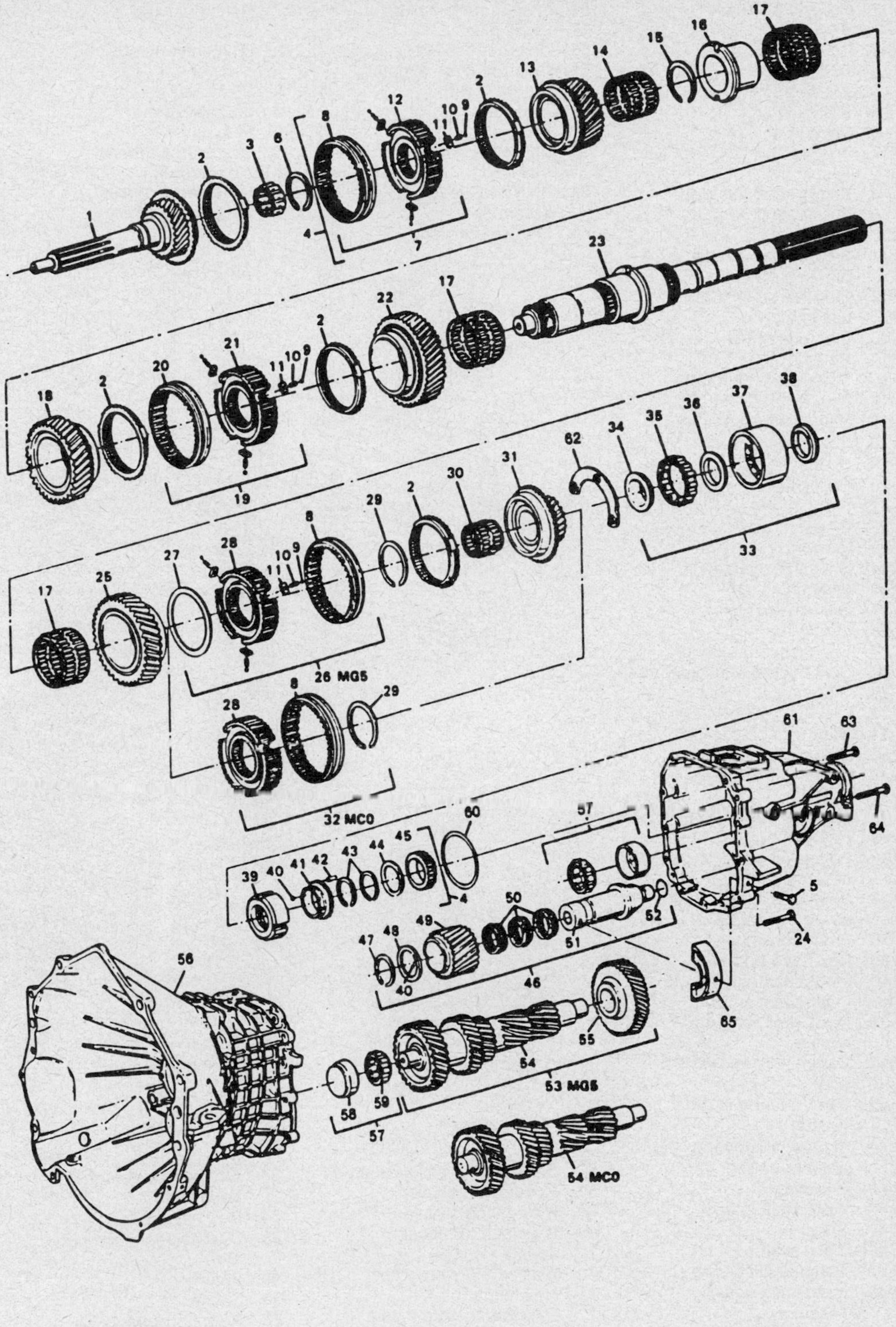
26 MG5
32 MCO
53 MG5
54 MCO

Gear components 2WD

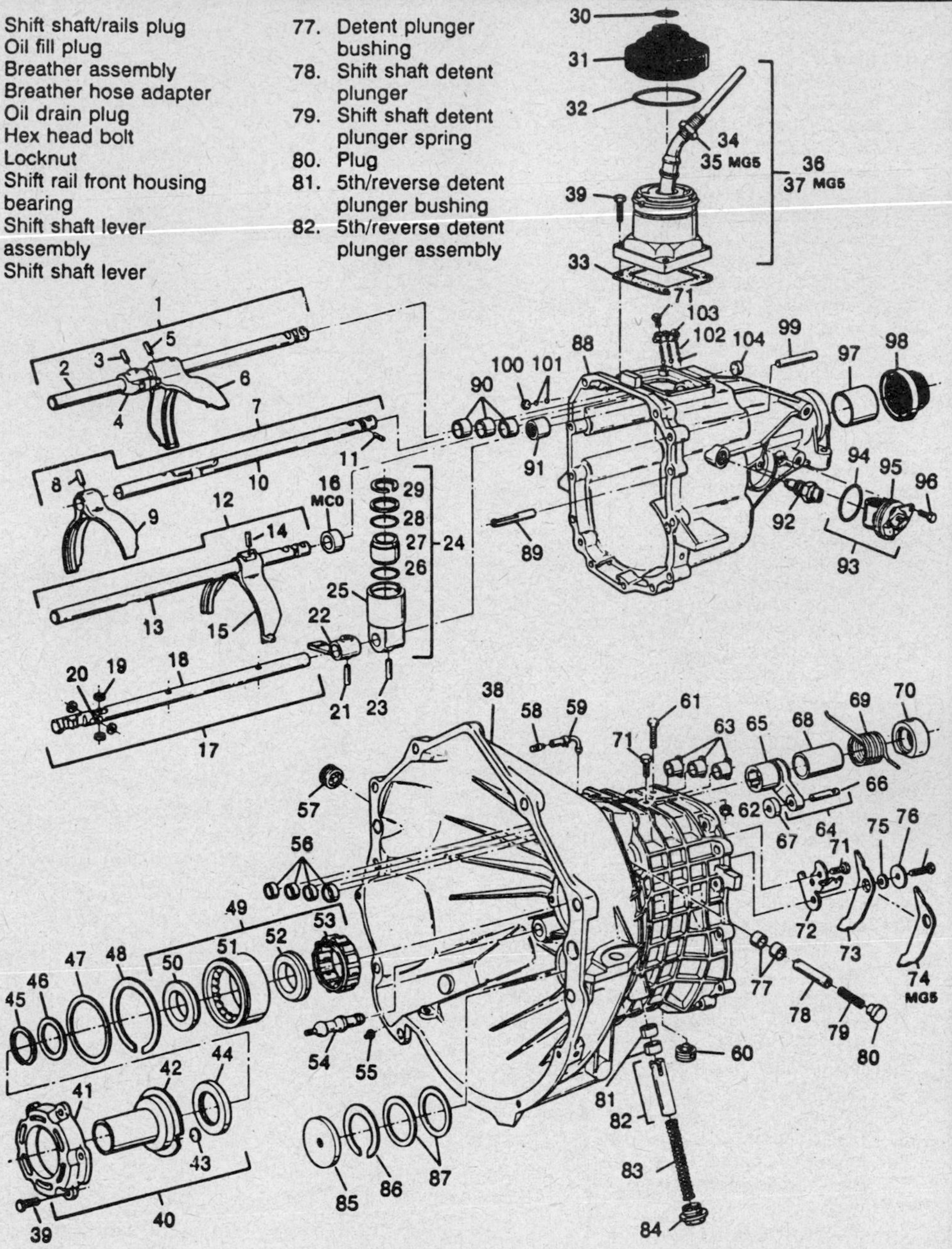

1. 1st/2nd shift rail assembly
2. 1st/2nd shift rail
3. Roll pin
4. 1st/2nd shift yoke
5. Roll pin
6. 1st/2nd shift fork
7. 3rd/4th shift rail assembly
8. 3rd/4th fork roll pin
9. 3rd/4th shift fork
10. 3rd/4th rail
11. 3rd/4th shift interlock pin
12. 5th/reverse shift rail assembly
13. 5th/reverse rail
14. Roll pin
15. 5th/reverse shift fork
16. Spacer block (MCO)
17. Shift shaft assembly
18. Shift shaft
19. Shift shaft roller
20. Roller pin
21. Roll pin
22. Finger
23. Roll pin
24. Shift shaft socket assembly
25. Shift shaft socket
26. O-ring
27. Shift socket bushing
28. Shift socket washer
29. Snapring
30. O-ring
31. Boot
32. O-ring
33. Shift lever housing seal
34. Hex jam nut (MCO)
35. Hex jam nut (MG5)
36. Shift lever housing assembly (MCO)
37. Shift lever housing assembly (MG5)
38. Front housing
39. Hex head bolt
40. Input shaft bearing retainer assembly
41. Input shaft bearing retainer
42. Clutch release bearing pilot
43. Input bearing retainer washer
44. Input shaft bearing retainer oil seal
45. Snapring
46. Input shaft spacer
47. Shim
48. Input shaft bearing assembly snapring
49. Input shaft bearing assembly
50. Outer ball bearing race
51. Outer bearing assembly race
52. Inner ball bearing race
53. Roller bearing
54. Clutch fork pivot assembly
55. Lube fitting
56. Shift shaft/rails plug
57. Oil fill plug
58. Breather assembly
59. Breather hose adapter
60. Oil drain plug
61. Hex head bolt
62. Locknut
63. Shift rail front housing bearing
64. Shift shaft lever assembly
65. Shift shaft lever
66. Shift shaft lever pin
67. Cover detent roller
68. Bias spring sleeve
69. Bias load torsional spring
70. Bias spring/sleeve seat
71. Hex head bolt
72. 5th/reverse detent cam support
73. 5th/reverse detent cam (MCO)
74. 5th/reverse detent cam (MG5)
75. Detent cam pivot sleeve
76. Retaining washer
77. Detent plunger bushing
78. Shift shaft detent plunger
79. Shift shaft detent plunger spring
80. Plug
81. 5th/reverse detent plunger bushing
82. 5th/reverse detent plunger assembly
83. 5th/reverse detent spring
84. Hex head plug
85. Countershaft bearing plug
86. Snapring
87. Shims
88. Rear housing (2WD)
89. Oil delivery tube assembly
90. Rear housing shift rail bearing
91. Rear housing shift shaft bearing
92. Backup light switch assembly
93. Electronic speed sensor assembly
94. O-ring
95. Electronic speed sensor
96. Hex head bolt
97. Rear extension bearing (2WD)
98. Slip yoke oil seal (2WD)
99. Dowel pin
100. Plug
101. Detent/interlock ball
102. Detent spring
103. Detent spring cover
104. Plug

Shift mechanism and case components 2WD

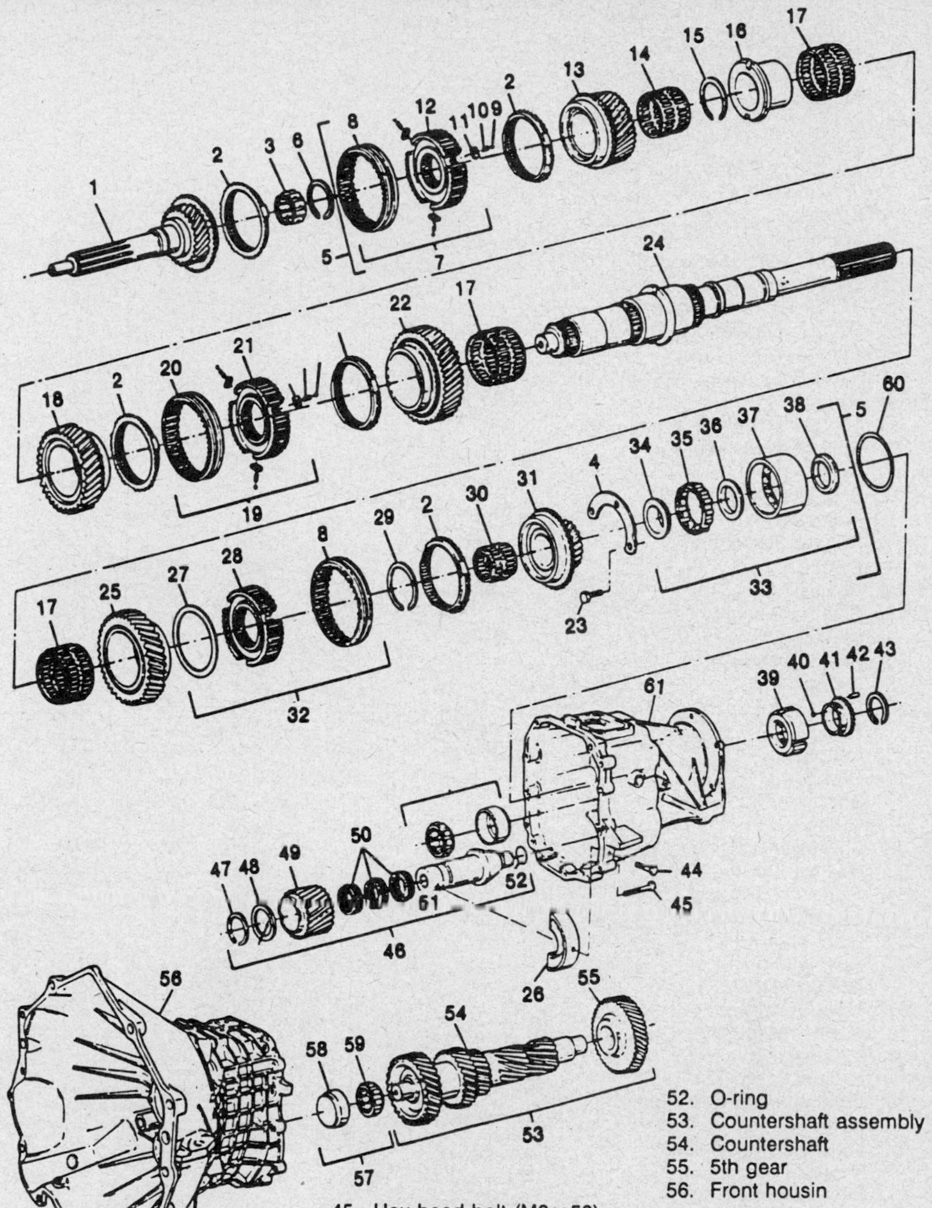

1. Input shaft
2. Synchronizer ring
3. Pilot bearing
4. Output shaft bearing retainer (4WD)
5. Mainshaft assembly (4WD)
6. Snapring
7. 3-4 synchronizer assembly
8. Synchronizer sleeve
9. Synchronizer detent ball
10. Synchronizer detent spring
11. Synchronizer key
12. 3-4 synchronizer hub
13. 3rd speed gear assembly
14. 3rd speed gear needle bearing assembly
15. Snapring
16. 2nd speed gear needle inner bearing race
17. Speed gear needle bearing assembly
18. 2nd speed gear assembly
19. 1-2 synchronizer assembly
20. 1-2 synchronizer sleeve
21. 1-2 synchronizer hub
22. 1st speed gear
23. Hex head bolt (M8 × 25)
24. Output shaft (4WD)
25. Reverse speed gear assembly
26. Idler shaft support
27. Spiral lock ring
28. 5th/reverse synchronizer hub
29. Snapring
30. 5th speed gear needle bearing assembly
31. 5th speed gear assembly
32. Reverse synchronizer assembly
33. Output shaft bearing assembly
34. Roller bearing assembly
35. Roller bearing
36. Inner ball bearing race
37. Outer bearing assembly race
38. Outer ball bearing race
39. Outer threaded thrust ring
40. 4mm ball
41. Inner threaded thrust ring
42. Roll spring pin
43. Snapring
44. Hex head bolt (M8 × 50)
45. Hex head bolt (M8 × 50)
46. Reverse idler assembly
47. Snapring
48. Thrust washer
49. Reverse gear
50. Needle bearing assembly
51. Reverse idler shaft
52. O-ring
53. Countershaft assembly
54. Countershaft
55. 5th gear
56. Front housin g
57. Countershaft bearing assembly
58. Countershaft bearing race
59. Countershaft bearing
60. Shim
61. Rear housing (4WD)

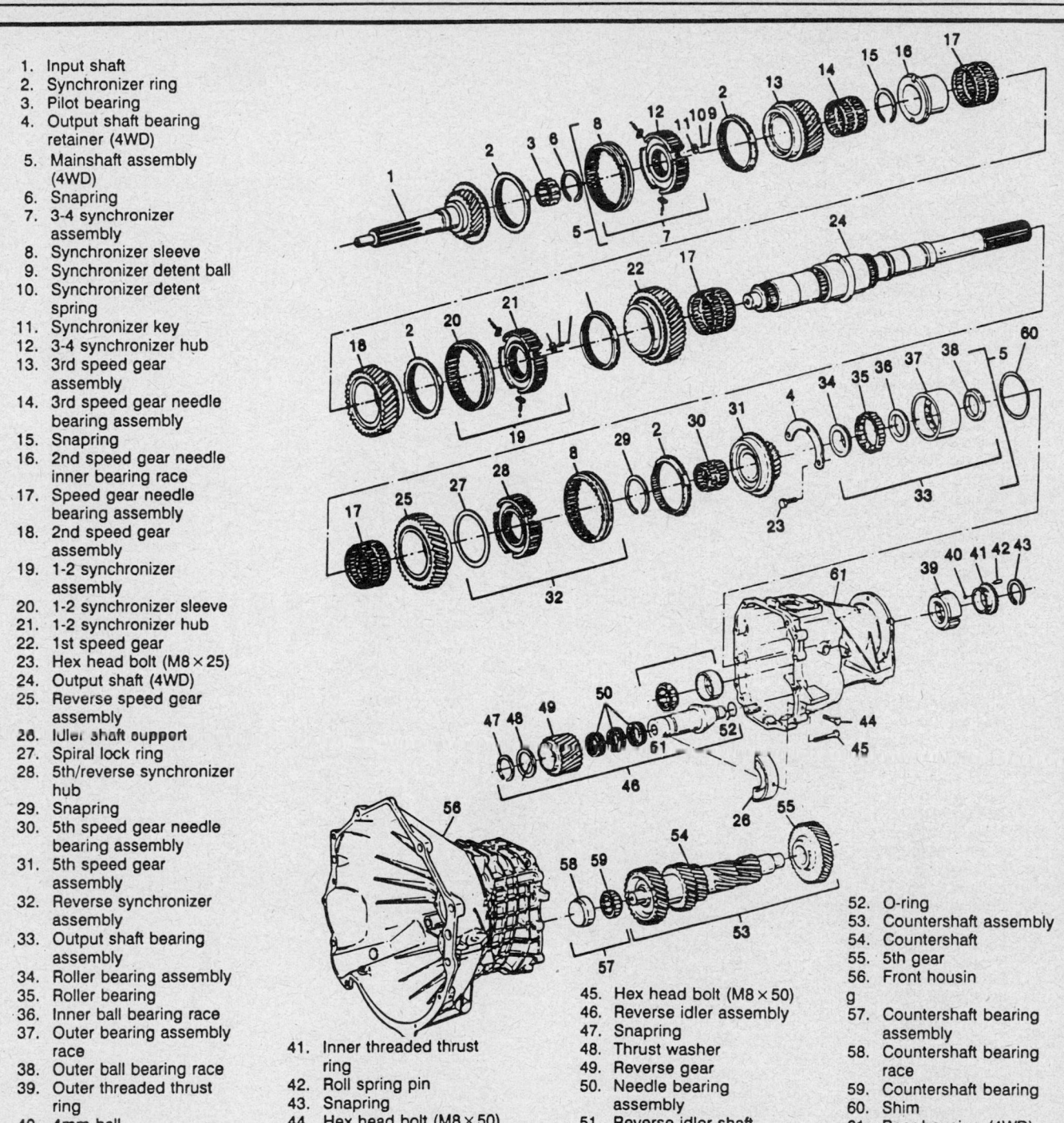

Gear components 4WD

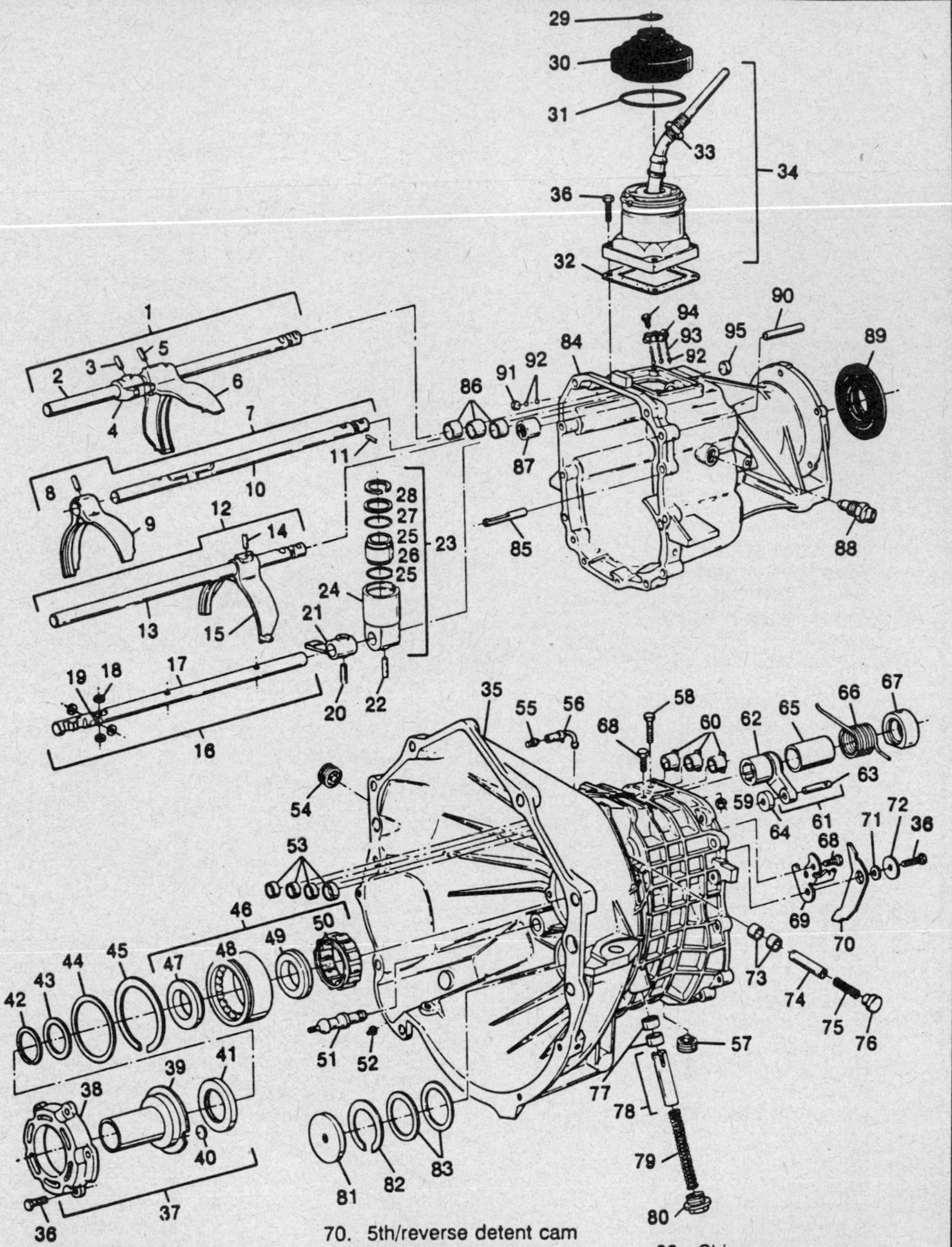

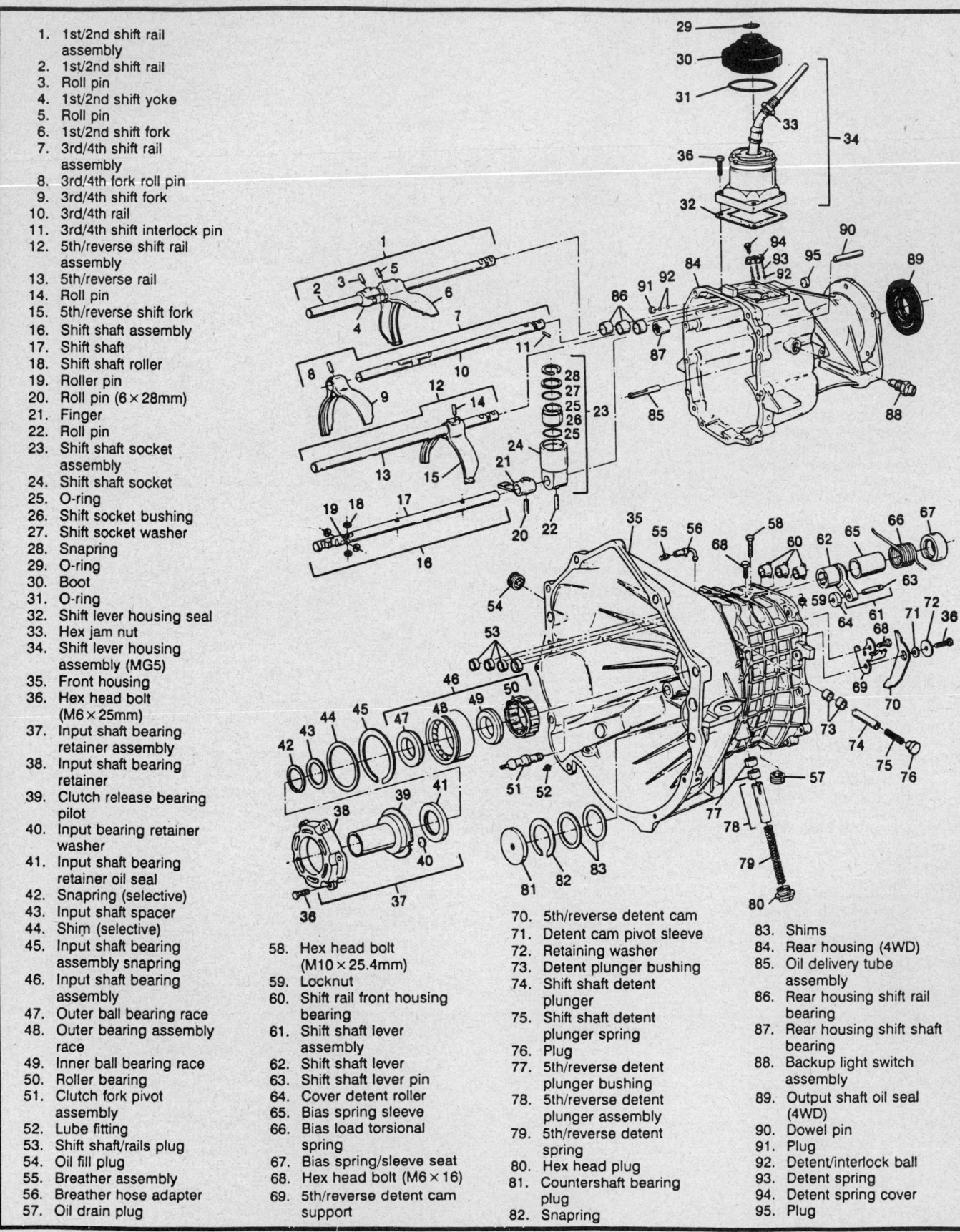

1. 1st/2nd shift rail assembly
2. 1st/2nd shift rail
3. Roll pin
4. 1st/2nd shift yoke
5. Roll pin
6. 1st/2nd shift fork
7. 3rd/4th shift rail assembly
8. 3rd/4th fork roll pin
9. 3rd/4th shift fork
10. 3rd/4th rail
11. 3rd/4th shift interlock pin
12. 5th/reverse shift rail assembly
13. 5th/reverse rail
14. Roll pin
15. 5th/reverse shift fork
16. Shift shaft assembly
17. Shift shaft
18. Shift shaft roller
19. Roller pin
20. Roll pin (6×28mm)
21. Finger
22. Roll pin
23. Shift shaft socket assembly
24. Shift shaft socket
25. O-ring
26. Shift socket bushing
27. Shift socket washer
28. Snapring
29. O-ring
30. Boot
31. O-ring
32. Shift lever housing seal
33. Hex jam nut
34. Shift lever housing assembly (MG5)
35. Front housing
36. Hex head bolt (M6×25mm)
37. Input shaft bearing retainer assembly
38. Input shaft bearing retainer
39. Clutch release bearing pilot
40. Input bearing retainer washer
41. Input shaft bearing retainer oil seal
42. Snapring (selective)
43. Input shaft spacer
44. Shim (selective)
45. Input shaft bearing assembly snapring
46. Input shaft bearing assembly
47. Outer ball bearing race
48. Outer bearing assembly race
49. Inner ball bearing race
50. Roller bearing
51. Clutch fork pivot assembly
52. Lube fitting
53. Shift shaft/rails plug
54. Oil fill plug
55. Breather assembly
56. Breather hose adapter
57. Oil drain plug
58. Hex head bolt (M10×25.4mm)
59. Locknut
60. Shift rail front housing bearing
61. Shift shaft lever assembly
62. Shift shaft lever
63. Shift shaft lever pin
64. Cover detent roller
65. Bias spring sleeve
66. Bias load torsional spring
67. Bias spring/sleeve seat
68. Hex head bolt (M6×16)
69. 5th/reverse detent cam support
70. 5th/reverse detent cam
71. Detent cam pivot sleeve
72. Retaining washer
73. Detent plunger bushing
74. Shift shaft detent plunger
75. Shift shaft detent plunger spring
76. Plug
77. 5th/reverse detent plunger bushing
78. 5th/reverse detent plunger assembly
79. 5th/reverse detent spring
80. Hex head plug
81. Countershaft bearing plug
82. Snapring
83. Shims
84. Rear housing (4WD)
85. Oil delivery tube assembly
86. Rear housing shift rail bearing
87. Rear housing shift shaft bearing
88. Backup light switch assembly
89. Output shaft oil seal (4WD)
90. Dowel pin
91. Plug
92. Detent/interlock ball
93. Detent spring
94. Detent spring cover
95. Plug

Shift mechanism and case components 4WD

TRANSMISSION MODIFICATIONS

There is no transmission modifications at the time of publication.

TROUBLE DIAGNOSIS

CHILTON'S THREE C'S TRANSMISSION DIAGNOSIS

Condition	Cause	Correction
Leaks lubricant	a) Lubricant level too high b) Main drive bearing retainer or gasket loose or damaged c) Side cover or gasket loose or damaged d) Rear extension seal damaged e) Countershaft loose in case	a) Drain to correct level b) Tighten or replace c) Tighten or replace d) Replace rear extension seal e) Replace case
Noisy shifting	a) Clutch worn or damaged b) Synchronizers or gears worn or damaged	a) Replace clutch b) Repair the transmission
Noisy neutral	a) Pilot bearing worn or damaged b) Main drive gear or countergear bearings worn or damaged	a) Replace pilot bearing b) Repair the transmission
Noisy operation	a) Lubricant level low b) Shift rods damaged c) Sychronizers worn or damaged d) Bearings worn or damaged e) Gears worn or damaged	a) Fill to correct level b) Replace shift rods c) Repair the transmission d) Repair the transmission e) Repair the transmission
Slips out of gear	a) Shift lever seal stiff b) Pilot bearing loose or damaged c) Dirt between the clutch housing and transmission d) Transmission loose e) Main drive gear retainer loose or damaged f) Transmission not aligned	a) Replace shift lever seal b) Replace pilot bearing c) Clean the mating surfaces d) Tighten transmission e) Tighten or replace f) Align transmission

ON CAR SERVICE

Services

REAR EXTENSION HOUSING SEAL

Removal and Installation

1. Raise the vehicle and support it safely.
2. Drain transmission oil.
3. Remove dirveshaft.
4. Using the proper tool, remove rear oil seal.
5. Apply locking compound on the outside of the new seal.
6. Install the new seal using a suitable tool.
7. Install driveshaft.
8. Refill with new transmission oil.
9. Lower the vehicle.

VEHICLE SPEED SENSOR (VSS)

Removal and Installation

1. Raise the vehicle and support it safely.
2. Remove electrical harness connector.
3. Remove VSS cover attaching screw and remove cover, if used.
4. Remove VSS and seal assembly.
5. Install a new seal on sensor. Coat the new seal with a thin flim of transmission oil.
6. Install the VSS assembly.
7. Install VSS cover and attaching screw, if used.
8. Reconnect electrical harness connector.
9. Lower the vehicle.

SHIFT CONTROL LEVER

Removal and Installation

1. Remove transfer case shift lever boot, if used.
2. Remove shift lever retainer attaching screw, retainer and boot.
3. Loosen shift lever jam nut and remove lever.
4. Position shift lever in place and tighten jam nut.

NOTE: The shift pattern on the knob must be horizontal to the driver.

5. Install the boot, retainer and attaching screws.
6. Install transfer case shift lever boot, if used.

REMOVAL AND INSTALLATION

TRANSMISSION REMOVAL

1. Disconnect battery negative cable.
2. Remove shift lever boot retainer attaching screws, retainer and boot.
3. Remove shift tower and seal from case. Cover the opening.
4. Raise the vehicle and support it safely. Drain the transmission lubricant.
5. Disconnect driveshaft.
6. Remove transfer case, if equipped.
7. Disconnect parking brake and controls.
8. Remove exhaust pipes.
9. Note the positions of any lines or wires before removing. Disconnect electrical harness connector from transmission.
10. Disconnect clutch slave cylinder and lay aside.
11. Remove inspection cover attaching screws and remove cover.
12. Support the transmission with a jack.

NOTE: Do not let the transmission hang from the clutch. Use guide pins to pull the transmission straight back on the clutch hub splines.

13. Remove crossmember and any other parts needed for clearance.
14. Remove the top 2 studs last and install guide pins. Remove screws, spring washers and studs as used.
15. Support the clutch release bearing and remove transmission.
16. Remove engine to transmission support plugs, if they are loose or damaged.

TRANSMISSION INSTALLATION

1. Install new engine to transmission support plugs, if needed.
2. Before installation, shift the transmission into high gear and coat the main drive gear splines with high temperature grease.
3. Install the top 2 studs first. Torque screws and studs to 37 ft. lbs. (50 Nm).

NOTE: Do not let the transmission hang from the clutch. Use guide pins to pull the transmission straight back on the clutch hub splines.

4. Install crossmember and remove jack.
5. Install any parts that were removed for clearance.
6. Install inspection cover and attaching screws.
7. Install the clutch slave cylinder.
8. Install electrical harness connectors, exhaust pipes, parking brake and controls.
9. Install transfer case, if equipped.
10. Install driveshaft.
11. Fill transmission with new lubricant.
12. Lower the vehicle and connect battery cable.
13. Install shift tower, a new seal, shift control lever and boot.

BENCH OVERHAUL

Before Disassembly

Thoroughly clean the exterior of the transmission before any attempt is made to disassemble it. This prevents dirt or other foreign materials from entering the transmission assembly or its internal parts.

NOTE: If steam cleaning is done to the exterior of the transmission, immediate disassembly should be done to avoid rusting, caused by condensation forming on the internal parts.

Transmission Disassembly

EXCEPT REAR HOUSING

1. Remove idler shaft support and 2 bottom bolts for special tools.
2. Support the transmission with holding fixtures J-8763-02, J-8763-21 and J-36824 or equivalent.
3. Remove backup lamp switch assembly.
4. Remove the electronic speed sensor assembly.
5. Remove detent plug, spring and plunger using removal tool J-36509 and J-23907 or equivalent.
6. Remove the detent spring cover 2 attaching bolts and remove cover, detent springs and 3 detent/interlock balls.

NOTE: It may be necessary to remove sealant from inside of holes to remove the balls.

7. On 2WD vehicles, remove the output shaft oil seal using tool J-36825 and J-23907 or equivalent. On 4WD vehicles, screw J-36825 into 1 of the 3 perforated holes in the seal and remove seal.
8. Rotate the transmission in a vertical position.
9. Remove the 6 input shaft bearing retainer attaching bolts and remove retainer assembly.
   a. Tap on clutch release bearing pilot with a rubber hammer.
   b. Save the input bearing retainer washer.

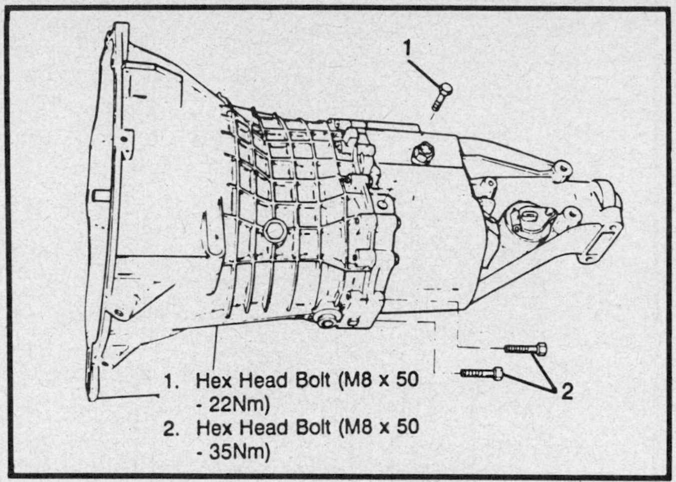
1
2
1. Hex Head Bolt (M8 x 50 - 22Nm)
2. Hex Head Bolt (M8 x 50 - 35Nm)

Bolt removal for special tool installation

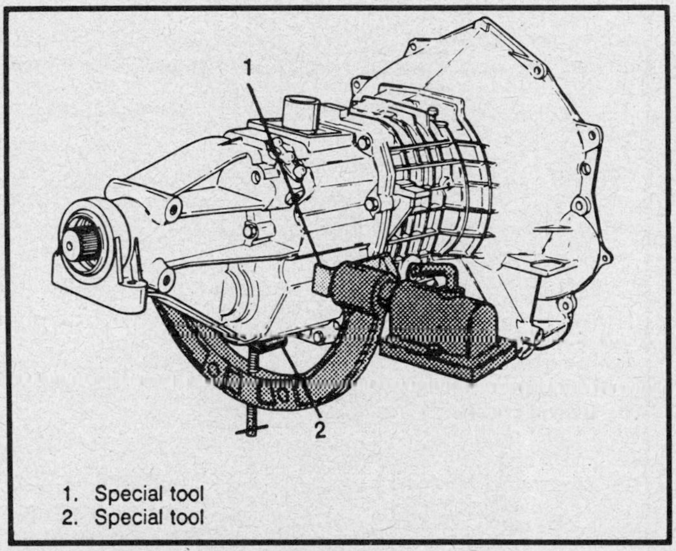
1
2
1. Special tool
2. Special tool

Transmission holding fixture

10. Remove snapring (selective), input shaft spacer, ball bearing inner race and shim.
11. Position the transmission horizontally.
12. Remove front housing to rear housing attaching bolts.
13. Drive dowels into front housing and remove housing.
14. Remove the countershaft bearing.
15. Remove bearing inner race and roller bearing.

NOTE: Degreasing with a liquid cleaner will make the bearing easier to remove. Grab the outer diameter edge of the inner race and remove with large pliers. Do not damage bearing cage while removing bearing inner race.

16. Remove idler shaft support.
17. Remove 4 rollers. Pull shift shaft forward and cock to detent cover side and remove roll pin. Support the shift shaft end while driving out finger roll pin.
18. Remove shift shaft socket assembly roll pin.
19. Remove shift shaft, shift shaft socket assembly and finger.
20. With a pair of diagonal cutters, pry out 3rd and 4th shift fork roll pin. If the roll pin breaks off, put the transmission into 3rd gear, cut the remainder of pin and drive through rail.
21. Remove the 3/4 rail plug by hitting on one side, allowing it to cock.

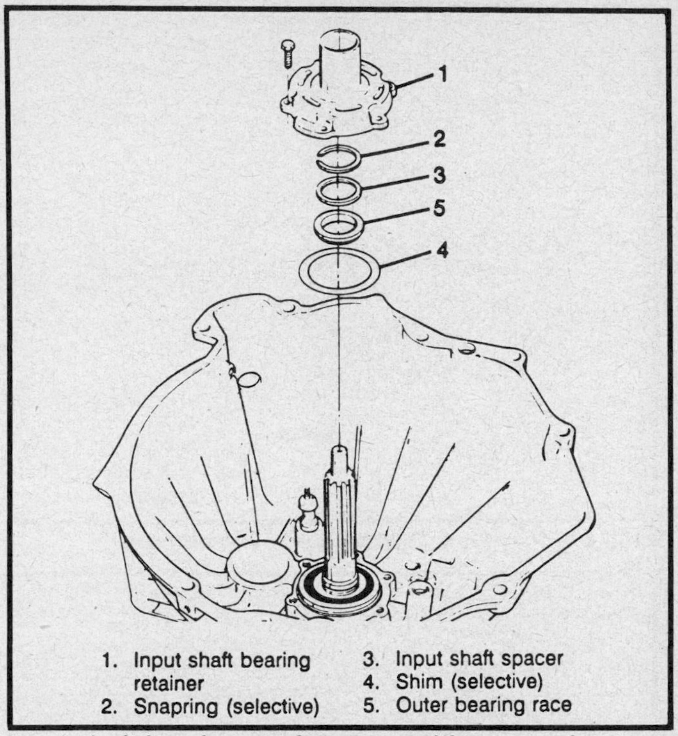
1
2
3
5
4
1. Input shaft bearing retainer
2. Snapring (selective)
3. Input shaft spacer
4. Shim (selective)
5. Outer bearing race

Input shaft bearing retainer assembly removal

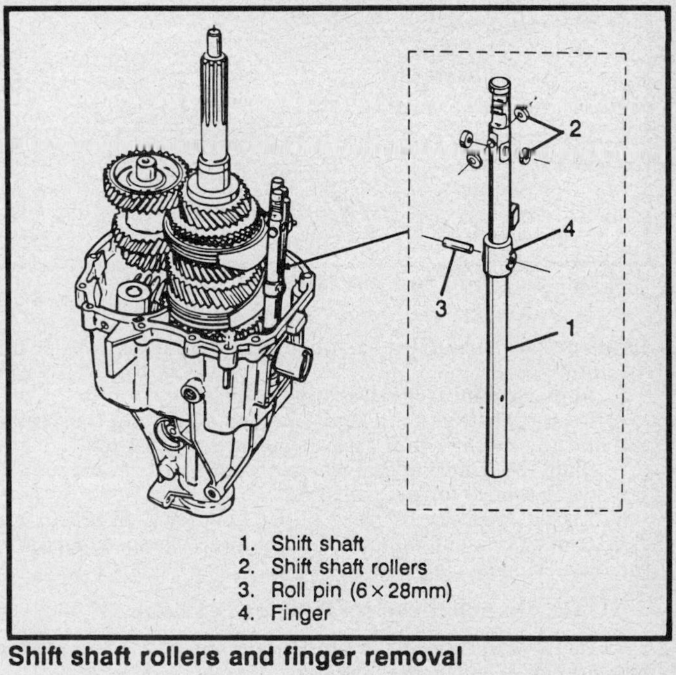
2
4
3
1
1. Shift shaft
2. Shift shaft rollers
3. Roll pin (6×28mm)
4. Finger

Shift shaft rollers and finger removal

22. Remove the 3/4 rail and detent/interlock balls.

REAR HOUSING

2WD VEHICLES

1. Support transmission assembly on tool J-36515 and J-36515-12 or equivalent. Remove holding fixtures J-36824, J-8763-02 and J-8763-21 or equivalent.
2. Remove rear housing mounting bolts. Alternately tap up-

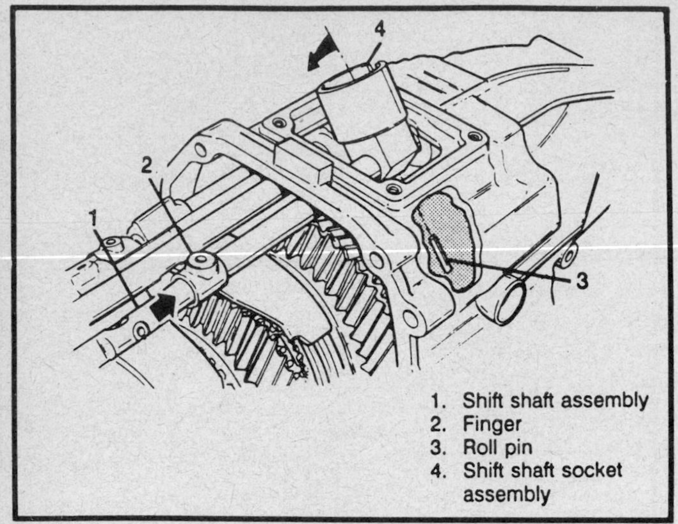
1
2
3
4
1. Shift shaft assembly
2. Finger
3. Roll pin
4. Shift shaft socket assembly

Shift shaft socket assembly roll pin removal

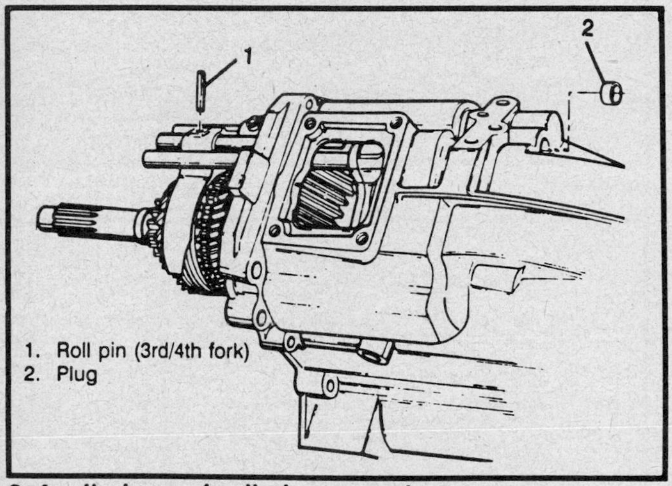
1
2
1. Roll pin (3rd/4th fork)
2. Plug

3–4 rail plug and roll pin removal

ward on rear housing with a rubber mallet and remove housing assembly.

3. Remove shim (selective) from housing.
4. Lock up assembly in 2nd/4th gear by sliding the 1st/2nd rail and the 3rd/4th shift fork downward against tool.
5. Slide the snapring and washer from speed sensor rotor to provide clearance for special tools.
6. Install speedometer gear puller adapter J–21427–01 and J–8205 or equivalent, on mainshaft and remove speed sensor rotor. Remove snapring and washer.

NOTE: Do not reuse rotor after removal.

7. Remove the sprial roll pin from mainshaft using spanner nut wrench J–36516 or equivalent.
   a. Position the black depth locating tang for the spiral roll pin remover/installer for clearance and drive downward.
   b. Position the black depth locating tang back into place with the rod going through it.
   c. Rotate spanner nut wrench J–36516 or equivalent clockwise until it stops.
8. Remove the threaded thrust ring (inner), ball and threaded thrust ring (outer) from mainshaft.
   a. Hold the countershaft against the mainshaft.
   b. Rotate spanner nut wrench J–36516 or equivalent clockwise until both threaded thrust rings are completely apart.

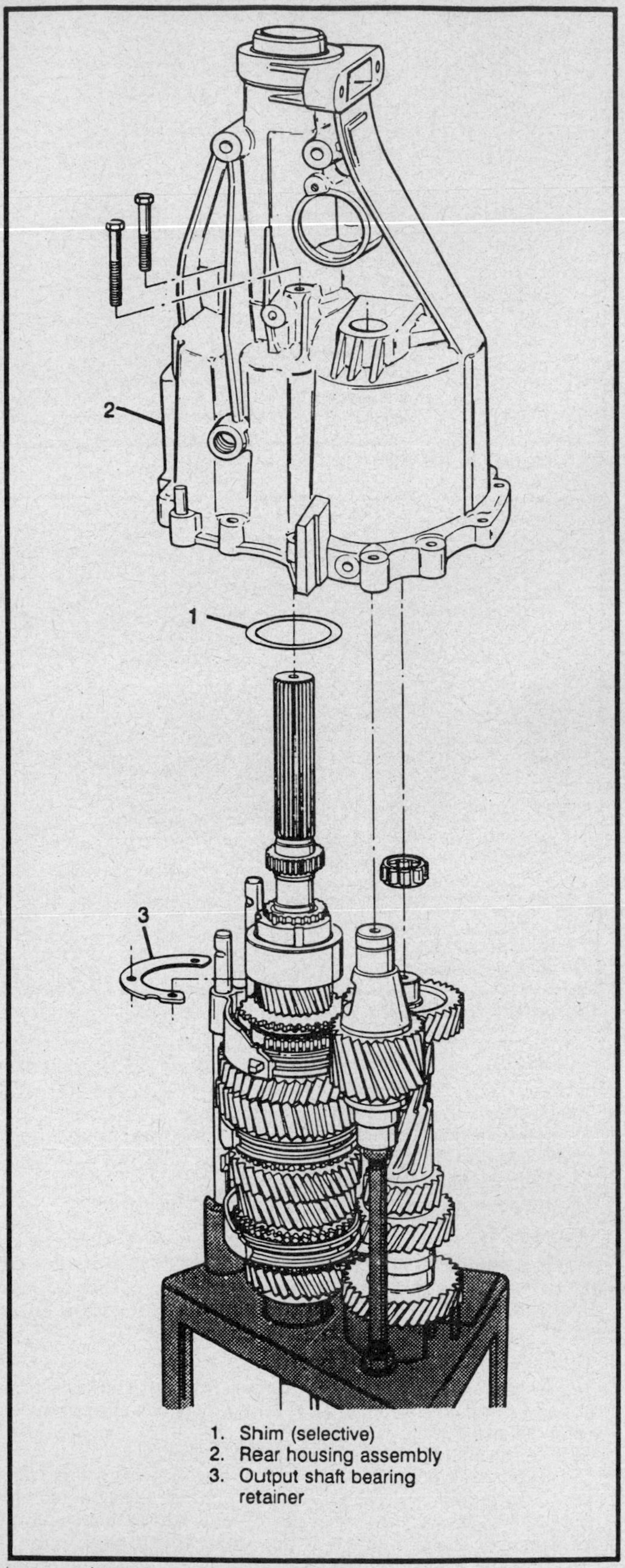
1
2
3
1. Shim (selective)
2. Rear housing assembly
3. Output shaft bearing retainer

Rear housing removal (2WD)

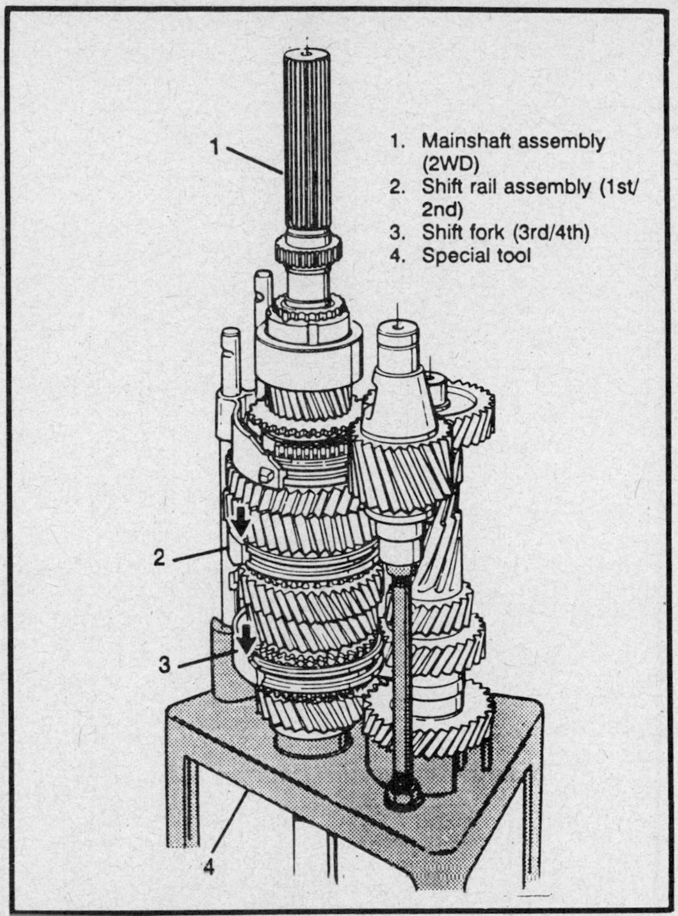
1. Mainshaft assembly (2WD)
2. Shift rail assembly (1st/2nd)
3. Shift fork (3rd/4th)
4. Special tool

Locking up transmission typical

9. Remove the reverse idler gear and countershaft.
10. Remove 1/2 shift rail assembly, 5/reverse shift rail assembly and 3/4 shift fork. On 4 speed transmissions, remove the spacer block.
11. Remove the mainshaft assembly.

NOTE: Leave the synchronizer ring on the 3/4 synchronizer assembly to prevent the synchronizer detent balls from popping out.

12. Remove the input shaft gear and pilot bearing.

4WD

1. Lock up the transmission in 2nd/4th gear by sliding the 1st/2nd shift rail assembly and 3rd/4th shift fork downward against tool.
2. Remove the spiral roll pin using spanner nut wrench J-36516 or equivalent.
   a. Position the black depth locating tang for the spiral roll pin remover/installer for clearance and drive downward.
   b. Position the black depth locating tang back into place with the rod going through it.
   c. Rotate spanner nut wrench J-36516 or equivalent clockwise until it stops.
3. Remove the threaded thrust ring (inner), ball and threaded thrust ring (outer) from mainshaft. Rotate spanner nut wrench J-36516 or equivalent clockwise until both threaded thrust rings are completely apart.
4. Remove the rear housing by alternately tapping upward on rear housing with a rubber mallet and remove housing assembly.
5. Save the ball bearing outer race that will be left in the rear housing.
6. Remove the countershaft bearing, countershaft and reverse idler assembly.
7. Remove 5th/reverse shift rail assembly, 1st/2nd shift rail assembly and 3rd/4th shift fork.

NOTE: Leave the synchronizer ring on the 3/4 synchronizer assembly to prevent the synchronizer detent balls from popping out.

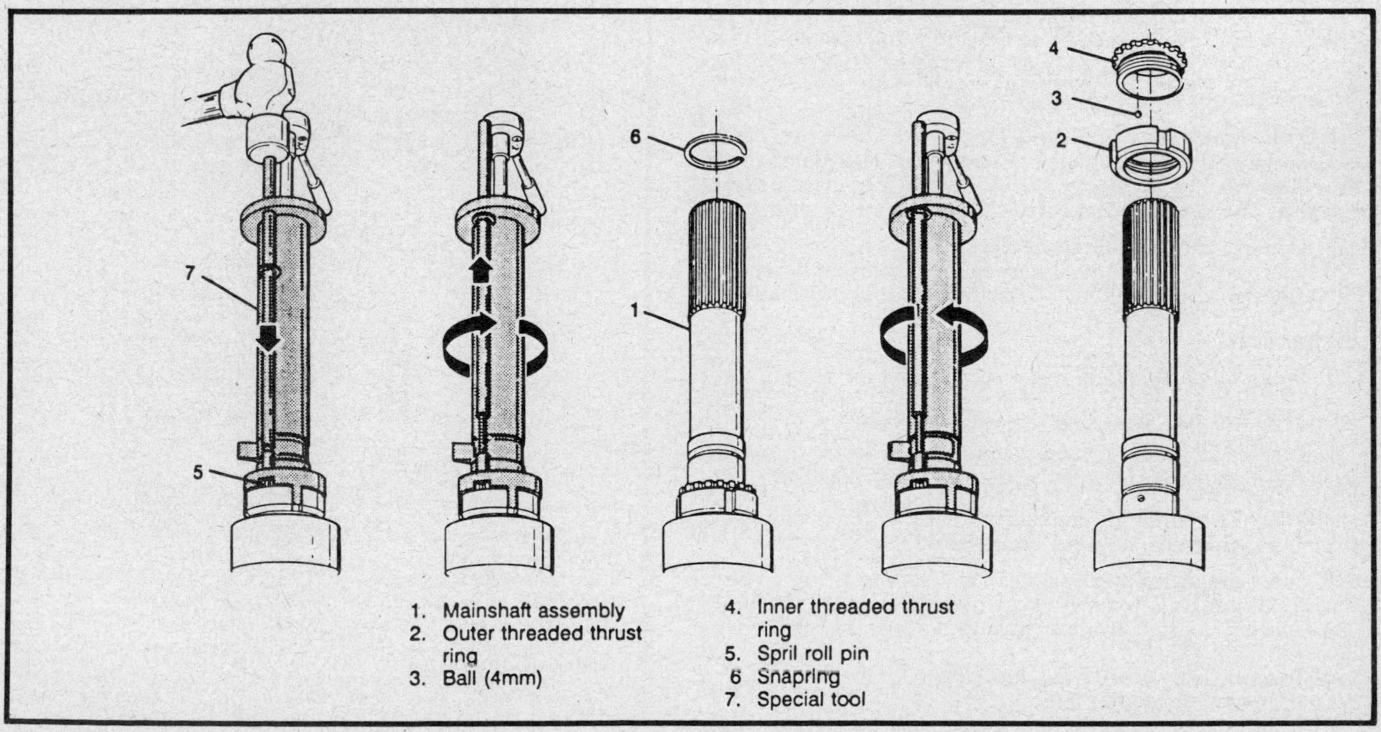
1. Mainshaft assembly
2. Outer threaded thrust ring
3. Ball (4mm)
4. Inner threaded thrust ring
5. Spril roll pin
6 Snapring
7. Special tool

Threaded thrust rings removal typical

8. Remove the input shaft gear and pilot bearing.

Unit Disassembly and Assembly

MAINSHAFT

Disassembly

1. Remove snapring (selective) from mainshaft.
2. Using bearing separator plate J-36513 or equivalent and a hydraulic press, remove 3/4 synchronizer assembly, synchronizer rings and 3rd speed gear.

NOTE: Mark the hub and sleeve so they can be installed in the same position. Also, leave the synchronizer rings to prevent the synchronizer detent balls from popping out.

3. Remove the 3rd gear bearing.
4. Remove the snapring (selective).
5. Using bearing separator plate J-36513 or equivalent and a hydraulic press, remove 2nd speed gear assembly and 2nd speed gear bearing race.
6. Remove the bearing assembly. Leave the synchronizer ring on the 1/2 synchronizer assembly to prevent the synchronizer detent balls from popping out.
7. Using bearing separator plate J-36513 or equivalent and a hydraulic press, remove the 1st speed gear and 1/2 synchronizer assembly.

NOTE: Mark the hub and sleeve so they can be installed in the same position. Also, leave the synchronizer rings on the 1/2 synchronizer assembly to prevent the synchronizer detent balls from popping out.

8. Remove the bearing assembly.
9. Using bearing separator plate J-36513 and a hydraulic press, remove the inner ball bearing race, roller bearing, roller bearing race and 5th speed gear.
10. Remove the 5th speed gear bearing.
11. Remove snapring (selective).
12. Using bearing separator plate J-36513 and a hydraulic press, remove reverse speed gear assembly and (5th speed) 5th/reverse synchronizer asembly or (4th speed) reverse synchronizer assembly.

NOTE: Mark the hub and sleeve so they can be installed in the same position. Also, leave the synchronizer rings on the 5th/reverse synchronizer assembly to prevent the synchronizer detent balls from popping out.

13. Remove the bearing assembly.
14. Place the 1/2, 3/4 and 5th speed synchronizers in separate shop towels, wrap assemblies and press against inner hub.

Inspection

1. Wash all parts in a suitable solvent and air dry.

NOTE: Do not spin dry the ball bearings.

2. Inspect all gears for cracks, chipped gear teeth, thrust face surfaces, bearing surface diameters and other damage.

NOTE: The black phosphate coating will develop wear patterns, this is a normal condition.

3. Inspect synchronizer sleeves for a sliding fit on synchronizer hubs and hubs to have a force fit on the mainshaft splines.
4. Inspect the synchronizer springs, keys and rings for wear or other damage.
5. Inspect the synchronizer clutching teeth for wear, scuffed, nicked, burred or broken teeth.
6. Inspect gear clutching cones for synchronizer ring metal transfer.

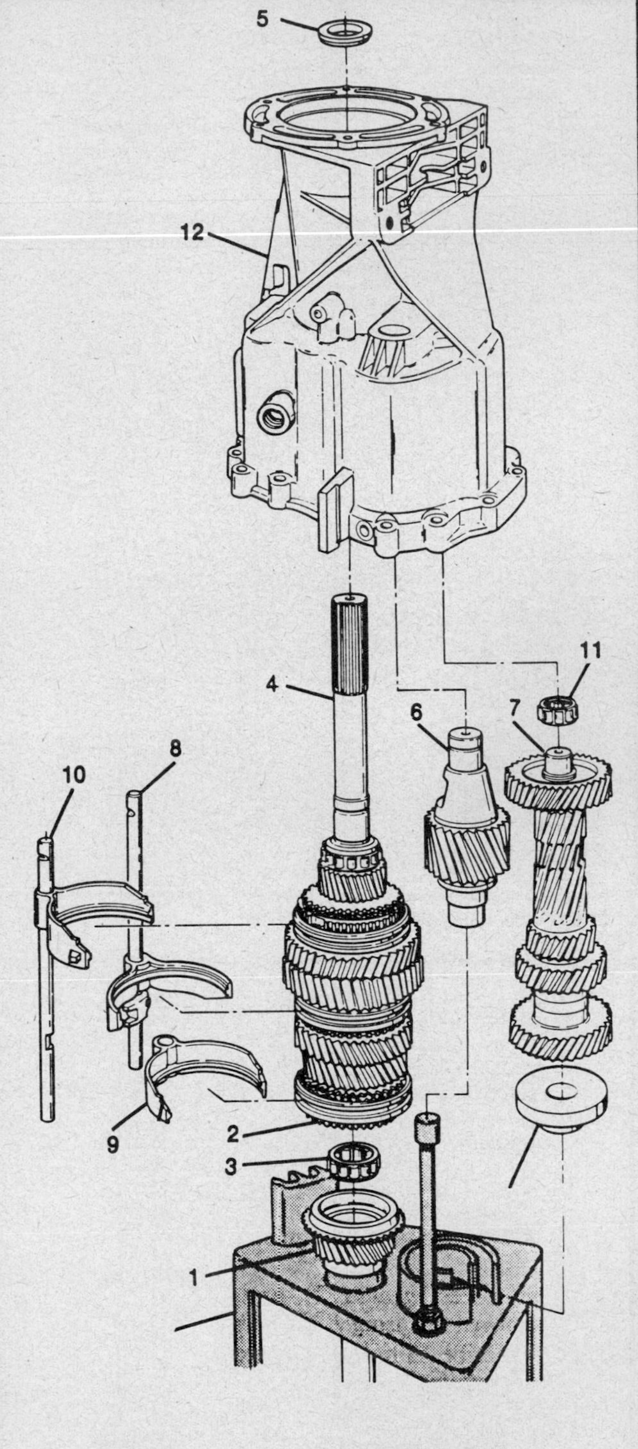
1
2
3
4
5
6
7
8
9
10
11
12

1. Input shaft
2. Synchronizer ring
3. Pilot bearing
4. Mainshaft assembly
5. Outer ball bearing race
6. Reverse idler assembly
7. Countershaft assembly
8. 1st/2nd shift rail assembly
9. 3rd/4th shift fork
10. 5th/reverse shift rail assembly
11. Countershaft bearing
12. Rear housing
13. Special tool
14. Special tool

Rear housing removal (4WD)

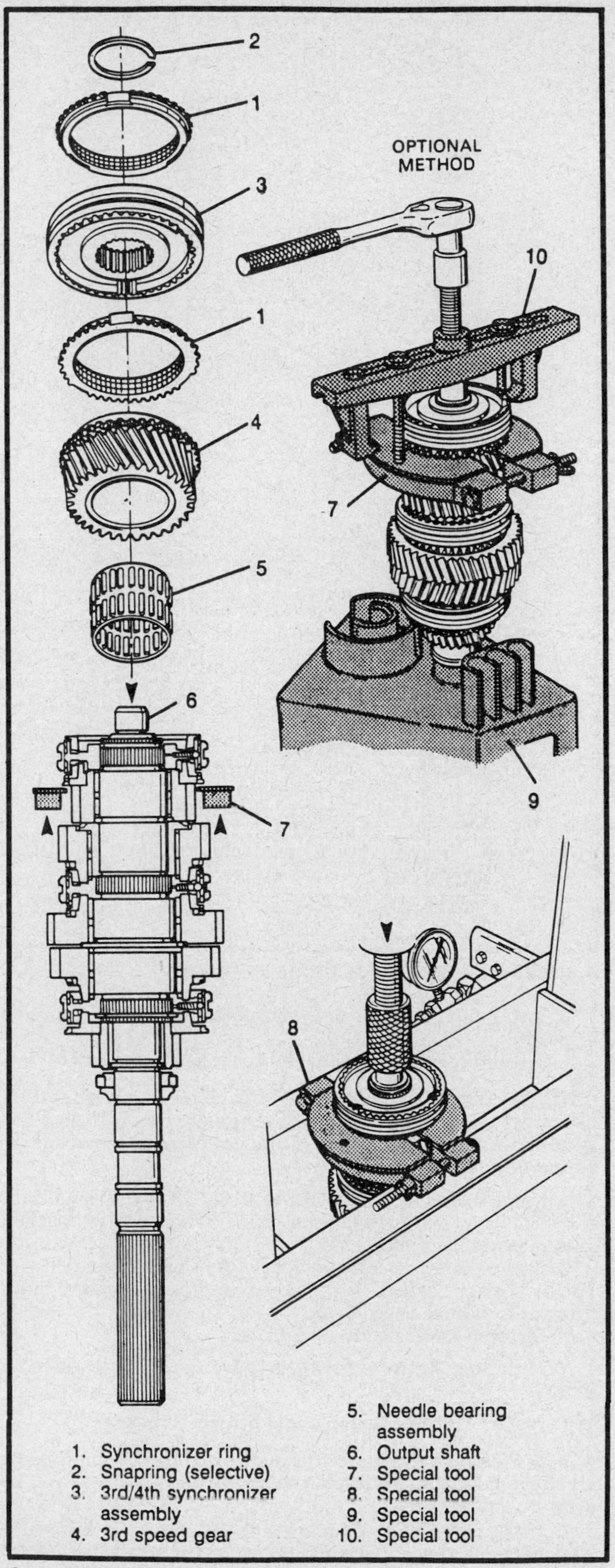
OPTIONAL METHOD
1
2
3
4
5
6
7
8
9
10
1. Synchronizer ring
2. Snapring (selective)
3. 3rd/4th synchronizer assembly
4. 3rd speed gear
5. Needle bearing assembly
6. Output shaft
7. Special tool
8. Special tool
9. Special tool
10. Special tool

3–4 gear components removal

7. Inspect the bearings and bearing surfaces for nicks, burrs, bent cages and wear.

9. Lubricate all bearings with light engine oil and check for rough rotation.

a. If scuffed, nicked, burred, scoring, or synchronizer ring metal transfer conditions cannot be removed with a soft cloth, replace the component and inspect mating parts.

b. Lubricate all components with 5W–30 oil or equivalent.

Assembly

NOTE: Prior to installation, heat the roller bearing race, ball bearing inner race and the speed sensor rotor

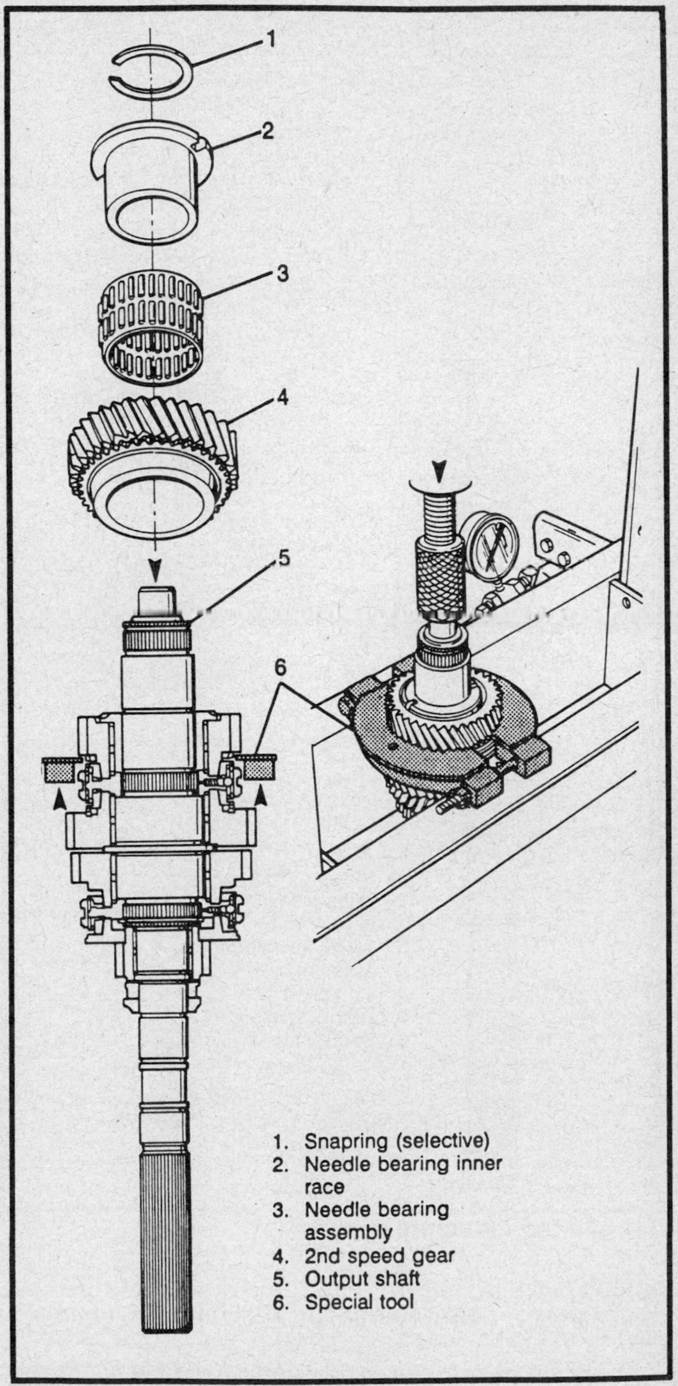
1
2
3
4
5
6
1. Snapring (selective)
2. Needle bearing inner race
3. Needle bearing assembly
4. 2nd speed gear
5. Output shaft
6. Special tool

2nd gear components removal

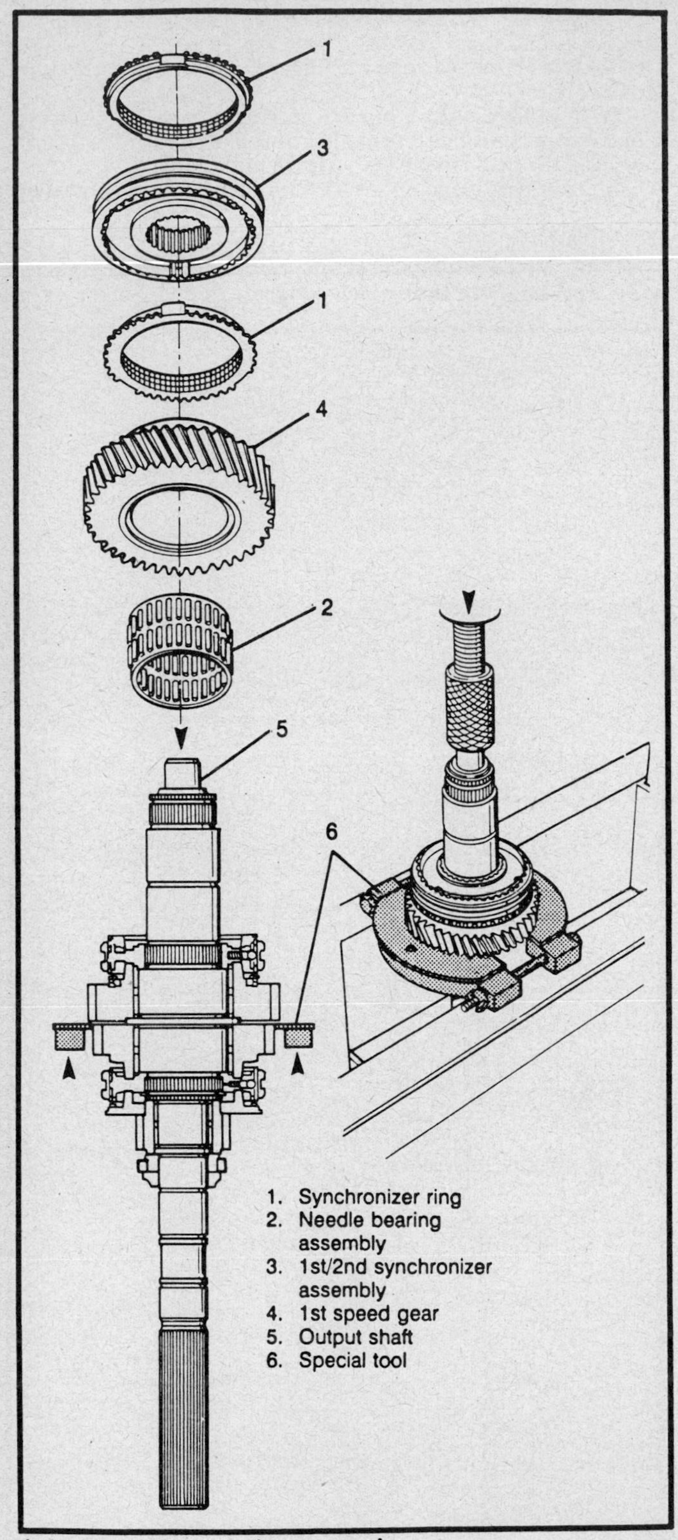
1
3
1
4
2
5
6
1. Synchronizer ring
2. Needle bearing assembly
3. 1st/2nd synchronizer assembly
4. 1st speed gear
5. Output shaft
6. Special tool

1st gear components removal

(2WD) for a period of 7–10 minutes at 250°F and 2nd speed gear race for a period of 20 minutes minimum, at 250°F.

1. Install 1st/2nd gear synchronizer assembly and 3rd/4th gear synchronizer assembly.

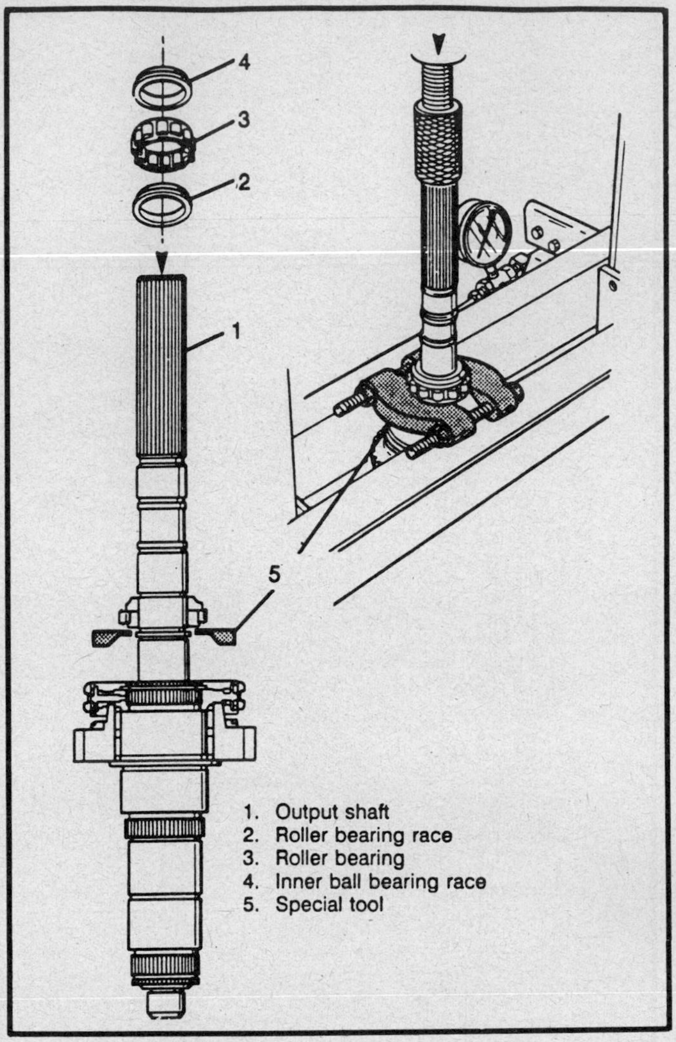
4
3
2
1
5
1. Output shaft
2. Roller bearing race
3. Roller bearing
4. Inner ball bearing race
5. Special tool

4 speed ball and roller races removal

2. Install 5th/reverse gear synchronizer assembly (4 speed or 5 speed).
3. Install bearing assembly and reverse speed gear.

NOTE: When pressing the 5th/reverse synchronizer assembly, manually align and engage the splines. Press synchronizer until seated. Make certain all shavings are removed.

4. Using tools J–36183, J–36184, J–36513 or equivalents and a hydraulic press, install 5th/reverse synchronizer assembly with synchronizer ring.
   a. Aling marks made previously, for correct positions.
   b. With spiral lock ring (5 speed) or groove (4 speed) in hub towards reverse speed gear.
5. Install a new selective snapring.

NOTE: Install the thickest snapring that will fit in the groove.

6. Install bearing assembly and 5th speed gear.

NOTE: Press the bearing races onto the mainshaft until there is no clearance between roller bearing race and stop on mainshaft.

7. Using tool J–6133–01 and bearing separator plate J–36513 or equivalent and a hydraulic press, install the roller bearing race. Shoulder down towards reverse gear.

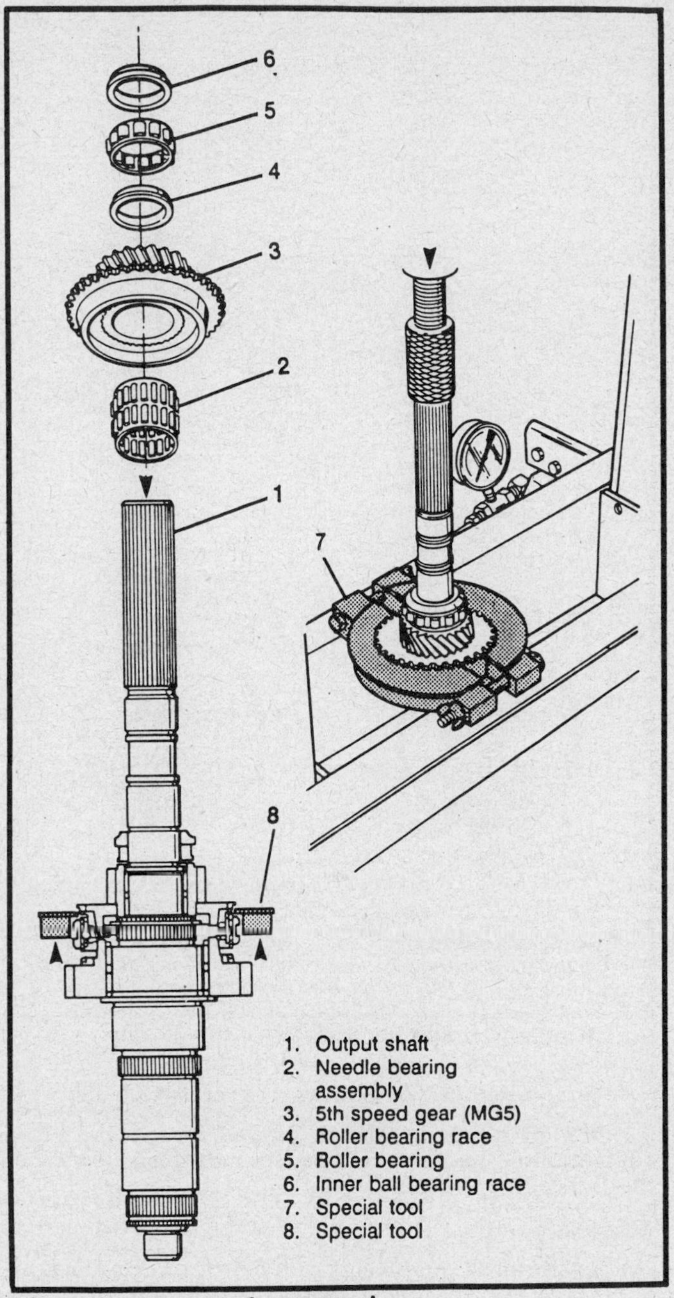
1. Output shaft
2. Needle bearing assembly
3. 5th speed gear (MG5)
4. Roller bearing race
5. Roller bearing
6. Inner ball bearing race
7. Special tool
8. Special tool

5th gear components removal

8. Using tool J–6133–01, J–36513 or equivalent and a hydraulic press, install roller bearing and inner ball bearing race.
   a. Apply grease to roller bearing.
   b. Smaller diameter of bearing cage up.
   c. Shoulder of inner ball bearing race down towards reverse gear.
9. Install bearing assembly and 1st speed gear.
   a. When pressing 1st/2nd synchronizer assembly, manually align and engage splines.
   b. Start pressing and stop before tangs engage.
   c. Lift and rotate gear to engage synchronizer ring.
   d. Press until seated.
   e. Make certain all shavings are removed.
10. Using tools J–36183, J–36184, J–22912–01 and a hydraulic press, install 1/2 synchronizer assembly with both synchronizer rings.

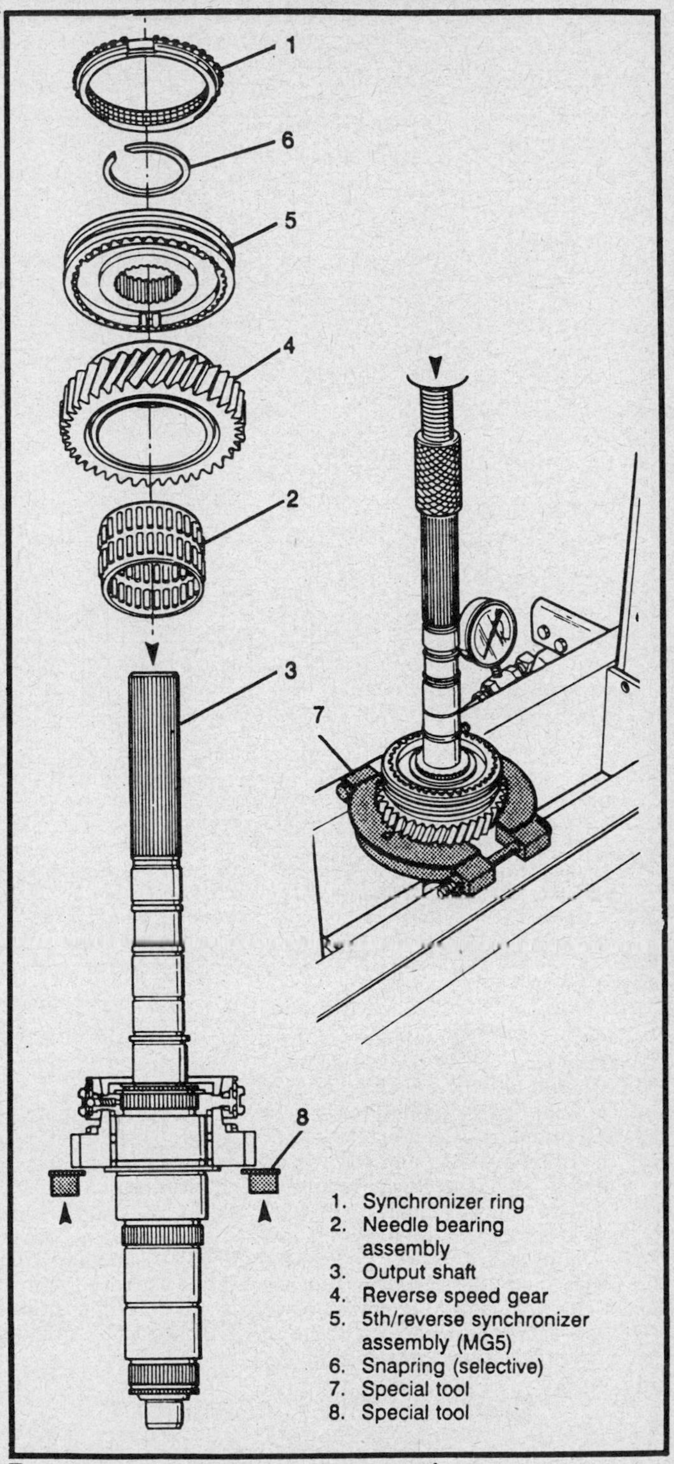
1. Synchronizer ring
2. Needle bearing assembly
3. Output shaft
4. Reverse speed gear
5. 5th/reverse synchronizer assembly (MG5)
6. Snapring (selective)
7. Special tool
8. Special tool

Reverse gear components removal

   a. Align marks made previously, for correct positions.
   b. Groove on outside of sleeve must go toward the 2nd speed gear to prevent gear clash during 1st and 2nd gear shifts (teeth on sleeve have different angles).
11. Install bearing assembly and 2nd speed gear. Make sure bearing cage is together.
12. Using tools J–36183, J–36184, J–22912–01 and a hydraulic press, install 2nd gear race (Heated).

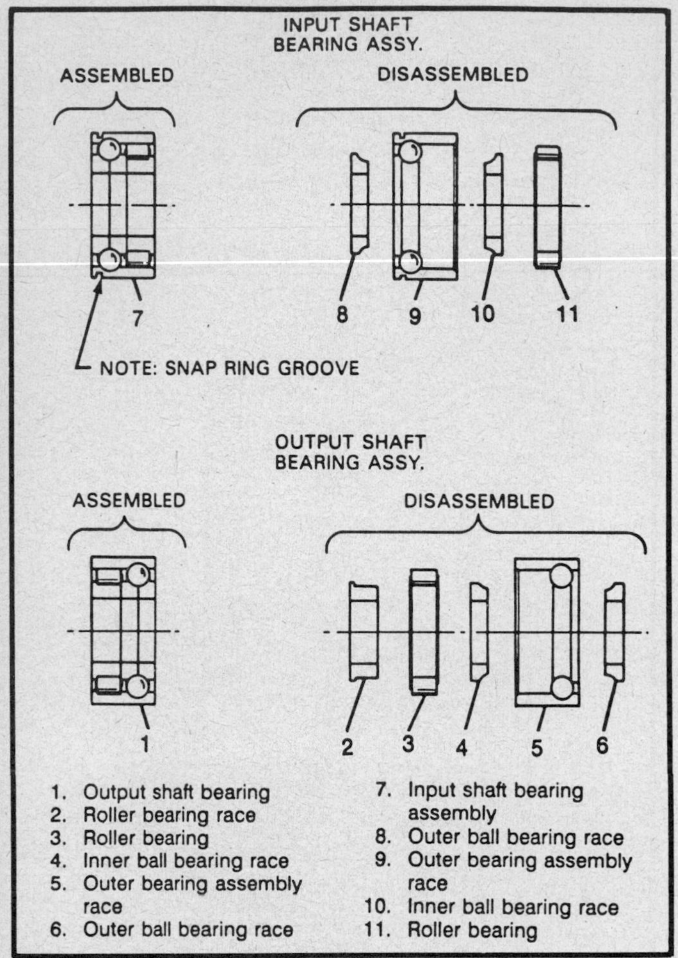
INPUT SHAFT
BEARING ASSY.
ASSEMBLED
DISASSEMBLED
7
8
9
10
11
NOTE: SNAP RING GROOVE
OUTPUT SHAFT
BEARING ASSY.
ASSEMBLED
DISASSEMBLED
1
2
3
4
5
6
1. Output shaft bearing
2. Roller bearing race
3. Roller bearing
4. Inner ball bearing race
5. Outer bearing assembly race
6. Outer ball bearing race
7. Input shaft bearing assembly
8. Outer ball bearing race
9. Outer bearing assembly race
10. Inner ball bearing race
11. Roller bearing

Main bearing components

13. Install a new selective snapring. Install the thickest snapring that will fit in the groove.
14. Install bearing assembly and 3rd speed gear.
    a. When pressing 3rd/4th synchronizer assembly, manually align and engage splines.
    b. Start pressing and stop before tangs engage.
    c. Lift and rotate gear to engage synchronizer ring.
    d. Press until seated.
    e. Make certain all shavings are removed.
15. Using tools J-36183, J-36184, J-22912-01 and a hydraulic press, install 3/4 synchronizer assembly with both synchronizer rings. Align marks made previously, for correct positions.
16. Install a new selective snapring. Install the thickest snapring that will fit in the groove.

REVERSE IDLER ASSEMBLY

Disassembly

1. Remove snapring.
2. Remove thrust washer, ball, reverse gear, 3 bearing assemblies and O-ring.

Inspection

1. Wash all parts in a suitable solvent.
2. Inspect gear teeth for scuffed, nicked, burred or broken teeth.
3. Inspect bearing assemblies for roughness while rotating, burred or pitted condition and gage damage.

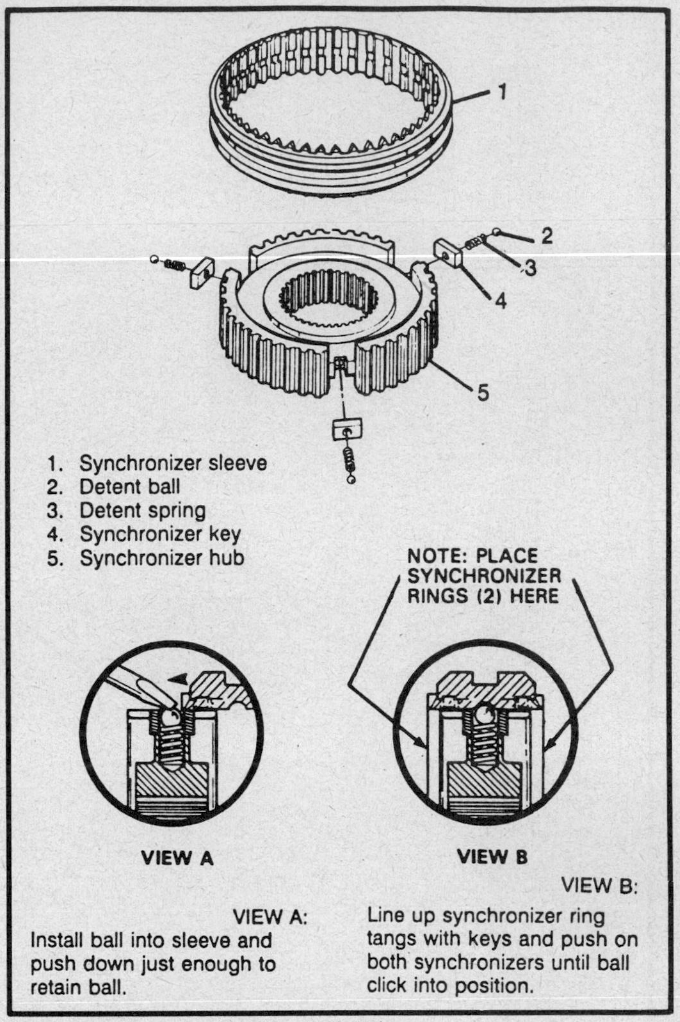
1
2
3
4
5
1. Synchronizer sleeve
2. Detent ball
3. Detent spring
4. Synchronizer key
5. Synchronizer hub
NOTE: PLACE SYNCHRONIZER RINGS (2) HERE
VIEW A
VIEW B
VIEW A:
Install ball into sleeve and push down just enough to retain ball.
VIEW B:
Line up synchronizer ring tangs with keys and push on both synchronizers until ball click into position.

Synchronizer assembly 3rd–4th

4. Inspect the shaft for scoring, wear or overheating.

NOTE: If scuffed, nicked, burred or scoring conditions cannot be removed with a soft stone or crocus cloth, replace the component.

Assembly

1. Lubricate all components with 5W-30 oil or equivalent.
2. Install the 3 bearing assembly.
3. Install the reverse gear.
    a. Extended part of the hub faces thrust washer.
4. Install the thrust washer and ball. Retain ball with petroleum jelly.
5. Install a new snapring and O-ring.

COUNTERSHAFT

Disassembly

If the 5th gear cannot be pressed off the countershaft, replace the complete assembly.

Inspection

1. Wash all parts in a suitable solvent.
2. Inspect countershaft for cracks.

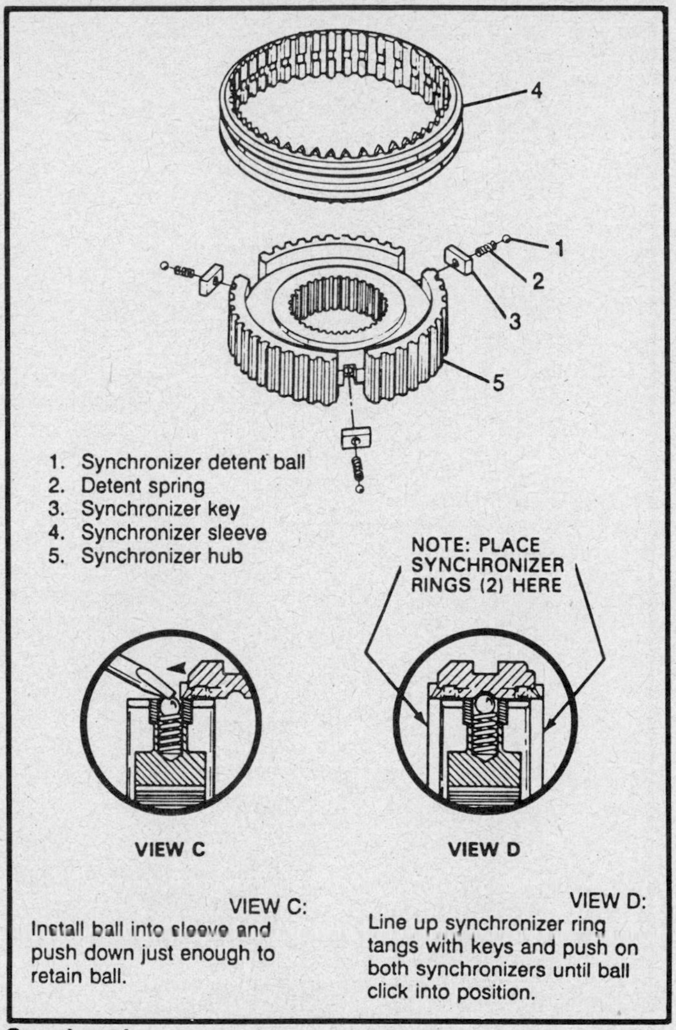
1
2
3
4
5
1. Synchronizer detent ball
2. Detent spring
3. Synchronizer key
4. Synchronizer sleeve
5. Synchronizer hub
NOTE: PLACE SYNCHRONIZER RINGS (2) HERE
VIEW C
VIEW D
VIEW C:
Install ball into sleeve and push down just enough to retain ball.
VIEW D:
Line up synchronizer ring tangs with keys and push on both synchronizers until ball click into position.

Synchronizer assembly 1st–2nd

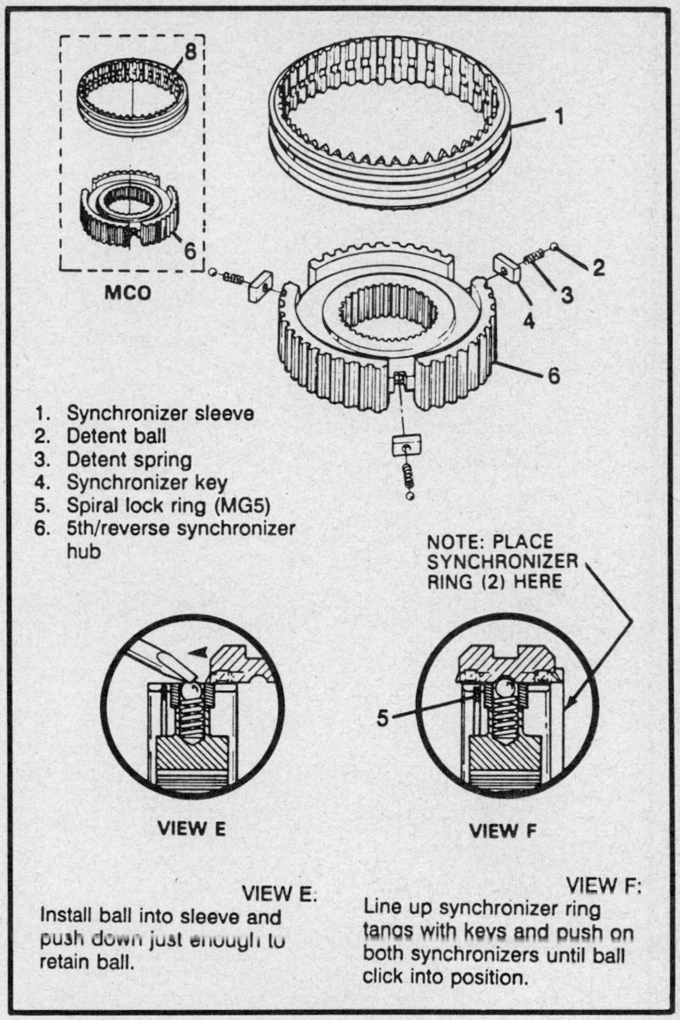
8
1
6
MCO
2
3
4
6
1. Synchronizer sleeve
2. Detent ball
3. Detent spring
4. Synchronizer key
5. Spiral lock ring (MG5)
6. 5th/reverse synchronizer hub
NOTE: PLACE SYNCHRONIZER RING (2) HERE
5
VIEW E
VIEW F
VIEW E:
Install ball into sleeve and push down just enough to retain ball.
VIEW F:
Line up synchronizer ring tangs with keys and push on both synchronizers until ball click into position.

Synchronizer assembly 5th–reverse

3. Inspect bearings for roughness while rotating, burred or pitted condition.
4. Inspect bearing races for scoring, wear or overheating.
5. Inspect gear teeth for scuffed, nicked, burred or broken teeth.

NOTE: If scuffed, nicked, burred or scoring conditions cannot be removed with a soft stone or crocus cloth, replace the component.

Assembly

1. Install 5th gear to countershaft assembly or replace assembly, if necessary.
2. Lubricate assembly with 5W–30 oil or equivalent.

Endplay Adjustment

NOTE: This procedure must be performed if the countershaft bearings, front housing countershaft bearing race or the front or rear housing was replaced. If a gear rattle noise complaint (not clutch disc related) is experienced, it is recommended the countershaft endplay be checked.

1. Remove bearing race using a brass drift.
2. Remove the countershaft bearing plug.
3. Remove the snapring, shims and oil fill plug.
4. Inspect bearing and countershaft.
   a. Smaller diameter of bearing cage into bearing race (202).
   b. Install bearings in bearing races prior to assembly and retain with petroleum jelly.
5. Install tools J–8001 and J–25025–A or equivalent, to front housing in one of the input shaft bearing retainer bolt holes.
   a. Use a pry bar, pry the countershaft upward, noting the dial indicator travel. Use the fill plug hole to reach the third gear part of the countershaft for prying.
   b. Measure off gear tooth.
   c. Make certain tool J–8001 stays on gear tooth while measuring endplay.
   d. Allow the countershaft to lower to its original position, noting the dial indicator travel. Total travel should be 0.005–0.009 in. (0.13–0.23mm).
6. Move bearing race with tool J–8092 and J–36799 or equivalent to achieve specified endplay.
7. Use the least number of shims to retain specified endplay and install snapring.
8. Seat the countershaft bearing race using tool J–8092 and J–36799 or equivalent.
9. Recheck Step 5a–7 to make sure endplay is still correct after seating countershaft bearing race.
10. Install oil fill plug.
11. Install a new countershaft bearing plug. Apply gasket maker GM part number 1052943 or equivalent to outer edge of plug.

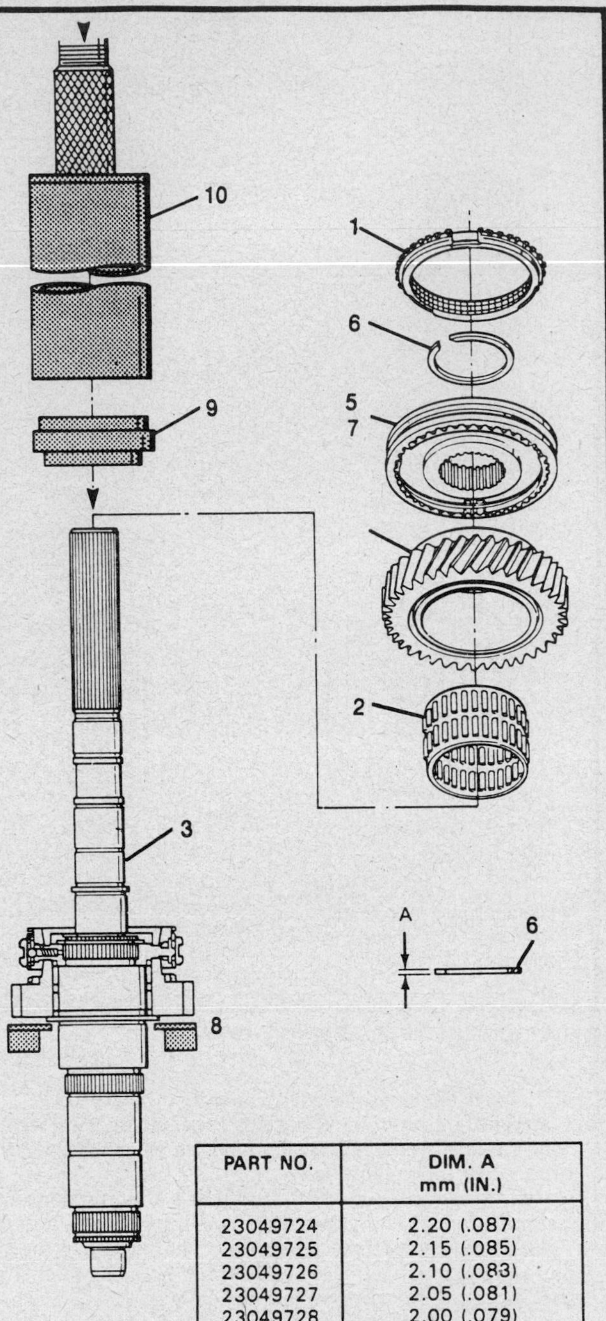
10
1
6
5
7
9
2
3
A
6
8

PART NO.	DIM. A mm (IN.)
23049724	2.20 (.087)
23049725	2.15 (.085)
23049726	2.10 (.083)
23049727	2.05 (.081)
23049728	2.00 (.079)

1. Synchronizer ring
2. Needle bearing assembly
3. Output shaft
4. Reverse gear assembly
5. 5th/reverse synchronizer assembly (MG5)
6. Snapring (selective)
7. Reverse synchronizer assembly (MCO)
8. Special tool
9. Special tool
10. Special tool

Reverse gear components installation

NOTE: The countershaft bearing plug will require cooling for 20 minutes minimum, at 32°F (0°C) prior to installation during assembly.

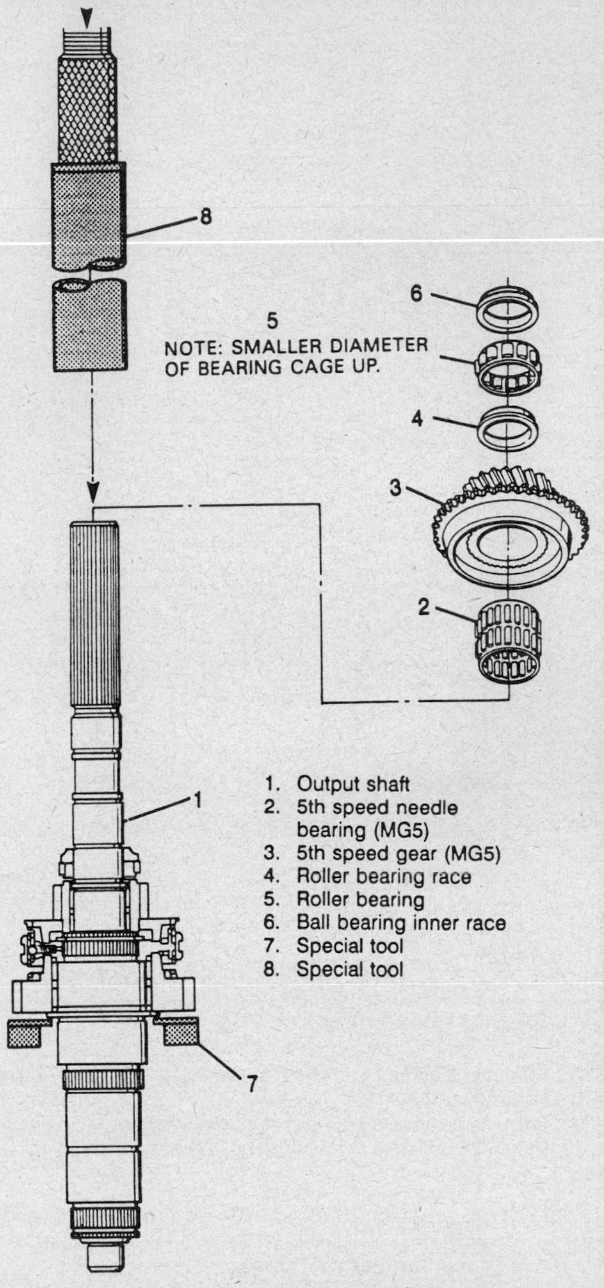
8
6
5
NOTE: SMALLER DIAMETER
OF BEARING CAGE UP.
4
3
2
1
7

1. Output shaft
2. 5th speed needle bearing (MG5)
3. 5th speed gear (MG5)
4. Roller bearing race
5. Roller bearing
6. Ball bearing inner race
7. Special tool
8. Special tool

Bearing race and 5th gear installation

FRONT HOUSING ASSEMBLY

Disassembly

1. Using a brass drift, remove ball bearing assembly outer race.
2. Using tools J-36800 and J-23907 or equivalent, remove the 3 shift rail front housing bearings.
3. Using tools J-36509 and J-23907 or equivalent, remove clutch fork pivot assembly.
4. Use a punch and drive out shift shaft/rail plugs.
5. Remove countershaft bearing plug.
   a. Destake plug first.

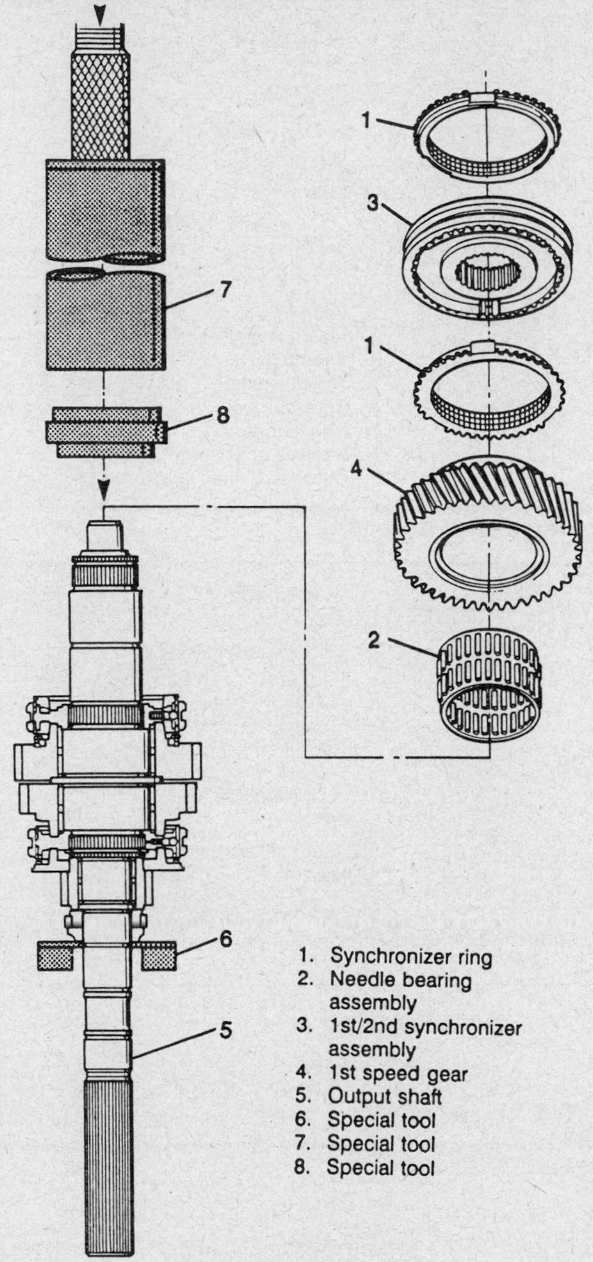
1
3
1
4
2
7
8
6
5

1. Synchronizer ring
2. Needle bearing assembly
3. 1st/2nd synchronizer assembly
4. 1st speed gear
5. Output shaft
6. Special tool
7. Special tool
8. Special tool

1st gear components installation

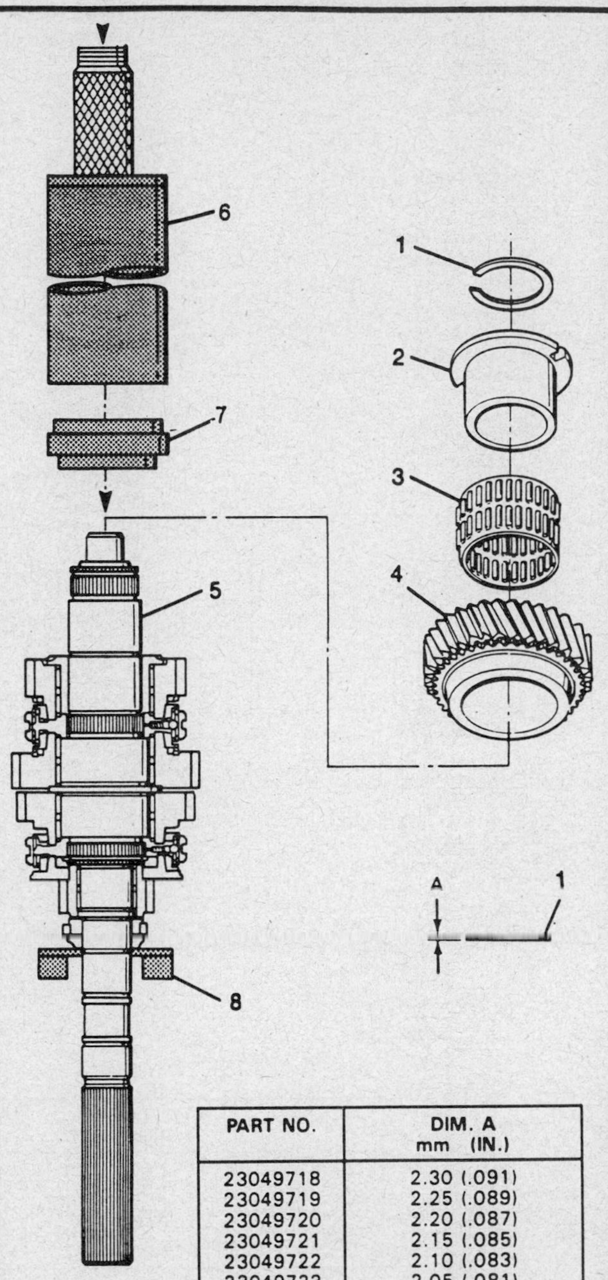
6
1
2
7
3
4
5
8
A
1

PART NO.	DIM. A mm (IN.)
23049718	2.30 (.091)
23049719	2.25 (.089)
23049720	2.20 (.087)
23049721	2.15 (.085)
23049722	2.10 (.083)
23049723	2.05 (.081)
23049790	2.00 (.077)

1. Snapring (selective)
2. Inner needle bearing race
3. Needle bearing assembly
4. 2nd speed gear
5. Output shaft
6. Special tool
7. Special tool
8. Special tool

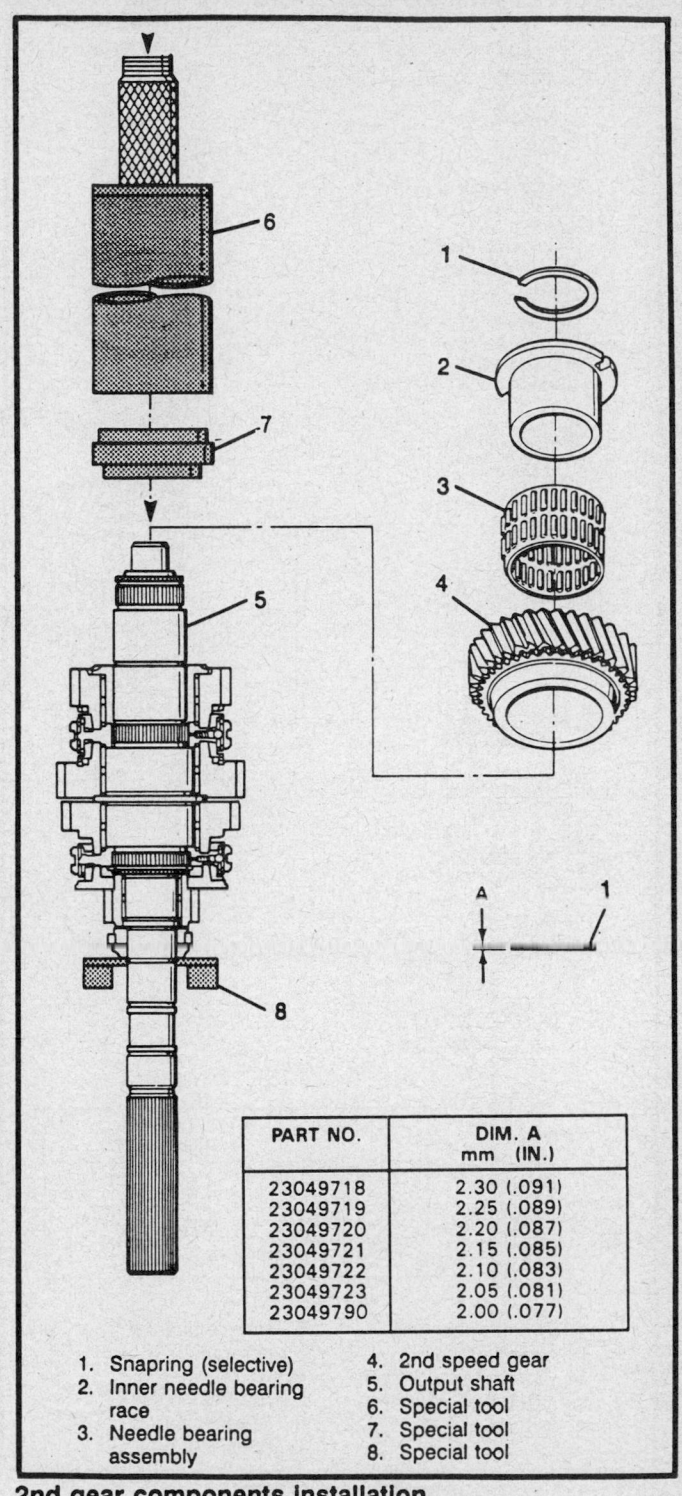

2nd gear components installation

b. Screw a 3/8 × 16 bottom tap into the plug and remove with tool J–6725 or equivalent.

6. Using tools J–36799 and J–8092 or equivalent, remove snapring, shims and countershaft bearing race.

a. Tap in countershaft bearing race before removing snapring.

b. Check countershaft endplay adjustment, if countershaft bearing race is removed.

c. Measure the countershaft bearing race bore in 2 locations diagonally, 0.157 in. (4mm) in from inside of transmission housing. If the housing bore is not within 2.045–2.046 in. (51.946–51.965mm), replace the housing.

7. Remove 5th/reverse detent hex head plug, 5th/reverse detent spring and 5th/reverse detent plunger assembly.

8. Remove detent cam bolt, washer, reverse detent cam (4 speed) or 5th/reverse cam (5 speed) and bushing.

a. Using a flat blade type tool, release tension of bias spring from end of shift lever pin.

NOTE: Spring is under high tension.

9. Remove bias spring bolt, bias spring and sleeve seat, bias

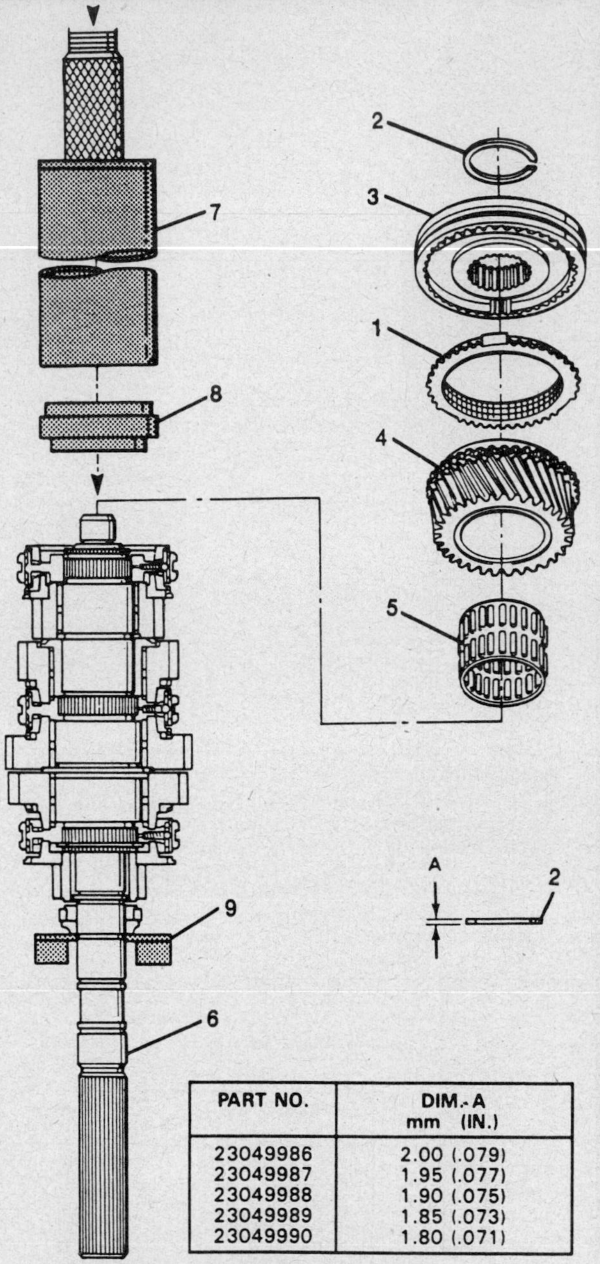
2
3
1
4
5
7
8
9
6
A
2

PART NO.	DIM. A mm (IN.)
23049986	2.00 (.079)
23049987	1.95 (.077)
23049988	1.90 (.075)
23049989	1.85 (.073)
23049990	1.80 (.071)

1. Synchronizer ring
2. Snapring (selective)
3. 3rd/4th synchronizer assembly
4. 3rd speed gear
5. Needle bearing assembly
6. Output shaft
7. Special tool
8. Special tool
9. Special tool

3rd gear components installation

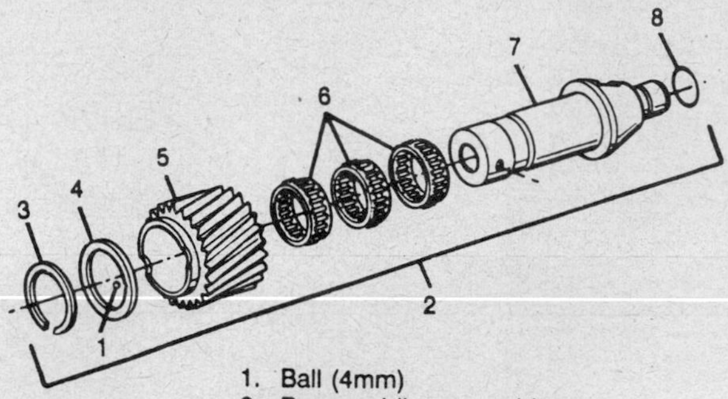
8
7
6
5
4
3
1
2

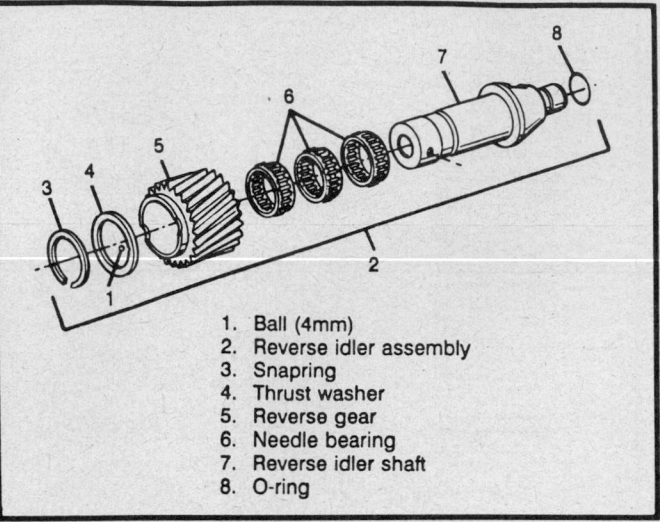

1. Ball (4mm)
2. Reverse idler assembly
3. Snapring
4. Thrust washer
5. Reverse gear
6. Needle bearing
7. Reverse idler shaft
8. O-ring

Reverse idler assembly

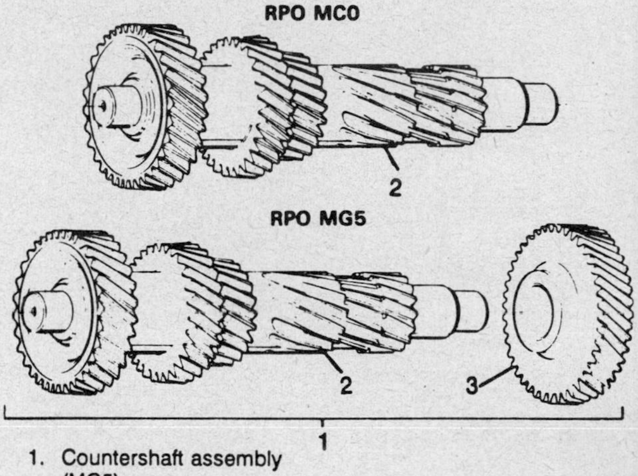
RPO MCO
2
RPO MG5
2
3
1

1. Countershaft assembly (MG5)
2. Countershaft assembly (MCO)
3. 5th gear (MG5)

Countershaft assembly

spring, bias spring sleeve, shift shaft lever assembly and detent roller.

10. Remove detent cam support bolt and remove detent cam support.

11. Using tool J-36506, remove the 2 5th/reverse detent bushings.

12. Using tool J-36507, remove the 2 detent bushings. Remove 1 bushing at a time.

13. Remove the breather assembly. Metal tube must remain in case.

14. Remove drain and fill plugs using tool J-36511 or equivalent.

Inspection

1. Clean all gasket material from case using liquid gasket remover.

2. Clean all parts in a suitable solvent and air dry.

3. Inspect the bearing race bore for wear, scratches or grooves. If countershaft bearing race is worn or damaged, the rear housing must be replaced.

4. Inspect bushings for scores, burrs, roundness or evidence of overheating.

5. Inspect the case for cracks, damaged threads, mounting surfaces for nicks, burrs, or scratches. If the case is crack, it must be replaced.

6. Using a straight edge, check machined surfaces for flatness.

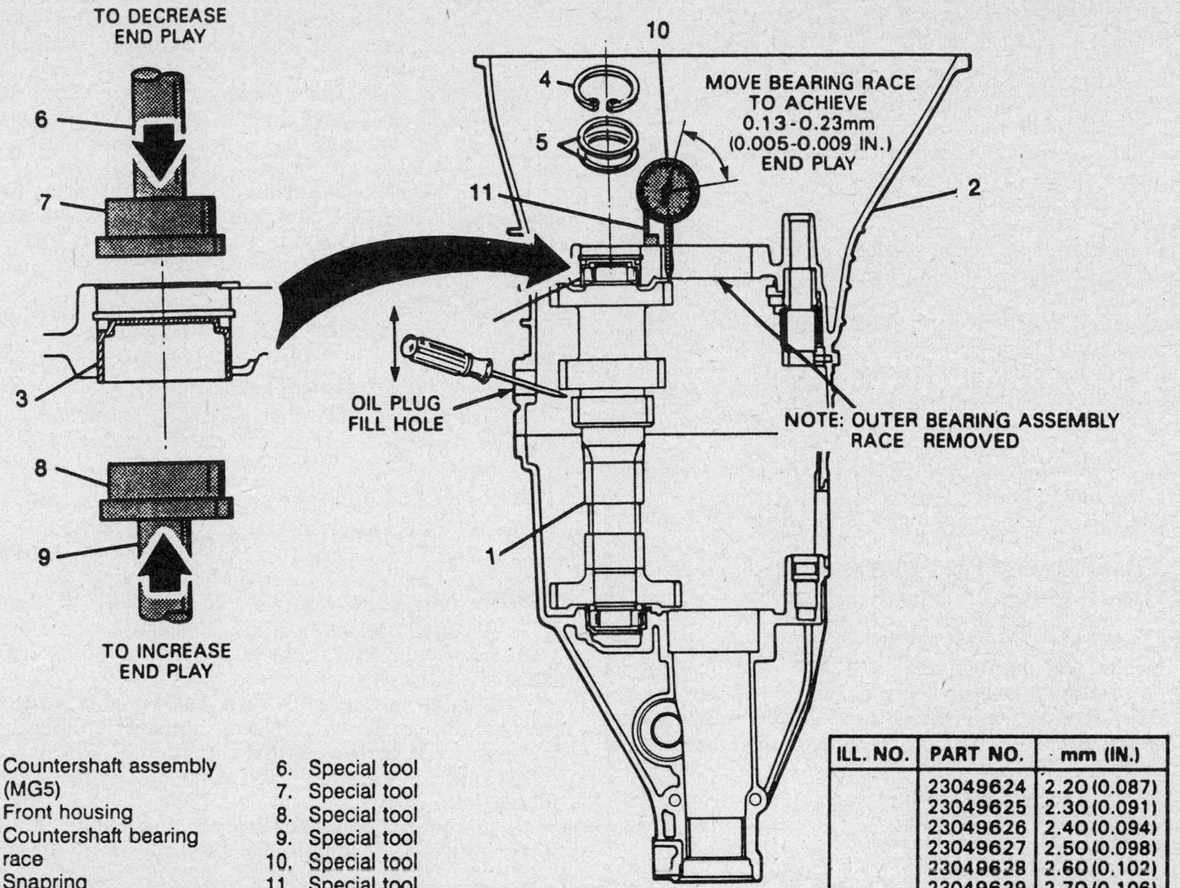
TO DECREASE
END PLAY
TO INCREASE
END PLAY
MOVE BEARING RACE
TO ACHIEVE
0.13-0.23mm
(0.005-0.009 IN.)
END PLAY
OIL PLUG
FILL HOLE
NOTE: OUTER BEARING ASSEMBLY
RACE REMOVED

1. Countershaft assembly (MG5)
2. Front housing
3. Countershaft bearing race
4. Snapring
5. Shim
6. Special tool
7. Special tool
8. Special tool
9. Special tool
10. Special tool
11. Special tool

ILL. NO.	PART NO.	mm (IN.)
	23049624	2.20 (0.087)
	23049625	2.30 (0.091)
	23049626	2.40 (0.094)
	23049627	2.50 (0.098)
	23049628	2.60 (0.102)
	23049629	2.70 (0.106)
	23049630	0.40 (0.016)
	23049631	0.50 (0.020)

Countershaft endplay adjustment

NOTE: If scuffed, nicked, burred or scoring conditions cannot be removed with a soft stone or crocus cloth, replace the component. Clean-up damaged threads with the correct size tap.

Assembly

NOTE: If the countershaft bearing plug was removed, it will require cooling for a period of 20 minutes minumum, at 32°F (0°C) before installation.

1. Using a brass driff, install the ball bearing outer race.
2. Using tools J-36798-1 and J-36190 or equivalent, install 3 shift rail front housing bearings. Install until flush and stake using tools J-36798-1, J-36798-2 and J-36190 or equivalent. Do not stake the tabs on the bushings.
3. Using tools J-36190 and J-36510 or equivalent, install the clutch fork pivot assembly. Grease assembly after installation through lube fitting.
4. Using tool J-36507 or equivalent, install 2 detent plunger bushings.
   a. Install 1 bushing at a time.
   b. Install the first bushing until the second scribe mark on the tool lines up with the housing.
   c. Install the second bushing until the first scribe mark on the tool lines up with the housing.
5. Using tools J-36799 and J-8092 or equivalent, install countershaft bearing race. Align lube slot in race with groove in the housing. Install shims and snapring.

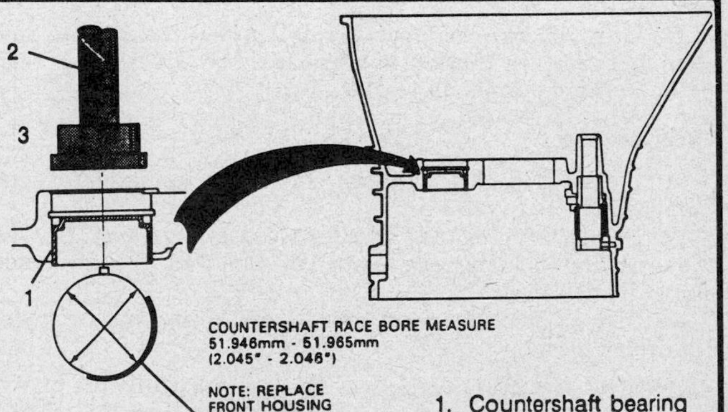
COUNTERSHAFT RACE BORE MEASURE
51.946mm - 51.965mm
(2.045" - 2.046")
NOTE: REPLACE
FRONT HOUSING
IF THERE IS
EXCESSIVE
METAL TRANSFER

1. Countershaft bearing race
2. Special tool
3. Special tool

Countershaft bearing race bore measurement

NOTE: If the countershaft bearing races, countershaft bearings, countershaft or front or rear housing are replaced, the countershaft endplay must be checked.

6. Using tools J-36799 and J-8092 or equivalent, install a new countershaft bearing plug. Apply gasket maker GM part

number 1052943 or equivalent, to the outside edge of the plug and stake in 3 places evenly apart.

7. Using tool J-36506, install both 5th/reverse detent bushings.

a. Install 1 bushing at a time.

b. Install the first bushing until the second scribe mark on the tool lines up with the housing.

c. Install the second bushing until the first scribe mark on the tool lines up with the housing.

8. Install detent cam support and bolt. Torque bolt to 7 ft. lbs. (8.5 Nm).

9. Install reverse detent cam (4 speed) or 5th/reverse cam (5 speed), bushing, washer and bolt. Torque bolt to 7 ft. lbs. (8.5 Nm).

10. Install detent roller on shift shaft lever assembly and install into housing.

11. Install bias spring sleeve, bias load torsional spring, bias spring and sleeve seat and bolt. Torque bolt to 7 ft. lbs. (8.5 Nm).

12. Install bias load torsional spring end back onto shift shaft lever pin.

13. Install 5th/reverse plunger assembly, 5th/reverse detent spring and hex head plug.

a. Make certain slot in plunger is lined up with reverse cam.

b. Use pipe sealant GM part number 1052080 or equivalent to threads.

c. Torque hex head plug to 46 ft. lbs. (60 Nm).

14. Install breather assembly, oil fill plug and drain plug. Apply pipe sealant GM part number 1052080 or equivalent on threads and torque drain plug to 46 ft. lbs. (60 Nm).

15. Install shift shaft/rails plugs until flush. Apply gasket maker GM part number 1052943 or equivalent to the edge of plugs.

REAR HOUSING

Disassembly

1. Using bushing remover J-36800 or equivalent and a slide hammer, remove the 3 rear housing shift rail bearings.

2. Using a slide hammer with a pilot bearing remover, remove rear housing shift shaft bearing. Position pilot bearing remover legs behind ball bearing cage.

3. Remove 3 bolts, output shaft bearing retainer (4WD) and bearing assembly outer race using a brass drift then remove shims.

4. Using oil seal remover J-26941 or equivalent and a slide hammer, remove the slip yoke oil seal (2WD vehicles).

5. Using a punch, drive out plug.

Inspection

1. Clean all gasket material from case using liquid gasket remover.

2. Inspect bearing race bore for wear, scratches or grooves, if countershaft bearing race is worn or damaged, the rear housing must be replaced.

3. Inspect bushings for scores, burrs, roundness or evidence of overheating.

NOTE: On 2WD vehicles, the rear extension bushing cannot be serviced. Replace the rear housing, if necessary.

4. Inspect the case for cracks, nicks on surfaces, burrs, scratches or damage thread holes. If the case is crack, it must be replaced.

5. Using a stright edge, check machined mating surfaces for flatness.

NOTE: If scuffed, nicked, burred or scoring conditions cannot be removed with a soft stone or crocus cloth, replace the component. Clean-up damaged threads with the correct size tap.

Assembly

1. Using plug driver tool J-36190 and staking tool J-36798-1 or equivalent, install the 3 rear housing shift rail bearings and stake as necessary.

2. Using bushing installer tool J-36506, install rear housing shift shaft bearing until flush.

3. On 2WD vehicles, use a brass drift to install shims and bearing assembly outer race.

4. On 4WD vehicles, install output shaft bearing retainer. Apply threadlocker GM part number 12345382 or equivalent to the 3 threaded bolts and install. Torque bolts to 17 ft. lbs. (22 Nm).

NOTE: The output bearing selective shim procedure must be performed before installing a new bearing assembly.

5. Install the 2 dowel pins until flush.

6. Using oil seal installer J-36503, install slip yoke oil seal. Fill between the seal lips with chassis grease (2WD vehicles).

7. Apply gasket maker GM part number 1052943 or equivalent to plug and install plug.

SHIFT SHAFT/RAIL AND FORK

Inspection

1. Wash all parts in a suitable solvent and air dry.

2. Inspect 1st/2nd shift rail, 3rd/4th rail, 5th/reverse rail and shift shaft for wear or scoring.

3. Inspect 1st/2nd shift fork, 3rd/4th shift fork and 5th/reverse shift fork for wear, scoring or distortion. Black coloring on the edge of forks is a normal condition.

4. Inspect 1st/2nd shift yoke and finger for wear or distortion.

5. Inspect both pins for wear or distortion.

6. Inspect shift shaft socket aassembly for wear or distortion.

NOTE: Wear, scoring or distortion requires replacement of assembly and inspection of mating parts.

SELECTIVE SHIM PROCEDURE

Input Shaft Bearing Retainer

1. Measure the distance between the sealing flange of the retainer and the flanged part of the release guide. Record this measurement.

2. Measure the height of the input bearing outer race from the sealing surface of the front housing. Record this measurement.

3. Subtract the bearing race height from the retainer depth and select a shim the same (or next smaller) size as the difference in measurements.

Output Shaft Bearing

1. Measure the distance between the bearing retainer surface and the bottom of the bearing bore in the rear housing. Record this measurement.

2. Measure the width of the mainshaft rear bearing. Record this measurement.

3. Subtract the bearing width from the housing bore depth. Select a shim which is the same (or next smaller) size as the difference in measurements.

Transmission Assembly

REAR HOUSING

2WD

1. Install the pilot bearing into input shaft with the smaller

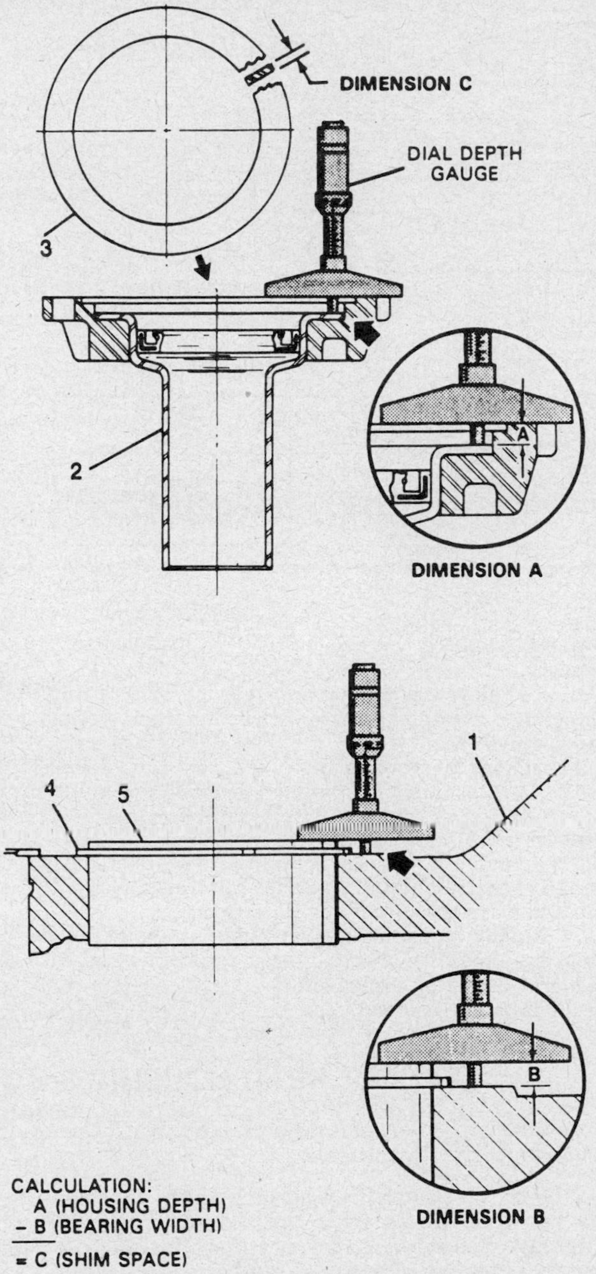
DIMENSION C
DIAL DEPTH GAUGE
3
2
A
DIMENSION A
1
4
5
B
DIMENSION B
CALCULATION:
A (HOUSING DEPTH)
– B (BEARING WIDTH)
= C (SHIM SPACE)

ILL. NO.	PART NO.	DIM. C mm (IN.)
	23049825 23049826 23049827	0.30 (0.012) 0.40 (0.016) 0.50 (0.020)

1. Front housing
2. Input shaft bearing retainer
3. Shim (selective)
4. Snapring
5. Outer bearing race

Shim selection input shaft bearing retainer

diameter of bearing cage toward input shaft. Retain with petroleum jelly.

2. Assemble the input shaft, pilot bearing, synchronizer ring and mainshaft assembly on fixture J-36515 or equivalent.
3. Assemble adapter J-36515-12 or equivalent on countershaft and install assembly onto fixture J-36515 or equivalent.
4. Install the reverse idler assembly using a new O-ring.
5. Install bearing assembly outer race and ball bearing outer race.
6. Install on the mainshaft assembly:
   a. 3/4 shift fork with taper on fork towards 3rd gear.
   b. 1/2 shift rail assembly and spacer block (4 speed only).
   c. Lock up the transmission by sliding the 3/4 shift fork and 1/2 shift rail assembly downward towards fixture J-36515 or equivalent.
7. Install ball to output shaft and retain with petroleum jelly.
8. Install the inner threaded thrust ring and outer threaded thrust ring. Make certain the old spiral roll pin is removed from the outer threaded thrust ring.
   a. Completely screw the rings together, then back off of rings until both identification slots for the ball line up.
   b. Slide the assembled rings over the ball that is retained on the output shaft.
   c. Slide the rod up on spanner nut wrench J-36516 or equivalent, turn the black depth locating tang over, slide the rod through so the roll pin will be installed to the correct depth.
   d. Allow clearance for the snapring by screwing the rings together counterclockwise with spanner nut wrench J-36516 or equivalent.
9. Install a new snapring.
10. Turn spanner nut wrench J-36516 or equivalent clockwise to obtain a torque of 12 ft. lbs. (15 Nm), then advance to the next sprial roll pin notch.
11. Install a new spiral roll pin into spanner nut wrench J-36516 or equivalent and retain with petroleum jelly. Lined up roll pin and drive pin into outer thrust ring.
12. Install the snapring and speedo gear thrust washer.
13. Install the speed sensor rotor. Rotor must be heated.
14. Install shim onto output shaft bearing assembly and retain with pertroleum jelly.

NOTE: The rear housing output shaft bearing assembly bore must be heated for 3–5 minutes before assembly.

15. Install retainer alignment cables J-36515-10 or equivalent, through bolt and holes in rear housing.
   a. Screw into the output shaft bearing retainer.
   b. Notch in retainer towards oil delivery tube.
16. Install bearing in bearing race of housing and retain with petroleum jelly. The smaller diameter of bearing cage into bearing race.

NOTE: Press each roller towards the race to secure them for easier assembly.

17. Install the rear housing assembly.
   a. Make certain the reverse idler shaft is lined up with the hole in the case.
   b. Rotate back and forth while pulling down.
   c. Pull up on retainer alignment cables J-36515-10 or equivalent, while installing rear housing.

NOTE: Bring the housing straight down. If resistance is felt at about ¼ in., the rollers are cocked. Repeat procedure above. Do not force housing down.

18. Remove tool J-36515-10 or equivalent and install bolts. Apply pipe sealant GM part number 1052080 or equivalent to bolt holes of rear housing. Apply threadlocker GM part number 12345382 or equivalent to bolt threads. Torque bolts to 17 ft. lbs. (22 Nm).

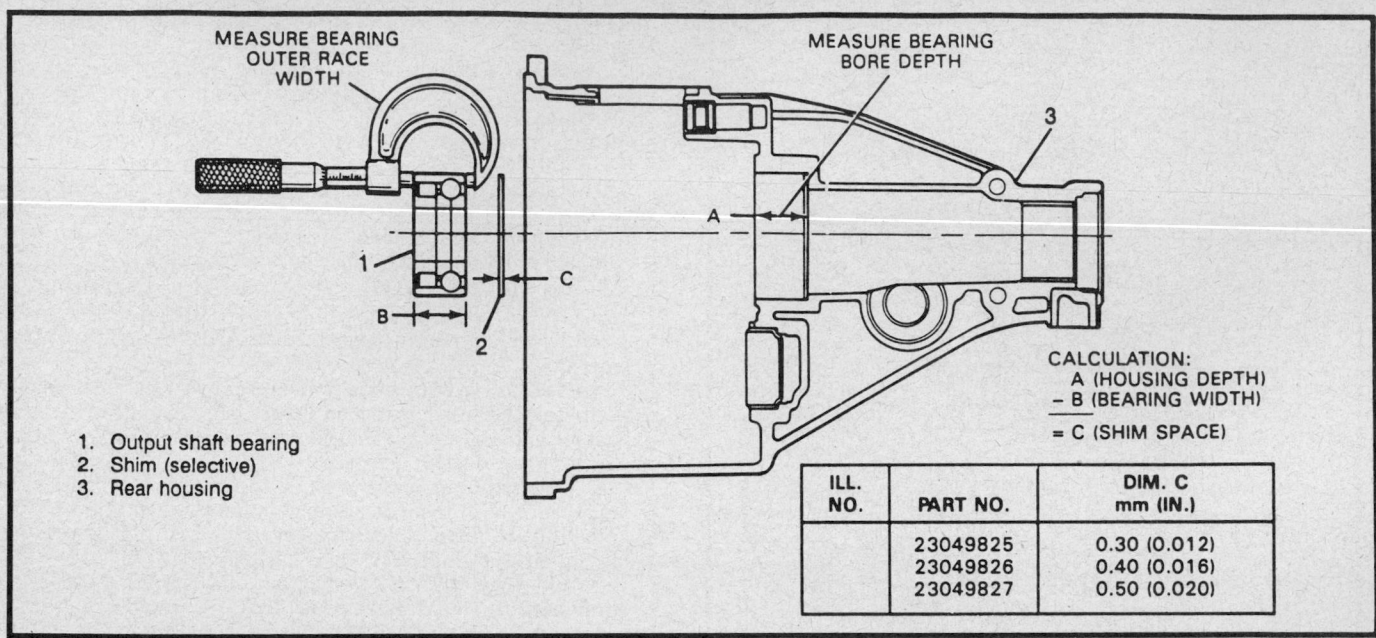

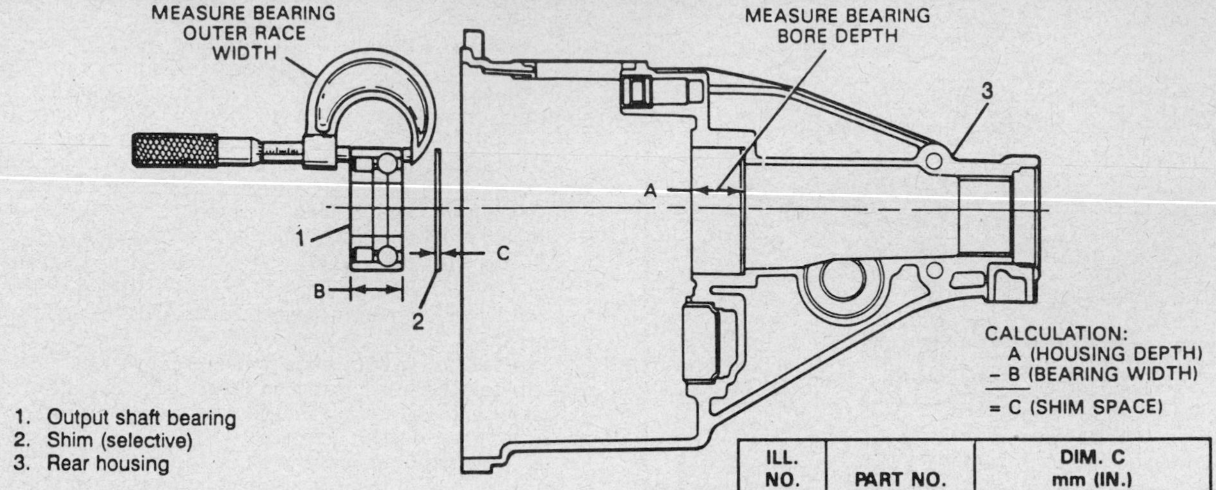
MEASURE BEARING
OUTER RACE
WIDTH
MEASURE BEARING
BORE DEPTH
3
A
1
C
B
2
CALCULATION:
A (HOUSING DEPTH)
– B (BEARING WIDTH)
= C (SHIM SPACE)
1. Output shaft bearing
2. Shim (selective)
3. Rear housing

ILL. NO.	PART NO.	DIM. C mm (IN.)
	23049825 23049826 23049827	0.30 (0.012) 0.40 (0.016) 0.50 (0.020)

Shim selection output shaft bearing

4WD

1. Install the pilot bearing into input shaft with the smaller diameter of bearing cage toward input shaft. Retain with petroleum jelly.
2. Assemble the input shaft, pilot bearing, synchronizer ring and mainshaft assembly on fixture J-36515 or equivalent.
3. Assemble adapter J-36515-12 or equivalent on countershaft and install assembly onto fixture J-36515 or equivalent.
4. Install the reverse idler assembly using a new O-ring.
5. Install bearing assembly outer race and ball bearing outer race.
6. Install on the mainshaft assembly:
   a. 3/4 shift fork with taper on fork towards 3rd gear.
   b. 1/2 shift rail assembly and spacer block (4 speed only).
   c. Lock up the transmission by sliding the 3/4 shift fork and 1/2 shift rail assembly downward towards fixture J-36515 or equivalent.
7. Install bearing in bearing race of housing and retain with petroleum jelly. Smaller diameter of bearing cage into bearing race.

NOTE: Press each roller towards the race to secure them for easier assembly.

8. Install the rear housing assembly. Make certain the reverse idler shaft is lined up with hole in the case. Rotate back and forth while pulling down.

NOTE: Bring the housing straight down. If resistance is felt at about ¼ in., the rollers are cocked. Repeat procedure above. Do not force housing down.

9. Install the outer bearing race and ball. Retain ball with petroleum jelly.
10. Install the inner threaded thrust ring and outer threaded thrust ring. Make certain the old spiral roll pin is removed from the outer threaded thrust ring.
   a. Completely screw the rings together, then back off of rings until both identification slots for the ball line up.
   b. Slide the assembled rings over the ball that is retained on the output shaft.
   c. Slide the rod up on spanner nut wrench J-36516 or equivalent, turn the black depth locating tang over, slide the rod through so the roll pin will be installed to the correct depth.
   d. Allow clearance for the snapring by screwing the rings together counterclockwise with spanner nut wrench J-36516 or equivalent.
11. Install a new snapring.
12. Turn spanner nut wrench J-36516 or equivalent clockwise to obtain a torque of 12 ft. lbs. (15Nm), then advance to the next sprial roll pin notch.
13. Install a new spiral roll pin into spanner nut wrench J-36516 or equivalent and retain with petroleum jelly. Line up roll pin and drive pin into outer thrust ring.
14. Install seal protector onto output shaft. Fill the output shaft seal lips with chassis grease and install seal with seal installer J-36502 or equivalent.
15. Remove seal protector.

EXCEPT REAR HOUSING

1. Install the idler shaft support. Line up bolt threads in idler shaft support with bolt hole.

NOTE: The machined surface on the face casting must be installed down into case because the bolt hole is slightly off center. Incorrect installation will cause incorrect reverse gear tooth pattern under load.

   a. Apply pipe sealant GM part number 1052080 or equivalent to bolt holes of rear housing.
   b. Apply threadlocker GM part number 12345382 or equivalent to bolt threads.
   c. Hold reverse idler shaft against idler support while torqueing.
   d. Torque bolts to 17 ft. lbs. (22 Nm).
2. Assemble transmission holding fixture J-36824 or equivalent to transmission case.
3. Place all forks in neutral position. Coat the 2 interlock check balls with petroleum jelly and insert balls through plug hole using a small magnetic. Push 1 ball into 1/2 shift rail side and the other into 3/4 shift rail side.
4. Install the 3/4 shift rail with interlock pin. Retain pin with petroleum jelly.

a. The detent slots in 3/4 shift rail faces up.
b. Install through 3/4 shift fork and into the rear housing shift rail bearing.

NOTE: The 1/2 and 5th/reverse shift rail must be in NEUTRAL position or the interlock system will not allow the 3/4 rail to engage.

5. Install the roll pin to a depth where a maximum of 12.5mm of the roll pin remains. Measure from the edge of the 3/4 rail to the top of the roll pin.

NOTE: If the roll pin is not installed to its proper depth, it may rub against the front housing and cause the 3rd or 4th gear to hop-out.

a. Test the interlock system to be certain the interlock balls are in place, by trying to move 2 shift rails.
6. Apply gasket maker GM part number 1052943 or equivalent to plug and install plug until flush.
7. Install the shift shaft, finger and shift shaft socket assembly. The detent slots must faces the idler support side of the transmission and the finger extension must be on the underside of the 5/reverse shift rail.
8. Install roll pins.
9. Install the 3 detent/interlock balls and springs.
10. Apply gasket maker GM part number 1052943 or equivalent to the outside of bolt holes of the detent spring cover and install cover and attaching bolts. Torque bolts to 7 ft. lbs. (8.5 Nm).
11. Install the 4 rollers to shift shaft. Retain with petroleum jelly.
12. Install the countershaft bearing in bearing race of front housing. The smaller diameter of the bearing cage toward bearing race.
13. Install roller bearing and ball bearing outer race.
14. Apply gasket maker GM part number 1052943 or equivalent to the outside of bolt holes of the rear housing and install front housing.

NOTE: Bring the housing straight down. If resistance is felt at about ¼ in., the rollers are cocked. Repeat procedure above. Do not force housing down.

15. Install the dowel pins and attaching bolts. Do not tighten bolts at this time.
16. Install ball bearing outer race, input shaft spacer and a new selective snapring.
a. Install the thickest snapring that fits into the groove. It may be necessary to pull out on the input shaft to install the selective snapring.
17. Apply gasket maker GM part number 1052943 or equivalent to the inside cover bolt hole pattern.

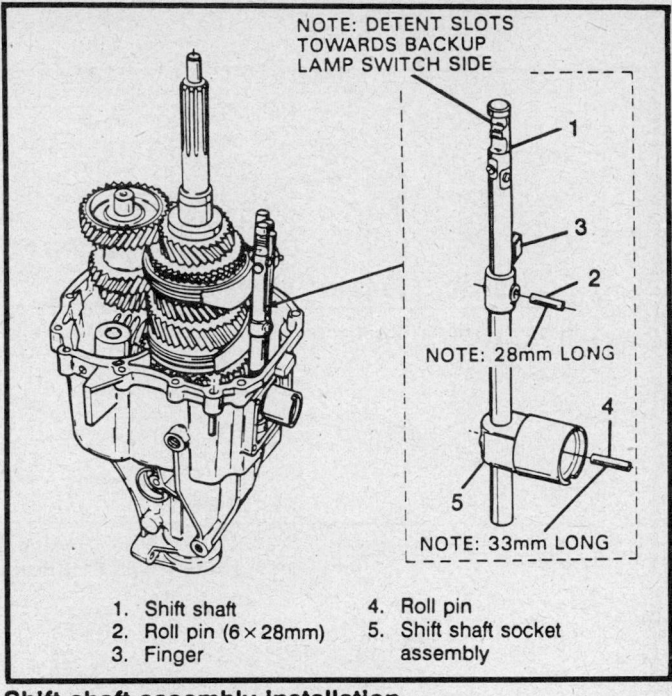

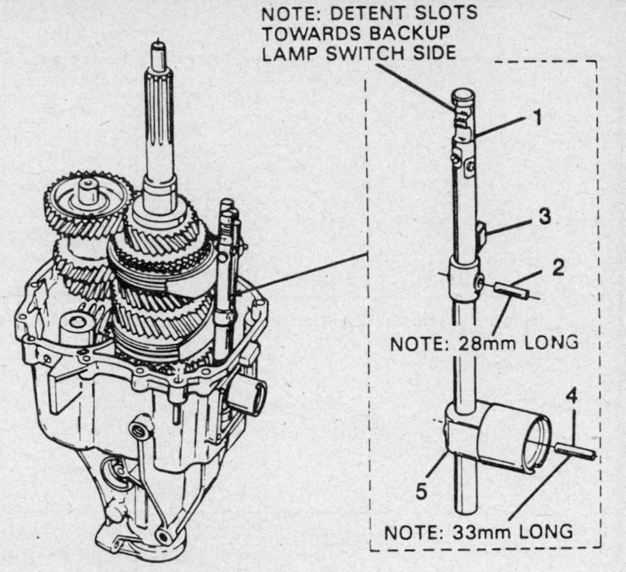
NOTE: DETENT SLOTS TOWARDS BACKUP LAMP SWITCH SIDE
1
2
3
4
5
NOTE: 28mm LONG
NOTE: 33mm LONG

1. Shift shaft
2. Roll pin (6×28mm)
3. Finger
4. Roll pin
5. Shift shaft socket assembly

Shift shaft assembly installation

NOTE: Do not apply too much gasket maker around the oil drain back hole. It may clog the hole causing a low fluid flow through the bearing resulting in premature bearing failure.

18. Install the shim into the input shaft bearing retainer assembly. Make certain the input bearing retainer washer is in place. Retain with petroleum jelly.
19. Align the oil drain back hole of the input shaft bearing retainer assembly with the hole in the housing and install attaching bolts. Torque bolts to 7 ft. lbs. (8.5 Nm).
20. Install the shift shaft detent plunger, shift shaft detent spring and plug using a brass drift. Apply pipe sealant GM part number 1052080 or equivalent to the plug.
21. On 2WD vehicles, install a new O-ring on the electronic speed sensor and coat with a thin flim of transmission oil. Install the electronic speed sensor.
22. Install back-up lamp switch. Apply pipe sealant GM part number 1052080 or equivalent to the threads.
23. Tighten housing attaching bolts to 27 ft. lbs. (35 Nm).

SPECIFICATIONS

TORQUE SPECIFICATIONS

Description	Ft. lbs.	Nm
Inner/outer threaded thrust ring	12	15
Input shaft bearing retainer bolts	7	8.5
Oil fill plug	46	60
Oil drain plug	46	60
5th/reverse rail deflection bolt	27	35
5th/reverse detent hex head plug	46	60

Description	Ft. lbs.	Nm
Outer shaft bearing retainer bolts	17	22
Idler shaft support bolts	17	22
Front/rear housing bolts	27	35
Electronic speed sensor bolt	7	9
Shift lever housing assembly bolts	7	8.5

SPECIAL TOOLS

Holding Fixture

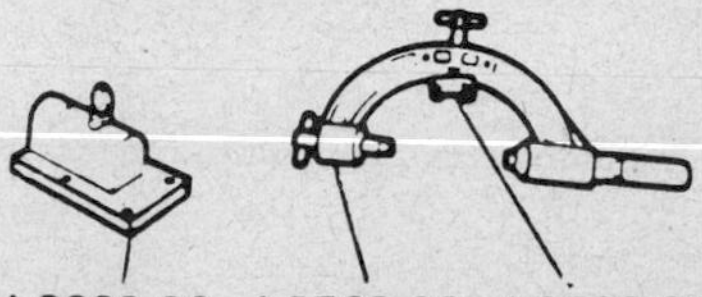

J 3289-20, J 8763-02, J 8763-21

Speedometer Gear Puller Adapter

J 21427-01

Input Shaft Remover/Installer Press Tube with Cap

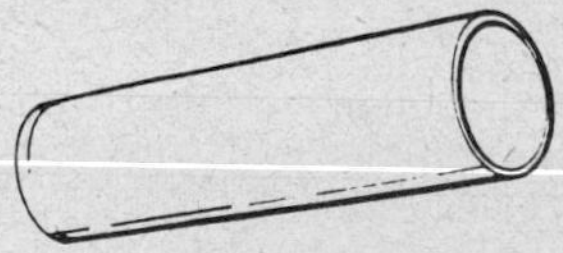

J 36183

Slide Hammer With Adapter

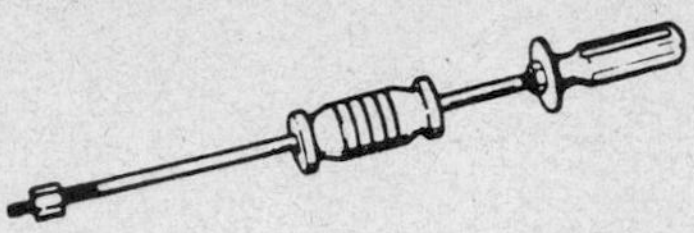

J 6125-1B

Separator Plate or Equivalent

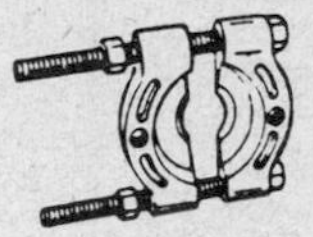

J 22912-01

Adapter Press Tube Reducer

J 36184

Bearing Race Installer

J 6133-01

Slide Hammer With Pilot Bearing Puller

J 23907

Universal Driver Handle

J 36190

Dial Indicator Set

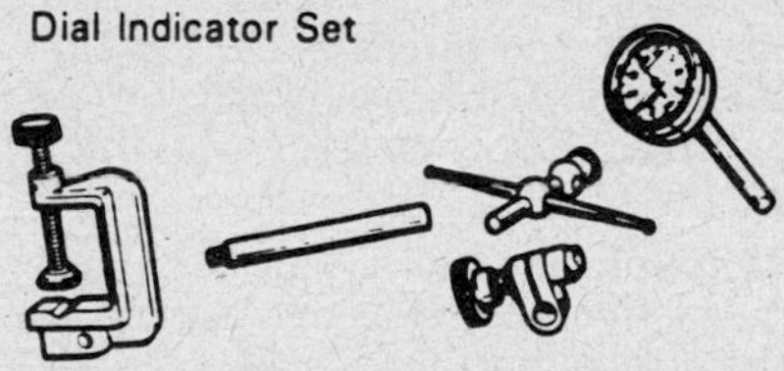

J 8001

Dial Indicator Stand and Guide Pin Set

J 25025-B

4WD Output Shaft Oil Seal Installer

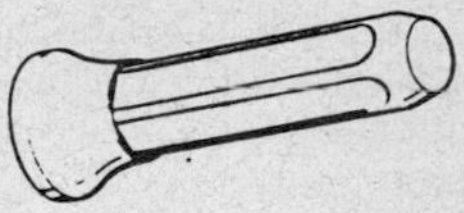

J 36502

Universal Driver Handle

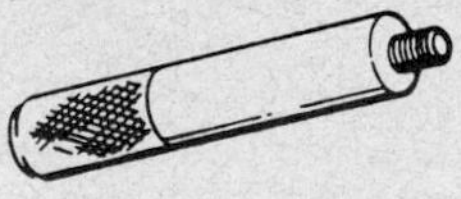

J 8092

Heat Gun or Equivalent

J 25070

2WD Output Shaft Oil Seal Installer

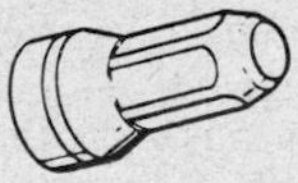

J 36503

Gear Puller or Equivalent

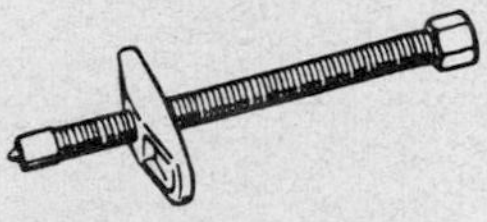

J 8105

2WD Output Shaft Oil Seal Remover

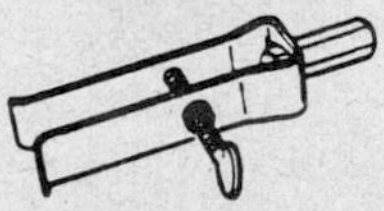

J 26941

Input Shaft Bearing Retainer Oil Seal Installer

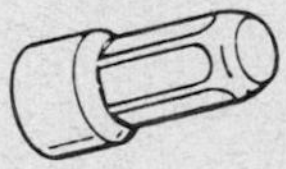

J 36504

Speedometer Gear Puller

J 8433

Input Shaft Bearing Retainer Oil Seal Remover

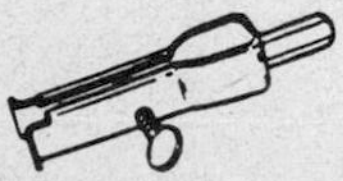

J 29369-2

5th and Reverse Detent Bushing Remover/Installer

J 36506

Section 7

T-4 and T-5 Transmissions
Borg Warner

APPLICATION

1984–86 CJ–7
1984–88 Eagle
1984–85 Scrambler
1984–89 Cougar, Mustang, Thunderbird
1984–86 Capri
1985–89 Astro, Safari
1984–89 Camaro, Firebird, S–10 and S–15 Trucks
1984–87 Chevette and 1000

GENERAL DESCRIPTION

The Borg Warner T–4 is a 4 speed transmission and the T–5 is a 5 speed transmission. Both transmissions provide fully synchronized forward speeds and a single non-synchronized reverse gear. All forward gears are helical cut and are in constant mesh. These transmissions are manufactured with metric measurements and all replacement parts must be the correct metric dimension. The transmission case, cover and adapter/extension housing are aluminum. AMC/Jeep/Eagle use both transmissions and simply labels them as T–4 and T–5. General Motors vehicles refer to it as the 77mm transmission.

Metric Fasteners

Metric bolt sizes and thread pitches are used for all fasteners on the T–4 and T–5 transmission. The use of metric tools is mandatory in the service of this transmission.

Do not attempt to interchange metric fasteners for inch system fasteners. Mismatched or incorrect fasteners can result in damage to the transmission unit through malfunctions, breakage or possible personal injury. Care should be taken to reuse the fasteners in the same location as removed, whenever possible. Due to the large number of alloy parts used and the aluminum casing, torque specifications should be strictly observed. Before installing bolts into aluminum parts, always dip threads into oil, or sealant (if specifically required), to prevent the screws from galling the aluminum threads and to prevent seizing.

Capacities

Checking Fluid Level

The vehicle must be level before an accurate reading can be obtained. The fluid level considered full when the lubricant is to the bottom of the fill hole.

FLUID CAPACITY

All capacities given in quarts

Year	Vehicle	Transmission	Fluid Type	Overhaul Capacity
1984–86	CJ–7	T–4	8983–000–000	2.0
1984–88	Eagle	T–4	8983–000–000	2.0
1984–85	Scrambler	T–4	8983–000–000	2.0
1984–86	CJ–7	T–5	8983–000–000	2.3
1985–88	Eagle	T–5	8983–000–000	2.3
1984–86	Scrambler	T–5	8983–000–000	2.3
1984–86	Capri	T–5	Dexron®II	2.8
1984–89	Cougar, Mustang and Thunderbird	T–5	Dexron®II	2.8
1984–87	Camaro and Firebird	T–5	5W–30	3.3
1988–89	Camaro and Firebird	T–5	Dexron®II	2.9
1985–89	Astro and Safari	T–5	Dexron®II	2.2
1984–87	Chevette and 1000	T–5	Dexron®II	2.2
1984–89	S–10 and S–15	T–5	Dexron®II	2.2

TRANSMISSION MODIFICATIONS

There were no modifications for AMC/Jeep/Eagle or Ford at the time of this publication. The following modification only apply to General Motors vehicles with the 77mm transmission.

Deceleration Rattle

Some 1985 S–10 and S–15 vehicles equipped with the 2.5L engine and T–5 transmission may require a clutch driven disc, part number 15550410, to reduce gear rattle during deceleration.

Fluid Leak

Some 1984 Camaro and Firebird vehicles with the T–5 transmission may require a speedometer driven gear housing, part number 3869912, to prevent a fluid leak.

3rd/4th Gear Clash

Some 1984 Camaro and Firebird vehicles may require synchronizer keys to repair gear clash during shifts into 3rd and 4th speeds.

Deceleration Rattle

Some 1984–85 Camaro and Firebird vehicles with 2.5L engines may require a pressure plate, part number 14091980, to reduce gear rattle. The driven disc alone will reduce deceleration gear rattle.

Crack Shift Boots

Some 1985–87 Camaro and Firebird vehicles may have gear shift boots that crack in cold weather. A replacement boot, part number 10095205, should be used.

TROUBLE DIAGNOSIS

CHILTON'S THREE C'S TRANSMISSION DIAGNOSIS
Borg Warner T-4 and T-5—4WD

Condition	Cause	Correction
Transmission shifts hard	a) Clutch adjustment incorrect	a) Adjust clutch.
	b) Clutch linkage or cable binding	b) Lubricate or repair as necessary.
	c) Shift rail binding	c) Check for mispositioned selector arm roll pin, loose cover bolts, worn shift rail bores, worn shift rail, distorted oil seal, or extension housing not aligned with case. Repair as necessary.
	d) Internal bind in transmission caused by shift forks, selector plates, or synchronizer assemblies	d) Remove, disassemble and inspect transmission. Replace worn or damaged components as necessary.
	e) Clutch housing misalignment	e) Check runout at rear face of clutch housing.
	f) Incorrect lubricant	f) Drain and refill transmission.
	g) Block rings and/or cone seats worn	g) Blocking ring to gear clutch tooth face clearance must be 0.030 in. or greater. If clearance is correct, it may still be necessary to inspect blocking rings and cone seats for excessive wear. Repair as necessary.
Gear clash when shifting from one gear to another	a) Clutch adjustment incorrect	a) Adjust clutch
	b) Clutch linkage or cable binding	b) Lubricate or repair as necessary.
	c) Clutch housing misalignment	c) Check runout at rear of clutch housing.
	d) Lubricant level low or incorrect lubricant	d) Drain and refill transmission and check for lubricant leaks if level was low. Repair as necessary.
	e) Gearshift components, or synchronizer assemblies worn or damaged	e) Remove, disassemble and inspect transmission. Replace worn or damaged components as necessary.

CHILTON'S THREE C'S TRANSMISSION DIAGNOSIS
Borg Warner T-4 and T-5—4WD

Condition	Cause	Correction
Transmission noisy	a) Lubricant level low or incorrect lubricant	a) Drain and refill transmission. If lubricant level was low, check for leaks and repair as necessary.
	b) Clutch housing-to-engine, or transmission-to-clutch housing bolts loose	b) Check and correct bolt torque as necessary.
	c) Dirt, chips, foreign material in transmission	c) Drain, flush, and refill transmission.
	d) Gearshift mechanism, transmission gears, or bearing components worn or damaged	d) Remove, disassemble and inspect transmission. Replace worn or damaged components as necessary.
	e) Clutch housing misalignment	e) Check runout at rear face of clutch housing.
Jumps out of gear	a) Clutch housing misalignment	a) Check runout at rear face of clutch housing.
	b) Gearshift lever loose	b) Check lever for worn fork. Tighten loose attaching bolts.
	c) Offset lever nylon insert worn or lever attaching nut loose	c) Remove gearshift lever and check for loose offset lever nut or worn insert. Repair or replace as necessary.
	d) Gearshift mechanism, shift forks, selector plates, interlock plate, selector arm, shift rail, detent plugs, springs or shift cover worn or damaged	d) Remove, disassemble and inspect transmission cover assembly. Replace worn or damaged components as necessary.
	e) Clutch shaft or roller bearings worn or damaged	e) Replace clutch shaft or roller bearings as necessary.
	f) Gear teeth worn or tapered, synchronizer assemblies worn or damaged, excessive end play caused by worn thrust washers or output shaft gears	f) Remove, disassemble, and inspect transmission. Replace worn or damaged components as necessary.
	g) Pilot bushing worn	g) Replace pilot bushing.
Will not shift into one gear	a) Gearshift selector plates, interlock plate, or selector arm, worn, damaged, or incorrectly assembled	a) Remove, disassemble, and inspect transmission cover assembly. Repair or replace components as necessary.
	b) Shift rail detent plunger worn, spring broken, or plug loose	b) Tighten plug or replace worn or damaged components as necessary.
	c) Gearshift lever worn or damaged	c) Replace gearshift lever.
	d) Synchronizer sleeves or hubs, damaged or worn	d) Remove, disassemble and inspect transmission. Replace worn or damaged components.
Locked in one gear—cannot be shifted out	a) Shift rail(s) worn or broken, shifter fork bent, setscrew loose, center detent plug missing or worn	a) Inspect and replace worn or damaged parts.
	b) Broken gear teeth on countershaft gear, clutch shaft, or reverse idler gear	b) Inspect and replace damaged part.
	c) Gearshift lever broken or worn, shift mechanism in cover incorrectly assembled or broken, worn damaged gear train components	c) Disassemble transmission. Replace damaged parts or assemble correctly.

CHILTON'S THREE C'S TRANSMISSION DIAGNOSIS
Borg Warner T-4 and T-5—4WD

Condition	Cause	Correction
Transfer case difficult to shift or will not shift into desired range	a) Vehicle speed too great to permit shifting.	a) Stop vehicle and shift into desired range. Or reduce speed to 2–3 mph (3–4 km/h) before attempting to shift.
	b) If vehicle was operated for extended period in 4H mode on dry paved surface, driveline torque load may cause difficult shifting.	b) Stop vehicle, shift transmission to neutral, shift transfer case to 2H mode and operate vehicle in 2H on dry paved surfaces.
	c) Transfer case external shift linkage binding.	c) Lubricate or repair or replace linkage, or tighten loose components as necessary.
	d) Insufficient or incorrect lubricant.	d) Drain and refill to edge of fill hole with SAE 85W–90 gear lubricant only.
	e) Internal components binding, worn, or damaged.	e) Disassemble unit and replace worn or damaged components as necessary.
Transfer case noisy in all drive modes	a) Insufficient or incorrect lubricant.	a) Drain and refill to edge of fill hole with SAE 85W–90 gear lubricant only. Check for leaks and repair if necessary. Note: If unit is still noisy after drain and refill, disassembly and inspection may be required to locate source of noise.
Noisy in—or jumps out of four wheel drive low range	a) Transfer case not completely engaged in 4L position.	a) Stop vehicle, shift transfer case in Neutral, then shift back into 4L position.
	b) Shift linkage loose or binding.	b) Tighten, lubricate, or repair linkage as necessary.
	c) Shift fork cracked, inserts worn, or fork is binding on shift rail.	c) Disassemble unit and repair as necessary.
Lubricant leaking from output shaft seals or from vent	a) Transfer case overfilled.	a) Drain to correct level.
	b) Vent closed or restricted.	b) Clear or replace vent if necessary.
	c) Output shaft seals damaged or installed incorrectly.	c) Replace seals. Be sure seal lip faces interior of case when installed. Also be sure yoke seal surfaces are not scored or nicked. Remove scores, nicks with fine sandpaper or replace yoke(s) if necessary.
Abnormal tire wear	a) Extended operation on dry hard surface (paved) roads in 4H range.	a) Operate in 2H on hard surface (paved) roads.

POWER FLOW

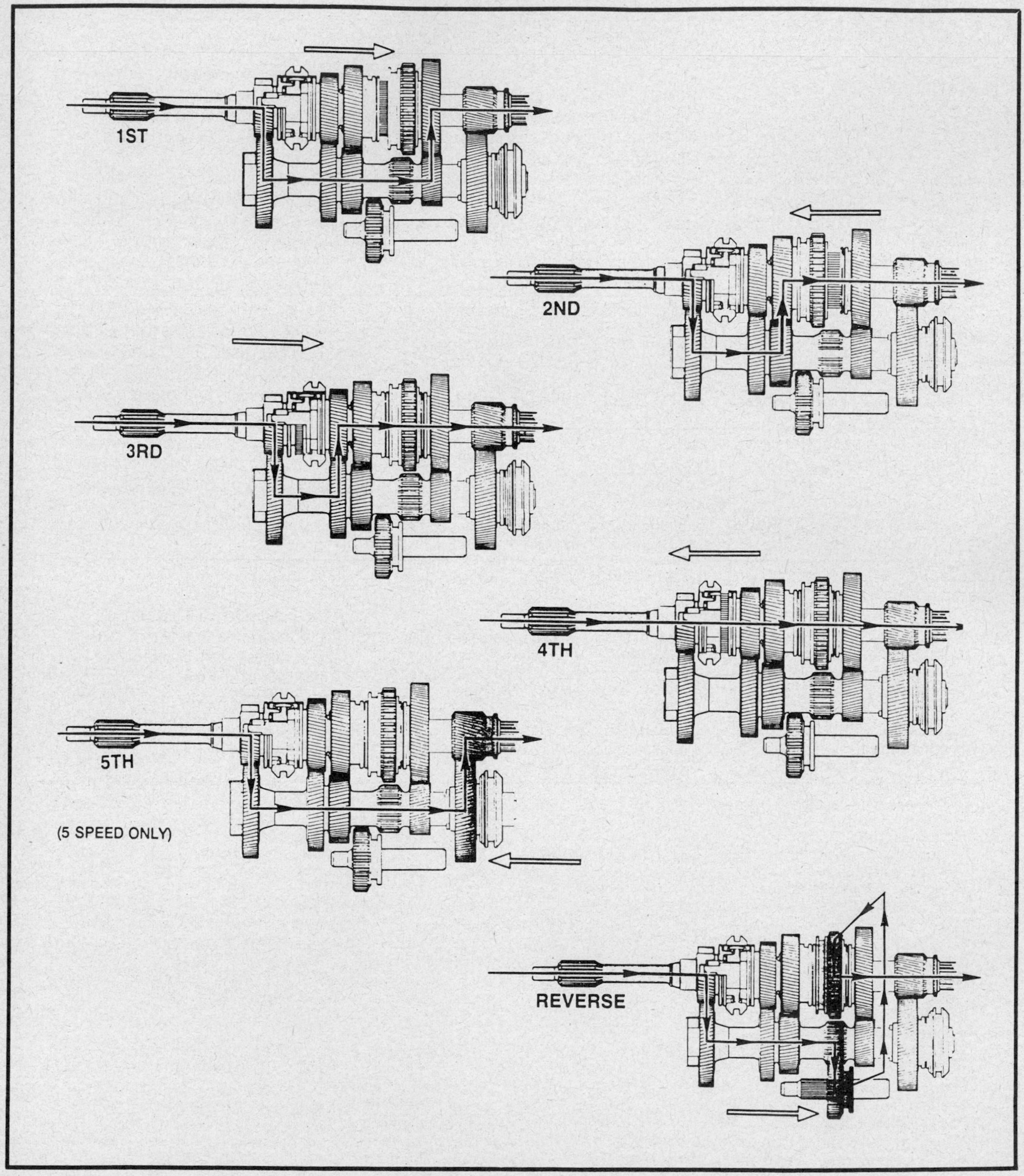
1ST
2ND
3RD
4TH
5TH
(5 SPEED ONLY)
REVERSE

Powerflow for T–4 and T–5

ON CAR SERVICES

Services

FLUID CHANGES

1. Raise and support vehicle safely.
2. Remove drain plug, if equipped or siphon transmission fluid into a pan.
3. Remove filler plug.
4. Install drain plug.
5. Fill to the level of the filler plug hole with the proper type and amount of automatic transmission fluid.
6. Install fill plug and lower vehicle.

REAR SEAL

Removal and Installation

1. Raise and support vehicle safely.
2. Remove drain plug, if equipped or siphon transmission fluid into a pan.
3. Remove filler plug.
4. Install drain plug.
5. Mark driveshaft to yoke for reassembly and remove driveshaft.
6. Pry out old seal using an appropriate tool.
7. Install new seal, coated with locking compound, using tool J–21426 or equivalent.
8. Install the driveshaft, making certain to align marks and torque bolts to 15 ft. lbs. (20 Nm).
9. Fill to the level of the filler plug hole with the proper type and amount of automatic transmission fluid.
10. Install fill plug and lower vehicle.

REMOVAL AND INSTALLATION

TRANSMISSION REMOVAL

AMC/JEEP/EAGLE

1. Remove the screws attaching the shift lever boot to the floorpan and slide boot up. Remove the bolts attaching the shift lever housing to the transmission and remove the lever and housing
2. Disconnect the negative battery cable.
3. Raise and support the vehicle safely. Remove the skid plate, if equipped.
4. Mark the driveshaft to output yoke for assembly alignment reference.
5. Disconnect the driveshaft and support it to vehicle with wire.
6. Support engine, by placing a safety stand place at the clutch housing. Remove the crossmember.
7. On 1984–88 Eagle, it is necessary to remove the catalytic converter bracket from the transfer case and the brace rod from the bracket.
8. Mark and disconnect the speedometer adapter (if equipped) and cable.
9. Remove the back-up light wiring, drive indicator wiring, if equipped and vent hoses, if equipped.
10. Mark the front driveshaft to output yoke for assembly alignment reference.
11. Disconnect the driveshaft and support it to vehicle with wire.
12. Remove the transfer case shift shaft retaining nut.
13. Remove the cotter pin that retains the shift control link pins in the shift rods and remove the pins.
14. Remove the shift shaft and disengage the shift lever from the shift control links.
15. Slide the lever upward in the boot to move the lever out of the way.
16. Support the transmission/transfer case assembly with a transmission jack.
17. Remove the bolts attaching the transmission to the clutch housing and remove the transmission/transfer case assembly.
18. Remove the transfer case from the transmission.
19. Remove the pilot bushing lubricating wick from the bushing and soak the wick in engine oil.

FORD MOTOR CO.

1. Remove the screws attaching the shift lever boot to the floorpan and slide boot up. Remove the bolts attaching the shift lever housing to the transmission and remove the lever and housing.
2. Disconnect the negative battery cable.
3. Raise and support the vehicle safely.
4. Mark the driveshaft to output yoke for assembly alignment reference.
5. Disconnect the driveshaft and support it to vehicle with wire.
6. Support engine, by placing a safety stand under the clutch housing. Remove the crossmember.
7. Remove the catalytic converter and if necessary the inlet pipe.
8. Mark and disconnect the speedometer adapter (if equipped) and cable.
9. Remove the back-up light and neutral sensing wiring.
10. Support transmission with transmission jack and remove remove crossmember if not already removed.
11. Remove the bolts attaching the transmission to the clutch housing and remove the transmission.

GENERAL MOTORS

These procedures will vary between 2WD and 4WD vehicles and may vary slightly between the different vehicle models.

1. Disconnect the negative battery cable.
2. Remove the shift lever boot.
3. Shift transmission into **N** and remove shift lever.
4. Raise and safely support vehicle.
5. Mark the driveshaft to output yoke for assembly alignment reference.
6. Disconnect and remove the driveshaft from vehicle.
7. Support engine, by placing a safety stand under the clutch housing. Remove the crossmember.
8. Remove the catalytic converter bracket, the brace rod from the bracket and any other parts needed for clearance.
9. Remove the speedometer cable or speed sensor and speedometer adapter, if used. If a speedometer adapter is used mark it to the transmission or transfer case for reference during reassembly.

10. Remove the back-up light wiring, drive indicator wiring, if equipped and vent hoses, if equipped.
11. Remove exhaust pipes as needed.
12. Remove slave cylinder.
13. If a 2WD vehicle, carefully remove the transmission by pulling straight back. If a 4WD vehicle, continue with Step 14.
14. Mark the front driveshaft to output yoke for assembly alignment reference.
15. Disconnect and support the driveshaft to the vehicle with wire, or remove it.
16. Remove the transfer case shift shaft retaining nut.
17. Remove the cotter pin that retains the shift control link pins in the shift rods and remove the pins.
18. Remove the shift shaft and disengage the shift lever from the shift control links.
19. Slide the lever upward in the boot to move the lever out of the way.
20. Support the transmission/transfer case assembly with a transmission jack.
21. Remove the bolts attaching the transmission to the clutch housing and remove the transmission/transfer case assembly.
22. Remove the transfer case from the transmission.

TRANSMISSION INSTALLATION

AMC/JEEP/EAGLE

1. Install the pilot bushing lubricating wick and align the throwout bearing with the splines in the driven plate hub.
2. Shift the transmission into gear, using a long appropriate tool.
3. Mount the transmission on the transmission jack, raise and install transmission.
4. Make certain transmission is installed properly and torque bolts to 55 ft. lbs. (75 Nm).
5. Apply Permatex® No. 3, or equivalent, to both sides of the transmission-to-transfer case gasket and install gasket.
6. Place transfer case in transmission jack, raise and install onto transmission.
7. Make certain transfer case is installed properly and torque the attaching bolts to 30 ft. lbs. (41 Nm).
8. Install the transfer case shift lever.
9. Install the shifter shaft, link pins and control assembly.
10. Connect the front driveshaft, making certain to align the marks and torque the clamp strap bolt to 15 ft. lbs. (20 Nm).
11. Connect the vent hose, backup wiring and drive indicator wiring.
12. Install the speedometer adapter, if equipped, making certain to align marks and speedometer cable. Torque adapter to 100 inch lbs. (11 Nm).
13. Install the brace rod, if equipped and rear crossmember and torque attaching bolts to 30 ft. lbs. (41 Nm).
14. Remove engine support.
15. On the 1984–88 Eagle, attach the catalytic converter bracket to transfer case and torque to 26 ft. lbs. (35 Nm).
16. Connect the rear driveshaft, making certain to align marks and torque the clamp strap to 15 ft. lbs. (20 Nm).
17. Check transmission and transfer case lubrication levels, install skid plate, if equipped and torque bolts to 30 ft. lbs. (41 Nm).
18. Lower the vehicle and connect the negative battery cable.
19. Apply RTV—type sealant to the gear shift lever mounting cover and install cover. Torque bolts to 13 ft. lbs. (18 Nm).
20. Install the gearshift lever on the mounting cover. Make certain the the lever is engaged with the shift rail before torque bolts to 18 ft. lbs. (24 Nm).
21. Position the shift lever boot on the floorpan and install boot attaching screws.

FORD MOTOR CO.

1. Align the throwout bearing with the splines in the driven plate hub.
2. Shift the transmission into gear, using a long appropriate tool.
3. Mount the transmission on the transmission jack, raise and install transmission.
4. Make certain transmission is installed properly and torque bolts to 55 ft. lbs. (75 Nm).
5. Install speedometer adapter, if used, making certain to align marks on case. Install the speedometer cable or speed sensor.
6. Install the brace rod(s), if equipped and rear crossmember and torque attaching bolts to 35 ft. lbs. (47 Nm).
7. Remove engine support.
8. Install the catalytic converter and torque the attaching bolts to 20–30 ft. lbs. (27–41 Nm).
9. Connect the driveshaft, making certain to align marks and torque the U-bolts to 42–57 ft. lbs. (56–77 Nm).
10. Check transmission lubrication level.
11. Lower the vehicle and connect the negative battery cable.
12. Apply RTV—type sealant to the gear shift lever mounting cover and install cover. Torque bolts to 13 ft. lbs. (18 Nm).
13. Install the gearshift lever on the mounting cover. Make certain the the lever is engaged with the shift rail before torque bolts to 18 ft. lbs. (24 Nm).
14. Position the shift lever boot on the floorpan and install boot attaching screws.

GENERAL MOTORS

These procedures will vary between 2WD and 4WD vehicles and may vary slightly between different vehicle models.

1. Shift transmission into gear and lubricate main drive gear splines with a light coat of high temperature gear.
2. Mount the transmission on the transmission jack, raise and install transmission.
3. Make certain transmission is installed properly and torque bolts to 55 ft. lbs. (75 Nm).

NOTE: On Chevette, torque the bolts to 30 ft. lbs. (40 Nm), if the Chevette damper was removed, torque the damper bolts to 30 ft. lbs. (40 Nm).

4. If a 4WD vehicle continue with Step 5. If a 2WD vehicle continue with Step 12.
5. Apply Permatex® No. 3, or equivalent, to both sides of the transmission-to-transfer case gasket and install gasket.
6. Place transfer case in transmission jack, raise and install onto the transmission.
7. Make certain transfer case is installed properly and torque the attaching bolts to 38 ft. lbs. (52 Nm).

NOTE: If the adapter plate was removed from the transmission, torque adapter-to-transmission bolts to 32 ft. lbs. (43 Nm).

8. Install the transfer case shift lever.
9. Install the shifter shaft, link pins and control assembly.
10. Connect the front driveshaft, making certain to align the marks and torque the bolts to 15 ft. lbs. (20 Nm).
11. Connect the vent hose, backup wiring and drive indicator wiring.
12. Install speedometer adapter, if used, making certain to align marks on case. Install the speedometer cable or speed sensor.
13. Install the brace rod(s), if equipped and rear crossmember and torque attaching bolts to 35 ft. lbs. (47 Nm).
14. Remove engine support.
15. Install the catalytic converter bracket, if equipped to transfer case and torque to 26 ft. lbs. (35 Nm).
16. Connect the rear driveshaft, making certain to align marks and torque the bolts to 15 ft. lbs. (20 Nm).
17. Install the slave cylinder, if equipped and torque nuts to 13 ft. lbs. (18 Nm), or install clutch cable.
18. Check transmission and transfer case lubrication levels, install skid plate, if equipped and torque bolts to 30 ft. lbs. (41 Nm).

19. Lower the vehicle and connect the negative battery cable.
20. Apply RTV—type sealant to the gear shift lever mounting cover and install cover. Torque bolts to 13 ft. lbs. (18 Nm).
21. Install the gearshift lever on the mounting cover. Make certain the the lever is engaged with the shift rail before torque bolts to 18 ft. lbs. (24 Nm).
22. Position the shift lever boot on the floorpan and install boot attaching screws.

BENCH OVERHAUL

Before Disassembly

Clean the exterior of the transmission assembly before any attempt is made to disassembly it, in order to prevent dirt or other foreign materials from entering the transmission assembly or its internal parts.

NOTE: If steam cleaning is done to the exterior of the transmission, immediate disassembly should be done to avoid rusting, caused by condensation forming on the internal parts.

Transmission Disassembly

4 SPEED

These procedures may vary slightly on different manufactured vehicles, 2WD or 4WD and from model to model.

1. Drain the transmission lubricant. Some models are not equipped with a drain plug; the fluid must be siphoned from the transmission.

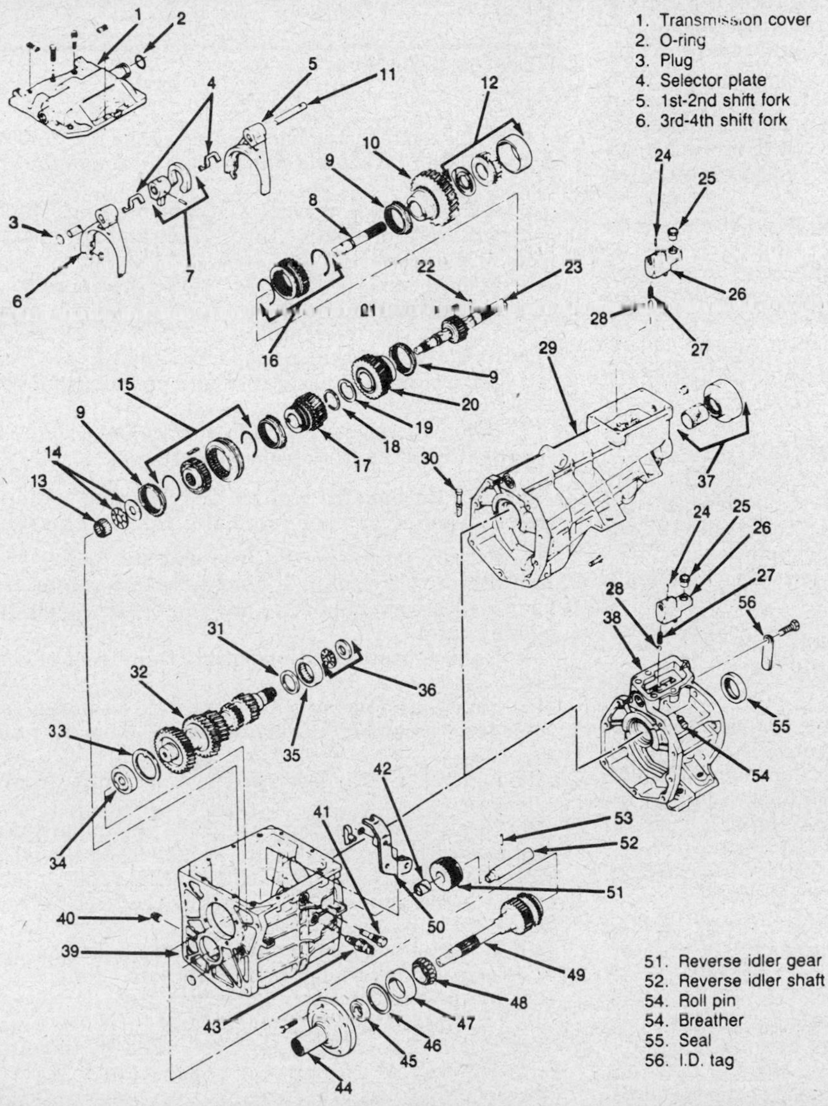

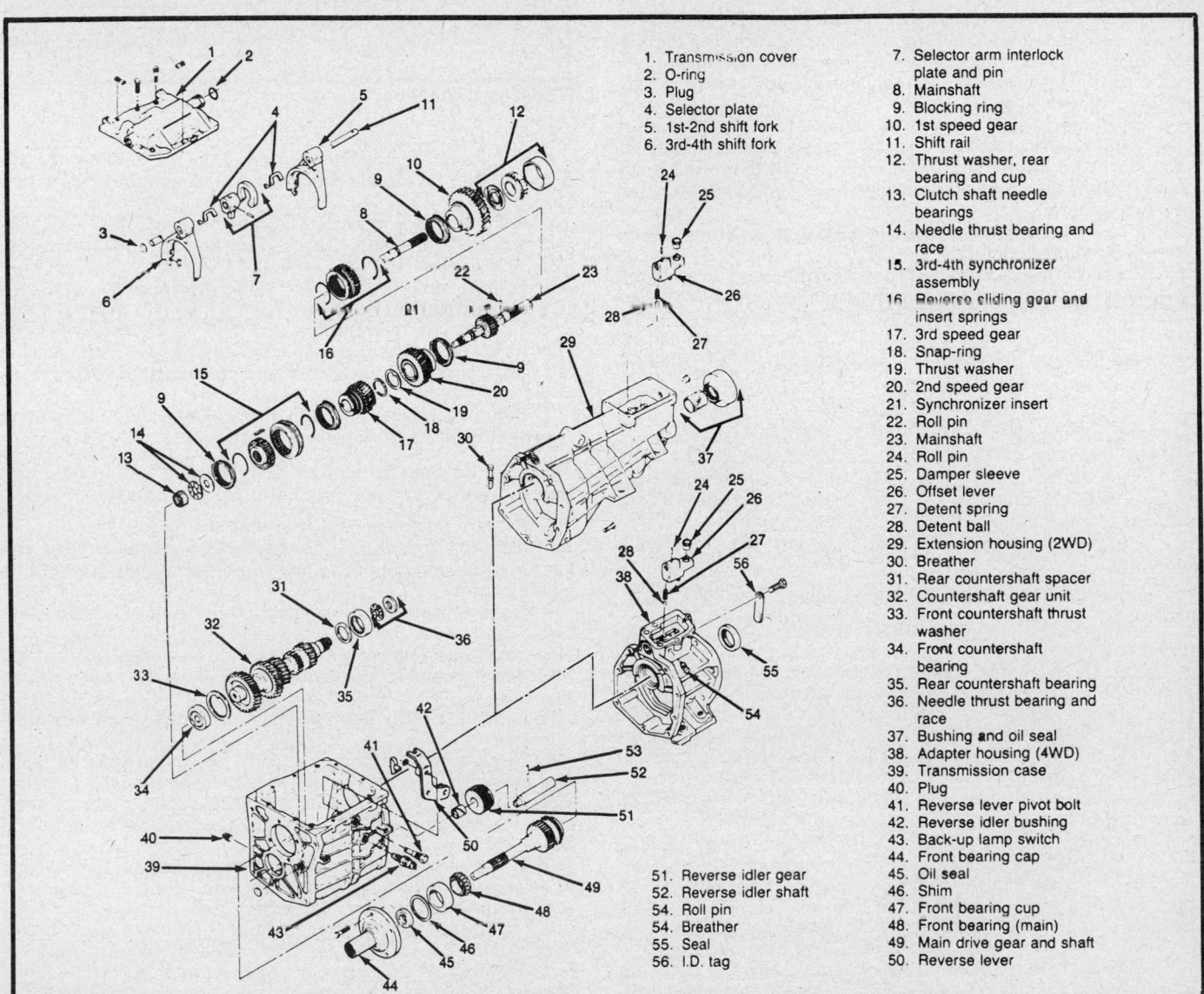

1. Transmission cover
2. O-ring
3. Plug
4. Selector plate
5. 1st-2nd shift fork
6. 3rd-4th shift fork
7. Selector arm interlock plate and pin
8. Mainshaft
9. Blocking ring
10. 1st speed gear
11. Shift rail
12. Thrust washer, rear bearing and cup
13. Clutch shaft needle bearings
14. Needle thrust bearing and race
15. 3rd-4th synchronizer assembly
16. Reverse sliding gear and insert springs
17. 3rd speed gear
18. Snap-ring
19. Thrust washer
20. 2nd speed gear
21. Synchronizer insert
22. Roll pin
23. Mainshaft
24. Roll pin
25. Damper sleeve
26. Offset lever
27. Detent spring
28. Detent ball
29. Extension housing (2WD)
30. Breather
31. Rear countershaft spacer
32. Countershaft gear unit
33. Front countershaft thrust washer
34. Front countershaft bearing
35. Rear countershaft bearing
36. Needle thrust bearing and race
37. Bushing and oil seal
38. Adapter housing (4WD)
39. Transmission case
40. Plug
41. Reverse lever pivot bolt
42. Reverse idler bushing
43. Back-up lamp switch
44. Front bearing cap
45. Oil seal
46. Shim
47. Front bearing cup
48. Front bearing (main)
49. Main drive gear and shaft
50. Reverse lever
51. Reverse idler gear
52. Reverse idler shaft
54. Roll pin
54. Breather
55. Seal
56. I.D. tag

Borg Warner T-4 transmission—exploded view

2. Use a pin punch and hammer to remove the offset lever-to-shift rail roll pin.
3. Remove the extension housing (2WD) or the adapter (4WD). Remove the housing and the offset lever as an assembly.
4. Remove the detent ball and spring from the offset lever. Remove the roll pin from the extension housing or adapter.
5. Remove the countershaft rear thrust bearing and race.
6. Remove the transmission cover and shift fork assembly. Take notice that 2 of the transmission cover bolts are alignment type dowel pins. Mark their location so that they may be reinstalled in their original locations.
7. Remove the reverse lever to reverse lever pivot bolt C-clip.
8. Remove the reverse lever pivot bolt. Remove the reverse lever and fork as an assembly.
9. Mark the position of the front bearing cap to case, then remove the bearing cap bolts and cap.
10. Remove the front bearing race and the shims from the bearing cap. Use a small pry bar and remove the front seal from the bearing cap.
11. Rotate the main drive gear shaft until the flat portion of the gear faces the countershaft, then remove the main drive gear shaft assembly.
12. Remove the thrust bearing and 15 roller bearings from the clutch shaft. Remove the output shaft bearing race. Tap the output shaft with a plastic hammer to loosen it if necessary.
13. Tilt the output shaft assembly upward and remove the assembly from the case.
14. Carefully pull off the countershaft rear bearing, using the proper puller after marking the position for reinstallation.
15. Move the countershaft rearward and tilt it upward to remove it from the transmission case. Remove the countershaft bearing spacer.
16. Remove the reverse idler shaft roll pin, then remove the reverse idler shaft and gear.
17. Press off the countershaft front bearing. Use the appropriate pullers and remove the bearing from the main drive gear shaft.
18. Remove the extension housing or adapter oil seal and remove the back-up light switch from the case.

5 Speed

These procedures will vary slightly on different manufactured vehicles, 2WD or 4WD and from model to model. The Ford, 1988–89 Camaro and Firebird use a similar design which is slightly different than the T-5 transmission that is used on the other General Motors, AMC and Jeep/Eagle vehicles.
1. Remove drain bolt on transmission case and drain lubricant.
2. Thoroughly clean the exterior of the transmission assembly.
3. Using pin punch and hammer, remove roll pin attaching offset lever to shift rail.
4. Remove extension housing to transmission case bolts and remove housing and offset lever as an assembly. Do not attempt to remove the offset lever while the extension housing is still bolted in place. The lever has a positioning lug engaged in the housing detent plate which prevents moving the lever far enough for removal.
5. Remove detent ball and spring from offset lever and remove roll pin from extension housing or offset lever.
6. On older models, remove plastic funnel, thrust bearing race and thrust bearing from rear of countershaft. The countershaft rear thrust bearing, bearing washer and plastic funnel may be found inside the extension housing. On newer models, pry the oiling funnel from the rear of the countershaft. This may be a very snug fit and the funnel may become damaged.
7. Remove bolts attaching transmission cover and shift fork assembly and remove cover. Take notice that 2 of the transmission cover attaching bolts are alignment-type dowel bolts. Note the location of these bolts for assembly reference.

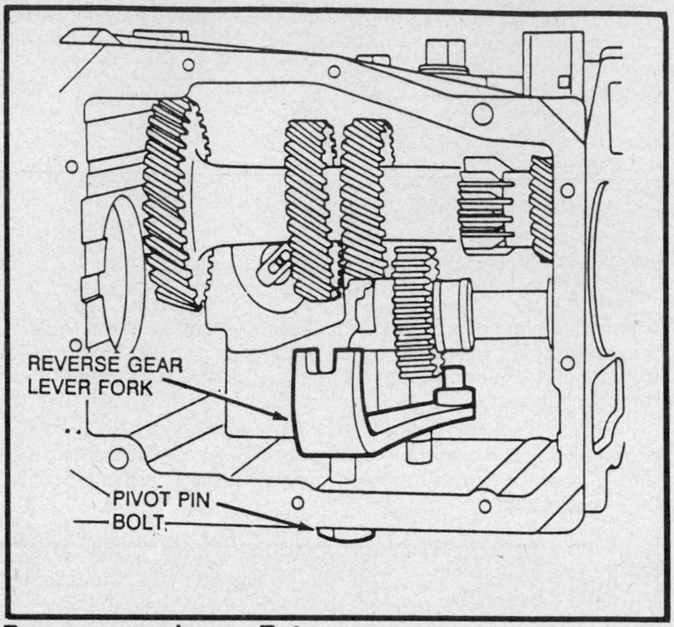
REVERSE GEAR
LEVER FORK
PIVOT PIN
BOLT
Reverse gear lever—T-4

8. Using a punch and hammer, drive the roll pin from the 5th gearshift shift fork while supporting the end of the shaft with a block of wood.
9. Remove 5th synchronizer gear snapring, shift fork, 5th gear synchronizer sleeve, blocking ring and 5th speed drive gear from rear of countershaft.
10. Remove snapring from 5th speed driven gear.
11. Using a hammer and punch, mark both bearing cap and case for assembly reference.
12. Remove front bearing cap bolts and remove front bearing cap. Remove front bearing race and endplay shims from front bearing cap.
13. Rotate drive gear until flat surface faces countershaft and remove drive gear from transmission case.

NOTE: Be careful not to drop the 15 roller bearing, thrust washer or race from the rear of the input shaft.

14. Remove reverse lever C-clip and pivot bolt.
15. Remove mainshaft rear bearing race and then tilt mainshaft assembly upward and remove assembly from transmission case.
16. Unhook overcenter link spring from front of transmission case.
17. Rotate 5th/reverse gear shift rail to disengage rail from reverse lever assembly. Remove shift rail from rear of transmission case.
18. Remove reverse lever and fork assembly from transmission case.
19. On all vehicles except Ford, 1988–89 Camaro and Firebird, continue with Step 20. On all Ford, 1988–89 Camaro and Firebird, bend back the countershaft bearing retainer lock tabs, remove the 4 retaining bolts, remove the bearing retainer and shims.
20. Using hammer and punch, drive roll pin from forward end of reverse idler shaft and remove reverse idler shaft, rubber O-ring and gear from the transmission case.
21. On Ford, 1988–89 Camaro and Firebird, continue with Step 25. On all vehicles except Ford, 1988–89 Camaro and Firebird, remove rear countershaft snapring and spacer.
22. Insert a brass drift through drive gear opening in front of transmission case and, using an arbor press, carefully press countershaft rearward to remove rear countershaft bearing.

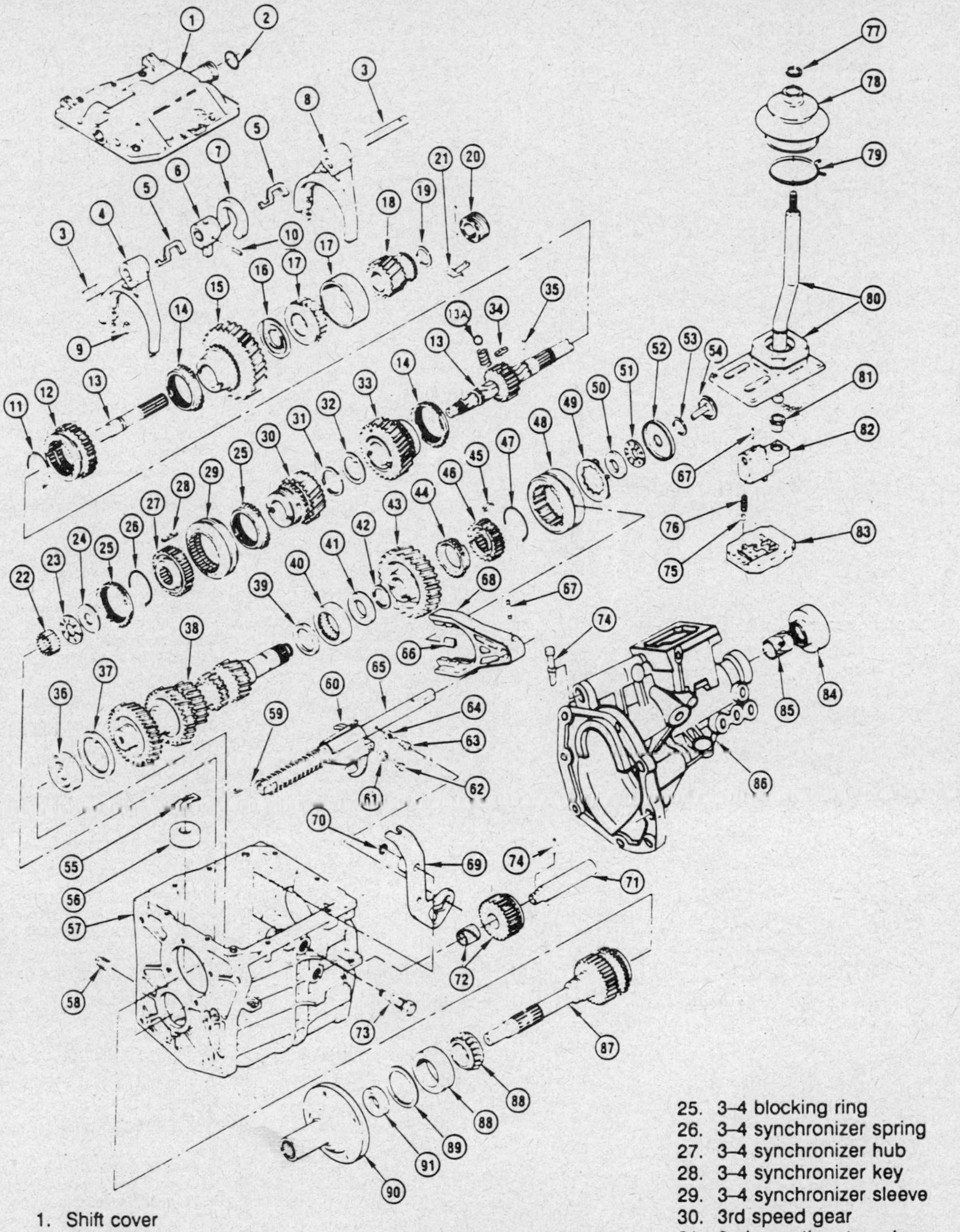

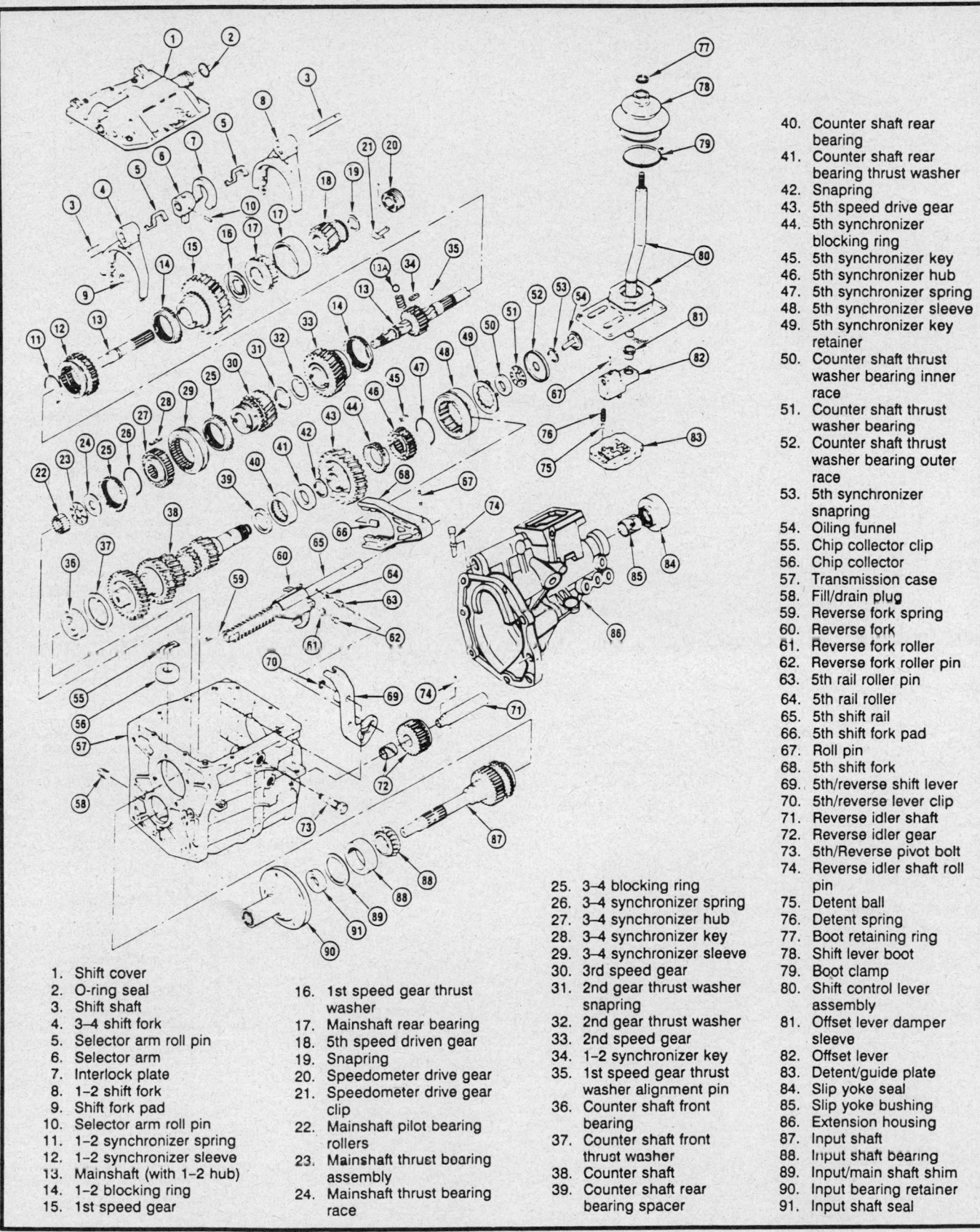

1. Shift cover
2. O-ring seal
3. Shift shaft
4. 3–4 shift fork
5. Selector arm roll pin
6. Selector arm
7. Interlock plate
8. 1–2 shift fork
9. Shift fork pad
10. Selector arm roll pin
11. 1–2 synchronizer spring
12. 1–2 synchronizer sleeve
13. Mainshaft (with 1–2 hub)
14. 1–2 blocking ring
15. 1st speed gear
16. 1st speed gear thrust washer
17. Mainshaft rear bearing
18. 5th speed driven gear
19. Snapring
20. Speedometer drive gear
21. Speedometer drive gear clip
22. Mainshaft pilot bearing rollers
23. Mainshaft thrust bearing assembly
24. Mainshaft thrust bearing race
25. 3–4 blocking ring
26. 3–4 synchronizer spring
27. 3–4 synchronizer hub
28. 3–4 synchronizer key
29. 3–4 synchronizer sleeve
30. 3rd speed gear
31. 2nd gear thrust washer snapring
32. 2nd gear thrust washer
33. 2nd speed gear
34. 1–2 synchronizer key
35. 1st speed gear thrust washer alignment pin
36. Counter shaft front bearing
37. Counter shaft front thrust washer
38. Counter shaft
39. Counter shaft rear bearing spacer
40. Counter shaft rear bearing
41. Counter shaft rear bearing thrust washer
42. Snapring
43. 5th speed drive gear
44. 5th synchronizer blocking ring
45. 5th synchronizer key
46. 5th synchronizer hub
47. 5th synchronizer spring
48. 5th synchronizer sleeve
49. 5th synchronizer key retainer
50. Counter shaft thrust washer bearing inner race
51. Counter shaft thrust washer bearing
52. Counter shaft thrust washer bearing outer race
53. 5th synchronizer snapring
54. Oiling funnel
55. Chip collector clip
56. Chip collector
57. Transmission case
58. Fill/drain plug
59. Reverse fork spring
60. Reverse fork
61. Reverse fork roller
62. Reverse fork roller pin
63. 5th rail roller pin
64. 5th rail roller
65. 5th shift rail
66. 5th shift fork pad
67. Roll pin
68. 5th shift fork
69. 5th/reverse shift lever
70. 5th/reverse lever clip
71. Reverse idler shaft
72. Reverse idler gear
73. 5th/Reverse pivot bolt
74. Reverse idler shaft roll pin
75. Detent ball
76. Detent spring
77. Boot retaining ring
78. Shift lever boot
79. Boot clamp
80. Shift control lever assembly
81. Offset lever damper sleeve
82. Offset lever
83. Detent/guide plate
84. Slip yoke seal
85. Slip yoke bushing
86. Extension housing
87. Input shaft
88. Input shaft bearing
89. Input/main shaft shim
90. Input bearing retainer
91. Input shaft seal

Borg Warner T-5—Ford, 1988–89 Camaro and Firebird

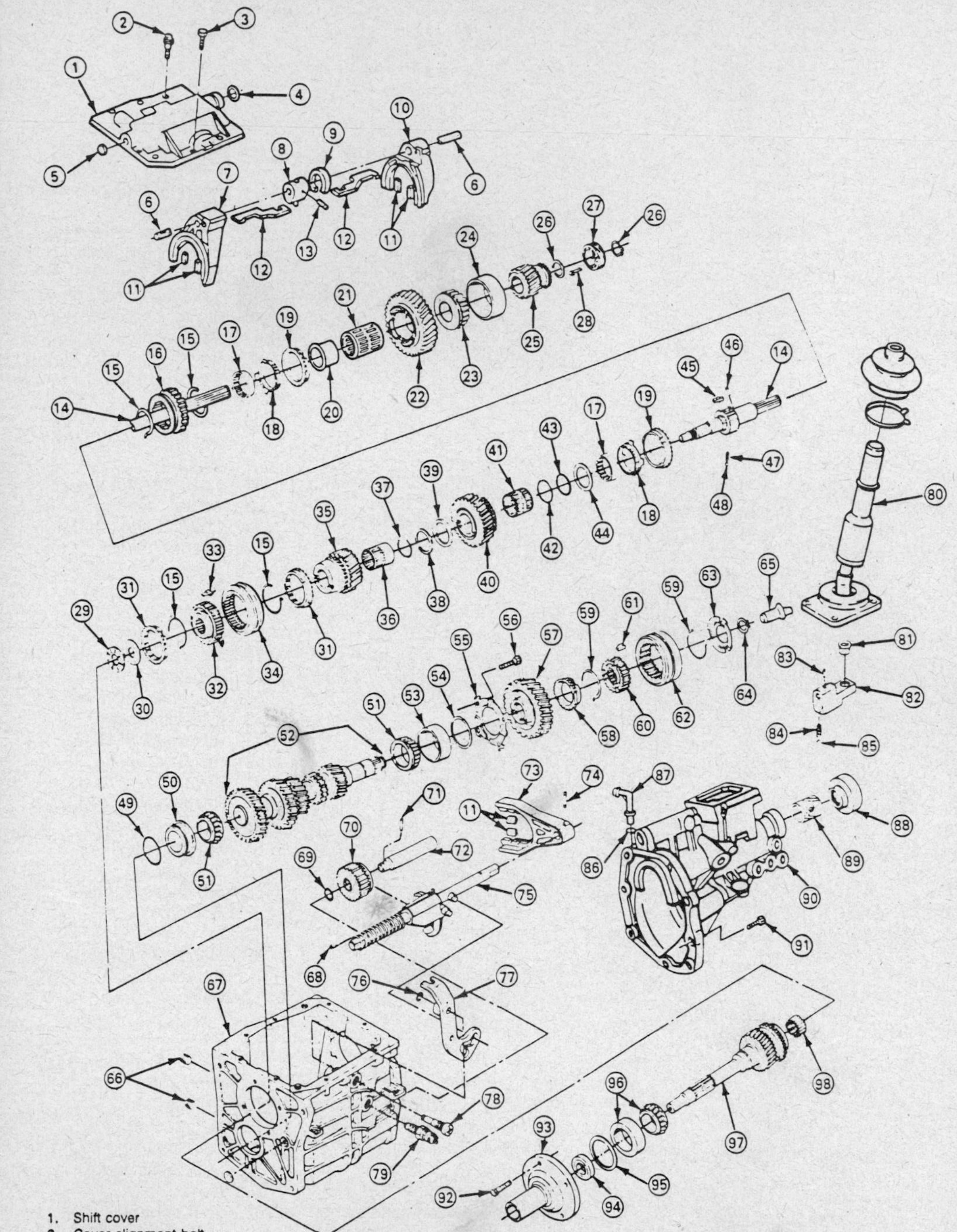

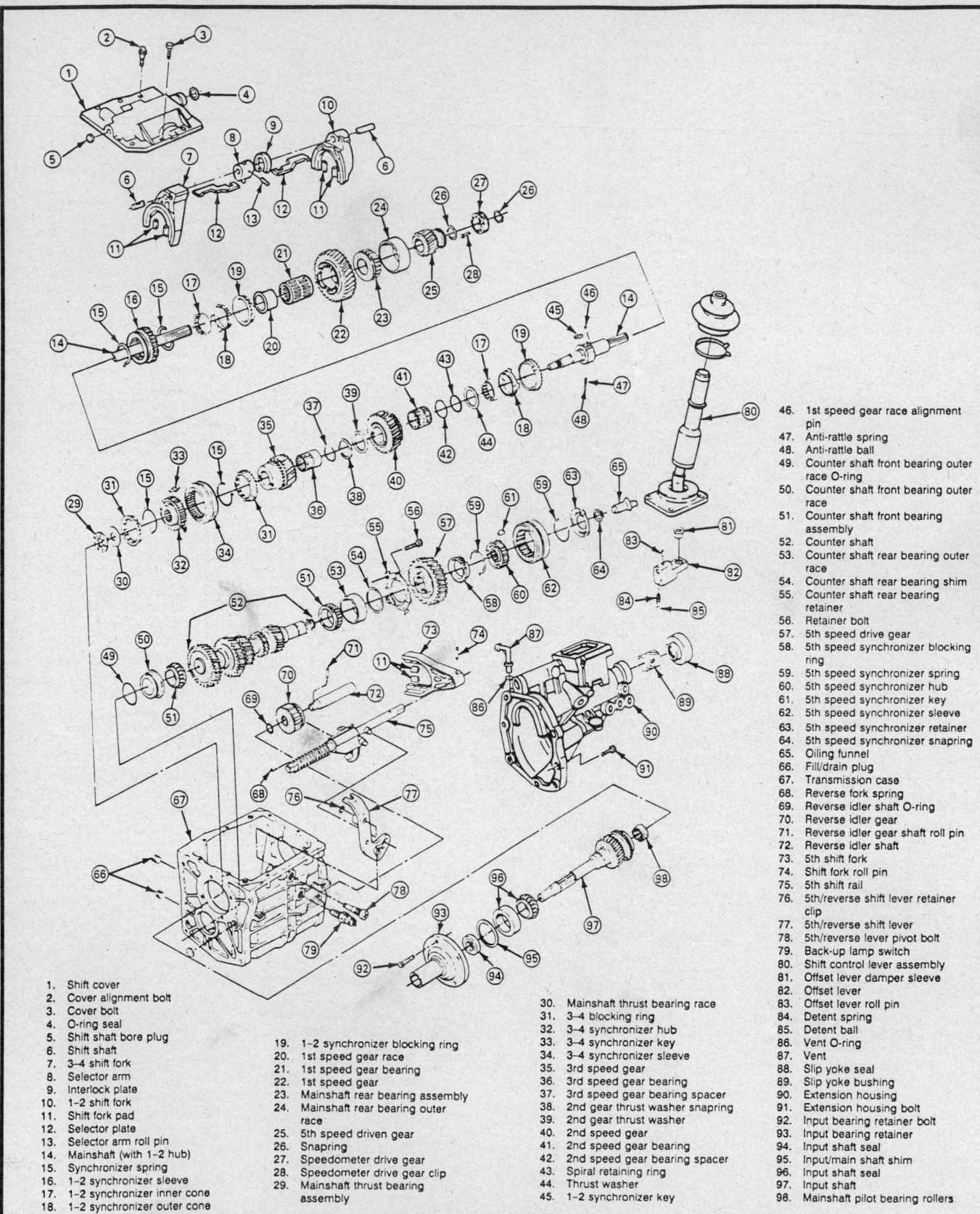

1. Shift cover
2. Cover alignment bolt
3. Cover bolt
4. O-ring seal
5. Shift shaft bore plug
6. Shift shaft
7. 3–4 shift fork
8. Selector arm
9. Interlock plate
10. 1–2 shift fork
11. Shift fork pad
12. Selector plate
13. Selector arm roll pin
14. Mainshaft (with 1–2 hub)
15. Synchronizer spring
16. 1–2 synchronizer sleeve
17. 1–2 synchronizer inner cone
18. 1–2 synchronizer outer cone
19. 1–2 synchronizer blocking ring
20. 1st speed gear race
21. 1st speed gear bearing
22. 1st speed gear
23. Mainshaft rear bearing assembly
24. Mainshaft rear bearing outer race
25. 5th speed driven gear
26. Snapring
27. Speedometer drive gear
28. Speedometer drive gear clip
29. Mainshaft thrust bearing assembly
30. Mainshaft thrust bearing race
31. 3–4 blocking ring
32. 3–4 synchronizer hub
33. 3–4 synchronizer key
34. 3–4 synchronizer sleeve
35. 3rd speed gear
36. 3rd speed gear bearing
37. 3rd speed gear bearing spacer
38. 2nd gear thrust washer snapring
39. 2nd gear thrust washer
40. 2nd speed gear
41. 2nd speed gear bearing
42. 2nd speed gear bearing spacer
43. Spiral retaining ring
44. Thrust washer
45. 1–2 synchronizer key
46. 1st speed gear race alignment pin
47. Anti-rattle spring
48. Anti-rattle ball
49. Counter shaft front bearing outer race O-ring
50. Counter shaft front bearing outer race
51. Counter shaft front bearing assembly
52. Counter shaft
53. Counter shaft rear bearing outer race
54. Counter shaft rear bearing shim
55. Counter shaft rear bearing retainer
56. Retainer bolt
57. 5th speed drive gear
58. 5th speed synchronizer blocking ring
59. 5th speed synchronizer spring
60. 5th speed synchronizer hub
61. 5th speed synchronizer key
62. 5th speed synchronizer sleeve
63. 5th speed synchronizer retainer
64. 5th speed synchronizer snapring
65. Oiling funnel
66. Fill/drain plug
67. Transmission case
68. Reverse fork spring
69. Reverse idler shaft O-ring
70. Reverse idler gear
71. Reverse idler gear shaft roll pin
72. Reverse idler shaft
73. 5th shift fork
74. Shift fork roll pin
75. 5th shift rail
76. 5th/reverse shift lever retainer clip
77. 5th/reverse shift lever
78. 5th/reverse lever pivot bolt
79. Back-up lamp switch
80. Shift control lever assembly
81. Offset lever damper sleeve
82. Offset lever
83. Offset lever roll pin
84. Detent spring
85. Detent ball
86. Vent O-ring
87. Vent
88. Slip yoke seal
89. Slip yoke bushing
90. Extension housing
91. Extension housing bolt
92. Input bearing retainer bolt
93. Input bearing retainer
94. Input shaft seal
95. Input/main shaft shim
96. Input shaft seal
97. Input shaft
98. Mainshaft pilot bearing rollers

Borg Warner T-5—except Ford, 1988–89 Camaro and Firebird

23. Move countershaft assembly rearward, tilt countershaft upward and remove from case. Remove countershaft front thrust washer and rear bearing spacer.
24. Remove countershaft front bearing from transmission case using an arbor press.
25. On Ford, 1988–89 Camaro and Firebird, remove the rear bearing outer race by pushing it rearward. Remove the bearing assembly from shaft using a puller and move the countershaft assembly rearward, tilt countershaft upward and remove from case.

Unit Disassembly and Assembly

MAINSHAFT GEAR TRAIN

Disassembly

4 SPEED

1. Remove the thrust bearing washer from the front of the output shaft.
2. Scribe matchmarks on the hub and sleeve of the 3rd/4th synchronizer so that these parts may be reassembled properly.
3. Remove the 3rd/4th synchronizer blocking ring, sleeve and hub as an assembly.
4. Remove the insert springs and the inserts from the 3rd/4th synchronizer and separate the sleeve from the hub.
5. Remove the 3rd speed gear from the shaft.
6. Remove the 2nd speed gear to output shaft snapring, the tabbed thrust washer and the 2nd speed gear from the shaft.
7. Use an appropriate puller and remove the the output shaft bearing.
8. Remove the 1st gear thrust washer, the roll pin, the 1st speed gear and the blocking ring.
9. Scribe matchmarks on the 1st/2nd synchronizer sleeve and the output shaft.
10. Remove the insert spring and the inserts from the 1st/reverse sliding gear, then remove the gear from the output hub.

5 SPEED

NOTE: Do not attempt to remove the 1st/2nd reverse hub from mainshaft. The hub and shaft are assembled and machined as a matched set. Note the order of disassembly of all components, transmission's component disassembly may vary from model to model.

AMC/Jeep/Eagle

1. Remove thrust bearing washer from front end of mainshaft.
2. Scribe reference mark on 3rd/4th synchronizer hub and sleeve for reassembly.
3. Remove 3rd/4th synchronizer blocking ring, sleeve, hub and 3rd gear as an assembly from mainshaft.
4. Remove snapring, tabbed thrust washer and 2nd gear from mainshaft.
5. Remove 5th gear with tool J-22912-01 or equivalent and arbor press. Slide rear bearing of mainshaft.
6. Remove 1st gear thrust washer, roll pin, 1st gear and synchronizer ring form mainshaft.
7. Scribe reference mark on 1st/2nd synchronizer hub and sleeve for reassembly.
8. Remove synchronizer spring and keys from 1st/reverse sliding gear and remove gear from mainshaft hub.

Ford Motor Co. and General Motors

1. Remove thrust bearing washer from front end of mainshaft.
2. Scribe reference mark on 3rd/4th synchronizer hub and sleeve for reassembly.
3. Using a press and puller plate, remove 3rd/4th synchronizer blocking ring, sleeve, hub and 3rd gear as an assembly from mainshaft.

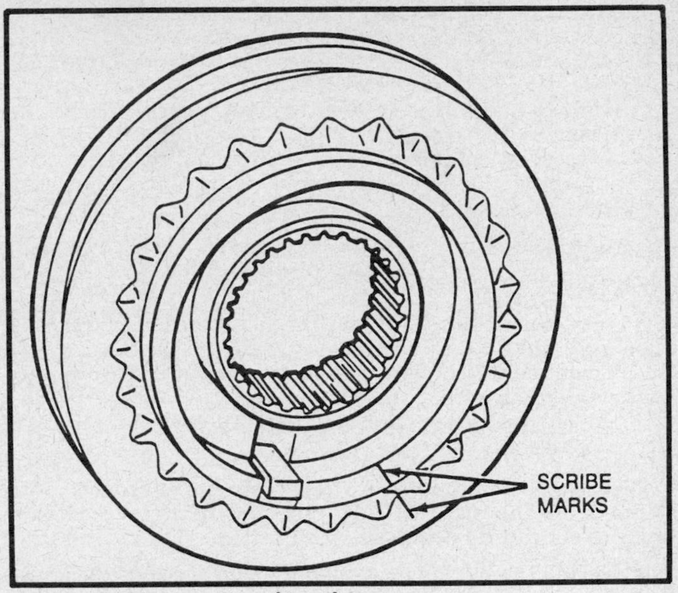
SCRIBE
MARKS
Scribe marks on synchronizer

4. On vehicles except Ford, 1988–89 Camaro and Firebird, remove the snapring, tabbed thrust washer and 2nd gear and blocking ring from the mainshaft. On Ford, 1988–89 Camaro and Firebird, remove the 3rd speed gear bearing and spacer, the 2nd speed gear snapring, thrust washer, gear, bearing spacer, spiral retaining ring and 3 piece 2nd speed blocking ring assembly.
5. Scribe matchmarks on the 1st/2nd synchronizer hub and sleeve for reassembly purposes.
6. Remove the 1st/2nd synchronizer sleeve from its hub, including the anti-rattle ball and spring (not included on all models), keys and springs.
7. Using an arbor press and the appropriate driver, remove the 5th speed gear from the mainshaft.
8. Slide the rear mainshaft bearing off of the mainshaft.
9. Remove the 1st gear thrust washer, roll pin, if equipped, 1st speed gear and blocking ring, bearing and inner race, setting them in order for reassembly reference.

NOTE: Close attention must be given to the order in which the 1st gear assembly is removed. The 1st gear and blocking ring assembly are not installed in the same order in all vehicles.

Inspection

NOTE: This inspection procedure includes counter shaft and reverse idler components. AMC recommends replacement of countershaft if any mainshaft gears are defective, to reduce gear noise and maintain proper gear mesh.

1. Clean all components, except plastic or nylon, is solvent.
2. Inspect for broken, chipped or worn gear teeth.
3. Inspect for bent or broken inserts.
4. Check for damaged roller thrust or needle bearings and inspect the bearing bores for cracks or damage.
5. Check for worn or loss of surface metal from the ccountershaft and hub, clutch shaft or reverse idler gear shaft.
6. Check the thrust washers for damage and wear.
7. Check for nicked, broken or worn output or clutch shaft splines.

8. Check for bent, weak or distorted snaprings.
9. Check the reverse idler gear busing for wear.
10. Inspect front and rear bearings for roughness, wear and damage.
11. Inspect the shift rails, selector arms, plates, interlock for worn, bent or damaged parts.
12. Inspect synchronizers damage, wear and rough surfaces.
13. If any of the above are found, replace the appropriate part(s).

Assembly

4 SPEED

1. Coat the output shaft and the gear bores with transmission lubricant.
2. Align the matchmarks and install the 1st/2nd synchronizer sleeve on the output shaft hub.
3. Install the 3 inserts and 2 springs into the 1st/reverse synchronizer sleeve.

NOTE: The tanged end of each spring should be positioned on the same insert but the open face of each spring should be opposite each other.

4. Install the blocking ring and the 2nd speed gear onto the output shaft.
5. Install the tabbed thrust washer and 2nd gear snapring in the output shaft; be sure that the washer is properly seated in the notch.
6. Install the blocking ring and the 1st speed gear onto the output shaft, then install the 1st gear roll pin.
7. Press the rear bearing onto the shaft.
8. Install the remaining components onto the output shaft: The 1st gear thrust washer. The 3rd speed gear. The 3rd/4th synchronizer hub inserts and the sleeve (the hub offset must face forward). The thrust bearing washer on the rear of the countershaft.

5 SPEED

AMC/Jeep/Eagle

1. Lubricate the mainshaft and gear boxes with a liberal coating of automatic transmission fluid.
2. Align and install the 1st/2nd synchronizer sleeve on the mainshaft, using the matchmarks made during disassembly.
3. If removed, install the synchronizer inserts and springs into the 1st/2nd synchronizer sleeve. Note the tanged end of each spring should be positioned on the same insert but the open face of each spring should be opposite the other.
4. Install the blocking ring and 2nd speed gear onto the mainshaft.
5. Install the tabbed thrust washer and the 2nd speed gear retaining snapring on the mainshaft. Be sure that the washer tab is fully seated into the notch of the mainshaft.
6. Install the blocking ring and the 1st speed gear onto the mainshaft.
7. Carefully drive the 1st gear roll pin into place.
8. Press the rear bearing onto the mainshaft using an arbor press and the appropriate driver.
9. Install the 1st gear thrust washer.
10. Install the 3rd speed gear, 3rd/4th synchronizer inserts and sleeve onto the mainshaft. The offset of the hub must face forward.
11. Install the thrust bearing washer onto the forward end of the mainshaft.

Ford Motor Co. and General Motor

1. Assemble the synchronizers.
2. Coat mainshaft and gear bores with transmission lubricant.
3. On all vehicles except Ford, 1988–89 Camaro and Firebird, install the 1st speed blocking ring, gear, roll pin, if equipped and thrust washer. On Ford, 1988–89 Camaro and Firebird, install the 3 piece blocking ring assembly, roll pin and 1st gear bearing race and the 1st gear and 1st gear bearing.
4. Install the mainshaft rear bearing assembly.
5. Install 5th speed gear on mainshaft using an arbor press. Install snapring on mainshaft. If snapring doesn't fit the 5th gear is not on far enough.
6. On all vehicles except Ford, 1988–89 Camaro and Firebird, proceed to Step 9. On Ford, 1988–89 Camaro and Firebird, install the 3 piece blocking ring assembly for the 2nd gear, thrust washer for blocking ring inner core, spiral retaining ring, 2nd speed gear spacer, bearing and gear.
7. Install the 2nd gear thrust washer and snapring.
8. Install the 3rd speed gear spacer, bearing, the 3rd speed gear and the blocking ring.
9. install the 2nd speed blocking ring, tabbed thrust washer, snapring, 3rd speed gear and blocking ring.
10. Install the 3-4 synchronizer hub. Using a arbor press, install the hub onto the mainshaft. Make certain to align the blocking ring with the synchronizer keys while installing the hub.

COVER AND FORKS

Disassembly

NOTE: Mark the position of the parts so that they may be properly installed. The procedures may vary slightly between different manufacturers, note the order of components as disassembled for reference during assembly.

1. Place the selector arm plates and the shift rail centered in the **N** position.
2. Rotate the shift rail counterclockwise until the selector arm disengages from the selector arm plates; the selector arm roll pin should now be accessible.
3. Pull the shift rail rearward until the selector contacts the 1st/2nd shift fork.
4. Use a $^3/_{16}$ in. pin punch and remove the selector arm roll pin and the shift rail.
5. Remove the shift forks, the selector arm, the roll pin and the interlock plate.
6. Remove the shift rail oil seal and O-ring.
7. Remove the nylon inserts and the selector arm plates from the shift forks.

Inspection

1. Clean all components, except plastic or nylon, is solvent.
2. Inspect for cracks in the bores, sides, bosses or bolt holes.
3. Inspect for stripped threads in the bolt holes.
4. Check for nicks, burrs or rough surfaces in the shaft bores ro on the gasket surfaces.
5. Check for worn shift fork inserts and for broken, cracked or worn shift forks.
6. Inspect the shift rails, selector arms, plates, interlock for worn, bent or damaged parts, if not already done.
7. If any of the above are found, replace the appropriate part(s).

Assembly

1. Attach the nylon inserts to the selector arm plates and through the shift forks.
2. If removed, coat the edges of the shift rail plug with sealer and install the plug.
3. Coat the shift rail and the rail bores with petroleum jelly, then slide the shift rail into the cover until the end of the rail is flush with the inside edge of the cover.
4. Position the 1st/2nd shift fork into the cover; with the offset of the shift fork facing the rear of the cover. Push the shift rail through the fork. The 1st/2nd fork is the larger of the forks.

5. Position the selector arm and the C-shaped interlock plate into the cover, then push the shift rail through the arm. The widest part of the interlock plate must face away from the cover and the selector arm roll pin must face downward, toward the rear of the cover.
6. Position the 3rd/4th shift fork into the cover with the fork offset facing the rear of the cover. The 3rd/4th shift selector arm plate must be positioned under the 1st/2nd shift fork selector arm plate.
7. Push the shift rail through the 3rd/4th shift fork and into the front cover rail bore.
8. Rotate the shift rail until the forward selector arm plate faces away from parallel to the cover.
9. Align the roll pin holes of the selector arm and the shift rail and install the roll pin. The roll pin must be installed flush with the surface of the selector arm to prevent selector arm plate to pin interference.
10. Install the O-ring into the groove of the shift rail oil seal, then install the oil seal carefully after lubricating it.

Transmission Assembly

4 SPEED

NOTE: If any replacement fastener must be used, make certain that it matches the original exactly. Metric fasteners are used in this transmission.

1. Apply a coating of Loctite® 601 (or its equivalent) to the outer cage of the front countershaft bearing. Press the bearing fully into its bore; it should be flush with the transmission case.
2. Apply a coating of petroleum jelly to the tabbed countershaft thrust washer. Install the thrust washer so that its tab engages the corresponding depression in the transmission case.
3. Tip the transmission case on end and install the countershaft into the front bearing bore.
4. Install the countershaft rear bearing spacer and coat the rear countershaft bearing with petroleum jelly. Install the rear countershaft bearing using the appropriate special tools (number J-29895 installer and J-33032 sleeve protector), or equivalents.

NOTE: When properly installed, the rear bearing will extend 0.125 in. beyond the transmission case surface.

5. Position the reverse idler gear into the transmission case (shift lever groove facing rearward) and install the reverse idler shaft from the rear of the case.
6. Install the shaft retaining pin.
7. Install the assembled output shaft into the transmission case.
8. If removed, install the main drive gear bearing onto the main drive gear shaft, using tool number J-2995, or equivalent and an arbor press.
9. Coat the main drive gear roller bearing (15) with petroleum jelly and install them into the recess of the main drive gear.
10. Install the thrust bearing and race into the recess of the main drive gear.
11. Install the 4th gear blocking ring onto the output shaft.
12. Install the rear output shaft bearing race.
13. Install the main drive gear assembly into the transmission case, engaging the 3rd/4th synchronizer blocking ring.
14. Evenly and carefully tap a new front bearing cap seal into place.
15. Install a new oil seal into the adapter housing (4WD) in the same manner as in Step 14.
16. Install the front bearing race into the front bearing cap. Do not yet install the front bearing cap shims.
17. Temporarily install the front bearing cap, without sealer.
18. Install the reverse lever, pivot pin (coat the threads with non-hardening sealer) and the retaining C-clip.

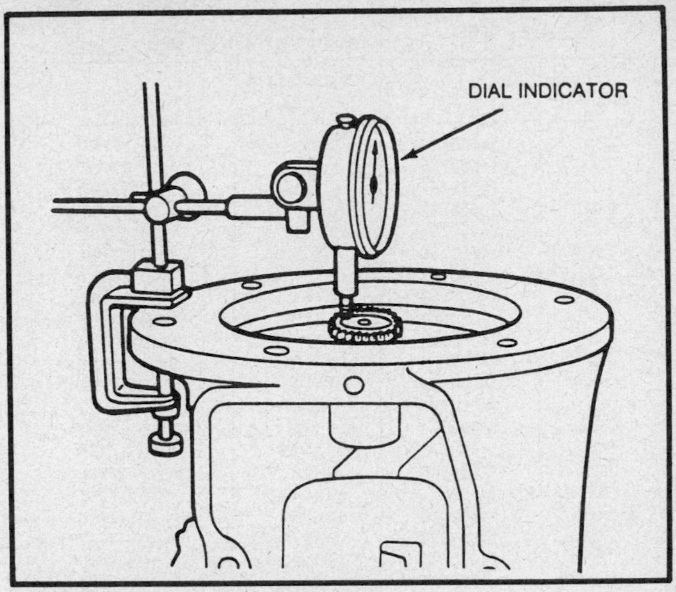
DIAL INDICATOR
Measuring mainshaft endplay

NOTE: Be sure that the reverse lever fork is engaged with the reverse idler gear.

19. Coat the countershaft rear bearing race and the thrust bearing with petroleum jelly. Install these parts into the extension/adapter housing.
20. Temporarily install the extension,/adapter housing, without sealer. Tighten, but do not final torque the bolts.
21. Turn the transmission case on end. Mount a dial indicator on the extension/adapter housing so that the indicator needle contacts the end of the output shaft.
22. Rotate the main drive gear shaft and the output shaft, then zero the dial indicator.
23. Pull upward on the output shaft to remove the endplay. Read the indicator and record the reading.

NOTE: To completely eliminate the total endplay, the bearings must be preloaded from 0.001–0.005 in.

24. Select a shim pack which measures 0.001–0.005 in. thicker than the endplay reading obtained during Step 23.
25. Move the transmission so that it sits horizontally and remove the front bearing cap and bearing race. Install the shim pack and reinstall the bearing race.
26. Apply a 1/8 in. bead of RTV sealer on the case mating surface of the front bearing cap. Align the case and cap matchmarks which were made during disassembly. Install the bearing cap and torque the bolts to 15 ft. lbs. (20 Nm).
27. Recheck the endplay; no endplay should exist.
28. Remove the extension/adapter housing.
29. Move the shift forks and the synchronizer sleeves to their Neutral positions.
30. Apply a 1/8 in. bead of RTV sealer to the cover mating surface of the transmission. While aligning the shift forks with the synchronizer sleeves, carefully lower the cover assembly into place on the transmission.
31. Center the cover in order to engage the reverse relay lever. Install the 2 alignment-type (dowel) cover attaching bolts. Torque all cover bolts to 9 ft. lb. (13 Nm).

NOTE: The offset lever-to-shift rail roll pin hole must be positioned vertically; if it is not, repeat Steps 29–31.

32. Apply a 1/8 in. bead of RTV sealer to the extension/adapter housing mating surface of the transmission and install the housing over the output shaft.

Input/Main Shaft		Counter Shaft	
Thickness (inches)	Part Number	Thickness (inches)	Part Number
0.012	14050783	0.1005	10117781
0.014	14050780	0.102	10117782
0.016	14050781	0.1035	10117783
0.018	14050782	0.105	10117784
0.020	14050784	0.1065	10117785
0.022	14050786	0.108	10117786
0.023	14050787	0.1095	10117787
0.024	14069868	0.111	10117788
0.025	14069869	0.1125	10117789
0.026	14069870	0.114	10117790
0.027	14069871	0.1155	10117791
0.028	14069872	0.117	10117792
0.029	14069873	0.1185	10117793
0.030	14069874	0.120	10117794
0.031	14069875	0.1215	10117795
0.032	14069876	0.123	10117796
0.033	14069877	0.1245	10117797
0.034	14069878	0.126	10117798
0.035	14069879	0.1275	10117799
0.036	14050785	0.129	10117800
0.037	14069880	0.1305	10117801
0.038	14069881	0.132	10117802
0.039	14069882	0.1335	10117803
0.040	14069883	0.135	10117804
0.041	14069884	0.1365	10117805
0.042	14069885	0.138	10117806
0.043	14069886	0.1395	10117807
0.044	14069887	0.141	10117808
		0.1425	10117809
		0.144	10117810
		0.1455	10117811
		0.147	10117812
		0.1485	10117813
		0.150	10117814
		0.1515	10117815
		0.1530	10117816

Shim selection

NOTE: The shift rail must be positioned so that it just enters the shift cover opening.

33. Install the detent spring into the offset lever and place the steel ball into the neutral guide plate detent. Apply pressure to the steel ball with the detent spring and the offset lever, then slide the offset lever on the shift rail and seat the extension/adapter housing against the transmission case. Tighten the housing retaining bolts to 25 ft. lbs. (32 Nm).
34. Install the roll pin into the offset lever and shift rail.
35. Install the damper sleeve in the offset lever. Coat the back-up lamp switch threads with RTV sealer and install the switch into the transmission case. Torque the switch to 15 ft. lbs. (20Nm).

5 SPEED

1. Install the countershaft into the case. On all vehicles except Ford, 1988–89 Camaro and Firebird models, make certain the front thrust washer is in place. If necessary use heavy petroleum jelly to hold washer in case slot.
2. Using a press, install the countershaft rear bearing into the case. This bearing is press onto the shaft on some models and into the case on other models.
3. On AMC/Jeep/Eagle vehicles, install the countershaft rear bearing spacer and coat the rear countershaft bearing with petroleum jelly. Install the rear countershaft bearing using the appropriate special tools J–29895 installer and J–33032 sleeve protector, or equivalents) and proceed to Step 6. On all Ford, 1988–89 Camaro and Firebird, install the countershaft rear bearing outer race and retainer (without shims). Tighten the retaining bolts to 15 ft. lbs. (20 Nm).

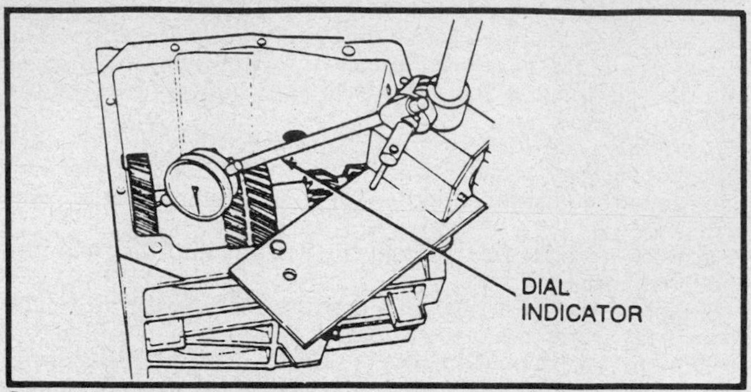
DIAL
INDICATOR

Checking counter shaft endplay—Ford method

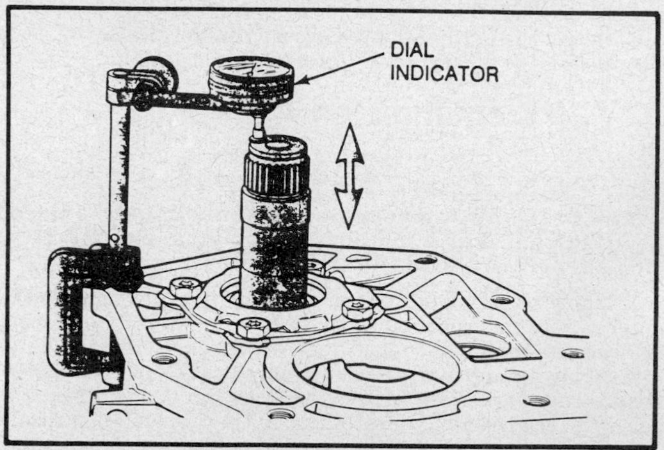
DIAL
INDICATOR

Checking counter shaft endplay—General Motors method

NOTE: When properly installed, the rear bearing will extend 0.125 in. beyond the transmission case surface.

4. Mount a dial indicator on the case and set it up to measure countershaft endplay. Select a shim which is the same thickness as the dial indicator reading, or up to 0.004 in. less than the reading.
5. Remove the countershaft rear bearing retainer and outer race.
6. Install the reverse idler shaft, gear and O-ring. Use a $^{3}/_{16}$ in. diameter pin punch to install the reverse idler roll pin. On all AMC/Jeep/Eagle vehicles and General Motors except 1988–89 Camaro and Firebird, proceed with Step 8.
7. On all Ford, 1988–89 Camaro and Firebird, install the countershaft rear bearing outer race, shims (already selected) and retainer. Torque the bolts to 15 ft. lbs. (20 Nm) and bend the lock tabs on the retainer using a hammer and punch.
8. Install the reverse fork and spring. On older models, install the 5th/reverse lever along with the fork.
9. Install the mainshaft into the case.
10. Install the mainshaft rear bearing outer race into the rear of the case.
11. Install the 4th speed blocking ring into the front of the mainshaft.
12. Install the input shaft onto the front of the mainshaft.
13. Install the retainer (without shims) onto the front of the case.
14. Torque the retaining bolts to 15 ft. lbs. (20 Nm).
15. Install the 5th speed drive gear and its blocking ring on the rear of the countershaft.
16. Assemble the 5th synchronizer and rail/fork. Guide the rail through the reverse fork and spring into the front of the case. Push the synchronizer assembly onto the splines of the ccountershaft.

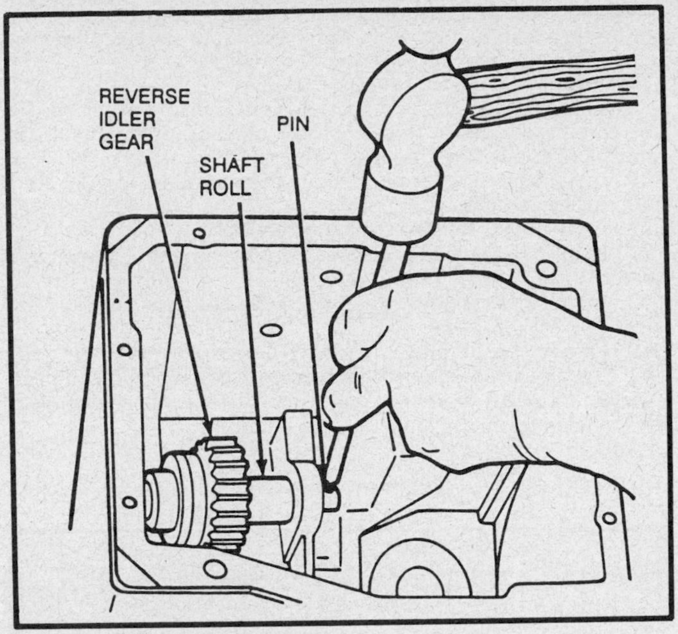
REVERSE
IDLER
GEAR
PIN
SHAFT
ROLL

Reverse idler gear

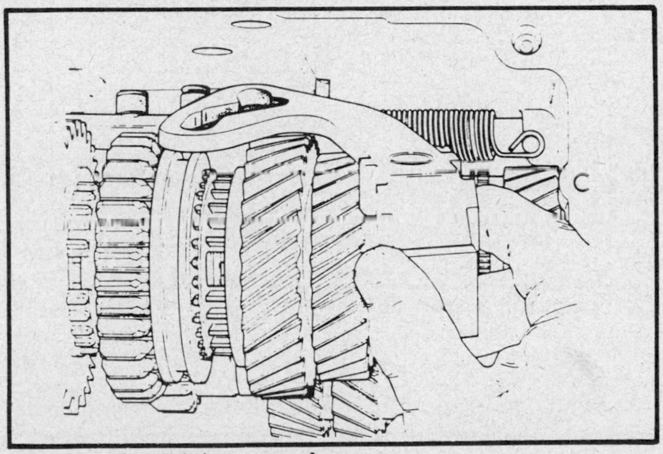

Aligning the 5th/reverse lever

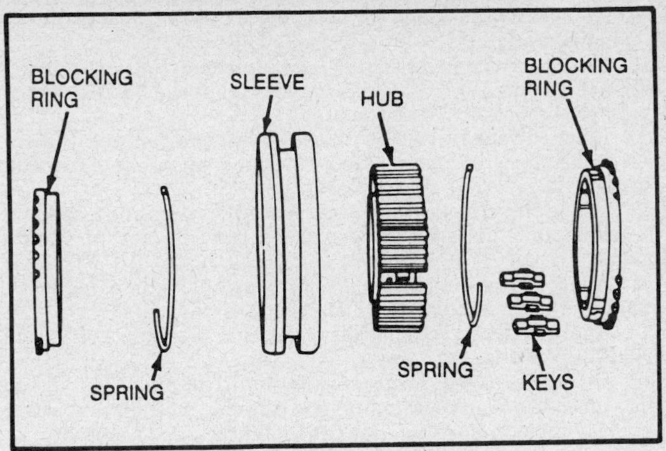
BLOCKING
RING
SLEEVE
HUB
BLOCKING
RING
SPRING
SPRING
KEYS

3rd/4th synchronizer assembly—typical

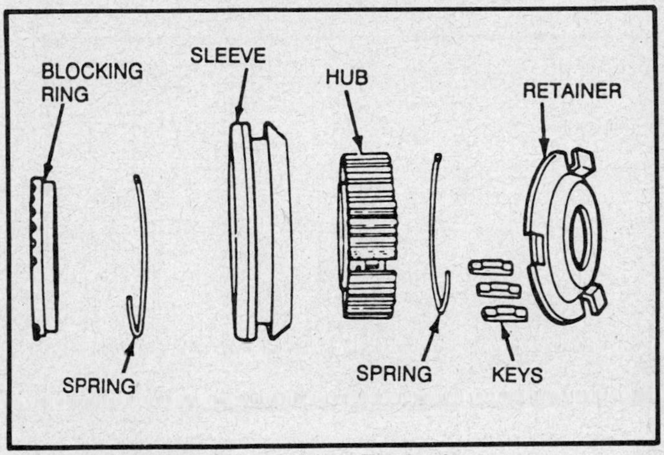
BLOCKING
RING
SLEEVE
HUB
RETAINER
SPRING
SPRING
KEYS

5th synchronizer assembly—Typical

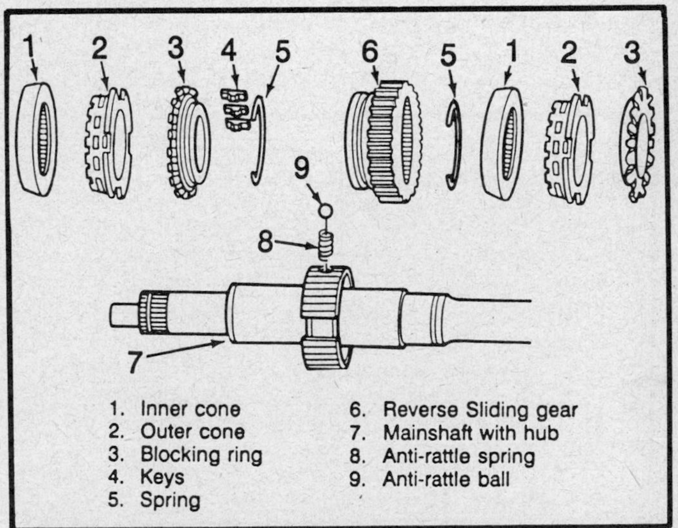
1
2
3
4
5
6
5
1
2
3
9
8
7
1. Inner cone
2. Outer cone
3. Blocking ring
4. Keys
5. Spring
6. Reverse Sliding gear
7. Mainshaft with hub
8. Anti-rattle spring
9. Anti-rattle ball

1st/2nd synchronizer assembly—typical

17. On all vehicles except Ford, 1988–89 Camaro and Firebird, install the countershaft thrust bearing inner race, snapring, thrust bearing assembly, outer race and oiling funnel. Use petroleum jelly to hold these items in place. On all Ford, 1988–89 Camaro and Firebird, install the 5th synchronizer snapring and oiling funnel, if equipped.

18. On newer models, install the 5th/reverse lever into the case. Align the slots of the lever with the rollers of the reverse fork and 5th shift rail.

19. Coat the threads of the pivot bolt with sealer and install it into the case. Make certain the pivot bolts is properly aligned and torque the bolt to 20 ft. lbs. (27 Nm).

20. Install the 5th/reverse lever clip.

21. Install the reverse fork spring to its proper position and check the operation of the 5th/reverse mechanism.

22. If 2WD, install the speedometer drive gear onto the rear of the mainshaft, making certain the speedometer gear retaining clip is fully seated.

23. Install the slip yoke snapring onto the end of the mainshaft.

24. Apply a ⅛ in. bead of RTV sealant on the surface of the shift cover.

25. Make certain that the 1st/2nd and 3rd/4th synchronizer sleeves and the 5th/reverse lever are in **N**.

26. Lower the cover onto the case, allowing the 1–2 and 3–4 forks to slide onto their sleeves. Slide the cover toward the 5th/reverse lever side just before the cover contacts the case.

27. Install the cover bolts and torque to 10 ft. lbs. (14 Nm).

28. Apply a 1/8 in. bead of RTV sealant on the surface of the extension housing.

29. Lubricate the detent/guide plate in the extension housing with lithium grease. Install the detent ball in the **3RD/4TH** position of the detent pattern plate.

30. Place the offset lever (with the detent spring) in the extension housing detent/guide plate area and push the extension against the case and shift cover.

31. Guide the offset lever onto the shift shaft while installing the extension housing and compressing the detent spring against its ball.

32. Apply sealer to the threads of the extension housing bolts, install and torque bolts to 23 ft. lbs. (27 Nm).

33. Apply sealer to the threads of the back-up lamp switch and drain plug and install.

34. Install the offset lever-to-shift shaft roll pin.

35. Place the transmission on end and mount a dial indicator on the extension housing. Set the dial indicator to measure input/main shaft endplay. Move input/main shaft up and down.

36. Select a shim that is within 0.001 in. of the endplay measure. This will give the required 0 endplay.

37. Remove the input bearing retainer.

38. Install the shims behind the input bearing outer race in the retainer. If more than a single shim is used, install the smaller shims before the larger shims.

39. Install the retainer and torque bolts to 15 ft. lbs. (20 Nm).

40. Recheck endplay, if endplay is not less than 0.002 change shims as needed. Remove bearing retainer.

41. Apply a 1/8 in. bead of RTV sealant to the retainer and apply sealer to the bolt threads.

42. Install the retainer and torque the bolts to 15 ft. lbs. (20 Nm).

43. Loosen the fill plug.

44. Install the speedometer driven gear assembly, if equipped.

45. Install the transmission and fill to proper level with Dexron®II automatic transmission fluid.

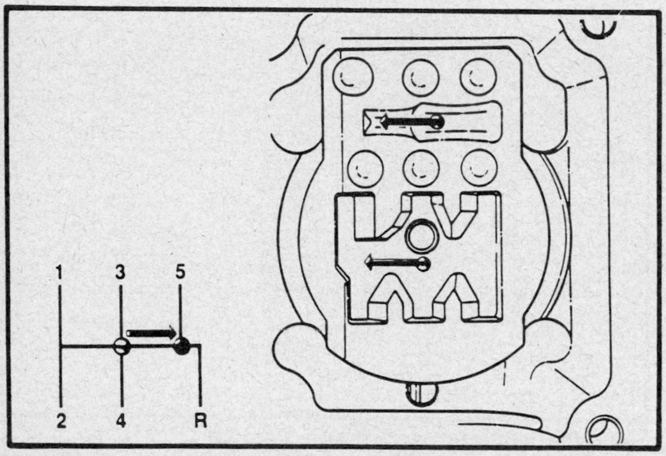
1
3
5
2
4
R
Detent guide action in 5th—shift feel T-5

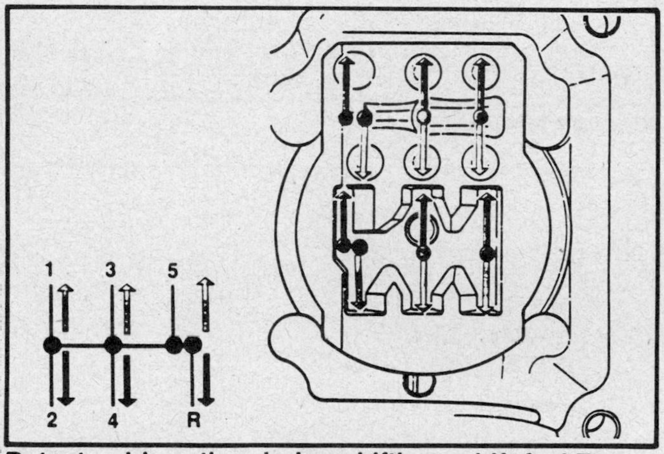
1
3
5
2
4
R
Detent guide action during shifting—shift feel T-5

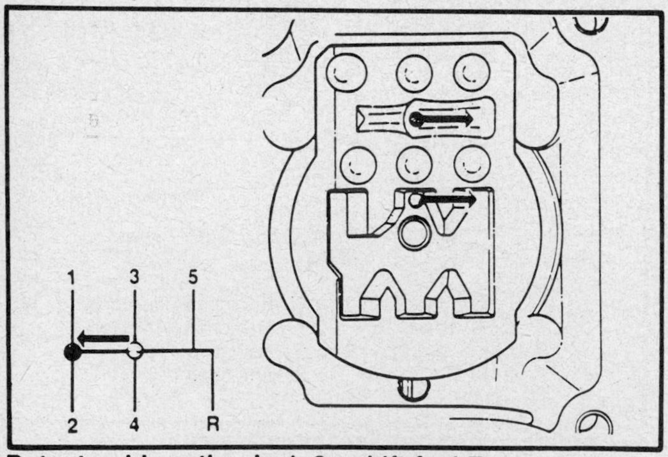
1
3
5
2
4
R
Detent guide action in 1-2—shift feel T-5

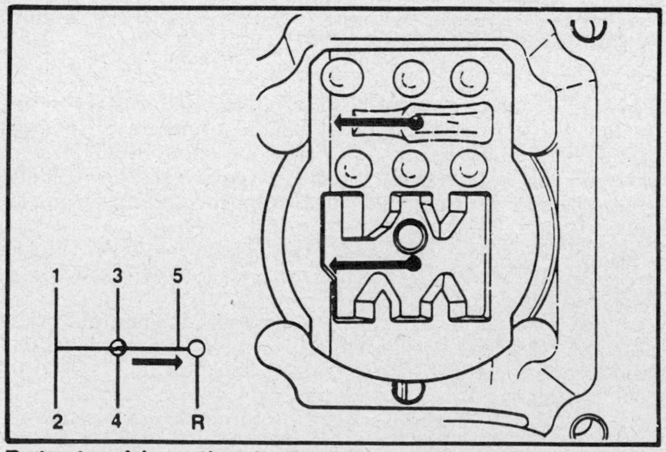
1
3
5
2
4
R
Detent guide action in Reverse—shift feel T-5

SPECIFICATIONS

TORQUE SPECIFICATIONS

Description	ft. lbs.	Nm
Bearing retainer	11–20	15–27
Shift cover	6–11	8–15
Extension housing	20–45	27–61
Turret cover	11–15	15–20
Drain plug	15–30	20–41
Transmission to flywheel housing	35–55	61–88
Shift tower to extension housing	23–32	31–43
Speedometer cable retaining screw	3–5	4–6
Transmission support	36–50	48–68

Description	ft. lbs.	Nm
Transmission extension housing bolts	28–36	38–48
Driveshaft U-bolts	42–57	56–77
Catalytic converter attaching bolts	20–30	27–41
Shift lever to transmission	23–32	31–43
Shift boot to floor pan	3–7	4–9.5
Cluster gear rear bearing retainer	10–15	15–20
Top gear sensing switch	12–18	17–24
Backup lamp switch	12–18	17–24
Neutral sensing switch	12–18	17–24

SPECIAL TOOLS

Tool Number	Description
J-3289-20	Fixture base
J-6133-A	Press tube
J-8092	Driver handle
J-8433	Puller (with bolts)
J-21426	Seal installer
J-22912-01	Puller plate
J-23064-14	Bushing driver
J-23103	(J-37372) Press tube
J-23907	Slide hammer
J-24127	Speedometer puller
J-25234	Press tube
J-29369-2	Seal/bearing pullers
J-29369-3	Seal/bearing pullers
J-29895	Press tube

Tool Number	Description
J-34162	Holding fixture
J-36842	Holding fixture
J-37357	Bearing driver
J-37358	Case support/pilot
J-37359	Puller plate adapters
J-37360	Guide sleeve
J-37375	Seal installer
J-37372	Press tube
J-5590	Press tube
	MEASURING TOOLS
J-8001	Dial indicator set
J-29600-1	Zero to 1 in. micrometer
J-7872	Dial indicator set (with magnetic base)

SPECIAL TOOLS

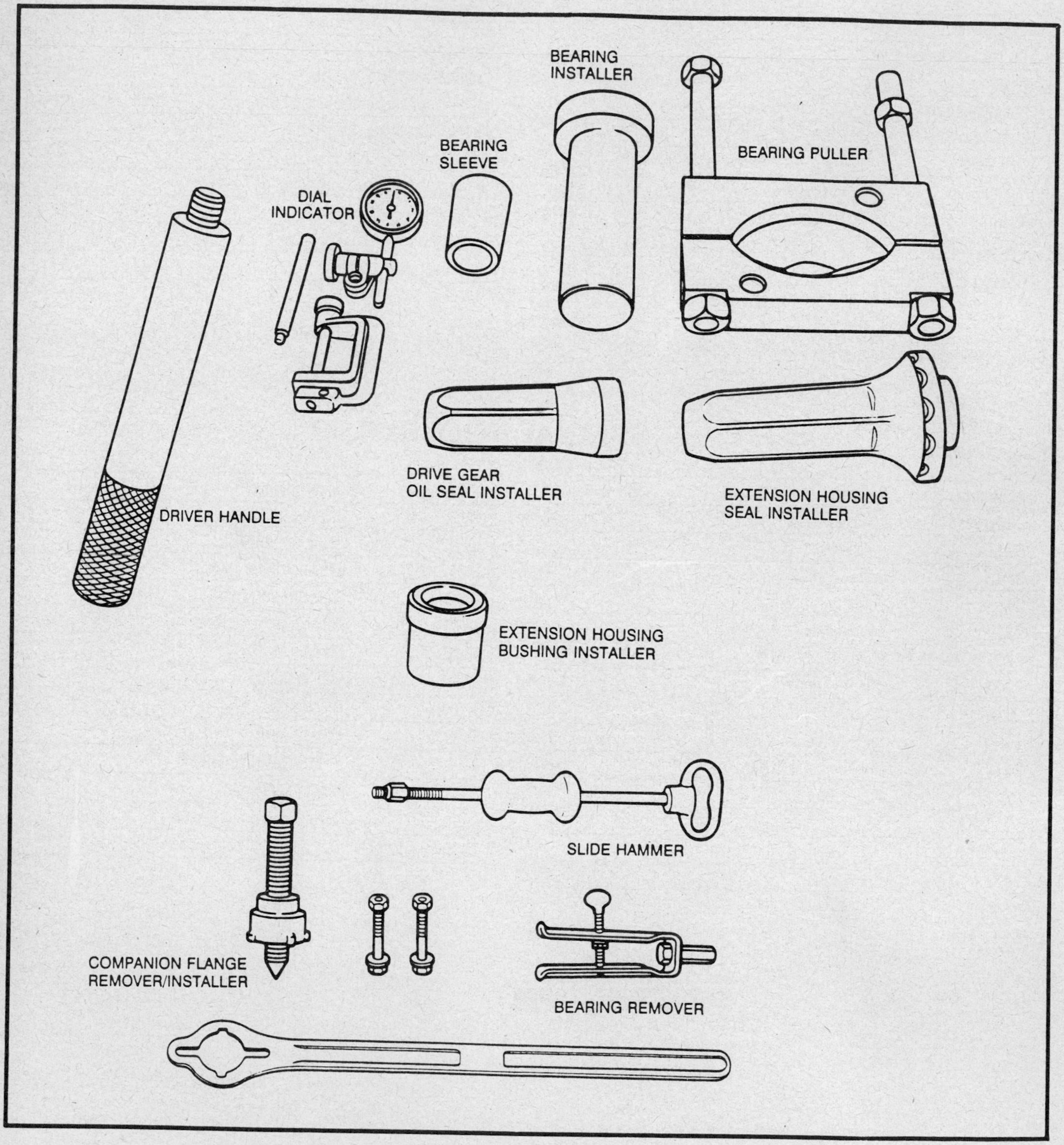
BEARING INSTALLER
BEARING SLEEVE
BEARING PULLER
DIAL INDICATOR
DRIVE GEAR OIL SEAL INSTALLER
EXTENSION HOUSING SEAL INSTALLER
DRIVER HANDLE
EXTENSION HOUSING BUSHING INSTALLER
SLIDE HAMMER
COMPANION FLANGE REMOVER/INSTALLER
BEARING REMOVER

Section 7

T-176 Transmission
Borg Warner

APPLICATION

T-176

Year	Vehicle	Engine	VIN Code
1984	Jeep/CJ-7	4.2L	C
	Jeep/Scrambler	4.2L	C
	Jeep/Wagoneer	4.2L	C
	Jeep/J-10 Truck	4.2L	C
	Jeep/J-10 Truck	5.9L	N
1985	Jeep/CJ-7	4.2L	C
	Jeep/Scrambler	4.2L	C
	Jeep/Wagoneer	4.2L	C
	Jeep/J-10 Truck	4.2L	C
	Jeep/J-10 Truck	5.9L	N
1986	Jeep/CJ-7	4.2L	C
	Jeep/Scrambler	4.2L	C

Year	Vehicle	Engine	VIN Code
	Jeep/Wagoneer	4.2L	C
	Jeep/J-10 Truck	4.2L	C
	Jeep/J-10 Truck	5.9L	N
1987	Jeep/Wagoneer	4.2L	C
	Jeep/J-10 Truck	4.2L	C
	Jeep/J-10 Truck	5.9L	N
1988	Jeep/Wagoneer	4.2L	C
	Jeep/J-10 Truck	4.2L	C
	Jeep/J-10 Truck	5.9L	N
1989	Jeep/Wagoneer	4.2L	C
	Jeep/J-10 Truck	4.2L	C
	Jeep/J-10 Truck	5.9L	N

GENERAL DECRIPTION

The Warner T-176 transmission is a 4 speed constant mesh unit, providing synchronized engagement in all forward gears. The shift mechanism is located within the shift control housing which also serves as the transmission top cover. The shift mechanism does not require adjustment and can be serviced independently of the transmission.

A spring and plunger type backup lamp switch is used with this transmission. The switch is located in the transmission case and is actuated by the reverse shift rail. The switch does not require adjustment and is serviced as an assembly.

Transmission Identification

The transmission identification tag is bolted to the shift control lever housing near the left rear corner. The information on this tag is necessary to obtain the correct replacement parts. Be sure the tag is securely attached to the original location after completing all service operations.

Metric Fasteners

The metric fastener dimensions are very close to the dimensions of the familiar inch system fasteners. For this reason, replacement fasteners must have the same measurement and strength as those removed.

Do not attempt to interchange metric fasteners for inch system fasteners. Mismatched or incorrect fasteners can result in damage to the transmission unit through malfunctions, breakage or possible personal injury.

Care should be taken to re-use the fasteners in the same locations as removed.

Common metric fastener strength property classes are 9.8 and 12.9 with class identification embossed on the head of each bolt. The inch strength classes range from grade 2–8 with the line identification embossed on each bolt head. The markings correspond to 2 lines less than actual grade (for example grade 8 bolt will exhibit 6 embossed lines on the bolt head). Some metric nuts will be marked with a single digit strength identification numbers on the nut face.

Capacities

The transmission lubricant capacity is 2.0 quarts (1.7L). The correct recheck or top-off level is to the bottom edge of the fill plug hole. The recommended lubricant for the T-176 transmission is AMC/Jeep manual gear lubricant or an equivalent SAE 75W-90, API grade GL5 gear lubricant.

Checking Fluid Level

1. Raise and support the vehicle safely.
2. Remove the transmission oil fill plug.
3. The correct level should be to the bottom edge of the fill plug hole.
4. Fill the proper level with the recommended gear oil and reinstall the oil fill plug.
5. Lower the vehicle.

Transmission Modifications

There are no transaxle modifications for the T-176 transmission at the time of this publication.

TROUBLE DIAGNOSIS

CHILTON'S THREE C'S TRANSMISSION DIAGNOSIS

T-176 4 Speed Manual Transmission

Condition	Cause	Correction
Transmission shifts hard	a) Clutch adjustment incorrect	a) Adjust clutch
	b) Clutch linkage or cable binding	b) Lubricate or repair as necessary
	c) Shift rail binding	c) Check for mispositioned selector arm roll pin, loose cover bolts, worn shift rail bores, worn shift rail, distorted oil seal, or extension housing not aligned with case. Repair as necessary
	d) Internal bind in transmission caused by shift forks, selector plates, or synchronizer assemblies	d) Remove, disassemble and inspect transmission. Replace worn or damaged components as necessary
	e) Clutch housing misalignment	e) Check runout at rear face of clutch housing
	f) Incorrect lubricant	f) Drain and refill transmission
	g) Block rings and/or cone seats worn	g) Blocking ring to gear clutch tooth face clearance must be 0.030 inch or greater. If clearance is correct it may still be necessary to inspect blocking rings and cone seats for excessive wear. Repair as necessary
Gear clash when shifting from one gear to another	a) Clutch adjustment incorrect	a) Adjust clutch
	b) Clutch linkage or cable binding	b) Lubricate or repair as necessary
	c) Clutch housing misalignment	c) Check runout at rear of clutch housing
	d) Lubricant level low or incorrect lubricant	d) Drain and refill transmission and check for lubricant leaks if level was low. Repair as necessary
	e) Gearshift components, or synchronizer assemblies worn or damaged	e) Remove, disassemble and inspect transmission. Replace worn or damaged components as necessary
Transmission Noisy	a) Lubricant level low or incorrect lubricant	a) Drain and refill transmission. If lubricant level was low, check for leaks and repair as necessary
	b) Clutch housing-to-engine, or transmission-to-clutch housing bolts loose	b) Check and correct bolt torque as necessary
	c) Dirt, chips, foreign material in transmission	c) Drain, flush, and refill transmission
	d) Gearshift mechanism, transmission gears, or bearing components worn or damaged	d) Remove, disassemble and inspect transmission. Replace worn or damaged components as necessary
	e) Clutch housing misalignment	e) Check runout at rear face of clutch housing
Jumps out of gear	a) Clutch housing misalignment	a) Check runout at rear face of clutch housing
	b) Gearshift lever loose	b) Check lever for worn fork. Tighten loose attaching bolts
	c) Offset lever nylon insert worn or lever attaching nut loose	c) Remove gearshift lever and check for loose offset lever nut or worn insert. Repair or replace as necessary
	d) Gearshift mechanism, shift forks, selector plates, interlock plate, selector arm, shift rail, detent plugs, springs or shift cover worn or damaged	d) Remove, disassemble and inspect transmission cover assembly. Replace worn or damaged components as necessary
	e) Clutch shaft or roller bearings worn or damaged	e) Replace clutch shaft or roller bearings as necessary

CHILTON'S THREE C'S TRANSMISSION DIAGNOSIS
T-176 4 Speed Manual Transmission

Condition	Cause	Correction
Jumps out of gear	f) Gear teeth worn or tapered, synchronizer assemblies worn or damaged, excessive end play caused by worn thrust washers or output shaft gears	f) Remove, disassemble, and inspect transmission. Replace worn or damaged components as necessary
	g) Pilot bushing worn	g) Replace pilot bushing
Will not shift into one gear	a) Gearshift selector plates, interlock plate, or selector arm, worn, damaged, or incorrectly assembled	a) Remove, disassemble, and inspect transmission cover assembly. Repair or replace components as necessary
	b) Shift rail detent plunger worn, spring broken, or plug loose	b) Tighten plug or replace worn or damaged components as necessary
	c) Gearshift lever worn or damaged	c) Replace gearshift lever
	d) Synchronizer sleeves or hubs, damaged or worn	d) Remove, disassemble and inspect transmission. Replace worn or damaged components
Locked in one gear—can not be shifted out	a) Shift rail(s) worn or broken, shifter fork bent, setscrew loose, center detent plug missing or worn	a) Inspect and replace worn or damaged parts
	b) Broken gear teeth on countershaft gear, clutch shaft, or reverse idler gear	b) Inspect and replace damaged part
	c) Gearshift lever broken or worn, shift mechanism in cover incorrectly assembled or broken, worn, damaged gear train components	c) Disassemble transmission. Replace damaged parts or assemble correctly

Services

FLUID CHANGE

1. Raise and support the vehicle safely.
2. Remove the transmision oil fill plug and allow the gear oil to drain into a suitable drain pan.
3. The correct level should be to the bottom edge of the fill plug hole.
4. Fill to the proper level with the recommended gear oil and reinstall the oil fill plug.
5. Lower the vehicle.

REMOVAL AND INSTALLATION

TRANSMISSION REMOVAL

1. Remove the screws attaching the transmission shift lever boot to the floor pan and slide the boot upward onto the lever.
2. Press and turn the transmission shift lever retainer counterclockwise to release the lever. Remove the lever, boot, spring and seat as an assembly.
3. Raise and support the vehicle safely.
4. Mark the driveshaft and transfer case yoke for assembly alignment references.
5. Disconnect the rear driveshaft at the transfer case yoke. Move the shaft aside and secure it to the underbody with wire.
6. Disconnect the front parking brake cable equalizer. Remove the clip that retains the rear cable to the rear crossmember and move the cable aside.
7. Position a safety stand under the clutch housing to support the engine.
8. Remove the nuts and bolt attaching the rear crossmember to the frame rails and rear support cushion and remove the crossmember.
9. Disconnect the speedometer cable. Disconnect the back-up lamp switch wire. Disconnect the 4WD indicator switch wire.
10. Disconnect the transfer case vent hole at the transfer case. Mark the front driveshaft and transfer case yoke for assembly alignment reference.
11. Disconnect front driveshaft from the transfer case yoke. Move the shaft aside and secure it to the underbody with a piece of wire.
12. On the CJ-7 and Scrambler, remove the shifter shaft retaining nut. Remove the cotter pin and washer that connect the shift control link pins in the shift rods and remove the pins. Remove the shifter shaft and disengage the shift lever from the shift control links.
13. On all other vehicles, remove the cotter pin and washer that connect the link to the shift lever and disconnect the link from the shift lever.
14. Support the transmission/transfer case assembly with a suitable transmission jack. Use a safety chain to secure the assembly on the jack.
15. Remove the bolts attaching the transmission to the clutch housing and remove the transmission/transfer case assembly.
16. Remove the bolts attaching the transfer case to the transmission and remove the transfer case. Clean the old gasket material and sealer from the mating surfaces of the transmission and transfer case.
17. Remove the pilot bushing lubricating wick from the bushing and soak the wick in clean engine oil. Use a long needle nose pliers to remove the wick from the bushing.

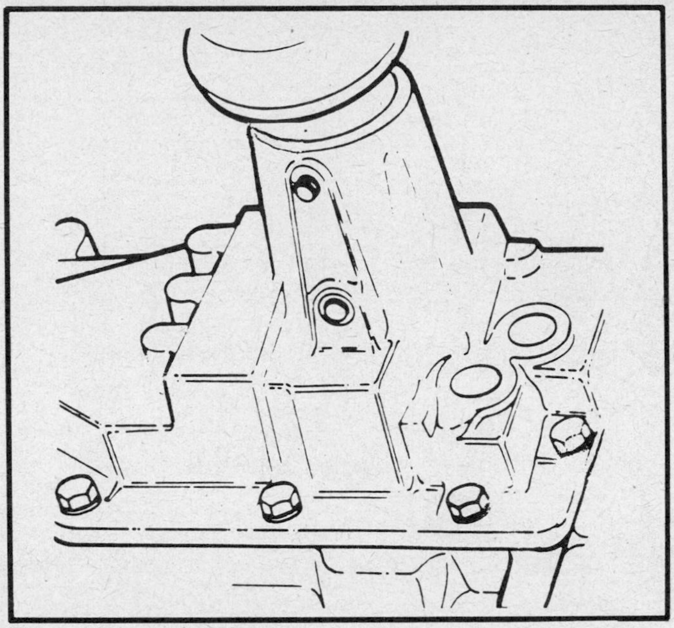

Removing the shift lever

TRANSMISSION INSTALLATION

1. Install the pilot bushing lubricating wick and align the throwout bearing with the splines in the driven plate hub.
2. Shift the transmission into gear using the shift lever. This prevents the clutch shaft from rotating during installation and makes the clutch shaft to driven plate spline alignment easier.
3. Mount the transmission onto the transmission jack. Raise the transmission and align the transmission clutch shaft with the splines in the driven plate hub.
4. Install the transmission. When the transmission is seated on the clutch housing, install and torque the transmission to clutch housing bolts to 55 ft. lbs.
5. Apply a suitable sealer to both sides of the replacement gasket for the transmission to transfer case. Position the gasket on the transfer case.
6. Mount the transfer case onto the transmission jack. Raise the transfer case and align the transmission output shaft and transfer case input shaft spilnes.
7. Install the transfer case to transmission retaining bolts/ nuts and torque them to 40 ft. lbs.
8. On the CJ-7 and Scrambler, install the shifter shaft and engage the shift lever into the shift control links. Install the cotter pin and washer that connect the shift control link pins in the shift rods and install the pins. Install the shifter shaft retaining nut.
9. On all other vehicles, connect the shift lever link to the operating lever on the transfer case.
10. Connect the front driveshaft to the transfer case yoke. Torque the clamp strap bolts to 15 ft. lbs. Be sure that the shaft and yoke are aligned according to the reference marks made at disassembly.
11. Connect the vent hose to the transfer case. Connect the wire to the 4WD indicator switch. Connect the speedometer.
12. Install the rear crossmember. Torque the crossmember attaching nuts and bolts to 30 ft. lbs.
13. Remove the safety stand used to support the engine. Connect the parking brake rear cable to the clip that retains the cable to the crossmember and connect the front cable to equalizer.
14. Connect the rear driveshaft to the transfer case yoke. Torque the clamp strap bolts to 15 ft. lbs. Be sure that the shaft and yoke are aligned according to the reference marks made at disassembly.
15. Check and correct the transmission and transfer case lubricant levels, if necessary.
16. Lower the vehicle.
17. Install the shift lever assembly. Seat the lever in the shift housing, press and turn the lever retainer clockwise to lock the lever in the housing and install the lever boot onto the housing.
18. Position the shift lever boot on the floorpan and install the boot attaching screws.

BENCH OVERHAUL

Before Disassembly

Cleanliness during disassembly and assembly is necessary to avoid further transmisson trouble after overhaul. Before removing any of the transmission subassemblies, plug all the openings and clean the outside of the transmission thoroughly. Steam cleaning or car wash type high pressure equipment is preferable. During disassembly, clean all parts in suitable solvent and dry each part. Do not use cloth or paper towels to dry parts. Use compressed air only.

Transaxle Disassembly

1. Remove the transfer case from the rear of the transmission.
2. Remove the shift control housing. Mark the location of the two dowel bolts in the housing.
3. Drain the lubricant from the transmission, if not previously done. Remove the rear adapter housing.
4. With a dummy countershaft tool, remove the countershaft from the transmission, front to rear. Allow the cluster gear to lay on the bottom of the case.

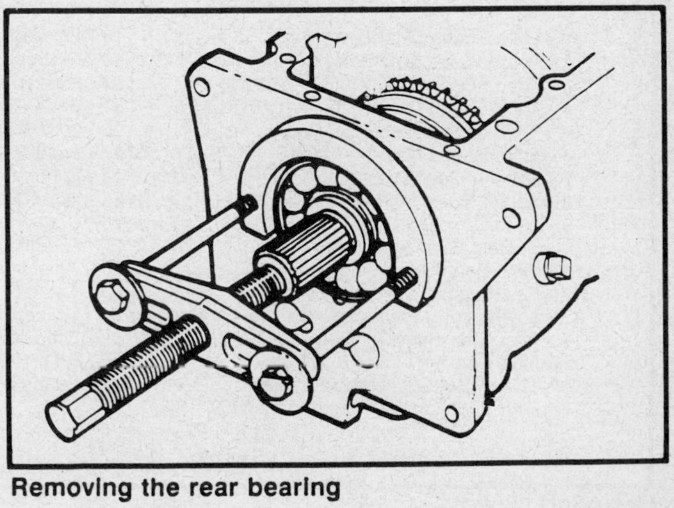

Removing the rear bearing

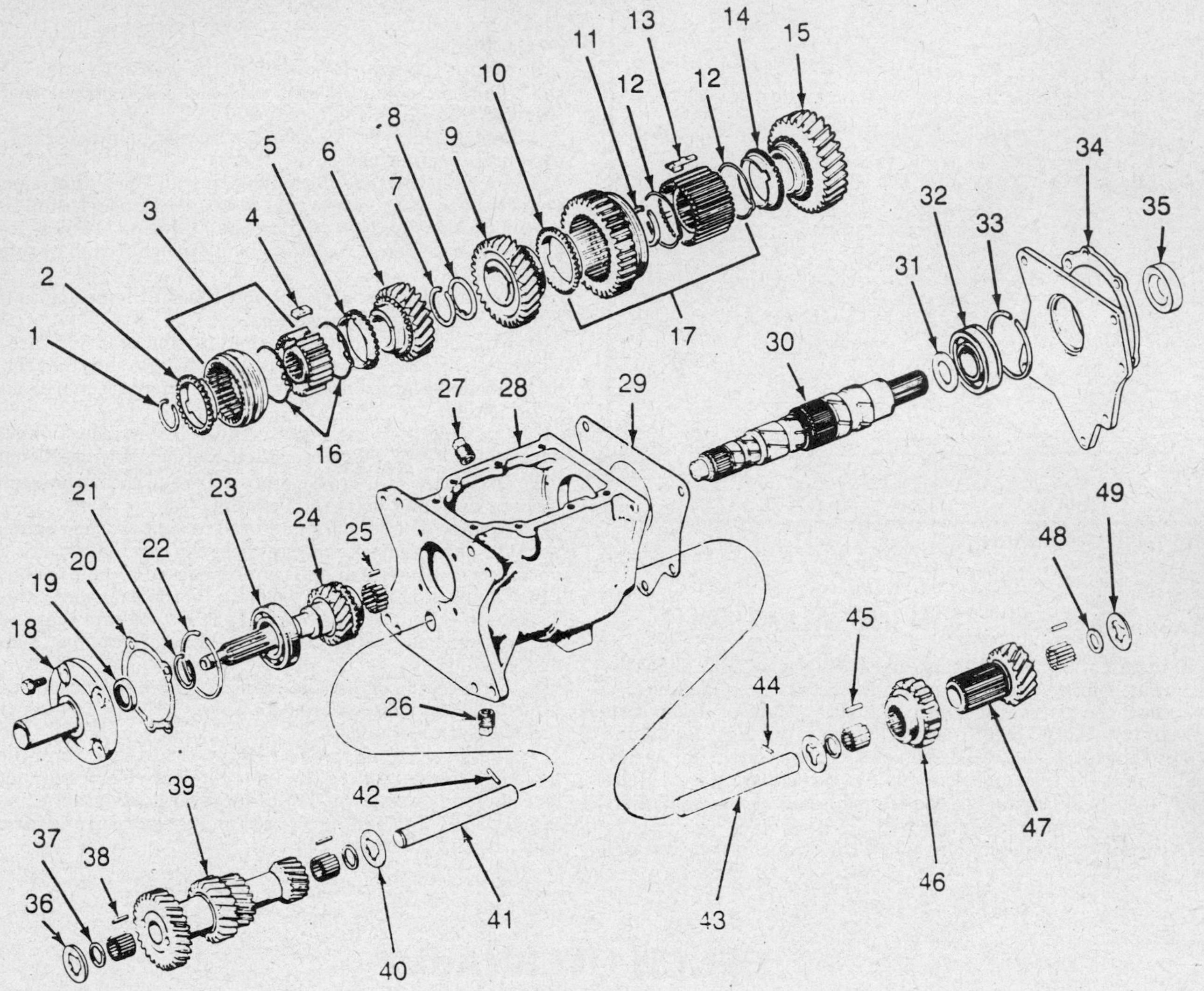
1
2
3
4
5
6
7
8
9
10
11
12
13
12
14
15
16
17
18
19
20
21
22
23
24
25
26
27
28
29
30
31
32
33
34
35
36
37
38
39
40
41
42
43
44
45
46
47
48
49

1 - Third-Fourth Gear Snap Ring
2 - Fourth Gear Synchronizer Ring
3 - Third-Fourth Gear Clutch Assembly
4 - Third-Fourth Gear Plate
5 - Third Gear Synchronizer Ring
6 - Third Speed Gear
7 - Second Gear Snap Ring
8 - Second Gear Thrust Washer
9 - Second Speed Gear
10 - Second Gear Synchronizer Ring
11 - Main Shaft Snap Ring
12 - First-Second Synchronizer Spring
13 - Low-Second Plate
14 - First Gear Synchronizer Ring
15 - First Gear
16 - Third-Fourth Synchronizer Spring
17 - First-Second Gear Clutch Assembly
18 - Front Bearing Cap
19 - Oil Seal
20 - Gasket
21 - Snap Ring
22 - Lock Ring
23 - Front Ball Bearing
24 - Clutch Shaft
25 - Roller Bearing
26 - Drain Plug
27 - Fill Plug
28 - Case
29 - Gasket
30 - Spline Shaft
31 - First Gear Thrust Washer
32 - Rear Ball Bearing
33 - Snap Ring
34 - Adapter Plate
35 - Adapter Seal
36 - Front Countershaft Gear Thrust Washer
37 - Roller Washer
38 - Rear Roller Bearing
39 - Countershaft Gear
40 - Rear Countershaft Thrust Washer
41 - Countershaft
42 - Pin
43 - Idler Gear Shaft
44 - Pin
45 - Idler Gear Roller Bearing
46 - Reverse Idler Sliding Gear
47 - Reverse Idler Gear
48 - Idler Gear Washer
49 - Idler Gear Thrust Washer

Exploded view

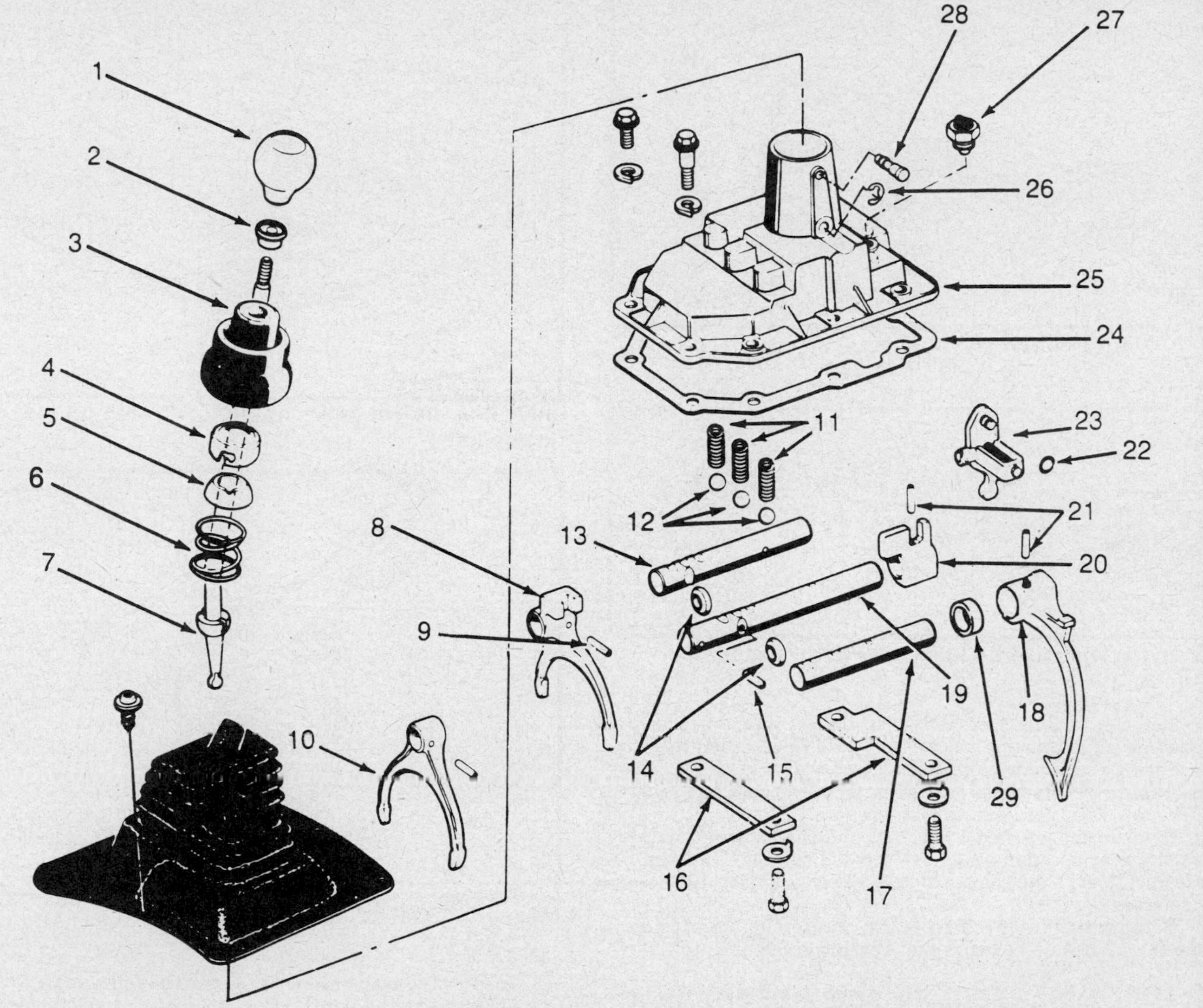
1
2
3
4
5
6
7
8
9
10
11
12
13
14
15
16
17
18
19
20
21
22
23
24
25
26
27
28
29

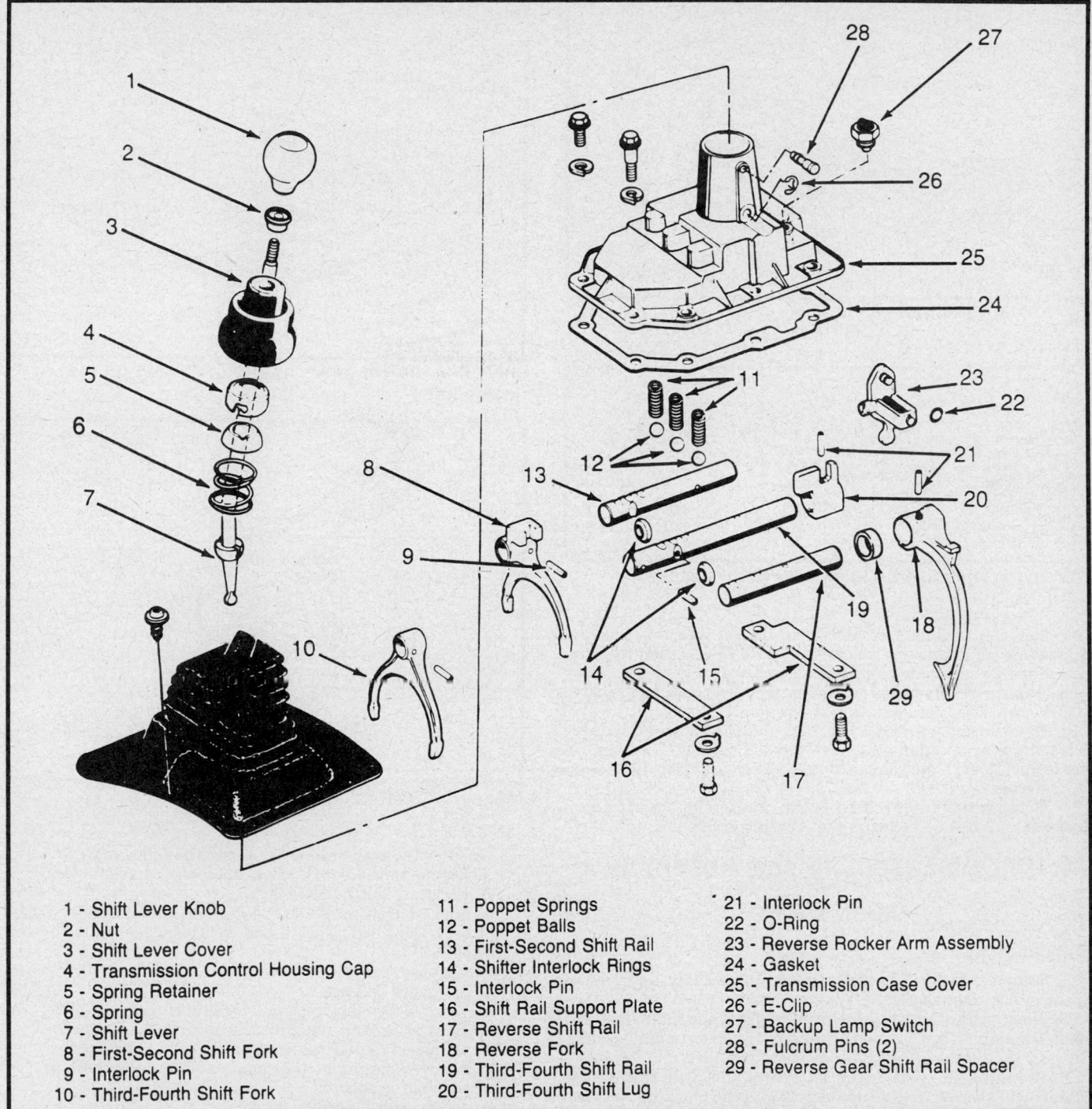

1 - Shift Lever Knob
2 - Nut
3 - Shift Lever Cover
4 - Transmission Control Housing Cap
5 - Spring Retainer
6 - Spring
7 - Shift Lever
8 - First-Second Shift Fork
9 - Interlock Pin
10 - Third-Fourth Shift Fork
11 - Poppet Springs
12 - Poppet Balls
13 - First-Second Shift Rail
14 - Shifter Interlock Rings
15 - Interlock Pin
16 - Shift Rail Support Plate
17 - Reverse Shift Rail
18 - Reverse Fork
19 - Third-Fourth Shift Rail
20 - Third-Fourth Shift Lug
21 - Interlock Pin
22 - O-Ring
23 - Reverse Rocker Arm Assembly
24 - Gasket
25 - Transmission Case Cover
26 - E-Clip
27 - Backup Lamp Switch
28 - Fulcrum Pins (2)
29 - Reverse Gear Shift Rail Spacer

Exploded view of the shift control housing

5. Remove the rear bearing locating and retaining snaprings. Remove the rear bearing with a bearing remover tool or equivalent.

6. Matchmark the front bearing retainer to the case for easier installation, remove the retaining bolts and the retainer. Remove and discard the bearing oil cap seal.

7. Remove the locating and retaining snaprings from the front bearing. Remove the front bearing and the input (clutch) shaft using a puller tool or equivalent.

8. Remove the 4th gear synchronizer ring from the clutch shaft or synchronizer hub.

9. Remove the mainshaft pilot bearing rollers from the input (clutch) shaft pocket.

10. Remove the mainshaft and gear train assembly by moving the 3rd/4th synchronizer sleeve to the neutral position. Tilt the rear end of the shaft downward and lift the front end of the shaft upward and out of the case. Remove the countershaft gear assembly.

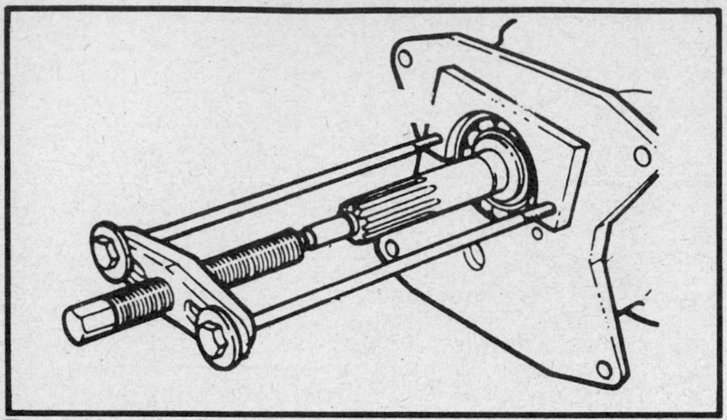

Removing the input assembly

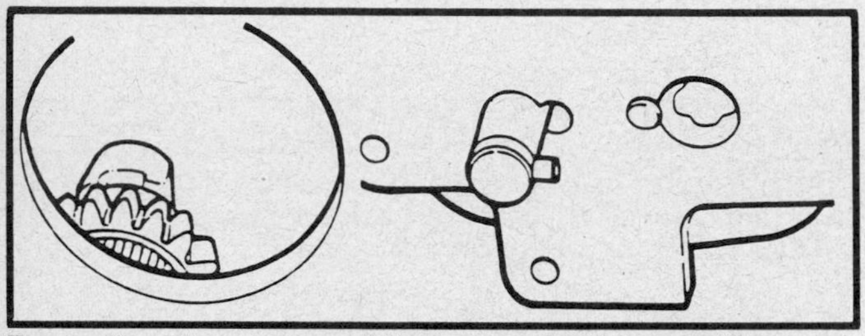

Removing the reverse idler gear shaft assembly

11. Remove the countershaft gear thrust washers and any mainshaft pilot bearing rollers that may have fallen into the case during the input (clutch) shaft removal.
12. Tap the reverse idler gear shaft from the case and remove the reverse idler gear and thrust washers.
13. Remove the 44 needle bearings and the bearing retainers from the gear assembly. Remove the sliding gear from the reverse idler gear. Note the position of the sliding gear for assembly reference.
14. Remove the arbor tool from the countershaft gear and remove the 42 needle bearings and bearing retainers.

Unit Disassembly and Assembly

MAINSHAFT

Disassembly

1. Remove the 3rd/4th speed synchronizer snapring from the front of the mainshaft.
2. Remove the 3rd/4th synchronizer from the mainshaft and slide the hub from the sleeve. Remove the inserts and springs. Inspect the blocking rings for wear and damage.
3. Remove the 3rd speed gear and the 2nd speed gear snapring. Remove the 2nd speed gear and the blocking ring. Remove the tabbed thrust washer.
4. Remove the snapring from the 1st/2nd synchronizer hub. Remove the hub and the reverse gear with sleeve as an assembly. Matchmark the hub and sleeve for assembly references. Remove the inserts and springs as the sleeve is removed.
5. Remove the 1st speed gear thrust washer from the rear of the shaft and remove the 1st speed gear and the blocking ring.

Inspection

1. Clean and inspect the main shaft and gear assemblies.
2. Replace any and all worn or damaged components as necessary.

Installing the 1st gear and blocking ring on the mainshaft

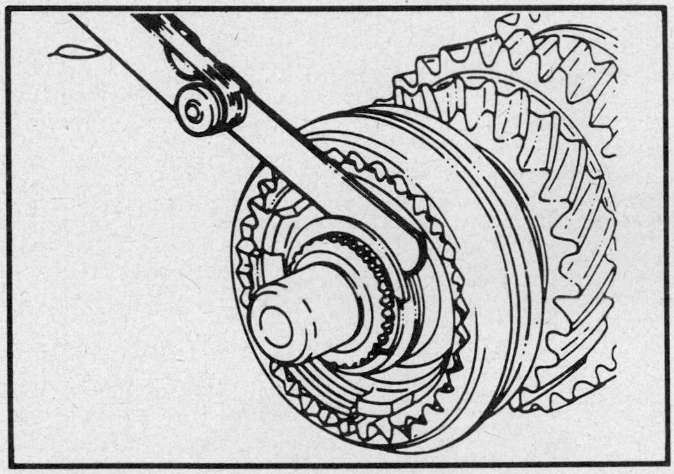

Measuring the mainshaft endplay

3. Inspect the transmission case for the following:
   a. Cracks in the bores, sides, bosses or at bolt holes.
   b. Stripped bolt hole threads.
   c. Nicks, burrs, roughness on gasket or shaft bore surfaces.
4. Check the gears, shafts and synchronizer units for the following:
   a. Chipped, broken or worn gear teeth.
   b. Damaged splines.
   c. Worn or broken teeth or blocking rings.
   d. Bent or broken synchronizer inserts or springs.
   e. Damaged needle bearings or operating surfaces.
   f. Wear or galling of the mainshaft, countershaft, clutch shaft or idler gear shaft.
   g. Worn or broken thrust washers.
   h. Bent, distorted, broken or weak snaprings.
   i. Rough, galled, worn or broken front or rear bearing.

Assembly

1. Lubricate the mainshaft, synchronizer assemblies and gear bores with transmission lubricant.
2. Assemble the 1st/2nd synchronizer hub and reverse gear sleeve. install the gear and sleeve on the hub and place the assembly on a workbench.
3. Drop the inserts into the hub slots. Install the insert springs. Position the loop end of the spring in one insert, compress the spring ends and insert the spring ends under the lips

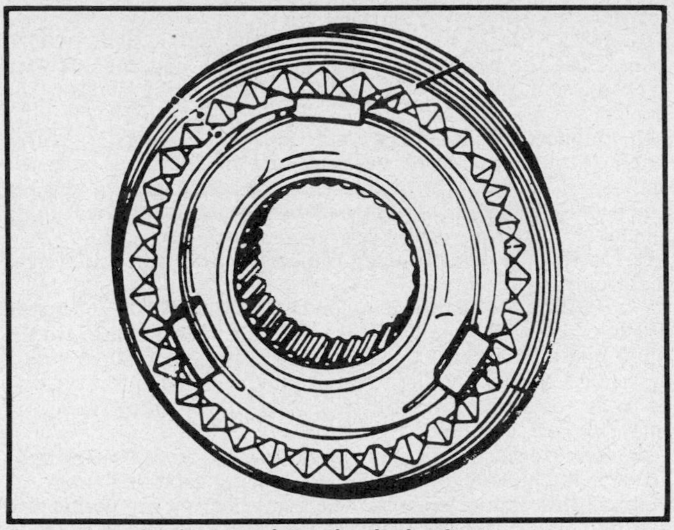

Dropping the inserts into the hub slots

of the remaining 2 inserts. Be sure the spring is under the lip of each insert. Turn the assembly over and install the remaining insert spring. Be sue to install the spring so that the open end faces 180 degrees opposite the first spring.

4. Install the assembled 1st/2nd speed synchronizer hub and the reverse gear with sleeve, on the mainshaft. Secure with a new snapring.

5. Install the 1st speed gear and blocking ring on the rear of the mainshaft and install the 1st gear thrust washer.

6. Install a new tabbed thrust washer on the mainshaft with the tab seated in the mainshaft tab bore.

7. Install the 2nd speed gear and the blocking ring on the mainshaft and secure with a new snapring.

8. Install the 3rd speed gear and blocking ring on the mainshaft.

9. Assemble the 3rd/4th speed synchronizer hub. Install the sleeve on the synchronizer hub, Use the reference marks to align the marks. Place the assembled hub and sleeve flat on a workbench and drop the inserts into the hub slots.

10. Install the insert springs. Position the loop end of the spring in one insert, compress the spring ends and insert the spring ends under the lips of the remaining 2 inserts. se sure the spring is under the lip of each insert. Turn the assembly over and install the remaining insert spring. Be sure to install the spring so that the open end faces 180 degrees opposite the first spring.

11. Install the assembled 3rd/4th speed synchronizer on the mainshaft and secure with a new snapring.

12. The measured endplay between the snapring and the 3rd/4th speed synchronizer should be 0.004–0.014 in.

SHIFT CONTROL HOUSING

Disassembly

1. Remove the shift lever cover, control housing cap and retainer.

2. Remove the shift lever and spring.

3. Position the transmission case cover in a vise so the shift forks are facing upward. Use wood blocks to protect the cover from the vise jaws and do not overtighten the vise.

4. Place all the shift forks in the neutral position.

5. Remove the shift rail support plate attaching bolts and tabbed washers.

6. Remove the support plate, 1st/2nd shift rail, 3rd/4th shift rail, shift lug, interlock pins, reverse shift rail, poppet balls, shifter interlock rings, poopet springs, fulcrum pins and reverse rocker arm assembly.

7. Remove the cover from the vise.

Inspection

1. Inspect all components. Replace any components that are nicked, cracked, broken or excessively worn.

2. Do not discard the spacer on the reverse gear shift rail. This spacer was added to prevent reverse gear overtravel and must remain in place on the shift rail.

Assembly

1. Clamp the transmission case cover in a vise using protective wooden blocks and install the fulcrum pins in the cover.

2. Install the replacement O-ring seal on the reverse rocker arm assembly.

3. Install the assembly and clip. Lubricate the shift rails and shift rail grooves in the cover with petroleum jelly or equivalent.

4. Install the poppet springs in the transmission case cover bores.

5. Install the poppet balls, 1 in each spring.

6. Position the reverse gear shift rail and fork on the reverse rocker arm in the transmission cover.

NOTE: Be sure the notch on the shift rail is positioned over the reverse poppet ball and that the reverse rocker arm is engaged in the reverse fork slot.

7. Install the 3rd/4th shift rail and shift fork in the transmission case cover.

NOTE: Be sure the interlock pin is in position in the shift rail before further assembly.

8. Install the 1st/2nd shift rail and fork assembly. Be sure the shift rail notch is over the poppet ball in the transmission case cover.

9. Install the shifter interlock rings in the cover and between the poppet balls.

10. Press downward on the shift rails to compress the poppet balls and springs. Use a wood black that is long enough to contact all 3 shift rails to press the rails downward evenly.

11. While holding the shift rails downward, position the shift rail retaining plates on the housing and install the plate attaching bolts and tabbed washers finger-tight.

12. Remove the wood block and torque the shift rail retaining bolts to 12–15 ft. lbs. Be sure the tabbed washers are in the correct position before bending the washer tabs.

13. Check the shift rail operation. Each rail must slide smoothly in the cover groove. Be sure it is not possible to overshift into another gear position. After checking the shift operation, place the forks in the 3rd gear position.

14. Install the shift lever, spring, spring retainer and control housing cap. Push the cap downward and turn the lever retainer clockwise to install and seat.

Transaxle Assembly

1. Lubricate the reverse idler gear shaft bore and sliding gear with transmission gear oil. Install the sliding gear on the reverse idler gear.

2. Install 22 needle bearings and 1 bearing retainer at each end of the gear.

3. Coat the reverse idler gear thrust washer surfaces with petroleum jelly and install the thrust washers in the case.

NOTE: The thrust washers have flats on them. Be sure to install the washers so that these flats will face the mainshaft. Also, be sure to engage the thrust washer locating tabs in the case locating slots.

4. Install the reverse idler gear assembly. Align the gear bore, thrust washers and case bores. Install the reverse idler gear shaft from the rear of the case, Be sure to seat the roll pin in the shaft, align the roll pin with the counterbore in the case and push the shaft into the rear of the case.

5. Measure the reverse idler gear endplay by inserting a feeler gauge between the thrust washer and gear. The endplay should be 0.004–0.018 in. If the endplay exceeds 0.018 in., remove the idler gear and replace the thrust washer.

6. Coat the countershaft gear bore, needle bearings and bearing bores in the gear with petroleum jelly. Install the 21 needle bearings and 1 retainer in each end of the gear using the dummy countershaft as a bearing holder.

7. Coat the countershaft gear thrust washer surfaces with petroleum jelly and position the thrust washers in the case. Be sure to engage the locating tabs on the thrust washers in the locating slots in the case.

8. Insert the countershaft into the rear case bore just far enough to hold the rear thrust washer in position displacing the dummy countershaft out the front case bore hole. Do not completely remove the dummy countershaft. This will prevent the washer from being displaced when the countershaft gear is installed.

9. Install the countershaft gear. Align the gear bore, thrust washers and bores in the case and install the countershaft part way into the case.

10. Measure the countershaft gear endplay by inserting a feeler gauge between the washer and the gear. The endplay should be 0.004–0.018 in. If the endplay exceeds 0.018 in., remove the gear and replace the thrust washers. After the correct endplay is obtained, allow the countershaft gear to remain at the bottom of the case. Leave the countershaft in the rear case bore to hold the rear thrust washer in place.

NOTE: The countershaft gear must remain at the bottom of the case to provide sufficient clearance for installation of the mainshaft and clutch shaft assemblies.

11. Install the mainshaft geartrain assembly into the case. Be sure the synchronizers are in the neutral position so the sleeves will clear the top of the case when the assembly is installed.

NOTE: Do not use chassis grease or a similar heavy grease in the clutch shaft bore. Use petroleum jelly only. Heavy grease will plug the lubrication holes in the shaft and prevent proper lubrication of the roller bearings.

12. Install the locating snapring on the front bearing and install the front bearing part way onto the clutch shaft. Do not install the bearing completely at this time as the shaft will not clear the countershaft.

13. Coat the bearing bore in the clutch shaft and mainshaft pilot roller bearings with petroleum jelly.

14. Install the 15 roller bearings in the clutch shaft bearing bore. Coat the blocking ring surface of the clutch shaft with transmission gear oil and position the blocking ring on the shaft.

15. Support the mainshaft assembly and insert the clutch shaft through the front bearing bore in the case. Seat the mainshaft pilot hub in the clutch shaft roller bearing and tap the front bearing and clutch shaft in the case using a soft faced mallet.

16. Install the front bearing cap and tighten the bolts finger tight only.

17. Position the rear bearing on the mainshaft. Do not install the bearing locating ring at this time. Start the bearing into the shaft and into the case bore. Complete the bearing installation by using a soft face mallet to tap it into place. When the bearing is fully seated on the shaft, install the bearing retaining snapring.

18. Remove the front bearing cap, seat the front bearing fully on the clutch shaft and install the bearing retaining snapring. Apply a thin film of sealer to the front bearing cap gasket and position the gasket on the case. Be sure the gasket notch is aligned with the oil return hole in the case.

19. Install a new front bearing cap oil seal. Install the front bearing cap and torque the bolts to 12 ft. lbs.

20. Install the locating ring on the rear bearing. If necessary, reseat the bearing in the case use a soft face mallet.

NOTE: When installing the countershaft, be careful not to damage the thrust washers. Be sure they are aligned with the case bores and gear bores before tapping the countershaft into place.

21. Install the countershaft as follows:

a. Turn the transmission case on end, positioning the case at the edge of the workbench with the clutch shaft pointing downward. Be sure the countershaft bore in the front of the case is accessible. It may be necessary to have an assistant hold the case.

b. Align the countershaft gear bores with the thrust washers and case bores and tap the shaft into place and displace the dummy countershaft out the front of the case. Do not allow the dummy shaft to drop to the floor.

c. Shift the synchronizer sleeves into all the gear positions and check their operation. If the clutch shaft and main shaft appear to bind in the neutral position, check the synchronizer rings sticking on the tappered portion of the gears.

22. Fill the transmission to the proper level with the recommended gear oil and reinstall the oil fill plug. Position the new shift control housing on the case and install the control housing. Torque the housing bolts to 12 ft. lbs.

SPECIFICATIONS

TORQUE SPECIFICATIONS

Component	ft. lbs.	Nm
Backup lamp switch	15	20
Drain and fill plugs	15	20
Front bearing cap bolts	13	18
Shift housing-to-transmission case bolts	13	18
Support plate bolts	18	24
Detent retainer bolt	23	31
Drain and fill plugs	35	47
Front/rear yoke nuts	120	163
Indicator switch	18	24
Operating lever locknut	18	24

Component	ft. lbs.	Nm
Rear case-to-front case bolts	23	31
Rear retainer bolts	23	31
Detent retainer bolt	23	31
Drain and fill plugs	18	24
Front/rear yoke nuts	120	163
Operating lever locknut	18	24
Rear retainer bolts	23	31
Transfer case-to-transmission adapter nuts	26	35
Universal joint strap bolt-to-transfer case	170①	19

① inch lbs.

Section 7

T-18 Transmission
Borg Warner

APPLICATION

1984–87 Ford E, F Series and Bronco

GENERAL DESCRIPTION

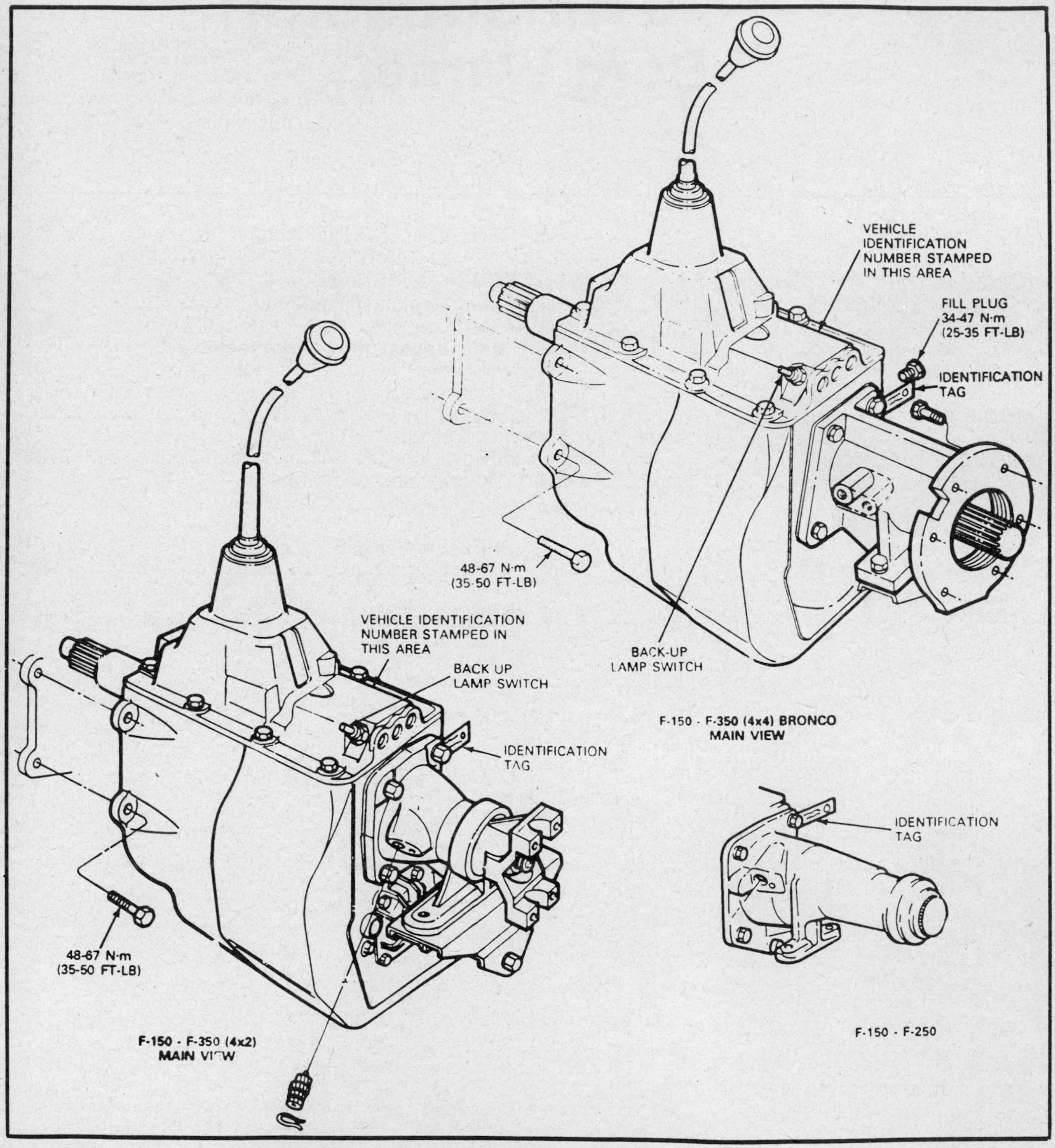
VEHICLE IDENTIFICATION NUMBER STAMPED IN THIS AREA
FILL PLUG 34-47 N·m (25-35 FT-LB)
IDENTIFICATION TAG
48-67 N·m (35-50 FT-LB)
BACK-UP LAMP SWITCH
F-150 - F-350 (4x4) BRONCO MAIN VIEW
VEHICLE IDENTIFICATION NUMBER STAMPED IN THIS AREA
BACK UP LAMP SWITCH
IDENTIFICATION TAG
IDENTIFICATION TAG
48-67 N·m (35-50 FT-LB)
F-150 - F-350 (4x2) MAIN VIEW
F-150 - F-250

T–18 transmission and related components

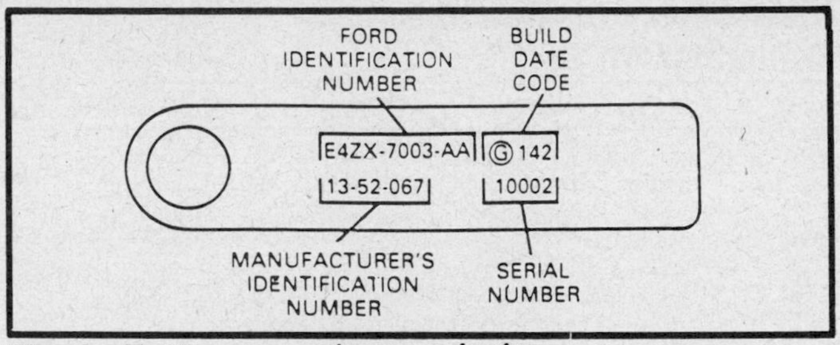
FORD IDENTIFICATION NUMBER
BUILD DATE CODE
E4ZX-7003-AA
Ⓖ 142
13-52-067
10002
MANUFACTURER'S IDENTIFICATION NUMBER
SERIAL NUMBER

Warner T–18 manaual transmission

The Warner T–18 is a semi-synchronized 4 speed transmission. The 1st and reverse gears are of the spur type and the 3rd and 4th gears are of the helical type and are synchronized for the ease of shifting. The T–18 transmission also incorporates a floor mounted shifter.

Transmission Identification

Transmission identification codes are found on the Vehicle Safety Compliance Certification label located on the driver's door lock pillar.

Manual transmissions have service identification tags to identify assemblies for service purposes. The Warner transmissions have an identification tag located on the left middle bolt retaining the cover to the case.

Metric Fasteners

The Warner T–18 transmission uses standard SAE fasteners. Do not attempt to interchange standard SAE fasteners for metric system fasteners. Mismatched or incorrect fasteners can result in damage to the transmission through malfunctions, breakage or possible personal injury. Care should be taken to reuse the fasteners in the same location as removed, whenever possible. Before installing cap screws into aluminum parts, dip theminto oil to prevent the screws from galling the aluminum threads and to prevent seizing.

Capacities

The Warner T–18 transmission requires a total capacity of 3.5 qts. (1.6L) of DOAZ–19C547–A SAE 80 gear oil, or equilvalent.

Checking Fluid Level

1. Raise the vehicle and support it safely.
2. Remove the fill plug located on the side of the case and check the level. The fluid level should be at the lower edge of the fill plug hole, if not add fluid meeting specified recomendations.
3. Install the fill plug and lower the vehicle.

TRANSMISSION MODIFICATIONS

Rattle, Ringing Noise From Transmission Area

Some transmissions built before December 15, 1985 and equipped with 4.9L engine, a 10 in. clutch and have a gross vehicle weight under 8500 lbs., may have a rattle or ringing noise that can be heard during shifting coming from the transmission area. This condition is caused by clutch disc vibration. The clutch disc must be removed and replaced with a different designed disc, part number E5TZ–7550–B.

TROUBLE DIAGNOSIS

CHILTON'S THREE C'S TRANSMISSION DIAGNOSIS
Borg Warner T-18 Manual Transmission

Condition	Cause	Correction
Shift lever—improper operation	a) Shift rods out of adjustment	a) Check crossover operation. Hand lever must move freely through neutral gate without catch or drag. Adjust shift rods
	b) Steering column shift tube out of alignment	b) Check lever position for proper transmission gear selection. Replace steering column shift tube if necessary
Transmission shifts hard	a) Clutch pedal free-travel out of adjustment	a) Adjust clutch pedal free-travel
	b) Clutch does not completely release	b) Inspect complete clutch system
	c) Transmission fluid low or improper type	c) Add lubricant or change lubricant as required
	d) Shift lever binding or worn	d) Remove cap from shift tower. Eliminate binding condition or replace components as required

CHILTON'S THREE C'S TRANSMISSION DIAGNOSIS
Borg Warner T-18 Manual Transmission

Condition	Cause	Correction
Transmission shifts hard	e) Worn or damaged internal shift mechanism	e) Remove transmission cover. Check internal shift mechanism by shifting into and out of all gears. Repair or replace as required
	f) Binding of sliding gears and/or synchronizers	f) Check for free movement of gears and synchronizers. Repair or replace as required
	g) Housings and/or shafts out of alignment	g) Remove transmission and check for binding condition between input shaft and engine crankshaft pilot bearing or bushing. Check flywheel housing alignment. Repair or replace as required
Noisy in forward gears	a) Lubricant level low, or improper type	a) Add lubricant, or refill with specified lubricant
	b) Components grinding on transmission	b) Check for screws, bolts, etc., of cab or other components grounding out. Correct as required
	c) Component housing bolts loose	c) Check torque on transmission to flywheel housing bolts, output shaft flange nut and flywheel housing to engine block bolts. Tighten bolts to specification
	d) Flywheel housing to engine crankshaft alignment	d) Check and align flywheel housing to engine crankshaft
	e) Noisy bearings or gears	e) Remove and disassemble transmission. Inspect input, output and countershaft bearings. Inspect speedometer gear and gear teeth for wear or damage. Replace as required
Gears clash when shifting from one forward gear to another	a) Engine idle speed too high	a) Adjust engine idle speed
	b) Clutch pedal free-travel incorrect	b) Check clutch for pedal free-travel, complete release and spin time. Adjust as required
	c) Improper manual shift linkage	c) Adjust and repair manual shift linkage as required
	d) Pilot bearing binding	d) Remove transmission and check for a binding condition between input shaft and engine crankshaft pilot bearing. Replace as required
	e) Damaged gear teeth and/or synchronizer	e) Disassemble transmission, repair or replace as required
Transmission jumps out of gear	a) Manual shift linkage binding, out of adjustment. Stiff shift boot	a) Adjust linkage. If necessary align steering column to vehicle body
	b) Improper fit of form isolation pad	b) Replace shift boot if exceptionally stiff. Replace or rework pad to provide clearance
	c) Loose transmission to engine mounting bolts, or loose levers	c) Tighten transmission to flywheel housing, and flywheel housing to engine block bolts to specifications. Loosen all bolts and reseat flywheel housing. Tighten all bolts. Tighten levers if necessary
	d) Flywheel housing to engine crankshaft out of line	d) Shim or replace housing as required
	e) Crankshaft pilot bearing worn	e) Replace bearing

CHILTON'S THREE C'S TRANSMISSION DIAGNOSIS
Borg Warner T-18 Manual Transmission

Condition	Cause	Correction
Transmission jumps out of gear	f) Interior components damage	f) Disassemble transmission. Inspect the synchronizer sleeves for free movement on their hubs. Inspect the synchronizer blocking rings for widened index slots, rounded clutch teeth and smooth internal surface. Check countershaft cluster gear for excessive endplay. Check shift forks for worn or loose mounting on shift rails. Check fork pads for excessive wear. Inspect synchronizer sliding sleeve and gear clutch teeth for wear or damage. Repair or replace as required
	g) Worn gear teeth due to partial engagement	g) Replace worn or damaged gears
Transmission is locked in one gear. It cannot be shifted out of that gear	a) Manual-shift linkage out of adjustment, binding or damaged	a) Disconnect the problem shift rod from transmission shift lever. Try to shift the transmission lever into and out of gear position. If OK, repair or replace linkage parts
	b) Internal components	b) Remove transmission. Inspect problem gears, shift rails and forks and synchronizer for wear or damage. Repair as required. Check for broken fork slot tabs on TOD
	c) Loose fork on rail	c) Check shift rail interlock system
Transfer case makes noise	a) Incorrect tire inflation pressures and/or incorrect size tires and wheels	a) Assure that all tires and wheels are the same size, and that inflation pressures are correct
	b) Excessive tire tread wear	b) Check tire tread wear to see if there is more than .06 in. difference in tread wear between front and rear. Interchange one front and one rear wheel. Reinflate tires to specifications
	c) Internal components	c) Operate vehicle in all transmission gears with transfer case in 2HI, or Hi range. If there is noise in transmission in neutral gear, or in some gears and not in others, remove and repair transmission. If there is noise in all gears, operate vehicle in all transfer case ranges. If noisy in all ranges or HI range only, disassemble transfer case. Check input gear, intermediate and front output shaft gear for damage. Replace as required. If noisy in LO range only, inspect intermediate gear and sliding gears for damage. Replace as required
4WD transfer case jumps out of gear	a) Incomplete shift linkage travel	a) Adjust linkage to provide complete gear engagement
	b) Loose mounting bolts	b) Tighten mounting bolts
	c) Front and rear driveshaft slip yokes dry or loose	c) Lubricate and repair slip yokes as required. Tighten flange yoke attaching nut to specifications
	d) Internal components	d) Disassemble transfer case. Inspect sliding clutch hub and gear clutch teeth for damage. Replace as required

CHILTON'S THREE C'S TRANSMISSION DIAGNOSIS
Borg Warner T-18 Manual Transmission

Condition	Cause	Correction
Transmission will not shift into one gear—all others OK	a) Manual linkage out of adjustment	a) Adjust manual linkage
	b) Manual shift linkage damaged or worn	b) Lubricate, repair or replace parts as required
	c) Back-up switch ball frozen	c) If reverse is problem, check back-up switch for ball frozen in extended position (if so equipped)
	d) Internal components	d) Remove transmission. Inspect shift rail and fork system synchronizer system and gear clutch teeth for restricted travel. Repair or replace as required

ON CAR SERVICE

Services

FLUID CHANGE

1. Raise and support the vehicle safely.
2. Remove the transmision oil fill plug and allow the gear oil to drain into a suitable drain pan.
3. The correct level should be to the bottom edge of the fill plug hole.
4. As required, fill to the proper level with the recommended gear oil and reinstall the oil fill plug.
5. Lower the vehicle.

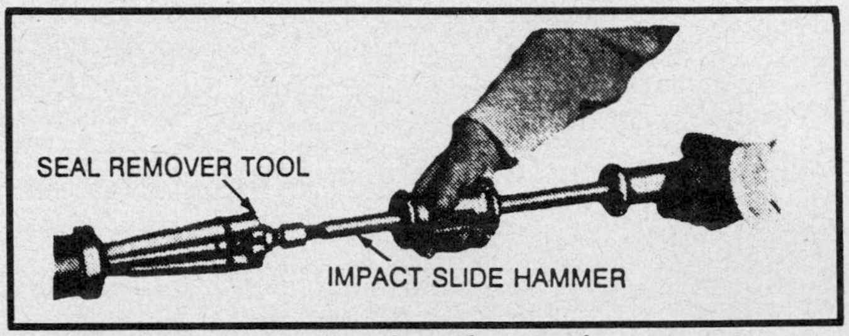
SEAL REMOVER TOOL
IMPACT SLIDE HAMMER
Removing the extension housing seal

REAR SEAL

Removal and Installation

1. Raise the vehicle and suppport it safely.
2. Remove the driveshaft or coupling shaft.
3. Remove the seal from the extension housing using a suitable seal puller tool.
4. Install a new seal in the extension housing using a suitable installer tool. Lubricate the seal with a muti-purpose lubricant.
5. Install the driveshaft and lower the vehicle.

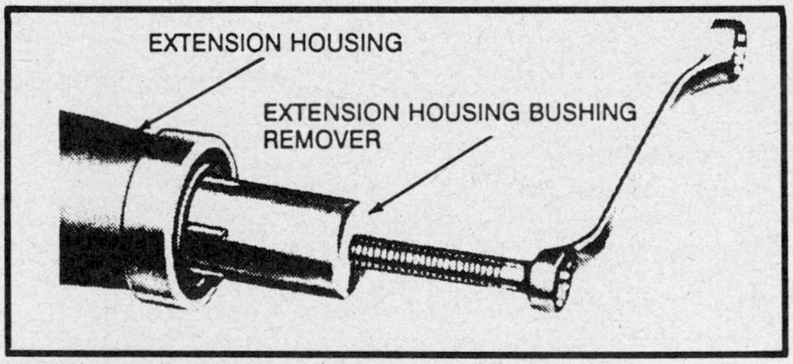
EXTENSION HOUSING
EXTENSION HOUSING BUSHING REMOVER
Removing the extension housing bushing

REAR BUSHING AND SEAL

Removal and Installation

1. Raise the vehicle and support it safely.
2. Disconnect the driveshaft or coupling shaft from the transmission.
3. Remove the seal using a suitable seal remover tool.
4. Insert a suitable bushing remover tool into the extension housing until it grips on the front side of the bushing.
5. Turn the bushing remover screw clockwise until the bushing is free of the housing.
6. Lubricate the new bushing with a multi-purpose lubricant and drive it into the extension housing.
7. Install a new seal and lubricate it with a multi-purpose lubricant.
8. Connect the driveshaft or coupling shaft to the transmission. Lower the vehicle.

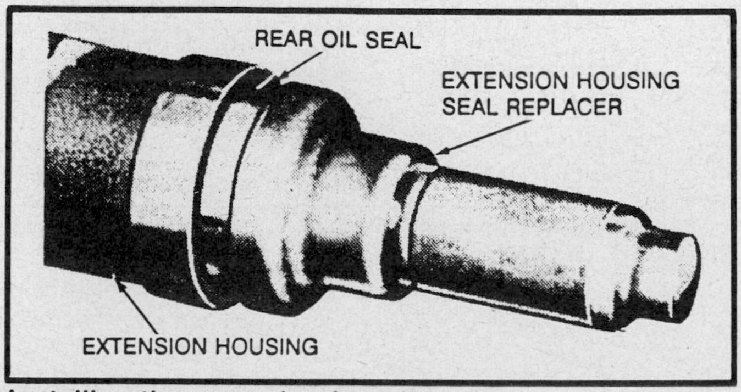
REAR OIL SEAL
EXTENSION HOUSING SEAL REPLACER
EXTENSION HOUSING
Installing the extension housing seal

REMOVAL AND INSTALLATION

TRANSMISSION REMOVAL

F-150 AND F-350 2WD

1. Disconnect the negative battery cable. Remove the floor mat, the body floor pan cover, the gearshift lever shift ball and boot as an assembly. Remove the isolator pad.
2. Raise the vehicle and support it safely.
3. Position a transmission jack under the transmission and secure it in place.
4. Disconnect the speedometer cable.
5. Disconnect the back-up lamp switch located at the rear of the gear shift housing cover.
6. Disconnect the driveshaft or coupling shaft from the transmission.
7. Disconnect the clutch linkage from the transmission and wire it to one side.
8. Remove the transmission-to-clutch housing attaching bolts.
9. Move the transmission to the rear until the input shaft clears the clutch housing. Lower the transmission from the vehicle.

F-150, F-350 4WD AND BRONCO

1. Disconnect the negative battery cable. Remove the screws retaining the floor mat to the floor and remove the mat.
2. Remove the screws retaining the access cover to the floor pan. Place the shift lever in the reverse position and remove the cover.
3. Remove the insulator and dust cover.
4. Remove the transfer case shift lever, shift ball and boot as an assembly.
5. Remove the transmission case shift lever, shift ball and boot as an assembly.
6. Raise the vehicle and support it safely.
7. Remove the drain plug and drain the transmission fluid.
8. Disconnect the rear driveshaft from the transfer case and wire it out of the way.
9. Disconnect the front driveshaft from the transfer case and wire it out of the way.
10. Remove the shift link from the transfer case.
11. Remove the speedometer cable from the transfer case.
12. Position a transmission jack under the transfer case. Remove the bolts retaining the transfer case to the transmission and lower the transfer case from the vehicle.
13. Remove the rear support bracket-to-transmission bolts.
14. Position a transmision jack under the transmission and remove the rear support bracket and brace.
15. Remove the transmission-to-bell housing bolts. Lower the transmision from the vehicle.

TRANSMISSION INSTALLATION

F-150 AND F-350 2WD

1. Position the transmission on a transmission jack and raise it into position behind the housing.
2. Install a guide pin in each lower mounting bolt hole of the flywheel housing.
3. Start the input shaft through the release bearing. Align the splines on the input shaft with the splines in the clutch disc.
4. Move the transmission forward on the guide pins until the input shaft pilot enters the bearing or bushing in the crankshaft. If the transmission front bearing retainer binds up on the clutch release bearing hub, work the release bearing lever until the hub slides onto the transmission front bearing retainer.
5. Remove the guide pin studs and install the lower attaching bolts.
6. Connect the speedometer cable and driven gear and clutch linkage.
7. Install the bolts attaching the front U-joint of the coupling shaft to the transmission output shaft flange.
8. Connect the back-up lamp switch.
9. Install the shift lever, boot and shift ball as an assembly and lubricate the spherical ball seat with a multi-purpose lubricant.
10. Install the isolator pad. Install the floor pan cover and floor mat.

F-150, F-350 4WD AND BRONCO

1. Place the transmission on a transmission jack and install it in the vehicle. Install guide pin studs in the bell housing top holes, to guide the transmission into position.
2. Install the lower bolts. Remove the guide pin studs and install the upper bolts.
3. Place the rear support bracket in position and install the remaining transmission-to-bell housing bolts.
4. Install the rear support bracket bolts and the rear support bracket. Remove the jack.
5. Position the transfer case on the transmission jack and install the retaining bolts and gasket. Position the transfer case on the transmission and tighten the bolts.
6. Install the transfer case shift link.
7. Position and install the speedometer cable.
8. Remove the wire holding the front driveshaft to the side and connect the driveshaft.
9. Remove the wire holding the rear driveshaft to the side and connect the driveshaft.
10. Fill the transfer case with Motorcraft XT-2-QDX, Dexron®II automatic tranmission fluid or equilvalent.
11. Fill the manual transmission with standard transmission lubricant SAE 80 W D8DZ-19C547-A or equilvalent.
12. Lower the vehicle.
13. Position the gasket and the shift cover in place.
14. Install the pilot bolts and install the remaining shift cover retaining bolts.
15. Install the transfer case shift lever, shift ball and boot as assembly .
16. Install the dust cover and the insulator.
17. Install the access cover to the floor pan screws.
18. Install the floor mat screws.
19. Install the shift boot area screws.

BENCH OVERHAUL

Before Disassembly

When servicing the unit, it is recommended that as each part is disassembled, it is cleaned in solvent and dried with compressed air. Disassembly and reassembly of this unit and its parts must be done on a clean work bench. Also, before installing bolts into aluminum parts, always dip the threads into clean transmission oil. Anti-seize compound can also be used to prevent bolts from galling the aluminum and seizing. Always use a torque wrench to keep from stripping the threads. Take care with the seals when installing them, especially the smaller O-rings. The slightest damage can cause leaks. Aluminum parts are very susceptible to damage so great care should be exercised when handling them. The internal snaprings should be expanded and the external snaprings compressed if they are to be reused. This will help

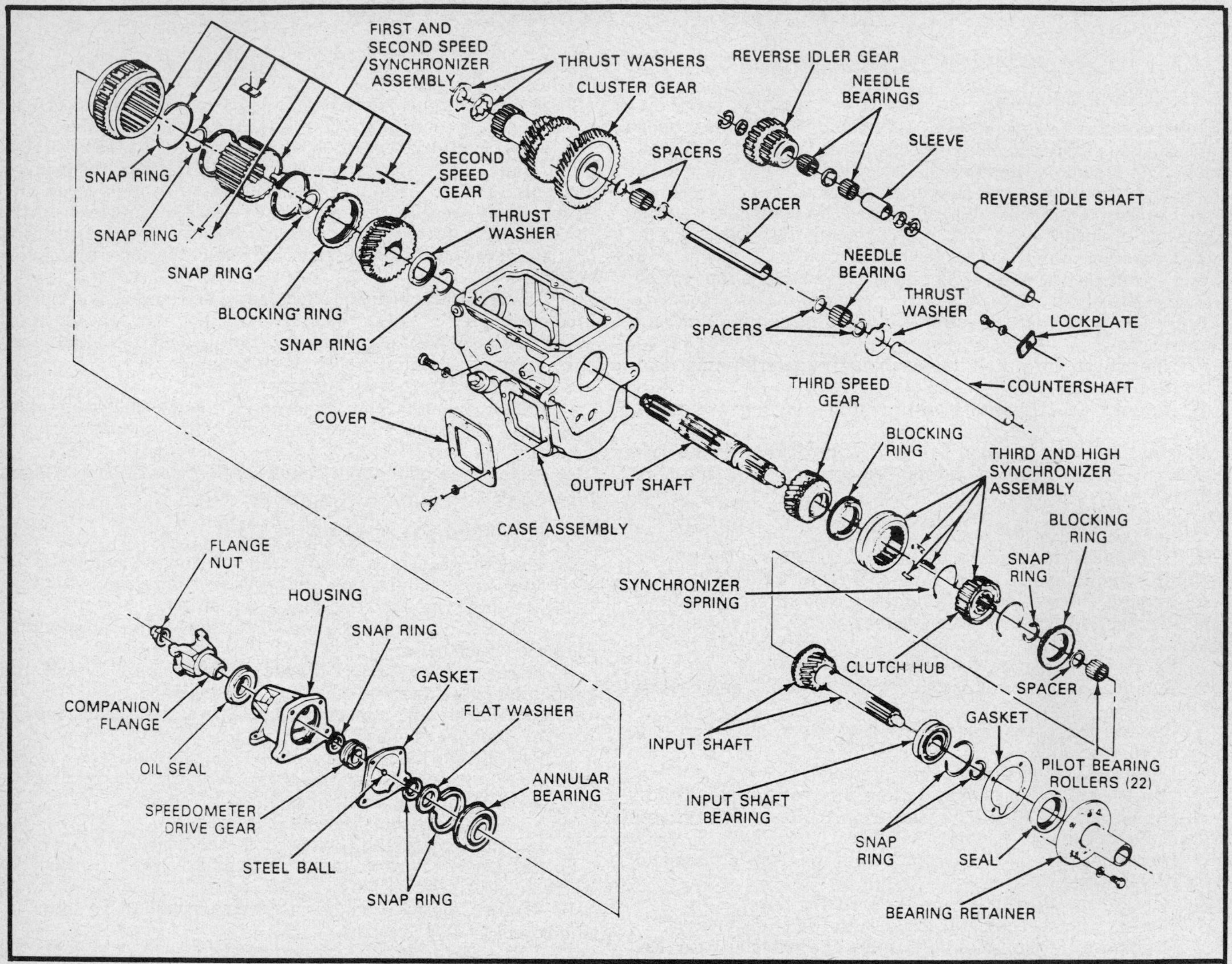
FIRST AND SECOND SPEED SYNCHRONIZER ASSEMBLY
THRUST WASHERS
REVERSE IDLER GEAR
CLUSTER GEAR
NEEDLE BEARINGS
SLEEVE
SPACERS
SECOND SPEED GEAR
SNAP RING
SNAP RING
SPACER
REVERSE IDLE SHAFT
THRUST WASHER
SNAP RING
NEEDLE BEARING
BLOCKING RING
THRUST WASHER
SNAP RING
SPACERS
LOCKPLATE
THIRD SPEED GEAR
COUNTERSHAFT
COVER
BLOCKING RING
THIRD AND HIGH SYNCHRONIZER ASSEMBLY
OUTPUT SHAFT
CASE ASSEMBLY
BLOCKING RING
FLANGE NUT
SNAP RING
SYNCHRONIZER SPRING
HOUSING
SNAP RING
CLUTCH HUB
GASKET
SPACER
COMPANION FLANGE
FLAT WASHER
GASKET
INPUT SHAFT
OIL SEAL
ANNULAR BEARING
PILOT BEARING ROLLERS (22)
SPEEDOMETER DRIVE GEAR
INPUT SHAFT BEARING
STEEL BALL
SNAP RING
SEAL
SNAP RING
BEARING RETAINER
Exploded view—Warner T-18 manual transmission

insure proper seating when installed. Be sure to replace any O-ring, gasket, or seal that is removed.

Transmission Disassembly

1. Position the transmission assembly in a suitable holding fixture. Drain the transmission and shift the unit into 2nd gear before removing the housing assembly.
2. Lock the transmission in gear, remove the U-joint flange and oil seal.
3. Remove the speedometer driven gear and bearing assembly.
4. Remove the output shaft bearing retainer or extension housing.
5. Remove the speedometer drive gear snapring retainer. Slide the speedometer drive gear off the output shaft.
6. Remove the output shaft bearing snapring retainers from the output shaft and from the bearing. Remove the bearing spacer.
7. Install tools T75L–7025–B, F, H and T84T–7025–B or equivalents on the output shaft and over the output shaft bearing. Remove the output shaft bearing.
8. Remove the input shaft bearing retainer. Remove the input shaft bearing snapring from the input shaft and from the bearing.
9. Install the tool on the input shaft and over the input shaft bearing. Remove the input shaft bearing.
10. Remove the output shaft assembly from the case.
11. Remove the input shaft assembly from the case. Do not lose the 22 pilot bearing rollers from the inner end of the shaft.
12. Remove the reverse idler gear shaft and the countershaft retainer from the end of the transmission case.
13. Remove the reverse idler gear shaft, using the tools T50T–100–A and T50T–7140–C or equilvalents.
14. Remove the reverse idler gear from the case.
15. Remove the countershaft with the same tools used to remove the reverse idler gear shaft.
16. Install a suitable dummy shaft tool in the countershaft. Remove the countershaft gear assembly from the case. Guide the countershaft assembly with the dummy shaft tool installed, out of the case so that the roller bearings and spacers that remain in the countershaft are not lost.

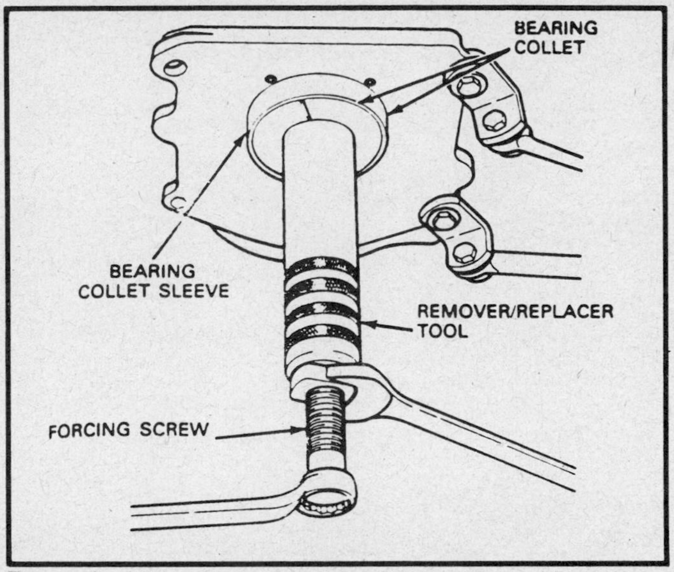
BEARING
COLLET
BEARING
COLLET SLEEVE
REMOVER/REPLACER
TOOL
FORCING SCREW

Removing the output shaft bearing—Warner T-18 manual transmission

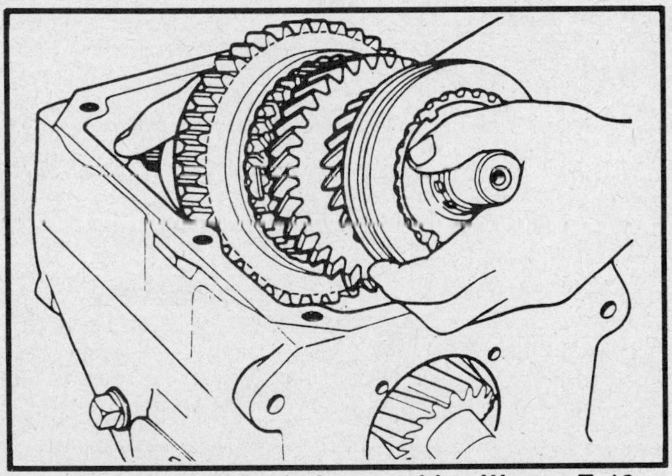

Removing the output shaft assembly—Warner T-18 manual transmission

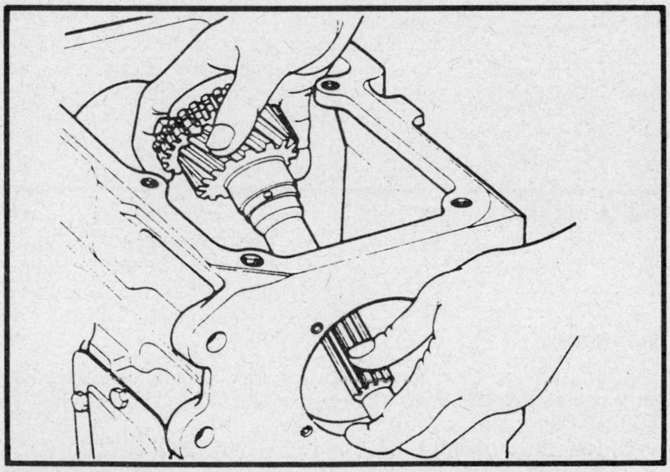

Removing the input shaft shaft assembly—Warner T-18 manual transmission

Unit Disassembly and Assembly

SHIFT LEVER

Disassembly

NOTE: Remove the shift ball only if the shift ball, boot or lever is to be replaced. If either the ball, boot or lever is not being replaced, remove the ball, boot and lever as an assembly.

1. Remove the plastic insert from the shift ball. Warm the ball with a heat gun to 140–180°F (60–82°C) knock the ball off the lever with a block of wood and a hammer, taking care not to damage the finish on the shift lever.

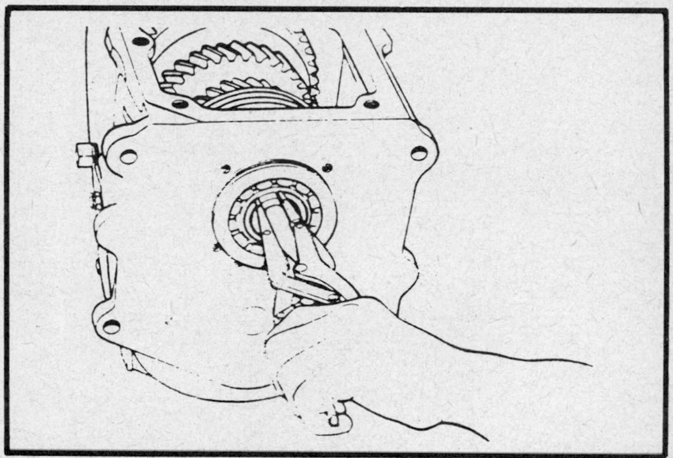

Removing the input shaft bearing snapring—Warner T-18 manual transmission

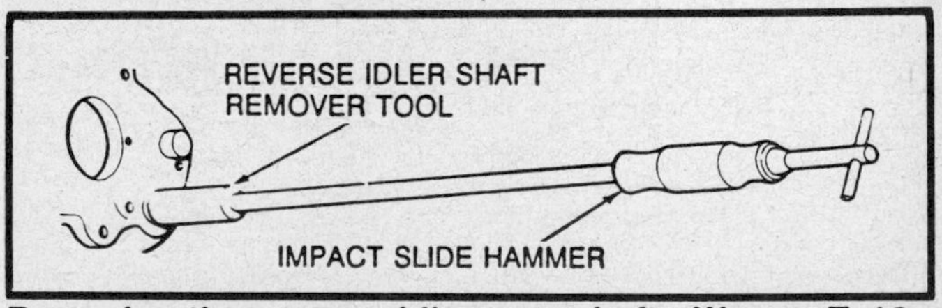
REVERSE IDLER SHAFT
REMOVER TOOL
IMPACT SLIDE HAMMER

Removing the reverse idler gear shaft—Warner T-18 manual transmission

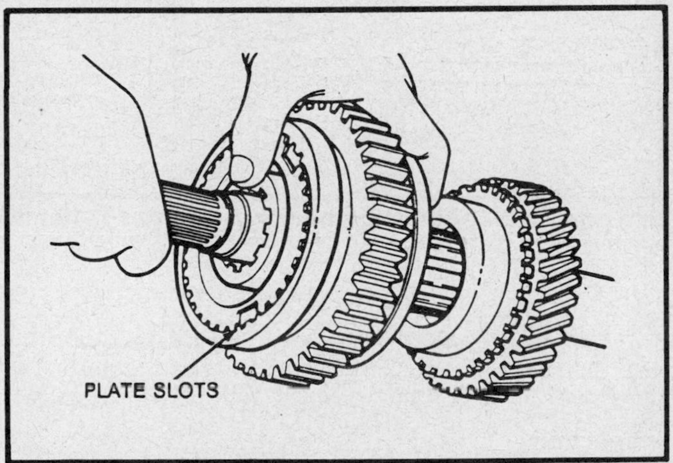
PLATE SLOTS

Removing the 2nd gear synchronizer assembly—Warner T-18 manual transmission

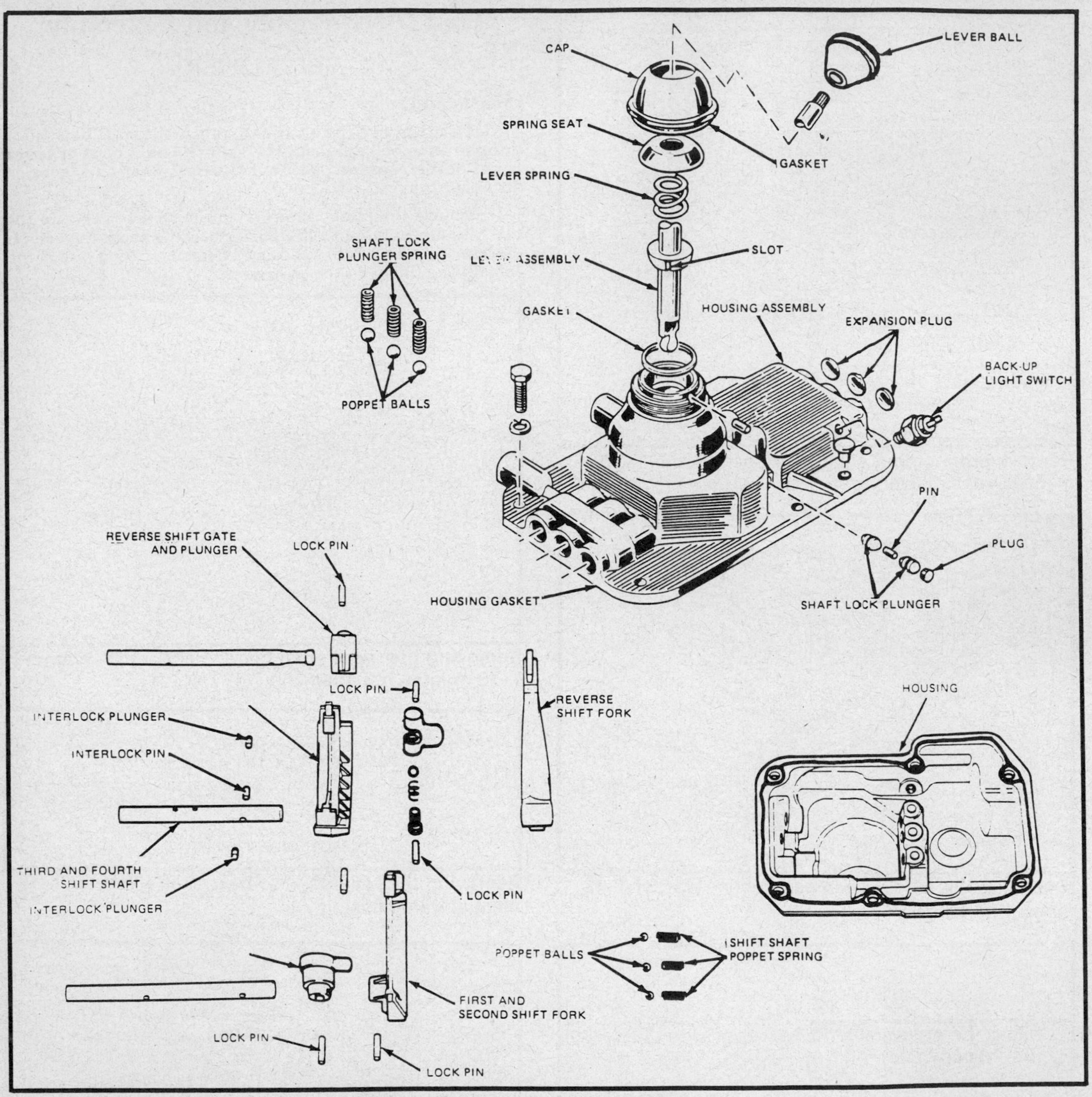
CAP
LEVER BALL
SPRING SEAT
GASKET
LEVER SPRING
SHAFT LOCK
PLUNGER SPRING
SLOT
HOUSING ASSEMBLY
EXPANSION PLUG
BACK-UP
LIGHT SWITCH
POPPET BALLS
PIN
PLUG
REVERSE SHIFT GATE
AND PLUNGER
LOCK PIN
HOUSING GASKET
SHAFT LOCK PLUNGER
LOCK PIN
REVERSE
SHIFT FORK
HOUSING
INTERLOCK PLUNGER
INTERLOCK PIN
THIRD AND FOURTH
SHIFT SHAFT
INTERLOCK PLUNGER
LOCK PIN
POPPET BALLS
SHIFT SHAFT
POPPET SPRING
FIRST AND
SECOND SHIFT FORK
LOCK PIN
LOCK PIN

Exploded view of the shift housing—Warner T-18 manual transmission

2. Remove the rubber boot and floor pan cover.
3. Shift the unit into 2nd gear, remove the lock pin and remove the shift lever from the shifter housing.

Inspection

1. Carefully check the shift lever for distortion, wear or looseness.
2. Replace worn or defective parts.

Assembly

1. Install the shift lever in the shifter housing, making sure that the slot in the lever aligns with the tab in the housing. Install the lock pin.
2. Install the rubber boot and floor pan cover.
3. Warm the ball with a heat gun to 140–180°F (60–82°C) and tap the ball on the lever with a suitable tool and mallet. Install the plastic shift pattern insert.

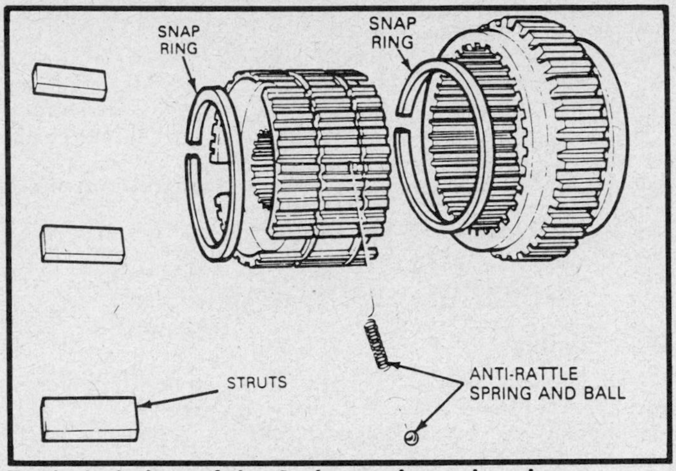
SNAP RING
SNAP RING
STRUTS
ANTI-RATTLE SPRING AND BALL

Exploded view of the 2nd speed synchronizer—Warner T-18 manual transmission

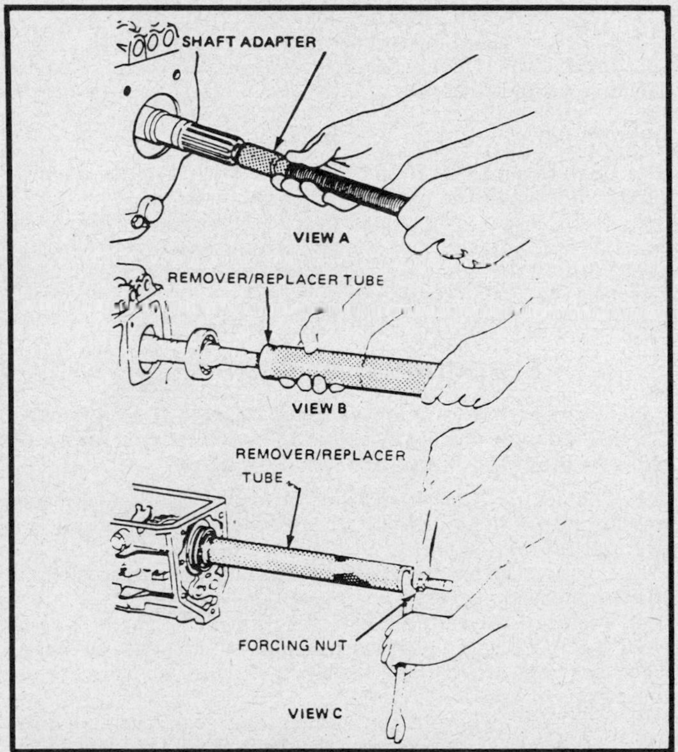
SHAFT ADAPTER
VIEW A
REMOVER/REPLACER TUBE
VIEW B
REMOVER/REPLACER TUBE
FORCING NUT
VIEW C

Installing the output shaft bearing—Warner T-18 manual transmission

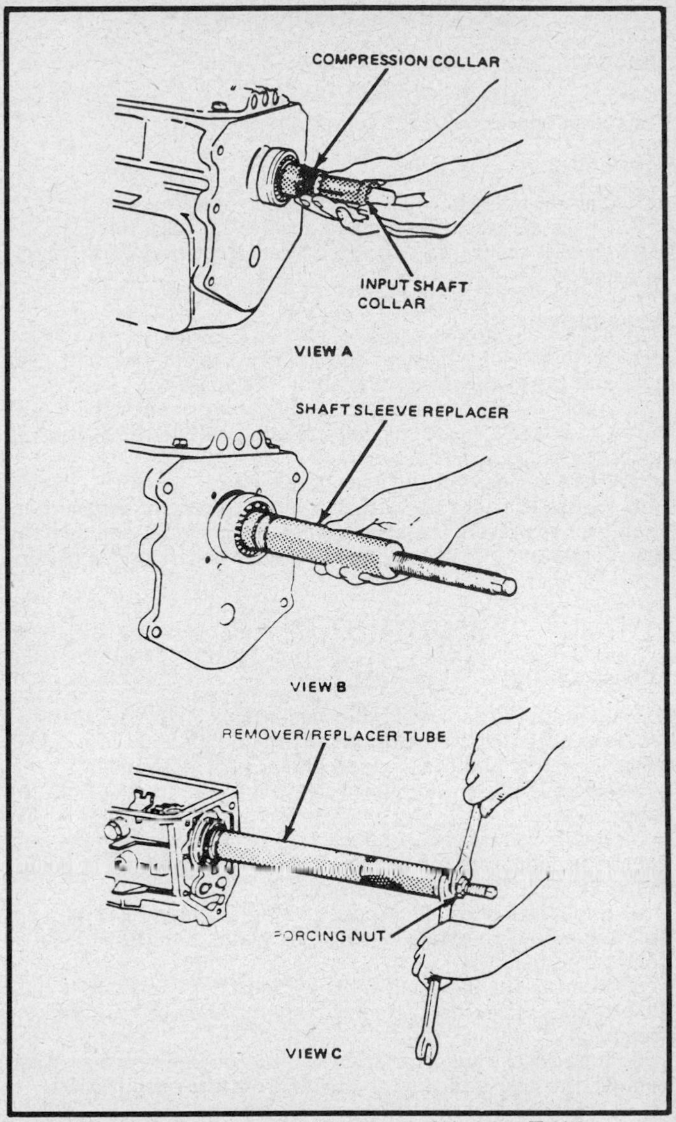
COMPRESSION COLLAR
INPUT SHAFT COLLAR
VIEW A
SHAFT SLEEVE REPLACER
VIEW B
REMOVER/REPLACER TUBE
FORCING NUT
VIEW C

Installing the input shaft bearing—Warner T-18 manual transmission

SYNCHRONIZERS

Disassembly

1. Remove the synchronizer sleeve and the inserts from the hub.
2. Before removing the snaprings from the ends of the hub, check the endplay of the 2nd speed gear. There should be 0.005–0.024 in. (0.127–0.609mm) of endplay.
3. Slide the low and 2nd speed gear off the hub. Be careful not to lose any of the balls, springs, or plates nor the anti-rattle spring and ball.
4. Remove the snapring from behind the synchronizer hub. Pull synchronizer hub from the shaft. Remove the blocking ring.

Inspection

1. Check that the ring and gear engage smoothly.
2. Check for worn teeth or tapered surface.
3. Inspect the synchronizer hub for worn or damaged splines, synchronizer key groove or end surface.
4. Check for smooth hub and sleeve engagement when it is installed.
5. Check the hub sleeve splines and sleeve fork groove for wear or damage.

Assembly

1. Install the blocking ring and synchronizer hub from the shaft.
2. Install the snapring in front of the synchronizer hub.
3. Carefully place the balls, springs, plates and the anti-rattle spring and ball in place in the synchronizer.
4. Slide the low and 2nd speed gear over the hub. Install the snaprings at both ends.
5. Slide the synchronizer sleeve and the inserts over the hub.

COUNTERSHAFT GEAR

Disassembly

Remove the dummy shaft, bearing rollers, bearing spacers and the center spacer from the countershaft gear.

Inspection

1. Inspect the roller bearing for wear or damage.
2. Check all gears for chipped, broken or worn teeth.
3. Inspect the countershaft for wear. Replace defective parts as needed.

Assembly

1. Slide the long bearing spacer into the countershaft gear bore and insert the dummy shaft in the spacer.
2. Apply a film of petroleum jelly to the countershaft gear bore and install 1 of the bearing spacers. Position the 22 bearing rollers in the gear bore.
3. Place a spacer in the gear bore.
4. Hold a large thrust washer against the end of the countershaft gear to prevent the rollers from dropping out and turn the assembly over. Install bearing spacer, 22 rollers and a spacer.

OUTPUT SHAFT

Disassembly

1. Remove the 3rd and high-speed synchronizer hub snapring from the output shaft and slide the 3rd and high-speed synchronizer assembly and the 3rd speed gear off the shaft.
2. Remove the synchronizer sleeve and the inserts from the hub.
3. Before removing the 2 snaprings from the ends of the hub, check the endplay of the 2nd speed gear. There should be 0.005–0.024 in. (0.127–0.609mm) of endplay.
4. Slide the low and 2nd speed gear off the hub. Be careful not to loose any of the balls, springs, or plates nor the anti-rattle spring and ball.
5. Remove the snapring from behind the synchronizer hub. Pull synchronizer hub from the shaft. Remove the blocking ring.
6. Remove the snapring from behind the 2nd speed gear and remove the gear and thrust washer from the output shaft.

Inspection

1. Inspect the bearings for wear or damage.
2. Check the teeth, splines and journals of the outputshaft for damage.
3. Check all gears for chipped, broken or worn teeth.
4. Check synchronizer sleeves for free movement on their hubs.
5. Inspect the synchronizer blocking rings for wear marks.

Assembly

1. Place output shaft with the threaded end up in a soft-jawed vise.
2. Place an output shaft ring in the 3rd groove from the threaded end of the shaft. Place the recessed side of the 2nd speed gear thrust washer over the snapring.
3. Place the 2nd speed gear against the washer and assemble the snapring in groove behind the gear.
4. Place the blocking ring on 2nd speed gear.
5. Assemble the 2nd speed synchronizer assembly over the splines of main shaft, aligning the 3 blocking ring cut-outs with shifting plates. The low and 2nd gear shift fork groove should be located to rear of transmission.
6. Place a snapring in the mainshaft groove behind clutch hub.
7. Turn output shaft over and assemble 3rd speed gear against output shaft shoulder.
8. Place blocking ring on 3rd speed gear.
9. Assemble 3rd and high-speed synchronizer assembly over output shaft splines. Align the 3 blocking ring slots with shifting plates and position the end of the hub which has the long chamfer to the front of the transmission.
10. Place the snapring in output shaft groove in front of 3rd and high-speed synchronizer assembly.
11. Assemble the spacer on output shaft.

REVERSE IDLER GEAR SHAFT

Disassembly

1. Remove the reverse idler gear shaft and the countershaft retainer from the end of the transmission case.
2. Remove the reverse idler gear shaft, using the tools, T50T–100–A and T50T–7140–C or equivalents.
3. Remove the reverse idler gear from the case.

Inspection

1. Inspect the roller bearing for wear or damage.
2. Check all gears for chipped, broken or worn teeth.
3. Inspect the idler shaft and trustwashers for wear. Replace defective parts as needed.

Assembly

1. Coat the inside off the idler gear with a multi-purpose grease and install the needle bearings and sleeve.
2. Position the reverse idler gear assembly in the transmission case and install the reverse idler gear shaft.
3. Position the slot in the rear of the shaft so that it can be engaged by the shaft retainer.

Transmission Assembly

NOTE: If a bearing is not used, be sure that a protector is used over the 3/4 synchronizer to prevent jamming the 4th blocking ring onto the cone seat.

1. Coat all parts, especially the bearings, with specified transmission lubricant to prevent scoring when the transmission is first put into operation.
2. Position the countershaft gear assembly thrust washers in the transmission case.
3. Position the countershaft gear assembly (with dummy shaft tool installed) in the transmission case. Use care so that no roller bearings are lost and so that the thrust washers are not moved out of position.
4. Carefully, drive out the countershaft gear assembly dummy shaft by installing the countershaft from the rear of the transmission case. Position the slot in the rear of the countershaft so that it can be engaged by the shaft retainer.
5. Position the reverse idler gear assembly in the transmission case and install the reverse idler gear shaft. Position the slot in the rear of the shaft so that it can be engaged by the shaft retainer.
6. Install the countershaft and reverse idler gear shaft retainer.
7. Load the 22 pilot bearing rollers in the inner end of the input shaft (use petroleum jelly to keep the pilot bearings in position). Position the input shaft assembly in the transmission case and install the blocking ring on the input shaft.
8. Install the output shaft assembly in the transmission case. Use care so that the pilot bearing rollers are not permitted to drop out of the input shaft.
9. Install a dummy bearing tool T75L–7025–Q or equivalent on the transmission input shaft. This tool is necessary to keep

the input and output shafts in alignment when installing the output shaft bearing.

10. Assemble the locating snapring to the outer race of the output shaft bearing in the groove provided.

11. Install the output shaft bearing using tools, T75L–7025–B, L or equivlents.

12. Install the flatwasher against the rearward face of the output shaft bearing. The properly installed washer will be external to the main body of the transmission.

13. Install a snapring at the rearward surface of the washer in the output shaft groove provided.

14. Install the input shaft bearing using tools, T75L–7025–B, K, R, S or equivalents and install the snapring. Use the thickest select fit snaprings which will fit on the bearing.

15. Install the input shaft bearing spacer and retainer gasket and retainer. Tighten the bolts to 30–36 ft. lbs. (41–48 Nm).

16. Position the speedometer drive gear and spacer, if used on the output shaft over lock ball and install the speedometer drive gear retaining snapring.

17. Using a new gasket install the output shaft bearing retainer or extension housing. Tighten the bolts to 42–50 ft. lbs. (57–67 Nm).

18. Lubricate the extension housing bushing and seal and the U-joint flange with multi-purpose lubricant.

19. Install the U-joint flange. Lock the transmission in 2 gears and tighten the retaining nuts.

20. Install the gear shift housing assembly with standard transmission lubricant SAE 80W D8DZ–19C547–A, ESP–MC83C or equivalent.

21. Fill the transmission to the proper level with standard transmission lubricant SAE 80W D8DZ–19C547–A, ESP–MC83C or equivalent. Add ½ pt. (¼L) of lubricant through the speedometer cable hole in the rear transmission extension housing.

22. Install the transmission into the vehicle.

SPECIFICATIONS

TORQUE SPECIFICATIONS

Item	ft. lbs.	Nm
Back-Up light switch	15–25	20–47
Clutch housing to transmission mounting bolts	35–50	47–67
Case cover	25–35	34–47
Countershaft rear retainer	25–35	34–47
Drain plug	25–40	34–54
Filler plug	25–40	34–54
Output shaft flange nut	75–110	102–149
Mainshaft rear retainer	25–35	34–47
Mainshaft rear retainer	40–50	54–67
Power take off cover bolt	25–35	34–47
Reverse idler shaft/countershaft locking bolt	25–35	34–47
Front bearing retainer to case	10–15	14–20
Clutch housing to engine block	40–50	54–67

SPECIAL TOOLS

Number	Description
T50T-100-A	Impact slide hammer
T57L-500-B	Bench mounted holding fixture
T00L-1175-AC	Seal remover
T75L-4201-A	Clutch housing alignment adapter
T75L-4201-B	Clutch housing alignment adapter
D78P-4201-B	Dial indicator with base
T00L-4201-C	Dial indicator with bracketry
T75L-6392-A	Clutch housing alignment tool
D79L-7000-A	Retaining ring pliers
T72J-7025	Mainshaft bearing cone replacer
T71P-7025-A	Output shaft bearing replacer
T75L-7025-B	Remover/replacer tube
T75L-7025-F	Bearing collet
T75L-7025-H	Bearing collet sleeve
T84T-7025-B	Forcing screw
T75L-7025-K	Shaft sleeve replacer
T75L-7025-L	Shaft adapter—replacing
T77L-7025-L	Manual transmission bearing set
T75L-7025-Q	Dummy bearing
T75L-7025-R	Compression collar
T75L-7025-S	Impact shaft collet
T50T-7140-C	Reverse idler shaft remover

Section 7

T-19 Transmission
Borg Warner

APPLICATION

Year	Vehicle	Engine
1984	F250	6.9L
		7.5L
	F350	6.9L
		7.5L
1985	F250	6.9L
		7.5L
	F350	6.9L
		7.5L
1986	F250	6.9L
	F350	6.9L
1987	F250	6.9L
	F350	6.9L

GENERAL DESCRIPTION

The 4 speed fully synchronized T19 manual transmissions are equipped with a center, floor mounted gear shift lever. The reverse gear and the 1st, 2nd, 3rd, and 4th speed gears are helical cut. The input shaft is supported by a ball bearing. The front end of the output shaft is supported by a pilot bearing installed in the input shaft. The rear end of the output shaft is supported by a ball bearing. The ball bearing and shaft are retained in the case by a snapring.

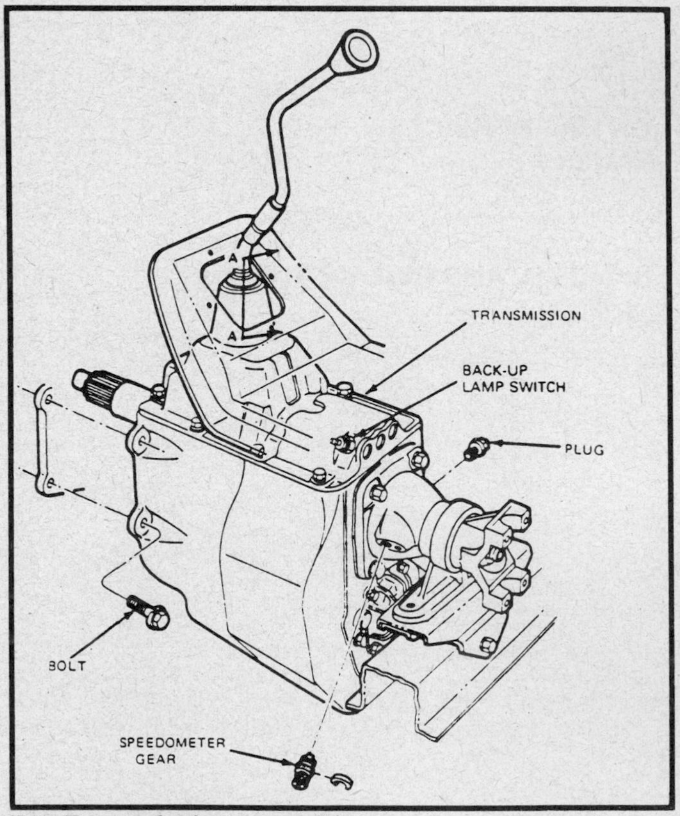
TRANSMISSION
BACK-UP LAMP SWITCH
PLUG
BOLT
SPEEDOMETER GEAR
T19 Transmission assembly

Capacities

All vehicles equipped with the T19 transmission require 7.0 pts. (3.3L) of transmission gear lubricant Ford specification D8DZ-19C547-A (SAE 80W)or equivalent.

Checking Fluid Level

The vehicle must be level before an accurate reading can be obtained. The fluid level considered full when the lubricant is to the bottom of the fill hole.

TRANSMISSION MODIFICATIONS

There were no modifications for the T19 transmissions at the time of this publication.

TROUBLE DIAGNOSIS

CHILTON'S THREE C'S TRANSMISSION DIAGNOSIS
Warner T19

Condition	Cause	Correction
Shift lever—improper operation	a) Shift rods out of adjustment	a) Check crossover operation. Hand lever must move freely through neutral gate without catch or drag. Adjust shift rods
	b) Steering column shift tube out of alignment	b) Check lever position for proper transmission gear selection. Replace steering column shift tube if necessary
Transmission shifts hard	a) Clutch pedal free-travel out of adjustment	a) Adjust clutch pedal free-travel
	b) Clutch does not completely release	b) Remove and disassemble clutch assembly. Replace as required
	c) Transmission fluid low or improper type	c) Add lubricant or change lubricant as required
	d) Shift lever binding or worn	d) Remove cap from shift tower. Eliminate binding condition or replace components as required
	e) Worn or damaged internal shift mechanism	e) Remove transmission cover. Check internal shift mechanism by shifting into and out of all gears. Repair or replace as required
	f) Binding of sliding gears and/or synchronizers	f) Check for free movement of gears and synchronizers. Repair or replace as required
	g) Housings and/or shafts out of alignment	g) Remove transmission and check for binding condition between input shaft and engine crankshaft pilot bearing or bushing. Check flywheel housing alignment. Repair or replace as required
Noisy in forward gears ①	a) Lubricant level low, or improper type	a) Add lubricant, or refill with specified lubricant
	b) Components grinding on transmission	b) Check for screws, bolts, etc., of cab or other components grounding out. Correct as required
	c) Component housing bolts loose	c) Check torque on transmission to flywheel housing bolts, output shaft flange nut and flywheel housing to engine block bolts. Tighten bolts to specification
	d) Flywheel housing to engine crankshaft alignment	d) Check and align flywheel housing to engine crankshaft
	e) Noisy bearings or gears	e) Remove and disassemble transmission. Inspect input, output and countershaft bearings. Inspect speedometer gear and gear teeth for wear or damage. Replace as required
Gears clash when shifting from one forward gear to another	a) Engine idle speed too high	a) Adjust engine idle speed
	b) Clutch pedal free-travel incorrect	b) Check clutch for pedal free-travel, complete release and spin time. Adjust as required
	c) Improper manual-shift linkage	c) Adjust and repair manual-shift linkage as required
	d) Pilot bearing binding	d) Remove transmission and check for a binding condition between input shaft and engine crankshaft pilot bearing. Replace as required
	e) Damaged gear teeth and/or synchronizer	e) Disassemble transmission, repair or replace as required

CHILTON'S THREE C'S TRANSMISSION DIAGNOSIS
Warner T19

Condition	Cause	Correction
Transmission jumps out of gear	a) Manual-shift linkage binding, out of adjustment. Stiff shift boot	a) Adjust linkage. If necessary align steering column to vehicle body
	b) Improper fit of form isolation pad	b) Replace shift boot if exceptionally stiff. Replace or rework pad to provide clearance
	c) Loose transmission to engine mounting bolts, or loose levers	c) Tighten transmission to flywheel housing, and flywheel housing to engine block bolts to specifications. Loosen all bolts and reseat flywheel housing. Tighten all bolts. Tighten levers if necessary
	d) Flywheel housing to engine crankshaft out of line	d) Shim or replace housing as required
	e) Crankshaft pilot bearing worn	e) Replace bearing
	f) Interior components damage	f) Disassemble transmission. Inspect the synchronizer sleeves for free movement on their hubs. Inspect the synchronizer blocking rings for widened index slots, rounded clutch teeth and smooth internal surface. Check countershaft cluster gear for excessive endplay. Check shift forks for worn or loose mounting on shift rails. Check fork pads for excessive wear. Inspect synchronizer sliding sleeve and gear clutch teeth for wear or damage. Repair or replace as required
	g) Worn gear teeth due to partial engagement	g) Replace worn or damaged gears
Transmission will not shift into one gear—all others OK	a) Manual linkage out of adjustment	a) Adjust manual linkage
	b) Manual-shift linkage damaged or worn	b) Lubricate, repair or replace parts as required
	c) Back-up switch ball frozen	c) If reverse is problem, check back-up switch for ball frozen in extended position (if so equipped)
	d) Internal components	d) Remove transmission. Inspect shift rail and fork system synchronizer system and gear clutch teeth for restricted travel. Repair or replace as required
Transmission is locked in one gear. It cannot be shifted out of that gear	a) Manual-shift linkage out of adjustment, binding or damaged	a) Disconnect the problem shift rod from transmission shift lever. Try to shift the transmission lever into and out of gear position. If OK, repair or replace linkage parts
	b) Internal components	b) Remove transmission. Inspect problem gears, shift rails and forks and synchronizer for wear or damage. Repair as required. Check for broken fork slot tabs on TOD
	c) Loose fork on rail	c) Check shift rail interlock system
Transfer case makes noise	a) Incorrect tire inflation pressures and/or incorrect size tires and wheels	a) Assure that all tires and wheels are the same size, and that inflation pressures are correct
	b) Excessive tire tread wear	b) Check tire tread wear to see if there is more than .06 inch difference in tread wear between front and rear. Interchange one front and one rear wheel. Reinflate tires to specifications

CHILTON'S THREE C'S TRANSMISSION DIAGNOSIS
Warner T19

Condition	Cause	Correction
Transfer case makes noise	c) Internal components	c) Operate vehicle in all transmission gears with transfer case in 2HI, or Hi range. If there is noise in transmission in neutral gear, or in some gears and not in others, remove and repair transmission. If there is noise in all gears, operate vehicle in all transfer case ranges. If noisy in all ranges or HI range only, disassemble transfer case. Check input gear, intermediate and front output shaft gear for damage. Replace as required. If noisy in LO range only, inspect intermediate gear and sliding gears for damage. Replace as required
4-wheel drive transfer case jumps out of gear	a) Incomplete shift linkage travel	a) Adjust linkage to provide complete gear engagement
	b) Loose mounting bolts	b) Tighten mounting bolts
	c) Front and rear driveshaft slip yokes dry or loose	c) Lubricate and repair slip yokes as required. Tighten flange yoke attaching nut to specifications
	d) Internal components	d) Disassemble transfer case. Inspect sliding clutch hub and gear clutch teeth for damage. Replace as required
① While verifying the condition, determine whether the noise is gear roll-over noise, release bearing rub or some other transmission related noise.	Gear roll-over noise, inherent in manual transmissions, is caused by the constant mesh gears turning at engine idle speed, while the clutch is engaged and the transmission is in neutral; and release bearing rub are sometimes mistaken for mainshaft bearing noise.	Gear roll-over noise will disappear when the clutch is disengaged or when the transmission is engaged in gear. Release bearing rub will disengage when the clutch is engaged in the event that a bearing is damaged, the noise is more pronounced while engaged in gear under load or coast than in neutral.

ON CAR SERVICE

Services

FLUID CHANGE

Procedure

1. Raise and support vehicle safely.
2. Remove the drain plug if equipped, or siphon transmission fluid into a pan.
3. Remove filler plug.
4. Install drain plug.
5. Fill to the level of the filler plug hole with transmission gear lubricant Ford specification D8DZ–19C547–A (SAE 80W) or equivalent.
6. Install fill plug and lower vehicle.

SHIFT LEVER

Removal and Installation

NOTE: Remove the shift ball only if the shift ball, boot or lever is to be replaced. If either the ball, boot or lever is not being replaced, remove the ball, boot and lever as an assembly.

1. Remove the plastic insert from the shift ball. Warm the ball with a heat gun to 140–180°F (60–82°C) knock the ball off the lever with a block of wood and a hammer, taking care not to damage the finish on the shift lever.
2. Remove the rubber boot and floor pan cover.
3. Shift the unit into 2nd gear, remove the lock pin and remove the shift lever from the shifter housing.
4. Install the shift lever in the shifter housing, making sure that the slot in the lever aligns with the tab in the housing. Install the lock pin.
5. Install the rubber boot and floor pan cover.
6. Warm the ball with a heat gun to 140–180°F (60–82°C) and tap the ball on the lever with a $^{7}/_{16}$ inch socket and mallet. Install the plastic shift pattern insert.

REMOVAL AND INSTALLATION

TRANSMISSION REMOVAL

F-250–F-350 EXCEPT 4WD

1. Remove the floor mat, the body floor pan cover, the gearshift lever shift ball and boot as an assembly. Remove the isolator pad assembly.
2. Raise the vehicle and position safely on jackstands. Position a transmission jack under the transmission and disconnect the speedometer cable.
3. Disconnect the back-up lamp switch located at the rear of the gear shift housing cover.
4. Disconnect the driveshaft or coupling shaft and clutch linkage from the transmission and wire it to one side.
5. Remove the transmission rear insulator and lower retainer. Remove the crossmember.
6. Remove the transmission attaching bolts.
7. Move the transmission to the rear until the input shaft clears the clutch housing. Lower the transmission.

F-250–F-350 WITH 4WD

1. Remove the screws holding the floor mat.
2. Remove the screws holding the access cover to the floor pan.
3. Place the shift lever in the reverse position and remove the cover.
4. Remove the insulator and dust cover.
5. Remove the transfer case shift lever, shift ball and boot as an assembly.
6. Remove transmission shift lever, shift ball and boot as an assembly.
7. Raise the vehicle and support safely.
8. Remove the drain plug and drain the transmission.
9. Disconnect the rear driveshaft from the transfer case and wire it out of the way.
10. Disconnect the front driveshaft from the transfer case and wire it out of the way.
11. Remove the shift link from transfer case.
12. Remove the speedometer cable from the transfer case.
13. Position a transmission jack under the transfer case. Remove the 6 bolts holding the transfer case to the transmission and lower the transfer case from the vehicle.
14. Position a transmission jack under the transmission and remove the crossmember.
15. Remove the 4 bolts that hold the transmission to the bell housing.
16. Remove the transmission from the vehicle.

TRANSMISSION INSTALLATION

F-250–F-350 EXCEPT 4WD

1. Place the transmission on a transmission jack, install guide studs in the clutch housing and raise the transmission until the input shaft splines are aligned with the clutch disc splines. The clutch release bearing and hub must be properly positioned in the release lever fork.
2. Slide the transmission forward on the guide studs until it is in position on the clutch housing. Install the attaching bolts and tighten them to 35–50 ft. lbs. (48–67 Nm). Remove the guide studs and install the 2 lower attaching bolts.
3. Install the crossmember. Position the insulator and retainer between the transmission and crossmember. Install bolts and tighten to 45–60 ft. lbs. (60–80 Nm). Install the nut retaining the insulator and retainer to crossmember then tighten to 50–70 ft. lbs. (68–94 Nm). Remove the transmission jack.
4. Connect the speedometer cable and driven gear and clutch linkage.
5. Install the bolts attaching the front u-joint of the coupling shaft to the transmission output shaft flange. Tighten the bolts and nuts to specifications. Install the transmission rear support and upper and lower absorbers.
6. Connect the back-up lamp switch.
7. Install the shift lever, boot and shift ball as an assembly and lubricate the spherical ball seat with multi-purpose lubricant.
8. Install the isolator pad assembly. Install the floor pan cover, and floor mat.

F-250–F-350 WITH 4WD

1. Place the transmission on a transmission jack and install it in the vehicle installing 2 guide studs in the bell housing top holes, to guide the transmission into position.
2. Install the 2 lower bolts. Remove the guide studs and install the upper bolts.
3. Place the rear support bracket in position and install the 8 retaining bolts.
4. Install the crossmember. Remove the transmission jack.
5. Position the transfer case on the transmission jack and install the 6 retaining bolts and gasket. Position the transfer case on the transmission and tighten the bolts.
6. Install the transfer case shift link.
7. Position and install the speedometer cable.
8. Remove wire and connect front driveshaft.
9. Remove wire and connect rear driveshaft.
10. Fill the transfer case with DEXRON®II, automatic transmission fluid or equivalent and fill the manual transmission with standard transmission lubricant (SAE 80W), D8DZ 19C547–A (ESP–M2C83–C) lubricant or equivalent.
11. Lower the vehicle.
12. Clean and prepare the gasket area.
13. Position gasket and shift cover.
14. Install 2 pilot bolts, then install remaining shift cover retaining bolts.
15. Install transfer case shift lever, shift ball and boot as an assembly and transmission shift lever, shift ball and boot as an assembly.
16. Install dust cover and insulator.
17. Install access cover to floor pan screws.
18. Install the floor mat screws.
19. Install the boot area screws.

BENCH OVERHAUL

Before Disassembly

Clean the exterior of the transmission assembly before any attempt is made to disassemble it in order to prevent dirt or other foreign materials from entering the transmission assembly or its internal parts.

NOTE: If steam cleaning is done to the exterior of the transmission, immediate disassembly should be done to avoid rusting, caused by condensation forming on the internal parts.

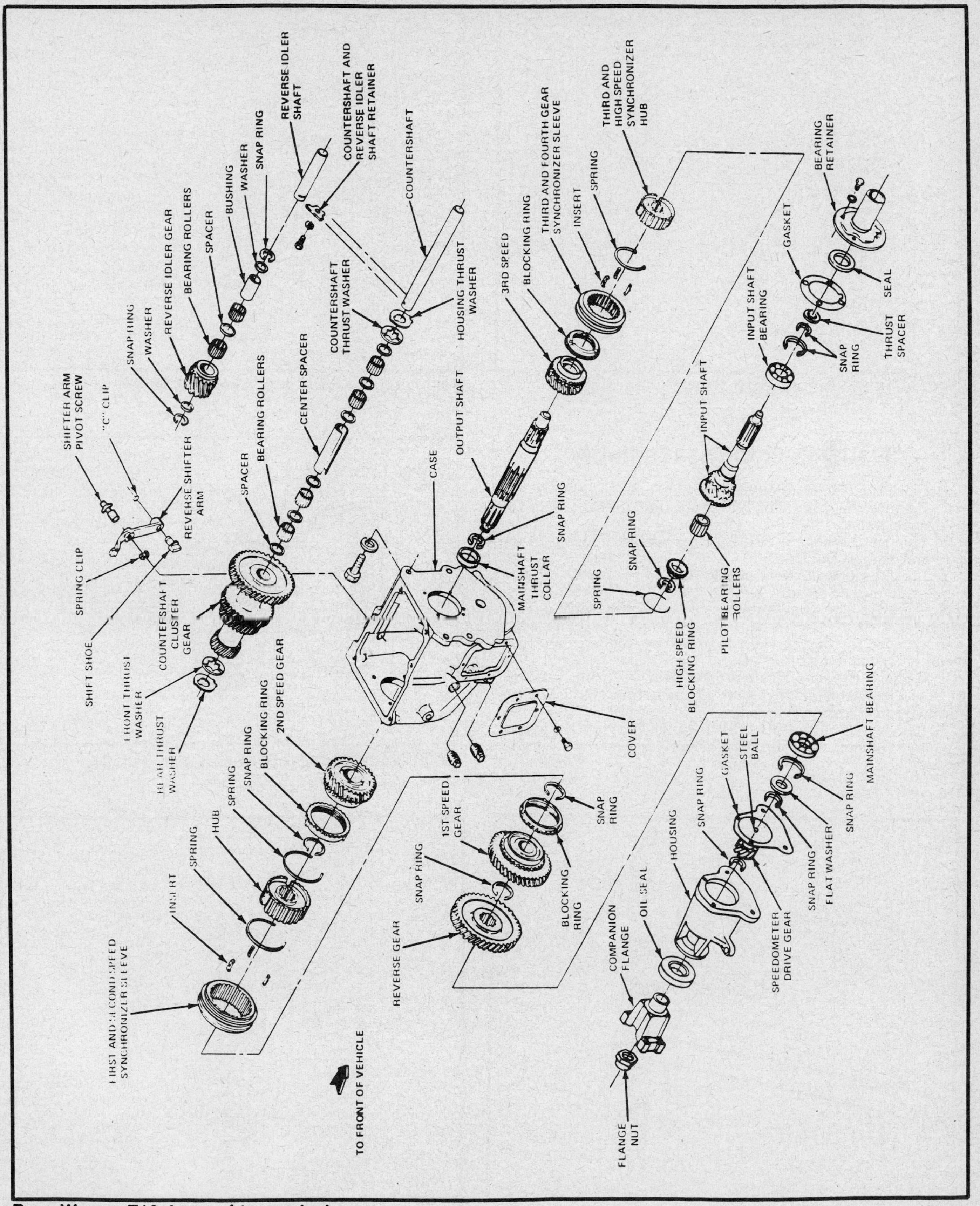
REVERSE IDLER SHAFT
COUNTERSHAFT AND REVERSE IDLER SHAFT RETAINER
COUNTERSHAFT
SNAP RING
WASHER
BUSHING
SPACER
BEARING ROLLERS
REVERSE IDLER GEAR
WASHER
SNAP RING
SHIFTER ARM PIVOT SCREW
"C" CLIP
REVERSE SHIFTER ARM
SPRING CLIP
SHIFT SHOE
COUNTERSHAFT CLUSTER GEAR
FRONT THRUST WASHER
REAR THRUST WASHER
SPACER
BEARING ROLLERS
CENTER SPACER
COUNTERSHAFT THRUST WASHER
HOUSING THRUST WASHER
CASE
OUTPUT SHAFT
SNAP RING
MAINSHAFT THRUST COLLAR
3RD SPEED
BLOCKING RING
THIRD AND FOURTH GEAR SYNCHRONIZER SLEEVE
INSERT
SPRING
THIRD AND HIGH SPEED SYNCHRONIZER HUB
BEARING RETAINER
GASKET
SEAL
THRUST SPACER
SNAP RING
INPUT SHAFT BEARING
INPUT SHAFT
PILOT BEARING ROLLERS
SNAP RING
SPRING
HIGH SPEED BLOCKING RING
COVER
2ND SPEED GEAR
BLOCKING RING
SNAP RING
SPRING
HUB
SPRING
INSERT
FIRST AND SECOND SPEED SYNCHRONIZER SLEEVE
1ST SPEED GEAR
SNAP RING
SNAP RING
BLOCKING RING
REVERSE GEAR
TO FRONT OF VEHICLE
MAINSHAFT BEARING
SNAP RING
GASKET
STEEL BALL
SNAP RING
HOUSING
SNAP RING
FLAT WASHER
SPEEDOMETER DRIVE GEAR
OIL SEAL
COMPANION FLANGE
FLANGE NUT

Borg Warner T19 4 speed transmission

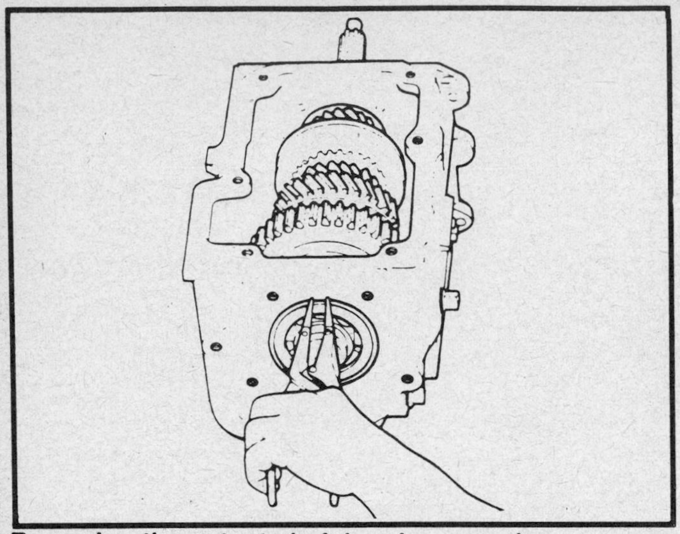
Removing the output shaft bearing snapring

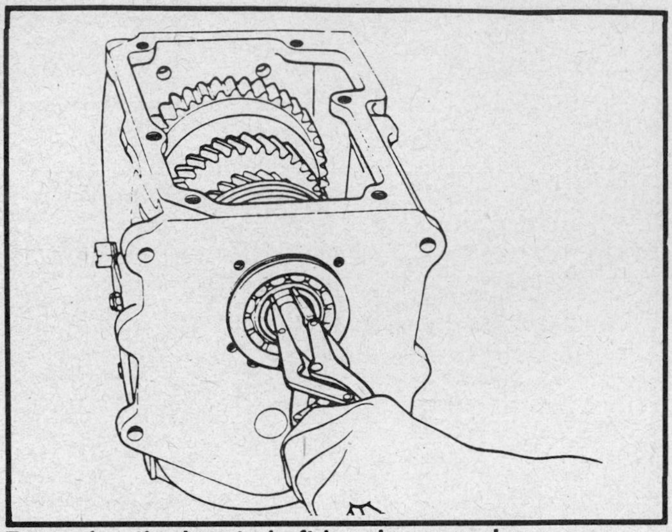
Removing the input shaft bearing snapring

Transmission Disassembly

1. Place the transmission in a suitable holding fixture.
2. Remove the drain plug and drain the lubricant from the transmission.
3. Remove the bolts attaching the gearshift housing to the case and remove the housing.
4. Lock the transmission in 2 gears and remove the u-joint flange and output oil seal.
5. Remove the speedometer driven-gear and bearing assembly.
6. Remove the output shaft rear bearing retainer or extension housing and gasket.
7. Remove the speedometer drive gear snapring retainer. Slide the speedometer drive gear off the output shaft. Remove the speedometer gear drive ball.
8. Remove the output shaft bearing snapring from the output shaft and the snapring from the bearing. Remove bearing spacer washer.

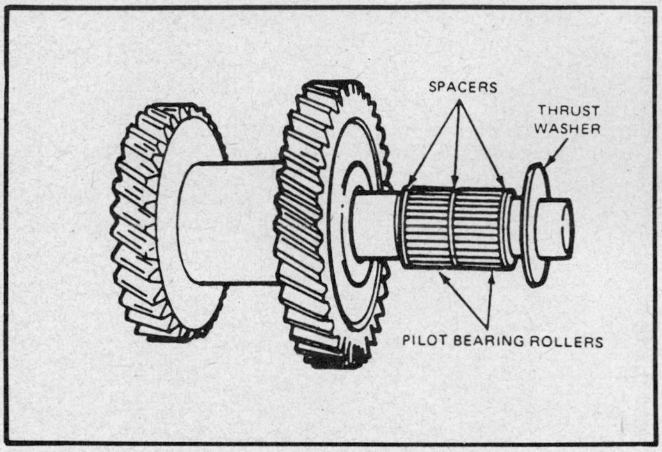
SPACERS
THRUST WASHER
PILOT BEARING ROLLERS
Countershaft gear roller bearing installation

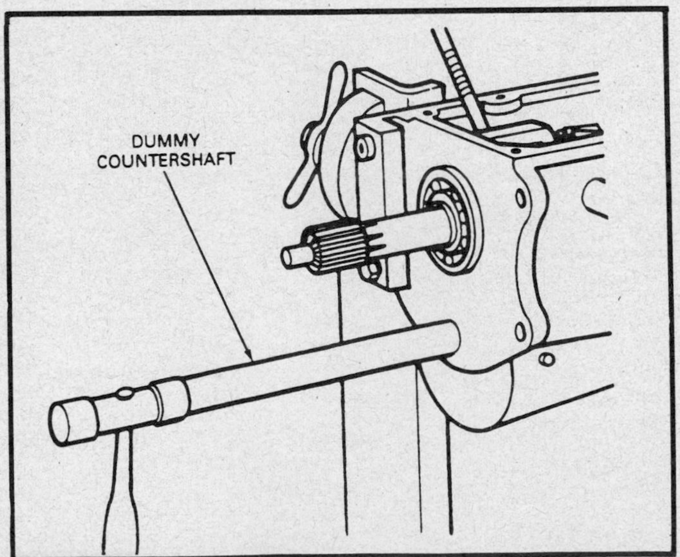
DUMMY COUNTERSHAFT
Dummy countershaft tool

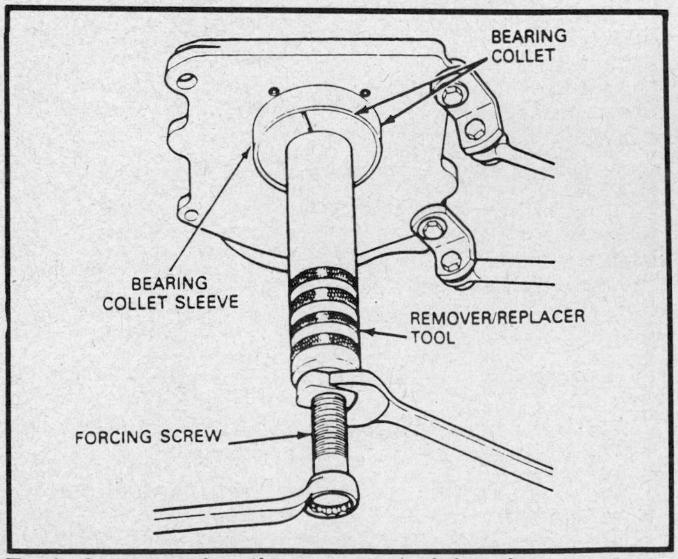
BEARING COLLET
BEARING COLLET SLEEVE
REMOVER/REPLACER TOOL
FORCING SCREW
Tools for removing the output shaft bearing

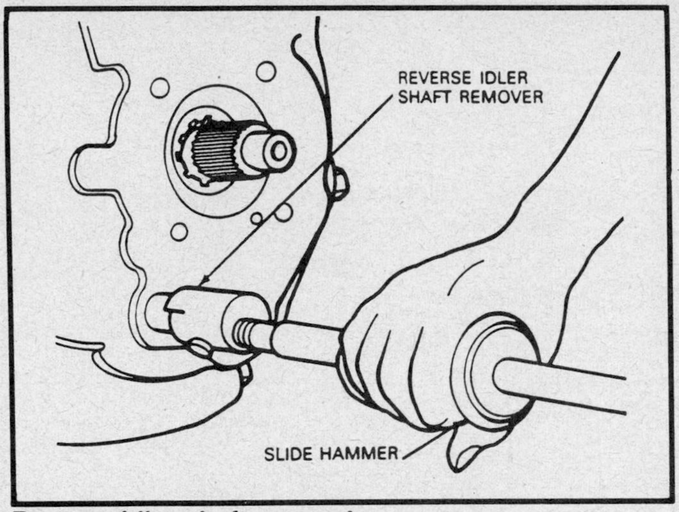
REVERSE IDLER
SHAFT REMOVER
SLIDE HAMMER
Reverse idler shaft removal

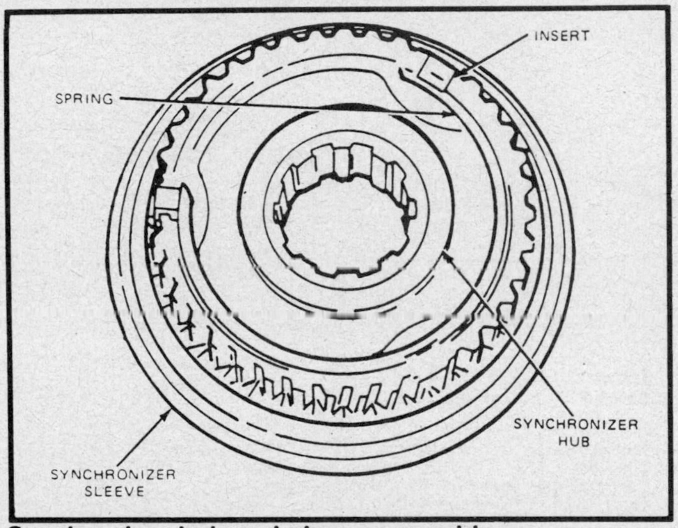
INSERT
SPRING
SYNCHRONIZER
HUB
SYNCHRONIZER
SLEEVE
Synchronizer hub and sleeve assembly

9. Install tools T75L–7025–B, F, H and T84T–7025–B on the output shaft and over the output shaft bearing. Remove the output shaft bearing.
10. Remove the input shaft bearing retainer and gasket. Remove the input shaft bearing snapring from the input shaft and the snapring from the bearing.
11. Install the tool on the input shaft and over the input shaft bearing. Remove the input shaft bearing.
12. Remove the spring clip from the reverse shifter arm pivot screw and remove the pivot screw from the shifter arm, and case. Lift the shifter arm and shoe assembly out of the case by prying with a suitable tool.
13. Remove the input shaft (with the flat facing upward) assembly from the case. Do not lose the 22 pilot bearing rollers from the inner end of the shaft. Remove the thrust spacer.
14. Remove the output shaft and gear assemblies from the case.
15. Use the dummy countershaft tool T83T–7111–B or equivalent to drive out the countershaft (from the front).

NOTE: Keep the dummy shaft in contact with the countershaft to avoid dropping the rollers. The dummy shaft should be 9.379–9.380 in. (23.82–23.83 cm) long and 1.133–1.135 in. (2.878–2.882 cm) in diameter.

Remove the countershaft cluster gear. Make sure the front and rear thrust washers are removed from the case. Be careful not to lose any rollers.
16. Remove the idler shaft (from the front of the transmission) using the tools T50T–100–A and T50T–7140–C or equivalent. Exercise caution so as not to damage the transmission while using the tool to remove the idler shaft.
17. Remove the reverse idler gear, and thrust washers, being careful not to lose any of the rollers.

Unit Disassembly and Assembly

INPUT SHAFT

Disassembly

1. Remove the thrust spacer and pilot bearing rollers from the gear bore.
2. Using a suitable puller, remove the input shaft ball bearing.

Inspection

Replace the input shaft if it is worn, bent, or twisted, if the gear has chipped, nicked, worn or missing teeth, or if the cone surface is damaged. If the pilot bearing bore is scored, replace the gear and gear rollers.

Assembly

1. Press the ball bearing onto the input shaft and against the gear.
2. Coat the bore of the gear with grease, and place the 22 pilot bearing rollers in the bore.
3. Install the thrust spacer in the bore against the rollers. Use grease to hold it in position.

OUTPUT SHAFT

Disassembly

1. Remove the snapring. Slide the 3rd and 4th speed synchronizer assembly, blocking ring and the 3rd speed gear off the shaft.
2. Place the output shaft in an arbor press and press the reverse gear off the shaft or pull the reverse gear off the shaft with 2-jaw puller tool D80L–1002–L or 2–3 jaw puller, D80L–1013–A.
3. Remove the 1st speed selective snapring and slide the 1st speed gear and blocker ring off the output shaft.
4. Remove the 1st and 2nd speed synchronizer snapring. Slide the synchronizer off the shaft.
5. Remove the snapring from the rear of the 2nd gear. Remove the blocking ring and synchronizer gear from the output shaft.

Inspection

1. Clean all components, except plastic or nylon, in solvent.
2. Inspect for broken, chipped or worn gear teeth.
3. Inspect for bent or broken inserts.
4. Check for damaged roller thrust or needle bearings and inspect the bearing bores for cracks or damage.
5. Check for worn or loss of surface metal from the countershaft and hub, clutch shaft or reverse idler gear shaft.
6. Check the thrust washers for damage and wear.
7. Check for nicked, broken or worn output or clutch shaft splines.
8. Check for bent, weak or distorted snaprings.
9. Check the reverse idler gear busing for wear.
10. Inspect front and rear bearings for roughness, wear and damage.

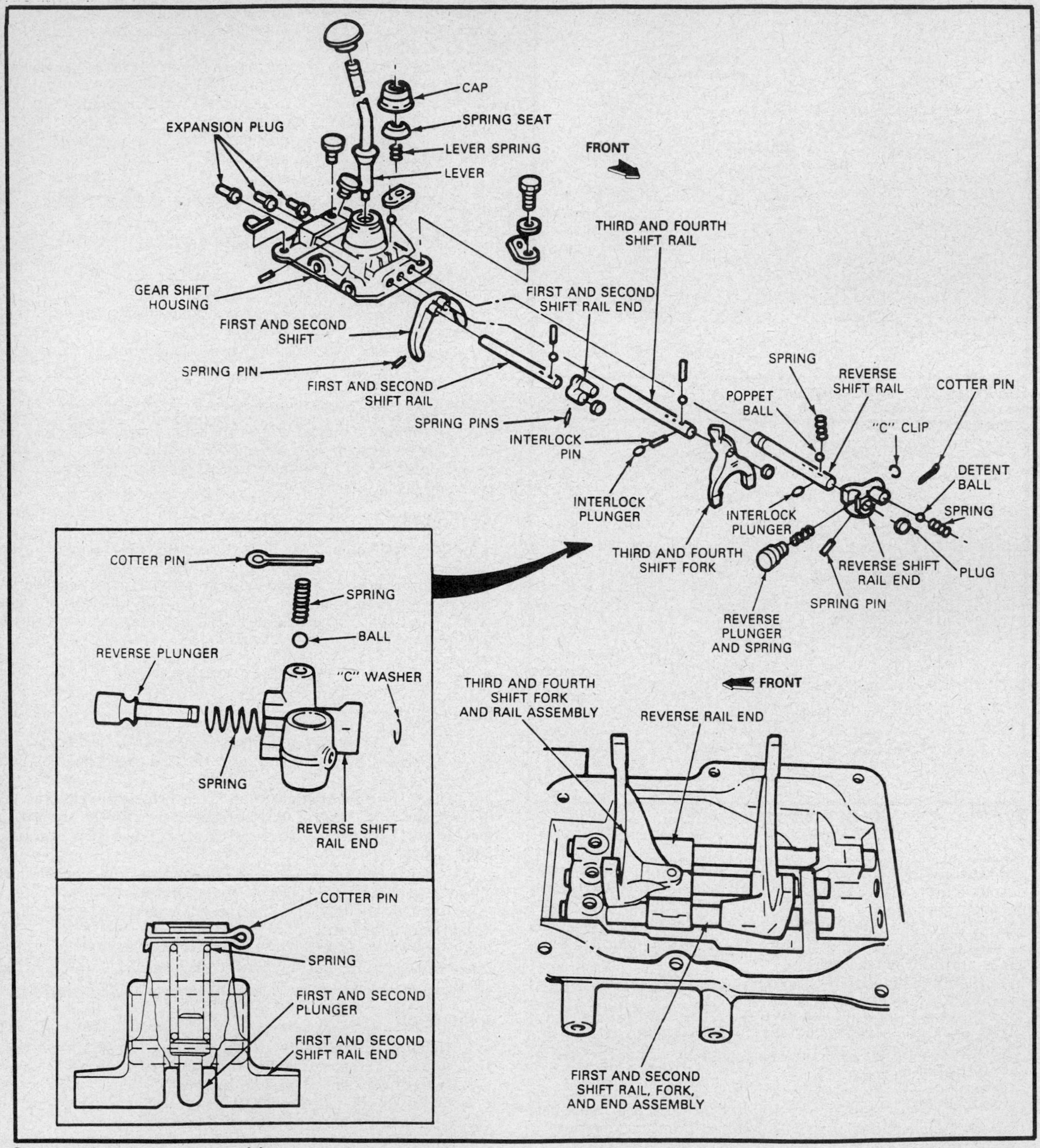
CAP
SPRING SEAT
EXPANSION PLUG
LEVER SPRING
FRONT
LEVER
THIRD AND FOURTH SHIFT RAIL
GEAR SHIFT HOUSING
FIRST AND SECOND SHIFT RAIL END
FIRST AND SECOND SHIFT
SPRING PIN
SPRING
REVERSE SHIFT RAIL
COTTER PIN
FIRST AND SECOND SHIFT RAIL
POPPET BALL
SPRING PINS
"C" CLIP
INTERLOCK PIN
DETENT BALL
SPRING
INTERLOCK PLUNGER
INTERLOCK PLUNGER
THIRD AND FOURTH SHIFT FORK
REVERSE SHIFT RAIL END
PLUG
SPRING PIN
REVERSE PLUNGER AND SPRING
COTTER PIN
SPRING
BALL
REVERSE PLUNGER
"C" WASHER
SPRING
REVERSE SHIFT RAIL END
COTTER PIN
SPRING
FIRST AND SECOND PLUNGER
FIRST AND SECOND SHIFT RAIL END
FRONT
THIRD AND FOURTH SHIFT FORK AND RAIL ASSEMBLY
REVERSE RAIL END
FIRST AND SECOND SHIFT RAIL, FORK, AND END ASSEMBLY

Gear shift housing assembly

11. Inspect the shift rails, selector arms, plates, interlock for worn, bent or damaged parts.
12. Inspect synchronizers damage, wear and rough surfaces.
13. If any of the above are found, replace the appropriate part(s).

Assembly

1. Hold the output shaft in a vertical position with the front of the shaft down, and slide the 2nd speed gear onto the shaft (gear cone toward rear).

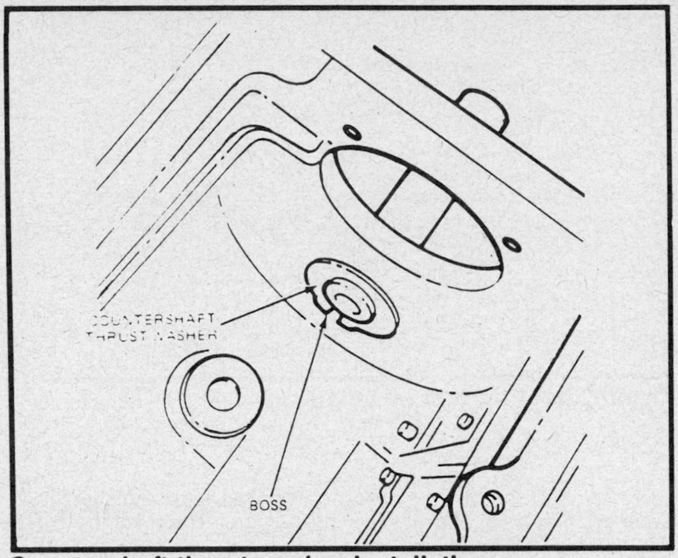
COUNTERSHAFT
THRUST WASHER
BOSS

Countershaft thrust washer installation

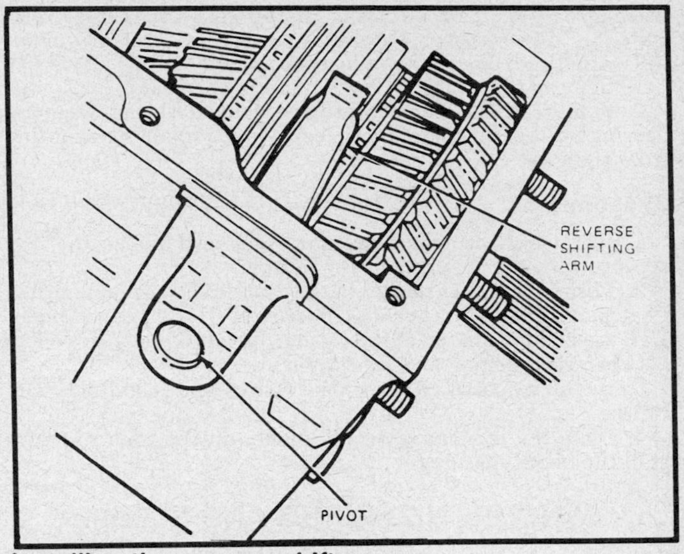
REVERSE
SHIFTING
ARM
PIVOT

Installing the reverse shifter arm

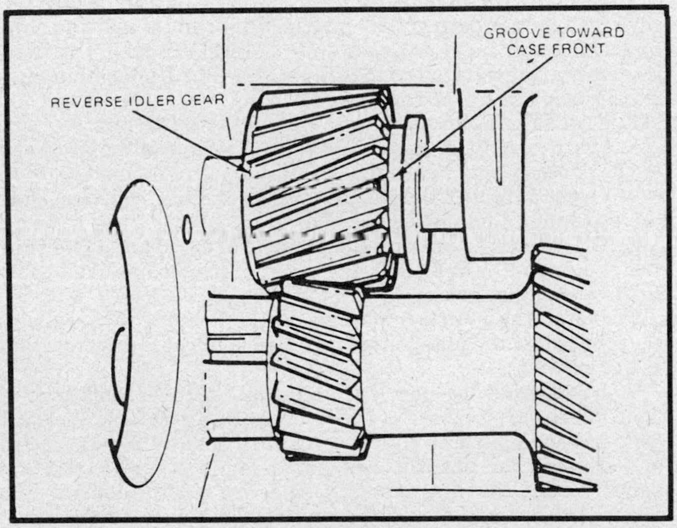
GROOVE TOWARD
CASE FRONT
REVERSE IDLER GEAR

Installing the reverse idler gear

2. Install the snapring 0.92–0.94 in. (2.34–2.39mm) thick on the output shaft at the rear of the 2nd-speed gear.
3. Place a blocking ring in the 1st and 2nd speed synchronizer assembly next to the side of the hub with the counterbore. Make sure that the ring slots are aligned with the insert.
4. Hold the blocking ring in the 1st and 2nd speed synchronizer assembly (with ring slots aligned with inserts) on the side with the hub counterbore must be toward the 2nd speed gear. Install the snapring. (Use thickest selective snapring that can be fit into the groove.)
5. Install the 2nd blocking ring in the synchronizer assembly, making sure that the ring slots are aligned with the inserts.
6. Install the 1st speed gear (coned portion toward the 1st and 2nd speed synchronizer assembly) and 0.101–0.103 in. (2.57–2.62mm) thick snapring on the shaft.
7. Mount the output shaft in a press and press the reverse gear (longer hub toward 1st speed gear) on the shaft. Press the rear bearing cone on the output shaft. If the output shaft bearing is to be replaced, be sure it is a maximum load-rated bearing.
8. Remove the output shaft from the press. Install the 3rd speed gear (cone toward front).
9. Place a blocking ring in the 3rd and 4th speed synchronizer assembly on the side with the larger hub diameter. Make sure that the slots are aligned with the inserts.
10. Hold the blocking ring in position, and slide the synchronizer assembly onto the output shaft. Blocking ring and large hub diameter must be toward the 3rd speed gear.
11. Install the thrust race on the shaft against the synchronizer hub. The flange must be toward the front. Install the thrust bearing against the race.
12. Apply grease to the face of the blocking ring and install in on the shaft and in the 3rd and 4th speed synchronizer assembly.

COUNTERSHAFT GEAR

Disassembly

Remove the dummy shaft, bearing rollers, bearing spacers and center spacer (sleeve) from the countershaft gear.

Assembly

1. Slide the long bearing spacer (sleeve) into the countershaft gear bore and insert the dummy shaft in the spacer.

NOTE: The dummy shaft should be coated liberally with grease.

2. Hold the gear in a vertical position and install one of the bearing spacers. Position the 22 pilot bearing rollers in the gear bore.
3. Place a spacer on top of the rollers and install 22 more rollers and another spacer.
4. Coat the face of the large thrust washer with grease, and hold a large thrust washer against the end of the countershaft gear to prevent the rollers from dropping out and turn the assembly over. Install the rollers, spacers and thrust washers in the other end of the gear.

REVERSE IDLER GEAR

Disassembly

Check the idler gear roller bearings for roughness by holding the bushing to prevent its turning while rotating the gear. The gear should then be installed on the shaft to check for roughness between the shaft and bushing. If the gear turns freely and smoothly, disassembly of the unit is not necessary. If any roughness is noticed, disassemble the unit as follows.

1. Remove the snapring from 1 end of the gear.
2. Remove the idler gear bearing rollers, thrust washers, bearing spacer, bushing (idler sleeve) and remaining snapring from the gear.

Assembly

1. Install a snapring in 1 end of the idler gear and set the gear on end, with the snapring at the bottom.
2. Position a thrust washer in the gear on top of the snapring. Coat the outside of the bushing (idler sleeve) with grease and install the bushing on top of the washer. Insert 37 bearing rollers between the bushing and the gear bore.
3. Install a spacer on top of the rollers and install 37 more rollers.
4. Place the remaining thrust washer on the rollers and install the other snapring.

SYNCHRONIZER HUB AND SLEEVE

Disassembly

1. Remove the spring from each side of the assembly, and remove the 3 inserts.
2. Slide the hub out of the sleeve.

Assembly

1ST AND 2ND SPEED SYNCHRONIZER

1. Install the 1st and 2nd speed clutch hub in the sleeve. The hub counterbore should be on the same side as the sleeve chamfer.
2. Place the 3 inserts in the hub slots.
3. Hook the end of a spring under an insert and position the spring around the hub and under each of the inserts.
4. Turn over the assembly and hook the end of the second spring over the other end of the insert used for hooking the first spring. Position the spring around the hub and under each insert, but in the opposite direction of the first spring.

3RD AND HIGH SPEED SYNCHRONIZER

Assembly is the same as that for the 1st and 2nd speed synchronizer with the following note: The 2 grooves on the chamfered portion of the clutch sleeve can be assembled in either direction on the hub.

1. Install the 1st and 2nd speed clutch hub in the sleeve. The hub counterbore should be on the same side as the sleeve chamfer.
2. Place the 3 inserts in the hub slots.
3. Hook the end of a spring under an insert and position the spring around the hub and under each of the inserts.
4. Turn over the assembly and hook the end of the second spring over the other end of the insert used for hooking the first spring. Position the spring around the hub and under each insert, but in the opposite direction of the first spring.

GEAR SHIFT HOUSING

Disassembly

1. Remove the gear shift lever housing cap and lift the lever out of the housing. Be sure all shafts are in neutral before disassembly.
2. Remove the spring pins from the shift forks and shift rail ends. Remove the expansion plugs from the ends of the housing.
3. Tap the shift rails out of the housing while holding one hand over the holes in the housing to prevent loss of the poppet springs and balls. Remove the shift rail ends and forks. Lift the 2 shaft interlock plungers and the pin out of the housing.

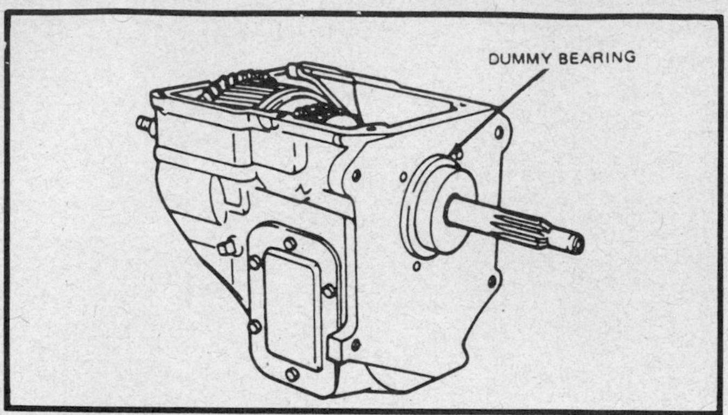
DUMMY BEARING

Dummy bearing tool installed

4. To disassemble the reverse shift rail end, remove the C-washer to release the plunger and spring. Remove the cotter pin, spring and ball.

Assembly

1. Position the notched end of the 1st and 2nd shift rail through the rear bore of the housing. Slide the 1st and 2nd shift fork (use outer hole) onto the shift rail. Do not slide the shaft into its bore at the front end at this time. The 3 poppet notches should face the top of the housing.
2. Slide the 1st and 2nd shift rail end into the rail.
3. Place the poppet spring and ball in the hole at the front end of the cover.
4. Use a suitable tool to depress the ball and spring and slide the rail into its bore over the ball.
5. Drive a spring pin through the hole in the 1st and 2nd shift rail end and into the hole in the rail. Secure the shift rail to its neutral position (center poppet).
6. Install the interlock plunger in the housing making sure that the end of the plunger is in the side notch of the 1st and 2nd shift rail.
7. Install the 3rd and 4th shift rail (notched end toward front) in the center bore of the housing, and assemble the shift fork, interlock pin, and poppet spring and ball. Note that the 3rd and 4th shift rail passes through a 2nd hole in the 1st and 2nd shift fork and that the poppet notches are toward housing top. Secure the 3rd and 4th shift fork to the rail with a spring pin. Slide the rail to neutral position (center poppet).
8. Install the interlock plunger and make sure it is positioned in the notch in the 3rd and 4th shift rail.
9. Assemble the reverse plunger and spring in the reverse shift rail end. Retain with a C-washer inserted in the plunger groove. Assemble the ball, spring and cotter pin in the shift rail end.
10. Install the reverse shift rail (notched end toward front), reverse shift rail end, and poppet spring and ball in the housing.

Transmission Assembly

1. Coat all parts, especially the bearings, with the specified transmission lubricant to prevent scoring when the transmission is first operated.
2. Start the countershaft (small end first) into its bore at the rear of the case. Insert the shaft just enough to position the rear countershaft steel thrust washer on the shaft and against the case. Apply grease to the washer to hold it in position. Using the reverse idler shaft as a temporary holding tool, insert the small end of the shaft into the front countershaft bore just enough to hold the front countershaft steel thrust washer in position. Install the thrust washer.

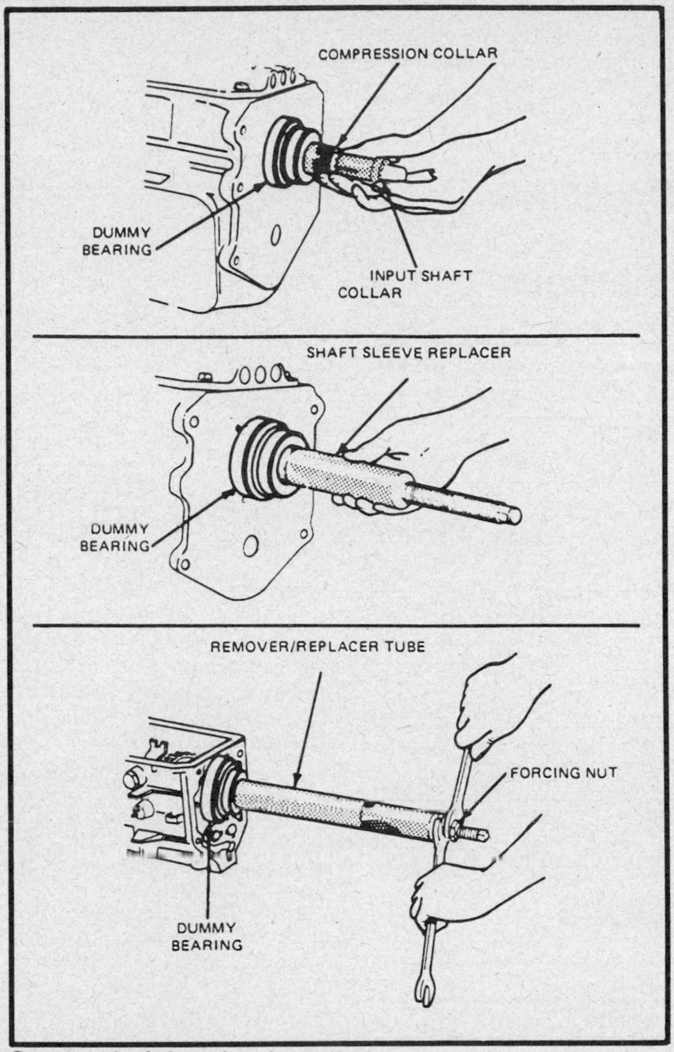
COMPRESSION COLLAR
DUMMY BEARING
INPUT SHAFT COLLAR
SHAFT SLEEVE REPLACER
DUMMY BEARING
REMOVER/REPLACER TUBE
FORCING NUT
DUMMY BEARING

Output shaft bearing installation

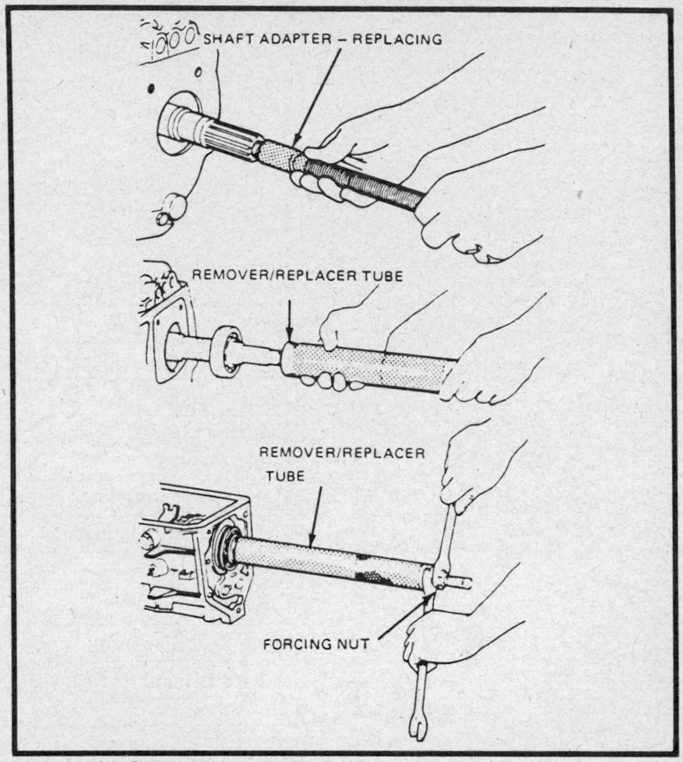
SHAFT ADAPTER – REPLACING
REMOVER/REPLACER TUBE
REMOVER/REPLACER TUBE
FORCING NUT

Input shaft bearing installation

NOTE: Make sure that the notch in the thrust washers are aligned with the boss at each end of the case.

3. Position the countershaft cluster gear assembly in the case. Do not lose any rollers.
4. Slide out the reverse idler shaft and countershaft gear dummy shaft by installing the countershaft from the rear. Keep shaft ends in contact so that rollers cannot drop out of position. Do not drive the countershaft completely into the press fit at the rear of the case at this time.
5. Position the reverse idler gear assembly in the case and install the idler (small end of the shaft toward the front). Shift fork groove of the gear should be toward the front of the case. Do not drive the shaft completely into position.
6. Make sure that the countershaft and reverse idler gear shaft are properly aligned so that the retainer can be positioned in the shaft slots. Drive the shafts into position in the case. Install the retainer and bolt.
7. Install the output shaft assembly in the case.
8. With the output shaft shifted slightly to the right, position the reverse shifter arm and shoe assembly on the reverse idler gear. Install the pivot screw through the hole in the left side of the case and into the shifter arm hole. Install the spring clip to retain the shifter arm to the pivot screw. Center the output shaft to the case bore.
9. Install the input shaft through the front bore with the flats on the shaft facing upward. When past the countershaft, turn the input shaft so the flats are facing downward. Guide the input shaft onto the output shaft.
10. Install a dummy bearing tool T75L–7025–Q or equivalent on the transmission input shaft. This tool is necessary to keep the input and output shafts in alignment when installing the output shaft bearing.
11. Assemble the locating snapring to the outer race of the output shaft bearing in the groove provided.
12. Install the output shaft bearing using the tools T75L–7025–B, L or equivalent.
13. Install the spacer washer against the rearward face of the input shaft bearing. (The properly installed washer will be external to the main body of the transmission.)
14. Install a snapring at the rearward surface of the washer in the output shaft groove provided.
15. Remove the dummy bearing from the input shaft.
16. Install the input shaft bearing using the Tools T75L–7025B, K, Q, R, S and install the snapring. Use the thickest select fit snaprings which will fit on the bearing.
17. Install the input shaft bearing thrust spacer and retainer gasket and retainer. Tighten the bolts to 15–25 ft. lbs. (21–33 Nm).
18. Position the speedometer drive gear (and spacer, if used) on the output shaft over lock ball and install the speedometer drive gear retaining snapring.
19. Using a new gasket install the output shaft bearing retainer (or extension housing). Tighten the bolts to 34–45 ft. lbs. (48–61 Nm).
20. Lubricate the retainer, bushing and seal and the u-joint flange with multi-purpose lubricant.
21. Install the U-joint flange. Lock the transmission in 2 gears and tighten the retaining nut to 75–115 ft. lbs. (102–155 Nm).
22. Install the gear shift housing assembly (with housing assembly and unit shifted into 2nd gear) on the transmission and tighten the cover bolts to 25–35 ft. lbs. (34–47 Nm).

23. Fill the transmission to the proper level with standard transmission lubricant D8DZ–19C547–A (SAE 80W) or equivalent. Add ½ pint of lubricant through the speedometer cable hole in the rear transmission extension housing.

SPECIFICATIONS

TORQUE SPECIFICATIONS

Item	ft. lbs.	Nm
Transmission to flywheel housing	37–42	51–56
Gear shift housing to case	25–35	34–47
Speedometer cable retainer to output shaft bearing retainer	3–4.5	4.5–6
Output shaft bearing retainer to case	34–45	48–61
Flywheel housing to engine	40–50	55–67
Filler plug	25–40	34–54
Drain plug	25–40	34–54
Output shaft flange nut	75–115	102–155
Countershaft and reverse idler shaft retainer bolt	25–35	34–47
Power take off cover	25–35	34–47
Input shaft bearing retainer	15–25	21–33

SPECIAL TOOLS

Number	Description
T50T-100-A	Impact slide hammer
T57L-500-B	Bench mounted holding fixture
D80L-1002-A	Two-jaw puller
D80L-1013-A	Two-three jaw puller
T75L-7025-B	Remover replacer tube
T75L-7025-F	Bearing collet
T75L-7025-H	Bearing collet sleeve
T84T-7025-B	Forcing screw
T75L-7025-K	Shaft sleeve replacer
T75L-7025-L	Threaded shaft adapter replacer
T75L-7025-Q	Dummy bearing
T75L-7025-R	Compression collar
T75L-7025-S	Input shaft collar
T83T-7111-B	Dummy countershaft
T83T-7137-A	Clutch aligner
T50T-7140-C	Reverse idler shaft remover

Section 7

NP-2500 Transmission
New Process

APPLICATION

NP2500 TRANSMISSION

Year	Vehicle	Engine	VIN Code
1987	Dakota	2.2L, 3.9L	C, M
1988	Dakota	2.2L, 3.9L	C, M
	D-100	3.9L, 5.2L, 5.9L	X, Y, W
	D-150	3.9L, 5.2L, 5.9L	X, Y, W
1989	Dakota	2.5L, 3.9L	K, X
	D-100	3.9L	X
	D-150	3.9L	X
	D-250	5.9L	8
	D-350	5.9L	8
	W-350	5.9L	8

GENERAL DESCRIPTION

The NP2500 5 speed manual transmission was designed and built by Chrysler Motors. The transmission is fully synchronized with an overdrive 5th gear. The single-unit top-mounted shifter includes a 3-rail shift system. The reverse gear is also in constant mesh and is located directly behind the overdrive in the shift pattern. A reverse blocker has been incorporated to prevents accidental shifts from overdrive to reverse.

Transmission Identification

An identification plate is affixed to the left side of the transmission above the fill plug. This plate provides model and vehicle identification numbers.

Metric Fasteners

The metric fastener dimensions are very close to the dimensions of the familiar inch system fasteners. For this reason, replacement fasteners must have the same measurement and strength as those removed.

CAUTION

Do not attempt to interchange metric fasteners for inch system fasteners. Mismatched or incorrect fasteners can result in damage to the transmission unit through malfunctions, breakage or possible personal injury.

Care should be taken to re-use the fasteners in the same locations as removed.

Common metric fastener strength property classes are 9.8 and 12.9 with class identification embossed on the head of each bolt. The inch strength classes range from grade 2 to 8 with the line identification embossed on each bolt head. Markings correspond to 2 lines less than actual grade (for example grade 8 bolt will exhibit 6 embossed lines on the bolt head). Some metric nuts will be marked with a single digit strength identification numbers on the nut face.

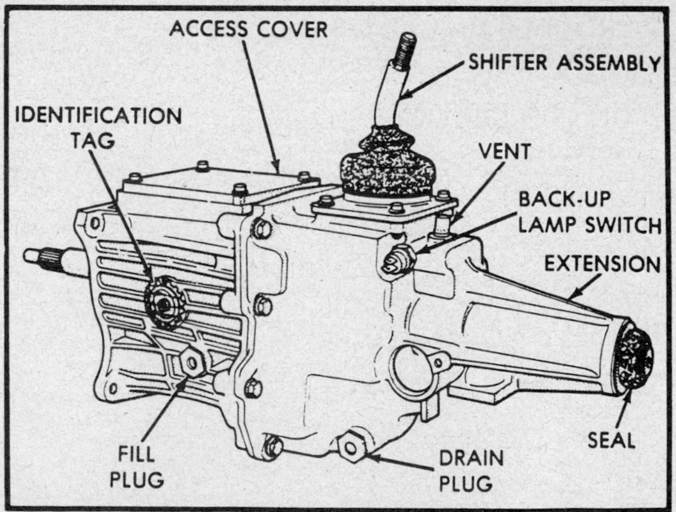
ACCESS COVER
SHIFTER ASSEMBLY
IDENTIFICATION TAG
VENT
BACK-UP LAMP SWITCH
EXTENSION
SEAL
FILL PLUG
DRAIN PLUG
NP2500 5 speed manual transmission

Capacities

The NP2500 transmission is filled with SAE 10W-30 engine oil at the factory and should be used for transmission fluid changes. The fluid capacity of this unit is 2 qts.

Checking Fluid Level

Check the fluid level in the transmission by removing the fill plug located on the left side under the identification tag. The fluid level should be at the bottom of the fill opening.

TRANSMISSION MODIFICATIONS

Clutch Housing Mis-Installation

1988 Ram Wagon/Van, Ram Charger 4WD, Ram Pick-Up 2WD, Ram Chassis Cab and Dakota vehicles have experienced excessive spin time, clutch chatter, failure of transmission drive pinions and pilot bushing, and/or noise and vibration complaints. These symptoms are attributed to mis-installation of the clutch housing to the strut block. If any of these conditions exist, the clutch housing and struts must be loosened, repositioned and tightened by performing the following procedure:

1. Loosen the clutch housing-to-engine block fasteners.
2. Install the clutch housing to the engine block and torque all housing-to-block fasteners to the following values: 3/8 in. bolts to 30 ft. lbs. (41 Nm), 7/16 in. bolts to 50 ft. lbs. (68 Nm) and 12mm bolts to 70 ft. lbs. (96 Nm).
3. Torque the strut-to-housing and strut-to-clutch fasteners as described in Step 2.

POWER FLOW

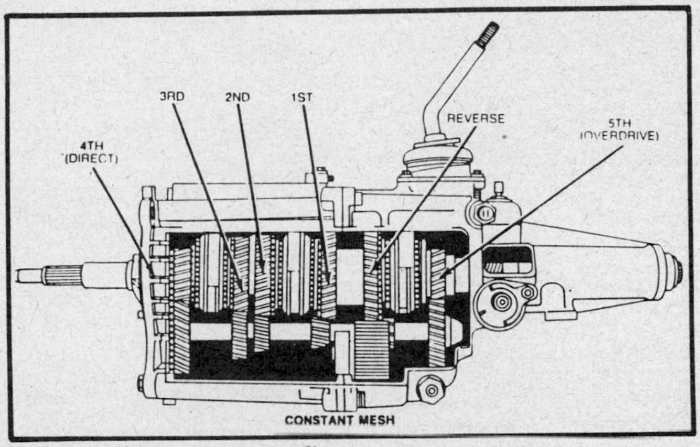
3RD
2ND
1ST
REVERSE
5TH (OVERDRIVE)
4TH (DIRECT)
CONSTANT MESH

Constant mesh power flow

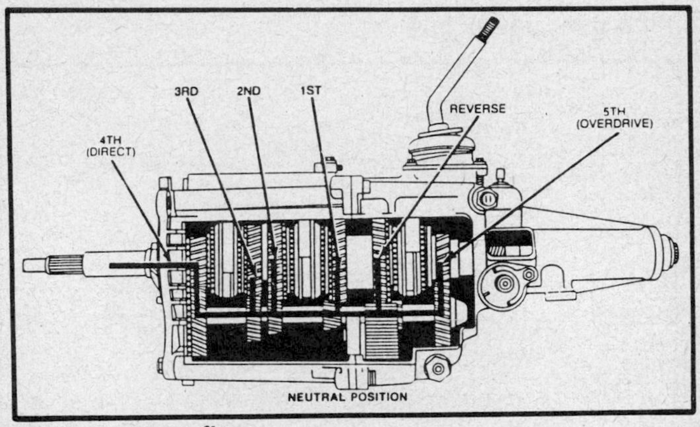
3RD
2ND
1ST
REVERSE
5TH (OVERDRIVE)
4TH (DIRECT)
NEUTRAL POSITION

Neutral power flow

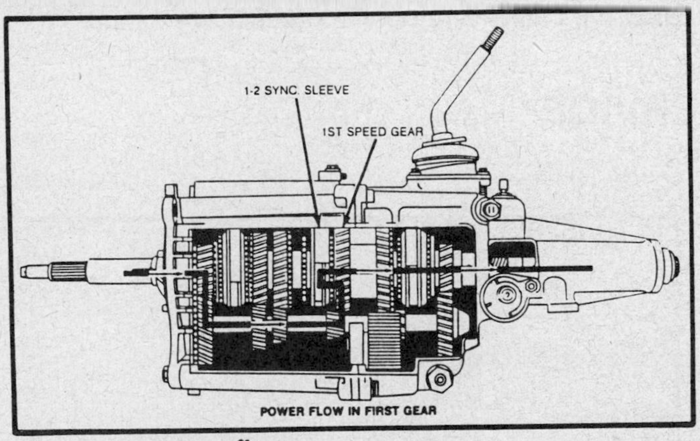
1-2 SYNC. SLEEVE
1ST SPEED GEAR
POWER FLOW IN FIRST GEAR

First gear power flow

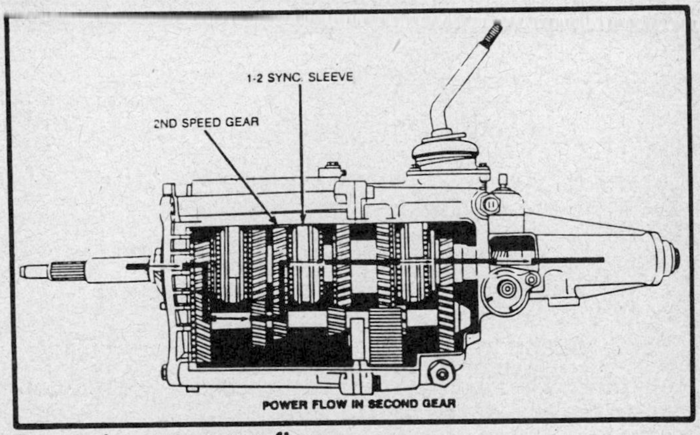
1-2 SYNC. SLEEVE
2ND SPEED GEAR
POWER FLOW IN SECOND GEAR

Second gear power flow

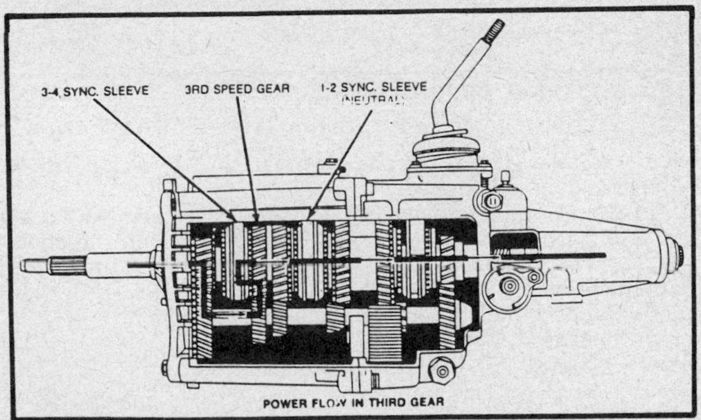
3-4 SYNC. SLEEVE
3RD SPEED GEAR
1-2 SYNC. SLEEVE
POWER FLOW IN THIRD GEAR

Third gear power flow

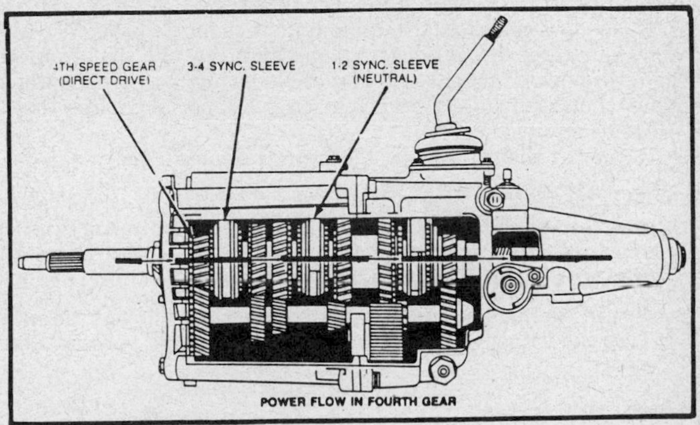
SPEED GEAR (DIRECT DRIVE)
3-4 SYNC. SLEEVE
1-2 SYNC. SLEEVE (NEUTRAL)
POWER FLOW IN FOURTH GEAR

Fourth gear power flow

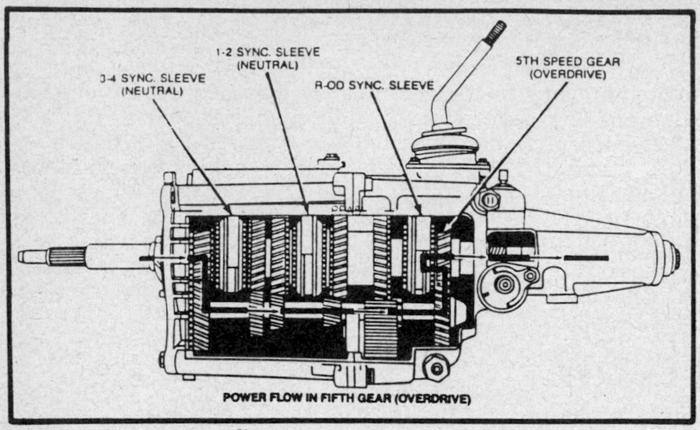
1-2 SYNC. SLEEVE
(NEUTRAL)
3-4 SYNC. SLEEVE
(NEUTRAL)
R-OD SYNC. SLEEVE
5TH SPEED GEAR
(OVERDRIVE)
POWER FLOW IN FIFTH GEAR (OVERDRIVE)

Fifth gear power flow

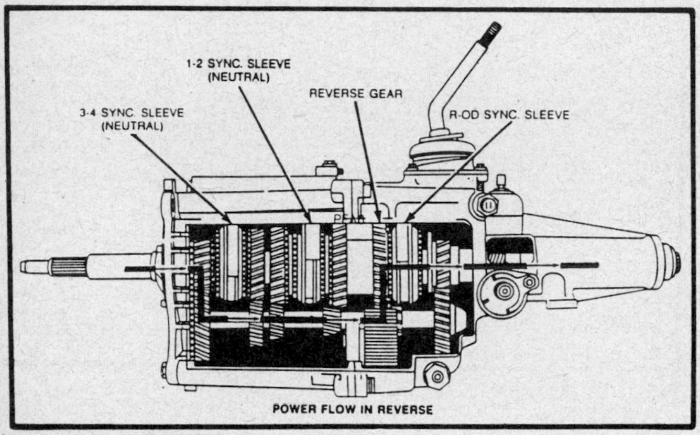
1-2 SYNC. SLEEVE
(NEUTRAL)
REVERSE GEAR
3-4 SYNC. SLEEVE
(NEUTRAL)
R-OD SYNC. SLEEVE
POWER FLOW IN REVERSE

Reverse power flow

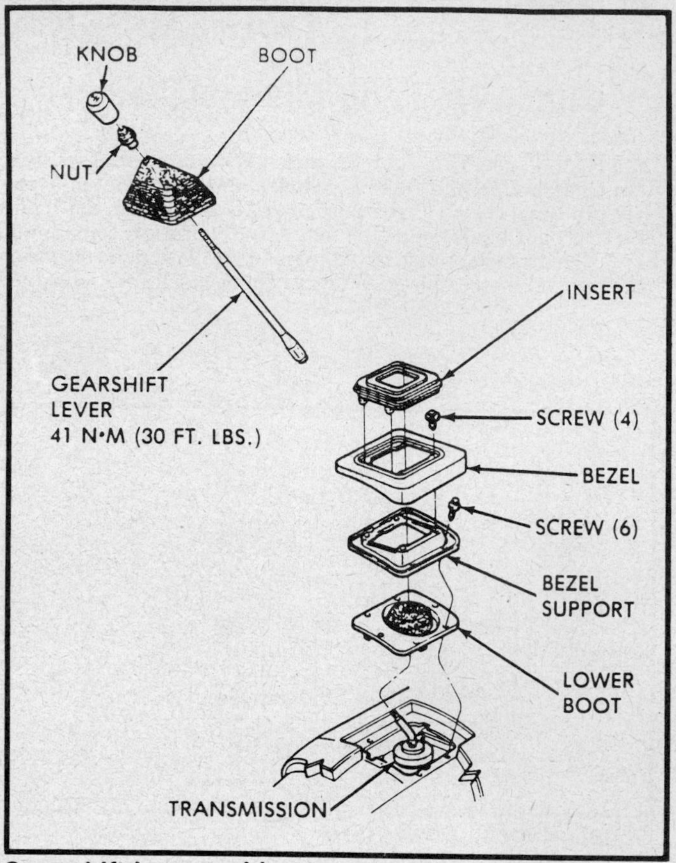
KNOB
BOOT
NUT
INSERT
GEARSHIFT
LEVER
41 N•M (30 FT. LBS.)
SCREW (4)
BEZEL
SCREW (6)
BEZEL
SUPPORT
LOWER
BOOT
TRANSMISSION

Gear shift lever and boot components – Dakota

ON CAR SERVICE

Services

SPEEDOMETER PINION GEAR

Removal and Installation

NOTE: The rear axle gear ratio and the tire size determine the size of the pinion gear.

1. Remove the bolt and retainer attaching the speedometer pinion to the extension housing.
2. If oil is found in the cable housing, replace the oil seal as follows: Remove the lockring and pry the old seal from the adapter bore. Start the seal and the lockring into the adapter by hand then seat the seal and lockring with special adpater tool C–4004 or equivalent.
3. Count the number of teeth on the pinion gear.

NOTE: Before the adapter and pinion are installed, the adapter flange and the extension housing mating surafaces must be cleaned thoroughly. Dirt or sand will cause mis-alignment of the gear and cause damage.

4. Rotate the speedometer pinion gear and adapter assembly so that the identification number on the adapter, corresponding to the number of teeth on the gear, is in the 6 o'clock position as the assembly is installed.

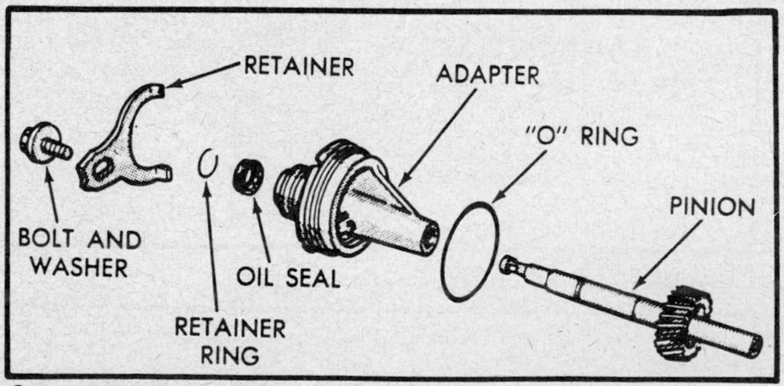
RETAINER
ADAPTER
"O" RING
PINION
BOLT AND
WASHER
OIL SEAL
RETAINER
RING

Speedometer pinion and adapter

5. Install the retainer so that the tangs are aligned with the adapter positioning slots. Install the bolt and tap the adapter firmly into the extension housing. Torque the retainer bolt to 100 inch lbs.
6. Connect the speedometer cable to the adapter.
7. Fill the transmission to the proper level.

REMOVAL AND INSTALLATION

TRANSMISSION REMOVAL

1. Raise and support the vehicle safely.
2. Remove the skid plate, if installed. Drain lubricant from the transmission and transfer case, if equipped.
3. Disconnect the speedometer cable.
4. Disconnect and matchmark the front and rear driveshafts as required. Suspend each shaft from a convenient place; do not allow them to hang free.
5. Disconnect the shift rods at the transfer case, if equipped.
6. Remove the rear driveshaft. Matchmark the driveshaft and rear U-joints before removing the driveshaft.
7. Support the transfer case properly, if equipped and remove the extension-to-transfer case mounting bolts.
8. Move the transfer case rearward to disengage the front input shaft spline.
9. Lower and remove the transfer case as required.
10. Disconnect the back-up light switch electrical connector.
11. Support the engine safely.
12. Support the transmission safely.
13. Remove the transmission crossmember.
14. Remove the transmission-to-clutch housing bolts.
15. Slide the transmission rearward until the mainshaft clears the clutch disc.
16. Position the transmission on a suitable holding fixture.

TRANSMISSION INSTALLATION

NOTE: The transmission pilot bushing in the end of the crankshaft requires high-temperature grease. Multipurpose grease should be used. Do not lubricate the end of the mainshaft, clutch splines, or clutch release levers.

1. Remove the transmission from the holding fixture and raise into position under the vehicle.
2. Slide the transmission forward until the mainshaft engages the clutch disc, then push it all the way forward.
3. Install the transmission-to-clutch housing bolts. Torque the bolts to 50 ft. lbs. (68 Nm).
4. Install the and bolt the transmission crossmember to the frame.
5. Connect the back-up light switch.
6. Raise the transfer case into position and support safely.
7. Move the transfer case rearward to disengage the front input shaft spline, if equipped.
8. Install the extension-to-transfer case mounting bolts, if equipped. Torque the bolts to 40 ft. lbs. (54 Nm).
9. Install the rear driveshaft.
10. Connect the shift rods at the transfer case and adjust the linkage.
11. Connect the front driveshafts, as required.
12. Connect the speedometer cable.
13. Install the skid plate, if removed. Fill the transmission and transfer case to the proper level.
14. Lower the vehicle.

BENCH OVERHAUL

Before Disassembly

Clean the transmission case with a suitable solvent and dry with compressed air. Inspect the case for cracks and stripped threads in the various bolt holes. Check the machined mating surfaces for burrs, nicks or any other surface imperfections that would make the transmission case unfit for service in the vehicle. The front mating surface should be smooth. Any minor burrs may be dressed with a fine mill file. Repair damaged threaded holes with an appropriate insert or by tapping and re-threading the hole.

Transmision Disassembly

1. Position the transmission on a suitable holding fixture.
2. Remove the shifter assembly bolts.
3. Pry the shifter assembly loose. Clean all RTV sealant from both mating surfaces.
4. Remove the access cover bolts.
5. Pry the access cover loose. Clean all RTV sealant from both mating surfaces.
6. Remove the detent springs and bullets with a magnet. Note where each spring and bullet goes.
7. Remove the extension housing-to-case bolts.
8. Pry the extension housing loose. When pulling off the housing, hold the reverse/overdrive shift rail stationary. Clean all RTV sealant from both mating surfaces.
9. Remove the reverse/overdrive shift rail and blocker spring.
10. Remove the tapered pins from the shift forks.
11. Remove the 1st/2nd shift rail and retrieve the interlock plates.
12. Remove the countershaft overdrive gear snapring.

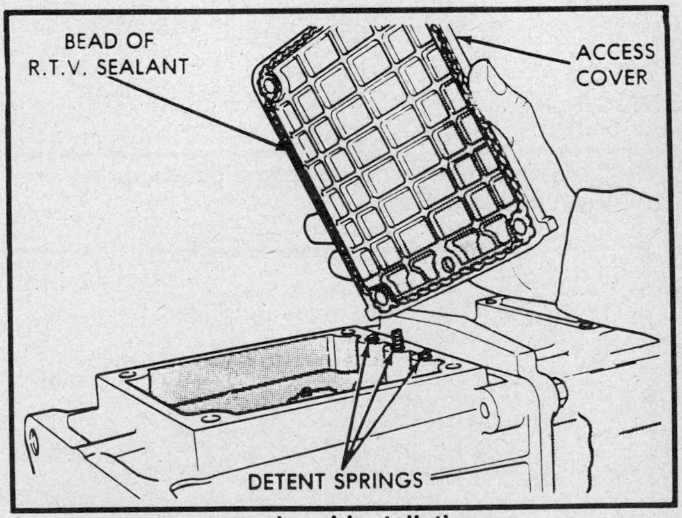
BEAD OF
R.T.V. SEALANT
ACCESS
COVER
DETENT SPRINGS
Access cover removal and installation

13. Remove the countershaft overdrive gear using special puller tool C-4982 or equivalent.
14. Remove the mainshaft overdrive gear snapring.
15. Remove the mainshaft overdrive gear thrust washer and anti-spin pin.
16. Remove the mainshaft overdrive gear.
17. Remove the reverse/overdrive hub snapring.
18. Remove the reverse/overdrive synchronizer, fork, and rail assembly.

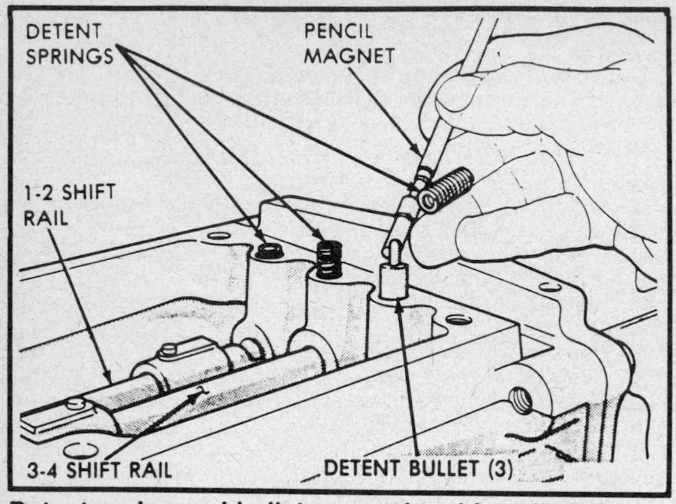
DETENT SPRINGS
PENCIL MAGNET
1-2 SHIFT RAIL
3-4 SHIFT RAIL
DETENT BULLET (3)

Detent spring and bullet removal and installation

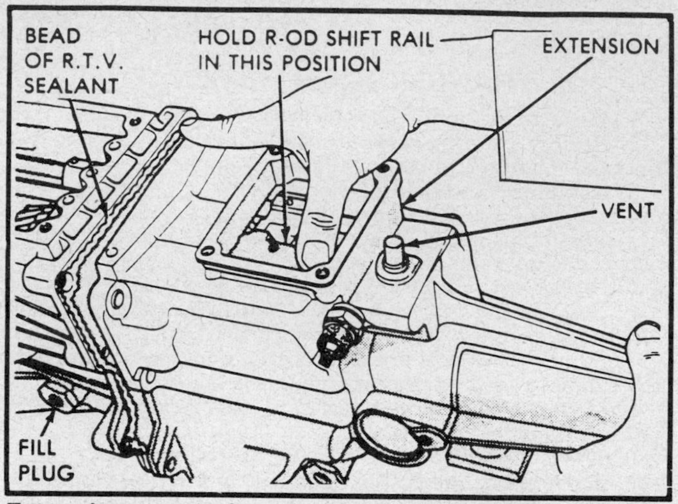
BEAD OF R.T.V. SEALANT
HOLD R-OD SHIFT RAIL IN THIS POSITION
EXTENSION
VENT
FILL PLUG

Extension removal and installation. To remove, hold reverse-overdrive shift rail as shown

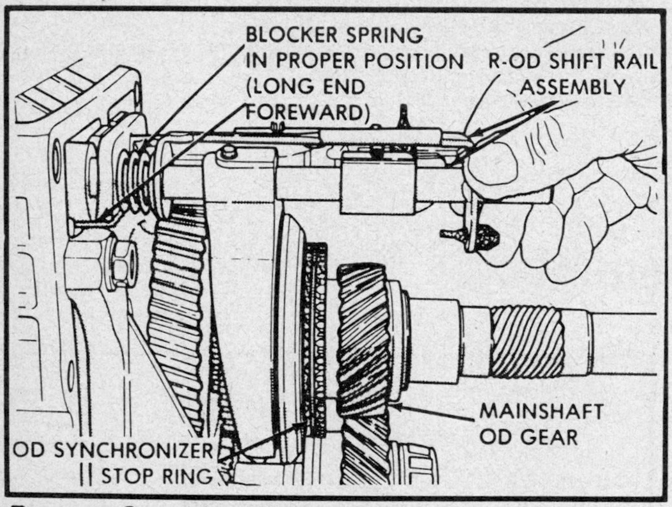
BLOCKER SPRING IN PROPER POSITION (LONG END FOREWARD)
R-OD SHIFT RAIL ASSEMBLY
MAINSHAFT OD GEAR
OD SYNCHRONIZER STOP RING

Reverse-Overdrive shift rail and blocker spring removal and installation

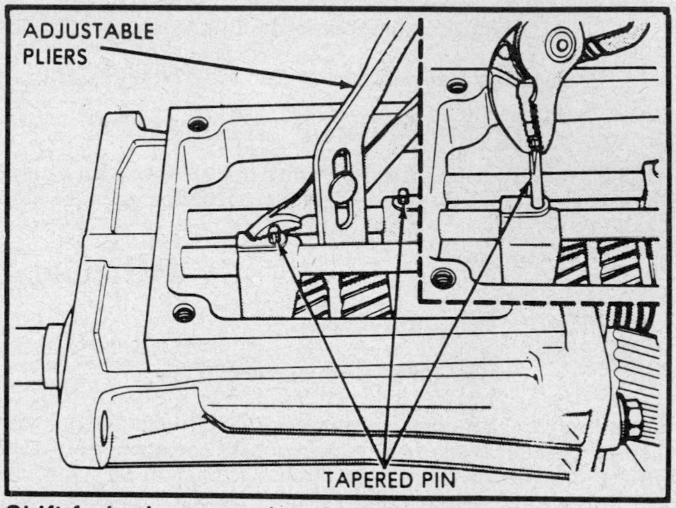
ADJUSTABLE PLIERS
TAPERED PIN

Shift fork pin removal and installation

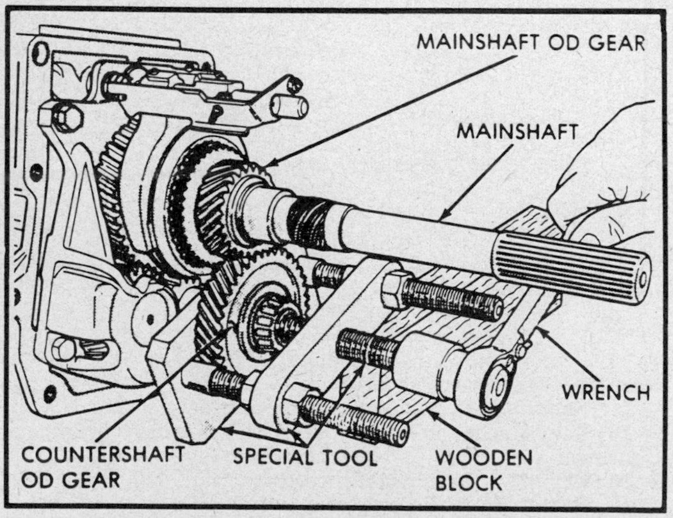
MAINSHAFT OD GEAR
MAINSHAFT
WRENCH
COUNTERSHAFT OD GEAR
SPECIAL TOOL
WOODEN BLOCK

Removal of countershaft overdrive with the use of special puller tool

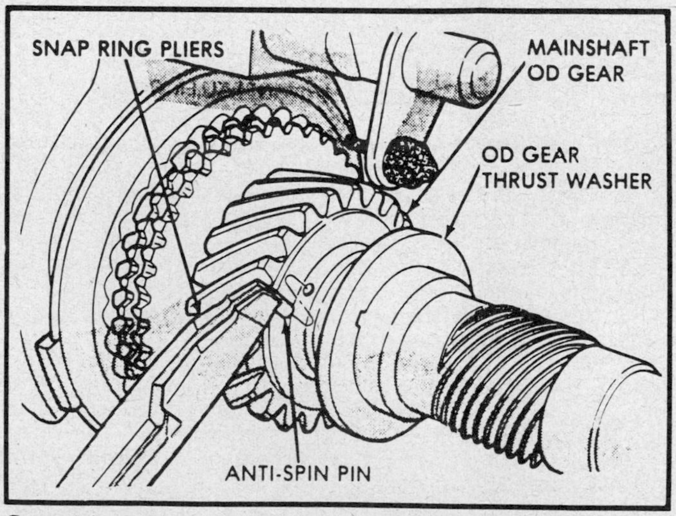
SNAP RING PLIERS
MAINSHAFT OD GEAR
OD GEAR THRUST WASHER
ANTI-SPIN PIN

Overdrive gear thrust washer and anti-spin pin removal and installation

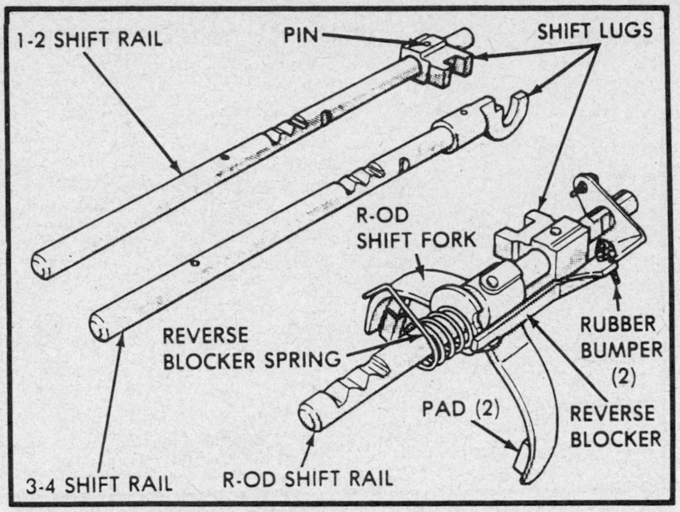
1-2 SHIFT RAIL
PIN
SHIFT LUGS
R-OD
SHIFT FORK
REVERSE
BLOCKER SPRING
RUBBER
BUMPER
(2)
PAD (2)
REVERSE
BLOCKER
3-4 SHIFT RAIL
R-OD SHIFT RAIL

Shift rail components

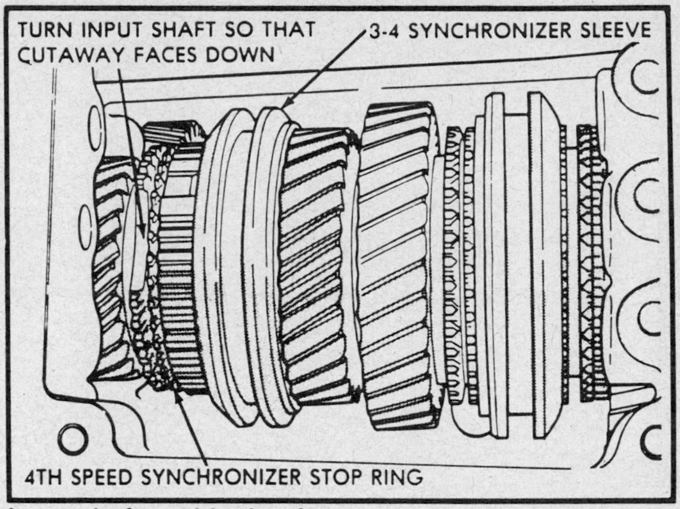
TURN INPUT SHAFT SO THAT
CUTAWAY FACES DOWN
3-4 SYNCHRONIZER SLEEVE
4TH SPEED SYNCHRONIZER STOP RING

Input shaft positioning for removal and installation

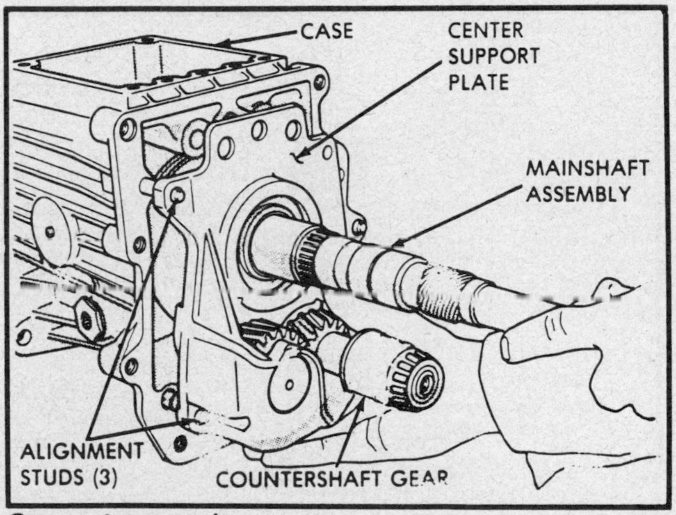
CASE
CENTER
SUPPORT
PLATE
MAINSHAFT
ASSEMBLY
ALIGNMENT
STUDS (3)
COUNTERSHAFT GEAR

Gear set removal

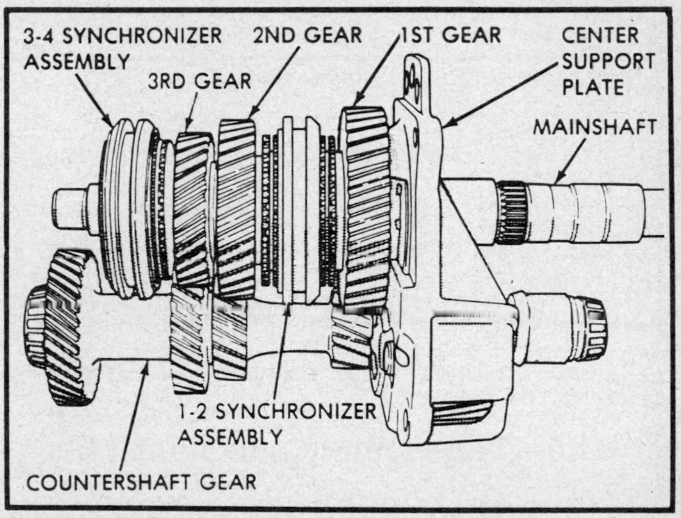
3-4 SYNCHRONIZER
ASSEMBLY
3RD GEAR
2ND GEAR
1ST GEAR
CENTER
SUPPORT
PLATE
MAINSHAFT
1-2 SYNCHRONIZER
ASSEMBLY
COUNTERSHAFT GEAR

Countershaft gear removal and installation

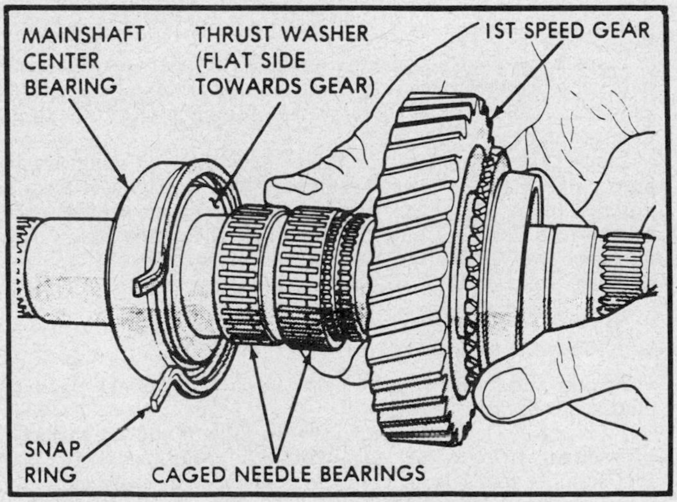
MAINSHAFT
CENTER
BEARING
THRUST WASHER
(FLAT SIDE
TOWARDS GEAR)
1ST SPEED GEAR
SNAP
RING
CAGED NEEDLE BEARINGS

1st speed gear removal and installation

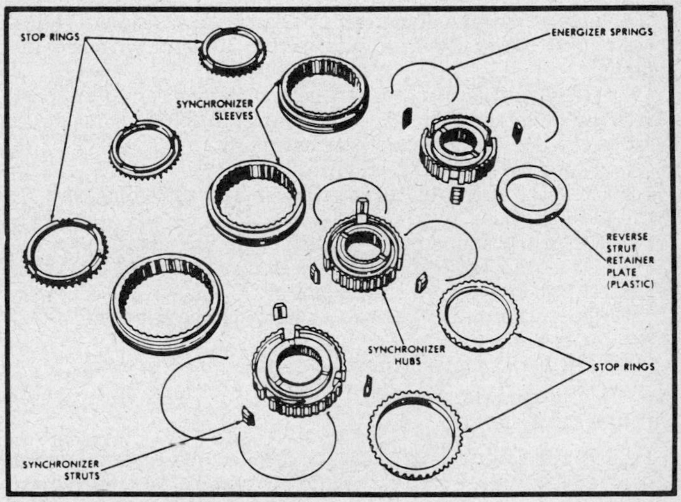
STOP RINGS
ENERGIZER SPRINGS
SYNCHRONIZER
SLEEVES
REVERSE
STRUT
RETAINER
PLATE
(PLASTIC)
SYNCHRONIZER
HUBS
STOP RINGS
SYNCHRONIZER
STRUTS

Synchronizer components

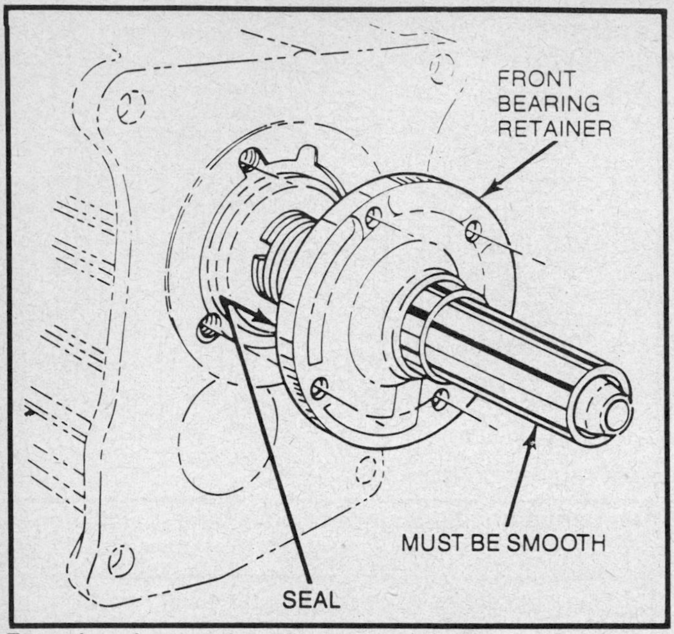
FRONT
BEARING
RETAINER
MUST BE SMOOTH
SEAL
Front bearing retainer inspection points

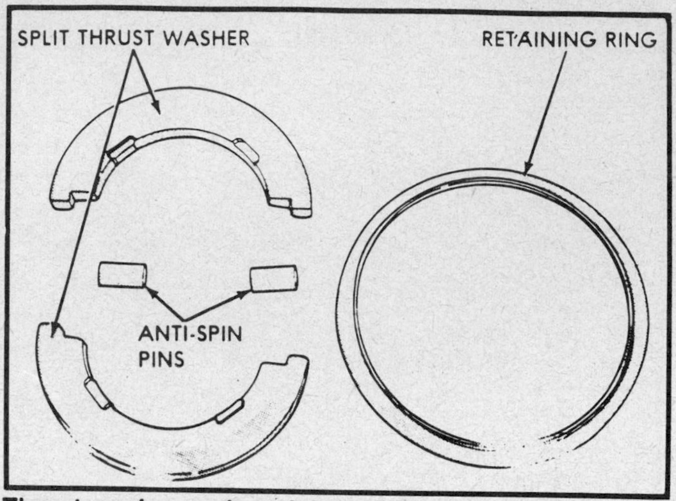
SPLIT THRUST WASHER
RETAINING RING
ANTI-SPIN
PINS
Thrust washer and retaining ring

19. Remove the 3rd/4th shift rail and retrieve the interlock plate.
20. Remove the 1st/2nd and 3rd/4th shift forks.
21. Remove the 1st/2nd and 3rd/4th shift rails.
22. Remove the reverse gear and thrust washer from the mainshaft.
24. Remove the center support plate bolts.
25. Position the drive pinion (on the input shaft) so that the cut-away on the pinion faces that countershaft.
26. Remove the gear set/support plate assembly from the transmission housing.

Unit Disassembly and Assembly

GEAR SET

Disassembly

1. Open the mainshaft center bearing snapring and tap the mainshaft assembly from the center support with a rubber mallet.
2. Remove the countershaft gear from the center support.
3. Remove the 3rd/4th synchronizer hub snapring and remove the 3rd/4th synchronizer assembly.
5. Remove the 3rd speed gear.
6. Remove the split thrust washer and retaining ring from the mainshaft.
7. Open the snapring and remove the 2nd speed gear.
8. Open the 1st/2nd synchronizer hub snapring and remove the 1st/2nd synchronizer assembly.
9. Open the mainshaft center bearing snapring and remove the 1st speed gear.
10. The synchronizers may be disassembled by removing the energizing springs.

Inspection

Inspect the countershaft gear for chipped and broken teeth and evidence of wear. Small nicks and burrs must be removed using the proper method. Replace the gear if damaged beyond the scope of minor repair. Inspect the countershaft bearing cones for wear, damage or failure. Replace the bearing cones as necessary.

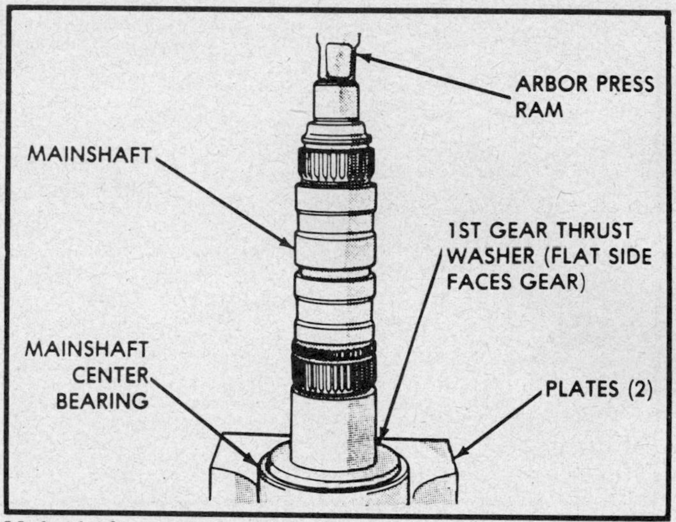
ARBOR PRESS
RAM
MAINSHAFT
1ST GEAR THRUST
WASHER (FLAT SIDE
FACES GEAR)
MAINSHAFT
CENTER
BEARING
PLATES (2)
Mainshaft center bearing installation

Inspect the drive pinion gear teeth. If the teeth are excessively worn chipped or broken, replace the gear. If the oil seal contact area is rusted, pitted or scored, a new pinion is recommended to ensure maximum seal life.

Inspect the spline teeth on the synchronizer clutch gears and stop rings. Replace all components showing signs of wear or chipped teeth. Make sure the synchronizer sleeve slides easily on the hub. Inspect the stop rings for cracks and wear. If the rings are cracked or show signs of wear in the area of the threaded bore, replace them. Check the struts for wear and damage.

NOTE: Service the synchronizers as an assembly. Do not interchange parts.

Inspect the shift forks for wear in the area of the pads and shafts. Replace as necessary.

Thoroughly clean the bearings with solvent and dry with compressed air. Inspect the bearings for roughness by slowly turning them by hand. Lubricate the bearing with clean engine oil prior to installation.

On all needle type roller bearings and spacers, look for flat spots, brinneling, wear and galling. Inspect the roller contact surfaces of the gear and shaft.

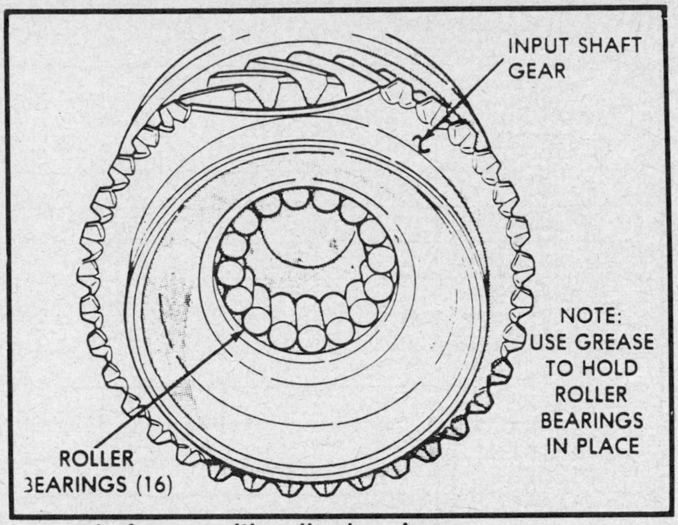
INPUT SHAFT GEAR
NOTE: USE GREASE TO HOLD ROLLER BEARINGS IN PLACE
ROLLER BEARINGS (16)

Input shaft gear with roller bearings

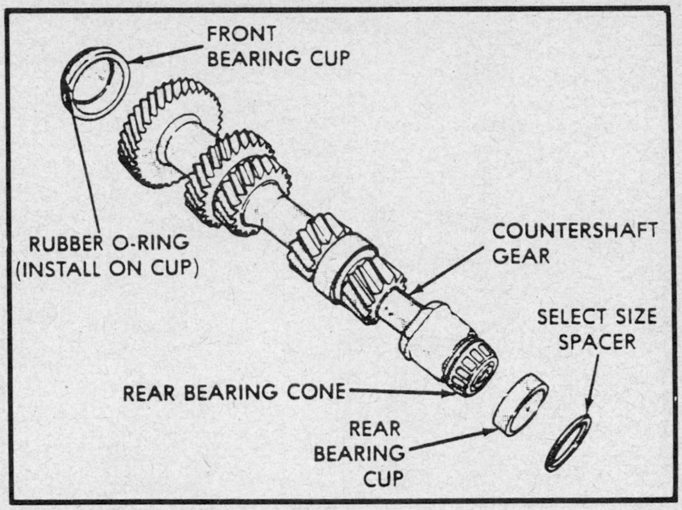
FRONT BEARING CUP
RUBBER O-RING (INSTALL ON CUP)
COUNTERSHAFT GEAR
SELECT SIZE SPACER
REAR BEARING CONE
REAR BEARING CUP

Countershaft gear assembly

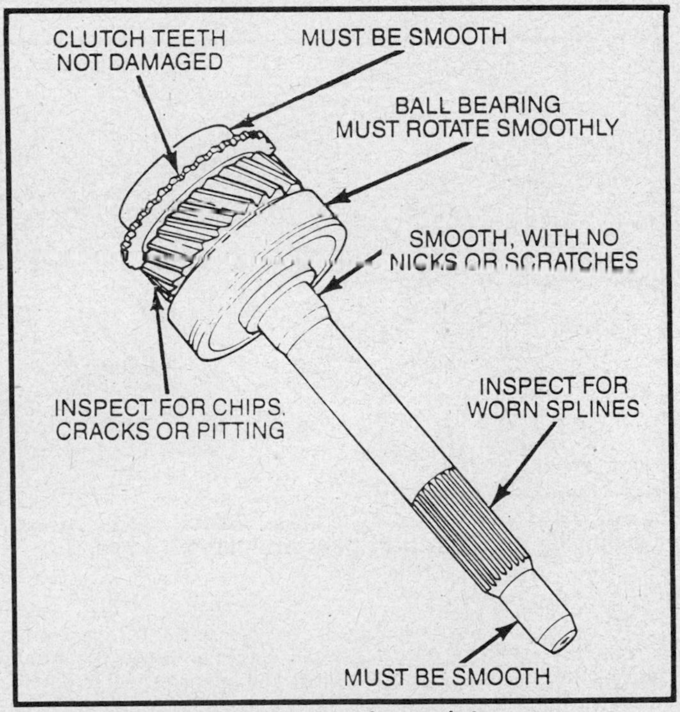
CLUTCH TEETH NOT DAMAGED
MUST BE SMOOTH
BALL BEARING MUST ROTATE SMOOTHLY
SMOOTH, WITH NO NICKS OR SCRATCHES
INSPECT FOR CHIPS, CRACKS OR PITTING
INSPECT FOR WORN SPLINES
MUST BE SMOOTH

Input shaft assembly inspection points

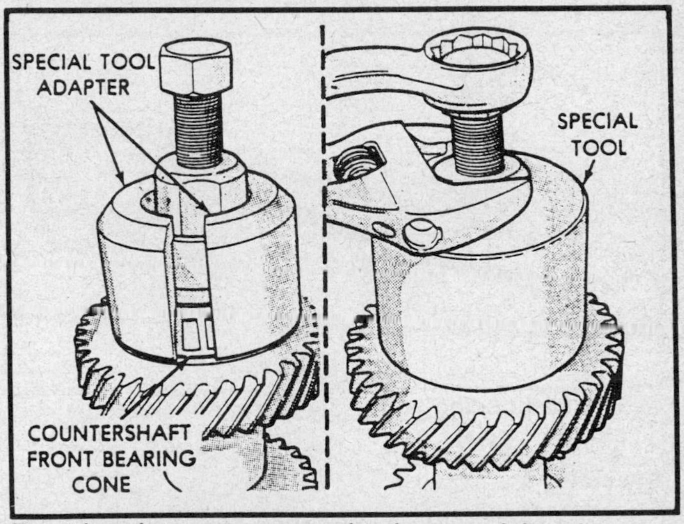
SPECIAL TOOL ADAPTER
SPECIAL TOOL
COUNTERSHAFT FRONT BEARING CONE

Front bearing cone removal using special tools

Assembly

1. Assemble the synchronizers in the reverse order of disassembly.
2. Install the needle bearings and snapring with the chamfered side of the snapring facing 1st gear.
3. Install the mainshaft bearing snapring, loosely.
4. Install the 1st speed gear and thrust washer so that the flat side of the washer faces the gear. Lock the gear with the snapring.
5. Install the 1st/2nd synchronizer assembly.
6. Install the 1st/2nd synchronizer hub snapring.
7. Install the 2nd speed gear.
6. Install the split thrust washer and retaining ring on the mainshaft.
8. Install the 3rd speed gear.
9. Install the 3rd/4th synchronizer assembly.
10. Install the 3rd/4th synchronizer hub snapring.
11. Install the countershaft gear through the center support.
12. Open the mainshaft center bearing snapring and using a plastic mallet on the center support, tap the mainshaft assembly onto the center support.

MAINSHAFT

Disassembly

1. Remove the mainshaft center bearing thrust washer and snapring.
2. Place the mainshaft in an arbor press and press off the mainshaft center bearing.

Inspection

Inspect the mainshaft gear and journal bearing mating surfaces for galling and excessive wear. They must be smooth and free of defects. Inspect the snapring grooves for burred edges. Polish the grooves with a fine file or crokus cloth as required. Inspect the splines on the synchronizer hub shaft for wear. Replace the mainshaft as necessary.

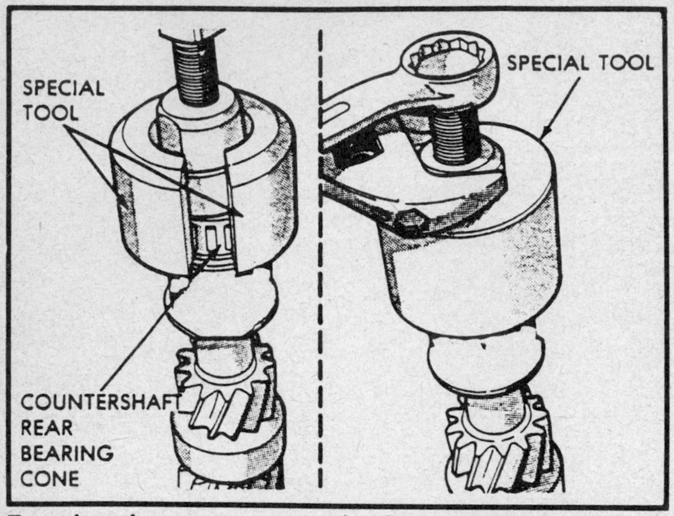
SPECIAL TOOL
SPECIAL TOOL
COUNTERSHAFT REAR BEARING CONE

Rear bearing conre removal using special tools

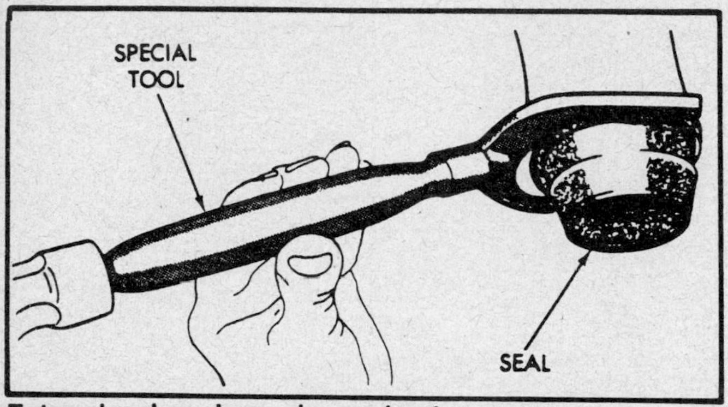
SPECIAL TOOL
SEAL

Extension housing yoke seal using special tool

Assembly

1. Place the mainshaft in an arbor press and press on the mainshaft center bearing.
2. Install the thrust washer so that the flat side is facing the gear.
3. Install the mainshaft center bearing snapring.

INPUT SHAFT

Disassembly

1. Remove the input shaft bearing retainer bolts.
2. Remove the retainer and discard the gasket.
3. Remove the large input bearing snapring.
4. Remove the input shaft from the case.

NOTE: There should be 16 roller bearings in the bore of the input shaft.

5. Remove the small input shaft bearing retaining snapring.
6. Using an arbor press, remove the input bearing.
7. Drive oil seal from the bearing retainer.

Inspection

Starting at the front end of the input shaft, inspect the end of the shaft that rides in the crankshaft pilot bushing for smoothness. The splines should not be worn. Inspect the oil seal seating area for smoothness. A scratch or ding in this area could cause the oil seal to fail.

SELECT SHIM CHART

PART NO.	THICKNESS MILLIMETERS	THICKNESS INCHES
4338275	1.37-1.39	.0539
4338276	1.46-1.48	.0579
4338277	1.55-1.57	.0614
4338278	1.64-1.66	.0650
4338279	1.73-1.75	.0685
4338280	1.82-1.84	.0720
4338281	1.91-1.93	.0756
4338282	2.00-2.02	.0791
4338283	2.09-2.11	.0827
4338284	2.18-2.20	.0862
4338285	2.27-2.29	.0898

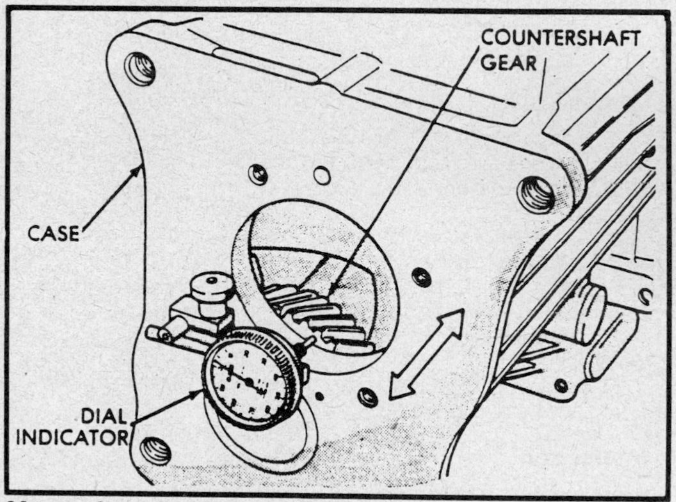
COUNTERSHAFT GEAR
CASE
DIAL INDICATOR

Measuring countershaft gear endplay

The front bearing retainer must be free of cracks, burrs and gouges on the case mating surface. The long sleeve-like portion must be smooth to ensure a satisfactory sliding surface for the throw-out bearing. Also, the retainer gasket surface must be completely free of the old gasket material.

Remove the roller bearings from the rear end of the input shaft and make sure the bore is smooth and not pitted. Inspect the rollers in the same manner.

Assembly

1. Install a new input shaft retainer oil seal.
2. Using an arbor press and special tool C–4965 or equivalent, install the input bearing.
3. Install the small input shaft bearing retaining snapring.
4. Install the input shaft in the case.
5. Install the large input bearing snapring.
6. Install the retainer and new gasket. Torque the bolts to 21 ft. lbs. (35 Nm).
7. Assemble the 16 roller bearings in the bore of the input shaft. Hold them in place by coating them with heavy grease.

CENTER SUPPORT PLATE

Disassembly

1. Remove the reverse idler shaft snapring from the center support plate.
2. Remove the reverse idler gear and shaft.
3. Remove the countershaft center bearing snapring.
4. Using an arbor press, remove the countershaft center bearing.

Inspection

Look for cracks in the center support plate and replace it if any are found. Inspect the counter gear bearing for pitting or any other signs of failure. Replace the bearing if necessary.

Assembly

1. Using an arbor press and special tool C–4171 or equivalent, install the countershaft center bearing into the support plate.
2. Install the countershaft center bearing snapring.
3. Assemble and install the reverse idler gear and shaft.
4. Install the reverse idler shaft snapring on the center support plate.

COUNTERSHAFT GEAR AND EXTENSION HOUSING

Disassembly

1. Using special puller tool C–4983–4 and adapter or equivalents, remove the front bearing cone from the countershaft.
2. Using an arbor press remove the front bearing cup from the case.
3. Using special tool C–4983–5 and adapter or equivalents, remove the rear bearing cone from the countershaft
4. Using tools L–4454 and L–4518 or equivalents, remove the countershaft rear bearing cup from the extension housing.
5. Remove the snapring and remove the extension housing ball bearing.
6. Remove the extension housing yoke seal.
7. Drive out the extension housing bushing.

Inspection

All countershaft gears should be free of piting, cracks, nicks and broken teeth. The center support journal must be free of imperfections. Check the bearing cones for damage or signs of failure and replace as necessary.

Look for cracks in the extension and correct or replace the extension as required. Re-tap all damaged threaded holes. Check the bushing in the rear of the extension for damage. During the bench overhaul, the extension seal should be replaced.

Remove the ball bearing and check it for smooth and quiet operation. Noisy bearings can be quieted with a few drops of clean engine oil. Check the housing mating surfaces for nicks and gouges and correct as necessary. Make sure the plastic plug on the right side of the extension (near the shifter assembly) is installed properly. Inspect the counter gear bearing cup for damage and wear and replace it as necessary.

Assembly

1. Drive in a new extension housing bushing.
2. Install a new extension housing yoke seal.
3. Install the extension housing ball bearing and snapring.
4. Using an arbor press and special tools C–4171 and C–4973 or equivalents, install the countershaft rear bearing cup in the extension housing.
5. Using an arbor press and tool C–4966 or equivalent, install the rear bearing cone on the countershaft.
6. Using an arbor press and tool C–4171 or equivalent, install the front bearing cup in the case.
7. Using an arbor press and tool C–4967 or equivalent, install the front bearing cone on the countershaft.
10. Using a dial indicator, check the countershaft gear endplay with the assembly positioned in the case. Endplay should be from 0.001–0.005 in. Use the proper size select-fit shim to adjust the endplay to within specifications.

Transmission Assembly

1. If the magent was removed from the transmision case, install it.
2. Install the gear set/support plate assembly. Fabricate alignment studs to aide in positioning the center support plate.
3. Turn the input shaft so that the drive pinion gear cut-away is facing down.
4. Install the center support plate bolts. Torque the bolts to 40 ft. lbs. (54 Nm).
5. Install the square anti-spin pin into the mainshaft locating hole and slide the reverse gear thrust washer onto the shaft.
6. Install the reverse gear.
7. Install the 1st/2nd and 3rd/4th shift forks.
8. Insert the 3rd/4th shift rail through the center support place and install the interlock plate.
9. Install the reverse/overdrive synchronizer, fork, and rail assembly.
10. Install the reverse/overdrive hub snapring.
11. Install the mainshaft overdrive gear.
12. Install the mainshaft overdrive gear thrust washer and anti-spin pin.
13. Install the mainshaft overdrive gear snapring.
14. Install the countershaft overdrive gear using special tool C–4982 with adapter tools C–4982–2 or their equivalents.
15. Install the countershaft overdrive gear snapring.
16. Install the 1st/2nd shift rail with the two interlocks.
17. Install the tapered pins on the shift forks.
18. Install the reverse/overdrive shift rail and blocker spring. The long end of the blocker spring faces forward.
19. Seal the extension housing with a bead of RTV and install the extension housing. Torque the bolts to 40 ft. lbs. (54 Nm).
20. Install the detent springs and bullets.
21. Seal the the access cover with a bead of RTV. Install the access cover bolts and torque to 21 ft. lbs.
22. Install the shifter assembly. The shifter assembly is sealed with a bead of RTV sealant instead of a gasket.
23. Install the shifter assembly bolts. Torque the bolts to 21 ft. lbs. (35 Nm).
24. Remove the transmission from the holding fixture and install in the vehicle.

SPECIFICATIONS

TORQUE SPECIFICATIONS

Item	ft. lbs.	Nm
Access cover bolts	21	28
Back-up lamp switch	15	20
Center support to case	40	54
Extension housing to case bolts	40	54
Gearshift lever	30	41
Input shaft bearing retaining bolts	21	28
Shifter assembly to extension bolts	21	28
Speedometer drive clamp screw	8	11
Transmission to clutch housing bolts	50	68
Transmission drain plug	30	41
Transmission fill plug	30	41

Section 7

NP-435 Transmission
New Process

APPLICATION

FORD MOTOR CO.

Year	Vehicle	Engine
1984	E350 Chassis Cab	All
	F150 4×4	All
	F250 4×4	All
	F250 4×2	300, 351W
	F350 4×4	All
	Bronco	All
1985	E350 Chassis Cab	All
	F150 4×4	All
	F250 4×4	All
	F250 4×2	300, 351W
	F350 4×4	All
	Bronco	All
1986	E350 Chassis Cab	All
	F150 4×4	All
	F250 4×4	All
	F250 4×2	300, 351W
	F350 4×4	All
	Bronco	All
1987	E350 Chassis Cab	All
	F150 4×4	All
	F250 4×4	All
	F250 4×2	300, 351W
	F350 4×4	All
	Bronco	All

CHRYSLER CORPORATION

Year	Vehicle	Engine	Vin Code
1984	D-100	3.7L, 5.2L, 5.9L	H, T, W
	D-150	3.7L, 5.2L, 5.9L	H, T, W
	D-250	3.7L, 5.2L, 5.9L	H, T, W
	D-350	5.9L	W
	AW-150	5.2L, 5.9L	T, W
	W-100	3.7L, 5.2L, 5.9L	H, T, W
	W-150	3.7L, 5.2L, 5.9L	H, T, W
	W-250	3.7L, 5.2L, 5.9L	H, T, W
	W-350	5.9L	W

CHRYSLER CORPORATION

Year	Vehicle	Engine	Vin Code
1985	D-100	3.7L, 5.2L, 5.9L	H, T, W
	D-150	3.7L, 5.2L, 5.9L	H, T, W
	D-250	5.2L, 5.9L	T, W
	D-350	5.9L	W
	AW-150	5.2L, 5.9L	T, W
	W-100	3.7L, 5.2L, 5.9L	H, T, W
	W-150	3.7L, 5.2L, 5.9L	H, T, W
	W-250	5.2L, 5.9L	T, W
	W-350	5.9L	W
1986	D-100	3.7L, 5.2L, 5.9L	H, T, W
	D-150	3.7L, 5.2L, 5.9L	H, T, W
	D-250	5.2L, 5.9L	T, W
	D-350	5.9L	W
	AW-150	5.2L, 5.9L	T, W
	W-100	3.7L, 5.2L, 5.9L	H, T, W
	W-150	3.7L, 5.2L, 5.9L	H, T, W
	W-250	5.2L, 5.9L	T, W
	W-350	5.9L	W
1987	D-150	3.7L, 5.2L, 5.9L	H, T, W
	D-250	3.7L, 5.2L, 5.9L	H, T, W
	D-350	5.9L	W
	W-150	3.7L, 5.2L, 5.9L	H, T, W
	W-250	5.2L, 5.9L	T, W
	Ramcharger (2 WD/4WD)	5.2L, 5.9L	T, W
1988	D-100	3.9L, 5.2L, 5.9L	X, Y, W
	D-150	5.9L	W
	D-250	3.9L, 5.2L, 5.9L	X, Y, W
	D-350	5.9L	W
	W-100	5.2L, 5.9L	Y, W
	W-150	5.2L, 5.9L	Y, W
	W-250	5.2L, 5.9L	Y, W
	W-350	5.9L	W
	Ramcharger (2 WD/4WD)	5.2L, 5.9L	Y, W
1989	D-150	5.2L	Y
	Ramcharger (D-100/150)	5.2L	Y
	Ramcharger (W-100/150)	5.2L, 5.9L	Y, W

GENERAL DESCRIPTION

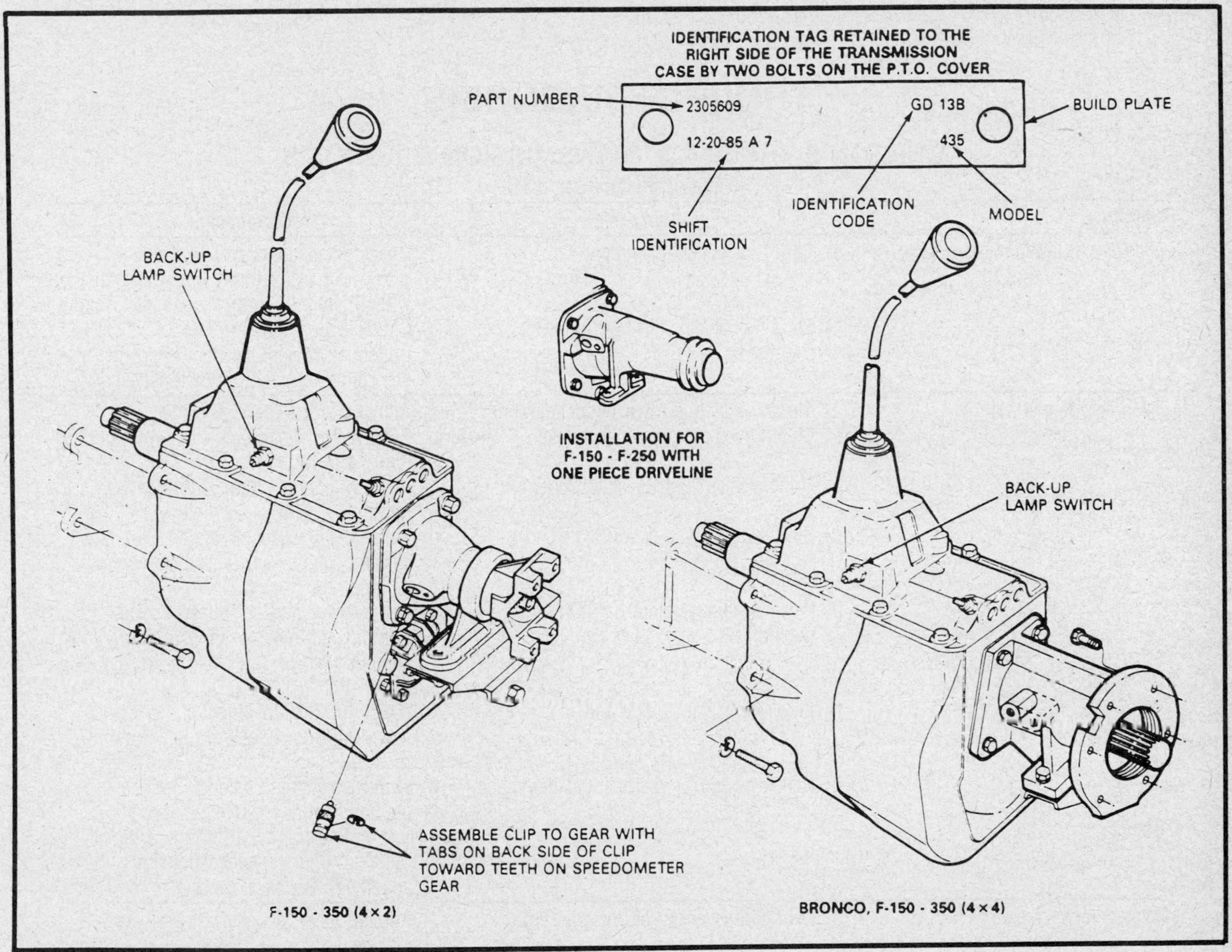
IDENTIFICATION TAG RETAINED TO THE RIGHT SIDE OF THE TRANSMISSION CASE BY TWO BOLTS ON THE P.T.O. COVER
PART NUMBER
2305609
GD 13B
12-20-85 A 7
435
BUILD PLATE
SHIFT IDENTIFICATION
IDENTIFICATION CODE
MODEL
BACK-UP LAMP SWITCH
INSTALLATION FOR F-150 - F-250 WITH ONE PIECE DRIVELINE
BACK-UP LAMP SWITCH
ASSEMBLE CLIP TO GEAR WITH TABS ON BACK SIDE OF CLIP TOWARD TEETH ON SPEEDOMETER GEAR
F-150 - 350 (4 × 2)
BRONCO, F-150 - 350 (4 × 4)

The New Process NP435 transmission

The New Process 435 4 speed transmission is provided with a floor-mounted shift lever. The 1st and reverse gears are spur cut. The 2nd, 3rd and 4th gears are helical cut and are synchronized to permit easier shifting. The input shaft is supported by a tapered roller bearing. The input shaft endplay is controlled by the thickness of gaskets between the case and the front bearing retainer. The front end of the output shaft is supported by a pilot bearing that is located in the input shaft. The rear end of the output shaft is supported by a ball bearing that is pressed onto the shaft. The bearing and shaft are held in the case by a snapring and a bearing retainer. Retention of the output shaft in the case may vary, depending on application, with the various extension housings that are used to fit several vehicle models. The 3rd and 4th speed synchronizer is mounted on the forward end of the output shaft and is held in place by the gear end of the input shaft and the 3rd speed gear. The 3rd and 2nd speed gears are located between the shoulder and the 3rd and 4th speed synchronizer on the output shaft. The 2nd speed synchronizer and the 1st speed sliding gear are located toward the rear of the output shaft. A spacer is provided to prevent the 1st speed sliding gear from contacting the case. The countershaft gear is supported by caged roller bearings installed at both ends of the gear. A roller-type thrust bearing and a bearing race are provided at the rear end of the gear. A thrust washer is also provided at the front of the gear. The reverse idler uses a roller bearing-type gear. The bearings are installed between the gear and a sleeve on the reverse idler shaft. The sets of bearings are separated by a snapring and bearing race retainer spacer. Snaprings, installed in the center of the reverse idler gear, hold the entire assembly in position.

Capacities

All vehicles equipped with the NP435 transmission require 7.0 pts. (3.3L) of transmission gear lubricant. Ford specification is D8DZ–19C4547–A (SAE 80W) or equivalent. Chrysler specification is multipurpose gear oil meeting API specification GL-5 or engine oil labeled for API service SF.

Checking Fluid Level

The vehicle must be level before an accurate reading can be obtained. The fluid level considered full when the lubricant is to the bottom of the fill hole.

TRANSMISSION MODIFICATIONS

There were no modifications for the NP435 transmission at the time of this publication.

TROUBLE DIAGNOSIS

CHILTON'S THREE C'S TRANSMISSION DIAGNOSIS
New Process 435

Condition	Cause	Correction
Shift lever—improper operation	a) Shift rods out of adjustment	a) Check crossover operation. Hand lever must move freely through neutral gate without catch or drag. Adjust shift rods
	b) Steering column shift tube out of alignment	b) Check lever position for proper transmission gear selection. Replace steering column shift tube if necessary
Transmission shifts hard	a) Clutch pedal free-travel out of adjustment	a) Adjust clutch pedal free-travel
	b) Clutch does not completely release	b) Remove and disassemble clutch assembly. Replace as required
	c) Transmission fluid low or improper type	c) Add lubricant or change lubricant as required
	d) Shift lever binding or worn	d) Remove cap from shift tower. Eliminate binding condition or replace components as required
	e) Worn or damaged internal shift mechanism	e) Remove transmission cover. Check internal shift mechanism by shifting into and out of all gears. Repair or replace as required
	f) Binding of sliding gears and/or synchronizers	f) Check for free movement of gears and synchronizers. Repair or replace as required
	g) Housings and/or shafts out of alignment	g) Remove transmission and check for binding condition between input shaft and engine crankshaft pilot bearing or bushing. Check flywheel housing alignment. Repair or replace as required
Noisy in forward gears ①	a) Lubricant level low, or improper type	a) Add lubricant, or refill with specified lubricant
	b) Components grinding on transmission	b) Check for screws, bolts, etc., of cab or other components grounding out. Correct as required
	c) Component housing bolts loose	c) Check torque on transmission to flywheel housing bolts, output shaft flange nut and flywheel housing to engine block bolts. Tighten bolts to specification
	d) Flywheel housing to engine crankshaft alignment	d) Check and align flywheel housing to engine crankshaft
	e) Noisy bearings or gears	e) Remove and disassemble transmission. Inspect input, output and countershaft bearings. Inspect speedometer gear and gear teeth for wear or damage. Replace as required
Gears clash when shifting from one forward gear to another	a) Engine idle speed too high	a) Adjust engine idle speed
	b) Clutch pedal free-travel incorrect	b) Check clutch for pedal free-travel, complete release and spin time. Adjust as required
	c) Improper manual-shift linkage	c) Adjust and repair manual-shift linkage as required

CHILTON'S THREE C'S TRANSMISSION DIAGNOSIS
New Process 435

Condition	Cause	Correction
Gears clash when shifting from one forward gear to another	d) Pilot bearing binding	d) Remove transmission and check for a binding condition between input shaft and engine crankshaft pilot bearing. Replace as required
	e) Damaged gear teeth and/or synchronizer	e) Disassemble transmission, repair or replace as required
Transmission jumps out of gear	a) Manual-shift linkage binding, out of adjustment. Stiff shift boot	a) Adjust linkage. If necessary align steering column to vehicle body
	b) Improper fit of form isolation pad	b) Replace shift boot if exceptionally stiff. Replace or rework pad to provide clearance
	c) Loose transmission to engine mounting bolts, or loose levers	c) Tighten transmission to flywheel housing, and flywheel housing to engine block bolts to specifications. Loosen all bolts and reseat flywheel housing. Tighten all bolts. Tighten levers if necessary
	d) Flywheel housing to engine crankshaft out of line	d) Shim or replace housing as required
	e) Crankshaft pilot bearing worn	e) Replace bearing
	f) Interior components damage	f) Disassemble transmission. Inspect the synchronizer sleeves for free movement on their hubs. Inspect the synchronizer blocking rings for widened index slots, rounded clutch teeth and smooth internal surface. Check countershaft cluster gear for excessive endplay. Check shift forks for worn or loose mounting on shift rails Check fork pads for excessive wear. Inspect synchronizer sliding sleeve and gear clutch teeth for wear or damage. Repair or replace as required
	g) Worn gear teeth due to partial engagement	g) Replace worn or damaged gears
Transmission will not shift into one gear—all others OK	a) Manual linkage out of adjustment	a) Adjust manual linkage
	b) Manual-shift linkage damaged or worn	b) Lubricate, repair or replace parts as required
	c) Back-up switch ball frozen	c) If reverse is problem, check back-up switch for ball frozen in extended position (if so equipped)
	d) Internal components	d) Remove transmission. Inspect shift rail and fork system synchronizer system and gear clutch teeth for restricted travel. Repair or replace as required
Transmission is locked in one gear. It cannot be shifted out of that gear	a) Manual-shift linkage out of adjustment, binding or damaged	a) Disconnect the problem shift rod from transmission shift lever. Try to shift the transmission lever into and out of gear position. If OK, repair or replace linkage parts
	b) Internal components	b) Remove transmission. Inspect problem gears, shift rails and forks and synchronizer for wear or damage. Repair as required. Check for broken fork slot tabs on TOD
	c) Loose fork on rail	c) Check shift rail interlock system

CHILTON'S THREE C'S TRANSMISSION DIAGNOSIS
New Process 435

Condition	Cause	Correction
Transfer case makes noise	a) Incorrect tire inflation pressures and/or incorrect size tires and wheels	a) Assure that all tires and wheels are the same size, and that inflation pressures are correct
	b) Excessive tire tread wear	b) Check tire tread wear to see if there is more than .06 inch difference in tread wear between front and rear. Interchange one front and one rear wheel. Reinflate tires to specifications
	c) Internal components	c) Operate vehicle in all transmission gears with transfer case in 2HI, or Hi range. If there is noise in transmission in neutral gear, or in some gears and not in others, remove and repair transmission. If there is noise in all gears, operate vehicle in all transfer case ranges. If noisy in all ranges or HI range only, disassemble transfer case. Check input gear, intermediate and front output shaft gear for damage. Replace as required. If noisy in LO range only, inspect intermediate gear and sliding gears for damage. Replace as required
4-wheel drive transfer case jumps out of gear	a) Incomplete shift linkage travel	a) Adjust linkage to provide complete gear engagement
	b) Loose mounting bolts	b) Tighten mounting bolts
	c) Front and rear driveshaft slip yokes dry or loose	c) Lubricate and repair slip yokes as required. Tighten flange yoke attaching nut to specifications
	d) Internal components	d) Disassemble transfer case. Inspect sliding clutch hub and gear clutch teeth for damage. Replace as required

① While verifying the condition, determine whether the noise is gear roll-over noise, release bearing rub or some other transmission related noise.
Gear roll-over noise, inherent in manual transmissions, is caused by the constant mesh gears turning at engine idle speed, while the clutch is engaged and the transmission is in neutral; and release bearing rub are sometimes mistaken for mainshaft bearing noise.
Gear roll-over noise will disappear when the clutch is disengaged or when the transmission is engaged in gear.
Release bearing rub will disengage when the clutch is engaged in the event that a bearing is damaged, the noise is more pronounced while engaged in gear under load or coast than in neutral.

ON CAR SERVICE

Services

FLUID CHANGE

Procedure

1. Raise and support vehicle safely.
2. Remove the drain plug if equipped, or siphon transmission fluid into a pan.
3. Remove filler plug.
4. Install drain plug.
5. Fill to the level of the filler plug hole with the proper transmission gear lubricant.
6. Install fill plug and lower vehicle.

SHIFT LEVER

Removal and Installation

NOTE: Remove the shift ball only if the shift ball, boot or lever is to be replaced. If either ball, boot or lever is not being replaced, remove the ball, boot and lever as an assembly.

1. Remove the plastic insert from the shift ball. Warm the ball with a heat gun to 140–180°F (60–82°C) knock the ball off the lever with a block of wood and a hammer, taking care not to damage the finish on the shift lever.
2. Remove the rubber boot and floor pan cover.
3. Shift the unit into 2nd gear, remove the lock pin and re-

move the shift lever from the shifter housing.

4. Install the shift lever in the shifter housing, making sure that the slot in the lever aligns with the tab in the housing. Install the lock pin.

5. Install the rubber boot and floor pan cover.

6. Warm the ball with a heat gun to 140–180°F (60–82°C) and tap the ball on the lever with a $^{7}/_{16}$ inch socket and mallet. Install the plastic shift pattern insert.

REMOVAL AND INSTALLATION

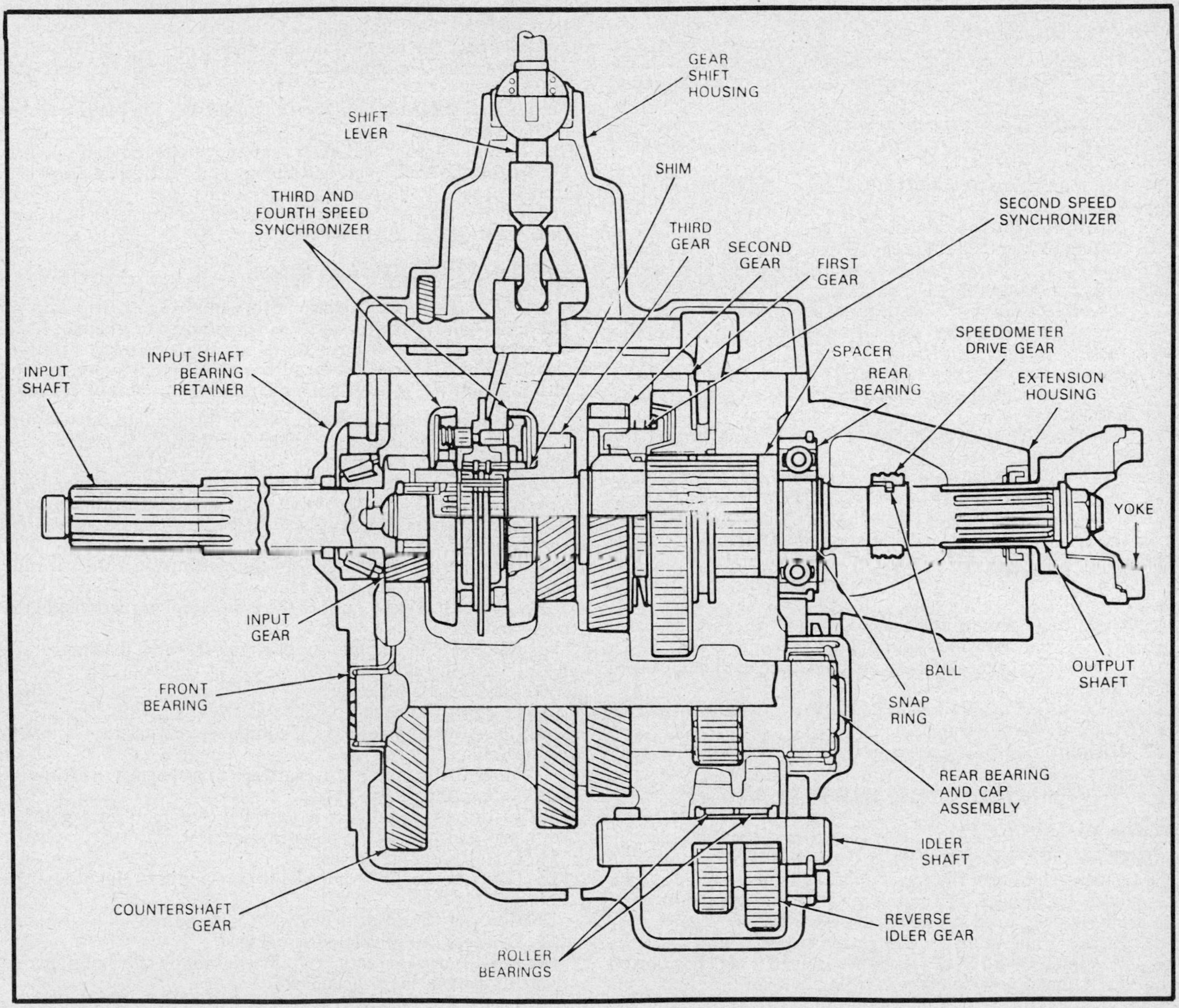
GEAR SHIFT HOUSING
SHIFT LEVER
SHIM
THIRD AND FOURTH SPEED SYNCHRONIZER
THIRD GEAR
SECOND GEAR
FIRST GEAR
SECOND SPEED SYNCHRONIZER
SPEEDOMETER DRIVE GEAR
SPACER
REAR BEARING
EXTENSION HOUSING
INPUT SHAFT
INPUT SHAFT BEARING RETAINER
YOKE
INPUT GEAR
BALL
OUTPUT SHAFT
FRONT BEARING
SNAP RING
REAR BEARING AND CAP ASSEMBLY
IDLER SHAFT
COUNTERSHAFT GEAR
REVERSE IDLER GEAR
ROLLER BEARINGS

NP435 transmission—cut-away view

TRANSMISSION REMOVAL

FORD MOTOR CO.

1. Remove the floor mat. Remove the shift lever, boot and ball as an assembly. On 4WD vehicles, remove the transfer case shift lever, rubber boot, and ball as an assembly.

2. Remove the floor pan transmission cover. Remove the weather pad on F-150—F-350. It may be necessary to remove the seat assembly.

3. Remove the gearshift lever and knob by first removing the inner cap with the use of tool T73T–7220–A or equivalent. Then remove the spring seat and spring. Remove the gearshift lever from the housing.

4. Disconnect the back-up lamp switch located on the left hand side of the gearshift housing cover.

5. Raise the vehicle and position safely. Disconnect the speedometer cable. Remove the rear driveshaft and wire it out of the way. On 4WD vehicles follow Steps 6 through 11.

6. Drain the transfer case.

7. Remove the front driveshaft from transfer case and wire out of the way.

8. Remove the cotter key holding the shift link and remove the shift link.

9. Remove the speedometer cable from the transfer case.
10. Remove 3 bolts holding support bracket to transfer case.
11. Position a transmission jack under the transfer case. Remove the 6 bolts holding the transfer case to the transmission. Remove the transfer case from the vehicle.
12. Place a transmission jack under the transmission, remove the rear support.
13. Remove the transmission attaching bolts at the clutch housing and remove the transmission.

CHRYSLER CORPORATION

1. The shift lever retainer is removable by pressing down, rotating the retainer counterclockwise a small amount and then releasing.
2. Raise and support the vehicle safely.
3. if equipped, remove the skid plate and associated components.
4. Drain lubricant from the transmission and transfer case, if equipped.
5. Disconnect the speedometer cable.
6. Matchmark and disconnect the front and rear driveshafts as required. Suspend each shaft from a convenient place;do not allow them to hang free.
7. Disconnect the shift rods at the transfer case, if equipped.
8. Support the transfer case, if equipped, with a suitable transmission jack.
9. Remove the extension-to-transfer case mounting bolts.
10. Move the transfer case rearward to disengage the front input shaft spline.
11. Lower and remove the transfer case from under the vehicle.
12. Disconnect the back-up light switch electrical connector.
13. Install engine support fixture C–3487A with special adapter tool DD–1279 or equivalents so that they are firmly attached over the frame rails. The support ends must be flush with the underside of the oil pan flange.
14. Support the transmission with a suitable transmission jack.
15. Unbolt and remove the transmission crossmember.
16. Remove the transmission-to-clutch housing bolts.
17. Slide the transmission rearward until the mainshaft clears the clutch disc.
18. Once the mainshaft is clear, lower the transmission and remove from under the vehicle.
19. Position the transmission on a suitable holding fixture.

TRANSMISSION INSTALLATION

FORD MOTOR CO.

1. Place the transmission on a transmission jack and raise the transmission until the input shaft splines are aligned with the clutch disc splines. The clutch release bearing and hub must be properly positioned in the release lever fork.
2. Install guide studs in the clutch housing and slide the transmission forward on the guide studs until it is in position on the clutch housing. Install the attaching bolts or nuts and tighten them to specification. Remove the guide studs and install the 2 lower attaching bolts. On vehicles equipped with 4WD, follow Steps 3–7.
3. Place the transfer case on a transmission jack. Install the 6 retaining bolts and gasket. Position the transfer case on the transmission and tighten bolts to specification.
4. Install transfer case shift link and cotter key.
5. Position and install speedometer cable.
6. Position and install support bracket with 3 retaining bolts. Tighten to specification.
7. Remove wires and connect front driveshaft.
8. Remove wires and connect rear driveshaft. Install the transmission rear support.
9. Connect the back-up lamp switch.
10. On vehicles equipped with 4WD, fill the transfer case with lubricant.
11. Install transmission cover plate. Install the seat assembly if it was removed.
12. Install weather pad, floor mat and rubber boot as applicable. Install the shift lever, ball and boot on the transmission and transfer case (if equipped) as an assembly.
13. Fill the transmission with standard transmission lubricant D8DZ–19C547–A (SAE 80W) or equivalent.

CHRYSLER CORPORATION

NOTE: The transmission pilot bushing in the end of the crankshaft requires high-temperature grease. Multipurpose grease should be used, such as part number 3410863 or its equivalent. Do not lubricate the end of the mainshaft, clutch splines, or clutch release levers.

1. Remove the transmission from the holding fixture and raise into position under the vehicle until the drive pinion is engaged with the clutch housing bore.
2. Roll the transmission forward until the mainshaft engages the clutch disc. Turn the pinion shaft to align the splines then push the transmission forward until it is fully seated against the clutch housing.
3. Install the transmission-to-clutch housing bolts. Torque the bolts to 105 ft. lbs. (142 Nm).
4. Attach the rear mount to the crossmember then install the crossmember.
5. Remove the engine support fixture and disengage the hooks from the side rails.
6. Install the shift lever.
7. Connect the back-up light switch.
8. Raise the transfer case, if equipped, into position and support safely.
9. Move the transfer case rearward to disengage the front input shaft spline, if equipped.
10. Install the extension-to-transfer case mounting bolts, if equipped and torque the bolts to specification.
11. Install the rear driveshaft.
12. Connect the shift rods at the transfer case and adjust the linkage.
13. Connect the front driveshafts, as required.
14. Connect the speedometer cable.
15. Install the skid plate, if equipped. Fill the transmission and transfer case to the proper level.
16. Lower the vehicle.
17. Perform a road test to make sure that the transmission shifts smoothly and quietly.

BENCH OVERHAUL

Before Disassembly

Clean the exterior of the transmission assembly before any attempt is made to disassemble it in order to prevent dirt or other foreign materials from entering the transmission assembly or its internal parts.

NOTE: If steam cleaning is done to the exterior of the transmission, immediate disassembly should be done to avoid rusting, caused by condensation forming on the internal parts.

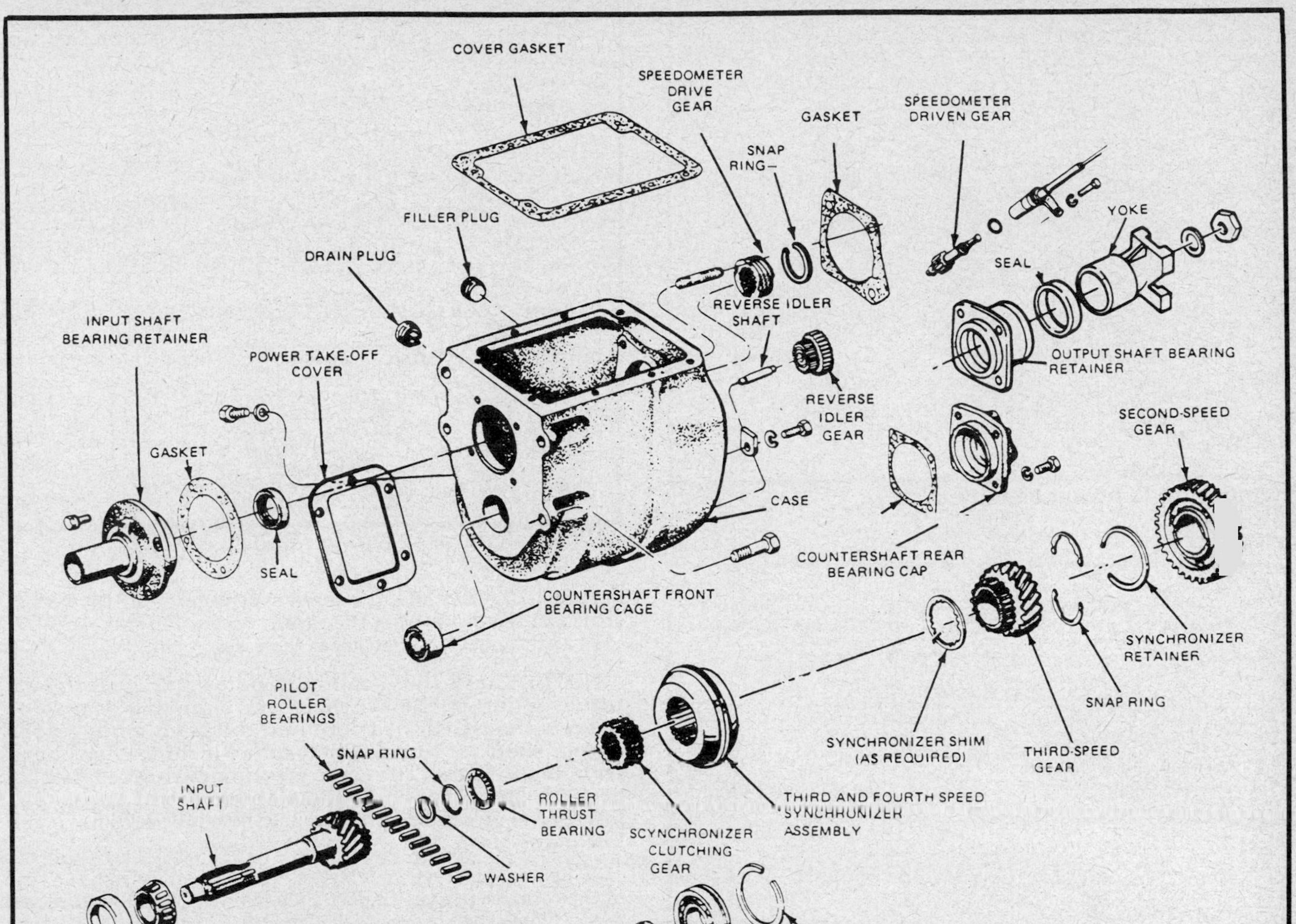
COVER GASKET
SPEEDOMETER DRIVE GEAR
GASKET
SPEEDOMETER DRIVEN GEAR
SNAP RING—
FILLER PLUG
YOKE
DRAIN PLUG
SEAL
REVERSE IDLER SHAFT
INPUT SHAFT BEARING RETAINER
POWER TAKE-OFF COVER
OUTPUT SHAFT BEARING RETAINER
REVERSE IDLER GEAR
SECOND-SPEED GEAR
GASKET
CASE
SEAL
COUNTERSHAFT REAR BEARING CAP
COUNTERSHAFT FRONT BEARING CAGE
SYNCHRONIZER RETAINER
PILOT ROLLER BEARINGS
SNAP RING
SYNCHRONIZER SHIM (AS REQUIRED)
THIRD-SPEED GEAR
SNAP RING
INPUT SHAFT
ROLLER THRUST BEARING
THIRD AND FOURTH SPEED SYNCHRONIZER ASSEMBLY
SCYNCHRONIZER CLUTCHING GEAR
WASHER
SNAP RING
RACE
BEARING
SPRING
REVERSE IDLER GEAR COMPONENTS
OUTPUT SHAFT BEARING
FIRST-SPEED GEAR STOP
REVERSE IDLER GEAR SHAFT
OUTPUT SHAFT
FIRST-SPEED GEAR—
SPRING
NEEDLE BEARING
SECOND SPEED SYNCHRONIZER BRAKE AND SPRING
NEEDLE THRUST BEARING AND RACE
RETAINER
COUNTERSHAFT GEAR
THRUST WASHER

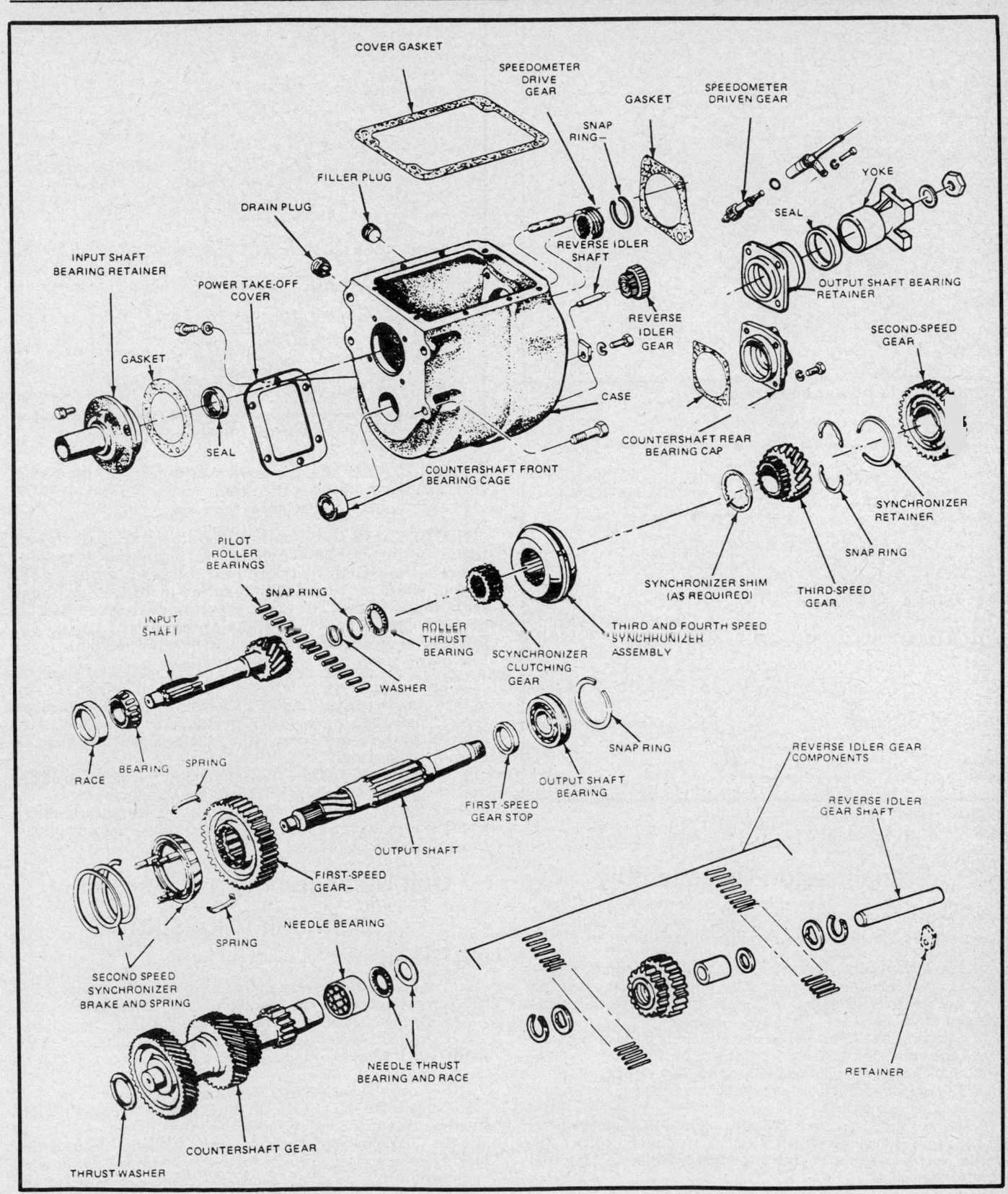

NP435 transmission—exploded view

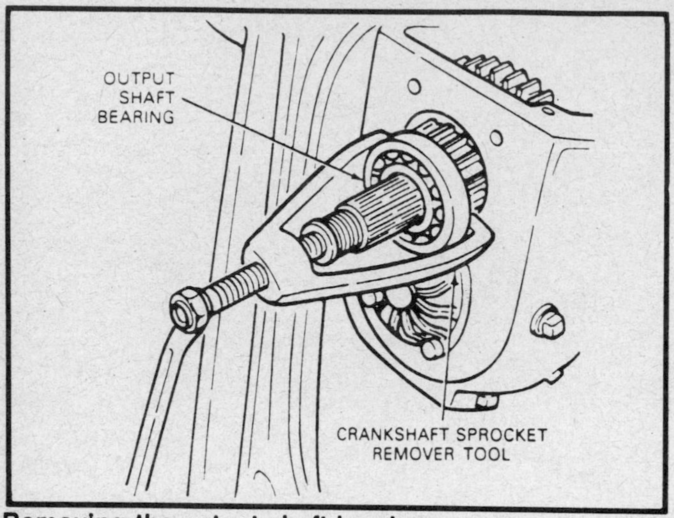
OUTPUT SHAFT BEARING
CRANKSHAFT SPROCKET REMOVER TOOL
Removing the output shaft bearing

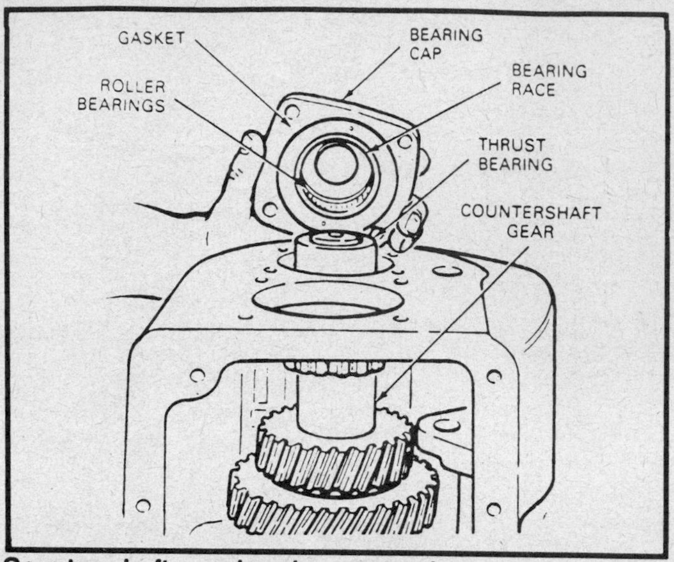
GASKET
BEARING CAP
ROLLER BEARINGS
BEARING RACE
THRUST BEARING
COUNTERSHAFT GEAR
Counter shaft rear bearing removal

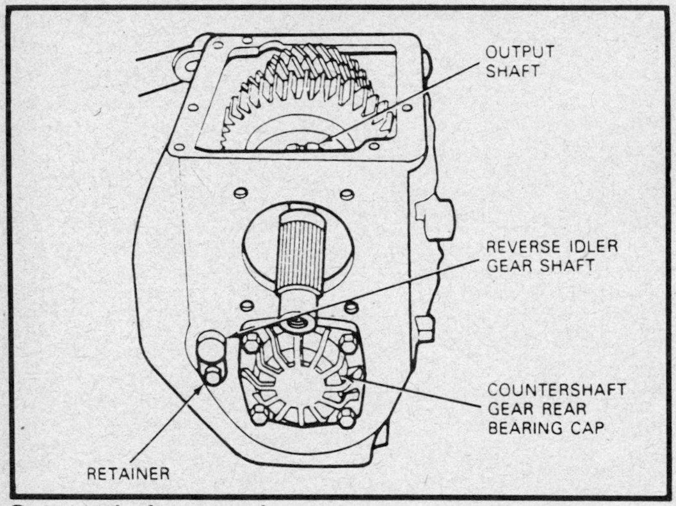
OUTPUT SHAFT
REVERSE IDLER GEAR SHAFT
COUNTERSHAFT GEAR REAR BEARING CAP
RETAINER
Output shaft removal

Transmission Disassembly

1. Place the transmission in a suitable holding stand.
2. Remove the drain plug and drain the lubricant from the transmission.
3. Place the gearshift lever in the neutral position, remove the gearshift housing attaching bolts, and remove the housing.
4. Lock the transmission in 2 gears, and remove the output flange nut.
5. Remove the extension housing attaching bolts or nuts. Remove the extension housing from the output shaft.
6. Slip the speedometer drive gear off the output shaft.
7. Remove the front bearing retainer attaching bolts. Remove the retainer and gasket.
8. Rotate the input shaft gear as required to align the notch in the input gear clutch teeth with the countershaft drive gear teeth. Remove the input gear and tapered roller bearing from the transmission.
9. Remove the output shaft bearing, snapring and bearing using tool T58T-6306-A or equivalent.
10. Remove the roller-type thrust bearing from the front of the output shaft.
11. Slide the 3rd and 4th speed synchronizer off the output shaft and remove it from the case.
12. Lift the output shaft from the case.

NOTE: Check the location of paint marks on the external spline of the synchronizer clutching gear (side nearest the thrust bearing) and the paint mark on the output shaft (at the external spline leading edge chamfer). Reposition to the same location during reassembly or 180 degrees apart. Production parts will have a yellow, green or orange paint mark; service parts have yellow paint mark.

13. Remove the reverse idler gear shaft retainer from the rear of the case. Secure the impact slide hammer, T50T-100-A and reverse idler shaft remover T50T-7140-C or equivalent to the shaft. Hold the gear to prevent it from dropping, and then remove the idler shaft.
14. With the front of the transmission in the downward position, remove the attaching bolts from the countershaft gear rear bearing cap. Tap the cap with a soft-faced hammer to free it from the case. Remove the cap, race thrust bearing, and gasket. Lift the gear from the case.

Unit Disassembly and Assembly

GEARSHIFT HOUSING

Disassembly

1. The gearshift housing should be disassembled only if it is necessary to replace a shift fork, shift rail, or the cover itself.
2. Slide the boot off the cap, if so equipped.
3. Using shift lever retaining cap tool T73T-7220-A or equivalent, turn the cap counterclockwise and remove the lever from the cover.
4. Remove the back-up lamp switch from the housing.
5. Remove the spiral pin from the 1st and 2nd speed shift fork and the gate with screw extractor. To insert the screw extractor into the spiral pin, tap lightly on fabricated handle while slowly turning counterclockwise. When extractor grips the pin, stop tapping an continue turning until spiral pin is removed.
6. Push the shifter shaft out through the rear to force the plug out of the housing. Cover the detent ball access hole to prevent the ball and spring from flying out as the shaft clears the hole. Remove the shift fork and gate.

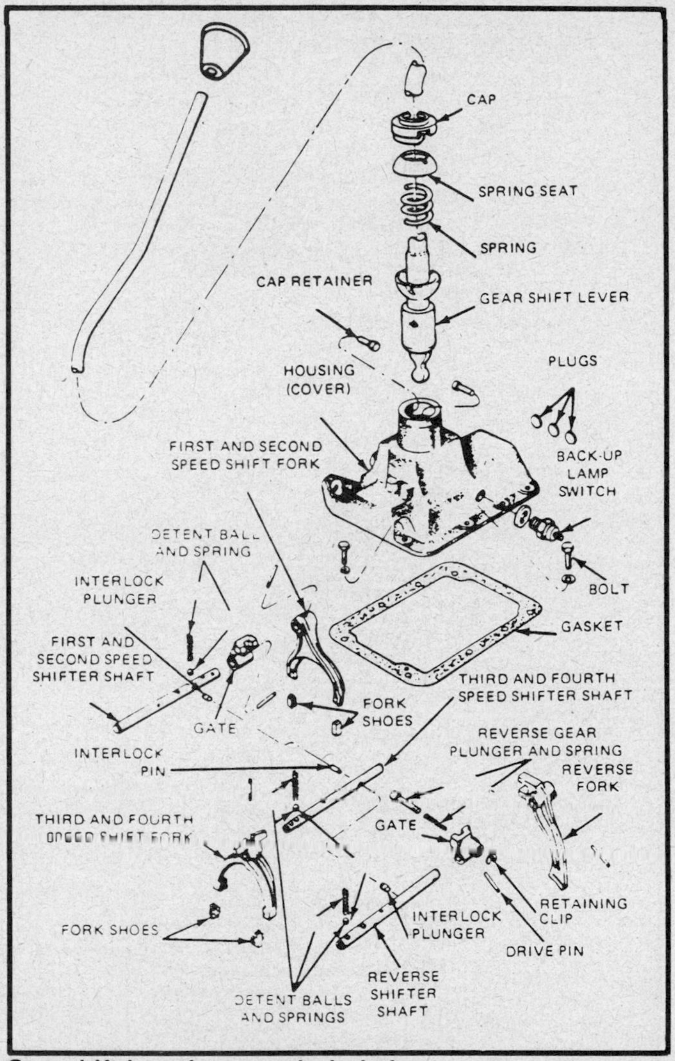
CAP
SPRING SEAT
SPRING
CAP RETAINER
GEAR SHIFT LEVER
HOUSING (COVER)
PLUGS
FIRST AND SECOND SPEED SHIFT FORK
BACK-UP LAMP SWITCH
DETENT BALL AND SPRING
BOLT
INTERLOCK PLUNGER
GASKET
FIRST AND SECOND SPEED SHIFTER SHAFT
THIRD AND FOURTH SPEED SHIFTER SHAFT
FORK SHOES
GATE
REVERSE GEAR PLUNGER AND SPRING
INTERLOCK PIN
REVERSE FORK
THIRD AND FOURTH
GATE
RETAINING CLIP
INTERLOCK PLUNGER
FORK SHOES
DRIVE PIN
REVERSE SHIFTER SHAFT
DETENT BALLS AND SPRINGS

Gearshift housing—exploded view

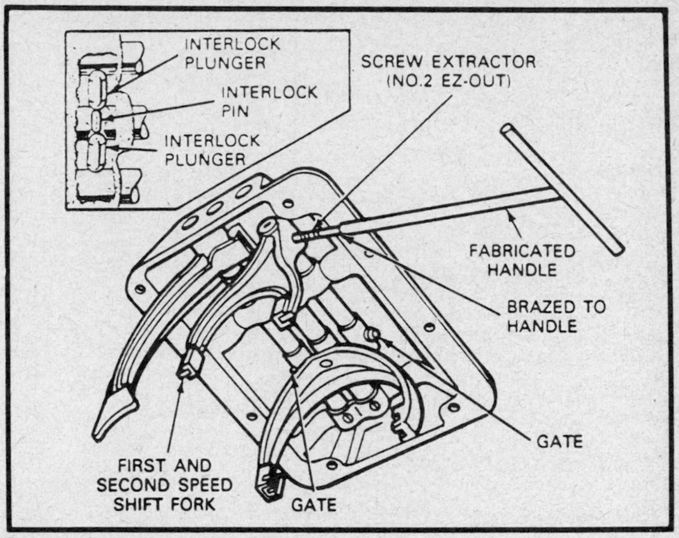
INTERLOCK PLUNGER
SCREW EXTRACTOR (NO.2 EZ-OUT)
INTERLOCK PIN
INTERLOCK PLUNGER
FABRICATED HANDLE
BRAZED TO HANDLE
GATE
FIRST AND SECOND SPEED SHIFT FORK
GATE

Removing the shift fork spiral pin

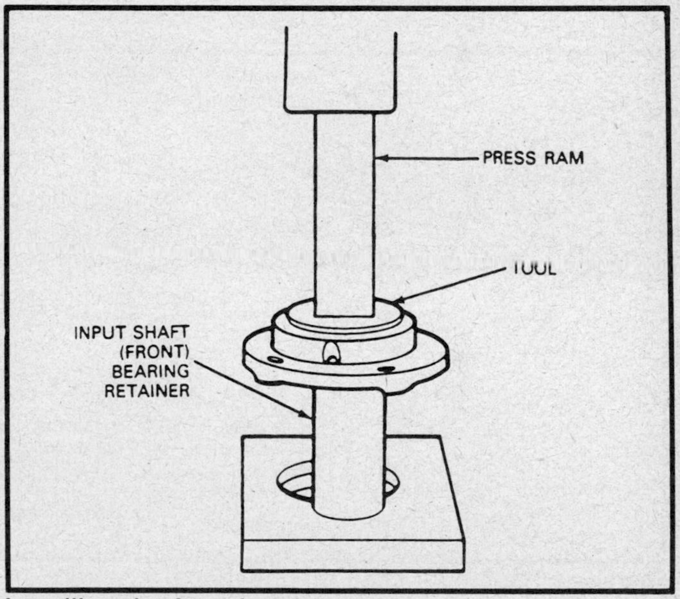
PRESS RAM
TOOL
INPUT SHAFT (FRONT) BEARING RETAINER

Installing the front bearing race

7. Remove the 3rd and 4th speed shifter shaft in the same manner, then the reverse shifter shaft.
8. Compress the reverse gear plunger and remove the retaining clip. Remove the plunger and the spring from the gate.

Assembly

1. Assemble the spring on the reverse gear plunger and hold it in the reverse shift gate. Compress it in the gate and install the retaining clip.
2. Enter the reverse shifter shaft in the cover and place the detent spring and ball in position. Depress the ball and slide the shifter shaft over it.
3. Install the gate and fork on the shaft. Install new spiral pins in the gate and fork.
4. Apply a film of sealer in the plug seat at the rear of the cover. Install a new plug in the reverse shifter shaft bore.
5. Place the reverse fork in the neutral position.
6. Install the 2 interlock plungers in the bores.
7. Insert the interlock pin in the 3rd and 4th speed shifter shaft. Install the shaft in the same manner as the reverse shifter shaft mechanism.
8. Install the 1st and 2nd speed shifter shaft in the same manner, making sure that the interlock plunger is in place.
9. Lubricate the spherical ball seat with multi purpose lubricant, C1AZ–19590–B or equivalent. Position the shift lever and cap in place.
10. Install the back-up lamp switch, and tighten to 20–30 ft. lbs.(28–40 Nm).

COUNTERSHAFT GEAR FRONT BEARING

Disassembly

Drive the countershaft gear front bearing cage from the case. The bearing can be driven out from the outside of the case.

Assembly

Carefully press the bearing cage into the case until it is flush with the front of the case.

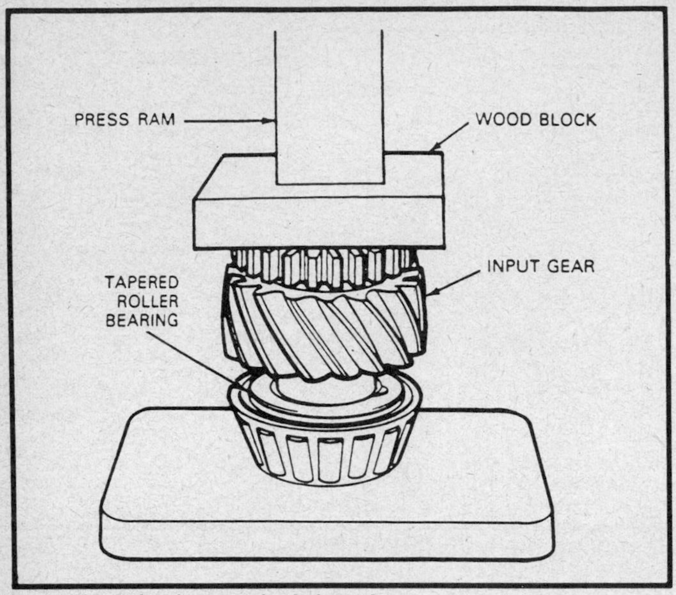
PRESS RAM
WOOD BLOCK
TAPERED ROLLER BEARING
INPUT GEAR

Installing the input shaft bearing

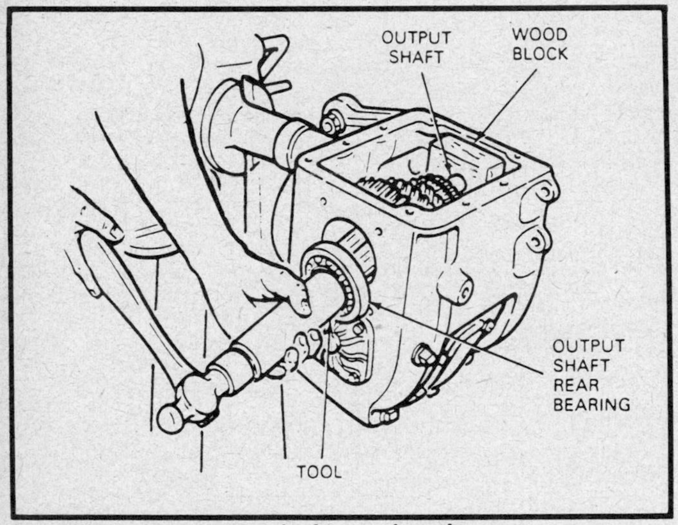
OUTPUT SHAFT
WOOD BLOCK
OUTPUT SHAFT REAR BEARING
TOOL

Installing the output shaft rear bearing

COUNTERSHAFT GEAR REAR BEARING

Disassembly and Assembly

1. If the countershaft gear rear bearing was removed, position the transmission so the front of the case is facing downward.
2. If uncaged bearings are re-used, the loose rollers (or needles) can be held in place with suitable grease. (This is not required with the caged-type bearing.) Position the bearings in the cap.
3. Position the race thrust bearing and the cap. Tighten the attaching bolts to 20–40 ft. lbs (28–54 Nm).

INPUT SHAFT BEARING RACE

Disassembly

Pull the bearing race from the front bearing retainer with impact slide hammer tool T50T-100-A or equivalent and internal puller tool, D80L-943-A or equivalent.

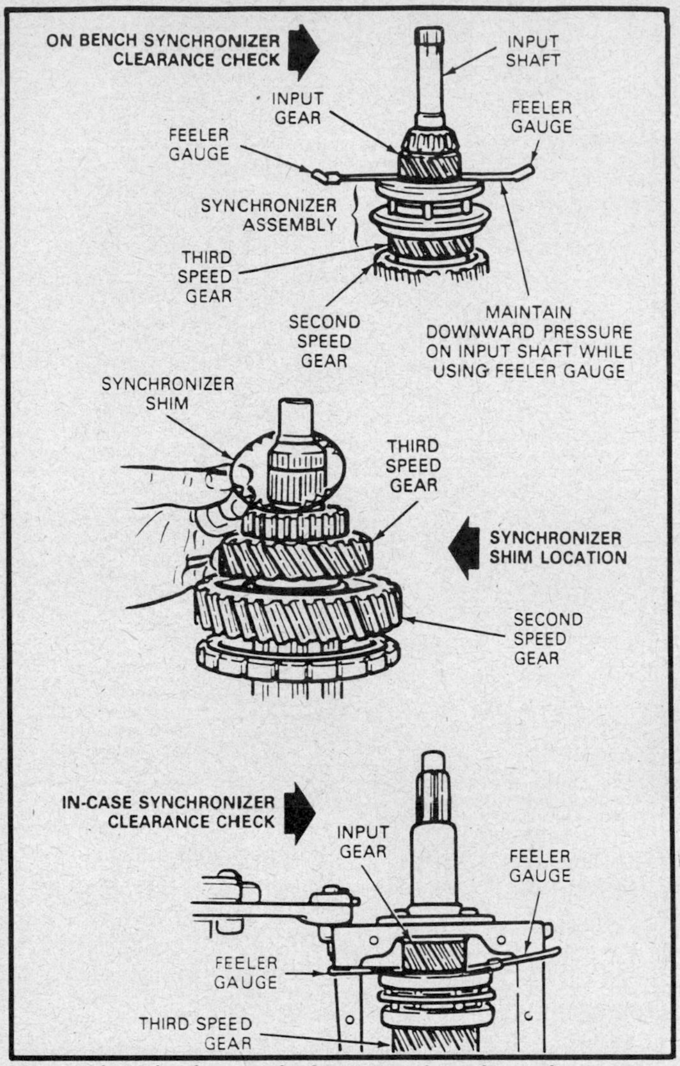
ON BENCH SYNCHRONIZER CLEARANCE CHECK
INPUT SHAFT
INPUT GEAR
FEELER GAUGE
FEELER GAUGE
SYNCHRONIZER ASSEMBLY
THIRD SPEED GEAR
SECOND SPEED GEAR
MAINTAIN DOWNWARD PRESSURE ON INPUT SHAFT WHILE USING FEELER GAUGE
SYNCHRONIZER SHIM
THIRD SPEED GEAR
SYNCHRONIZER SHIM LOCATION
SECOND SPEED GEAR
IN-CASE SYNCHRONIZER CLEARANCE CHECK
INPUT GEAR
FEELER GAUGE
FEELER GAUGE
THIRD SPEED GEAR

Measuring the input shaft-to-synchronizer clearance

Assembly

Press a new race into the front bearing retainer with a suitable tool.

INPUT SHAFT SEAL

Disassembly

Pull the input shaft seal with puller attachment tool T58L-101-A and slide hammer tool T50T-100-A or equivalent.

Assembly

Press a new seal into place in the retainer making sure that the lip of the seal is toward the mounting surface.

INPUT SHAFT BEARING

Disassembly

Remove the tapered roller bearing from the input shaft.

Assembly

Position axle bearing/seal plate T75L-1165-B or equivalent in

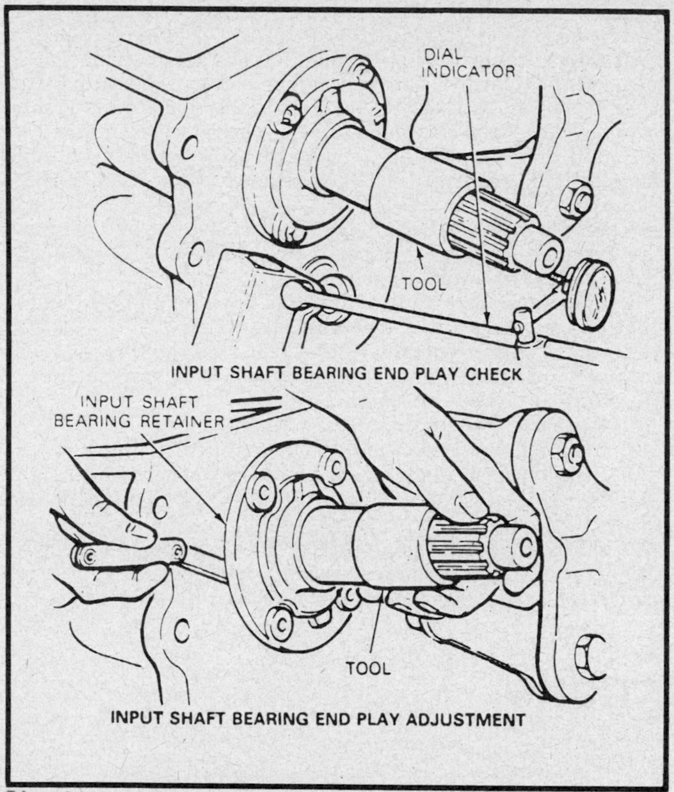
DIAL INDICATOR
TOOL
INPUT SHAFT BEARING END PLAY CHECK
INPUT SHAFT BEARING RETAINER
TOOL
INPUT SHAFT BEARING END PLAY ADJUSTMENT

Checking the input shaft endplay

an arbor press. Place a wood block on the input gear and press the gear into the bearings until it contacts the bearing inner race.

INPUT SHAFT ROLLER BEARINGS

Disassembly

1. Remove the snapring and retaining washer from the shaft.
2. Remove the bearings from the input shaft bore.

Assembly

1. Coat the bore of the input shaft gear with a thin film of multi purpose lubricant grease.
2. Install the 14 roller bearings in the bore of the input shaft. Slide the final roller axially into place. Secure the bearings with the retaining washer and the snapring.

REVERSE IDLER GEAR BEARINGS

The reverse idler gear and bearings are serviced only as an assembly. Once the bearing lock ring snaps into place, it cannot be removed without destroying either the gear or the bearing.

1ST SPEED GEAR

Disassembly

1. Remove both the sliding 1st speed gear and the output shaft from the transmission.
2. Mark the gear and output shaft so the gear and shaft splines can be assembled in the same position. Remove gear from output shaft. Do not lose the spline springs.

Assembly

1. If a new gear and/or shaft is being installed, select-fit the gear to the shaft for the best sliding action. Do the same if the 1st speed gear has been sticking on the shaft splines. Position the spline springs on the coast side of the spline teeth of the 1st speed gear. To facilitate assembly, use a daub of multi purpose lubricant grease to retain the springs on the sides of the spline teeth. Two notches on the hub of the 1st speed gear identify the location of the relieved splines.
2. Place the new 1st speed gear assembly in its position on the output shaft.

2ND SPEED GEAR

Disassembly

1. Slide the 3rd speed gear, 1st speed gear stop, and the 1st speed gear off the output shaft. Do not lose the slider gear springs.
2. Carefully pry up on 1 end of the 2-piece snapring and remove it from the shaft. If any damage to the rings results from this, it should not be re-used.
3. Slide the 2nd speed gear and synchronizer brake off the shaft.
4. Remove the snapring from the synchronizer brake pins. Separate the brake and spring from the 2nd speed gear.

Assembly

1. Assemble the spring and synchronizer brake to the 2nd speed gear. Secure the brake with the snapring, making sure that the snapring tangs are away from the gear.
2. Slide the 2nd speed gear onto the front of the output shaft, being sure that the synchronizer brake is toward the rear. Secure the gear to the shaft with the 2-piece snapring.
3. Slide the spring loaded 1st speed gear and the gear stop onto the rear of the shaft.
4. Slide the 3rd speed gear, synchronizer shim, if required, the 3rd and 4th speed synchronizer, and the roller-type thrust bearing onto the front of the shaft.

NOTE: The synchronizer clutching gear and the output shaft should be assembled so that the paint marks are located 180 degrees apart.

Transmission Assembly

NOTE: Coat all parts with the specified transmission lubricant to prevent scoring when the transmission is first operated.

1. Place the input shaft, 3rd/4th speed synchronizer assembly clutch gear and thrust bearing on the output shaft and secure the complete assembly in a vise.
2. On F-150—F-350 and Bronco, check the distance between the synchronizer and input shaft gear. If the distance is more than 0.063–0.081 in. (1.60–2.05mm) install the necessary thickness of shims between the 3rd speed gear and the synchronizer outer stop ring. After the proper thickness of shims has been established, remove the input shaft from the output shaft.
3. On F-150—F-350 and Bronco, the 3rd and 4th speed synchronizer is a 1 piece assembly consisting of a sliding clutch hub, 3 energizing pins, 2 energizing springs, 3 blocking ring pins, and 2 aluminum inner stop rings assembled as an integral unit. In addition, there are 2 outer stop rings.

NOTE: The 2 oil slots in the clutching gear must be installed facing the 3rd speed gear. The oil slot must not face against the thrust bearing.

4. On F-150—F-350 and Bronco, when replacing the 3rd and 4th speed synchronizer, use the latest assembly. Also, be sure to use the new outer stop rings, which are necessary to protect the inner stop rings against premature failure. Never mix old and new parts nor use old-style or worn outer stop rings.

5. Position the transmission case with the front in the downward position.

6. Apply a thin film of multi purpose lubricant grease on the front thrust washer and position it in the front of the case. The thrust washer is bored off-center, therefore, make sure that the tangs match the slots in the case boss.

7. Place the countershaft gear in the case.

8. Place the rear thrust bearing and then the bearing race on the rear of the countershaft gear.

9. Install a new gasket on the bearing cap. Install the bearing cap on the rear of the case. Tighten the attaching bolts to 20–40 ft. lbs. (28–54 Nm).

10. Install the reverse idler gear in the case with the larger gear toward the rear of the case. Coat the rear of the reverse idler gear shaft with perfect seal sealing compound, B5A–19554–A (ESR–M18P2–A and ESE–M4G115–A) or equivalent, sealer before installing the retainer. Press idler shaft into position. Secure the idler shaft retainer with a bolt.

11. Position the output shaft assembly in the case.

12. Place a wood block in the front of the case. Drive the bearing onto the rear of the output shaft while holding the front of the shaft against the block using tool T71P–7025–A.

13. Install the rear extension housing on the transmission.

14. With the cutaway portion of the clutch teeth in the downward position on the input shaft, install the gear in the case.

15. Install the input shaft bearing retainer with no gasket or capscrews.

Using a suitable tool to hold the shaft and retainer concentric, measure the clearance between the retainer and the case. Install a gasket shim pack 0.010–0.015 in.(0.254–0.381mm) greater than the measured clearance between the retainer and the case to obtain the required 0.007–0.014 in. (0.177–0.355mm) input shaft endplay. Tighten the front retainer bolts, then re-check the endplay.

When the input shaft endplay has been established, re-check the synchronizer clearance. It should be 0.070–0.095 in. (1.77–2.41mm). Adjust, if required.

16. Lubricate the extension housing bushing and seal and u-joint flange with multi purpose lubricant.

17. Install the speedometer drive gear, and the flange attaching nut. Lock the transmission in gear and tighten the nut to 75–110 ft. lbs. (102–149 Nm).

18. Place the transmission gears in neutral.

19. Install a new gasket and the gear shift housing.

20. Install 2 cover aligning bolts and lockwashers finger tight.

21. Install the rest of housing attaching bolts. Tighten all bolts to specifications.

22. Fill the transmission through the speedometer cable attachment opening in the rear bearing retainer, until the lubricant reaches the lower lever of the regular filler opening.

SPECIFICATIONS

TORQUE SPECIFICATIONS

Description	ft. lbs.	Nm
Back-up lamp switch	20–30	28–40
Bell housing mounting bolts	70–110	95–149
Case cover	20–40	28–54
Countershaft rear retainer	20–40	28–54
Drain plug	25–35	34–47
Filler plug	25–35	34–47

TORQUE SPECIFICATIONS

Description	ft. lbs.	Nm
Flange nut	75–110	102–149
Mainshaft rear retainer	35–45	48–61
Power take off cover bolt	12–18	17–24
Reverse idler shaft locking bolt	20–40	28–54
Input shaft retainer to case	25–35	34–47

SPECIAL TOOLS

Number	Description
T50T-100-A	Impact slide hammer—2½ lbs.
T59L-100-B	Impact slide hammer—2½ lbs.
D79P-100-A	Impact slide hammer—5 lbs.
T58L-101-A	Puller attachment
T75L-500-B	Bench mounted holding fixture
D80L-943-A	Internal puller
T75L-1165-B	Axle bearing/seal plate
TOOL-1175-AC	Seal remover
T75L-4201-A	Clutch housing alignment adapter
T75L-4201-B	Clutch housing alignment adapter

Number	Description
D78P-4201-B	Dial indicator magnetic base
TOOL-4201-C	Dial indicator with bracketry
T58T-6306-A	Sprocket remover
T75L-6392-A	Clutch housing alignment tool
D79L-7000-A	Retaining ring pliers
T72J-7025	Mainshaft bearing cone replacer
T71P-7025-A	Output shaft bearing replacer
T50T-7140-C	Reverse idler shaft removal tool
T73T-7220-A	Shift lever cap retaining tool

Section 8

Transfer Case

BORG WARNER 1345 TRANSFER CASE

Application

1984–89 Ford F-150, F-250 and F-350

General Description

The Borg Warner 1345 transfer case is a 2 piece all aluminum part time transfer case. The unit lubrication is done by a positive displacement oil pump. This oil pump, channels the oil flow through drilled holes in the rear output shaft. The oil pump turns with the rear output shaft and allows towing of the vehicle for extended distances without having to disconnect the driveshaft.

Trouble Diagnosis

CHILTON'S THREE C'S TRANSFER CASE DIAGNOSIS
Borg Warner 1345 Transfer Case

Condition	Cause	Correction
Slips out of gear	a) Shifting poppet spring weak	a) Replace the shifting poppet spring
	b) Bearing broken or worn	b) Replace the bearing
	c) Shifter fork bent	c) Replace shifter fork
	d) Improper control rod adjustment	d) Adjust the control rod
Hard shifting	a) Lack of lubricant	a) Add the recommended fluid to the proper level
	b) Shift lever binding on shaft	b) Repair and adjust the shift lever
	c) Shifting poppet ball scored	c) Replace the shifting poppet ball
	d) Shifter fork bent	d) Replace shifter fork
	e) Low tire pressure	e) Inflate tires to the correct pressure
Backlash	a) Companion yoke loose	a) Torque the companion yoke to specifications
	b) Transfer case loose on mounts	b) Torque the mount bolts to specifications
	c) Internal parts excessively worn	c) Rebuild the transfer case
Noisy	a) Low lubricant level	a) Add the recommended fluid to the proper level
	b) Bearings improperly adjusted or excessively worn	b) Adjust or replace the bearing
	c) Gears worn or damaged	c) Replace the gear in question
	d) Improper alignment of driveshaft or U-joint	d) Realign the driveshaft and U-joint
Oil leakage	a) Excessive amount of oil in the case	a) Drain the fluid to the proper level
	b) Oil vent clogged	b) Clean oil vent
	c) Gaskets or seals leaking	c) Replace leaking gaskets or seals
	d) Bearings improperly adjusted or excessively worn	d) Adjust or replace the bearing
	e) Driveshaft yoke mating surfaces scored	e) Machine the driveshaft yoke mating surface
Overheating	a) Excessive amount of oil in the case	a) Drain the fluid to the proper level
	b) Low lubricant level	b) Add the recommended fluid to the proper level
	c) Bearings adjustment too tight	c) Adjust or replace the bearing

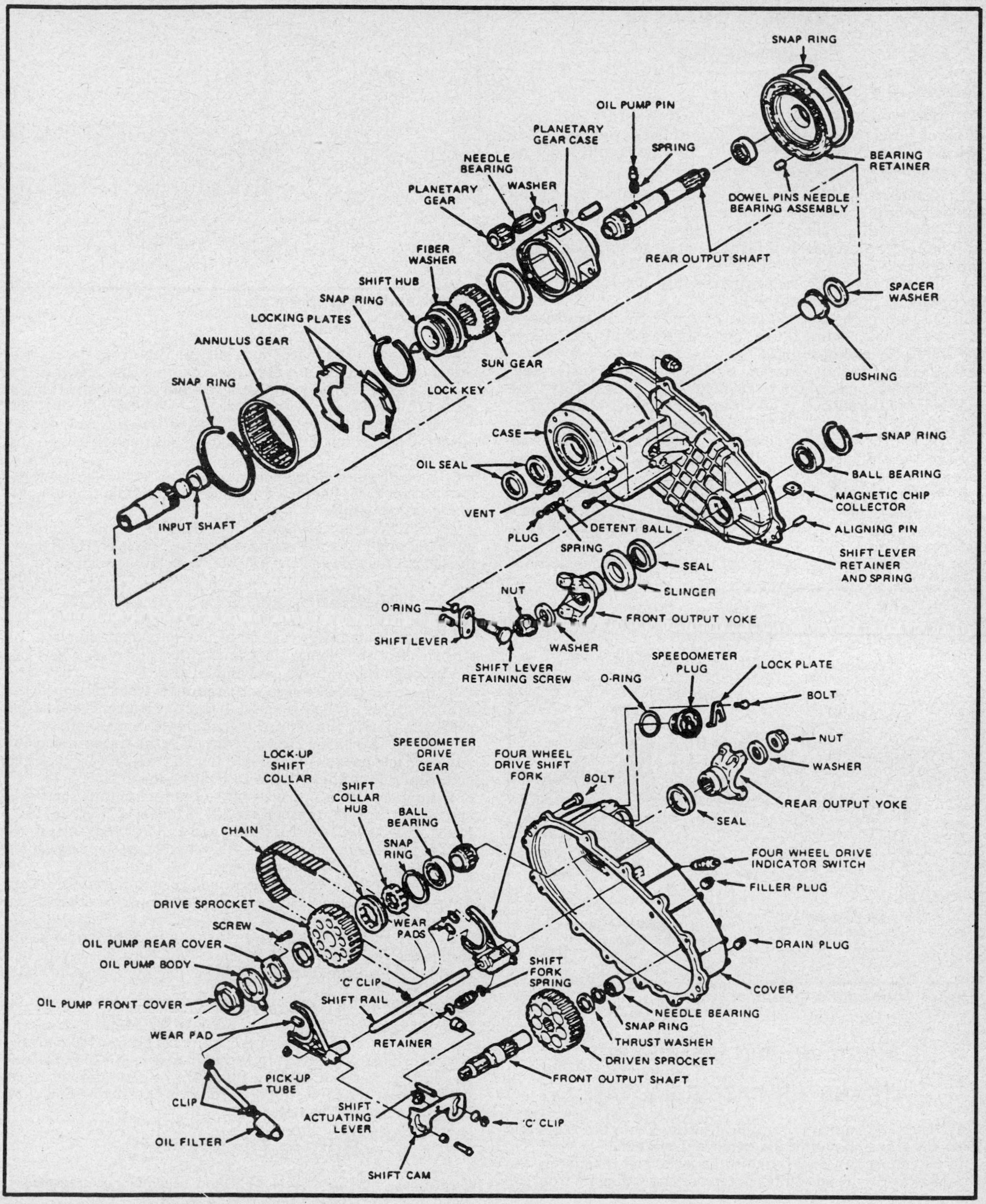
SNAP RING
OIL PUMP PIN
PLANETARY GEAR CASE
SPRING
BEARING RETAINER
NEEDLE BEARING
WASHER
PLANETARY GEAR
DOWEL PINS NEEDLE BEARING ASSEMBLY
FIBER WASHER
REAR OUTPUT SHAFT
SHIFT HUB
SNAP RING
SPACER WASHER
LOCKING PLATES
ANNULUS GEAR
SUN GEAR
BUSHING
SNAP RING
LOCK KEY
CASE
SNAP RING
OIL SEAL
BALL BEARING
MAGNETIC CHIP COLLECTOR
VENT
INPUT SHAFT
DETENT BALL
ALIGNING PIN
PLUG
SPRING
SHIFT LEVER RETAINER AND SPRING
SEAL
NUT
SLINGER
O-RING
FRONT OUTPUT YOKE
SHIFT LEVER
WASHER
SHIFT LEVER RETAINING SCREW
SPEEDOMETER PLUG
O-RING
LOCK PLATE
BOLT
SPEEDOMETER DRIVE GEAR
NUT
LOCK-UP SHIFT COLLAR
FOUR WHEEL DRIVE SHIFT FORK
WASHER
BOLT
SHIFT COLLAR HUB
REAR OUTPUT YOKE
BALL BEARING
SEAL
CHAIN
SNAP RING
FOUR WHEEL DRIVE INDICATOR SWITCH
FILLER PLUG
DRIVE SPROCKET
WEAR PADS
SCREW
OIL PUMP REAR COVER
DRAIN PLUG
OIL PUMP BODY
SHIFT FORK SPRING
'C' CLIP
COVER
OIL PUMP FRONT COVER
SHIFT RAIL
NEEDLE BEARING
SNAP RING
WEAR PAD
THRUST WASHER
RETAINER
DRIVEN SPROCKET
FRONT OUTPUT SHAFT
PICK-UP TUBE
CLIP
SHIFT ACTUATING LEVER
'C' CLIP
OIL FILTER
SHIFT CAM

Exploded view of the 1345 transfer case

On Vehicle Services

SHIFT LEVER

Removal and Installation

NOTE: Remove the shift ball only if the shift ball, boot or lever have to be replaced. If any of these parts are not being replaced, remove the shift ball, boot and lever as an assembly.

1. Remove the plastic insert from the shift ball. Warm the ball with a heat gun or equivalent until it reaches approximately 140–180°F. Using a block of wood and a hammer, knock the shift ball off the lever. Be careful not to damage the finish on the shift lever.
2. Remove the rubber boot with the floor pan cover. Disconnect the vent hose from the shift lever.
3. Disconnect the transfer case shift rod from the shift lever. Remove the bolt holding the shift lever to transfer case. Remove the shift lever and bushings.
4. Position the shift lever and bushings on the transfer case. Install the bolt and tighten to 70–90 ft. lbs. (95–122 Nm). Coat the pivots on the shift lever and transfer case lever with a multipurpose lubricant C1AZ–19590–B or equivalent.
5. Install the shift rod on the pivots and connect the vent hose to the shift lever. Install the rubber boot and floor pan cover.
6. Warm the ball with a heat gun or equivalent until it reaches approximately 140–180°F. Using a 7/16 in. socket and a mallet, tap the ball on to the lever and install the plastic shift pattern insert.
7. Check the transfer case for proper shifting and operation.

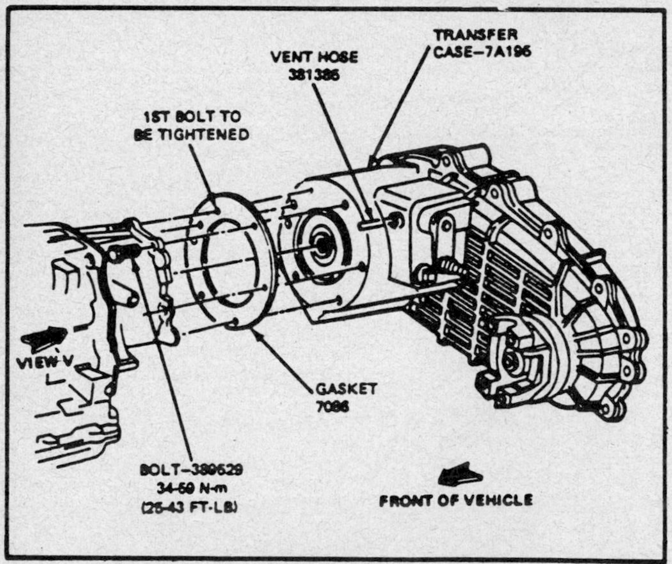
TRANSFER
CASE–7A196
VENT HOSE
381386
1ST BOLT TO
BE TIGHTENED
VIEW V
GASKET
7086
BOLT–389629
34-59 N-m
(25-43 FT-LB)
FRONT OF VEHICLE

Proper shift cam engagement

Removal and Installation

TRANSFER CASE REMOVAL

1. Raise and support the vehicle safely. Drain the gear fluid from the transfer case into a suitable drain pan.
2. Disconnect the 4WD indicator switch connector at the transfer case and if so equipped, remove the skid plate from the frame.
3. Disconnect the front driveshaft from the front output yoke. Disconnect the rear driveshaft from the rear output yoke.

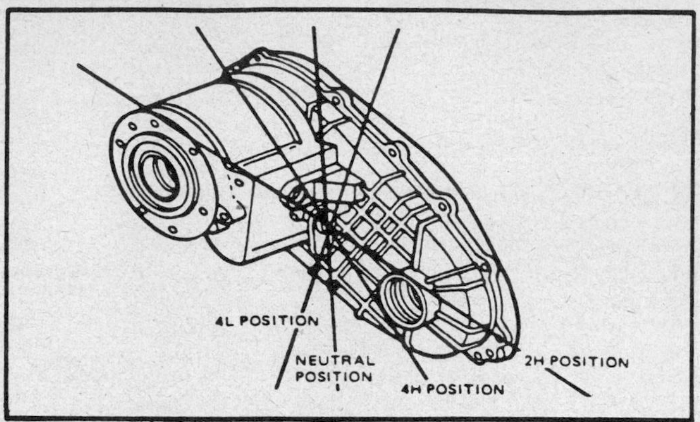
4L POSITION
NEUTRAL
POSITION
4H POSITION
2H POSITION

Proper shift lever installation

4. Disconnect the speedometer driven gear from the transfer case bearing retainer. Remove the retaining clips and shift rod from the transfer case control lever and remove the transfer case shift lever.
5. Disconnect the vent hose from the transfer case and remove the heat shield from the engine mount bracket and transfer case.
6. Using a suitable transmission jack or equivalent, support the transfer case. Remove the bolts that hold the transfer case to the transmission.
7. Slide the transfer case rearward off the transmission output shaft and lower the transfer case from the vehicle. Remove the old gasket between the transfer case and the adapter.

TRANSFER CASE INSTALLATION

1. Place the transfer case on asuitable tranmission jack. Install the heat shield onto the transfer case and place a new gasket between the transfer case and adapter.
2. Raise the transfer case with a suitable transmission jack or equivalent, raise it high enough so that the transmission output shaft aligns with the splined transfer case input shaft.
3. Slide the transfer case forward on to the transmission output shaft and onto the dowel pin. Install transfer case retaining bolts and torque them to 26–43 ft. lbs. (35–58 Nm).
4. Connect the rear driveshaft to the rear output shaft yoke and torque the retaining bolts to 20–28 ft. lbs. (27–38 Nm) (8–15 ft. lbs. (11–20 Nm) on the F-150 and F-350 4WD). Attach the shift rod to the transfer case shift lever and transfer case control rod and using the retaining rings.
5. Connect the speedometer driven gear to the transfer case. Connect the 4WD indicator switch wire connector at the transfer case.
6. Connect the front driveshaft to the front output yoke and torque the yoke nut to 8–15 ft. lbs. (11–20 Nm). Attach the heat shield to the engine mounting bracket and mounting lug on the transfer case.
7. Install the skid plate to the frame. Install the transfer case drain plug and torque the plug to 6–14 ft. lbs. (8–19 Nm).
8. Fill the transfer case with 6.5 pints of Dexron®II transmission fluid or equivalent. Torque the fill plug to 15–25 ft. lbs. (20–34 Nm). Lower the vehicle. Start the engine and check the transfer case for correct operation. Stop the engine and check the fluid level, add as necessary.

Before Disassembly

Cleanliness during disassembly and assembly is necessary to avoid further transfer case trouble after overhaul. Before removing any of the transfer case subassemblies, plug all the openings and clean the outside of the of the transfer case thoroughly. Steam cleaning or car wash type high pressure equipment is preferable. During disassembly, clean all parts in suitable solvent and dry each part. Do not use cloth or paper towels to dry parts. Use compressed air only.

Transfer Case Disassembly

1. Remove the transfer case from the vehicle. Drain the fluid from the case by removing the filler plug from the case half. Remove the speedometer cover.
2. Remove both output shaft yoke nuts and washers. Remove the front and rear output yokes. Remove the 4WD indicator switch.
3. Separate the cover from the case by removing the attaching bolts. Pry the case and cover apart using a pry bar. Remove the magnetic chip collector from the boss in the bottom of the case half.
4. Slide the shift collar hub off the rear output shaft and compress the shift fork spring. Remove the upper and lower spring retainers from the shaft.
5. As an assembly lift out from the case the 4 wheel lockup shift collar. Be careful not to lose the nylon wear pads on the lockup fork. Note the location of the holes on the nylon wear pad and lockup fork. Lift the output shaft from the case.
6. Remove the snapring from the front output shaft and remove the thrust washer. Grip the chain and both sprockets and lift straight up to to remove the drive sprockets, driven sprocket and chain from the output shafts. Remove the thrust washer from the rear output shaft.
7. Lift the front output shaft out from the case. Remove the 4 oil pump attaching bolts and remove the oil pump rear cover, pickup tube filter and pump body, 2 pump pins, pump spring and oil pump front cover from the rear output shaft. Disconnect the oil pump pickup tube from the pump body.

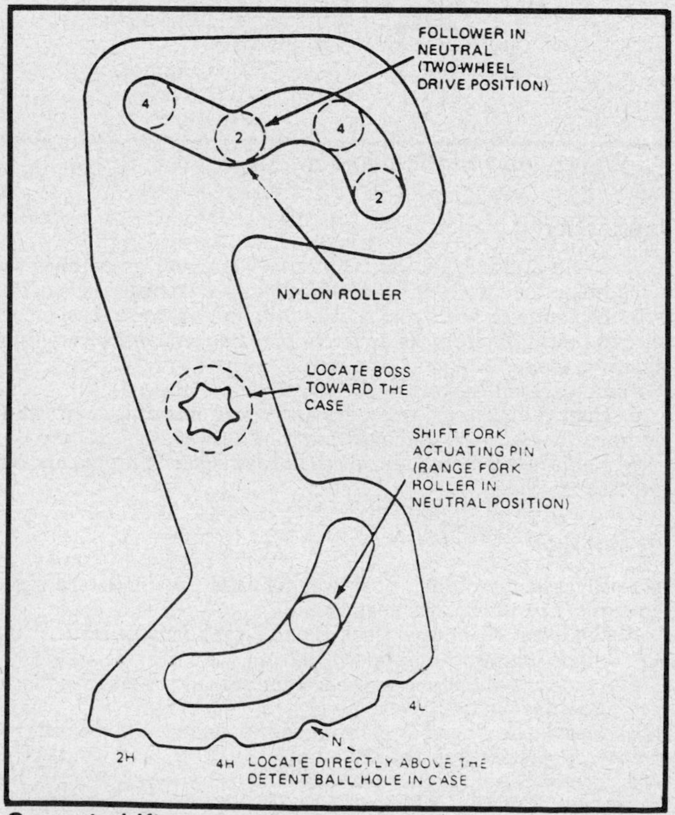
FOLLOWER IN NEUTRAL (TWO-WHEEL DRIVE POSITION)
NYLON ROLLER
LOCATE BOSS TOWARD THE CASE
SHIFT FORK ACTUATING PIN (RANGE FORK ROLLER IN NEUTRAL POSITION)
4L
N
2H
4H LOCATE DIRECTLY ABOVE THE DETENT BALL HOLE IN CASE

Correct shift cam engagement

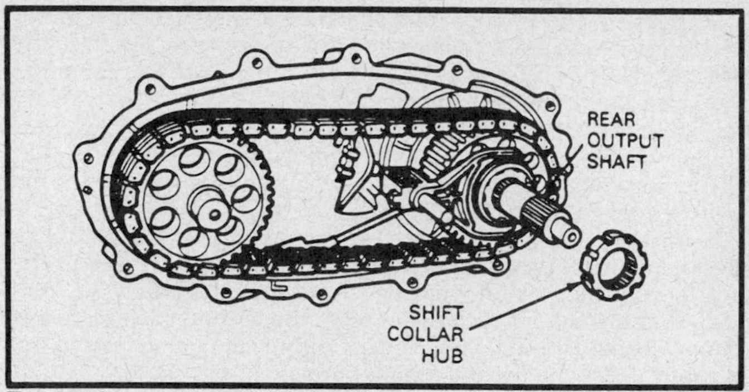
REAR OUTPUT SHAFT
SHIFT COLLAR HUB

Removing the shift collar hub

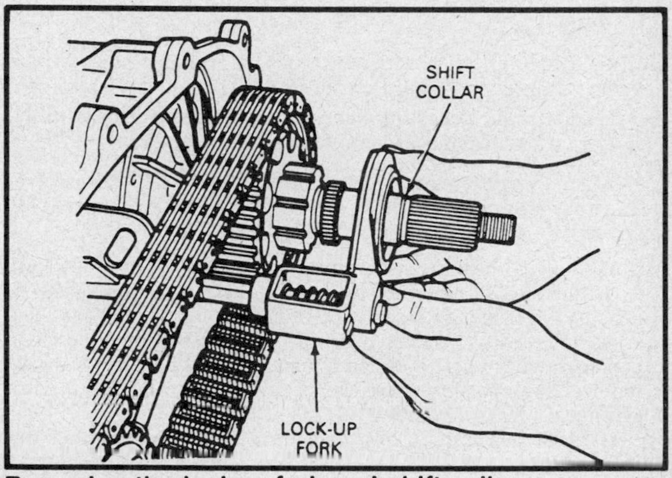
SHIFT COLLAR
LOCK-UP FORK

Removing the lockup fork and shift collar

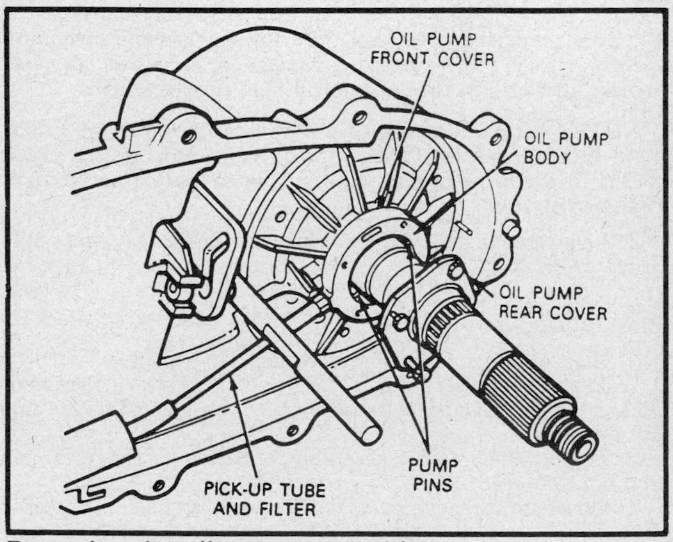
OIL PUMP FRONT COVER
OIL PUMP BODY
OIL PUMP REAR COVER
PUMP PINS
PICK-UP TUBE AND FILTER

Removing the oil pump assembly

8. Remove the snapring that holds the bearing retainer inside the case. Lift the rear output shaft while tapping on the bearing retainer with a plastic or soft mallet. Lift the rear output shaft and bearing retainer from the case.

NOTE: 2 dowel pins will fall into the case when the retainer is removed, be sure not to lose them.

9. Remove the rear output shaft from the bearing retainer and if necessary, press the needle bearing assembly out from the bearing retainer. Remove the C-clip that holds the shift cam to the shift actuating lever inside the case.
10. Remove the shift lever retaining screw and remove the shift lever from the case. When removing the lever, the shift cam will disengage from the shift lever and may release the detent ball and spring from the case.
11. As an assembly, remove the planetary gear set, shift rail, shift cam, output shaft and shift forks from the case. Be careful not to lose the 2 nylon wear pads on the shift fork.
12. Remove the spacer washer from the bottom of the case and remove the bushing. Using a drift pin or equivalent, drive out the plug from the detent spring bore.

CLEANING AND INSPECTION

Cleaning

During overhaul, all components of the transfer case (except bearing assemblies) should be thoroughly cleaned with solvent and dried with air pressure prior to inspection and reassembly.
1. Clean the bearing assemblies as follows:

NOTE: Proper cleaning of bearings is of utmost importance. Bearings should always be cleaned separately from other parts.

a. Soak all bearing assemblies in clean solvent or fuel oil. Bearings should never be cleaned in a hot solution tank.
b. Slush bearings in solvent until all old lubricant is loosened. Hold races so that bearings will not rotate; then clean bearings with a soft bristled brush until all dirt has been removed. Remove loose particles of dirt by tapping bearing flat against a block of wood.
c. Rinse bearings in clean solvent; then blow bearings dry with air pressure.

CAUTION

Do not spin bearings while drying.

d. After drying, rotate each bearing slowly while examining balls or rollers for roughness, damage, or excessive wear. Replace all bearings that are not in first class condition.

NOTE: After cleaning and inspecting bearings, lubricate generously with recommended lubricant, then wrap each bearing in clean paper until ready for reassembly.

2. Remove all portions of old gaskets from parts, using a stiff brush or scraper.

Inspection

1. Inspect all parts for discoloration or warpage.
2. Examine all gears and splines for chipped, worn, broken or nicked teeth. Small nicks or burrs may be removed with a fine abrasive stone.
3. Inspect the breather assembly to make sure that it is open and not damaged.
4. Check all threaded parts for damaged, stripped, or crossed threads.
5. Replace all gaskets, oil seals and snaprings.
6. Inspect housings, retainers and covers for cracks or other damage. Replace the damaged parts.
7. Inspect keys and keyways for condition and fit.
8. Inspect shift forks for wear, distortion or any other damage.
9. Check detent ball springs for free length, compressed length, distortion or collapsed coils.
10. Check bearing fit on their respective shafts and in their bores or cups. Inspect bearings, shafts and cups for wear.

NOTE: If either bearings or cups are worn or damaged, it is advisable to replace both parts.

11. Inspect all bearing rollers or balls for pitting or galling.
12. Examine detent balls for corrosion or brinneling. If shift bar detents show wear, replace them.
13. Replace all worn or damaged parts. When assembling the transfer case, coat all moving parts with recommended lubricant.

Unit Disassembly and Assembly

PLANETARY GEAR SET

Disassembly

1. Slide the input shaft rearward out of the planetary gear set.
2. Remove the snapring from the annulus gear. Remove the shift hub and planetary gear case from the annulus gear.
3. Unlock the 2 locking plates from the hub. Remove the shift hub snapring and T-shaped lock key. Lift the shift hub from the assembly.
4. Remove the outer fiber washer, sun gear and inner fiber washer. Rotate the inner fiber washer slightly upon removal to allow positioning tabs to clear the planetary gears. Replace the fiber washers upon assembly.

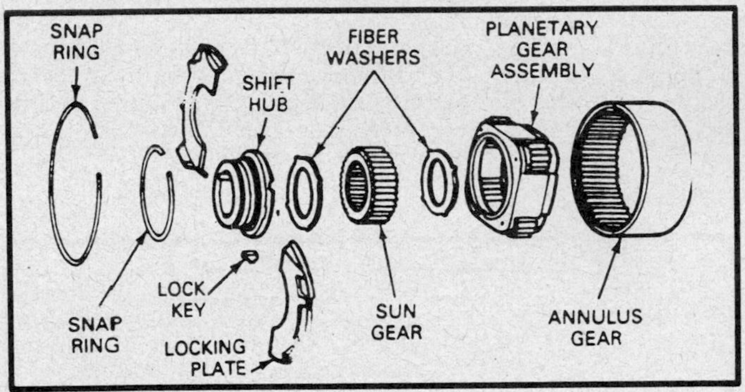
SNAP RING
FIBER WASHERS
PLANETARY GEAR ASSEMBLY
SHIFT HUB
LOCK KEY
SNAP RING
LOCKING PLATE
SUN GEAR
ANNULUS GEAR

Exploded view of the planetary gear set

Inspection

1. Clean and inspect the input shaft and gear assemblies.
2. Inspect all parts for discoloration or warpage.
3. Examine all gears and splines for chipped, worn, broken or nicked teeth. Small nicks or burrs may be removed with a fine abrasive stone.
4. Replace all gaskets, oil seals and snaprings.
5. Inspect housings, retainers and covers for cracks or other damage. Replace the damaged parts.
6. Replace any and all worn or damaged components as necessary.

Assembly

1. Place a new inner fiber washer into the planetary gear housing. Install the sun gear.
2. Coat the new outer fiber washer with petroleum jelly or equivalent. Install the outer washer on the hub.
3. Place the planetary gear cage and install the t-shaped lock kek and snapring.
4. Install the ring gear plate, with the dished side toward the planetary gear set, on the shift hub.
5. Lower the planetary assembly into the annulus gear. Be sure the tabs on the locking plate engage the annulus gear teeth. Install the snapring.

COVER

Disassembly

1. Remove the snapring retainer from the rear output shaft ball bearing assembly in the cover.
2. Turn the cover over and remove the rear output shaft seal with a suitable slide hammer and seal removal tool.
3. Remove the speedometer drive gear. Press the rear output shaft ball bearing out from the cover.
4. Remove the speedometer gear adapter. Remove the front output shaft inner needle bearing from the cover with a suitable slide hammer.

Inspection

1. Clean and inspect the output shaft and bearing assemblies.
2. Inspect all parts for discoloration or warpage.
3. Replace all gaskets, oil seals and snaprings.
4. Inspect housings, retainers and covers for cracks or other damage. Replace the damaged parts.
5. Replace any and all worn or damaged components as necessary.

Assembly

1. Press a new needle bearing into the cover using a suitable output bearing installation tool.
2. Press a new ball bearing assembly into the cover. Install the snapring. Turn the cover over and install the speedometer drive gear.
3. Install the new output shaft seal into position. Install the speedometer gear adapter.

CASE

Disassembly

1. Remove the snapring retaining front output shaft ball bearing assembly in the case.
2. Remove the output shaft seal and both input shaft seals. Press the front output shaft bearing and input shaft bushing from the case.

Inspection

1. Clean and inspect the output shaft and ball bearing assemblies.
2. Inspect all parts for discoloration or warpage.
3. Replace all gaskets, oil seals and snaprings.
4. Inspect housings, retainers and covers for cracks or other damage. Replace the damaged parts.
5. Replace any and all worn or damaged components as necessary.

Assembly

1. Press the new input shaft bushing into the case. Be sure the lug is in the downward position.
2. Install the new output shaft ball bearing. Install the snapring.
3. Press both input shaft seal into the case. Press the front output shaft seal into the case.

TRANSFER CASE ASSEMBLY

NOTE: Before starting the assembly procedure, lubricate all the internal parts, with Dexron®II transmission fluid or equivalent.

1. Assemble the planetary gear set, shift rail, shift cam, input shaft and shift fork together as a unit. Be sure that the boss on the shift cam is installed toward the case. Install the spacer washer on the input shaft.

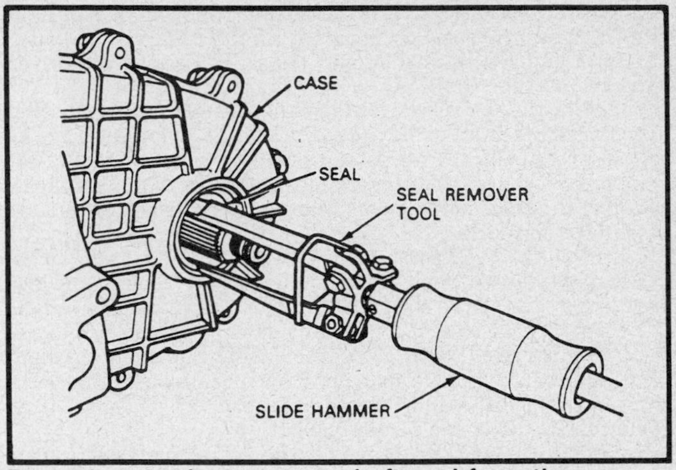
CASE
SEAL
SEAL REMOVER TOOL
SLIDE HAMMER

Removing the front output shaft seal from the case

2. Place the rear output shaft in the planetary gear set, being sure that the shift cam engages the shift fork actuating pin. Lay the case on its side and insert the rear output shaft with planetary gear set into the case. Be sure that the spacer washer remains on the input shaft.
3. Install the shift rail into the hole in the case. Install the outer roller bushing into the guide in the case. Remove the rear output shaft and position the shift fork in neutral.
4. Place the shift control lever shaft through the cam and install the clip ring. Be sure that the shift control lever is pointed downward and is parallel to the front face of the case in the neutral position.
5. Check the shift fork and planetary gear engagement. The unit should move freely with out any binding. Press a new needle bearing into the needle bearing retainer using output bearing replacer tool T80T-7127-C or equivalent.
6. Insert the output shaft through the bearing retainer from the bottom side outward. Insert the rear output shaft pilot into the input shaft rear bushing. Align the dowel holes and lower the bearing into position.
7. Install the dowel pins. Install the snapring that retains the bearing retainer in the case. Insert the detent ball and spring in the detent bore in the case half.
8. Coat the seal plug with RTV sealant or equivalent. Drive the plug into the case until the lip of the plug is $^1/_{32}$ in. below the surface of the case. Peen the case over the plug in 2 places.
9. Install the oil pump front cover over the output shaft with the flanged side down. The word **TOP** must be facing the top of the transfer case as the position the case is installed in the vehicle.
10. Install the oil pump spring and 2 pump pins with the flat side outward in the hole in the output shaft. Push both pins in to install the oil pump body, pickup tube and filter. The rear markings on the pump must be facing upward. Be sure to prime the pump with Dexron®II transmission fluid or equivalent.
11. Place the oil pump rear cover on the output shaft with the flanged side outward. The word **TOP REAR** is positioned toward the the top of the transfer case in the position the transfer case is installed in the vehicle. Apply Loctite® or equivalent to the oil pump bolts and install the pump cover. Torque the bolts to 3–4 ft. lbs. (4–5 Nm) and be sure to rotate the pump while tightening.

NOTE: When the oil pump is correctly installed, it will rotate freely on the output shaft.

12. Install the thrust washer on the rear output shaft next to the oil pump. Install the chain on the drive sprocket and driven sprocket. Lower the chain and sprockets into position in the

case. The driven sprocket is installed over the front output shaft and the drive the sprocket is placed on the rear output shaft.

13. Assemble the washer and snapring behind the driven sprocket. Engage the 4WD fork on the shift collar.

14. Slide the shift fork over the shift shaft and the shift collar over the rear output shaft. Be sure that the nylon pads are installed on the shift fork tips and the necked down part of the shift collar lockup fork are assembled correctly. Note that the location of the holes in the nylon wear pad and the lockup fork are assembled correctly.

15. Push the 4WD shift spring downward and install the upper spring retainer. Push the spring upward and install the lower retainer. Install the shift collar hub on the rear output shaft.

16. Apply a bead of RTV sealant or equivalent on the case mounting surface. Lower the cover over the rear output shaft and align the shift rail to its blind hole in the cover. Be sure the front output shaft is fully seated in its support bearing. Install the attaching bolts and torque the bolts to 40–45 ft. lbs. (54–61 Nm). Allow 1 hour curing time for the gasket material prior to the operating vehicle.

17. Install the 4WD indicator switch and torque the bolts down at 8–12 ft. lbs. (11–16 Nm). Press an oil slinger on the front yoke.

18. Install the front and rear output shaft yokes. Install the anti-spill oil seal. Coat the faces of the yoke nuts and output shaft threads with a suitable thread sealer. Torque the yoke nuts to 100–130 ft. lbs. (136–176 Nm). Install the speedometer assembly.

19. Refill the transfer case with 6.5 pints of Dexron® II transmission fluid or equivalent. Torque the level plug and the drain plugs to 6–14 ft. lbs. (8–19 Nm). Torque the fill plug to 15–22 ft. lbs. (20–32 Nm).

20. Install the transfer case.

21. Start the engine and check the transfer case for correct operation. Stop the engine and check the fluid level, add as necessary.

22. Fluid should drip from the level hole. If the fluid flows out level hole in a stream, the pump may not be operating properly.

Special Tools

Number	Description
T50T-100-A	Impact slide hammer—2½ lb.
D80L-100-A	Blind hole puller set
D80L-100-T	Blind hole puller
D80L-100-H	Actuator pin
TOOL-1175-AC	Seal remover
T80T-7127-B	Output bearing replacer—front
T80T-7127-C	Output bearing replacer—rear

Specifications

Description	N-m	Ft-Lbs
Case Half Attaching Bolts	48-54	35-40
Four Wheel Drive Indicator Switch	11-16	8-12
Front and Rear Output Yokes to Transfer Case	163-203	120-150
Drain Plug	13-24	14-22
Fill Plug	21-33	15-25
Transfer Case to Transmission Adapter	34-58	25-43
Heat Shield to Transfer Case	54-61	40-45
Skid Plate to Frame	20-27	15-20
Front Driveshaft to Front Output Yoke	163-203	120-150
Rear Driveshaft to Rear Output Yoke	163-203	120-150

BORG WARNER 1350 MANUAL SHIFT TRANSFER CASE

Application

1984–89 Ford Ranger

General Description

The Borg Warner 1350 is a 3 piece aluminum part time transfer case. It transfers power from the transmission to the front drive axle and the rear axle and when actuated. The unit is lubricated by a positive displacement oil pump that channels oil flow through drilled holes in the rear output shaft. The pump turns with the rear output shaft and allows towing of the vehicle at maximum legal road speeds for extended distances without disconnecting the front and/or rear driveshaft.

Trouble Diagnosis

CHILTON'S THREE C'S TRANSFER CASE DIAGNOSIS

Condition	Cause	Correction
Slips out of gear	a) Shifting poppet spring weak b) Bearing broken or worn c) Shifter fork bent d) Improper control rod adjustment	a) Replace the shifting poppet spring b) Replace the bearing c) Replace shifter fork d) Adjust the control rod
Hard shifting	a) Lack of lubricant b) Shift lever binding on shaft c) Shifting poppet ball scored d) Shifter fork bent e) Low tire pressure	a) Add the recommended fluid to the proper level b) Repair and adjust the shift lever c) Replace the shifting poppet ball d) Replace shifter fork e) Inflate tires to the correct pressure
Backlash	a) Companion yoke loose b) Transfer case loose on mounts c) Internal parts excessively worn	a) Torque the companion yoke to specifications b) Torque the mount bolts to specification c) Rebuild the transfer case
Noisy	a) Low lubricant level b) Bearings improperly adjusted or excessively worn c) Gears worn or damaged d) Improper alignment of driveshafts or U-joint	a) Add the recommended fluid to the proper level b) Adjust or replace the bearing c) Replace the gear in question d) Realign the driveshaft and U-joint
Oil leakage	a) Excessive amount of oil in the case b) Oil vent clogged c) Gaskets or seals leaking d) Bearings improperly adjusted or excessively worn e) Driveshaft yoke mating surfaces scored	a) Drain the fluid to the proper level b) Clean oil vent c) Replace leaking gaskets or seals d) Adjust or replace the bearing e) Machine the driveshaft yoke mating surface
Overheating	a) Excessive amount of oil in the case b) Low lubricant level c) Bearing adjustment too tight	a) Drain the fluid to the proper level b) Add the recommended fluid to the proper level c) Adjust or replace the bearing

On Vehicle Services

SHIFT LINKAGE

Adjustment

If partial or incomplete engagement of the transfer case shift lever detent position occurs, or should the transfer case controll assembly ever require removal, proceed as follows:

1. Raise the shift boot to expose the top surface of the cam plates.
2. Loosen the 1 large and 1 small bolt, approximately 2 turns. Move the transfer case shift lever to the **4L** position.
3. Move the cam plate rearward until the bottom chamfered corner of the neutral lug just contacts the forward right edge of the shift lever.
4. Hold the cam plate in this position and torque the larger bolt first to 70–90 ft. lbs. (95–122 Nm) and torque the smaller bolt to 31–42 ft. lbs. (42–57 Nm).
5. Move the transfer case in cab shift lever to all shift positions to check for positive engagement. There should be a clearance between the shift lever and the cam plate in the **H** front and **4H** rear (clearance not to exceed 0.13 in.) and **4L** shift positions.
6. Install the shift boot assembly.

SHIFT LEVER

Removal and Installation

NOTE: Remove the shift ball only if the shift ball, boot or lever is to be replaced. If the ball, boot or lever is not

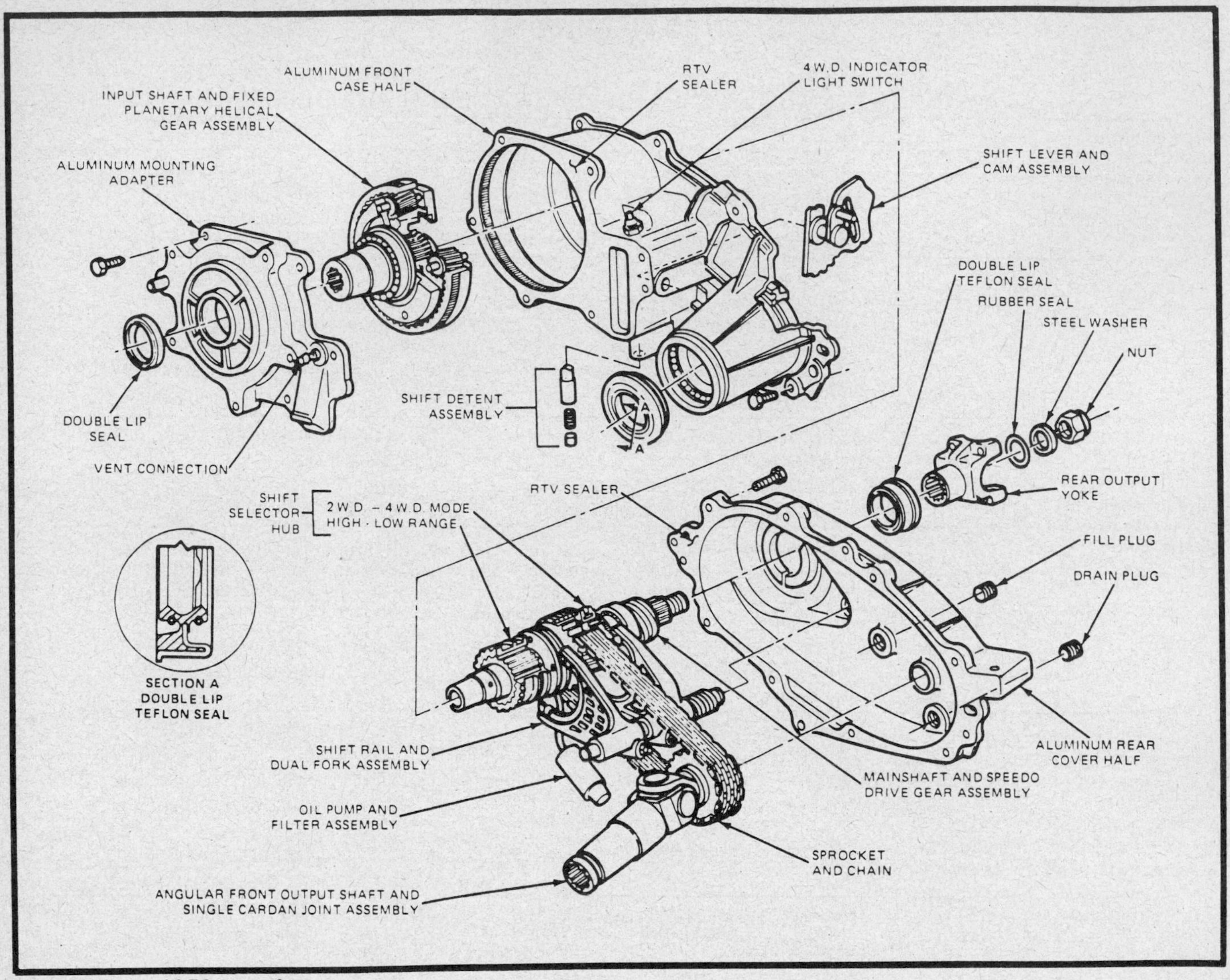
ALUMINUM FRONT CASE HALF
RTV SEALER
4 W.D. INDICATOR LIGHT SWITCH
INPUT SHAFT AND FIXED PLANETARY HELICAL GEAR ASSEMBLY
ALUMINUM MOUNTING ADAPTER
SHIFT LEVER AND CAM ASSEMBLY
DOUBLE LIP TEFLON SEAL
RUBBER SEAL
STEEL WASHER
NUT
SHIFT DETENT ASSEMBLY
DOUBLE LIP SEAL
A
VENT CONNECTION
RTV SEALER
REAR OUTPUT YOKE
SHIFT SELECTOR HUB
2 W.D. - 4 W.D. MODE
HIGH - LOW RANGE
FILL PLUG
DRAIN PLUG
SECTION A DOUBLE LIP TEFLON SEAL
SHIFT RAIL AND DUAL FORK ASSEMBLY
ALUMINUM REAR COVER HALF
MAINSHAFT AND SPEEDO DRIVE GEAR ASSEMBLY
OIL PUMP AND FILTER ASSEMBLY
SPROCKET AND CHAIN
ANGULAR FRONT OUTPUT SHAFT AND SINGLE CARDAN JOINT ASSEMBLY

Borg-Warner 1350 transfer case

being replaced, remove the ball, boot and lever as an assembly.

1. Remove the plastic insert from the shift ball. Warm the ball with a heat gun to 60–87°C (140–180°F) and knock the ball off the lever with a block of wood and a hammer. Be careful not to damage the finish on the shift lever.
2. Remove the rubber boot and floor pan cover.
3. Disconnect the vent hose from the control lever.
4. Unscrew the shift lever from the control lever.
5. Remove the bolts retaining the shifter to the extension housing. Remove the control lever and bushings.
6. Move the transfer case shift lever to the **4L** position.
7. Install the shifter assembly with the bolts finger tight and move the cam plate rearward until the bottom chamfered corner of the neutral lug just contacts the forward right edge of the shift lever.
8. Hold the cam plate in this position and torque the larger bolt first to 70–90 ft. lbs. (95–122 Nm) and torque the smaller bolt to 31–42 ft. lbs. (42–57 Nm).
9. Move the transfer case in cab shift lever to all shift positions to check for positive engagement. There should be a clearance between the shift lever and the cam plate in the **H** front and **4H** rear (clearance not to exceed 0.13 in.) and **4L** shift positions.
10. Install the shift lever to the control lever and torque the bolts to 27–37 ft. lbs. (37–50 Nm). Install the vent assembly so that the white marking on the house is in the position in the notch in the shifter.

NOTE: The upper end of the vent hose should be 2 in. above the top of the shifter and positioned inside of the shift lever boot.

11. Install the rubber boot and floor pan cover.
12. Warm the ball with a heat gun or equivalent until it reaches approximately 140–180°F. Using a $^7/_{16}$ in. socket and a mallet, tap the ball on to the lever and install the plastic shift pattern insert. The end of the shift ball should be to the end of the knurl on the upper portion of the shift lever.
13. Check the transfer case for proper shifting and operation.

FRONT OUTPUT SHAFT OIL SEAL

Removal and Installation

1. Raise and support the vehicle safely.

2. Place a drain pan under transfer case, remove the drain plug and drain fluid from the transfer case.
3. Disconnect the front driveshaft from the axle input yoke.
4. Loosen the clamp retaining the front driveshaft boot to the transfer case, and pull the driveshaft and front boot assembly out of the transfer case front output shaft.
5. Remove the oil seal from the front output housing bore with a suitable seal remover and slide hammer.
6. Make sure that the housing face and bore are free from nicks and burrs. Coat the seal with a suitable grease.
7. Position the oil seal into the front output housing bore, making sure the oil seal is not cocked in the bore. Drive the new seal into the housing with a suitable seal installer.
8. Clean the transfer case front output female spline and apply a suitable grease. Insert the front driveshaft male spline.
9. Connect the front driveshaft to the front output yoke and torque the yoke nut to 8–15 ft. lbs. (11–20 Nm). Attach the heat shield to the engine mounting bracket and mounting lug on the transfer case.
10. Install the skid plate to the frame. Install the transfer case drain plug and torque the plug to 14–22 ft. lbs. (19–32 Nm).
11. Fill the transfer case with 6.5 pints of Dexron®II transmission fluid or equivalent. Torque the fill plug to 14–22 ft. lbs. (19–32 Nm). Lower the vehicle. Start the engine and check the transfer case for correct operation. Stop the engine and check the fluid level, add as necessary.

Removal and Installation

TRANSFER CASE REMOVAL

The catalytic converter is located beside the transfer case. Be careful when working around the catalytic converter because of the extremely high temperatures generated by the converter.

1. Raise and support the vehicle safely.
2. If so equipped, remove the skid plate from frame.
3. Place a drain pan under transfer case, remove the drain plug and drain fluid from the transfer case.
4. Disconnect the 4WD indicator switch wire connector at the transfer case.
5. Disconnect the front driveshaft from the axle input yoke.
6. Loosen the clamp retaining the front driveshaft boot to the transfer case, and pull the driveshaft and front boot assembly out of the transfer case front output shaft.
7. Disconnect the rear driveshaft from the transfer case output shaft yoke.
8. Disconnect the speedometer driven gear from the transfer case rear cover.
9. Disconnect the vent hose from the control lever.
10. Loosen or remove the large bolt and the small bolt retaining the shifter to the extension housing. Pull on the control lever until the bushing slides off the transfer case shift lever pin. If necessary, unscrew the shift lever from the control lever.
11. Remove the heat shield from the transfer case.
12. Support the transfer case with a transmission jack.
13. Remove the 5 bolts retaining the transfer case to the transmission and the extension housing.
14. Slide the transfer case rearward off the transmission output shaft and lower the transfer case from the vehicle. Remove the gasket from between the transfer case and extension housing.

TRANSFER CASE INSTALLATION

1. Install the heat shield onto the transfer case and place a new gasket between the transfer case and adapter.
2. Raise the transfer case with a suitable transmission jack or equivalent, raise it high enough so that the transmission output shaft aligns with the splined transfer case input shaft.
3. Slide the transfer case forward on to the transmission output shaft and onto the dowel pin. Install transfer case retaining bolts and torque them to 26–43 ft. lbs. (36–58 Nm).
4. Connect the rear driveshaft to the rear output shaft yoke and torque the retaining bolts to 20–28 ft. lbs. (27–38 Nm). Attach the shift rod to the transfer case shift lever and transfer case control rod and attach with retaining rings.
5. Connect the speedometer driven gear to the transfer case. Connect the 4WD indicator switch wire connector at the transfer case.
6. Connect the front driveshaft to the front output yoke and torque the yoke nut to 8–15 ft. lbs. (11–20 Nm). Attach the heat shield to the engine mounting bracket and mounting lug on the transfer case.
7. Install the skid plate to the frame. Install the transfer case drain plug and torque the plug to 14–22 ft. lbs. (19–32 Nm).
8. Fill the transfer case with 6.5 pints of Dexron®II transmission fluid or equivalent. Torque the fill plug to 14–22 ft. lbs.(19–32 Nm).
9. Lower the vehicle. Start the engine and check the transfer case for correct operation. Stop the engine and check the fluid level, add as necessary.

Before Disassembly

Cleanliness during disassembly and assembly is necessary to avoid further transfer case trouble after overhaul. Before removing any of the transfer case subassemblies, plug all the openings and clean the outside of the of the transfer case thoroughly. Steam cleaning or car wash type high pressure equipment is preferable. During disassembly, clean all parts in suitable solvent and dry each part. Do not use cloth or paper towels to dry parts. Use compressed air only.

Transfer Case Disassembly

1. Place the transfer case into a suitable holding fixture.
2. Remove the transfer case drain plug with a 3/8 in. drive ratchet and drain the fluid.
3. Remove the 4WD indicator switch and the breather vent.
4. Remove the rear output shaft yoke by removing the 30mm nut, steel washer and rubber seal from the output shaft.
5. Remove the 9 bolts which retain the front case to the rear cover. Insert a ½ in. drive breaker bar between the 3 pry bosses and separate the front case from the rear cover. Remove all traces of RTV gasket sealant from the mating surfaces of the front case and rear cover.

NOTE: When removing RTV sealant, take care not to damage the mating surface of the aluminum case.

6. If the speedometer drive gear or ball bearing assembly is to be replaced, first, drive out the output shaft oil seal from either the inside of the rear cover with a brass drift and hammer or from the outside by bending and pulling on the curved-up lip of the oil seal. Remove and discard the oil seal. Remove the speedometer drive gear assembly (gear, clip and spacer). Note that the round end of the speedometer gear clip faces the inside of the rear cover.
7. Remove the internal snapring that retains the rear output shaft ball bearing in the bore. From the outside of the case, drive out the ball bearing using the output shaft bearing replacer tool T83T–7025–B and drive handle tool T80T–4000–W or equivalent.
8. If required, remove the front output shaft caged needle bearing from the rear cover with puller collet tool D80L–100–S and impact slide hammer tool T50T–100–A or equivalent.
9. Remove the 2WD–4WD shift fork spring from the boss in the rear cover.
10. Remove the shift collar hub from the output shaft. Remove the 2WD–4WD lockup assembly and the 2WD–4WD shift fork together as an assembly. Remove the 2WD–4WD fork from the 2WD–4WD lockup assembly. If required, remove the external

clip and remove the roller bushing assembly (bushing, shaft and external clip) from the 2WD–4WD shift fork.

11. If required to disassemble the 2WD–4WD lockup assembly, remove the internal snapring and pull the lockup hub and spring from the lockup collar.

12. Remove the external snapring and thrust washer that retains the drive sprocket to the front output shaft.

13. Remove the chain, driven sprocket and drive sprocket as an assembly.

14. Remove the collector magnet from the notch in the front case bottom.

15. Remove the output shaft and oil pump as an assembly.

16. If required to disassemble the oil pump, remove the 4 bolts from the body. Note the position and markings of the front cover, body, pins, spring, rear cover, and pump retainer as removed.

17. Pull out the shift rail.

18. Slip the high-low range shift fork out of the inside track of the shift cam. If required, remove the external clip and remove the roller bushing assembly (bushing, shaft and external clip) from the high-low range shift fork.

19. Remove the high-low shift hub from out of the planetary gearset in the front case.

20. Push and pull out the anchor end of the torsion spring from the locking post in the front case half. Remove the torsion spring and roller out of the shift cam, if so equipped.

21. Turn the front case over and remove the 6 bolts retaining the mounting adapter to the front case. Remove the mounting adapter, input shaft and planetary gearset as an assembly.

22. If required, remove the ring gear from the front case using a press. Note the relationship of the serrations to the chamfered pilot diameter during removal.

23. Expand the tangs of the large snapring in the mounting adapter and pry under the planetary gearset and separate the input shaft and planetary gearset from the mounting adapter.

24. If required, remove the oil from the mounting adapter with seal remover tool 1175–AC and impact slide hammer tool T50T–100–A or equivalent.

25. Remove the internal snapring from the planetary carrier and separate the planetary gearset from the input shaft assembly.

26. Remove the external snapring from the input shaft. Place the input shaft assembly in a press and remove the ball bearing from the input shaft using bearing splitter tool D79L–4621–A or equivalent. Remove the thrust washer, thrust plate and sun gear off the input shaft.

27. Move the shift lever by hand until the shift cam is in the 4 wheel high detent position (4WD-HI) and mark a line on the outside of the front case using the side of the shift lever and a grease pencil.

28. Remove the 2 set screws from the front case and from the shift cam.

29. Turn the front case over and remove the external clip. Pry the shift lever out of the front case and shift cam.

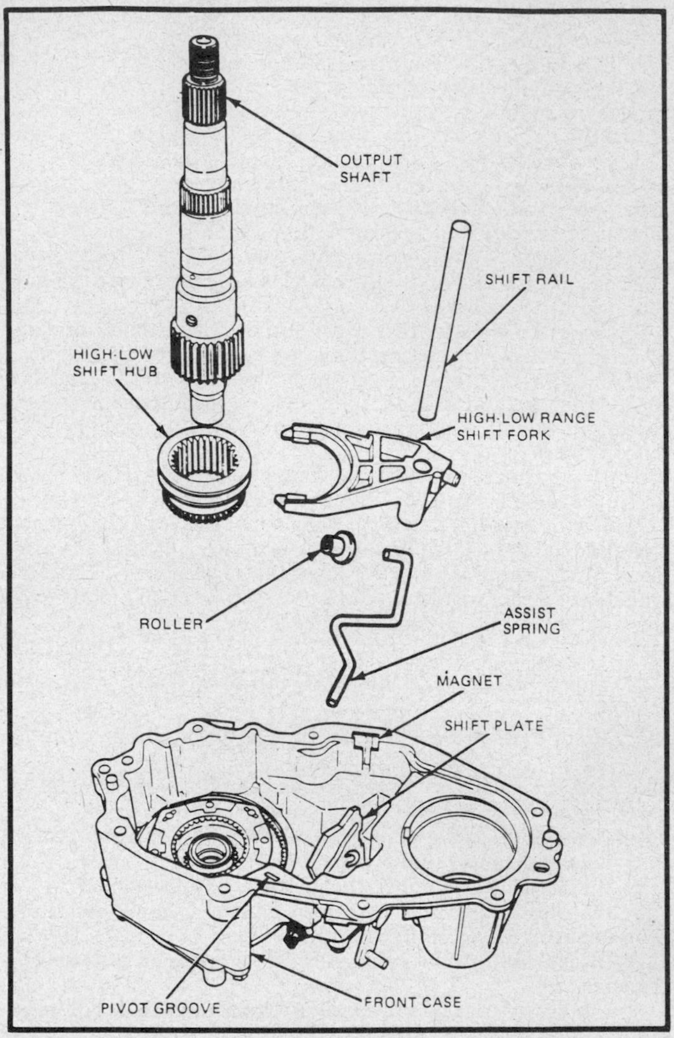
OUTPUT SHAFT
SHIFT RAIL
HIGH-LOW SHIFT HUB
HIGH-LOW RANGE SHIFT FORK
ROLLER
ASSIST SPRING
MAGNET
SHIFT PLATE
PIVOT GROOVE
FRONT CASE

High-low range shift assembly and output shaft

NOTE: Do not pound on the external clip during removal. Removal of 4WD indicator switch will ease removal of the shift lever and shift cam assembly.

30. Remove the O-ring from the second groove in the shift lever shaft.

31. Remove the detent plunger and compression spring from the inside of the front case.

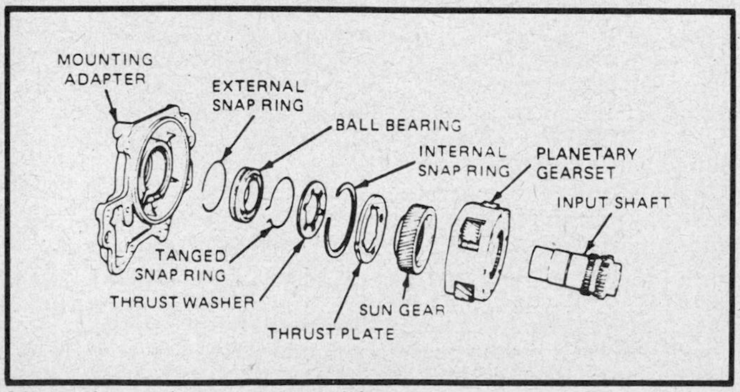
MOUNTING ADAPTER
EXTERNAL SNAP RING
BALL BEARING
INTERNAL SNAP RING
PLANETARY GEARSET
INPUT SHAFT
TANGED SNAP RING
THRUST WASHER
SUN GEAR
THRUST PLATE

Mounting adapter assembly

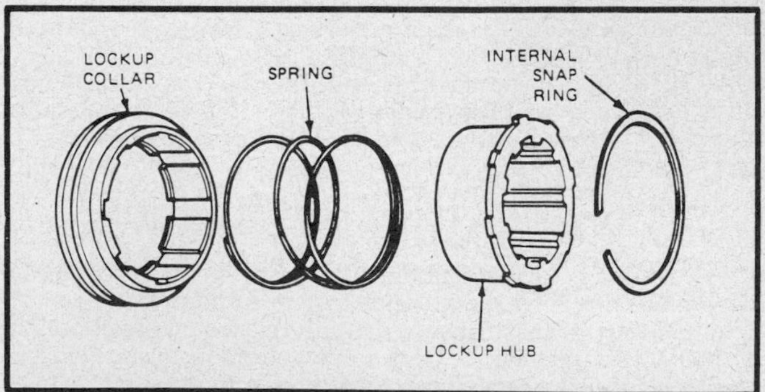
LOCKUP COLLAR
SPRING
INTERNAL SNAP RING
LOCKUP HUB

Exploded view of the 2WD and 4WD lockup assembly

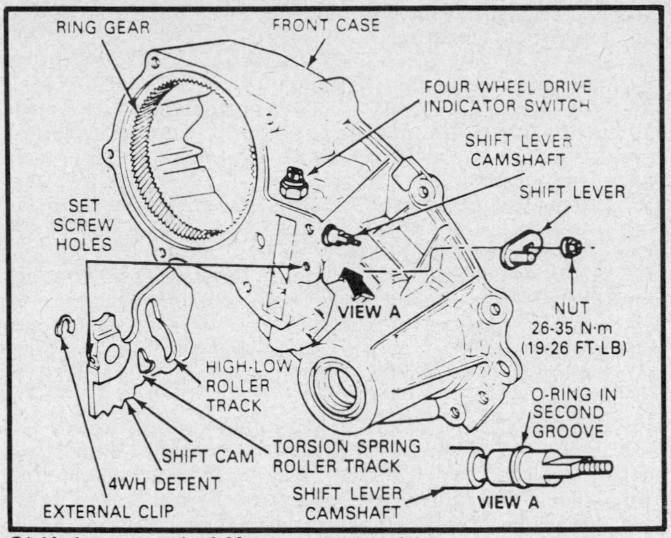
RING GEAR
FRONT CASE
FOUR WHEEL DRIVE INDICATOR SWITCH
SHIFT LEVER CAMSHAFT
SHIFT LEVER
SET SCREW HOLES
VIEW A
NUT 26-35 N·m (19-26 FT-LB)
HIGH-LOW ROLLER TRACK
O-RING IN SECOND GROOVE
TORSION SPRING ROLLER TRACK
SHIFT CAM
4WH DETENT
SHIFT LEVER CAMSHAFT
EXTERNAL CLIP
VIEW A

Shift lever and shift cam assembly

32. Remove the internal snapring and remove the ball bearing retainer from the front case by tapping on the face of the front output shaft and U-joint assembly with a plastic hammer. Remove the internal snapring and drive the ball bearing out of the bearing retainer using output shaft bearing replacer tool T83T–7025–B and driver handle tool T80T–4000–W or equivalent.

NOTE: The clip is required to prevent the bearing retainer from rotating. Do not discard the clip.

33. Remove the front output shaft and U-joint assembly from the front case. If required, remove the oil seal with seal remover tool–1175–AC and impact slide hammer tool T50T–100–A or equivalent. If required, remove the internal snapring and drive the ball bearing out of the front case bore using output shaft replacer tool T83T–7025–B and driver handle tool T80T–4000–W or equivalent.

34. If required, place the front output shaft and U-joint assembly in a vise, being careful not to damage the assembly. Use copper or wood vise jaws.

35. Remove the internal snaprings that retain the bearings in the shaft.

36. Position the U-joint tool T74P–4635–C or equivalent, over the shaft ears and press the bearing out. If the bearing cannot be pressed all the way out, remove it with vise grip or channel lock pliers.

37. Re-position the U-joint tool on the spider in order to remove the opposite bearing.

38. Repeat the above procedure until all bearings are removed.

CLEANING AND INSPECTION

Cleaning

During overhaul, all components of the transfer case (except bearing assemblies) should be thoroughly cleaned with solvent and dried with air pressure prior to inspection and reassembly.

1. Clean the bearing assemblies as follows:

NOTE: Proper cleaning of bearings is of utmost importance. Bearings should always be cleaned separately from other parts.

a. Soak all bearing assemblies in clean solvent or fuel oil. Bearings should never be cleaned in a hot solution tank.

b. Slush bearings in solvent until all old lubricant is loosened. Hold races so that bearings will not rotate; then clean

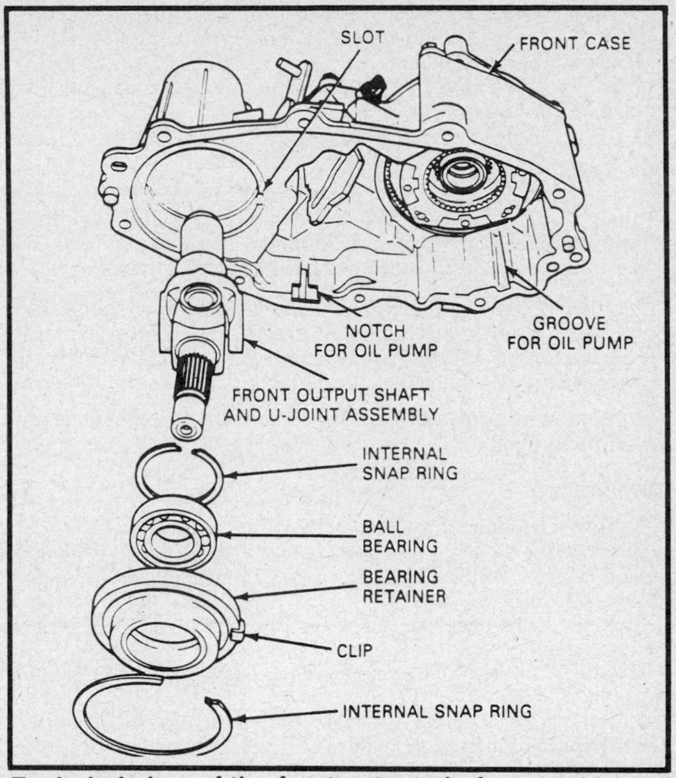
SLOT
FRONT CASE
NOTCH FOR OIL PUMP
GROOVE FOR OIL PUMP
FRONT OUTPUT SHAFT AND U-JOINT ASSEMBLY
INTERNAL SNAP RING
BALL BEARING
BEARING RETAINER
CLIP
INTERNAL SNAP RING

Exploded view of the front output shaft assembly

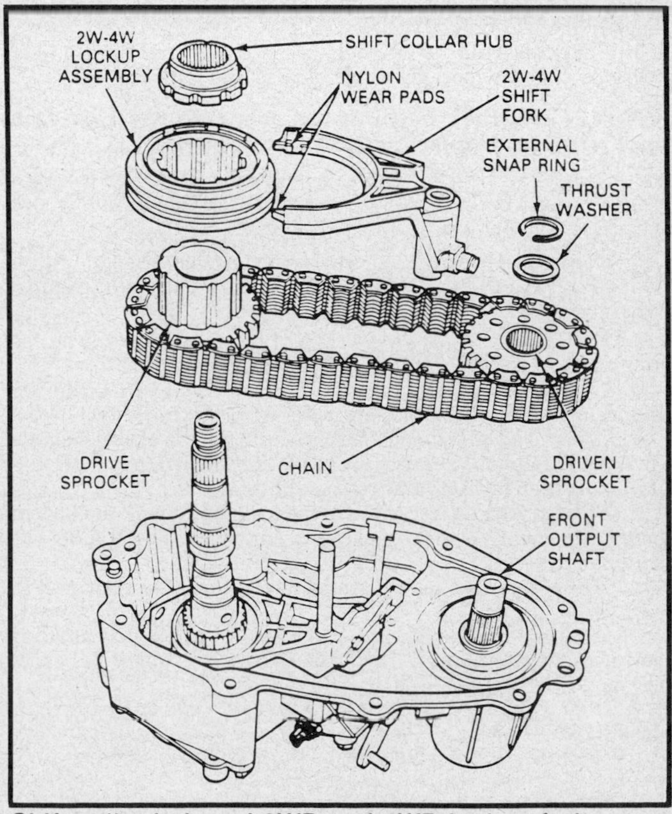
2W-4W LOCKUP ASSEMBLY
SHIFT COLLAR HUB
NYLON WEAR PADS
2W-4W SHIFT FORK
EXTERNAL SNAP RING
THRUST WASHER
DRIVE SPROCKET
CHAIN
DRIVEN SPROCKET
FRONT OUTPUT SHAFT

Shift collar hub and 2WD and 4WD lockup fork assemblies

bearings with a soft bristled brush until all dirt has been removed. Remove loose particles of dirt by tapping bearing flat against a block of wood.

c. Rinse bearings in clean solvent; then blow bearings dry with air pressure.

CAUTION

Do not spin bearings while drying.

d. After drying, rotate each bearing slowly while examining balls or rollers for roughness, damage, or excessive wear. Replace all bearings that are not in first class condition.

NOTE: After cleaning and inspecting bearings, lubricate generously with recommended lubricant, then wrap each bearing in clean paper until ready for reassembly.

2. Remove all portions of old gaskets from parts, using a stiff brush or scraper.

Inspection

1. Inspect all parts for discoloration or warpage.
2. Examine all gears and splines for chipped, worn, broken or nicked teeth. Small nicks or burrs may be removed with a fine abrasive stone.
3. Inspect the breather assembly to make sure that it is open and not damaged.
4. Check all threaded parts for damaged, stripped, or crossed threads.
5. Replace all gaskets, oil seals and snaprings.
6. Inspect housings, retainers and covers for cracks or other damage. Replace the damaged parts.
7. Inspect keys and keyways for condition and fit.
8. Inspect shift forks for wear, distortion or any other damage.
9. Check detent ball springs for free length, compressed length, distortion or collapsed coils.
10. Check bearing fit on their respective shafts and in their bores or cups. Inspect bearings, shafts and cups for wear.

NOTE: If either bearings or cups are worn or damaged, it is advisable to replace both parts.

11. Inspect all bearing rollers or balls for pitting or galling.
12. Examine detent balls for corrosion or brinneling. If shift bar detents show wear, replace them.
13. Replace all worn or damaged parts. When assembling the transfer case, coat all moving parts with recommended lubricant.

Transfer Case Assembly

Before assembly, lubricate all parts with Dexron®II ATF.

1. If removed, start a new bearing into an end of the shaft ear. Support the output shaft in a vise equipped with copper or wood jaws, in order not to damage the shaft.
2. Position the spider into the bearing and press the bearing below the snapring groove using U-joint tool, T74P–4635–C or equivalent.
3. Remove the tool and install a new internal snapring on the groove.
4. Start a new bearing into the opposite side of the shaft ear and using the tool, press the bearing until the opposite bearing contacts the snapring.
5. Remove the tool and install a new internal snapring in the groove.
6. Re-position the front output shaft assembly and install the other 2 bearings in the same manner.
7. Check the U-joint for freedom of movement. If a binding condition occurs due to misalignment during the installation procedure, tap the ears of both shafts sharply to relieve the bind. Do not install the front output shaft assembly if the U-joint shows any sign of binding.
8. If removed, drive the ball bearing into the front output case bore using output shaft bearing replacer tool T83T–7025–B and drive handle tool T80T–4000–W or equivalent. Drive the ball bearing in straight, making sure that it is not cocked in the bore. Install the internal snapring that retains the ball bearing to the front case.
9. If removed, install the front output oil seal in the front case bore using output shaft seal installer tool T83T–7065–B and driver handle tool T80T–4000–W or equivalent.
10. If removed, install the ring gear in the front case. Align the serrations on the outside diameter of the ring gear to the serations previously cut in the front case bore. Using a press, start the piloted chamfered end of the ring gear first and press in until it is fully seated. Make sure the ring gear is not cocked in the bore.
11. If removed, install the ball bearing in the bearing retainer bore. Drive the bearing into the retainer using output shaft bearing replacer tool T83T–7025–B and driver handle tool T80T–4000–W or equivalent. Make sure the ball bearing is not cocked in the bore. Install the internal snapring that retains the ball bearing to the retainer.
12. Install the front output shaft and U-joint assembly through the front case seal. Position the ball bearing and retainer assembly over the front output shaft and install in the front case bore. Make sure the clip on the bearing retainer aligns with the slot in the front case. Tap the bearing retainer into place with a plastic hammer. Install the internal snapring that retains the ball bearing and retainer assembly to the front case.
13. Install the compression spring and the detent plunger into the bore from the inside of the front case.
14. Install a new O-ring in the second groove of the shift lever shaft. Coat the shaft and O-ring with multi-purpose long-life lubricant.

NOTE: Use a rubber band to fill the first groove so as not to cut the O-ring. Discard the rubber band.

15. Position the shift cam inside the front case with the 4WH detent position over the detent plunger. Holding the shift cam by hand, push the shift lever shaft into the front case to engage the shift cam aligning the side of the shift lever with the mark previously scribed on the front case. Install the external clip on the end of the shift lever shaft.
16. Install the 2 set screws in the front case and in the shift cam. Tighten the screws to 5–7 ft. lbs. (7–10 Nm). Make sure the set screw in the front case is in the first groove of the shift lever shaft and not bottomed against the shaft itself. The shift lever should be able to move freely to all detent positions.
17. Slide the sun gear, thrust plate, thrust washer, and press the ball bearing over the input shaft. Install the external snapring to the input shaft.

NOTE: The sun gear recessed face and ball bearing snapring groove should be toward the rear of the transfer case. The stepped face of the thrust washer should face towards the ball bearing.

18. Install the planetary gear set to the sun gear and input shaft assembly. Install the internal snapring to the planetary carrier.
19. Drive the oil seal into the bore of the mounting adapter with input shaft seal installer tool T83–T–7065–A and driver handle tool T80T–4000–W or equivalent.
20. Place the tanged snapring in the mounting adapter groove. Position the input shaft and planetary gearset in the mounting adapter and push inward until the planetary assembly and input shaft assembly are seated in the adapter. When properly seated, the tanged snapring will snap into place. Check installation by holding the mounting adapter by hand and tapping the face of the input shaft against a wooden block to ensure that the

snapring is engaged.

21. Remove all traces of RTV gasket sealant from the mating surfaces of the front case and mounting adapter. Install a bead of RTV gasket sealant on the surface of the front case.

22. Position the mounting adapter on the front case. Install 6 bolts and tighten to 25–30 ft. lbs. (31–41 Nm).

23. Position the roller on the 90 degree bent tang of the torsion spring. The larger diameter end of the spring must be installed first.

24. Install the roller into the torsion spring roller track of the shift cam while locating the center spring in the pivot groove in the front case. Push the anchor end of the torsion spring behind the locking post adjacent to the ring gear face.

25. Position the high-low shift hub into the planetary gearset. Slip the high-low shift fork bushing into the high-low roller track of the shift cam and the groove of the high-low shift hub.

NOTE: Make sure the nylon wear pads are installed on the shift fork. Make sure the dot on the pad is installed in the fork hole.

26. Install the shift rail through the high-low fork and make sure the shift rail is seated in the bore in the front case.

27. Install the oil pump front cover over the output shaft with the flanged side down. The word **TOP** must be facing the top of the transfer case as the position of the case as installed in the vehicle.

28. Install the oil pump spring and 2 pump pins with the flat side outward in the hole in the output shaft. Push both pins in to install the oil pump body, pickup tube and filter. The rear markings on the pump must be facing upward. Be sure to prime the pump with Dexron®II transmission fluid or equivalent.

29. Place the oil pump rear cover on the output shaft with the flanged side outward. The word **TOP REAR** is positioned toward the the top of the transfer case in the position the transfer case is installed in the vehicle. Apply Loctite® or equivalent to the oil pump bolts and install the pump cover. Torque the bolts to 3–4 ft. lbs. (4–5 Nm) and be sure to rotate the pump while tightening.

NOTE: When the oil pump is correctly installed, it will rotate freely on the output shaft.

30. Install the thrust washer on the rear output shaft next to the oil pump. Install the chain on the drive sprocket and driven sprocket. Lower the chain and sprockets into position in the case. The driven sprocket is installed over the front output shaft and the drive the sprocket is placed on the rear output shaft.

31. If disassembled, assemble the 2WD–4WD shift fork to the 2WD–4WD lockup assembly. Install the spring in the lockup collar. Place the lockup hub over the spring and engage the lockup hub in the notches in the lockup collar. Retain the lockup hub to the lockup collar with an internal snapring.

32. Install the 2WD–4WD shift fork to the 2WD–4WD lockup assembly. If removed, make sure the nylon wear pads are installed on the fork. The dot on the pad must be installed in the hole in the fork. Install the 2WD–4WD lockup collar and hub assembly over the the output shaft and onto the shift rail. If removed, install the shaft, bushing and external clip to the 2WD-4WD lockup fork.

33. Install the shift collar hub to the output shaft.

34. If removed, drive the gauged needle bearing into the rear cover bore with the needle bearing replacer tool T83T–7127–A and driver handled tool T80T–4000–W or equivalent.

35. If removed, install the ball bearing in the rear cover bore. Drive the bearing into the rear cover bore with output shaft bearing replacer tool T83T–7025–B and driver handled tool T80–4000–W or equivalent. Make sure the ball bearing is not cocked in the bore. Install the internal snapring that retains the ball bearing to the rear cover.

36. Install the speedometer drive gear assembly into the rear cover bore with round end of the speedometer gear clip facing towards the inside of the rear cover. Drive the oil seal into the rear cover bore with output shaft seal installer tool T83T–7065–B and driver handle tool T80T–4000–W or equivalent.

37. Install the 2WD–4WD shift fork spring on the inside boss of the rear cover.

38. Prior to final assembly of the rear cover to front case half, the transfer case shift lever assembly should be shifted into **4H** detent position to assure positioning of the shift rail to the rear cover.

39. Coat the mating surface of the front case with a bead of Loctite® sealant or equivalent.

40. Position the rear cover on the front case, making sure that the 2WD–4WD shift fork spring engages the shift rail and does not fall off the rear cover boss. Install the nine bolts, starting with the bolts on the rear cover and torque the bolts to 23–30 ft. lbs. (21–41 Nm).

NOTE: If the rear cover assembly does not seat properly, move the rear cover up and down slightly to permit the end of the shift rail to enter the shift rail hole in the rear cover boss.

41. Install the front and rear output shaft yokes. Install the anti-spill oil seal. Coat the faces of the yoke nuts and output shaft threads with a suitable thread sealer. Torque the yoke nuts to 120–150 ft. lbs. (163–203 Nm).

42. Install the 4WD indicator switch and torque the bolts down at 23–35 ft. lbs. (21–48 Nm).

43. Refill the transfer case with 3.0 pints of Dexron®II transmission fluid or equivalent. Torque the level plug and the drain plugs to 14–22 ft. lbs. (19–32 Nm).

44. Install the transfer case.

45. Start the engine and check the transfer case for correct operation. Stop the engine and check the fluid level, add as necessary.

46. Fluid should drip from the level hole. If the fluid flows out level hole in a stream, the pump may not be operating properly.

Special Tools

Number	Description
T50T-100-A	Impact slide hammer
D80L-100-S	Puller collet
Tool-1175-AC	Seal remover
T80T-4000-W	Driver handle
D79L-4621-A	Bearing splitter
T74P-4635-C	U-Joint tool

Specifications

TORQUE SPECIFICATIONS

Description	Nm	Ft. Lb.
Breather Vent	8–19	6–14
Case to Cover Bolts	31–41	23–30
Drain and Fill Plug	19–30	14–22
Four-Wheel Drive Indicator Switch	34–47	25–35
Front and Rear Driveshaft Bolts	16–20	12–15
Shift Control Bolts—Large	95–122	70–90
Shift Control Bolts—Small	42–57	31–42
Shift Shaft and Shift Cam Set Screw	6.8–9.5	5–7
Skid Plate to Frame Bolt	30–41	22–30

TORQUE SPECIFICATIONS

Description	Nm	ft. lbs.
Transfer Case to Transmission Adapter	34-47	25-35
Upper Shift Control Lever and Heat Shield Bolts	37-50	27-37
Yoke Nut	163-203	120-150
	N•m	In. Lb.
Oil Pump Bolts	4.0-4.5	36-40
Speedometer Screw	2.3-2.8	20-25

NOTE: The output shaft must turn freely within the oil pump. If binding occurs, loosen the four bolts and re-tighten again.

BORG WARNER 1350 ELECTRONIC SHIFT TRANSFER CASE

Application

1986–89 Ford Ranger and Bronco II

General Description

The Borg Warner 1350 is a 3 piece aluminum part time transfer case. It transfers power from the transmission to the rear axle and when electronically actuated, also to the front drive axle. The unit is lubricated by a positive displacement oil pump that channels oil flow through drilled holes in the rear output shaft. The pump turns with the rear output shaft and allows towing of the vehicle at maximum legal road speeds for extended distances without disconnecting the front and/or rear driveshaft.

The electronic system consists of a push button control, an electronic control module, an electric shift motor with an integral shift position sensor and a speed sensor. The push button control is located in the overhead console. The electronic control module is located to the left of the right hand speaker in the instrument panel. The electronic control module, controls the operation of the transfer case in response to inputs to the push button control by the vehicle operator.

The speed sensor, is mounted to the rear of the transfer case, the senor tells the control module the proper speed to shift the transfer case. The shift position sensor, is an integral part of the electric shift motor, tells the control module the shift position of the transfer case.

The electronic shift motor mounted externally at the rear of the transfer case, drives a rotary helical cam which moves the 2WD–4WD shift fork and **4H** and **4L** reduction shift fork to the selected vehicle drive position.
The transfer case is equipped with a magnetic clutch, similar to an air conditioning compressor clutch, which is located inside the transfer case adjacent to the 2WD–4WD shift collar. The clutch is used to spin up the front drive system from zero to vehicle speed in milliseconds. This spin up allows the shift between 2 high and 4 high to be made at any vehicle speed. When the transfer case rear and front output shafts reach synchronous speed, the spring loaded shift collar mechanically engages the mainshaft hub to the chain drive sprocket and the magnetic clutch is deactivated.

Shifts between **4H** and **4L** can only occur with the clutch interlock or transmission safety switches closed. The vehicle's speed must also be within specified limits as determined by the transfer case speed sensor (3 mph or under).

When the operator selects the drive combination through the push button control, an electric motor turns a helical cam, which is linked to the high low and 2WD–4WD shift forks through fork mounted roller bushing assemblies. As the electric motor turns the helical cam, the high low fork bushing rides in a slotted lobe in the cam to make low high or high low range changes; and the 2WD–4WD for bushing rides on lobes at the end of the cam to make the 2WD–4WD or 4WD–2WD shift.

Trouble Diagnosis

CHILTON'S THREE C'S TRANSFER CASE DIAGNOSIS

Condition	Cause	Correction
Slips out of gear	a) Shifting poppet spring weak	a) Replace the shifting poppet spring
	b) Bearing broken or worn	b) Replace the bearing
	c) Shifter fork bent	c) Replace shifter fork
	d) Improper control rod adjustment	d) Adjust the control rod
Hard shifting	a) Lack of lubricant	a) Add the recommended fluid to the proper level
	b) Shift lever binding on shaft	b) Repair and adjust the shift lever
	c) Shifting poppet ball scored	c) Replace the shifting poppet ball
	d) Shifter fork bent	d) Replace shifter fork
	e) Low tire pressure	e) Inflate tires to the correct pressure

CHILTON'S THREE C'S TRANSFER CASE DIAGNOSIS

Condition	Cause	Correction
Backlash	a) Companion yoke loose b) Transfer case loose on mounts c) Internal parts excessively worn	a) Torque the companion yoke to specifications b) Torque the mount bolts to specification c) Rebuild the transfer case
Noisy	a) Low lubricant level b) Bearings improperly adjusted or excessively worn c) Gears worn or damaged d) Improper alignment of driveshafts or U-joint	a) Add the recommended fluid to the proper level b) Adjust or replace the bearing c) Replace the gear in question d) Realign the driveshaft and U-joint
Oil leakage	a) Excessive amount of oil in the case b) Oil vent clogged c) Gaskets or seals leaking d) Bearings improperly adjusted or excessively worn e) Driveshaft yoke mating surfaces scored	a) Drain the fluid to the proper level b) Clean oil vent c) Replace leaking gaskets or seals d) Adjust or replace the bearing e) Machine the driveshaft yoke mating surface
Overheating	a) Excessive amount of oil in the case b) Low lubricant level c) Bearing adjustment too tight	a) Drain the fluid to the proper level b) Add the recommended fluid to the proper level c) Adjust or replace the bearing

Electronic Controls

DIAGNOSIS AND TESTING

The battery feed circuit, through a circuit breaker provides memory capability for the electronic control module. The ignition **RUN** and **ACC** feed circuits through a fuse, provide power for the switches and the electric shift motor. The side marker lamp feed circuit provides the power for night time illumination of the overhead roof console vehicle graphics.

CONTROL MODULE

Self Test

The electronic control module has a diagnostic capability of its own circuitry. The self test procedures are as follows:

1. Remove the 5 wire connector and the 8 wire connector from the electronic control module.
2. Turn the ignition switch to the **RUN** position.
3. Activate the self test switch and note the results.
4. A flashing indicator lamp (approximately 1 flash per second) indicates that the control module id functioning properly.
5. A steady indicator light indicates that the control module is inoperative and must be replaced.

CONTROL MODULE CIRCUIT

There are 3 wiring harness connected to the electronic control module, the 8 wire pigtail harness connector, the 5 wire harness connector and the 8 wire harness connector. To check the integrity of these circuits, disconnect the harness from the electronic control module and perform the following test.

8 Wire Pigtail Harness Connector Test

1. Remove the 8 wire pigtail harness connector and connect a suitable voltmeter between terminal 8 and the ground. The voltmeter should indicate battery (12 volts) voltage at all times.

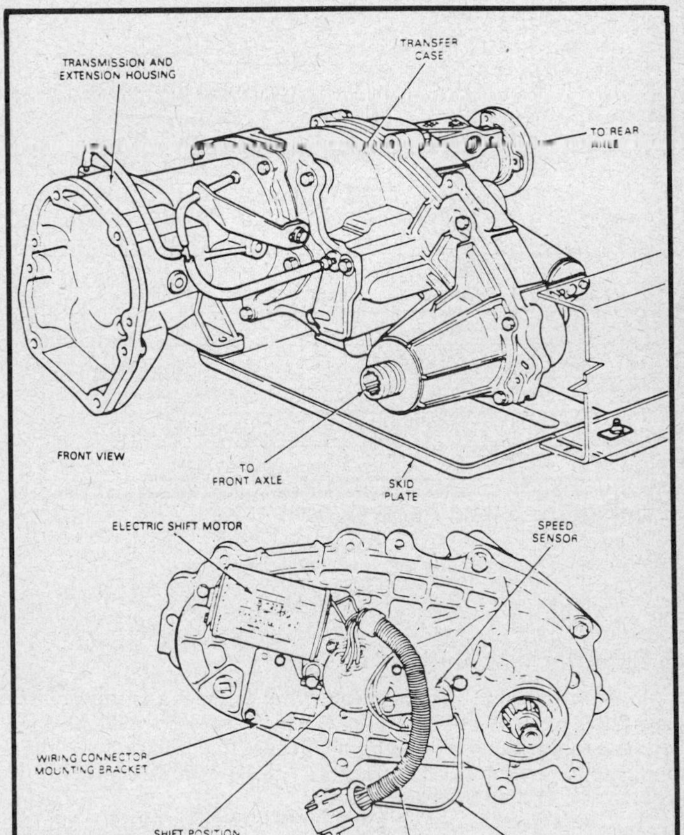
TRANSMISSION AND EXTENSION HOUSING
TRANSFER CASE
TO REAR
FRONT VIEW
TO FRONT AXLE
SKID PLATE
ELECTRIC SHIFT MOTOR
SPEED SENSOR
WIRING CONNECTOR MOUNTING BRACKET
SHIFT POSITION SENSOR
REAR VIEW
WIRING HARNESS
WIRE FEED FOR THE MAGNETIC CLUTCH

Front and rear view of the 1350 electronic shift transfer case

2. Connect the voltmeter between terminal 7 and ground. Turn the ignition switch to the **RUN** position. The voltmeter should indicate battery (12 volts) voltage.

NOTE: In the following test where the usage of an ohmmeter is specified, always remember that an ohmmeter should never be connected into a live or powered circuit. If an ohmmeter is connected to a powered circuit, severe damage will be done to the instrument. The vehicle's battery should be disconnected before disconnecting any circuit with an ohmmeter to prevent any accidental damage to the instrument.

3. Connect an ohmmeter between terminal 6 and ground. The ohmmeter should indicate a low resistance valve of less than 10 ohms.
4. Connect an ohmmeter between terminal 4 and 5. The ohmmeter should indicate a low resistance valve of less than 10 ohms.
5. Connect an ohmmeter between terminal 3 and ground. The ohmmeter should indicate a 0 ohms.
6. Connect an ohmmeter between terminal 2 and ground. The ohmmeter should indicate a 0 ohms.

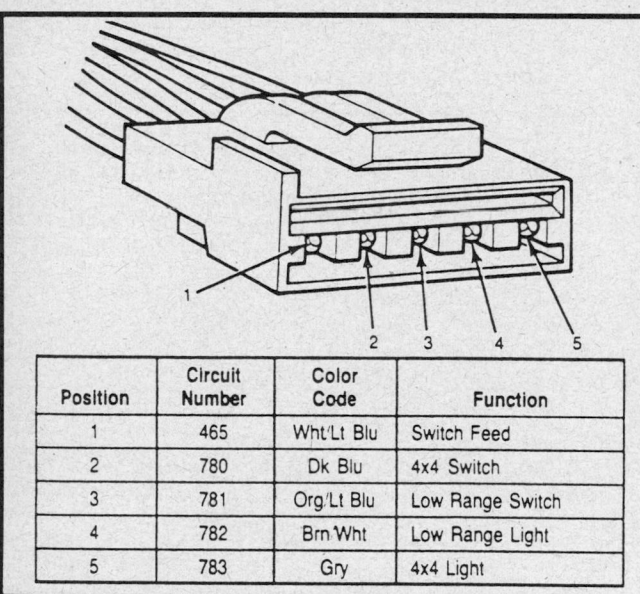

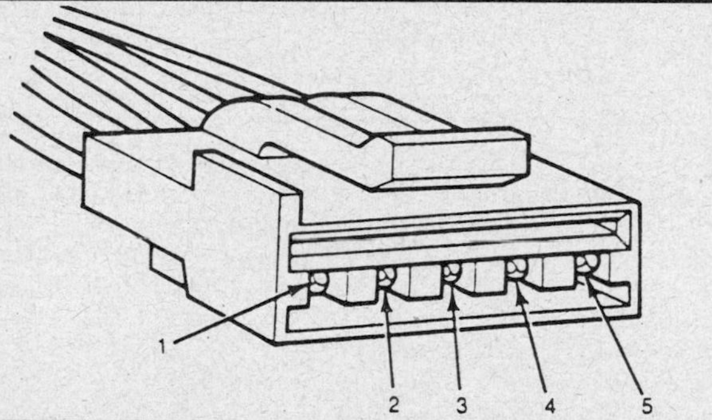
1
2
3
4
5

Position	Circuit Number	Color Code	Function
1	465	Wht/Lt Blu	Switch Feed
2	780	Dk Blu	4x4 Switch
3	781	Org/Lt Blu	Low Range Switch
4	782	Brn/Wht	Low Range Light
5	783	Gry	4x4 Light

Testing the 5 wire harness connector

5 Wire Harness Connector Test

1. Remove the 5 wire connector and connect a suitable ohmmeter between terminal 1 and 2. Then depress the 4WD (2H–4H) switch in the overhead roof console. The ohmmeter should indicate a very low resistance valve (less than 50 ohms) while the switch is being depressed.
2. Connect a suitable ohmmeter between terminal 1 and 3. Then depress the low range switch in the overhead roof console. The ohmmeter should indicate a very low resistance valve (less than 50 ohms) while the switch is being depressed.
3. Connect a test lead between terminal 4 and ground. Then turn the ignition switch to the **RUN** position and observe the overhead roof console. The light in the console low range bar should illuminate.
4. Connect a test lead between terminal 5 and ground. Then turn the ignition switch to the **RUN** position and observe the overhead roof console. The light in the console 4WD bar should illuminate.

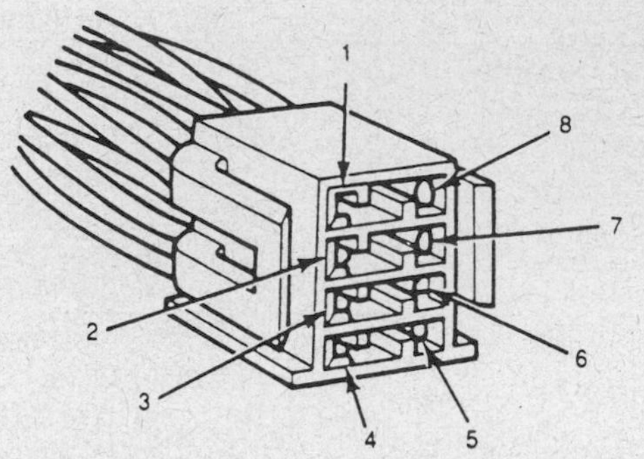
1
8
7
2
6
3
4
5

Position	Circuit Number	Color Code	Function
1	OPEN		
2	57	Blk	Ground
	57A	Blk	Ground
3	396	Blk/Org	Logic Ground
4	778	Org	Transfer Case Motor Control (Clockwise) 2H-4H-4L
5	777	Yel	Transfer Case Motor Control (Counterclockwise) 4L-4H-2H
6	779	Brn	Electro-Magnetic Clutch (Feed)
7	296	Wht/Ppl	Ignition Run and Accessory Feed (Fused)
8	517	Blk/Wht	Battery Feed (Circuit Breaker)

Testing the 8 wire pigtail harness connector

8 Wire Harness Connector Test

1. Remove the 8 wire connector and connect a suitable ohmmeter between terminal 1 and ground. Then depress the clutch pedal and observe the ohmmeter. The ohmmeter should indicate a very low resistance valve (less than 50 ohms) while the clutch pedal is being depressed.
2. If the vehicle is equipped with a automatic transmission, connect a suitable ohmmeter between terminal 1 and ground. Then shift the transmission into the **N** position and observe the ohmmeter. The ohmmeter should indicate a very low resistance valve (less than 50 ohms) while in the **N** position.
3. Connect a suitable ohmmeter between terminal 2 and 3. The ohmmeter should indicate a very low resistance valve (200–350 ohms).
4. This will check the continuity of the speed sensor that is located in the transfer case. The speed sensor picks up the rotating speed of the output shaft from 2 notches that are cut in opposite sides of the outer ring of the clutch housing assembly.
5. Connect an ohmmeter between terminal 8 and terminals 4, 5, 6 and 7, respectively. Use the chart provided for the ohmmeter reading in each transfer case position.

OHMMETER READINGS FOR SHIFT MOTOR POSITION SENSOR

Ohmmeter Connection	Transfer Case Gear Position		
	2 High	4 High	4 Low
Meter Reading From Terminal #8 to #4	Short	Open	Short
Meter Reading From Terminal #8 to #5	Open	Open	Short
Meter Reading From Terminal #8 to #6	Short	Short	Open
Meter Reading From Terminal #8 to #7	Open	Short	Open

NOTE: SHORT is a "low" resistance reading on the ohmmeter (zero ohms).
OPEN is a "high" resistance reading on the ohmmeter (infinity).

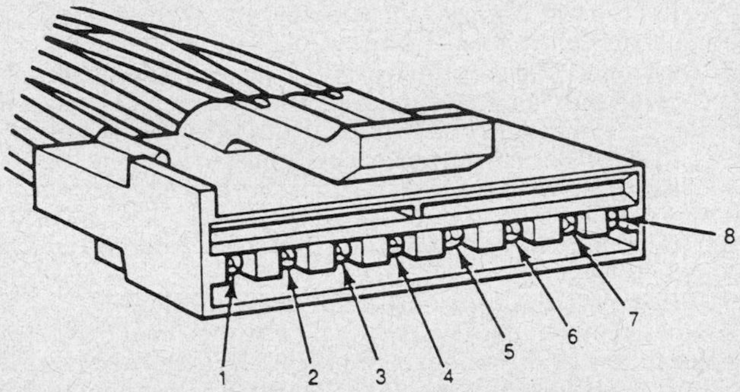
1
2
3
4
5
6
7
8

Position	Circuit Number	Color Code	Function
1	32	Red Lt Blu	Manual Transmission Clutch Interlock Switch
	463	Red Wht	Automatic Transmission Neutral Safety Switch
2	774	Lt Grn	Speed Sensor (Feed)
3	772	Lt Blu	Speed Sensor Return
4	771	Violet	Wire #5, Contact Plate Position Sensor in Transfer Case
5	770	Wht	Wire #4, Contact Plate Position Sensor in Transfer Case
6	764	Brn Wht	Wire #3, Contact Plate Position Sensor in Transfer Case
7	763	Org Wht	Wire #2, Contact Plate Position Sensor in Transfer Case
8	762	Yel Wht	Wire#1, Contact Plate Position Sensor in Transfer Case

Testing the 8 wire harness connector

On Vehicle Services

OVERHEAD CONSOLE

Removal and Installation

1. Disconnect the negative battery cable. Insert a suitable tool in the notch at the top of each light lens and snap out both lenses to expose 2 mounting screws.
2. Remove the 2 mounting screws and pull down the rear of the console approximately 1 in. Push forward to disengage the console from the front mounting clips.
3. Disconnect the 2 pin and 10 pin connectors and remove the console assembly.
4. Position the console assembly up to the overhead roof area and connect the 2 pin and 10 pin connectors.
5. Engage the forward end of the console into the 2 mounting clips. Pull rearward to assure engagement.
6. Install the 2 console mounting screws. Install the 2 light lenses by snapping then into place.

4WD SHIFT SWITCH

Removal and Installation

1. Disconnect the negative battery cable. Remove the overhead console assembly.
2. Remove 1 switch mounting screw and remove the switch from the console.
3. When installing the switch, position the switch in the console and retain it with 1 mounting screw.
4. Install the overhead assembly into the overhead roof area.

CONTROL MODULE

Removal and Installation

1. Disconnect the negative battery cable. Remove the instrument panel pad.
2. Remove the 2 module to instrument panel attaching screws.
3. Lift out the module and disconnect 3 wiring connectors. Remove the module assembly.
4. Position the new module to the instrument panel opening. Connect the 3 wiring connectors to the module.
5. Place module into the instrument panel opening and in stall the 2 mounting screws.
6. Install the instrument panel pad.

SHIFT LEVER

Removal and Installation

NOTE: Remove the shift ball only if the shift ball, boot or lever is to be replaced. If the ball, boot or lever is not being replaced, remove the ball, boot and lever as an assembly.

1. Remove the plastic insert from the shift ball. Warm the ball with a heat gun to 140–180°F (60–87°C) and knock the ball off the lever with a block of wood and a hammer. Be careful not to damage the finish on the shift lever.
2. Remove the rubber boot and floor pan cover.
3. Disconnect the vent hose from the control lever.
4. Unscrew the shift lever from the control lever.
5. Remove the bolts retaining the shifter to the extension housing. Remove the control lever and bushings.
6. Move the transfer case shift lever to the **4L** position.
7. Install the shifter assembly with the bolts finger tight and move the cam plate rearward until the bottom chamfered corner of the neutral lug just contacts the forward right edge of the shift lever.
8. Hold the cam plate in this position and torque the larger bolt first to 70–90 ft. lbs. (95–122 Nm) and torque the smaller bolt to 31–42 ft. lbs. (42–57 Nm).
9. Move the transfer case in cab shift lever to all shift positions to check for positive engagement. There should be a clearance between the shift lever and the cam plate in the **H** front and **4H** rear (clearance not to exceed 0.13 in.) and **4L** shift positions.

10. Install the shift lever to the control lever and torque the bolts to 27–37 ft. lbs. (37–50 Nm). Install the vent assembly so that the white marking on the house is in the position in the notch in the shifter.

NOTE: The upper end of the vent hose should be 2 in. above the top of the shifter and positioned inside of the shift lever boot.

11. Install the rubber boot and floor pan cover.
12. Warm the ball with a heat gun or equivalent until it reaches approximately 140–180°F. Using a $^7/_{16}$ in. socket and a mallet, tap the ball on to the lever and install the plastic shift pattern insert. The end of the shift ball should be to the end of the knurl on the upper portion of the shift lever.
13. Check the transfer case for proper shifting and operation.

FRONT OUTPUT SHAFT OIL SEAL

Removal and Installation

1. Raise and support the vehicle safely.
2. Place a drain pan under transfer case, remove the drain plug and drain fluid from the transfer case.
3. Disconnect the front driveshaft from the axle input yoke.
4. Loosen the clamp retaining the front driveshaft boot to the transfer case, and pull the driveshaft and front boot assembly out of the transfer case front output shaft.
5. Remove the oil seal from the front output housing bore with a suitable seal remover and slide hammer.
6. Make sure that the housing face and bore are free from nicks and burrs. Coat the seal with a suitable grease.
7. Position the oil seal into the front output housing bore, making sure the oil seal is not cocked in the bore. Drive the new seal into the housing with a suitable seal installer.
8. Clean the transfer case front output female spline and apply a suitable grease. Insert the front driveshaft male spline.
9. Connect the front driveshaft to the front output yoke and torque the yoke nut to 8–15 ft. lbs. (11–20 Nm). Attach the heat shield to the engine mounting bracket and mounting lug on the transfer case.
10. Install the skid plate to the frame. Install the transfer case drain plug and torque the plug to 14–22 ft. lbs. (19–32 Nm).
11. Fill the transfer case with 6.5 pints of Dexron®II transmission fluid or equivalent. Torque the fill plug to 14–22 ft. lbs. (19–32 Nm). Lower the vehicle. Start the engine and check the transfer case for correct operation. Stop the engine and check the fluid level, add as necessary.

REAR OUTPUT SHAFT OIL SEAL

Removal and Installation

1. Raise and support the vehicle safely.
2. Place a drain pan under transfer case, remove the drain plug and drain fluid from the transfer case.
3. Disconnect the rear driveshaft from the axle output shaft flange. Wire the driveshaft out of the way.
4. Remove the output shaft flange by removing the 30mm nut, steel washer and rubber seal from the rear of the output shaft and remove the flange.
5. Remove the oil seal from the rear output housing bore with a suitable seal remover and slide hammer.
6. Make sure that the housing face and bore are free from nicks and burrs. Coat the seal with a suitable grease.
7. Position the oil seal into the rear output housing bore, making sure the oil seal is not cocked in the bore. Drive the new seal into the housing with a suitable seal installer.
8. Install the yoke, rubber seal, steel washer and nut on the output shaft. Torque the nut to 150–180 ft. lbs. (203–244 Nm).
9. Connect the rear driveshaft to the transfer case output shaft flange. Torque the bolts to 61–87 ft. lbs. (83–118).
10. Fill the transfer case with 6.5 pints of Dexron®II transmission fluid or equivalent. Torque the fill plug to 14–22 ft. lbs. (19–32 Nm). Lower the vehicle. Start the engine and check the transfer case for correct operation. Stop the engine and check the fluid level, add as necessary.

Removal and Installation

TRANSFER CASE REMOVAL

The catalytic converter is located beside the transfer case. Be careful when working around the catalytic converter because of the extremely high temperatures generated by the converter.

1. Raise and support the vehicle safely.
2. If so equipped, remove the skid plate from frame.
3. Place a drain pan under transfer case, remove the drain plug and drain fluid from the transfer case.
4. Remove the wire connector from the feed wire harness at the rear of the transfer case. Be sure to squeeze the locking tabs, then pull the connectors apart.
5. Disconnect the front driveshaft from the axle input yoke.
6. Loosen the clamp retaining the front driveshaft boot to the transfer case, and pull the driveshaft and front boot assembly out of the transfer case front output shaft.
7. Disconnect the rear driveshaft from the transfer case output shaft yoke.
8. Disconnect the speedometer driven gear from the transfer case rear cover.
9. Disconnect the vent hose from the control lever.
10. Loosen or remove the large bolt and the small bolt retaining the shifter to the extension housing. Pull on the control lever until the bushing slides off the transfer case shift lever pin. If necessary, unscrew the shift lever from the control lever.
11. Remove the heat shield from the transfer case.
12. Support the transfer case with a transmission jack.
13. Remove the 5 bolts retaining the transfer case to the transmission and the extension housing.
14. Slide the transfer case rearward off the transmission output shaft and lower the transfer case from the vehicle. Remove the gasket from between the transfer case and extension housing.

TRANSFER CASE INSTALLATION

1. Install the heat shield onto the transfer case and place a new gasket between the transfer case and adapter.
2. Raise the transfer case with a suitable transmission jack or equivalent, raise it high enough so that the transmission output shaft aligns with the splined transfer case input shaft.
3. Slide the transfer case forward on to the transmission output shaft and onto the dowel pin. Install transfer case retaining bolts and torque them to 26–43 ft. lbs. (35–58 Nm).
4. Connect the rear driveshaft to the rear output shaft yoke and torque the retaining bolts to 20–28 ft. lbs. (27–38 Nm). Attach the shift rod to the transfer case shift lever and transfer case control rod and attach with retaining rings.
5. Connect the speedometer driven gear to the transfer case. Connect the 4WD indicator switch wire connector at the transfer case.
6. Connect the front driveshaft to the front output yoke and torque the yoke nut to 8–15 ft. lbs. (11–20 Nm). Attach the heat shield to the engine mounting bracket and mounting lug on the transfer case.
7. Install the skid plate to the frame. Connect the the wire connectors on the rear of the transfer case, making sure the retaining tabs lock.
8. Fill the transfer case with 6.5 pints of Dexron®II transmission fluid or equivalent. Torque the fill plug to 14–22 ft. lbs. (19–32 Nm). Lower the vehicle. Start the engine and check the transfer case for correct operation. Stop the engine and check the fluid level, add as necessary.

Before Disassembly

Cleanliness during disassembly and assembly is necessary to avoid further transfer case trouble after overhaul. Before removing any of the transfer case subassemblies, plug all the openings and clean the outside of the of the transfer case thoroughly. Steam cleaning or car wash type high pressure equipment is preferable. During disassembly, clean all parts in suitable solvent and dry each part. Do not use cloth or paper towels to dry parts. Use compressed air only.

Transfer Case Disassembly

1. Place the transfer case into a suitable holding fixture.
2. Remove the transfer case drain plug with a 3/8 in. drive ratchet and drain the fluid.
3. Remove the 4WD indicator switch and the breather vent.
4. Remove the rear output shaft yoke by removing the 30mm nut, steel washer and rubber seal from the output shaft. Remove the wire connector assembly from the mounting bracket on the rear cover. If required, remove the 2 bolts and remove the bracket.
5. Form a small hook at the end of a paper clip or safety pin. Remove the locking sleeve from the wire connector by hooking the paper clip or safety pin and pulling it up from the bottom. Be sure not to damage the wire connector locking sleeve.
6. Remove the brown wire from the No. 1 center position in the connector. If required, remove the speed sensor green wire from the No. 4 connector position and blue wire from the No. 5 connector position.
7. Remove the speed sensor bracket retaining bolt and bracket. If required, remove the bracket and speed sensor.
8. Remove the 9 bolts which retain the front case to the rear cover. Insert a 1/2 in. drive breaker bar between the 3 pry bosses and separate the front case from the rear cover. Remove all traces of RTV gasket sealant from the mating surfaces of the front case and rear cover.

NOTE: When removing RTV sealant, take care not to damage the mating surface of the aluminum case.

9. Remove the remaining 3 bolts retaining the motor to the rear cover and remove the motor. Note the position of the triangular shaft extending out of the rear cover and the triangular slot in the motor.

NOTE: The motor is serviced as a complete assembly. Do not remove the screws that secure the rear cover to the motor gear housing.

10. If the speedometer drive gear or ball bearing assembly is to be replaced, first, drive out the output shaft oil seal from either the inside of the rear cover with a brass drift and hammer or from the outside by bending and pulling on the curved-up lip of the oil seal. Remove and discard the oil seal. Remove the speedometer drive gear assembly (gear, clip and spacer). Note that the round end of the speedometer gear clip faces the inside of the rear cover.
11. Remove the internal snapring that retains the rear output shaft ball bearing in the bore. From the outside of the case, drive out the ball bearing with output shaft bearing replacer tool T83T-7025-B and drive handle tool T80T-4000-W or equivalent.
12. If required, remove the front output shaft caged needle bearing from the rear cover with puller collet tool D80L-100-S and impact slide hammer tool T50T-100-A or equivalent. Remove the nuts retaining the clutch coil assembly to the rear cover. Pull the assembly along with the O-rings and brown wire from the cover.
13. Remove the 2WD-4WD shift fork spring from the boss in the rear cover.
14. Remove the shift collar hub from the output shaft. Remove the 2WD-4WD lockup assembly and the 2WD-4WD shift fork together as an assembly. Remove the 2WD-4WD fork from the 2WD-4WD lockup assembly. If required, remove the external clip and remove the roller bushing assembly (bushing, shaft and external clip) from the 2WD-4WD shift fork.
15. If required to disassemble the 2WD-4WD lockup assembly, remove the internal snapring and pull the lockup hub and spring from the lockup collar.
16. Remove the helical cam assembly from the front case. If required, remove the helical cam, torsion spring and sleeve from the shaft. Remove the external snapring and thrust washer that retains the drive sprocket to the front output shaft.
17. Remove the chain, driven sprocket and drive sprocket as an assembly.
18. Remove the collector magnet from the notch in the front case bottom.
19. Remove the output shaft and oil pump as an assembly.
20. If required to disassemble the oil pump, remove the 4 bolts from the body. Note the position and markings of the front cover, body, pins, spring, rear cover, and pump retainer as removed.
21. Pull out the shift rail.
22. Slip the high-low range shift fork out of the inside track of the shift cam. If required, remove the external clip and remove the roller bushing assembly (bushing, shaft and external clip) from the high-low range shift fork.
23. Remove the high-low shift hub from out of the planetary gearset in the front case.
24. Push and pull out the anchor end of the torsion spring from the locking post in the front case half. Remove the torsion spring and roller out of the shift cam, if so equipped.
25. Turn the front case over and remove the 6 bolts retaining the mounting adapter to the front case. Remove the mounting adapter, input shaft and planetary gearset as an assembly.
26. If required, remove the ring gear from the front case using a press. Note the relationship of the serrations to the chamfered pilot diameter during removal.
27. Expand the tangs of the large snapring in the mounting adapter and pry under the planetary gearset and separate the input shaft and planetary gearset from the mounting adapter.
28. If required, remove the oil from the mounting adapter with seal remover, tool 1175-AC and impact slide hammer tool T50T-100-A or equivalent.
29. Remove the internal snapring from the planetary carrier and separate the planetary gearset from the input shaft assembly.
30. Remove the external snapring from the input shaft. Place the input shaft assembly in a press and remove the ball bearing from the input shaft using bearing splitter tool D79L-4621-A or equivalent. Remove the thrust washer, thrust plate and sun gear off the input shaft.
31. Move the shift lever by hand until the shift cam is in the 4 wheel high detent position (4WH) and mark a line on the outside of the front case using the side of the shift lever and a grease pencil.
32. Remove the 2 set screws from the front case and from the shift cam.
33. Turn the front case over and remove the external clip. Pry the shift lever out of the front case and shift cam.

NOTE: Do not pound on the external clip during removal.

34. Remove the O-ring from the second groove in the shift lever shaft.
35. Remove the detent plunger and compression spring from the inside of the front case.
36. Remove the internal snapring and remove the ball bearing retainer from the front case by tapping on the face of the front output shaft and U-joint assembly with a plastic hammer. Remove the internal snapring and drive the ball bearing out of the

bearing retainer using output shaft bearing replacer tool T83T–7025–B and driver handle tool T80T–4000–W or equivalent.

NOTE: The clip is required to prevent the bearing retainer from rotating. Do not discard the clip.

37. Remove the front output shaft and U-joint assembly from the front case. If required, remove the oil seal with seal remover tool 1175–AC and impact slide hammer tool T50T–100–A or equivalent. If required, remove the internal snapring and drive the ball bearing out of the front case bore using output shaft replacer tool T83T–7025–B and driver handle tool T80T–4000–W or equivalent.
38. If required, place the front output shaft and U-joint assembly in a vise, being careful not to damage the assembly. Use copper or wood vise jaws.
39. Remove the internal snaprings that retain the bearings in the shaft.
40. Position the U-joint tool tool T74P–4635–C or equivalent, over the shaft ears and press the bearing out. If the bearing cannot be pressed all the way out, remove it with vise grip or channel lock pliers.
41. Re-position the U-joint tool on the spider in order to remove the opposite bearing.
42. Repeat the above procedure until all bearings are removed.

CLEANING AND INSPECTION

Cleaning

During overhaul, all components of the transfer case (except bearing assemblies) should be thoroughly cleaned with solvent and dried with air pressure prior to inspection and reassembly.

1. Clean the bearing assemblies as follows:

NOTE: Proper cleaning of bearings is of utmost importance. Bearings should always be cleaned separately from other parts.

a. Soak all bearing assemblies in clean solvent or fuel oil. Bearings should never be cleaned in a hot solution tank.

b. Slush bearings in solvent until all old lubricant is loosened. Hold races so that bearings will not rotate; then clean bearings with a soft bristled brush until all dirt has been removed. Remove loose particles of dirt by tapping bearing flat against a block of wood.

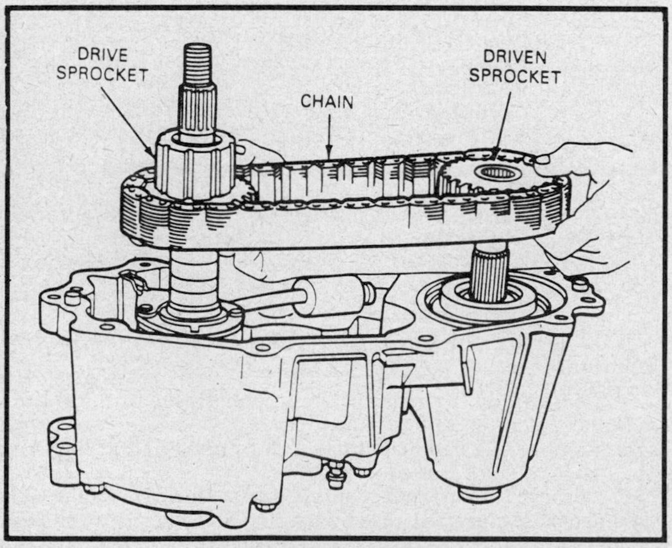
DRIVE SPROCKET
CHAIN
DRIVEN SPROCKET

Removing the chain drive assembly

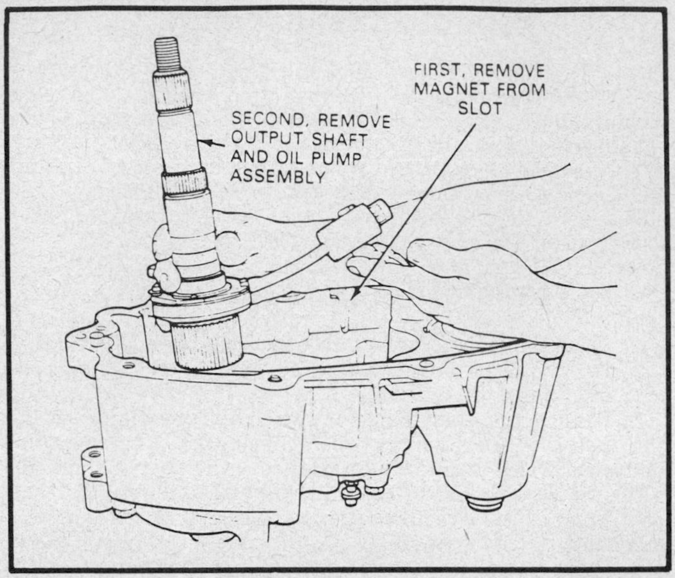
FIRST, REMOVE MAGNET FROM SLOT
SECOND, REMOVE OUTPUT SHAFT AND OIL PUMP ASSEMBLY

Removing the oil pump assembly

c. Rinse bearings in clean solvent; then blow bearings dry with air pressure.

CAUTION

Do not spin bearings while drying.

d. After drying, rotate each bearing slowly while examining balls or rollers for roughness, damage, or excessive wear. Replace all bearings that are not in first class condition.

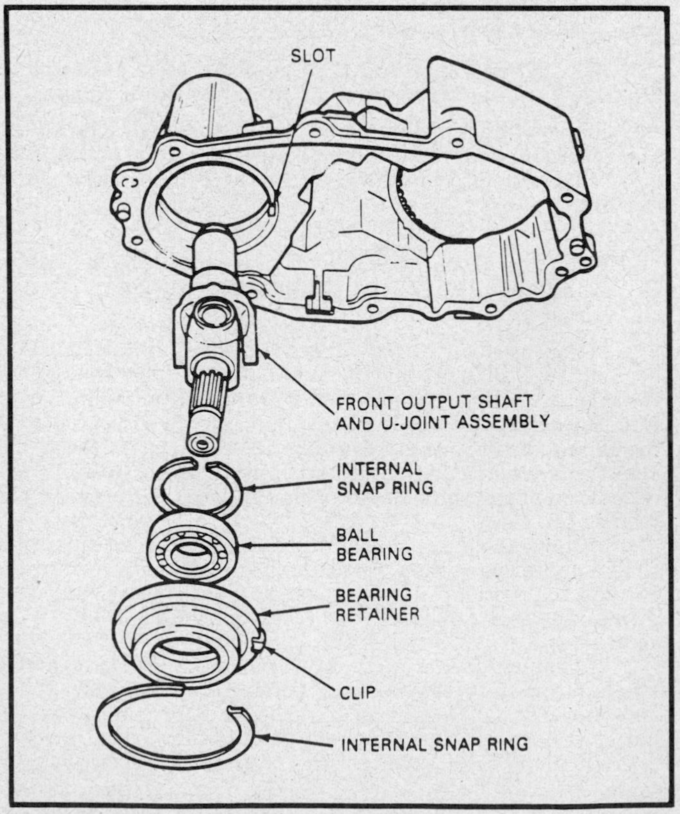
SLOT
FRONT OUTPUT SHAFT AND U-JOINT ASSEMBLY
INTERNAL SNAP RING
BALL BEARING
BEARING RETAINER
CLIP
INTERNAL SNAP RING

Removing the front output shaft assembly

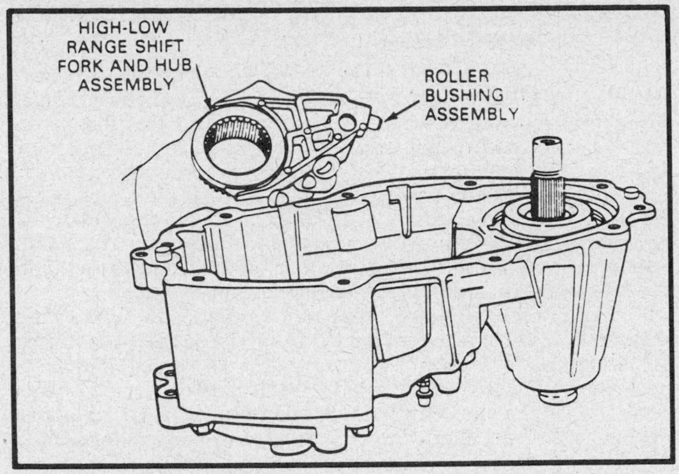
HIGH-LOW RANGE SHIFT FORK AND HUB ASSEMBLY
ROLLER BUSHING ASSEMBLY

Removing the high-low range shift fork assembly

NOTE: After cleaning and inspecting bearings, lubricate generously with recommended lubricant, then wrap each bearing in clean paper until ready for reassembly.

2. Remove all portions of old gaskets from parts, using a stiff brush or scraper.

Inspection

1. Inspect all parts for discoloration or warpage.
2. Examine all gears and splines for chipped, worn, broken or nicked teeth. Small nicks or burrs may be removed with a fine abrasive stone.
3. Inspect the breather assembly to make sure that it is open and not damaged.
4. Check all threaded parts for damaged, stripped, or crossed threads.
5. Replace all gaskets, oil seals and snaprings.
6. Inspect housings, retainers and covers for cracks or other damage. Replace the damaged parts.
7. Inspect keys and keyways for condition and fit.
8. Inspect shift forks for wear, distortion or any other damage.
9. Check detent ball springs for free length, compressed length, distortion or collapsed coils.
10. Check bearing fit on their respective shafts and in their bores or cups. Inspect bearings, shafts and cups for wear.

NOTE: If either bearings or cups are worn or damaged, it is advisable to replace both parts.

11. Inspect all bearing rollers or balls for pitting or galling.
12. Examine detent balls for corrosion or brinneling. If shift bar detents show wear, replace them.
13. Replace all worn or damaged parts. When assembling the transfer case, coat all moving parts with recommended lubricant.

Transfer Case Assembly

Before assembly, lubricate all parts with Dexron®II, automatic transmission fluid.

1. If removed, start a new bearing into an end of the shaft ear. Support the output shaft in a vise equipped with copper or wood jaws, in order not to damage the shaft.
2. Position the spider into the bearing and press the bearing below the snapring groove using U-joint tool, T74P–4635–C or equivalent.
3. Remove the tool and install a new internal snapring on the groove.

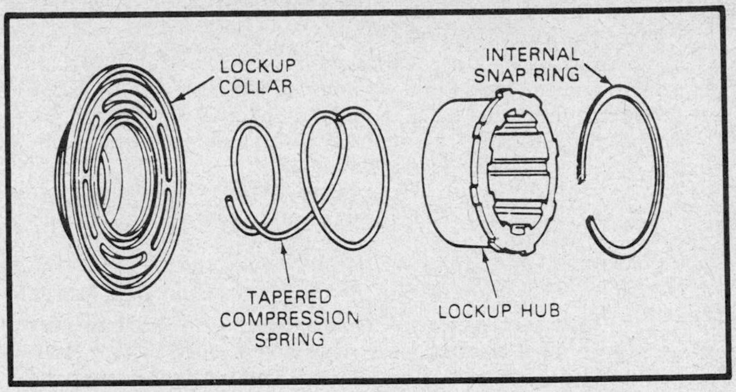
LOCKUP COLLAR
INTERNAL SNAP RING
TAPERED COMPRESSION SPRING
LOCKUP HUB

Exploded view of the disassembled 2WD–4WD lockup assembly

4. Start a new bearing into the opposite side of the shaft ear and using the tool, press the bearing until the opposite bearing contacts the snapring.
5. Remove the tool and install a new internal snapring in the groove.
6. Re-position the front output shaft assembly and install the other 2 bearings in the same manner.
7. Check the U-joint for freedom of movement. If a binding condition occurs due to misalignment during the installation procedure, tap the ears of both shafts sharply to relieve the bind. Do not install the front output shaft assembly if the U-joint shows any sign of binding.
8. If removed, drive the ball bearing into the front output case bore using output shaft bearing replacer tool T83T–7025–B and drive handle tool T80T–4000–W or equivalent. Drive the ball bearing in straight, making sure that it is not cocked in the bore. Install the internal snapring that retains the ball bearing to the front case.
9. If removed, install the front output oil seal in the front case bore using output shaft seal installer tool T83T–7065–B and driver handle tool T80T–4000–W or equivalent.
10. If removed, install the ring gear in the front case. Align the serrations on the outside diameter of the ring gear to the serations previously cut in the front case bore. Using a press, start the piloted chamfered end of the ring gear first and press in until it is fully seated. Make sure the ring gear is not cocked in the bore.
11. If removed, install the ball bearing in the bearing retainer bore. Drive the bearing into the retainer using output shaft bearing replacer tool T83T–7025–B and driver handle tool T80T–4000–W or equivalent. Make sure the ball bearing is not cocked in the bore. Install the internal snapring that retains the ball bearing to the retainer.
12. Install the front output shaft and U-joint assembly through the front case seal. Position the ball bearing and retainer assembly over the front output shaft and install in the front case bore. Make sure the clip on the bearing retainer aligns with the slot in the front case. Tap the bearing retainer into place with a plastic hammer. Install the internal snapring that retains the ball bearing and retainer assembly to the front case.
13. Install the compression spring and the detent plunger into the bore from the inside of the front case.
14. Install a new O-ring in the second groove of the shift lever shaft. Coat the shaft and O-ring with multi-purpose long-life lubricant.

NOTE: Use a rubber band to fill the first groove so as not to cut the O-ring. Discard the rubber band.

15. Position the shift cam inside the front case with the 4WH detent position over the detent plunger. Holding the shift cam by hand, push the shift lever shaft into the front case to engage the shift cam aligning the side of the shift lever with the mark

previously scribed on the front case. Install the external clip on the end of the shift lever shaft.

16. Install the 2 set screws in the front case and in the shift cam. Tighten the screws to 5–7 ft. lbs. (7–10 Nm). Make sure the set screw in the front case is in the first groove of the shift lever shaft and not bottomed against the shaft itself. The shift lever should be able to move freely to all detent positions.

17. Slide the sun gear, thrust plate, thrust washer, and press the ball bearing over the input shaft. Install the external snapring to the input shaft.

NOTE: The sun gear recessed face and ball bearing snapring groove should be toward the rear of the transfer case. The stepped face of the thrust washer should face towards the ball bearing.

18. Install the planetary gear set to the sun gear and input shaft assembly. Install the internal snapring to the planetary carrier.

19. Drive the oil seal into the bore of the mounting adapter with input shaft seal installer tool T83–T–7065–A and driver handle tool T80T–4000–W or equivalent.

20. Place the tanged snapring in the mounting adapter groove. Position the input shaft and planetary gearset in the mounting adapter and push inward until the planetary assembly and input shaft assembly are seated in the adapter. When properly seated, the tanged snapring will snap into place. Check installation by holding the mounting adapter by hand and tapping the face of the input shaft against a wooden block to ensure that the snapring is engaged.

21. Remove all traces of RTV gasket sealant from the mating surfaces of the front case and mounting adapter. Install a bead of RTV gasket sealant on the surface of the front case.

22. Position the mounting adapter on the front case. Install 6 bolts and tighten to 25–30 ft. lbs. (31–41 Nm).

23. Position the roller on the 90 degree bent tang of the torsion spring. The larger diameter end of the spring must be installed first.

24. Install the roller into the torsion spring roller track of the shift cam while locating the center spring in the pivot groove in the front case. Push the anchor end of the torsion spring behind the locking post adjacent to the ring gear face.

25. Position the high-low shift hub into the planetary gearset. Slip the high-low shift fork bushing into the high-low roller track of the shift cam and the groove of the high-low shift hub.

NOTE: Make sure the nylon wear pads are installed on the shift fork. Make sure the dot on the pad is installed in the fork hole.

26. Install the shift rail through the high-low fork and make sure the shift rail is seated in the bore in the front case.

27. Install the oil pump front cover over the output shaft with the flanged side down. The word **TOP** must be facing the top of the transfer case as the position the case id installed in the vehicle.

28. Install the oil pump spring and 2 pump pins with the flat side outward in the hole in the output shaft. Push both pins in to install the oil pump body, pickup tube and filter. The rear markings on the pump must be facing upward. Be sure to prime the pump with Dexron®II transmission fluid or equivalent.

29. Place the oil pump rear cover on the output shaft with the flanged side outward. The word **TOP REAR** is positioned toward the the top of the transfer case in the position the transfer case is installed in the vehicle. Apply Loctite® or equivalent to the oil pump bolts and install the pump cover. Torque the bolts to 3–4 ft. lbs. (4–5 Nm) and be sure to rotate the pump while tightening.

NOTE: When the oil pump is correctly installed, it will rotate freely on the output shaft.

30. Install the thrust washer on the rear output shaft next to the oil pump. Install the chain on the drive sprocket and driven sprocket. Lower the chain and sprockets into position in the case. The driven sprocket is installed over the front output shaft and the drive the sprocket is placed on the rear output shaft.

31. If disassembled, assemble the 2WD–4WD shift fork to the 2WD–4WD lockup assembly. Install the spring in the lockup collar. Place the lockup hub over the spring and engage the lockup hub in the notches in the lockup collar. Retain the lockup hub to the lockup collar with an internal snapring.

32. Install the 2WD–4WD shift fork to the 2WD–4WD lockup assembly. If removed, make sure the nylon wear pads are installed on the fork. The dot on the pad must be installed in the hole in the fork. Install the 2WD–4WD lockup collar and hub assembly over the the output shaft and onto the shift rail. If removed, install the shaft, bushing and external clip to the 2WD-4WD lockup fork.

33. Slide the spring spacer on the camshaft and position it beneath the drive tang. Place the torsion spring on the camshaft. Position the first spring tang to the left of the camshaft drive tang. Wind the second spring tang clockwise past the drive tang. Push the torsion spring and sleeve in as far as it will go. This will seat the second spring tang on the right side of the drive tang. Install the helical cam and slide the drive tang between the torsion spring tangs as far as it will go.

34. Install the tang end of the camshaft assembly of the alignment pin pressed in the front of the case. Position the torsion spring tangs and camshaft tang so that they are pointing toward the top side of the transfer case and are touching the high low shift fork assembly.

35. Lift the 2WD–4WD shift fork slightly while holding the shift rail down and rotate the helical cam track into the high low and 2WD–4WD for roller bushings by turning the camshaft assembly. The triangular shaft will be in the 2 wheel high **2H** position at final assembly.

36. Install the shift collar hub to the output shaft.

37. If removed, drive the gaged needle bearing into the rear cover bore with the needle bearing replacer tool T83T–7127–A and driver handled tool T80T–4000–W or equivalent.

38. If removed, install the ball bearing in the rear cover bore. Drive the bearing into the rear cover bore with output shaft bearing replacer tool T83T–7025–B and driver handled tool T80–4000–W or equivalent. Make sure the ball bearing is not cocked in the bore. Install the internal snapring that retains the ball bearing to the rear cover.

39. Install the speedometer drive gear assembly into the rear cover bore with round end of the speedometer gear clip facing towards the inside of the rear cover. Drive the oil seal into the rear cover bore with output shaft seal installer tool T83T–7065–B and driver handle tool T80T–4000–W or equivalent.

40. Install the 2WD–4WD shift fork spring on the inside boss of the rear cover.

41. Prior to final assembly of the rear cover to front case half, the transfer case shift lever assembly should be shifted into **4H** detent position to assure positioning of the shift rail to the rear cover.

42. Coat the mating surface of the front case with a bead of Loctite® sealant or equivalent.

43. Position the rear cover on the front case, making sure that the 2WD–4WD shift fork spring engages the shift rail and does not fall off the rear cover boss. Install the 9 bolts, starting with the bolts on the rear cover and torque the bolts to 23–30 ft. lbs. (21–41 Nm).

NOTE: If the rear cover assembly does not seat properly, move the rear cover up and down slightly to permit the end of the shift rail to enter the shift rail hole in the rear cover boss.

44. Using a pair of pliers equipped with copper or wood jaws, rotate the triangular shaft so that it aligns with the triangular slot in the motor. Install the motor and torque the retaining bolts to 6–8 ft. lbs. (8–11 Nm).

NOTE: If the shaft will not stay in 4H position, rotate the shaft to the 2H position. Install the motor and rotate the counterclockwise until the motor is aligned with the mounting holes.

45. Install the speed sensor and bracket and torque the bracket bolts to 6–8 ft. lbs. (8–11 Nm).

46. Install the brown clutch coil wire to the No. 1 center terminal and if removed, the speed sensor green wire to the No. 4 connector position and the blue wire to the No. 5 connector position. Install the locking sleeve.

47. Position the wire connector mounting bracket on the rear cover. Install the bolts and torque them to 5–7 ft. lbs. (6–10 Nm). Install the wiring connector to the mounting bracket.

48. Install the front and rear output shaft yokes. Install the anti-spill oil seal. Coat the faces of the yoke nuts and output shaft threads with a suitable thread sealer. Torque the yoke nuts to 120–150 ft. lbs. (163–203 Nm).

49. Install the 4WD indicator switch and torque the bolts down at 23–35 ft. lbs. (21–48 Nm).

50. Refill the transfer case with 3.0 pints of Dexron®II transmission fluid or equivalent. Torque the level plug and the drain plugs to 14–22 ft. lbs. (19–32 Nm).

51. Install the transfer case. Lower the vehicle.

52. Start the engine and check the transfer case for correct operation. Stop the engine and check the fluid level, add as necessary.

53. Fluid should drip from the level hole. If the fluid flows out level hole in a stream, the pump may not be operating properly.

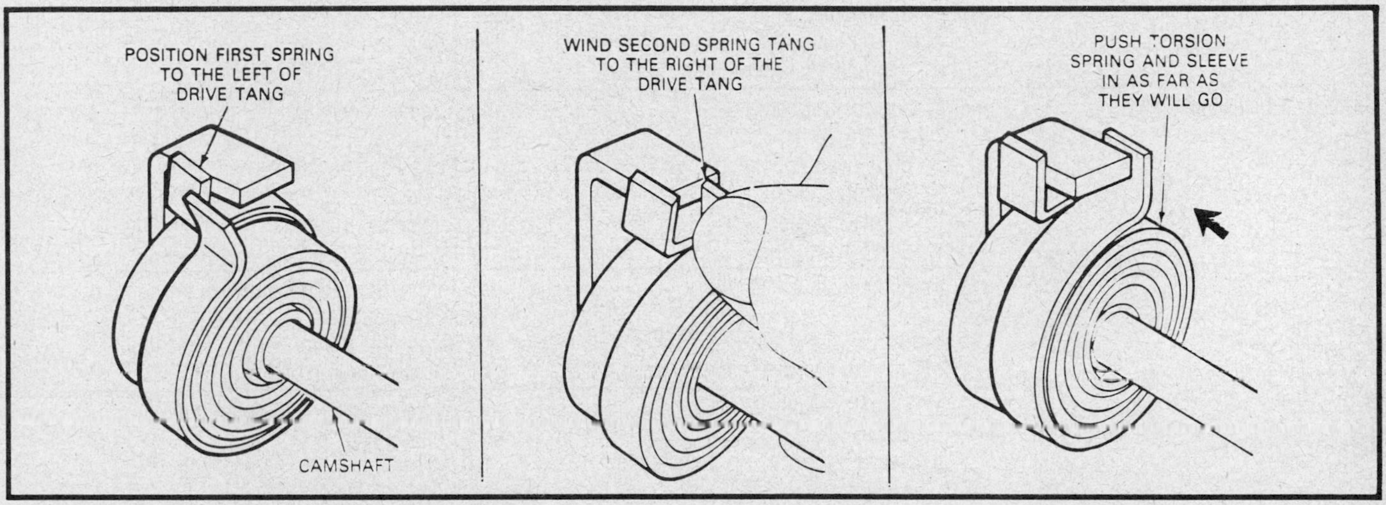
POSITION FIRST SPRING
TO THE LEFT OF
DRIVE TANG
CAMSHAFT
WIND SECOND SPRING TANG
TO THE RIGHT OF THE
DRIVE TANG
PUSH TORSION
SPRING AND SLEEVE
IN AS FAR AS
THEY WILL GO

Sliding the spring spacer onto the camshaft

Specifications

TORQUE SPECIFICATIONS
1350 Electronic Shift Transfer Case

Description	ft. lbs.	Nm
Breather vent	6–14	8–19
Case to cover bolts	23–30	31–41
Drain and fill plug	14–22	19–30
Front driveshaft yoke bolts	12–15	16–20
Skid plate to frame bolt	22–30	30–41
Transfer case to transmission adapter	25–43	34–58
Yoke nut	150–180	203–244
Motor mount, motor bracket and clutch coil nut	6–8	8.1–10.8
Wire connector bracket	5–7	6.8–9.5
Rear driveshaft flange bolts	61–87	83–118
Oil pump bolts	36–40①	4.0–4.5
Speedometer screw	20–25①	2.3–2.8

① Inch lbs.

Special Tools

Number	Description
T50T-100-A	Impact slide hammer
D80L-100-S	Puller collet
Tool-1175-AC	Seal remover
T80T-4000-W	Driver handle
D79L-4621-A	Bearing splitter
T74P-4635-C	U-Joint tool

BORG WARNER 1356 MANUAL SHIFT TRANSFER CASE

Application

1987–89 Ford Bronco, F-150, F-250 and F-350

General Description

The Borg Warner 1356 transfer case is a 3 piece all magnesium design (except for the manual shift unit equipped with a power take off which will have a magnesium case half and an aluminum case half), part time transfer case. Lubrication is accomplished by a positive displacement oil pump. This oil pump, channels the oil flow through drilled holes in the rear output shaft. The oil pump turns with the rear output shaft and allows towing of the vehicle for extended distances without having to disconnect the driveshaft.

Trouble Diagnosis

CHILTON'S THREE C'S TRANSFER CASE DIAGNOSIS

Condition	Cause	Correction
Slips out of gear	a) Shifting poppet spring weak	a) Replace the shifting poppet spring
	b) Bearing broken or worn	b) Replace the bearing
	c) Shifter fork bent	c) Replace shifter fork
	d) Improper control rod adjustment	d) Adjust the control rod
Hard shifting	a) Lack of lubricant	a) Add the recommended fluid to the proper level
	b) Shift lever binding on shaft	b) Repair and adjust the shift lever
	c) Shifting poppet ball scored	c) Replace the shifting poppet ball
	d) Shifter fork bent	d) Replace shifter fork
	e) Low tire pressure	e) Inflate tires to the correct pressure
Backlash	a) Companion yoke loose	a) Torque the companion yoke to specifications
	b) Transfer case loose on mounts	b) Torque the mount bolts to specification
	c) Internal parts excessively worn	c) Rebuild the transfer case
Noisy	a) Low lubricant level	a) Add the recommended fluid to the proper level
	b) Bearings improperly adjusted or excessively worn	b) Adjust or replace the bearing
	c) Gears worn or damaged	c) Replace the gear in question
	d) Improper alignment of driveshafts or U-joint	d) Realign the driveshaft and U-joint
Oil leakage	a) Excessive amount of oil in the case	a) Drain the fluid to the proper level
	b) Oil vent clogged	b) Clean oil vent
	c) Gaskets or seals leaking	c) Replace leaking gaskets or seals
	d) Bearings improperly adjusted or excessively worn	d) Adjust or replace the bearing
	e) Driveshaft yoke mating surfaces scored	e) Machine the driveshaft yoke mating surface
Overheating	a) Excessive amount of oil in the case	a) Drain the fluid to the proper level
	b) Low lubricant level	b) Add the recommended fluid to the proper level
	c) Bearing adjustment too tight	c) Adjust or replace the bearing

On Vehicle Services

SHIFT LEVER

Removal and Installation

NOTE: Remove the shift ball only if the shift ball, boot or lever have to be replaced. If any of these parts are not being replaced, remove the shift ball, boot and lever as an assembly.

1. Remove the plastic insert from the shift ball. Warm the ball with a heat gun or equivalent until it reaches approximately 140–180°F. Using a block of wood and a hammer, knock the shift ball off the lever. Be careful not to damage the finish on the shift lever.
2. Remove the rubber boot with the floor pan cover. Disconnect the vent hose from the shift lever.
3. Disconnect the transfer case shift rod from the shift lever. Remove the bolts holding the shift lever to transfer case. Remove the shift lever and bushings.
4. Before installing the shifter assembly, move the transfer case lever to the **4L** position.
5. Install the shifter assembly with the bolts finger tight, and move the cam plate rearward until the bottom chamfered corner of the neutral lug just contact the forward right edge of the shift lever.

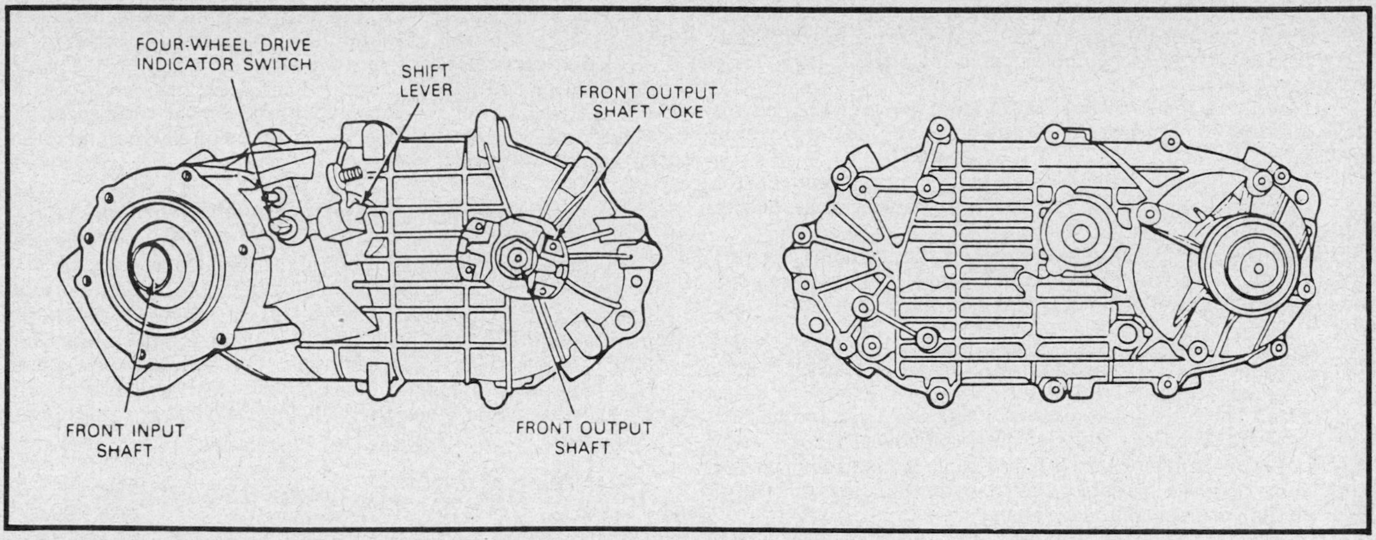
FOUR-WHEEL DRIVE INDICATOR SWITCH
SHIFT LEVER
FRONT OUTPUT SHAFT YOKE
FRONT INPUT SHAFT
FRONT OUTPUT SHAFT

Borg Warner 1356 transfer case—Manual shift shown, electric shift similar

6. Hold the cam plate in the proper position and tighten the bolts to 71–90 ft. lbs. (96–122 Nm).
7. Move the transfer case in-cab shift lever to all positions to check for positive engagement. There should be clearance between the shift lever and the cam plate in the **2H** front, and **4H** rear (clearance should not exceed 2.0mm), and the **4L** shift positions.
8. Attach the shift lever to the control lever and tighten to 23–32 ft. lbs. (21–43 Nm).
9. Install the vent assembly so the white marking on the housing is in the position in the notch in the shifter. Install the rubber boot and the floor pan cover.
10. Warm the ball with a heat gun or equivalent until it reaches approximately 140–180°F. Using a $^{7}/_{16}$ in. socket and a mallet, tap the ball on to the lever and install the plastic shift pattern insert.
11. Check the transfer case for proper shifting and operation.

REAR OR FRONT OUTPUT SHAFT WITH FIXED YOKES OIL SEAL

Removal and Installation

1. Raise and support the vehicle safely.
2. Place a drain pan under transfer case, remove the drain plug and drain fluid from the transfer case.
3. Disconnect the front or rear driveshaft from the transfer case output shaft yoke. Wire the driveshaft out of the way.
4. Remove the oil seal by removing the 30mm nut, steel washer and rubber seal from the front or rear output shaft and remove the yokes.
5. Remove the oil seal from the front of the rear output housing bore with a suitable seal remover and slide hammer. Remove the oil slinger from each yoke.
6. Make sure that the housing face and bore are free from nicks and burrs. Coat the seal with a suitable lubricant. Position the oil seal into the front or rear output housing bore, making sure the oil seal is not cocked in the bore. Drive the new seal into the housing with a suitable seal installer. Install a new oil seal slinger on each yoke.
7. Install the yoke, rubber seal, steel washer and locknut on the front or rear output shafts. Torque the locknut to 150–180 ft. lbs. (203–244 Nm).
8. Connect the front or rear driveshaft to the output yoke and torque the bolts to 61–87 ft. lbs. (83–118 Nm).

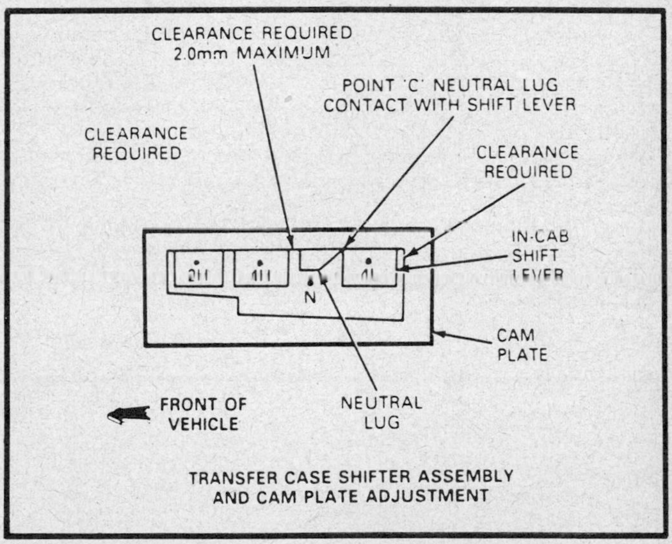
CLEARANCE REQUIRED
2.0mm MAXIMUM
POINT 'C' NEUTRAL LUG
CONTACT WITH SHIFT LEVER
CLEARANCE REQUIRED
CLEARANCE REQUIRED
IN-CAB SHIFT LEVER
2H
4H
N
4L
CAM PLATE
FRONT OF VEHICLE
NEUTRAL LUG
TRANSFER CASE SHIFTER ASSEMBLY AND CAM PLATE ADJUSTMENT

Transfer case shifter and cam plate adjustment

9. Fill the transfer case with 6.5 pints of Dexron®II transmission fluid or equivalent. Torque the fill plug to 14–22 ft. lbs. (19–32 Nm). Lower the vehicle. Start the engine and check the transfer case for correct operation. Stop the engine and check the fluid level, add as necessary.

Removal and Installation

TRANSFER CASE REMOVAL

The catalytic converter is located beside the transfer case. Be careful when working around the catalytic converter because of the extremely high temperatures generated by the converter.

1. Raise and support the vehicle safely. Drain the gear fluid from the transfer case into a suitable drain pan.
2. Disconnect the 4WD indicator switch connector at the transfer case and if so equipped, remove the skid plate from the frame.

3. Disconnect the front driveshaft from the front output yoke. Disconnect the rear driveshaft from the rear output yoke.
4. Disconnect the speedometer cable from the transfer case bearing retainer.
5. Disconnect the vent hose from the transfer case and remove the heat shield from the engine mount bracket and transfer case.
6. Using a suitable transmission jack, support the transfer case. Remove the bolts that hold the transfer case to the transmission.
7. Slide the transfer case rear ward off the transmission output shaft and lower the transfer case from the vehicle. Remove the old gasket between the transfer case and the adapter.

TRANSFER CASE INSTALLATION

1. Install the heat shield onto the transfer case and place a new gasket between the transfer case and adapter.
2. Raise the transfer case with a suitable transmission jack or equivalent, raise it high enough so that the transmission output shaft aligns with the splined transfer case input shaft.
3. Slide the transfer case forward on to the transmission output shaft and onto the dowel pin. Install transfer case retaining bolts and torque them to 26–43 ft. lbs. Remove the transmission jack from the transfer case.
4. Connect the rear driveshaft to the rear output shaft yoke and torque the retaining bolts to 20–25 ft. lbs. (27–35 Nm) for the Bronco, and 8–15 ft. lbs. (11–20 Nm) for the F series trucks. Attach the shift rod to the transfer case shift lever and transfer case control rod and attach with retaining rings.
5. Connect the speedometer driven gear to the transfer case. Connect the 4WD indicator switch wire connector at the transfer case.
6. Connect the front driveshaft to the front output yoke and torque the yoke nut to 8–15 ft. lbs. (11-20 Nm). Attach the heat shield to the engine mounting bracket and mounting lug on the transfer case.
7. Install the skid plate to the frame. Install the transfer case drain plug and torque the plug to 6–14 ft. lbs. (8-19 Nm).
8. Fill the transfer case with 6.5 pints of Dexron®II transmission fluid or equivalent. Torque the fill plug to 15–25 ft. lbs. (20 – 35 Nm). Lower the vehicle. Start the engine and check the transfer case for correct operation. Stop the engine and check the fluid level, add as necessary.

Before Disassembly

Cleanliness during disassembly and assembly is necessary to avoid further transfer case trouble after overhaul. Before removing any of the transfer case subassemblies, plug all the openings and clean the outside of the of the transfer case thoroughly. Steam cleaning or car wash type high pressure equipment is preferable. During disassembly, clean all parts in suitable solvent and dry each part. Do not use cloth or paper towels to dry parts. Use compressed air only.

Transfer Case Disassembly

1. Place the transfer case in a suitable holding fixture. Drain the fluid from the case by removing the filler plug from the case half. Remove the speedometer cover.
2. Remove both output shaft yoke nuts and washers. Remove the front and rear output yokes. Remove the 4WD indicator switch.
3. Remove the front and rear output shaft yoke seals using special tool T74P–77248–A and T50L–100–A or equivalents.
4. Remove the input shaft seals using the tools mentioned above.
5. Remove the 4 No. 50 Torx® head bolts securing the rear bearing retainer to the cover. Pry the rear bearing retainer from the cover using a ½ in. drive breaker bar between the pry bosses and separate and remove the bearing retainer from the cover. Remove all traces of RTV gasket sealant from the mating surfaces of the cover and the bearing retainer.

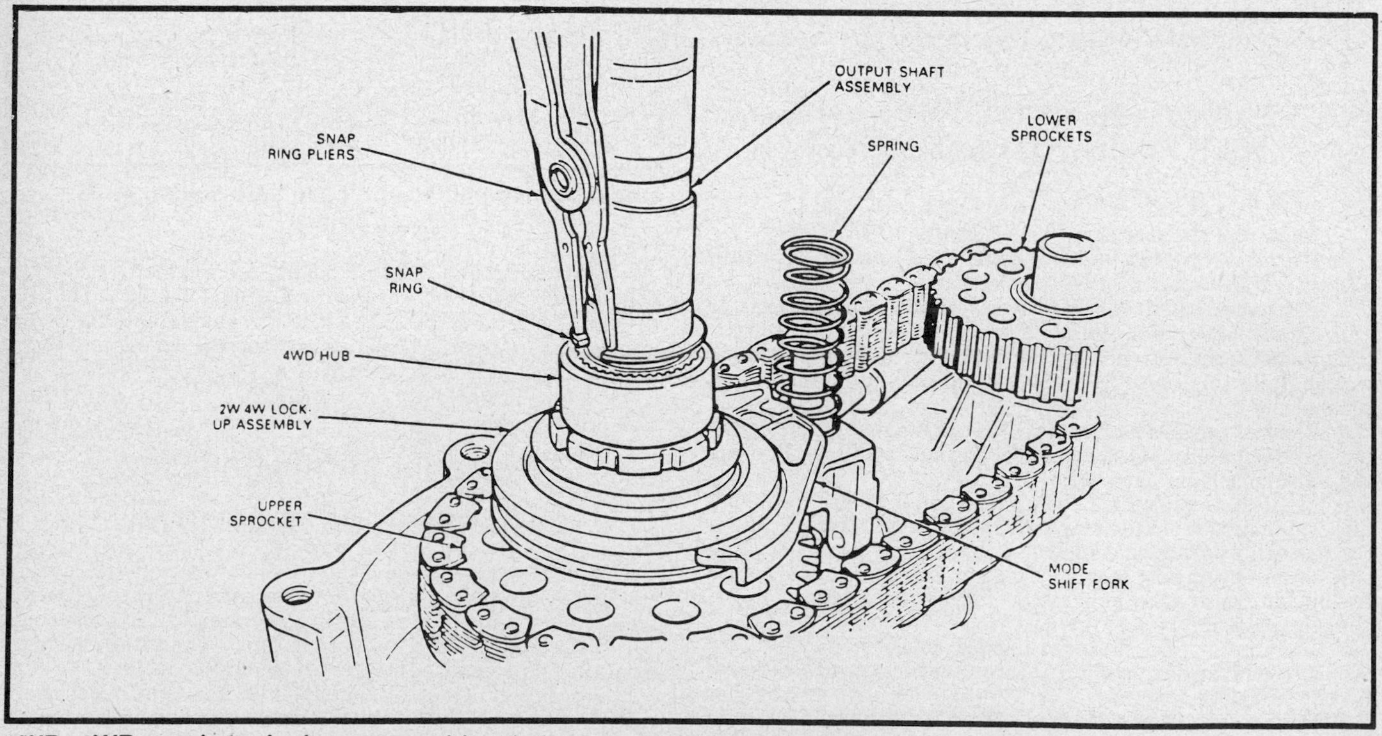
OUTPUT SHAFT ASSEMBLY
LOWER SPROCKETS
SPRING
SNAP RING PLIERS
SNAP RING
4WD HUB
2W 4W LOCK-UP ASSEMBLY
UPPER SPROCKET
MODE SHIFT FORK

2WD, 4WD sprokets, lock-up assembly, chain, upper and lower sprockets

NOTE: When removing the RTV sealer, be careful not to damage the mating surfaces of the magnesium cases.

6. Lift the rear output shaft and using a small prybar, remove the speedometer gear retaining clip.
7. Slide the speedometer gear forward and remove the ball with a small magnet. The speedometer gear can now be removed off of the rear output shaft.
8. Remove the snapring on the output shaft retaining the upper rear ball bearing using a suitable tool.
9. Remove the 12 No. 50 Torx® bolts that retain the front case to the rear cover. Insert a ½ in. drive breaker bar between the pry bosses and separate. Lift the front case from the rear cover. Remove all traces of RTV gasket sealant from the mating surfaces of the cover and the bearing retainer.

NOTE: When removing the RTV sealer, be careful not to damage the mating surfaces of the magnesium cases.

10. Remove the front output shaft inner needle bearing from the rear cover with special slide hammer tool T50T–100–A and D80L–100–T, collet from D80L–100–A, blind hole puller set or equivalents.
11. Drive out the rear output shaft bearing from the inside of the case using the appropriate tools.
12. Remove the snapring on the output shaft securing the clutch hub. Slide the 4WD hub off of the output shaft.
13. Remove the spring from the shift rail and lift the mode shift fork complete with the shifting collar from the upper sprocket spline.
14. Dissassemble the 2WD–4WD lockup assembly by removing the internal snapring and pull the lockup hub and spring from the collar.
15. Remove the snapring retaining the lower sprocket to the lower output shaft. Grasp the upper and lower sprocket complete with the chain and lift them at the same time from the upper and lower output shafts.
16. Remove the snapring retaining the lower sprocket complete with the chain and lift them at the same time from the upper and lower output shafts.
17. Remove the shift rail by sliding it straight out from the shift fork.
18. Remove the high and low shift fork by first rotating it until the roller is free from the cam then sliding out of the engagement from the shift hub.
19. Remove the chip collecting magnet from its slot in the case.
20. Lift out the pump screen and remove the output shaft assembly with the pump assembled on it. If the pump is to be disassembled, remove the 4 bolts from the pump body. Note the position of the pump front body, pins, spring, rear cover and pump retainer as removed.
21. Remove the high low shift hub.
22. Remove the front output shaft from the case.
23. Turn the front case over and remove the front oil seal from the case using tool T74P–77248–A and T50T–100–A. or equivalents.
24. Reaching through the front opening with a pair of snapring pliers, expand the snapring on the input shaft allowing it to drop out of the bearing. The carrier assembly, including the input shaft is serviced as an assembly only. If the bearing or bushing is to be replaced, drive out both of them through the input spline using suitable tools.
25. Remove the ring gear by prying out the internal snapring and lift out the gear.
26. Remove the power take off drive gear from the input shaft carrier assembly, if equipped using appropriate tools.
27. Remove the internal snapring securing the input shaft bearing to the case and drive it out from the outside of the case using tool T73T–1202–A and T80T–4000–W or equivalents.
28. Remove the internal snapring securing the front output shaft bearing in the magnesium housing and drive the bearing out from the front of the case using tool T73T–1202–B and T80T–4000–W or equivalents.
29. Remove the shift cam by removing the retaining clip and sliding the shift shaft out of the case.
30. Remove the shift shaft seal by carefully prying it out of the case, being careful not to damage the case.
31. Remove the shift cam, assist spring, and the assist spring bushing from the case.

CLEANING AND INSPECTION

Cleaning

During overhaul, all components of the transfer case (except bearing assemblies) should be thoroughly cleaned with solvent and dried with air pressure prior to inspection and reassembly.
1. Clean the bearing assemblies as follows:

NOTE: Proper cleaning of bearings is of utmost importance. Bearings should always be cleaned separately from other parts.

a. Soak all bearing assemblies in clean solvent or fuel oil. Bearings should never be cleaned in a hot solution tank.
b. Slush bearings in solvent until all old lubricant is loosened. Hold races so that bearings will not rotate; then clean bearings with a soft bristled brush until all dirt has been removed. Remove loose particles of dirt by tapping bearing flat against a block of wood.
c. Rinse bearings in clean solvent; then blow bearings dry with air pressure.

CAUTION

Do not spin bearings while drying.

d. After drying, rotate each bearing slowly while examining balls or rollers for roughness, damage, or excessive wear. Replace all bearings that are not in first class condition.

NOTE: After cleaning and inspecting bearings, lubricate generously with recommended lubricant, then wrap each bearing in clean paper until ready for reassembly.

2. Remove all portions of old gaskets from parts, using a stiff brush or scraper.

Inspection

1. Inspect all parts for discoloration or warpage.
2. Examine all gears and splines for chipped, worn, broken or nicked teeth. Small nicks or burrs may be removed with a fine abrasive stone.
3. Inspect the breather assembly to make sure that it is open and not damaged.
4. Check all threaded parts for damaged, stripped, or crossed threads.
5. Replace all gaskets, oil seals and snaprings.
6. Inspect housings, retainers and covers for cracks or other damage. Replace the damaged parts.
7. Inspect keys and keyways for condition and fit.
8. Inspect shift forks for wear, distortion or any other damage.
9. Check detent ball springs for free length, compressed length, distortion or collapsed coils.
10. Check bearing fit on their respective shafts and in their bores or cups. Inspect bearings, shafts and cups for wear.

NOTE: If either bearings or cups are worn or damaged, it is advisable to replace both parts.

11. Inspect all bearing rollers or balls for pitting or galling.
12. Examine detent balls for corrosion or brinneling. If shift bar detents show wear, replace them.
13. Replace all worn or damaged parts. When assembling the transfer case, coat all moving parts with recommended lubricant.

Transfer Case Assembly

NOTE: Before starting the assembly procedure, lubricate all the internal parts, with Dexron®II transmission fluid or equivalent.

1. Install the input shaft and the front output shaft bearings in the case using the appropriate tools. Install the internal snaprings retaining the bearings in the case.
2. Drive the front output shaft seal into the case unit until it is fully seated against the case using tool T86T–7034–CH or equivalent.
3. Install the front output shaft through the lower bearing. The front output shaft is held in place in the case by the front output yoke and oil seal slinger assembly. Install the front yoke assembly onto the front output shaft then the rubber seal, flat washer and 30mm locknut. Torque the yoke locknut to 130–180 ft. lbs. (176–244 Nm).
4. Press the power take-off drive gear onto the input shaft assembly if it was removed.
5. Press the needle bearing and bronze bushing into the input shaft with the appropriate tools.
6. Install the ring gear into the slots in the case and retain it with the large internal snapring making sure that it is fully seated.
7. Install the input shaft and carrier assembly in the case through the input shaft bearing being careful not to damage the gear teeth when aligning them with the ring gear teeth.
8. While supporting the carrier assembly in position, install a new spring on the front side of the input bearing making sure that it is fully seated in the snapring groove of the input shaft.
9. Install the upper input shaft oil seal into the case using an appropriate tool until it is fully seated against the case.
10. Install a new shifter shaft seal into the case using an appropriate tool.
11. Assemble the shift cam assembly into the case by sliding the shift shaft and lever assembly through the case and seal into engagement with the shift cam. Secure the shift cam with the retaining clip.
12. Install the shift cam assist spring in position in the bushing of the shift cam and in the recess in the case.
13. Assemble the pump and output shaft as follows: Place the oil pump cover with the word **TOP** facing the front of the front case. Install the 2 pins (with the flats facing upwards) with the spring between the pins and place the assembly in the oil pump bore in the output shaft. Place the oil pump body and the pick up tube over the shaft and make sure that the pins are riding against the inside of the pump body. Place the oil pump rear cover with the words **TOP REAR** facing the rear of the case. The word **TOP** on the front cover and the rear cover should be on the same side. Install the pump retainer with the tabs facing the front of the transfer case. Install the 4 retaining bolts and rotate the output shaft while tightening the bolts to prevent the pump from binding. Tighten the bolts to 36–40 inch lbs. Lubricate the assembly with automatic transmission fluid.

NOTE: The output shaft must turn freely within the oil pump. If binding occurs, loosen the 4 bolts and retighten again.

14. Install the high low shift hub. Install the high low shift fork by engaging it with the shift hub flange and rotating it until the roller is engaged with the lower groove of the cam.
15. Install the shift rail through the high low fork bore and into the rail bore in the case.
16. Install the output shaft and oil pump assembly in the input shaft. Make certain that the external splines of the output shaft engage the internal splines of the high low shift hub. Make sure that the oil pump retainer and oil filter leg are in the groove and notch of the front case. Install the collector magnet in the notch in the front case.
17. Assemble the upper and lower sprockets with the chain and place them as an assembly over the upper and lower output shafts. Install the washer and snapring which retain the lower sprocket to the front output shaft.
18. Assemble the 2WD–4WD lockup assembly by installing the tapered compression spring in the lockup collar with the small end installed first. Place the lockup hub over the spring and compress the spring while installing the internal snapring which holds the lockup assembly together.

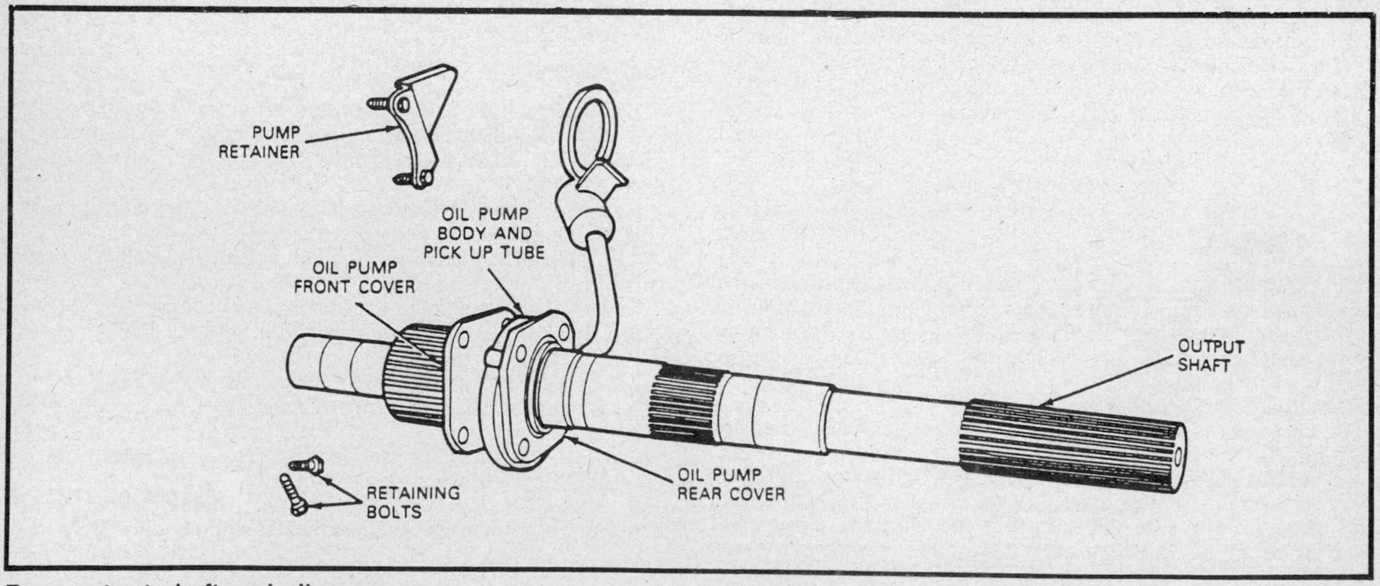
PUMP RETAINER
OIL PUMP BODY AND PICK UP TUBE
OIL PUMP FRONT COVER
OUTPUT SHAFT
RETAINING BOLTS
OIL PUMP REAR COVER

Rear output shaft and oil pump

19. Install the lockup assembly and its shift fork over the external splines of the upper sprocket and the shift rail with the long boss of the shift rail facing foreward.

20. Assemble the 4WD return spring over the shift rail and against the shift fork.

21. Place the 4WD hub over the external splines of the output shaft and secure with the appropriate snapring. Make sure that the snapring is fully seated in the snapring groove.

22. Press the lower output needle bearing in its bore in the rear cover using an appropriate tool.

23. Press the rear output shaft bearing into position in the cover. Install the bearing snapring retainer in the cover.

24. Install the rear output shaft oil seal in the bearing retainer using the appropriate tool making sure it is fully seated.

25. Coat the mating surface of the front case with a bead of RTV.

26. Place the cover on the case making sure that the lower output shaft, shift shaft and the shift rail are aligned. Install and torque the 12 No. 50 Torx® head case to cover bolts to 22–36 ft. lbs. (29–49 Nm).

27. Install the bearing retainer snapring on the output shaft making sure that the snapring is fully seated in the groove on the shaft.

28. Place the speedometer drive gear over the shaft with the slot aligned with the hole for the drive ball. The gear should go completely against the snapring which retains the output shaft. Place the ball in the hole and pull the speedometer gear over the ball. Snap the retaining clip between the snapring and the speedometer gear.

29. Apply a bead of RTV to the face of the rear bearing retainer or to the rear slip yoke extension housing.

30. Place the rear bearing retainer or the rear slip yoke extension housing in its position and secure with the 4 torx head bolts.

31. On the transfer case with the slip yoke bearing retainer housing remove the extension oil seal using tool T74P–77248–A and T50T–100–A or equivalents. Remove the extension housing bushing using tool T85–7034–AH or equivalent. Install a new bushing using tool T85T–7034–BH and T80T–4000–W or equivalents. Install a new seal using tool T61L–7657–B or equivalent.

32. Install the rear output shaft yoke and slinger assembly onto the rear splines of the output shaft. Install the rubber seal, flat steel washer and 30mm locknut on the output shaft and torque to 150–180 ft. lbs. (203–244 Nm).

33. Install the drain plug and tighten to 14–22 ft. lbs. (19–32 Nm).

34. Install the 4WD indicator light switch and aluminum washer into the case.

35. Place a ⅜ in. drive ratchet in the fill plug and remove the plug. Fill the transfer case with 64 oz. of Dexron®II transmission fluid.

36. Install the fill plug and tighten to 14–22 ft. lbs. (19-32 Nm).

37. Install the transfer case. Lower the vehicle.

38. Start the engine, check the transfer case for proper operation. Stop the engine and check the fluid level. The fluid should drip out of the level hole. If the fluid flows out of the level hole, the oil pump may not be funtioning properly.

Specifications

TORQUE SPECIFICATIONS
1356 Manual Shift Transfer Case

Description	ft. lbs.	Nm
Case half attaching bolts	35–40	48–54
Four wheel drive indicator switch	8–12	11–16
Front and real output yokes to transfer case	120–150	163–203
Drain plug	14–22	19–30
Fill plug	14–22	19–30
Transfer case to transmission adapter	25–43	34–58
Heat shield to transfer case	40–45	54–61
Skid plate to frame	15–20	20–27
Front driveshaft to front output yoke	8–15	11–20
Rear driveshaft to rear output yoke—bolt—Bronco	20–28	28–33
Rear driveshaft to rear output yoke—nut—F150-F350 4×4	8–15	11–20

Special Tools

Number	Description
T50T-100-A	Impact slide hammer—2½ lb.
D80L-100-A	Blind hole puller set
D80L-100-T	Blind hole puller
D80L-100-H	Actuator pin
TOOL-1175-AC	Seal remover
T80T-7127-B	Output bearing replacer—front
T80T-7127-C	Output bearing replacer—rear
T80T-4000-P	Bearing installer
T74P-77248-A	Oil seal remover
T61L-7657-B	Oil seal installer
T86L-7034-AH	Extension housing bushing remover
T85T-7034-BH	Extension housing bushing installer
T80T-4000-W	Bearing installer
T86T-7034-CH	Oil seal installer
T73T-1202-B	Bearing installer
T73T-1202-A	Bearing installer

BORG WARNER 1356 ELECTRONIC SHIFT TRANSFER CASE

Application

1987–89 Ford Bronco with automatic transmission

General Description

The Borg Warner 1356 electronic shift transfer case is a 3 piece magnesium part time transfer case. It transfers power from the transmission to the rear axle and when electronically actuated, also to the front drive axle. The unit is lubricated by a positive displacement oil pump that channels oil flow through drilled holes in the rear output shaft. The pump turns with the rear output shaft and allows towing of the vehicle at maximum legal road speeds for extended distances without disconnecting the front and/or rear driveshaft.

The electronic system consists of a 2 rocker switch control system, an electronic control module, an electric shift motor with an integral shift position sensor and a speed sensor. The 2 rocker control switches are located on the lower right hand coner of the instrument panel for fingertip shift control. The electronic control module is located on the right hand cowl side. The electronic control module, controls the operation of the transfer case in response to inputs to the push button control by the vehicle operator.

The speed sensor, is mounted to the rear of the transfer case, the sensor tells the control module the proper speed to shift the transfer case. The shift position sensor, is an integral part of the electric shift motor, tells the control module the shift position of the transfer case.

The electronic shift motor mounted externally at the rear of the transfer case, drives a rotary helical cam which moves the 2WD–4WD shift fork and **4H–4L** reduction shift fork to the selected vehicle drive position.

The transfer case is equipped with a magnetic clutch, similar to an air conditioning compressor clutch, which is located inside the transfer case adjacent to the 2WD–4WD shift collar. The clutch is used to spin up the front drive system from zero to vehicle speed in milliseconds. This spin up allows the shift between **2H** and **4H** to be made at any vehicle speed. When the transfer case rear and front output shafts reach synchronous speed, the spring loaded shift collar mechanically engages the mainshaft hub to the chain drive sprocket and the magnetic clutch is deactivated.

Shifts between **4H** and **4L** can only occur with the clutch interlock or transmission safety switches closed. The vehicle's speed must also be within specified limits as determined by the transfer case speed sensor (3 mph or under).

When the operator selects the drive combination through the push button control, an electric motor turns a helical cam, which is linked to the high low and 2WD–4WD shift forks through fork mounted roller bushing assemblies. As the electric motor turns the helical cam, the high low fork bushing rides in a slotted lobe in the cam to make low high or high low range changes; and the 2WD–4WD for bushing rides on lobes at the end of the cam to make the 2WD–4WD or 4WD–2WD shift.

Trouble Diagnosis

CHILTON'S THREE C'S TRANSFER CASE DIAGNOSIS

Condition	Cause	Correction
Slips out of gear	a) Shifting poppet spring weak	a) Replace the shifting poppet spring
	b) Bearing broken or worn	b) Replace the bearing
	c) Shifter fork bent	c) Replace shifter fork
	d) Improper control rod adjustment	d) Adjust the control rod
Hard shifting	a) Lack of lubricant	a) Add the recommended fluid to the proper level
	b) Shift lever binding on shaft	b) Repair and adjust the shift lever
	c) Shifting poppet ball scored	c) Replace the shifting poppet ball
	d) Shifter fork bent	d) Replace shifter fork
	e) Low tire pressure	e) Inflate tires to the correct pressure
Backlash	a) Companion yoke loose	a) Torque the companion yoke to specifications
	b) Transfer case loose on mounts	b) Torque the mount bolts to specification
	c) Internal parts excessively worn	c) Rebuild the transfer case
Noisy	a) Low lubricant level	a) Add the recommended fluid to the proper level
	b) Bearings improperly adjusted or excessively worn	b) Adjust or replace the bearing
	c) Gears worn or damaged	c) Replace the gear in question
	d) Improper alignment of driveshafts or U-joint	d) Realign the driveshaft and U-joint
Oil leakage	a) Excessive amount of oil in the case	a) Drain the fluid to the proper level
	b) Oil vent clogged	b) Clean oil vent
	c) Gaskets or seals leaking	c) Replace leaking gaskets or seals
	d) Bearings improperly adjusted or excessively worn	d) Adjust or replace the bearing
	e) Driveshaft yoke mating surfaces scored	e) Machine the driveshaft yoke mating surface

CHILTON'S THREE C'S TRANSFER CASE DIAGNOSIS

Condition	Cause	Correction
Overheating	a) Excessive amount of oil in the case b) Low lubricant level c) Bearing adjustment too tight	a) Drain the fluid to the proper level b) Add the recommended fluid to the proper level c) Adjust or replace the bearing

Electronic Controls

DIAGNOSIS AND TESTING

The battery feed circuit, through a circuit breaker provides memory capability for the electronic control module. The ignition **RUN** and **ACC** feed circuits through a fuse, provide power for the switches and the electric shift motor. The headlamp dimmer circuit provides the power for night time illumination of the overhead roof console vehicle graphics.

CONTROL MODULE

Self Test

The electronic control module has a diagnostic capability of its own circuitry. The self test procedures are as follows:

1. Remove the 5 wire connector and the 8 wire connector from the electronic control module.
2. Turn the ignition switch to the **RUN** position.
3. Activate the self test switch and note the results.
4. A flashing indicator lamp (approximately one flash per second) indicates that the control module is functioning properly.
5. A steady indicator light indicates that the control module is inoperative and must be replaced.

CONTROL MODULE CIRCUIT

There are 3 wiring harness connected to the electronic control module, the 8 wire pigtail harness connector, the 5 wire harness connector and the 8 wire harness connector. To check the integrity of these circuits, disconnect the harness from the electronic control module and perform the following test.

8 Wire Pigtail Harness Connector Test

1. Remove the 8 wire pigtail harness connector and connect a suitable voltmeter between terminal 8 and the ground. The voltmeter should indicate battery (12 volts) voltage at all times.
2. Connect the voltmeter between terminal 7 and ground. Turn the ignition switch to the **RUN** position. The voltmeter should indicate battery (12 volts) voltage.

NOTE: In the following test where the usage of an ohmmeter is specified, always remember that an ohmmeter should never be connected into a live or powered circuit. If an ohmmeter is connected to a powered circuit, severe damage will be done to the instrument. The vehicle's battery should be disconnected before disconnecting any circuit with an ohmmeter to prevent any accidental damage to the instrument.

3. Connect an ohmmeter between terminal 6 and ground. The ohmmeter should indicate a low resistance valve of less than 10 ohms.
4. Connect an ohmmeter between terminal 4 and 5. The ohmmeter should indicate a low resistance valve of less than 10 ohms.
5. Connect an ohmmeter between terminal 3 and ground. The ohmmeter should indicate a 0 ohms.
6. Connect an ohmmeter between terminal 2 and ground. The ohmmeter should indicate a 0 ohms.

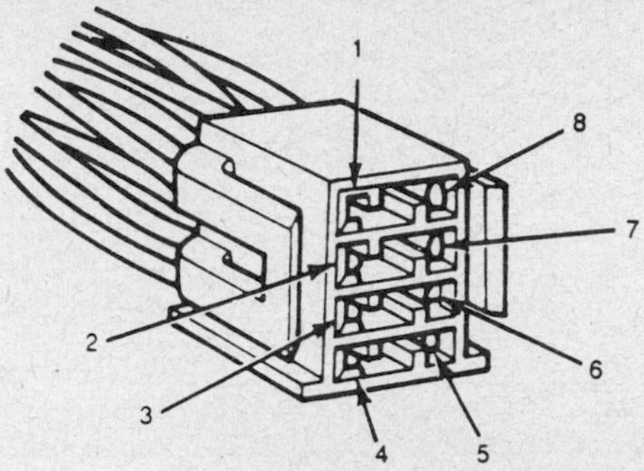
1
8
7
2
6
3
4
5

Position	Circuit Number	Color Code	Function
1	OPEN		
2	57	Blk	Ground
	57A	Blk	Ground
3	·396	Blk/Org	Logic Ground
4	778	Org	Transfer Case Motor Control (Clockwise) 2H-4H-4L
5	777	Yel	Transfer Case Motor Control (Counterclockwise) 4L-4H-2H
6	79	Brn	Electro-Magnetic Clutch (Feed)
7	296	Wht/Ppl	Ignition Run and Accessory Feed (Fused)
8	517	Blk/Wht	Battery Feed (Circuit Breaker)

Testing the 8 wire pigtail harness connector

5 Wire Harness Connector Test

1. Remove the 5 wire connector and connect a suitable ohmmeter between terminal 1 and 2. Then depress the 4WD (2H–4H) switch in the overhead roof console. The ohmmeter should indicate a very low resistance valve (less than 50 ohms) while the switch is being depressed.
2. Connect a suitable ohmmeter between terminal 1 and 3. Then depress the low range switch in the overhead roof console. The ohmmeter should indicate a very low resistance valve (less than 50 ohms) while the switch is being depressed.
3. Connect a test lead between terminal 4 and ground. Then turn the ignition switch to the **RUN** position and observe the overhead roof console. The light in the console low range bar should illuminate.
4. Connect a test lead between terminal 5 and ground. Then turn the ignition switch to the **RUN** position and observe the overhead roof console. The light in the console 4WD bar should illuminate.

8 Wire Harness Connector Test

1. Remove the 8 wire connector and connect a suitable ohmmeter between terminal 1 and ground. Then depress the clutch pedal and observe the ohmmeter. The ohmmeter should indicate a very low resistance valve (less than 50 ohms) while the clutch pedal is being depressed.

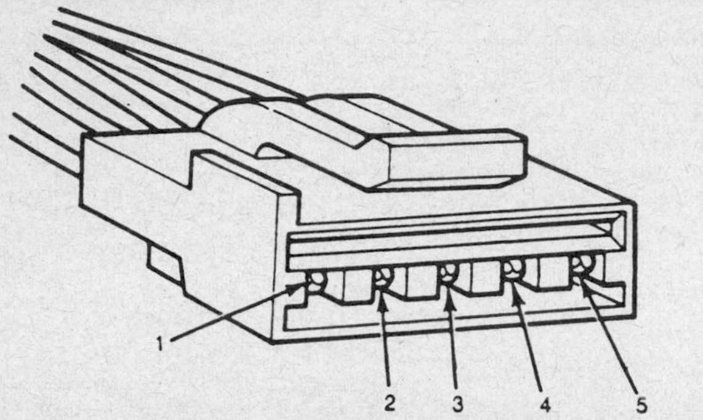
1
2
3
4
5

Position	Circuit Number	Color Code	Function
1	465	Wht/Lt Blu	Switch Feed
2	780	Dk Blu	4x4 Switch
3	781	Org/Lt Blu	Low Range Switch
4	782	Brn/Wht	Low Range Light
5	783	Gry	4x4 Light

Testing the 5 wire harness connector

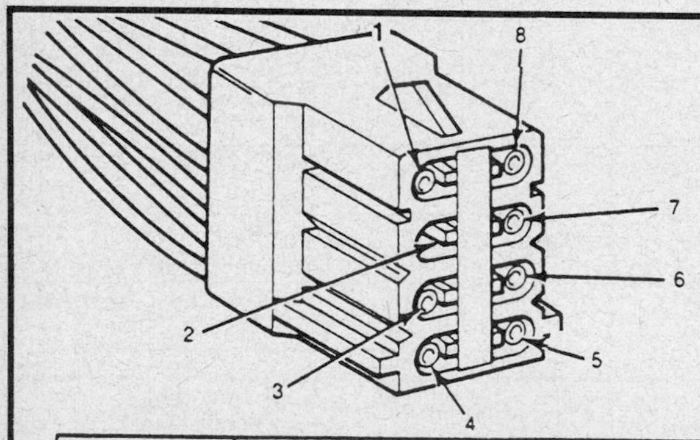
1
8
7
6
2
5
3
4

Position	Circuit Number	Color Code	Function
1	782	Red/Yel	Ignition Run and Accessory Feed (Fused)
2 (Used Only on Auto Trans)	463	Red/Wht	Automatic Transmission Neutral Safety Switch
3	296	Wht/Ppl	Ignition Run and Accessory Feed (Fused)
4	57	Blk	Ground
5	780	Lt Blu/Red	Side Marker Lamp (Feed)
6	783	Lt Grn/Yel	Dome Lamp (Feed)
7 (Used Only on Manual Trans)	463	Red/Lt Blu	Manual Transmission Clutch Interlock Switch
8	517	Blk/Wht	Battery Feed (Circuit Breaker)

Main feed connector terminal pin identification

2. If the vehicle is equipped with a automatic transmission, connect a suitable ohmmeter between terminal 1 and ground. Then shift the transmission into the **N** position and observe the ohmmeter. The ohmmeter should indicate a very low resistance valve (less than 50 ohms) while in the **N** position.

3. Connect a suitable ohmmeter between terminal 2 and 3. The ohmmeter should indicate a very low resistance valve (200–350 ohms).

4. This will check the continuity of the speed sensor that is located in the transfer case. The speed sensor picks up the rotat-

OHMMETER READINGS FOR SHIFT MOTOR POSITION SENSOR

Ohmmeter Connection	Transfer Case Gear Position		
	2 High	4 High	4 Low
Meter Reading From Terminal #8 to #4	Short	Open	Short
Meter Reading From Terminal #8 to #5	Open	Open	Short
Meter Reading From Terminal #8 to #6	Short	Short	Open
Meter Reading From Terminal #8 to #7	Open	Short	Open

NOTE: SHORT is a "low" resistance reading on the ohmmeter (zero ohms).
OPEN is a "high" resistance reading on the ohmmeter (infinity).

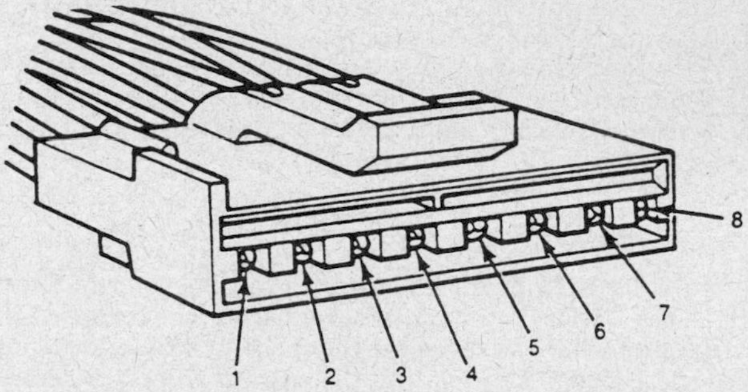
8
7
6
5
1
2
3
4

Position	Circuit Number	Color Code	Function
1	32	Red/Lt Blu	Manual Transmission Clutch Interlock Switch
	463	Red/Wht	Automatic Transmission Neutral Safety Switch
2	774	Lt Grn	Speed Sensor (Feed)
3	772	Lt Blu	Speed Sensor Return
4	771	Violet	Wire #5, Contact Plate Position Sensor in Transfer Case
5	770	Wht	Wire #4, Contact Plate Position Sensor in Transfer Case
6	764	Brn/Wht	Wire #3, Contact Plate Position Sensor in Transfer Case
7	763	Org/Wht	Wire #2, Contact Plate Position Sensor in Transfer Case
8	762	Yel/Wht	Wire#1, Contact Plate Position Sensor in Transfer Case

Testing the 8 wire harness connector

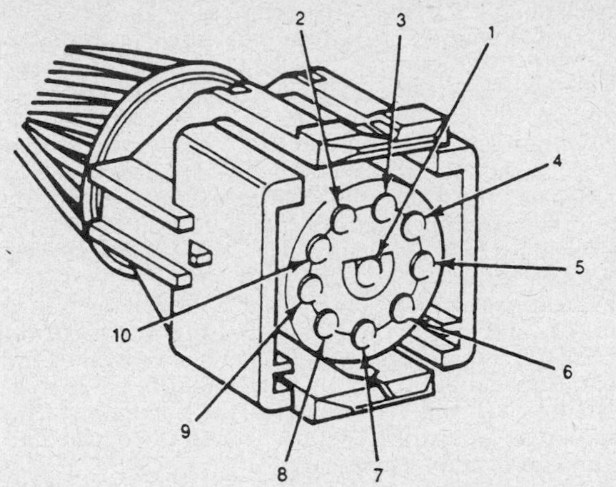
2
3
1
4
5
10
6
9
8
7

Position	Circuit Number	Color Code	Function
1	779	Brn	Electro-Magnetic Clutch (Feed)
2	778	Org	Transfer Case Motor Control (Clockwise) 2H-4H-4L
3	777	Yel	Transfer Case Motor Control (Counterclockwise) 4L-4H-2H
4	774	Lt Grn	Speed Sensor (Feed)
5	772	Lt Blu	Speed Sensor (Return)
6	771	Violet	Wire #5, Shift Position Sensor in Transfer Case (Output to Module)
7	770	Wht	Wire #4, Shift Position Sensor in Transfer Case (Output to Module)
8	764	Brn/Wht	Wire #3, Shift Position Sensor in Transfer Case (Output to Module)
9	763	Org/Wht	Wire #2, Shift Position Sensor in Transfer Case (Output to Module)
10	762	Yel/Wht	Wire #1, Shift Position Sensor in Transfer Case (Input from Module)

Electronic transfer case feed 10 pin connector terminal pin identification

ing speed of the output shaft from 2 notches that are cut in opposite sides of the outer ring of the clutch housing assembly.

5. Connect an ohmmeter between terminal 8 and terminals 4, 5, 6 and 7, respectively. Use the chart provided for the ohmmeter reading in each transfer case position.

On Vehicle Services

4WD SHIFT SWITCH

Removal and Installation

1. Disconnect the negative battery cable. Remove the instrument panel assembly.
2. Remove switch mounting screw and remove the switch from the panel.
3. When installing the switch, position the switch in the panel and retain it with the mounting screw.
4. Install the instrument panel assembly.

CONTROL MODULE

Removal and Installation

The control module is located on the inside right hand cowl panel.

1. Disconnect the negative battery cable.
2. Remove the right hand cowl panel kick panel.
3. Remove the 2 module to instrument panel attaching screws.
4. Lift out the module and disconnect the 3 wiring connectors. Remove the module assembly.
5. Position the new module to the kick panel opening. Connect the 3 wiring connectors to the module.
6. Place module into the kick panel opening and install the 2 mounting screws.
7. Install the kick panel pad.

SHIFT LEVER

Removal and Installation

NOTE: Remove the shift ball only if the shift ball, boot or lever have to be replaced. If any of these parts are not being replaced, remove the shift ball, boot and lever as an assembly.

1. Remove the plastic insert from the shift ball. Warm the ball with a heat gun or equivalent until it reaches approximately 140–180°F. Using a block of wood and a hammer, knock the shift ball off the lever. Be careful not to damage the finish on the shift lever.
2. Remove the rubber boot with the floor pan cover. Disconnect the vent hose from the shift lever.
3. Disconnect the transfer case shift rod from the shift lever. Remove the bolts holding the shift lever to transfer case. Remove the shift lever and bushings.
4. Before installing the shifter assembly, move the transfer case lever to the **4L** position.
5. Install the shifter assembly with the bolts finger tight, and move the cam plate rearward until the bottom chamfered corner of the neutral lug just contact the forward right edge of the shift lever.
6. Hold the cam plate in the position mentioned above and tighten the bolts to 71–90 ft. lbs. (95–122 Nm).
7. Move the transfer case in-cab shift lever to all positions to check for positive engagement. There should be clearance between the shift lever and the cam plate in the **2H** front, and **4H** rear (clearance should not exceed 2.0mm), and the **4L** shift positions.
8. Attach the shift lever to the control lever and tighten to 23–32 ft. lbs. (31–43 Nm).
9. Install the vent assembly so the white marking on the housing is in the position in the notch in the shifter. Install the rubber boot and the floor pan cover.
10. Warm the ball with a heat gun or equivalent until it reaches approximately 140–180°F. Using a $^{7}/_{16}$ in. socket and a mallet, tap the ball on to the lever and install the plastic shift pattern insert.
11. Check the transfer case for proper shifting and operation.

REAR OR FRONT OUTPUT SHAFT WITH FIXED YOKES OIL SEAL

Removal and Installation

1. Raise and support the vehicle safely.
2. Place a drain pan under transfer case, remove the drain plug and drain fluid from the transfer case.
3. Disconnect the front or rear driveshaft from the transfer case output shaft yoke. Wire the driveshaft out of the way.
4. Remove the oil seal by removing the 30mm nut, steel washer and rubber seal from the front or rear output shaft and remove the yokes.
5. Remove the oil seal from the front of the rear output housing bore with a suitable seal remover and slide hammer. Remove the oil slinger from each yoke.
6. Make sure that the housing face and bore are free from nicks and burrs. Coat the seal with a suitable lubricant. Position the oil seal into the front or rear output housing bore, making sure the oil seal is not cocked in the bore. Drive the new seal into the housing with a suitable seal installer. Install a new oil seal slinger on each yoke.
7. Install the yoke, rubber seal, steel washer and locknut on the front or rear output shafts. Torque the locknut to 150–180 ft. lbs. (203–244 Nm).
8. Connect the front or rear driveshaft to the output yoke and torque the bolts to 61–87 ft. lbs. (83–118 Nm).
10. Fill the transfer case with 6.5 pints of Dexron®II transmission fluid or equivalent. Torque the fill plug to 14–22 ft. lbs. (19–32 Nm). Lower the vehicle. Start the engine and check the transfer case for correct operation. Stop the engine and check the fluid level, add as necessary.

Removal and Installation

TRANSFER CASE REMOVAL

The catalytic converter is located beside the transfer case. Be careful when working around the catalytic converter because of the extremely high temperatures generated by the converter.

1. Raise and support the vehicle safely. Drain the gear fluid from the transfer case into a suitable drain pan.
2. Disconnect the 4WD indicator switch connector at the transfer case and if so equipped, remove the skid plate from the frame.
3. Remove the wire connector from the feed wire harness at the rear of the transfer case. Be sure to squeeze the locking tabs, then pull the connectors apart.
4. Disconnect the front driveshaft from the front output yoke. Disconnect the rear driveshaft from the rear output yoke.
5. Disconnect the speedometer driven gear from the transfer case rear cover. Disconnect the vent hose from the mounting bracket.
6. Using a suitable transmission jack, support the transfer case. Remove the bolts that hold the transfer case to the transmission.
7. Slide the transfer case rearward off the transmission output shaft and lower the transfer case from the vehicle. Remove the old gasket between the transfer case and the adapter.

TRANSFER CASE INSTALLATION

1. Install the heat shield onto the transfer case and place a new gasket between the transfer case and adapter.
2. Raise the transfer case with a suitable transmission jack or equivalent, raise it high enough so that the transmission output shaft aligns with the splined transfer case input shaft.
3. Slide the transfer case forward on to the transmission output shaft and onto the dowel pin. Install transfer case retaining bolts and torque them to 26–43 ft. lbs. (36–58 Nm). Remove the transmission jack from the transfer case.
4. Install the vent hose so that the white marking on the hose aligns with the mounting bracket. Connect the speedometer driven gear to the transfer case rear cover. Torque the screw to 20–25 inch lbs. (27–35 Nm).
5. Connect the rear driveshaft to the rear output shaft yoke and torque the retaining bolts to 20–28 ft. lbs.(27–38 Nm).
6. Connect the front driveshaft to the front output yoke and torque the yoke nut to 20–28 ft. lbs. (27–38 Nm). Attach the heat shield to the engine mounting bracket and mounting lug on the transfer case.
7. Install the skid plate to the frame. Install the transfer case drain plug and torque the plug to 6–14 ft. lbs. (8–19 Nm).
8. Fill the transfer case with 6.5 pints of Dexron®II transmission fluid or equivalent. Torque the fill plug to 15–25 ft. lbs. (11–35 Nm). Lower the vehicle. Start the engine and check the transfer case for correct operation. Stop the engine and check the fluid level, add as necessary.

Before Disassembly

Cleanliness during disassembly and assembly is necessary to avoid further transfer case trouble after overhaul. Before removing any of the transfer case subassemblies, plug all the openings and clean the outside of the of the transfer case thoroughly. Steam cleaning or car wash type high pressure equipment is preferable. During disassembly, clean all parts in suitable solvent and dry each part. Do not use cloth or paper towels to dry parts. Use compressed air only.

Transfer Case Disassembly

1. Remove the transfer case from the vehicle.
2. Remove the transfer case drain plug with a 3/8 in. drive ratchet and drain the fluid.
3. Remove the rear output shaft yoke by removing the 30mm nut, steel washer and rubber seal from the output shaft. Remove the front and rear output shaft yoke seals using special tool T74P–77248–A and T50L–100–A or equivalents. Remove the input shaft seals using the same tools.
4. Remove the wire connector assembly from the mounting bracket on the rear cover. If required, remove the 2 bolts and remove the bracket.

Form a small hook at the end of a paper clip or safety pin. Remove the locking sleeve from the wire connector by hooking the paper clip or safety pin and pulling it up from the bottom. Be sure not to damage the wire connector locking sleeve.

5. Remove the brown wire from the No. 1 center position in the connector. If required, remove the speed sensor green wire from the No. 4 connector position and blue wire from the No. 5 connector position.
6. Remove the 4 No. 50 Torx® head bolts securing the rear bearing retainer to the cover. Pry the rear bearing retainer from the cover using a 1/2 in. drive breaker bar between the pry bosses and separate and remove the bearing retainer from the cover. Remove all traces of RTV gasket sealant from the mating surfaces of the cover and the bearing retainer.

NOTE: When removing the RTV sealer, be careful not to damage the mating surfaces of the magnesium cases.

7. Lift the rear output shaft and using a small prybar, remove the speedometer gear retaining clip. Slide the speedometer gear forward and remove the ball with a small magnet. The speedometer gear can now be removed off of the rear output shaft. Remove the speed sensor from the rear cover.
8. Remove the 4 bolts retaining the shift motor to the rear cover and remove the motor. Note the position of the triangular shaft extending out of the rear cover and the triangular slot in the motor.

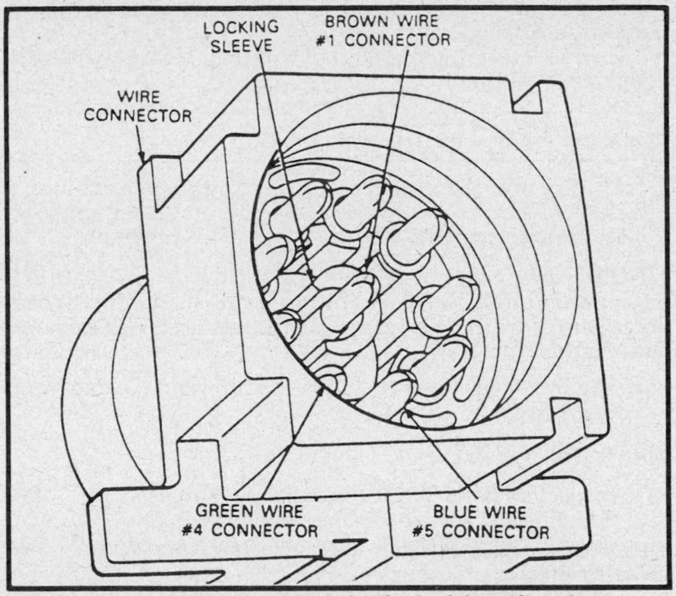
LOCKING SLEEVE
BROWN WIRE #1 CONNECTOR
WIRE CONNECTOR
GREEN WIRE #4 CONNECTOR
BLUE WIRE #5 CONNECTOR

Rear wiring connector terminal pin identification

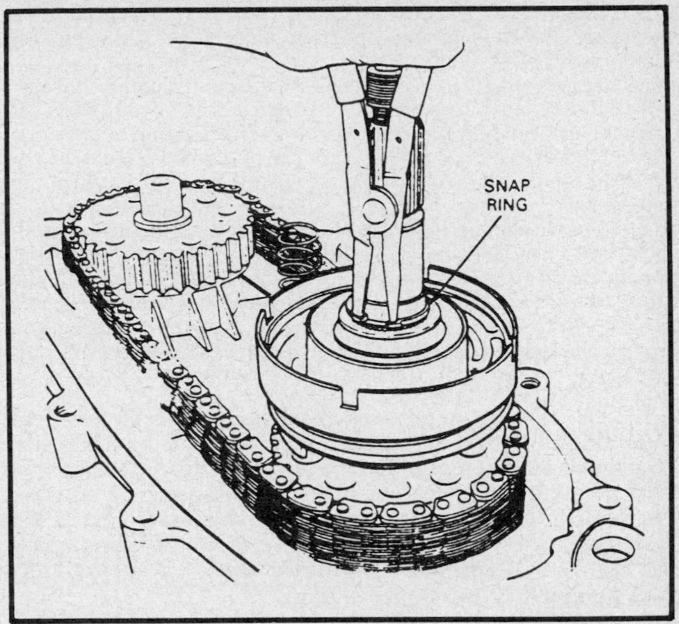
SNAP RING

Removing the clutch housing retaining snapring

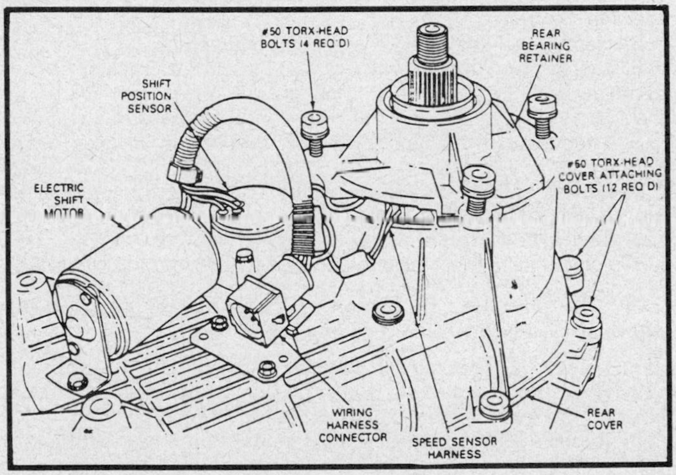
#50 TORX-HEAD BOLTS (4 REQ'D)
REAR BEARING RETAINER
SHIFT POSITION SENSOR
#60 TORX-HEAD COVER ATTACHING BOLTS (12 REQ D)
ELECTRIC SHIFT MOTOR
WIRING HARNESS CONNECTOR
SPEED SENSOR HARNESS
REAR COVER

Removing the rear bearing retainer

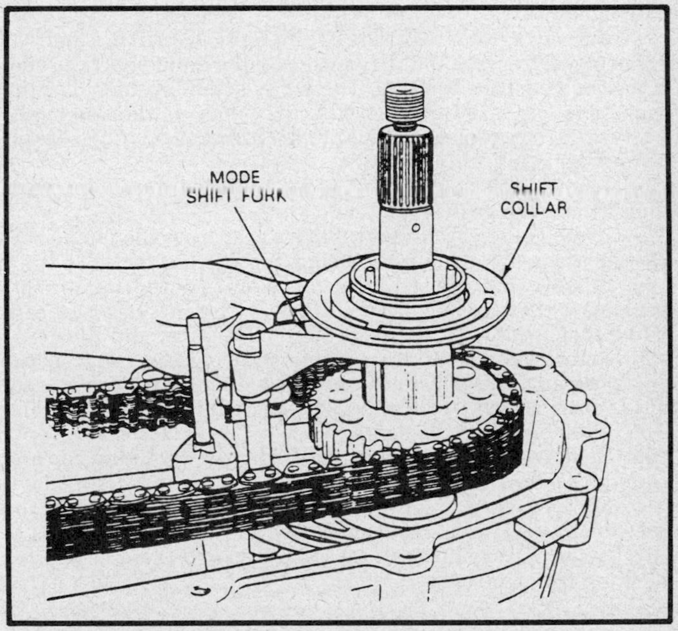
MODE SHIFT FORK
SHIFT COLLAR

Removing the mode shift fork and shift collar

NOTE: The motor is serviced as a complete assembly. Do not remove the screws that secure the rear cover to the motor gear housing.

9. Remove the snapring on the output shaft retaining the upper rear ball bearing using a suitable tool.
10. Remove the 12 No. 50 Torx® bolts that retain the front case to the rear cover. Insert a ½ in. drive breaker bar between the pry bosses and separate. Lift the front case from the rear cover. Remove all traces of RTV gasket sealant from the mating surfaces of the cover and the bearing retainer.

NOTE: When removing the RTV sealer, be careful not to damage the mating surfaces of the magnesium cases.

11. Remove the front output shaft inner needle bearing from the rear cover with special slide hammer tool T50T–100–A and D80L–100–T, collet from D80L–100–A blind hole puller set or equivalents.
12. Drive out the rear output shaft bearing from the inside of the case using the appropriate tools.
13. Remove the snapring on the output shaft securing the clutch hub. Slide the 4WD hub off of the output shaft.
14. Remove the spring from the shift rail and lift the mode shift fork complete with the shifting collar from the upper sprocket spline.
15. Dissassemble the 2WD–4WD lockup assembly by removing the internal snapring and pull the lockup hub and spring from the collar.
16. Remove the snapring retaining the lower sprocket to the lower output shaft. Grasp the upper and lower sprocket complete with the chain and lift them at the same time from the upper and lower output shafts.
17. Remove the snapring retaining the lower sprocket complete with the chain and lift them at the same time from the upper and lower output shafts.
18. Remove the shift rail by sliding it straight out from the shift fork.

19. Remove the high and low shift fork by first rotating it until the roller is free from the cam then sliding out of the engagement from the shift hub.
20. Remove the chip collecting magnet from its slot in the case.
21. Lift out the pump screen and remove the output shaft assembly with the pump assembled on it. If the pump is to be disassembled, remove the 4 bolts from the pump body. Note the position of the pump front body, pins, spring, rear cover and pump retainer as removed.
22. Remove the high low shift fork by rotating it until the roller is free from the cam then sliding out the engagement from the shifting hub. Remove the helical cam assembly from the front case. If required, remove the helical cam, torsion spring and sleeve from the shaft. Remove the external snapring and thrust washer that retains the drive sprocket to the front output shaft.

NOTE: If it is necessary to disassemble the helical cam assembly care should be execised as the cam is slid rearward to disengage it from the spring. The spring is energized and can release violently. The spring must be removed from the helical cam and the shaft finger. Do not get your fingers in the way of the disengaging spring. It will rotate to the point that the spring ends will be roughly 180 degrees apart.

23. Remove the front output shaft from the case.
24. Turn the front case over and remove the front oil seal from the case using tool T74P-77248-A and T50T-100-A or equivalents.
25. Reaching through the front opening with a pair of snapring pliers, expand the snapring on the input shaft allowing it to drop out of the bearing. The carrier assembly, including the input shaft is serviced as an assembly only. If the bearing or bushing is to be replaced, drive out both of them through the input spline using suitable tools.
26. Remove the ring gear by prying out the internal snapring and lift out the gear.
27. Remove the power take off drive gear from the input shaft carrier assembly, if equipped using appropriate tools.
28. Remove the internal snapring securing the input shaft bearing to the case and drive it out from the outside of the case using tool T73T-1202-A and T80T-4000-W or equivalents.
29. Remove the internal snapring securing the front output shaft bearing in the magnesium housing and drive the bearing out from the front of the case using tool T73T-1202-B and T80T-4000-W or equivalents.
30. Remove the shift cam by removing the retaining clip and sliding the shift shaft out of the case.
31. Remove the shift shaft seal by carefully prying it out of the case, being careful not to damage the case.
32. Remove the shift cam, assist spring, and the assist spring bushing from the case.

CLEANING AND INSPECTION

Cleaning

During overhaul, all components of the transfer case (except bearing assemblies) should be thoroughly cleaned with solvent and dried with air pressure prior to inspection and reassembly.
1. Clean the bearing assemblies as follows:

NOTE: Proper cleaning of bearings is of utmost importance. Bearings should always be cleaned separately from other parts.

a. Soak all bearing assemblies in clean solvent or fuel oil. Bearings should never be cleaned in a hot solution tank.
b. Slush bearings in solvent until all old lubricant is loosened. Hold races so that bearings will not rotate; then clean bearings with a soft bristled brush until all dirt has been removed. Remove loose particles of dirt by tapping bearing flat against a block of wood.
c. Rinse bearings in clean solvent; then blow bearings dry with air pressure.

CAUTION

Do not spin bearings while drying.

d. After drying, rotate each bearing slowly while examining balls or rollers for roughness, damage, or excessive wear. Replace all bearings that are not in first class condition.

NOTE: After cleaning and inspecting bearings, lubricate generously with recommended lubricant, then wrap each bearing in clean paper until ready for reassembly.

2. Remove all portions of old gaskets from parts, using a stiff brush or scraper.

Inspection

1. Inspect all parts for discoloration or warpage.
2. Examine all gears and splines for chipped, worn, broken or nicked teeth. Small nicks or burrs may be removed with a fine abrasive stone.
3. Inspect the breather assembly to make sure that it is open and not damaged.
4. Check all threaded parts for damaged, stripped, or crossed threads.
5. Replace all gaskets, oil seals and snaprings.
6. Inspect housings, retainers and covers for cracks or other damage. Replace the damaged parts.
7. Inspect keys and keyways for condition and fit.
8. Inspect shift forks for wear, distortion or any other damage.
9. Check detent ball springs for free length, compressed length, distortion or collapsed coils.
10. Check bearing fit on their respective shafts and in their bores or cups. Inspect bearings, shafts and cups for wear.

NOTE: If either bearings or cups are worn or damaged, it is advisable to replace both parts.

11. Inspect all bearing rollers or balls for pitting or galling.
12. Examine detent balls for corrosion or brinneling. If shift bar detents show wear, replace them.
13. Replace all worn or damaged parts. When assembling the transfer case, coat all moving parts with recommended lubricant.

Transfer Case Assembly

NOTE: Before starting the assembly procedure, lubricate all the internal parts, with Dexron®II transmission fluid or equivalent.

1. Install the input shaft and the front output shaft bearings in the case using the appropriate tools. Install the internal snaprings retaining the bearings in the case.
2. Drive the front output shaft seal into the case unit until it is fully seated against the case using tool T86T-7034-CH or equivalent.
3. Install the front output shaft through the lower bearing. The front output shaft is held in place in the case by the front output yoke and oil seal slinger assembly. Install the front yoke assembly onto the front output shaft then the rubber seal, flat washer and 30mm locknut. Torque the yoke locknut to 130-180 ft. lbs. (176-244 Nm).
4. Press the power take-off drive gear onto the input shaft assembly if it was removed.
5. Press the needle bearing and bronze bushing into the input shaft with the appropriate tools.

6. Install the ring gear into the slots in the case and retain it with the large internal snapring making sure that it is fully seated.

7. Install the input shaft and carrier assembly in the case through the input shaft bearing being careful not to damage the gear teeth when aligning them with the ring gear teeth.

8. While supporting the carrier assembly in position, install a new spring on the front side of the input bearing making sure that it is fully seated in the snapring groove of the input shaft.

9. Install the upper input shaft oil seal into the case using an appropriate tool until it is fully seated against the case.

10. Install a new shifter shaft seal into the case using an appropriate tool.

11. Assemble the shift cam assembly into the case by sliding the shift shaft and lever assembly through the case and seal into engagement with the shift cam. Secure the shift cam with the retaining clip.

12. Install the shift cam assist spring in position in the bushing of the shift cam and in the recess in the case.

13. Assemble the pump and output shaft as follows: Place the oil pump cover with the word **TOP** facing the front of the front case. Install the 2 pins (with the flats facing upwards) with the spring between the pins and place the assembly in the oil pump bore in the output shaft. Place the oil pump body and the pick up tube over the shaft and make sure that the pins are riding against the inside of the pump body. Place the oil pump rear cover with the words **TOP REAR** facing the rear of the case. The word **TOP** on the front cover and the rear cover should be on the same side. Install the pump retainer with the tabs facing the front of the transfer case. Install the 4 retaining bolts and rotate the output shaft while tightening the bolts to prevent the pump from binding. Tighten the bolts to 36–40 inch lbs. Lubricate the assembly with automatic transmission fluid.

NOTE: The output shaft must turn freely within the oil pump. If binding occurs, loosen the 4 bolts and retighten again.

14. Install the high low shift hub. Install the high low shift fork by engaging it with the shift hub flange and rotating it until the roller is engaged with the lower groove of the cam.

15. Install the shift rail through the high low fork bore and into the rail bore in the case.

16. Install the output shaft and oil pump assembly in the input shaft. Make certain that the external splines of the output shaft engage the internal splines of the high low shift hub. Make sure that the oil pump retainer and oil filter leg are in the groove and notch of the front case. Install the collector magnet in the notch in the front case.

17. Assemble the upper and lower sprockets with the chain and place them as an assembly over the upper and lower output shafts. Install the washer and snapring which retain the lower sprocket to the front output shaft.

18. Assemble the 2WD–4WD lockup assembly by installing the tapered compression spring in the lockup collar with the small end installed first. Place the lockup hub over the spring and compress the spring while installing the internal snapring which holds the lockup assembly together.

19. Install the lockup assembly and its shift fork over the external splines of the upper sprocket and the shift rail with the long boss of the shift rail facing foreward.

20. Assemble the 4WD return spring over the shift rail and against the shift fork.

21. Place the 4WD hub over the external splines of the output shaft and secure with the appropriate snapring. Make sure that the snapring is fully seated in the snapring groove.

22. Press the lower output needle bearing in its bore in the rear cover using an appropriate tool.

23. Press the rear output shaft bearing into position in the cover. Install the bearing snapring retainer in the cover.

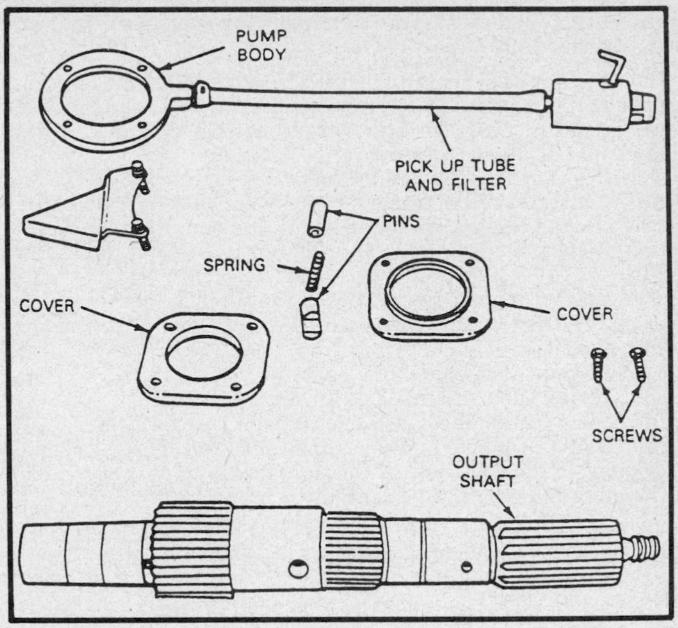
PUMP
BODY
PICK UP TUBE
AND FILTER
PINS
SPRING
COVER
COVER
SCREWS
OUTPUT
SHAFT

Exploded view of the rear output shaft and oil pump assembly

24. Install the rear output shaft oil seal in the bearing retainer using the appropriate tool making sure it is fully seated.

25. Coat the mating surface of the front case with a bead of RTV.

26. Place the cover on the case making sure that the lower output shaft, shift shaft and the shift rail are aligned. Install and torque the 12 No. 50 Torx® head case to cover bolts to 22–36 ft. lbs. (30–49 Nm).

27. Install the bearing retainer snapring on the output shaft making sure that the snapring is fully seated in the groove on the shaft.

28. Place the speedometer drive gear over the shaft with the slot aligned with the hole for the drive ball. The gear should go completely against the snapring which retains the output shaft. Place the ball in the hole and pull the speedometer gear over the ball. Snap the retaining clip between the snapring and the speedometer gear. Install the speed sensor in its bore in the cover.

29. Apply a bead of RTV to the face of the rear bearing retainer or to the rear slip yoke extension housing.

30. Place the rear bearing retainer or the rear slip yoke extension housing in its position and secure with the 4 Torx® head bolts.

31. On the transfer case with the slip yoke bearing retainer housing remove the extension oil seal using tool T74P–77248–A and T50T–100–A or equivalents. Remove the extension housing bushing using tool T85–7034–AH or equivalent. Install a new bushing using tool T85T–7034–BH and T80T–4000–W or equivalent. Install a new seal using tool T61L–7657–B. or equivalent.

32. Install the rear output shaft yoke and slinger assembly onto the rear splines of the output shaft. Install the rubber seal, flat steel washer and 30mm locknut on the output shaft and torque to 150–180 ft. lbs. (203–244 Nm).

33. Using a pair of pliers equipped with copper or wood jaws, rotate the triangular shaft so that it aligns with the triangular slot in the motor. Install the motor and torque the retaining bolts to 6–8 ft. lbs. (8–11 Nm).

NOTE: If the shaft will not stay in 4H position, rotate the shaft to the 2H position. Install the motor and rotate

the counterclockwise until the motor is aligned with the mounting holes.

34. Install the brown clutch coil wire to the No. 1 center terminal and if removed, the speed sensor green wire to the No. 4 connector position and the blue wire to the No. 5 connector position. Install the locking sleeve.
35. Position the wire connector mounting bracket on the rear cover. Install the bolts and torque them to 5–7 ft. lbs. (6–10 Nm). Install the wiring connector to the mounting bracket.
36. Install the drain plug and tighten to 14–22 ft. lbs. (19–32 Nm).
37. Place a ⅜ in. drive ratchet in the fill plug and remove the plug. Fill the transfer case with 64 oz. of Dexron®II transmission fluid.
38. Install the fill plug and tighten to 14–22 ft. lbs.
39. Install the transfer case. Lower the vehicle.
40. Start the engine, check the transfer case for proper operation. Stop the engine and check the fluid level. The fluid should drip out of the level hole. If the fluid flows out of the level hole, the oil pump may not be funtioning properly.

Specifications

TORQUE SPECIFICATIONS
1356 Electronic Shift Transfer Case

Description	ft. lbs.	Nm
Case half attaching bolts	35–40	48–54
Four wheel drive indicator switch	8–12	11–16
Front and rear output yokes to transfer case	120–150	163–203
Drain plug	14–22	19–30
Fill plug	14–22	19–30
Transfer case to transmission adapter	25–43	34–58
Heat shield to transfer case	40–45	54–61
Skid plate to frame	15–20	20–27
Front driveshaft to front output yoke	8–15	11–20

TORQUE SPECIFICATIONS
1356 Electronic Shift Transfer Case

Description	ft. lbs.	Nm
Rear driveshaft to rear output yoke—bolt—Bronco	20–28	28–33
Rear driveshaft to rear output yoke—Nut—F150-F350 4×4	8–15	11–20

Special Tools

Number	Description
T50T-100-A	Impact slide hammer—2½ lb.
D80L-100-A	Blind hole puller set
D80L-100-T	Blind hole puller
D80L-100-H	Actuator pin
TOOL-1175-AC	Seal remover
T80T-7127-B	Output bearing replacer—front
T80T-7127-C	Output bearing replacer—rear
T80T-4000-P	Bearing installer
T74P-77248-A	Oil seal remover
T61L-7657-B	Oil seal installer
T86L-7034-AH	Extension housing bushing remover
T85T-7034-BH	Extension housing bushing installer
T80T-4000-W	Bearing installer
T86T-7034-CH	Oil seal installer
T73T-1202-B	Bearing installer
T73T-1202-A	Bearing installer

BORG WARNER 1359 TRANSFER CASE

Application

1987–89 Bronco II

General Description

The Borg Warner 1359 is a in-line 3 piece aluminum transfer case. It transfers power from the transmission to the rear axle.

The input shaft assembly and the output shaft assembly are both supported by prelubricated, double sealed ball bearing requiring no periodic lubrication. The input shaft and output shaft are connected together by a coupling sleeve providing direct drive between the 2 shafts. The mounting adapter, front case, and rear cover are assembled together without RTV type sealer required for sealing purposes. The case assembly contains no lubricant and none should be added.

Trouble Diagnosis

CHILTON'S THREE C'S TRANSFER CASE DIAGNOSIS

Condition	Cause	Correction
Slips out of gear	a) Shifting poppet spring weak b) Bearing broken or worn c) Shifter fork bent d) Improper control rod adjustment	a) Replace the shifting poppet spring b) Replace the bearing c) Replace shifter fork d) Adjust the control rod
Hard shifting	a) Lack of lubricant b) Shift lever binding on shaft c) Shifting poppet ball scored	a) Add the recommended fluid to the proper level b) Repair and adjust the shift lever c) Replace the shifting poppet ball

CHILTON'S THREE C'S TRANSFER CASE DIAGNOSIS

Condition	Cause	Correction
Hard shifting	d) Shifter fork bent e) Low tire pressure	d) Replace shifter fork e) Inflate tires to the correct pressure
Backlash	a) Companion yoke loose b) Transfer case loose on mounts c) Internal parts excessively worn	a) Torque the companion yoke to specifications b) Torque the mount bolts to specification c) Rebuild the transfer case
Noisy	a) Low lubricant level b) Bearings improperly adjusted or excessively worn c) Gears worn or damaged d) Improper alignment of driveshafts or U-joint	a) Add the recommended fluid to the proper level b) Adjust or replace the bearing c) Replace the gear in question d) Realign the driveshaft and U-joint
Oil leakage	a) Excessive amount of oil in the case b) Oil vent clogged c) Gaskets or seals leaking d) Bearings improperly adjusted or excessively worn e) Driveshaft yoke mating surfaces scored	a) Drain the fluid to the proper level b) Clean oil vent c) Replace leaking gaskets or seals d) Adjust or replace the bearing e) Machine the driveshaft yoke mating surface
Overheating	a) Excessive amount of oil in the case b) Low lubricant level c) Bearing adjustment too tight	a) Drain the fluid to the proper level b) Add the recommended fluid to the proper level c) Adjust or replace the bearing

On Vehicle Services

REAR OUTPUT SHAFT OIL SEAL

Removal and Installation

1. Raise and support the vehicle safely.
2. Place a drain pan under transfer case, remove the drain plug and drain fluid from the transfer case.
3. Disconnect the rear driveshaft from the transfer case output shaft yoke. Wire the driveshaft out of the way.
4. Remove the oil seal by removing the 30mm nut, steel washer and rubber seal from the front or rear output shaft and remove the yokes.
5. Remove the oil seal from the rear output housing bore with a suitable seal remover and slide hammer.
6. Make sure that the housing face and bore are free from nicks and burrs. Coat the seal with a suitable lubricant. Position the oil seal into the rear output housing bore, making sure the oil seal is not cocked in the bore. Drive the new seal into the housing with a suitable seal installer.
7. Install the yoke, rubber seal, steel washer and locknut on the front or rear output shafts. Torque the locknut to 150–180 ft. lbs. (203–244 Nm).
8. Connect the front or rear driveshaft to the output yoke and torque the bolts to 61–87 ft. lbs. (83–118 Nm).
9. Lower the vehicle. Start the engine and check the transfer case for correct operation.

Removal and Installation

TRANSFER CASE REMOVAL

The catalytic converter is located beside the heat shield. Be careful when working around the converter because of the extremely high temperatures generated by the converter.

1. Raise the vehicle and support safely.
2. Disconnect the rear driveshaft from the transfer case output shaft flange.

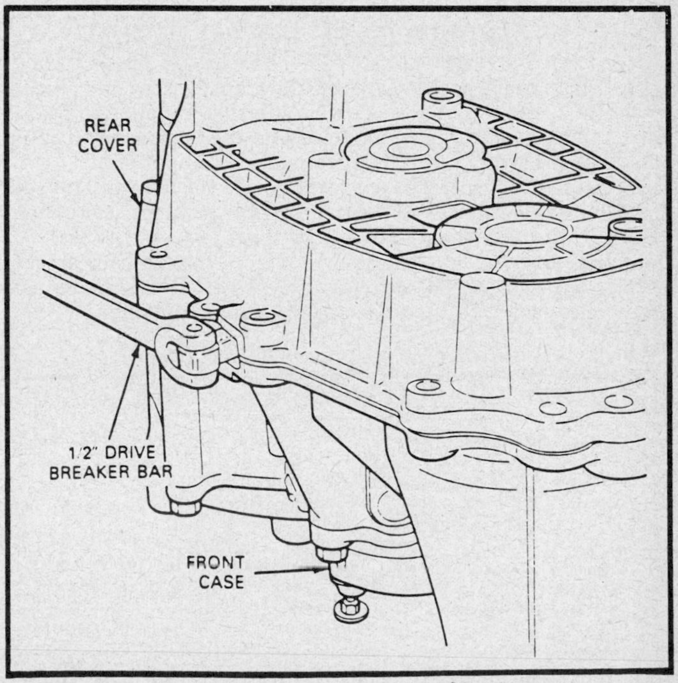
REAR
COVER
1/2" DRIVE
BREAKER BAR
FRONT
CASE

Separating the rear cover from the front case

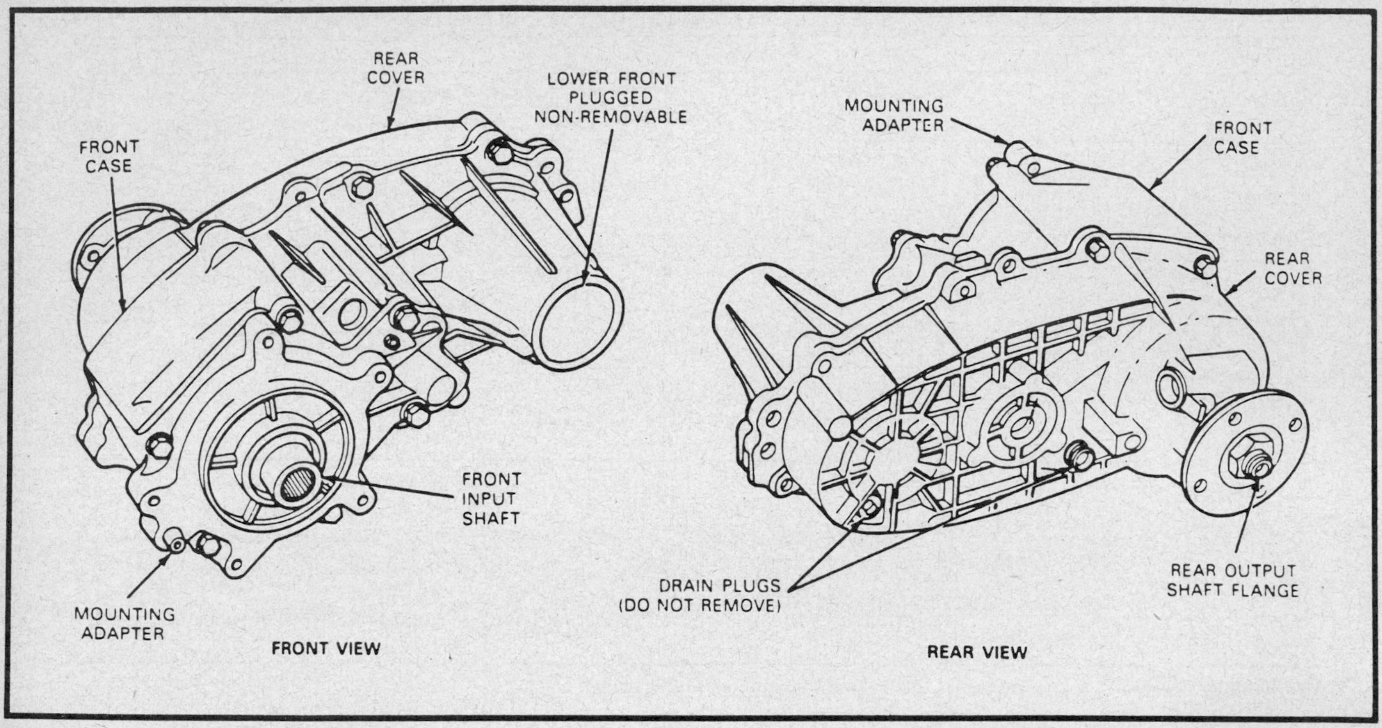
REAR COVER
LOWER FRONT PLUGGED NON-REMOVABLE
FRONT CASE
FRONT INPUT SHAFT
MOUNTING ADAPTER
FRONT VIEW
MOUNTING ADAPTER
FRONT CASE
REAR COVER
DRAIN PLUGS (DO NOT REMOVE)
REAR OUTPUT SHAFT FLANGE
REAR VIEW

Borg Warner 1359 transfer case

3. Disconnect the speedometer driven gear from the transfer case rear cover.
4. Support the transfer case with a transmission jack.
5. Remove the 5 bolts retaining the transfer case to the transmission and the extension housing.
6. Slide the transfer case rearward off the transmission output shaft and lower the transfer case from the vehicle. Remove the gasket from between the transfer case and extension housing.

TRANSFER CASE INSTALLATION

1. Place a new gasket between the transfer case and adapter.
2. Raise the transfer case with a suitable transmission jack or equivalent, raise it high enough so that the transmission output shaft aligns with the splined transfer case input shaft.
3. Slide the transfer case forward on to the transmission output shaft and onto the dowel pin. Install transfer case retaining bolts and torque them to 25–43 ft. lbs.
4. Remove the transmission jack from the transfer case.
5. Connect the speedometer cable assembly to the transfer case rear cover. Tighten the screw to 20–25 inch lbs.

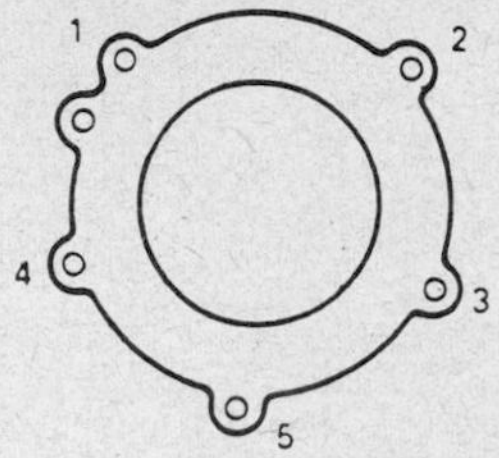
1
2
3
4
5

Transmission adapter to Transfer case torque sequence

6. Connect the rear driveshaft to the output shaft flange and torque the yoke nut to 61–87 ft. lbs. (83–118 Nm).
7. Lower the vehicle.

Before Disassembly

Cleanliness during disassembly and assembly is necessary to avoid further transfer case trouble after overhaul. Before removing any of the transfer case subassemblies, plug all the openings and clean the outside of the of the transfer case thoroughly. Steam cleaning or car wash type high pressure equipment is preferable. During disassembly, clean all parts in suitable solvent and dry each part. Do not use cloth or paper towels to dry parts. Use compressed air only.

Transfer Case Disassembly

1. Place the transfer case into a suitable holding fixture.
2. Remove the rear output shaft yoke by removing the 30mm nut, steel washer and rubber seal from the output shaft. Remove the flange from the output shaft.
3. Remove the rear output shaft oil seal from the rear cover using tool T74P-77248-A and T50T-100-A or equivalents.
4. Remove the 9 bolts which retain the front case to the rear cover. Insert a ½ in. drive breaker bar between the 3 pry bosses and separate the front case from the rear cover.
5. Remove the speedometer drive gear assembly (gear, clip and spacer). Note that the round end of the speedometer gear clip faces the inside of the rear cover.
6. Remove the rear spacer collar from the rear output shaft.
7. Remove the rear output shaft and snapring assembly by lifting it out of the locking coupling and front input shaft in the front case.
8. Remove the locking coupling from the external splined end of the input shaft assembly.
9. Remove the internal snapring that retains the rear output shaft ball bearing in the bore. From the outside of the case, drive out the ball bearing with output shaft bearing replacer tool

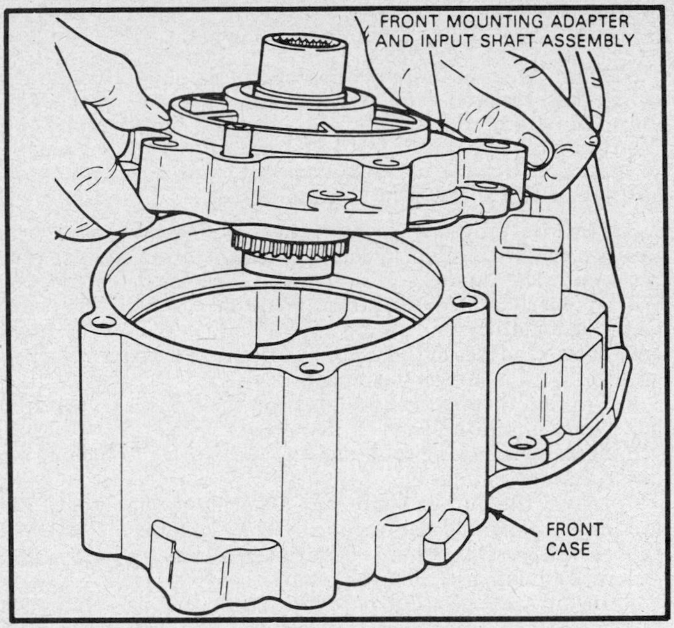
FRONT MOUNTING ADAPTER
AND INPUT SHAFT ASSEMBLY
FRONT
CASE

Removing the front mounting adapter

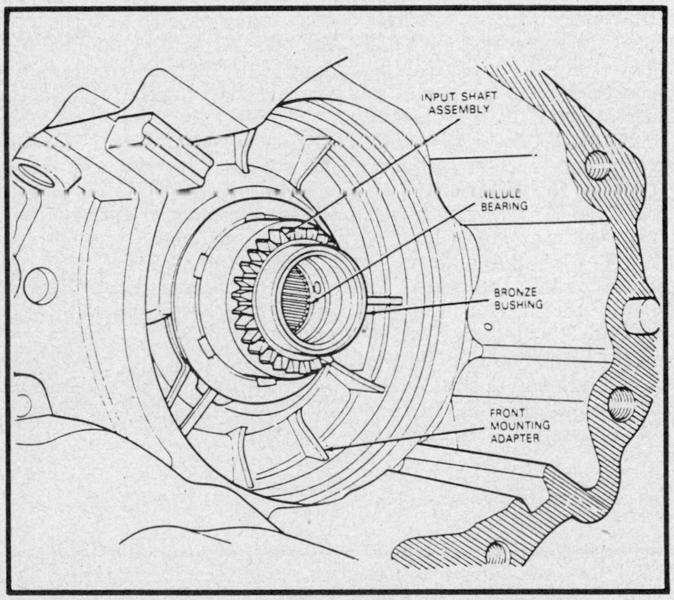
INPUT SHAFT
ASSEMBLY
NEEDLE
BEARING
BRONZE
BUSHING
FRONT
MOUNTING
ADAPTER

Removing the input shaft needle bearing and bushing

T83T-7025-B and drive handle tool T80T-4000-W or equivalents.

10. Turn the front case over and remove the 6 bolts retaining the mounting adapter to the front case. Remove the mounting adapter, input shaft bearing, thrust plate and sleeve as an assembly.

11. Place the mounting adapter in a suitable holding fixture and remove the output shaft pilot and needle bearing and bushing from the rear bore of the input shaft.

12. Expand the tangs of the large snapring in the mounting adapter and pry the snapring apart and remove the input shaft assembly from the mounting adapter.

13. Remove the external snapring from the input shaft. Place the input shaft assembly in a press and remove the ball bearing from the input shaft using bearing splitter tool D79L–4621–A or

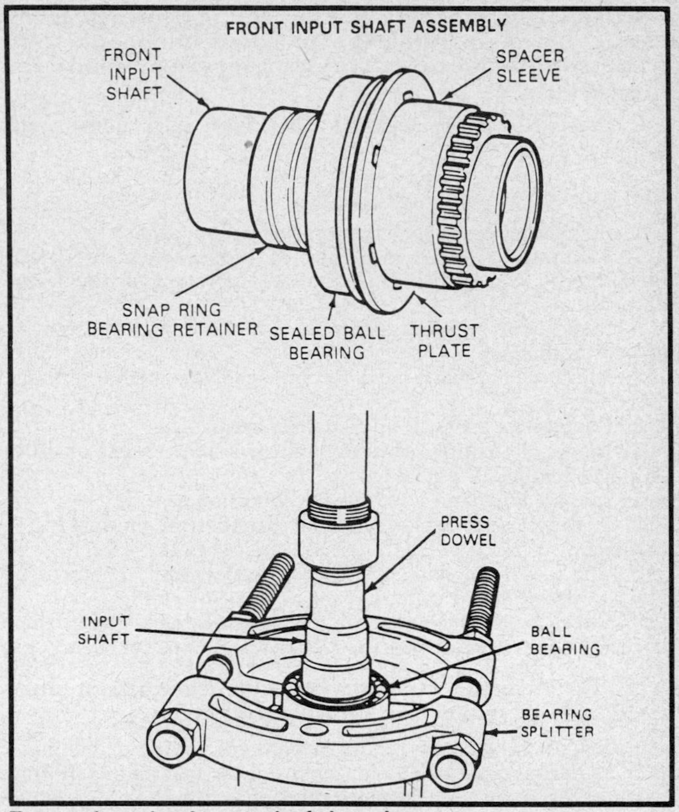
FRONT INPUT SHAFT ASSEMBLY
FRONT
INPUT
SHAFT
SPACER
SLEEVE
SNAP RING
BEARING RETAINER
SEALED BALL
BEARING
THRUST
PLATE
PRESS
DOWEL
INPUT
SHAFT
BALL
BEARING
BEARING
SPLITTER

Removing the input shaft bearing

equivalent. Remove the thrust plate and sleeve off the input shaft.

14. Remove the seal from the mounting adapter with seal remover tool T77248–A and impact slide hammer tool T50T–100–A or equivalent.

CLEANING AND INSPECTION

Cleaning

During overhaul, all components of the transfer case (except bearing assemblies) should be thoroughly cleaned with solvent and dried with air pressure prior to inspection and reassembly.

1. Clean the bearing assemblies as follows:

NOTE: Proper cleaning of bearings is of utmost importance. Bearings should always be cleaned separately from other parts.

a. Soak all bearing assemblies in clean solvent or fuel oil. Bearings should never be cleaned in a hot solution tank.

b. Slush bearings in solvent until all old lubricant is loosened. Hold races so that bearings will not rotate; then clean bearings with a soft bristled brush until all dirt has been removed. Remove loose particles of dirt by tapping bearing flat against a block of wood.

c. Rinse bearings in clean solvent; then blow bearings dry with air pressure.

CAUTION

Do not spin bearings while drying.

d. After drying, rotate each bearing slowly while examining balls or rollers for roughness, damage, or excessive wear. Replace all bearings that are not in first class condition.

NOTE: After cleaning and inspecting bearings, lubricate generously with recommended lubricant, then wrap each bearing in clean paper until ready for reassembly.

2. Remove all portions of old gaskets from parts, using a stiff brush or scraper.

Inspection

1. Inspect all parts for discoloration or warpage.
2. Examine all gears and splines for chipped, worn, broken or nicked teeth. Small nicks or burrs may be removed with a fine abrasive stone.
3. Inspect the breather assembly to make sure that it is open and not damaged.
4. Check all threaded parts for damaged, stripped, or crossed threads.
5. Replace all gaskets, oil seals and snaprings.
6. Inspect housings, retainers and covers for cracks or other damage. Replace the damaged parts.
7. Inspect keys and keyways for condition and fit.
8. Inspect shift forks for wear, distortion or any other damage.
9. Check detent ball springs for free length, compressed length, distortion or collapsed coils.
10. Check bearing fit on their respective shafts and in their bores or cups. Inspect bearings, shafts and cups for wear.

NOTE: If either bearings or cups are worn or damaged, it is advisable to replace both parts.

11. Inspect all bearing rollers or balls for pitting or galling.
12. Examine detent balls for corrosion or brinneling. If shift bar detents show wear, replace them.
13. Replace all worn or damaged parts. When assembling the transfer case, coat all moving parts with recommended lubricant.

Transfer Case Assembly

Before assembly, lubricate all parts with Dexron®II, Automatic Transmission Fluid.

1. If removed, install the front output oil seal in the front case bore using output shaft seal installer tool T83T-7065-B and driver handle tool T80T-4000-W or equivalents.
2. Slide the spacer sleeve, thrust plate, and press the ball bearing onto the input shaft assembly. Install the external snapring to the input shaft to retain the ball bearing.

NOTE: The snapring groove of the ball bearing should be facing toward the rear of the transfer case. The stepped face of the thrust plate should face toward the front of the case, against the bearing. The spacer sleeve is installed between the thrust plate and the external splines of the input shaft assembly.

3. Place the tanged snapring in the mounting adapter groove. Position the input shaft in the mounting adapter and push inward until the input shaft assembly is seated in the adapter. When properly seated, the tanged snapring will snap into place. Check installation by holding the mounting adapter by hand and tapping the face of the input shaft against a wooden block to ensure that the snapring is engaged.
4. Position the mounting adapter on the front case. Install 6 bolts and tighten to 25–30 ft. lbs. (31–41 Nm).
5. Install the locking coupling onto the external splines of the input shaft assembly.
6. Install the output shaft, and snapring assembly into the locking coupling and the input shaft. Make sure the external splines of the output shaft engauge the internal splines of the locking coupling and the input shaft.
7. Install the spacer hub onto the output shaft assembly with the square teeth side facing toward the front of the transfer case.
8. If removed, install the ball bearing in the rear cover bore. Drive the bearing into the rear cover bore with output shaft bearing replacer tool T83T-7025-B and driver handled tool T80-4000-W or equivalents. Make sure the ball bearing is not cocked in the bore. Install the internal snapring that retains the ball bearing to the rear cover.
9. Install the speedometer drive gear assembly into the rear cover bore with round end of the speedometer gear clip facing towards the inside of the rear cover. Pack the speedometer gear cavity in the rear cover with long life lubricant. Drive the oil seal into the rear cover bore with output shaft seal installer tool T83T-7065-B and driver handle T80T-4000-W or equivalents.
10. Position the rear cover on the front case. Install the nine bolts, starting with the bolts on the rear cover and torque the bolts to 23–30 ft. lbs. (31–41 Nm).
11. Install the rear output shaft flange. Coat the faces of the yoke nuts and output shaft threads with a suitable thread sealer. Torque the yoke nuts to 150–180 ft. lbs. (203–244 Nm).
12. Install the transfer case.

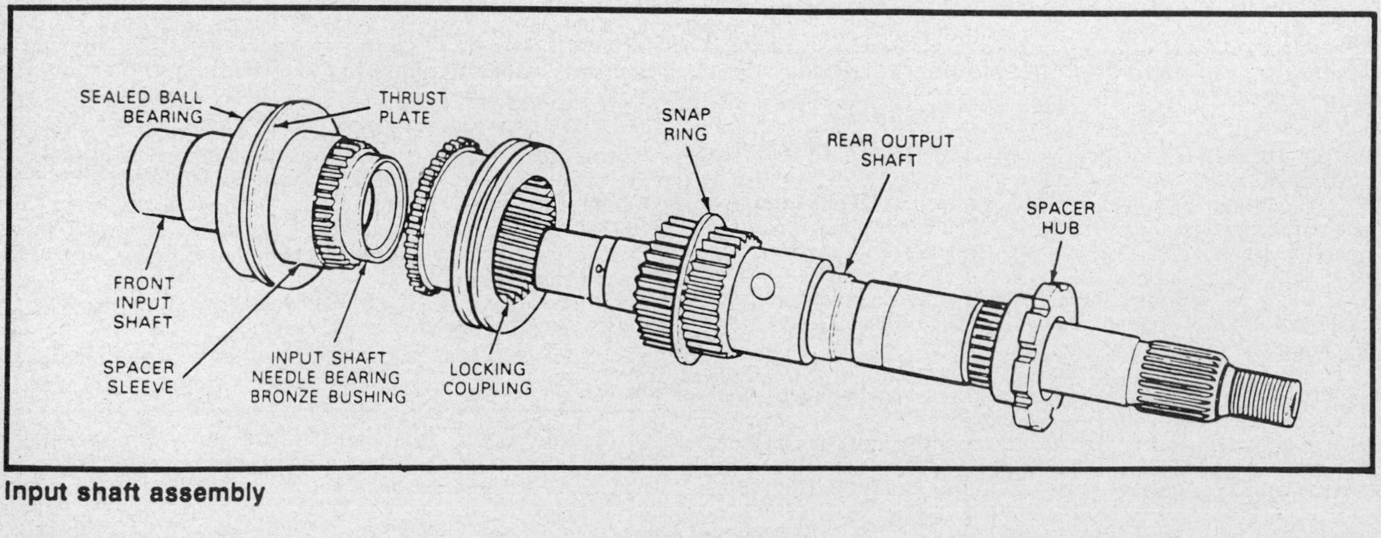
SEALED BALL BEARING
THRUST PLATE
SNAP RING
REAR OUTPUT SHAFT
SPACER HUB
FRONT INPUT SHAFT
SPACER SLEEVE
INPUT SHAFT NEEDLE BEARING BRONZE BUSHING
LOCKING COUPLING

Input shaft assembly

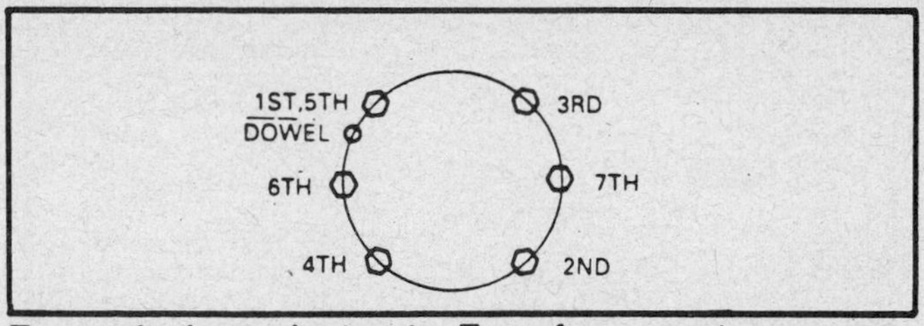
1ST,5TH
DOWEL
3RD
6TH
7TH
4TH
2ND

Transmission adapter to Transfer case torque sequence

Special Tools

Number	Description
T50T-100-A	Impact slide hammer
Tool-1175-AC	Seal remover
T80T-4000-W	Driver handle
D79L-4621-A	Bearing splitter
T74P-77248-A	Seal remover
T83T-7025-B	Output shaft bearing replacer
T83T-7025-C	Input shaft bearing replacer
T83T-7065-A	Input shaft seal installer
T83T-7065-B	Output shaft seal installer

Specifications

TORQUE SPECIFICATIONS

Description	Torque N•m	Torque Ft. Lb.
Case to Cover Bolts	31–41	23–30
Rear Driveshaft Bolts	83–118	61–87
Skid Plate to Frame Bolt	30–41	22–30
Transfer Case to Transmission Adapter	34–47	25–35
Yoke Nut	203–244	150–180
Speedometer Screw	2.3–2.8	20–25

DANA 300 TRANSFER CASE

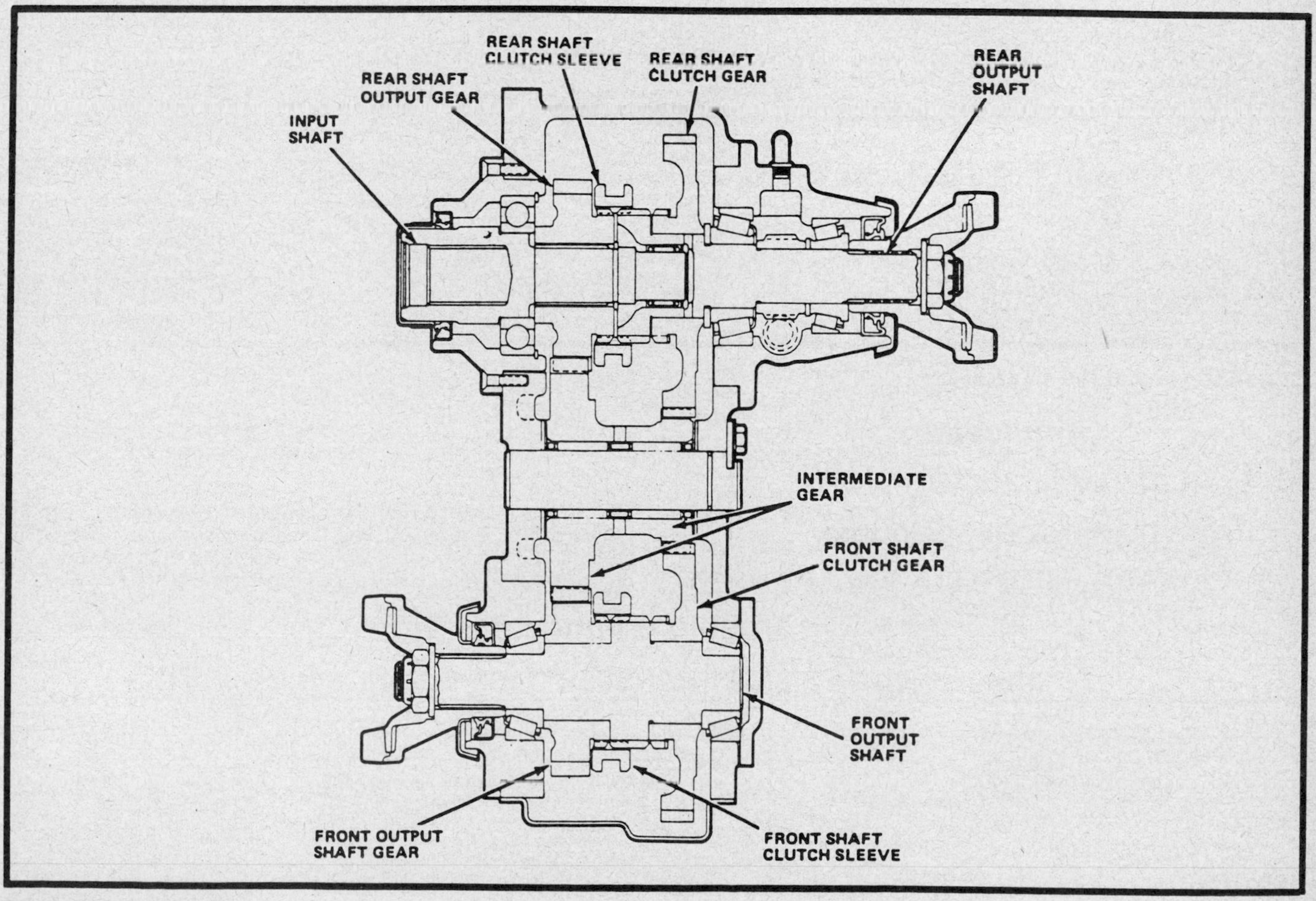
REAR SHAFT CLUTCH SLEEVE
REAR SHAFT CLUTCH GEAR
REAR OUTPUT SHAFT
REAR SHAFT OUTPUT GEAR
INPUT SHAFT
INTERMEDIATE GEAR
FRONT SHAFT CLUTCH GEAR
FRONT OUTPUT SHAFT
FRONT OUTPUT SHAFT GEAR
FRONT SHAFT CLUTCH SLEEVE

Dana 300 power flow

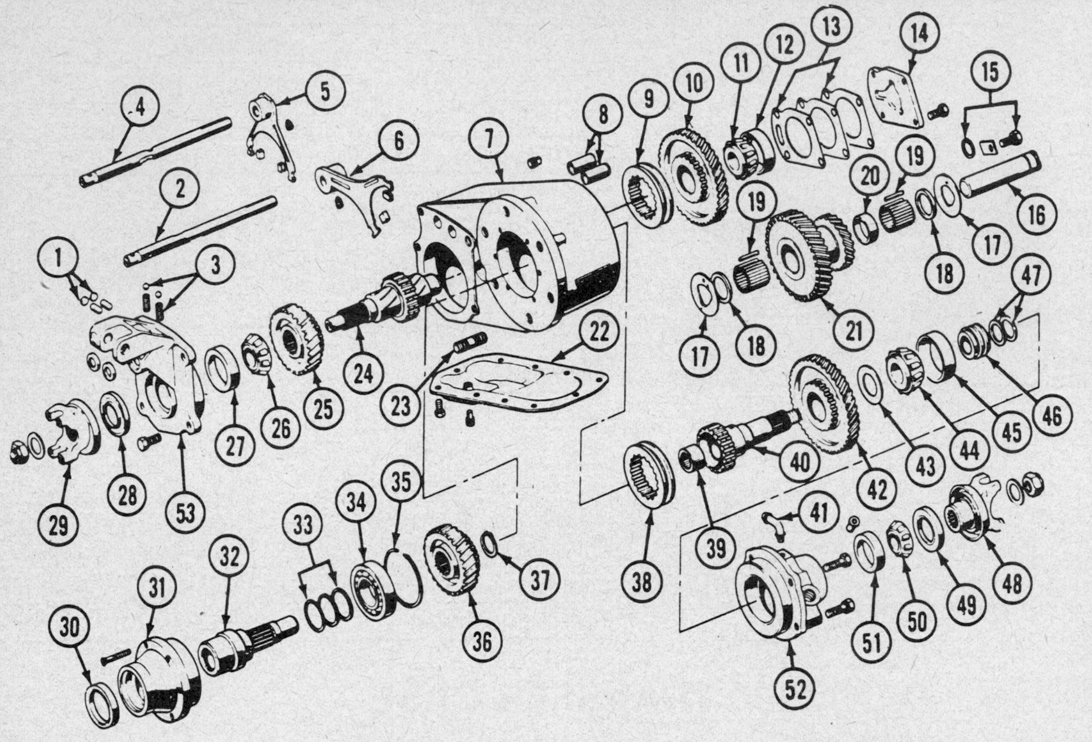

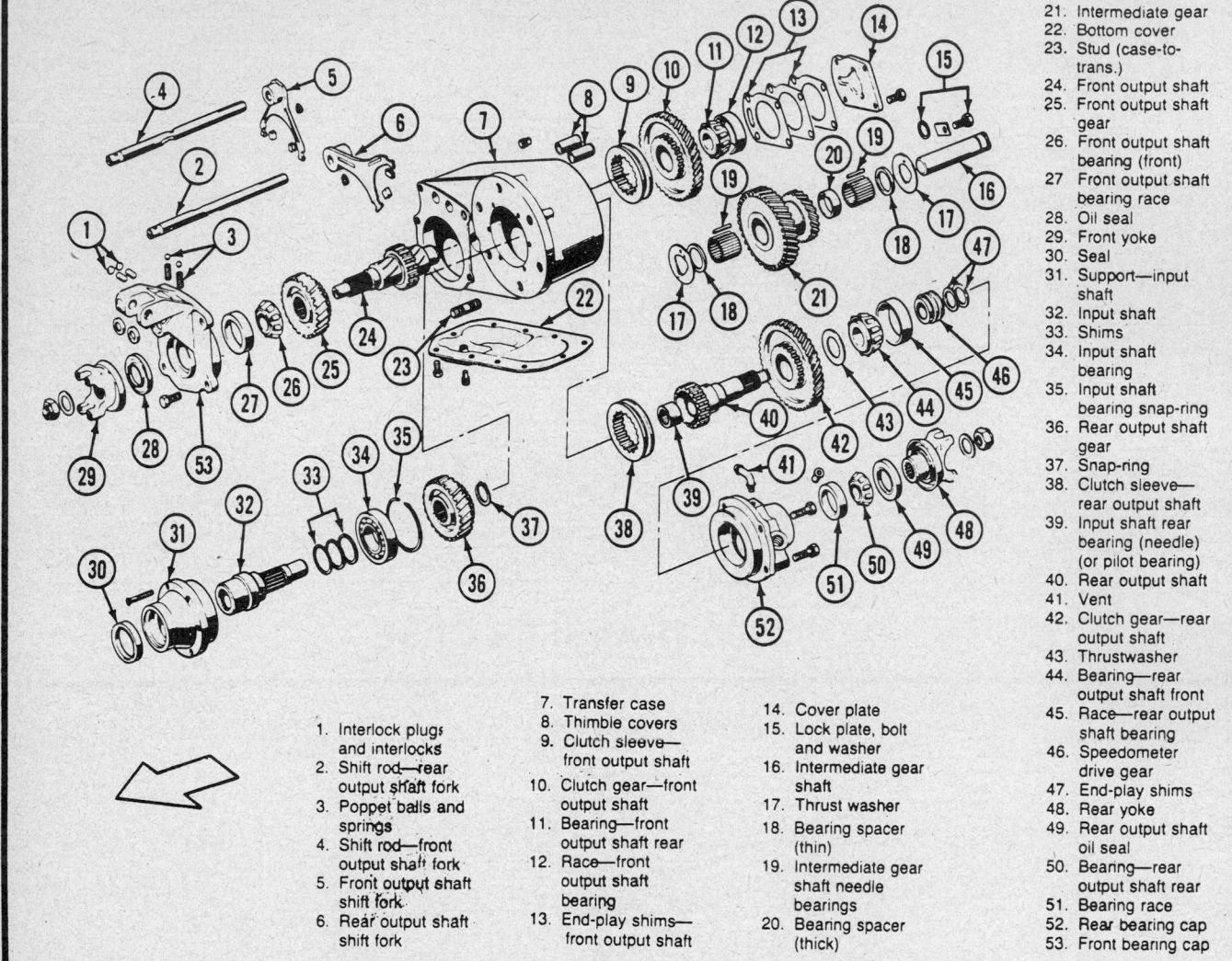

1. Interlock plugs and interlocks
2. Shift rod—rear output shaft fork
3. Poppet balls and springs
4. Shift rod—front output shaft fork
5. Front output shaft shift fork
6. Rear output shaft shift fork
7. Transfer case
8. Thimble covers
9. Clutch sleeve—front output shaft
10. Clutch gear—front output shaft
11. Bearing—front output shaft rear
12. Race—front output shaft bearing
13. End-play shims—front output shaft
14. Cover plate
15. Lock plate, bolt and washer
16. Intermediate gear shaft
17. Thrust washer
18. Bearing spacer (thin)
19. Intermediate gear shaft needle bearings
20. Bearing spacer (thick)
21. Intermediate gear
22. Bottom cover
23. Stud (case-to-trans.)
24. Front output shaft
25. Front output shaft gear
26. Front output shaft bearing (front)
27. Front output shaft bearing race
28. Oil seal
29. Front yoke
30. Seal
31. Support—input shaft
32. Input shaft
33. Shims
34. Input shaft bearing
35. Input shaft bearing snap-ring
36. Rear output shaft gear
37. Snap-ring
38. Clutch sleeve—rear output shaft
39. Input shaft rear bearing (needle) (or pilot bearing)
40. Rear output shaft
41. Vent
42. Clutch gear—rear output shaft
43. Thrustwasher
44. Bearing—rear output shaft front
45. Race—rear output shaft bearing
46. Speedometer drive gear
47. End-play shims
48. Rear yoke
49. Rear output shaft oil seal
50. Bearing—rear output shaft rear
51. Bearing race
52. Rear bearing cap
53. Front bearing cap

Exploded view of the Dana 300

Application

1984–86 Jeep CJ-7 and Scrambler

General Description

The Dana 300 transfer case has a cast iron case, 4 gear positions and employs an external floor mounted gearshift linkage for range control. It is a part time, 2 speed unit with undifferentiated high and low ranges. It is used with both manual and automatic transmission. Low range reduction is 2.6:1.

Trouble Diagnosis

CHILTON'S THREE C'S TRANSFER CASE DIAGNOSIS

Condition	Cause	Correction
Transfer case difficult to shift or will not shift into desired range	a) Vehicle speed too great to permit shifting	a) Stop vehicle and shift into desired range. Or reduce speed to 2–3 mph (3–4 km/h) before attempting to shift
	b) If vehicle was operated for extended period in 4H mode on dry paved surface, driveline torque load may cause difficult shifting	b) Stop vehicle, shift transmission to neutral, shift transfer case to 2H mode and operate vehicle in 2H on dry paved surfaces

CHILTON'S THREE C'S TRANSFER CASE DIAGNOSIS

Condition	Cause	Correction
Transfer case difficult to shift or will not shift into desired range	c) Transfer case external shift linkage binding d) Insufficient or incorrect lubricant e) Internal components binding, worn, or damaged	c) Lubricate or repair or replace linkage, or tighten loose components as necessary d) Drain and refill to edge of fill hole e) Disassemble unit and replace worn or damaged components as necessary
Transfer case noisy in all drive modes	a) Insufficient or incorrect lubricant	a) Drain and refill to edge of fill hole. Check for leaks and repair if necessary. Note: If unit is still noisy after drain and refill, disassembly and inspection may be required to locate source of noise.
Noisy in—or jumps out of four wheel drive low range	a) Transfer case not completely engaged in 4L position b) Shift linkage loose or binding c) Shift fork cracked, inserts worn, or fork is binding on shift rail	a) Stop vehicle, shift transfer case in Neutral, then shift back into 4L position b) Tighten, lubricate, or repair linkage as necessary c) Disassemble unit and repair as necessary
Lubricant leaking from output shaft seals or from vent	a) Transfer case overfilled b) Vent closed or restricted c) Output shaft seals damaged or installed incorrectly	a) Drain to correct level b) Clear or replace vent if necessary c) Replace seals. Be sure seal lip faces interior of case when installed. Also be sure yoke seal surfaces are not scored or nicked. Remove scores, nicks with fine sandpaper or replace yoke(s) if necessary
Abnormal tire wear	a) Extended operation on dry hard surface (paved) roads in 4H range	a) Operate in 2H on hard surface (paved) roads

On Vehicle Services

TRANSFER CASE SHIFT LINKAGE

Removal and Installation

1. Remove the screws that attach the shift lever boot to the floorpan.
2. Remove the shift lever knob and slide the boot up and off the lever.
3. Raise and support the vehicle. Remove the shifter shaft retaining nut.
4. Remove the cotter pins that retain the link pins in the shift rods and remove the link pins. Discard ther old cotter pins.
5. Remove the shifter shaft from thr shift lever.

NOTE: On some vehicles, the shifter shaft is threaded onto the shift lever and must be unthreaded to be removed. On other vehicles, the shaft is removed simply by sliding it out of the lever and front cover bosses.

6. Remove the shift control links from the shift rods. Clean and inspect the linkage components. Replace any components that are broken, bent cracked or excessively worn or scored.
7. Install the shift control links. Install the shift lever.
8. Install the shifter shaft in the front cover bosses and shift lever.
9. Install and tighten the shifter shaft retainer nut. Install the link pins into the shift rods. Secure the pins with new cotter pins.
10. Lower the vehilce. Install the boot on the shift lever. Install the knob on the shift lever. Position the boot onto the floor pan and install the boot attaching screws.

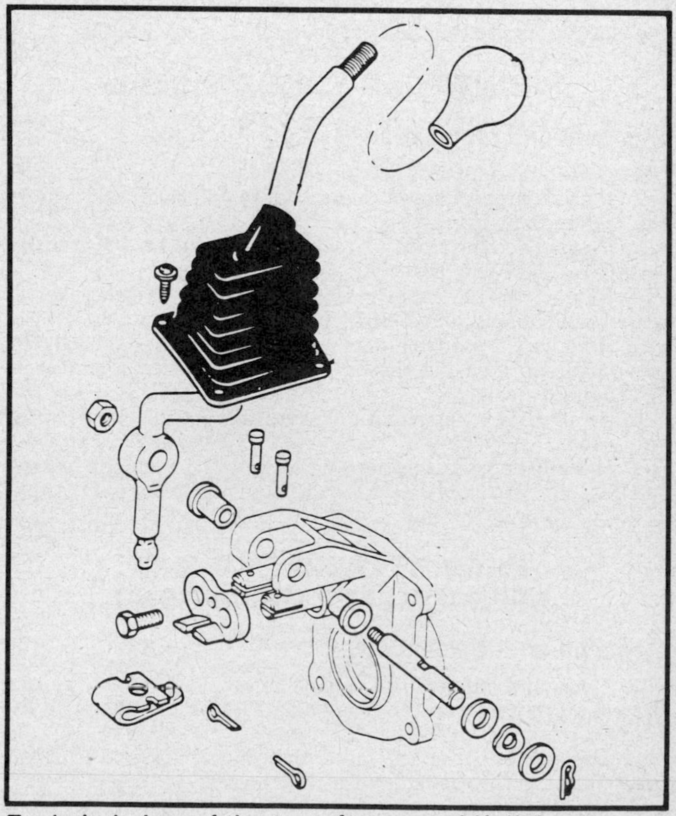

Exploded view of the transfer case shift linkage

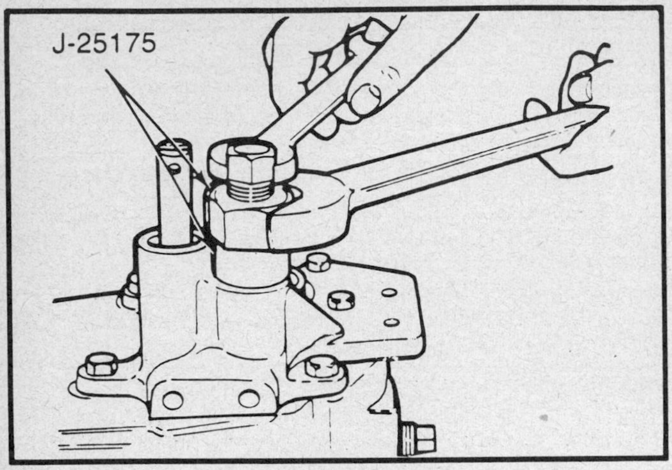
J-25175

Removing the shift rod oil seal

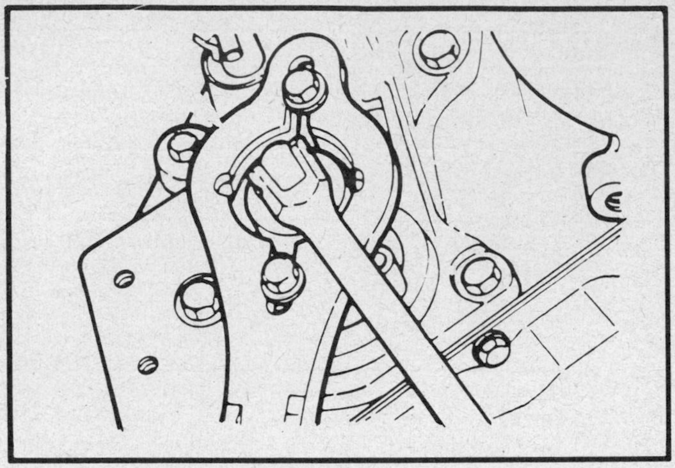

Output shaft yoke nut removal

SHIFT ROD OIL SEAL

Removal and Installation

1. If the left side shift rod seal is to replaced, shift the transfer case into the **4L** position.
2. Raise and support the vehicle safely.
3. Remove the clevis pins connecting the control links to the transfer case shift rods.
4. Remove the shift rod oil seal, using seal remover tool J-25175 or equivalent.
5. Install the replacement seal using a thimble and driver tool J-25167 or equivalent.
6. Install the clevis pins connecting the control links to the transfer case shift rods. Use replacement cotter pins to secure the clevis pins. Lower the vehicle.

FRONT/REAR YOKE OIL SEAL

Removal and Installation

1. Raise and support the vehicle safely.
2. Place a support stand under the transmission and remove the rear crossmember.
3. Disconnect the front or rear driveshaft from the transfer case yoke. Wire the driveshaft out of the way.
4. Remove the oil seal by removing the transfer case yoke nut and washer, using tool J-8614-01 or equivalent.
5. Remove the transfer yoke using tool J-8614-01, 02 and 03 or equivalent. Remove the oil seal with seal remover tool J-25180 or equivalent.
6. Install the replacement seal with tool J-25160 or equivalent.
7. Install the yoke, washer and locknut. Torque the locknut to 120 ft. lbs. Use tool J-8614-01 or equivalent to hold the yoke while tightening the nut.

REAR BEARING CAP AND SPEEDOMETER DRIVE GEAR

Removal and Installation

1. Raise and support the vehicle safely. Disconnect the rear driveshaft at the transfer case yoke. Tie the driveshaft to the frame with wire.
2. Disconnect the speedometer driven gear sleeve and driven gear. Remove the transfer case vent hose.
3. Remove the transfer yoke using tool J-8614-01, 02 and 03 or equivalents.
4. Remove the bearing cap to transfer case bolts and remove the bearing cap.

NOTE: The bearing cap has been coated with a sealant. Use a putty knife to break the seal and work the knife around the bearing cap to loosen it and remove it.

5. Remove the shims and speedometer drive gear from the output shaft. Be sure to keep the shims together for use in assembly.
6. Remove the speedometer driven gear bushing from the bearing cap, if necessary.
7. Install the speedometer driven gear bushing if the bushing was removed. Install the speedometer drive gear and shims onto the shaft.
8. Apply a bead of thread sealer to the mating surface of the cap and install the cap. Use 2 cap screws to align the bolt holes. Use a plastic mallet to tap the cap into position. Torque the bearing cap bolts to 35 ft. lbs. (48 Nm).
9. Install the output shaft yoke and torque the locknut to 120 ft. lbs. (163 Nm) Be sure to use tool J-8614-01 or equivalent to hold the yoke while tightening the nut.
10. Check the rear output shaft endplay as follows:
    a. Attach the dial indicator to the bearing cap and position the indiactor stylus agianst the output shaft.
    b. Pry the output shaft back and forth to check the endplay, the endplay should be 0.001-0.005 in. (0.025-0.127mm).
    c. If the endplay is not correct, remove or add shims between the speedometer drive gear and the output shaft rear bearing.
11. Install the transfer case vent hose. Install the speedometer driven gear sleeve and driven gear.
12. Install the speedometer cable. Install the driveshaft and torque the retaining bolts to 16 ft. lbs.

Removal and Installation

TRANSFER CASE REMOVAL

1. On vehicles equipped with manual transmissions, remove the shift lever knob, trim ring and boot from the transmission and transfer shift levers.
2. Remove the floor covering, if so equipped and remove the transmission access cover from the floorpan.
3. Raise and support the vehicle safely and drain the lubricant from the transfer case.

4. Position a support stand under the clutch housing to support the engine and transmission and remove the rear crossmember.
5. Disconnect the front and rear driveshafts at the transfer case. Disconnect the speedometer cable at the transfer case.
6. If necessary, disconnect the parking brake cable at the equalizer. Disconnect the exhaust pipe support bracket at the transfer case, if so equipped.
7. Remove the the bolts attaching the transfer case to the transmission and remove the transfer case.

TRANSFER CASE INSTALLATION

1. Shift the transfer case into the **4L** position.
2. Rotate the transfer case output shaft by turning the yoke until the transmission outyput shaft gear engages the transfer case input shaft. Move the transfer case forward until the case seats against the transmission.

NOTE: Be sure that the transfer case is flush against the transmission. Severe damage to the transfer case will result if the attaching bolts are tighten while the transfer case is cocked or in a bind.

3. Install the transfer case to transmission attaching bolts and torque them to 30 ft. lbs.
4. Fill the transfer case with the recommended lubricant to the proper level. Connect the speedometer driven gear to the transfer case.
5. Connect the transfer case shift lever and control links to the transfer case shift rods.
6. Connect the front and rear driveshafts to the transfer case yokes. Torque the bolts to 16 ft. lbs.
7. Install the rear crossmember and remove the support stand from under the clutch housing.
8. Connect the parking brake cable to the equalizer and connect the exhaust pipe support bracket to the transfer case if disconnected.
9. Lower the vehicle. Install the transmission access cover plate on the floor pan. Install the floor covering, if so equipped. Install the boots, trim rings and shift knobs.

Before Disassembly

Cleanliness during disassembly and assembly is necessary to avoid further transfer case trouble after overhaul. Before removing any of the transfer case subassemblies, plug all the openings and clean the outside of the of the transfer case thoroughly. Steam cleaning or car wash type high pressure equipment is preferable. During disassembly, clean all parts in suitable solvent and dry each part. Do not use cloth or paper towels to dry parts. Use compressed air only.

Transfer Case Disassembly

1. Drain the unit and remove the shift lever assembly.
2. Remove the bottom cover.

NOTE: The bottom cover has been coated with a sealant. Use a putty knife to break the seal and work the knife around the bottom of the cover to break it loose. Don't try to wedge the cover off.

3. With a puller, remove the front and rear yokes.
4. Unbolt and remove the input shaft support from the case. The rear output shaft gear and input shaft will come with it as an assembly.

NOTE:The support has been coated with sealant. Use a putty knife to break the seal and work the knife around the bottom of the cover to break it loose. Don't try to wedge the cover off.

5. Remove the rear output shaft clutch sleeve from the case.
6. Remove and discard the snapring retaining the rear output shaft gear on the input shaft and remove the gear.
7. Remove and discard the input bearing snapring.
8. Remove the input shaft bearing from the support. Tap the end of the shaft with a soft mallet to aid removal.
9. Remove the input shaft bearing and endplay shims from the shaft with an arbor press.
10. Remove the input shaft oil seal from the support.
11. Unbolt and remove the intermediate shaft lockplate.
12. Remove the intermediate shaft. Tap the shaft out of the case using a brass punch and plastic mallet.
13. Remove and discard the intermediate shaft O-ring seal.
14. Remove the intermediate gear assembly and thrust washers.

NOTE: The thrust washers have locating tabs which must fit into notches in the case at assembly.

15. Remove the needle bearings and spacers from the intermediate gear. There are 48 needle bearings and 3 spacers.
16. Remove the rear bearing cap attaching bolts and remove the cap. A plastic mallet will aid in removal.

NOTE: The rear bearing cap has been coated with sealant.

17. Remove the endplay shims and speedometer drive gear from the rear output shaft.
18. Remove and discard the rear output shaft oil seal. Remove the bearings and races from the rear cap.
19. Unbolt and remove the front and rear output shaft shift forks from the shift rods.
20. Remove the shift rods. Insert a punch through the clevis pin holes in the rods and rotate the rods while pulling them out of the case.

NOTE: The shift rods are free of the case, take care to avoid losing the shift rod poppet balls and springs.

21. Remove the shift forks from the case.
22. Remove the bolts attaching the front cap to the case and remove the cap.

NOTE: The front cap has been coated with sealant.

23. Remove the front output shaft and shift rod oil seals from the front cap.
24. Remove the bearing race from the front cap.
25. Remove the cover plate bolts and remove the plate and endplay shims from the case. Keep the shims together for assembly.
26. Move the front output shaft toward the front of the case.
27. Remove the front output shaft rear bearing race.
28. Remove the rear output shaft front bearing. Position the case on wood blocks. Seat the clutch gear on the case interior surface and tap the shaft out of the bearing with a soft mallet.

NOTE: If the bearing is difficult to remove, an arbor press may have to be used.

29. Remove the rear output shaft front bearing, thrust washer, clutch gear and output shaft from the case.
30. Remove the front output shaft rear bearing with an arbor press.

CAUTION

Be sure to support the case with wood blocks positioned on either side of the case bore.

31. Remove the case from the press and remove the output shaft, clutch gear and sleeve and the shaft rear bearing.
32. Remove the front output shaft front bearing with an arbor press and tool J-22912-01 or its equivalent.
33. Remove the front output shaft from the gear.

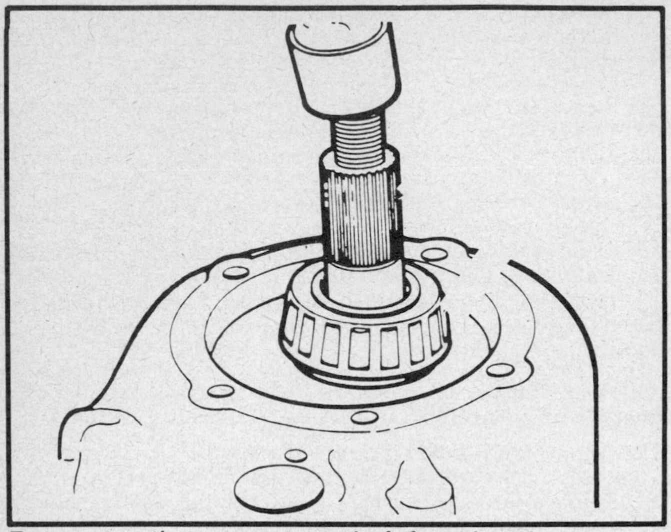
Removong the rear output shaft front bearing

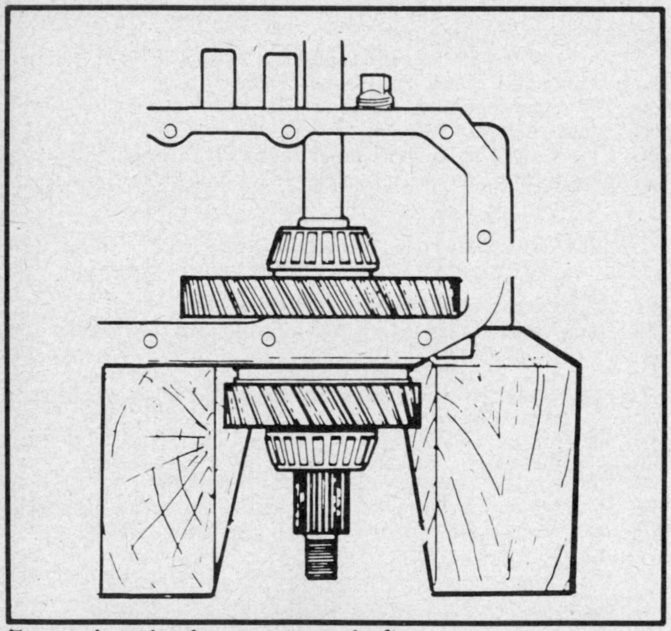
Removing the front output shaft

34. Remove the input shaft rear needle bearing from the rear output shaft using tool J-29369-1 or its equivalent. Support the shaft in a vise during removal.
35. Using a $^3/_8$ in. drive, $^7/_{16}$ in. socket, remove the shift rod thimbles from the case.

CLEANING AND INSPECTION

Cleaning

During overhaul, all components of the transfer case (except bearing assemblies) should be thoroughly cleaned with solvent and dried with air pressure prior to inspection and reassembly.

1. Clean the bearing assemblies as follows:

NOTE: Proper cleaning of bearings is of utmost importance. Bearings should always be cleaned separately from other parts.

a. Soak all bearing assemblies in clean solvent or fuel oil. Bearings should never be cleaned in a hot solution tank.
b. Slush bearings in solvent until all old lubricant is loosened. Hold races so that bearings will not rotate; then clean bearings with a soft bristled brush until all dirt has been removed. Remove loose particles of dirt by tapping bearing flat against a block of wood.
c. Rinse bearings in clean solvent; then blow bearings dry with air pressure.

CAUTION

Do not spin bearings while drying.

d. After drying, rotate each bearing slowly while examining balls or rollers for roughness, damage, or excessive wear. Replace all bearings that are not in first class condition.

NOTE: After cleaning and inspecting bearings, lubricate generously with recommended lubricant, then wrap each bearing in clean paper until ready for reassembly.

2. Remove all portions of old gaskets from parts, using a stiff brush or scraper.

Inspection

1. Inspect all parts for discoloration or warpage.
2. Examine all gears and splines for chipped, worn, broken or nicked teeth. Small nicks or burrs may be removed with a fine abrasive stone.
3. Inspect the breather assembly to make sure that it is open and not damaged.
4. Check all threaded parts for damaged, stripped, or crossed threads.
5. Replace all gaskets, oil seals and snaprings.
6. Inspect housings, retainers and covers for cracks or other damage. Replace the damaged parts.
7. Inspect keys and keyways for condition and fit.
8. Inspect shift forks for wear, distortion or any other damage.
9. Check detent ball springs for free length, compressed length, distortion or collapsed coils.
10. Check bearing fit on their respective shafts and in their bores or cups. Inspect bearings, shafts and cups for wear.

NOTE: If either bearings or cups are worn or damaged, it is advisable to replace both parts.

11. Inspect all bearing rollers or balls for pitting or galling.
12. Examine detent balls for corrosion or brinneling. If shift bar detents show wear, replace them.
13. Replace all worn or damaged parts. When assembling the transfer case, coat all moving parts with recommended lubricant.

Transfer Case Assembly

Coat all parts with SAE 85W-90 oil before assembly.

1. Apply Loctite® 220 or its equivalent to the thimbles and install them in the case.
2. Install the front output shaft gear on the front output shaft. Be sure that the clutch teeth on the gear face the shaft gear teeth.
3. Install the front bearing on the front output shaft using an arbor press. Be sure that the bearing is seated against the gear.
4. Install the front output shaft in the case and install the clutch sleeve and gear on the shaft.

5. Install the front output shaft rear bearing using an arbor press.

NOTE: Install an old yoke nut on the shaft to avoid damage to the threads.

6. Install the input shaft needle bearings in the rear output shaft with tool J-29179 or its equivalent.
7. Position the rear output shaft clutch gear in the case and insert the rear output shaft into the gear.
8. Install the thrust washer and front bearing on the rear output shaft using an arbor press.
9. Install the shims and bearing on the input shaft using an arbor press.
10. Install a new input shaft seal.
11. Using a new snapring, install the input shaft and bearing in the support.
12. Install the rear output shaft gear on the input gear and install a new gear retaining ring.
13. Measure the clearance between the input gear and the gear retaining snapring using a feeler gauge. Clearance should not exceed 0.003 in. If clearance is beyond tolerance, add shims between the input shaft and bearing.
14. Install the clutch sleeve on the rear output shaft.
15. Apply Loctite® 515 or equivalent to the mating surfaces of the input shaft support and install the support assembly, shaft and gear in the case. Use 2 support bolts to align the support on the case and tap the support into position with a soft mallet. Torque the support bolts to 10 ft. lbs.
16. Install the rear bearing cap front bearing race.
17. Install the rear bearing cap rear bearing race.
18. Position the rear output shaft rear bearing in the rear bearing cap.
19. Install the rear output shaft yoke oil seal.
20. Install the speedometer gear and endplay shims on the rear output shaft.
21. Apply Loctite® 515 or equivalent to the mating surfaces of the cap and install the rear bearing cap. Use 2 cap bolts to align the cap and tap it into place with a soft mallet.
22. Tighten the cap bolts to 35 ft. lbs.
23. Install the rear output shaft yoke. Torque a new locknut to 120 ft. lbs.
24. Clamp a dial indicator on the rear output shaft bearing cap. Position the indicator stylus so that it contacts the end of the shaft.
25. Pry the shaft back and forth to check endplay. Endplay should be 0.001–0.005 in. If play is not correct, remove or add shims between the speedometer drive gear and the output shaft rear bearing.

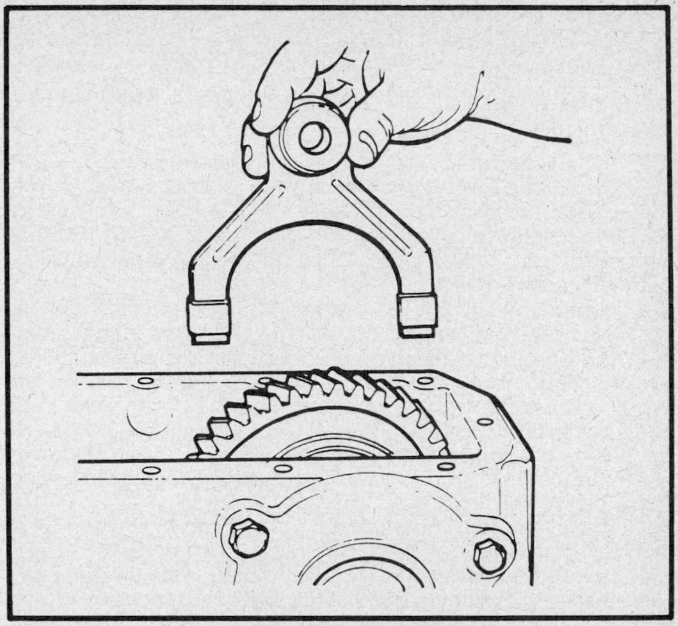
Installing the front output shaft shift fork

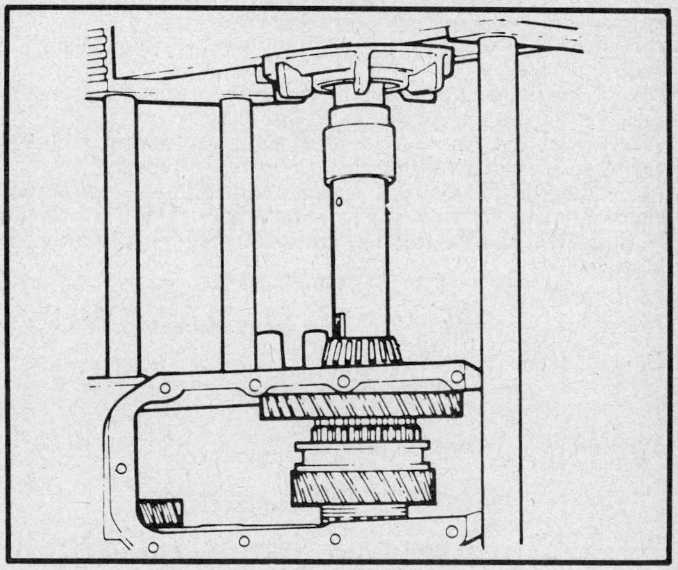
Installing the front output shaft bearing

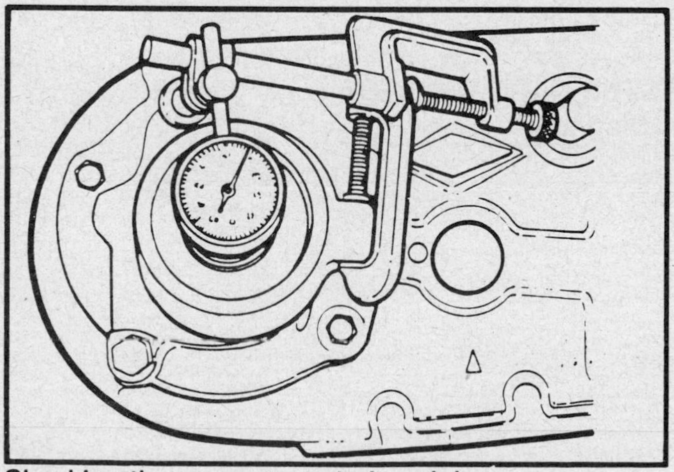
Checking the rear output shaft endplay

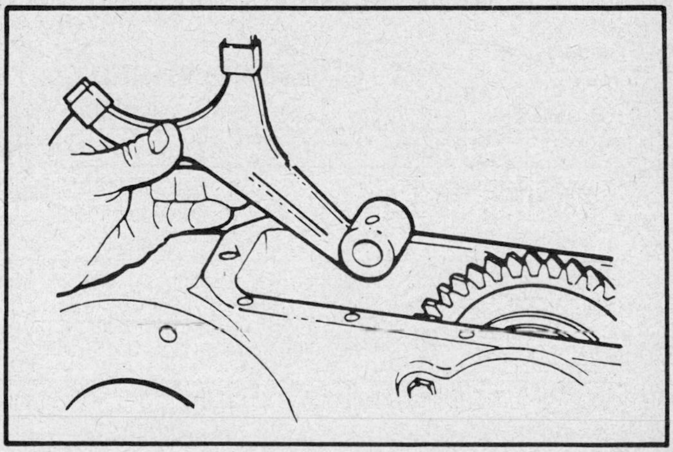
Installing the rear output shaft shift fork

26. Install the front output shaft rear bearing race.
27. Install the front output shaft endplay shims and cover plate. Tighten the cover plate bolts to 35 ft. lbs.

NOTE: Apply Loctite® 220 to the bolts before installation.

28. Install the front output shaft front bearing race.
29. Install the front output shaft yoke oil seal.
30. Install the shift rod oil seals.
31. Install the front bearing cap, using Loctite® 515 on the mating surfaces. Use 2 bolts to align the cap and tap it into position with a soft mallet.
32. Install and tighten the bearings cap bolts to 35 ft. lbs.
33. Seat the rear bearing cup against the cover plate by tapping the end of the front output shaft with a plastic mallet. Mount a dial indicator on the front bearing cap and position the stylus against the end of the output shaft. Pry the shaft back and forth to check endplay. Endplay should be 0.001–0.005 in. If the play is not correct, add or remove shims between the cover plate and case. If shims are added seat the rear bearing cup again before checking.
34. Install the front output shaft yoke. Tighten the new locknut to 120 ft. lbs.
35. Install the front and rear output shaft shift forks.
36. Install the front output shaft shift rod poppet ball and spring in the front bearing cap.
37. Compress the poppet ball and spring and install the front output shaft shift rod part way in the case.
38. Insert the front output shaft shift rod through the shift fork.
39. Align the setscrew hole in the shift fork and rod. Install and tighten the setscrew to 14 ft. lbs.
40. Install the rear output shaft shift rod poppet ball and spring in the front bearing cap.
41. Compress the ball and spring and install the rear output shaft shift rail part way. The front output shaft shift rod should be in neutral and the interlocks seated in the front bearing cap bore.
42. Insert the rear output shaft shift rod through the shift fork.
43. Align the setscrew holes in the fork and rod. Torque the setscrew to 14 ft. lbs.
44. Insert tool J-25142 or equivalent in the intermediate gear and install the needle bearings and spacer.
45. Install the intermediate gear thrust washers in the case. Make sure that the tangs are aligned with the grooves in the case. The thrust washers may be held in place with petroleum jelly.
46. Install a new O-ring seal on the intermediate shaft.
47. Position the intermediate gear in the case.
48. Install the intermediate shaft in the case bore. Tap the shaft into the gear until the shaft forces the tool out of the case.
49. Install the intermediate shaft lock plate and bolt. Torque the bolt to 23 ft. lbs.
50. Install the bottom cover, applying Loctite® 515 or equivalent to the mating surfaces. Install and torque the bolts to 15 ft. lbs.
51. Fill the case with 4 pints of SAE 85W–90W gear oil.

Specifications

TORQUE SPECIFICATIONS
Dana 300 Transfer Case

Component	ft. lbs.	Nm
Bottom cover bolts	15	20
Cover plate bolts	35	47
Front bearing cap bolts	35	47
Front/rear yoke locknuts	120	163
Input shaft support screws	10	14
Lockplate bolts	23	31
Shift fork setscrews	14	19

DANA (AWD) TRANSFER CASE

Application

1988–89 Ford Tempo and Mercury Topaz

General Description

The transfer case is actuated by an electrically controlled vacuum servo system. When the the all-wheel drive switch is turned **ON** a relay activates the 4WD solenoid valve. The 4WD solenoid valve allows vacuum to be created in the lefthand chamber of the vacuum servo. The vacuum moves the servo rod and sliding collar into engagement with the transfer case output gears, driveshaft and rear axle. When the 2WD switch is turned **ON** a relay activates the 2WD solenoid valve. Vacuum is created in the righthand chamber of the vacuum servo disengaging the transfer case, driveshaft and rear axle output gears. The transfer case lubrication is integral with the transaxle. The transaxle/transfer case assembly requires 8.3 qts. of Mercon® automatic transmission fluid.

CHILTON'S THREE C's TRANSFER CASE DIAGNOSIS
AWD—Vacuum Diagnosis

Condition	Cause	Correction
Insufficient vacuum	a) Damaged or clogged manifold fitting b) Damaged hoses c) Damaged or worn check valve	a) Service or replace fitting b) Service as required c) Replace/service
Reservoir not maintaining vacuum	a) Worn or damaged reservoir	a) Check for leak by installing a vacuum gauge at rubber tee (input to dual solenoids). Gauge should rear (16–20 inches) 54–67 kPa vacuum
Dual solenoid assembly inoperative	a) Damaged or worn solenoid assembly	a) Check for vacuum at solenoids as outlined
No AWD engagement	a) Insufficient vacuum at vacuum servo	a) Disconnect vacuum harness at single to double connector and install a vacuum

CHILTON'S THREE C's TRANSFER CASE DIAGNOSIS
AWD—Vacuum Diagnosis

Condition	Cause	Correction
No AWD engagement		gauge. With engine running and AWD switch in proper position, check for vacuum
	b) Damaged or worn vacuum servo	b) Place transaxle in neutral. Raise vehicle on a hoist and disconnect vacuum harness at single to double connector. Install a hand vacuum pump onto red tube connector and block off black connector. Apply (16–20 inches) 54–67 kPa vacuum at servo end of harness. While rotating front wheels, note that rear wheels also rotate. If rear wheels do not rotate, replace vacuum servo

CHILTON'S THREE C's TRANSFER CASE DIAGNOSIS
AWD—Electrical Diagnosis

Condition	Cause	Correction
AWD system inoperative	a) Blown fuse b) Connector at fuse panel disengaged	a) Replace fuse b) Install connector firmly into fuse panel
AWD switch indicator inoperative	a) Loose connection at switch b) Worn or damaged switch	a) Push connector firmly into switch b) Replace switch
AWD relay inoperative	a) Poor connection at relay b) Open or short in harness c) Worn or damaged relay	a) Check connection at relay b) Service or replace harness as necessary c) Replace relay
AWD dual solenoid inoperative	a) Open or short in harness	a) Service or replace harness

Removal and Installation

TRANSFER CASE REMOVAL

1. Disconnect the negative battery cable. Raise and support the vehicle safely.
2. Remove cup bore plug and drain transfer case.
3. Remove the line retaining bracket.
4. Remove the driveshaft front retaining bolts and caps. Disengage the front driveshaft from the drive yoke.
5. Check backlash before removal in order to reset to existing backlash.

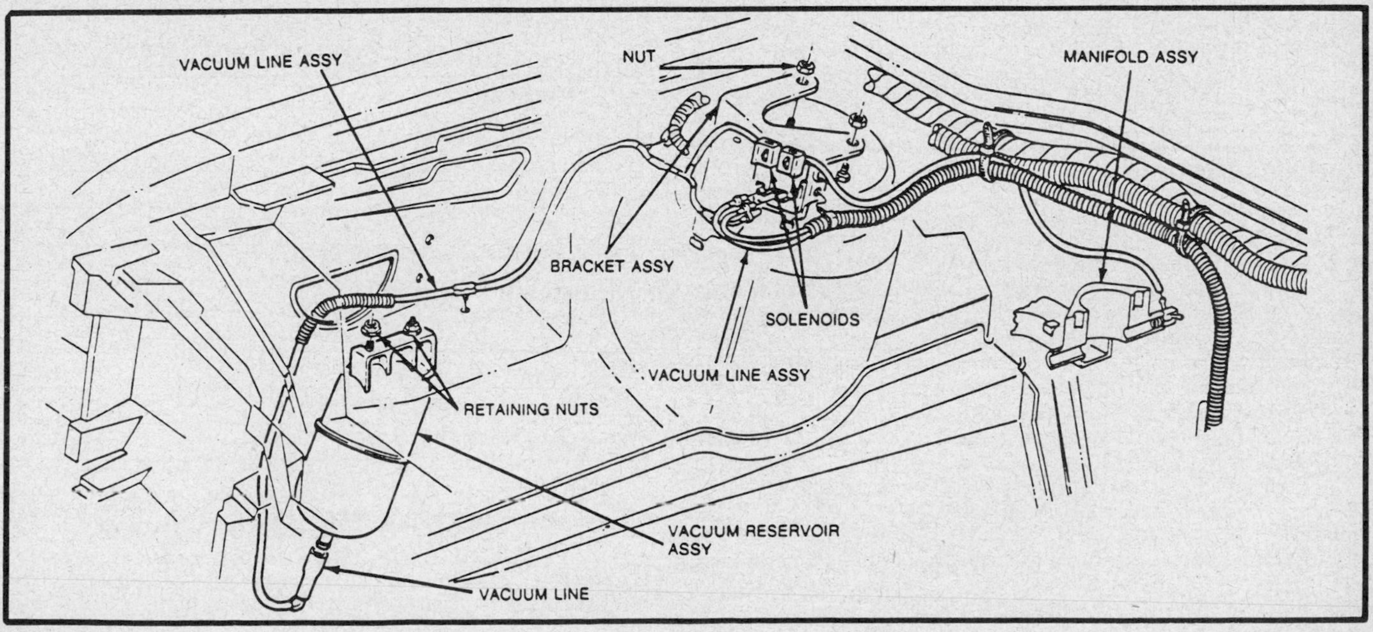
VACUUM LINE ASSY
NUT
MANIFOLD ASSY
BRACKET ASSY
SOLENOIDS
VACUUM LINE ASSY
RETAINING NUTS
VACUUM RESERVOIR ASSY
VACUUM LINE

Vacuum controls

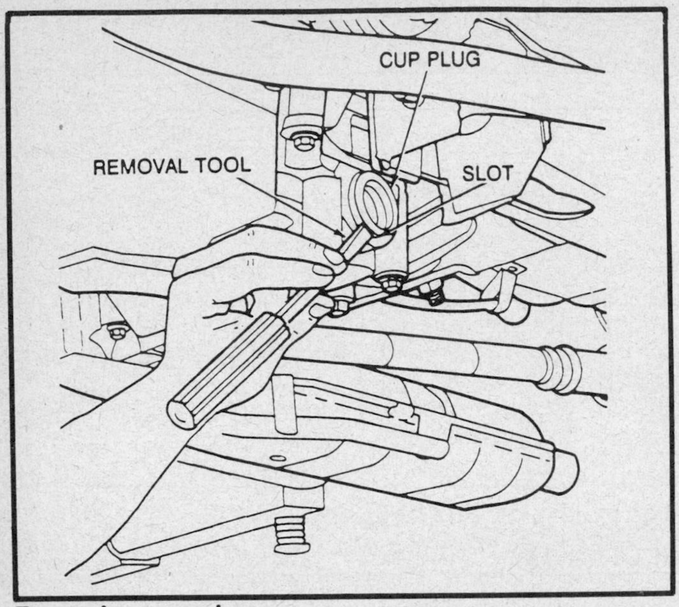
CUP PLUG
REMOVAL TOOL
SLOT

Removing cap plug

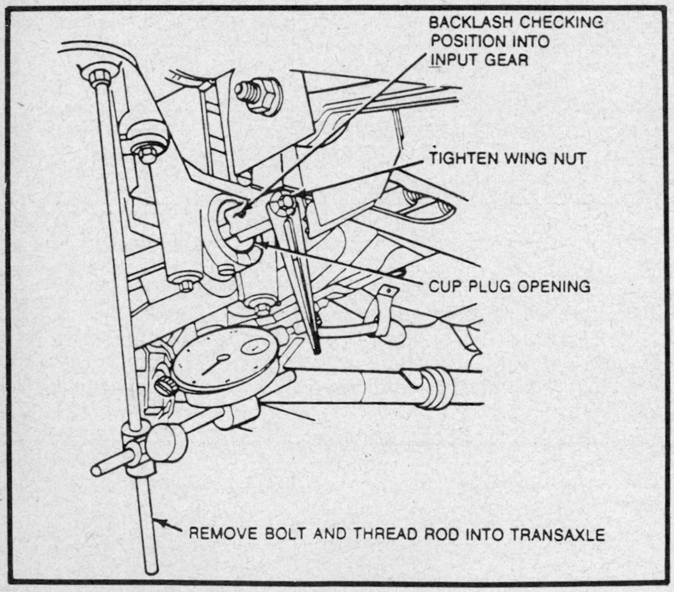
BACKLASH CHECKING POSITION INTO INPUT GEAR
TIGHTEN WING NUT
CUP PLUG OPENING
REMOVE BOLT AND THREAD ROD INTO TRANSAXLE

Checking backlash

6. Install backlash measuring gauge T87P–4020–B, or equivalent through the cup plug opening into the input gear.
7. Make certain the transaxle is in **P**.
8. Install a rod on the transaxle panrail. Secure a dial indicator to rod. Rotate both front wheels together, until park gear is wedged tight against the park pawl.
9. Position stylus of dial indicator on end of backlash checking tool. Maintain load on park gear, park pawl and wheels while reading backlash.
10. Push backlash measuring gauge upward and zero dial indicator.
11. Pushing down on backlash measuring gauge, measure backlash. Correct backlash should be 0.012–0.024 in. (0.3–0.6mm).
12. Remove 3 bolts retaining vacuum motor shield and remove the shield.
13. Remove the vacuum lines from the servo.
14. Remove the 13 bolts retaining the transfer case to the transaxle and remove the transaxle.

NOTE: Mark the bolts as to location removed for installation reference. The length of Bolts differ so bolts must be install in the same location removed.

TRANSFER CASE INSTALLATION

1. Discard all old gasket material from mating surfaces.
2. Select new maximum thickness gasket, 7A191–H.
3. Install transfer case and torque bolts to 15–19 ft. lbs. (21–25 Nm) making certain to install bolts in same locations removed from.
4. Install backlash measuring gauge T87P–4020–B, or equivalent through the cup plug opening into the input gear.
5. Make certain the transaxle is in **P**.
6. Install a rod on the transaxle panrail. Secure a dial indicator to rod. Rotate both front wheels together, until park gear is wedged tight against the park pawl.
7. Position stylus of dial indicator on end of backlash checking tool. Maintain load on park gear, park pawl and wheels while reading backlash.
8. Push backlash measuring gauge upward and zero dial indicator.
9. Pushing down on backlash measuring gauge, measure backlash. Correct backlash should be 0.012–0.024 in. (0.3–0.6mm).
10. If measurement is not within specification, remove transfer case, select correct gasket and reinstall transfer case. Recheck the backlash and change gasket size if necessary.
11. Install transfer case and install bolts in proper locations. Torque bolts to 15–19 ft. lbs. (21–25 Nm).
12. Install vacuum motor supply hose connector.
13. Install vacuum motor shield and torque bolts to 7–12 ft. lbs. (9–16 Nm).
14. Position driveshaft to drive yoke. Apply Loctite® 242, or equivalent, to bolt threads. Install bolts and torque to 15–17 ft. lbs. (21–23 Nm).

NOTE: Do not apply high strength threadlock to bolts or retaining bolts make become impossible to remove.

15. Install vacuum line retaining bracket. Torque bolt to 7–12 ft. lbs. (9–16 Nm).
16. Lower vehicle and connect the negative battery cable.
17. Fill transaxle with 8.3 qts. of Mercon® automatic transmission fluid.

SELECT GASKET CHART

Measurement Obtained		Select Gasket Required
in.	mm	
.012–.020	0.30–0.50	7A191-H
.021–.024	0.51–0.62	7A191-G
.025–.030	0.63–0.76	7A191-F
.031–.035	0.77–0.90	7A191-E
.036–.042	0.91–1.06	7A191-D
.043–.048	1.07–1.21	7A191-C
.049–.054	1.22–1.38	7A191-B
.055–.064	1.39–1.62	7A191-A

Before Disassembly

Clean the exterior of the transfer case assembly before any attempt is made to disassemble it, in order to prevent dirt or other foreign materials from entering the transfer case assembly or its internal parts.

NOTE: If steam cleaning is done to the exterior of the transfer case, immediate disassembly should be done to avoid rusting, caused by condensation forming on the internal parts.

Transfer Case Disassembly

1. Drain and remove the transfer case.
2. Remove the transfer case side cover bolts.
3. Clean gasket material from transfer case and cover.
4. Remove the housing retaining bolts and remove the gear housing subassembly.
5. Remove the O-ring and shims. Wire the shim stack together for reassembly.
6. Remove snaprings from vacuum servo shaft and shift fork.

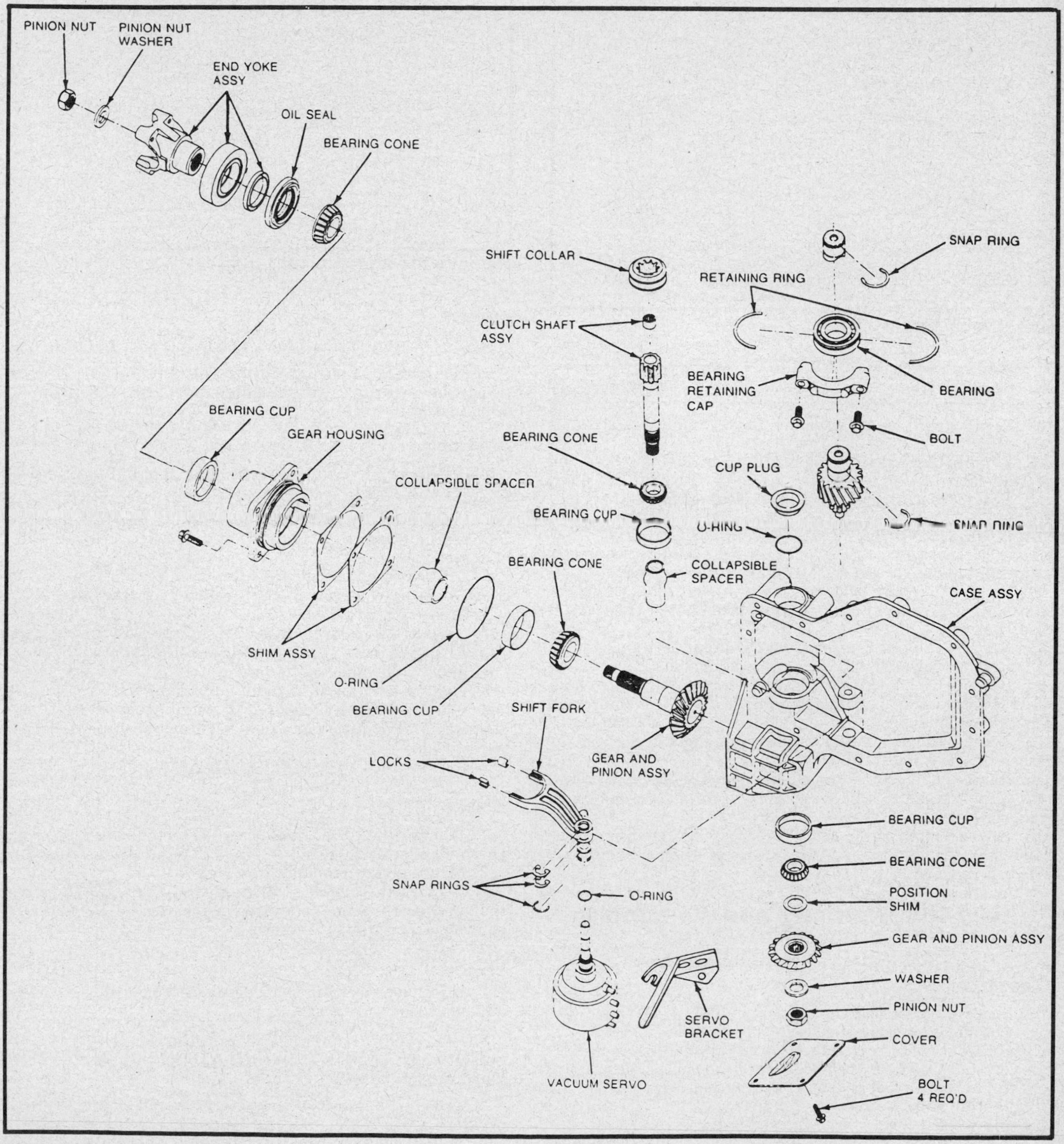
PINION NUT
PINION NUT WASHER
END YOKE ASSY
OIL SEAL
BEARING CONE
SHIFT COLLAR
RETAINING RING
SNAP RING
CLUTCH SHAFT ASSY
BEARING RETAINING CAP
BEARING
BOLT
BEARING CUP
GEAR HOUSING
BEARING CONE
CUP PLUG
COLLAPSIBLE SPACER
BEARING CUP
O-RING
SNAP RING
BEARING CONE
COLLAPSIBLE SPACER
CASE ASSY
SHIM ASSY
O-RING
BEARING CUP
SHIFT FORK
LOCKS
GEAR AND PINION ASSY
BEARING CUP
BEARING CONE
SNAP RINGS
O-RING
POSITION SHIM
GEAR AND PINION ASSY
WASHER
PINION NUT
SERVO BRACKET
COVER
VACUUM SERVO
BOLT 4 REQ'D

All-wheel drive—Tempo/Topaz

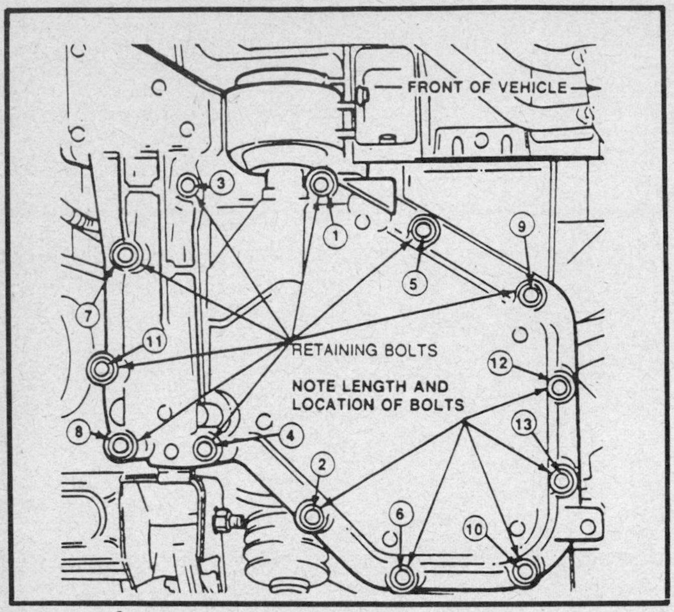
FRONT OF VEHICLE
RETAINING BOLTS
NOTE LENGTH AND LOCATION OF BOLTS
1
2
3
4
5
6
7
8
9
10
11
12
13

Bolt torque sequence

CAUTION

Eye protection must be worn when removing or install snaprings.

7. Remove shift motor assembly. Remove shift fork and shift fork clips.
8. Remove the transfer case bearing cap retaining bolts and bearing cap.
9. Rotate bearing and remove 2 piece snapring from the bearing.
10. Using a thin prybar, remove the inner snapring which positions the input gear to the ball bearing. Slide the bearing toward the input gear and remove the outer snapring.
11. Remove cup plug. Slide input gear toward ball bearing until the input gear and bearing can be lifted out of the transfer case.
12. Remove the ball bearing from the input gear.
13. Remove the shift collar from the clutch shaft.
14. Remove the pinion nut and washer from the clutch shaft. Use a breaker bar and holding tool T87P–7120–A, or equivalent. Tap the clutch shaft from the transfer case, with a soft drift.
15. Remove the pinion gear, outer bearing and shims from the transfer case. Wire the shims together.
16. Remove and discard the clutch shaft collapsible spacer.
17. Mount transfer case in holding fixture T57L–500–B, or equivalent.
18. Install clutch shaft inner bearing cup remover T87P–7120–D. Remove inner and outer bearing, using slide hammer into cup remover.

Unit Disassembly and Assembly

DRIVE GEAR HOUSING

Disassembly

1. Place the gear housing subassembly in a soft-jawed vise. Remove pinion nut, yoke end and washer.
2. Tap in drive gear with a soft-faced hammer to remove from the gear housing. Remove and discard the collapsible spacer.
3. Remove the inner bearing cone from the drive gear, using a press and pinion bearing cone remover.
4. Mount the drive in a soft-jawed vise.
5. Remove drive gear housing oil seal, using a roll head prybar.

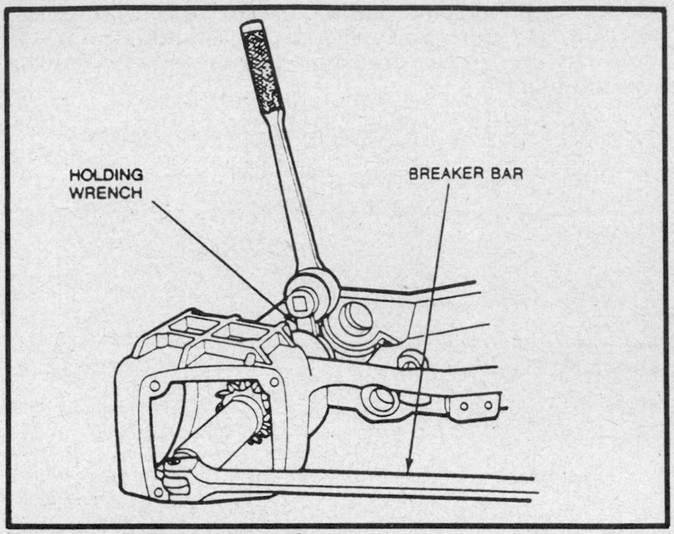
HOLDING WRENCH
BREAKER BAR

Remove the clutch shaft

6. Remove the inner and outer drive gear bearing cups, using a brass drift and hammer.

NOTE: Remove any burrs and wipe bores clean.

7. Remove gear housing from vise. Install new inner and outer drive bearing cups, using bearing cup replacer T87P–4616–A, or equivalent.

Assembly

NOTE: Install the nut on the end of the drive gear to protect the shaft.

1. Clean the drive gear in solvent. Install a new inner bearing cone assembly using pinion bearing cone replacer T62F–4621, or equivalent.
2. Lubricate and install new outer drive bearing cone. Install a new oil seal, using install T87P–7065–B, or equivalent.
3. Grease the end of the seal.
4. Install a new collapsible spacer on drive gear stem.5. Install drive gear into gear housing.
6. Install end yoke, washer and nut.
7. Tighten pinion nut in small increments until rotation effort is 15–32 inch lbs. (1.7–3.6 Nm) with new bearings. Do not exceed this specification or a new collapsible spacer will have to be installed.

CLUTCH SHAFT

Disassembly

1. Remove the clutch shaft inner bearing using a press and bearing puller attachment D84L–1123–A, or equivalent.
2. Mount the clutch shaft in a vise.
3. Remove the clutch shaft needle bearings, which centers the input gear, using a pilot bearing replacer T87P–7120–C, or equivalent and a slide hammer.

Assembly

1. Install a new clutch shaft needle bearing, using a hammer and pilot bearing replacer.

NOTE: When install the needle bearing into the clutch shaft, install it with the tapered end down (toward clutch shaft).

2. Pack the bearing with grease to maintain proper needle position.
3. Install the clutch shaft inner bearing cone, using a press and bearing puller attachment D84L–1123–A, or equivalent.

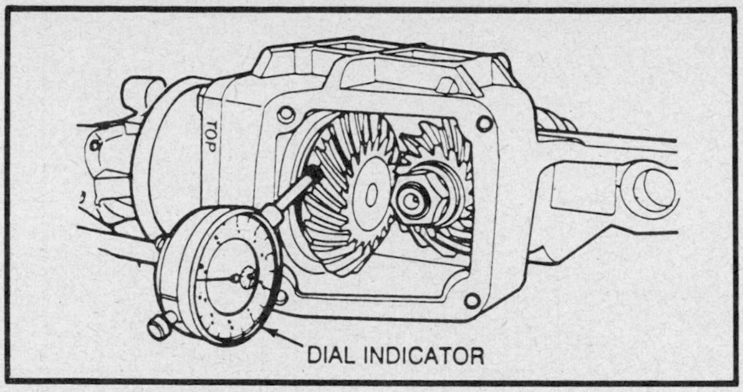
TOP
DIAL INDICATOR

Checking backlash

TRANSFER CASE ASSEMBLY

1. Wipe bearing bores clean. Install inner and outer bearing cups, using bearing cup replacer T87P–7120–B, or equivalent.
2. Install a new collapsible spacer on clutch shaft.
3. Install clutch shaft into transfer case. Assemble original shim and pinion gear.
4. Assemble washer and pinion nut. Torque nut using breaker bar and holding wrench until rotational effort is 4.0–8.0 inch lbs. (0.45–0.9 Nm) with new bearings. Do not exceed this specification of a new collapsible spacer will be required to obtain proper preload.
5. Position shims and a new O-ring onto the gear housing. Lubricate O-ring.
6. Install gear housing subassembly to transfer case and torque bolts to 8.0–12.0 ft. lbs. (11–16 Nm).
7. Check backlash between drive and pinion gear. Correct backlash should be 0.004–0.006 in. (0.10–0.15mm).

NOTE: Check gear contact tooth pattern. If a gross pattern error is detected with backlash correct, adjust the drive pinion gear shim stack. Increasing the shim stack should move the contact pattern on the drive (pull) side of gear toward toe of tooth.

8. Install shift collar onto clutch shaft.
9. Slide the ball bearing onto the input gear.
10. Install the input gear into the transfer case. Slide the small end of the input gear into clutch shaft.
11. Install snapring onto outer end of shaft.
12. Slide the bearing outboard and install snapring onto the inner end of the input shaft. Make certain the snaprings are completely seated in grooves.
13. Install 2 piece snapring into groove for ball bearing transfer case.
14. Install bearing cap and retainer bolts. Tighten bolts to 18–24 ft. lbs. (24–33 Nm).
15. Inspect shift fork clips and replace if necessary. Install shift fork onto clutch collar.
16. Install a new O-ring onto the vacuum servo shaft. Lubricate O-ring with automatic transmission fluid.
17. Install vacuum servo assembly into transfer case. Install the snapring, making certain it is fully seated in groove.
18. Install the shift fork snaprings.
19. Apply a bead of silicone rubber sealer on cover surface. Install the transfer case side cover and torque bolts to 7–12 ft lbs. (9–16 Nm).
20. Install transfer case, checking backlash, onto transaxle.

SPECIFICATIONS

Description	ft. lbs.	Nm
Vacuum solenoids-to-shock tower	21–30	29–40

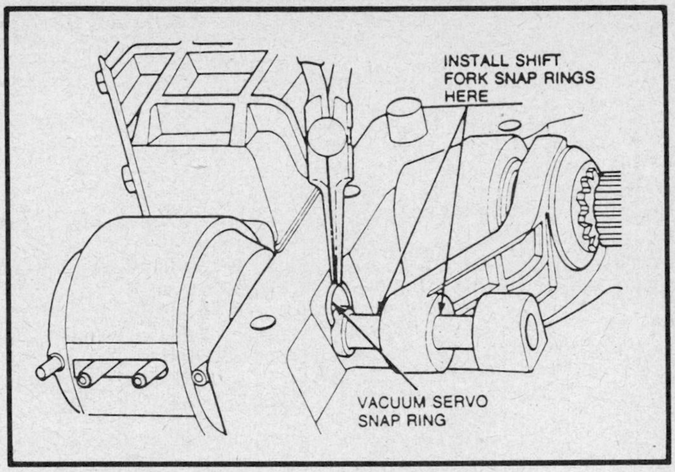
INSTALL SHIFT FORK SNAP RINGS HERE
VACUUM SERVO SNAP RING

Servo snapring location—Tempo/Topaz

Specifications

Description	ft. lbs.	Nm
Vacuum reservoir retaining nuts	34–38 ①	3.8–4.3
Vacuum servo line bracket	7–12	9–16
Vacuum servo shield-to-transfer case	7–12	9–16
Transfer case retaining bolts	15–19	21–25
Driveshaft-to-drive yoke	15–17	21–23
Gear housing-to-transfer case	8–12	11–16
Bearing cap retaining bolts	18–24	24–33
Transfer case side cover retaining bolts	7–12	9–16

① Inch lbs.

SPECIAL TOOLS

Tool Number	Description
T57L-500-B	Bench mounted holding fixture
D79L-4621-A	Pinion bearing cone remover
D83L-7059-B	Vacuum pump
D84L-1123-A	Bearing puller attachment
T62F-4621-A	Pinion bearing cone replacer
T75L-1165-B	Seal plate
T80T-4000-W	Driver handle
T87P-4020-B	Backlash measuring gauge
T87P-4616-A	Bearing cup replacer
T87P-7065-B	Output seal replacer
T87P-7120-A	Holding wrench
T87P-7120-B	Transfer drive bearing cup replacer
T87P-7120-C	Pilot bearing replacer
T87P-7120-D	Bearing cup remover
TOOL-4201-C	Dial indicator

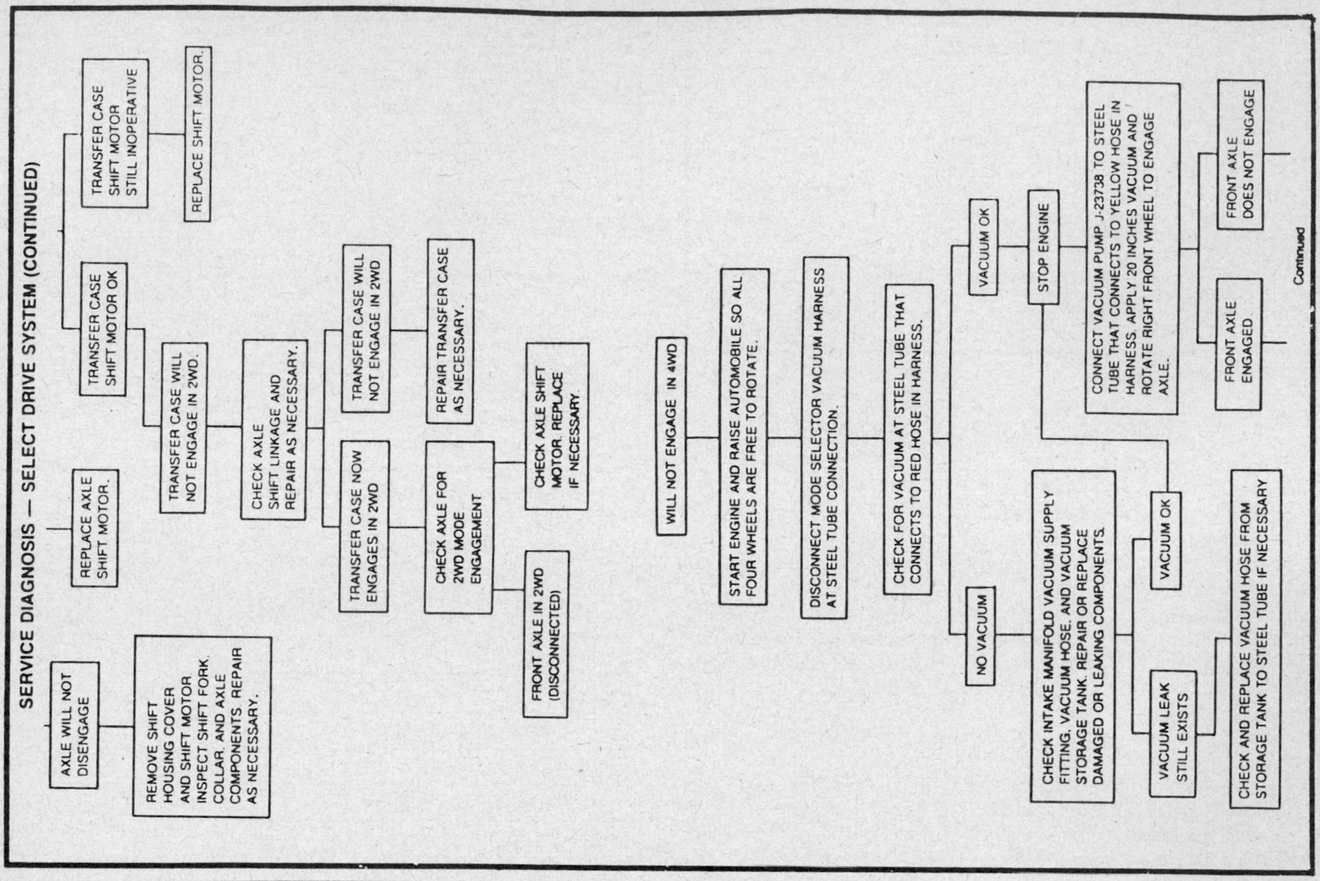
SERVICE DIAGNOSIS — SELECT DRIVE SYSTEM (CONTINUED)
TRANSFER CASE SHIFT MOTOR STILL INOPERATIVE
REPLACE SHIFT MOTOR.
TRANSFER CASE SHIFT MOTOR OK
REPLACE AXLE SHIFT MOTOR.
TRANSFER CASE WILL NOT ENGAGE IN 2WD.
CHECK AXLE SHIFT LINKAGE AND REPAIR AS NECESSARY.
TRANSFER CASE WILL NOT ENGAGE IN 2WD
REPAIR TRANSFER CASE AS NECESSARY.
TRANSFER CASE NOW ENGAGES IN 2WD
CHECK AXLE FOR 2WD MODE ENGAGEMENT
CHECK AXLE SHIFT MOTOR. REPLACE IF NECESSARY.
FRONT AXLE IN 2WD (DISCONNECTED)
AXLE WILL NOT DISENGAGE
REMOVE SHIFT HOUSING COVER AND SHIFT MOTOR INSPECT SHIFT FORK, COLLAR, AND AXLE COMPONENTS. REPAIR AS NECESSARY.
WILL NOT ENGAGE IN 4WD
START ENGINE AND RAISE AUTOMOBILE SO ALL FOUR WHEELS ARE FREE TO ROTATE.
DISCONNECT MODE SELECTOR VACUUM HARNESS AT STEEL TUBE CONNECTION.
CHECK FOR VACUUM AT STEEL TUBE THAT CONNECTS TO RED HOSE IN HARNESS.
VACUUM OK
STOP ENGINE
CONNECT VACUUM PUMP J-23738 TO STEEL TUBE THAT CONNECTS TO YELLOW HOSE IN HARNESS. APPLY 20 INCHES VACUUM AND ROTATE RIGHT FRONT WHEEL TO ENGAGE AXLE.
FRONT AXLE DOES NOT ENGAGE
FRONT AXLE ENGAGED.
NO VACUUM
CHECK INTAKE MANIFOLD VACUUM SUPPLY FITTING, VACUUM HOSE, AND VACUUM STORAGE TANK. REPAIR OR REPLACE DAMAGED OR LEAKING COMPONENTS.
VACUUM OK
VACUUM LEAK STILL EXISTS
CHECK AND REPLACE VACUUM HOSE FROM STORAGE TANK TO STEEL TUBE IF NECESSARY
Continued

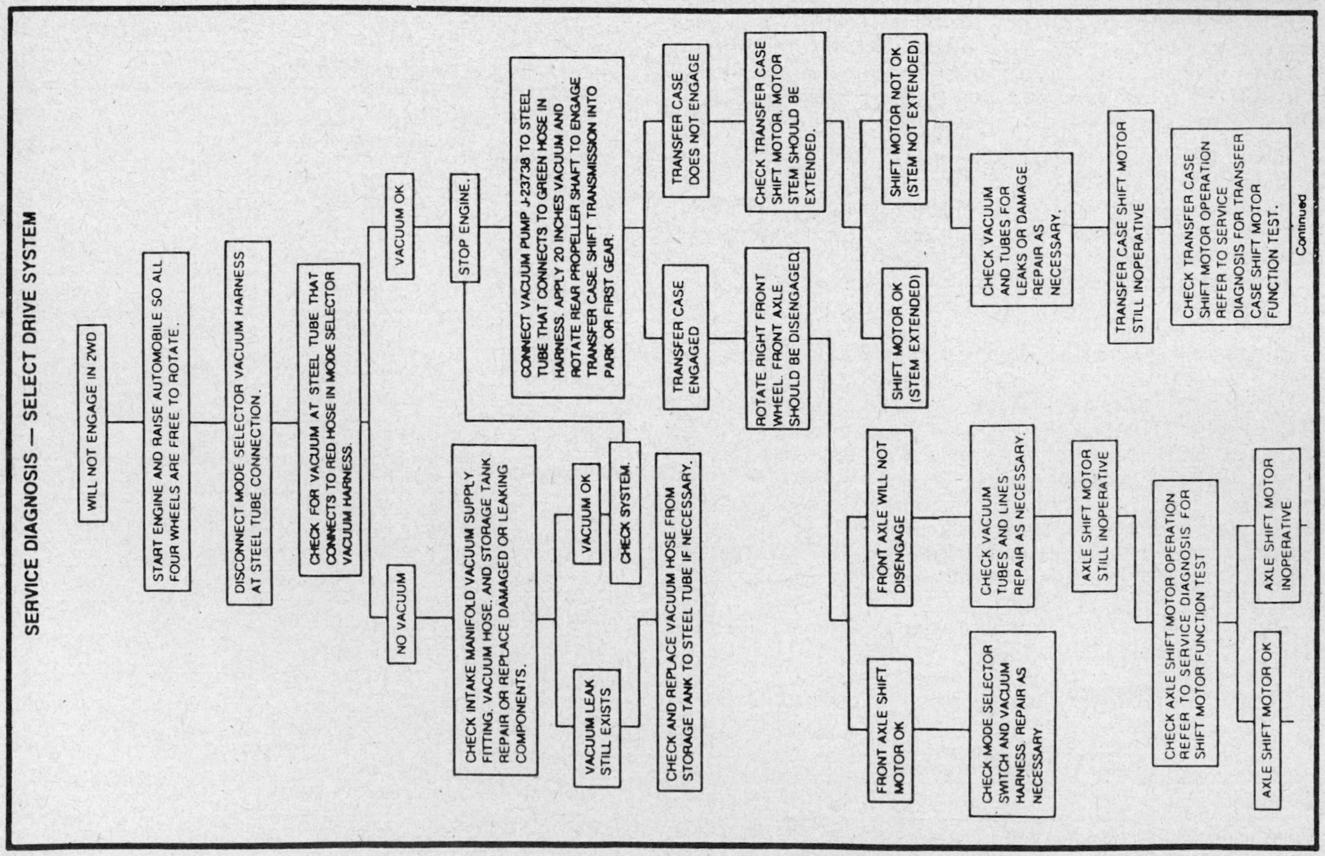
SERVICE DIAGNOSIS — SELECT DRIVE SYSTEM
WILL NOT ENGAGE IN 2WD
START ENGINE AND RAISE AUTOMOBILE SO ALL FOUR WHEELS ARE FREE TO ROTATE.
DISCONNECT MODE SELECTOR VACUUM HARNESS AT STEEL TUBE CONNECTION.
CHECK FOR VACUUM AT STEEL TUBE THAT CONNECTS TO RED HOSE IN MODE SELECTOR VACUUM HARNESS.
VACUUM OK
STOP ENGINE.
CONNECT VACUUM PUMP J-23738 TO STEEL TUBE THAT CONNECTS TO GREEN HOSE IN HARNESS. APPLY 20 INCHES VACUUM AND ROTATE REAR PROPELLER SHAFT TO ENGAGE TRANSFER CASE. SHIFT TRANSMISSION INTO PARK OR FIRST GEAR.
TRANSFER CASE DOES NOT ENGAGE
CHECK TRANSFER CASE SHIFT MOTOR. MOTOR STEM SHOULD BE EXTENDED.
SHIFT MOTOR NOT OK (STEM NOT EXTENDED)
CHECK VACUUM AND TUBES FOR LEAKS OR DAMAGE REPAIR AS NECESSARY.
TRANSFER CASE SHIFT MOTOR STILL INOPERATIVE
CHECK TRANSFER CASE SHIFT MOTOR OPERATION REFER TO SERVICE DIAGNOSIS FOR TRANSFER CASE SHIFT MOTOR FUNCTION TEST.
SHIFT MOTOR OK (STEM EXTENDED)
TRANSFER CASE ENGAGED
ROTATE RIGHT FRONT WHEEL. FRONT AXLE SHOULD BE DISENGAGED.
FRONT AXLE WILL NOT DISENGAGE
CHECK VACUUM TUBES AND LINES REPAIR AS NECESSARY.
AXLE SHIFT MOTOR STILL INOPERATIVE
CHECK AXLE SHIFT MOTOR OPERATION REFER TO SERVICE DIAGNOSIS FOR SHIFT MOTOR FUNCTION TEST
AXLE SHIFT MOTOR INOPERATIVE
AXLE SHIFT MOTOR OK
FRONT AXLE SHIFT MOTOR OK
CHECK MODE SELECTOR SWITCH AND VACUUM HARNESS. REPAIR AS NECESSARY
NO VACUUM
CHECK INTAKE MANIFOLD VACUUM SUPPLY FITTING, VACUUM HOSE, AND STORAGE TANK REPAIR OR REPLACE DAMAGED OR LEAKING COMPONENTS.
VACUUM OK
CHECK SYSTEM.
VACUUM LEAK STILL EXISTS
CHECK AND REPLACE VACUUM HOSE FROM STORAGE TANK TO STEEL TUBE IF NECESSARY.
Continued

SERVICE DIAGNOSIS — SELECT DRIVE SYSTEM

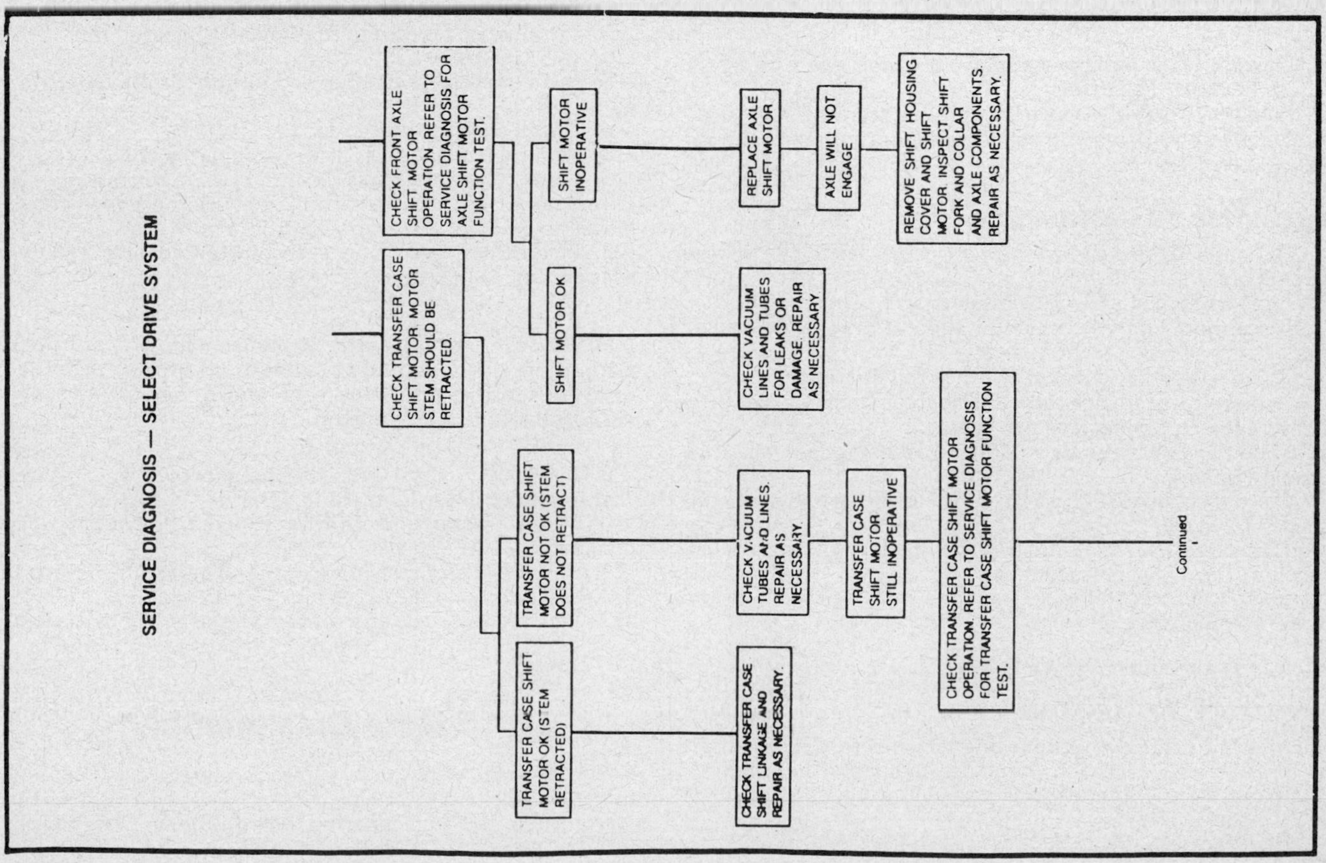
CHECK TRANSFER CASE SHIFT MOTOR MOTOR STEM SHOULD BE RETRACTED.
TRANSFER CASE SHIFT MOTOR OK (STEM RETRACTED)
CHECK TRANSFER CASE SHIFT LINKAGE AND REPAIR AS NECESSARY.
TRANSFER CASE SHIFT MOTOR NOT OK (STEM DOES NOT RETRACT)
CHECK VACUUM TUBES AND LINES. REPAIR AS NECESSARY
TRANSFER CASE SHIFT MOTOR STILL INOPERATIVE
CHECK TRANSFER CASE SHIFT MOTOR OPERATION. REFER TO SERVICE DIAGNOSIS FOR TRANSFER CASE SHIFT MOTOR FUNCTION TEST.
Continued
CHECK FRONT AXLE SHIFT MOTOR OPERATION REFER TO SERVICE DIAGNOSIS FOR AXLE SHIFT MOTOR FUNCTION TEST.
SHIFT MOTOR OK
CHECK VACUUM LINES AND TUBES FOR LEAKS OR DAMAGE. REPAIR AS NECESSARY
SHIFT MOTOR INOPERATIVE
REPLACE AXLE SHIFT MOTOR
AXLE WILL NOT ENGAGE
REMOVE SHIFT HOUSING COVER AND SHIFT MOTOR. INSPECT SHIFT FORK AND COLLAR AND AXLE COMPONENTS. REPAIR AS NECESSARY.

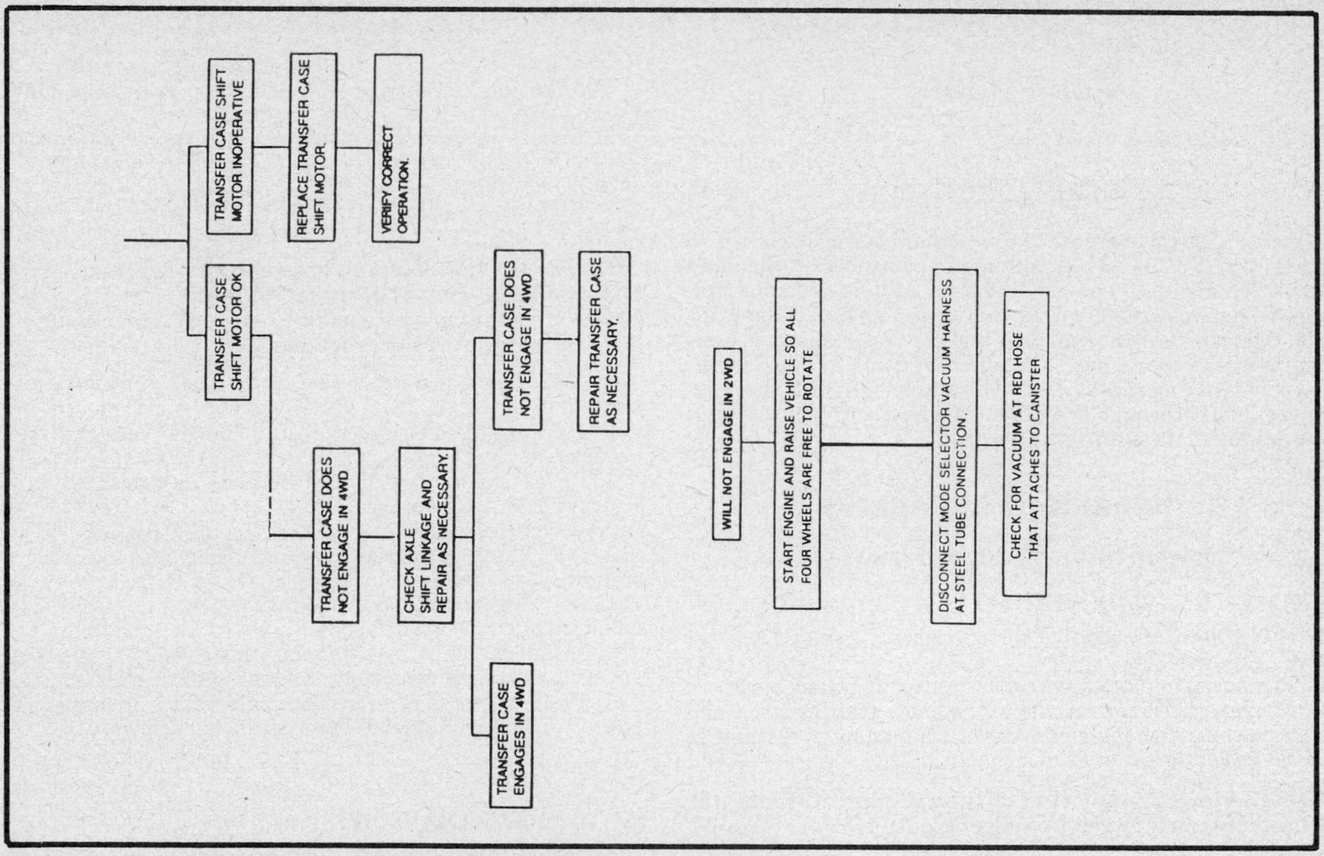
TRANSFER CASE SHIFT MOTOR OK
TRANSFER CASE DOES NOT ENGAGE IN 4WD
CHECK AXLE SHIFT LINKAGE AND REPAIR AS NECESSARY.
TRANSFER CASE ENGAGES IN 4WD
TRANSFER CASE DOES NOT ENGAGE IN 4WD
REPAIR TRANSFER CASE AS NECESSARY.
TRANSFER CASE SHIFT MOTOR INOPERATIVE
REPLACE TRANSFER CASE SHIFT MOTOR.
VERIFY CORRECT OPERATION.
WILL NOT ENGAGE IN 2WD
START ENGINE AND RAISE VEHICLE SO ALL FOUR WHEELS ARE FREE TO ROTATE
DISCONNECT MODE SELECTOR VACUUM HARNESS AT STEEL TUBE CONNECTION.
CHECK FOR VACUUM AT RED HOSE THAT ATTACHES TO CANISTER

NEW PROCESS 129 TRANSFER CASE

Application

1984–88 AMC Eagle

General Description

The model 129 transfer case is used on AMC Eagle vehicles. It requires 3 qts. of Dexron®II automatic transmission fluid and anti-friction additives must not be used. It uses a viscous coupling and pinion assembly, that is not serviceable. In the **2WD** mode, the transfer case shift fork and clutch sleeve are not engaged with the spline gear and sprocket carrier. This disconnects the drive sprocket from the mainshaft and prevents torque transfer through the drive train to the driven sprocket, front output shaft and front driveshaft.

Removal and Installation

TRANSFER CASE REMOVAL

MANUAL TRANSMISSION

1. Disconnect the negative battery cable. Raise and support the vehicle safely.
2. Remove skid plate and rear brace rod at transfer case.
3. Remove speedometer adapter retainer attaching bolt and remove retainer, adapter and cable. Plug adapter opening in transfer case to prevent excessive oil spillage.

NOTE: Mark the position of the speedometer adapter for assembly reference before removing it.

4. Mark driveshaft shafts and axle yokes for assembly alignment reference and disconnect driveshaft shafts at transfer case.
5. On vehicles so equipped, remove transfer case shift motor vacuum harness.
6. Support transfer case with a suitable transmission jack.
7. Remove nuts from transfer case mounting studs and remove transfer case.

AUTOMATIC TRANSMISSION

1. Disconnect the negative battery cable. Raise the vehicle and support it safely.
2. Support engine and transmission with transmission jack.
3. Disconnect catalytic converter support bracket at adapter housing.
4. Remove skid plate and rear brace rod at transfer case.
5. Remove speedometer cable and adapter from transfer case. Discard adapter O-ring, it is not reusable.
6. Matchmark driveshaft shafts and transfer case yokes for assembly reference.
7. Disconnect driveshaft shafts at yokes. Secure shafts to underside of vehicle.
8. Disconnect gearshift and throttle linkage at transmission.
9. Lower the rear crossmember.
10. Remove all transfer case to adapter housing stud nuts and remove transfer case.

TRANSFER CASE INSTALLATION

MANUAL TRANSMISSION

1. Align transmission output and transfer case input shafts and install transfer case on transmission adapter housing.
2. Install and tighten transfer case mounting stud nuts to 33 ft. lbs. torque.
3. Remove jack used to support transfer case.
4. Align and connect driveshaft shafts to axle yokes. Tighten clamp strap bolts to 15 ft. lbs. torque.
5. Install replacement O-ring on speedometer adapter and install adapter and cable and retainer. Tighten retainer bolt to 100 inch lbs. torque.

NOTE: Do not attempt to reuse the original adapter O-ring. The O-ring is designed to swell in service to improve its sealing qualities and it could be cut or torn during installation if reuse is attempted.

6. Install skid plate and rear brace rod. Torque retaining bolts to 30 ft. lbs.
7. On vehicles so equipped, install transfer case shift motor vacuum harness.
8. Check and correct lubricant levels in transmission and transfer case, if necessary.
9. Lower the vehicle and connect the negative battery cable.
10. Install nut on transfer case mounting stud located inside transmission adapter housing. Torque nut to 33 ft. lbs.
11. Install gearshift lever mounting cover on transmission adapter housing.
12. Install gearshift lever on mounting cover. Be sure lever is engaged with shift rail before tightening lever attaching bolts.
13. Position gearshift lever boot and bezel on floorpan or console, if so equipped. Install bezel attaching screws.

AUTOMATIC TRANSMISSION

1. Install transfer case on adapter housing. Be careful not to damage output shaft splines during installation.
2. Install transfer case to adapter housing stud nuts. Torque nuts to 33 ft. lbs.
3. Install rear crossmember and torque the attaching nuts to 30 ft. lbs.
4. Install the rear brace rod.
5. Remove transmission jack or support stand.
6. Connect gearshift and throttle linkage to transmission.
7. Connect driveshaft shafts. Torque clamp strap bolts to 15 ft. lbs.
8. Install new O-ring on speedometer adapter and install adapter and cable in transfer case.

NOTE: Do not attempt to reuse the old adapter O-ring. O-ring is designed to swell in service to provide improved sealing qualities and could be cut or torn if reinstallation is attempted.

9. Install skid plate and stiffening brace, if so equipped. Tighten retaining bolts to 30 ft. lbs.
10. Connect catalytic converter support bracket to adapter housing.
11. Check transfer case lubricant level and transmission linkage adjustments to make certain they are correct.
12. Lower the vehicle and connect the negative battery cable.

Before Disassembly

Clean the exterior of the transfer case assembly before any attempt is made to disassemble it, in order to prevent dirt or other foreign materials from entering the transfer case assembly or its internal parts.

NOTE: If steam cleaning is done to the exterior of the transfer case, immediate disassembly should be done to avoid rusting, caused by condensation forming on the internal parts.

Transfer Case Disassembly

1. Remove the drain plug and drain the lubricant from the transfer case.
2. Remove the nut and bolt which attaches the shift motor bracket to the transfer case.
3. Remove the motor and bracket as an assembly.
4. Remove the nuts which attach the yokes. Discard the sealing washers. Remove the yokes.
5. Mount the transfer case on wood blocks which have V-notches cut into them to clear the front transfer case mounting studs.
6. Mark the relationship between the rear retainer and the case. Remove the retainer attaching bolts.
7. Pry the retainer off of the case, using a pair of pry bars placed into the slots provided in the retainer for this purpose.
8. Remove the differential shim(s) and the speedometer gear from the rear output shaft.
9. Remove the front case-to-rear case bolts and pry the cases apart with a pair of prybars.

NOTE: Slots are provided at each end of the rear case for this purpose. Do not attempt to wedge the halves apart.

10. Remove the thrust bearing and races from the front output shaft. Note their relationships so that these parts may be reinstalled properly.
11. Remove the oil pump from the rear output shaft, noting its position for reassembly.
12. Remove the rear output shaft from the viscous coupling.
13. Remove the pilot bearing rollers from the shaft or coupling. Set the rollers aside in a group.
14. Remove the mainshaft O-ring from the end of the shaft.
15. Remove the viscous coupling from the mainshaft and side gear.
16. Lift the front output shaft, sprocket and chain upward. Tilt the front output shaft toward the mainshaft. Slide the chain off of the mainshaft drive sprocket and remove the assembly.
17. Remove the front thrust bearing assembly. The bearing will be positioned on either the front output shaft or the case.
18. Remove the drive chain from the front output shaft and sprocket.
19. Remove the driven sprocket-to-front output shaft snapring. Mark the sprocket and shaft so that they may be reassembled properly. Remove the sprocket from the shaft.
20. Remove the mainshaft, side gear, clutch gear, drive sprocket and spline gear as an assembly.
21. Remove the range fork, rail and clutch sleeve as an assembly. Mark the sleeve and fork so that they may be reassembled properly. Remove the sleeve from the fork.
22. Remove the pin to separate the fork and the rail, if necessary.
23. Inspect the rail, bracket and fork for excessive wear, scoring, distortion, etc. Replace any part which is damaged.
24. Slide the rail through the range fork. Install the retaining pin.
25. Remove the mainshaft thrust washer from the input gear. Remove the input gear, thrust bearing and race.
26. Remove the detent ball, spring and bolt.
27. Remove the retaining nut and washers from the range sector shaft. Tap the sector shaft with a plastic mallet to remove it from the case.
28. Remove the O-ring seal and seal retainer from the sector shaft bore in the case.

MAINSHAFT
FRONT
OUTPUT
SHAFT
DRIVE
CHAIN
WOOD
BLOCKS
FRONT
CASE

Removing the front outputshaft, driven sprocket and drive chain

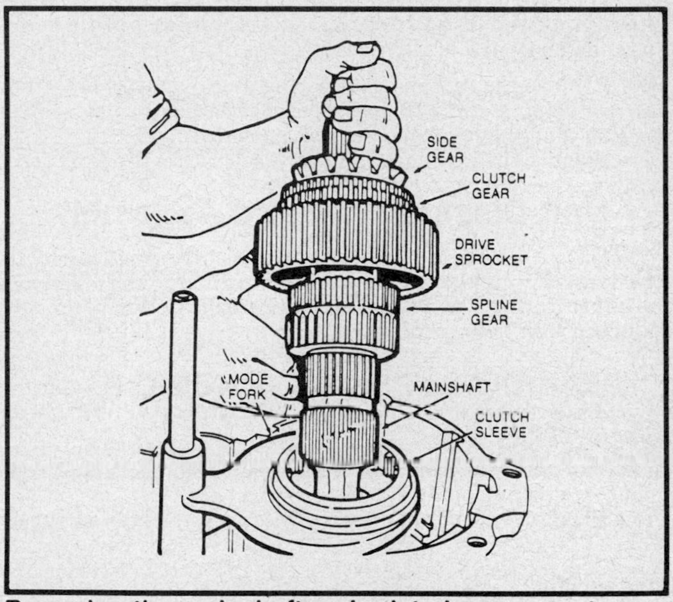
SIDE
GEAR
CLUTCH
GEAR
DRIVE
SPROCKET
SPLINE
GEAR
MODE
FORK
MAINSHAFT
CLUTCH
SLEEVE

Removing the mainshaft and related components as an assembly

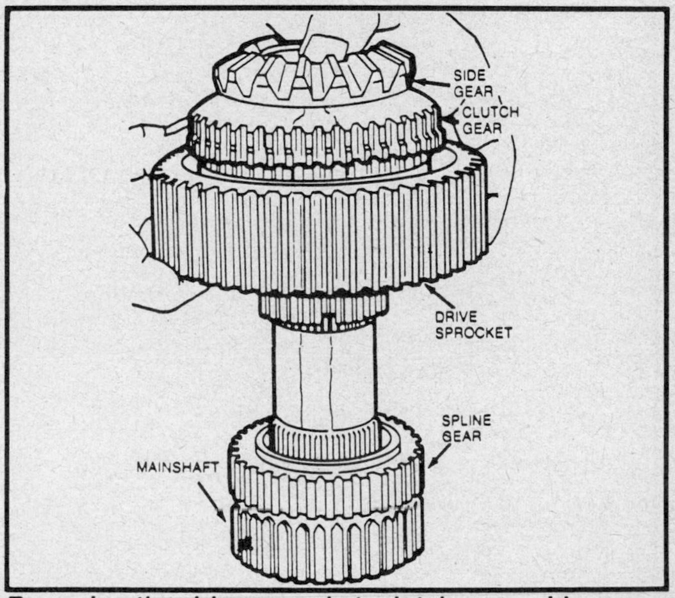
SIDE
GEAR
CLUTCH
GEAR
DRIVE
SPROCKET
SPLINE
GEAR
MAINSHAFT

Removing the drive sprocket, clutch gear, side gear and sprocket carrier

29. Pull the drive sprocket, clutch gear and side gear upward and off of the mainshaft.
30. Remove the 82 needle bearings and 2 bearing spacers from the mainshaft. Note the position of the spacers so that they may be reinstalled properly.
31. Remove the spline gear and thrust washer from the mainshaft.
32. Remove the side gear, clutch gear and thrust washer from the sprocket carrier and sprocket.
33. Remove the clutch gear and thrust washer from the side gear.
34. Remove the sprocket carrier snapring. Remove the drive sprocket from the carrier. Mark the sprocket and the carrier so that they may be reassembled in their proper relationship.
35. Remove the 3 bearing spacers and the sprocket carrier 120 needle bearings from the carrier.

NOTE: Do not intermix the mainshaft needle bearings from Step 30 with the sprocket carrier needle bearings, as they are of different sizes.

36. Remove the rear output bearing and rear yoke seal from the rear retainer. Note that a side of the bearing is shielded. The bearing must be reinstalled in the same direction.
37. Remove the input gear and yoke seals from the front case.

Unit Disassembly and Assembly

NOTE: All bearings must be correctly positioned in the transfer case to avoid oil feed hole blockage. Always be certain that the feed holes are not blocked after any bearing has been replaced.

REAR OUTPUT SHAFT BEARING RACE

Disassembly

1. Pull the race from the transfer case bore, using a slide hammer and an appropriate adapter.
2. Using a small pry bar, carefully pry out the rear output lip seal.

Assembly

1. Install a new output lip seal.
2. Carefully drive a new bearing race into place, using a bearing driver.
3. Remove the tool and make certain that the oil feed hole is clear.

FRONT OUTPUT SHAFT FRONT BEARING

This bearing may be removed and installed with a bearing driver. The driver must contact the bearing squarely. Make certain that the oil feed hole is not blocked after the bearing is in place.

FRONT OUTPUT SHAFT REAR BEARING

This bearing may be removed and installed with a bearing driver. The driver must contact the bearing squarely. Make certain that the oil feed hole is not blocked after the bearing is in place. The bearing must be seated flush with the edge of the bore in the case to allow room for the thrust bearing.

INPUT GEAR FRONT AND REAR BEARINGS

1. Drive both bearings out at the same time, using an appropriate bearing driver.
2. Drive the new bearings into place, rear bearing first.
3. After installation, check that the oil feed holes are not blocked and that the bearings are flush with the case bore surface.
4. Carefully drive a new oil seal into place.

MAINSHAFT PILOT BUSHING

1. Position the input gear on an opened vise (bushing facing downward). The vise must be opened enough for the bushing to be clear of the vise jaws when pulled downward.
2. Using a slide hammer-type bushing puller, remove the bushing.
3. Drive the new bushing into place, making sure that the oil feed hole is properly aligned.

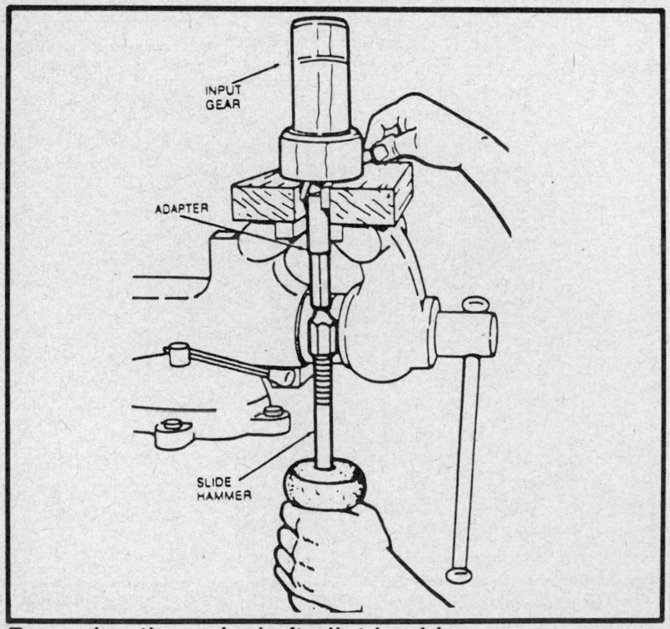
INPUT GEAR
ADAPTER
SLIDE HAMMER
Removing the mainshaft pilot bushing

REAR RETAINER BEARING AND SEAL

1. Remove the bearing, using a brass drift and a hammer. The seal is removed in the same manner.
2. Drive the new bearing into place, making sure that the shielded side of the bearing faces the interior of the transfer case.
3. Carefully drive a new seal into the retainer.

BENCH TESTING THE VISCOUS COUPLING

This torque bias test should be performed while the transfer case is disassembled for any reason, as the viscous coupling is the key to the operation of the 4WD system.
1. Install the clutch gear onto the side gear.
2. Install the clutch gear/side gear assembly into the viscous coupling.
3. Mount the coupling and gear assembly in a vise, with wood blocks between the side gear and vise jaws. Clamp the side gear firmly.
4. Make sure that the clutch gear is firmly engaged in the coupling and install the rear output shaft in the viscous coupling.
5. Install the yoke on the rear output shaft and attach with the retaining nut.
6. Attach a socket (of the same size as the yoke nut) to a torque wrench. With the socket engaged to the yoke nut, rotate the rear output shaft and note the torque reading obtained with the torque wrench. The minimum acceptable rotational torque reading is 25 ft. lbs. If the reading is at or above 25 ft. lbs., the

coupling is good. If a lower torque reading is obtained, the coupling is defective and must be replaced.

7. Remove the yoke nut and yoke from the rear output shaft. Remove the coupling assembly from the vise.

Transfer Case Assembly

NOTE: All parts should be lubricated prior to assembly, with either the specified lubricant Dexron®II automatic transmission fluid or petroleum jelly if stated within the procedure. Do not use any type of heavy grease (i.e. chassis lubricant) during assembly of the transfer case, as lubricants of this nature can block oil passages.

1. Install new yoke oil seals.
2. Install a new O-ring and retainer into the range sector shaft bore of the case.
3. Install the range sector and locknut on the sector shaft. Install the O-ring seal, retainer, range lever, washer and locknut on the sector shaft.
4. Torque the sector shaft locknut to 17 ft. lbs.
5. Install the thrust bearing and race on the input gear. Install the gear into the front of the case.
6. Install the mainshaft thrust washer into the input gear.
7. Assemble the range fork, rail and clutch sleeve. Install the assembly in to the case. Make certain that the rail is fully seated in the case bore.

NOTE: The rail bore of the front case must be perfectly dry. A small amount of oil in the bore will prevent proper seating of the rail.

8. Install a thrust washer and a new O-ring on the mainshaft.
9. Coat the mainshaft needle bearing surface with petroleum jelly and install the needle bearings and spacers in the following order:
   a. Install the short bearing spacer on the shaft.
   b. Install the first 41 needle bearings.
   c. Install the long bearing spacer.
   d. Install the remaining 41 needle bearings.

When installing the spacers, be careful not to disturb the needle bearings. If necessary, use additional petroleum jelly to hold the bearings in place.

10. Install the splined gear on the mainshaft, being careful not to disturb the bearings.
11. Install the sprocket carrier in the drive sprocket, being sure to align the carrier-to-sprocket reference marks made during disassembly.

NOTE: The tapered carrier teeth must be positioned on the same side as the deep races of the drive sprocket.

12. Install the sprocket carrier snaprings.
13. Install the sprocket carrier needle bearings and spacers in the following manner.
   a. Coat both the sprocket carrier recess and the needle bearings with petroleum jelly.
   b. Install the center spacer.
   c. Install 60 needle bearings into each end of the sprocket carrier.
   d. Install the remaining 2 spacers, 1 at each side of the carrier.
14. Install the sprocket carrier and drive sprocket assembly onto the mainshaft, being careful not to disturb the bearings. Note that the recessed side of the drive sprocket must face upward.
15. Position the clutch gear thrust washer on the thrust surface of the sprocket carrier.
16. Install the clutch gear on the side gear, with the tapered edge of the clutch gear facing the side gear teeth.
17. Install the side gear and clutch gear assembly onto the mainshaft, being careful not to disturb the bearings. The side gear must be fully seated in the sprocket carrier.

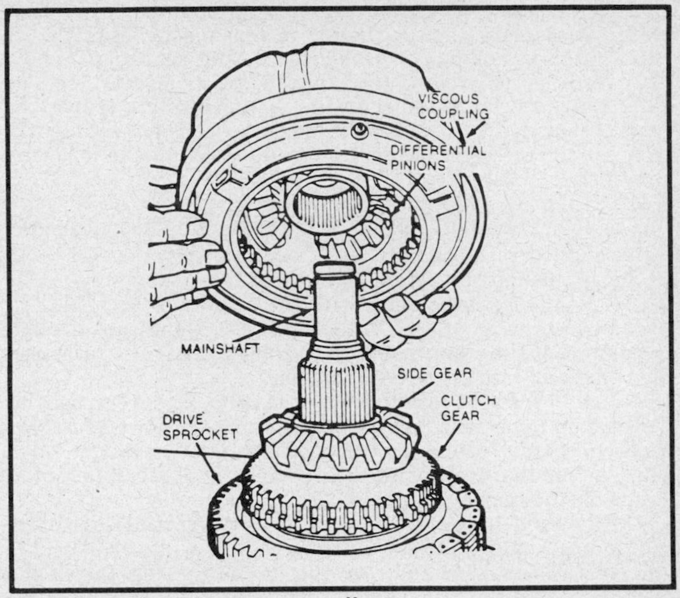
VISCOUS COUPLING
DIFFERENTIAL PINIONS
MAINSHAFT
SIDE GEAR
CLUTCH GEAR
DRIVE SPROCKET

Installing the viscous coupling

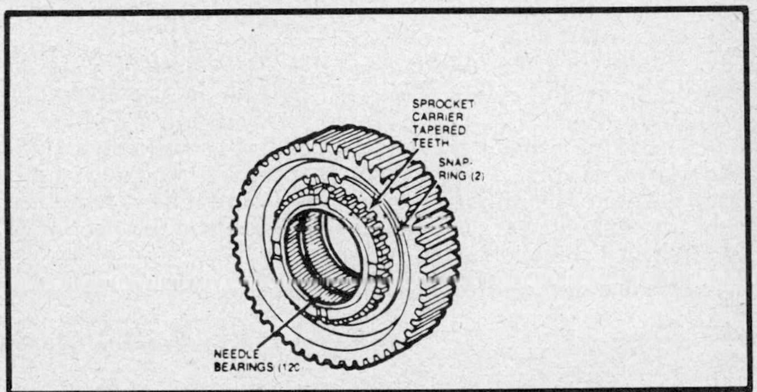
SPROCKET CARRIER TAPERED TEETH
SNAP. RING (2)
NEEDLE BEARINGS (120

Sprocket carrier and drive sprocket assembly

18. Install the mainshaft and gear assembly into the case, being sure that the mainshaft is fully seated in the input gear.
19. Install the driven sprocket on the front output shaft, being sure to align the sprocket-to-shaft reference marks which were made during disassembly. Install the sprocket snapring.
20. Install the thick front thrust bearing race into the case, followed by the bearing and the thin race.
21. Install the drive chain on the driven sprocket.
22. Raise and tilt the driven sprocket and chain in order to attach the opposite end of the chain to the drive sprocket.
23. Align the front output shaft with the bore of the case and install the shaft. Make sure that the front shaft thrust bearing assembly is fully seated in the case.
24. Install the thin race of the front output shaft rear thrust bearing, followed by the bearing and the thick rear thrust bearing race.
25. Install the viscous coupling on the side gear and clutch gear. The coupling must be fully seated on the clutch gear, the clutch gear must be flush with the coupling and the gear teeth should not be visible.
26. Coat the pilot bearing surface of the mainshaft and all of the pilot roller bearings with petroleum jelly. Install the pilot roller bearings on the shaft, using additional petroleum jelly to hold the bearings in place, if necessary.
27. Install the rear output shaft on the mainshaft and into the viscous coupling. Be careful not to disturb the bearings during

shaft installation. The shaft must be fully seated in the coupling; if necessary, tap the shaft with a plastic mallet to seat it.

28. Install the oil pump on the rear output shaft.

29. Install a new rear output shaft bearing oil seal.

30. Apply a bead of sealer to the mating surface of the rear case. If removed, reinstall the case magnet. If the rear case will not seat completely into the front case, check for the following conditions:

a. Oil present in the range fork rail bore.

b. Rear thrust bearing assembly of the front output shaft is not aligned with the rear case.

c. Mainshaft not completely seated.

d. Rear case not aligned with the oil pump.

31. Install the front case to the rear case, being sure to align the dowels at the front case with the bolt holes of the rear case. Seat the rear case onto the front case.

32. Install the rear case-to-front case bolts. Be sure to use flat washers on the bolts at the case ends where the alignment dowels are located. Torque the bolts to 23 ft. lbs.

33. Install the speedometer drive gear and the differential shims on the rear output shaft.

34. Align and temporarily install the rear retainer. Tighten, but do not final-torque the bolts.

35. Install the front and rear output shaft yokes. Install the original yoke nuts, finger-tight only.

36. Mount a dial indicator on the rear retainer so that the stylus contacts the rear yoke nut. The stylus must be in line with the rear output shaft.

37. Rotate the front output shaft 10–20 revolutions. Zero the dial indicator and rotate the front shaft an additional revolution. Note the dial indicator reading which should be 0.002–0.010 in. If the endplay is correct, proceed to the next step. If the endplay is not correct, remove the rear retainer and add or subtract differential shims as required. Reinstall the rear retainer, yoke and nut. Recheck the endplay. Repeat the adjustment procedure until the endplay is correct.

38. Remove the front and rear yokes. Discard the original yoke nuts.

39. Remove the rear retainer. Apply sealer to the retainer mating surface and all of the retainer bolt threads. Install the retainer and the bolts. Torque the retainer bolts to 23 ft. lbs.

40. Install the front and rear yokes, using new sealing washers and yoke nuts. Tighten the yoke nuts to 120 ft. lbs.

41. Install the detent ball, spring and bolt. Apply sealer to the bolt threads and tighten the bolt to 23 ft. lbs.

42. Install the drain plug and washer.

43. Fill the transfer case with the 3 qts. of Dexron®II automatic transmission fluid.

44. Install the fill plug and washer, tighten both the drain and fill plugs to 25 ft. lbs.

45. If removed, install the plug and washer in the front case. Tighten to 18 ft. lbs.

46. Install the shift motor and bracket.

47. Install the transfer case.

TORQUE SPECIFICATIONS

Component	ft. lbs.	Nm
Detent retainer bolt	20–25	27–34
Drain and fill plugs	15–20	20–27
Front and rear yoke nuts	90–130	122–176
Indicator switch	15–20	20–27
Operating lever locknut	14–20	19–27
Rear case-to-front case bolts—all	20–25	27–34
Rear retainer bolts	20–25	27–34
Skid plate bolts	25–35	34–47

NEW PROCESS 205 TRANSFER CASE

Application

1984–89 Dodge/Plymouth Pick Up/Ramcharger
1984–89 General Motors Blazer/Jimmy, Suburban/Pick Up

General Description

The New Process model 205 transfer case is a 2 speed gearbox mounted between the main transmission and the rear axle. The gearbox transmits power from the transmission and engine to the front and rear driving axles. It requires 2.6 qts. of Dexron®II automatic transmission fluid.

CHILTON'S THREE C's TRANSFER CASE DIAGNOSIS

Condition	Cause	Correction
Excessive noise	a) Lubricant level-low b) Worn or damaged bearings c) Misalignment of drive shafts or universal joints d) Yoke bolts loose e) Loose adapter bolts	a) Fill as required b) Replace c) Align d) Torque to specs. e) Torque to specs.
Shifter lever difficult to move	a) Binding inside transfer case	a) Repair as required
Shifter lever disengages from position	a) Gears worn or damaged b) Shift rod bent c) Missing detent ball or spring	a) Replace b) Replace c) Replace
Lubricant leaking	a) Excessive lubricant in case b) Leaking seals or gaskets c) Loose bolts d) Scored yoke in seal contact area	a) Adjust level b) Replace c) Tighten d) Refinish or replace

Removal and Installation

TRANSFER CASE REMOVAL

1. Disconnect the negative battery cable, raise and safely support the vehicle..
2. Remove the skid plate, if equipped. Drain the lubricant from the transfer case. Remove lower front output case bolt to drain fluid, on transfer cases not equipped with a drain plug.
3. Mark the transfer case front and rear output shaft yokes and driveshafts for assembly alignment reference.
4. Disconnect the speedometer cable or speed sensor and indicator switch wires.
5. Disconnect the shift lever link from operating lever.
6. Pace and support stand under transmission and remove the rear crossmember.
7. Disconnect the front and rear driveshaft shafts at the transfer case yokes. Secure shafts to frame rails with wire. Do not allow shafts to hang.
8. Disconnect the parking brakel cable guide from the pivot located on the right frame rail, if necessary.
9. Remove the bolts attaching exhaust pipe support bracket to transfer case, if necessary.
10. Remove the transfer case-to-transmission bolts.
11. Move the transfer case assembly rearward until free of the transmission output shaft and remove assembly.
12. Remove all gasket material from the rear of the transmission adapter housing.

TRANSFER CASE INSTALLATION

1. Place transfer case on a transmission jack and install onto transmission.

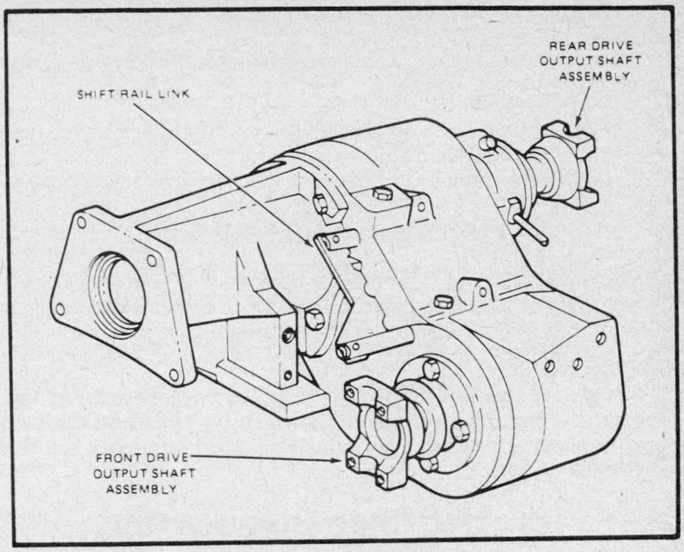
REAR DRIVE
OUTPUT SHAFT
ASSEMBLY
SHIFT RAIL LINK
FRONT DRIVE
OUTPUT SHAFT
ASSEMBLY

Front view of a model 205 transfer case

NOTE: Do not install any transfer case attaching bolts until the transfer case is completely seated against the transmission.

2. Torque the transfer case attaching bolts to 40 ft. lbs. (54 Nm). Torque the shift lever locknut to 18 ft. lbs. (24 Nm).
3. Connect the shift rods.

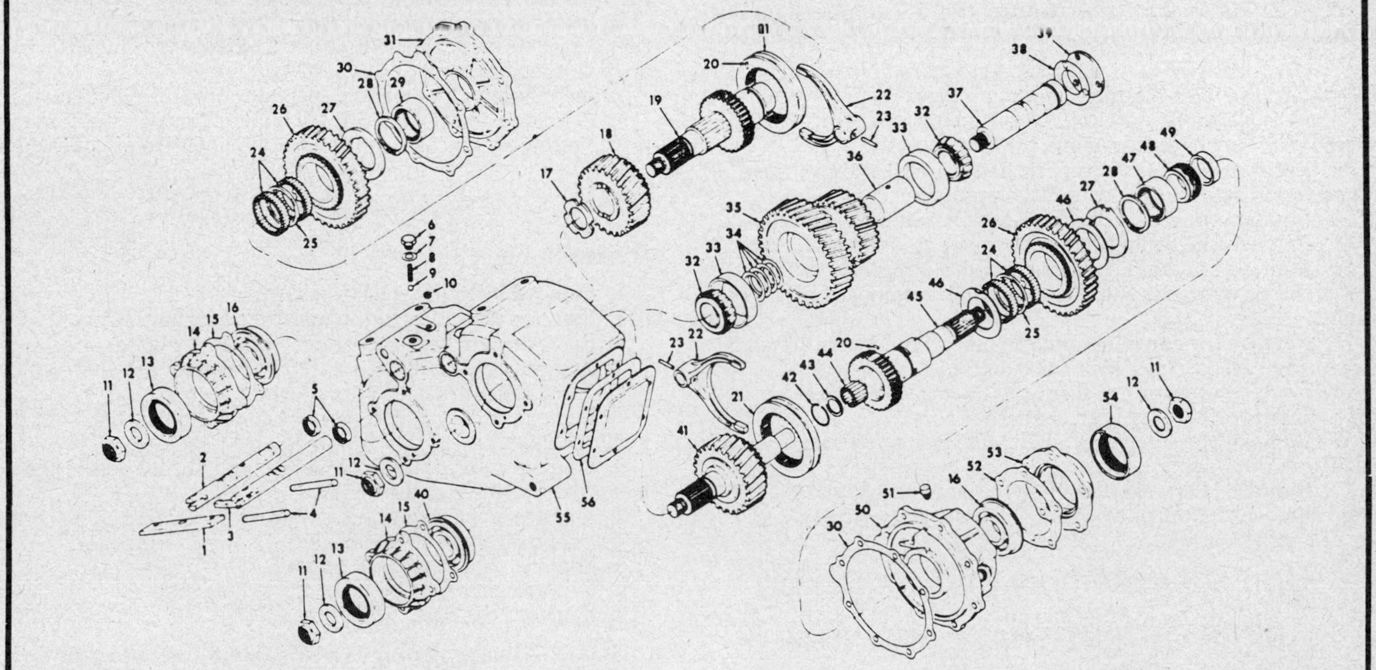

1. Shift lever link
2. Bar
3. Bar
4. Plunger
5. Seal
6. Screw
7. Gasket
8. Spring
9. Ball
10. Plug
11. Nut
12. Washer
13. Seal
14. Retainer
15. Gasket
16. Bearing
17. Washer
18. Gear
19. Shaft
20. Pin
21. Clutch
22. Fork
23. Pin
24. Bearing
25. Spacer
26. Gear
27. Washer
28. Ring
29. Bearing
30. Gasket
31. Retainer
32. Cone
33. Cup
34. Shim set
35. Gear
36. Spacer
37. Shaft
38. Gasket
39. Cover
40. Bearing
41. Shaft
42. Ring
43. Washer
44. Bearing
45. Gear
46. Washer
47. Bearing
48. Gear
49. Spacer
50. Retainer
51. Breather
52. Gasket
53. Retainer
54. Seal
55. Case
56. Gasket

Exploded view of a new process 205 transfer case

4. Install the front and rear driveshafts, making certain to align marks.
5. Install speedometer cable or speed sensor and connect any electrical wiring.
6. Install the skid plate, if removed.
7. Fill the transfer case with to fill hole with Dexron®II type automatic transmission fluid.
8. Lower the vehicle and connect the negative battery cable.

Before Disassembly

Clean the exterior of the transfer case assembly before any attempt is made to disassemble it, in order to prevent dirt or other foreign materials from entering the transfer case assembly or its internal parts.

NOTE: If steam cleaning is done to the exterior of the transfer case, immediate disassembly should be done to avoid rusting, caused by condensation forming on the internal parts.

Transfer Case Disassembly

1. Clean the exterior of the case.
2. Remove the nuts from the universal joint flanges.
3. Remove the front output shaft rear bearing retainer, front bearing retainer and drive flange.
4. Tap the front output shaft assembly from the case with a soft hammer. Remove the sliding clutch, front output high gear, washer and bearing from the case.
5. Remove the rear output shaft housing attaching bolts and remove the housing, output shaft, bearing retainer and speedometer gear.
6. Slide the rear output shaft from the housing.

NOTE: Be careful not to lose the 15 needle bearings that will be loose when the rear output shaft is removed.

7. Drive the two ¼ in. shift rail pin access hole plugs into the transfer case with a punch and hammer.
8. Remove the 2 shift rail detent nuts and springs from the case. Use a magnet to remove the detent balls.
9. Position both shift rails in neutral and remove the shift fork retaining roll pins with a long punch.
10. Remove the clevis pin from a shift rail and rail link.
11. Remove the range shift rail, then the 4WD shift rail.
12. Remove the shift forks and and sliding clutch from the case. Remove the input shaft bearing retainer, bearing and shaft.
13. Remove the cup plugs and rail pins, if they were driven out, from the case.
14. Remove the locknut from the idler gear shaft.
15. Remove the idler gear shaft rear cover.
16. Remove the idler gear shaft, using a soft hammer and a drift.
17. Roll the idler gear assembly to the front output shaft hole and remove the assembly from the case.

Unit Disassembly and Assembly

REAR OUTPUT SHAFT AND YOKE

Disassembly

1. Loosen rear output shaft yoke nut.
2. Remove shaft housing bolts, then remove the housing and retainer assembly.
3. Remove retaining nut and yoke from the shaft, then remove the shaft assembly.
4. Remove and discard snapring.
5. Remove thrust washer and pin.
6. Remove tanged bronze washer. Remove gear needle bearings, spacer and second row of needle bearings.
7. Remove tanged bronze thrust washer.
8. Remove pilot rollers, retainer ring and washer.
9. Remove oil seal retainer, ball bearing, speedometer gear and spacer. Discard gaskets.
10. Press out bearing.
11. Remove oil seal from the retainer.

Assembly

1. Install 2 rows of needle bearings into the output low gear, retaining them with grease.

NOTE: Each row consists of 32 needle bearings and the 2 rows are separated by a spacer.

2. Install thrust washer (with tang down in clutch gear groove) onto the rear output shaft.
3. Install output low gear onto shaft with clutch teeth facing downward.
4. Install thrust washer over gear with tab pointing up and away. Install washer pin.
5. Install large thrust washer over shaft and pin. Turn washer until tab fits into slot located approximately 90 degrees away from pin.
6. Install snapring and measure shaft endplay. If endplay is not 0.002–0.027 adjust as necessary.
7. Grease pilot bore and install needle bearings.

NOTE: There are 15 pilot needle bearings.

8. Install thrust washer and new snapring in pilot bore.
9. Press new bearing into retainer housing.
10. Install housing on output shaft assembly.
11. Install spacer and speedometer gear. Install rear bearing.
12. Install rear bearing retainer seal.
13. Install bearing retainer assembly on housing, using additional gaskets to achieve specified clearance. Torque attaching bolts to specifications.
14. Install yoke, washer and locknut on output shaft.
15. Position range rail in HI, then install output shaft and retainer assembly on case. Torque housing bolts to specifications.

FRONT OUTPUT SHAFT

Disassembly

1. Remove locknut, washer and yoke.
2. Remove attaching bolts and front bearing retainer.
3. Remove rear bearing retainer attaching bolts.
4. Tap output shaft with a soft-faced hammer and remove shaft, gear assembly and rear bearing retainer.
5. Remove sliding clutch, gear, washer and bearing from output high gear.
6. Remove sliding clutch from the high output gear; then remove gear, washer and bearing.
7. Remove gear retaining snapring from the shaft, using large snapring picks. Discard ring.
8. Remove thrust washer and pin.
9. Remove gear, needle bearings and spacer.
10. Replace rear bearing, if necessary.

NOTE: Always replace the bearing and retainer as an assembly. Do not try to press a new bearing into an old retainer.

Removal

1. Install 2 rows of needle bearings in the front low output gear and retain with grease.

NOTE: Each row consists of 32 needle bearings and the 2 rows are separated by a spacer.

2. Position front output shaft in a soft-jawed vise, with spline end down. Place front low gear over shaft with clutch gear facing down; then install thrust washer pin, thrust washer and new snapring. Position snapring gap opposite the thrust washer pin.
3. Place front drive high gear and washer in case. Install sliding clutch in the shift fork, then put fork and rail into 4WD-HI position, meshing front drive high gear and clutch teeth.
4. Align washer, high gear and sliding clutch and bearing bore. Insert front output shaft and low gear assembly through the high gear assembly.
5. Install front output bearing and retainer with a new seal in the case.
6. Clean and grease rollers in front output rear bearing retainer. Install on case with a single gasket and bolts coated with sealant. Torque bolts to specifications.
7. Install front output yoke, washer and locknut. Torque locknut to specifications.

SHIFT RAILS AND FORKS

Disassembly

1. Remove the 2 poppet nuts, springs. Remove the poppet balls, using a magnet.
2. Remove cup plugs on top of case, using a ¼ in. punch.
3. Position both shift rails in neutral, then remove fork pins with a long handled screw extractor.
4. Remove clevis pins and shift rail link.
5. Lower the upper rail, then the lower rail.
6. Remove shift forks and sliding clutch.
7. Remove the front output high gear, washer and bearing. Remove the shift rail cup plugs.

Assembly

1. Press the 2 rail seals into the case.

NOTE: Install seals with metal lip outward.

2. Install interlock pins from inside case.
3. Insert slotted end of front output drive shift rail (with poppet notches up) into back of case.
4. While pushing rail through to neutral position, install shift fork (long end inward).
5. Install input shaft and bearing into case.
6. Install end of range rail (with poppet notches up) into front of case.
7. Install sliding clutch on fork, then place over input shaft in case.
8. Push range rail, while engaging sliding clutch and fork, through to neutral position.
9. Drive new lockpins into forks through holes at top of case.

NOTE: Tilt case on power take-off opening to install range rail lockpin.

IDLER GEAR

Disassembly

1. Remove idler gear shaft nut.
2. Remove rear cover.
3. Tap out idler gear shaft, using a soft-faced hammer and a drift approximately the same diameter as the shaft.
4. Remove idler gear through the front output shaft hole.
5. Remove the 2 bearing cups from the idler gear.

Assembly

1. Press the bearing cups in the idler gear.
2. Assemble the 2 bearing cones, spacer, shims and idler gear on a dummy shaft, with bore facing up. Endplay should not be greater than 0.002 in..
3. Install idler gear assembly (with dummy shaft) into the case, large end first, through the front output shaft bore.
4. Install idler shaft from large bore side, driving it through with a soft-faced hammer or mallet.
5. Install washer and new locknut. Check for free rotation and measure endplay. Torque locknut to specifications.
6. Install idler shaft cover and new gasket. Torque cover bolts to specifications.

NOTE: Flat side of cover must be positioned towards front output shaft rear cover.

Transfer Case Assembly

1. Assemble the idler shaft gears, bearings, spacer and shims, and bearings on a dummy shaft tool and install the assembly into the case through the front output shaft bore, large end first.
2. Install the idler shaft from the large bore side, using a soft-faced hammer to drive it through the bearings, spacer, gears and shims.
3. Install a washer and new locknut on the end of the idler shaft. Check to make sure the idler gear rotates freely. Tighten the locknut to specification.
4. Install the idler shaft cover with a new gasket so the flat side faces the rear bearing retainer of the front output shaft. Install and tighten the retaining screws to the proper torque.
5. Install the interlock pins into the interlock bore through the front of the output shaft opening.
6. Start the 4WD shift rail into the front of the case, solid end of the rail first, with the detent notches facing up.
7. Position the shift fork onto the shift rail with the long end facing inward. Push the rail through the fork and into the neutral position.
8. Position the input shaft and bearing in the case.
9. Start the range shift rail into the case from the front, with the detent notches facing up.
10. Position the sliding clutch to the shift fork. Place the sliding clutch on the input shaft and align the fork with the shift rail. Push the rail through the fork into the neutral position.
11. Install the roll pins that lock the shift forks to the shift rails with a long punch.
12. Position the front wheel drive high gear and its thrust washer in the case. Position the sliding clutch in the shift fork. Shift the rail and fork into the front wheel drive 4WD-HI position, while at the same time, meshing the clutch with the mating teeth on the front wheel drive high gear.
13. Align the thrust washer, high gear and sliding clutch with the bearing bore in the case and insert the front output shaft and low gear into the high gear assembly.
14. Install a new seal in the front bearing retainer of the front output shaft. Install the bearing and retainer and new gasket in the case. Tighten the bearing retainer cap screws to the proper torque.
15. Lubricate the roller bearing in the front output shaft rear bearing retainer, which is the aluminum cover. Install it over the front output shaft and to the case. Install and tighten the retaining screws to the proper torque.
16. Move the range shift rail to the HI position and install the rear output shaft and retainer assembly to the housing and input shaft. Use additional new gaskets, as required, to adjust the clearance on the input shaft pilot. Install the rear output shaft housing retaining bolts and tighten to specification.
17. Using a punch and sealing compound, install the shift rail pin access plugs.
18. Install the fill and drain plugs and the cross-link clevis pin.
19. Install power take-off cover and gasket. Torque attaching bolts to specifications.
20. Install cup plugs at rail pin holes and seal the cup plugs.

Specifications

ENDPLAY SPECIFICATIONS

	in.
Idler gear	0.000–0.002
Rear output shaft	0.002–0.027

TORQUE SPECIFICATIONS

	ft. lbs.
Idler shaft locknut	150
Idler shaft cover	20
Front output shaft front bearing retainer	30–35
Front output shaft yoke locknut	130–150
Rear output shaft bearing retainer and housing	30–35

TORQUE SPECIFICATIONS

	ft. lbs.
Rear output shaft yoke locknut	130–150
Power take off cover	15
Front output shaft rear bearing retainer	30–35
Filler and drain plugs	30
Case to frame	130
Case to adapter	25
Adapter mount	75
Case bracket to frame—Upper	30
Case bracket to frame—Lower	65
Adapter to transmission—Manual transmission	30–35
Adapter to transmission—Automatic transmission	30–35

NEW PROCESS 207 TRANSFER CASE

Application

1984–88 General Motors – S-10/S-15
1986–87 Dodge – Dakota
1986–88 Jeep – Comanche/Wrangler

General Description

The 207 transfer case is an aluminum case, chain drive, 4 position unit providing 4WD **HI** and **LO** ranges, a 2WD **HI** range and a **N** position. It is a parttime 4WD unit. Torque input in 4WD **HI** and **LO** ranges is undifferentiated. The range positions on the 207 transfer case are selected by a floor mounted gearshift lever. Dexron®II automatic transmission fluid, or equivalent, is the recommend lubricant.

The 207 case is a 2 piece aluminum case containing front and rear output shafts, 2 drive sprockets, a shift mechanism and a planetary gear assembly. The drive sprockets are connected and operated by the drive chain. The planetary assembly which consists of a 3 pinion carrier and an annulus gear provide the 4WD drive **LO** range when engaged.

On Vehicle Service

LINKAGE

Adjustment

1. Place the transfer case in **2WD**.
2. Insert a ⅛ in. (3mm) spacer between the gate and lever.
3. Hold the lever in this position.
4. Place the transfer case lever in **2WD**.
5. Adjust the link adjuster on the shifter shaft, to provide a free pin at the transfer case outer lever.

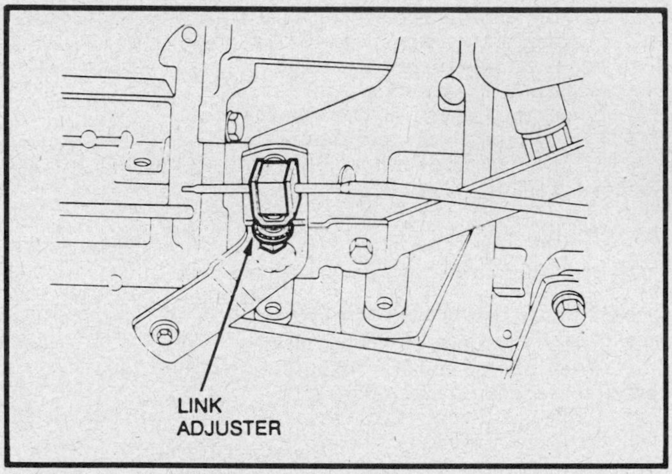
LINK
ADJUSTER

Linkage adjustment – Model 207

Removal and Installation

TRANSFER CASE REMOVAL

GENERAL MOTORS S-SERIES AND JEEP WRANGLER/COMANCHE

1. Disconnect the negative battery cable. Shift the transfer case into the **4WD-HI** position.
2. Raise and support the vehicle safely. If equipped, remove the skid plate. Drain the fluid from the transfer case.
3. Matchmark the transfer case front output shaft yoke and driveshaft for reassembly. Disconnect the driveshaft from the transfer case.
4. Matchmark the rear axle yoke and the driveshaft for reassembly. Remove the driveshaft.
5. Disconnect the speedometer cable. Disconnect the vacuum harness and all electrical connections at the transfer case. Remove the catalytic converter hanger bolts at the converter assembly.
6. Raise the transmission and transfer case assembly. Remove the transmission mount retaining bolts. Remove the mount and the catalytic converter hanger. Lower the transmission and transfer case assembly.
7. Properly support the transfer case assembly. Remove the transfer case retaining bolts.
8. On vehicles equipped with automatic transmission, it will be necessary to remove the shift lever bracket mounting bolts from the transfer case adapter in order to remove the upper left transfer case retaining bolt.
9. Separate the transfer case from its mounting and remove it from the vehicle.

DODGE DAKOTA

1. Disconnect the negative battery cable, raise and safely support the vehicle..

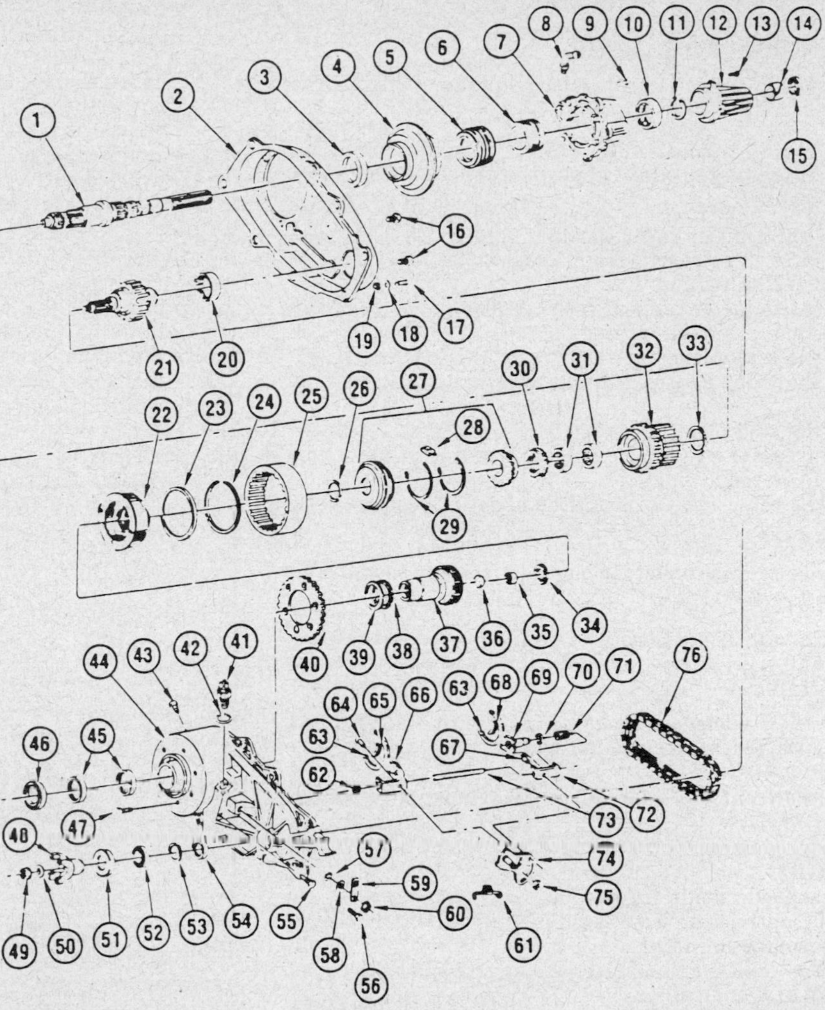

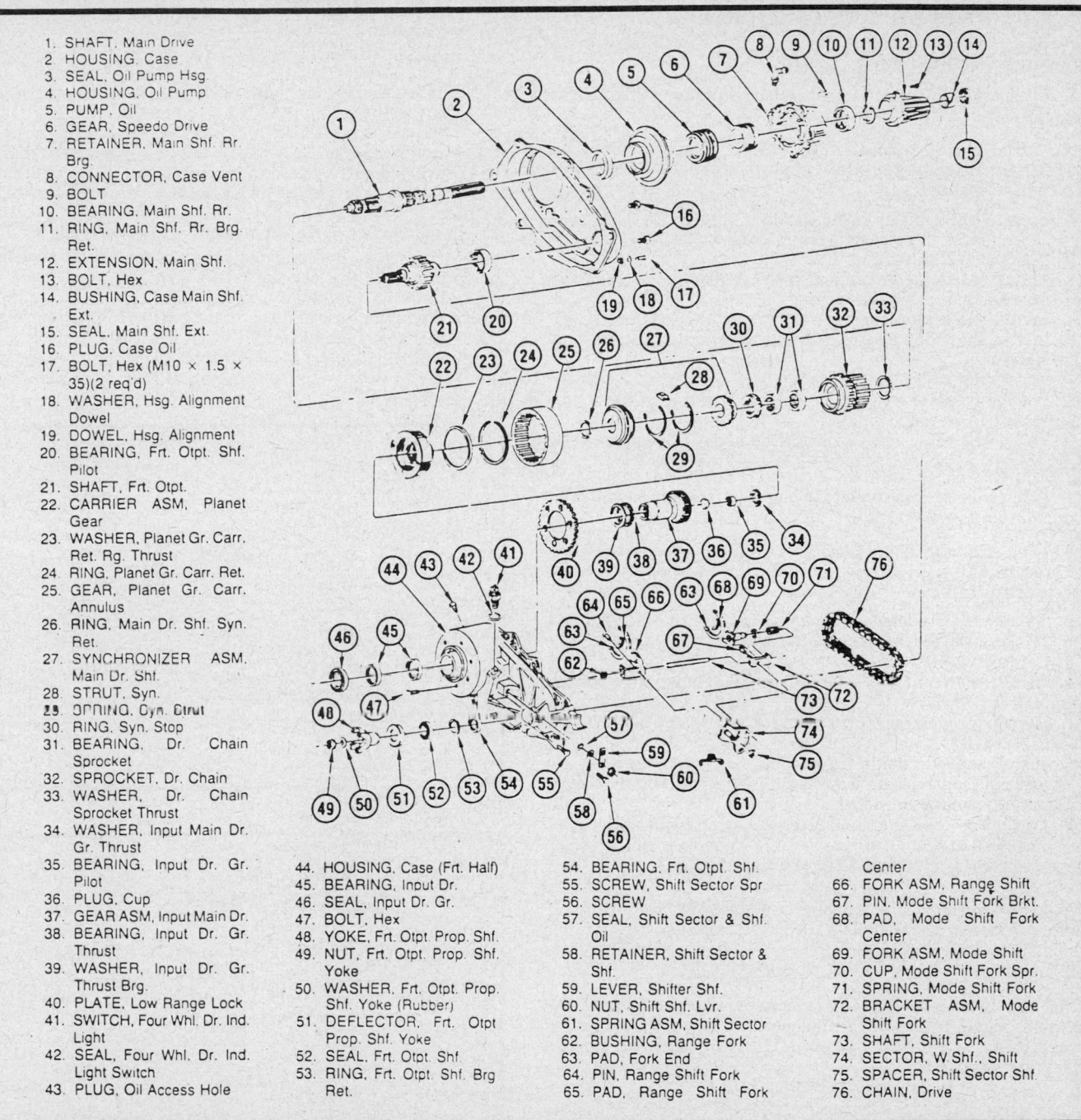

1. SHAFT, Main Drive
2. HOUSING, Case
3. SEAL, Oil Pump Hsg.
4. HOUSING, Oil Pump
5. PUMP, Oil
6. GEAR, Speedo Drive
7. RETAINER, Main Shf. Rr. Brg.
8. CONNECTOR, Case Vent
9. BOLT
10. BEARING, Main Shf. Rr.
11. RING, Main Shf. Rr. Brg. Ret.
12. EXTENSION, Main Shf.
13. BOLT, Hex
14. BUSHING, Case Main Shf. Ext.
15. SEAL, Main Shf. Ext.
16. PLUG, Case Oil
17. BOLT, Hex (M10 × 1.5 × 35)(2 req'd)
18. WASHER, Hsg. Alignment Dowel
19. DOWEL, Hsg. Alignment
20. BEARING, Frt. Otpt. Shf. Pilot
21. SHAFT, Frt. Otpt.
22. CARRIER ASM, Planet Gear
23. WASHER, Planet Gr. Carr. Ret. Rg. Thrust
24. RING, Planet Gr. Carr. Ret.
25. GEAR, Planet Gr. Carr. Annulus
26. RING, Main Dr. Shf. Syn. Ret.
27. SYNCHRONIZER ASM, Main Dr. Shf.
28. STRUT, Syn.
29. SPRING, Syn. Strut
30. RING, Syn. Stop
31. BEARING, Dr. Chain Sprocket
32. SPROCKET, Dr. Chain
33. WASHER, Dr. Chain Sprocket Thrust
34. WASHER, Input Main Dr. Gr. Thrust
35. BEARING, Input Dr. Gr. Pilot
36. PLUG, Cup
37. GEAR ASM, Input Main Dr.
38. BEARING, Input Dr. Gr. Thrust
39. WASHER, Input Dr. Gr. Thrust Brg.
40. PLATE, Low Range Lock
41. SWITCH, Four Whl. Dr. Ind. Light
42. SEAL, Four Whl. Dr. Ind. Light Switch
43. PLUG, Oil Access Hole
44. HOUSING, Case (Frt. Half)
45. BEARING, Input Dr.
46. SEAL, Input Dr. Gr.
47. BOLT, Hex
48. YOKE, Frt. Otpt. Prop. Shf.
49. NUT, Frt. Otpt. Prop. Shf. Yoke
50. WASHER, Frt. Otpt. Prop. Shf. Yoke (Rubber)
51. DEFLECTOR, Frt. Otpt Prop. Shf. Yoke
52. SEAL, Frt. Otpt. Shf.
53. RING, Frt. Otpt. Shf. Brg Ret.
54. BEARING, Frt. Otpt. Shf.
55. SCREW, Shift Sector Spr.
56. SCREW
57. SEAL, Shift Sector & Shf. Oil
58. RETAINER, Shift Sector & Shf.
59. LEVER, Shifter Shf.
60. NUT, Shift Shf. Lvr.
61. SPRING ASM, Shift Sector
62. BUSHING, Range Fork
63. PAD, Fork End
64. PIN, Range Shift Fork
65. PAD, Range Shift Fork Center
66. FORK ASM, Range Shift
67. PIN, Mode Shift Fork Brkt.
68. PAD, Mode Shift Fork Center
69. FORK ASM, Mode Shift
70. CUP, Mode Shift Fork Spr.
71. SPRING, Mode Shift Fork
72. BRACKET ASM, Mode Shift Fork
73. SHAFT, Shift Fork
74. SECTOR, W Shf., Shift
75. SPACER, Shift Sector Shf.
76. CHAIN, Drive

Exploded view of a new process 207 transfer case

2. Remove the skid plate, if equipped. Drain the lubricant from the transfer case. Remove lower front output case bolt to drain fluid, on transfer cases not equipped with a drain plug.
3. Mark the transfer case front and rear output shaft yokes and driveshafts for assembly alignment reference.
4. Disconnect the speedometer cable or speed sensor and indicator switch wires.
5. Disconnect the shift lever link from operating lever.
6. Place and support stand under transmission and remove the rear crossmember.
7. Disconnect the front and rear driveshaft shafts at the transfer case yokes. Secure shafts to frame rails with wire. Do not allow shafts to hang.
8. Disconnect the parking brake cable guide from the pivot located on the right frame rail, if necessary.
9. Remove the bolts attaching exhaust pipe support bracket to transfer case, if necessary.
10. Remove the transfer case-to-transmission bolts.
11. Move the transfer case assembly rearward until free of the transmission output shaft and remove assembly.
12. Remove all gasket material from the rear of the transmission adapter housing.

TRANSFER CASE INSTALLATION

GENERAL MOTORS S-SERIES AND JEEP WRANGLER/COMANCHE

1. Place transfer case on a transmission jack and install onto transmission.

NOTE: Do not install any transfer case attaching bolts until the transfer case is completely seated against the transmission.

2. Torque the transfer case attaching bolts to 25 ft. lbs. (34 Nm). Torque the shift lever locknut to 18 ft. lbs. (24 Nm).
3. Connect the shift rods.
4. Install speedometer cable or speed sensor and connect any vacuum lines and electrical wiring.
5. Install the catalytic converter hanger and assembly.
6. Install the front and rear driveshafts, making certain to align marks.
7. Install the skid plate, if removed.
8. Fill the transfer case with to fill hole with Dexron®II type automatic transmission fluid.
9. Lower the vehicle and connect the negative battery cable.

DODGE DAKOTA

1. Place transfer case on a transmission jack and install onto transmission.

NOTE: Do not install any transfer case attaching bolts until the transfer case is completely seated against the transmission.

2. Torque the transfer case attaching bolts to 40 ft. lbs. (54 Nm). Torque the shift lever locknut to 18 ft. lbs. (24 Nm).
3. Connect the shift rods.
4. Install the front and rear driveshafts, making certain to align marks.
5. Install speedometer cable or speed sensor and connect any electrical wiring.
6. Install the skid plate, if removed.
7. Fill the transfer case with to fill hole with Dexron®II type automatic transmission fluid.
8. Lower the vehicle and connect the negative battery cable.

Before Disassembly

Clean the exterior of the transfer case assembly before any attempt is made to disassemble it, in order to prevent dirt or other foreign materials from entering the transfer case assembly or its internal parts.

NOTE: If steam cleaning is done to the exterior of the transfer case, immediate disassembly should be done to avoid rusting, caused by condensation forming on the internal parts.

Transfer Case Disassembly

1. Remove fill and drain plugs.
2. Remove front yoke. Discard yoke seal washer and yoke nut.
3. Turn transfer case on end and position front case on wood blocks.
4. Shift transfer case to **4 LO**.
5. Remove extension housing attaching bolts. Using a hammer, tap the shoulder on the extension housing to break sealer loose.
6. Remove the snapring for the rear bearing from the mainshaft and discard.
7. Remove the rear retainer attaching bolts. Using a hammer, tap the shoulder on the retainer to break sealer loose.
8. Remove the rear retainer and pump housing from the transfer case.
9. Remove the pump seal from the pump housing and discard.
10. Remove the speedometer drive gear from the mainshaft.
11. Remove the pump gear from the mainshaft.
12. Remove the bolts attaching the rear case to the front case and remove rear case. To separate the case, insert a prybar into the slots casted in the case ends and pry upward. Do not attempt to wedge the case halves apart at any point on the mating surfaces.
13. Remove the front output shaft and drive chain as an assembly. It may be necessary to raise the mainshaft slightly for the output shaft to clear the case.
14. Pull up on the mode fork rail until rail clears range fork and rotate mode fork and rail and remove from transfer case.
15. Pull up on the mainshaft until it separates from the planetary assembly. Remove the mainshaft from the transfer case.
16. Remove the planetary assembly with the range fork from the transfer case.

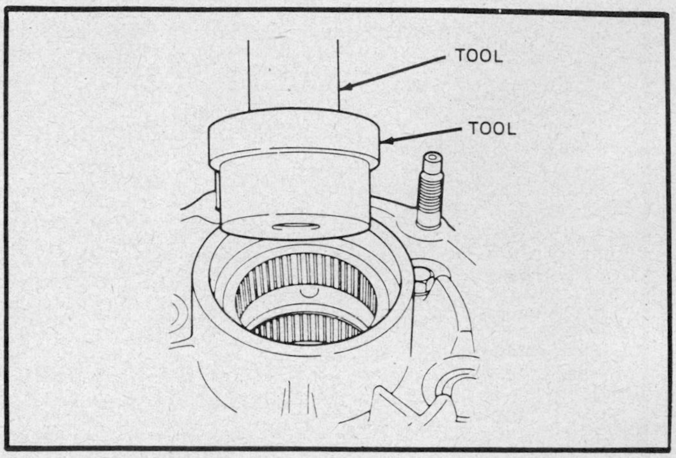
TOOL
TOOL

Input gear removal

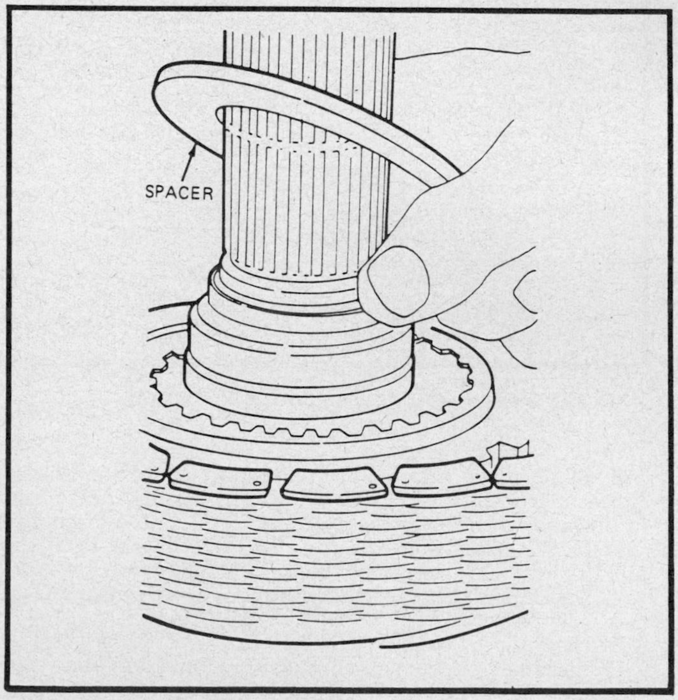
SPACER

Drive sprocket thrust washer

17. Remove the planetary thrust washer, input gear thrust bearing and front thrust washer from the transfer case.
18. Remove the shift sector detent spring and retaining bolt.
19. Remove the shift sector, shaft and spacer from the transfer case.
20. Remove the locking plate retaining bolts and lock plate from the transfer case.
21. Remove the input gear pilot bearing, using J–29369–1 or equivalent with a slide hammer.
22. Remove the front output shaft seal, input shaft seal and the rear extension seal, using a brass drift.
23. Using J–33841 with J–8092 or equivalents, press the 2 caged roller bearings for the front input shaft gear from the transfer case.
24. Using J–29369–2 with J–33367, or a slide hammer, remove the rear bearing for the front output shaft.
25. Using a hammer and drift, remove the rear mainshaft bearing from the rear retainer.
26. Using an awl, remove the snapring retaining the front output shaft bearing. Using a hammer and drift, remove the bearing from the case.
27. Remove the bushing from the extension housing using J–33839 with J–8092 or equivalent. Press bushing from the extension housing.

Unit Disassembly and Assembly

MAINSHAFT

Disassembly

1. Remove the speedometer gear.
2. Using an awl, pry off the pump gear from the mainshaft.
3. Remove the snapring retaining the synchronizer hub from the mainshaft.
4. Using a brass hammer, tap the synchronizer hub from mainshaft.
5. Remove the drive sprocket.
6. Using J–33826 and J–8092 or equivalent, press 2 caged roller bearings from the drive sprocket.
7. Remove synchronizer keys and retaining rings from the synchronizer hub.

Inspection

1. Clean and inspect all parts.
2. Replace any parts if they show evidence of excessive wear, distortion or damage.

Assembly

1. Using J–33828 and J–8092 or equivalent, install the front drive sprocket bearing. Press bearing until tool bottoms. Bearing should be flush with front surface. Reverse tool on J–8092 or equivalent and press rear bearing into sprocket until tool bottoms. The rear bearing should be recessed after installation.
2. Install thrust washer on the mainshaft.
3. Install drive sprocket on the mainshaft.
4. Install blocker ring and synchronizer hub on the mainshaft. Seat hub on main shaft and install a new snapring to retain.
5. Install pump gear on the mainshaft. Tap the gear with a hammer to seat on mainshaft.
6. Install speedometer gear on the mainshaft.

PLANETARY GEAR

Disassembly

1. Remove the snapring retaining the planetary gear in the annulus gear.
2. Remove outer thrust ring and discard.
3. Remove planetary assembly from the annulus gear.
4. Remove inner thrust ring from the planetary assembly and discard.

Inspection

1. Clean and inspect parts.
2. Replace any parts if they show evidence of excessive wear, distortion or damage.

Assembly

1. Install the inner thrust ring on planetary assembly.
2. Install the planetary assembly into the annulus gear.
3. Install the outer thrust ring and then the snapring.

Transfer Case Assembly

All of the bearings used in the transfer case must be correctly positioned to avoid covering the bearing oil feed holes. After installation of bearings, check the bearing position to be sure the feed hole is not obstructed or blocked by a bearing.

1. Install the lock plate in the transfer case. Coat case and lock plate surfaces around bolt holes with Loctite®515 or equivalent.
2. Position the lock plate to the case and align bolt holes in lock plat with case. Install attaching bolts and torque to specification.
3. Install the roller bearings for the input shaft into the transfer case, using J–33830 and J–8092 or equivalent. Press bearings until tool bottoms in bore.
4. Install the front output shaft rear bearing, using J–33832 and J–8092 or equivalent. Press bearing until tool bottoms in case.
5. Install the front output shaft front bearing using J–33833 and J–8092 or equivalent. Press bearing until tool bottoms in bore.
6. Install the snapring that retains the front output shaft bearing in case.
7. Install the front output shaft seal using J–33834 or equivalent.
8. Install the input shaft seal using J–33831 or equivalent.
9. Install spacer on shift sector shaft and install sector in transfer case. Install shift lever and retaining nut. Torque to specification.
10. Install shift sector detent spring and retaining bolt.
11. Install the pilot bearing into the input gear, using J–33829 and J–8092 or equivalent. Press bearing until tool bottom.
12. Install the input gear front thrust bearing and input gear in transfer case.
13. Install the planetary gear thrust washer on the input gear. Position range fork on planetary assembly and install planetary assembly into the transfer case.
14. Install the mainshaft into the transfer case. Make sure the thrust washer is aligned with the input gear and planetary assembly before installing mainshaft.
15. Install mode fork on synchronizer sleeve and rotate until mode fork is aligned with range fork. Slide mode fork rail down through range fork until rail is seated in bore of transfer case.
16. Position drive chain on front output shaft and install chain on drive sprocket. Install front output shaft in the transfer case. It may be necessary to slightly raise the mainshaft to seat the output shaft in the case.
17. Install the magnet into pocket of transfer case.
18. Apply ⅛ in. bead of Loctite®515 or equivalent to the mating surface of the front case. Install rear case on the front case aligning dowel pins. Install bolts and torque to 20–25 ft. lbs. Install the 2 bolts with washers into the dowel pin holes.
19. Install the output bearing into the rear retainer, using J–33833 and J–8092 or equivalent. Press bearing until seated in bore.

20. Install pump seal in pump housing, using J-33835 or equivalent. Apply petroleum jelly to pump housing tabs and install housing in rear retainer.
21. Apply ⅛ in. bead of Loctite®515 or equivalent to mating surface of rear retainer. Align retainer to case and install retaining bolts. Torque bolts to specification 15–20 ft. lbs.
22. Using a new snapring, install snapring on mainshaft. Pull up on mainshaft and seat snapring in its groove.
23. Install bushing in extension housing, using J-33826 and J-8092 or equivalent. Press bushing until tool bottoms in bore.
24. Install a new seal in the extension housing, using J-33843 or equivalent.
25. Apply ⅛ in. bead of Loctite®515 or equivalent to mating surface of extension housing. Align extension housing to the rear retainer and install attaching bolts. Torque bolts to specification 20–25 ft. lbs.
26. Install front yoke on output shaft. Install a new yoke seal washer with a new nut and torque to specification.
27. Install drain plug and torque to specification. Install fill plug.

Specifications

TORQUE SPECIFICATIONS
Model 207

Description	N•m	Ft. Lb.
Bolt Locking Plate to Transfer Case	27-40	20-30
Nut-Front Output Yoke	122-176	90-130
Switch Vacuum	20-34	15-25
Nut-Shift Lever	20-27	15-20
Bolt-Transfer Case	27-34	20-25
Bolt-Rear Retainer	20-27	15-20
Bolt-Extension Housing	27-34	20-25
Bolt-Drain-Fill	40-54	30-40
Bolt-Adapter to Transfer Case	26-40	19-29
Bolt-Shift Bracket	65-85	47-62
Bolt-Shift Lever Pivot	120-140	88-103
Bolt-Shift Lever Adjusting	34-48	25-35

NEW PROCESS 208 TRANSFER CASE

Application

1984–89 General Motors Blazer/Jimmy, Suburban/Pick Up
1984–89 Dodge Pick Up/Ramcharger
1984–87 Ford F150/F250/Bronco
1984–88 Jeep Grand Wagoneer

General Description

The NP208 is a part time unit with a 2 piece aluminum housing. On the front case half, the front output shaft, front input shaft, 4WD indicator switch and shift lever assembly are located. On the rear case half, the rear output shaft, bearing retainer and drain and fill plugs are located. It requires 5-5.5 qts. of Dexron®II automatic transmission fluid.

CHILTON'S THREE C's TRANSFER CASE DIAGNOSIS

Condition	Cause	Correction
Transfer case difficult to shift or will not shift into desired range	a) Vehicle speed too great to permit shifting	a) Stop vehicle and shift into desired range. Or reduce speed to 2–3 mph (3–4 km/h) before attempting to shift
	b) If vehicle was operated for extended period in 4H mode on dry paved surface, driveline torque load may cause difficult shifting	b) Stop vehicle, shift transmission to neutral, shift transfer case to 2H mode and operate vehicle in 2H on dry paved surfaces
	c) Transfer case external shift linkage binding	c) Lubricate or repair or replace linkage, or tighten loose components as necessary
	d) Insufficient or incorrect lubricant	d) Drain and refill to edge of fill hole with Dexron®II only
	e) Internal components binding, worn or damaged	e) Disassemble unit and replace worn or damaged components as necessary
Transfer case noisy in all drive modes	a) Insufficient or incorrect lubricant	a) Drain and refill to edge of fill hole with Dexron®II only. Check for leaks and repair if necessary. Note: If unit is still noisy after drain and refill, disassembly and inspection may be required to locate source of noise
Noisy in—or jumps out of four wheel drive low range	a) Transfer case not completely engaged in 4L position	a) Stop vehicle, shift transfer case in Neutral, then shift back into 4L position
	b) Shift linkage loose or binding	b) Tighten, lubricate or repair linkage as necessary
	c) Range fork cracked, inserts worn, or fork is binding on shift rail	c) Disassemble unit and repair as necessary
	d) Annulus gear or lockplate worn or damaged	d) Disassemble unit and repair as necessary

CHILTON'S THREE C's TRANSFER CASE DIAGNOSIS

Condition	Cause	Correction
Lubricate leaking from output shaft seals or from vent	a) Transfer case overfilled b) Vent closed or restricted c) Output shaft seals damaged or installed incorrectly	a) Drain to correct level b) Clear or replace vent if necessary c) Replace seals. Be sure seal lip faces interior of case when installed. Also be sure yoke seal surfaces are not scored or nicked. Remove scores, nicks with fine sandpaper or replace yoke(s) if necessary
Abnormal tire wear	a) Extended operation on dry hard surface (paved) roads in 4H range	a) Operate in 2H on hard surface (paved) roads

On Vehicle Service

LINKAGE

Adjustment

1. Place transfer case in **4WD-HI**.
2. Push lower shifter lever forward to the **4WD-HI** stop.
3. Install the rod swivel in the shift lever hole.
4. Hang a 0.200 in. thick gauge behind swivel.
5. Tighten rear rod nut against the gauge with the shifter against the **4WD-HI** stop.
6. Remove the gauge and push swivel rearward against the rear rod nut.
7. Tighten the front rod nut against the swivel.
8. Check for proper operation.

Removal and Installation

TRANSFER CASE REMOVAL

GENERAL MOTORS BLAZER/JIMMY, SUBURBAN AND 1982-87 PICK-UP

1. Disconnect the negative battery cable. Shift the transfer case into the **4WD-HI** position.
2. Raise and support the vehicle safely. Drain the fluid from the transfer case. Remove the cotter pin from the shift lever swivel.
3. Matchmark the transfer case front output shaft yoke and driveshaft for reassembly. Disconnect the driveshaft from the transfer case.
4. Matchmark the rear axle yoke and the driveshaft for reassembly. Remove the driveshaft.

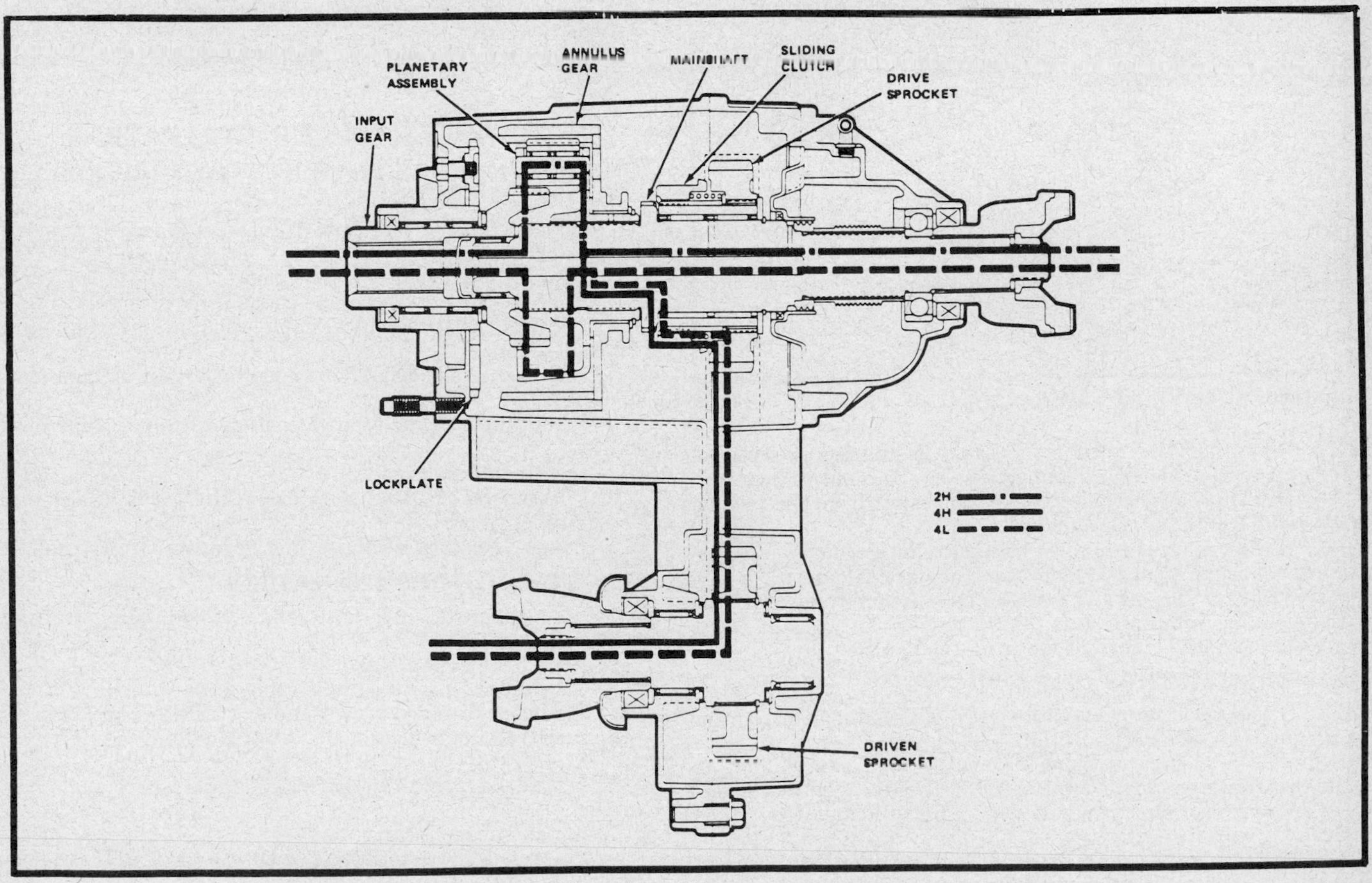
PLANETARY ASSEMBLY
ANNULUS GEAR
SLIDING CLUTCH
DRIVE SPROCKET
INPUT GEAR
LOCKPLATE
2H
4H
4L
DRIVEN SPROCKET

Power flow new process 208 transfer case

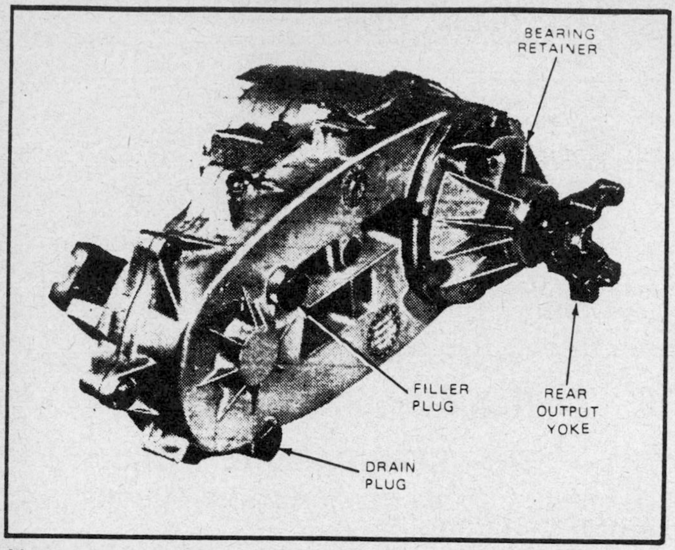
BEARING
RETAINER
FILLER
PLUG
REAR
OUTPUT
YOKE
DRAIN
PLUG

New process 208 rear case view

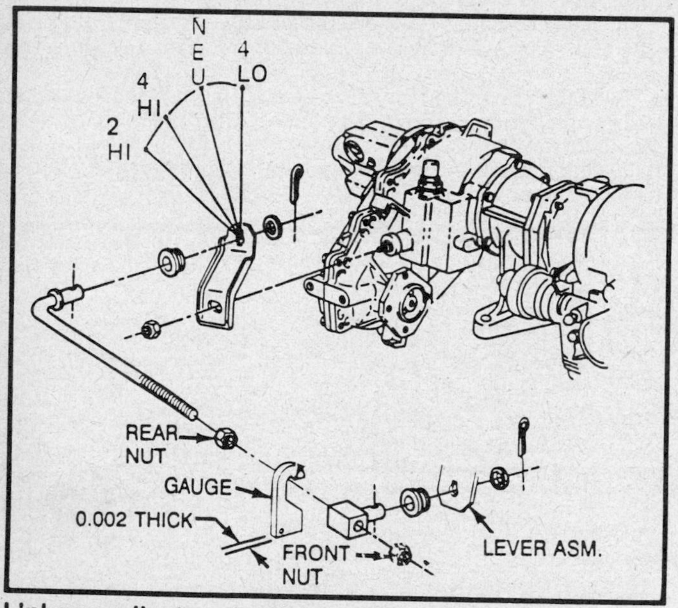
N
E
U
4
LO
4
HI
2
HI
REAR
NUT
GAUGE
0.002 THICK
FRONT
NUT
LEVER ASM.

Linkage adjustment – Model 208

5. Disconnect the speedometer cable. Disconnect the vacuum harness and all electrical connections at the transfer case.
6. Disconnect the parking brake cable guide from the pivot on the right frame rail, as required.
7. If the vehicle is equipped with automatic transmission, remove the right strut rod from the transfer case assembly.
8. Properly support the transfer case assembly. Remove the transfer case retaining bolts.
9. Remove the transfer case from the vehicle.

GENERAL MOTORS 1988–89 PICK-UP

1. Disconnect the negative battery cable. Shift the transfer case into the **4WD-HI** position.
2. Raise and support the vehicle safely. Drain the fluid from the transfer case. Remove the the shift lever swivel. Disconnect the speed sensor electrical wire. Disconnect the indicator switch electrical wire.
3. Matchmark the transfer case front output shaft yoke and driveshaft for reassembly. Disconnect the driveshaft from the transfer case.
4. Matchmark the rear axle yoke and the driveshaft for reassembly. Remove the driveshaft.
5. Disconnect the parking brake cable guide from the pivot on the right frame rail, as required.
6. Properly support the transfer case assembly. Remove the transfer case retaining bolts.
7. Remove the skid shield and plate. Remove the transfer case from the vehicle.

EXCEPT GENERAL MOTORS

These procedures may vary slightly between different manufacturers.

1. Disconnect the negative battery cable. Disconnect the floor covering access cover, if equipped.
2. Raise the vehicle and support it safely. Drain the fluid from the transfer case.
3. Disconnect the 4WD indicator switch wire connector at the transfer case.
4. Disconnect the speedometer driven gear from the transfer case rear bearing retainer.
5. Remove the nut retaining the transmission shift lever assembly to the transfer case.
6. Remove the skid plate from the frame, if so equipped.
7. Remove the heat shield from the frame.
8. Support transfer case with a transmission jack or equivalent.
9. Disconnect the front driveshaft from the front output shaft yoke.
10. Support the transmission with suitable stand.
11. Disconnect the rear driveshaft from the rear output shaft yoke.
12. On some vehicles it may be necessary to disconnect the parking brake cable and exhaust support bracket.
13. Remove the bolts retaining the transfer case to the transmission adapter.
14. Lower the transfer case from the vehicle.

TRANSFER CASE INSTALLATION

GENERAL MOTORS BLAZER/JIMMY, SUBURBAN AND 1982-87 PICK-UP

1. Install transfer case onto transmission, using transmission jack.
2. Install the strut rod, if removed.
3. Connect the parking brake cable, if necessary.
4. Connect the speedometer cable, vacuum lines and electrical connections.
5. Install the rear driveshaft, making certain to align match marks.
6. Install the front driveshaft, making certain to align match marks.
7. Install shifter lever.
8. Fill transfer case to correct level with Dexron®II automatic transmission fluid.
9. Lower vehicle and connect the negative battery cable.

GENERAL MOTORS 1988–89 PICK-UP

1. Install transfer case onto transmission, using transmission jack.
2. Install the skid plate.
3. Connect the parking brake cable, if necessary.
4. Connect the speed sensor and electrical connections.
5. Install the rear driveshaft, making certain to align match marks.
6. Install the front driveshaft, making certain to align match marks.
7. Install shifter lever.
8. Fill transfer case to correct level with Dexron®II automatic transmission fluid.
9. Lower vehicle and connect the negative battery cable.

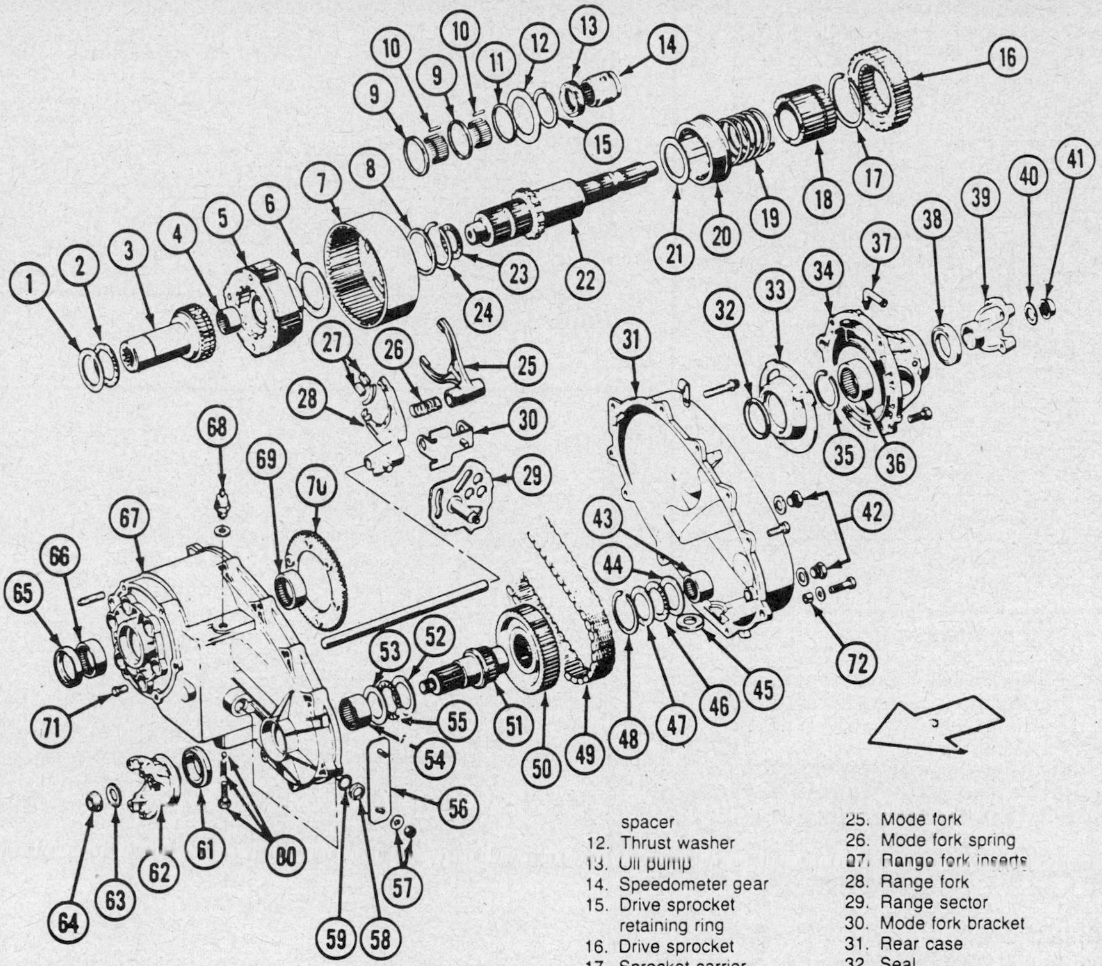

1. Input gear thrust washer
2. Input gear thrust bearing
3. Input gear
4. Mainshaft pilot bearing
5. Planetary assembly
6. Planetary thrust washer
7. Annulus gear
8. Annulus gear thrust washer
9. Needle bearing spacers
10. Mainshaft needle bearings (120)
11. Needle bearing spacer
12. Thrust washer
13. Oil pump
14. Speedometer gear
15. Drive sprocket retaining ring
16. Drive sprocket
17. Sprocket carrier stop ring
18. Sprocket carrier
19. Clutch spring
20. Sliding clutch
21. Thrust washer
22. Mainshaft
23. Mainshaft thrust bearing
24. Annulus gear retaining ring
25. Mode fork
26. Mode fork spring
27. Range fork inserts
28. Range fork
29. Range sector
30. Mode fork bracket
31. Rear case
32. Seal
33. Pump housing
34. Rear retainer
35. Rear output bearing
36. Bearing snap-ring
37. Vent tube
38. Rear seal
39. Rear yoke
40. Yoke seal washer
41. Yoke nut
42. Drain and fill plugs
43. Front output shaft rear bearing
44. Front output shaft rear thrust bearing race (thick)
45. Case magnet
46. Front output shaft rear thrust bearing
47. Front output shaft rear thrust bearing race (thin)
48. Driven sprocket retaining ring
49. Drive chain
50. Driven sprocket
51. Front output shaft
52. Front output shaft front thrust bearing race (thin)
53. Front output shaft front thrust bearing race (thick)
54. Front output shaft front bearing
55. Front output shaft front thrust bearing
56. Operating lever
57. Washer and locknut
58. Range sector shaft seal retainer
59. Range sector shaft seal
60. Detent ball, spring and retainer bolt
61. Front seal
62. Front yoke
63. Yoke seal washer
64. Yoke nut
65. Input gear oil seal
66. Input gear front bearing
67. Front case
68. Lock mode indicator switch and washer
69. Input gear rear bearing
70. Lockplate
71. Lockplate bolts
72. Case alignment dowels

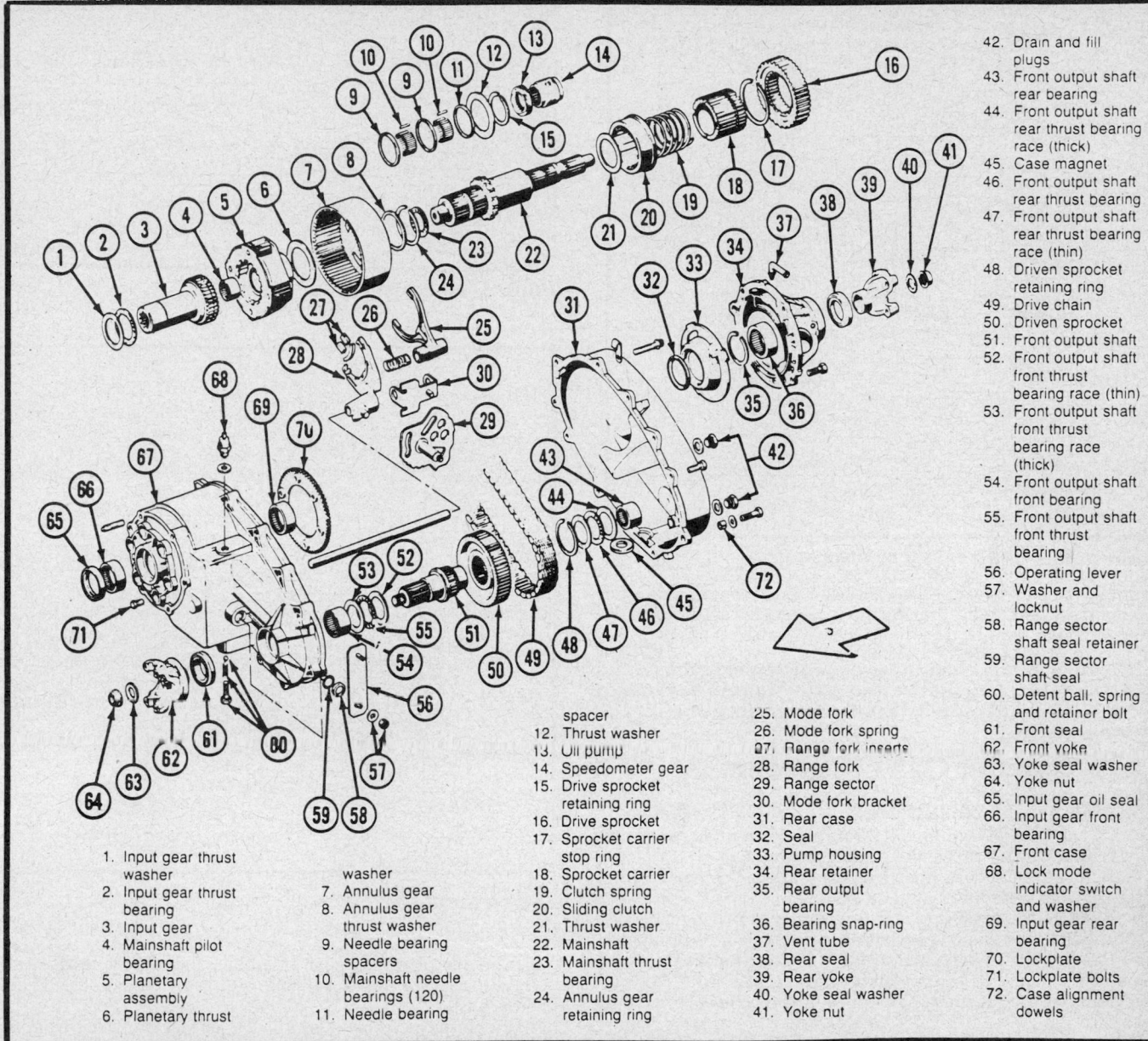

Exploded view of the new process 208 transfer case

EXCEPT GENERAL MOTORS

These procedures may vary slightly between different manufacturers.

1. When installing place a new gasket between the transfer case and the adapter.
2. Raise the transfer case with a transmission jack so the transmission output shaft aligns with the splined transfer case input shaft.
3. Install the bolts retaining the case to the adapter.
4. Connect the rear driveshaft to the rear output shaft yoke.
5. Connect the front driveshaft to the front output yoke.
6. Remove the transmission jack from the transfer case.
7. Position the heat shield to the frame crossmember and mounting lug to the transfer case and install and tighten the bolts and screw.
8. Install the skid plate to the frame.
9. Install the shift lever to the transfer case and tighten the retaining nut.
10. Install the speedometer driven gear to the transfer case.
11. Connect the 4WD indicator switch wire to the transfer case.
12. Install the drain plug. Remove the filler plug and install approximately 5.5 pts. of Dexron®II or equivalent type transmission fluid. The transfer case is full when the lubricant is at or slightly below the fill plug hole.
13. Install the exhaust support bracket and parking brake cable, if removed.
14. Lower the vehicle and connect the negative battery cable.

Before Disassembly

Clean the exterior of the transfer case assembly before any at-

FRONT OUTPUT SHAFT REAR
THRUST BEARING ASSEMBLY

Front output shaft rear thrust washer

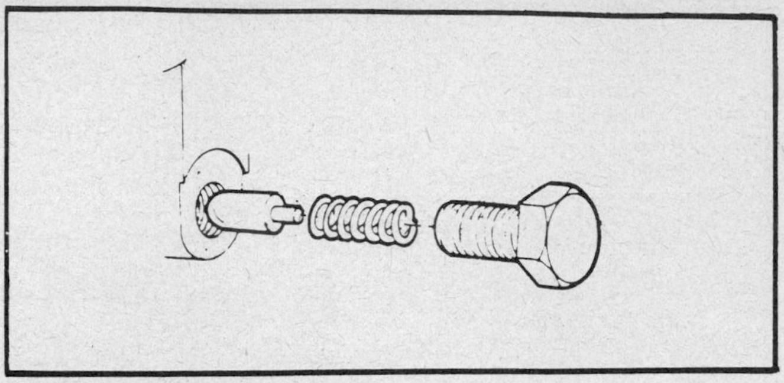

Poppet, spring and bolt

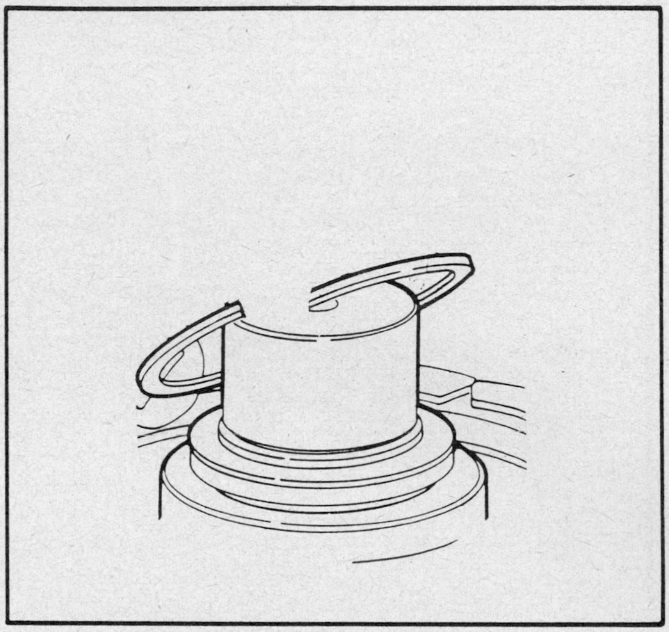

Driven sprocket retaining snap-ring

tempt is made to disassemble it, in order to prevent dirt or other foreign materials from entering the transfer case assembly or its internal parts.

NOTE: If steam cleaning is done to the exterior of the transfer case, immediate disassembly should be done to avoid rusting, caused by condensation forming on the internal parts.

Transfer Case Disassembly

1. Drain the fluid from the case.
2. Remove the attaching nuts from the front and rear output yokes. Remove the yokes and sealing washers.
3. Remove the 4 bolts and separate the rear bearing retainer from the rear case half.
4. Remove the retaining ring, speedometer drive gear nylon oil pump housing and oil pump gear from the rear output shaft.
5. Remove the 11 bolts and separate the case halves by inserting a screw driver in the pry slots on the case.
6. Remove the magnetic chip collector from the bottom of the rear case half.
7. Remove the thick thrust washer, thrust bearing and thin thrust washer from the front output shaft assembly.
8. Remove the drive chain by pushing the front input shaft inward and by angling the gear slightly to obtain adequate clearance to remove the chain.
9. Remove the output shaft from the front case half and slide the thick thrust washer, thrust bearing and thin thrust washer off the output side of the front output shaft.
10. Remove the screw, poppet spring and check ball from the front case half.
11. Remove the 4WD indicator switch and washer from the front case half.
12. Position the front case half on its face and lift out the rear output shaft, sliding clutch and clutch shift fork and spring.
13. Place a shop towel on the shift rail. Clamp the rail with locking pliers so that they lay between the rail and the case edge. Position a pry bar under the pliers and pry out the shift rail.
14. Remove the snapring and thrust washer from the planetary gear set assembly in the front case half.
15. Remove the annulus gear assembly and thrust washer from the front case half.
16. Lift the planetary gear assembly from the front case half.
17. Lift out the thrust bearing, sun gear, thrust bearing and thrust washer.
18. Remove the 6 bolts and lift the gear locking plate from the front case half.
19. Remove the nut retaining the external shift lever and washer. Press the shift control shaft inward and remove the shift selector plate and washer from the case.
20. From the rear output shaft, remove the snapring and thrust washer retaining the chain drive sprocket and slide the sprocket from the drive gear.
21. Remove the retaining ring from the sprocket carrier gear.
22. Carefully slide the sprocket carrier gear from the rear output shaft. Remove the 2 rows of 60 loose needle bearings. Remove the 3 separator rings from the output shaft.

Unit Disassembly and Assembly

LOCK PLATE

Disassembly

1. Remove and discard lock plate attaching bolts.
2. Remove the lock plate from the case.
3. Clean mating surfaces.

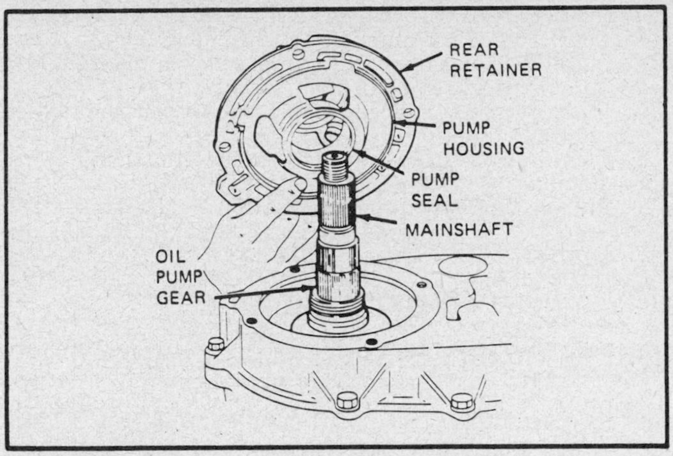
REAR
RETAINER
PUMP
HOUSING
PUMP
SEAL
MAINSHAFT
OIL
PUMP
GEAR

Rear retainer

Assembly

1. Coat case and lock plate surfaces around bolt holes with sealant.
2. Position new lock plate in the case and align bolt holes.
3. Coat new lock plate bolts with Loctite® or equivalent sealant.
4. Install and torque bolts to 30 ft. lbs. (41 Nm).

REAR OUTPUT BEARING AND SEAL

Disassembly

1. Drive bearing out of retainer, using brass drift.
2. Remove rear seal, using brass drift.

Assembly

1. Install new bearing, using tool J–7818 or equivalent. Make certain the shielded side of the bearing faces the interior of the case.
2. Install the bearing retaining snapring.
3. Install a new seal, using tool J–29162 or equivalent.

FRONT OUTPUT SHAFT BEARINGS

Disassembly

1. Remove the front bearing, using tool J–8092 and J–29168 or equivalent.
2. Remove the rear bearing, using tool J–26941 and a slide hammer.

Assembly

1. Install new front bearing, using tool J–8092 and J–29167 or equivalent.
2. Remove installer and check bearing position to be sure oil feed hole is not covered.
3. Install the new rear bearing, using tool J–8092 and J–29163 or equivalent.
4. Remove installer and check bearing position to be sure oil feed hole is not covered. Make certain the bearing is seated flush with the edge of case bore to allow room for the thrust bearing assembly.

INPUT GEAR BEARINGS

Disassembly

1. Proper secure input shaft on tool J–8092.

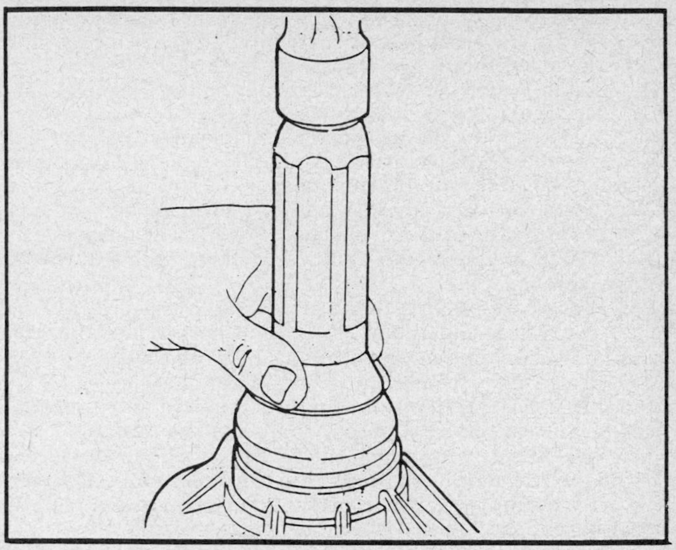

Rear seal installation

2. Remove both bearing simultaneously, using tool J–8092 and J–29170.

Assembly

1. Install rear bearing, using tool J–8092 and J–29169 or equivalent.
2. Install front bearing, using same tool.
3. Check that oil feed holes are not covered and bearings are flush with case bore surfaces.

MAINSHAFT PILOT BEARING

Disassembly

1. Remove bearing, by hand.
2. If bearing can not be removed by handed, use slide hammer and an internal type blind hole bearing puller.

Assembly

1. If necessary, install new bearing using a driver and installer.
2. Check that oil feed holes are not covered and bearings are flush with edge of oil hole.

Transfer Case Assembly

1. Slide the thrust washer against the gear on the rear output shaft.
2. Place the 3 spacer rings in position on the rear output shaft. Liberally coat the shaft with petroleum jelly and install the 2 rows of 60 needle bearings in position on the rear output shaft.
3. Carefully slide the sprocket gear carrier over the needle bearings. Be careful not to dislodge any of the needles.
4. Install the retaining ring on the sprocket gear.
5. Slide the chain drive sprocket onto the sprocket carrier gear.
6. Install the thrust washer and snapring on the rear output shaft.
7. Install the shift selector plate and washer through the front of the case.
8. Place the shift lever assembly on the shift control shaft and torque the nut to 14–20 ft. lbs.
9. Place the locking plate in the front case half and torque the bolts to 25–35 ft. lbs.

10. Place the thrust bearing and washer over the input shaft of the sun gear. Insert the input shaft through the front case half from the inside and insert the thrust bearing.
11. Install the planetary gear assembly so the fixed plate and planetary gears engage the sun gear.
12. Slide the annulus gear and clutch assembly with the shift fork assembly engaged, over the hub of the planetary gear assembly. The shift fork pin must engage the slot in the shift selector plate. Install the thrust washer and snapring.
13. Position the shift rail through the shift fork hub in the front case. Tap lightly with a soft-faced hammer to seat the rail in the hole.
14. Position the sliding clutch shift fork on the shift rail and place the sliding clutch and clutch shift spring into the front case half. Slide the rear output shaft into the case.
15. On the output side of the front output shaft, assemble the thin thrust washer, thrust bearing and thick thrust washer and partially insert the front output shaft into the case.
16. Place the drive chain on the rear output shaft drive gear. Insert the rear output shaft into the front case half and engage the drive chain on the front output shaft drive gear. Push the front output shaft into position in the case.
17. Assemble the thin thrust washer, thrust bearing and thick thrust washer on the inside of the front output shaft drive gear.
18. Position the magnetic chip collector into position in the front case half.
19. Place a bead of RTV sealant completely around the face of the front case half and assemble the case halves being careful that the shift rail and forward output shafts are properly retained.
20. Alternately tighten the bolts to 20–25 ft. lbs.
21. Slide the oil pump gear over the input shaft and slide the spacer collar into position.
22. Engage the speedometer drive gear onto the rear output shaft and slide the retaining ring into position.
23. Use petroleum jelly to hold the nylon oil pump housing in position at the rear bearing retainer. Apply a bead of RTV sealant around the mounting surface of the retainer and carefully position the retainer assembly over the output shaft and onto the rear case half. The retainer must be installed so that the vent hole is vertical when the case is installed.
24. Torque the retainer bolts alternately to 20–25 ft. lbs.
25. Place a new thrust washer under each yoke and install the yokes on their respective shafts. Place the oil slinger under the front yoke. Torque the nuts to 90–130 ft. lbs.
26. Install the poppet ball, spring and screw in the front case half. Torque the screw to 20–25 ft. lbs.
27. Install the 4WD indicator switch and washer and tighten to 15–20 ft. lbs.
28. Fill the unit with 5.5 pints of Dexron®II automatic transmission fluid.
29. Recheck fluid level after road test.

Specifications

TORQUE SPECIFICATIONS

	ft. lbs.	Nm
Nut, shift lever-to-shifter assembly	17	23
Nut, knob assembly-to-shift lever	24	33
Bolt, shifter assembly-to-transfer case	96	130
Nut, shift arms-to-case	13	17
Screw, shift lever boot retainer	24 ①	2.7
Bolt, detent retainer	23	31
Switch, indicator	111	149
Bolt, adapter-to-transmission	24	33
Bolt, adapter-to-transmission case	24	33
Filler plug	35	47
Nut, skid plate-to-crossmember	47	63
Bolt, support strut rod—Transmission end	35	47
Bolt, support strut rod—Transfer case end	129	175
Bolts, yoke steps		
Transfer case, front	55	75
Transfer case, rear	11	15

① Inch Pounds

NEW PROCESS 228 TRANSFER CASE

Application

1984–88 Jeep Comanche

General Description

The NP228 transfer case is very similar to the NP229 full time transfer case used on previous Jeep vehicles. The main difference in these transfer cases, is that the NP228 uses a differential unit in place of the viscous coupling. It also uses modified shift collars to preclude engine run away during a delayed shift.

The NP228 is used with the Selec-Trac system. A vacuum shift motor disconnect system is connected to the transfer case to disengage the front differential. It is a 3 position, dual range, full time/part time unit, with integral **LO** range and a **N** position. The NP228 requires 3.5 qts. of Dexron®II automatic transmission fluid.

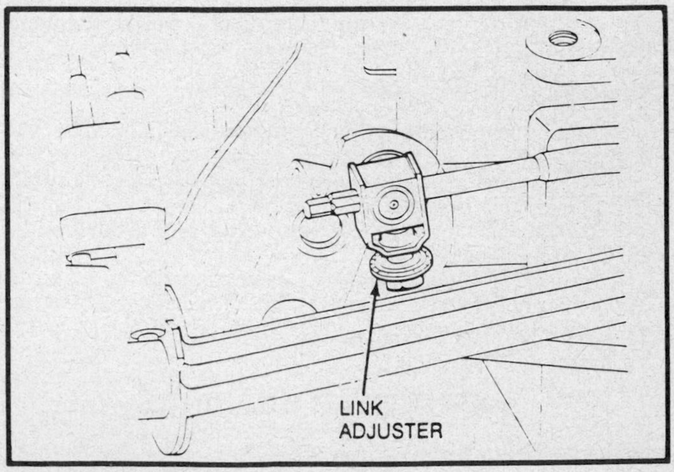
LINK
ADJUSTER

Linkage adjustment—Model 228

SERVICE DIAGNOSIS — SELECT DRIVE SYSTEM

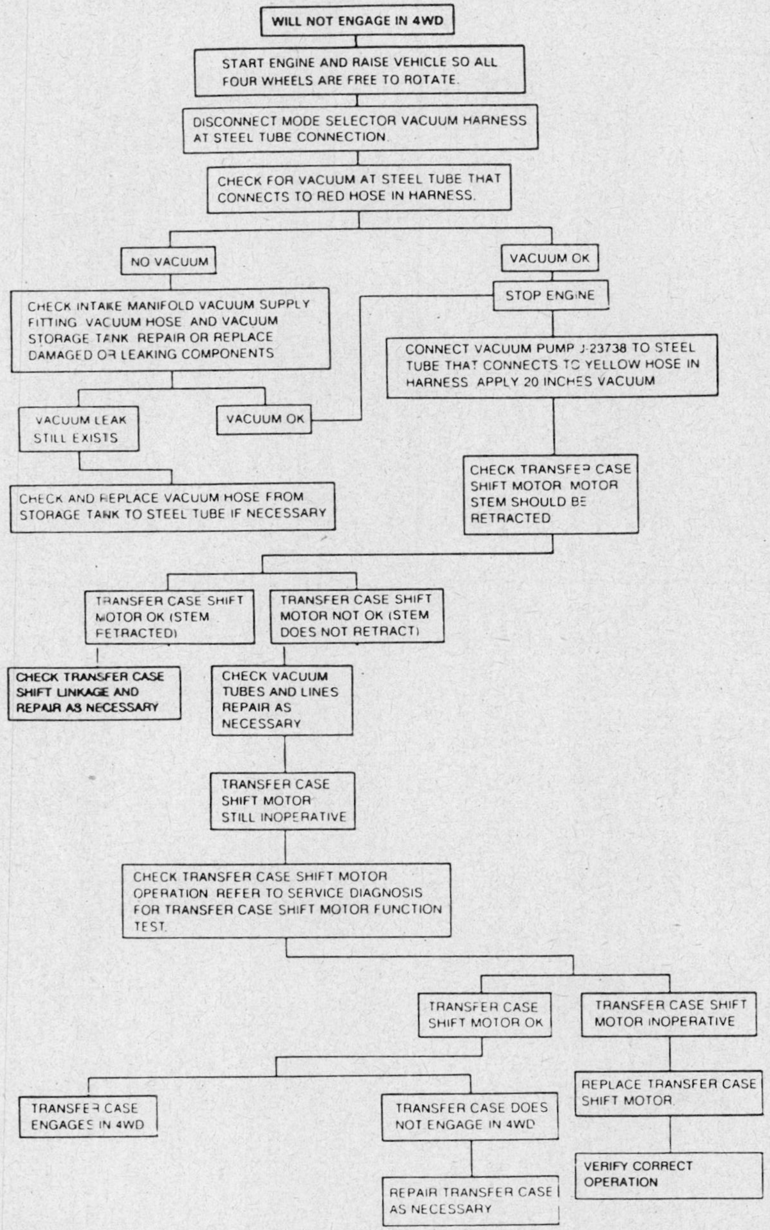
WILL NOT ENGAGE IN 4WD
START ENGINE AND RAISE VEHICLE SO ALL FOUR WHEELS ARE FREE TO ROTATE.
DISCONNECT MODE SELECTOR VACUUM HARNESS AT STEEL TUBE CONNECTION.
CHECK FOR VACUUM AT STEEL TUBE THAT CONNECTS TO RED HOSE IN HARNESS.
NO VACUUM
VACUUM OK
CHECK INTAKE MANIFOLD VACUUM SUPPLY FITTING VACUUM HOSE AND VACUUM STORAGE TANK REPAIR OR REPLACE DAMAGED OR LEAKING COMPONENTS
STOP ENGINE
VACUUM LEAK STILL EXISTS
VACUUM OK
CONNECT VACUUM PUMP J-23738 TO STEEL TUBE THAT CONNECTS TO YELLOW HOSE IN HARNESS APPLY 20 INCHES VACUUM
CHECK AND REPLACE VACUUM HOSE FROM STORAGE TANK TO STEEL TUBE IF NECESSARY
CHECK TRANSFER CASE SHIFT MOTOR MOTOR STEM SHOULD BE RETRACTED
TRANSFER CASE SHIFT MOTOR OK (STEM RETRACTED)
TRANSFER CASE SHIFT MOTOR NOT OK (STEM DOES NOT RETRACT)
CHECK TRANSFER CASE SHIFT LINKAGE AND REPAIR AS NECESSARY
CHECK VACUUM TUBES AND LINES REPAIR AS NECESSARY
TRANSFER CASE SHIFT MOTOR STILL INOPERATIVE
CHECK TRANSFER CASE SHIFT MOTOR OPERATION REFER TO SERVICE DIAGNOSIS FOR TRANSFER CASE SHIFT MOTOR FUNCTION TEST.
TRANSFER CASE SHIFT MOTOR OK
TRANSFER CASE SHIFT MOTOR INOPERATIVE
TRANSFER CASE ENGAGES IN 4WD
TRANSFER CASE DOES NOT ENGAGE IN 4WD
REPLACE TRANSFER CASE SHIFT MOTOR.
REPAIR TRANSFER CASE AS NECESSARY
VERIFY CORRECT OPERATION

SERVICE DIAGNOSIS — SELECT DRIVE SYSTEM (CONTINUED)

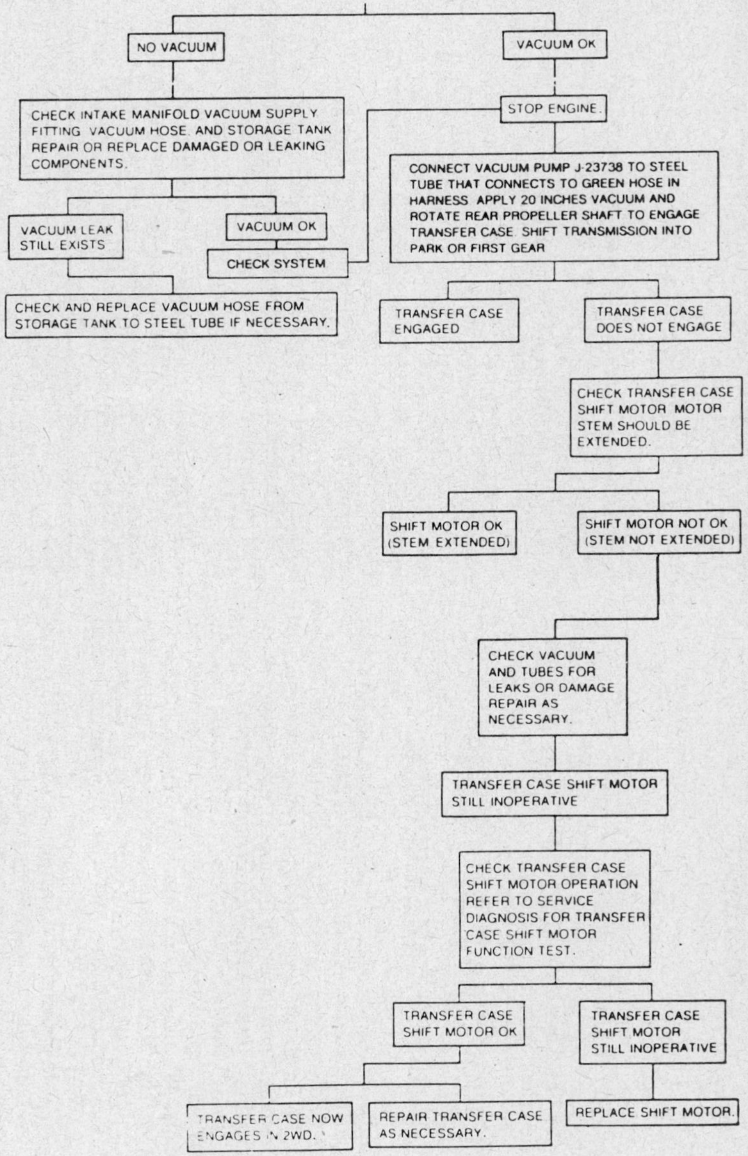
NO VACUUM
VACUUM OK
CHECK INTAKE MANIFOLD VACUUM SUPPLY FITTING VACUUM HOSE AND STORAGE TANK REPAIR OR REPLACE DAMAGED OR LEAKING COMPONENTS.
STOP ENGINE.
CONNECT VACUUM PUMP J-23738 TO STEEL TUBE THAT CONNECTS TO GREEN HOSE IN HARNESS APPLY 20 INCHES VACUUM AND ROTATE REAR PROPELLER SHAFT TO ENGAGE TRANSFER CASE. SHIFT TRANSMISSION INTO PARK OR FIRST GEAR
VACUUM LEAK STILL EXISTS
VACUUM OK
CHECK SYSTEM
CHECK AND REPLACE VACUUM HOSE FROM STORAGE TANK TO STEEL TUBE IF NECESSARY.
TRANSFER CASE ENGAGED
TRANSFER CASE DOES NOT ENGAGE
CHECK TRANSFER CASE SHIFT MOTOR MOTOR STEM SHOULD BE EXTENDED.
SHIFT MOTOR OK (STEM EXTENDED)
SHIFT MOTOR NOT OK (STEM NOT EXTENDED)
CHECK VACUUM AND TUBES FOR LEAKS OR DAMAGE REPAIR AS NECESSARY.
TRANSFER CASE SHIFT MOTOR STILL INOPERATIVE
CHECK TRANSFER CASE SHIFT MOTOR OPERATION REFER TO SERVICE DIAGNOSIS FOR TRANSFER CASE SHIFT MOTOR FUNCTION TEST.
TRANSFER CASE SHIFT MOTOR OK
TRANSFER CASE SHIFT MOTOR STILL INOPERATIVE
TRANSFER CASE NOW ENGAGES IN 2WD.
REPAIR TRANSFER CASE AS NECESSARY.
REPLACE SHIFT MOTOR.

On Vehicle Service

LINKAGE

Adjustment

1. Place the transfer case in **2WD**.
2. Insert a 1/8 in. (3mm) spacer between the gate and lever.
3. Hold the lever in this position.
4. Place the transfer case lever in **2WD**.
5. Adjust the link adjuster on the shifter shaft, to provide a free pin at the transfer case outer lever.

Removal and Installation

TRANSFER CASE REMOVAL

1. Raise and support the vehicle safely. Drain the lubricant out of the transfer case into a suitable drain pan.
2. Disconnect the speedometer cable and vent hose. Disconnect the transfer case shift lever link at the opening lever. Disconnect the catalytic converter or bracket if necessary.
3. Place a suitable jack stand or equivalent, under the transmission and remove the rear crossmember. Mark the transfer case front and rear output shafts at the transfer case yokes and driveshaft shafts for installation alignment reference.
4. Disconnect the front and rear driveshafts at the transfer case yokes and secure the shafts along the frame. Disconnect the shift motor vacuum hoses and the transfer case shift linkage.
5. Place a suitable transmission jack under the transfer case. Remove the transfer case to transmission mounting bolts and move the transfer case assembly rearward until it clears the transmission output shaft.
6. Lower the jack with the transfer case on it and remove the transfer case from the vehicle. Remove all gasket material from the rear of the transmission adapter housing.

TRANSFER CASE INSTALLATION

1. Install transfer case onto transmission, using transmission jack.

NOTE: Do not install any mounting bolts until all parts are aligned. Make sure that all splined shafts mesh properly before tightening bolts.

2. Install bolts and torque the transfer case to transmission bolts to 40 ft. lb. and the driveshaft yoke nuts to 10 ft. lb.
3. Connect the catalytic converter or bracket, if removed. Connect the parking brake cable, if removed.
4. Connect the speedometer cable or sensor.
5. Install shifter linkage.
6. Install the rear driveshaft, making certain to align match marks.
7. Install the front driveshaft, making certain to align match marks.
8. Install the vacuum lines and electrical connections.
9. Fill transfer case to correct level with Dexron®II automatic transmission fluid.
10. Install the skid plate, if equipped.
11. Lower vehicle and connect the negative battery cable.

Before Disassembly

Clean the exterior of the transfer case assembly before any attempt is made to disassemble it, in order to prevent dirt or other foreign materials from entering the transfer case assembly or its internal parts.

NOTE: If steam cleaning is done to the exterior of the transfer case, immediate disassembly should be done to avoid rusting, caused by condensation forming on the internal parts.

Transfer Case Disassembly

1. Drain all the lubricant from the transfer case and remove the front and rear yoke nuts along with their seal washers. Discard the seal washers.
2. Mark the front and rear yokes for easy installation alignment reference and remove the front and rear yokes. It may be necessary to use tool J–8614–01 or equivalent to remove the yokes.
3. Place the transfer case on wooden blocks. Cut V-notches in the blocks of wood so there is clearance for the front case mounting studs. Mark the retainer and the rear case for easy assembly reference.
4. Remove the rear retainer bolts and pry off the retainer with a suitable prybar. Remove the differential shim(s) and speedometer drive gear from the rear output shaft.
5. Remove the bolts that attach the rear transfer case half to the front case half. Be sure to get the washer that are used with the bolts on each end of the transfer case.

NOTE: Insert 2 small prybars into the slots at each end of the rear of the transfer case half to loosen it. Do not attempt to wedge the transfer case halves apart or the case mating surface will be damages.

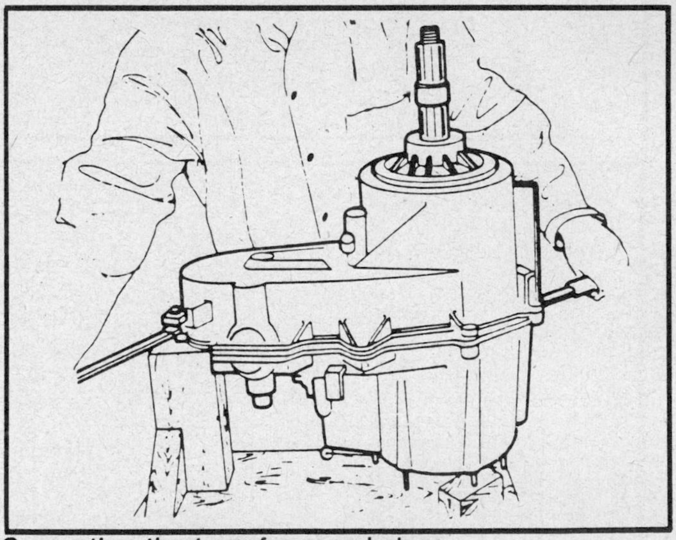

Separating the transfer case halves

6. Remove the rear transfer case half from the front case. Remove the thrust bearing and races from the front output shaft and be to note the order of the bearing and races for easy assembly reference.
7. Remove the oil pump from the rear output shaft, note the position of the pump for easy assembly reference. The recessed side of the pump faces the case interior.
8. Remove the rear output shaft from the mainshaft and remove the 15 main needle bearing rollers from the shaft or coupling. Remove the mainshaft O-ring from the end of the shaft. Remove the differential from the mainshaft and side gear.
9. Remove the front output shaft, driven sprocket and drive chain assembly. Lift the front shaft, sprocket and chain upward. Tilt the front shaft toward the mainshaft. Slide the chain off the drive sprocket and remove the assembly.
10. Remove the front output shaft and front thrust bearing as-

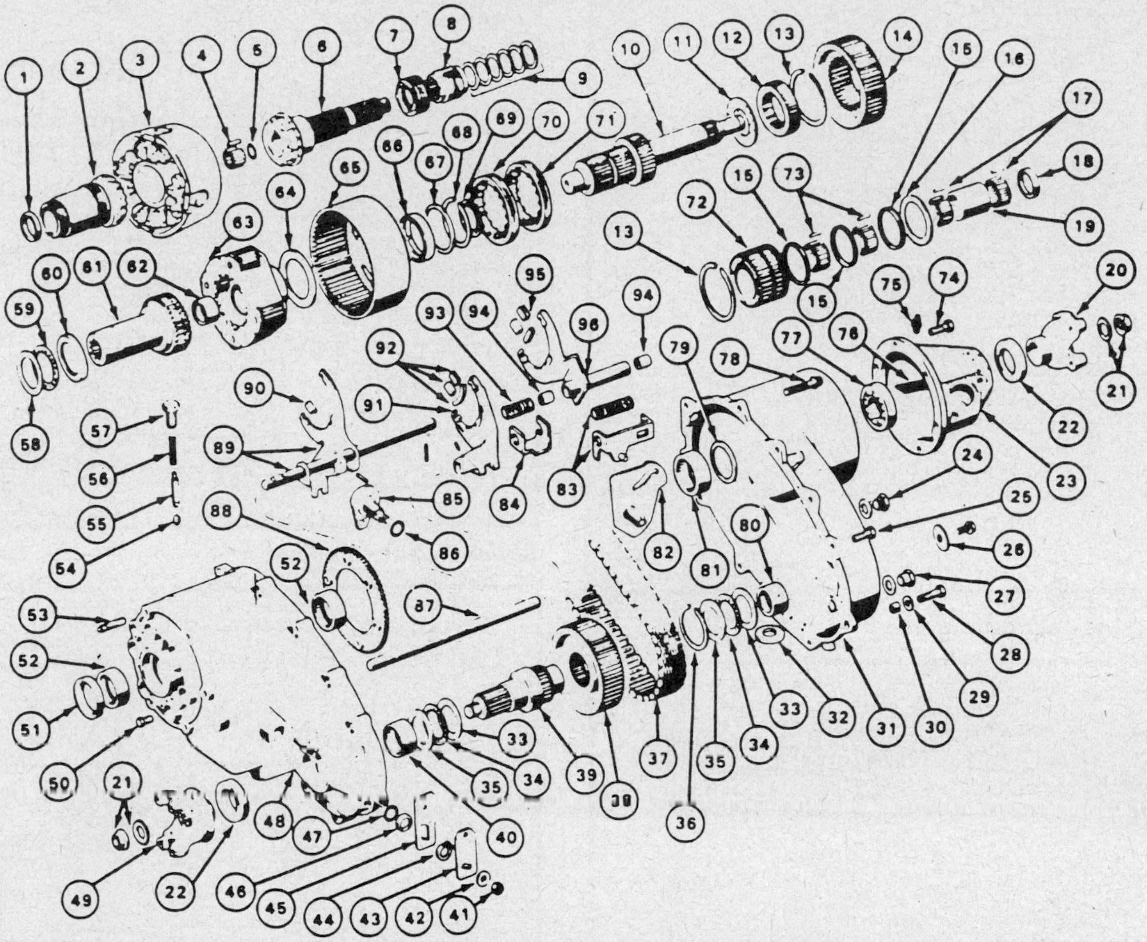

1. Spacer
2. Side gear
3. Differential
4. Pilot bearing rollers (15)
5. O-ring seal
6. Rear output shaft
7. Oil pump
8. Speedometer drive gear
9. Shim kit
10. Mainshaft
11. Mainshaft thrust washer
12. Spline gear
13. Retaining ring
14. Sprocket
15. Spacer
16. Sprocket thrust washer
17. Side gear roller (82)
18. Spacer (short)
19. Spacer (long)
20. Rear yoke
21. Nut and seal washer
22. Seal
23. Rear retainer
24. Plug assembly
25. Bolt
26. Identification tag
27. Plug assembly
28. Dowel bolt
29. Dowl bolt washer
30. Case half dowell
31. Rear half case
32. Magnet
33. Front output shaft bearing assembly race (thick)
34. Front output shaft bearing assembly thrust
35. Front output shaft bearing assembly race (thin)
36. Retaining ring
37. Chain
38. Driven sprocket
39. Front output shaft
40. Front output front bearing
41. Nut
42. Washer
43. Mode lever
44. Snap ring
45. Range lever
46. O-ring retainer
47. O-ring seal
48. Front half case
49. Front output yoke
50. Low range plate bolt
51. Input shaft oil seal
52. Input shaft bearing
53. Stud
54. Ball
55. Plunger
56. Plunger spring
57. Screw
58. Input race
59. Input thrust bearing
60. Input race (thick)
61. Input shaft
62. Input bearing
63. Planetary gear assembly
64. Input gear thrust washer
65. Annulus gear assembly
66. Annulus bushing
67. Thrust washer
68. Retaining ring
69. Thrust bearing
70. High range sliding clutch sleeve
71. Mode sliding clutch sleeve
72. Carrier
73. Carrier rollers (120)
74. Rear retainer bolt
75. Vent
76. Vent seal
77. Output bearing
78. Bolt
79. Seal
80. Front output rear bearing
81. Output shaft inner bearing
82. Range sector
83. Range bracket (outer) and spring
84. Range bracket (inner)
85. Mode sector
86. O-ring seal
87. Range rail
88. Low range lockout plate
89. Mode fork, rail and pin
90. Mode fork pad
91. Range fork
92. Range fork pads
93. Range bracket spring (inner)
94. Locking fork bushing
95. Locking fork pads
96. Locking fork

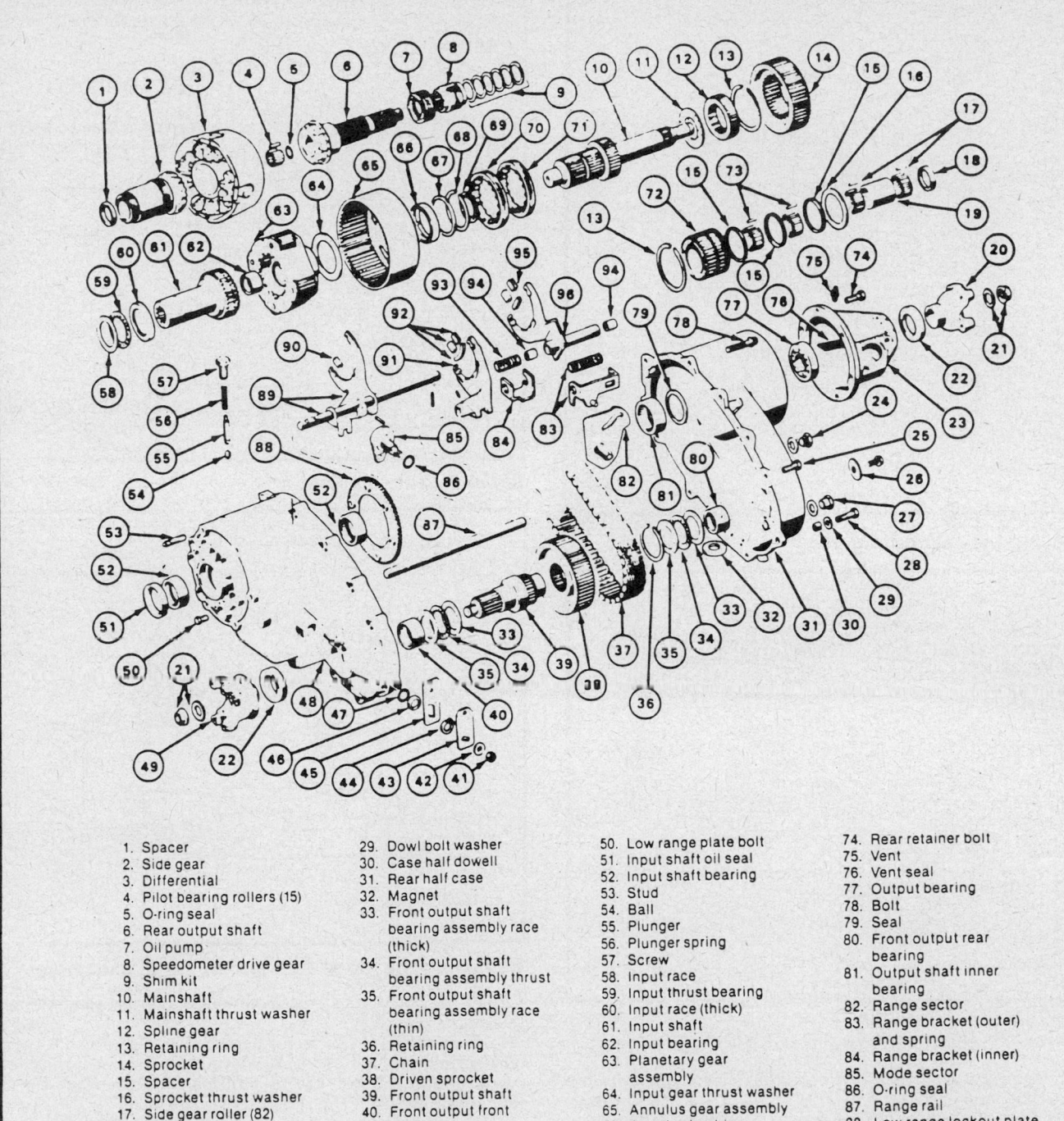

Exploded view of the 228 transfer case

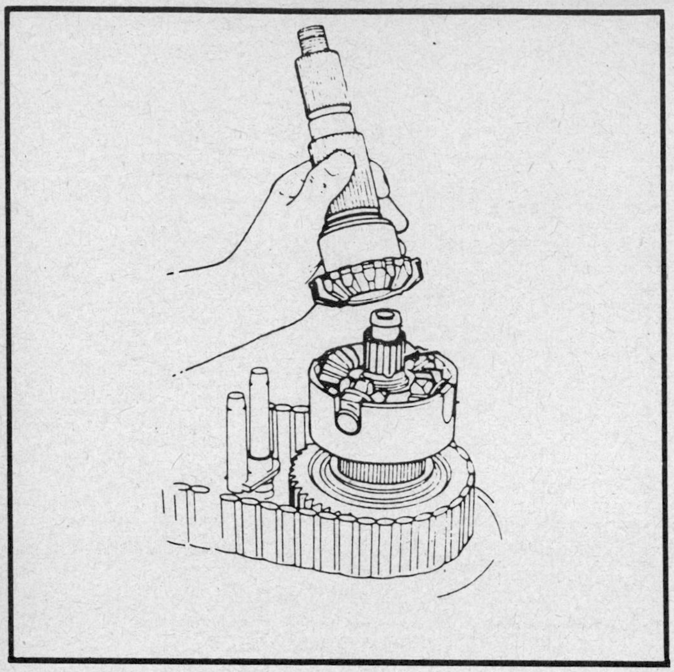

Removing the rear output shaft

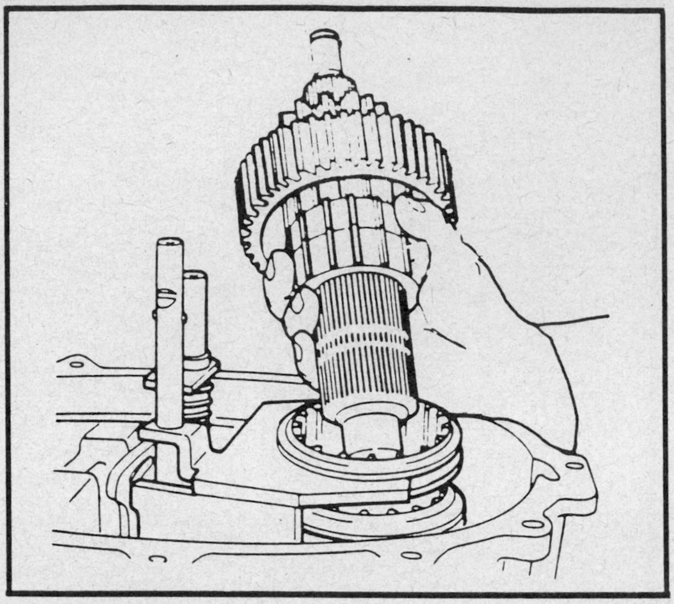

Removing the mainshaft

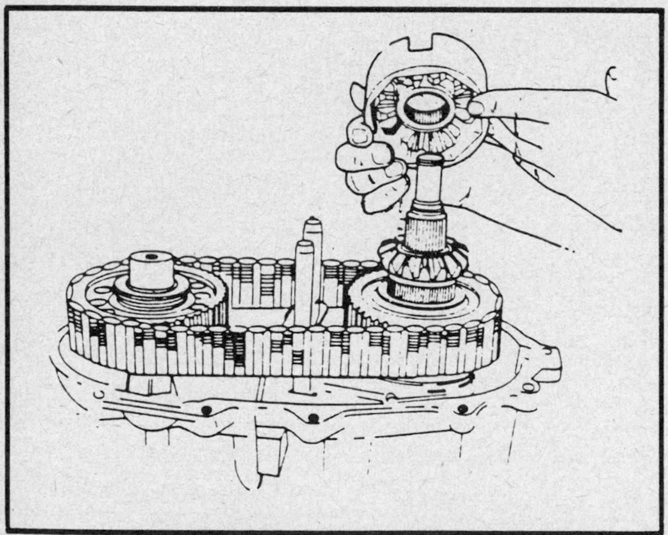

Removing the differential

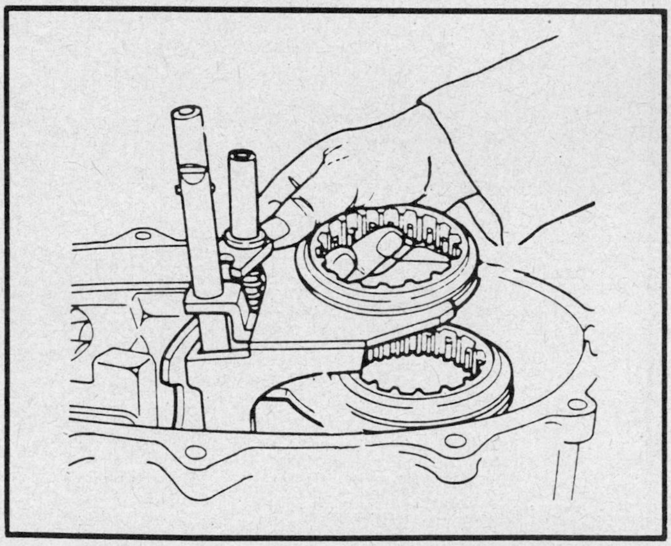

Removing the mode fork and shift rail assembly

sembly from the front case. Remove the drive chain from the output shaft and sprocket.

11. Remove the snapring that retains the driven sprocket on the front output shaft. Mark the sprocket and shaft for assembly reference and remove the sprocket from the shaft.

12. Remove the mainshaft, side gear, drive sprocket and spline gear as an assembly.

13. Remove the mode fork, shift rail and mode sliding clutch sleeve as an assembly. Mark the sleeve and fork for easy assembly reference and remove the sleeve from the fork.

NOTE: The mode fork and rail are pinned together so that they will operate as a unit. Remove the pin to separate the 2 components if necessary.

14. Remove the locking fork, high range sliding clutch sleeve, fork brackets and fork spring as an assembly. Be sure to take note of the position of these components for an easy assembly reference.

15. Remove the range sector detent screw and remove the detent spring, plunger and ball. Move the range operating lever downward to the last detent position and disengage the low range fork lug from the range sector slot.

16. Remove the retaining snapring from the annulus gear and remove the thrust washer. Remove the annulus gear, range fork and rail as an assembly and separate the components for cleaning and inspection.

17. Remove the planetary thrust washer from the planetary assembly hub. Remove the planetary assembly, by grasping the planetary hub and lifting the assembly upward.

18. Remove the mainshaft thrust bearing from the input shaft. Remove the input shaft, input shaft thrust bearing and race.

19. Remove the range sector and operating lever attaching nut and lockwasher. Remove the lever.

20. Remove the range sector and shaft from the front case and remove the range sector O-ring and retainer.

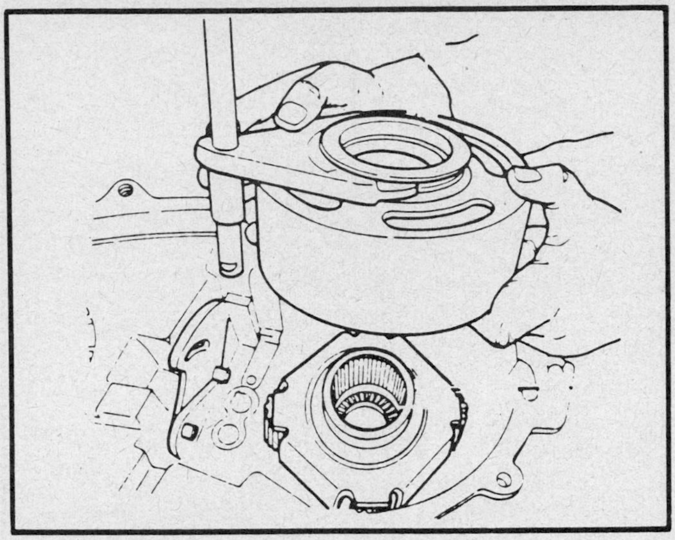

Removing the annulus gear

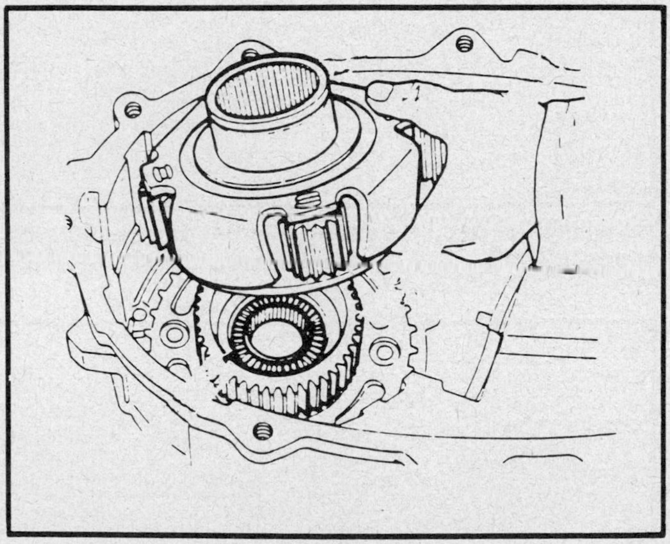

Removing the planetary assembly

Transfer Case Assembly

NOTE: During the assembly, lubricate all of the transfer case internal components with Dexron®II transmission fluid or petroleum jelly as indicated in the procedure. Do not use chassis lubricant or similar thick lubricants.

1. Install a replacement input shaft and rear output shaft bearing oil seals. Set the seals flush with the edge of the seal bore or in the seal groove in the transfer case. Coat the seal lips with petroleum jelly after installation.
2. Install the input shaft thrust bearing race in the transfer case counterbore. Install the input gear thrust bearing on the input shaft and install the shaft and bearing in the transfer case.
3. Install the mainshaft thrust bearing in the bearing recess in the input shaft. Install the planetary assembly on the input shaft. Ensure that the planetary pinion teeth mesh fully with the input shaft.
4. Install the planetary thrust washer on the planetary hub.

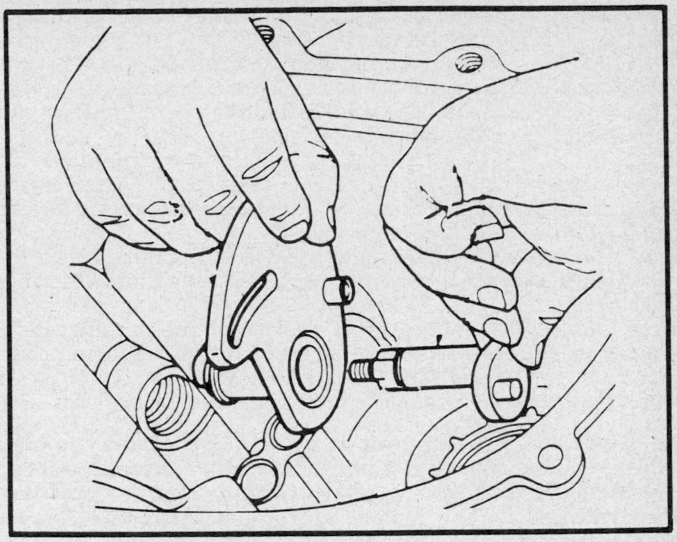

Installing the mode sector shaft into the range sector

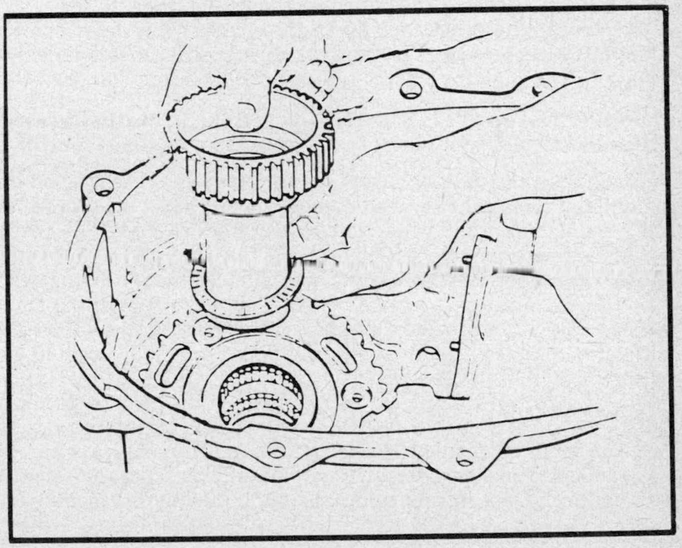

Removing the input shaft assembly

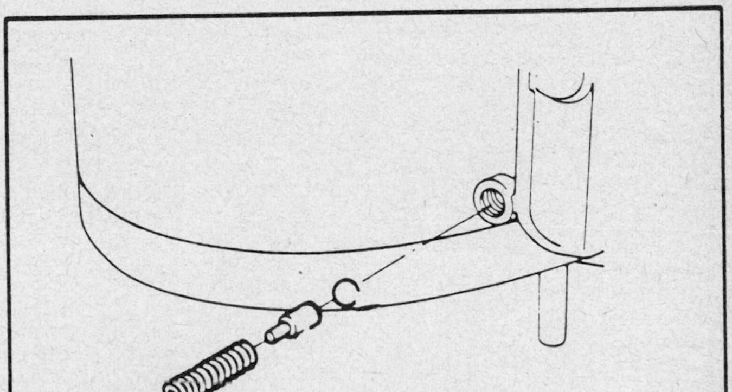

Installing the detent ball assembly

Install a replacement sector shaft O-ring and install the retainer in the shaft bore in the transfer case.

5. Install the O-ring on the mode sector shaft and insert the mode sector through the range sector. Install the range sector in the front of transfer case half. Install the operating lever and the snapring on the range sector shaft.

6. Install the lever, attaching washer and locknut on the mode sector shaft. Torque the locknut to 17 ft. lbs. Assemble the annulus gear, range fork and rail. Install the assembled fork on and over the planetary assembly.

7. Be sure that the annulus gear is fully meshed with the planetary pinions. Engage the range sector lug into the range sector.

8. Install the annulus thrust washer and the annulus retaining ring onto the annulus gear hub. Install the detent ball, plunger, spring and retaining screw in the front transfer case half detent bore. Torque the retaining screw to 22 ft. lbs.

NOTE: The locking mode clutch sleeve and the high range clutch sleeve are not interchangeable. The sleeve splines are different. So be sure that the correct sleeve is installed in the proper shift fork. Also, the sleeves must be replaced as a set.

9. Assemble and install the locking fork, fork bracket, fork springs and high range clutch sleeves. Be sure that the lug on the fork is seated in the range sector detent slot.

10. Install the range fork lug in the range sector detent notch. Move the range sector to the HI range position. Assemble and install the range fork, shift rail and mode clutch sleeve.

NOTE: Steps 11–16 are to be used if the mainshaft was disassembled, when the transfer case was disassembled.

11. Install the thrust washer and a replacement O-ring on the mainshaft. Install the needle bearings and bearing spacers on the mainshaft. Coat the shaft bearing surface and all needle bearings with petroleum jelly.

12. Install the first 41 needle bearings, the long bearing spacer, the remaining 41 needle bearings and the remaining short spacer. Be careful to avoid displacing the bearing when the spacers are installed. Apply additional petroleum jelly to hold the bearing in place if necessary.

13. Install the spline gear on the mainshaft, be careful not to displace the bearing while installing the gear. Install the sprocket carrier in the drive sprocket and install the sprocket carrier snaprings. Make sure that the carrier and sprockets are aligned according to the reference marks made during disassembly.

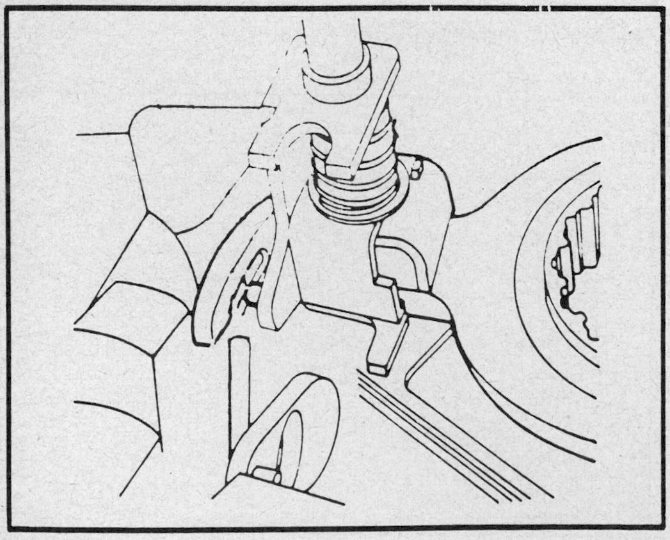

Installing the locking fork

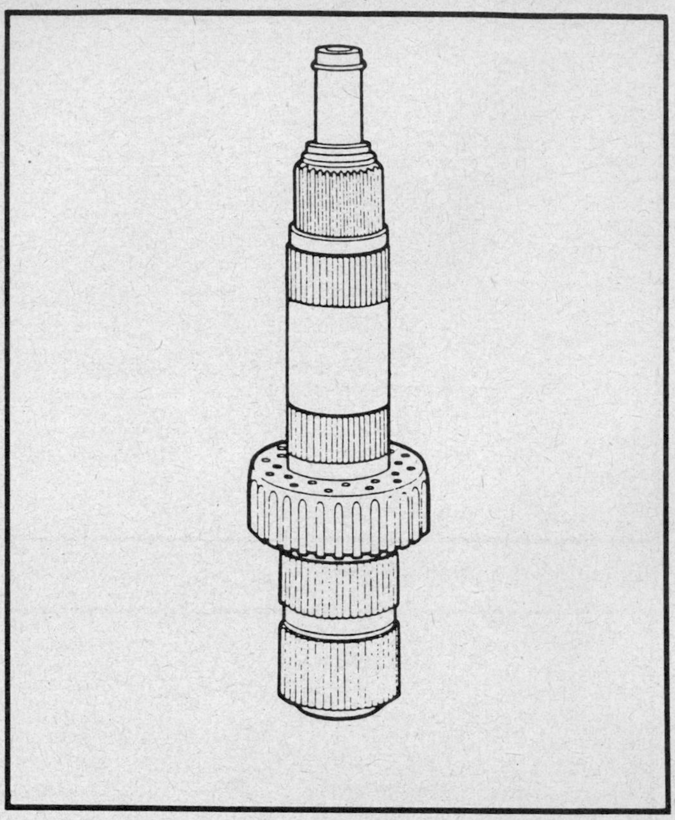

Installing the needle bearing and bearing spacers on the mainshaft

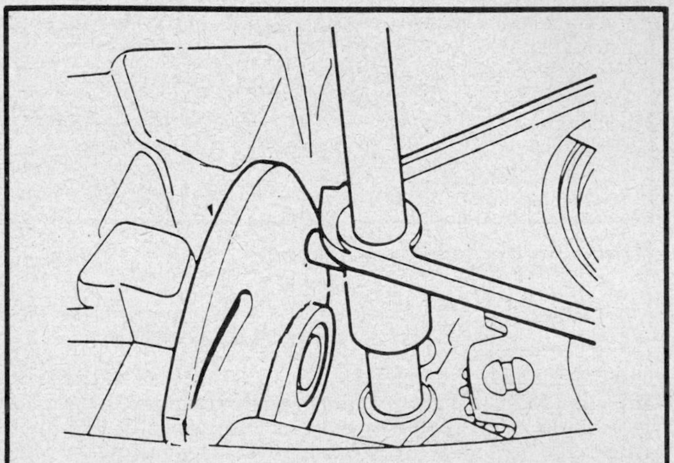

Engage the range sector lug into the range sector

NOTE: The sprocket carrier teeth are tapered on a side and the drive sprocket has a deep recess on a side. Be sure that these components are assembles so that the carrier tapered teeth and sprocket recess are on the same side.

14. Install the sprocket carrier bearings and spacers. Coat the

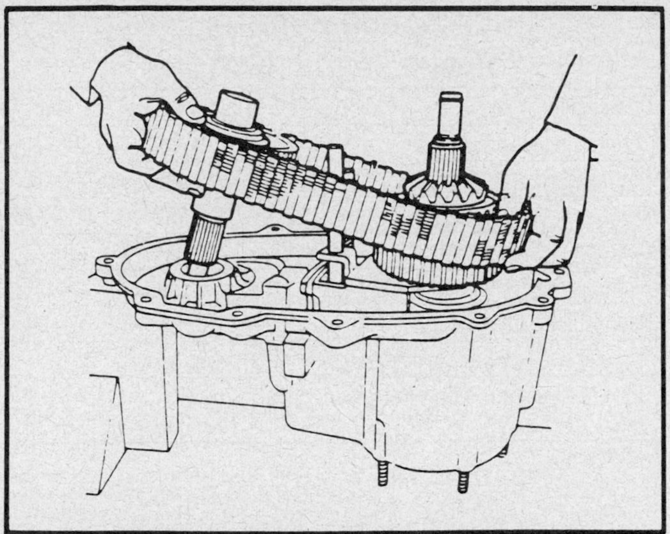

Installing the drive chain, front output shaft and driven socket

carrier bore and all the 120 carrier needle bearings with petroleum jelly. Install the center spacer.

15. Install the 60 needle bearings in each end of the carrier and install the remaining 2 spacers, 1 at each side of the carrier. Apply additional petroleum jelly to hold the bearings in place if necessary.

16. Install the assembled sprocket carrier and drive sprocket on the mainshaft. Do not displace the mainshaft bearing during installation. Be sure that the recessed side of the drive sprocket is facing downward.

17. Install the trust washer in the mainshaft, position the washer on the sprocket carrier. Install the side gear on the mainshaft and be sure that the side gear is fully seated in the sprocket carrier. Be careful not to displace any of the carrier or mainshaft needle bearings.

18. Install the mainshaft and gear assembly in the case, making sure that the mainshaft is fully seated in the input gear. Install the driven sprocket on the front output shaft and install the sprocket retaining snapring. Be sure that the sprocket is installed according to the reference marks made during disassembly.

19. Install the front output shaft front thrust bearing assembly in the transfer case front half. Install the thick race in the transfer case and then install the bearing and the thin race.

20. Install the drive chain, front output shaft and driven sprocket. Install the chain on the driven sprocket. Raise and tilt the driven sprocket and chain and install the opposite end of the chain on the drive sprocket.

21. Align the front output shaft with the shaft bore in the transfer case front half and install the shaft in the transfer case. Be sure that the front shaft thrust bearing assembly is seated in the transfer case.

22. Install the front output shaft rear thrust bearing assembly on the front output shaft. Install the tin race, then install the bearing and the thick race. Install the differential on the side gear, making sure that the differential is fully seated.

23. Coat the mainshaft pilot bearing surface and all 15 needle bearings with petroleum jelly and install the bearing on the shaft. Apply additional petroleum jelly to hold the bearings in place if necessary.

24. Install the rear output shaft on the mainshaft and into the differential, making sure that the shaft is completely seated. If necessary tap the shaft with a plastic mallet or equivalent to seat the shaft. Do not displace the pilot bearing during shaft installation.

Checking the end play on the rear output shaft

25. Install the oil pump and the rear output shaft. Install the oil pump with the recessed side facing down. Install a replacement rear output shaft bearing seal in the rear transfer case half.

26. Apply a bead of Loctite®515 sealant or equivalent, to the mating surface of the rear transfer case half. Install the magnet in the case and attach the rear transfer case half to the front transfer case half. Be sure that the alignment dowels at the front case half ends are aligned with the bolt holes in the rear case half and mate the rear case half with the front case half.

NOTE: If the rear transfer case half will not mate completely with the front case. Inspect the following; oil in the range fork rail bore, the front output shaft rear thrust bearing assembly is not aligned with the rear case half, the mainshaft is not completely seated or the rear case half is not aligned with the oil pump.

27. Install the rear case half to the front case half bolts. Torque the bolts to 23 ft. lbs. Be sure that the flat washers are used on the bolts at the case end where the alignment dowels are located. Install the speedometer drive gear on the rear output shaft.

28. Measure the thickness of the shim pack and record. Install a 0.030 in. shim on the rear output shaft. Align the rear retainer on the rear transfer case half and install the retainer. Install the retainer bolts and tighten them securely, do not torque to specifications.

29. Install the front and rear output shaft yokes and the original yoke nuts. Tighten the yoke nuts finger tight and check the differential endplay.

30. Set the shift lever in the 4WD-HI range position. Place a dial indicator on the rear retainer and position the indicator stylus so that it contacts the rear yoke nut.

31. Pull upward on the rear output yoke, note the dial indicator pointer position and record it. Remove the retainer and add or subtract differential shims as necessary to correct the endplay. The endplay should be between 0.002–0.010 in. The recommended endplay is 0.006 in.

32. After adjusting the endplay, remove the front and rear yokes. Discard the original yoke nuts. Apply a bead of Loctite®515 sealant or equivalent, to the retainer mating sur-

face and install the retainer. Apply the sealer to the retaining bolts and install the bolts. Torque the bolts to 23 ft. lbs.

33. Position the front and rear yokes and install the replacement yoke seal washers and nuts. Using tool J–8614–01, or equivalent hold the yokes in place and torque the yoke nuts to 120 ft. lbs.

34. Install the detent ball, spring and bolt if these were not installed previously. Apply sealer to the bolt before installing it and torque the bolt to 23 ft. lbs.

35. Install the drain plug and washer. Fill the transfer case with 7 pints of Dexron®II transmission fluid or equivalent. Install the fill plug and washer and torque the drain and fill plugs to 18 ft. lbs.

36. Install the plug and washer in the front transfer case half, if removed and torque the plug to 18 ft. lbs.

Specifications

TORQUE SPECIFICATIONS

Component	ft. lbs.	Nm
Detent retainer bolt	20–25	27–34
Drain and fill plugs	15–25	20–34
Front/rear yoke nuts	90–130	122–176
Operating lever locknut	15–20	20–27
Rear case-to-front case bolts (All)	20–25	27–34
Rear retainer bolts	20–25	27–34
Transfer case-to-transmission adapter nuts	20–30	27–41
Universal joint strap bolt-to-transfer case yoke	140–200 ①	16–23

① Inch lbs.

NEW PROCESS 229 TRANSFER CASE

Application

1984–88 Jeep Cherokee/Wagoneer
1984–89 Jeep Pick Up/Grand Wagoneer

General Description

The model 229 is used with the Selec-Trac® system. There are 3 shift forks used, the mode and low range forks are pinned to the rail and operate as a unit. The models 219 and 229 are similar in appearance, but the internal components are different in configuration, oil passages and dimensions. It requires 3 qts. of Dexron®II automatic transmission fluid and anti-friction additives must not be used. It uses a viscous coupling and pinion assembly, that is not serviceable.

On Vehicle Service

MODE ROD

Adjustment

1. Place transfer case in **2WD-HI**.
2. Both range lever and mode lever must be aligned on the same centerline prior to mode rod adjustment
3. Rotate output shaft as needed to assist positioning of mode lever.
4. Rotate the rear driveshaft, while applying a load on the mode lever to fully engage into **2WD-HI**.
5. Adjust the mode rod to approximately 5.9 in. (149.9mm) to elimate freeplay.
6. Check all vacuum connections and test drive vehicle, shifting into **4WD** and back to **2WD**.
7. Recheck mode lever position after road test. If the lever is not aligned with the range lever, length rod 1 turn and repeat Steps 1–7.

NOTE: With the transfer case vacuum motor shaft fully extended and the transfer case in 2WD-HI adjust the mode rod so that the pin move freely in the hole through the vacuum motor shaft and mode rod.

RANGE ROD

Adjustment

1. Adjust the mode rod.
2. Road test and recheck mode rod positioning.
3. Loosen the locknuts on range rod and transfer case end.
4. Adjust linkage to position the range lever ½–1.0 in. (13–25mm) above the floor, when in the **2WD-HI** range.
5. Check for sufficient travel for complete range engagement.
6. Tighten the locknuts.

Removal and Installation

TRANSFER CASE REMOVAL

1. Raise and support the vehicle safely. Drain the lubricant out of the transfer case into a suitable drain pan.
2. Disconnect the speedometer cable and vent hose. Disconnect the transfer case shift lever link at the opening lever.
3. Place a suitable jack stand or equivalent, under the transmission and remove the rear crossmember. Mark the transfer case front and rear output shafts at the transfer case yokes and driveshaft shafts for installation alignment reference.
4. Disconnect the front and rear driveshafts at the transfer case yokes and secure the shafts along the frame. Disconnect the shift motor vacuum hoses and the transfer case shift linkage. Disconnect the catalytic converter or bracket if necessary.
5. Place a suitable transmission jack under the transfer case. Remove the transfer case to transmission mounting bolts and move the transfer case assembly rearward until it clears the transmission output shaft.
6. Lower the jack with the transfer case on it and remove the transfer case from the vehicle. Remove all gasket material from the rear of the transmission adapter housing.

TRANSFER CASE INSTALLATION

1. Install transfer case onto transmission, using transmission jack.

NOTE: Do not install any mounting bolts until all parts are aligned. Make sure that all splined shafts mesh properly before tightening bolts.

2. Install bolts and torque the transfer case to transmission bolts to 40 ft. lbs. and the driveshaft yoke nuts to 10 ft. lbs.

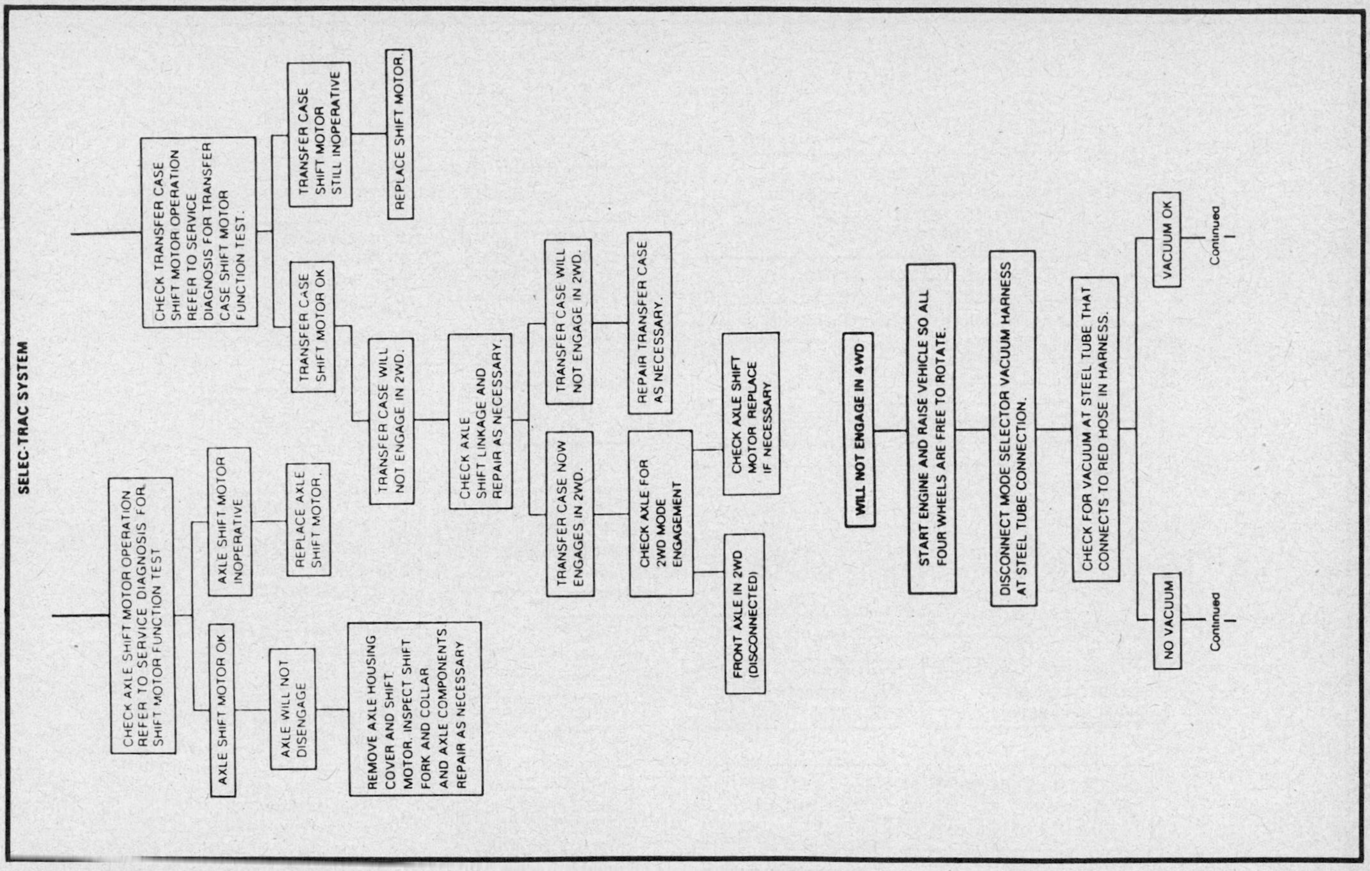

SELEC-TRAC SYSTEM
CHECK AXLE SHIFT MOTOR OPERATION REFER TO SERVICE DIAGNOSIS FOR SHIFT MOTOR FUNCTION TEST
AXLE SHIFT MOTOR OK
AXLE SHIFT MOTOR INOPERATIVE
AXLE WILL NOT DISENGAGE
REPLACE AXLE SHIFT MOTOR.
REMOVE AXLE HOUSING COVER AND SHIFT MOTOR. INSPECT SHIFT FORK AND COLLAR AND AXLE COMPONENTS REPAIR AS NECESSARY
CHECK TRANSFER CASE SHIFT MOTOR OPERATION REFER TO SERVICE DIAGNOSIS FOR TRANSFER CASE SHIFT MOTOR FUNCTION TEST.
TRANSFER CASE SHIFT MOTOR OK
TRANSFER CASE SHIFT MOTOR STILL INOPERATIVE
REPLACE SHIFT MOTOR.
TRANSFER CASE WILL NOT ENGAGE IN 2WD.
CHECK AXLE SHIFT LINKAGE AND REPAIR AS NECESSARY.
TRANSFER CASE NOW ENGAGES IN 2WD.
TRANSFER CASE WILL NOT ENGAGE IN 2WD.
REPAIR TRANSFER CASE AS NECESSARY.
CHECK AXLE FOR 2WD MODE ENGAGEMENT
FRONT AXLE IN 2WD (DISCONNECTED)
CHECK AXLE SHIFT MOTOR REPLACE IF NECESSARY
WILL NOT ENGAGE IN 4WD
START ENGINE AND RAISE VEHICLE SO ALL FOUR WHEELS ARE FREE TO ROTATE.
DISCONNECT MODE SELECTOR VACUUM HARNESS AT STEEL TUBE CONNECTION.
CHECK FOR VACUUM AT STEEL TUBE THAT CONNECTS TO RED HOSE IN HARNESS.
NO VACUUM
VACUUM OK
Continued
Continued

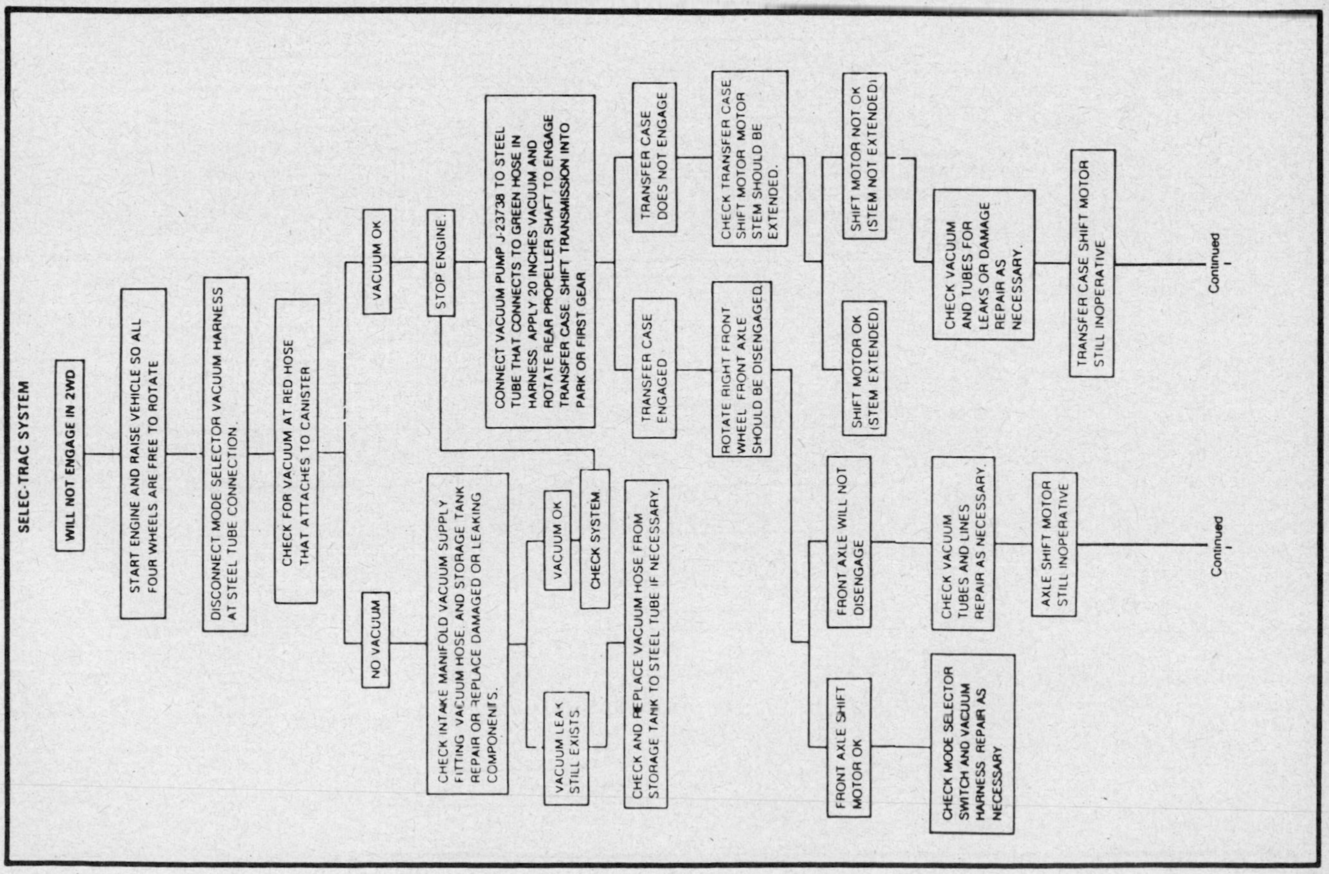

SELEC-TRAC SYSTEM
WILL NOT ENGAGE IN 2WD
START ENGINE AND RAISE VEHICLE SO ALL FOUR WHEELS ARE FREE TO ROTATE
DISCONNECT MODE SELECTOR VACUUM HARNESS AT STEEL TUBE CONNECTION.
CHECK FOR VACUUM AT RED HOSE THAT ATTACHES TO CANISTER
NO VACUUM
VACUUM OK
CHECK INTAKE MANIFOLD VACUUM SUPPLY FITTING VACUUM HOSE, AND STORAGE TANK REPAIR OR REPLACE DAMAGED OR LEAKING COMPONENTS.
VACUUM LEAK STILL EXISTS.
VACUUM OK
CHECK SYSTEM.
CHECK AND REPLACE VACUUM HOSE FROM STORAGE TANK TO STEEL TUBE IF NECESSARY.
STOP ENGINE.
CONNECT VACUUM PUMP J-23738 TO STEEL TUBE THAT CONNECTS TO GREEN HOSE IN HARNESS. APPLY 20 INCHES VACUUM AND ROTATE REAR PROPELLER SHAFT TO ENGAGE TRANSFER CASE. SHIFT TRANSMISSION INTO PARK OR FIRST GEAR
TRANSFER CASE ENGAGED
TRANSFER CASE DOES NOT ENGAGE
ROTATE RIGHT FRONT WHEEL. FRONT AXLE SHOULD BE DISENGAGED.
CHECK TRANSFER CASE SHIFT MOTOR MOTOR STEM SHOULD BE EXTENDED.
FRONT AXLE SHIFT MOTOR OK
FRONT AXLE WILL NOT DISENGAGE
SHIFT MOTOR OK (STEM EXTENDED)
SHIFT MOTOR NOT OK (STEM NOT EXTENDED)
CHECK MODE SELECTOR SWITCH AND VACUUM HARNESS. REPAIR AS NECESSARY
CHECK VACUUM TUBES AND LINES REPAIR AS NECESSARY.
CHECK VACUUM AND TUBES FOR LEAKS OR DAMAGE REPAIR AS NECESSARY.
AXLE SHIFT MOTOR STILL INOPERATIVE
TRANSFER CASE SHIFT MOTOR STILL INOPERATIVE
Continued
Continued

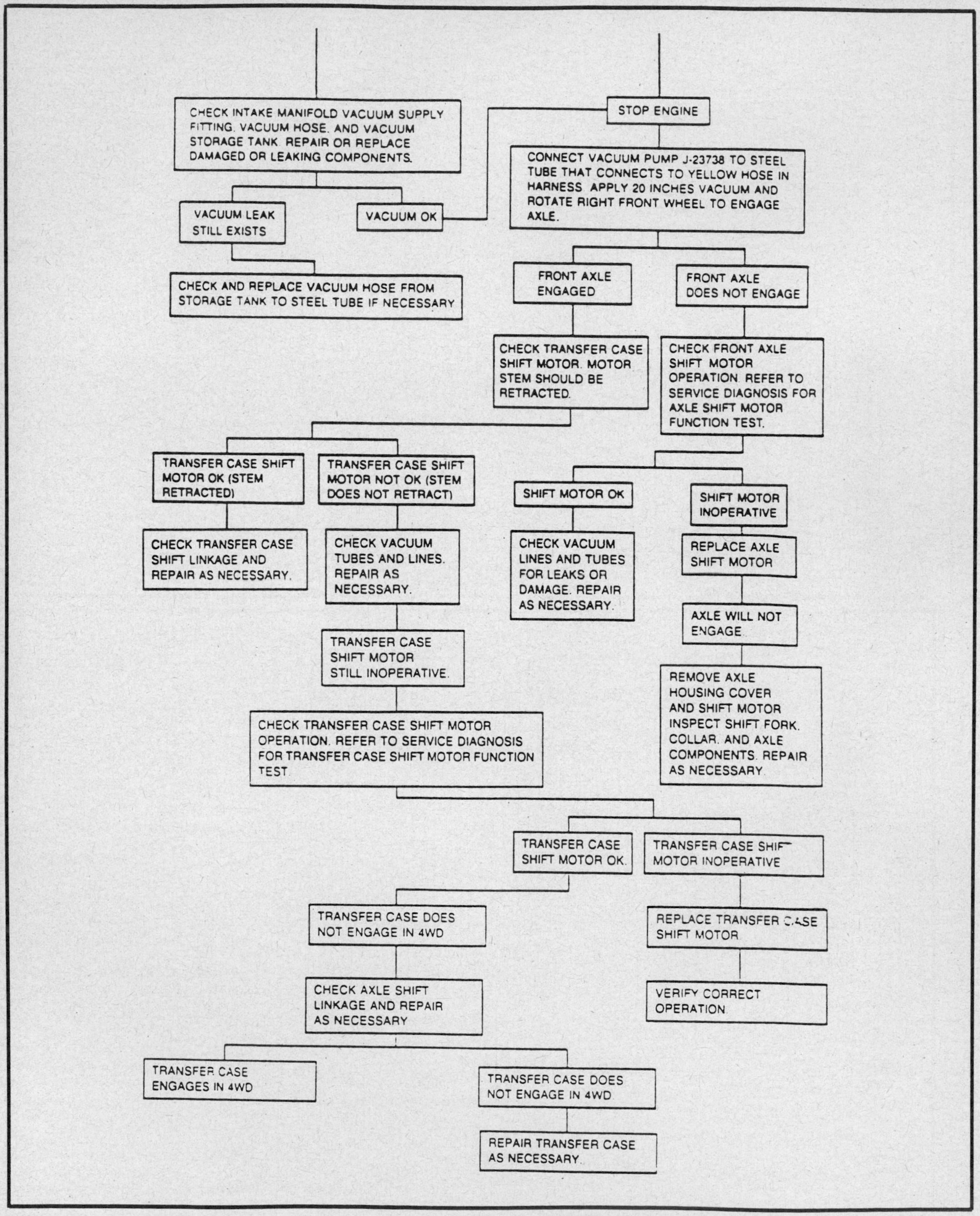
CHECK INTAKE MANIFOLD VACUUM SUPPLY FITTING, VACUUM HOSE, AND VACUUM STORAGE TANK. REPAIR OR REPLACE DAMAGED OR LEAKING COMPONENTS.
STOP ENGINE
CONNECT VACUUM PUMP J-23738 TO STEEL TUBE THAT CONNECTS TO YELLOW HOSE IN HARNESS. APPLY 20 INCHES VACUUM AND ROTATE RIGHT FRONT WHEEL TO ENGAGE AXLE.
VACUUM LEAK STILL EXISTS
VACUUM OK
CHECK AND REPLACE VACUUM HOSE FROM STORAGE TANK TO STEEL TUBE IF NECESSARY
FRONT AXLE ENGAGED
FRONT AXLE DOES NOT ENGAGE
CHECK TRANSFER CASE SHIFT MOTOR. MOTOR STEM SHOULD BE RETRACTED.
CHECK FRONT AXLE SHIFT MOTOR OPERATION. REFER TO SERVICE DIAGNOSIS FOR AXLE SHIFT MOTOR FUNCTION TEST.
TRANSFER CASE SHIFT MOTOR OK (STEM RETRACTED)
TRANSFER CASE SHIFT MOTOR NOT OK (STEM DOES NOT RETRACT)
SHIFT MOTOR OK
SHIFT MOTOR INOPERATIVE
CHECK TRANSFER CASE SHIFT LINKAGE AND REPAIR AS NECESSARY.
CHECK VACUUM TUBES AND LINES. REPAIR AS NECESSARY.
CHECK VACUUM LINES AND TUBES FOR LEAKS OR DAMAGE. REPAIR AS NECESSARY.
REPLACE AXLE SHIFT MOTOR
AXLE WILL NOT ENGAGE.
TRANSFER CASE SHIFT MOTOR STILL INOPERATIVE.
REMOVE AXLE HOUSING COVER AND SHIFT MOTOR INSPECT SHIFT FORK, COLLAR, AND AXLE COMPONENTS. REPAIR AS NECESSARY.
CHECK TRANSFER CASE SHIFT MOTOR OPERATION. REFER TO SERVICE DIAGNOSIS FOR TRANSFER CASE SHIFT MOTOR FUNCTION TEST.
TRANSFER CASE SHIFT MOTOR OK.
TRANSFER CASE SHIFT MOTOR INOPERATIVE
TRANSFER CASE DOES NOT ENGAGE IN 4WD
REPLACE TRANSFER CASE SHIFT MOTOR
CHECK AXLE SHIFT LINKAGE AND REPAIR AS NECESSARY
VERIFY CORRECT OPERATION.
TRANSFER CASE ENGAGES IN 4WD
TRANSFER CASE DOES NOT ENGAGE IN 4WD.
REPAIR TRANSFER CASE AS NECESSARY.

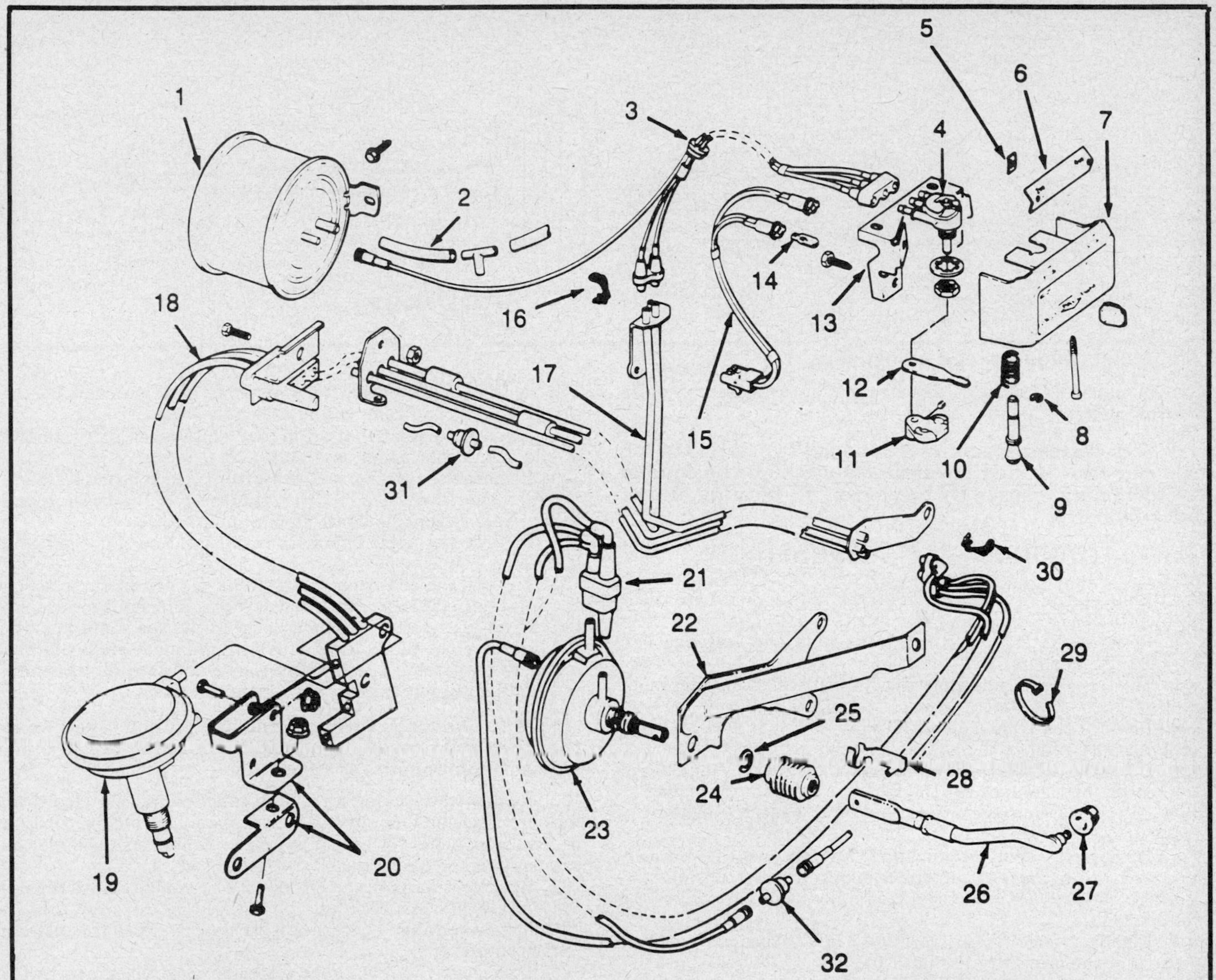
1
2
3
4
5
6
7
8
9
10
11
12
13
14
15
16
17
18
19
20
21
22
23
24
25
26
27
28
29
30
31
32

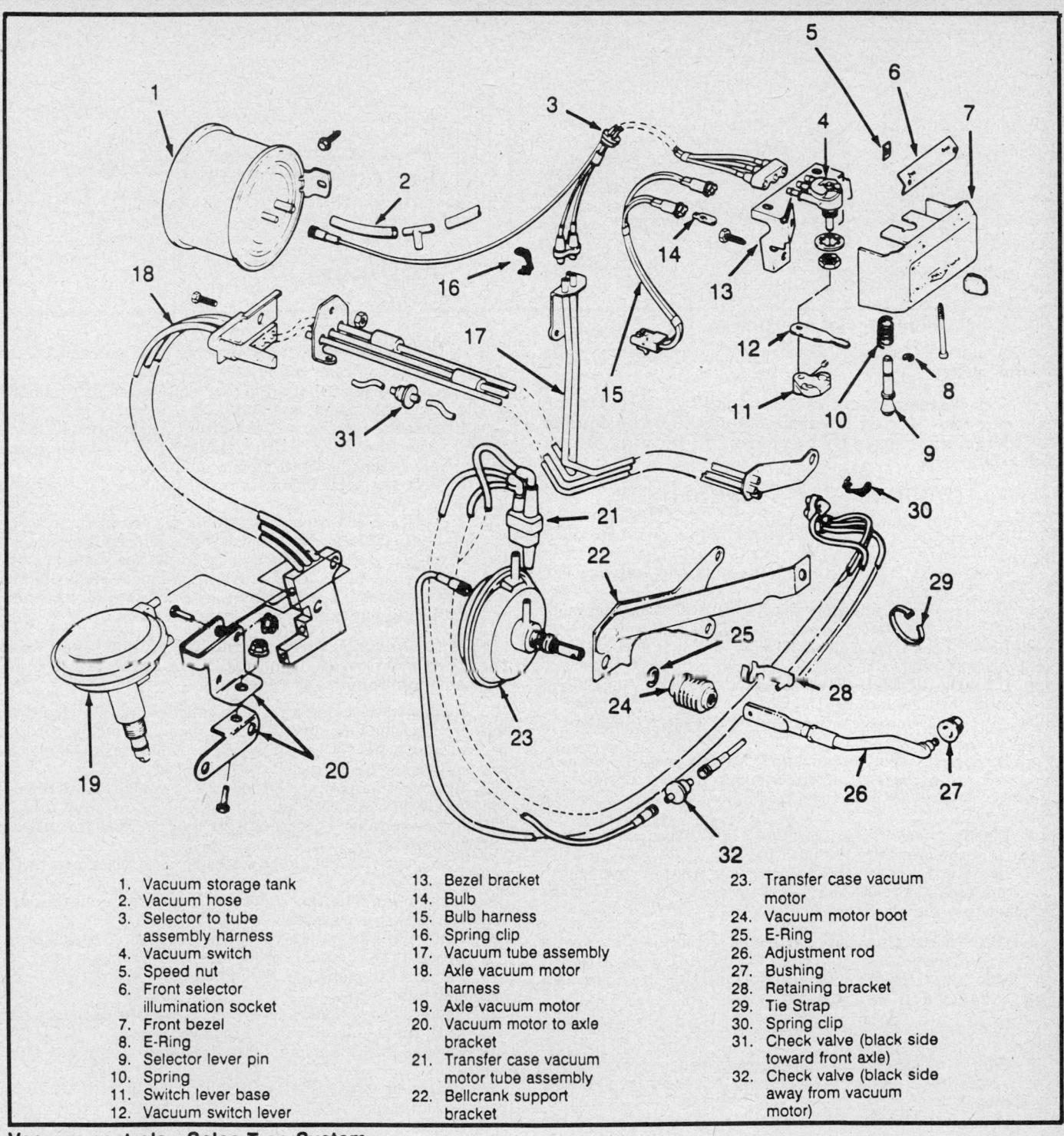

1. Vacuum storage tank
2. Vacuum hose
3. Selector to tube assembly harness
4. Vacuum switch
5. Speed nut
6. Front selector illumination socket
7. Front bezel
8. E-Ring
9. Selector lever pin
10. Spring
11. Switch lever base
12. Vacuum switch lever
13. Bezel bracket
14. Bulb
15. Bulb harness
16. Spring clip
17. Vacuum tube assembly
18. Axle vacuum motor harness
19. Axle vacuum motor
20. Vacuum motor to axle bracket
21. Transfer case vacuum motor tube assembly
22. Bellcrank support bracket
23. Transfer case vacuum motor
24. Vacuum motor boot
25. E-Ring
26. Adjustment rod
27. Bushing
28. Retaining bracket
29. Tie Strap
30. Spring clip
31. Check valve (black side toward front axle)
32. Check valve (black side away from vacuum motor)

Vacuum controls – Selec-Trac System

3. Connect the catalytic converter or bracket, if removed. Connect the parking brake cable, if removed.
4. Connect the speedometer cable or sensor.
5. Install shifter linkage.
6. Install the rear driveshaft, making certain to align matchmarks.
7. Install the front driveshaft, making certain to align matchmarks.
8. Install the vacuum lines and electrical connections.
9. Fill transfer case to correct level with Dexron®II automatic transmission fluid.
10. Install the skid plate, if equipped.
11. Lower vehicle and connect the negative battery cable.

Before Disassembly

Clean the exterior of the transfer case assembly before any attempt is made to disassembly it, in order to prevent dirt or other

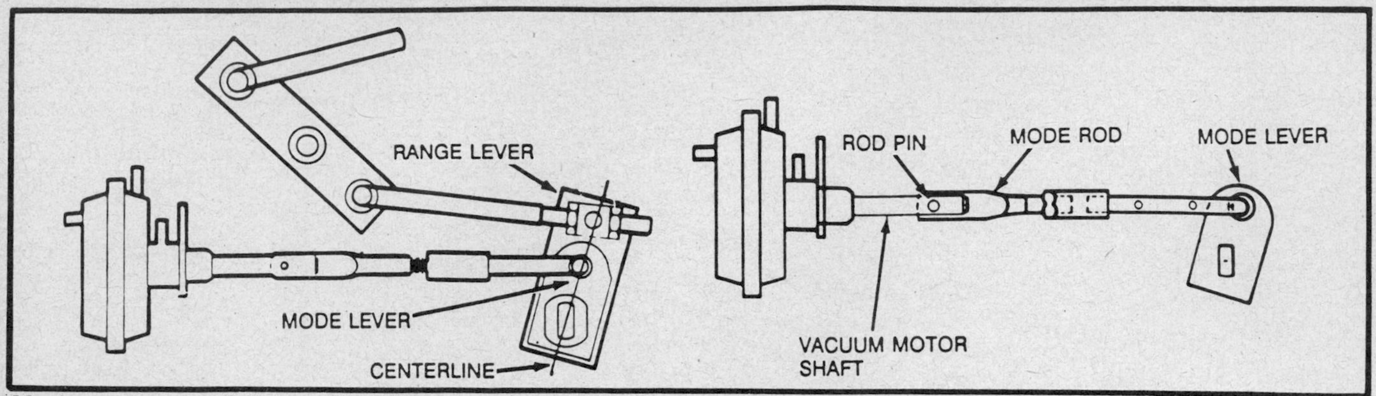
RANGE LEVER
MODE LEVER
CENTERLINE
ROD PIN
MODE ROD
MODE LEVER
VACUUM MOTOR
SHAFT

Mode and range linkage adjustment

foreign materials from entering the transfer case assembly or its internal parts.

NOTE: If steam cleaning is done to the exterior of the transfer case, immediate disassembly should be done to avoid rusting, caused by condensation forming on the internal parts.

Transfer Case Disassembly

1. Remove the drain plug and drain the lubricant from the transfer case.
2. Remove the front and rear yoke nuts and seal washers. Discard the washers.
3. Mark the front and rear yokes for installation alignment reference.
4. Remove the front and rear yokes. Use tool J–8614–01 or equivalent to remove the yokes if necessary.
5. Place the transfer case on wooden blocks. Cut V-notches in the blocks for clearance for the front case mounting studs.
6. Mark the rear retainer and rear case for assembly reference.
7. Remove the rear retainer bolts and remove the retainer. Use a pair of prybars to pry the retainer off the transfer case. Position the prybars in slots in the retainer and case to pry the retainer loose.
8. Remove the differential shim(s) and speedometer drive gear from the rear output shaft.
9. Remove the bolts attaching the rear transfer case half to the front case half. Note that the bolts used at each end of the transfer case require flat washers.

NOTE: Insert a pair of prybars in the slots at each end of the rear transfer case half to loosen it. Do not attempt to wedge the transfer case halves apart or the case mating surfaces will be damaged.

10. Remove the rear transfer case half from the front case half using a pair of prybars.
11. Remove the thrust bearing and races from the front output shaft. Note the position of the bearing and races for assembly reference.
12. Remove the oil pump from the rear output shaft. Note the position of the pump for assembly reference. The recessed side of the pump faces the case interior.
13. Remove the rear output shaft from the viscous coupling.
14. Remove the 15 mainshaft pilot bearing rollers from the shaft or coupling (if the rollers dropped off during removal of the rear output shaft).
15. Remove the mainshaft O-ring from the end of the shaft.
16. Remove the viscous coupling from the mainshaft and side gear.
17. Remove the front output shaft, driven sprocket and drive chain assembly. Lift the front shaft, sprocket and chain upward. Tilt the front shaft toward the mainshaft. Slide the chain off the drive sprocket and remove the assembly.
18. Remove the mainshaft, side gear, clutch gear, drive sprocket and spline gear as an assembly.
19. Remove the front output shaft front thrust bearing assembly from the front case, or from the shaft, if the bearing and races remained on the shaft during removal.
20. Remove the drive chain from the front output shaft and sprocket.
21. Remove the snapring that retains the driven sprocket on the front output shaft. Mark the sprocket and shaft for assembly reference and remove the sprocket from the shaft.
22. Remove the mode fork, shift rail and mode sliding clutch sleeve as an assembly. Mark the sleeve and fork for assembly reference and remove the sleeve from the fork.

NOTE: The mode fork and rail are pinned together so that they will operate as a unit. Remove the pin to separate the components if necessary.

23. Remove the locking fork, high range sliding clutch sleeve, fork brackets and fork springs as an assembly. Note the position of the components for assembly reference and disassemble the components for cleaning and inspection.
24. Remove the range sector detent screw and remove the detent spring, plunger and ball.
25. Move the range operating lever downward to the last detent position.
26. Disengage the low range fork lug from the range sector slot.
27. Remove the retaining snapring from the annulus gear and remove the thrust washer.
28. Remove the annulus gear, range fork and rail as an assembly. Separate the components for cleaning and inspection.
29. Remove the planetary thrust washer from the planetary assembly hub.
30. Remove the planetary assembly. Grasp the planetary hub and lift the assembly upward to remove it.
31. Remove the mainshaft thrust bearing from the input shaft.
32. Remove the input shaft and remove the input shaft thrust bearing and race.
33. Remove the range sector and operating lever attaching nut and lockwasher. Remove the lever.
34. Remove the range sector and shaft from the front case.
35. Remove the range sector O-ring and retainer.

Unit Disassembly and Assembly

MAINSHAFT

Disassembly

1. Grasp the drive sprocket and lift the sprocket clutch gear and side gear upward and off the mainshaft.

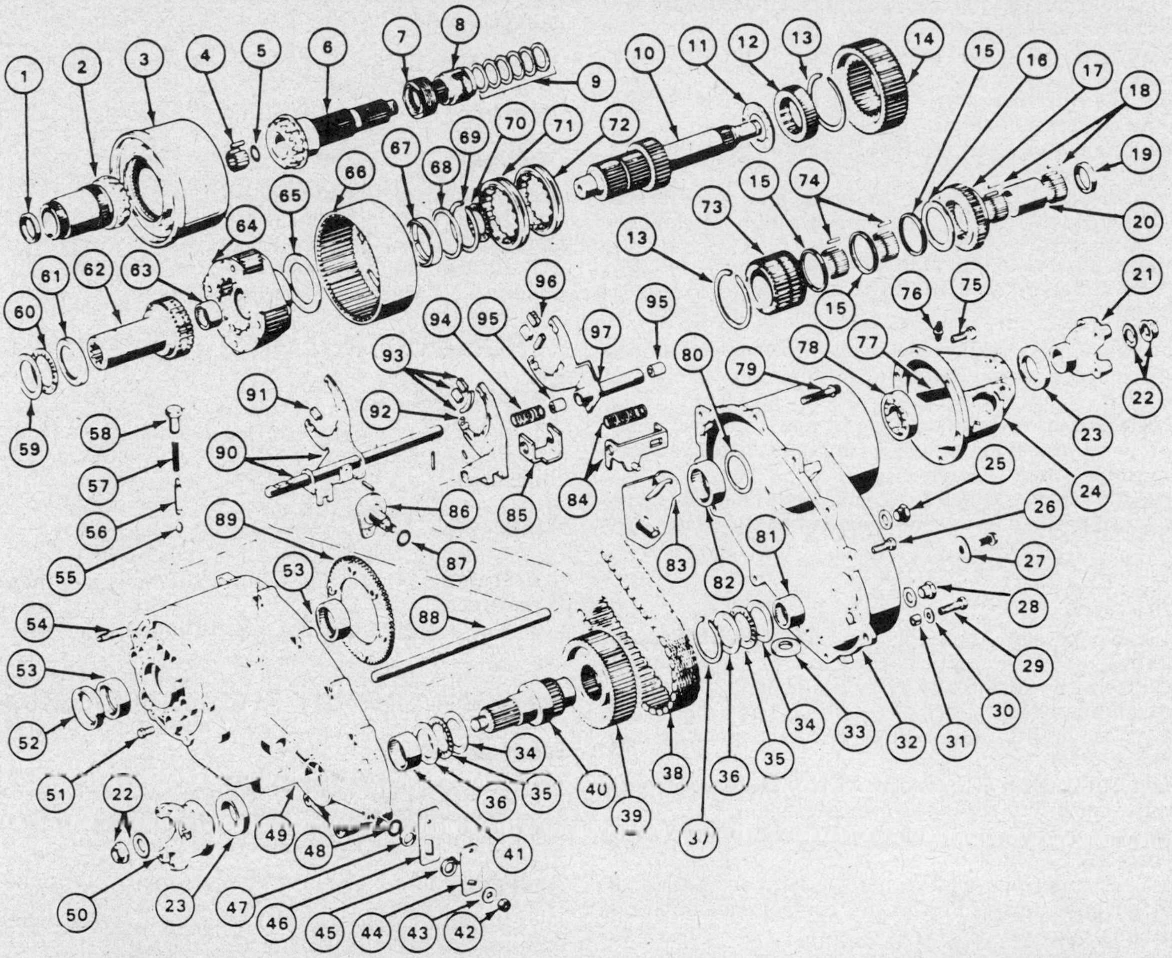

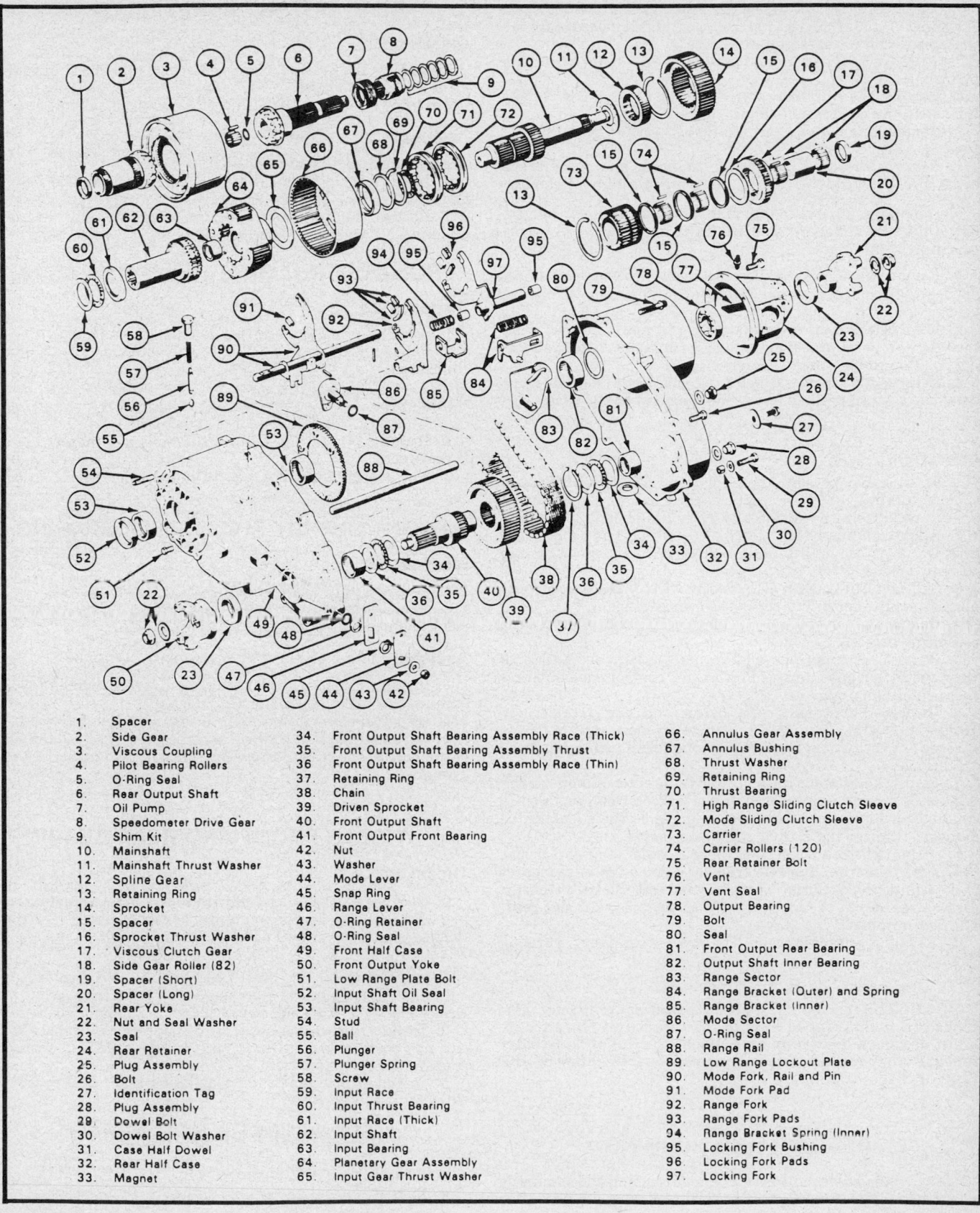

1. Spacer
2. Side Gear
3. Viscous Coupling
4. Pilot Bearing Rollers
5. O-Ring Seal
6. Rear Output Shaft
7. Oil Pump
8. Speedometer Drive Gear
9. Shim Kit
10. Mainshaft
11. Mainshaft Thrust Washer
12. Spline Gear
13. Retaining Ring
14. Sprocket
15. Spacer
16. Sprocket Thrust Washer
17. Viscous Clutch Gear
18. Side Gear Roller (82)
19. Spacer (Short)
20. Spacer (Long)
21. Rear Yoke
22. Nut and Seal Washer
23. Seal
24. Rear Retainer
25. Plug Assembly
26. Bolt
27. Identification Tag
28. Plug Assembly
29. Dowel Bolt
30. Dowel Bolt Washer
31. Case Half Dowel
32. Rear Half Case
33. Magnet
34. Front Output Shaft Bearing Assembly Race (Thick)
35. Front Output Shaft Bearing Assembly Thrust
36. Front Output Shaft Bearing Assembly Race (Thin)
37. Retaining Ring
38. Chain
39. Driven Sprocket
40. Front Output Shaft
41. Front Output Front Bearing
42. Nut
43. Washer
44. Mode Lever
45. Snap Ring
46. Range Lever
47. O-Ring Retainer
48. O-Ring Seal
49. Front Half Case
50. Front Output Yoke
51. Low Range Plate Bolt
52. Input Shaft Oil Seal
53. Input Shaft Bearing
54. Stud
55. Ball
56. Plunger
57. Plunger Spring
58. Screw
59. Input Race
60. Input Thrust Bearing
61. Input Race (Thick)
62. Input Shaft
63. Input Bearing
64. Planetary Gear Assembly
65. Input Gear Thrust Washer
66. Annulus Gear Assembly
67. Annulus Bushing
68. Thrust Washer
69. Retaining Ring
70. Thrust Bearing
71. High Range Sliding Clutch Sleeve
72. Mode Sliding Clutch Sleeve
73. Carrier
74. Carrier Rollers (120)
75. Rear Retainer Bolt
76. Vent
77. Vent Seal
78. Output Bearing
79. Bolt
80. Seal
81. Front Output Rear Bearing
82. Output Shaft Inner Bearing
83. Range Sector
84. Range Bracket (Outer) and Spring
85. Range Bracket (Inner)
86. Mode Sector
87. O-Ring Seal
88. Range Rail
89. Low Range Lockout Plate
90. Mode Fork, Rail and Pin
91. Mode Fork Pad
92. Range Fork
93. Range Fork Pads
94. Range Bracket Spring (Inner)
95. Locking Fork Bushing
96. Locking Fork Pads
97. Locking Fork

Jeep Select Trac 229

2. Remove the mainshaft needle bearings and bearing spacers from the mainshaft, a total of 82 bearings are used, note the spacer position for assembly reference.
3. Remove the spline gear and thrust washer from the mainshaft.
4. Remove the side gear, clutch gear and clutch gear thrust washer from the sprocket carrier and sprocket.
5. Remove the clutch gear and thrust washer from the side gear.
6. Remove the sprocket carrier snapring and remove the drive sprocket from the carrier, mark for assembly reference.

NOTE: The sprocket carrier and mainshaft needle bearings are different in size. Take care to avoid intermixing them.

7. Remove the 3 bearing spacers and all sprocket carrier needle bearings from the carrier, a total of 120 needle bearings are used.
8. Remove the rear output bearing and rear yoke seal from the rear retainer, the bearing is shielded on a side, note the bearing position for assembly reference.
9. Remove the input gear and front yoke seals from the front case, use a small pry bar to pry the seals out of the case.

Inspection

1. Wash all components thoroughly in clean solvent. Ensure that all lubricant, metallic particles, dirt and foreign material are removed from the surfaces of every component.
2. Apply compressed air to each oil supply port and channel in each transfer case half to remove any obstructions or cleaning solvent residue.
3. Inspect all gear teeth for excessive wear or damage. Inspect all gear splines for burrs, nicks, wear or damage.
4. Remove minor nicks or scratches. Replace any component exhibiting excessive wear or damage.
5. Inspect all snaprings and thrust washers for excessive wear, distortion and damage. Replace any component exhibiting these conditions.
6. Inspect the transfer case halves and rear retainer for cracks, porosity, damaged mating surfaces, stripped bolt threads and distortion. Replace any component exhibiting these conditions.
7. Inspect the viscous coupling and differential pinions. If the pinions or carrier are damaged or worn excessively, replace the coupling as an assembly. If the coupling is cracked, leaking, or damaged, replace the coupling as an assembly.
8. Inspect the condition of all needle, roller, ball and thrust bearings in the front and rear transfer case halves. Also inspect to determine the condition of the bearing bores in both transfer case halves and in the input gear, rear output shaft, side gear, and rear retainer.
9. Replace any component that is excessively worn or damaged.

NOTE: The front output shaft thrust bearing race surfaces are heat treated during manufacture. Heat treatment causes a brown or blue discoloration of these surfaces. Do not replace a front output shaft because of this type of discoloration.

BEARING AND BUSHING

All of the bearings used in the transfer case must be correctly positioned to avoid blocking the bearing oil supply holes. After replacing any bearing, check the bearing position and ensure that the supply hole is not obstructed by the bearing.

REAR OUTPUT SHAFT BEARING

Disassembly

1. Remove the bearing, using remover tool J-26941 or equivalent and slide hammer.
2. Remove the rear output lip seal using a small awl.

Assembly

1. Install a replacement lip seal.
2. Install a replacement bearing, using driver handle J-8092 and installer tool J-29166 or equivalent.
3. Remove the tools and inspect the oil supply hole. The bearing must not obstruct the supply hole.

FRONT OUTPUT SHAFT FRONT BEARING

Disassembly

1. Secure output shaft in tool J-8092, or equivalent.
2. Remove the bearing, using tools J-8092 and J-29168 or equivalent.

Assembly

1. Install new bearing, remove the tools and inspect the oil supply hole.
2. The bearing must not obstruct the supply hole.

FRONT OUTPUT SHAFT REAR BEARING

Disassembly

1. Secure output shaft in tool J-8092, or equivalent.
2. Remove the bearing, using remover tool J-26941 or equivalent and slide hammer.

Assembly

1. Install a replacement bearing, using driver handle J-8092 and installer tool J-29163 or equivalent.
2. Inspect the bearing position to ensure the oil supply hole is not obstructed and that the bearing is seated flush with the edge of the bore in the case to allow clearance for the thrust bearing assembly.

INPUT GEAR FRONT AND REAR BEARINGS

Disassembly

1. Secure input gear assembly in tool J-8092, or equivalent.
2. Remove both bearings simultaneously, using driver handle J-8092 and remover tool J-29170 or equivalent.

Assembly

1. Install the rear bearing, then install the front bearing. Use driver handle J-8092 and installer tool J-29169 or equivalent.
2. Inspect the bearing position to ensure the oil supply holes are not obstructed and that the bearings are flush with the transfer case bore surfaces.
3. Install a replacement oil seal, using seal installer tool J-29162 or equivalent.

MAINSHAFT PILOT BUSHING

Disassembly

1. Secure assembly in soft-jawed vise.
2. Remove the bushing, using slide hammer J-2619-01 and remover tool J-29369-1 or equivalent.

Assembly

1. Install a replacement bearing, using driver handle J–8092 and installer tool J–29174 or equivalent.
2. Inspect bushing position to ensure that the oil supply hole is not obstructed.

ANNULUS GEAR BUSHING

Disassembly

1. Secure assembly in soft-jawed vise.
2. Remove the bushing, using driver handle J–8092 and remover/installer tool J–29185 or equivalent.

Assembly

1. Install a replacement bushing, using tools J–8092 and J–29185–2 or equivalent.
2. Remove any chips generated by the bushing removal and/or installation.

REAR OUTPUT BEARING AND YOKE SEAL

Disassembly

1. Remove the bearing, using a brass drift and hammer.
2. Remove the seal from the retainer, using a brass drift and hammer.

Assembly

NOTE: The rear output bearing is shielded on a side. Ensure that the shielded side faces the transfer case interior after installation.

1. Install a replacement bearing, using driver handle J–8092 and installer tool J–7818 or equivalent.
2. Install a replacement seal in the retainer, using tool J–29162 or equivalent.

Transfer Case Assembly

1. Install a replacement input shaft and rear output shaft oil seals. Seat the seals flush with the edge of the seal bore or in the seal groove of the transfer case. Coat the seal lips with petroleum jelly after installation.
2. Install the input shaft thrust bearing race in the transfer case counterbore.
3. Install the input shaft thrust bearing on the input shaft and install the shaft and bearing in the transfer case.
4. Install the mainshaft thrust bearing in the bearing recess

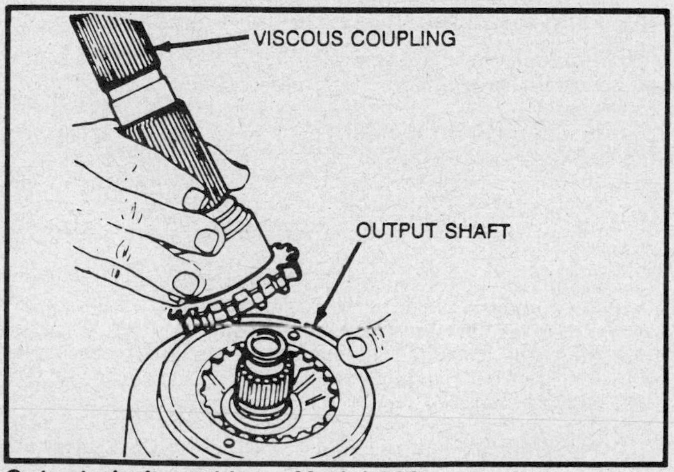
VISCOUS COUPLING
OUTPUT SHAFT
Output shaft position—Model 229

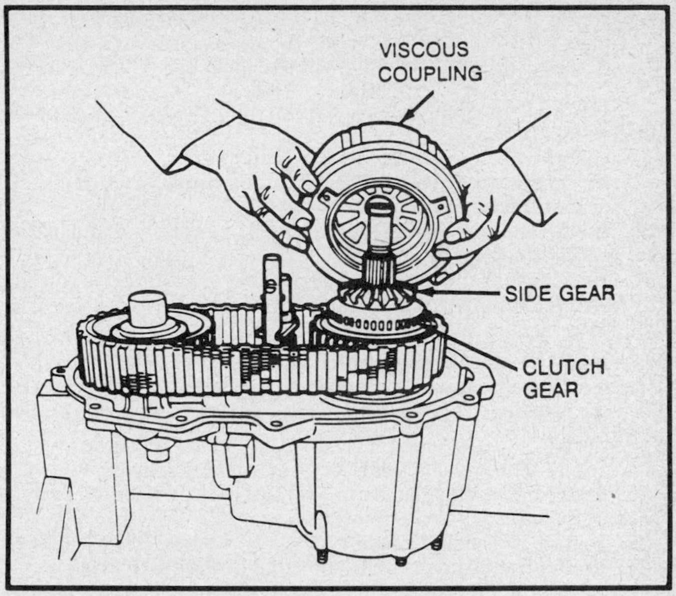
VISCOUS
COUPLING
SIDE GEAR
CLUTCH
GEAR
Viscous Coupling position—Model 229

in the input shaft.

5. Install the planetary assembly on the input shaft. Make certain the planetary pinion teeth mesh fully with the input shaft.
6. Install the planetary thrust washer on the planetary hub.
7. Install a replacement sector shaft O-ring and install the retainer in the shaft bore in the transfer case.
8. Install the O-ring on the mode sector shaft and insert the mode sector through the range sector.
9. Install the range sector in the front transfer case half. Install the operatin lever and the snapring on the range sector shaft.
10. Install the lever, attaching washer and locknut on the mode sector shaft. Tighten the locknut to 17 ft. lbs. of torque.
11. Assemble the annulus gear, range fork and rail.
12. Install the assembled fork on and over the planetary pinions.
13. Ensure the range sector lug into the range sector.
14. Engage the range sector lug into the range sector.
15. Install the annulus thrust washer and the annulus retaining ring on the annulus gear hub.
16. Install the detent ball, plunger, spring and retaining screw in the front transfer case half detent bore.
17. Tighten the bolt to 22 ft. lbs. of torque.

NOTE: The locking mode clutch sleeve and the high range clutch sleeve are not interchangeable. The sleeve splines are different. Ensure that the correct sleeve is installed in the corresponding shift fork.

18. Assemble and install the locking fork, fork bracket, fork springs and high range clutch sleeves.
19. Ensure that the lug on the fork is seated in the range selector detent slot.
20. Install the range fork lug in the range sector detent slot.
21. Move the range fork lug in the range sector detent notch.
22. Assemble and install the range fork, shift rail and mode clutch sleeve.
23. Install the thrust washer and replacement O-ring on the mainshaft.
24. Install the needle bearings and bearing spacers on the mainshaft.
25. Coat the shaft bearing surface and all needle bearings with petroleum jelly.
26. Install the first 41 needle bearings.

27. Install the long bearing spacer, the remaining 41 needle bearings and the remaining short spacer.
28. Be careful to avoid displacing the bearings when the spacers are installed.
29. Use additional petroleum jelly to hold the bearing in place, if necessary.
30. Install the spline gear on the mainshaft.
31. Be careful to avoid displacing the bearings when the spacers are installed.
32. Install the sprocket carrier in the drive sprocket and install the sprocket carrier snaprings.
33. Ensure that the carrier and sprocket are aligned according to the reference marks made during disassembly.

NOTE: The sprocket carrier teeth are tapered on a side and the drive sprocket has a deep recess on a side. Ensure that these components are assembled so that the carrier tapered teeth and sprocket recess are on the same side.

34. Install the sprocket carrier bearing and spacers.
35. Coat the carrier bore and the 120 carrier needle bearings with petroleum jelly.
36. Install the center spacer.
37. Install the 60 bearings in each end of the carrier and install the remaining 2 spacers at each side of the carrier.
38. Use additional petroleum jelly to hold the bearing in place, if necessary.
39. Install the assembled sprocket carrier and drive sprocket on the mainshaft. Be careful to avoid displacing the bearings when installed.
40. Ensure that the recessed side of the drive sprocket is facing downward.
41. Install the clutch gear thrust washer in the mainshaft.
42. Position the washer on the sprocket carrier.
43. Install the clutch gear on the side gear.
44. Ensure that the tapered edge of the clutch gear faces the side gear teeth.
45. Install the assembled side gear and clutch gear on the mainshaft. Ensure that the side gear is fully seated in the sprocket carrier.
46. Take care to avoid displacing the bearings when installed.
47. Install the mainshaft and gear assembly in the case.
48. Ensure that the mainshaft is fully seated in the input gear.
49. Install the driven sprocket on the front output shaft and install the sprocket retaining snapring. Ensure that the sprocket is install according to the reference marks made during disassembly.

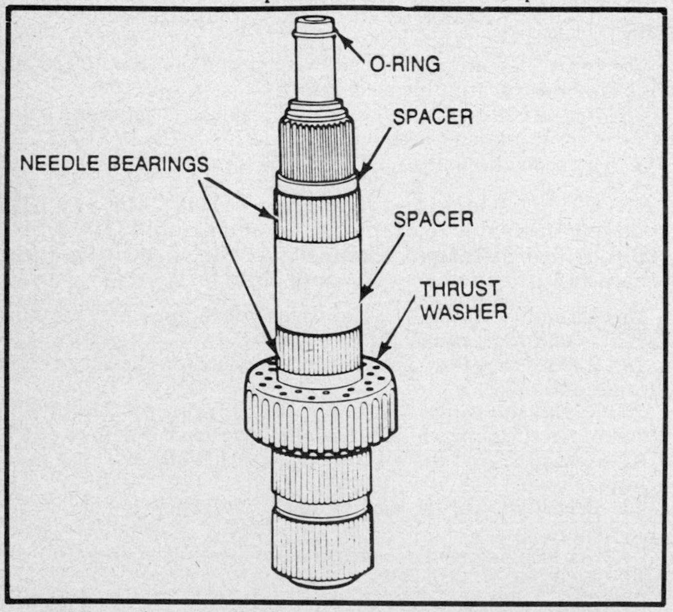
O-RING
SPACER
NEEDLE BEARINGS
SPACER
THRUST
WASHER

Installing the needle bearings and spacers on the mainshaft

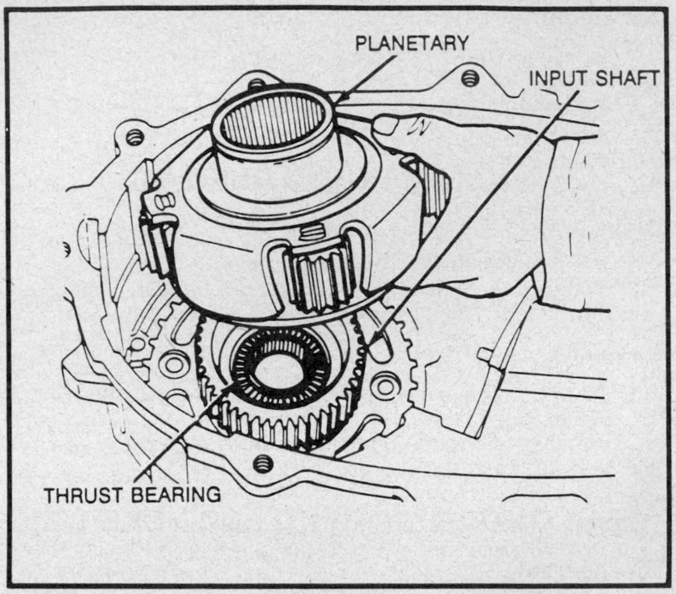
PLANETARY
INPUT SHAFT
THRUST BEARING

Installing the planetary assembly—Model 229

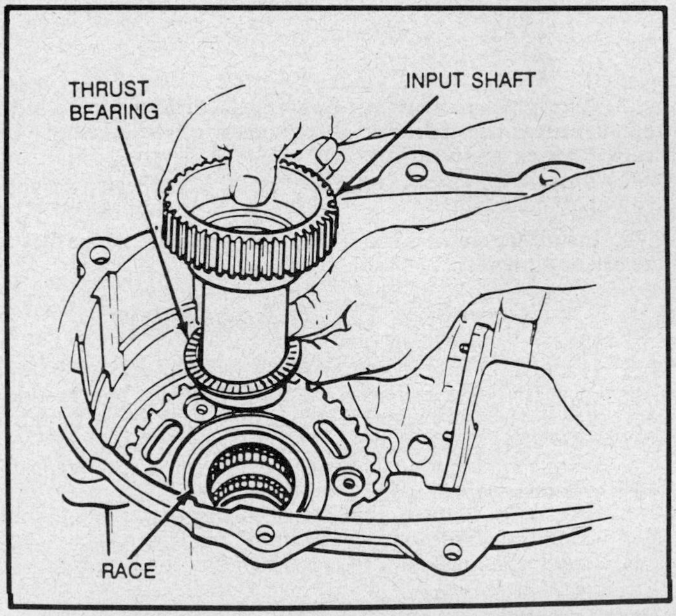
THRUST
BEARING
INPUT SHAFT
RACE

Installing the input shaft assembly—Model 229

50. Install the front output shaft's front thrust washer bearing assembly in the transfer case front half.
51. Install the thick race in the transfer case and then install the bearing and the thin race.
52. Install the drive chain, front output shaft and driven sprocket.
53. Install the chain on the driven sprocket.
54. Raise and tilt the driven sprocket and chain and install the opposite end of the chain on the drive sprocket.
55. Align the front output shaft with the shaft bore in the transfer case front half and install the shaft in the transfer case.
56. Ensure that the front shaft thrust bearing assembly is seated in the transfer case.
57. Install the front output shaft rear thrust bearing assembly on the front output shaft.

58. Install the thin race, then install the bearing and thick race.
59. Install the viscous coupling on the side gear and clutch gear.
60. Ensure that the coupling is fully seated on the clutch gear. The clutch gear should be flush with the coupling and the gear teeth should be visible.
61. Coat the mainshaft pilot bearing surface and all 15 pilot bearings with petroleum jelly and install the bearing on the shaft.
62. Use additional petroleum jelly to hold the bearing in place, if necessary.
63. Install the rear output shaft on the mainshaft and into the viscous coupling. Ensure that the shaft is completely seated in the coupling.
64. Tap the shaft with a plastic mallet or brass punch to seat it, if necessary.
65. Do not displace the pilot bearing during installation of the shaft.
66. Install a replacement rear output shaft bearing seal in the rear transfer case half.
67. Apply a bead of Loctite® 515, or equivalent, to the mating surface of the rear transfer case half.
68. Install the magnet in the case, if removed.
69. Attach the rear transfer case half to the front transfer case half. Ensure that the alignment dowels at the front case half ends are aligned with the bolt holes in the rear case half and mate the rear case half with the front case half.

NOTE: If the rear transfer case half with not mate completely with the front case, inspect for the following: oil in the range fork rail bore, the front output shaft rear thrust bearing assembly is not aligned with the rear case half, the mainshaft is not completely seated, the rear case half is not aligned with the oil pump.

70. Install the rear case half to the front case half bolts. Tighten the bolts with 23 ft. lbs. of torque. Ensure that the flat washers are used on both bolts at the case end where the alignment dowels are located.
71. Install the speedometer drive gear on the rear output shaft.
72. Measure and record the thickness of the shim pack.
73. Install a 0.030 in. shim on the rear output shaft.
74. Align the rear retainer on the rear transfer case half and install the retainer. Install the retainer bolts. Tighten bolts securely but not to specified torque.
75. Install the front and rear output shaft yokes and original yoke nuts. Tighten nuts finger tight and check differential endplay.
76. Set the shift lever in the **4WD–HI** range.
77. Position a dial indicator on the rear retainer and position the indicator stylus so it contacts the rear yoke nut.
78. Support the transfer case to prevent the front output yoke from turning.
79. Slowly turn the rear output shaft while maintaining moderate inward pressure on the rear yoke. Turn the rear output shaft at least 2 full turns to determine the maximum run-out of the shaft.

NOTE: A wrench should be used to turn the yoke to provide the leverage needed to turn the viscous coupling in the transfer case.

80. Set the shaft at its maximum run-out point and zero the dial indicator.
81. Pull upward on the rear output yoke and record the dial indicator measurement.
82. Remove the retainer. Add or subtract differential shims as necessary to correct the endplay to 0.002 – 0.010 in.
83. Remove front and rear yokes and discard the old nuts.
84. Apply a bead of Loctite® 515, or equivalent, to the retainer mating surface.
85. Apply sealer to the retainer bolts and install the bolts. Torque the bolts to 23 ft. lbs.
86. Position the front and rear yokes. Install the replacement yoke seal washers and nuts.
87. Torque the yoke nuts to 120 ft. lbs., while holding yokes in place.
88. Install the detent ball, spring and bolt if not previously installed. Apply sealer to bolt, install and torque to 23 ft. lbs.
89. Install the drain plug and washer.
90. Fill the transfer case to bottom of fill hole, with approximately 6 pts. of Dexron®II automatic transmission fluid.
91. Install the fill plug. Torque the fill and drain plugs to 18 ft. lbs.
92. Install the transfer case into vehicle.

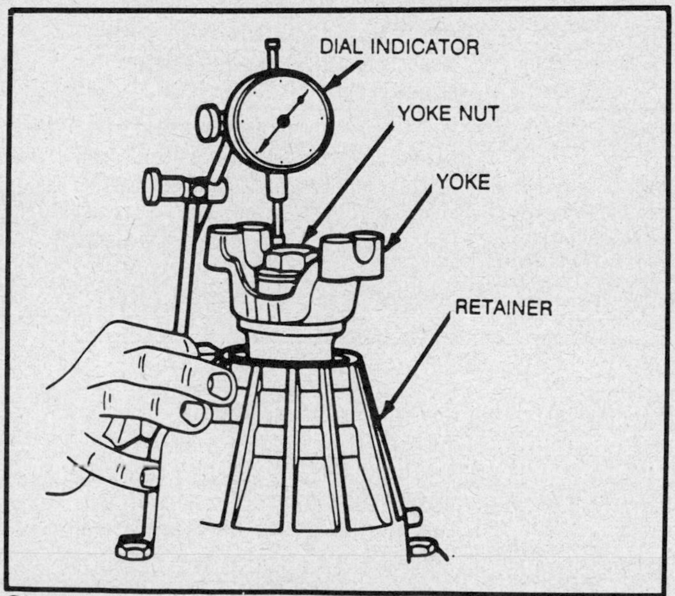
DIAL INDICATOR
YOKE NUT
YOKE
RETAINER

Checking endplay – Model 229

Specifications

TORQUE SPECIFICATIONS Model 229

Description	ft. lbs.	Nm
Detent retainer bolt	20–25	27–34
Drain and fill plugs	15–20	20–34
Front/rear yoke nuts	90–130	122–176
Operating lever locknut	14–20	19–27
Rear case-to-front case bolts (All)	20–25	27–34
Rear retainer bolts	20–25	27–34
Transfer case-to-transmission adapter nuts	22–30	38–41
Universal joint strap bolt-to-transfer case	140–200 ①	16–23

① Inch lbs.

NEW PROCESS 231 TRANSFER CASE

Application

1988–89 Dodge Dakota
1987-89 General Motors S-10 and S-15 Series Trucks
1987-89 Jeep Cherokee, Comanche, Wagoneer and Wrangle

General Description

The NP231 is a part time unit with a built-in low range reduction gear system. It has 3 operating ranges plus a Neutral position. The low-range system provides a 2.72:1 gear reduction ratio for increased low-speed torque capacity. The unit has a 2 piece aluminum housing assembly. On the front case half, the front output shaft, front input shaft, 4WD indicator switch and shift lever assembly are located. On the rear case half, the rear output shaft, bearing retainer and drain and fill plugs are located.

Trouble Diagnosis

CHILTON'S THREE C's TRANSFER CASE DIAGNOSIS

Condition	Cause	Correction
Transfer case difficult to shift or will not shift into desired range	a) Vehicle speed too great to permit shifting	a) Stop vehicle and shift into desired range. Or reduce speed to 2–3 mph (3–4 km/h) before attempting to shift
	b) If vehicle was operated for extended period in 4H mode on dry paved surface, driveline torque load may cause difficult shifting	b) Stop vehicle, shift transmission to neutral, shift transfer case to 2H mode and operate vehicle in 2H on dry paved surfaces
	c) Transfer case external shift linkage binding	c) Lubricate or repair or replace linkage, or tighten loose components as necessary
	d) Insufficient or incorrect lubricant	d) Drain and refill to edge of fill hole with Dexron®II only
	e) Internal components binding, worn or damaged	e) Disassemble unit and replace worn or damaged components as necessary
Transfer case noisy in all drive modes	a) Insufficient or incorrect lubricant	a) Drain and refill to edge of fill hole with Dexron®II only. Check for leaks and repair if necessary. Note: If unit is still noisy after drain and refill, disassembly and inspection may be required to locate source of noise
Noisy in—or jumps out of four wheel drive low range	a) Transfer case not completely engaged in 4L position	a) Stop vehicle, shift transfer case in Neutral, then shift back into 4L position
	b) Shift linkage loose or binding	b) Tighten, lubricate or repair linkage as necessary
	c) Range fork cracked, inserts worn, or fork is binding on shift rail	c) Disassemble unit and repair as necessary
	d) Annulus gear or lockplate worn or damaged	d) Disassemble unit and repair as necessary
Lubricate leaking from output shaft seals or from vent	a) Transfer case overfilled	a) Drain to correct level
	b) Vent closed or restricted	b) Clear or replace vent if necessary
	c) Output shaft seals damaged or installed incorrectly	c) Replace seals. Be sure seal lip faces interior of case when installed. Also be sure yoke seal surfaces are not scored or nicked. Remove scores, nicks with fine sandpaper or replace yoke(s) if necessary
Abnormal tire wear	a) Extended operation on dry hard surface (paved) roads in 4H range	a) Operate in 2H on hard surface (paved) roads

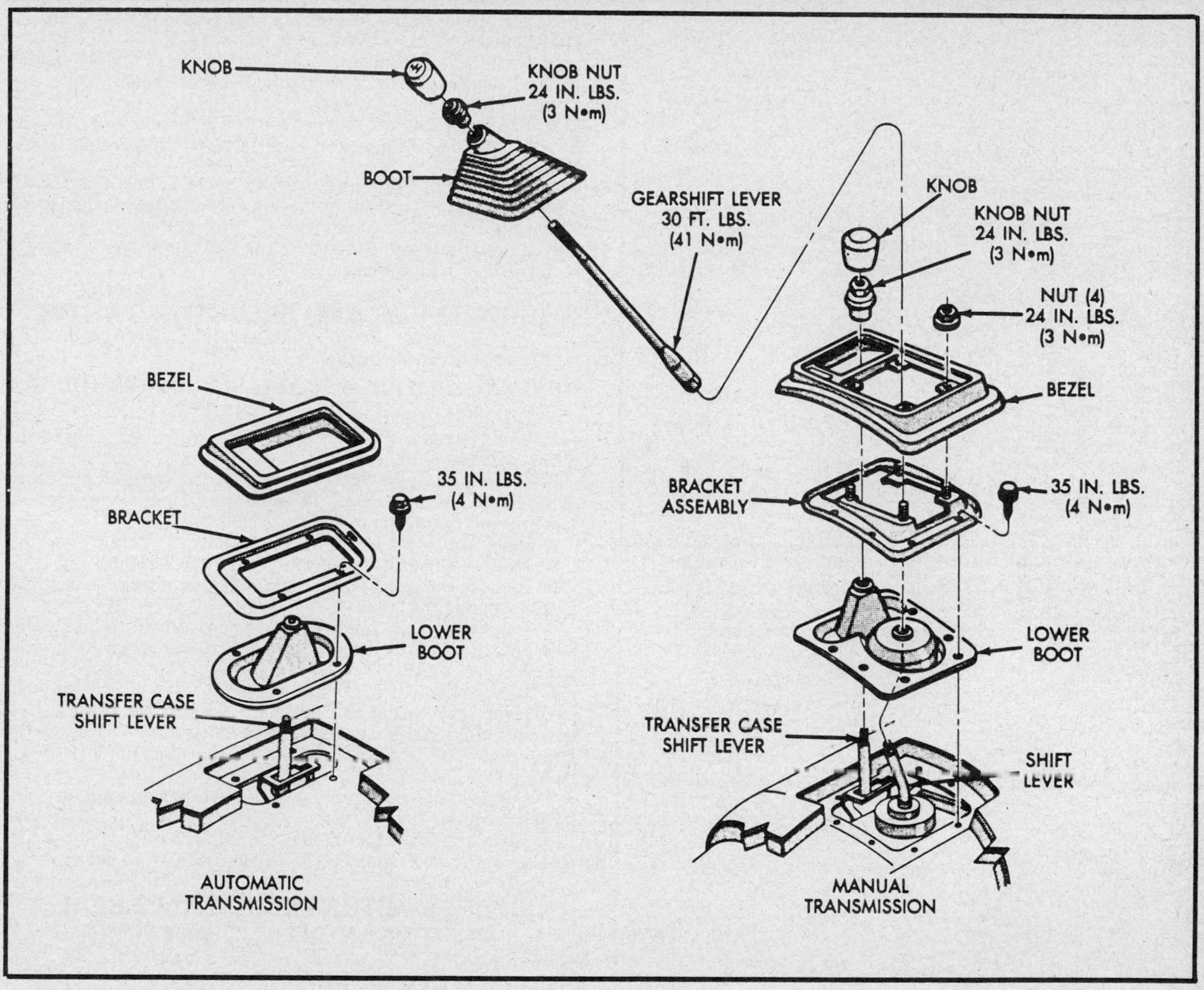
KNOB
KNOB NUT
24 IN. LBS.
(3 N•m)
BOOT
GEARSHIFT LEVER
30 FT. LBS.
(41 N•m)
KNOB
KNOB NUT
24 IN. LBS.
(3 N•m)
NUT (4)
24 IN. LBS.
(3 N•m)
BEZEL
BEZEL
35 IN. LBS.
(4 N•m)
BRACKET
BRACKET
ASSEMBLY
35 IN. LBS.
(4 N•m)
LOWER
BOOT
LOWER
BOOT
TRANSFER CASE
SHIFT LEVER
TRANSFER CASE
SHIFT LEVER
SHIFT
LEVER
AUTOMATIC
TRANSMISSION
MANUAL
TRANSMISSION

Exploded view of the transfer case shift linkage – Dodge Dakota

On Vehicle Services

SHIFT LINKAGE

Adjustment

DODGE DAKOTA

1. Place the shift lever in the **4H** position.
2. Remove the screws attaching the shift lever boot to the floorpan. Slide the boot upward to provide access to the shift gate.
3. Insert a ⅛ in. (3mm) spacer between the shift lever and the forward edge of the shift lever gate. Secure the lever and spacer in position with tape or wire.
4. Raise and support the vehicle safely. Loosen the adjusting link enough to allow the linkage to slide freely in the link.
5. Move the tranfer case range lever to the **4H** position. Position the linkage so it is a free fir in the range lever. Tighten the setscrew, in adjusting link, securely.
6. Lower the vehicle. Remove the shift lever spacer and install the boot to the floorpan.

GENERAL MOTORS S-10 AND S-15 SERIES TRUCKS

1. Remove the console assembly. Raise the upper boot on the shift lever.
2. Loosen the shift assembly lock bolt. Loosen the pivot bolt. Place the transfer case shifter lever in the **4HI** position.
3. Insert a $^{5}/_{8}$ (8mm) drill bit through the shifter into the bracket.
4. Install a bolt at the lower transfer case shift lever. This will lock the transfer case in the **4HI** position.
5. Remove the bolt from the lower transfer case shift lever. Remove the drill bit.
6. Install the shift assembly lock bolt and torque it to 30 ft. lbs. Torque the shift lever through bolt with the grease fitting on it to 96 ft. lbs.
7. Install the upper boot and the console assembly.

JEEP CHEROKEE, COMANCHE AND WAGONEER

1. Remove the shift lever boot. Place the shift lever in the **4L** position.
2. Insert a 0.157 in. (4mm) spacer between the shift lever and

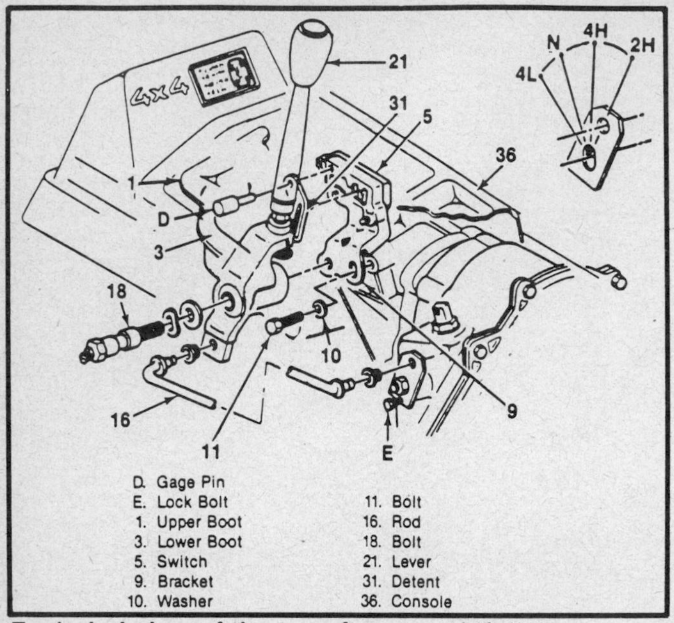
4x4
N
4H
2H
4L
21
31
5
36
1
D
3
18
10
9
16
11
E
D. Gage Pin
E. Lock Bolt
1. Upper Boot
3. Lower Boot
5. Switch
9. Bracket
10. Washer
11. Bolt
16. Rod
18. Bolt
21. Lever
31. Detent
36. Console

Exploded view of the transfer case shift linkage—General Motors S-10, S-15 and light trucks

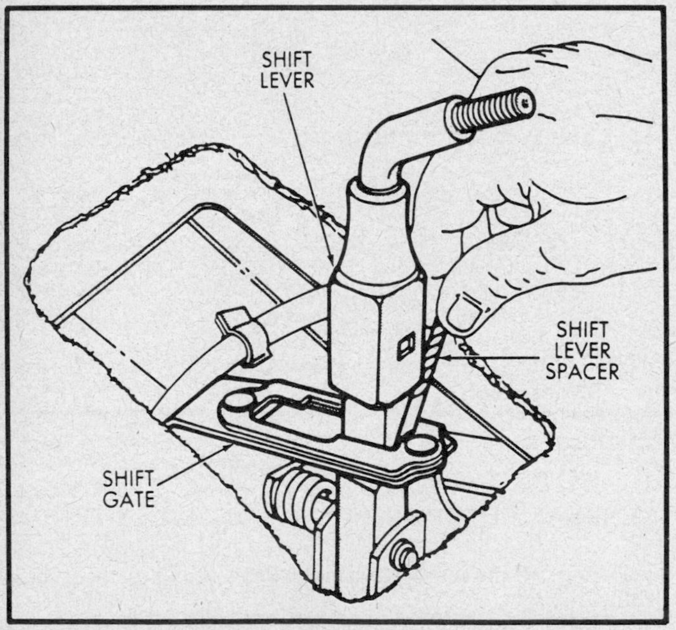
SHIFT
LEVER
SHIFT
LEVER
SPACER
SHIFT
GATE

Installing the shift lever spacer—Jeep Wrangler

the forward edge of the shift lever gate. Secure the lever and spacer in position with tape or wire.

3. Raise and support the vehicle safely. Loosen the trunnion lock bolt. The linkage rod should now slide freely in the trunnion.

4. Be sure that the tranfer case range lever is still in the **4L** position. Position the linkage so it is a free fit in the range lever. Tighten the trunnion locknut.

5. Lower the vehicle. Remove the shift lever spacer and install the boot and bezel.

JEEP WRANGLER

1. Remove the transfer case shift knob locknut and remove the knob. Remove the shift lever boot. Place the shift lever in the **4L** position.

2. Insert a ⅛ in. (3mm) spacer between the shift lever and the forward edge of the shift lever gate. Secure the lever and spacer in position with tape or wire.

3. Raise and support the vehicle safely. Loosen the trunnion lock bolt. The linkage rod should now slide freely in the trunnion.

4. Be sure that the tranfer case range lever is still in the **4L** position. Position the linkage so it is a free fit in the range lever. Tighten the trunnion locknut.

5. Lower the vehicle. Remove the shift lever spacer and install the boot and shift knob.

SHIFT LEVER AND SELECTOR SWITCH

Removal and Installation

GENERAL MOTORS S-10 AND S-15 SERIES TRUCKS

1. Disconnect the negative battery cable.
2. Remove the console assembly and shift boot. Loosen the jam nut at the shift lever.
3. Disconnect the selector switch wiring harness and front axle switch wiring harness. Remove the selector switch retaining bolt and remove the switch.
4. Raise and support the vehicle safely.
5. Remove the shift rod from the shifter assembly.
6. Remove the pivot bolt, adjusting bolt and remove the shift lever assembly.
7. Install the shift lever assembly, pivot bolt and adjusting bolt. Install the shift rod into the shifter assembly. Adjust the shift linkage.
8. Lower the vehicle. Install the shift lever by screwing the shift lever down until the pawl clears the bracket. Then turn it down an additional 1½ turns. Install the jam nut and torque the jam nut to 18 ft. lbs. Install the selector switch and retaining bolt.
9. Connect the front axle switch wire harness and selector switch harness.
10. Install the shift boot and retaining screws. Install the console assembly and reconnect the negative battery cable.

SPEEDOMETER GEAR, SHAFT SEAL, REAR BEARING AND RETAINER

Removal and Installation

The front and rear output shaft seals, extension, rear retainer, bearing and speedometer drive gear can all be serviced with the transfer case in the vehicle. The following combined procedure outlines removal and installation of these components.

1. Raise and support the vehicle safely.
2. Place a drain pan under transfer case, remove the drain plug and drain fluid from the transfer case.
3. Remove the driveshaft and secure it to the under side of the vehicle.
4. Remove the extension seal with a suitable seal removal tool. Install a new seal with a seal installer tool.
5. Remove the extension housing. Remove the speedometer gear from the rear retainer. Mark the retainer for assembly alignment reference.
6. Remove the retainer attaching bolts and remove the retainer. Pry the retainer with a suitable tool to remove it.
7. Completely clean off all the old RTV sealant. If the retainer or bearing are to be replaced, remove the bearing retainer snapring from the rear retainer and remove the bearing.
8. Install the rear output bearing in the rear retainer. Apply a ⅛ in. bead or RTV sealant to the mating surface of the rear retainer. Align the retainer case reference marks and install the rear retainer on the case. Torque the rear retainer bolts to 18 ft. lbs.

9. Install the output shaft seal. Install the speedometer driven gear.
10. Fill the transfer case to the edge of the fill plug opening with the recommended lubricant. Install the speedometer driven gear. Install the drain plug to 35 ft. lbs.
11. Install the driveshaft and tighten the clamp strap bolts to 170 inch lbs. (19 Nm). Lower the vehicle.

Removal and Installation

TRANSFER CASE REMOVAL

EXCEPT GENERAL MOTORS S–10 AND S–15 SERIES TRUCKS

1. Raise the vehicle and support safely.
2. Remove the drain plug and drain the fluid from the transfer case.
3. Mark the transfer case front and rear output shaft yokes and driveshafts for assembly alignment and reference.
4. Disconnect the speedometer cable and vacuum switch hoses.
5. Disconnect the shift lever link from the operating lever.
6. Place a support stand under the transmission and remove the rear crossmember.
7. Mark the transfer case front and rear output shaft yokes and driveshafts for assembly alignment reference.
8. Disconnect the front and rear driveshafts at the transfer case yokes. Secure the shafts to the frame rails with wire.

NOTE: Do not allow the shafts to hang.

9. Remove the bolts attaching the exhaust pipe support bracket to the transfer case, if necessary.
10. Remove the transfer case-to-transmission nuts.
11. Move the transfer case assembly rearward until free of the transmission output shaft and remove the assembly.
12. Remove all the gasket material from the rear of the transmission adapter housing.

GENERAL MOTORS S–10 AND S–15 SERIES TRUCKS

1. Shift the vehicle into the **4HI** position. Raise the vehicle and support safely.
2. Remove the drain plug and drain the fluid from the transfer case.
3. Mark the transfer case front and rear output shaft yokes and driveshafts for assembly alignment and reference.
4. Remove the skid plate. Remove the front and rear driveshafts and secure them to the frame.
5. Disconnect the speedometer cable, electrical connections and vacuum harness.
6. Remove the shift lever from the case. Remove the catalytic converter hanger bolts at the converter.
7. Place a suitable transmission jack under the transmission and transfer case assembly. Remove the transmission mounting bolts and raise the transmission assembly.
8. Remove the catalytic converter hanger. Lower the transmission assembly.
9. Suport the transfer case with a suitable stand and remove the transfer case to transmission case bolts. On vehicles with automatic transmissions, remove the shift lever bracket mounting bolts from the adapter to permit the removal of the upper left transfer attaching bolts.
10. Remove the transfer case from the adpater or the the extension housing.

TRANSFER CASE INSTALLTION

EXCEPT GENERAL MOTORS S–10 AND S–15 SERIES TRUCKS

1. Apply a suitable sealant to both sides of the transfer case-to-transmission gasket and position the gasket on the transmission.
2. Align and install the transfer case assembly on the transmission. Be sure the transfer case input gear splines are aligned with the transmission output shaft. Align the splines by rotating the transfer case rear output shaft yoke as necessary.

NOTE: Do not install any transfer case attaching nuts until the transfer case is completely seated against the transmission.

3. Align and install the transfer case attaching nuts. Tighten the nuts to 35 ft. lbs (48 Nm).
4. Install the rear crossmember and remove the transmission support stand.
5. Attach the exhaust pipe support bracket to the transfer case, if removed.
6. Align and connect the driveshafts.
7. Connect the speedometer cable and vacuum switch hoses.
8. Connect the shift lever to the operating lever. Tighten the locknut to 18 ft. lbs. (23 Nm).
9. Fill the transfer case with the recommended lubricant to the proper level.
10. Lower the vehicle.

GENERAL MOTORS S–10 AND S–15 SERIES TRUCKS

1. Apply a suitable sealant to both sides of the transfer case-to-transmission gasket and position the gasket on the transmission.
2. Align and install the transfer case assembly on the transmission. Be sure the transfer case input gear splines are aligned with the transmission output shaft. Align the splines by rotating the transfer case rear output shaft yoke as necessary.

NOTE: Do not install any transfer case attaching nuts until the transfer case is completely seated against the transmission.

3. Align and install the transfer case attaching nuts. Tighten the nuts to 23 ft. lbs.
4. Install the transmission mount and the catalytic converter hanger. Torque the tranmission mount bolts to 22 ft. lbs. (29 Nm) and the converter hanger bolts to 40 ft. lbs. (54 Nm).
5. Install the shift lever. Reconnect the speedometer, electrical connections and vacuum harness.
6. Install the front and rear driveshafts and torque the bolts to 15 ft. lbs. (20 Nm).
7. Fill the transfer case with the recommended lubricant to the proper level.
8. Install the skid plate and torque the bolts to 21 ft. lbs. (28 Nm). Lower the Vehicle. Install the negative battery and road test the vehicle.

Before Disassembly

Cleanliness during disassembly and assembly is necessary to avoid further transfer case trouble after overhaul. Before removing any of the transfer case subassemblies, plug all the openings and clean the outside of the of the transfer case thoroughly. Steam cleaning or car wash type high pressure equipment is preferable. During disassembly, clean all parts in suitable solvent and dry each part. Do not use cloth or paper towels to dry parts. Use compressed air only.

Transfer Case Disassembly

1. Remove the transfer case from the vehicle.
2. Remove the attaching nuts from the front and rear output yokes. Remove the yokes and sealing washers. Move the transfer case selector lever rearward to the **4L** position.
3. Remove the bolts and tap the extension housing off of the

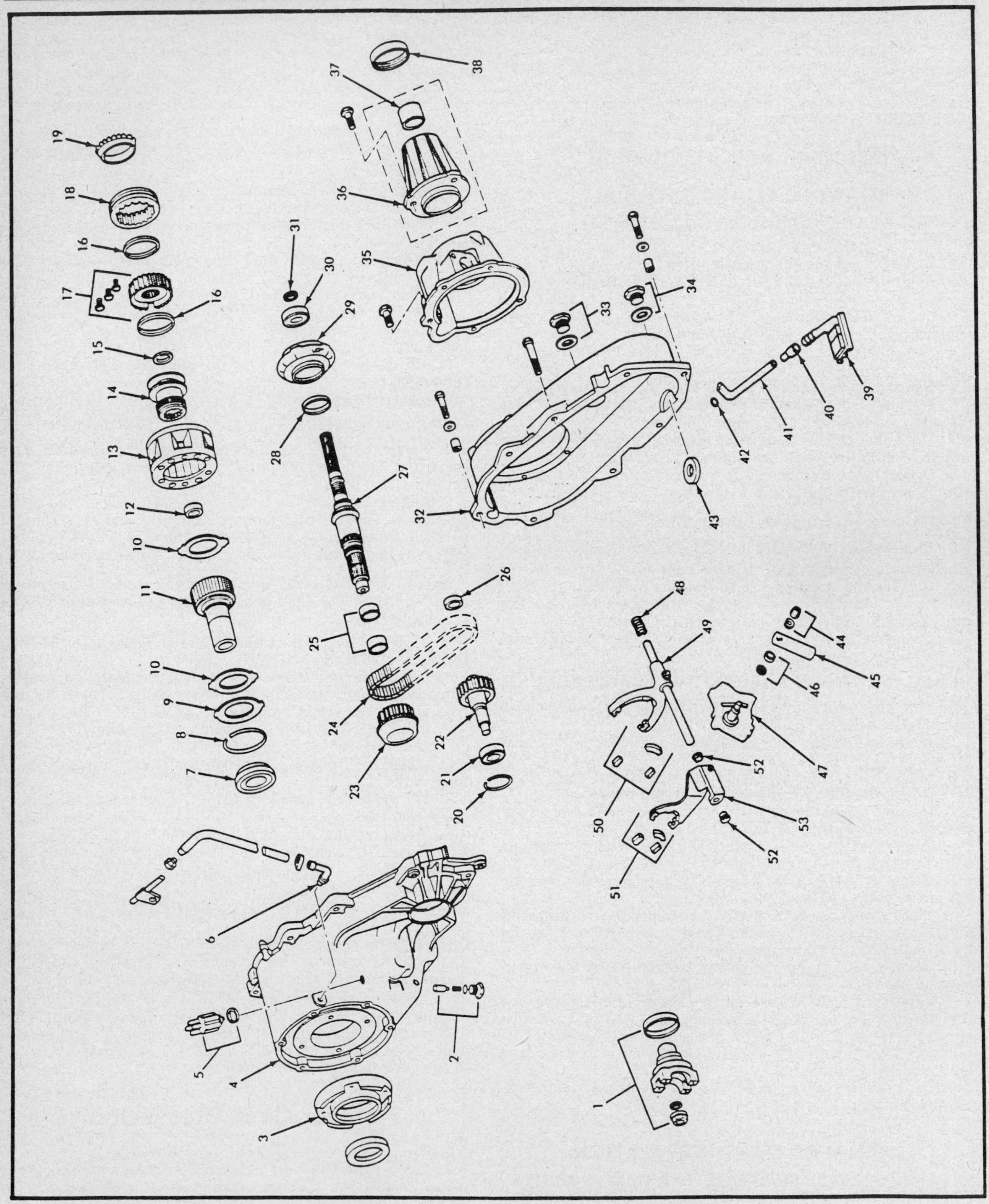
1
2
3
4
5
6
7
8
9
10
10
11
12
13
14
15
16
16
17
18
19
20
21
22
23
24
25
26
27
28
29
30
31
32
33
34
35
36
37
38
39
40
41
42
43
44
45
46
47
48
49
50
51
52
52
53

Exploded view of the NP231 Transfer case

1. Front yoke nut, seal washer, yoke, and oil seal
2. Shift detent plug, spring, and pin
3. Front retainer and seal
4. Front case
5. Vacuum switch and seal
6. Vent assembly
7. Input gear bearing and snap ring
8. Low range gear snap ring
9. Input gear retainer
10. Low range gear thrust washers
11. Input gear
12. Input gear pilot bearing
13. Low range gear
14. Range fork shift hub
15. Synchronizer hub snap ring
16. Synchronizer hub springs
17. Synchronizer hub and inserts
18. Synchronizer sleeve
19. Synchronizer stop ring
20. Snap ring
21. Output shaft front bearing
22. Output shaft (front)
23. Drive sprocket
24. Drive chain
25. Drive sprocket bearings
26. Output shaft rear bearing
27. Mainshaft
28. Oil seal
29. Oil pump assembly
30. Rear bearing
31. Snap ring
32. Rear case
33. Fill plug and gasket
34. Drain plug and gasket
35. Rear retainer
36. Extension housing
37. Bushing
38. Oil seal
39. Oil pickup screen
40. Tube connector
41. Oil pickup tube
42. Pickup tube O-ring
43. Magnet
44. Range lever nut and washer
45. Range lever
46. O-ring and seal
47. Sector
48. Mode spring
49. Mode fork
50. Mode fork inserts
51. Range fork inserts
52. Range fork bushings
53. Range fork

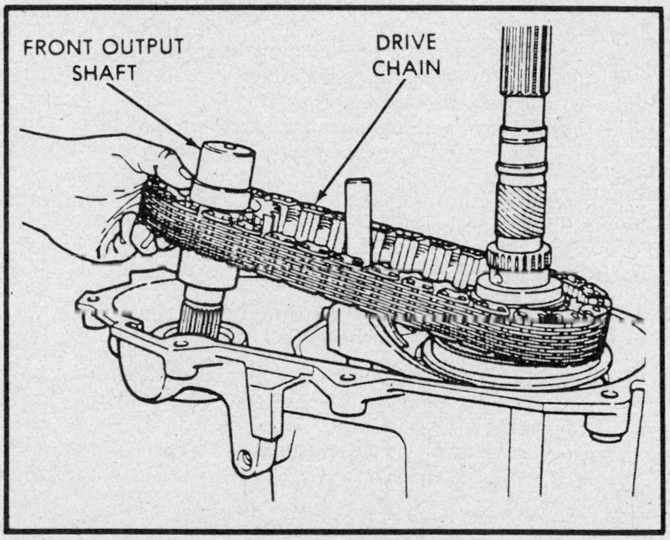
FRONT OUTPUT SHAFT
DRIVE CHAIN

Drive chain removal and installation

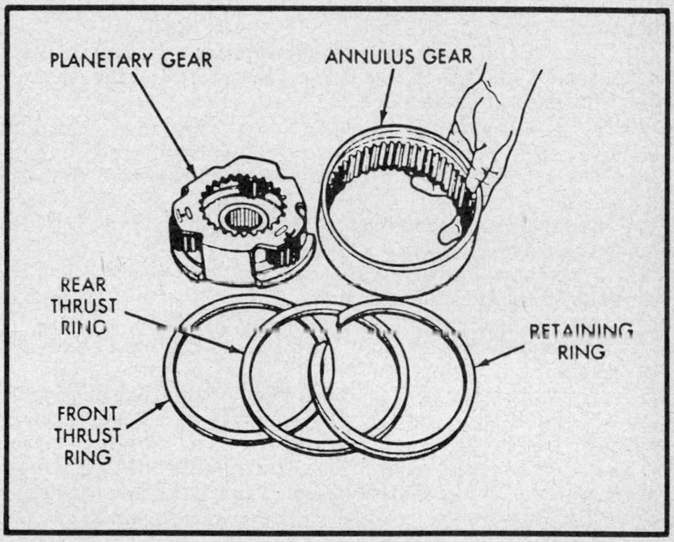
PLANETARY GEAR
ANNULUS GEAR
REAR THRUST RING
RETAINING RING
FRONT THRUST RING

Planetary assembly components

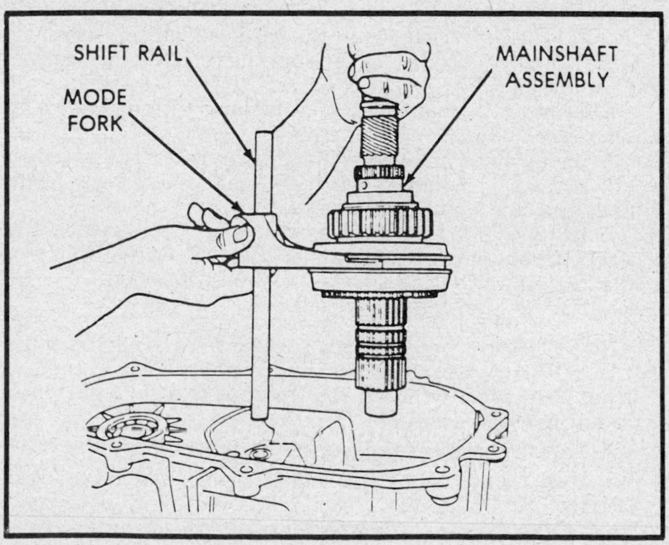
SHIFT RAIL
MODE FORK
MAINSHAFT ASSEMBLY

Mainshaft assembly removal

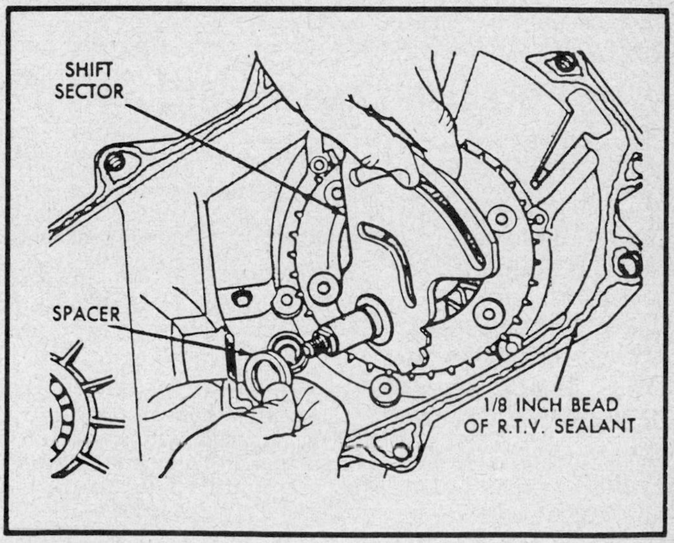
SHIFT SECTOR
SPACER
1/8 INCH BEAD OF R.T.V. SEALANT

Shift selector removal

rear retainer. Tap the extension housing in the clockwise direction to break the sealer bead, then remove the housing.

NOTE: To avoid damaging the sealing surfacee of the extension housing and rear retainer, do not attempt to pry or wedge the housing off of the retainer.

4. Remove the snapring from the rear bearing, then, remove the 4 bolts and separate the rear bearing retainer from the rear case half.
5. Remove the rear retainer, position a pry bar under each of the tabs on the retainer housing and carefully pry the retainer upward and off the rear case. Remove the bolts attaching the rear case to the front case. Separate the rear case from the front case.
6. Remove the oil pump and rear case as an assembly. Slide the oil screen out of the case pocket. Disconnect the screen from the pickup tube and remove the screen. Remove the pickup tube from the oil pump and remove the oil pump from the rear case.
7. Remove the pickup tube O-ring from the oil pump and mark the oil pump housing for assembly reference.
8. Remove the screws that attach the 2 halves of the pump. Remove the feed housing from the gear housing. Be sure to note the position of the pump gears and remove them from the housing.
9. Remove the mode spring from the shift rail. Tap the front output shaft upward to free it from the shaft bearing. Remove the front output shaft and drive chain as an assembly.
10. Remove the mainshaft, mode fork and spring rail assembly from the front case half. Remove the mode fork and shift rail from the synchro sleeve. Remove the synchro sleeve from the mainshaft.
11. Remove the synchro hub snapring, hub and stop ring. Remove the drive sprocket.
12. Slide the range fork pin out of the shift selector. Remove the range fork and shift hub. Remove the transfer case range lever from the sector shaft. Remove the shift sector.
13. Remove the sector shaft bushing and O-ring. Remove the shift detent pin, spring and plug.
14. Turn the front case over and remove the front bearing retainer bolts. Remove the front bearing retainer. Remove the input gear snapring.
15. Press the input low range gear assembly out of the input bearing with a suitable arbor press. Remove the low range gear snapring. Remove the retainer, thrust washers and input gear from the low range gear.
16. Remove the oil seals from the retainer, rear extension housing, oil pump feed housing and case halves. Remove the magnet from the front of the case.
17. Remove the speedometer driven gear, seals and adaptor.

CLEANING AND INSPECTION

Cleaning

During overhaul, all components of the transfer case (except bearing assemblies) should be thoroughly cleaned with solvent and dried with air pressure prior to inspection and reassembly.
1. Clean the bearing assemblies as follows:

NOTE: Proper cleaning of bearings is of utmost importance. Bearings should always be cleaned separately from other parts.

a. Soak all bearing assemblies in clean solvent or fuel oil. Bearings should never be cleaned in a hot solution tank.
b. Slush bearings in solvent until all old lubricant is loosened. Hold races so that bearings will not rotate; then clean bearings with a soft bristled brush until all dirt has been removed. Remove loose particles of dirt by tapping bearing flat against a block of wood.
c. Rinse bearings in clean solvent; then blow bearings dry with air pressure.

CAUTION

Do not spin bearings while drying.

d. After drying, rotate each bearing slowly while examining balls or rollers for roughness, damage, or excessive wear. Replace all bearings that are not in first class condition.

NOTE: After cleaning and inspecting bearings, lubricate generously with recommended lubricant, then wrap each bearing in clean paper until ready for reassembly.

2. Remove all portions of old gaskets from parts, using a stiff brush or scraper.

Inspection

1. Inspect all parts for discoloration or warpage.
2. Examine all gears and splines for chipped, worn, broken or nicked teeth. Small nicks or burrs may be removed with a fine abrasive stone.
3. Inspect the breather assembly to make sure that it is open and not damaged.
4. Check all threaded parts for damaged, stripped, or crossed threads.
5. Replace all gaskets, oil seals and snaprings.
6. Inspect housings, retainers and covers for cracks or other damage. Replace the damaged parts.
7. Inspect keys and keyways for condition and fit.
8. Inspect shift forks for wear, distortion or any other damage.
9. Check detent ball springs for free length, compressed length, distortion or collapsed coils.
10. Check bearing fit on their respective shafts and in their bores or cups. Inspect bearings, shafts and cups for wear.

NOTE: If either bearings or cups are worn or damaged, it is advisable to replace both parts.

11. Inspect all bearing rollers or balls for pitting or galling.
12. Examine detent balls for corrosion or brinneling. If shift bar detents show wear, replace them.
13. Replace all worn or damaged parts. When assembling the transfer case, coat all moving parts with recommended lubricant.

Transfer Case Assembly

Before assembly, lubricate all parts with the recommended lubricant. The bearing bores in the various transfer case components contain oil feed holes. Be sure the replace bearings do not block these feed holes.
1. Remove the input gear pilot bearing with a suitable slide hammmer and seal removal tool.
2. Install a new input gear pilot bearing into the input gear.
3. Assemble the low range gear, input gear thrust washers, input gear and input gear retainer.
4. Install the low range gear snapring. Lubricate the input gear with automatic transmission fluid. Start the input gear into the front bearing. Press the input gear into the front bearing.

NOTE: Use a proper size tool to press the input gear into the front bearing. An incorrect tool could push the input gear pilot bearing too far into the gear bore. Also, do not press against the end faces of the low range gear. The gear cae and thrust washers could be damaged.

5. Install a new input gear snapring. Install a new oil seal in the front bearing retainer. Apply a ⅛ in. (3mm) wide bead of RTV sealant to the front bearing retainer seal surface. Install the front bearing retainer on the front case. Torque the retainer bolts to 16 ft. lbs. (21 Nm).

6. Install a new sector shaft O-ring and bushing. Install the shift sector in the case. Install the range lever and lever attaching nut on the shift sector. Torque the attaching nut to 22 ft. lbs. (32 Nm).

7. Install the detent pin, spring and plug. Torque the plug to 15 ft. lbs. (20 Nm). Install new pads and shift rail bushings in the range fork. Assemble the range fork and shift hub. Engage the range fork pin in the sector slot.

8. If the drive sprocket bearings are to be replaced install them as follows:

a. Press both bearings out of the sprockets simultaneously.

b. Install a new front bearing by pressing the bearing flush with the edge of the bore.

c. Install the rear bearing by pressing the bearing in until it is $^{3}/_{16}$ in. (4.6mm) below the edge of the bore.

NOTE: Do not press the bearings any farther into the sprocket than specified. The bearings could block the mainshaft oil feed hole pressed too deeply into the sprocket.

9. Install the synchro hub springs struts and spring. Lubricate the drive sprocket bearings, the stop ring and synchro hub with the recommended fluid. Install the sprocket, stop ring and synchro hub on the mainshaft. Be sure to seat the hub struts on the stop ring lugs.

10. Install the synchro hub snapring. Install the sleeve on the synchro hub. Be sure the sleeve is installed with the beveled spline ends facing the stop ring. Install the new pads on the mode fork and install the shift rail in the fork.

11. Engage the mode fork in the synchro sleeve. Install the mode fork mainshaft assembly in the case. Be sure the mode fork rail is seated in both of the range fork bushings.

12. Assemble and install the output shaft and drive chain. Lift the mainshaft slightly to ease the chain and shaft installation. Install the mode spring on the shift rail.

13. Install a new output shaft rear bearing. Remove the bearing with an internal type puller and slide hammer. Seat the new bearing in the case with a suitable seal installer.

14. Install a new seal in the oil pump feed housing. Assemble the oil pump. Lubricate and install the 2 gears in the gear housing. Align and install the feed housing on the gear housing. Install and tighten the oil pump screws to 14 inch lbs.

15. Install the pickup tube O-ring in the oil pump. Insert the oil pickup tube in the oil pump. Attach the oil screen and connecting hose to the pickup tube. Install thre assembled oil pump, pickup tube and screen in the rear case. Be sure that the screen isseated in the case slot. Install the magnet in the front case.

16. Apply a ⅛ in. (3mm) wide bead of RTV sealant to the sealing surface of the front case. Align and install the rear case on the front case. Be sure the case locating dowels are in place and the mainshaft splines are engaged in the oil pump inner gear.

17. Install and torque the front case to rear case attaching bolts to 22 ft. lbs. (32 Nm). Be sure to install a washer under each of the bolts use at each case dowel locations.

18. Apply a ⅛ in. (3mm) wide bead of RTV sealant to the flange surface of the rear retainer. Install the locating dowel in the rear retainer and install the retainer on the case. Torque the retainer bolts to 18 ft. lbs. (23 Nm). Install a new rear bearing snapring. Lift the mainshaft slightly to seat the snapring in the shaft groove.

19. Install a new seal in the extension housing. Apply a ⅛ in. (3mm) wide bead of RTV sealant to the flange surface of the extension housing. Install the exstension housing on the case, torque the retaining bolt to 30 ft. lbs. (41 Nm).

20. Install the front yoke. Secure the yoke with a replacement seal washer and nut. Torque the nut to 110 ft. lbs. (149 Nm).

21. Install a replacement gasket on the vacuum switch and install the switch in the case.

22. Install the speedometer driven gear, seals and adapter. Refill the transfer case with the recommended fluid to the proper level. Install and torque the drain plug to 35 ft. lbs. (48 Nm).

Specifications

TORQUE SPECIFICATIONS
New Process 231 Transfer Case

Description	Torque Ft. Lbs.	Torque N•m
Bolt, Low Range Lock Plate	25	34
Vacuum Switch	20	27
Nut, Range Lever	18	24
Bolt, Front Case-to-Rear Case	22	30
Bolt, Rear Retainer	18	24
Bolt, Extension Housing	22	30
Plug, Drain/Fill	35	47
Bolt, Adapter-to-Transfer Case	24	33
Nut, Transfer Case-to-Adapter	26	35
U-Joint Clamp Strap Bolts	14	19

NEW PROCESS 241 TRANSFER CASE

Application

1988–89 Dodge Pick-Up and Ramcharger
1987–89 General Motors CK Series Pick-Up

General Description

The NP241 is a part time unit with a 2 piece aluminum housing. The case contains a front and rear output shaft, 2 drive sprockets, a shift mechanism and a planetary gear assembly. The drive sprockets are connected and operated by the drive chain. The planetary assembly which consists of a 4 pinion carrier an an annulus gear provide the 4WD low range when engaged. The reduction ratio is 2.72:1 in this range.

On Vehicle Services

SHIFT LINKAGE

Adjustment

DODGE PICK-UP AND RAMCHARGER

1. Place the shift lever in the **4H** position.
2. Remove the screws attaching the shift lever boot to the floorpan. Slide the boot upward to provide access to the shift gate.
3. Insert a ⅛ in. (3mm) spacer between the shift lever and the forward edge of the shift lever gate. Secure the lever and spacer in position with tape or wire.
4. Raise and support the vehicle safely. Loosen the adjusting link enough to allow the linkage to slide freely in the link.
5. Move the tranfer case range lever to the **4H** position. Position the linkage so it is a free fir in the range lever. Tighten the setscrew, in adjusting link, securely.
6. Lower the vehicle. Remove the shift lever spacer and install the boot to the floorpan.

GENERAL MOTORS CK SERIES PICK-UP

1. Remove the console assembly. Raise the upper boot on the shift lever.
2. Loosen the shift assembly lock bolt. Loosen the pivot bolt. Place the transfer case shifter lever in the **4HI** position.
3. Insert a ⅝ (8mm) drill bit through the shifter into the bracket.
4. Install a bolt at the lower transfer case shift lever. This will lock the transfer case in the **4HI** position.
5. Remove the bolt from the lower transfer case shift lever. Remove the drill bit.
6. Install the shift assembly lock bolt and torque it to 30 ft. lbs. (41 Nm). Torque the shift lever through bolt with the grease fitting on it to 96 ft. lbs. (128 Nm).
7. Install the upper boot and the console assembly.

SPEEDOMETER GEAR, SHAFT SEAL, REAR BEARING AND RETAINER

Removal and Installation

The front and rear output shaft seals, extension, rear retainer, bearing and speedometer drive gear can all be serviced with the transfer case in the vehicle. The following combined procedure outlines removal and installation of these components.

1. Raise and support the vehicle safely.
2. Place a drain pan under transfer case, remove the drain plug and drain fluid from the transfer case.
3. Remove the driveshaft and secure it to the under side of the vehicle.
4. Remove the extension seal with a suitable seal removal tool. Install a new seal with a seal installer tool.
5. Remove the extension housing. Remove the speedometer gear from the rear retainer. Mark the retainer for assembly alignment reference.
6. Remove the retainer attaching bolts and remove the retainer. Pry the retainer with a suitable tool to remove it.
7. Completely clean off all the old RTV sealant. If the retainer or bearing are to be replaced, remove the bearing retainer snapring from the rear retainer and remove the bearing.
8. Install the rear output bearing in the rear retainer. Apply a ⅛ in. bead or RTV sealant to the mating surface of the rear retainer. Align the retainer case reference marks and install the rear retainer on the case. Torque the rear retainer bolts to 18 ft. lbs. (23 Nm).
9. Install the output shaft seal. Install the speedometer driven gear.
10. Fill the transfer case to the edge of the fill plug opening with the recommended lubricant. Install the speedometer driven gear. Install the drain plug to 35 ft. lbs. (46 Nm).
11. Install the driveshaft and tighten the clamp strap bolts to 170 inch lbs. (19 Nm). Lower the vehicle.

Removal and Installation

TRANSFER CASE REMOVAL

DODGE PICK-UP AND RAMCHARGER

1. Raise the vehicle and support safely.
2. Remove the drain plug and drain the fluid from the transfer case.
3. Mark the transfer case front and rear output shaft yokes and driveshafts for assembly alignment and reference.
4. Disconnect the speedometer cable and vacuum switch hoses.
5. Disconnect the shift lever link from the operating lever.
6. Place a support stand under the transmission and remove the rear crossmember.
7. Mark the transfer case front and rear output shaft yokes and driveshafts for assembly alignment reference.
8. Disconnect the front and rear driveshafts at the transfer case yokes. Secure the shafts to the frame rails with wire.

NOTE: Do not allow the shafts to hang.

9. Remove the bolts attaching the exhaust pipe support bracket to the transfer case, if necessary.
10. Remove the transfer case-to-transmission nuts.
11. Move the transfer case assembly rearward until free of the transmission output shaft and remove the assembly.
12. Remove all the gasket material from the rear of the transmission adapter housing.

GENERAL MOTORS CK SERIES PICK-UP

1. Shift the vehicle into the **4HI** position. Raise the vehicle and support safely.
2. Remove the drain plug and drain the fluid from the transfer case.
3. Mark the transfer case front and rear output shaft yokes and driveshafts for assembly alignment and reference.
4. Remove the skid plate. Remove the front and rear driveshafts and secure them to the frame.
5. Disconnect the speedometer cable, electrical connections and vacuum harness.

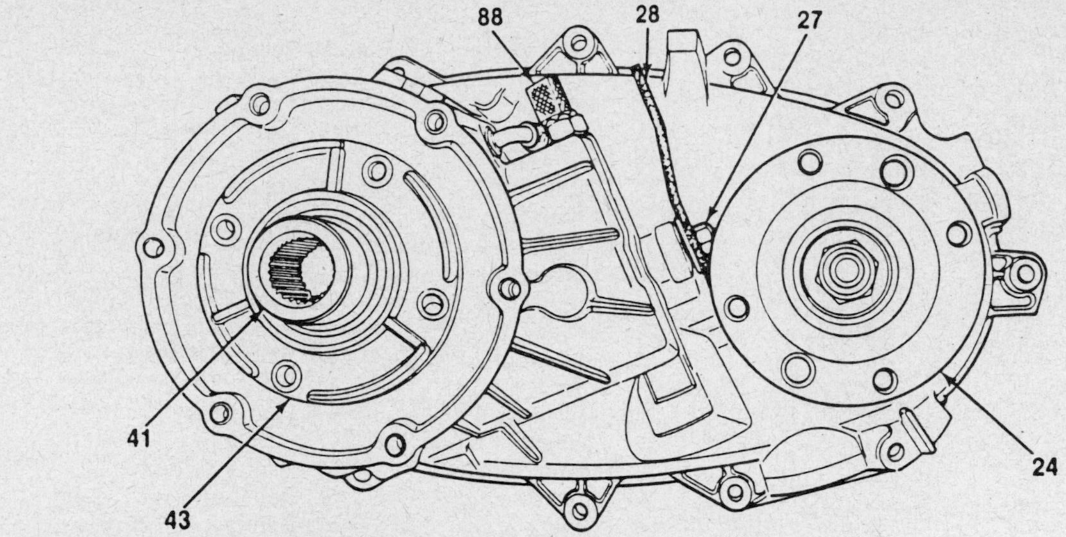
88
28
27
41
43
24

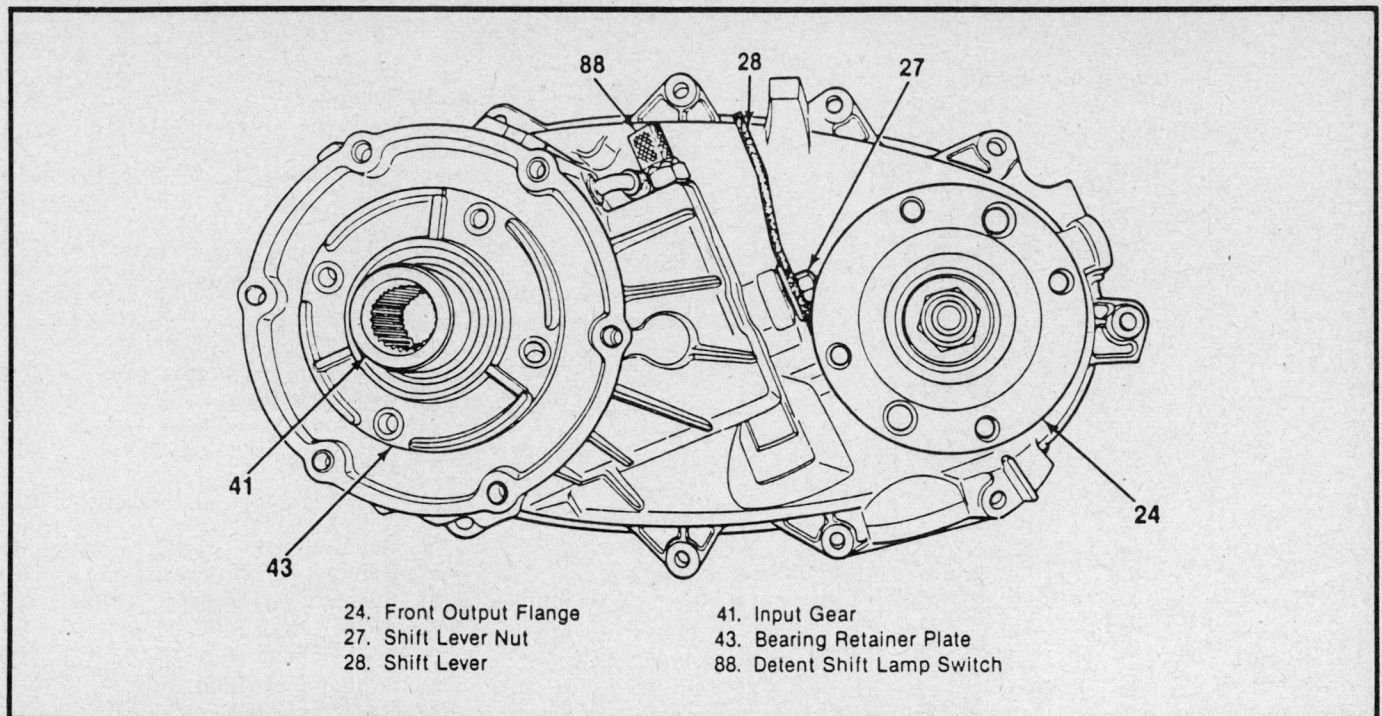

24. Front Output Flange
27. Shift Lever Nut
28. Shift Lever
41. Input Gear
43. Bearing Retainer Plate
88. Detent Shift Lamp Switch

NP241 Transfer case—front view

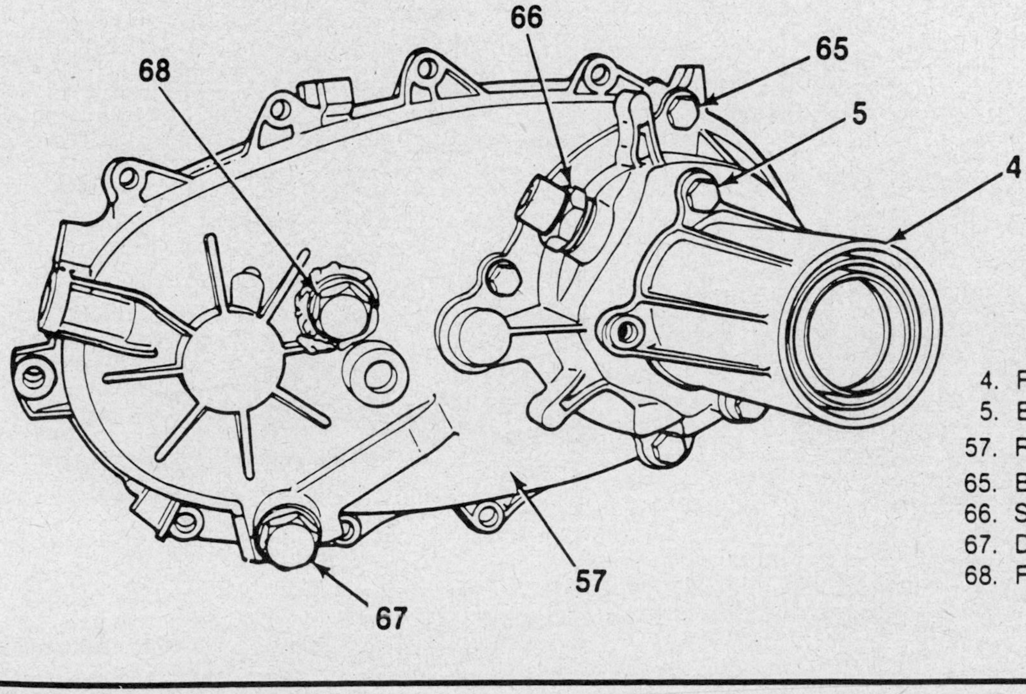
66
68
65
5
4
57
67

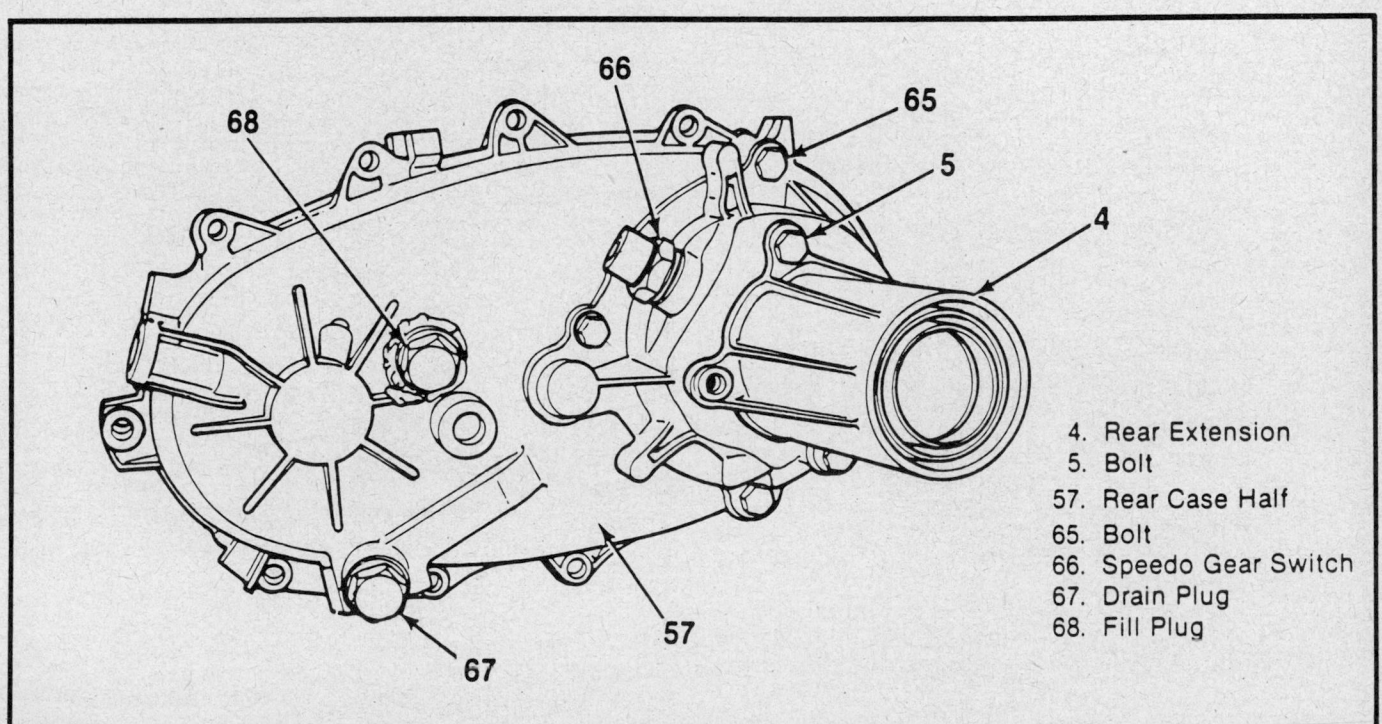

4. Rear Extension
5. Bolt
57. Rear Case Half
65. Bolt
66. Speedo Gear Switch
67. Drain Plug
68. Fill Plug

NP241 Transfer case—rear view

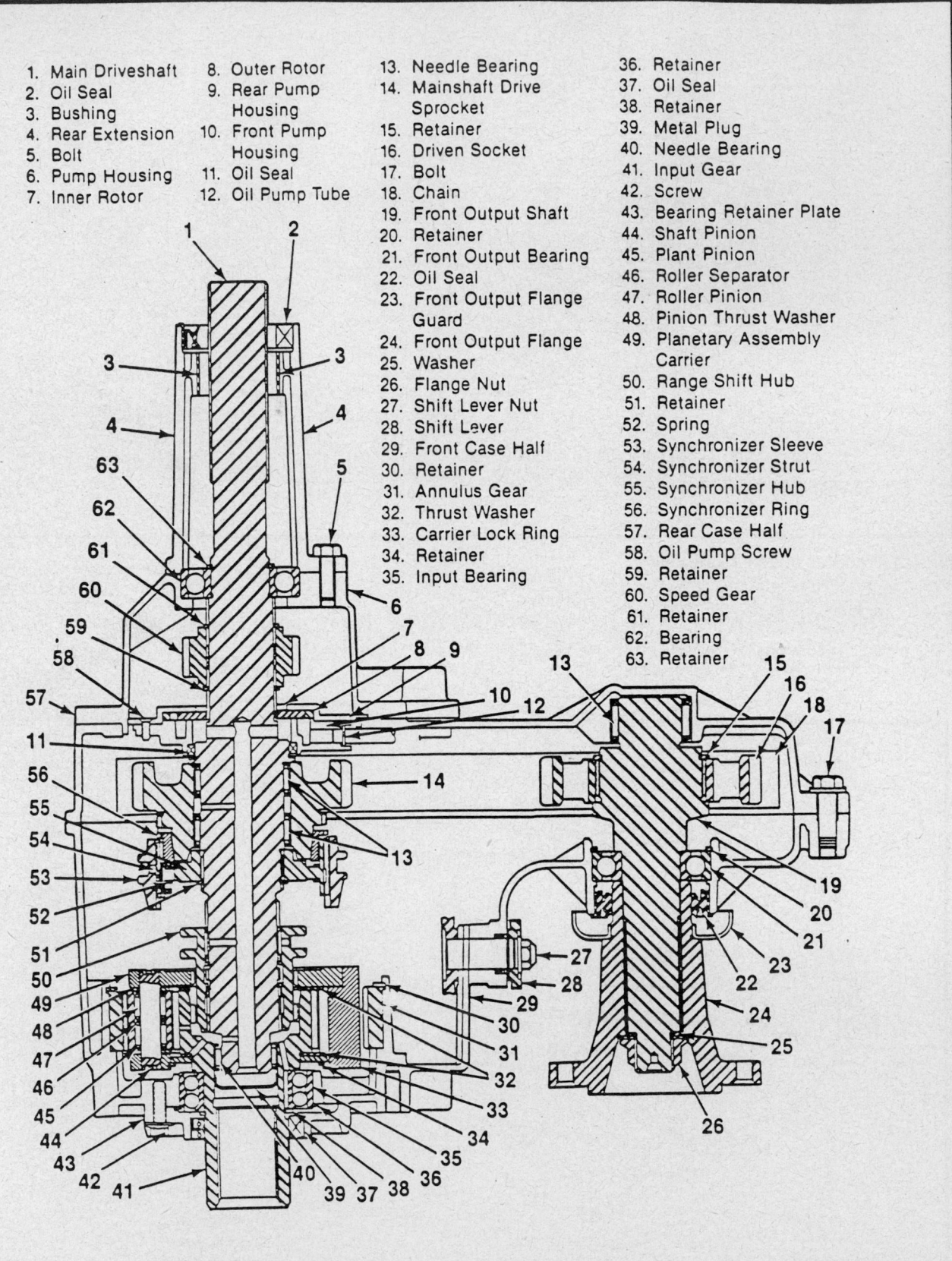

1. Main Driveshaft
2. Oil Seal
3. Bushing
4. Rear Extension
5. Bolt
6. Pump Housing
7. Inner Rotor
8. Outer Rotor
9. Rear Pump Housing
10. Front Pump Housing
11. Oil Seal
12. Oil Pump Tube
13. Needle Bearing
14. Mainshaft Drive Sprocket
15. Retainer
16. Driven Socket
17. Bolt
18. Chain
19. Front Output Shaft
20. Retainer
21. Front Output Bearing
22. Oil Seal
23. Front Output Flange Guard
24. Front Output Flange
25. Washer
26. Flange Nut
27. Shift Lever Nut
28. Shift Lever
29. Front Case Half
30. Retainer
31. Annulus Gear
32. Thrust Washer
33. Carrier Lock Ring
34. Retainer
35. Input Bearing
36. Retainer
37. Oil Seal
38. Retainer
39. Metal Plug
40. Needle Bearing
41. Input Gear
42. Screw
43. Bearing Retainer Plate
44. Shaft Pinion
45. Plant Pinion
46. Roller Separator
47. Roller Pinion
48. Pinion Thrust Washer
49. Planetary Assembly Carrier
50. Range Shift Hub
51. Retainer
52. Spring
53. Synchronizer Sleeve
54. Synchronizer Strut
55. Synchronizer Hub
56. Synchronizer Ring
57. Rear Case Half
58. Oil Pump Screw
59. Retainer
60. Speed Gear
61. Retainer
62. Bearing
63. Retainer

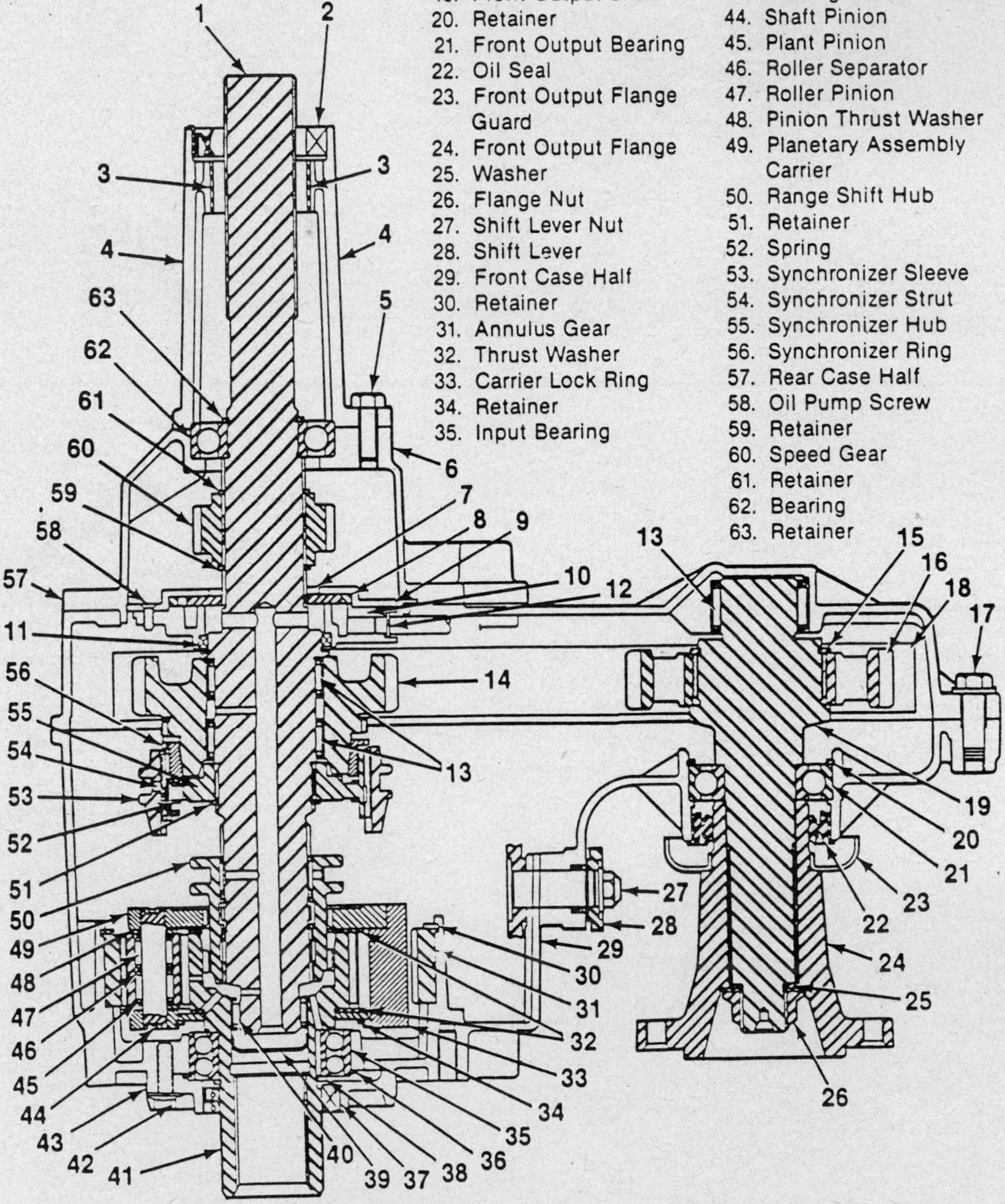

NP241 Transfer case—phantom view

Trouble Diagnosis

CHILTON'S THREE C'S TRANSFER CASE DIAGNOSIS

Condition	Cause	Correction
Transfer case difficult to shift or will not shift into desired range	a) Vehicle speed too great to permit shifting	a) Stop vehicle and shift into desired range. Or reduce speed to 2–3 mph (3–4 km/h) before attempting to shift
	b) If vehicle was operated for extended period in 4H mode on dry paved surface, driveline torque load may cause difficulty	b) Stop vehicle, shift transmission to neutral, shift transfer case to 2H mode and operate vehicle on 2H on dry paved surfaces
	c) Transfer case external shift linkage binding	c) Lubricate, repair or replace linkage, or tighten loose components as necessary
	d) Insufficient or incorrect lubricant	d) Drain and refill to edge of fill hole
	e) Internal components binding, worn or damaged	e) Disassemble unit and replace worn or damaged components as necessary
Transfer case noisy in all drive modes	a) Insufficient or incorrect lubricant	a) Drain and refill to edge of fill hole. If unit is still noisy after drain and refill, disassembly and inspection may be required to locate source of noise
Noisy in—or jumps out of the four-wheel-drive low range	a) Transfer case not completely engaged in 4L position	a) Stop vehicle, shift transfer case to Neutral, then shift back into 4L position
	b) Shift linkage loose or binding	b) Tighten, lubricate, or repair linkage as necessary
	c) Range fork cracked, inserts worn, or fork is binding on shift rail	c) Disassemble unit and repair as necessary
	d) Annulus gear or lockplate worn or damaged	d) Disassemble unit and repair as necessary
Lubricant leaking from output shaft seals or from vent	a) Transfer case overfilled	a) Drain to correct level
	b) Vent closed or restricted	b) Clear or replace vent if necessary
	c) Output shaft seals damaged or installed incorrectly	c) Replace seals. Be sure seal lip faces interior of case when installed. Also be sure yoke seal surfaces are not scored or nicked. Remove scores and nicks with fine sandpaper or replace yoke(s) if necessary
Abnormal tire wear	a) Extended operation on dry hard surface (paved) roads in 4H range	a) Operate in 2H on hard surface (paved) roads

6. Remove the shift lever from the case. Remove the catalytic converter hanger bolts at the converter.
7. Place a suitable transmission jack under the transmission and transfer case assembly. Remove the transmission mounting bolts and raise the transmission assembly.
8. Remove the catalytic converter hanger. Lower the transmission assembly.
9. Suport the transfer case with a suitable stand and remove the transfer case to transmission case bolts. On vehicles with automatic transmissions, remove the shift lever bracket mounting bolts from the adapter to permit the removal of the upper left transfer attaching bolts.
10. Remove the transfer case from the adpater or the the extension housing.

TRANSFER CASE INSTALLATION

DODGE PICK-UP AND RAMCHARGER

1. Apply a suitable sealant to both sides of the transfer case-to-transmission gasket and position the gasket on the transmission.
2. Align and install the transfer case assembly on the transmission. Be sure the transfer case input gear splines are aligned with the transmission output shaft. Align the splines by rotating the transfer case rear output shaft yoke as necessary.

NOTE: Do not install any transfer case attaching nuts until the transfer case is completely seated against the transmission.

3. Align and install the transfer case attaching nuts. Tighten the nuts to 35 ft. lbs. (46 Nm).
4. Install the rear crossmember and remove the transmission support stand.
5. Attach the exhaust pipe support bracket to the transfer case, if removed.
6. Align and connect the driveshafts.
7. Connect the speedometer cable and vacuum switch hoses.
8. Connect the shift lever to the operating lever. Tighten the locknut to 18 ft. lbs. (23 Nm).
9. Fill the transfer case with the recommended lubricant to the proper level.
10. Lower the vehicle.

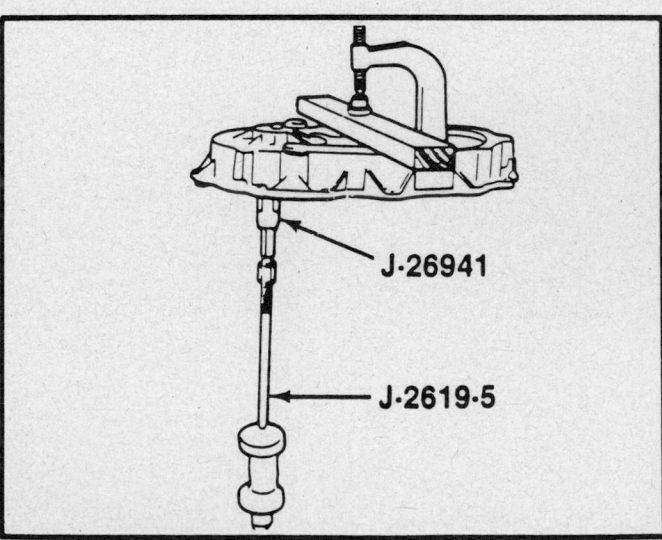

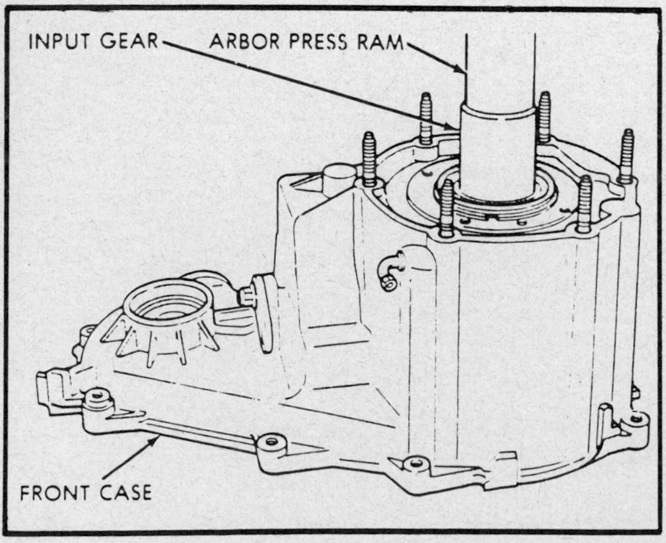
INPUT GEAR
ARBOR PRESS RAM
FRONT CASE
Removing the input gear

GENERAL MOTORS CK SERIES PICK-UP

1. Apply a suitable sealant to both sides of the transfer case-to-transmission gasket and position the gasket on the transmission.
2. Align and install the transfer case assembly on the transmission. Be sure the transfer case input gear splines are aligned with the transmission output shaft. Align the splines by rotating the transfer case rear output shaft yoke as necessary.

NOTE: Do not install any transfer case attaching nuts until the transfer case is completely seated against the transmission.

3. Align and install the transfer case attaching nuts. Tighten the nuts to 23 ft. lbs. (31 Nm).
4. Install the transmission mount and the catalytic converter hanger. Torque the tranmission mount bolts to 22 ft. lbs. and the converter hanger bolts to 40 ft. lbs. (54 Nm).
5. Install the shift lever. Reconnect the speedometer, electrical connections and vacuum harness.
6. Install the front and rear driveshafts and torque the bolts to 15 ft. lbs. (20 Nm).
7. Fill the transfer case with the recommended lubricant to the proper level.
8. Install the skid plate and torque the bolts to 21 ft. lbs. (28 Nm). Lower the vehicle. Install the negative battery and road test the vehicle.

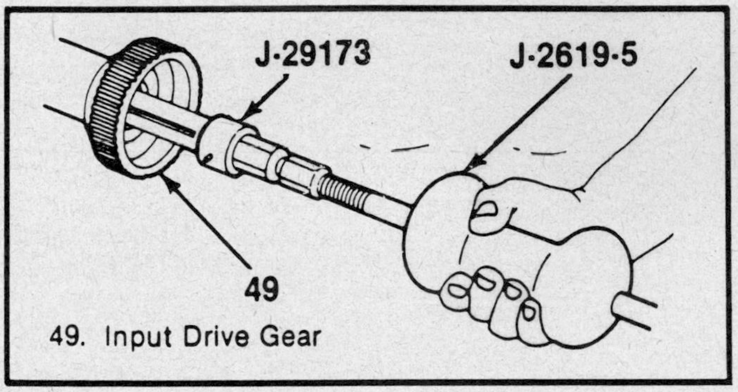
J-29173
J-2619-5
49
49. Input Drive Gear
Input gear bearing removal

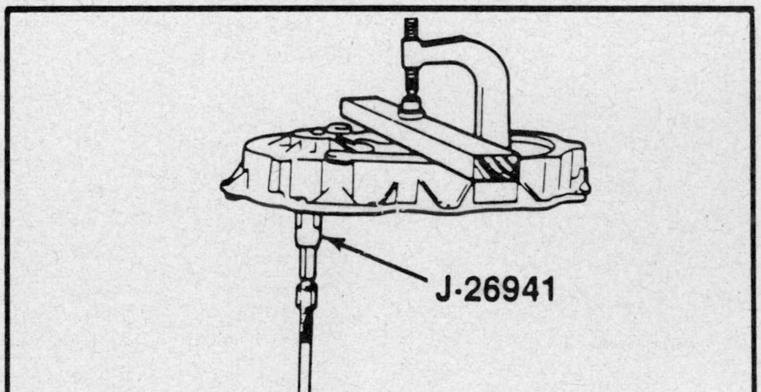
J-26941
J-2619-5
Output shaft bearing removal

Before Disassembly

Cleanliness during disassembly and assembly is necessary to avoid further transfer case trouble after overhaul. Before removing any of the transfer case subassemblies, plug all the openings and clean the outside of the of the transfer case thoroughly. Steam cleaning or car wash type high pressure equipment is preferable. During disassembly, clean all parts in suitable solvent and dry each part. Do not use cloth or paper towels to dry parts. Use compressed air only.

Transfer Case Disassembly

1. Drain the fluid from the case if not already drained.
2. Remove the attaching nuts from the front and rear output yokes. Remove the yokes and sealing washers.
3. Remove the indicator switch and seal, then, remove the speedometer switch and seal.
4. Remove the poppet screw, spring and the range selection plunger.
5. Remove the bolts and disconnect the mainshaft extension housing from the rear case half.
6. Remove the retaining ring, speedometer drive gear nylon, oil pump housing, and the oil pump gear from the rear output shaft.
7. Remove the case bolts and separate the case halves by inserting a small pry bar in the pry slots on the case.
8. Remove the fork shift spring.
9. Remove the oil pump pickup tube and the magnetic chip collector.
10. Remove the retainer from the driven socket.
11. Remove the mainshaft, chain and driven sprocket as a unit from the front case half.
12. Remove the synchronizer assembly retainer, then remove the syncro. assembly, sleeve, thrust washer, hub and ring.
13. Remove the range fork, range selector, mode fork and range shift hub. Remove the shift lever nut, washer, and shift lever.
14. Remove the input shaft bearing retainer plate and seal.
15. Remove the input shaft bearing retainer, bearing, and the planetary assembly. Remove the bearing from the input gear with special tool J-22912-1 or equivalent.
16. Remove the retainer, lock ring and thrust washer from the annulus gear assembly.
17. Remove the needle bearing from the input gear as follows:
   a. Insert needle bearing tool J-29369-1 and adapter with J-2619-5 slide hammer or equivalents.
   b. Using the tools mentioned above, hammer the bearing from the input gear.
18. Remove the drive sprocket from the main driveshaft.
19. Remove the needle bearings from the drive sprocket as follows:
   a. Insert needle bearing tool J-29369-2 and adapter with J-2619-5 slide hammer or equivalents.
   b. Using the tools mentioned above, hammer the bearing from the drive sprocket.
20. Remove the retainer and bearing from the front output shaft. Use bearing remover tool J-33832 and driver J-8092 or equivalents to drive the bearing from the case.
21. Remove the seal from the mainshaft extension housing and the seal from the front input bearing retainer.
22. Remove the needle bearing from the rear case half as follows:
   a. Insert needle bearing tool J-29369-2 and adapter with J-2619-5 slide hammer or equivalents.
   b. Using the tools mentioned above, hammer the bearing from the case.

CLEANING AND INSPECTION

Cleaning

During overhaul, all components of the transfer case (except bearing assemblies) should be thoroughly cleaned with solvent and dried with air pressure prior to inspection and reassembly.

1. Clean the bearing assemblies as follows:

NOTE: Proper cleaning of bearings is of utmost importance. Bearings should always be cleaned separately from other parts.

   a. Soak all bearing assemblies in clean solvent or fuel oil. Bearings should never be cleaned in a hot solution tank.
   b. Slush bearings in solvent until all old lubricant is loosened. Hold races so that bearings will not rotate; then clean

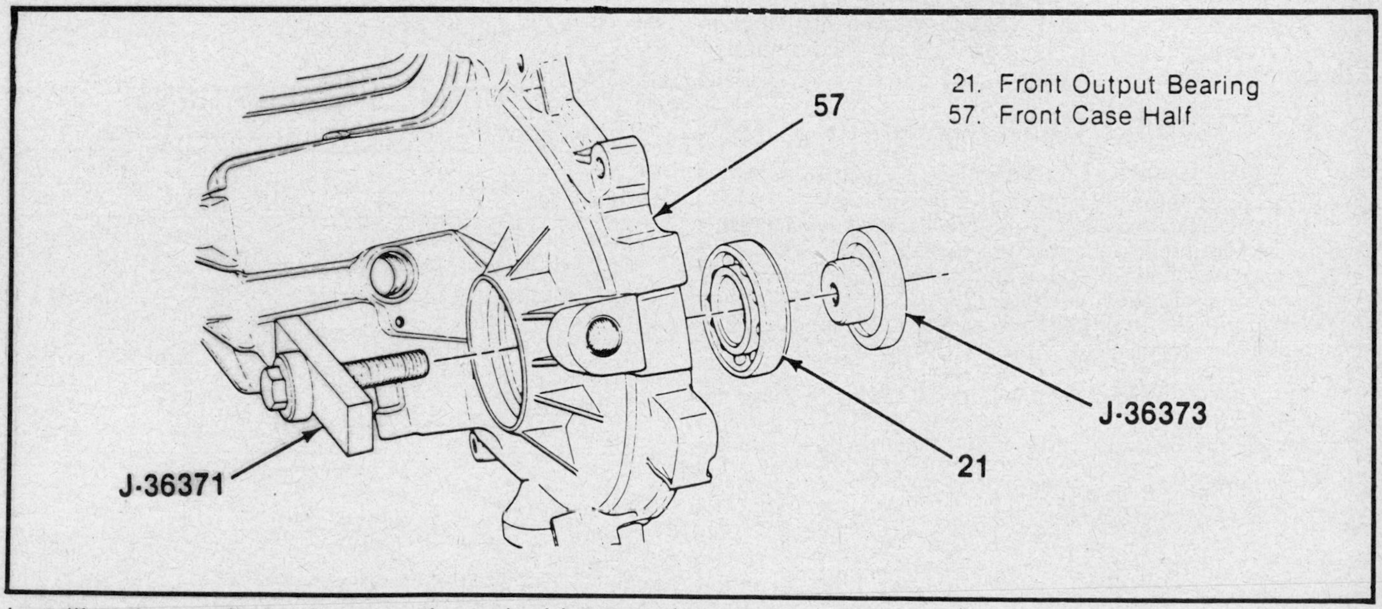
21. Front Output Bearing
57. Front Case Half
57
J-36373
21
J-36371

Installing the needle bearings on the main drive sprocket

bearings with a soft bristled brush until all dirt has been removed. Remove loose particles of dirt by tapping bearing flat against a block of wood.

c. Rinse bearings in clean solvent; then blow bearings dry with air pressure.

CAUTION

Do not spin bearings while drying.

d. After drying, rotate each bearing slowly while examining balls or rollers for roughness, damage, or excessive wear. Replace all bearings that are not in first class condition.

NOTE: After cleaning and inspecting bearings, lubricate generously with recommended lubricant, then wrap each bearing in clean paper until ready for reassembly.

2. Remove all portions of old gaskets from parts, using a stiff brush or scraper.

Inspection

1. Inspect all parts for discoloration or warpage.
2. Examine all gears and splines for chipped, worn, broken or nicked teeth. Small nicks or burrs may be removed with a fine abrasive stone.
3. Inspect the breather assembly to make sure that it is open and not damaged.
4. Check all threaded parts for damaged, stripped, or crossed threads.
5. Replace all gaskets, oil seals and snaprings.
6. Inspect housings, retainers and covers for cracks or other damage. Replace the damaged parts.
7. Inspect keys and keyways for condition and fit.
8. Inspect shift forks for wear, distortion or any other damage.
9. Check detent ball springs for free length, compressed length, distortion or collapsed coils.
10. Check bearing fit on their respective shafts and in their bores or cups. Inspect bearings, shafts and cups for wear.

NOTE: If either bearings or cups are worn or damaged, it is advisable to replace both parts.

11. Inspect all bearing rollers or balls for pitting or galling.
12. Examine detent balls for corrosion or brinneling. If shift bar detents show wear, replace them.
13. Replace all worn or damaged parts. When assembling the transfer case, coat all moving parts with recommended lubricant.

Transfer Case Assembly

1. Use tools J-36370 and J-8092 or equivalent to drive the needle bearings onto the main shaft drive sprocket.
2. With special tools J-36372 and J-8092 or equivalent, drive the needle bearing into the rear case half.
3. With tools J-36373 and J-8092 or equivalent, drive the needle bearing into the input gear.
4. Install the bearing into the front case half, using J-36371 and J-36373 or equivalent to insert the bearing. Install the bearing.
5. Use J-36371 or equivalent and install the bearing into the oil pump housing.
6. Install the bearing to the input gear, then, install the thrust washer, carrier lock ring and the retainer. Install the bearing to the input gear with tool J-36372 or equivalent.
7. Install the input gear, bearing and planetary assembly into the annulus ring. Use a hammer and a brass drift to seat the bearing.
8. Install the retainer to the input shaft bearing.
9. Install the retainer to the input gear.
10. Install the input shaft bearing retainer, seal and bolts. Tighten the bolts to 14 ft. lbs.
11. Install the range shift hub, mode fork, range selector and range fork. Install the shift lever, washer and nut and tighten to 20 ft. lbs. (27 Nm).
12. Install the drive sprocket and needle bearings to the main driveshaft.
13. Assembly the synchronizer assembly; ring, hub, thrust washer and sleeve, then install the retainer.
14. Install the mainshaft, chain and driven sprocket as a unit into the front case half.
15. Install the retainer to the driven sprocket. Install the shift fork spring.
16. Install the oil pump pickup, filler and magnetic washer to the rear case half.
17. Apply a bead of Loctite® 515 sealer or equivalent to the case matinng surfaces, then connect the rear and front case halves. Install the bolts and tighten to 23 ft. lbs. (31 Nm).
18. Install the speedometer gear retainer, speedometer gear, then the 2nd retainer.
19. Apply a bead of Loctite® 515 sealer or equivalent to the mating surfaces of the pump housing, then connect the housing. Install the bolts and tighten to 30 ft. lbs. (41 Nm).
20. Install the bearing retainer to the mainshaft.
21. Apply a bead of Loctite® 515 sealer or equivalent to the mating surfaces of the extension housing, then connect the housing. Install the bolts and tighten to 23 ft. lbs. (31 Nm).
22. Install the range selector plunger, spring and poppet screw.
23. Install the speedometer pickup switch and seal. Tighten the switch to 23 ft. lbs.
(31 Nm).
24. Install the indicator lamp switch and seal. Tighten the switch to 17 ft. lbs. (22 Nm).
25. Install the front output flange, washer and nut. Tighten the nut to 110 ft. lbs. (169 Nm).
26. Install the transaxle into the vehicle.
27. Fill the unit with 4.6 pints of Dexron®II transmission fluid.

TORQUE SPECIFICATIONS
New Process 241 Transfer Case

Description	Torque N•m	Torque Ft. Lbs.
Input Shaft Retainer Bolts	19	14
Shift Selector Lever Nut	27	20
Shift Selector Light Switch	24	17
Case Half Bolts	31	23
Pump Housing Bolts	41	30
Mainshaft Extension Housing Bolts	31	23
Speedometer Pick-up Switch	31	23
Front Propeller Shaft Flange Bolts	149	110

SPECIAL TOOLS

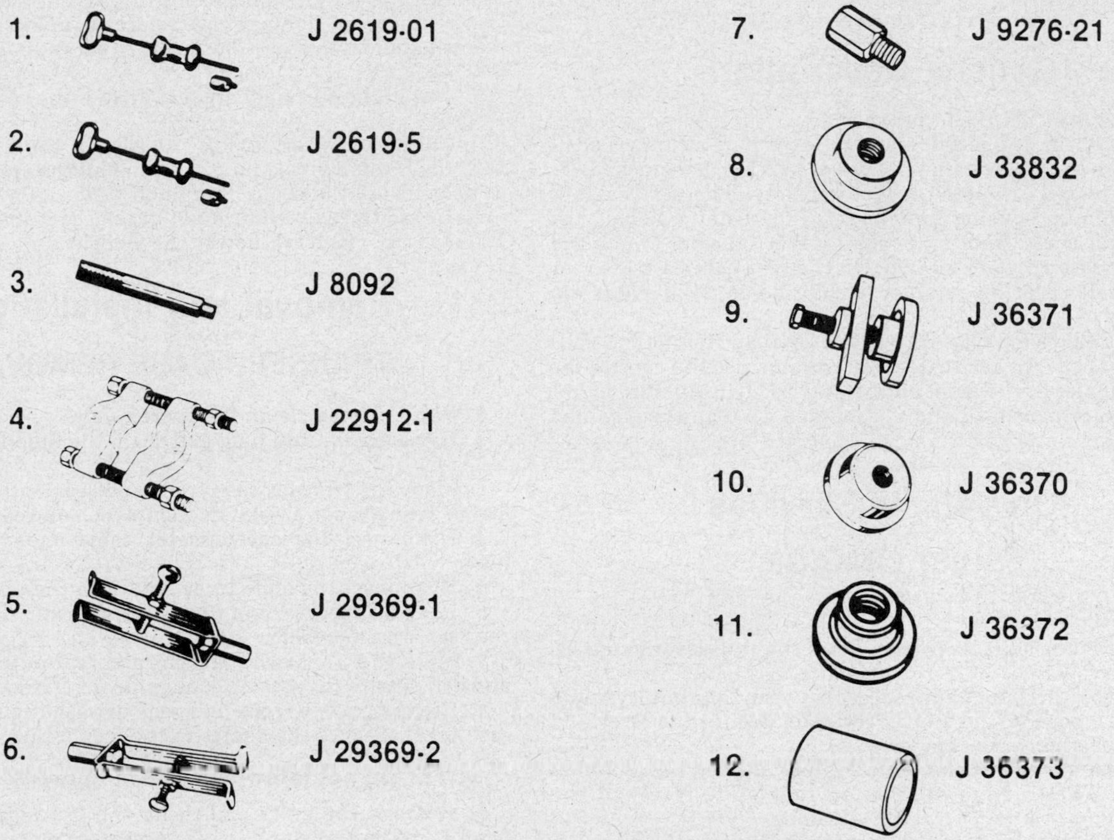
1. J 2619-01
2. J 2619-5
3. J 8092
4. J 22912-1
5. J 29369-1
6. J 29369-2
7. J 9276-21
8. J 33832
9. J 36371
10. J 36370
11. J 36372
12. J 36373

1. Slide Hammer
2. Slide Hammer
3. Driver Handle
4. Bearing Remover
5. Output Shaft Bearing Remover
6. Output Shaft Bearing Remover
7. Slide Hammer Adapter
8. Bearing Installer/Remover
9. Front and Rear Output Shaft Bearing Installer
10. Driver Sprocket Needle Bearing Installer
11. Input Gear Needle Bearing Installer
12. Input Gear Ball Bearing Installer

F-05437

NEW PROCESS 242 TRANSFER CASE

Application

1989 Jeep Cherokee and Wagoneer

General Description

The NP242 is a part time unit with a built-in low range reduction gear system. It has 4 operating ranges plus a Neutral position. The low-range system provides a 2.72:1 gear reduction ratio for increased low-speed torque capacity. The unit has a 2 piece aluminum housing assembly. On the front case half, the front output shaft, front input shaft, 4WD indicator switch and shift lever assembly are located. On the rear case half, the rear output shaft, bearing retainer and drain and fill plugs are located.

The NP242 transfer case provides a 2WD and a full time 4WD drive operation. An interaxle differential is used to control the torque transfer to the front and rear axles. The differential has a locking mechanism to undifferentiated 4WD in the high and low ranges.

On Vehicle Services

SHIFT LINKAGE

ADJUSTMENT

1. Remove the shift lever boot. Place the shift lever in the **4L** position.
2. Insert a 0.157 in. (4mm) spacer between the shift lever and the forward edge of the shift lever gate. Secure the lever and spacer in position with tape or wire.
3. Raise and support the vehicle safely. Loosen the trunnion lock bolt. The linkage rod should now slide freely in the trunnion.
4. Be sure that the transfer case range lever is still in the **4L** position. Position the linkage so it is a free fit in the range lever. Tighten the trunnion locknut.
5. Lower the vehicle. Remove the shift lever spacer and install the boot and bezel.

SPEEDOMETER GEAR, SHAFT SEAL, REAR BEARING AND RETAINER

Removal and Installation

The front and rear output shaft seals, extension, rear retainer, bearing and speedometer drive gear can all be serviced with the transfer case in the vehicle. The following combined procedure outlines removal and installation of these components.

1. Raise and support the vehicle safely.
2. Place a drain pan under transfer case, remove the drain plug and drain fluid from the transfer case.
3. Remove the driveshaft and secure it to the under side of the vehicle.
4. Remove the extension seal with a suitable seal removal tool. Install a new seal with a seal installer tool.
5. Remove the extension housing. Remove the speedometer gear from the rear retainer. Mark the retainer for assembly alignment reference.
6. Remove the retainer attaching bolts and remove the retainer. Pry the retainer with a suitable tool to remove it.
7. Completely clean off all the old RTV sealant. If the retainer or bearing are to be replaced, remove the bearing retainer snapring from the rear retainer and remove the bearing.
8. Install the rear output bearing in the rear retainer. Apply a ⅛ in. bead or RTV sealant to the mating surface of the rear retainer. Align the retainer case reference marks and install the rear retainer on the case. Torque the rear retainer bolts to 18 ft. lbs. (23 Nm).
9. Install the output shaft seal. Install the speedometer driven gear.
10. Fill the transfer case to the edge of the fill plug opening with the recommended lubricant. Install the speedometer driven gear.Install the drain plug to 35 ft. lbs. (48 Nm).
11. Install the driveshaft and tighten the clamp strap bolts to 170 inch lbs. (19 Nm). Lower the vehicle.

Removal and Installation

TRANSFER CASE REMOVAL

1. Raise the vehicle and support safely.
2. Remove the drain plug and drain the fluid from the transfer case.
3. Mark the transfer case front and rear output shaft yokes and driveshafts for assembly alignment and reference.
4. Disconnect the speedometer cable and vacuum switch hoses.
5. Disconnect the shift lever link from the operating lever.
6. Place a support stand under the transmission and remove the rear crossmember.
7. Mark the transfer case front and rear output shaft yokes and driveshafts for assembly alignment reference.
8. Disconnect the front and rear driveshafts at the transfer case yokes. Secure the shafts to the frame rails with wire.

NOTE: Do not allow the shafts to hang.

9. Remove the bolts attaching the exhaust pipe support bracket to the transfer case, if necessary.
10. Remove the transfer case-to-transmission nuts.
11. Move the transfer case assembly rearward until free of the transmission output shaft and remove the assembly.
12. Remove all the gasket material from the rear of the transmission adapter housing.

TRANSFER CASE INSTALLATION

1. Apply a suitable sealant to both sides of the transfer case-to-transmission gasket and position the gasket on the transmission.
2. Align and install the transfer case assembly on the transmission. Be sure the transfer case input gear splines are aligned with the transmission output shaft. Align the splines by rotating the transfer case rear output shaft yoke as necessary.

NOTE: Do not install any transfer case attaching nuts until the transfer case is completely seated against the transmission.

3. Align and install the transfer case attaching nuts. Tighten the nuts to 26 ft. lbs. (35 Nm).
4. Install the rear crossmember and torque the rear crossmember bolts to 30 ft. lbs. (41 Nm). Remove the transmission support stand.
5. Attach the exhaust pipe support bracket to the transfer case, if removed.
6. Align and connect the driveshafts.
7. Connect the speedometer cable and vacuum switch hoses.

8. Connect the shift lever to the operating lever. Tighten the locknut to 18 ft. lbs. (23 Nm).
9. Fill the transfer case with the recommended lubricant to the proper level.
10. Lower the vehicle.

Before Disassembly

Cleanliness during disassembly and assembly is necessary to avoid further transfer case trouble after overhaul. Before removing any of the transfer case subassemblies, plug all the openings and clean the outside of the of the transfer case thoroughly. Steam cleaning or car wash type high pressure equipment is preferable. During disassembly, clean all parts in suitable solvent and dry each part. Do not use cloth or paper towels to dry parts. Use compressed air only.

Transfer Case Disassembly

1. Place the transfer case into a suitable holding fixture.
2. Remove the attaching nuts from the front and rear output yokes. Remove the yokes and sealing washers. Move the transfer case selector lever rearward to the **4L** position.
3. Remove the bolts and tap the extension housing off of the rear retainer. Tap the extension housing in the clockwise direction to break the sealer bead, then remove the housing.

NOTE: To avoid damaging the sealing surface of the extension housing and rear retainer, do not attempt to pry or wedge the housing off of the retainer.

4. Remove the snapring from the rear bearing, then, remove the 4 bolts and separate the rear bearing retainer from the rear case half.
5. Remove the rear retainer, position a pry bar under each of the tabs on the retainer housing and carefully pry the retainer upward and off the rear case. Remove the bolts attaching the rear case to the front case. Separate the rear case from the front case.
6. Remove the oil pump and rear case as an assembly. Slide the oil screen out of the case pocket. Disconnect the screen from the pickup tube and remove the screen. Remove the pickup tube from the oil pump and remove the oil pump from the rear case.
7. Remove the pickup tube O-ring from the oil pump and mark the oil pump housing for assembly reference.
8. Remove the screws that attach the 2 halves of the pump. Remove the feed housing from the gear housing. Be sure to note the position of the pump gears and remove them from the housing.
9. Remove the mode spring from the shift rail. Tap the front output shaft upward to free it from the shaft bearing. Remove the front output shaft and drive chain as an assembly.
10. Remove the transfer case shift lever nut and lever. Remove the shift detent plug, spring and pin.
11. Remove the seal plug from the low range fork lockpin access hole. Move the shift selector to align the low range fork lockpin with the access hole. Remove the range fork lockpin with a size No. 1 easy out[®] tool or equivalent. Grip the easy out with a locking pliers and remove the pin with a counterclockwise, twist and pull motion. Remove the shift rail by pulling it strait up and out of the fork.
12. Remove the mainshaft, mode fork and spring rail assembly from the front case half. Remove the mode fork and shift rail from the synchro sleeve. Remove the synchro sleeve from the mainshaft.

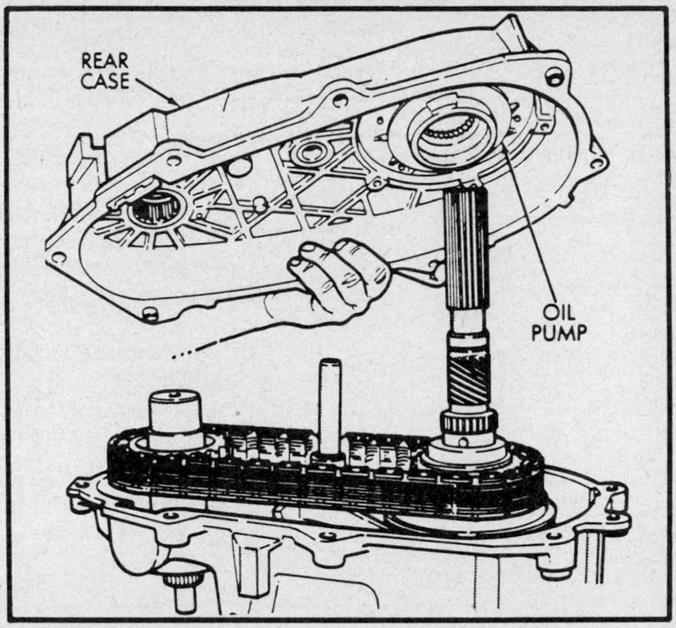
REAR CASE
OIL PUMP
Removing the rear case and oil pump assembly

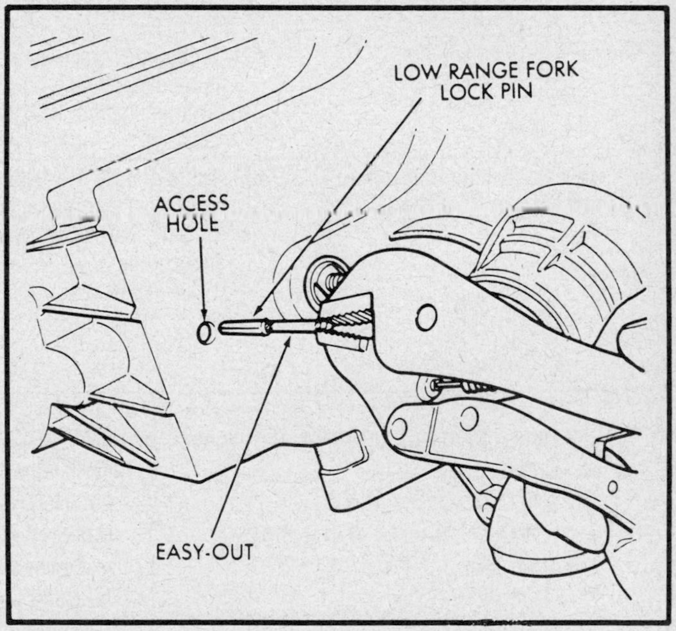
LOW RANGE FORK LOCK PIN
ACCESS HOLE
EASY-OUT
Removing the low range fork lockpin

13. Remove the intermediate clutch shaft snapring, the clutch shaft thrust ring, intermediate clutch shaft, differential snapring and remove the differential.
14. Remove the differential needle bearings and the 2 needle bearing thrust washers from the mainshaft.
15. Slide the range fork pin out of the shift selector. Remove the range fork and shift hub. Remove the transfer case range lever from the sector shaft. Remove the shift sector.

Trouble Diagnosis

CHILTON'S THREE C'S TRANSFER CASE DIAGNOSIS

Condition	Cause	Correction
Transfer case difficult to shift or will not shift into desired range	a) Vehicle speed too great to permit shifting	a) Stop vehicle and shift into desired range. Or reduce speed to 2–3 mph (3–4 km/h) before attempting to shift
	b) If vehicle was operated for extended period in 4H mode on dry paved surface, driveline torque load may cause difficulty	b) Stop vehicle, shift transmission to neutral, shift transfer case to 2H mode and operate vehicle on 2H on dry paved surfaces
	c) Transfer case external shift linkage binding	c) Lubricate, repair or replace linkage, or tighten loose components as necessary
	d) Insufficient or incorrect lubricant	d) Drain and refill to edge of fill hole
	e) Internal components binding, worn or damaged	e) Disassemble unit and replace worn or damaged components as necessary
Transfer case noisy in all drive modes	a) Insufficient or incorrect lubricant	a) Drain and refill to edge of fill hole. If unit is still noisy after drain and refill, disassembly and inspection may be required to locate source of noise
Noisy in—or jumps out of the four-wheel-drive low range	a) Transfer case not completely engaged in 4L position	a) Stop vehicle, shift transfer case to Neutral, then shift back into 4L position
	b) Shift linkage loose or binding	b) Tighten, lubricate, or repair linkage as necessary
	c) Range fork cracked, inserts worn, or fork is binding on shift rail	c) Disassemble unit and repair as necessary
	d) Annulus gear or lockplate worn or damaged	d) Disassemble unit and repair as necessary
Lubricant leaking from output shaft seals or from vent	a) Transfer case overfilled	a) Drain to correct level
	b) Vent closed or restricted	b) Clear or replace vent if necessary
	c) Output shaft seals damaged or installed incorrectly	c) Replace seals. Be sure seal lip faces interior of case when installed. Also be sure yoke seal surfaces are not scored or nicked. Remove scores and nicks with fine sandpaper or replace yoke(s) if necessary

1 FRONT BEARING RETAINER AND SEAL
2 FRONT CASE
3 SHIFT SECTOR
4 LOW RANGE FORK AND INSERTS
5 SHIFT RAIL
6 SHIFT BRACKET
7 SLIDER BRACKET
8 BUSHING AND SPRING
9 MODE FORK AND INSERTS
10 BUSHING
11 FORK SPRING
12 BUSHING
13 VENT TUBE ASSEMBLY
14 INPUT GEAR BEARING AND SNAP RING
15 LOW RANGE GEAR SNAP RING
16 RETAINER, LOW RANGE GEAR
17 THRUST WASHER, LOW RANGE GEAR
18 INPUT GEAR
19 REAR CASE
20 DRAIN/FILL PLUGS
21 REAR BEARING RETAINER
22 EXTENSION HOUSING
23 BUSHING AND OIL SEAL
24 VACUUM SWITCH
25 MAGNET
26 THRUST RING
27 SNAP RING
28 SHIFT SLEEVE
29 LOW RANGE GEAR
30 PILOT BUSHING (INPUT GEAR/MAINSHAFT)
31 FRONT OUTPUT SHAFT FRONT BEARING AND SNAP RING
32 INTERMEDIATE CLUTCH SHAFT
33 SHIFT SLEEVE
34 SNAP RING
35 MAINSHAFT
36 DIFFERENTIAL ASSEMBLY
37 OIL PUMP TUBE O-RING
38 OIL PUMP PICKUP TUBE AND SCREEN
39 MAINSHAFT BEARING ROLLERS
40 DRIVE SPROCKET
41 DRIVE CHAIN
42 SNAP RING
43 OIL PUMP SEAL
44 OIL PUMP
45 REAR BEARING AND SNAP RING
46 FRONT OUTPUT SHAFT REAR BEARING
47 SNAP RING
48 DRIVEN SPROCKET
49 FRONT OUTPUT SHAFT
50 MAINSHAFT BEARING SPACERS
51 SHIFT LEVER WASHER AND NUT
52 SHIFT LEVER
53 SECTOR O-RING AND SEAL
54 DETENT PIN, SPRING AND PLUG
55 SEAL PLUG
56 FRONT YOKE NUT, SEAL WASHER, YOKE, SLINGER AND OIL SEAL

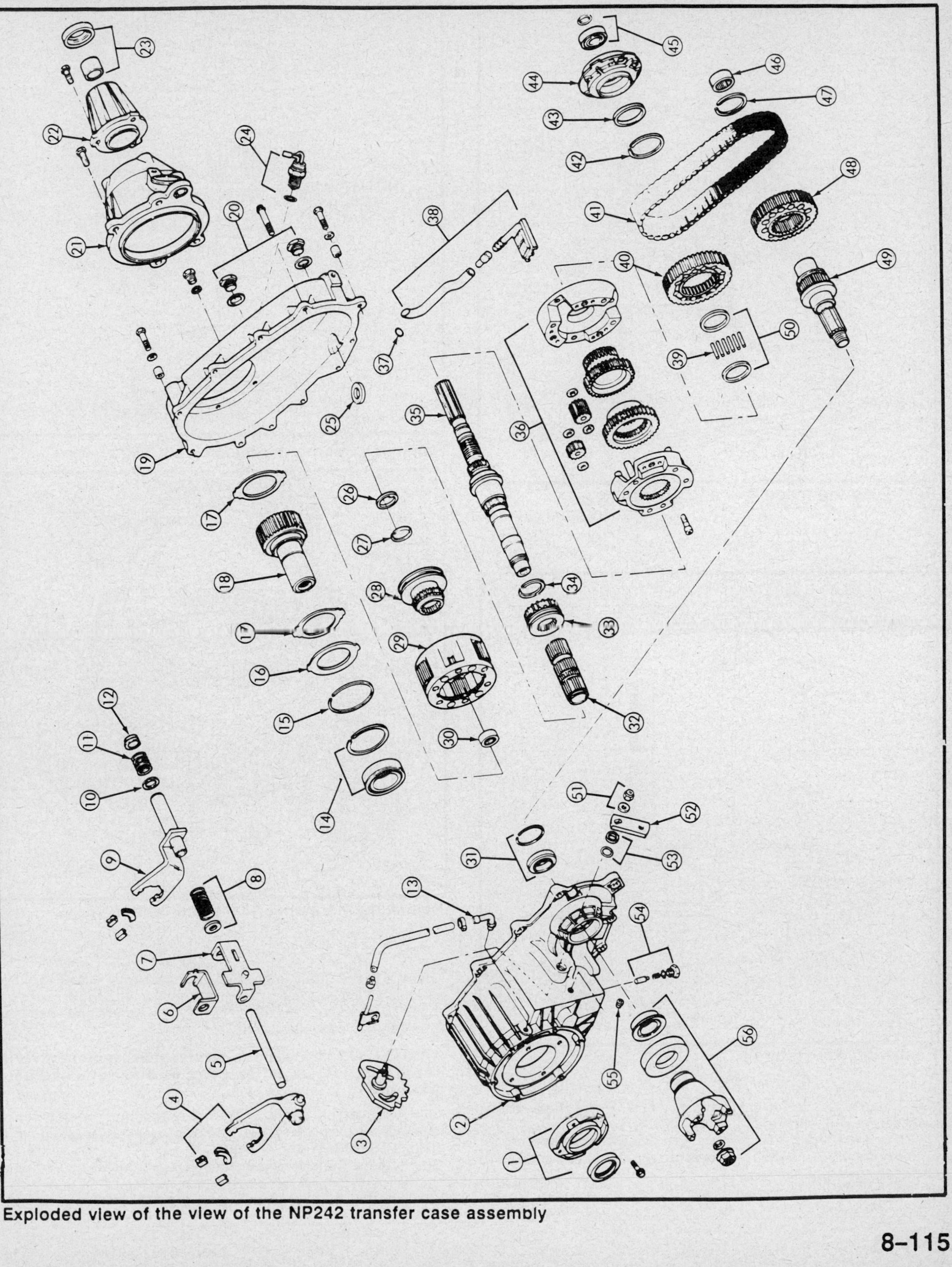
1
2
3
4
5
6
7
8
9
10
11
12
13
14
15
16
17
18
17
19
20
21
22
23
24
25
26
27
28
29
30
31
32
33
34
35
36
37
38
39
40
41
42
43
44
45
46
47
48
49
50
51
52
53
54
55
56

Exploded view of the view of the NP242 transfer case assembly

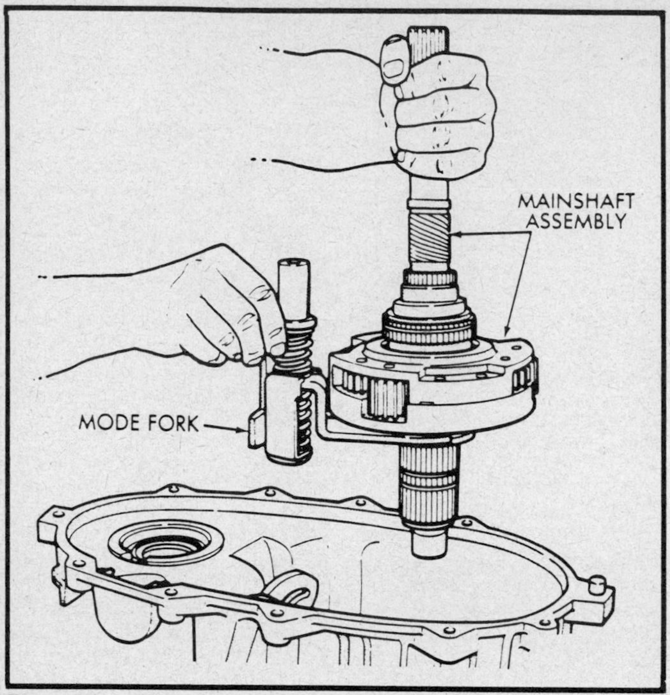
MAINSHAFT
ASSEMBLY
MODE FORK
Removing the mode fork and mainshaft

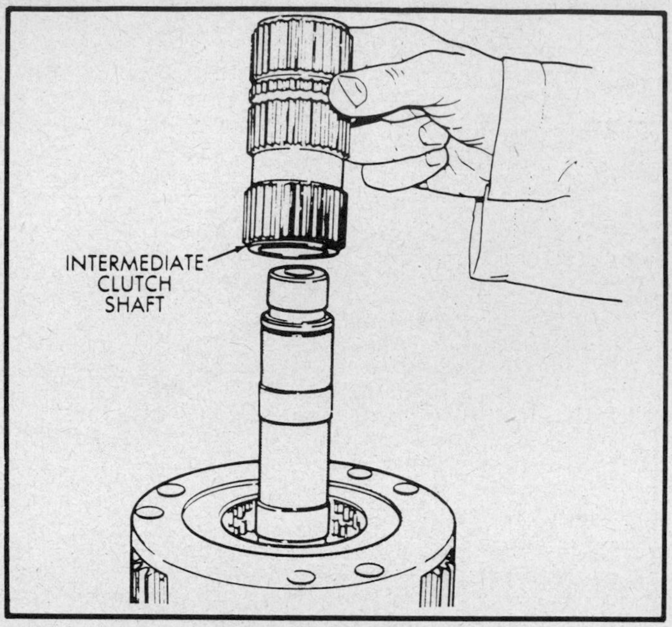
INTERMEDIATE
CLUTCH
SHAFT
Removing the intermediate clutch shaft

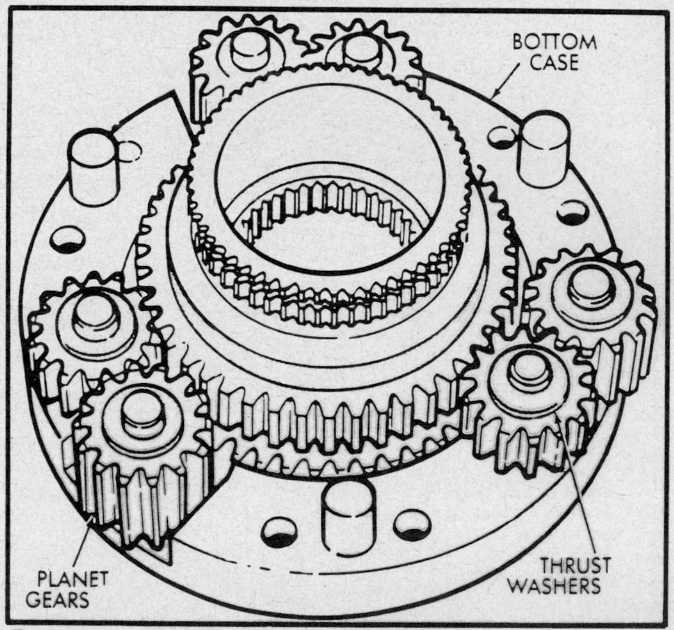
BOTTOM
CASE
PLANET
GEARS
THRUST
WASHERS
Removing the planet gears and carrier

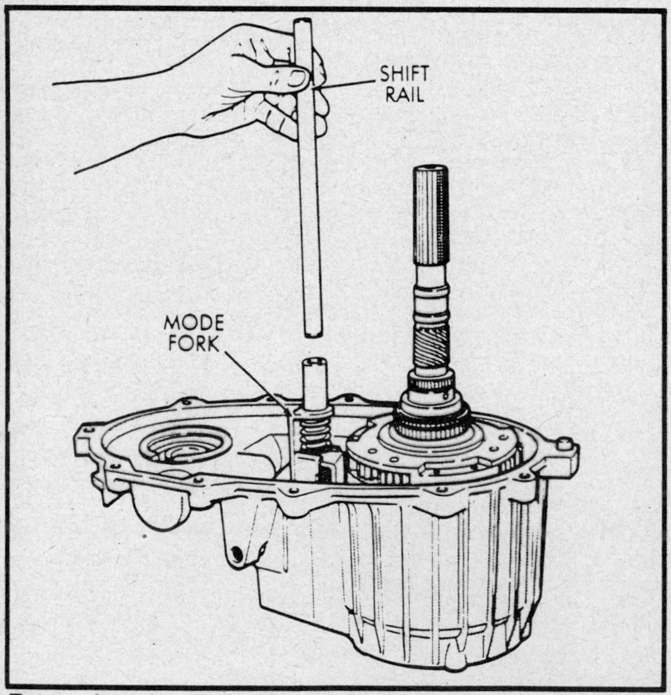
SHIFT
RAIL
MODE
FORK
Removing the shift rail

16. Remove the sector shaft bushing and O-ring. Remove the shift detent pin, spring and plug.
17. Turn the front case over and remove the front bearing retainer bolts. Remove the front bearing retainer. Remove the input gear snapring.

18. Press the input low range gear assembly out of the input bearing with a suitable arbor press. Remove the low range gear snapring. Remove the retainer, thrust washers and input gear from the low range gear.

NOTE: The gear is not a serviceable component. If it is damaged, replace the gear and front case as an assembly.

19. Remove the oil seals from the retainer, rear extension housing, oil pump feed housing and case halves. Remove the magnet from the front of the case.
20. Remove the speedometer driven gear, seals and adaptor.

CLEANING AND INSPECTION

Cleaning

During overhaul, all components of the transfer case (except bearing assemblies) should be thoroughly cleaned with solvent and dried with air pressure prior to inspection and reassembly.

1. Clean the bearing assemblies as follows:

NOTE: Proper cleaning of bearings is of utmost importance. Bearings should always be cleaned separately from other parts.

a. Soak all bearing assemblies in clean solvent or fuel oil. Bearings should never be cleaned in a hot solution tank.

b. Slush bearings in solvent until all old lubricant is loosened. Hold races so that bearings will not rotate; then clean bearings with a soft bristled brush until all dirt has been removed. Remove loose particles of dirt by tapping bearing flat against a block of wood.

c. Rinse bearings in clean solvent; then blow bearings dry with air pressure.

CAUTION

Do not spin bearings while drying.

d. After drying, rotate each bearing slowly while examining balls or rollers for roughness, damage, or excessive wear. Replace all bearings that are not in first class condition.

NOTE: After cleaning and inspecting bearings, lubricate generously with recommended lubricant, then wrap each bearing in clean paper until ready for reassembly.

2. Remove all portions of old gaskets from parts, using a stiff brush or scraper.

Inspection

1. Inspect all parts for discoloration or warpage.
2. Examine all gears and splines for chipped, worn, broken or nicked teeth. Small nicks or burrs may be removed with a fine abrasive stone.
3. Inspect the breather assembly to make sure that it is open and not damaged.
4. Check all threaded parts for damaged, stripped, or crossed threads.
5. Replace all gaskets, oil seals and snaprings.
6. Inspect housings, retainers and covers for cracks or other damage. Replace the damaged parts.
7. Inspect keys and keyways for condition and fit.
8. Inspect shift forks for wear, distortion or any other damage.
9. Check detent ball springs for free length, compressed length, distortion or collapsed coils.
10. Check bearing fit on their respective shafts and in their bores or cups. Inspect bearings, shafts and cups for wear.

NOTE: If either bearings or cups are worn or damaged, it is advisable to replace both parts.

11. Inspect all bearing rollers or balls for pitting or galling.
12. Examine detent balls for corrosion or brinneling. If shift bar detents show wear, replace them.
13. Replace all worn or damaged parts. When assembling the transfer case, coat all moving parts with recommended lubricant.

Transfer Case Assembly

Before assembly, lubricate all parts with the recommended lubricant. The bearing bores in the various transfer case components contain oil feed holes. Be sure the replace bearings do not block these feed holes.

1. Remove the input gear pilot bearing with a suitable slide hammmer and seal removal tool.
2. Install a new input gear pilot bearing into the input gear.
3. Assemble the low range gear, input gear thrust washers, input gear and input gear retainer.
4. Install the low range gear snapring. Lubricate the input gear with automatic transmission fluid. Start the input gear into the front bearing. Press the input gear into the front bearing.

NOTE: Use a proper size tool to press the input gear into the front bearing. An incorrect tool could push the input gear pilot bearing too far into the gear bore. Also, do not press against the end faces of the low range gear. The gear cae and thrust washers could be damaged.

5. Install a new input gear snapring. Install a new oil seal in the front bearing retainer. Apply a ⅛ in. (3mm) wide bead of RTV sealant to the front bearing retainer seal surface. Install the front bearing retainer on the front case. Torque the retainer bolts to 16 ft. lbs. (21 Nm).
6. Install a new sector shaft O-ring and bushing. Install the shift sector in the case. Install new pads in the low range fork. Install the range lever and lever attaching nut on the shift sector. Torque the attaching nut to 22 ft. lbs. (30 Nm).
7. Install the detent pin, spring and plug. Torque the plug to 15 ft. lbs. (20 Nm). Install new pads and shift rail bushings in the range fork. Assemble the range fork and shift hub. Engage the range fork pin in the sector slot.
8. Lubricate the differential components with the recommended lubricant. Install the sprocket gear in the differential bottom case. Install the differential planet gears and new thrust washers. Be sure the thrust washers are installed at the top and bottom of each planet gear.
9. Install the differential mainshaft gear. Align using a scribe marks made at disassembly. Install the first mainshaft bearing spacer on the mainshaft. Install the bearing rollers on the mainshaft. Be sure to coat the bearing rollers with a petroleum jelly to hold them in place.
10. Install the remaining bearing spacer on the mainshaft. Do not display any bearings while installing the spacer. Install the differential. Do not displace the mainshaft bearing when installing the differential.
11. Install the differential snapring, intermediate clutch shaft, clutch shaft thrust washer and clutch shaft snapring.
12. Inspect the mode fork assembly. Replace the pads and bushing if necessary. Replace the fork tube if the bushings inside the tube are worn or damaged. Also check the springs and slider bracket. Replace all worn or damaged parts.
13. Install the mode sleeve in the mode fork. Install the assembled sleeve and fork on the mainshaft. Be sure the mode sleeve splines are engaged in the differential splines.
14. Install the mode fork and mainshaft assembly in the case. Rotate the manishaft slightly to engage the shaft with the low range gears. Rotate the mode fork pin into the shift sector slot.
15. Install thr shift rail, be sure that the rail is seated in both shift forks. Rotate the shift sector to align the lockpin hole in the low range fork with the access hole in the case. Insert an easy out or equivalent in the range fork lockpin to hole it securely for installation. The lockpin is slightly tappered on 1 end. Insert the tappered end into the fork and rail.
16. Insert the lockpin through access hole and into the shift fork. Then remove the easy out and seat the pin with a pin punch. Install the plug in the lockpin access hole.
17. Install the transfer case shift lever and attaching nut. Torque the nut to 22 ft. lbs.
18. Install the detent plunger, detent spring and detent plug in the case. Install the front output shaft.
19. Install the drive chain. Lift the mainshaft slightly to ease

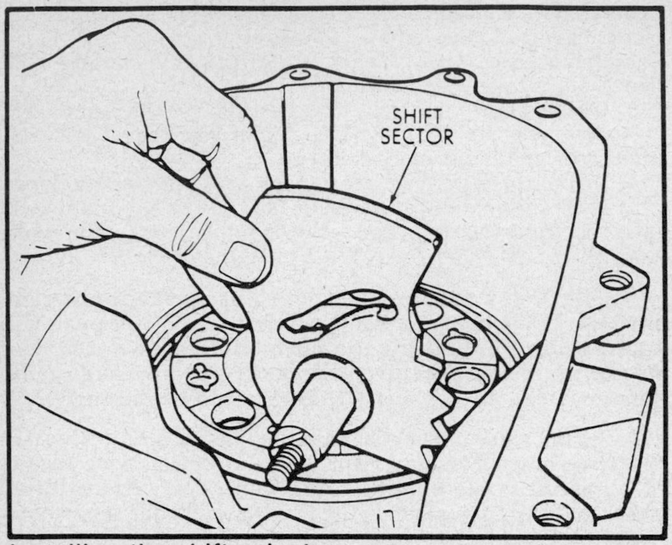
SHIFT
SECTOR
Installing the shift selector

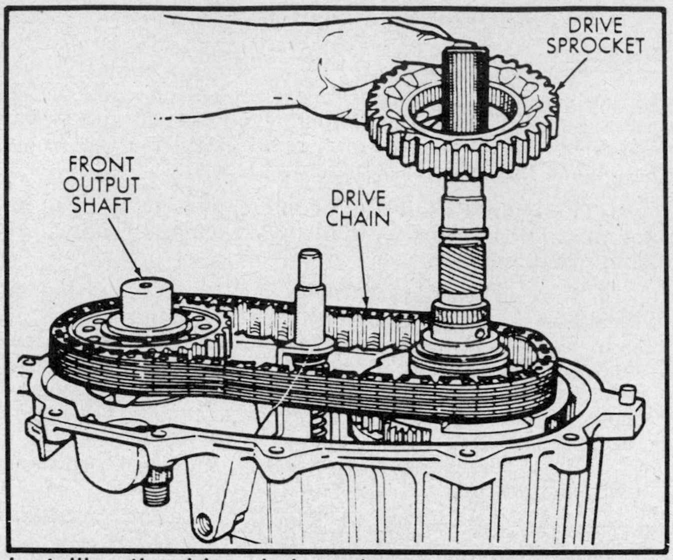
DRIVE
SPROCKET
FRONT
OUTPUT
SHAFT
DRIVE
CHAIN
Installing the drive chain and sprocket assembly

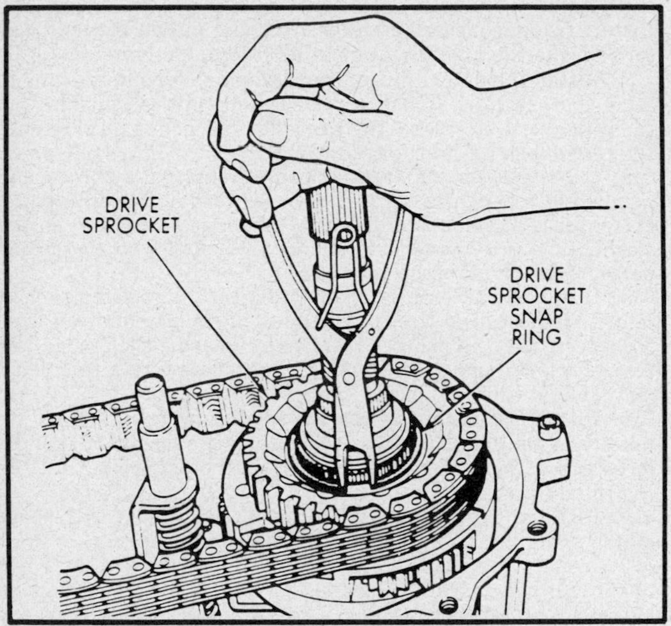
DRIVE
SPROCKET
DRIVE
SPROCKET
SNAP
RING
Installing the drive sprocket snapring

the chain and shaft installation. Install the drive sprocket. Engage the drive sprocket with the chain. Engage the sprocket splines with the mainshaft splines. Install a new drive socket snapring.

20. Install a new output shaft rear bearing. Remove the bearing with an internal type puller and slide hammer. Seat the new bearing in the case with a suitable seal installer.

21. Install a new seal in the oil pump feed housing. Assemble the oil pump. Lubricate and install the 2 gears in the gear housing. Align and install the feed housing on the gear housing. Install and tighten the oil pump screws to 14 inch lbs.

22. Install the pickup tube O-ring in the oil pump. Insert the oil pickup tube in the oil pump. Attach the oil screen and connecting hose to the pickup tube. Install thre assembled oil pump, pickup tube and screen in the rear case. Be sure that the screen isseated in the case slot. Install the magnet in the front case.

23. Apply a ⅛ in. (3mm) wide bead of RTV sealant to the sealing surface of the front case. Align and install the rear case on the front case. Be sure the case locating dowels are in place and the mainshaft splines are engaged in the oil pump inner gear.

24. Install and torque the front case to rear case attaching bolts to 30 ft. lbs. (41 Nm).Be sure to install a washer under each of the bolts use at each case dowel locations.

25. Apply a ⅛ in. (3mm) wide bead of RTV sealant to the flange surface of the rear retainer. Install the locating dowel in the rear retainer and install the retainer on the case. Torque the retainer bolts to 30 ft. lbs. (41 Nm). Install a new rear bearing snapring. Lift the mainshaft slightly to seat the snapring in the shaft groove.

26. Install a new seal in the extension housing. Apply a ⅛ in. (3mm) wide bead of RTV sealant to the flange surface of the extension housing. Install the exstension housing on the case, torque the retaining bolt to 30 ft. lbs. (41 Nm).

27. Install the front yoke. Secure the yoke with a replacement seal washer and nut. Torque the nut to 110 ft. lbs. (169 Nm).

28. Install a replacement gasket on the vacuum switch and install the switch in the case. Torque the switch to 20 ft. lbs. (27 Nm).

29. Install the speedometer driven gear, seals and adapter. Refill the transfer case with the recommended fluid to the proper level. Install and torque the drain plug to 35 ft. lbs. (48 Nm).

Specifications

TORQUE SPECIFICATIONS
NP242 Transfer Case

Component	ft. lbs.	Nm
Extension housing bolt	30	41
Rear retainer bolt	30	41
Front case-to-rear case bolt	30	41
Front yoke nut	110	149
Front bearing retainer bolt	16	21
Differential housing bolt	N/A	N/A
Shift lever nut	22	30
Switch	20	27
Detent spring cover	15	20
Drain and fill plug	35	47
Oil pump screw	14	1.6

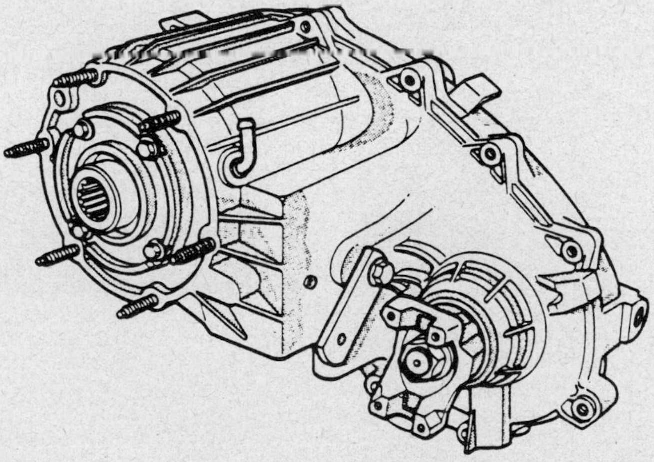

Model 242 Transfer Case

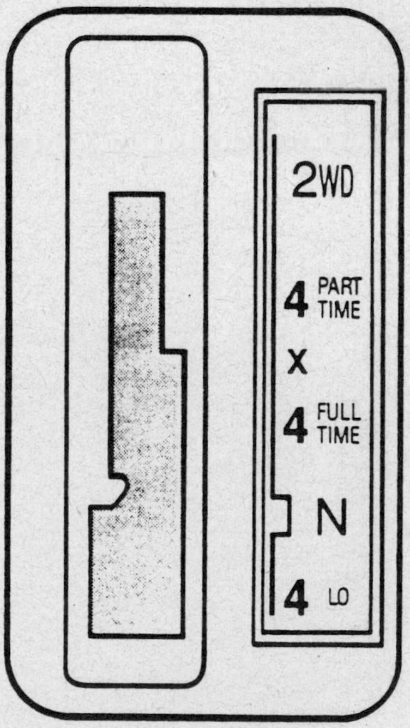
2WD
4 PART TIME
X
4 FULL TIME
N
4 LO

Shift Pattern

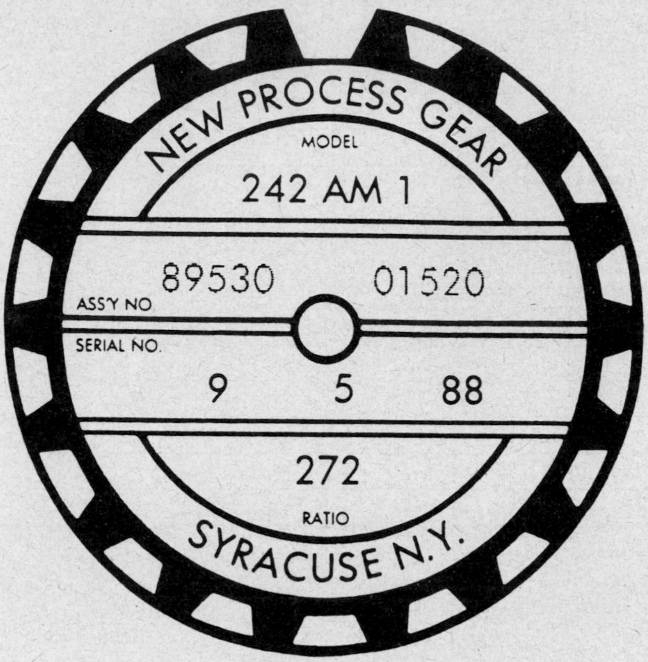
NEW PROCESS GEAR
MODEL
242 AM 1
ASS'Y NO.
89530
01520
SERIAL NO.
9
5
88
272
RATIO
SYRACUSE N.Y.

Transfer Case I.D. Tag

Section 9

Oil Flow Circuits

PARK—A500

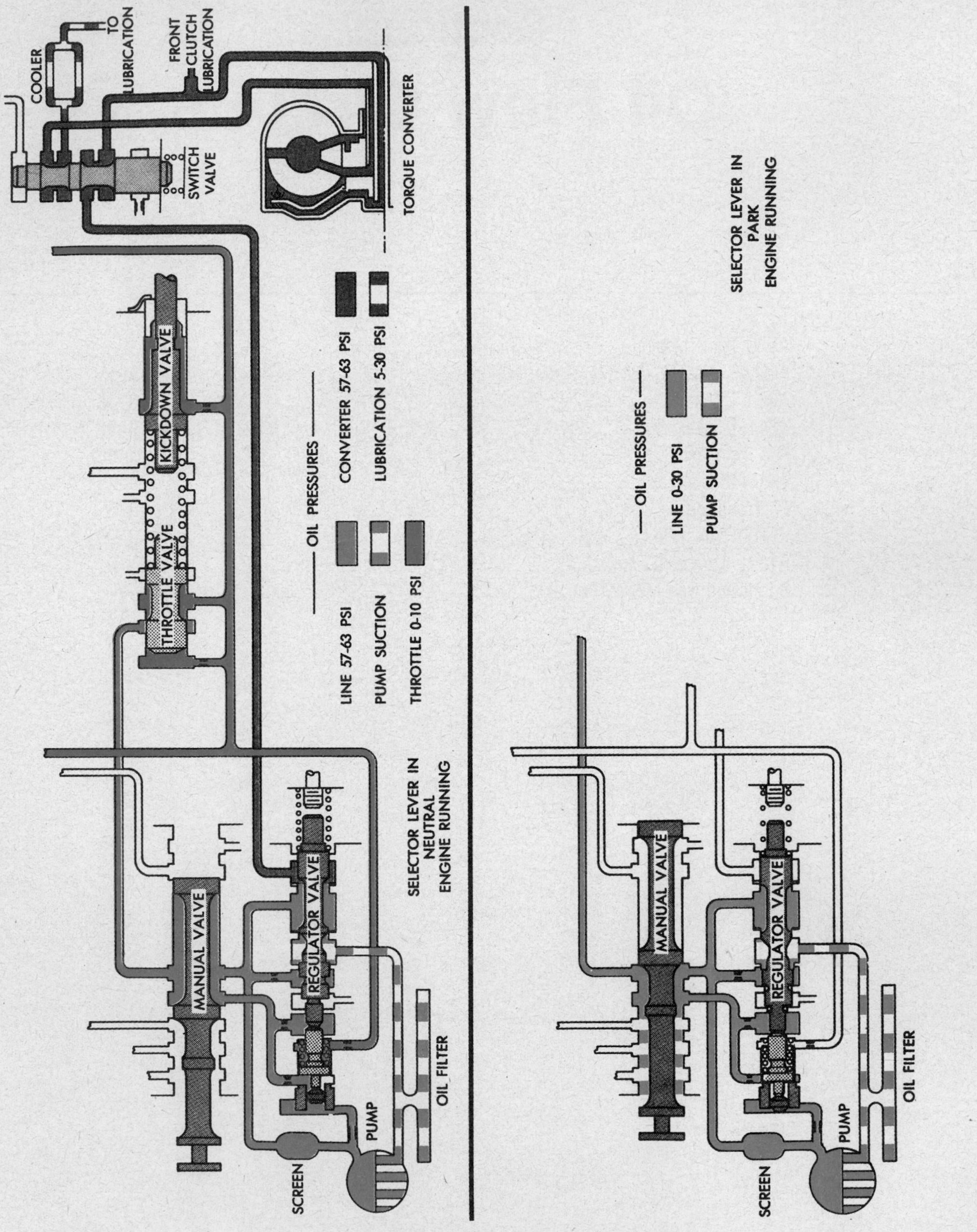
COOLER
TO LUBRICATION
FRONT CLUTCH LUBRICATION
SWITCH VALVE
TORQUE CONVERTER
KICKDOWN VALVE
THROTTLE VALVE
— OIL PRESSURES —
LINE 57-63 PSI
PUMP SUCTION
THROTTLE 0-10 PSI
CONVERTER 57-63 PSI
LUBRICATION 5-30 PSI
MANUAL VALVE
REGULATOR VALVE
SELECTOR LEVER IN NEUTRAL ENGINE RUNNING
SCREEN
PUMP
OIL FILTER
SELECTOR LEVER IN PARK ENGINE RUNNING
— OIL PRESSURES —
LINE 0-30 PSI
PUMP SUCTION
MANUAL VALVE
REGULATOR VALVE
SCREEN
PUMP
OIL FILTER

DRIVE (BREAKAWAY) HALF THROTTLE—A500

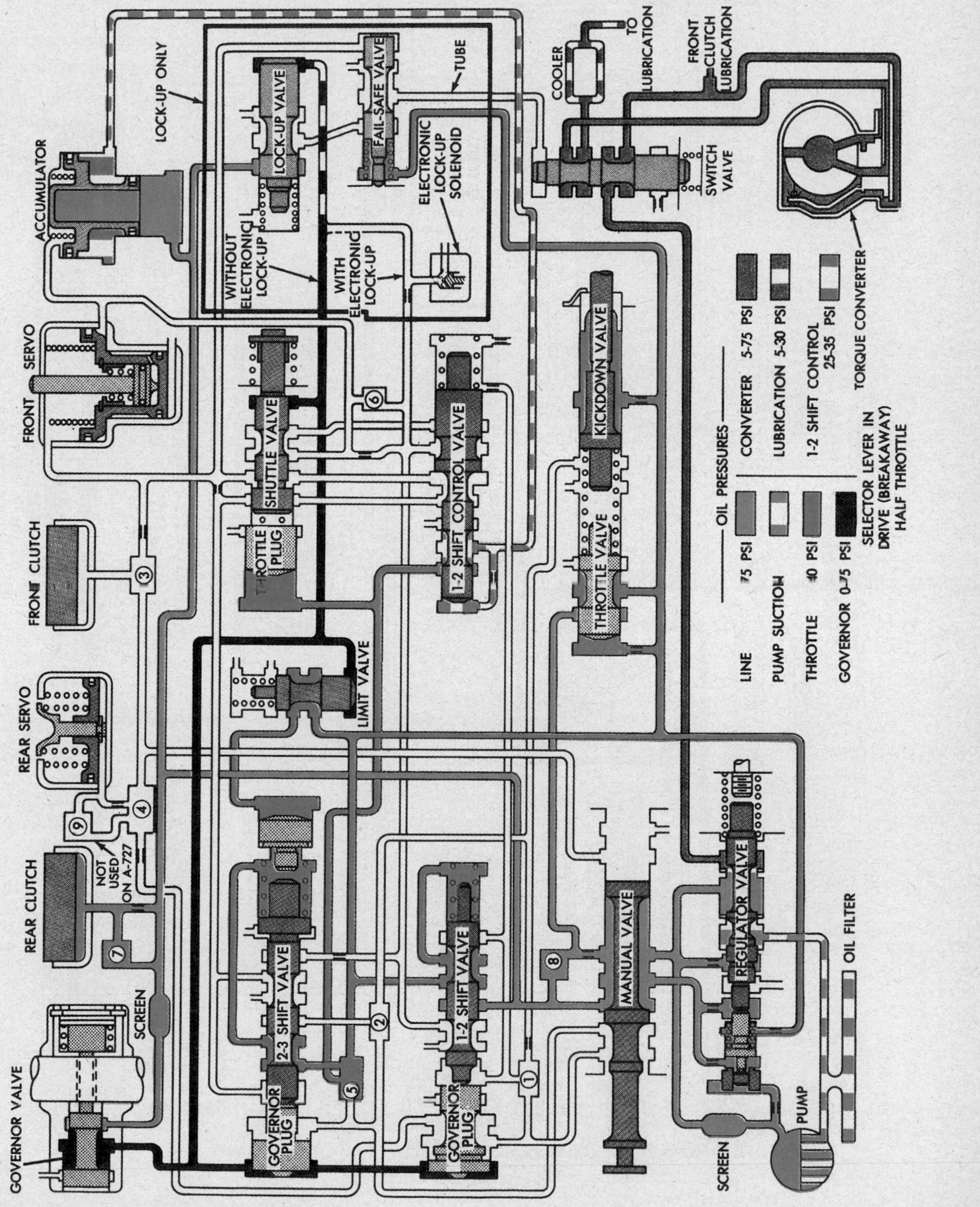
LOCK-UP ONLY
ACCUMULATOR
LOCK-UP VALVE
FAIL-SAFE VALVE
TUBE
ELECTRONIC LOCK-UP SOLENOID
WITHOUT ELECTRONIC LOCK-UP
WITH ELECTRONIC LOCK-UP
COOLER
TO LUBRICATION
FRONT CLUTCH LUBRICATION
SWITCH VALVE
TORQUE CONVERTER
FRONT SERVO
SHUTTLE VALVE
THROTTLE PLUG
1-2 SHIFT CONTROL VALVE
KICKDOWN VALVE
THROTTLE VALVE
OIL PRESSURES
LINE 75 PSI
PUMP SUCTION
THROTTLE 40 PSI
GOVERNOR 0-75 PSI
CONVERTER 5-75 PSI
LUBRICATION 5-30 PSI
1-2 SHIFT CONTROL 25-35 PSI
SELECTOR LEVER IN DRIVE (BREAKAWAY) HALF THROTTLE
FRONT CLUTCH
REAR SERVO
LIMIT VALVE
REAR CLUTCH
NOT USED ON A-727
SCREEN
GOVERNOR VALVE
2-3 SHIFT VALVE
GOVERNOR PLUG
1-2 SHIFT VALVE
MANUAL VALVE
REGULATOR VALVE
OIL FILTER
PUMP
1
2
3
4
5
6
7
8
9

DRIVE (2ND) HALF THROTTLE–A500

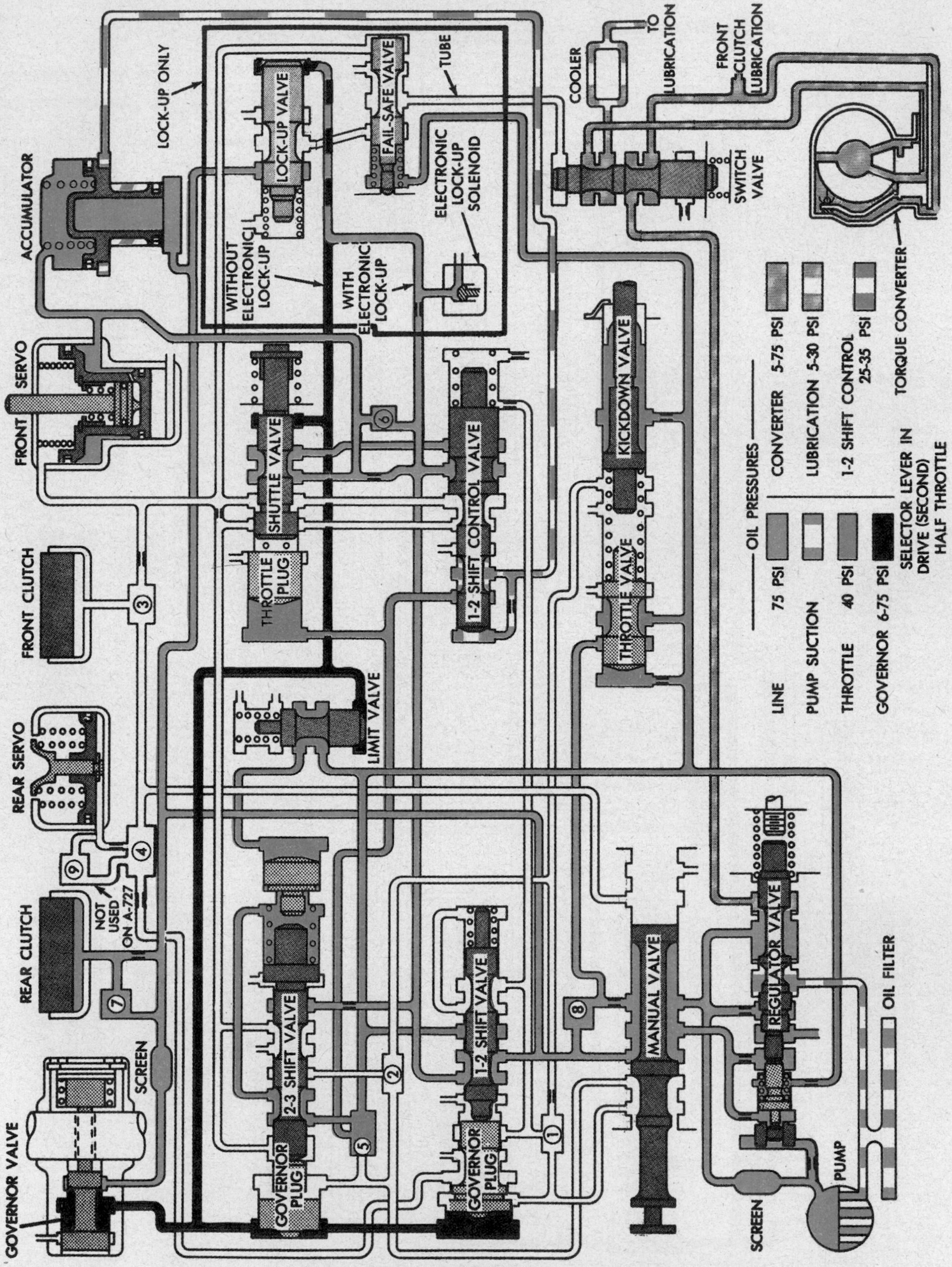
LOCK-UP ONLY
LOCK-UP VALVE
FAIL-SAFE VALVE
TUBE
COOLER
TO LUBRICATION
FRONT CLUTCH LUBRICATION
ACCUMULATOR
ELECTRONIC LOCK-UP SOLENOID
SWITCH VALVE
WITHOUT ELECTRONIC LOCK-UP
WITH ELECTRONIC LOCK-UP
TORQUE CONVERTER
FRONT SERVO
SHUTTLE VALVE
1-2 SHIFT CONTROL VALVE
KICKDOWN VALVE
THROTTLE PLUG
THROTTLE VALVE
FRONT CLUTCH
OIL PRESSURES
LINE 75 PSI
PUMP SUCTION
THROTTLE 40 PSI
GOVERNOR 6-75 PSI
CONVERTER 5-75 PSI
LUBRICATION 5-30 PSI
1-2 SHIFT CONTROL 25-35 PSI
SELECTOR LEVER IN DRIVE (SECOND) HALF THROTTLE
LIMIT VALVE
REAR SERVO
NOT USED ON A-727
REAR CLUTCH
SCREEN
2-3 SHIFT VALVE
1-2 SHIFT VALVE
MANUAL VALVE
REGULATOR VALVE
OIL FILTER
GOVERNOR VALVE
GOVERNOR PLUG
PUMP

DRIVE (DIRECT) HALF THROTTLE—A500

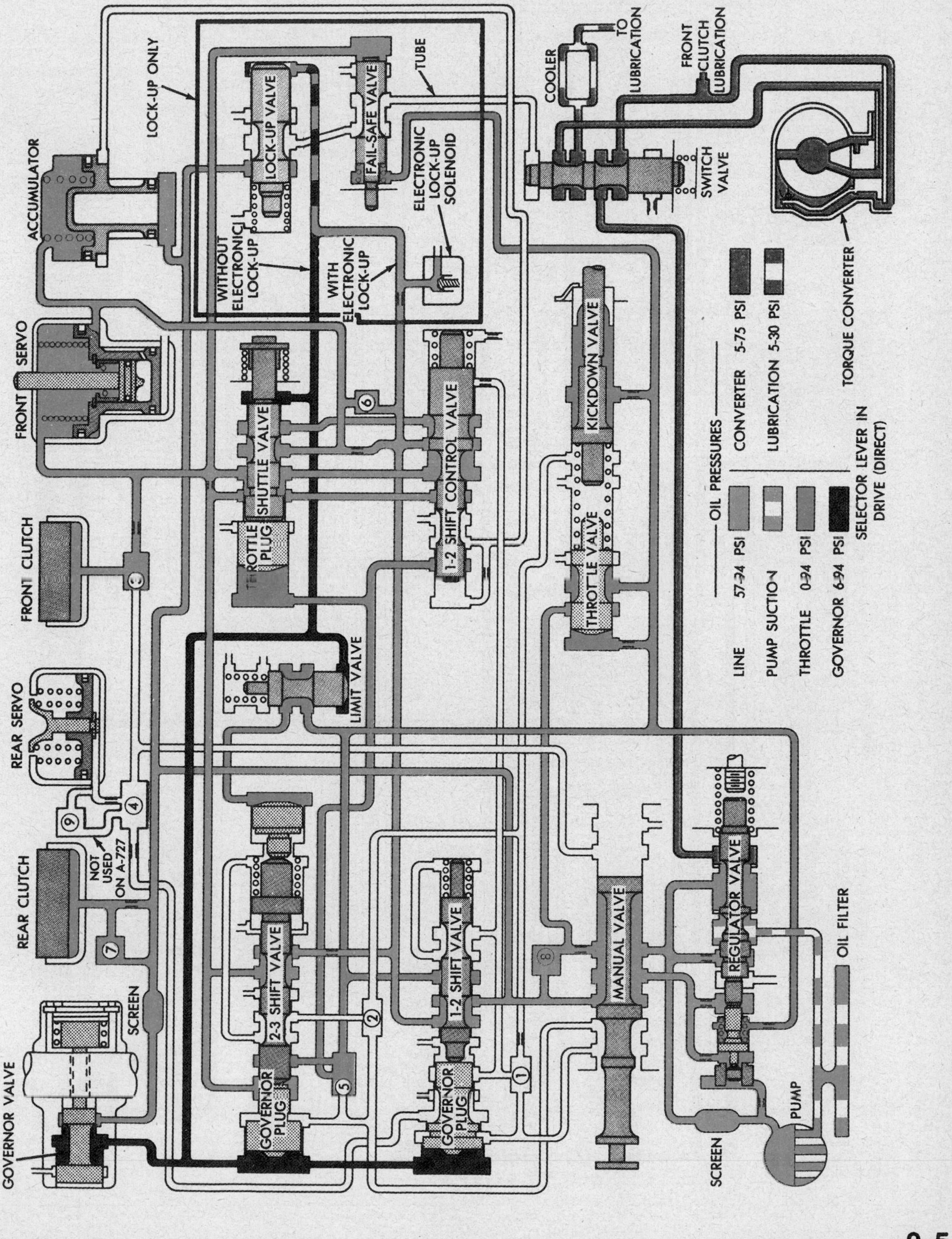
TO LUBRICATION
FRONT CLUTCH LUBRICATION
COOLER
TUBE
LOCK-UP ONLY
LOCK-UP VALVE
FAIL-SAFE VALVE
ELECTRONIC LOCK-UP SOLENOID
SWITCH VALVE
ACCUMULATOR
WITHOUT ELECTRONIC LOCK-UP
WITH ELECTRONIC LOCK-UP
TORQUE CONVERTER
FRONT SERVO
KICKDOWN VALVE
SHUTTLE VALVE
SHIFT CONTROL VALVE
1-2 SHIFT CONTROL VALVE
THROTTLE PLUG
FRONT CLUTCH
THROTTLE VALVE
OIL PRESSURES
LINE 57-94 PSI
PUMP SUCTION
THROTTLE 0-94 PSI
GOVERNOR 6-94 PSI
CONVERTER 5-75 PSI
LUBRICATION 5-30 PSI
SELECTOR LEVER IN DRIVE (DIRECT)
LIMIT VALVE
REAR SERVO
NOT USED ON A-727
REAR CLUTCH
2-3 SHIFT VALVE
1-2 SHIFT VALVE
MANUAL VALVE
REGULATOR VALVE
OIL FILTER
SCREEN
GOVERNOR VALVE
GOVERNOR PLUG
PUMP

DRIVE (LOCKUP) HALF THROTTLE—A500

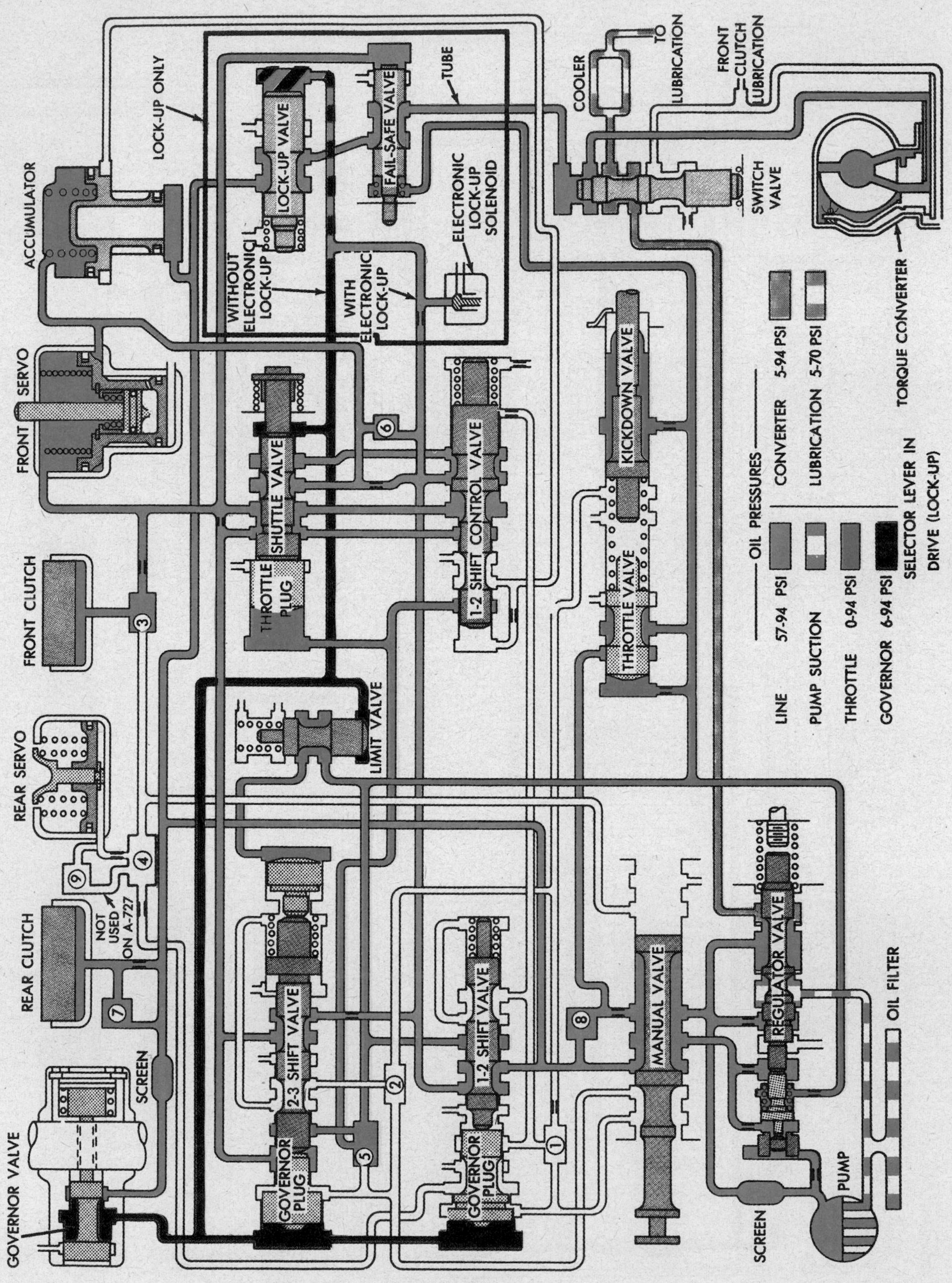
LOCK-UP ONLY
TUBE
COOLER
TO LUBRICATION
FRONT CLUTCH LUBRICATION
ACCUMULATOR
LOCK-UP VALVE
FAIL-SAFE VALVE
ELECTRONIC LOCK-UP SOLENOID
SWITCH VALVE
WITHOUT ELECTRONIC LOCK-UP
WITH ELECTRONIC LOCK-UP
FRONT SERVO
SHUTTLE VALVE
1-2 SHIFT CONTROL VALVE
KICKDOWN VALVE
TORQUE CONVERTER
THROTTLE PLUG
FRONT CLUTCH
THROTTLE VALVE
LIMIT VALVE
REAR SERVO
NOT USED ON A-727
REAR CLUTCH
REGULATOR VALVE
2-3 SHIFT VALVE
1-2 SHIFT VALVE
MANUAL VALVE
SCREEN
OIL FILTER
GOVERNOR VALVE
GOVERNOR PLUG
GOVERNOR PLUG
PUMP
SCREEN
OIL PRESSURES
LINE 57-94 PSI
PUMP SUCTION
THROTTLE 0-94 PSI
GOVERNOR 6-94 PSI
CONVERTER 5-94 PSI
LUBRICATION 5-70 PSI
SELECTOR LEVER IN DRIVE (LOCK-UP)

DRIVE (PART THROTTLE KICKDOWN) BELOW 40 MPH—A500

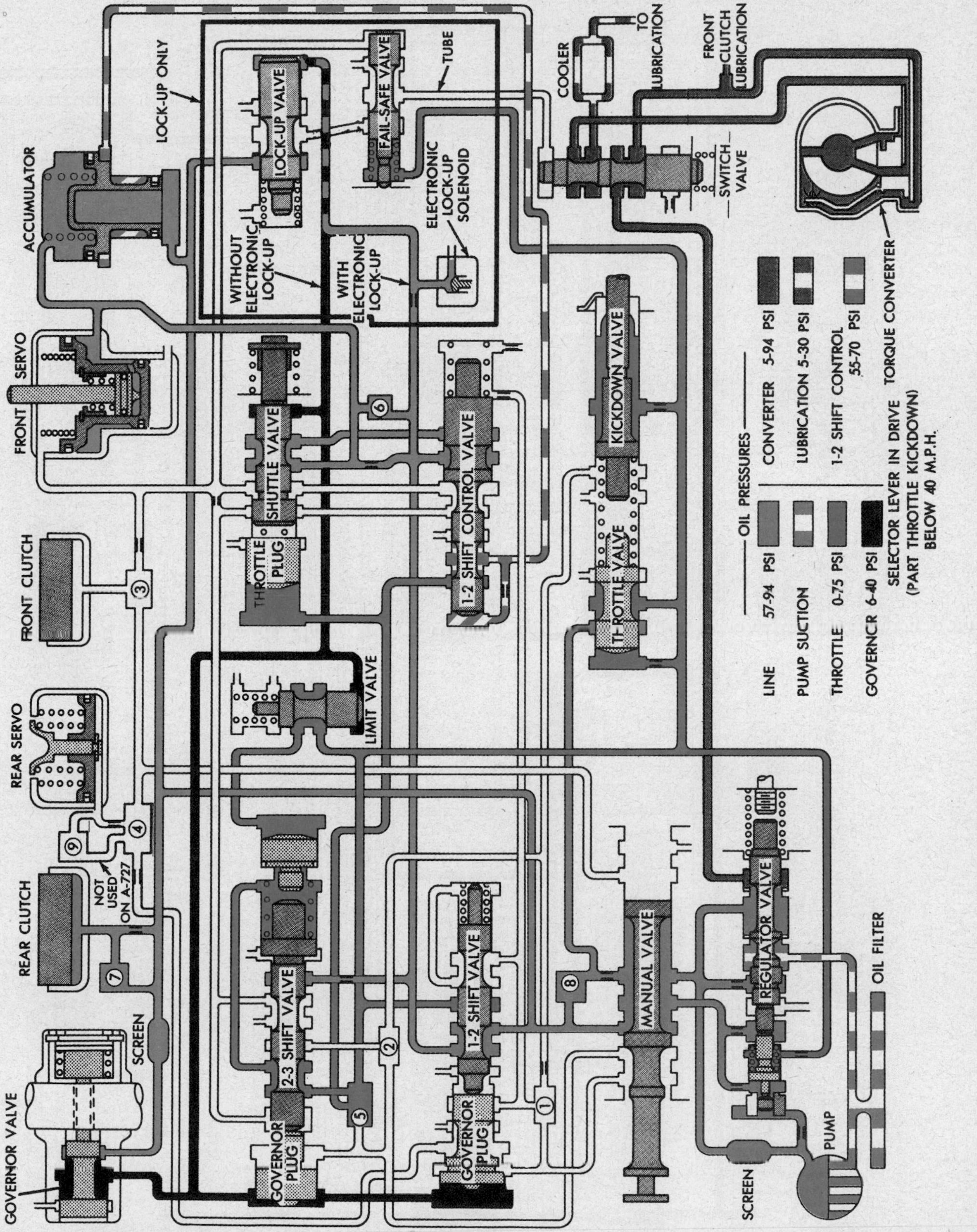
LOCK-UP ONLY
LOCK-UP VALVE
FAIL-SAFE VALVE
TUBE
TO LUBRICATION
COOLER
FRONT CLUTCH LUBRICATION
ACCUMULATOR
WITHOUT ELECTRONIC LOCK-UP
WITH ELECTRONIC LOCK-UP
ELECTRONIC LOCK-UP SOLENOID
SWITCH VALVE
TORQUE CONVERTER
FRONT SERVO
SHUTTLE VALVE
1-2 SHIFT CONTROL VALVE
KICKDOWN VALVE
THROTTLE PLUG
THROTTLE VALVE
FRONT CLUTCH
LIMIT VALVE
REAR SERVO
NOT USED ON A-727
REAR CLUTCH
SCREEN
GOVERNOR VALVE
2-3 SHIFT VALVE
GOVERNOR PLUG
1-2 SHIFT VALVE
MANUAL VALVE
REGULATOR VALVE
OIL FILTER
PUMP
SCREEN
OIL PRESSURES
LINE 57-94 PSI
PUMP SUCTION
THROTTLE 0-75 PSI
GOVERNOR 6-40 PSI
CONVERTER 5-94 PSI
LUBRICATION 5-30 PSI
1-2 SHIFT CONTROL 55-70 PSI
SELECTOR LEVER IN DRIVE (PART THROTTLE KICKDOWN) BELOW 40 M.P.H.

DRIVE (FULL THROTTLE KICKDOWN)—A500

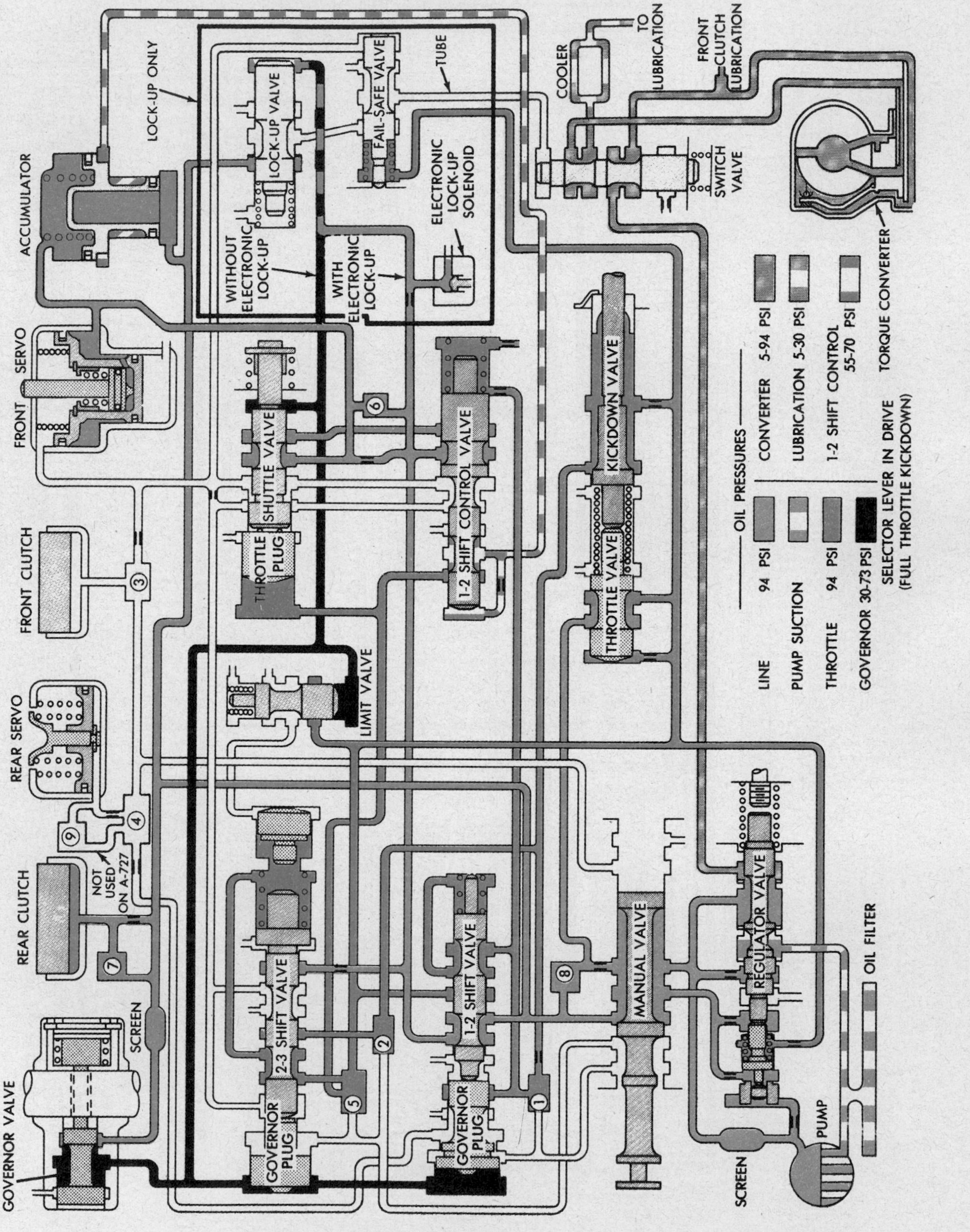
LOCK-UP ONLY
LOCK-UP VALVE
FAIL-SAFE VALVE
TUBE
TO LUBRICATION
COOLER
FRONT CLUTCH LUBRICATION
ACCUMULATOR
ELECTRONIC LOCK-UP SOLENOID
SWITCH VALVE
WITHOUT ELECTRONIC LOCK-UP
WITH ELECTRONIC LOCK-UP
TORQUE CONVERTER
FRONT SERVO
SHUTTLE VALVE
1-2 SHIFT CONTROL VALVE
KICKDOWN VALVE
FRONT CLUTCH
THROTTLE PLUG
THROTTLE VALVE
OIL PRESSURES
LINE 94 PSI
PUMP SUCTION
THROTTLE 94 PSI
GOVERNOR 30-73 PSI
CONVERTER 5-94 PSI
LUBRICATION 5-30 PSI
1-2 SHIFT CONTROL 55-70 PSI
SELECTOR LEVER IN DRIVE (FULL THROTTLE KICKDOWN)
LIMIT VALVE
REAR SERVO
NOT USED ON A-727
REAR CLUTCH
2-3 SHIFT VALVE
1-2 SHIFT VALVE
MANUAL VALVE
REGULATOR VALVE
OIL FILTER
SCREEN
GOVERNOR VALVE
GOVERNOR PLUG
PUMP
SCREEN

2—MANUAL 2ND (CLOSED THROTTLE)—A500

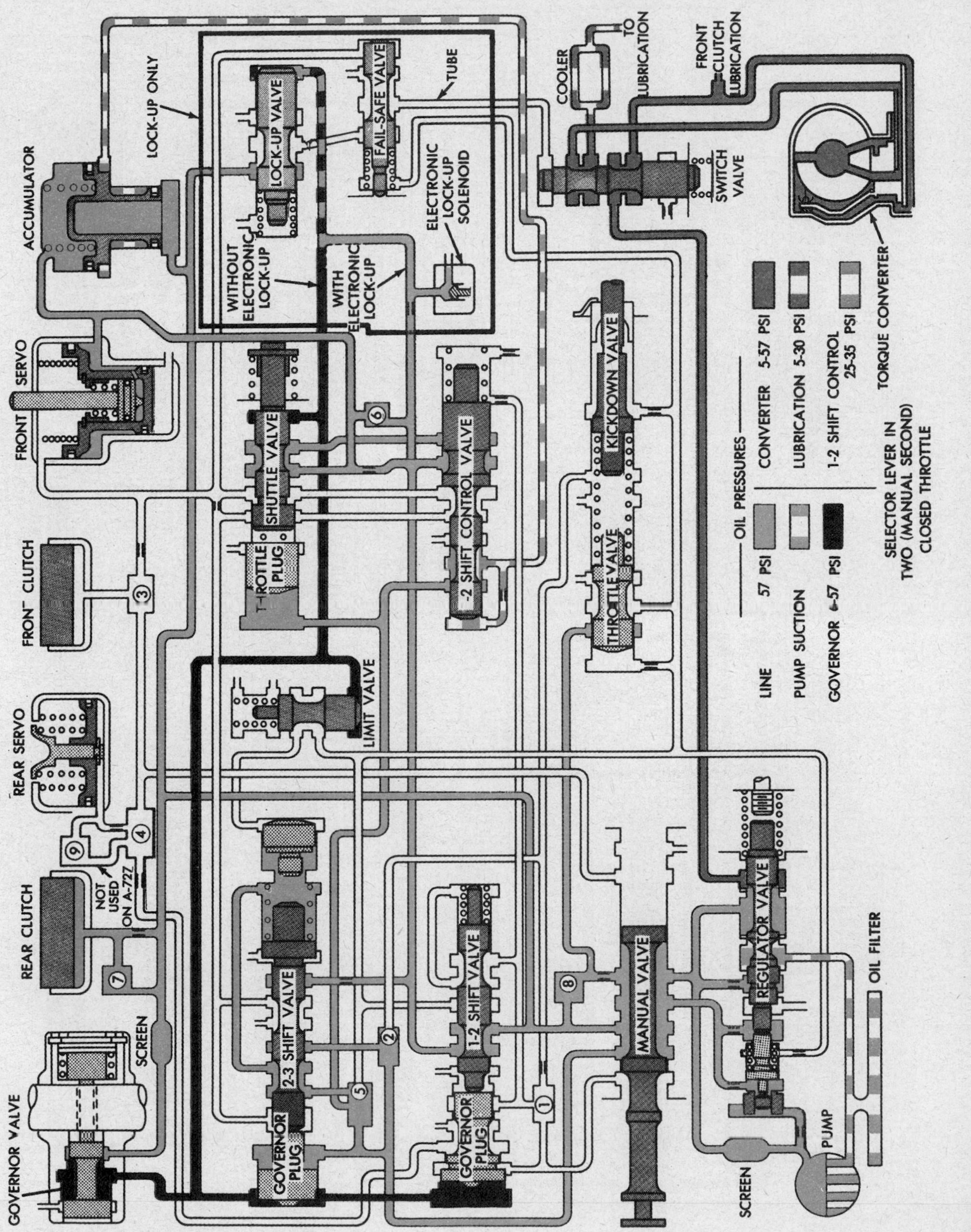
ACCUMULATOR
LOCK-UP ONLY
LOCK-UP VALVE
FAIL-SAFE VALVE
TUBE
COOLER
TO LUBRICATION
FRONT CLUTCH LUBRICATION
SWITCH VALVE
ELECTRONIC LOCK-UP SOLENOID
WITHOUT ELECTRONIC LOCK-UP
WITH ELECTRONIC LOCK-UP
TORQUE CONVERTER
FRONT SERVO
SHUTTLE VALVE
1-2 SHIFT CONTROL VALVE
KICKDOWN VALVE
THROTTLE PLUG
THROTTLE VALVE
FRONT CLUTCH
LIMIT VALVE
REAR SERVO
NOT USED ON A-727
REAR CLUTCH
SCREEN
GOVERNOR VALVE
2-3 SHIFT VALVE
GOVERNOR PLUG
1-2 SHIFT VALVE
GOVERNOR PLUG
MANUAL VALVE
REGULATOR VALVE
OIL FILTER
PUMP
SCREEN
OIL PRESSURES
LINE 57 PSI
PUMP SUCTION
GOVERNOR 6-57 PSI
CONVERTER 5-57 PSI
LUBRICATION 5-30 PSI
1-2 SHIFT CONTROL 25-35 PSI
SELECTOR LEVER IN TWO (MANUAL SECOND) CLOSED THROTTLE

1—MANUAL LOW (CLOSED THROTTLE)—A500

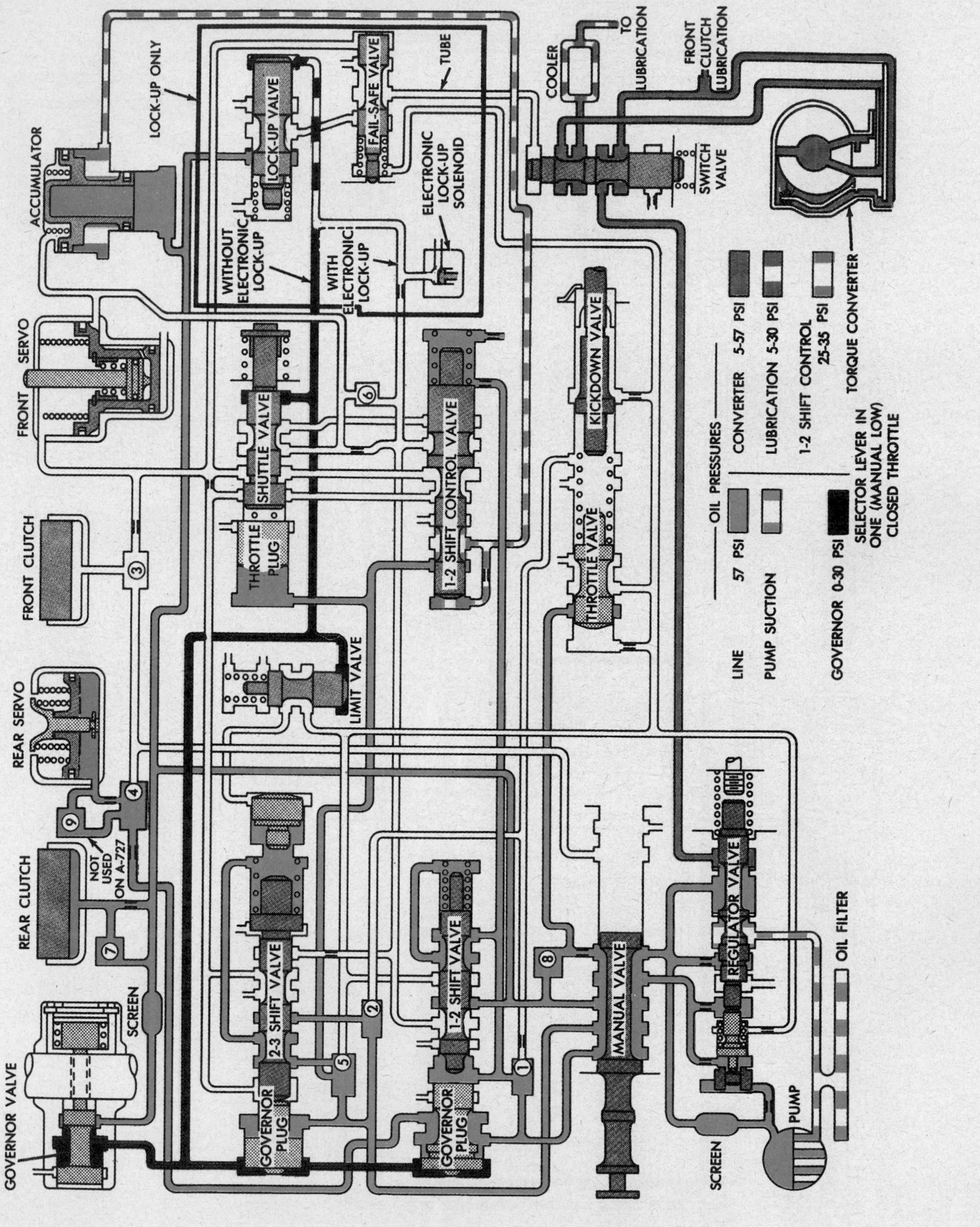
ACCUMULATOR
LOCK-UP ONLY
LOCK-UP VALVE
FAIL-SAFE VALVE
TUBE
ELECTRONIC LOCK-UP SOLENOID
COOLER
TO LUBRICATION
FRONT CLUTCH LUBRICATION
SWITCH VALVE
WITHOUT ELECTRONIC LOCK-UP
WITH ELECTRONIC LOCK-UP
FRONT SERVO
SHUTTLE VALVE
THROTTLE PLUG
1-2 SHIFT CONTROL VALVE
KICKDOWN VALVE
THROTTLE VALVE
FRONT CLUTCH
LIMIT VALVE
REAR SERVO
NOT USED ON A-727
REAR CLUTCH
SCREEN
GOVERNOR VALVE
2-3 SHIFT VALVE
GOVERNOR PLUG
1-2 SHIFT VALVE
GOVERNOR PLUG
MANUAL VALVE
REGULATOR VALVE
OIL FILTER
PUMP
SCREEN
OIL PRESSURES
LINE 57 PSI
PUMP SUCTION
GOVERNOR 0-30 PSI
CONVERTER 5-57 PSI
LUBRICATION 5-30 PSI
1-2 SHIFT CONTROL 25-35 PSI
SELECTOR LEVER IN ONE (MANUAL LOW) CLOSED THROTTLE
TORQUE CONVERTER
1
2
3
4
5
6
7
8
9

REVERSE—A500

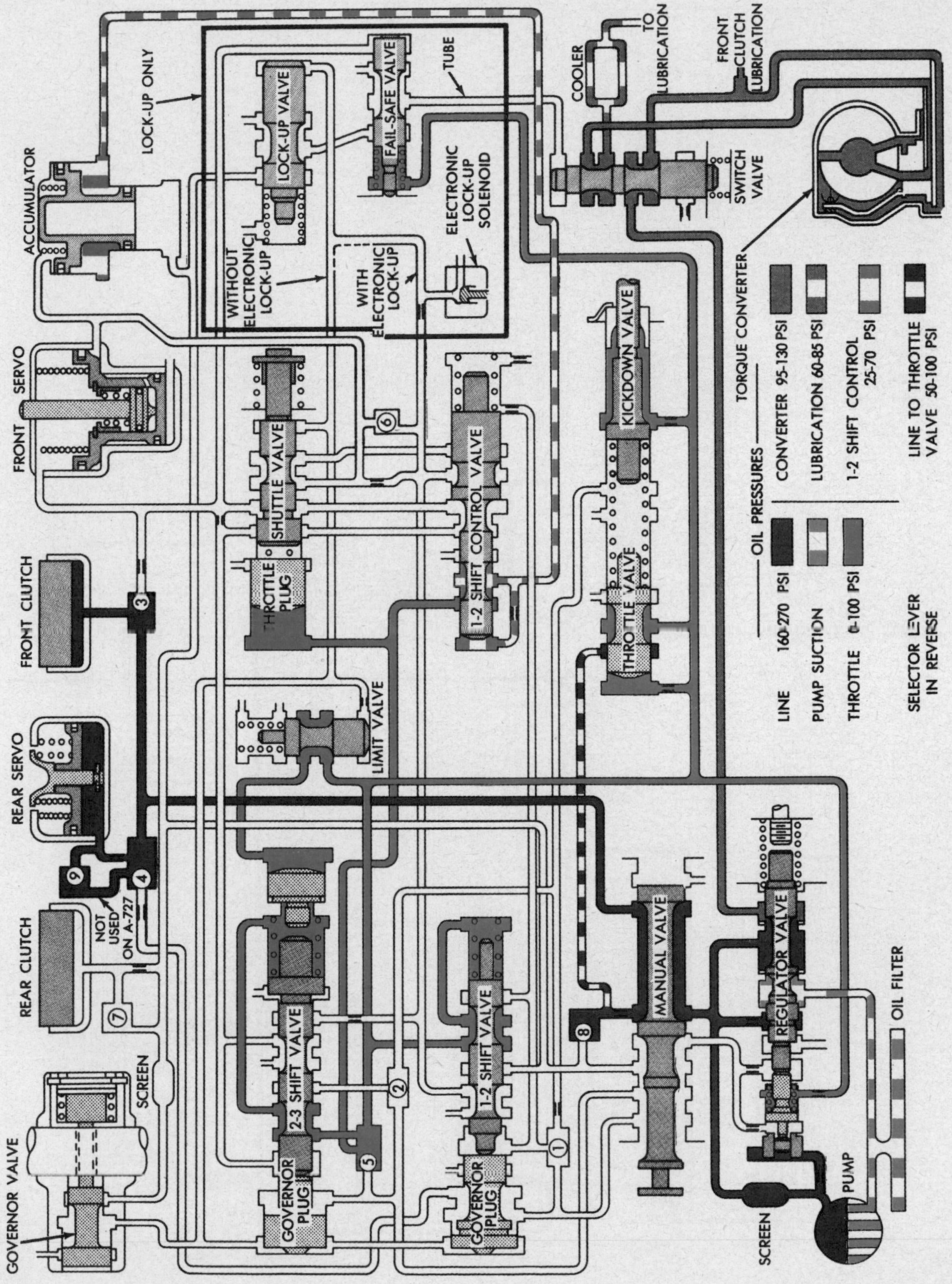
LOCK-UP ONLY
ACCUMULATOR
LOCK-UP VALVE
FAIL-SAFE VALVE
TUBE
COOLER
TO LUBRICATION
FRONT CLUTCH LUBRICATION
SWITCH VALVE
ELECTRONIC LOCK-UP SOLENOID
WITHOUT ELECTRONIC LOCK-UP
WITH ELECTRONIC LOCK-UP
FRONT SERVO
SHUTTLE VALVE
THROTTLE PLUG
1-2 SHIFT CONTROL VALVE
KICKDOWN VALVE
THROTTLE VALVE
TORQUE CONVERTER
OIL PRESSURES
LINE 160-270 PSI
PUMP SUCTION
THROTTLE 0-100 PSI
CONVERTER 95-130 PSI
LUBRICATION 60-85 PSI
1-2 SHIFT CONTROL 25-70 PSI
LINE TO THROTTLE VALVE 50-100 PSI
SELECTOR LEVER IN REVERSE
FRONT CLUTCH
REAR SERVO
LIMIT VALVE
NOT USED ON A-727
REAR CLUTCH
SCREEN
GOVERNOR VALVE
2-3 SHIFT VALVE
GOVERNOR PLUG
1-2 SHIFT VALVE
MANUAL VALVE
REGULATOR VALVE
OIL FILTER
PUMP
SCREEN

DRIVE (2ND) HALF THROTTLE COMPLETE WITH OVERDRIVE UNIT—A500

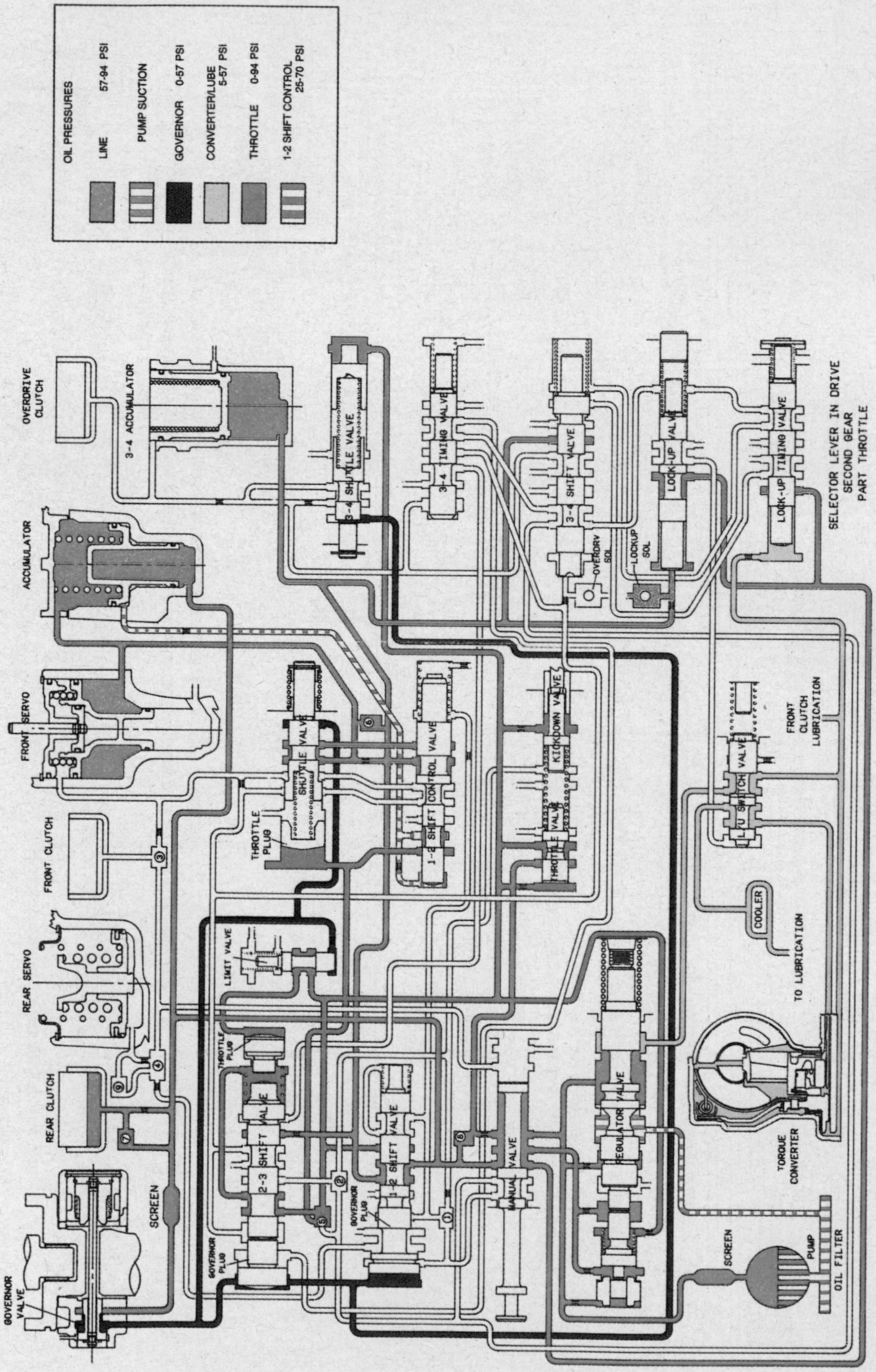
OIL PRESSURES
LINE 57-94 PSI
PUMP SUCTION
GOVERNOR 0-57 PSI
CONVERTER/LUBE 5-57 PSI
THROTTLE 0-94 PSI
1-2 SHIFT CONTROL 25-70 PSI
OVERDRIVE CLUTCH
3-4 ACCUMULATOR
3-4 SHUTTLE VALVE
3-4 TIMING VALVE
3-4 SHIFT VALVE
OVERDRV SOL
LOCKUP SOL
LOCK-UP VALVE
LOCK-UP TIMING VALVE
SELECTOR LEVER IN DRIVE
SECOND GEAR
PART THROTTLE
ACCUMULATOR
FRONT SERVO
SHUTTLE VALVE
THROTTLE PLUG
1-2 SHIFT CONTROL VALVE
KICKDOWN VALVE
THROTTLE VALVE
FRONT CLUTCH LUBRICATION
L/U SWITCH VALVE
FRONT CLUTCH
COOLER
TO LUBRICATION
REAR SERVO
LIMIT VALVE
REAR CLUTCH
2-3 SHIFT VALVE
1-2 SHIFT VALVE
MANUAL VALVE
REGULATOR VALVE
TORQUE CONVERTER
SCREEN
GOVERNOR PLUG
GOVERNOR VALVE
PUMP
OIL FILTER

DRIVE (DIRECT) COMPLETE WITH OVERDRIVE UNIT AND TORQUE CONVERTER UNLOCKED—A500

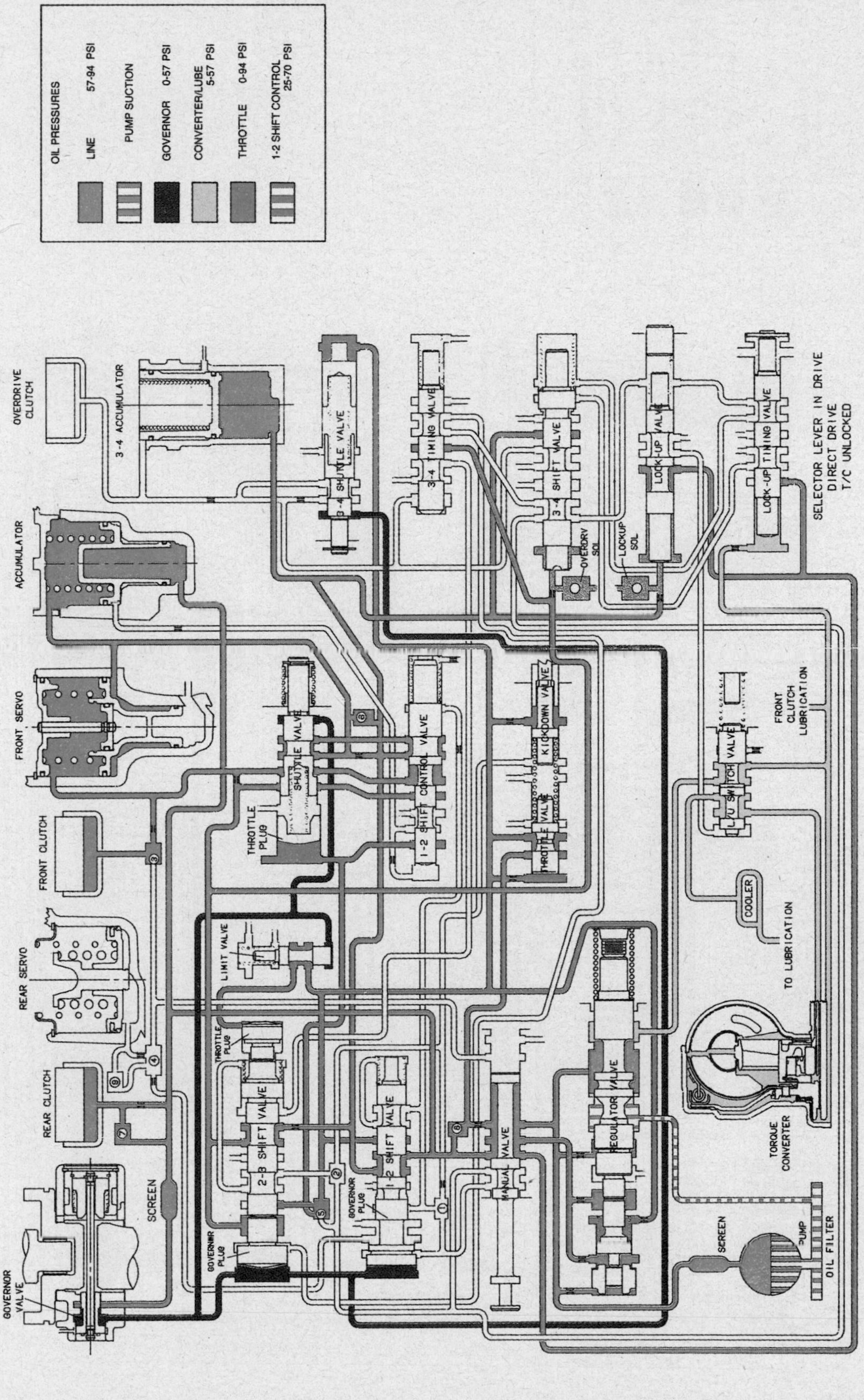
OIL PRESSURES
LINE 57-94 PSI
PUMP SUCTION
GOVERNOR 0-57 PSI
CONVERTER/LUBE 5-57 PSI
THROTTLE 0-94 PSI
1-2 SHIFT CONTROL 25-70 PSI
OVERDRIVE CLUTCH
3-4 ACCUMULATOR
3-4 SHUTTLE VALVE
3-4 TIMING VALVE
3-4 SHIFT VALVE
OVERDRV SOL
LOCKUP SOL
LOCK-UP VALVE
LOCK-UP TIMING VALVE
SELECTOR LEVER IN DRIVE
DIRECT DRIVE
T/C UNLOCKED
ACCUMULATOR
FRONT SERVO
SHUTTLE VALVE
THROTTLE PLUG
1-2 SHIFT CONTROL VALVE
KICKDOWN VALVE
THROTTLE VALVE
L/U SWITCH VALVE
FRONT CLUTCH LUBRICATION
FRONT CLUTCH
COOLER
TO LUBRICATION
REAR SERVO
LIMIT VALVE
THROTTLE PLUG
2-3 SHIFT VALVE
1-2 SHIFT VALVE
MANUAL VALVE
REGULATOR VALVE
TORQUE CONVERTER
REAR CLUTCH
SCREEN
GOVERNOR PLUG
GOVERNOR PLUG
GOVERNOR VALVE
SCREEN
PUMP
OIL FILTER

DRIVE (OVERDRIVE GEAR) WITH TORQUE CONVERTER UNLOCKED – A500

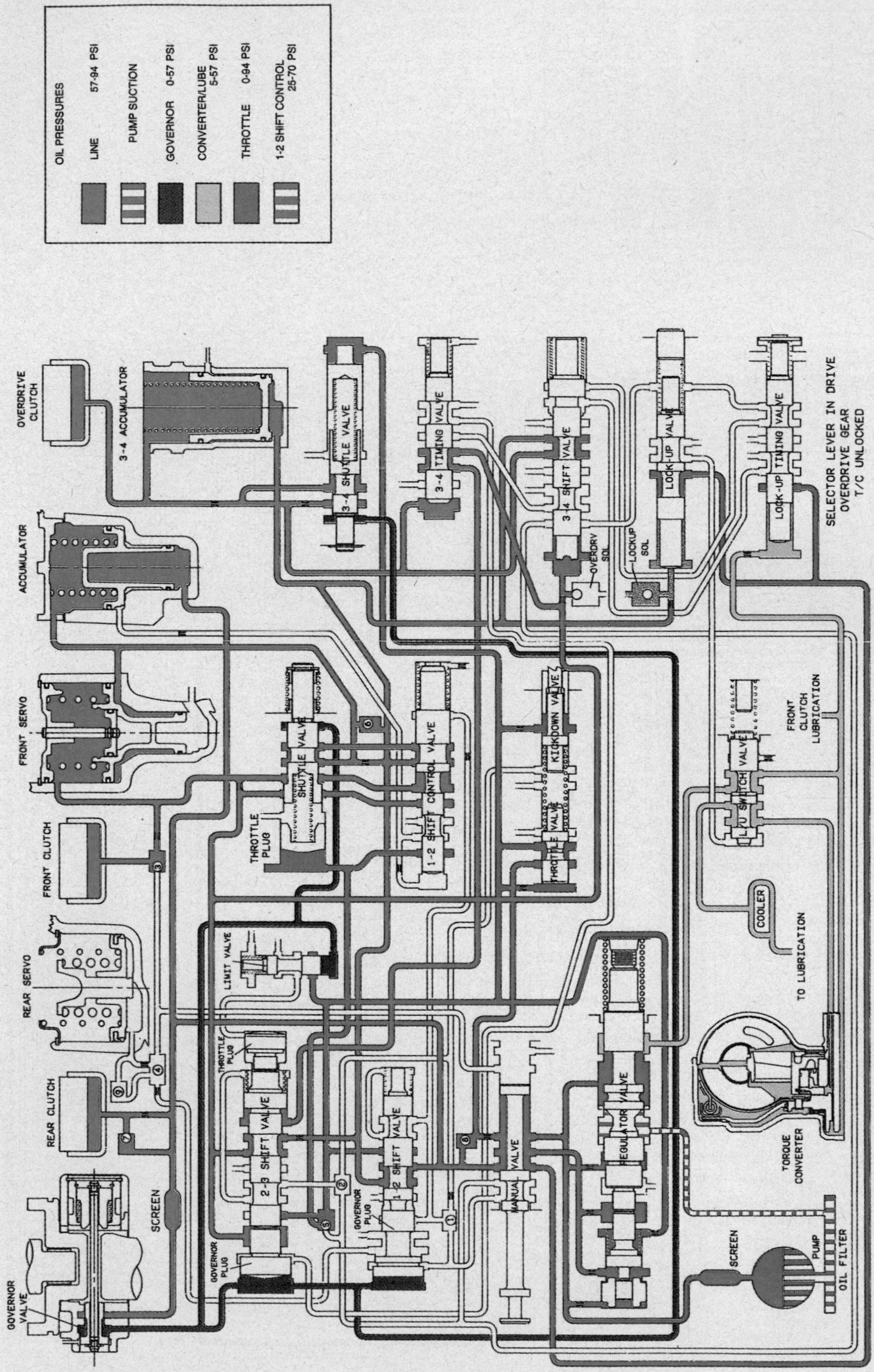
OIL PRESSURES
LINE 57-94 PSI
PUMP SUCTION
GOVERNOR 0-57 PSI
CONVERTER/LUBE 5-57 PSI
THROTTLE 0-94 PSI
1-2 SHIFT CONTROL 25-70 PSI
OVERDRIVE CLUTCH
3-4 ACCUMULATOR
ACCUMULATOR
FRONT SERVO
FRONT CLUTCH
REAR SERVO
REAR CLUTCH
GOVERNOR VALVE
SCREEN
3-4 SHUTTLE VALVE
3-4 TIMING VALVE
3-4 SHIFT VALVE
OVERDRV SOL
LOCK-UP VALVE
LOCKUP SOL
LOCK-UP TIMING VALVE
SELECTOR LEVER IN DRIVE
OVERDRIVE GEAR
T/C UNLOCKED
SHUTTLE VALVE
THROTTLE PLUG
1-2 SHIFT CONTROL VALVE
THROTTLE VALVE
KICKDOWN VALVE
L/U SWITCH VALVE
FRONT CLUTCH LUBRICATION
COOLER
TO LUBRICATION
LIMIT VALVE
THROTTLE PLUG
2-3 SHIFT VALVE
1-2 SHIFT VALVE
GOVERNOR PLUG
GOVERNOR PLUG
MANUAL VALVE
REGULATOR VALVE
TORQUE CONVERTER
SCREEN
PUMP
OIL FILTER

NEUTRAL—KM148 AND AW132

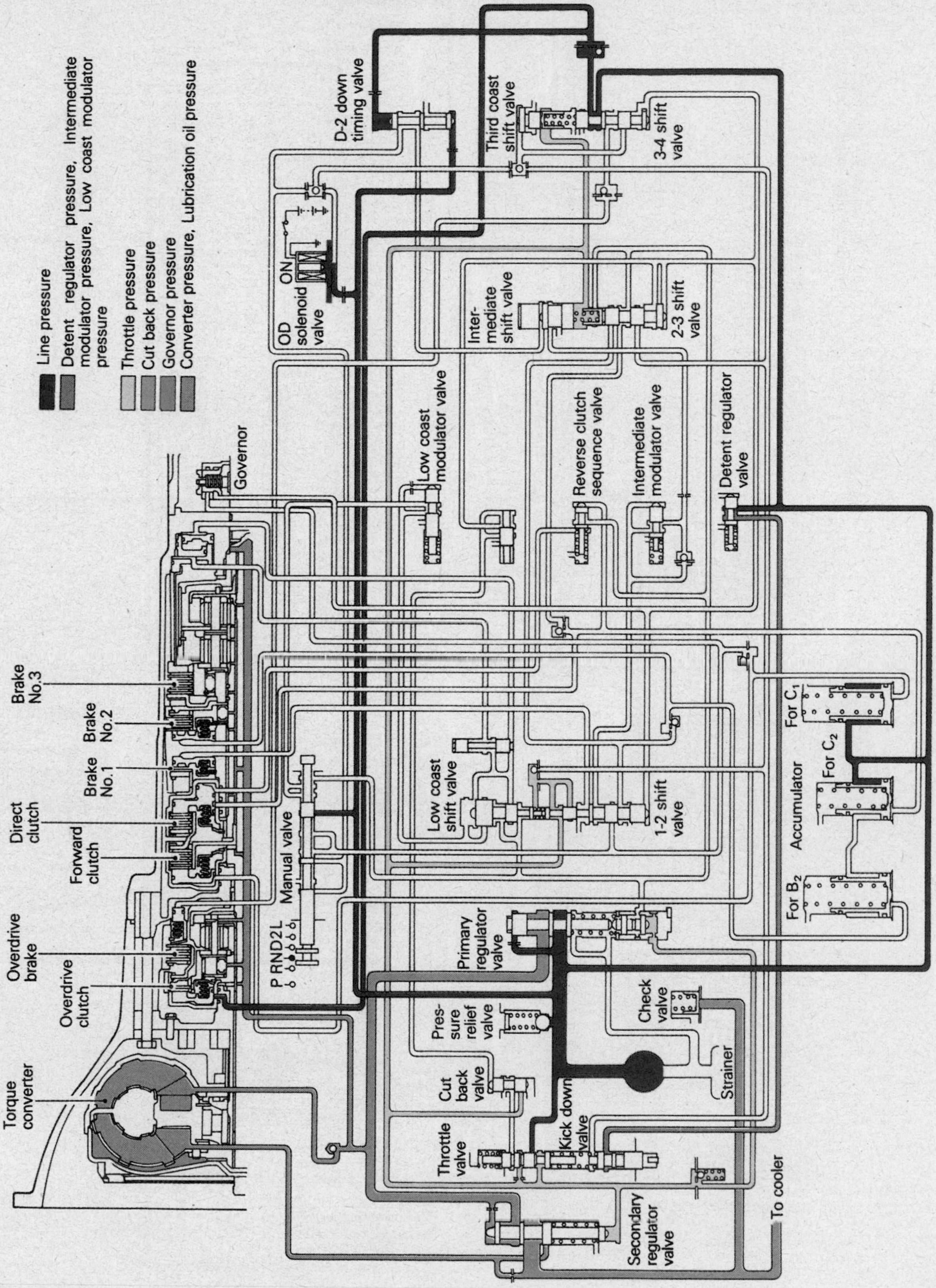
Line pressure
Detent regulator pressure, Intermediate modulator pressure, Low coast modulator pressure
Throttle pressure
Cut back pressure
Governor pressure
Converter pressure, Lubrication oil pressure
D-2 down timing valve
Third coast shift valve
3-4 shift valve
OD solenoid valve
ON
Inter-mediate shift valve
2-3 shift valve
Governor
Low coast modulator valve
Reverse clutch sequence valve
Intermediate modulator valve
Detent regulator valve
Brake No.3
Brake No.2
Brake No.1
Direct clutch
Forward clutch
Overdrive brake
Overdrive clutch
Torque converter
Manual valve
P RND2L
Low coast shift valve
1-2 shift valve
For C1
For C2
Accumulator
For B2
Primary regulator valve
Pres-sure relief valve
Check valve
Strainer
Cut back valve
Throttle valve
Kick down valve
Secondary regulator valve
To cooler

PARK—KM148 AND AW132

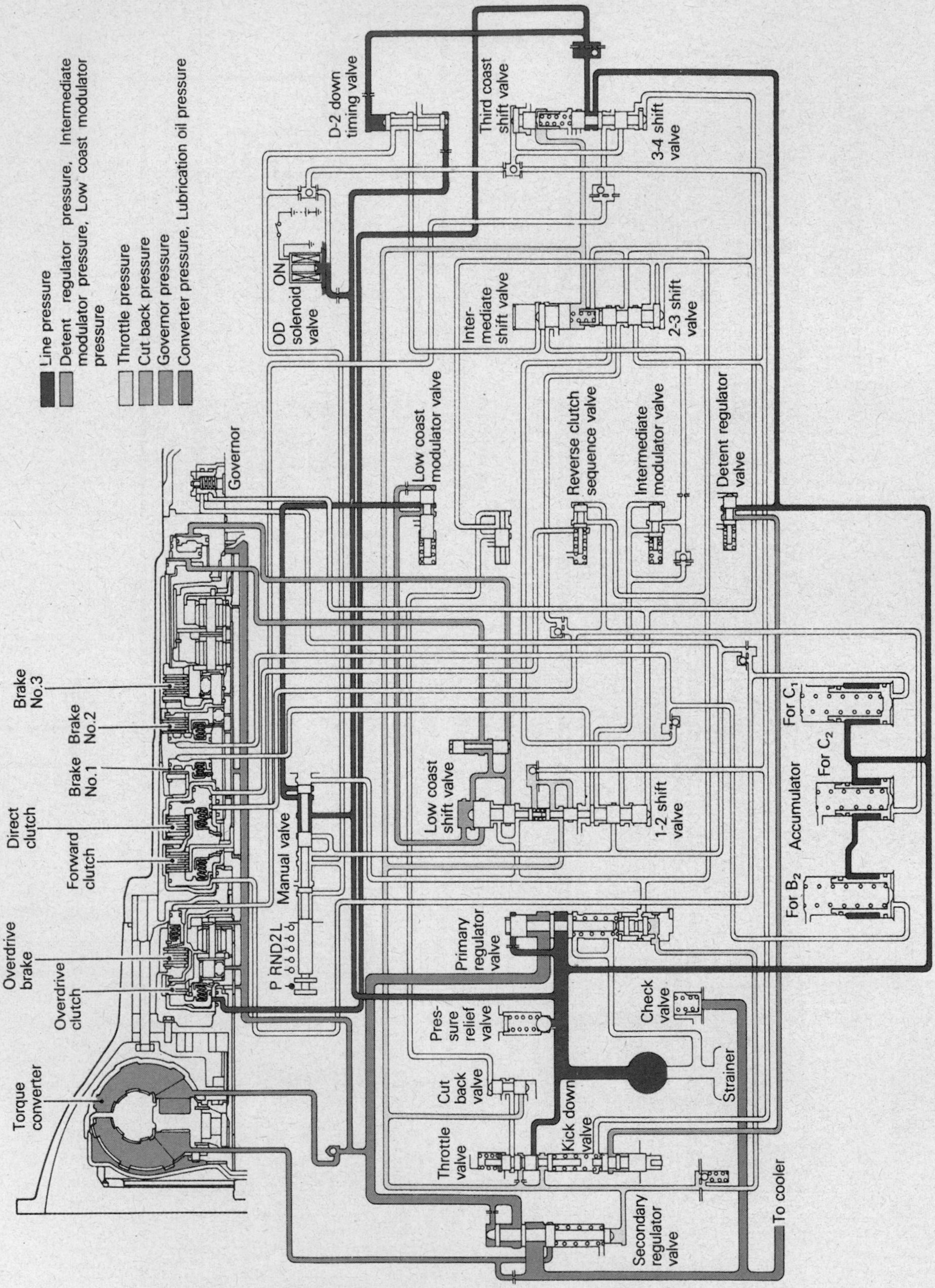
Line pressure
Detent regulator pressure, Intermediate modulator pressure, Low coast modulator pressure
Throttle pressure
Cut back pressure
Governor pressure
Converter pressure, Lubrication oil pressure
D-2 down timing valve
Third coast shift valve
3-4 shift valve
OD solenoid valve
ON
Inter-mediate shift valve
2-3 shift valve
Low coast modulator valve
Reverse clutch sequence valve
Intermediate modulator valve
Detent regulator valve
Governor
Brake No.3
Brake No.2
Brake No.1
Direct clutch
Forward clutch
Overdrive brake
Overdrive clutch
Torque converter
Low coast shift valve
1-2 shift valve
Manual valve
P RND2L
For C1
For C2
Accumulator
For B2
Primary regulator valve
Pres-sure relief valve
Check valve
Strainer
Cut back valve
Kick down valve
Throttle valve
To cooler
Secondary regulator valve

DRIVE—1ST GEAR—KM148 AND AW132

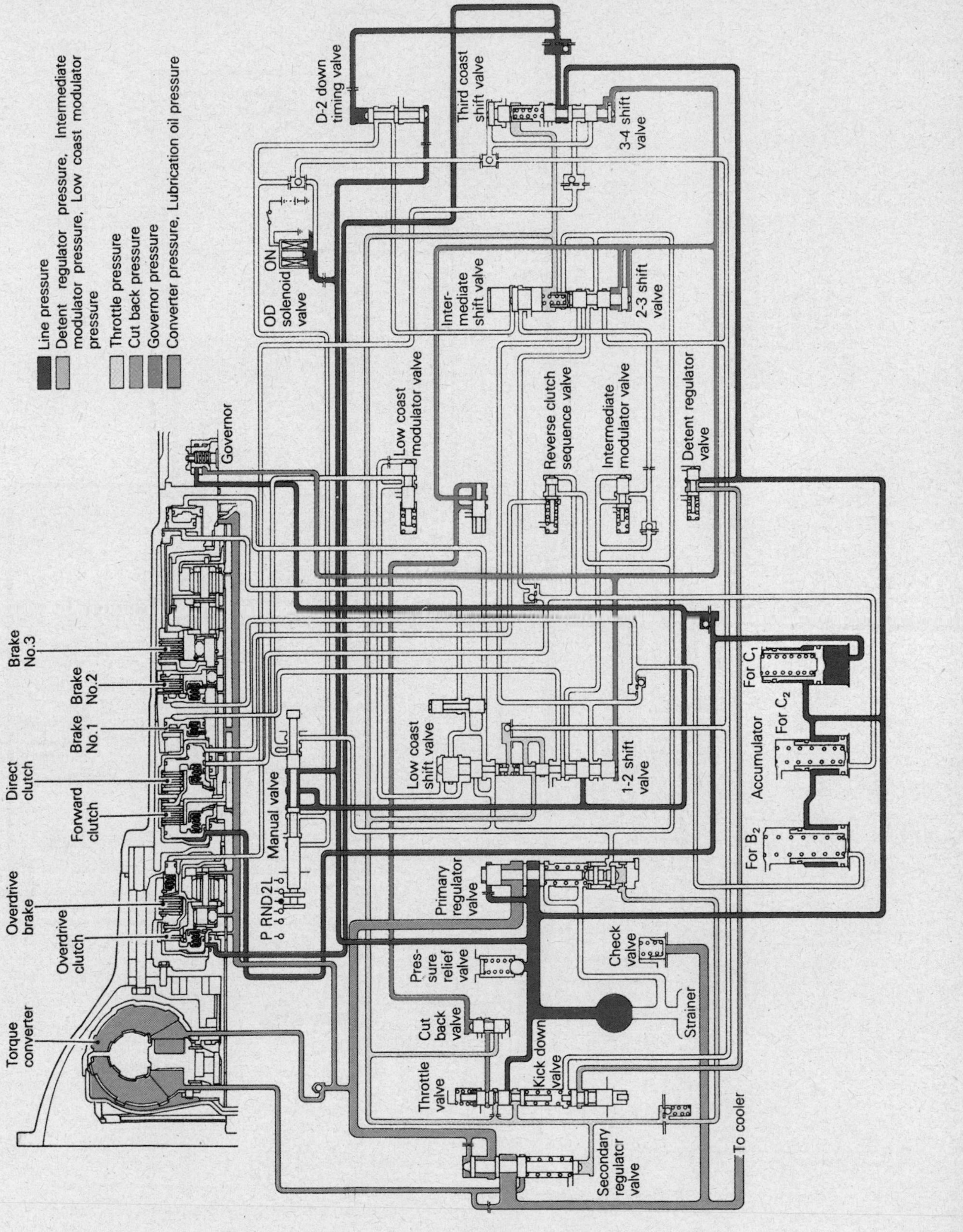
Line pressure
Detent regulator pressure, Intermediate modulator pressure, Low coast modulator pressure
Throttle pressure
Cut back pressure
Governor pressure
Converter pressure, Lubrication oil pressure
D-2 down timing valve
Third coast shift valve
3-4 shift valve
OD solenoid valve
ON
Inter-mediate shift valve
2-3 shift valve
Governor
Low coast modulator valve
Reverse clutch sequence valve
Intermediate modulator valve
Detent regulator valve
Brake No.3
Brake No.2
Brake No.1
Direct clutch
Forward clutch
Overdrive brake
Overdrive clutch
Torque converter
Manual valve
P RND2L
Low coast shift valve
1-2 shift valve
For C1
For C2
Accumulator
For B2
Primary regulator valve
Pres-sure relief valve
Check valve
Strainer
Cut back valve
Kick down valve
Throttle valve
To cooler
Secondary regulator valve

DRIVE–2ND GEAR–KM148 AND AW132

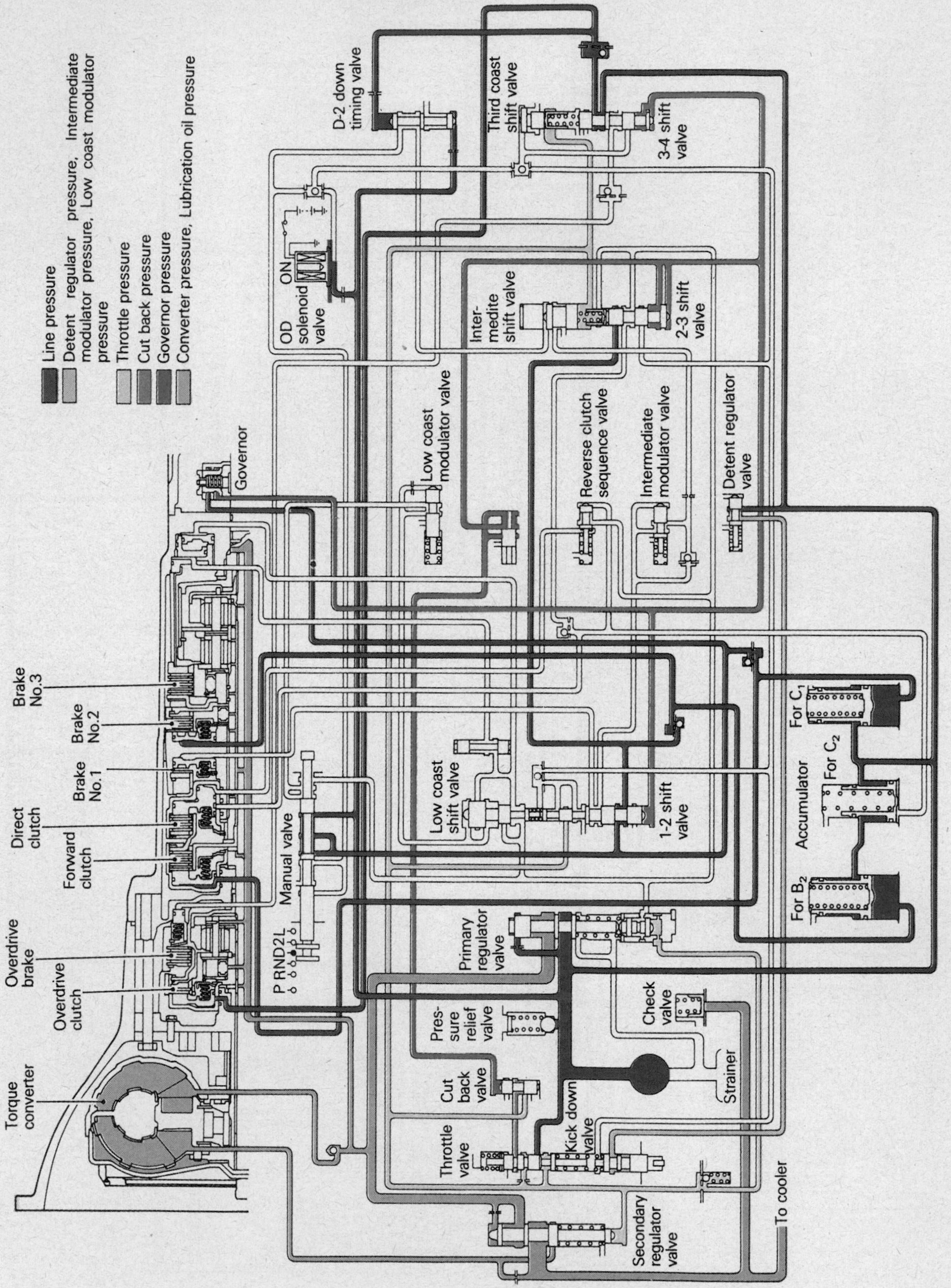
Line pressure
Detent regulator pressure, Intermediate modulator pressure, Low coast modulator pressure
Throttle pressure
Cut back pressure
Governor pressure
Converter pressure, Lubrication oil pressure
D-2 down timing valve
Third coast shift valve
3-4 shift valve
OD solenoid valve
ON
Inter-medite shift valve
2-3 shift valve
Governor
Low coast modulator valve
Reverse clutch sequence valve
Intermediate modulator valve
Detent regulator valve
Brake No.3
Brake No.2
Brake No.1
Direct clutch
Forward clutch
Overdrive brake
Overdrive clutch
Torque converter
Manual valve
P R N D 2 L
Low coast shift valve
1-2 shift valve
For C1
For C2
Accumulator
For B2
Primary regulator valve
Pres-sure relief valve
Check valve
Strainer
Cut back valve
Throttle valve
Kick down valve
Secondary regulator valve
To cooler

DRIVE—3RD GEAR—KM148 AND AW132

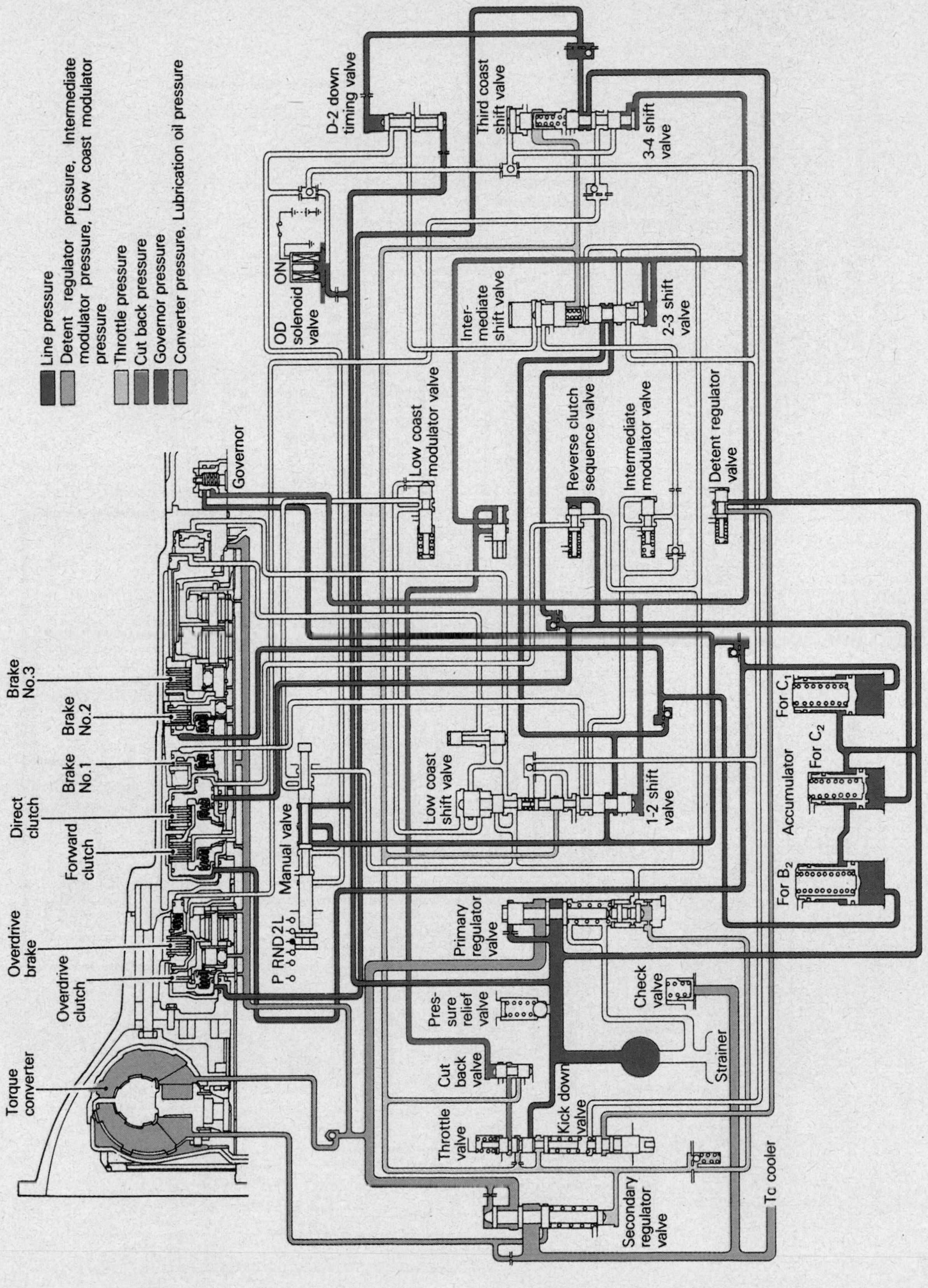
Line pressure
Detent regulator pressure, Intermediate modulator pressure, Low coast modulator pressure
Throttle pressure
Cut back pressure
Governor pressure
Converter pressure, Lubrication oil pressure
D-2 down timing valve
Third coast shift valve
3-4 shift valve
OD solenoid valve
ON
Inter-mediate shift valve
2-3 shift valve
Low coast modulator valve
Reverse clutch sequence valve
Intermediate modulator valve
Detent regulator valve
Governor
Brake No.3
Brake No.2
Brake No.1
Direct clutch
Forward clutch
Overdrive brake
Overdrive clutch
Torque converter
Manual valve
P RND2L
Low coast shift valve
1-2 shift valve
For C_1
For C_2
Accumulator
For B_2
Primary regulator valve
Pres-sure relief valve
Check valve
Strainer
Cut back valve
Throttle valve
Kick down valve
Secondary regulator valve
Tc cooler

DRIVE—3RD GEAR (OD SOLENOID VALVE OFF)—KM148 AND AW132

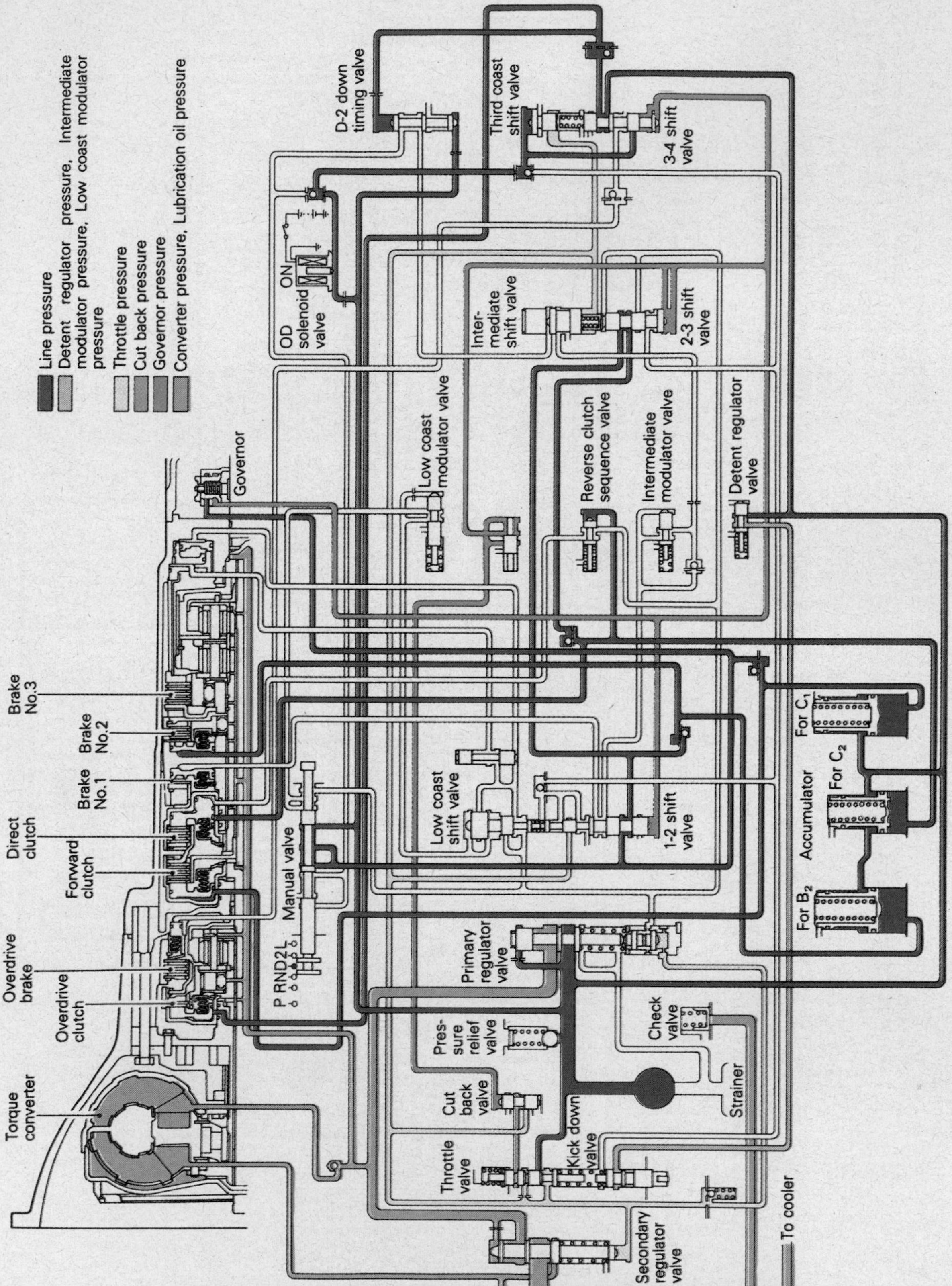
Line pressure
Detent regulator pressure, Intermediate modulator pressure, Low coast modulator pressure
Throttle pressure
Cut back pressure
Governor pressure
Converter pressure, Lubrication oil pressure
D-2 down timing valve
Third coast shift valve
3-4 shift valve
OD solenoid valve
ON
Inter-mediate shift valve
2-3 shift valve
Governor
Low coast modulator valve
Reverse clutch sequence valve
Intermediate modulator valve
Detent regulator valve
Brake No.3
Brake No.2
Brake No.1
Direct clutch
Forward clutch
Overdrive brake
Overdrive clutch
Torque converter
Manual valve
P RND2L
Low coast shift valve
1-2 shift valve
For C_1
For C_2
Accumulator
For B_2
Primary regulator valve
Pres-sure relief valve
Check valve
Strainer
Cut back valve
Throttle valve
Kick down valve
Secondary regulator valve
To cooler

DRIVE—4TH GEAR—KM148 AND AW132

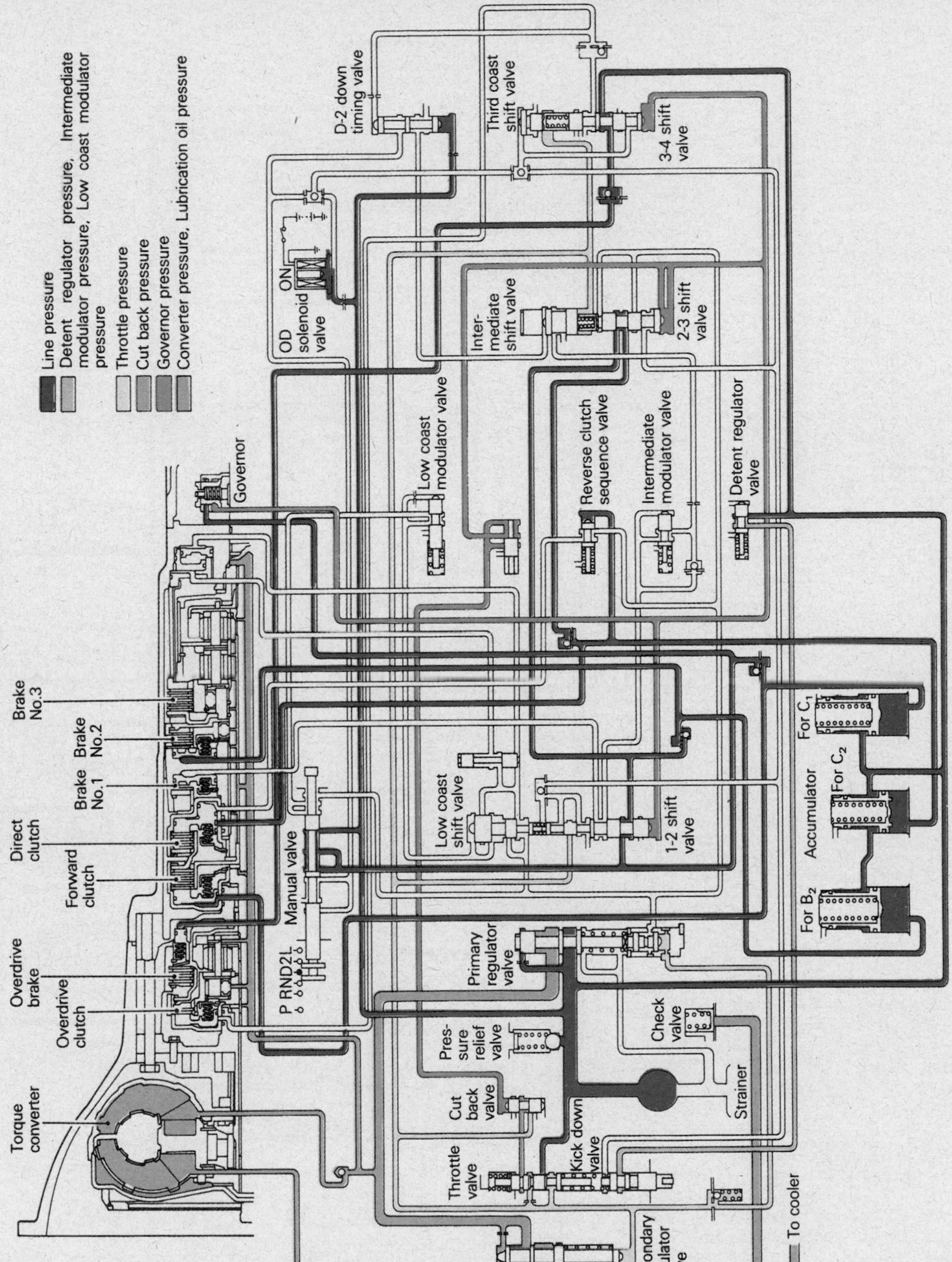
Line pressure
Detent regulator pressure, Intermediate modulator pressure, Low coast modulator pressure
Throttle pressure
Cut back pressure
Governor pressure
Converter pressure, Lubrication oil pressure
OD solenoid valve
ON
D-2 down timing valve
Third coast shift valve
3-4 shift valve
Inter-mediate shift valve
2-3 shift valve
Low coast modulator valve
Reverse clutch sequence valve
Intermediate modulator valve
Detent regulator valve
Governor
Brake No.3
Brake No.2
Brake No.1
Direct clutch
Forward clutch
Overdrive brake
Overdrive clutch
Torque converter
Low coast shift valve
1-2 shift valve
Manual valve
P RND2L
For C_1
For C_2
Accumulator
For B_2
Primary regulator valve
Pres-sure relief valve
Check valve
Cut back valve
Strainer
Throttle valve
Kick down valve
To cooler
Secondary regulator valve

DRIVE—KICKDOWN GEAR (4TH TO 3RD)—KM148 AND AW132

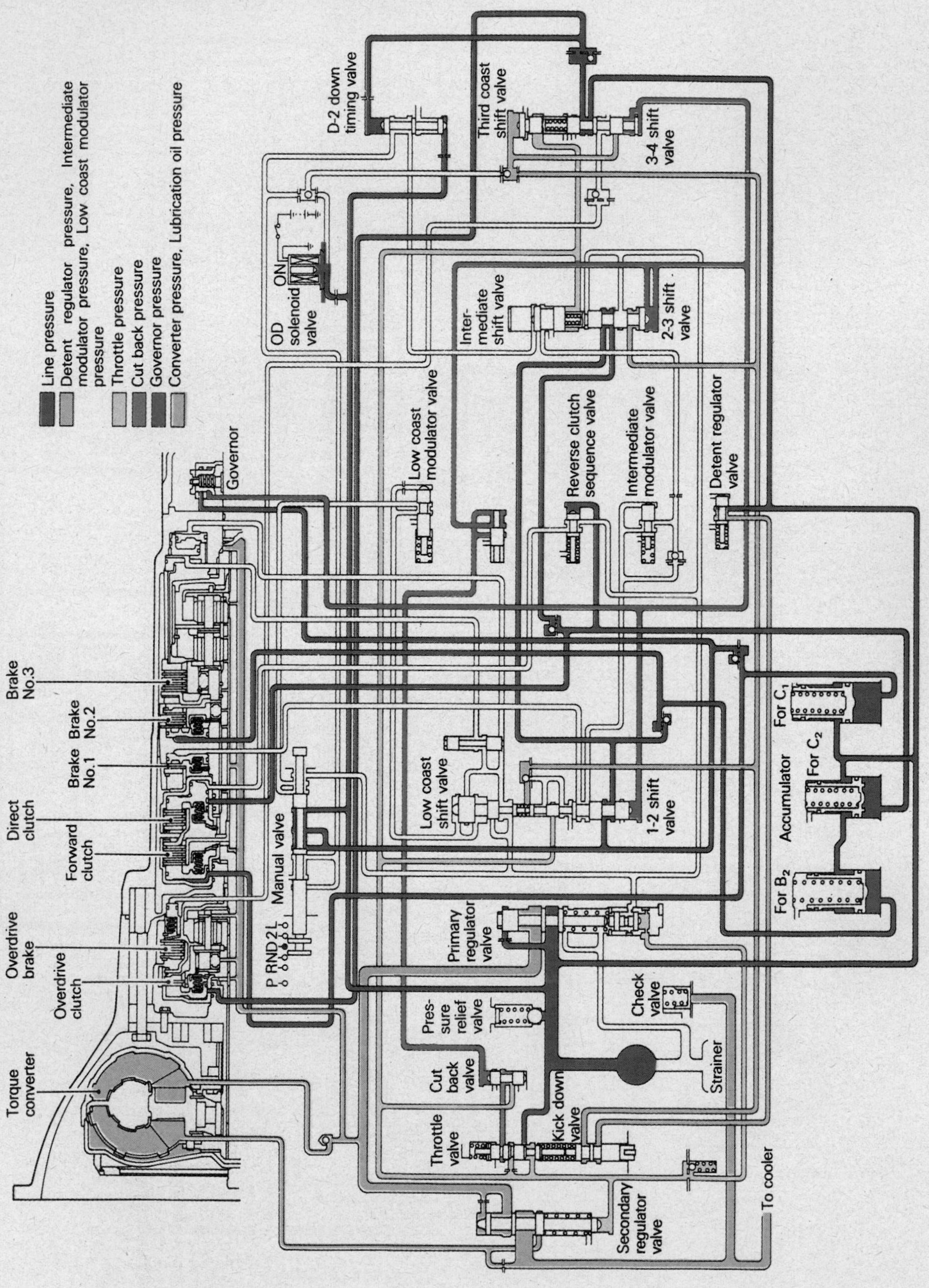
Line pressure
Detent regulator pressure, Intermediate modulator pressure, Low coast modulator pressure
Throttle pressure
Cut back pressure
Governor pressure
Converter pressure, Lubrication oil pressure
D-2 down timing valve
Third coast shift valve
3-4 shift valve
OD solenoid valve
ON
Inter-mediate shift valve
2-3 shift valve
Governor
Low coast modulator valve
Reverse clutch sequence valve
Intermediate modulator valve
Detent regulator valve
Brake No.3
Brake No.2
Brake No.1
Direct clutch
Forward clutch
Overdrive brake
Overdrive clutch
Torque converter
Manual valve
P R N D 2 L
Low coast shift valve
1-2 shift valve
For C1
For C2
Accumulator
For B2
Primary regulator valve
Pres-sure relief valve
Check valve
Strainer
Cut back valve
Throttle valve
Kick down valve
Secondary regulator valve
To cooler

D2—2ND GEAR—KM148 AND AW132

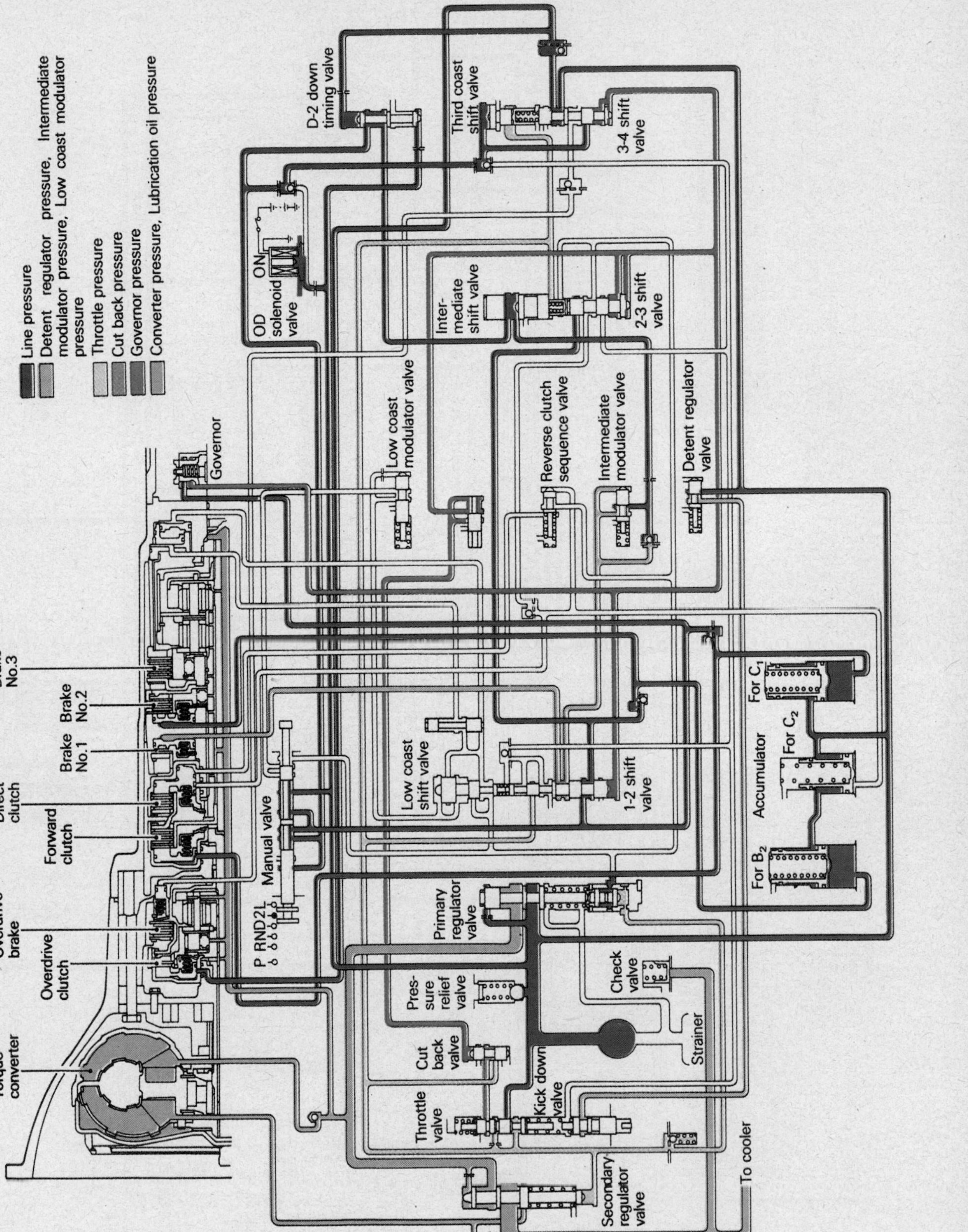

Line pressure
Detent regulator pressure, Intermediate modulator pressure, Low coast modulator pressure
Throttle pressure
Cut back pressure
Governor pressure
Converter pressure, Lubrication oil pressure
D-2 down timing valve
Third coast shift valve
3-4 shift valve
OD solenoid valve
ON
Inter-mediate shift valve
2-3 shift valve
Low coast modulator valve
Reverse clutch sequence valve
Intermediate modulator valve
Detent regulator valve
Governor
Brake No.3
Brake No.2
Brake No.1
Direct clutch
Forward clutch
Overdrive brake
Overdrive clutch
Torque converter
Manual valve
P RND2L
Low coast shift valve
1-2 shift valve
For C_1
For C_2
Accumulator
For B_2
Primary regulator valve
Pres-sure relief valve
Check valve
Strainer
Cut back valve
Kick down valve
Throttle valve
To cooler
Secondary regulator valve

DRIVE—LOCKUP GEAR—KM148 AND AW132

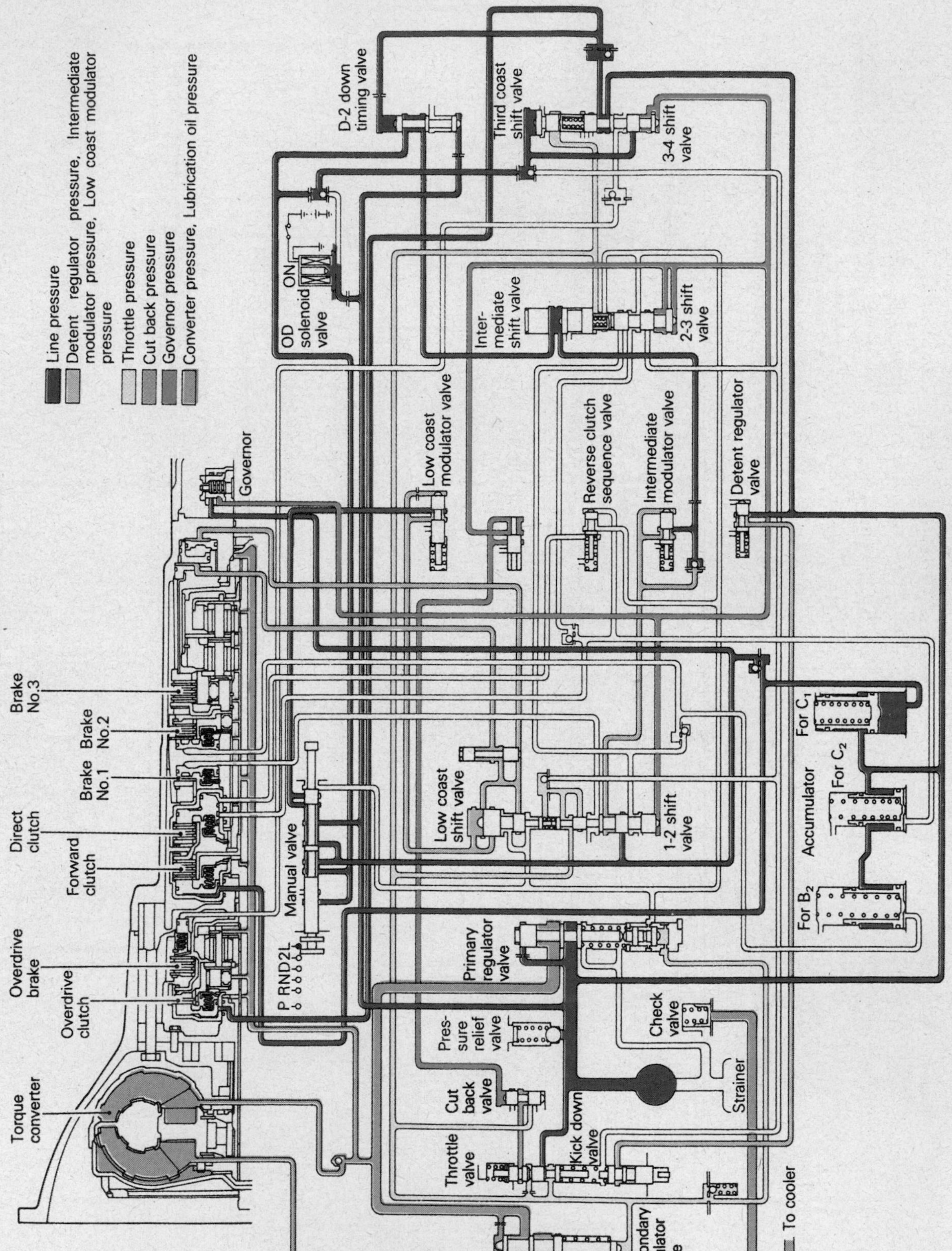
Line pressure
Detent regulator pressure, Intermediate modulator pressure, Low coast modulator pressure
Throttle pressure
Cut back pressure
Governor pressure
Converter pressure, Lubrication oil pressure
D-2 down timing valve
Third coast shift valve
3-4 shift valve
ON
OD solenoid valve
Inter-mediate shift valve
2-3 shift valve
Governor
Low coast modulator valve
Reverse clutch sequence valve
Intermediate modulator valve
Detent regulator valve
Brake No.3
Brake No.2
Brake No.1
Direct clutch
Forward clutch
Overdrive brake
Overdrive clutch
Torque converter
Low coast shift valve
1-2 shift valve
Manual valve
P RND2L
For C1
For C2
Accumulator
For B2
Primary regulator valve
Pres-sure relief valve
Check valve
Strainer
Cut back valve
Kick down valve
Throttle valve
To cooler
Secondary regulator valve

REVERSE—KM148 AND AW132

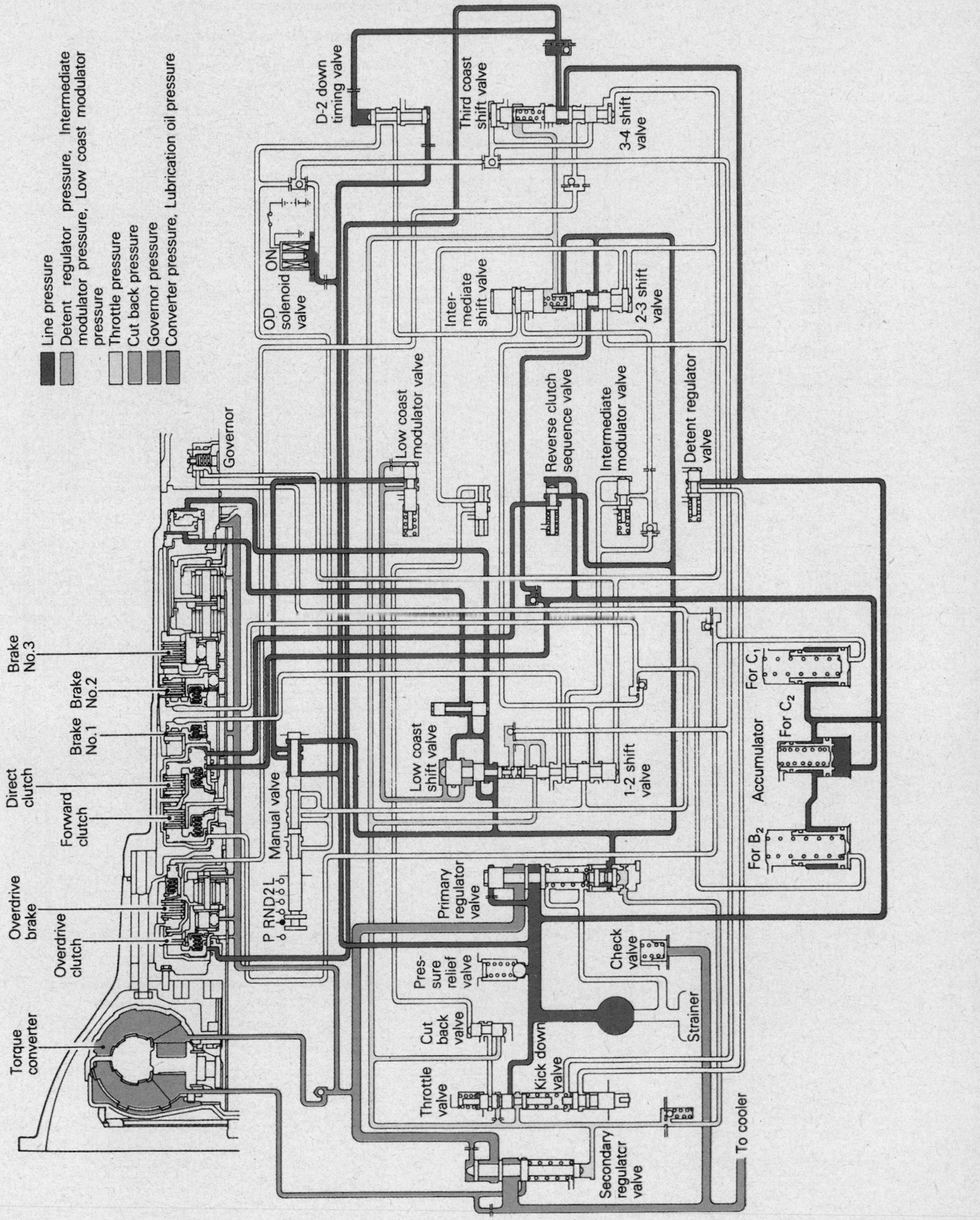

Line pressure
Detent regulator pressure, Intermediate modulator pressure, Low coast modulator pressure
Throttle pressure
Cut back pressure
Governor pressure
Converter pressure, Lubrication oil pressure
D-2 down timing valve
Third coast shift valve
3-4 shift valve
OD solenoid valve
ON
Inter-mediate shift valve
2-3 shift valve
Governor
Low coast modulator valve
Reverse clutch sequence valve
Intermediate modulator valve
Detent regulator valve
Brake No.3
Brake No.2
Brake No.1
Direct clutch
Forward clutch
Overdrive brake
Overdrive clutch
Torque converter
Manual valve
P RND2L
Low coast shift valve
1-2 shift valve
For C1
For C2
Accumulator
For B2
Primary regulator valve
Pres-sure relief valve
Check valve
Strainer
Cut back valve
Kick down valve
Throttle valve
To cooler
Secondary regulator valve

NEUTRAL GEAR—AW132L

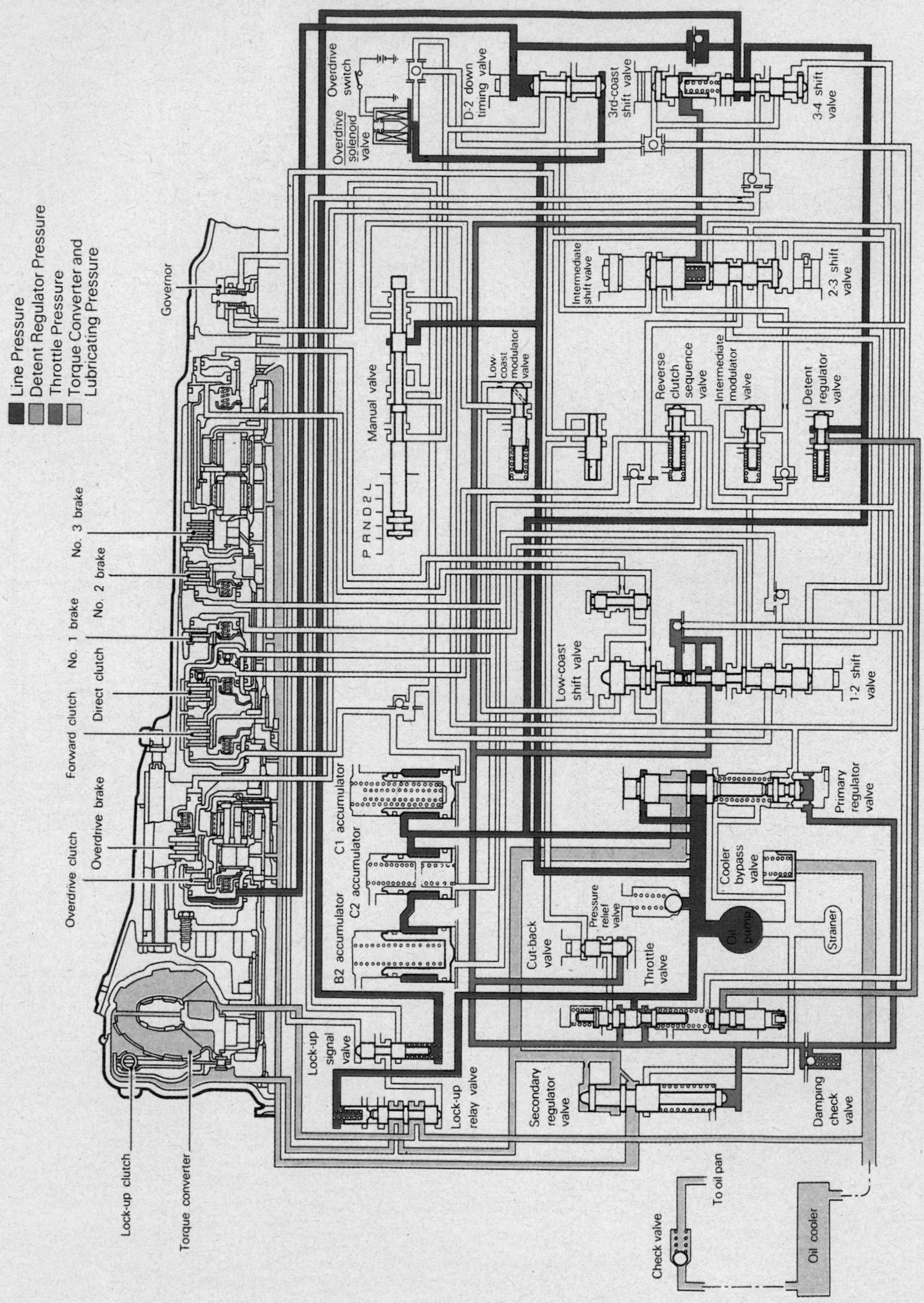
Line Pressure
Detent Regulator Pressure
Throttle Pressure
Torque Converter and Lubricating Pressure
Overdrive switch
Overdrive solenoid valve
D-2 down timing valve
3rd-coast shift valve
3-4 shift valve
Governor
Intermediate shift valve
2-3 shift valve
Manual valve
P R N D 2 L
Low-coast modulator valve
Reverse clutch sequence valve
Intermediate modulator valve
Detent regulator valve
No. 3 brake
No. 2 brake
No. 1 brake
Direct clutch
Forward clutch
Overdrive brake
Overdrive clutch
Low-coast shift valve
1-2 shift valve
B2 accumulator
C1 accumulator
C2 accumulator
Primary regulator valve
Cooler bypass valve
Pressure relief valve
Cut-back valve
Throttle valve
Oil pump
Strainer
Lock-up signal valve
Lock-up relay valve
Secondary regulator valve
Damping check valve
Lock-up clutch
Torque converter
To oil pan
Check valve
Oil cooler

PARK—AW132L

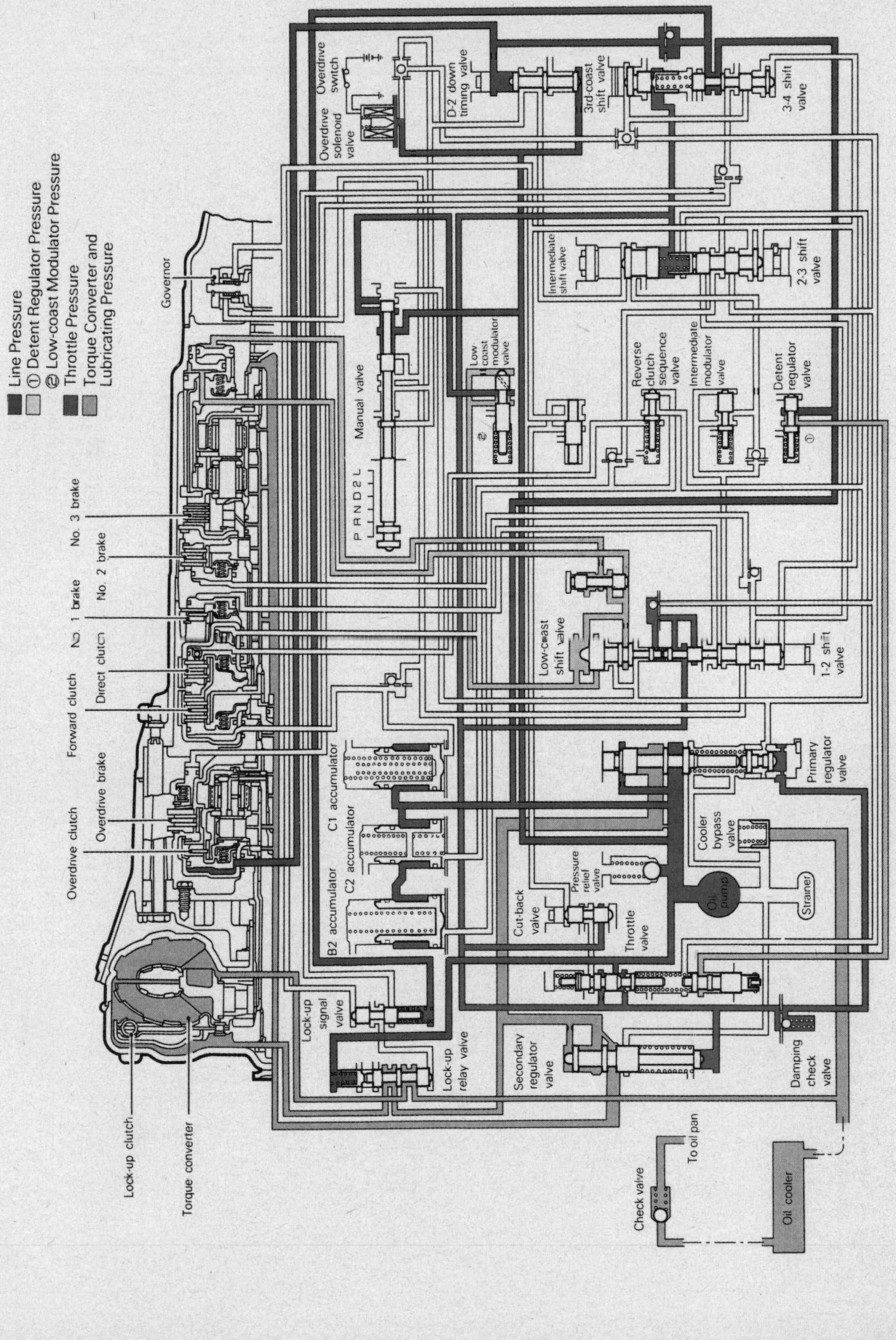
Line Pressure
① Detent Regulator Pressure
② Low-coast Modulator Pressure
Throttle Pressure
Torque Converter and Lubricating Pressure
Governor
Overdrive solenoid valve
Overdrive switch
D-2 down timing valve
3rd-coast shift valve
3-4 shift valve
Intermediate shift valve
2-3 shift valve
Manual valve
P R N D 2 L
Low coast modulator valve
Reverse clutch sequence valve
Intermediate modulator valve
Detent regulator valve
No. 3 brake
No. 2 brake
No. 1 brake
Direct clutch
Forward clutch
Overdrive brake
Overdrive clutch
Low-coast shift valve
1-2 shift valve
C1 accumulator
C2 accumulator
B2 accumulator
Primary regulator valve
Cooler bypass valve
Pressure relief valve
Oil pump
Strainer
Cut-back valve
Throttle valve
Lock-up signal valve
Lock-up relay valve
Secondary regulator valve
Damping check valve
Lock-up clutch
Torque converter
Check valve
To oil pan
Oil cooler

DRIVE—1ST GEAR—AW132L

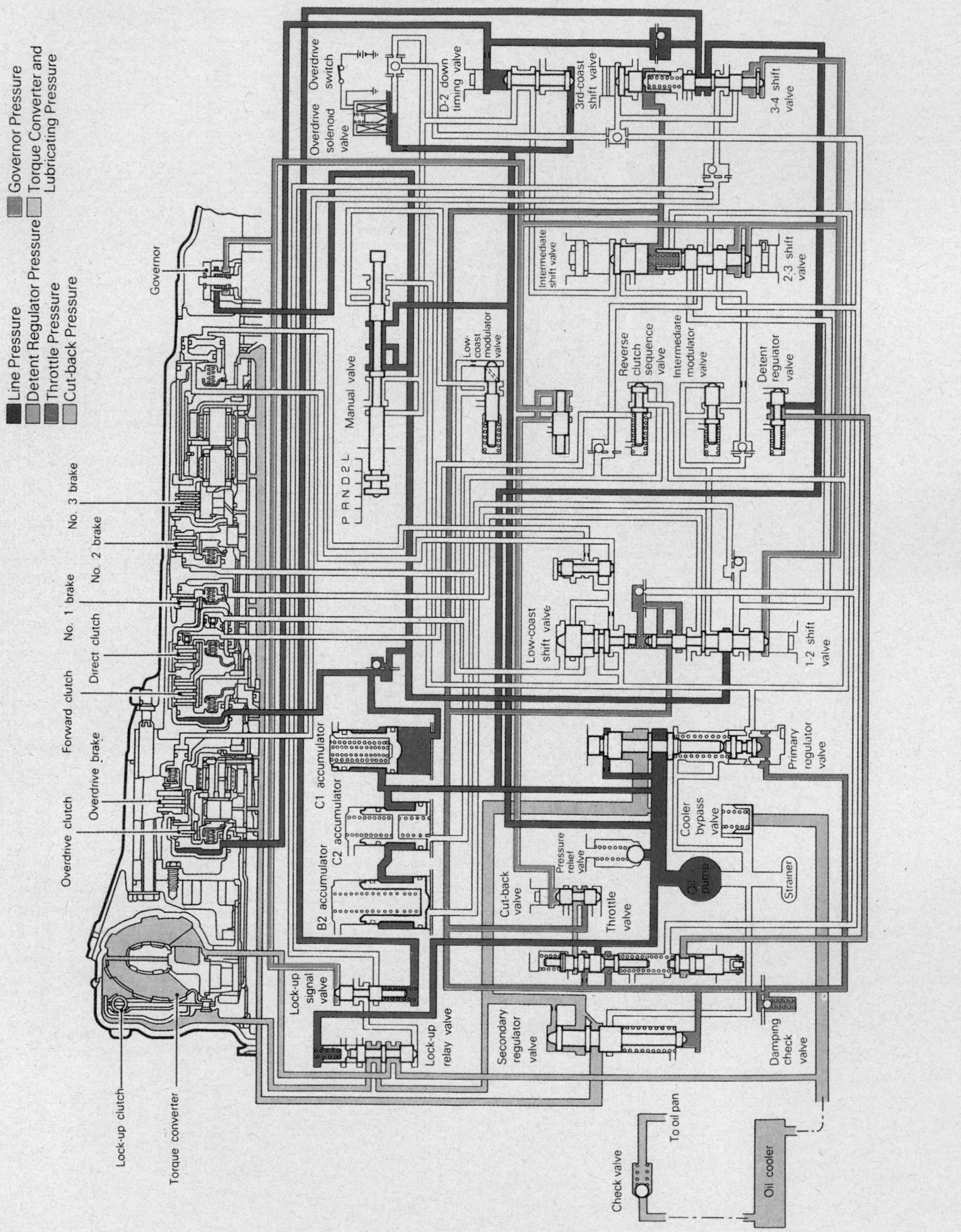
Line Pressure
Detent Regulator Pressure
Throttle Pressure
Cut-back Pressure
Governor Pressure
Torque Converter and Lubricating Pressure
Governor
No. 3 brake
No. 2 brake
No. 1 brake
Direct clutch
Forward clutch
Overdrive brake
Overdrive clutch
Lock-up clutch
Torque converter
Overdrive solenoid valve
Overdrive switch
D-2 down timing valve
3rd-coast shift valve
3-4 shift valve
Intermediate shift valve
2-3 shift valve
Manual valve
P R N D 2 L
Low-coast modulator valve
Reverse clutch sequence valve
Intermediate modulator valve
Detent regulator valve
Low-coast shift valve
1-2 shift valve
C1 accumulator
C2 accumulator
B2 accumulator
Primary regulator valve
Cooler bypass valve
Pressure relief valve
Oil pump
Strainer
Cut-back valve
Throttle valve
Lock-up signal valve
Lock-up relay valve
Secondary regulator valve
Damping check valve
To oil pan
Check valve
Oil cooler

DRIVE—2ND GEAR—AW132L

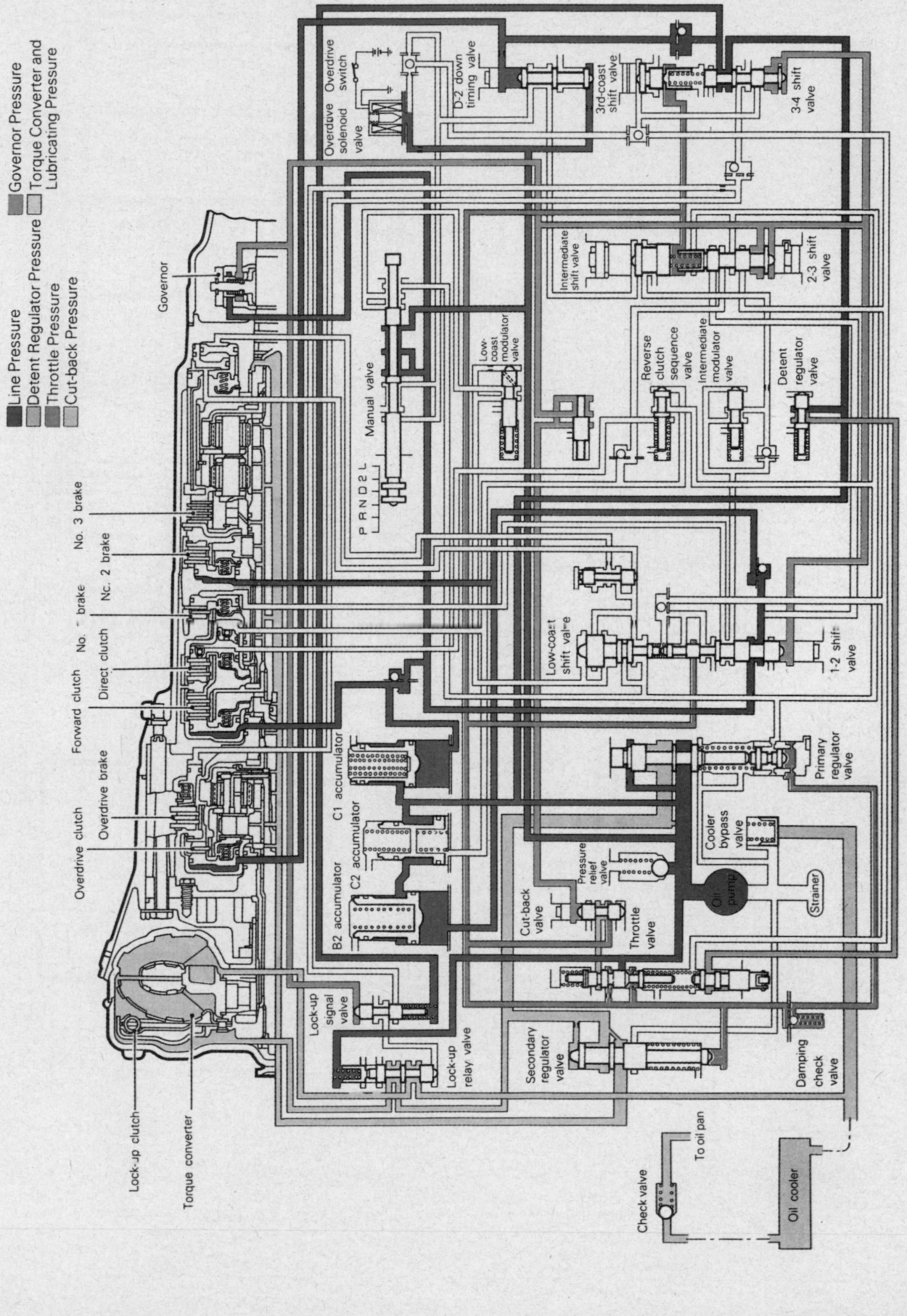
Line Pressure
Detent Regulator Pressure
Throttle Pressure
Cut-back Pressure
Governor Pressure
Torque Converter and Lubricating Pressure
Governor
No. 3 brake
No. 2 brake
No. 1 brake
Direct clutch
Forward clutch
Overdrive brake
Overdrive clutch
Lock-up clutch
Torque converter
Overdrive solenoid valve
Overdrive switch
D-2 down timing valve
3rd-coast shift valve
3-4 shift valve
Intermediate shift valve
2-3 shift valve
Manual valve
P R N D 2 L
Low-coast modulator valve
Reverse clutch sequence valve
Intermediate modulator valve
Detent regulator valve
Low-coast shift valve
1-2 shift valve
C1 accumulator
C2 accumulator
B2 accumulator
Primary regulator valve
Cooler bypass valve
Pressure relief valve
Oil pump
Strainer
Cut-back valve
Throttle valve
Lock-up signal valve
Lock-up relay valve
Secondary regulator valve
Damping check valve
Check valve
To oil pan
Oil cooler

DRIVE—3RD GEAR—AW132L

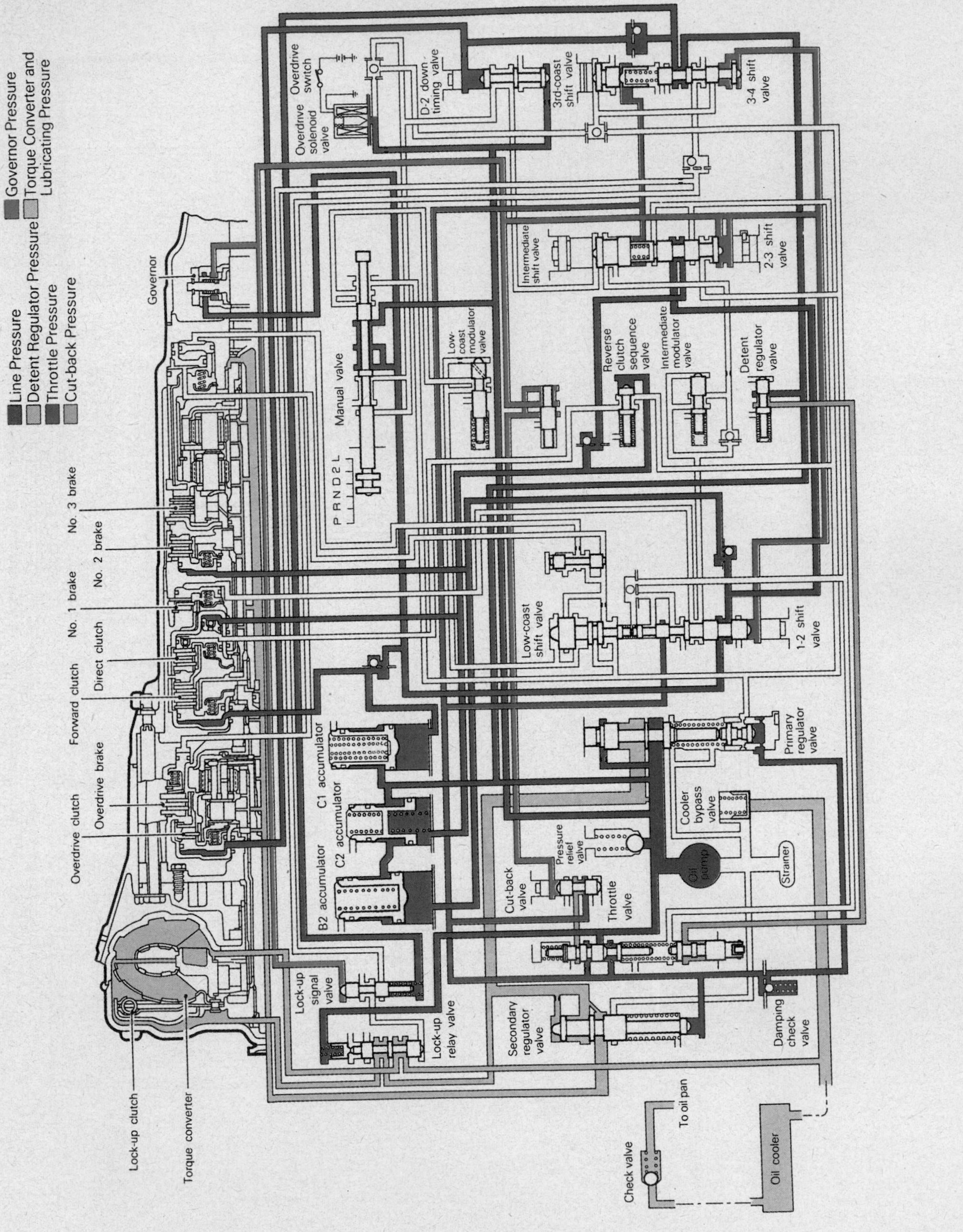
Line Pressure
Detent Regulator Pressure
Throttle Pressure
Cut-back Pressure
Governor Pressure
Torque Converter and Lubricating Pressure
Governor
No. 3 brake
No. 2 brake
No. 1 brake
Direct clutch
Forward clutch
Overdrive brake
Overdrive clutch
Lock-up clutch
Torque converter
Overdrive switch
Overdrive solenoid valve
D-2 down-timing valve
3rd-coast shift valve
3-4 shift valve
Intermediate shift valve
2-3 shift valve
Manual valve
P R N D 2 L
Low-coast modulator valve
Reverse clutch sequence valve
Intermediate modulator valve
Detent regulator valve
Low-coast shift valve
1-2 shift valve
Primary regulator valve
C1 accumulator
C2 accumulator
B2 accumulator
Cooler bypass valve
Pressure relief valve
Oil pump
Strainer
Cut-back valve
Throttle valve
Lock-up signal valve
Lock-up relay valve
Secondary regulator valve
Damping check valve
To oil pan
Check valve
Oil cooler

DRIVE—4TH GEAR (LOCKUP CLUTCH OFF)—AW132L

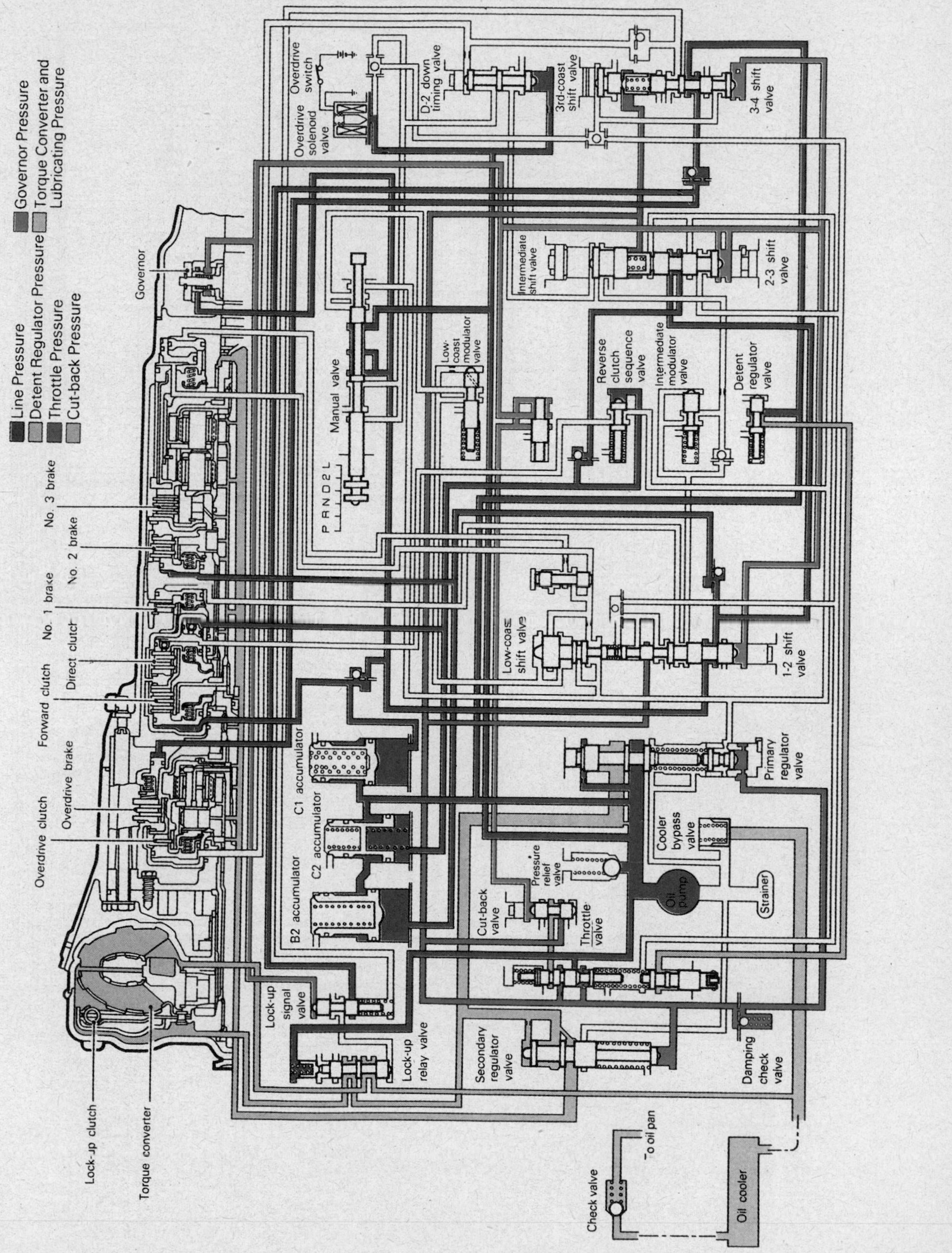
Line Pressure
Detent Regulator Pressure
Throttle Pressure
Cut-back Pressure
Governor Pressure
Torque Converter and Lubricating Pressure
Overdrive switch
Overdrive solenoid valve
D-2 down timing valve
3rd-coast shift valve
3-4 shift valve
Governor
Intermediate shift valve
2-3 shift valve
Low-coast modulator valve
Manual valve
P R N D 2 L
Reverse clutch sequence valve
Intermediate modulator valve
Detent regulator valve
No. 3 brake
No. 2 brake
No. 1 brake
Direct clutch
Forward clutch
Overdrive brake
Overdrive clutch
Low-coast shift valve
1-2 shift valve
C1 accumulator
C2 accumulator
B2 accumulator
Primary regulator valve
Cooler bypass valve
Pressure relief valve
Oil pump
Strainer
Cut-back valve
Throttle valve
Lock-up signal valve
Lock-up relay valve
Secondary regulator valve
Damping check valve
Lock-up clutch
Torque converter
To oil pan
Check valve
Oil cooler

DRIVE—4TH GEAR (LOCKUP CLUTCH ON)—AW132L

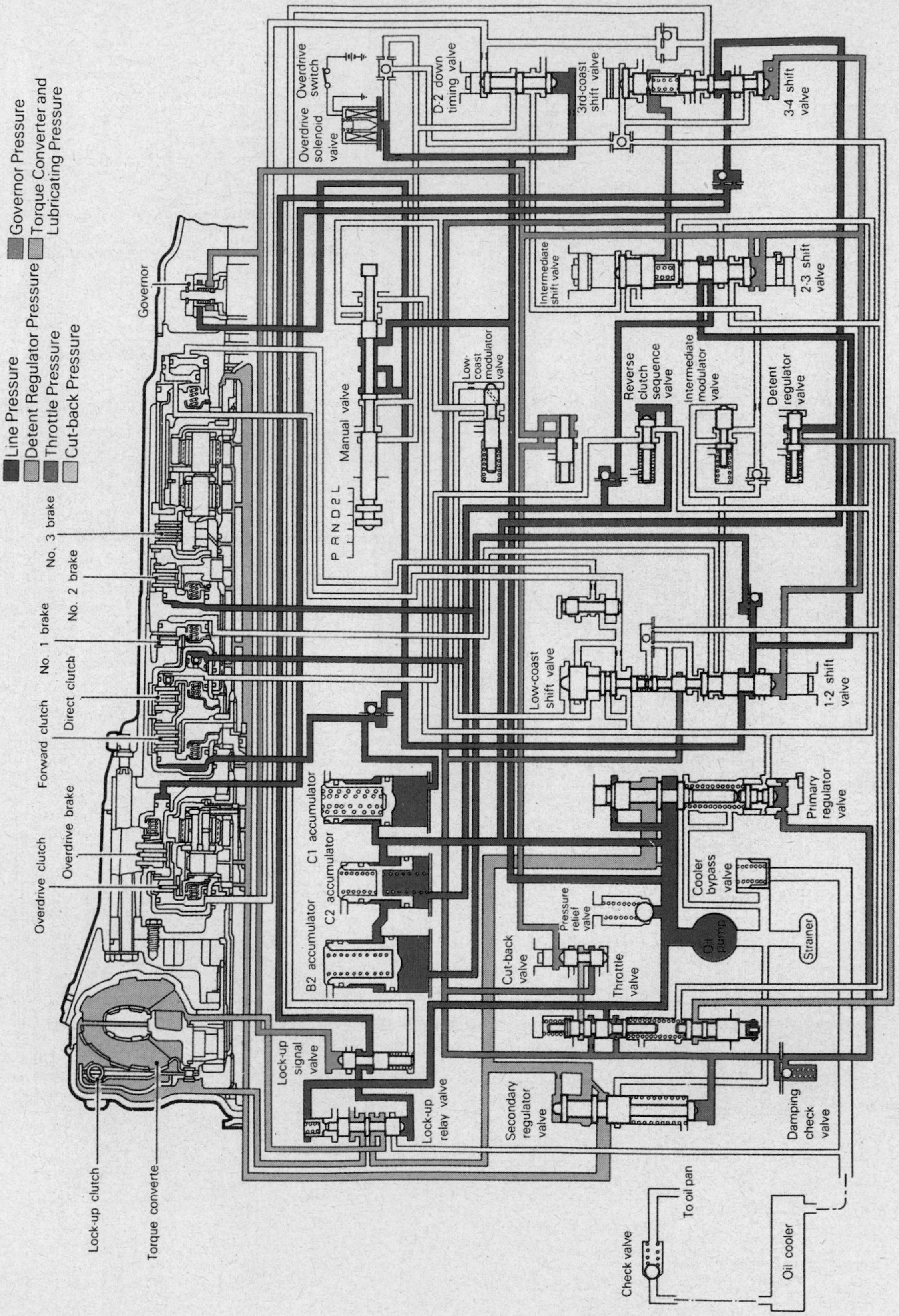
Line Pressure
Detent Regulator Pressure
Throttle Pressure
Cut-back Pressure
Governor Pressure
Torque Converter and Lubricating Pressure
Governor
No. 3 brake
No. 2 brake
No. 1 brake
Direct clutch
Forward clutch
Overdrive brake
Overdrive clutch
Lock-up clutch
Torque converter
Overdrive switch
Overdrive solenoid valve
D-2 down timing valve
3rd-coast shift valve
3-4 shift valve
Intermediate shift valve
2-3 shift valve
Manual valve
P R N D 2 L
Low-coast modulator valve
Reverse clutch sequence valve
Intermediate modulator valve
Detent regulator valve
Low-coast shift valve
1-2 shift valve
C1 accumulator
C2 accumulator
B2 accumulator
Primary regulator valve
Cooler bypass valve
Pressure relief valve
Oil pump
Strainer
Cut-back valve
Throttle valve
Lock-up signal valve
Lock-up relay valve
Secondary regulator valve
Damping check valve
To oil pan
Check valve
Oil cooler

DRIVE—KICKDOWN GEAR (4TH TO 3RD)—AW132L

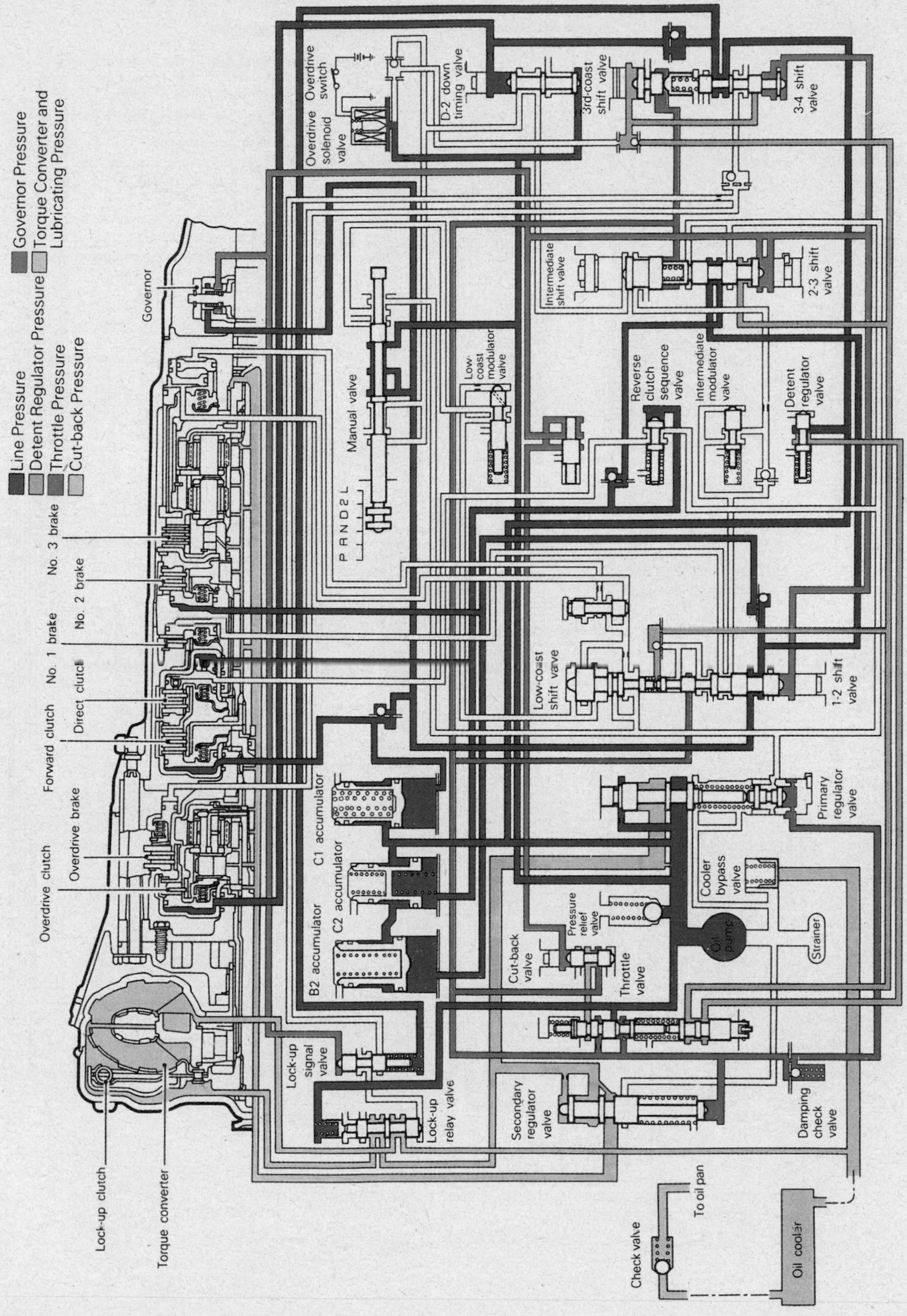
Line Pressure
Detent Regulator Pressure
Throttle Pressure
Cut-back Pressure
Governor Pressure
Torque Converter and Lubricating Pressure
Governor
No. 3 brake
No. 2 brake
No. 1 brake
Direct clutch
Forward clutch
Overdrive brake
Overdrive clutch
Lock-up clutch
Torque converter
Overdrive solenoid valve
Overdrive switch
D-2 down timing valve
3rd-coast shift valve
3-4 shift valve
Intermediate shift valve
2-3 shift valve
Manual valve
P R N D 2 L
Low-coast modulator valve
Reverse clutch sequence valve
Intermediate modulator valve
Detent regulator valve
Low-coast shift valve
1-2 shift valve
C1 accumulator
C2 accumulator
B2 accumulator
Primary regulator valve
Cooler bypass valve
Pressure relief valve
Oil pump
Strainer
Cut-back valve
Throttle valve
Lock-up signal valve
Lock-up relay valve
Secondary regulator valve
Damping check valve
Check valve
To oil pan
Oil cooler

DRIVE—3RD GEAR (OD SWITCH OFF)—AW132L

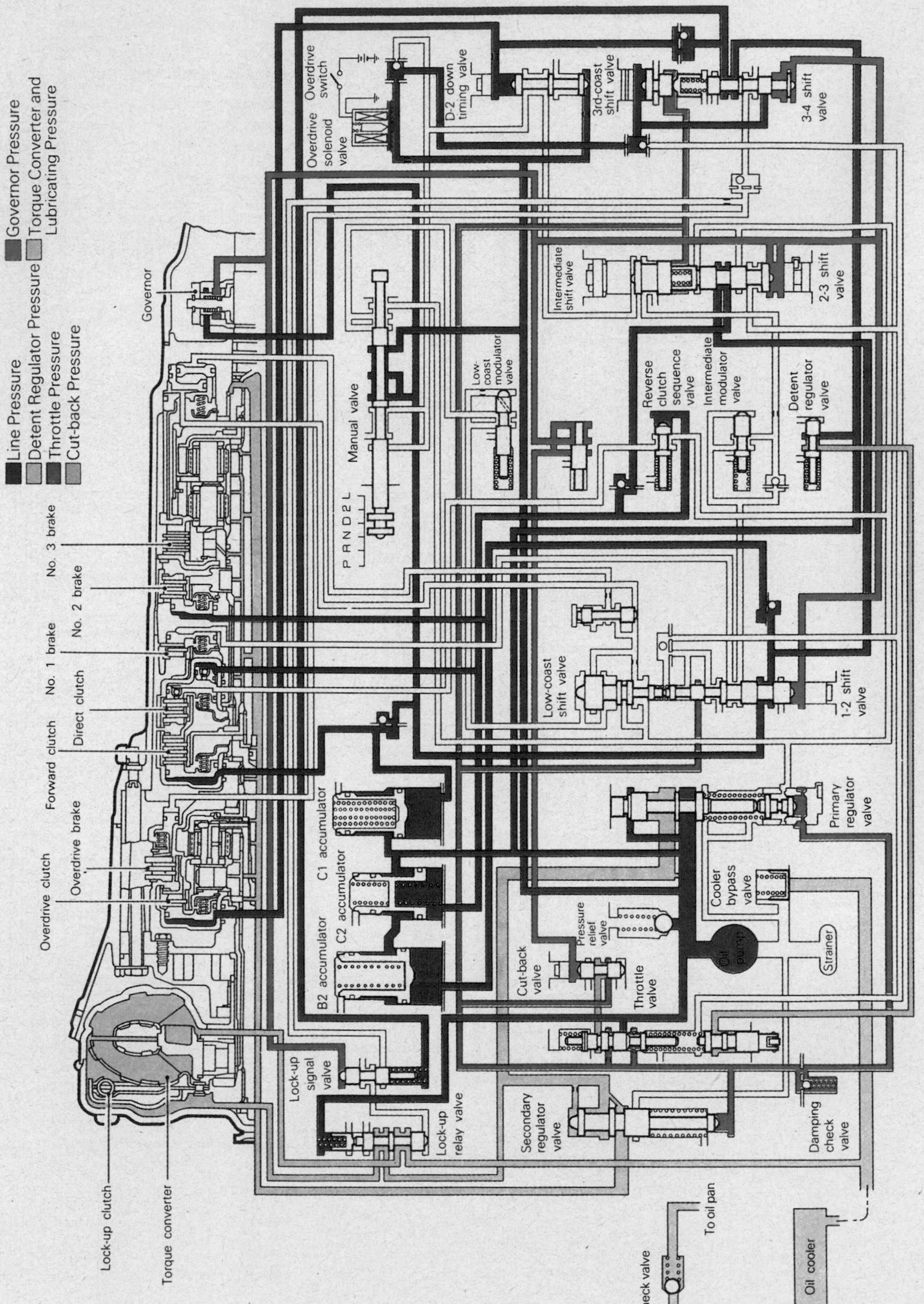
Line Pressure
Detent Regulator Pressure
Throttle Pressure
Cut-back Pressure
Governor Pressure
Torque Converter and Lubricating Pressure
Governor
No. 3 brake
No. 2 brake
No. 1 brake
Direct clutch
Forward clutch
Overdrive brake
Overdrive clutch
Lock-up clutch
Torque converter
Overdrive solenoid valve
Overdrive switch
D-2 down timing valve
3rd-coast shift valve
3-4 shift valve
Intermediate shift valve
2-3 shift valve
Manual valve
P R N D 2 L
Low-coast modulator valve
Reverse clutch sequence valve
Intermediate modulator valve
Detent regulator valve
Low-coast shift valve
1-2 shift valve
C1 accumulator
C2 accumulator
B2 accumulator
Primary regulator valve
Cooler bypass valve
Pressure relief valve
Oil pump
Strainer
Cut-back valve
Throttle valve
Lock-up signal valve
Lock-up relay valve
Secondary regulator valve
Damping check valve
Check valve
To oil pan
Oil cooler

DRIVE—2ND GEAR—AW132L

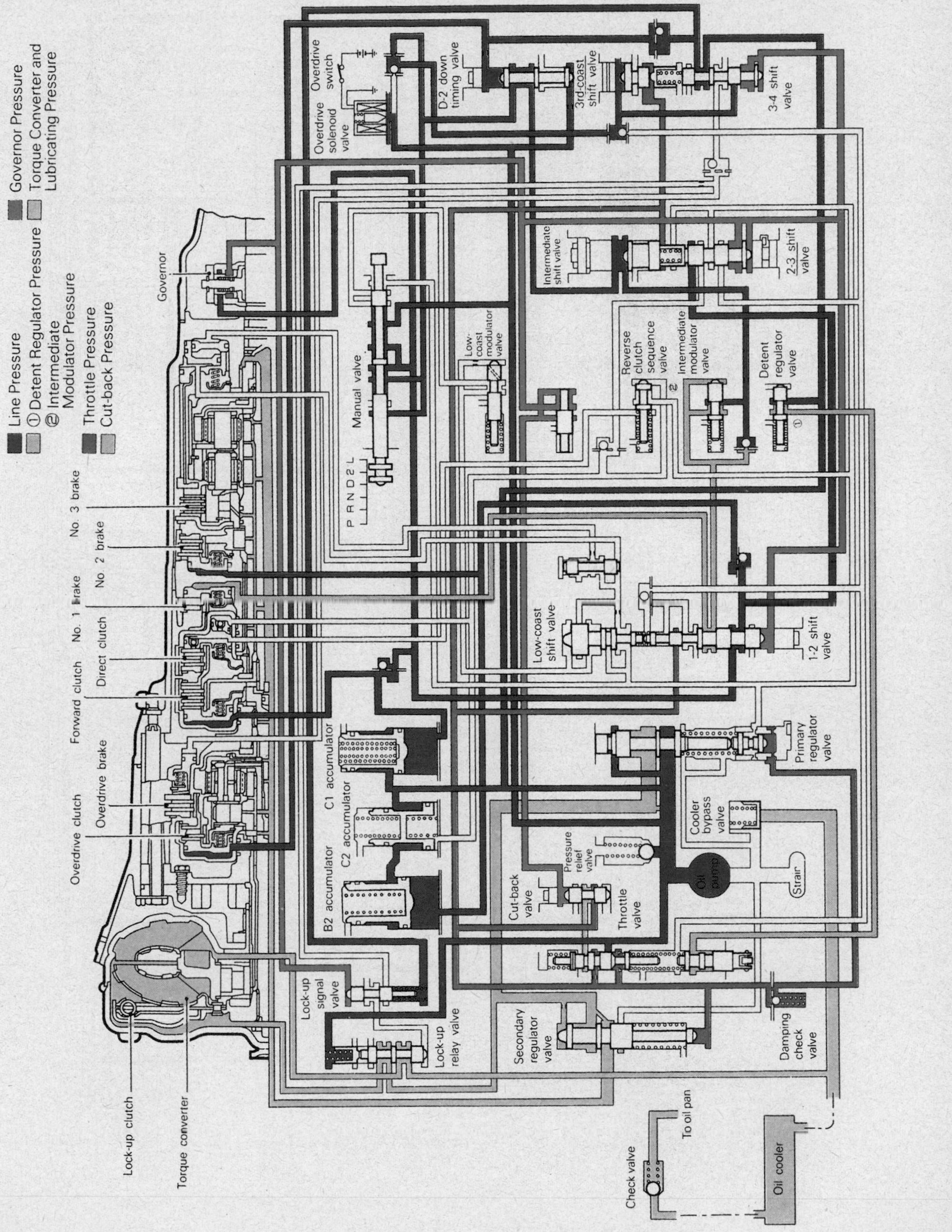
Line Pressure
① Detent Regulator Pressure
② Intermediate Modulator Pressure
Throttle Pressure
Cut-back Pressure
Governor Pressure
Torque Converter and Lubricating Pressure
Governor
No. 3 brake
No. 2 brake
No. 1 brake
Direct clutch
Forward clutch
Overdrive brake
Overdrive clutch
Lock-up clutch
Torque converter
Overdrive solenoid valve
Overdrive switch
D-2 down timing valve
3rd-coast shift valve
3-4 shift valve
Intermediate shift valve
2-3 shift valve
Manual valve
P R N D 2 L
Low-coast modulator valve
Reverse clutch sequence valve
Intermediate modulator valve
Detent regulator valve
Low-coast shift valve
1-2 shift valve
C1 accumulator
C2 accumulator
B2 accumulator
Primary regulator valve
Cooler bypass valve
Pressure relief valve
Oil pump
Strainer
Cut-back valve
Throttle valve
Lock-up signal valve
Lock-up relay valve
Secondary regulator valve
Damping check valve
Check valve
To oil pan
Oil cooler

DRIVE—LOCKUP GEAR—AW132L

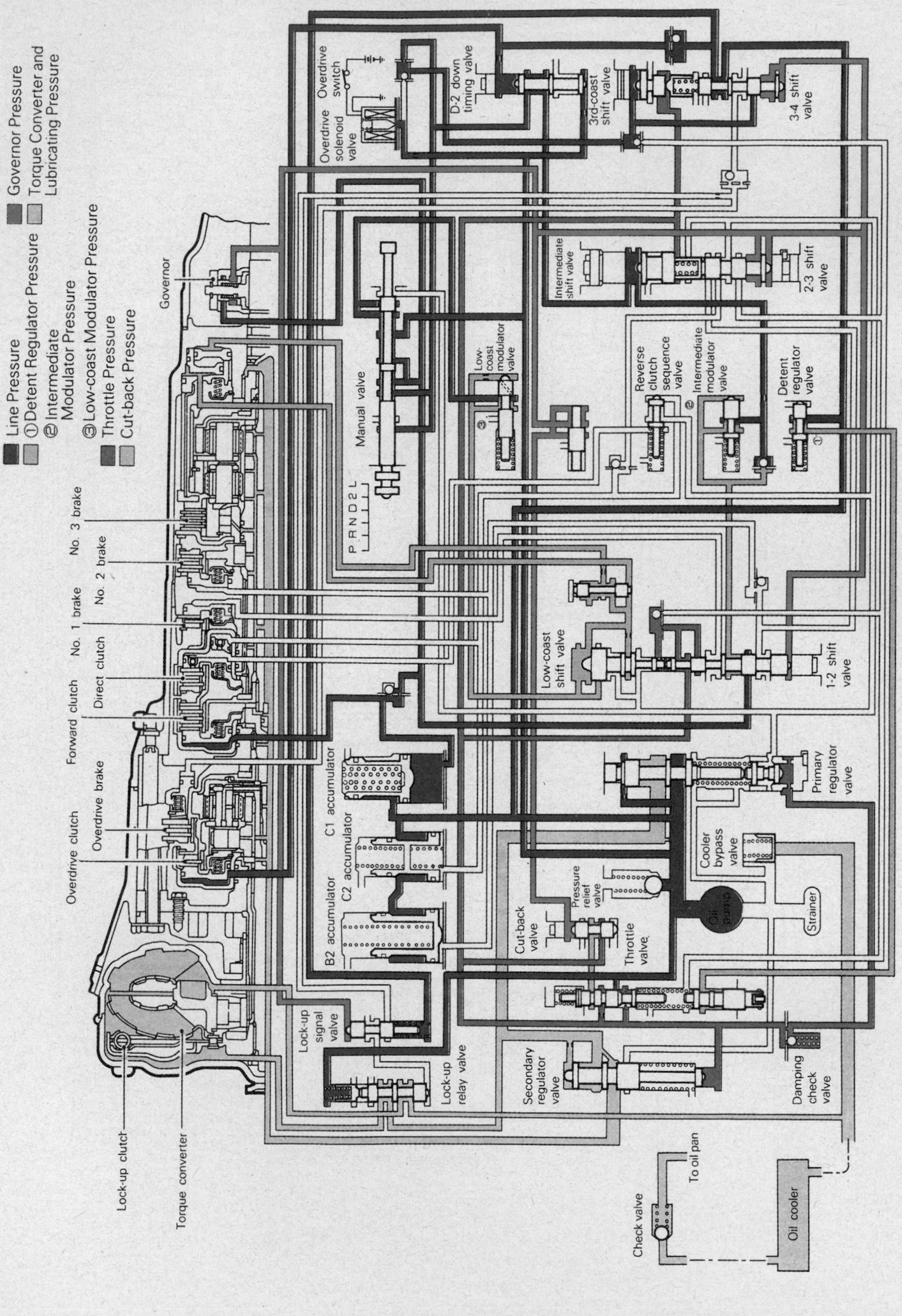
Line Pressure
① Detent Regulator Pressure
② Intermediate Modulator Pressure
③ Low-coast Modulator Pressure
Throttle Pressure
Cut-back Pressure
Governor Pressure
Torque Converter and Lubricating Pressure
Governor
No. 3 brake
No. 2 brake
No. 1 brake
Direct clutch
Forward clutch
Overdrive brake
Overdrive clutch
Lock-up clutch
Torque converter
Overdrive solenoid valve
Overdrive switch
D-2 down timing valve
3rd-coast shift valve
3-4 shift valve
Intermediate shift valve
2-3 shift valve
Manual valve
P R N D 2 L
Low-coast modulator valve
Reverse clutch sequence valve
Intermediate modulator valve
Detent regulator valve
Low-coast shift valve
1-2 shift valve
C1 accumulator
C2 accumulator
B2 accumulator
Primary regulator valve
Cooler bypass valve
Pressure relief valve
Oil pump
Strainer
Cut-back valve
Throttle valve
Lock-up signal valve
Lock-up relay valve
Secondary regulator valve
Damping check valve
Check valve
To oil pan
Oil cooler

REVERSE—AW132L

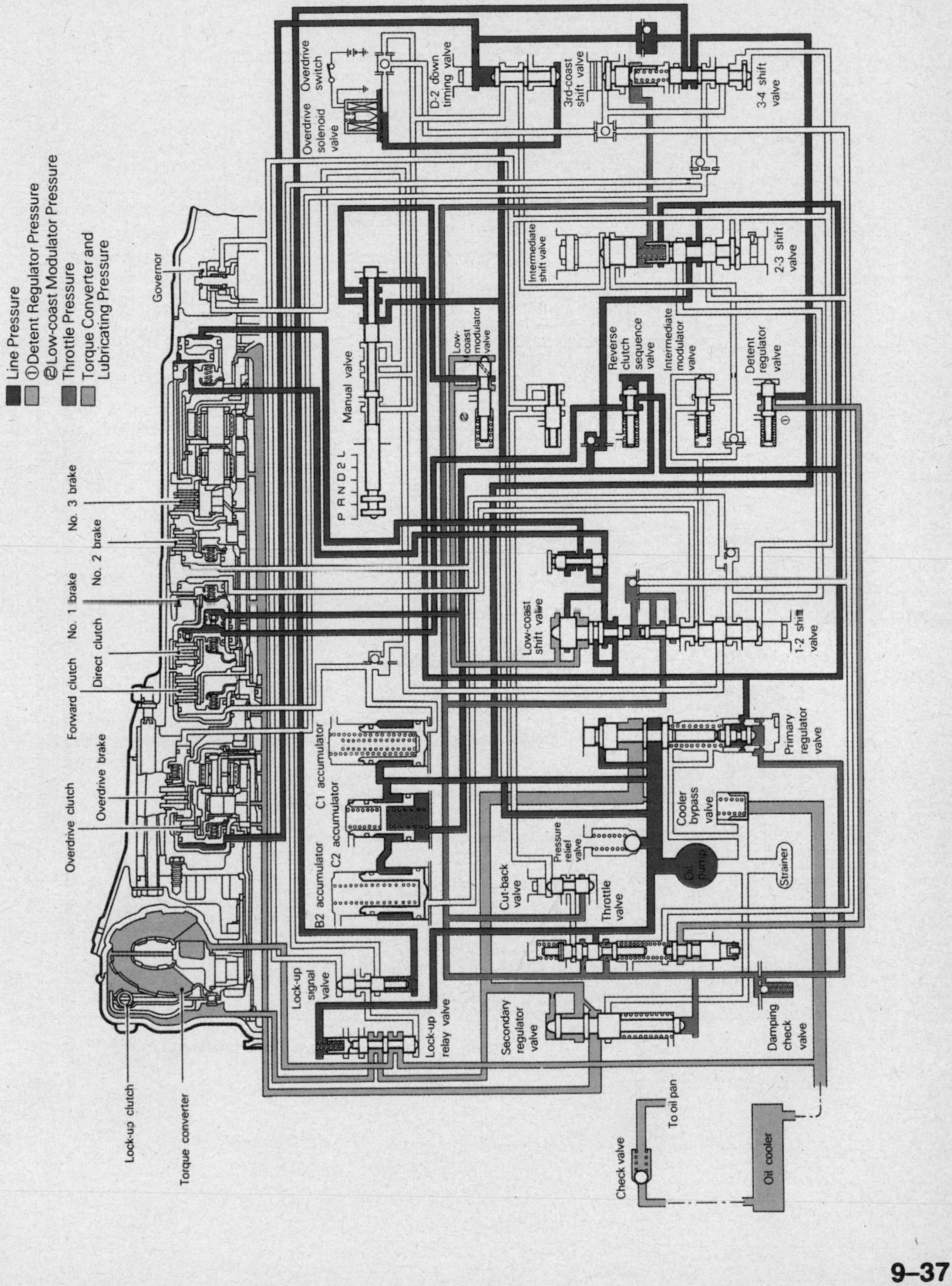
Line Pressure
①Detent Regulator Pressure
②Low-coast Modulator Pressure
Throttle Pressure
Torque Converter and Lubricating Pressure
Governor
No. 3 brake
No. 2 brake
No. 1 brake
Direct clutch
Forward clutch
Overdrive brake
Overdrive clutch
Lock-up clutch
Torque converter
Overdrive solenoid valve
Overdrive switch
D-2 down timing valve
3rd-coast shift valve
3-4 shift valve
Intermediate shift valve
2-3 shift valve
Manual valve
P R N D 2 L
Low-coast modulator valve
Reverse clutch sequence valve
Intermediate modulator valve
Detent regulator valve
Low-coast shift valve
1-2 shift valve
C1 accumulator
C2 accumulator
B2 accumulator
Primary regulator valve
Cooler bypass valve
Pressure relief valve
Cut-back valve
Throttle valve
Oil pump
Strainer
Lock-up signal valve
Lock-up relay valve
Secondary regulator valve
Damping check valve
Check valve
To oil pan
Oil cooler

PARK/NEUTRAL—SPEED UNDER 8 MPH—A604

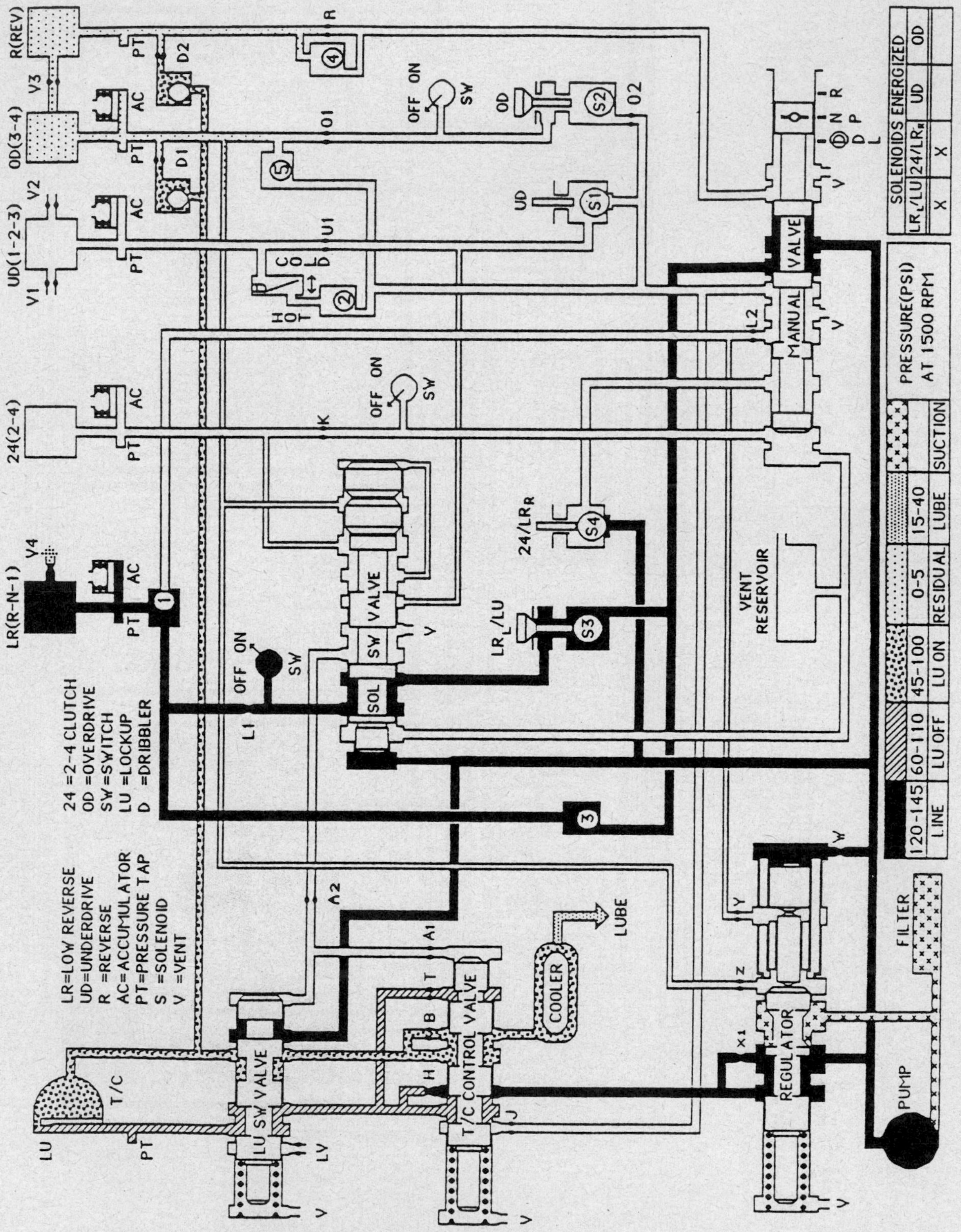
LR=LOW REVERSE
UD=UNDERDRIVE
R =REVERSE
AC=ACCUMULATOR
PT=PRESSURE TAP
S =SOLENOID
V =VENT
24 =2-4 CLUTCH
OD =OVERDRIVE
SW=SWITCH
LU =LOCKUP
D =DRIBBLER
R(REV)
OD(3-4)
UD(1-2-3)
24(2-4)
LR(R-N-1)
MANUAL VALVE
SW VALVE
SOL
T/C CONTROL VALVE
LU SW VALVE
REGULATOR
COOLER
LUBE
PUMP
FILTER
VENT RESERVOIR
T/C
PRESSURE(PSI) AT 1500 RPM
120-145 LINE
60-110 LU OFF
45-100 LU ON
0-5 RESIDUAL
15-40 LUBE
SUCTION
SOLENOIDS ENERGIZED
LR/LU
24/LR
UD
OD

NEUTRAL – SPEED OVER 8 MPH – A604

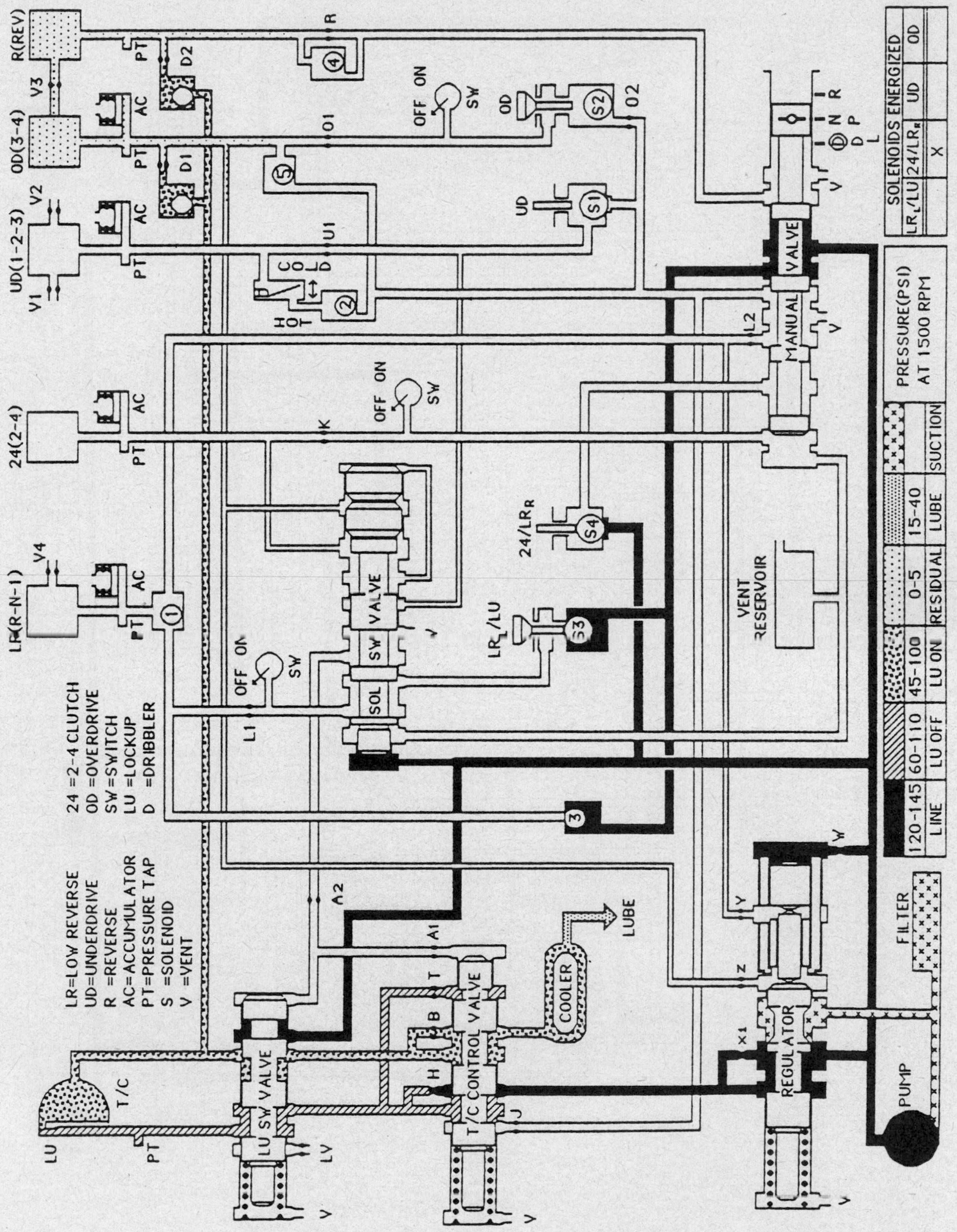
LR=LOW REVERSE
UD=UNDERDRIVE
R =REVERSE
AC=ACCUMULATOR
PT=PRESSURE TAP
S =SOLENOID
V =VENT
24 =2-4 CLUTCH
OD =OVERDRIVE
SW=SWITCH
LU =LOCKUP
D =DRIBBLER
LR(R-N-1)
24(2-4)
UD(1-2-3)
OD(3-4)
R(REV)
MANUAL VALVE
SOL SW VALVE
LU SW VALVE
T/C CONTROL VALVE
REGULATOR
PUMP
FILTER
COOLER
LUBE
VENT RESERVOIR
T/C
PRESSURE(PSI) AT 1500 RPM
120-145 LINE
60-110 LU OFF
45-100 LU ON
0-5 RESIDUAL
15-40 LUBE
SUCTION
SOLENOIDS ENERGIZED
LR/LU
24/LR
UD
OD
X

REVERSE—A604

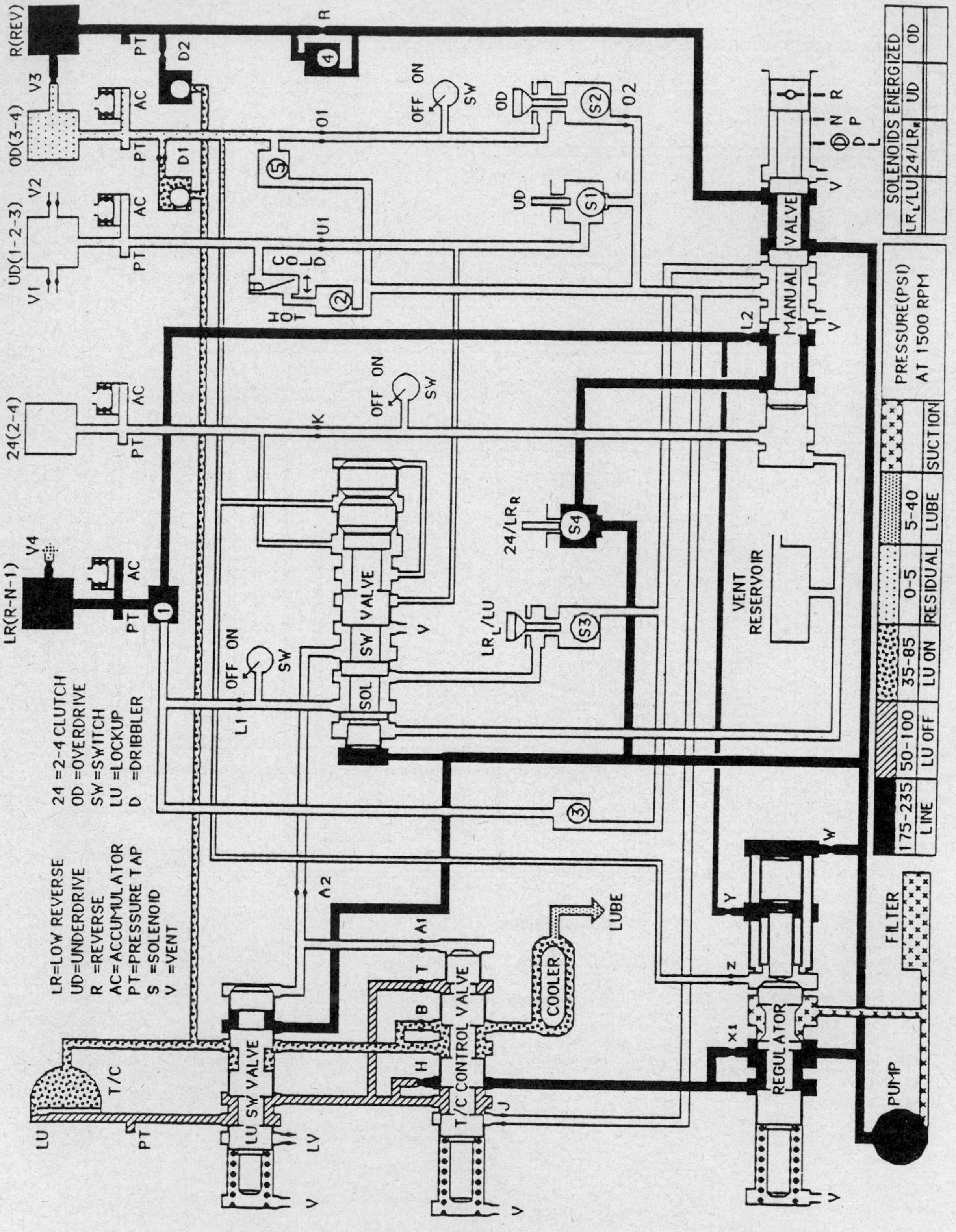
LR=LOW REVERSE
UD=UNDERDRIVE
R =REVERSE
AC=ACCUMULATOR
PT=PRESSURE TAP
S =SOLENOID
V =VENT
24 =2-4 CLUTCH
OD =OVERDRIVE
SW=SWITCH
LU =LOCKUP
D =DRIBBLER
R(REV)
OD(3-4)
UD(1-2-3)
24(2-4)
LR(R-N-1)
T/C
LU
COOLER
LUBE
FILTER
PUMP
REGULATOR
T/C CONTROL VALVE
LU SW VALVE
SOL SW VALVE
MANUAL VALVE
VENT RESERVOIR
SOLENOIDS ENERGIZED
LR_L/LU
24/LR_R
UD
OD
PRESSURE(PSI) AT 1500 RPM
175-235 LINE
50-100 LU OFF
35-85 LU ON
0-5 RESIDUAL
5-40 LUBE
SUCTION

REVERSE BLOCK—SHIFT TO REVERSE WITH SPEED OVER 8 MPH—A604

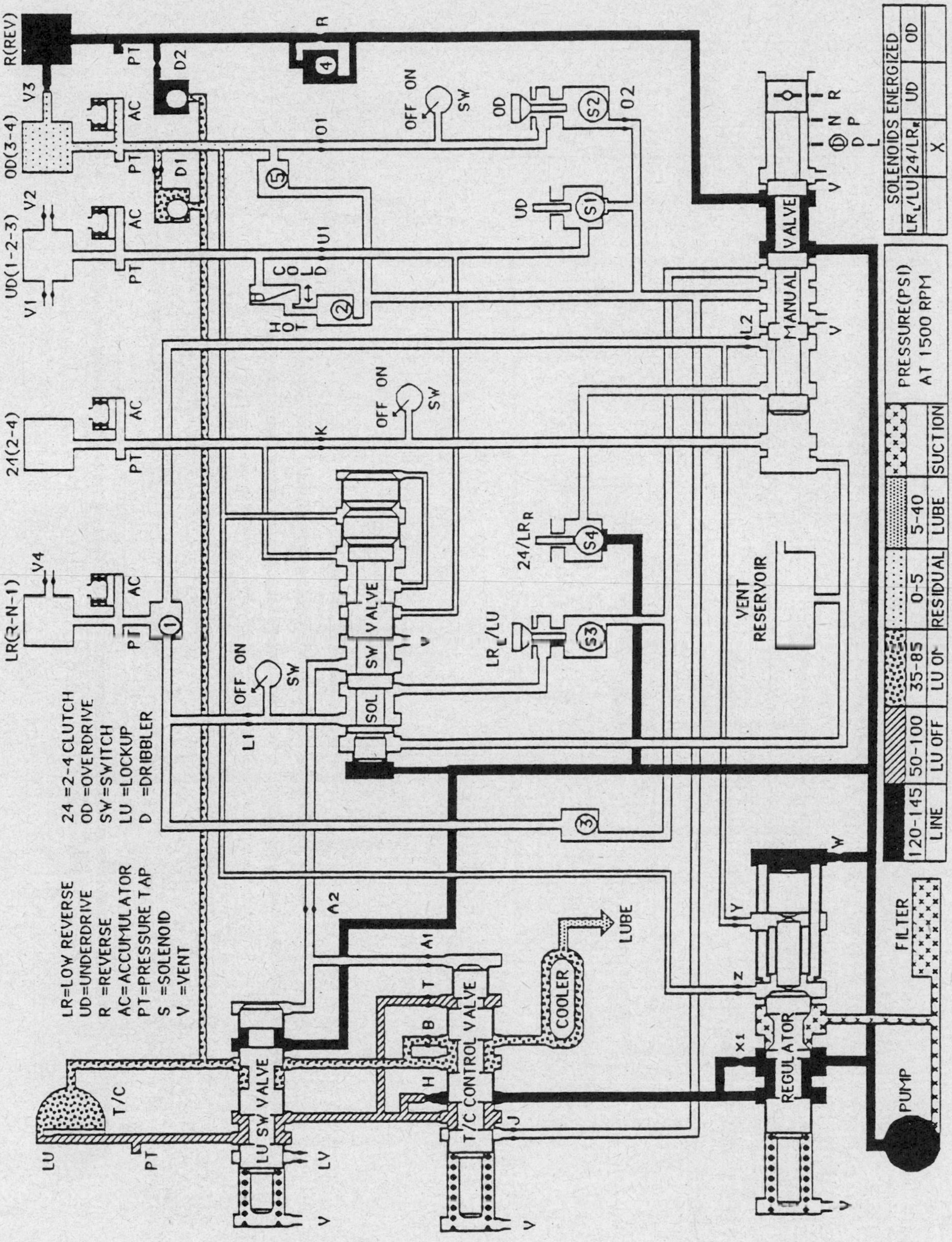
LR=LOW REVERSE
UD=UNDERDRIVE
R =REVERSE
AC=ACCUMULATOR
PT=PRESSURE TAP
S =SOLENOID
V =VENT
24 =2-4 CLUTCH
OD =OVERDRIVE
SW=SWITCH
LU =LOCKUP
D =DRIBBLER
R(REV)
OD(3-4)
UD(1-2-3)
24(2-4)
LR(R-N-1)
SOLENOIDS ENERGIZED
LR_L/LU 24/LR_R UD OD
X
PRESSURE(PSI) AT 1500 RPM
120-145 LINE
50-100 LU OFF
35-85 LU ON
0-5 RESIDUAL
5-40 LUBE
SUCTION
MANUAL VALVE
SOL SW VALVE
VENT RESERVOIR
REGULATOR
PUMP
FILTER
COOLER
LUBE
T/C CONTROL VALVE
LU SW VALVE
T/C

1ST GEAR—A604

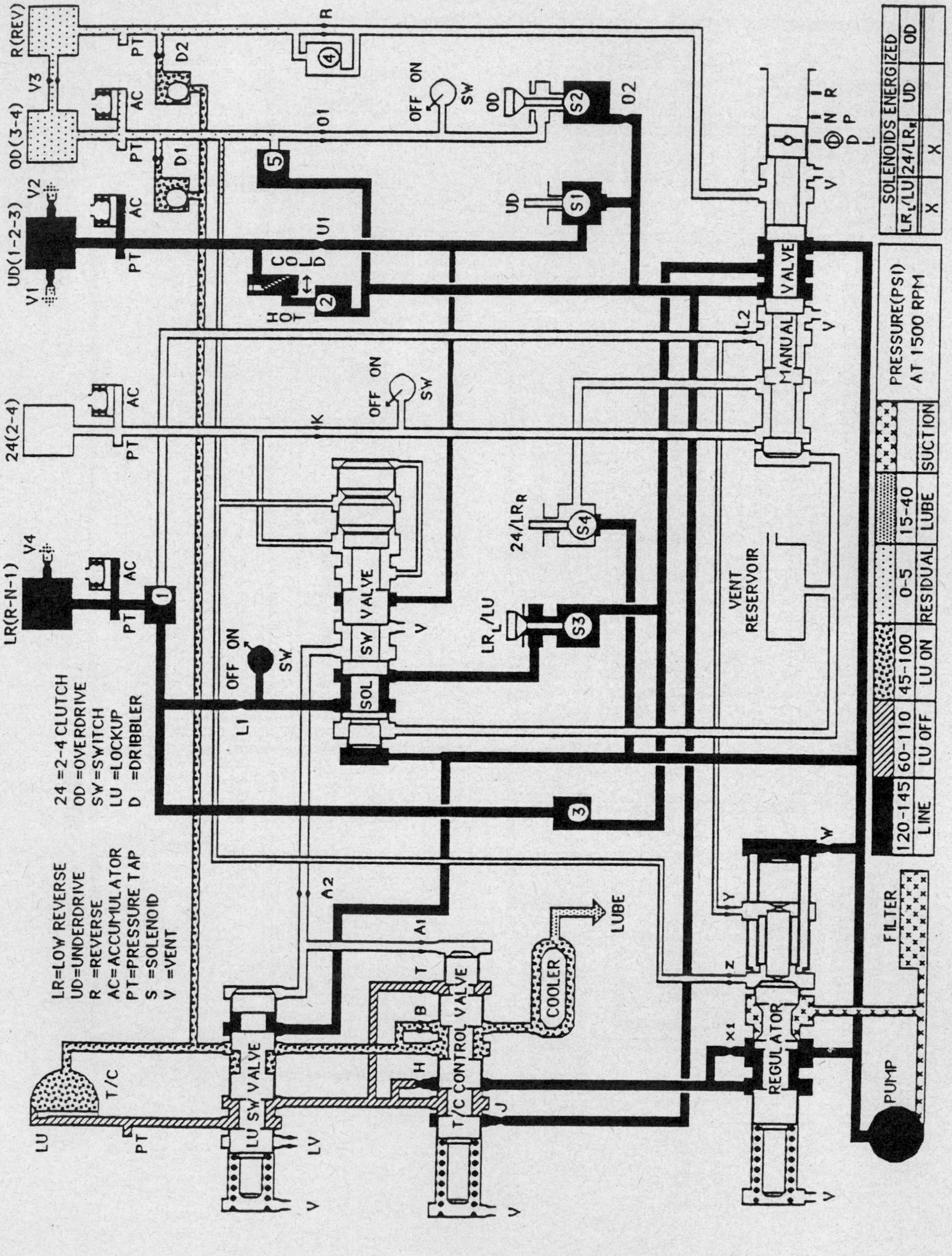
LR=LOW REVERSE
UD=UNDERDRIVE
R =REVERSE
AC=ACCUMULATOR
PT=PRESSURE TAP
S =SOLENOID
V =VENT
24 =2-4 CLUTCH
OD =OVERDRIVE
SW=SWITCH
LU =LOCKUP
D =DRIBBLER
R(REV)
OD(3-4)
UD(1-2-3)
24(2-4)
LR(R-N-1)
MANUAL VALVE
SOL SW VALVE
LU SW VALVE
T/C CONTROL VALVE
REGULATOR
COOLER
LUBE
FILTER
PUMP
VENT
RESERVOIR
SOLENOIDS ENERGIZED
LR$_L$/LU
24/LR$_R$
UD
OD
PRESSURE(PSI) AT 1500 RPM
120-145 LINE
60-110 LU OFF
45-100 LU ON
0-5 RESIDUAL
15-40 LUBE
SUCTION

2ND GEAR–A604

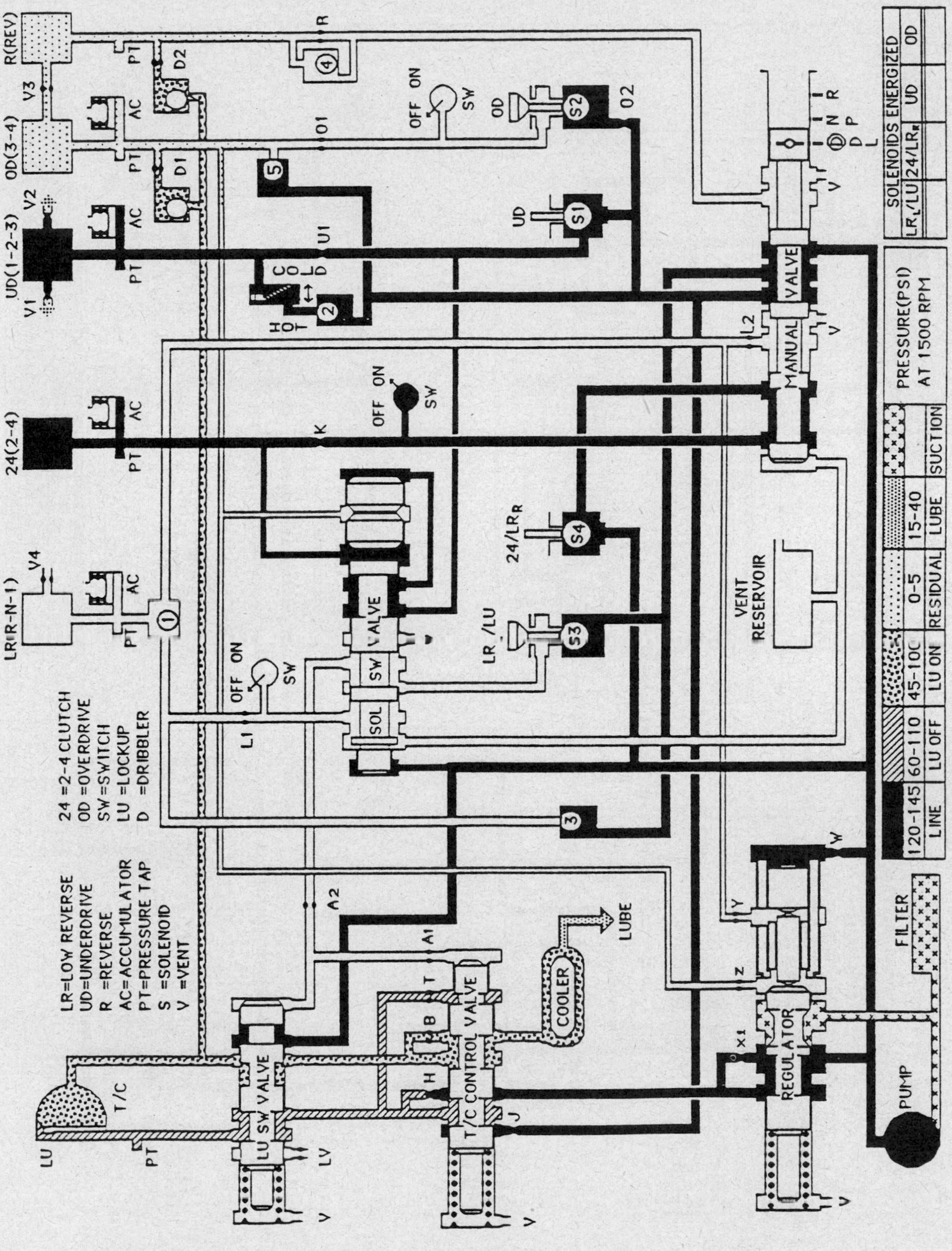
LR=LOW REVERSE
UD=UNDERDRIVE
R =REVERSE
AC=ACCUMULATOR
PT=PRESSURE TAP
S =SOLENOID
V =VENT
24 =2-4 CLUTCH
OD =OVERDRIVE
SW=SWITCH
LU =LOCKUP
D =DRIBBLER
R(REV)
OD(3-4)
UD(1-2-3)
24(2-4)
LR(R-N-1)
SOLENOIDS ENERGIZED
LR/LU
24/LR
UD
OD
PRESSURE(PSI) AT 1500 RPM
120-145 LINE
60-110 LU OFF
45-100 LU ON
0-5 RESIDUAL
15-40 LUBE
SUCTION
MANUAL VALVE
SW VALVE
SOL
VENT RESERVOIR
REGULATOR
T/C CONTROL VALVE
LU SW VALVE
COOLER
LUBE
FILTER
PUMP
T/C

2ND GEAR—PARTIAL LOCKUP—A604

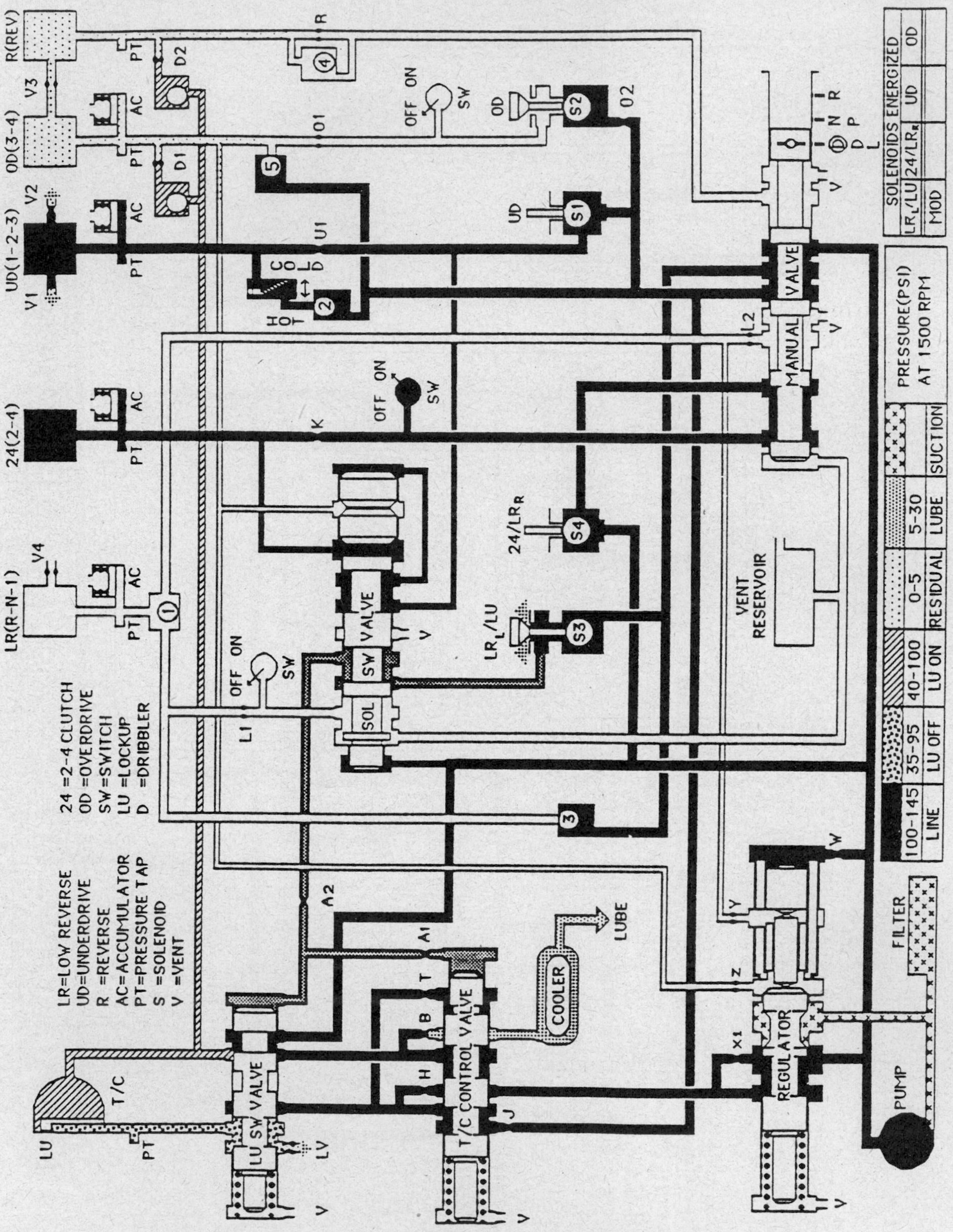
LR=LOW REVERSE
UD=UNDERDRIVE
R =REVERSE
AC=ACCUMULATOR
PT=PRESSURE TAP
S =SOLENOID
V =VENT
24 =2-4 CLUTCH
OD =OVERDRIVE
SW=SWITCH
LU =LOCKUP
D =DRIBBLER
R(REV)
OD(3-4)
UD(1-2-3)
24(2-4)
LR(R-N-1)
V1
V2
V3
V4
D1
D2
O1
O2
U1
K
L1
L2
A1
A2
T
B
H
J
W
Y
Z
X1
SOLENOIDS ENERGIZED
LR/LU
24/LR
UD
OD
MOD
PRESSURE(PSI) AT 1500 RPM
100-145 LINE
35-95 LU OFF
40-100 LU ON
0-5 RESIDUAL
5-30 LUBE
SUCTION
MANUAL VALVE
SW VALVE
SOL
LU SW VALVE
T/C CONTROL VALVE
REGULATOR
COOLER
LUBE
FILTER
PUMP
VENT RESERVOIR
T/C
COLD
HOT
OFF ON SW
S1
S2
S3
S4
24/LR
LR/LU
R N D D L

DIRECT GEAR—A604

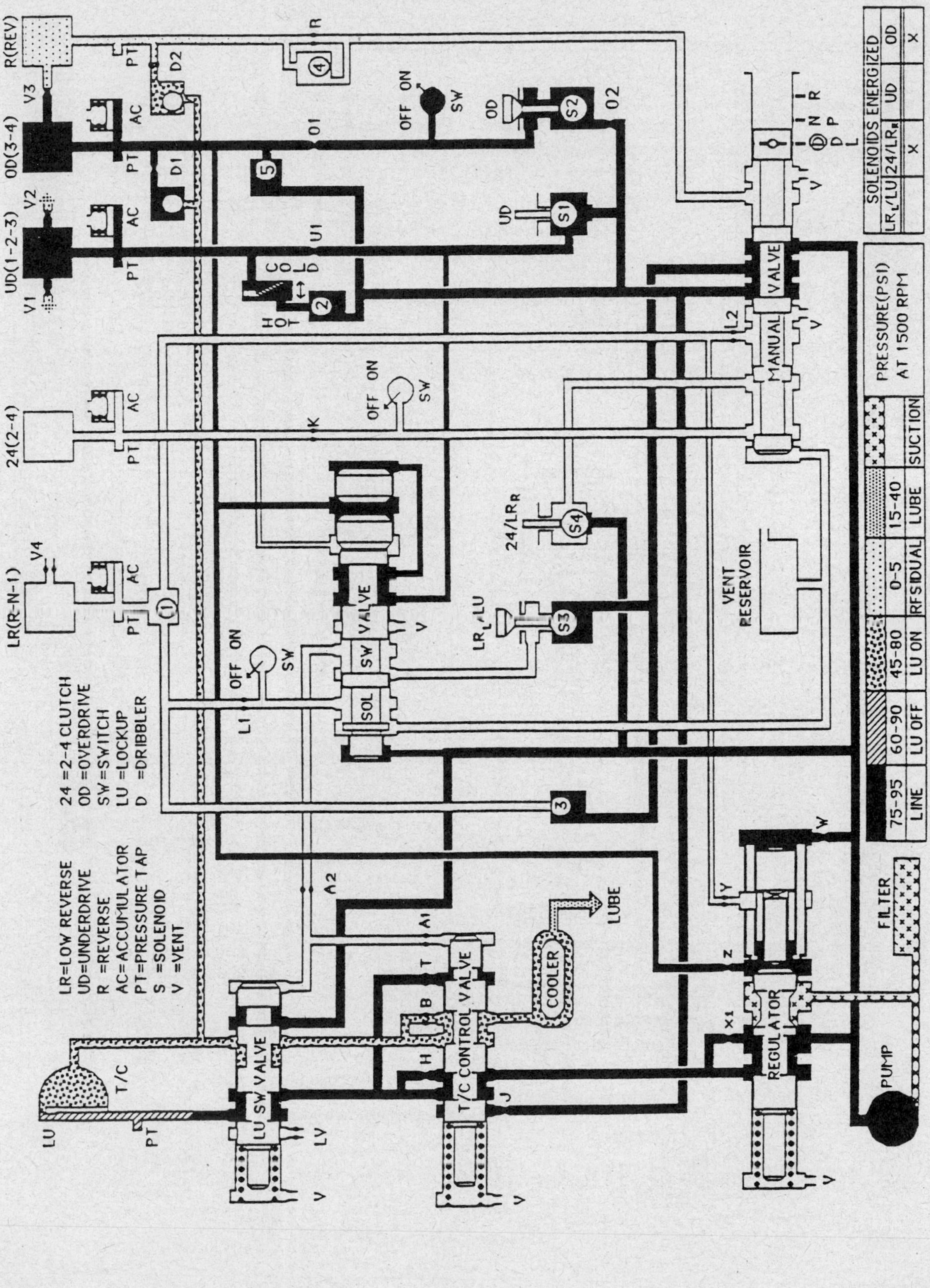
LR=LOW REVERSE
UD=UNDERDRIVE
R =REVERSE
AC=ACCUMULATOR
PT=PRESSURE TAP
S =SOLENOID
V =VENT
24 =2-4 CLUTCH
OD =OVERDRIVE
SW=SWITCH
LU =LOCKUP
D =DRIBBLER
R(REV)
OD(3-4)
UD(1-2-3)
24(2-4)
LR(R-N-1)
V1
V2
V3
V4
AC
PT
D1
D2
01
02
U1
OFF ON
SW
OD
UD
S1
S2
S3
S4
24/LRR
LRL LU
HOT
COLD
MANUAL VALVE
SOL
SW VALVE
VENT RESERVOIR
L1
L2
K
A1
A2
T
B
H
J
T/C
LU
LV
LU SW VALVE
T/C CONTROL VALVE
COOLER
LUBE
REGULATOR
PUMP
FILTER
W
Y
Z
X1
R N D D L
SOLENOIDS ENERGIZED
LRL/LU
24/LRR
UD
OD
PRESSURE(PSI) AT 1500 RPM
75-95 LINE
60-90 LU OFF
45-80 LU ON
0-5 RESIDUAL
15-40 LUBE
SUCTION

DIRECT GEAR—PARTIAL LOCKUP—A604

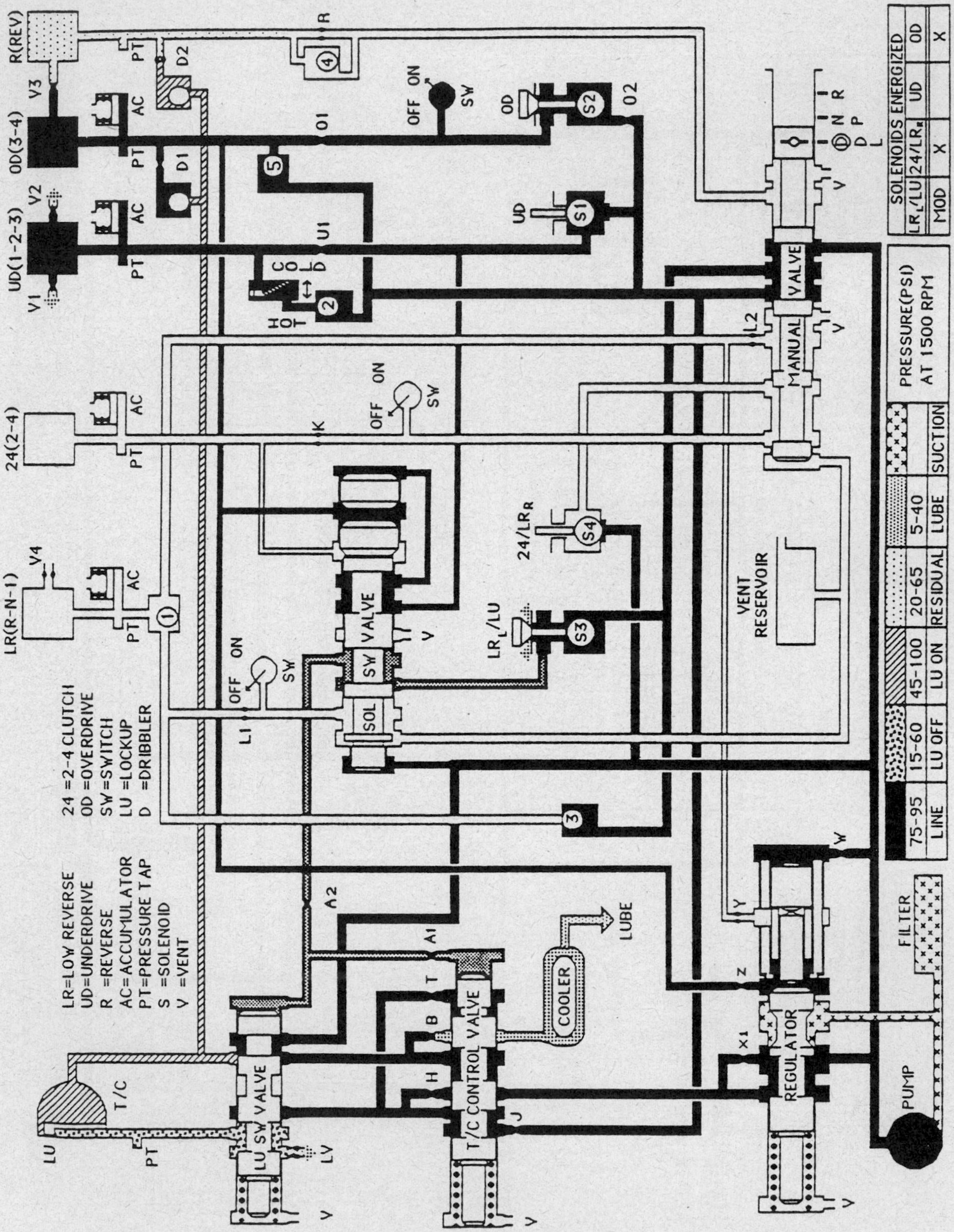
LR=LOW REVERSE
UD=UNDERDRIVE
R =REVERSE
AC=ACCUMULATOR
PT=PRESSURE TAP
S =SOLENOID
V =VENT
24 =2-4 CLUTCH
OD =OVERDRIVE
SW=SWITCH
LU =LOCKUP
D =DRIBBLER
SOLENOIDS ENERGIZED
MOD
PRESSURE(PSI) AT 1500 RPM
75-95 LINE
15-60 LU OFF
45-100 LU ON
20-65 RESIDUAL
5-40 LUBE
SUCTION
MANUAL VALVE
REGULATOR
PUMP
FILTER
COOLER
LUBE
T/C CONTROL VALVE
LU SW VALVE
SW VALVE
SOL
VENT RESERVOIR
T/C
R(REV)
OD(3-4)
UD(1-2-3)
24(2-4)
LR(R-N-1)

DIRECT GEAR—FULL LOCKUP—A604

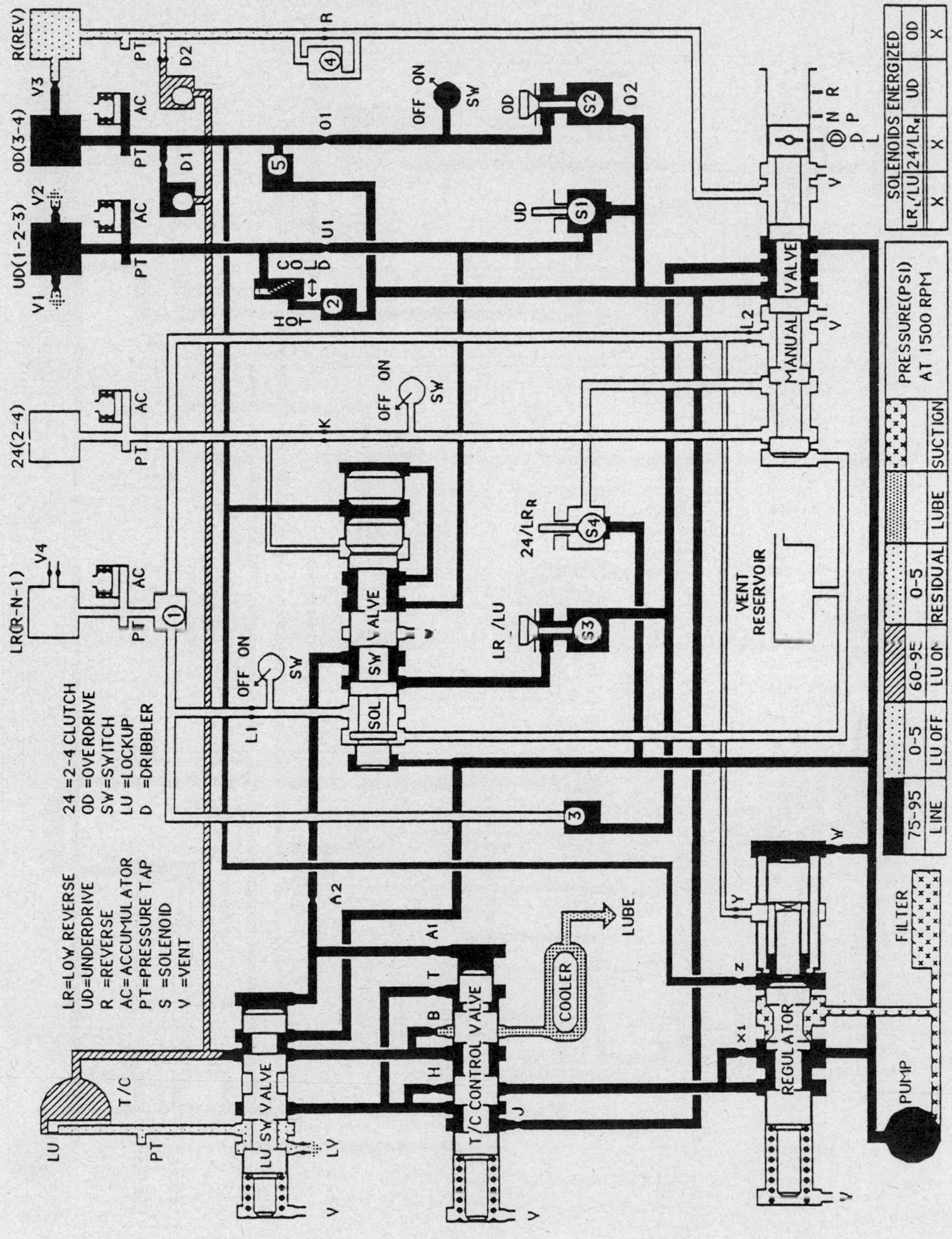
LR=LOW REVERSE
UD=UNDERDRIVE
R =REVERSE
AC=ACCUMULATOR
PT=PRESSURE TAP
S =SOLENOID
V =VENT
24 =2-4 CLUTCH
OD =OVERDRIVE
SW=SWITCH
LU =LOCKUP
D =DRIBBLER
R(REV)
OD(3-4)
UD(1-2-3)
24(2-4)
LR(R-N-1)
T/C
LU SW VALVE
T/C CONTROL VALVE
COOLER
LUBE
REGULATOR
PUMP
FILTER
MANUAL VALVE
SOL SW VALVE
VENT RESERVOIR
SOLENOIDS ENERGIZED
LR/LU
24/LR
UD
OD
PRESSURE(PSI) AT 1500 RPM
75-95 LINE
0-5 LU OFF
60-95 LU ON
0-5 RESIDUAL
LUBE
SUCTION

OVERDRIVE—A604

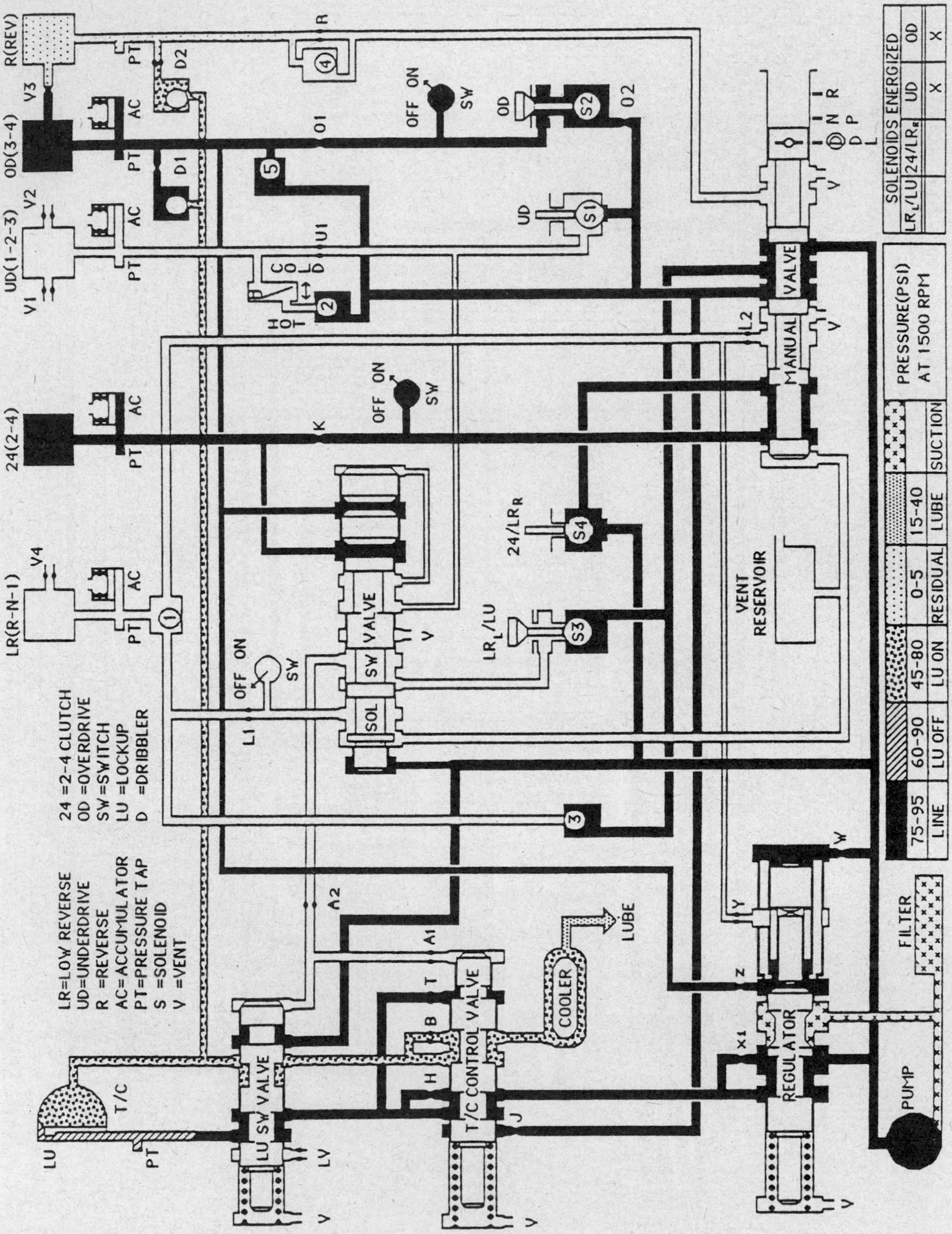
LR=LOW REVERSE
UD=UNDERDRIVE
R =REVERSE
AC=ACCUMULATOR
PT=PRESSURE TAP
S =SOLENOID
V =VENT
24 =2-4 CLUTCH
OD =OVERDRIVE
SW=SWITCH
LU =LOCKUP
D =DRIBBLER
R(REV)
OD(3-4)
UD(1-2-3)
24(2-4)
LR(R-N-1)
T/C
LU
LU SW VALVE
T/C CONTROL VALVE
COOLER
LUBE
REGULATOR
PUMP
FILTER
VENT
RESERVOIR
MANUAL VALVE
SW VALVE
SOL
S1
S2
S3
S4
LR_L/LU
24/LR_R
OFF ON
SOLENOIDS ENERGIZED
LR_L/LU
24/LR_R
UD
OD
X
X
PRESSURE(PSI) AT 1500 RPM
75-95 LINE
60-90 LU OFF
45-80 LU ON
0-5 RESIDUAL
15-40 LUBE
SUCTION

OVERDRIVE—FULL LOCKUP—A604

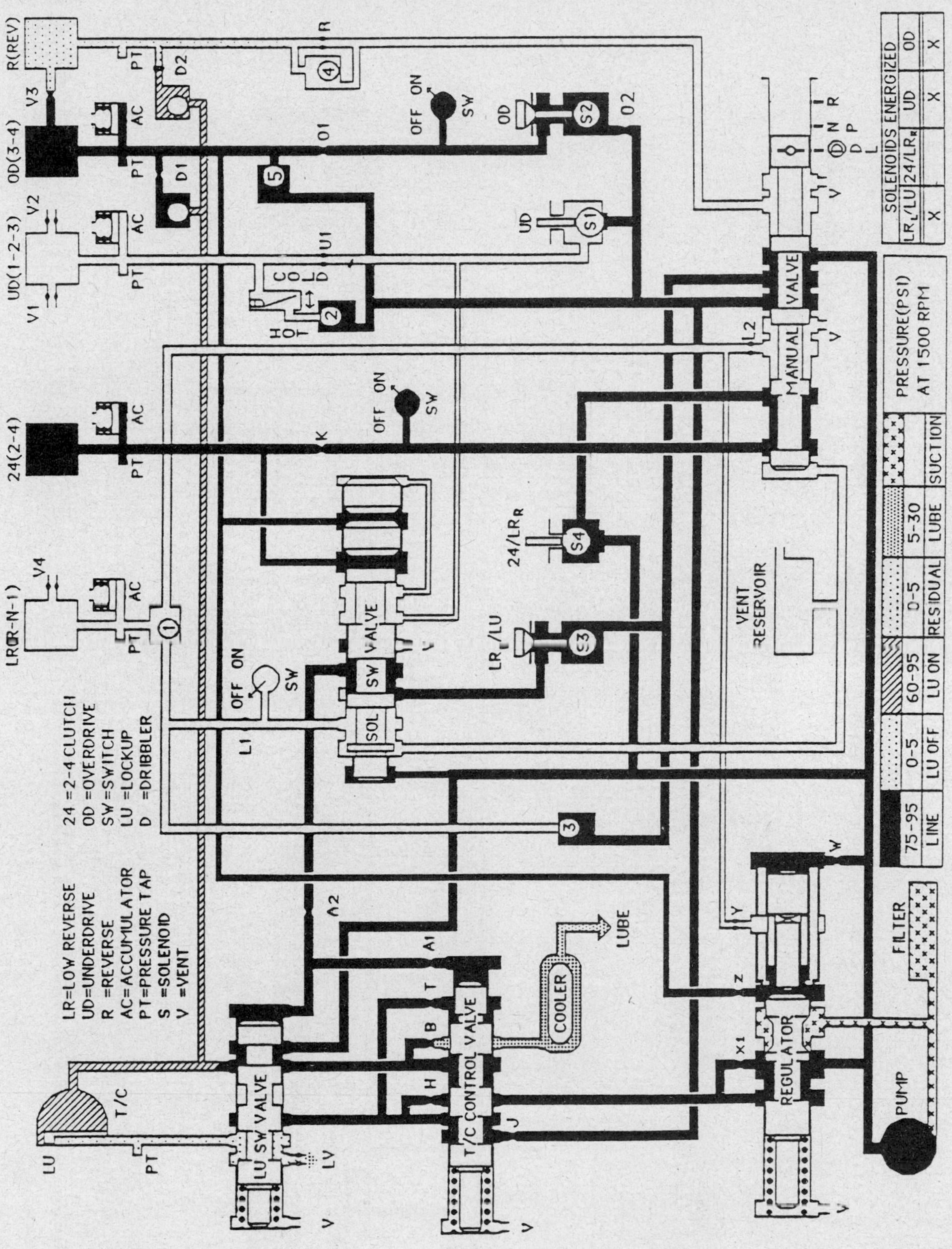
LR=LOW REVERSE
UD=UNDERDRIVE
R =REVERSE
AC=ACCUMULATOR
PT=PRESSURE TAP
S =SOLENOID
V =VENT
24 =2-4 CLUTCH
OD =OVERDRIVE
SW=SWITCH
LU =LOCKUP
D =DRIBBLER
SOLENOIDS ENERGIZED
LR$_L$/LU
24/LR$_R$
UD
OD
X
X
X
X
PRESSURE(PSI) AT 1500 RPM
75-95 LINE
0-5 LU OFF
60-95 LU ON
0-5 RESIDUAL
5-30 LUBE
SUCTION
FILTER
PUMP
REGULATOR
T/C CONTROL VALVE
LU SW VALVE
COOLER
LUBE
MANUAL VALVE
SOL
SW VALVE
VENT RESERVOIR
R(REV)
OD(3-4)
UD(1-2-3)
24(2-4)
LR(R-N-1)
T/C
LU
LV

PARK—4EAT

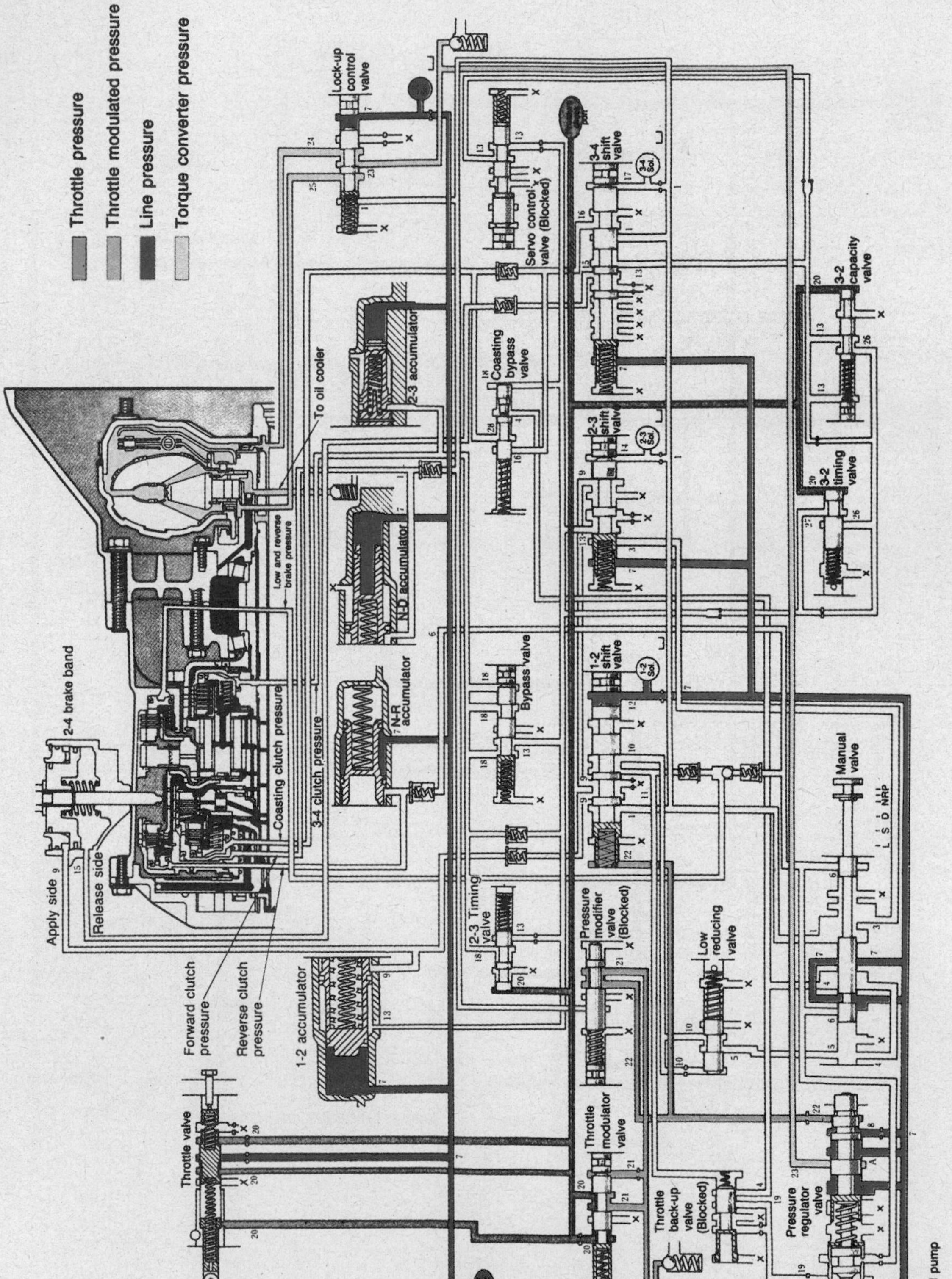
Throttle pressure
Throttle modulated pressure
Line pressure
Torque converter pressure
Lock-up control valve
To oil cooler
2-3 accumulator
N-D accumulator
N-R accumulator
Low and reverse brake pressure
Coasting clutch pressure
3-4 clutch pressure
2-4 brake band
Apply side
Release side
Forward clutch pressure
Reverse clutch pressure
1-2 accumulator
Throttle valve
Servo control valve (Blocked)
Coasting bypass valve
Bypass valve
2-3 Timing valve
Pressure modifier valve (Blocked)
Throttle modulator valve
3-4 shift valve
2-3 shift valve
1-2 shift valve
Low reducing valve
Throttle back-up valve (Blocked)
3-2 capacity valve
3-2 timing valve
Manual valve
L S D NRP
Pressure regulator valve
Oil pump

REVERSE—4EAT

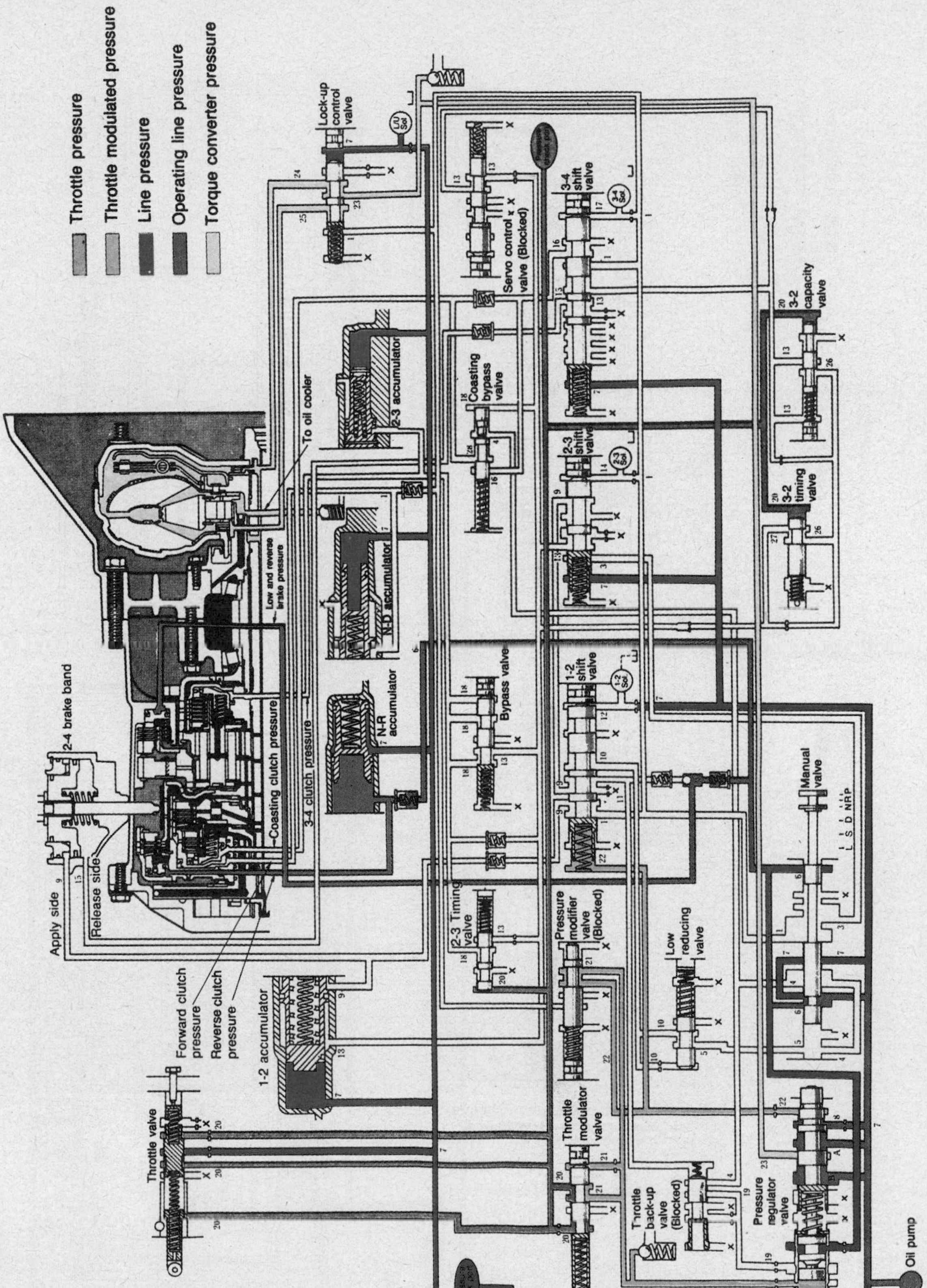
Throttle pressure
Throttle modulated pressure
Line pressure
Operating line pressure
Torque converter pressure
Lock-up control valve
L/U Sol
Servo control valve (Blocked)
3-4 shift valve
3-4 Sol
3-2 capacity valve
2-3 accumulator
Coasting bypass valve
2-3 shift valve
2-3 Sol
3-2 timing valve
To oil cooler
N-D accumulator
Low and reverse brake pressure
2-4 brake band
Coasting clutch pressure
3-4 clutch pressure
N-R accumulator
Bypass valve
1-2 shift valve
1-2 Sol
Manual valve
L S D N R P
Apply side
Release side
2-3 Timing valve
Pressure modifier valve (Blocked)
Low reducing valve
Forward clutch pressure
Reverse clutch pressure
1-2 accumulator
Throttle valve
Throttle modulator valve
Throttle back-up valve (Blocked)
Pressure regulator valve
Oil pump

NEUTRAL BELOW 11 MPH—4EAT

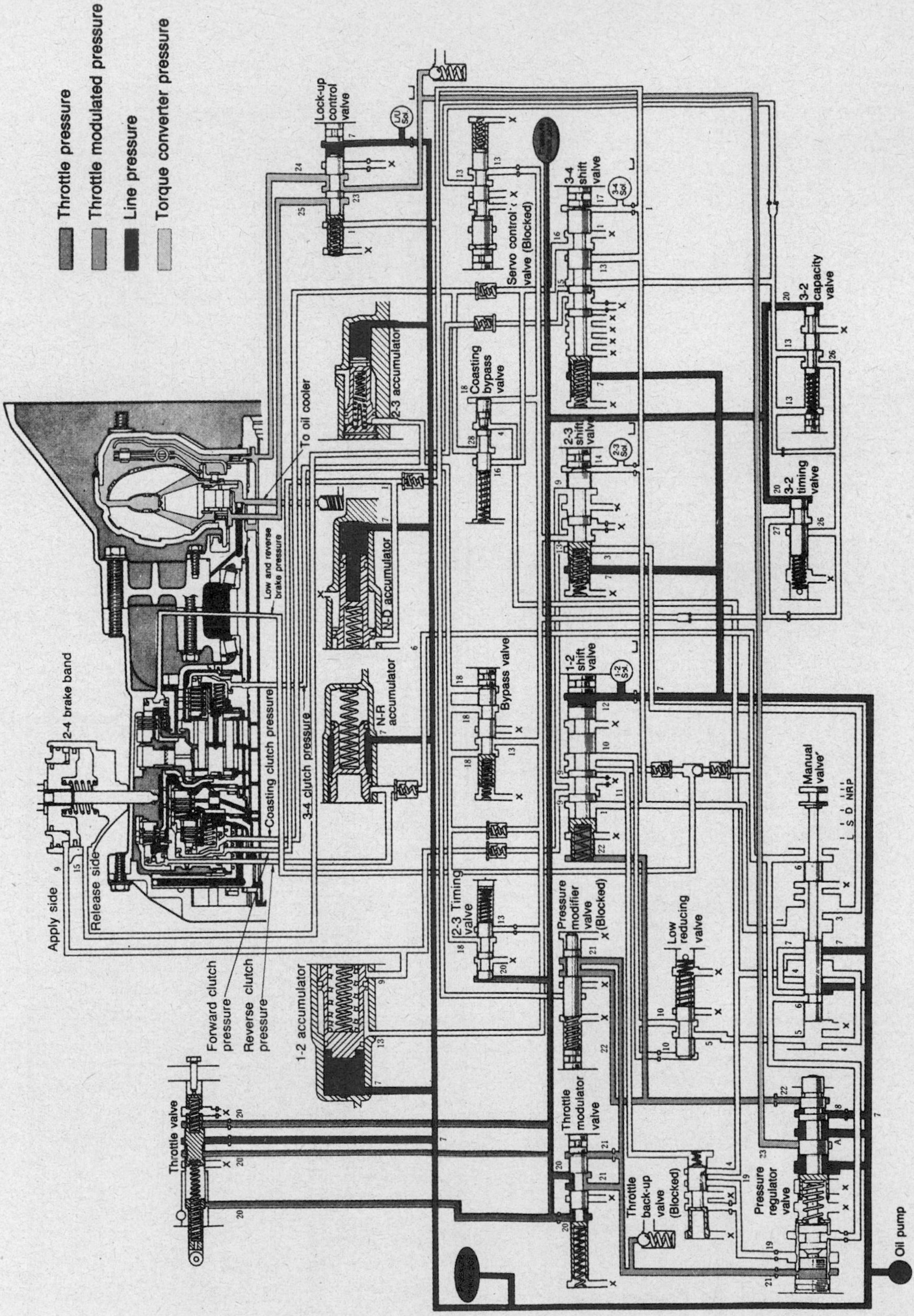
Throttle pressure
Throttle modulated pressure
Line pressure
Torque converter pressure
Lock-up control valve
L/U Sol
3-4 shift valve
3-4 Sol
Servo control valve (Blocked)
3-2 capacity valve
2-3 accumulator
Coasting bypass valve
2-3 shift valve
2-3 Sol
3-2 timing valve
To oil cooler
Low and reverse brake pressure
N-D accumulator
1-2 shift valve
1-2 Sol
Bypass valve
N-R accumulator
2-4 brake band
Coasting clutch pressure
3-4 clutch pressure
Manual valve
L S D N R P
Apply side
Release side
2-3 Timing valve
Pressure modifier valve (Blocked)
Low reducing valve
Forward clutch pressure
Reverse clutch pressure
1-2 accumulator
Throttle valve
Throttle modulator valve
Throttle back-up valve (Blocked)
Pressure regulator valve
Oil pump

NEUTRAL ABOVE 11 MPH—4EAT

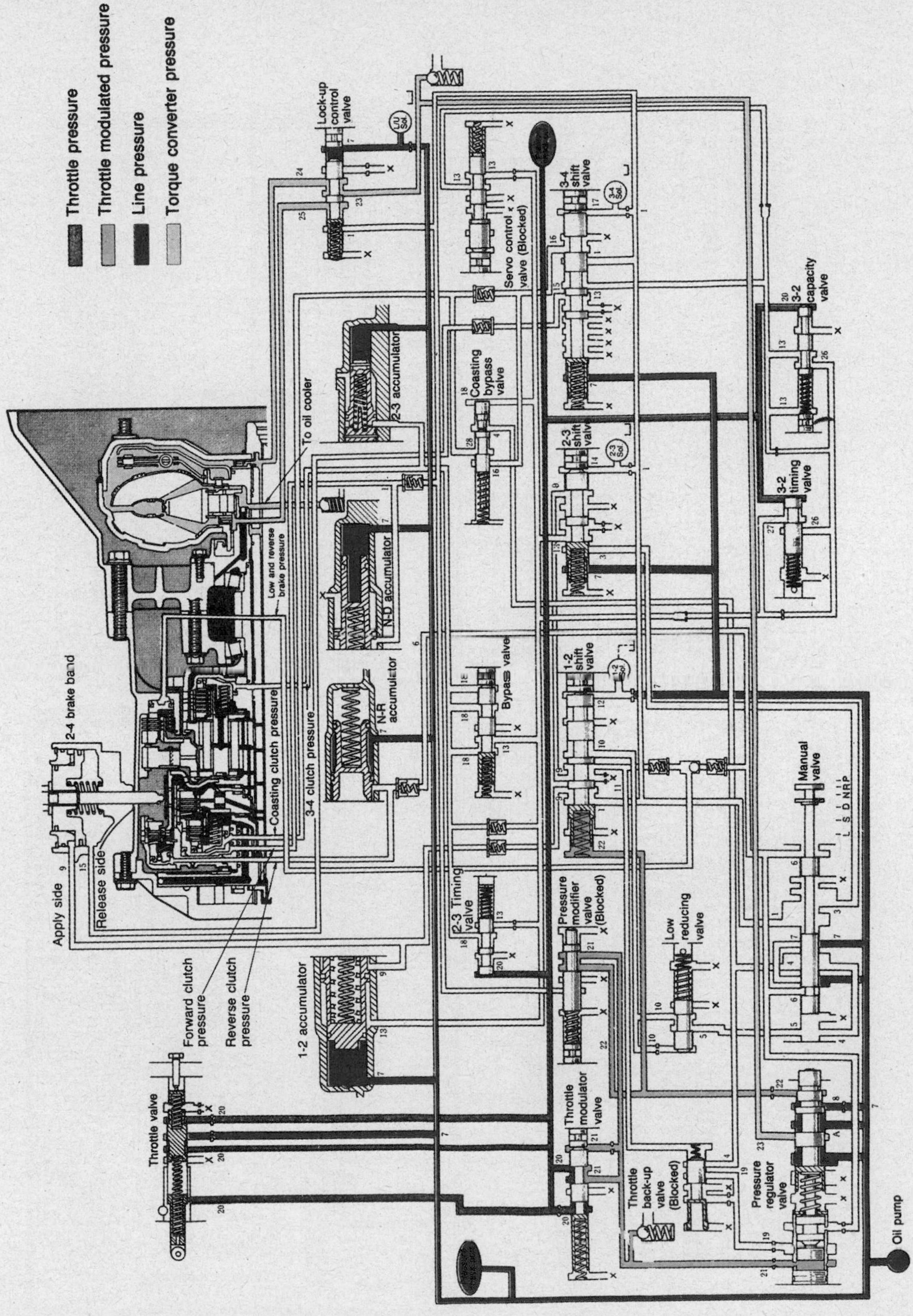
Throttle pressure
Throttle modulated pressure
Line pressure
Torque converter pressure
Lock-up control valve
L/U Sol
3-4 shift valve
3-4 Sol
Servo control valve (Blocked)
3-2 capacity valve
Coasting bypass valve
2-3 accumulator
To oil cooler
2-3 shift valve
2-3 Sol
3-2 timing valve
N-D accumulator
Low and reverse brake pressure
2-4 brake band
Coasting clutch pressure
3-4 clutch pressure
N-R accumulator
Bypass valve
1-2 shift valve
1-2 Sol
Manual valve
L S D N R P
Apply side
Release side
2-3 Timing valve
Pressure modifier valve (Blocked)
Low reducing valve
Forward clutch pressure
Reverse clutch pressure
1-2 accumulator
Throttle valve
Throttle modulator valve
Throttle back-up valve (Blocked)
Pressure regulator valve
Oil pump

DRIVE—1ST GEAR—4EAT

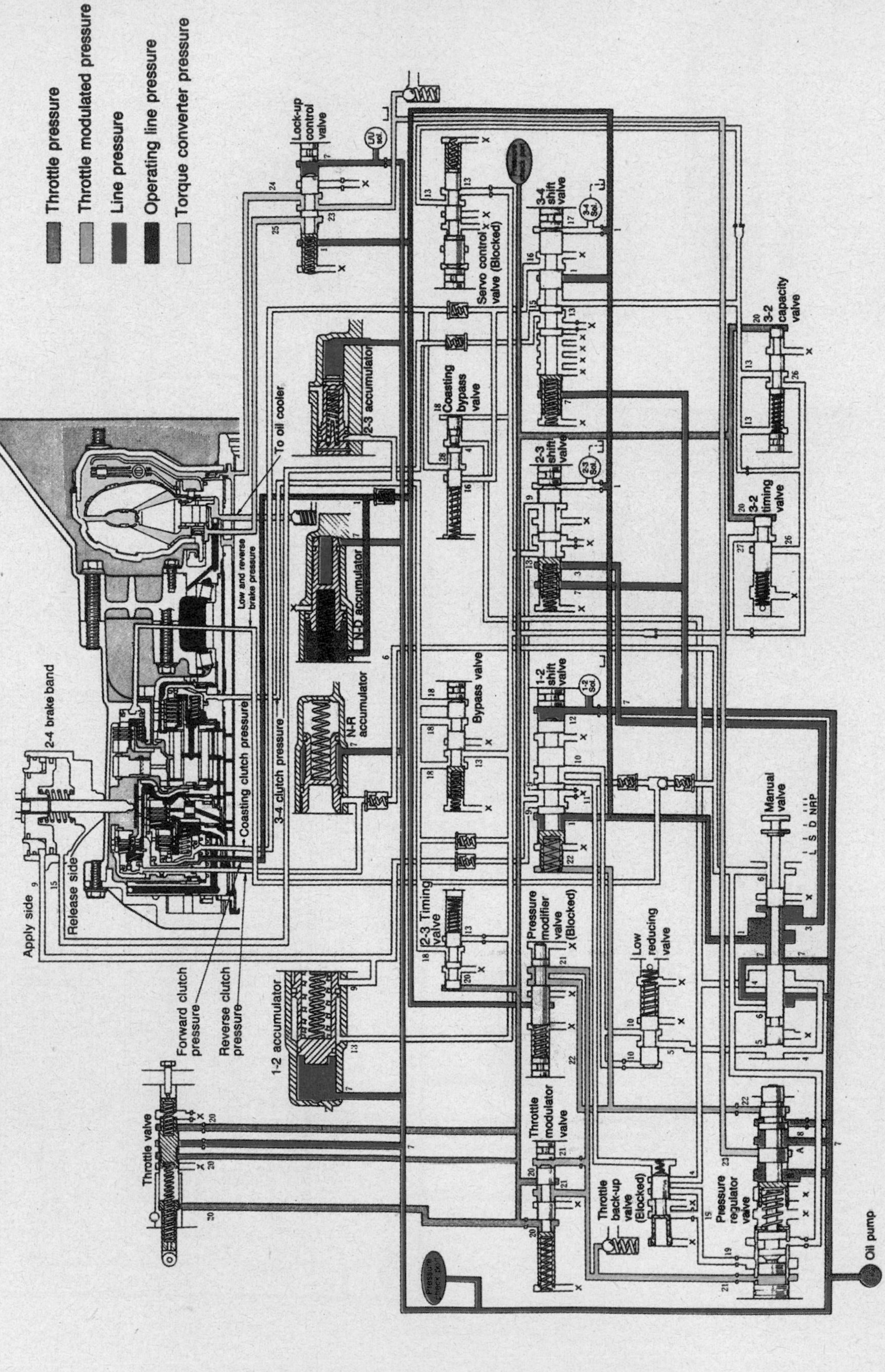
Throttle pressure
Throttle modulated pressure
Line pressure
Operating line pressure
Torque converter pressure
Lock-up control valve
Servo control valve (Blocked)
3-4 shift valve
3-2 capacity valve
Coasting bypass valve
2-3 accumulator
To oil cooler
2-3 shift valve
3-2 timing valve
N-D accumulator
Low and reverse brake pressure
N-R accumulator
Bypass valve
1-2 shift valve
Manual valve
2-4 brake band
Coasting clutch pressure
3-4 clutch pressure
Apply side
Release side
2-3 Timing valve
Pressure modifier valve (Blocked)
Low reducing valve
Forward clutch pressure
Reverse clutch pressure
1-2 accumulator
Throttle valve
Throttle modulator valve
Throttle back-up valve (Blocked)
Pressure regulator valve
Oil pump

DRIVE—2ND GEAR—4EAT

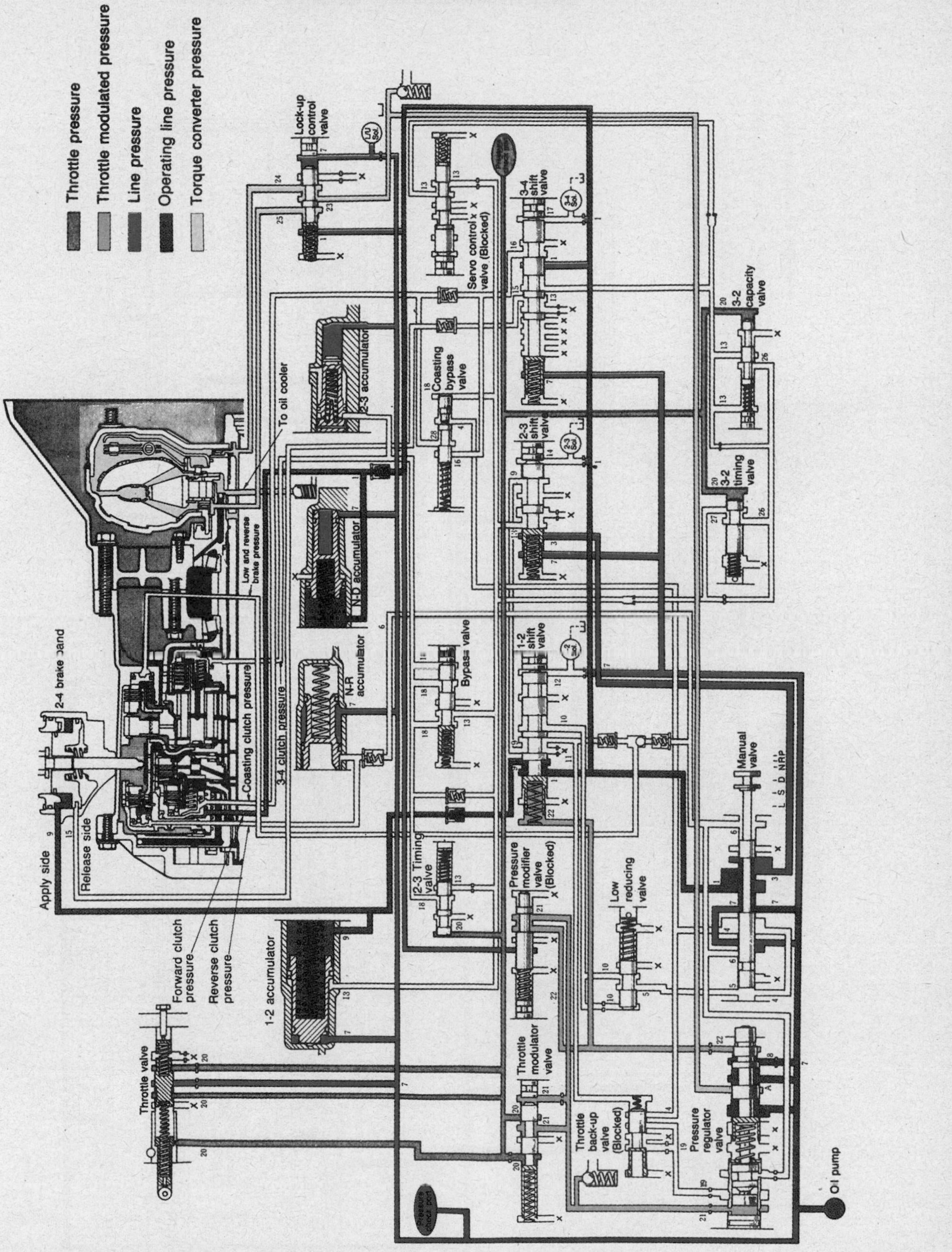
Throttle pressure
Throttle modulated pressure
Line pressure
Operating line pressure
Torque converter pressure
Lock-up control valve
Servo control valve (Blocked)
3-4 shift valve
3-2 capacity valve
To oil cooler
2-3 accumulator
Coasting bypass valve
2-3 shift valve
3-2 timing valve
Low and reverse brake pressure
N-D accumulator
N-R accumulator
Bypass valve
1-2 shift valve
2-4 brake band
Coasting clutch pressure
3-4 clutch pressure
Manual valve
Apply side
Release side
2-3 Timing valve
Pressure modifier valve (Blocked)
Low reducing valve
Forward clutch pressure
Reverse clutch pressure
1-2 accumulator
Throttle valve
Throttle modulator valve
Throttle back-up valve (Blocked)
Pressure regulator valve
Oil pump

DRIVE—3RD GEAR BELOW 25 MPH—4EAT

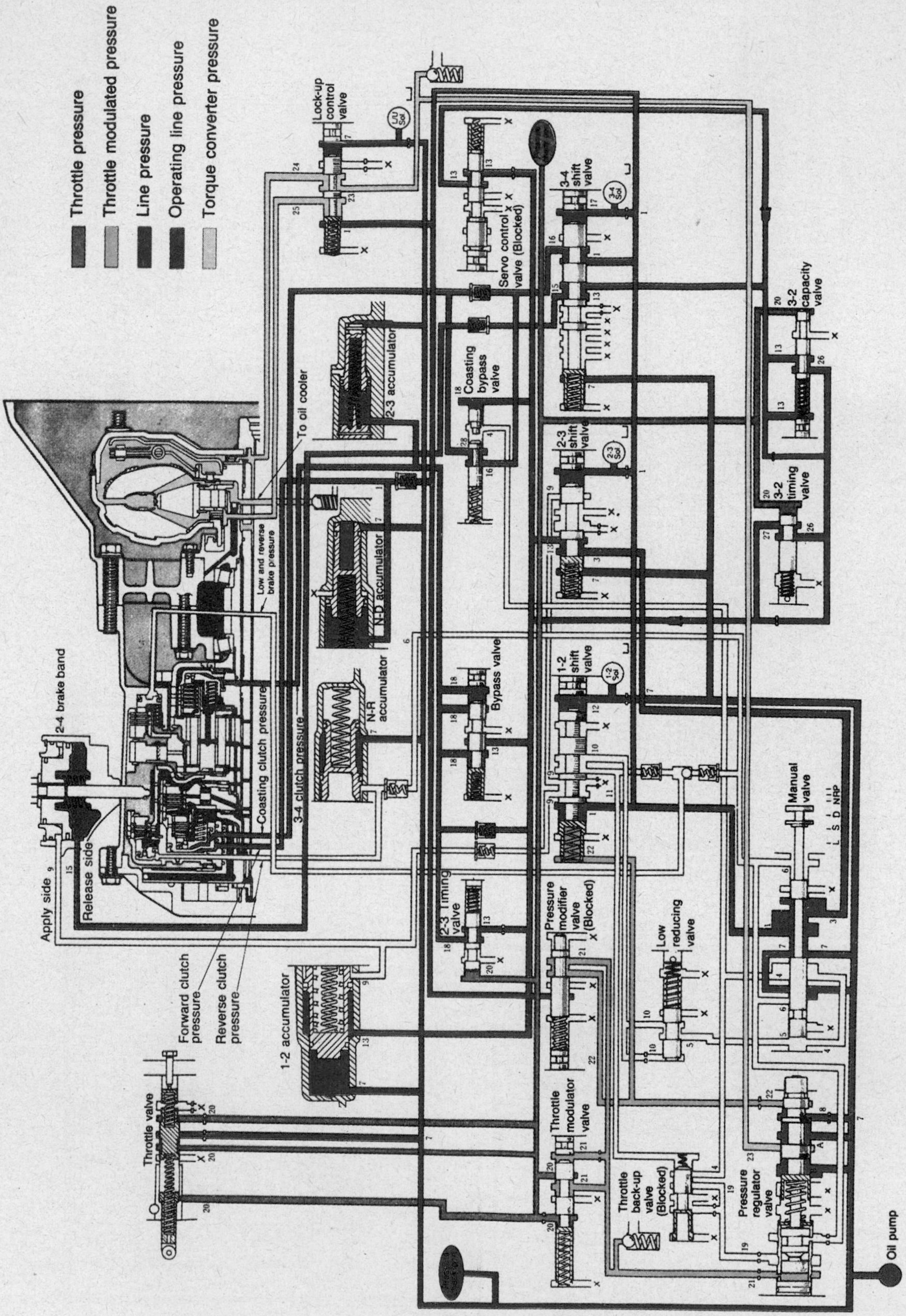
Throttle pressure
Throttle modulated pressure
Line pressure
Operating line pressure
Torque converter pressure
Lock-up control valve
Servo control valve (Blocked)
3-4 shift valve
3-2 capacity valve
3-2 timing valve
2-3 shift valve
Coasting bypass valve
2-3 accumulator
To oil cooler
N-D accumulator
Low and reverse brake pressure
N-R accumulator
Bypass valve
1-2 shift valve
2-4 brake band
Coasting clutch pressure
3-4 clutch pressure
Apply side
Release side
Forward clutch pressure
Reverse clutch pressure
1-2 accumulator
2-3 Timing valve
Pressure modifier valve (Blocked)
Low reducing valve
Manual valve
L S D NRP
Throttle valve
Throttle modulator valve
Throttle back-up valve (Blocked)
Pressure regulator valve
Oil pump

DRIVE—3RD GEAR ABOVE 25 MPH LOCKUP ON—4EAT

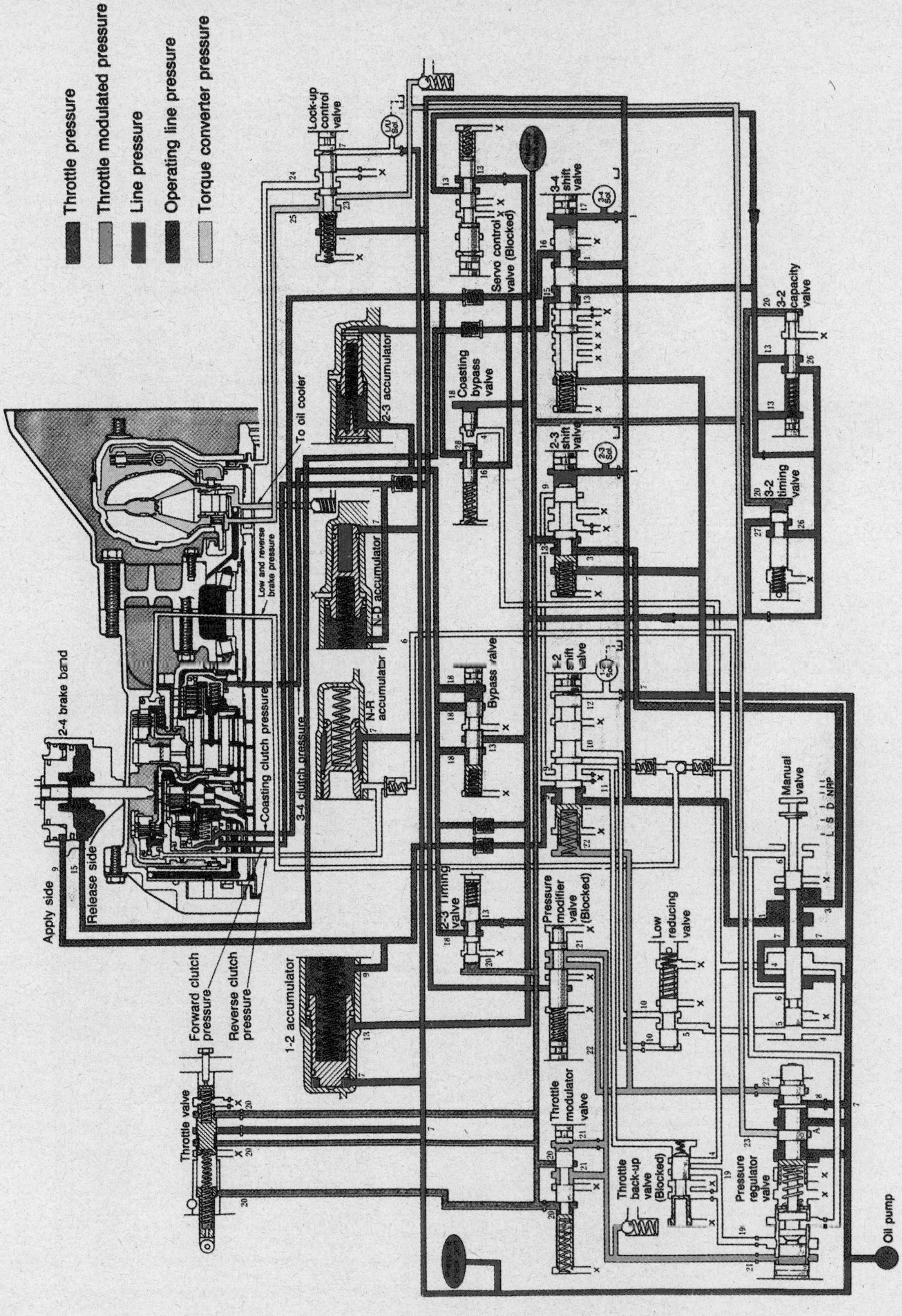
Throttle pressure
Throttle modulated pressure
Line pressure
Operating line pressure
Torque converter pressure
Lock-up control valve
Servo control valve (Blocked)
3-4 shift valve
3-2 capacity valve
Coasting bypass valve
2-3 accumulator
To oil cooler
2-3 shift valve
3-2 timing valve
N-D accumulator
Low and reverse brake pressure
Bypass valve
1-2 shift valve
N-R accumulator
2-4 brake band
Coasting clutch pressure
3-4 clutch pressure
Manual valve
Apply side
Release side
2-3 Timing valve
Pressure modifier valve (Blocked)
Low reducing valve
Forward clutch pressure
Reverse clutch pressure
1-2 accumulator
Throttle valve
Throttle modulator valve
Throttle back-up valve (Blocked)
Pressure regulator valve
Oil pump

DRIVE—OVERDRIVE LOCKUP ON—4EAT

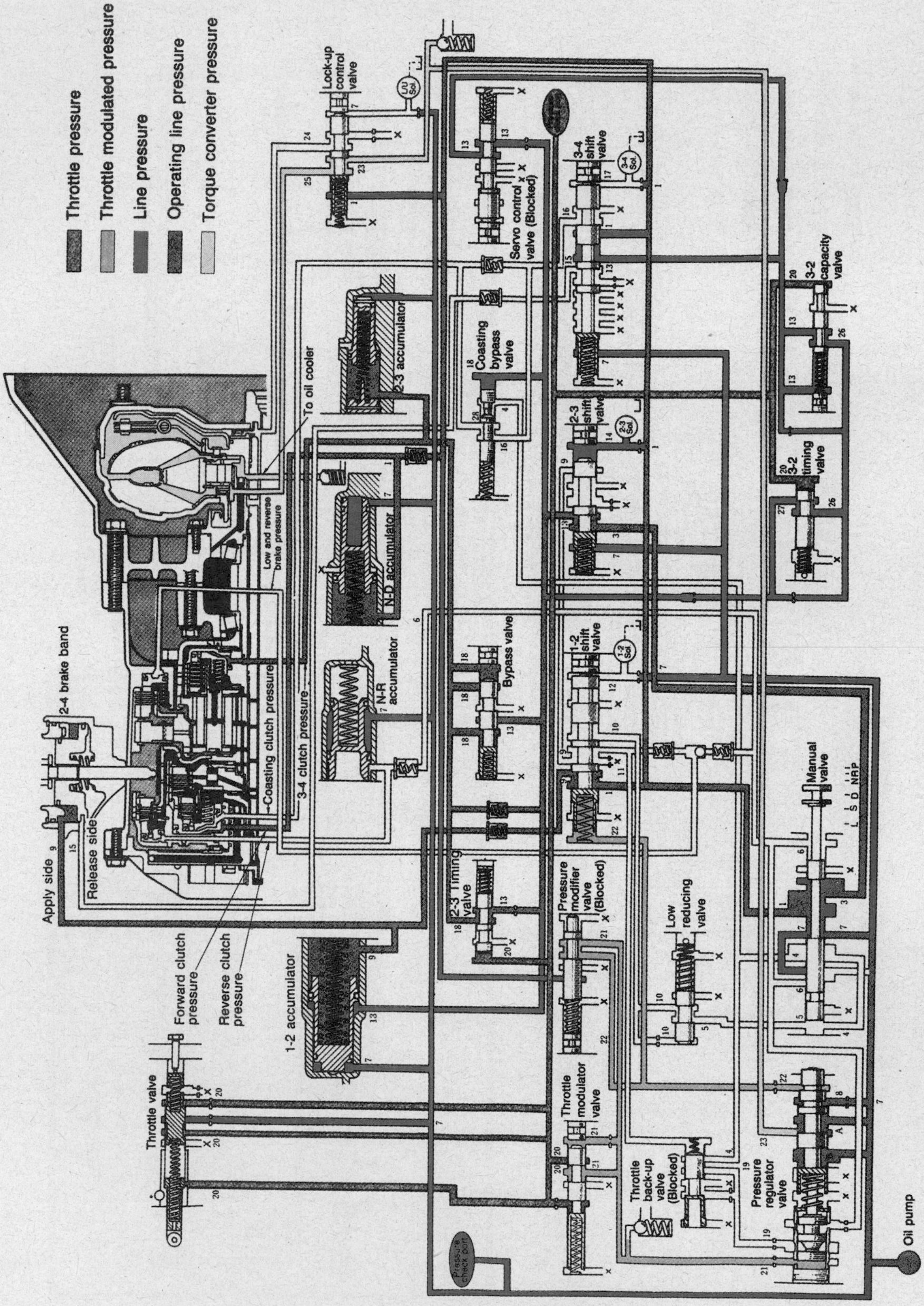
Throttle pressure
Throttle modulated pressure
Line pressure
Operating line pressure
Torque converter pressure
Lock-up control valve
L/U Sol
Servo control valve (Blocked)
3-4 shift valve
3-4 Sol
3-2 capacity valve
2-3 accumulator
Coasting bypass valve
To oil cooler
2-3 shift valve
2-3 Sol
3-2 timing valve
Low and reverse brake pressure
N-D accumulator
N-R accumulator
1-2 shift valve
1-2 Sol
Bypass valve
2-4 brake band
Coasting clutch pressure
3-4 clutch pressure
Manual valve
L S D N R P
Apply side
Release side
2-3 timing valve
Pressure modifier valve (Blocked)
Low reducing valve
Forward clutch pressure
Reverse clutch pressure
1-2 accumulator
Throttle valve
Throttle modulator valve
Throttle back-up valve (Blocked)
Pressure regulator valve
Oil pump

S—1ST GEAR—4EAT

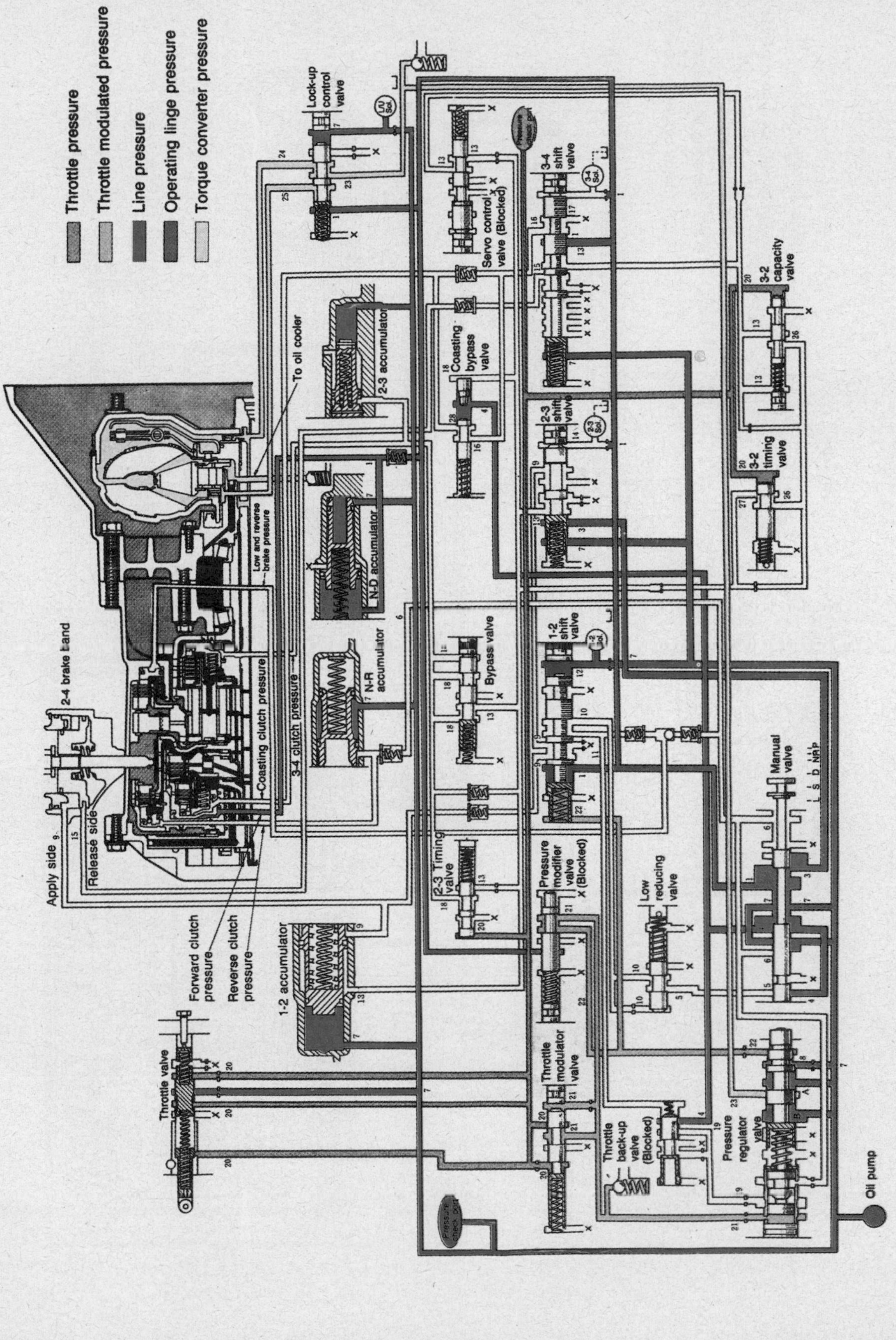
Throttle pressure
Throttle modulated pressure
Line pressure
Operating linge pressure
Torque converter pressure
Lock-up control valve
Servo control valve (Blocked)
3-4 shift valve
3-2 capacity valve
To oil cooler
2-3 accumulator
Coasting bypass valve
2-3 shift valve
3-2 timing valve
Low and reverse brake pressure
N-D accumulator
N-R accumulator
1-2 shift valve
Bypass valve
2-4 brake band
Coasting clutch pressure
3-4 clutch pressure
Apply side
Release side
2-3 Timing valve
Pressure modifier valve (Blocked)
Low reducing valve
Manual valve
Forward clutch pressure
Reverse clutch pressure
1-2 accumulator
Throttle valve
Throttle modulator valve
Throttle back-up valve (Blocked)
Pressure regulator valve
Oil pump

S—2ND GEAR—4EAT

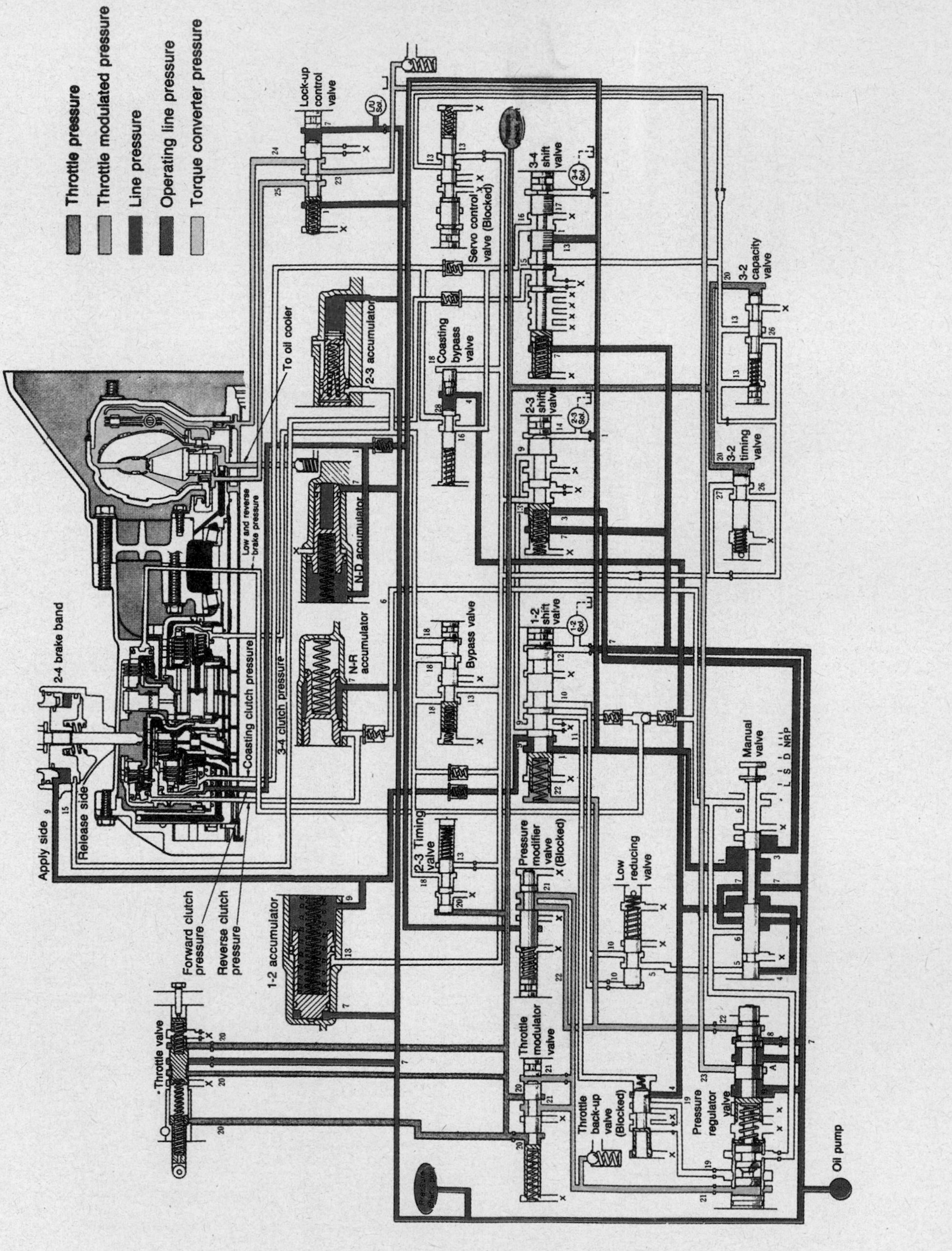
Throttle pressure
Throttle modulated pressure
Line pressure
Operating line pressure
Torque converter pressure
Lock-up control valve
Servo control valve (Blocked)
3-4 shift valve
3-2 capacity valve
To oil cooler
2-3 accumulator
Coasting bypass valve
2-3 shift valve
3-2 timing valve
Low and reverse brake pressure
N-D accumulator
N-R accumulator
Bypass valve
1-2 shift valve
2-4 brake band
Coasting clutch pressure
3-4 clutch pressure
Apply side
Release side
Manual valve
L S D NRP
2-3 Timing valve
Pressure modifier valve X (Blocked)
Low reducing valve
Forward clutch pressure
Reverse clutch pressure
1-2 accumulator
Throttle valve
Throttle modulator valve
Throttle back-up valve (Blocked)
Pressure regulator valve
Oil pump

S—2ND GEAR HOLD—4EAT

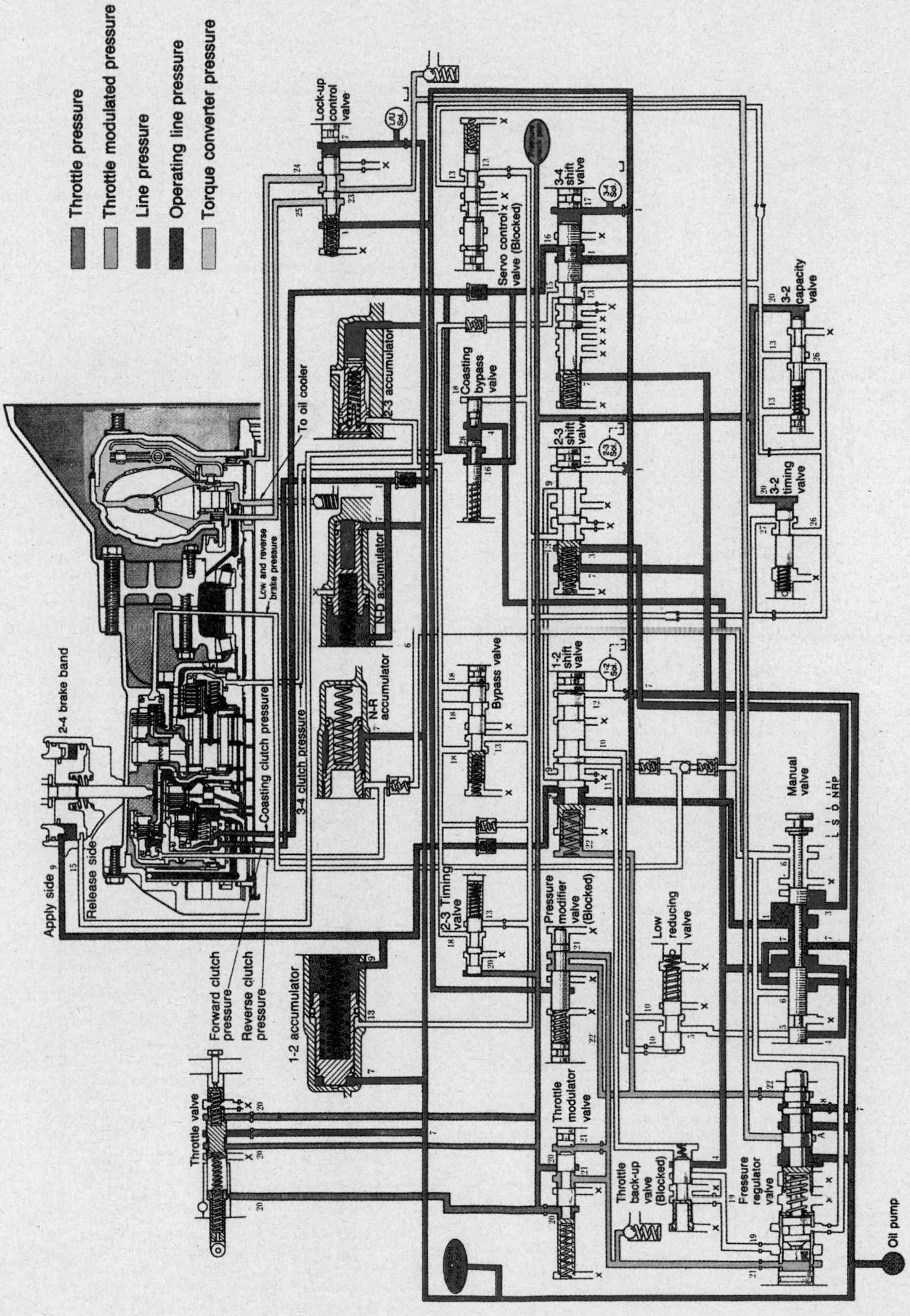
Throttle pressure
Throttle modulated pressure
Line pressure
Operating line pressure
Torque converter pressure
Lock-up control valve
L/U Sol.
Servo control valve (Blocked)
3-4 shift valve
3-4 Sol.
3-2 capacity valve
Coasting bypass valve
2-3 accumulator
To oil cooler
2-3 shift valve
2-3 Sol.
3-2 timing valve
N-D accumulator
Low and reverse brake pressure
1-2 shift valve
1-2 Sol.
Bypass valve
N-R accumulator
2-4 brake band
Coasting clutch pressure
3-4 clutch pressure
Manual valve
L S D N R P
Apply side
Release side
2-3 Timing valve
Pressure modifier valve (Blocked)
Low reducing valve
Forward clutch pressure
Reverse clutch pressure
1-2 accumulator
Throttle valve
Throttle modulator valve
Throttle back-up valve (Blocked)
Pressure regulator valve
Oil pump

S—3RD GEAR BELOW 25 MPH—4EAT

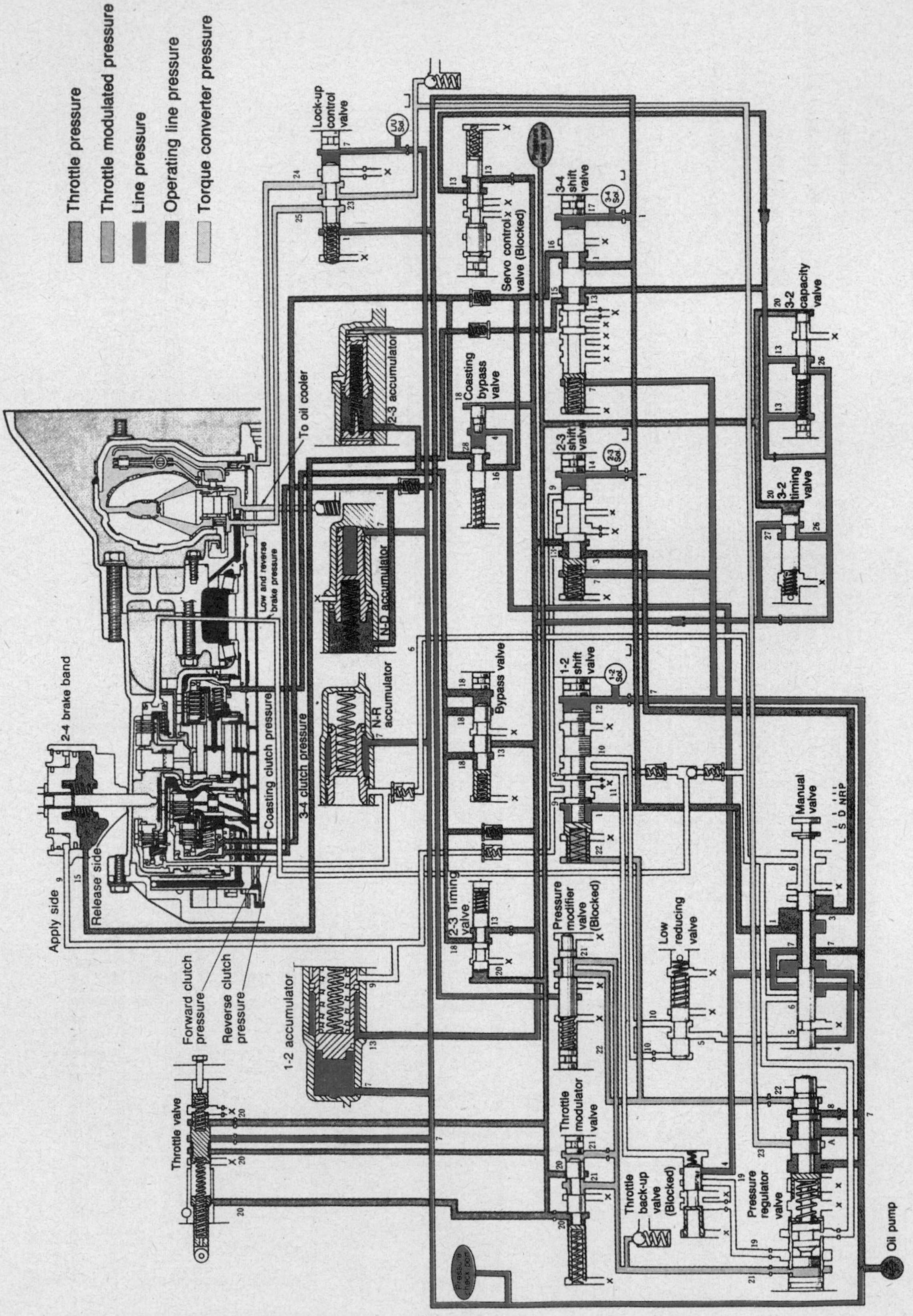
Throttle pressure
Throttle modulated pressure
Line pressure
Operating line pressure
Torque converter pressure
Lock-up control valve
Servo control valve (Blocked)
3-4 shift valve
3-2 capacity valve
2-3 accumulator
To oil cooler
Coasting bypass valve
2-3 shift valve
3-2 timing valve
Low and reverse brake pressure
N-D accumulator
N-R accumulator
Bypass valve
1-2 shift valve
2-4 brake band
Coasting clutch pressure
3-4 clutch pressure
Manual valve
L S D N R P
Apply side
Release side
2-3 Timing valve
Pressure modifier valve (Blocked)
Low reducing valve
Forward clutch pressure
Reverse clutch pressure
1-2 accumulator
Throttle valve
Throttle modulator valve
Throttle back-up valve (Blocked)
Pressure regulator valve
Oil pump

S—3RD GEAR ABOVE 25 MPH—4EAT

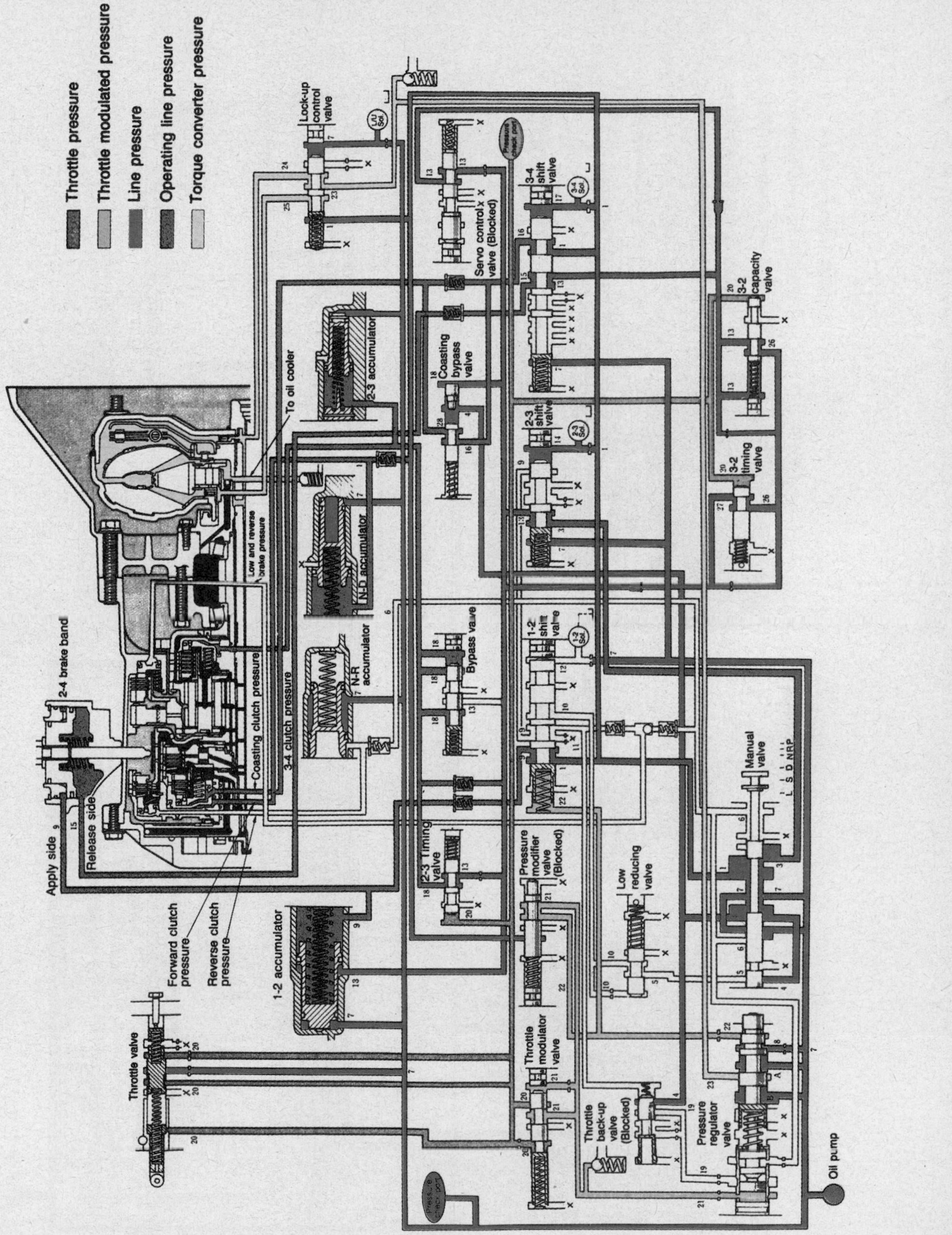
Throttle pressure
Throttle modulated pressure
Line pressure
Operating line pressure
Torque converter pressure
Lock-up control valve
L/U Sol.
Servo control valve (Blocked)
Pressure check port
3-4 shift valve
3-4 Sol.
3-2 capacity valve
Coasting bypass valve
2-3 accumulator
To oil cooler
2-3 shift valve
2-3 Sol.
3-2 timing valve
N-D accumulator
Low and reverse brake pressure
2-4 brake band
N-R accumulator
Bypass valve
1-2 shift valve
1-2 Sol.
Coasting clutch pressure
3-4 clutch pressure
Manual valve
L S D N R P
Apply side
Release side
2-3 Timing valve
Pressure modifier valve (Blocked)
Low reducing valve
Forward clutch pressure
Reverse clutch pressure
1-2 accumulator
Throttle valve
Throttle modulator valve
Throttle back-up valve (Blocked)
Pressure regulator valve
Oil pump

L—1ST GEAR—4EAT

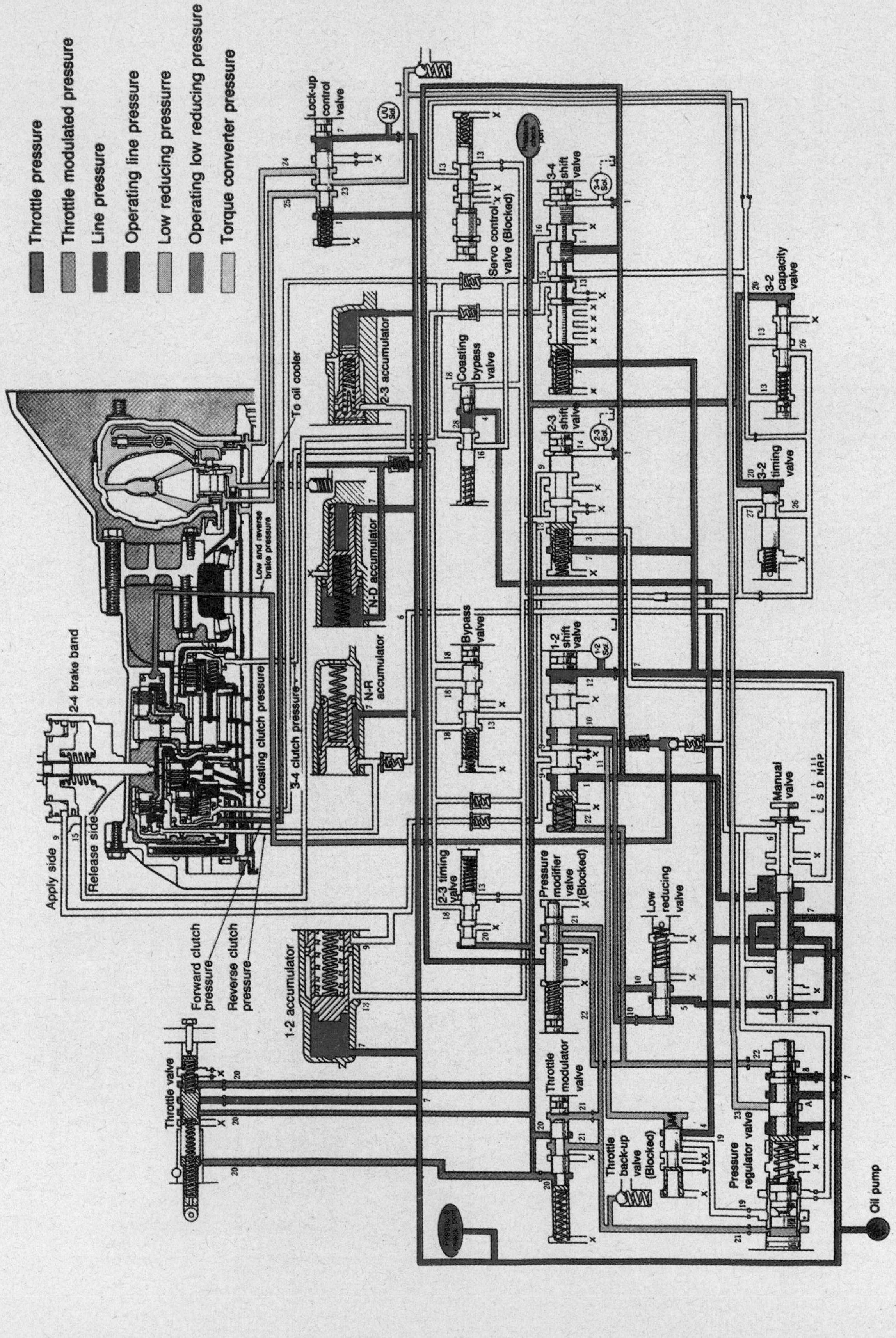
Throttle pressure
Throttle modulated pressure
Line pressure
Operating line pressure
Low reducing pressurre
Operating low reducing pressure
Torque converter pressure
Lock-up control valve
Servo control valve (Blocked)
3-4 shift valve
3-2 capacity valve
2-3 accumulator
Coasting bypass valve
2-3 shift valve
3-2 timing valve
To oil cooler
N-D accumulator
Low and reverse brake pressure
Bypass valve
1-2 shift valve
N-R accumulator
2-4 brake band
Coasting clutch pressure
3-4 clutch pressure
Manual valve
L S D N R P
Apply side
Release side
2-3 timing valve
Pressure modifier valve (Blocked)
Low reducing valve
Forward clutch pressure
Reverse clutch pressure
1-2 accumulator
Throttle valve
Throttle modulator valve
Throttle back-up valve (Blocked)
Pressure regulator valve
Oil pump

L—1ST GEAR HOLD—4EAT

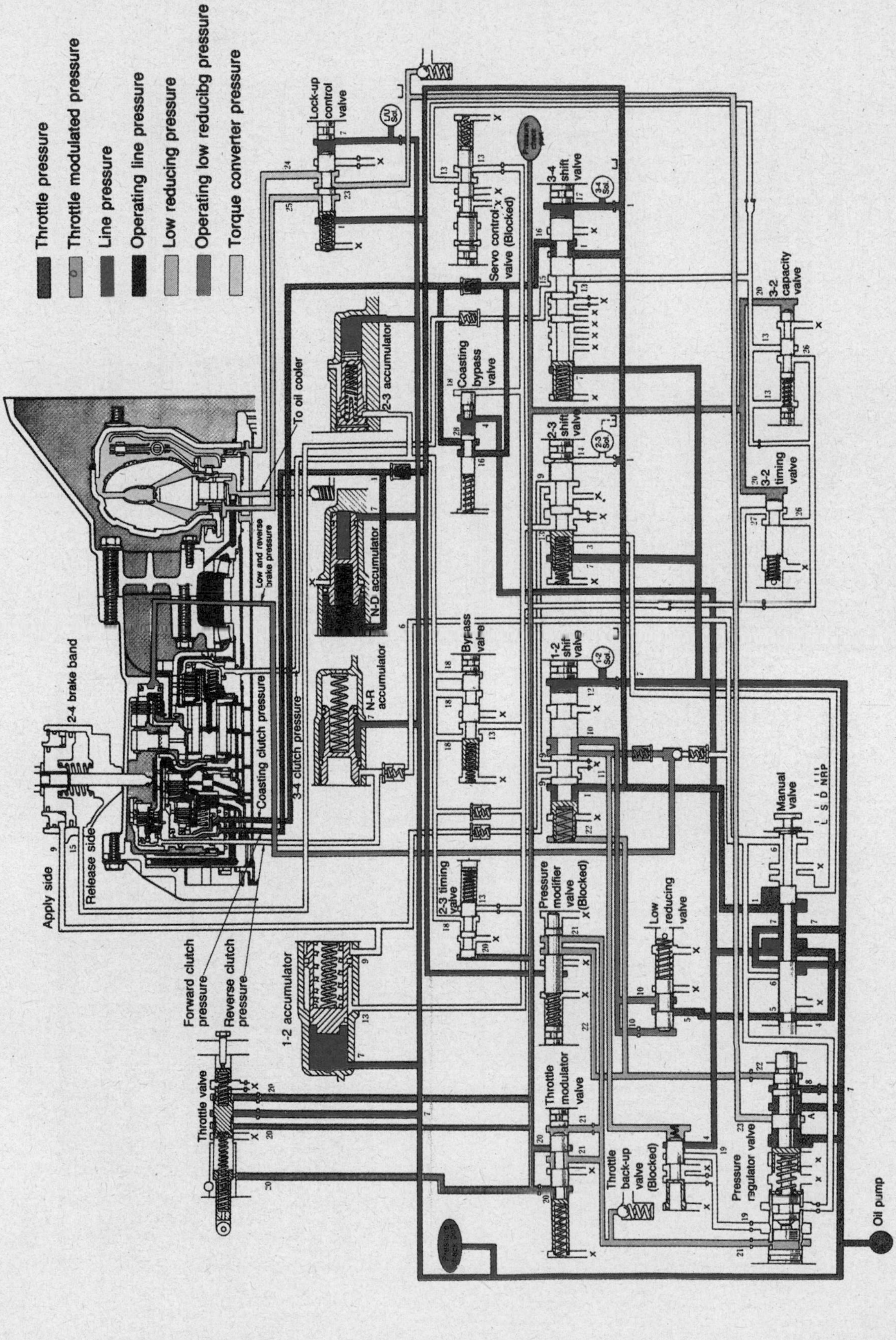
Throttle pressure
Throttle modulated pressure
Line pressure
Operating line pressure
Low reducing pressure
Operating low reducibg pressure
Torque converter pressure
Lock-up control valve
Servo control valve (Blocked)
3-4 shift valve
3-2 capacity valve
2-3 accumulator
Coasting bypass valve
2-3 shift valve
3-2 timing valve
To oil cooler
N-D accumulator
Low and reverse brake pressure
Bypass valve
1-2 shift valve
2-4 brake band
N-R accumulator
Coasting clutch pressure
3-4 clutch pressure
Apply side
Release side
Manual valve
L S D NRP
2-3 timing valve
Pressure modifier valve (Blocked)
Low reducing valve
Forward clutch pressure
Reverse clutch pressure
1-2 accumulator
Throttle valve
Throttle modulator valve
Throttle back-up valve (Blocked)
Pressure regulator valve
Oil pump

L—2ND GEAR BELOW 68 MPH—4EAT

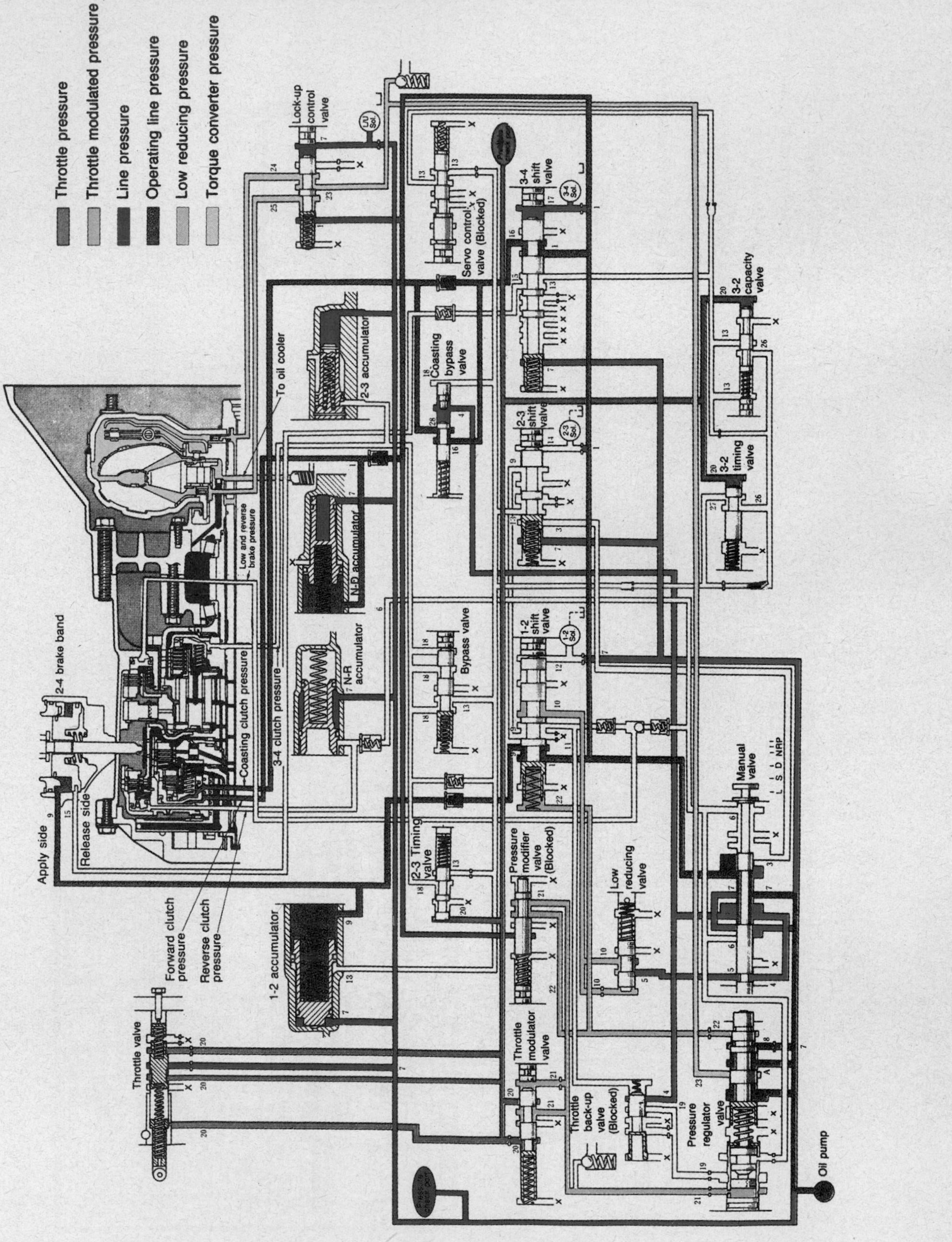
Throttle pressure
Throttle modulated pressure
Line pressure
Operating line pressure
Low reducing pressure
Torque converter pressure
Lock-up control valve
L/U Sol.
Servo control valve (Blocked)
3-4 shift valve
3-4 Sol.
3-2 capacity valve
3-2 timing valve
2-3 shift valve
2-3 Sol.
1-2 shift valve
1-2 Sol.
Coasting bypass valve
Bypass valve
2-3 accumulator
N-D accumulator
N-R accumulator
To oil cooler
Low and reverse brake pressure
2-4 brake band
Coasting clutch pressure
3-4 clutch pressure
Apply side
Release side
Forward clutch pressure
Reverse clutch pressure
1-2 accumulator
2-3 Timing valve
Pressure modifier valve (Blocked)
Low reducing valve
Manual valve
L S D NRP
Throttle valve
Throttle modulator valve
Throttle back-up valve (Blocked)
Pressure regulator valve
Oil pump

L—2ND GEAR ABOVE 68 MPH—4EAT

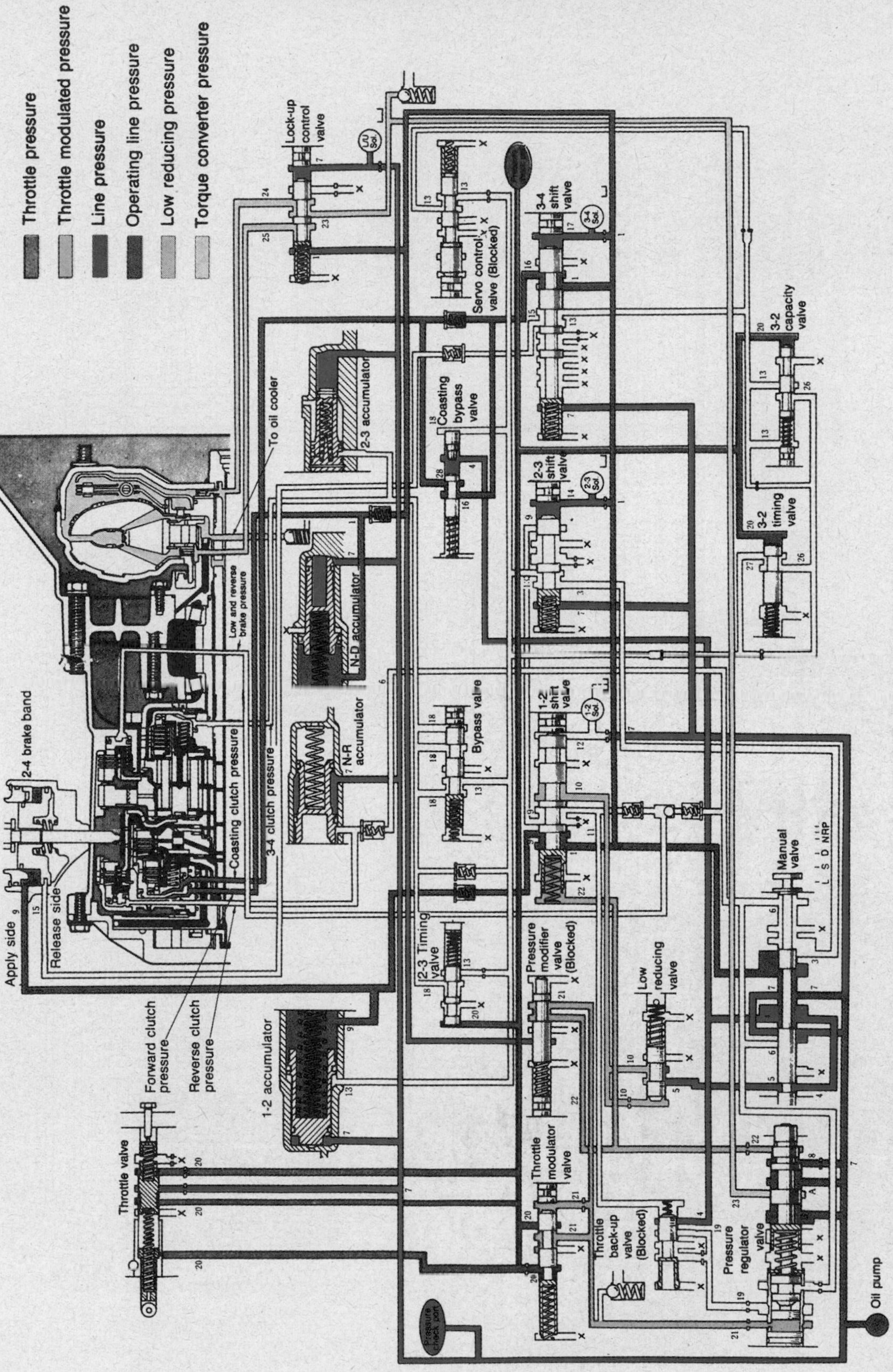
Throttle pressure
Throttle modulated pressure
Line pressure
Operating line pressure
Low reducing pressure
Torque converter pressure
Lock-up control valve
L/U Sol
Servo control valve (Blocked)
3-4 shift valve
3-4 Sol
3-2 capacity valve
To oil cooler
2-3 accumulator
Coasting bypass valve
2-3 shift valve
2-3 Sol
3-2 timing valve
Low and reverse brake pressure
N-D accumulator
2-4 brake band
Coasting clutch pressure
3-4 clutch pressure
N-R accumulator
Bypass valve
1-2 shift valve
1-2 Sol
Manual valve
L S D N R P
Apply side
Release side
2-3 Timing valve
Pressure modifier valve (Blocked)
Low reducing valve
Forward clutch pressure
Reverse clutch pressure
1-2 accumulator
Throttle valve
Throttle modulator valve
Throttle back-up valve (Blocked)
Pressure regulator valve
Oil pump
Pressure check port

NEUTRAL—F3A

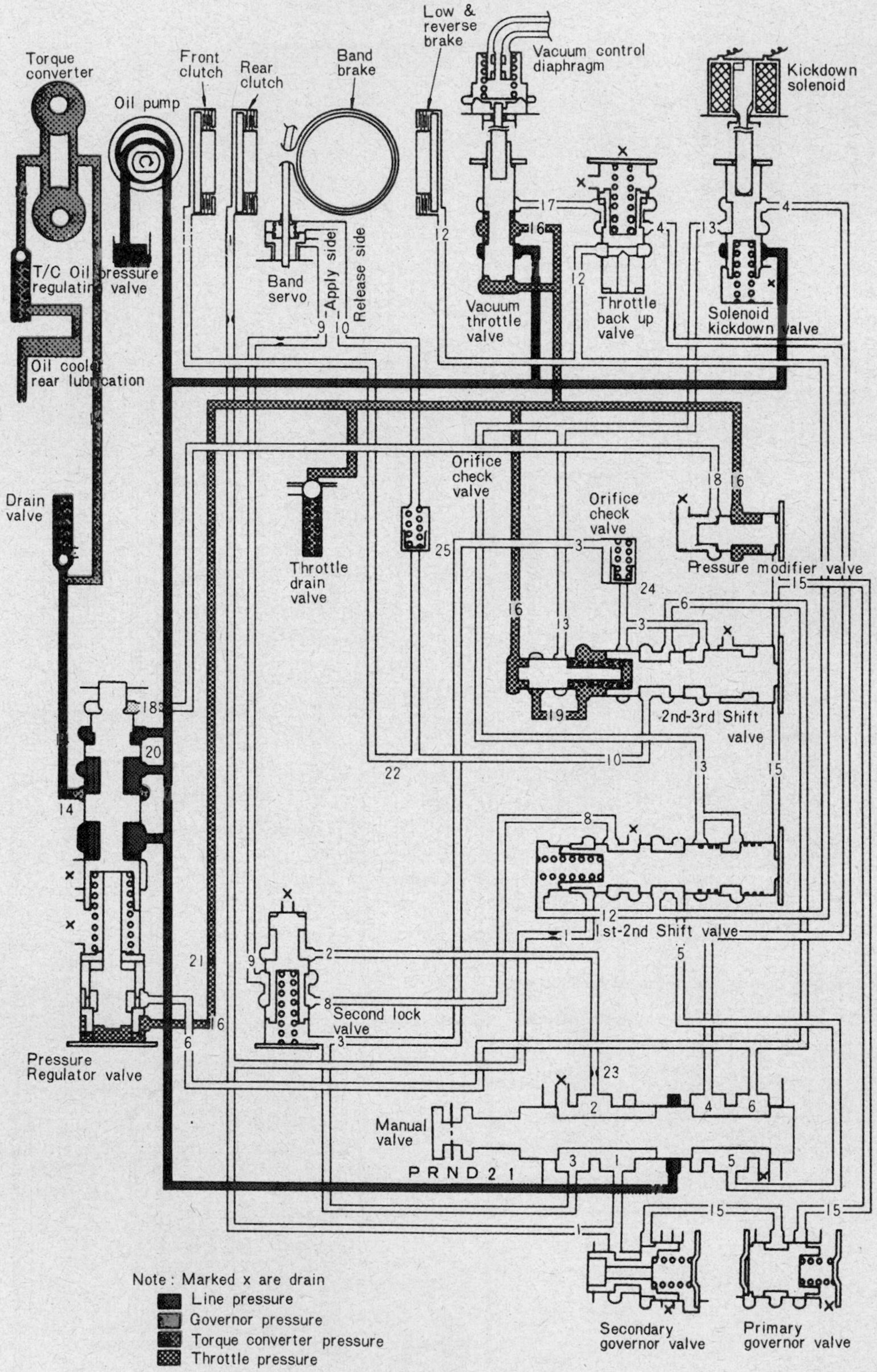
Torque converter
Oil pump
Front clutch
Rear clutch
Band brake
Low & reverse brake
Vacuum control diaphragm
Kickdown solenoid
T/C Oil pressure regulating valve
Band servo
Apply side
Release side
Vacuum throttle valve
Throttle back up valve
Solenoid kickdown valve
Oil cooler rear lubrication
Drain valve
Throttle drain valve
Orifice check valve
Orifice check valve
Pressure modifier valve
2nd-3rd Shift valve
1st-2nd Shift valve
Second lock valve
Pressure Regulator valve
Manual valve
P R N D 2 1
Secondary governor valve
Primary governor valve
Note: Marked x are drain
Line pressure
Governor pressure
Torque converter pressure
Throttle pressure

D1—1ST GEAR—F3A

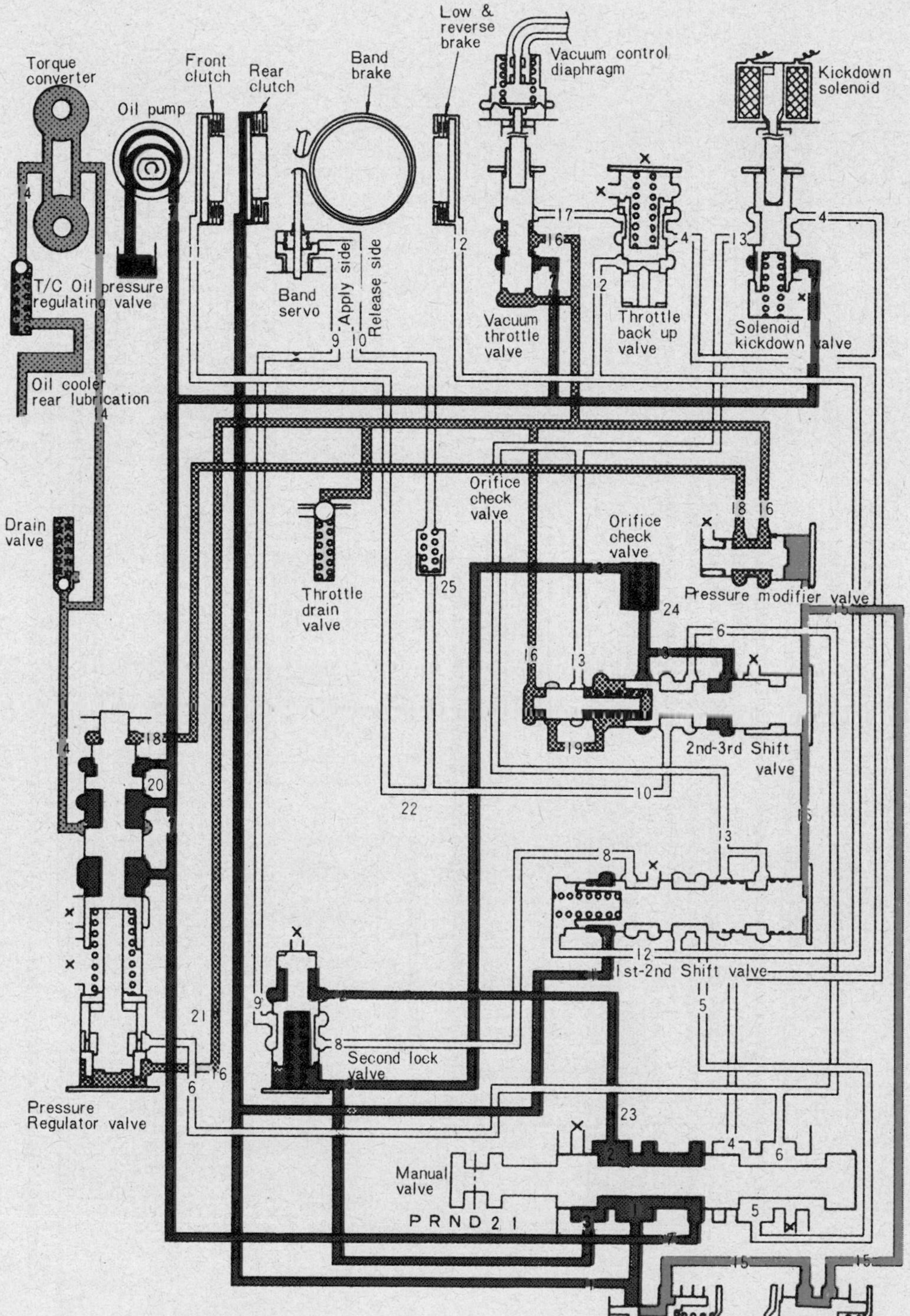
Torque converter
Oil pump
Front clutch
Rear clutch
Band brake
Low & reverse brake
Vacuum control diaphragm
Kickdown solenoid
T/C Oil pressure regulating valve
Band servo
Apply side
Release side
Vacuum throttle valve
Throttle back up valve
Solenoid kickdown valve
Oil cooler rear lubrication
Drain valve
Throttle drain valve
Orifice check valve
Orifice check valve
Pressure modifier valve
2nd-3rd Shift valve
1st-2nd Shift valve
Second lock valve
Pressure Regulator valve
Manual valve
P R N D 2 1
Secondary governor valve
Primary governor valve

Note: Marked x are drain
- Line pressure
- Governor pressure
- Torque converter pressure
- Throttle pressure

D2—2ND GEAR—F3A

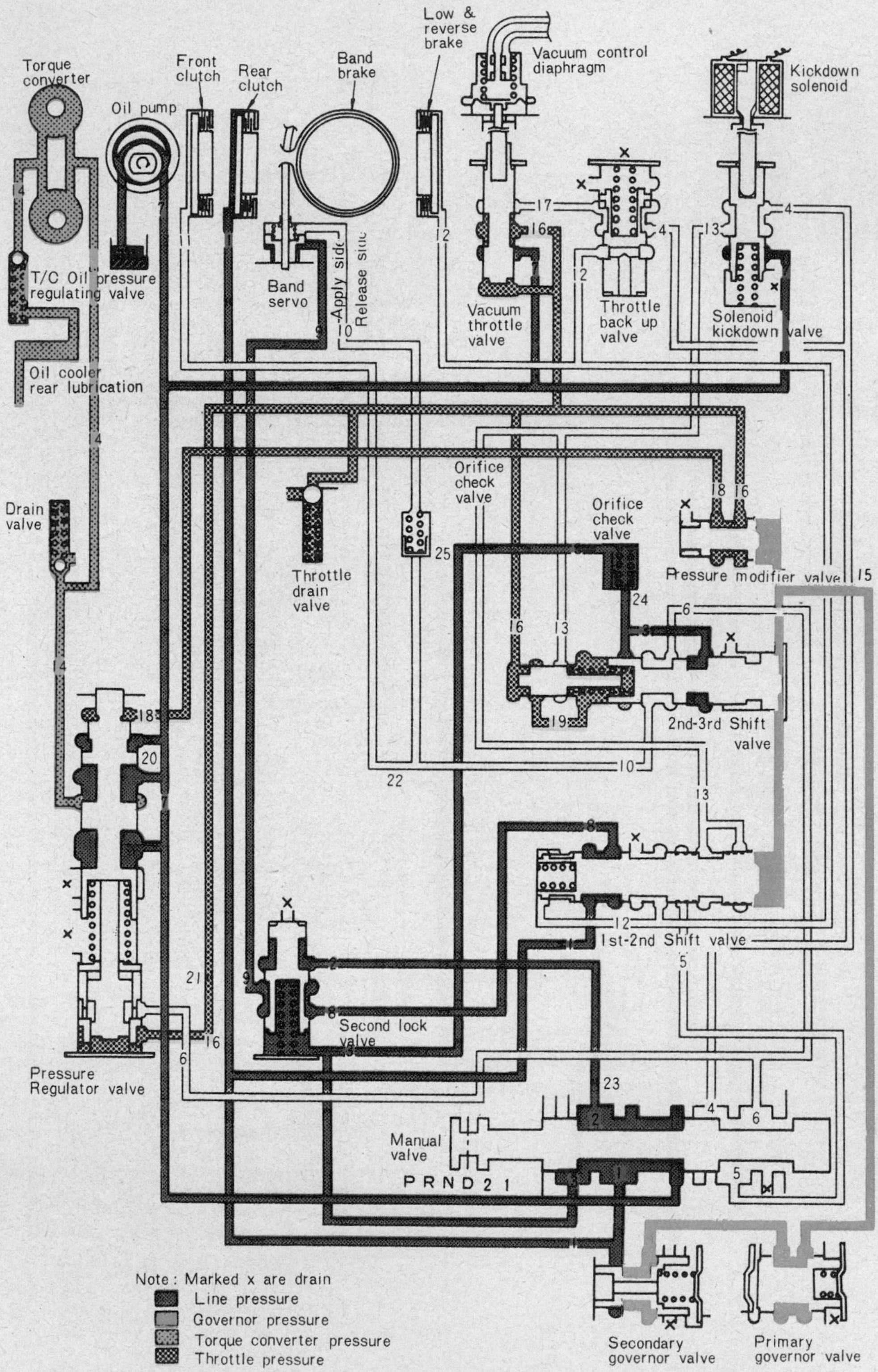
Torque converter
Oil pump
Front clutch
Rear clutch
Band brake
Low & reverse brake
Vacuum control diaphragm
Kickdown solenoid
T/C Oil pressure regulating valve
Band servo
Apply side
Release side
Vacuum throttle valve
Throttle back up valve
Solenoid kickdown valve
Oil cooler rear lubrication
Orifice check valve
Orifice check valve
Drain valve
Throttle drain valve
Pressure modifier valve
2nd-3rd Shift valve
1st-2nd Shift valve
Second lock valve
Pressure Regulator valve
Manual valve
P R N D 2 1
Secondary governor valve
Primary governor valve
Note: Marked x are drain
Line pressure
Governor pressure
Torque converter pressure
Throttle pressure

D3—3RD GEAR—F3A

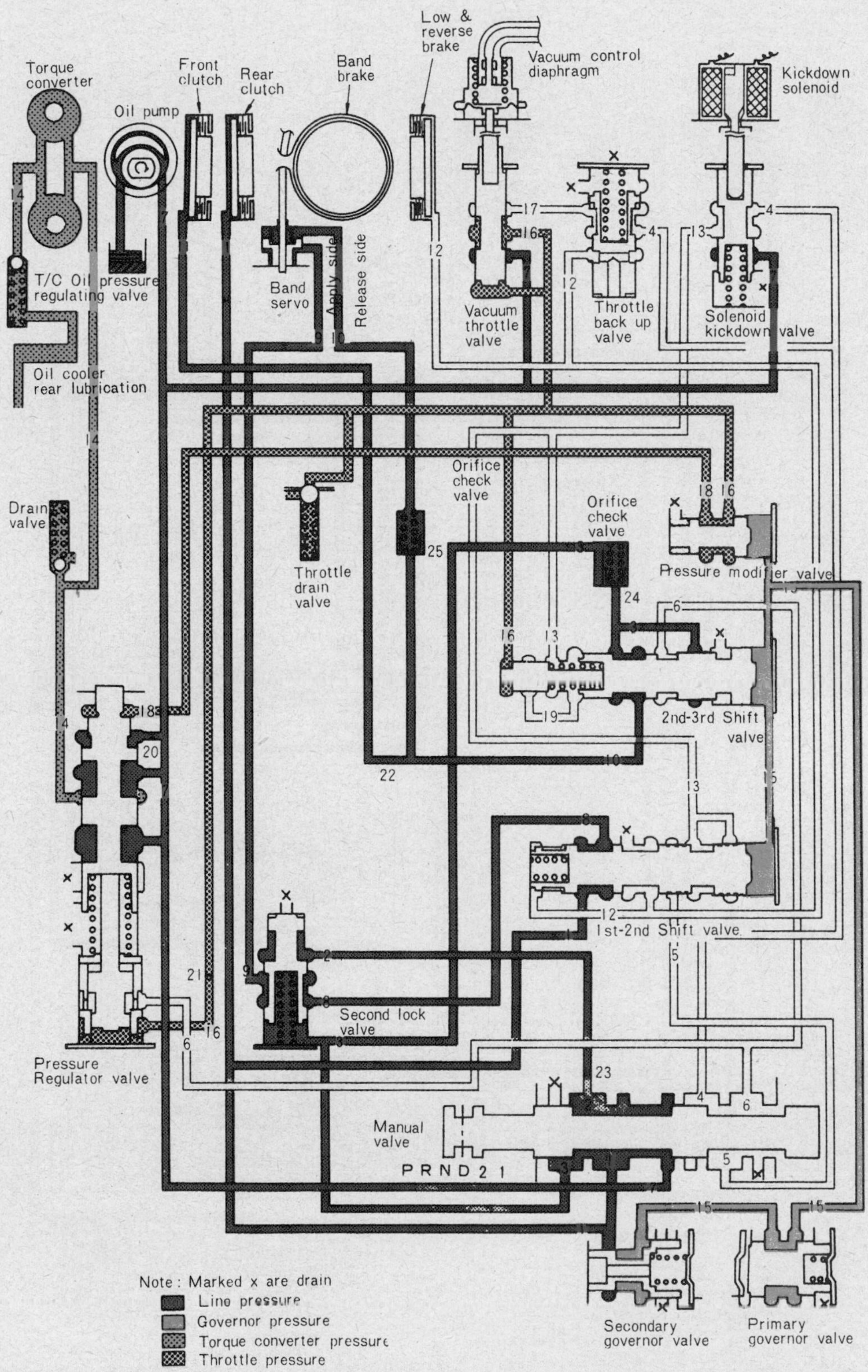
Low & reverse brake
Vacuum control diaphragm
Kickdown solenoid
Torque converter
Front clutch
Rear clutch
Band brake
Oil pump
Apply side
Release side
Band servo
T/C Oil pressure regulating valve
Vacuum throttle valve
Throttle back up valve
Solenoid kickdown valve
Oil cooler rear lubrication
Orifice check valve
Orifice check valve
Drain valve
Throttle drain valve
Pressure modifier valve
2nd-3rd Shift valve
1st-2nd Shift valve
Second lock valve
Pressure Regulator valve
Manual valve
P R N D 2 1
Secondary governor valve
Primary governor valve
Note: Marked x are drain
Line pressure
Governor pressure
Torque converter pressure
Throttle pressure

2–2ND GEAR—F3A

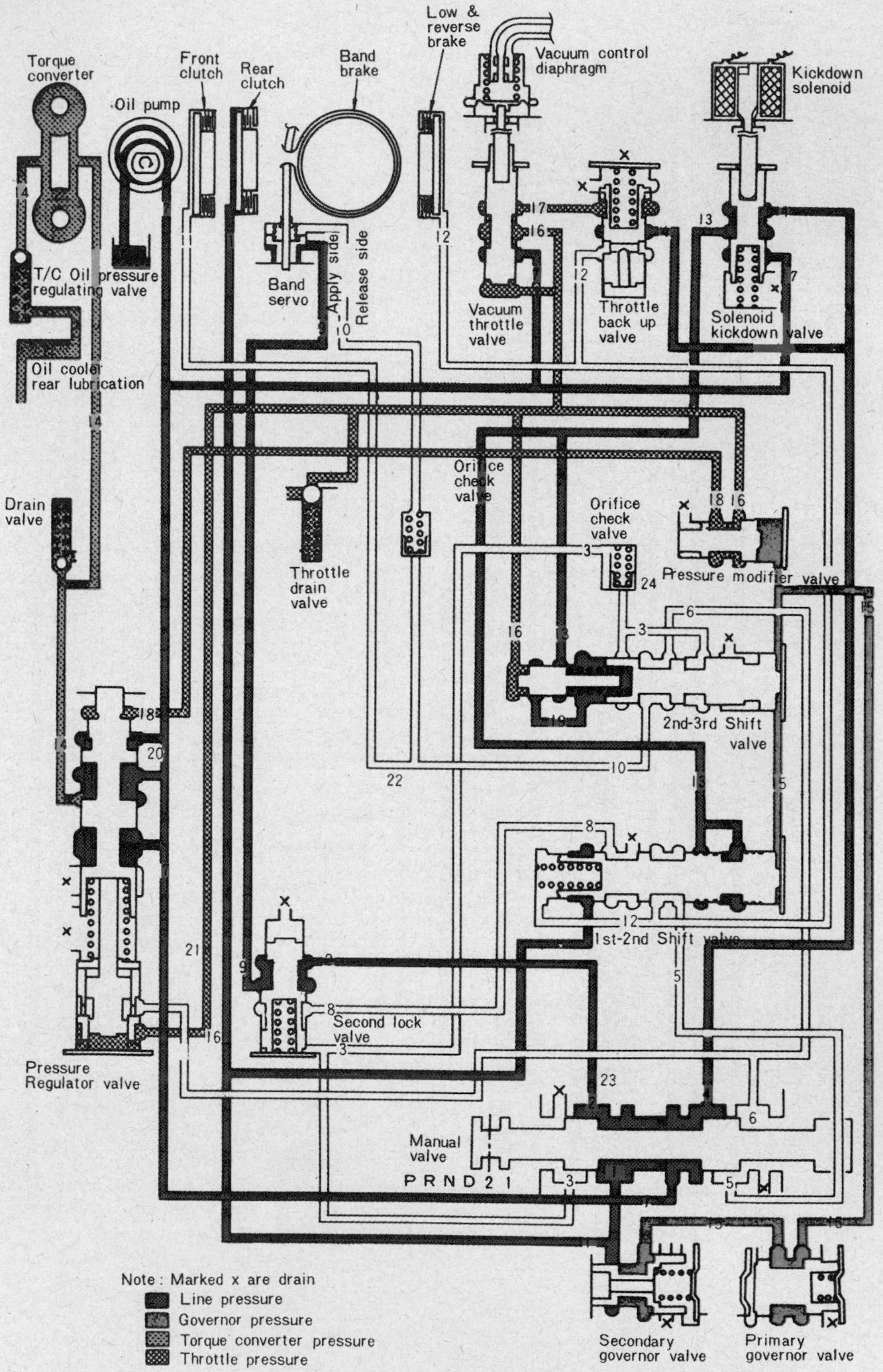
Torque converter
Oil pump
Front clutch
Rear clutch
Band brake
Low & reverse brake
Vacuum control diaphragm
Kickdown solenoid
T/C Oil pressure regulating valve
Band servo
Apply side
Release side
Vacuum throttle valve
Throttle back up valve
Solenoid kickdown valve
Oil cooler rear lubrication
Orifice check valve
Orifice check valve
Drain valve
Throttle drain valve
Pressure modifier valve
2nd-3rd Shift valve
1st-2nd Shift valve
Second lock valve
Pressure Regulator valve
Manual valve
P R N D 2 1
Secondary governor valve
Primary governor valve
Note: Marked x are drain
Line pressure
Governor pressure
Torque converter pressure
Throttle pressure

1—1ST GEAR—F3A

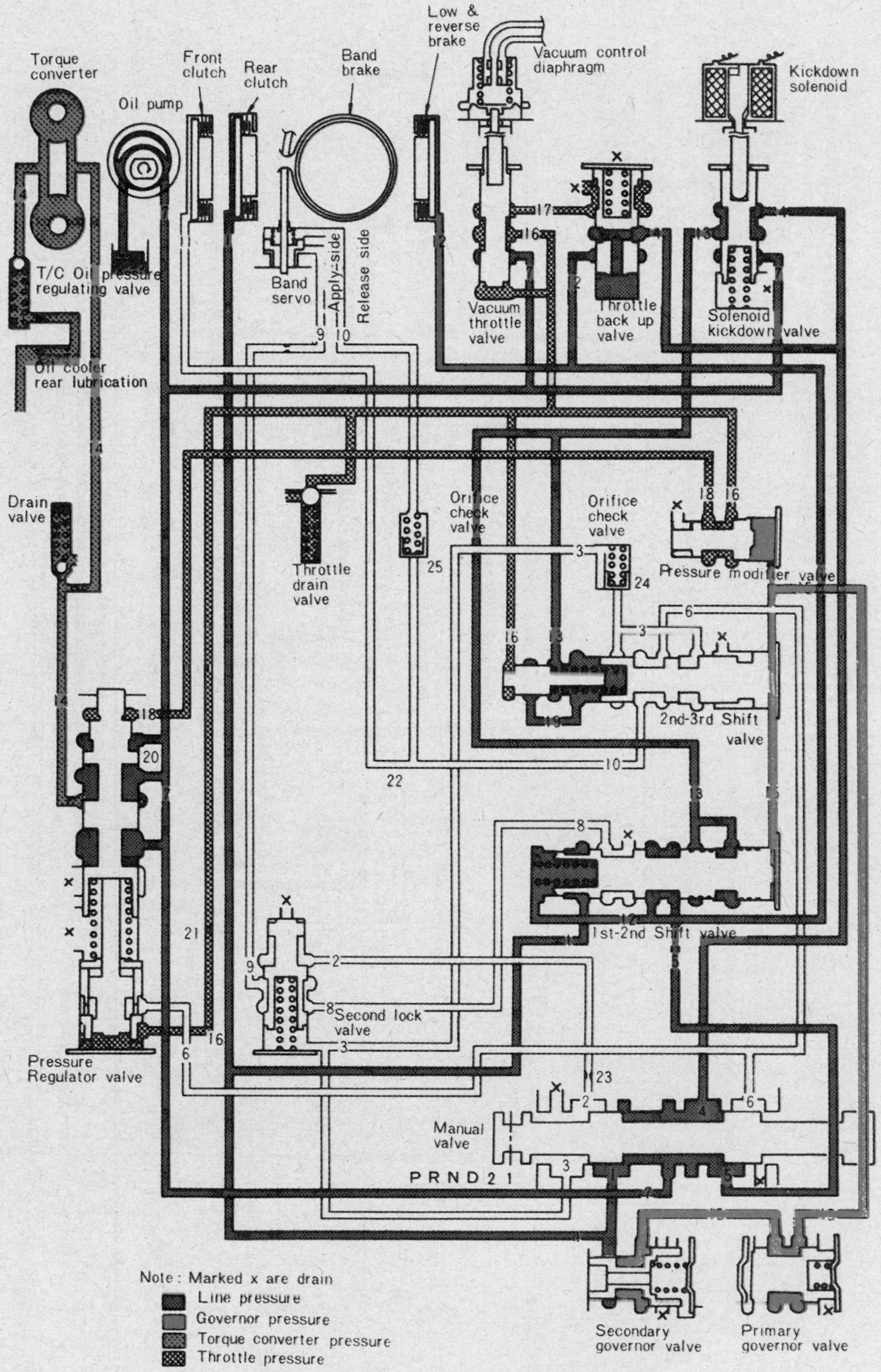
Torque converter
Oil pump
Front clutch
Rear clutch
Band brake
Low & reverse brake
Vacuum control diaphragm
Kickdown solenoid
T/C Oil pressure regulating valve
Band servo
Apply-side
Release side
Vacuum throttle valve
Throttle back up valve
Solenoid kickdown valve
Oil cooler rear lubrication
Drain valve
Throttle drain valve
Orifice check valve
Orifice check valve
Pressure modifier valve
2nd-3rd Shift valve
Pressure Regulator valve
1st-2nd Shift valve
Second lock valve
Manual valve
P R N D 2 1
Secondary governor valve
Primary governor valve
Note: Marked x are drain
Line pressure
Governor pressure
Torque converter pressure
Throttle pressure

REVERSE—F3A

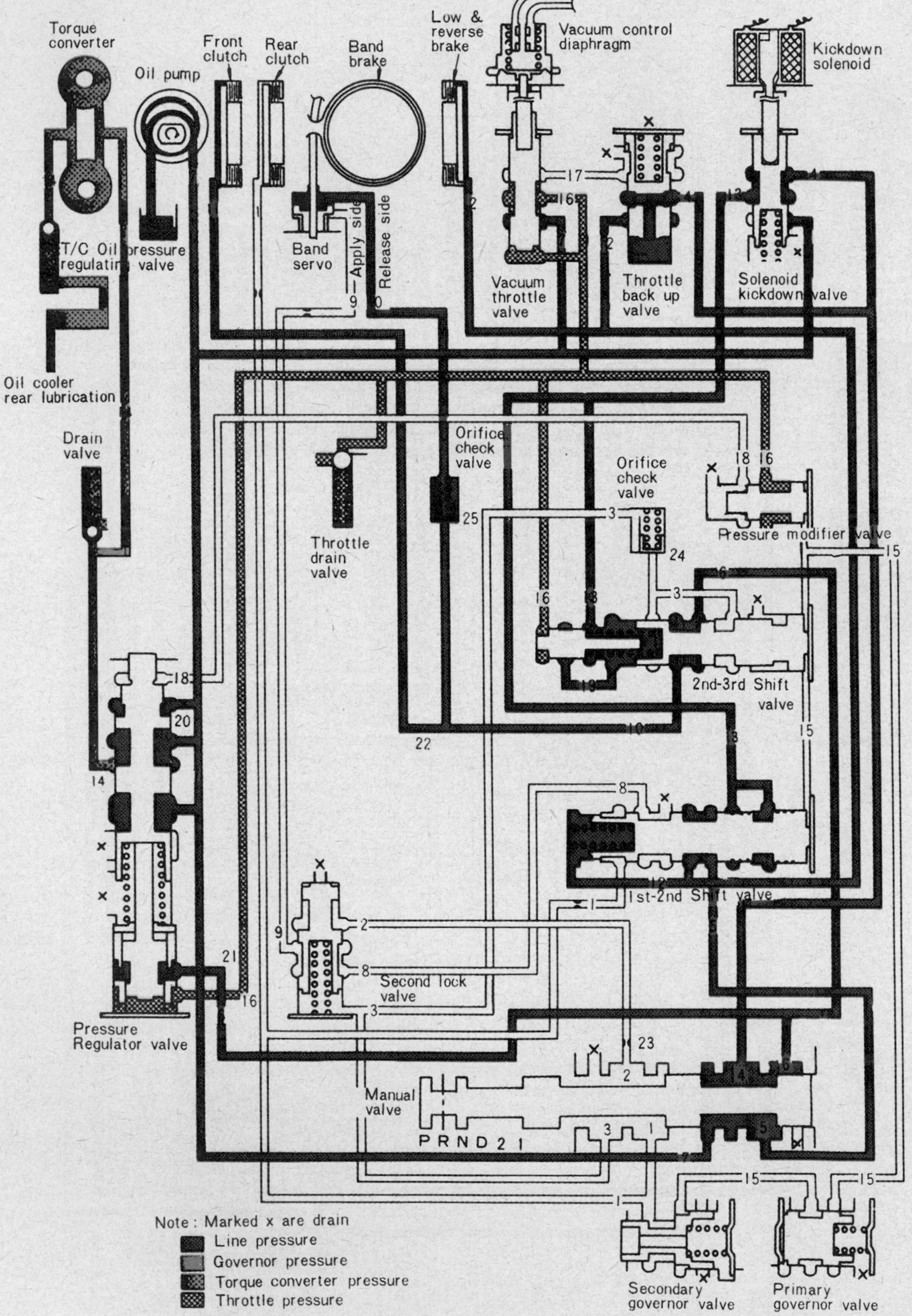
Torque converter
Oil pump
Front clutch
Rear clutch
Band brake
Low & reverse brake
Vacuum control diaphragm
Kickdown solenoid
T/C Oil pressure regulating valve
Band servo
Apply side
Release side
Vacuum throttle valve
Throttle back up valve
Solenoid kickdown valve
Oil cooler rear lubrication
Drain valve
Orifice check valve
Orifice check valve
Throttle drain valve
Pressure modifier valve
2nd-3rd Shift valve
1st-2nd Shift valve
Second lock valve
Pressure Regulator valve
Manual valve
P R N D 2 1
Secondary governor valve
Primary governor valve
Note: Marked x are drain
Line pressure
Governor pressure
Torque converter pressure
Throttle pressure

PARK—440–T4

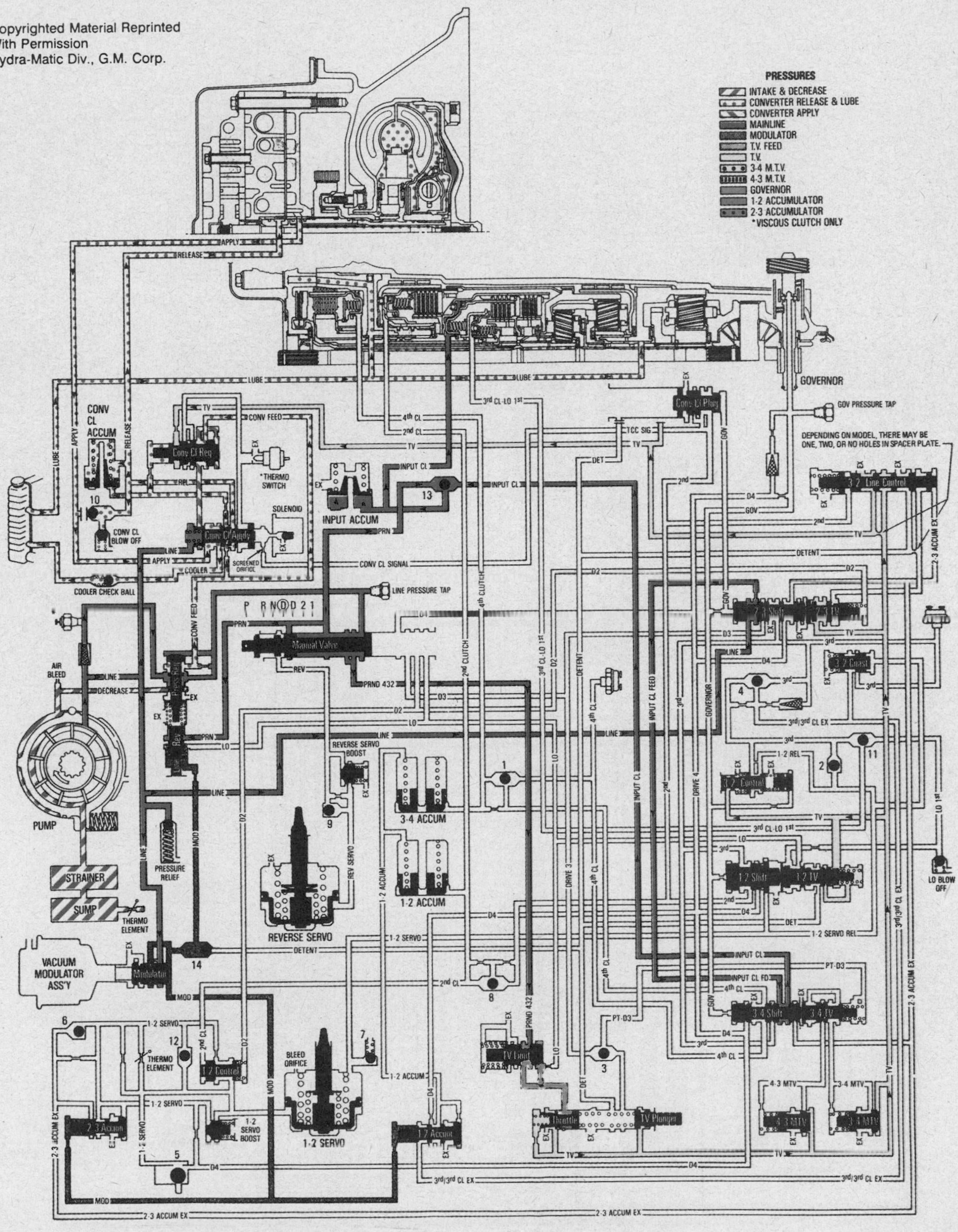
PRESSURES
INTAKE & DECREASE
CONVERTER RELEASE & LUBE
CONVERTER APPLY
MAINLINE
MODULATOR
T.V. FEED
T.V.
3-4 M.T.V.
4-3 M.T.V.
GOVERNOR
1-2 ACCUMULATOR
2-3 ACCUMULATOR
*VISCOUS CLUTCH ONLY
APPLY
RELEASE
LUBE
GOVERNOR
GOV PRESSURE TAP
DEPENDING ON MODEL, THERE MAY BE ONE, TWO, OR NO HOLES IN SPACER PLATE.
CONV CL ACCUM
Conv Cl Req
*THERMO SWITCH
SOLENOID
Conv Cl Apply
CONV CL BLOW OFF
SCREENED ORIFICE
COOLER
COOLER CHECK BALL
CONV FEED
INPUT ACCUM
INPUT CL
CONV CL SIGNAL
LINE PRESSURE TAP
P R N D D 2 1
Manual Valve
Conv Cl Plug
3-2 Line Control
2-3 Shift
2-3 TV
3-2 Coast
3-2 Control
1-2 Shift
1-2 TV
3-4 Shift
3-4 TV
AIR BLEED
DECREASE
PUMP
STRAINER
SUMP
THERMO ELEMENT
PRESSURE RELIEF
REVERSE SERVO BOOST
REVERSE SERVO
3-4 ACCUM
1-2 ACCUM
VACUUM MODULATOR ASS'Y
Modulator
1-2 Control
2-3 Accum
1-2 SERVO BOOST
BLEED ORIFICE
1-2 SERVO
1-2 Accum
TV Limit
Throttle
TV Plunger
4-3 MTV
3-4 MTV
LO BLOW OFF
2-3 ACCUM EX
3rd/3rd CL EX

DRIVE—1ST GEAR—440–T4

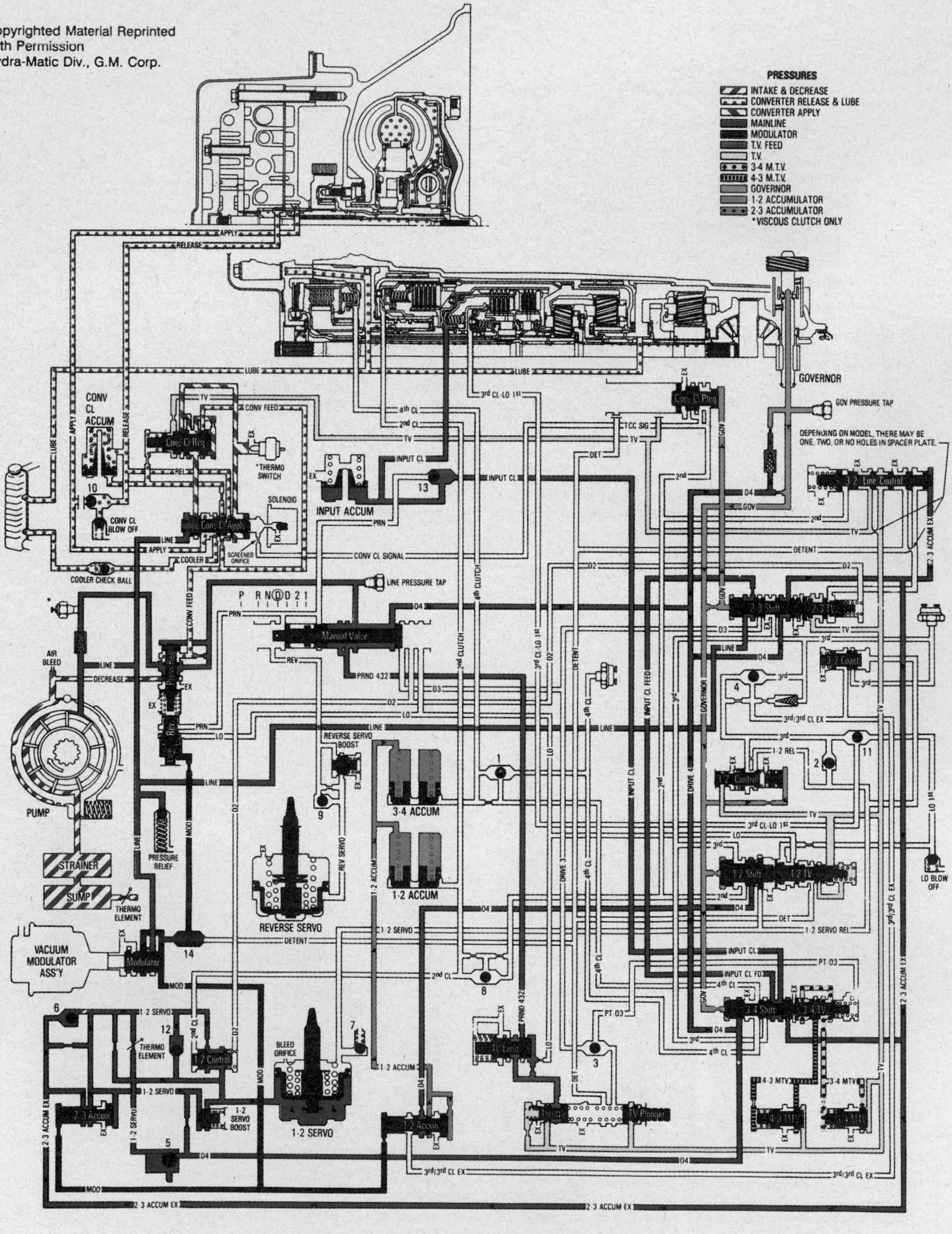
PRESSURES
INTAKE & DECREASE
CONVERTER RELEASE & LUBE
CONVERTER APPLY
MAINLINE
MODULATOR
T.V. FEED
T.V.
3-4 M.T.V.
4-3 M.T.V.
GOVERNOR
1-2 ACCUMULATOR
2-3 ACCUMULATOR
*VISCOUS CLUTCH ONLY
APPLY
RELEASE
LUBE
GOVERNOR
GOV PRESSURE TAP
DEPENDING ON MODEL, THERE MAY BE ONE, TWO, OR NO HOLES IN SPACER PLATE.
CONV CL ACCUM
Conv Cl Reg
*THERMO SWITCH
SOLENOID
Conv Cl Apply
CONV CL BLOW OFF
SCREENED ORIFICE
COOLER CHECK BALL
INPUT ACCUM
Conv Cl Plug
3-2 Line Control
LINE PRESSURE TAP
P R N D D 2 1
Manual Valve
AIR BLEED
PUMP
STRAINER
SUMP
THERMO ELEMENT
PRESSURE RELIEF
REVERSE SERVO BOOST
REVERSE SERVO
3-4 ACCUM
1-2 ACCUM
2-3 Shift
2-3 TV
3-2 Coast
1-2 Control
1-2 Shift
1-2 TV
LO BLOW OFF
VACUUM MODULATOR ASS'Y
Modulator
1-2 Control
2-3 Accum
1-2 SERVO BOOST
BLEED ORIFICE
1-2 SERVO
1-2 Accum
TV Limit
Throttle
TV Plunger
3-4 Shift
3-4 TV
4-3 MTV
3-4 MTV
2-3 ACCUM EX
3rd/3rd CL EX

DRIVE—2ND GEAR—440-T4

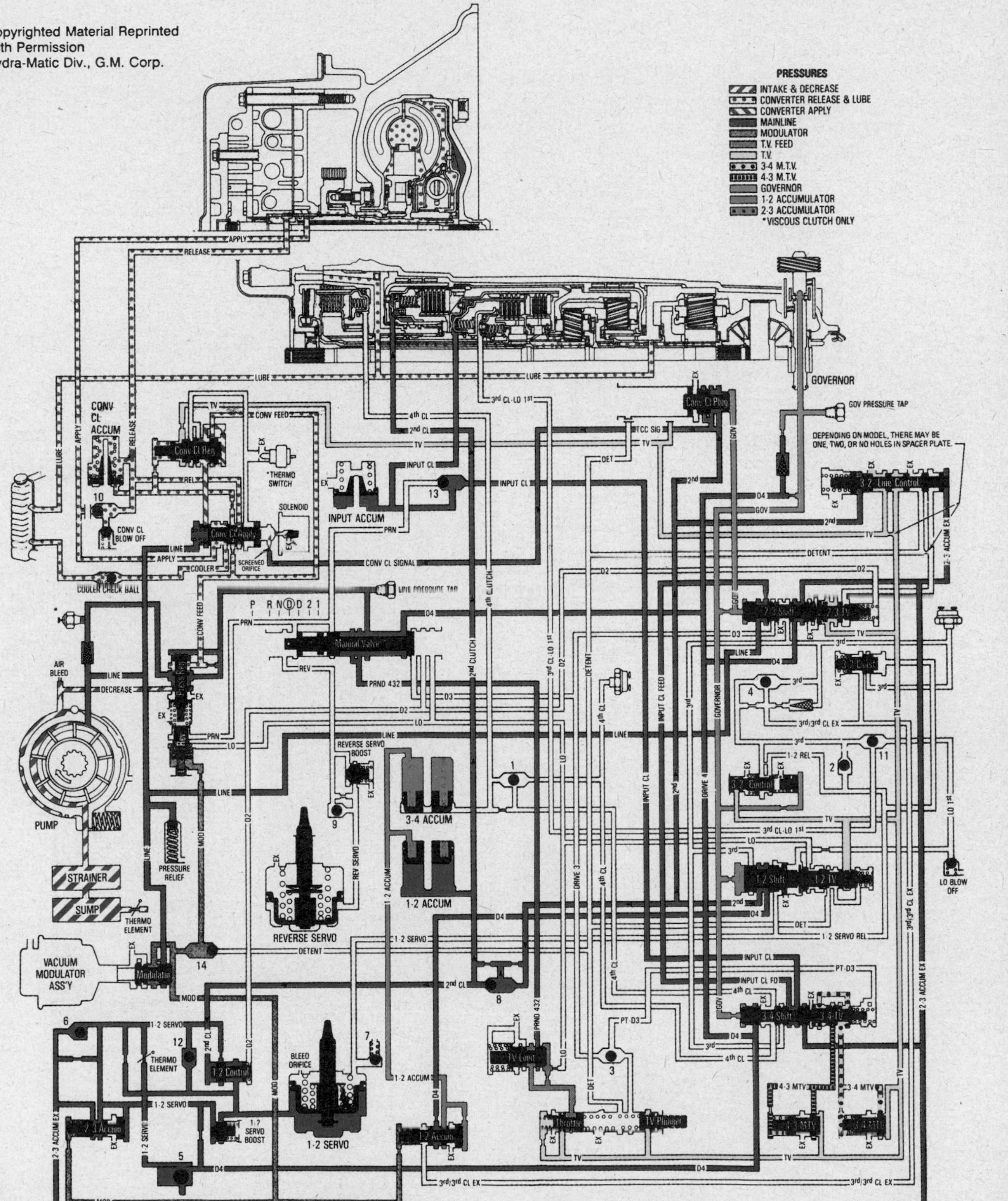
PRESSURES
INTAKE & DECREASE
CONVERTER RELEASE & LUBE
CONVERTER APPLY
MAINLINE
MODULATOR
T.V. FEED
T.V.
3-4 M.T.V.
4-3 M.T.V.
GOVERNOR
1-2 ACCUMULATOR
2-3 ACCUMULATOR
*VISCOUS CLUTCH ONLY
APPLY
RELEASE
LUBE
GOVERNOR
GOV PRESSURE TAP
DEPENDING ON MODEL, THERE MAY BE ONE, TWO, OR NO HOLES IN SPACER PLATE.
CONV CL ACCUM
*THERMO SWITCH
SOLENOID
INPUT ACCUM
CONV CL BLOW OFF
COOLER CHECK BALL
SCREENED ORIFICE
CONV CL SIGNAL
LINE PRESSURE TAP
P R N D D 2 1
AIR BLEED
PUMP
STRAINER
SUMP
THERMO ELEMENT
PRESSURE RELIEF
VACUUM MODULATOR ASS'Y
REVERSE SERVO BOOST
3-4 ACCUM
1-2 ACCUM
REVERSE SERVO
BLEED ORIFICE
1-2 SERVO
1-2 SERVO BOOST
LO BLOW OFF

DRIVE—3RD GEAR—440-T4

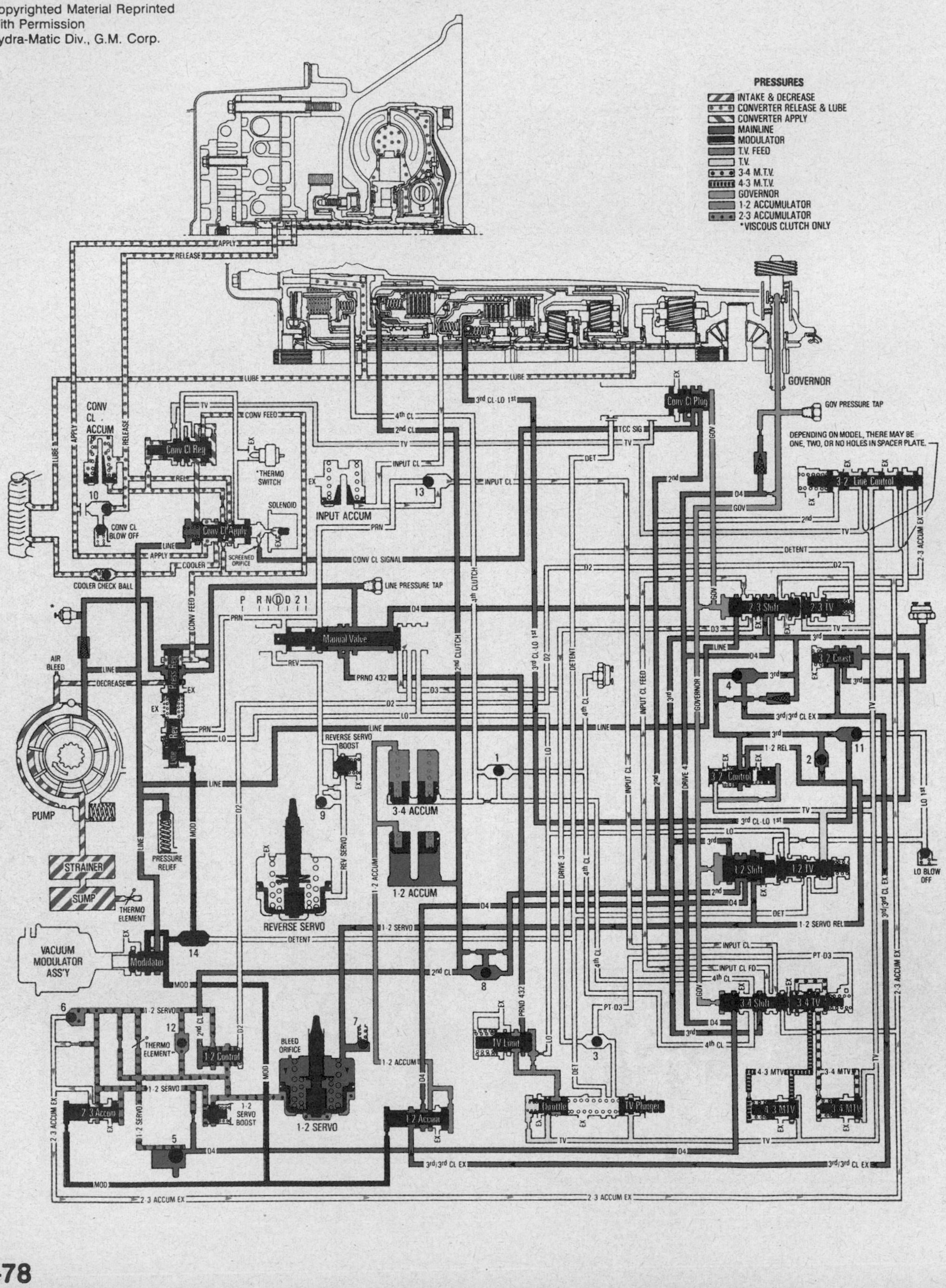
PRESSURES
INTAKE & DECREASE
CONVERTER RELEASE & LUBE
CONVERTER APPLY
MAINLINE
MODULATOR
T.V. FEED
T.V.
3-4 M.T.V.
4-3 M.T.V.
GOVERNOR
1-2 ACCUMULATOR
2-3 ACCUMULATOR
*VISCOUS CLUTCH ONLY
APPLY
RELEASE
LUBE
GOVERNOR
GOV PRESSURE TAP
DEPENDING ON MODEL, THERE MAY BE ONE, TWO, OR NO HOLES IN SPACER PLATE.
CONV CL ACCUM
Conv Cl Reg
CONV FEED
*THERMO SWITCH
SOLENOID
Conv Cl Apply
SCREENED ORIFICE
CONV CL BLOW OFF
COOLER CHECK BALL
COOLER
INPUT ACCUM
INPUT CL
Conv Cl Plug
TCC SIG
3-2 Line Control
CONV CL SIGNAL
LINE PRESSURE TAP
P R N D D 2 1
Manual Valve
PRND 432
2-3 Shift
2-3 TV
3-2 Coast
AIR BLEED
Press Reg
Rev
PUMP
STRAINER
SUMP
THERMO ELEMENT
PRESSURE RELIEF
REVERSE SERVO BOOST
REVERSE SERVO
3-4 ACCUM
1-2 ACCUM
3-2 Control
1-2 Shift
1-2 TV
LO BLOW OFF
VACUUM MODULATOR ASS'Y
Modulator
1-2 Control
2-3 Accum
1-2 SERVO BOOST
BLEED ORIFICE
1-2 SERVO
1-2 Accum
TV Limit
Throttle
TV Plunger
3-4 Shift
3-4 TV
4-3 MTV
3-4 MTV
2-3 ACCUM EX
3rd/3rd CL EX

DRIVE – CONVERTER CLUTCH APPLY – 440–T4

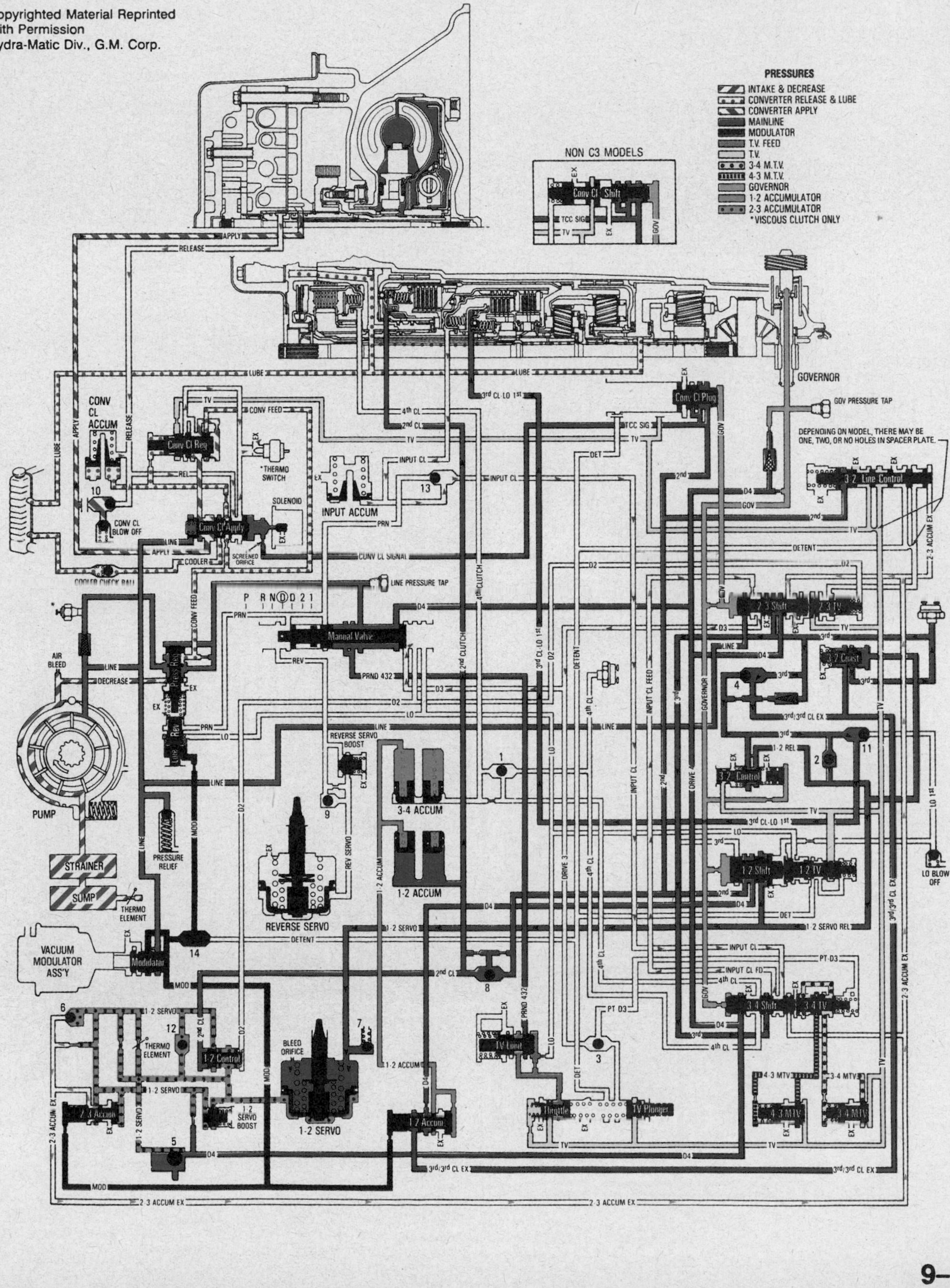
PRESSURES
INTAKE & DECREASE
CONVERTER RELEASE & LUBE
CONVERTER APPLY
MAINLINE
MODULATOR
T.V. FEED
T.V.
3-4 M.T.V.
4-3 M.T.V.
GOVERNOR
1-2 ACCUMULATOR
2-3 ACCUMULATOR
*VISCOUS CLUTCH ONLY
NON C3 MODELS
Conv Cl Shift
TCC SIG
TV
EX
GOV
APPLY
RELEASE
LUBE
GOVERNOR
GOV PRESSURE TAP
DEPENDING ON MODEL, THERE MAY BE ONE, TWO, OR NO HOLES IN SPACER PLATE.
3-2 Line Control
CONV CL ACCUM
Conv Cl Req
CONV FEED
*THERMO SWITCH
SOLENOID
Conv Cl Apply
CONV CL BLOW OFF
SCREENED ORIFICE
COOLER
COOLER CHECK BALL
INPUT ACCUM
CONV CL SIGNAL
Conv Cl Plug
LINE PRESSURE TAP
P R N D D 2 1
Manual Valve
AIR BLEED
Press Reg
PUMP
STRAINER
SUMP
THERMO ELEMENT
PRESSURE RELIEF
VACUUM MODULATOR ASS'Y
Modulator
REVERSE SERVO BOOST
3-4 ACCUM
1-2 ACCUM
REVERSE SERVO
2-3 Shift
2-3 TV
3-2 Coast
3-2 Control
1-2 Shift
1-2 TV
LO BLOW OFF
3-4 Shift
3-4 TV
TV Limit
Throttle
TV Plunger
4-3 MTV
3-4 MTV
1-2 Control
THERMO ELEMENT
2-3 Accum
1-2 SERVO BOOST
BLEED ORIFICE
1-2 SERVO
1-2 Accum
2-3 ACCUM EX

DRIVE—4TH GEAR—440-T4

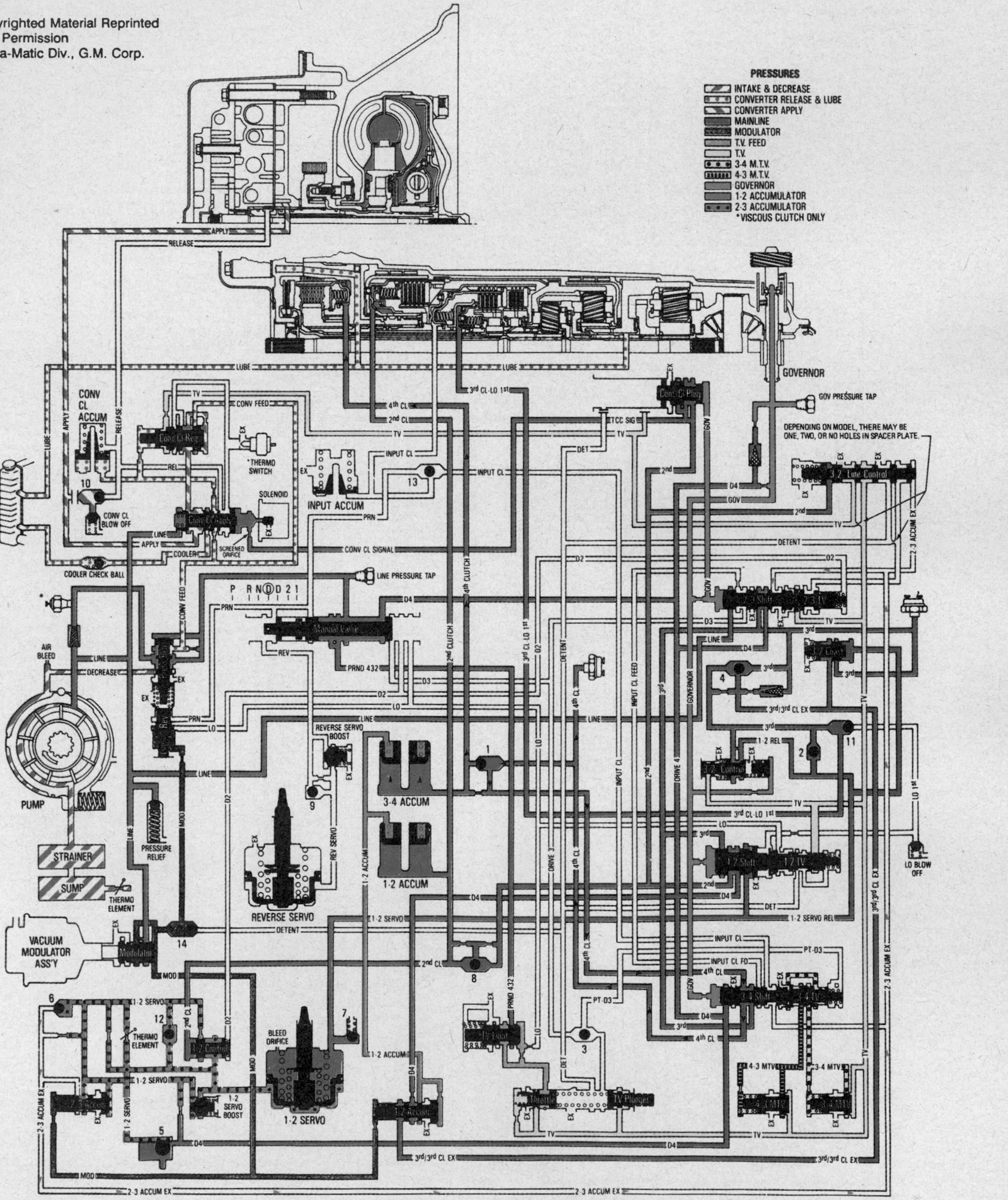
PRESSURES
INTAKE & DECREASE
CONVERTER RELEASE & LUBE
CONVERTER APPLY
MAINLINE
MODULATOR
T.V. FEED
T.V.
3-4 M.T.V.
4-3 M.T.V.
GOVERNOR
1-2 ACCUMULATOR
2-3 ACCUMULATOR
*VISCOUS CLUTCH ONLY
APPLY
RELEASE
LUBE
GOVERNOR
GOV PRESSURE TAP
DEPENDING ON MODEL, THERE MAY BE ONE, TWO, OR NO HOLES IN SPACER PLATE.
3-2 Line Control
CONV CL ACCUM
CONV FEED
*THERMO SWITCH
SOLENOID
CONV CL BLOW OFF
COOLER CHECK BALL
SCREENED ORIFICE
CONV CL SIGNAL
INPUT ACCUM
LINE PRESSURE TAP
P R N D D 2 1
Manual Valve
AIR BLEED
DECREASE
PUMP
STRAINER
SUMP
THERMO ELEMENT
PRESSURE RELIEF
REVERSE SERVO BOOST
REVERSE SERVO
3-4 ACCUM
1-2 ACCUM
VACUUM MODULATOR ASS'Y
Modulator
BLEED ORIFICE
1-2 SERVO
1-2 SERVO BOOST
2-3 Shift
3-2 Control
1-2 Shift
3-4 Shift
4-3 MTV
3-4 MTV
Throttle
TV Plunger
LO BLOW OFF
2-3 ACCUM EX
3rd/3rd CL EX

PART THROTTLE 4-3 AND MODULATED DOWNSHIFT (VALVES SHOWN IN 3RD GEAR POSITION)—440-T4

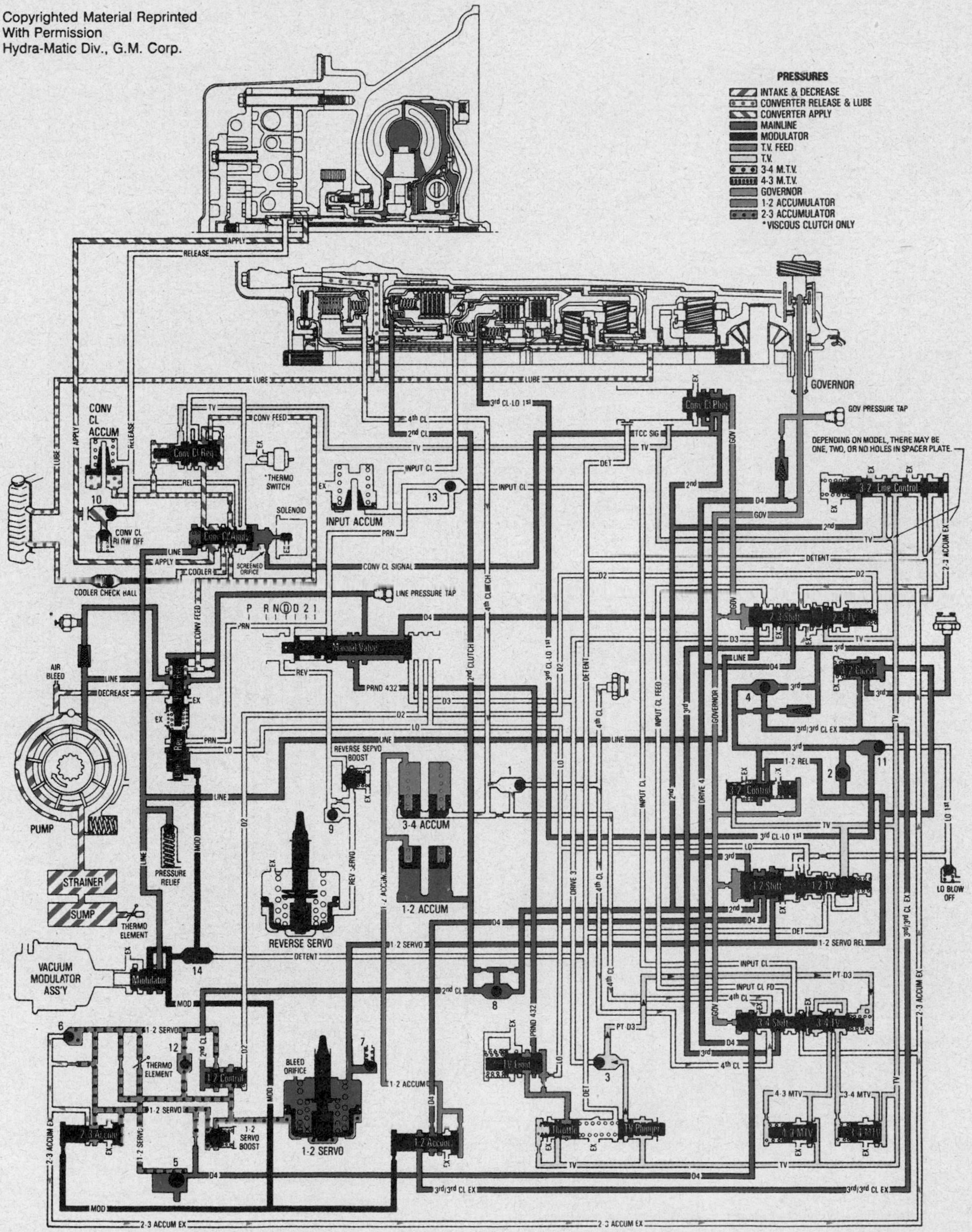
PRESSURES
INTAKE & DECREASE
CONVERTER RELEASE & LUBE
CONVERTER APPLY
MAINLINE
MODULATOR
T.V. FEED
T.V.
3-4 M.T.V.
4-3 M.T.V.
GOVERNOR
1-2 ACCUMULATOR
2-3 ACCUMULATOR
*VISCOUS CLUTCH ONLY
GOVERNOR
GOV PRESSURE TAP
DEPENDING ON MODEL, THERE MAY BE ONE, TWO, OR NO HOLES IN SPACER PLATE.
CONV CL ACCUM
*THERMO SWITCH
SOLENOID
INPUT ACCUM
CONV CL BLOW OFF
COOLER CHECK BALL
LINE PRESSURE TAP
P R N D D 2 1
Manual Valve
AIR BLEED
PUMP
STRAINER
SUMP
THERMO ELEMENT
PRESSURE RELIEF
REVERSE SEPVO BOOST
3-4 ACCUM
1-2 ACCUM
REVERSE SERVO
VACUUM MODULATOR ASS'Y
Modulator
BLEED ORIFICE
1-2 SERVO
1-2 SERVO BOOST
1-2 Control
2-3 Accum
1-2 Accum
TV Limit
Throttle
TV Plunger
3-2 Line Control
2-3 Shift
2-3 TV
1-2 Coast
3-2 Control
1-2 Shift
1-2 TV
3-4 Shift
3-4 TV
4-3 MTV
3-4 MTV
Conv Cl Plug
Conv Cl Reg
Conv Cl Apply
LO BLOW OFF
LUBE
APPLY
RELEASE
TV
CONV FEED
4th CL
2nd CL
INPUT CL
3rd CL-LO 1st
TCC SIG
DET
DETENT
CONV CL SIGNAL
D4
D3
D2
LINE
PRN
REV
PRND 432
LO
DECREASE
MOD
2-3 ACCUM EX
3rd/3rd CL EX
1-2 SERVO REL
INPUT CL FD
PT-D3
4-3 MTV
3-4 MTV

DETENT 3–2 DOWNSHIFT (VALVES SHOWN IN 2ND GEAR POSITION) 440–T4

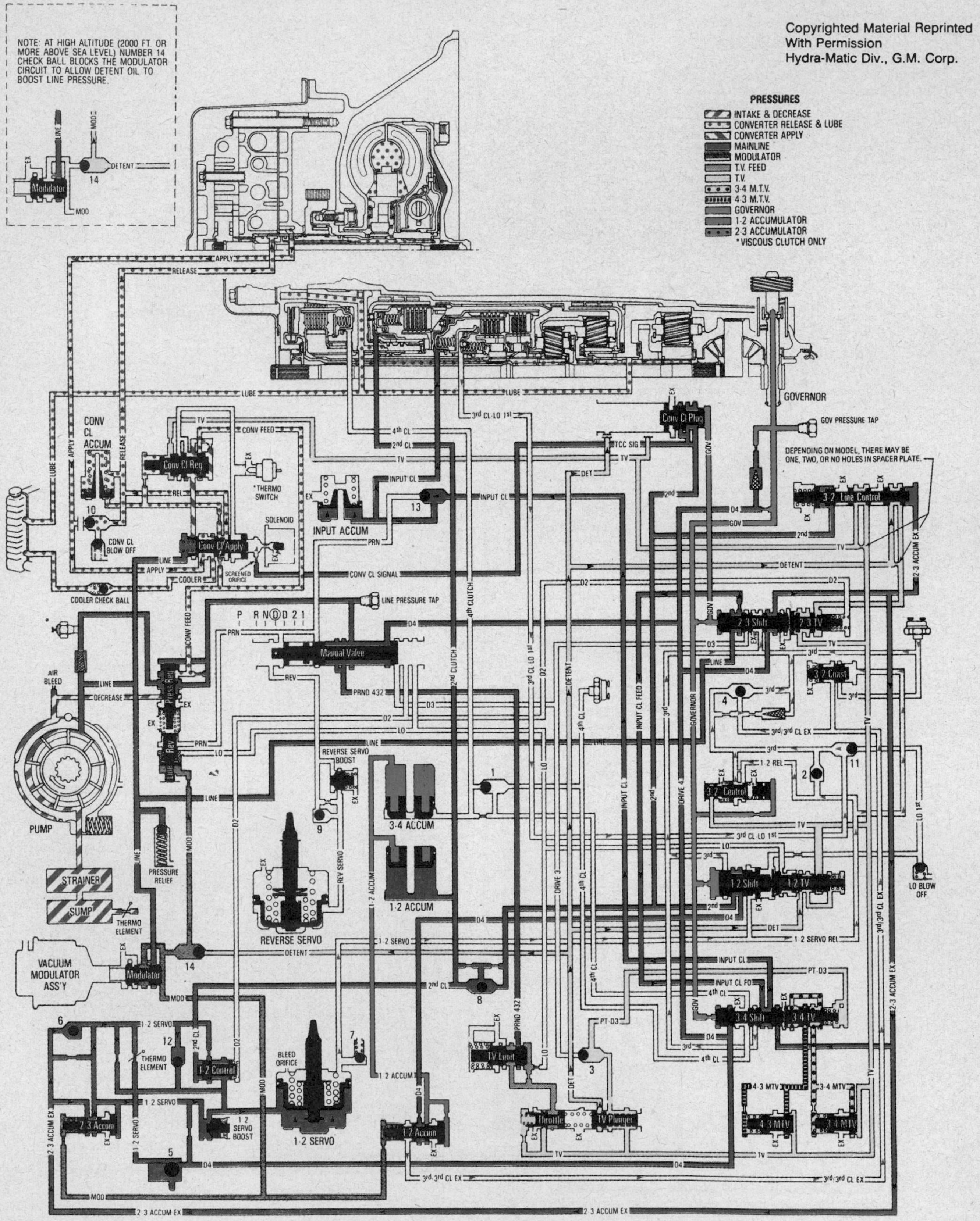
NOTE: AT HIGH ALTITUDE (2000 FT. OR MORE ABOVE SEA LEVEL) NUMBER 14 CHECK BALL BLOCKS THE MODULATOR CIRCUIT TO ALLOW DETENT OIL TO BOOST LINE PRESSURE.
PRESSURES
INTAKE & DECREASE
CONVERTER RELEASE & LUBE
CONVERTER APPLY
MAINLINE
MODULATOR
T.V. FEED
T.V.
3-4 M.T.V.
4-3 M.T.V.
GOVERNOR
1-2 ACCUMULATOR
2-3 ACCUMULATOR
*VISCOUS CLUTCH ONLY
GOVERNOR
GOV PRESSURE TAP
DEPENDING ON MODEL, THERE MAY BE ONE, TWO, OR NO HOLES IN SPACER PLATE.
3-2 Line Control
CONV CL ACCUM
Conv Cl Reg
*THERMO SWITCH
SOLENOID
Conv Cl Apply
CONV CL BLOW OFF
COOLER CHECK BALL
SCREENED ORIFICE
Conv Cl Plug
INPUT ACCUM
LINE PRESSURE TAP
P R N D D 2 1
Manual Valve
AIR BLEED
PUMP
STRAINER
SUMP
THERMO ELEMENT
PRESSURE RELIEF
VACUUM MODULATOR ASS'Y
Modulator
REVERSE SERVO BOOST
3-4 ACCUM
1-2 ACCUM
REVERSE SERVO
2-3 Shift
2-3 TV
3-2 Coast
3-2 Control
1-2 Shift
1-2 TV
LO BLOW OFF
3-4 Shift
3-4 TV
TV Limit
Throttle
TV Plunger
4-3 MTV
3-4 MTV
1-2 Control
2-3 Accum
1-2 SERVO BOOST
1-2 SERVO
BLEED ORIFICE
1-2 Accum
2-3 ACCUM EX

MANUAL 3RD—440-T4

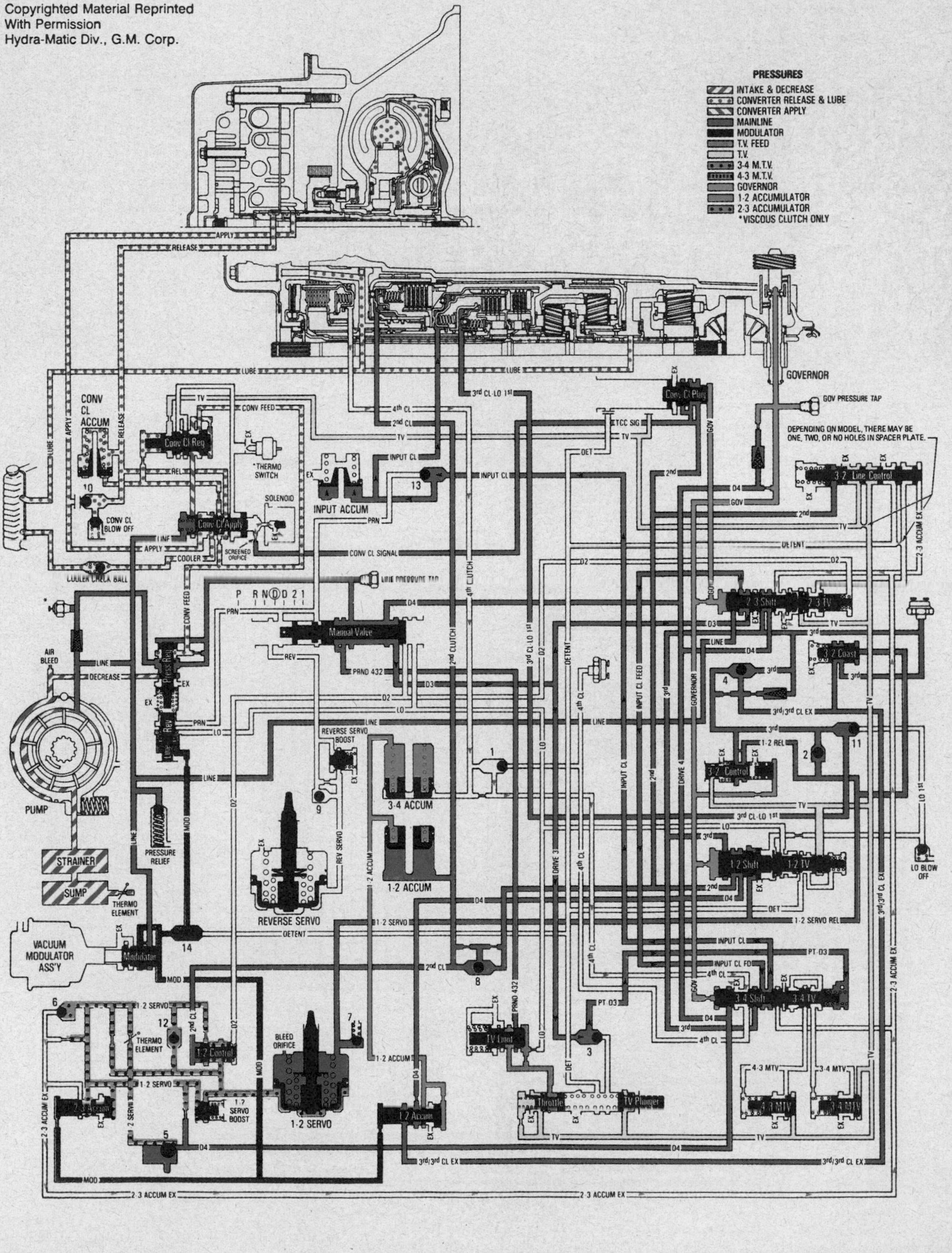
PRESSURES
INTAKE & DECREASE
CONVERTER RELEASE & LUBE
CONVERTER APPLY
MAINLINE
MODULATOR
T.V. FEED
T.V.
3-4 M.T.V.
4-3 M.T.V.
GOVERNOR
1-2 ACCUMULATOR
2-3 ACCUMULATOR
*VISCOUS CLUTCH ONLY
APPLY
RELEASE
LUBE
GOVERNOR
GOV PRESSURE TAP
DEPENDING ON MODEL, THERE MAY BE ONE, TWO, OR NO HOLES IN SPACER PLATE.
CONV CL ACCUM
Conv Cl Reg
CONV FEED
*THERMO SWITCH
SOLENOID
Conv Cl Apply
SCREENED ORIFICE
CONV CL BLOW OFF
COOLER
COOLER CHECK BALL
INPUT ACCUM
Conv Cl Plug
3-2 Line Control
CONV CL SIGNAL
LINE PRESSURE TAP
P R N D D 2 1
Manual Valve
AIR BLEED
LINE
DECREASE
Press Reg
Rev
PUMP
STRAINER
SUMP
THERMO ELEMENT
PRESSURE RELIEF
VACUUM MODULATOR ASS'Y
Modulator
REVERSE SERVO BOOST
3-4 ACCUM
1-2 ACCUM
REVERSE SERVO
2-3 Shift
2-3 TV
3-2 Coast
3-2 Control
1-2 Shift
1-2 TV
LO BLOW OFF
3-4 Shift
3-4 TV
TV Limit
Throttle
TV Plunger
4-3 MTV
3-4 MTV
1-2 Control
2-3 Accum
1-2 SERVO BOOST
BLEED ORIFICE
1-2 SERVO
1-2 Accum
2-3 ACCUM EX
3rd/3rd CL EX

MANUAL 2ND – 440–T4

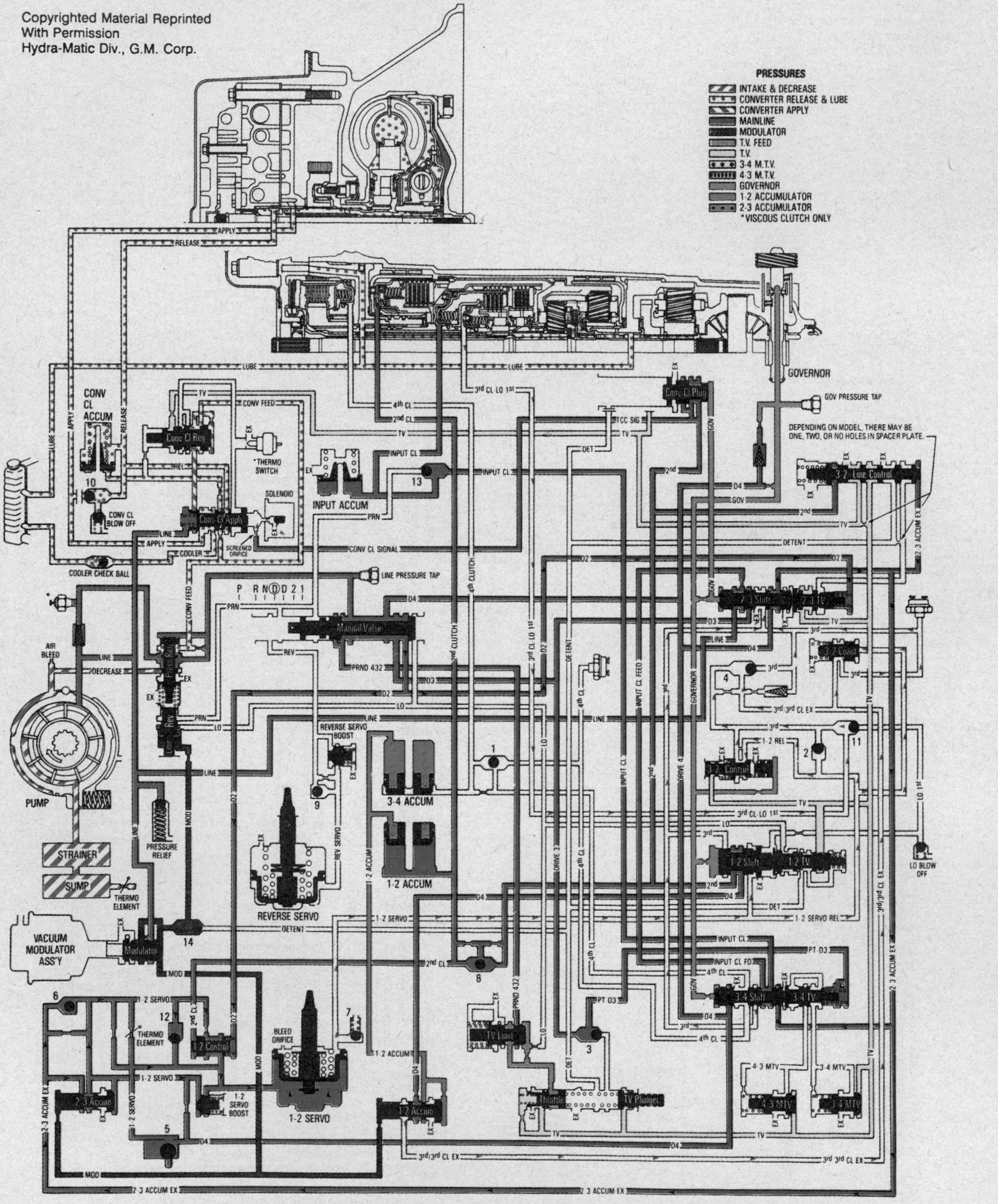
PRESSURES
INTAKE & DECREASE
CONVERTER RELEASE & LUBE
CONVERTER APPLY
MAINLINE
MODULATOR
T.V. FEED
T.V.
3-4 M.T.V.
4-3 M.T.V.
GOVERNOR
1-2 ACCUMULATOR
2-3 ACCUMULATOR
*VISCOUS CLUTCH ONLY
APPLY
RELEASE
LUBE
GOVERNOR
GOV PRESSURE TAP
DEPENDING ON MODEL, THERE MAY BE ONE, TWO, OR NO HOLES IN SPACER PLATE.
CONV CL ACCUM
Conv Cl Reg
CONV FEED
*THERMO SWITCH
SOLENOID
Conv Cl Apply
CONV CL BLOW OFF
SCREENED ORIFICE
COOLER CHECK BALL
COOLER
INPUT ACCUM
INPUT CL
Conv Cl Plug
TCC SIG
3-2 Line Control
DETENT
CONV CL SIGNAL
LINE PRESSURE TAP
P R N D D 2 1
Manual Valve
PRND 432
AIR BLEED
LINE
DECREASE
Press Reg
Rev
REVERSE SERVO BOOST
PUMP
PRESSURE RELIEF
STRAINER
SUMP
THERMO ELEMENT
3-4 ACCUM
1-2 ACCUM
REVERSE SERVO
2-3 Shift
2-3 TV
3-2 Coast
3-2 Control
1-2 Shift
1-2 TV
LO BLOW OFF
VACUUM MODULATOR ASS'Y
Modulator
1-2 SERVO
1-2 Control
BLEED ORIFICE
1-2 SERVO BOOST
2-3 Accum
1-2 Accum
TV Limit
Throttle
TV Plunger
3-4 Shift
3-4 TV
4-3 MTV
3-4 MTV
2-3 ACCUM EX
3rd 3rd CL EX
MOD

MANUAL LOW—440-T4

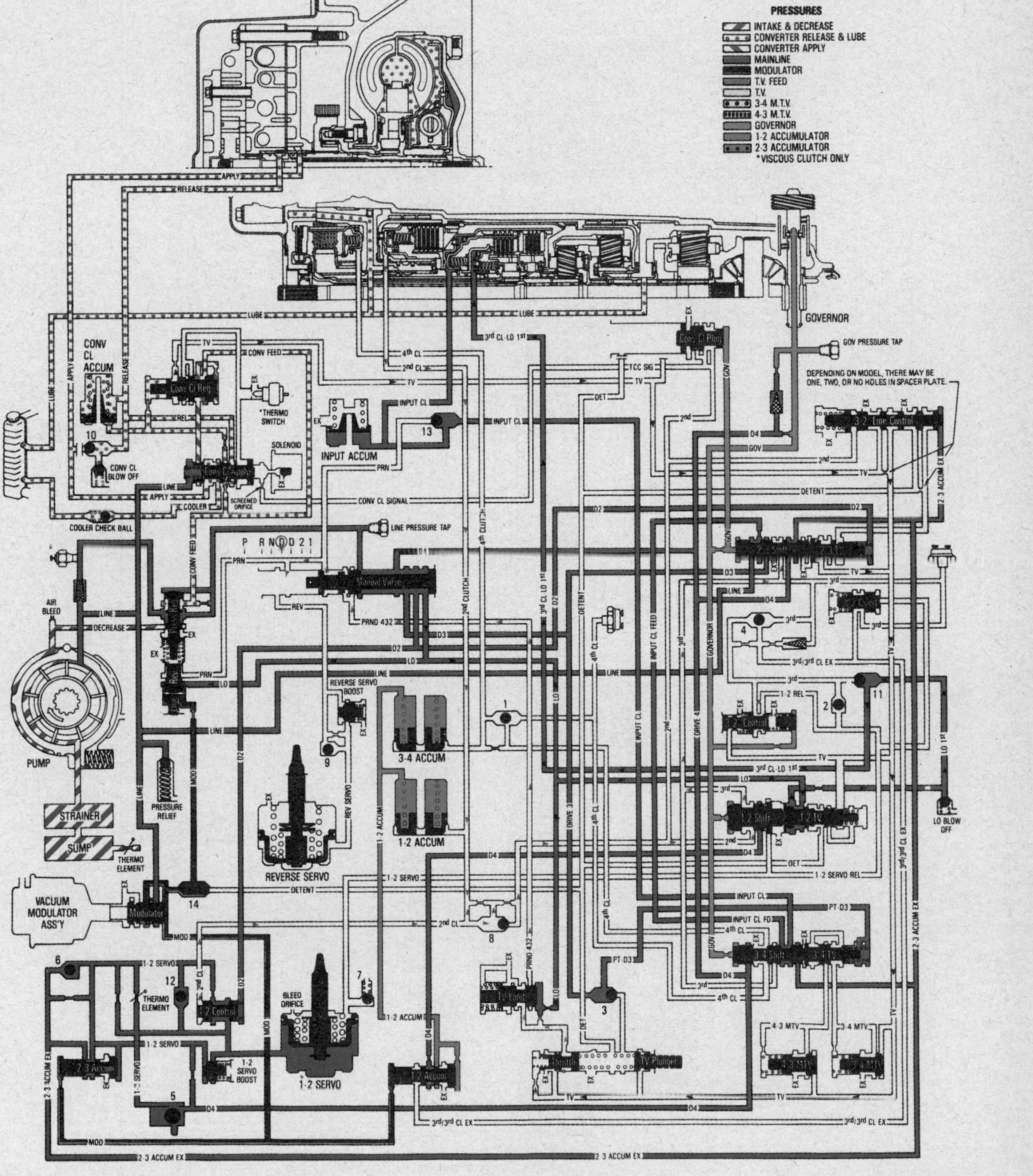
PRESSURES
INTAKE & DECREASE
CONVERTER RELEASE & LUBE
CONVERTER APPLY
MAINLINE
MODULATOR
T.V. FEED
T.V.
3-4 M.T.V.
4-3 M.T.V.
GOVERNOR
1-2 ACCUMULATOR
2-3 ACCUMULATOR
*VISCOUS CLUTCH ONLY
APPLY
RELEASE
LUBE
GOVERNOR
GOV PRESSURE TAP
DEPENDING ON MODEL, THERE MAY BE ONE, TWO, OR NO HOLES IN SPACER PLATE.
CONV CL ACCUM
Conv Cl Reg
*THERMO SWITCH
SOLENOID
Conv Cl Apply
CONV CL BLOW OFF
SCREENED ORIFICE
COOLER CHECK BALL
INPUT ACCUM
CONV CL SIGNAL
LINE PRESSURE TAP
P R N D D 2 1
Manual Valve
3-2 Line Control
AIR BLEED
PUMP
STRAINER
SUMP
THERMO ELEMENT
PRESSURE RELIEF
REVERSE SERVO BOOST
3-4 ACCUM
1-2 ACCUM
REVERSE SERVO
3-2 Control
1-2 Shift
1-2 TV
LO BLOW OFF
VACUUM MODULATOR ASS'Y
Modulator
2-3 Shift
3-4 TV
2-3 Accum
1-2 SERVO BOOST
BLEED ORIFICE
1-2 SERVO
TV Plunger
4-3 MTV
3-4 MTV

REVERSE—440-T4

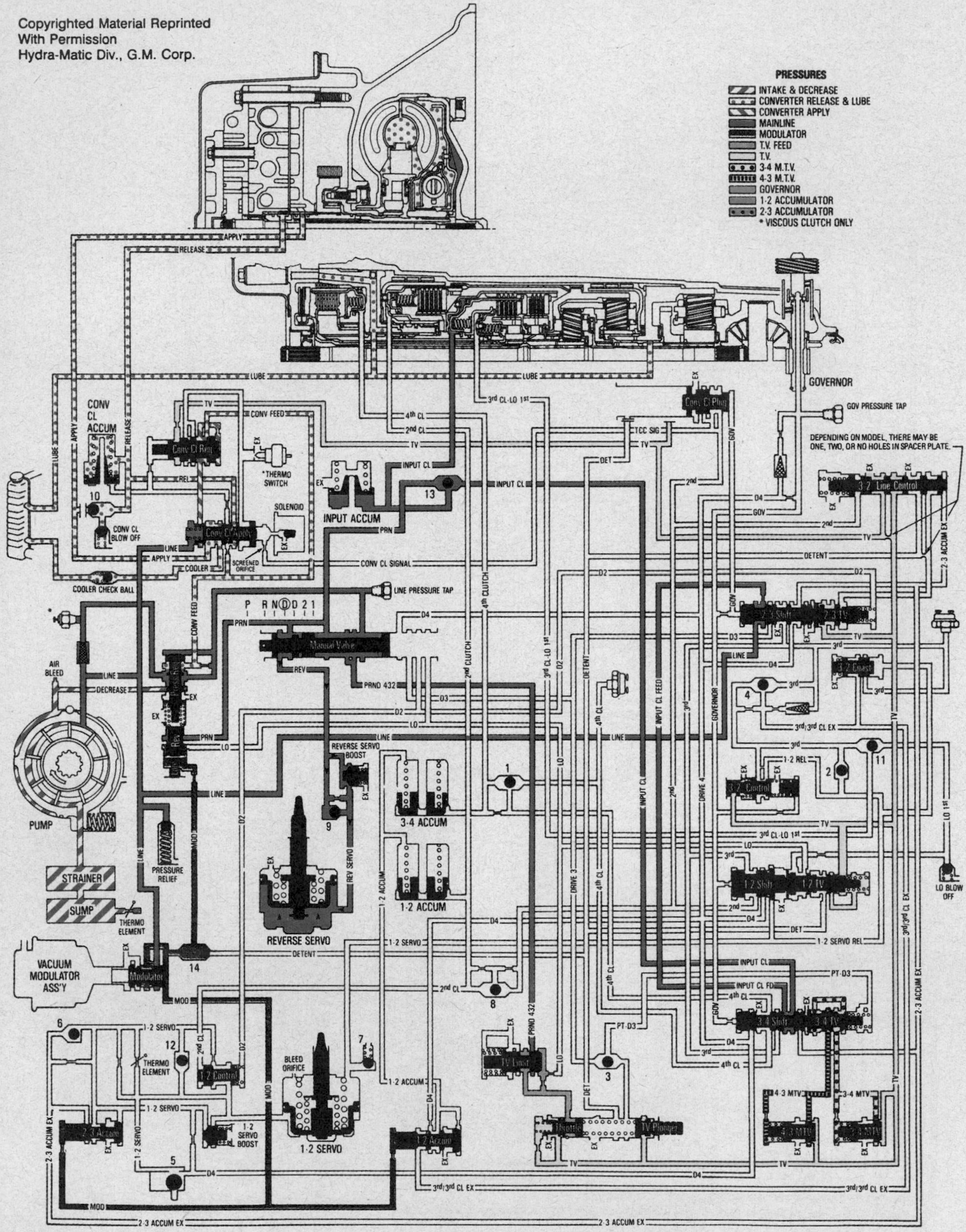
PRESSURES
INTAKE & DECREASE
CONVERTER RELEASE & LUBE
CONVERTER APPLY
MAINLINE
MODULATOR
T.V. FEED
T.V.
3-4 M.T.V.
4-3 M.T.V.
GOVERNOR
1-2 ACCUMULATOR
2-3 ACCUMULATOR
* VISCOUS CLUTCH ONLY
APPLY
RELEASE
LUBE
GOVERNOR
GOV PRESSURE TAP
DEPENDING ON MODEL, THERE MAY BE ONE, TWO, OR NO HOLES IN SPACER PLATE.
CONV CL ACCUM
CONV FEED
Conv Cl Reg
*THERMO SWITCH
SOLENOID
CONV CL BLOW OFF
Conv Cl Apply
SCREENED ORIFICE
COOLER
COOLER CHECK BALL
INPUT ACCUM
INPUT CL
CONV CL SIGNAL
LINE PRESSURE TAP
P R N D D 2 1
Manual Valve
Conv Cl Plug
3-2 Line Control
2-3 Shift
2-3 TV
3-2 Coast
3-2 Control
1-2 Shift
1-2 TV
3-4 Shift
3-4 TV
4-3 MTV
3-4 MTV
LO BLOW OFF
AIR BLEED
PUMP
STRAINER
SUMP
THERMO ELEMENT
PRESSURE RELIEF
REVERSE SERVO BOOST
REVERSE SERVO
3-4 ACCUM
1-2 ACCUM
VACUUM MODULATOR ASS'Y
Modulator
1-2 Control
2-3 Accum
1-2 SERVO BOOST
BLEED ORIFICE
1-2 SERVO
1-2 Accum
TV Limit
Throttle
TV Plunger
2-3 ACCUM EX
3rd/3rd CL EX
MOD

PARK—F7

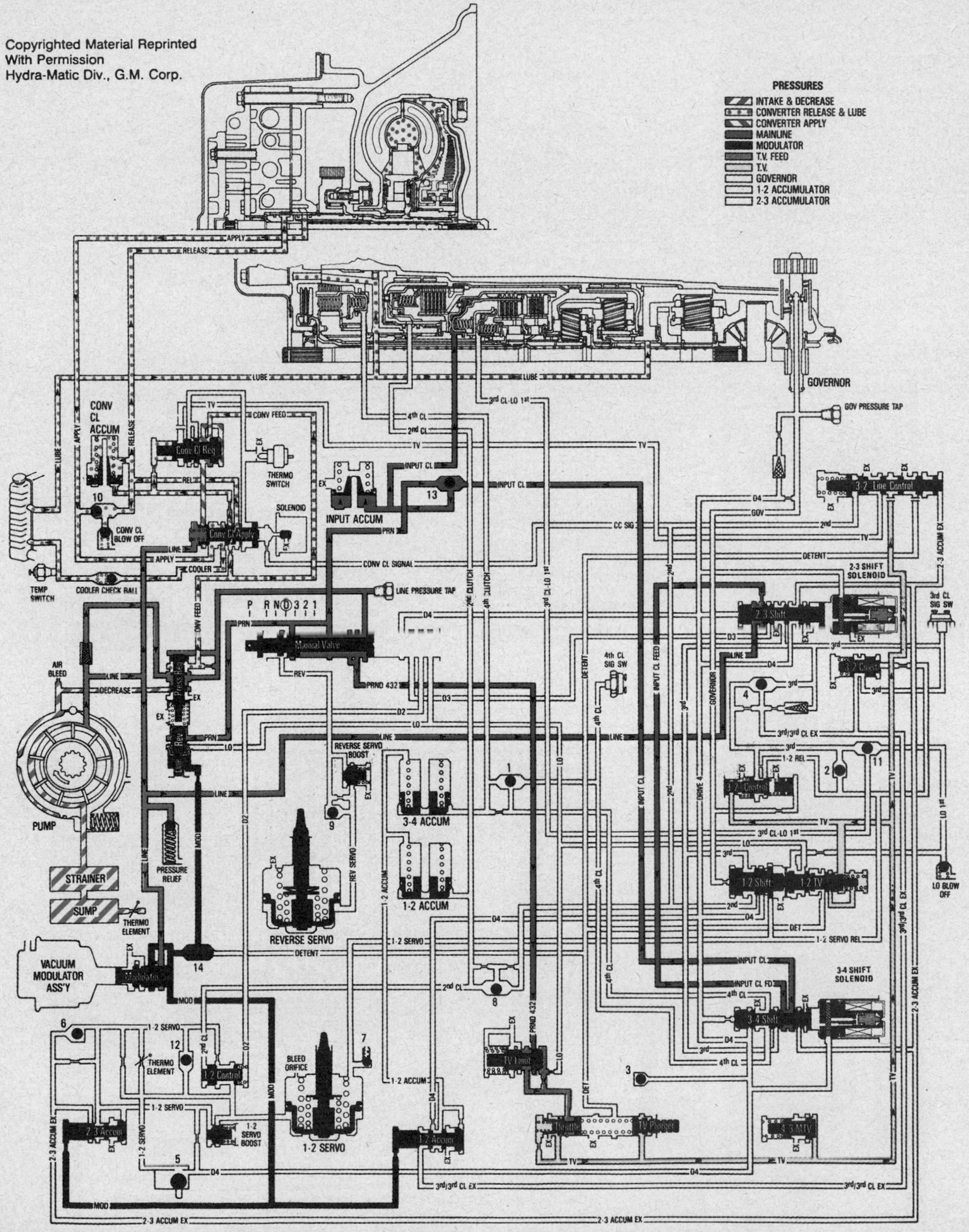
PRESSURES
INTAKE & DECREASE
CONVERTER RELEASE & LUBE
CONVERTER APPLY
MAINLINE
MODULATOR
T.V. FEED
T.V.
GOVERNOR
1-2 ACCUMULATOR
2-3 ACCUMULATOR
APPLY
RELEASE
LUBE
GOVERNOR
GOV PRESSURE TAP
CONV CL ACCUM
Conv Cl Reg
CONV FEED
THERMO SWITCH
SOLENOID
INPUT ACCUM
INPUT CL
3-2 Line Control
CONV CL BLOW OFF
Conv Cl Apply
TEMP SWITCH
COOLER CHECK BALL
COOLER
CONV CL SIGNAL
LINE PRESSURE TAP
P R N D 3 2 1
Manual Valve
2-3 SHIFT SOLENOID
3rd CL SIG SW
2-3 Shift
3-2 Coast
4th CL SIG SW
AIR BLEED
DECREASE
Press Reg
PUMP
REVERSE SERVO BOOST
3-4 ACCUM
1-2 ACCUM
1-2 Control
1-2 Shift
1-2 TV
LO BLOW OFF
STRAINER
SUMP
THERMO ELEMENT
PRESSURE RELIEF
REVERSE SERVO
VACUUM MODULATOR ASS'Y
Modulator
3-4 SHIFT SOLENOID
3-4 Shift
TV Limit
BLEED ORIFICE
1-2 SERVO
1-2 SERVO BOOST
2-3 Accum
1-2 Accum
Throttle
TV Plunger
4-3 MTV
2-3 ACCUM EX
3rd/3rd CL EX

DRIVE RANGE—1ST GEAR-F7

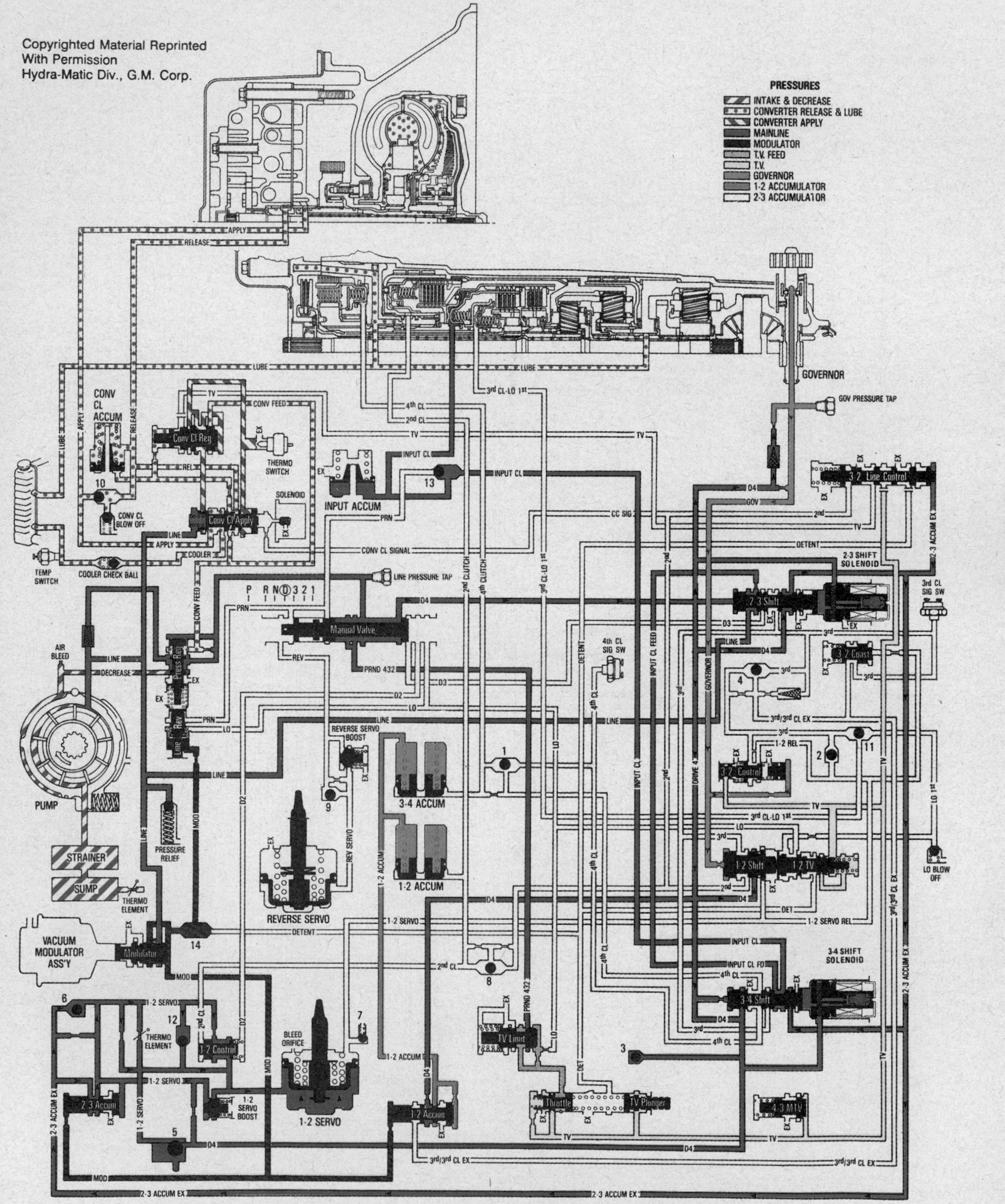
PRESSURES
INTAKE & DECREASE
CONVERTER RELEASE & LUBE
CONVERTER APPLY
MAINLINE
MODULATOR
T.V. FEED
T.V.
GOVERNOR
1-2 ACCUMULATOR
2-3 ACCUMULATOR
GOVERNOR
GOV PRESSURE TAP
CONV CL ACCUM
Conv Cl Reg
THERMO SWITCH
SOLENOID
CONV CL BLOW OFF
Conv Cl Apply
TEMP SWITCH
COOLER CHECK BALL
INPUT ACCUM
LINE PRESSURE TAP
Manual Valve
P R N D 3 2 1
AIR BLEED
PUMP
STRAINER
SUMP
THERMO ELEMENT
PRESSURE RELIEF
VACUUM MODULATOR ASS'Y
Modulator
REVERSE SERVO BOOST
REVERSE SERVO
3-4 ACCUM
1-2 ACCUM
3-2 Line Control
2-3 SHIFT SOLENOID
2-3 Shift
3rd CL SIG SW
3-2 Coast
4th CL SIG SW
3-2 Control
1-2 Shift
1-2 TV
LO BLOW OFF
3-4 SHIFT SOLENOID
3-4 Shift
1-2 Control
BLEED ORIFICE
1-2 SERVO
1-2 SERVO BOOST
2-3 Accum
1-2 Accum
TV Limit
Throttle
TV Plunger
4-3 M1V
2-3 ACCUM EX

DRIVE RANGE—2ND GEAR—F7

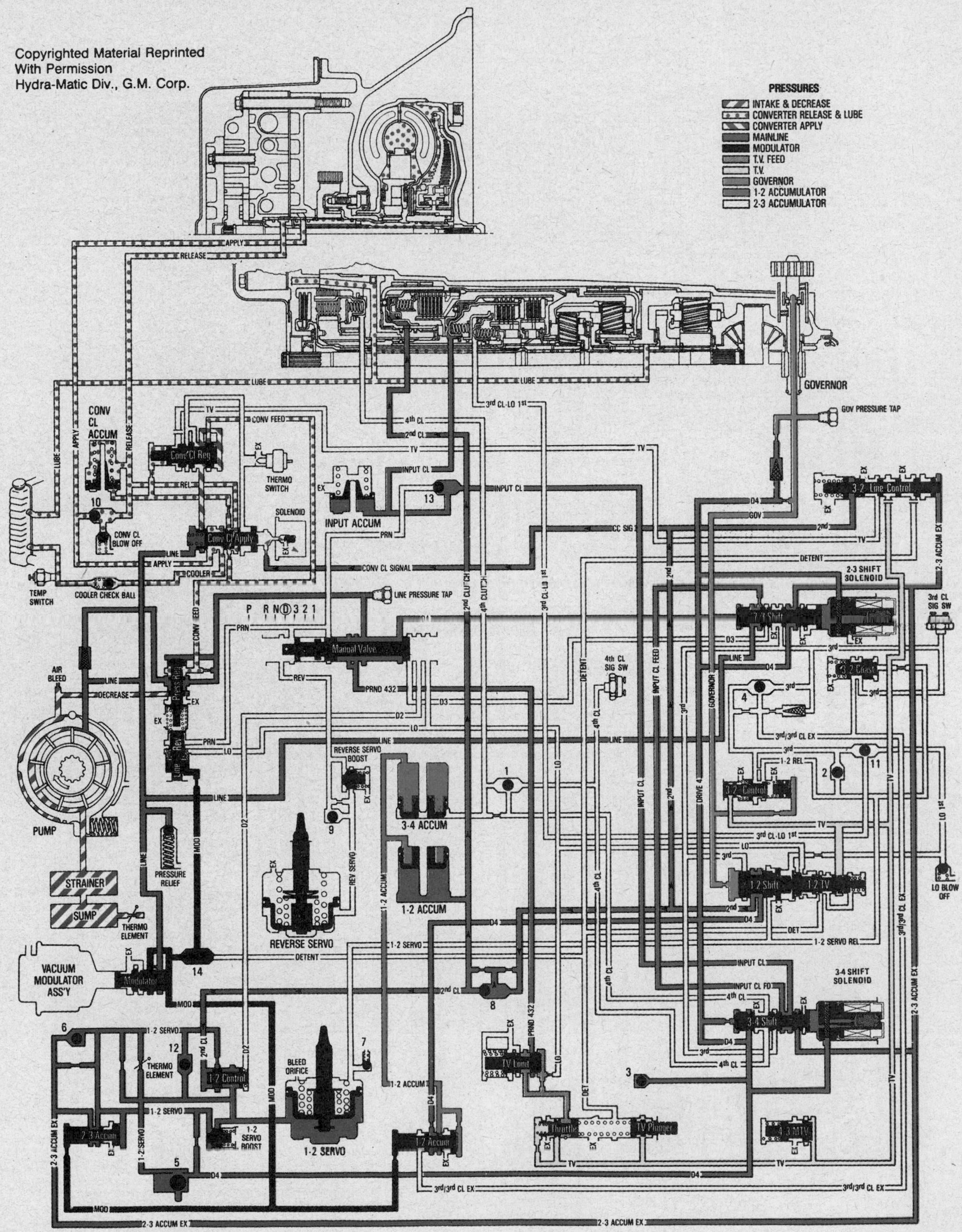
PRESSURES
INTAKE & DECREASE
CONVERTER RELEASE & LUBE
CONVERTER APPLY
MAINLINE
MODULATOR
T.V. FEED
T.V.
GOVERNOR
1-2 ACCUMULATOR
2-3 ACCUMULATOR
APPLY
RELEASE
LUBE
GOVERNOR
GOV PRESSURE TAP
CONV CL ACCUM
Conv Cl Reg
CONV FEED
THERMO SWITCH
SOLENOID
Conv Cl Apply
CONV CL BLOW OFF
COOLER
COOLER CHECK BALL
TEMP SWITCH
INPUT ACCUM
INPUT CL
CONV CL SIGNAL
LINE PRESSURE TAP
P R N D 3 2 1
Manual Valve
3-2 Line Control
2-3 SHIFT SOLENOID
3rd CL SIG SW
2-3 Shift
3-2 Coast
4th CL SIG SW
3-2 Control
1-2 Shift
1-2 TV
LO BLOW OFF
3-4 SHIFT SOLENOID
3-4 Shift
AIR BLEED
DECREASE
Press Reg
Rev
PUMP
STRAINER
SUMP
THERMO ELEMENT
PRESSURE RELIEF
REVERSE SERVO BOOST
3-4 ACCUM
1-2 ACCUM
REVERSE SERVO
VACUUM MODULATOR ASS'Y
Modulator
1-2 Control
BLEED ORIFICE
2-3 Accum
1-2 SERVO BOOST
1-2 SERVO
1-2 Accum
TV Load
Throttle
TV Plunger
4-3 MTV
2-3 ACCUM EX
3rd/3rd CL EX

DRIVE RANGE—3RD GEAR—F7

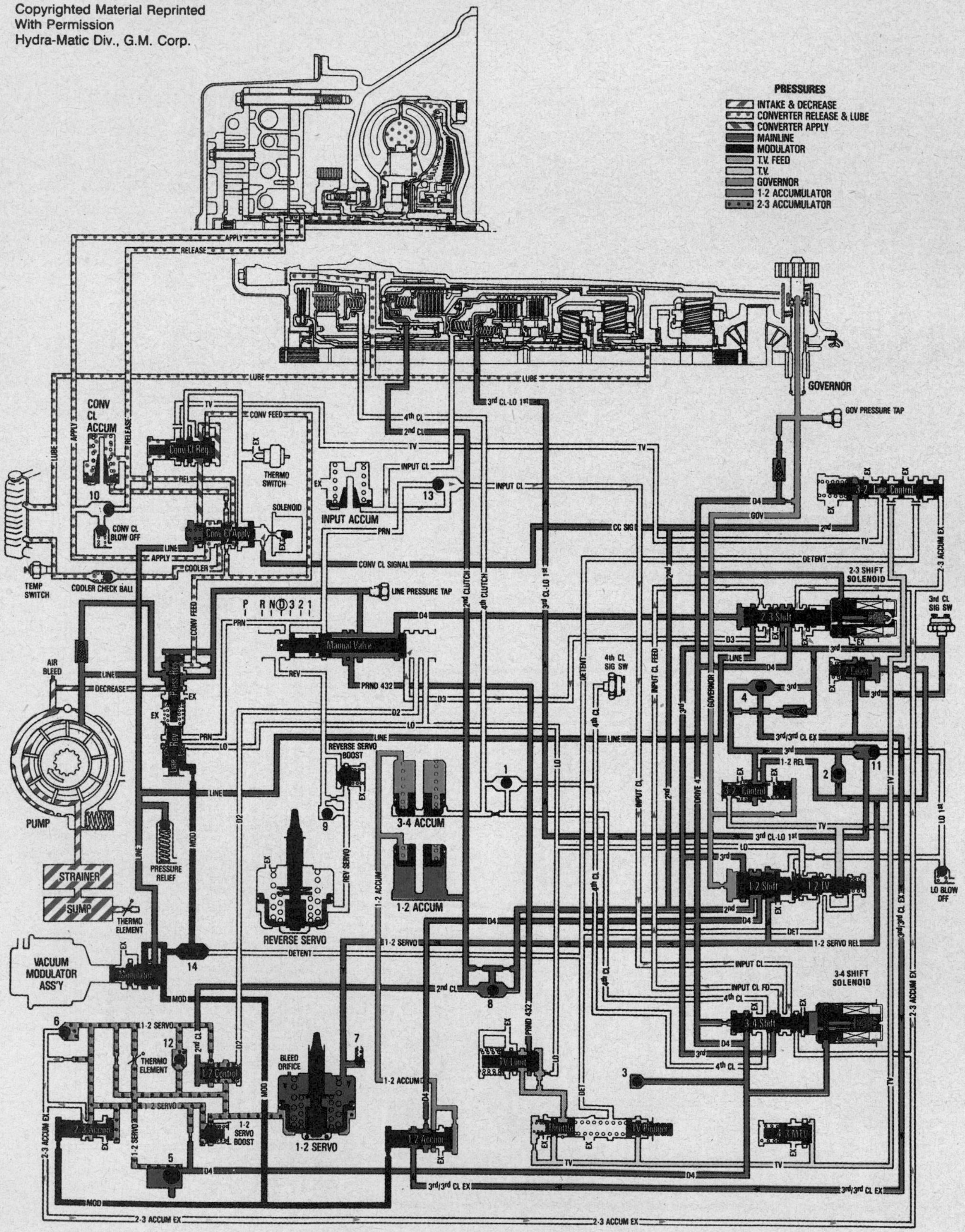
PRESSURES
INTAKE & DECREASE
CONVERTER RELEASE & LUBE
CONVERTER APPLY
MAINLINE
MODULATOR
T.V. FEED
T.V.
GOVERNOR
1-2 ACCUMULATOR
2-3 ACCUMULATOR
APPLY
RELEASE
LUBE
GOVERNOR
GOV PRESSURE TAP
CONV CL ACCUM
Conv Cl Reg
THERMO SWITCH
SOLENOID
INPUT ACCUM
Conv Cl Apply
CONV CL BLOW OFF
TEMP SWITCH
COOLER CHECK BALL
LINE PRESSURE TAP
P R N D 3 2 1
Manual Valve
3-2 Line Control
2-3 SHIFT SOLENOID
3rd CL SIG SW
2-3 Shift
3-2 Boost
4th CL SIG SW
AIR BLEED
PUMP
STRAINER
SUMP
THERMO ELEMENT
PRESSURE RELIEF
REVERSE SERVO BOOST
3-4 ACCUM
1-2 ACCUM
REVERSE SERVO
3-2 Control
1-2 Shift
1-2 TV
LO BLOW OFF
3-4 SHIFT SOLENOID
3-4 Shift
VACUUM MODULATOR ASS'Y
Modulator
1-2 Control
2-3 Accum
1-2 SERVO BOOST
BLEED ORIFICE
1-2 SERVO
1-2 Accum
TV Limit
Throttle
TV Plunger
4-3 MTV
2-3 ACCUM EX

CONVERTER CLUTCH APPLIED—3RD GEAR—F7

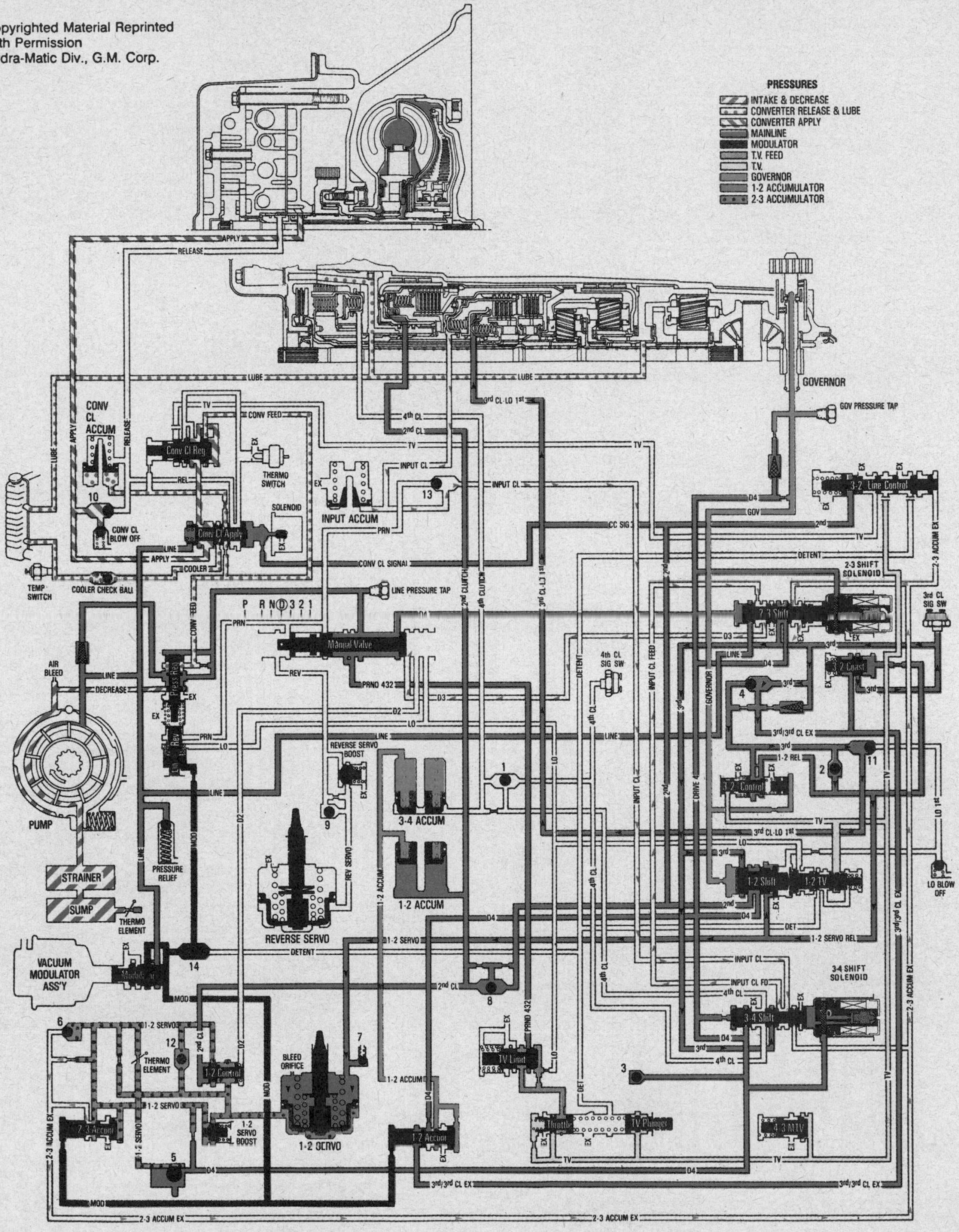
PRESSURES
INTAKE & DECREASE
CONVERTER RELEASE & LUBE
CONVERTER APPLY
MAINLINE
MODULATOR
T.V. FEED
T.V.
GOVERNOR
1-2 ACCUMULATOR
2-3 ACCUMULATOR
APPLY
RELEASE
LUBE
GOVERNOR
GOV PRESSURE TAP
CONV CL ACCUM
Conv Cl Reg
CONV FEED
THERMO SWITCH
SOLENOID
Conv Cl Apply
CONV CL BLOW OFF
INPUT ACCUM
INPUT CL
COOLER
TEMP SWITCH
COOLER CHECK BALL
CONV CL SIGNAL
LINE PRESSURE TAP
P R N D 3 2 1
Manual Valve
3-2 Line Control
2-3 SHIFT SOLENOID
2-3 Shift
3rd CL SIG SW
3-2 Coast
4th CL SIG SW
AIR BLEED
DECREASE
Press Reg
PUMP
REVERSE SERVO BOOST
3-4 ACCUM
1-2 ACCUM
3-2 Control
PRESSURE RELIEF
STRAINER
SUMP
THERMO ELEMENT
REVERSE SERVO
1-2 Shift
1-2 TV
LO BLOW OFF
VACUUM MODULATOR ASS'Y
Modulator
3-4 SHIFT SOLENOID
3-4 Shift
1-2 Control
BLEED ORIFICE
TV Limit
2-3 Accum
1-2 SERVO BOOST
1-2 SERVO
1-2 Accum
Throttle
TV Plunger
4-3 MTV
2-3 ACCUM EX

DRIVE RANGE—OVERDRIVE

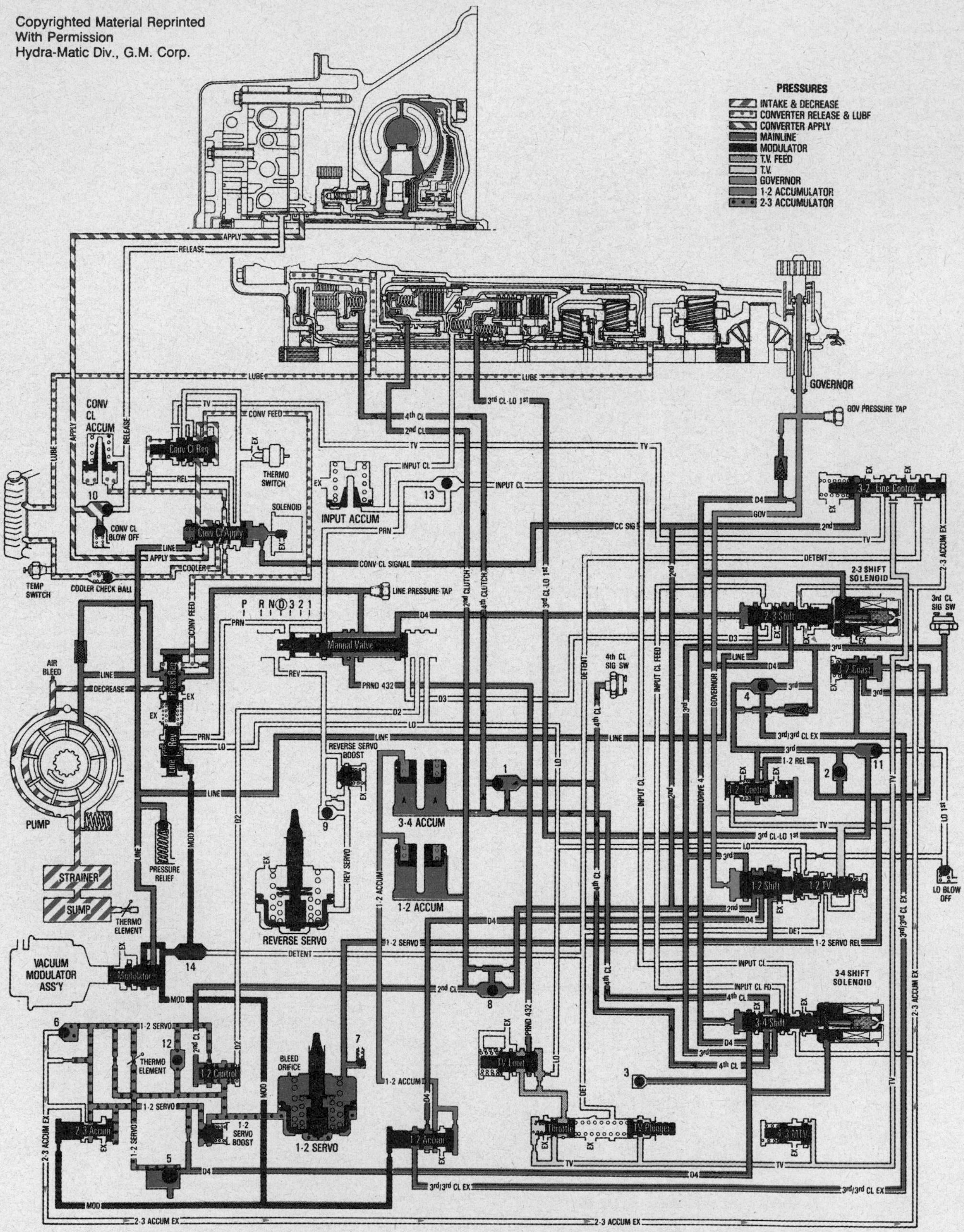
PRESSURES
INTAKE & DECREASE
CONVERTER RELEASE & LUBE
CONVERTER APPLY
MAINLINE
MODULATOR
T.V. FEED
T.V.
GOVERNOR
1-2 ACCUMULATOR
2-3 ACCUMULATOR
APPLY
RELEASE
LUBE
GOVERNOR
GOV PRESSURE TAP
CONV CL ACCUM
Conv Cl Reg
CONV FEED
THERMO SWITCH
SOLENOID
Conv Cl Apply
CONV CL BLOW OFF
COOLER
COOLER CHECK BALL
TEMP SWITCH
INPUT ACCUM
CONV CL SIGNAL
LINE PRESSURE TAP
P R N D 3 2 1
Manual Valve
3-2 Line Control
2-3 SHIFT SOLENOID
2-3 Shift
3rd CL SIG SW
3-2 Coast
4th CL SIG SW
AIR BLEED
DECREASE
Press Reg
Rev Boost
Line Boost
PUMP
REVERSE SERVO BOOST
3-4 ACCUM
1-2 ACCUM
REVERSE SERVO
PRESSURE RELIEF
STRAINER
SUMP
THERMO ELEMENT
VACUUM MODULATOR ASS'Y
Modulator
3-2 Control
1-2 Shift
1-2 TV
LO BLOW OFF
3-4 SHIFT SOLENOID
3-4 Shift
TV Limit
Throttle
TV Plunger
4-3 MTV
2-3 Accum
1-2 Control
1-2 SERVO BOOST
BLEED ORIFICE
1-2 SERVO
1-2 Accum
2-3 ACCUM EX
3rd/3rd CL EX

PART THROTTLE 4-3 AND MODULATED DOWNSHIFT (VALVES SHOWN IN 3RD GEAR POSITION)—F7

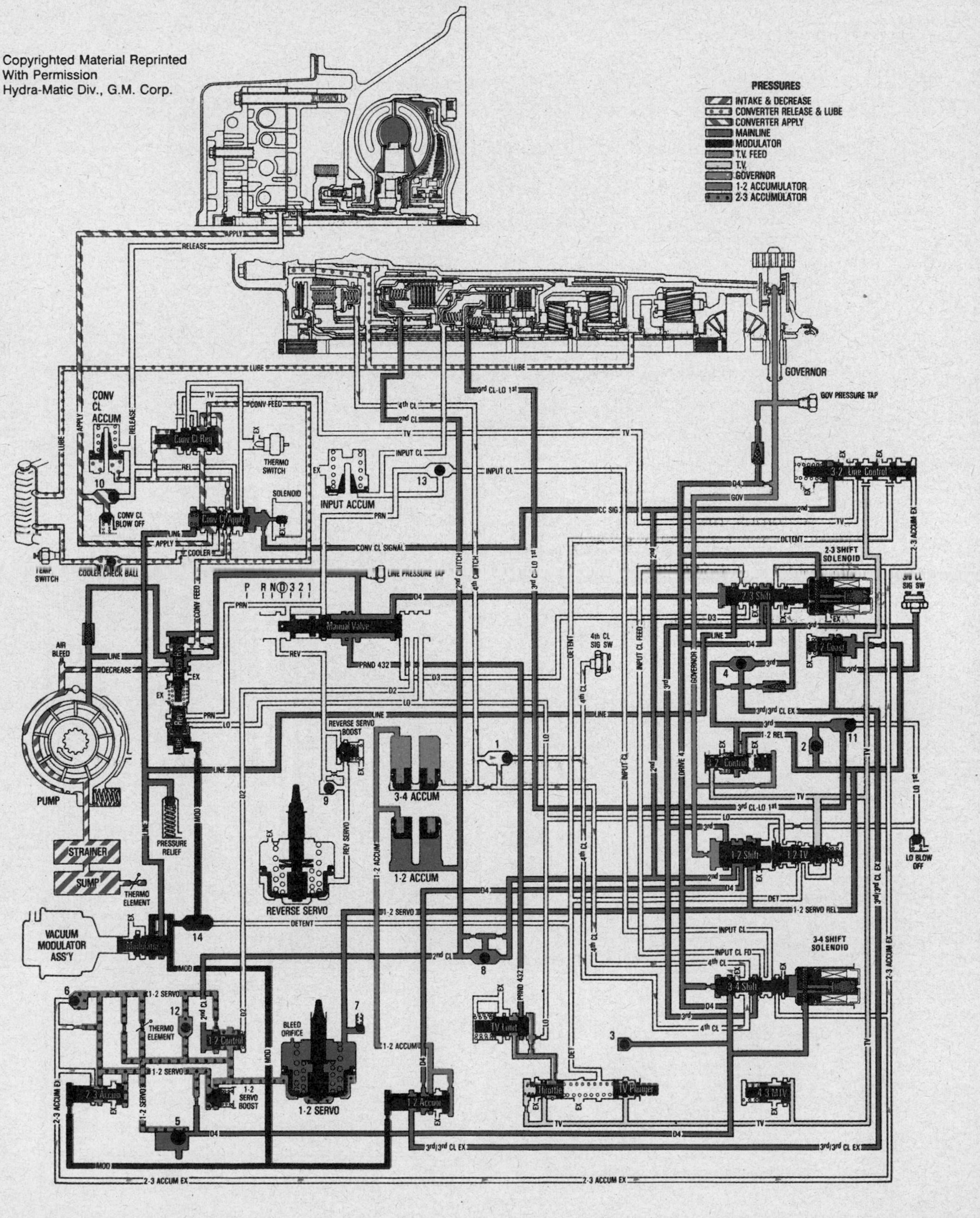
PRESSURES
INTAKE & DECREASE
CONVERTER RELEASE & LUBE
CONVERTER APPLY
MAINLINE
MODULATOR
T.V. FEED
T.V.
GOVERNOR
1-2 ACCUMULATOR
2-3 ACCUMULATOR
APPLY
RELEASE
LUBE
GOVERNOR
GOV PRESSURE TAP
CONV CL ACCUM
Conv Cl Reg
CONV FEED
THERMO SWITCH
SOLENOID
Conv Cl Apply
CONV CL BLOW OFF
INPUT ACCUM
INPUT CL
CC SIG
CONV CL SIGNAL
COOLER
COOLER CHECK BALL
TEMP SWITCH
LINE PRESSURE TAP
P R N D 3 2 1
Manual Valve
3-2 Line Control
2-3 SHIFT SOLENOID
2-3 Shift
3rd CL SIG SW
3-2 Coast
4th CL SIG SW
AIR BLEED
LINE
DECREASE
Press Reg
PUMP
PRESSURE RELIEF
STRAINER
SUMP
THERMO ELEMENT
REVERSE SERVO BOOST
3-4 ACCUM
1-2 ACCUM
REVERSE SERVO
2-3 Control
LO BLOW OFF
1-2 Shift
1-2 TV
VACUUM MODULATOR ASS'Y
Modulator
3-4 SHIFT SOLENOID
3-4 Shift
TV Limit
1-2 Control
BLEED ORIFICE
1-2 SERVO
1-2 SERVO BOOST
2-3 Accum
1-2 Accum
Throttle
TV Plunger
3-4 MTV
2-3 ACCUM EX
3rd/3rd CL EX
MOD
DETENT
D4

DETENT DOWNSHIFTS (VALVES SHOWN IN 2ND GEAR POSITION)—F7

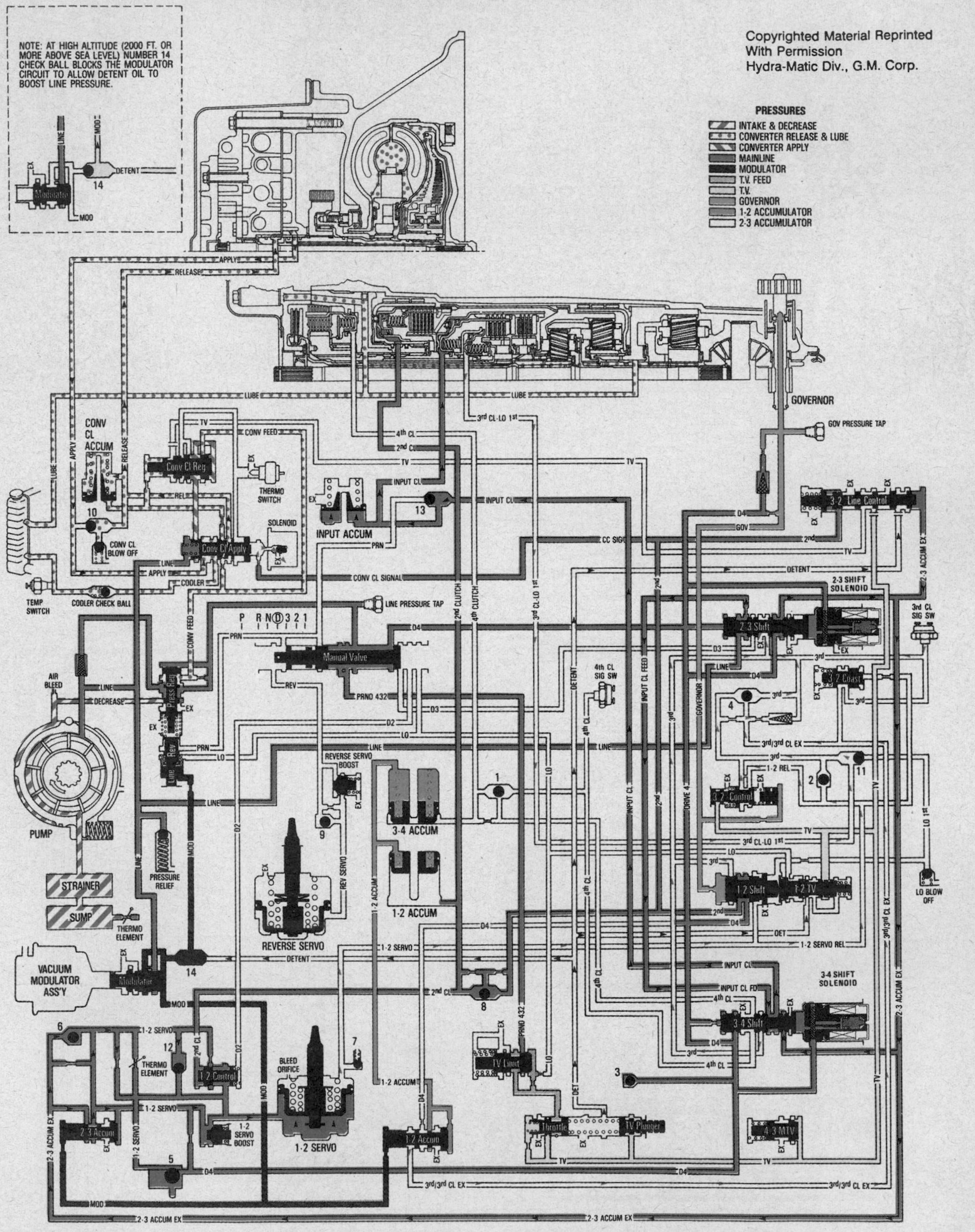
NOTE: AT HIGH ALTITUDE (2000 FT. OR MORE ABOVE SEA LEVEL) NUMBER 14 CHECK BALL BLOCKS THE MODULATOR CIRCUIT TO ALLOW DETENT OIL TO BOOST LINE PRESSURE.
PRESSURES
INTAKE & DECREASE
CONVERTER RELEASE & LUBE
CONVERTER APPLY
MAINLINE
MODULATOR
T.V. FEED
T.V.
GOVERNOR
1-2 ACCUMULATOR
2-3 ACCUMULATOR
GOVERNOR
GOV PRESSURE TAP
CONV CL ACCUM
Conv Cl Reg
THERMO SWITCH
SOLENOID
CONV CL BLOW OFF
Conv Cl Apply
TEMP SWITCH
COOLER CHECK BALL
INPUT ACCUM
3-2 Line Control
LINE PRESSURE TAP
P R N D 3 2 1
Manual Valve
2-3 SHIFT SOLENOID
2-3 Shift
3rd CL SIG SW
3-2 Coast
4th CL SIG SW
AIR BLEED
Press Reg
PUMP
REVERSE SERVO BOOST
3-4 ACCUM
1-2 ACCUM
3-2 Control
PRESSURE RELIEF
STRAINER
SUMP
THERMO ELEMENT
REVERSE SERVO
1-2 Shift
1-2 TV
LO BLOW OFF
VACUUM MODULATOR ASS'Y
Modulator
3-4 SHIFT SOLENOID
3-4 Shift
1-2 Control
BLEED ORIFICE
TV Limit
2-3 Accum
1-2 SERVO BOOST
1-2 SERVO
1-2 Accum
Throttle
TV Plunger
4-3 MTV

MANUAL 3RD—F7

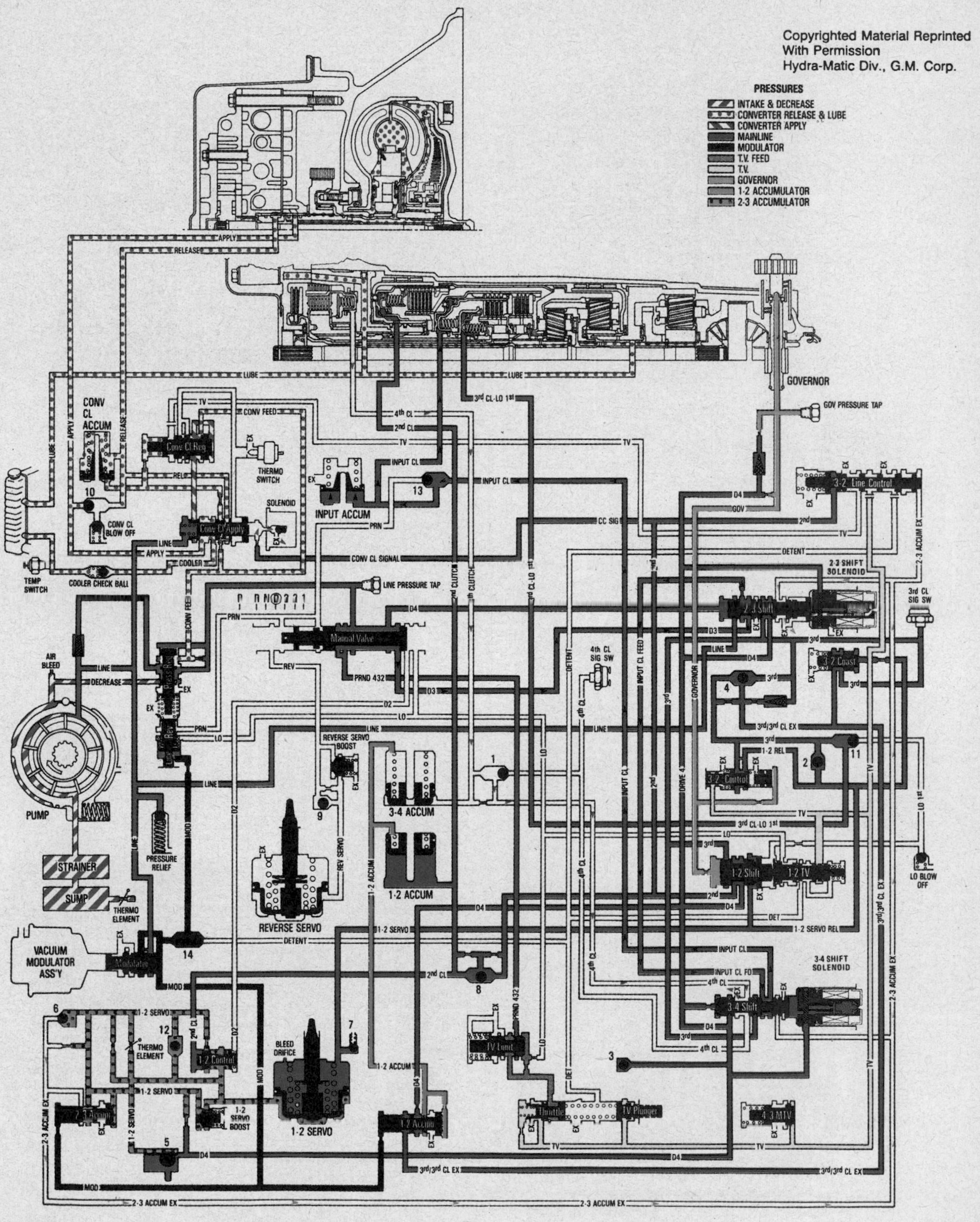
PRESSURES
INTAKE & DECREASE
CONVERTER RELEASE & LUBE
CONVERTER APPLY
MAINLINE
MODULATOR
T.V. FEED
T.V.
GOVERNOR
1-2 ACCUMULATOR
2-3 ACCUMULATOR
APPLY
RELEASE
LUBE
GOVERNOR
GOV PRESSURE TAP
CONV CL ACCUM
CONV FEED
Conv Cl Reg
THERMO SWITCH
SOLENOID
Conv Cl Apply
CONV CL BLOW OFF
TEMP SWITCH
COOLER CHECK BALL
INPUT ACCUM
INPUT CL
CONV CL SIGNAL
LINE PRESSURE TAP
Manual Valve
3-2 Line Control
2-3 SHIFT SOLENOID
2-3 Shift
3rd CL SIG SW
3-2 Coast
4th CL SIG SW
AIR BLEED
Press Reg
PUMP
REVERSE SERVO BOOST
3-4 ACCUM
1-2 ACCUM
REVERSE SERVO
PRESSURE RELIEF
STRAINER
SUMP
THERMO ELEMENT
VACUUM MODULATOR ASS'Y
Modulator
3-2 Control
1-2 Shift
1-2 TV
LO BLOW OFF
3-4 SHIFT SOLENOID
3-4 Shift
TV Limit
Throttle
TV Plunger
4-3 MTV
1-2 Control
2-3 Accum
1-2 SERVO BOOST
BLEED ORIFICE
1-2 SERVO
1-2 Accum
2-3 ACCUM EX

MANUAL 2ND—F7

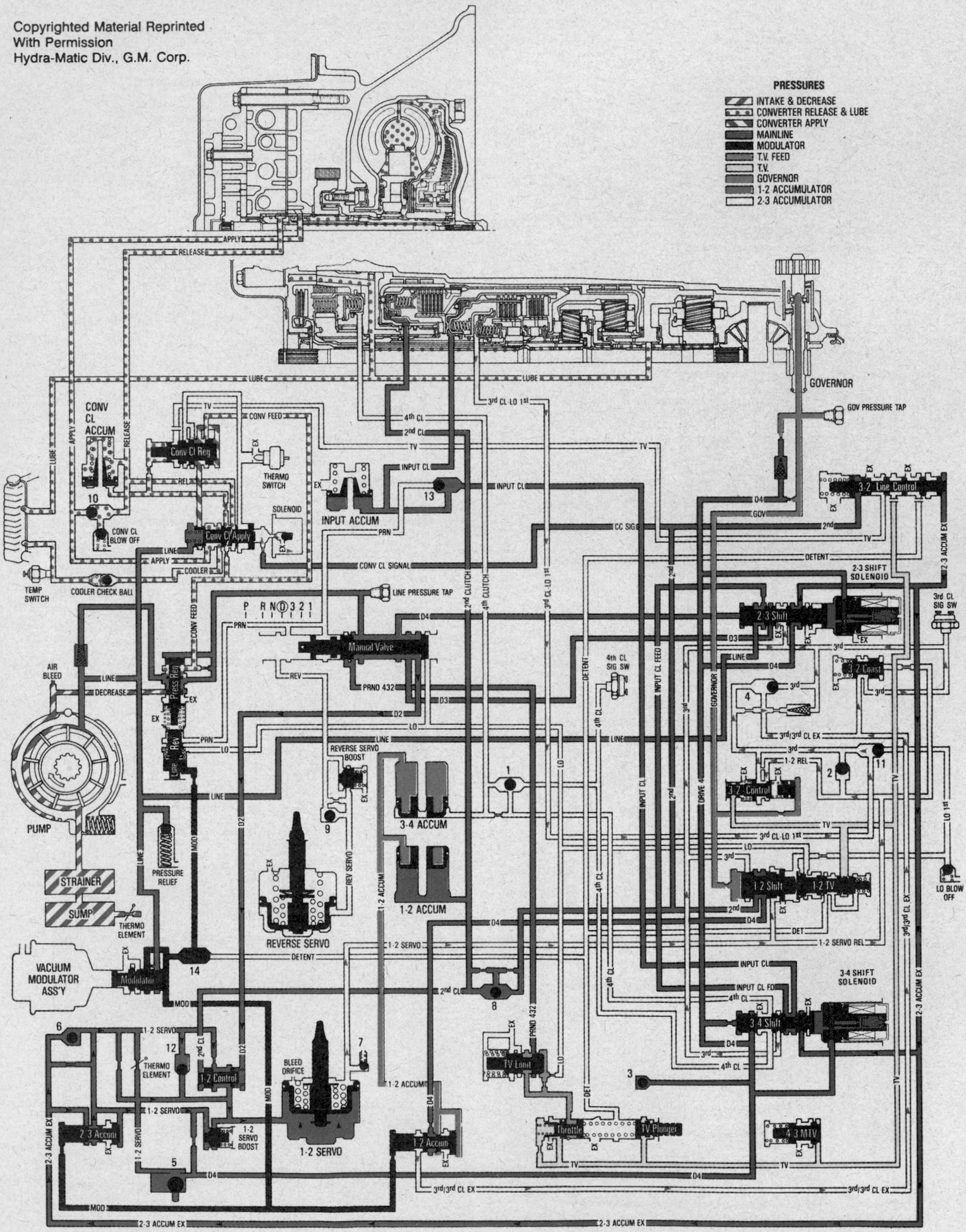
PRESSURES
INTAKE & DECREASE
CONVERTER RELEASE & LUBE
CONVERTER APPLY
MAINLINE
MODULATOR
T.V. FEED
T.V.
GOVERNOR
1-2 ACCUMULATOR
2-3 ACCUMULATOR
APPLY
RELEASE
LUBE
GOVERNOR
GOV PRESSURE TAP
CONV CL ACCUM
Conv Cl Reg
CONV FEED
THERMO SWITCH
SOLENOID
INPUT ACCUM
INPUT CL
4th CL
2nd CL
3rd CL-LO 1st
CONV CL BLOW OFF
Conv Cl Apply
CONV CL SIGNAL
CC SIG
TEMP SWITCH
COOLER CHECK BALL
COOLER
LINE PRESSURE TAP
P R N D 3 2 1
Manual Valve
3-2 Line Control
2-3 SHIFT SOLENOID
2-3 Shift
3rd CL SIG SW
3-2 Coast
4th CL SIG SW
DETENT
AIR BLEED
DECREASE
Press Reg
Rev
PUMP
STRAINER
SUMP
THERMO ELEMENT
PRESSURE RELIEF
REVERSE SERVO BOOST
REVERSE SERVO
3-4 ACCUM
1-2 ACCUM
3-2 Control
1-2 Shift
1-2 TV
LO BLOW OFF
1-2 SERVO REL
3-4 SHIFT SOLENOID
3-4 Shift
VACUUM MODULATOR ASS'Y
Modulator
1-2 Control
2-3 Accum
1-2 SERVO BOOST
BLEED ORIFICE
1-2 SERVO
1-2 Accum
TV Limit
Throttle
TV Plunger
4-3 MTV
3rd/3rd CL EX
2-3 ACCUM EX
MOD

MANUAL LOW—F7

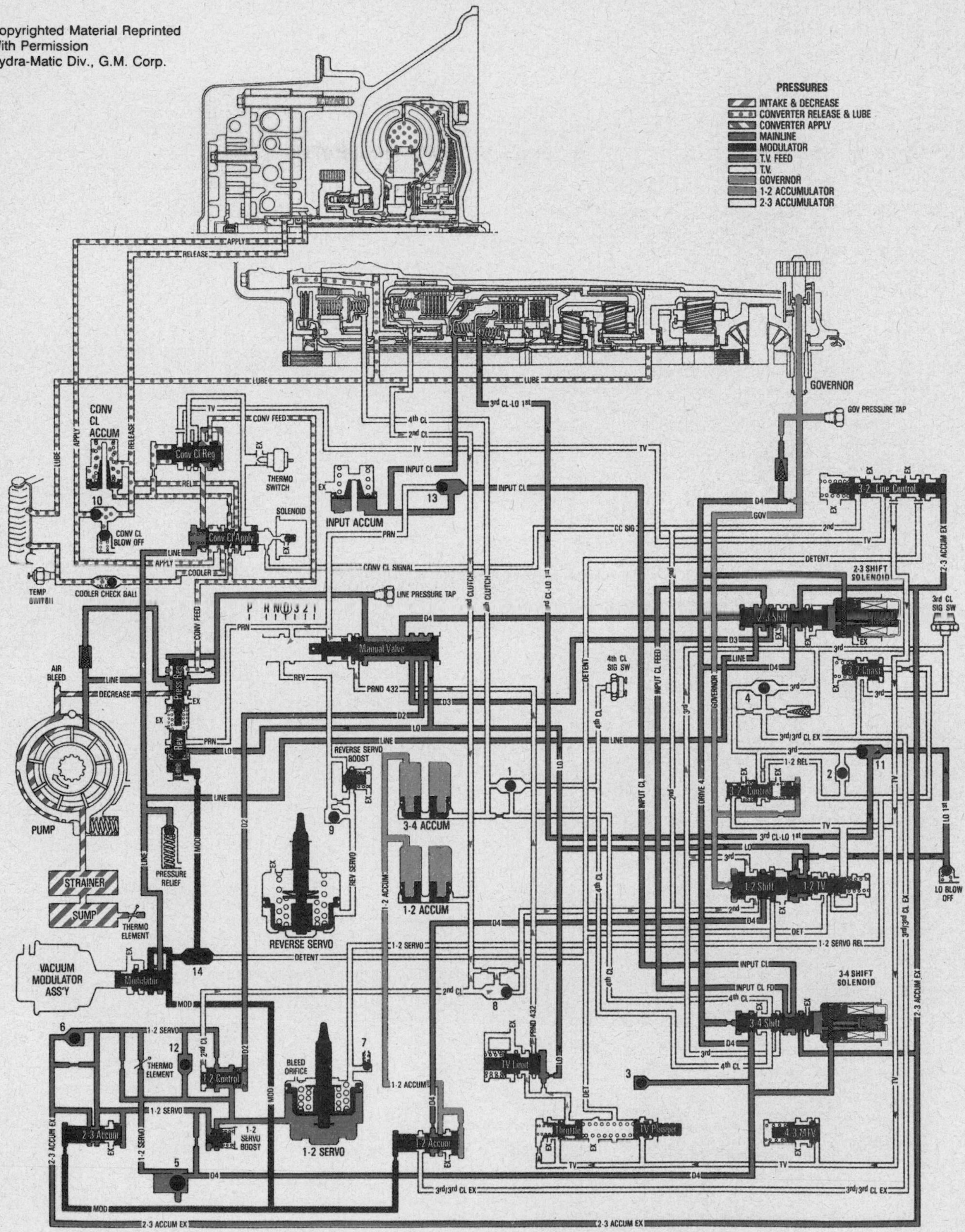
PRESSURES
INTAKE & DECREASE
CONVERTER RELEASE & LUBE
CONVERTER APPLY
MAINLINE
MODULATOR
T.V. FEED
T.V.
GOVERNOR
1-2 ACCUMULATOR
2-3 ACCUMULATOR
APPLY
RELEASE
LUBE
GOVERNOR
GOV PRESSURE TAP
CONV CL ACCUM
Conv Cl Reg
THERMO SWITCH
SOLENOID
Conv Cl Apply
CONV CL BLOW OFF
TEMP SWITCH
COOLER CHECK BALL
INPUT ACCUM
3-2 Line Control
2-3 SHIFT SOLENOID
2-3 Shift
3rd CL SIG SW
LINE PRESSURE TAP
Manual Valve
3-2 Coast
4th CL SIG SW
AIR BLEED
Press Reg
PUMP
REVERSE SERVO BOOST
3-4 ACCUM
3-2 Control
1-2 ACCUM
PRESSURE RELIEF
STRAINER
SUMP
THERMO ELEMENT
REVERSE SERVO
1-2 Shift
1-2 TV
LO BLOW OFF
VACUUM MODULATOR ASS'Y
Modulator
3-4 SHIFT SOLENOID
3-4 Shift
TV Limit
BLEED ORIFICE
1-2 Control
2-3 Accum
1-2 SERVO BOOST
1-2 SERVO
1-2 Accum
Throttle
TV Plunger
4-3 MTV
2-3 ACCUM EX

REVERSE—F7

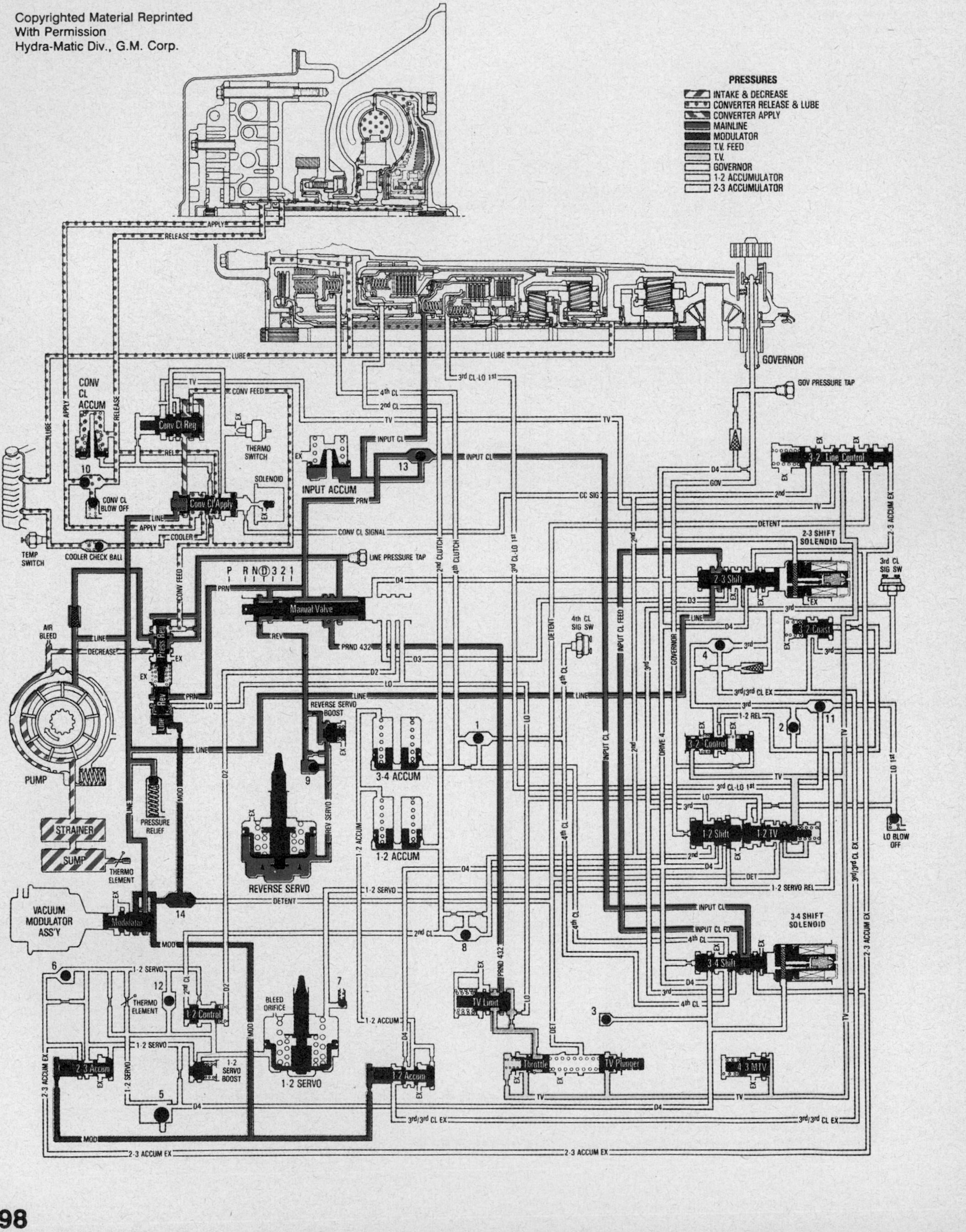
PRESSURES
INTAKE & DECREASE
CONVERTER RELEASE & LUBE
CONVERTER APPLY
MAINLINE
MODULATOR
T.V. FEED
T.V.
GOVERNOR
1-2 ACCUMULATOR
2-3 ACCUMULATOR
GOVERNOR
GOV PRESSURE TAP
CONV CL ACCUM
Conv Cl Reg
THERMO SWITCH
SOLENOID
INPUT ACCUM
Conv Cl Apply
CONV CL BLOW OFF
TEMP SWITCH
COOLER CHECK BALL
LINE PRESSURE TAP
Manual Valve
AIR BLEED
Press Reg
PUMP
STRAINER
SUMP
THERMO ELEMENT
PRESSURE RELIEF
REVERSE SERVO BOOST
REVERSE SERVO
3-4 ACCUM
1-2 ACCUM
VACUUM MODULATOR ASS'Y
Modulator
1-2 Control
BLEED ORIFICE
1-2 SERVO
2-3 Accum
1-2 SERVO BOOST
1-2 Accum
TV Limit
Throttle
TV Plunger
4-3 MTV
3-2 Line Control
2-3 SHIFT SOLENOID
2-3 Shift
3rd CL SIG SW
3-2 Coast
4th CL SIG SW
3-2 Control
1-2 Shift
1-2 TV
LO BLOW OFF
3-4 SHIFT SOLENOID
3-4 Shift

PARK—KF100

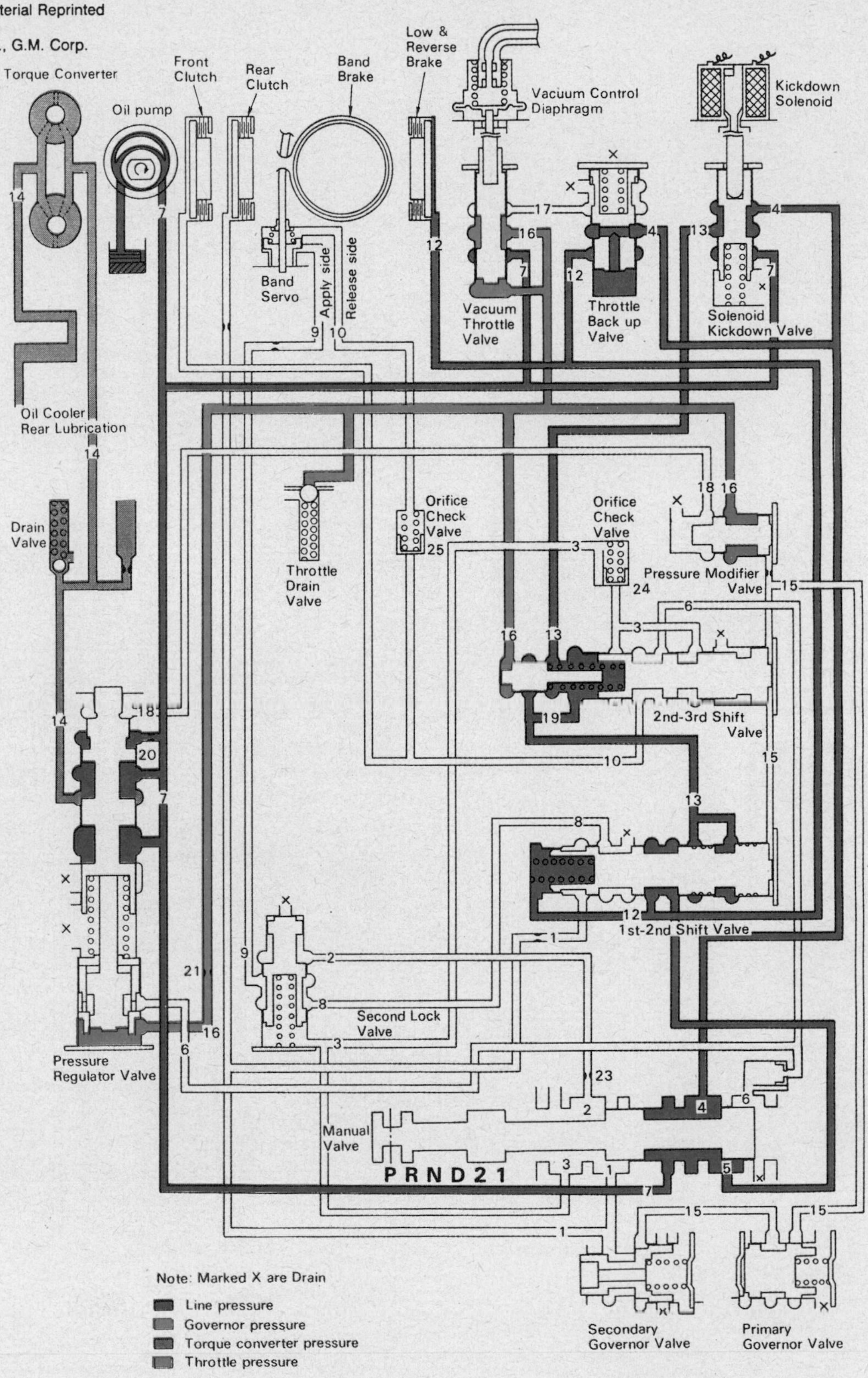
Torque Converter
Oil pump
Front Clutch
Rear Clutch
Band Brake
Low & Reverse Brake
Vacuum Control Diaphragm
Kickdown Solenoid
Band Servo
Apply side
Release side
Vacuum Throttle Valve
Throttle Back up Valve
Solenoid Kickdown Valve
Oil Cooler
Rear Lubrication
Drain Valve
Throttle Drain Valve
Orifice Check Valve
Orifice Check Valve
Pressure Modifier Valve
2nd-3rd Shift Valve
1st-2nd Shift Valve
Second Lock Valve
Pressure Regulator Valve
Manual Valve
P R N D 2 1
Secondary Governor Valve
Primary Governor Valve
Note: Marked X are Drain
Line pressure
Governor pressure
Torque converter pressure
Throttle pressure

REVERSE—KF100

Copyrighted Material Reprinted
With Permission
Hydra-Matic Div., G.M. Corp.

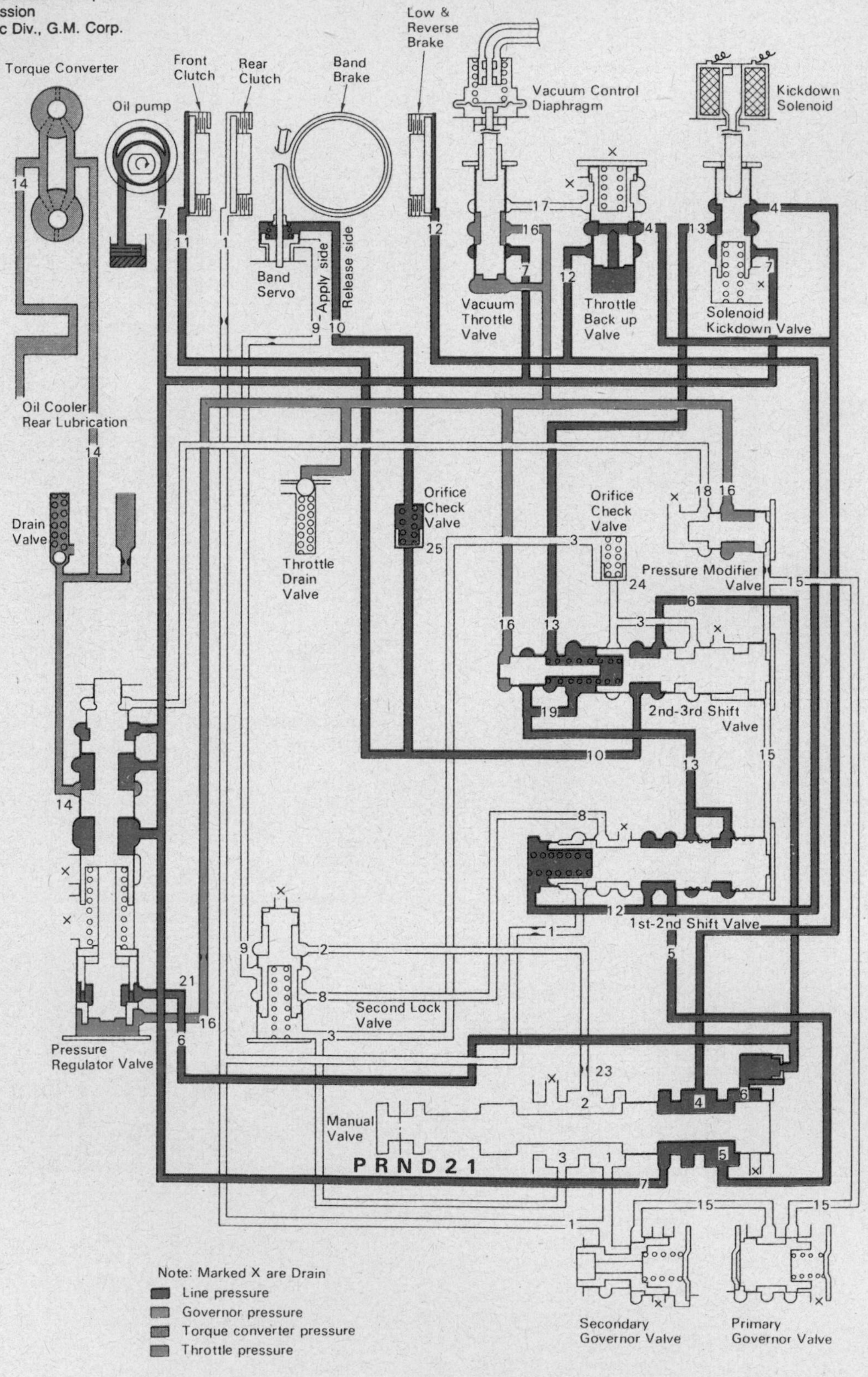
Low & Reverse Brake
Torque Converter
Front Clutch
Rear Clutch
Band Brake
Oil pump
Vacuum Control Diaphragm
Kickdown Solenoid
Band Servo
Apply side
Release side
Vacuum Throttle Valve
Throttle Back up Valve
Solenoid Kickdown Valve
Oil Cooler Rear Lubrication
Drain Valve
Throttle Drain Valve
Orifice Check Valve
Orifice Check Valve
Pressure Modifier Valve
2nd-3rd Shift Valve
1st-2nd Shift Valve
Second Lock Valve
Pressure Regulator Valve
Manual Valve
P R N D 2 1
Secondary Governor Valve
Primary Governor Valve
Note: Marked X are Drain
Line pressure
Governor pressure
Torque converter pressure
Throttle pressure

NEUTRAL—KF100

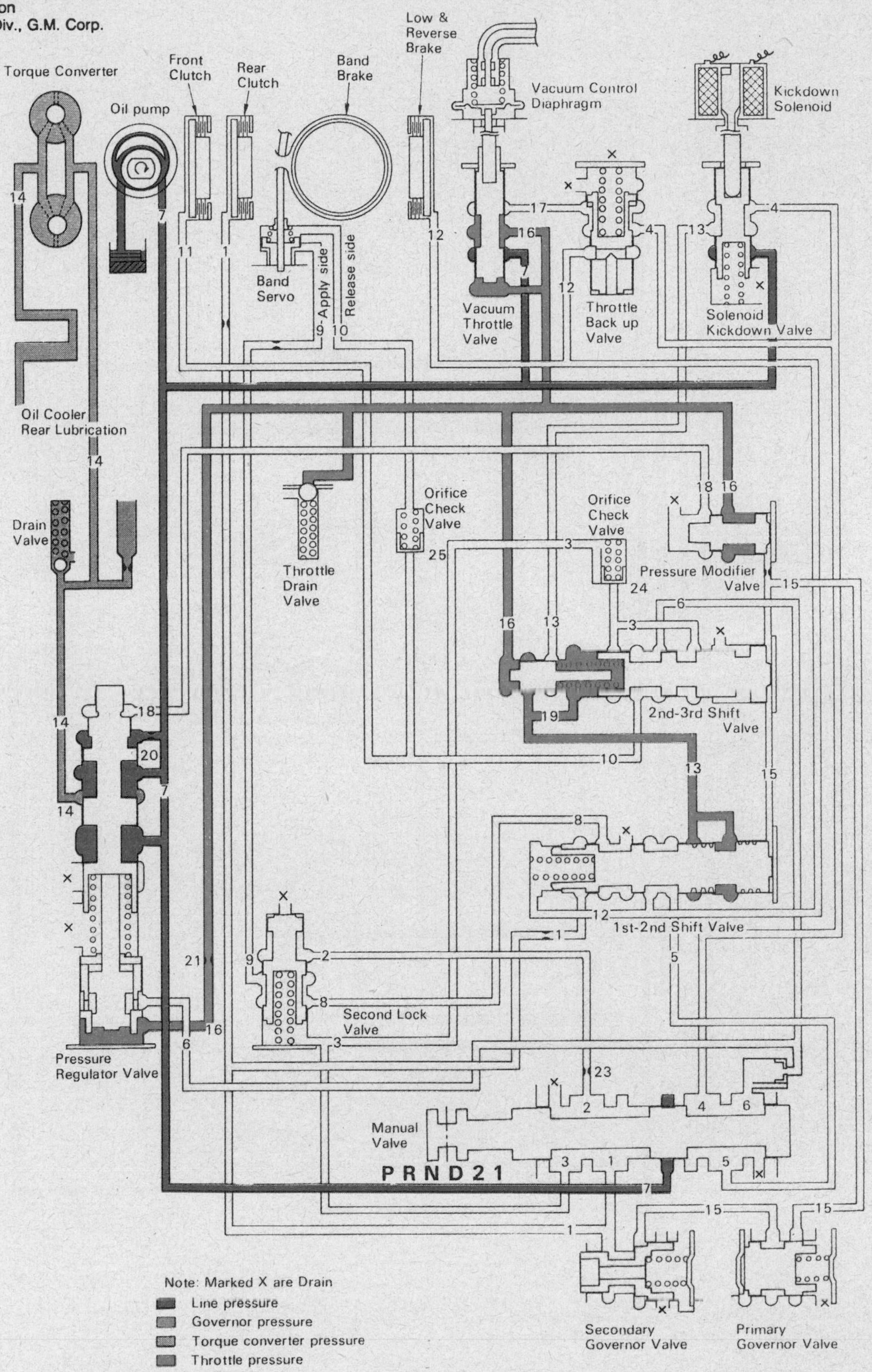
Torque Converter
Oil pump
Front Clutch
Rear Clutch
Band Brake
Low & Reverse Brake
Vacuum Control Diaphragm
Kickdown Solenoid
Band Servo
Apply side
Release side
Vacuum Throttle Valve
Throttle Back up Valve
Solenoid Kickdown Valve
Oil Cooler Rear Lubrication
Drain Valve
Throttle Drain Valve
Orifice Check Valve
Orifice Check Valve
Pressure Modifier Valve
2nd-3rd Shift Valve
1st-2nd Shift Valve
Second Lock Valve
Pressure Regulator Valve
Manual Valve
P R N D 2 1
Secondary Governor Valve
Primary Governor Valve
Note: Marked X are Drain
Line pressure
Governor pressure
Torque converter pressure
Throttle pressure

DRIVE—LOW GEAR—KF100

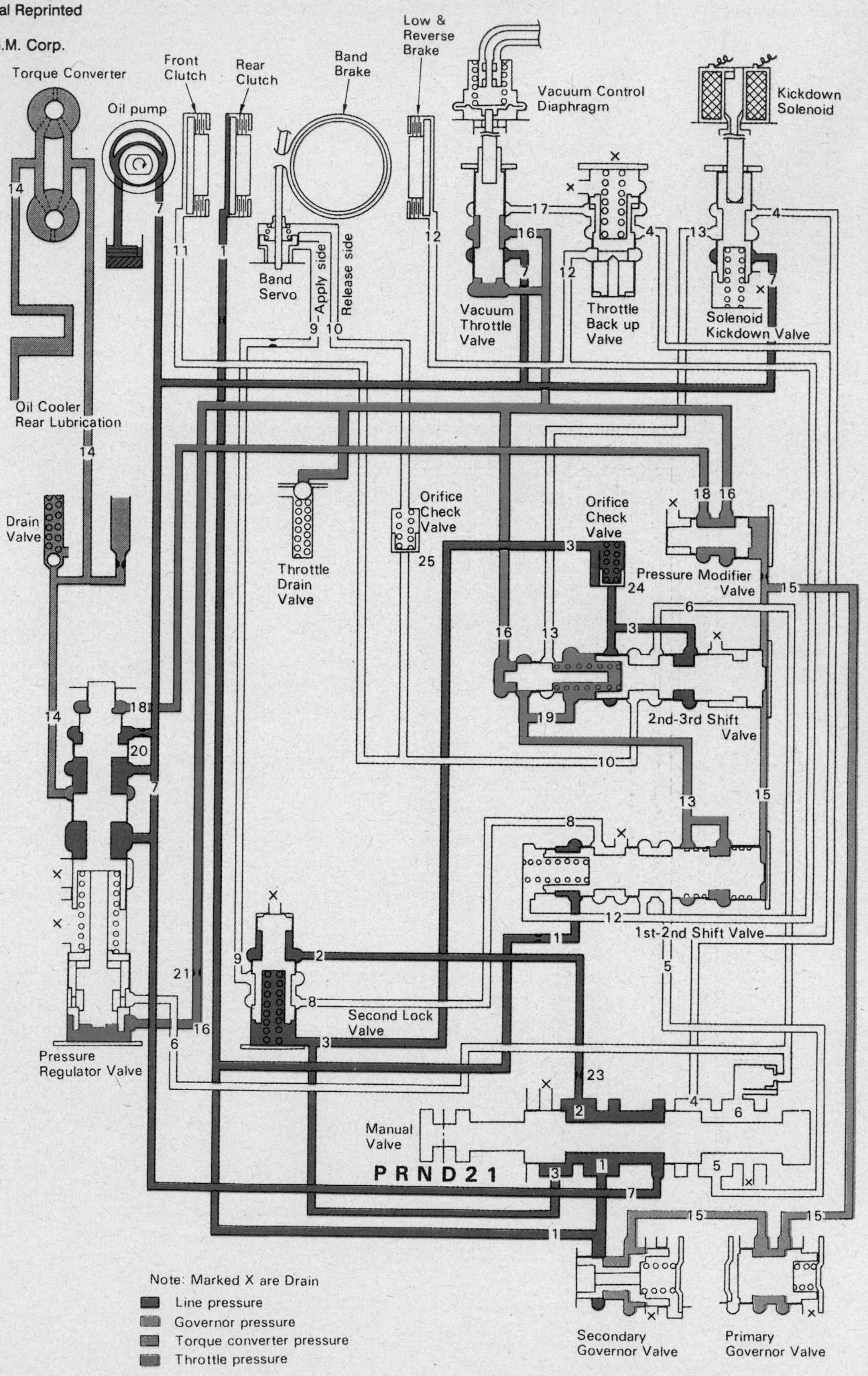
Torque Converter
Oil pump
Front Clutch
Rear Clutch
Band Brake
Low & Reverse Brake
Vacuum Control Diaphragm
Kickdown Solenoid
Band Servo
Apply side
Release side
Vacuum Throttle Valve
Throttle Back up Valve
Solenoid Kickdown Valve
Oil Cooler Rear Lubrication
Drain Valve
Throttle Drain Valve
Orifice Check Valve
Orifice Check Valve
Pressure Modifier Valve
2nd-3rd Shift Valve
1st-2nd Shift Valve
Second Lock Valve
Pressure Regulator Valve
Manual Valve
P R N D 2 1
Secondary Governor Valve
Primary Governor Valve
Note: Marked X are Drain
Line pressure
Governor pressure
Torque converter pressure
Throttle pressure

DRIVE—2ND GEAR—KF100

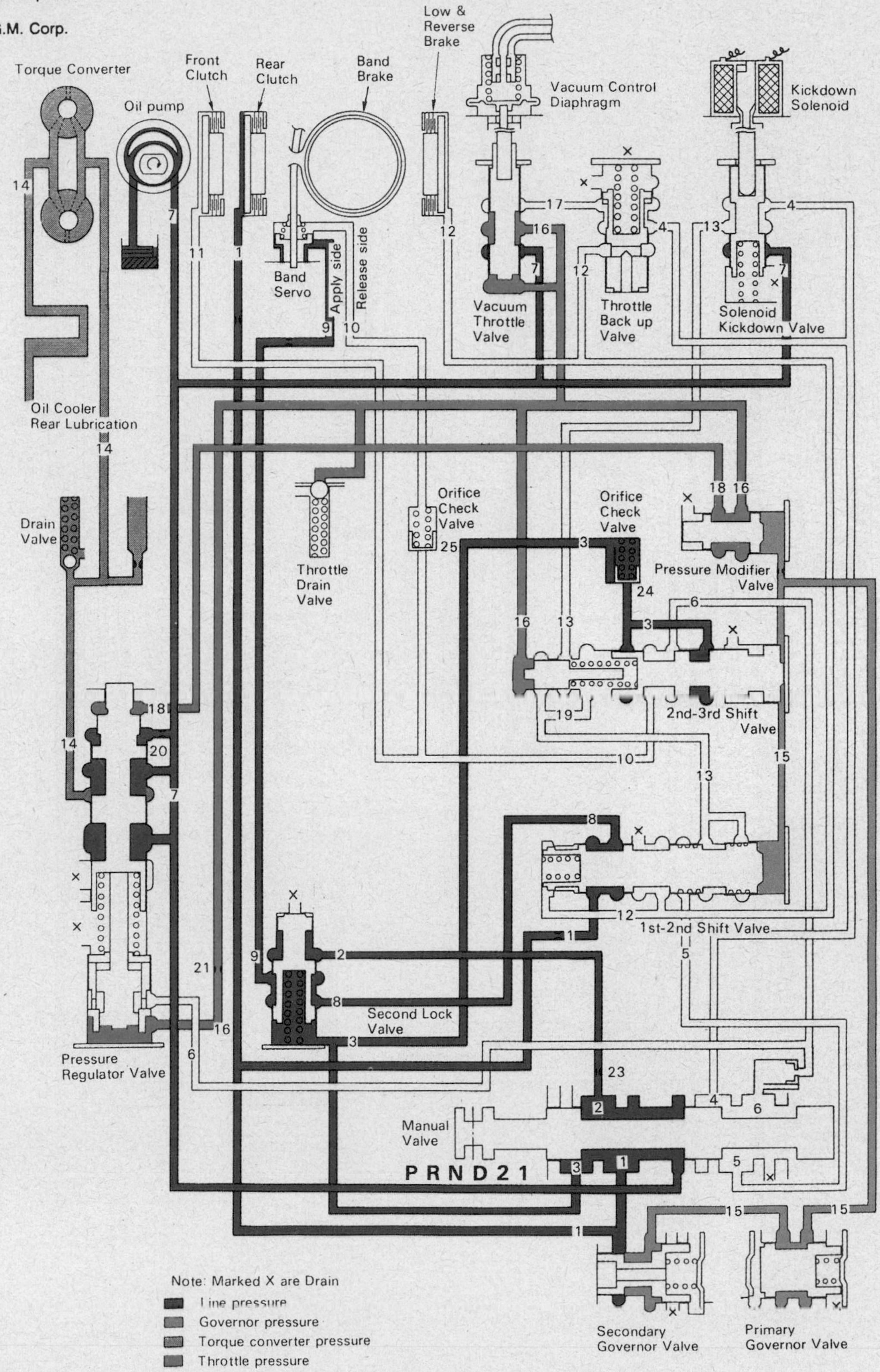
Torque Converter
Oil pump
Front Clutch
Rear Clutch
Band Brake
Low & Reverse Brake
Vacuum Control Diaphragm
Kickdown Solenoid
Band Servo
Apply side
Release side
Vacuum Throttle Valve
Throttle Back up Valve
Solenoid Kickdown Valve
Oil Cooler Rear Lubrication
Drain Valve
Throttle Drain Valve
Orifice Check Valve
Orifice Check Valve
Pressure Modifier Valve
2nd-3rd Shift Valve
1st-2nd Shift Valve
Second Lock Valve
Pressure Regulator Valve
Manual Valve
P R N D 2 1
Secondary Governor Valve
Primary Governor Valve
Note: Marked X are Drain
Line pressure
Governor pressure
Torque converter pressure
Throttle pressure

DRIVE—3RD GEAR—KF100

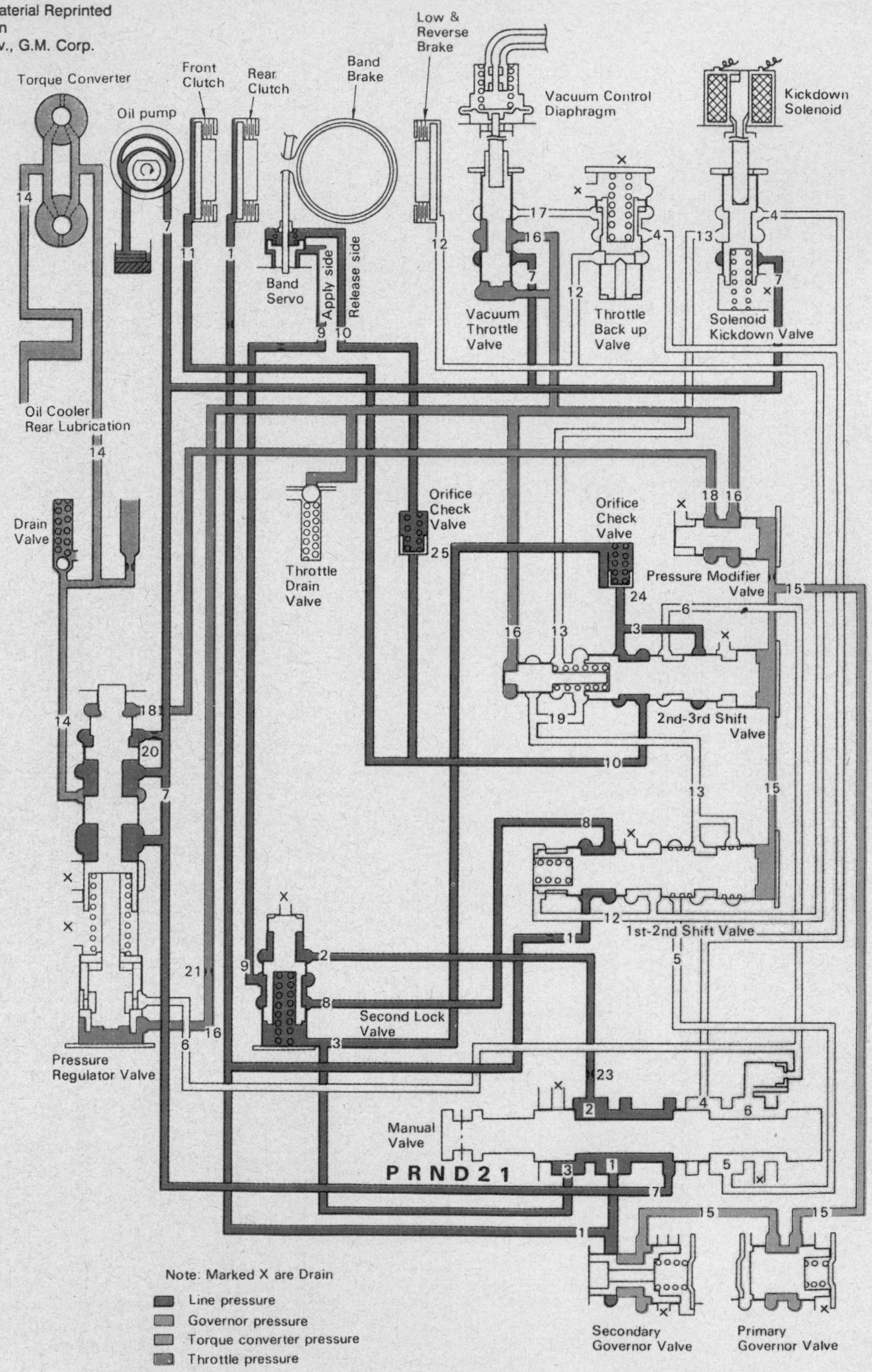
Torque Converter
Oil pump
Front Clutch
Rear Clutch
Band Brake
Low & Reverse Brake
Vacuum Control Diaphragm
Kickdown Solenoid
Band Servo
Apply side
Release side
Vacuum Throttle Valve
Throttle Back up Valve
Solenoid Kickdown Valve
Oil Cooler Rear Lubrication
Drain Valve
Throttle Drain Valve
Orifice Check Valve
Orifice Check Valve
Pressure Modifier Valve
2nd-3rd Shift Valve
1st-2nd Shift Valve
Second Lock Valve
Pressure Regulator Valve
Manual Valve
P R N D 2 1
Secondary Governor Valve
Primary Governor Valve
Note: Marked X are Drain
Line pressure
Governor pressure
Torque converter pressure
Throttle pressure

DRIVE—3-2 DOWNSHIFT—KF100

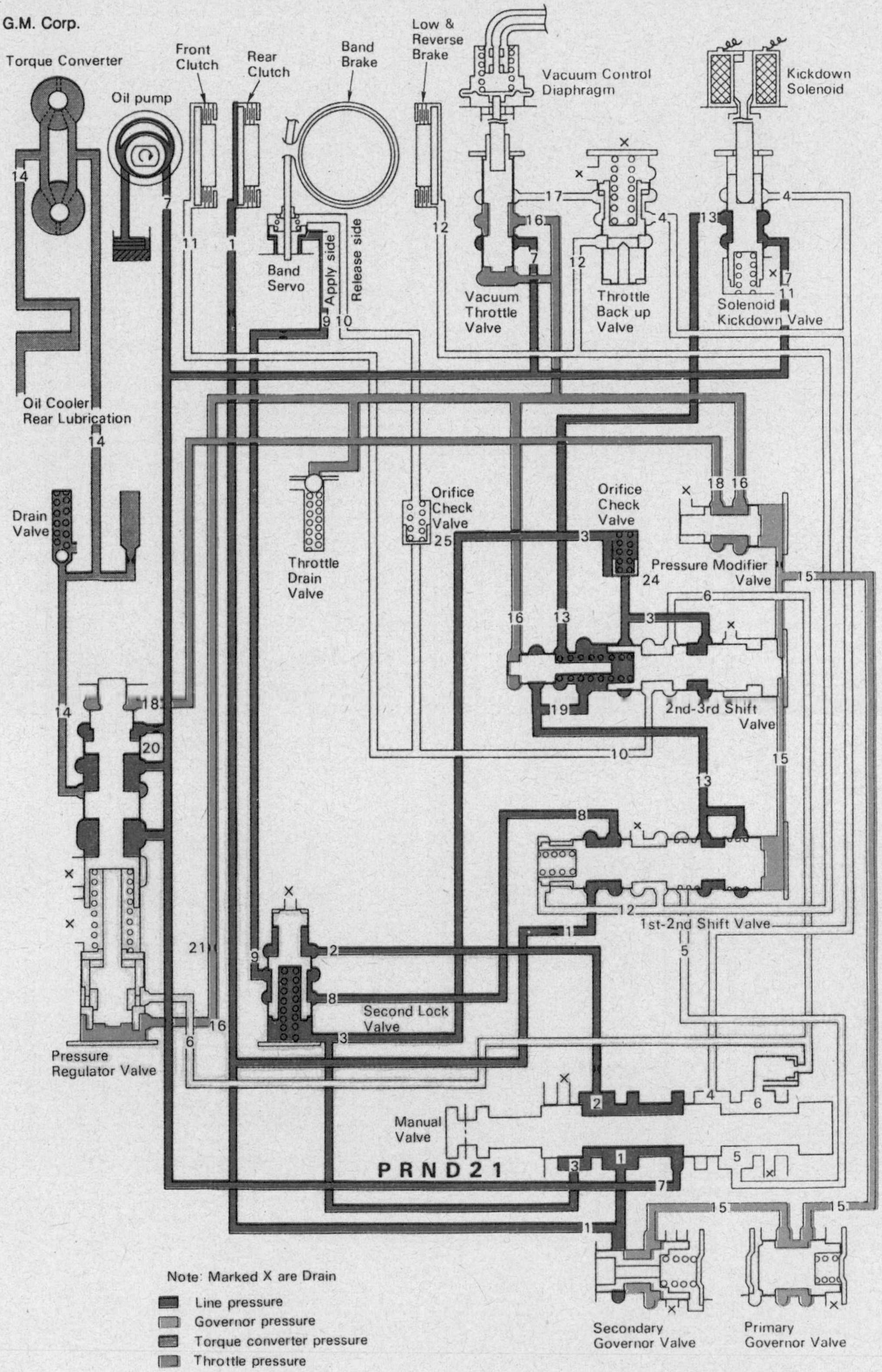
Torque Converter
Oil pump
Front Clutch
Rear Clutch
Band Brake
Low & Reverse Brake
Vacuum Control Diaphragm
Kickdown Solenoid
Band Servo
Apply side
Release side
Vacuum Throttle Valve
Throttle Back up Valve
Solenoid Kickdown Valve
Oil Cooler Rear Lubrication
Drain Valve
Throttle Drain Valve
Orifice Check Valve
Orifice Check Valve
Pressure Modifier Valve
2nd-3rd Shift Valve
1st-2nd Shift Valve
Second Lock Valve
Pressure Regulator Valve
Manual Valve
PRND21
Secondary Governor Valve
Primary Governor Valve
Note: Marked X are Drain
Line pressure
Governor pressure
Torque converter pressure
Throttle pressure

MANUAL 2—2ND GEAR—KF100

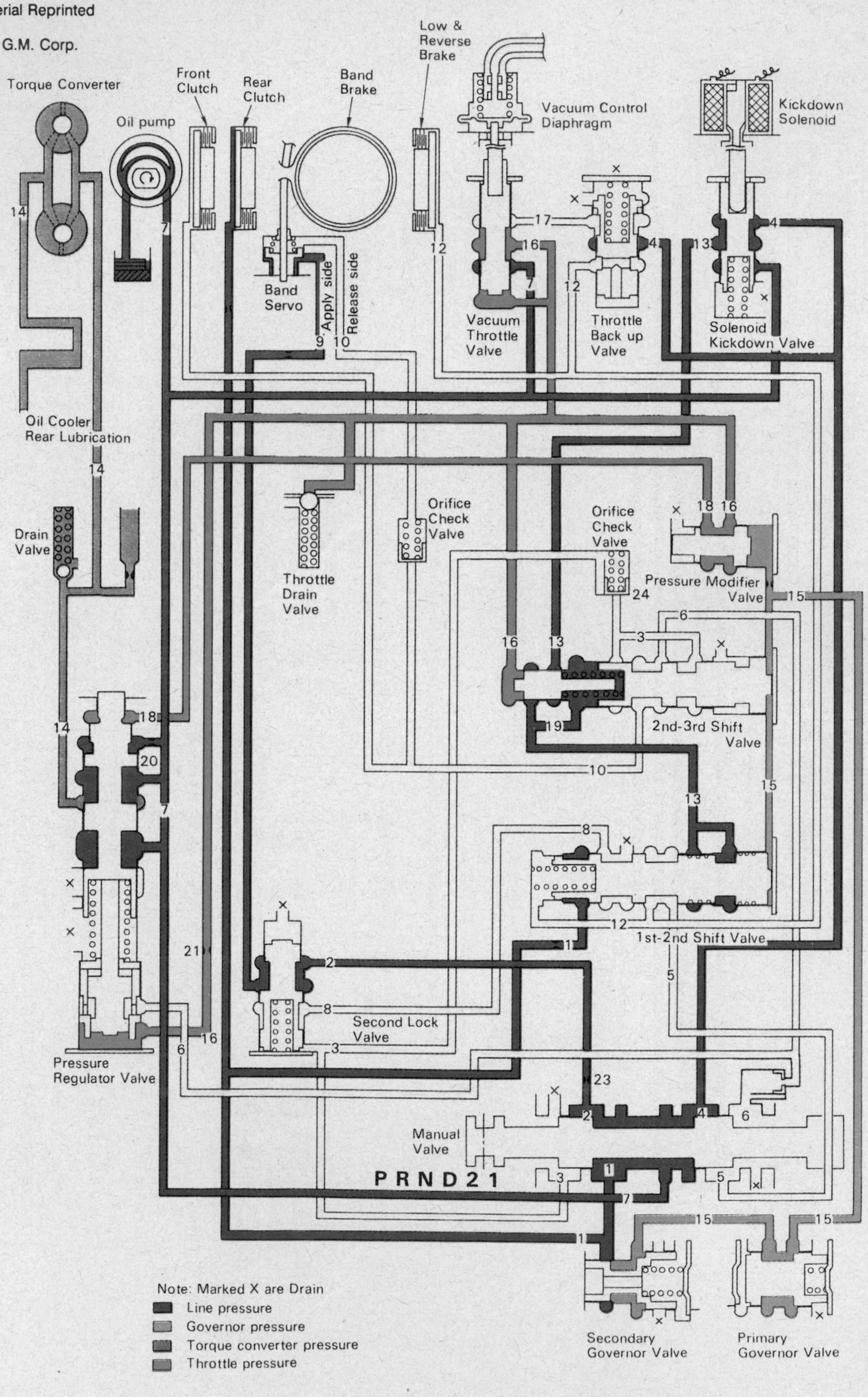
Torque Converter
Oil pump
Front Clutch
Rear Clutch
Band Brake
Low & Reverse Brake
Vacuum Control Diaphragm
Kickdown Solenoid
Band Servo
Apply side
Release side
Vacuum Throttle Valve
Throttle Back up Valve
Solenoid Kickdown Valve
Oil Cooler Rear Lubrication
Drain Valve
Throttle Drain Valve
Orifice Check Valve
Orifice Check Valve
Pressure Modifier Valve
2nd-3rd Shift Valve
1st-2nd Shift Valve
Second Lock Valve
Pressure Regulator Valve
Manual Valve
P R N D 2 1
Secondary Governor Valve
Primary Governor Valve
Note: Marked X are Drain
Line pressure
Governor pressure
Torque converter pressure
Throttle pressure

MANUAL 1—2ND GEAR—KF100

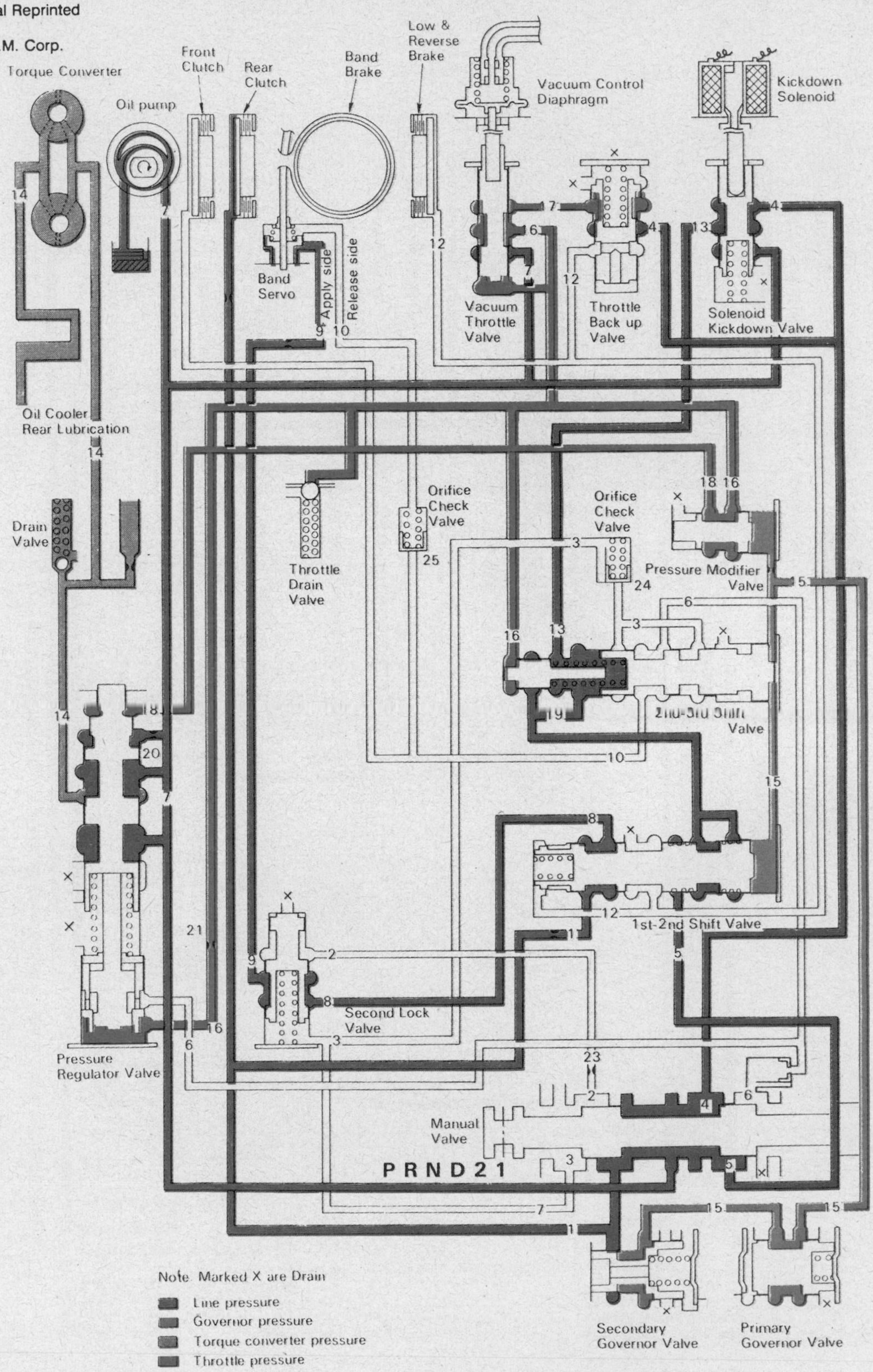
Torque Converter
Oil pump
Front Clutch
Rear Clutch
Band Brake
Low & Reverse Brake
Vacuum Control Diaphragm
Kickdown Solenoid
Band Servo
Apply side
Release side
Vacuum Throttle Valve
Throttle Back up Valve
Solenoid Kickdown Valve
Oil Cooler Rear Lubrication
Drain Valve
Throttle Drain Valve
Orifice Check Valve
Orifice Check Valve
Pressure Modifier Valve
2nd-3rd Shift Valve
1st-2nd Shift Valve
Second Lock Valve
Pressure Regulator Valve
Manual Valve
P R N D 2 1
Secondary Governor Valve
Primary Governor Valve

Note: Marked X are Drain

- Line pressure
- Governor pressure
- Torque converter pressure
- Throttle pressure

MANUAL 1—1ST GEAR—KF100

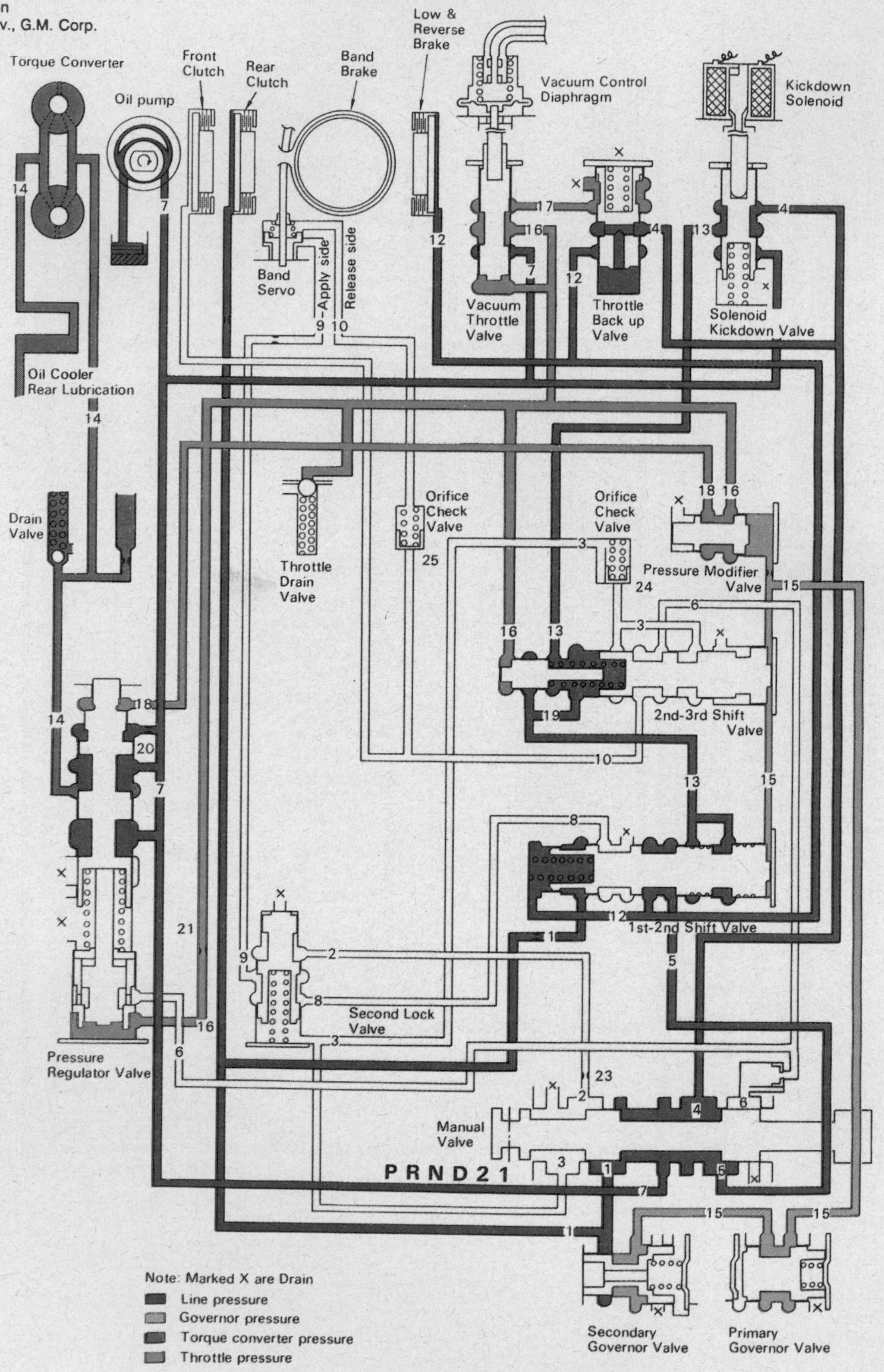
Torque Converter
Oil pump
Front Clutch
Rear Clutch
Band Brake
Low & Reverse Brake
Vacuum Control Diaphragm
Kickdown Solenoid
Band Servo
Apply side
Release side
Vacuum Throttle Valve
Throttle Back up Valve
Solenoid Kickdown Valve
Oil Cooler Rear Lubrication
Drain Valve
Throttle Drain Valve
Orifice Check Valve
Orifice Check Valve
Pressure Modifier Valve
2nd-3rd Shift Valve
1st-2nd Shift Valve
Second Lock Valve
Pressure Regulator Valve
Manual Valve
PRND21
Secondary Governor Valve
Primary Governor Valve
Note: Marked X are Drain
Line pressure
Governor pressure
Torque converter pressure
Throttle pressure

PARK OR NEUTRAL WITH ENGINE RUNNING—A131 L

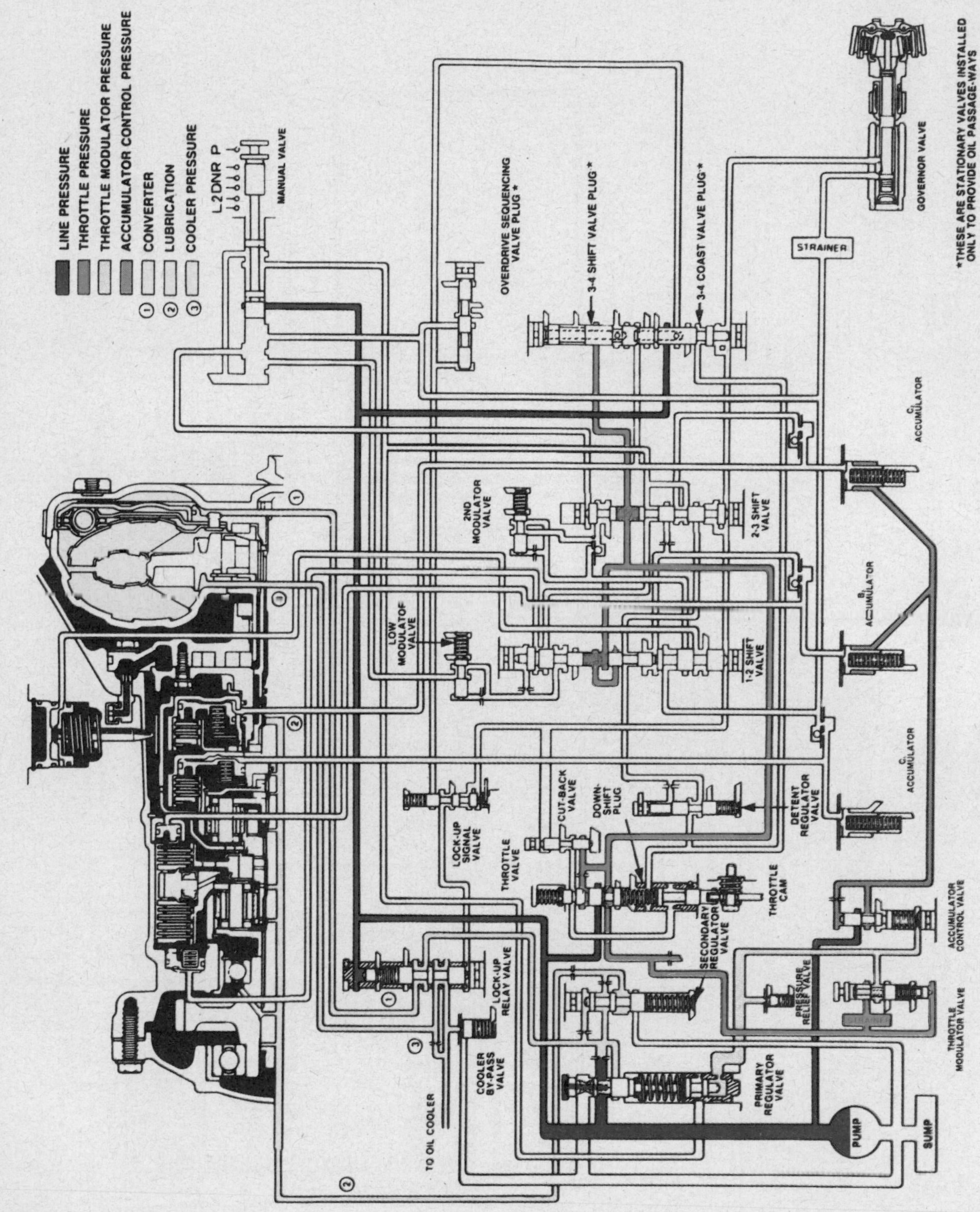
LINE PRESSURE
THROTTLE PRESSURE
THROTTLE MODULATOR PRESSURE
ACCUMULATOR CONTROL PRESSURE
① CONVERTER
② LUBRICATION
③ COOLER PRESSURE
L2DNR P
MANUAL VALVE
OVERDRIVE SEQUENCING VALVE PLUG *
3-4 SHIFT VALVE PLUG *
3-4 COAST VALVE PLUG *
STRAINER
GOVERNOR VALVE
*THESE ARE STATIONARY VALVES INSTALLED ONLY TO PROVIDE OIL PASSAGE-WAYS
ACCUMULATOR
2ND MODULATOR VALVE
2-3 SHIFT VALVE
ACCUMULATOR
LOW MODULATOR VALVE
1-2 SHIFT VALVE
ACCUMULATOR
CUT-BACK VALVE
DOWN-SHIFT PLUG
DETENT REGULATOR VALVE
LOCK-UP SIGNAL VALVE
THROTTLE VALVE
THROTTLE CAM
SECONDARY REGULATOR VALVE
ACCUMULATOR CONTROL VALVE
LOCK-UP RELAY VALVE
PRESSURE RELIEF VALVE
THROTTLE MODULATOR VALVE
COOLER BY-PASS VALVE
TO OIL COOLER
PRIMARY REGULATOR VALVE
PUMP
SUMP

REVERSE—A131L

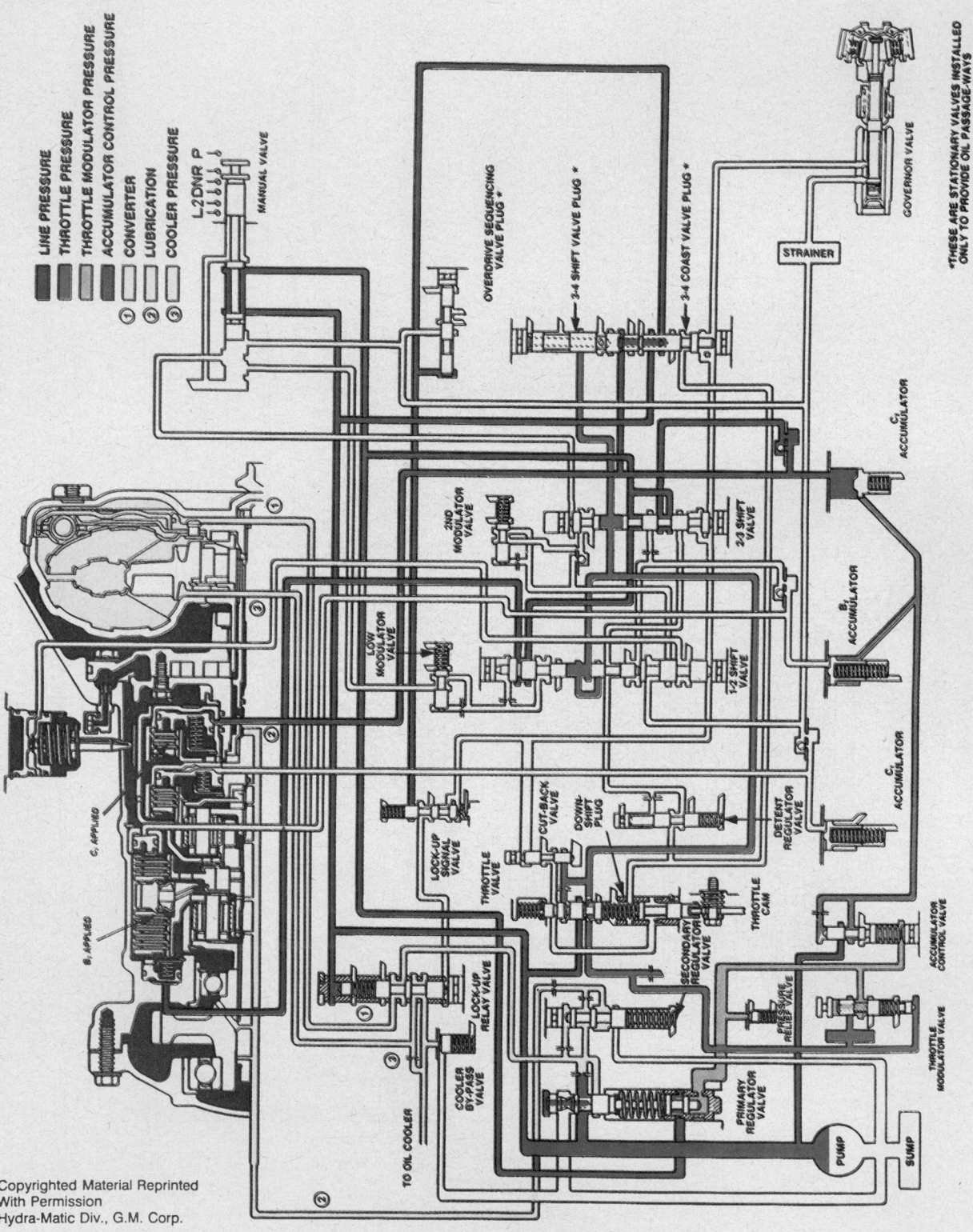

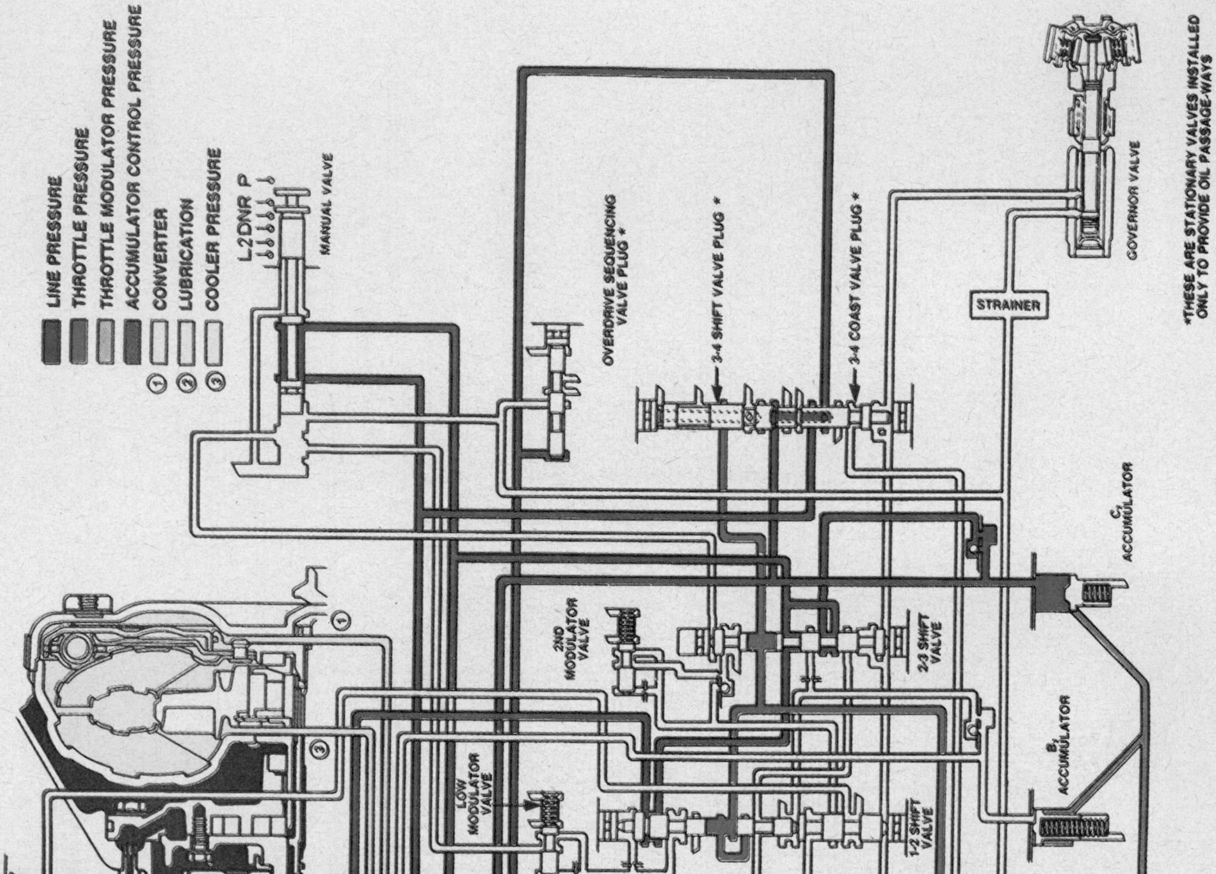
LINE PRESSURE
THROTTLE PRESSURE
THROTTLE MODULATOR PRESSURE
ACCUMULATOR CONTROL PRESSURE
① CONVERTER
② LUBRICATION
③ COOLER PRESSURE
L2DNR P
MANUAL VALVE
OVERDRIVE SEQUENCING VALVE PLUG *
3-4 SHIFT VALVE PLUG *
3-4 COAST VALVE PLUG *
STRAINER
GOVERNOR VALVE
*THESE ARE STATIONARY VALVES INSTALLED ONLY TO PROVIDE OIL PASSAGE-WAYS
ACCUMULATOR
2ND MODULATOR VALVE
2-3 SHIFT VALVE
ACCUMULATOR
LOW MODULATOR VALVE
1-2 SHIFT VALVE
ACCUMULATOR
APPLIED
APPLIED
LOCK-UP SIGNAL VALVE
CUT-BACK VALVE
DOWN-SHIFT PLUG
DETENT REGULATOR VALVE
THROTTLE VALVE
THROTTLE CAM
SECONDARY REGULATOR VALVE
ACCUMULATOR CONTROL VALVE
LOCK-UP RELAY VALVE
PRESSURE RELIEF VALVE
THROTTLE MODULATOR VALVE
COOLER BY-PASS VALVE
TO OIL COOLER
PRIMARY REGULATOR VALVE
PUMP
SUMP

DRIVE—1ST GEAR—A131L

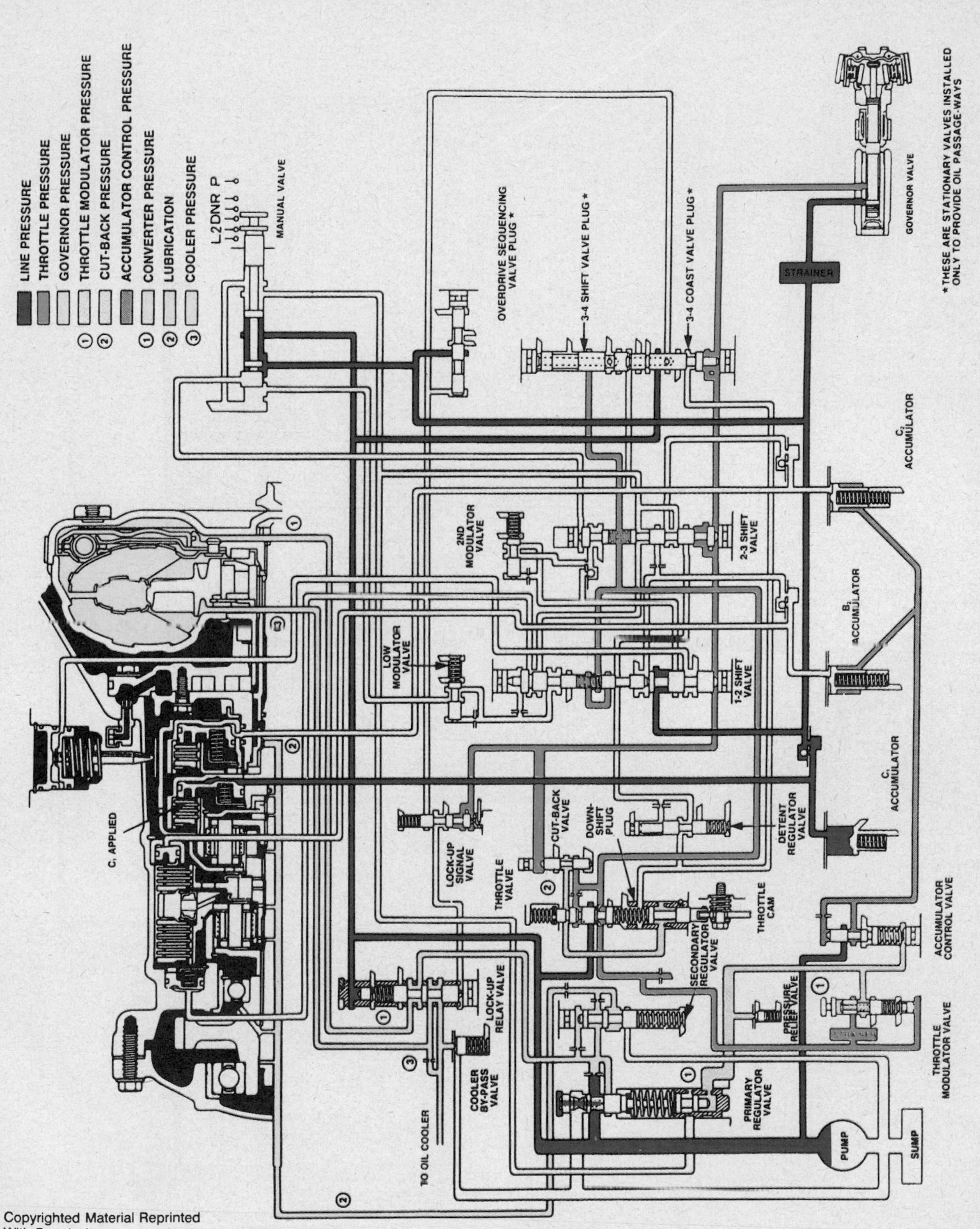

LINE PRESSURE
THROTTLE PRESSURE
GOVERNOR PRESSURE
① THROTTLE MODULATOR PRESSURE
② CUT-BACK PRESSURE
ACCUMULATOR CONTROL PRESSURE
① CONVERTER PRESSURE
② LUBRICATION
③ COOLER PRESSURE
L2DNR P
MANUAL VALVE
OVERDRIVE SEQUENCING VALVE PLUG *
3-4 SHIFT VALVE PLUG *
3-4 COAST VALVE PLUG *
STRAINER
GOVERNOR VALVE
*THESE ARE STATIONARY VALVES INSTALLED ONLY TO PROVIDE OIL PASSAGE-WAYS
C2 ACCUMULATOR
2ND MODULATOR VALVE
2-3 SHIFT VALVE
B2 ACCUMULATOR
LOW MODULATOR VALVE
1-2 SHIFT VALVE
C1 ACCUMULATOR
C1 APPLIED
CUT-BACK VALVE
DOWN-SHIFT PLUG
DETENT REGULATOR VALVE
LOCK-UP SIGNAL VALVE
THROTTLE VALVE
THROTTLE CAM
ACCUMULATOR CONTROL VALVE
SECONDARY REGULATOR VALVE
PRESSURE RELIEF VALVE
THROTTLE MODULATOR VALVE
LOCK-UP RELAY VALVE
COOLER BY-PASS VALVE
TO OIL COOLER
PRIMARY REGULATOR VALVE
PUMP
SUMP

DRIVE—2ND GEAR—A131L

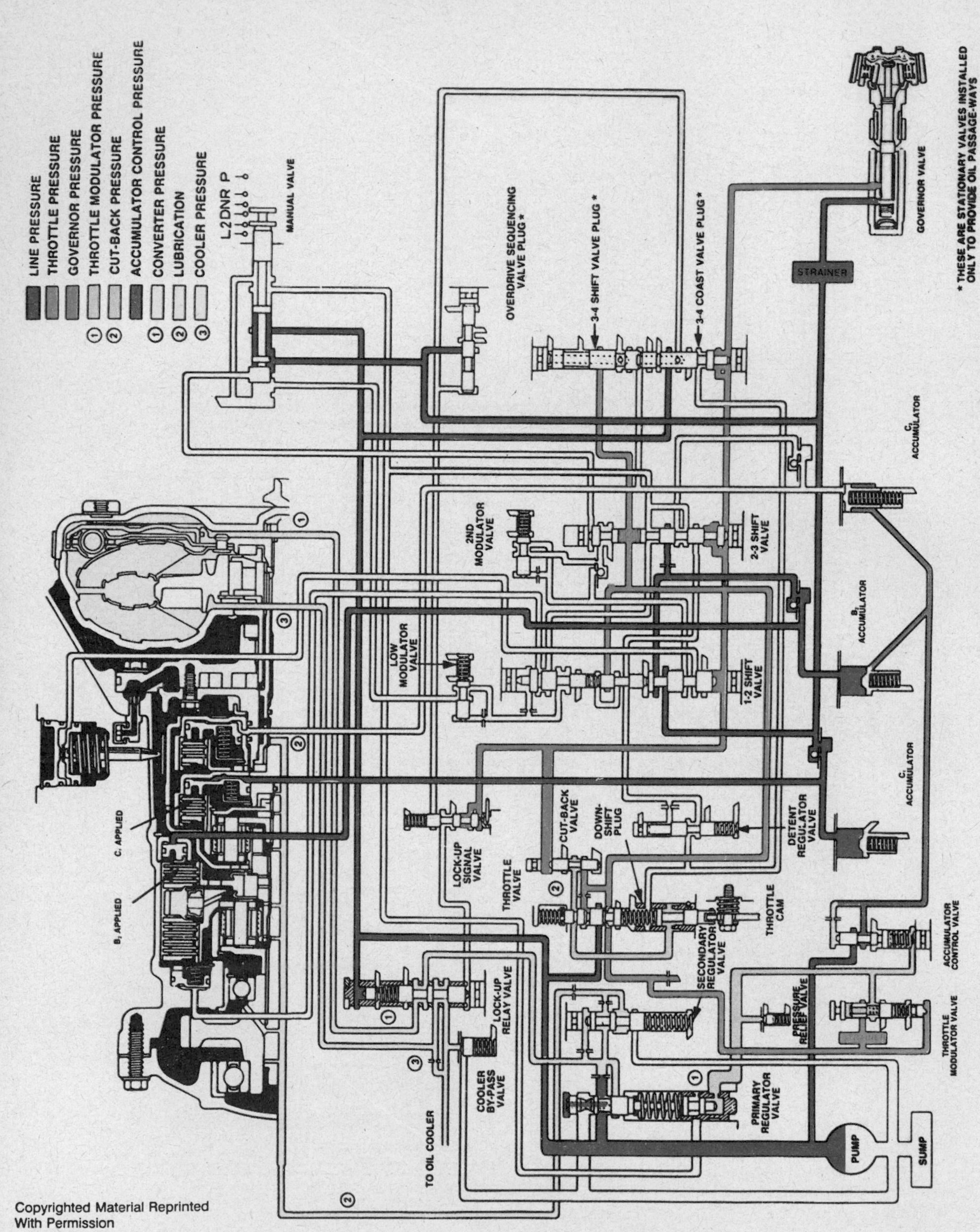
LINE PRESSURE
THROTTLE PRESSURE
GOVERNOR PRESSURE
① THROTTLE MODULATOR PRESSURE
② CUT-BACK PRESSURE
ACCUMULATOR CONTROL PRESSURE
① CONVERTER PRESSURE
② LUBRICATION
③ COOLER PRESSURE
L2DNR P
MANUAL VALVE
OVERDRIVE SEQUENCING VALVE PLUG *
3-4 SHIFT VALVE PLUG *
3-4 COAST VALVE PLUG *
STRAINER
GOVERNOR VALVE
* THESE ARE STATIONARY VALVES INSTALLED ONLY TO PROVIDE OIL PASSAGE-WAYS
C₂ ACCUMULATOR
2ND MODULATOR VALVE
2-3 SHIFT VALVE
B₂ ACCUMULATOR
LOW MODULATOR VALVE
1-2 SHIFT VALVE
C₁ ACCUMULATOR
C₁ APPLIED
B₂ APPLIED
CUT-BACK VALVE
DOWN-SHIFT PLUG
DETENT REGULATOR VALVE
LOCK-UP SIGNAL VALVE
THROTTLE VALVE
THROTTLE CAM
SECONDARY REGULATOR VALVE
ACCUMULATOR CONTROL VALVE
LOCK-UP RELAY VALVE
PRESSURE RELIEF VALVE
THROTTLE MODULATOR VALVE
COOLER BY-PASS VALVE
TO OIL COOLER
PRIMARY REGULATOR VALVE
PUMP
SUMP

DRIVE—3RD GEAR—A131L

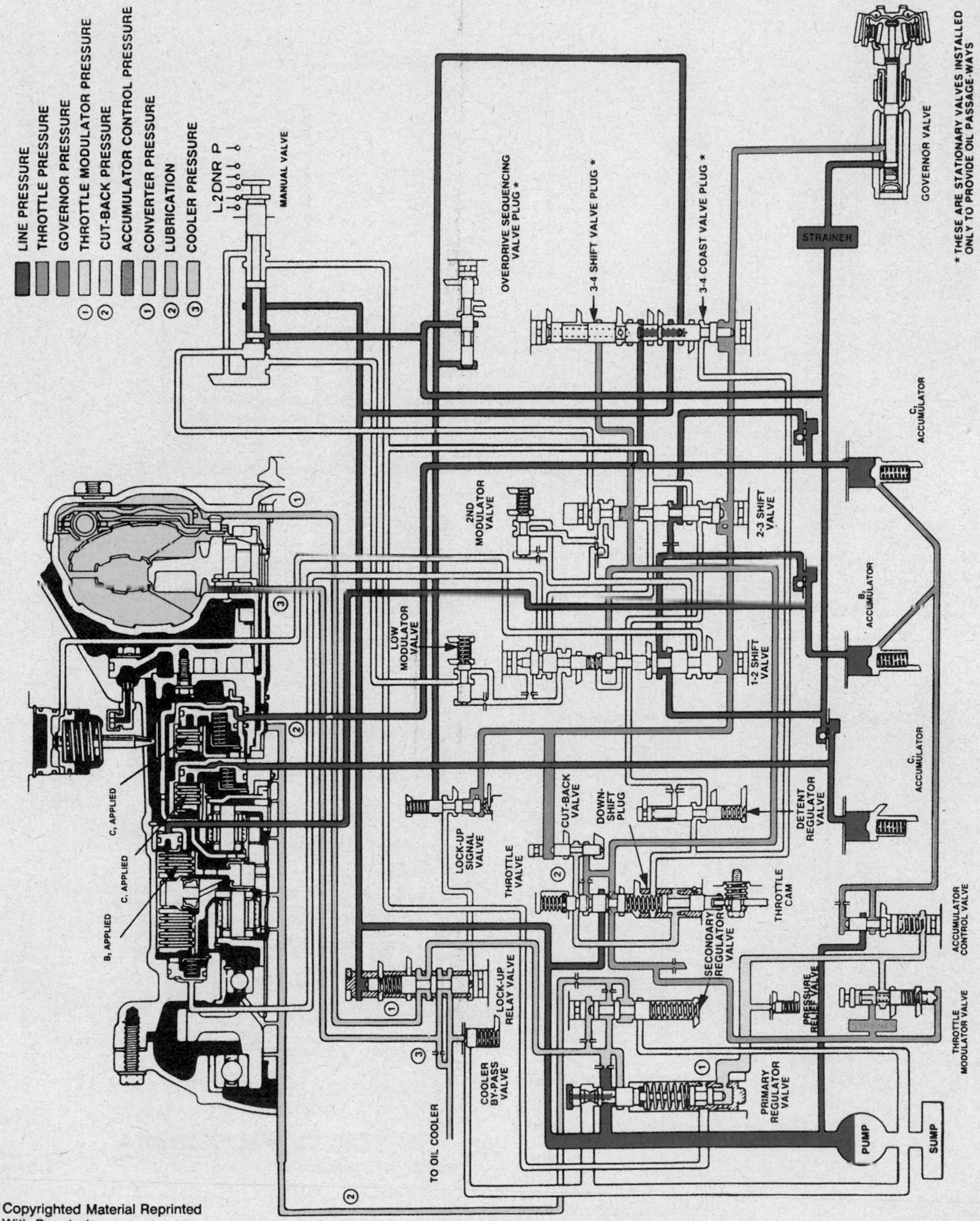
LINE PRESSURE
THROTTLE PRESSURE
GOVERNOR PRESSURE
① THROTTLE MODULATOR PRESSURE
② CUT-BACK PRESSURE
ACCUMULATOR CONTROL PRESSURE
① CONVERTER PRESSURE
② LUBRICATION
③ COOLER PRESSURE
L2DNR P
MANUAL VALVE
OVERDRIVE SEQUENCING VALVE PLUG *
3-4 SHIFT VALVE PLUG *
3-4 COAST VALVE PLUG *
STRAINER
GOVERNOR VALVE
* THESE ARE STATIONARY VALVES INSTALLED ONLY TO PROVIDE OIL PASSAGE-WAYS
C₁ ACCUMULATOR
2ND MODULATOR VALVE
2-3 SHIFT VALVE
B₁ ACCUMULATOR
LOW MODULATOR VALVE
1-2 SHIFT VALVE
C₂ ACCUMULATOR
CUT-BACK VALVE
DOWN-SHIFT PLUG
DETENT REGULATOR VALVE
LOCK-UP SIGNAL VALVE
THROTTLE VALVE
THROTTLE CAM
SECONDARY REGULATOR VALVE
ACCUMULATOR CONTROL VALVE
C₁ APPLIED
C₂ APPLIED
B₁ APPLIED
LOCK-UP RELAY VALVE
PRESSURE RELIEF VALVE
STRAINER
THROTTLE MODULATOR VALVE
COOLER BY-PASS VALVE
TO OIL COOLER
PRIMARY REGULATOR VALVE
PUMP
SUMP

DRIVE—3RD GEAR (TCC APPLIED)—A131L

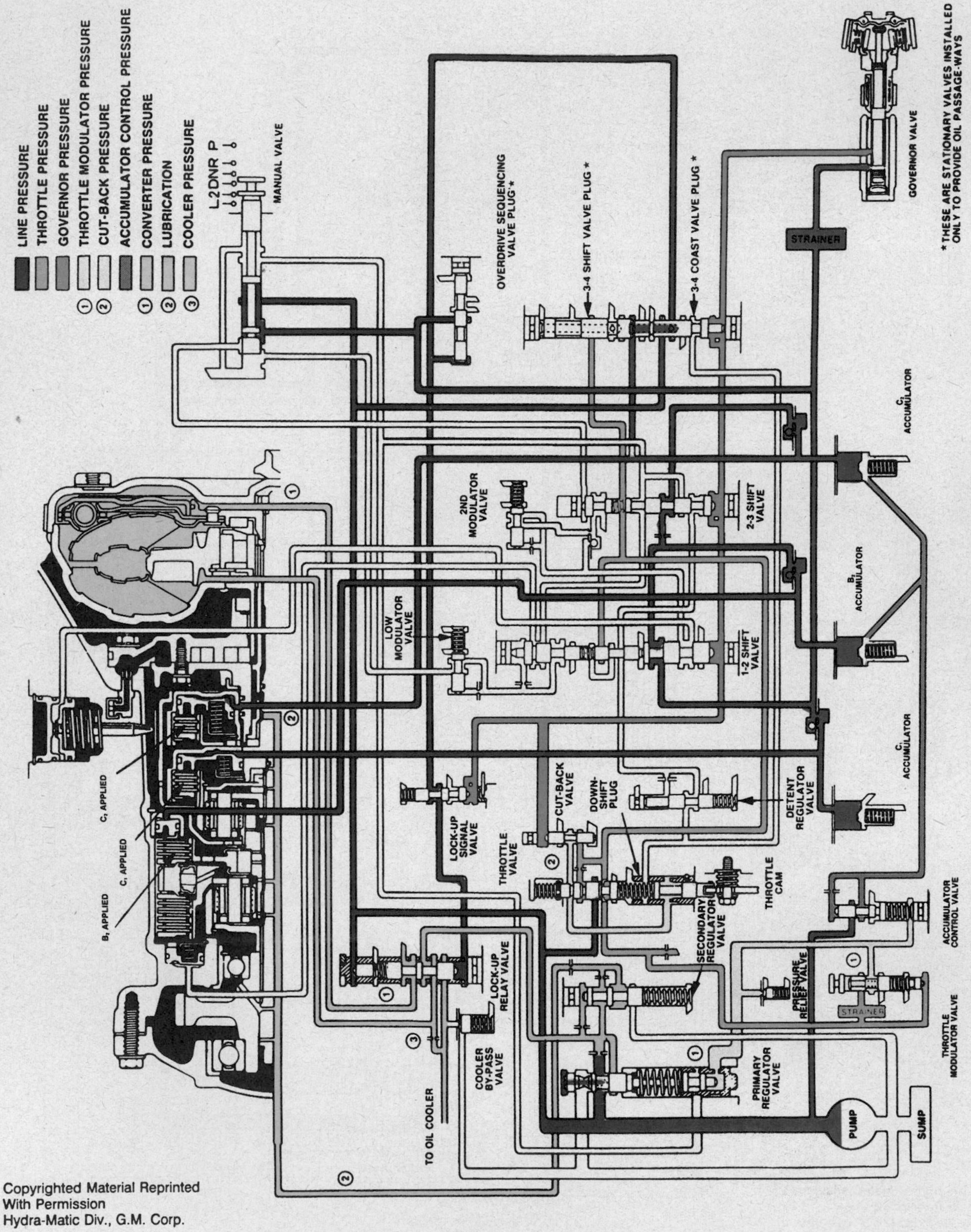
LINE PRESSURE
THROTTLE PRESSURE
GOVERNOR PRESSURE
① THROTTLE MODULATOR PRESSURE
② CUT-BACK PRESSURE
ACCUMULATOR CONTROL PRESSURE
① CONVERTER PRESSURE
② LUBRICATION
③ COOLER PRESSURE
L2DNR P
MANUAL VALVE
OVERDRIVE SEQUENCING VALVE PLUG*
3-4 SHIFT VALVE PLUG*
3-4 COAST VALVE PLUG*
STRAINER
GOVERNOR VALVE
*THESE ARE STATIONARY VALVES INSTALLED ONLY TO PROVIDE OIL PASSAGE-WAYS
C₂ ACCUMULATOR
2ND MODULATOR VALVE
2-3 SHIFT VALVE
B₂ ACCUMULATOR
LOW MODULATOR VALVE
1-2 SHIFT VALVE
C₁ ACCUMULATOR
C₂ APPLIED
C₁ APPLIED
B₂ APPLIED
LOCK-UP SIGNAL VALVE
THROTTLE VALVE
CUT-BACK VALVE
DOWN-SHIFT PLUG
DETENT REGULATOR VALVE
THROTTLE CAM
SECONDARY REGULATOR VALVE
ACCUMULATOR CONTROL VALVE
LOCK-UP RELAY VALVE
PRESSURE RELIEF VALVE
STRAINER
THROTTLE MODULATOR VALVE
COOLER BY-PASS VALVE
TO OIL COOLER
PRIMARY REGULATOR VALVE
PUMP
SUMP

Copyrighted Material Reprinted
With Permission
Hydra-Matic Div., G.M. Corp.

DRIVE—KICKDOWN—A131 L

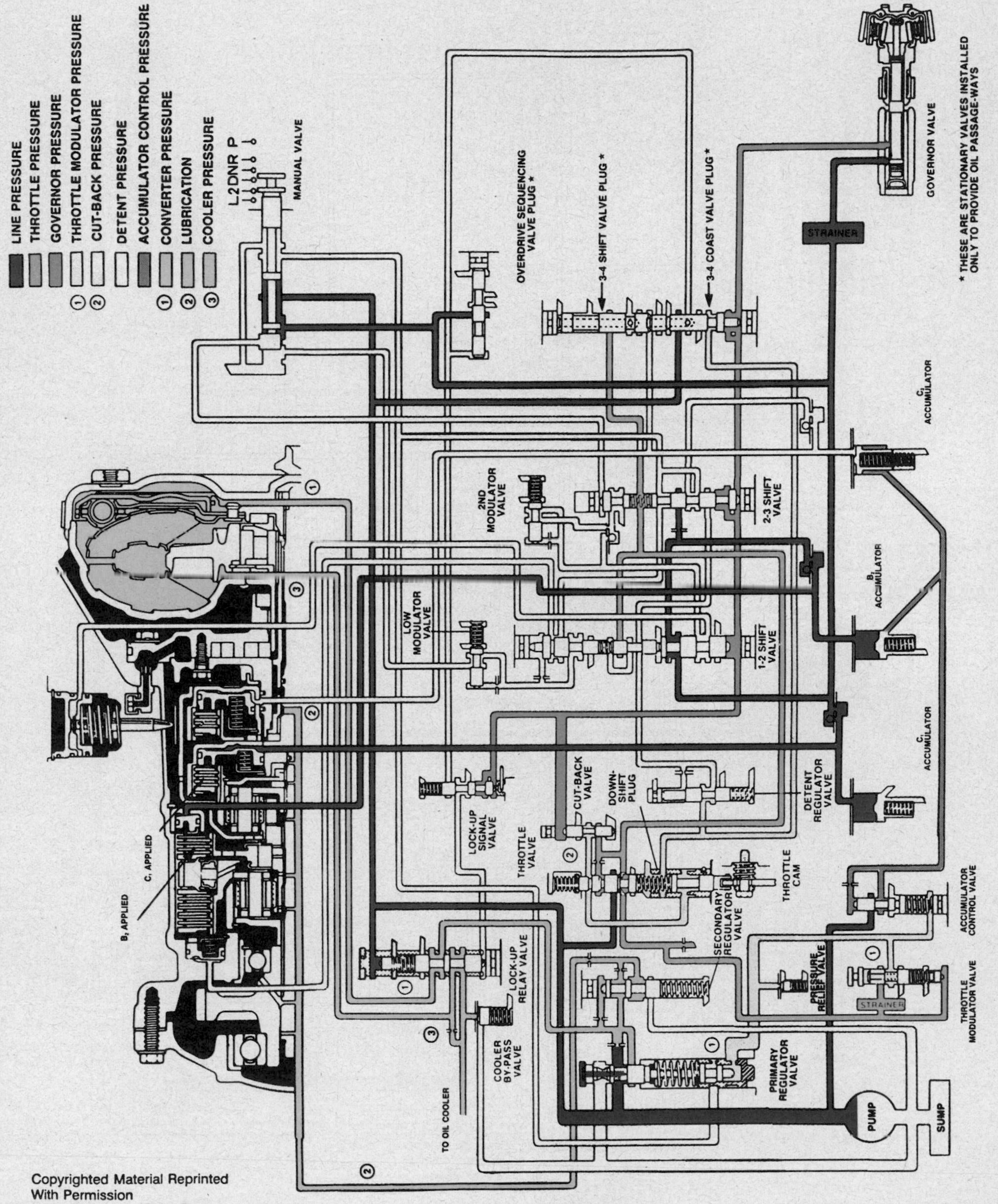
LINE PRESSURE
THROTTLE PRESSURE
GOVERNOR PRESSURE
① THROTTLE MODULATOR PRESSURE
② CUT-BACK PRESSURE
DETENT PRESSURE
ACCUMULATOR CONTROL PRESSURE
① CONVERTER PRESSURE
② LUBRICATION
③ COOLER PRESSURE
L2DNR P
MANUAL VALVE
OVERDRIVE SEQUENCING VALVE PLUG *
3-4 SHIFT VALVE PLUG *
3-4 COAST VALVE PLUG *
STRAINER
GOVERNOR VALVE
* THESE ARE STATIONARY VALVES INSTALLED ONLY TO PROVIDE OIL PASSAGE-WAYS
C_1 ACCUMULATOR
2ND MODULATOR VALVE
2-3 SHIFT VALVE
B_2 ACCUMULATOR
LOW MODULATOR VALVE
1-2 SHIFT VALVE
C_2 ACCUMULATOR
CUT-BACK VALVE
DOWN-SHIFT PLUG
DETENT REGULATOR VALVE
LOCK-UP SIGNAL VALVE
THROTTLE VALVE
THROTTLE CAM
SECONDARY REGULATOR VALVE
ACCUMULATOR CONTROL VALVE
B_2 APPLIED
C_1 APPLIED
LOCK-UP RELAY VALVE
PRESSURE RELIEF VALVE
STRAINER
THROTTLE MODULATOR VALVE
COOLER BY-PASS VALVE
TO OIL COOLER
PRIMARY REGULATOR VALVE
PUMP
SUMP

MANUAL 2—2ND GEAR—A131L

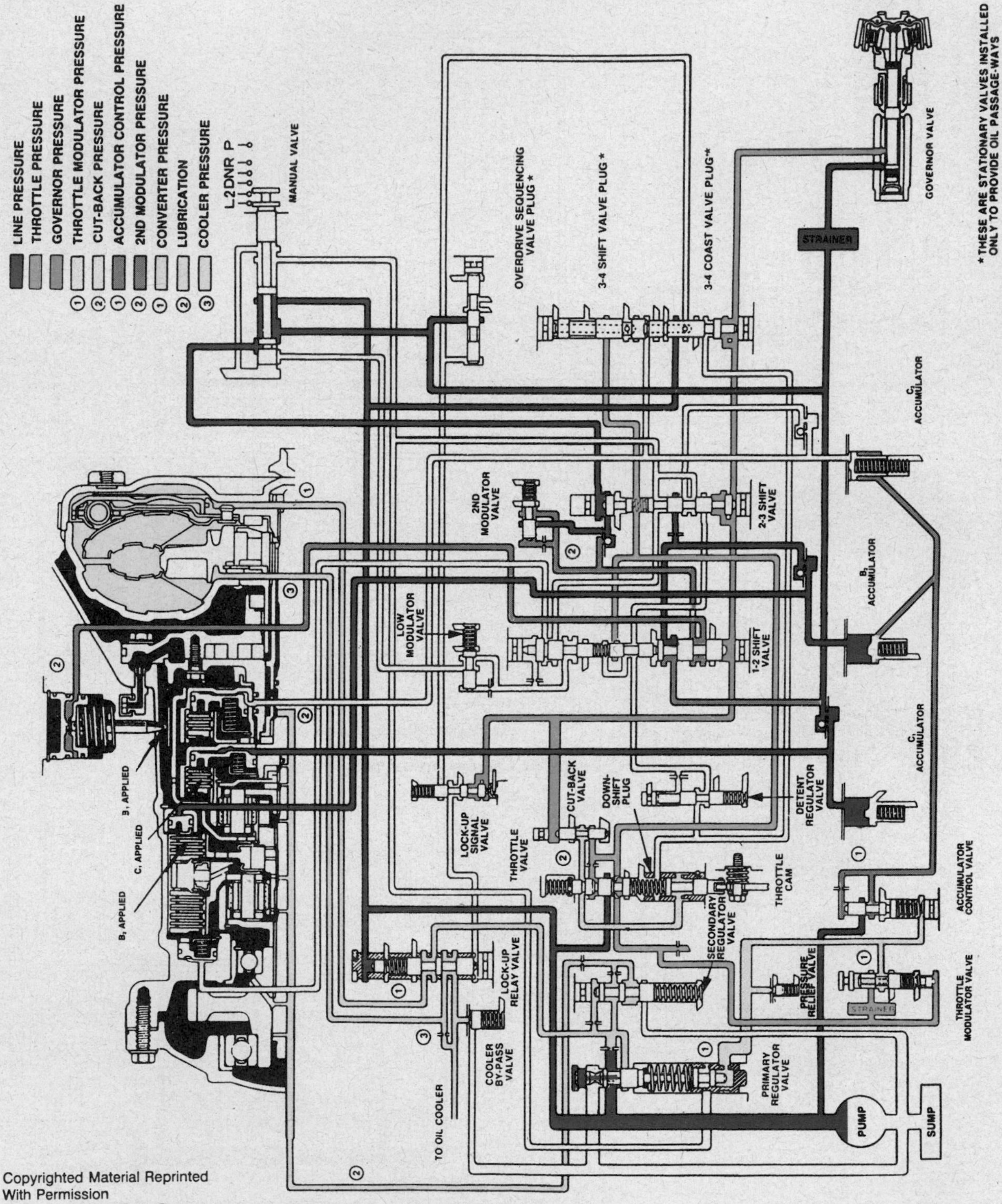
LINE PRESSURE
THROTTLE PRESSURE
GOVERNOR PRESSURE
① THROTTLE MODULATOR PRESSURE
② CUT-BACK PRESSURE
① ACCUMULATOR CONTROL PRESSURE
② 2ND MODULATOR PRESSURE
① CONVERTER PRESSURE
② LUBRICATION
③ COOLER PRESSURE
L2DNR P
MANUAL VALVE
OVERDRIVE SEQUENCING VALVE PLUG *
3-4 SHIFT VALVE PLUG *
3-4 COAST VALVE PLUG*
STRAINER
GOVERNOR VALVE
* THESE ARE STATIONARY VALVES INSTALLED ONLY TO PROVIDE OIL PASSAGE-WAYS
C₂ ACCUMULATOR
2ND MODULATOR VALVE
2-3 SHIFT VALVE
B₂ ACCUMULATOR
LOW MODULATOR VALVE
1-2 SHIFT VALVE
C₁ ACCUMULATOR
CUT-BACK VALVE
DOWN-SHIFT PLUG
DETENT REGULATOR VALVE
LOCK-UP SIGNAL VALVE
THROTTLE VALVE
THROTTLE CAM
B₂ APPLIED
C₁ APPLIED
B₁ APPLIED
ACCUMULATOR CONTROL VALVE
SECONDARY REGULATOR VALVE
PRESSURE RELIEF VALVE
LOCK-UP RELAY VALVE
THROTTLE MODULATOR VALVE
STRAINER
COOLER BY-PASS VALVE
PRIMARY REGULATOR VALVE
TO OIL COOLER
PUMP
SUMP

MANUAL L—1ST GEAR—A131L

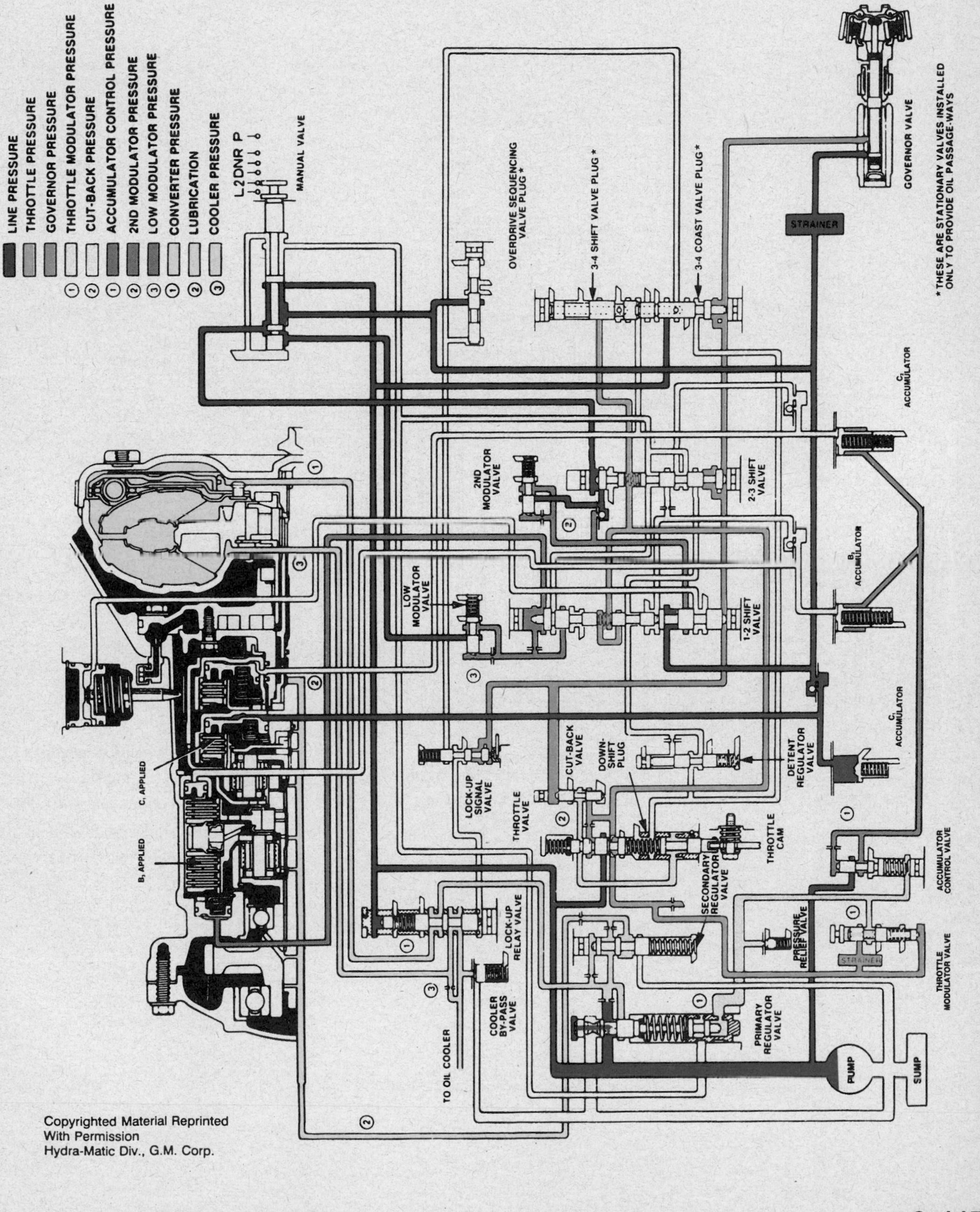
LINE PRESSURE
THROTTLE PRESSURE
GOVERNOR PRESSURE
① THROTTLE MODULATOR PRESSURE
② CUT-BACK PRESSURE
① ACCUMULATOR CONTROL PRESSURE
② 2ND MODULATOR PRESSURE
③ LOW MODULATOR PRESSURE
① CONVERTER PRESSURE
② LUBRICATION
③ COOLER PRESSURE
L 2 D N R P
MANUAL VALVE
OVERDRIVE SEQUENCING VALVE PLUG*
3-4 SHIFT VALVE PLUG*
3-4 COAST VALVE PLUG*
STRAINER
GOVERNOR VALVE
*THESE ARE STATIONARY VALVES INSTALLED ONLY TO PROVIDE OIL PASSAGE-WAYS
C2 ACCUMULATOR
2ND MODULATOR VALVE
2-3 SHIFT VALVE
B2 ACCUMULATOR
LOW MODULATOR VALVE
1-2 SHIFT VALVE
C1 ACCUMULATOR
CUT-BACK VALVE
DOWN-SHIFT PLUG
DETENT REGULATOR VALVE
LOCK-UP SIGNAL VALVE
THROTTLE VALVE
THROTTLE CAM
ACCUMULATOR CONTROL VALVE
C1 APPLIED
B1 APPLIED
SECONDARY REGULATOR VALVE
PRESSURE RELIEF VALVE
STRAINER
THROTTLE MODULATOR VALVE
LOCK-UP RELAY VALVE
COOLER BY-PASS VALVE
TO OIL COOLER
PRIMARY REGULATOR VALVE
PUMP
SUMP

PARK—SPRINT AND GEO TRACKER

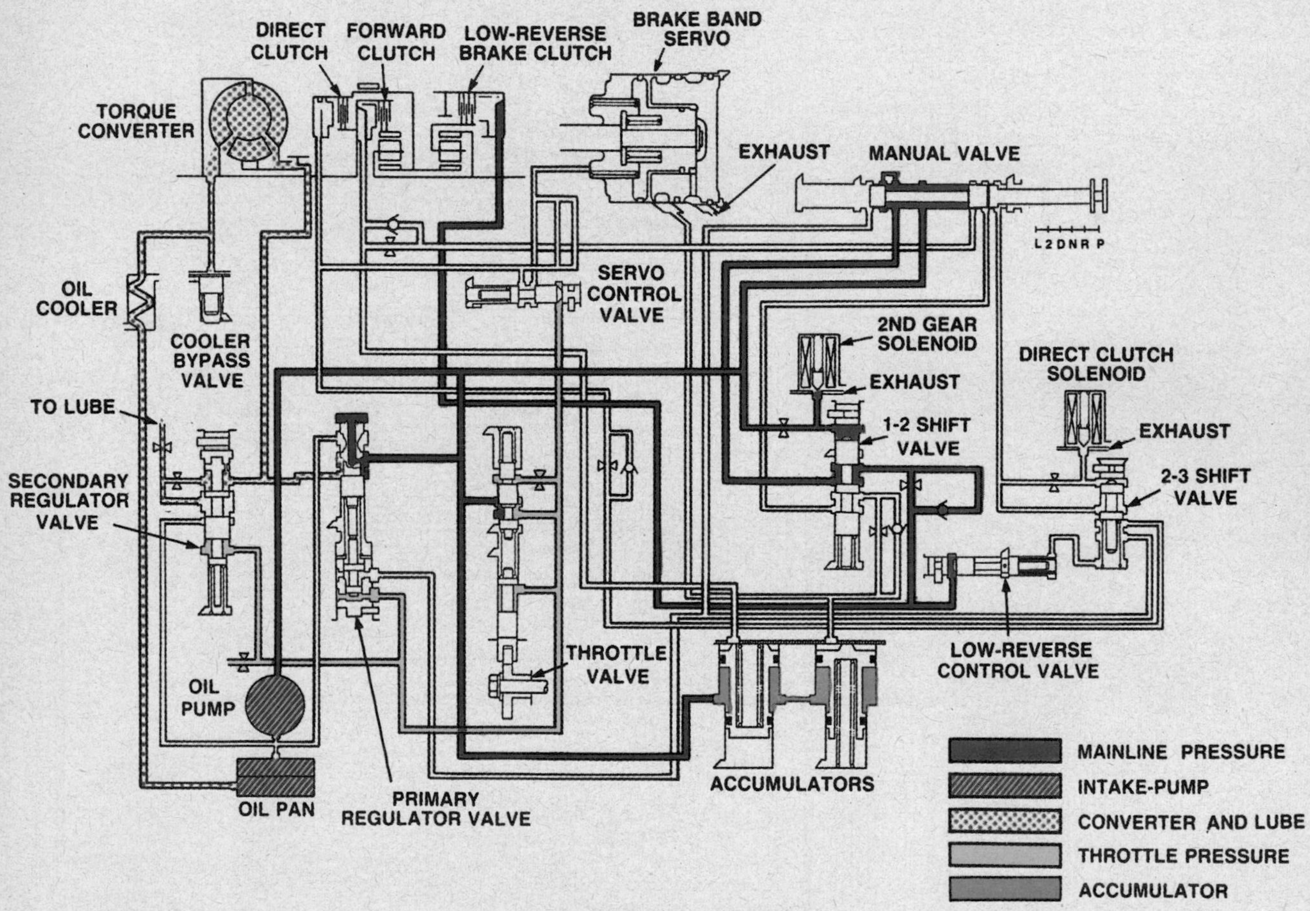
DIRECT CLUTCH
FORWARD CLUTCH
LOW-REVERSE BRAKE CLUTCH
BRAKE BAND SERVO
TORQUE CONVERTER
EXHAUST
MANUAL VALVE
L 2 D N R P
OIL COOLER
SERVO CONTROL VALVE
2ND GEAR SOLENOID
COOLER BYPASS VALVE
DIRECT CLUTCH SOLENOID
EXHAUST
TO LUBE
1-2 SHIFT VALVE
EXHAUST
SECONDARY REGULATOR VALVE
2-3 SHIFT VALVE
THROTTLE VALVE
LOW-REVERSE CONTROL VALVE
OIL PUMP
MAINLINE PRESSURE
INTAKE-PUMP
CONVERTER AND LUBE
THROTTLE PRESSURE
ACCUMULATOR
ACCUMULATORS
OIL PAN
PRIMARY REGULATOR VALVE

NEUTRAL – SPRINT AND GEO TRACKER

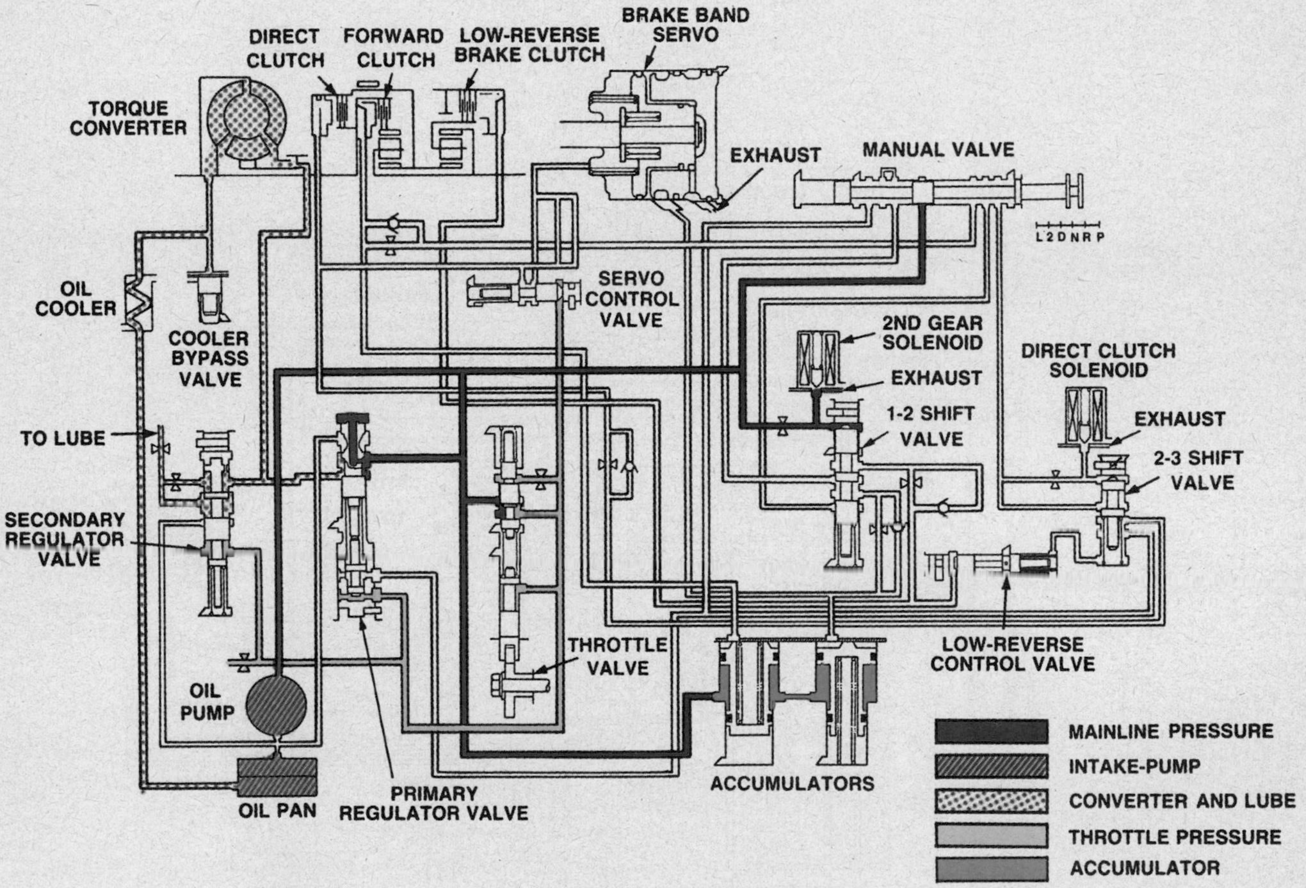
DIRECT CLUTCH
FORWARD CLUTCH
LOW-REVERSE BRAKE CLUTCH
BRAKE BAND SERVO
TORQUE CONVERTER
EXHAUST
MANUAL VALVE
L 2 D N R P
OIL COOLER
COOLER BYPASS VALVE
SERVO CONTROL VALVE
2ND GEAR SOLENOID
EXHAUST
DIRECT CLUTCH SOLENOID
1-2 SHIFT VALVE
EXHAUST
2-3 SHIFT VALVE
TO LUBE
SECONDARY REGULATOR VALVE
THROTTLE VALVE
LOW-REVERSE CONTROL VALVE
OIL PUMP
ACCUMULATORS
OIL PAN
PRIMARY REGULATOR VALVE
MAINLINE PRESSURE
INTAKE-PUMP
CONVERTER AND LUBE
THROTTLE PRESSURE
ACCUMULATOR

DRIVE OR 2ND RANGE—1ST GEAR—SPRINT AND GEO TRACKER

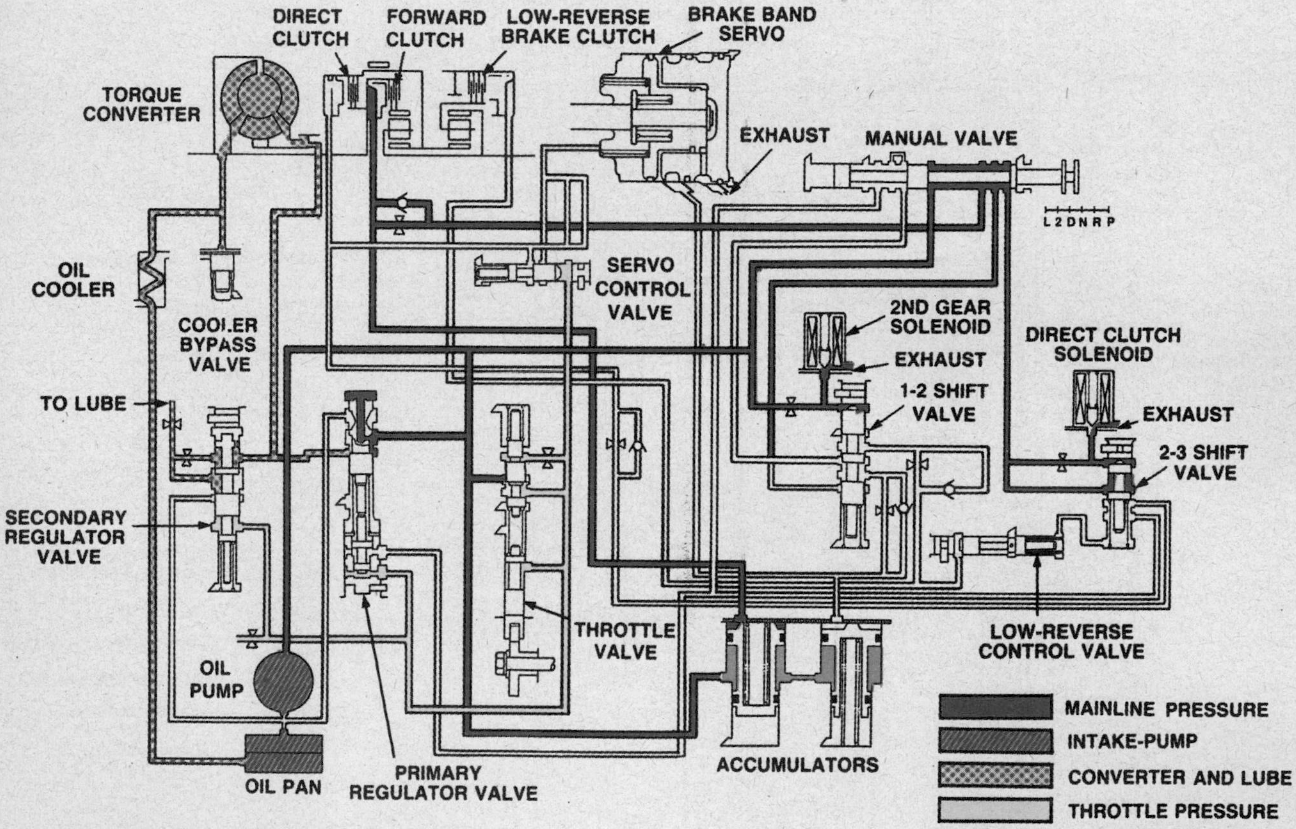
DIRECT CLUTCH
FORWARD CLUTCH
LOW-REVERSE BRAKE CLUTCH
BRAKE BAND SERVO
TORQUE CONVERTER
EXHAUST
MANUAL VALVE
L 2 D N R P
OIL COOLER
SERVO CONTROL VALVE
COOLER BYPASS VALVE
2ND GEAR SOLENOID
EXHAUST
DIRECT CLUTCH SOLENOID
TO LUBE
1-2 SHIFT VALVE
EXHAUST
2-3 SHIFT VALVE
SECONDARY REGULATOR VALVE
THROTTLE VALVE
LOW-REVERSE CONTROL VALVE
OIL PUMP
MAINLINE PRESSURE
INTAKE-PUMP
CONVERTER AND LUBE
THROTTLE PRESSURE
ACCUMULATOR
OIL PAN
PRIMARY REGULATOR VALVE
ACCUMULATORS

DRIVE OR 2ND RANGE—2ND GEAR—SPRINT AND GEO TRACKER

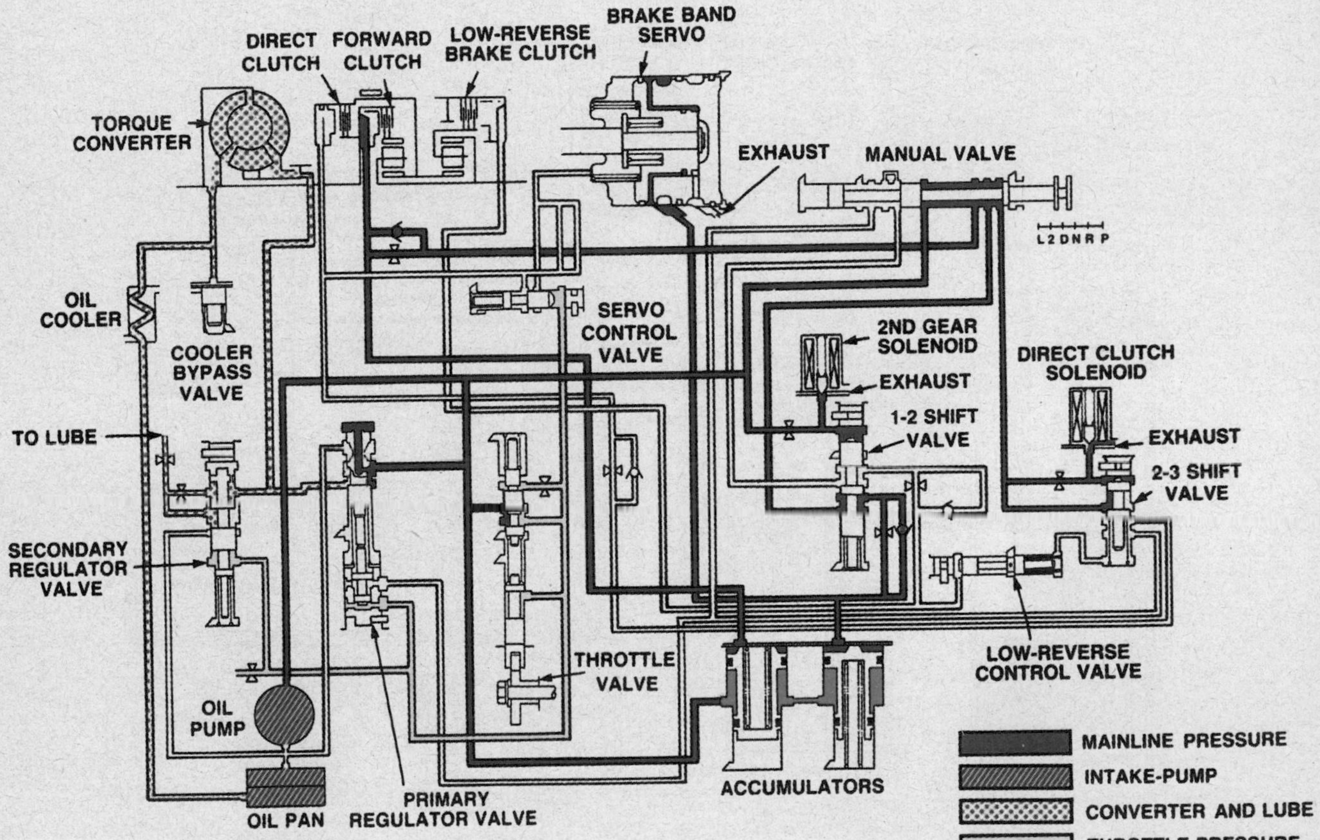
BRAKE BAND SERVO
DIRECT CLUTCH
FORWARD CLUTCH
LOW-REVERSE BRAKE CLUTCH
TORQUE CONVERTER
EXHAUST
MANUAL VALVE
L 2 D N R P
OIL COOLER
COOLER BYPASS VALVE
SERVO CONTROL VALVE
2ND GEAR SOLENOID
EXHAUST
1-2 SHIFT VALVE
DIRECT CLUTCH SOLENOID
EXHAUST
2-3 SHIFT VALVE
TO LUBE
SECONDARY REGULATOR VALVE
THROTTLE VALVE
LOW-REVERSE CONTROL VALVE
OIL PUMP
OIL PAN
PRIMARY REGULATOR VALVE
ACCUMULATORS
MAINLINE PRESSURE
INTAKE-PUMP
CONVERTER AND LUBE
THROTTLE PRESSURE
ACCUMULATOR

DRIVE RANGE – 3RD GEAR – SPRINT AND GEO TRACKER

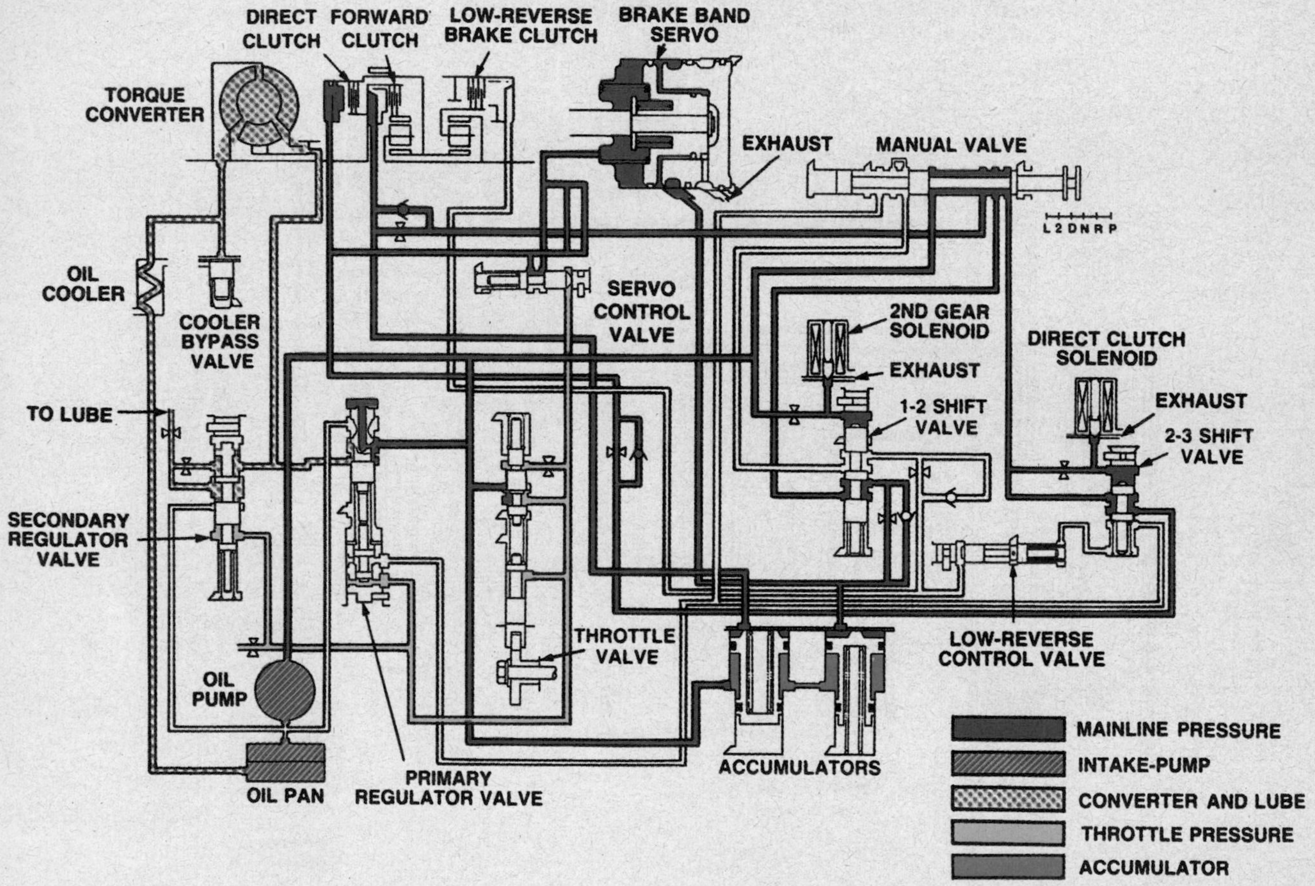
DIRECT CLUTCH
FORWARD CLUTCH
LOW-REVERSE BRAKE CLUTCH
BRAKE BAND SERVO
TORQUE CONVERTER
EXHAUST
MANUAL VALVE
L 2 D N R P
OIL COOLER
COOLER BYPASS VALVE
SERVO CONTROL VALVE
2ND GEAR SOLENOID
EXHAUST
DIRECT CLUTCH SOLENOID
TO LUBE
1-2 SHIFT VALVE
EXHAUST
2-3 SHIFT VALVE
SECONDARY REGULATOR VALVE
THROTTLE VALVE
LOW-REVERSE CONTROL VALVE
OIL PUMP
ACCUMULATORS
OIL PAN
PRIMARY REGULATOR VALVE
MAINLINE PRESSURE
INTAKE-PUMP
CONVERTER AND LUBE
THROTTLE PRESSURE
ACCUMULATOR

MANUAL L RANGE – 1ST GEAR – SPRINT AND GEO TRACKER

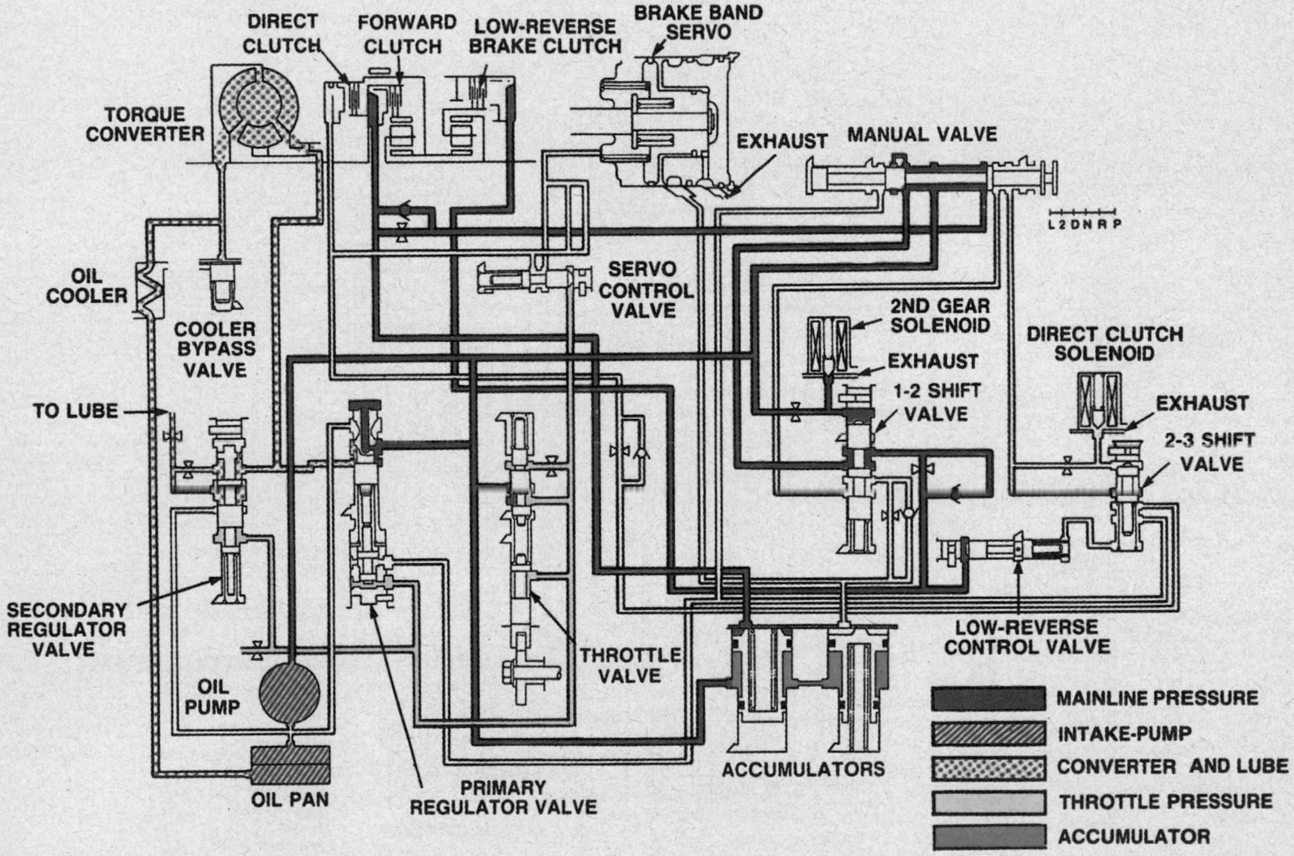
DIRECT CLUTCH
FORWARD CLUTCH
LOW-REVERSE BRAKE CLUTCH
BRAKE BAND SERVO
TORQUE CONVERTER
EXHAUST
MANUAL VALVE
L 2 D N R P
OIL COOLER
SERVO CONTROL VALVE
COOLER BYPASS VALVE
2ND GEAR SOLENOID
DIRECT CLUTCH SOLENOID
EXHAUST
1-2 SHIFT VALVE
EXHAUST
2-3 SHIFT VALVE
TO LUBE
SECONDARY REGULATOR VALVE
THROTTLE VALVE
LOW-REVERSE CONTROL VALVE
OIL PUMP
OIL PAN
PRIMARY REGULATOR VALVE
ACCUMULATORS
MAINLINE PRESSURE
INTAKE-PUMP
CONVERTER AND LUBE
THROTTLE PRESSURE
ACCUMULATOR

REVERSE—SPRINT AND GEO TRACKER

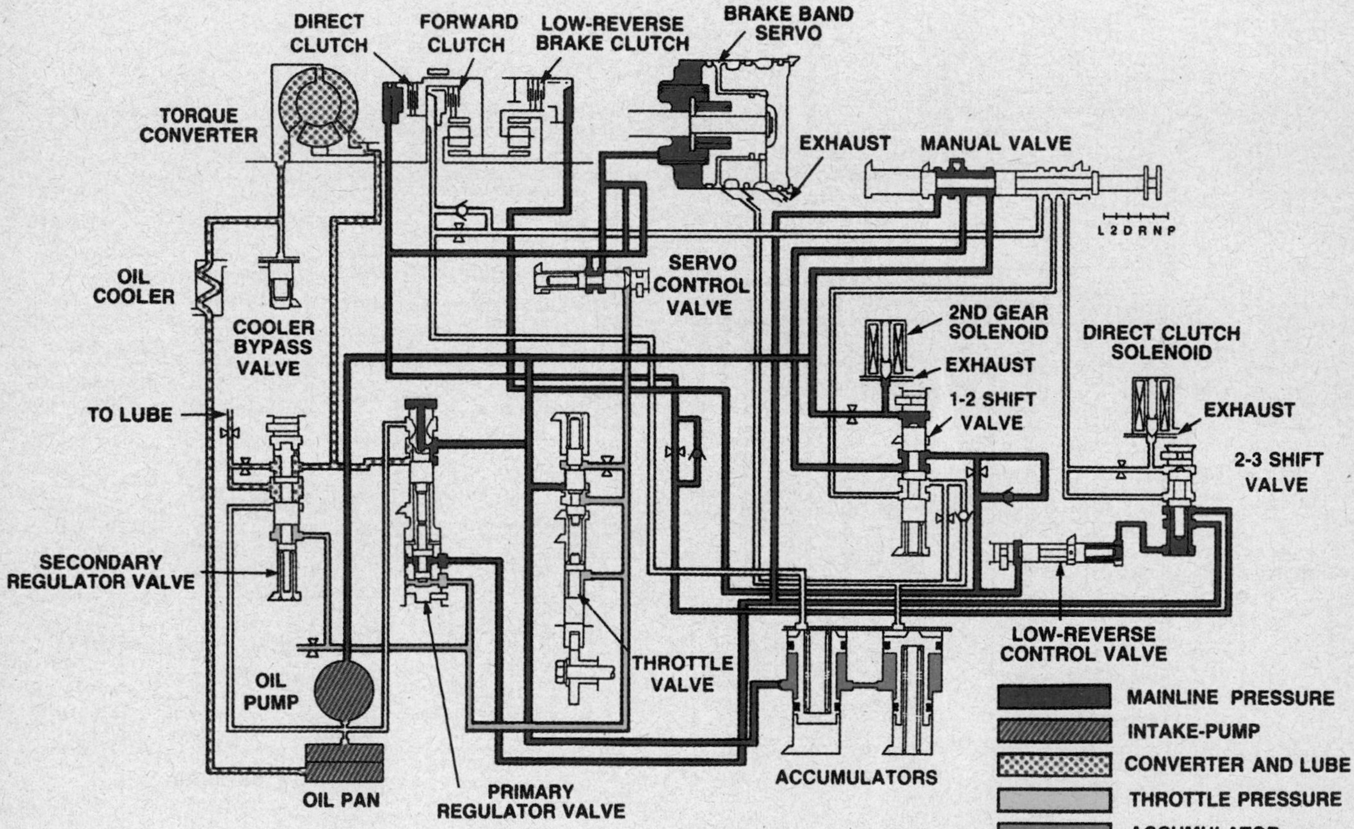
DIRECT CLUTCH
FORWARD CLUTCH
LOW-REVERSE BRAKE CLUTCH
BRAKE BAND SERVO
TORQUE CONVERTER
EXHAUST
MANUAL VALVE
L 2 D R N P
OIL COOLER
COOLER BYPASS VALVE
SERVO CONTROL VALVE
2ND GEAR SOLENOID
EXHAUST
DIRECT CLUTCH SOLENOID
1-2 SHIFT VALVE
EXHAUST
TO LUBE
2-3 SHIFT VALVE
SECONDARY REGULATOR VALVE
LOW-REVERSE CONTROL VALVE
OIL PUMP
THROTTLE VALVE
MAINLINE PRESSURE
INTAKE-PUMP
CONVERTER AND LUBE
THROTTLE PRESSURE
ACCUMULATOR
OIL PAN
PRIMARY REGULATOR VALVE
ACCUMULATORS

STANDARD TORQUE SPECIFICATIONS AND CAPSCREW MARKINGS

Newton-Meter has been designated as the world standard for measuring torque and will gradually replace the foot-pound and kilogram-meter torque measuring standard. Torquing tools are still being manufactured with foot-pounds and kilogram-meter scales, along with the new Newton-Meter standard. To assist the repairman, foot-pounds, kilogram-meter and Newton-Meter are listed in the following charts, and should be followed as applicable.

U.S. BOLTS

SAE Grade Number	1 or 2			5			6 or 7			8		
Capscrew Head Markings Manufacturer's marks may vary. Three-line markings on heads below indicate SAE Grade 5. 	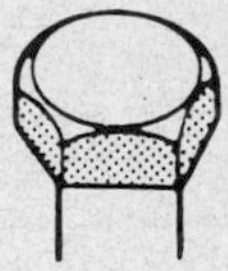						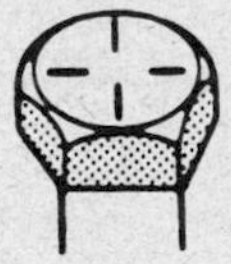					
Usage	Used Frequently			Used Frequently			Used at Times			Used at Times		
Quality of Material	Indeterminate			Minimum Commercial			Medium Commercial			Best Commercial		
Capacity Body Size	**Torque**			**Torque**			**Torque**			**Torque**		
(inches)-(thread)	Ft-Lb	kgm	Nm	Ft-Lb	kgm	Nm	Ft-Lb	kgm	Nm	Ft-Lb	kgm	Nm
1/4-20	5	0.6915	6.7791	8	1.1064	10.8465	10	1.3630	13.5582	12	1.6596	16.2698
-28	6	0.8298	8.1349	10	1.3830	13.5582				14	1.9362	18.9815
5/16-18	11	1.5213	14.9140	17	2.3511	23.0489	19	2.6277	25.7605	24	3.3192	32.5396
-24	13	1.7979	17.6256	19	2.6277	25.7605				27	3.7341	36.6071
3/8-16	18	2.4894	24.4047	31	4.2873	42.0304	34	4.7022	46.0978	44	6.0852	59.6560
-24	20	2.7660	27.1164	35	4.8405	47.4536				49	6.7767	66.4351
7/16-14	28	3.8132	37.9629	49	6.7767	66.4351	55	7.6065	74.5700	70	9.6810	94.9073
-20	30	4.1490	40.6745	55	7.6065	74.5700				78	10.7874	105.7538
1/2-13	39	5.3937	52.8769	75	10.3725	101.6863	85	11.7555	115.2445	105	14.5215	142.3609
-20	41	5.6703	55.5886	85	11.7555	115.2445				120	16.5860	162.6960
9/16-12	51	7.0533	69.1467	110	15.2130	149.1380	120	16.5960	162.6960	155	21.4365	210.1490
-18	55	7.6065	74.5700	120	16.5960	162.6960				170	23.5110	230.4860
5/8-11	83	11.4789	112.5329	150	20.7450	203.3700	167	23.0961	226.4186	210	29.0430	284.7180
-18	95	13.1385	128.8027	170	23.5110	230.4860				240	33.1920	325.3920
3/4-10	105	14.5215	142.3609	270	37.3410	366.0660	280	38.7240	379.6240	375	51.8625	508.4250
-16	115	15.9045	155.9170	295	40.7985	399.9610				420	58.0860	568.4360
7/8-9	160	22.1280	216.9280	395	54.6285	535.5410	440	60.8520	596.5520	605	83.6715	820.2590
-14	175	24.2025	237.2650	435	60.1605	589.7730				675	93.3525	915.1650
1-8	236	32.5005	318.6130	590	81.5970	799.9220	660	91.2780	894.8280	910	125.8530	1233.7780
-14	250	34.5750	338.9500	660	91.2780	849.8280				990	136.9170	1342.2420

METRIC BOLTS

Description	Torque ft-lbs. (Nm)			
Thread for general purposes (size x pitch (mm))	Head Mark 4		Head Mark 7	
6 x 1.0	2.2 to 2.9	(3.0 to 3.9)	3.6 to 5.8	(4.9 to 7.8)
8 x 1.25	5.8 to 8.7	(7.9 to 12)	9.4 to 14	(13 to 19)
10 x 1.25	12 to 17	(16 to 23)	20 to 29	(27 to 39)
12 x 1.25	21 to 32	(29 to 43)	35 to 53	(47 to 72)
14 x 1.5	35 to 52	(48 to 70)	57 to 85	(77 to 110)
16 x 1.5	51 to 77	(67 to 100)	90 to 120	(130 to 160)
18 x 1.5	74 to 110	(100 to 150)	130 to 170	(180 to 230)
20 x 1.5	110 to 140	(150 to 190)	190 to 240	(160 to 320)
22 x 1.5	150 to 190	(200 to 260)	250 to 320	(340 to 430)
24 x 1.5	190 to 240	(260 to 320)	310 to 410	(420 to 550)

CAUTION: Bolts threaded into aluminum require much less torque

ENGLISH TO METRIC CONVERSION: TORQUE

Torque is now expressed as either foot-pounds (ft./lbs.) or inch-pounds (in. lbs.). The metric measurement unit for torque is the Newton-meter (Nm). This unit—the Nm—will be used for all SI metric torque references, both the present ft./lbs. and in./lbs.

ft lbs	N-m	ft lbs	N-m	ft lbs	N-m	ft lbs	N-m
0.1	0.1	33	44.7	74	100.3	115	155.9
0.2	0.3	34	46.1	75	101.7	116	157.3
0.3	0.4	35	47.4	76	103.0	117	158.6
0.4	0.5	36	48.8	77	104.4	118	160.0
0.5	0.7	37	50.7	78	105.8	119	161.3
0.6	0.8	38	51.5	79	107.1	120	162.7
0.7	1.0	39	52.9	80	108.5	121	164.0
0.8	1.1	40	54.2	81	109.8	122	165.4
0.9	1.2	41	55.6	82	111.2	123	166.8
1	1.3	42	56.9	83	112.5	124	168.1
2	2.7	43	58.3	84	113.9	125	169.5
3	4.1	44	59.7	85	115.2	126	170.8
4	5.4	45	61.0	86	116.6	127	172.2
5	6.8	46	62.4	87	118.0	128	173.5
6	8.1	47	63.7	88	119.3	129	174.9
7	9.5	48	65.1	89	120.7	130	176.2
8	10.8	49	66.4	90	122.0	131	177.6
9	12.2	50	67.8	91	123.4	132	179.0
10	13.6	51	69.2	92	124.7	133	180.3
11	14.9	52	70.5	93	126.1	134	181.7
12	16.3	53	71.9	94	127.4	135	183.0
13	17.6	54	73.2	95	128.8	136	184.4
14	18.9	55	74.6	96	130.2	137	185.7
15	20.3	56	75.9	97	131.5	138	187.1
16	21.7	57	77.3	98	132.9	139	188.5
17	23.0	58	78.6	99	134.2	140	189.8
18	24.4	59	80.0	100	135.6	141	191.2
19	25.8	60	81.4	101	136.9	142	192.5
20	27.1	61	82.7	102	138.3	143	193.9
21	28.5	62	84.1	103	139.6	144	195.2
22	29.8	63	85.4	104	141.0	145	196.6
23	31.2	64	86.8	105	142.4	146	198.0
24	32.5	65	88.1	106	143.7	147	199.3
25	33.9	66	89.5	107	145.1	148	200.7
26	35.2	67	90.8	108	146.4	149	202.0
27	36.6	68	92.2	109	147.8	150	203.4
28	38.0	69	93.6	110	149.1	151	204.7
29	39.3	70	94.9	111	150.5	152	206.1
30	40.7	71	96.3	112	151.8	153	207.4
31	42.0	72	97.6	113	153.2	154	208.8
32	43.4	73	99.0	114	154.6	155	210.2

ENGLISH TO METRIC CONVERSION: PRESSURE

The basic unit of pressure measurement used today is expressed as pounds per square inch (psi). The metric unit for psi will be the kilopascal (kPa). This will apply to either fluid pressure or air pressure, and will be frequently seen in tire pressure readings, oil pressure specifications, fuel pump pressure, etc.

To convert pounds per square inch (psi) to kilopascals (kPa): multiply the number of psi by 6.89

Psi	kPa	Psi	kPa	Psi	kPa	Psi	kPa
0.1	0.7	37	255.1	82	565.4	127	875.6
0.2	1.4	38	262.0	83	572.3	128	882.5
0.3	2.1	39	268.9	84	579.2	129	889.4
0.4	2.8	40	275.8	85	586.0	130	896.3
0.5	3.4	41	282.7	86	592.9	131	903.2
0.6	4.1	42	289.6	87	599.8	132	910.1
0.7	4.8	43	296.5	88	606.7	133	917.0
0.8	5.5	44	303.4	89	613.6	134	923.9
0.9	6.2	45	310.3	90	620.5	135	930.8
1	6.9	46	317.2	91	627.4	136	937.7
2	13.8	47	324.0	92	634.3	137	944.6
3	20.7	48	331.0	93	641.2	138	951.5
4	27.6	49	337.8	94	648.1	139	958.4
5	34.5	50	344.7	95	655.0	140	965.2
6	41.4	51	351.6	96	661.9	141	972.2
7	48.3	52	358.5	97	668.8	142	979.0
8	55.2	53	365.4	98	675.7	143	985.9
9	62.1	54	372.3	99	682.6	144	992.8
10	69.0	55	379.2	100	689.5	145	999.7
11	75.8	56	386.1	101	696.4	146	1006.6
12	82.7	57	393.0	102	703.3	147	1013.5
13	89.6	58	399.0	103	710.2	148	1020.4
14	96.5	59	406.8	104	717.0	149	1027.3
15	103.4	60	413.7	105	723.9	150	1034.2
16	110.3	61	420.6	106	730.8	151	1041.1
17	117.2	62	427.5	107	737.7	152	1048.0
18	124.1	63	434.4	108	744.6	153	1054.9
19	131.0	64	441.3	109	751.5	154	1061.8
20	137.9	65	448.2	110	758.4	155	1068.7
21	144.8	66	455.0	111	765.3	156	1075.6
22	151.7	67	461.9	112	772.2	157	1082.5
23	158.6	68	468.8	113	779.1	158	1089.4
24	165.5	69	475.7	114	786.0	159	1096.3
25	172.4	70	482.6	115	792.9	160	1103.2
26	179.3	71	489.5	116	799.8	161	1110.0
27	186.2	72	496.4	117	806.7	162	1116.9
28	193.0	73	503.3	118	813.6	163	1123.8
29	200.0	74	510.2	119	820.5	164	1130.7
30	206.8	75	517.1	120	827.4	165	1137.6
31	213.7	76	524.0	121	834.3	166	1144.5
32	220.6	77	530.9	122	841.2	167	1151.4
33	227.5	78	537.8	123	848.0	168	1158.3
34	234.4	79	544.7	124	854.9	169	1165.2
35	241.3	80	551.6	125	861.8	170	1172.1
36	248.2	81	558.5	126	868.7	171	1179.0

ENGLISH TO METRIC CONVERSION: PRESSURE

The basic unit of pressure measurement used today is expressed as pounds per square inch (psi). The metric unit for psi will be the kilopascal (kPa). This will apply to either fluid pressure or air pressure, and will be frequently seen in tire pressure readings, oil pressure specifications, fuel pump pressure, etc.

To convert pounds per square inch (psi) to kilopascals (kPa): multiply the number of psi by 6.89

Psi	kPa	Psi	kPa	Psi	kPa	Psi	kPa
172	1185.9	216	1489.3	260	1792.6	304	2096.0
173	1192.8	217	1496.2	261	1799.5	305	2102.9
174	1199.7	218	1503.1	262	1806.4	306	2109.8
175	1206.6	219	1510.0	263	1813.3	307	2116.7
176	1213.5	220	1516.8	264	1820.2	308	2123.6
177	1220.4	221	1523.7	265	1827.1	309	2130.5
178	1227.3	222	1530.6	266	1834.0	310	2137.4
179	1234.2	223	1537.5	267	1840.9	311	2144.3
180	1241.0	224	1544.4	268	1847.8	312	2151.2
181	1247.9	225	1551.3	269	1854.7	313	2158.1
182	1254.8	226	1558.2	270	1861.6	314	2164.9
183	1261.7	227	1565.1	271	1868.5	315	2171.8
184	1268.6	228	1572.0	272	1875.4	316	2178.7
185	1275.5	229	1578.9	273	1882.3	317	2185.6
186	1282.4	230	1585.8	274	1889.2	318	2192.5
187	1289.3	231	1592.7	275	1896.1	319	2199.4
188	1296.2	232	1599.6	276	1903.0	320	2206.3
189	1303.1	233	1606.5	277	1909.8	321	2213.2
190	1310.0	234	1613.4	278	1916.7	322	2220.1
191	1316.9	235	1620.3	279	1923.6	323	2227.0
192	1323.8	236	1627.2	280	1930.5	324	2233.9
193	1330.7	237	1634.1	281	1937.4	325	2240.8
194	1337.6	238	1641.0	282	1944.3	326	2247.7
195	1344.5	239	1647.8	283	1951.2	327	2254.6
196	1351.4	240	1654.7	284	1958.1	328	2261.5
197	1358.3	241	1661.6	285	1965.0	329	2268.4
198	1365.2	242	1668.5	286	1971.9	330	2275.3
199	1372.0	243	1675.4	287	1978.8	331	2282.2
200	1378.9	244	1682.3	288	1985.7	332	2289.1
201	1385.8	245	1689.2	289	1992.6	333	2295.9
202	1392.7	246	1696.1	290	1999.5	334	2302.8
203	1399.6	247	1703.0	291	2006.4	335	2309.7
204	1406.5	248	1709.9	292	2013.3	336	2316.6
205	1413.4	249	1716.8	293	2020.2	337	2323.5
206	1420.3	250	1723.7	294	2027.1	338	2330.4
207	1427.2	251	1730.6	295	2034.0	339	2337.3
208	1434.1	252	1737.5	296	2040.8	240	2344.2
209	1441.0	253	1744.4	297	2047.7	341	2351.1
210	1447.9	254	1751.3	298	2054.6	342	2358.0
211	1454.8	255	1758.2	299	2061.5	343	2364.9
212	1461.7	256	1765.1	300	2068.4	344	2371.8
213	1468.7	257	1772.0	301	2075.3	345	2378.7
214	1475.5	258	1778.8	302	2082.2	346	2385.6
215	1482.4	259	1785.7	303	2089.1	347	2392.5